Chemistry of Coal Utilization

SECOND SUPPLEMENTARY VOLUME

Chemistry of Coal Utilization

SECOND SUPPLEMENTARY VOLUME

Prepared Under the Guidance of the Committee on Chemistry of Coal Utilization

Edited by

MARTIN A. ELLIOTT

Board on Mineral and Energy Resources
Commission on Natural Resources
National Research Council
National Academy of Sciences
National Academy of Engineering

A WILEY-INTERSCIENCE PUBLICATION

JOHN WILEY & SONS   New York • Chichester • Brisbane • Toronto

Notice

The project that is the subject of this volume was approved by the Governing Board of the National Research Council, whose members are drawn from the Councils of the National Academy of Sciences, the National Academy of Engineering, and the Institute of Medicine. The members of the Committee responsible for the volume were chosen for their special competences and with regard for appropriate balance.

This volume has been reviewed by a group other than the authors according to procedures approved by a Report Review Committee consisting of members of the National Academy of Sciences, the National Academy of Engineering, and the Institute of Medicine.

Library of Congress Cataloging in Publication Data

Main entry under title:

Chemistry of coal utilization.

"A Wiley-Interscience publication."

Supplements the work issued by the Committee on Chemical Utilization of Coal, National Research Council.

Includes index.

1. Coal. 2. Coal-tar products. I. Elliott, Martin Anderson, 1909– II. National Research Council. Committee on Chemistry of Coal Utilization. III. National Research Council. Committee on Chemical Utilization of Coal. Chemistry of coal utilization.

TP325.C443 662.6′2 80-13296
ISBN 0-471-07726-7

Printed in the United States of America

10 9 8 7 6 5 4 3 2 1

To Homer H. Lowry, an outstanding coal scientist, who brought into being the first three volumes of this work so extensively used and greatly appreciated by workers in coal science and coal technology

National Academy of Sciences Steering Committee on Chemistry of Coal Utilization

Contributors

Seymour B. Alpert
Electric Power Research Institute
Palo Alto, California

E. Aristoff
ARCO Chemical
Glenolden, Pennsylvania

Paul Averitt
U.S. Geological Survey
Denver, Colorado

Homer E. Benson
Benfield Corporation
Pittsburgh, Pennsylvania

Hans-Dieter Beyer
Bergbau-Forschung GmbH
Essen, Germany

Sidney A. Bresler
Bresler & Associates
New York, New York

Thomas G. Callcott
Broken Hill Proprietary
Company, Ltd.
New South Wales, Australia

F. J. Ceely
Foster Wheeler Energy Corporation
Livingston, New Jersey

Billy L. Crynes
Oklahoma State University
Stillwater, Oklahoma

Ernest L. Daman
Foster Wheeler Energy Corporation
Livingston, New Jersey

Albert W. Deurbrouck
U.S. Department of Energy
Pittsburgh, Pennsylvania

Ian G. C. Dryden
Editor, *Fuel*
Surrey, United Kingdom

Thomas F. Edgar
University of Texas at Austin
Austin, Texas

Werner Eisenhut
Bergbau-Forschung GmbH
Essen, Germany

Martin A. Elliott
Energy Consultant
Houston, Texas

Robert H. Essenhigh
Ohio State University
Columbus, Ohio

Carl W. Gehrs
Oak Ridge National Laboratory
Oak Ridge, Tennessee

Harold J. Gluskoter
Exxon Production Research Company
Houston, Texas

Everett Gorin
San Rafael, California

David W. Gregg
Lawrence Livermore Laboratory
Livermore, California

Herbert A. Grosick
Koppers Company, Inc.
Pittsburgh, Pennsylvania

Diethard Habermehl
Bergbau-Forschung GmbH
Essen, Germany

Dennis Hebden
British Gas Corporation
London, England

Stephen E. Herbes
Oak Ridge National Laboratory
Oak Ridge, Tennessee

Laszlo A. Heredy
Atomics International
Canoga Park, California

Jack B. Howard
Massachusetts Institute of Technology
Cambridge, Massachusetts

Richard E. Hucko
U.S. Department of Energy
Pittsburgh, Pennsylvania

James L. Johnson*
Institute of Gas Technology
Chicago, Illinois

Harald Jüntgen
Bergbau-Forschung GmbH
Essen, Germany

Karl Knoblauch
Bergbau-Forschung GmbH
Essen, Germany

Jürgen Klein
Bergbau-Forschung GmbH
Essen, Germany

J.E. Kovacic
Koppers Company, Inc.
Pittsburgh, Pennsylvania

James T. McCartney
U.S. Department of Energy
Pittsburgh, Pennsylvania

Donald McNeil
Technical Consultant
Bradford, England

Richard C. Neavel
Exxon Research and Engineering Co.
Baytown, Texas

Martin B. Neuworth
U.S. Department of Energy
Washington, D.C.

Fred Orywal
Bergbau-Forschung GmbH
Essen, Germany

Michael Perch
Koppers Company, Inc.
Verona, Pennsylvania

Harry Perry
Resources for the Future
Washington, D.C.

Fred B. Raymer
Bechtel Inc.
San Francisco, California

William T. Reid
Consultant, Energy Conversion
Columbus, Ohio

Robert W. Rieve
ARCO Chemical
Glenolden, Pennsylvania

Rodney R. Ruch
Illinois State Geological Survey
Urbana, Illinois

Eliahu J. Salmon
Resources for the Future
Washington, D.C.

Walter Schinzel
Bergbau-Forschung GmbH
Essen, Germany

*Deceased

Hans-Jürgen Schröter
Bergbau-Forschung GmbH
Essen, Germany

Joachim Schulze
Technical University of Berlin
Berlin, Germany

Leonard Seglin
Econergy Associates
New York, New York

Harold Shalit
ARCO Chemical
Glenolden, Pennsylvania

A.G. Sharkey, Jr.
U.S. Department of Energy
Pittsburgh, Pennsylvania

Neil F. Shimp
Illinois State Geological Survey
Urbana, Illinois

David S. Shriner
Oak Ridge National Laboratory
Oak Ridge, Tennessee

A.V. Slack
SAS Corporation
Sheffield, Alabama

Gregory B. Smith
Broken Hill Proprietary Company, Ltd.
New South Wales, Australia

Henry J.F. Stroud
British Gas Corporation
London, England

Terry L. Thompson
Bechtel Inc.
Houston, Texas

Ronald H. Wolk
Electric Power Research Institute
Palo Alto, California

Irving Wender
Pittsburgh Energy Research Center
Pittsburgh, Pennsylvania

G. Robert Yohe*
Illinois State Geological Survey
Urbana, Illinois

*Deceased

Preface

The first two volumes of this work, published in 1945, and the first supplemental volume, published in 1963, all under the editorship of H. H. Lowry, are recognized classics. Their comprehensive coverage and critical review of the literature on coal science and technology have been invaluable to workers at all levels of experience. Between 1963 and 1976, however, funding of coal research and development by government and industry in the United States increased from about $22 million to about $322 million per year. Throughout this period the expansion in the literature on coal science and technology and in the number of newcomers to the field made evident the need for a second supplementary volume to *Chemistry of Coal Utilization*.

Seymour Alpert, Electric Power Research Institute, and George R. Hill, University of Utah (formerly of Electric Power Research Institute), began to lay the groundwork for a second supplement as early as 1974. The new volume was not to replace the earlier volumes but was to bring up to date the review of the state of knowledge of the science and technology of coal utilization. The publisher agreed to reprint the first two volumes and to keep the first supplemental volume in print during the lifetime of the new supplement.

The earlier volumes had been prepared and published under the auspices of the National Research Council (NRC) of the National Academy of Sciences, and the Council was approached again about its interest in overseeing the current review. The NRC accepted the task and assigned it to the Board on Energy Studies under the NRC's Commission on Natural Resources. In September 1975 the board selected George R. Hill as chairman of a project steering committee, and I agreed to serve as Technical Editor. (Dr. Lowry died in 1971.) A contract providing financial support was signed between the U.S. Energy Research and Development Administration (ERDA) and the NRC in February 1976, and work under the contract began at once. In mid-February the following appointments to the Steering Committee were announced: George R. Hill (Chairman), University of Utah; Norbert Berkowitz, University of Alberta; Joan L. Coles, Utah Energy Conservation and Development Council; R. Tracy Eddinger, Cogas Development Company; Lionel S. Galstaun, Bechtel Incorporated; James R. Garvey, Bituminous Coal Research, Inc.; John M. Holmes, Oak Ridge National Laboratory; Hoyt C. Hottel, Massachusetts Institute of Technology; Henry R. Linden, Gas Research Institute; Johnstone S. Mackay, U.S. Steel Corporation; Douglas S. Montgomery, University of Alberta; Harry Perry, Resources for the Future; Eric H. Reichl, CONOCO Coal Development Company; Jack A. Simon, Illinois State Geological Survey; and Lawrence E. Swabb, Jr., Exxon Research and Engineering Company.

At its first meeting the Steering Committee approved the division of the task into chapters and suggested the names of authorities to write them. Sixty internationally recognized experts accepted the committee's invitation to write or collaborate on individual chapters, and the committee added four chapters dealing with topics not found in the earlier volumes: Chapter 1, The Coal Industry and Coal Research and Development in Perspective; Chapter 2, Coal Resources; Chapter 22, Control of Pollution from Combustion Processes; and Chapter 31, Environmental, Health and Safety Implications of Increased Coal Utilization. In addition, the broad areas of pyrolysis, combustion, gasification, and liquefaction were subdivided for treatment in three separate chapters on fundamentals, processes, and treatment of products. One committee member was assigned responsibility for the technical review of each of the 31 chapters. (The Acknowledgments that follow this Preface specify the designated members and their assignments.)

All the authors had been selected and all but one of the chapter outlines had been submitted before the second meeting of the Steering Committee in May 1976. At that time it was decided that I would suggest reviewers for the individual chapters, in consultation and agreement with the other committee members. (The Acknowledgments section lists the names of those who reviewed manuscripts and who contributed so substantially to the quality of the final product.)

The first manuscript was received in August 1976, the last in December 1978. Because of this unavoidable time spread, the cutoff date of the coverage of the literature varies from chapter to chapter just as it did in the first supplement. The current authors were asked to pick up review of their topics at the point where the antecedent chapter in the earlier volume left off and to bring the review down to the time of completion of their individual manuscripts. The completed manuscript was sent to the publisher in the spring of 1979.

The basic philosophy of the second supplemental volume is the same as that stated by Dr. Lowry in his preface to the first supplement. He wrote that the manuscripts were to

> include critical reviews of important world literature [through 1958]. Complete bibliographic coverage was to be considered less important than reference to the more significant publications. Both sides of controversial subjects were to be documented. Each chapter was to be reviewed by competent people in the field of activity covered, the reviewers being selected by consultation with members of the Committee on Chemistry of Coal Utilization. The review system was to be designed to avoid undue emphasis on particular subjects solely because of immediate personal interests of the authors . . . .

Authors of the current volume are solely responsible for the technical content and conclusions in their chapters. Certain minor changes in the text were made in the editorial process for the sake of uniformity. I take full responsibility for the following policy decisions.

Units The form to be followed in using units in the individual chapters was left to the discretion of the authors. In the chapters dealing with coal science, c.g.s. units were used consistently. In certain chapters dealing with coal technology, the form used by the authors of the referenced articles was followed. Thus both c.g.s. and f.p.s. units may appear in the same chapter.

maf and daf Throughout the book, ''maf'' is used consistently for ''moisture-and-ash-free'' basis. This decision was made before I learned that some of the authors had a strong preference for the use of ''daf'' for the ''dry-ash-free'' basis, which in certain instances is less ambiguous than maf. The point is brought out here to prevent confusion, since some articles in the literature by authors of chapters in this book advocate use of ''daf,'' yet the term ''maf'' appears in chapters they wrote for this book.

Repetition of subject matter I believe that some repetition of material from chapter to chapter is justified if it allows the discussion in each chapter to be more self-contained and minimizes or avoids interruptions by cross-references. As it turns out, there is comparatively little repetition among chapters, and where it does occur, it contributes to continuity of presentation.

Economics The Steering Committee decided that data on economics should not be incorporated into the technical discussions. The decision was made because of rapidly changing economic conditions as a result of inflation, and because differences in assumptions make economic estimates from different sources generally not comparable. In a few instances economic estimates were made by the same organization and used, not in the absolute sense, but for comparative purposes between technologies. In those few instances, I included economic comparisons to allow for more comprehensive coverage of the subject.

Glossary The bewildering array of acronyms that has come into use in the past decade, for government entities and other organizations as well as for processes and procedures, made a glossary indispensable.

One of the primary objectives of everyone participating in this effort was to make the finished product as widely available as possible. To that end we have made every effort commensurate with maintaining quality to hold down the list price of the published book. The NRC provided partial restitution of income foregone and out-of-pocket expenses incurred for only 11 of the many contributors. In the case of the earlier volumes, no royalties were paid by the publisher either to the NRC or to any of the authors. Recent National Academy of Sciences policy provides for royalties to be paid to the NRC on privately published sales of reports for which there may be high demand. In accord with that policy the contractual agreement concerning this volume provides for royalties to be paid to the NRC if the total number of copies sold exceeds 5000. Any money received under these terms will be placed in a special trust fund to support the preparation of the next volume in the *Chemistry of Coal Utilization* series.

The efforts of everyone involved in this ambitious undertaking will be more than amply rewarded if, as we hope, the wealth of scientific and technical knowledge contained here assists the nation in utilizing efficiently its vast coal resource.

MARTIN A. ELLIOTT

Technical Editor
Committee on Chemistry of Coal Utilization
National Academy of Sciences
National Research Council

Houston, Texas
January 1981

Acknowledgments

Obvious though it is, it must be stressed again that this book is the aggregate of the contributions of its authors. Their knowledge and experience and their skill at distilling what they had learned into the chapters that follow have, alone, made the present volume possible. Words cannot express my gratitude for the dedication of the authors to their task, or the pride I have taken in the privilege of serving as technical editor.

For an effort of this magnitude many individuals in addition to the authors made contributions, and it is a pleasure to acknowledge them.

Ernest E. Donath did a superb job of translating into English the four chapters written in German. The difficulties presented to the translator by the subtle distinctions in terminology cannot be overestimated. Dr. Donath surmounted them all.

The work of the Steering Committee and its chairman in advising and guiding me and in reviewing manuscripts was of inestimable value. The contribution of the committee members responsible for carrying out their special roles also deserves our particular notice. The committee members bore responsibility for chapters as follows: N. Berkowitz, Chapters 4, 5, 7, 8; J. Coles, Chapter 31; T. Eddinger, Chapters 12, 13, 14; L. Galstaun, Chapters 23, 24, 25; J. Garvey, Chapters 9, 10, 11, 20; J. Holmes, Chapters 6, 21; H. Hottel, Chapter 19; H. Linden, Chapter 1; J. Mackay, Chapters 26, 30; D. Montgomery, Chapters 15, 16, 17, 18; H. Perry, Chapter 22; J. Simon, Chapters 2, 3; L. Swabb, Chapters 27, 28, 29.

The reviewers listed below contributed generously and willingly of their time and abilities and played a crucial role in our efforts to maintain the high standards set by the earlier volumes. In alphabetical order they were Clayton G. Ball, Paul Weir Company, Inc.; Harold Beuther, Gulf Oil Research and Development; Charles Brandenburg, U.S. Energy Research and Development Administration; Irving Breger, U.S. Geological Survey; John Burchard, U.S. Environmental Protection Agency; Thomas G. Callcott, Broken Hill Proprietary Company, Ltd.; John Campbell, University of California; Richard Carpenter, East-West Environment and Policy Institute; Ralph L. Coates, Brigham Young University; Henry O. Cochran, Jr., Oak Ridge National Laboratory; Anthony Colucci, Electric Power Research Institute; Cyril Comar, Electric Power Research Institute; Arthur L. Conn, Amoco Research and Development Center; Richard Corey, U.S. Energy Research and Development Administration; William L. Crentz, U.S. Energy Research and Development Administration; Ernest Donath, Consultant; Sabri Ergun, Consultant; Harold Falkenberry, Tennessee Valley Authority; B. T. Fant, Exxon Research and Engineering Company; A. J. Forney, Consultant; L. D. Freidman, Cogas Development Company; J. F. Fryer, Alberta Research Council; Peter Given, Pennsylvania State University; Ralph Gray, U.S. Steel Corporation; Robert A. Graff, City University of New York; George Haddad, U.S. Steel Corporation; William J. Halvorsen, Consolidation Coal Company; Homer E. Harris, Eastern Associated Coal Corporation; Roy Helfenstine, Illinois Geological Survey; J. F. Hiorns, Consultant; Gerald A. Hollinden, Tennessee Valley Authority; Michael Holowaty, Inland Steel Corporation; Jack Howard, Massachusetts Institute of Technology; H. Raymond Hoy, British Coal Utilisation Research Association; Arthur Ingberman, Union Carbide Corporation; Clarence Karr, U.S. Energy Research and Development Administration; Raphael Kasper, National Research Council; Arthur Kohl, Rockwell International; George Land, Amax Coal Company; Bert Louks, Stanford Research Institute; Michael McCloskey, Sierra Club; Wayne A. McCurdy, U.S. Energy Research and Development Administration; D. G. Madley, Coal Research Establishment; Michael Massey, Carnegie-

Mellon University; Lowell Miller, U.S. Energy Research and Development Administration; Alexander G. Mills, U.S. Energy Research and Development Administration; Maurice F. R. Mulcahy, Mineral Research Laboratories, CSIRO; Milton Orchin, University of Cincinnati; Arthur A. Orning, Consultant; P. H. Pinchbeck, British Steel Corporation (Chemicals) Ltd.; Arthur L. Plumley, C-E Power Systems; Neil Rice, International Briquetting Association; Alan G. Roberts, British Coal Utilisation Research Association Ltd.; Bruce K. Schmid, Gulf Mineral Resources Company; Leo A. Schrider, U.S. Energy Research and Development Administration; A. A. Schuldt, Steel Company of Canada, Ltd.; Robert Saltsman, Bituminous Coal Research; William Spackman, Pennsylvania State University; John Stanton, British Coal Utilisation Research Association; I. McC. Stewart, University of New Castle; Robert Tardiff, National Research Council; Josephus Thomas, Illinois Geological Survey; Richard Thompson, Bethlehem Steel Corporation; Harold Turnbell, USS Chemicals, Division of U.S. Steel Corporation; Roger P. Van Driesen, Petroleum and Coal Process Marketing Combustion Engineering Inc.; Stanley Van Katesan, Powell Construction Company; Billy Walls, Carter Oil Company; Jack Walters, Pittsburgh Energy Research Center; Irving Wender, U.S. Energy Research and Development Administration; Edgar C. Winegartner, Exxon Research and Engineering Company; Wendell H. Wiser, University of Utah; David E. Wolfson, Consultant; Joseph Yerushalmi, City College of New York; Peter Zubovic, U.S. Geological Survey.

Many members of the NRC staff contributed in a variety of ways to the project. Winfred E. Berg, Special Projects Officer of the Commission on Natural Resources, initiated and assisted in establishing the project through negotiations with ERDA and, subsequently, the U.S. Department of Energy (DOE). Captain Berg also helped to set up the Steering Committee, and throughout the project we had the benefit of his unflagging interest and wise counsel.

I am deeply indebted to George White, Executive Secretary of the Committee on Chemistry of Coal Utilization. His administrative ability, dedication, competence, and diligence were major factors in ensuring the success of this project. Also indispensable was Erika Douglas, without whose competence, dedication, thoroughness, and diligence it would not have been possible to put the manuscript into final form for publication.

It is a particular pleasure to acknowledge the performances of these outstanding staff officers: Robert C. Rooney, Philippa Shepherd, Maggie Elliott, Claire Chow, Judith Cummings, and Anne Norman, editors at the Commission on Natural Resources, who assisted with the innumerable editorial tasks involved in readying the manuscript for the printer. The exhaustive subject index is the work of Bev Anne Ross. Estelle Miller, Barbara Brown, Ralph Davis, and David Savage, of the Manuscript Processing Unit, brought the computer to bear on our manuscript preparation needs for first draft to final copy. The Academy's Office of Publications consulted on details of production and contractual publishing arrangements.

Finally, to acknowledge the support without which the entire project could not have been undertaken or completed, I express my appreciation to ERDA/DOE and to their representatives responsible for contractual arrangements with the NRC, Ray Zahradnik, Martin Neuworth, and James Batchelor.

Dr. G. Robert Yohe, co-author of Chapter 1, "The Coal Industry and Coal Research and Development in Perspective," died in August 1979. Dr. Yohe had a long and distinguished career in the field of coal chemistry. Those associated with the preparation of this volume and his colleagues were saddened to see his productive career ended.

The Steering Committee, the Technical Editor, and the NRC staff were also saddened by the premature death of Dr. James L. Johnson, author of Chapter 23, "Fundamentals of Coal Gasification." Dr. Johnson's many outstanding contributions to science and technology had already received international recognition within his lifetime. It was our pleasure to cooperate with the Institute of Gas Technology in a memorial volume containing a collection of Dr. Johnson's writings by making available for inclusion the chapter written by him for the *Second Supplementary Volume to Chemistry of Coal Utilization*.

M.A.E.

Contents

Chemistry of Coal Utilization

SECOND SUPPLEMENTARY VOLUME

CHAPTER **1**

The Coal Industry and Coal Research and Development in Perspective

MARTIN A. ELLIOTT AND G. ROBERT YOHE*

Dr. ***Martin A. Elliott*** *received his B.E. in 1930 and his Ph.D. in 1933, in gas engineering, from Johns Hopkins University, and has been active in the field of fuel technology since 1928. From 1934 to 1938 he was an engineer in the manufactured gas plant of the Baltimore Gas and Electric Company. He joined the U.S. Bureau of Mines in Pittsburgh and eventually served as Assistant Chief of the Explosives Division, Assistant Chief and Chief of the Research and Development Branch of the Office of Synthetic Fuels. Dr. Elliott went to Illinois Institute of Technology, Chicago, in 1952 as Research Professor of Mechanical Engineering and eventually served as Director of the Institute of Gas Technology and finally as Academic Vice President. In 1967 he joined the Texas Eastern Transmission Corporation as Corporate Scientific Advisor, a position he held until 1975 when he retired and became a private energy consultant. His research interests through the years include: catalysis of steam-carbon reactions, studies of the hazards of using diesel engines underground, combustion in diesel engines, neutralizing the hazards of military and industrial explosives, coal hydrogenation, coal gasification, the Fischer–Tropsch synthesis, and the long-range producibility of fossil fuels.*

Dr. G. Robert Yohe *earned his B.A. degree at Cornell College, Iowa, in 1925 and his M.S. (1927) and Ph.D. (1929) degrees in chemistry at the University of Illinois. He was a research chemist at the Goodyear Tire and Rubber Company from 1929 to 1930 and taught organic chemistry at Ohio Wesleyan University from 1930 to 1937. From 1937 until his retirement in 1972, Dr. Yohe was engaged in research on the chemical nature of coal at the Illinois State Geological Survey in Urbana, Illinois.*

* Deceased.

1.1 PURPOSE AND SCOPE

It is the purpose of this chapter to bring together in one place enough of the historical information about coal and its nature and uses to give the reader an adequate introduction to the chapters that follow. As it is the aim of this volume to give the developments in coal science since the publication in 1963 of the first supplementary volume *Chemistry of Coal Utilization,* H. H. Lowry (Ed.), this chapter summarizes the developments up to about 1960. A detailed recounting of all events and data is both unnecessary and undesirable; the reader who wishes to become acquainted with such details can find them in Volumes I and II of the original publication[1] in the first supplement,[2] and in the references cited there.

[1] *Chemistry of Coal Utilization,* H. H. Lowry, Ed., two volumes, John Wiley and Sons, New York, 1945.

[2] *Chemistry of Coal Utilization,* H. H. Lowry, Ed., Supplementary Volume, John Wiley and Sons, New York, 1963.

1.2 COAL—ITS PROPERTIES AND USES IN BRIEF

What is coal? That question was the subject of a suit in Edinburgh, Scotland, in 1853[3] when a lessor of mineral rights tried to get some of the revenue from the lucrative sale of coal to make gas. As part of his instructions to the jury, the judge said,

> To find a scientific definition of coal after what had been brought to light within the last five days is out of the question. But was it coal in the common use of the word, as it must be understood to be used in language that does not profess to be the purest science, but in the ordinary acceptation of business transactions, reduced to writing? What is coal in that sense? . . .

Now (1980), 127 years after this suit, we are still seeking the scientific definition of coal in the sense that would allow a description of

[3] John Armstrong, *Carbonization Technology and Engineering,* Charles Griffin and Co., Ltd., London, 1929, 471 pp., p. 9.

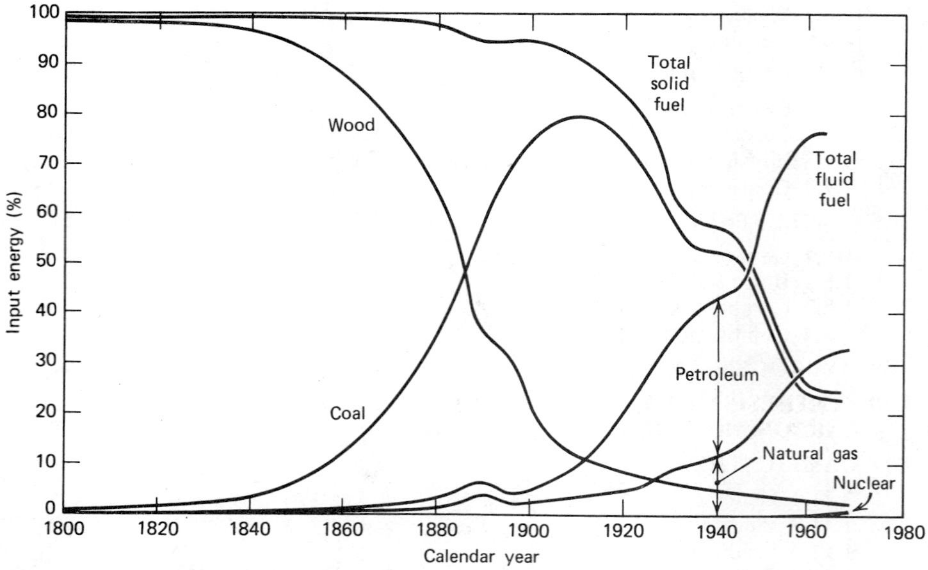

Figure 1.1 Contribution of each primary fuel to energy input, 1800–1975. Adapted from Refs. 4 and 5.

its molecular structure in extensive detail. In that time span we have learned much about the physical and chemical properties of coal so that we can do better than describe it in the sense of a "business transaction, reduced to writing." Today we can give an accurate technical definition of coal, in considerable depth if necessary. A brief version follows.

Coal is a chemically and physically heterogeneous mineral or rock consisting principally of carbon, hydrogen, and oxygen, with lesser amounts of sulfur and nitrogen. Other constituents are the ash-forming inorganic compounds distributed as discrete particles of mineral matter throughout the coal substance. Some coals melt and become plastic when heated and give off tars, liquors, and gases, leaving a residue of coke. Coals that do not melt also give off tars, liquors, and gases when heated, but leave a residue of a friable char instead of coke. Coal may be burned to generate steam (73% of U.S. domestic consumption was used by electric utilities in 1975) or it may be carbonized to produce coke (15% of U.S. domestic consumption in 1975). Coal may also be hydrogenated to produce liquid fuels, or hydrogasified to produce methane. Synthesis gases or gaseous fuels can be produced as the principal product by completely gasifying coal with oxygen and steam or air and steam. This thumbnail sketch of coal and its utilization is further amplified in this chapter and elsewhere in this book.

1.3 THE COAL INDUSTRY

Coal is the most abundant fossil fuel in both the world and in the United States, comprising about 75% of the total world and total U.S. resources of fossil fuels. Coal was the backbone of the world and U.S. energy systems until shortly after the beginning of this century. This state of affairs is clearly indicated by Fig. 1.1 which shows the contribution of each fuel source to the U.S. energy system from 1800 to 1975. The same qualitative relations apply also to the world energy system. Figure 1.1 shows the transition from an energy system based on wood to one based predominantly on coal, then to one based predominantly on the fluid fuels, petroleum and natural gas. It now appears that coal is on the threshold of once again becoming a major source of energy supply. We may also have seen the beginning of another transition with the emergence of nuclear-gener-

[4] P. C. Putnam, *Energy in the Future,* D. Van Nostrand and Co., New York, 1953, 556 pp.

[5] National Coal Association, *Coal Data,* 1975 Edition, Washington, D.C., 1976.

ated power as a contributor in the early 1960s.

Let us now examine the vicissitudes of coal utilization through the years. Coal mining is not discussed, because in spite of its relevance, its inclusion would broaden the scope of the book to the point of being unmanageable.

1.3.1 Early History

It is reported[6] that trade in coal started in England in A.D. 1215, although there is a record that coal had been given to the monks in the Abbey of Peterborough in A.D. 852 as an offering or in settlement of a claim on the people by the monastery. The latter date is of historical interest, but the former can be regarded as the beginning of the coal industry. But before that there is evidence "of the working of coal by the Britons of a so-called prehistoric time."[6] The Chinese used coal as early as 1000 B.C.[7] The Greeks and Romans knew of it around 400 B.C. but had little use for it. However in 238 B.C. Theophrastus, the favorite pupil of Aristotle, wrote of the nature, sources, and uses of coal, particularly by the smiths.[6] Finally, there is no doubt that the Romans engaged in the mining and utilization of coal in England in the early part of the Christian era.[6]

There is "a fairly clear historical account at least of the spread of coal mining in England from Newcastle to Durham and Yorkshire and finally to Lancashire, the Midlands and South Wales. . . . In 1662, £200,000 per annum was raised by the 'Hearth Tax,' imposed by Charles II on 'every fireplace or hearth in England.' "[6]

English ships were transporting coal to the European mainland in the seventeenth century, although for a time the people of the chief towns of France, Belgium, Holland, and Germany strongly objected to the use of coal. This attitude was reminiscent of an event that occurred in England during the reign of Edward I (1272–1301) when "the smoke from burning coal caused such annoyance to the prelates, nobles, and the gentry of the land that they were unable to stay in the town—because of 'the noisome smell and thick air caused by burning cole [sic].' A proclamation was therefore issued forbidding the use of coal."[6] This probably marks the beginning of the environmental problems associated with the mining and utilization of coal.

Developments in the eighteenth century laid the foundation for the "golden age" of coal, which began in the 1800s (see Fig. 1.1). Bone[8] states:

> The rise of the industrial supremacy which Great Britain enjoyed throughout the nineteenth century may be dated from the years 1730 to 1735 when the Abraham Darbys, father and son, rediscovered and revived the art of smelting iron with pit coal, which had been in abeyance since the days of Dud Dudley (1599–1684) who had originated it in 1621. The substitution of pit coal for wood charcoal as fuel in the blast furnace had an immediate and far-reaching effect upon the economic future of Great Britain and, through it, upon the welfare of mankind in general.

In fact, the increased use of coal for iron smelting and for fueling steam engines driving locomotives and supplying power to other machines was one of the important contributing factors to bringing about the Industrial Revolution. The use of coal in the blast furnace obviated the restrictions of an Act of Parliament during Elizabethan times prohibiting further expansion of iron smelting in England because of inroads on the forests.[8] For example,[9] one ton of iron could be produced with the coal from a seam of average thickness (probably 3–4 ft) and a superficial area of one square yard, whereas the older practice required wood equivalent to eight substantial trees, which would grow in a space of 500 square yards. Nef[10] estimates that 35,000 tons of metal (iron, lead, copper, and tin) may have been made annually in Great Britain, early in the seventeenth century; on the basis of this estimate, the annual

[6] James Tonge, *Coal*, Archibald Constable and Co, Ltd., London, 1907, pp. 1–4.

[7] R. Tracy Eddinger, "Coal—Nature's Black Diamond," *Prepr., Am. Chem. Soc., Div. Fuel Chem.*, **19** (4), 5 (1974).

[8] William A. Bone, *Coal and Its Scientific Uses*, Longmans Green and Co., London, New York, 1918, 491 pp.

[9] R. A. Mott, *The History of Cokemaking and of the Coke Oven, Coke Oven Managers' Association*, W. Heffner and Sons, Ltd., Cambridge, 1936, 139 pp.

[10] U. Nef, *The Rise of the British Coal Industry*, Vol. 1, London, p. 190., 1932.

production would consume the wood from 5000 acres, a very substantial woodland.

James Nielson's discovery in 1828 of the value of preheating air fed to the blast furnace[8] was the forerunner of all subsequent developments in improving fuel economy in iron smelting. This occurrence may well mark the start of the conservation ethic.

"Just as the need of pumping the mines was the chief incentive to the first pioneers of the steam engine, so the development of steam-power by British engineers during the eighteenth century greatly increased the commercial aspects of coal mining; . . . "[8] In fact, the increased production of coal was responsible for the construction of the first canal in England in 1755 and the first railroad in 1825.[8]

Nicol[11] states, "The conversion of coal into coke for use in iron smelting was first carried out on a large scale in this country [Great Britain], and in Germany about the middle of the eighteenth century." However there is a record of its use in the blast furnaces in England in 1718 and in malt drying and smiths' forges in the second half of the seventeenth century.

Coke was first made by partial combustion of small piles of coal in the hearth process,[9] in which coal was piled around a loose brick chimney that carried off the combustion gases. The "closed" oven or beehive oven was used for the distillation of wood in the seventeenth century[9] and was first used in England around 1750–1760[9] for coking coal.

In some beehive ovens arrangements were made to recover by-products, but the first practical by-product ovens were built at Sulzbach, near Saarbrucken, in 1770.[9] This development occurred more than 100 years after the Reverend John Clayton had distilled coal in a retort and obtained a "phlegm" (liquor), a black oil (tar), and a noncondensible spirit (coal gas) that could be stored in bladders and burnt.[9]

Exploitation of the illuminating power of coal gas led to the establishment of the manufactured gas industry. The first application of coal gas to lighting is credited to William Murdock, who in 1792 used coal gas produced in his backyard to light one of the rooms of his home.[8,9] In 1810 an Act of Parliament was obtained for forming a statutory company to supply coal gas to London.

The horizontally flued coke oven was the first satisfactory by-product recovery oven. One hundred of these ovens were installed at Dudweiler and Sulzbach, Saarbrucken, in 1851.[9] The gas retort and the by-product coke oven were designed to produce different end products. The development of the two types of processes went on more or less independently, with the coke oven reaching a relatively advanced stage of development in the late 1800s. This led to increased use of coke ovens in the early 1900s. It soon became apparent that it was more efficient to produce gas in the by-product coke oven than in the more labor-intensive gas retort. By 1930 there were no gas retorts operating in the United States, and less than 5% of the metallurgical coke was produced in beehive ovens (1.4% in 1974).

Developments in other fuel fields had far-reaching effects on coal conversion processes and on the demand for coal. For example, in 1855 Bunsen invented the atmospheric gas burner, which opened the way for gas markets to expand from lighting to a wide range of heating applications. A high illuminating power for gas was no longer required after Welsbach invented the incandescent mantle in 1884. The need would have been eliminated anyway in a few years when the incandescent electric light was invented, and the gas lighting load was eventually lost in the beginning of the twentieth century.

The advent of electric lighting opened up a great potential for coal combustion in the generation of steam for power generation. Thus the combustion of coal went through the stages of burning in open fireplaces in the early days, to burning in the fuel beds of small household furnaces and subsequently large industrial furnaces, and finally to burning as pulverized fuel in large central station furnaces. The last application was first tried in 1876, but was not successful in the United States until the factors affecting proper furnace design were elucidated in 1917.[12,13]

[11] E. W. L. Nicol, *Coke and Its Uses: In Relation to Smoke Prevention and Fuel Economy,* Ernest Benn Ltd., London, 1923, 184 pp.

[12] A. C. Fieldner and W. E. Rice, *Research and Progress in the Production and Use of Coal,* National Resources Planning Board, Washington, D.C., October 1941.

[13] P. Nicholls and J. F. Barkley, Section on "Combustion," Ref. 12, pp. 20–27.

As the use of coal gas developed, the production of gas for heating purposes was also developing. The first gas producer making low-Btu gas was built in 1832.[14] This principle was not widely used until it was applied by the Siemens brothers in connection with their invention of the open-hearth furnace in 1861.[14] The use of producer gas increased from that time, until at the turn of the century it was an important method for heating furnaces. But then its use declined, and in 1920 there were only 11,000 producers in use in the United States.[14]

Another development in the field of gas making stemmed from Fontana's discovery in 1780 of the production of blue gas when steam was passed over incandescent carbon.[15] This discovery was ignored until the period between 1823 and 1859, when additional experimental work was done on the steam-carbon reaction.[15] This eventually led to the development of processes for producing blue water gas between 1859 and 1875. The first successful commercial process was developed by Lowe in 1875 when he introduced the carburetted water gas set.[15] The application of this gasification technology increased until natural gas had supplanted manufactured gas.

The history of coal hydrogenation begins in 1913[16] when work on the Bergius concept of direct hydrogenation of coal under hydrogen pressure at an elevated temperature was undertaken in Germany. A 31-ton/day pilot plant was built in Germany in 1921.

The possibility of producing liquid hydrocarbons from water gas (Fischer-Tropsch synthesis) was conceived in 1913.[17] A report on the first experimental work on this synthesis, which is sometimes referred to as the indirect hydrogenation of coal, was published in 1923.[18]

Thus we come to the 1920s, when coal contributed almost 80% of the input to the U.S. energy system and nearly all the basic technology essential for the utilization of coal in a variety of applications had been developed. The further development of this technology to around 1960 is covered later in this chapter. Developments to 1975 are discussed in the remaining chapters of this book.

1.3.2 Coal Production

The consumption of world[4,19] and U.S. coal[5] in 10^{15} Btu/year is shown in Fig. 1.2. World coal consumption in this century has increased at an annual growth rate of about 1.9%, which is less than the annual growth rate of world energy consumption (3.3%). Coal consumption in the United States has not deviated too greatly from an average of about 12×10^{15} Btu/year ($\sim 500 \times 10^6$ short tons per year).

The portion of the total energy consumption (fossil fuel consumption plus hydropower) supplied by coal has decreased markedly both in the world and in the United States. Landsberg and Schurr[20] state,

> No single event or technological change can be cited to account for the relative decline in coal consumption during most of the 20th century. But two great shocks were undoubtedly the loss of the transportation field, when both railway locomotives and steamships switched rapidly to diesel power after World War II, and the more gradual elimination of coal as a household fuel.

This changing pattern in the U.S. markets served by coal is clearly shown in Figs. 1.3 and 1.4 for the period 1950–1975. The increase in the percentage of coal going to electric utilities is dramatic.

1.3.3 The Competitive Position of Coal Through the Years

Table 1.1 gives the comparative adjusted prices per million Btu at the point of production and certain points of consumption in the

[14] R. T. Haslam and R. P. Russell, *Fuels and Their Combustion*, McGraw-Hill Book Co., New York, 1926, 809 pp.

[15] J. J. Morgan, in Ref. 1, Vol. 2, Ch. 37: T. W. Stone, "History of Water Gas," presented at the 1926 Convention of the Canadian Gas Association.

[16] H. H. Storch, in Ref. 1, Vol. 2, Ch. 38.

[17] F. Fischer and R. Lessing, *The Conversion of Coals into Oils*, Ernest Benn, Ltd., London, 1925, 284 pp.

[18] H. H. Storch, N. Golumbic, and R. B. Anderson, *The Fischer-Tropsch and Related Syntheses*, John Wiley and Sons, New York, 1951, 610 pp.

[19] Martin A. Elliott and Nelson C. Turner, "Estimating the Future Rate of Production of the World's Fossil Fuels," *J. Inst. Fuel*, **47**, 55–61 (March 1974).

[20] H. H. Landsberg and S. H. Schurr, *Energy in the United States*, Random House, New York, 1960, p. 35.

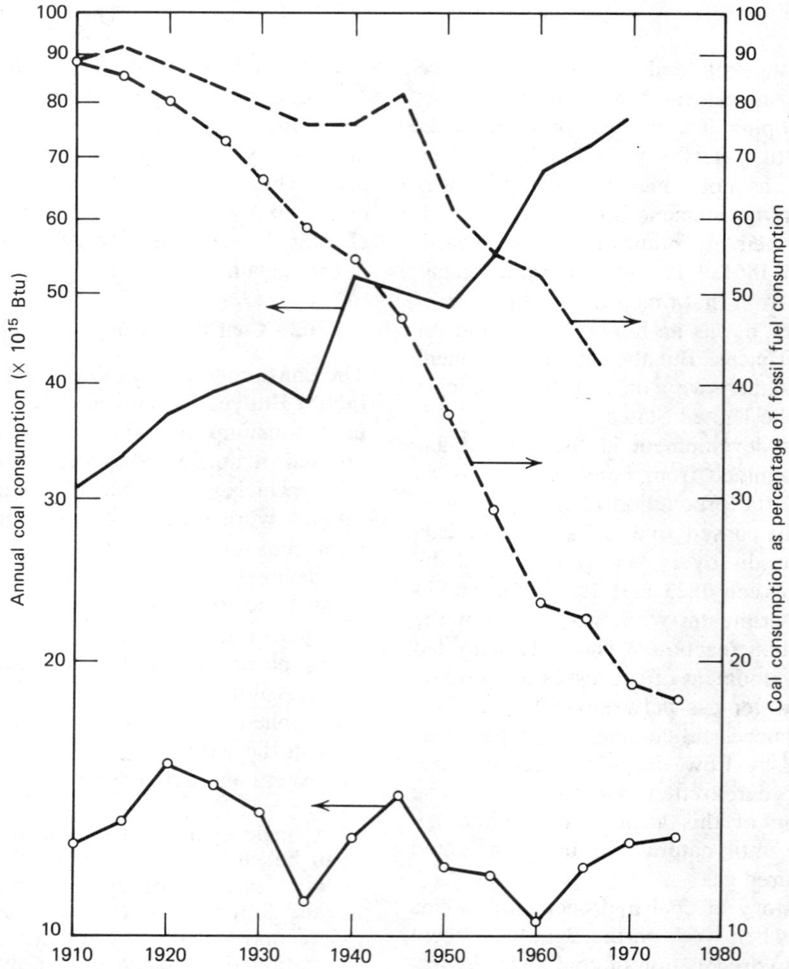

Figure 1.2 Consumption of world and U.S. (○) coal and contribution of coal to the energy system.

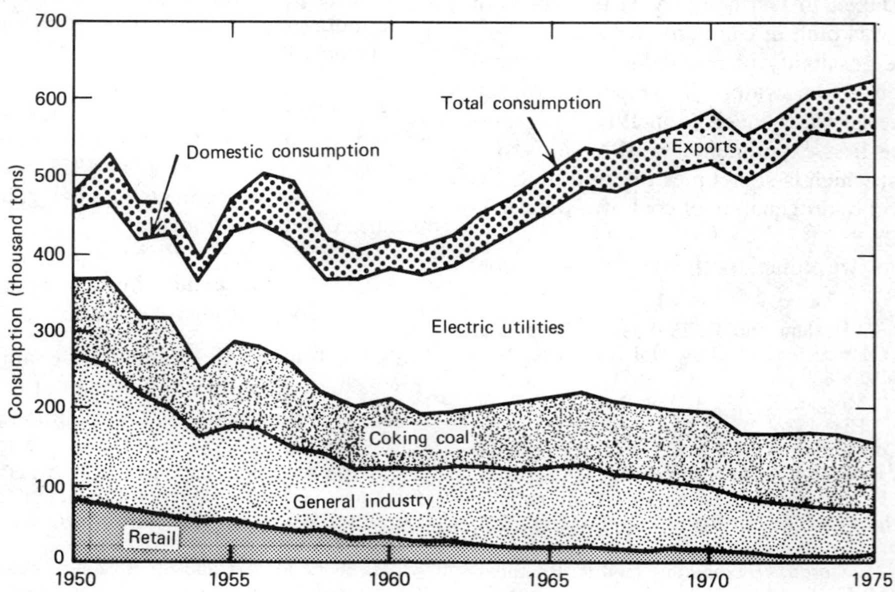

Figure 1.3 Coal consumption by category.[5]

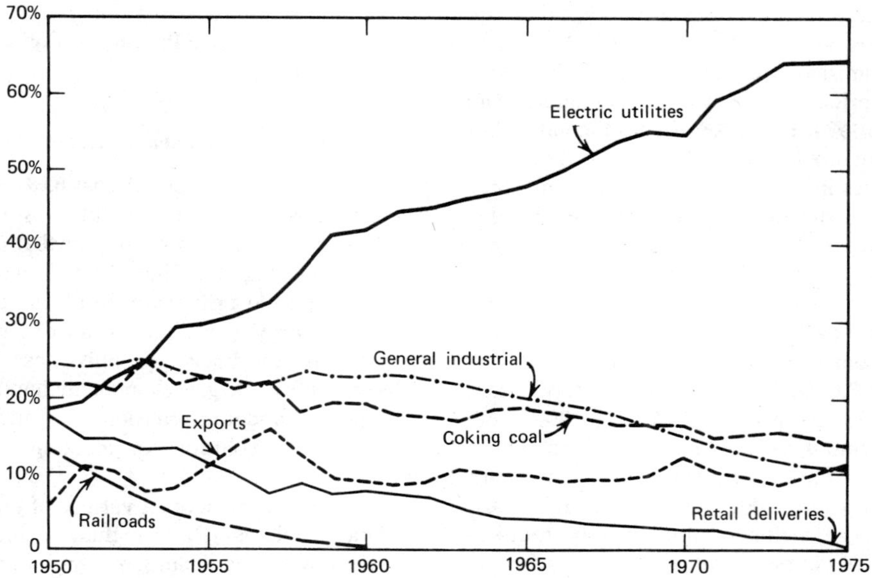

Figure 1.4 Coal consumption categories as a percentage of total consumption.[5]

Table 1.1 Average Prices of Fuels and Electricity, in the United States in Cents per MMB (All Prices in Constant 1972 Dollars)[21]

Fuel	1930	1940	1950	1960	1970	1975
Bituminous coal and lignite						
F.O.B. mines	20.5	25.1	34.5	26.6	28.2	65.5
Retail	—	93.9	100.0	96.8	—	—
Anthracite						
F.O.B. mines	63.3	54.0	65.3	45.9	47.5	99.8
"Coal," type unspecified, to power plants	—	—	50.2	37.9	34.2	67.9
Natural gas						
Wellhead	21.7	14.1	11.0	18.5	17.0	31.8
Average at point of consumption	64.1	71.0	46.7	69.5	61.9	90.1
Average at power plants	—	—	22.9	35.5	29.8	58.9
Crude petroleum at the wellhead	64.4	60.5	80.7	72.4	60.0	102.4
"oil" to power plants			59.3	49.9	43.6	157.3
Electricity						
Average for all customers	2451	2075	989	721	510	622
Dollar value index, 1972 = 100	314.3	345.0	186.5	145.6	109.5	78.6

United States for coal, natural gas, oil, and electricity, for selected years.

In spite of its premium qualities, natural gas was the lowest priced fuel at the point of production and at power plants for the periods shown. The price was low from 1930 to June 1954 because the interstate pipeline system was not extensive and gas was surplus in the producing areas. The price until the date of the *Phillips* decision,* June 1954,

[21] Dr. J. D. Parent and Dr. H. R. Linden, private communication.

* In the case of *Phillips Petroleum Company* v. *State of Wisconsin,* 347 U.S. 672 (1954) The Supreme Court extended the jurisdiction of the Federal Power Commission to natural gas sales made by independent producers to interstate pipeline transmission companies.

was a free-market price. Since that time the price has been regulated by the Federal Power Commission and has been artificially low.

The price of oil was relatively stable from 1930 to 1974. However in 1973 the nations of the Organization of Petroleum Exporting Countries imposed an oil embargo, and since that time the price of oil produced by this cartel (about 40% of world production) has been subject to periodic rises.

The average price of coal (in current dollars for a particular year) was relatively constant from 1950 to 1969, ranging $4.39–$5.08 per ton F.O.B. mines. In the successive years since 1969 the price was $6.26, $7.00, $7.60, $8.53, and finally $15.75 per ton.[5] These increases have occurred because of sharp decreases in productivity, resulting, in part, from more stringent mine safety requirements. Of course, inflation was also a factor.

The price of electricity decreased from 1930 to 1970 (see Table 1.1). This trend resulted chiefly from increases in conversion efficiencies and improvements in boiler and furnace designs. The increase in 1975 reflects higher fuel and capital costs to control sulfur oxides, oxides of nitrogen, and particulate emissions, as well as currency inflation.

1.3.4 Factors Affecting the Competitive Position of Coal Today

1.3.4.1 *Environmental Considerations*

As indicated in Fig. 1.4, almost 65% of U.S. coal is consumed by electric utilities. Coal is still the major supplier of fuel for that market as shown in Fig. 1.5; since 1965, however, it has lost ground principally to oil but also to nuclear energy as a direct result of stringent regulations on stack gas emissions. In the absence of stack gas cleanup technology in the period under discussion, low-sulfur oil was an easy solution for meeting requirements.

Coal mining presents a variety of environmental problems. These include maintaining a safe environment and working conditions in underground mines, restoring the terrain around strip mines, and prevention of stream pollution by acid mine water drainage and drainage from refuse banks.

In coal utilization we have mentioned the stack gas cleanup problem. However there are other potential air pollution problems, such as the discharge of trace elements. The

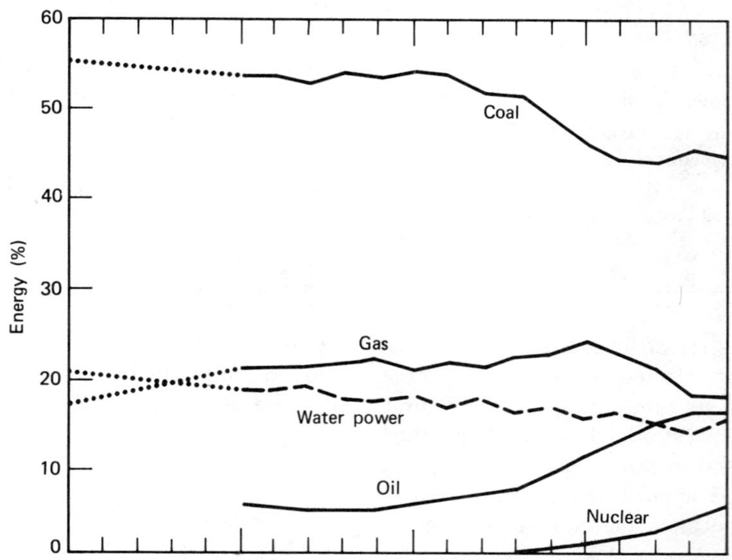

Figure 1.5 Source of electric energy production by electric utilities (percentage on a kilowatt-hour basis).[5]

solution of these problems and others is essential to the preservation of the environment, but implementation will increase the cost of coal or the cost of using coal, thus will adversely affect its competitive position in relation to fuels that present fewer environmental problems.

In addition, there are site-specific socioeconomic problems associated with the opening of new mines and the construction of coal conversion facilities. Such activities will frequently necessitate the establishment of new communities, the construction of roads, and the provision of community services. The provision of these facilities also results in such socioeconomic advantages as creation of jobs and increase of tax base.

1.3.4.2 *Factors Related to Technology*

The development of nuclear technology has added another fuel to compete with coal. The thermal reactor has already made inroads into the use of coal for power generation (see Fig. 1.4) although nuclear energy has not achieved the role predicted for it. Breeder and fusion reactors are under intensive development, and they are expected to be formidable competitors when they are commercialized. However this development work is going to take years, and in the interim coal will remain a major source of fuel for power generation. Coal will undoubtedly continue to be a significant contributor to our energy system, even after the advanced nuclear technology has been developed, because our energy requirements are so great.

Just as technological developments can increase competition of other fuels with coal, so can such developments improve the competitive position of coal. The more important areas where the development of technology can improve the competitive position of coal both economically and environmentally are mining, preparation, transportation, stack gas cleanup, fluidized-bed combustion in boilers, combined cycles, magnetohydrodynamics, solvent-refined coal, and the production of synthetic liquid and gaseous fuels.

Developments in these and other areas (except mining) are discussed in subsequent chapters. Finally, coal has a major long-term competitive advantage over other fossil fuels because it comprises about 75% of both the world and U.S. resources of fossil fuel. Inasmuch as U.S. petroleum and natural gas are already in short supply, and the same situation will exist in the world toward the end of this century, inevitably coal will come to supply a major fraction of our energy needs as it did in the first half of this century.

Coal Research and Development. As pointed out above, research and development on coal are vitally important to the economic and environmentally acceptable production and use of coal. In spite of this need, research on coal done by government in 1963 was funded at $11 million, only 3.3% of the total expenditures on energy research and only 1.8% of industry expenditures (Table 1.2). In fiscal year 1977 (Table 1.3), government research on coal increased over that done in 1963 by almost 35-fold: $380 million in the budget for the energy research categories listed in Table 1.3. Industry research in 1976 increased almost elevenfold over that done in 1963, and the percentage level increased significantly to 7.5%. The increased emphasis by the U.S. government and industry on coal

Table 1.2 Estimated[22] Expenditures for Energy Research and Development, 1963

	Expenditure[a] (thousands of dollars)	
Energy Source	Government	Industry
Coal	11,000	11,000
Oil and gas	40,000	336,000
Nuclear (fission)	210,000	90,000
Electricity:		
Generation	1,000	94,000
Transmission	250	62,000
Solar	3,000	500
Magnetohydrodynamics	9,000	4,000
Fuel cell	8,200	9,000
Thermionics	11,000	2,000
Thermoelectricity	11,000	2,600
Thermonuclear fusion	26,000	3,000
Total	330,450	614,100

[a] The figures of this table were compiled from a survey made by the energy study group. Industry numbers are order-of-magnitude expenditures, and any one may include research on another commodity.

[22] Executive Office of the President, *Energy R&D and National Progress*, Government Printing Office, Washington, D.C., 1964.

Table 1.3 Budgeted Expenditures for Energy Research and Development by ERDA for Fiscal Years 1976 and 1977 and by Industry for 1975 and 1976 (millions of dollars)[23]

	Federal Government		Industry			
			1975			1976 (Est.)
Type of Expenditure	FY 1976	FY 1977	Total	Federal	Company	Total
Energy						
Fossil fuel						
Oil	43	37	335	5	330	353
Gas	—	—	76	—[a]	—[a]	75
Shale	14	22	14	—[a]	—[a]	13
Coal						
Synthetic fuel	143	119	37	—[a]	—[b]	47
Mining			8	—[a]	—[a]	9
Other	166	261	53	15	38	60
Total	309	380	98	22	76	118
Other fossil fuel	1	1	26	—[a]	—[a]	31
Nuclear						
Fission	602	823	657	501	161	765
Fusion	250	392	42	36	6	51
Total	852	1215	699	537	161	816
Geo thermal	31	100[b]	7	—[a]	—[a]	10
Solar	115	160	18	—[a]	—[a]	35
Magnetohydrodynamics	30	37				
Conservation and utilization	75	120	54	9	45	60
All other	173	347	100	30	70	116
Basic energy sciences	210	227				
Total energy R&D	1853	2646	1427	620	807	1527
Pollution abatement						
Air			482	16	466	492
Water			71	( )	( )	76
Solid waste	206	219	22	( )	( )	( )
All other			76	23	53	74
Total pollution abatement	206	219	651	44	607	663
Grand total	2059	2865	2078	664	1414	2078

[a] Not available but included in total.
[b] Includes $50 million for loan guarantees.

research in the 1963–1977 period was a direct result of the energy crisis and the realization that it was essential to use our indigenous coal resources to a much greater extent and to develop technology for producing synthetic liquid and gaseous fuels from coal.

Tables 1.2 and 1.3 also show the research emphasis on coal's competitor, nuclear energy. Even when it is considered that pollution control technology and magnetohydrodynamics research could benefit coal, the funding for coal research is still overshadowed by that for nuclear research.

A particularly significant development of the expanded U.S. government research on coal has been the expansion of basic research and university research on coal.[24] Until comparatively recently a relatively small number of universities were doing coal research. The number active today—mostly in basic coal research—has increased dramatically, which augurs well for future advances in coal science and technology.

At the other end of the research and development spectrum is the commercialization of successful projects. Several projects in the

[23] U.S. Energy Research and Development Administration, Fiscal Year 1977 Budget to Congress—Statistical Highlights.

[24] National Science Foundation, Science Resources Highlights No. NSF 76-324, Washington, D.C., October 27, 1976.

United States are reaching the demonstration plant stage, but it will be many years before commercial plants based on this new technology can be built. However there is existing technology for commercial coal-gasification plants for producing substitute natural gas. It is extremely important to build one of these plants to get information on real-world economics and real-world materials and operating problems.

1.4 HISTORY AND EARLY DEVELOPMENTS IN THE UTILIZATION OF COAL

Most of the early developments in the utilization of coal were covered in Section 1.3.1 on the history of the coal industry. However related developments—such as the recognition of gas as a state of matter, studies of combustion and steam-carbon reactions, and the invention of the incandescent lamp—had a marked influence on developments of the utilization of coal. These are among the 99 items that follow, along with developments previously mentioned. The list was compiled by us[3,6,10,25] and supplemented by lists of significant developments submitted by authors of other chapters or by members of the Steering Committee.[26] Even though the list is completely subjective, it is a succinct summary of significant developments in coal utilization and related technologies; developments are listed chronologically and given consecutive numbers.

1.	Coal became a commercial commodity	1215
2.	Coke made from coal	1587
3.	Recovery of tar in the pile-coking process	1600
4.	Gas recognized as a state of matter	1620
5.	Coal used to smelt iron	1621
6.	Coke made by the hearth process	1640
7.	Beehive oven with tar recovery in wood distillation	1657
8.	Combustibility of coal mine gas (firedamp) discovered	1659
9.	Laboratory distillation of coal	1660
10.	Steam engine invented	1705
11.	Coke used in blast furnace	1718
12.	Combustion reactions studied	1760
13.	First practical by-product coke oven	1770
14.	Steam-carbon reactions studied	1780
15.	Desulfurization of coal gas with limestone	1785
16.	Coal gas produced for lighting	1792
17.	Coke ovens with openings on both ends	1798
18.	Water washing of coking coals	1810
19.	First commercial distillation of coal tar	1818
20.	Briquetting process for coking low-grade coals	1820
21.	Spreader stoker invented	1822
22.	Discovery of benzol in coal gas	1825
23.	Preheated air to blast furnace	1828
24.	Pulverized coal used in firing brick kilns and glass furnaces	1831
25.	First gas producer	1832
26.	Preheated combustion air to coke ovens	1832
27.	Underfeed stoker invented	1833
28.	Traveling grate stoker invented	1834
29.	First research on use of coal mixtures in coking	1835
30.	Creosote from coal tar used to preserve railroad ties	1838
31.	Sole-heated coke ovens invented	1839
32.	Isolation of benzene from coal-tar naphtha and conversion to nitrobenzene and aniline	1845
33.	Laming's material used for gas purification	1847
34.	First continuous hydraulically operated coal-charging machine for coke ovens	1848
35.	First reverberatory furnace	1851
36.	Invention of horizontal-chamber coke oven with vertical flues	1851
37.	Invention of horizontal-chamber coke oven with horizontal flues	1852
38.	Bunsen burner invented	1855

[25] M. A. Elliott, H. R. Batchelder, H. W. Nelson and G. A. Mills, *Fuels Chemistry —A Mid-Century Perspective*, Battelle Energy Program Report, Columbus, Ohio, November 1974.

[26] H. E. Benson, A. W. Deurbrouck, R. T. Eddinger, T. F. Edgar, W. Eisenhut, R. H. Essenhigh, L. S. Galstaun, T. L. Thompson, D. McNeil, W. T. Reid and L. E. Swabb, Jr., private communications.

39. Synthesis of mauveine and the beginning of the dyestuffs industry 1856
40. Invention of regenerative heating 1856
41. Heavy media washer developed 1858
42. Water gas process invented 1859
43. Open hearth process invented 1861
44. Coke oven with regenerators 1863
45. Underground gasification of waste and slack coal in mine was suggested 1868
46. Desulfurization of gas with ammonia water 1868
47. Experiments on use of pulverized coal in steam generation and in metallurgical furnaces 1869
48. Incandescent electric light invented 1879
49. Lux-Masse (iron oxide) used for gas purification 1881
50. Welsbach mantle invented 1884
51. Recuperative coke oven invented 1886
52. Discovery of coumarone 1890
53. Concentrating table invented 1890
54. Pumping of coal slurry patented 1891
55. Baum jig invented 1892
56. First briquetting machine 1897
57. First coke oven with silica walls 1900
58. Synthesis of methane from carbon monoxide and hydrogen 1902
59. Direct process for ammonia recovery 1903
60. Multiple retort underfeed stoker 1904
61. Indirect process for ammonia recovery 1904
62. Individual or cross-regenerative coke ovens 1904
63. First patent on underground gasification of coal 1909
64. Slagging gas producer 1912
65. Synthetic ammonia produced by the Haber-Bosch process 1913
66. Coal hydrogenated (Bergius process) 1913
67. Fischer-Tropsch synthesis conceived 1913
68. Coal pipeline constructed in England 1914
69. Pulverized coal firing in industrial plants 1914
70. Scientific studies of combustion in fuel beds 1916
71. Chance sand cone invented 1917
72. Pulverized coal firing in central station power plants 1917
73. Froth flotation of coal 1920
74. Application of petrography to coal 1920
75. Pott-Broche donor solvent extraction process invented 1925
76. Gasification at elevated pressures 1926
77. Slagging boiler furnaces 1929
78. Suspension (entrained) gasification of coal 1931
79. First experimental installation for preheating coal fed to coke oven 1931
80. Use of fluid heat carriers in coal gasification 1931
81. Fluidization invented 1931
82. First work on underground gasification of coal and lignite 1933
83. Hydrogasification invented 1936
84. Cyclone coal washer invented 1945
85. Cyclone furnace invented 1947
86. Pressurized fluid bed hydrogenation demonstrated in laboratory 1948
87. Carbon dioxide acceptor process studied 1949
88. Suspension gasification of coal at elevated pressure 1950
89. Use of solid heat carriers in coal gasification 1952
90. Fluid bed coal dryer developed 1955
91. Use of slag as a heat carrier in coal gasification 1956
92. First multistation long-distance coal pipeline 1957
93. Modern steam-iron process 1958
94. Electrogasification 1959
95. Introduction of unit train concept 1960
96. Molten salt process for coal gasification 1964
97. Molten iron process for coal gasification 1965
98. Combined cycle plant operated 1974
99. Fluidized-bed combustor for central station power plants 1976

The consecutive number of developments from this list are shown in relation to time on a semilog plot in Fig. 1.6. The number of developments increases exponentially with time at an annual average growth rate of

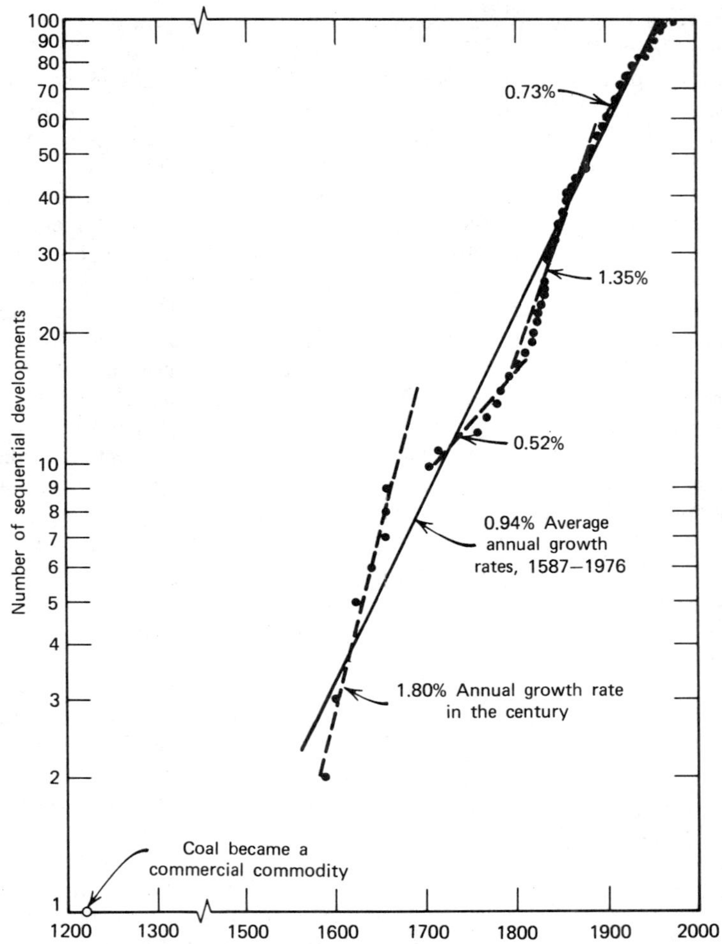

Figure 1.6 Sequential developments in coal utilization and related technologies, 1587–1976.

about 1%. The relation shown indicates that in the past the number of developments has doubled in about 69 years. Figure 1.6 shows also annual growth for individual centuries. Except for the very early developments, the nineteenth century was the most productive in new developments in coal utilization.

From our previous discussion it is evident that the technology for burning and gasifying coal and for converting it to coke had been developed by the beginning of the present century. However at that time the scientific study of coal was in its infancy. Great Britain, Germany, the United States, and France were the most active countries in coal research in the early 1900s.[27] Coal research, as measured by fuel abstracts per capita, increased in these countries and in Russia and Japan in this century, with Russia increasing its contributions significantly.[27]

Mine safety organizations played a major role in coal research at the beginning of this century, and they have continued this activity throughout the years. Notable among these and other coal research organizations are the Safety in Mines Research Board, the British Coal Utilization Research Association and the Fuel Research Board in Great Britain. The Kaiser Wilhelm Institute and the Bergbau Forschung in Germany, Charbonnages de France, the Dutch State Mines, and the U.S. Bureau of Mines (USBM), the U.S.

[27] Report of the Panel on Fuel Science and Technology of the Committee on Mineral Science and Technology, Division of Engineering, National Research Council–National Academy of Sciences, Washington, D.C., 1969.

Geological Survey, and the Illinois Geological Survey in the United States.

Until recently comparatively few universities were doing coal research, either abroad or in the United States. However because of increased government funding for coal research and development and because of the energy crisis, the interest of U.S. universities in coal research has increased markedly since 1973.

The urgency for making greater use of our indigenous coal resources has been responsible for increases in government funding of coal research and development. Thus we will see the continuous increase in significant developments, just as in the past. The acceleration of coal research and development has brought new people into the field. It is hoped that this supplemental volume to *Chemistry of Coal Utilization* will be helpful to them. It is also hoped that the forthcoming developments and the interest in utilizing coal will again make coal a major contributor to our energy system.

1.5 ORIGIN (COALIFICATION), PETROGRAPHY, AND CLASSIFICATION

Our knowledge of the origin,[28,31] petrography,[29,31] and classification[30,31] of coals for the time covered by this chapter has been presented so fully that it is quite unnecessary to report the details of these topics here. Only brief summaries are given.

1.5.1 Origin

For all practical purposes there is universal agreement that coals originated through the accumulation of plant debris that was later covered, compacted, and changed into the organic rock that we find today. "[As early] as 1598, Matthiolus, in correspondence with Klein, remarked on the origin of brown coal from wood."[32]

Most bituminous coal seams were deposited in swamps that were regularly flooded with nutrient-containing water that supported abundant peat-forming vegetation. The lower levels of the swamp waters were anaerobic and acidic; this environment promoted certain structural changes and biochemical decompositions of the plant remnants. This microbial and chemical alteration of the cellulose, lignin, and other plant substances, and later the increasing depth of burial, resulted in a decrease in the percentage of moisture and a gradual increase in the percentage of carbon content of the mass. This change from peat through the stages of lignite (brown coal), bituminous coal, and ultimately to anthracite (the process called "coalification") is characterized physically by decreasing porosity and increasing gelification and vitrification. Chemically there is a decrease in "volatile matter" content, as well as an increase in the percentage of carbon, a gradual decrease in the percentage of oxygen, and ultimately—as the anthracitic stage is approached—a marked decrease in the percentage content of hydrogen.

In common terminology the progressive changes involved in the coalification process are called an "advance in rank" of coal.

Some investigators have held that high pressures exerted by massive overburdening strata have promoted the chemical changes of coalification as well as the physical compaction. Others, however, believe that the time-temperature factors have been predominant, and that it is probably the effect of increasing temperature at increasing depth in the earth's crust that accounts for the general truth of Hilt's rule,[33] which states that in any region, coals of greater depth of burial will have greater fixed-carbon content (i.e., will be of more advanced rank).

Although most coals apparently formed where the parent plants grew and died, there are some deposits—especially many of the

[28] T. A. Hendricks, in Ref. 1, Vol. 1, Ch. 1, pp. 1–24.
[29] G. H. Cady, in Ref. 1, Vol. 1, Ch. 3, pp. 86–131.
[30] H. J. Rose, in Ref. 1, Vol. 1, Ch. 2, pp. 25–85.
[31] B. C. Parks, in Ref. 2, Ch. 1, pp. 1–34.
[32] M. C. Stopes and R. V. Wheeler, *Monograph on the Constitution of Coal,* HMSO, London, 1918, p. 1.
[33] C. Hilt, *Z. Verh. Dtsch. Ing.,* **17**, 194–202 (1873).

Gondwana coals (most coal seams in India, eastern Australia, and southern Africa)—that are not easily explained by this model. The Morwell brown coal of Australia is an excellent example: seven beds have been drilled, showing a total of 781 ft of coal, with the thickest seams being about 228 and 266 ft.[34] This tremendous accumulation is thought to have been brought about by water transportation of vegetable debris that for some reason was deposited in an area of some 50 square miles in what is now the Latrobe Valley.

Despite good general agreement among scientists about the origin and metamorphosis of coal, there have been a few controversial points. One of these involved the question of whether the cellulose or the lignin of the plant tissues became the principal coal-forming material. One school, typified by Franz Fischer and his collaborators (see Chapter 8), believed that the existence of benzene ring structures in coal pointed definitely to lignin as coal's progenitor, and this idea was further strengthened by the common observation that cellulose is more susceptible to destruction by rot than is lignin. Bergius[35] and Berl and his co-workers[36] concluded from their experiments on artificial coalification of lignin and cellulose that both these materials could and probably did contribute to the formation of coal. Yohe believes that most investigators consider lignin to have been the more important contributor to the coal substance, but no one is in a position to maintain that cellulose (and other plant ingredients) had no part in it.

Another controversy has been in regard to the origin of "mineral charcoal" or fusain, the friable, dust-forming ingredient that is found in most banded coals (see Ref. 28, pp. 19–21; Ref. 31, pp. 4–5). One group believed that this material, which resembles charcoal in appearance and properties, actually was charcoal that formed when fires burned in the forests or in the peaty accumulations of plant material. Others thought that some type of chemical or biochemical change occurred in the woody tissues prior to burial; this may have been a dry-rot kind of bacterial attack.

There is general agreement that the fusinization of woody fragments occurred very early in the coal formation process, and the preponderant opinion during the time involved in this review was that bacterial or chemical change other than fire was involved.

During the 1950–1960 decade it was noted that some coals contained appreciable amounts of uranium and that these coals evidently were of somewhat higher rank than might be expected from their age and geologic environment,[37] although low-rank coals seemed most likely to have relatively high uranium contents.[38] Breger[39] noted that high uranium contents were accompanied by decreases in organic hydrogen, volatile content, and Btu values for coalified wood, and suggested that dehydrogenation by alpha-particle bombardment might lead to such changes. Similar effects were reported for European coals that contain radioactive minerals.[40,41] Cross-linking as well as loss of hydrogen may have been brought about by the exposure to this radioactivity.

1.5.2 Petrography

1.5.2.1 *Beginning of Coal Petrography as a Science*

The banded or layered appearance of almost all coals has long attracted the attention of interested scientists, leading them to inquire into the differences between the bright bands and the dull layers and to wonder about the source materials and/or the coalification processes that led to their formation.

[34] W. A. Bone and G. W. Himus, *Coal, Its Constitution and Uses*, Longmans, Green and Co., London, 1936, p. 64.

[35] F. Bergius, *Naturwissenschaften*, **16**, 1–10 (1928).

[36] E. Berl and A. Schmidt, *Ann.*, **493**, 97–152 (1932); E. Berl and H. Keller, *Ann.*, **501**, 84–106 (1933).

[37] S. Ergun, W. F. Donaldson, and I. A. Breger, *Fuel*, **39**, 71–7 (1960)

[38] I. A. Breger, *Econ. Geol.*, **53**, 823–841 (1958).

[39] I. A. Breger, *Proc. 3rd Conf. Origin Constitution Coal*, Crystal Cliffs, Nova Scotia, June 20–22, 1956, pp. 356–380.

[40] M. Teichmüller and R. Teichmüller, *Geol. Mijnbouwkd.*, **20**, 41–66 (1958).

[41] E. Stach, *Brennstoff-Chemie*, **39**, 329–331 (1958).

Although White and Thiessen in their comprehensive bulletin *The Origin of Coal*[42] reported in detail on the microscopic examination of coals, the systematic organization of this kind of information, which constituted the real beginnings of coal petrography as a science, occurred in two important publications in 1919 and 1920.

The first report was published in England by Marie C. Stopes.[43] On the basis of a study of British bituminous coals, Stopes named four ingredients that were distinguishable without the aid of magnification: (1) fusain, the friable, charcoal-like "mother of coal"; (2) durain, the dull, compact, hard or tough layers; (3) clarain, the bright, often finely laminated material; and (4) vitrain, the bright, apparently structureless, glasslike portion that breaks with a conchoidal fracture. She pointed out that these ingredients differed from each other in their behavior on treatment with alcoholic potassium hydroxide and on oxidation with nitric acid followed by aqueous alkali treatment: the latter resulted in almost complete dissolving of vitrain to give a dark brown solution, less complete dissolving of clarain to give a lighter brown solution, still less attack on the durain, and almost no action on the fusain, which gave a nearly colorless solution.

Because the Stopes nomenclature was easily applied to macroscopic specimens, her terminology was readily adopted among coal investigators, particularly in Europe.

The second significant paper was the monumental work of Reinhardt Thiessen[44] in the United States. Thiessen gave a detailed history of the development of coal examination, citing the work and conclusions of numerous investigators in the field, starting with Henry Witham in 1831. Thiessen discussed the origin and deposition of peat in present-day bogs, the disintegration of plant materials, and the partial decay and the ultimate preservation of woody components, small and large. Many bituminous coals were described in detail, and there were photographs of coal pieces and photomicrographs of thin sections with the various ingredients labeled and identified. Thiessen's aim was to correlate the distinguishable components of coal with the plant materials from which they were derived. The bright portions derived from wood (the vitrain and the bright streaks in the clarain of Stopes) were called "anthraxylon." Dull coal was described as probably made up of microscopic fragments (megaspores, cuticles, spore exines, rodlets, anthraxylon chips, etc.), and Thiessen called these materials "attritus." The charcoal-like portion he mentioned as "mother coal," but he did not, at this time, use the term "fusain" employed by Stopes for this material.

[42] D. White and R. Thiessen, U. S. Bur. Mines Bull. No. 38, 1913, 390 pp.

[43] Marie C. Stopes, *Proc. R. Soc. London,* **90B,** 470–487 (1919).

[44] Reinhardt Thiessen, U. S. Bur. Mines Bull. No. 117, 1920, 296 pp.

1.5.2.2 *Modifications of Descriptions of Petrographic Constituents*

Both nomenclatures given in Section 1.5.2.1 were subsequently modified by their originators and by others. Thiessen found it convenient to consider attritus under two headings: that which is opaque (essentially the durain of Stopes), and that which is more or less transparent or translucent in thin section.[45] A few years later Stopes[46] revised and expanded her classification of coal ingredients and introduced the term "maceral" (supposedly analogous to "mineral" in crystalline rocks) for fragments seen under the microscope. Thus in the Stopes system the microscopic "vitrain" bands were "vitrinite," which included the macerals "ulminite" (usually woody or cortical tree tissues), "collinite" (any plant part that had been in solution and reprecipitated as a jelly), "periblinite" (cortical tissues that show some indication of cell structure), "suberinite" (corky tissues showing cell structure), "xylinite" (wood tissues), and "resinite." All these presumably have gone through a jellylike stage. The clarain of Stopes contains all these macerals plus others such as "cutinite," "exinite," and "fusinite" (cuticules, exines, fusain fragments, respectively), which are also components of durain. The complete chart given by Stopes[46] has been reproduced by Cady.[29]

Although the systems of Stopes and of Thiessen have been compared, the two sets of terms are not really equivalent, and a true correlation cannot be attained.

[45] R. Thiessen, *Trans. Am. Inst. Mining Metall. Eng. Coal Div.,* **88,** 644–672 (1930).

[46] Marie C. Stopes, *Fuel,* **14,** 4–13 (1935).

Quite naturally, similar problems were being considered in other coal-producing countries. Table 1.4 compares the nomenclature for the banded ingredients formerly used in different countries and shows their approximate relationships.[47]

Cady's detailed discussion of coal petrography[29] up to about 1940 should be examined by readers who are interested in more than a brief introduction to the topic.

1.5.2.3 *Status of Coal Petrography by 1960*

The two decades of the 1940s and 1950s brought to coal petrography new developments that have been described by Parks.[31]

The optical property of reflectance was used with much success by Hoffmann and Jenkner.[48] They made the important discovery of the correlation of the reflectance of vitrinite and the rank of the coal. The reflectance has been used by most coal petrographers since about 1947, and this property has become one of the most useful parameters for the microscopic study of coal.

This use of reflected light and of the oil immersion microscope lens was pursued vigorously in Germany, and the publication of several books[49–51] describing techniques made it possible for other workers to adopt the same methods. Laboratories for studying coal petrography were established in most of the world's coal-producing countries.

In the mid-1950s C. E. Marshall published a detailed and thoroughly documented paper, "Coal Petrology,"[52] which should be consulted by readers who want a more complete knowledge of the accomplishments during this period.

Applications of coal petrography are to be found in studies of the prediction of coke quality from petrographic data of coals and their blends, in the identification of coal seams, in oil and gas exploration, and in the evaluation of coal-cleaning processes. Schapiro and Gray[53] used a reflectance study in assessing the suitability of coals for the manufacture of metallurgical coke, and Burstlein[54] developed a process for control of the particle size of coke-oven feed based on differences in resistance to crushing of the vitrain, clarain, durain, and fusain of the coal. Partial crushing, sieving, recycling the coarser (and harder) fractions, and final recombination to give the desired blend are involved. Several patents were issued to Burstlein and his associates[55] for this process and modifications of it.

Before we leave the topic of coal petrography, it should be mentioned that the petrographic constituents undergo changes as the rank of a coal changes, and such changes are revealed in the chemical composition.

[47] E. Hoffmann, *Compt. Rend. 2me Congr. l'Avance, Études Strat. Carbonifère,* Heerlen, 1935, Vol. 1, (1937), pp. 313–325.

[48] E. Hoffman and A. Jenkner, "Die Inkohlung und ihre Erkennung im Mikrobild," *Glückauf,* **68,** 81–88 (1932).

[49] E. Stach, *Lehrbuch der Kohlenmikroskopie,* Glückauf, Kettwig, 1949, 285 pp.

[50] *Atlas für angewandte Steinkohlenpetrographie,* Glückauf, Essen, 1951, 329 pp.

[51] H. Freund, *Handbuch der Mikroskopie in der Technik,* Umschau Verlag, Frankfurt am Main, Vol. 2, 1952, 759 pp. (See especially pp. 3–63 for historical matter.)

[52] C. E. Marshall, *Econ. Geol.,* 50th Anniversary Volume, 757–834 (1955).

[53] N. Schapiro and R. J. Gray, *Proc. Ill. Min. Inst. 1960,* pp. 83–97.

[54] E. M. Burstlein, *Chal. Ind.,* **35,** 351–370 (1954); **36,** 14–28 (1955).

[55] British Pat. 694,197, 1953; U.S. Pats, 2,752-293, 1956, and 2,782,147, 1957.

Table 1.4 Comparison of the Nomenclature for Coal Types (Banded Ingredients) as Applied in Different Countries

Macerals	Germany	England		United States		France	Belgium	Holland
Fusit	*Faserkohle*	Mineral charcoal	Fusain	Mineral charcoal		*Fusain*	*Fusain*	*Fusit*
Telit	*Glanzkohle*	Bright coal	Vitrain	Bright coal	Anthraxylon	*Houille brillante*	*Charbon brillant*	*Telit*
Vitrit								*Vitrit*
Clarit			Clarain		Translucent attritus	*Houille semibrillante*	*Charbon semibrillant*	*Durit*
	Mattkohle	Dull coal		Semisplint	Opaque attritus			
Durit			Durain	Splint		*Houille matte*	*Charbon mat*	

Source: Originally published by Cady in Ref. 29. Reproduced from Ref. 1 (Vol. I, p. 99) with permission of John Wiley and Sons, Inc.

1.5.3 Classification

In view of the wide variation in the properties and compositions of coals and the many ways in which they are studied and used, it is not surprising that the problems of classifying them are complex. Each person involved would like to have the classification scheme that would be most useful to himself.

Some 30 years ago H. J. Rose made an elegant presentation of the topic,[30] and although he devoted 60 pages to it, he indicated that he had left much unsaid. Two of his comments bear repeating. "Practically every chemical and physical characteristic of coal, as well as every item of its commercial description, has been used or seriously proposed for classification purposes," and, in his conclusion: "An attempt has been made to give some attention to almost every phase of the subject and to show that there are as many methods of coal classification as there are viewpoints concerning the nature and uses of coal." The reader who wishes to learn all the details, documented by more than 300 references to the original literature, is referred to Dr. Rose's account.

Some of the more important bases suggested and/or used for classification are the following:

1. *Geologic period.* Cenozoic, Mesozoic, Paleozoic (Carboniferous and Devonian), and so on.
2. *Type.* This involves the nature and biochemical alteration of the original plant material.
3. *Grade.* This considers the particle size, calorific value, and the amount and nature of impurities.
4. *Visual features.*[56] Coarsely banded, finely banded, microbanded, nonbanded; splint, semisplint, cannel, boghead.
5. *Behavior toward organic solvents.*
6. *Behavior toward alkali extraction* (humic acid content).
7. *Adsorption of solvents, or of solutes from solution* (e.g., aqueous ferric chloride).
8. *Oxidizability* (permanganate number, yield of mellitic acid on oxidation, absorption of gaseous oxygen, weathering behavior).
9. *Behavior on hydrogenation.*
10. *Friability and grindability.*
11. *Color* ("streak test" on porcelain).
12. *Use*[57] (steam coals, gas coals, coking coals).
13. *Classification by rank on the basis of* (*a*) *Ultimate analysis:* carbon content, oxygen content, ratios of hydrogen to oxygen, carbon to hydrogen, and carbon to oxygen. (*b*) *Proximate analysis:* volatile matter (VM) or fixed carbon (FC) or FC : VM ratio. (*c*) *Calorific value,* or a combination of calorific and analytical values. (*d*) *Reflectance:* mean of random or of maximum measurements under oil immersion.
14. Sharpness of X-ray diffraction peaks.

[56] ASTM Standards, 1961, Part 8, Designation D493-39, p. 1347.

1.5.3.1 *Classification According to Use and Grade*

Although the American Society for Testing and Materials (ASTM) has reviewed specifications for gas and coking coals,[56,57] the classification of coals by use "must necessarily be left to the judgment of experienced operators."

The ASTM also set up specifications[58] for a code expressing certain quality characteristics of coals. However this specification was discontinued by ASTM in 1962 because it was used so infrequently.

1.5.3.2 *Classification by Rank*

Probably the most often used and most generally useful classification is that according to rank, which is essentially a means of designating the stage of metamorphosis a coal has reached in its slow, progressive transformation from oxygen-rich plant debris in the peat deposit to carbon-rich, oxygen-poor anthracitic coal.

As this transformation takes place, the

[57] ASTM Standards, 1961, Part 8, Designation D166-24, pp. 1235–1237.
[58] ASTM Standards 1961, Part 8, Designation D389-37, pp. 1233–1234.

following changes in properties may be noted, in general:

1. Moisture content decreases, more rapidly in the early stages, slowly in the late.
2. Percentage of carbon content increases, while percentage of oxygen decreases.
3. Percentage of hydrogen changes very little until late stages are reached; then it decreases markedly.
4. Percentage of volatile matter generally decreases.
5. Calorific value (moisture- and mineral-matter-free basis) increases, levels off, then drops slightly as the anthracitic stage is reached.

Thus neither the moisture content, the hydrogen content, nor the calorific value would be a suitable criterion for judging rank over the entire range, although each might be used for portions of it.

The percentage content of carbon (and/or oxygen) and the volatile matter (or fixed carbon) should be the most reliable bases for judging rank. Let us see what has developed.

Regnault has been given credit for performing the first reasonably accurate ultimate analyses of coal.[59,60] He concluded that coals of similar properties were similar in their analyses, and that the sum of oxygen and nitrogen percentages was a good indication of the nature of the coal. Years later Grüner[61] tabulated this information (Table 1.5) as a classification scheme that utilized both ultimate and proximate analysis data.

Later Seyler,[62] working mainly with coals from South Wales, developed his noteworthy system of classification based on ultimate analyses. This work was later extended to include data on a large number of European and British coals.[63] Despite the somewhat confusing and complex terminology, Seyler's scheme (Table 1.6) was an important step in bringing order to the classification of coal.

[59] V. Regnault, *Ann. Chim. Phys.* **66** *(2)* 337–365 (1837).

[60] V. Regnault, *Ann. Mines*, **12** (3) 161–240 (1837).

[61] L. Grüner, *Ann. Mines*, **4** (7) 169–207 (1874).

[62] C. A. Seyler, *Proc. S. Wales Inst. Eng.*, **21**, 483–526 (1899); **22**, 112–20 (1900); *Colliery Guardian*, **80**, 17–19, 80–82, 134–136 (1900).

[63] C. A. Seyler, *Fuel*, **3**, 15–26, 41–49, 79–83 (1924).

The ASTM Classification by Rank. During the 1920s and 1930s a number of investigators were working on systems that used both proximate analysis and calorific value,[64–69] and several of them (Stansfield, Thom) recognized the desirability of using a moisture-containing basis for the lower rank coals. Parr preferred the "unit coal" basis.

In 1927 the American Standards Association (now the American National Standards Institute) and the ASTM embarked on a project to devise a satisfactory and acceptable means of rank classification of coals. A large group of coal specialists from Canada and the United States worked together for some 10 years; their proposal was given "tentative standard" status in 1934, adopted as standard in 1937, revised slightly in 1938, and remained in that form through the period covered by this chapter.[70] (Some changes in the definition of ranks below low-volatile bituminous were made in 1966.) This classification made use of the moist, mineral-matter-free calorific value, supplemented by other characteristics such as weathering and agglomerating properties in the lower ranks, and depended on the dry, mineral-matter-free fixed carbon (or volatile matter) in the higher ranks.

This classification scheme has proved to be satisfactory from commercial as well as certain scientific viewpoints. See Ref. 1 of Chapter 3 for current ASTM standard. For example, if a coal is described as being in a certain rank group, some estimate of its heating value is readily made, and it may be noted that both the lowest (subbituminous) and highest (anthracitic) are indicated as "nonagglomerating" and would therefore not receive consideration as possibly coking coals. Figure 3.18 shows the approximate

[64] S. W. Parr, *J. Ind. Eng. Chem.*, **14**, 919–923 (1922).

[65] S. W. Parr, Univ. Illinois Eng. Exp. Sta., Bull. No. 180, 1928, 59 pp.

[66] S. W. Parr, *Fuel*, **16**, 52–58, 72–85 (1937).

[67] E. Stansfield, *Trans. Can. Inst. Min. Met.*, **28**, 252–266 (1925).

[68] F. Landsberg, *Glückauf*, **61**, 1427–1430 (1925).

[69] W. J. Thom, Jr., *Trans. Am. Inst. Min. Met. Pet. Eng.*, **71**, 282–288 (1925); **88** *(Coal Div.)*, 406–408 (1930).

[70] ASTM Standards, 1961, Part 8, Designation D388-38, pp. 1227–1232.

Table 1.5 The Regnault-Grüner Classification of Coals, as Revised by Bone and Himus[a]

Genus	Class	Chief Uses	Percentage Composition C	H	O + N + S	% Volatiles at 900° C	% Fixed Carbon	Character of Carbonaceous Residue
A, Lignites	Noncaking	—	60–75	~ 5.0	20–35	> 45	< 55	Noncoherent
B, Bituminous	1. Noncaking, long flame	Reverberatory furnaces	75–80	4.5–5.5	15–20	40–45	55–60	Noncoherent
	2. Caking, long flame	Gas making	80–85	~ 5.0	10–15	32–40	60–68	Very porous coke
	3. Hard coking	Coke manufacture	84–89	5.0–5.6	5.5–11.0	26–32	68–74	Dense coke
	4. Hard coking, short flame	Coke manufacture and steam raising	88–90	4.5–5.5	5.5–6.5	18–26	74–82	Very dense coke
BC, Semibituminous	Noncaking, short flame	Steam raising	90–92	4.0–4.5	4.0–5.5	15–20	80–85	Weakly caking or noncoherent
C, Anthracitic and anthracites	1. Anthracitic, noncaking	Steam raising	92–94	3.0–4.0	3.0–4.5	8–15	85–92	Pulverent
	2. Anthracites, noncaking	Domestic and central heating; malting kilns				< 8	> 92	Pulverulent

[a] All numerical data refer to the dry ashless coal.

Source: Reproduced from W. A. Bone and G. W. Himus, *Coal: Its Constitution and Uses*, Longmans, Green and Co., London, 1936, p. 41, with permission of Longman Group, Ltd.

Table 1.6 Seyler's Classification of Coals

Carbon Plane	Anthracitic: over 93.3%	Carbonaceous: 93.3–91.2%	Bituminous Meta-: 91.2–89.0%	Bituminous Ortho-: 89.0–87.0%	Bituminous Para-: 87.0–84.0%	Lignitous Meta-: 84–80%	Lignitous Ortho-: 80–75%
Perbituminous genus			Per-meta-bituminous: North Country steam coals	Per-ortho-bituminous: cannels	Per-para-bituminous: cannels	Perlignitous	
H, > 5.8% Vol, 30–44%			> 5.7 30–44	> 5.7 Seldom < 36	> 5.8 Seldom < 40	> 5.8 31–57	
Bituminous genus		(Pseudo-bituminous species)	Metabituminous: Welsh coking coals	Orthobituminous: Durham coking coals, smithy and gas coals	Parabituminous: many of the best gas coals, coking coals, and free-burning steam coals	Lignitous	
H 5.0–5.8%; Vol,—		5.0–5.8 > 23	4.9–5.7 23–30	5.0–5.7 23–30	5.0–5.8 30–40	4.7–5.8 31–57	
Semibituminous genus		Semibituminous species (ortho-semi-bituminous): generally coke; caking index increases with hydrogen content; steam coals	Sub-meta-bituminous: continental coking coals	Sub-ortho-bituminous: best Westphalian coking coals	Sub-para-bituminous	Sublignitous	
H; 4.5–5.0% Vol, 16–23%		4.45 14–24	4.5–4.9 1·6–23	4.5–5.0 16–23	5.0 and under 16–29		

Table 1.6 (*cont.*)

Carbon Plane	Anthracitic: over 93.3%	Carbonaceous: 93.3–91.2%	Bituminous Meta-: 91.2–89.0%	Bituminous Ortho-: 89.0–87.0%	Bituminous Para-: 87.0–84.0%	Lignitous Meta-: 84–80%	Lignitous Ortho-: 80–75%
Carbonaceous genus	Semianthracitic species: dry (noncoking) steam coals	Carbonaceous species (ortho-carbonaceous): Welsh smokeless steam coals, slightly coking	Pseudocarbonaceous (sub-meta-bituminous): steam coals	Pseudocarbonaceous (sub-ortho-bituminous)	Pseudocarbonaceous (sub-para-bituminous)	Sublignitous	
H, 4.0–4.5%	> 4	4.2–4.5	3.7–4.5	> 4.5			
Vol, 10–16%	9–15	10–14	10–16	> 16			
Anthracitic Genus	Orthoanthracite: true anthracites	Pseudoanthra - cite (subcarbonaceous): dry steam coals; bastard anthracites	Pseudoanthracitic (sub-meta-bituminous)	Pseudo-thracitic (sub-ortho-bituminous)	Pseudoanthracitic: (sub-para-bituminous)		
H, < 4.0%	< 4.0	< 4.2	< 3.7				
Vol, 10%	5–9	< 7.7	< 10				

Source: Reproduced from W. A. Bone and G. W. Himus, *Coal: Its Constitution and Uses*, Longmans, Green and Co., London, 1936, p. 42, with permission of Longman Group, Ltd.

correlation between different rank-related parameters.

1.6 CHEMICAL AND PHYSICAL PROPERTIES

1.6.1 Chemical Analysis

Since coals vary widely in composition and properties, and are used under widely varying conditions, various kinds of analytical and testing data are of scientific, technical, and commercial interest.

The ASTM was founded in 1898 as a scientific and technical organization for "the development of standards on characteristics and performance of materials, products, systems, and services; and the promotion of related knowledge." The establishment of standard methods for analyzing and testing coal and coke has been among the many activities of ASTM over the years, and these standard methods are frequently employed in the United States and in many other countries. The detailed procedures that comprise these methods are constantly scrutinized, frequently brought up to date, and published annually.[71] Because this chapter gives the status of coal studies up to about 1960, the 1961 edition of the Standards is cited here. Part 26 of the 1975 edition gives methods for coal and coke. In each annual edition the standards are generally listed in alphanumeric order of designation number for each main Technical Committee of ASTM.

Other countries, too, have developed standard methods for analyzing and testing coal. See, for example, the publications of the British Standards Institution (BSI), the Association Française de Normalisation (AFNOR), and the Deutscher Normenausschuss (DIN).

As coal became a commodity more and more heavily involved in world trade, the desirability of using the same standard methods everywhere became increasingly apparent, and after World War II the International Organization for Standardization (ISO), through its Technical Committee 27 on Solid Mineral Fuels, took up this task. By 1960 much progress had been made in studying and selecting methods to recommend as possible standards.

The status of the chemical analysis of coal in 1960 has been described by Ode[72] and is not given in detail here, except for a very brief description of ultimate analysis and determination of calorific value, proximate analysis, and the fundamental nature of coal (chemical constitution). Such items as grindability and plasticity are included in Section 1.6.6, "Physical Evaluation."

1.6.1.1 *Ultimate Analysis*

Ultimate analysis is the determination of the percentage composition in terms of carbon, hydrogen, sulfur, nitrogen, and ash, and the calculation of the oxygen content by difference. Determination of the moisture content is not by definition a part of ultimate analysis, but it must be done so that the "as determined" values can be calculated to the "as received," "dry," or other bases (see Ref. 71, pp. 1231, 1267–1268 for formulas). The methods of ultimate analysis were well developed many years ago, and no significant changes were made during the later years of the period covered by this chapter. Semimicro methods have been developed and used for the determination of nitrogen.[73,74] Another modification in this analytical procedure[75] has not been widely used.

1.6.1.2 *Ash*

Ash is the residue remaining when the coal sample (normally about 1 g) is ignited carefully under specified conditions (see Ref. 71, pp. 1244–1245) so that complete burning of the organic matter is attained without mechanical loss of material due to too rapid expulsion of volatile matter. It is usually done with the sample in a crucible heated to 700–750°C for coal, or 750–950°C for coke.

[71] See, e.g., ASTM Standards, 1961, Part 8, pp. 1227–1367.

[72] W. H. Ode, in Ref. 2, Ch. 5, pp. 202–231.

[73] A. E. Beet and R. Belcher, *Fuel,* **17,** 53–55 (1938).

[74] AFNOR, French Standard No., M03-018, 1954; Deutscher Normenausschuss, DIN 51722, 1954; BSI, British Standard No. 1016, Part 6, 1958.

[75] W. Mantel and W. Schreiber, *Glückauf,* **74,** 939–945 (1938).

1.6.1.3 *Oxygen*

The direct determination of oxygen has been under investigation for many years; oxidation methods involving a known excess of oxygen at the start,[76,77] and reduction procedures with hydrogen[78–80] or with carbon (the Schutze-Unterzaucher method[81,82] have been used.

The ASTM standards of 1961 indicate that none of the direct methods has been found to be satisfactory and recommend determining oxygen by difference: 100 − (% H + % C + % N + % S + % moisture + % ash).

1.6.1.4 *Calorific Value*

Selvig and Gibson[83] have given a full description of the determinations of the calorific value of coal, starting with the 1843 method of Berthier[84] (in which the heating value was considered to be proportional to the weight of a lead button obtained by heating 1 g of coal with litharge), continuing through the constant-pressure oxygen stream combustion, the potassium nitrate–potassium chlorate calorimeter of Lewis Thompson, the Parr sodium peroxide calorimeter, and constant-pressure oxygen combustion, and finally describing the constant-volume combustion with oxygen under pressure in a bomb. The oxygen bomb method was the ASTM standard in 1945, and was still the standard in 1961 (Ref. 71, pp. 1262–1267). This constant-volume calorimetry gives the "gross calorific value," in which water exists as a liquid at the completion of the process. The "net calorific value" considers that all water in the products exists as vapor at constant (atmospheric) pressure, and is obtained by subtracting 572 cal/g (1030 Btu/lb) of water formed from a unit weight of the fuel, from the gross value (see Ref. 71, p. 1348).

Early studies of the calculation of calorific value from the data of proximate analysis were reported in 1902 by Goutal,[85] who based his formula on fixed carbon, volatile matter, and a function of the moisture- and ash-free (maf) volatile matter. During the 1920s and 1930s there was a great deal of activity in devising and revising such formulas; Selvig and Gibson[83] give nearly a dozen such equations each for ultimate and for proximate analysis data. Typical of the well-known formulas are Dulong's for ultimate analysis (Btu/lb = $14{,}544C + 62{,}028\,(H - O/8) + 4050S$, where C, H, O, and S are the contents of carbon, hydrogen, oxygen, and sulfur expressed as decimal fractions), and Goutal's[85] for proximate analysis (cal/g = $82C + aV$, where C is fixed carbon, V is volatile matter in percent, and a is a factor that varies with V', the volatile matter content on the maf basis). The value of V' is found by

$$V' = \frac{100V}{C + V}$$

Goutal's relationship of a with V' was given:

V'	5	10	15	20	25	30	35	38	40
a	145	130	117	109	103	98	94	85	80

The relationship between the English and metric units for calorific values is cal/g × 1.8 = Btu/lb.

1.6.1.5 *Proximate Analysis*

Proximate analysis involves determining—in addition to moisture, ash, and sulfur—only the volatile matter and the fixed carbon.[86] The precision of the results obtained in proximate analysis depends on strict adherence to the empirical procedures; thus the direc-

[76] M. Dolch and H. Will, *Brennstoff-Chemie,* **12,** 141–146, 166–169 (1931).

[77] W. R. Kirner, *Ind. Eng. Chem., Anal. Ed.,* **8,** 57–61 (1936).

[78] H. ter Meulen, *Rec. Trav. Chim. Pays-Bas,* **43,** 899–904 (1924); **53,** 118–225 (1934).

[79] R. Wildenstein, *Chal. Ind.,* **32,** 293–298 (1951).

[80] H. Guerin and P. Marcel, *Bull. Soc. Chim. Fr.,* **1952,** 310–311.

[81] M. Schütze, *Z. Anal. Chem.,* **118,** 245–258 (1939).

[82] J. Unterzaucher, *Ber. Dtsch. Chem. Ges.,* **73B,** 391–404 (1940). See Ref. 2, p. 220, for numerous other references describing improvements of this method.

[83] W. A. Selvig and F. H. Gibson, in Ref. 1, Vol. 1, Ch. 4, pp. 132–144.

[84] P. Jarrier, *Fuel,* **8,** 486–491 (1929). Berthier's method is described in "notes by translator."

[85] M. Goutal, *Compt. Rend.,* **135,** 477–479 (1902); *J. Soc. Chem. Ind.,* **21,** 1267 (1902).

[86] S. W. Parr, *The Analysis of Fuel, Gas, Water and Lubricants,* McGraw-Hill Book Co., New York, 1932, Ch. 24, pp. 208–217. See also Ref. 71, pp. 1243–1247.

tions for the standard methods must be followed conscientiously by the analyst.

The data of proximate analysis may be of more value than those of ultimate analysis for judging the suitability of a given coal for a specific use: for example, the manufacture of metallurgical coke.

1.6.2 The Chemical Constitution of Coal

In the early years of organic chemistry, the discovery of isomerism (that two or more different compounds may have the same percentage composition and molecular weight) led to investigations of molecular structure. Chemists strove to understand the properties of carbon compounds in terms of the manner and order in which the constituent atoms were linked together.

Solving the problems of structural organic chemistry proceeded reasonably well as long as studies were limited to the rather low molecular weight compounds that were available in the pure state. But complex mixtures and compounds of high molecular weight gave much greater difficulty.

Inevitably, investigators raised the question, "What is the chemical constitution, the molecular architecture of coal?" And in that question, organic chemists have found one of their most difficult and challenging problems. This is because all coals differ, one from another; all are mixtures not readily separated into their constituents; and it is apparent that most if not all of these constituents are of high molecular weight.

The fact that a substance has a high molecular weight does not, in itself, preclude the gaining of knowledge of its structure. Rubber $(C_5H_8)_n$ is known to be a polymer of isoprene (methylbutadiene), and a great deal is known about its constitution and configuration. Cellulose $(C_6H_{10}O_5)_n$ is a "condensation polymer" of glucose, $C_6H_{12}O_6$, and it too has been well characterized. The key to solving the problems of chemical constitution in these cases has been the successful "depolymerization" and identification of the monomers or building units that are repeated over and over in the polymer.

But coal is not so accommodating. Attempts to depolymerize it by pyrolysis, hydrolysis, or other chemical treatment have never yielded an identifiable monomer, nor even a preponderance of simple structures that might be considered as building units in a copolymerization process. The idea of a repetitive structural group has been proposed,[87] but satisfactory evidence has not been found.

What, then, can be done to elucidate the chemical structure of coal? The answer lies in finding out as much as possible about the carbon skeleton, or the manner in which the carbon atoms are linked together; in learning how other atoms (mainly hydrogen, oxygen, nitrogen, and sulfur) are attached to or built into this carbon skeleton; in identifying and measuring the amounts of identifiable functional groups; and in studying the manner in which these values vary from coal to coal or change as the coal increases in rank.

1.6.2.1 *The Carbon Skeleton*

The data of ultimate analysis reveal little about the architecture of the molecules of coal substance, but they do make it possible to deduce what these molecules cannot be. Since bituminous coals usually contain about 5% hydrogen (see Table 1.7), we know, of course, that they are not predominantly composed of such saturated structures as aliphatic or alicyclic hydrocarbon derivatives, because the limiting composition of these (CH_2) contains more than 14% hydrogen. Neither are they extensive condensed-ring aromatic networks; coronene (Fig. 1.7) contains only 4% hydrogen, and similar structures made up of still more rings would contain an even lower percentage of hydrogen.

The idea that coal was an impure form of elemental carbon persisted in places well into the present century, even though more than 100 years ago Berthelot[89] concluded from his experiments on reduction with hydriodic acid that coal consisted of definite compounds of carbon rather than elemental carbon itself.

The importance of benzenoid ring units in the makeup of coal structures has been indicated by the results of various chemical and physical approaches, including oxida-

[87] Paul Erasmus, *Über die Bildung und den chemischen Bau der Kohlen,* Enke, Stuttgart, 1938, p. 10. See also W. Fuchs, *Die Chemie der Kohle,* Julius Springer, Berlin, 1931, pp. 325–326.

[89] M. Berthelot, *Ann. Chim. Phys.,* **20** (4), 530–531 (1870).

Table 1.7 Analytical Data for Some Coals of Different Ranks

Coal	Seam	County and State	As Received: Moisture (%)	As Received: Ash (%)	Ultimate Composition, maf (%): C	H	N	O	S	Methoxyl	Calorific Value (cal/g[a])
Lignite	Coteau	Ward, N.Dak.	32.5	5.0	72.38	5.30	1.12	20.53	0.67	2.22	4477
Subbituminous B	Unnamed	Weld, Col.	20.7	3.9	77.69	5.24	1.60	14.98	0.49	0.25	5818
High-volatile C bituminous	5	Fulton, Ill.	13.7	10.3	81.41	5.64	1.56	7.89	3.50	Trace	7005
High-volatile B (vitrain) bituminous	6	Franklin, Ill.	5.4	2.2	81.96	4.98	1.65	10.23	1.18	Nil	7611
High-volatile A bituminous	Pittsburgh	Allegheny, Pa.	1.7	7.8	82.92	5.91	1.80	6.34	3.03	—	8352
Medium-volatile bituminous	Sewell	Fayette, W. Va.	1.4	1.8	89.47	4.93	1.66	3.49	0.45	—	8559
Low-volatile bituminous	Pocahontas 3	McDowell, W. Va.	0.6	7.4	91.36	4.60	1.24	2.07	0.73	—	8697
Anthracite	Lykens	Luzerne, Pa.	4.2	2.4	94.88	1.83	0.67	1.78	0.84	—	7878

[a] Moist, mineral matter free. Multiply by 1.8 to get Btu/lb values used in ASTM classification by rank.

Source: Reprinted with permission from Ref. 88: G. R. Yohe and E. O Blodgett, *J. Am. Chem. Soc.*, **69,** 2646 (1947). Copyright by the American Chemical Society.

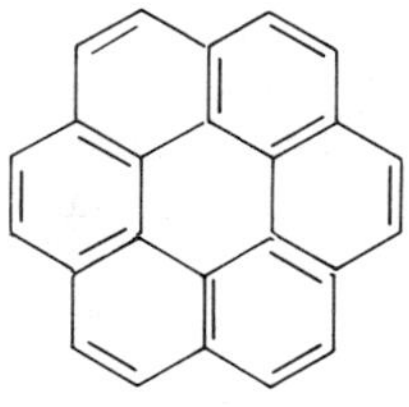

Figure 1.7. Coronene, $C_{24}H_{12}$.

tion, hydrogenation, hydrolysis, halogenation, X-ray diffraction, infrared spectroscopy, nuclear magnetic resonance, diamagnetic susceptibility, and the methods of statistical structural analysis.

1.6.2.2 *Oxidation*

Although many investigators have used various methods of oxidation in their studies of coal,[90] this review refers only to the works of several major groups of contributors.

In the early 1920s Franz Fischer and his co-workers at the Kaiser-Wilhelm Institute für Kohlenforschung in Germany published a series of papers[91] summarizing the work of earlier investigators on the oxidation of coal, and reporting their own results of oxidation studies with air at elevated temperatures and pressures and in the presence of an alkaline solution. The procedure was applied to cellulose, lignin, humic materials made from sucrose, peat, and coals of various ranks including bituminous. Benzene carboxylic acids were isolated from coal oxidation products, and this led "to the view that coals possess a predominantly aromatic structure." Fischer and his colleagues also concluded that the lignin rather than the cellulose of wood was the precursor of coal.

W. A. Bone and his co-workers at the Imperial College of Science and Technology, London University, pursued this attack still further, using alkaline potassium pemanganate as the oxidizing agent.[92] They measured the amounts of carbon dioxide, acetic acid, oxalic acid, and benzene carboxylic acids produced during the oxidation of lignin, peat, brown coal, bituminous coals, and anthracite, and concluded that their experiments "have completely proved, not only the benzenoid character of the main coal substance, but its progressive development in the lignin →peat→coal→anthracite series."

Similar oxidation of cellulose gave no benzenoid products, but from the various ranks of coals, at one time or another every benzene carboxylic acid except benzoic was obtained. The percentage of the carbon in the coal that was converted to benzene carboxylic acids ranged from 22% in the brown coals to 50% in an anthracite.

Under the leadership of H. C. Howard, the group at the Coal Research Laboratory of the Carnegie Institute of Technology in Pittsburgh, undertook an investigation of the aromatic acids formed from medium-volatile bituminous coal by oxidation with oxygen and sodium hydroxide at about 270°C and 900 psig.[93] Studies were made of the acids themselves, of their decarboxylation products, of their esters, and of the products obtained by hydrogenating the esters to convert the carboalkoxy groups to methyl groups. The aromatic nuclei thus identified, presumably parts of the original coal structure, included benzene, naphthalene, phenanthrene, biphenyl, diphenylene oxide, benzophenone, fluorenone, phthalan, pyridine, quinoline, and the phenylpyridines.

R. S. Montgomery and co-workers at the Dow Chemical Company in Midland, Michigan, studied samples of the same acids and found essentially the same aromatic nuclei.[93] They pointed out, however, that finding such nuclei as toluene and alkyl naphthalenes indicated that alkyl groups must exist in coal and contribute to its properties.[94]

1.6.2.3 *Hydrogenation*

Studies on the reduction of coal by various means have been summarized by Weiler[95] and by Dryden.[96] These experiments resulted

[90] See H. C. Howard, in Ref. 1, Vol. 1, Ch. 9, pp. 346–376; I. G. C. Dryden, in Ref. 2, pp. 272–289; R. A. Glenn, in Ref. 2, pp. 1088–1089, for summaries and original literature references.

[91] Franz Fischer et al., *Ges. Abh. Kennt. Kohle,* Vol.5, 1922, papers 17–27 incl. pp. 135–306.

[92] W. A. Bone and G. W. Himus, *Coal, Its Constitution and Uses,* Longmans, Green and Co., London, (1936), pp. 181–193.

[93] See I. G. C. Dryden, in Ref. 2, pp. 285–289, for many references to original papers.

[94] E. D. Holly and R. S. Montgomery, *Fuel,* **35,** 49–55 (1956).

[95] J. F. Weiler, in Ref. 1, Vol. 1, Ch. 10, pp. 377–417.

[96] I. G. C. Dryden, in Ref. 2, pp. 252–254.

in the isolation of multiring aromatic hydrocarbons such as naphthalene, anthracene, coronene, and their hydrogenated or partially hydrogenated derivatives. The identification of phenols in the reduction products also confirmed that benzene ring structures exist in coals.

Coals of lower ranks are more easily reduced and give higher percentage yields of low molecular weight, chloroform-soluble products than do coals of higher rank (Ref. 95, pp. 394 ff), indicating that the molecules of the coal substance increase in size and complexity and probably in the extent of the fused aromatic ring structural units as the rank increases.

More recent experiments carried out under milder conditions of electrolytic reduction by the polarographic technique[97] gave evidence that structural nuclei such as biphenyl, triphenylene, naphthalene, and phenanthrene were important in the makeup of coal extracts.[98]

Reduction of coal by the use of lithium in ethylene diamine[99] together with X-ray studies of the products[100] confirmed that aromatic structures had been hydrogenated.

Hydrogenolysis of a bituminous coal (85.6% C) followed by further hydrogenation and fractionation of the products (Ref. 96, p. 254) gave evidence that units of three to eleven condensed rings existed in the coal, but that large pericondensed structures such as coronene probably were absent.

1.6.2.4 *Hydrolysis*

Coals are inert to attack by dilute hydrochloric or sulfuric acid and undergo no appreciable change with dilute alkali, except under drastic conditions or in the presence of an oxidizing agent (e.g., air),[101] thus indicating that such hydrolyzable groups as ester, anhydride, and amide are absent. When concentrated sodium or potassium hydroxide was used at elevated temperatures, appreciable amounts of the coal substance were converted to alkali-soluble material, and this conversion was greater in extent with lower rank coals. Most of the alkali-soluble product was phenolic.

The hydrolysis of a bright coal from Wyoming (77% C, maf) with 5*N* sodium hydroxide at 250°C[102] gave phenols, aliphatic acids, hydrocarbons (including aromatic and partly hydrogenated cyclics of condensed-ring structures). It was postulated that ether and olefinic structures had been ruptured in the reaction.

1.6.2.5 *Halogenation*

The results of numerous experiments on the halogenation of coals have been summarized by Weiler[103] and by Dryden (Ref. 90, pp. 270–272). Fischer[104] concluded from his bromination studies that most of the bromine was used in addition reactions involving unsaturated structures, and only a little was used in substitution. More recent experiments[105] have shown that the bromine uptake of coals decreases as the rank increases, and the removal by aqueous sodium hydroxide of only 30–40% of the bromine suggests that bromine substitution involves mainly aromatic systems. Weiler's chlorination experiments[106] led him to conclude that about 85% of the carbon of a Pittsburgh Seam bituminous coal is in complex condensed-ring structures, "of which the nitrogen in the coal is probably an integral part."

1.6.3 Physical Methods of Investigation

The boundary line between the chemical and the physical characteristics of materials is not always easy to define. Quite often, the

[97] P. H. Given, V. Lupton and M. E. Peover, *Nature (London),* **181,** 1059–1060 (1958); *Proc. Inst. Fuel Conf.,* "Science in the Use of Coal," Sheffield, 1958, Paper No. 3, pp. A-38 to A-42. Summarized in Ref. 2, p. 253.

[98] P. H. Given and M. E. Peover, *Fuel,* **39,** 463–473 (1960).

[99] L. Reggel, R. Raymond, S. Friedman, R. A. Friedel and I. Wender, *Fuel,* **37,** 126–128 (1958).

[100] S. Ergun and I. Wender, *Fuel,* **37,** 503–506 (1958).

[101] H. C. Howard, in Ref. 1, Vol. 1, Ch. 11, pp. 418–424.

[102] F. G. Parker, J. P. Fugassi and H. C. Howard, *Ind. Eng. Chem.,* **47,** 1586–1592 (1955).

[103] J. F. Weiler, in Ref. 1, Vol. 1, Ch. 8, pp. 337–345.

[104] Ferd. Fischer, *Z. Angew. Chem.,* **12,** 764–767, 787–790 (1899).

[105] J. K. Brown, P. H. Given, V. Lupton and W. F. Wyss, *Proc. Inst. Fuel Conf.,* "Science in the Use of Coal," Sheffield, 1958, Paper No. 4, pp. A-43 to A-47.

[106] J. F. Weiler, *J. Am. Chem. Soc.,* **58,** 1112–1114 (1936); see also Ref. 103.

Table 1.8 Applications of Physical Methods

Measurement of	Yields Information on
X-ray diffraction	Size distribution of aromatic ring systems Average diameter of aromatic lamellae Mean bond length Average thickness of the packets of lamellae
Ultraviolet and visible absorption	Aromaticity Average aromatic ring size
Reflectance	Aromatic surface area
Refractive index (molecular refraction)	Optical anisotropy
Infrared absorption	Characteristic groups such as OH, CH_{ar}, CH_{al}, $(C=C)_{ar}$; the H_{ar}/H_{al} ratio
Electron spin resonance	Free radicals
Proton spin resonance	The H_{ar}/H_{al} ratio
Electrical conductivity	Changes in size of aromatic ring systems
Diamagnetic susceptibility (molar diamagnetic susceptibility)	Average number of aromatic rings
Dielectric constant	Dipole moment
Sound velocity	Aromaticity
Density (molar volume)	Aromaticity Ring condensation index

Source: Reproduced from H. H. Lowry, *Chemistry of Coal Utilization,* Supplementary Volume, John Wiley and Sons, 1963, p. 35, with permission of John Wiley and Sons, Inc.

data obtained in one of these domains are helpful in making interpretations in the other.

Among the newer tools available to chemists engaged in studying problems of chemical structure are a number of methods of instrumental analysis, including those that measure physical properties. Some of these, for example, X-ray diffraction, have been used for some decades, whereas others have undergone much of their development during the 1950s and later.

Tschamler and de Ruiter[107] have presented these methods as they were developed up to about 1960 in considerable detail; only very brief accounts of some of them are given here. Table 1.8, taken from the chapter by Tschamler and de Ruiter in the first Supplementary Volume, summarizes these methods and the type of information they yield.

1.6.3.1 *X-Ray Diffraction*

Although X-ray diffraction studies are most valuable in elucidating the physical nature of coals,[107] the patterns obtained are consistent with the ideas that bituminous coals contain both aliphatic and aromatic carbon structures[108] and condensed aromatic nuclei of average size not exceeding four fused rings.[109] The diffraction patterns show broad, diffuse bands characteristic of amorphous or colloidal materials rather than the sharp lines or spots given by powdered crystals or large single crystals, respectively. Yet the carbon-to-carbon spacings derived from these diffraction patterns suggest a graphitelike arrangement the layers of which are so limited in area and so poorly oriented in relation to one another that they cannot be called truly crystalline, although the term "crystallites" has been applied to these tiny groupings.

The average number of aromatic rings per cluster, which increases with increasing rank, has been estimated as about twelve for anthracites.[110]

It may be of interest to devote a few sentences to one of the ideas regarding the

[107] H. Tschamler and E. de Ruiter, in Ref. 2, Ch. 2, pp. 35–118.

[108] S. Ergun and V. H. Tiensuu, *Fuel,* **38,** 64–78 (1959).

[109] J. B. Nelson, *Fuel,* **33,** 153–75, 381–393 (1954); see also I. G. C. Dryden, in Ref. 2, pp. 254–257.

[110] P. B. Hirsch, *Proc. Inst. Fuel Conf.,* "Science in the Use of Coal," Sheffield, 1958, Paper No. 1, pp. A-29 to A-33.

possible arrangement of these essentially planar clusters of rings. Biscoe and Warren,[111] from their X-ray studies of carbon blacks, suggested the term "turbostratic" to describe their arrangement, and the use of an analogy may help to clarify the term. If a quantity of small disks (e.g., coins) is dumped into a box, they will be arranged in an essentially random manner, yet in some places they will tend to stack in small groups of essentially parallel units. If these disks are poured out on a table top, they will, in the main, take positions essentially parallel to each other and to the table top, but their arrangement in the other two dimensions will be random. It may be the same with these coal units, or lamellae: that is, they may be more completely random in the low-rank coals, attaining a greater degree of orientation as rank increases or as the coal is subjected to high temperature treatment as in carbonization. Also, as coal rank increases, the interlayer distance decreases.

1.6.3.2 *Infrared Absorption Spectra*

A number of interesting conclusions have been drawn from infrared absorption evidence. Brown[112] examined 12 vitrains of 78–94% carbon content and found that their structural skeletons were similar. Aromatic hydrogen increased and aliphatic hydrogen decreased with increasing rank. The strong absorption at 1600 cm^{-1} was attributed to aromatic ring vibrations, and the predominance of benzenoid rather than multiring nuclei has been suggested;[113] keto-enol tautomerism may contribute to absorption in this region.[114] Ethylenic double bonds and long-chain aliphatic units are absent.

1.6.3.3 *Nuclear Magnetic Resonance*

Studies of coal by nuclear magnetic resonance[115] have given values of aromaticity that are essentially in agreement with infrared data. Brown and co-workers[116] concluded that methylene bridges between aromatic rings were not present in coal extracts and probably not in the original coal.

1.6.3.4 *Diamagnetic Susceptibility*

The rather limited amount of work done on coal by the diamagnetic susceptibility method prior to 1960 gave results that were consistent with the ideas of the existence of aromatic clusters in the coal substance but differed somewhat in regard to the size of these units. Schuyer[117] reported that the number of rings in aromatic substances, including coals, could be determined with fair reliability by diamagnetic susceptibility measurements.

1.6.3.5 *Statistical Analysis*

We have mentioned already that the percentage of hydrogen in bituminous coals indicates that these materials are neither predominantly aliphatic structures, nor extensive condensed-ring aromatic networks.

Howard,[118] in summarizing the evidence for the cyclic structure of bituminous coals, called attention to the significance of the atomic C : H ratio and showed that this ratio for coal is not consistent with a highly aliphatic or linear structure. Some condensed cyclic structures must be present to satisfy the demands of the atomic C : H ratios found.

Van Krevelen and his co-workers[119] extended the statistical method and included other data such as density, refractive index, and heat of combustion. They were able to formulate equations for calculating a "ring condensation index" ($2R/C'$, where R is the number of rings in a cluster and C' the total number of carbon atoms in a cluster unit) and for deriving the fraction of aromatic carbon in a coal. The values of R found by Van Krevelen's methods are larger than those derived from X-ray data, for example, but

[111] J. Biscoe and B. E. Warren, *J. Appl. Phys.*, **13**, 364–371 (1942).

[112] J. K. Brown, *J. Chem. Soc.*, **1955**, 744–752.

[113] R. A. Friedel and J. A. Queiser, *Anal. Chem.*, **28**, 22–30 (1956).

[114] J. D. Brooks, R. A. Durie and S. Sternhell, *Aust. J. Appl. Sci.*, **9**, 63–80 (1958).

[115] See I. G. C. Dryden, in Ref. 2, pp. 258–259.

[116] J. K. Brown, W. R. Ladner and N. Sheppard, *Fuel*, **39**, 79–86 (1960); J. K. Brown and W. R. Ladner, *Fuel*, **39**, 87–97 (1960).

[117] J. Schuyer, *Naturwissenschaften*, **42**, 388 (1955).

[118] H. C. Howard, *Ind. Eng. Chem.*, **44**, 1083–1088 (1952).

[119] See I. G. C. Dryden, in Ref. 2, pp. 260–262, for references.

such discrepancies need not detract from the principal and important conclusion: clusters of condensed aromatic ring structures form an important part of the carbon skeleton of coal.

Howard[118] called attention to another characteristic of coals: they show none of the properties of linear high polymers, such as the capability of melting without decomposition when heated. One must entertain the idea that coals are of cross-linked structure, except possibly the vitrinites of the "highly fluid" coking coals, which do melt a little above 400°C, although usually with some decomposition.

1.6.4 Elements Attached to the Carbon Skeleton

1.6.4.1 *Hydrogen*

It seems quite obvious that most of the hydrogen in coals is bonded directly to carbon atoms. Because aromatic rings and condensed clusters of such rings form an important part of the carbon skeleton of the bituminous and anthracitic coals, it seems equally obvious that part of this hydrogen must be attached to these ring carbons or to the peripheral atoms of the condensed-ring clusters.

An increase in rank brings an increase in the aromatic and a decrease in the aliphatic hydrogen. Brown,[112] using infrared spectroscopy, found this change to be most marked between 89 and 94% carbon.

An interesting study by Kinney[120] on the correlation of yields of acetic acid on nitric acid–dichromate oxidation of coals with yields of methane on carbonization indicated that there are two structurally different sources of methane. One yields acetic acid on oxidation and is largely decomposed at carbonization temperatures up to 500°C; the second type of grouping does not oxidize to acetic acid and yields methane at carbonization temperatures of 500–1000°C. It was postulated that the acetic acid–yielding structures were mostly methyl groups, and the second type were more stable units such as methylene.

The amount of hydrogen split off from 10 Donets basin coals by heating them to 350°C with sulfur was between 1.00 and 1.30%; when anthracene oil or retene was used as a solvent, the hydrogen removed was more nearly uniform for most coal varieties, 2.4 to 2.9% at 350°C.[121]

Using an oxidation method, Mazumdar and Lahiri[122] concluded that the nonaromatic carbon must be either aliphatic or alicyclic; if the former, in chain lengths of not less than four carbons, or if the latter, not less than seven carbon systems. The ratio of aromatic hydrogen to total hydrogen for five coals was: 0.29 (coal of 71.5% C), 0.31 (coal of 80.7% C), 0.35 (coal of 85.2% C), 0.46 (coal of 90.3% C), and 0.79 (coal of 93.2% C).[123] Brown and Hirsch[124] examined a coal of 85% C by infrared absorption spectroscopy and reported that the "ratio of aliphatic to aromatic C—H bonds has been estimated to be approximately 4 : 1 for the coal."

The application of broad-line proton magnetic resonance spectroscopy to the study of coals[125] of 79, 83, and 93% carbon indicated that about two-thirds of the hydrogen in the lowest rank coal was in aliphatic groups; for the highest rank coal this value was about one-third. It was later found that tertiary aliphatic hydrogen behaved as aromatic in this method,[126] and the application of certain corrections brought these values for aromaticity into essential agreement with those derived from the infrared.[112]

Some hydrogen may be attached to atoms other than carbon; oxygen, nitrogen, and sulfur are the only such elements of importance in coal, and they are considered in the following paragraphs.

The near-constancy of the hydrogen contents of coals of the bituminous ranks and

[120] C. R. Kinney, *J. Am. Chem. Soc.*, **69**, 284–289 (1947).

[121] V. I. Zabavin and A. I. Khrisanfova, *Bull Acad. Sci. URSS, Cl. Sci. Tech.*, 1629–1634; (1946) *C.A.*, **41**, 3274i.

[122] B. K. Mazumdar and A. Lahiri, *Fuel*, **37**, 495–498 (1958).

[123] S. K. Chakrabartty, B. K. Mazumdar and A. Lahiri, *Fuel*, **37**, 498–501 (1958).

[124] J. K. Brown and P. B. Hirsch, *Nature (London)*, **175**, 229–233 (1955).

[125] P. C. Newman, L. Pratt and R. E. Richards, *Nature (London)*, **175**, 645 (1955); C. L. M. Bell, R. E. Richards and R. W. Yorke, *Brennstoff-Chemie*, **39**, *Sonderausgabe*, 30–31 (1958); *C.A.*, **52**, 15874f.

[126] I. G. C. Dryden, in Ref. 2, page 258, cites a private communication from R. E. Richards.

the abrupt decrease as anthracitic rank is attained are shown in Table 1.7.

1.6.4.2 *Oxygen*

All coals contain oxygen, and the percentage of this element present decreases as the rank of the coal increases. This trend also is apparent in Table 1.7. The anomalous low oxygen value for the high-volatile C bituminous coal may be due to its higher sulfur content.

Dryden has pointed out (Ref. 2, p. 262) that no appreciable amounts of alcoholic hydroxyl, carboxyl, or methoxyl groups occur in bituminous coals, although all these may be found in coal-forming materials (i.e., peats) and may persist in small amounts in the lower rank coals, such as brown coals, lignites, or even the subbituminous (see Table 1.7 for methoxyl). The absence of carboxyl groups that have been protected by esterification is indicated by the failure of attempts to hydrolyze coals with aqueous acids or alkalis under conditions that usually effect such hydrolysis. Thus the oxygen groups that remain to be considered are essentially narrowed down to phenolic hydroxyl, carbonyl (including quinone), and ether (including oxygen heterocycles).

Studies on the removal of oxygen from coal by reaction with hydrogen under pressure, at elevated temperature (430°C, 3 hr, maximum pressures 150–200 kg/cm^2 of H_2) in the presence of an equal weight of tetralin and 1% of tin sulfide[127–129] showed that some carbon dioxide was derived from the lowest rank coals, indicating the presence of some carboxyl groups. The main part of the oxygen was removed at rates that changed during the reaction, suggesting that about 60% of the oxygen was in a chemical combination of one type (removed most rapidly), whereas the remainder, removed more slowly, was combined in a different manner. No inference was made about the types of structures involved.

After summarizing the various chemical methods for estimating hydroxyl (sorption and titration with barium hydroxide and calcium acetate; neutralization with sodium hydroxide and sodium carbonate or potassium hydroxide and carbon dioxide; titration with sodium aminoethoxide; methylation with sodium hydroxide and dimethyl sulfate or with diazomethane; acetylation with acetic anhydride, ketene, or sodium acetate in acetic anhydride; and etherification with dinitrofluorobenzene and sodium bicarbonate or with hexamethyl-disilazane), Dryden[130] showed that coals have been reported to contain from 0 to as high as 11% hydroxyl oxygen, and stated that ". . . it is probably not far wrong to assume that about 60% of the total oxygen in bituminous coal exists as hydroxyl, although this amount is not strictly constant. Above 90% C, the percentage may drop to a much lower figure. . . ."

Yohe and Blodgett,[88] using a methylation procedure, and Brown[112] by means of infrared spectroscopy, both indicated that phenolic groups diminished as coals reached the higher ranks and were essentially lacking in coals of 89% and higher carbon content.

Dryden (Ref. 2, pp. 265–267) has also summarized the methods used for estimating carbonyl, and he concluded that Given's polarographic reduction methods[131,132] with coal extracts in dimethyl formamide with tetraethyl ammonium iodide were more reliable than hydrazone or oxime formation. Dryden estimates that carbonyl groups may account for from 20% to as much as 50% of the oxygen in coals. Given and Peover[133] have suggested that highly conjugated ketone groups chelated to hydroxyl may be the major form of carbonyl in the vitrains they examined.

Oxygen bound in ether structures, including heterocyclic rings, does not lend itself well to analytical determination, but its presence is inferred[112,134] from variation of in-

[127] H. H. Storch, C. H. Fisher, A. Eisner and L. Clarke, *Ind. Eng. Chem.*, **32,** 346–353 (1940).

[128] C. H. Fisher, G. C. Sprunk, A. Eisner, L. Clarke, M. L. Fein and H. H. Storch, *Fuel,* **19,** 132–138 (1940).

[129] J. F. Weiler, in Ref. 1, Vol. 1, Ch. 10, pp. 408–410.

[130] See Ref. 2, pp. 263–265.

[131] P. H. Given and M. E. Peover, *Proc. Inst. Fuel Conf.,* "Science in the Use of Coal," Sheffield, 1958, pp. A-33 to A-37.

[132] P. H. Given and J. M. Schoen, *J. Chem. Soc.,* 2680–2684 (1958).

[133] P. H. Given and M. E. Peover, *J. Chem. Soc.,* 394–400 (1960).

[134] M. M. Roy, *Fuel,* **36,** 249–250 (1957).

frared spectra with rank, or assumed as likely for the oxygen that is not otherwise accounted for.

1.6.4.3 *Nitrogen*

Very little is known about the manner in which nitrogen is combined in coal, except that it is part of the organic matter. Terres[135] investigated the formation of ammonia during coking and reported that in the usual procedure 10–14% of the nitrogen present was recoverable as ammonia, but that this amount could be increased to as much as 35% by slow distillation. He further investigated the destructive distillation of various types of nitrogen compound and found that only amino and substituted amino compounds yielded ammonia. His conclusion was that nitrogenous compounds in coal must have the character of proteins.

Shacklock and Drakeley[136] subjected coals of various ranks from peat to anthracite to extraction with boiling 33% sulfuric acid and found that the percentage of the total nitrogen extracted thus decreased markedly from 78.5% in peat to 5.4% in bituminous coal and 2.6% in anthracite.

Kinney and co-workers[137] made a study of the fate of nitrogen in the potassium permanganate oxidation of coal; their results suggested that "the basic nitrogen structures in the protein have not been greatly altered by the process of coalification," although proteins, as such, were not present in coal. It was also indicated that the nitrogen in coal became more predominantly of the tertiary amine type as the rank of the coal increased.

Part of the nitrogen in coals may have been derived from alkaloids[138] and related compounds, which contain the nitrogen in heterocyclic rings that are considered sufficiently stable to have been retained during the coalification process. This theory is in harmony with the findings of Beet[139] and of Entel.[140] Beet found nicotinic acid (3-pyridinecarboxylic acid) in incompletely digested Kjeldahl-type (sulfuric acid) reaction mixtures, and Entel isolated pyridine and related heterocyclics from products obtained by decarboxylating a mixture of acids formed by oxidizing coal with oxygen and alkali at elevated temperatures and pressure.

1.6.4.4 *Sulfur*

Sulfur is usually considered in two main categories as it exists in coal: organic, in which the sulfur atoms are a part of the coal structure, and inorganic, in which mineral sulfur compounds are physically admixed as impurities in the coal.

The inorganic sulfur is predominantly pyritic (iron pyrite, FeS_2), although in some cases of weathered or oxidized coals, significant amounts of sulfate (as calcium sulfate and/or ferrous sulfate) may be found. The importance of these mineral sulfur compounds and the problems posed by their presence for industrial users of coal have received a great deal of attention over the many years that man has been using coal. These questions have been ably discussed by Thiessen[141] and are not considered further here.

It is the organically bound sulfur that is really a part of the coal, and as such, of interest in this review.

The amount of organic sulfur in coal varies widely from a small fraction of a percent to a very unusual value of nearly 8%.[142] Kreulen,[143] describing what seems to be the same high-sulfur coal, says: "The Rasa coal of Istria (Croatia), a coal with a high (11 percent) organic sulfur and low oxygen content, exhibits a dual character." However his tabular data show the *total* sulfur content (maf) to be 10.9%; the oxygen content was 6.6%.

[135] E. Terres, *Jahrb. Gasbeleucht.*, **59**, 519–521 (1916); *J. Soc. Chem. Ind.*, **36**, 586 (1917); *C.A.*, **11**, 2727[3].

[136] C. W. Shacklock and T. J. Drakeley, *J. Soc. Chem. Ind.*, **46**, 478–481T (1927).

[137] C. R. Kinney, J. W. Eckerd, P. Rexford and H. B. Charmbury, *J. Am. Chem. Soc.*, **70**, 392–394 (1948).

[138] R. Thiessen, in discussion of paper by J. W. Cobb, *Trans. Am. Inst. Min. Metall. Eng.*, **71**, 214 (1925).

[139] A. E. Beet, *Fuel*, **19**, 108–109 (1940).

[140] J. Entel, *J. Am. Chem. Soc.*, **77**, 611–615 (1955).

[141] G. Thiessen, in Ref. 1, Vol. 1, Ch. 12, pp. 425–429.

[142] E. Donath and A. Indra, *Chem. Ztg.*, **36**, 1118–1120 (1912), reported on a coal from Istria: 1.09% pyritic, 0.02% sulfate, and 7.9% organic sulfur.

[143] D. J. W. Kreulen, *Fuel*, **31**, 462–467 (1952).

Organic sulfur is usually distributed throughout the coal and, of course, cannot be removed by coal-cleaning methods based on physical procedures. Only a far-reaching chemical degradation can release these sulfur atoms.

Very little is known from experimental evidence about the chemical structures involving the organic sulfur. Postowsky and Harlampovich[144] applied a methyl iodide reaction to three coals and reported that 18 to 32% of the sulfur present existed in thioether linkages. Kreulen[143] stated that a large part of the organic sulfur in the Istria coal was present in ring structures, and that sulfur-humic acids containing 8% sulfur were obtained by oxidizing the coal. He noted that the sum of the sulfur and oxygen percentages gave a value near what the oxygen content of such a coal might be expected to be.

It is tempting to speculate that since coals have lain for millions of years in a reducing environment, the sulfur they contain should be in the reduced state, that is, as C—S—H or C—S—C linkages. In this valence state sulfur occupies positions in molecular structures analogous to those held by oxygen. In this connection the observation by Kreulen cited in the preceding paragraph, and the oxygen and sulfur values of Table 1.7 are of interest.

In the absence of reports of mercaptan structures in coal, it would seem that organic sulfur must exist predominantly as thioethers, including heterocyclics.

The source of organic sulfur in coals is usually considered to be the sulfur-containing amino acid structures that are found in both plant and animal proteins.

1.6.5 Mineral Matter, Ash, and Rare Elements

Quite naturally, there has always been interest in the ash that remains when coal is burned, and in the mineral matter from which that ash is derived.

The inorganic (i.e., noncarbonaceous) materials that were part of the plant structures became the "inherent" mineral matter of the coal, and the amount of this is relatively small. The "adventitious" mineral matter is that which was introduced, probably as airborne dusts or waterborne silts and/or solutes during the peat bog or later stages of coal formation; this is ordinarily much greater in quantity, and quite varied in both quantity and composition.

The predominant minerals in coal are usually kaolinite, detrital clay, pyrite, and calcite, and they give rise to the oxides of silicon and other compounds, and oxides of aluminum, iron, and calcium in the ash. Lesser, but appreciable, quantities of compounds of magnesium, sodium, potassium, manganese, phosphorus, and sulfur are likewise found in coal ash generally.

Over the years, ash analyses, involving the determinations of these and perhaps in some cases a few more elements, have been done by the conventional wet methods developed for use in rock analysis. These methods and the information they have yielded have been discussed by Thiessen.[145]

With the advent of emission spectroscopy as a means of analysis, a new phase in the study of coal ash and coal mineral matter came about. There was a growing interest in the identification and determination of rarer elements that were present in small or even trace amounts.

Thiessen[145] also gives results of this type of analytical work done before about 1940; among elements that had been found in coal ashes were beryllium, boron, scandium, cobalt, nickel, zinc, gallium, germanium, arsenic, yttrium, zirconium, molybdenum, antimony, tin, lead, bismuth, silver, gold, rhodium, palladium, and platinum.[146] The summary by Ode[147] well conveys the growth of interest in the minor elements in coal and the proliferation of work on them during the next two decades.

It is beyond the scope of this chapter to reproduce all these data; the citation of a few examples will suffice. Headlee and Hunter[148] gave details of determining 38 elements in 16 of the major coal beds of West Virginia. A similar survey of Pennsylvania anthracites

[144] J. J. Postowsky and A. B. Harlampovich, *Fuel*, **15**, 229–232 (1936).

[145] G. Thiessen, in Ref. 1, Vol. 1, Ch. 14, pp. 485–495.

[146] V. M. Goldschmidt, *Ind. Eng. Chem.*, **27**, 1100–1102 (1935).

[147] W. H. Ode, in Ref. 2, Ch. 5, pp. 225–231.

[148] A. J. W. Headlee and R. G. Hunter, *Ind. Eng. Chem.*, **45**, 548–551 (1953); West Virginia Geol. Survey Bull. No. 13A, (1955), pp. 36–122.

was published by Nunn et al.[149] and of the low-rank coals of Texas, Colorado, and the Dakotas by Duel and Annell.[150]

Surveys of the same kind have been made in most if not all of the world's important coal-producing countries. The reader is referred to Ode's chapter[147] for the many references to these publications.

The methods of emission spectroscopy have been perhaps the most important in this work, but spectrophotometry, flame photometry, polarography, and chelametric titration have also been used.

The interested reader will want to consult Chapter 7 of this book for newer methods and data pertaining to this topic.

1.6.6 Physical Evaluation

Some of the uses of physical properties and their measurements have been mentioned as they relate to studies of the chemical constitution of coal. This section emphasizes what might more properly be called the mechanical and plastic properties, and the tests that have been devised to evaluate them.

Yancey and Geer[151] reviewed the status of knowledge on hardness, strength, and grindability of coal up to the early 1940s, and Brown and Hiorns[152] discussed developments in the succeeding 20 years in the area of mechanical properties, placing considerable emphasis on the fundamental studies and theoretical aspects of elastic properties and strength of the coal material.

1.6.6.1 *Hardness and Grindability*

A number of tests have been devised for evaluating coals with respect to their ability to withstand handling without excessive size degradation, and their amenability to the grinding that is required for charging coke ovens or for combustion in powdered-fuel units. Although various modifications of such tests have been proposed and used, this section deals primarily with methods that were generally accepted and used before about 1960; these are, essentially, the ASTM standards.

Two tests for breakage and/or degradation are the shatter test[153] and the tumbler test.[154]

When Yancey and Geer wrote their review,[151] systematic studies of grindability were in their early stages; the Hardgrove machine was relatively new,[155] and no standard procedure had been defined. In subsequent years methods of determining grindability were studied rather thoroughly, and the procedures deemed most suitable were the ball mill method, in which an estimate was made of the energy needed to pulverize a given amount of coal, and the Hardgrove machine method, in which the energy input was essentially constant and grindability was evaluated on the basis of the fraction of the sample that was reduced to suitable fineness. Discussion of these topics and the factors that affect grindability, including rank and petrographic composition, are given by Yancey and Geer[151] and by Brown and Hiorns.[152] In the meantime, the ASTM standard for determining grindability by means of the Hardgrove machine was adopted in 1951.[156]

The ASTM Standards also give specifications for equipment and procedures for screen analysis[157] and sieve analysis[158] of coal.

1.6.7 Plasticity

The problem of determining and evaluating the plasticity characteristics of coals is beset with numerous complexities. We are not dealing with the determination of melting point, density, or viscosity of a pure compound, but rather with attempts to describe and characterize in a useful manner a group of changes, both physical and chemical, that

[149] R. C. Nunn, H. L. Lovell and C. C. Wright, *Trans. Annu. Anthracite Conf. Lehigh Univ.*, **11**, 51–65 (1953).

[150] M. Duel and C. S. Annell, U.S. Geol. Survey Bull. No. 1036H, 1956, pp. 155–172.

[151] H. F. Yancey and M. R. Geer, in Ref. 1, Vol. 1, Ch. 5, pp. 145–159.

[152] R. L. Brown and F. J. Hiorns, in Ref. 2, Ch. 3, pp. 119–149.

[153] ASTM Standards, 1961, Part 8, Designation D440-49, pp. 1294–1298.

[154] ASTM Standards, 1961, Part 8, Designation D441-45, pp. 1340–1344.

[155] R. M. Hardgrove, *Trans. ASME, Fuels Steam Power*, **54**, 37–46 (1932).

[156] ASTM Standards, 1961, Part 8, Designation D409-51, pp. 1317–1320.

[157] ASTM Standards, 1961, Part 8, Designation D410-38, pp. 1328–1330.

[158] ASTM Standards, 1961, Part 8, Designation D311-30, pp. 1331–1332.

take place concurrently and sequentially, as the coal is heated.

Important among these changes are the development of a degree of fluidity, which may be very limited or quite extensive; the development of a stickiness or tackiness that enables the particles to cohere or to adhere to "inert" particles; the evolution of gas that may escape readily (e.g., in cases of high fluidity), or may be effectively trapped in a tough, viscous mass, causing considerable swelling; and the setting to a firm solid as higher temperatures are attained.

Thus the investigator may be interested in the temperatures at which fluidity develops and setting occurs; in the degree of fluidity attained; in the amount of swelling that occurs when no restraint is imposed, or in the pressure that develops if the mass of coal is held at essentially constant volume; in the strength of the solid residue that remains after setting has occurred, or the strength of the mass if an inert solid was admixed with the coal before heating; or in the variation of any or all of these properties with changes in rank of the coal, in the petrographic composition of the coal, or with changes in the test conditions, such as rate of heating.

In general the development of a plastic state is limited to the coals of bituminous ranks; the low-rank coals, such as lignites, and the high-rank anthracites, if pulverized and heated, remain as powders throughout the range of pyrolysis and gas evolution. These coals are not suitable for coke making. Of the petrographic constituents, fusain is devoid of these plastic properties throughout the range of ranks.

R. E. Brewer has given a complete and detailed description of the status of plasticity investigations in the early 1940s.[159] He points out the difficulties that stem from "the almost total lack of accepted standards and specifications," defines the terms in common use, describes apparatus and procedures used and the results obtained in numerous methods, and cites the uses that have been made of various data items. Only a very brief summary of these topics can be given here.

Some of the terms defined by Brewer follow:

Plasticity: "The plasticity of coal is a complex phenomenon which appears to be induced primarily by the pressure of gases causing surface flow at the moment when, under the action of heat, the molecules at the surface have attained a degree of freedom comparable with that obtaining in a liquid."[160]

Swelling: A volume change that takes place during heating wherein the coal can expand freely in a direction perpendicular to the heating surface.

Swelling Pressure: Pressure (force per unit area) that the coal mass exerts when obstructed from free expansion as it is heated.

Agglutinating Value: A measure of the binding qualities of a coal when heated in admixture with an inert material (e.g., sand or silicon carbide); usually measured by noting the strength of the resulting coke button by means of a compression test.

Agglomerating Value: The tendency to form coke when the coal sample is heated (as in a crucible) without admixture of an inert material. The form characteristics and strength of the coke button are usually noted.

Test Methods: Although most of the laboratory procedures that have been used to investigate coal plasticity have been designed to evaluate primarily a single characteristic (e.g., the fluidity that develops), usually other data items are noted at the same time, such as the temperatures at which fluidity is first observed and at which setting occurs.

Some tests were devised so that the observer might see what takes place, either by heating finely granulated coal under a microscope,[161] or by subjecting blocks (e.g., one-inch cubes) to a heating program in an inert atmosphere and observing the changes that had occurred.[162]

Other tests were designed to measure the swelling characteristics. In the "BSI swelling

[159] R. E. Brewer, in Ref. 1, Vol. 1, Ch. 6, pp. 160–309.

[160] R. A. Mott and R. V. Wheeler, *The Quality of Coke,* Chapman and Hall, London, 1939, pp. 11–12.

[161] P. Damm, *Fuel,* **8,** 167 (1929).

[162] R. A. Mott, *J. Soc. Chem. Ind.,* **46,** 85–92T (1927); *Fuel,* **6,** 217–231 (1927).

number''[163] test the coal sample was heated in a special crucible with no load imposed (free swelling), and the profile of the resulting button was compared with a set of standard profiles numbered 1–9; the number of the shape of closest fit was then assigned, as the swelling number. This procedure was later modified slightly and adopted as a standard for ''Free Swelling Index'' by the ASTM.[164]

Much work was done with the dilatometer test methods, in which a coal sample, usually a cylinder of pressed powdered coal, was heated in a vertical tube and allowed to push upward against a weighted piston device, the movement of which was indicated on a suitable scale, or by means of a dial (see Ref. 159, pp. 189–199 for detailed discussion and numerous references). This procedure made it possible to detect and measure contraction as well as expansion, and to note the temperature at which the changes occurred.

When the coal sample being heated is subjected to such confinement that it cannot swell (constant-volume tests), it is the ''swelling pressure'' that is measured. Such data are of interest in the coking industry because coke ovens are for all practical purposes constant-volume devices, and damage to the oven walls may result if excessive pressure develops. Brewer (Ref. 159, pp. 248–278) has described such tests and given many original references; he also included the larger scale test coke ovens that use coal charges weighing several hundred pounds and give results that are more readily translated into meaningful values for assessing the suitability of a coal for use in commercial coke ovens.

Several methods and many modifications have been developed for measuring the fluidity of the coal in its plastic state, and/or for measuring the temperatures at which such fluidity develops and the range of temperature through which it persists. Two that apparently are more useful in revealing temperature data than fluidity values are the penetrometer methods (see Ref. 159, pp. 199–203) and the gas flow procedures[165,166] (Ref. 159, pp. 203–220). Interest in these two methods has waned in recent years, as other more suitable methods have been developed.

Two methods that measure resistance to shear in a heated coal mass have been of considerable interest. Both involve a cylindrical charge of coal in which is embedded a coaxial stirrer having rabble arms. As the coal is heated, torque is applied to the coal charge or to the stirrer.

In the Davis plastometer[167,168] the coal charge is rotated at a constant speed of 1.5 rpm, and the changes in torque are recorded as the temperature is increased at 3°C/min through the range 300–500°C. Resistance to rotation is low in the unchanged charge, but when the coal begins to soften, resistance increases rapidly. It may then decrease if sufficient fluidity develops, then increasing until solidification occurs. The coke then breaks up, and resistance returns to a very low value, essentially as at the start of the test. Thus the softening temperature, the plastic range, and the solidification temperature are all obtained, as well as a measure of the fluidity and its changes as the test progresses.

The Gieseler plastometer[169,170] uses a coal charge that is static at the start of the test. A constant torque is applied to the stirrer, and the rate of turning is observed as the test progresses. The temperature of first movement of the stirrer, the maximum rate of stirring and its temperature, and the temperature when movement ceases give the initial softening temperature, the temperature of maximum fluidity, the plastic range, and a measure of the greatest fluidity attained. The rate at which the stirrer turns is indicated by a moving pointer on a dial that has 100 divisions for one revolution of the stirrer, and is reported as ''dial divisions per minute.'' This method was approved as a Tentative Standard by the ASTM in 1960, and a detailed description of the apparatus and procedure was published.[171] The values ob-

[163] BSI, Publication No. 804, 1938, 17 pp.
[164] ASTM Standards, 1961, Part 8, Designation D720-57, pp. 1312–1316.
[165] G. E. Foxwell, *J. Soc. Chem. Ind.*, **40,** 193–201T (1921); *Fuel*, **3,** 122–128 (1924).
[166] E. T. Layng and W. S. Hathorne, *Ind. Eng. Chem.*, **17,** 165–167 (1925).
[167] J. D. Davis, *Ind. Eng. Chem., Anal. Ed.*, **3,** 43–45 (1931).
[168] R. E. Brewer and R. G. Atkinson, *Ind. Eng. Chem., Anal. Ed.*, **8,** 443–449 (1936).
[169] K. Gieseler, *Glueckauf*, **70,** 178–183 (1934).
[170] R. E. Brewer and J. E. Triff, *Ind. Eng. Chem., Anal. Ed.*, **11,** 242–247 (1939).
[171] ASTM Standards, 1961, Part 8, Designation D1812-60T, pp. 1321–1327.

tained by this method are illustrated in data for Illinois coals used in coking tests, where initial softening temperatures ranged from 358 to 387°C, solidification temperatures from 438 to 459°C, and maximum fluidity from about 4 to about 350 dial divisions per minute. Eastern coking coals (eastern Kentucky, West Virginia, Virginia) gave initial softening temperatures of 354–437°C, solidification temperatures of 462–502°C, and maximum fluidity from about 10 to higher than could be read conveniently, and recorded as more than 15,000 dial divisions per minute.[172]

The application of extrusion methods (in which the softened, heated coal is forced through an orifice) to coal plasticity evaluation has been proposed, and some studies have been made (Ref. 159, pp. 234–240), but such procedures have never been widely used.

A number of tests have been used by those who are interested in coke making, not to measure the plastic characteristics as such, but rather to evaluate and compare coals as to their coke-making properties. These tests involve actually making coke and examining it. Among them, the test for agglomerating value and agglutinating value, both crucible tests, have already been mentioned. Another procedure that has proved useful is laboratory-scale coking in a horizontal tube,[173,174] accepted as a standard method in Great Britain as the Gray-King assay. In this test, which utilizes a 20-g coal sample, yields of coke, tar, liquor, and gas are determined as well as the quality of the coke. Calorific value of the gas may also be determined, so that the test may be a valuable guide to various aspects of by-product coking.

Baum and Heuser[175] discussed coal swelling, swelling pressure and its effects on coke ovens, and various methods then available for testing coals in this regard. These include the Otto Company's muffle test and the Koppers box test, in which the sample to be tested was inserted in one part of a two-compartment box, then placed in a commercial coke oven along with the regular charge. The extent to which the test sample moved the dividing partition of the box was noted after coking was complete. Koppers and Jenkner[176] also described the box test, as well as the Koppers laboratory hydraulic device for measuring pressure developed when a sample of coal is coked at constant volume.

Beginning in the 1930s and continuing for several decades, the USBM conducted an extensive survey of the coking and gas-making qualities of U.S. coals, using the USBM–American Gas Association (BM–AGA) cylindrical steel retort, 26 in. high and 13 in. in diameter, which took a charge of 75–100 lb of coal.[177]

Many of these investigations included studies on the variation of plastic behavior with particle size of the coal, moisture content, density of the charge, rank and petrographic composition of the coal, and the heating rate used in the test.

Brewer (Ref. 159, pp. 298–301) also discussed the relationship of plastic behavior of coal to combustion phenomena, particularly as it related to use in the underfeed stoker that was widely used in domestic heating during the 1930–1940 era.

Additional work that was done on the plastic properties of coal during the 1940s and 1950s has been summarized.[178] In general it consists of further studies of the methods previously available.

The plastic behavior of coal is affected by heating rate, particle size, nature of the coal,[179] pressure, oxidation, hydrogenation, and petrographic composition.

1.6.8 Density

Three categories are recognized in speaking of the density of coal: the "bulk" density, the true density, and the "apparent" density.

The bulk density is the weight per unit volume, measured by filling a box of known volume level full of crushed coal and weigh-

[172] F. H. Reed, H. W. Jackman, O. W. Rees, G. R. Yohe and P. W. Henline, Illinois State Geol. Survey Bull. No. 71, 1947, pp. 50–51.

[173] T. Gray and J. G. King, Fuel Research Board, Dept. Sci. Ind. Res., (U.K.) Fuel Research Tech. Paper No. 1, 1921; *Iron Coal Trades Rev.*, **102**, 362 (1921) *C.A.*, **15**, 1390.

[174] J. G. King, *J. Inst. Fuel*, **12**, 350–362 (1939).

[175] K. Baum and P. Heuser, *Fuel*, **10**, 51–64 (1931).

[176] H. Koppers and A. Jenkner, *Fuel*, **10**, 232–239 (1931).

[177] A. C. Fieldner, J. D. Davis and D. A. Reynolds, *Ind. Eng. Chem.*, **22**, 1113–1123 (1930).

[178] R. Loison, A. Peytavy, A. F. Boyer and R. Grillot, in Ref. 2, pp. 150–201.

[179] J. G. King, *Fuel*, **23**, 61–64 (1944).

ing it. Since the void spaces in the coal vary with the particle size and distribution and with the degree of compaction (and the degree of compaction varies with moisture content of the coal and the procedure used in filling the box), the values obtained for bulk density may differ considerably. The ASTM has published detailed directions for standard procedures for determining cubic foot weight of crushed bituminous coal,[180] both uncompacted and compacted by dropping it 6 ft into the standard one-cubic-foot box.

In writing of various aspects of moisture in coal, A. W. Gauger[181] has discussed the influence of moisture content on bulk density, showing that for crushed coal of less than 0.25 in. particle size, bulk density decreases as the moisture content increases. A minimum value is attained at about 6–10% moisture, then the values increase as further moisture is added. The difference between bulk densities of wet and dry coal was less for both finer and coarser coal than for intermediate sizes, and the addition of wetting agents to the water gave higher bulk densities than those obtained with water alone.

The methods for determining true density and apparent density have been summarized by Tschamler and de Ruiter.[182] They point out that since the sample weight can be measured with satisfactory precision, the determination of true density depends on precise measurement of the volume of the sample. Coals contain pores, the smallest of which are 5–10 Å in diameter.[183] Thus helium, with an atomic diameter of 1.78 Å, is a satisfactory fluid to use in measuring the "true volume" of the coal substance; the density values so obtained are often referred to as "helium densities."

Apparent density measurements involve the use of a fluid that does not penetrate the pore spaces completely. Thus the apparent density values for a given coal may differ slightly when density is measured in different fluids, depending on molecular size and the ability of the fluid to penetrate into the pores.

Both helium and apparent densities reach a minimum value (for bright coals) in the vicinity of 1.3 g/ml at carbon contents of 85–90%. Densities increase slowly as the carbon content drops below this range, and rather sharply at carbon contents above 90%.

Black durains have been reported as having the lowest density, followed by clarain, vitrain, and gray durain; fusain has the greatest density.[184] It should be remembered, however, that more or less mineral matter is present in all coal fractions, and if no correction for the heavier, inorganic substances is made, vitrain is often the least dense because it commonly contains the least mineral matter.

Tschamler and de Ruiter have discussed other topics related to density, including the effects of chemical treatment of the coal, the relationship of density to hydrogen content, and the use of density values in statistical methods of arriving at structural parameters (see Section 1.6.3.5).

Other physical measurements that have been used in the studies of coal have been shown in Table 1.8 in Section 1.6.3.

1.7 CARBONIZATION

1.7.1 General

The carbonization of coal involves the interaction of a wide range of chemical and physical properties of the coal itself in a high-temperature environment to produce a combustible gas, hydrocarbon and other liquids, tar, and coke. The nature of the tar and its yield, as well as the physical properties and reactivity of the coke and its yield, are affected by the coal used and by the carbonizing conditions. Thus an understanding of the complex coal carbonization process involves research on the plastic, agglutinating, agglomerating, and swelling properties of coals (discussed earlier in this chapter); dependence of yields of products on type of coal, temperature, and rate of heating; the chemical nature of coal tar; and, again, the physical properties and reactivity of coke.

Very few coals, when carbonized alone, will yield a metallurgical coke having the desired physical properties—particularly

[180] ASTM Standards, 1961, Part 8, Designation D291-60, pp. 1288–1293.

[181] A. W. Gauger, in Ref. 1, Vol. 1, pp. 620–622.

[182] H. Tschamler and E. de Ruiter, in Ref. 2, pp. 106–116.

[183] D. W. Van Krevelen, *Brennstoff-Chemie*, **35**, 257–262 (1954).

[184] A. M. Wandless and J. C. Macrae, *Fuel*, **13**, 4–15 (1934).

high strength. To achieve that goal, it is necessary to carbonize blends of coking coals. In spite of the work that has been done on blends of coals, it is still necessary to rely on empirical tests to determine their behavior in the coke oven and the strength of the coke produced.

Without going into detail in the areas just mentioned, we will try to indicate, through summaries and selected examples, some of the developments prior to 1960. We can note only such important related areas as the properties of coke-oven gas, light oil from coke-oven gas, removal of sulfur compounds and miscellaneous constituents from coke-oven gas, and the treatment of ammoniacal liquors.[1,2] Finally, the broad field of industrial coal carbonization can be covered only by the works[185,186] that review the status of the technology at various stages before 1960.

1.7.2 Carbonization Tests

In discussing the experimental carbonization of coals, Reynolds[187] states:

> A large number of assays for determining the coke- and by-product making properties of coal have been developed but these methods give little information concerning the quality of products because of the small amount of sample carbonized. In 1929 the Bureau of Mines, in cooperation with the American Gas Association, developed a carbonization test using charges weighing up to 200 pounds.[188] The yields of coke and by-products from this test are large enough to be tested by standard methods, thereby affording more complete data than are obtainable from tests on a smaller scale. Forty-eight coals have been carbonized by this method at high, medium, and low temperatures, and the qualities of the cokes and by-products determined. Petrographic, agglutinating, plastic, expanding, friability, slacking, and oxidizing properties of these coals have been studied also. The comparison of test results and plant data (available for some coals) is an important part of this work and the agreement has been close enough to fully justify confidence in the method of testing. Referring to the results of these investigations, Mulcahy[189] wrote: "Although much work on coal has been regarded as too fundamental or theoretical to be of practical value, the work of the Bureau of Mines in this report is of tremendous value in indicating the trends we can expect under different carbonizing conditions."

The subject of this section was also reviewed in detail by Davis.[190]

1.7.3 Physical Properties and Reactivity of Coke

The physical properties of coke such as density, size distribution, and strength as determined by the shatter, tumbler, and other tests, are significant in relation to the use of coke in fuel beds. Work in this area has been reviewed by Mayers.[191]

The "reactivity" of coke is "somewhat loosely used to express the ability of a coke to interact with, usually, oxygen, carbon dioxide or steam."[192] In his review of work on reactivity, Mayers[191] points out that it is not necessary to determine reactivities with various agents. By comparing a series of cokes of different reactivities, it has been shown that the reactivity with air is parallel to the ignition point[193]; that the reactivity to carbon dioxide[194] and steam[195] is parallel to changes in the ignition point; that the reactivities to air, to oxygen, and to carbon dioxide are parallel to changes in the ignition temperature[196–199]; and finally that reactivi-

[185] W. Gluud and D. I. Jacobson, *International Handbook of the By-Product Coke Industry,* Chemical Catalog Co., New York, 1932.

[186] Fred Denig, "Industrial Coal Carbonization," in Ref. 1, Vol. 1, Ch. 21.

[187] D. A. Reynolds, Section on "Carbonization," in Ref. 12, pp. 28–32.

[188] A. C. Fieldner and J. D. Davis, U.S. Bur. Mines, Monogr. No. 5, 1934, 164 pp.

[189] B. P. Mulcahy, *Foundry Coke, An Interpretive Discussion,* American Gas Association Production Conf., New York, May 23–25, 1938, 14 pp.

[190] J. D. Davis, in Ref. 1, Vol. 1, Ch. 22.

[191] M. A. Mayers, in Ref. 1, Vol. 1, Ch. 24.

[192] R. A. Mott and R. V. Wheeler, *The Quality of Coke,* Chapman & Hall, London, 1939, 464 pp.

[193] W. Melzer, *Glueckauf,* **66,** 1565–1576 (1930).

[194] T. J. Drakeley, *J. Soc. Chem. Ind.,* **50,** 317–330 (1931).

[195] K. Bunte and A. Giessen, *Gas Wasserfach,* **73,** 241–247 (1930).

[196] H. Bahr and F. Fallbohmer, *Gas Wasserfach,* **69,** 909–912, 929–932, 943–947 (1926).

[197] K. Bunte, *Z. Angew. Chem.,* **39,** 132–138 (1926).

[198] W. J. Muller and E. Jandl, *Brennstoff-Chemie,* **19,** 27–29 (1938).

[199] J. D. Davis and A. C. Fieldner, *Proc. Am. Gas Assoc.,* 636–659 (1935).

ties to air, to oxygen, to carbon dioxide, and to steam are all parallel.[200–202] No attempt is made here to review reactivity tests. Instead, certain data from such tests are considered in connection with the discussion of gasification reactions (see Section 1.11.2).

1.7.4 Coal Tar

Coal tar is one of the primary products of coal carbonization. It is a complex mixture containing principally aromatic compounds, many of which are polynuclear. Some 348 compounds were identified as coming from the carbonization of coal. Most of these appeared in the tar. Rhodes[203] and Weiler[204] also give information on the effect of the kind of coal and carbonization temperature on the characteristics and yield of coal tar, and more detailed information on certain of the individual chemical constituents of the tar.

1.8 COMBUSTION

1.8.1 General

Studies of the combustion of solid fuels have involved experiments with graphite, coal, and wood charcoal; with sizes ranging from particles smaller than 200 mesh to 1-in. spheres; with individual particles, clouds of particles, and filaments; and with both stationary and moving fuel beds. This wide range of conditions can be said to fall into one of two categories: basic studies and studies of combustion processes. The former includes studies with graphite in filaments or in simple geometric shapes. The latter embraces coal in the more complex systems. The variety of such systems exists because no single system is capable of burning efficiently the complete range of available coals. The aim of studies of combustion is to develop rational methods for designing equipment to suit the fuel available and to improve economic performance.

[200] J. D. Davis and J. W. Greene, *Proc. Am. Gas. Assoc.*, 1160–1164 (1926).
[201] J. D. Davis and D. A. Reynolds, *Ind. Eng. Chem.*, **20**, 617–621 (1928).
[202] B. Neumann, C. Kröger and E. Fingas, *Gas Wasserfach*, **74**, 565–572 (1931).
[203] E. C. Rhodes, in Ref. 1, Vol. 2, Ch. 31.
[204] J. F. Weiler, in Ref. 2, Ch. 14.

1.8.2 Basic Studies

In the latter part of the nineteenth century it was established that carbon monoxide and carbo dioxide were produced together in the reaction of carbon with oxygen.[205,206] The work of Rhead and Wheeler[207] indicated that carbon monoxide was the primary product at temperatures above 1000°C, whereas the work of Langmuir and others,[208–210] involving electrically heated filaments at low pressure, indicated carbon dioxide was the main product. In 1938 it was shown that low-pressure filament results were not valid because thermions upset the initial equilibrium of the primary reaction.[211,212] Confirmation of the results of Rhead and Wheeler[207] resulted from high-velocity experiments.[213]

According to Nicholls and Barkley[214] "the chemistry of combustion and its physics as regards to 'temperature, time and diffusion, were well understood in 1840,'[215] but such knowledge was not well disseminated." Thus it is interesting to note that in the 1920s it was recognized that the rate of combustion reactions was limited by the rate at which oxygen could be transported to the reacting surface.[216–218] This conclusion was supported by experimental work[219,220] on the rate of

[205] C. J. Baker, *J. Chem. Soc.*, **51**, 249–258 (1887).
[206] H. B. Baker, *Philos. Trans. R. Soc. London*, **179A**, 571–591 (1888).
[207] T. F. E. Rhead and R. V. Wheeler, *J. Chem. Soc.*, **101**, 846–856 (1912).
[208] I. Langmuir, *J. Am. Chem. Soc.*, **37**, 1139–1366 (1915).
[209] L. Meyer, *Z. Phys. Chem.*, **B17**, 385–404 (1932).
[210] V. Sihvonen, *Trans. Faraday Soc.*, **34**, 1062–1074 (1938).
[211] R. F. Strickland-Constable, *Trans. Faraday Soc.*, **34**, 1074–1080 (1938).
[212] R. F. Strickland-Constable, *Trans. Faraday Soc.*, **40**, 333–343 (1944).
[213] M. K. Grodzovskii and Z. F. Chukhanov *Fuel*, **15**, 321–328 (1936).
[214] P. Nicholls and J. F. Barkley, "Combustion," in Ref. 12, pp. 20–27.
[215] C. W. Williams, *Combustion of Coal*, Liverpool, England, 1840.
[216] W. Nusselt, *Z. Ver Dtsch. Ing.*, **68**, 124–128 (1924).
[217] P. Rosin, *Braunkohle*, **24**, 241–259 (1925).
[218] J. T. Ward and J. E. Hamblen, *Ind. Eng. Chem.*, **19**, 1025–1027 (1927).
[219] D. F. Smith and A. Gudmundsen, *Ind. Eng. Chem.*, **23**, 277–285 (1931).
[220] C. M. Tu, H. Davis and H. C. Hottel, *Ind. Eng. Chem.*, **26**, 749–757 (1934).

oxidation of spheres of graphite, which showed[221] "that the rate of the specific surface reaction depended strongly on the temperature at low temperatures[220] but less so at high ones[219,220]; that it was directly proportional to the oxygen concentration in the gas flowing over the carbon surface[220]; that it depended on approximately the 0.4 power of the mass velocity of the gas at high temperatures[219,220]; and that, for small particles, it was approximately inversely proportional to the diameter of the particle.[219] Similar effects were found for the reduction of carbon dioxide and steam[222] in work in which the monotonic effect of increasing gas velocity was used to estimate the absolute reaction rate by increasing the gas flow until no further increase in reaction rate, as measured by gas analysis, was found."

The impossibility of accounting for the dependence of reaction rates on gas velocity by a pure diffusion process led to the application of the concept of the stagnant film to account for this dependence.[220–223] This concept has been an important part of the development of our understanding of general heterogeneous reaction kinetics, which is extensively discussed by Thring and Essenhigh.[224] As discussed subsequently, the extensive application of the combustion of pulverized coal in large steam boilers led to the initiation of studies of burning particles in the size range of pulverized coal.

Tests with particles 60–90 mesh in size dropped through a heated tube showed that some coal particles ignited at about 1100°F, but a temperature of about 1300°F was require to ignite the major fraction of the particles.[225] Later tests showed that a temperature of about 1500°F was required to obtain uniform ignition of certain bituminous coals.[226] Combustion of individual particles of coal is complicated by the formation of cenospheres as a result of simultaneous softening and evolution of volatile matter, which when burned shows up as a luminous gas flame in photographic studies.[225] A theoretical analysis indicated that for a given quantity of excess air, the burning time of a solid particle is greater than that of a cenosphere containing the same weight of combustible.[227] The magnitude of the difference depends on the ratio of the size of the cenosphere to the size of the original particle. Burning times increase as the quantity of excess air decreases.

After reviewing basic and applied work on the combustion of pulverized coal, Orning[227] concluded in 1945:

> Considering all the data on the mechanism and rate of combustion, it appears that a transition must occur from diffusion to surface reactivity as the controlling factor. Immediately after the zone of ignition and burning of volatile matter, the particle temperature rises very rapidly. In this region the smaller particles burn to a low carbon content within a time proportional to the square of the radius of an equivalent solid sphere and to a function depending upon the influence of coal characteristics and flame conditions upon cenosphere formation.
>
> The larger particles will still contain a considerable portion of the total carbon after the rate of heat release has passed its peak. Continued rapid combustion of these particles depends upon their ability to maintain themselves above a certain temperature. With the combustion rate, and hence the heat release, varying as the particle diameter while the rate of loss by radiation is proportional to the surface, the largest particles will have the lowest temperature.
>
> The reactivity of the original fuel is no indication of the temperature at which the surface reaction begins to control. The high particle temperatures in the early stages of combustion transform the residue into a material whose reactivity is mainly a function of flame conditions. The characteristics of the original fuel appear in their effect on cenosphere formation, in a possible catalysis by ash, and in a hindrance by ash incrustations.

[221] M. A. Mayers, in Ref. 1, Vol. 2, Ch. 33.

[222] M. A. Mayers, *J. Am. Chem. Soc.*, **56,** 70–76, 1879–1881 (1934); **61,** 1053–1058 (1939).

[223] S. P. Burke and T. E. W. Schumann, *Ind. Eng. Chem.*, **23,** 406–412 (1931); *Proc. 3rd In. Conf. Bituminous Coal,* Vol. 2, 1931, pp. 489–509; *Ind. Eng. Chem.*, **24,** 451–453 (1932).

[224] M. W. Thring and R. H. Essenhigh, in Ref. 2, Ch. 17.

[225] H. E. Newall and F. S. Sinnatt, *Fuel,* **5,** 335–339 (1926).

[226] H. K. Griffin, J. R. Adams and D. F. Smith, *Ind. Eng. Chem.*, **22,** 808–815 (1929).

[227] A. A. Orning, in Ref. 1, Vol. 2, Ch. 34.

A complete calculation of burning times must consider the possibility that the surface reaction rate will begin to control beyond a certain degree of combustion. Lack of data on the reactivity and configuration of combustion residues and the dependence of particle temperature upon an equilibrium with furnace surroundings make it inadvisable to attempt a general solution.

Progress in this area between 1940 and 1963 is summarized by Thring and Essenhigh.[224]

1.8.3 Combustion in Fuel Beds

Combustion in fuel beds has been reviewed by Mayers,[221] Nicholls and Barkley,[214] Haslam and Russell,[14] Thring and Essenhigh,[224] and Sherman and Landry.[228] This type of combustion encompasses the open grate, the central house-heating furnace, and the various types of mechanical stoker.

At the beginning of this century extensive research had already been done to improve the convenience and efficiency of the coal fireplace.[229] Also extensive tests had been made by the USBM and other agencies on the combustion of various coals on a grate in a cylindrical furnace.[230–233] These tests provided information on the effect of rate of firing on gas composition in the fuel bed and on the temperature of the fuel bed. The tests showed conclusively that the quantity of coal burned or the capacity of the furnace depends entirely on the quantity of primary air coming into the furnace under the grate. The USBM also studied the combustion of gases over the fuel bed, and in this instance demonstrated conclusively that the quantity, distribution, and turbulence of secondary air determines the efficiency of the furnace.

Studies have been made on the combustion of various coals in house-heating boilers,[234] on the performance of domestic anthracite stokers,[235] on the fundamentals of combustion in small underfeed stokers,[236] and on burning various coals in domestic overfeed stokers.[237] Most of the work on various stokers involved improvements in mechanical design to secure a high degree of reliability and a high conversion of fuel into heat.[238–242]

Mayers[221] stated:

> The conditions within the beds of travelling-grate stokers have been determined experimentally.[243–245] These investigations show that the layer of coal entering the furnace is ignited at the top of the bed; and that the surface dividing ignited from raw coal moves down as the layer passes through the furnace. The ignition surface reaches the grate . . . where the dense flame of volatile matter stops; all the fuel has been ignited and coked at this point, and only the coked residue is burned out beyond it. As this process proceeds, a larger and larger proportion of the bed

[228] R. L. Sherman and B. A. Landry, in Ref. 2, Ch. 18.
[229] N. W. Fishenden, *The Coal Fire,* Fuel Research Board of England, Spec. Rep. No. 4, 1920.
[230] H. Kreisinger, F. K. Ovitz and C. E. Augustine, U.S. Bur. Mines Bull. No. 137, 1917.
[231] H. Kreisinger, C. E. Augustine and S. H. Katz, U.S. Bur. Mines Tech. Paper No. 139, 1918.
[232] H. Kreisinger, C. E. Augustine and F. K. Ovitz, U.S. Bur. Mines Bull. No. 135, 1917.
[233] M. E. Audibert, *Rev. Ind. Miner.,* **37,** 351 (1922).

[234] A. C. Willard, A. P. Kratz and V. S. Day, *Investigations of Warm-Air Furnaces and Heating Systems,* Univ. Illinois Eng. Exp. Sta. Bull. Nos. 120, 1921; 141, 1924; 188, 1929; 189, 1929; A. C. Willard, A. P. Kratz, S. Konzo, Univ. Illinois Eng. Exp. Sta. Bull. No. 246, 1932; A. P. Kratz and S. Konzo, Univ. Illinois, Eng. Exp. Sta. Bull. No. 266, 1934.
[235] P. A. Mulcey, "The Selection, Installation, and Performance of Domestic Anthracite Stokers," *Trans. 2nd Annu. Anthracite Conf.,* 1939 (with bibliography).
[236] C. A. Barnes, *Fundamentals of Combustion in Small Underfed Stokers,* Bituminous Coal Research Rep. No. 4 1938 (with bibliography).
[237] H. F. Yancey, K. A. Johnson, A. A. Lewis and J. B. Cordiner, Jr., *Burning of Various Coals Continuously and Intermittently on a Domestic Overfeed Stoker,* U.S. Bur. Mines, Rep. Invest. No. 3379, 1938, 33 pp.
[238] W. Schultes, *Arch. Waermewirtsch.,* **16,** 117–118 (1935).
[239] F. Schulte, *Z. Verh. Dtsch. Ing.,* **80,** 1237–1241 (1936).
[240] K. Cleve, *Feuerungstech.,* **25,** 317–322 (1937).
[241] A. R. Mayer, *Feuerungstech.,* **26,** 148–150, 201–211 (1938).
[242] K. Beck, *Arch. Waermewirtsch.,* **20,** 93–98, 301–316 (1939).
[243] H. Werkmeister, *Arch. Waermewirtsch.,* **12,** 225–232 (1931).
[244] J. D. Maughan, H. B. Spalding and B. M. Thornton, *Engineering,* **137,** 587–589, 656–658, 669–671 (1934); *J. Inst. Fuel,* **8,** 11–20 (1934).
[245] R. Loewenstein, *Waerme,* **57,** 97–101, 121–125 (1934).

becomes ashy refuse. The fire should be operated so that all the fuel is burned out of the refuse short of the end of the grate, but the air quantity passing through this end of the grate must be controlled to prevent excessive losses.

Spreader stokers were developed in the 1930s. The principles of operation are essentially similar, as far as the fuel bed is concerned, to those obtaining in handfiring, the stoker being merely a means of producing a continuous flow of a small quantity of fresh fuel into the fuel bed. Descriptions of such units and the results of tests made on them are given in Ref. 246 and by Barkley[247] and Sherman and Landry.[228]

1.8.4 Combustion of Pulverized Coal

The use of pulverized coal in the firing of large steam boilers came into practice shortly after the beginning of this century. For example, the boiler heating surface fired with pulverized coal increased from 2×10^5 ft^2 in 1918 to 2×10^6 ft^2 in 1925.[14] The progress of this development has been detailed in Orning,[227] Haslam and Russell,[14] Nicholls and Barkley,[214] and Sherman and Landry.[228]

The early developments in pulverized coal combustion have been summarized by Nicholls and Barkley[214]:

> In 1917 a report of the Bureau of Mines showed that a long travel of flame was required to give complete combustion unless the air and the combustible were mixed vigorously.[232] When furnaces for pulverized-coal firing were so built they gave satisfactory performance and good efficiences.[248] The cooperative work of the Bureau of Mines and Milwaukee Electric Railway and Light Co. resulted in the first successful application of pulverized fuel in the steam boiler field.
>
> In the earlier designs successful removal of the ash required that the particles be deposited in a dry or unmolten state. These dry-bottom furnaces were not so successful with coals having low-fusion ashes. The problem was solved by going to the other extreme of maintaining the ash in liquid state and tapping it intermittently or continuously; the wet-bottom furnace was the outcome.
>
> Operation of these furnaces depends largely on the properties of the coal ash and its slag. As relatively few data were available, the Bureau of Mines initiated in 1929 a study on the ash factor in such furnaces and the properties of coal-ash slags in general.[249] More recent work on coal ash and its effect on boiler availability was summarized by Ely and Barnhart.[250]

The initiation of the systematic study on coal ash slags followed closely the first successful application of the slagging-type furnace[251] and the first studies on slagged conditions in full-scale furnaces.[252] Other studies of conditions of temperature and gas composition in full-scale furnaces were directed toward the explanation of the striking difference in refractory service conditions in two similar furnaces[253] and toward the effect of different burner and heat absorption configurations.[254] Sherman and Landry[228] present data on the design of pulverized-coal burners and furnaces, on heat release rates in furnaces, and on the control of air pollution from combustion processes.

1.9 HYDROGENATION

Coal hydrogenation was practiced for the first time on a large scale in Germany in 1925 and subsequently in France, Great Britain, Japan, and the United States.[255] In com-

[246] Anon., *Power,* **80,** 476–477 (1936).

[247] J. F. Barkley, *Trans. ASME,* **59,** 259–265 (1937).

[248] Henry Kreisinger, John Blizard, C. E. Augustine and B. J. Cross, *An Investigation of Powdered Coal as Fuel for Power-Plant Boiler,* U.S. Bur. Mines Bull. No. 223, 1923, 91 pp.

[249] P. Nicholls and W. T. Reid, "Fluxing of Ashes and Slags as Related to the Slagging-Type Furnace," *Trans. ASME, Res. Pap.,* **54,** RP-54-9, 167–190 (1932); "Slags from Slag-Tap Furnaces and Their Properties," *Trans. ASME,* **56,** 447–465 (1934).

[250] F. G. Ely and D. H. Barnhart, in Ref. 2, Ch. 19.

[251] H. M. Cushing, *Trans. ASME,* **51,** 263–270 (1929).

[252] R. A. Sherman, P. Nicholls and E. Taylor, *Trans. ASME,* **51,** 339–410 (1929).

[253] R. A. Sherman and E. Taylor, *Trans. ASME,* 49–50, FSP 50-38, 125–140 (1927–1928).

[254] E. H. Tenney, *Proc. 3rd Int. Conf. Bituminous Coal,* Vol. 2, 1931, pp. 370–399.

[255] W. R. Wu and H. H. Storch, *Hydrogenation of Coal and Tar,* U.S. Bur. Mines Bull. No. 633, 1968, 195 pp.

menting on these early developments, Storch[256] states:

> Except for a relatively small amount of research published by the British Fuel Research Laboratory, little information on the fundamentals of coal hydrogenation is available, despite the voluminous patent literature, which is often more confusing than informative. Unquestionably much more is known by industrial interests, particularly in Germany and Great Britain, than appears in print. However, a study of the development of coal hydrogenation in these countries reveals that the intensive drive made during the past decade to achieve large-scale production has left little, if any, time and funds for fundamental research.

The literature on coal hydrogenation prior to 1931 was reviewed by Skinner[257] and Zerbe[258] and that between 1931 and 1938 by Fieldner.[259] These reviews, along with two books by Berthelot and Hot[260] and reviews by Storch[256] and Wu and Storch,[255] cover the basic work on coal and tar hydrogenation before 1943. Work done between then and 1963 was reviewed by Donath.[261]

Studies of the hydrogenation of a number of British coals in small bombs showed that the yield of oil boiling below 360°C at atmospheric pressure increased approximately linearly with the carbon content of the coal.[262] As a corollary, the reactivity of the coals to hydrogenation was approximately inversely proportional to their carbon content.

Similar studies on the hydrogenation of the mechanically separable constituents of American coals showed that the translucent constituents (spores, resins, oil algae, translucent attritus and anthraxylon) are liquefied completely. The opaque constituents (fusain and opaque attritus) contain about 75 to 95% and 20 to 60%, respectively, of materials resistant to hydrogenation.

On the basis of the foregoing results, it was suggested that "petrographic analysis would be a desirable, although only semiquantitative, tool for predicting liquefaction yield upon hydrogenation."[263–265] This possibility was confirmed by work done in the USBM experimental plant, which showed fair agreement between petrographic estimates of organic residue yield and yields obtained in the continuous operation of the plant.[266] These tests were made on nine coals ranging in rank from lignite to high-volatile bituminous A. Laboratory work on the effect of hydrogenating coal in the presence of a catalyst (stannous oxide), with and without a vehicle, showed that these two variables are not independent of each other.[267,268] For example, with small amounts of catalyst (< 0.1% of charge) the addition of a vehicle results in an appreciably greater yield of volatile products. With higher concentrations (> 0.5%) the effect is small and may even become negative.

The extensive work done on the effect of catalysts on the liquid phase hydrogenation of coal has been reviewed in considerable detail by Storch,[256] and Donath[261] has summarized work on the vapor phase hydrogenation.

In laboratory studies the catalytic effect of the walls of the converter cannot be neglected.[269] This effect is less important in large-

[256] H. H. Storch, in Ref. 1, Vol. 2, Ch. 38.

[257] D. G. Skinner, *Fuel*, **10**, 109–137 (1931).

[258] C. Zerbe, *Chem.-Ztg.*, **55**, 4, 18, 38, 94, 114, 126, 162 (1931).

[259] A. C. Fieldner, U. S. Bur. Mines *Minerals Yearbooks; 1933*, Nos. 433; *1934*, 627, *1935*, 691, *1936*, 625, *1937*, 935; U.S. Bur. Mines Tech. Paper No. 613, 1940, pp. 66–76.

[260] C. Berthelot and A. Hot, *Carburants de Synthese et de Remplacement*, Dunod, Paris, 1936, 350 pp.; *Technique et Economie Nouvelles des Carburants de Synthese*, Dunod, Paris, 1937, 369 pp.

[261] E. E. Donath, in Ref. 2, Ch. 22.

[262] Fuel Research Board, Dept. Sci. Ind. Research (U.K.) *Rept. for Year Ended March 31, 1936*, pp. 152–160.

[263] C. H. Fisher, G. C. Sprunk, A. Eisner, L. Clarke and H. H. Storch, *Ind. Eng. Chem.*, **31**, 190–195, 1155–1161, (1939); *Fuel*, **18**, 132–141, 196–203 (1939); **19**, 132–138, 162–172 (1940).

[264] C. C. Wright and G. C. Sprunk, *Pennsylvania State Coll.*, Mine. Ind. Exp. Sta. Bull., Nos. **26**, (1939), 32 pp; 28, 1939, 23 pp.

[265] C. H. Fisher, G. C. Sprunk, A. Eisner, H. J. O'Donnell, L. Clarke and H. H. Storch, U.S. Bur. Mines Tech. Paper No. 642, 1942, 162 pp.

[266] H. H. Storch, L. L. Hirst, C. H. Fisher and G. C. Sprunk, U. S. Bur. Mines Tech. Paper No. 622, 1941, 110 pp.

[267] L. Horton, J. G. King and F. A. Williams, *J. Inst. Fuel*, **7**, 85–97 (1933).

[268] T. E. Warren, R. W. Bowles and R. E. Gilmore, *Ind. Eng. Chem.*, **11**, 415–419 (1939).

[269] Fuel Research Board, Dept. Sci. Ind. Research (U.K.), *Rept. for the Year Ended March 1937*, pp. 149–151.

diameter industrial reactors, although it may be of considerable importance in engineering development work involving the use of small-diameter tubes.[270–272]

The effect of the halogens in enhancing the activity of stannous oxide, stannous hydroxide, ammonium molybdate, and nickel sesquioxide as liquid phase hydrogenation catalysts was demonstrated by the British[273,274] and the Japanese.[275,276] Results obtained with various catalysts on the liquid phase hydrogenation of coal are generally applicable to the liquid phase hydrogenation of tar.[277]

The preparation and performance of catalysts for the vapor phase hydrogenation of low-temperature tar and middle oil has been discussed by Storch.[256] Information on such catalysts as supported molybdenum oxide, pelleted molybdenum or tungsten disulfide, and many others, is also presented.

Tests on the effect of pressure on the liquid phase hydrogenation of coal showed little effect on the capacity of a small-scale plant in the 180–250 atm range.[278] However Pier[279] stated: "The conversion of oil coals or of asphalts of high molecular weight, which goes but slowly at 200 to 300 atmospheres, proceeds without difficulty and at a satisfactory rate for large scale production at considerably higher pressures, e.g., 700 atmospheres or more."

An increase in temperature in the 420–460°C range increases the yield of crude gasoline, middle oil, and hydrocarbon gas and decreases the yield of heavy oil (boiling at 300°C) and solid organic residue.[280,281] The hydrogen absorbed also increases with increasing temperature. The effect of temperature and contact time are analogous in that higher temperatures are equivalent to a longer contact time.[280]

Storch[256] points out:

> Coal is usually hydrogenated for the single purpose of producing hydrocarbon fuels. Consequently, there has been little incentive to study the characteristics of intermediate products such as the "middle" oil (boiling range 210 to 330°C) from liquid-phase coal hydrogenation. As a source of organic chemicals, however, this intermediate product is most interesting and merits examination. The middle oils from liquid-phase coal hydrogenation contain benzene, toluene, xylene, cyclohexane, methylcyclohexane, about 20 percent of tar acids (phenol, cresols, xylenols), and 2 to 5 percent of nitrogen bases. . . . [In addition] several patents mention the isolation of many polynuclear products. Sometimes the actual isolation is preceded by dehydrogenation. Products of this kind include pyrene, chrysene, retene, fluoranthene, carbazole, methyl- and dimethylpyrenes, 1, 12-benzoperylene, coronene, picene, fluorene, methylanthracene, and naphthalene.
>
> The products of the hydrogenation of coal are largely aromatic and naphthenic in nature, whereas appreciable yields (about 12 percent) of the paraffin hydrocarbons have been reported in the hydrogenation of German brown coals.

Tar may be hydrogenated under the same conditions as coal[261] except that no solid materials other than a small quantity of catalyst powder in the case of high-temperature tar must be introduced or removed from the pressure vessels. Low-temperature tar may be processed directly by mixing with recycle oil and hydrogen and passing the mixture

[270] F. Vallette, *Mem. Soc. Ing. Civ. Fr.*, **90,** 740–745 (1937).

[271] F. Vallette, *Bull. Soc. Encour. Ind. Natl.*, **135,** 353–365 (1936).

[272] G. T. Morgan and J. T. Veryard, *J. Soc. Chem. Ind.*, **57,** 152–162 (1938).

[273] Fuel Research Board, Dept. Sci. Ind. Research (U.K.) Tech. Papers No. 29, (1931), 37 pp.; No. 42, (1935), 58 pp.

[274] Fuel Research Board, Dept. Sci. Ind. Research (U.K.), *Rept. for the Year Ended March 1938,* pp. 206–207.

[275] R. Abe, S. Huzikawa, T. Kakutani, T. Okamura and K. Sugiyama, *J. Soc. Chem. Ind. Jap.*, **41,** Suppl. Bind., 417–421B (1938).

[276] S. Kurokawa, R. Hirota, M. Huziwara and N. Ascka, *J. Fuel Soc. Jap.*, **18,** 32–36A (1939).

[277] Fuel Research Board, Dept. Sci. Ind. Research (U.K.) Tech. Paper No. 40, 1935, 107 pp.

[278] N. Booth and F. A. Williams, *J. Inst. Fuel*, **11,** 493–502 (1938). Fuel Research Board, Dept. Sci. Ind. Research (U.K.), *Rept. for the Year Ended March 1937,* p. 155; *Report for the Year Ended March 1938,* p. 208.

[279] M. Pier, *Trans. Faraday Soc.*, **35,** 967–979 (1939).

[280] L. L. Hirst, et al., *Ind. Eng. Chem.*, **31,** 869–877 (1939).

[281] K. Gordon, *Trans. Inst. Min. Eng. (London)*, **82,** 348–359 (1931), *Proc. World Pet. Congr. London,* **2,** 317–321 (1933); *J. Inst. Fuel,* **9,** 69–84 (1935).

over a fixed catalyst bed in a high-pressure vessel.

The effect of process variables on the yield and distribution of products is essentially the same as that shown for coal in Storch's review[256] of tar hydrogenation, which also shows that this similarity exists in the nature of the products obtained from the two feedstocks, as well.

Developments in the hydrogenation of coal and tar in the liquid phase in commercial plants and in the USBM demonstration plant are reviewed by Donath,[261] who also discusses developments in vapor phase hydrogenation of primary products.

1.10 SYNTHESIS OF HYDROCARBONS FROM MIXTURES OF CARBON MONOXIDE AND HYDROGEN

The catalytic conversion of mixtures of carbon monoxide and hydrogen to hydrocarbons, which is referred to as the Fischer-Tropsch synthesis or the gas synthesis process, was developed and commercialized in the first part of this century. The details of laboratory research and industrial development have been reviewed by Storch.[282] A more comprehensive treatise on the Fischer-Tropsch and related syntheses was published in 1951.[283] The vast amount of work done on the Fischer-Tropsch synthesis in Germany, Great Britain, Japan, and the United States cannot be presented here. Instead, we quote the excellent brief summary of developments given in the book by Storch et al.[283]

> The Fischer-Tropsch synthesis had its origin in an observation made in 1923 by Franz Fischer and Hans Tropsch that alkalized iron turnings at 100–150 atm of hydrogen plus carbon monoxide and 400–450°C catalyzed the production of "synthol." The product consisted chiefly of oxygenated compounds and a very small quantity of hydrocarbon products at 7 atm and, later, at 1 atm Fischer found that the distribution of oxygenated and hydrocarbon products was reversed. The first Fischer-Tropsch catalyst was an iron–zinc oxide preparation which was reported by Fischer in 1925. It was tested at atmospheric pressure with a mixture containing 3 volumes of hydrogen and 1 volume of carbon monoxide, and at a temperature of 370°C. However, the development of iron catalysts was unsuccessful until 1937, which was several years after the Fischer-Tropsch was in commercial operation with a cobalt catalyst.
>
> An important advance in the development of Fischer-Tropsch catalysts was the precipitation of nickel-thoria on kieselguhr in 1931. This catalyst was followed in 1932 by the corresponding cobalt catalyst, 100 Co : 18 ThO_2 : 100 kieselguhr, parts by weight, which was the starting point in the development of the Ruhrchemie commercial catalyst. Subsequent changes in catalyst composition, concomitant with process development, resulted in a standard composition in 1937 of 100 Co : 18 ThO_2 : 200 kieselguhr and in 1938 of 100 Co : 5 ThO_2 : 8 MgO : 200 kieselguhr.
>
> The advantages of iron over cobalt, in being easily available from many sources and in imparting flexibility to the process, stimulated continuous research on iron catalysts. In 1937, alkalized, precipitated iron catalysts were successfully operated on a laboratory scale by the Kaiser-Wilhelm Institut für Kohlenforschung at 15 atm of water gas at temperatures of 235–250°C. . . .
>
> In the Fischer-Tropsch process as developed commercially by the Ruhrchemie A.G. in Germany in 1935–1940, synthesis gas containing 2 volumes of hydrogen per volume of carbon monoxide was compressed to about 7 atm and passed through a granular bed of cobalt catalyst at 185–205°C. The major products of the synthesis were wax, oil, water, gaseous hydrocarbons, and a minor amount of carbon dioxide. The hydrocarbons were largely straight-chain paraffins. The olefin content varied with pressure, contact time, and temperature. Improved modifications of the Ruhrchemie process involved the recycling of exit gas from the reactor and the use of iron to replace cobalt in catalyst preparations. These changes increased the olefin content of the product and the flexibility of the process, particularly in the control of the pro-

[282] H. H. Storch, in Ref. 1, Vol. 2, Ch. 39.

[283] H. H. Storch, N. Golumbic and R. E. Anderson, *The Fischer-Tropsch and Related Syntheses*, John Wiley and Sons, New York, 1951.

portions of gasoline, diesel oil, and alcohols.

1.11 GASIFICATION

1.11.1 General

As in the case of combustion, research and development work on coal gasification can be divided into basic studies and studies on the development of gasification processes. Work in the latter areas can be considered in three categories: (1) processes for making low-Btu or producer gas; (2) continuous processes for making gas of intermediate heating value that can be used directly for heating, as synthesis gas, or converted to high-Btu gas; and (3) intermittent processes for making blue water gas and carburetted water gas.

1.11.2 Basic Studies

In all gasification processes the endotherm of the steam-carbon reaction must be supplied by the combustion of some of the fuel with air or oxygen or by some other source of heat such as nuclear heat or electricity. Accordingly, the results of much of the basic work on combustion of carbon are also applicable to the gasification of carbon. However gasification also involves the reactions of carbon dioxide with carbon and steam with carbon.

At the beginning of this century considerable work had been done on the reaction of carbon dioxide with carbon to form carbon monoxide. This reaction is referred to as the "Boudouard reaction" because Boudouard was the first to study its equilibrium.[284] The reversal of this reaction can occur in the gas space above the fuel bed and is referred to as the "Neumann reversal" because Neumann was the first to direct attention to the possibility of its occurrence.[285] The carbon dioxide–carbon reaction and the steam-carbon reaction were studied extensively by Clement and co-workers[286] and by Rhead and Wheeler.[207,287] The former team presented data on the effect of temperature and time of contact on carbon dioxide or steam conversion, whereas Rhead and Wheeler published data on reaction rate constants and advanced the concept of a carbon surface covered with an adsorbed film of oxygen referred to as the "carbon-oxygen complex." Evidence was eventually presented of the existence of the carbon-oxygen complex[210,288–290] and its role in the mechanism of gasification reactions was considered.

Basic studies of gasification prior to 1963 have been thoroughly reviewed by Mayers,[221,291] Orning,[227] Haslam and Russell,[14] and von Fredersdorff and Elliott.[292] In addition to reviewing such studies, van der Hoeven[293] has summarized developments in the production of producer gas, and Morgan[294] has discussed developments in the production of blue water gas. Work on the gasification reactions done before 1933 has been reviewed by Logan[295] and Elliott,[296] who subsequently covered developments in the gasification field prior to 1940.[297]

As stated earlier, the reactivities of carbon to air, to oxygen, to steam, and to carbon dioxide are all parallel.[191] Therefore relative

[284] C. Boudouard, *Bull. Soc. Chem.*, **21,** 465–467, 713–715 (1899); **25,** 227–230, 833–840 (1901); *Ann. Chem. Phys.*, **24,** 1–85 (1901).

[285] K. Neumann, *Stahl Eisen*, **33,** 394–402 (1913); *J. Soc. Chem. Ind.*, **32,** 353, (1913); *Z. Verh. Dtsch. Ing.*, **58,** 1482–1484, 1501–1504 (1914).

[286] J. K. Clement, Univ. Illinois, Bull. No. 30, 1909, 47 pp.; J. K. Clement and H. A. Grine, U.S. Geol. Survey Bull. No. 393, 1910, pp. 15–27. J. K. Clement, I. H. Adams and C. N. Haskins, U.S. Bur. Mines Bull. No. 7, 1911, 58 pp.

[287] T. F. E. Rhead and R. V. Wheeler, *J. Chem. Soc*,. **103,** 461–489, 1210–1214 (1913).

[288] W. E. J. Broom and M. W. Travers, *Proc. R. Soc. London,* **A135,** 512–537 (1932).

[289] E. J. W. Verwey and J. H. deBoer, *Rec. Trav. Chem. Pays-Bas,* **55,** 674–687 (1936).

[290] A. Klemenc, R. Wechsberg and G. Wagner, *Z. Electrochem.*, **40,** 488–489 (1934).

[291] M. A. Mayers, *Chem. Rev.*, **14,** 31–53 (1934).

[292] C. G. von Fredersdorff and M. A. Elliott, in Ref. 2, Ch. 20.

[293] B. J. C. van der Hoeven, in Ref. 1, Vol. 2, Ch. 36.

[294] Jerome J. Morgan, "Water Gas," in Ref. 1, Vol. 2, Ch. 37.

[295] L. L. Logan, "An Investigation into the Manufacture of Water Gas with Especial Reference to the Decomposition of Steam," Ph.D. dissertation, Johns Hopkins University, 1929.

[296] M. A. Elliott, "The Catalytic Effect of Soda on the Reaction between Steam and Carbon from the Pyrolysis of low Grade Oils," Ph.D. dissertation, Johns Hopkins University, 1933.

[297] M. A. Elliott, "Gasification," in Ref. 12, pp. 33–44.

rates of reaction determined with one reactant are generally the same with any one of the other reactants.

Blakeley and Cobb[298] studied the effect of contact time, at two different temperatures, on the rate of reaction of carbon dioxide with a variety of cokes and other forms of carbon and also the effect of temperature at constant time of contact. Reactivity decreased with an increase in contact time because the more reactive parts of the surfaces had already been consumed. Similar effects would be expected with steam.

In a study of the influence of coke reactivity on the formation of water gas, Bunte and Giessen[195] observed that in the case of the more reactive cokes a large fraction of the steam was decomposed by the reaction, forming one mole of carbon monoxide and one mole of hydrogen per mole of carbon. They pointed out that the value obtained for the quantity of steam decomposed was not characteristic of the actual conditions in the fuel bed, but was influenced by the subsequent establishment of water gas shift equilibrium in the gas phase.

P. Dolch[299] analyzed the data of Bunte and Giessen[195] and of M. Dolch and Dietzel[300] and concluded that there was no experimental evidence that steam reacts with carbon to form carbon dioxide and hydrogen, but that the reaction forming carbon monoxide and hydrogen with the subsequent water gas shift reaction is of primary importance. In the experiments analyzed, water gas equilibrium was attained at a temperature of 800°C with reactive carbon materials such as lignite coke and wood charcoal. With less reactive cokes, such as gas coke, water gas equilibrium was not attained until temperatures approached 1000°C.

An extensive review of the detailed mechanisms of the carbon-oxygen, the carbon-oxygen-steam, the carbon–carbon dioxide, and carbon-steam reactions has been presented by von Fredersdorff and Elliott.[292]

In earlier studies of the steam-carbon reaction, the catalytic effect of ash on the water gas reaction was observed.[300,301] The first studies on the magnitude of the effect of catalyzing the steam-carbon reaction were made by Taylor and Neville[302] in 1921. They found that potassium and sodium carbonates were the most effective of all the catalysts investigated, and they clearly demonstrated that their effect was indeed catalytic. They believed that the catalysts accelerated the decomposition of the surface complex of carbon and oxygen suggested by Rhead and Wheeler,[207] thereby exposing a clean carbon surface that was reactive to carbon dioxide. Carbon dioxide produced in the water gas shift reaction then interacted with the carbon to form additional carbon monoxide, which in the presence of excess steam reacted as above to form additional quantities of carbon dioxide and hydrogen, thus increasing the net increase of gas from the steam-carbon reaction.

Marson and Cobb[303] discovered that sodium carbonate was the most effective additive in gasifying laboratory coke at 1000°C. It was found that additions of potassium carbonate and sodium sulfate were just as effective in catalyzing the steam-carbon reaction as was sodium carbonate. In the same investigation it was observed that the additives that produced a maximum effect on the steam-carbon reaction exhibited an analogous effect on the reduction of carbon dioxide by carbon.

Additives were effective in the range of concentrations from 0.5 to 5.0%. Sutcliffe and Cobb[304] studied the effect of temperature on coke containing 5% additives of various constituents. The accelerated effect was considerably greater at 600°C than at 1000°C. Similar studies were made by other investigators.[305]

Studies[306] of the mechanism of the catalysis of the carbon gasification reactions showed that the acceleration is brought about by the alternate reduction and reformation of the

[298] T. H. Blakeley and J. W. Cobb, *Inst. Gas Eng., Commun.*, **104,** *Gas World,* **101,** 452–455, 498–499, 612–613 (1934).

[299] P. Dolch, *Gas Wasserfach,* **75,** 807–811 (1932).

[300] M. Dolch and E. Dietzel, *Braunkohle,* **30,** 455, 467 (1931).

[301] J. Gwosdz, *Z. Angew. Chem.*, **31,** 137 (1918).

[302] H. S. Taylor and H. A. Neville, *J. Am. Chem. Soc.*, **43,** 2055–2071 (1921).

[303] C. B. Marson and J. W. Cobb, *Gas J.*, **175,** 882–891 (1926).

[304] J. A. Sutcliffe and J. W. Cobb, *Gas J.*, **178,** 895–901 (1927).

[305] B. Neumann, C. Kröger, and E. Fingas, *Z. Anorg. Allg. Chem.*, **197,** 321–328 (1931).

[306] D. A. Fox and A. H. White, *Ind. Eng. Chem.*, **23,** 259 (1931).

alkali carbonate in the following manner. The sodium carbonate first reacts with carbon, yielding carbon monoxide and sodium in a molecular ratio of 3 : 2. The sodium thus transported into the gaseous phase is at liberty to react with carbon dixoide, with the subsequent formation of carbon monoxide and sodium oxide. The sodium oxide may then further react with carbon dioxide to form sodium carbonate, which is deposited on the carbon surface and is therefore capable of further reaction.

The steam-carbon reaction may also be catalyzed when carbon is deposited on a refractory impregnated with a sodium-containing salt.[296]

In 1937 Dent[307] and co-workers laid the foundation for the hydrogasification of coal in studies of the kinetics of methane formation in the reaction of cokes and coals with hydrogen at elevated temperatures and at pressures ranging from 1 to 100 atm. von Fredersdorff and Elliott[292] also reviewed work on the mechanism and kinetics of the carbon-hydrogen reaction and work on the early development of the hydrogasification process.

1.11.3 Producer Gas

Several investigators studied the reactions occurring in the fuel bed of a gas producer.[14,308,309] From bottom to top, the zones are designated: ash zone; oxidation zone; primary reduction zone, in which the steam-carbon and carbon dioxide–carbon reactions occur; secondary reduction zone, in which the water gas shift and carbon dioxide reduction reactions occur; distillation zone, in which the volatile matter in the fuel is added to the gas; and gas space, in which a decrease in heating value may occur as a result of the so-called Neumann reversal of the Boudouard reaction.[14]

The effect of steam in the air blast of a producer was well known early in this century. Subsequent work established that the efficiency of the producer reached a maximum between 0.3 and 0.4 lb of steam per pound of coal.[310,311] However this range may have to be exceeded in practice to avoid clinker formation.

Work on the performance of full-scale producers has been summarized.[14,292,293] Van der Hoeven[293] discusses a variety of theoretical, thermodynamic, and stoichiometric analyses of the producer gas process.

Development work on the slagging gas producer was done early in this century.[293] Data from tests on those producers[293,312] and on some of their further developments and operating characteristics are described by Guillon.[313]

1.11.4 Continuous Production of Intermediate Heating Value Gas

The use of oxygen in the blast of a producer was advocated in the early 1920s.[314–317] Experimental work done first at atmospheric pressure led eventually to the development of the Lurgi process, which operated at 20 atm.[318] Data on the effect of pressure in increasing the heating value of gas from the Lurgi process were presented by Hubmann[319] and Danulat.[320]

[307] F. J. Dent, W. H. Blackburn and H. C. Millett, *Inst. Gas Eng., Commun.*, **167,** (1937), 47 pp.; **190** (1938), 69 pp.

[308] W. A. Bone and R. V. Wheeler, *J. Iron Steel Inst.*, **73,** 126–160 (1907).

[309] W. A. Bone and R. V. Wheeler, *J. Iron Steel Inst.*, **73** 126–160, 1907; **78** 206–233 (1908).

[310] R. T. Haslam, I. T. Ward, and R. F. Mackie, "Reactions in the Fuel Bed of a Gas Producer. III. Effects of Steam-Coal Ratio," *Ind. Eng. Chem.*, **19** (1), 141–144 (1927).

[311] J. K. Clement, *J. Iron Steel Inst.*, **107,** 97 (1923).

[312] A. Heczko, *12th Congr. Chem. Ind.*, Prague, 1932, pp. 25–29; *Chim. Ind.*, Special No., June 1933, pp. 420–426.

[313] H. Guillon, *Chal. Ind.*, **16,** 483–484, 518–524, 560–562 (1935), **17** *153–156, 191–197, 230–234, 281–285, 451–454 (1936),* **19,** *344–348, 533–540 (1938).*

[314] H. J. Hodsman and J. W. Cobb, *Trans. Inst. Gas Eng.*, **18,** 431–452 (1919–1920); *Gas J.*, **150,** 640–647 (1920).

[315] H. J. Hodsman, *Gas World*, **72,** 492–502 (1920); *Gas J.*, **153,** 92 (1921).

[316] E. A. W. Jefferies, *Gas Age*, **47,** 145–150 (1921); *Gas Rec.*, **19,** 45–55 (1921).

[317] L. J. Willien, *Proc. Am. Gas Asso.* **1923,** 969–974.

[318] R. Drawe, *Gas Wasserfach*, **69,** 1013–1015 (1926), **76,** 541–545 (1933), **80,** 806–810 (1937); *Gas J.*, **203,** 95 (1933); *Arch. Waermewirtsch.*, **19,** 201–203 (1938).

[319] O. Hubmann, *Metallges. Period Rev.*, **8,** 9–15 (1934).

[320] F. Danulat, *Mitt. Metallges.*, No. 13, 14–22 (1938).

The Lurgi process, commercialized in 1936, made possible the continuous production of gas having a heating value as high as 450 Btu/ft^3. The results of tests with various fuels in large gasifiers were summarized in a review of gasification processes.[292]

Development work on slagging gasifiers operating at elevated pressure was begun in the late 1950s.[321,322]

The Winkler generator, which used oxygen or oxygen-enriched air and steam to gasify lignite or semicoke, was developed and used commercially in the 1930s.[323,324] In this process a fluidized bed was used for the first time in gasifying solid fuel.

In their review of coal gasification, von Fredersdorff and Elliott[292] summarized development work on entrained or suspension gasification by the USBM, the Institute of Gas Technology, and Ruhrgas. Information was also given on the two commercial suspension gasification processes: Koppers-Totzek, operating at atmospheric pressure, and Texaco, operating at 400–500 psig. Other contacting systems and processes were described, including the Leuna and Thyssen-Galoczy slagging generators, the Winkler fluidized-bed gasifier, the Panindco process, the Rummel slag bath gasifier, and the Flesch-Demag process.

1.11.5 Carburetted Water Gas

The production of carburetted water gas reached a peak in 1926 when it represented about 60% of the total manufactured gas in the United States.[294] To put this situation into perspective, the energy in this quantity of gas represents about 0.5% of the energy in the natural gas produced in 1976. Since by 1977 carburetted water gas was not being produced in the United States, developments are mentioned only briefly. These have been discussed[294–296,325] and include the application of automatic controls, the use of mechanical grates, the substitution of heavy oil for gas oil, the application of the back-run process, the re-forming of refinery and natural gas in water gas generators, and the production of high-Btu gas that could be mixed with or even substituted for natural gas.

[321] D. Hebden and R. F. Edge, Gas Council (U.K.) Res. Commun. No. GC 50, 1958, 29 pp.; No. GC 50/44, 1959, 8 pp.

[322] K. I. Kolodtsov and V. I. Babii, *Teploenergetika*, **5**: (5), 25–31 (1959).

[323] Anon., *Z. Verh. Dtsch. Ing.*, **79**, 1094 (1935).

[324] C. Bosch, *Chem. Fabrik*, **7**, 1–10 (1934).

[325] J. J. Morgan, *Manufactured Gas*, Vol. 1, National Process Co. New York, 1926, 518 pp.

1.12 CONCLUDING STATEMENT

This chapter has covered briefly the history of the coal industry, except for coal mining, and many aspects of research and development work on the nature of coal and on its utilization, from the inception of the industry up to about 1960. The subsequent chapters cover, in considerable detail, research and development work on coal and coal utilization for the period from about 1960 to around mid-1976 for most chapters, mid-1977 for the remainder. Thus with its brief coverage of work done before about 1960, this supplement gives a picture of the present (mid-1976 to mid-1977) status of research and development work on coal and its utilization and on how that status was reached. For a comprehensive understanding of how we arrived at the current state of research and development on coal, Volumes I and II and the first supplementary volume of this series should be consulted.

As stated earlier, research and development work on coal has been accelerating for the past several years. Many companies, institutions, and people without previous association with coal have entered the field. The need to make available to such people and organizations an authoritative, up-to-date review of the status of coal research and development was the major factor responsible for the decision to publish this second supplement to *Chemistry of Coal Utilization*, and the Steering Committee felt that the overview material presented in this chapter would be useful.

The "first golden age" of coal began in the middle of the nineteenth century and reached its peak in the first quarter of the twentieth century. We are most certainly on the threshold of the "second golden age" of coal for the following reasons:

1. Coal is the only fossil fuel resource available in sufficient quantity to make up for declining U.S. and eventually world resources of petroleum and natural gas.

2. Future energy demands are so great that coal will supply a significant fraction of those demands even when advanced nuclear technologies are developed.
3. The currently expanding research and development work on coal will provide the technology for the conversion of coal to environmentally acceptable fuels and for the direct use of coal in an environmentally acceptable manner.

Why is an expanded research and development program necessary when there were commercial-size coal liquefaction plants in Germany during World War II and when manufactured gas at one time constituted a major portion of the gas distributed to towns and cities? Much of the older technology cannot be used today for environmental, technical, and economic reasons. For example, the high-Btu oil gas process, which made a gas that was interchangeable with natural gas, was a cyclic process. The generator was blown with air, which was discharged to the atmosphere, then fed with steam and oil to be cracked during the gas-making period. Cleaning the blast gas in a cyclic process is a formidable problem. In addition, the high price of oil and its long-range availability make the use of this technology uneconomic. The carburetted water gas process suffers from the foregoing problems; in addition, it requires expensive coke for part of its feed, and its product is not interchangeable with natural gas.

The technology under development will culminate in processes that operate continuously, require coal only as a feed, and produce fuel gases or synthesis gases that can be converted to chemicals or substitute natural gas or other fluid fuels.

When the technology for producing synthetic liquid fuels from coal was employed in Germany during World War II, the aim was to approach 100% conversion of coal to liquids. To minimize the use of catalysts, extremely high pressures were needed. This requirement would make such a technology uneconomic today.

The technology under development does not try to achieve 100% conversion. Accordingly, it is able to operate at less extreme pressures and seeks to reach a level of conversion such that the residue can be gasified to supply the hydrogen required for the process.

The new technology for the "second golden age" of coal has just been mentioned. This technology and the concomitant basic research will probably be the basis for the third supplement to *Chemistry of Coal Utilization*. These developments are progressing at such a rapid pace that the period covered by the next supplement is likely to be significantly shorter than the period covered by this one.

CHAPTER 2

Coal Resources

PAUL AVERITT

Paul Averitt *was employed by the U.S. Geological Survey from 1938 to 1976. He was active in work on coal and coal resources from the late 1940s until his retirement. He has written many reports on coal and several on the coal resources of the United States, published in 1975 and earlier. As a form of occupational therapy, he has also worked from time to time on the stratigraphy and structure of southwestern Utah and, more recently, in 1977 and early 1978, he served as Interim Science Editor for the Geological Society of America.*

2.1 INTRODUCTION

Coal is widespread and abundant in most parts of the United States, and like petroleum and natural gas, it has contributed significantly to our industrial and economic growth. Of the three fossil fuels, coal is by far the most abundant. On the basis of an adjusted and weighted analysis of data available on resources of fossil fuel in the United States as of January 1, 1974, the recoverable resources of coal contain about 10 times more heat value than the combined recoverable resources of petroleum and natural gas. Use of coal, however, has lagged behind use of petroleum and natural gas because those fuels have been cleaner to burn, easier to handle, and in the case of natural gas and certain fuels derived from petroleum, lower in cost. Nevertheless production of coal since 1940 has remained in the range of 400–600 million tons annually. In 1974 and 1975, 610 and 654 million tons, respectively, were produced, and production is certain to increase in the future because of the decreased availability and concomitant increased cost of petroleum and natural gas.

The basic information on coal in the United States is contained in about 1500 detailed geologic reports published by the U.S. Geological Survey, and in about the same number of reports published by other agencies and organizations, including state Geological Surveys, the U.S. Bureau of Mines, and professional societies. Additional information is contained in technical journals and in records of coal mining companies, railroads, and land-holding companies. From these various sources, most states have prepared summary reports on the geology and occurrence of coal, including estimates of coal resources.

This chapter, which is a condensation of a longer report[1] based in part on the state summary reports, is a discussion and analysis of total U.S. coal resources. Resource terms are defined in the glossary, Section 2.12.

2.2 CLASSIFICATION OF COAL

2.2.1 Rank of Coal

Rank is a way of expressing the progressive metamorphism of coal from lignite (low in rank) to meta-anthracite (high in rank). It is based on relative heat value, which is calculated on a moist, mineral-matter-free basis for low rank coals; and on percentage of fixed carbon, which is calculated on a dry, mineral-matter-free basis for higher rank coals.[2] As shown in Fig. 2.1, the heat value and the percentage of fixed carbon increase from lignite to low-volatile bituminous coal as the percentages of moisture and volatile matter decrease. The decrease in heat value in semianthracite and coal of higher ranks is caused by the pronounced decrease in volatile matter shown in the bottom part of Fig. 2.1.[1] The changes in rank result mainly from the weight of overlying sedimentary rock, the heat produced by depth of burial, time, and structural deformation, all of which contribute to progressive compaction and devolatilization. Rank is quite independent of grade, which is a way of expressing quality. Since coals of different ranks are adapted to

[1] Paul Averitt, *Coal Resources of the U.S., January 1, 1974,* U.S. Geol. Survey Bull. No. 1412, 1975 (reprinted 1976), 131 pp.

[2] ASTM, Designation D 388-66, "Standard Specifications for Classification of Coals by Rank," reapproved 1972, in *Gaseous Fuels, Coal, and Coke; Atmospheric Analyses,* ASTM, Pt. 26, 1974, pp. 54–58.

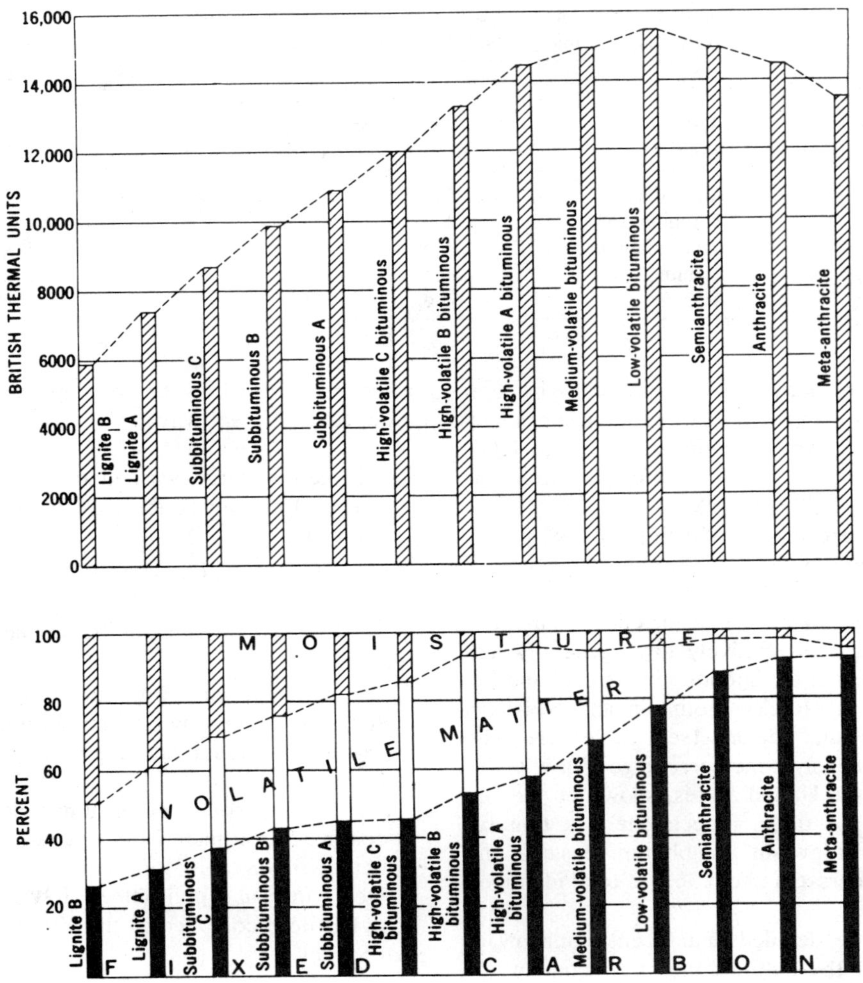

Figure 2.1 Comparison on moist, mineral-matter- free basis of heat values and proximate analyses of coal of different ranks. Source: Ref. 1, fig. 3, p. 17.

different uses, rank is a major basis of differentiation in coal resource calculations.

2.2.2 Grade of Coal

Coal is classified by grade largely according to the ash content, the ash fusion temperature, the sulfur content, and the presence of other deleterious constituents. Thus far in work on coal resources, a preliminary classification on the basis of sulfur content has been made, but classification on the basis of ash content has not been made because the content of ash is more variable than that of sulfur.

In recent years information on trace elements in coal has increased significantly, but classification according to the content of trace elements is not yet possible.

2.2.2.1 *Sulfur*

Sulfur is an undesirable element in coal. It lowers the quality of coke and of the resulting iron and steel products. It contributes to corrosion, to the formation of boiler deposits, and to air pollution. Its presence in spoil banks inhibits the growth of vegetation, and as sulfuric acid it is the main deleterious compound in the acid mine waters that contribute to stream pollution.

The sulfur content of coal in the United States ranges from 0.2 to about 7.0%, but the average in all coal is 1.0–2.0%. Most of the sulfur, perhaps 40–80%, occurs as a constituent of pyrite and marcasite (FeS_2). The

Table 2.1 Distribution of Identified U.S. Coal Resources According to Rank and Sulfur Content[4]

	Sulfur Content		
Rank	Low: 0–1%	Medium: 1.1–3.0%	High: 3+%
Anthracite	97.1	2.9	—
Bituminous coal	29.8	26.8	43.4
Subbituminous coal	99.6	0.4	—
Lignite	90.7	9.3	—
All ranks	65.0	15.0	20.0

remainder occurs as organic sulfur in combination with the coal-forming vegetal material, as hydrous ferrous sulfate ($FeS_4 \cdot 7H_20$) derived by the weathering of pyrite, and as gypsum ($CaSO_4 \cdot 2H_2O$).[3]

The percentages of sulfur and of pyritic sulfur are highest in bituminous coals of Pennsylvanian age in the Appalachian, Illinois, and Western Interior coal basins. The percentage is relatively low—generally less than 1%—in the subbituminous coal and lignite of the Rocky Mountain and Northern Great Plains regions (see Table 2.1). The conspicuously large percentage of low-sulfur coal in the United States, shown on the last line of the table, reflects the fact that resources of low-sulfur subbituminous coal and lignite represent about 56% of total identified resources.

A more detailed and recent summary of data on the sulfur content of coal in the demonstrated coal reserve base is contained in two reports of the U.S. Bureau of Mines.[5,6]

2.2.2.2 *Minor Elements*

Coal contains small quantities of virtually all metallic and nonmetallic elements, which were introduced into the coal bed in one or all of four different ways:

1. As inert material washed or blown into the coal swamp at the time of plant accumulation.
2. As biochemical precipitates from the swamp water.
3. As minor constituents of the original plant cells.
4. As later additions, introduced after coal formation, primarily by groundwater moving downward and laterally.

An excellent summary of minor elements in coal has been made by Sun et al.,[7] and a selected bibliography on minor and trace elements in coal, applicable mainly to U.S. coal, has been compiled by the U.S. Geological Survey.[8]

2.3 DISTRIBUTION OF COAL IN THE UNITED STATES

Coal-bearing rocks underlie about 13% of the land area of the 50 states and about 14% of the land area of the 48 conterminous states (Figs. 2.2 and 2.3[9]). The rocks are present in

[3] F. E. Walker and F. E. Hartner, *Forms of Sulfur in U.S. Coals,* U.S. Bur. Mines Inf. Circ. No. 8301, 1966, 51 pp.

[4] J. A. De Carlo, E. T. Sheridan, and Z. E. Murphy, *Sulfur Content of United States Coals,* U.S. Bur. Mines Inf. Circ. No. 8313, 1966, 44 pp.

[5] U.S. Bureau of Mines, *The Reserve Base of U.S. Coals by Sulfur Content* (2 parts), Pt. 1, *The Eastern States,* U.S. Bur. Mines Inf. Circ. No. 8980, 1975, 537 pp.

[6] U.S. Bureau of Mines, *The Reserve Base of U.S. Coal by Sulfur Content* (2 parts), Pt. 2, *The Western States,* U.S. Bur Mines Inf. Circ. No. 8693, 1975, 322 pp.

[7] S. C. Sun, H. Vasquez-Rosas, and D. Augenstein, *Pennsylvania Anthracite Refuse—A Literature Survey on Chemical Elements in Coal and Coal Refuse,* Pennsylvania State Univ. Spec. Res. Rep. No. SR-82, 1971, 72 pp.

[8] U.S. Geological Survey, *Minor and Trace Elements in Coal—A Selected Bibliography of Reports in English, January 1976,* U.S. Geol. Survey Open-File Rep. No. 76-481, 1976, 16 pp.

[9] F. F. Barnes, Coal Fields of the United States—Sheet 2, Alaska, U.S. Geol. Survey Map, scale 1:5,000,000, 1961.

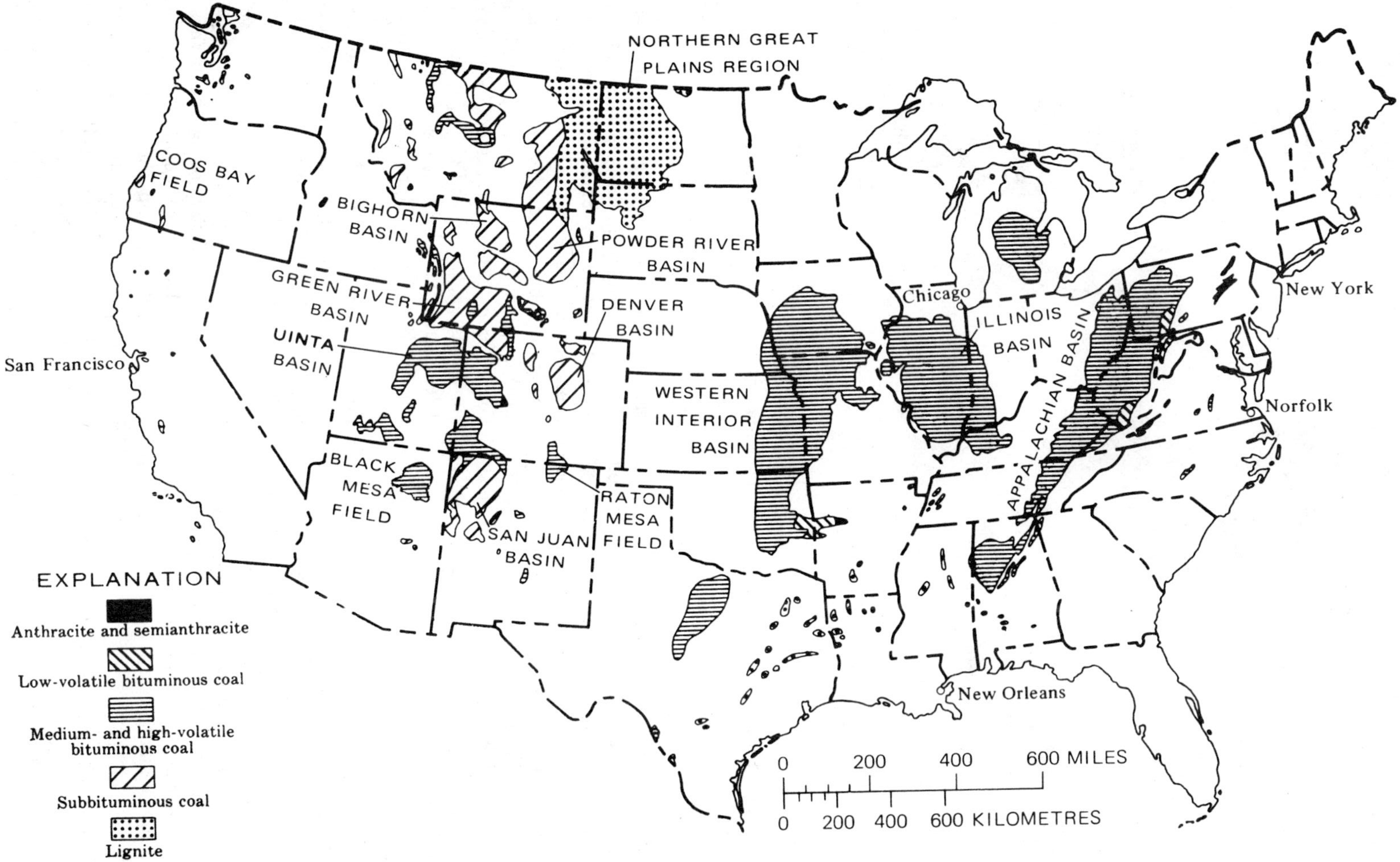

Figure 2.2 Coalfields of the conterminous United States. Source: Ref. 1, fig. 1, p. 5.

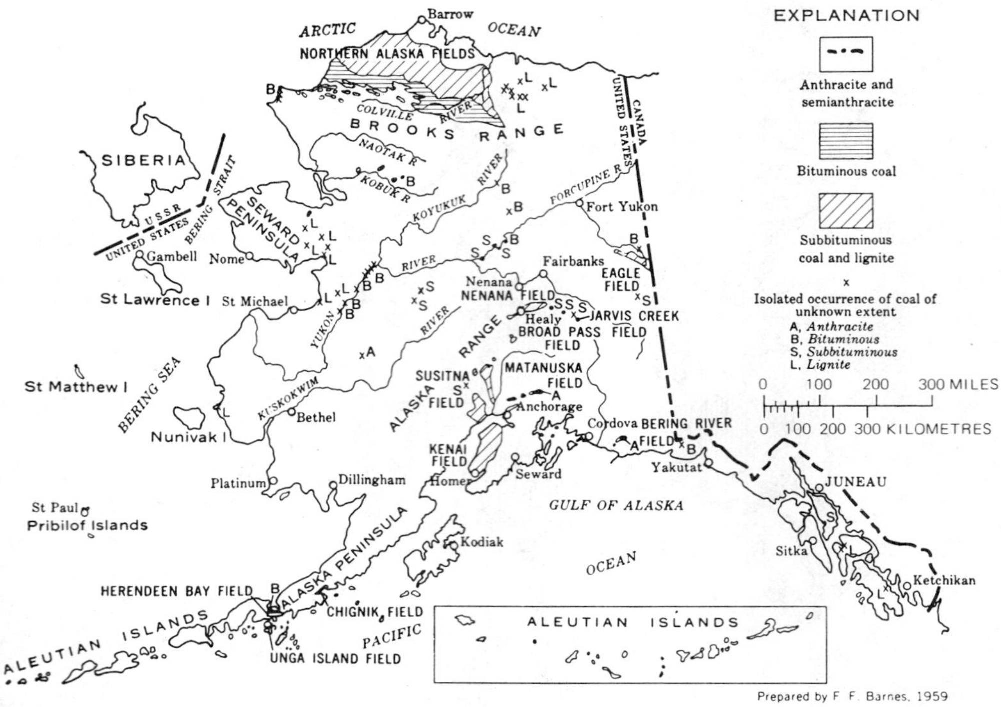

Figure 2.3 Coalfields of Alaska. (Prepared by Barnes as noted thereon.) Source: Ref. 1, fig. 2, p. 6.

37 states; in many states the coal-bearing areas represent a substantial percentage of the total area of the state, and in a few states, like Illinois and West Virginia, the percentage is more than half. Coal-bearing rocks range in thickness from a few hundred feet to somewhat more than 10,000 ft, but in most coal-bearing areas they are typically less than 3000 ft thick.

2.3.1 Number of Coal Beds

Coal beds are distributed irregularly but in substantial numbers throughout the sequences of coal-bearing rock. Of the 15 states where coal bed correlations are sufficiently well established to permit statewide application of coal bed names, five states have 60–117 named and described coal beds, five have 36–53, and five have 2–24.

2.3.2 Main Structural Basins

In most coalfields the coal-bearing rocks and their enclosed coal beds lie in structural basins or synclines, the largest of which are broad and shallow. In the Appalachian basin most of the coal is generally less than 3000 ft below the surface. In the Illinois and Western Interior basins, most of it is less than 2000 ft below the surface. In the Northern Great Plains region of eastern Montana, North Dakota, and South Dakota, all the coal is less than 1500 ft below the surface. In the Powder River basin of northeastern Wyoming, almost all the coal to the base of the Fort Union Formation is less than 2000 ft below the surface. In the San Juan basin of northwestern New Mexico and southwestern Colorado, most of the coal is less than 4000 ft below the surface. In the Raton Mesa field of Colorado and New Mexico, most of the coal is less than 2000 ft below the surface.

Other coal basins, particularly those in the Rocky Mountain region and in the Pacific Northwest, are very deep. In the Green River basin of southwest Wyoming, the coal-bearing rocks are as much as 15,000 ft below the surface; and in the Wind River and Bighorn basins of central and northern Wyoming, and in the Uinta basin of Utah and Colorado, they are as much as 20,000 ft below the surface in the deepest parts of the basins.

In these and other deep basins, the coal-bearing rocks dip steeply basinward, and only a few miles from the outcrops they are more than 6000 ft below the surface. In such structural settings, resource estimates, which must be restricted to narrow bands of accessible coal near the outcrops, tend to be conservative.

The occurrence of coal-bearing rocks in the United States in two diverse structural settings—many large, shallow basins, and a few very deep ones—accounts for the concentration of coal resources in the shallower overburden categories.

2.4 COAL RESOURCES IN THE UNITED STATES

The coal resources in the United States are classified in two major categories: identified resources and hypothetical resources. Identified resources include coal in beds that have been mapped by tracing surface exposures or explored by drilling. Tonnage is calculated in the measured, indicated, and inferred resource categories, and in additional categories according to rank, thickness of bed, and thickness of overburden (see Glossary).

Hypothetical resources include coal in unmapped and unexplored parts of known coal-field areas. The approximate magnitude of hypothetical resources has been estimated on a state-by-state basis by extrapolating from information on nearby areas of identified resources. The extrapolation is based on the area and thickness of coal-bearing rock and the tonnage of coal in that rock in nearby areas of identified resources. This information is then applied to the area and thickness of coal-bearing rock in the unmapped and unexplored areas.

2.4.1 Resources Versus Reserves

"Resources" is an umbrella term comprising many separate categories of coal in the ground within specified minimum limits for thickness of beds and maximum limits for thickness of overburden. The term has no economic connotation.

The "demonstrated reserve base" comprises coal in the ground considered suitable for mining by current methods. It represents a small, choice part of identified resources, and it is the only category that has an economic connotation.

As determined by analysis and summation of information from many sources, the remaining U.S. coal resources as of January 1, 1974, totaled 3968 billion short tons, distributed in several major resource categories (Table 2.2).

The data in Table 2.2, and in the more detailed Tables 2.3–2.6, are more significant in the trillions and billions columns and less significant in the millions columns. In general, the grand total of 3968 billion tons may be expressed as 4 trillion tons with no apprecia-

Table 2.2 Estimated Remaining (in the ground) Coal Resources of the United States, January 1, 1974

Category	Billions (10^9) of Short Tons
1. Identified (measured, indicated and inferred) resources	
Demonstrated reserve base (Table 2.5)	434
Additional identified resources	1297
Total identified resources (Table 2.3)	1731
2. Hypothetical resources	
0–3000 ft overburden (Table 2.4)	1849
3000–6000 ft overburden (Table 2.4)	388
Total hypothetical resources	2237
3. Total remaining resources (Table 2.4)	3968

ble loss of accuracy; but the total of 3968 billion tons and the other totals and subtotals presented in this chapter have arithmetic value because they facilitate accounting and aid in preserving and identifying the many smaller individual figures on which they are based.

2.4.2 Identified Resources

The estimate of 1731 billion short tons for remaining identified resources as of January 1, 1974, is given by states and by ranks of coal in Tables 2.3 and 2.4. Most of the estimates in Table 2.3 were obtained from summary reports on coal in the individual states, as cited in the right-hand column of the table. These reports present data on the occurrence and distribution of coal in many resource categories, and they contain information on the stratigraphy of coal-bearing rocks, and on the thickness, continuity, and composition of individual coal beds. Most of them include bibliographies for sources of more detailed information. Summary reports are invaluable in the initial or comprehensive phases of study of coal in any state, but they are not substitutes for the larger number of detailed reports on which they are based.

The estimates by states in Table 2.3 are based primarily on mapped coal beds and on measurements of coal thicknesses along coal outcrops, supplemented by information in mine workings and drill holes downdip from the outcrops. The information is concentrated in the 0–1000 ft overburden category, but in some relatively small areas data on coal at greater depths are available.

Estimates of identified resources are of great importance for several reasons: (1) they are based on fact, (2) they include accessible coal of current economic interest (see Section 2.4.3.1, "Demonstrated Reserve Base"), (3) they aid in selecting areas favorable for further exploration and development and in planning industrial expansion, and (4) they provide data on which estimates of coal in the deeper and less accessible parts of the coal basins may be obtained by extrapolation.

Although they are based on detailed information accumulated slowly by the laborious processes of mapping outcrops of coal beds and drilling holes to test coal thickness, the estimates of identified resources in Table 2.3 are still minimum estimates, and they may increase in the future as mapping, prospecting, and development continue.

2.4.3 Distribution of Identified Resources in Selected Categories

To serve different interests and needs, coal in each identified category has been divided into various subcategories. They are: Demonstrated Reserve Base, Region, Rank, Thickness of Overburden, Reliability of Estimates, Thickness of Beds, and a combined category of Overburden, Reliability, and Thickness of Beds.

2.4.3.1 *Demonstrated Reserve Base*

The demonstrated reserve base is a selected portion of the identified resources deemed suitable for mining by 1974 methods. The coal in the demonstrated reserve base is: (1) in the measured and indicated (demonstrated) resource category, (2) in beds 28 or more in. thick for bituminous coal and anthracite, and 60 in. or more thick for subbituminous coal and lignite, and (3) in the 0–120 ft overburden category for lignite, which is deemed suitable only for strip mining, and in the 0–1000 ft overburden category for the higher ranks of coal, which is deemed suitable for strip, auger, and underground mining methods. The reserve base may also include coal that does not fit these parameters, if it is being mined locally or is considered to be commercially minable.[10] The demonstrated coal reserve base of the United States by states and by method of mining is shown in Table 2.5.

The data in Table 2.5 are for coal in the ground, at least 50% of which is recoverable. This portion is termed "reserves" as distinguished from the reserve base. To avoid any possible ambiguity, "reserves" may also be termed "recoverable reserves."

A substantial amount of information on the sulfur content of coal in the demonstrated reserve base has been published by the U.S. Bureau of Mines (Section 2.2.2, "Grade of Coal").

[10] U.S. Bureau of Mines, *Demonstrated Coal Reserve Base of the U.S. on January 1, 1974*, U.S. Bur. Mines Min. Ind. Surveys, June 1974, 6 pp. (see also Ref. 33).

Table 2.3 Identified Coal Resources of the United States, January 1, 1974

In millions (10^6) of short tons. Estimates include beds of bituminous coal and anthracite generally 14 in. or more thick, and beds of subbituminous coal and lignite generally 2½ ft. or more thick to an overburden depth of 3000 ft. Figures are for resources in the ground. Of the reported tonnage, 91% is less than 1000 ft below the surface, and most of the remainder is 1000–2000 ft below the surface. Dashes mean negligible, or not pertinent.

			Resources Depleted to Jan. 1, 1974			
State	Type of Estimate[a]	Estimated Original or Remaining Identified Resources	Production[b]	Production Plus Loss in mining[c]	Remaining Identified Resources, Jan. 1, 1974	Source of Estimate
					Bituminous Coal	
Alabama	R-1958	13,754	246[d]	492	13,262	Culbertson (1964)
Alaska	Orig.	19,429	8	16	19,413	Barnes (1967)
Arizona	Orig.	21,250[e]	8	16	21,234	Peirce et al. (1970)
Arkansas	Orig.	1,816	89	178	1,638	Haley (1960)
Colorado	Orig.	110,003	443	886	109,117	Holt *(in press)*[f]
Georgia	R-1945	24	—	—	24	Johnson (1946)[f]
Illinois	R-1972	146,255[g]	127[d]	254	146,001	Hopkins and Simon (1974; written commun., 1974)[f]
Indiana	R-1965	33,238	185[d]	370	32,868	Wier (1973)
Iowa	Orig.	7,237	366	732	6,505	Landis (1965)
Kansas	R-1957	18,706	19[d]	38	18,668	Schoewe (1958)[f]
Kentucky						
Eastern	Orig.	33,440	2,607	5,214	28,226	Huddle et al. (1963)
Western	Orig.	38,878	1,379	2,758	36,120	Huddle et al. (1963)
Maryland	R-1950	1,200	24[d]	48	1,152	Averitt (1974)
Michigan	Orig.	297	46	92	205	Cohee et al. (1950)
Missouri	Orig.	31,860[h]	338	676	31,184	Robertson (1971; 1973)
Montana	Orig.	2,363	32	64	2,299	Combo et al. (1949)
New Mexico	Orig.	10,948	100	200	10,748	Read et al. (1950)
North Carolina	Orig.	112	1	2	110	Reinemund (1949; 1955)
Ohio	Orig.	46,488	2,661	5,322	41,166	Brant and DeLong (1960)[f]
Oklahoma	R-1974	7,117	0	0	7,117	Friedman (1975)[f]
Oregon	Orig.	50	—	—	50	R. S. Mason (written commun., 1965)[f]
Pennsylvania	R-1970	64,552[i]	306[l]	612	63,940	Edmunds (1972)[f]
Tennessee	R-1959	2,748	109[l]	218	2,530	Luther (1959; written commun., 1965)
Texas	Orig.	6,100	26	52	6,048	Mapel (1967)
Utah	R-1971	23,216[j]	15[d]	30	23,186	Doelling (1973); Doelling and Graham (1972a,b)

Table 2.3 (*cont.*)

	Type of Estimate[a]	Estimated Original or Remaining Identified Resources	Resources Depleted to Jan. 1, 1974: Production[b]	Resources Depleted to Jan. 1, 1974: Production Plus Loss in mining[c]	Remaining Identified Resources, Jan. 1, 1974	Source of Estimate
Virginia	Orig.	11,696	1,240	2,480	9,216	Andrew Brown et al. (1952)
Washington	R-1960	1,869	1[d]	2	1,867	Beikman et al. (1961)
West Virginia	Orig.	116,618	8,234	16,468	100,150	Headlee and Nolting (1940)
Wyoming	Orig.	13,235	266	532	12,703	Berryhill et al. (1950)
Other states[k]	Orig.	610	—	—	610	
Total		785,109	18,876	37,752	747,357	
				Subbitumious Coal		
Alaska	Orig.	110,696[l]	15	30	110,666	Barnes (1967)
Arizona	Orig.	—[e]	—	—	—	Peirce et al. (1970)
Colorado	Orig.	19,985	126	252	19,733	Holt (in press)[f]
Montana	Orig.	177,151[m]	166	332	176,819	Combo et al. (1949)[f]
New Mexico	Orig.	50,801	81	162	50,639	Read et al. (1950)[f]
Oregon	Orig.	290	3	6	284	R. S. Mason (written commun., 1965)[f]
Utah	R-1971	173[j]	—	—	173	Doelling (1973)
Washington	R-1960	4,194	7[d]	14	4,180	Beikman et al. (1961)
Wyoming	Orig.	123,656[l]	208	416	123,240	Berryhill et al. (1950): Glass (1972)
Other states[n]	Orig.	40	4	8	32	
Total		486,986	610	1,220	485,766	
				Lignite		
Alabama	Orig.	2,000[o]	—	—	2,000	Daniel (1973)
Alaska	Orig.	—[l]	—	—	—	Barnes (1967)
Arkansas	Orig.	350	—	—	350	Haley (1960)
Colorado	Orig.	20	—	—	20	Soister (1974)
Kansas	Orig.	—[p]	—	—	—	Schoewe (1952; 1958)
Montana	Orig.	112,533[m]	6	12	112,521	Combo (1949)[f]

North Dakota	Orig.	350,910	154	308	350,602	Brant (1953)
Oklahoma	Orig.	—[p]	—	—	—	Trumbull (1957)
South Dakota	Orig.	2,187[q]	1	2	2,185	M.J. Tipton (written commun., 1973); D. M. Brown (1952)
Texas	Orig.	10,426[r]	85	133	10,293	Kaiser (1974)
Washington	R-1960	117	—	—	117	Beikman et al. (1961)
Wyoming	Orig.	—[t]	—	—	—	Berryhill et al. (1950)
Other states[s]	Orig.	50	2	4	46	
Total		478,593	248	459	478,134	
Anthracite and Semianthracite						
Alaska	Orig.	—[t]	—	—	—	Barnes (1951)
Arkansas	Orig.	456	14	28	428	Haley (1960)
Colorado	Orig.	90	6	12	78	Landis (1959)
New Mexico	Orig.	6	2	2	4	Read et al. (1950)
Pennsylvania	R-1965	19,000	94[d]	188	18,812	Arndt et al, (1968)
Virginia	Orig.	355	10	20	335	Andrew Brown et al. (1952)
Washington	R-1960	5	—	—	5	Beikman et al. (1961)
Total		19,912	125	250	19,662	
Total all ranks		1,770,600	19,859	39,681	1,730,919	

Note: For References for Tables 2.3 and 2.4, see pp. 88 and 89.

[a] R, remaining resources in the ground as of January 1 of the year indicated; Orig., original resources in the ground before the advent of mining.

[b] Production, 1800–1885, from Eavenson (1942); 1886–1923, from U.S. Geological Survey (1886–1923); 1924–1972 from U.S. Bureau of Mines (Minerals Yearbooks, 1924–1972); 1973, from U.S. Bureau of Mines (1974a). For a few states, production records of state coal mine inspectors were used.

[c] With a few minor exceptions, past losses in mining assumed to equal past production.

[d] Production from year remaining resources were estimated through 1973.

[e] All tonnage is in the Black Mesa field. Some coal in the Dakota Formation is near the rank boundary between bituminous and subbitiminous coal. Does not include small resources of thin and impure coal in the Deer Creek and Pinedale fields.

[f] See other summary reports on coal resources in individual states as follows: Colorado (Hornbaker and Holt, 1973; Landis, 1959; Shomaker and Holt, 1973); Georgia (Butts and Gildersleeve, 1948; Sullivan, 1942); Illinois (Cady, 1952); Indiana (Spencer, 1953); Kansas (Abernathy et al., 1947); eastern Montana (Averitt, 1965); New Mexico (Fassett and Hinds, 1971; Shomaker and Holt, 1973); Ohio (Struble et al., 1971); Oklahoma (Trumbull, 1957); Oregon (Mason and Erwin, 1955; Mason, 1969); Pennsylvania anthracite (Ashley, 1945; Ashmead, 1926; Rothrock, 1950); Pennsylvania bituminous coal (Reese and Sisler, 1928); Texas bituminous coal (Evans, 1974); and Texas lignite (Fisher, 1963; Perkins and Lonsdale, 1955).

Table 2.3 (*cont.*)

[g] Includes beds to a minimum thickness of 18 in. for overburden of 0–150 ft, and beds to a minimum thickness of 28 in. for overburden in excess of 150 ft.

[h] Includes 6719 million tons in beds 12–18 in. thick. Much of this thin coal is strippable.

[i] Minimum bed thickness: Beaver, Butler, and Lawrence counties, 14 in.; all other counties, 18 in.

[j] Excludes coal in beds less than 4 ft thick.

[k] California, Idaho, Nebraska, and Nevada.

[l] Small resources and production of lignite included with subbituminous coal.

[m] Original subbituminous coal resources of 132,151 million tons as estimated by Combo et al. (1949), increased 45,000 million tons to 177,151 million tons; and original lignite resources of 87,533 million tons increased 25,000 million tons to 112,533 million tons, on basis of more recent geologic mapping and exploratory drilling in eastern Montana.

[n] California and Idaho.

[o] Single discontinuous bed 0–40 ft thick; maximum overburden, 250 ft.

[p] Small resources of lignite in western Kansas and western Oklahoma in beds generally less than 30 in. thick.

[q] Original lignite resources of 2033 million tons as estimated by D. M. Brown (1952), increased to 2187 million tons to accord with new data on stripping coal provided by M. J. Tipton (written commun., 1973).

[r] Overburden, 0–200 ft; lignite beds, 3–10 ft thick.

[s] California, Idaho, Louisiana, Mississippi, and Nevada.

[t] Small resources of anthracite in the Bering River field believed to be too badly crushed and faulted to be economically recoverable (Barnes, 1951).

[u] Less than total recorded cumulative production of about 42.3 billion tons. See also footnotes *a* and *d*.

Table 2.4 Total Estimated Remaining Coal Resources of the United States, January 1, 1974

In millions (10^6) of short tons. Estimates include beds of bituminous coal and anthracite generally 14 in. or more thick, and beds of subbituminous coal and lignite generally 2½ ft or more thick, to overburden depths of 3000 and 6000 ft. Figures are for resources in the ground.

	Overburden 0–3000 ft								
	Remaining Identified Resources, Jan. 1, 1974 (from Table 2.3)								
State	Bituminous Coal	Subbituminous Coal	Lignite	Anthracite and Semi-anthracite	Total	Estimated Hypothetical Resources in Unmapped and Unexplored Areas[a]	Estimated Total Identified and Hypothetical Resources Remaining in the Ground	Overburden 3000–6000 ft: Estimated Additional Hypothetical Resources in Deeper Structural Basins[a]	Overburden 0–6000 ft: Estimated Total Identified and Hypothetical Resources Remaining in the Ground
Alabama	13,262	0	2,000	0	15,262	20,000	35,262	6,000	41,262
Alaska	19,413	110,666	—[b]	—[c]	130,079	130,000	260,079	5,000	265,079
Arizona	21,234[d]	—[d]	0	0	21,234	0	21,234	0	21,234
Arkansas	1,638	0	350	428	2,416	4,000[e]	6,416	0	6,416
Colorado	109,117	19,733	20	78	128,948	161,272	290,220	143,991	434,211
Georgia	24	0	0	0	24	60	84	0	84
Illinois	146,001	0	0	0	146,001	100,000	246,001	0	246,001
Indiana	32,868	0	0	0	32,868	22,000	54,868	0	54,868
Iowa	6,505	0	0	0	6,505	14,000	20,505	0	20,505
Kansas	18,668	0	—[f]	0	18,668	4,000	22,668	0	22,668
Kentucky									
Eastern	28,226	0	0	0	28,226	24,000	52,226	0	52,226
Western	36,120	0	0	0	36,120	28,000	64,120	0	64,120
Maryland	1,152	0	0	0	1,152	400	1,552	0	1,552
Michigan	205	0	0	0	205	500	705	0	705
Missouri	31,184	0	0	0	31,184	17,489	48,673	0	48,673
Montana	2,299	176,819	112,521	0	291,639	180,000	471,639	0	471,639
New Mexico	10,748	50,639	0	4	61,391	65,556[g]	126,947	74,000	200,947
North Carolina	110	0	0	0	110	20	130	5	135
North Dakota	0	0	350,602	0	350,602	180,000	530,602	0	530,602
Ohio	41,166	0	0	0	41,166	6,152	47,318	0	47,318
Oklahoma	7,117	0	—[f]	0	7,117	15,000	22,117	5,000[h]	27,117

Table 2.4 (*cont.*)

State	Overburden 0–3000 ft: Remaining Identified Resources, Jan. 1, 1974 (from Table 2.3): Bituminous Coal	Subbituminous Coal	Lignite	Anthracite and Semianthracite	Total	Estimated Hypothetical Resources in Unmapped and Unexplored Areas[a]	Estimated Total Identified and Hypothetical Resources Remaining in the Ground	Overburden 3000–6000 ft: Estimated Additional Hypothetical Resources in Deeper Structural Basins[a]	Overburden 0–6000 ft: Estimated Total Identified and Hypothetical Resources Remaining in the Ground
Oregon	50	284	0	0	334	100	434	0	434
Pennsylvania	63,940	0	0	18,812	82,752	4,000[i]	86,752	3,600[j]	90,352
South Dakota	0	0	2,185	0	2,185	1,000	3,185	0	3,185
Tennessee	2,530	0	0	0	2,530	2,000	4,530	0	4,530
Texas	6,048	0	10,293	0	16,341	112,100[k]	128,441	—[k]	128,441
Utah	23,186[l]	173	0	0	23,359	22,000[m]	45,359	35,000	80,359
Virginia	9,216	0	0	335	9,551	5,000	14,551	100	14,651
Washington	1,867	4,180	117	5	6,169	30,000	36,169	15,000	51,169
West Virginia	100,150	0	0	0	100,150	0	100,150	0	100,150
Wyoming	12,703	123,240	—[b]	0	135,943	700,000	835,943	100,000	935,943
Other states[n]	610	32[o]	46[p]	0	688	1,000	1,688	0	1,688
Total	747,357	485,766	478,134	19,662	1,730,919	1,849,649	3,580,568	387,696	3,968,264

Note: For References for Tables 2.3 and 2.4, see pp. 88 and 89.

[a] Acknowledgments: The geologists listed below following names of individual states provided the estimates of hypothetical resources used in this table. Most of these geologists are authors of reports on identified resources cited in Table 2.3; geologists who prepared estimates of hypothetical resources for the remaining states have engaged in recent intensive work on coal in these states: Alabama: W. C. Culbertson; Arkansas, B. R. Haley; Colorado, R. D. Holt; Illinois, M. E. Hopkins and J. A. Simon; Indiana, C. E. Wier; Iowa, E. R. Landis; Kentucky, K. J. Englund; Missouri, C. E. Robinson; Montana, R. E. Mason; New Mexico, J. E. Fassett and J. S. Hinds; North Dakota, R. A. Brant; Ohio, H. R. Collins and D. O. Johnson; Oklahoma, S. A. Friedman; Oregon, R. S. Mason; Pennsylvania anthracite, H. H. Arnt; Pennsylvania bituminous coal, W. E.

Edmunds; Tennessee, E. T. Luther; Texas lignite, W. R. Kaiser; Virginia, K. J. Englund; Utah, H. H. Doelling; Washington, H. M. Beikman; and Wyoming, N. M. Denson, G. B. Glass, W. R. Keefer, and E. M. Schell. Estimates for the few remaining states were prepared by the author.

[b] Small resources of lignite included under subbituminous coal.

[c] Small resources of anthracite in the Bering River field believed to be too badly crushed and faulted to be economically recoverable (Barnes).

[d] All tonnage is in the Black Mesa field. Some coal in the Dakota Formation is near the rank boundary between bituminous and subbituminous coal. Does not include small resources of thin and impure coal in the Deer Creek and Pinedale fields.

[e] Lignite.

[f] Small resources of lignite in western Kansas and western Oklahoma in beds generally less than 30 in. thick.

[g] After Fassett and Hinds, who reported 85,222 million tons "inferred by zone" to an overburden depth of 3000 ft in the Fruitland Formation of the San Juan basin. Their figure has been reduced by 19,666 million tons as reported by Read for coal in all categories also to an overburden depth of 3000 ft in the Fruitland Formation of the San Juan basin. The figure of Read was based on measured surface sections and is included in the identified tonnage recorded in Table 2.3.

[h] Includes 100 million tons inferred below 3000 ft.

[i] Bituminous coal.

[j] Anthracite.

[k] Lignite, overburden 200–5000 ft; identified and hypothetical resources undifferentiated. All beds assumed to be 2 ft thick, although many are thicker.

[l] Excludes coal in beds less than 4 ft thick.

[m] Includes coal in beds 14 in. or more thick, of which 15,000 million tons is in beds 4 ft or more thick.

[n] California, Idaho, Nebraska, and Nevada.

[o] California and Idaho.

[p] California, Idaho, Louisiana, and Mississippi.

Table 2.5 Demonstrated Coal Reserve Base of the United States, January 1, 1974, by State and by Method of Mining

In millions (10^6) of short tons. Figures are for coal in the ground.

State	Potential Mining Method		
	Underground	Surface	Total
Alabama	1,798	1,184	2,982
Alaska	4,246	7,399	11,645
Arizona	—[a]	350	350
Arkansas	402	263	665
Colorado	14,000	870	14,870
Georgia	1	—	1
Illinois	53,442	12,223	65,665
Indiana	8,949	1,674	10,623
Iowa	2,885	—[b]	2,885
Kansas	—[a]	1,388	1,388
Kentucky, eastern	9,467	3,450	12,917
Kentucky, western	8,720	3,904	12,624
Maryland	902	146	1,048
Michigan	118	1	119
Missouri	6,074	3,414	9,488
Montana	65,165	42,562	107,727
New Mexico	2,136	2,258	4,394
North Carolina	31	—[b]	31
North Dakota	—	16,003	16,003
Ohio	17,423	3,654	21,077
Oklahoma	860	434	1,294
Oregon	1	—[b]	1
Pennsylvania	29,819	1,181	31,000
South Dakota	—	428	428
Tennessee	667	320	987
Texas	—	3,272	3,272
Utah	3,780	262	4,042
Virginia	2,971	679	3,650
Washington	1,446	508	1,954
West Virginia	34,378	5,212	39,590
Wyoming	27,554	23,674	51,228
Total	297,235	136,713	433,948

[a] Data insufficient to establish reserve base.
[b] Less than 1 million tons.

Source: U.S. Bur Mines, Ref. 10, p. 4.

2.4.3.2 *Region*

The distribution of identified resources according to eight major coal basins or comparable large regions is given in Table 2.6. These subdivisions provide a natural breakdown of data, and they can be considered separately or combined in various ways for study and analysis. Region 1, for example, represents coal convenient for use by the densely populated, highly industrialized northeastern states. Regions 1 and 2 combined represent the Appalachian coal basin, which provides coal to the eastern seaboard and coal that is exported to Japan, Canada, Western Europe, and elsewhere. Regions 1, 2, 3, and 4 combined represent all coal east of the Mississippi River, whereas regions 5, 6, 7, and 8 combined represent all coal west of the Mississippi. Regions 1, 2, 3, 4, and 5 lie east of, and regions 6, 7, and 8 lie west of an imaginary northeast-trending line extending from the Texas Panhandle to Minnesota, which marks an important division of regions and resources according to age and rank of coal. Regions 6 and 7 combined represent the Rocky Mountain and Northern Great Plains provinces. Region 8 represents the west coast and Alaska.

The tonnage figures in column 1 of Table 2.6 are for the demonstrated coal reserve base as presented by states in Table 2.5. The figures in column 2 express the same information in percentages. The tonnage figures in column 4 are for total remaining identified resources as of January 1, 1974, as presented by rank and by states in Tables 2.3 and 2.4. Column 3 gives the difference between column 4 and column 1.

The amount of coal included in the demonstrated reserve base (Table 2.6, column 1) is much larger in some regions than in others because of differences in the thickness and number of coal beds and differences in the structure and topography of the major coal-bearing basins.

The demonstrated reserve base in region 1, the Northern Appalachian basin, is large as compared with the markedly smaller reserve base in region 2, the Southern Appalachian basin, because the center of coal deposition was in the northern part of the Appalachian basin. Hence in region 1 coal beds are thicker, more continuous, and more numerous than those in region 2. Also, the bulk of the coal-bearing sequence in the Northern Appalachian basin is preserved in a large syncline, whereas in the Southern Appalachian basin the entire upper part of the coal-bearing sequence was eroded in post–Pennsylvanian time.

The large demonstrated reserve base in

Table 2.6 Distribution, by Basin or Region of the Demonstrated Coal Reserve Base and of Total Remaining Identified Coal Resources of the United States, January 1, 1974

In billions (10^9) of short tons—indicate negligible amount of coal. Figures are for reserves and resources in the ground. At least half of the reserve base is recoverable.

	Overburden 0–3000 ft			
	Demonstrated Reserve Base, 0–1000 ft Overburden[a] (from Table 2.5)		Resources in Thin Beds and Inferred Resources. 0–1000 ft Overburden, and Identified Resources in All Beds 1000–3000 ft Overburden	Total Remaining Identified Resources (from Table 2.3, rounded)
Basin or Region	Tons (1)	Percent (2)	(3)	(4)
1. Northern Appalachian basin (Pa., Ohio, W. Va., and Md.)	93	21	132	225
2. Southern Appalachian basin (eastern Ky., Va., Tenn., N.C., Ga., and Ala.)	20	5	36	56
3. Michigan basin	—	—	—	—
4. Illinois basin (Ill., Ind., and western Ky.)	89	20	126	215
5. Western Interior basin (Iowa, Kansas, Mo., Okla., Ark., and Texas)	19	4	63	82
6. Northern Rocky Mountains (N. Dak., S. Dak., Mont., Wyo., and Idaho)	175	41	606	781
7. Southern Rocky Mountains (Colo., Utah, Ariz., and N. Mex.)	24	6	211	235
8. West coast (Alaska, Wash., Oreg., and Calif.)	14	3	123	137
Total	434	100	1297	1731

[a] Includes coal in the measured and indicated (demonstrated) category in beds 28 in. or more thick for bituminous coal and anthracite, and 5 ft or more thick for subbituminous coal and lignite. Maximum overburden is 1000 ft for subbituminous coal, bituminous coal, and anthracite, and 120 ft for lignite. May include coal outside these parameters if such coal is being mined or is considered to be commercially minable.

Source for cols. (1) and (2): U.S. Bureau of Mines, 1974, Demonstrated Coal reserve base of the United States on January 1, 1974: U.S. Bur. Mines Mins. Ind. Surveys, June, 1974, 6 pp. *Source for cols (3) and (4):* Ref. 1., p. 9.

region 4, the Illinois basin, results because the Illinois basin is relatively shallow and the topography is relatively flat; thus the coal is less than 1000 ft below the surface over thousands of square miles. However, much of this coal can be reached only by vertical or inclined shafts.

The demonstrated reserve base in region 5, the Western Interior basin, is relatively small because the coal-bearing rocks are thin, and the coal beds are few and generally thinner than beds in the Illinois basin.

The very large demonstrated reserve base in region 6, the Northern Rocky Mountains, represents 41% of the total in column 1. Tonnage and percentage are large because the coal beds are very thick, numerous, and closely spaced; the coal-bearing rocks lie nearly flat; and the topography is relatively flat over thousands of square miles in North Dakota, eastern Montana, and northeastern Wyoming. Much of the coal included in the reserve base of region 6 is within reach by strip mining methods.

The demonstrated reserve base in region 7, the Southern Rocky Mountains, is modest compared with that in region 6, because in most of region 7 the coal-bearing rocks are on the edges of moderately to steeply dipping structural basins. In parts of the region, particularly in the Wasatch Plateau and Book Cliffs of central Utah, the moderately dipping coal crops out at the bases of nearly vertical cliffs, thus passing below 1000 ft of overburden a short distance from the outcrops. All the coal occurring in this topographic setting can be reached by drift mines, and even larger tonnages with overburden more than 1000 ft thick can be reached conveniently through the same openings.

The small demonstrated reserve base in region 8, the west coast, reflects the location of the coal in Washington; most of it lies on the flanks of steeply dipping basins, thus also passes below 1000 ft of overburden a short distance from the outcrops. In addition, most of the coal in Alaska is classified as inferred; the region actually contains larger total resources than is suggested by the small figure for the demonstrated reserve base.

2.4.3.3 *Rank*

United States coal is distributed unequally among the four categories of rank. As shown numerically in Tables 2.3 and 2.4, and graphically in Fig. 2.4*a*, 43.1% of the original identified resources is bituminous coal, including 1.1% low-volatile bituminous coal. By comparison, 28.1% is subbituminous coal, 27.7% is lignite, and only 1.1% is anthracite. It should be noted that the comparison in Fig. 2.4*a* is based on weight. A comparison based on the contained heat value of the coal shows a marked percentage increase for bituminous coal, a modest percentage increase for anthracite, and progressive decreases for subbituminous coal and lignite, as indicated by short tick marks to the right of the respective columns.

The geographic distribution of resources of the different ranks of coal is also unequal. In the conterminous United States, about 83% of the identified resources of bituminous coal and anthracite lies east of an imaginary line trending toward the northeast from the Texas Panhandle to Minnesota (Fig. 2.2). About 99% of the subbituminous coal and lignite lies west of the line, an unequal geographic distribution largely related to differences in geologic age. Nearly all the coal in states east of the imaginary line is of Pennsylvanian age, whereas nearly all the coal in states west of the line is of a more recent age—Cretaceous or Tertiary. The younger, western coal attains high rank only where it has been deformed and altered by the forces that accompanied mountain building and by the intrusion of igneous rock.

The resources of subbituminous coal and lignite of the western states are lower in heat value and are somewhat more difficult to ship and store than the more widely used bituminous coal of the eastern states. The low-rank coals of the western states, however, are well suited for the production of electric power and for the production of synthetic gas and liquid fuels, and in many parts of the west can be mined efficiently by stripping methods. Because of these advantages, the low-rank coals in the west have received increased attention since the late 1960s.

2.4.3.4 *Thickness of Overburden*

Figure 2.4*b* shows the percentage distribution of classified resources in three categories according to thickness of overburden as follows: 0–1000, 1000–2000, and 2000–3000 ft. It is noteworthy that 91% of the classified resources is less than 1000 ft below the

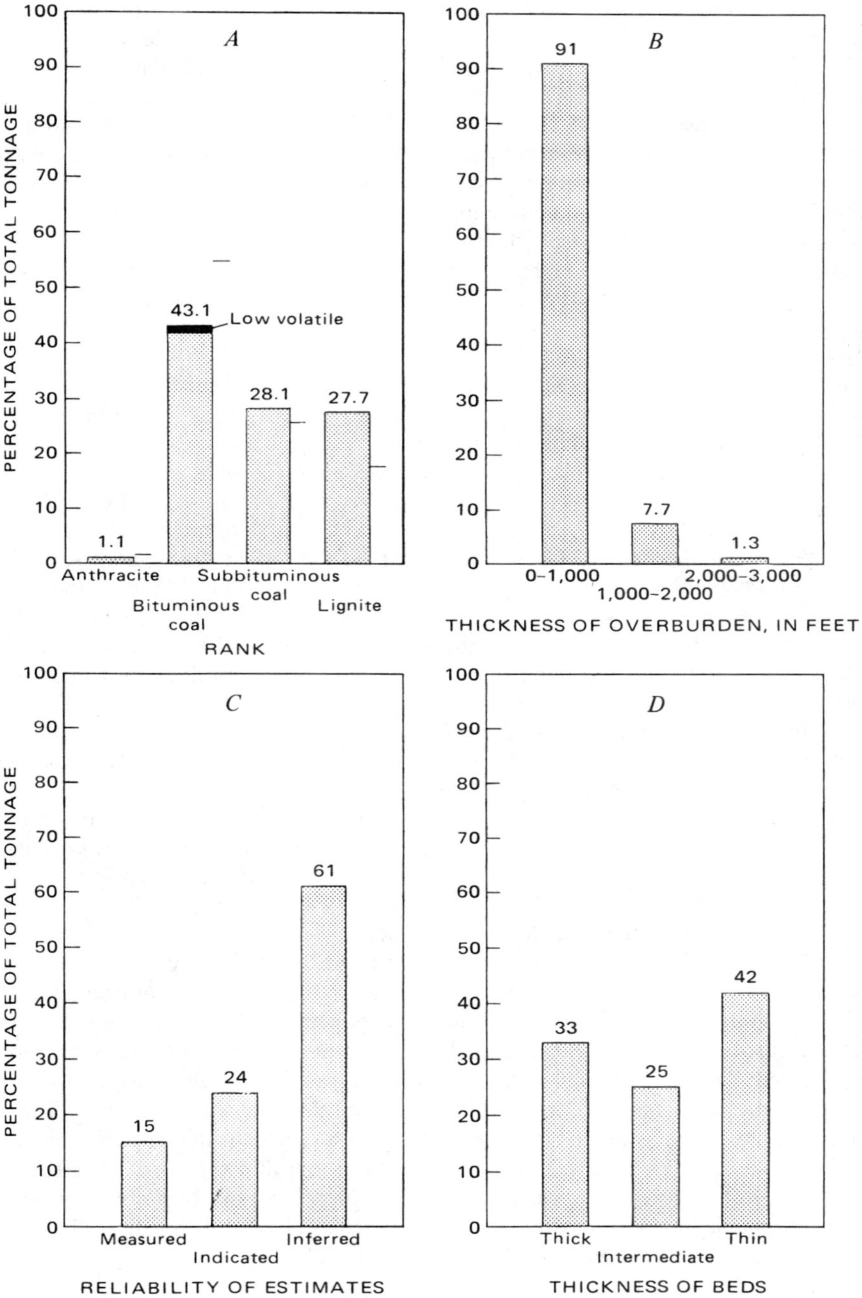

Figure 2.4 Percentage distribution by weight of estimated original identified coal resources: (*a*) in states according to rank and in 21 states according to (*b*) thickness of overburden, (*c*) reliability of all estimates, and (*d*) thickness of beds. Ticks beside or above bars in *a* indicate comparison by rank based on contained heat value of the coal. Source: Ref. 1, fig. 5, pp. 37.

surface and that only 7.7 and 1.3%, respectively, are present in the 1000–2000 and the 2000–3000 ft overburden categories.

The reason for the impressive concentration of the classified resources in the 0–1000 ft overburden category is that in most parts of the United States, coal-bearing rocks lie near the surface, and exploration and mining are concentrated in this shallow overburden zone. As a result, less attention has been given to coal more than 1000 ft below the surface.

Only a small amount of coal in the United States is mined from beds 1000–2000 ft below the surface, and no appreciable amount is mined from beds more than 2000 ft below the surface. Mining below 1000 ft has been, or is being, carried out in the Pennsylvania Anthracite region, in the Coosa and Cahaba fields of Alabama, in the Book Cliffs of Utah, and in several fields in Washington. In Great Britain, Belgium, Germany, and Poland, however, mining has been extended to depths of 4000 ft. As exploration and mine development are extended to greater depths in the United States, it is certain that the identified resources will increase considerably by the addition of tonnage in the deeper overburden categories.

2.4.3.5 *Reliability of Estimates*

Figure 2.4*c* shows the percentage distribution of classified resources in the 21 states in the measured, indicated, and inferred categories. Of the large tonnage thus classified, 15% is classed as measured, 24% as indicated, and 61% as inferred. The 15% classed as measured is somewhat large compared to the 24% classed as indicated, according to stated definitions for these categories. This minor disparity is the result of intensive exploratory drilling for strippable coal in the early 1970s, which delineated large tonnages of measured coal in areas with no more than 200 ft of overburden. A reexamination of the known strippable coal areas from the point of view of total resources would undoubtedly result in an increase in the tonnage and percentage of coal that could be classified as "indicated."

The figure of 61% for inferred resources is large because of the lack of data in areas remote from outcrops. It does, however, express the approximate amount of coal that can be inferred to be present in areas remote from outcrops on the basis of current geologic information. Additional geologic mapping, exploratory drilling, and study in areas of inferred resources would undoubtedly increase the percentage of measured and indicated resources and would decrease the percentage of inferred resources. On the other hand, by providing new data, such exploration might actually increase the tonnage of total identified resources.

2.4.3.6 *Thickness of Beds*

The terms "thick," "intermediate," and "thin," as used in Fig. 2.4*d*, refer to beds of coal in three categories that differ for the different ranks of coal. Defined as "thick" are beds of bituminous coal and anthracite more than 42 in. thick, and beds of subbituminous coal and lignite more than 10 ft thick. Defined as "intermediate" are beds of bituminous coal and anthracite 28–42 in. thick, and beds of subbituminous coal and lignite 5–10 ft thick. Defined as "thin" are beds of bituminous coal and anthracite 14–28 in. thick, and beds of subbituminous coal and lignite 2½–5 ft thick.

As noted in Figure 2.4*d*, coal in thick beds makes up 33% of the total, coal in beds of intermediate thickness makes up 25%, and coal in thin beds makes up 42%. The percentage of resources in beds of intermediate thickness is relatively low in part because of the large amount of coal in thick beds delineated during the recent period of intensive exploration for strippable coal and in part because of a human tendency to assign minimum thicknesses to beds in the inferred category, which increases the percentage of coal in the thin category. Thus the emphasis on thick coal and thin coal has been at the expense of the intermediate thickness category.

2.4.3.7 *Combined Categories of Overburden, Reliability, and Thickness of Beds*

Figure 2.5 summarizes the distribution of resources in the three major categories presented in Figs. 2.4*b*, 2.4*c*, and 2.4*d*. Figure 2.5 clearly shows (1) the preponderance of resources in the 0–1000 ft category, (2) the disproportionate relationship between meas-

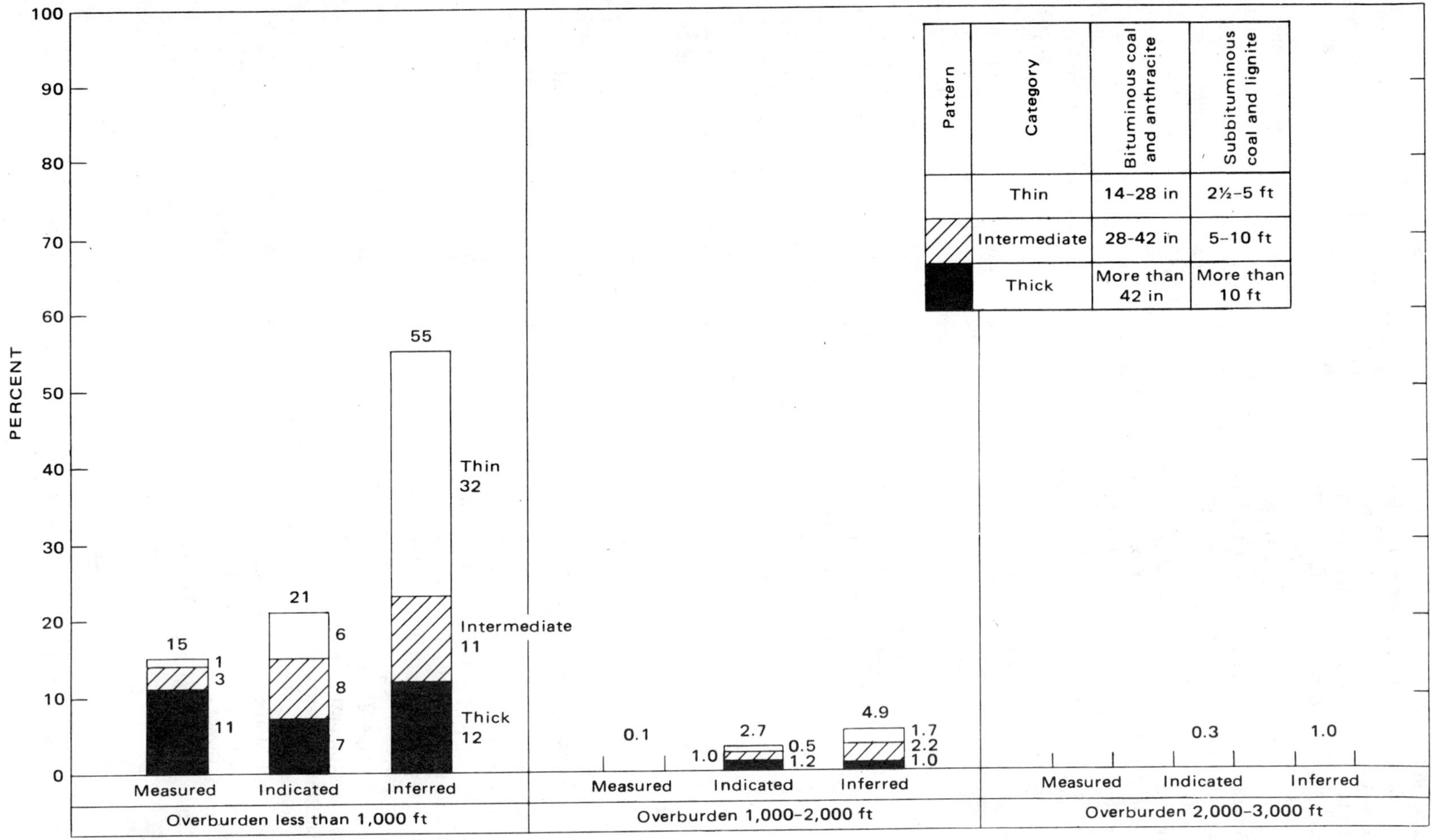

Figure 2.5 Summary of percentage distribution by major resource categories of estimated original identified coal resources in 21 states recorded in Fig. 2.4. Source: Ref. 1, fig. 6, p. 40.

ured and indicated resources, and (3) the disproportionate relationship between measured resources and indicated resources in thick beds. Resources are present in each of the categories in Fig. 2.5, except the one that represents measured resources in thin beds, 2000–3000 ft below the surface. The amounts in several categories are less than 1% of the total and could not be shown on a diagram at this scale.

As in Figs. 2.4*b*, 2.4*c*, and 2.4*d*, Fig. 2.5 shows the conservative character of the estimates of identified resources as discussed below. The large percentages of resources in the indicated and inferred categories and the small percentage in the measured category reflect a lack of data rather than a lack of coal. The small percentage of coal in the 1000–2000 ft overburden category compared to that in the 0–1000 ft category is also due primarily to lack of data. The deeper overburden categories obviously contain additional coal that could not be included in estimates of identified resources (see Section 2.6, "Hypothetical Resources").

2.5 COAL RECOVERED IN PAST MINING

2.5.1 Percentage of Coal Recovered

Most studies of recoverability in past coal mining are based on comparisons between the amount of coal produced and the amount of coal originally present, or the amount left behind, in the block of ground being studied. These comparisons require coal production records and maps of individual mines; for this reason, comparisons are generally focused on the performance of an individual mine or on mines under ownership of one company. Such studies usually show a higher percentage recovery than the expected recovery on a broader regional or national basis. However several regional studies of recovery have been made in the Appalachian and Illinois basins, and several individuals with long experience in the coal-mining industry have provided subjective opinions on recoverability that are of interpretative value.

2.5.1.1 *Underground Mining*

In nine regional studies recovery in underground mining ranged from 33 to 60%, and the median recovery experience was about 50%.[11]

It should be noted, however, that these studies do not uniformly take into account coal left in barrier pillars, in restricted areas around oil and gas wells and fields, under towns, railroads, highways, streams, and reservoirs, in top and lower benches, in rider beds, and in local areas of faulting and folding. None takes into account coal in higher, unmined beds damaged by earlier mining of lower beds, and none takes into account coal lost in the cleaning process. Therefore the average recovery in past underground mining is likely to be slightly lower that 50%. The rounded figure of 50%, however, is convenient and meaningful for use in discussions of past and near-future average recovery in underground mining, and it is accepted for use in this chapter.

Many underground mines, particularly those that use the longwall mining method, achieve a recovery higher than 50% and the gradual introduction of more efficient mining methods will probably result eventually in a higher national average recovery.

2.5.1.2 *Strip Mining*

Under favorable conditions, recovery in strip mining may be as high as 90% of the coal originally in the ground. However most strip mine operators agree that the average recovery is closer to 80%, and this figure is used in preparing many estimates of recoverable strip mining reserves.

2.5.1.3 *Auger Mining*

The maximum possible recovery in auger mining is about 75%, but when the many operations involved in this mining mode are considered, the actual average recovery is probably no more than about 50%. Actual recovery in auger mining is less than the possible maximum because (1) the auger holes are generally smaller in diameter than the thickness of the bed being mined, (2)

[11] See Ref. 1. pp. 29–31.

spaces ranging in width from a few inches to 1 ft or more are routinely left between adjacent auger holes, and (3) some holes collapse and cannot be drilled to the planned depth.

2.6 HYPOTHETICAL RESOURCES

The preceding analysis of data on the distribution of identified resources (Section 2.4.3) provides convincing evidence that unmapped and unexplored areas in known coalfields contain substantial additional resources, but they must be classed as hypothetical. Their approximate magnitude has been estimated at 2237 billion tons by a process of extrapolation from nearby areas of identified resources. Estimates for each state are presented in separate columns in Table 2.4, and the evidence on which the estimates are based is summarized in the paragraphs that follow.

In most states for which modern estimates of identified coal resources have been prepared, substantial areas of coal-bearing rock were omitted from consideration because of the lack of specific information about the occurrence and thickness of the coal. In Colorado, for example, 75% of the coal-bearing area was omitted; in eastern Kentucky, 13% was omitted; in Montana, 9.3%; in North Dakota, 1.7%; in Washington, 66%; and in Wyoming, 53.5%. A part of the estimated tonnage of hypothetical resources is present in such areas.

Because most exploration and mining in the United States is concentrated along outcrops, the amount of detailed information on coal decreases rapidly with the distance from the outcrop. With a few conspicuous exceptions, such as the Illinois basin, only general information is available about coal in the centers of the large coal basins, and for that reason most of the identified resources are confined to a narrow zone a few miles wide parallel to the outcrops of individual coal beds. This is illustrated by Fig. 2.4*b*, which shows that 91% of the classified resources are less than 1000 ft below the surface.

Many coal-bearing areas, particularly those remote from current means of transportation or centers of use, have been mapped or examined only in reconnaissance. In such areas points of information are widely spaced and are confined to the thicker, more exposed beds. As a result, resource estimates tend to be small. The estimated hypothetical resources include an allowance for additional coal that should be discovered when detailed geologic mapping is extended into such areas.

In areas covered by reconnaissance mapping, and in many others as well, data on the coal-bearing rocks and on individual coal beds are generally insufficient to establish correlations between coal beds in all parts of the areas. Where correlations cannot be established, the estimated resources are restricted to the vicinity of known outcrops. Where correlations can be established, resources can be inferred to exist at greater distances between outcrops, and the total estimated resources tend to be larger. The estimated hypothetical resources include an allowance for coal that may be delineated as a result of improved knowledge of stratigraphy and coal bed correlations.

From the foregoing discussion and from the distribution pattern of identified resources shown in Fig. 2.5, it is apparent that the bulk of the estimated hypothetical resources is in the 1000–2000 ft overburden zone, and that smaller amounts are present in other overburden zones. Figure 2.6 gives the probable distribution of the total estimated coal resources of the United States in the identified and hypothetical categories combined, according to thickness of overburden.

The estimated hypothetical resources are, of course, only an approximation, based primarily on extrapolation from the more reliable and more useful estimates of identified resources. Although the estimated hypothetical resources are large, for the most part they are relatively inaccessible to mining at present, and a more exact delineation of the magnitude, distribution, and utility of such resources can be ascertained only by future detailed geologic mapping, exploration, and study. Nevertheless the estimated hypothetical resources constitute an important part of total coal resources that need to be considered in future planning for the use of energy.

2.7 COKING COAL RESOURCES

Coke is usually manufactured from blends of two or more coals of different rank and

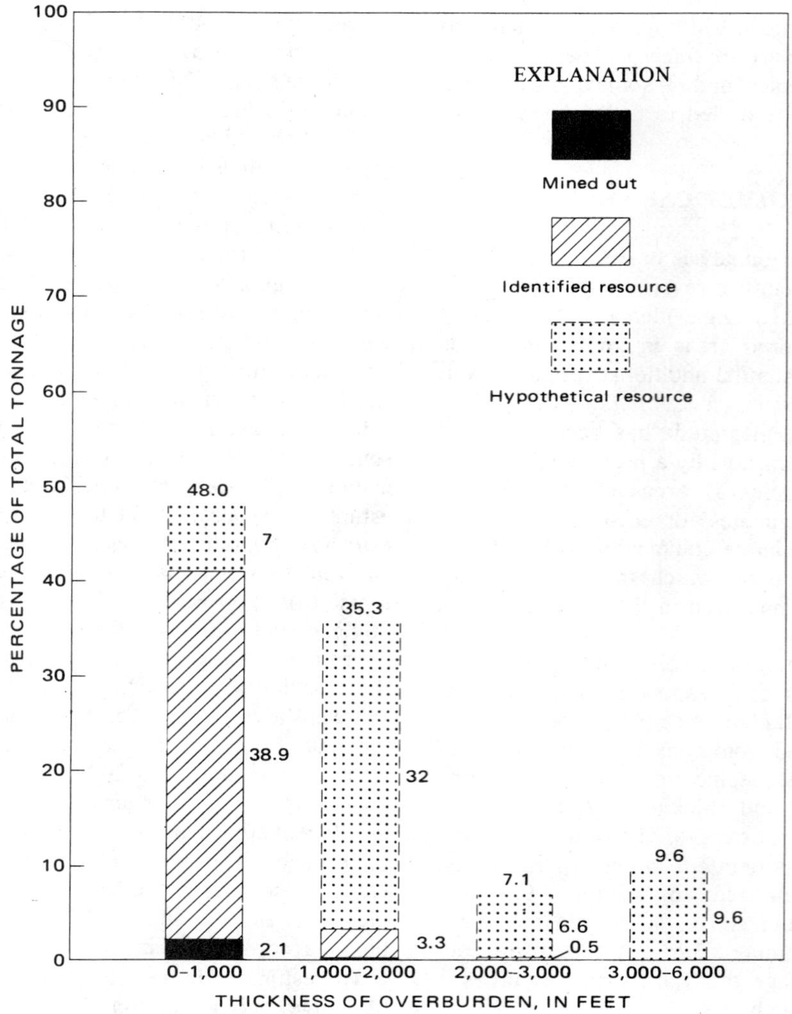

Figure 2.6 Probable percentage distribution of total U.S. coal resources according to thickness of overburden. Source: Ref. 1, fig. 7, pp. 45.

composition and may incorporate small amounts of other ingredients, such as anthracite fines, petroleum coke, or low-temperature char. The term "coking coal," therefore, typically refers to a variety of coals and only rarely to a single coal with unique properties. Although a single coal of medium-volatile bituminous rank that is low in ash, sulfur, and phosphorus will produce a satisfactory metallurgical coke, resources of such coal are small, and the properties desired in a coke are more readily obtained and standardized by a blending procedure. Blending also permits use of coals that individually do not yield a satisfactory coke.

In 1973 low-volatile bituminous coal, which is strongly coking, constituted 20.2% of the tonnage of coal received by oven-coke plants in the United States; medium-volatile coal, which is moderately coking, constituted 12.6%; and high-volatile coal, which is weakly coking, constituted 67.2%. The average volatile-matter content of coking coal blends prepared during the year was 30.2%, on a moisture-free basis.[12]

In addition to rank, the original plant constituents of coal are a factor in determining coking properties, as are the deleterious con-

[12] U.S. Bureau of Mines, *Minerals Yearbooks* (annual volumes for the years indicated), Government Printing Office, Washington, D.C., 1932–1973.

stituents—ash, sulfur, and phosphorus. With many variables that must be taken into account, modern coking coal blends have become complex mixtures of carbonaceous materials.

Most areas of high-rank, high-quality coal best suited for the manufacture of coke and coke chemicals are in the northern part of the Appalachian basin, principally in West Virginia, Pennsylvania, eastern Kentucky, and Virginia. Substantial amounts of coal suitable for the manufacture of coke are also present in Alabama at the southern end of the Appalachian basin. Bibliographies accompanying summary reports on individual Appalachian basin states, as cited in Table 2.3, contain information on the occurrence and composition of coking coal in the respective states. Additional information is contained in reports by Dowd et al.,[13] Wallace et al.,[14] Williams et al.,[15] Hershey et al.,[16] Blaylock et al.,[17] Travis et al.,[18] Lowe et al.,[19] Provost et al.,[20] and Tavenner et al.[21]

Coal in the Illinois basin is weakly coking, but in steel manufacturing centers at the southern end of Lake Michigan and near East St. Louis, about 4.5 million tons of Illinois coal is used annually in coking coal blends because of proximity and low cost. Since 1970 the Illinois basin has supplied more than 25% of the coal carbonized in

[13] J. J. Dowd, J. M. Provost, R. F. Abernethy, and D. A. Reynolds, *Estimate of Known Recoverable Reserves of Coking Coal in Ohio County, West Virginia,* U.S. Bur. Mines Rep. Invest. No. 5171, 1955, 14 pp; *Estimate of Known Recoverable Reserves of Coking Coal in Marshall County, West Virginia,* U.S. Bur. Mines Rep. Invest. No. 5207, 1956, 21 pp.; J. J. Dowd, J. M. Provost, R. G. Travis, R. F. Abernethy, and D. A. Reynolds, *Estimate of Known Recoverable Reserves of Coking Coal in Brooke County, West Virginia,* U.S. Bur. Mines Rep. Invest. No. 5160, 1955, 17 pp.; J. J. Dowd, A. L. Toenges, R. F. Abernethy, and D. A. Reynolds, *Estimate of Known Recoverable Reserves of Coking Coal in Jefferson County, Pennsylvania,* U.S. Bur. Mines Rep. Invest. No. 4840, 1952, 18 pp.; *Estimate of Known Recoverable Reserves of Coking Coal in Raleigh County, West Virginia,* U.S. Bur. Mines Rep. Invest. No. 4893, 1952, 37 pp.; *Estimate of Known Recoverable Reserves of Coking Coal in Knott County, Kentucky,* U.S. Bur. Mines Rep. Invest. No. 4897, 1952, 20 pp.; J. J. Dowd, L. A. Turnbull, A. L. Toenges, R. F. Abernethy, and D. A. Reynolds, *Estimate of Known Recoverable Reserves of Coking Coal in Pike County, Kentucky,* U.S. Bur. Mines Rep. Invest. No. 4792, 1951, 34 pp.; *Estimate of Known Recoverable Reserves of Coking Coal in Armstrong County, Pennsylvania,* U.S. Bur. Mines Rep. Invest. No. 4801, 1951, 16 pp.; *Estimate of Known Recoverable Reserves of Coking Coal in Westmoreland County, Pennsylvania,* U.S. Bur. Mines Rep. Invest. No. 4803, 1951, 16 pp.; *Estimate of Known Recoverable Reserves of Coking Coal in Fayette County, Pennsylvania,* U.S. Bur. Mines Rep. Invest. No. 4807, 1951, 19 pp.; *Estimate of Known Recoverable Reserves of Coking Coal in Floyd County, Kentucky,* U.S. Bur. Mines Rep. Invest. No. 4813, 1951, 16 pp.; J. J. Dowd, L. A. Turnbull, A. L. Toenges, H. M. Cooper, R. F. Abernethy, D. A. Reynolds, and W. L. Crentz, *Estimate of Known Recoverable Reserves of Coking Coal in Indiana County, Pennsylvania,* U.S. Bur. Mines Rep. Invest. No. 4757, 1950, 22 pp.; J. J. Dowd, L. A. Turnbull, A. L. Toenges, H. M. Cooper, R. F. Abernethy, D. A. Reynolds, and Thomas Fraser, *Estimate of Known Recoverable Reserves of Coking Coal in Cambria County, Pennsylvania,* U.S. Bur. Mines Rep. Invest. No. 4734, 1950, 25 pp.

[14] J. J. Wallace, J. J. Dowd, J. A. Bowsher, R. F. Abernethy, and D. A. Reynolds, *Estimate of Known Recoverable Reserves of Coking Coal in Somerset County, Pennsylvania,* U.S. Bur. Mines Rep. Invest. No. 4998, 1953, 20 pp.; J. J. Wallace, J. J. Dowd, J. M. Provost, R. F. Abernethy, and D. A. Reynolds, *Estimate of Known Recoverable Reserves of Coking Coal in Allegheny County, Pennsylvania,* U.S. Bur. Mines Rep. Invest. No. 5003, 1953, 16 pp.; *Estimate of Known Recoverable Reserves of Coking Coal in Washington County, Pennsylvania,* U.S. Bur. Mines Rep. Invest. No. 5109, 1955, 23 pp.; *Estimate of Known Recoverable Reserves of Coking Coal in Greene County, Pennsylvania,* U.S. Bur. Mines Rep. Invest. No. 5143, 1955, 22 pp.; J. J. Wallace, J. J. Dowd, W. H. Tavenner, R. F. Abernethy, and D. A. Reynolds, *Estimate of Known Recoverable Reserves of Coking Coal in Wyoming County, West Virginia,* U.S. Bur. Mines Rep. Invest. No. 4966, 1953, 39 pp.; *Estimate of Known Recoverable Reserves of Coking Coal in Mercer County, West Virginia,* U.S. Bur. Mines Rep. Invest. No. 5077, 1954, 20 pp.; J. J. Wallace, J. J. Dowd, R. G. Travis, R. F. Abernethy, and D. A. Reynolds, *Estimate of Known Recoverable Reserves of Coking Coal in Letcher County, Kentucky,* U.S. Bur. Mines Rep. Invest. No. 5016, 1953, 26 pp.; *Estimate of Known Recoverable Reserves of Coking Coal in Harlan County, Kentucky,* U.S. Bur. Mines Rep. Invest. No. 5037, 1954, 26 pp.; *Estimate of Known Recoverable Reserves of Coking Coal in Perry County, Kentucky,* U.S. Bur. Mines Rep. Invest. No. 5083, 1954, 26 pp.; J. J. Wallace, J. J. Dowd, Lloyd Williams, R. F. Abernethy, and D. A. Reynolds, *Estimate of Known Recoverable Reserves of Coking Coal in Allegany County, Maryland,* U.S. Bur. Mines Rep. Invest. No. 4970, 1953, 18 pp.

[15] Lloyd Williams, R. F. Abernethy, B. W. Gandrud, D. A. Reynolds, and D. E. Wolfson, *Estimate of Known Recoverable Reserves and the*

midwestern coke plants. In blends using Illinois coal, the required larger amount of more strongly coking coal is obtained from the Appalachian basin.[22]

In a few areas in the West, principally in Colorado, Utah, Oklahoma, Arkansas, Washington, and New Mexico, coal is produced that is satisfactory for the manufacture of coke when used in blends. The most important areas are the Raton Mesa region, Colorado–New Mexico; the Sunnyside field, Utah; and the Somerset–Crested Butte–Carbondale region, Colorado. These areas are prominent in plans for the industrial development of the West. Summary information about resources of coking coal in the West is contained in reports by Averitt,[23] Haley,[24] R. B. Johnson,[25] Landis,[26] and Trumbull.[27]

Because of the almost limitless possibilities

Preparation and Carbonizing Properties of Coking Coal in Overton County, Tennessee, U.S. Bur. Mines Rep. Invest. No. 5131, 1955, 27 pp.; Lloyd Williams, H. K. Gibbs, W. L. Crentz, and J. W. Miller, *Estimate of Known Recoverable Reserves and Preparation Characteristics of Coking Coal in Scott County, Tennessee,* U.S. Bur. Mines Rep. Invest. No. 5232, 1956, 36 pp.; Lloyd Williams, H. K. Gibbs, W. L. Crentz, J. W. Miller, and D. A. Reynolds, *Estimate of Known Recoverable Reserves and Preparation and Carbonizing Properties of Coking Coal in Campbell County, Tennessee,* U.S. Bur. Mines Rep. Invest. No. 5258, 1956, 78 pp.; Lloyd Williams and R. E. Hershey, *Estimate of Known Recoverable Reserves of Coking Coal in Bledsoe County, Tennessee,* U.S. Bur. Mines Rep. Invest. No. 5234, 1956, 18 pp.; Lloyd Williams, R. E. Hershey, R. F. Abernethy, B. W. Gandrud, and D. A. Reynolds, *Estimate of Known Recoverable Reserves and the Preparation and Carbonizing Properties of Coking Coal in Sequatchie County, Tennessee,* U.S. Bur. Mines Rep. Invest. No. 5136, 1955, 28 pp.; Lloyd Williams, R. E. Hershey, and B. W. Gandrud, *Estimate of Known Recoverable Reserves and the Preparation and Carbonizing Properties of Coking Coal in Marion County, Tennessee,* U.S. Bur. Mines Rep. Invest. No. 5159, 1955, 30 pp.; Lloyd Williams, C. W. James, W. L. Crentz, and J. W. Miller, *Estimate of Known Recoverable Reserves and the Preparation Characteristics of Coking Coal in Morgan County, Tennessee,* U.S. Bur. Mines Rep. Invest. No. 5266, 1956, 43 pp.; Lloyd Williams C. W. James, B. W. Gandrud, and D. A. Reynolds, *Estimate of Known Recoverable Reserves and the Preparation and Carbonizing Properties of Coking Coal in Anderson County, Tennessee,* U.S. Bur. Mines Rep. Invest. No. 5185, 1955, 52 pp.; Lloyd Williams, R. W. Lowe, L. A. Turnbull, E. P. Carman, W. L. Crentz, D. A. Reynolds, and R. F. Abernethy, *Estimate of Known Recoverable Reserves and the Preparation and Carbonizing Properties of Coking Coal in Putnam County, Tennessee,* U.S. Bur. Mines Rep. Invest. No. 5029, 1954, 21 pp.

[16] R. E. Hershey, Lloyd Williams, W. L. Crentz, and J. W. Miller, *Estimate of Known Recoverable Reserves and Preparation Characteristics of Coking Coal in Van Buren County, Tennessee,* U.S. Bur. Mines Rep. Invest. No. 5208, 1956, 25 pp.; *Estimate of Known Recoverable Reserves and the Preparation Characteristics of Coking Coal in Hamilton County, Tennessee,* U.S. Bur. Mines Rep. Invest. No. 5263, 1956, 27 pp.; R. E. Hershey, Lloyd Williams, W. L. Crentz, J. W. Miller, and D. A. Reynolds, *Estimate of Known Recoverable Reserves and Preparation and Carbonizing Properties of Coking Coal in Claiborne County, Tennessee,* U.S. Bur. Mines Rep. Invest. No. 5229, 1956, 44 pp.; R. E. Hershey, Lloyd Williams, and B. W. Gandrud, *Estimate of Known Recoverable Reserves of Coking Coal in Grundy County, Tennessee,* U.S. Bur. Mines Rep. Invest. No. 5148, 1955, 16 pp.

[17] D. W. Blaylock, J. J. Dowd, R. F. Abernethy, and D. A. Reynolds, *Estimate of Known Recoverable Reserves of Coking Coal in Clearfield County, Pennsylvania,* U.S. Bur. Mines Rep. Invest. No. 5166, 1955, 36 pp.; *Estimate of Known Recoverable Reserves of Coking Coal in Clarion County, Pennsylvania,* U.S. Bur. Mines Rep. Invest. No. 5231, 1956, 27 pp.

[18] R. G. Travis, J. J. Dowd, R. F. Abernethy, and D. A. Reynolds, *Estimate of Known Recoverable Reserves of Coking Coal in Monongalia County, West Virginia,* U.S. Bur. Mines Rep. Invest. No. 5233, 1956, 30 pp.

[19] R. W. Lowe, W. L. Crentz, and J. W. Miller, *Estimate of Known Recoverable Reserves and Preparation Characteristics of Coking Coal in Fentress County, Tennessee,* U.S. Bur. Mines Rep. Invest. No. 5235, 1956, 34 pp.

[20] J. M. Provost, J. J. Dowd, R. F. Abernethy, and D. A. Reynolds, *Estimate of Known Recoverable Reserves of Coking Coal in Marion County, West Virginia,* U.S. Bur. Mines Rep. Invest. No. 5267, 1956, 30 pp.

[21] W. H. Tavenner, J. J. Dowd, R. F. Abernethy, and D. A. Reynolds, *Estimate of Known Recoverable Reserves of Coking Coal in Logan County, West Virginia,* U.S. Bur. Mines Rep. Invest. No. 5259, 1956, 50 pp.

[22] H. W. Jackman and R. J. Helfinstine, *A Survey of the Coking Properties of Illinois Coals,* Illinois Geol. Survey Circ. No. 412, 1967, 27 pp.

[23] Paul Averitt, *Coking-Coal Deposits of the Western United States,* U.S. Geol. Survey Bull. No. 1222-G, 1966, 48 pp.

[24] B. R. Haley, *Coal Resources of Arkansas, 1954,* U.S. Geol. Survey Bull. No. 1072-P, 1960, pp. 795–831.

[25] R. B. Johnson, *Coal Resources of the Trinidad Coal Field in Huerfano and Las Animas Counties, Colorado,* U.S. Geol. Survey Bull. No. 1112-E, 1961, pp. 129–180.

of blending coals and hydrocarbons in the manufacture of coke, and because of the certainty that the acceptable amounts of impurities in coke will be allowed to increase and coking properties to decrease as the higher rank and higher grade bituminous coals are depleted, it is likely that lower rank and lower quality bituminous coal will be beneficiated (Chapter 10) for use in the future. Thus it can be said that the resources of such coal are likely to be very large. Of the identified bituminous coal resources remaining in the ground as of January 1, 1974 (Table 2.3), about 35%, or about 260 billion tons, is suitable on the basis of rank, quality, and composition as a major or minor component in coking coal blends.

2.7.1 Low-Volatile Bituminous Coal

Low-volatile bituminous coal is high in heat value, low in volatile matter, and generally low in ash and sulfur content. Of all coal used in the manufacture of coke, low-volatile bituminous coal is the most important because (1) it is strongly coking and can be used in coking coal blends to upgrade the much larger resources of high-volatile bituminous coal, which is less strongly coking, (2) most areas of low-volatile bituminous coal are on the eastern edge of the Appalachian coal basin near centers of population and industry on the eastern seaboard, and (3) for uses other than the manufacture of coke, low-volatile bituminous coal contributes less to air pollution than do lower ranks of coal.

Low-volatile bituminous coal is mined extensively for the manufacture of metallurgical coke used in the United States, and it constitutes a substantial portion of the coal and coke exported to Japan, Canada, and Western Europe. It is also used on a smaller and steadily decreasing scale by nearby manufacturing industries and electric utilities because of its high heat value and low ash and sulfur content.

This choice fuel is in relatively short supply. An analysis of data on the occurrence of low-volatile bituminous coal in the summary reports on the states of Pennsylvania, West Virginia, Maryland, Virginia, Alabama, Oklahoma, Arkansas, and Colorado suggests that the original resources of low-volatile bituminous coal in the ground totaled about 20,000 million tons. The amount is about 1.1% of the originally identified total coal resources of the United States. The proportion will not change significantly because any change in the amount of a coal comprising such a low percentage of the total resource results in a comparatively small change in the percentage for that resource.

In many areas of less desirable and less readily accessible coal in the United States, the remaining resources are very nearly equal to the original resources, because little mining has been done. The areas containing low-volatile bituminous coal, on the other hand, are being mined very rapidly, and the remaining resources of this coal are now less than 1% of the remaining identified resources of the United States. Because only a limited supply of low-volatile bituminous coal is available, its use for purposes other than the manufacture of coke is a waste of a national asset, and coking operations and metallurgical processes must ultimately be adjusted to permit increased use of lower rank coal.

2.8 STRIPPABLE COAL RESOURCES

The amount of coal mined and potentially minable by strip-mining methods has increased steadily throughout the years, along with increases in the number, size, and efficiency of strip-mining machines. In 1917 strip mining accounted for only 1% of the total United States production of bituminous coal, subbituminous coal, and lignite, as compared to 54.7% in 1975. By the end of 1975 strip mining accounted for 14% of the total cumulative United States coal production. During 1975 almost the entire production of 10 states—Alaska, Indiana, Kansas, Missouri, Montana, North Dakota, Oklahoma, South Dakota, Texas, and Wyoming—was achieved by strip-mining methods.

In 1917 the largest steam shovel in operation had a capacity of only a few cubic yards. By 1957 the largest shovel in operation had a capacity of 70 cubic yards, or 105 tons. In

[26] E. R. Landis, *Coal Resources of Colorado,* U.S. Geol. Survey Bull. No. 1072-C, 1959, pp. 131–232.

[27] J. V. A. Trumbull, *Coal Resources of Oklahoma,* U.S. Geol. Survey Bull. No. 1042-J, 1957, pp. 307–382.

succeeding years still larger shovels were constructed, and in 1972 the largest shovel in operation had a capacity of 180 cubic yards, or 265 tons, and was capable of removing 16,000 tons of overburden an hour. Shovels of 200-cubic-yard capacity are a future possibility.

In 1968 a walking dragline with a 250-ft boom and a 145-cubic-yard bucket of 215-ton capacity was placed in operation in Indiana. The dragline can handle overburden to a maximum depth of 96 ft.[28] In 1969 in southern Ohio a walking dragline with a 310-ft boom and a 220-cubic-yard bucket of 325-ton capacity was placed in operation handling overburden to a maximum depth of 185 ft.[29] Several draglines with 360-ft booms are on order.

The increase in the size and efficiency of strip-mining machinery has permitted a steady increase in the average and maximum thickness of overburden removed, and as a result, the ratio of average thickness of overburden to average thickness of recovered coal has also increased (see Table 2.7).

The averages presented in Table 2.7 include several noteworthy extremes. In the famous Wyodak mine, Wyoming, for example, a 90-ft bed of coal is recovered by removing 25–40 ft of overburden, a ratio of 3.6 : 1 to 2.2 : 1. In Alaska the average thickness of overburden removed in 1965 was nearly 67 ft, and the average thickness of coal recovered was nearly 43 ft; these figures yield a favorable statewide ratio of 1.4 : 1.

In marked contrast, the average thickness of overburden removed in Oklahoma in 1965 was 43 ft, and the average thickness of coal recovered was 1.5 ft, for a statewide ratio of 29 : 1.[31] In one outstanding operation in Alabama, overburden ranging in thickness from a few feet to nearly 100 ft, and averaging about 60 ft, was removed to recover a bed of high-quality metallurgical coal 22 in. thick.[32] The ratio was 22 : 1. In at least one operation in Kansas 45 ft of overburden was removed to recover 1.5 ft of high-quality coal, a ratio of 30 : 1. In Illinois ratios larger than 30 : 1 have been handled and are being planned in parts of large-scale stripping projects where the coal is 28–36 in. thick.

These examples suggest that within the lift and swing limits of existing machinery, the 30 : 1 ratio is technically feasible as a local maximum for current and near-future strip mining. In the current highly competitive energy market, however, the success of each strip-mining operation depends on many factors in addition to the ratio between thickness of overburden and thickness of coal. These factors include thickness and quality of the coal; density and hardness of the overburden, stability of the highwall, and capacity of the machinery; size of property and amount of coal available; selling price of coal from competing sources; distance to transportation facilities and markets; availability of electric power, labor, and supporting facilities; and many environmental considerations. Because of the continued availability of coal with more favorable overburden ratios, the average ratio nationwide will continue to be less than 30 : 1 for many years (see the average ratios for recent years in Table 2.7).

[28] "Giant Digs Deeper at Dugger," *Coal Age*, **73**(12), 54–60 (1968).
[29] "Big Muskie—King of the Giants," *Coal Age*, **74**(12), 50–61 (1969).
[30] W. H. Young, *Thickness of Bituminous Coal and Lignite Seams Mined in 1965*, U.S. Bur. Mines Inf. Circ. No. 8345, 1967, 18 pp.
[31] W. H. Young, see Ref. 30, p. 18.
[32] "Thin-Seam Strip Mining in Alabama," *Coal Age*, **73**(6), 68–72 (1968).

Table 2.7 Average and Maximum Thickness of Overburden Removed and Average Thickness of Bituminous Coal and Lignite Recovered by Strip Mining in the United States for Selected Years

	1946	1950	1955	1960	1965	1970 (Est.)
Average thickness of overburden removed (ft)	32	39	42	46	50	55
Maximum thickness of overburden removed (ft)	—	—	70+	100	125	185
Average thickness of coal recovered (ft)	5.2	5.1	4.9	5.1	5.2	5.0
Ratio of average overburden thickness to average coal thickness	.6 : 1	8 : 1	8.5 : 1	9 : 1	10 : 1	11 : 1

Source: Modified from Ref. 30, p. 18.

The remaining strippable coal reserve base of the United States as of January 1, 1974, totals 137 billion tons, as shown by states in column 2 of Table 2.5. Based on an earlier study by the U.S. Bureau of Mines,[33] about 80% of this total, or 110 billion tons, is within reach by available machinery and current methods of mining, and 50% of it, or 55 billion tons, is economically recoverable.

The figure of 55 billion tons is too large to be appreciated except by comparison with smaller, more comprehensible numbers. It is, for example, nearly 10 times the cumulative strip coal production in the United States from the beginning of strip mining to January 1, 1974, and it is 200 times the production of strip coal in 1974 alone. These comparisons are not intended to suggest life expectancy of the economically recoverable strippable coal reserves, because the rate of production and the estimated size of the strippable coal reserve base are certain to change in the future.

About 70% of the economically recoverable strippable coal reserves contains 1% or less sulfur, and this low percentage is a major factor in the current interest in such coal and its increased production.

2.8.1 Reclamation of Strip-Mined Land

The abandoned spoil banks of past strip-mining operations in the United States are usually cited as a major objection to future strip mining. It should be noted, however, that the older abandoned spoil banks are the product of a time when the amount of disturbed land was relatively small, coal mining was a highly competitive, low-profit business, the value of the land before mining was low, and there was no strong public pressure for reclamation. The practices of the past, however, need not be the practices of the future: attitudes have changed, land reclamation laws have been passed, and highly advanced levels of land reclamation have been achieved in many parts of the United States. Actually, reclamation of strip-mined land is easier and less costly than almost any other environmental objective. It can be accomplished in large part by the same men and machinery that first removed the coal, and no new or expensive technology is required. When reclamation is carried on concurrently with strip mining, the cost of returning the land to a pleasing contour with a surface that will support vegetation should be in the range of \$2000–\$6000 per acre.[34] That these figures may be higher than the original value of the land is not significant, because the usefulness of the land will be restored, environmental objectives will be achieved, and the value of the land can be expected to appreciate.

When the cost of reclamation is expressed in terms of the coal recovered, it is found to be surprisingly low. Table 2.8 shows the cost associated with various thicknesses of coal recovered at four levels of estimated reclamation costs per acre. The cost is much lower in the West than in the East because western coal beds are thicker than eastern beds, and the acreage disturbed by strip mining in the West is small in relation to the amount of coal recovered. The western coals and associated rock are also low in sulfur, and the disturbed overburden in the West is less acidic than that in the East.

The estimated reclamation costs shown in Table 2.8 range from less than a penny per ton of coal recovered to \$1.44 per ton. Because western coal will probably contribute substantially to future strip mine production, it is likely that over the near and intermediate terms the national average will be in the lower part of this range, or roughly 25¢ per ton of coal recovered.

2.9 PEAT RESOURCES

Although peat is the precursor of coal, it is not classified as coal and usually is considered separately. Peat is an important fuel in Europe and elsewhere, but only small quantities have been used as fuel in the United States because of the abundance of other fuels. However the United States has substantial deposits of peat, and it is produced commercially for a variety of nonfuel purposes. Air-dried peat is a source of concen-

[33] U.S. Bureau of Mines, *Strippable Reserves of Bituminous Coal and Lignite in the United States*, U.S. Bur. Mines Inf. Circ. No. 8531, 1971, 148 pp. (see also Ref. 10).

[34] R. E. Matson, Montana Bur. Mines Geol., personal communication, April 1975.

Table 2.8 Costs of Reclaiming Strip-Mined Land as Related to Thickness of Underlying Coal

Costs of reclamation are expressed in dollars and cents per ton of coal recovered by strip mining at four assumed levels of cost per acre for reclamation of the strip-mined land.

Thickness of Coal[a] (ft)	Estimated Recovery per Acre[b] (tons)	Assumed Costs of Reclamation per Acre $1000	$2000	$3000	$4000
2	2,800	$0.36	$0.72	$1.08	$1.44
3	4,300	0.23	0.46	0.69	0.92
4	5,700	0.18	0.35	0.53	0.70
5	7,100	0.14	0.28	0.42	0.56
10	14,200	0.07	0.14	0.21	0.28
15	21,300	0.05	0.095	0.14	0.19
20	28,400	0.035	0.07	0.105	0.14
25	35,400	0.028	0.056	0.084	0.112
50	70,800	0.014	0.028	0.042	0.056
100	141,600	0.007	0.014	0.021	0.028

[a] Thin beds, 2–5 ft. typical of strip mining in the eastern and central United States: thick beds, 10–100 ft. typical of Northern Rocky Mountain and Great Plains regions.
[b] Assuming specific gravity of coal to be 1.3 and recovery to be 80%.

Source: Ref. 1, Table 7, p. 57.

trated organic matter that contains about 2% nitrogen. Because of these properties, it is used in the United States primarily as a soil conditioner. In 1973, for example, 87% of the peat consumed in the United States was used directly as an admixture to soil, and the remainder was used primarily in potting mixtures and fertilizers, and for packing flowers, shrubs, and bulbs. Small amounts were used in the culture of mushrooms and earthworms.

There may be a change in trends in the use of peat in the United States. The Minnesota Gas Company is exploring the possibility of applying coal gasification technology to the manufacture of pipeline gas from Minnesota peat in a study financed by a $1.2 million federal contract. The utility is also seeking a 25-year lease on state-owned lands containing 200,000 acres of peat bogs.[35]

During 1973 U.S. production of peat totaled 634,503 tons, and imports, primarily from Canada, totaled 323,501 tons.[36]

The peat resources of the conterminous United States were described in considerable detail by Soper and Osbon,[37] who estimated that the original peat resources totaled 13,827 million tons, calculated on an air-dried basis. Of this total, only about 11 million tons was mined between 1922 and 1973.

The peat resources occur primarily in local deposits distributed throughout two general regions. The northern peat region, which contains about 80% of the total resources, comprises Minnesota, Wisconsin, Michigan, eastern South Dakota, the northern parts of Iowa, Illinois, Indiana, Ohio, and Pennsylvania, and New York, New Jersey, and the New England states. The Atlantic coastal region, which contains approximately 19% of the total resources, comprises the southern part of Delaware, the eastern parts of Maryland, Virginia, North Carolina, South Carolina, and Georgia, and all Florida. Peat is also present in quantity in many parts of Alaska. Small deposits of peat occur in a narrow belt of land adjoining the Gulf of Mexico; in the valleys of the Sacramento and San Joaquin rivers, California; in Siskiyou, Los Angeles, Orange, and San Bernardino

[35] *Wall Street Journal,* July 29, 1976, p. 1.

[36] U.S. Bureau of Mines, *1973 Minerals Yearbook,* Vol. 1, p. 890, Government Printing Office. Washington, D.C.

[37] E. K. Soper and C. C. Osbon, *The Occurrence and Uses of Peat in the United States,* U.S. Geol. Survey Bull. No. 728, 1922, 207 pp.

counties in California; and in the basins of lakes and rivers in Oregon, Washington, and the Rocky Mountain states.

Several recent detailed studies of the occurrence and origin of peat in the eastern United States have been published by Cameron.[38]

2.10 WORLD COAL RESOURCES

Estimates of the coal resources of the world may be based on either of two predominant points of view. The first lists resources by individual countries to arrive at a total for the world. The second considers the world as a whole, notes the distribution of coal-bearing rock, and on the basis of existing data extrapolates from known to unknown areas. The first approach is typified by a report of the World Energy Conference,[39] which contains a 6-page table presenting invaluable information on resources and reserves of coal, by rank, in individual countries, and providing references to sources of more detailed information. (See also the monthly magazine *World Coal.*[40]) This table yields a figure of 10,753,884 megatonnes (11,851 billion short tons) for total world coal resources, of which 1,420,274 megatonnes (1565 billion short tons) represents total world coal reserves in place. On the basis of these figures, world coal reserves in place represent 13.2% of total resources. By way of comparison, the demonstrated reserve base in the United States represents 25% of identified world resources and 11% of total resources.

Solicitations for resource information used in the World Energy Conference Report table (Ref. 39; p.50) specified a minimum bed thickness of 0.30 m (12 in.) and a maximum overburden depth of 1200 m (3937 ft) for anthracite, bituminous coal, and subbituminous coal; and a minimum bed thickness of 0.30 m (12 in.) and a maximum overburden depth of 500 m (1640 ft) for lignite and brown coal. Many of the estimates used in the table are based on different parameters, however, mainly because the available estimates themselves were based on them. The estimates differ also because of differences in the points of view of the individual estimators, and because of differences in the amount of geologic and exploratory information available.

There are differences in points of view because coal is an abundant bulk commodity in most parts of the world, and annual production is typically only a small part of the available potential in the ground. Economic interest is thus centered on the thicker and more accessible beds, but long-range national planning and good resource management require consideration of thinner and less accessible beds that may be needed in the future. For certain countries, particularly the highly industrialized countries that make extensive use of coal, estimates are available for resources in several categories according to thickness of coal and overburden and several other points of view. For some countries this figure is comparable to the figure for identified resources as defined in this chapter; for others, it is comparable to the figure for total resources (identified plus hypothetical resources) as defined here.

After making reasonable adjustments for differences in points of view, minimum thickness of coal included in estimates, and maximum thickness of overburden considered, and after taking into account the worldwide distribution and thickness of coal-bearing rock, I concluded that world coal resources total about 16,620 billion short tons divided by continent and coal resource category as presented in Table 2.9.

The figures for identified resources in column 1 of Table 2.9 are based reliably on factual data and are conservative. The figures for hypothetical resources (column 2) and for total resources (column 3) are less reliable, but they are based in part on published estimates and in part on extrapolation from the figures in column 1.

[38] C. C. Cameron, "Peat," in *Mineral Resources of the Appalachian Region,* U.S. Geol. Survey Prof. Paper No. 580, 1968, pp. 136–145; *Peat Deposits of Northeastern Pennsylvania,* U.S. Geol. Survey Bull. No. 1317-A, 1971, 90 pp.; *Peat Deposits of Southeastern New York,* U.S. Geol. Survey Bull. No. 1317-B, 1970, 32 pp.; "Peat Resources of the Unglaciated Uplands Along the Allegheny Structural Front in West Virginia, Maryland, and Pennsylvania," in *Geological Survey Research 1970,* U.S. Geol. Survey Prof. Paper No. 700-D, 1970, pp. D153–D161.

[39] World Energy Conference, *The 1974 Survey of Energy Resources,* Engineers Joint Council, New York, 1974, pp. 74–79.

[40] *World Coal Magazine,* **1**(9), 37 (1975).

Table 2.9 Estimated Total Original Coal Resources (billions of short tons) of the World, by Continents[a]

Continent	Identified Resources (1)	Estimated Hypothetical Resources (2)	Estimated Total Resources (3)
Asia[b]	3,000[c]	8,000	11,000[d]
North America	1,900	2,500	4,400
Europe[e]	300	500	800
Africa	90	160	250
Oceania[f]	70	60	130
South and Central America	30	10	40
Total	5,390	11,230	16,620

[a] Original resources in the ground in beds 12 in. or more thick, and generally less than 4,000 ft below surface but includes small amounts between 4000 and 6000 ft.
[b] Includes European U.S.S.R.
[c] Includes about 2,300 billion short tons in the U.S.S.R. (see Ref. 41).
[d] Includes about 9,500 billion shor tons in the U.S.S.R. (see Ref. 41).
[e] Includes Turkey.
[f] Australia, New Zealand, New Caledonia.
Source: Ref. 1, Table 9, p. 91.

For comparison, the figure of 1565 billion short tons for world coal reserves in place, taken from the World Energy Conference table, represents 29% of world identified resources presented in column 1 of Table 2.9, and 9% of total resources presented in column 3 of Table 2.9. These percentages are not greatly different from those determined for the United States.

Most of the tonnage shown in column 1 of Table 2.9 lies between 0 and 2000 ft below the surface, and only a small amount lies between 2000 and 4000 ft. The bulk of that listed in column 3 also lies between 0 and 2000 ft, but larger amounts are present between 2000 and 4000 ft, and a small additional amount lies between 4000 and 6000 ft. Because most of the coal in the world occurs in shallow structural basins, the amount potentially present decreases with each 1000-ft increase in depth, and the amount potentially present below 3000 or 4000 ft is small compared to the larger amounts at shallow depths.

Some of the figures used in obtaining the continent totals in column 1 are for coal resources remaining in the ground as of various dates in the past; others are for original resources. Most of the data used to obtain the continent totals in column 3 are for original resources. The bulk of the tonnage in Table 2.9 is properly classified as original resources.

The data for the United States, as recorded in Table 2.4, are included in the totals for North America in Table 2.9. On this basis the United States has about one-fourth of the world's identified resources and about one-fifth of the world's total resources.

Table 2.9 shows clearly that Asia contains most of the world's potential coal resources. The tonnage is concentrated in the U.S.S.R. and in the People's Republic of China, both of which are important coal-producing countries. Other continents lag behind Asia in a sequence of rapidly decreasing tonnages. In Europe the coal resources have been well documented by many years of detailed geologic mapping and extensive exploration, and economic parameters have been applied with increasing frequency to coal in the identified category. As a result, much of the tonnage classed as hypothetical is in beds too thin or too deeply buried to be mined economically.

[41] N. V. Mel'nikov, "The Role of Coal in the Energy Fuel Resources of the USSR," *Can. Min. Metall. Bull.*, **65**(722), 77–82 (1972).

In Africa most coal in the hypothetical category is too impure to be mined economically; but much coal-bearing rock in Africa is concealed by younger rock, and estimates of resources in all categories are likely to increase in the future.

2.11 CONCLUDING STATEMENT

As an abundant, widespread, and economical source of heat and energy, coal is certain to be used in increasing quantities in the immediate future. It will share in long-term energy growth and, in particular, will contribute substantially to the generation of electricity and to the manufacture of steel.

As a remarkably versatile high-Btu chemical compound, coal is also certain to become a future major source of synthetic gas, liquid fuels, and lubricants, as well as thousands of hydrocarbon compounds used by the chemical and manufacturing industries.

2.12 GLOSSARY OF COAL RESOURCE TERMS

Demonstrated Reserve Base. A selected portion of coal in the ground in the measured and indicated (demonstrated) category; restricted primarily to coal in thick and intermediate beds less than 1000 ft below the surface and deemed to be economically and legally available for mining at the time of the determination.

Demonstrated Resources. The combined tonnage in the measured and indicated resource categories as defined above.

Hypothetical Resources. The tonnage of coal estimated to be in the ground in unmapped and unexplored parts of known coal basins to an overburden depth of 6000 ft; determined by extrapolation from the nearest areas of identified resources. Not otherwise classified. Future exploration to determine thickness, continuity, and quality of beds, and a more accurate estimate of tonnage will permit reclassification as identified resources. If data permit, some tonnage may be reclassified and added to the reserve base.

Identified Resources. Combined tonnage of coal in the measured, indicated, and inferred resource categories (see below); all coal in the identified category is further classified according to rank, thickness of bed, and thickness of overburden.

Identified Subeconomic Resources. The tonnage in the identified category minus the tonnage in the reserve base; some of this remaining tonnage may later be reclassified and added to the reserve base as a result of better information or changed economic or legal conditions.

Indicated Resources. The tonnage of coal in the ground based partly on specific observations and partly on reasonable geological projections; the points of observation and measurement are about 1 mile apart but may be 1½ miles apart for beds of known continuity.

Inferred Resources. The tonnage of coal in the ground based on an assumed continuity of coal beds downdip from and adjoining areas containing measured and indicated resources. Generally inferred coal lies 2 miles or more from outcrops or from points of precise information.

Measured Resources. The computed tonnage of coal in the ground based on assured coal bed correlations and on observations spaced about one-half mile apart; tonnage is judged to be accurate within 20% of the true tonnage.

Original Resources. Resources in the ground before the advent of mining.

Recoverability Factor. The percentage of coal in the reserve base that can be recovered by established mining practices.

Remaining Resources. Resources remaining in the ground as of a stated date; obtained by subtracting from original resources production and estimated losses during mining, or by eliminating mined-out areas as of a stated date.

Reserve. The tonnage that can be recovered from the reserve base by application of the recoverability factor; may be termed the "recoverable reserve."

Resources. The total quantity of coal in the ground within specified limits of thickness of bed overburden; includes identified and hypothetical resources.

Speculative Resources. A category for discussion of possible areas of coal occurrence outside known coalfields and coal basins in the United States, as currently defined for coal resource studies; examples: coal more than 6000 ft below the surface in deep Rocky Mountain coal basins, or coal on the continental shelves; no estimate is prepared for coal in this category.

REFERENCES FOR TABLES 2.3 AND 2.4

G. E. Abernathy, J. M. Jewett, and W. H. Schoewe, *Coal Reserves in Kansas,* Kansas Geol. Survey Bull. No. 70, Pt. 1, 1947, 20 pp.

H. H. Arndt, Paul Averitt, James Dowd, D. J. Frendzel, and P. A. Gallo "Coal," in *Mineral Resources of the Appalachian Region,* U.S. Geol. Survey Prof. Paper No. 580, 1968, pp. 102–133.

G. H. Ashley, *Anthracite Reserves,* Pennsylvania Topog. Geol. Survey Prog. Rep. No. 130, 1945, 13 pp.

D. C. Ashmead, *Anthracite Losses and Reserves in Pennsylvania,* Pennsylvania Geol. Survey 4th Ser. Bull. No. M8, 1926, 71 pp.

Paul Averitt, "Coal Deposits of Eastern Montana," in *Mineral Potential of Eastern Montana—A Basis for Future Growth,* U.S. Senate, 89th Cong., 1st sess., Senate Doc. No. 12, 1965; pp. 9–25. (Also distributed by the Montana Bureau of Mines and Geology as Spec. Publ. No. 33, and reproduced as Appendix B in Spec. Publ. No. 36, pp. 69–80.)

F. F. Barnes, *A Review of the Geology and Coal Resources of the Bering River Coalfield, Alaska,* U.S. Geol. Survey Circ. No. 146, 1951, 11 pp.

F. F. Barnes, *Coal Resources of Alaska,* U.S. Geol. Survey Bull. No. 1242-B, 1967, 36 pp.

H. M. Beikman, H. D. Gower, and T. A. M. Dana, *Coal Reserves of Washington,* Washington Div. Mines Geol. Bull. No. 47, 1961, 115 pp.

H. L. Berryhill, Jr., D. M. Brown, Andrew Brown, and D. A. Taylor, *Coal Resources of Wyoming,* U.S. Geol. Survey Circ. No. 81, 1950, 78 pp.

R. A. Brant, *Lignite Resources of North Dakota,* U.S. Geol. Survey Circ. No. 226, 1953, 78 pp.

R. A. Brant, and R. M. DeLong, *Coal Resources of Ohio,* Ohio Div. Geol. Survey Bull. No. 58, 1960, 245 pp.

Andrew Brown, H. L. Berryhill, Jr., D. A. Taylor, and J. V. A. Trumbull, *Coal Resources of Virginia,* U.S. Geol. Survey Circ. No. 171, 1952, 57 pp.

D. M. Brown, *Lignite Resources of South Dakota,* U.S. Geol. Survey Circ. No. 159, 1952, 18 pp.

Charles Butts and Benjamin Gildersleeve, *Geology and Mineral Resources of the Paleozoic Area in Northwest Georgia,* Georgia Geol. Survey Bull. No. 54, 1948, 176 pp.

G. H. Cady, *Minable Coal Reserves of Illinois,* Illinois Geol. Survey Bull. No. 78, 1952, 138 pp.

G. V. Cohee, R. N. Burns, Andrew Brown, R. A. Brant, and Dorothy Wright, *Coal Resources of Michigan,* U.S. Geol. Survey Circ. No. 77, 1950, 56 pp.

J. X. Combo, D. M. Brown, H. F. Pulver, and D. A. Taylor, *Coal Resources of Montana,* U.S. Geol. Survey Circ. No. 53, 1949, 28 pp.

W. C. Culbertson, *Geology and Coal Resources of the Coal-Bearing Rocks of Alabama,* U.S. Geol. Survey Bull. No. 1182-B, 1964, 79 pp.

T. W. Daniel, Jr., *A Strippable Lignite Bed in South Alabama,* Geol. Survey of Alabama Bull. No. 101, 1973, 101 pp.

H. H. Doelling, *Central Utah Coal Fields—Sevier-Sanpete, Wasatch Plateau, and Emery,* Utah Geol. Mineral. Survey Monogr. No. 3, 1973, 570 pp.

H. H. Doelling, and R. L. Graham, *Southwestern Utah Coal Fields—Alton, Kaiparowits Plateau, and Kolob-Harmony,* Utah Geol. Mineral. Survey Monogr. No. 1, 1972, 332 pp.

H. H. Doelling, and R. L. Graham, *Eastern and Northern Utah Coal Fields—Vernal, Henry Mountains, Sego, La Sal–San Juan, Tabby Mountain, Coalville, Henrys Fork, Goose Creek, and Lost Creek,* Utah Geol. Mineral. Survey Monogr. No. 2, 1972, 409 pp.

H. N. Eavenson, *The First Century and a Quarter of the American Coal Industry,* privately printed, Waverly Press, Inc., Baltimore, 1942, 701 pp.

T. H. Evans, *Bituminous Coal in Texas,* Texas Bur. Econ. Geol. Handbook No. 4, 1974, 65 pp. (see Mapel, 1967).

W. E. Edmunds, *Coal Reserves of Pennsylvania—Total, Recoverable, and Strippable, January 1, 1970,* Pennsylvania Bur. Topogr. Geol. Survey Inf. Circ. No. 72, 1972, 40 pp.

J. E. Fassett, and J. S. Hinds, *Geology and Fuel Resources of the Fruitland Formation and Kirtland Shale of the San Juan Basin, New Mexico and Colorado,* U.S. Geol. Survey Prof. Paper No. 676, 1971, 76 pp.

W. L. Fisher, *Lignites of the Texas Gulf Coastal Plain,* Texas Univ. Bur. Econ. Geol. Rep. Invest. No. 50, 1963, 164 pp. (see Perkins and Lonsdale, 1955; Kaiser, 1974).

S. A. Friedman, "Investigation of the Coal Reserves in the Ozarks Section of Oklahoma and Their Potential Uses," *Final Report to the Ozarks Regional Commission, July 10, 1975,* Oklahoma Geol. Survey, 1975, 117 pp.

G. B. Glass, *Midyear Review of Wyoming Coal Fields, 1972,* Wyoming Geol. Survey, 1972, 43 pp.

B. R. Haley, *Coal Resources of Arkansas, 1954,* U.S. Geol. Survey Bull. No. 1072-P, 1960, pp. 795–831.

A. J. W. Headlee and J. P. Nolting, Jr., *Characteristics of Minable Coals of West Virginia,* West Virginia Geol. Survey Rep., Vol. 13, 1940, 272 pp.

R. D. Holt, *Coal Resources in Colorado,* Colorado Geol. Survey Bull. No. 34-B, in press.

M. E. Hopkins and J. A. Simon, *Coal Resources of Illinois,* Illinois Geol. Survey Miner. Note No. 53, 1974, 24 pp.

A. L. Hornbaker and R. D. Holt, *1972 Summary*

of Coal Resources in Colorado, Colorado Geol. Survey Spec. Publ. No. 3, 1973, 15 pp.

J. W. Huddle, E. J. Lyons, H. L. Smith, and J. C. Ferm, *Coal Reserves of Eastern Kentucky,* U.S. Geol. Survey Bull. No. 1120, 1963, 274 pp.

V. H. Johnson, *Coal Deposits on Sand and Lookout Mountains, Dade and Walker Counties, Georgia,* U.S. Geol. Survey Coal Invest. Map, scale 1 : 48,000, 1946.

W. R. Kaiser, *Texas Lignite—Near-Surface and Deep-Basin Resources,* Texas Bur. Econ. Geol. Rep. Invest. No. 79, 1974, 70 pp.

E. R. Landis, *Coal Resources of Colorado,* U.S. Geol. Survey Bull. No. 1072-C, 1959, pp. 131–232.

E. R. Landis, *Coal Resources of Iowa,* Iowa Geol. Survey Tech. Paper No. 4, 1965, 141 pp.

E. T. Luther, *The Coal Reserves of Tennessee,* Tennessee Div. Geol. Bull. No. 63, 1959, 294 pp.

W. J. Mapel, *Bituminous Coal Resources of Texas,* U.S. Geol. Survey Bull. No. 1242-D, 1967, 28 pp. (see Evans, 1974).

R. S. Mason, "Coal," in *Mineral and Water Resources of Oregon,* Committee on Interior and Insular Affairs, U.S. Senate, 90th Cong., 2d Sess., Comm. print, 1969, pp. 272–280.

R. S. Mason and M. I. Erwin, *Coal Resources of Oregon,* U.S. Geol. Survey Circ. No. 362, 1955, 7 pp.

H. W. Peirce, S. B. Keith, and J. C. Wilt, *Coal, Oil, Natural Gas, Helium, and Uranium in Arizona,* Arizona Bur. Mines Bull. No. 182, 1970, 289 pp.

J. M. Perkins and J. T. Lonsdale, *Mineral Resources of the Texas Coastal Plain* (prelim. rep.), Texas Univ. Bur. Econ. Geol. Miner. Resour. Circ. No. 38, 1955, 65 pp. (see Fisher, 1963; Kaiser, 1974).

C. B. Read, R. T. Duffner, G. H. Wood, and A. D. Zapp; *Coal Resources of New Mexico,* U.S. Geol. Survey Circ. No. 89, 1950, 24 pp.

J. F. Reese and J. D. Sisler, *Coal Resources, Introductory Volume,* Pt. 3 of *Bituminous Coal Fields of Pennsylvania,* Pennsylvania Geol. Survey, 4th Ser., Bull. No. M-6, 1928, 153 pp.

J. A. Reinemund, *Geology of the Deep River Coal Field, Chatham, Lee, and Moore Counties, North Carolina,* U.S. Geol. Survey Coal Invest. Map, scale 1 : 24,000, 1949.

J. A. Reinemund, *Geology of the Deep River Coal Field, North Carolina,* U.S. Geol. Survey Prof. Paper No. 246, 1955, 159 pp.

C. E. Robertson, *Evaluation of Missouri's Coal Resources,* Missouri Geol. Survey Rep. Invest. No. 48, 1971, 92 pp.

C. E. Robertson, *Minable Coal Reserves of Missouri,* Missouri Geol. Survey Rep. Invest. No. 54, 1973, 71 pp.

H. E. Rothrock, "A Summary and Critique of the Estimates of Pennsylvania Anthracite Reserves," in *Pennsylvania Academy of Science Symp. Mineral Resources in Pennsylvania, Pa. Acad. Sci. Proc.,* **24,** 192–197 (1950).

W. H. Schoewe, *Coal Resources of the Cretaceous System (Dakota Formation) in Central Kansas,* Kansas Geol. Survey Bull. No. 96, Pt. 2, 1952, pp. 69–156.

W. H. Schoewe, "Economic Geography; Mineral Resources," Pt. 4 of *The Geography of Kansas, Kans. Acad. Sci. Trans.,* **61:**4, 359–468 (1959).

J. W. Shomaker and R. D. Holt, *Coal Resources of Southern Ute and Ute Mountain Ute Indian Reservations, Colorado and New Mexico,* New Mexico Bur. Mines Miner. Resour. Circ. No. 134, 1973, 22 pp.

F. D. Spencer, *Coal Resources of Indiana,* U.S. Geol. Survey Circ. No. 266, 1953, [1954], 42 pp.

P. E. Soister, *A Preliminary Report on a Zone Containing Thick Lignite Beds, Denver Basin, Colorado,* U.S. Geol. Survey Open-File Rep. No. 74-27, 1974. Planimetric map, scale 1 : 250,000; text, 46 pp.

R. A. Struble, H. R. Collins, and D. L. Kohout, *Deep-Core Investigation of Low-Sulfur Coal Possibilities in Southeastern Ohio,* Ohio Div. Geol. Survey Rep. Invest. No. 81, 1971, 29 pp.

J. W. Sullivan, *The Geology of the Sand-Lookout Mountain Area, Northwest Georgia,* Georgia Geol. Survey Inf. Circ. No. 15, 1942, 68 pp.

J. V. A. Trumbull, *Coal Resources of Oklahoma,* U.S. Geol. Survey Bull. No. 1042-J, 1957, pp. 307–382 (see Friedman, 1975).

U.S. Bureau of Mines, *Mineral Resources of the U.S.* (annual volumes for the years indicated), Government Printing Office, Washington, D.C., 1924–1931.

U.S. Bureau of Mines, *Minerals Yearbooks* (annual volumes for the years indicated), Government Printing Office, Washington, D.C., 1932–1973.

U.S. Bureau of Mines, Weekly Coal Report No. 2950, Pittsburgh, U.S. Bur. Mines, March 29, 1974, 8 pp.

U.S. Geological Survey, *Mineral Resources of the U.S.* (annual volumes for the years indicated), Government Printing Office, Washington, D.C., 1886–1923.

C. E. Wier, *Coal Resources of Indiana,* Indiana Geol. Survey Bull. No. 42-I, 1973, 40 pp. (esp. pp. 32–33).

CHAPTER 3

Origin, Petrography, and Classification of Coal

RICHARD C. NEAVEL

Richard C. Neavel *is a Scientific Advisor with the Baytown (Texas) Coal Research Laboratory of Exxon Research and Engineering Company, having been with that company since 1966. After receiving his B.A. degree in geology from Temple University, he began his studies of the origin, constitution, and utilization of coal as a graduate student at the Pennsylvania State University. He earned M.S. and Ph.D. degrees in geology from Penn State. Dr. Neavel served as a coal petrologist for the Indiana Geological Survey and consulted for numerous organizations. He was Chairman of the 1975 Fuels Science Gordon Research Conference, and Program Chairman for the Fuel Chemistry Division of the American Chemical Society in 1978–1979. His 25-year professional research career has focused principally on relating preparation, coking, gasification, and liquefaction responses of coals to their chemical, physical, and petrographic properties.*

3.1 DEFINITIONS, BASIC TERMINOLOGY, AND AN OVERVIEW OF CLASSIFICATION

Coal is a sedimentary rock accumulated as peat and composed principally of macerals, subordinately of minerals, and containing water and gases in submicroscopic pores. Macerals are organic substances derived from plant tissues and exudates that have been incorporated in sedimentary strata, variably subjected to decay, then compacted, hardened, and chemically altered by natural (geological) processes. The discipline wherein coal is studied as a rock constituted principally of macerals is called *coal petrology* (*petros* = rock; *logos* = the science of), and the description or classification of coal as a rock is referred to as *coal petrography* (-*graphy* = description of). Coal petrologists have commonly been trained as geologists, mineralogists, or botanists; therefore they are frequently concerned with questions about the geological and botanical origin of coals. Consequently many petrologically oriented studies and classification systems have a decided mineralogical or genetic orientation, as well. The language of the petrographer frequently confuses and discourages the chemist or engineer who attempts to understand the significance of the petrographic constitution of coal. This is unfortunate, because the value and validity of any study of the "chemistry" or of the utility of coal is enhanced by the recognition of the rocklike character of coal.

Coal is not a uniform mixture of carbon, hydrogen, oxygen, sulfur, and other elements; nor is it, as is often implied, simply a uniform, polyaromatic, "polymeric" substance. Rather it is an aggregate of microscopically distinguishable, physically distinctive, and chemically different substances called *macerals* and *minerals*.

In many respects coal resembles a fruitcake, for like a fruitcake coal is formed initially as a mixture of diverse ingredients that are then altered (*metamorphosed*) in a natural "baking" process leading to a final product that is visibly heterogeneous. The principal source ingredients of coal were various plant materials that were preserved and accumulated in the moist environment of peat swamps, many of which extended over wide areas, and from which mineral matter was substantially barred.

The heterogeneity that is evident even in a hand specimen of coal (Fig. 3.1) is inherited from the diversity of source materials. As a simplified overview, Fig. 3.2 shows schematically the various substances that accumulated as peat deposits. The dominant material under typical conditions was "wood," and several fates are shown for it: humification (incorporation essentially intact), partial decomposition, and "charring" by partial combustion. However not only "woods" of variable character, but other chemically and physically distinctive plant substances also accumulated in typical peats. The metamorphosed remains of these plant-derived materials are recognizable in coals, as the photomicrograph of a polished surface of coal shows (Fig 3.3). The distinctive areas recognizable in the photomicrograph represent discrete volumes of different substances, each referable to a specific class of maceral substances.

Adding further to the complexity of the source materials were inorganic substances that entered the swamp as particulate mineral grains and as dissolved ions. Many of the dissolved ions, especially those of sulfur, either combined with the organic fraction or were precipitated in place to form discrete mineral grains.

Different coals, or even samples of coal taken from different places in a given coal seam, can be described on the basis of the relative proportions and the textural or spatial associations of the macerals and minerals comprising them. On the basis of these features, different *types* of coal are recognizable. However specification of the coal type (maceral and mineral content and texture) is inadequate for classifying coals definitively, since similar types can be found metamorphosed to different ranks. The baking process that altered these accumulated plant substances after burial by thick layers of sediments is termed metamorphism; the degree of "doneness" is referred to as the rank of the coal. The interrelationships between rank and type are shown in Fig. 3.4. The simplest constituents comprising any coal are moisture, organic material (macerals), and mineral matter.*

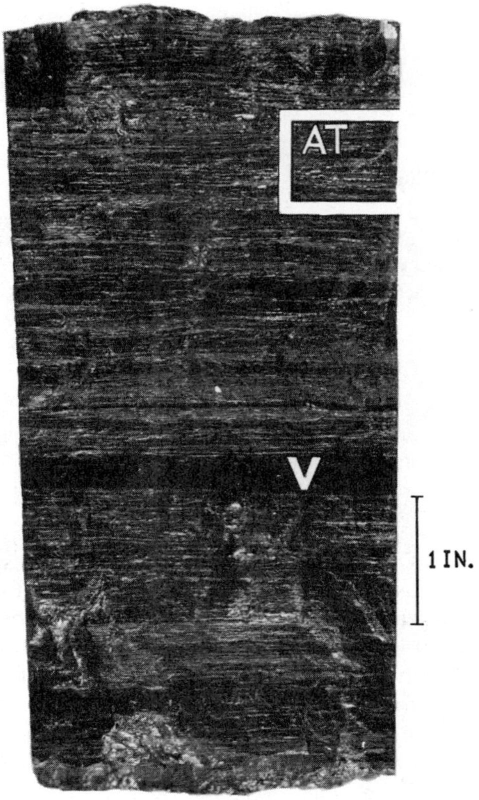

Figure 3.1 Natural broken surface of a block of bituminous coal: oriented as it would be found in the seam. The banding or striation reflects the heterogeneity of the coal resulting from progressive accumulation of different plant substances in the peat swamp.

*Coal does not contain ash. Ash, the residue left after combustion, is derived from pyrolyzed, oxidized, and commonly sulfated inorganic (mineral) material.

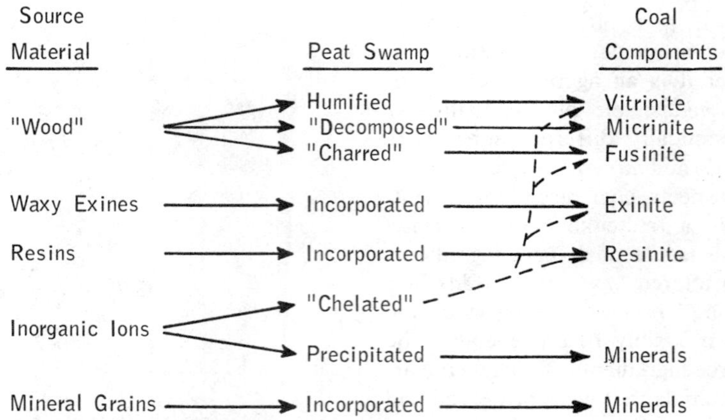

Figure 3.2 Physically discrete and chemically distinctive plant materials that accumulated along with inorganic substances in the peat swamp retain their identity as distinctive macerals in the coal.

Coals are porous, and the pores are saturated with water. With increasing rank to the anthracitic stage, porosity decreases; hence moisture-holding capacity decreases (the trend is reversed slightly in the highest rank anthracites). For convenience of presentation, the hypothetical compositions in Fig. 3.4 are shown on a moisture-free basis.

For a given coal, the relative proportions of organic (maceral) material and mineral matter were established in the swamp (or, in some instances, shortly after peat burial when epigenetic minerals may have formed). Furthermore, the relative proportions of the progenitors of different macerals were also set in the swamp. These proportions varied from time to time and place to place even in a single swamp. In the hypothetical example shown in Fig. 3.4, vitrinite comprises 55% of the lignite; each of the other constituents (fusinite, exinite, mineral matter) comprises about 15% of the lignite. The progenitors of each of these constituents occurred in the originating peat in essentially the same proportions. If the originating peat had been metamorphosed to the bituminous rank or to

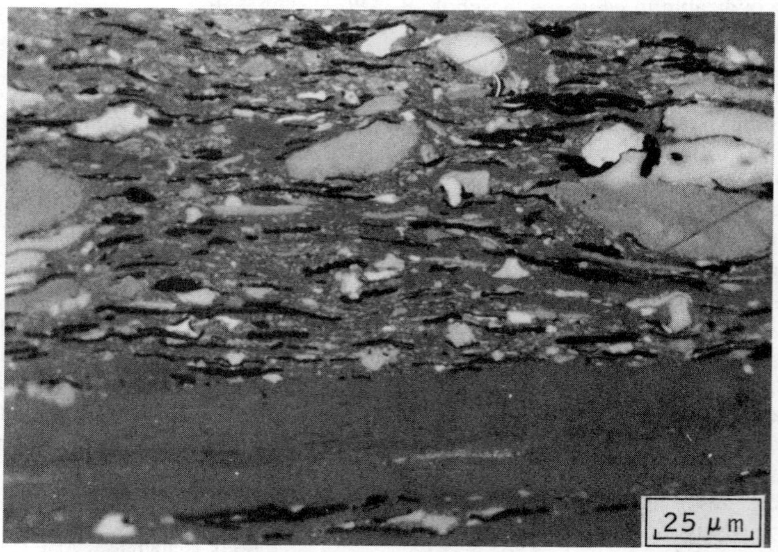

Figure 3.3 Reflected light photomicrograph of polished coal surface. Gray matrix is vitrinite, black bodies are exinite (coalified spore exines), and light (bright) material is inertinite.

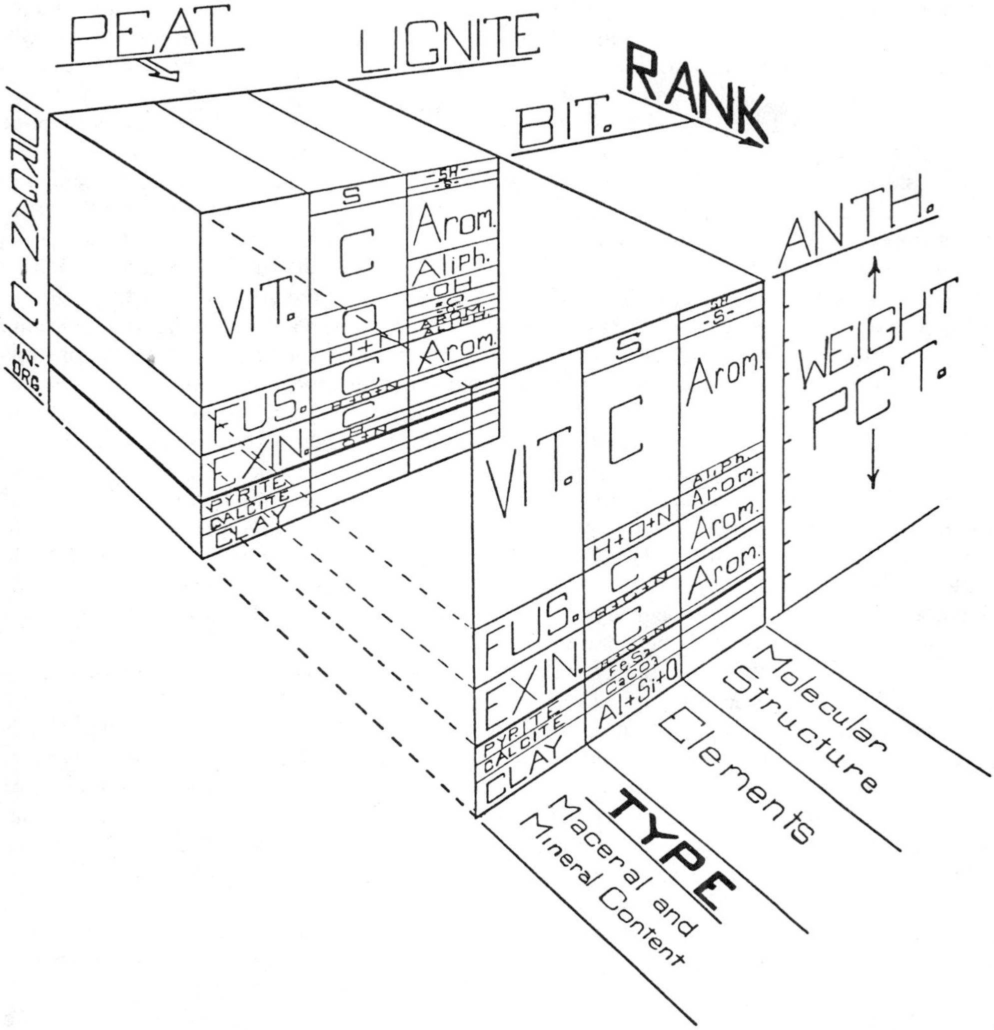

Figure 3.4 Interrelationship of type and rank; depicting the compositional changes that would occur to a specific, hypothetical peat as a result of metamorphism. The proportions of constituent macerals and minerals remain relatively constant irrespective of rank (near column). However, other properties, such as elemental compositions and molecular structure, change progressively (middle and far columns). Similar figures for other peats would exhibit different maceral (and mineral) proportions, although the elemental composition and molecular structure of a given maceral at a given rank would be essentially similar to those shown.

the anthracitic rank (as shown by the "slices" through the block diagram), the relative proportions of those constituents would not be changed (except for some differential loss of volatile products). Consequently, from the hypothetical peat, the relative proportions of the major classes of macerals and minerals, (thus the coal type), would be essentially the same, no matter how severely the peat was to be metamorphosed (at least to the low-volatile bituminous rank, where the properties of some macerals become similar). However the extent of the metamorphic alteration would affect the elemental composition and molecular structure of the various organic constituents. Thus wood-derived vitrinite can be recognized microscopically in all ranks of coal, but its physical properties and chemical composition differ from rank to rank.

From a chemical standpoint, maturation (coalification) is predominantly a phenomenon whereby oxygen (initially) and hydrogen (subsequently) in functional groups are stripped from the cyclic carbon "backbone" and evolved from the coal seam as low molecular weight substances (H_2O, CO_x, CH_4). As a result, the relative proportion of carbon in the residual, metamorphosed coal increases with increasing rank. For this reason, carbon content is often used as a measure of rank. However each of the classes of maceral material originated as chemically dissimilar substances and each responds differently to the metamorphic process. Therefore two different types of coal that were subjected to equally severe metamorphic processes (e.g., two distinctive layers in one seam) can differ significantly in carbon contents, even on a mineral-free basis. Similarly, other kinds of analyses of different types of coal of the same rank can yield significantly different results, representing, as they do, the weighted average compositions or properties of the constituent macerals. Thus a chemical analysis reflects not only the rank, but also the type of the coal. But because typical coals are made up predominantly of wood-derived vitrinite, analyses of the carbon content or of the volatile matter yield of whole coals do provide comparatively useful information about the rank of the coal.

The rank classification system used in North America is that specified in ASTM Designation D388,[1] wherein volatile matter (dry, mineral-matter-free basis) and Btu per pound (moist, mineral-free) are employed as classifying criteria (Chapter 2). On the basis of arbitrary classes defined by these parameters, coals of progressively higher rank (more severely metamorphosed) are referred to as lignitic, subbituminous, bituminous, and anthracitic.

Ideally the classification of a coal according to petrographic composition (type) and the extent of metamorphic alteration (rank) should be sufficient to allow one to predict the response of that coal in any processing situation. However limited availability of correlations between type and rank on the one hand and the operational responses on the other hand restricts the application of this concept. Consequently the value or "grade" of coals is appraised most commonly by one or more empirical tests that apply specifically to the use to which the coal is to be put. A coal of high grade for one application (e.g., steam raising) may be a poor grade for another application (e.g., coke making: see Chapter 2). Yet the fundamental properties of rank and type remain the same no matter how the coal is to be used.

Rank and type are fundamental properties established by the conditions of formation of the coal, whereas the grade or economic value is established by the interrelationships between the fundamental properties, including those of associated mineral matter, and the use to which the coal will be put. Just as studies of the "chemistry" of coals can be put into a meaningful context only when the rank and type (petrographic constitution) are recognized, so the empirically defined grade can be understood and related to other coals only when rank and type are defined adequately.

To summarize, coals are sedimentary rocks composed principally of macerals with subordinate mineral concentrations. Coals may be classified as to rank, type, and grade. Coal types are distinguished on the basis of megascopic or microscopic observations that recognize constitutional differences resulting from variations in proportions and distributions of the various macerals and minerals. Coal rank classification is based on measurement of properties that are sensitive to geologically induced metamorphism. Hence rank reflects the extent or severity of metamorphic alteration to which the plant-derived macerals have been subjected. Coals may be graded according to their value for a specific use, employing whatever empirically applicable test parameters are appropriate to that use, including but not limited to ash content, moisture content, calorific value, rank, and petrographic composition (type).

Consider an investigator who wishes to select a coal to be used as a feedstock for the commercialization of a liquefaction process. What properties should he or she evaluate in the coals that are potentially available? Given that there are presently available only the most rudimentary correlations between the liquefaction properties of coals and the fundamental parameters of rank and type, the investigator must empirically test samples of the coals that are potentially mineable at

[1] ASTM Standards, 1977, Part 27, Designation D388-77, Standard Specifications for Classification of Coals by Rank.

a competitive price. Would not the search be much more efficient if there were available a correlation such as shown in Fig. 3.5? Ideally the investigator would need only to assess the rank and type of the available coals and then, given the correlation shown, predict their relative values as liquefaction feedstocks. Unfortunately, no such correlation exists, but this chapter attempts to bring together the information relating rank and type to grade that is now available. It represents a summary of the present thinking regarding the parameters that are important to the practical classification of coals, first defining the causes of the variations in coal properties, then discussing the significance of these variations in practical situations.

3.2 COAL PETROLOGY: MACROSCOPIC AND MICROSCOPIC EVALUATION OF COAL TYPES

3.2.1 Petrographic Constitution

As noted in the preceding section, the heterogeneity of coal is evident from a visual examination (Figs. 3.1 and 3.3). In most coal seams vertical zones or benches can be distinguished on the basis of gross differences in luster, texture, or fracture pattern (Fig. 3.6). Variations in appearance can be traced to variations in the depositional environment, such as water level and velocity, water chemistry, and nature of the plant communities. Different depositional environments led to the formation of different types of coal.

Investigations of coal type based on megascopic appearance can yield information useful for seam correlation, for predicting minability (including roof and floor stability), for predicting lateral changes in coal properties, and for providing guidance to optimize preparation procedures. Megascopic observations, however, are adequate only for qualitatively predicting or correlating behavior of the coal during utilization. A brief description of the megascopic characterization technique follows.

In many coal seams there are especially distinctive layers or benches, called partings, that are rich in mineral matter (predominantly clays). When the coal is mined, these partings, ranging from less than an inch to several feet thick, are mined, along with the purer associated coal. Hence some pieces of mined "coal" may consist of almost pure mineral matter. If the coal is cleaned or beneficiated, much of this material will report to the refuse pile; however commercial cleaning operations are never 100% efficient, and even a "clean" coal contains pieces of mineral-rich material.

The discrete mineral-rich partings were formed during transient periods when flood-

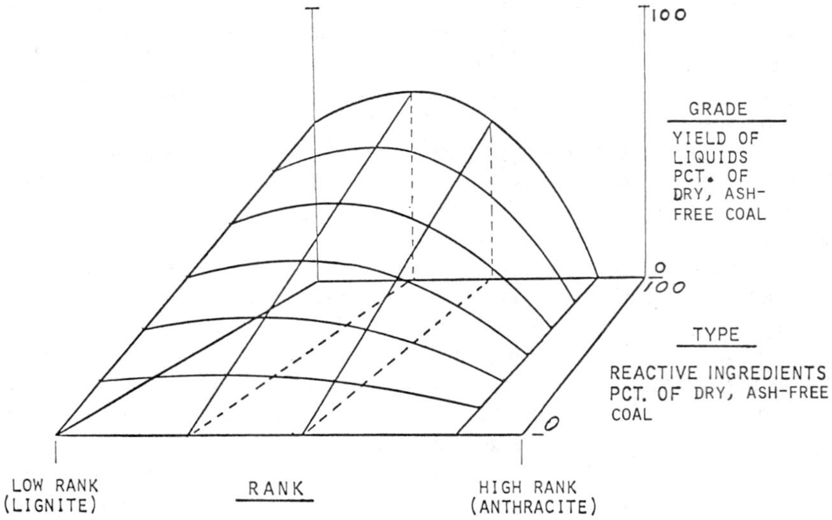

Figure 3.5 Coal liquefaction response surface indicating effects of rank and type of grade (i.e., liquid yield). Figure is illustrative and does not accurately depict yields.

Figure 3.6 A coal seam face. Benches reflect variations in depositional environments. (Photo courtesy of W. Spackman)

ing transported detrital minerals (clay, silt) across the swamp. More subtle alterations in the swamp environment (i.e., minor changes in water level or nutrient supply) resulted in changes in coal properties that were less dramatic than the formation of a mineral-rich parting, but just as important from the standpoint of the coal use. For instance, in a given swamp, environmental conditions might have remained relatively stable during the accumulation of one meter of peat. The layer of coal formed from the peat would then be relatively uniform in its maceral content and texture. Changing environmental conditions might then have resulted in the accumulation of a layer composed of a different mix of macerals. Each of these layers would consist of a different coal type, and the types in each bench might respond dissimilarly to utilization processes.

Most typical coal seams, then, are comprised of benches differentiated on the basis of the nature and appearance of the organic matter itself. The most common coals are referred to as banded coals, well exemplified by the coal in Fig. 3.1. The smooth, homogeneous black band (really a segment of a broad, flattened lens) labeled *V* is called vitrain. Not only does it look homogeneous at the macroscopic level, but even when viewed microscopically it appears to be relatively homogeneous. Microscopically visible traces of relict plant cell structure in many vitrain bands indicate that they consist of the flattened, coalified remains of single pieces of woody material. This wood-derived material constituting vitrain bands is referred to as the maceral vitrinite.

Numerous other thinner vitrain bands can be distinguished on the face of the coal shown in Fig. 3.1. Where individual vitrain bands are so thin that they are barely distinguishable, or where they are virtually absent, the coal is referred to as attrital coal or attritus. Attritus is comprised of fine particulate material and may be dull or bright depending on the size of the particles and the proportions of the macerals and minerals comprising it. Individual benches of coal can be described on the basis of the relative proportion and thickness of the vitrain bands and the brightness (or shininess) of the attritus (compared to the vitrain). Generally speaking, the brighter the coal (for a given rank), the more of the maceral vitrinite it contains. Bright, vitrinite-rich coal is sometimes referred to as clarain, whereas duller, vitrinite-deficient coal is called durain.

Another macroscopically visible, organic ingredient of coal is fusain, which is often found accumulated in thin, horizontal layers. Fusain is commonly porous, soft, and friable;

therefore the coal breaks preferentially along planes where it is concentrated. Commonly, silky, fibrous chips of fusain are visible on the top and bottom surfaces of pieces of coal (Fig. 3.7). Occasionally the pores of fusain are filled with pyrite or other minerals. Just as the maceral substance comprising vitr*ain* bands is called vitr*inite,* the macroscopically recognizable fus*ain* is composed of the maceral fus*inite,* which, in lignitic and bituminous rank coals, is of higher carbon content than the associated vitrinite.

From the foregoing brief description of the megascopic characteristics of coal, it should be obvious that different pieces of coal, even from a bench that is relatively uniform in terms of its vitrain banding and attrital coal characteristics, can and do differ in their maceral contents. Furthermore, as already implied, most coal seams are not uniform in terms of their banded characteristics. In different layers or benches, greater or lesser amounts of the different macerals (the number and properties of which are more complex than so far discussed—see following sections) leads to the production of particles of coal that not only are internally heterogeneous but differ significantly from one another. Thus different pieces of coal from the same seam may react quite dissimilarly in certain processing situations, and the investigator of, for example, reaction rate phenomena, should keep in mind that the subject reactant is highly diverse, not only at the molecular level, but also at the maceral and particle levels.

3.2.1.1 *Purposes and Advantages of Microscopy*

The microscopic examination and analysis of coal is the only technique presently used to characterize coal type quantitatively. Several decades ago Wilfrid Francis[2] developed a complex extraction technique referred to as a "rational analysis" for estimating the proportions of (1) resins and hydrocarbons, (2) ulmins, (3) cuticular material, (4) opaque matter, and (5) fusain. These components correspond approximately to microscopically discernible (1) extractable waxes, (2) vitrinite, (3) liptinite (exinite), (4) semifusinite (+ micrinite?), and (5) fusinite. The rational analysis procedure is potentially useful as a "chemical" petrographic analysis technique; however it is complex and probably more labor intensive than the microscopic technique. Thus rational analysis appears to offer

[2] W. Francis, *Coal, Its Formation and Composition,* Edward Arnold (Publishers), Ltd., London, 1961.

Figure 3.7 Chips of charcoallike fusain on a bedding plane. (Photo courtesy of W. Spackman)

few advantages over microscopy, whereas microscopy is applicable to mixtures of coals of different ranks (and is useful for distinguishing rank). Furthermore, it provides the capability of recognizing oxidation and contaminants (i.e., coke, chars, pieces of coal from other sources) that are frequently found in coal supplies.

Macerals and minerals occur as volumetric segments of coal seams, segments of sufficient size to be resolved at microscopic magnifications of about 100 diameters and greater. By examining correctly prepared samples, it is possible to determine the relative proportions of the various constituent macerals and minerals, their associations and juxtaposition with each other (texture), and certain optical features that correlate with specific chemical properties. The most important application of the study of optical properties is the correlation that has been developed between the reflectance of vitrinite and its carbon content (or H/C ratio) which, in turn, correlates with rank.

3.2.1.2 *Microscopic Techniques*

The earliest microscopic studies of coal were carried out on very thin slices, and for certain paleobotanical investigations this is still the preferred technique. However for applied petrographic investigations, the microscopy of polished surfaces offers at least four advantages over the thin-section technique: (1) preparation of the sample is faster and requires less skill; (2) polished surfaces can be used for making quantitative reflectance and hardness measurements useful for assessing coal rank; (3) because of poor light transmittance, thin sections are unusable for the higher rank coals; and (4) quantitative analyses of the proportions of the macerals is more accurate on the plane, polished surfaces than on thin sections having a discrete thickness that can lead to overlapping or masking.

3.2.1.3 *Sample Preparation*

Detailed procedures for sample preparation and microscopic analysis of coal are described in ASTM Standards D2796–D2799.[3–6] Also, deVries et al.[7] and Chistyakova[8] describe procedures in some detail. The reader is referred to these publications for descriptions of techniques; however a summary discussion, along with some cautions and observations, is presented here.

Oriented blocks of coal can be polished and studied microscopically, and are often used in geological studies such as exploration and seam correlation. For evaluating a coal to be used in processing, however, a representative split of crushed coal is molded into a block with epoxy or polyester binder, after which the block is polished. Sampling should be carried out according to procedures that assure a representative sample at all stages of subdivision, preferably by riffling or careful use of the cone-and-quartering procedure.[9] Because petrographic components differ in hardness and friability, specific ones commonly occur preferentially in certain sizes of crushed coal. It is especially important, therefore, to be sure that all sizes are represented in the final sample. This caution is as applicable to crushing and sampling for chemical analyses or reactivity tests as it is for microscopic analyses.

Procedures for preparing samples for petrographic examination are described in ASTM D2797,[4] which is summarized below. A representative sample of air-dried coal is crushed to pass a No. 4 (4.75 mm) screen, minimizing the production of fines in the process. About 250 g is riffled off and crushed to pass a No. 20 (0.85 mm) sieve, intermit-

---

[3] ASTM Standards, 1977, Part 27, Designation D2796-72, Standard Definitions of Terms Relating to Lithologic Classes and Physical Components of Coal.

[4] ASTM Standards 1977, Part 27, Designation D2797-72, Standard Method of Preparing Coal Samples for Microscopical Analysis by Reflected Light.

[5] ASTM Standards 1977, Part 27, Designation D2798-72, Standard Method for Microscopical Determination of the Reflectance of the Organic Components in a Polished Specimen of Coal.

[6] ASTM Standards, 1977, Part 27, Designation D2799-72, Standard Method for Microscopical Determination of Volume Percent of Physical Components of Coal.

[7] H. A. W. deVries, P. J. Habets, and C. Bokhoven, *Brennstoff-Chemie,* **49**(1), 15–21 (1968); *C. A.,* **68,** 61570 (1968).

[8] A. S. Chistyakova, *Tr. Vses. Nauchn.-Issled. Geol. Inst.* **132,** 285–289 (1968); *C. A.,* **70,** 70023 (1969).

[9] ASTM Standards, 1977, Part 27, Designation D2013-72, Standard Method of Preparing Coal Samples for Analysis.

tently screening off the material that will pass the screen and crushing the oversize in a mortar and pestle (stage crushing). Some investigators recommend removing excess mineral matter by employing a sink-float separation in a liquid of 1.9 specific gravity.[10] Since this could result in a biased sample when mineral matter is preferentially associated with specific macerals, however, the procedure is not recommended when accurate quantitative analyses are desired.

About 10 g of the −20 mesh coal is mixed well with about 4 g of an epoxy resin to make a paste that is placed in a cylindrical mold of 1-in. I.D. (wall about 1/4 in. thick, height about 1 1/2 in.), the inside of which is coated with a release agent. Sliding-fit plugs seal the bottom and top, and a hydraulic press is used to compact the mass and expel air. The epoxy is allowed to cure, and the briquette is pressed from the mold. When the pellet is to be used in an automated microscope, the relative coal and binder volumes are carefully measured.

I have found a very satisfactory alternative method for use when only a very small quantity of sample is available. The coal is mixed on a sheet of foil with an excess of epoxy so that the mixture will flow like a viscous liquid. The slurry is evacuated in a dessicator to remove air bubbles and to forcefully impregnate porous particles. The mixture is then poured into a small glass vial, coated with release agent, about 7–10 mm diameter and 20–25 mm high. After the epoxy has cured, the glass is broken away from the coal/epoxy cylinder, the cylinder is split lengthwise, and the two pieces are laid flat side down in the bottom of one of the 1-in. molds and weighted down with a piece of lead. Epoxy is then poured on top to fill the mold. When cured, the bottom of the briquette offers lengthwise sections of the initial cylindrical pellet, and though some differential settling may have occurred during the initial embedding process, a representative sample is exposed to view.

The briquettes are ground and polished on a rotary lap (see ASTM D2797)[4] using water-slurried abrasives of successively finer particle size. Normally the final polish is made with submicron-sized alumina on a cloth-covered lap. In a commonly employed technique an apparatus is used that holds six briquettes and rotates the briquettes as the lap is turned. When proceeding from one abrasive size to the next finer size, particular care should be exercised to remove all the coarser abrasive, lest residual particles leave scratches on the surface. An ultrasonic cleaner with water containing a wetting agent or detergent is highly recommended for the cleaning operation. A final wipe with a mild detergent solution on a cotton swab followed by a flush with distilled water will remove any last remaining abrasive particles. The finished surface should be plane and scratch-free, exposing polished cross sections of the individual coal grains. Figure 3.8 is a low-magnification photomicrograph of a typical polished briquette.

Normally the polished surfaces should exhibit a minimum of the relief that results from differential hardness of the constituents. Relief is generally minimized by using low-nap cloths and by keeping polishing time at each step as short as possible, consistent with cutting enough material away to prepare the surface for the next finer abrasive. For special investigations, however, it is sometimes useful to develop relief deliberately, so that selected constituents do stand out. Similarly, surfaces can be etched, commonly by thermal oxidation or oxidizing solutions such as a solution of 25 g of potassium permanganate and 5 g of sulfuric acid in 70 ml of water.[11] Etching often exposes details of relict botanical structure not visible on a conventionally polished surface.

3.2.1.4 *The Microscope*

A microscope with a vertical illuminator (Fig. 3.9) is employed to examine the polished surfaces. For analysis of maceral content requiring maximum resolving power the reflector is a partially reflecting glass plate. However for measurement of reflectance, a Berek prism is employed (see Section 3.2.1.6, Reflectance Analysis"). For relatively low magnification (<200 diameters), dry objectives are commonly used; however at higher

[10] M.-Th. Mackowsky, in *Textbook of Coal Petrology,* E. Stach, Ed., Gebrüder Borntraeger, Berlin and Stuttgart, 1975, p. 252.

[11] See Ref. 10, p. 281.

Figure 3.8 Photomicrograph of polished surface of crushed coal briquette. Dark matrix is epoxy binder.

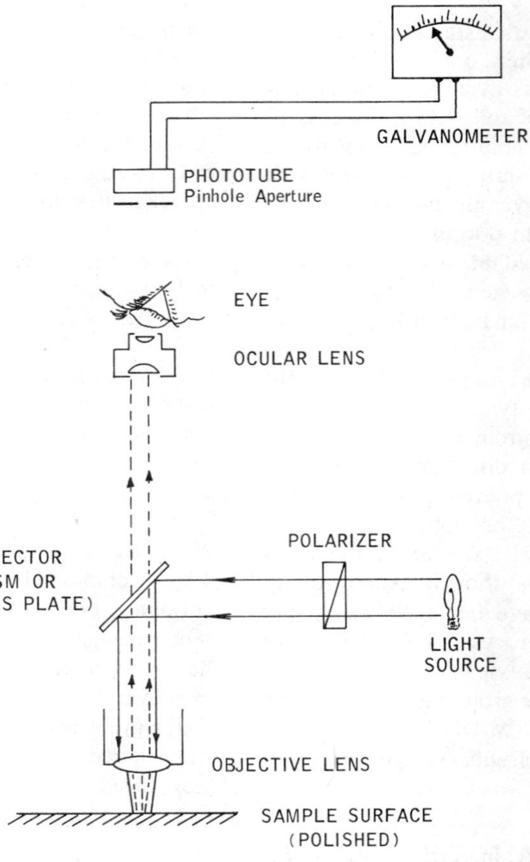

Figure 3.9 Diagram of reflected-light microscope for visual observation or reflectance measurement.

magnifications, oil immersion objectives are used, and special microscope immersion oil is employed as a medium between the objective and the specimen. The oil serves to enhance the contrast between macerals of different reflecting properties. Figure 3.10 contrasts the same field of view using a dry objective and oil immersion. Methylene iodide (with proper objective and adequate safety precautions—principally ventilation) can be used as an immersion medium to enhance contrast even more than conventional immersion oil.

3.2.1.5 *Maceral Analysis*

The quantitative, microscopic assessment of the relative proportions of different macerals comprising a sample is referred to as a petrographic analysis. A later section of this chapter defines the commonly recognized macerals (see Table 3.1, Section 3.2.2). One of the eyepieces of the microscope is fitted with a cross hair or a graticule (lined grid) to serve as a reference for selecting specific locations on the polished surface. The microscope is also fitted with a stage that permits moving the sample by fixed increments in two perpendicular directions. It is common practice to employ a stage that is electrically activated by switches connected to a series of counters, each of which accumulates the "counts" representing the encounter of the eyepiece reference point ("locator") and a specific maceral. A magnification of at least 500 diameters and oil immersion are recommended.

The unpolished (bottom) surface of the briquette is mounted with a small piece of modeling clay on a glass or metal plate that fits the holding mechanism of the mechanical stage. The polished (top) surface is rendered

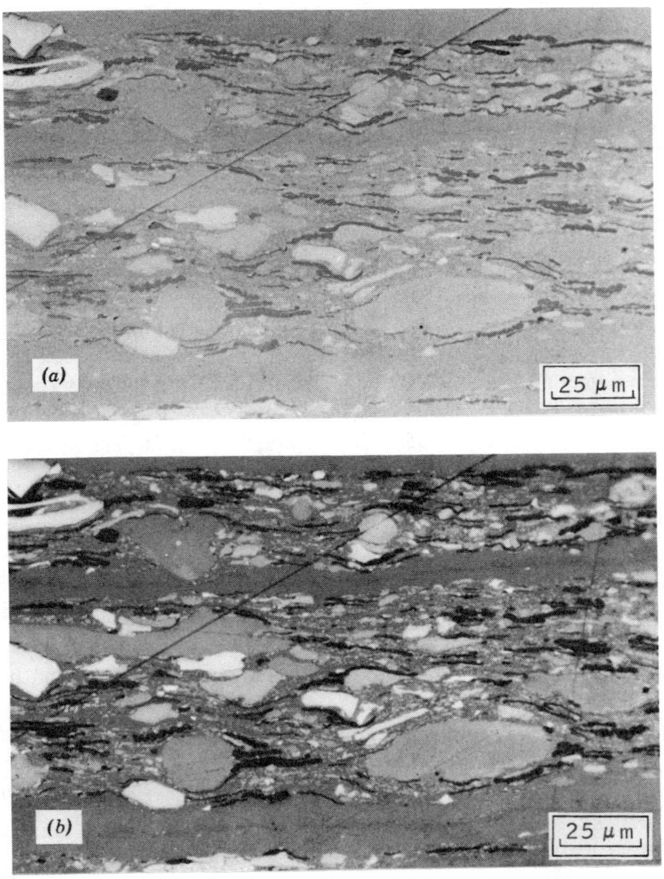

Figure 3.10 Comparison of a microscope field under air (a) and immersion oil (b). Note the enhanced contrast between constituents that results from the use of immersion oil.

parallel to the plate (thereby parallel to the microscope stage) by pressing between two parallel plattens, the clay serving to compensate for any nonparallelism of the top and bottom of the briquette. Alternatively, specialized substages can be used to level the sample. The mounted pellet is then placed on the stage, a drop of immersion oil is placed on the surface, and the microscope is carefully focused.

The stage must be moved in such a way that the surface is "sampled" in an unbiased manner. This is accomplished by moving the stage along equally spaced, parallel transects and stopping at equally spaced intervals. That is, the surface of the briquette is imagined to be overlain by a rectangular grid, and the analyst determines the maceral that lies at each intersection. The spacing between sample locations depends on the eyepiece locator (cross hair or graticule) used. ASTM (D2799)[6] recommends the use of either a simple cross hair (in which case sample "grid spacing" is 0.5 mm) or a graticule, four "corner" points of which are designated as "locators" ("grid spacing" of 1mm).

If an electrically operated stage is used, the locator must be a cross hair according to the ASTM procedure. Because the cross hair falls on epoxy binder 10 to 30% of the time, this technique necessitates moving the stage many times without identifying any material. An alternative method is to employ an eyepiece grid, the center intersection of which is designated as the locator. If the locator falls on binder, an adjacent grid point is used as a locator, progressing in a consistent spiral fashion until a grid point is found that overlays identifiable material. Figure 3.11 illustrates the concept: if point 1 lies on binder, then point 2 is utilized as a locator; Should point 2 also fall on binder, point 3 is utilized, and so forth, until a point falls on identifiable material. This technique maintains the requirement of unbiased sampling and minimizes the number of stage movements needed in an analysis.

No universally accepted procedures have been devised to handle situations of the locator falling on the intersection of two substances. The choice of a procedure to resolve this dilemma lies with the analyst; however it must be a procedure that does not lead to a bias. The simplest procedure is to identify always the material that falls on the left side

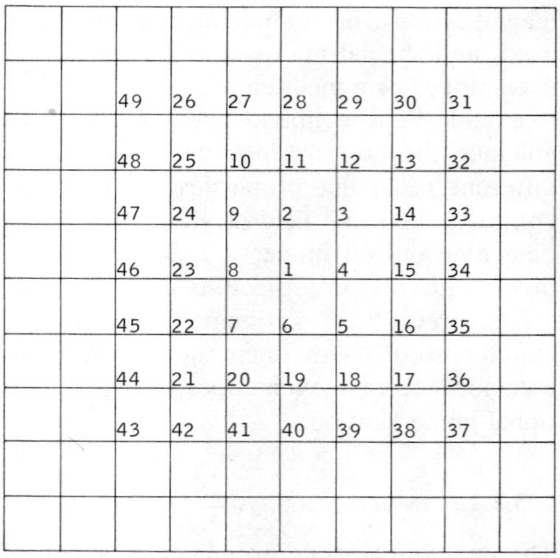

Figure 3.11 Representation of an ocular grid used as a "cross-hair" locator. Numbers (not present on actual grid) designate the order of intersections used to select a location for identification. See text for description of technique.

of the locator, or else to ignore the point completely and progress to the next location.

A record is kept of the substances (macerals, minerals) identified, either on the totalizers connected to the electrically operated stage or on a separate counter or data sheet. After 500 points on a briquette have been counted, the data are calculated to a percentage basis. ASTM D2799[6] recommends counting 500 points on each of two briquettes. Minerals are not normally identified unless there is a special reason for doing so: therefore the data represent volumetric percentages of the identified macerals on a mineral-free basis. These can be calculated to a mineral-containing basis by estimating mineral content (volumetric basis) according to the following formula:

$$MM = \frac{100\,(PM/2.8)}{[\,(100 - PM)/1.35] + [PM/2.8]} \tag{3.1}$$

where MM = mineral matter (vol %)
PM = mineral matter (w%, Parr formula)
= (1.08 × % ash) + (0.55 × % sulfur)

Though some investigators quantify minerals microscopically, microscopic determinations are seldom as accurate as estimation of mineral content from ash and sulfur analyses. If desired, data may be calculated to a weight percentage basis by using estimated specific gravities for the macerals and minerals. Specific gravities differ in different ranks of coal. Unless definite values are known, values of 1.2 for exinite, 1.3 for vitrinite, and 1.4 for inertinite are assumed,[12] and the weight percentage mineral matter is estimated directly from the Parr formula.

3.2.1.6 *Reflectance Analysis*

In the reflected-light microscope the light impinges normally (perpendicularly) on the polished specimen surface. Most of the light is absorbed by the coal, but some is reflected back through the objective lens to the eyepiece. Since the proportion of incident light that is reflected is controlled by the chemical composition and molecular structure of the specimen, measurement of the reflectance can be used as a guide to the chemical properties of very small areas (volumes) of coal. McCartney and Ergun[13] discuss in detail the physics that control reflectance. The reflectance r (i.e., the fraction of incident light that is reflected back through the objective) is related to the index of refraction of the specimen n, the index of refraction of the medium between the objective and the specimen n_m, and the absorption index of the specimen k, according to the Fresnel-Beer equation:

$$r = \frac{(n - n_m)^2 + n^2 k^2}{(n + n_m)^2 + n^2 k^2} \tag{3.2}$$

Reflectance r increases with increasing index of refraction n. The reflectance normally is reported as a percentage value ($\%R = r \times 100$).

Through the determination of reflectance using two different media with different indices of refraction (n_{m1}, n_{m2}), it is possible to evaluate the magnitude of n and k for the coal sample. Routinely, however, R is determined in only one medium, usually immersion oil, and the symbol for reflectance becomes $\%R_0$ (reflectance, percentage in oil). Its absolute value is arrived at by comparison with a substance of known reflectance, usually a polished piece of glass with a known n. Although the reflectance can be determined for all the macerals encountered in a coal sample, commonly it is determined only on the vitrinite.

Some investigators, particularly in Europe,[14] determine reflectance in air. In the United States reflectance measurements on vitrinite are made using immersion oil ($n_m = 1.518$) and polarized light made approximately monochromatic (green, 546 nm) with an interference filter or a combination of Kodak Wratten filters (Nos. 58 and 77). The incident light is polarized to minimize variations that can result from variable orientations of the molecular structure of the sample particles in the briquettes.

The molecular structure of vitrinite in coals of bituminous rank and higher is partially oriented, meaning that polycondensed aromatic "units" tend to have a statistically preferred orientation as a result of the unidirectional confining pressure exerted by overlying rocks during metamorphism. Except for a few rare instances, vitrinite affects light as though the vitrinite were a uniaxial negative crystalline substance. Cook et al.[15] report that vitrinite in some anthracites reflects light as though it were biaxial, but such instances are apparently rare. As a consequence of this pseudocrystallinity (optical anisotropy), the index of refraction varies depending on the direction of impingement of the light rays with respect to the "crystal" axes. By using polarized light and a rotating microscope stage, there will be a sample position (rotational position) that yields a maximum reflectance value no matter how the crystallographic axes of the particle under study are oriented with respect to the polished surface.

The reasoning behind this can be seen by referring to the reflectance ellipsoid shown in Fig. 3.12, which represents the effect of coal on a beam of polarized light impinging normally on a polished section of the coal and being partially reflected from it. Any polished section through a particle of coal can be

[12] See Ref. 10, p. 255.
[13] J. T. McCartney and S. Ergun, U.S. Bur. Mines Bull. No. 641, 1967, pp. 1–49.

[14] N. N. Ulanov, *Khim. Tverd. Topl.*, **8**(6), 30–36 (1974).
[15] A. C. Cook, D. G. Murchison, and E. Scott, *Fuel*, **51**(3), 180–184 (1972).

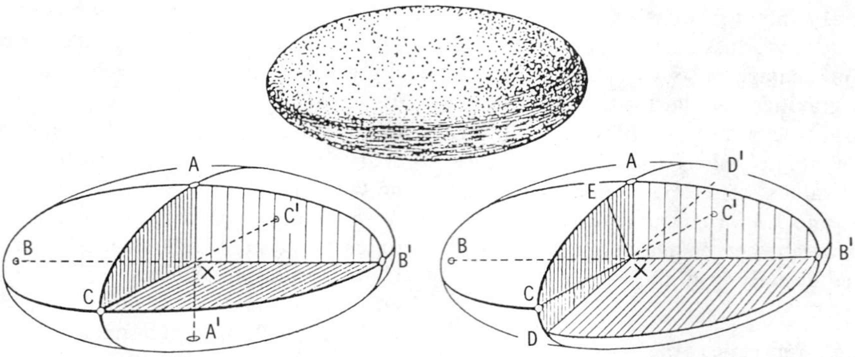

Figure 3.12 The manner in which coal affects polarized light can be represented by a reflectance ellipsoid (analogous to index ellipsoid of optical crystallography). The effect on light of any polished section of coal can be represented by an appropriate cross section (e.g., *A'- B'- A- B*) of the ellipsoid through the center *X*. The length of any line (e.g., *B-B'*) through *X* on this plane represents the magnitude of the reflectance when a beam of light impinges normally to the plane (i.e., *C- X*) and the polarization vibration direction coincides with the line (i.e., *B-B'*).

represented by an appropriate cross section through the ellipsoid. By convention, the light ray is conceived as impinging normally on that surface at *X*. Polarized light, which is always employed, can be conceived of as vibrating normally to the direction of propagation. The relative length of any line radiating from *X* on the ellipsoid cross section represents the relative reflectance of the light when the polarization vibration direction coincides with that line. Except for a few unusual instances, the optical properties of the coal are such that the ellipsoid is oriented so that plane *B–C–B'–C'* coincides with the bedding plane of the coal.

For instance, if the particle is polished along the bedding plane (i.e., in the plane parallel to the earth's surface), the reflectance properties would be represented by plane *B–C–B'–C'*. Axes *B–B'* and *C–C'* are equal and represent diameters of a sphere normal to *A–A'*. Line *A*–X then represents a light ray impinging on this plane. No matter what the direction of polarization (e.g., along *B–B'*, or *C–C'*, or any other line in the plane and through X), the reflectance would be the same and would be the maximum value obtainable from that coal particle.

If the particle is polished normally to the bedding, the plane cross section *A–B–A'–B'* would be the appropriate diagrammatic representation. This figure is an ellipse with principal axes *A–A'* and *B–B'*. The light ray travels from *C* to *X* and impinges normally on the plane. When the polarization direction coincides with *B–B'*, reflectance will be a maximum for this surface and will be equivalent to the reflectance of a bedding plane surface ("true maximum"). When the vibration direction coincides with *A–A'*, reflectance will be a minimum for that surface and will be the lowest value obtainable for that coal particle ("true minimum"). When the polarization direction coincides with any other line in the *A–B–A'–B'* plane and through *X*, the reflectance will be proportional to the length of that line and will, therefore, be some value between maximum and minimum.

The statistical probability of sectioning a particle in the bedding plane or normally to it is small: hence most coal particles in a pellet will be sectioned (polished) in some other plane, for instance, *B–D–B'–D'*. In this example the light ray follows line *E–X*, which is normal to plane *B–D–B'–D'*. When the polarization direction coincides with *B–B'*, the same maximum reflectance is obtained as in any other section. When it is vibrating along *D–D'*, however, a minimum reflectance value for that section will be observed. This value lies somewhere between the true maximum (*B–B'*) and the true minimum (*A–A'*). Thus it can be seen that by rotating the microscope stage, for all sections except those in the bedding plane, the reflectance will vary from a maximum value (equivalent to the true maximum) to a minimum value (only equivalent to the true minimum when the section is normal to the bedding). Only a small proportion of the randomly oriented particles in a briquettte of crushed coal will yield a true minimum reflectance value, but all will yield a true maximum value.

Details of the procedure most commonly employed for conducting an analysis of reflectance values are provided in ASTM D2798.[5] A sensitive photomultiplier tube detects the light reflected from a small area of the sample. Generally the area, as measured on the sample surface, is about 2–10 μm diameter, limited by some means such as an aperture placed in the focal plane of an eyepiece that projects the image onto the phototube. The signal from the photomultiplier is amplified and displayed digitally or on a meter. The system is calibrated using polished glass standards of known index of refraction. Because standardization is not carried out simultaneously with reflectance determinations on the coal, it is important that the sensing system and the light source be stabilized so that once standardized, sensitivity and light intensity remain constant during the actual analysis. It is recommended that the system linearity be checked by comparing at least two standards. Standardization should be checked every 15 to 30 min or after every 25 reflectance determinations.

The glass standards are chosen to exhibit reflectances that fall within the range from about 0.3% reflectance to about 1.9% reflectance, a range that encompasses most coals of commercial interest. The glass is mounted as a prism in epoxy binder and polished in a manner similar to that used for a coal sample. It is important to mount the prism in its binder in such a way as to prevent internally reflected light from emerging coincidently with the desired light reflected from the polished surface. This is accomplished as shown in Fig. 3.13. The reflectance of the standard glass is calculated according to Equation 3.3:

$$\%R_0 = \frac{(n_{\text{glass}} - n_{\text{oil}})^2 \times 100}{(n_{\text{glass}} + n_{\text{oil}})^2} \tag{3.3}$$

It is also important to align the polarization plane of the polarizer at a 45° angle to the principal axes of the microscope. If this is not done, elliptically polarized light results and one will not obtain a true maximum reflectance. For the prevention of elliptical polarization, it is also important to use a Berek prism rather than a glass plate as the

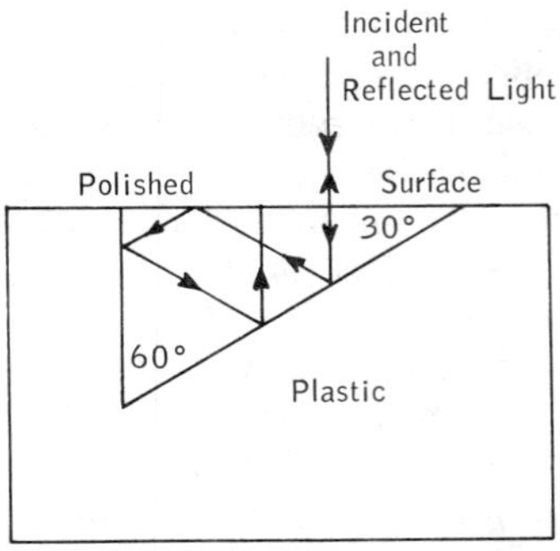

Figure 3.13 Shape of the glass prism used for standardizing a microscope system for reflectance measurements.

microscope reflector. Mackowsky[16] recommends that parasitic reflections in the optical system be assessed by determining the current generated in the display when a completely absorptive substance is placed under the objective. This value then is routinely subtracted from the analytical results.

Harrison[17] first reported, and Thomas and Damberger[18] confirmed, that moisture in pores of vitrinite of high-volatile bituminous coals can affect the reflectance, decreasing $\%R_0$ by as much as 0.1% below that of moisture-free samples. Harrison therefore recommended storing polished briquettes in a desiccator for a minimum of 15 hr before making a reflectance analysis. Mason[19] found no effect of moisture, though he showed by calculation that there should be an effect. It would appear that for maximum consistency and comparability, the practice of storage in a desiccator should be encouraged, particularly for coals below high-volatile A bituminous rank.

The investigator must constantly be on the alert for the presence of air bubbles in the immersion oil between the objective and the sample surface. Often these bubbles cannot be seen in the microscope field, but their presence can significantly reduce (or, less commonly, increase) the light intensity. During a routine analysis wherein a number of measurements are being made on different particles of the same coal, the operator

[16] See ref. 10, p. 269.
[17] J. A. Harrison, *Fuel,* **44,** 225–228 (1965).
[18] J. Thomas and H. H. Damberger, Illinois State Geol. Survey Circ. No. 493, 1976, pp. 1–38.
[19] D. Mason, *Am Chem. Soc., Div. Fuel Chem., Prepr.,* **14**(1), 75–80 (1970).

should suspect the encroachment of a bubble if the reflectance values suddenly begin to run to lower (or, rarely, higher) values. Bubbles can often be seen by using a Bertand lens or by removing the eyepiece and viewing the rear focal plane of the objective conoscopically.

In routine analyses the maximum reflectance is determined for each of 100 specimens of vitrinite in a coal briquette. The briquette surface is sampled in a manner that provides for the selection of particles to be analyzed in some routine, unbiased fashion. Commonly, on selecting a specific piece of vitrinite to be analyzed, it is necessary to adjust its position slightly with respect to the location of the sensed spot to avoid scratches. Here too, it is necessary to take maximum precautions against biasing the location of the analysis spot, since all too frequently it is tempting to select the smoothest, brightest area within the vitrinite. Once the location has been selected, the reflected light is diverted to the photometer and the stage is rotated through 360°. A maximum value will be observed at each of two rotational positions 180° apart, and the average of these two (provided they are within ±5% of each other) is recorded as the maximum reflectance for that location. If the maxima in any given pair differ by more than ±5% (relative), the pair should be rejected. Such differences likely arise from relief or from inaccurate centering of the lens.

The initial results of an analysis are represented by a series of meter scale division numbers, the magnitude of which depend on the graduation system of the meter. Each of these values is converted to reflectance values according to Equation 3.4:

$$R_V = R_S \times \frac{M_V}{M_S} \qquad (3.4)$$

where R_V = reflectance of vitrinite
R_S = reflectance of standard
M_V = meter reading for vitrinite
M_S = meter reading for standard

Invariably, the values from a single coal will span a range. This is largely because of slight differences in reflectance properties of vitrinite from different layers of the seam, and partly because of simple experimental error. An arithmetic mean is generally reported (so-called mean maximum reflectance). Dispersion of the data can be reported as a variance or standard deviation, or, more commonly, according to a procedure devised by Schapiro and Gray[20] in which the proportions of the analysis results that fall in classes defined by 0.1% reflectance intervals (e.g., 0.70–0.799, 0.80–0.899) are reported. Such classes are often referred to as "V-types" or "V-classes." The vitrinite with reflectance falling in the class 0.70–0.799 is referred to as V_7, that in the class 0.80–0.899 as V_8, etc.

Many European investigators do not use polarized light, do not rotate the stage, and therefore do not determine the maximum reflectance at each sampled location. Their data are commonly reported as mean reflectance (R_M), which should not be confused with the statistical average of reflectance maxima (mean maximum reflectance) commonly reported by U.S. investigators.

3.2.1.7 *Automated Microscopy*

Subjective discrimination of macerals in bituminous rank coals is accomplished partly on the basis of observed reflectance differences. In theory, then, with the proper scanning technique and discriminator circuits, it should be possible to automate the microscope to perform a maceral analysis. A number of partially successful systems have been constructed. One of the first, designed and operated by Bomberger and Duel,[21] utilized a belt drive to advance the microscope stage and an electronic discriminator circuit to tabulate the proportion and the particle sizes of pyrite grains in coal samples. Subsequent development of the machine was carried out by W. F. Berry of Bituminous Coal Research, Inc., and L. P. Vitt of FEMCo, Inc., and by Denton, Bayer, and Hassel of Youngstown Steel Company.[22] These machines were limited to distinguishing between substances of fairly disparate reflectances; therefore their maximum use was in distinguishing pyrite, with its very high reflectance, from the lower reflecting coal substances, and in defining the

[20] N. Schapiro and R. J. Gray, *Proc. Ill. Min. Inst.*, 83–97 (1960).

[21] R. D. Bomberger and M. Duel, *Proc. Soc. Min. Eng.*, 65–69 (March 1965).

[22] G. H. Denton, J. L. Bayer, and R. E. Hassel, *J. Met.*, **19**, 88 (1967).

macropore structure of cokes (pores low reflectance, coke high reflectance).

McCartney and his co-workers at the Bureau of Mines[23,24] used a ball-disk integrator to drive a stage in a spiral motion, thereby avoiding sudden changes in speed and direction. With this drive, they could scan a transect of 636 mm/hr. Their discrimination circuit was interfaced with a digital computer that could display a frequency histogram of reflectances for all the components in the samples examined. The first reported use was to analyze pyrite contents and particle sizes. The automated technique yielded results averaging 92% of the pyrite that was determined chemically, and 94% of that determined visually. Size determinations were not as accurate as those determined visually because the automated equipment could not discriminate "defects" in single grains; thus it reported a partially "broken" grain as two or more separate grains. Subsequent application of the equipment to analyze petrographic composition yielded reasonably good results if certain assumptions were made about the reflectance boundaries for the constituents. However the automated results did not produce maceral percentages that coincided specifically with those assessed visually.

Davis and Vastola[25] refer to several innovations in automated microscopy reported by Kojima at the Nippon Steel Corporation. These authors also describe a similar automated system being developed at the Pennsylvania State University. Samples are scanned linearly using a motor-driven stage, and the reflectances of sequentially adjacent areas are assessed. The photomultiplier response data are digitized and fed to a computer. The computer is programmed to provide a cumulative reflectance histogram (so-called reflectogram) that can be interpreted in terms of maceral contents. In addition, vitrinite reflectance values can be calculated and pyrite concentrations, and particle sizes (chord lengths) can be determined. R. J. Gray at the Monroeville, Pennsylvania, research laboratory of U.S. Steel is developing a similar system that is referred to as a photon-counting system. R. R. Thompson and L. G. Benedict of Bethlehem Steel are also evaluating an automated system being developed by E. Leitz, Inc., which estimates, in a microscopic field, the volumetric proportions of materials that fall within arbitrarily defined reflectance levels.

Though any automated microscopic equipment would likely be relatively expensive to construct, it is probable that it could pay for itself by reducing labor costs for making the analyses when many samples must be analyzed routinely. However several problems besides the cost have retarded the development of a truly universal automated system for coal petrographic analyses. It is essentially impossible to prepare a briquette surface so that there is no relief (viz., sloping contacts) between components of differing hardness. Wherever a surface is not perpendicular to the light beam, the light is reflected away from the objective, and the detector system senses a low-reflectance substance. Binder (epoxy) has a low reflectance, as do liptinite and some mineral matter, and an automated system is not generally capable of distinguishing such materials from each other. And, of course, an automated system is not capable of "seeing" the form of a material or its surroundings, features that are commonly employed in visual analyses to identify some maceral materials.

No automated system has been reported that can take into account optical anisotropy of vitrinites; hence only an average reflectance can be assessed. Finally, an automated system would not be usable for mixtures of high- and low-rank coals, since it would not be able to distinguish the vitrinite in the higher rank coal (relatively higher reflectance) from semifusinite of similar reflectance in the lower rank coal.

In spite of these difficulties, it is reasonable to expect that useful data could be developed by automated procedures. Though some of the subtleties that are evaluated visually might be sacrificed, differentiation of components into high-reflecting and low-reflecting classes could prove valuable for screening of coals as liquefaction feedstocks, for instance. It seems certain that there will be additional developments in this area in the future.

[23] J. T. McCartney and S. Ergun, U.S. Bur. Mines Rep. Invest. No. 7256, 1969, pp. 1–15.

[24] J. T. McCartney, H. J. O'Donnel, and S. Ergun, *Fuel*, **50**(3), 226–235 (1971).

[25] A. Davis and F. J. Vastola, *J. Microsc.*, **109**(1), 3–12 (1977).

3.2.1.8 *Fluorescence Microscopy*

Although first investigated more than 40 years ago, the use of fluorescence microscopic techniques has only recently received the attention it deserves. Some components of coal, in particular those with higher hydrogen contents, fluoresce when illuminated with ultraviolet (UV) light. Special dichroic mirrors are used in the vertical illuminator, replacing the usual glass plate or Berek prism, or else special filters are employed. Relatively high-intensity light sources rich in UV must also be employed (i.e., high-pressure mercury or xenon lamps). Stach[26] and Ottenjann et al.[27] describe the procedures in some detail, as does the 1975 edition of *International Handbook of Coal Petrography.*[28]

Several substances that have been identified simply as liptinite (exinite) in the past have been shown to be autoemitters (fluorescent substances), and these probably have distinctive chemical properties compared to other, nonfluorescing spore exine substances.[27]

Besides the use in differentiating certain maceral substances subjectively, several objective measuring techniques have been developed. These involve measuring the intensity of fluorescent illumination throughout the spectral range of about 400–700 nm. Such techniques require spectrophotometric equipment. It is reported that the wavelength of maximum fluorescence of sporinites increases with increasing rank[27,29] and with increasing temperature of heat treatment.[30] Also, the so-called red/green quotient

$$Q = \frac{I \text{ @ } 650 \text{ nm}}{I \text{ @ } 500 \text{ nm}} \tag{3.5}$$

shows similar positive correlation with rank.

3.2.2 Component Classification and Properties

A perusal of coal petrological literature discloses that there are hundreds of appellations reputing to refer to distinctive macerals. This abundance of terms probably leads more to confusion than to clarification, and results largely from a failure of many coal petrologists to come to grips with a definition of the term "maceral." The most technologically appropriate definition is that proposed in 1958 by William Spackman, who defined macerals as "organic *substances* or optically homogeneous aggregates of organic substances that possess distinctive physical and chemical properties."[31]

This definition is appropriate because the the technologist is interested foremost in the substances that constitute a coal sample—specifically, in the physical and chemical properties of the substances, and in how these substances will react in some specific process. The coal petrologist, then, to offer assistance and insight to the coal technology community, must identify and quantify specific substances principally on the basis of microscopically discernible criteria.

With the microscope, one observes that the polished surface of a coal sample, such as that shown in Fig. 3.14, can be differentiated into numerous "areas" ranging from essentially white, through various shades of gray, to black. Distinctive areas of relatively uniform tone are defined or delineated by relatively abrupt changes in tone. The more of the incident light that an area reflects (hence the higher its reflectance), the brighter it appears. The uniformity of reflectance within a definable area, such as that designated by the bracket labeled *V*-1, implies a relative uniformity of chemical and physical properties of the substance comprising the area (and, by extension, a volume of material of which the area is a cross section). This optical homogeneity constitutes the essence of Spackman's definition of macerals.

In practice, the petrographer first recognizes an area of uniform reflectance, commonly disregarding minor variations in re-

[26] E. Stach, in *Textbook of Coal Petrology,* E. Stach, Ed., Gebrüder Borntraeger, Berlin and Stuttgart, 1975, p. 292.

[27] K. Ottenjann, M. Teichmüller, and M. Wolf, *Fortschr. Geol. Rheinl. and Westfalen,* **24,** *1–36 (1974).*

[28] International Committee for Coal Petrology, *International Handbook of Coal Petrography,* Centre National de Recherche Scientifique, Paris, 1963; Supplements, 1971, 1975.

[29] P. van Gijzel, "Review of the UV-Fluorescence of Fresh and Fossil Exines and Exosporia," in *Sporopollenin,* J. D. Brooks, Ed., Academic Press, New York, 1971, p. 659.

[30] F. T. C. Ting and H. B. Lo, *Fuel,* **54,** 201–204 (July 1975).

[31] W. Spackman, *Trans. N.Y. Acad. Sci., Ser. II,* **20**(5), 411–423 (1958).

Figure 3.14 Photomicrograph of polished surface of a high volatile bituminous coal showing liptinite (*L*), inertinite (*I*), and vitrinite (*V*) with different reflectances. Black areas are holes.

flectance. Using the criteria of the reflectance relative to associated areas and the form, disposition, or shape of the cross-sectional area, the investigator then attempts to identify the substance. Areas with similar reflectances and modes of occurrence are assumed to be representative of the same substance.

In all ranks of coal below low-volatile bituminous, three major classes of substances are recognized on the basis of their relative reflectances. These are the major maceral classes referred to as inertinite macerals (bright), liptinite or exinite macerals (dark), and vitrinite macerals (intermediate). Representatives of these classes are designated by *I*, *L*, and *V* in Fig. 3.14, which depicts a typical bituminous coal.

It is evident in Fig. 3.14 that particles of bright inertinite and dark liptinite are surrounded by (enclosed within) a matrix of vitrinite (intermediate gray) that exhibits slight variations in reflectance. It is further evident that there are particles (or areas) of materials that exhibit reflectances spanning the range from the gray of the dominant vitrinitic material to absolute white. These variations in reflectance result from minor differences in physical and chemical properties of the substances; therefore the concept of there being only three classes of macerals must be slightly modified. For even in a small portion of one coal, such as that shown in Fig. 3.14, there appear to be a substantial number of different substances based on differences in reflectance. Nevertheless, even though there are materials gradational between vitrinite and the highest reflectant inertinite, and even though materials of discernibly different shades of gray are classified together as vitrinite, the tripartite classification is useful. Further discussion of the gradational nature of the maceral classes is found in Section 3.2.2.2, "Inertinite."

The range of variations in the vitrinite properties within a coal is commonly characterized by the spread of reflectance values. Not only are there variations in the vitrinite within one coal, which are minor, but more important, the vitrinitic material in a lignite is chemically quite dissimilar from the vitrinitic material in a bituminous coal. Thus reference to *the maceral* "vitrinite," though common (and even employed in this chapter) is technically inaccurate. For the term "vitrinite" refers to a large suite of materials identified by their form or disposition (implying a common genesis) and by their reflectance relative to other materials in the same sample.

The names of all maceral classes contain the suffix "-inite." Spackman[31] proposed that macerals be classified hierarchically and suggested that the suffix "-inoid" be employed at one level of classification. Although the suffix is often used, Spackman's concept of how it should be applied is frequently ignored. Hence vitrin*oid* is often used synonymously with vitrin*ite*.

As previously mentioned, hundreds of so-called maceral names can be found in the literature. Almost all refer to subclasses of the vitrinite, liptinite, or inertinite classes.

Some of these subclasses appear to be justifiable on technological grounds, but it is likely that these are only a small proportion of the many proposed. Much of this nomenclatural proliferation can be attributed to the botanical or geological training of many of the petrologists who are investigating questions of coal origin and stratigraphy. Clues that bear on these questions can often be obtained by recognizing botanical derivations of some maceral forms. When the objective is to infer information about the peat swamp environment, there is a strong tendency to be a "splitter" and to differentiate (with the application of a classifying word) as many things as one can, never knowing when such differences might yield the key to a particular study.

From the technological standpoint, however, differentiation or subclassification should be made only when it can be demonstrated that this is chemically or physically justified. That this injunction is so often ignored simply reflects a failure on the part of many investigators to recognize that the microscopic identification of macerals should be a means to an end, not the end itself.

In particular, I believe that the differentiation of "macerals" solely on the basis of size, shape, or form as viewed microscopically is unjustified and unnecessary unless it can be demonstrated that the materials so classified are chemically distinctive. If coalified spore exines are large or small, they still consist of liptinite and should be recognized as varieties of liptinite only if their reflectances (or other objectively measured properties) are significantly different. Likewise, regardless of whether different pieces of inertinite have holes (cell lumina) in them, they should be considered representatives of the same maceral if their reflectances are the same. If inertinite with holes contribute to, for instance, specific crushing responses or to specific differences in the combustion rate when compared to inertinite without holes, it would be most appropriate to refer to these specimens, respectively, as cellular inertinite and noncellular inertinite. Largely on the basis of this philosophy, in this chapter I have purposely not provided definitions for the many so-called macerals that can be encountered in the literature.

Vitrinite and liptinite are referred to collectively as "reactives" and inertinite macerals are referred to as "inerts." However these terms must be interpreted in the context of relative reactivity rather than in any absolute sense. The reactive macerals are those that decompose when heated to about 400°C in an oxygen-free atmosphere, yielding a relatively high proportion of volatile substances compared to associated inert macerals, or they convert to a plastic phase and, with continued heating and some devolatization, solidify to produce a cellular coke. The inert macerals are those that do not become plastic and that evolve less volatile matter than do associated reactive macerals. Except in the higher rank coals, where the properties of the macerals differ little and the organic petrography is of less significance, the inertinite macerals of a coal are carbon rich compared to vitrinite in the same coal, whereas liptinite macerals are comparatively hydrogen rich. Compared to vitrinite in the same coal, inertinite has a higher reflectance, whereas liptinite has a lower reflectance. These differences are shown in Fig. 3.14.

Now to add a dollop of confusion to a neat, simple, and comprehensible classification, it must be pointed out that in coals of high (anthracitic) rank, the substances that are recognizable microscopically as having been reactive macerals in lower rank coals are relatively inert (in the sense already defined). With increasing rank, the reactive macerals become more highly condensed cyclic materials, while the already carbon-rich inert macerals are altered very little by metamorphism. Therefore most of the distinctions between them ultimately blur. Similarly in coals of very low rank (lignites), substances that are distinctive and recognizable as "inert" macerals (because they exhibit higher reflectance than the associated vitrinite) may exhibit a reasonable reactivity in certain processes. For instance, some semifusinite in lignite, although it may not become plastic when heated, is sufficiently reactive that it can be converted to liquids in the hydrogen donor process.

The most common classes of macerals are shown in Table 3.1. These categories are applicable to all ranks of coal, though they are most applicable to coals of bituminous rank. Different nomenclature has been used or suggested for coals ranked higher and lower than bituminous (e.g., vitrinite in lignites has been called huminite). Table 3.1

Table 3.1 Classification of Macerals, Especially in Bituminous Coals

Group	Subcategory
Vitrinite	Vitrinite bands (telinite, telocollinite)
	Attrital vitrinite
	Pseudovitrinite[a]
Liptinite (exinite)	Sporinite, cutinite
	Suberinite
	Alginite
	Resinite
	"Oily" macerals
Inertinite	Sclerotinite
	Fusinite
	Semifusinite
	Macrinite
	Micrinite

[a] Some investigators would prefer classifying pseudovitrinite in the inertinite group.[38] Though it may be relatively inert (nonfusible) in the coking process, pseudovitrinite is reactive in, for instance, liquefaction.[40]

gives only the subcategories of the vitrinite, liptinite, and inertinite group macerals that have been most commonly recognized. Other, microscopically differentiable subcategories *may* be significantly different technologically, but there are few data to justify their separation into these other categories.

3.2.2.1 *Vitrinite*

Vitrinite originates principally from plant structural elements, which are composed of lignin and cellulose. It is gray in reflected light, the reflectance varying slightly with location in a single seam, but varying significantly with the rank of the coal. Vitrain bands (see Fig. 3.1) are composed of vitrinite, and the matrix or groundmass of common attrital coal layers is vitrinite. Because most coals contain more than 50% (commonly 75–80%) vitrinite, the gross properties of coals reflect the properties of vitrinite. Consequently the many properties of coals defined and described in other chapters are, with few exceptions, principally the properties of the vitrinite in those coals. No attempt is made here to describe the many properties of vitrinite, but Section 3.3.1 surveys the variations in properties with rank.

The reflectance of vitrinite is related to the rank of the coal, higher reflectance being associated with more severely metamorphosed coals. In a single coal, however, the reflectance of the vitrinite is seldom completely uniform, generally spanning a range of about 0.3 percent. All vitrinites with the same reflectance are not identical; therefore reflectance, though a valuable index, is not an infallible guide to vitrinite properties.[32] Because vitrinites of different reflectances within a coal sample cannot be separated and concentrated readily, relating the subtle reflectance differences to subtle chemical differences is difficult.

However three varieties of vitrinite have been recognized and have been shown to be sufficiently different to make justifiable their differentiation in a petrographic analysis. The three varieties are vitrinite occurring as bands or lenses (referred to variously as telocollinite,[33] vitrinite A,[34] banded,[35] telinite[36]), vitrinite occurring as a groundmass of attrital coal (vitrodetrinite,[37] desmocollinite,[37] matrix vitrinite,[35] vitrinite B[34], collinite[36]), and pseudovitrinite.[38]

Vitrinite occurring as bands or lenses (vitrain) typically has higher reflectance than associated attrital vitrinite. Because bands can be handpicked to concentrate samples, it is possible to study them chemically, although handpicked vitrinite may not represent all lenticular vitrinite. This lenticular form of vitrinite has less hydrogen and yields less volatile matter than the attrital vitrinite of the same coal.[34] Also, it has been reported that the oxygen content is lower.[39] In some

[32] C. R. Binder, L. J. Duffy, and P. H. Given, *Prepr., Am. Chem. Soc., Div. Fuel Chem.*, **7**(2), 145–152 (1963).

[33] M. Teichmüller, in *Textbook of Coal Petrology*, E. Stach, Ed., Gebrüder Borntraeger, Berlin and Stuttgart, 1975, p. 187.

[34] H. R. Brown, A. C. Cook, and G. H. Taylor, *Fuel*, **43**(2), 111–124 (1964).

[35] A. R. Cameron, *Proc. 1st Geol. Conf. W. Canadian Coal 1972*, Research Council of Alberta, Inf. Ser. No. 60, pp. 31–45.

[36] P. A. Hacquebard, "Value of Quantitative Separation of Vitrinite into Telinite and Collinite for Petrography of Coking Coals," *Proc. Int. Comm. Coal Petrology*, No. 3, 1960, pp. 131–141.

[37] See Ref. 33, p. 193.

[38] L. G. Benedict, R. R. Thompson, J. J. Shigo, and R. P. Aikman, *Fuel*, **47**(2), 125–143 (1968).

[39] A. V. Lapo, *Tr. Vses. Nauchn.-Issled. Geol. Inst.*, **132**, 91–103 (1968); *C. A.*, **70**, 39586 (1969).

coals lenticular vitrinite appears to be more responsible for swelling and the development of plastic properties than the attrital vitrinite.[36]

Pseudovitrinite may occur as either the lenticular or the attrital form. It is recognized in grain-mount preparations by its slightly higher reflectance, slitted structure, remnant cell structure, and the presence of serrated edges. First recognized as essentially unfused particles in coke, it was then sought and found in coal.[38] Commonly it comprises 15–25% of vitrinite in coal and may be as much as 50%. Vitrinite of bituminous rank coals is the fusible component that yields porous coke on pyrolysis, whereas pseudovitrinite, although only slightly different in appearance from vitrinite, ranges from fusible to nonfusible (inert). Although it is claimed that some pseudovitrinite is inert in the coking process, and is therefore considered a variety of inertinite by some investigators,[38] it is apparently reactive (i.e., converts to liquids) in liquefaction processes.[40] Hence to include it in the inertinite class would be misleading. In fact, it is unfortunate that the word "inertinite" is used, since macerals that are "inert" in coking may not be inert in other senses. "Nonplastic" or "nonfusible" would be just as accurate and much less confusing.

[40] A. Davis, W. Spackman, and P. H. Given, *Energy Sources*, **3**(1), 55–81 (1976).

3.2.2.2 *Inertinite*

If reflectance measurements are made on a large sampling of all the components in a bituminous coal, and if the data are then arranged as a frequency histogram, they will form a continuum such as that shown in Fig. 3.15. Liptinite will be represented by a cluster of data at the low reflectance end offset from the remainder of the data. The remaining material is readily divisible into two major categories. At the lower reflectance end of the nonliptinite spectrum is vitrinite. It is the most abundant material in typical coals, commonly forming the matrix in which other constituents are embedded. Because of its abundance, it is represented by an obvious peak in the histogram (Fig. 3.15). In a higher rank coal (below metaanthracitic rank) the position of the vitrinite and inertinite data would be displaced toward higher reflectance values, but there would still be an obvious vitrinite peak.

A complete reflectance spectrum such as that shown in Fig. 3.15 is not obtained in routine petrographic analyses. Therefore the division between vitrinite and inertinite is not normally based on reflectance measurements. In practice, the vitrinite is readily

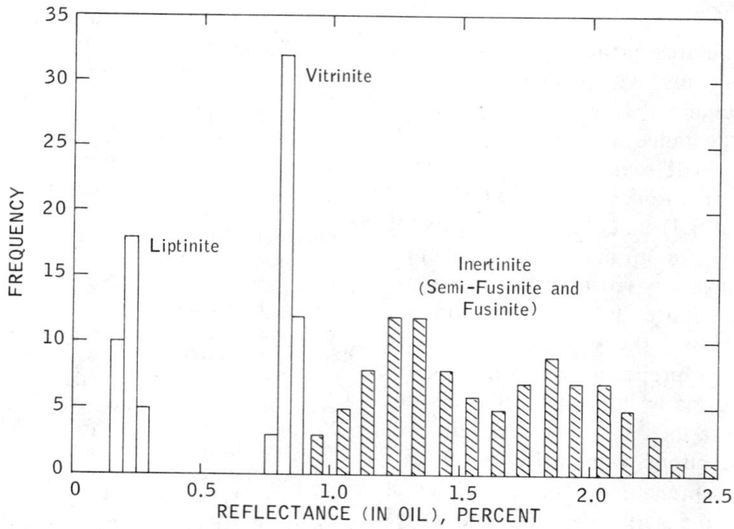

Figure 3.15 Distribution of reflectances of components in a high volatile A bituminous coal, Chilton Bed, West Virginia.[13]

recognized because of its abundance, its relatively gray appearance, and its form (bands, matrix). The inertinite occurs commonly as discrete particles that except for a small proportion with reflectance near that of the vitrinite, contrast strikingly with the vitrinite. In a petrographic analysis, the classification of the transitional material as vitrinite or inertinite would have little influence on the utility of the data, since this substance comprises only a small part of the coal.

Thus except for unusual and relatively rare coals, there is little difficulty in classifying material as either vitrinite or inertinite. However subdivision of the inertinite class is a much more subjective activity. Two criteria are employed to make this subdivision: reflectance and form. Five classes of inertinite macerals are commonly recognized: sclerotinite, fusinite, semifusinite, macrinite, and micrinite (see Table 3.2). Whether this subdivision is justifiable and necessary on technological grounds is highly debatable; because these subclasses are widely recognized, however, they are described in the following paragraphs.

Sclerotinite. This component is relatively rare in coals of Carboniferous age (Appalachia, Interior basins), but relatively common in coals of Mesozoic and Tertiary age (Great Plains, Rocky Mountains). In its most common form sclerotinite occurs as round or ovoid bodies, commonly more than 20 μm in diameter, and more reflectant than associated vitrinite, but seldom exceeding a reflectance of about 1.5%. Often the bodies exhibit an internal cellular structure. They are the remains of fungal bodies that were chitinous, resistant to decay, and relatively reflectant in their natural state. Long, thin strips of moderately high reflecting material derived from fungal hyphae are also classified as sclerotinite by some petrologists. Little is known about the properties of sclerotinite, except that its reflectance would imply that it has a high carbon content relative to associated vitrinite. See Stach[41] for further discussion of sclerotinite.

Fusinite, Semifusinite. The material comprising the inertinite fraction of the sample characterized in Fig. 3.15 is classifiable as fusinite and semifusinite. In recognition of the diversity of this material, at least in terms of its reflectance properties, coal petrologists have operationally recognized a low-reflectance component (semifusinite) and a high-reflectance component (fusinite). Unfortunately the boundary between these subdivisions is drawn subjectively by each petrologist according to personal whims; hence one person's semifusinite will be another's fusinite. Schapiro et al.[42] suggested that the boundary be set at 2.5% reflectance; but because in routine analyses reflectance is not measured on inertinite, this is routinely impractical. In contrast, Bennett,[43] who made a comprehensive study of the subject, contends that essentially all material with reflectance above that of the vitrinite is infusible, therefore is essentially equivalent from a coke-making standpoint.

From a technological point of view, it

[41] E. Stach, in *Textbook of Coal Petrology,* E. Stach, Ed., Gebrüder Borntraeger, Berlin and Stuttgart, 1975.

[42] N. Schapiro, R. J. Gray, and G. R. Eusner, *Blast Furn., Coke Oven Raw Mater. Proc.,* **20,** 89–112 (1961).

[43] A. J. R. Bennett, *Fuel,* **47**(1), 51–62 (1968).

Table 3.2 Classification of Inertinite

Form	Max. Dimension	Reflectance Low	Reflectance High
With cell lumina		Semifusinite	Fusinite
No cell lumina			
	>10 μm	Macrinite[3]	Macrinite[3]
	<10 μm	Micrinite[3]	Micrinite[3]
	<30 μm–(but >1–2 μm)	Inertodetrinite[41]	Inertodetrinite[41]
May or may not have cell lumina	20–300 μm	Sclerotinite[3,41]	Sclerotinite[3,41]

appears to be unsatisfactory to accept the presently employed subjective division, or the arbitrary 2.5% reflectance boundary proposed by Schapiro et al., or the concept of treating all inertinite as a uniform substance.

It has never been possible both to isolate physically and to analyze chemically material from one coal (other than the vitrinite or the liptinite) that falls within a narrow range of the reflectance spectrum; therefore no direct correlations of reflectance with chemical properties of the fusinitic materials have been made. However maceral material found concentrated in fusain bands has been analyzed. Such samples almost always consist of material with a range of reflectances and generally even include some contaminating vitrinite and exinite.The chemical and physical properties of fusinites have been deduced from these extremely heterogeneous samples. Invariably such samples have lower hydrogen and higher carbon contents than vitrinite picked from vitrain bands in the same coal. It is most probable that the carbon/hydrogen ratio increases with increasing reflectance.

If volatile yield and reactivity properties (other than fusibility, which is definable only on a fusible-infusible scale) do vary progressively with reflectance, then for critical uses of petrographic analyses, such as in prediction of liquefaction potential or volatile product yield, a direct measurement of the reflectance distribution of the fusinitic material (including semifusinite) would be desirable.

It seems reasonable to characterize inertinite according to reflectance where its properties may be critical. However conventional classification procedures also use form or morphology. For instance, according to ASTM Specification D2796[3] and other investigators (e.g., Stach[44]), the classification of microscopically assessed material as fusinite and semifusinite (in contrast to other inertinite macerals), requires that some relict cell structure still be visible. However as discussed below, existence of such features does not appear to be a particularly valid criterion for the differentiation of chemicophysically distinct substances.

Particulate fusinite, which commonly occurs as angular shards and does not exhibit cell structure (see Fig. 3.16), should be identical in composition to larger pieces of similar reflectance that show relict plant cell structure. Therefore if the concept of macerals as classes of substances is to be adhered to, fusinite is fusinite regardless of whether it has holes (cell lumina) in it.

It may be useful, however, to recognize different "sizes" of fusinite. For example, larger pieces of fusinite can act as loci for cracks in coke, thereby decreasing the overall coke strength. On the other hand, finer particles of fusinite can be incorporated into the cell walls of the coke, thereby strengthening the coke walls. The dividing line between beneficial and deleterious particle sizes is likely to depend on the rank of the coal, because the coke cell wall thickness, therefore the incorporability of fusinite, will depend on the properties of vitrinite, which are correlative with rank.[45]

In coal liquefaction processes highly reflectant fusinite is not converted to a liquid. Therefore it emerges from such a process as particles of the same size as were fed to the process. If an operator is concerned about the potential particle size of undissolved material, some predictive insight can be gained by a knowledge of the particle size range of fusinite (and other nonliquefiable materials—i.e., mineral matter—also inert to liquefaction).

It is common practice in the United States to refer to small fusinite particles as massive micrinite or macrinite, although unfortunately both terms imply a relationship between these substances and micrinite (see following section). Some European investigators refer to fine particles of fusinite (<30 μm) as inertodetrinite, just as they refer to particulate vitrinite as vitridetrinite and to particulate liptinite as liptodetrinite. Such terminology emphasizes the detrital (particulate) form rather than the composition of these materials. To emphasize composition, it would be preferable to use the terms detrivitrinite, detrifusinite, and detriliptinite.

Macrinite, Micrinite. Particulate materials with reflectance greater than that of associated vitrinite and not exhibiting cellular structure have been referred to by a number

[44] Stach, E., in *Textbook of Coal Petrology,* E. Stach, Ed., Gebrüder Borntraeger, Berlin and Stuttgart, 1975, p. 95.

[45] R. J. Gray, *Ill. Min. Inst.,* 20–47 (1967).

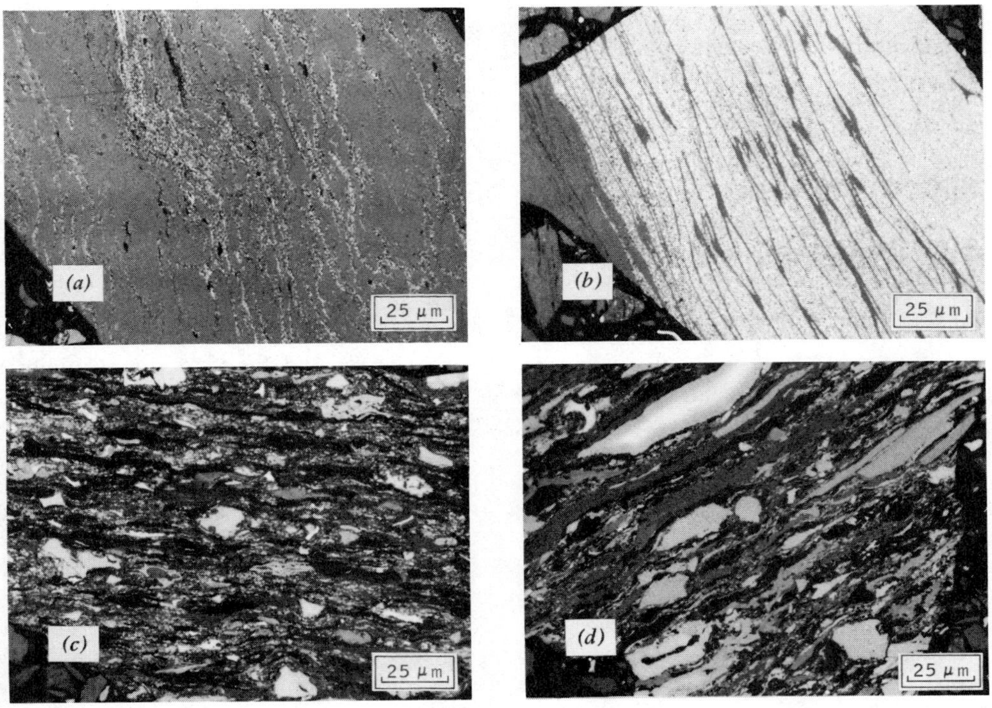

Figure 3.16 Photomicrographs showing distinction between micrinite (*sensu stricto*) and particulate fusinite (detrifusinite), sometimes also referred to incorrectly as micrinite. (a) Thin layers of micrinite (bright) in vitrinite. (b) Concentrated micrinite (bright) enclosing thin strips of vitrinite. (c) Complex mixture of vitrinite, liptinite, and inertinite. Larger bright particles are detrifusinite, very fine (approximately 1μm) bright particles are micrinite. (d) Detrifusinite (bright) with vitrinite (dark gray) and liptinite (black).

of names. According to definitions found in ASTM D2796,[3] if the maximum dimension of these materials is greater than 10 μm, they are called *macrinite;* and if less than 10 μm, they are called *micrinite.*

From the standpoint of identifying and classifying *substances*, this arbitrary size boundary is unfortunate. Furthermore, the distinction between materials that exhibit relict cell structure and those that do not (see previous discussion of fusinite) is also unfortunate, for it seems certain that composition is unlikely to be related to size or to morphological distinctions. As already discussed, it would be far more logical to make a division based on measured reflectance differences, at least for the coarser (macrinite) materials.

Material of small particle size and high reflectance has been called micrinite. Yet this term has been used to refer to substances that are probably dissimilar. "Micrinite" should be used to refer only to material with reflectance higher than that of associated vitrinite and occurring as bodies about 1 μm in maximum dimension. Close examination at high microscopic resolution indicates that such bodies are spherical.[13] Micrinite may occur dispersed through a matrix of vitrinite, as in Fig. 3.16*A*, or else concentrated, often to the exclusion of other macerals (Fig. 3.16B). In the latter form it has been referred to as the microlithotype *microite*.

Micrinite has been classified as one of the inert macerals principally on the basis of its lack of plasticity during pyrolysis. However it has been reported that micrinite found in a polished coal surface disappeared when the surface was heated in vacuum with a heat lamp.[46] Although the temperature was not measured, it could not have been high (200–300°C?). Therefore it was considered remarkable that the presumably inert micrinite could have "reacted." It is possible that what was assumed to be micrinite actually

[46] B. N. Nandi and D. S. Montgomery, *Fuel,* **46**(5), 394–398 (1967).

consisted of small (1 μm) holes in vitrinite, a phenomenon fairly commonly encountered in high-volatile C bituminous coal. Such holes, each acting like a concave mirror, look deceivingly like small, bright particles. If holes were indeed observed, they could have "melted" together with the application of heat. Some other evidence, however, suggests that micrinite is in fact fairly reactive,[47] a conclusion implied by its scarcity in higher rank coals.[48]

Some investigators have concentrated and analyzed samples of material that they refer to as micrinite.[47,49] It seems likely that the material with which they worked comprised, in fact, relatively fine particles of fusinite (and semifusinite), now more commonly referred to as macrinite. Petrographers have not always adhered to the distinction between such material and the 1-μm material (micrinite, *sensu stricto*). Detrifusinite (or macrinite) particles may range from very fine (presumably 1 μm or less), at which size they would be difficult to distinguish from true micrinite, to particles many millimeters in maximum dimension. These distinctions are shown in Fig. 3.16. Although occasionally even the experienced petrographer could fail to differentiate very fine (1-μm) fusinite particles from true micrinite, there can be no controversy about the differentiation of only slightly coarser fusinite from true micrinite. Therefore the different classes (micrinite, fusinite) are well justified, and referring to particles of material obviously representing fragments of fusinite as "micrinite" can only lead to further confusion.

3.2.2.3 *Liptinite (Exinite)*

In bituminous and lower rank coals liptinites (exinites) are low-reflectance substances of high hydrogen content. Four major categories of liptinite have been recognized, but their differentiation may be more justifiable on the basis of origin than on the basis of differences in chemical and physical properties. These categories are referred to as sporinite (plus cutinite), suberinite, resinite, and alginite.

Strictly speaking, sporinite and cutinite are the only true exinite macerals, although "exinite" has been commonly used to refer synonymously to all liptinite macerals. Sporinite and cutinite exhibit similar low reflectances, and available chemical data on concentrates of these materials also indicate their basic similarity. Sporinite is derived from the exines of spore and pollen grains and is seen microscopically as thin lenses typically measuring about 3–5 by 5–25 μm. Cutinite is derived from chemically similar material (cuticle) that coats leaves, needles, and some fruiting bodies; it commonly appears in microscopic preparations as elongate strips with serrated edges. Both sporinite and cutinite have highly polymerized, alicyclic structures, with a significantly higher proportion of hydroaromatic hydrogen than is found in associated vitrinite.[47] Infrared (IR) spectroscopic studies have indicated absorption bands characteristic of resins and waxes.[50] These true exines are highly resistant to oxidation, a property that is exploited in the maceration process to recover concentrates of spore exines from coal for botanical examination. Sporinite and cutinite have been reported to decompose thermally at temperatures lower than[51,52] and higher than[53] associated vitrinite. There is general agreement that when pyrolyzed, they yield significantly greater proportions of volatile products (gases, oils, and tars) than does associated vitrinite. Sporinite and cutinite are considerably tougher than the more brittle vitrinite with which they are associated.

During metamorphism sporinite and cutinite are little altered until around the subbituminous/bituminous boundary, where, it has

[47] P. H. Given, M. E. Peover, and W. F. Wyss, *Fuel,* **44**(6), 425–435 (1965).

[48] L. A. Flueckinger, R. R. Dutcher, and A. R. Cameron, *J. Geol.,* **80,** 237–247 (1972).

[49] D. W. van Krevelen, *Coal,* Elsevier, Amsterdam, 1961.

[50] T. Ratajczak, *Bull. Acad. Pol. Sci., Ser. Sci. Geol. Geogr.,* **16**(2), 55–59 (1968); *C. A.,* **70,** 89434 (1969).

[51] J. L. Bayer, G. H. Denton, and M. C. Chang, *Prepr., Am. Chem. Soc., Div. Fuel Chem.,* **7**(2),94 (1963).

[52] E. I. Toryanik, R. G. Dyukanov, and V. G. Zashkvara, *Khim. Tverd. Topl.* (No. 4), 49–54 (1967); *C. A.,* **68,** 14674 (1968).

[53] R. R. Thompson and L. G. Benedict, "Relationship Between Cokability and Petrographic Composition of Agglomerating Coals," *Proc. Coal Agglom. Conversion Symp.,* West Virginia Geol. Econ. Survey, Morgantown, May 5–6, 1975 (published 1976).

been suggested, some forms of these macerals are decarboxylated and converted to petroleumlike substances. This stage is referred to as the first coalification jump by M. Teichmüller, who defines a second coalification jump at the high-volatile/medium-volatile bituminous rank boundary.[54] There exinites are converted to substances with reflectance equivalent to or possibly higher than the associated vitrinite.

Suberinite is recognized in Tertiary and Mesozoic coals only. It is formed from cork cell walls that are commonly impregnated with wax. Consequently it exhibits properties similar to cutinite and sporinite. It is recognized microscopically by its low reflectance and its characteristic cellular texture.

Alginite originates from oil-rich algae. It is relatively rare, being found only in coal layers formed under very wet conditions, and it is a typical component of sapropelic or boghead coals (see Section 3.2.3.2, "Coal Types as a Reflection of Depositional Environments"). Algae contain abundant fatty substances, principally high carbon number alkanes and olefins. The IR spectra of alginites are similar to those of resinite, with strong nonaromatic carbon-hydrogen bands and little evidence of aromaticity.[55] Alginite exhibits low reflectance and is distinguished from other liptinite macerals by its form, occurring as ovoid bodies with a crenulated border. Algae form not only identifiable alginite, but may also contribute significantly to the content of aliphatic constituents dispersed in coal (seldom recognizable microscopically).

Resinite, as its name implies, is derived from resinous source materials. Resinites are materials of relatively low reflectance that occur as spherical or ovoid bodies, or as irregularly shaped masses. Because the resinous source materials differ, it has been suggested that resulting resinites should be differentiated. M. Teichmüller[56] has divided the resinites into three genetic groups: terpene resinite, lipid resinite, and secondary resinite (exsudatinite). All share a common origin from hydrogen-rich substances, and retain high hydrogen/carbon ratios in coals through the bituminous rank (commonly having a higher H/C than associated sporinite).[57]

The terpene resins occur in secretion cells and ducts of plants, especially in conifers; in this form they are fairly readily recognized either still in the cell lumina or isolated following the destruction of the cell walls. However terpene resins may also be exuded from the plant, forming large, irregular masses. Lipid resins originate from fatty oils and waxes and commonly do not exhibit any identifiable form. Teichmüller suggests that they may serve as progenitors for petroleumlike substances in coal.

Exsudatinite (secondary resinite) has been described and named by M. Teichmüller.[58] This low-reflecting substance has infiltrated cracks and other cavities in some coal seams; it is found predominantly in subbituminous and high-volatile bituminous coals, apparently having been emplaced through the migration of tarry or bituminous material at the metamorphic stage characteristic of subbituminous coal. Some micrinite may be material left after migration of the mobile phase that forms exsudatinite.[58] Murchison had earlier noted the differences between resinite in low-rank coals and higher rank coals.[59] In a later report[60] he supports, on the basis of IR absorption spectra, Teichmüller's hypothesis of the differentiation of primary and secondary resinites.

Two hydrogen-rich substances recently described by M. Teichmüller[58] do not fit into the three major groups of macerals; these are fluorinite and bituminite, which can be referred to appropriately (along with exsudatinite) as "oily macerals." Both are recognized principally by their fluorescence in blue or UV light. Fluorinite is a high-hydrogen, low-reflectance (but highly fluorescent) substance that apparently originated from plant oils. It is found as lenses and is responsible for the

[54] M. Teichmüller, in *Textbook of Coal Petrology*, E. Stach, Ed., Gebrüder Borntraeger, Berlin and Stuttgart 1975, p. 198.

[55] R. Millais and D. G. Murchison, *Fuel*, **47**(3), 247–258 (1969).

[56] See Ref. 54, p. 205.

[57] D. G. Murchison and J. M. Jones, *Fuel*, **42**, 141–158 (1963).

[58] M. Teichmüller, "Generation of Petroleum-like Substances in Coal Seams as Seen Under the Microscope," *Advances in Organic Geochemistry*, Editions Techniques, Paris, 1973, pp. 379–407.

[59] D. G. Murchison, "Properties of Coal Macerals. Infrared Spectra of Resinites and Their Carbonized and Oxidized Products," in Am. Chem. Soc. Adv. Chem. Ser. No. 55, (1966), p. 307.

[60] D. G. Murchison, *Fuel*, **55**(1), 79–83 (1976).

Table 3.3 Properties of Macerals Compared to Vitrinite in Same Coal (Subbituminous and High-Volatile Bituminous Only).

Magnitude of property greater than (>), less than (<), or equal to (=) that of associated vitrinite.

	Inertinite			
	Semifusinite	Fusinite	Micrinite	Exinite
Optical properties				
Reflectance	>	$\gg$	>	<
Fluorescence				>
Chemical structure				
Basic carbon structure[62,63]				=
Molecular weight[63]				>
H/C ratio[58,63]	<	$\ll$	=	>
H aliphatic/H total;—CH_2—; hydroaromaticity[64]				>
Fraction aromatic C; rings/unit[62,64,65]		>	>	<
$Oxygen_{OH}/oxygen_{total}$[58,63]		<		<
Unpaired spins by ESR[66,68]		>		<
Unpaired spins of fusinite with same carbon as vitrinite[68]		>		
Reactivity				
Methane sorption[67]		>		
Decomposition temperature[52]				<
Oxidizability[63]				<
Reduction with lithium in ethylenediamine[63]		<	=	<

"smeared" appearance of polished surfaces of some low-rank bituminous coals. Bituminite is a formless, low-reflectance material with weak brown fluorescence. It may be mistaken for shredded spore exines, but its lack of distinct shape, slightly higher fluorescence, and slightly lower reflectance serve to distinguish it from spore exines. It apparently represents a decomposition product of algae, animal plankton, bacterial lipids, and so on. Bituminite is the common organic material in oil source rocks. Spackman et al.[61] have also discussed and illustrated the oily macerals exsudatinite, bituminite, and fluorinite.

3.2.2.4 *Properties of Macerals from Bituminous Rank Coals*

Table 3.3 summarizes the properties of the major categories of macerals. Rather than presenting specific values, the magnitude of the selected property is designated as being greater than, equal to, or less than the magnitude of that property in vitrinite of coal of

[61] W. Spackman, A. Davis, and G. D. Mitchell, *Brigham Young Univ. Geol. Stud.*, **22**(3), 59–75 (1976).

[62] W. Ruland and V. Sandor, *Fuel*, **44**(1), 39–44 (1965).

[63] P. H. Given, M. E. Peover, and W. F. Wyss, *Fuel*, **39**, 323–340 (1960).

[64] T. Tschamler and E. de Ruiter, "Comparative Study of Exinite, Vitrinite, and Micrinite," in Am. Chem. Soc. Adv. Chem. Ser. No. 55, (1966), pp. 332–343.

[65] S. K. Mookherjea and B. K. Mazumdar, *Fuel*, **48**(3), 277–283 (1969).

[66] Yu. B. Tyutyunnikov, L. G. Sintserova, and I. S. Romodanov, *Khim. Tverd. Topl.*, No. 5, 61–68 (1967); *C. A.*, **68**, 61571 (1968).

[67] I. Ettinger, I. Eremin, B. Zimakov, and M. Yonovskaya, *Fuel*, **45**(4), 267–275 (1966).

[68] D. E. G. Austen, D. J. E. Ingram, P. H. Given, C. P. Binder, and L. W. Hill, "Electron Spin Resonance Study of Pure Macerals," in Am. Chem. Soc. Adv. Chem. Ser. No. 55, (1966), 344–362.

the same rank. The absence of a symbol means that no data are available for that property and substance.

3.2.2.5 *Macerals of Lignitic Rank Coals*

Very little attention has been paid to the petrographic character of lignitic (and subbituminous) rank coals in the United States during the last several decades. In contrast, European investigators have produced a substantial body of literature on the subject. Most petrological studies of these low-rank coals have leaned heavily toward botanically oriented descriptions because botanical structures of many of the components are readily apparent in the lower rank coals. A distinctive terminology has emerged; however little information is available relating the technological properties to the descriptive evaluations.

A thorough discussion of the terminology employed in petrological studies of lignite and brown coal is not warranted here because of the large numbers of terms and the often conflicting uses of terms. Synonyms or subcategories of vitrinite are listed below simply to make the reader aware of their existence.

Terms Synonymous with Vitrinite or Representing Subcategories of Vitrinite in Lignitic and Brown Coals

Huminite	Xylinite
Humotelinite	Xylotelinite
Humodetrinite	Desminite
Humocollinite	Desmocollinite
Textinite	Eugelinite
Ulminite	Gelocollinite
Attrinite	Porigelinite
Densinite	Euulminite
Gelinite	Textoulminite
Corpohuminite	Humoresinite
Dopplerite	Phlobaphinite
Detrinite	

It is possible that the group maceral classification will not be adequate for characterizing the petrographic composition of low-rank coals that might be employed as conversion feedstocks. Whether the terms and concepts so far employed by other investigators for describing low-rank coals will prove useful in this endeavor remains to be seen. For further elucidation of these terms, the reader is referred to *International Handbook of Coal Petrography* and its supplements[28] and to Stach's *Textbook of Coal Petrology*.[41]

3.2.2.6 *Inorganic Material*

In addition to the organic elements (carbon, hydrogen, oxygen, nitrogen, and sulfur) that comprise the macerals, coals contain inorganic elements in variable proportions. On combustion of the coal, most of the inorganic elements are left in their oxidized form as an ash residue. Carbon, hydrogen, oxygen, and sulfur may also occur in inorganic form. These may or may not remain as elemental components of the ash depending on their initial form and on the combustion conditions.

From a utilitarian standpoint, the inorganic elements almost always impart undesirable properties to the coal, and some of these are discussed in Section 3.4, "Classification of Coal According to Grade." From a petrological standpoint, they can be discussed most effectively on the basis of their form in the coal. Their form affects how they respond in specific utilization processes, including the relative ease or difficulty of their removal during coal "cleaning."

The inorganic elements are found in three principal forms: dissolved in the pore water, combined chemically with the organic materials (either ion exchanged or as metallo-organics), and as discrete mineral particles.

The most common elements found dissolved in the pore water are sodium and chlorine, their concentration often being similar to the concentration in connate brine waters of the associated strata.[69] Though these dissolved elements can be removed by aqueous extraction,[70] they are commonly left in the coal when it is dried before being analyzed or utilized.

Alkali and alkaline earth elements, especially sodium and calcium, are found replacing protons on oxygenated functional groups of the organic matter. These ion-exchanged

[69] H. J. Gluskoter, *Econ. Geol.*, **60**, 614–620 (1965).

[70] R. C. Neavel, K. K. Koh, and N. C. Nahas, *Trans. AIME, Soc. Min. Eng.* **262**(3), 263–267 (September 1977).

species occur principally in the lower rank coals (lignitic and subbituminous) in which the vitrinites are characteristically rich in oxygenated functional groups.

Numerous elements, especially those present in trace amounts, may be present as metallo-organic substances. Very little investigative effort has been made in this area. Further discussion can be found in Section 3.2.3.4, "Effects of Environments on Inorganic Components."

Inorganic elements occur in coal predominantly as discrete particles of mineral matter. The most abundant minerals in typical coals are the detrital minerals, principally silicates (e.g., quartz, clays) that were carried into the peat swamps by flowing water. Because of the low velocity of streams flowing into and through peat swamps, the detrital minerals are generally fine grained. Silicate minerals are often found concentrated in distinctive layers formed during periods of flooding, although they also may be found intermixed with the organic (maceral) matrix. Detrital minerals other than silicates are frequently encountered,[2] though they are seldom as abundant as the silicates. Essentially all detrital minerals exhibit low reflectance and are often difficult to distinguish microscopically from the similarly low-reflectance liptinite macerals. The distinctive form of liptinite is generally the major distinguishing feature.

Clays and calcite (or dolomite) may fill vertical cracks (cleats) in some coal seams. Because these deposits are often thin and poorly cemented to the organic material, thin flakes are liberated when the coal is crushed.

Iron disulfide (FeS_2), as the mineral pyrite (or marcasite), is found in varying proportions in almost all coal seams. It occurs in localized concentrations ranging from micron-sized crystals to boulder-sized aggregates. It is readily recognized microscopically by its exceptionally high reflectance. Often pyrite occurs preferentially associated with specific maceral components in a particular portion of a seam. For instance, it may be found filling the cell lumens in fusain, or as fine crystals in vitrinite layers. Frequently it is found in or closely associated with exinite. Like some clays and calcite, pyrite may also be found filling vertical cleats. Pyrite and the occurrence of sulfur generally, are discussed in more detail in Section 3.2.3.4, "Effects of Environments on Inorganic Components."

Mineral matter is generally not quantified in routine petrographic (microscopic) analyses, though its occurrence may be noted and recorded. Its nature and quantity can be readily assessed by other analytical techniques. Microscopic observation can, however, effectively assess the distribution of mineral matter, which in turn influences its liberation and separation potential in a "cleaning process." Such processes normally separate materials on the basis of size-hindered settling or specific gravity, and a washability test is often more effective when combined with a microscopic examination of separated fractions.

When coals are subjected to a separation such as cleaning, certain macerals may tend to report selectively to one or the other of the product streams.[71] Therefore where the coal type appears to be critical to a particular use, it is advisable to evaluate the petrographic characteristics of both the raw coal and the cleaned coal.

3.2.3 Coal Types: Formation and Significance

3.2.3.1 *Chemistry of Transformation of Plants to Coal*

The transformation of plant substances into coal is generally considered to involve two stages: the biochemical (peat) stage, which is instrumental in establishing coal type, and the dynamochemical (geochemical, metamorphic) stage, which establishes coal rank. For a coal seam to exist, the plant substances from which it formed had to be preserved from oxidative and microbial destruction. The preservation and accumulation took place in swamps where relatively quiescent water covered or saturated the fallen plant parts (but was not deep enough to preclude plant growth). In most areas topographic subsidence (relative to sea level) maintained

[71] B. K. Johnston and A. C. Cook, discussion to "Some Coal Washing Developments in Australia," by J. J. Post, in *Australian Black Coal*, A. C. Cook, Ed., Australasian Institute of Mining and Metallurgy, Illawara Branch, 1975, pp. 210–211.

the water level at the surface of the peat. Not only did the water level have to balance peat accumulation during the active growth stage, but peat accumulation had to be terminated by relatively rapid burial under waterborne sediments (clays, sand), to seal the peat from oxidative or erosional destruction. Coal seams represent, then, the products of this delicate balance, making the formation of coals several hundred feet thick seem quite extraordinary.

Because the plant components that ultimately form liptinite are highly resistant to decay, they were incorporated into peat essentially intact. However the volumetrically dominant woody cellular tissues are more or less susceptible to chemical alteration during the peat-accumulating stage; hence any consideration of the biochemical stage involves principally the transformations undergone by lignin and cellulose.

"Wood," as the term is used here, is highly complex: that is, the structural elements of the plants are referred to as wood, including the leaf tissues (exclusive of cuticle), roots, bark (exclusive of periderm), and trunk-stem structures. The chemical constituents comprising wood are predominantly classifiable as celluloses (including hemicelluloses) and lignin, which are variable in their composition and molecular structures. Lignocellulosic material is the source both of vitrinite and of most inertinite.

In the formation of vitrinite the cell wall materials (cellulose, lignin) and cell contents (proteins and some waste substances) were altered, principally by biochemical action, to humic substances generally recognized to be the progenitors of the vitrinitic macerals in coal. The magnitude of the contribution of each of the cell components, and the physical and chemical nature of the resulting maceral material, depended on the chemical conditions within the peat.[72,73] Microscopic studies of woody tissues from a lignite indicated that at least 13 microscopically different substances relatable to the macerals vitrinite, micrinite, and resinite in higher rank coals could be identified.[74] This observation indicates the difficulty of attempting to summarize the biochemical stage of coalification succinctly.

Cellulose and lignin are intimately intermixed in the cell walls, their relative proportions depending on the specific layer of the wall, the specific tissue, and the plant species. In paper making the lignin is extracted from the wood using caustic, whereas in peat the cellulose seems to be preferentially (though not completely) attacked by microorganisms that prefer it as an energy source. Some fungi, on the other hand, utilize the lignin.[75] Proteins are readily decomposed biologically.[76] Phenolic tannins can impregnate the wood of some species of plants, rendering it highly resistant to microbial decay.[75] Thus depending on the plants and the conditions within the peat deposit, certain parts of the cell walls may have been preferentially destroyed by being converted to water, carbon dioxide, and microbial protein.[77]

The problem that seems to have received the most investigative consideration in the area of biochemical coalification is the origin of the cyclical compounds that predominate in the vitrinite of coal. Polycyclic aromatics are absent in plants, yet are found in extracts of peat,[78] and a number of hypotheses have been proposed to explain the chemistry of their formation. It is generally presumed that they were formed by microbial activities, for instance, through a shikimic acid intermediate[79] or through condensation of benzoquinones.[77] It has also been suggested that benzoquinones can act as catalysts as well as reactants, particularly with the aid of co-

[72] A. Davis and W. Spackman, *Fuel*, **43**(3), 215–224 (1964).

[73] L. Belau, *Bergbautechnik*, **19**(4), 201–206 (1969); C. A., **71**, 62884 (1969).

[74] W. Spackman and E. S. Barghoorn, "Coalification of Woody Tissue as Deduced from a Petrographic Study of Brandon Lignite," in Am. Chem. Soc. Adv. Chem. Ser. No. 55, (1966), pp. 695–707.

[75] See Ref. 54, p. 183.

[76] W. Flaig, *Landbau-Forsch. Voelkenrode*, **71**(1), 1–16, 17–26 (1967).

[77] W. Flaig, "Biochemical Factors in Coal Formation," in *Coal and Coal Bearing Strata*, D. G. Murchison and T. S. Westoll, Eds., Oliver and Boyd, Edinburgh, 1968, pp. 197–232.

[78] N. G. Golovanov and V. V. Borsoglebski, *Khim. Tverd. Topl.*, No. (6), 163–168 (1968); *C. A.*, **70**, 70021 (1969).

[79] M. H. Rogoff, I. Wender, and R. B. Anderson, U. S. Bur. Mines Inf. Circ. No. 8075 (1962), pp. 1–55.

catalyzing enzymes from microbes.[77] The products of the cyclization that obviously did occur in the peat are mixtures of humic acids, a term referring to multifunctional, polycondensed, aromatic acids, generally of indeterminate, but high, molecular weight. Aerobic decay is claimed to favor the formation of humic acids.[75,80] Through the lignitic rank, significant quantities of humic acids can be extracted with aqueous caustic.

It is known that not all lignin and cellulose were converted to humic acids (or their nonacidic analogues humins or humates), because material identified microscopically as cellulose has been found in low-rank lignites[74]; none has been reported from higher rank coals, however. Both nitrogen and sulfur appear to be fixed during the biochemical stage, the nitrogen originating from plant and microbial proteins, the sulfur from dissolved SO_2^{2-} (Section 3.2.3.4, "Effects of Environments on Inorganic Components" has further discussion of sulfur). It has been proposed that as available nitrogen was fixed into the decaying vegetation, microbiological activity diminished.[76] The biochemical stage was essentially terminated when the peat was buried to a depth of 10m or more.[81] Articles by Flaig[76,77,82] and a monograph by Rogoff et al.[79] discuss the microbiology of coal in detail.

In addition to the biochemical alteration of the cellular tissues and cell contents, more or less physical disintegration of the tissues also occurred. Some cellular tissues, though chemically altered, retained their integrity as physically intact entities varying in size from logs to chips; some were thoroughly broken down into dispersed molecules of humic material. Thus wood derivatives may be recognizable as vitrain bands, as attrital vitrinite, and even as coalified gels formed in place or deposited from colloidal dispersions. It has been suggested that in coal from a peat layer where severe disintegration was the norm, the coalified vitrinite exhibits a low reflectance (and, by inference, a higher H/C ratio.)[83] Certainly when the lignin and cellulose were converted to carbon oxides and water, only the resistant waxy (exinitic) materials remained, and the peat differed significantly from more normal humic peats. On the other hand, where severe oxidative reactions resulted in a form of *in situ* pyrolysis, inertinite macerals (fusinite, semifusinite) and possibly pseudovitrinite may have been formed from the cellular tissues.

Although in typical peat swamps the dominant fate of lignocellulose was to form vitrinite progenitors, some was converted to inertinite.[84] Most fusinite/semifusinite appears to have been formed predominantly by a charring process that operated before deposition in the peat. Among other studies, electron spin resonance measurements[68] and analyses of specific heat and electrical conductivity[85] (Chapter 4) have led investigators to conclude that fusain specimens studied had been subjected to temperatures well above those that could possibly have affected the seam as a whole. The most logical source for such high temperatures (700–800°C have been deduced)[85] is simply a fire; hence fusinite is just charcoal.[84] Though one investigator claims that the absence of ash associated with fusain layers belies this hypothesized origin,[86] it is likely that fusinite-rich fusain layers represent aqueous "placer deposits" rather than layers of fusinite formed *in situ*. In such layers the presence of chips of vitrinite and spore exines, intimately mixed with fusinite exhibiting a broad range of reflectances, tends to support the placer deposit hypothesis.

It has been suggested that some fusinite may have been formed by biochemical activity and by dehydrogenation (of lignocellulose) "by some means or other."[85] It has also been suggested that some fusinite might have originated from carbon-rich structures of some plants ("primary fusinite").[87] Furthermore there is some evidence that meta-

[80] N. K. Larina, G. N. Meshkova, K. T. Cherinskaya, Z. V. Grigor'eva, and V. I. Kasatochkin, *Khim. Tverd. Topl.*, No. 2, 37–44 (1968); *C. A.*, **69**, 29165 (1968).

[81] M. Teichmüller and R. Teichmüller, in *Textbook of Coal Petrology*, E. Stach, Ed., Gebrüder Borntraeger, Berlin and Stuttgart, 1975, p. 25.

[82] W. Flaig, "Chemistry of Humic Substances in Relation to Coalification," in Am. Chem. Soc. Adv. Chem. Ser. No. 55, (1966), pp. 58–68.

[83] See Ref. 54, p. 193.

[84] J. A. Bell and D. G. Murchison, *Fuel*, **45**(5), 407 (1966).

[85] A. Banerjee, B. K. Mazumdar, and A. Lahiri, *Fuel*, **45**(2), 145–150 (1966).

[86] J. M. Schopf, *Econ. Geol.*, **66**, 1153–1156 (1971).

[87] See Ref. 54, p. 222.

morphism might have affected different source materials differentially, forming from some substances a coalified material referable to fusinite (so-called rank fusinite).[88,89]

McCartney[90] determined the reflectance distribution and chemical composition of fusain samples from four coals of different ranks. The reflectance values for fusinite from higher rank coals were higher, indicating that the inert macerals may have been affected by metamorphism.

Micrinite apparently originated in several ways. M. Teichmüller[58] has concluded that some micrinite represents a residue after a fractionation process that resulted from metamorphic alteration of hydrocarbon materials. It is her contention that when heated in the geological milieux to a temperature corresponding to that which forms bituminous coal from lignite, certain lipoid or waxy substances formed a petroleumlike substance (exsudatinite) that migrated freely into "traps" in the coal where it solidified, leaving behind a more carbonaceous residue (micrinite). Thus according to Teichmüller, most micrinite is a secondary maceral found only in higher rank coals. However, Spackman and Barghoorn,[74] among others, recognized in lignite finely granular, opaque substances referable to micrinite that were derived from degradation of plant cell walls. Thus at least some micrinite appears to have been derived from lignocellulosic plant substances.

3.2.3.2 *Coal Types as a Reflection of Depositional Environments*

Physical conditions, principally water depth and related water movement, affected chemical conditions in the accumulating peat. The chemical conditions, especially hydrogen ion concentration (pH) and oxidation/reduction (redox) potential, in turn affected the microbiological activities. These interrelated physical, chemical, and microbiological conditions interacted with the indigenous plant materials to form specific types of peat, hence specific coal types. Though widespread environmental influences such as the proximity of marine conditions might have influenced the gross characteristics of a peat, localized conditions such as the presence of pools and isolated plant communities resulted in variations in peat properties from place to place in contemporaneously deposited layers. Thus from time to time and from place to place, the progenitors of different types of coal could have been formed in a continuously accumulating peat deposit.

Spackman et al.[91] describe in considerable detail the kinds of localized environmental variation found in a modern-day peat swamp of the Florida Everglades.

Coal from a seam is not uniform; hence single pieces cannot be used to characterize the coal from one seam even in a given location. Furthermore, the common practice of defining a sample solely by designating the seam from which it is derived, rather than specifically defining the location and/or properties of a sample, is virtually useless for conveying technological information.

The variations in coal properties affected by depositional environments may have been widespread or localized. Regional trends that are independent of specific seams have been recognized. For instance, often ash and sulfur contents exhibit regional patterns that persist through superimposed seams, reflecting a persistent areal geography of the swamp-bearing area.[92,93] Alternatively, it is reported that in British Columbia and Alberta vitrinite decreases—while fusinite and semifusinite increase—in progressively deeper, older seams; changes that appear to correlate with increasing coarseness of sediments in the enclosing strata.[35] Coals from the central floodplains of paralic (near-sea) basins are reported to be vitrinite rich, with considerable exinite, whereas coals from the landward basin margins are rich in resinite and inertinite. Coals deposited in lagoonal areas are reported to contain abundant attrital vitrin-

[88] See Ref. 54, p. 220.

[89] G. D. Petrovskii, *Tr. Vses. Nauchn.-Issled. Geol. Inst.*, **132,** 16–24 (1968).

[90] J. T. McCartney, *Fuel,* **56**(2), 223–224 (1977).

[91] W. Spackman, C. P. Dolsen, and W. Riegel, *Paelontographica,* **117(B),** 135–152 (1966).

[92] S. P. Babu, J. A. Barlow, L. L. Craddock, R. V. Hidalgo, and E. Friel, West Virginia Geol. Econ. Survey, Coal Geol. Bull. No. 1, 1973, 32 pp.

[93] B. G. Gaberman and A. I. Dydyshko, *Tr. Khim.-Met. Inst. Akad. Nauk Kaz. SSR,* **2,** 110–121 (1968); *C. A.,* **69,** 88652 (1968).

ite.[94] Even the properties of kerogen in sedimentary rocks vary depending on the depositional proximity to marine conditions, being more hydrogen rich, the greater the marine influence.[95] Characteristic minerals within seams have also been related to depositional environments.[96]

There are often broad variations in coal type within a single seam, as exemplified by the properties of the Harbour Seam of Nova Scotia, studied petrographically by Hacquebard et al.[97] Finely striated, vitrinite-rich coal (clarain) formed in reed moors, impure coal accumulated in more open areas, and coals with thick vitrain characterized forest-moor areas. In his studies of several seams in Pennsylvania, Koppe[98,99] found that the petrographic composition changed significantly in a direction normal to present-day anticlines (that probably were topographically high during coal deposition), but was relatively uniform along the strike of the structures. Higher fusain contents were found in the Upper Freeport Seam where it crossed structural highs.

One of the best reasoned and lucidly written accounts of the relationships of depositional environments and coal types is that by Marlies and Rolf Teichmüller in which they summarize the geology and depositional history of the Cenozoic-Mesozoic coals in Germany.[100] The articles by Hacquebard and coworkers are recommended as exemplary studies of the effect of Carboniferous coal/peat environments on coal type.[94,97,101] Cook discusses the relationships between paleogeography, paleotectonism, and coal types in Australian coals.[102]

Although there are generally recognized classification systems for coal macerals, there is far less unanimity in the classification of coal types. The most comprehensive, megascopic description technique, briefly summarized earlier in this chapter, was published by Schopf in 1960.[103] Other investigators commonly define coal types on the basis of the percentages of constituent macerals determined microscopically.[97,101,104] Two major coal types are fairly widely recognized: the humic coals and the sapropelic coals. An intermediate, unnamed type is also fairly commonly encountered. Humic coals are vitrinite rich and are often referred to as clarains or clarites. These are differentiated further on the basis of the thickness and amount of vitrain and of the characteristics of the attrital fraction, properties that appear to be closely related to depositional environment. The sapropelic coals, consisting predominantly of the hydrogen-rich macerals (liptinites), are presumed to have been accumulated in deeper water where humic material was sparse. Cannel coals (spore-exine-rich), boghead coals (alginite-rich), and even some oil source rocks fall into the category of sapropelic deposits. Sapropelic coal types may occur as benches in otherwise humic coals, and seldom exceed 0.5 m in thickness. Moore offers a review of sapropelic deposits in a 1968 publication.[105]

[94] P. A. Hacquebard, T. F. Birmingham, and J. R. Donaldson, *Proc. Symp. Science Technol. Coal,* Canadian Institute of Mining Metallurgy and Canadian Advisory Commission on Coal Research, Ottawa, March 29–31, 1967, published by Mines Branch, Dept. Energy, Mines, Resources, Ottawa, (1967), pp. 84–97.

[95] I. A. Breger and A. Brown, *Science,* **137,** 221–224 (1962).

[96] C. P. Rao and H. J. Gluskoter, Illinois Geol. Survey Circ. No. 476, (1973), pp. 1–56.

[97] P. A. Hacquebard, A. R. Cameron, and J. R. Donaldson, Geol. Survey Canada Paper No. 65-15, (1965), pp. 1–31.

[98] E. F. Koppe, "Petrographic Continuity of Pennsylvania Coals," in Am. Chem. Soc. Adv. Chem. Ser., No. 55, (1966), pp. 69–79.

[99] E. F. Koppe, *Trans. Am. Inst. Min. Metall. Pet. Eng.,* **217,** 185–189 (1960).

[100] M. Teichmüller and R. Teichmüller, "Cenozoic and Mesozoic Coal Deposits of Germany," in *Coal and Coal Bearing Strata,* D. G. Murchison and T. S. Westoll, Eds., Oliver and Boyd, Edinburgh, 1968, pp. 347–379.

[101] P. A. Hacquebard and M. S. Barss, "Paleogeography and Facies Aspects of the Minto Coal Seam, New Brunswick, Canada," *C. R. 6th Cong. Intr. Strat. Geol. Carbonif.,* Sheffield, 1967, Vol. 3, 1970, pp. 861–872.

[102] A. C. Cook, "The Spatial and Temporal Variation of the Type and Rank of Australian Coals," in *Australian Black Coal,* A. C. Cook, Ed., Australasian Institute of Mining and Metallurgy, Illawara Branch, 1975, pp. 63–84.

[103] J. M. Schopf, U.S. Geol. Survey Bull. No. 1111-B, 1960.

[104] I. E. Val'ts, A. I. Ginzburg, and N. M. Krylonova, *Khim. Tverd. Topl.,* No. (3), 9–21 (1968); *C. A.,* **69,** 53426 (1968).

[105] L. Moore, "Cannel Coals, Bogheads and Oil Shales," in *Coal and Coal Bearing Strata,* D. G. Murchison and T. S. Westoll, Eds., Oliver and Boyd, Edinburgh, 1968, pp. 19–29.

Intermediate between the vitrinite-rich humic coals and the fine-grained, vitrinite-poor sapropelic coals are the durainic coals or durites. These are generally only sparsely banded with vitrain; hence they are finely textured and appear dull. They apparently formed under conditions of either extreme dryness or extreme wetness where lignin and cellulose were sparsely available or were destroyed by oxidative decay, or were converted to carbon-rich inertinite macerals. Durains are commonly inertinite rich, and often contain considerable liptinite.

Depositional environments can often be inferred by studies of the spore exines that can be isolated from the coals and identified, on the basis of ornamental features, as being derived from specific environment-indicating plants. Several interesting studies relating petrographic types and spore exine characteristics have been published by Smith.[106,107]

3.2.3.3 *Microlithotypes*

Not only are there coal types that can be recognized and classified on the basis of their megascopic appearances, it has also been found operationally useful to recognize certain associations of macerals that occur in particles of crushed coal. These are referred to as microlithotypes. Mackowsky describes thoroughly the recognition criteria and procedures for microscopic analysis according to microlithotypes.[108]

The microlithotype concept is predicated on the observations that few coal particles are monomaceralic, and that the reactions of individual coal particles in technological situations are often influenced by the interactions of their constituent macerals and minerals. Where whole particles or portions of particles more than 50 μm thick (normal to the banding) are relatively uniform in their textural and compositional characteristics, the particle or zone can be identified and classified as a specific "type" of coal. The classification of microlithotypes according to Mackowsky is shown in Table 3.4. Commonly the relative proportions of the macerals within a particle or zone are quantified microscopically by using a 20-point grid, each point of which represents 5 vol %.[109]

Whether the microlithotype concept as described above or some variation is employed, the identification of the maceral (and mineral) associations in coal particles is potentially useful. Especially in the areas of pyrolysis and liquefaction, the proximity of certain macerals could be as important as the overall (modal) maceral composition. I have observed, for instance, that vitrinite enclosed within relatively inert materials (e.g., mineral

[106] A. H. V. Smith, "Seam profiles and seam characters," in *Coal and Coal Bearing Strata,* D. G. Murchison and T. S. Westoll, Eds., Oliver and Boyd, Edinburgh, (1968) pp. 31–42.

[107] A. H. V. Smith, *Geol Mag.,* **94**(5), 345–362 (1957).

[108] M.-Th. Mackowsky, in *Textbook of Coal Petrology,* E. Stach, Ed., Gebrüder Borntraeger, Berlin and Stuttgart, 1975, p. 256.

[109] M.-Th. Mackowsky, in *Textbook of Coal Petrology,* E. Stach, Ed., Gebrüder Borntraeger, Berlin and Stuttgart, 1975, p. 258.

Table 3.4 Classification of Microlithotypes[109]

Microlithotype	Maceral	Volume %
Monomaceralic		
Vitrite	Vitrinite (V)	>95%
Liptite	Liptinite (L)	>95%
Inertite	Inertinite (I)	>95%
Bimaceralic		
Clarite	V + L	>95%
Vitrinertite	V + I	>95%
Durite	I + L	>95%
Trimaceralic		
Duroclarite	V + I + L	V > (I + L)
Clarodurite	V + I + L	I > (L + V)
Vitrinertoliptite	V + I + L	L > (I + V)
Carbominerite	V,L,I and mineral matter (MM)	MM > 20%, < 60%

matter, inertinite) is not readily liquefied because of poor contact with the potential hydrogen-donating solvent or vehicle. Cameron and Babu[110] relate swelling and plasticity characteristics to specific coal types and discuss the effects of these properties on the coke-making characteristics of western Canadian coals. Finally, it is worth repeating that certain coal types may be found more commonly in specific particle size ranges of crushed coal because of unique properties (hardness, toughness, friability).[111] Therefore, it must never be assumed that a sample of a biased size fraction from broken coal, or worse yet, a single lump, represents the total pile or total seam.

3.2.3.4 *Effects of Environments on Inorganic Components*

A previous section classified the minerals found in coal seams in terms of their occurrence. Minerals may also be classified, on the basis of their origin, as detrital, syngenetic, and epigenetic. Detrital minerals are those that were transported into the swamp as particles borne either by water or wind. Syngenetic minerals were crystallized in place from dissolved ions at essentially the same time that the peat was deposited. Pyrite is the most common syngenetic mineral. Epigenetic minerals are those that were formed after the peat accumulated. Calcite ($CaCO_3$), dolomite [$Ca,Mg(CO_3)_2$], kaolinite [$Al_2Si_2O_5(OH)_4$], and some pyrite (FeS_2) are the most common epigenetic minerals.

Essentially all detrital and syngenetic inorganic material found in coal was transported into the peat swamp by water, but a minimal transport role was also played by wind. Hence the commonly encountered attribution of origin of some inorganic elements found in coal to the plants growing in the peat swamp can be somewhat misleading. It might be expected, however, that certain elements passing through the peat swamp could have been selectively concentrated by the plants growing there, and that plants could have selectively leached certain elements from the floor sediments. Similarly, certain elements might have reacted with the organic material while it accumulated as peat, thereby resulting in their concentration in coal seams in proportions greater than are characteristic for the enclosing sedimentary rocks. Sulfur, in its "organic" form, is one of the elements so concentrated.

There is sulfur in the protoplasmic substances of the growing plants, and sulfur is found in the organic macerals of almost all coal. Because of its extreme variability from seam to seam, however, it is justifiable to think of sulfur as an extraneous (if not solely inorganic) material. The sulfur is found in coal seams principally as pyritic sulfur (FeS_2) and organic sulfur (as an elemental component of the molecular structure of the macerals).

The regionally averaged, total sulfur content of a coal seam reflects the gross geographic environment in which the peat accumulated. The concentration of sulfate ions in the water of the peat swamp was very influential in this regard, and peats accumulated in marine-influenced environments tended to become high-sulfur coals. Dorofeev[112] concluded that the typical sulfur contents of Carboniferous age coals of the Donetz basin showed the following relationships with depositional environments:

Depositional Location	Typical Sulfur Content (%)
Coastal swamps on continental margin	1.6
As above, weak marine influence	2.2
Coastal swamps on marine side	3.4
Swamps with prolonged marine exposure	3.9

Williams and Keith[113] found a statistically positive correlation between sulfur content

[110] A. R. Cameron and S. K. Babu, Geol. Survey Canada Paper No. 68-35, 1968, pp. 1–57.

[111] W. Spackman, W. F. Berry, R. R. Dutcher, and A. H. Brisse, *Yearb. American Iron and Steel Inst.*, 403–449 (1960).

[112] A. P. Dorofeev, *Khim. Tverd. Topl.*, **8**(3), 54–59 (1974).

[113] E. G. Williams and M. L. Keith, *Econ. Geol.*, **58**, 720–729 (1963).

of coals of central Pennsylvania and the presence of overlying marine shales. Gluskoter and Hopkins[114] asserted that sulfur content was lower in an Illinois coal seam where a gray shale covered the seam, thereby (they hypothesized) excluding sulfate-bearing marine waters from percolating down into the seam. It seems as likely, however, that any relationship found between sulfur content and the lithology of the overburden reflects the effects of some persistent topographic feature. Several studies indicate that sulfur content is established in the peat swamp and does not appear to be affected significantly by subsequent deposition or metamorphism.[115–117]

Both organic sulfur and syngenetic pyritic sulfur owe their emplacement in the seam to the activities of sulfate-reducing bacteria. *Desulfovibrio desulfuricans* enzymatically utilizes the oxygen of sulfate ions for its metabolism, thereby producing sulfide (S^{2-}) ions. These combine with hydrogen and evolve as hydrogen sulfide (H_2S), react with iron to form pyrite, or react with oxygenated functional groups of organic material to form organic sulfur. Thus all organic sulfur and most pyritic sulfur are probably syngenetic with peat accumulation,[116,117] and concentrations of both forms tend to be positively correlated. Some pyrite formed in cracks or cleats, probably from infiltrating solutions in the late peat or early lignite stage.[102] Because of the very low solubility of pyrite, it is difficult to envision how Fe^{2+} and S^{2-} ions could be transported together and be deposited as pyrite in the cleats. More likely, iron-bearing water circulated through the cleats where it contacted sulfide sulfur (possibly as H_2S) that migrated from the peat. Troilite, a progenitor of pyrite, would then grow in the cleats.

Variations in sulfur content across short distances both vertically and laterally reflect localized availability of sulfate ion or the effects of localized redox conditions on bacterial activity, the redox potential being most influenced by localized water movement reflecting topographic variations.[116] In addition, the availability of iron also could affect the concentration of pyrite wherever available sulfate ions and bacterial activity produced abundant sulfide ions.[96,116] Because iron most likely entered the peat swamp adsorbed on clays, pyrite tends to be more abundant adjacent to clay-rich zones; hence pyrite is commonly more abundant at the top and bottom of seams as well as below clay-rich partings.[116,118]

The organic sulfur content of fusinite is invariably lower than that of associated vitrinite, though fusinite cell lumina may be filled with pyrite.[116,118] In one study using an electron microprobe, the organic sulfur content of liptinite was found to be both higher and lower than that in associated vitrinite, but averaged somewhat higher in the liptinite than in the vitrinite.[116]

The transformation, recrystallization, and mobility of other inorganic substances is significantly affected by the pH of the peat swamp.[119] Zubovic[120] discusses thoroughly the preferred relationship of certain elements with the organic fractions of the coal. Both salinity and pH appear to be important determinants of the elements that will be found in various positions within coal seams. The elements boron, chlorine, chromium, copper, germanium, nickel, ruthenium, sodium, and vanadium are considered to be saline indicators.[121,122] Elements that have been reported as ion-exchanged in coals, where pH is important,[123] are gallium,[124] calcium, magne-

[114] H. J. Gluskoter and M. E. Hopkins, *Ill. Geol. Survey Guideb.*, **8,** 89–95 (1970).

[115] Z. Balazs and A. Juhasz, *Banyasz. Kohasz Lapok, Banyasz.*, **101**(6), 408–414 (1969); *C. A.*, **71,** 83391 (1969).

[116] R. C. Neavel, Ph.D. thesis, Pennsylvania State University, (1966).

[117] W. C. Grady, AIME, Soc. Min. Eng. Prepr. No. 76-F-315, 1976.

[118] H. J. Gluskoter and J. A. Simon, Illinois Geol. Survey Circ. No. 432, (1968), pp. 1–28.

[119] M. Duel, Am. Chem. Soc., Div. Gas Fuel Chem., Boston, April 1959.

[120] P. Zubovic, "Physiocochemical Properties of Certain Minor Elements as Controlling Factors in Their Distribution in Coal." in Am. Chem. Soc. Adv. Chem. Ser. No. 55, (1966), pp. 221–231.

[121] P. P. Timofeev, L. I. Bogolybova, L. V. Miesserova, and N. P. Federovskaya, *Khim. Tverd. Topl.*, No. 5, 96–102 (1967); *C. A.*, **68,** 61567 (1968).

[122] E. Skipsey, *Fuel,* **54**(2), 121–125 (1975).

[123] V. F. Razdoroznyi, *Geokhimiya,* No. 1, 99–105 (1968).

[124] Gr. Eskenazy, *Fuel,* **46**(3), 187–191 (1967).

sium (especially around the rims of peat-accumulating basins),[125] titanium (favored by higher pH),[126] sodium,[127] nickel, and beryllium.[123]

The commonly determined[128] and most abundant elements (Si, Al) are found principally as detrital mineral grains that were transported into the peat swamp. Their concentration varies from layer to layer and from place to place because of variations in the flow of sediment-bearing streams into and through the swamp.

Chapter 7 provides further discussions of mineral matter and its relationship to depositional environments.

3.3 METAMORPHISM AND COAL RANK

Following the biochemical (diagenetic) stage of active peat accumulation, additional physical and chemical changes took place during the metamorphic stage under the confining pressure of overlying sediments. Pressure compacted, cemented, and hardened the organic materials, while subsurface heat (from deep-seated geothermal processes) drove off oxygen and hydrogen as water, oxides of carbon, and methane. Heat and pressure thereby converted the organic material progressively to substances more and more similar to graphite. Almost all properties of the organic substances that can be measured or assayed changed in response to progressively more severe metamorphism. Coals are ranked according to how severely they were metamorphosed, hence where they are located along this continuum from peat to graphite.

3.3.1 Measurement of Rank

Because vitrinite precursors (humates, humic acids) predominated in typical peats, the story of metamorphism and its effects is principally the story of the changes in properties of vitrinite. Some of the more important properties of vitrinite that did change with metamorphism are listed in Table 3.5. Metamorphism also affected the other macerals, but the relationships between the severity of metamorphism and the magnitude of change are different from those of vitrinite. The classical example of this appears in Fig. 3.17, which shows the "coalification tracks" of the major maceral materials.

The greatest value of the concept of the development of rank is that under normal circumstances the many characteristic properties of the vitrinite changed in a progressive fashion. As a consequence, by determining the magnitude of some metamorphism-sensitive property of the vitrinite, thereby classifying the coal according to rank, other properties of the vitrinite can be inferred with reasonable accuracy from available correlations. Some properties of the other macerals can also be inferred, but because of difficulties associated with concentrating other macerals, fewer data are available for them than for vitrinite.

Table 3.5 Properties of Vitrinite Affected Progressively by Metamorphism

Increase with increasing rank
Reflectance (microscopic) and optical anistropy
Carbon content
Aromaticity (viz., $f_a = C_{arom.}/C_{total}$)
Condensed-ring fusion
Parallelization of molecular moieties
Calorific value (slight decrease at very high rank)
Decrease with increasing rank
Volatile matter (especially oxygenated compounds)
Oxygen content (especially as functional groups)
Oxidizability
Solubility (especially in aqueous alkalis and polar hydrocarbons)
Increase initially to a maximum, then decrease
Hardness (minor increase at very high rank)
Plastic properties
Hydrogen content
Decrease initially to a minimum, then increase
Surface area
Porosity (and moisture-holding capacity)
Density (in helium)

[125] V. G. Klinger, I. F. Largin, and M. V. Popov, *Prir. Bolot. Metody Ikh. Issled., Mater., Vses. Sovesch.*, "Sovrem. Puti Metody Issled. Bolot," 254–259 (1964) (published 1967), *C. A.*, **70**, 49260.

[126] Gr. Eskenazy, *Fuel,* **51**(3), 221–223 (1972).

[127] L. E. Paulson and W. W. Fowkes, U.S. Bur. Mines Rep. Invest. No. 7176, 1968, pp. 1–18.

[128] ASTM Standards, 1977, Part 27, Designation D2795-69 (reapproved 1974) Standard Methods of Analysis of Coal and Coke Ash.

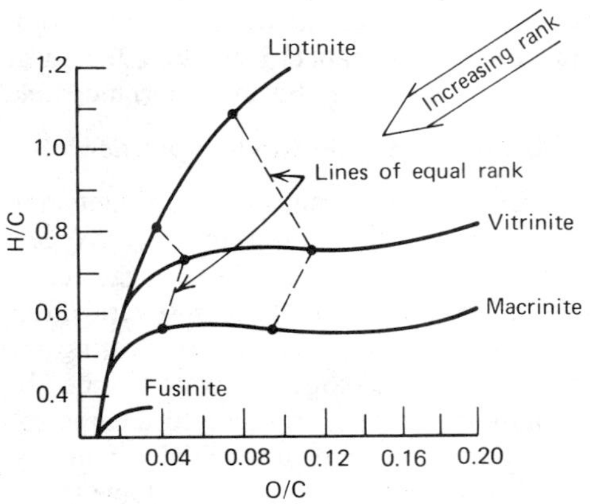

Figure 3.17 Coalification tracks of different macerals based on H/C and O/C atomic ratios. From Teichmüller and Teichmüller[129] after van Krevelen.[49] Although van Krevelen's initial diagrams referred to "micrinite," the designation was changed to "macrinite" in the Teichmüller figure. The lines represent typical materials, but specific data points diverge significantly from the regressions shown.

Several very important reservations must be attached to the foregoing generalization:

1. Data employed to establish rank and employed in correlations are most meaningful when obtained from vitrinite-rich coals (or, more meaningfully, from vitrinite only).
2. Although there are broadly generalizable correlations between properties, there is significant scatter even in data from carefully purified vitrinites because of variations in plant-source materials, variability of diagenetic alterations, and variability of relationships between temperature and pressure during metamorphism.

Although minerals may be altered by metamorphic processes, it is the organic material that is most sensitive to these processes.[130] Therefore analyses designed to elucidate coal rank should be carried out on organic-rich fractions, or else the results should be calculated to an ash-free (or preferably mineral-free) basis. Isolated samples of vitrinite represent the best samples for determination of rank.

The most useful rank-defining properties, each with its own special attractiveness, are (1) elemental carbon content, (2) volatile matter yield, (3) moisture-holding capacity, (4) functional group properties, (5) calorific value, and (6) microscopic reflectance of vitrinite. The ideal property would be easily and objectively measured, and sensitively responsive throughout the transition from plant substance to graphite. No single property meets these criteria.

3.3.1.1 *Carbon Content*

The elemental carbon content is probably the chemical property most validly used in scientific investigations to express coal rank.[131] Unfortunately it is difficult to isolate vitrinite routinely to assess the carbon content. Some equivocal definitions of coal rank result when carbon content (or any other chemical property) is determined on petrographically heterogeneous, whole-coal samples. For instance, a low-rank coal with high fusinite content will have a carbon content higher than more typical vitrinite-rich coals of that rank. Furthermore, to a first approximation, sulfur atoms may be molecular substitutes for carbon atoms, and it has been suggested that the sum of carbon and organic sulfur contents may be a more meaningful rank parameter than carbon content alone.[132]

3.3.1.2 *Volatile Matter*

When coal is heated in an inert atmosphere or in a vacuum to temperatures above about 350°C, the molecular structure decomposes. Low molecular weight decomposition products are thereby evolved (volatile matter), leaving a residue of carbon-rich char (fixed carbon) and inorganic materials. The yield of volatile matter evolved under standardized heating conditions is probably the most commonly employed (though far from the "best") measure of coal rank (especially for bituminous and higher rank cals—see ASTM

[129] M. Teichmüller and R. Teichmüller, in *Textbook of Coal Petrology,* E. Stach, Ed., Gebrüder Borntraeger, Berlin and Stuttgart, 1975, pp. 35–54.

[130] J. R. Castaño and D. M. Sparks, *Interpretation of Vitrinite Reflectance Measurements in Sedimentary Rocks,* Geol. Soc. America Spec. Paper No. 153, (1974), pp. 31–52.

[131] M. F. Kessler, *Fuel,* **52**(3), 191–197 (1973).

[132] R. M. Elofson and K. F. Schulz, *Prepr., Am. Chem. Soc., Div. Fuel Chem.,* **11**(4), 259–263 (1967).

D388[1]). The volatile yield decreases with increasing coal rank. Although volatile matter yield is easy to assess, its use as a rank parameter suffers some of the same deficiencies as the use of carbon content. In addition, it has several deficiencies of its own. Originally, the volatile matter/fixed carbon test was employed as a measure of the amount of coke that could be expected from a coal. Different standardized techniques for the determination of volatile matter yield lead to different results, principally because volatile matter is the product of pyrolytic decomposition, and different heating rates and end temperatures result in differences in decomposition.[133,134] The test is especially sensitive to variations in petrographic composition (coal type). For instance, volatile matter yields ranging from 42 to 50% (dry, ash-free) were obtained from petrographically distinctive layers of a single lignite from Saskatchewan,[135] and these differences certainly could not have reflected differences in the extent of metamorphism (rank).

3.3.1.3 *Moisture-Holding Capacity*

With increasing level of metamorphism to anthracitic rank, porosity decreases. A reasonably effective measure of porosity is obtained by determining the moisture-holding capacity of the coal. Saturation of the pores can be achieved by soaking a sample of crushed coal in water, then equilibrating the sample in an atmosphere of 97% relative humidity until all surface moisture is removed (see ASTM D1412[136] for equilibrating procedure). The equilibrated sample is then dried quantitatively at 105°C, and the loss is considered to be pore-held moisture. This property has been especially useful for determining the rank of mildly metamorphosed coals where volatile matter and carbon content have not yet been affected significantly by metamorphism. Moisture-holding capacity is also affected by the functional group characteristics, and in coals where cationic elements replace protons on acid functional groups, moisture-holding capacity is reduced.[137] As with most other properties, moisture-holding capacity can be affected by coal type as well as by metamorphic rank.

3.3.1.4 *Functional Group Properties*

For assessment of rank of mildly metamorphosed coals, the measurement of oxygenated functional groups has been suggested.[138] This can be carried out through several direct analysis procedures and by measuring the yield of carbon dioxide and water formed by decomposition.[139,140] Because to a large extent metamorphism results initially in deoxygenation, the concept appears to have merit, although it has not been accepted as a standardized technique.

3.3.1.5 *Calorific Value*

The calorific value of the organic fraction increases progressively with the severity of metamorphism (largely as a result of oxygen loss), hence is widely used as a measure of rank (there is a slight drop in calorific value at the anthracitic stage). The ASTM Classification of Coals by Rank[1] employs the calorific value expressed on a mineral-free and bed-moist basis to differentiate the low-rank coals. This parameter (moist, mineral-free Btu/lb) has the advantage of reflecting for progressively higher rank coals both the increased calorific value of the organic material and the decreased moisture-holding capacity.

3.3.1.6 *Reflectance of Vitrinite*

The rank-measuring parameter receiving the most interest in the last several decades is the microscopically determined reflectance of vitrinite (ASTM D2798[5]; Soviet GOST 12113). The greatest advantage of this technique (described in Section 3.2.1.6) is that it measures a rank-sensitive property on only one petrographic constituent; therefore it is applicable even where the coal type is unusual or atypical. With increasing rank, both aromaticity and molecular cluster size increase. These properties influence the index of refraction (n) and index of absorption (k) respectively, and as discussed previously, the

[133] H. Pichler, A. Stoeckel, and K. H. Lauer, *Gas Wasserfach,* **109**(35), 959–965 (1968).
[134] W. Duhameau, Bull. Tech. Houille Deriv. INICHAR, No. 37, 1968, pp. 1–20.
[135] A. R. Cameron and T. F. Birmingham, Geol. Survey Canada Paper No. 71-8, (1971), pp. 1–20.
[136] ASTM Standards, 1977, Part 27, Designation D1412-74, Standard Test Method for Equilibrium Moisture of Coal.
[137] E. I. Rukin and T. P. Gorskaya, *Khim. Tverd. Topl.,* **8**(5), 28–30 (1974).
[138] N. N. Ulanov and I. D. Druchina, *Khim. Tverd. Topl.,* **8**(2), 25–30 (1974).
[139] N. N. Ulanov, *Khim. Tverd. Topl.,* **8**(5), 3–9, (1974).
[140] V. I. Zabavin, *Khim. Tverd. Topl.* (5), 38–43 (1968); *C. A.,* **70,** 13259 (1969).

reflectance increases as a function of increased n and k. Excellent correlations have been established between reflectance and volatile matter[141] and the ratio of atomic hydrogen to carbon.[13] In the latter study the relationship

$$\%R_0 = 15.64 \exp\left(-3.6\frac{H_{atom}}{C_{atom}}\right) \quad (3.6)$$

was found for a series of hand picked vitrinites. The authors suggested that the specific reflectance ($\%R_0$/density) might be an even better parameter because it extrapolates to the value for graphite at 100% carbon (H/C = 0). However routine measurement of the density of vitrinite from a sample would be difficult.

The reflectance technique appears to be one of the more universally definitive rank-measuring tests, as evidenced by its applicability to petrographically dissimilar Gondwana (Southern Hemisphere) and European coals.[142] Besides correlations with volatile matter and H/C ratio, reflectance has been correlated with calorific value, free swelling index, yields of coke, light oil, gas, and tar from coking, and coke gas properties.[143] The technique can be applied to organic fragments isolated from sedimentary rocks, yielding values comparable to those obtained from stratigraphically adjacent coals.[144] Reflectance is not affected by hydrofluoric acid used to dissolve silicates and free the vitrinite particles from host rocks,[130] and it is affected in only a minor way by low-temperature oxidation, which makes the technique particularly applicable for assessing the rank of weathered, outcrop coals[143,145] (see also Section 3.3.3.).

Although the reflectance is controlled predominantly by the stage of regional metamorphism, several other phenomena influence it. The reflectance of vitrinites from a single seam spans a narrow range of values as a result of minor differences in vitrinite properties. Peat swamp redox potential, source plant types, and extent of diagenetic (biochemical) decay influenced the nature of the vitrinite progenitors and thereby affected the path of reflectance through the rank series.[146] Brown et al.[34] suggest that the reflectance of vitrinite from vitrain bands is the best guide to rank, but Thompson and Benedict[147] caution that the reflectance of pseudovitrinite (which often occurs as lenticular bands) is a poor guide to rank. Since the reflectance can be influenced by pressure during heating[148] and by the duration of heating,[149] reflectance relates to metamorphism in a complex fashion. Reflectance anisotropy is affected more by pressure than by the heat that is the principal metamorphic agency; therefore bireflectance may be an additional clue to the nature of the metamorphic process in a specific area. The use of reflectance is best confined to coals higher in rank than lignites, because the early metamorphic changes, though often physicochemically significant, exerted little influence on reflectance.

3.3.1.7 *Correlation of Properties*

The relationships between the various rank-measuring parameters are summarized in Fig. 3.18; note that the correlating tie points are only approximate. Each of the parameters is most useful within a specific range of the rank sequence as shown by the cross-hatched bars. The "scientific rank stages" as defined in Fig. 3.18 are based principally on the

[141] J. T. McCartney and M. Teichmüller, *Fuel,* **51**(1), 64–68 (1972).

[142] D. Chandra and G. H. Taylor, in *Textbook of Coal Petrology,* E. Stach, Ed., Gebrüder Borntraeger, Berlin and Stuttgart, 1975.

[143] L. G. Benedict and W. F. Berry, "Further Applications of Coal Petrography," in Am. Chem. Soc. Adv. Chem. Ser. No. 55, (1966), 577–601.

[144] A. Hood and J. R. Castaño, *United Nations ESCAP, CCOP Tech. Bull.,* **8,** 85–118 (1974).

[145] D. Chandra, *Fuel,* **41,** 185–193 (1962).

[146] I. I. Ammosov, *Khim. Tverd. Topl.,* **8**(4), 3–12 (1974).

[147] R. R. Thompson and L. G. Benedict, "Vitrinite Reflectance as an Indicator of Coal Metamorphism for Cokemaking," Geol. Soc. America Spec. Paper No. 153, (1974), pp. 95–108.

[148] D. Chandra, *Fuel,* **44**(3), 171–175 (1965).

[149] H. A. W. DeVries, P. J. Habets, and C. Bokhoven, *Brennstoff-Chemie,* **49**(4), 105–110 (1968).

[150] M. Teichmüller and R. Teichmüller in *Textbook of Coal Petrology,* E. Stach, Ed., Gebrüder Borntraeger, Berlin and Stuttgart, 1975; "Geological Aspects of Coal Metamorphism," in *Coal and Coal Bearing Strata,* D. G. Murchison and T. S. Westoll, Eds., Oliver and Boyd, Edinburgh, 1968, pp. 233–267.

[151] B. C. Parks and H. J. O'Donnell, *Petrography of American Coals,* U.S. Bur. Mines Bull. No. 550, 1956.

[152] A. Hood and J. R. Castaño, *United Nations ESCAP, CCOP Tech. Bull.,* **8,** 85–118, 1974.

[153] W. H. Ode and W. H. Frederic, *The International Systems of Hard-Coal Classification and Their Application to American Coals,* U.S. Bur Mines Rep. Invest. No. 5435, 1958.

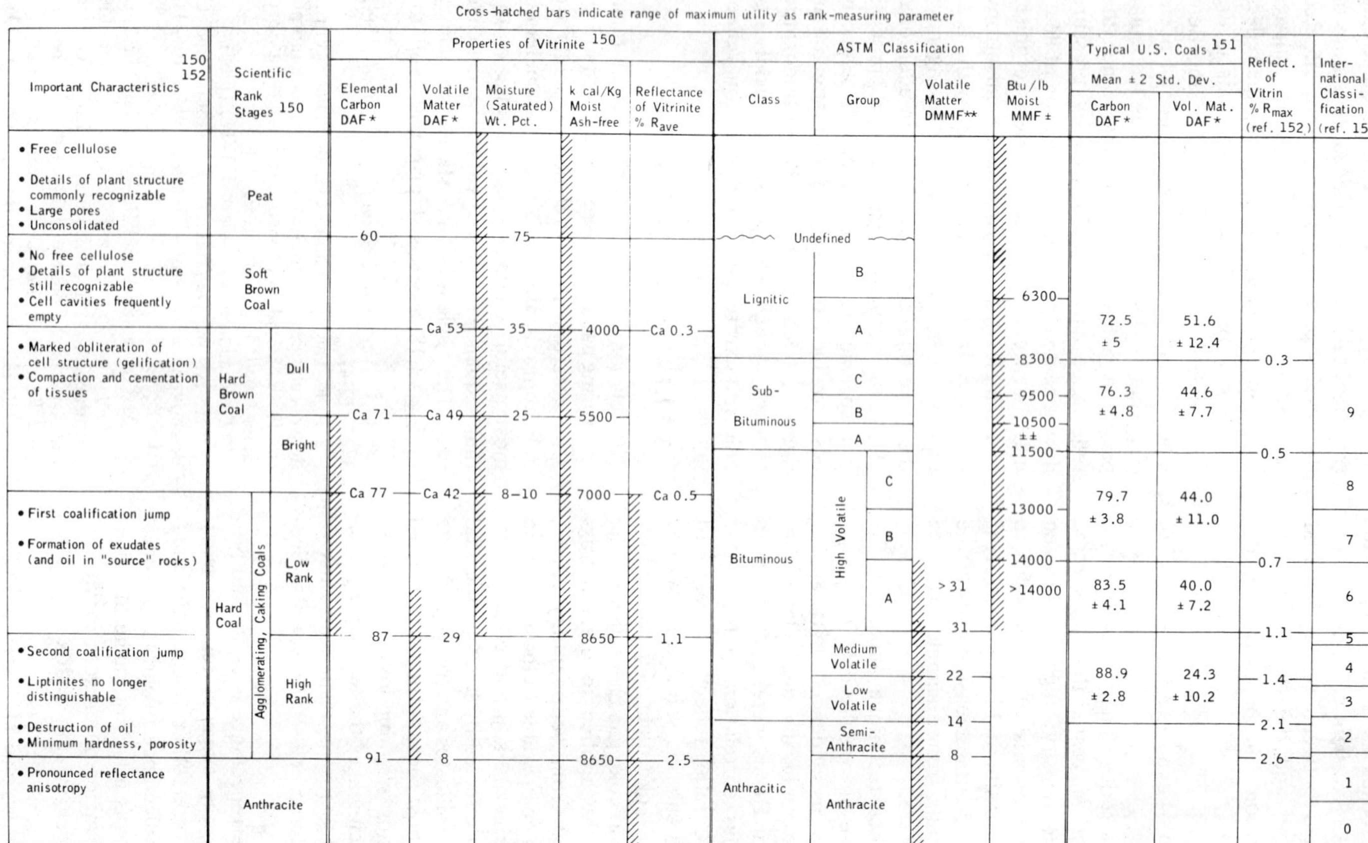

* Dry, ash-free basis, weight percentage
** Dry, mineral-free basis, weight percentage
± Mineral-free, but with saturated pore moisture content
±± Bituminous if agglomerating, subbituminous if non-agglomerating

Figure 3.18 Correlation of coal ranks (approximate).

German rank classification system.[150] On this basis, low-rank coals are called brown coals; they exhibit no tendency to fuse (cake, agglomerate) when heated. Intermediate- (hard or bituminous) rank coals do fuse and swell, forming a vesicular coke, the physical properties of which vary with rank and type. The high-rank coals (anthracites) are, like the low-rank coals, nonagglomerating, but for a different reason. In the low-rank coals high oxygen content and a virtual absence of mobile, bituminous substances (bitumen) preclude fusion, whereas in high-rank coals bitumen has been metamorphically destroyed and the molecular structure has been highly polycondensed.[154] Correlations between reflectance and some chemical properties of vitrinites are plotted in Fig. 3.19. References to some of the published correlations between the principal rank-measuring parameters and other properties are listed in Table 3.6.

[154] R. C. Neavel, "Coal Plasticity Mechanism: Inferences from Liquefaction, Studies," *Proc. Coal Agglom. Conversion Symp.*, West Virginia Geol. Econ. Survey, Morgantown, May 5–6, 1975 (published 1976).

3.3.2 Causes of Metamorphism

The agency or agencies that cause metamorphism operate on a regional basis, and with only rare exceptions, coal rank is relatively uniform or else changes only gradually with geographic-geologic distance. Only four agencies have been seriously considered as possible causes of coal metamorphism, namely, pressure, heat, tectonism (shear), and gamma-ray irradiation. Artificial coalification experiments should have resolved any controversy about the relative importance of each of these processes, but in general such experiments have been far from conclusive. The one factor that cannot be duplicated in the laboratory is the long time span over

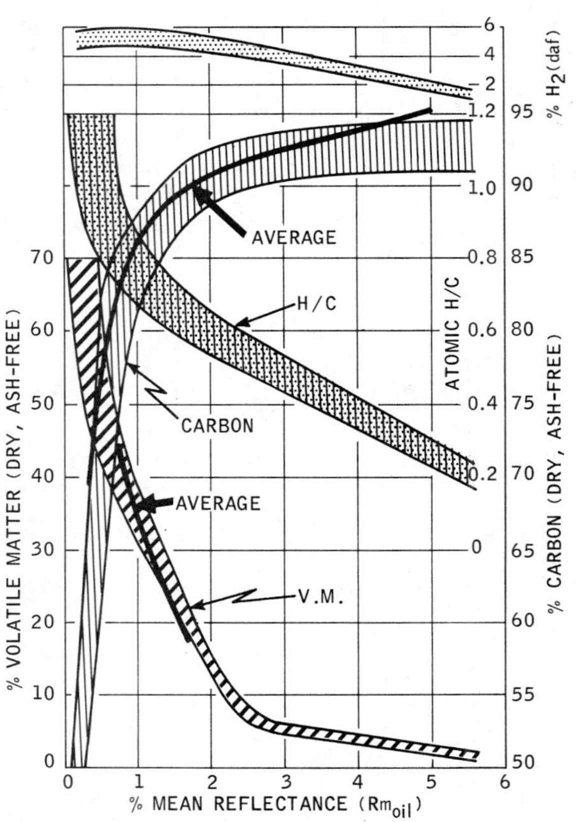

Figure 3.19 Relationship between vitrinite reflectance and different chemical rank parameters.[129] Average lines for volatile matter and for carbon from van Krevelen.[49]

Table 3.6 Correlations of Rank-related Parameters Found in Literature[a]

	Elemental Carbon	Reflectance	Volatile Matter	Moisture
Reflectance	49, 155			
Volatile matter	156, 157	141, 143, 147, 158, 159		
Moisture	156	159, 160		
Hydrogen	156	159		
Oxygen	49, 156			
Functional groups	49			
Calorific value	156			161
Aromaticity	49			
Density	49			
Hardness	49, 162			
Surface area	49, 160, 163	160	160	164
Porosity	49		165	160
Atomic H/C	166	141, 159		
Swelling	156		156	

[a] Numbers designate footnoted references.

which the natural metamorphic process operated. Hence present theories about the process of metamorphism have been generated principally from field studies.

In a classic series of papers, Patteisky and M. Teichmüller[157] presented data relating vitrinite composition to the depth of burial for more than 300 samples secured from boreholes. Coals ranked from lignite to anthracite were encountered, and data from these coals served to establish the correlations shown in Fig. 3.19. That rank increases with depth of burial has now been unequivocally established.[14,158,167–171] That the fundamental cause is the increased temperature associated with increased depth seems to be well established. The best arguments favoring this hypothesis are those of Marlies and Rolf Teichmüller as presented thoroughly in 1966[172]; however they are more readily accessible in some of the authors' other works.[159,170,171,173] In addition, Damberger has written several excellent syntheses of the relationships between rank

[155] M. Teichmüller, *Erdoel Kohle, Erdgas, Petrochem.*, **24**, 69–76 (1971).

[156] H. R. Brown, G. H. Taylor, and A. C. Cook, *Fuel*, **43**(1), 43–54 (1964).

[157] von K. Patteisky and M. Teichmüller, *Brennstoff-Chemie*, **41**(3), 79–84; **41**(4), 97–104; **41**(5), 133–137 (1960).

[158] A. S. Stankevich, A. E. Proskuryakov, E. G. Raskina, and D. N. Sidel'nikov, *Khim. Tverd. Topl.*, **8**(6), 22–29 (1974).

[159] M. Teichmüller, in *Textbook of Coal Petrology*, E. Stach, Ed., Gebrüder Borntraeger, Berlin and Stuttgart, 1975.

[160] J. Thomas Jr. and H. H. Damberger, Illinois State Geol. Survey Circ. No. 493, 1976, pp. 1–38.

[161] H. Damberger, *Econ. Geol.*, **66**, 488–494 (1971).

[162] A. Battacharyya, *Sci. Cult.*, **34**(5), 221–222 (1968).

[163] H. Gan, S. P. Nandi, and P. L. Walker, Jr., *Fuel*, **51**(4), 272–277 (1972).

[164] J. Thomas and H. H. Damberger, Illinois State Geol. Survey Circ. No. 493, (1976), 1–38.

[165] S. P. Nandi and P. L. Walker, Jr., "Diffusion of Argon from Coals of Different Rank," in Am. Chem. Soc. Adv. Chem. Ser. No. 55, (1966), pp. 379–385.

[166] V. I. Baryshev, *Khim. Tverd. Topl.*, **8**(5), 47–52 (1974).

[167] H. Damberger, *Brennstoff-Chemie*, **49**(3), 73–77 (1968).

[168] P. A. Hacquebard and J. R. Donaldson, *Rank Studies of Coals in the Rocky Mountains and Inner Foothills Belt, Canada*, Geol. Soc. America Spec. Paper No. 153, 1974, pp. 75–94.

[169] M. Sasaki, *Nenryo Kyokai-shi*, **46**(488), 864–871 (1967).

[170] M. Teichmüller and R. Teichmüller, "Diagenesis of Coal," in *Diagenesis in Sediments*, G. Larsen and G. V. Chilinger, Eds., Elsevier, Amsterdam, 1967, pp. 391–415.

[171] M. Teichmüller and R. Teichmüller, "Geological Causes of Coalification," in Am. Chem. Soc. Adv. Chem. Ser. No. 55, 1966, pp. 133–155, 213–217.

[172] M. Teichmüller and R. Teichmüller, *Z. Dtsch. Geol. Ges.*, **117**(1), 243–279 (1966).

[173] M. Teichmüller and R. Teichmüller, "Geological Aspects of Coal Metamorphism," in *Coal and Coal Bearing Strata*, D. G. Murchison and T. S. Westell, Eds., Oliver and Boyd, Edinburgh, 1968, pp. 233–267.

and thermal processes as inferred from mid-continent and eastern U.S. coals.[174,175]

Principally on the basis of the data collected from deep boreholes in Europe, the Teichmüllers conclude that geothermal gradients typical of most areas in the earth (about 1°C/25 m of depth) were adequate to create high-rank coals. Where geothermal gradients were greater—for instance, above high-temperature magmatic zones—coal ranks are higher for a given depth. Initially, burial resulted only in the compaction of peat into brown coal, but with prolonged exposure to temperatures of about 50–200°C, chemical changes occurred. These especially involved deoxygenation, and resulted in decreased volatile matter yields and "increased" carbon and hydrogen contents (hydrogen less dramatically, since it was lost—along with the oxygen—as water, whereas carbon was evolved as carbon oxides). Though these chemical changes also occur during mild heating of peat or lignite in the laboratory, such processing does not produce a material with the caking or agglomerating properties typical of bituminous rank coals. Thus heat is necessary, but it is not sufficient. Apparently pressure serves to retain relatively low molecular weight hydrocarbons (bitumens) that escape as volatile products if peat or low-rank coal is heated in an unconfined system. Bitumen appears to be a necessary component of bituminous coals that do become plastic when heated.[154] It has been proposed that pressure may even promote depolymerization to produce bitumen.[176] It has also been suggested that pressure retards chemical alteration.[170] On the basis of thermodynamic principles it is difficult to envision how pressure could promote (or retard) depolymerization; however it could alter the course of the reactions, especially by retaining "reactants" in the system.

Following the formation of caking bituminous coals from lignite, greater metamorphic severity leads to the production of noncaking, high-carbon coals (anthracite). Predominantly, the reaction is one of dehydrogenation by way of methane production. At about the level of 87% carbon (in vitrinite), a position in the rank scale referred to as the second coalification jump, liptinite reflectance increases to (and possibly beyond) that of associated vitrinite, largely because of dehydrogenation.[177] Methane formed by the dehydrogenation of liptinite and vitrinite is often found in the coal pores and in associated sedimentary strata, its derivation being inferred from the $^{13}C/^{12}C$ ratio.[178,179]

Folding of the coal-containing strata, and associated tectonic processes (i.e., shear forces), appear to exert little influence on metamorphic development. This conclusion, based on several lines of evidence,[180,181] is significant in light of the common geographic congruence of folding and high rank that has been observed, leading to the presumption of a causal relationship. However, as the Teichmüllers point out,[170] burial is commonly deepest and later folding is commonly most severe in geosynclinal foredeeps; thus although folding and high rank may be statistically correlative, there appears to be no causal relationship.

The duration of heating during metamorphism also influences the final rank; where peat/coal deposits were not buried deeply enough to have been heated, however, time alone was inadequate to exert any metamorphic effect.[170] Friedel et al. demonstrated that coallike materials were formed from cellulose and sawdust when they were heated to 200°C in a confined system for 2 years, whereas

[174] H. Damberger, *Econ. Geol.*, **66**, 488–494 (1971).

[175] H. Damberger, *Coalification Patterns of Pennsylvanian Coal Basins of the Eastern United States*, Geol. Soc. America Spec. Paper No. 153, 1974, pp. 53–74.

[176] S. Karaivanov and Z. Petrounova, *Fuel*, **46**(4–5), 239–245 (1967).

[177] E. Stach, "Basic Principles of Coal Petrology," in *Coal and Coal Bearing Strata*, D. G. Murchison and T. S. Westoll, Eds., Oliver and Boyd, Edinburgh, 1968, pp. 3–17.

[178] U. Columbo, F. Gazzarrini, R. Gonfiantini, G. Kneuper, M. Teichmüller, and R. Teichmüller, "Das C^{12}/C^{13} Verhältnis von Kohlen und Kohlenbürtigem Methan," *Proc. 6th Int. Cong. Strat. Geol. Carbonif.*, Sheffield, Vol. 2, (1967), pp. 557–574.

[179] H. U. Friedrich and H. Juentgen, *Erdoel, Kohle, Erdgas, Petrochem. Brennst.-Chem.*, **26**(11), 636–639 (1973).

[180] H. Juentgen, M. Teichmüller, and D. Juendorf, *Brennstoff-Chemie*, **50**(2), 40–45 (1969).

[181] L. E. Shterenberg, K. T. Cherkinskaya, I. L. Ettinger, M. M. Ellison, and V. I. Kasatvchkin, *Dokl. Akad. Nauk SSSR*, **180**(1), 214–217 (1968); *C. A.*, **69**, 29162 (1968).

little change was observed after only short heating.[182]

All these observations are consistent with classical thermodynamic and kinetic theories. Only the time scale is different from that commonly employed experimentally, and so lower temperatures exerted much greater effects than commonly are observed in experimental situations. The reactions may also be catalyzed by minerals in the coal seams. According to Hood and Castãno,[144] metamorphic reactions seem to conform to the standard kinetic dictum that reaction time is halved for each 10°C temperature increase. One author contends[183] that exact kinetic expressions for upranking cannot be defined for each of the many coal-bearing areas, but several investigators note that the reactions do seem to follow first-order reaction kinetics, at least in tectonically uncomplex areas.[184]

Abnormal heat sources such as intrusive volcanic magma (lava) have altered coals in some places. The distance over which intrusive heat is effective varies with the rank already achieved by the coal,[185] but does not seem to exceed about 2 times the thickness of the intrusive body.[184] Such localized alterations lead to abnormal coals by virtue of excessively high temperatures acting over relatively short times. As a generalization, the effect is more similar to a charring or coking process that might be conducted in the laboratory than it is to the molecular reorganization that results from typical (viz., regional) metamorphism.[186,187] Occasionally localized heating by igneous rocks has been sufficiently slow and under sufficient confining pressure to produce typical metamorphic effects rather than charring.[188]

Irradiation from radioactive decay can alter coal properties,[177,189] but this phenomenon occurs only very locally (e.g., as halos around radioactive minerals).

Different coal types are influenced somewhat differently by equivalent metamorphic processes. Coals deposited under more calcareous conditions appear to be higher in bitumen content for a given rank.[190] Data from Japanese coals do not plot on regressions that depict the progressive metamorphic changes of typical coals from other areas. Density, H/C, and aromaticity (f_a) especially fail to plot against carbon content in the same way as the majority of the world's coals.[191] It is possible that other examples of such hydrogen-rich coals will be found as coal characterization efforts become more sophisticated.

3.3.3 Oxidation

The vitrinite in coals of high-volatile bituminous rank and lower is susceptible to oxidation at relatively mild (i.e., ambient) conditions. Although still imperfectly understood, the mechanism of mild oxidation is generally assumed to involved initially the chemisorption of oxygen to the coal throughout the internal pore surface (if the surface is "available," i.e., not blocked by water in the pores). Even the chemisorption of less than 1% by weight of oxygen can significantly affect the properties and usefulness of some coals.

The effects of oxidation have been found to be deleterious in several areas of coal utilization. For instance, oxidation reduces the yield of liquids that can be obtained in coal liquefaction processes.[192] Increasing interest has recently been focused on the ef-

[182] R. A. Friedel, J. A. Queiser, and H. L. Retcofsky, *Prepr., Am. Chem. Soc., Div. Fuel Chem.*, **13**(1), 1–7 (1969).

[183] V. N. Nagornyi and Yu. N. Nagornyi, *Khim. Tverd. Topl.*, **8**(4), 37–44 (1974).

[184] W. G. Dow, "The Effects of Time and Temperature on the Maturation of Organic Matter and Generation of Petroleum," *Symp. Thermal Alteration Organic Materials,* 8th Annual Meeting American Association of Stratigraphic Palynologists, Houston, October 29, (1974), p. 9.

[185] O. I., Gavulova, *Vop. Metamorf. Uglei Epigeneza Umeschayushchikh Porod.*, 163–182 (1968).

[186] J. M. Schopf and W. E. Long, "Coal Metamorphism and Igneous Associations in Antarctica," in Am. Chem. Soc. Adv. Chem. Ser. No. 55, (1966), pp. 156–195.

[187] N. Schapiro and R. J. Gray, "Physical Variations in Highly Metamorphosed Antarctic Coals," in Am. Chem. Soc. Adv. Chem. Ser. No. 55, 1966, pp. 196–217.

[188] A. C. Cook and R. G. Wilson, *Australas. Inst. Min. Metall. Proc.*, No. 232, 27–39 (1969).

[189] J. Jedwab, "Significance and Use of Optical Phenomena in Uraniferous Caustobioliths," in Am. Chem. Soc. Adv. Chem. Ser. No. 55, (1966), pp. 119–131.

[190] See Ref. 81, p. 28.

[191] S. Fujii and H. Tsuboi, *Fuel,* **46**(5), 361–366 (1967).

[192] R. C. Neavel, *Fuel,* **55**(3), 237–242 (1976).

fects of oxidation on the coking properties of vitrinite. Even relatively mild oxidation can reduce or destroy the capability of some vitrinites to become plastic; this phenomenon is especially evident in vitrinites from low-rank coals.[193,194] Goodarzi and Murchison[195] conducted an especially noteworthy study of this phenomenon, employing microscopic observation of cokes formed from oxidized coals.

Heating coal in free oxygen results in exothermic reactions that pyrolyze the coal, creating high reflectance "skins" on coal particles.[196] However low-temperature oxidation either decreases reflectance slightly[197] or has a negligible effect on this property.[198] Therefore reflectance of vitrinite can be used as a guide to the rank of weathered or oxidized coals, and available correlations of rank-sensitive properties with reflectance can then be used to infer properties of the coal in an unoxidized state. No other analytical parameter appears to be as useful for assessing the rank of oxidized coal.

Benedict and Berry[143] concluded that for coals with vitrinite reflectance of between 0.4 and 1.6%, oxidation could be detected microscopically by several criteria. The volatile matter (VM) content of fresh (unoxidized) coals was found to correlate (correlation coefficient, $r = 0.96$) with the reflectance according to Equation 3.7:

$$\%\ \text{VM (dry, ash-free)} = 60.65 - (25.14 \times \%\ R_0) \qquad (3.7)$$

Any coals that diverge significantly from this relationship (and for which there does not seem to be a valid explanation in terms of coal type) should be suspected of being oxidized. To estimate the properties of such coals in their unoxidized state, Benedict and Berry suggested projecting a line from the datum point (defined by VM and R_0) to the above-defined regression such that it intersects the regression line at a right angle. The intersection could be assumed to represent the VM and R_0 of an unoxidized sample of that coal. No justification for this suggestion was presented, nor has its validity been confirmed.

Gray et al.[193] discuss thoroughly the effects and detection of oxidation. In contrast to fresh coals, they found the following effects of oxidation: (1) free-swelling index decreases, (2) froth flotability decreases, (3) solidification temperature in Gieseler plastometer increases, (4) the pH of a water slurry decreases, (5) solubility in aqueous alkali increases, (6) oxidized coal acquires a negative charge when it slides down a grounded plate, (7) oxidized coals absorb Safranio "O" (red stain) from a solution in potassium hydroxide, then appear green by reflected light microscopy.

3.4 CLASSIFICATION OF COAL ACCORDING TO GRADE

The grade of a coal establishes its economic value for a specific use; it is a complex function of the interaction of the process system and both rank and type. Since most commonly referenced coal classifications are utilization oriented, they are classifications according to grade. A typical example is the international classification, discussed thoroughly by van Krevelen and compared to the ASTM classification by Ode and Frederic.[154] The classification is based on a three-digit code in which the first digit refers to the class [sic] of the coal defined by volatile matter or calorific value. (In reality, the first digit is principally an indication of rank; higher digits refer to lower ranks.) The second digit indicates the group [sic] as defined by *caking* properties evaluated either by the free-swelling index or by the Roga index. The third digit defines a subgroup based on *coking* properties (dilatometer or Gray-King assay).

[193] R. J. Gray, A. H. Rhoades, and D. T. King, AIME, Soc. Min. Eng., Prepr. No. 74F63, 1974.

[194] R. J. Gray and J. P. McGinnes, "Effect of Oxidized Coal on Coke Plant Operations," presented to Eastern States Blast Furnace and Coke Oven Operators, February 5–6, 1976.

[195] F. Goodarzi and D. G. Murchison, *Fuel*, **55**(2), 141–147 (1976).

[196] D. M. Mason and F. C. Schora, Jr., *Am. Chem. Soc., Div. Fuel Chem., Prepr.*, **10**(4), 140–150 (1966).

[197] R. J. Gray and K. C. Krupinski, "Use of Microscopic Procedures to Determine the Extent of Destruction of Agglomerating Properties of Coal," *Proc. Coal Agglom. Conversion Symp.*, West Virginia Geol. Econ. Survey, Morgantown, May 5–6, 1975 (published 1976).

[198] D. Chandra, in *Textbook of Coal Petrology*, E. Stach, Ed., Gebrüder Borntraeger, Berlin and Stuttgart, 1975, pp. 159–165.

As can be seen, this classification is strongly oriented toward coke making.

Other than this so-called international system and the ASTM classification according to rank, there are no classification schemes that are truly universal. In fact, there is really no classification that allows assignment of a coal to a neat, all-encompassing pigeonhole from which all technological properties could be inferred. Therefore, except for the rank classification and Seyler's graphic correlation of properties,[199] all so-called classifications are little more than an arrangement of coals according to one or two properties of specific technological interest. In other words, most coals are "classified" according to nothing more sophisticated than how well they meet the criteria for some specific use. The following sections of this chapter discuss the relative impacts of the scientific classifying parameters of rank and type on the grade for specific processes and uses.

3.4.1 Mining and Preparation

Mining and crushing operations are affected by hardness, friability, and toughness, properties that in turn are influenced by both coal rank and coal type. The hardness of vitrinite, as measured by Vickers or Knoop microhardness tests, is related to rank in a complex fashion, achieving minimal values at about 70 and 90% carbon.[49] The Hardgrove Grindability Index (HGI), a measure of the work required to grind coal, increases with increasing rank to about 1.4% vitrinite reflectance (23% volatile matter, approximately 90% C), and then decreases (note that higher HGI values indicate less work required). Superimposed on this fundamental influence of rank on hardness and HGI are effects of coal type.[200] HGI increases with increasing mineral matter content in the friable, low-volatile bituminous rank coals, but is not markedly affected by mineral content in high-volatile coals where the organic material itself is tough. On the other hand, higher macrinite, micrinite, and liptinite concentrations make high volatile coals tougher (lower HGI). It is reported that low-volatile Gondwana coals with less than 40% vitrinite are also difficult to grind.[201]

There is an interaction of coal types with different mechanical miners. Layers especially rich in pyrite exert significant shock on the miner. One study showed that the current required by a mechanical miner varied irregularly as the head progressed from top to bottom of the seam.[202] These variations reflected the varying properties of the coal types comprising the different benches of the seam.

Fracturing, either at the face or during subsequent crushing, is affected significantly by the textural arrangement of petrographic components. Breakage along the bedding plane (horizontal planes of lamination) occurs preferentially at layers of fusain (often only thin films). Breakage along vitrain bands is essentially proportional to their abundance in the coal, whereas breakage along bands of attrital coal is less frequent than would be expected from their abundance.[203] Where they are present, thin clay laminae often serve as planes of breakage, and severe disintegration can result where they are abundant.[204]

Differential hardness of various types of coal found in different layers can result in the formation of uniquely different kinds of particles in specific size ranges,[205] a phenomenon that has been exploited in a procedure to separate vitrinite-rich and vitrinite-poor fractions.[206,207] Fusinite and vitrinite tend to accumulate in the finer fractions, often being

[199] B. C. Parks, "Origin, Petrography, and Classification of Coal," in *Chemistry of Coal Utilization,* Supplementary Volume, H. H. Lowry, Ed., John Wiley and Sons, New York, 1963, pp. 1–34.

[200] A. A. Terchick, R. W. Shoenberger, B. Perlic, and L. F. DeRusha, *Prepr., Am. Chem. Soc., Div. Fuel Soc. Chem.* 7(2), 95–109 (1963).

[201] A. C. Cook, personal communication, University of Wollongong, New South Wales, Australia, 1977.

[202] J. T. Peters, N. Schapiro, and R. J. Gray, *Trans. AIME, Soc. Min. Eng.,* **223**(1), 1–6 (1962).

[203] J. A. Harrison, *Trans. AIME, Soc. Min. Eng.,* **226,** 346–357 (1963).

[204] R. C. Neavel, *Trans. AIME, Soc. Min. Eng.,* **217,** 438–442 (1960).

[205] W. Beimann, H. S. Auvil, and D. C. Coleman, "High-Temperature Carbonization," in *Chemistry of Coal Utilization*, Supplementary Volume, H. H. Lowry, Ed., John Wiley and Sons, New York, 1963, p. 473.

[206] C. R. Jackson and M. O. Holowaty, *Iron Steel Eng.,* **46**(1), 80–83 (1969).

[207] V. I. Skorokhodov and D. V. Sycheva, *Koks Khim.,* No. 6, 3–8 (1968); *C. A.,* **69,** 5342 (1968).

enriched in a narrow size range, whereas liptinite—because it imparts a toughness to the coal—is sometimes found concentrated in coarser sizes.[203] In western Canadian coals vitrinite concentrated in particles larger than ⅛ in. (3.2 mm), whereas fusinite (contrary to usual experiences and possibly because of mineralization), concentrated in intermediate sizes.[208] Significant variations in mineral content are commonly encountered in different particle sizes.

During cleaning or coal preparation, microscopic observations often can serve as an aid to understanding the response of a coal, thereby suggesting methods to improve efficiency. For instance, controlled pulverization guided by microscopic observations has been found useful for optimum liberation of pyrite.[209] In one study,[210] sink-float removal of pyrite was directly related to pyrite grain size as determined microscopically (correlation coefficient, $r = 0.89$ for coal particles <1.4 mm; $r = 0.79$ for <38 mm). Note that whenever size measurements are carried out microscopically on cross-sectioned particles, a correction factor must be employed to compensate for sections that do not pass through the middle of the particle.

Fusinite, vitrinite, dull attrital coal (durain),[211] and oxidized coal[193] are reported to be separable in electrostatic cleaning operations. The final moisture content of a wet-cleaned coal is a function of both the extraneous water and the water in the pores, which is rank-related. For lower rank coals, the tendency toward explosivity of dust is reported to be positively correlated with the sum of the liptinite, vitrinite, and pyrite contents.[212] The tendency for spontaneous combustion in storage piles can be a serious problem with low-rank coals.[213] It has been reported that spontaneous combustion tendency is correlative with fusinite content.[214] Because of the intrinsically low reactivity of fusinite, however, it seems likely that associated properties such as pyrite in fusinite or physical decrepitation along fusain layers is responsible.

3.4.2 Combustion

Spicer[215] presented a comprehensive survey of coal characteristics important to combustion. The properties related to rank and type that he considers important can be categorized as follows:

1. *Calorific value:* Directly related to rank, indirectly to mineral (ash) content; higher in liptinite-rich coal types.
2. *Grindability:* Related to rank and to type as discussed in the previous section.
3. *Combustibility:* Controlled by volatile matter and caking characteristics, which are principally rank-related. Swelling, which can affect combustibility, has been reported to be independent of rank in furnaces,[216] whereas microscopic studies of particles in a hot stage indicate that vitrinite ignites faster than fusinite, and resinite expands explosively.[217]
4. *Ash properties:* Related principally to mineral composition, which was primarily controlled by deposition.

Benedict and Berry[143] developed a nomograph relating the design capacity for stokers, in pounds per hour per square foot of grate, to the free-swelling index, ash fusion properties, particle size, and rank (assessed by vitrinite reflectance). Combustion tempera-

[208] A. R. Cameron and J. C. Botham, "Petrography and Carbonization Characteristics of Some Western Canadian Coals," in Am. Chem. Soc. Adv. Chem. Ser. No. 55, 1966, pp. 564–576.

[209] R. J. Gray, N. Schapiro, and G. D. Coe, *Trans. AIME, Soc. Min. Eng.* **226** (June 1963).

[210] R. A. Glenn and R. D. Harris, "Liberation of Pyrite from Steam Coals," *American Power Conf.*, Chicago, March 21–23, Printed by Bituminous Coal Research, Inc., 1961.

[211] F. Fraas, U.S. Bur. Mines Bull. No. 603, 1962, pp. 125–127.

[212] G. V. Nenarokov, G. N. Krikunov, and E. G. Kessler, *Rasved. Okhr. Nedr.*, **34**(10), 13–14 (1968); *C. A.*, **70**, 13277 (1969).

[213] L. E. Paulson, S. A. Cooley, and R. C. Elman, U.S. Bur. Mines Inf. Circ. No. 8650, (1974), pp. 49–75.

[214] G. N. Krikunov, L. A. Polyanskaya, T. K. Svanidze, and V. F. Yakovenko, *Khim. Tverd. Topl.*, No. 3, 48–54 (1969).

[215] T. S. Spicer, Am. Inst. Min. Metall. Pet. Eng., San Francisco, February, Prepr. No. TP59F85, 1959.

[216] R. H. Essenhigh and G. C. Yorke, *Prepr., Am. Chem. Soc., Div. Fuel Chem.*, **8**(2) 172–184 (1964).

[217] M. J. Shibaoka, *J. Inst. Fuel*, **42**(337), 59–66 (1969).

ture is reported to be independent of rank, but sensitive to ash and moisture contents.[218]

In pilot-scale combustion tests of western Canadian coals, it was found that combustion efficiencies of the coals were inversely related to the inert maceral contents of the coals. Fusinite, semifusinite, and oxidized vitrinite persisted through the flame in a partially reacted form, producing high levels of unburned carbon in the fly ash.[219]

3.4.3 Pyrolysis (Excluding Coke-Making) and Plasticity

When coal is heated to about 350°C and higher in the absence of free oxygen, it decomposes into volatile substances and a residual char or coke. Industrial processes that capitalize on this pyrolytic phenomenon are generally designed either to maximize the production of liquid products (condensable tars and oils) or to optimize the quality of the coke product. The value of coals for pyrolytic production of liquids is significantly influenced by two characteristics; namely, caking properties and yield pattern.

In most pyrolytic processes, coal is fed in a crushed form to a hot reactor (often containing a fluidized bed of char). When heated rapidly, the vitrinite of some coals softens and particles can agglomerate or "cake" into a coherent mass, hence such coals are referred to as plastic, agglomerating, or caking coals (see Neavel[154] for a discussion of the mechanism of plastic development). Unless special precautions are taken when caking coals are fed to a hot reactor, the reactor inlet plugs or the fluidization of the bed is adversely affected. Since caking or agglomeration can also affect the feeding of coal to gasification reactors, the discussion is relevant to that area as well.

Gray and Krupinski[197] conducted tests in a small fluidized bed at 750°C and found that potentially deleterious agglomeration characteristics were confined to coals in which the vitrinite reflectance ranged from 0.6 to 1.9% (mean maximum). Preoxidation at a nominal bed temperature of 400°C eliminated agglomeration tendencies for vitrinite with reflectance of 0.7%, but not for vitrinites with R_0 of 0.9% or above. In the higher rank (higher reflectance) coals, microscopically detectable skins of high reflectance were formed on coal particles oxidized at 400°C in air, but these skins ruptured when the coal was subsequently heated to 750°C, allowing plastic material to exude. Mason and Schora[196] reported similar results from a similar microscopically oriented study. It is likely that the high-reflectance "skin" results from pyrolysis caused by high temperatures on the particle surface as a result of heat generated by oxidation. Reflectance is known to increase progressively with the temperature of pyrolysis when oxygen is excluded, and this property can be used as a measure of the temperature to which the coal was exposed.[220] It has also been found that optical changes occur when coal is heated for a long period (up to 32 weeks) at temperatures lower than those considered to be nominal decomposition temperatures.[221]

For typical vitrinite-rich coals, plastic and caking properties are predominantly correlative with rank. However type factors can exert an influence. Liptinite produces more volatile matter than vitrinite from the same coal, and where it is sufficiently abundant, this production can lead to abnormally severe swelling (vesicle development).[222,223] Furthermore, because of its higher hydrogen content, liptinite could serve as a hydrogen donor, thereby promoting fluidity. Pyrite may be a dehydrogenating material that reduces fluidity.[224] Other minerals and the inert macerals, including most semifusinite[43] and pseudovitrinite, serve as nonreactive diluents that would tend to decrease plastic properties of otherwise caking coals.

[218] P. G. Sal'kov, *Teploenergetika*, **15**(8), 90–91 (1968); *C. A.*, **69**, 79102 (1968).

[219] B. N. Nandi, T. D. Brown, and G. K. Lee, *Fuel*, **56**(2), 125–130 (1977).

[220] D. Chandra and R. L. Bond, "Reflectance of Carbonized Coals," *Int. Comm. Coal Petrology Proc.*, No. 2, Liege, May 23–25, published by R. Louis, Ixelles-Brusels, (1956), pp. 47–51.

[221] F. Goodarzi and D. G. Murchison, *Fuel*, **56**(1), 89–96 (1977).

[222] I. M. Glushchenko and E. M. Zinchenko, *Khim. Tverd. Topl.*, No. 1, 140–142 (1968).

[223] P. A. Hacquebard and T. E. Tibbetts, "Practical Significance of Coal Petrography to Coking," *Proc. 12th Dominion–Provincial Coal Res. Conf.*, Ottawa, October 12–15, 1960, pp. 13–57.

[224] A. V. Polubentsev, I. K. Klopotov, V. A. Uykhovanets, and V. Ya. Shevchenko, *Izv. Nauchn.-Issled. Inst. Nefte-Uglekhim. Sin. Irkutsk. Univ.*, **8**, 49–59 (1966); *C. A.*, **70**, 21683 (1969).

Solutions to the problems engendered by agglomeration that have been tried or suggested include engineering approaches (e.g., stirring, staged-temperature increase, very high temperature, dilution with inerts) and coal pretreatment (e.g., oxidation, alkali impregnation,[225] other chemical additives or treatments[226]).

The product yields and properties from pyrolysis processes, hence the return on investment (ROI), are significantly affected by the properties of the coal. For five coals studied in the FMC-COED process (Chapter 13) estimated ROI ranged from 8 to 23% solely as a response to differences in coal properties.[227] Though no correlation was made between ROI and specific coal properties, bituminous rank coals were better feeds than subbituminous or lignitic coals. Within bituminous ranks, higher hydrogen coals yielded oil that was easier to hydrogenate catalytically.[228]

Benedict and Berry[143] found that yields of coke, oil, tar, gases (and gas properties), and liquor from high-temperature carbonization of more than 200 samples could be correlated with the reflectance of the vitrinite. Correlation coefficients all exceeded 0.9 except those for liquid yield (0.72) and Btu per standard cubic foot of gas (0.87). Since yields could be predicted to within about ±10% (of the yield) based solely on vitrinite reflectance, it appears that for the coals studied, coal rank was responsible for as much as 50–90% of the variability in pyrolysis product yields. However the coals from which Benedict and Berry secured their data were limited to those used commercially for the production of metallurgical coke (metcoke). Coals of rank or type not suited for metcoke production were excluded, and inclusion of these coals might be expected to alter the variance of the correlation between rank and yield.

Comberiati et al.[229] also reported that the properties of pyrolysis tars (e.g., proportions of tar acids, olefins, certain hydrocarbons) varied with rank of the coal from which they were derived, and noted that liptinite-rich coal gave an abnormally high yield of aliphatic hydrocarbons.

3.4.4 Metallurgical Coke

As already discussed, when particles of caking coals are heated in the absence of free oxygen, they soften, decompose, swell, agglomerate, and ultimately resolidify into a porous, carbon-rich solid (coke). Coals that form cokes that are usable for the reduction of iron ores in the blast furnace are referred to as coking coals. All coking coals are caking coals, but not all caking coals are coking coals. The processing of coal into metcoke must be carefully controlled in terms of both the conditions attendant to the coke making and the characteristics of the coal employed. It is likely that through the years, more research effort has been devoted to optimizing coke making than was devoted to any other use of coal. Chapters 14, 15, and 17 give details of the technology. Here I summarize briefly the more recent findings relating coal grade to coke quality, but it should be kept in mind that there is no universal agreement on the criteria that should be used to determine the relative quality of cokes.[230] Day-to-day uniformity of the feed and of the coke properties is almost as important as the absolute level of any quality criterion.[231]

Figure 3.20 shows the principal variables associated with the manufacture of metcoke. The dependent variables are influenced most by the grade of feed coal employed; however the processing conditions can be varied within rather narrow limits to optimize the dependent variables for a given feedstock.

The chemical quality of coke is a direct reflection of the quality of the feed coals. Essentially all inorganic elements remain in the coke; hence the ash content of the coke is directly proportional to the ash or mineral

[225] J. W. Patrick and F. H. Shaw, *Fuel*, **51**(1), 69–75 (1972).

[226] A. C. Cunningham and W. F. Wyss, *Fuel*, **46**(2), 137–143 (1967).

[227] Anon., *Chem. Eng. News*, **46**(51), 41 (1968).

[228] H. E. Jacobs, J. F. Jones, and R. T. Eddinger, *Prepr., Am. Chem. Soc., Div. Fuel Chem.*, **14**(4), 42–57 (1970).

[229] J. R. Comberiati, C. Karr, Jr., E. E. Childers, J. O. Mapstone, Jr., and P. A. Estep, *J. Inst. Fuel*, **43**(352), 147–150 (1970).

[230] N. Schapiro and R. J. Gray, *Blast Furn. Steel Plant*, 273–280 (April 1963).

[231] M. C. Chang and G. H. Denton, "Quality Considerations for Future Blast Furnace Coke," *Ironmaking Conf. Metall, Soc. AIME*, Buffalo, New York, April 1963.

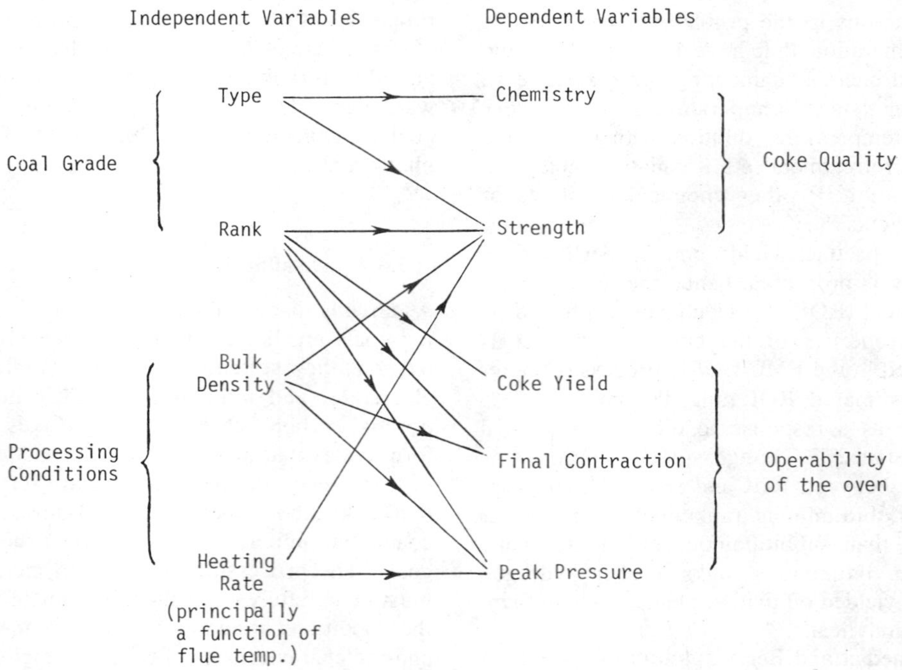

Figure 3.20 Interrelationships between variables in metcoke production.

content of the coal, as shown in Equation 3.8:

$$\text{ash (wt\% in coke)} = \text{ash (wt\% in coal)} \times \frac{100}{100 - 0.87\%\text{VM}} \tag{3.8}$$

Sulfur content is also a direct function of sulfur content of the coal, but not all sulfur remains behind in the coke. Typically, the sulfur in coke (as weight percentage of the coke) is equal to about 0.8–0.9 times the weight percentage of sulfur in coal (e.g., if coal sulfur is 1.0%, coke sulfur is about 0.85%).[209] Although there are no absolute restrictions on ash or sulfur contents of acceptable coking coals, most commercial installations prefer cokes with less than 10% ash and less than 1% sulfur.

Strength is the most important property of cokes made from coals with satisfactory ash and sulfur contents: it is the ability of the coke to maintain its physical integrity as it descends with the burden in a blast furnace. Strength is most commonly tested using some variation of a drop-shatter test (ASTM D440[232]) or a tumbler test (ASTM D3402[233]). The tumbler stability index [percentage of coke fragments larger than 1 in. (25 mm) remaining after 2 × 3 in. coke is tumbled for 1400 revolutions in a 36-in. drum with lifters] is the single most widely used coke strength parameter. (See Chapter 15).

Through the years technologists have evaluated and employed with varying degrees of success many parameters to grade coals as coke-making feedstocks. Standard proximate and ultimate analytical parameters can indicate whether a coal is approximately suitable for coke making, but they cannot predict with sufficient accuracy the strength characteristics of the coke that would be produced.

As a consequence of the failure of standard coal analyses to provide such predictive information, numerous empirical tests of the physical response of coal to heating have

[232] ASTM Standards, 1977, Part 27, Designation D440-49, Standard Method of Drop Shatter Test for Coal.

[233] ASTM Standards, 1977, Part 27, Designation D3402-76, Standard Method of Tumbler Test for Coke.

been widely employed as guides to predict coke strength. In this category, covered in Chapter 6, are tests of plasticity or fluidity [Gieseler plastometer (ASTM D2639,[234] ASTM D1812[235]), Audibert-Arnu dilatometer,[49] Sapozhnikov plastometer (Soviet GOST-1186)] and agglomerating capacity [Roga index,[49] Gray-King analysis,[49] free-swelling index (ASTM D720[236]), inert binding capacity[237]]. These test methods have proved useful for comparing various coals to each other in a qualitative fashion, and for indicating whether a coal should even be considered for coking purposes. However the values obtained from these tests have failed to correlate directly with coke strength,[237,238] hence the tests have been useful as quantitative predictive tools only within limited ranges of coal rank and type.

The feed to essentially all coke ovens is a blend of two or more coals, most commonly a high-volatile bituminous coal for bulk and a medium- or low-volatile bituminous coal for strength. Unfortunately none of the empirical tests of agglomeration characteristics and plasticity is capable of yielding information that allows one to predict in what way coals should be blended to produce better coke.

Ammosov et al.[239] published one of the first truly scientific approaches to the problem of how to employ analytical parameters to predict the strength of coke from either a single coal or a blend of several coals. At about the same time, Spackman and his coworkers at Penn State were finding that when some poor coking coals were added to some good grades of coking coal, a superior coke could be produced.[111] They concluded that the optimal blending of petrographic constituents was responsible. Schapiro and Gray at the Monroeville, Pennsylvania, laboratory of the U.S. Steel Corporation conducted extensive investigations into the relationship between coke strength and the coal properties of rank and type (petrographic composition). Their research resulted in the development of a technique similar to Ammosov's for predicting the strength of coke (stability index) from microscopic analyses of the coal.[20,42,240] This technique has met with wide acceptance in the United States, either in its original form or as modified to suit the requirements of the user.

The underlying principle is that bituminous rank coals consist of materials that will fuse when heated (reactive macerals) and materials that will not fuse (minerals and inert macerals). In typical coals vitrinite is the dominant reactive maceral, and it will of itself produce coke, the strength of which depends on the rank of the coal. For some ranks of coal a stronger coke can be made if an optimum proportion of inert materials, acting like aggregate in concrete, is admixed with the reactive vitrinite.

Figure 3.21 illustrates the dependence of coke strength on the joint effects of coal rank and the proportions of inert and reactive macerals (petrographic composition) for single coals. Coke strength (the dependent variable) is represented by lines of equal tumbler stability index, and the independent variables are expressed as the coordinates of rank (vitrinite reflectance) and petrographic composition (proportion of inerts or reactives).

For the low-volatile coals with reflectance greater than about 1.4%, the addition of any inerts is deleterious to coke strength, and strength decreases with increased inert content. However for the lower rank coals (medium- and high-volatile bituminous), the addition of 20–30% inerts strengthens the coke considerably.

The isostability lines in Fig. 3.21 were calculated according to the procedures devised by Schapiro and Gray, who based their conclusions on data from many 500-lb coke-

[234] ASTM Standards, 1977, Part 27, Designation D2639-74, Standard Test Method for Plastic Properties of Coal by the Constant Torque Gieseler Plastometer.

[235] ASTM Standards, 1977, Part 27, Designation D1812-69, Standard Test Method for Plastic Properties of Coal by the Gieseler Plastometer.

[236] ASTM Standards, 1977, Part 27, Designation D720-67, Standard Test Method for Free-Swelling Index of Coal.

[237] H. B. Zaalberg, H. N. M. Dormans, F. J. Huntjens, and G. Kardaun, *Fuel,* **44**(2), 95–108 (1965).

[238] J. L. Bayer, G. H. Denton, and M. C. Chang, *Prepr., Am. Chem. Soc., Div. Fuel Chem.,* **7**(2), 54 (1963).

[239] I. I. Ammosov, I. V. Eremin, S. I. Sukhenko, and L. S. Oshurkova, *Koks Khim.,* No. 12, 9–12 (1957).

[240] N. Schapiro and R. J. Gray, *Inst. Fuel,* 234–242 (June 1964).

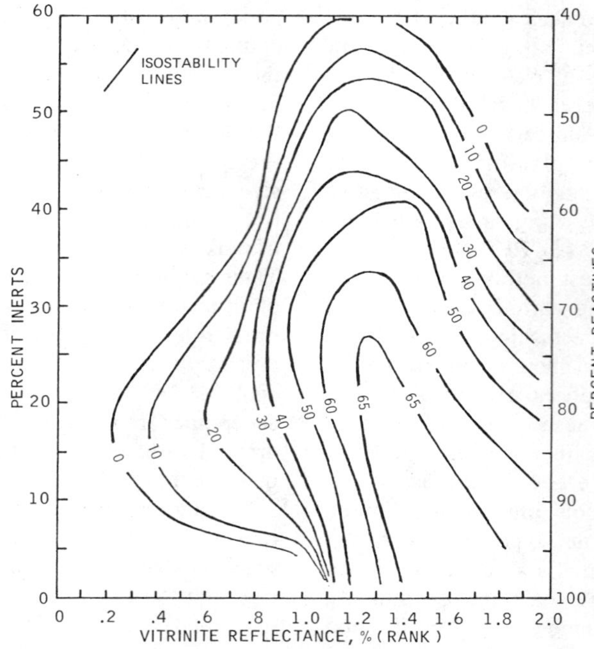

Figure 3.21 Influence of rank (vitrinite reflectance) and type (percentage of inert and reactive macerals) on coke strength. Contours are lines of equal coke strength (tumbler stability) calculated by the author according to the Schapiro and Gray procedure as described in the text.

oven tests of various coal blends, all standardized (actually or by calculation) to a feed with 2% moisture, less than 12% ash, and 80% less than ⅛ in. size. A charged bulk density of about 50 lb/ft^3 and heating rate of 1 in./hr were assumed. They obtained a linear correlation coefficient of 0.98 when estimated coke strengths were compared to experimental values; however this correlation was achieved with the data from which the procedure was derived.[240] Though the validity of the calculation procedure has been questioned, as discussed below, it has been widely applied, and has been described in detail in a publication produced privately by Bituminous Coal Research, Inc.[241]

Procedure for Estimating Coke Stability from Petrographic Analysis of Coal, Based on Work of Schapiro and Gray

1. Microscopically determine volumetric proportions of vitrinite, liptinite, semifusinite, fusinite, and micrinite according to procedures specified in ASTM D2799.[6]

[241] R. Moses, *Use of Petrographic Analysis for the Production of Coke Stability,* Bituminous Coal Research, Inc., 350 Hochberg Rd., Monroeville, Pa., 1976, 19 pp.

2. Estimate volumetric proportion of mineral matter from ash and sulfur contents (Section 3.2.1.5, "Maceral Analyses"). Normalize maceral data to mineral-containing basis.
3. Determine reflectance of 100 particles of vitrinite according to procedures specified in ASTM D2798.[5]
4. Apportion vitrinite into V-types according to reflectance (e.g., V_5 = vitrinite with reflectance between 0.50 and 0.59).
5. Normalize all data in volume percent as follows:
   (*a*) Total inerts = % fusinite + % micrinite + % mineral matter + (% semifusinite × ⅔ below 1.4% R_0 and 4/5 above 1.4% R_0).
   (*b*) Total reactives = 100 − total inerts.
   (*c*) Weight V-types: multiply percentages of each type determined in reflectance analysis by total reactives divided by 100.
6. Place data in form shown as Fig. 3.22, in column labeled "vol % reactives."
7. For each V-type determine strength factor from Fig. 3.23, using total inerts as the independent variable (abscissa).

Vitrinite Type	Optimum Inerts	Volume Percent Reactives	Strength Factor	Optimum Inert Index	Optimum Strength Index
2	5.2	______	______	______	______
3	4.7	______	______	______	______
4	4.3	______	______	______	______
5	3.8	______	______	______	______
6	3.5	______	______	______	______
7	3.1	______	______	______	______
8	2.8	______	______	______	______
9	2.6	______	______	______	______
10	2.4	______	______	______	______
11	2.7	______	______	______	______
12	3.2	______	______	______	______
13	4.0	______	______	______	______
14	5.2	______	______	______	______
15	7.0	______	______	______	______
16	9.5	______	______	______	______
17	12.3	______	______	______	______
18	14.8	______	______	______	______
19	17.1	______	______	______	______
TOTAL		______		______	______

Inerts = 100 - reactives = ____________

$$\text{Balance Index} = \frac{\%\ \text{inerts in coal}}{\Sigma \text{OII}} = \_\_\_\_\_\_\_\_\_\_$$

$$\text{Strength Index} = \frac{\Sigma \text{OSI}}{\%\ \text{reactives in coal}} = \_\_\_\_\_\_\_\_\_\_$$

Tumbler stability index = ____________

Figure 3.22 Data form for estimation of coke strength from petrographic data.

8. For each V-type calculate optimum inert index (OII) with Equation 3.9:

$$\text{OII} = \frac{\%\ \text{of coal in V-type (col. 3)}}{\text{optimum inerts (col. 2)}} \tag{3.9}$$

Note: Optimum inerts is the ratio of reactives to inerts that yields maximum strength coke for that V-type. These values were derived empirically by Gray and Schapiro and are constants.

9. For each V-type, use Equation 3.10 to calculate optimum strength index (OSI):

$$\text{OSI} = \%\ \text{of coal in V-type} \times \text{strength factor} \tag{3.10}$$

10. Sum OII values for all V-types and calculate balance index (BI), using Equation 3.11:

$$\text{BI} = \frac{\%\ \text{inerts in coal}}{\Sigma \text{OII}} \tag{3.11}$$

11. Sum OSI values for all V-types and calculate strength index (SI) with Equation 3.12:

$$\text{SI} = \frac{\Sigma \text{OSI}}{\%\ \text{reactives in coal}} \tag{3.12}$$

12. From Fig. 3.24 estimate anticipated coke strength (tumbler stability index) using balance index and strength index parameters.

The procedure is applicable to one coal or a blend of two or more coals. Where petrographic data are available for several coals, the coals may be "blended on paper" by weighting the data in any desired proportions, depending on desired strength and cost of the coals. In the example presented as Fig. 3.25, an estimate is made of the coke strength that would be expected if two hypothetical coals were to be blended in a 70 : 30 proportion.

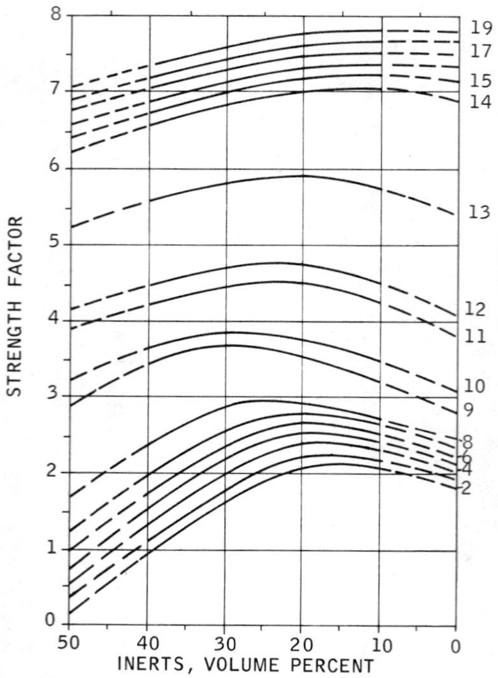

Figure 3.23 Relative coke strength (strength factor) varies as a function of volume percentage of inert macerals and the properties of the reactive macerals (designated by V-type derived from vitrinite reflectance). From Moses[241] after Schapiro, Gray, and Eusner.[42]

In the predictive technique, liptinite and one-third or four-fifths of the semifusinite are treated as reactive components essentially equivalent to vitrinite. Yet even Schapiro and Gray indicate that coals with high liptinite content do not fit their correlation.[240] Several other authors also indicate that liptinites do not act like vitrinite when heated.[242–244] At the other extreme, in one study semifusinite with reflectance only slightly higher than vitrinite in the same coal was found to be inert.[245] Thus although the technique works as a predictive tool, certain fundamental questions about the role of various macerals remain unanswered. In addition, as described below, the technique has been found to be deficient in some respects.

On the basis of experimental data obtained at the Illinois State Geological Survey laboratory, Harrison et al.[246] adjusted the isostability line positions of Fig. 3.25 to provide somewhat better estimates of coke stability from very low-rank bituminous coals such as are found in Illinois. It has been reported[247] that the finely dispersed mineral matter and micrinite in Gondwana (Southern Hemisphere) coals lead to inaccuracies in estimating coke strength using the procedure of Schapiro and Gray.

The procedure leads to inaccuracies for western Canadian coals also, likely because their oxygen contents (especially as −OH) are higher for a given reflectance than those of older, Appalachian coals.[248] Ammosov et al.[239] noted that variations in the degree of reduction and oxidation could influence fusibility, and this seems to be an area worthy of additional research. It has been reported that Gieseler plastometer fluidities of western Canadian coals correlated with the concentration of the microlithotype vitrite rather than with the vitrinite content.[208]

Brown et al.[156] contend that the Schapiro-Gray procedure does not work effectively for Australian coals, holding that the coke strength data reported by Schapiro and Gray could be correlated directly with the coal rank (based on mean maximum reflectance of vitrinite) irrespective of the petrographic composition. However Schapiro and Gray showed in a paper published in the same year[240] that coals with similar reflectance do in fact yield different strength cokes as a result of variable inert/reactive ratios. Brown et al. dispute that there is a fixed optimum inert content for maximizing coke strength from a specific rank coal, contending that different forms of inerts (i.e., sizes, associations) will interact differently with the vitrin-

[242] L. G. Ol'shanetskii and A. K. Vetrova, *Khim. Tverd. Topl.*, **8**(2), 9–15 (1974).

[243] E. I. Toryanik and A. S. Zhidko, *Koks Khim.*, No. 7, 6–11 (1968); *C. A.*, **69**, 68814 (1968).

[244] A. S. Stankevich, M. V. Romanovskaya, and V. V. Artamonov, *Nauchn. Tr. Kuznetskii Nauchn.-Issled. Proekt.-Konstr. Inst. Ugleobogashch*, No. 4, 208–217 (1968); *C. A.*, **70**, 79802 (1969).

[245] G. H. Taylor, M.-Th. Mackowsky, and B. Alpern, *Fuel*, **46**(6), 431–440 (1967).

[246] J. A. Harrison, H. W. Jackman, and J. A. Simon, Illinois State Geol. Survey Circ. No. 366, 1964, pp. 1–20.

[247] N. N. Chatterjee, B. N. Nandi, and D. S. Montgomery, *Can. Mining Met. Bull.*, **59**, 615–625 (1966).

[248] B. Ignasiak and N. Berkowitz, *Can Inst. Min. Bull.*, **July**, 72–76 (1974).

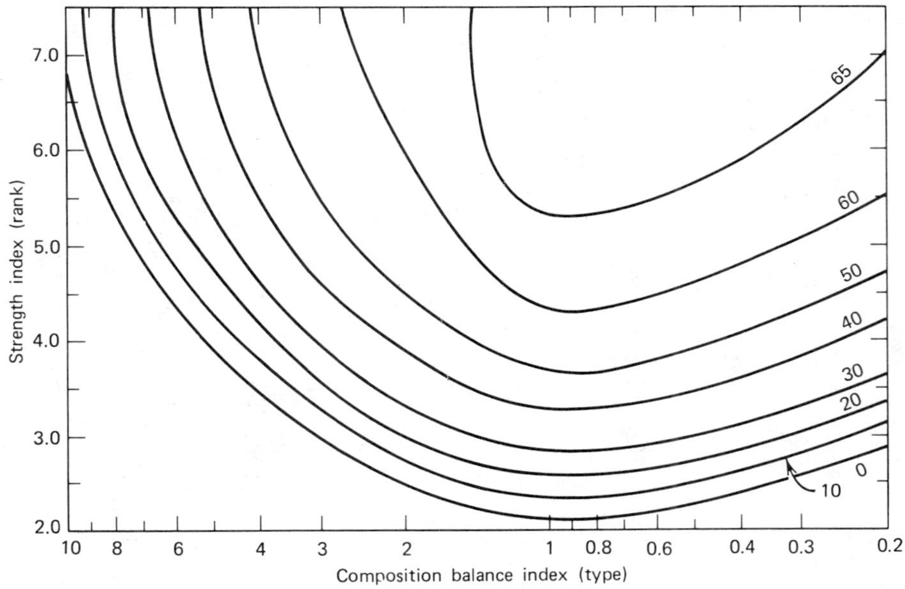

Figure 3.24 Correlation chart for estimating tumbler stability index of coke and blending potential from petrographic data. From Moses[241] after Schapiro, Gray and Eusner.[42]

ite. Mackowsky[249] also makes this point. Brown et al.[156] showed that swelling (British Standard Swelling Index), volatile matter, and coke strength for Australian coals were all related to rank (expressed as percentage of carbon, either from direct analysis or estimated from vitrinite reflectance) and to petrographic composition as defined by the percentage of the microlithotypes vitrite and clarite (see Section 3.2.3.3).

Coking tests of some Australian coals by Cook and Edwards[250] and Cook and Wilson[188] further confirmed that the correlation of Schapiro and Gray correlation to be inapplicable to some Australian coals. Some of their data appear in Fig. 3.26, which is identical to Fig. 3.21 except for the added data points. It can be seen that the cokes made from high-inertinite Australian coals from the Bulli Seam are much stronger than the formulas of Schapiro and Gray would predict. Conversely, the cokes from the Wongwalli, Paparo, and Stockton seams are weaker than would be predicted. Only vitrinite content on a mineral-matter-free basis is reported for the Australian coals, and the liptinite and mineral matter contents could displace the points up or down by as much as 10–15%. However it is unlikely that this adjustment would bring the data in line with the isostability lines calculated according to the Schapiro-Gray procedure.

Thompson and his associates at the Homer Research Laboratories of the Bethlehem Steel Corporation also conducted extensive experiments relating coal petrography to coke making. They reported that not all vitrinite, even in coking-range coals, is reactive, naming the nonreactive vitrinite "pseudovitrinite."[38] Their investigations led to the observation that there were several discontinuities in the relationships between rank (vitrinite reflectance) and coke strength, volatile matter, and coking pressure. As shown graphically in Fig. 3.27, Thompson and Benedict conclude that coals can be separated into three populations based on the reflectance of vitrinite excluding pseudovitrinite ($<0.7\%\ R_0$, 0.7–$1.35\%\ R_0$, $>1.35\%\ R_0$). They offer the hypothesis that the discontinuities represent fundamental changes in coal (vitrinite) composition and should be relatable to specific stages in the coalification process.

At the Bethlehem Steel Corporation, evaluation of the coke-making potential of coals is based on the techniques devised by

[249] See Ref. 108, p. 351.
[250] A. C. Cook and G. E. Edwards, *Fuel,* **50,** 41–52 (1970).

1. Basic Petrographic Data

Constituent	Volume % in Constituent in Coal		
	Coal A(70%)	Coal B(30%)	Composite
Vitrinite	75.6	90.8	80.2
Liptinite	4.6	1.4	3.6
Semifusinite	6.1	0.6	4.5
Fusinite	8.4	2.7	6.7
Micrinite	2.1	2.2	2.1
Mineral Matter	3.2	2.3	2.9
TOTAL	100.0	100.0	100.0
Inerts	17.8	7.6	14.7
Reactives	82.2	92.4	85.3

2. Reflectance Distribution

V-Type	Volume % in V-Type Class			
	Vitrinite			Reactives*
	Coal A	Coal B	Composite	Composite
6	17.2	--	12.0	10.2
7	61.3	--	42.9	36.6
8	21.5	--	15.1	12.9
9	--	--	--	--
10	--	4.3	1.3	1.1
11	--	18.7	5.6	4.8
12	--	75.9	22.8	19.4
13	--	1.1	0.3	0.3
TOTAL	100.0	100.0	100.0	85.3

3. Calculation of Indexes

V-Type	Optimum Inerts	Vol.% Reactives	Strength Factor	Optimum Inert Index	Optimum Strength Index
6	3.5	10.2	2.6	2.9	26.5
7	3.1	36.6	2.7	11.8	98.8
8	2.8	12.9	2.8	4.6	36.1
9	2.6	--	--	--	--
10	2.4	1.1	3.6	0.5	4.0
11	2.7	4.8	4.4	1.8	21.1
12	3.2	19.4	4.6	6.1	89.2
13	4.0	0.3	5.8	0.1	1.7
TOTAL		85.3		27.8	277.4

Balance Index = $\frac{14.7}{27.8} = 0.53$ Strength Index = $\frac{277.4}{85.3} = 3.25$

Estimated Stability Index = 35

*Composite vitrinite X 0.853 (i.e., fraction of reactives)

Figure 3.25 Worked example of estimation of coke strength from petrographic data. Estimate is for anticipated strength of coke if coals A and B were to be blended in 70:30 proportions.

Thompson and his associates.[38,251] A series of curves relates coke stability from single coals to effective inert content for various levels of vitrinite reflectance (Fig. 3.28). They claim that in blends, the composition (rank level) and the percentage of the low-volatile coal establishes the strength of the coke. Gin et al. had earlier come to a similar conclusion.[252]

In practice, the strength of cokes from blends of coals with less than 1.4% R_0 can

[251] L. G. Benedict, R. R. Thompson, and R. C. Wenger, *Blast Furn. Steel Plant* (1968).

[252] T. T. Gin, C. L. Dahl, and D. G. Wilson, "Petrographic Evaluation of Coking Coals," AISI Regional Meeting, San Francisco, November 8, (1963).

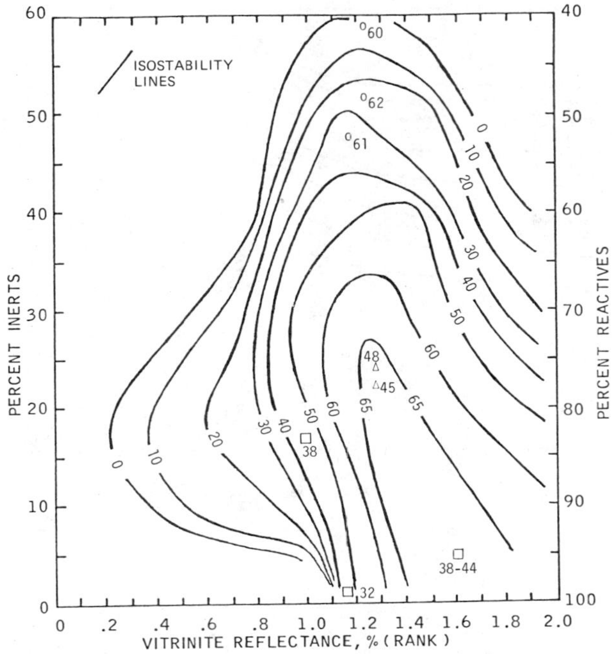

Figure 3.26 Coke strength data (stability index) from Australian coals[188] plotted on figure derived from strength estimating procedure of Schapiro and Gray. These Australian coals respond differently from those used by Schapiro and Gray to establish their estimating system. ○ = Bulli Seam, <12% ash △ = Wongwalli Seam, <20% ash □ = Paparo and Stockton seams, ash content unspecified.

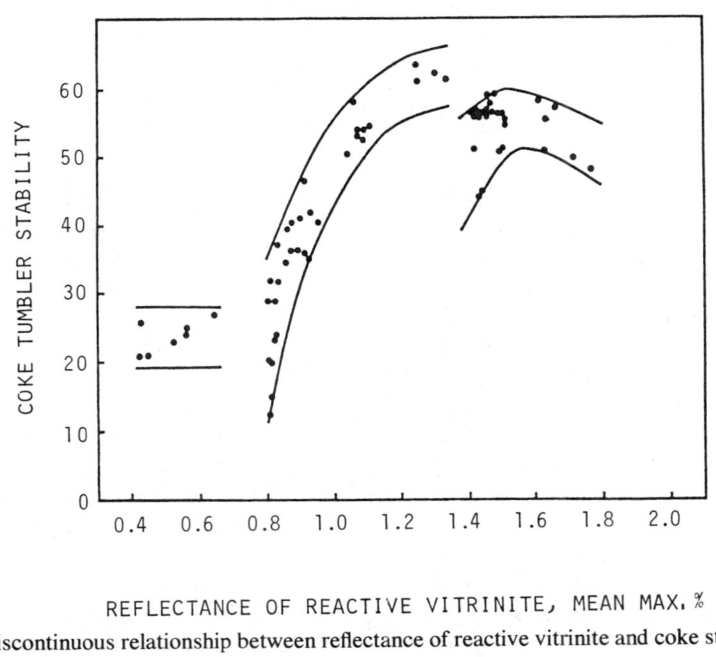

Figure 3.27 Discontinuous relationship between reflectance of reactive vitrinite and coke strength (tumbler stability).[147]

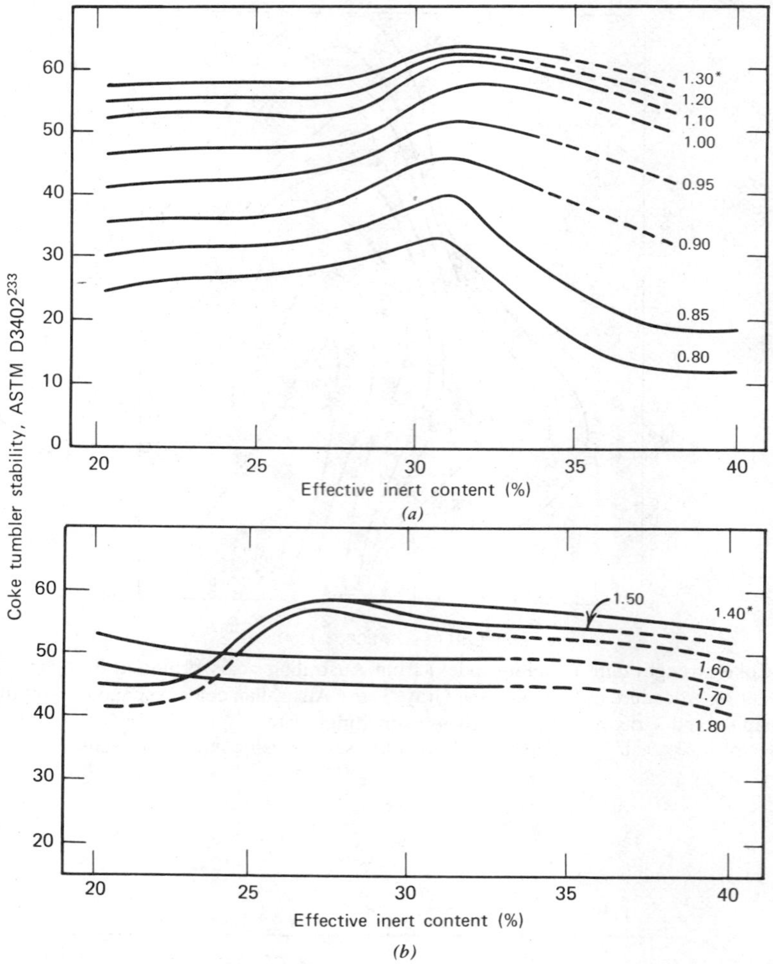

Figure 3.28 Effect of rank (vitrinite reflectance) and type (inert content) on stability of coke from single coals.[251] Numbers with asterisks refer to reflectance of reactive vitrinite (mean maximum percent). (a) Curves for high and medium-volatile coals (<1.4% R_o). (b) Curves for low-volatile coals (>1.4% R_o).

be predicted by calculating a weighted average of the stabilities for the individual coals using the curves in Fig. 3.28*a*. When coals with greater than 1.4% R_0 are added to the lower reflectance coals, the strength of the coke from the blend is estimated by using Fig. 3.29. The strength of the coke from the high- and medium-volatile coals (<1.4% R_0) is estimated from Fig. 3.28*a* and is plotted to the appropriate height above the zero abscissa point in Fig. 3.29. Then the strength of the coke from the low-volatile coals (>1.4% R_0) is estimated from Fig. 3.28*b* and is plotted above the 100% point in Fig. 3.29. A curve is drawn in essentially the shape shown on the left-hand side of Fig. 3.29, intersecting a horizontal line from the right-hand point at a position vertically above the 23% abscissa point. The blend coke strength is estimated from this curve by employing the percentage of low-volatile coal in the mix as the independent variable.

Soviet investigators have published numerous articles on the subject of classifying coals for metcoke production. In general, their procedures are somewhat more involved than the Schapiro-Gray technique. As an example, one investigator suggests using a combination of petrographic composition, a rank parameter, reducibility, microlithotype characteristics, and the extent of coincidence of the plastic temperatures of the

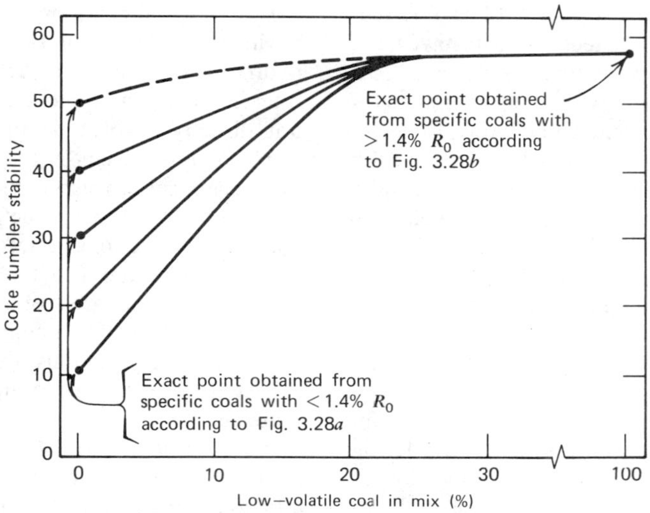

Figure 3.29 Combined correlation system for estimating coke strength from coal blend containing high- and low-volatile coals. Modified from Thompson.[253]

different coals, to predict coke strength.[254] In the Soviet Union considerable reliance is still placed on the plastometer as a guide to estimating coking coal behavior.[255]

Japanese practice, as represented by the work of Miyazu et al.,[256] employs both Gieseler plasticity and vitrinite reflectance to evaluate the coke-making properties of coals. To obtain maximum coke strength, coals are blended such that the average (maximum) vitrinite reflectance of the coals in the blend falls between 1.2 and 1.3%, and the average Gieseler fluidity of the blend is between 200 and 1000 dial divisions per minute. Also, according to Matsuoka,[257] it is common practice to control the volatile matter content of the blend to values between 29 and 33%, and to blend coals such that the composition balance index falls between 0.8 and 1.5 and the strength index is about 3.7–4.3.

Marshall[258] has proposed a classification of coking coals according to a code represented by a four-digit number XX.YY, where XX is the mean maximum reflectance times 10 (i.e., for R_0 of 1.3–1.4, X = 13), and YY is dilatation according to the Ruhr dilatometer divided by 10 (i.e., for dilatation of 130–140%, YY = 13). No information is presented to relate this classification system to coke strengths, however.

In addition to coke stability, petrographic analyses have been used to predict contraction properties, coke-oven pressures, and reactivity of the coke. Contraction is reported to increase with increased content of inerts in medium-volatile coals.[240] Coals with vitrinite reflectance above 1.35% can exert excessive pressure on oven walls during coking.[53] For this reason, although the higher rank coals make strong cokes when coked alone, they are most commonly blended with lower rank coals. Graphical correlation of vitrinite reflectance and oven pressure is given by Thompson and Benedict.[147]

If a coke is too highly reactive, it could be

[253] R. R. Thompson,Bethlehem Steel Corp. personal communication of unpublished work, 1977.

[254] A. S. Stankevich, *Nauchn. Tr. Kuznetskii Nauchn.-Issled. Proekt.-Konstr. Inst. Ugleobogashch,* No. 4, 154–177 (1968).

[255] T. A. Stekleneva, *Nauchn. Tr. Kuznetskii Nauchn.-Issled. Proekt.-Konstr. Inst. Ugleobogashch,* No. 4, 208–217 (1968); *C. A.,* **70,** 79799 (1969).

[256] T. Miyazu, Y. Okuyama, T. Fukuyama, N. Suzuki, and T. Mori, "Evaluation and Design of Blends Using Many Kinds of Coal for Coke Making," Int. Iron Steel Congr., Düsseldorf, Paper No. 1.2.2.1, 1974, pp. 1–18.

[257] H. Matsuoka, "Requirements for Coals in Japanese Coking Blends," in *Australian Black Coal,* A. C. Cook, Ed., Australasian Institute of Mining and Metallurgy, Illiwara Branch, (1975), pp. 251–263.

[258] R. J. Marshall, *Fuel,* **55**(4), 346–347 (1976).

consumed too rapidly in the blast furnace. Therefore some coke technologists have been concerned with evaluating the effect of coal properties on reactivity characteristics of coke.[259] Schapiro and Gray[230] and Benedict and Thompson[260] correlated reactivity with coke structures assessed microscopically. Reactivity is principally related to coke porosity, which in turn is related to coal rank. Lower rank coals, alkalis, and inertinite macerals produce more highly reactive cokes.

3.4.5 Liquefaction

Hydrocarbon liquids can be produced from coal by condensing the vapors formed during pyrolysis. However a greater proportion of the coal can be converted to liquids by heating pulverized coal slurried in an appropriate liquid hydrocarbon vehicle to about 400–450°C under pressures up to several hundred atmospheres. Under these conditions the vitrinite and liptinite of coals ranked bituminous or lower decompose, and much of the decomposition product is solubilized in the vehicle. Subsequent processing can then separate the undissolved solids (mineral matter and inert macerals) and heavy molecules from the liquids. A portion of the liquids is then recycled as a vehicle to slurry fresh coal (the vehicle may or may not be hydrogenated between passes).

This process is often incorrectly referred to as a hydrogenation process. Though it is common for there to be several hundred atmospheres of hydrogen pressure in the system, liquefaction can be carried out with no molecular hydrogen present.[261] In the liquefaction reaction, hydrogenation in the sense of creating, for example, cyclohexane from benzene, or decalin from napthalene, is negligible. Most hydrogen appears to enter the reaction by way of combination with free radicals created by the coal decomposition (Chapter 27).

Numerous process configurations are presently being investigated in this area (Chapter 28). All have in common the thermal decomposition-dissolution of the coal in a liquid hydrocarbon vehicle, a step that is correctly referred to as the liquefaction step. It does not directly involve a supported catalyst, even if one is present in the reactor. As with most chemical processes, yields and product properties from coal liquefaction are influenced by interactions between the processing conditions and the feedstock properties. Because this is an area of emerging—rather than established—technology, however, only the grossest correlations are available relating coal constitution to liquefaction yield patterns.

Present practice in the technical literature is to provide data on the proximate and ultimate (elemental) composition of coals employed in various liquefaction research programs. However if results are to be compared from laboratory to laboratory (or even from sample to sample in the same laboratory), it will be necessary to provide petrographic compositional data as well, for of all the uses of coal, liquefaction appears to be most sensitive to coal type (petrographic composition). The grading of coals according to their liquefaction potential will almost certainly have to be done on the basis of *both* rank and type criteria.

Only lignites, subbituminous, and high-volatile bituminous coals are potentially applicable to liquefaction, and within those ranks only coals that consist predominantly of vitrinite and liptinite will prove to be commercially attractive. The yield of liquid products decreases with increasing rank for coals ranked higher than high-volatile bituminous.[262] In one study, no matter how the processing conditions were varied, no conversion was obtained for coals containing more than 92% carbon (organic basis).[263] About 88% elemental carbon (27% volatile matter, vitrinite R_0 = 1.3–1.4%) seems to be a more realistic upper rank limit. Coals ranked high-volatile C bituminous and lower are equally convertible (if petrographically equivalent),[261] but from progressively lower rank coals, less liquid and more gas is pro-

[259] R. R. Thompson, A. F. Mantione, and R. P. Aikman, presented at Pittsburgh Regional Technical Meeting of AISI, November 12, 1970.

[260] L. G. Benedict and R. R. Thompson, presented at the Annual Meeting, Northeast Section, Geological Society of America, Allentown, Pa., March 22, 1973.

[261] R. C. Neavel, *Fuel*, **55**(3), 237–242 (1976).

[262] P. H. Given, D. C. Cronauer, W. Spackman, H. L. Lovell, A. Davis, and B. Biswas, *Fuel*, **54**(1), 34–41 (1975).

[263] E. Rusin, *Koks, Smola, Gaz*, **17**(9), 271–276 (1972).

duced.[262] Liquid products from lower rank coals are more highly oxygenated, whereas higher viscosity is exhibited by liquids from progressively higher rank coals.[262] There appears to be no relation between feed coal rank and the deactivation by carbon of catalysts used for liquid upgrading.[264] Sodium that may be found exchanged onto oxygenated functional groups in low-rank coals may serve as a natural catalyst in the reaction; products from sodium-rich coals have been found to have lower viscosity than products from sodium-poor coals of similar rank.[262] Calcium, which also may occur ion-exchanged on low-rank coals, is known to produce calcium carbonate in the form of deleterious reactor wall scale and spherical concretions.

Vitrinite and liptinite from coals of satisfactory rank appear potentially to be completely convertible to a dissolved substance.[261,262] However under commercial conditions 100% of these macerals is not converted to usable liquid products. Some gas is produced, and though the remainder may "dissolve" completely, some high molecular weight product appears to be produced even under the best of conditions. Pseudovitrinite appears to be as reactive as vitrinite. Alginite, as might be expected because of its naturally high H/C ratio, is exceptionally reactive.[262] When reacted in a hydrogen-deficient vehicle (anthracene oil), liptinite is reported to give a higher yield than associated vitrinite.[265]

Of the so-called inertinite macerals, semifusinite and macrinite (detrifusinite) may be partially reactive. A macrinite-rich coal from Indiana (No. 1 Block Coal) yielded more liquids than could have been produced from vitrinite and liptinite in the sample; hence it was concluded that some macrinite had been converted.[266] Similarly the same investigation revealed that some semifusinite in an Australian coal was apparently converted, and residues from semifusinite were observed microscopically to exhibit signs of having become plastic.

[264] C. C. Kang and E. S. Johanson, *Prepr., Am. Chem. Soc., Div. Fuel Chem.*, **21**(5), 32–42 (1976).

[265] J. Rauk, A. Rutkowski, and M. Rutkowski, *Pr. Nauk. Inst. Chem. Technol. Nafty Wegla Politech. Wroclaw,* **11**, 28–42 (1972).

[266] G. D. Mitchell, A. Davis, and W. Spackman, *Prepr., Am. Chem. Soc., Div. Fuel Chem.*, **21**(5), 189–198 (1976).

In another investigation samples of lignite from the Savage Mine (Richland County, Montana) were liquefied at 425°C in small tubing bombs, using tetralin as a vehicle (see Neavel[261] for procedure). Pyridine-insoluble residues decreased progressively in samples from 5, 30, and 80 min residence, and most of the material converted in the longer runs was low-reflectance fusinite. The data (Fig. 3.30), indicate that after 80 min residence time a significant proportion of the fusinite with reflectance less than 1.4% had been converted.

Unfortunately, there has been insufficient investigation of the precise relationships between liquefaction and coal constitution. Therefore it is not possible yet to predict liquid yield on the basis of a petrographic analysis (including vitrinite reflectance).[262] To some extent this obstacle may result from the confounding influence of catalysis from mineral matter, especially iron sulfide.[267,268] Also, great care must be taken to minimize oxidation of the coal before liquefaction, because even slight oxidation appears to be deleterious to conversion.[261,269] Texture may

[267] A. R. Tarrer, J. A. Guin, W. S. Pitts, J. P. Henley, J. W. Prather, and G. A. Styles, *Prepr., Am. Chem. Soc., Div. Fuel Chem.*, **21**(5), 59–77 (1976).

[268] C. H. Wright and D. E. Severson, *Prepr., Am. Chem. Soc., Div. Fuel Chem.*, **16**(2), 68–93 (1975).

[269] A. A. Krichko and T. A. Titova, *Khim. Tverd. Topl.*, No. 4, 70–75 (1972).

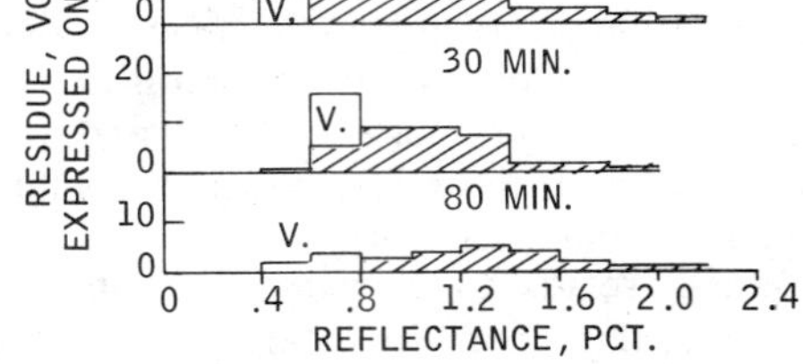

Figure 3.30 Conversion of fusinite in Savage Mine lignite to benzene solubles with increased residence time in liquefaction system (425°C, tetralin vehicle, no H_2). From previously unpublished data by R. C. Neavel.

also be important, since I have noticed in examining residues from liquefaction that vitrinite enclosed by clayey mineral matter (bone coal) does not convert, likely because it is protected from the hydrogen-donating vehicle by the mineral matter.

3.4.6 Gasification

Whereas liquefaction is a relatively mild process that can be influenced subtly by petrographic composition, gasification is a severe process that utilizes coal principally as a source of carbon to be gasified by steam, oxygen, or hydrogen. Though moisture, mineral matter, and elements other than carbon are all diluents, it does not follow that the highest grade coals for gasification are high-rank coals (those with high carbon contents). The high-carbon coals (anthracites and low-volatile bituminous rank) are poorly reactive in gasification, and for some purposes even the bituminous rank coals are unsatisfactory feedstocks because of their propensity to cake. For hydrogasification processes it is generally desirable to produce as much methane as possible directly from pyrolysis, thereby utilizing the coal-hydrogen most efficiently. Therefore bituminous rank coals, though they introduce problems because of caking, are favored for hydrogasification because they evolve more methane than do other ranks. Thus the selection of coals for gasification must balance a number of contradictory requirements, and grade cannot be defined in simple terms.

From laboratory studies of the gasification of 36 coal samples with hydrogen (hydrogasification) and with steam in combination with hydrogen, Johnson[270] established an inverse relationship between the reactivity (of devolatilized coal, viz. chars) and the coal rank, based on carbon content of the raw coal (daf basis). Attempts to relate the reactivity to porosity characteristics and to intrinsic carbon reactivity were not successful, and it is likely that both these rank-sensitive factors are equally important.

Directly related to the rank is the capacity of the coal to contain catalytic elements distributed throughout the coal structure as the cationic components of organic acid salts. Lignites in particular, with their high oxygen content, often contain alkali and alkaline earth elements ion-exchanged to the vitrinite molecules. Because they are well dispersed, these elements are especially effective as catalysts for both steam gasification and hydrogasification.[270] Experiments designed to relate reactivity to rank or type independently of inorganic catalysis should be made on samples from which such ion-exchanged metals have been removed through substitution of protons.

Of the 36 samples tested by Johnson, 9 were handpicked fusains, 11 were handpicked vitrains, and 16 were whole coals. Irrespective of their petrographic character, all samples conformed to the same regression, relating reactivity to carbon content of the coal sample before pyrolysis. Mason,[271] on the other hand, noted that inertinite was unreactive during the first stage of the hydrogasification process, and liptinite was destroyed by oxidative pretreatment at 800°F (427°C).

The variability in petrographic character of particles fed to a fluid-bed gasifier can influence the characteristics of a steady state bed. Because particles of pure mineral matter will not fluidize in the same way as organic-rich char, investigators need to be aware of the particle-to-particle differences in their feed coal samples.

In all existing or projected gasification processes, crushed or pulverized coal is fed into a reactor wherein the temperature exceeds 600°C. As already described in the section on pyrolysis, this can result in agglomeration of particles of caking (bituminous rank, vitrinite-rich) coals.

The tendency for lignites to decrepitate as a result of differential shrinkage when dried or pyrolyzed can create problems in some systems. This phenomenon could be more or less severe with different types of lignites. Briquetting has been employed as a solution to this problem, and the petrographic character of the coal significantly affects briquet quality.[272–276]

[270] J. L. Johnson, *Prepr., Am. Chem. Soc., Div. Fuel Chem.*, **20**(4), 85–101 (1975).

[271] D. Mason, *Prepr., Am. Chem. Soc., Div. Fuel Chem.*, **12**(3), 83–94 (1968).

[272] J. Fica, *Freiberger Forschungsh, A* **429**, 101–109 (1968).

[273] L. H. King, Dept. Mines Tech. Survey, Mine Branch, Canada, Tech. Bull. No. 4, 1959.

Finally, the slagging or fusion properties of the mineral components can be significant in some processes.[277] Though few data have been reported on the subject, it seems certain that standard ash fusion tests will not be adequate to characterize the slagging or agglomerating properties of the minerals. In fact, it is possible that no bulk test of the fusion properties will detect the propensity of one or a few mineral species to "melt" at relatively low temperatures (especially in unusual atmospheres), thereby adhering to each other and to reactor surfaces. Considerable research effort would be desirable in this area.

[274] T. Kruszewski, *Kwart. Geol.*, **12**(1), 199–205 (1968).
[275] St. Velchev, *Vuylishta*, **23**(2), 22–23 (1968); *C. A.*, **69**, 37736 (1968).
[276] M. F. Woods, G. M. Habbersam, K. Elsworth, and S. Bennett, *Fuel*, **46**(3), 193–209 (1967).
[277] W. C. Corder, H. R. Batchelder, and W. A. Goldberger, *5th Synthetic Pipeline Gas Symp.* (Chicago), (1973), 21 pp.

3.5 SUMMARY

Coal can be defined succinctly as a sedimentary rock composed principally of macerals and subordinantly of minerals, containing water and gases in submicroscopic pores. Coals originated as peat deposits formed in swamps through the accumulation of physically discrete and chemically distinctive plant substances (and minor proportions of inorganic substances). The distinctive organic components of peat were converted (metamorphosed) to distinctive coal constituents (macerals) by compaction from subsequently deposited detrital sediments and by heat from subterranean sources.

Wood was the dominant plant material in typical peat swamps; therefore most coals consist predominantly of the wood-derived maceral vitrinite. All coals can be ranked by measuring one or more of the vitrinite properties (e.g., reflectance, carbon content, volatile matter), which responded to the severity of the metamorphic processes that affected the region containing the coals. Identification of the rank of a coal provides a key, through established correlations, to many of the phys-

Table 3.7 Relationships of Principal Grade Criteria to Fundamental Rank and Type Properties

		Controlling Fundamental Coal Property[a]	
Process	Principal Grade Criteria	Rank	Type
Mining, preparation	Grindability, hardness, abrasiveness	1	2
	Formation of unique particles		1
Combustion	Calorific value	1	2
	Combustibility, swelling	1	2
	Ash properties		1
Pyrolysis (excluding metcoke)	Yield structure	1	2
	Agglomeration	1	2
Metcoke	Coke strength	1	2
	Coke yield	1	2
	Contraction, pressure	1	2
	Impurities in coke		1
Liquefaction	Gas + liquids yield	1	2
	Gas/liquids ratio	1	
	Liquid properties	1	2
	Autocatalysis, exchanged ions	1	
	Autocatalysis, minerals		1
Gasification	Reactivity	1	2
	Autocatalysis, exchanged ions	1	

[a] 1 = dominant; 2 = secondary.

ical and chemical properties of the vitrinite. Characterization of the coal type according to its content of vitrinitic and nonvitrinitic macerals and of minerals provides insight into the expectable deviations of whole-coal properties from the properties of vitrinite of that rank.

Thus to compare and to contrast (i.e., to classify) coals optimally, both the *coal rank,* reflecting the stage of metamorphic alteration, and the *coal type,* reflecting the heterogeneity inherited from the diverse source materials, must be recognized. Coals may also be compared according to their economic value as raw materials, an activity properly referred to as classification according to *grade*. A coal rated as high grade for one application may be virtually worthless for another: therefore there is no simple classification according to grade. Instead, coals must be graded according to criteria that are applicable to the specific use situation.

The utility (grade) of coal for any use is reduced in direct proportion to its content of included impurities (moisture and ash-forming inorganic material). Additionally, for each process in which coal is used, there are specific criteria (other than purity) by which the coal is valued or graded. As Table 3.7 indicates, each of these grade criteria is affected directly by the rank, by the type (petrographic constitution), or by both rank and type.

As Table 3.7 implies, a substantial proportion of the variability in coal properties critical to grade (other than content of nonorganic impurities) is attributable to variations in rank. This is largely because most coals are vitrinite rich, and vitrinite properties principally reflect the extent of metamorphism or rank. However Table 3.7 also shows that variations attributable to depositionally related coal type factors can influence coal grade. For some applications, even where the rank is right, the type can significantly affect value and may preclude the use of a coal completely.

Hypothetically, the classification of a coal according to criteria that define rank and type adequately should provide the capability for grading that coal. That today such an ideal is only partially achievable reflects principally a failure of many coal technologists to recognize the validity of the concept and to test accordingly.

CHAPTER **4**

Physical Properties of Coal and Its Products

A. G. SHARKEY, JR. AND JAMES T. McCARTNEY

A. G. Sharkey, Jr. *is Deputy Director of the Pittsburgh Energy Technology Center. He joined the U.S. Bureau of Mines in 1946 and has been associated with the Bruceton, Pennsylvania, laboratory during the transition to ERDA in 1974 and then to the Department of Energy in 1977. Mr. Sharkey's major research interest has been the application of spectral analysis techniques to coal and coal-derived fuels. He has more than 200 publications to his credit, mostly in the field of mass spectrometry and including investigations of coal, coal extracts, and gasification and liquefaction products. He received the Department of Interior Distinguished Service Award in 1970 and is a Past Chairman of the Pittsburgh Conference on Analytical Chemical and Applied Spectroscopy.*

James T. McCartney *was formerly a research physicist in the Chemical and Instrumental Analysis Division of the Pittsburgh Energy Technology Center and is currently affiliated with the Lawrence Berkeley Laboratory. From 1935 until 1979 he was with the U.S. Bureau of Mines and subsequently ERDA and the Department of Energy. His research interests have been the physical properties of coals and carbons, coal petrography, and electron microscopy; he has authored more than 40 publications in these fields. Mr. McCartney was awarded the Distinguished Service Medal of the Department of the Interior in 1968.*

4.1 INTRODUCTION

A key question in preparing a chapter on the physical properties of coal and its products is the definition of physical properties. In probing for an answer, several leading researchers were queried and a variety of responses were obtained. Many of the replies showed that the overlap between chemical and physical properties is such that a clear-cut definition cannot be given. To provide a useful extension of the material in the first supplement, it was therefore decided to follow the same format: namely, to organize the material in terms of the type instrumentation used to measure the physical properties.

Two major factors have determined the size of the effort in studies of the physical properties of coal since 1958. First, there has been a tremendous increase in the number of types of analytical instruments and in their sophistication; and second, there has been a sharp decrease in the number of laboratories concerned with basic studies of the physical properties of coal. In the last 5 years in particular, there has been a major increase in process studies as related to coal, but most of the advances are based on studies of the physical and chemical properties of coal carried out many years ago.

This lack of new information concerning the properties of coal is evident not only in process design for synthetic fuels (synfuels), but also in coal mining. The information available and the current research effort are minimal when one considers the enormity of the problems involved in coal extraction and related problems of health effects in mining. The basis for environmental assessment consists of data about the physical and chemical properties of coal. Yet even the important environmental concerns have failed to generate adequate support for basic research into the physical properties of coal.

Use of several techniques has expanded, including magnetic resonance and infrared absorption, for studies of characteristic groupings and various physical parameters. The use of ultraviolet and visible absorption approaches has declined markedly in the past few years. Several techniques described in the previous supplement, including sound velocity, dielectric constant, and diamagnetic susceptibility, have played only a minor role in the past few years. Mass spectrometry was not included in the previous supplement, yet in the last 10 years it has been one of the most popular techniques for studying coal and its products. Many of the analytical techniques applicable to coal-derived liquids are most effective after separation according to functionality or other factors. Since separation techniques have been used in many of the investigations described, a section concerning sample selection and separation has been included. Only the analytical techniques not included in the previous supplement are described in detail in the current chapter. Table 4.1 summarizes all the methods and the information derived from them. There have been few review articles to summarize the many results of the past 18 years, and

Table 4.1 Summary of Analytical Techniques for Coal Examined in Chapter 4

Technique	Information Derived
Optical properties	True rank criterion Size and ordering of aromatic layers Graphitic structure and anisotropy
Density and pore structure	Reactivity Diffusion of gases Absorbing power Molecular sieve material
Electron microscopy	Ultrafine granular structure Pore structure Graphitic crystallization and layering Element distribution in fine components
X-Ray diffraction	Size distribution of aromatic ring systems Diameter and thickness of lamellae Mean bond length
Electrical properties	Free radical concentrations Ring structure and sizes Energy gap–semiconductor properties Graphitic structure and anisotropy
Thermal analysis	Rate of change of physical state
Coal liquids—sample separations	Solvent separation Polarity Functionability
Mass spectrometry	Molecular weight and formulas Hydrocarbon type analysis Carbon number distribution Trace elements
Nuclear magnetic resonance	H_{ar}/H, C_{ar}/C ratios Distribution of isotopic tracers
Electron spin resonance	Free radicals, charge carriers
Infrared absorption	Functional groups (e.g., OH, CH_{ar},CH_{ali}, CH_3) Minerals
Ultraviolet-visible absorption	Aromaticity Aromatic ring size
Optical fluorescence	Identification of coal-derived components
Electron spectroscopy for chemical analysis (ESCA)	Elemental distribution and chemical state of surfaces
Emission spectroscopy	Quantitative and qualitative elemental analysis Single element and survey analysis Major, minor, and trace element analysis for metals and semimetals
X-Ray fluorescence	Inorganic element analysis
Atomic absorption	Precise quantitative determinations for metals and semimetals Single element analysis for minor and trace constituents
Mössbauer spectroscopy	Iron compounds and association with sulfur

this emphasizes the urgent need for the present work. Although a detailed evaluation of all the information available on physical properties is outside the scope of this chapter, a cursory assessment is made in individual sections and in the conclusion.

4.2 REFLECTANCE AND OPTICAL CONSTANTS

4.2.1 Introduction

Many significant deductions concerning the origins of coals, their structure, and their behavior in various processes have been based on determinations of the reflectance of the vitrinite, the major physical component of most coals.[1] The principal optical constants of the vitrinite can be derived from the reflectance measurements in two media by use of Equation 4.1:

$$R = \frac{(n - n_m)^2 + n^2K^2}{(n + n_m)^2 + n^2 K^2} \qquad (4.1)$$

where R is the reflectance, n and n_m the indices of refraction of the vitrinite and the observational medium, respectively, and K the absorption index of the vitrinite. The variations in R, n, and K that occur during metamorphosis of coal deposits and during physical and chemical changes in processing have served as bases for measurement and interpretation of the makeup and behavior of coals.

4.2.2 Methods for Reflectance Measurement

Early measurements of coal reflectance were made with a Berek microphotometer, which employed a double beam and visual comparison settings. This method has been superseded by photoelectric techniques that largely eliminate subjectivity and increase the speed of measurement. Several reports of the early 1960s describe variations of earlier photoelectric methods.[2–4] Jones described an unusual variation of the Berek method using modulated light signals and electronic detection to reduce errors due to matching and power fluctuations.[5] Murchison reviewed methods of reflectance determination developed by 1964.[6] Descriptions of reflectance equipment and techniques used in the Netherlands and in the Soviet Union were published in the late 1960s.[7,8] Harrison[9,10] discussed possible errors in reflectance measurement of lower rank coals of high internal surface area due to absorbed moisture. He recommended drying for about 30 hr before measurement. Routine methods of specimen preparation and reflectance measurement in the United States were published as tentative standards by the American Society for Testing and Materials (ASTM) in 1969 and were approved as standards in 1972.[11] The basic equipment for current reflectance measurements is shown in Fig. 4.1,[12] excluding the driven stage and the electronic recording components.

Automated techniques of reflectance measurement were first described in 1955[13] and have increased in sophistication to the present time. Automation of reflectance on vitrinite alone is limited by the necessity for visually observing each field for the presence of vitrinite rather than another component. Kotter and Langner[14] and Harrison[15] described equipment for this purpose in which automation was confined to recording and plotting the distribution of reflectance values in the vitrinite. Bayer et al.[16] described a system

[1] *Chemistry of Coal Utilization,* Supplementary Volume, H. H. Lowry, Ed., John Wiley and Sons, New York, 1963, p. 49.

[2] K. Kotter, *Brennstoff-Chemie,* **41,** 263 (1960).

[3] W. Pickhardt and K. Robock, *Brennstoff-Chemie,* **46,** 44 (1965).

[4] L. Soos, *Acta Geol. Acad. Sci. Hung.,* **8,** 1 (1964).

[5] J. M. Jones, *Econ. Geol.,* **57,** 42 (1962).

[6] D. G. Murchison, *Bull. Inst. Min. Met.,* **689,** 479 (1964).

[7] H. A. W. de Vries, P. J. Habets, and C. Bokhoven, *Brennstoff-Chemie,* **49,** 11 (1968).

[8] L. I. Sarbeeva, *Vop. Metamorf. Uglei Epigenza Umeschayuschikh Porod,* G. A. Ivonov, Ed., Izv. "Nauk" Leningrad. Otd., Leningrad, U.S.S.R., 1968, p. 50.

[9] J. A. Harrison, *Fuel,* **44,** 225 (1965).

[10] J. A. Harrison and J. Thomas, Jr., *Fuel,* **45,** 501 (1966).

[11] ASTM Standards, 1974, Part 26, Designations D2797 and D2798, p. 551.

[12] J. T. McCartney and S. Ergun, U.S. Bur. Mines Bull. No. 641, 1967, 49 pp.

[13] J. T. McCartney and L. J. E. Hofer, *Anal. Chem.,* **27,** 1320 (1955).

[14] K. Kotter and W. D. Langner, *Brennstoff-Chemie,* **42,** 380 (1961).

[15] J. A. Harrison, *Microscope,* **18,** 223 (1970)

[16] J. L. Bayer, G. H. Denton, and R. E. Hassel, ASTM Spec. Tech. Publ. No. 430, 1968, p. 118.

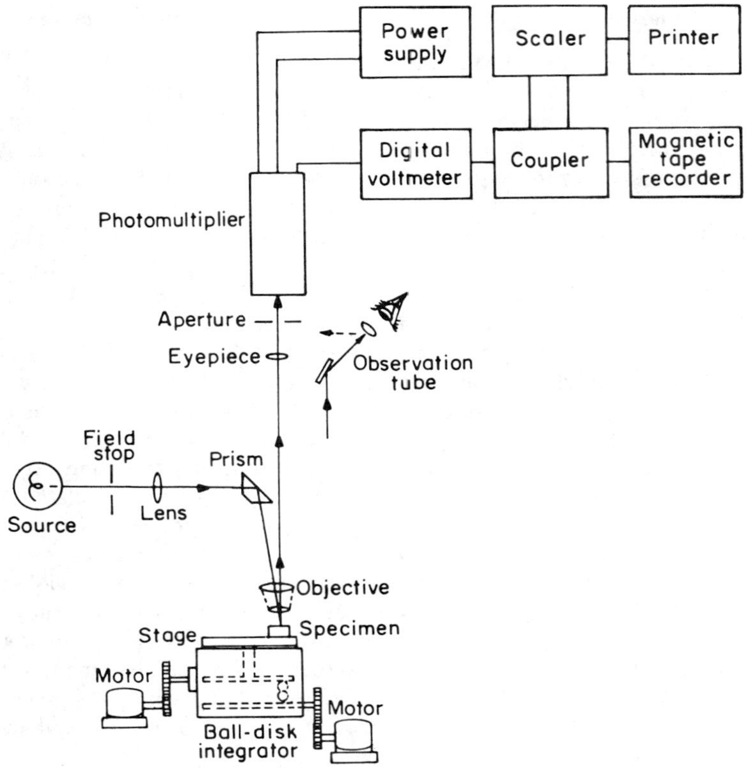

Figure 4.1 Schematic diagram of the automated reflectance scanning microscope system.[12]

for completely automated reflectance readings on a solid or pulverized coal sample by continuously recording reflected light intensities from areas as small as a fraction of a micrometer in diameter at a rate of 1000 points/s. These readings were stored in a number of channels of reflectance level with thresholds set according to a calibrated standard level, and they determined ranges of reflectance of components of interest in the coal sample. Size distributions of the components could also be determined. Such methods are open to error due to limited resolution of surface detail, boundary effects, and polishing defects. However applications described for materials somewhat less complex than coal show reasonably reproducible results fairly comparable to those obtained visually. McCartney and Ergun[17] described a similar technique of automated reflectance determination, but they recorded the reflected intensities on magnetic tape and analyzed the results by computer programming (Fig. 4.1). Application of the equipment to determination of pyrite content and size distribution in pulverized coal samples resulted in values for pyrite content within 15% of those determined by chemical analysis, not quite as good agreement as that between chemical and visual determinations. This equipment was also applied to determination of petrographic components, but the results were not promising.[18] However correlation of such information with behavior of certain coals in specific utilization processes may be useful. Davis and Vastola[19] have described one of the latest developments in automated microscopy of coal. Results with their equipment are generally similar to the foregoing, but the data can be gathered at much higher speeds.

[17] J. T. McCartney and S. Ergun, U.S. Bur. Mines Rep. Invest. No. 7256, 1969, 15 pp.

[18] J. T. McCartney, H. J. O'Donnell, and S. Ergun, *Fuel,* **50,** 226 (1971).

[19] A. Davis and F. J. Vastola, *J. Microsc.,* **109,** Pt. 1, 3 (1977).

4.2.3 Optical Constants

In the 1960s interest continued in the derivation of optical constants and consequent structure deductions from reflectance measurements on the vitrinite of coals. Piller and von Gehlen[20] discussed errors involved in reflectance measurement and in the derivation of refractive index *n* and absorption (extinction) coefficient *k* from reflectance. They concluded that reflectance measurement has several sources of significant error, which usually ranges from 1 to 5% and could reach 10% for low-reflecting materials such as coal. Errors in *n* and *k* would probably be comparable for the ranges found in coals. McCartney and Ergun[12] compared *n* and *k* values derived by two investigators and by reflection and transmission (Table 4.2). They found an average difference of 23% in *k* between reflection and transmission results and 36% between results of two sets of investigators. Some of the results for these comparisons were obtained by studies of ultrathin (0.05–0.10 μm) coal sections with an interference microscope.[21]

[20] H. Piller and K. von Gehlen, *Am. Miner.*, **49**, 867 (1964).
[21] J. T. McCartney and S. Ergun, *J. Opt. Soc. Am.*, **52**, 197 (1962).

4.2.4 Spectral Variation of Optical Constants

Most of the early studies of reflectance in coals were conducted in green light equivalent to the mercury line at 546 nm. Later work has dealt with variations of *R*, *n*, and *k* in the ultraviolet (UV) and visible spectra. Ergun and McCartney[22] found that the specific reflectances (reflectance in air divided by density *d*) of four vitrinites ranged between those of graphite and diamond, in order of decreasing rank. Those of graphite and anthracite increased with visible wavelength (440–867 nm), whereas those of three bituminous coals and diamond fell slightly or remained nearly the same. At 546 nm the *R*/*d* ratios were linearly related to H/C ratios and the linear relation extended to the *R*/*d* of graphite. Gilbert[23] made similar measurements over an extended UV spectral range (235–600 nm). He found that the reflectance of diamond increased continuously with decreasing wavelength in the UV, that of graphite peaked at 250 nm, and those of seven coal vitrains decreased. He found no evidence for

[22] S. Ergun and J. T. McCartney, *Fuel*, **39**, 449 (1960).
[23] L. A. Gilbert, *Fuel*, **39**, 393 (1960).

Table 4.2 Extinction Coefficients, Refractive Indices, and Absorption Indices of Vitrinites of Differing Ranks and of Graphite at 5460 Å[12]

Sample				Absorption Index		
					From Reflectance[c]	
Rank	State	Extinction Coefficient,[a] *k*	Refractive index,[b] *n*	From *n* and *k*	McCartney and Ergun	Huntjens and van Krevelen
Lignite	North Dakota	0.066	1.70	0.039		0.03
Subbituminous	Wyoming	0.093	1.72	0.054		0.04
hvCb	Illinois	0.12	1.76	0.068	0.04	0.05
hvAb	Pennsylvania	0.18	1.80	0.10	0.08	0.06
Mvb	West Virginia	0.24	1.90	0.13	0.12	0.07
lvb	West Virginia	0.22	1.85	0.12	0.13	0.08
Subanthracite	West Virginia	0.48	2.05	0.23	0.12–0.16[d]	0.11–0.12[d]
Anthracite	Pennsylvania	0.54	2.05	0.26	0.27–0.31[d]	0.16–0.26[d]
Graphite		1.12	2.20	0.51	0.66	0.56

[a] By transmission through ultrathin sections.
[b] By interferometry of ultrathin sections.
[c] Interpolated from reflectance-rank relationship.
[d] Range of minimum to maximum values.

Source: J. T. McCartney and S. Ergun, U.S. Bur. Mines Bull. No. 641, 1967, 49 pp.

absorption by individual condensed hydrocarbons in the vitrains. In a subsequent note[24] he reported similar and confirming results with three more vitrains.

Gilbert[25] later determined variations of *n* and *k* in six vitrinites in the spectral range 240–600 nm by measuring ratios of reflectance for light plane-polarized in and perpendicular to the plane of incidence, for two fixed angles of incidence (Fig. 4.2). Values of *n* and *k* were read from graphs constructed from these data. The bituminous coals showed marked peaks of *k* at about 270 nm, and an anthracite had continuously high absorption into the visible region. It was concluded that all the coals contain significant proportions of aromatic material, but that the degree of conjugation in anthracite is much greater. McCartney et al.[26] also determined variations of *n* and *k* in the UV-visible range for several vitrains, but used the two-media technique, which unlike Gilbert's, did not neglect possible anisotropy. Their results generally agreed with those of Gilbert, but showed less pronounced peaks of *k* in the UV and greater absorption in the visible region.

Ergun et al.[27] studied the absorption properties of vitrinites in the UV-visible by measuring transmission through ultrathin sections (50–220 nm) rather than by reflectance. Results were similar to those obtained in reflectance studies. The relation of absorption in coals to that in graphite was shown. It was concluded that coalification proceeds by two different mechanisms. In the subbituminous and bituminous stages, small aromatic nuclei are formed without much increase in size; in the anthracite stage, aromatic units coalesce to form much larger layers. However X-ray studies reveal that the ultimate transition to three-dimensional graphite crystallinity probably does not occur until the layers reach a size of about 3 nm.

Three investigations were conducted to extend determinations of *n* and *k* into the infrared (IR) spectral region. McCartney and Ergun[28] found that *n* and *k* for metaanthracite increased appreciably up to 2 μm as did that of graphite. The constants for an anthracite and lower rank coals, of increased magnitude with rank, varied little in the near IR. Foster and Howarth[29] extended the studies to 10 μm. Blokh and Burak[30] measured *n* and *k* for an anthracite in the 0.8–5.5 μm range. They found that *n* increased up to 5 μm and *k* up to 3 μm, both remaining constant subsequently.

4.2.5 Optical Anisotropy

It has been well established that the optical constants of vitrinites in high-rank coals depend on the orientation of the bedding planes of the coal with the impingent radiation. Most studies have shown that these coals are optically uniaxial, with the optic axis perpendicular to the bedding. Maximum values of the constants occur when the vibrating plane of the radiation is parallel to the bedding and minimum values occur when it is perpendicular to the bedding (Fig. 4.3). However one study[32] has claimed that anthracites are slightly biaxial, although not enough to affect significantly the use of maximum reflectance as a measure of rank.

De Vries et al.[31] discussed the anisotropy of vitrinites and pointed out one advantage of measuring maximum reflectance rather than average reflectance. They showed that maximum reflectance measurements on a pulverized coal in briquets vary less than those of average reflectance, and also that the maximum changes more strongly with rank than the average. As a consequence, they demonstrated that maximum *R* could be more useful than average *R* for distinguishing several coals in a blend.

Ergun et al.[33] studied the optical properties

[24] L. A. Gilbert, *Fuel,* **40,** 72 (1961).
[25] L. A. Gilbert, *Fuel,* **41,** 351 (1962).
[26] J. T. McCartney, J. B. Yasinsky, and S. Ergun, *Fuel,* **44,** 349 (1965).
[27] S. Ergun, J. T. McCartney, and R. E. Walline, *Fuel,* **40,** 109 (1961).
[28] J. T. McCartney and S. Ergun, *Sympo. Science Technology Coal,* Ottawa, March 29–31, 1967, Dept. Energy, Mines, Resources, Canada, p. 79.
[29] P. J. Foster and C. R. Howarth, *Carbon (Oxford),* **6,** 719 (1968).
[30] A. G. Blokh and L. D. Burak, *Inzh.-Fiz. Zh.,* **23,** 1054 (1972).
[31] H. A. W. de Vries, P. J. Habets, and C. Bokhoven, *Brennstoff-Chemie,* **49,** 47 (1968); also in *Geol. Mijnbouw.,* **47,** 423 (1968).
[32] A. C. Cook, D. G. Murchison, and E. Scott, *Fuel,* **51,** 180 (1972).
[33] S. Ergun, J. B. Yasinsky, and J. R. Townsend, *Carbon (Oxford),* **5,** 403 (1967).

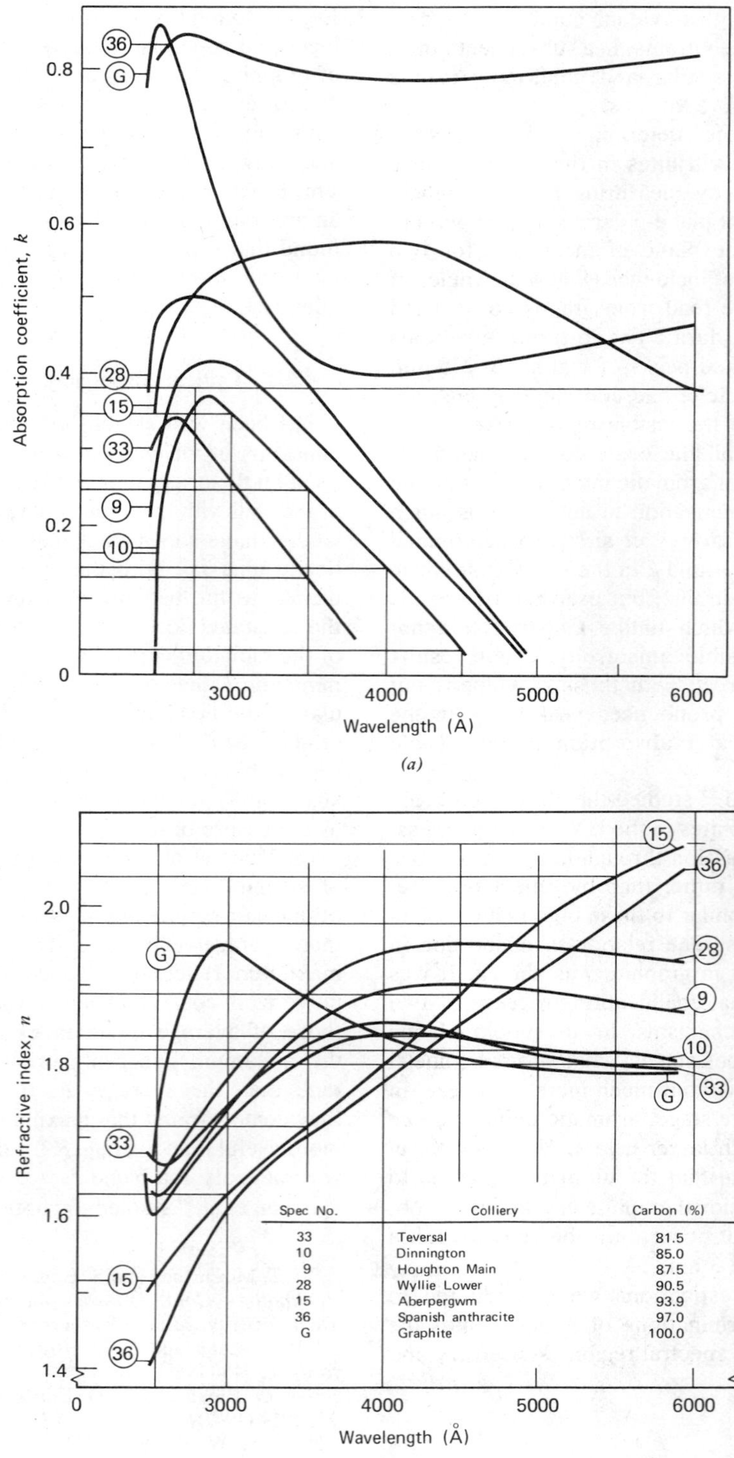

Figure 4.2 Variation of optical constants of vitrinites in the UV-visible spectrum: (*a*) absorption coefficient versus wavelength and (*b*) refractive index versus wavelength.[25]

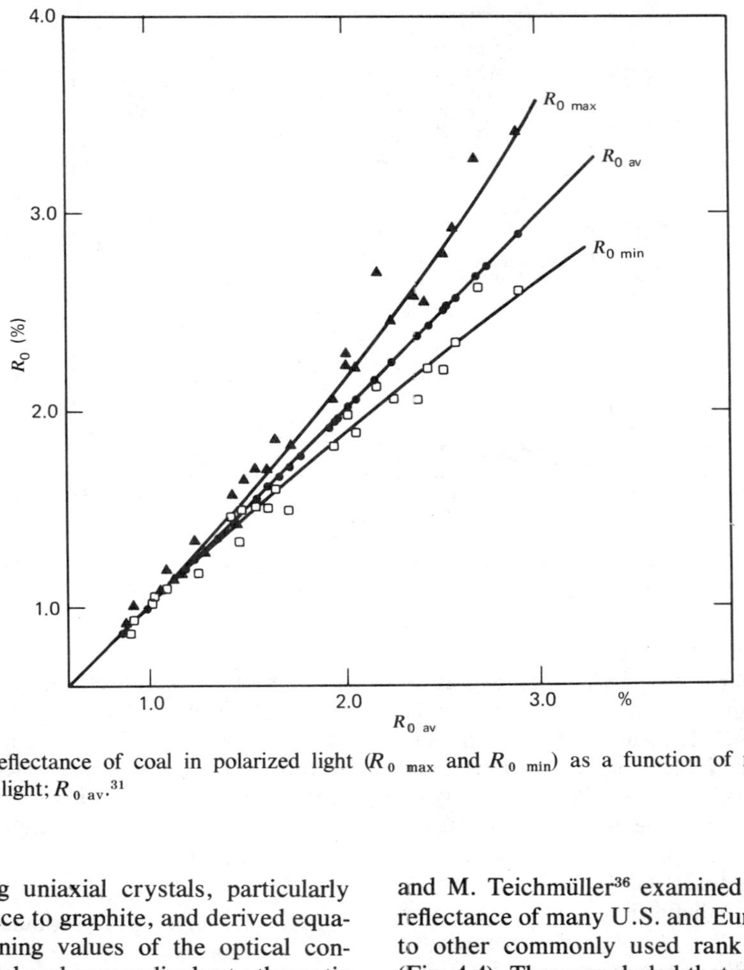

Figure 4.3 Reflectance of coal in polarized light ($R_{0\ max}$ and $R_{0\ min}$) as a function of reflectance in nonpolarized light; $R_{0\ av}$.[31]

of absorbing uniaxial crystals, particularly with reference to graphite, and derived equations governing values of the optical constants parallel and perpendicular to the optic axis. A review of the work of Soviet geologists on the optical anisotropy of brown coal was published.[34] Malan[35] discussed reflectance, luminescence, and anisotropy of lignite.

4.2.6 Application of Coal Reflectance to Geology and Metamorphism

The continuous increase of vitrinite reflectance with the rank of coal has been discussed by a number of investigators. It usually is ascribed to the growth of aromatic layer clusters and to greater ordering of these parallel to the bedding planes. McCartney and M. Teichmüller[36] examined relations of reflectance of many U.S. and European coals to other commonly used rank parameters (Fig. 4.4). They concluded that reflectance is the best single criterion of rank in bituminous coals with less than 30% volatile matter and is probably the best for anthracites. For high-volatile bituminous coals reflectance seems to be of nearly equal value to moisture content or calorific value. Its use for judging rank in subbituminous coals and lignites is questionable. The authors list the reflectance ranges established for present rank categories in the United States and propose that they be included in standard classification systems as supplementary criteria of rank.

Ponomareva et al.[37] studied the reflectance characteristics of some 60 Donets basin

[34] P. Mandev, *God. Sofiskiya Univ., Geol.-Geograf., Fak., Kniga, 1-Geol,* **57,** 241 (1964).

[35] O. Malan, *Freiberger Forschungsh.,* **A253,** 5 (1962).

[36] J. T. McCartney and M. Teichmüller, *Fuel,* **51,** 64 (1972).

[37] M. N. Ponomareva, M. M. Lifshits, and G. P. Vyrich, *Issled. Klassifik. Uglei, Donetskii-Nauchn.-Issled. Ugol'n Inst.* No. 25, Moscow, Sb. 1962, p. 52.

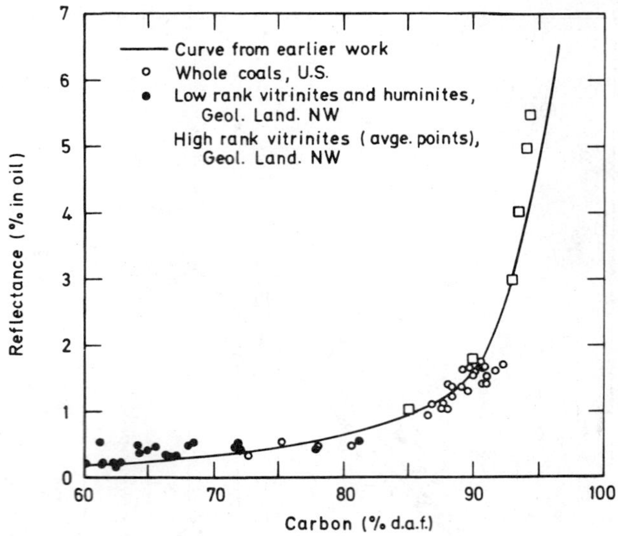

Figure 4.4 Reflectance-carbon relations for European and U.S. whole coals or vitrinites.[36]

coals. They later showed how maximum reflectance characterizes the true degree of metamorphism of coals from six basins in the Soviet Union.[38] They also examined the use of n and k and indices of metamorphism and concluded that n is suitable for classifying coals from cannel to lean, but inadequate for anthracites. Absorption coefficient k is applicable to lean coals and those of high rank.[39] Dobranravov[40] defined a volume reflection index $R_v = \frac{2}{3}R_{\max} + \frac{1}{3}R_{\min}$, as the mean for three mutually perpendicular planes. This characterizes all three factors affecting metamorphism—temperature, pressure, and time. The average of measurements in unpolarized light on a large number of random grains in a briquet is $R_v = R_{av}$. The isothermal reflectance, expressed in Equation 4.2, is an index of the temperature of coalification.

$$R_t = \frac{R_{\max} + R_{\min}}{2} \qquad (4.2)$$

Chaudhuri et al.[41] traced a relation between variation of reflectance with rank and variations in the content of oxygen functional groups. They attempted to correlate apparent points of inflection in the rate-change curve of reflectance versus carbon content with changes in oxygen distribution in terms of carboxyl, hydroxyl, carbonyl, and unaccounted oxygen.

4.2.7 Geologic and Miscellaneous Properties

Reflectance has been used by a number of investigators to characterize coals of Japan,[42] Antarctica,[43] France,[44] New Zealand,[45] Portugal,[46] the Himalayas,[47] India,[48,49] Bulgaria,[50] Poland,[51] and Czechoslovakia.[52]

[38] N. M. Dvuzhil'naya, M. N. Ponomareva, and G. P. Vyrich, *Issled., Ispol'z. Stand. Uglei., Donetskii-Nauchn.-Issled. Ugol'n. Inst.* 1965, p. 3.
[39] M. N. Ponomareva, M. M. Lifshits, and G. P. Vyrich, in Ref. 38, p. 20.
[40] V. F. Dobranravov, *Khim. Tverd. Topl.* **9,** 11 (1975).
[41] S. G. Chaudhuri, A. K. Gupta, G. S. De, M. K. Basu, S. Tumuluri, and K. C. Lahiri, *J. Mines, Met., Fuels,* 131, (April 1976).

[42] H. Sugimura and M. Shiwa, *Nenryo Kyokaishi,* **46** (488), 911 (1967).
[43] N. Schapiro and R. J. Gray, in Am. Chem. Soc. Adv. Chem. Ser. No. 55, (1966) p. 196.
[44] E. Meriaux, *Ann. Soc. Geol. Nord (Lille),* **84,** 19 (1964).
[45] D. Chandra, *Econ. Geol.,* **60,** 1041 (1965).
[46] M. J. Lemos DeSousa, Separata do Tomo 5S, *Communicacoes does Lervicos Geologicas de Portugal,* Lisbon, 1971, p. 181.
[47] T. K. Ghosh, *Fuel,* **49,** 226 (1970).
[48] T. K. Ghosh, *Econ. Geol.,* **64,** 683 (1969).
[49] M. K. Rishi, *Econ. Geol.,* **65,** 700 (1970).
[50] G. Shishkov, *Dokl. Bolg. Akad. Nauk,* **24,** 79 (1971).
[51] S. Jasienko and K. Bratek, *Pr. Nauk Inst. Chem. Tech. Nofty Wegla, Politech. Wroclaw,* **25,** 41 (1975).
[52] M. Kosina et al., *Uhli,* **23,** 65 (1975).

Jedwab[53] discussed optical anomalies, such as high reflectance and anisotropic halos, in coals affected by radioactivity of adjacent minerals. Soos[54] studied optical properties of coal macerals. Ammosov and Malinin[55] related reflectance to metamorphism and the structures of related rocks. Ammosov also analyzed the ranges and causes of reflectance variations in vitrinites of all levels of metamorphism.[56] He pointed out the greater ranges of reflectance in higher rank coals and attributed them to variations in ordering of the molecular structures in different areas of the vitrinite. Variations in lower rank vitrinites are attributed to differences in the precursory plant material. Alpern discussed the reflectance-metamorphism relation[57] and the usefulness of coal and bitumen reflectance in petroleum prospecting.[58] Zahn et al.[59] studied optical activity in oils derived from coals. Olaf and Robock[60] developed the theory of light scattering by coal and mineral particles.

4.2.8 Effects of Temperature or Carbonization

Previously it had been shown that the reflectances of coals increase as they are carbonized at increasing temperatures, and that the increase becomes sharp beyond a temperature in the range 300–500°C, depending on the rank of the coal. Ghosh[61] had found that the ultimate reflectance depends on the rank, carbonizing temperature, and rate and time of heating, and that it is possible in turn to deduce something about these factors from a study of the reflectance. De Vries et al.[62] presented a critical reexamination of changes in reflectance at increasing temperature (Fig. 4.5). Their conclusions essentially agreed with those of earlier investigators concerning rates of increase of R in various temperature ranges and also concerning the importance of "soak" at a particular temperature, which causes R to continue to rise. They also discussed the development of random anisotropic grain structure in the plastic state and the increase of this effect with temperature, probably due to condensation of the aromatic clusters.

Marshall and Murchison[63] studied three vitrinites of differing rank carbonized at 450, 600, and 750°C and recorded the variations of the optical parameters with wavelength in the visible spectrum. They found that the dispersion trends for the carbonized samples up to 600°C were similar to those reported for naturally occurring vitrinites from high-volatile to anthracitic rank. Carbonization at 750°C caused further changes not observed in natural vitrinites. These included a continued rise in reflectance and anisotropy and changes in dispersions of the constants that correlate with increased aromatic layer diameters and improved stacking order. Chandra[64] and Marshall and Murchison[65] engaged in some rather minor discussions concerning the latter's paper. Goodarzi and Murchison[66] made a more detailed study by carbonizing three vitrinite-rich coals at 25°C intervals over the range 20–950°C. These experiments at small temperature intervals emphasized that refractive index is more sensitive than reflectance as an indicator of structural changes, particularly in correlating with crystallite height in the aromatic systems, as indicated by X-ray diffraction studies. This work also emphasized the effect of the plastic state during heating of coking coals, as contrasted with lower and higher rank noncoking coals, in promoting structure with better ordered crystallite orientation. These observations conform to the finding that the plastic state favors the development of "graphitizing" carbons that eventually form well-or-

[53] J. Jedwab, in Am. Chem. Soc. Adv. Chem. Ser. No. 55, (1966), p. 119.
[54] L. Soos, *Acta Chim. Acad. Sci. Hung.*, **47**, 67 (1966).
[55] I. I. Ammosov and S. I. Malinin, *Geol. Uglei Sibiri Dal'nego Vostoka, Akad. Nauk SSSR, Sibirsk. Otd., Inst. Geol. Geofiz.*, 1965. p. 5.
[56] I. I. Ammosov, *Khimi. Tverd. Topl.*, **8**, 3 (1974). English translation by Allerton Press, Inc., Philadelphia, p. 1.
[57] B. Alpern, *Ann. Soc. Geol. Nord.*, **89**, 143 (1969).
[58] B. Alpern, *C. R. Acad. Sci., Ser. D.*, **272**, 1717 (1971).
[59] C. Zahn, B. D. Blaustein, G. Pantages, and I. Wender, U.S. Bur. Mines Rep. Invest. No. 6525, 1964, 32 pp.
[60] J. Olaf and K. Robock, *Staub*, **21**, 495 (1961).
[61] T. K. Ghosh, *Econ. Geol.*, **63**, 182 (1968).
[62] H. A. W. de Vries, P. J. Habets, and C. Bokhoven, *Brennstoff-Chemie*, **49**, 105 (1968).

[63] R. J. Marshall and D. G. Murchison, *Fuel*, **50**, 4 (1971).
[64] D. Chandra, *Fuel*, **51**, 88 (1972).
[65] R. J. Marshall and D. G. Murchison, *Fuel*, **51**, 252 (1972).
[66] F. Goodarzi and D. G. Murchison, *Fuel*, **51**, 322 (1972).

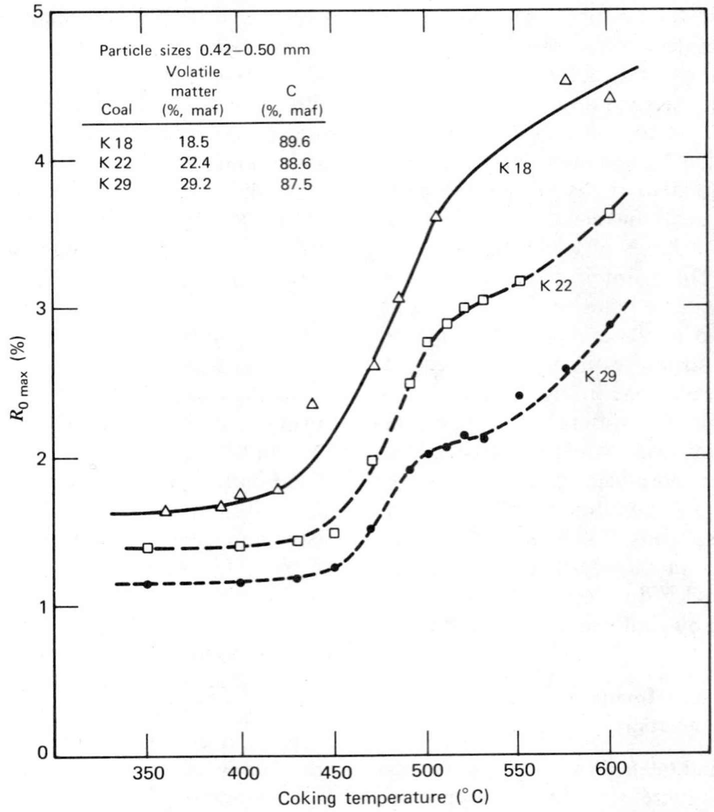

Figure 4.5 Variations of reflectance of vitrinites as a function of temperature.[62]

dered graphite. Finally, Goodarzi and Murchison[67] investigated optical properties of carbonized vitrinites after preoxidation for 14 days at 105°C. Preoxidation produced only relatively small changes in the trends of optical parameters of fresh vitrinites with temperature. The most noticeable difference was the reduction in contrast between the refractive index track for the coking coal vitrinite and that for the lower and higher rank vitrinites.

Brown and Taylor[68] studied the optical properties and other characteristics of unusual Antarctic coals. They found that the reflectance and anisotropy were higher than normal for all six coals and that two had values higher than any reported in the literature for anthracites, approaching those of graphite. Other characteristics, such as X-ray diffraction parameters, confirmed this unusual behavior. They concluded that all the coals had been subject to thermal metamorphosis at 600 to 1000°C. Chandra[69] also found that the maximum reflectances of thermally metamorphosed and laboratory-carbonized coals were greater than those of normal coals at any given carbon content. Chandra[70] later claimed that the temperatures to which thermally metamorphosed or carbonized coals were subjected, with freedom for escape of decomposition products, could be estimated by measuring their reflectance. Finally, Chandra[71] studied reflectances of ball coals (separable rounded entities found in coal beds) and found that some were thermally affected, possibly by nearby igneous intrusions, although visibly of the same composition as the surrounding coal.

[67] F. Goodarzi and D. G. Murchison, *Fuel,* **52,** 164 (1973).
[68] H. R. Brown and G. H. Taylor, *Fuel,* **40,** 211 (1961).
[69] D. Chandra, *Fuel,* **42,** 69 (1963).
[70] D. Chandra, *Fuel,* **44,** 171 (1965).
[71] D. Chandra, *Fuel,* **48,** 271 (1969).

4.2.9 Oxidation

Chandra[72] reported that mild oxidation of two coals in the laboratory had resulted in no significant changes in reflectance. He later examined weathered coals from several seams and found no change in reflectance.[73] Later he exposed eight coals with carbon content ranging from 77 to 93% to the atmosphere for 10 years and again found no effect on reflectance.[74] Work by other investigators has shown an increase in reflectance near the surfaces of oxidized coal particles. It was suggested that oxidation above 170°C was required to produce these "oxidation rims."[75]

4.3 PORE STRUCTURE, SURFACE AREA, AND DENSITY*

Coals are highly porous materials, and a significant proportion of their total open pore volume is located in very small pores. Porosity of coal and its dependence on rank are of great practical importance. Pore volume and pore size distribution of coals determine the ease with which methane can diffuse out of the internal pore structure during mining. The preparation of coals for marketing by removal of mineral matter depends on the difference in the specific gravities of the coal and the minerals; the specific gravity of the organic fraction of the coal is partly dependent on its porosity. The porosity and pore size distribution of coals have their greatest significance in the utilization processes. A number of carbonaceous adsorbents used commercially in gas and water purification are prepared by heating coal in the absence of air to produce a char, then activating the char in an oxidizing atmosphere. Although the porosity of the char, not that of the coal precursor, is used in these materials, a knowledge of the porosity of the coal makes it possible to predict whether it will give a suitable char for a specific application. In the use of metallurgical coke and in the gasification and liquefaction conversion processes, chemical reactions must occur between the reactant(s) and the surface of the solids, much of which is located within pores. Also, the reaction products must be able to diffuse out rapidly to permit access of fresh reactants. Since all these processes involve some thermal treatment, some kind of char is always an intermediate product of the process. Again, the rank of the coal and the nature of its pore structure permit a prediction of the kind of porosity that will be developed during processing and, hence, an assessment of the coal for the above processes.

The term "ultrafine structure" of coals has been used by Dryden[76] to define holes whose sizes range from gaps a few angstroms wide to comparatively larger channels up to 100 Å wide. In addition, large capillaries and cracks are present in the coal structure.[76] In the case of porous adsorbents, Dubinin[77] has suggested use of the following classification, which is now also generally accepted for coals, for distinguishing between pores of different sizes:

1. *Macropores:* Pores that have diameters greater than 200 Å.
2. *Transitional pores:* Pores that are between 20 and 200 Å in diameter.
3. *Micropores:* Pores that have diameters less than 20 Å.

Dryden[76] has referred to the methods used for the quantitative characterization of the internal pore structure of coals. Briefly, they are (1) helium and apparent densities, which can give total open pore volumes and volumes accessible to adsorbates of different molecular sizes; (2) heat of wetting, from which surface areas can be obtained; and (3) vapor and gas adsorption, from which both volume and area distributions can be derived.

4.3.1 Surface Area

Heat of wetting in methanol has been used extensively in the past for determining surface areas of coals. The relevant literature

[72] D. Chandra, *Econ. Geol.*, **53**, 102 (1958).
[73] D. Chandra, *Fuel*, **41**, 185 (1962).
[74] D. Chandra, *Econ. Geol.*, **61**, 754 (1966).
[75] B. Alpern and F. Maume, *Rev. Ind. Miner.*, **51**, 979 (1969).
* Contributed by Om P. Mahajan and Philip L. Walker, Jr.

[76] *Chemistry of Coal Utilization*, Supplementary Volume, H. H. Lowry, Ed., John Wiley and Sons, New York, 1963, p. 232.
[77] M. M. Dubinin, in *Chemistry and Physics of Carbon*, Vol. 2, P. L. Walker, Jr., Ed., Marcel Dekker, New York, 1966, p. 51.

has been adequately reviewed by Dryden[76] and Marsh.[78] The method is now only of historical importance. There is overwhelming evidence[76,78–80] that adsorption of methanol on carbonaceous solids involves specific interactions between the hydroxyl group of methanol and oxygen-containing functional groups present on the carbon surfaces. Therefore it is not surprising that heat of wetting in methanol per unit surface area is a function of the concentration of oxygen functional groups, hence of coal rank.[81,82] Furthermore, Fugassi et al.[83] have suggested that the structure of coal gel is changed irreversibly by the sorption of methanol. They showed that the sorption of methanol and *n*-pentane on a Pittsburgh Seam coal was isothermally reversible. However following desorption of methanol, subsequent adsorption of *n*-pentane increased by about 100%.

Adsorption of nitrogen (N_2) at liquid N_2 temperature (−196°C) has conventionally been used for measuring the Brunauer-Emmett-Teller (BET) surface areas of porous and nonporous solids.[84] However N_2 adsorption on coals measures surface areas that are considerably lower than those expected. Maggs[85] showed that in the temperature range −196 to −93°C, N_2 adsorption on coals exhibited a positive temperature dependence. Since adsorption is an exothermic process, the extent of physical adsorption should actually decrease with increasing temperature. It has been suggested[86,87] that at −196°C the system of fine pores in coals is inaccessible to N_2 molecules because of an activated diffusion process and/or shrinkage of pores.

Ganguli et al.[88] have shown that at −196°C enhanced adsorption of argon (Ar) on coals is obtained when this gas is introduced at room temperature and the samples are then cooled slowly to −196°C, rather than when the samples are first cooled to −196°C, then exposed to Ar. Korta et al.[89] studied the adsorption of N_2 on a coal already cooled to −196°C and on a coal that was simultaneously exposed to N_2 and cooled to −196°C. The latter method gave a higher value of surface area. Nandi and Walker[90] studied diffusion of N_2 from coals at temperatures between 25 and 140°C. Diffusion was activated; once the activation energy was known, the diffusion parameter at −196°C could be estimated. They concluded that the low N_2 areas of coals measured at −196°C are due to very low diffusion rates of N_2 and that the physical shrinkage of pores is not of primary importance. Anderson et al.[91] have suggested that N_2 adsorption measures the external area of coals. Walker and Geller[86] concluded that the N_2 surface area represents the macropore area of coals.

There is another drawback associated with the use of N_2 adsorption at −196°C for surface area measurements. Lamond and Marsh[92] suggested that when the diameters of pores in a carbon molecule are only a few times larger than the diameter of a N_2 molecule, the N_2 at −196°C fills these pores at very low relative vapor pressures, resulting in reversible capillary condensation before the apparent monolayer capacity is reached. Obviously this can result in unrealistically high values of the BET areas.

Walker and co-workers[93,94] have determined the N_2 surface areas of a number of coals varying in rank from anthracite to lignite. Figure 4.6 plots these areas, expressed

[78] H. Marsh, *Fuel,* **44,** 253 (1965).

[79] D. H. T. Spencer, in *Porous Carbon Solids,* R. L. Bond, Ed., Academic Press, New York, 1967, p. 87.

[80] B. R. Puri, in *Chemistry and Physics of Carbon,* Vol. 6, P. L. Walker, Jr., Ed., Marcel Dekker, New York, 1970, p. 191.

[81] L. Robert and S. Pregermain, *Fuel,* **42,** 389 (1963).

[82] L. Robert and H. Brusset, *Fuel,* **44,** 309 (1965).

[83] P. Fugassi, R. Hudson, and G. Ostapchenko, *Fuel,* **37,** 25 (1958).

[84] S. Brunauer, *The Adsorption of Gases and Vapours,* Oxford University Press, Oxford, and Princeton University Press, Princeton, N.J., 1945.

[85] F. A. P. Maggs, *Research,* **6,** 13S (1953).

[86] P. L. Walker, Jr., and I. Geller, *Nature (London),* **178,** 1001 (1956).

[87] R. B. Anderson, W. K. Hall, J. A. Leckey, and K. C. Stein, *J. Phys. Chem.,* **60,** 1548 (1956).

[88] N. C. Ganguli, P. N. Mukherjee, and A. Lahiri, *Fuel,* **40,** 525 (1961).

[89] A. Korta, M. Lason, J. Kawecka, and J. Klosinska-Drwalova, *Arch. Gorn.,* **1959,** 219; cf. *Fuel Abstracts Current Titles,* Vol. 2, Abstr. No. 81, 1961

[90] S. P. Nandi and P. L. Walker, Jr., *Fuel,* **43,** 385 (1964).

[91] R. B. Anderson, J. Bayer, and L. J. E. Hofer, *Fuel,* **44,** 443 (1965).

[92] T. G. Lamond and H. Marsh, *Carbon (Oxford),* **1,** 281 (1964).

[93] H. Gan, S. P. Nandi, and P. L. Walker, Jr., *Fuel,* **51,** 272 (1972).

[94] S. P. Nandi and P. L. Walker, Jr., *Fuel,* **50,** 345 (1971).

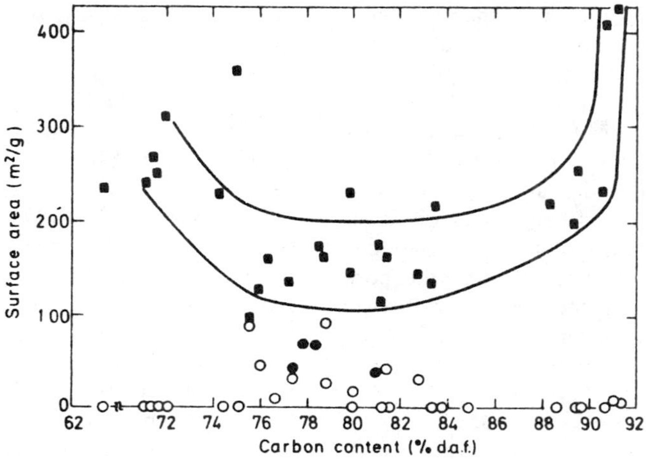

Figure 4.6 Variation of N_2 and CO_2 surface areas of coals with carbon content: ● = N_2 (from Ref. 93): ○ = N_2, ■ = CO_2 (from Ref. 94).

on a mineral-matter-containing basis, versus carbon contents, expressed on a moisture- and ash-free (maf) basis. In general, coals with N_2 areas greater than 10 m^2/g fall in the carbon content range of about 76–82%. Surface areas of most coals on both sides of this carbon content range are less than 1 m^2/g. However anthracites are exceptions to this general behavior and have surface areas of 5–8 m^2/g.

Surface areas of mineral matter (measured by nitrogen adsorption) obtained from the low-temperature ashing of coals can be as high as 10 m^2/g.[95] Since in some cases the specific surface of the whole coal has been found to be less than 1 m^2/g (Fig. 4.6), it has been suggested[93] that finely divided mineral particles distributed in the coal organic matrix are not accessible to N_2 at −196°C.

Because of the inherent limitations associated with the use of adsorption of N_2 at −196°C, it was realized that to get meaningful surface areas of coals, it was essential to measure the adsorption of gases at the highest temperature possible. Bond and Spencer[96] used the extent of neon (Ne) adsorption on a carbon black (Neo-Spectra) at 25°C and 760 torr as a basis for estimating surface areas of coals and chars. That is, it was assumed that the amount of Ne adsorbed per unit surface of carbon black could be taken to convert Ne uptake on coals at 25°C and 760 torr to surface area. However Marsh and Wynne-Jones[97] observed that under the same experimental conditions used by Bond and Spencer,[96] the amounts of N_2 and oxygen adsorbed per unit surface of carbons varied markedly with the nature of the starting material and the carbonization temperature. The Ne adsorption method is empirical and should be valid only for adsorbents associated with adsorption potentials similar to those involved in the adsorption on the Neo-Spectra carbon black. Since adsorption of Ne was studied much above its critical temperature, the fraction of surface covered at 760 torr is expected to be very small. It is known[92,98,99] that at low relative vapor pressures (i.e., low surface coverages), the extent of physical adsorption is governed primarily by the pore size of the adsorbent rather than by the magnitude of its surface. Lamond and Marsh[92] have shown that on increasing activation of a given carbon, which results in enlarging the pore size, the quantity of a substance adsorbed at lower pressures decreases, even though the surface area of the sample increases. It has been suggested[92]

[95] J. V. O'Gorman, "Mineral Matter and Trace Elements in U.S. Coals," Ph.D. thesis, Pennsylvania State University, 1971.

[96] R. L. Bond and D. H. T. Spencer, *Conf. Industrial Carbon Graphite*, Soc. Chem. Ind., London, 1957, p. 231.

[97] H. Marsh and W. F. K. Wynne-Jones, *Carbon (Oxford)*, **1**, 269 (1964).

[98] O. P. Mahajan and P. L. Walker, Jr., *J. Colloid Interface Sci.*, **29**, 129 (1969).

[99] O. P. Mahajan, M. Morishita, and P. L. Walker, Jr., *Carbon (Oxford)*, **8**, 167 (1970).

that as the pore size increases, the probability of interaction of the adsorbate molecule with more than one pore wall decreases. As a result, the heat of adsorption, hence the extent of adsorption per unit surface, decreases with increase in pore size.

Adsorption of hydrocarbons at or near room temperature has been used for determining surface areas of coals.[100–102] Juentgen and Langhoff[102] have shown that the pore area, calculated by the Langmuir theory, decreases with increasing volatile matter of the coal and is expressed by Equation 4.3:

$$A_{CH_4}(m^2/g) = 105.9 - 2.1086\,VM + 0.0213\,VM^2 \quad (4.3)$$

where *VM* is the percent of volatile matter. However it has been suggested[103,104] that hydrocarbon gases be used with caution in making surface area measurements because of the possibility of dissolution of the gases in the coal structure. Furthermore, it has been reported[83] that adsorption of hydrocarbons on coals brings about irreversible changes in their structure.

Surface areas of coals have been determined from the adsorption of krypton (Kr) and xenon (Xe) at low temperatures.[104–107] The BET surface areas calculated from the adsorption of Kr at −110 and −90°C are significantly higher than those obtained from the adsorption of Ar or N_2 at −196°C.[104] Kini[106] suggested that at −110°C, Kr does not completely penetrate the coal pore structure. He used adsorption of Kr at −78°C and of Xe at 0°C for surface area measurements.[107] However he observed some molecular sieving effects with Xe in some coals and chars.

Carbon dioxide adsorption at −78°C has been used extensively for surface area measurements of coals.[86,91,92,97,108,109] Adsorption of CO_2 at −78°C greatly exceeds the adsorption of N_2 at −196°C. Anderson et al.[108] have suggested that reliable surface areas, not complicated by diffusion or imbibition, can be obtained from the CO_2 isotherms at −78°C. Nandi and Walker[90] studied the diffusion of CO_2 and N_2 from four coals varying in rank from anthracite to high-volatile bituminous. They found that N_2 has a higher activation energy for diffusion than does CO_2. They estimated that for an equilibration time of 30 min (per sorption point), about 85% of the pore volume of the anthracite would be filled with CO_2 at −78°C, whereas only 32% of the pore volume of a bituminous coal having a carbon content (maf) of 83% would be filled with CO_2. However at 25°C essentially the entire pore volume of all four coals investigated would be filled with CO_2. Bearing this in mind, Walker and Kini[110] used CO_2 adsorption at 25°C to calculate surface areas of coals. For the sake of comparison, they also measured surface areas from adsorption of N_2 (−196°C), Kr (−78°C), CO_2 (−78°C), and Xe (0°C). To use the BET equation, it is necessary to measure sorption up to a relative vapor pressure of about 0.2.[84] Since CO_2 at 25°C and Xe and Kr at the selected adsorption temperatures have saturation vapor pressures far in excess of atmospheric pressure, it was necessary to use a high-pressure system for adsorption. The surface area values obtained by Walker and Kini are listed in Table 4.3. These results show that adsorption of CO_2 at −78°C and of Xe at 0°C should *usually* measure essentially the total surface of coals, whereas CO_2 adsorption at 25°C should *invariably* measure essentially the total surface area of coals. However adsorption of CO_2 at −78°C for surface area measurements should be used with caution. For calculating the molecular area of adsorbed CO_2 at −78°C, CO_2 is usually assumed to be in the supercooled liquid state. However Walker and Kini[110] argue that this may not be true when adsorption occurs

[100] J. Van der Sommen, P. Zwietering, B. J. M. Eillebrecht, and D. W. van Krevelen, *Fuel,* **34,** 444 (1955).

[101] S. J. Gregg and M. I. Pope, *Fuel,* **38,** 501 (1959).

[102] H. Juentgen and J. Langhoff, *Brennstoff-Chemie,* **45,** 50 (1964).

[103] K. A. Kini and A. Lahiri, *Fuel,* **39,** 515 (1960).

[104] L. D. Ahuja, S. P. Nandi, K. A. Kini, and A. Lahiri, *J. Sci. Ind. Res. (India),* **20B,** 295 (1961).

[105] S. P. Nandi, K. A. Kini, and A. Lahiri, *Proc. Symp. Nature Coal,* Jealgora, India, Central Fuel Research Institute, New Delhi, 1959, p. 119.

[106] K. A. Kini, *Fuel,* **42,** 103 (1963).

[107] K. A. Kini, *Fuel,* **43,** 173 (1964).

[108] R. B. Anderson, L. J. E. Hofer, and J. Bayer, *Fuel,* **41,** 559 (1962).

[109] J. Thomas, Jr., and H. Damberger, Illinois State Geol. Survey, Circ. No. 493, (1976), p. 1.

[110] P. L. Walker, Jr., and K. A. Kini, *Fuel,* **44,** 453 (1965).

Table 4.3 Surface Areas of Coals as Measured by Different Adsorbates and at Different Temperatures[110]

	Surface area (m^2/g)				
C (%, maf)	N_2, (−196°C)	Kr, (−78°C)	CO_2, (−78°C)	Xe, (0°C)	CO_2, (25°C)
95.2	34	176	246	226	224
90.0	Nil	96	146	141	146
86.2	Nil	34	107	109	125
83.6	Nil	20	80	62	104
79.2	11	17	92	84	132
72.7	22	84	198	149	139

Source: P. L. Walker, Jr., and K. A. Kini, *Fuel,* 44, 453 (1965).

in very fine pores. They have concluded that for such systems the true state of adsorbed CO_2 at −78°C is intermediate between a perfect bulk liquid and a perfect bulk solid.

Marsh and Siemieniewska[111] suggested the use of the Polanyi-Dubinin (P-D) equation, given below, to calculate surface area of coals from CO_2 adsorption at 0 and 20°C:

$$\log V = \log V_0 - \frac{0.434BT^2}{\beta^2}\left(\log\frac{p_s}{p}\right)^2 \quad (4.4)$$

where V = amount adsorbed at equilibrium pressure p

V_0 = micropore capacity

p_s = saturation vapor pressure of the adsorbate

B = a constant that is a measure of the micropore size

T = temperature of adsorbate in K

β = affinity coefficient of the adsorbate relative to N_2 or benzene

The micropore capacity, V_0, is obtained from the intercept of the plot between log V and $[\log (p_s/p)]^2$. The micropore capacity can be converted to a micropore surface area if the area occupied by the adsorbate molecule is known. The unique feature of the P-D equation is that it permits the calculation of surface area from adsorption data taken below 1 atm pressure and obtained in a conventional volumetric apparatus. For a number of coals and chars, excellent agreement has been reported between the surface areas calculated by the BET and P-D equations from the adsorption of CO_2 at 25°C measured in two different pressure ranges.[112] That is, the micropore surface area (from P-D) very closely approximates the total surface area (from BET) in coals because of the large percentage of microporosity existing in coals.

The data of Marsh and Siemieniewska[111] show that higher rank coals have higher CO_2 areas than lower rank coals, and that a shallow minimum in area occurs at about 89% carbon. They suggest that, since surface areas of coals of different rank are not far from 200 m^2/g, the observed variations in the areas with rank might be caused by differences in accessibility rather than by differences in true surface area. Thomas and Damberger agree with this.[109] Gan et al.[93] measured CO_2 areas (25°C) for a number of coals varying in rank from anthracite to lignite. The CO_2 areas of different samples are plotted against carbon contents of the coals in Fig. 4.6. There is a much wider spread in the CO_2 areas than that reported by Marsh and Siemieniewska.[111] The area decreases, in general, with a decrease in carbon content down to about 83%. It remains more or less constant down to about 76% carbon content and increases thereafter. The CO_2 areas for coals of similar carbon content fall within a band rather than on a line. In general, coals with higher N_2 areas have lower CO_2 areas (this is not true for Illinois hvAb coals).

It has been argued[91,113] that the adsorption of CO_2 may be influenced by the quadrupole

[111] H. Marsh and T. Siemieniewska, *Fuel,* **44,** 355 (1965).

[112] P. L. Walker, Jr., and R. I. Patel, *Fuel,* **49,** 91 (1970).

[113] V. R. Deitz, F. G. Carpenter, and R. B. Arnold, *Carbon (Oxford),* **1,** 245 (1964).

moment of the molecule. Deitz et al.[113] studied the adsorption of CO_2 on various carbons and observed that the amount of CO_2 adsorbed per square meter of surface area (as measured from N_2 isotherms) and the heat of adsorption of CO_2 increased with increase in the fraction of surface covered by hydroxyl groups. Anderson et al.[91] have reported that on carbon blacks the amount of CO_2 adsorbed (−78°C) relative to N_2 (−196°C) increased with increasing oxygen content of the carbon blacks per unit surface area, as estimated from the N_2 isotherms. The surface area of a sample of lampblack, as measured from CO_2 adsorption, was found to be about 4 times as high as that measured by N_2 adsorption. The enhanced adsorption of CO_2 was attributed to the chemical interaction of CO_2 with the oxygen-containing groups present on the carbon surfaces. But Ramsey[114] showed that although gamma irradiation of an anthracite in air increased the oxygen content by about 40%, the CO_2 area increased by only 5%. He concluded that the adsorption of CO_2 can be used for determining surface areas of not only anthracites but also of highly oxygenated coals. Marsh[78] has suggested that because of enhanced adsorption potential in microporous coals, the additional effect of surface oxygen groups on CO_2 adsorption is slight. Marsh's conclusion appears to be supported by data presented in Table 4.3. That is, Xe surface areas measured at 0°C are in reasonably close agreement with those measured in CO_2 at −78 and 25°C. Fictitiously large CO_2 surface areas, because of chemical interaction of CO_2 with functional groups on the surface of the coals, are not in evidence.

To evaluate absolute surface area from monolayer capacity, a suitable value for the cross-sectional area of the adsorbate molecule must be used. In pores of molecular dimensions, the molecular area could be up to 4 times that found from adsorption on a flat surface. It was suggested that pores in coals are slit shaped.[115,116] Therefore molecular areas in such pores are likely to be 2–3 times greater than those found on a flat surface. In this context Marsh and O'Hair[117] have argued that increasing the measured CO_2 area by a factor of 2 raises the surface area of high rank coals to 600 m^2/g. This value is close to the area (665 m^2/g) determined by small-angle X-ray scattering for an anthracite coal carbonized at 400°C.[118] Since anthracites do not undergo significant structural changes on heat treatment to 400°C, the correlation between these two results suggests[117] that an agreement, though not unequivocal, exists between the adsorption and small-angle X-ray scattering methods.

There has been extensive deliberation on the validity and justification of reporting the surface areas of microporous adsorbents. For one thing, it is not possible to assign a definite value to the cross-sectional area of the adsorbate molecules, as discussed above. Furthermore, Dubinin[77] argues that because there is an adsorption force field in the entire volume of micropores, adsorption results in volume filling of the micropores. Because of this, Dubinin emphasizes that the concept of layer-by-layer filling (as is envisaged in the derivation of the BET equation[84]) and of area of micropores is meaningless. According to Spencer and Bond,[119] surface area of coals has no physical significance, therefore should not be reported. They suggested reporting monolayer volumes or total volumes of sorbate uptake.

4.3.2 Densities and Pore Volumes

Tschamler and de Ruiter[120] have discussed the definition of true and apparent densities of coals and the experimental methods for determining them.

4.3.2.1 *Densities from Displacement of Helium and Wetting Liquids*

Helium (He) is the smallest atom available. Thus it has the best chance of penetrating all the porosity in coals. It was earlier be-

[114] J. W. Ramsey, *Fuel,* **44,** 277 (1965).

[115] P. L. Walker, Jr., L. G. Austin, and S. P. Nandi, in *Chemistry and Physics of Carbon,* Vol. 2, P. L. Walker, Jr., Ed., Marcel Dekker, New York, 1966, p. 257.

[116] R. L. Bond, *Nature (London),* **178,** 104 (1956).

[117] H. Marsh and T. E. O'Hair, *Fuel,* **45,** 301 (1966).

[118] P. Chiche, S. Durif, and S. Pregermain, *Fuel,* **44,** 5 (1965).

[119] D. H. T. Spencer and R. L. Bond, in Am. Chem. Soc. Adv. Chem. Ser. No. 55, 1966, p. 724.

[120] *Chemistry of Coal Utilization,* Supplementary Volume, H. H. Lowry, Ed., John Wiley and Sons, New York, 1963, p. 35.

lieved[121] that coal structures have no closed-off pores that are inaccessible to He atoms. However X-ray studies on anthracites have shown the presence of some porosity that is closed to He.[122] Therefore the He density of a coal would be lower than its true density (i.e., the density of the solid phase).

When determining densities of carbonaceous solids by He displacement, it has been assumed that the gas is not adsorbed at room temperature. However this is not always true.[79,123–127] Maggs et al.[123] observed that the apparent dead space of a bulb containing porous carbon adsorbents increased significantly with a decrease in the temperature of measurement over the range 307 to −196°C. This was attributed to occurrence of adsorption, the extent of adsorption increasing with decrease in temperature. Kini and Stacy[125] showed that at 25°C adsorption of He on carbons varied with the nature of the material and increased, in general, with increasing internal surface area and "true" density of the sample. They cautioned that the accuracy of the He densities of high-density carbons, such as cokes, can be adversely affected by adsorption of He. Kipling et al.[126] and Kipling and McEnaney[127] also reported the adsorption of He on porous carbons. However Kipling et al.[126] also found that for some carbons associated with very small pores, He densities measured at lower temperatures were lower than those measured at higher temperatures. In view of the possibility of the adsorption of He on carbons at room temperature, it has been recommended[126] that measurement of He density for highly adsorptive carbons be made at 300°C. This is not recommended for raw coals, which will undergo structural changes at that temperature. Given the discussion above and the possibility of adsorption, He densities of coals reported in the literature may be too high. However Maggs et al.[123] reported that for raw coals, He adsorption at room temperature is not appreciable enough to cause significant errors in the densities. Close agreement between He and water densities in coals, to be discussed shortly, supports this finding.

Determination of He density requires an elaborate vacuum system. Furthermore, the experimental procedure is cumbersome and time-consuming. Attempts have been made in the past[120] to determine densities of coals by the simple pycnometer method, using liquids of small molecular dimensions such as methanol and water. Methanol densities of coal are invariably higher than the corresponding He densities, for reasons discussed elsewhere.[120] For lower rank coals, water densities—like the methanol densities—are higher than the He densities, whereas for higher rank coals the water densities are lower.[120] The lower values have been attributed to the extensive hydrophobic character of higher rank coals,[120,128] as a result of which air present in the pores is not completely displaced by water.

Ettinger and Zhupakhina[128] suggested that water penetration into the hydrophobic coal structure can be improved with the help of a wetting agent composed of a mixture of polyethylene glycol and monoalkylphenyl ethers. The molecules of the wetting agent, which have hydrophobic hydrocarbon radicals attached to hydrophilic polar groups, undergo orientation in water. It has been suggested that in the case of hydrophobic coals, the polar groups of the wetting agent would be turned into the water and the nonpolar groups toward the coal surface. Therefore subject to penetration by the wetting agent itself, water will penetrate into the coals quickly and more completely.

Ettinger and Zhupakhina[128] used the following procedure for determining the water densities of coals. After the addition of the wetting agent–water mixture to the coal sample (−60 mesh) contained in a pycnometer, the suspension was heated in a boiling water bath for 30 min. The pycnometer was then

[121] R. E. Franklin, *Trans. Faraday Soc.*, **25**, 274 (1949).

[122] W. V. Kotlensky and P. L. Walker, Jr., *Proc. 4th Carbon Conf.*, Pergamon Press, New York, 1960, p. 423.

[123] F. A. P. Maggs, P. H. Schwabe, and J. H. Williams, *Nature (London)*, **186**, 956 (1960).

[124] J. B. Donnet, P. Couderc, and E. Papirer, *J. Chim. Phys.*, **64**, 1656 (1967).

[125] K. A. Kini and W. O. Stacy, *Carbon (Oxford)*, **1**, 17 (1963).

[126] J. J. Kipling, J. N. Sherwood, P. V. Shooter, N. R. Thompson, and R. N. Young, *Carbon (Oxford)*, **4**, 5 (1966).

[127] J. J. Kipling and B. McEnaney, *2nd Conf. Industrial Carbon Graphite*, Soc. Chem. Ind., London, 1966, p. 380.

[128] I. L. Ettinger and E. Zhupakhina, *Fuel*, **39**, 387 (1960).

cooled for 30 min in a water thermostat maintained at 20°C and finally weighed. Water and He densities for a number of coals are given in Table 4.4. Agreement between the two sets of values is excellent. With three exceptions, the difference in the densities determined by the two methods does not exceed 1%.

The lower water densities of hydrophobic higher rank coals (when no wetting agent is used) have been attributed, as discussed above, to the inability of water molecules to displace the air present in the pores. This view is not supported by the recent work of Walker et al.[129] They determined the He, water, and methanol densities of a series of 13 coal chars of 40 × 100 mesh particle size. The chars were prepared in an inert atmosphere at 855 and 1000°C; subsequently carbon was deposited on the char surfaces from the cracking of methane.[130] All samples were devoid of essentially all oxygen-containing functional groups. As a result, the char surfaces were hydrophobic. Walker et al.[129] allowed sufficient contact times for the water and methanol densities of all samples to attain limiting constant values. After 5 days, methanol and water densities, as measured by the pycnometer method, were in close agreement with helium densities.

[129] A. M. Youssef, O. P. Mahajan, and P. L. Walker, Jr., unpublished results.
[130] M. Kamishita, O. P. Mahajan, and P. L. Walker, Jr., *Fuel,* **56,** 444 (1977).

Table 4.4 Helium and Water Densities for Coals of Varying Rank[128]

VM (%, maf)	Density (g/cm³) With Helium	Density (g/cm³) With Water	Difference (%)
44.5	1.31	1.31	0.0
34.8	1.36	1.36	0.0
32.9	1.27	1.28	+0.8
29.6	1.30	1.29	−0.8
26.6	1.28	1.28	0.0
24.9	1.34	1.33	−0.7
24.6	1.38	1.45	+5.1
23.7	1.31	1.32	+0.8
17.0	1.34	1.33	−0.7
14.0	1.45	1.46	+0.7
12.9	1.40	1.37	−2.1
11.7	1.39	1.45	+4.3
3.4	1.57	1.57	0.0
1.7	1.58	1.58	0.0

Source: I. L. Ettinger and E. Zhupakhina, *Fuel,* **39**, 387 (1960).

4.3.2.2 *Densities from Displacement of Mercury and Gas Flow Methods*

Particle density is normally determined from the volume of mercury that the sample displaces under a given pressure. In the determination of mercury density of powders, there is some uncertainty about the minimum pressure needed to fill the interparticle void volume. The coals usually possess a polymodal distribution of pore sizes consisting of micro-, transitional, and macropores.[93] The voids between solid particles behave as macropores (i.e., voids greater than 300 Å in diameter). Gan et al.[93] used the following approach to estimate the pressure corresponding to complete filling of the void volume between particles. They determined this filling pressure on the same size fraction of nonporous glass powders as the coal powders used for mercury density measurements. For the 40 × 70 mesh glass spheres, significant amounts of mercury were forced into the voids between particles at pressures below 20 psi. Between 20 and 60 psi, mercury intrusion was much less; above 60 psi, further mercury penetration was negligible.

Gan et al.[93] conducted a similar run with a 40 × 70 mesh fraction of an anthracite that was known to have negligible macroporosity. It was observed that mercury penetration occurred up to 60 psi and was negligible between 60 and 500 psi. It was concluded that for the 40 × 70 mesh fraction of coal particles, the interparticle voids would be filled at 60 psi and that penetration at higher pressures would be due to the filling of intraparticle pores.

Ettinger and Zhupakhina[128] have suggested that the particle density of coal can also be determined by "silanization," which involves coating the coal surface with a very thin transparent film of an organosilicon compound. The thin film is impervious to water and has practically no influence on the weight or volume of the coal particles. A pycnometric determination of particle density is made on the silanized coal using a very dilute solution of pyridine in water. It was found that, for 14 coals having volatile matter contents (maf) in the 2.1–48.6% range, the difference in particle density by the silanization method and by geometric measurements on cubes of the coals was less than 1%.

Ergun[131] measured particle densities on a series of coal cokes, using the principle of gas flow through beds of particles packed to different bulk densities. From measurements of pressure drop through the beds as a function of gas flow rates, and by use of equations developed earlier by Ergun and Orning,[132] particle densities of sieve fractions ranging from 8-16 to 100-140 mesh were determined. The particle density increased monotonically with a decrease in particle size, as progressively smaller voids were removed from the particles as a result of continuing comminution.

4.3.2.3 *Change of Density and Pore Volume with Rank*

The variation of He density (mineral-matter-free basis, assuming the density of mineral matter to be 2.7 g/cm^3)[120] with carbon content of American coals is shown in Fig. 4.7. The curve indicates a shallow minimum at about 81% carbon and rises sharply at about 90% carbon. Franklin[121] and Fugii and Tsuboi[133] have also reported similar trends for British and Japanese coals, respectively, although the minimum in each case occurred at somewhat higher carbon contents than those observed for the American coals. These results thus indicate that for a given coal rank, geological history has a marked effect on coal structure. In this context Hatami et al.[134] have reported that surface areas of Japanese coals are lower than those of the same rank of foreign coals.

A plot of the variation of mercury density with carbon content of coals has a shape similar to that for He density.[93]

Toda[135] measured the densities of vitrain-enriched fractions of several coals of different rank in methanol, *n*-hexane, and mercury. These densities, as well as the He densities, showed higher values in the lower and higher rank coals and minima at about 87% carbon content. Toda found it convenient to consider the specific volume (V) of fluid displaced by coals (cm^3 fluid/g coal). The specific volume is the reciprocal of density measured in that fluid. The total open pore (specific) volume (i.e., $V_{Hg} - V_{He}$) was essentially independent of coal rank. However the volume accessible to *n*-hexane (i.e., $V_{Hg} - V_{n-hex}$) increased with coal rank up to about 87% carbon content and decreased thereafter. Toda compared the micropore volumes W_0, obtained from CO_2 adsorption data using the P-D equation, with $V_{Hg} - V_{He}$ and $V_{Hg} - V_{MeOH}$. As seen in Fig. 4.8, the $W_0 - (V_{Hg} - V_{He})$ values decrease with increase in coal rank, show a minimum at about 87% carbon con-

[131] S. Ergun, *Anal. Chem.*, **23,** 151 (1951).
[132] S. Ergun and A. A. Orning, *Ind. Eng. Chem.*, 1179 (1949).
[133] S. Fujii and H. Tsuboi, *Fuel,* **46,** 361 (1967).
[134] M. Hatami, Y. Toda, and Y. Yoshida, *Nenryo Kyokai-shi,* **48,** 235 (1969).
[135] Y. Toda, *Fuel,* **51,** 108 (1972).

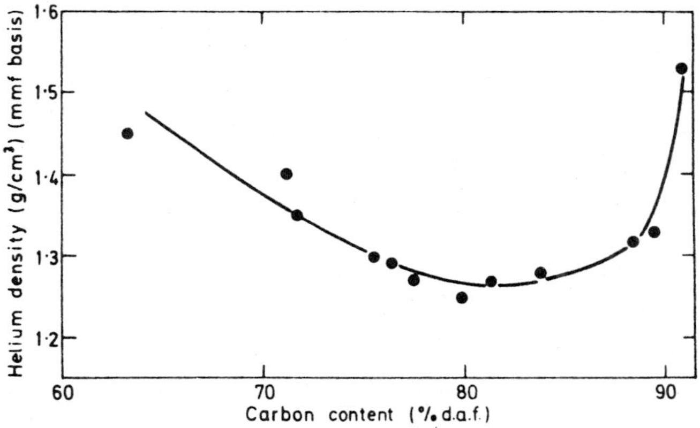

Figure 4.7 Variation of helium density of coals with carbon content.[93]

tent, and increase thereafter. From the positive magnitude $W_0 - (V_{Hg} - V_{He})$ values for many coals, Toda concluded that the CO_2 molecule could penetrate into pores that were inaccessible to He; the number of such pores being greater in the lower and higher rank coals. Since the minimum dimension of a CO_2 molecule is larger than that of a He atom, it is difficult to conceive how some pores can be accessible to CO_2 but not to He. Total open pore volume is a measure of the volume of micro-, transitional, and macropores, whereas W_0[77,136] gives the volume of micropores. Since coals of all ranks have some proportion of their total open pore volume contained in transitional and macropores, it is not understood why the total open pore volume should be less than the micropore volume.

Toda[135] also observed that the $V_{Hg} - V_{MeOH}$ values for various coals were invariably higher than the corresponding W_0 values. It was inferred that coals contain pores that are accessible to methanol but are inaccessible to CO_2. Furthermore, a plot of $(V_{Hg} - V_{MeOH}) - W_0$ against the hydroxyl contents of coals gave a straight line passing through the origin, indicating that the volume of pores accessible to methanol but inaccessible to CO_2 is proportional to the concentration of hydroxyl groups. Toda attributed the observed size-penetration discrepancy to the inbibition of methanol by the coal structure.

4.3.2.4 *Relation Between Specific Volume and Hydrogen Content*

Mentser and Ergun[137] found that the specific volumes of anthracitic and bituminous coals calculated from their He densities were a linear function of their hydrogen contents on an atomic basis. The specific volume at zero hydrogen content agreed with that of graphite. The specific volume was expressed by Equation 4.5:

$$V_{He} = 0.44 + 0.84X_H \qquad (4.5)$$

where 0.44 = specific volume of graphite (cm^3/g)
0.84 = slope of the straight line
X_H = hydrogen atom percentage

This relation is plotted in Fig. 4.9. From the conformity of anthracites and metaanthracites to this relationship, it was concluded that both the hydrogen content and V_{He} are excellent criteria to be used for evaluating the degree of coalification of higher rank coals.

[136] Y. Toda, M. Hatami, S. Toyoda, Y. Yoshida, and H. Honda, *Carbon (Oxford)*, **8**, 565 (1970).

[137] M. Mentser and S. Ergun, *Fuel*, **39**, 509 (1960).

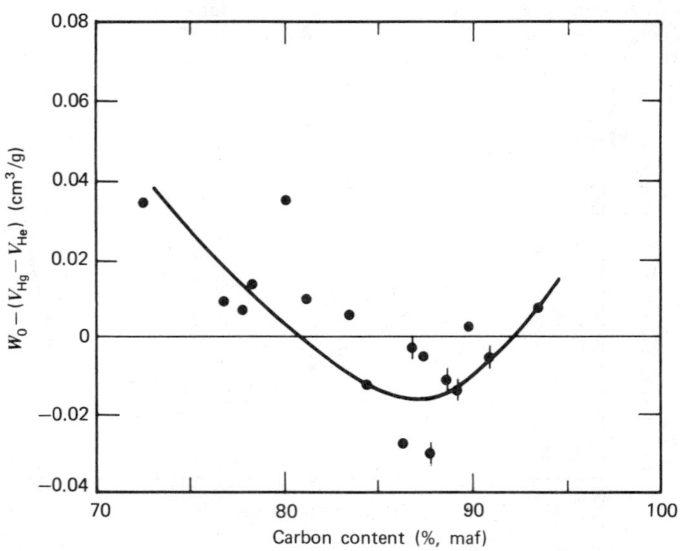

Figure 4.8 Relation between $W_0 - (V_{Hg} - V_{He})$ and coal rank: ● = Japanese coals, ⧫ = other coals.[135]

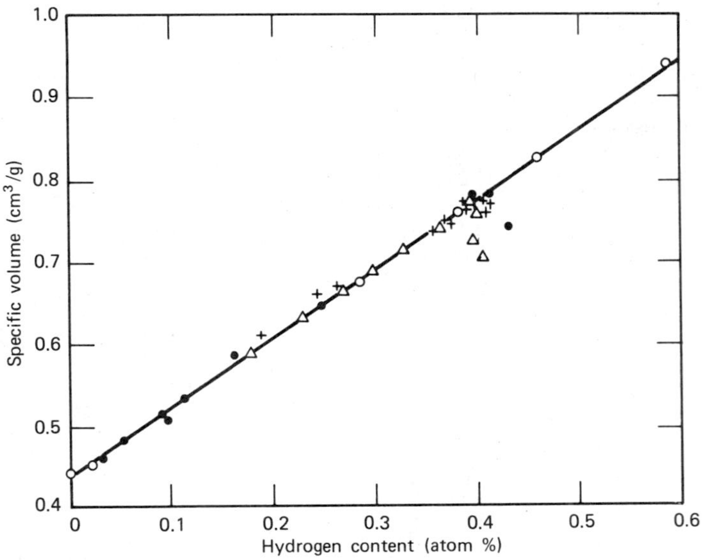

Figure 4.9 Change in specific volume of coals as measured by helium displacement as a function of atom percent hydrogen.[137] Data of Franklin (+); van Krevelen and Chermin (△); Mentser and Ergun (●); Ergun, Mentser, and Howard (○).

4.3.2.5 *Distribution of Total Pore Volume in Coals*

Gan et al.[93] have made a detailed study of the porosities of coals varying widely in rank. They estimated the pore volumes contained in the following different pore diameter ranges:

1. Total open pore volume V_T accessible to He, as estimated from the He and mercury densities.
2. Pore volume V_1 contained in pores greater in diameter than 300 Å, as estimated from penetration of mercury under a pressure of 5800 psi.
3. Pore volume V_2 contained in pores in the diameter range 300 to about 12 Å, as estimated from the analysis of the absorption branch of the N_2 isotherms (-196°C) by the Cranston and Inkley method.[138]
4. Pore volume V_3 contained in pores smaller in diameter than 12 Å as estimated from $V_3 = V_T - (V_1 + V_2)$.

The relevant results for different coals are given in Table 4.5. The proportion of V_3 is significant for all coals. Its value is a maximum in the anthracite sample (PSOC-80) and a minimum in the lignite sample (PSOC-89). These results thus demonstrate that essentially all American coals, irrespective of their rank, show molecular sieve properties. From the results in Table 4.5 it may be concluded that (1) porosity in coals with carbon contents less than about 75% is primarily due to the presence of macropores, (2) porosity in coals with about 75–84% carbon content is predominantly due to the presence of micro- and transitional pores, and (3) for coals ranging in carbon content from about 85 to 91%, microporosity predominates.

A higher CO_2 area is associated with the presence of a larger proportion of micropores. However it is surprising that even though the three lignite samples have only about 41% or less of their total open pore volume in pores with diameters less than 300 Å, their CO_2 areas are higher than those of the hvCb samples, which have almost 80% of their total open pore volume in pores smaller in diameter than 300 Å (Table 4.6). Furthermore, because of the pronounced macroporosity in lignites, their N_2 surface areas would be expected to be larger than those found experimentally. It appears, therefore, that the low N_2 surface areas of the lignites may be due to physical shrinkage

[138] R. W. Cranston and F. A. Inkley, in *Advances in Catalysis,* Vol. 9, D. D. Eley et al., Eds., Academic Press, New York, 1957, p. 143.

Table 4.5 Gross Open Pore Distributions in Coals[93]

Sample	Rank	C (%, maf)	V_T (cm³/g)	V_1 (cm³/g)	V_2 (cm³/g)	V_3 (cm³/g)	V_3 (%)	V_2 (%)	V_1 (%)
PSOC-80	Anthracite	90.8	0.076	0.009	0.010	0.057	75.0	13.1	11.9
PSOC-127	lvb	89.5	0.052	0.014	0.000	0.038	73.0	Nil	27.0
PSOC-135	mvb	88.3	0.042	0.016	0.000	0.026	61.9	Nil	38.1
PSOC-4	hvAb	83.8	0.033	0.017	0.000	0.016	48.5	Nil	51.5
PSOC-105A	hvBb	81.3	0.144	0.036	0.065	0.043	29.9	45.1	25.0
Rand	hvCb	79.9	0.083	0.017	0.027	0.039	47.0	32.5	20.5
PSOC-26	hvCb	77.2	0.158	0.031	0.061	0.066	41.8	38.6	19.6
POC-197	hvBb	76.5	0.105	0.022	0.013	0.070	66.7	12.4	20.9
PSOC-190	hvCb	75.5	0.232	0.040	0.122	0.070	30.2	52.6	17.2
PSOC-141	Lignite	71.7	0.114	0.088	0.004	0.022	19.3	3.5	77.2
PSOC-87	Lignite	71.2	0.105	0.062	0.000	0.043	40.9	Nil	59.1
PSOC-89	Lignite	63.3	0.073	0.064	0.000	0.009	12.3	Nil	87.7

Source: H. Gan, S. P. Nandi, and P. L. Walker, Jr., *Fuel,* **51,** 272 (1972).

of pores at −196°C, and the low N_2 areas of other coals are due to an activated diffusion process.

4.3.3 Effect of Heat Treatment of Coal on Internal Structural Properties

When coals are thermally heated, volatile matter is released from the coal skeleton framework, creating additional porosity and surface area. Some of the additional porosity comes from the opening up of previously closed porosity; the remainder comes from the enlargement of pores that were previously open. Concurrent with the loss of volatile matter, some cross links are broken, which permits the aromatic regions to better align, resulting in a loss of open porosity and surface area. At lower temperatures the loss of volatiles predominates and results in an overall opening up of the structure. But as the temperature is further raised and volatile release becomes small, breakage of cross links and crystallite alignment are dominant. Open pore volume and surface area loss become significant. Thus it is to be expected that over a certain temperature range, the surface area and open pore volume will go through a maximum. The magnitude of the changes in surface area and open porosity on heat treatment of coals is expected to vary as a function of such parameters as heating rate, maximum temperature, soak time at maximum temperature, atmosphere under which heating occurred, particle size, inorganic impurities present and their distribution in coal, maceral composition, and whether the coal behaves as a thermoplastic or thermosetting precursor.

Table 4.6 Nitrogen and Carbon Dioxide Areas of Coals[93]

Sample	Rank	Surface area (m²/g) N_2	CO_2
PSOC-80	Anthracite	7.0	408
PSOC-127	lvb	< 1.0	253
PSOC-135	mvb	< 1.0	214
PSOC-4	hvAb	< 1.0	213
PSOC-105A	hvBb	43.0	114
Rand	hvCb	17.0	147
PSOC-26	hvCb	35.0	133
PSOC-190	hvCb	83.0	96
PSOC-141	Lignite	2.3	250
PSOC-87	Lignite	< 1.0	268
PSOC-89	Lignite	< 1.0	238

Source: H. Gan, S. P. Nandi, and P. L. Walker, Jr., *Fuel,* **51,** 272 (1972).

Toda[139] studied changes in the micropore structure of a number of coals heated to a series of temperatures up to 1200°C by measuring densities in He, *n*-hexane, and methanol. The carbon contents of the coals investigated ranged between 77.8 and 89.8% (maf). For a given heat treatment temperature (HTT), the specific volumes of the three fluids displaced by all the samples heat treated up to about 700 to 800°C (i.e., cm³ fluid/g

[139] Y. Toda, *Fuel,* **52,** 99 (1973).

coal) varied in the order: $V_{n\text{-hex}} > V_{He} > V_{MeOH}$. At higher temperatures the specific volumes for all the samples in methanol were larger than those in He. Since the specific volumes displaced by various coals changed only slightly on heat treatment of the coals up to 350°C, it was concluded that apart from dehydration, no significant changes in coal structure occurred up to this temperature. The specific volume of each fluid displaced decreased rapidly for coals heated above about 350°C. For the lower rank coals, the specific volumes of methanol and He displaced showed minima at about 800 and 900°C, respectively, whereas no such minima were observed for the higher rank coals. That is, for the higher rank coals specific volumes of methanol and He displaced continued to decrease up to the maximum HTT of 1200°C. These results reflect differences in the thermal stability of the open porosity in coals of varying rank when they are taken to elevated temperatures, since increasing fluid displacement can be taken to coincide primarily with the conversion of open porosity to closed porosity.

Figure 4.10 shows changes in the volumes of micropores that are accessible to methanol but inaccessible to n-hexane (i.e., $V_{n\text{-hex}} - V_{MeOH}$) as a function of HTT for different coals. These volumes show a maximum at about 350°C, then decrease to a minimum at 400°C for the caking coals. However the corresponding values for the noncaking coals remain essentially constant up to 400°C. Further heat treatment produces a pronounced maximum in $V_{n\text{-hex}}\,V_{MeOH}$ for a HTT of about 600°C, before the difference in specific volumes falls close to zero for all coals heat treated to 1200°C. These results dramatically show that coals can be converted to molecular sieve materials by their heat treatment—the character of the molecular sieving depending on the rank of the starting coal and its HTT.

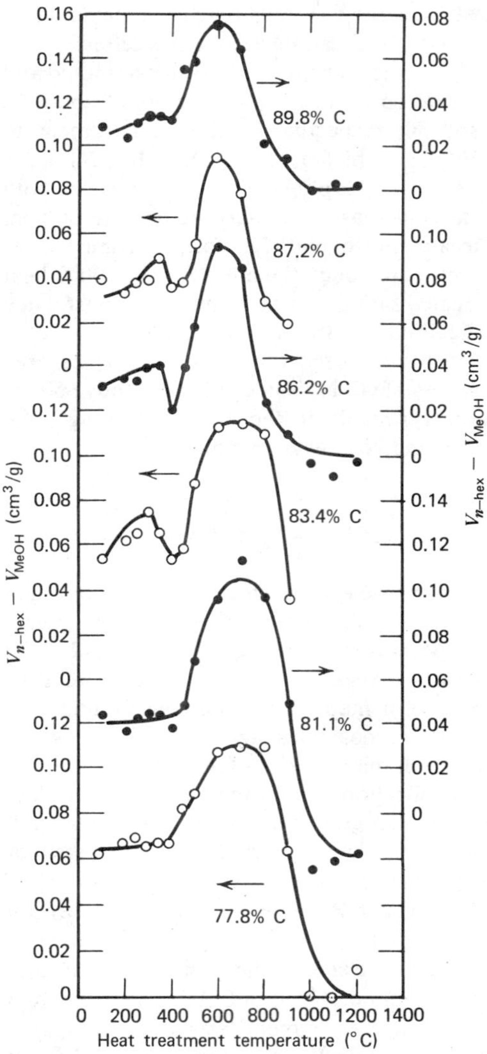

Figure 4.10 Changes in ($V_{n\text{-hex}} - V_{MeOH}$) with HTT for a series of Japanese coals.[139]

Changes in surface areas of coals occurring during heat treatment have been investigated by several workers.[136,140–145] In general, surface area increases with increasing HTT, passes through a maximum, and decreases thereafter. The temperature at which the maximum occurs depends on the starting coal. For example, N_2 and CO_2 areas of anthracites reach maxima at about 700°C and decrease sharply at 900°C.[140] For an hvA bituminous coal, N_2 surface area remains essentially constant (<1 m^2/g) up to 900°C,

[140] S. P. Nandi, V. Ramdass, and P. L. Walker, Jr., *Carbon (Oxford)*, **2**, 199 (1964).

[141] R. G. Jenkins, S. P. Nandi, and P. L. Walker, Jr., *Fuel*, **52**, 288 (1973).

[142] T. Siemieniewska, *Koks, Smola, Gaz*, **13**, 57, 231, 309 (1968).

[143] R. I. Razouk, F. Z. Saleeb, and A. M. Youssef, *Carbon (Oxford)*, **6**, 325 (1968).

[144] M. S. Orenbakh, *Izv. Sib. Otd. Akad. Nauk SSSR, Ser. Tekh. Nauk*, **1**, 143 (1968).

[145] P. Chiche, *J. Chim. Phys.*, **60**, 792, 799, 825 (1963).

whereas the CO_2 area reaches a maximum at about 600°C and decreases thereafter.[141] The N_2 surface areas of hvB bituminous and lignite coals increase by factors of about 15 and 30, respectively, on heat treatment to 600°C; at higher temperature the N_2 areas decrease sharply.[141] The CO_2 areas in both cases increase by a factor of about 3 on heat treatment to 600°C before starting to decrease, although the areas for the 900°C heat treated samples in each case are almost twice the areas of the starting coals. For brown coals carbonized up to 400°C, surface area accessible to N_2 is negligible.[142] Above 500°C and especially in the 700–900°C range, the CO_2 and N_2 areas are almost equal.

4.4 ELECTRON MICROSCOPY

Perhaps the earliest attempt to study ultrafine structures in coal with an electron microscope was reported by Preston and Cuckow.[146] However they did little more than present a method of preparing thin-film replicas of coal cleavage surfaces and show electron micrographs of such replicas at low magnification. McCartney[147] presented electron micrographs of microspores and cell structures in vitrinite. They were taken of replicas of polished and etched coal surfaces. Alpern and Pregermain[148] observed several kinds of fine granular structure in replicas of polished surfaces. The grains in a vitrinite were around 200 to 300 Å in diameter, and those in the "ciment" surrounding spores (probably granular micrinite) measured about 1000 Å. A still different granularity, of unspecified size, was observed in massive micrinite. Interpretations of these granular structures were not attempted. Mackowsky and Nemetschek[149] subjected vitrite particles to intense bombardment by the electron beam and observed the residues to be porous aggregates of particles 100–300 Å in diameter. Electron diffraction patterns indicated that graphitization had occurred.

In 1959 Pregermain and Guillemot[150] first described a technique for cutting ultrathin sections of coal (500–2000 Å) using a microtome with a diamond knife. These proved to be good specimens for electron microscopy. Structures were observed in vitrinites consisting of grains of average size from 250 Å in a lignite to 100 Å in a semianthracite. Such granular details were not seen in graphite fragments. It was supposed that a relationship existed between these granular structures and the porous texture indicated by sorption methods; however no quantitative relation was traced. McCartney et al.[151,152] prepared ultrathin sections following Pregermain's method and presented electron micrographs of apparent cell structures and striations of vitrinites. They cautioned that these might be preparation artifacts. Brown and Taylor[153] discussed this latter work and stated that the structures shown were definitely artifacts. They also pointed out the general improvement in resolution of familiar structures by electron microscopy.

Taylor[154] later described two types of vitrinite as revealed by electron micrographs. Vitrinite A is homogeneous, and vitrinite B consists of vitrinite A plus sheets of exinite approximately 1000 Å thick. Taylor also identified exinite, micrinite, and semifusinite in ultrathin sections and ultrafine pore structures in vitrinite. McCartney et al.[155] made additional studies of ultrathin sections of petrologic components of coals of different rank. They observed ultrafine structures in two general size ranges, one of the order of hundreds of angstroms, the other less than a hundred angstroms. Particles having the general form of spheroids, curved cylinders, and round and polygonal platelets were observed. A later work[156] resulted in clearer micro-

[146] G. D. Preston and F. W. Cuckow, *Ultra Fine Structure of Coals and Cokes,* BCURA, London, 1944, p. 334.

[147] J. T. McCartney, *Econ. Geol.,* **44,** 617 (1949).

[148] B. Alpern and S. Pregermain, *Bull. Microscop. Appl.,* **6,** 16 (1956).

[149] M. Th. Mackowsky and Th. Nemetschek, *Naturwissenschaften,* **43,** 442 (1956).

[150] S. Pregermain and C. Guillemot, *4th Int. Conf. Electron Microscopy,* Vol. 1, Springer-Verlag, Berlin, 1960, p. 384.

[151] J. T. McCartney, S. Ergun, and R. E. Walline, *Nature (London),* **191,** 1361 (1961).

[152] J. T. McCartney, R. E. Walline, and S. Ergun, U.S. Bur. Mines Rep. Invest. No. 5885, 1961, 5 pp.

[153] H. R. Brown and G. H. Taylor, *Nature (London),* **193,** 1146 (1962).

[154] G. H. Taylor, in Am. Chem. Soc. Adv. Chem. Ser. No. 55, 1966, p. 274.

[155] J. T. McCartney, H. J. O'Donnell, and S. Ergun, in Am. Chem. Soc. Adv. Chem. Ser. No. 55, 1966, p. 261.

[156] J. T. McCartney, *Fuel,* **49,** 409 (1970).

graphs and more definite deductions concerning ultrafine particles of granular micrinite. It was concluded that their sizes ranged from 100 to 1000 Å (10–100 nm), most being in the 400–600 Å range. They appeared to have spheroidal or disk shapes, and it was not clear which was predominant. Taylor[154] also described particles in this size range, which he believed were similar to those just noted. Earlier suggested evidence for incipient three-dimensional crystallinity in these particles was not confirmed. McCartney and Ergun[157] found that a vitrain band in a metaanthracite was composed largely of ovoid or polygonal platelets averaging 800 Å in diameter. This approximation of crystallite size was also derived from X-ray diffraction studies. Electron micrographs and electron diffraction patterns proved that the platelets are oriented parallel to the bedding planes.

Waisman et al.[158] showed electron micrographs of replicas of natural cleavage surfaces of coal (coal components unspecified). They observed pores as small as 200 Å (20 nm) in diameter and channels connecting the pores. Harris and Yust[159] observed porosity in the exinite and inertinite constituents of a high-volatile bituminous coal, using a high-voltage (1-MeV) electron microscope (Fig. 4.11). Some of the pores in exinite were described as channellike and in the size range 200–100 nm; those in inertinite lay in the range of 2–50 nm, with a large proportion less than 10 nm. This porosity was interpreted as being continuous and having a high permeability. Small particles (~25 nm in diameter) were seen in most of the exinite pores and may act as catalysts of pore formation. Pregermain[160] reported studies of ultrathin sections of semicoke carbonized at 500°C. She observed granular structures similar to those in raw coal and containing pores 20–50 Å (2–5 nm) in diameter. These sizes were compared favorably with those ob-

[157] J. T. McCartney and S. Ergun, *Nature (London)*, **205**, 962 (1965).

[158] B. A. Waisman, M. D. Krivitzkii, and F. E. Krigman, *Dokl. Akad. Nauk SSSR*, **97**, 1031 (1954).

[159] L. A. Harris and C. S. Yust, *Fuel*, **55**, 233 (1976).

[160] S. Pregermain, *European Regional Conf. Electron Microscopy*, Vol. 1, Nederlandse Vereeniging voor Electronemicroscopie, Delft, 1961, p. 516.

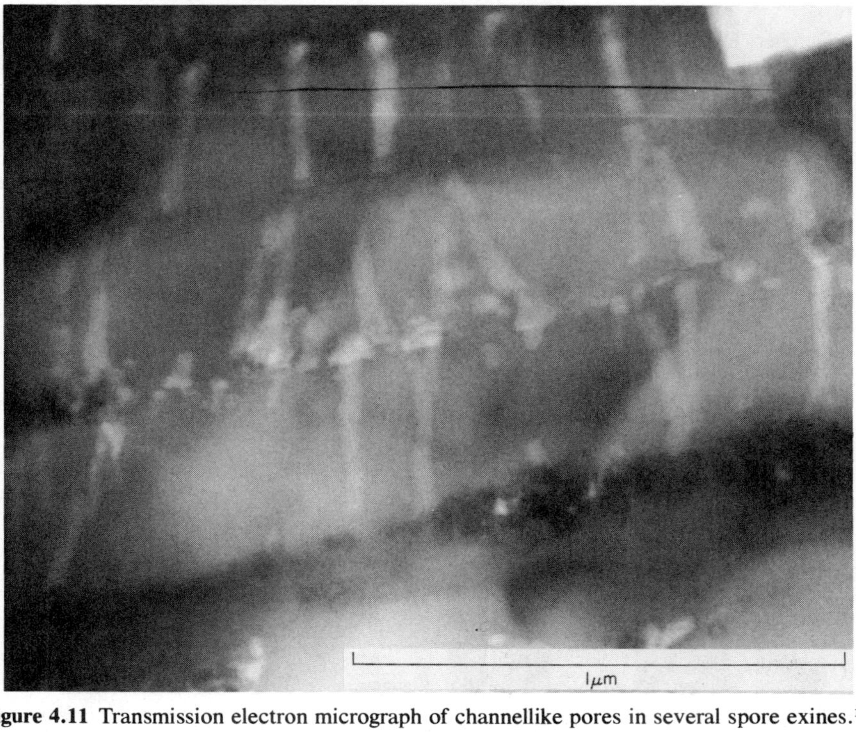

Figure 4.11 Transmission electron micrograph of channellike pores in several spore exines.[159]

tained by Bond and Spencer[161] in absorption measurements. Augustyn et al.[162] presented 40 scanning electron micrographs (SEM) and 9 energy-dispersive X-ray diagrams of mid-European brown coals, showing topographical features, woody structures, and elemental compositions of inorganic constituents. They suggested that these techniques should be useful in petrographic classification. Boateng and Phillips[163] examined coal surfaces with optical and scanning electron microscopes and an electron microprobe. They found by the latter that organic sulfur was fairly evenly distributed. Dutcher et al.[164] studied the distribution of mineral elements in coals by the X rays generated by an electron microprobe and by electron backscatter patterns. Sutherland[165] investigated the possibility of determining organic sulfur by electron microprobe in a brief study limited to two coals. Lishtvan and Terentev[166] investigated solvation products (bitumens) of brown coals and peats by electron microscopy and electron diffraction. They appeared to be similar to paraffin waxes from petroleum. Ergun and McCartney[167] recorded electron diffraction patterns of coals of different rank by transmission through ultrathin sections.

McCartney[168] presented SEMs of metallurgical cokes and residues from rapid pyrolysis of coals (Fig. 4.12). Pores in the cokes ranged from 10 to 150 μm. Bubble structures ranging from 5 to 25 μm were observed in the higher temperature cokes, and bubblelike projections seen on these larger bubbles ranged in size from 0.5 to 1.3 μm. Frigge and Wolff[169] compared the structures seen with a scanning electron microscope in an anisotropic coke from a fat coal with isotropic coke from a high-volatile coal. Addair et al.[170] studied by SEM the changes in a blast furnace coke on gasification. They saw that surface pitting and fissure formation occurred rather than a smooth erosion, and suggested that the pitting may occur at locations of catalytic impurities.

[161] R. L. Bond and D. H. T. Spencer, *Industrial Carbon and Graphite,* Soc. Chem. Ind., London, 1958, p. 231.

[162] D. Augustyn, M. Iley, and H. Marsh, *Fuel,* **55,** 25 (1976).

[163] D. A. D. Boateng and C. R. Phillips, *Fuel,* **55,** 318 (1976).

[164] R. R. Dutcher, E. W. White, and W. Spackman, *Proc. 22nd Ironmaking Conf., Iron & Steel Div.,* Metall. Soc. AIME, New York, 1964, p. 463.

[165] J. K. Sutherland, *Fuel,* **54,** 132 (1975).

[166] I. I. Lishtvan and A. A. Terentev, *Khim. Tverd. Topl.,* **9,** 74 (1975).

[167] S. Ergun and J. T. McCartney, *Science,* **134,** 1620 (1961).

[168] J. T. McCartney, *Fuel,* **50,** 457 (1971).

[169] J. Frigge and E. M. Wolff, *Fuel,* **49,** 436 (1970).

[170] R. R. Addair, E. H. Boult, and H. Marsh, *Fuel,* **51,** 57 (1972).

4.5 X-RAY DIFFRACTION*

High-rank coals produce diffraction patterns with a strong line at about 3.5 Å and weak lines at about 2.1 and 1.76 Å, and in some

* Contributed by Sidney S. Pollack.

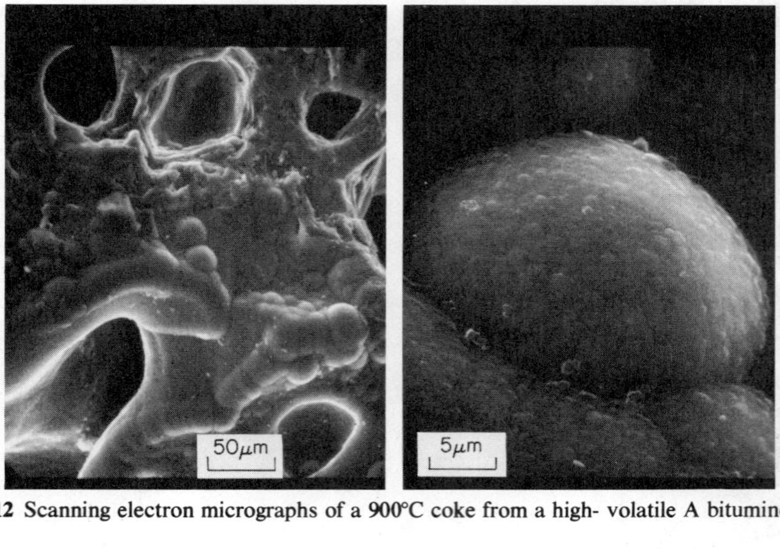

Figure 4.12 Scanning electron micrographs of a 900°C coke from a high- volatile A bituminous coal.[168]

cases a peak at about 20 Å. With decrease in rank, the 3.5 spacing increases to about 3.9 Å. Exinite and some coals also produce an additional line at about 4.5 Å, called the gamma band, which may be much stronger than the 3.5 line, as in exinites, or much weaker.

Figure 4.13 contains the diffraction patterns of three vitrains, showing the increase in sharpness of the lines with increasing carbon content and higher rank. Figure 4.14 compares patterns of a lignite, coal spores,

[171] L. Cartz and P. B. Hirsch, *Philos. Trans. R. Soc. London,* **A252,** 557 (1960).

[172] S. Ergun, U.S. Bur. Mines Bull. No. 648, 1968, 38 pp.

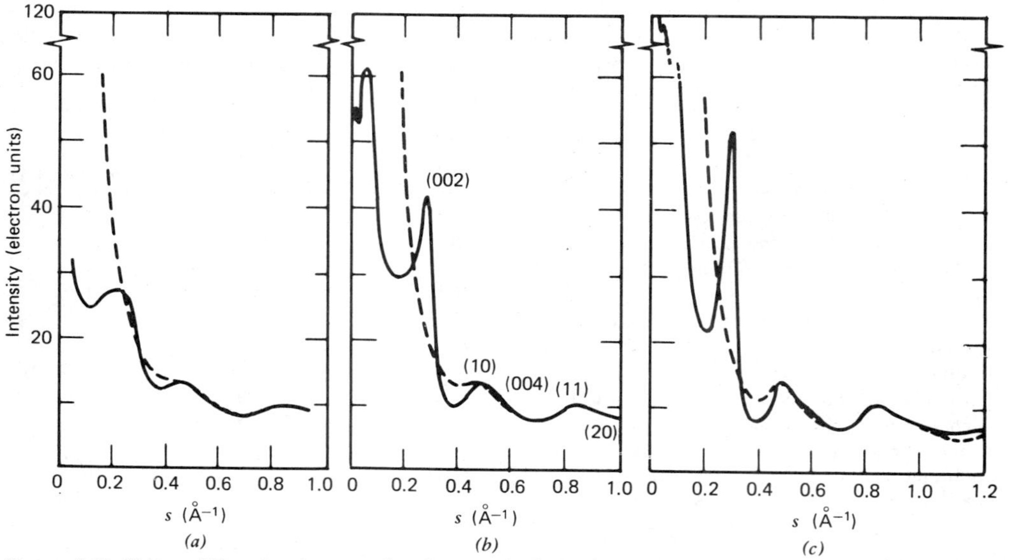

Figure 4.13 X-Ray diffraction patterns for three typical vitrains of (*a*) 78% carbon, (*b*) 89% carbon, and (*c*) 94% carbon. The intensity is plotted as a function of s ($= 2 \sin \theta/\lambda$).[171]

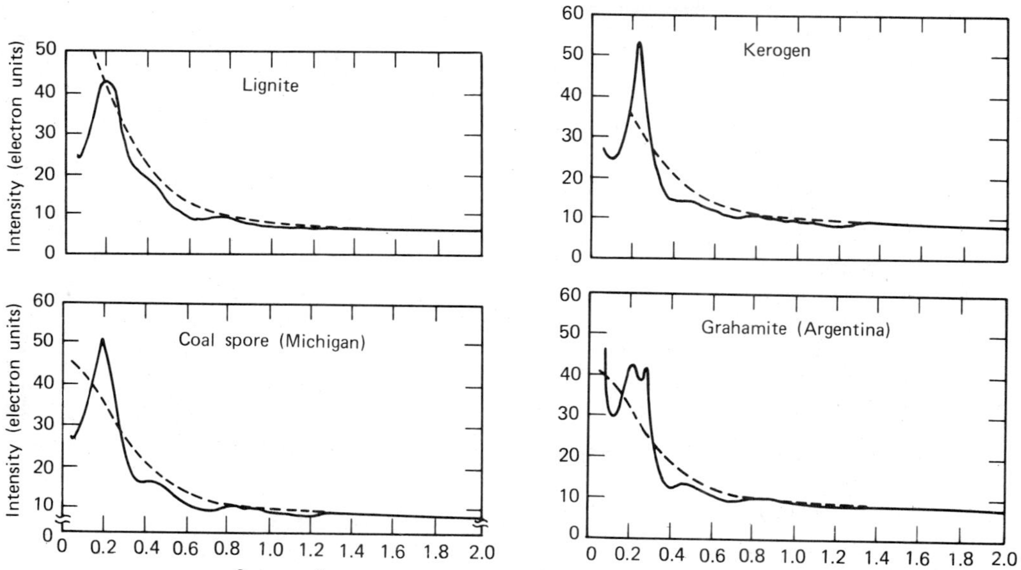

Figure 4.14 X-Ray scattering from a lignite, kerogen, coal spores, and a grahamite.[172]

kerogen (the organic matter in oil shale), and a grahamite, showing both 3.5- and 4.5-Å peaks. The chemical composition and the *d*-spacing of the strongest peak for vitrains from coals of various ranks and other organic materials are listed in Tables 4.7 and 4.8. The increase in the *d*-spacing of the strongest peak with decreasing carbon for the coal samples in Table 4.7 is shown in Fig. 4.15. There is no such relationship for the noncoal samples.

Although most coals produce patterns with only a few diffuse lines, high-rank anthracites and metaanthracites show additional lines identical to those of graphite.[173] The similar-

[173] M. Mentser, H. O'Donnell, and S. Ergun, *Proc. 5th Carbon Conf.*, Vol. 2, Pergamon Press, 1962, p. 493.

Table 4.7 Chemical Composition and *d*-Spacings of Strongest Diffraction Line for Demineralized Vitrinites from Coals of Differing Rank[172]

	Ultimate Analysis (wt %, maf)					
Sample	C	H	O	N	S	*d*-Spacing (Å)
Anthracite, Luzerne County, Pa.	92.8	2.7	2.9	1.0	0.6	3.51
Semianthracite, Northumberland County, Pa.	90.5	3.9	3.4	1.5	0.7	3.57
Low-volatile bituminous, Wyoming County, W. Va.	90.8	4.6	3.3	0.7	0.6	3.57
Medium-volatile bituminous, Wyoming County, W. Va.	89.1	5.0	3.6	1.7	0.6	3.64
High-volatile A bituminous, Allegheny County, Pa.	84.9	5.6	6.9	1.6	1.0	3.68
High-volatile B bituminous, Saline County, Ill.	81.9	5.1	10.5	1.9	0.6	3.77
High-volatile C bituminous, St. Clair County, Ill.	77.3	4.9	14.3	1.2	2.3	3.80
Subbituminous A, Sweetwater County, Wyo.	78.5	5.3	13.9	1.5	0.8	3.70
Subbituminous B, Weld County, Col.	72.3	4.7	21.0	1.7	0.3	3.70
Subbituminous C, Campbell County, Wyo.	70.6	4.8	23.3	0.7	0.6	3.85
Lignite, Burke County, N. Dak.	70.6	4.7	23.4	0.7	0.6	3.92

Table 4.8 Chemical Composition and *d*-Spacings of Strongest Diffraction Line for Sedimented Organic Materials[172]

	Ultimate Analysis (wt %, maf)					
Sample	C	H	O	N	S	*d*-Spacing (Å)
1. Kerogen from oil shale	80.5	11.1	4.4	2.7	1.3	4.5
2. Trona acids	75.8	9.8	10.9	1.9	1.6	4.5
3. Asphaltene	83.0	10.8	2.3	3.1	0.8	4.2
4. Coal spores	78.9	7.2	11.6	1.0	1.3	4.6
5. Ambrite	82.1	10.4	6.0	0.1	1.4	5.9
6. Coal resin	85.6	10.4	3.2	0.4	0.4	6.0
7. Gilsonite	85.5	10.3	1.1	2.9	0.2	4.8
8. Impsonite	89.2	5.4	1.6	2.0	1.8	3.5
9. Elaterite	78.2	10.3	6.7	1.0	3.8	4.6
10. Grahamite, Pennsylvania[a]	87.6	7.9	1.7	1.2	1.6	4.4 3.6
11. Grahamite, Argentina[a]	86.6	7.6	2.4	0.8	2.6	4.4 3.5
12. Grahamite, Peru[a]	81.8	7.7	2.9	1.5	6.1	4.4 3.5

[a] Sample has composite peaks.

Source: S. Ergun, U.S. Bur. Mines Bull. No. 648, 1968, 38 pp.

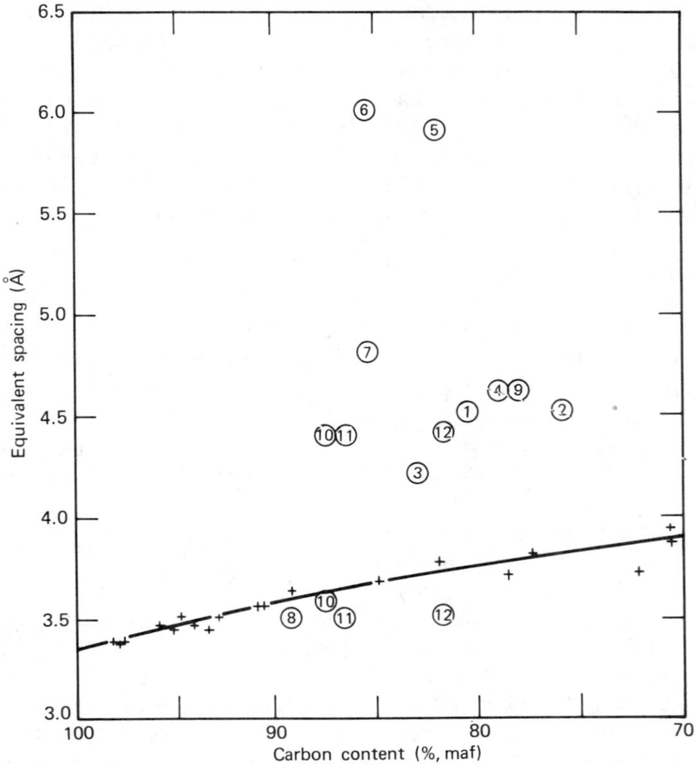

Figure 4.15 Spacings of the strongest diffraction line from coals (+) and sedimented organic matter (numbers). See Table 4.7 for description of noncoal samples.[172]

ities to graphite have led to the interpretation that coal consists of very small, graphitelike layers packed turbostratically. This means that successive layers are displaced and rotated parallel, but otherwise more or less randomly, with respect to each other. The two-dimensional (10), (11), and (20) and three-dimensional (002) and (004) indices are marked in Fig. 4.13*b*. The gamma band is not related to the graphite structure, and its origin is not understood. Possible explanations are alicyclic structures[174] with parallel stacking, the packing of aliphatic side chains,[175] or the irregular packing of aromatic layers.[171] The gamma band is the most prominent feature of the exinite pattern.[172]

Coal structures are much too small for single-crystal studies; therefore all X-ray diffraction interpretation is based on powder techniques and procedures. Peak half-width measurements, radial distribution functions, and comparison with theoretical scattering curves have all been used to estimate the structural parameters of coal. Kwan and Yen[176] have reported carbon aromaticities for four coals of different rank by comparing the intensities of the (002) and gamma bands. The method requires careful deconvolution of the diffraction patterns.

Theoretical studies of three-dimensional[177] and two-dimensional[178] crystallites showed that peak half-widths can be used to measure crystallites as small as 10–30 Å if appropriate corrections are made and if no imperfections or distortions are present. Since the "crystallites" in coal are imperfect,[179] the widths may provide useful information for compar-

[174] S. Ergun and I. Wender, *J. Appl. Chem.*, **10**, 189 (1960).

[175] V. I. Kasatochkin, *Izv. Akad. Nauk, Otd. Tekh. Nauk*, 1321 (1951).

[176] J. T. Kwan and T. F. Yen, *Prepr., Am. Chem. Soc., Div. Fuel Chem.*, **21**(7), 67 (1976).

[177] S. Ergun, V. H. Tiensuu, and L. E. Alexander, *J. Appl. Phys.*, **35**, 1718 (1964).

[178] B. E. Warren and P. Bodenstein, *Acta Crystallogr.*, **18**, 282 (1965).

[179] R. Diamond, *Philos. Trans. R. Soc. London*, **A252**, 193 (1960).

ative studies but do not reflect an accurate size measurement. In studies comparing radial distribution analysis[180] and electron microscopic direct lattice imaging[181] with peak half-width measurements, the latter gave smaller crystallite sizes for carbons and heated coals. A Fourier transform method[182,183] for analyzing (*hk*) profiles, which separates size and distortion effects, has been developed for turbostratic carbons but has not been applied to coals.

A detailed procedure[179] for estimating layer diameters in coal, developed in the late 1950s, was described in detail in the first Supplementary Volume of *Chemistry of Coal Utilization*.[184] The presence of imperfections in the layers makes it impossible to deduce an unambiguous structure, but a model of highly perfect small units linked together to form a large imperfect unit was developed.[171] For low-rank bituminous coals the vitrains consist of one to three aromatic rings; vitrains with about 90% carbon contain two to five rings. The more perfect regions are linked together by C—C linkages and aliphatic bonds to form either larger imperfect layers that are buckled or layers in which adjacent aromatic rings are rotated relative to each other about an axis normal to the plane of the sheet. The diameter and the thickness of the aromatic layer are about 7–8 Å, and the imperfect layers may be 20 Å or larger for a vitrain with about 90% carbon.

The use of Diamond's technique[179] for the study of coals was criticized[185] on three grounds: (1) the amorphous carbon determined in the analysis is simply an adjustment term in curve fitting, (2) only carbon atoms are considered in analyzing the scattered intensities, even though an ash-free coal may contain as little as 75% C, and (3) alicyclic structures that are present in coal produce peaks in the same positions as the (*hk*) bands of aromatic structures.

It has been shown that allowing for the presence of impurity atoms reduced significantly the amorphous fraction found in carbonized coals[179] and an anthracite,[186] but still left 20–30% nonaromatic carbon in a vitrinite[187] and its benzene- and pyridine-soluble extracts and residues.

A model of statistical disorder in parallel layer stacks in which the spacing between layers was not constant was developed.[187] The interlayer spacing calculated in this way increases the measured value by about 10%. There is good agreement between these values and those calculated from helium density measurement.[188]

4.5.1 Pitches and Chemically Treated Coals

Several studies of pitch tars found the crystallites to be 7–9 Å[171,189] using Diamond's procedure, and up to 20 Å[190,191] when half-width measurements were used. Examination of the Fourier transform of the (002) band of vitrains before and after oxidation in air[192] showed an increase in parallel layer groups with two and three layers and a decrease in single layers.

Coals were irradiated[172] with a ^{60}Co 2.8-MeV source with 10^9 R for 24 hr. High-volatile A and low-volatile coals showed no change, but a lignite and high-volatile B coals showed slight changes in their diffraction patterns.

Carbon blacks made from coal[193] had crystallite sizes of 6–17 Å for L_a and 12–14 Å for L_c, dimensions comparable to commercial blacks.

[180] S. Ergun, *Carbon (Oxford)*, **6**, 141 (1968).
[181] E. L. Evans and J. L. Jenkins, *Carbon (Oxford)*, **10**, 637 (1972).
[182] W. Ruland, *Acta Crystallogr.* **22**, 615 (1967).
[183] W. Ruland and R. Perret, *J. Appl. Crystallogr.*, **1**, 257 (1968).
[184] *Chemistry of Coal Utilization*, Supplementary Volume, H. H. Lowry, Ed., John Wiley and Sons, New York, 1963, pp. 39, 43, 48.
[185] S. Ergun, in *Chemistry and Physics of Carbon*, Vol. 3, P. L. Walker, Jr., Ed., Marcel Dekker, New York, 1968, p. 211.
[186] M. Short and H. McKinstry, *Proc. 5th Carbon Conf.*, Vol. 2, Pergamon Press, New York, 1962, p. 507.
[187] W. Ruland, *Proc. 5th Carbon Conf.*, Vol. 1, Pergamon Press, New York, 1962, p. 429.
[188] W. Ruland and V. Sandor, *Fuel*, **44**, 39 (1965).
[189] H. Honda, H. Kimura, and Y. Sanada, *Carbon (Oxford)*, **8**, 181 (1970).
[190] M. Shiraishi and K. Kobayashi, *Bull. Chem. Soc. Jap.* **46**, 2575 (1973).
[191] S. S. Pollack and L. E. Alexander, *J. Chem. Eng. Data*, **5**, 87 (1960).
[192] S. Z. Ali, *Fuel*, **39**, 369 (1960).
[193] J. Bayer and S. Ergun, *Carbon (Oxford)*, **5**, 107 (1966).

4.5.2 Carbonization and Graphitization

Carbonization studies of coals are discussed in detail in the first supplementary Volume of *Chemistry of Coal Utilization,*[184] and there are numerous articles on graphitization studies of carbons in the series *Chemistry and Physics of Carbon.*[194]

4.5.3 Small-Angle Scattering

The small-angle-scattering curve for coals usually shows its highest intensity at the lowest measurable angle and decreases as the angle increases. Many coals show a maximum at about 20 Å. Possible explanations of this peak were discussed in the first Supplementary Volume of *Chemistry of Coal Utilization* on page 36 and following. Kumura[195] measured a spacing of 35 to 40 Å in coals with 83 and 81.5% carbon.

The surface area of the coals can be determined from the continuous portion of the small-angle scattering curve if the slope at the tail or highest angle portion of the curve reaches −3 or −4, depending on whether infinitely high slits or pinhole collimation is used. Since small-angle scattering is produced by a difference in electron density, such as that of pores in a solid, and X rays can easily pass through thin specimens of coal, this technique can measure surface area of enclosed pores as well as those connected to larger channels. Kröger and Mües[196] found that vitrinite, micrinite, and exinite have surface areas between 24 and 40 m^2/g, which decrease by about one-third on coking at 1000°C. Durif[197] examined six vitrinites that had surface areas from 3 to 480 m^2/g. The samples were heated in steps from 400 to 3000°C. Some showed a maximum surface area at about 700°C and others at about 1500°C.

[194] *Chemistry and Physics of Carbon,* 12 vols., P. L. Walker, Jr., Ed., Marcel Dekker, New York, 1961–1976.

[195] T. Kumura, *Kogyo Kataku Zasshi,* **74,** 2202 (1971).

[196] C. Kröger and G. Mües, *Brennstoff-Chemie,* **42,** 78 (1961).

[197] S. Durif, *J. Chim. Phys.,* **60,** 816 (1963).

4.5.4 Crystalline Materials

An exhaustive survey of the minerals in coals and their reactions on thermal decomposition has been made by the British Coal Utilization Research Association (BCURA).[198,199] The minerals obtained by radiofrequency, low-temperature ashing have also been reported.[200]

Bitumens[201] of oil-saturated rocks, coals, and organic substances in rocks show diffraction patterns typical of solid solutions of paraffins.

4.6 ELECTRICAL CONDUCTIVITY

A good basic discussion of the electrical conductivity of coal and related materials was presented in the previous supplementary volume.[202] Techniques of measurement on both solid coal pieces and pulverized samples were discussed, and effects of pressure and moisture content were pointed out. (An analytical relation was later derived for expressing the electrical resistance of a powder as a function of the ρ of the substance, pressure, grain size, and mean diameter of the surface of contact.[203]) Coal was classed as a semiconductor, with specific resistances in the general range of $1–10^{14}$ Ω-cm. It was generally observed that specific resistance decreased with increase in both rank and temperature, the latter behavior being characteristic of semiconductors. Also discussed was the anisotropy of conductivity in layered carbonaceous substances such as graphite and high rank coals. The ρ of graphite is of the order of 10^{-4} Ω-cm parallel to the layer planes, but 2 Ω-cm perpendicular to the layers.

[198] J. D. Watt, BCURA Survey, Part I, Leatherhead, Surrey, 1968.

[199] Part II of Ref. 198, 1969.

[200] R. S. Mitchell and H. J. Gluskoter, *Fuel,* **55,** 90 (1976).

[201] C. I. Nosov, K. F. Rodionova, V. I. Dobovik, and V. V. Il'inskaya, *Tr. Vses. Neft. Nauchn. Issled. Geologorazved. Inst.* **68,** 71 (1969).

[202] *Chemistry of Coal Utilization,* Supplementary Volume, H. H. Lowry, Ed., John Wiley and Sons, New York, 1963, p. 95.

[203] A. A. Popova and K. I. Gaevaya, *Khim. Tverd. Topl.,* **8,** 151 (1974).

Pope and Gregg[204] determined specific conductivities for a series of vitrains of differing rank, using solid and pulverized samples (Fig. 4.16). Differences in results from the two sample forms were explained on the basis of absorbed moisture and contact resistances in the pulverized samples. A slow increase of conductivity with rank occurred up to about 90% carbon, after which the increase became progressively more rapid. The relation is closely similar to that between mean aromatic layer diameter and carbon as shown by X-ray analysis. It was deduced from this that all coals are intrinsic semiconductors. De Ruiter et al.[205] also measured changes in resistance of powdered coal samples with rank. They found that the electron activation energy (energy gap) calculated from the relationship between resistance and temperature, effectively a measure of the energy difference between the valence and conduction bands,

$$\Delta\epsilon = 0.397\,\frac{\log \rho_2 - \log \rho_1}{1/T_2 - 1/T_1}, \qquad (4.6)$$

[204] M. I. Pope and S. J. Gregg, *Fuel*, **40**, 123 (1961).

[205] E. de Ruiter, R. Leutner, and H. Tschamler, *Fuel*, **41**, 118 (1962).

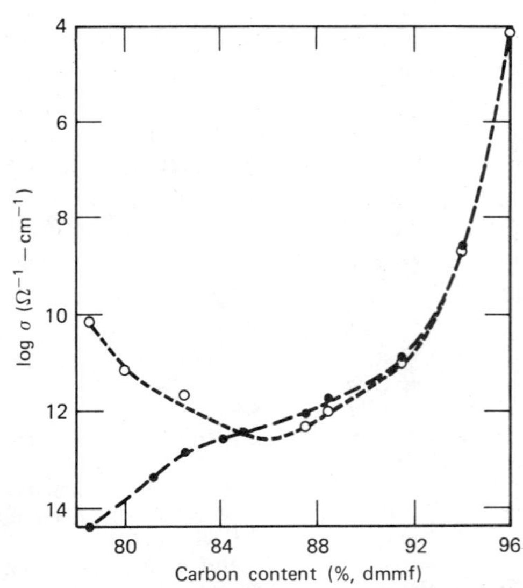

Figure 4.16 The specific electrical conductivity of undried compacts (○) and dried, machine-cut cuboids of vitrain (●) plotted against carbon content of the dry mineral- and moisture-free material.[204]

is nearly constant for vitrains from 77 to 89% carbon but falls abruptly with higher C. These findings were considered in relation to several postulates concerning aromatic cluster size and type of bonding between clusters. Similar conductivity behavior was found for coal oils and their extraction products.[206]

Ouchi[207] determined the specific resistance and energy gap of coals and carbonized coals. The marked difference observed between coalification and carbonization effects may be due to low molecular weight compounds having insulating effects in coals. After volatilization of these, all the carbonized products had the same $\Delta\epsilon$ and log R values at given HTTs. Ohuchi et al.[208] measured electrical conductivities of Japanese coals while heating to 800°C at 10°C/min. Temperature dependence and irreversible changes of electrical conductivity, and relations between $\Delta\epsilon$ and carbon content were explained for unit structural models of coals. Waters[209] found that the resistances of cokes from 11 coals, measured up to 900°C, depended on carbonization temperature only. He, too, inferred that coke behaves as an intrinsic semiconductor, with π electrons of aromatic layers being the dominant current carriers. Studies of the effects of temperature on resistance were extended to 3000°C for weakly coking and gas coals and anthracites, among other carbonaceous products.[210] Specific resistances decreased rapidly up to 1000°C, leveled off in the precrystalline stages at 1500–2000°C, and decreased during homogeneous crystallization above 2000°C. These changes corresponded to structural variations during graphitization (increase in thickness of the C layer from <80 Å at 1600°C to 400 Å at 3000°C and decrease in interlayer distances from 3.42 to 3.35 Å).

It was previously reported[202] that fusains have relatively low specific resistances of 10^2–10^4 Ω-cm and that vitrains, clarains, and durains have comparatively high values

[206] E. de Ruiter, R. Leutner, and H. Tschamler, *Rec. Trav. Chim. Pays-Bas*, **81**, 5 (1962).

[207] K. Ouchi, *Fuel*, **46**, 71 (1967).

[208] H. Ohuchi, T. Isobe, and J. Sohma, *Nenryo Kyokai-shi*, **47**(500), 856 (1968).

[209] P. L. Waters, *Proc. 5th Conf. Carbon*, Vol. 2, Pergamon Press, New York, 1962, p. 131.

[210] Z. V. Grigor'eva, T. M. Khrenkova, and V. I. Kasatochkin, *Tr. Inst. Goryuch. Iskop. Moscow*, **23**(2), 19 (1967).

(>10^8 Ω-cm). It was also reported that ρ is less dependent on rank for fusains than for the other components. Tonkonogov and Veksler[211] found that ρ increased with content of vitrain and decreased with fusain in a series of coals at 20 and 400°C. Oh and Choi[212] correlated ρ and CO_2 reactivity of anthracites and graphite at 800 and 900°C. Reactivity increased with increased ρ. The value of ρ could be used to differentiate graphite from Korean anthracite.

4.7 DIELECTRIC CONSTANT

The fundamentals of the application of dielectric constant concepts to coal, and consequent deductions concerning structure and behavior, were discussed in the previous supplementary volume.[213] Dielectric constants (ϵ) are usually measured by inserting coal—either in solid pieces or powdered dispersions in organic liquids or paraffin—into a condenser as the dielectric medium and measuring changes in capacitance by several techniques, depending on the use of high or low frequencies. At high frequencies the static dielectric constant ϵ must be replaced by a complex constant equal to $\epsilon' - \epsilon''$, the real part minus the imaginary part, which is dependent on energy losses. Although ϵ'' is stated to have significant values in the region from about 10^3 to 10^6 H_z, above this, particularly above 10^9 H_z, ϵ'' is practically zero and $\epsilon' = \epsilon \sim n^2$, the usual relation between dielectric constant and refractive index for insulators.

Variations of ϵ were shown for a series of vitrains, as measured at 10^6 H_z: ϵ decreased from 5 at 70% C to near the value of n^2 (3.5) at about 87% C, then increased rapidly in higher rank coals. Similar results were obtained in at least two studies. The higher values of ϵ in the lower rank coals were explained by the presence of many polar groups and those in the higher rank coals by increased electrical conductivity. Evidence for semiconducting character was seen in high-rank English, American, and Indian coals, but not in Japanese coals—an indication that large aromatic systems are present in the latter to a much smaller extent. In measurements on cokes ϵ' was found to increase slowly with pyrolysis temperature up to 600°C and rapidly at higher temperatures.

Sen et al.[214] determined dielectric constants of Indian coals of 78–84% C and of their acetylated products as a function of temperature of carbonization up to 600°C. A sudden increase of ϵ of the original coals at about 600°C was attributed to an increase in electrical conductivity. On acetylation, ϵ of the coals decreased at normal temperature, and the sudden increase was observed at less than 600°C in acetylated coals, particularly in the lower rank coals. This behavior in acetylation was attributed to a decrease in polar interactions between coal units that permitted coke formation at a lower temperature than was possible with the original coal. Hydrogenated coals exhibited a similar behavior. Tonkonogov and Veksler[215] found that ϵ of dry coals along the layers in the sonic frequency range (< 10^3 H_z) is far higher than across the strata, but at high frequencies the difference disappears. Dielectric losses along the layers were also found to be higher than across them. The temperature dependence of the loss tangent indicated a very loose coal structure along the layers and a high correlation of loss with electrical conductivity.

Measurements of dielectric properties were reported for coals of Russia,[216] Hungary,[217] Rumania,[218] and Japan.[219] Because of the great effect of moisture content on dielectric constants of coals, a number of applications for monitoring moisture content by

[211] M. P. Tonkonogov and V. A. Veksler, *Izv. Vyssh. Uchebn. Zaved., Fiz.*, **12**(2), 119 (1969).

[212] S. S. Oh and C. S. Choi, *Kungnip Kongap Yonguso Pogo,* **14,** 128 (1964).

[213] *Chemistry of Coal Utilization,* Supplementary Volume, H. H. Lowry, Ed., John Wiley and Sons, New York, 1963, p. 101.

[214] D. K. Sen, S. K. Sen Gupta, K. A. Kini, and A. Lahiri, *J. Sci. Ind. Res. (India),* **19B,** 184 (1960).

[215] M. P. Tonkonogov and V. A. Veksler, *Izv. Vyssh. Uchebn. Zaved. Fiz.*, **11**(1), 152 (1968).

[216] E. I. Arsh, L. A. Krasin, and G. R. Nosov, *Fiz.-Tekh. Probl. Razrab. Polez. Iskop.*, No. 1, 83 (1969).

[217] A. Miklos, *Banyasz. Kut. Int. Kozlemen,* **9**(1), 243 (1964).

[218] A. Luca et al., *Bul. Inst.-Politek-Iasi, Sect. II, Chim. Ing., Chim.*, **21**(1–2), 17 (1975).

[219] K. Chitoku and K. Higasi, *Bull. Chem. Soc. Jap.*, **34,** 320 (1961).

measuring ϵ were described, especially during conveying or processing.[220–226]

4.8 MAGNETIC SUSCEPTIBILITY

The application of magnetic susceptibility characteristics and determinations to deductions concerning the structure of coal has been treated thoroughly.[227] It was deduced that the organic portions are essentially diamagnetic, but they also reveal traces of paramagnetism, perhaps due to unpaired electrons or free radicals. Some of the associated mineral matter may be ferromagnetic and should be chemically removed, if possible, before measurements. Correction for the paramagnetic contribution is possible by establishing the effect of temperature; then extrapolating to infinite temperature where paramagnetism essentially vanishes. Diamagnetic susceptibility of coals has been measured by determining with a Gouy balance the force exerted by a magnetic field, by microscopic measurements of deflection in a field, or by pendulum period measurements with and without a field. Magnetic susceptibilities (χ) of demineralized coals, vitrains, clarains, and durains were found to lie between -0.5 and -0.6×10^{-6} cgs units. An anthracite showed $\chi = -1.15 \times 10^{-6}$ and exhibited anisotropy parallel and perpendicular to the layers.

A slight dependence of true diamagnetic susceptibility of coals on carbon content has been demonstrated.[228] Such results have been compared to the calculated sums of the atomic susceptibilities of the component elements, with modifications for carbon in various ring structures. From such comparisons and studies on pure aromatic hydrocarbons, it was concluded that coals up to 88% C contain one to four aromatic rings per structural unit and that larger rings are present above 90% C. Variations of χ with temperature appear to be small up to about 1300°C, after which χ increases rapidly, approaching the value for graphite (-7.5×10^{-6} cgs units) above 2000°C.

Little research on magnetic susceptibility of coals has been reported since 1959. Bivins and Ergun[229] published results of research on a range of U.S. bituminous coals and anthracites. They found no consistent variation with rank and very slight anisotropy up to anthracite. Values ranged from -0.42 to -0.77×10^{-6} cgs units. Berkovitz et al.[230, 231] measured the diamagnetic component of the susceptibility of coal chars and their free radical concentrations. The value of χ_{dia} fell unexpectedly in the 700–750°C range, possibly because of development of triplet electron states and/or chemical reactions favoring bond localization. A difference between free radical estimates from electron spin resonance and susceptibility data was tentatively ascribed to π-σ interactions in the Mrozowski model. The behavior of susceptibility and the diamagnetic component χ_{dia} during heating at several temperatures was shown to be primarily associated with changes in molecular or lamellar shape. Effects of lamellar size increases on χ were confined to temperatures above 700°C. It was thought likely that variations in χ and χ_{dia} during short heat treatment arose from ring-opening reactions. Since χ and χ_{dia} changed essentially similarly with heating time, it appeared that the "free radicals" in coal chars are probably due to charge-transfer processes and σ-σ bond rupture.

4.9 THERMAL ANALYSIS*

Thermal analysis methods differ from most of the other methods used to determine the physical properties of coals in that each

[220] V. M. Smirnow and L. M. Mashkevich, *Avtoma. Dorogi,* **23**(3), 13 (1960).

[221] R. Imaeda, H. Oikawa, H. Fujiwara, and T. Nozaki, *Tanken 1960,* 15th Anniversary Issue, p. 69.

[222] H. J. Kersting, *Braunkohle,* **17**(6), 217 (1965).

[223] S. Hornig, *Glueckauf,* **101**(11), 660 (1965).

[224] R. M. Green, "Humidity Moisture," *Papers Int. Symp.,* Washington, D.C., 1965, Vol. 4, p. 141.

[225] H. J. Kersting, *Braunkohle,* **19**(9), 321 (1967).

[226] A. A. Agroskin et al., *Ugol Ukr.,* **18,** 46 (1974).

[227] *Chemistry of Coal Utilization,* Supplementary Volume, H. H. Lowry, Ed., John Wiley and Sons, New York, 1963, p. 90.

[228] H. Honda and K. Ouchi, *Fuel,* **36,** 159 (1957).

[229] D. Bivins and S. Ergun, *Science,* **159,** 83 (1968).

[230] N. Berkovitz, P. A. Cavell, and R. M. Elofson, *Fuel,* **40,** 279 (1961).

[231] H. M. Brown and N. Berkovitz, *Fuel,* **45,** 207 (1966).

* Contributed by Jack G. Walters.

method is itself responsible for producing chemical and physical changes in the sample under investigation. Most of the uses of coal result in thermal treatment in some manner, and many test methods have been developed to determine the behavior and characteristics of coal when thermally treated. The tests are generally empirical and were formulated to obtain specific information about a particular process or method of thermal treatment. Therefore many of the physical changes that occur in coal during thermal treatment are addressed in the chapters dealing with plastic properties, pyrolysis techniques, low- and high-temperature carbonization, and the products of these processes.

Two general methods of thermal analysis that have been thoroughly investigated are differential thermal analysis (DTA) and thermogravimetric analysis (TGA). The DTA produces a deviation, ΔT, from the temperature of an unreactive reference sample; thus there are indications of endothermic or exothermic reactions or changes occurring in the sample under investigation. Because coal is a heterogeneous material, the deviation noted by this method cannot be construed to be from a single cause, but is rather an average value obtained from all the causes. Therefore DTA for coals has limited meaning without explicit data.[232] It cannot be recommended for accurately assessing any chemical or physical properties or attributes of coal, but only as a rough criterion for distinguishing the thermal behavior of coals of different ranks. It has limited potential for evaluating coals for specific processes because coals ideally suited for a specific process may have a DTA identical to that of a coal that is totally unsuited for the process.[233]

Thermogravimetric analysis involves the determination of weight loss from a sample as a function of rate of heating and temperature or the rate of weight loss as related to time at a specific temperature. Physical changes are observed only indirectly and then only if they involve a change in the release of volatile matter, which results in a weight change. The weight loss accompanying the release of volatile matter, if not a surface reaction, involves the transfer of the volatile matter to the surface and is analogous to the vaporization of a sorbed substance from inside a porous solid. Corrales and van Krevelen[234] found that when carbon tetrachloride or toluene was absorbed on activated carbon, the maximum rate of weight loss, due to evaporation of the solvent, occurred at temperatures 180°C above the boiling point of the sorbed material.

Such behavior can be explained by applying the theory of physical sorption. In the studies of the decomposition of coal by TGA, however, it is necessary to account for all the physical desorption, physical mass transfer, and diffusion that occur within individual particles.

In fractional TGA, an interesting technique developed by Waters[235,236] at the Commonwealth Scientific and Industrial Research Organization (CSIRO) in Australia, the volatile products of solids are analyzed by condensation, absorption, or chemical conversion. Variations in the rates of evolution of the tar, liquid, and gases from coal or similar compounds can be followed during thermal treatment.

4.9.1 Heat Content and Specific Heat

Specific heat and heat content of coal and associated materials (ash, coke, and volatile products) at various temperatures are significant in that their determination is necessary for the following purposes.

1. Design and development of any process that includes the thermal processing of coal, such as heat balances for coke ovens and various gasification processes.
2. The complete understanding of the fundamental properties of the material.

The specific heats of coals have been the subject of many investigations, but there are numerous uncertainties in the literature data due to sample uncertainty, variations in methods of determination, and failure to ad-

[232] H. Luther, W. Eisenhut, and O. Abel, *Brennstoff-Chemie,* **47,** 258 (1966).

[233] M. F. Kesseler and H. Romovackova, *Fuel,* **40,** 161 (1961).

[234] J. A. Corrales and D. W. van Krevelen, *J. Inst. Fuel,* **33,** 10 (1960).

[235] P. L. Waters, CSIRO, Coal Research Div. Tech. Commun. No. 34, 1960, 35 pp, CSIRO, Chatswood, New South Wales, Australia.

[236] P. L. Waters, *Anal. Chem.,* **32,** 852, (1960).

equately describe experimental conditions. For the investigations that permit the establishment of a common ground for comparison, there is consistency in the data for coke, coal volatiles, and related hydrocarbons.[237] However there is a very noticeable lack of data in the plastic range (300–550°C), where rapid decomposition and physical change occur. An interesting investigation by Agroskin and Goncharov[238] provides information on the specific heats of coal during carbonization to 900°C. Of particular interest is the relationship shown in Fig. 4.17, where the increase in specific heat that peaks at 550°C correlates with the large endotherm determined by DTA. In fact, the specific heat curve is almost a mirror image of the DTA thermogram. Kirov[237] has related mean specific heats of coals to volatile matter—those of coke to temperature of formation, and those of ash to temperature—and has studied the relation of moisture to specific heat determinations. Equations that predict many of the published experimental results to within 5% were developed for the specific heats of mineral impurities, moisture, residual coke, gaseous and liquid hydrocarbons, and volatile matter of coal. Mean specific heat values[237] were used to compute the total energy required for coking coals of various volatile matter content. It was found in general that variations due to volatile matter and moisture and ash contents are comparatively small, and that an average total heat may be computed from Equation 4.7:

$$Q = 82 + 0.286T \pm 7\% \qquad (4.7)$$

Therefore the energy required to coke a coal in the temperature range 20–950°C is 354 ±25 cal/G, which is within the range determined by other methods. Mean specific heat data for subbituminous coals and lignites were determined by Gomez et al.[239] using ice calorimetric techniques. Heat contents were also determined; the heat of carbonization for a coking coal, to 990°C, was determined to be 345 cal/g, which agrees remarkably with the data of Kirov.[237]

[237] N. Y. Kirov, *BCURA Bull.*, **29,** 33 (1965).
[238] A. A. Agroskin and E. I. Goncharov, *Coke Chem.*, **11,** 16 (1965).

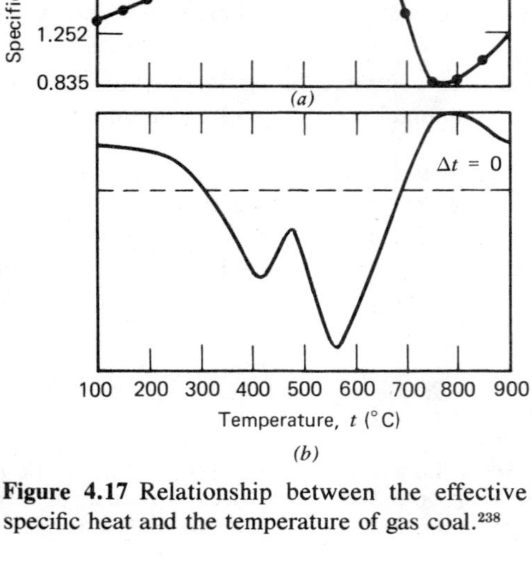

Figure 4.17 Relationship between the effective specific heat and the temperature of gas coal.[238]

4.10 COAL LIQUIDS: SAMPLE SELECTION AND SEPARATION*

A well-defined coal product characterization should take into consideration both sample selection and the solvent extraction procedures that provide the physically and/or chemically defined subfractions for future analysis. A principal concern and a major part of a separation procedure is choosing a representative sample. This concern has always been present among those working with whole coal, but has proved to be of greater importance when coal liquids are being sampled from a process stream. Present-day technology can separate an oil into hundreds of individual components or subfractions, but only a relatively small number of such subfractions are usually used to reflect the operating conditions and mechanism of an entire process.

The sample to be analyzed must be well-documented and chosen to best represent

[239] M. Gomez, J. B. Gayle, and A. R. Taylor, Jr., U.S. Bur. Mines Rep. Invest. No. 6607, (1965), 45 pp.

* Contributed by Frank K. Schweighardt.

that part of the system under study. For whole coals this usually means a petrographic analysis as well as a geographic description. For coal-derived liquids and residues from a process stream, one should report analysis of the feed coal, processing conditions, site of sampling, time of sampling, time of analysis after sampling, sample handling and storage conditions, and the conventional experimental parameters of materials and instrumentation.

These factors are important because after a solvent separation there is a tendency to forget that the final subfractions, which may be only a few milligrams, are subsequently analyzed by instrumentation that can detect nanogram quantities. If a sample is poorly chosen, a precise measurement made for a single component could be inaccurate as it relates to the total process. It is therefore important that the sample represent the system well, and that the separation procedure neither change nor react irreversibly with any component in the sample.

4.10.1 Coal Extraction

Solvent extraction by the Soxhlet method is by far the most popular method for initiating a separation of whole coal. Solvents most commonly used are pyridine,[240,241] carbon disulfide,[242] and benzene. A review of solvent extractions of coals by Roy and Lahiri[243] summarizes 84 investigations of different mechanisms of extraction and their structural implications.

In general, once a whole coal or a petrographically selected fraction has been chosen, it should be carefully ground and completely sieved to less than 200 mesh. The coal particles are then dried under vacuum (10^{-2} torr), mixed with prewashed Celite (1 : 1) to increase active surface area and allow for swelling, and loaded into a prewashed, fine, double-thickness cellulose thimble. Extraction is begun with a freshly distilled dry solvent. To avoid side reactions and agglomeration of extracts, the solvent should be changed after the first 3–5 hr. After extraction, the most troublesome problem is solvent removal, especially with pyridine. Schweighardt and Retcofsky[244] have found that a solvent-free extract can be otained by rotovaping the pyridine extract under vacuum (10^{-1} torr) at 100°C until a concentrated pyridine solution (25–50 ml) remains, then adding benzene (2 : 1, v/v) and freeze drying under high vacuum (10^{-2} torr) for 8 hr. Solvent removal is confirmed by IR analysis.

Recently Bartle et al.[245] and Whitehead and Williams[246] reported using a supercritical gas (toluene) to volatilize high-boiling materials from a nonvolatile residue.[247] They applied this method to coal undergoing mild pyrolysis to extract the liquids that are being formed. They reported that since the extraction takes place at temperatures below those at which volatile matter is evolved by destructive distillation, thermal degradation is minimized. Figure 4.18 is a flow diagram for gas extraction of heated coals.

Vahrman and co-workers have completed companion studies of solvent extractions of coal at atmospheric[248] and at high pressure[249] at 200–300°C. Their studies were directed toward finding aliphatic hydrocarbons in coal. Such hydrocarbons, which were present in quantities in excess of 2–3%, were found to be the main progenitors of the hydrocarbons in primary extracted tars at atmospheric pressure.

Another method of separating hydrocarbon material from coal is the use of ultrasonic solvation in quinoline as described by Kessler et al.[250] They were able to dissolve almost 50% of a Pittsburgh Seam hvAb coal in 24 hr

[240] F. Vecerikova and M. F. Kessler, *Acta Mont,* **35,** 93 (1975).
[241] S. Yokoyama, N. Ounisi, and G. Takeya, *Nenryo Kyokai-shi,* **52,** 906 (1973).
[242] H. L. Retcofsky and R. A. Friedel, *Fuel,* **47,** 487 (1968).
[243] J. Roy and A. Lahiri, *Proc. Symp. Chem. Oil Coal,* Central Fuel Research Institute, Dhnabad, India, 1972, p. 447.
[244] F. K. Schweighardt and H. L. Retcofsky, private communication.
[245] K. D. Bartle, T. G. Martin, and D. F. Williams, *Fuel,* **54,** 226 (1975).
[246] J. C. Whitehead and D. F. Williams, *J. Inst. Fuel,* **48,** 182 (1975).
[247] P. F. M. Paul and W. S. Wise, *The Principles of Gas Extraction,* Mills and Born, London, 1971.
[248] J. A. Spence and M. Vahrman, *Fuel,* **49,** 395 (1970).
[249] M. Rahman and M. Vahrman, *Fuel,* **50,** 318 (1971).
[250] T. Kessler, R. A. Friedel, and A. G. Sharkey, Jr., *Fuel,* **49,** 222 (1970).

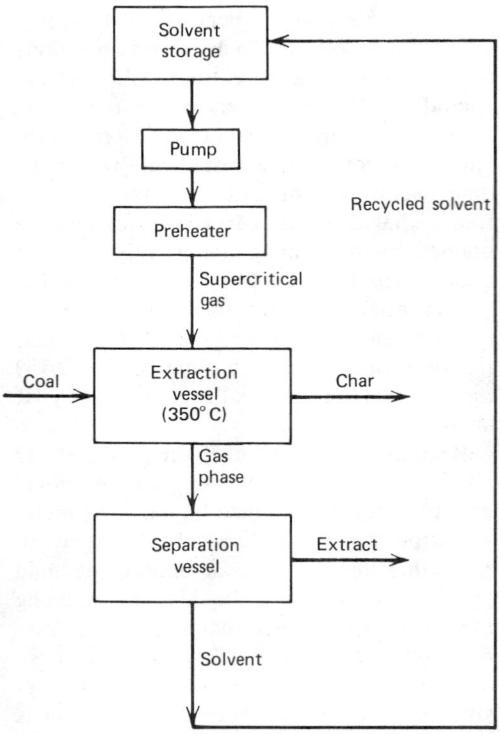

Figure 4.18 Gas extraction of heated coals.[245]

at 36°C. Romovacek and Kuhat,[251] on the other hand, developed a method of rapid pyrolysis of coal. A preweighed sample of coal encapsulated in tin is dropped into a bath of molten tin and pyrolyzed in less than 50 s. A carrier gas is used to remove the products rapidly to a gas chromatograph for detection and to largely eliminate secondary degradation. They report that the pyrolysis products accurately reflect the chemical nature of the original coal. A similar method was devised by Girling[252] to separate the more volatile C_4–C_{10} hydrocarbons from whole coal over a temperature range of room temperature to 700°C.

4.10.2 Coal-Derived Liquids

Experience has taught that in most cases the properties of the whole are the sum of its individual parts. Whether this axiom applies to coal liquids and their component parts has yet to be proved. Because of the great complexity of such coal-derived oils, a synergistic effect or the converse may well be important in relating the physical-chemical properties of liquid fractions to those of the whole liquid product.

The first problem in the separation of coal-derived liquids is not choosing a method, but deciding how many fractions the product should be separated into. Sternberg et al.[253] have stated that new and useful information obtained about the nature of a product increases with the number of fractions, but very quickly reaches a maximum. Thereafter useful information that further defines the system decreases with an increase in the number of fractions (Fig. 4.19). The methods described here are some of the better examples of judicious choosing of the number of subfractions related to the amount of useful information obtained.

A classic first approach to coal-liquid separation has been solvent solubility studies[254] in benzene, pentane, and pyridine. In 1951 Weller et al.[255] proposed a simple solvent extraction scheme (Fig. 4.20). More recently Mima et al.[256] published a solvent analysis method that has been used for more than 20 years at the Pittsburgh Energy Technology

[251] J. Romovacek and S. Kuhat, *Anal. Chem.*, **40,** 1119 (1968).
[252] G. W. Girling, *J. Appl. Chem.*, **13,** 77 (1963).

[253] H. W. Sternberg, R. Raymond, and F. K. Schweighardt, *Prepr., Am. Chem. Soc. Div. Pet. Chem.*, **21**(1), 198 (1976).
[254] M. L. Fein, A. Eisner, H. M. Cooper, and C. H. Fisher, *Anal. Chem.*, **11,** 432 (1939).
[255] S. Weller, M. G. Pelipetz, and S. Friedman, *Ind. Eng. Chem.*, **43,** 1572 (1951).
[256] M. J. Mima, H. Schultz, and W. E. McKinstry, ERDA/PERC RI-76/6, U.S. ERDA, Washington, D.C., 1976.

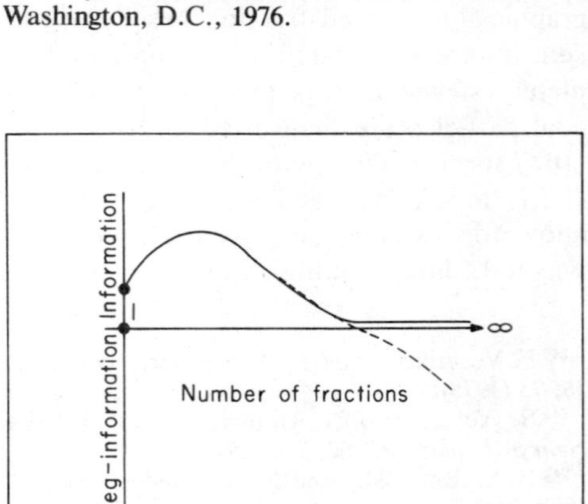

Figure 4.19 Information versus number of fractions.[253]

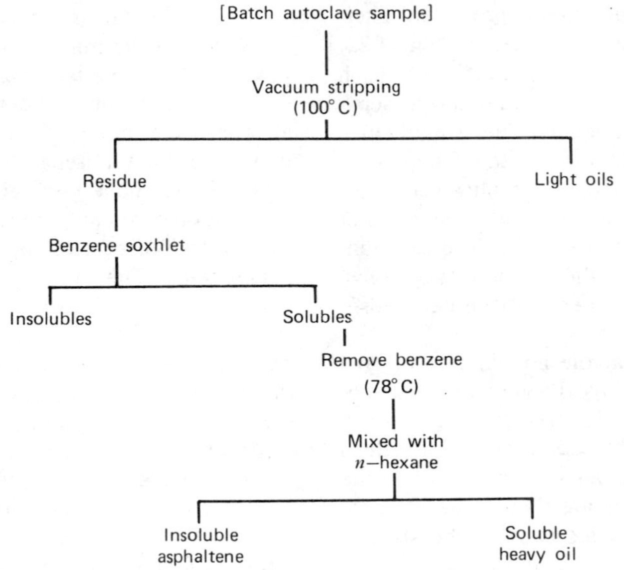

Figure 4.20 Scheme of Weller et al.[255] for solvent separation of coal-derived liquids.

Center on coal liquefaction products (Fig. 4.21). Their method defines the separation of a coal liquefaction product into oils (pentane-soluble), asphaltenes (benzene-soluble, pentane-insoluble), and benzene insolubles. Sternberg et al.[253] have extended this separation by treating the benzene insolubles with pyridine or tetrahydrofuran to give a pyridine-soluble (preasphaltene) fraction and leave a coal-derived residue. A separation scheme used by Schwager and Yen[257] provides five fractions from coal liquids: oils, resins, asphaltenes, carboids, and carbenes. At the Pittsburgh Energy Technology Center, Schultz et al.[258] are investigating a number of solvent analysis methods to standardize such separation procedures. Ruberto et al.[259] have proposed a separation scheme for coal-derived liquid fuel based on their experience

[257] I. Schwager and T. F. Yen, *Prepr., Am. Chem. Soc., Div. Fuel Chem.,* **21,** 199 (1976).

[258] H. Schultz, M. J. Mima, and F. K. Schweighardt, private communication.

[259] R. G. Ruberto, D. M. Jewell, R. K. Jensen, and D. C. Cronauer, *Prepr., Am. Chem. Soc., Div. Fuel Chem.,* **19**(2), 258 (1974).

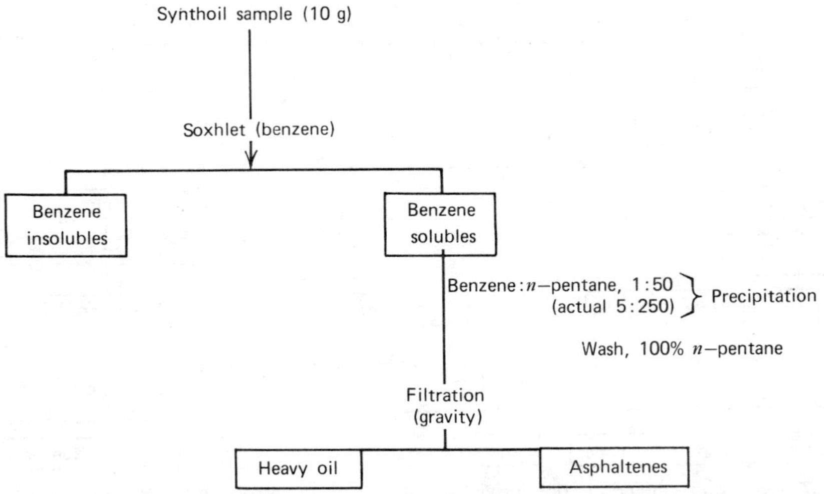

Figure 4.21 Scheme of Mima et al.[256] for solvent separation of coal-derived liquids.

with heavy end petroleum products. Their separation scheme is outlined in Fig. 4.22.

Farcasiu[260] and Callen et al.[261] have each established a more direct approach to separating coal-derived liquids. Their liquid chromatography procedure consists of sequential elution[262] of the sample from silica gel adsorbent with specific solvents and a minimum of mixing between each solvent. The resultant chemical classes of the fractions are shown in Fig. 4.23 and are described in detail elsewhere.[263]

If we accept that the liquefaction of coal can be represented by the generalized mechanism in Fig. 4.24, then the separation schemes outlined provide a first approach to understanding coal-derived products and the processes that generate them. The ultimate mechanism that relates back to the starting coal may be far more complex and may involve both the mineral matter and general petrography of the feed coal.

Fractionation of coal-derived liquids may also be made on an analytical scale (1–1000 mg) rather than a preparative scale (2–10 g). One of the most popular methods is gel permeation chromatography (GPC), using the total tetrahydrofuran- or pyridine-soluble components. The use of μ-Sytrogel[265] glass beads or coated silica under pressure (500–3000 psi) gives a separation based in part on molecular dynamic volume, chemical functionality, and secondary solvent–solute–column packing interactions.[266] Once a distribution is achieved, intermediate fractions are collected as a function of elution volume (time) or specific detector response. Yokayama et al.[267] have reported GPC separations of coal extracts and hydrogenated coal liquids, and Kulczycka et al.[268] have reviewed the use of gel chromatography for separating coal chemical products.

[260] M. Farcasiu, *Fuel,* **56,** 9 (1977).

[261] R. B. Callen, J. G Bendoraitis, C. A. Simpson, and S. E. Voltz, *Ind. Eng. Chem., Prod. Res. Dev.,* **15,** 222 (1976).

[262] R. P. W. Scott and P. Kucera, *Anal. Chem.,* **45,** 749 (1973).

[263] D. D. Whitehurst, M. Farcasiu, and T. O. Mitchell, *The Nature and Origin of Asphaltenes in Processed Coals,* EPRI Rep. No. AF-252, February 1976.

[264] F. K. Schweighardt and A. G. Sharkey, Jr., "Spectral Analysis of Effluents from Coal Liquefaction Processes," Preprints, 1976 Coal Chemistry Workshop, Stanford Research Institute, Menlo Park, Calif., August 1976, p. 72.

[265] R. Hendrik and J. Lijzenga, *Anal. Chem.,* **47,** 2160 (1975).

[266] C. E. Dougan, *Molecular Size Distribution of Asphalt as Determined by Gel Permeation Chromatography,* Bur. Highways, Connecticut Dept. Transportation, No. PB192978, 1970.

[267] S. Yokoyama, N. Ouinshi, H. Itoh, and G. Takeya, *Nenryo Kyokai-shi,* **53,** 1033 (1974).

[268] J. Kulczycka, A. Rusin, and A. Tobiczyk, *Koks, Smola, Gaz,* **19,** 133 (1974).

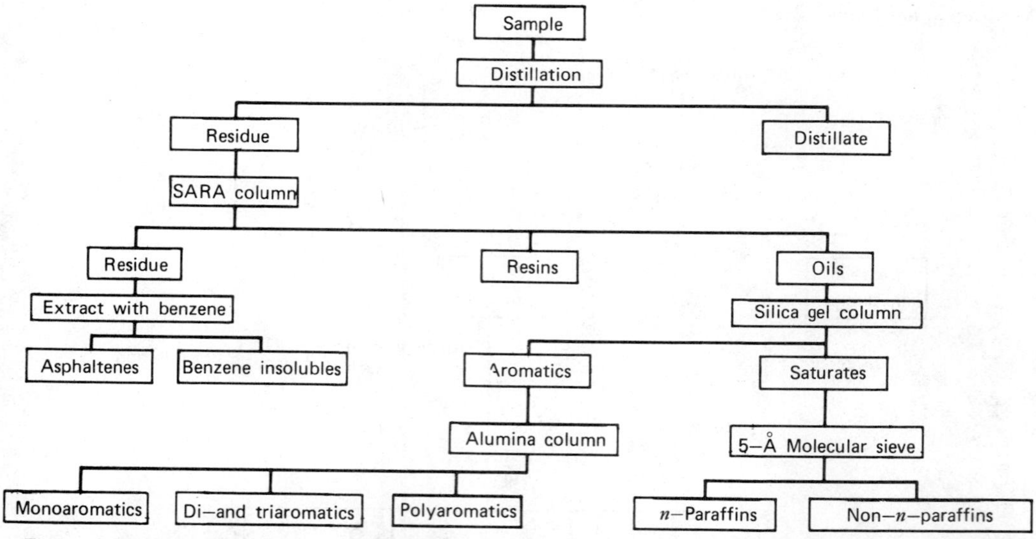

Figure 4.22 Separation scheme of coal-derived liquids based on petroleum heavy end products.[259] SARA = *S*aturates, *A*romatics, *R*esins, and *A*sphaltenes.

Classical Description	Fraction	Elution Solvent	Major Compounds
	1.	Hexane	Saturates
Oils, 1–3	2.	Hexane–15% benzene	Aromatics
	3.	Chloroform	Polar aromatics; nonbasic N, O, S heterocyclics
Asphaltenes, 3–5	4.	Chloroform–10% Et_2O	Simple phenols
	5.	Et_2O–3% EtOH	Basic nitrogen heterocyclics
	6.	McOH	Highly functional molecules (>10 wt % heteroatoms)
Multifunctional compounds, 5–9	7.	Chloroform–3% EtOH	Polyphenols
	8.	THF	Increasing O content and increasing basicity of nitrogen
	9.	Pyridine	
	10.		Noneluted, unknown materials

Figure 4.23 Sequential solvent elution scheme.[260]

It is important to mention that separations involving high-performance liquid chromatography (HPLC) are specific with respect to detector—as well as column packing—and solvent. The closest to a universal detector for high-boiling, coal-derived liquids that is independent of solvent yet carbon specific is the moving wire, flame ionization detector.[269] Other detectors such as the variable wavelength ultraviolet (VUW) spectrophotometer are dependent on a molecular extinction coefficient, and in such cases saturated hydrocarbons usually go undetected. Refractive index detection is usually limited to a single solvent, gradient elution being very difficult to control in a reference flow cell. Therefore the choice of the detector can greatly affect the separation and the resulting interpretation of the data.

Another form of separation is reversible chemical derivatization of specific functional groups in coal-derived liquids. Such deriva-

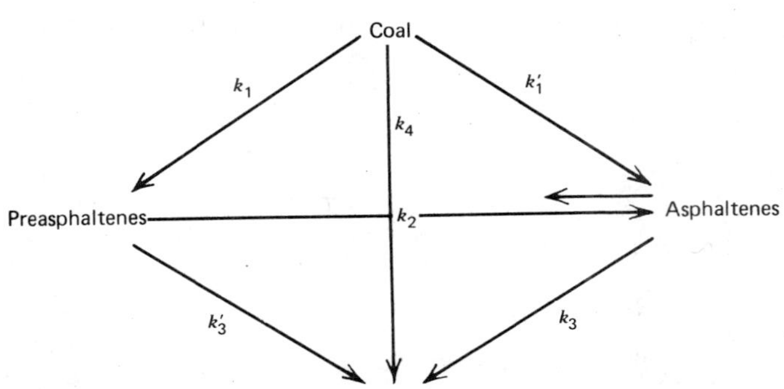

Figure 4.24 Reaction pathways of coal liquefaction.[264]

[269] J. A Yancey and T. J. Filipi, "The Use of First Run LC and Its Application to Complex and Unknown Mixtures," presented at the Pittsburgh Conference on Analytical Chemistry and Applied Spectroscopy, Cleveland, March 3–7, 1975.

tives have recently been used to selectively separate nitrogen heterocyclics from asphaltene material. Sternberg et al.[270] used gaseous hydrogen chloride (HCl) to form HC1 adducts with pyridinelike nitrogen heterocyclics and separated these from benzene solution of coal-derived asphaltenes (Fig. 4.25).

Shultz et al.[271] have extended the use of this technique to include the separation of nitrogen heterocyclics from the pentane-soluble oils and the total coal liquefaction products. The results indicate that the separation of nitrogen heterocyclics by HCl adduct formation can be made preparative and requires short workup times, and that side reactions are minimized and limited to HCl addition to the acid/neutral components. Strong base extraction, such as Claisen base,[272] may react irreversibly with hydroxyl moieties at high pH.

A summary[273,274] of separation methods, as they apply to coal-derived liquids, has recently been published; it includes detailed experimental conditions for column and thin-layer chromatography, thin-layer electrophoresis, solvent separation analysis, and HCl adduct formation.

[270] H. W. Sternberg, R. Raymond, and F. K. Schweighardt, *Science,* **188,** 49 (1975).

[271] J. L. Shultz, C. M. White, F. K. Schweighardt, and A. G. Sharkey, Jr., ERDA/PERC/RI-77/7, U.S. ERDA, Washington, D.C., 1977.

[272] R. Husak and C. Golumbic, *J. Am. Chem. Soc.,* **73,** 1567 (1951).

[273] F. K. Schweighardt, ERDA/PERC/RI-77/3, U.S. ERDA, Washington, D.C., 1977.

[274] F. K. Schweighardt, H. L. Retcofsky, and R. A. Friedel, *Fuel,* **55,** 313 (1976).

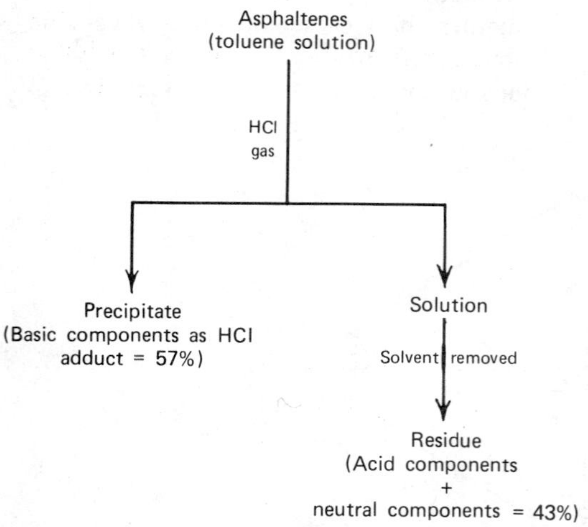

Figure 4.25 Acid-base separation scheme of asphaltenes.[270]

4.11 MASS SPECTROMETRY*

Trace elements as well as organic material associated with coal and its products can be investigated by mass spectrometry. Several excellent summaries of applications of mass spectrometers to organic and inorganic analyses are in the literature.[275,276] Mass spectrometers equipped with electron bombardment sources are the most popular type for studies of organic material. These instruments require volatile species having a vapor pressure of at least a few micrometers of mercury (μm Hg) for analysis. Combined gas chromatographs–mass spectrometers are very popular, since separation prior to analysis by mass spectrometry can be obtained in a single operation by mating the two units with an appropriate interface.[277] Direct probe introduction of samples into the ion source permits examination of coal and many of the heavier fractions from coal-derived oils and process samples. High-resolution mass spectrometry provides precise masses of individual components in a complex mixture from which molecular formulas can be determined. A particular isomeric form cannot be determined by this technique, but formulas for as many as 800 components in coals and coal extracts have been reported. Mass spectrometry is unique in that the molecular weights of individual components in a mixture can be determined in many instances.

Spark-source mass spectrometry (SSMS) has been used to survey for as many as 60 trace elements in coal and materials derived from coal. A major advantage of this tech-

* Contributed by Janet L. Shultz and Marion S. Edmunds.

[275] Mynard C. Hamming and Norman G. Foster, *Interpretation of Mass Spectra of Organic Compounds,* Academic Press, New York and London, 1972.

[276] Arthur J. Ahern, *Trace Analysis by Mass Spectrometry,* Academic Press, New York, 1972.

[277] William McFadden, *Techniques of Combined Gas Chromatography/Mass Spectrometry Applications in Organic Analysis,* John Wiley and Sons, New York, 1973.

nique is that survey data for a large number of trace elements can be obtained from a single spectrum. With most other techniques, either chemical processing must be done for individual determinations or the instrumentation must be altered for each element. SSMS has been applied to coal, coal ash, and many process samples. The technique has the added advantage of high sensitivity; in many instances concentrations down to the ppb range can be determined.

4.11.1 Coal and Its Petrographic Components

Coal and its petrographic components are being studied by conventional mass spectrometry. Because of the high molecular weights involved, however, it is difficult to obtain representative samples in vapor phase at the conditions of temperature and pressure in the mass spectrometer. It is usually necessary to use special techniques such as pyrolysis and extraction, which may alter the original sample and/or limit the analysis to portions of the original coal. Nonetheless, the spectra yield useful information.

Coal vitrain was used to test a device for volatilizing substances near the ionizing beam of a mass spectrometer.[278] The mass spectrum from 200 to 370°C showed loosely bound aliphatic material of a wide range of molecular weights, alkyl-substituted benzenes, naphthalenes, and higher aromatics. Above this temperature the detection of methane indicated decomposition. Homologous series of substituted phenols and naphthols were also detected.

In a second study at a temperature range of 80–450°C, similar products were evolved from five coals with carbon content ranging from 82 to 94%.[279] Both studies attempted to determine how material is held in the coal matrix. The release of alkyl aromatics between 170 and 300°C suggested that they are bound physically rather than chemically in the coal and that their appearance was a result of vaporization from the micropore structure of the coal.

A study of high molecular weight compounds, each containing an aromatic nucleus, was made to help elucidate the structure of coal and its products.[280] Spectra of coal and coal extracts showed evidence for the presence of methylene units, which may be part of an aliphatic chain or an alicyclic ring. Results indicated that in coal there is a repeating unit with a molecular weight of about 180.

When bombarded by rare gas ions of moderate energy, a number of solid surfaces, including coal, emitted secondary particles that were examined by mass spectrometry.[281] The coal samples were characterized by unusually large acetylene peaks in the neutral and negative ion spectra.

In an investigation of whole coal, extracts of coal, and pyrolysis products, it was determined that aromatic hydrocarbons obtained by vacuum pyrolysis of coal below 300°C were possibly sorbed material rather than decomposition products.[282] Material extracted from coal at room temperature is altered even by mild thermal treatment (Fig. 4.26).

A bituminous coal was analyzed by a combined method of ozonolysis, gas chromatography, and mass spectrometry.[283] Products expected from the oxidation of structures proposed for bituminous coals were found. At any stage of the oxidative degradation, the residual coal retained virtually the same properties as the original, so the oxidation might be visualized as a removal layer by layer.

SSMS was used to determine the amounts of minor and trace elements in coal particles of various sizes, including the respirable range (<5 μm).[284] The organic material did not interfere with the determinations. Fifty-six elements were determined in samples from mines in the Pittsburgh and Upper Freeport seams.

A survey of the applications of SSMS showed it to be particularly useful in the

[278] H. W. Holden and J. C. Robb, *Nature (London)*, **182**, 340 (1958).
[279] H. W. Holden and J. C. Robb, *Fuel*, **39**, 39 (1960).

[280] R. I. Reed, *Fuel*, **39**, 341 (1960).
[281] R. E. Honig, *Proc. 2nd Conf. Adv. Mass Spectrosc.*, Oxford, Vol. 2, 1963, p. 25. Author at RCA, Princeton, N.J.
[282] A. G. Sharkey, Jr., J. L. Shultz, and R. A. Friedel, U.S. Bur. Mines Rep. Invest. No. 6318, 1963, 32 pp.
[283] Sister M. Carol Bitz and Bartholomew Nagy, *Anal. Chem.*, **39**, 1310 (1967).
[284] T. Kessler, A. G. Sharkey, Jr., and R. A. Friedel, U.S. Bur. Mines Tech. Prog. Rep. No. 42, September 1971.

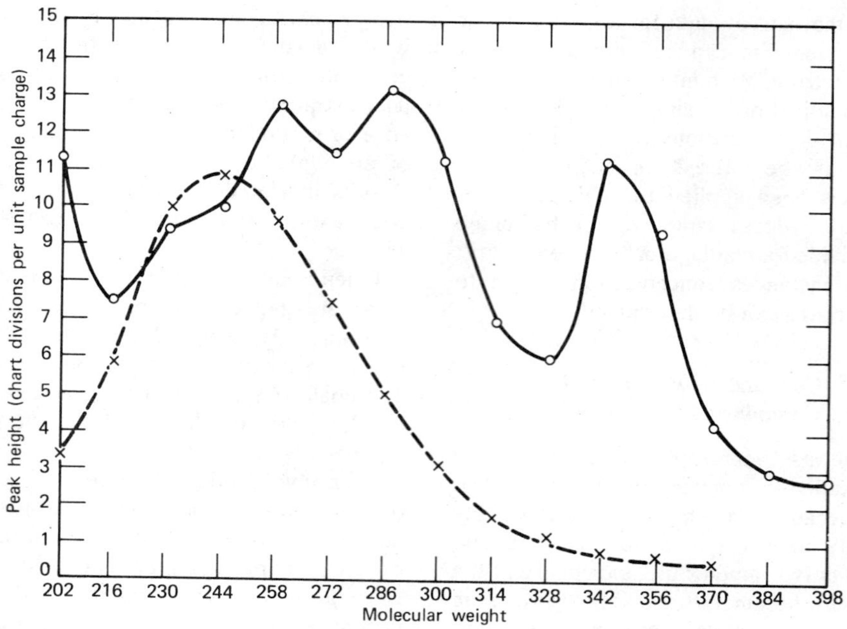

Figure 4.26 Distribution of four-ring, pericondensed aromatic hydrocarbons in pyridine extract from Wyoming subbituminous coal: ○ = unheated extract; × = extract heated 4 hr at 450°C.[282]

determination of trace and ultratrace elements in samples that are difficult to analyze.[285] It can be used to conveniently analyze coal, fly ash, respirable coal, lung tissue, and so on, which makes it especially suitable for assessing environmental hazards. Figure 4.27 plots the relative concentrations of certain elements in lung samples and respirable particulate matter from coal mines; remarkable correlation is shown.

A method of quantitative analysis for six trace elements was reported, using atomic absorption values for comparison.[286] The agreement was good and indicated that SSMS may be used to determine elements that are difficult to determine by atomic absorption, with the added advantage of eliminating the dissolution step in the sample preparation.

As part of a respirable dust mine research program, SSMS was used for semiquantitative determination of trace elements (≤200 ppm) and minor elements in raw coals.[287] One anthracite, two medium-volatile bituminous, four low-volatile, and six high-volatile A bituminous coals, representing ten coal seams located in Pennsylvania, West Virginia, Utah, and Colorado, were examined. Sixty-four elements ranging from 0.01 to 41,000 ppm weight were determined. The range of concentrations is shown in Fig. 4.28.

Several analytical techniques—neutron activation, atomic absorption, SSMS, optical emission, anodic stripping voltammetry, and X-ray fluorescence—were compared for the determination of trace elements in coal, fly ash, fuel oil, and gasoline.[288] Twenty-eight elements were determined with the various procedures. Order of magnitude agreement was found for only seven: silicon, calcium, potassium, strontium, iron, chromium, and nickel. Since emissions of pollutants into the air depend on both the quantity of material combusted or processed and the concentration of pollutants in the material, measurements must be precise and accurate. This study showed that total reliance on a single analytical method for trace analysis may con-

[285] R. Brown, M. L. Jacobs, and H. E. Taylor, *Am. Lab.*, 4(11), 29 (1972).

[286] Richard J. Guidoboni, *Anal. Chem.*, **45**(7), 1275 (1973).

[287] T. Kessler, A. G. Sharkey, Jr., and R. A. Friedel, U.S. Bur. Mines Rep. Invest. No. 7714, 1973.

[288] Darryl J. Von Lehmden, Robert H. Jungers, and Robert E. Lee, Jr., *Anal. Chem.*, **46**, 239 (1974).

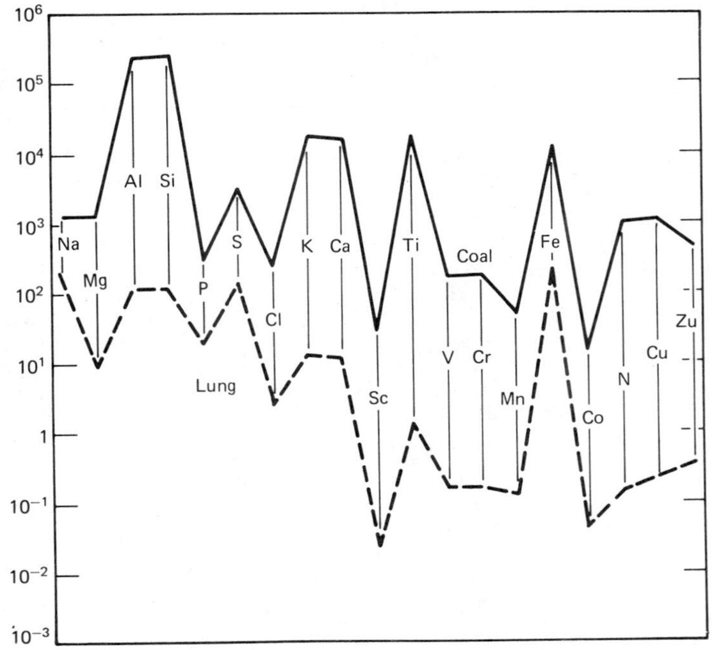

Figure 4.27 Relative concentrations of elements in coal dust (solid lines) and dry lung tissue (dashed lines).[285]

tribute to misleading results for a number of elements.

The continuing modifications of environmental impact surveys have necessitated the development of analytical techniques capable of acquiring data on many elements in a single sample.[289] Spark-source techniques are applicable for essentially each element in the periodic table in common environmental matrices (i.e., coal, fly ash, soil, and water). Data for elements not currently considered important are obtained as well, and can be stored for future consideration.

In addition to coal dust the atmosphere in a mine can contain other materials such as rock dust ($CaCO_3$), slate, and organic contaminants; therefore respirable dust may have a different composition from coal of a similar size from the same mine.[290] For samples from eight coal seams, the elemental compositions of the respirable dust and the 3.3–5.5 μm coal fraction were similar. The elements showing differences in concentration greater than the factor of 2, which is expected in duplicate spark-source determinations, were silver, cadmium, copper, zirconium, rubidium, calcium, chlorine, phosphorus, and bromine. In a majority of the samples these values were higher in respirable dust than in the corresponding coal fraction.

Spark-source mass spectrometry and thermal emission mass spectrometry (TEMS) were used to detect ppb quantities of elements in energy sources such as coal, fuel oil, and gasoline.[291] Toxic metals—cadmium, mercury, lead, and zinc—may be determined by SSMS with an estimated precision of ±5%, and metals that ionize thermally may be determined by TEMS with an estimated ±1% precision using the isotope dilution technique. These techniques were used for trace element balance of a coal-fired steam plant. Procedures for the analysis of lead in gasoline and uranium in coal and fly ash by TEMS are presented.

An investigation of organic compounds in respiratory coal dust by high-resolution mass spectrometry showed that the organic com-

[289] H. E. Taylor and R. Brown, *Instrum. Min. Metall. Ind.*, **2,** 21 (1974).

[290] A. G. Sharkey, Jr., T. Kessler, and R. A. Friedel, in Am. Chem. Soc. Adv. Chem. Ser. No. 141, 1975, p. 48.

[291] J. A. Carter, R. L. Walker, and J. R. Sites, in Am. Chem. Soc. Adv. Chem. Ser. No. 141, 1975, p. 74.

Weight (ppm)

H N.D.								
Li 4–163	Be 0.4–3							
Na 100–1000	Mg 500–3500							
K 300–6500	Ca 800–6100	Sc 3–30	Ti 200–1800	V 2–77	Cr 26–400	Mn 5–240	Fe 1400–12,000	Co 1–90
Rb 1–150	Sr 17–1000	Y 3–25	Zr 28–300	Nb 5–41	Mo 1–5	Tc N.D.	Ru <0.1	Rh <0.1
Cs 0.2–9	Ba 20–1600	La 0.3–29	Hf <0.3–4	Ta <0.1–8	W <0.1–0.4	Re <0.2	Os <0.2	Ir <0.2
Fr N.D.	Ra N.D.	Ac N.D.	Ce 1–30	Pr 1–8	Nd 4–36	Pm N.D.	Sm 1–6	Eu <0.1–0.4
			Th <0.1–5	Pa N.D.	U <0.1–1			

Figure 4.28 Range of concentrations of the elements in 13

ponents differed (1) in concentration but not in composition with particle size, (2) with rank of coal, (3) in coals of the same rank from different coal seams, and (4) in respirable mine dust and respirable coal dust (prepared) from the same mine.[292] The eight dusts were from coal seams in the Appalachian region, where the incidence of coal worker's pneumoconiosis is relatively high. The molecular weights of the major organic components varied greatly; three showed strong concentrations of polynuclear aromatic hydrocarbons. A further study of the organic material in respiratory size coal dust (prepared) and mine dust (collected in coal mines) was made to identify and classify the samples from mines having either high or low incidence of pneumoconiosis, from coals of various ranks, and from different geographic areas.[293] The investigation of the organic components in 21 respirable mine dusts and six 3.3–5.5 μm coal dusts (prepared) produced the following findings:

1. Respirable mine dusts are similar in composition to respirable coal dusts except for the molecular weight range, which is greater in the coal dusts.
2. Many respirable mine dusts are contaminated with a highly saturated hydrocarbon material, probably originating from lubricating oils and greases associated with mining machinery.
3. The mass spectral data for respirable mine dusts correlate with rank and geographical origin.
4. No correlation is obvious between the mass spectra of the organic components in respirable mine dusts and the incidence of pneumoconiosis for the 21 mine dusts studied.

4.11.2 Extracts

Solvent extraction, a long-established method for the study of the nature of coal, has been applied to coal and to products of coal

[292] J. L. Shultz, R. A. Friedel, and A. G. Sharkey, Jr., U.S. Bur. Mines Tech. Prog. Rep. No. 61, October 1972.

[293] J. L. Shultz, R. A. Friedel, and A. G. Sharkey, Jr., PERC/RI-75/4, U.S. ERDA, Washington, D.C., October 1975.

			B 1–230	C N.D.	N N.D.	O N.D.	F 1–110
			Al 3000– 23,000	Si 5000– 41,000	P 6–310	S 700– 10,000	Cl 10– 1500
Ni 3–60	Cu 3–180	Zn 3–80	Ga 0.3–10	Ge 0.03–1	As 1–10	Se 0.04–0.3	Br 1–23
Pd <0.1	Ag <0.01–3	Cd <0.01–0.7	In Std.	Sn 1–47	Sb <0.1–2	Te <0.1–0.4	I <0.1–4
Pt <0.3	Au <0.1	Hg <0.3–0.5	Tl <0.1–0.3	Pb 1–36	Bi <0.1–0.2	Po N.D.	At N.D.
Gd <0.1–3	Tb <0.1–2	Dy <0.1–5	Ho <0.1–0.4	Er <0.1–0.4	Tm <0.1	Yb <0.1–0.5	Lu <0.1–0.3

raw coals analyzed by spark-source mass spectrometry.[287]

from pyrolysis, carbonization, liquefaction, and so on. The extracts vary considerably in yield and composition, depending on the coal and its preparation, the solvent or solvent mixture, and the conditions of extraction (i.e., time, temperature, pressure, particle size, atmosphere, and pretreatment). Useful information is obtained when such information is considered in the interpretation of the mass spectral data obtained from extracts.

A study by mass spectrometric techniques of extracts from Grandville Marquis coal and from its pyrolysis products involved a variety of solvents, including carbon tetrachloride, acetone, and pyridine.[294] Complex spectra led to further attempts at fractionation using chromatography, treatment with acids and bases, extraction with successive solvents, and vacuum distillations. Information was obtained about the aliphatic content and the reactivity of components of the extracts. The investigations showed the advantages of combining fractionation with the mass spectrometric analyses.

[294] H. W. Holden and J. C. Robb, *Fuel*, **39**, 485 (1960).

The mass spectra of a vacuum pyrolysis condensate at 450°C and a methanol-benzene extract of Pittsburgh Seam (hvAb) coal were compared[295] and showed similar peak distributions in the two materials and a similar molecular weight range. Low ionizing voltage analyses are shown in Table 4.9. The major difference was that the extract contained much less oxygenated material than the condensate.

To aid in the analyses of coal extracts and pyrolysis products, mass spectra were obtained of pyrolyzates of several aromatic structures identified in pyridine extracts of coal.[296] Twenty hydrocarbons having from one to four aromatic rings and molecular weights from 116 (indene) to 228 (chrysene) were pyrolyzed and their liquid phase pyrolysis products determined. Thermal treatment of binary mixtures of model aromatic compounds indicated extensive interaction to form larger molecules up to seven condensed

[295] A. G. Sharkey, Jr., J. L. Shultz, and R. A. Friedel, *Fuel*, **40**, 423 (1961).
[296] A. G. Sharkey, Jr., J. L. Shultz, and R. A. Friedel, *Carbon (Oxford)*, **4**, 365 (1966).

Table 4.9 Mass Spectrometric Analyses (vol %) of Extract and Vacuum Pyrolysis Condensate from Pittsburgh Seam 84% Carbon Coal[295]

Compound Types Including Alkyl Derivatives[a]	Extract[b]	Condensate[c]	Compound Types Including Alkyl Derivatives[a]	Extract[b]	Condensate[c]
Indans, tetralins[d]	2.2	5.1	Phenols	4.1	16.4
Naphthalenes[d]	21.1	13.4	Dihydric, alkoxyphenols	0.7	3.3
Acenaphthenes, biphenyls[d]	8.6	9.5	Naphthols	1.3	4.8
Acenaphthylenes, fluorenes	7.7	10.0	Indanols, tetralols	0.0	6.9
Anthracenes, phenanthrenes	8.3	7.8	Benzenes	26.3	—[e]
Phenylnaphthalenes	4.4	4.8	Unknown (molecular weight 310 → 450) ≈ 10%		
Pyrenes, fluoranthenes	3.5	4.8			
Chrysenes, tetrazenes	1.8	3.2			

[a] Based on materials being primarily aromatic and phenolic.
[b] Solvent 60% benzene, 40% methanol; room temperature extraction.
[c] Removed from pyrolysis tube and analyzed in benzene-methanol solvent.
[d] Also includes neutral oxygenated compounds having same molecular weight as these hydrocarbons (i.e., benzofuran and indan).
[e] Lower members of homologous series in gases from vacuum pyrolysis.

Source: A. G. Sharkey, Jr., J. L. Shultz, and R. A. Friedel, *Fuel*, **40**, 423 (1961).

rings. Products from the low-temperature carbonization of coal generally have molecular weights below 400. The explanation may be that major structural types in a room temperature extract of coal produce components resulting primarily from the recombination of radicals. The effect of liquid phase pyrolysis on a pyridine extract from Pittsburgh Seam coal is illustrated in Fig. 4.29. The mass spectrometer is a useful tool in following the sequence of reactions that produce a heavy residue such as a coal tar.

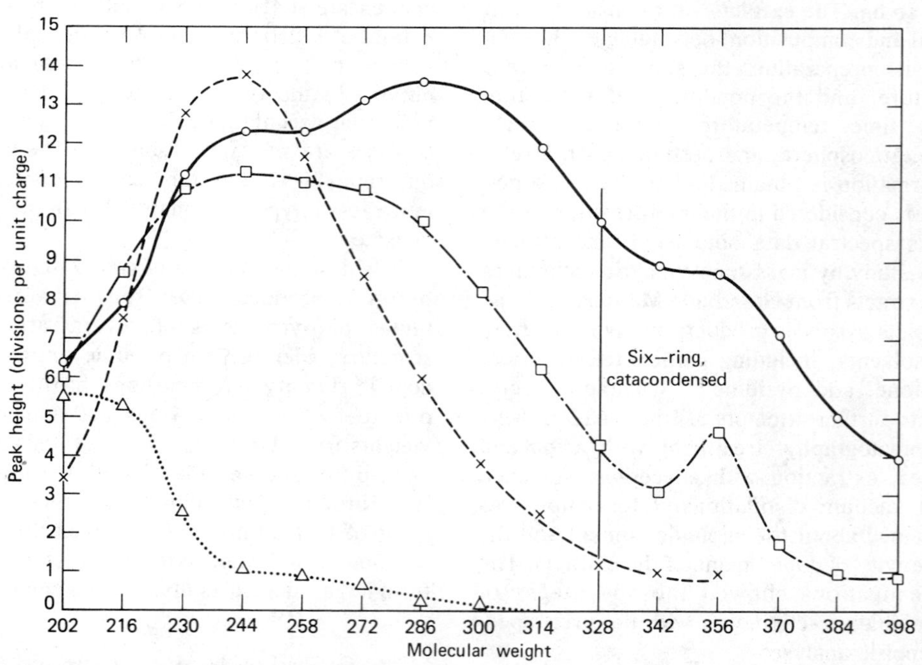

Figure 4.29 Partial mass spectra of pyridine extracts of Pittsburgh Seam (hvAb) coal. Four-ring pericondensed aromatics: ○ = unheated extract, × = extract heated at 450°C, 4 hr; □ = extract heated at 450°C, seconds; △ = extract heated at 500°C, 4 hr.[296]

A progressive extraction of coal with fresh batches of pyridine has shown that different constituents predominate at different stages of the extraction.[297] The first extract contained the alkyl benzenes and phenols, while polynuclear aromatics were extracted later; the higher concentration of phenols in the initial 5-min extract was a direct indication that the basicity of pyridine is an important factor in coal extractions.

In preparing extracts for examination by mass spectrometry, investigations were carried out to determine factors required for the solvation of coal. Quinoline was found to be the best of the five solvents studied.[298] The solubilization of coal by ultrasonic irradiation of coal-quinoline systems at ambient temperature was then studied. The amount of coal solvated was found to be a function of time and particle size.[299] Neither temperature nor ambient atmosphere affected the solvation results. The mass spectrometric analyses of pyridine extracts of reduced and untreated coals (Fig. 4.30) indicated a greater concentration of oxygen-containing compounds (except for dibenzofurans) in the reduced vitrain, an eightfold decrease in sulfur as shown by the ultimate analyses, a lower concentration of higher molecular weight hydrocarbons, and an increase in the amount of hydroaromatic compounds.[300]

Molecular weights of chloroform extracts of coals and chars determined by mass spectrometric analysis differed very little for a wide range of coals; this indicated that the same type of material was being extracted from the coals examined.[301] The structural type analyses of the extracts supported this indication because the percentage of each type of structure was relatively constant. Chloroform is a selective solvent and seems to dissolve the portion of coal that is a major

[297] R. A. Friedel, J. L. Shultz, and A. G. Sharkey, Jr., *Fuel,* **47,** 403 (1968).
[298] T. Kessler, R. A. Friedel, and A. G. Sharkey, Jr., *Fuel,* **49,** 222 (1970).
[299] T. Kessler, R. A. Friedel, and A. G. Sharkey, Jr., *J. Appl. Chem.,* **20,** 245 (1970).
[300] T. Kessler, R. Raymond, and A. G. Sharkey, Jr., *Fuel,* **48,** 179 (1969).
[301] J. G. Walters, J. L. Shultz, and J. A. Gloenzer, U.S. Bur. Mines Rep. Invest. No. 6973, July 1967.

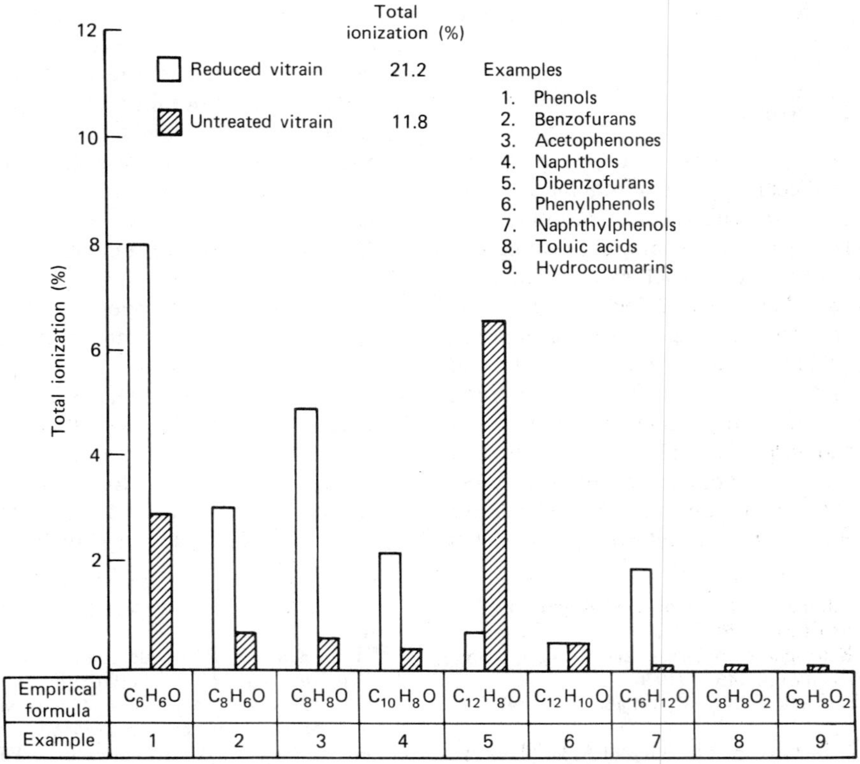

Figure 4.30 Oxygen compounds in pyridine extracts of Pittsburgh Seam vitrain.[300]

contributor to the development of a plastic state during thermal treatment.

A gas flame coal and a 95% vitrinite were extracted with chloroform in a nitrogen atmosphere.[302] A fraction of the neutral portion was examined by gas chromatography–mass spectrometry and it was possible to identify *n*-paraffins, monomethyl paraffins, phytol derivatives such as phytane and pristane, as well as various methyl-, ethyl- and butylbenzenes, naphthalene, phenanthrene, anthracene, fluoranthene, and pyrene.

Organic matter from two samples of coal-graphite from the Ketilidian of southwest Greenland was examined by solvent extraction, gas chromatography, and mass spectrometry.[303] The samples were separated by column chromatography followed by urea inclusion. The resulting fractions were analyzed by gas chromatography–mass spectrometry. The compounds identified included straight-chain alkanes from C_{12} to C_{21}, isoalkanes C_{14}–C_{19}, alkyl cyclohexanes, alkylbenzenes and alkylnaphthalenes, 15 monoterpenoid compounds, normal fatty acids from C_8 to C_{18}, seven branched-chain fatty acids, and two unsaturated C_{16} and C_{18} fatty acids. These organic compounds, together with the fossils, indicated biologic activity in this Precambrian coal-graphite layer.

4.11.3 Pyrolysis

The emphasis of studies of the pyrolytic products of coal has shifted from the standard low- and high-temperature carbonization products to products from rapid heating and quenching techniques such as flash and laser irradiation. Flash irradiation with a xenon lamp can produce temperatures of approximately 1000°C *in vacuo* in milliseconds, and temperatures up to 3000°C have been achieved under the same conditions with laser irradiation. Several studies, using coal as a target, were designed to investigate the temperatures and pressures achieved by laser irradiation.[304,305] An investigation of products from the high-temperature irradiation and pyrolysis of coal indicated that the products obtained from flash and laser irradiations of coal are different from those produced in high-temperature carbonization.[306] Results are shown in Table 4.10. The gaseous products are richer in unsaturates, with acetylene predominating. A higher percentage of acetylene was obtained with the laser irradiations than with the flash, probably indicating that the temperature was higher in the laser experiments. The products of high temperature irradiation were gas and a solid residue, with little evidence of the formation of liquids or tarry components; vaporization was probably in excess of 60%, using focused energy from a 2–10 J source. Based on these experiments, the distribution of products appears to be a function of the energy of irradiation, atmosphere, and probably particle size of the sample. The products were a simple mixture of low molecular weight gases, and little could be deduced concerning the chemical structure of coal.

The gaseous products from laser pyrolysis of coals of various ranks were studied by mass spectrometry.[307] Lower rank coals produced approximately 4 times as much gas as the higher rank coals, such as anthracite, and also showed a higher concentration of acetylene. Gases from laser irradiation in general had a lower concentration of saturated species such as methane and a higher concentration of acetylene than gases from low-temperature processing.

Coals of various ranks were also studied using a laser–mass spectrometer system with the samples irradiated in the ionization chamber of a time-of-flight spectrometer.[308] All gave large peaks at mass/charge *(m/e)* ratios of 50 and 74 and a smaller peak at 26, possibly due to diacetylene, triacetylene, and acetylene, respectively. The ratio of intensities, 50/74, varied little with coal rank, indicating a common origin for the substances responsible. This is believed to be the fragmentation of benzene rings in monocyclic

[302] Leonhard Birkofer and Wolfgang Pauly, *Brennstoff-Chemie,* **50,** 376 (1969).

[303] K. Raunsgaard Pedersen and Joergen Lam, *Medd. Greenland,* **185,** 1 (1970).

[304] F. S. Karn and J. M. Singer, *Fuel,* **47** 235 (1968).

[305] F. S. Karn, R. A. Friedel, and A. G. Sharkey, Jr., *Fuel,* **48,** 297 (1969).

[306] A. G. Sharkey, Jr., J. L. Shultz, and R. A. Friedel, U.S. Bur. Mines Rep. Invest. No. 6868, 1966.

[307] F. S. Karn, R. A. Friedel, and A. G. Sharkey, Jr., *Carbon (Oxford),* **5,** 25 (1967).

[308] Peter H. Given, in *Adv. Org. Geochem., Proc. 4th Int. Meeting on Organic Geochemistry, Amsterdam, 1968,* P. A. Schenck, Ed., Pergamon Press, Oxford, 1969, p. 427.

Table 4.10 Gaseous Products from Laser and Flash Irradiations and High-Temperature Carbonization of Coal[306]

Compound (mole %)[a]	Energy Source or Treatment: Ruby Laser, ~2-J Output, Sample Form, Cube	Xenon Lamp, ~4000-J Input Sample Form, <200 Mesh	Carbonization at 900°C
Hydrogen	45.4	69.2	55.6
Carbon monoxide	14.4	16.1	7.4
Methane	4.8	4.1	31.5
Acetylene	20.9	7.8	0.05[b]
Ethylene	4.9	0.05[b]	3.4
Ethane	0.6	0.1	1.2
Hydrogen cyanide	1.4	1.7	—
Carbon dioxide	2.0	0.5	0.4
Total hydrocarbons above C_2	5.6	0.4	0.5
Foregoing gases as weight-percent of coal	60[c]	18[d]	15[e]

[a] Water-free analysis.
[b] Maximum.
[c] Estimation based on volume of craters in surface following irradiation.
[d] Estimated from weight of coal on inner surface of tube.
[e] Carbonization product supplied by J. G. Walters, U.S. Bureau of Mines, Pittsburgh Coal Research Center.

Source: A. G. Sharkey, Jr., J. L. Shultz, and R. A. Friedel, Bureau of Mines Rept. Invest. 6868, 1966.

and polycyclic aromatic structures. In addition to peaks believed to be due to homologous series of alkyl benzenes and naphthalenes, there were groups tentatively attributed to mono- and dihydroxy benzenes. Dihydroxy peaks at masses 110 and 124 were most prominent in the spectra of the lowest rank coals and decreased in intensity relative to the monohydroxy and hydrocarbon peaks with higher rank coals. These data constituted the first evidence of multiple OH groups in single ring aromatic components.

Products evolved when coal particles were heated in the source of the time-of-flight spectrometer were classified from their spectra into three groups.[309]

1. *Radicals and atoms:* The peak at mass 15 (CH_3) was thought to come from methyl radicals because the ionizing voltage was too low to rupture the C—CH_3 or C—H bonds. From the size of mass 1 and mass 12 peaks, it is thought that hydrogen and carbon atoms were generated during the pyrolysis.
2. *Polyacetylenes:* The principal hydrocarbon product was acetylene, and peaks at masses 50, 74, and 98 were assigned to di-, tri-, and tetraacetylene, respectively.
3. *Ions:* Peaks at mass 7, 23, and 39 were attributed to ions of the alkali metals lithium, sodium, and potassium.

A further study of the products of rapid pyrolysis using laser heating and a time-of-flight mass spectrometer was done on coal, aromatic compounds, and plastics.[310] Only at high laser energies were radicals, polyacetylenes, and ions produced from coal. At low energies the principal products were water, methane, and ethylene.

The laser combined with time-of-flight mass spectrometry enhances the examination of the structure of coal.[311] Rapid pulses of short duration heat the surface rather than the coal bulk. By focusing the laser output it is possible to pyrolyze the petrographic constituents of coal *in situ*. A spectrum at 50 eV showed both aromatic and aliphatic substituted aromatic compounds and abnormal mass distributions indicative of free radicals, results similar to those reported above.[309]

[309] W. K. Joy, W. R. Ladner, and E. Pritchard, *Nature (London)*, **217**(5129), 640 (1968).

[310] Walter K. Joy, W. R. Ladner, and E. Pritchard, *Fuel*, **49**, 26 (1970).

[311] F. J. Vastola and A. J. Pirone, *Prepr. Am. Chem. Soc., Div. Fuel Chem.*, **10**(2), C-53 (1966).

A mass spectrometric study of the gaseous products from laser irradiation of Pittsburgh Seam coal in various atmospheres showed that the acetylene content of the gas was decreased by the presence of helium and increased in the presence of nitrogen and argon, reaching a maximum at argon pressures of 80–400 torr.[312] Table 4.11 shows gaseous composition in various atmospheres at constant pressure. High temperature and rapid quenching using laser irradiation favor the production of acetylene, which can exist in equilibrium with carbon and hydrogen at temperatures above 3000 K. The role of argon or nitrogen in enhancing acetylene production is probably related to the thermal properties of the gases.

When macerals from high-volatile bituminous coals were irradiated by 6-J pulses from a ruby laser, the total product gas increased in the order fusinite, micrinite, vitrinite, and exinite, which is the same order shown for increasing volatile matter content of the macerals.[313] The major gases evolved were hydrogen, carbon monoxide, and acetylene. Relative concentrations of the products varied little.

Although laser pyrolysis yields a variety of gaseous and solid products, there is little evidence for tar and pitch, which are characteristic of coal carbonization.[314] High-energy irradiation from a pulsed ruby laser produced a gas high in hydrogen, acetylene, and carbon monoxide and a solid with an elemental composition similar to the original coal. Low-energy irradiation with a continuous CO_2 laser produced a complex mixture of hydrogen, carbon monoxide, and hydrocarbon gases with little or no acetylene, and a solid with a lower C/H ratio than the parent coal. The solid had very low density, high solubility in benzene, and an ultimate analysis similar to ash-free coal. High-resolution mass spectrometry showed it to be a complex mixture of organic molecules with many of the characteristics of coal.

Gases arising from the rapid pyrolysis of coal vitrains were examined by mass spectrometric analysis.[315] The major components of the gas were hydrogen, methane, and carbon monoxide, with lesser amounts of CO_2 and hydrocarbons up to C_6. Hydrocarbons are present both as saturates and unsaturates, with the exception of acetylene, the absence of which is most likely attributable to the lower temperatures and lower heating rates employed relative to flash and laser irradiation.

Coal was pyrolyzed by exposure to a mil-

[312] J. L. Shultz and A. G. Sharkey, Jr., *Carbon (Oxford),* **5,** 57 (1967).
[313] F. S. Karn and A. G. Sharkey, Jr., *Fuel,* **47,** 193 (1968).
[314] F. S. Karn, R. A. Friedel, and A. G. Sharkey, Jr., *Fuel,* **51,** 113 (1972).
[315] M. Mentser, H. J. O'Donnell, S. Ergun, and R. A. Friedel, in Am. Chem. Soc. Adv. Chem. Ser. No. 131, 1974, p. 1.

Table 4.11 Composition of Gaseous Products (wt %) from Laser Irradiation of Pittsburgh Seam (hvAb) Coal in Various Atmospheres[312]

Atmosphere	Pressure of Gas, 20 torr					
	Vacuum	Ar	N_2	H_2O	H_2	He
Carbon dioxide	3.7	14.3	19.3	18.0	17.8	43.4
Carbon monoxide	41.2	26.3	33.8	34.6	37.2	15.6
Hydrogen	6.0	4.3	3.9	6.1	8.3[a]	2.6
Methane	8.5	3.1	4.9	6.1	5.9	7.1
Acetylene	25.9	36.7	32.6	22.0	25.5	6.7
Ethane	3.0	0.8	0.7	1.1	1.0	2.4
Diacetylene	2.7	11.1	1.6	1.0	0.4	0.5
Total hydrocarbons above C_2	4.1	1.1	0.9	7.4	1.5	21.5
Hydrogen cyanide	4.9	2.3	2.3	3.7	2.4	0.2

[a] Hydrogen produced by the laser pyrolysis.

Source: J. L. Shultz and A. G. Sharkey, Jr. *Carbon* **5,** 57 (1967).

lisecond high-intensity light flash.[316] Only gaseous products were formed, which were analyzed by a mass spectrometer. Small molecules such as hydrogen, carbon monoxide, methane, and acetylene predominated. Changes in the energy of the flash and in the pyrolysis atmosphere caused variations in the product distribution.

Mass spectrometric analysis of gases produced when a sample of subbituminous B coal was exposed to the radiant flux in an arc-imaging furnace showed a composition differing radically from that found in conventional carbonization (coking) of coal.[317] Only gases were found.

4.11.4 Carbonization

The analysis of coal carbonization products by high-temperature, high-resolution, and low ionizing voltage mass spectrometry has greatly increased the data base for these products, confirming earlier component identifications achieved by tedious chemical and spectroscopic studies and extending the number of components identified to include alkyl derivatives and heteroatom structures. Mass spectrometric methods cannot distinguish among isomers; thus a minimum number of components have been identified by molecular weight or formula.

Analysis by mass spectrometry of gases formed during low-temperature carbonization of coal indicated that decomposition began below 175°C with the main products being CO and CO_2.[318] As the temperature was increased, paraffins and olefins with more than three carbons were found. With a further rise in temperature, these hydrocarbons decreased and hydrogen, methane, ethane, and ethylene increased. Benzene and its derivatives, hydrocarbons with more than six carbon atoms, and O and N heterocyclics were not found.

An investigation of the low molecular weight alkanes from high-temperature coal carbonization revealed their similarity to those from crude oil and Fischer-Tropsch synthesis product.[319] The naphthene isomers from crude oil and high-temperature coal carbonization also have similar distributions.

Several investigations were concerned with the characterization of coal-tar distillates. Low ionizing voltage mass spectrometric analyses were obtained for naphthalene and anthracene oils, heavy creosote, and three fractions of a pitch.[320] Table 4.12 summarizes the average molecular weight, aromaticity, mean structural unit, and ring distribution for each fraction. The investigation of these fractions of coal tar resulted in semiquantitative data for 38 structural types and carbon number distribution data for their alkyl derivatives. The molecular weights of 10 of these structural types indicate ring systems not previously reported in products from high-temperature carbonization of coal. Two of these structures are polynuclear aromatic hydrocarbons, two are oxygen-containing compounds, one contains sulfur, and five are nitrogen-containing compounds. High-resolution mass spectrometry has been very useful in studying organic species present in complex mixtures, particularly in providing information concerning heteroatom species usually present in lower concentrations than polynuclear aromatic hydrocarbons. Qualitative analysis of the 80–85°C softening point pitch yielded molecular formulas for 239 compounds corresponding to a minimum of 108 structural types and their alkyl derivatives.[321] These data are summarized in Fig. 4.31; structures containing CH, CHO, and CHN are the most numerous, but species containing S, N_2, ON, O_2, and O_2N_2 are also present.

A review of several studies, including the fractionation of a coal-tar pitch directly in the mass spectrometer ion source, a comparison of the mass spectra of fresh and weathered road tar, and the investigation of "average stability" and "low-reactivity" coke-oven pitch has shown the usefulness of high-resolution mass spectrometry for character-

[316] E. Rau and L. Seglin, *Fuel,* **43,** 147 (1964).
[317] E. Rau and R. T. Eddinger, *Fuel,* **43,** 246 (1964).
[318] Yoshio Kawana, Yasushi Kobayashi, Hidemasa Honda, and Kozo Hirota, *Nippon Kogaku Zasshi,* **76,** 1228 (1955).
[319] R. A. Friedel and A. G. Sharkey, Jr., in Am. Chem. Soc. Adv. Chem. Ser. No. 55, 1966, p. 32.
[320] J. L. Shultz, R. A. Friedel, and A. G. Sharkey, Jr., U.S. Bur. Mines Rep. Invest. No. 7000, August 1967.
[321] J. L. Shultz, T. Kessler, R. A. Friedel, and A. G. Sharkey, Jr., *Fuel,* **51,** 242 (1972).

Table 4.12 Summary of Data Derived from Mass Spectrometric Analyses of Products from High-Temperature Carbonization of Coal[320]

	Naphthalene Oil	Anthracene Oil	Heavy Creosote	Pitch Sample		
				A	B	C
Boiling range (°C)	200–230	300–360	>360	—	—	—
Softening point (°C)	—	—	—	80–85	100–110	140–150
Average molecular weight	134	195	218	242	252	267
Aromaticity, f_a	0.86	0.89	0.93	0.94	0.95	0.94
Mean structural unit R_{ar}	1.6	2.9	3.7	4.2	4.5	4.8
Analysis %[a]						
1 Aromatic ring	39.6	0.9	—	—	—	—
2 Aromatic rings	53.9	18.2	2.0	2.2	0.9	0.6
3 Aromatic rings	3.4	46.3	27.6	14.9	8.3	4.8
4 Aromatic rings	0.5	19.5	30.8[b]	21.0[b]	23.3[b]	14.3[b]
5 Aromatic rings	—	1.2	6.6	11.6	16.4	15.4
6 Aromatic rings	—	—	1.8	7.0	9.3	10.9
7 Aromatic rings	—	—	0.3	1.5	0.9	1.6
Oxygen, sulfur, and nitrogen compounds[c]	2.6	8.2	21.6	12.7	12.3	10.4
Residue[d]	—	5.7	9.3	29.1	28.6	42.0

[a] Naphthalene and anthracene oil are in volume percent; creosote and pitch samples in weight percent.
[b] Includes the five-ring structure, benzo [*ghi*] fluoranthene.
[c] Data include oxygen-containing compounds with molecular weights corresponding to those of phenols, dihydric phenols, and the quinones. If present, other oxygenated structures containing ring oxygens (e.g., dibenzofuran) are included in the data for the aromatic hydrocarbons having the same molecular weight.
[d] Not volatile at 300°C.

Source: J. L. Shultz, R. A. Friedel, and A. G. Sharkey, Jr., Bureau of Mines Rept. Invest. 7000, Aug 1967.

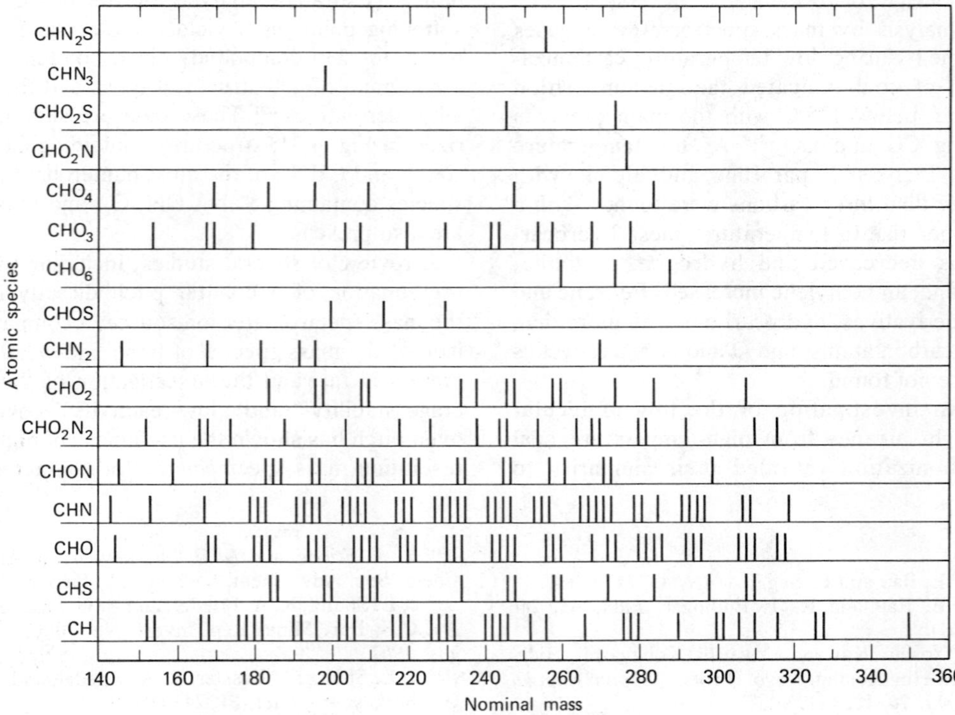

Figure 4.31 Molecular formula map, coal-tar pitch.[321]

izing complex coal carbonization products.[322] By the use of similar techniques, it was possible to correlate the electrode binder pitch quality index and the overall ranking of the pitches with the ratio of unsubstituted to alkyl-substituted aromatics.[323]

A high-temperature coal tar (Saar coal-tar distillate, b.p. <700°C) was separated into acid, base, and neutral fractions and analyzed by high-resolution mass spectrometry.[324,325] About 500 compounds were detected, based on 116 aromatic hydrocarbon types including the following combinations of elements: CHO, CHO_2, CHO_3, CHN, CHN_2, CHON, CHS, CHOS, and CHNS.

Carbon number distributions were obtained by low-voltage mass spectrometry for tar acid fractions from five low-temperature tars resulting from the carbonization of western United States low-rank coals.[326] The data support the view that low-temperature tars contain the same chemical compounds but in different proportions. Similar analysis of a coal-tar phenolic fraction obtained by medium-temperature pyrolysis of hard coals showed 27% alkyl phenols, 42% indanols and hydroxytetralins, 15.8% naphthol, 7.5% indenols, and 7.5% dihydric phenols.[327]

4.11.5 Coal Liquids: Process Samples

Many instrumental techniques, such as hydrocarbon type analysis by mass spectrometry, have provided analytical data in petroleum laboratories for many years, and it is only natural that an attempt is being made to extend these established techniques to coal-derived liquids. Currently available hydrocarbon type-analysis methods, in which the total concentration of various structural types such as paraffins, naphthenes, and aromatics are determined in petroleum products, cannot be applied directly to coal-derived oils because of the differences in the organic material.[282,328–330] Some of the major differences between the coal-derived liquids and conventional petroleum fractions are the higher aromaticity of the coal-derived liquids, the shorter chain substituents on ring structures in coal liquids, and differences in the concentrations of the heteroatom species including oxygen, nitrogen, and sulfur. Individual investigations as well as cooperative studies have shown that prior separation by compound class is necessary. The higher aromaticity of the coal-derived liquids is an advantage, particularly in terms of mass spectrometric analysis, since the polynuclear aromatics and other large ring systems lead to intense molecular ions, thus facilitating analysis.[331,332]

Five types of information can be derived from mass spectrometric analysis of organic components in coal-derived liquids:

1. Determination of total hydrocarbons.
2. Analysis for hydrocarbon types.
3. Carbon number distribution, particularly for aromatic structures.
4. Compound identification.
5. Structure elucidation.

Information from these analyses ranges from the least specific to the most specific that can be derived by mass spectrometry.

A group type-analysis technique for 17 hydrocarbon components, shown in Table 4.13, was developed using data for the major aromatic hydrocarbons observed in a number of coal liquids.[330] Group type analysis was performed with high-energy electrons, and the compound types were identified by low ionizing voltage mass spectrometry. In a concurrent investigation a procedure was developed to provide comparative information on

[322] A. G. Sharkey, Jr., J. L. Shultz, T. Kessler, and R. A. Friedel, *Spectrometry of Fuels*, Plenum Press, New York, 1970, p. 1.

[323] J. L. Shultz, R. A. Friedel, and A. G. Sharkey, Jr., *Fuel*, **49**, 175 (1970).

[324] Helmut Pichler and Albert Herlan, *Erdoel Kohle, Erdgas, Petrochem. Brennst.-Chem.*, **26**, 338 (1973).

[325] Helmut Pichler and Albert Herlan, *Erdoel Kohle, Erdgas, Petrochem. Brennst.-Chem.*, **26**, 401 (1973).

[326] Manuel Gomez, W. S. Landers, Janet L. Shultz, and A. G. Sharkey, Jr., U.S. Bur. Mines Rep. Invest. No. 6586, 1965.

[327] V. G. Lavrent'ev, A. P. Lapan, and A. A. Polyakova, *Khim. Tverd. Topl.*, **3**, 144 (1975).

[328] R. J. Clerc, A. Hood, and M. J. O'Neil, *Anal. Chem.*, **27**, 868 (1955).

[329] S. H. Hastings, *Anal. Chem.*, **28**, 1243 (1956).

[330] J. T. Swansiger, F. E. Dickson, and H. T. Best, *Anal. Chem.*, **46**, 730 (1974).

[331] A. G. Sharkey, Jr., Gorton Wood, and R. A. Friedel, *Chem. Ind. (London)*, 833 (1958).

[332] A. G. Sharkey, Jr., Gorton Wood, Janet L. Shultz, I. Wender, and R. A. Friedel, *Fuel*, **38**, 315 (1959).

Table 4.13 Total Aromatic Fraction[330]

No.	Compound Type	Group Sum	Partial Volume	% Based on 100
−28	Benzopyrenes	219.3	565.9	2.0
−24	Chrysenes	408.7	1092.9	3.9
−12	Decahydropyrenes	447.2	2214.9	7.9
−16	Hexahydropyrenes	976.7	4250.9	15.3
−18	Tetrahydrofluoranthenes	571.5	1271.9	4.6
−22	Pyrenes/fluoranthenes	2349.1	4490.4	16.1
−20	Dihydropyrenes	1618.2	2863.7	10.3
−10	Octahydrophenanthrenes	875.3	2661.6	9.6
−16	Hexahydrophenanthrenes	394.4	−51.9	0.0
−14	Tetrahydrophenanthrenes	1208.8	5106.5	18.3
−18	Phenanthrenes	1434.9	1729.2	6.2
−16	Fluorenes/dihydrophenanthrenes	1743.5	−240.7	0.0
−14	Acenaphthenes/biphenyls	1090.9	−333.3	0.0
−8	Tetralins	895.4	1374.5	4.9
−10	Tetrahydroacenaphthenes	882.3	197.5	0.7
−12	Naphthalenes	1295.2	−398.3	0.0
−6	Benzenes	588.4	44.9	0.2
	Totals	17,089.8	27,864.8	100.0

[a] Estimated accuracy, ±2%.

Source: J. T. Swansiger, F. E. Dickson, and H. T. Best, *Anal. Chem.*, 46, 730 (1974).

coal liquids using mass spectrometry as one of the major techniques for characterization.[333]

Survey data indicating the major structural types in all process streams from coal liquefaction and gasification have been obtained by mass spectrometry. Major structural types in heavy oil and asphaltene fractions of coal hydroliquefaction products have been analyzed by low ionizing voltage technqiues.[334] High-resolution mass spectrometry plus computerized data acquisition and reduction has been used for characterization of coal liquids, and this combination is a powerful tool for characterization.[335,336] Studies of coal liquids have indicated up to 99 different structural types, with approximately 500 components identified. Compounds containing O, O_2, O_3, N, S, SO, and SO_2 were identified.

A study was made of the organic sulfur compounds in coal hydrogenation products. This investigation, using high-resolution mass spectrometry, gave insight into several of the organic sulfur compounds in coal liquids and also determined the most stable structural types. In samples from the noncatalyzed hydroliquefaction of coal, 13 of the 14 organic sulfur compounds detected (Table 4.14) contained the thiophene nucleus.[337]

A study of the composition and character of a synthetic crude derived from coal was carried out using conventional petroleum characterization techniques. The sample was first separated into three distillates and a residue; each distillate was further separated by extraction and chromatographic methods, as outlined in Fig. 4.32. The various fractions were then characterized by several spectral

[333] R. G. Ruberto and D. M. Jewell, "Characterization of Coal Liquids," *Symp. Supplemental Fuels from Coal, New Analytical Needs,* 26th Pittsburgh Conf., Cleveland, March 3–7, 1975, Spectroscopy Society of Pittsburgh, Society for Analytical Chemistry, Pittsburgh.

[334] A. G. Sharkey, Jr., J. L. Shultz, C. E. Schmidt, and R. A. Friedel, PERC/RI-75/5, U.S. ERDA, Washington, D.C., November 1975.

[335] Thomas Aczel, J. O. Foster, and J. H. Karchmeyer, *Am. Chem. Soc., Div. Fuel Chem. Prepr.*, **13,** 8 (1969).

[336] Thomas Aczel, *Rev. Anal. Chem.*, **1,** 226 (1972).

[337] Sayeed Akhtar, A. G. Sharkey, Jr., J. L. Shultz, and P. M. Yavorsky, presented at 167th National Meeting American Chemical Society, Los Angeles, March 31–April 5, 1974.

Table 4.14 Structural Formulas of the Organic Sulfur Compounds[337]

m/e[a]	Formula	Structure	Compound
98	C_5H_6S	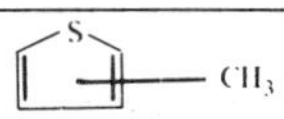	Methylthiophene
134	C_8H_6S	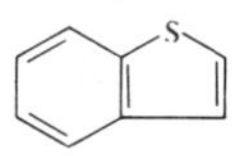	Benzothiophene
138	$C_8H_{10}S$	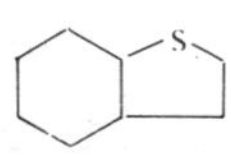	Tetrahydrobenzothiophene
148	C_9H_8S	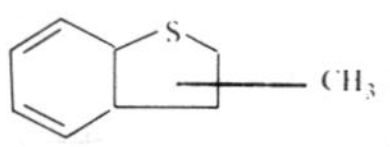	Methylbenzothiophene
162	$C_{10}H_{10}S$	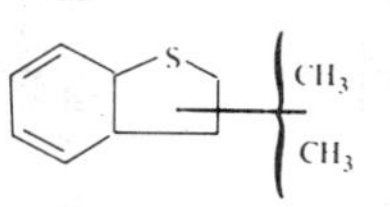	Dimethylbenzothiophene
174	$C_{11}H_{10}S$	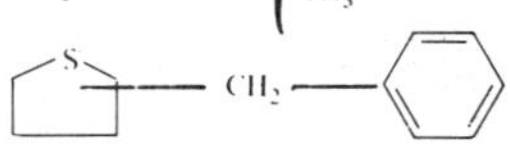	Benzylthiophene
184	$C_{12}H_8S$	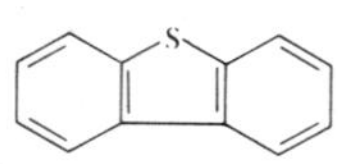	Dibenzothiophene
198	$C_{13}H_{10}S$	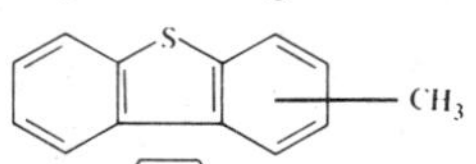	Methyldibenzothiophene
208	$C_{14}H_8S$	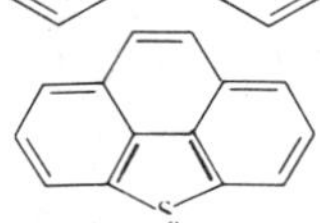	Benzo[*def*]dibenzothiophene
234	$C_{16}H_{10}S$	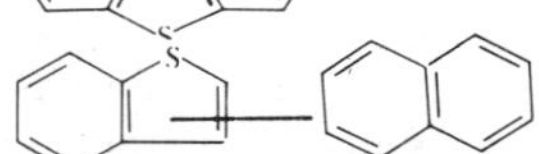	Naphthobenzothiophene
248	$C_{17}H_{12}S$	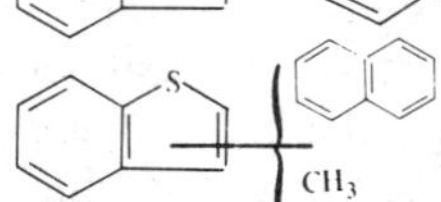	Methylnaphthobenzothiophene
284	$C_{20}H_{12}S$	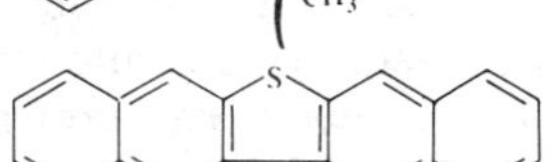	Dinaphthothiophene
298	$C_{21}H_{14}S$	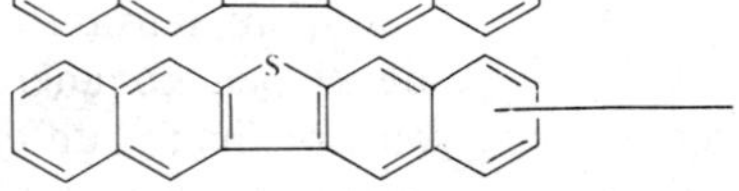	Methyldinaphthothiophene

[a]mass to charge ratio

Source: S. Akhtar, A. G. Sharkey, Jr., J. L. Shultz, and P. M. Yavorsky, presented at 167th National Meeting American Chemical Society, Los Angeles, March 31–April 5, 1974.

techniques.[338] The analyses provided quantitative compositional data that will be useful in developing efficient refining processes and suggest that much of the present petroleum technology will be applicable to processing synthetic crudes.

Condensate waters and other waters derived from the treatment and processing of coal have been investigated primarily to determine organic contaminants.[339] Methylene chloride extracts of the waters were investigated by high-resolution mass spectrometry,

[338] P. W. Woodward, G. B. Sturm, Jr., D. W. Vogh, S. A. Holmes, and J. E. Dooley, BERC/RI-76/2, U.S. ERDA, Washington, D.C., January 1976.

[339] C. E. Schmidt, A. G. Sharkey, Jr., and R. A. Friedel, U.S. Bur. Mines Tech. Prog. Rep. No. 86, 1974.

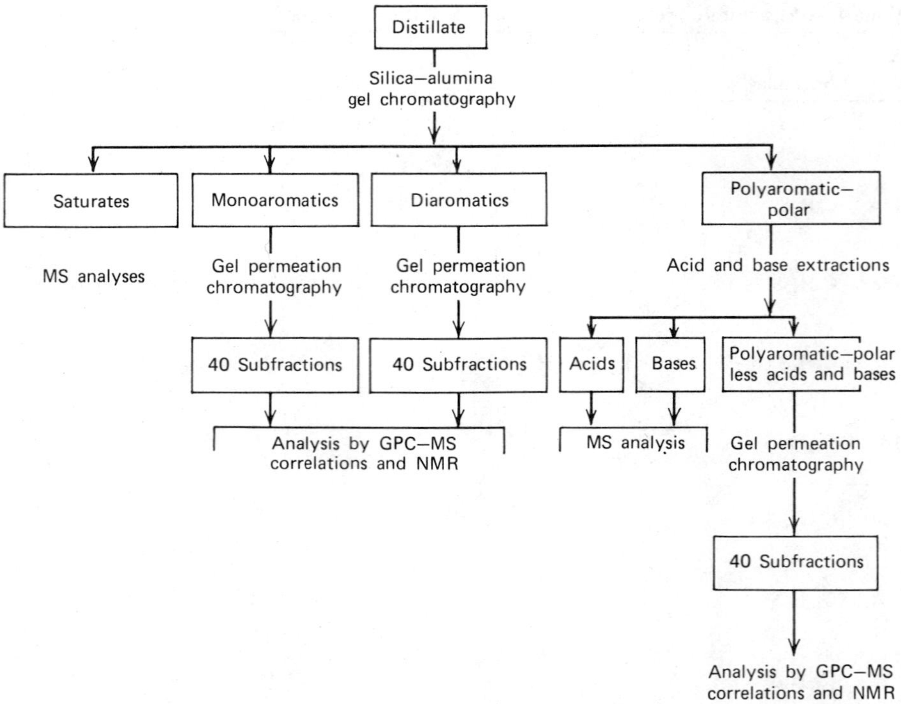

Figure 4.32 Analysis scheme for the 207–363°C and 363–531°C distillates.[338]

combined gas chromatography–mass spectrometry, and related techniques. Extracts of process waters from coal gasification contain 60–80% phenolic material. Some nitrogen and sulfur compounds were also detected. Many structural types in the phenolic portion of coal-derived liquids can be determined using trimethylsilyl ether derivatives.[340] Carbon number distribution data were obtained for compounds having up to 22 carbon atoms in a methanol extract of low-temperature tar from lignite.[341]

4.12 NUCLEAR MAGNETIC RESONANCE*

The applications of nuclear magnetic resonance (NMR) spectrometry in coal research have increased markedly since the publication of the first Supplementary Volume.[342] Only 14 references to NMR investigations of coals and coal-derived materials were cited in the first supplement. As in nearly all fields of study involving the determination of molecular structure, the use of NMR in coal research has grown because more reliable and more sensitive spectrometers have been developed commercially. Additionally, new and distinctly different methods for better detecting nuclear magnetic resonances have been introduced. Notable among these is the pulse Fourier transform (FT) technique,[343] which significantly reduces the time needed to obtain high-quality NMR spectra and has several other advantages over more conventional continuous wave (CW) techniques. The primary difference between CW and pulse FT NMR is that radiofrequency (rf) power is applied to the sample continuously while the CW spectrum is being obtained,

[340] S. H. Langer, S. Connell, and I. Wender, *J. Org. Chem.*, **23**, 50 (1958).
[341] A. G. Sharkey, Jr., Janet L. Shultz, and R. A. Friedel, *Fuel*, **41**, 359 (1962).
* Contributed by Herbert L. Retcofsky.

[342] *Chemistry of Coal Utilization*, Supplementary Volume, H. H. Lowry, Ed., John Wiley and Sons, New York, 1968, pp. 85–89, 117.
[343] T. C. Farrar and E. D. Becker, *Pulse and Fourier Transform NMR—Introduction to Theory and Methods*, Academic Press, New York, 1971.

whereas FT NMR spectra are obtained by Fourier transformation of the signals arising from the nuclear spin system after the sample has been subjected to short rf pulses of discrete frequency. FT NMR spectrometry, now a relatively common analytical tool in many laboratories, is largely responsible for the rapid growth of NMR applications to nuclei that do not possess the intrinsically high NMR sensitivity of the proton (^1H). Carbon-13 (^{13}C), which has a natural abundance of only 1.1% in addition to a relatively low magnetogyric ratio, is a prime example. Perhaps even more important in coal research was the development of multiple-pulse[344] and cross-polarization[345] techniques, which for many materials permit high-resolution NMR studies in the *solid* state.

Because of space limitations, detailed discussions of the theoretical and experimental aspects of the many NMR methods that have found utility in coal research cannot be given here. Liberal use of references to appropriate texts and treatises as well as to the original scientific literature is made to help alleviate any problems that this may impose on the reader.

4.12.1 A Brief History of NMR in Coal Research

NMR was first applied to coal in 1955,[346] just 9 years after reports[347,348] of the discovery of the phenomenon in bulk matter. The resonances observed were those of the protons in the coals and were detected by what is now commonly referred to as broadline NMR. The width and shape of the spectra were governed primarily by proton dipolar interactions. Spectral second moments were later used to estimate the ratio of protons in aromatic groups and in methyl groups undergoing hindered motion about their C_3 axes to protons in methylene and perimethyl groups.[349]

Unlike the broad resonances generally observed in NMR spectra of solids, liquids or materials in solution give rise to much more highly resolved spectra. The first high-resolution ^1H NMR spectrum of a coal derivative was published in 1959.[350] The spectrum was that of the carbon disulfide–soluble material from a coal hydrogenation asphaltene and consisted essentially of three envelopes of resonances. The band envelopes could be assigned to aromatic and phenolic hydrogens ($H_{ar+\phi}$), hydrogens on saturated carbons adjacent to aromatic rings (H_α), and hydrogens on saturated carbons β and further removed from aromatic rings (H_0).[350,351] Brown and Ladner,[352] in a classic paper published in 1960, developed a set of formulas that is still commonly used to deduce information concerning the carbon atom distribution from high-resolution ^1H NMR data. The method requires a few a priori assumptions concerning the molecular composition of the coal derivative and a knowledge of its elemental composition.

Most of these early ^1H NMR studies were limited to neat liquids or materials soluble in a rather narrow range of NMR solvents such as carbon disulfide and methylene chloride. Applications of NMR in coal research were considerably expanded after deuterated solvents became available. Of particular significance was the introduction in 1961 of deuterated pyridine as an NMR solvent.[353] This permitted determination of the hydrogen distribution in coal extracts representing a relatively large proportion of the parent coal.

Other advances in ^1H NMR studies of coals and related materials include the use of pulse techniques to measure proton relaxation times in coal (1968)[354] and the use of a superconducting solenoid to obtain extremely well-resolved spectra of soluble materials

[344] J. S. Waugh and L. M. Huber, *J. Chem. Phys.*, **47**, 1862 (1967).

[345] A. Pines, M. G. Gibby, and J. S. Waugh, *J. Chem. Phys.*, **56**, 1176 (1972).

[346] P. C. Newman, L. Pratt, and R. E. Richards, *Nature (London)*, **175**, 645 (1955).

[347] E. M. Purcell, H. C. Torrey, and R. V. Pound, *Phys. Rev.*, **69**, 37 (1946).

[348] F. Bloch, W. W. Hansen, and M. Packard, *Phys. Rev.*, **69**, 127 (1946).

[349] C. L. M. Bell, R. E. Richards, and R. W. Yorke, *Brennstoff-Chemie*, **39**, S32 (1958).

[350] R. A. Friedel, *J. Chem. Phys.*, **31**, 280 (1959).

[351] J. K. Brown, W. R. Ladner, and N. Sheppard, *Fuel*, **39**, 79 (1960).

[352] J. K. Brown and W. R. Ladner, *Fuel*, **39**, 87 (1960).

[353] R. A. Friedel and H. L. Retcofsky, *Proc. 5th Carbon Conf.*, Pergamon Press, New York, 1963, p. 149.

[354] H. L. Retcofsky and R. A. Friedel, *Fuel*, **47**, 391 (1968).

from coal (1969).[355] More recently (1977),[356] a report was published that presented the exciting possibility that multiple-pulse NMR experiments, after suitable refinement, may allow researchers to obtain high-resolution ^1H NMR spectra of *solid* coals. Preliminary experiments[356] resulted in narrowing the proton spectral line widths of bituminous coals from ~30 kHz to ~700 Hz.

In addition to those of the proton, NMR studies of the naturally occurring ^{13}C in coals and coal-derived materials have also received considerable attention. Although the full impact of ^{13}C NMR spectrometry as a tool for determining molecular structure was only recently realized, its potential in coal research has long been recognized. Carbon-13 NMR spectra of materials derived from coal were published as early as 1966.[357] To observe the very weak signals, rapid-passage dispersion mode techniques[358] were employed. These resulted in spectra consisting of asymmetric peaks and other distortions; nevertheless, some degree of quantitative integrity was maintained. These early spectra were referred to as high resolution, although by present standards they would probably be called medium-resolution spectra. The quantitative reliability of ^{13}C NMR in coal research was greatly improved after the introduction of signal averaging[359] to improve spectral signal/noise ratios. By signal averaging 1000 absorption mode ^{13}C NMR spectral scans of a coke-oven pitch sample, Hammel and Smith[360,361] were able to obtain what is thought to be the first unambiguous value of the carbon aromaticity of a material derived from coal. It was not until the mid-1970s that the more modern NMR techniques, correlation NMR[362] and FT NMR, were used to observe ^{13}C NMR resonances in soluble materials from coal.[363–367]

Carbon-13 NMR has also been applied to whole coals. In 1971[368] broadline ^{13}C NMR spectrometry was used to confirm the high aromaticity of anthracite. The spectrum consisted of a single broad peak devoid of any structure. The problem of resolution in ^{13}C NMR spectra of solid coals was overcome to a great extent some 5 years later[369] by use of the cross-polarization (CP) NMR technique often called proton-enhanced nuclear induction spectroscopy.[345] CP NMR spectra of coals consist of two overlapping resonances, which can be assigned to aromatic and nonaromatic carbons. The spectra can provide *estimates* of the carbon aromaticity (f_a). Further improvements in resolution of CP NMR spectra were attained by combining the CP experiment with spinning of the coal sample at the "magic angle."[370] "Magic angle" spinning reduces to zero the anisotropic chemical shift contribution to the line width.[371]

NMR spectra of nuclei other than ^1H and ^{13}C in coal have not been published yet. The potential of ^{33}S NMR in coal research has been explored (1972)[372] and found to be somewhat less than promising. A reinvestigation using more sophisticated instrumentation would be warranted. Deuterium (^2H) NMR was applied as a tracer tool in studies

[355] K. D. Bartle and D. W. Jones, *Fuel,* **48,** 21 (1969).

[356] B. C. Gerstein, C. Chow, R. G. Pembleton, and R. C. Wilson, *J. Phys. Chem.*, **81,** 565 (1977).

[357] R. A. Friedel and H. L. Retcofsky, *Chem. Ind. (London),* **1966,** 455.

[358] P. C. Lauterbur, in *Determination of Organic Structures by Physical Methods,* Vol. 2, F. C. Nachod and W. D. Phillips, Eds., Academic Press, New York, 1962, pp. 472–475.

[359] M. P. Klein and G. W. Barton, *Rev. Sci. Instrum.*, **34,** 754 (1963).

[360] H. L. Retcofsky and R. A. Friedel, in *Spectrometry of Fuels,* R. A. Friedel, Ed., Plenum Press, New York, 1970, p. 99.

[361] The actual spectrum is shown in Fig. 12 of Retcofsky and Friedel, Ref. 360. It appears that Messrs. Hammel and Smith were not properly credited.

[362] J. Dadok and R. F. Sprecher, *J. Mag. Resonance,* **13,** 243 (1974).

[363] F. K. Schweighardt, H. L. Retcofsky, and R. A. Friedel, *Fuel,* **55,** 313 (1976).

[364] H. L. Retcofsky and R. A. Friedel, *Fuel,* **55,** 363 (1976).

[365] K. D. Bartle, T. G. Martin, and D. F. Williams, *Chem. Ind. (London),* **1975,** 313.

[366] Y. Maekawa, S. Ueda, Y. Hasegawa, Y. Nakata, S. Yokoyama, and Y. Yoshida, *Prepr. Am. Chem. Soc., Div. Fuel Chem.*, **20,** 1 (1975).

[367] J. A. Franz, J. R. Morrey, J. A. Campbell, G. L. Tingey, R. J. Pugmire, and D. M. Grant, *Prepr., Am. Chem. Soc., Div. Fuel Chem.*, **20,** 12 (1975).

[368] H. L. Retcofsky and R. A. Friedel, *Anal. Chem.*, **43,** 485 (1971).

[369] D. L. VanderHart and H. L. Retcofsky, *Fuel,* **55,** 202 (1976).

[370] J. Schaefer and E. O. Stejskal, Poster Paper, 17th Experimental NMR Conf., Pittsburgh, April 25–29, 1976.

[371] E. R. Andrews, *Prog. Nuclear Magn. Resonance Spectosc.*, **8,** 1 (1971).

[372] H. L. Retcofsky and R. A. Friedel, *J. Am. Chem. Soc.*, **94,** 6579 (1972).

of coal liquefaction in 1976.[373] An interesting application of ^{14}N NMR in coal research was published in the same year.[374] Inferences on hydrogen bonding between the acid and base components in coal-derived asphaltenes were made by following the changes in the chemical shift of the ^{14}N resonance of pyridine with the addition of acid components separated from the asphaltenes.

4.12.2 NMR Studies of Coals and Related Materials in the Solid State

4.12.2.1 *Broadline ^1H NMR*

The experimental aspects of broadline ^1H NMR spectrometry were reviewed in the first edition. Data have been reported for whole coals,[346,347,375,376] vitrains,[377–380] vitrinites,[376,381–384] exinites,[376,384,385] micrinites,[376,385] fusinites,[384] carbonized coals,[386] brominated coals,[387] coal extracts,[376,381,383] and coal-tar pitches.[380,388] A series of papers proposing the use of broadline NMR to measure the moisture contents of coals has also appeared.[389–392]

Coal and Its Petrographic Components. The low-temperature proton second moments for coals, vitrains, and vitrinites, as reported by several investigators, are plotted as a function of the carbon contents of the materials in Fig. 4.33. With the exception of one high result (21.2 G^2),[376] which may be due to incomplete removal of moisture from the sample,[377] the agreement among data obtained by different investigators is quite good. As Fig. 4.33 indicates, for high rank coals the second moment decreases as the carbon content of the coal increases. Data for macerals other than vitrinite are quite limited; available data are summarized in Table 4.15. It appears that the second moments of the macerals decrease in the order exinite > vitrinite > fusinite > micrinite. It has also been shown[384] that the second moments of exinites and fusinites decrease with increasing carbon content, as was found for vitrinites. Again, the amount of data was small, consisting only of values for five exinites and three fusinites.

The usual method for deducing information on the chemical structure of coal from broadline ^1H NMR measurements is to assume that the second moments for the various proton-containing groupings in the coal are additive:

$$\Delta H^2 = \Sigma\, \Delta H_x^2\left(\frac{H_x}{H}\right), \qquad x = 1 \rightarrow N \qquad (4.8)$$

where H_x/H is the fraction of hydrogen in form x and ΔH^2_x is the contribution to the second moment for that form. Studies of pure hydrocarbons[376] and phenols[375] reduce this equation to:

$$\Delta H^2 = 9.7\left(\frac{H_{ar} + H_{OH} + H_{CH_3} + H_{CH}}{H}\right) + 27.5\left(\frac{H_{CH_2} + H_{CH_3}^{peri}}{H}\right) \qquad (4.9)$$

[373] F. K. Schweighardt, B. C. Bockrath, R. A. Friedel, and H. L. Retcofsky, *Anal. Chem.*, **48,** 1255 (1976).
[374] F. K. Schweighardt, R. A. Friedel, and H. L. Retcofsky, *Appl. Spectosc.* **30,** 291 (1976).
[375] W. R. Ladner and A. E. Stacey, *Fuel,* **43,** 13 (1964).
[376] R. E. Richards and R. W. Yorke, *J. Chem. Soc.,* 2489 (1960).
[377] J. Smidt, W. van Raayen, and D. W. Van Krevelen, *Fuel,* **38,** 539 (1959).
[378] W. R. Ladner and A. E. Stacey, *Fuel,* **40,** 295 (1961).
[379] Y. Sanada and H. Honda, *Fuel,* **41,** 437 (1963).
[380] J. Smidt, W. van Raayen, and D. W. Van Krevelen, *Fuel,* **41,** 527 (1962).
[381] J. F. M. Oth and H. Tschamler, *Fuel,* **40,** 119 (1961).
[382] H. Tschamler and E. de Ruiter, *Brennstoff-Chemie,* **43,** 28 (1962).
[383] J. F. M. Oth and H. Tschamler, *Fuel,* **42,** 467 (1963).
[384] W. R. Ladner and A. E. Stacey, *Fuel,* **42,** 75 (1963).
[385] H. Tschamler and E. de Ruiter, in *Coal Science,* P. H. Given, Ed., Am. Chem. Soc. Adv. Chem. Ser., No. 55, 1966, p. 332.
[386] W. R. Ladner and A. E. Stacey, *Fuel,* **44,** 71 (1965).
[387] A. C. Cunningham, W. R. Ladner, R. Wheatley, and W. F. Wyss, *Fuel,* **45,** 61 (1966).
[388] R. M. Pearson, *Fuel,* **52,** 80 (1973).
[389] W. R. Ladner and A. E. Stacey, *Br. J. Appl. Phys.,* **13,** 136 (1962).
[390] I. F. Galbraith, W. R. Ladner, and A. E. Stacey, *Proc. 1963 Int. Symp. Humidity and Moisture,* Reinhold, New York, 1965, p. 163.
[391] W. R. Ladner, *Instru. Pract.,* **18,** 1039, (1964).
[392] W. R. Ladner and R. Wheatley, *J. Inst. Fuel,* **39,** 280 (1966).

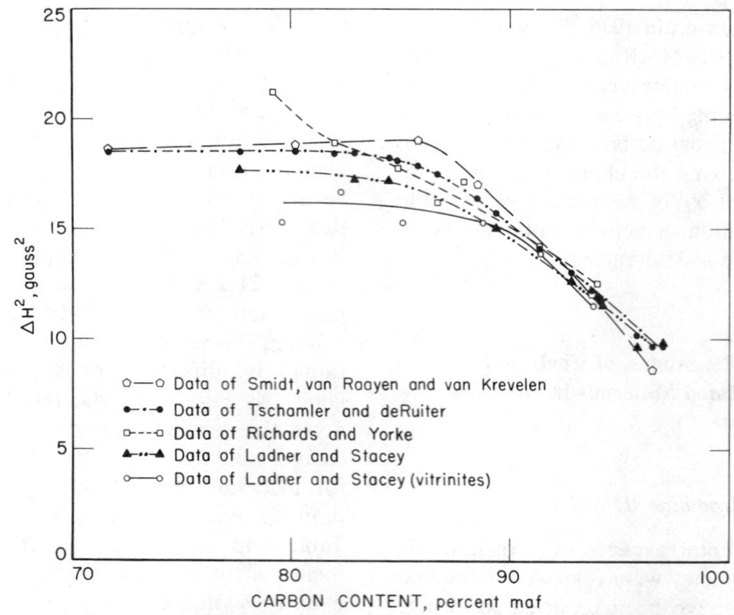

Figure 4.33 Low-temperature proton NMR second moments of vitrains and vitrinites as a function of carbon content as determined by various workers. (Data from Refs. 377, 382, 376, 378, and 384 respectively.)

By use of appropriate approximations or by evaluating some of the parameters by other means, a number of structural parameters can be evaluated, including the following:

$$\left(\frac{H_{ar} + H_{CH_3}}{H_{CH_2} + H_{CH_3}^{peri}}\right)^{376} \qquad (4.10)$$

$$\left(\frac{H_{CH_2}}{H}\right)^{375} \qquad (4.11)$$

and limiting values of

$$\left(\frac{H_{ar}}{H_{CH_2} + H_{CH_3}}\right)^{376} \qquad (4.12)$$

Another method for utilizing the second-moment data is to use *model substances* such as soluble fractions from coal rather than *model compounds* for calibration materials. For example, high-resolution NMR spectra of soluble materials from low-temperature carbonization of coals have been used to evaluate the contribution of aromatic protons to their broadline spectra.[378] The proportion of aromatic hydrogen was then estimated, using this calibration, for a range of coals from their second-moment values (Table 4.16). A number of investigators[393–395] have

[393] P. H. Given, *Fuel,* **39,** 147 (1960).
[394] W. R. Ladner and A. E. Stacey, *Fuel,* **40,** 452 (1961).
[395] I. G. C. Dryden, *Fuel,* **41,** 301 (1962).

Table 4.15 Proton NMR Second Moments for Macerals from Coal

		ΔH^2 (G^2)				
Coal	C (%, maf)[a]	Vitrinite	Exinite	Fusinite	Micrinite	Ref.
Cannock wood	79.6	15.3	19.0	12.2	—	384
Markham Main	82.3	16.7	19.1	11.4	—	384
Dinnington Main	85.1	15.3	18.3	—	—	384
American Bright coal	84.5	14.5	20.0	—	12.7	376
Dull coal	83.9	14.2	18.5	—	12.1	376 385

[a] Carbon content of the vitrinite component.

Table 4.16 Estimates of the Ratios of Aromatic to Aliphatic Hydrogen in Coals from Broadline ^{1}H NMR Measurements[378]

C (%, maf)	H_{ar}/H_{ali}
77.7	0.06
82.9	0.13
84.5	0.13
89.4	0.47
92.8	1.20

Source of data: W. R. Ladner and A. E. Stacey, *Fuel,* **40,** 295, 1961.

used broadline NMR data, often in association with other data, to deduce possible coal structures and changes in coal structure with rank.

It is emphasized that the second moments discussed above were all derived from spectra of coals at low temperatures. At higher temperatures the second moment of a hydrogen-containing substance generally decreases because of increased molecular motion within the sample. Spectra of coals at room temperature sometimes exhibit a narrow peak superposed on a broad one.[379,380,383] In undried coals the main contributor to the narrow band is almost certainly moisture in the coal. Some thoroughly dried samples also exhibit this narrow band in their spectra. The narrow component in the latter coals is thought to be due to groupings in the coal that undergo rapid molecular motion. The temperature dependence of the second moments of coals has been measured by two groups of investigators.[379,380] The second moment of an anthracite (96.5%C)[380] changes very little with temperature, indicating that the protons in the sample are bound to very rigid units. Coals of lower carbon content show a more pronounced decrease in second moment with increasing temperature.[380] A Japanese coal (84.9% C)[379] and two European coals (85.8 and 80.2% C)[380] show the appearance of the narrow component ~ 350–400 K, suggesting the presence of molecular units having large differences in mobility. In contrast to these, an American lignite (71.7% C)[380] shows a strong decrease of the second moment over the temperature range 77–450 K, while the spectrum becomes constantly narrower without the appearance of a second narrow peak. The spectral behavior of this coal suggests that the protons are highly mobile, which is in accord with the large number of nonaromatic protons thought to be present in the sample.

Moisture in Coal. In a series of papers[389–392] it was proposed that broadline ^{1}H NMR could be used to determine both inherent and free moisture in coal. The method was reportedly reliable to ±1% and independent of the rank, mineral matter content, and particle size of the coal. The total moisture contents of continuous coal feeds were measured with an accuracy similar to that obtained for stationary samples.[389] The method is based on the observation that the spectra of moist coals consist of a combination of two spectra: a broad spectrum resulting from the hydrogen in the solid coal, and a narrow spectrum due to the more mobile hydrogen in water associated with the coal. To the best of our knowledge the method is no longer in use.

Coal Extracts. Only a few broadline NMR studies of coal extracts have been published.[376,381,383] Early studies of extracts obtained by progressive extraction of a vitrinite (84.0% C) with pyridine, trichlorobenzene, benzene, and carbon tetrachloride showed that there was little difference in second moment between the extracted samples and the original vitrinite.[376] Later studies of products from a vitrinite of quite similar carbon content extracted with pyridine, benzene, or petroleum ether indicated that each of the second moments of the extraction products was slightly less than that of the vitrinite.[383] Figure 4.34 illustrates the close similarity of the spectra of the vitrinite and its benzene extract. It was deduced, using reasonable assumptions, that at least one methyl group is present for every four methylene groups in the vitrinite and that the methyl content of the products from the stepwise extraction increases progressively.[383]

Treated Coals. A broadline ^{1}H NMR study of coals before and after carbonization at temperatures up to 900°C has been reported.[386] The plot of the temperature dependence of the second moment for each of the three coals studied (maf C = 83.6, 89.4, and 92.8%) shows three distinct regions: an initial region of constant second moment, followed

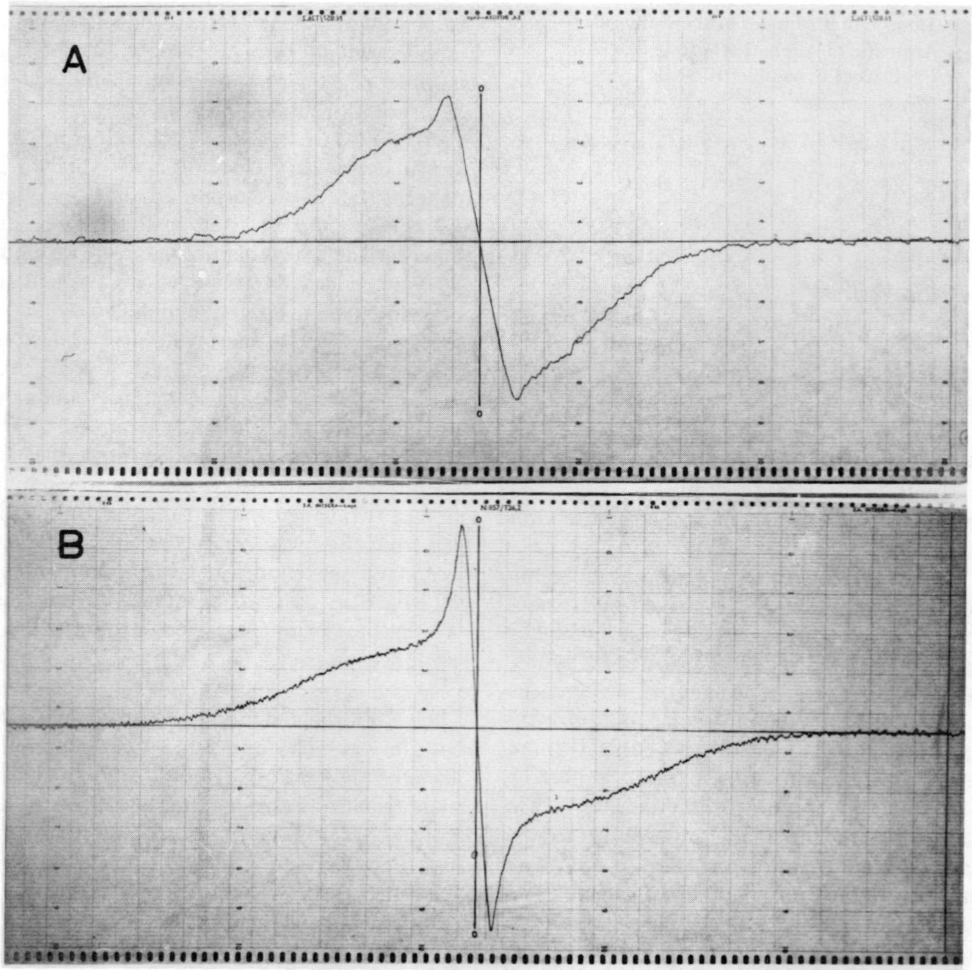

Figure 4.34 Room temperature broadline ^1H NMR spectra of a vitrinite (*a*) and its benzene extract (*b*).[383]

by a region in which the second moment decreases steadily, and finally a region in which the second moment continues to decrease but at a rate independent of the starting material. The plot of second moment versus temperature bears a remarkable similarity to that of the hydrogen content of the coal versus temperature. The decrease in second moment at the higher temperature is attributed to thermal decomposition of the aliphatic structures in the coals.

Broadline ^1H NMR was applied to brominated coals and heat-treated brominated coals to estimate the amount of methylene hydrogen.[387] The NMR results show that more than half the methylene groups remain in the coal after bromination and all the methylene hydrogen is eliminated from both coals and brominated coals by heating to 600°C.

Coal-Tar Pitches. Broadline ^1H NMR has been used to estimate the concentration of aromatic plus alpha-aliphatic protons in whole pitches.[388] The method uses a calibration curve based on the second moments of chloroform extracts of pitch and the concentration of aromatic + alpha-aliphatic protons as determined by high-resolution NMR of the extracts in deuterated chloroform. Broadline NMR measurements on a series of pitch fractions as a function of temperature (77.2–450 K) showed that the mobility in the fractions changes little with rising temperature until the softening point is reached.[380] The investigators point out that the spectra of pitches differ from those of semicrystalline polymers in that the latter generally show a superposition of a broad resonance due to relatively immobile protons and a narrow one due to mobile protons.

4.12.2.2 *Pulse and Multiple-Pulse ¹H NMR*

Measurement of Nuclear Relaxation Times. Among other information, pulse and multiple-pulse NMR experiments can provide values for the nuclear relaxation times T_1 and T_2. The spin lattice relaxation time T_1 is the time required for the difference between the excess spin population in the upper magnetic energy states and its equilibrium value to be reduced by the factor e. It involves that component of the nuclear magnetization along the direction of the applied magnetic field M_z. It can alternatively be defined as the half-life as M_z approaches its equilibrium value M_0. The spin-spin relaxation time T_2 is the characteristic time of the decay of the transverse components of the magnetization M_x and M_y. In nonmagnetic solids such as coals, homonuclear (^1H–^1H) dipolar interactions constitute the dominant mechanism for spin-spin relaxation. Sometimes T_1 and T_2 are referred to as the longitudinal and transverse relaxation times, respectively.

The most commonly used pulse methods for measuring spin-lattice relaxation times involve application of a 180° pulse to invert the magnetization followed by a 90° pulse after a time τ. After waiting at least 5 T_1' s, the sequence is repeated for a different value of τ. The decay rate of M_z can now be established and T_1 deduced, since

$$M_z = M_0 (1 - 2e^{-\tau/T^1}) \qquad (4.13)$$

Early measurements on a vitrain (82.6% C) from a high-volatile A bituminous coal gave a room temperature T_1 value of 50 μs.[354] More recent measurements[356] on a variety of coals of different ranks gave considerably larger values ranging from 8.7 to 425 ms. In the case of two of the coals examined, multiple T_1 values were reported.[356]

The spin-spin relaxation time can be deduced from the free induction decay (FID) following a 90° pulse. For coals having carbon contents between 75 and 90.3%, T_2 values of 5–120 μs have been reported.[354,356] Multiple T_2 values were found for several of the coals.[356] It should be noted that if certain conditions are met, the Fourier transform of the FID gives the true NMR spectrum of the material being studied. Thus the NMR second moment can be derived not only from the broadline spectrum discussed earlier, but also by appropriate mathematical treatment of the results of pulse NMR experiments.

Determination of Total Hydrogen in Coal. It is of interest that pulse ¹H NMR spectrometry has recently been used to determine the hydrogen contents of coals.[396] Initial values of FIDs were shown to provide a rapid, nondestructive quantitative measure of the protons in coals, provided certain experimental conditions were met. Table 4.17 compares hydrogen contents obtained by this NMR method with those obtained by combustion analysis.

Higher Resolution in Solids. Preliminary applications of multiple-pulse NMR experiments[397,398] to remove homonuclear line

[396] B. C. Gerstein and R. G. Pembleton, *Anal. Chem.*, **49**, 75 (1977).
[397] W. K. Rhim, D. P. Elleman, and R. W. Vaughan, *J. Chem. Phys.*, **58**, 1772 (1973).
[398] C. R. Dybowski and R. W. Vaughan, *Macromolecules*, **8**, 50 (1975).

Table 4.17 Comparison of Quantitative Analysis of Hydrogen in Coals by NMR and Combustion Analysis[396]

	Hydrogen (%)	
Coal	NMR	Combustion Analysis
Pocahontas No. 4 vitrain	4.72	4.12
Powellton vitrain	4.87	4.80
Star vitrain	5.63	5.53
Star vitrain	6.31	5.53
Star vitrain	5.60	5.53
Lovilia vitrain	5.88	5.54
Upper Michigan vitrain	5.05	5.32
Lovilia channel	5.29	5.40
Michigan run of mine	6.20	5.63

Source: B. C. Gerstein and R. G. Pembleton, *Anal. Chem.*, **49**, 75 (1977).

broadening and chemical shift anisotropy effects from high-resolution ^{1}H NMR spectra of coals have been reported.[356] The investigators were successful in reducing the proton spectral line widths in coals by a factor of 40. The resulting line shape unfortunately exhibited no structure indicative of separate peaks for aromatic and aliphatic protons. Comparison of these results with results obtained from studies on model compounds led the authors to estimate the free radical–proton line broadening in coals to be about 400 Hz. The use of multiple-pulse NMR techniques in coal research will undoubtedly receive a great deal of attention in the future.

4.12.2.3 *Carbon-13 NMR*

Broadline ^{13}C NMR. Very few applications of ^{13}C NMR spectrometry to solid coals have been reported. This situation is expected to change in the near future, however, given recent advances in line-narrowing techniques. Broadline ^{13}C NMR spectra of coals,[368,399] diamonds,[399] and graphite[399–401] have been published. Although the spectra of coals consisted of a single broad peak, the results, when compared to those for diamonds and graphite, suggested an increasing carbon aromaticity with increasing coal rank, leading ultimately to the formation of graphitelike structures.[399]

Proton-Enhanced Nuclear Induction Spectroscopy. The introduction of proton-enhanced nuclear induction spectroscopy[345] as a method for obtaining high-resolution NMR spectra of dilute spins in solids has led to renewed interest in ^{13}C NMR in coal research. This recently developed technique, which involves cross-polarization and high-power proton decoupling, overcomes the problem of dipolar line broadening by protons and is potentially a powerful method for quantitative determination of the carbon aromaticity of coal. The method differs from conventional high-resolution ^{13}C NMR in the following ways:

[399] H. L. Retcofsky and R. A. Friedel, *J. Phys. Chem.*, **77**, 68 (1973).
[400] A. Abragam, A. Landesman, and J. M. Winter, *C.R. Acad. Sci., Paris*, **247**, 1852 (1958).
[401] G. P. Carver, *Phys. Rev., B*, **2**, 2248 (1970).

1. Sufficient proton decoupling power must be present to decouple the direct dipolar coupling rather than the indirect dipolar coupling that is seen in liquid samples. This requires the use of power levels about 100 times greater than those used to obtain high-resolution proton decoupled ^{13}C spectra of liquids.
2. The carbon magnetization is generated exclusively from the proton polarization by polarization transfer. The signal-to-noise ratio in this experiment is greatly improved relative to conventional ^{13}C NMR because (*a*) the lines are narrower due to proton decoupling, (*b*) a modest signal enhancement is realized in the cross-polarization experiment, and (*c*) experiments may be repeated on the order of the *proton* rather than the typically longer ^{13}C relaxation times. In the initial work the technique was applied to vitrains from an lvb and an hvAb coal. Spectra of the solid vitrains are shown in Fig. 4.35. Each spectrum consists of two overlapping resonances. The resonances are broad because of anisotropic chemical shift effects as well as chemical shift dispersion due to differences in the isotropic chemical shifts of the aromatic or nonaromatic carbons present. Line shape analysis of the two spectra yielded estimates of the carbon aromaticities for the two coals. Figure 4.36 compares the resulting aromaticities by this new technique with coal aromaticities obtained by a variety of other methods.[402–405]

4.12.3 High-Resolution NMR Studies of Soluble Materials from Coal

NMR spectra of materials in solution nearly always show considerably more detail than spectra of solids. This results largely from the greater degree of translational and rotational freedom experienced by molecules in

[402] H. N. M. Dormans, F. J. Huntjens, and D. W. Van Krevelen, *Fuel*, **36**, 321 (1957).
[403] D. W. Van Krevelen, H. A. G. Chermin, and J. Schuyer, *Fuel*, **38**, 483 (1959).
[404] S. K. Chakrabartty and H. O. Kretschmer, *Fuel*, **53**, 132 (1974).
[405] H. L. Retcofsky, *Appl. Spectosc.* **31**, 116 (1977).

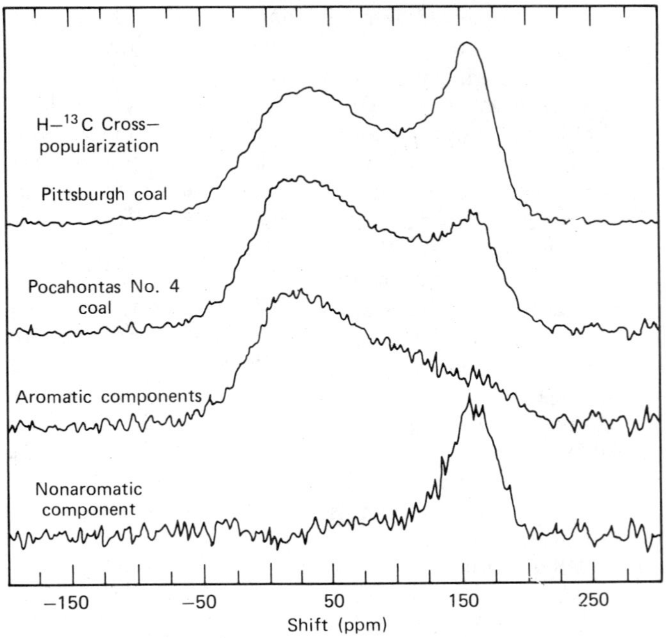

Figure 4.35 Proton-enhanced ^{13}C NMR spectra of solid coals at −95°C (shifts with respect to CS_2).[369]

the liquid state. High-resolution NMR spectra of soluble materials from coal are of special interest because they provide quantitative data for the distribution of specific nuclei among various organic structures. ^1H and ^{13}C are the nuclei that have received the most attention.

4.12.3.1 *Experimental Techniques*

Most published high-resolution ^1H NMR spectra of materials from coal have been obtained using conventional CW techniques. A discussion of the experimental aspects of CW NMR was presented in the first Supple-

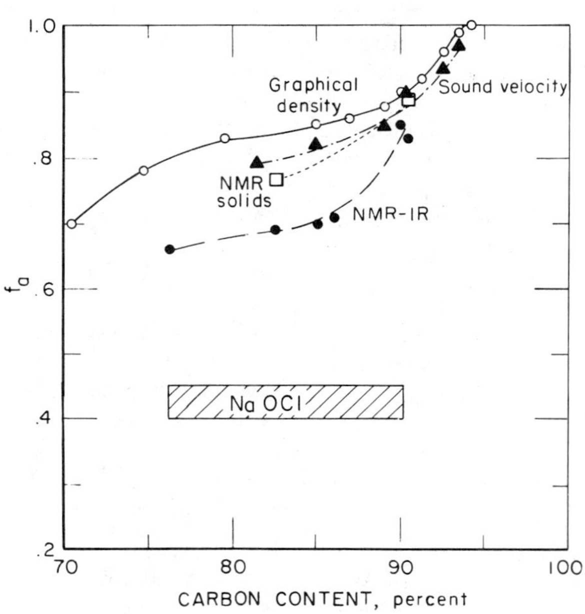

Figure 4.36 Aromaticities of coals as determined by various methods.

mentary Volume[342] and is not repeated here. An excellent report treating the quantitative aspects has appeared.[406] Although the early ^{13}C NMR applications to materials from coal also involve CW techniques, more recent work and probably most future work will be done using FT[343] or perhaps correlation spectroscopy.[362] The FT technique is by far the more commonly used of the two. Quantitative analysis by FT NMR has been discussed for both the proton[407] and ^{13}C[408,409] cases. Since one of the principal reasons for obtaining the ^{13}C NMR spectrum of a material from coal is to measure unambiguously the carbon aromaticity (f_a), it is important that the data be quantitatively reliable.

Two important problems face the experimenter who attempts to obtain reliable quantitative ^{13}C NMR data using FT methods. The first is the low NMR sensitivity of ^{13}C ($\sim$ 1/5700 that of the proton), and the second is the long T_1' s generally associated with the nuclei. To overcome the first problem, it is common to decouple the protons, which reduces spin multiplets due to coupling with protons to single lines, and to accumulate FIDs following several hundred or more pulses in a digital computer before transforming to the spectrum. The long T_1' s, however, place severe limitations on the pulse repetition rate that can be used. One way to remedy this situation is to introduce a delay of several T_1' s between pulses. For many materials this can result in very long accumulation times. Studies of coal hydrogenation products suggest that short repetition times might be used, since the naturally occurring free radicals in the materials probably reduce the ^{13}C T_1' s.[366] The reader is cautioned that one must perform the appropriate experiments to determine the most reliable pulse repetition rate. To do otherwise *can* lead to unreliable data.

Another method for overcoming the T_1 problem is to dissolve a paramagnetic material such as chromium acetylacetonate in the sample to make all T_1' s short compared to the pulse repetition rate. This method has not yet found wide use in studies of materials from coal. Sufficient data to determine its reliability or its need in coal research are not available. It is conceivable that the free radicals that are normally present in most coal-derived materials may produce the same effect.

The collapse of NMR spin multiplets in ^{13}C NMR by proton decoupling frequently introduces another problem, which may lead to NMR peak intensities that are not directly proportional to the number of carbons responsible for the resonances. Decoupling of protons in a ^{13}C experiment results in an enhanced population in the lower ^{13}C energy level relative to that of the normal Boltzmann distribution. This effect, termed the nuclear Overhauser effect (NOE), may be different for different carbons in the same sample. The problem of variable NOEs is most simply overcome by "gating" the proton decoupler.[410] The method, which has been applied effectively to materials from coal liquefaction,[411] involves turning the decoupling field on during the data acquisition time, and off for a relatively long period of time before the sequence is repeated. The excess ^{13}C spin population will decay during the latter time period.

It is emphasized again that it is the responsibility of the researcher to be aware of the precautions that must be taken to ensure the quantitative reliability of NMR data. Unwary experimenters are almost certain to encounter a large number of problems. Unfortunately they are not likely to recognize many of them.

4.12.3.2 *Treatment of the Data*

The method of Brown and Ladner[352] is the method most frequently used to deduce structural information from ^1H NMR spectra of soluble materials from coal. For purposes of illustration, the ^1H spectrum of a pyridine extract of coal is reproduced in Fig. 4.37.

[406] *Radio-Frequency Spectroscopy,* Vol. 3, Varian Associates, Palo Alto, Calif., 1960, p. 1.
[407] B. Thiault and M. Mersseman, *Org. Magn. Resonance,* **7,** 575 (1975).
[408] B. Thiault and M. Mersseman, *Org. Magn. Resonance,* **8,** 28 (1976).
[409] J. N. Shoolery and W. C. Jankowski, Application Note No. NMR-73-4, Varian Associates, Palo Alto, Calif., 1973.
[410] R. Freeman, H. D. Hill, and R. J. Kapten, *J. Magn. Resonance,* **7,** 327 (1972).
[411] D. L. Wooten, H. C. Dorn, L. T. Taylor, and W. M. Coleman, *Fuel,* **55,** 224 (1976).
[412] H. L. Retcofsky and R. A. Friedel, in *Spectrometry of Fuels,* R. A. Friedel, Ed., Plenum Press, New York, 1970, p. 70.

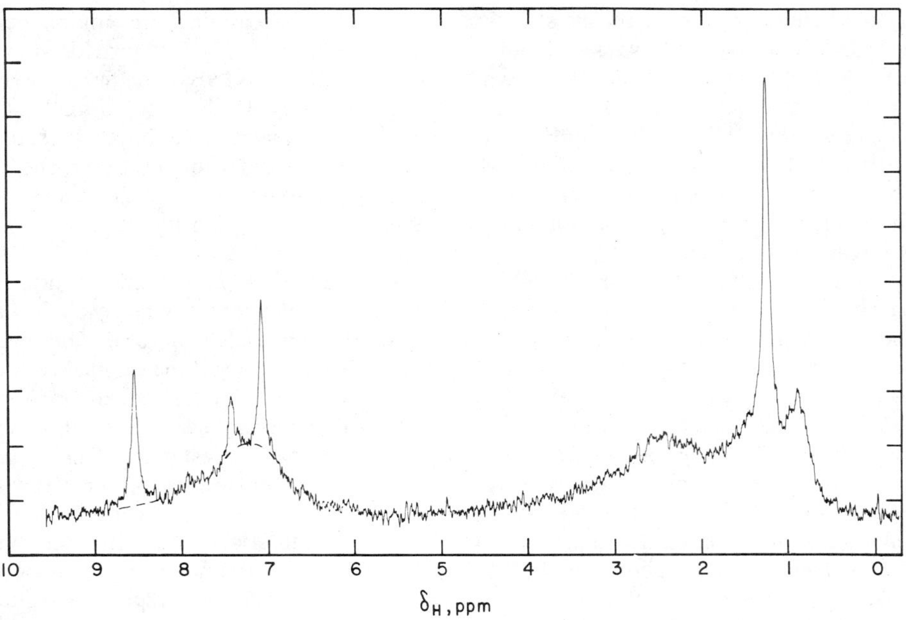

Figure 4.37 High-resolution ^{1}H NMR spectrum of pyridine extract from Adaville subbituminous coal.[412]

Chemical shifts (δ, in units of ppm of applied radiofrequency) are referred to those of tetramethylsilane (δ, 0 ppm). The solvent employed was deuterated pyridine, and the residual protons in the solvent are responsible for the three relatively sharp peaks occurring between δ, 6.5 and δ, 9.0 ppm. The broad resonances centered near 7.2, 2.4, and 1.2 ppm are assigned to $H_{ar+\phi}$, H_α, and H_0, respectively. Brown and Ladner suggested that the ^{1}H NMR data could be used to provide considerably more information about the chemical structure if the atomic ratio of hydrogen to carbon in the nonaromatic systems were known. If this ratio is x for the α-carbon groupings and y for the indirectly attached or unattached groupings, then—from knowledge of the type of oxygen structure present—it is possible to estimate three important structural parameters, f_a, σ, and H_{aru}/C_{ar}:

$$f_a = \frac{\frac{C}{H} - \frac{H^*_\alpha}{x} - \frac{H^*_0}{y}}{\frac{C}{H}} \qquad (4.14)$$

$$\sigma = \frac{\frac{H^*_\alpha}{x} + \frac{O}{H}}{\frac{H^*_x}{x} + \frac{O}{H} + H^*_{ar}} \qquad (4.15)$$

$$\frac{H_{aru}}{C_{ar}} = \frac{\frac{H^*_x}{x} + H^*_{ar} + \frac{O}{H}}{\frac{C}{H} - \frac{H^*_\alpha}{x} - \frac{H^*_0}{y}} \qquad (4.16)$$

where f_a = the aromaticity, defined as the ratio of the aromatic carbon to the total

σ = the degree of substitution of the aromatic systems, that is, the fraction of the available aromatic edge atoms occupied by substituents

H_{aru}/C_{ar} = the atomic hydrogen-to-carbon ratio of the hypothetical unsubstituted aromatic material

C/H, O/H = atomic ratios determined directly from the elementary analysis

$H^*_\alpha = H_\alpha/H$

$H^*_0 = H_0/H$

$H^*_{ar} = H_{ar}/H$

Brown and Ladner point out that these equations require that all the oxygen be attached directly to aromatic systems and not be shared between them (e.g., as in aromatic ethers) and that the aromatic rings not be linked by C—C bonds, as in a polyaryl system. For most materials from coal, x and y are given the value 2 on the assumption that the aliphatic structures are predominantly methylene groups. A recent study[413] showed that unambiguous f_a values obtained from ^{13}C NMR measurements on a variety of coal-derived materials were in excellent agreement with those deduced from ^{1}H NMR data using the Brown and Ladner treatment. The Brown and Ladner equations, of course, can be modified to include terms for methylene bridge protons, methyl groups, or other proton types. With the advent of ^{13}C NMR, f_a can now be measured directly (if suitable solvents are available). The denominator of Equation 4.16 can also be determined directly. In addition H_{ali}/C_{ali}, which is the true weighted average of x and y, can be obtained.

In addition to obtaining the H/C ratios for various organic structures present in coal-derived materials, ^{13}C and ^{1}H NMR data in conjunction with other information can be used to deduce "mean structural units." A possible average molecule for the "asphaltol" component of a solvent-refined coal (SRC) is shown in Fig. 4.38. It should be noted that treatment of the data in this fashion seldom provides a *unique* representation of the average structure.

Although the principal use of ^{13}C NMR is to determine f_a, the detail frequently observed in spectra of materials from coal can be used to estimate the number of carbons in specific structural groups (e.g., aromatic bridgehead carbons, phenolic carbon, aromatic carbons bonded directly to hydrogen, etc.[365]).

[413] H. L. Retcofsky, F. K. Schweighardt, and M. Hough, *Anal. Chem.*, **49**, 585 (1977).
[414] M. Farcasiu, T. O. Mitchell, and D. D. Whitehurst, *Prepr., Am. Chem. Soc., Div. Fuel Chem.*, **21**, 11 (1976).

A report describing the complementary use of ^{1}H and ^{13}C NMR to provide atomic H/C ratios for coal-derived materials was published recently.[410] The method makes use of dioxane as an internal intensity standard. Preliminary results for the hexamethylphosphoramide-soluble material from a sample of SRC agreed well with the value obtained by combustion analysis.

The remaining part of this section is devoted to applications of high-resolution NMR to soluble materials from coal. The reader is encouraged to consult the original references for more detailed explanations, and in particular, to obtain a better description of the actual materials investigated. This is particularly important because all coal investigators do not use the same names to describe the same materials, and different authors sometimes use the same name to describe different materials. It is especially disturbing that individual authors are sometimes inconsistent in the use of names to describe material from coal even in the same paper.

4.12.3.3 *Coal Extracts*

High-resolution ^{1}H NMR spectral data have been reported for pyridine,[353,383,405,412,415–419] benzene,[360,381,383,415,416,420] carbon disul-

[415] J. F. M. Oth, E. de Ruiter, and H. Tschamler, *Brennstoff-Chemie*, **42**, 22 (1961).
[416] H. Tschamler and E. de Ruiter, *Brennstoff-Chemie*, **43**, 20 (1962).
[417] G. Takeya, M. Itoh, A. Suzuki, and S. Yokoyama, *Bull. Chem. Soc. Jap.*, **36**, 1222 (1963).
[418] G. Takeya, M. Itoh, A. Suzuki, and S. Yokoyama, *J. Fuel Soc. Jap.*, **43**, 837 (1964).
[419] R. A. Durie, Y. Shewchyk, and S. Sternhell, *Fuel*, **45**, 99 (1966).

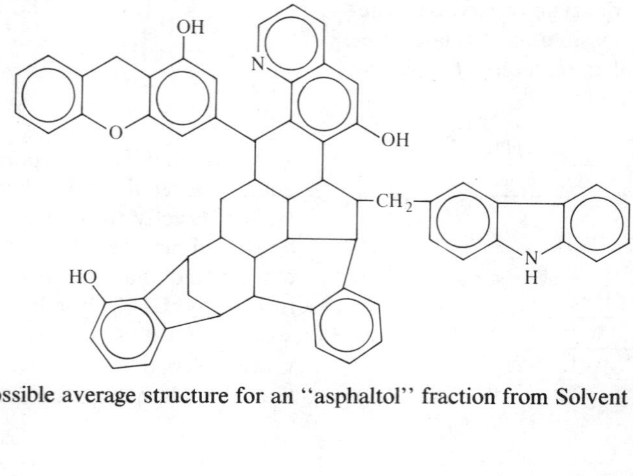

Figure 4.38 Possible average structure for an "asphaltol" fraction from Solvent Refined Coal.[414]

fide,[405,412,421] chloroform,[419,422] and petroleum ether[383,416,420] extracts of coal. Most of the extractions were carried out at the boiling point of the solvent, although in at least one case the extract was obtained by mechanical shaking of the coal and solvent at room temperature.[421] Typically, all extracts except those of pyridine were obtained by progressive extraction.

^{1}H NMR data for pyridine extracts of several U.S.[405] and Japanese[418] coals are summarized in Table 4.18. Also included are the mean structural parameters estimated by treatment of the data by the Brown and Ladner[352] method. For the Japanese coals a value of 2 was chosen for both x and y, whereas x and y values for the U.S. coals were based on the true H_{ali}/C_{ali} ratios obtained from complementary use of ^{1}H and ^{13}C NMR. The results of Table 4.18 indicate that both the hydrogen and carbon aromaticities of pyridine extracts increase with increasing rank of the starting coal. The corresponding decrease in H_{aru}/C_{ar} with coal rank suggests that the number of rings per polynuclear condensed aromatic cluster increases with increasing rank.

Figure 4.39 summarizes the carbon and hydrogen distributions in the pyridine and carbon disulfide extracts of six coals ranging in rank from subbituminous to low-volatile coal. It should be noted that the atom distributions for the carbon disulfide extracts are essentially unambiguous ones derived from both ^{1}H and ^{13}C NMR, whereas the carbon distributions for the pyridine extracts are estimates on the basis of ^{1}H NMR data.

The solvent extracts of coal described above were obtained using what are generally referred to as "mild" conditions. Extract yields are generally increased if the extraction is carried out in an autoclave at high temperature or under supercritical conditions. Generally thermal degradation of the coal occurs under these conditions. NMR has been applied to study the composition of products from the extraction of Yubari coal with quinoline in an autoclave[423] and from the extraction of Markham Main coal with supercritical toluene.[365,424]

The Yubari coal was extracted with quinoline at temperatures of 300, 350, 375, and 400°C. The pyridine-soluble material from the quinoline extracts was dissolved in deuterated pyridine and examined by ^{1}H NMR. The data indicated an increase of f_a of the

[420] J. F. M. Oth and H. Tschamler, *Brennstoff-Chemie*, **43**, 35 (1962).
[421] H. L. Retcofsky and R. A. Friedel, *Fuel*, **47**, 487 (1968).
[422] H. H. Oelert, *Z. Anal. Chem.*, **231**, 105 (1967).
[423] H. Kanai, A. Suzuki, M. Ito, and G. Takeya, *J. Fuel Soc. Jap.* **45**, 859 (1966).
[424] K. D. Bartle, T. G. Martin, and D. Williams, *Fuel*, **54**, 226 (1975).

Table 4.18 Hydrogen Distribution and Mean Structural Parameters for Pyridine Extracts of Coals[405,418]

Coal	C (%, maf)	Extract Yield (%)	Hydrogen Distribution			Structural Parameters			Ref.
			H^*_{ar}	H^*_α	H^*_0	f_a	σ	H_{aru}/C_{ar}	
Tokachi	61.5	13.8	0.07	0.12	0.75	0.41	0.74	0.93	418
Sohya	70.3	16.6	0.18	0.20	0.56	0.61	0.55	0.69	418
Sumiyoshi	75.5	15.8	0.21	0.20	0.53	0.62	0.52	0.72	418
Adaville	76.3	6.7	0.20	0.30	0.44	0.64	0.59	0.76	405
Taiheiyo	76.7	16.7	0.10	0.21	0.64	0.53	0.67	0.60	418
Bibai	80.7	—	0.27	0.22	0.45	0.70	0.45	0.65	418
Pittsburgh	82.6	21.4	0.35	0.26	0.36	0.73	0.37	0.68	405
Yubari	84.0	18.5	0.30	0.25	0.43	0.69	0.41	0.67	418
Powellton	85.1	20.9	0.27	0.29	0.39	0.72	0.47	0.59	405
Lower Banner	86.1	19.3	0.32	0.28	0.37	0.73	0.37	0.57	405
Pocahontas No. 3	90.0	2.8	0.55	0.31	0.13	0.85	0.27	0.63	405
Pocahontas No. 4	90.4	2.5	0.50	0.30	0.19	0.83	0.26	0.57	405

Sources of data: H. L. Retcofsky, *Appl Spectros.*, 31, 116 (1977); G. Takeya, M. Itoh, A. Suzuki, and S. YoKoyama, *J. Fuel Soc. Jap.*, **43**, 837 (1964).

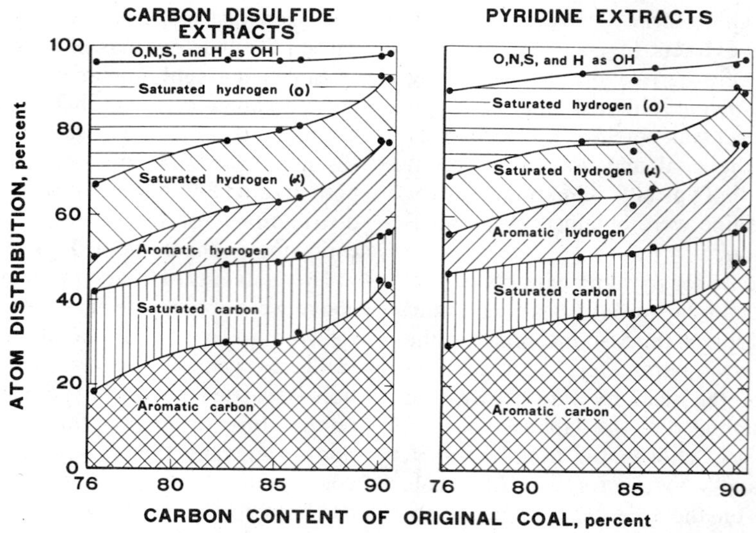

Figure 4.39 Distribution of carbon and hydrogen in pyridine and carbon disulfide extracts of selected coals as determined by high-resolution NMR spectrometry.[405]

extractable material with increasing temperature of extraction.

The chemical nature of the supercritical gas extraction of Markham Main coal at 350°C was probed by both ^{1}H and ^{13}C NMR.[365,424] In an attempt to obtain additional information on the carbon distribution, the researchers undertook a more detailed analysis of the ^{13}C spectra of the benzene and methanol eluates. The results are summarized in Table 4.19 and ^{13}C NMR spectra are reproduced in Fig. 4.40. Although the authors

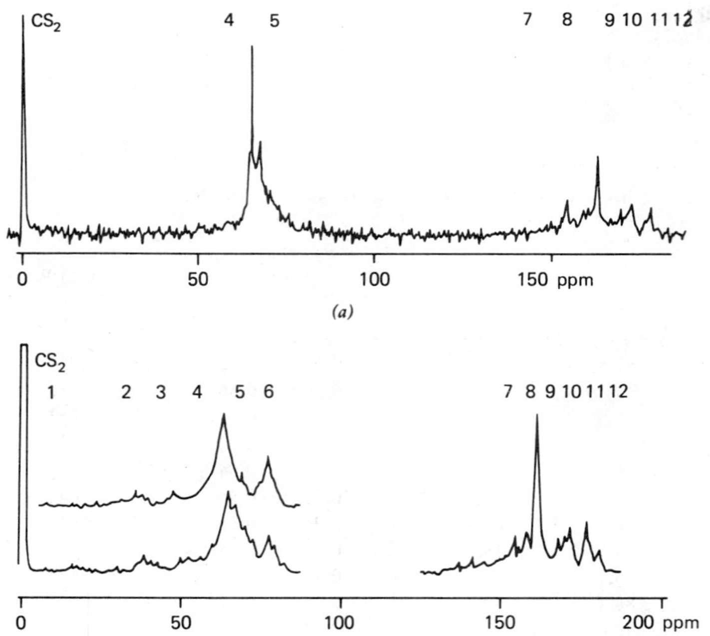

Figure 4.40 Broad-band proton-decoupled pulse ^{13}C NMR spectra of fractions of a supercritical fluid extract of coal. (*a*) Benzene eluate, 22.63 MHz, 27,100 scans, and (*b*) methanol eluate, 25.14 MHz = 28,000 scans (upper), 12,100 scans (lower). Numbers refer to carbon types in Table 4.19.[365]

Table 4.19 ^{13}C **NMR Spectroscopy of Silica Gel Fractions of Petroleum Ether Solubles of a Supercritical Gas Extract of Coal**[424]

No.	Carbon Type	^{13}C Shift Range (ppm from $^{13}CS_2$)	% Total Carbon: Benzene Eluate	% Total Carbon: Methanol Eluate	Benzene Eluate: From ^{13}C Spectrum		Benzene Eluate: From ^{1}H Spectrum		Methanol Eluate: From ^{13}C Spectrum			Methanol Eluate: From ^{1}H Spectrum	
1	Quinone C ═ 0	0–15	0	0	0				0				
4	Aromatic[13] C·alkyl, C at aromatic/saturated ring junctions; aromatic ^{13}C—Ph	45–55	9	11 ± 4[a]	2·0	7·5	5·1	7·8	2·4 ± 0·7		9·2 ± 0·9	$5{\cdot}3_5$	$7{\cdot}2_5$
5	Aromatic bridgehead C	55–64	24	26 ± 3	5·5		2·7		5·8 ± 0·6	6·8 ± 0·6			
2	Aromatic ether[13] C—O	28–35	0	4 ± 1	0				1·0 ± 0·2			1.9	
3	Phenolic C—OH	35–45	0	5 ± 1	0				1·0 ± 0·2				
6	Aromatic C—H	64–75	34	16 ± 3	7·7		$7{\cdot}4_5$		3·5 ± 0·7	5·9 ± 0·8		$6{\cdot}4_5$	
—	^{13}C—H ortho to C—O[b]	75–85	0	6 ± 2	0				1·4 ± 0·3				
	Methylene bridge C and bridgehead C of naphthenes	140–156	6	10 ± 2	1·4		0·4		2·2 ± 0·4			0·35	
8	C α to aromatic ring (other than CH_3)	156–183	11	12 ± 2	2·5		2·0		2·7 ± 0·3			2·6	
9	Certain C β to ring	163–170	7	5 ± 1	1·6	7·7	$1{\cdot}8_5$	7·5	1·1 ± 0·2	7·6 ± 0·6		2·3	8·8
10	Ch_3 α to ring	170–174	6	4 ± 1	1·4		1·6		0·9 ± 0·2			$1{\cdot}3_5$	
11	CH_3 of ethyl aromatics	174–176	2	1 ± 1	0·5		$0{\cdot}7_5$		0·3 ± 0·2			$1{\cdot}0_5$	
12	CH_3 γ or further from ring	176–180	2	2 ± 1	0·5		$0{\cdot}9_5$		0·5 ± 0·2			1·1	

[a] Standard deviation based on four spectra recorded on two spectrometers.
[b] Also includes olefinic C.

Source: K. D. Bartle, T. G. Martin, and D. Williams, *Fuel,* 226 (1975).

point out that some overlap of chemical shift ranges for different types of carbons is likely, their work represents one of the first attempts to fully utilize the ^{13}C NMR data.

4.12.3.4 *Reduced Coals*

^1H NMR spectrometry has been used to determine the hydrogen distribution in solvent extracts of electrochemically reduced coal[360,425] and of coals reduced chemically with lithium-ethylenediamine[360] and with lithium-ethylamine.[426] Table 4.20 shows the decrease in H^*_{ar} of the extracts as more hydrogen is added to the coal.

4.12.3.5 *Coal Carbonization Products*

NMR studies of products from the carbonization of coal have received considerable attention since the early 1960s. Unfortunately details of the carbonization conditions are not often given, and this makes the literature difficult to review adequately. In addition, the problem of nomenclature alluded to above compounds the difficulty.

^1H NMR spectra of a coke-oven pitch and a pitch from low-temperature coal carbonization appear in Fig. 4.41. Figure 4.42 shows the ^{13}C NMR spectrum of a coal tar, and the wealth of information to be derived from the high-field (51.5 kG) ^1H NMR spectrum of a coke-oven anthracene oil is indicated in Fig. 4.43.

A number of investigators have reported ^1H NMR studies of coal-tar pitches.[353,355,427–431] A peak assigned to methylene bridge protons appears in the spectra of high-temperature pitches. The presence of methylene bridges requires some slight modification[427] of Equations 4.15 and 4.16. ^1H NMR data indicate f_a values in excess of 0.9 for carbon disulfide extracts of pitches from high-temperature coal carbonization.[353,427] The high aromaticity of one such pitch has been confirmed by complementary ^{13}C NMR studies.[353,413,427] The ^{13}C and ^1H NMR results collectively suggest the presence of very large polynuclear condensed aromatic ring structures in the materials. In contrast to the extremely high f_a values for high-temperature pitches, the ^1H NMR spectrum of a low-temperature pitch suggested a much lower value of 0.67.

[425] H. W. Sternberg, C. L. Delle Donne, R. E. Markby, and I. Wender, *Fuel,* **45,** 469 (1966).
[426] C. R. Binder, L. J. Duffy, and P. H. Given, *Prepr., Am. Chem. Soc., Div. Fuel Chem.,* **7,** 145 (1963).
[427] C. W. DeWalt, Jr., and M. S. Morgan, *Prepr., Am. Chem. Soc., Div. Fuel Chem.,* Atlantic City, N.J., September 9–14, 1962, p. 33.
[428] H. S. Rao, G. S. Murty, and A. Lahiri, *Fuel,* **39,** 263 (1960).
[429] K. A. Kini and G. S. Murty, *Fuel,* **53,** 204 (1974).
[430] T. Furuta, M. Shiraishi, and Y. Sanada, *J. Jap. Pet. Inst.,* **18,** 1086 (1975).
[431] Y. Yamada, T. Furuta, and Y. Sanada, *Anal. Chem.,* **48,** 1637 (1976).

Table 4.20 ^1H NMR Data and Other Information for Solvent Extracts of Reduced Coals

Method of Reduction	Coal	H Atoms Added/100 C Atoms	Solvent	Yield (%)	H/C	H^*_{ar}	Ref.
—	Pocahontas No. 3	0	Pyridine	3	0.70	0.55	412
—	Pocahontas No. 3	0	Benzene	1	0.77	0.48	360
Electrochemical	Pocahontas No. 3	4	Benzene	3	1.02	0.19	360,425
Electrochemical	Pocahontas No. 3	18	Benzene	11	1.09	0.08	360,425
Electrochemical	Pocahontas No. 3	27	Benzene	8	1.12	0.07	360,425
Electrochemical	Pocahontas No. 3	34	Benzene	18	1.17	0.05	360,425
Electrochemical	Pocahontas No. 3	53	Benzene	21	1.26	0.03	360,425
Lithium-ethylamine	Lower Freeport	11	Chloroform	18	—	0.07[a]	426
—	Pittsburgh	0	Pyridine	21	0.88	0.35	360,412
Lithium-ethylenediamine	Pittsburgh	7.5	Pyridine	—	1.06	0.14	360
Lithium-ethylenediamine	Pittsburgh	16	Pyridine	39	1.07	0.11	360
Lithium-ethylenediamine	Pittsburgh	16	Benzene	13	—	0.07	360

[a] NMR data qualitative; H^*_{ar} estimated from IR spectrum

Source: Data from References 360, 412, 425, 426.

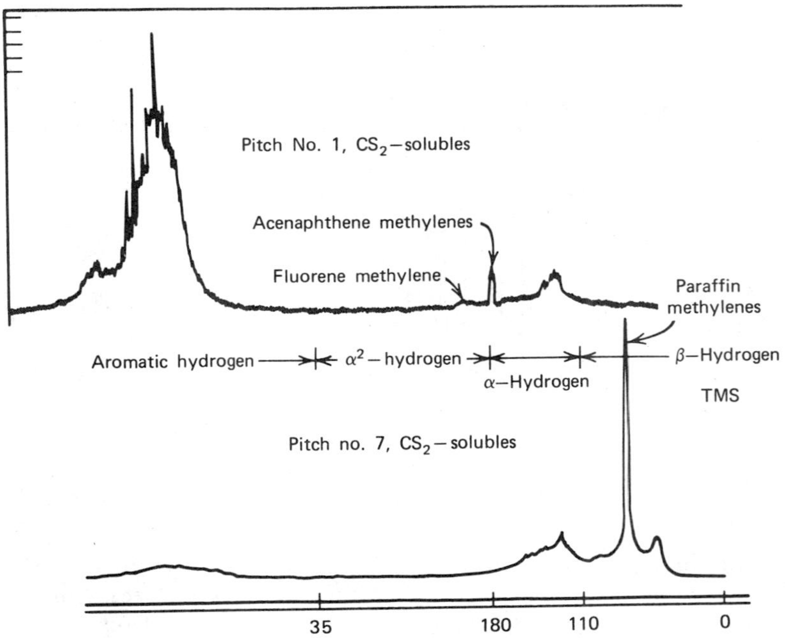

Figure 4.41 High-resolution ^{1}H NMR spectra of the carbon disulfide–soluble fractions from a coke-oven pitch (No. 1) and pitch produced by low-temperature coal carbonization (No. 7).[427]

Proton NMR spectrometry has also been applied in chemical structure studies of neutral oils from both low-[353,355] and high-temperature[413] carbonization, low[432] and high[433] molecular fractions of road tars, tar and other soluble materials from low-temperature carbonization,[351,352,378,434] anthracene oil,[355] and pitches and tars from vertical retorts.[355,432,433] The neutral oils from low-[357] and high-temperature[413] carbonization and the anthracene oil[357] were also the subject of ^{13}C NMR studies.

[432] K. D. Bartle and J. A. S. Smith, *Fuel,* **44,** 109 (1965).

[433] K. D. Bartle and J. A. S. Smith, *Fuel,* **46,** 29 (1967).

[434] J. D. Brooks and J. W. Smith, *Fuel,* **43,** 125 (1964).

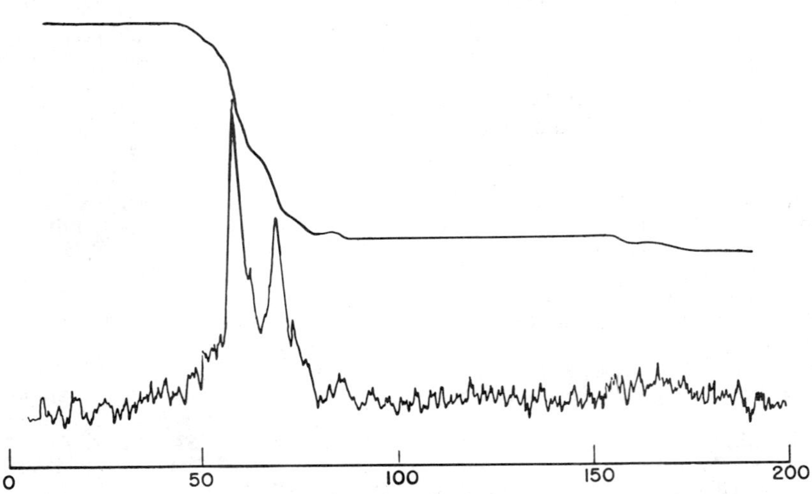

Figure 4.42 High resolution ^{13}C NMR spectrum of a coal-tar pitch fraction.[360,361]

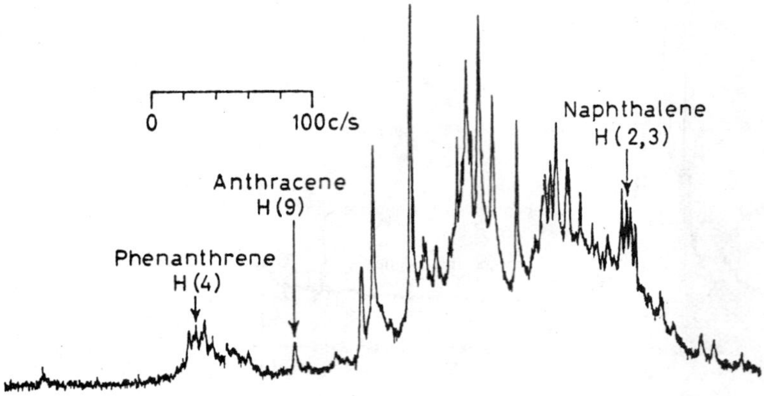

Figure 4.43 The aromatic region of the 220-MHz high-resolution ^1H NMR spectrum of a coke-oven anthracene oil.[355]

4.12.3.6 *Coal Liquefaction Products*

The energy shortages being experienced by many nations have prompted renewed interest in research and development activities to produce environmentally acceptable fuels from coal. Many of these processes produce liquid or solid fuels that are complex mixtures of hydrocarbons and other organic compounds. These processes are referred to generically as coal hydrogenation, hydrodesulfurization, or liquefaction, although the actual reaction mechanisms are not at all well understood. Many of these processes have advanced to the PDU (process development unit) or pilot-plant stage. Examples of such processes currently under development in the United States include H-coal,[435] SRC,[436] Synthoil,[437] and COED.[438] NMR has been an excellent tool to aid in the characterization of products from these processes. It is emphasized that the coal liquefaction products described below are from *experimental* units. Thus they do not necessarily represent the ultimate product from a plant that has reached maturation.

Table 4.21 gives a rather comprehensive list of references to NMR studies of various coal liquefaction products. The NMR nuclei of principal concern are ^{13}C and ^1H,[363,366,374,411–414,439–453] although some atten-

[435] C. A. Johnson, M. C. Chervenak, E. S. Johanson, and R. H. Wolk, in *Coal Processing Technology,* prepared by the editors of *Chemical Engineering Progress,* American Institute of Chemical Engineers, New York, 1974, p. 22.

[436] V. L. Brandt and B. K. Schmid, *Chem. Eng. Prog.,* **65**(12), 55 (1969).

[437] P. M. Yavorsky, S. Akhtar, J. J. Lacey, M. Weintraub, and A. A. Reznik, *Chem. Eng. Prog.,* **71,** 79 (1975).

[438] R. C. Merrill, L. J. Scotti, L. Ford, and D. J. Domina, in *Coal Processing Technology,* Vol. 2, prepared by the editors of *Chemical Engineering Progress,* American Institute of Chemical Engineers, New York, 1975, p. 88.

[439] R. B. Callen, C. A. Simpson, J. G. Bendoraitis, and S. E. Voltz, *Ind. Eng. Chem. Prod. Res. Dev.,* **15,** 223 (1976).

[440] R. P. Anderson, in *Coal Processing Technology,* Vol. 2, prepared by editors of *Chemical Engineering Progress,* American Institute of Chemical Engineers, New York, 1975, p. 130.

[441] I. Schwager and T. F. Yen, *Prepr., Am. Chem. Soc., Div. Fuel Chem.,* **21,** 199 (1976).

[442] N. Woolsey, R. Baltisberger, K. Klabunde, V. Stenberg, and R. Kaba, *Prepr., Am. Chem. Soc., Div. Fuel Chem.,* **21,** 33 (1976).

[443] M. Farcasiu, *Fuel,* **56,** 9 (1977).

[444] H. W. Sternberg, R. Raymond, and F. K. Schweighardt, *Science,* **188,** 49 (1975).

[445] C. M. White and J. O. H. Newman, *Chromatographic Investigation of a Light Oil Produced by the SYNTHOIL Process,* PERC/RI-76/3, U.S. ERDA, Washington, D.C., 1976.

[446] H. L. Retcofsky and T. A. Link, "Carbon-13 Magnetic Resonance Studies of the SYNTHOIL Coal Liquefaction Product," presented at 26th Pittsburgh Conf. Anal. Chem. Appl. Spect., Cleveland, March 1977.

[447] P. W. Woodward, G. P. Sturm, Jr., J. W. Vogh, S. A. Holmes, and J. E. Dooley, *Compositional Analyses of SYNTHOIL from West Virginia Coal,* BERC/RI-76/2, U.S. ERDA, Washington, D.C., 1976.

[448] S. R. Taylor, G. Galya, B. J. Brown, and N. C. Li, *Spectosc. Lett.,* **9,** 733 (1976).

Table 4.21 Coal Liquefaction Products or Fractions Thereof That Have Been Examined by High-Resolution NMR Spectrometry

Liquefaction Process	Nucleus Observed	Ref.
H-Coal	^{1}H	413
H-Coal	^{13}C	413, 439
SRC	^{1}H	411, 413, 414, 440–443
SRC	^{13}C	411, 413, 414, 439, 443
SRL	^{1}H	413, 442
SRL	^{13}C	413
Synthoil	^{1}H	363, 374, 413, 441, 444, 445, 447, 448
Synthoil	^{13}C	363, 374, 413, 439, 446
Synthoil	^{14}N	374
Synthoil	^{2}H	373
Laboratory autoclave experiments	^{1}H	350, 366, 449–453
Laboratory autoclave experiments	^{13}C	366, 451

tion has been given to ^{2}H tracer studies[373] and ^{14}N hydrogen-bonding studies.[374] Proton NMR data of liquefaction products are generally used to measure H^*_{ar}, H^*_{α}, and H^*_{O}, whereas f_a values are obtained from the ^{13}C spectra. The changes in these parameters, as liquefaction conditions such as temperature, hydrogen pressure, and recycle oil are modified, provide insight into reaction mechanisms and may provide information as to optimum conditions for pilot-plant operation.

Because of the complexity of these products, analysis by NMR and by other means is generally carried out on fractions from samples separated by solvent or chromatographic techniques. Representative ^{13}C NMR spectra of several fractions separated from products of the Synthoil process are shown in Fig. 4.44. The rich detail of these spectra suggests that considerably more information than simply a measurement of f_a could be deduced from them. The chemical shift–functional group relationships of Table 4.19 may be useful in this respect. One of the problems encountered in making chemical shift assignments is the lack of spectral data on pure compounds similar to those present in liquefaction products. Nevertheless, even cursory examination of Fig. 4.44 reveals the lack of aromatic carbons in the saturate fraction (no peaks at 130 ppm), the high aromatic content and the presence of bridgehead carbons in the aromatic fraction, and the presence of aromatic C—O groupings in the acid fraction (peaks at low field). Values of f_a for a variety of coal liquefaction products are shown graphically in Fig. 4.45; a few petroleum products are included for comparison. Average structures of benzene-insoluble material and chromatographic fractions from SRC, as suggested by ^{1}H and ^{13}C NMR data, are shown in Figs. 4.38 and 4.46 respectively.[414,443]

A topic of considerable interest in studies of coal liquefaction is the chemical nature of the asphaltene components. This interest was largely brought about by the observation that asphaltenes may affect the viscosity of the product.[454] NMR studies of asphaltenes from the Synthoil process[363,364,444,448] suggested that the asphaltenes consist of hydrogen-bonded complexes of the type depicted in Fig. 4.47. The illustration merely indicates the type of hydrogen bonding that may occur between acidic and basic components; it does not imply that asphaltenes consist of acidic and basic components in a 1/1 ratio.

[449] R. Yoshida, Y. Maekawa, T. Ishii, and G. Tekeya, *Fuel*, **55**, 341 (1976).
[450] S. Yokoyama, D. M. Bodily, and W. H. Wiser, *Prepr., Am. Chem. Soc., Div. Fuel Chem.*, **21**, 77 (1976).
[451] S. Yokoyama, D. M. Bodily, and W. H. Wiser, *Prepr., Am. Chem. Soc., Div. Fuel Chem.*, **21**, 84 (1976).
[452] R. G. Ruberto, D. C. Cronauer, D. M. Jewell, and K. S. Seshadri, *Fuel*, **56**, 17 (1977).
[453] R. G. Ruberto, D. C. Cronauer, D. M. Jewell, and K. S. Seshadri, *Fuel*, **56**, 25 (1977).
[454] H. W. Sternberg, R. Raymond, and S. Akhtar, *Prepr., Am. Chem. Soc., Div. Pet. Chem.*, **20**, 711 (1975).

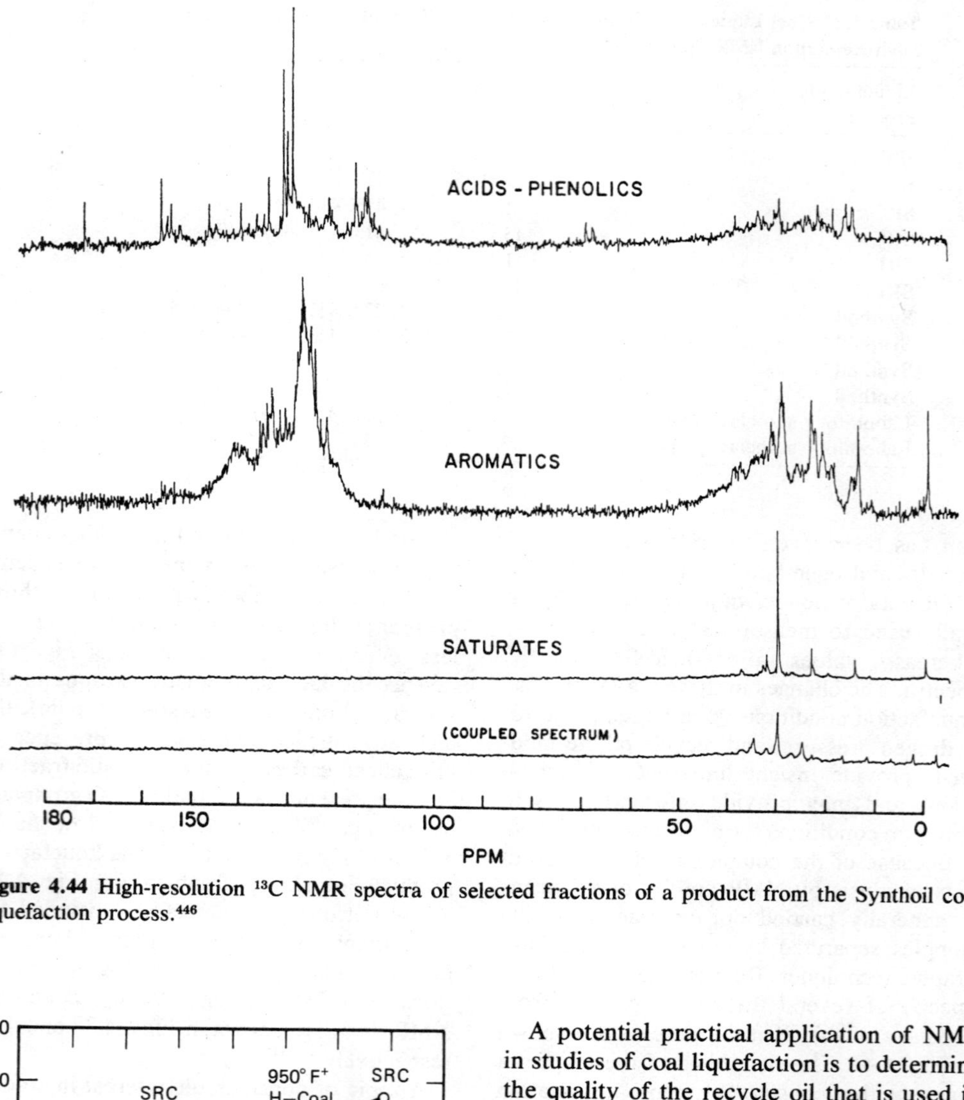

Figure 4.44 High-resolution ^{13}C NMR spectra of selected fractions of a product from the Synthoil coal liquefaction process.[446]

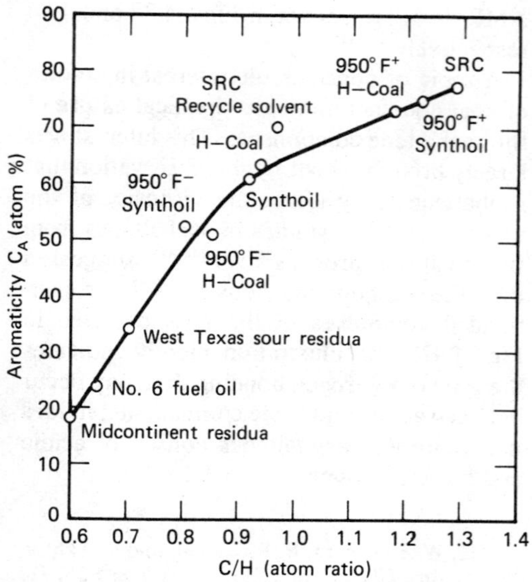

Figure 4.45 Aromaticities of coal liquefaction products and petroleum products versus C/H.[439]

A potential practical application of NMR in studies of coal liquefaction is to determine the quality of the recycle oil that is used in most liquefaction processes. NMR spectra of SRC recycle solvents during successive passes through a laboratory-scale continuous reactor showed a gradual increase in aromatic structures and a decrease in aliphatic side chains, suggesting that the solvent has improved in its hydrogen transfer capability.[440] The NMR spectra also showed the presence of materials that are not useful in the hydrogen transfer process; this material was present even at the end of the run, when the initial anthracene oil recycle solvent was for all practical purposes replaced with process-derived solvent.

There is little doubt that NMR and other analyses of coal liquefaction products will receive considerable attention throughout the 1980s.

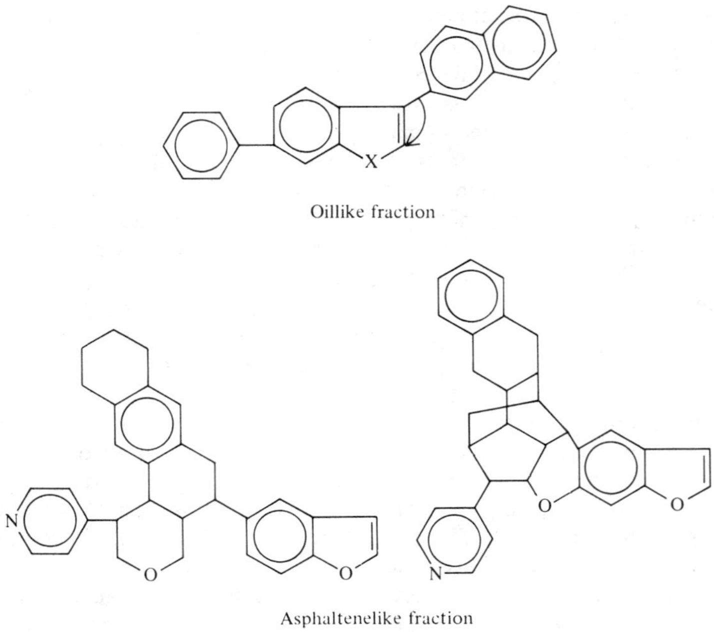

Figure 4.46 Possible average structures of selected fractions from solvent refined coal, derived via sequential elution by solvents chromatography.[443]

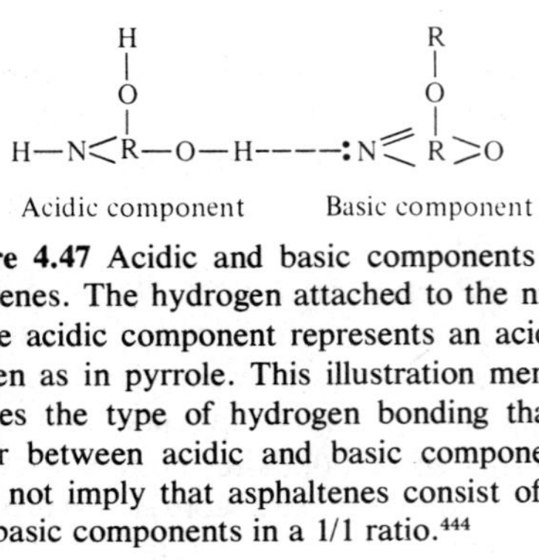

Figure 4.47 Acidic and basic components of asphaltenes. The hydrogen attached to the nitrogen in the acidic component represents an acidic hydrogen as in pyrrole. This illustration merely indicates the type of hydrogen bonding that may occur between acidic and basic components; it does not imply that asphaltenes consist of acidic and basic components in a 1/1 ratio.[444]

4.12.3.7 *Depolymerized Coal*

Several papers have been published in which NMR spectrometry was used to characterize the soluble products from the depolymerization of coals[367,455–457] and oxidized coals.[458] The principal objectives of the work on coals were to depolymerize coals under "mild" conditions, thus to recover unaltered monomeric units, and to investigate the possible existence of methylene bridges in the coals. Studies of the depolymerization products from treatment of six bituminous coals with phenol-boron trifluoride led to the conclusion that bituminous coals contain large structural units connected to smaller units by methylene bridges. Carbon-13 NMR spectra of Knife River lignite depolymerized with phenol and *p*-toluene sulfonic acid have also been reported. The spectra show a complex aromatic structure with a much smaller complement of aliphatic structure. Preliminary interpretation of the spectra indicated the possible presence of methylene and methine bridge structures in the soluble material.

[455] L. A. Heredy and M. B. Neuworth, *Fuel,* **41,** 221 (1962).

[456] L. A. Heredy, A. E. Kostyo, and M. Neuworth, in *Coal Science,* Am. Chem. Soc. Adv. Chem. Ser. No. 55, 1966, p. 493.

[457] L. A. Heredy, A. E. Kostyo, and M. B. Neuworth, *Fuel,* **44,** 125 (1965).

[458] L. J. Darlage, J. P. Weidner, and S. S. Block, *Fuel,* **53,** 54 (1974).

4.13 ELECTRON SPIN RESONANCE*

Electron spin resonance (ESR) spectrometry, sometimes called electron paramagnetic resonance (EPR) or simply electron magnetic resonance (EMR), is a means of detecting direct transitions between electron Zeeman levels. The phenomenon of electron spin

* Contributed by Herbert L. Retcofsky.

resonance is observed only for atomic or molecular systems having net electron spin angular momentum, that is, materials containing one or more unpaired electrons. Materials that meet this criterion include organic free radicals, molecules in their triplet state, charge transfer complexes, transition metal ions, semiconductors, metals, inorganic molecules such as oxygen that contain partially filled molecular orbitals, and crystals having certain point defects such as *F* centers.

The first successful ESR experiments were those of Zavoisky[459] in 1945 and Cummerow and Halliday[460] in 1946. ESR spectra are obtained by measuring the absorption of electromagnetic radiation—generally, but not necessarily, in the microwave region—by materials containing unpaired electrons in an appropriately strong magnetic field. The experiment is usually carried out at a fixed frequency while the magnetic field is slowly scanned. Microwave frequencies of 9–10 GHz are typically used in commercially available spectrometers. To detect organic free radicals at these frequencies, magnetic fields of about 0.35 tesla (T) (i.e., 3500 G) are required. Only the aspects of ESR spectrometry necessary for a basic understanding of applications in coal research are discussed here; more detailed treatment of the theoretical and experimental aspects of ESR can be found in several excellent treatises[461–466] or in the first Supplementary Volume to *Chemistry of Coal Utilization.*[467]

[459] E. Zavoisky, *J. Phys. USSR,* **9,** 211 (1945).

[460] R. L. Cummerow and D. Halliday, *Phys. Rev.,* **70,** 433 (1946).

[461] C. P. Poole, Jr., *Electron Spin Resonance—A Comprehensive Treatise on Experimental Techniques,* Interscience, New York, 1967, 922 pp.

[462] J. E. Wertz and J. R. Bolton, *Electron Spin Resonance—Elementary Theory and Practical Applications,* McGraw-Hill Book Co., New York, 1972, 497 pp.

[463] M. Bersohn and J. C. Baird, *An Introduction to Electron Paramagnetic Resonance,* W. A. Benjamin, New York, 1966, 274 pp.

[464] D. J. E. Ingram, *Spectroscopy at Radio and Microwave Frequencies,* Plenum Press, London, 1967, 534 pp.

[465] R. S. Alger, *Electron Paramagnetic Resonance: Techniques and Applications,* Interscience, New York, 1968, 580 pp.

[466] P. B. Ayscough, *Electron Spin Resonance in Chemistry,* Methuen, London, 1967, 451 pp.

[467] *Chemistry of Coal Utilization,* Supplementary Volume, H. H. Lowry, Ed., John Wiley and Sons, New York, 1963, p. 78.

4.13.1 Measurement of Spectra

4.13.1.1 *ESR Spectral Parameters*

The three most useful parameters that can be extracted from ESR spectra of coals and related carbonaceous materials are the spectral intensity, spectral line width, and *g* value. Other parameters, such as the electron relaxation times, can sometimes be measured or estimated, but generally these have had only limited use in coal research. Electron relaxation times of coals were reviewed in the first Supplementary Volume.[467] Electron-nuclear hyperfine splittings, usually observed in the ESR spectra of organic free radicals, have not been resolved in spectra of coals.

The concentration of unpaired electrons, presumably free radical electrons in the case of most coals, can be estimated by comparing the integrated ESR intensity of the coal sample with that of some standard substance such as 1,1-diphenyl-2-picrylhydrazyl (DPPH) obtained under the same experimental conditions. Since ESR spectra are nearly always recorded as first derivatives of absorption curves, double integrations are required. It is emphasized that it is not the number of free spins that is obtained directly from the ESR spectra, but rather the static spin susceptibility X_0 of the sample under study (see section 4.13.12).

The ESR spectral line width of a coal sample is generally determined by measuring the distance in gauss between the maximum and minimum points of the first derivative curve. This corresponds to the line width at maximum slope of the actual absorption curve. It should be noted that the line widths reported in early ESR studies of coals often were measured at half-intensity of the absorption curve; thus are larger than the peak-to-peak widths reported in more recent work.

The spectroscopic splitting factor (Lande *g* value) is given by Equation 4.17:

$$g = \frac{h\nu}{\beta H_0} \tag{4.17}$$

where β is the Bohr magneton, h is Planck's constant, and ν/H_0 is the ratio of the frequency to the magnetic field at resonance. Generally the frequency is measured using a wavemeter or high-speed electronic counter and the magnetic field is determined by recording the spectrum of some substance of known g value. Hall probes, Gauss meters, or NMR oscillators are sometimes used to determine H_0.

4.13.1.2 *A Cautionary Note*

The foregoing section suggests that the extraction of meaningful data from the ESR spectra of coals is a relatively straightforward matter. Unfortunately, this is not always the case. Singer,[468] in an excellent review article, discusses the physics of the ESR phenomena and points out various pitfalls that may confront the unwary experimenter, especially one involved in studies of coals and carbons. Although space does not permit an in-depth discussion of Singer's review, two areas—the measurement of spin concentration and g values—are of special concern to coal chemists and deserve some mention.

Assuming that the intensity of the ESR signal has been properly determined (i.e., saturation is absent, modulation amplitude used is small compared to the spectral line width, electrical conductivity effects on spectra are minimized, etc.), there remains the problem of relating the spin susceptibility (X_0) to the number of spins (N). For coals, as well as for most substances, DPPH is used as a standard material to compute N. The choice of any material as an ESP intensity standard, however, requires a knowledge of the temperature dependence of the susceptibility. For stable free radicals with almost free spins, X_0 is given by the familiar Curie law expression:

$$X_0 = \frac{Ng^2 \beta^2 S (S + 1)}{3kT} \tag{4.18}$$

where S is the spin quantum number. For coals, the usual procedure has been to choose $S = ½$ and $g = 2$ and to compute N relative to another material (e.g., DPPH,) that obeys Equation 4.18 rather well.

Although it has been reported that the ESR intensities of coals increase with decreasing temperature as required by Curie's law,[469,470] extensive studies on a wide variety of coals have apparently not been carried out. Recent studies of the temperature dependence of the ESR intensity for chars and carbons[468,471–473] that show non–Curie law behavior suggest the need for further studies on coals. It is also conceivable that such a study *may* provide information regarding the recent hypothesis that charge transfer complexes[474] rather than stable free radicals are responsible for the ESR absorption in coals.

Singer also discusses the necessity for the use of proper dispersal techniques and low microwave power levels when measuring ESR g values for highly conducting materials.[475] He points out that both electrical contact between particles and thermal contact of the sample with its surroundings may lead to incorrect g values. Electrical contact for high-temperature carbons and very high rank coals[476] may lead to asymmetric line shapes, thus to incorrect g values. For a polycrystalline graphite rod, a large difference in g value was observed when the sample was exposed to air; this resulted from resistive heating of the carbon by the microwaves in the cavity. These effects are unimportant for most coals; they can, however, affect measurements made on metaanthracites and heat-treated coals.

A thorough study of Singer's review article is recommended for any researcher involved in ESR studies of coals and carbons. It is believed that disagreements between the ESR results of different workers frequently are due to lack of attention to both the basic ESR phenomena and the microwave techniques necessary for their meaningful observation.

4.13.2 Recent Advances in Instrumentation

Although the advent of solid state circuitry and more reliable microwave sources has added to the sophistication of ESR spectrom-

[468] L. S. Singer, *Proc. 5th Carbon Conf.*, Vol. 2, Pergamon Press, New York, 1963, p. 37.

[469] D. E. G. Austen, D. J. E. Ingram, and J. G. Tapley, *Trans. Faraday Soc.*, **5K,** 400 (1958).

[470] J. van Gerven, A. van Itterbeeck, and E. De Wolf, *J. Phys. Radium*, **17,** 140 (1965).

[471] J. Kommander and L. S. Singer, unpublished results.

[472] L. S. Singer and G. Wagoner, *Proc. 5th Carbon Conf.*, Vol. 2, Pergamon Press, New York, 1963, p. 65.

[473] G. M. Arnold, *Carbon (Oxford)*, **5,** 33 (1967).

[474] R. M. Elofson and K. F. Schultz, *Prepr., Am. Chem. Soc., Div. Fuel Chem.*, **11,** 513 (1967).

[475] L. S. Singer and G. Wagoner, *J. Chem. Phys.*, **37,** 1812 (1962).

[476] H. L. Retcofsky, unpublished results.

eters, radical changes in the method of obtaining spectra comparable to the introduction of pulse FT techniques in NMR have not occurred. Nevertheless, sensitivity of instruments has improved over the years. Instruments capable of detecting 2×10^{10} ΔH spins are now commercially available; the use of signal averages can increase this sensitivity many fold. Most commercial spectrometers operate at X-band (9.5 GHz); however Q-band (35 GHz) instruments are also available. Obtaining spectra at two different frequencies (and magnetic fields) is an excellent way to determine whether multiple signals in an ESR spectrum are due to nuclear hyperfine interactions or to two different radical species. Spectra of a sample of vitrain-rich Pittsburgh coal obtained at these two frequencies[477] exhibited two field-dependent signals, which were assigned to two different types of radical. Other improvements in instrumentation introduced subsequent to publication of the first supplement include dual cavities for more rapid and more precise determination of spectra, field frequency control units to enhance spectrometer reproducibility, and commercial spectrometers to observe electron-nuclear double resonance (ENDOR).

4.13.3 Sample Preparation

Since it is known that oxygen and other paramagnetic gases affect the ESR line width and spectral intensity of some coals and chars,[467] it is important that ESR measurements be made on evacuated samples. Generally these effects are reversible and result from the paramagnetic properties of the gases rather than any particular chemical property of the coal. Anthracites,[478,479] heated coals,[469–478] fusains,[477–480] and fusinites[479] are particularly susceptible to the effects of paramagnetic gases. Measurements reported for a series of five Japanese coals having maf carbon contents of 70.6–85.4% showed an irreversible oxygen effect of particular significance.[481] The coals studied were taken from deep within the seams in such a manner that exposure to oxygen was avoided. In each case the intensity of the ESR signal of the virgin coals increased irreversibly when air was introduced into the sample tube, the increase in intensity being most pronounced for the coals of lower rank. To the best of our knowledge, this is the only ESR investigation of coals in which prior exposure of the coalified material to oxygen was avoided.

As indicated previously, coals and treated coals exhibiting electrical properties in the range of semiconductors should be ground to particle sizes less than the microwave skin depth and dispersed in a nonconducting medium such as silicon dioxide or potassium bromide before ESR measurements are made. Failure to make these preparations may result in incorrect spin concentrations, line shapes, and g values. In addition, the microwave power level incident on the sample must be low enough to prevent both saturation and resistive heating.

4.13.4 ESR Results for Coals and Their Petrographic Components

Since the discovery of ESR absorption in natural carbons by Uebersfeld et al.[482] and Ingram et al.,[483] investigators too numerous to list here have reported ESR data for a hundred or more coals, suggesting that stable free radicals or other species containing unpaired electrons are present in *all* coals. Although the exact nature of these species has not been established, ESR measurements have given considerable insight into the immediate environment of the free spins.

4.13.4.1 *Whole Coals, Vitrains, and Vitrinites*

ESR studies of whole coals, vitrains, and vitrinites give quite similar results, thus the three are discussed collectively. The effect of

[477] H. L. Retcofsky, J. M. Stark, and R. A. Friedel, *Anal. Chem.*, **40**, 1699 (1968).

[478] J. Smidt and D. W. van Krevelen, *Fuel*, **38**, 355 (1959).

[479] D. E. G. Austen, D. J. E. Ingram, P. H. Given, C. R. Binder, and L. W. Hill, in *Coal Science*, Am. Chem. Soc. Adv. Chem. Ser., No. 55, 1966, p. 344.

[480] E. de Ruiter, *Fuel*, **41**, 63 (1962).

[481] H. Ohuchi, M. Shiotani, and J. Sohma, *Fuel*, **48**, 187 (1969).

[482] J. Uebersfeld, A. Etienne, and J. Combrisson, *Nature (London)*, **174**, 614 (1954).

[483] D. J. E. Ingram, J. G. Tapley, R. Jackson, R. L. Bond, and A. R. Murnahgan, *Nature (London)*, **174**, 797 (1954).

petrographic composition on the ESR spectra of coals is presented later in this chapter.

As reported in the first supplement,[467] the free radical concentrations of coals increase approximately exponentially with increasing carbon content up to a carbon content of 94%, after which a rapid decrease in the number of free spins is observed. A number of more recent papers[477,479,481,484–489] contain ESR intensity data for additional coals that support or at least partially support the earlier findings. Typical free spin concentrations range from 1–6 $\times$ 10^{18} g^{-1} for peats and lignites to 5.3 $\times$ 10^{19} g^{-1} for an anthracite having a carbon content of 92.7%.[477] A concentration of 5 $\times$ 10^{17} free spins/g has been reported for a metaanthracite.[477]

The variation of ESR line widths of coals with rank or carbon content is also well documented.[467,477,479,484,488–490] For coals having carbon contents between 55 and 90%, peak-to-peak line widths, as measured using the derivative of the absorption curve, range from about 5.2 to 8.6 G. A gradual increase in line width with increasing carbon content is first observed; this trend is reversed at a carbon content of 80%. The rate of decrease becomes very rapid for coals having carbon contents between 90 and 95%; line widths of only a few tenths of a gauss have been measured for several such coals. Coals of the very highest rank (i.e., metaanthracites), with carbon contents generally in excess of 95%, often exhibit very broad resonances. This general trend in spectral line width (and in other spectral parameters) observed for coals is also exhibited by heat-treated coals and low-temperature organic chars in plots of ESR data as a function of pyrolysis temperature.[469] For a series of 53 coals representing all coal ranks except very high rank metaanthracites, a near-rectilinear relation between the ESR line widths ΔH and the hydrogen contents H of the coals was reported.[490] The correlation coefficient of 0.91 and the standard error of estimate value of 0.8 G, which is approximately twice the experimental error in the line width determination, indicated that the relation between ΔH and H is well represented by the regression line as shown in Equation 4.19:

$$\Delta H = 1.6H - 1.8 \tag{4.19}$$

It is again emphasized that the line widths of coals reported in most of the earlier ESR studies[467] were measured at half-height on the absorption curve rather than peak-to-peak on the first derivative curve.

The Landé g value, a very important ESR spectral parameter that is sensitive to the molecular environment of an unpaired electron, was treated only casually in early coal studies. The utility of this parameter in coal research became apparent as more precise values became available. Systematic studies of the ESR g values of a large variety of coals, including ones from the United States,[476,477,491] Canada,[489] Pakistan,[476,489] Yugoslavia,[476,489] Austria,[476] Antarctica,[476] and Japan,[484,488] have appeared. The reader is cautioned that any reported g value of coal or treated coal that is less than the free electron value (2.0023) is open to question (see Section 4.13.1.2). A recent plot of the ESR g values of coals as a function of carbon content[476] is reproduced in Fig. 4.48. Anisotropies in the g values of anthracitic coals have been reported.[490]

A 1959 report indicated that free radical concentrations and spectral line widths of coals from the same mine were independent of the depth from which the sample was taken.[492] More recent ESR measurements made on a Colorado coal that had been subjected to igneous intrusion showed considerable dependence on sampling depth.[493] These observations collectively suggest large

[484] S. Toyoda, S. Sugawara, and H. Honda, *J. Fuel Soc. Jap.*, **45,** 876 (1966).

[485] Y. B. Tyutyunnikov, I. S. Romodanov, L. G. Sinterserova, and A. V. Grebenchuk, *Khim. Tverd. Topl.*, **4,** 133 (1968).

[486] A. I. Skrigan, L. P. Dmitrieva, and V. S. Verner, *Dokl. Akad. Nauk Beloruss. SSR*, **12,** 628 (1968).

[487] H. Ohuchi, M. Shiotani, and J. Sohma, *Kogyo Kagaku Zasshi*, **71,** 1191 (1968).

[488] S. Toyoda and H. Honda, *Carbon (Oxford)*, **3,** 527 (1966).

[489] R. M. Elofson and K. F. Schulz, in *Spectrometry of Fuels*, R. A. Friedel, Ed., Plenum Press, New York, 1970, p. 202.

[490] H. L. Retcofsky, G. P. Thompson, R. Raymond, and R. A. Friedel, *Fuel*, **54,** 126 (1975).

[491] H. L. Retcofsky, J. M. Stark, and R. A. Friedel, *Chem. Ind. (London)*, 1327 (1967).

[492] G. N. Sarkar, A. Mukherji, R. N. Chatterjee, and U. S. Ghosh, *Indian J. Phys.*, **33,** 117 (1959).

[493] P. H. Given and C. R. Binder, in *Advances in Organic Geochemistry 1964*, Pergamon Press, Oxford, 1966, p. 147.

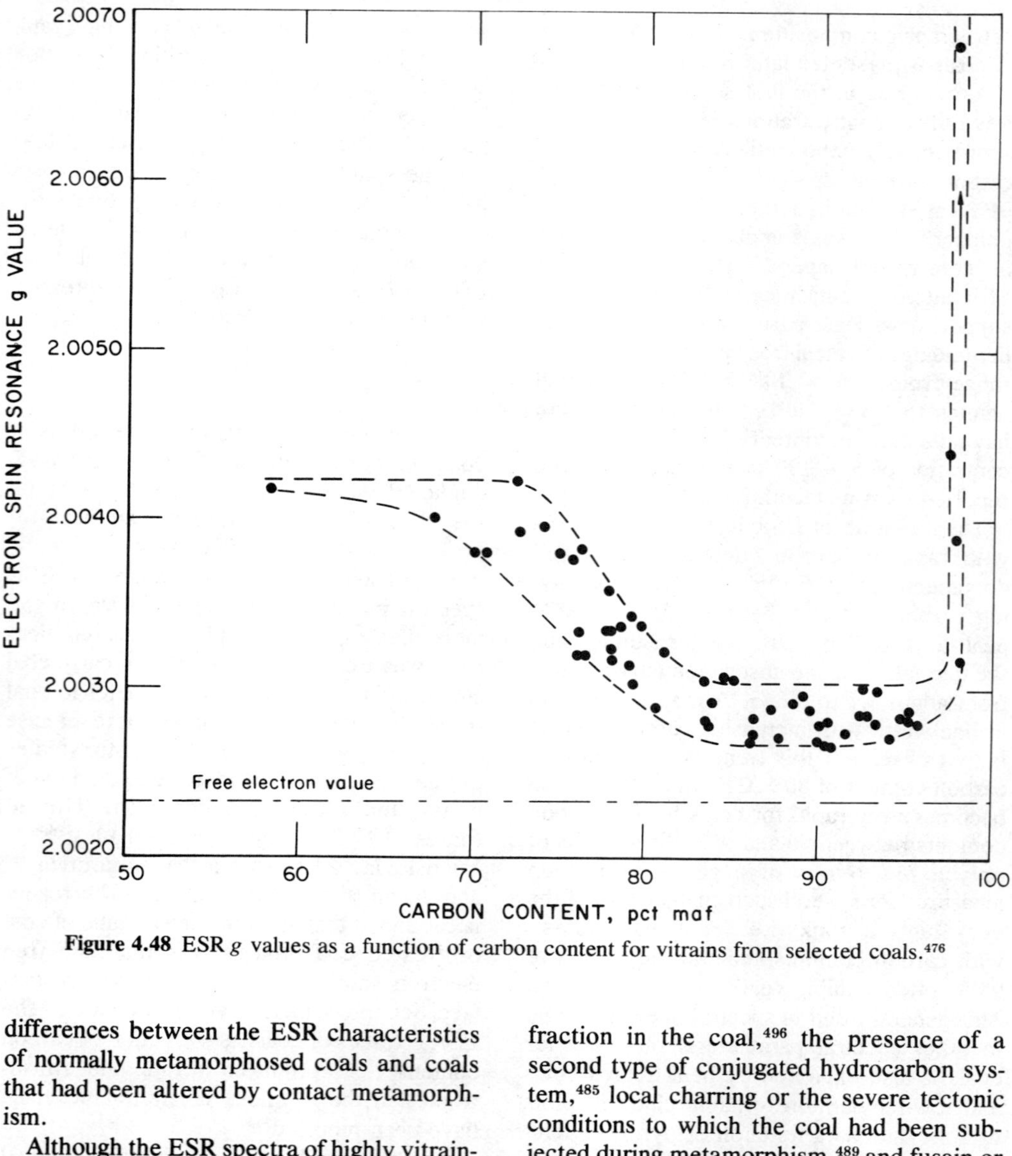

Figure 4.48 ESR g values as a function of carbon content for vitrains from selected coals.[476]

differences between the ESR characteristics of normally metamorphosed coals and coals that had been altered by contact metamorphism.

Although the ESR spectra of highly vitrainous coals generally consist of a single peak devoid of any fine structure, spectra of whole coals and "impure" vitrinites sometimes show dual peaks.[477,484,485,489,494,495] The narrower of the two signals has a peak-to-peak line width of 1–2 G and is very sensitive to oxygen. The origin of the narrow component is not well understood and may be different for different coals. It has been variously attributed to the presence of a primary tar fraction in the coal,[496] the presence of a second type of conjugated hydrocarbon system,[485] local charring or the severe tectonic conditions to which the coal had been subjected during metamorphism,[489] and fusain or fusinite intermingled with the vitrain in the sample.[476,497,498]

4.13.4.2 *Other Petrographic Components*

Although a considerable amount of ESR data is available for vitrains, vitrinites, and whole coals, other petrographic components have received far less attention. Some ESR data,

[494] H. Honda, K. Egi, S. Toyoda, Y. Sanada, and T. Furuta, *Carbon (Oxford),* **1,** 155 (1964).

[495] S. Toyoda and H. Honda, *J. Fuel Soc. Jap.,* **44,** 697 (1965).

[496] A. S. Sidorov and V. F. Anufrienko, *Kinet. Katal.,* **7,** 1060 (1966).

[497] C. Yokokawa, *J. Fuel Soc. Jap.,* **47,** 702 (1968).

[498] C. Yokokawa, *Fuel,* **48,** 29 (1969).

quite limited in scope, have been reported for fusains and fusinites,[476,477,479,480,499] micrinites,[479,500,501] and exinites.[479,500,501] Fusain, fusinite, and micrinite generally exhibit much higher spin concentrations and narrower line widths than vitrinite from the same coal, whereas the ESR characteristics of exinite are quite similar to those of the corresponding vitrinite. Independent measurements by two laboratories[500,501] on the vitrinite, exinite, and micrinite macerals from four coals having carbon contents between 83.5 and 88.8% indicate that the free radical contents of vitrinites and exinites increase with increasing coal rank, but the ESR signals of the two macerals are nearly indistinguishable for coals having carbon contents exceeding 89%. The radical contents of the micrinites decrease with increasing coal rank. The line widths of the vitrinite, exinite, and micrinite are nearly invariant with rank.[501] These generalizations must be viewed with some reservation because of the small number of samples investigated. ESR data for fusains,[476,479,499] taken collectively, show no clear-cut rank dependence. The small line widths reported for fusains from medium- and high-rank coals support the theory that fusinization of plant remains has an early inception and progresses rapidly, mostly in the peat and lignitic stages, after which the metamorphic changes are almost imperceptible.[502] The effect of petrographic composition of coals on their ESR spectra has been discussed[503] and the possibility of ambiguities in the ESR spectra of coals of unknown petrographic makeup has been noted.[479]

4.13.4.3 *Effect of Heat Treatment*

It was reported in the first supplement[467] that vitrinites having carbon contents in the 75.9–93.7% range showed an increase in radical concentration with increasing pyrolysis temperature, the concentration reaching a maximum between 450 and 600°C. Radical concentrations of coals of higher or lower rank than these were not noticeably changed by heat treatment to temperatures approaching 600°C. Line widths remained relatively constant with the initial increase in pyrolysis temperature, then decreased and passed through a minimum at approximately the temperature of the maximum radical concentration. Above 600°C the concentration of radicals decreased, frequently to a value below the sensitivity of the ESR instrument. ESR signals began to reappear in the case of coals pyrolyzed at temperatures above 1500°C. A number of more recent investigations[488,494,504–509] of whole coals and vitrains for all practical purposes confirm the earlier findings.

It was pointed out as early as 1959[510] that the concentration of free radicals in heat-treated coals is a function not only of the carbonization temperature but also of the heating rate. For a vitrain of 86.9% C, the maximum value for the free radical content was reached at a slightly lower temperature when the sample was pyrolyzed at a heating rate of 1°C/min than at a rate of 3°C/min. Several ESR studies of much more rapidly heated coals have since been reported.[511–514] In experiments using heating rates comparable to those encountered during the combustion of pulverized coal (°C × 10^4/s), the maximum free radical concentrations for two coals were found to occur at tempera-

[499] L. M. Vasil'eva, K. I. Bochkareva, K. N. Shiryaeva, A. A. Shklyaev, and V. F. Anufrienko, *Khim. Tverd. Topl.*, 48 (1972).
[500] C. Kröger, *Brennstoff-Chemie*, **39**, 29 (1958).
[501] E. de Ruiter and H. Tschamler, *Brennstoff-Chemie*, **42**, 290 (1961).
[502] J. M. Schopf, *Econ. Geol.*, **43**, 207 (1948).
[503] G. S. Berger, L. G. Kislitsina, and S. K. Akhmetov, *Khim. Tverd. Topl.*, **2**, 144 (1973).
[504] K. I. Kononenko and E. A. Kostyuk, *Ukr. Fiz. Zh.*, **12**, 1440 (1967).
[505] A. A. Sidorov and V. F. Anufrienko, *Izv. Sibrirsk. Otd. Akad. Nauk SSSR, Ser. Khim. Nauk*, **1**, 155 (1966).
[506] A. K. Dzizenko, V. V. Opritov, V. N. Sushin, and T. V. Shustova, *Khim. Tverd. Topl.*, **2**, 142 1969.
[507] S. Toyoda, *Shigen Gijutsu Shikensho Hokoku*, No. 74, 1969, 117 pp.
[508] K. I. Kononenko and E. A. Kostyuk, *Radiospektrosk. Mol. Kristallogr. Klitin*, 72 (1973).
[509] K. I. Kononenko and E. A. Kostyuk, *Radiospektrosk. Mol. Kristallogr. Klitin*, 81 (1973).
[510] J. Smidt and D. W. van Krevelen, *Nature (London)*, **184**, 719 (1959).
[511] R. Chauvin, P. Chiche, M. F. Quinton, and J. Uebersfeld, *C.R. Acad. Sci., Paris, Ser. C*, **265**, 1212 (1967).
[512] R. L. Bond, W. R. Ladner, and R. Wheatley, *Fuel*, **47**, 213 (1968).
[513] R. Chauvin, P. Chiche, M. F. Quinton, and J. Uebersfeld, *Carbon (Oxford)*, **7**, 307 (1969).
[514] R. J. R. Hayward, W. R. Ladner, and R. Wheatley, *Fuel*, **49**, 223 (1970).

tures approximately 175–250°C higher than for coals subjected to slow heat treatment (1¼°C/min).[512] Typical results are shown in Fig. 4.49. It was concluded that coals heated to a high temperature in a short period of time produce chars with free radical contents similar to those obtained from longer heating at a lower temperature. Using even faster heating rates (> °C × 10⁶/s) during flash-heating experiments, coals produced essentially electrically *nonconducting* chars at temperatures of about 2000°C.[514] In contrast, chars produced at temperatures of 800°C during conventional pyrolysis exhibit semiconductivity.

A very comprehensive study of the effect of heat treatment on various macerals from selected British and American coals has been reported.[479] The vitrinite components of the coal had carbon contents between 79.6 and 88.6%. The changes in the concentration of free spins and in the spectral line width for vitrinite, exinite, and fusinite from Chislet No. 5 (88.6% maf C) coal are shown in Fig. 4.50. Typically, on pyrolysis in nitrogen at temperatures up to 550°C, the signals given

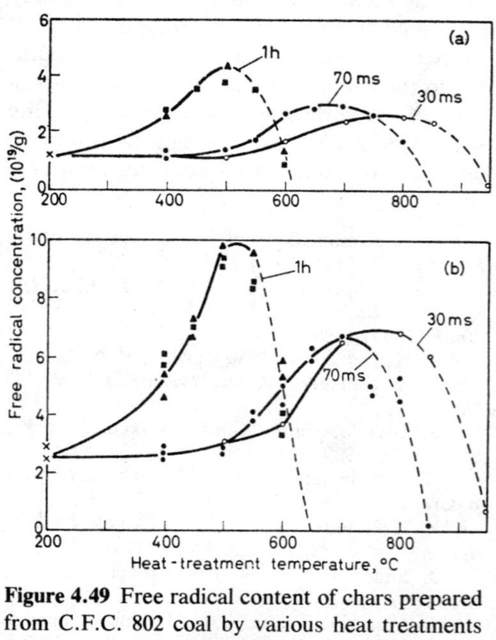

Figure 4.49 Free radical content of chars prepared from C.F.C. 802 coal by various heat treatments in the range of 400°–950°C: (*a*) measurements made in the air and (*b*) measurements made *in vacuo*; x = raw coal; ○, ● = rapid heating to temperature, residence times at temperature 30 and 70 ms, respectively; ▲ = slow heating to temperature (1¼°C/min), held at temperature for 1 hr; ■ = subsamples of ● reheated slowly.[512]

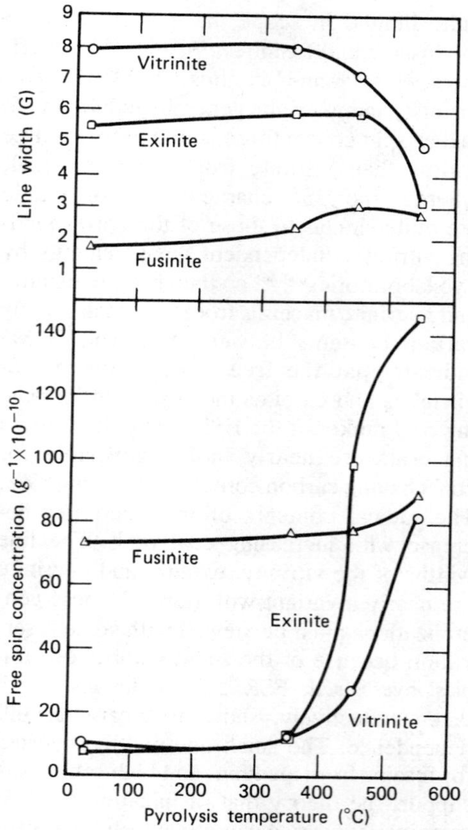

Figure 4.50 Free spin concentration and line width for macerals from Chislet No. 5 coal.[479]

by vitrinites and exinites increase in height and sharpness, whereas those given by fusinites undergo little change in either characteristic. Only one micrinite was studied; the behavior of the sample was similar to that of the fusinites.

4.13.5 Irradiated Coals

ESR studies of coals that had been subjected to various types of high-energy radiation including gamma rays,[515–520] neutrons,[518,521]

[515] I. Wender, C. Zahn, and R. F. Stewart, U.S. Bur. Mines Inf. Circ. No. 8377, 1968, 39 pp.

[516] R. Husack, G. A. Brady, and J. W. Eckerd, U.S. Bur. Mines Rep. Invest. No. 6391, 1964, 28 pp.

[517] R. McBrian, *Min. Congr. J.*, **44,** 82 (1958).

[518] A. Weinstein and P. L. Walker, Jr., *Fuel,* **40,** 357 (1961).

[519] J. Duchesne, J. Depireux, and J. M. van der Kaa, *Geochim. Cosmochim. Acta,* **23,** 209 (1961).

[520] S. Toyoda, *Fuel,* **42,** 431 (1963).

[521] R. A. Friedel and I. A. Breger, *Science,* **130,** 1762 (1959).

and electrons[522,523] have been reported. Although results from different laboratories may appear to be conflicting, all must be viewed in light of studies[519,520,523] indicating that radiation-induced radicals in coal are short-lived species. Frequently the ESR measurements were made long after the actual radiation experiments had been carried out.

4.13.5.1 *Gamma-Irradiated Coals*

Samples of lignite, hvAb, and lvb coals showed little if any difference in the concentrations of free radicals after exposure to gamma rays (1000 Mrad).[515] Many months had elapsed between the time of irradiation of the samples and the time of the ESR measurements reported. A 1964 study on irradiation of high-volatile anthracite[516] could not confirm an earlier report that irradiation of anthracite increased the free radical content.[517] ESR studies of a Japanese caking (85.2% maf C) and a noncaking (79.8% maf C) coal after exposure to ^{60}Co gamma irradiation indicated the presence of narrow lines newly produced by the irradiation.[520] The measurements were made one month after the irradiation experiments; the narrow lines could not be detected in 4-month-old samples. A Belgian coal, irradiated with ^{60}Co gamma rays at a flux of 0.35 Mrad/hr for a total dosage of 84 Mrad, showed a 27% increase in ESR intensity when measured a short time after irradiation.[519] The signal decreased rapidly in intensity during the next 50 hr to a constant value approximately 10% greater than that of the nonirradiated coal.

4.13.5.2 *Pile Irradiation*

Five coals ranging in rank from lignite to low-volatile bituminous coal were exposed to an integrated flux of 2.29×10^{19} slow neutrons per square centimeter for 120 days. Little or no change due to the irradiation was found in free radical content for the three coals of lowest rank (maf C = 69.9, 77.8, 82.4%), whereas the two highest rank coals (maf C = 89.3, 90.5%) showed increases of 81 and 32%, respectively.[521] No significant changes in free radical concentration of seven bituminous coals receiving up to 350 Mrad of radiation by gamma rays and fast neutrons in the presence of ammonia or under vacuum were observed.[518]

[522] D. J. Brown and G. J. Pitt, *Fuel,* **43,** 445 (1964).

[523] H. L. Retcofsky, A. G. Sharkey, Jr., and R. A. Friedel, *Fuel,* **46,** 109 (1967).

4.13.5.3 *Electron Irradiation*

ESR spectra of two English coals (maf C = 81.3 and 85.7%) were found to be essentially identical to the spectra of the corresponding coals after bombardment with 2 MeV electrons at dosages of 300 and 1200 Mrad using a van de Graaff generator. Vitrains from four American coals were irradiated directly in the cavity of an ESR spectrometer by directing the beam from a 6-MeV, 1-kW linear electron accelerator through an axial hole in one pole-piece of the ESR electromagnet.[523] Newly formed radicals were produced in a lignite that received 52 Mrad and in a low-volatile bituminous coal that received 49 Mrad. The spectra of the electron-irradiated coals resembled those of the gamma-irradiated Japanese coals discussed above. The radicals produced by electron irradiation had lifetimes of several minutes to perhaps a few hours. No detectable new radicals were observed during electron irradiation of vitrains from hvAb and hvCb coals.

4.13.6 Chemically Treated Coals

4.13.6.1 *Reduction*

A series of vitrains, ranging in maf carbon content from 82.5 to 93.1%, were reduced with lithium-ethylenediamine and subsequently examined by ESR.[524] The radical content of each of the reduced vitrains, except for the anthracite sample, was lower than that of the unreacted vitrain. A decrease in radical content was also observed for an lvb coal after electrochemical reduction.[476] The results are summarized in Table 4.22.

4.13.6.2 *Catalytic Dehydrogenation*

To investigate the hypothesis that unresolved electron-nuclear hyperfine interactions are prime contributors to the observed ESR line

[524] L. Reggel, R. Raymond, W. A. Steiner, R. A. Friedel, and I. Wender, *Fuel,* **40,** 339 (1961).

Table 4.22 Data for Reduced Vitrains[476]

	Vitrain				
	Pocahontas No. 3	Pittsburgh	Powellton	Sewell	Pocahontas No. 3
Carbon content of starting vitrain (%)	89.41	82.52	84.93	89.35	89.83
Method of reduction	Electrochemical	Lithium ethylene-diamine	Lithium ethylene-diamine	Lithium ethylene-diamine	Lithium ethylene-diamine
Hydrogen atoms added per 100 carbon atoms	9.8	16.7	19.1	29.1	44.7
Increase in atomic H/C (%)	17.2	36.1	39.2	56.2	83.4
Decrease in free radical content (%)	34	64	76	82	82

Source: H. L. Retcofsky, unpublished results.

widths of coals, vitrains from 10 coals were examined before and after catalytic dehydrogenation with palladium.[490] The line width of the dehydrogenated vitrain was smaller than that of the original sample in each case. The percentage decrease in the line widths of the coals after dehydrogenation appears as a function of the carbon contents of the original vitrains in Fig. 4.51. The plot is strikingly similar to that of the gas evolution data.[525] No special significance should be attached to the manner in which the experimental points are joined on the graph.

[525] L. Reggel, I. Wender, and R. Raymond, *Fuel,* **47,** 373 (1968).

4.13.6.3 *Sulfur-Treated Coals*

A vitrinite (83.9% maf C), its pyridine extract, and the extraction residue showed essentially the same increase in free radical concentration when heated in the presence of sulfur.[526] Studies on 15 aromatic compounds showed that the radical concentration found after sulfur treatment strongly depends on both the type of aromatic system and the type of substitution. The similar behavior of the vitrinite and its extraction products suggests a high similarity of the structural features of all three samples.

[526] E. de Ruiter, *Fuel,* **44,** 49 (1965).

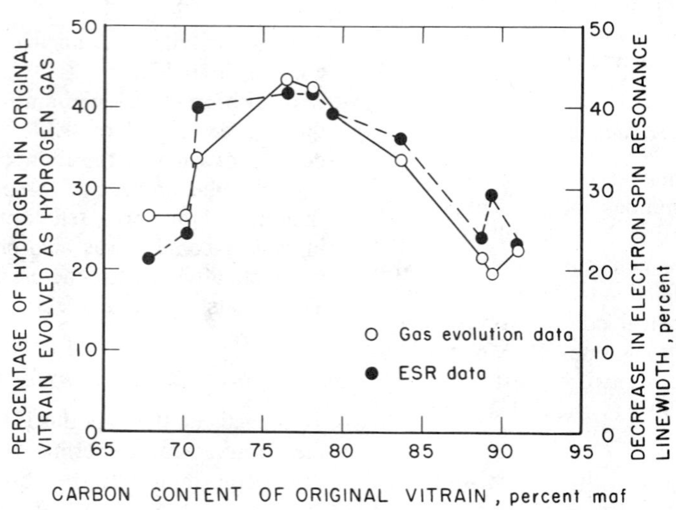

Figure 4.51 Hydrogen evolution and change in ESR line width of vitrains after catalytic dehydrogenation.[490]

4.13.6.4 *Iodine-Treated Coals*

The number of free radicals in coals (maf C = 84–87%), previously treated with HCl and HF or hydrolyzed, first decreased, then went through a maximum, and decreased as increasing amounts of iodine were adsorbed at 25°C.[527]

4.13.6.5 *Treatment with Concentrated Sulfuric Acid*

It was reported in the first Supplementary Volume (p. 85) that when a vitrain of 84% maf carbon, its pyridine extract, or the extract residue was treated with concentrated sulfuric acid at room temperature, the number of free radicals increased more than tenfold while the half-width remained constant. Similar results have been reported for a more extensive set of vitrains from 11 coals having carbon contents in the range 80.1–91.0-.[528] Exinites and micrinites were included in the more recent study. Exinites behaved similarly to the vitrinites. For the micrinites, a fivefold increase in free radical content was observed after treatment with sulfuric acid. This was accompanied by a 50% decrease in line width.

4.13.6.6 *Coals Treated with Various Solvents*

Seventeen vitrains from coals of different ranks were heated at temperatures between 120 and 260°C with various solvents including ethylenediamine, pyridine, cyclohexanone, and quinoline.[498] Spin concentrations of individual coals decreased considerably after treatment except for samples of brown coal and anthracite. The percentage decrease in the number of free spins was rank dependent, with the maximum change occurring for coals having carbon contents of 83–86%. ESR spectra of Southern Bell, Moura, Mickel, New River, and Itmann coals, after treatment with ethylenediamine at 140°C, exhibited dual peaks reminiscent of those frequently observed in the spectra of whole coals. It was suggested that ethylenediamine replaces the oxygen adsorbed on the spin centers in the coal.

4.13.7 *Other Materials from Coal*

ESR measurements for a number of other materials from coal have also been reported. Examples include coal tars,[529,530] coal-tar pitches,[531] SRC,[532] and the reductive degradation products of coals.[533] Although tars and their heptane-, methanol-, and benzene-soluble fractions give strong ESR spectra,[530] the concentrations of paramagnetic species in low molecular fractions of coke-oven and continuous-vertical-retort refined tars are low.[529] A study of nine coal-tar pitches and their quinoline-soluble and quinoline-insoluble fractions showed a general nonlinear increase in spin concentrations with increasing quinoline-insoluble content of the pitches.[531] ESR absorption was observed in solid SRC and in the recycled SRC liquid.[532] In addition to the "free radical" signal at g 2, signals at g 4 due to paramagnetic metal-containing species—possibly high-spin iron(III)—were present in the spectra of the recycled SRC liquid and a frozen tetrahydrofuran solution of the SRC. It should be noted that a dilute solid mixture of pitch in potassium chloride is supplied with most commercial ESR spectrometers as a sensitivity standard.

4.13.8 Artificial Coalification

It has been suggested that indirect studies of the structure of coal and the nature of the coalification process can be carried out through spectral studies, principally IR[534–536]

[527] S. Angelov, D. Mekhandzhiev, A. Terebenina, and M. Mikhailov, *Dokl. Bolg. Akad. Nauk,* **23,** 1517 (1970).

[528] E. de Ruiter and H. Tschamler, *Brennstoff-Chemie,* **42,** 290 (1961).

[529] K. D. Bartle and J. A. S. Smith, *Fuel,* **44,** 109 (1965).

[530] K. D. Bartle and J. A. S. Smith, *Fuel,* **46,** 29 (1967).

[531] K. A. Kini, *J. Indian Chem. Soc.,* **50,** 616 (1973).

[532] D. L. Wooton, H. C. Dorn, L. T. Taylor, and W. M. Coleman, *Fuel,* **55,** 224 (1976).

[533] I. M. Krivovyaz, K. Tuichiev, D. T. Zabramnyi, and N. A. Parpeiv, *Uzbeksk. Khim. Zh.,* **10,** 42 (1966).

[534] R. A. Friedel, *Appl. Opt.,* **2,** 1109 (1963).

[535] R. A. Friedel and H. L. Retcofsky, in *Spectrometry of Fuels,* R. A. Friedel, Ed., Plenum Press, New York, 1970, p. 46.

[536] R. A. Friedel, J. A. Queiser, and H. L. Retcofsky, *J. Phys. Chem.,* **74,** 908 (1970).

and ESR[536,537] spectrometry, of chars of model substances. ESR studies of heat-treated cellulose,[536,538] wood,[535,536] and very low rank coals[536] indicate changes in spin concentrations, line widths, and g values similar to those observed in plots of the ESR paramaters of coals as a function of rank (or carbon content). The possible importance of both time and temperature in coalification has been demonstrated by studying chars produced by slow pyrolysis at low temperatures comparable to those encountered during normal metamorphic processes.[535]

4.13.9 Interpretations of the ESR Results

It is unlikely that the present state of knowledge of the structure and chemistry of coal will allow a single explanation encompassing all the ESR results described above. If, however, one accepts the premise that the species responsible for the ESR absorption in coals and coallike substances (other than those that exhibit electrical semiconductivity) are free radicals, certain generalizations are possible. Most of the following discussion is based on that premise.

It is generally thought that coal metamorphism involves a progressive increase in the size and number of condensed aromatic ring structures at the expense of nonaromatic materials. The increase in radical concentrations of coals with increasing rank is best explained on the basis of resonance stabilization of the radicals that form during coal metamorphism. The formation of radicals during the slow pyrolysis of coals and model substances suggests that the increase in the number of radicals with increasing coal rank involves thermal reactions, although at least one group of researchers[519] suggested that radioactivity is more important than geothermal factors in the formation of free radicals in lignites. The decrease in the free radical contents of coal after chemical reduction supports the view that the free radicals in coal are associated with aromatic structures.

The variation of ESR line widths is best explained on the basis of unresolved hyperfine interactions between the unpaired electrons and protons in the coals. The near-rectilinear relationship between the line widths and the hydrogen contents of the coals as well as the observation that line widths are reduced after dehydrogenation (Fig. 4.38) is strong, albeit not conclusive, evidence that this is true.

The changes in the ESR g values of coals as coalification progresses (Fig. 4.48) suggest that the free radical electrons in peats and low-rank coals are at least partially localized on oxygen atoms. The decrease in g value during progressive coalification up to the medium-rank bituminous stage is undoubtedly the result of the decreasing oxygen contents of the coals. The plateau in the curve (~84 to ~95% maf C) indicates the presence of a more "hydrocarbon-type radical" for these more highly metamorphosed coals. The reported g value anisotropy in anthracitic coals[490] and the very large g values observed for metaanthracites suggest the formation of large condensed-ring structures that evolve into graphitelike structures. It has been suggested that sulfur and nitrogen structures, in addition to those containing oxygen, may also contribute to the observed g values of low-rank coals.[539]

Studies of the effect of heat treatment on the ESR spectra of coal macerals (see Fig. 4.50 for representative results) present the exciting possibility of using ESR behavior as a geothermal indicator.[493] Within certain limitations, the behavior of a carbon does not change until the specimen is heated to a temperature higher than that to which it had been exposed during its previous history. A comparative study of macerals from a select set of coals gave results consistent with the view that fusinites were subjected to some kind of pyrolysis to perhaps 400–660°C before incorporation into the organic sediment.[479]

4.14 INFRARED ABSORPTION*

4.14.1 Introduction

Because of the increased availability of commercial IR spectroscopic instrumentation and the yield of "some" information on almost

[537] C. Yokokawa, H. Yoshimoto, and Y. Kamitani, *Nenryo Kyokai-shi,* **49,** 874 (1970).

[538] B. Milsch, W. Windsch, and H. Heinzelmann, *Carbon (Oxford),* **6,** 807 (1968).

[539] T. F. Yen and S. R. Sprang, *Prepr., Am. Chem. Soc., Div. Petrol. Chem.,* **15,** A65 (1970).

*Contributed by Fred R. Brown.

all samples analyzed by this technique, IR analysis has become a routine laboratory measurement. This proliferation of instrumentation and the applicability of the technique to so many different sample types are evidenced by the vast literature existing on the IR analysis of coal and coal reaction products.

This literature contains several reviews of the application of IR spectroscopy, and in some cases other instrumental techniques such as UV-visible and NMR, to coal and coal-derived materials.[540–544] As noted previously,[545] as the material to be analyzed becomes more complex on a molecular level, either as a homogeneous material consisting of large molecular units or as a heterogeneous material consisting of large numbers of different molecular units, the amount and the research value of the data obtained tend to decrease. Notwithstanding this lessening in the absolute value of IR data when the technique is applied to materials like coal, it does appear that more meaningful information can be acquired on the materials obtained from the thermal and/or chemical treatment of coal, (e.g., analysis of chromatographic fractions of coal hydrogenation products[546]).

4.14.2 Techniques of Measurement

4.14.2.1 *Sampling Techniques*

The pressed window technique has become the most popular method of preparing samples for IR analysis. It consists of dispersing a ground sample of the solid to be analyzed in an IR transmitting matrix, such as potassium bromide (KBr), and pressing the matrix into a window. This method is much simpler than preparing thin sections of coal; it also has the advantage of no spectral interferences, which normally occur when the mull technique (suspending the ground solid sample in a paraffin or halogenated-paraffin oil) is used. KBr pellets of coal have been used to study gas-solid reactions; the adsorption of water vapor by coal has been monitored.[547] The reproducibility of coal absorption bands using the KBr technique has been examined.[548] Though some results show spectral changes based on the sampling technique used,[549] most reported spectra are independent of sampling technique.[550] The problem of spectral changes caused by oxidation of samples subjected to long-term grinding or pulverization can be acute.[551] This is important in the analysis of extremely intractable materials such as graphite and activated carbons, but it does not appear to be such a problem with coal samples, though long-term periods of grinding should be avoided.

4.14.2.2 *Instrumental Techniques*

Most IR instrumentation used in coal analysis is of the double-beam type. That is, the energy from the IR radiation source is optically split, with one component becoming the sample beam and the other component the reference beam. The instrument determines the difference in IR absorption between the sample in the sample beam and any material placed in the reference beam. Thus if instrumental balance is maintained, gases or vapors from ambient air present in the instrument are compensated for and do not contribute to the recorded spectrum. Similarly, any contamination in relatively low concentration in the matrix, mulling agent, or window material used can be eliminated from the spectrum.[552] An evaluation of the relationship among in-

[540] Shuya Fujii, *Nenryo Kyokai-shi,* **38,** 442 (1959).

[541] R. A. Friedel, H. L. Retcofsky, and J. A. Queiser, U.S. Bur. Mines Bull. No. 640, 1967, 48 pp.

[542] Hans H. Oelert and Ernst A. Hemmer, *Erdoel Kohle, Erdgas, Petrochem.,* **23**(2), 87 (1970).

[543] Aleksandra Rusin and Amalia Tobiczyk, *Koks, Smola, Gaz,* **17**(2), 51 (1972).

[544] J. H. Le Roux, W. Zenker, and M. J. Heal, *Proc. Symp. Mol. Spectrosc., 1973,* Council of Scientific and Industrial Research Pretoria, South Africa, 16 pp.

[545] *Chemistry of Coal Utilization,* Supplementary Volume, H. H. Lowry, Ed., John Wiley and Sons, New York, 1963, p. 61.

[546] Fred R. Brown and J. A. Queiser, unpublished results.

[547] Robert A. Friedel, J. A. Queiser, and Eneo C. Moroni, *Anal. Chem.,* **43**(6), 761 (1971).

[548] V. Vecerikova and F. M. Kessler, *Sb. Pr. Horn. Ustavu Csav,* **4,** 208 (1961).

[549] N. A. Kekin, M. G. Sklyar, and T. S. Palaguta, *Proizvod. Koksa,* **2,** 10 (1973).

[550] *Chemistry of Coal Utilization,* Supplementary Volume, H. H. Lowry, Ed., John Wiley and Sons, New York, 1963, p. 65.

[551] D. M. Smith, J. J. Griffin, and E. D. Goldberg, *Anal. Chem.,* **47**(2), 233 (1975).

[552] N. A. Kekin, *Sb. Nauchn. Tr. Ukr. Nauchn.-Issled. Uglekhim. Inst.,* **17,** 94 (1965).

strumental parameters for coal analysis using the KBr window method has been made with the goal of optimizing these parameters.[553]

The attenuated total reflection (ATR) technique has been used to record the IR spectra of a coal extract,[554] a whole coal (94% C, anthracite),[554] the volatile products obtained from a shock-heated coal,[555] and the interaction products (acid-base) of coal and pyridine.[555] ATR is based on the reflection of radiation at the interface of the sample and an IR transmitting prism of higher refractive index. Good optical contact must be made between the sample and the prism, and with most samples of whole coals the difficulties associated with making this contact limit the application of the technique.[554] Since optical penetration of the sample by the source beam is only slight, any spectrum obtained by ATR is associated only with the composition of the surface of the sample in most intimate contact with the prism. Depending on the information sought, this can either be an advantage or disadvantage.

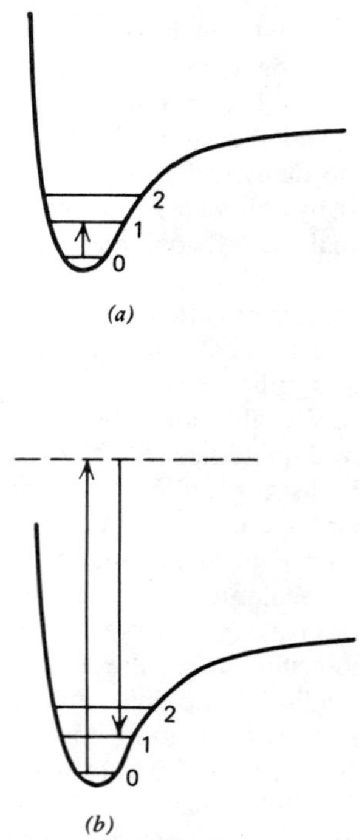

Figure 4.52 Fundamental processes associated with (*a*) IR and (*b*) Raman spectroscopies.[556]

4.14.3 Raman Spectroscopy

Raman spectroscopy is another technique that measures the vibrational energy spacings of molecular species. Whereas IR spectroscopy measures these levels directly through an absorption process, Raman spectroscopy measures the vibrational spacings through a scattering process. The fundamental processes involved are shown in Fig. 4.52. Raman spectra are recorded by measuring the differences in frequency of the monochromatic exciting line and the inelastically scattered photons (Raman bands). Formal selection rules for the two processes do exist, but for molecular complexes like coal, all vibrational transitions are formally allowed because of the lack of symmetry of the molecular species. However bands assignable to certain transitions may not occur because of low magnitudes of their respective intensity tensors.

Raman spectroscopy has been applied to whole coal samples.[557–560] All coals yield essentially the same spectrum (two bands at 1585 and 1360 cm^{-1}), with overall quality of the spectra increasing with coal rank (Fig. 4.53). Since the Raman spectra were recorded with a finely focused, intense laser beam as the source of monochromatic energy, the possibility that the recorded spectra are of carbonized or oxidized coal does exist.[557–560] One set of authors[560] prepared and recorded the Raman spectra of freshly obtained coal surfaces under an atmosphere of helium in an effort to prevent oxidation. Spectra similar to those for samples analyzed in air were obtained. In addition, samples were analyzed with the source laser beam defocused on the surface of the coal samples to minimize large

[553] V. Parausanu, I. Hogea, Fr. Barca, and M. Caramalau, *Chim. Anal. (Bucharest)*, **1**, 207 (1971).

[554] R. Bent and W. R. Ladner, *Fuel*, **44**, 243 (1965).

[555] R. A. Friedel and J. A. Queiser, *Prepr., Am. Chem. Soc., Div. Fuel Chem.*, **10**, 1 (1966).

[556] Fred R. Brown and Leo E. Makovsky, unpublished results.

[557] R. A. Friedel and G. L. Carlson, *Chem. Ind. (London)*, **40**, 1128 (1971).

[558] R. A. Friedel, J. A. Queiser, and G. L. Carlson, *Prepr., Am. Chem. Soc., Div. Fuel Chem.*, **15**, 123 (1971).

[559] R. A. Friedel and G. L. Carlson, *Fuel*, **51**, 194 (1972).

[560] L. E. Makovsky, P. Waldstein, and W. H. Edwards, *Nature (London)*, **231**, 154 (1971).

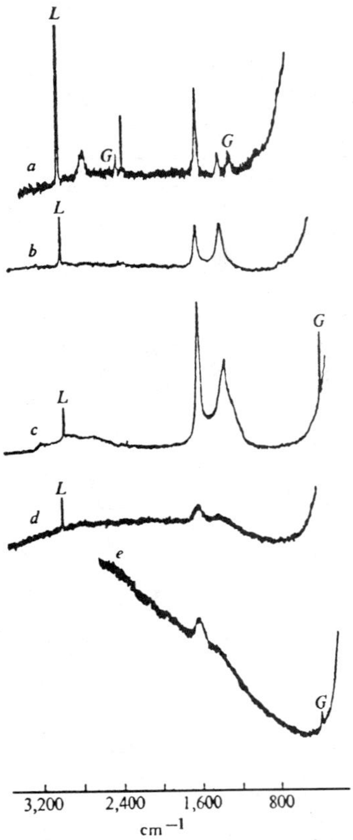

Figure 4.53 Raman spectra of graphite and coals: *a*, Graphite; *b*, metaanthracite; *c*, anthracite; *d*, bituminous; *e*, lignite. The exciting beam (4880 Å) is focused and the slit width is 9 cm^{-1}; scan rate, 50 cm^{-1}–min^{-1}; and time constant, 3 s.[560] L = emission line from laser; G = grating ghost.

temperature increases. Only an anthracite coal gave bands of sufficient intensity to monitor.

4.14.4 IR Spectra of Coals and Their Petrographic Components

The ease with which an IR spectrum of coal can be obtained is evidenced by the large number of papers concerning IR data on whole coals and/or their petrographic components. Table 4.23, which lists references on IR data of coals based on one parameter—geographic origin—indicates the wide use of IR spectroscopy.

As noted in the first Supplementary Volume to *Chemistry of Coal Utilization,*[561] most IR spectra of coal samples of the same rank are essentially the same. The range of spectral investigation in the IR has been expanded beyond that covered in the earlier work. Investigations into the lower mid-IR region have shown absorption from mineral matter present in coal,[541,562–565] but no discrete absorption assignable to organic structures. An investigation of a Pittsburgh vitrain in the far-IR region (25–125 μm; 400–80 cm^{-1}) resulted in a spectrum with a general absorption, but with no indication of specific bands.[541] As shown in Fig. 4.54[566], spectral changes as to band intensities do occur if coals of different ranks are compared. Qualitative explanations of these changes have been made.[567]

[561] *Chemistry of Coal Utilization,* Supplementary Volume, H. H. Lowry, Ed., John Wiley and Sons, New York, 1963, p. 65.

[562] Erich Preu, *Freiberger Forschungsh.,* **373A,** 91 (1965).

[563] Isac Blum, Fr. Barca, and Anca Bidulescu, *Bul. Inst. Politeh. "Gheorghe Gheorghiu-Dej" Bucuresti,* **30,** 59 (1968).

[564] Isac Blum, Fr. Barca, G. Musca, and Anca Bidulescu, *Freiberger Forschungsh.,* **473,** 73 (1969).

[565] Patricia A. Estep, John J. Kovach, Clarence Karr, Jr., Edward E. Childers, and Arthur L. Hiser, *Am. Chem. Soc., Div. Fuel Chem., Prepr.,* **13,** 18 (1969).

[566] S. Fujii and F. Yokoyama, *Fuel Soc. Jap.* **37,** 643 (1958).

[567] *Chemistry of Coal Utilization,* Supplementary Volume, H. H. Lowry, Ed., John Wiley and Sons, New York, 1963, p. 70.

Table 4.23 Coal Samples Subjected to IR Spectroscopic Analyses[a]

Sample Origin	Ref.
Japan	575, 576, 605, 611, 618
North Korea	599
Hungary	601, 613
India	617
East Germany	610
Rumania	553, 564, 620
Poland	577, 615, 619
Canada	616
U.S.S.R.	549, 552, 569, 583, 606–608
Australia	582
Thailand	568, 572
England	598, 612
United States	541

[a]This table is not an exhaustive survey of all coals analyzed by IR spectroscopy in the more than 150 papers dealing with the subject area. Rather, it indicates the wide application of the technique.

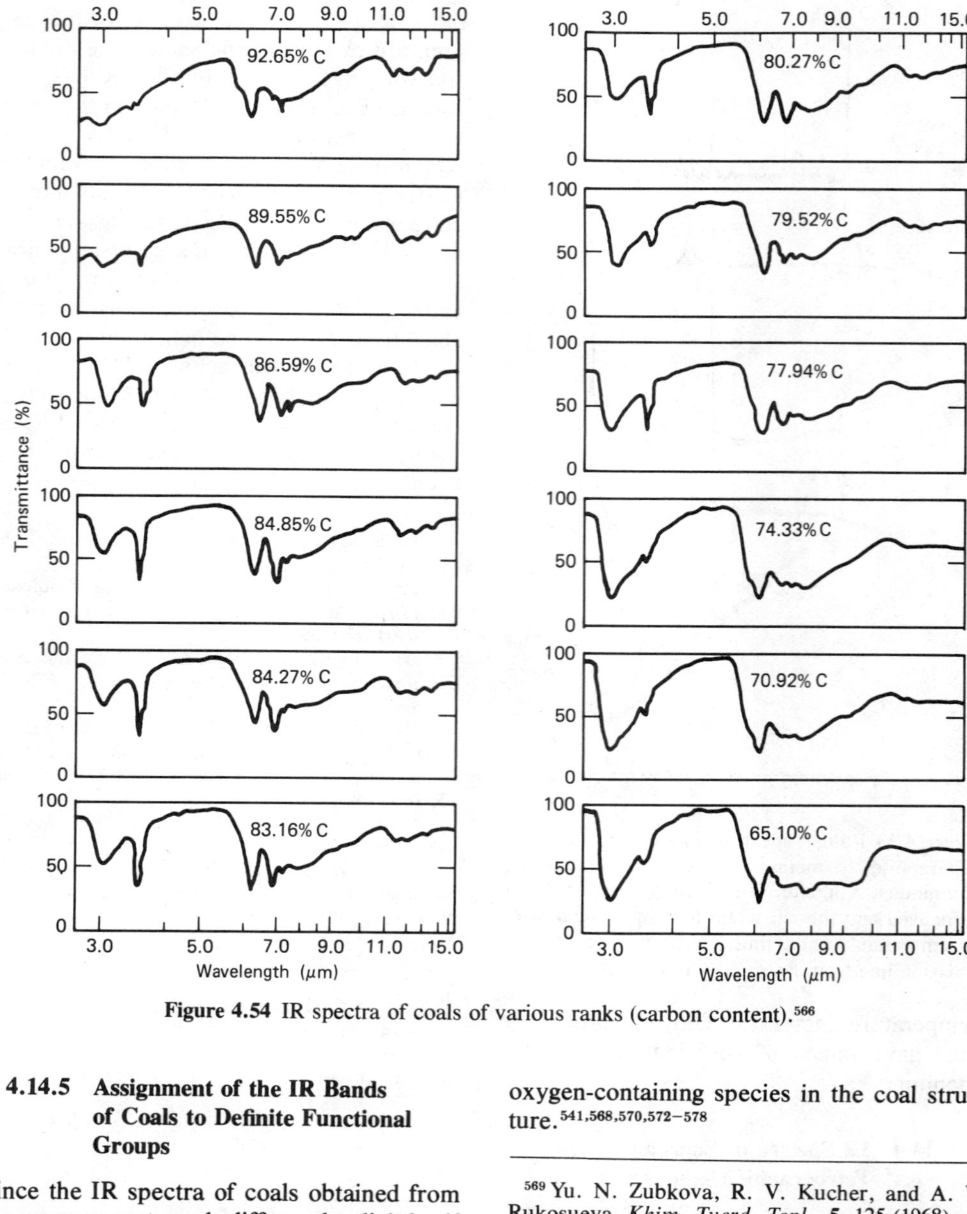

Figure 4.54 IR spectra of coals of various ranks (carbon content).[566]

4.14.5 Assignment of the IR Bands of Coals to Definite Functional Groups

Since the IR spectra of coals obtained from the more recent work differ only slightly, if at all, from the earlier work, it should not be surprising that the basic functional group assignments summarized in the first supplement have not materially changed. One controversy does remain: the assignment of the strong, specific absorption centered near 1600 cm^{-1}.

The alternatives for the assignment of this band are either that the band is due almost solely to the aromatic carbon skeleton[568–571] or that a significant contribution arises from oxygen-containing species in the coal structure.[541,568,570,572–578]

[568] H. S. Rao, P. L. Gupta, Mrs. F. Kaiser, and A. Lahiri, *Fuel,* **41,** 417 (1962).

[569] Yu. N. Zubkova, R. V. Kucher, and A. V. Rukosueva, *Khim. Tverd. Topl.,* **5,** 125 (1968).

[570] E. de Ruiter and H. Tschamler, *Brennstoff-Chemie,* **47,** 52 (1966).

[571] Yoshihiro Osawa, Jung Wei Shih, and Tetsunosuke Tsunoda, *Nenryo Kyokai-shi,* **49,** 898 (1970).

[572] L. Czuchajowski and G. J. Lawson, *Fuel,* **42,** 131 (1963).

[573] Shuya Fujii, *Fuel,* **42,** 341 (1963).

[574] Shuya Fujii, *Fuel,* **42,** 17 (1963).

[575] Yoshihiro Osawa, Hidehiko Sugimura, and Shuya Fujii, *Nenryo Kyokai-shi,* **48,** 703 (1969).

[576] Shuya Fujii, Yoshihiro Osawa, and Hidehiko Sugimura, *Fuel,* **49,** 68 (1970).

[577] Leszek Czuchajowski, *Arch. Gorn.,* **6,** 257 (1961).

[578] Leszek Czuchajowski, *Rocz. Chem.,* **36,** 747 (1962).

Results of the reduction of coal samples with lithium aluminum hydride ($LiAlH_4$) and the resulting influence on the intensity of the 1600 cm^{-1} band have been used as evidence in support of both assignments. The absence of any significant change in band intensity on the $LiAlH_4$ reduction or on carbonization is supportive of the lack of carbonyl contribution to the band.[568] But other work on coal reduction with $LiAlH_4$ did show a change in band intensity, which is supportive of a carbonyl contribution.[574] This change in band intensity results in a 10–30%[574] or 50%[573] contribution assignable to carbonyl structures. Since the amount of contribution from carbonyls varied over such a large range for various coals, and because the experimental parameters (most important, the solvent used) were not identical in the two sets of experiments, the results may not be inconsistent, although the generalized interpretations are.

Any carbonyl contribution to the 1600-cm^{-1} band must be unique because of the abnormally low frequency of the band. To reconcile the low frequency with a carbonyl-stretching assignment, hydrogen-bonded carbonyls[573–576,578,579] or, more formally, chelated rings[541,572,577] have been proposed. Model compound studies on hydroxyanthraquinones and on polycondensates of 5-hydroxy-1,4-naphthoquinone prepared in an alkaline medium possess an intense band at 1600 cm^{-1}.[572] These studies have been used to propose that chelated rings of more than six members contribute to the 1600-cm^{-1} band. Polycondensates of HCHO and model compounds that yielded strong 1600-cm^{-1} bands were used to conclude that hydrogen-bonded quinoid structures are present in coal.[577,578] Simple Hückel molecular orbital calculations have been used with a bond order concept to show that free carbonyls cannot contribute to the intensity of the band.[579] The calculations further indicate that strongly hydrogen-bonded carbonyls with bond orders of 0.81–0.85 can be expected to absorb in the 1600-cm^{-1} region. Molecular orbital calculations (by means of linear combination of atomic orbitals) using quinones and semiquinones have led others to conclude that semiquinone radicals contribute to the 1600-cm^{-1} band.[580,581]

Most IR investigations of coal use the KBr pellet sampling technique. Water intrinsic to the coal or present in the powdered KBr has been shown to contribute to the intensity of the 1600-cm^{-1} band.[582] This water can account for 15–20% of the total observed intensity.

A contribution to the 1600-cm^{-1} band from graphitic structures has been proposed.[558] This suggestion is based on the similarity of Raman spectra between high-rank coals and graphite. This work noted that the coal samples degraded under laser irradiation; therefore it is possible that these proposed "graphite structures" were the result of sample pyrolysis.

The "correct" assignment of the 1600-cm^{-1} band will undoubtedly show a contribution from various hydrocarbon and heteroatom functional groups. Aromatic structures, hydrogen-bonded carbonyls, physically and chemically adsorbed water, and possibly graphitic structures probably all contribute to the great intensity and broadness of the band. The relative contributions from the specific functional groups will vary with rank. The determination of these relative contributions in a given coal is basic to an understanding of the coal structure and its chemical reactivity.

4.14.6 IR Spectra of the Products and Residues of Thermally Treated Coals

IR measurements made on generic products obtained from the nonoxidative thermal treatment of coals, gases,[583] volatile liquids (oils, tars, and pitches),[555,584–594] and residues[571,577,595–598] have all been studied.

Gaseous products obtained at various tem-

[579] S. Fujii and H. Tsuboi, *Fuel,* **45,** 369 (1966).

[580] Lesjek Czuchajowski and Aleksander Erndt, *Bull. Acad. Pol. Sci., Ser. Sci. Chem.,* **18,** 35 (1970).

[581] Leszek Czuchajowski and Aleksander Erndt, *Rocz. Chem.,* **43,** 1451 (1969).

[582] R. A. Durie, Y. Shewchyk, and R. A. Friedel, *Fuel,* **46,** 53 (1967).

[583] I. V. Makarenko, B. V. Nelyubin, and G. P. Alaev, *Izv. Nauchn.-Issled. Inst. Nefte-Uglekhim. Sin. Irkutsh Univ.,* **12,** 79 (1970).

[584] F. P. Turenko, V. G. Sakharovskii, L. M. Turenko, and A. D. Baranskii, *Fiz.-Khim. Issled. Vzaimodeistviya Solei Shchelochnykh Metal. Rasplavakh Prod. Destruktsii Sapropelitov.,* 106 (1972).

[585] A. P. Bronshtein, S. S. Gitis, V. V. Platonov, and L. D. Khudoteplaya, *Sin., Anal. Strukt. Org. Soedin.,* **3,** 176 (1971).

peratures up to 900°C from whole coal and its petrographic components have been quantitatively determined by IR spectroscopy.[583] The productions of methane, water, and carbon dioxide were monitored. Rates of production, the temperatures of maximum rates of individual component production, and total yields were measured. The production of water over the range 300–900°C showed that at least two, and possibly three, successive processes were operative.

Analysis of the residues resulting from the thermal treatment of coals over a variety of temperature ranges yielded the following general conclusions:

1. More spectral changes occur at lower temperatures for lower rank coals.[571] Thus the 3300-cm^{-1} band, assigned to hydrogen-bonded —OH groups, disappeared in a vitrain and a lignite at 500 and 360°C, respectively.[577] Also, the intensity of the bands at 2860 and 2925 cm^{-1}, assigned to aliphatic C—H stretches, decreased more rapidly for lignite than for coals.[597]
2. The ratio of aromatic absorption at 3030 cm^{-1} to aliphatic absorption at 2920 cm^{-1}, measured either by integrated intensities[596] or by optical densities[571] (peak heights), increases with increasing temperature of carbonization. Similar qualitative results are obtained for resinites,[598] and a carbon disulfide extract of a pyridine extract of a heat-treated hard coal.[599]
3. This relative decrease of aliphatic C—H bonds also corresponds to an increase in absorption in the 700–900 cm^{-1} range.[571,597] Bands in this region are assigned to the HCC deformation (rocking) modes of aromatic systems.[600]
4. The 1600-cm^{-1} band persists in all coals, even at pyrolysis temperatures of 800°C.[571,577,596,597]

Coal-tar products obtained from the carbonization of coal can be analyzed directly by IR spectroscopy. The spectra obtained do not differ materially from those of the starting coals.[601] The IR absorption at 2 μm (5000 cm^{-1}) has been used to evaluate coal-tar pitches for coking properties[602] and the compressive strength of electrodes made from the pitch.[587] This background absorption was found to be related to the quinoline-insoluble fraction of the pitch, which determines many of its properties. IR spectra of pitches obtained from coal tar and from asphalt were measured.[588] Based on the ratio of aromatic to aliphatic bonds, it was concluded that the former contained primarily aromatic compounds, and the latter almost entirely aliphatic compounds.

The real utility of IR spectroscopy in coal science is most appropriately shown when the technique is used in conjunction with other instrumental techniques and with distillation, solvent, and chromatographic techniques for the characterization of liquids obtained from either thermal or chemical treatment of coal. Qualitative IR analyses for compound types have been performed on coal tars separated into fractions using silica gel column chromatography.[585,586] IR combined with more detailed separation techniques can provide a method of identification

---

[586] Z. A. Rumyantseva, E. P. Buryakova, Z. M. Perednikova, I. N. Grigina, and N. M. Karavaev, *Khim. Tverd. Topl.*, **3**, 94 (1973).

[587] K. A. Kini, *J. Indian Chem. Soc.*, **50**, 616 (1973).

[588] A. B. Moustafa, *J. Oil Colour Chem. Assoc.*, **57**, 44 (1974).

[589] I. A. Zehcheiko and I. A. Shaks, *Tr. Vses. Neft. Nauchn.-Issled. Geologorazved. Inst.*, **345**, 79 (1973).

[590] F. P. Turenko, N. A. Kuz'mina, and I. D. Shevchenko, *Khim. Tverd. Topl.*, **5**, 106 (1975).

[591] Z. I. Noskova and L. F. Lipatova, *Khim. Prod. Koksovaniya Uglei Vostoka SSSR*, **6**, 237 (1970).

[592] Horst Thielemann, *Mikrochim. Acta*, **5**, 717 (1971).

[593] Miroslav Janik, *Chem. Prum.*, **20**, 467 (1970).

[594] D. I. Tsekhovol'skaya, *Khim. Tverd. Topl.*, **5**, 125 (1970).

[595] R. M. Elofson, *Can. J. Chem.*, **35**, 926 (1957).

[596] N. A. Kekin and A. M. Miroshnichenko, *Khim. Tverd. Topl.*, **2**, 134 (1970).

[597] I. Blum and V. Popescu, *Cercet. Met., Inst. Cercet. Met., Bucharest*, **11**, 991 (1970).

[598] Duncan G. Murchison, in Am. Chem. Soc. Adv. Chem. Ser., No. 55, 1966, p. 307.

[599] Beum Sun Li and Chang Kyung Boo, *Chosun Kwahakwon Tongbo*, **5**, 34 (1964).

[600] *Chemistry of Coal Utilization*, Supplementary Volume, H. H. Lowry, Ed., John Wiley and Sons, New York, 1963, p. 69.

[601] *Chemistry of Coal Utilization*, Supplementary Volume, H. H. Lowry, Ed., John Wiley and Sons, New York, 1963, p. 73.

[602] P. L. Walker, Jr., and D. O. Baumbach, *Proc. 5th Carbon Conf.*, Vol. 2, Pergamon Press, New York, 1963, p. 175.

and quantification of the specific compounds pyrene and fluoranthene, present in coal-tar fractions.[591] Similarly, IR and thin-layer and gas-liquid chromatography were used to identify ten specific phenols present in the 228–280°C fraction of a medium-temperature coal tar,[603] and nine phenols (of which six were determined quantitatively) in the xylenol fraction from a coal-tar distillation.[592]

4.14.7 IR Spectra of Coal Extracts

Table 4.24 lists the solvent systems typically used for the preparation of coal extracts for instrumental analyses. Table 4.23 gave some of the coals, by geographic origin, that have been solvent extracted and analyzed.

In general, the larger the percentage of extraction obtained with a given solvent system, the more closely the IR spectrum compares with that of the precursor coal. Thus pyridine extracts, which can represent up to 30% of the original coal sample, give spectra very similar to those of the original coal, though some spectral differences in the CH-stretching region have been noted.[541] Even though similar spectra are obtained for whole coals and their pyridine or other high-yield solvent extracts, it is obvious that since these yields are still low, care must be taken in any attempt to analyze quantitatively the data obtained on the coal extracts. This similarity of spectra despite low overall yields is another reflection of the limitations of quantitative IR spectroscopy in the analysis of complex materials. At best, these extracts may yield data indicating certain structural types that are present in coal, but stating that "the extracts [from high-yield solvents] . . . may be taken as representative of the organic matter present in the original coal"[604] may be unwise.

[603] A. P. Lapan, V. G. Lavrent'ev, T. G. Paramonova, M. V. Efimova, and I. P. Titov, *Khim. Pererab. Tverd. Topl.*, 156 (1973).

Table 4.24 Solvent Systems Typically Used in Preparing Coal Extracts for Instrumental Analyses

Solvent	Ref.
Chloroform	606–608, 612
Pyridine	604, 605, 611, 614, 618
Benzene + ethanol	604, 613, 620
Basic extraction	604
Benzene + ethanol + base	620

The "solutions" obtained from the extraction process may not be optically clear; they also tend to be unstable.[604] Extract materials freed from solvent may not redissolve on readdition of solvent. Whether these properties result from undissolved materials being carried over in the extraction process or from oxidative or other chemical reactions of the materials, they again reflect the problems associated with extrapolating the analytical results back to the original coal.

A qualitative IR analysis of the pyridine extracts of a variety of coals[605] yielded the following results:

1. The intensity of the hydrogen-bonded —OH stretching band at 3300 cm^{-1} decreased linearly with increasing carbon content of the original coal.
2. The intensity of the aliphatic CH deformation modes at 1450 cm^{-1} decreased with increasing carbon content. Other bands resulting from aliphatic components did not change.
3. The amount of aromatic hydrogen increased with increasing carbon content.
4. The intensity of the 1600-cm^{-1} band decreased with increasing carbon content.

Similar results can be obtained for chloroform extracts.[606,607] These spectral changes that result for different ranks of coal have been related to distinct metamorphic changes through the series.[608]

Many studies have been directed toward quantifying the IR results for whole coals[570,609–611] and coal extracts. The total

[604] *Chemistry of Coal Utilization,* Supplementary Volume, H. H. Lowry, Ed., John Wiley and Sons, New York, 1963, p. 75.

[605] Yoshihiro Osawa, Jung Wei Shih, and Tetsunosuke Tsunoda, Nenryo Kyokai-shi, **50,** 31 (1971).

[606] E. A. Glebovskaya and T. N. Mel'tsanskaya, *Khim. Tverd. Topl.*, **4,** 103 (1967).

[607] I. A. Shaks, A. I. Ginzburg, and T. N. Mel'tanskaya, *Khim. Tverd. Topl.,* **5,** 22 (1970).

[608] E. A. Glebovskaya, V. V. Ivantsova, N. I. Babinkova, M. I. Kalmykov. T. N. Mel'tsanskaya, and L. I. Sarbeeva, *Tr. Vses. Neft. Nauchn.-Issled. Geologorazved. Inst.*, **353,** 26 (1974).

[609] V. Vecerikova and M. F. Kessler, *Freiberger Forschungsh.*, **388,** 223 (1967).

[610] Henning H. Oelert, *Erdoel Kohle,* **18,** 876 (1965).

[611] Ryoichi Yoshida, Yousuke Maekawa, Susumu Yokoyama, and Gen Takeya, *Nenryo Kyokai-shi,* **54,** 332 (1975).

hydrogen in bright coal present in methyl groups was determined to be 15–25% based on the IR intensity of the 1375-cm^{-1} band of chloroform extracts and from oxidation studies.[612] Qualitatively it was noted that total methyl group content decreased with increasing rank. The total yield of tar obtained by extracting lignite samples with 1 : 1 benzene/ethanol was directly related to the IR absorbance at 2925–2930 cm^{-1}.[613]

Quantitative results for the distribution of aromatic hydrogens have been obtained for vitrinite extracts using the absorption bands at 10.8–14.7 μm (925–680 cm^{-1}).[614] The vitrinite, a pyridine extract, a benzene extract of the pyridine extract, a petroleum ether extract of the benzene extract, and all residues were studied. All samples gave three bands of varying intensity in the spectral region of interest. Average integral absorption coefficients were established for these samples from the spectra of more than 350 model compounds.

Combining the results from IR spectroscopy with those from proton NMR has provided methods for determining the ratio of aromatic to aliphatic hydrogens,[611,615–618] the distribution of aliphatic hydrogen into CH, CH_2, and CH_3 groups,[541,617] and the aromaticity[618] of coal extracts. The NMR data provide more accurate calibration data for quantifying the IR results.

The extracts of coal extracted at elevated temperatures[619] and elevated pressures[620] using solvent containing tetralin have been examined by IR. Because of the known hydrogen-transferring ability of tetralin at elevated temperature and pressure, it is most probable that the extracts contained large quantities of coal hydrogenation products. That tetralin did chemically react with the coal was noted by the decrease in tetralin and the increase in naphthalene following extraction.[620]

4.14.8 IR Spectra of Mineral Matter

Mineral matter present in whole coals and various float-and-sink fractions can be monitored by IR spectroscopy.[545,562–565] Spectra of pure minerals, and their relationship to the IR spectra of coals have been discussed.[562–565]

Quantitative IR methods have been established for the determination of quartz content, which is a known health hazard, in respirable coal dust.[621–625] Either of the quartz absorption bands at 780 and 795 cm^{-1}, or a combination of both, is used for this determination. Sensitivity to $\leqslant 10\ \mu g$ of quartz is obtainable.[622] Other minerals for which quantitative measurements can be made are kaolinite, using its 913-cm^{-1} band, and muscovite, using its 555-cm^{-1} band.[621]

4.14.9 IR Spectra of Irradiated Coals

The IR spectra of whole coal and KBr pellets of whole coal, on exposure to gamma rays (480 Mrad) in an argon atmosphere, showed increased intense absorption at 1420–1460, 848, and 690 cm^{-1}.[626] These bands were assigned to *cis*-disubstituted and trisubstituted vinyl groups. The intensity of these bands decreased, and concurrently bands assigned to OH groups increased on exposure of the samples to air.

[612] R. Bent, W. K. Joy, and W. R. Ladner, *Fuel*, **43,** 5 (1964).

[613] Sandor Sipos and Eva Sipos-Kedves, *Szegedi Tanarkepzo Foiskola Tud. Kozlem,* 157 (1971).

[614] H. Tschamler and E. de Ruiter, *Brennstoff-Chemie,* **44,** 280 (1963).

[615] J. F. M. Oth, E. de Ruiter, and H. Tschamler, *Brennstoff-Chemie,* **42,** 378 (1961).

[616] H. Tschamler and E. de Ruiter, *Brennstoff-Chemie,* **43,** 212 (1962).

[617] J. F. M. Oth and H. Tschamler, *Fuel,* **42,** 467 (1963).

[618] Susumu Yokoyama, Mitsuomi Ito, and Gen Takeya, *Nenryo Kyokai-shi,* **54,** 340 (1974).

[619] Emanuel Rusin, *Pr. Nauk Inst. Chem. Technol. Nafty Wegla Politech. Wroclaw.,* **18,** 3 (1973).

[620] G. J. Kulcsar, J. Vodnar, L. Janku, J. Kohan, and E. Hamburg, *Stud. Univ. Babes-Bolyai Cluj, Ser. Chem.,* **8,** 457 (1963).

[621] M. Gade and M. Reisner, "Pneumoconiosis," in *Proc. 3rd Int. Conf.,* H. A. Shapiro, Ed., Oxford University Press, Capetown, South Africa, 1970, p. 636.

[622] Robert W. Freedman, Sam Z. Toma, and Helen W. Lang, *Am. Ind. Hyg. Assoc. J.,* **35,** 411 (1974).

[623] Donald J. Larsen, Laura J. Von Doenhoff, and John V. Crable, *Am. Ind. Hyg. Assoc. J.,* **33,** 367 (1972).

[624] Samuel A. Goldberg and Murray Jacobson, U.S. Bur. Mines Inf. Circ. No. 8545, 1972, 10 pp.

[625] B. A. Maguire and D. Barker, Saf. Mines Res. Estab. (U.K.) Res. Rep. No. 278, (1971), 16 pp.

[626] O. I. Egorova, *Khim. Tverd. Topl.,* **2,** 28 (1975).

4.14.10 IR Spectra of Chemically Treated Coals and Coal Extracts

IR spectra of the oxidation products of coal have continued to be of interest to numerous investigators. Spectral data for coal (or products derived from coal) oxidized by air,[627–630] air–ammonia,[631] oxygen,[632,633] hydrogen peroxide,[629,634,635] nitric acid,[630,634,635] alkaline nitrobenzene,[636] and alkaline potassium permanganate[637] have been obtained.

The resulting spectra invariably showed increased carbonyl absorption at about 1700 cm^{-1} associated with the formation of carboxylic acid groups. Other structural changes deduced from the spectra were the formation of phenolic and/or alcoholic hydroxy groups[629,631,635,636] and ethers.[629,632] The lignites oxidized with a 3 : 1 ammonia-to-air volumetric ratio indicated that nitrogen was incorporated as amide groups in stable heterocyclics.[631]

Other chemical treatments of coal or humic acids investigated include methylation and acetylation,[638] sulfomethylation,[639] and acid (HCl-WF) treatments[640] for removal of mineral matter.

Hydrogenation products of coals have been studied.[630,641–642] Primarily because of the possibility of their use as a substitute for petroleum products, liquids derived from the hydrogenation of coal will undoubtedly be the subject of more extensive investigations. Thus asphaltenes (benzene-soluble, hexane-insoluble materials) and their acidic and basic components have been studied.[643] An IR investigation of 13 subfractions obtained by solvent and chromatographic separation techniques of a coal hydrogenation product more adequately shows the utility of IR spectroscopy in coal science.[556]

4.15 ULTRAVIOLET-VISIBLE SPECTROSCOPY*

UV-visible spectroscopy has been an invaluable tool in the study of unsaturated and, in particular, aromatic hydrocarbons. It is widely used not only in obtaining structural information but also in qualitative and quantitative analysis. Since coal and coal-derived materials are usually highly aromatic, one would expect UV-visible spectroscopy to be very useful in the study of such materials. This is indeed true of the chemicals ultimately derived from coal, but the applicability of the technique to whole coals and subfractions without considerable separation has not been particularly useful.

The most difficult problem in interpreting the UV and visible spectra of coal and related materials is posed by the extreme complexity of the mixtures involved. As a consequence, coals, coal extracts, and products derived from the oxidation, carbonization, and hydrogenation of coal exhibit UV-visible spectra that are essentially continuous throughout the 200–750 nm region. The spectra of all these materials indicate an absorbance that monotonically increases with decreasing wavelength and exhibits little if any structure.

Several attempts have been made to reconstruct the UV spectra of coal as a sum of

[627] Shuya Fujii, *Kogyo Kagaku Zasshi,* **62,** 1574 (1959).

[628] N. A. Kekin, M. A. Stepaneko, and N. I. Matusyak, *Sb. Nauchn. Tr. Ukr. Nauchn.-Issled. Uglekhim. Inst.,* **23,** 146 (1971).

[629] L. N. Nefedova, N. K. Larina, and V. I. Kasatochkin, *Khim. Tverd. Topl.,* **5,** 59 (1970).

[630] Peter Heinz Berthold and Hans Koehler, *Freiberger Forschungsh.,* **365A,** 37 (1965).

[631] Antoni Andrzejak and Helena Wachowska, *Poznan. Tow. Przyj. Nauk, Pr. Kom. Mat-Przyr., Pr. Chem.,* **12**(6), 105 (1971).

[632] M. G. Reznik and B. Ya. Chekhovskoi, *Khim. Tverd. Topl.,* **5,** 35 (1970).

[633] Yves Grillet and Henri Guerin, *C.R. Acad. Sci., Paris, C,* **271**(19), 1156 (1970).

[634] L. Czuchajowski, *Fuel,* **39,** 377 (1960).

[635] Leszek Czuchajowski, *Arch. Gorn.,* **5,** 359 (1960).

[636] Bedri Bozkurt, Alec F. Gaines, Ayten Ogut, Yuda Yurum, and Aral Olcay, *Fuel,* **51**(1), 81 (1973).

[637] D. T. Zabramnyi, S. Nasritdinov, and O. I. Pobedorosteseva, *Uzb. Khim. Zh.,* **16**(3), 65 (1972).

[638] S. P. Mathur, *Soil Sci.,* **113**(2), 136 (1972).

[639] E. Spiros, *Fuel,* **49**(3), 336 (1970).

[640] G. Bergman, G. Huck, J. Karweil, and H. Luther, *Brennstoff-Chemie,* **38,** 193 (1957).

[641] S. Fujii, *Nenryo Kyokai-shi,* **43**(443), 181 (1964).

[642] J. Egbal and S. Sarkar, *Indian J. Technol.,* **9**(8), 313 (1971).

[643] Fred R. Brown, Sidney Friedman, Leo E. Makovsky, and F. K. Schweighardt, *Appl. Spectrosc.,* **31**(3), 241 (1977).

* Contributed by Dennis H. Finseth.

spectra of a suitably chosen set of model compounds.[644–646]

In the limit of noninteracting absorbers, the absorbance of an n-component mixture will be the sum of the individual absorbances of the components:

$$A_{\text{total}} = \sum_{i=1}^{n} A_i \qquad (4.20)$$

This expression is widely used in the analysis of simple mixtures in solution, but it is not expected to be rigorously valid at high concentrations. The relationship is particularly suspect in strongly interacting mixtures such as whole coals.

A number of workers have used this spectrum reconstruction method to estimate the aromatic content of coals and coal-derived materials.[646,646–648] In these investigations both the KBr disk[647,648] and the thin-section[649] sampling techniques were used. Qualitatively, the spectra obtained using these two techniques are essentially the same. The absorbance smoothly increases with decreasing wavelength, and the only structure evident is a consistent shoulder near 265 nm. Similar spectra can be obtained by dispersing the finely ground solid in a polyethylene matrix.[650,651] The UV spectrum of a Pittsburgh vitrain (84.0% C) in a KBr disk was interpreted as being consistent with a low concentration (< 25%) of two- to four-ring aromatics.[647] This conclusion was reached by comparing the coal spectrum with spectra of 14 two- to four-ring aromatic model compounds. A spectrum of the same Pittsburgh vitrain, obtained using the ultrathin (500–2000 Å) section sampling technique, yielded a specific extinction coefficient that was higher by a factor of 3–4 than the KBr disk result.[652] This finding is naturally consistent with an estimated aromatic content that is higher by the same factor. In another study the specific absorption coefficient of a Dinnington vitrain (85.0% C) was measured using a polarized light technique.[653] A value of the specific absorption coefficient consistent with that of the ultrathin section measurement above was determined.

A spectral study of a low-temperature carbonization product of black durains (83.1 and 85.2% C) run in chloroform solution concludes that the UV-visible spectrum is consistent with a high aromatic content.[644] In this work the spectrum of the carbonization product is compared with the composite spectrum of 23 aromatics with an average of four rings. The specific extinction coefficient found in this study is in reasonably good agreement with that found for the ultrathin section of Pittsburgh vitrain.[652] The result on the black durain is also in general agreement with results obtained on carbon disulfide extracts of six coals of varying rank with carbon contents (maf) ranging from 76.3 to 90.4%. In this work it was concluded that for five of the six coal extracts, the UV-visible spectrum was consistent with an extremely high aromatic content (in excess of 100% in some cases).[654]

The discrepancies among conclusions on aromatic content using this spectrum reconstruction method may well be due to differences in sampling technique. The differences among the specific extinction coefficients determined using KBr disks, thin sections, and solutions are not yet resolved; but from the point of view of quantitative accuracy and reproducibility, the dilute solution method is probably far superior. This method unfortunately suffers from the disadvantage of not being applicable to whole coals. In general, the estimation of aromatic character from the UV-visible spectrum should be viewed with appropriate caution. In a strongly interacting system like coal, it would not be at all unexpected to find that the absorbance of the

[644] R. Bent and W. R. Ladner, *Fuel,* **39,** 479 (1960).

[645] V. Sandor, E. de Ruiter, and H. Tschamler, *Erdoel Kohle,* **15,** 713 (1962).

[646] R. A. Friedel and J. A. Queiser, *Fuel,* **38,** 369 (1959).

[647] R. A. Friedel, H. L. Retcofsky, and J. A. Queiser, U.S. Bur. Mines Bull. No. 640, 1967.

[648] R. A. Friedel, *Brennstoff-Chemie,* **44**(1), 23 (1963).

[649] S. Ergun, J. T. McCartney, and R. E. Walline, *Nature (London),* **187,** 1014 1960.

[650] R. A. Friedel, *Proc. 5th Conf. Carbon,* Pergamon Press, New York, 1963, p. 149.

[651] F. R. McDonald and G. L. Cook, *Appl. Spectrosc.,* **15,** 110 (1961).

[652] J. T. McCartney and S. Ergun, U.S. Bur. Mines Bull. No. 641 (1967).

[653] L. A. Gilbert, *Fuel,* **41,** 351 (1962).

[654] H. L. Retcofsky and R. A. Friedel, in *Spectrometry of Fuels,* R. A. Friedel, Ed., Plenum Press, New York, 1970. p. 37.

mixture is not equal to the sum of the absorbances of the individual components. This procedure for estimating aromatic content neglects the absorbance due to species other than aromatics; therefore at best it can only set an upper bound on the aromatic content. The absorbance due to free radicals, metal complexes, and large substituted heteroaromatics, to name a few, must be considered. Unfortunately the contributions of such components cannot be quantified, since their identity in coal is largely unknown.

UV-visible spectra have also been used in another way to estimate the size of aromatic clusters. When extracts of coal and various chromatographic fractions thereof are dissolved in concentrated sulfuric acid, an intense, broad absorption maximum appears in the visible range near 485 nm.[655] This absorption is absent in solutions in other solvents and is reversible on addition of water. It has been attributed to the formation of aromatic carbonium ions and radical ions. The similarity of the wavelength maxima, intensity, and band width of this new band in various extracts and subfractions has been interpreted as evidence for the similarity of the aromatic cluster sizes in all these extracts.

The ultrathin section technique has been applied not only to whole coal, but to graphite as well. A UV maximum was observed in graphite near 250 nm.[656] This absorption band is assigned to the transition of an electron from the valence band to the conduction band in the solid, and this experimental result agrees well with the theoretically predicted energy for this transition.[657] The band observed in graphite is in the same spectral region as the 265-nm shoulder observed in coal spectra, which is commonly explained as being characteristic of aromatic systems.

The presence of high concentrations of free radical species has been suggested as an alternative explanation for the visible absorption of coal. In support of this hypothesis, a correlation has been observed between the free spin concentration and absorptivity of asphaltene fractions from coal hydrogenation products.[647] The intense absorption of many model radical compounds together with the high free spin concentration observed via ESR in many coal-derived products suggest that the contribution of such systems to the overall absorptivity of coal may be large. Further investigation of this possibility is needed.

Other possible contributors to the UV and visible spectra of coal are metal complexes and molecular complexes of the charge transfer type.[647] Little experimental work has been done to establish what contribution, if any, these make to the observed absorptivity.

UV-visible spectroscopy has been used analytically in a large number of laboratories to characterize coal extracts[655,658–661] and materials derived from coal by way of carbonization,[644] oxidation,[662–664] and hydrogenation.[665–668]

In most cases these materials as produced are very complex mixtures and exhibit spectra that are qualitatively very similar to those of coal; again UV-visible spectra are not very useful. However after separation of the mixtures, usually chromatographically, the UV has been very useful in identifying and quantifying individual compounds and molecular classes. Many of the pure compounds separated from coal-derived materials have been identified from their characteristic UV spectra.

The analysis of multicomponent aromatic mixtures derived from coal is routinely accomplished by way of the use of UV-visible

[655] E. de Ruiter and H. Tschamler, *Fuel,* **41,** 491 (1962).

[656] S. Ergun, J. T. McCartney, and R. E. Walline, *Fuel,* **40,** 109 (1961).

[657] C. A. Coulson and R. Taylor, *Proc. Phys. Soc. (London),* **A65,** 815 (1952).

[658] A. Halleux and H. de Greef, *Fuel,* **42,** 185 (1963).

[659] M. M. Roy, *Brennstoff-Chemie,* **47,** 65 (1966).

[660] A. V. Terebenina, *Khim. Tverd. Topl.* **6,** 40 (1971).

[661] H. L. Retcofsky and R. A. Friedel, *Fuel,* **37,**, 487 (1968).

[662] F. Micheel, J. Bernsmann, G. Draeger, and J. P. Riebe, *Brennstoff-Chemie,* **45**(12), 360 (1964).

[663] M. Bitz and B. Nagy, *Anal. Chem.,* **39**(11), 1310 (1967).

[664] S. M. Manskaya and L. A. Kodina, *Probl. Eqvolyutsionnoi Tekhn. Biokhim., Akad. Nauk. SSSR, Inst. Biokhim.,* 101 (1964).

[665] A. Rusin, *Chem. Anal. (Warsaw),* **19**(1), 145 (1974).

[666] A. Rusin, *Chem. Anal. (Warsaw),* **19**(5), 1021 (1974).

[667] A. Rusin and J. Kulczycka, *Chem. Anal. (Warsaw),* **20**(6), 1177 (1975).

[668] J. Kulczycka and A. Rusin, *Koks Smola Gaz,* **20**(2), 34 (1975).

spectroscopy.[669–671] This method of analysis is applicable only for mixtures having a small number of components, since the error quickly becomes large as the number of components increases.

UV-visible spectroscopy has been applied to investigations of coal-derived humic acids.[664,672–677] It was found that spectroscopic methods for the determination of stability constants in these systems yield values consistent with those determined by other methods.[676]

UV-visible spectroscopy will probably find its greatest utility in the characterization of specific compounds and molecular classes that are separated from coal or coal-derived materials. The application of UV-visible spectroscopy in conjunction with high-performance chromatography and other separation methods undoubtedly will be used widely in identifying and quantifying individual components of coal-derived materials.

A number of reviews that discuss the applications of UV-visible spectroscopy in coal science have appeared since 1958.[678–680] The interested reader may find these useful.

[669] R. J. Goodyear and F. Vernon, *Anal. Lett.*, **3**(6), 299 (1970).

[670] H. A. Bonnett and A. Bortoli, *Anal. Chem.*, **32**, 1153 (1960).

[671] A. P. Kolesov, M. P. Grebenshehikova, S. G. Bogomolov, and E. I. Gosbunova, *Materialy Tret'ego Ural'sk. Soveshch. po Spektroskopii*, Inst. Fiz. Metal. Acad. Nauk SSSR, Komis. po Spektroskopii, Sverdlovsk, 1960, p. 157.

[672] D. T. Zabramnyi, S. Nasritdinov, and O. I. Pobedonostseva, *Uzb. Khim. Zh.*, **16**(3), 65 (1972).

[673] M. G. Korniolva and L. A. Medvedeva, *7th Ural Konf. Spektrosk.*, No. 2, 1971, p. 149.

[674] I. D. Komissarov, L. F. Loginov, and I. N. Streltsova, *Nauchn. Tr. Tynmen. Selskokhoz Inst.*, No. 14, 75 (1970).

[675] M. Susaki, H. Oka, and Y. Onon, *Kitami Kogyo Tanki Daigaku, Kenkyu Hokoku*, **2**(1), 73 (1967).

[676] F. D. Ovcharenko, S. A. Gordienko, and T. F. Glushchenko, *Agrokem. Talajtan*, **20**(1–2), 3 (1971).

[677] M. M. Kononova, *Soil and Organic Matter*, Pergamon Press, New York, 1961.

[678] H. H. Oelert and E. A. Hemmer, *Erdoel Kohle, Erdgas, Petrochem.*, **23**(2), 87 (1970).

[679] R. A. Friedel, in *Encyclopedia of Spectroscopy*, G. L. Clark, Ed., Reinhold, New York, 1960, p. 414.

[680] N. K. Larina and V. I. Kasatochkin, *Khim. Tverd. Topl.*, No. 4, 49 (1969).

4.16 OPTICAL FLUORESCENCE*

Although fluorescence spectroscopy is, in general, very useful for the analysis of large aromatic compounds, the technique has been used very little in investigations of coal and coal-derived materials. Since fluorescence involves two independently variable experimental parameters—excitation and emission wavelength—it may be, in principle, a more selective technique than absorption spectroscopy. This improved discrimination has been used in studies of relatively simple mixtures but has not been thoroughly investigated in coal-derived systems.

Whole coal itself does not fluoresce to any appreciable extent,[681] but coal extracts and coal hydrogenation products have been found to fluoresce very strongly in solution. The fluorescence spectra of pyridine extracts of five coals ranging from 76.3 to 90.4% C were obtained together with spectra of coal hydrogenation asphaltenes and neutral oil, and a coal-tar pitch.[682] The spectra of these materials were compared with spectra of 11 model aromatic hydrocarbons, using emission wavelength maximum, band width, and fluorescent intensity. The spectra of four of the five extracts as well as the asphaltene and the pitch were very similar, having a broad emission band centered near 440 nm (315-nm excitation) and a high fluorescent intensity. The extract of the coal with the lowest carbon content (76.3%) exhibited a significantly lower intensity and a shorter wavelength emission maximum (390 nm). The neutral oil spectrum also indicated a lower intensity and a much shorter emission wavelength maximum (340 nm). These data on the oil were interpreted as indicating a high concentration of three- and four-ring aromatics.

In this study the fluorescent intensity of the pyridine extracts was found to increase approximately linearly with increasing carbon content of the parent coals. An attempt was made to use this intensity to estimate the aromaticity of the extracts, but the result was inconsistent with aromaticities determined by other means. Other workers have

*Contributed by Dennis H. Finseth.

[681] R. A. Friedel and H. L. Gibson, *Nature (London)*, **211**, 404 (1966).

[682] H. L. Retcofsky, T. J. Brendel, and R. A. Friedel, *Nature, Phys. Sci.*, **240**(97), 18 (1972).

used the fluorescence of benzene extracts of various coals for classification purposes.[683] In this study the fluorescence of the extract was greatest for bituminous coals and least for brown coals.

An unusual, long-wavelength fluorescence of coal hydrogenation products has been observed in the IR when using blue-green excitation.[682] The powdered asphaltene from a 400°C hydrogenation of Pittsburgh coal and the same asphaltene in benzene solution were found to fluoresce strongly in the IR with intensities comparable to that of the mineral greenockite (native cadmium sulfide). The source of this fluorescence is as yet undetermined, but it is suggested that either free radicals or charge transfer species may be involved.

Fluorescence spectroscopy has not been extensively applied to the analysis of coal-derived materials, but several workers have reported on its application to petroleum-derived materials of very similar composition, in particular high-boiling (400–500°C) petroleum distillates.[684,685] These materials contain many compound types in common with coal hydrogenation products. In these studies fluorescence spectroscopy was applied not to the analysis of the whole material, but rather to chromatographic fractions that were designed to yield compound types. The utility of fluorescence spectroscopy was demonstrated for the determination of molecular types, and good agreement was obtained between the spectra of model compounds and those of the chromatographic fractions. Characteristic spectra of carbazoles, 11*H*-carbazoles, 7*H*-carbazoles, chrysenes, coronenes, and benzo[*ghi*]perylenes were reported.[684]

The extremely high sensitivity of fluorescent methods has been used to determine quantitatively small concentrations of coal-derived polynuclear aromatics in a variety of systems. This high sensitivity is especially important in the detection and quantification of compounds such as benzo[*a*]pyrene, which have been shown to be health hazards. Sensitivities for the detection of benzo[*a*]pyrene have been reported in the range 10^{-9}–10^{-11} g/ml.[686,687] The sensitivity and selectivity of the fluorescence technique make it an attractive one, and in the future fluorescence spectroscopy may find greater application in the analysis of separated fractions of coal-derived products. Improved experimental techniques such as the direct analysis of thin-layer chromatographic plates may also extend the utility of the technique.

[683] A. Lissner, *Ber. Verh. Saechs. Akad. Wiss. Leipzig, Math-Naturw. Kl.*, **104**(2) (1960), 19 pp.
[684] J. F. McKay and D. R. Latham, *Anal. Chem.*, **44**(13), 2132 (1972).
[685] H. V. Drushel and A. L. Sommers, *Anal. Chem.*, **38**, 19 (1966).

4.17 ELECTRON SPECTROSCOPY FOR CHEMICAL ANALYSIS*

Bombardment of a solid material with soft (e.g., Al Kα – 1486.6 eV) X rays can result in electrons being ejected from the constituent atoms of the solid. Energy analysis of primary ejected electrons (the electrons ejected as a result of the direct absorption of the X ray by the electron) is the basis for electron spectroscopy for chemical analysis (ESCA).[688]

The process can be represented as follows:

$$A + h\nu \xrightarrow[\text{(X-ray source)}]{} A^{+*} + e^-_{\text{(primary)}} \quad (4.21)$$

[686] G. Bergmann, *Erdoel Kohle*, **15**, 612 (1962).
[687] F. P. Turenko, L. D. Nikolenko, N. D. Krasnitskaya, P. P. Dikum, A. D. Baranskii, and I. P. Titov, *Khim. Tverd. Topl.*, No. 1, 134 (1972).
* Contributed by Fred R. Brown.
[688] Secondary electrons can also be emitted from the constituent elements of a solid. After an ESCA process (i.e., the ejection of a core electron), an ion with a positive hole in its core electronic structure results. One process that can occur to lower the overall energy of the ion is for another electron from an outer orbital to fill this positive hole. This process is also energy releasing. This quantity of energy can be released by either of two competing processes: (1) the energy can be released as an X ray, the analysis of which is the basis for X ray fluorescence, discussed in Section 4.1.20 below. This can be represented as $A^{+*} \rightarrow A^+ + h\nu$ (X-ray fluorescence); (2) the energy can be released to a third electron of the ion. This ejected electron is called an Auger electron. The energy analysis of these *secondary* ejected electrons is the basis of Auger electron spectroscopy; that is, $A^{+*} \rightarrow A^+ + e^-$ (secondary). The fundamentals of ESCA are detailed in K. Siegbahn, C. Nordling, A. Fahlman, R. Nordberg, K. Hamrin, J. Hedman, G. Johansson, T. Bergmark, S. E. Karlsson, I. Lindgren, and B. Lindberg, *ESCA—Atomic, Molecular and Solid State Structures Studied by Means of Electron Spectroscopy*, Almqvist and Wiesells, Upsala, Sweden, 1967.

This energy analysis directly yields the kinetic energy of an ejected electron. Knowing this quantity and also the energy of the source (soft X rays) permits one to determine the orbital binding energy of the electron, as shown in Equation 4.22:

$$E_b = h\nu - E_k \tag{4.22}$$

where E_b is the orbital binding energy, $h\nu$ is the energy of the incident X ray, and E_k is the kinetic energy of the ejected electron.

The orbital binding energy is characteristic of the element from which the electron was emitted. Thus the elemental composition of the solid being analyzed can be determined. More important, the binding energy of the electron is also dependent on the chemical environment of its parent atom. Thus a chemical shift occurs for electrons ejected from atoms of the same element but present in different chemical species or environments.

Although penetration of the solid by the source X rays is great, the escape depth of the ejected electrons is small. This escape depth varies from sample to sample, but values of 5–50 Å (0.5–5.0 nm) are typical. ESCA thus can provide elemental and chemical information on the first several monolayers of the surface of a sample. Commercial instruments are generally equipped with a method (e.g., an Ar^+ gun) to initially "clean" the sample surface and subsequently remove surface layers so that a depth profile of the sample can be obtained.

4.17.1 ESCA Spectra of Coals

ESCA spectra of whole coals have been recorded in two independent investigations.[689,690] Primary emphasis of both works was on the characterization of sulfur moieties in the coal samples. For peak assignment purposes, both studies included the ESCA analysis of sulfur-containing minerals typically found in coal. A summary of these results appears in Table 4.25. The quality of spectra in the sulfur-2p region for coal samples is shown in Figs. 4.55 and 4.56. Basic differences in the interpretation of the resulting data occur for the minerals studied, and by analogy to these standards, for the coals. One set of authors (Schultz and Proctor[689]) assigned the two peaks obtained for pyrite to two chemically inequivalent sulfur species in the iron sulfide. A similar interpretation is given to the two peaks in the other iron sulfide, marcasite. The other set of authors (Frost et al.[690]) assigned the peak of lower binding energy in

[689] H. D. Schultz and W. G. Proctor, *Appl. Spectrosc.*, **27**, 347 (1973).
[690] D. C. Frost, W. R. Leeder, and R. L. Tapping, *Fuel*, **53**, 206 (1974).

Table 4.25 Summary of ESCA Results for Whole Coals and Minerals Typically Found in Coals

Sample	Investigation A,[689] Sulfur 2p (eV)		Investigation B,[690] Sulfur 2p (eV)	
Pyrite	170.6	164.4	170.9[a]	165.0[a]
Marcasite	171.1	165.1	170.3	164.2
$FeSO_4$	NA		171.0	—
$CaSO_4$ (anhydrite)	171.5	—	NA	
Dibenzothiophene		161.4	NA	
Coal[b]	169.7	162.0		
I	171.9	163.0	NA	
	—	164.0		
II	NA		169.7	163.7

[a] Statistical average of six "fresh" and four "weathered" samples.
[b] Numerous coal samples were analyzed by each set of authors. Values in the table are typical of the results obtained.

NA = Not analyzed.

Sources of data: H. D. Shultz and W. C. Proctor, *Appl Spectr.*, **27**, 347 (1973); D. C. Frost, W. R. Leeder, and R. L. Tapping, *Fuel*, **53**, 206 (1974).

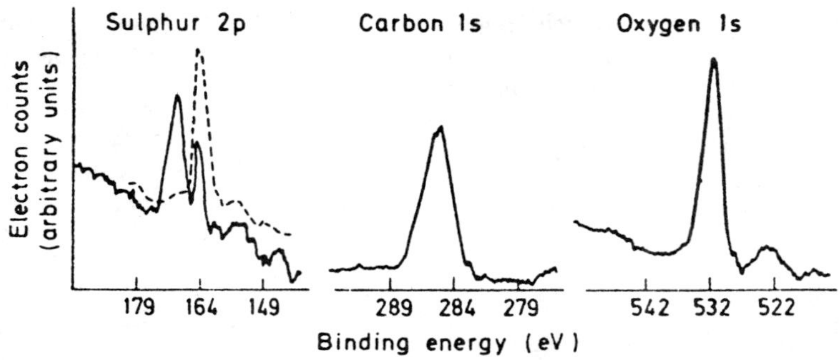

Figure 4.55 Typical spectra of coal showing the three orbital electron binding energy regions. The Sulfur 2*p* region shows freshly ground sample (upper trace) and weathered sample (lower trace).[690]

pyrite to the iron sulfide structure, while assigning the peak at higher binding energy (~171 eV) to iron sulfate, which has formed on the surface of the pyrite by air oxidation. The later assignments were substantiated by comparing the spectrum obtained for a pyrite sample freshly ground under nitrogen with the spectrum of a sample subsequently exposed to air. Little or no discrete electron emission was found at the higher binding energy position for the samples prepared under nitrogen; but the air-exposed samples yielded spectra that showed electron emission at this position, gained at the expense of the lower binding energy peak.

Extrapolating results for the sulfur-containing minerals, plus results for sulfur-containing organic structures, led Schultz and Proctor to conclude that in favorable coal samples, peaks assignable to three types of sulfur-containing species could be distinguished: sulfates, sulfides, and organic sulfur species. Accordingly, these authors listed the sulfur-2*p* peaks appearing in Fig. 4.55 and tabulated in Table 4.25 as follows: 163.0 eV, organic sulfur; 164.0 eV, pyritic sulfur; 171.9 eV, sulfate sulfur; 162.0 and 169.7 eV, unassigned. The interpretation of results for iron minerals, combined with literature values for organic sulfur compounds, led Frost et al. to conclude that no discrete organic sulfur emission could be found in coal. It was postulated that such electron emission was obscured by that from the pyritic sulfur and that this overlap would affect the quantifying of the pyrite content of the coals. Other elements in major concentrations in coal were also detected.[690] Attempts to evaluate these results quantitatively were only partially successful because of surface contamination from hydrocarbons and oxygen in the sampling area.

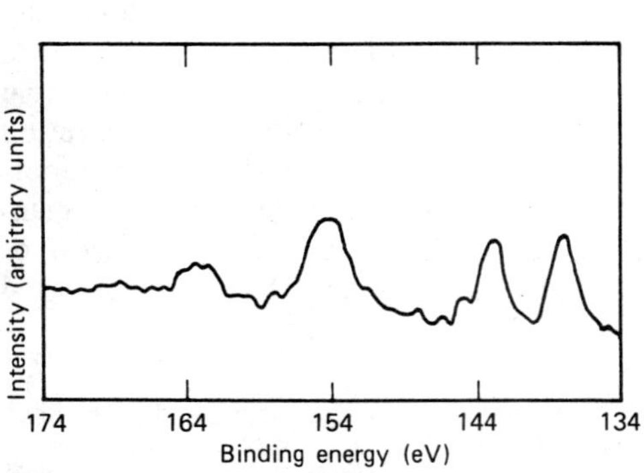

Figure 4.56 Electron emission spectra of bituminous coal, Christian County, Illinois.[689]

4.18 OPTICAL EMISSION SPECTROGRAPHY*

4.18.1 Introduction

Optical emission spectrography (OES) is a well-established technique for the determination of elemental abundances in coal ashes.[691] Quantitative, semiquantitative, or qualitative analyses of major, minor, and trace constituents of coal ash can be obtained using this technique.

The technique has some inherent disadvantages. OES cannot be used to analyze whole coal, so that some ashing procedure must be employed before analyses. The small quantity of coal ash normally employed in OES can result in nonrepresentative sampling.

* Contributed by Hyman Schultz.

[691] W. Fuchs, *Ind. Eng. Chem.*, **27** (1935).

Since the sample is exposed to high temperatures during the analysis, OES is not generally applicable to elements that are volatile. The method is destructive, and the results reflect elemental abundances and give no information about states of chemical combination.

4.18.2 Use as a Survey Technique

Since OES produces spectral information about a large number of elements simultaneously, its major use has been as a survey technique. Zubovic et al.[692] studied 15 minor elements in column, channel, and auger samples collected in five regions of the United States. Coal was ashed at 450°C, and 30 mg of the ash was used in each case. The overall coefficient of variation of the element contents was reported as ±15%.

Abernethy et al.[693], using a method developed by Peterson and Zing,[694] surveyed 827 U.S. commercial coals for 29 elements. The method was characterized as "semiquantitative" and involved ashing the coal at 450 and 800°C before analysis. Tables 4.26 and 4.27 summarize the results that were obtained.

Ruch et al.[695] and Dreher and Schlercher[696] used OES and other analytical techniques to characterize whole coal, "high-" and "low-" temperature ash, and coal float-sink fractions. They employed both direct-reading and photographic detection spectrographs, and on the basis of their results, recommended which elements could be optimally determined by each technique. They also noted that ashing of coal in a radiofrequency-excited stream of oxygen led to nonreproducible results, possibly because large amounts of unaltered mineral matter were present in the ash.

[692] P. Zubovic, T. Stadnichenko, and N. B. Sheffey, U.S. Geol. Survey Bull. No. 1117, 1961.

[693] R. E. Abernethy, M. J. Peterson, and F. H. Gibson, U.S. Bur. Mines Rep. Invest. No. 7281, 1969.

[694] M. J. Peterson and J. B. Zink, U.S. Bur. Mines Rep. Invest. No. 6496, 1964.

[695] R. R. Ruch, H. J. Gluskoter, and N. F. Shimp, Environmental Protection Technology Series, EPA-650/2-74-054, July 1974.

[696] G. B. Dreher and J. A. Schlercher, in *Trace Elements in Fuel*, S. P. Babu, Ed., Am. Chem. Soc. Adv. Chem. Ser. No. 141, (1975), p. 35.

4.18.3 Use for the Determination of Individual Elements

With the development of newer analytical techniques, the use of OES for the determination of individual elements in coal ash has declined. Stadnichenko et al.[697] studied the distribution of beryllium in 1385 coals from various parts of the United States. The coals were from column samples, some of which were washed. The absolute concentrations reported for beryllium content appear to be too high, but the ratios of the beryllium content of coals from different regions of the country may be useful.

Cory et al.[698] used OES and wet chemical methods to study the occurrence and distribution of germanium in coal ash from power plants. The investigation was conducted to determine whether coal ash from power plants could be used as an economical commercial source of germanium. Limits of detection for the OES method employed were cited as 50 ppm for quantitative analyses and less than 20 ppm for qualitative estimate. Results obtained in this study indicated that the germanium in coal tends to concentrate in the fly ash when the coal is burned in steam generators.

4.18.4 Standards

In late 1974 the U.S. National Bureau of Standards made available Standard Reference Material (SRM) 1632, "Trace Elements in Coal," which was certified for 14 trace element concentrations and uncertified for an additional seven trace element concentrations. In the same year the National Bureau of Standards also made available SRM 1633, "Trace Elements in Coal Fly Ash," whose composition was certified for 12 trace elements and uncertified for seven other trace elements. These SRMs were the first coal matrix standards available for calibration purposes. Before these reference materials were introduced, each laboratory either manufactured its own "standard" coal ash or used standard materials that were not based on a coal matrix.

[697] T. Stadnichenko, P, Zubovic, and N. B. Sheffey, U.S. Geol. Survey Bull. No. 1084-K, 1961.

[698] R. C. Corey, J. W. Myers, C. H. Schwartz, F. H. Gibson, and P. J. Colbassani, U.S. Bur. Mines Bull. No. 575, 1959.

4.18.5 New Developments

During the period covered by this supplement, no improvements were made in the instrumentation or theory of OES that had significant effects on the analysis of coal, coal ash, or the products of coal conversion.

The use of inductively coupled plasmas[699] as sources in OES has not been extensively applied to the analysis of the materials of concern here and therefore remains an unknown. If the use of plasma sources should prove applicable, they may offer greater sensitivity and precision in trace element analysis by means of OES.

4.19 X-RAY FLUORESCENCE SPECTROMETRY*

X-Ray emission spectrometry is concerned with the excitation and measurement of characteristic X rays from a sample for qualitative and quantitative analysis of the elements present. The X rays may be generated by electrons, radioisotopes (usually gamma rays), heavy ions, or other X rays. An instrument that uses X rays to excite the sample and measures the resultant radiation is called an X-ray fluorescence spectrometer. The emitted radiation can be measured using a wavelength dispersive detector system (Fig. 4.57) or an energy dispersive system.

In Fig. 4.57,[700] *A* represents a specimen held in some type of holder (*B*) in a specimen compartment (*C*). Primary X rays from the tube strike the sample and excite secondary X rays, which pass through a Soller slit collimator (*D*) to reach the analyzing crystal (*E*). The crystal is mounted on a stage that can be rotated, the axis of the crystal being the goniometer axis. As the crystal rotates, the angle θ varies and each wavelength in the beam is sequentially diffracted according to Bragg's law, $\lambda = 2d \sin \theta$, where λ is the wavelength and d is the interplanar spacing of the crystal. A second collimator (*F*) and detector are mounted on a common arm (*G*) that rotates on the same axis as the crystal. The arrangement is such that as the crystal moves through an angle θ, the detector moves 2θ. In the detector each photon is converted into a pulse of electric current, which is amplified and passed through a pulse height selector, then into ratemeter-recorder and scaler time circuits. Some units receive the output of printers or feed the data into computers for calculation of concentrations. For large numbers of analyses on many samples, computer-controlled multispecimen instruments are made.

In addition to the wavelength dispersive instruments just described, energy dispersive instruments are also commonly used. In these instruments the excitation source may be X rays, radioisotopes, or high-energy ions; the detector is stationary, and the emitted X rays are measured using a multichannel analyzer that separates the X rays according to their photon energies. Most work on coal until now has been carried out using the wavelength dispersive detectors, but energy dispersive units will see wider use because of their simplicity, economy, and absence of moving parts. An investigation of air particulates on filters, using instruments of both types,[701] concluded that major concentrations can be determined by either technique, but that concentrations approaching the detection limits can be determined only by the wavelength dispersive spectrometers. Work at the Illinois Geological Survey[702] showed X-ray fluorescence spectrometry to be a rapid, simple, and reasonably accurate technique for the determination of 21 minor and trace elements in coal; it may be the best method for analyzing large numbers of coal samples. The aluminum, iron, titanium, calcium, potassium, chlorine, sulfur, and silicon in Australian brown coals[703,704] and the vanadium, iron, nickel, copper, zinc, arsenic, bromine, and lead in 20 noncoal organic mineral materials[705] have been determined by

[699] V. A. Fassel and R. N. Kniseley, *Anal. Chem.*, **46**, 1155A (1974).

* Contributed by Sidney S. Pollack.

[700] E. P. Bertin, *Principles and Practice of X-Ray Spectrometric Analysis*, Plenum Press, New York, 1975, p. 94.

[701] J. V. Gilfrich, P. G. Burkhalter, and L. S. Birks, *Anal. Chem.*, **45**, 2002 (1973).

[702] J. F. Kuhn, W. F. Harst, and N. F. Shimp, in *Trace Elements in Fuel*, S. P. Babu, Ed., Am. Chem. Soc. Adv. Chem. Ser. No. 141, 1975, p. 66.

[703] T. R. Sweatman, K. Norris, and R. A. Durie, CSIRO, Div. Coal Res., Misc. Rep. No. 177, (1963), 30 pp.

[704] L. T. Kiss, *Anal. Chem.*, **38**, 1731 (1966).

[705] J. Otteman and G. Friese, *Geochim. Cosmochim. Acta*, **26**, 599 (1962).

Table 4.26 Average[a] Trace Element Content (%) in Ash of Coal from Three Areas[693]

			Eastern Province		Interior Province		Western States	
Element	Crustal Abundance[b]	Approx. Lower Limit of Detection[c]	Frequency of Detection	Average Trace Element Content of Ash	Frequency of Detection	Average Trace Element Content of Ash	Frequency of Detection	Average Trace Element Content of Ash
Barium	0.0425	0.002	100	0.0676	100	0.0399	100	0.1467
Beryllium	0.00028	0.0001	100	0.0012	100	0.0014	100	0.0006
Boron	0.0010	0.0002	100	0.0265	100	0.0731	100	0.0529
Chromium	0.0100	0.0001	100	0.0230	100	0.0224	100	0.0066
Cobalt	0.0025	0.0020	100	0.0184	98	0.0193	98	0.0097
Copper	0.0055	0.0001	100	0.0128	100	0.0089	100	0.0047
Gallium	0.0015	0.0002	100	0.0071	100	0.0039	100	0.0033
Germanium	0.00015	0.0003	99	0.0048	100	0.0104	95	0.0017
Lanthanum	0.0030	0.01	92	0.0145	86	0.0131	81	0.0128
Lead	0.0013	0.0001	100	0.0055	100	0.0131	100	0.0029
Lithium	0.0020	0.0001	100	0.0584	100	0.0235	100	0.0168
Manganese	0.0950	0.0001	100	0.0260	100	0.0325	100	0.0212
Molybdenum	0.00015	0.0001	99	0.0082	99	0.0073	100	0.0020
Nickel	0.0075	0.0001	100	0.0209	100	0.0262	100	0.0054
Scandium	0.0022	0.002	100	0.0089	100	0.0069	97	0.0052
Strontium	0.0375	0.001	100	0.1052	100	0.0658	100	0.1456
Tin	0.0002	0.0001	100	0.0019	99	0.0019	100	0.0017
Vanadium	0.0135	0.0001	100	0.0336	100	0.0325	100	0.0152
Ytterbium	0.00034	0.0001	100	0.0007	100	0.0005	100	0.0003

Yttrium	0.0033	0.001	100	0.0142	100	0.0118	100	0.0076
Zinc	0.0070	0.005	98	0.0230	100	0.0743	93	0.0258
Zirconium	0.0165	0.005	100	0.0704	100	0.0825	100	0.0850
Arsenic	0.00018	0.005	67	0.0159	41	0.0119	16	0.0073
				(0.0107)		(0.0049)		(0.0012)
Bismuth	0.00002	0.0001	82	0.0002	77	0.0001	83	0.0001
				(0.0002)		(0.0001)		(0.0001)
Cerium	0.0060	0.02	31	0.0238	11	0.0214	13	0.0238
				(0.0074)		(0.0024)		(0.0031)
Neodymium	0.0028	0.01	29	0.0213	10	0.0183	15	0.0295
				(0.0062)		(0.0018)		(0.0044)
Niobium (columbium)	0.0020	0.001	73	0.0053	88	0.0055	85	0.0053
				(0.0039)		(0.0048)		(0.0045)
Rubidium	0.0090	0.001	97	0.0239	100	0.0276	58	0.0064
				(0.0232)		(0.0276)		(0.0037)
Thallium	0.00005	0.0005	43	0.0019	49	0.0008	9	0.0005
				(0.0008)		(0.0004)		(0.00005)
Average ash (% dry coal)	—	—	—	9.3	—	10.5	—	9.8
Number of samples	—	—	—	600	—	123	—	104

[a] Averages calculated for number of samples in which element was detected, except that averages in parentheses were calculated for all the samples tested using zero for element contents below limit of detection.

[b] Brian Mason, *Principles of Geochemistry,* 3rd ed., John Wiley & Sons, New York, 1966, pp. 45–46.

[c] M. J. Peterson and J. B. Zink, *A Semiquantitative Spectrochemical Method for Analysis of Coal Ash,* U.S. Bur. Mines Rep. Invest. No. 6496 (1964) pp. 8–10.

Source: R. E. Abernethy, M. J. Peterson, and F. H. Gibson, U.S. Bur. Mines Rep. Inves. No. 7281 (1969).

Table 4.27 Percentage of Samples in Each Range of Trace Element Content of Ash[693]

Element and Area[a]	Analysis of Ash (%)										
	Not Detected	<0.0001	0.0001–0.0002	0.0003–0.0007	0.0008–0.002	0.003–0.007	0.008–0.02	0.03–0.07	0.08–0.2	0.3 to 0.7	0.8–2.0
	Elements Occurring in Most Samples										
Barium											
1	—	—	—	—	—	1	13	30	53	2	1
2	—	—	—	—	—	2	40	46	12	—	—
3	—	—	—	—	—	—	9	28	49	11	3
Beryllium											
1	—	4	10	30	50	5	1	—	—	—	—
2	—	2	3	26	60	8	1	—	—	—	—
3	—	13	25	42	17	2	1	—	—	—	—
Boron											
1	—	—	—	—	1	12	55	24	8	—	—
2	—	—	—	—	—	2	10	40	48	—	—
3	—	—	—	—	—	7	33	29	28	3	—
Chromium											
1	—	—	—	—	1	3	72	23	1	—	—
2	—	—	—	—	1	1	72	26	—	—	—
3	—	—	2	1	19	38	40	—	—	—	—
Cobalt											
1	—	—	—	—	—	8	77	14	1	—	—
2	2	—	—	—	1	9	77	6	5	—	—
3	2	—	—	1	1	33	61	2	—	—	—
Copper											
1	—	—	—	—	3	21	72	4	—	—	—
2	—	—	—	—	9	29	62	—	—	—	—
3	—	—	2	4	33	39	22	—	—	—	—
Gallium											
1	—	—	—	—	13	48	38	1	—	—	—
2	—	—	—	2	33	56	9	—	—	—	—
3	—	—	2	—	48	43	7	—	—	—	—

Germanium											
1	1	—	—	7	38	29	24	1	—	—	—
2	—	—	—	—	13	24	60	3	—	—	—
3	5	—	1	30	51	10	3	—	—	—	—
Lanthanum											
1	8	—	—	—	—	—	88	4	—	—	—
2	14	—	—	—	—	—	81	5	—	—	—
3	19	—	—	—	—	—	75	6	—	—	—
Lead											
1	—	—	—	2	27	41	29	—	1	—	—
2	—	—	—	—	31	21	38	8	2	—	—
3	—	—	2	1	56	35	6	—	—	—	—
Lithium											
1	—	—	—	—	—	6	35	26	33	—	—
2	—	—	—	—	2	40	39	10	9	—	—
3	—	—	—	4	10	36	34	12	4	—	—
Manganese											
1	—	—	—	—	2	13	59	16	10	—	—
2	—	—	—	—	1	5	53	31	10	—	—
3	—	—	—	—	7	20	49	18	6	—	—
Molybdenum											
1	1	—	1	2	27	24	44	1	—	—	—
2	1	—	—	—	16	41	41	1	—	—	—
3	—	—	5	17	65	9	4	—	—	—	—
Nickel											
1	—	—	—	—	—	4	80	14	2	—	—
2	—	—	—	—	—	6	71	14	9	—	—
3	—	—	7	1	25	38	29	—	—	—	—
Scandium											
1	—	—	—	—	1	40	58	1	—	—	—
2	—	—	—	—	6	59	33	2	—	—	—
3	3	—	—	—	14	63	20	—	—	—	—
Strontium											
1	—	—	—	—	—	1	8	21	68	2	—
2	—	—	—	—	—	1	20	42	35	2	—
3	—	—	—	—	—	—	8	14	65	13	—

Table 4.27 Percentage of Samples in Each Range of Trace Element Content of Ash[693] (Continued)

Element and Area[a]	Analysis of Ash (%)										
	Not Detected	<0.0001	0.0001–0.0002	0.0003–0.0007	0.0008–0.002	0.003–0.007	0.008–0.02	0.03–0.07	0.08–0.2	0.3 to 0.7	0.8–2.0
Elements Occurring in Most Samples											
Tin											
1	—	1	1	25	59	10	4	—	—	—	—
2	1	—	2	30	51	13	2	1	—	—	—
3	—	—	2	29	59	6	4	—	—	—	—
Vanadium											
1	—	—	—	—	—	1	48	44	7	—	—
2	—	—	—	—	—	2	43	53	2	—	—
3	—	—	—	—	2	18	71	8	1	—	—
Ytterbium											
1	—	—	13	45	42	—	—	—	—	—	—
2	—	1	39	43	16	1	—	—	—	—	—
3	—	1	59	34	6	—	—	—	—	—	—
Yttrium											
1	—	—	—	—	1	15	77	7	—	—	—
2	—	—	—	—	4	35	55	5	1	—	—
3	—	—	—	—	9	42	49	—	—	—	—
Zinc											
1	2	—	—	—	—	16	59	17	6	—	—
2	—	—	—	—	—	2	12	48	31	5	2
3	7	—	—	—	—	14	54	17	8	—	—
Zirconium											
1	—	—	—	—	—	—	10	40	50	—	—
2	—	—	—	—	—	—	2	31	67	—	—
3	—	—	—	—	—	—	6	20	74	—	—
Additional Elements Found in Many Samples											
Arsenic											
1	33	—	—	—	—	15	43	7	2	—	—
2	59	—	—	—	—	11	28	1	1	—	—
3	84	—	—	—	1	9	6	—	—	—	—

Bismuth												
1	15	2	78	1	1	—	—	—	—	—	—
2	23	6	71	—	—	—	—	—	—	—	—
3	17	12	70	1	—	—	—	—	—	—	—
Cerium												
1	69	—	—	—	—	—	23	8	—	—	—
2	89	—	—	—	—	—	10	1	—	—	—
3	87	—	—	—	—	—	10	3	—	—	—
Neodymium												
1	71	—	—	—	—	1	21	6	1	—	—
2	90	—	—	—	—	—	10	—	—	—	—
3	85	—	—	—	—	—	12	3	—	—	—
Niobium												
1	27	—	—	—	9	54	10	—	—	—	—
2	12	—	—	—	7	66	15	—	—	—	—
3	15	—	—	—	16	55	14	—	—	—	—
Rubidium												
1	3	—	—	—	4	24	48	13	8	—	—
2	—	—	—	—	1	14	57	16	11	—	1
3	42	—	—	—	10	29	19	—	—	—	—
Thallium												
1	57	—	—	7	29	5	2	—	—	—	—
2	51	—	—	33	15	1	—	—	—	—	—
3	91	—	—	8	1	—	—	—	—	—	—

[a] Area 1 = eastern province, 600 samples.
Area 2 = interior province, 123 samples.
Area 3 = western states, 104 samples.

Source: R. E. Abernethy, M. J. Peterson, and F. H. Gibson, U.S. Bur. Mines Rep. Inves. No. 7281 (1969).

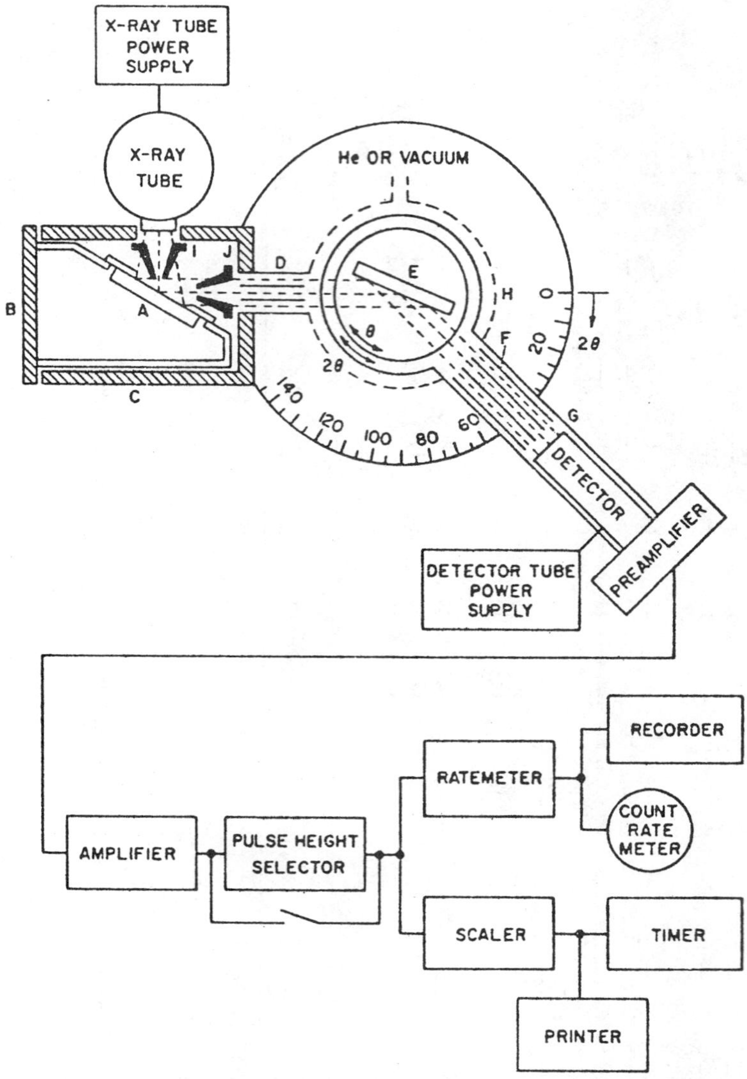

Figure 4.57 X-Ray secondary-emission (fluorescence) spectrometer.[700]

X-ray fluorescence spectrometry. The technique has been used also as a method for determining total sulfur in coals[706–708] and in liquid fuels[709] and as a way of distinguishing between sulfate, sulfide, and organic sulfur in coals.[710,711] It has also been used for the analysis of selenium[712] and mercury[713] where the minimum detectable concentration has been found to be 2–3 ppm under favorable conditions. Portable radioisotope units[714,715] for monitoring ash in coals have been developed. Using 40-MeV oxygen ions as an ex-

[706] BCRA, Coke Res. Rep. No. 71, (1972), 6 pp.

[707] J. Frigge, *Erdoel Kohle, Erdgas, Petrochem. Brennst.-Chem.*, **25,** 447 (1972).

[708] M. Berman and S. Ergun, U.S. Bur. Mines Rep. Invest. No. 7124, (1968), 20 pp.

[709] K. H. Koch, L. Schmitz, and W. Loose, *Arch. Eisenhuttenw.*, **42,** 883 (1971).

[710] Y. Gohshi and K. Yanagase, *Nenryo Kyokaishi,* **52,** 926 (1973).

[711] R. G. Hurley and F. W. White, *Anal. Chem.*, **46,** 2234 (1974).

[712] T. L. Belopol'skaya and I. V. Serikov, *Khim. Tverd. Topl.*, **6,** 73 (1969).

[713] R. L. Mylkebust, M. M. Darr, and K. F. J. Heinrich, U.S. Nat. Bur. Stand. Rep. No. COM-75-10686, (1974), 21 pp.

[714] J. R. Rhodes, U.S. At. Energy Comm., ORNL-ITC-5, (1965), p. 206.

[715] C. G. Clayton, *Bull. Inf. Assoc. Tech. Energ. Nucl. Suppl.*, **93,** 4 (1972).

citation source,[716] detection limits are about 10 ppm.

4.20 ATOMIC ABSORPTION SPECTROSCOPY*

Since atomic absorption spectroscopy (AAS) was introduced as a technique with commercially available instrumentation during the early 1960s, it has gained wide acceptance. The application of AAS to the analysis of coal ash was reported in the literature[717] as early as 1963. Since the first use of AAS for the analysis of coal and coal-related materials, the literature on the subject has expanded enormously. The current literature is voluminous, and only a small portion is cited here.

4.20.1 Description of the Technique

A number of books[718–720] and descriptive articles[721–723] that discuss in detail the practice and theory of ASS have been published in recent years. These should be consulted for a more complete description than space permits in this volume.

It has been known since Fraunhofer that atoms of an element in their ground state will absorb light at very specific wavelengths. The energies of the absorbed light quanta are equal to the difference in electronic energies between the atomic ground state and higher energy states. The decrease in intensity of a light beam passing through a cloud of ground state atoms at these absorbing or resonance wavelengths is proportional to the number of absorbing atoms in the light path. By measuring light intensities at specific wavelengths, quantitative estimates can be made of the concentration of chemical elements introduced into the light path. Atomic absorption can be observed using continuum light sources; in practice, however, a light source is used that employs the specific element of interest such as a hollow cathode lamp or an electrodeless discharge lamp.

A practical restriction on the use of AAS is the wavelength of the resonance effect. All metals and most semimetals have resonance wavelengths above about 180 nm, whereas nonmetals absorb at wavelengths below 180 nm. The practical difficulties involved in producing intrumentation to detect nonmetal atomic absorption has precluded the direct use of AAS in the case of nonmetals (Fig. 4.58).

Both flame and nonflame methods are employed to introduce atoms into the light path of a spectrometer. The most common procedure used in flame methods is the aspiration of a solution of the element of interest into a chamber where the solution is nebulized and mixed with fuel and oxidant gases. The most widely used fuel is acetylene, and the most common oxidizers are air and nitrous oxide. The gas-nebulized solution mixture is introduced into a burner head, usually of a laminar flow design, where it is burned. The flame is aligned in the light path so that an optimum absorption signal is produced. The flame serves the purpose of evaporating the solvent from the solution droplets and decomposing the compounds of the element of interest into ground state atoms. Figure 4.59 shows the components of an atomic absorption spectrometer.

In nonflame methods atomization is achieved by applying a small amount of solution to a nonreactive surface such as carbon or tantalum and raising the temperature of the surface by passing an electric current through it. Mercury is a special case, since the vapor pressure of elemental mercury at room temperature is high enough to produce detectable atomic absorption signals. Since mercury compounds in solution can be chemically reduced, and all mercury compounds also decompose at relatively low tempera-

[716] L. Shabuson, B. L. Cohen, G. H. Wedberg, and K. C. Chan, *J. Appl. Phys.*, **44**, 4749 (1973).

* Contributed by Hyman Schultz.

[717] R. A. Durie, H. N. S. Schafer, and D. J. Swaine, CSIRO Div. Coal Res., Misc. Rep. No. 178, (1963), p. 13.

[718] W. Slavin, *Atomic Absorption Spectroscopy*, Interscience, New York, 1968.

[719] G. F. Kirkbright and M. Sargent, *Atomic Absorption and Fluorescence Spectroscopy*, Academic Press, New York, 1974.

[720] H. Moenk, *Modern Trace Analysis*, Vol. 1, *Atomic Spectroscopy in Trace Analysis; Overview of the Methods, Apparatus, Assumptions, and Areas of Application*, Geest and Portig, Leipzig, East Germany, 1974.

[721] C. Woodward. *Annual Reports on Analytical Spectroscopy. 1974*, Vol. 4, Society for Analytical Chemistry, London, 1975.

[722] Galen W. Ewing, Ed., *Topics in Chemical Instrumentation*, Chemical Education Publishing Co., Easton, Pa., 1971, p. 1.

[723] W. W. Meinke and B. F. Schribner, Eds., U.S. Natl. Bur. Stand. Monog. No. 100, Washington, D.C., 1967, p. 266.

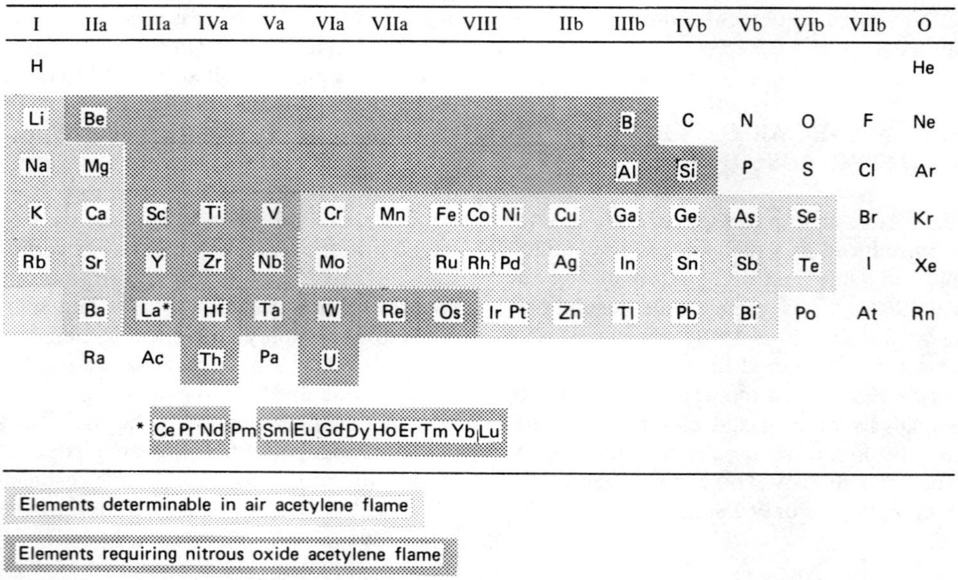

Figure 4.58 Elements determined by atomic absorption spectrometry (shaded areas—the other elements are not determinable by AAS).[724]

tures to elemental mercury, either mode of decomposition can take place outside the light path and the mercury vapor produced can be carried into the light path at room temperature.[726,727]

When carried into a flame[728] or a hot

[724] W. S. Price, *Analytical Atomic Absorption Spectrometry,* Heyden and Sons Ltd., London, 1972, frontispiece.

[725] W. S. Price, *Analytical Atomic Absorption Spectrometry,* Heyden and Sons Ltd., London, 1972, p. 18.

[726] W. R. Hatch and W. L. Ott, *Anal. Chem.*, **40,** 2085 (1970).

[727] W. W. Vaughn, U.S. Geol. Survey Circ. No. 540, 1967.

[728] H. L. Kahn and J. E. Schallis, *At. Absorpt. Newsl.*, **7,** 5 (1968).

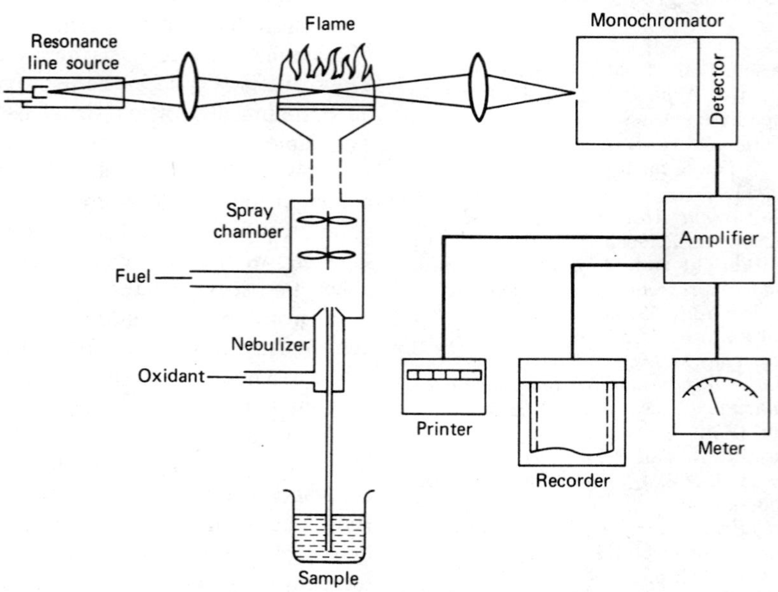

Figure 4.59 Practical system for atomic absorption spectrometry.[725]

tube,[729,730] volatile hydrides such as arsine and hydrogen selenide decompose into their respective elements and make these amenable to quantitative analysis by AAS.

4.20.2 Application of AAS to Coal and Related Materials

The development of AAS and other techniques during the 1960s and early 1970s coincided with increased interest in the determination of trace metals in coal and the products of coal utilization. The renewed interest was caused primarily by environmental concerns. The use of AAS, along with other techniques that were brought into use during the same period, allowed the detection and quantification of metallic elements at levels below those previously attainable. Table 4.28[731] shows typical detection limits achievable using AAS. When used with care, AAS also offered levels of precision in trace element analyses that were not easily attainable in prior years.

When mercury was determined in coal and concern was expressed about the quantity of mercury discharged from coal-fired power plants,[732,733] the analytical method used was flameless AAS. A round-robin study of methods for the determination of mercury[734] led to AAS being one of the two recommended methods. Indeed, follow-up studies on mercury emissions from coal-fired power plants used AAS[735–737] as the analytical method.

As with all analytical procedures, AAS is subject to interferences of various sorts, and in addition the sensitivity of AAS varies from element to element. In light of these considerations, AAS is not necessarily the method of choice for the determination of all metals and semimetals in all matrices. Ruch et al.[738]

[729] C. R. Chu, G. P. Barron, and P. A. W. Baumgarner, *Anal. Chem.*, **44**, 1476 (1972).

[730] J. Y. Hwang, P. A. Ullucci, C. J. Mokeler, and S. B. Smith, Jr., *Am. Lab.*, March 1973, p. 43.

[731] *Analytical Methods for Atomic Absorption Spectrophotometry*, Perkin-Elmer Corp., Norwalk, Conn.

[732] O. I. Joensuu, *Science*, **172**, 1027 (1971).

[733] J. V. O'Gorman, N. H. Suhr, and P. L. Walker, Jr., *Appl. Spectrosc.*, **26**(1), 44 (January 1972).

[734] M. D. Schlesinger and H. Schultz, U.S. Bur. Mines Rep. Invest. No. 7609, 1972, 11 pp.

[735] R. C. Diehl, E. A. Hattman, H. Schultz, and R. J. Haren, U.S. Bur. Mines Tech. Paper No. 54, May 1972.

[736] G. W. Kalb, in *Trace Elements in Fuels*, S. P. Babu, Ed., Am. Chem. Soc. Adv. Chem. Ser. No. 141, 1975, p. 154.

[737] W. L. Anderson and K. E. Smith, *Environ. Sci. Technol.*, **11**, 75 (1977).

Table 4.28 Atomic Absorption Detection Limits (Model 603)[731]

Element	Limit (μg/ml)	Element	Limit (μg/ml)
Ag	0.002	Nb[a,c]	1
Al[a]	0.02	Nd[a,c]	1
As[b]	0.2	Ni	0.005
Au	0.01	Os[a]	0.08
B[a]	0.7	P[a,b]	50
Ba[a,c]	0.008	Pb[b]	0.01
Be[a]	0.001	Pd	0.02
Bi	0.025	Pr[a,c]	5
Ca	0.0005	Pt	0.05
Ca[b]	0.001	Rb[c,d]	0.002
Co	0.01	Re[a]	0.5
Cr	0.003	Rh	0.004
Cs[c,d]	0.005	Ru	0.07
Cu	0.002	Sb	0.04
Dy[a,c]	0.05	Sc[a,c]	0.02
Er[a,c]	0.04	Se[b]	0.2
Eu[a,c]	0.02	Si[a]	0.02
Fe	0.005	Sm[a,c]	2
Ga[a]	0.05	Sn[a,b]	0.07
Gd[a,c]	1.2	Sr[c]	0.002
Ge[a,b]	0.1	Ta[a]	1
Hf[a]	2	Tb[a,c]	0.6
Hg[b]	0.25	Te[b]	0.03
Ho[a,c]	0.04	Ti[a]	0.04
In	0.02	Tl[b]	0.01
Ir	0.6	Tm[a,c]	0.01
K	0.002	U[a]	30
La[a,c]	2	V[a]	0.04
Li	0.0003	W[a]	1
Lu[a,c]	0.7	Y[a,c]	0.05
Mg	0.0001	Yb[a,c]	0.005
Mn	0.002	Zn	0.001
Mo[a]	0.02	Zr[a]	1
Na	0.0002		

[a] Nitrous oxide–acetylene flame.

[b] Electrodeless discharge lamp.

[c] 1000 μg/ml of potassium added to control ionization.

[d] Flame emission with air-acetylene flame.

Source: Analytical Methods for Atomic Absorption Spectrophotometry, Perkin-Elmer Corp., Norwalk, Conn.

[738] R. R. Ruch, H. J. Gluskoter, and N. F. Shimp, *Occurrence and Distribution of Potentially Volatile Trace Elements in Coal,* EPA-650/2-74-054, July 1974.

found AAS to be the method of choice for the determination of cadmium, zinc, lead, copper, and nickel in coals and the float-sink fractions from those coals. For the individual investigator, the use of AAS would depend on a number of factors such as matrix, precision required, sensitivity desired, and elements to be studied.

AAS can be applied to the analysis of coal,[733,738] the fate of trace elements from coal in power plants,[739,740] and the distribution of trace elements in coal conversion processes.[741,742]

4.20.3 Major and Minor Elements

Although the greatest use of AAS has been in the area of trace element determinations, its inherent sensitivity makes the technique applicable to the determination of major and minor elemental constituents in coal and the products of coal utilization. The advantage gained by the use of AAS is that several elements can be analyzed sequentially in the same solution. The methodology employed usually involves the ashing of the coal and the dissolution of the coal ash[743] prior to AAS analysis.

In addition to its general utility, AAS is being considered as a replacement for the current ASTM method for determining pyritic iron.[744] The proposed AAS method is faster and more precise than the current ASTM method and is being used in the U.S. Bureau of Mines Coal Analysis Laboratory and several other coal analysis laboratories.

[739] N. E. Bolton, J. A. Carter, J. F. Emery, C. Feldman, W. Fulkerson, L. D. Hulett, and W. S. Lyon, in *Trace Elements in Fuels,* S. P. Babu, Ed., Am. Chem. Soc. Adv. Chem. Ser. No. 141, 1975, p. 175.

[740] H. Schultz, E. A. Hattman, and H. B. Booher, in *Trace Elements in Fuels,* S. P. Babu, Ed., Am. Chem. Soc. Adv. Chem. Ser. No. 141, (1975), p. 139.

[741] H. Schultz, G. A. Gibbon, E. A. Hattman, H. B. Booher, and J. W. Adkins, PERC/RI-77/2, Pittsburgh Energy Research Center, ERDA, Pittsburgh, February 1977.

[742] A. Attari, EPA Report No. 650/2-73-004, Institute of Gas Technology, Chicago, August 1973.

[743] R. B. Muter and L. L. Nice, in *Trace Elements in Fuels,* S. P. Babu, Ed., Am. Chem. Soc. Adv. Chem. Ser. No. 141, (1975), p. 57.

[744] ASTM Standards, 1975, ASTM Designation D2492, Gaseous Fuels, Coal and Coke; Atmospheric Analysis.

4.20.4 Conclusions

AAS is an analytical technique that offers some distinct advantages when applied to the analysis of coal and the products of coal utilization. The method is sensitive and specific when used for the determination of most metals and semimetals. AAS can be applied to the analysis for major, minor, or trace elements because the range of concentrations over which it is useful is generally large. The amount of sample required is relatively small, ranging from microliter amounts of sample solution with the use of a carbon furnace to several cubic centimeters of sample solution with the normal aspiration methods. Since the sample is introduced into the instrument in solution in almost all cases, the aliquot used for analysis should be representative of the material dissolved (which in the case of a heterogeneous material such as coal can be made large enough to alleviate the sampling problems associated with small samples).

Although AAS offers many advantages to the analyst, there are a number of disadvantages that one should be aware of before selecting the method. AAS is used to determine one element at a time, which limits its application as a survey technique. As stated previously, it is not normally applicable to nonmetals, and the sensitivity and ease of application vary from element to element and from sample matrix to sample matrix. There are interferences of various sorts associated with AAS, but probably the most important when applied to coal is matrix interference. Matrix effects are related to the element being determined and its chemical environment. Since coals are heterogeneous and vary in chemical composition, the matrix interference effects—when they occur—are liable to vary, making the choice of calibration standards difficult if not impossible. Where matrix interference is significant, a technique such as the method of standard additions must be applied if accurate analyses are to be obtained using AAS.

4.21 MÖSSBAUER SPECTROSCOPY*

The relatively new technique of Mössbauer spectroscopy is based on the resonant absorption of gamma rays by an atomic nucle-

* Contributed by Dennis H. Finseth.

us.[745–747] The method probes the environment of the nucleus by observing the very small effects of the chemical environment on nuclear properties and therefore must be a very high resolution technique. The spectrum is scanned by moving the gamma-ray source at varying rates relative to the absorber, thus imparting slightly different energies to the gamma-ray photon via the Doppler shift. This technique is applicable to a limited number of elements, with iron being by far the most important.

Mössbauer spectroscopy has been used to investigate the nature of iron in coal.[748] The iron in coal can be classified as HCl-soluble, pyritic, and "organic." The HCl-soluble portion is that which is soluble in 25% HCl; it consists primarily of carbonates, oxides, sulfates, and so on. The pyritic portion is the iron that is not soluble in 25% HCl but is soluble in 25% HNO_3. This fraction is primarily iron bound in pyrites. The "organic" iron is not acid soluble under the above-described conditions, and its chemical nature is largely unknown.

Mössbauer spectroscopy is well suited to the study of iron in a matrix of light elements, as in the case of coal, because the gamma rays are transmitted through the bulk of the sample with little matrix absorption. The advantage of Mössbauer spectroscopy for these investigations is that the study is accomplished without altering the chemical environment of the sample. The Mössbauer work on coal was specifically initiated to elucidate the nature of the organic iron.

This work was performed using nine coal samples ranging from 72.2 to 92.7% C, which had a range of total iron content from 0.07 to 4.18%. The distribution of total iron between HCl-soluble, pyritic, and "organic" iron varied from sample to sample, with the highest "organic" content being 1.16%.

The Mössbauer spectra of these samples indicated that two distinct forms of iron were present. These were characterized by their isomer shift, quadrupole splitting, and temperature dependence. One of the iron resonances observed was identified as being characteristic of iron pyrites. The second Mössbauer doublet observed was distinct from the pyrite absorption and was most evident in samples analyzed as having high "organic" iron content. Quantitative agreement between Mössbauer absorption intensity and chemical analysis was not achieved in this work, however.

The spectrum obtained for this "organic" iron is characteristic of iron(II), but the spectral parameters determined differed from those of any known iron Mössbauer spectrum. The conclusion of this work was that the "organic" iron spectrum is consistent with a high-spin iron(II) species in octahedral symmetry, possibly bound to heterocyclic nitrogen compounds. The data do not, however, rule out the existence of iron in a claylike silicate mineral or gel.

Mössbauer spectroscopy was used in another study to characterize the oxidation of pyritic surfaces important in the formation of acid mine drainage.[749] This work indicated that the oxidation products were $Fe(OH)_3$ and $Fe_2(SO_4)_3$.

4.22 SUMMARY

Expanded use of new and more sophisticated analytical methods has produced much additional data concerning the physical properties of coal and its products.

Results of continued reflectance determinations on vitrinites have confirmed the development of small aromatic nuclei in low-rank coals and a rapid increase in size and ordering of aromatic layers in high-rank coals. Usefulness of reflectance as a more fundamental measurement of rank has been demonstrated. Reflectance measurements on trace occurrences of vitrinite in rock strata have aided in petroleum and gas prospecting. Automated reflectance analysis of coals has been demonstrated as a possible utilization criterion.

X-Ray diffraction studies have shown that high-rank metaanthracites closely approach graphite, but that the poorly crystalline low-rank coals contain too many imperfections

[745] E. Fluck, *Adv. Inorg. Chem. Radiochem.*, **6**, 433 (1964).

[746] J. F. Duncan and R. M. Golding, *Q. Rev. Chem. Soc. (London)*, **19**, 36 (1965).

[747] G. K. Wertheim, *Mössbauer Effect, Principles and Applications*, Academic Press, New York, 1964.

[748] John F. Lefelhocz, Robert A. Friedel, and Truman P. Kohman, *Geochim. Cosmochim. Acta*, **31**, 2761 (1967).

[749] S. Baker and R. Andrews, Water Pollut. Control Res. Ser. No. 1401F1103/71, 1971, 31 pp.

for accurate determinations of aromatic cluster sizes, spacing, and ordering.

Authorities seem to agree that surface area concepts of the pore structure of coals have little physical significance. A critical review of coal density measurements has concluded that porosity in coals of less than 75% C is due mostly to macropores (>200 Å), that porosity in coals of 75–84% C is due to micropores (<20 Å) and transitional (20–200 Å) pores, and that porosity in coals of 85–91% C is largely in micropores.

Electron microscopy has given direct evidence of pore structure in coals and of development of graphitic crystallites in very high rank coals, correlating with X-ray evidence. Ultrafine coke structure has also been revealed.

Additional studies of electrical properties of coal—conductivity, dielectric constant, and magnetic susceptibility—have supported postulates of semiconductor properties and approach of high-rank coals to the graphite layer structure. The dielectric constant has served as a monitor of moisture content in coals.

Studies have continued on the thermal behavior of coals by laboratory-scale methods of differential thermal analysis, thermogravimetric analysis, and specific heat determination. Total heats of coking coals to 950°C are around 350 cal/g.

Information concerning the distribution of carbon and hydrogen in aromatic and nonaromatic structures, the molecular weight distribution of components in extracts and coal-derived liquids, and the detailed carbon number distribution data for all major aromatic structures in coal and products from the high-temperature reactions of coal are but a few of the new data obtained by mass spectrometry in the past 15 years. Investigations of certain fractions, primarily for process development purposes, have given insight into the lower molecular weight components.

Similarity between light components volatilized from coal and light ends in certain process liquids has been established by mass spectrometry. Investigations of products from reduced and alkylated coals have given new insights into not only structural parameters but also factors leading to the increased solubility of coal. Application of high-resolution mass spectrometry has led to determinations of molecular formulas of many additional components associated with the coal structure. Trace element determinations are reviewed extensively in Chapter 7. However determinations related to the physical structure and composition of coal have been made by spark-source mass spectrometry.

Recent advances in NMR spectrometry have greatly expanded its usefulness in coal research. High-resolution NMR studies of soluble materials from coal have been extended to include ^{13}C NMR in addition to the more conventional ^{1}H NMR. Thus for soluble material such as coal estracts and coal-derived liquids, the distribution of hydrogen and carbon among various organic functional groups can now be determined quantitatively. These data, taken collectively, permit the deduction of the gross aspects of mean structural units. Even more exciting is the extension of high-resolution NMR to the solid state. Reasonably well-resolved resonances in ^{13}C spectra of solid coals can now be obtained by means of proton-enhanced nuclear induction spectral techniques, and progress is being made on developing techniques for obtaining ^{1}H NMR spectra of comparable quality. Although considerable controversy still remains over the chemical structure of coals, most NMR studies support the classical views that coals are highly aromatic substances, that the aromaticity of coal increases with increasing rank, and that the increase in aromaticity is accompanied by increases in the number of aromatic rings per condensed polynuclear aromatic ring system.

Many of the ESR results are consistent with the premise that free radicals give rise to the ESR absorption in coals. The increase in radical concentrations of coals with increasing rank is best explained on the basis of resonance stabilization of the radicals which form during coal metamorphism. These data are supported by the formation of radicals observed during the slow pyrolysis of coals. The changes in the ESR g values of coals as coalification progresses suggest that the free radical electrons in peats and low-rank coals are at least partially localized on oxygen atoms.

IR studies of whole coals, petrographic components, extracts, and treated coals have continued. Changes in the bank intensities as a function of coal rank have been observed and explanations proposed. The distribution of carbon and hydrogen in coal and coal-derived materials has been determined when IR data were combined with NMR and mass spectrometry results. Insight into the mineral matter in coals has been obtained by IR.

UV-visible absorption spectroscopy is very useful in the qualitative and quantitative characterization of separated fractions of coal-derived materials. Fluorescence spectroscopy shows some promise for qualitative and quantitative characterization of chemical classes present in coal-derived materials.

Many problems related to the physical properties of coal and its products remain to be solved. Little is known concerning the form of many elements in coal, that is, how particular elements are bound in the organic and inorganic material in the coal. The identity of the structures with which certain heteroatoms such as nitrogen, oxygen, and sulfur are bound in the heavy ends of the organic material has yet to be resolved. One of the encouraging signs at the moment is the deep involvement of petroleum laboratories in studies of coal. Indeed, much of the technology and analytical expertise available in petroleum laboratories should be brought to bear on many of the problems remaining concerning the physical and also the chemical properties of coal.

CHAPTER 5

Mechanical Properties of Coal

THOMAS G. CALLCOTT AND GREGORY B. SMITH

Thomas G. Callcott *is Fuel Research Manager at the Central Research Laboratories of the Broken Hill Proprietary Company, Ltd., Shortland, New South Wales, Australia. His principal current research interests include coal transportation and cleaning, coking including formcoke, new gasification methods, generation of SRC and synthetic coking components, pneumatic and fluidized-bed processes including combustion, particulate properties, energy utilization in the steel industry, coal exploration and evaluation, chemical and physical characteristics of coal and derived products. Since 1951 Dr. Callcott has worked on a series of problems associated with the physical, mechanical, and size reduction properties of raw and manufactured materials.*

Gregory B. Smith *is a Fuel Research Officer at the Central Research Laboratories of the Broken Hill Proprietary Company, Ltd., Shortland, New South Wales, Australia. His major research interests are currently in the chemical and physical upgrading of coal, especially for cokemaking. Mr. Smith's postgraduate studies have been in the area of coal comminution, especially by impact breakage.*

5.1 INTRODUCTION

Coals are encountered as virgin seams during exploration and mining, and as broken material in all kinds of use after mining. *In situ* coals are aggregates of diverse materials ranging from small isolated particles to substantial layers of minerals or organic matter. The organic matter varies greatly, ranging from charcoallike materials to gelified colloidal matter. The components of coal seams generally are disposed in lenticular arrays in essentially the same order in which the deposits built up during seam formation. Some coals are highly banded, whereas other coals show little or no macroscopic banding. Many of these features are described in macroscopic and microscopic petrographic studies reviewed elsewhere (e.g., Stach[1]) and in this volume (Chapter 3).

Gelified coal matter, such as bands of vitrain, usually exhibits distinct fractures that have become obvious in metabituminous and orthobituminous coals. Other lithotypes, such as durains, are more obviously composites and include more readily deformable exinitic macerals that may have had a significant part in accommodating deformations due to earth stresses, thereby avoiding the generation of the obvious fractures of vitrain. Das[2] has presented relevant examples of contortional structures.

The mechanical properties of individual coal macerals are poorly known, and we have had to rely on complex empirical tests, such as microhardness, to gain some idea of the relative mechanical properties of macerals and microlithotypes. Among the critical test conditions are factors such as the moisture content of the sample, the gas adsorbed within the coal, and the temperature at which tests are conducted. These test factors can alter apparent mechanical properties considerably, although all coals become brittle if the temperature of testing is low enough. At normal temperatures many coals deform nonelastically, especially if the specimen is small or in a condition corresponding to its freshly mined state. The type of stress system is as important as the properties of the bulk material.

Hard coals are commonly considered to behave as brittle materials, whereas brown coals and lignites often display brittleness only when the rates of stressing are extremely high. Thus, for example, comminution of brown coals is not satisfactory in machines that stress slowly.

[1] E. Stach et al., *Textbook of Coal Petrology,* Gebrüder Borntraeger, Berlin and Stuttgart, 1975.

[2] B. Das, *Int. J. Rock Mech. Min. Sci,* **11,** 453 (1974).

Brown and Hiorns[3] examined the situation at a period when efforts were being made to develop a general mathematical model for coals, but the model paid a minimum of attention to their highly heterogeneous composite character. Retrospectively it is clear that much of the work was influenced by notions of coal constitution that were current at the time, which led essentially to seeing coals as special kinds of polymeric macromolecular material and indeed to a widespread sympathy for the notion that a "coal molecule" existed. The anticipation of a physical model of coals has not been fulfilled—at least no model of the mechanical properties of coal thus far presented has been widely accepted. As with other technical areas of study, computers have enabled us to give more attention to problems of real material through the adoption of more complex and realistic mathematical descriptions. There is also little doubt that the growing general interest in composites[4] and the growing understanding of the emergence of special characteristics due to the peculiar structure of specific composites provide a better foundation for the study and use of mechanical properties of coal than did previous notions.

This chapter differs from its counterpart in the first Supplementary Volume[3] in two ways. First, we believe that the amount of progress made since 1963 does not warrant reworking the detailed evaluation studies of the previous volume; second, we emphasize here the significance of the mechanical properties of coal for utilization.

5.2 PRINCIPAL MECHANICAL PROPERTIES OF COAL

The principal mechanical properties of rocks (coals) are given in Table 5.1.[5–22] The significance of the various properties to the study of coals is indicated.

[3] R. L. Brown and F. J. Hiorns, *Chemistry of Coal Utilization,* in Supplementary Volume, H. H. Lowry, Ed., John Wiley and Sons, New York, 1963, pp. 119–149.

[4] K. Kendall, *Proc. R. Soc. London,* **344** (1637), 287 (1975).

[5] H. Ko and K. H. Gerstle, *Int. J. Rock Mech. Min. Sci.,* **13,** 81 (1976).

[6] D. W. Hobbs, *Strain Anal.,* **2**(4), 307 (1967).

The properties are not directly related to the heterogeneity of the specimens used in that they theoretically apply to isolated homogeneous particles as well as to heterogeneous ones. The interpretation depends on the degree of heterogeneity. For coals, the orientation of bedding planes of the specimens relative to applied stresses is the simplest expression of the major importance of heterogeneity. Furthermore, few of the properties have been or can be measured on homogeneous specimens taken from coals, because of the large specimen size required for tests relative to the smallness of any obviously homogeneous entity in coal specimens. Thus only inferences are used to assess individual homogeneous components of coal.

Fundamental properties of coal specimens are not easy to determine because of difficulties in preparing specimens and in conducting tests. For example, direct tensile tests, even with a satisfactory specimen, are not readily performed because it is so difficult to clamp a specimen without damaging it. Static properties are usually easier to determine than dynamic ones, and close agreement between these measurements of me-

[7] Y. G. Shliomovichus and G. G. Sennikov, *Sov. Min. Sci.,* **1,** 120 (1970).

[8] Australian Standard ASK 184-1969.

[9] R. H. Atkinson and H. Y. Ko, *Int. J. Rock Mech. Min. Sci.,* **10,** 351 (1973).

[10] C. H. Sholz, *J. Geophys. Res.,* **73**(4), 1417 (1968).

[11] D. R. Cooling, *Month. Bull. BCURA,* **29**(12), 409 (1965).

[12] M. Hobler, *Colliery Guardian,* **222**(4), 114 (1974).

[13] J. R. Hearst et al., *Int. J. Rock Mech. Min. Sci.,* **13,** 37 (1976).

[14] B. M. Butcher and A. L. Stevens, *Int. J. Rock Mech. Min. Sci.,* **12,** 147 (1975).

[15] A. Kidybinski, *Int. J. Rock Mech. Min. Sci.,* **3,** 279 (1966).

[16] B. M. Parish, *Br. J. Appl. Phys.,* **18**(2), 233 (1967).

[17] P. English and F. J. Hiorns, *Trans. Inst. Min. Met., Sect. C,* **75** (712), C78–C94, (1966).

[18] C. W. Goddard and A. F. Duzy, Preprint ASME No. 67 - WA/FU-1, November 1967.

[19] B. Das, *Fuel,* **52**(2), 118 (1973).

[20] B. Das, *Int. J. Rock Mech. Min. Sci.,* **9,** 783 (1972).

[21] F. J. Hiorns and B. M. Parish, *J. Inst. Fuel,* **39,** 126 (1966).

[22] B. M. Parish, *J. Inst. Fuel,* **40,** 76 (1967).

Table 5.1 Principal Mechanical Properties of Coal, Excluding Powders, Assemblies, Aggregates, and so on

Properties	Measurement Technique	Significance for Appreciation of Coal	Ref.
	Fundamental Properties		
Rupture moduli (ultimate strengths			
Static strengths			
Compression			
Uniaxial	Cylindrical or cubical specimens stressed between platens along axis	High	5
Biaxial			
Triaxial	As above, but with surrounding fluid at pressure	High	6, 7
Tension			
Direct	Cylindrical or shaped specimen extended along axis	Nil	
Indirect	1. Cylinder compressed diametrically between platens		5, 6
	2. Cylinder with hole along axis stressed diametrically		6
	3. Doublepoint loading with hemispherical indenters		5, 6
	4. Stressing square discs with pair of indenters	High	6
	5. Bending a beam	Low	5, 6
Shear			
Simple	Shear cells for cubes or square prisms	Low	
Torsion	Torque applied to solid or hollow cylinders	Nil	
Dynamic strengths	cyclic		
Moduli			
Static			
Compliance matrix	Multiaxial test cell for cubes	High	5, 8, 9
Young's modulus	Uniaxial loading, bending beams		5
Modulus of rigidity	Uniaxial loading, bending beams		5
Poisson ratio	Uniaxial loading, bending beams		5
Bulk modulus	Uniaxial loading, bending beams		5
Volumetric strain	In compression using special gauges		10
Dynamic			
Young's modulus	Fundamental flexural vibration		5, 11
Modulus of rigidity	Fundamental flexural vibration		5
Stress wave velocity			12, 13, 14
Rheology			
Creep behavior	Uniaxial loading, loaded beams	High	5, 15
Plasticity	Cubes between platens or rolls	Low	16
Surface energy			
Specific fracture energy	Fracture by bending or cleavage	High	17
Friction			
Internal friction	Fundamental flexural vibration, triaxial compression		7, 16
Sliding friction	Velocity of surface slippage inclined to line of compressive force		9
Interparticulate friction	Shear cell leg Jenike or annular cell, angle of repose		
	Semiempirical Properties		
Hardness			
Scratch	Mohs scale, surface area of standard scratch, etc.	Low	11
Indentation, static			
Macro	Spherical indenters (Brinell, Meyer, Rockwell) conical (Ludwick Rockwell, NCB), pyramidal (Knoop, Vickers)	Low	11,18
Micro	Vickers and Knoop using pyramidal indenters	High	11, 19, 20

Table 5.1 (*Continued*)

Properties	Measurement Technique	Significance for Appreciation of Coal	Ref.
	Empirical Properties		
Specific energy (cutability)	Drilling Rig (MRDE)		18
Grindability	Hardgrove, Laboratory Mills	High	
Impact strength (arbitrary stress rates)	Shatter tests	High	
Abrasivity	Drill wear, ball, roll and surface wear	High	18, 21, 22

chanical properties cannot be expected with such variable materials as coals.

The most common tests are empirical ones, using assemblies of particles. The tests usually include some degree of comminution as well as an appraisal of effects through estimates of changes in particle size. There is no clear relationship between this group of empirical tests and the fundamental properties, although empirical correlations can be developed.

It is evident that the condition of a sample and its previous mechanical history—as manifested in crack frequency—have a profound effect on specific results and make direct application of results in a wider context difficult. Therefore this chapter discusses three broad categories: seams, individual pieces, and assemblies of particles. However a number of inherent or fundamental coal properties that are characteristic of the coal mass or its surface are discussed first.

5.2.1 Elasticity

In common with many other rocks, Hard Coal [gross specific energy > 27.1 MJ/kg (maf)][8] exhibits a stress-strain curve in compression with an initial nonlinear portion associated with closure of cracks followed by an essentially linear elastic range.[5,23] Evans and Pomeroy[23] have suggested that the intercept on the strain axis of the extrapolated linear portion of the graph may be considered to be a measure of the open cracks and/or layers of very soft material in the specimen. For coal *in situ* the existence of the initial nonlinear range may be questioned.[5] A viscous effect due to frictional sliding along some closed cracks could be associated with the linear range, as evidenced by hysteresis effects on unloading.[24] Evans and Pomeroy[23] have shown, using Barnsley "Hards" (a colloquial name for this durain) coal, that the viscous component of deformation does not affect the measured elastic moduli to any appreciable extent, provided the strain is measured within a few seconds of application of load.

The deformation behavior of a solid exhibiting linear elasticity is described completely by the material stiffness matrix or its inverse, the compliance matrix, relating principal strains and stresses. Ko and Gerstle[5] have examined cubical rock specimens with a multiaxial compression cell, which can be used to measure deformations at any principal stress state. They used the cell along with a triaxial test cell to determine the compliance matrices for two high-volatile (32 and 39%) coals from the United States. They found that although the principal compliances were sufficiently similar to suggest isotropic behavior, the shear compliances were much larger than those expected for an isotropic elastic material, suggesting that coal could reasonably be modeled as an orthotropic material for engineering approximations.

Evans and Pomeroy,[23] on the other hand, found that of the principal dynamic elastic moduli for Barnsley "Hards" coal, the modulus perpendicular to the bedding plane was significantly lower than that parallel to the bedding plane. However the moduli parallel to the bedding plane appeared to be independent of direction of axis relative to the

[23] I. Evans and C. D. Pomeroy, *Strength, Fracture and Workability of Coal,* Pergamon Press, Oxford, 1966.

[24] B. T. Brady, *Int. J. Rock Mech. Min. Sci.,* **10,** 291 (1973).

cleat planes, and Evans and Pomeroy concluded that coal tended toward transverse isotropy.

Determinations of dynamic Young's modulus on Barnsley "Hards" coal have given 4.0 and 3.3 GPa, parallel and perpendicular to the bedding plane, respectively.[23] This compares with an average static value of 4.0 GPa perpendicular to the bedding plane for large *in situ* specimens in South African collieries,[25] where the data also showed an independence of specimen size. Using a triaxial test cell, Hobbs[6] found that Young's modulus for sedimentary rock, including coal, increases with confining pressure. Evans and Pomeroy found that for any particular coal the static modulus was lower than the dynamic.

One method of measuring dynamic elastic moduli, as noted in the previous supplement, is to measure the velocity of sound through the specimen for which the dependence is:

$$\left(\frac{E_\alpha}{\rho}\right)^{1/2} = \frac{l}{t} \tag{5.1}$$

where E_α = dynamic Young's modulus
ρ = density
l = path length
t = time

The velocity of longitudinal waves has been measured by various workers[12–14] as about 2 km/s in the direction of bedding planes. Shear wave velocity has been reported as being 1.3 km/s.

The limited tensile data of Evans and Pomeroy suggest that although Young's modulus in tension is about the same as in compression, the Poisson ratio tends to be lower for Barnsley "Hards" coal stressed parallel to the bedding plane.

5.2.1.1 *Time-Dependent Deformation*

Hard Coal is a viscoelastic material that can be represented by the Burgers model.[5,15] By the model, the application of a compressive stress σ to coal will result in a strain ϵ_B, which is given by[5]:

$$\epsilon_B = \sigma\left[\frac{t}{\eta_1} + \frac{l}{E_1} + \frac{1 - \exp(-tE_2/\eta_2)}{E_2}\right] \tag{5.2}$$

where t is time.

[25] W. L. Van Heerden, *J. S. Afr. Inst. Min. Metall.*, **75**(8), 207 (1975).

The component $\sigma t/\eta_1$ represents irrecoverable strain due to viscous flow, where η_1 is the effective viscosity when all transient components have disappeared. This viscosity is given by Evans and Pomeroy as 10^{15} Pa-s for Barnsley "Hards" coal. The recoverable component σ/E_1 is the instantaneous strain, and the remaining term represents recoverable elastic aftereffects.

Kidybinski[15] has determined the constants in Equation 5.2 for a cannel coal using a loaded beam as E_1 = 4.6 GPa, E_2 = 32.4 GPa, $\eta_2 = 5.7 \times 10^{15}$ Pa-s, and $\eta_1 = 1.9 \times 10^{17}$ Pa-s. A retardation time ($t = \eta_2/E_2$) of 50 hr was also determined. This figure is some 70 times larger than the value reported by Terry and Morgans[14] for Barnsley "Hards" coal. The difference in values may be attributed to differences in constitution of cannels and "Hards." Kidybinski[15] also postulated a more complicated rheological model for cannel coal, sandstone, and mudstone and proposed strength indexes for both delayed and instantaneous elastic strains.

5.2.2 Strength

Hard Coal is a brittle material in that failure is signified by a sudden and devastating change in structure. The ultimate stress at which such failure occurs is the strength, and in most cases it is clearly defined.

Breakage of a solid requires a certain minimum energy to create new surface. In practice, experimental determination of surface energy must include other energy consumed in crack propagation. The specific fracture energy has been measured by English and Hiorns[17] at approximately 10 J/m² using controlled cleavage fracture of bar specimens and the brittle fracture of bent beams.

The dependence of strength on specimen size, especially in compression, has been widely investigated and is treated in detail in Section 5.3.2. Although there has been some disagreement regarding the order of the dependence of compressive strength σ_c on size (e.g., length of the cube side H), the form of the dependence is generally agreed to be:

$$\sigma_c = kH^{-r} \tag{5.3}$$

The parameter k is a strength characteristic of the coal (and, to some extent, the test), but it has been shown[26] to be independent of

[26] W. A. Hustrulid, *Rock Mech.* **8**(2), 115 (1976).

size of specimen. The widespread adoption of k appears to have the advantage of avoiding ambiguity in reporting strength figures. In the SI system the units of k are Pa m^r. The exponent r is normally in the range 0–0.5.

The high degree of nonuniformity of coal with respect to the distribution of cleats and cracks and components of various strengths might be expected to produce a high degree of scatter of strengths. Evans and Pomeroy[23] presented a comprehensive account of their compression tests on English coals. They concluded that the crushing strength of coal cubes of a particular size showed a wide variation. However in Fig. 5.1 their results are included in a plot of standard deviation versus mean compressive strength determinations on a wide range of rocks, as reported by Coates.[27] The linearity of the relation on a log-log plot shows that the number of measurements necessary to determine the average strength within ±20% for any rock depends simply on the average strength. An important qualification here is that the data of Evans and Pomeroy are classified according to the orientation of the specimen with respect to bedding planes and cleat direction. If the specimens were randomly oriented, a wider scatter would be observed. A similar curve is observed with tensile strength, although there are different constants in the regression equation.

The triaxial compressive strength of coal has been of particular interest in relation to the confined breakage of coal in the seam. A detailed experimental analysis of the triaxial compression of two English coals as conducted by Murrell[28] was covered in the first Supplementary Volume.[3] Shliomovichus and Sennikov[7] have found that the relationship between confining pressure σ_3 and compressive strength σ_c of a Russian coal was rectilinear up to about $\sigma_3 = 12$ MPa. Hobbs[6] has found a slightly nonlinear relation up to about 35 MPa. Shliomovichus and Sennikov gave the relation:

[27] D. F. Coates, *Rock Mechanics Principles*, Canadian Mines Branch Monogr. No. 374, 1965.

[28] *Mechanical Properties of Non-Metallic Brittle Materials*, W. H. Walton, Ed., Butterworths Scientific Publications, London, 1958, pp. 123–156.

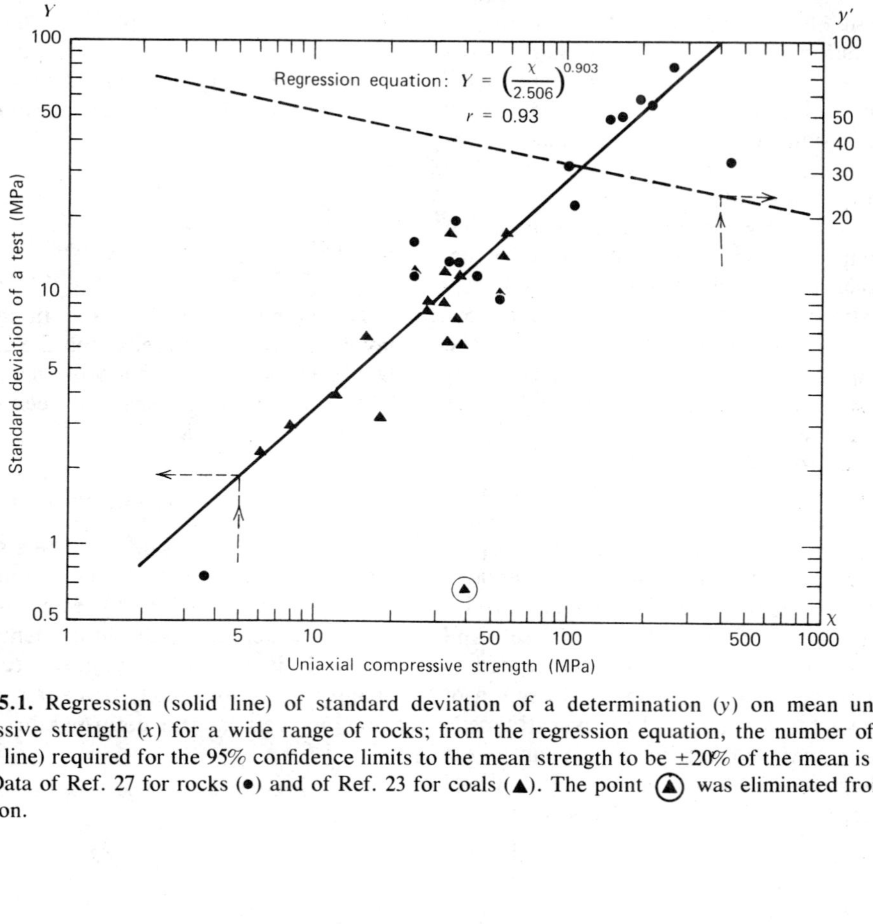

Figure 5.1. Regression (solid line) of standard deviation of a determination (y) on mean uniaxial compressive strength (x) for a wide range of rocks; from the regression equation, the number of tests (dashed line) required for the 95% confidence limits to the mean strength to be ±20% of the mean is given by y'. Data of Ref. 27 for rocks (●) and of Ref. 23 for coals (▲). The point (▲) was eliminated from the regression.

$$\sigma_c = \frac{2\cos\rho}{1-\sin\rho}K + \frac{1+\sin\rho}{1-\sin\rho}\sigma_3 \qquad (5.4)$$

The coefficient of cohesion, K, and angle of internal friction, ρ, can be found from experiment by the relations:

$$K = \frac{1-\sin\rho}{2\cos\rho}A \quad \text{and} \quad \sin\rho = \frac{B-1}{B+1} \qquad (5.5)$$

where A and B are the intercept and gradient, respectively, of the plot σ_c versus σ_3. Values of $K = 4.4$ MPa and $\rho = 38°$ were found for Volkoviskii Seam coal, for which no rank parameters were given.

Very little attention has been given to the measurement and understanding of tensile and shear failure in coal. The major contribution sighted was that of Evans and Pomeroy,[23] who noted that even when compressive stresses are applied to coal, the breakage appears to be essentially tensile in nature.

Evans and Pomeroy found that the determination of tensile strength of coal by bending tests gave a higher result than other tensile tests. This was attributed to a modification of the surface of the specimen in preparation, because the maximum tensile stresses are set up at the surface. The investigators advocated a method of diametrical compression of a disk or, alternatively, indentation testing, which were described in the previous supplement.

A disadvantage of both these tests is that the tensile stress is associated with an orthogonal compressive stress. The compressive stress alone rarely leads to failure, but the combination of compressive and tensile principal stresses means that large shear stresses are set up in the material. Thus testing coals of low shear strength by these methods may not be suitable.

Evans and Pomeroy provided a detailed account of the variation in tensile strength with orientation of the specimen with respect to cleats and bedding planes. Not enough coals were tested to enable reliable conclusions on the relative effects of rank and petrography. The tensile strength tended to be greatest for stressing in the direction of least visible fractures. The lowest tensile strength was 480 kPa perpendicular to the bedding plane of a medium volatile coal and the largest value was 4070 kPa parallel to the bedding plane and main cleat for a higher volatile coal.

The researchers also demonstrated that tensile stress depended on specimen size, using disks of Barnsley "Hards" coal with tensile stress acting parallel to bedding planes. The thickness of the disks ranged between about 8 and 50 mm. It was found that tensile strength decreased with increasing specimen size according to the power law:

$$\sigma_t \propto (\text{thickness})^{-0.23 \pm 0.03} \qquad (5.6)$$

It is interesting to note that this exponent is similar to that determined for cubes of the same coal crushed perpendicular to the bedding planes. The similarity tends to support Evans and Pomeroy's notion that failure in crushing tests is the result of tensile stresses. Evans and Pomeroy also presented earlier data of Pomeroy and Foote[29] on the ratio of compressive strength to tensile strength for a range of coals. In line with the abovementioned hypothesis on crushing, they compared the respective strengths determined at right angles. They found:

$$\frac{\sigma_c}{\sigma_t} = 28$$

where σ_c taken parallel to bedding
σ_t taken perpendicular to bedding

$$\frac{\sigma_c}{\sigma_t} = 17$$

where σ_c taken perpendicular to bedding
σ_t taken parallel to bedding

The compressive test used in this case involved dry steel platens, and the strengths would probably have been of the order of 30% lower if the platens had been suitably lubricated.

5.2.3 Hardness and Abrasiveness

Hardness can be considered to be a semiempirical property in that like compressive strength, it is measured by empirical tests, but it is closely related to fundamental properties. It is a convenient measure of the liability of a substance to cause abrasive wear as well as the deformability of the

[29] C. D. Pomeroy and P. Foote, *Colliery Eng.*, **37**, 146–154 (1960).

material. Although coal itself is not particularly hard, some of its associated minerals are hard enough to produce heavy abrasion, as indicated by Cooling.[11]

Cooling's excellent review described the common hardness tests including those of most relevance to coal: rebound and indentation hardness. The use of portable scleroscopes, such as the Schmidt hammer, which measure dynamic hardness by measuring the rebound of an indenter, is convenient for *in situ* measurements, although their application is usually confined to flat horizontal faces. Das[19] concluded that rebound strength is a complex function of ultimate strength, elastic limit stress, Young's modulus, elastic and ultimate strain, and brittleness. He noted that above a certain drop height a body is stressed beyond its elastic limit, and plastic failure begins to take over. In this case the share of energy returned in rebound energy is reduced, and correlations between rebound hardness and other mechanical properties are more complex. Therefore the drop height of the Shore scleroscope (247 mm) should be reduced for lignites and coking coals. Das's results of Shore hardness on a range of vitrinites are scattered, but a general trend can be observed. As with many mechanical and other properties (including microhardness), an inflection (minimum rebound strength) is observed at the 88–90% carbon level (dmmf).

The static indentation hardness test is the most commonly used. It is usually expressed as the ratio of the load applied to the contact area of the permanent indentation. The microindentation tests are particularly suitable for testing coal macerals and minerals, because they are among the few tests that can be performed on a wide range of small entities. The loads are such that the relatively brittle materials in coals can be tested without fracture occurring.

Cooling[11] has stressed the care necessary for the construction and operation of equipment to achieve meaningful results. One major source of apparent discrepancies between various studies was attributed to the different loads used. Cooling noted that for Vickers and Knoop tests, which use pyramidal indenters, the geometrical similarity of the impressions over the whole range of penetration should mean that hardness is independent of load. At loads below 1 kg, however, this was not necessarily the case. It is apparent from the definition below of Vickers hardness H_V that for this independence to apply the load must vary as the square of the diagonal d, where d is in millimeters:

$$H_V = \frac{L}{A} = \frac{2L}{d^2}\sin\left(\frac{\theta}{2}\right) \tag{5.7}$$

where A = actual area of impression (mm^2)
θ = angle between opposite faces of pyramid (136°)
L = load (kg)

On investigating the effect of load on Vickers indentation hardness of coal, Das found that values of N in the relation $L \propto d^N$ were close to 2. He recommended, however, that the load used always be quoted along with microhardness values.

Cooling[11] has concluded that despite numerous values in the literature of microhardness for coal and its associated minerals, experimental conditions such as load and orientation of specimen have not been well enough reported to enable other investigators to reproduce the work readily. The hardness of coal is generally in the range 10–70 kg/mm^2, depending on rank, with a minimum near 20% VM (~90% C, dmmf). However some anthracites and individual macerals may be as hard as 200 kg/mm^2 or even so elastic that the indentation is not preserved (e.g., as with cutinite). Cooling also stated that no anisotropy has been observed in the hardness of coal specimens.

Since coal is considerably softer than steel, abrasiveness depends almost exclusively on the type, quantity, and distribution of hard minerals in the coal. The hardest minerals commonly associated with coal or with the rocks surrounding the coal measure are quartz and pyrites, which have Vickers hardness numbers of 1100–1260 and 840–1130, respectively, compared with about 600–700 for a hard steel (see Hiorns and Parish[21]). Parish[30] investigated the effect of mineral matter type and content on coal abrasiveness using the Yancey-Geer-Price[31] test machine. These results suggested a linear relation between wear rate and mineral content. Quartz caused a three- to fourfold increase over

[30] B. M. Parish, BCURA Res. Rep. No. 363, February 1970, 113 pp.

[31] *Trans. AIME, Soc. Min. Eng.*, 262–268 (March 1951).

siderite, which was in turn about 3 times more abrasive than calcite.

During the period of this review a number of empirical tests have been developed to measure abrasiveness. The most relevant for pulverizers, which probably involve the most damaging environments, appears to be that of Hiorns and Parish.[21] In their test a sample of coal was passed between two test rollers of about 100 mm diameter. The gap between the rolls, the load on the rolls, and the differential speed of the rolls could be varied. The metal lost from the rolls was measured (to the order of 10^{-5} g) by a radioactive count on the pulverized coal, and one of the rolls was made of a section of steel containing a radioactive isotope of iron. Parish's[22] study with low-ash (1.6–3.7%, adb) English coals showed the expected parabolic relation of wear to percentage volatile matter. The hardness of coal did not correlate with wear. As expected from their similar correlations with volatile matter, grindability and compressive strength were related to wear.

5.2.4 Friction

The frictional properties of coal surfaces do not appear to have been discussed or measured in any recent literature cited in *Fuel Abstracts* [later *Fuel Abstracts and Current Titles* (FACTS) published by the Institute of Fuel, U.K.], the major initial source of references for this review.

However the work done at the Mines Research Establishment of the National Coal Board (Great Britain), briefly mentioned in the first supplement,[3] has been treated in more detail by Evans and Pomeroy.[23] They report the coefficients of surface friction μ, as determined against steel and brass. When tested against steel, it was found that μ increased with sliding distance up to about 300 mm (1 ft). This increase was shown to result from a change in the surface of the coal specimen, which was explained by Brown and Hiorns[3] as flattening of asperites. The dependence of μ on volatile matter for some British coals was such that μ started at about 0.5 for a volatile matter of 5% (maf), falling to about 0.4 at 20% (maf), and rising to about 0.6 at 40% (maf). Here μ was measured over the first 6 mm (¼ in.) of sliding. It was also found that μ decreased with increasing load according to:

$$\mu \propto (\text{load})^{d_f}$$

where $d_f \simeq 0.1$.

The hardness of the steel was found to have no effect on μ when a mean pressure of 345 kPa (50 psi) was employed. Wetted surfaces led to a μ of about 70% of the dry value. Evans and Pomeroy considered that this reduction was due not to a change in the mechanical properties of the coal but to the effect of a surface film of water molecules on the sliding surface.

The angle of internal friction (or slope of the Mohr envelope) for hard coal is generally in the range 35–40°.[7,32]

5.3 MECHANICAL PROPERTIES AND FORM OF THE COAL SAMPLE

5.3.1 Coal Seams

Very little attention seems to have been given to measuring and describing the overall mechanical properties of virgin coal seams. Such measurements would have to depend largely on dynamic methods (e.g., sound waves and radar).

Schwaetzer and Desbrandes[33] have discussed problems of reconciling laboratory and field measurements of the speed of sound in coal and associated rocks. For seams, the observed speeds were higher than laboratory results. The anisotropy of sonic velocity (relative to the seam) was larger for coal samples than for the most common coal measure rocks; but the results were imprecise, and the velocity of sound in coal was half that of the rock samples (approximately 1.5 : 3 km/s). The relation of velocity V_v (m/s) across the bedding for the Lorraine Seam to depth Z (m) was given by:

$$V_v = 2180 + 0.33Z \text{ (m/s)} \qquad (5.8)$$

and the effect of depth was considered most important. Schwaetzer and Desbrandes were optimistic about the applicability of sonic measurement to seam appraisal.

Mathematical descriptions of seams in highly simplified terms regularly appear in

[32] A. H. Wilson and D. P. Ashwin, *Min. Eng.*, No. 141, 409–417 (June 1972).

[33] T. Schwaetzer and R. Desbrandes, *Rev. Inst. Fr. Pet.*, **20**(1), 3 (1965).

the literature,[34,35] but observation of mechanical properties *in situ* does not. Similarly complex contortional patterns[2] have been noted for seams, but again analysis in terms of mechanical properties has not.

In situ effects of confining pressures have been inferred to be similar to those on laboratory specimens (e.g., biaxial, triaxial tests). However the feasibility of direct *in situ* measurements is low, even though they have been considered essential (Bieniawski[36]).

Holland and Olsen[37] inferred from laboratory studies that *in situ* coal properties are affected by interfacial friction (moisture effects). The inference seems to be in accord with current mining opinion, but the effects have not been observed directly.

Indentation (penetration) tests on seam faces (or bore walls) have been tried.[23,38] There appears to be no doubt that these methods have practical value to miners, but the tests need much more development.

Overall it can be stated that direct measurements of mechanical properties of coal *in situ* have not been extensive, and the few examples noted above only testify to the absence of a comprehensive treatment.

5.3.2 Single Pieces

A piece or particle is considered to be an isolated mass of coal bounded by a continuous surface or interface. Normally it is stressed by external objects and/or a contiguous medium, and the points of application of stress will be found on the surface of the piece. A particle possesses the inherent properties of the parent material, which may have been modified by its mechanical history, but a particle's behavior in a given stress situation depends on the locus of its boundary. Particle boundaries can be described by the basic properties of shape, size, and aspect. Because of anisotropy, the orientation or aspect of a coal particle with respect to the direction of the applied stress affects the response.

[34] B. V. Vlasenko and G. I. Gritsko, *Sov. Min. Sci., Part 6*, 607 (1965).

[35] B. V. Vlasenko and F. M. Erlikhmann, *Sov. Min. Sci.*, **8**(5) 491 (1972).

[36] Z. T. Bieniawski, *Eng. Geol.*, **2**(5) 325 (1968).

[37] C. T. Holland and D. A. Olsen, *Trans. AIME, Soc. Min. Eng.*, **241,** 323 (1968).

[38] B. Sh. Vinokur and G. D. Morozzov, *Sov. Min. Sci.*, **8**(5) 491 (1972).

The response of a piece of coal to a given stress situation is affected by the locus of its boundary in three ways:

1. The points of application of stress may depend on the position of the boundary.
2. Dynamic processes within the particle, such as wave propagation, are influenced by the particle surface.
3. The size of the particle may affect the mode of response and the ultimate stress at which the response is observed.

When a particle of coal or any other material contacts a solid object, the magnitude and pattern of stress established in the particle depend, in part, on its shape. Highly localized stress concentrations are more likely to occur at the point of contact if the particle has an irregular surface. Other points of stress concentration occur at notches on the surface as well as at the tips of cracks within the particle.

In the case of dynamic stressing, the shape of the particle also determines the path of the stress wave within it. When the surrounding medium is air, most of the compressive wave is reflected as a more destructive tensile wave. The compressive strength of coal is an order greater than its tensile strength. The proportion reflected and transmitted depends on the angle of incidence. In a composite such as coal the course of the stress waves is also modified by reflection and refraction at boundaries of its component entities. If at any time the particle ruptures, the wave propagation is modified.

The composite nature of coal means that pieces generated during the normal winning and utilization activities are of irregular shape and widely varying sizes. The advent of computer techniques for finite element analysis has meant that problems such as the calculation of stress distribution in a static situation[39] and the propagation of stress waves in a dynamic situation can be analyzed, if we know the material properties and have a mathematical description of the locus of the boundary of the object. For irregular objects such a description is a tedious and complex exercise[40] that is impractical at pres-

[39] S. S. Peng, *Int. J. Rock Mech. Min. Sci. Geomech. Abstr.*, **13**,97-10 (1976).

[40] K. Gotoh and J. L. Finney, *Powder Technol.* **12,** 125–130 (1975).

ent. In any case analytical techniques are probably not powerful enough to handle such irregular shapes. Normally descriptions of the shape of irregular particles appear to be limited to measures of their deviation from regular shapes such as spheres, ellipsoids, and prisms. However the size of a coal particle can be variously described by its volume, its projected area, or any of a number of characteristic dimensions that are loosely termed "diameters." In the strictest sense the diameter is equal to the maximum distance between parallel supporting planes.[41]

As a result of the complications introduced by irregular particle shape, standard tests for mechanical properties stipulate regular specimens. Because coal is such a difficult material to shape, tests have often been performed with irregular or at best roughly shaped specimens. Various investigators have developed empirical correlations between strengths of irregular and regular specimens for different rock materials.[42–44]

Hiramatsu and Oka[45] used three-dimensional photoelastic experiments on a sphere, a cube, and a rectangular prism of epoxy resin to show that the stress states within the vicinity of the axis of compressive loading are much the same in each case. They deduced that provided the loaded area and the Poisson ratio of the specimen are within certain limits, and provided the specimen breaks in tension, the ultimate tensile strength (S_t) is a function of the load at failure (F_0) and the distance between the points of loading (d) given by:

$$S_t = \frac{0.9F_0}{d^2} \tag{5.9}$$

However such a relation may not be applicable to coal and other anisotropic rocks.

The effect of size on the mechanical behavior of a piece of coal has been reasonably well investigated, and correlations based on experiment have been developed, in particular for failure in compression. In the previous review the relation between cube size and compressive strength was given as:

$$\sigma_c = \frac{k}{\sqrt{H}} \tag{5.10}$$

where σ_c = compressive strength of a cube having a side of length H
k = constant for each particular coal

Note that this relation is that of Equation 5.3 when $r = 0.5$.

Evans and Pomeroy[23] replaced the exponent $-\frac{1}{2}$ in Equation 5.10 with the values -0.17 and -0.32 for cubes of Deep Duffryn and Barnsley "Hards" coal, respectively. They noted similar exponents when testing irregular specimens, and they appear to have conducted one of the few studies providing statistically significant results in terms of the number of specimens tested (see Section 5.2). In a theoretical consideration of the compressive failure of cubes, Evans et al. concluded that the exponent of Equation 5.3 increases from $-\frac{1}{2}$ (when preexisting cracks are commensurate with the size of the specimen) as the size of flaws decreases relative to the specimen, up to the limit of zero (meaning no size dependence) when cracks are absent. A further possibility has been raised by Gonano and Brown,[46] namely, the less brittle the failure, the closer the exponent is to zero.

Bieniawski,[47] when testing large coal blocks underground, concluded that the decline in compressive strength does not continue indefinitely with increasing size. He proposed 1.5 m as the length of the limiting cube.

Hustrulid[26] reviewed comprehensively the size-strength problem and concluded that Equation 5.3 is valid for cubes up to a limiting size of about 1 m. He attributed the notable difference in the results of Evans and Pomeroy[23] to the difficulties of preparing and testing small cubes (3 mm and larger).

[41] I. M. Yaglom and V. G. Boltyanskii, *Convex Figures* (Trans. P. J. Kelly and L. G. Walters), Holt, Rinehart and Winston, New York, 1961.
[42] D. P. Singh and J. N. Singh, *J. Mines Met. Fuels,* 18–22 (January 1974).
[43] S. K. Bordia, *Met. Miner. Rev.,* **5** (April 1967).
[44] L. S. Srivastava and S. Behari, *J. Soil Mech. Found. Eng.,* **7**(1), 15–26 (January 1968).
[45] Y. Hiramatsu and Y. Oka, *Int. J. Rock Mech. Min. Sci.,* **3**(2), 89–99 (1966).
[46] L. P. Gonano and E. T. Brown, *5th Aust. Conf. Mechanics Structures Materials,* Melbourne, 1975, pp. 205–208.
[47] Z. T. Bieniawski, *Int. J. Rock Mech. Min. Sci.,* **5**, 325–335 (1968).

Schoenert[48] summarized the factors that lead to a continuing increase in strength as particle size decreases. The first factor, as was recognized by Weibull,[49] is that the strength of a particle is determined by the most effective flaw present, and that the occurrence of a flaw of a certain high effectiveness becomes less probable as particle size decreases. Effective flaws originally present are depleted as larger particles are crushed. Schoenert claimed that in addition to flaw depletion, two other aspects are important, namely, capacity for storing elastic energy and the influence of plastic deformation. The capacity for storing elastic energy is a volume property; however the energy consumed by a geometrically similar crack is proportional to the square of specimen diameter. Therefore small specimens have less energy available for fracturing and consequently require higher stress levels for crushing.

The possibility that differences in the relations arise from different mechanical histories of the specimens, including possible compensation of flaw depletion by the introduction of new flaws during specimen preparation, has not been excluded in the studies cited.

That the apparently brittle compressive failure of coal specimens changes to plastic deformation when the particles are small enough has been long known.[3] Parish[16] has studied these phenomena theoretically and experimentally. She derived the critical size d_c at which cubes adopt plastic failure:

$$d_c = x\left[\frac{C_x}{H(1+f(\mu))}\right]^{1/r} \tag{5.11}$$

where r is the empirical exponent of Equation 5.3 and μ is the coefficient of friction between coal and platen,

where $f(\mu) = \frac{1}{3}\sqrt{3}$ if $\mu \geq \frac{1}{3}$
$= \frac{1}{3}$ if $\mu < \frac{1}{3}$
H = microhardness (indentation)
C_x = compressive strength of cube of side x

[48] K. Schoenert, *Trans. AIME, Soc. Min. Eng.*, **252,** 21–26 (March 1972).
[49] W. Weibull, *Ing. Vet. Akad.*, **151,** 1 (1939).

Tests with irregular particles showed that none failed plastically for sizes larger than d_c and that the proportion deforming plastically increased as particle size decreased below d_c. Parish's results can be fitted by the log normal probability distribution, and they accord with her explanation that the apparent strength of irregular particles is reduced because loading occurs at a few points compared to theoretical loading of cubes.

Austin et al.[50] measured shape factors for 16 American coals from anthracite to high volatile in rank for the products of grinding in a Hardgrove machine. The shape factor variation with volatile matter (maf) was the typical "parabola" of bright American coals, with medium volatile coals being most nearly cubical. Similar shape factors were observed for tests with a disk and ball mill.

The literature on single specimens of coal is thus scanty compared to that for other materials, and it does not include comparison of coals with other material in ways that enable us to identify a distinctive analogue of coal.

5.3.3 Particle Assemblies

In the extreme, any lump of coal can be considered to be an assembly of particles, but the cohesiveness between contiguous entities appears to be of the same order as that of the entities themselves. One limit of an assembly is thus set as those collections of particles with interparticle bonds about an order weaker than the intraparticle bonds.

"Bulk materials," as they are commonly called, are included as assemblies of particles, as are all assemblies for which the mechanical properties are not merely those of the individual particles. Thus fluidized beds, slurries, and both cohesive and incoherent bulk materials are included within the class.

Coals in the various assemblies that we encounter generally do not behave differently from other solids.[51,52] The low density of coal relative to water, however, and the generally

[50] L. G. Austin, R. P. Gardiner and P. L. Walker, Jr., *Fuel*, **42,** 319 (1963).
[51] G. Herdan, *Small Particle Statistics*, Butterworths Scientific Publications, London, 1960.
[52] *The Rheology of Particulate Systems*, Symposium, Power Advisory Centre, Harrogate, U.K., March 1973.

hydrophobic nature of coal matter confer some special characteristics on coal assemblies in relation to water. In addition, the influence of the minerals contained in coal assemblies can alter profoundly their behavior under stress.

Coal assemblies are the result of the moving of broken coal. Indeed, all static coal assemblies are formed from moving ones. For this reason the character of mobile coal assemblies is indicated first.

During flow, coal tends to segregate so that larger particles collect together, as do smaller ones. This is evident both in the downflow in chutes and in the heap formed as the flowing coal comes to rest. Similarly, segregation develops for most coal assemblies during pneumatic (including fluidized) and hydraulic transportation unless the size distribution is tightly graded.

Although the rheology of coal suspensions is not usually Newtonian, most engineering design work assumes continuum properties and simple mathematical stress-deformation relations; that is, the designs do not allow (hypothetically) for the segregation of particle sizes in preferred zones.

Segregation during the formation of a static bed, as in a bin, originates in the accumulation of a "wall" of fines around the impact area. The larger particles roll down the heap, but the accumulated fines roll only at intermittent intervals that depend on the size distribution and the feed rate. Because of the cohesive effects of moisture, its increase in content in the flowing coal can reduce segregation. In general, the motion of coal assemblies does not tend toward the formation of an assembly of minimum voidage.

Segregated or not, static assemblies exhibit resistance to tensile and shear stresses. General reviews[53] cite many equations relating tensile strength to properties of the assembly: voidage, particle density, and cohesive-adhesive forces. The ratio of solids to voids seems to be the most significant single factor for the looser types of assembly, whereas the strength of the interparticle bond (e.g., a water bridge) becomes dominant if the proportion of such bonds is large. The effects of the bridges can be calculated from considerations of capillarity (e.g., calculations of strength of pellets). Linear Mohr-Coulomb relations are usually used to analyze tests, but with the help of a computer the use of nonlinear relations is not difficult.[54]

Flow from the static condition depends on both the shape and the size of the discharge aperture. Circular shapes pass more particulate material per unit area than do other shapes. Flow also depends on the cohesion of the assembly. Toyama[55] has shown that the inner core of solids flowing down a tube can maintain constant relative positions of particles. But particles in the boundary zone (4–8 particle diameters) do move relative to each other, and abrasion occurs in that zone.[55] Such phenomena are likely to occur in funnel flow from bunkers.

Jenike[56] discussed the transition between mass and funnel flow (Chapter 9). His graph shows approximately that if the angle of internal friction exceeded 40° (¾ semicone angle of bin), funnel flow develops. Mass flow bunkers are preferred in most industrial installations because they avoid some ill effects of segregation and useless dead space.

Oiling of coal is often practiced at coke ovens to improve the flow and increase the bulk density. It seems most likely that these effects are the result of elimination of pendular bridges of moisture as the oil displaces moisture from the surface of the coal matter. To say that the particles are "lubricated" by oiling is merely another way of expressing the decrease of cohesion.

The soil mechanics of spoil dumps (tips) have been much studied since the Aberfan disaster. Thomson and Rodin[57] and Taylor[58] described British work that showed that the fine component of the tips tended to contain twice as much coal matter as the coarse components. Clays and moisture, particularly water flows, dominated the development of instability, which could be calculated from analyses of tests using the linear Mohr-Coulomb relations. The studies of Roberts[59] ranged widely over soil and rock mechanics as they relate to mining, emphasizing assem-

[53] H. Schubert, *Powder Technol.*, **11**, 107 (1975).

[54] M. R. Kingston and A. J. M. Spencer, *J. Mech. Phys. Solids*, **18**, 233 (1970).

[55] S. Toyma, *Powder Technol.*, **4**, 214 (1971).

[56] A. W. Jenike, *Powder Technol.*, **1**, 327 (1967).

[57] C. McK. Thomson and S. Rodin, *Colliery Spoil Tips After Aberfan*, Institution of Civil Engineers, London, 1972.

[58] R. K. Taylor, *Geotech.*, **24**, 683 (1974).

[59] A. Roberts, *Min. Eng.*, **127**, 621 (1968).

blies of particulate materials as well as the rock properties themselves.

Loose and even rather cohesive assemblies of particulate materials "fracture" easily in tension or shear. Indeed it is often difficult to apply such stresses without fracture. Compression of assemblies is usually easy to achieve. It is extremely common in the mills used to reduce the size of coals, as with medium-speed rotating table mills, for example. As heavy compression is applied to a packed bed, even one that does not have confining walls to restrain outward movement of particles during stress application, particles break after dilatant expansion has been overcome.

An assembly can take on the character of a compact mass along with a decrease in voidage and the formation of usually weak interparticle bonds, in some cases involving plastic deformation of asperites. (In briquetting, the stresses applied suffice to eliminate most voids but not to fracture the coal extensively.) Release of the load for dried noncoherent materials allows an expansion of the bed to a looser state, which can be moved more easily. Under compression in confined spaces large single lumps can be changed into assemblies of broken particles. It is this character of compressive loading of assemblies that in medium-speed mills seems to be the most common source of multiple fracture (i.e., the breakage of progeny particles after the primary particle has first undergone rupture).

To sum up, assemblies of particles differ in mechanical properties from their constituent particles because of the contacts between particles and the presence of secondary substances on the surfaces, or of other surface adhesive properties. The assembly thereby acquires specific stress-deformation characteristics as internal friction, cohesive-adhesive qualities, higher resistance to breakage, and the occurrence of repeated fracturing of particles before stress is relieved. Other features of assemblies are variations in voidage and segregation that develop as the assemblies move or come to rest.

Coals themselves do not have special features that are not encountered in other materials, but in practice coals and their products contain materials that reduce the apparent strength of an assembly. Of these, water and swelling clays (e.g., montmorillonite-illite) are the most significant.

5.3.4 Coal Mineral Matter

Particles of minerals (often also called "ash") present in coals possess the same properties they exhibit when formed in rocks free of coal. The clay minerals are generally soft and absorb water to degrees that cause mechanical properties to vary. Ankerites (i.e., various carbonates) tend to be stronger and harder than coal. Thus the wear in coal-using equipment has been attributed to minor amounts of abrasive (hard) minerals,[11] but progress in evaluations of abrasion of coals has not advanced significantly since Cooling's reviews (see also the first Supplementary Volume).

Publications on the effects of mineral matter on coal grindability have not significantly altered established findings that free mineral particles break independently of coal particles, that some mineral matter is easier to comminute than others and some is more difficult and that observations are necessary in practical cases.

5.4 MECHANICAL PROPERTIES AND COALS' NONMECHANICAL CHARACTERISTICS

The nonmechanical characteristics of coals are used normally to classify coals. In particular, characteristics related to rank, such as carbon content, volatile matter, reflectance of vitrinite, gross specific energy, and petrographic compositions (e.g., maceral composition), have become the more important characteristics for describing coal matter. Mineral matter content [or its imperfect approximation, ash, MM = β(ash); $1.19 > \beta > 1.02$ for most coals] has a significant value in the assessment of whole coal. However none of the characteristics above correlates closely with any of the others over the range of all known coals. It is not surprising, therefore, that correlations of mechanical properties with each other or with nonmechanical characteristics are quite markedly scattered around any discernible trends.

This situation was well established by the first Supplementary Volume, and no change in the trends could have been expected on the basis of later studies (e.g., that of Meintjis[60]). None of the recent studies has been at all comprehensive, and there is no theoretical basis relating nonmechanical characteristics of coals to their mechanical properties.

Perhaps the property most likely to have some general usefulness in the immediate future will be microhardness, which correlates with coal characteristics (e.g., rank, petrography, or friction). There is some evidence that microhardness correlates with rank in the form:

$$Y = xe^{-x} + bx^m \tag{5.12}$$

where x = an appropriate function of rank parameters, such as carbon content (dmmf) C_0 or reflectance R_0

b, m = constants such that m is relatively large (e.g., 9) and b is small (e.g., 10^{-4})

Y = an appropriate function of microhardness

Correlations of both indentation and rebound hardness with rank have this form. However before much use could be made of such correlations, considerable extension of the results to many more coals from every nation would be needed, as well as rigorous standardization of testing practices. Complex testing problems exist where the petrographic entities are very small. Cindered coals (i.e., coals thermally metamorphosed by intrusion rather than by the slow "normal" coalification processes) probably differ from "normal" coals in their microhardness–volatile matter correlation, and one could also expect this to apply to air-oxidized coals (or at least oxidized vitrains and clarains). Such correlations bear close comparison with work on the properties of manufactured carbonaceous materials where observations suggest that maximum microhardness may follow heating to about 900°C, or more exactly the development of well-defined d_{002} lattice spacings.

Parish found that small coal particles often undergo plastic deformation; this can be seen in the formation of aggregates during comminution.[16] However her results were drawn from too small a range of properties to justify acceptance of the maxima and minima in her calculations. Similarly, data for her correlations with Hardgrove grindability were inadequate for generalization. Nevertheless, her work is the most significant contribution to the subject. It could also explain laboratory results in the late 1940s on "plastic deformation" by squashing, which included many observations of brittle failure rather than plastic deformation. New results for heat-treated coals have not been noted.

The structure of coal is undoubtedly significant in the mechanical reduction of coals, but it is more evident in size reduction because of coal's imbibition of swelling fluids.[61] Such "spontaneous" disintegration of coals occurs in many liquids, and moreover many liquids weaken coals (e.g., methanol rinsing raises the HGI). The products of spontaneous fracture after immersion in liquids differ from ground products with respect to size distribution. The former appear to have a mixed Rosin-Rammler distribution; for particles smaller than approximately 0.4 mm, $n = 2$, whereas for larger particles $n = n$ of run of mine coal.

Spontaneous fracture is also liable to occur in supercritical gas extraction, solvent refining, and hydrogenation. It is presumably time dependent, but if spontaneous fracture is maximized for liquids whose Hilderbrand-Scott solubility parameter lies between 18 and 30 $J^{1/2}/cm^{3/2}$, then many such liquids may be found in liquefaction processing slurries. Inhibition of spontaneous fracture by water (presumably other liquids can also inhibit spontaneous fracture) has also been demonstrated.

Only one major study has tried to correlate (empirically) the HGI with several coal characteristics.[62] These U.S. results involved complex statistical problems. The study used the interesting device of fitting xe^{-x} (where x is the independent variable) in addition to simple polynomial expressions and interactions to account for clusters of data that were well removed (step jumped) from the normal trend (note $xe^{-x}|_{\max}$ at $x = 1$). This approach aided the correlation of the skewed data sets.

[60] A. A. Meintjis, *J. S. Afr. Inst. Min. Metall.*, **65,** 629 (1965).

[61] D. V. Keller and C. D. Smith, *Fuel,* **55,** 273 (1976).

[62] J. Paone, U. S. Bur. Mines, Rep. Invest. No. 6880.

The scatter about the eventual regression was large, ± 10 HGI units. The work included some petrographic-based correlations, but it did not discuss why HGI varied with petrographic composition.

The size distribution of products of breakage processes is the obvious expression of a coal's response to mechanical forces. Neither rank nor petrography nor a combination of them is a good predictor of run of mine or other size distributions. A simple correlation, albeit scattered, has been used in Australia[63]:

$$n = \frac{4}{3} - \frac{\text{HGI}}{120} \quad (5.13)$$

where n is the Rosin-Rammler distribution coefficient and HGI is the Hardgrove Grindability Index. From this equation a size distribution for a little-known coal can be estimated if the 95% passing size is also given from practical experience.

5.5 MECHANICAL PROPERTIES AND THE UTILIZATION OF COAL

5.5.1 Exploration

We can cite no publication specifically dealing with the significance of the mechanical properties of coal for exploration.

Exploration for coal has increased steadily during the last decade. Diamond drilling methods have improved in terms of core recovery for slim holes (about 50 mm diameter) and for the rapid recovery of large cores (about 200 mm diameter), particularly in the shallower intersections potentially suited to open-cut mining operations. Also significant for coal exploration and analysis of associated strata[64] are recovery and the assessment of gas and water encountered during drilling.

Success in exploration for coal depends on the mechanical properties of coal in two senses:

1. The recovery of unbroken drill core with only minor losses depends on the interaction of drilling practice with seam conditions.
2. Seismic and similar dynamic methods of assessing the geophysics of coal seams provide first-order information about seam thickness and geological structures (e.g., faults) resulting from differences in coal and rock properties.[65]

A more or less arbitrary distinction can be made between exploration substantially remote from workings and exploration closely associated with workings. In the former case the principal purposes are assessment of reserves-resources, general seam correlations and properties, first-order estimates of mining, and beneficiation and utilization problems. For exploration closely associated with workings, the principal purposes are to assess problems involving such matters as rock bursts or faults that could develop with further extension of the workings.

Thus the significance of mechanical properties for coal exploration is largely indirect, apart from wear and tear on drilling equipment, which is low in coal itself relative to associated strata. However mechanical properties of the coal are of importance in that information gained during the first phase of exploration is used to design subsequent exploration phases.

5.5.2 Winning

Greater reliance on coal as a source of fuel and chemicals feedstock in the future will be possible only if the production rate can keep pace with the rapidly expanding demand. Duplication of current winning practices is unlikely to prove to be the answer in many parts of the world. We must develop innovative methods to achieve higher productivity, greater overall recovery, lower costs, less uncertainty with respect to safety, and easier access to deposits in difficult locations.

Mine design and planning is moving away from reliance on the trial-and-error methods of the past. As the pressure increases on resources, facilities, and production, added emphasis will be placed on better knowledge of the mechanical properties of coal and its associated strata.

The mechanical properties of coals are

[63] Discussion to Paper No. D2, *7th Int. Coal Preparation Conf.*, Australia, 1976.

[64] K. H. R. Moelle and J. D. Young, *Eng. Geol.*, **4**, 65 (1970).

[65] Anon., *World Min.*, **27** (10), 23–25 (1974).

significant for coal mining in two major respects: first in ensuring stable and safe mine design and operation, and second in ensuring that coal is freed from the seam most effectively.

5.5.2.1 *Mine Stability*

In designing an underground mine, the openings must be located in such a way that overall yield and ease of access to the face are maximized while stability is ensured, at least in the active areas. Since strata control involves the maintenance of either stable or controlled collapse of overlying rock measures, the physical state of associated coal measure rocks is often more important than the coal itself. However the response of the coal under the prevailing stresses has a profound effect on the stability of pillars, roadways, faces, and other elements supporting the coal. Indeed in some cases[66] where an unstable roof exists because laminates occur, failure due to the prevailing tensile stress can be avoided by leaving a layer of dull coal, where it exists, against the roof. This is not possible where friable bands (e.g., vitrain) occur near the roof, since these do not possess the structural strength needed to provide a bridge holding up the weaker overlying strata.

In extending an operating mine or opening a mine in a geologically well-known area, much emphasis is given to prior experience. Practical experience may be supported by analysis of the sedimentary and structural features of the coal seam and surrounding strata, since the strata exhibit mechanical anisotropies that may be met during exploration and development. Diessel and Moelle[66] have demonstrated that directional mining based on petrofabric analysis can significantly reduce mining instability.

When unexpected problems are encountered or new mines are planned, we have recourse to physical or mathematical models. Models have become popular because full-scale underground tests can rarely be conducted, even though pillar design relies to some extent on empirical formulas based on such tests. It should be noted that a prerequisite of modeling studies is knowledge of the magnitude of the principal virgin stress components.

The most common physical modeling for the study of instability in coal mines involves scale modeling with equivalent materials. A scale model of a section of strata is fabricated from materials of appropriate mechanical and physical properties and a compensating load is applied to account for the superincumbent strata not included in the model. The results of making openings (scaled) in the coal seam are then studied.

The coal properties to be considered in choosing a modeling material should depend on the problem under investigation. For a comprehensive dimensional analysis of modeling, Thakur[67] used the following mechanical properties: modulus of elasticity E, Poisson's ratio ν, angle of internal friction ϕ, coefficient of dynamic viscosity η, creep period t_0, "natural specific gravity" γ, and the appropriate strength characteristic σ. Thakur's study also showed the relations between scale factors for geometry, applied stress, and deformation time. He concluded that in the case of sedimentary rocks, such as coal measures, the properties of the modeling materials should be identical to those of the prototype materials in ν and ϕ, that the specific gravities of each of the modeling and prototype materials should be about the same, and that the following equation should hold:

$$\frac{E_m}{E_p} = \frac{\sigma_m}{\sigma_p} = \frac{L_m}{L_p} \tag{5.14}$$

where subscripts m and p refer to model and prototype materials, respectively, and L is a characteristic linear measurement such as seam thickness. The reader is referred to the original article for Thakur's complicated modeling of η and t_0.

In practice, the studies we cite do not by any means consider all the coal properties listed above. Australian model studies[68] have shown deformation and rupture patterns similar to those observed in mine practice when the major criterion for the modeling materials was the compressive strength. In this case the models were constructed on a scale of

[66] C. F. K. Diessel and K. H. R. Moelle, *8th Commonwealth Mining Metallurgical Conf.*, Australia and New Zealand, Paper No. 36, 1965.

[67] D. N. Thakur, *Int. J. Rock Mech. Min. Sci.*, **5**, 355–369 (1968).

[68] R. T. Hall, ACIRL Rep. No. 73-4.

1 : 24, using various mixtures of sand and portland cement.

In reality physical models are useful as a check once a tentative design has been selected. ACIRL (Australian Coal Industry Research Laboratories, Bellambi, N.S.W.) workers have used models to assess the effect of small or complicated geometrical features of seams. Some models have been demonstrative with respect to the safety and utility of innovative or unusual mining methods.

Electrical resistance analogues have been used to determine stresses and displacements in the strata surrounding tabular mining excavations such as coal workings.[69–71] These studies have made use of the mathematical analogy that exists between the elastic laws and equations governing the steady flow of electricity. The convergence distribution of the roof and floor of a deep tabular seam in an elastic rock mass is described by the Laplace equation. The method involves solving the equation by measuring the potential distribution in the case of steady state current flow in a homogeneous medium, which also happens to satisfy the Laplace equation. In practice an electrolytic or resistance network analogue is used, and the relevant mechanical properties are Young's modulus and the Poisson ratio.

Because of the structural and petrological complexity of some geological formations, it seems virtually impossible to formulate the stress-deformation problems as a closed-form mathematical solution. Consequently one must resort to numerical methods. The finite element method, which has been described in detail,[72] has proved to be a flexible means of analysis. In a two-dimensional analysis a grid usually composed of triangles or quadrilaterals is set up around the opening. The accuracy of the method depends on the extent of the grid and the fineness of the elements. The boundary stresses are set at measured or calculated values of the virgin stress, and the stresses at every node are calculated, assuming plane strain. The approach is equivalent to minimizing the strain energy of the system. Where Hooke's law is assumed, the coal and other rock properties considered are normally as follows: (*a*) stiffness or compliance matrix in the plane under consideration, (*b*) Poisson ratio—normal and transverse, (*c*) bulk modulus, (*d*) tensile strength, (*e*) Mohr envelope, and (*f*) density. More recently the method has proved to be capable of handling nonlinear elasticity, joint nonidealities, rupture events, and extension to three-dimensional analyses. The significant joint properties include normal stiffness, shear stiffness, and cohesion. Where the coal is incapable of sustaining tensile stresses because of joints and fissures, a "no tension" state and a zero (or very small) elastic modulus in the direction of tension have been introduced.[73]

For meaningful results from modeling studies, the stress field and the constitutive relations of the *in situ* material properties must be known. The stress field can be measured directly or estimated from the depth-pressure relationship, which is on the order of 30 kPa/m for the vertical stress component. Horizontal stresses are said to vary from one-third to three times the vertical component, depending on the locality. The measurement of *in situ* material properties has been exceedingly difficult except for a limited number of tests that can be performed close to the face. The use of values determined on small specimens in the laboratory has often been unsatisfactory. This is especially the case with coal, because it fractures and generates flaws when *in situ* stresses are released. A serious bias is also developed by the need to discriminate against weaker coal sections in choosing a specimen suited to testing. A further factor is the relation, previously discussed, between strength and size of test piece.

Kidybinski and Babcock have developed a promising approach to computing the effective material properties, involving an inverse approximation technique.[74] Their finite ele-

[69] M. D. G. Salamon et al., *J. S. Afr. Inst. Min. Metall.*, **65**, 115–37 (1964).

[70] M. D. G. Salamon et al., *Proc. 2nd Cong. ISRM*, Belgrade, pp. 4–18, 1970.

[71] M. D. G. Salamon et al., Part I, *J. S. Afr. Inst. Min. Metall.*, **65**, 319–338 (1964).

[72] O. C. Zienkiewicz and Y. K. Cheung, *The Finite Element Method in Structural and Continuum Mechanics*, McGraw-Hill Book Co., New York, 1967, p. 274.

[73] O. C. Zienkiewicz, S. Valliappan and I. P. King, *Geotechnique*, **18**, (1), 56–66 (March 1968).

[74] A. Kidybinski and C. O. Babcock, *Rock Mech. Suppl. 2*, **5**(1), 1–19 (1973).

ment computation was based on the moduli of elasticity as determined in the laboratory. Successive approximations to the effective moduli for coal and overlying strata were evaluated for model displacements matching those from underground measurements. In other words, the effective moduli for coal were taken as those that give a curve of coal seam compaction (vertical deformation) ahead of the face, as observed in reality. The effective modulus of elasticity further than 6 m from the face turned out to be half the laboratory determination. Closer to the face, the effective moduli were even lower.

One of the most common and most lethal instabilities in coal mining consists of the phenomena of instantaneous outbursts of coal and/or gas from working faces. In the case of a gas outburst the cause of the failure is a "pocket" of gas that remains at elevated pressure while the surrounding coal is relieved of confining pressure (and presumably weakened). Coal between the pocket and the working face is subjected to rock pressures and gas pressure, which have components at right angles. The critical gas pressure for failure can be calculated as a function of layer thickness and rock stresses; the pertinent coal properties are modulus of elasticity, Poisson ratio, and tensile strength.[75] Bursts occur most commonly at faults and adjacent zones where crushed coal contains gas at high pressure.

Other rock bursts can result from the sudden release of the strain energy stored in adjacent rock masses. According to a Polish study,[76] laboratory loading-unloading tests can be used to assess the "liability to rock burst" W_{ET}, which was defined as follows:

$$W_{ET} = \frac{\Phi_{sp}}{\Phi_{st}} \tag{5.15}$$

where Φ_{sp} = recoverable elastic energy accumulated in the sample on stressing and destressing

Φ_{st} = energy lost in permanent strain, loosening, and dissipation on stressing and destressing (a measure of hysteresis)

[75] Z. H. Erzhanov, A. S. Saginov and Yu A. Veksler, *Sov. Min. Sci.*, **9**(4) 349–351 (1973).

[76] Z. Szecowka, *Pr. Gl. Inst. Gorn. Komun.*, Part 568, 3–35 (1972).

Water infusion into the coal face has been used to reduce the risk of bursts. Szecowka[76] has shown that the effect of increasing moisture content ΔW on liability to rock burst is as follows:

$$W_{ET} = 6.65e^{-1.04\Delta W} \tag{5.16}$$

Other methods of counteracting seam rock bursts involve destressing the coal by blasting or drilling to move the peak abutment pressure away from the immediate vicinity of the face. These methods have not always solved the problem.

Often it is necessary to choose a mining geometry such that the zone of peak stress is not surrounded by yielding coal, thus avoiding dramatic redistribution of stress as the final element of coal yields.[77] Longwall and yielding-pillar mining systems have been effective in controlling coal bumps because ordinarily the zone of peak abutment pressure is not surrounded by destressed coal.

The sizing of coal pillars for roof support has been based largely on empirical formulas arising out of work on the relation between compressive strength and specimen size (see Section 5.2.2).

Hustrulid[26] has hypothesized that k, the proportionality constant of Equation 5.3, can be determined on specimens of the order of 50–100 mm. Thus the ultimate strength σ_c of a cube having an edge dimension equal to the seam height H can be predicted from Equation 5.3. The average stress on the coal pillar can be estimated from the following equation:

$$S = \frac{dT}{1 - e} \tag{5.17}$$

where S = average stress on coal pillar (kPa)

T = overburden thickness (m)

d = assumed increase in vertical stress component per meter of overburden (30 kPa/m)

e = extraction ratio

Since Equation 5.17 gives only the stress on the average pillar, it seems suited to preliminary calculations only. Indeed, individual pillar stresses vary according to the position of each pillar with respect to overall mine

[77] S. Crouch and C. Fairhurst, *Trans. AIME*, **256**(4), 317–323 (December 1974).

layout. For a better estimate of pillar stresses, the electrical analogue method has proved to be reasonably accurate.[78]

Finally, Hustrulid estimated the least width W of a rectangular sectioned pillar from the relation:

$$\frac{\sigma_p}{\sigma_c} = 0.778 + 0.222\frac{W}{H} \quad (5.18)$$

He then applied an unstated safety factor.

Equation 5.18, which was originally determined on sandstone prisms in 1876, has been shown by Hustrulid to fit the great majority of published data, including the more recent studies by Holland,[79] Salamon and Munro,[80] and Bieniawski.[36] In an analysis of *in situ* test data from three South African collieries, including the same work of Bieniawski, Van Heerdan[25] concluded that a relationship of the same form but with constants 0.64 and 0.36, respectively, is appropriate. However Hustrulid asserts that these constants are due to the use of an inappropriate cube size in normalizing the results of Bieniawski.

These formulas do not seem to have been rigorously tested by statistically meaningful underground observations. Moreover such simple approximations with respect to mechanical properties may fail to account for the peculiarities of a particular mining situation.

The coefficient k of Equation 5.3 has been shown to vary from coal to coal but to be independent of specimen size. It has also been shown[35,81] that k depends on the coefficient of friction between coal specimen and adjacent rock used as a platen. However for a number of coal measure rocks (sandstone, limestone, and two shales) Holland and Olsen[37] showed that the frictional effect did not vary significantly, the coefficient of friction averaging 0.47 dry and 0.43 wet. Holland and Olsen make the point, however, that clays and shales, especially those containing swelling clay minerals such as montmorillonite, easily turn to mud when under water, and this mud can become a lubricant affecting pillar strength.

Wilson and Ashwin[32] recognized that the empirical formulas quoted above apply to small pillars and, in general, cannot be extrapolated to large (wide) pillars. In considering the stress distribution in the broken and elastic zones of a large pillar, they proposed a plausible hypothesis concerning the load-bearing capacity, which is capable of further development. To keep the equations manageable, however, the effect of variations in mechanical properties of coal was neglected. Thus the formulas proposed should be treated with caution. Wilson and Ashwin appear to have erred on the conservative side when their data are compared with Bieniawski's data[36] for larger pillars.

5.5.2.2 *Extraction of the Seam*

Work on the mechanical properties of coal in relation to extraction of the seam includes considerations of blasting and various methods of attacking the face.

Disintegration follows the stress relief and tension waves that are created by a blast. By setting charges close to the face, advantage can be taken of the relatively low tensile strength of coal. In this case the longitudinal compressive stress waves generated by the explosion on reflection from the free surface cause scabbing, which is characteristic of tensile failure. The tensile strength, the compressional wave velocity (approximately 2 km/s for coal), the elastic moduli, and preexisting cracks and discontinuities affect the disintegration on blasting. The water content of the coal has a significant effect on wave velocity. Hobler[12] has shown that a greater volume of broken coal per mass of charge is obtainable when the coal is wetted, an accompanying phenomenon being the production of larger lumps.

Although release wave or tensile scabbing is important in shotfiring and other blasting operations, other mechanisms of breakage have been considered significant. Cook et al.[82] and Pifer and Ko[81] have reported that fracturing from release of compressive loading was the most destructive mechanism. It

[78] B. B. Dhar and D. F. Coates, *Int. J. Rock Mech. Min. Sci.*, **9** (6), 789–802 (November 1972).
[79] C. T. Holland, *Proc. 6th Symp. Rock Mechanics,* University of Missouri, Rolla, 1964, pp. 450–466.
[80] M. D. G. Salamon and A. H. Munro, *J. S. Afr. Inst. Min. Metal.*(September 1967), 68–78.
[81] D. A. Pifer and K. C. Ko, *Trend Eng.*, **19**(4), 23–26 (1967).
[82] M. A. Cook et al., *Trans. AIME, Soc. Min. Eng.*, **235**, 383–392 (December 1966).

occurs along with a minor contribution from radial fracture close to the borehole caused by lower velocity shear waves. The compression zone occurred in the intermediate area between charge and face. Such fracturing seems likely to be significant only if the charge is, first, severe enough and, second, far enough from the face.

Experimental field blasting studies by Hearst et al.[13] and Butkovich[83] have assessed to some extent the validity of computer models of confined coal breakage. The experimenters used the models of Shatz[84] and Burton,[85] which rely on constitutive relations between prevailing forces and displacements—including the pressure-volume curve, the failure envelope, and the elastic constants—to calculate the propagation of the stress wave and the resulting changes in density and velocity fields.

Apart from blasting, the other commonly used method to separate coal from the face of a seam involves attacking the coal mass with cutting bits. Warner[86] has stated that other methods, including high-pressure water jets, laser and electron beams, ultrasonics, and mechanical vibration, have been tested on coal and other rocks. None of these practices is widespread commercially, although hydraulic winning has been practiced in several countries.

The mechanism of disintegration has been the subject of many studies, such as those for aiding in tool design.[5,86,87] It is well documented that in general, the deeper the bit penetrates, the less dust it produces, apparently because a smaller portion of the broken coal originates from the region adjacent to the bit where local compressive fracture leads to fines. Pomeroy[88] showed that an increase in fines production was accompanied by an increase in energy expended per unit mass of coal extracted.

Investigations by Pomeroy and Foote[29] showed that of the various inherent mechanical properties of coal, tensile strength, and hardness (penetration)—as opposed to compressive or impact strength—dominate in determining coal "ploughability." Roxborough and Whitaker,[89] on the other hand, have shown that rebound hardness at the seam (an easily measured parameter) is related linearly to the production rate. Evans and Pomeroy[23] studied extensively the mechanism of breakage of coal by wedge action. They observed that during the penetration of wedges normal to the surface, in certain types of coal, cracks attributed to tensile breakage radiate from the tip of the wedge. Despite the existence of these cracks, the mode of entry of the wedge appears to be primarily by the crushing of coal against the surface of the wedge, and the force required is primarily related to the compressive strength of the coal. In general, therefore, the cracks observed are not fundamental to the penetration, but their extent would be expected to influence the make of coal per cut.

When coal is removed from the seam face, the overriding mechanical considerations pertain to the structure of the seam, especially the distribution and frequency of partings, cleats, and joints, and the petrological profile.[90] Pomeroy[88] showed quantitatively that the cutting force required was greatest when cutting at 90° to the main cleat direction, and that maximum coal yield per unit energy expended occurred at a cutting angle of about 45° to the cleat. When trepanners or shearers are used, cleat orientation has little effect because much of the cutting is done perpendicular to the bedding plane.

Apart from the fissures existing in the virgin field, which result from tectonic forces and forces internal to the coal seam (due to such factors as evaporation or metamorphism), mining also leads to dislocations induced by processes of stress redistribution. Conceivably a certain amount of control could be exerted on these induced disloca-

[83] T. R. Butkovitch, *Int. J. Rock Mech. Min. Sci.*, **13**(2) 45–51 (February 1976).

[84] J. F. Schatz, *SOC 73, A One-Dimensional Wave Propagation Code for Rock Media,* UCRL-51689, Lawrence Livermore Laboratory, 1974.

[85] D. E. Burton and J. F. Schatz, *Rock Modelling in TENSOR 74, a Two-Dimensional Lagrangian Shock Propagation Code,* UCID-16719, Lawrence Livermore Laboratory, 1975.

[86] E. M. Warner, *Min. Congr. J.*, **56**(8), 35–43, 56 (August 1970).

[87] I. Evans, "The Force Required to Cut Coal with Blunt Wedges," *Int. J. Rock Mech. Min. Sci.*, **2**, 1–12, (1965).

[88] C. O. Pomeroy, *Colliery Guardian,* **207**, 642, 672 (1963).

[89] F. F. Roxborough and R. D. R. Whitaker, "Roof Control and Coal Hardness," *Colliery Eng.*, **42**(491), 19–24 (January 1965).

[90] V. Hucka and B. Das, *West. Miner,* **48**(3), 35–36, 38, 40 (1975).

tions, to facilitate the extraction of coal at the face.

The petrographic profile of the seam defines, to some extent, first the strength profiles, then the optimum position of cut and preferred floor and roof horizons. The degree of mineralization and the distribution of stone bands are other significant factors in extraction strategy.

5.5.3 Transportation

There are four classifications of transportation of coals:

1. Carriage as static particulate assemblies in containers (e.g., rail wagons or ships) or on belt conveyors.
2. Transfer under gravity within and from containers (bins) or at transfer points where the particles of assemblies are subject to frequent collisions with each other and/or severe collision with heaped particles or rigid walls.
3. Carriage as aqueous suspensions (e.g., pipelining).
4. Carriage as gaseous suspensions [e.g., pulverized fuel (p. f.) firing].

In general the levels of stress increase in the order listed. The most relevant mechanical properties are those governing the motion of assemblies, size degradation, and wear on the containing equipment.

5.5.3.1 *Static Transfer*

The principal property controlling statically conveyed coal is the coefficient of friction, which can be determined from the angle of rest of the heaped coal, and decreases with decreasing moisture content, increasing particle size (e.g., Rosin-Rammler mean size, $\bar{x}$), and Rosin-Rammler distribution coefficient (n). We can readily generalize from Sligar's[91] analysis of the surface profile of a coal heap on a rotating table to estimate the effective coefficient of friction for transition of static to moving coal assemblies. In static situations, or in transfer at very slow rates, coal assemblies are strong enough not to degrade in size. The distribution on and transmission of forces to bin walls by an assembly is similar to that of other materials and is not peculiar to coal.[56]

5.5.3.2 *Gravity Transfer*

The gravitational discharge of coal from bins through apertures (Chapter 9) is adequately described by general correlations for particulate materials and does not require introduction of special properties. Given reasonably free-flowing coal (e.g., air dry or wet-oiled), apertures can be used to control discharge rates; but the development of cohesive forces from such factors as excess surface moisture makes calculation and control unreliable.

Breakage during transfer tends to be minimal in that the size distribution coefficients $\bar{x}$ and n change little, but dust can be raised as a result of breakage and aeration of dust already present. (The dust may not be representative of the coal, but neither does it always contain relatively high concentrations of specific mineral or maceral components). Matrix methods are often convenient for the analysis of breakage due to transfer of coal from one relatively static situation to another. Broadbent and Callcott's[92] method for determining size-dependent breakage, as modified by Berenbaum[93] for "survival" of pieces, has been used for such analyses, including the analogous shatter test. The analyses were recently reviewed and extended by Bull et al.[94]

Various feeders may be included in this transportation category, such as screw feeders, rotary star feeders, or lock hoppers.

Ferretti's work[95,96] on feeders reported wear as a major problem in new coal processes, but he did not identify the materials responsible for it. The work of Carleton et al.[97] on screw feeders–conveyors indicated that the coefficients of friction between blades and particles and within the particle

[91] J. Sligar, Ph.D. Thesis, Dept. Chem. Eng., University of Newcastle, New South Wales, Australia, 1977.

[92] S. R. Broadbent and T. G. Callcott, *Phil. Trans. Soc. London, Ser. A,* **249,** 99 (1956).

[93] R. Benenbaum, *J. Inst. Fuel,* **34,** 367 (1961).

[94] F. A. Bull et al., *J. Inst. Fuel,* **48,** 197 (1975).

[95] E. J. Ferretti, *Coal Min. Process.* **11**(10), 38, 52 (1974).

[96] E. J. Ferretti, *Chem. Eng.,* 113 (December 1974).

[97] A. J. Carleton, J. E. P. Miles and F. H. Valentin, ASME 1968, Paper No. 68-MH22.

assembly (as well as the cohesiveness of the assembly) were significant factors. The analysis should apply to coals.

The cohesive and adhesive properties of coal supplies are borderline mechanical properties, significant to bulk handling in that they arise from particle size, shape, and superficial moisture content. Sharkov[98] provided a clear account of the following factors:

1. Wet spheres of coal will adhere to surfaces on contact, provided the diameter in meters is less than

$$d = 0.154\,\rho^{-1/2} \tag{5.19}$$

where ρ = true density of fuel (kg/m^3)

With the slightest compression, the water layers holding the particles become thinner, and adhesion increases because of the action of the capillary forces. In this case the size of coal particles that are held by water layers increases 2 to 3 times compared with the size given by Equation 5.19.

2. Adhesion and cohesion of various coals were determined by Sharkov using Deryagin's general law of friction, which expresses the resistance to displacement of two surfaces relative to each other as follows:

$$T = \mu(F + Sp) \tag{5.20}$$

where μF = resistance due to friction forces
μ = friction factor
F = compressive force (N)
μSp = resistance due to cohesion of surfaces
S = contact area (m^2)
p = specific molecular pressure (N/m^2)

Sharkov measured μSp directly to assess experimental results. Anthracite, a Moscow coal, and peat showed similar nonlinear increases in adhesion and cohesion (N/m^2) for increases in moisture content, for increases of compression of the bulk assembly, and for duration of compression. Decreases in adhesion and cohesion, however, were less sensitive to particle size ($d^{-0.136}$) than theoretically predicted (d^{-1}). Huff's[99] earlier work was less general, but it also indicated a cohesion index increasing as moisture content increased for, again, several size distributions of anthracite, bituminous coals, and lignites. The index also increased (erratically) with particle size. A correlation between packed-bed bulk density and moisture content might have affected results.

Pietsch[100] comprehensively surveyed adhesion and agglomeration of solids during storage, flow, and handling, using Rumpf's classification of binding mechanisms (see Table 1 of Ref. 100). The mechanism of greatest importance for raw and cleaned coal supplies is Rumpf's mechanism III, in which binding is due to interfacial forces and capillary pressure at freely movable liquid surfaces. Of less importance for transportation are molecular forces [mechanism IV(1)], which account for the caking properties of dry coal fines, the adhesion and cohesion forces in not-freely movable binders (mechanism II), and the deposition of suspended colloidal particles [mechanism I(5b)].

5.5.4 Preparation

Coal preparation, as considered here, does not include deliberate size reduction (see Section 5.5.5, "Comminution"), although the size distribution of the raw coal, which is altered by degradation during winning, transportation, and comminution, has a strong influence on beneficiation processes (Chapters 9 and 10). The size distribution coefficients of run-of-mine coals correlate with the HGI, but the mean size is markedly affected by mining techniques. The breakdown of raw coal (run-of-mine) and associated rock during conventional coal washing affects the yields, not only with respect to the size grading of the product, but also in the yield of clean coal. These aspects of preparation were discussed in pilot plant work by Le Page and Pollard.[101] They concluded that in gauging the probable behavior of a particular coal for a given washery practice, the laboratory test methods used must take into account the influence of water on the degradation due to

[98] A. T. Sharkov, *Thermal Eng.*, **6**, 30 (1970).

[99] W. R. Huff, *Combustion*, **40**(2) 28 (1968).

[100] W. B. Pietsch, *Trans. AIME, Soc. Min. Eng.*, Paper No. 68 MH-21, 1968.

[101] A. Le Page and F. Pollard, *Dev. Coal Prep. Res. Mine Quarry*, **5**(6), 32, 34, 39, 40 (June 1976).

mechanical shear and abrasion forces in the plant, particularly during pumping. A Soviet analysis[102] of a washery stated that maximum degradation occurred when the coal was loaded into bunkers, handled on conveyors, and dewatered in centrifuges.

The cost of maintaining and replacing worn equipment in coal preparation plants is a significant component of the operating cost. Abrasion is the major cause of wear in dry areas of the plant, and the life of the equipment is often increased by using hard or abrasion-resistant steel in construction. Erosion resulting from interparticle impact can be reduced by avoiding sharp bends in flumes and pipelines and by minimizing the angle of discharge from chutes and belts.[103]

In wet areas of the plant, which constitute a major portion of preparation operations, corrosion can be the primary factor in wear, but abrasion accelerates corrosion by keeping metal surfaces exposed. Studies on rates of wear[104] have demonstrated the superiority of stainless steel in areas of high wear. Other benefits of stainless steel included better "slidability" and fewer holdup problems in chutes and hoppers because of low coefficients of friction between coal and smooth, slick surfaces that are not pitted from corrosion.

The selective preparation of coal to upgrade coking quality has become a commercial practice. The preparation processes use differences in the properties and distribution of various coal lithotypes in a feed to discriminate between the "fusible" and "infusible" coking constituents. In some operations the content of less friable durainous inerts is depleted in a coking product by directing them to steaming or other uses. Some coking practices[105,106] reduce the size of inert components relative to fusible materials to improve dissemination of the inerts in the end coke.

[102] S. V. Ovchinnikov and V. P. Kurbatov, *Coke Chem. USSR*, No. 1, 5–8 (January 1970).

[103] J. D. Swan, "The Control of Abrasion in Coal Preparation Plants," AIME, Soc. Min. Eng., Fall Meeting, October 21–23, 1970.

[104] American Iron & Steel Institute, *Stainless Steel: Effective Abrasion and Corrosion Control in Coal Handling and Preparation Equipment*, April 1976.

[105] P. C. Saha, "The Selective Preparation of Coal for Coke Production," *TISCO*, **18**(2), 56–62 (1971).

[106] Anon., "Selective Preparation of Coals for Coking Using Sovaco Screens," *FRI News*, **16**, 71–76 (1966).

5.5.5 Comminution

Industrial reduction of coal size involves a complex of processes including stress application, handling, separating, drying, and fluid flow techniques. Only a few aspects of these processes are notably affected by the mechanical properties of coals and associated rock. The principal theoretical and practical problems of size reduction involve the energy consumed for comminution, changes in size distributions, flow of particles (with special emphasis on size separation during transport), drying, wear on machines, and mechanical problems in constructing increasingly large pulverizers. The last aspect is significant because the capacity of plants has been increasing more rapidly than that of mills. Most studies of comminution have emphasized energy consumption and changes of size using simplified models of size reduction processes.[107]

The simplest approach to comminution has been to relate a size reduction characteristic of the mill to the energy used and later to correlate that relation with a laboratory test measure that could be further correlated with other properties of the coal. Bogot and Hamilton[108] continued this approach for pulverizer capacity and HGI, but as with much of the earlier information, experimental observations to verify the diagrams are lacking.

The next simplest approach has been to attempt to predict changes, due to milling, in the size distribution of coal. A mathematical model of a simple character is set up containing variables that can be correlated with a mill's characteristic variables. In most studies, however, fundamental relationships between the fragmentation of particles, the stress involved, and the machine conditions have been obscured by disagreement over details of the phenomenological model. Some investigators do not seem to recognize that they have used a model of experimental processes, and that the model amalgamates the parameters of material and machinery, perhaps inextricably.

[107] V. K. Gupta and T. C. Kapur, in *4th European Symp. Comminution*, Nuremburg, 1975. (Dechema, **79**, 447.)

[108] A. Bogot and T. B. Hamilton, *Am. Power Conf.*, **36**, 474 (1974).

The use of breakage and selection functions (matrices) in models of comminution processes is now well established, but only the rare investigator[107] notes that these functions lack independence.[109,110]

Gupta and Kapur's term "selekage" has much to recommend it for use with batch-grinding models, but its more general applicability would require study. Thus the kind of comminution process and its relation to more basic operations need examination.

Bown[111,112] showed that using the concept of "associated" energy to estimate the efficiency of milling is unsatisfactory. Nevertheless Stairmand[113] later continued to accept it as the appropriate method. It is a cumbersome technique, but it has the merit of emphasizing the relatively small proportion of energy used directly by mills in stressing to failure the material being broken. The same conclusion has been reached from other directions, as Chandler's[114] comparison of grindability tests clearly summarizes. Sligar's[91] (see also Sligar et al.[115]) detailed study of Lopulco mills demonstrated the complexity of mill mechanics and the numerous energy-using processes. It could be argued that the very complexity of milling plants precludes the adoption of simplistic analogue tests unless they give rise to indices directly related to the mechanical properties of coals.

Maruyama[116] related the mean compressive strength of Japanese coals to a fractional power a of particle volume. Plotting his results gives:

$$a = 0.467 - 0.0021 \text{ HGI} \qquad (5.21)$$

He also correlated mean compressive strength linearly with HGI (negatively). Maruyama's experimental results seem to agree with Chandler's data.[114] Bown[111,112] showed that the amount of energy required to break cubes of coal by slow compression per unit of fines produced is similar to the amount of energy needed to break by impact, and that the stored energy at breakage depends critically on the stress conditions. The stored energy at failure under uniaxial compression was an order smaller than for biaxial or triaxial compression.

Such results suggest that the energy required for coal breakage in special mechanical tests and in simulations of large comminution plants may well correlate, but departure from standardized conditions could render correlations inexact. Moreover, mathematical models with the simplistic inclusion of terms such as compression strength would also be inadequate to account for nonstandardized conditions.

Several workers have suggested that the friction of coal on mill surfaces is a significant factor in comminution. Sligar[91] made interesting use of wear patterns from the rubbing of coal on mill metal to assess the flow of particles within a mill. Rose and English[117] suggested that the friction of coal and coke reduces the capacity of a jaw crusher relative to calculated values generally, and to such materials as granite. Kapelson and Unterberger,[118] concluded that friction along beaters contributed to size reduction.

Bown's analysis[119] of slow, free crushing between platens predicted that friction between broken fragments will account for a considerable proportion of the energy supplied. Hiorns[120] proposed that the apparent concordance of many reported experimental data on size reduction with Rittinger's law is a result of the major energy requirement in grinding, namely, to overcome frictional resistance between particles. He showed that because frictional resistance is proportional to surface produced and to pressure, the minimum total work required for a given size reduction by compression is simply related to the reciprocal of product size.

[109] T. G. Callcott, *Trans. Inst. Min. Metall.*, **76**(729), C3-C33, 1967.
[110] T. G. Callcott, *Aust. Chem. Proc. Eng.*, 1969.
[111] R. W. Bown, *Trans. Inst. Min. Metall.* **75**(715), C163 (1966).
[112] R. W. Bown, BCURA IC 312–315, private communication.
[113] C. J. Stairmand, in *4th European Symp. Comminution,* Nuremburg, 1975.
[114] R. L. Chandler, BCURA Monthly Bull. Nos. 248, 1965.
[115] J. Sligar, T. G. Callcott, and I. McC. Stewart, in *3rd European Symp. Comminution*, 1971.
[116] T. Maruyama, *J. Fuel Soc. Jap.* **45,** 843 (1966).
[117] H. E. Rose and J. E. English, *Trans. Inst. Min. Metall.,* C76, C32 (1967).
[118] I. G. Kapelson and O. G. Unterberger, *Coke Chem. USSR,* **6,** 41 (1967).
[119] R. W. Bown, *Trans. Inst. Min. Metall.,* **C75** C173 (1966).
[120] F. J. Hiorns, in *2nd European Symp. Comminution,* 1966. (Dechema, **57,** 165.)

The correlation of directly determined mechanical properties of coals with expressions of comminution is rare and represents a negligible part of the literature of size reduction.

Many workers[107] have expressed the notion that a selection function, estimated relative to each worker's specific model, will express both the strength of the materials and the chances of particles being so positioned that breakage forces will fragment them. Regularities in the selection functions can be seen to be similar to the readily observed regularities in size distributions, but the mathematical operations enforce such correlations. In this way coal strength, product size distributions, and selection functions are interchangeable expressions of the coal's characteristics.

We can conclude that there is as yet no satisfactory way of relating the mechanical properties of coals to the behavior of coals in mills. Indeed the limited visual evidence[110] from high-speed photographic studies of coals being broken by the impact of hammers on lumps or lumps onto rigid walls suggests that the mechanical properties of coals are such that coals cannot resist the forces without fracture. The principal factor in achieving a high degree of size reduction by impact is the number of stress applications rather than their intensity.

The dearth of data from well-documented work on mechanical properties of coal macerals and maceral size distribution is notable. The view that some macerals generate smaller sizes than others (hence are weaker) has industrial value because it is widely accepted and is taken advantage of when maceral composition varies with particle size. Contrary cases have received little attention. Nevertheless, one must in general accept the conclusion that maceral concentrating phenomena do occur in industry as a function of initial coal size and maceral distribution.

Definitive studies of the relationship between mechanical properties of coal-associated minerals and size reduction processes have not been published. We continue to encounter mineral concentration in fines or in coarse fractions but without any clear explanation based on measured mechanical properties.

The more complex problems of breaking coal-mineral intergrown or bonded particles have not been given adequate study, although undocumented assertions of selective behavior have been made.

5.5.6 Agglomeration

Size enlargement of coal is a rapidly expanding practical and scientific subject. Agglomeration of coal shows considerable promise for improving the recovery of coal after hydraulic transport and the liberation of mineral matter by so-called selective agglomeration is an aid in beneficiation. In general, the forces that bond pellets or agglomerates rely on surface chemical effects, resulting from the physical proximity of the coal particles in the presence of more or less liquid. Such processes do not rely on the application of significant stresses or on a resultant particle-particle deformation. Briquetting processes, however, do involve such stress.

The actual mechanism or phenomenon of briquetting depends to a great extent on the mechanical properties of the solid particles. As discussed in Section 5.3.2, Hard Coal particles under essentially static stressing do not exhibit plastic deformation except at small sizes on the order of 100-μm cubes.[16] Using scanning electron microscopy, Hardman and Lilley[121] observed that at lower pressures the compaction of coal and of sucrose is dominated by particle fracture, with a movement of particle fragments into positions of closer proximity. At high pressures unlike sucrose, coal fragments tended to remain wedged in the form of chain bridges that maintained a more open structure in the compact. Hardman and Lilley also concluded that initial compaction at low pressures was not governed by the sliding of particle into new equilibrium positions. This, however, may not be the case when a liquid binder of some sort is present to lubricate the interparticle surfaces.

Based on empirical considerations, Rammler and Weber[122] proposed a quadratic dependence of compressive strength on molding pressure for binderless briquetting of peat and coal. Subsequent studies by

[121] J. S. Hardman and B. A. Lilley, "Mechanisms of Compaction of Powdered Materials," *Proc. R. Soc. London, Ser. A*, **333**, 183–199 (1973).

[122] E. Rammler and M. Weber, "Mathematical Dependence of Briquetting Strength on Moulding Pressure in Binderless Briquetting," *Freiburg Forschung* **A515**, 75–89 (1973).

Hardman and Lilley[123] showed that greater volume reduction could be achieved by applying a shear stress in addition to compression. They suggested that the reduction was the result of a repeated sequence involving the breakdown of existing chain bridges followed by further particle fracture. Their electron microscopy did not reveal any evidence to support the earlier theories of Gregory[124] and Rhys Jones[125] that by virtue of their plasticity the fine particles produced in such a process glued the larger particles together. Nevertheless, Gregory[124] did produce strong briquetting of Hard Coals at moderate pressures using shear strain during compression.

Based on probability considerations, Bockstiegel[126] verified the form of the empirical equation found by Habberjam[127] for hot coal briquetting:

$$P = \frac{M}{V_0}\left(\frac{V_0 - V}{V}\right) \tag{5.22}$$

where P = external uniform pressure
M = a constant with the dimension of stress
V = "material-related void volume"
V_0 = "initial material–related void volume"

Bockstiegel's probability theory gives:

$$M = V_0\sigma_c\frac{C}{6}\left(\frac{1+\nu}{2-2\nu}\right) \tag{5.23}$$

where σ_c = the compressive yield stress
ν = the Poisson ratio
C = a constant of the order of unity

The strength of briquettes or other coal agglomerates can be determined by almost any means applicable to solid objects. The most common properties quoted are ultimate compressive, tensile, and cohesive strengths; indentation; and shatter or impact strengths. Since briquettes are normally of regular shapes and uniform sizes, the drop test developed by Lyall[128] gives an appropriate measure of strength for green or semicoked briquettes during transport within a formcoke producer. Formcokes are most conveniently tested in the same tumblers as conventional cokes.

5.5.7 Chemical Processing

5.5.7.1 *Coking*

A compact review of coking and cokes has been made by Loison et al.[129] The review shows essentially that mechanical properties themselves are of minor direct significance in the thermal, chemical, and physical phenomena that occur during coking.

The only significance of the mechanical properties of coal is in the way they affect the distributions of size and components of coking blends. In this respect it has long been apparent that complex (and disputed) relationships exist between coking process phenomena and size-maceral-mineral distributions. In simplest terms, coke strength is favored, provided the charge can be fed to the coke ovens, by increasing the fineness of the charge, by increasing its dry bulk density, and by maximizing the intimacy of mixing of all distinguishable coal entities. In practice problems are encountered largely because of the qualification, "provided the charge can be fed to the coke ovens," a qualification that involves techniques, equipment, and economic aspects. Thus attention is directed to the mechanical properties of coke itself.

Metallurgical cokes are manufactured[129] by heating large amounts (20–40 tons) of approximately 5-mm blends of coals in narrow (c. 450 mm) slot ovens. Although new methods [130,131] for making substitutes for the clas-

[123] J. S. Hardman and B. A. Lilley, in *Proc. 1st Int. Conf. Compaction Consolidation Particulate Matter,* Brighton, England, October 1972, p. 115.

[124] H. R. Gregory, *Trans. Inst. Chem. Eng.*, **40,** 241 (1962).

[125] D. C. Rhys Jones, in *Chemistry of Coal Utilization,* Supplementary Volume, H. H. Lowry, Ed., John Wiley and Sons, New York, 1963, pp. 675–753.

[126] G. Bockstiegel, in *Proc. 1st Int. Conf. Compaction Consolidation Particulate Matter,* Brighton, England, October 1972, p. 51.

[127] G. M. Habberjam, *J. Appl. Phys.*, **15,** 1233–1242 (1964).

[128] K. D. Lyall, *Powder Technol.* **4** 165–167 (1970–1971).

[129] R. Loison, P. Foch and A. Boyer. "*Le Coke*"—*Principles de la Fabrication Recherche de la Qualitée,* Dunod, Paris, 1970.

[130] N. M. Potter, *J. Inst. Fuel,* 497 (1970), 313 (1972).

[131] T. G. Callcott, "Technical Development of Formcoke," IISI WG Presentation, Brussels, April 4, 1974, private communication.

sical cokes are being developed, they are not yet commercially significant and are excluded from this discussion, along with other specialized minor products.

Coke is a complex matrix of bonded particulate materials enclosing numerous vesicles. The porosity is high, around 50%. Cokes normally are encountered as relatively large, irregularly shaped lumps. Concavities abound, but they are largely ignored in coke size analyses or shape factors. Elongation along the direction of heat penetration is common, and pyramidal to blocky shapes are typical. Cokes also are permeated by fractures varying from large visible fissures to microcracks sometimes clustered around materials that did not fuse during coking. (Theories[129] about how fissures form during coking are outside the scope of this chapter.)

We can cite no theoretical analyses of the mechanical properties of cokes that relate closely to their known structures and chemical constitution, but efforts to measure static and dynamic properties have been persistent.[132,133] It is difficult, however, to prepare satisfactory test specimens and to prepare them in numbers sufficient for the results to be even moderately precise.

Most investigators have cut specimens from "fissure-free" coke material, and in so doing they have selected for study a small proportion of the coke that ostensibly is the strongest. Discussions of the consequent bias have not been recorded, but certainly compressive and tensile strengths have been measured on materials that are not representative of the great bulk of cokes used in metallurgical industries.

Coke test specimens are also difficult to prepare with normal standards of precision because of the abrasive and brittle character of this porous composite material. Indirect tensile tests (e.g., Brazilian) have been used, but there appear to be no detailed analyses of the phenomena occurring during diametral compression. Failure from central fractures in accordance with simple theory can be observed in some specimens, whereas in others failure seems to originate in wedges formed at the compressed periphery. Loison et al.[129] have not dealt with this critical difference in types of failure.

British Coke Research Association (BCRA, now the British Carbonisation Research Association) workers[134,135] described results of tensile tests of cokes and showed that individual values of a few megaPascals above or below the mean have a standard deviation on the order of 1/2–1/3 of the mean (i.e., some hundreds of results are required to obtain the mean within confidence limits of a few percent). Patrick and Stacey[136] recorded scattered negative correlations of median tensile strength with porosity and cell radius, and—not unexpectedly—positive correlation for cell thickness. Similarly, Miyagawa and Fujishima,[137] using two cokes of good but not exceptional strength, found that indirect tensile strength increased with apparent relative density.

Uncertainty in the interpretation of compressive strength is similar to that for tensile strength and, as indicated in the section on comminution, measures of the mechanical properties of cokes have not been strictly related to indices obtained by tumbling tests (i.e., autogenous grinding tests).

The mechanical properties of cokes are often considered to include results of empirical tests of the "strength" of blast furnace cokes. The autogenous grinding character of coke tumbling tests has been demonstrated by the well-known variation in size distribution with the number of revolutions of the drum used. (Bimodal distributions develop during all coke tumbling tests: the coarse mode has mean sizes on the order of 10–100 mm, whereas the fine mode has mean sizes on the order of 0.1 mm. Few particles have sizes in transition between the two modes. Refer also to Sanna and Pasletti.[138]) In essentials, coke tumbler tests differ in the size of initial coke lumps (which are dried to eliminate moisture), dimensions and configuration of the drum tumbler, and numbers of revolutions effected before the product size anal-

[132] N. S. Gryazbiv, Yu. A. Nechaev and A. I. Zolotukhin, *Coke Chem. USSR,* 7 20 (1972).

[133] R. Baruzdina, *Dokl. Akad. Nauk SSR,* **156,** 935 (1964).

[134] H. E. Blayden, *COMA Yearb.,* 197 (1966).

[135] Anon., BCRA Coke Res. Rep. Nos. 53, 59, 1969; 62, 1970.

[136] J. W. Patrick and A. E. Stacey, *Fuel,* **54,** 213 (1975).

[137] T. Miyagawa and I. Fujishima, *J. Fuel Soc. Jap.* **54,** 983 (1975).

[138] D. Sanna and R. Pasletti, *CDS Circ. Inf. Tech.,* **32,** 1347 (1975).

ysis is determined. A factor of some significance is the change in shape of cokes from elongated to more spherical as tumbling progresses, because coke size analysis remains essentially a single lump appraisal of the aperture (square or round) through which a lump can just be passed. Thus after prolonged tumbling, size analyses using round hole and square aperture screens (of equal diameter and length) are essentially the same, whereas for normal coke the equivalent round hole screen aperture is 1.16 times that of a square hole screen.[139] So far, all coke strength indices based on tumbler tests have been primitive, and they reflect only limited understanding of comminution by tumbling or the effects of particle shape, size, and autogenous breakage. Nevertheless correlations between indices are well known,[140] even though each correlation is imprecise. The set of inexact equations[141,142] in Table 5.2 approximates the linear relations between pairs of coke strength indices, where the imprecision of estimating y from x or x from y is indicated by $\pm L$, respectively. These relations apply broadly to familiar cokes.

[139] T. G. Callcott, *8th Commonwealth Min. Met. Congr.*, Melbourne, Australia, Paper No. 106 (1965).

[140] A. Milson, *COMA Yearb.* 360 (1962).

[141] T. G. Callcott and N. A. Brown, private communication.

[142] V. D. Frishberg et al., *Koks Khim.*, No. 8, 15 (1975).

Table 5.2 Simplified Trend Lines for Coke Tests[a]

$y = ax + b$	$\pm L_y$	$\pm L_x$
From Callcott-Brown[141]		
$S = 1.6I_{20} - 70$	± 10	± 10
$S = 0.9M_{40} - 14$	± 8	± 10
$H = -I_{10} + 87$	± 5	± 5
$H = 84 - 2M_{10}$	± 5?	$2\frac{1}{2}$?
$DI^{30}_{15} = 0.52\, I_{20} + 54$	$\pm 1\frac{1}{2}$	± 4
$DI^{150}_{15} = -1.5\, I_{20} + 112$	± 2	± 3
$DI^{150}_{15} = I_{20} + 6$	$\pm 2\frac{1}{2}$	$\pm 2\frac{1}{2}$
$I_{20} = 0.285M_{40} + 60$	(*Very inexact*)	
$M_{10} = 0.4I_{10}$	± 1	± 2
Recalculated from Frishberg et al.[142]		
Residue (kg) $= 8.283\ DI^{30}_{15} - 441$ (U.S.S.R.)	± 15	$\pm 1\frac{1}{2}$

[a] Symbols:
S = ASTM Stability Factor
H = ASTM Hardness Factor
I_{20}, I_{10} = Irsid Test indices
M_{40}, M_{10} = Micum Test indices
$DI^{30}_{15}, DI^{150}_{15}$ = Japanese Drum Test indices
a, b = constants
y, x = indices
L_y, L_x = approximate confidence limits

The decrease in coke size during prolonged tumbler testing has been correlated in a number of ways, but all show curvature in plots except where the fraction R oversize is correlated with the number of drum revolutions (i.e., duration) as shown in Equation 5.24.[131]

$$R_x = \exp\left(-\left(\frac{N}{\overline{N}_x}\right)^{k_x}\right) \tag{5.24}$$

where $\overline{N}_x$ is a constant for a specific coke and size x, and k_x also is a constant.

The values of $\overline{N}_x$ seem to be much the same for Micum and ASTM tumbler test results. They spanned from around 5000 for $x = 6$ mm to 1000 at $x = 30$ mm and 240 at $x = 40$ mm, for results given in a BCRA report.[143] The respective values of $\overline{N}_x$ for JIS tumbler of the same cokes were 1700, 1350, and 16 revolutions. The values of k_x for these cokes decreased with size from about 0.6 to 0.5, as size x increased from 6 to 40 mm. This relationship was found to be applicable to all reported size analyses from tumbler tests,[125] and we can conclude that all drum tests involve similar autogenous "disappearance kinetics." Despite the wide range over which the relationships have been consistent (up to 4000 revolutions), the analysis of coke degradation in tumblers needs further study if the large scatter in results throughout the industry is to be reduced by greater control over conditions and if cross-test relations are to be better understood.

Shatter test results after repeated drops were reported[131] to have the form above with $\overline{N}$ possibly replaced by $0.67\overline{N}$. The relationships have also been shown to apply to the degradation of formcokes.

The apparently inescapable conclusion is that measures of coke strength based on tumbling (and drop-shatter tests) are poorly correlated, inexactly described, and deserving of further study. Nevertheless there are those who differ from this conclusion. Among them is Milson,[140] who emphasized the existence of statistically significant cor-

[143] Anon., BCRA Carbonisation Res. Rep. No. 5, 1974.

relations as sufficient evidence for the view that correlated coke strength indices measure similar coke properties. Milson also stressed the large scatter, which implies the presence of other factors.

Breakage of Coke by Random Division. Experiment has shown that the theory of random division as applied by Manning[144] to the fracture of a solid may also be applied with good agreement to the breakage of coke, provided the "magnitude" to be divided at random is the lump's volume and not its characteristic linear dimension.[110,145]

If a piece of coke is dropped onto a hard flat surface and it breaks into N fragments, the theory predicts that the smallest particle will have a volume equal to $1/N^2$ of the original volume; the second smallest $[1/N + 1/(N - 1)]/N$ of the original volume; and so on, up to the largest, whose fraction of the original volume will be $[1/N + 1/(N - 1) + \ldots + 1/3 + 1/2 + 1]/N$.

Although the number of fragments N cannot presently be predicted, the theory of random division will predict closely the relative sizes of the particles that will form on dropping individual lumps of coke (ranging 100–50 mm) from heights of about 1 m onto a flat steel plate. A small percentage (i.e., 1–2%) of dust will also form. Rarely are more than 10 fragments produced by individual lumps.

5.5.7.2 *Gasification, Combustion, and Liquefaction*

How well coal conversion plants work and how reliable they prove, will depend crucially on the designers' ability to prevent excessive wear from corrosion, erosion, and sliding. Such wear is common when transporting coal and associated solids at high velocities in gases or liquids that are corrosive, especially under conditions of high temperature and pressure. Although there is knowledge of-some of the relevant material properties in associated industries, Schulz,[146] Silence,[147] and others have concluded that knowledge of the peculiar situations encountered in the emerging coal conversion industry is lacking.

The mechanical properties of the raw coal are of most importance in the comminution and handling activities that have been treated in this chapter. However there are transfer systems that are peculiar to the processing industry (e.g., extrusion, screw feeding, lock hoppers, and related valves). Other areas of relevance include breakage in fluidized beds which, although minor, may be important in gasification and combustion processes. Derived properties such as particle size and bed voidage are relevant because of effects on flow of fluids and mass-heat transfer. The possible significance of disintegration of particles by imbibition of specific liquids was mentioned earlier.

5.6 CONCLUSION

The most notable feature of our examination of the literature is the absence of any accepted and widely used theory on the mechanical properties of coal. Allied to this is the absence of any center devoted largely to the study of these properties and their significance for coal utilization, even though the 2.3 billion tonnes of coal produced each year would seem to warrant at least one such central facility. It is also notable that to a larger extent, the mechanical phenomena of coals are so loosely related to coal's general characteristics and so little used that none of them has been proposed for use classification.

[144] A. B. Manning, *J. Inst. Fuel,* **25,** 31 (1952).
[145] T. G. Callcott, BHP Bull. No. 26, 26, 1966.

[146] C. A. Schulz, *Materials Problems and Research Opportunities in Coal Conversion,* Columbus, Ohio, April 16–18, 1974, Paper No. 41 p. 441.
[147] W. L. Silence, *Materials Problems and Research Opportunities in Coal Conversion,* Columbus, Ohio, April 16–18, 1974, Paper No. 42, p. 449.

CHAPTER **6**

Plastic Properties of Coal

DIETHARD HABERMEHL, FRED ORYWAL, AND HANS-DIETER BEYER

Diethard Habermehl *became a Diplom-Physiker (graduated physicist) in nuclear physics from the University of Münster, West Germany, in 1968. He received his doctor's degree in natural science from the Institute of Technology in Aachen, West Germany, in 1972. In 1968 he joined the scientific staff of Bergbau-Forschung, GmbH, Essen, West Germany, where he is now head of the full-scale coking test plant. Dr. Habermehl's current research interests include investigations on the measurement of the shrinkage of coals during coking, optimization of preheating techniques (Precarbon® process), optimization of programmed coke ovens underfiring (CODECO process), development of coking techniques, and controlling the industrial coke oven process.*

Fred Orywal *received his Diplom-Chemiker (graduated chemist) degree in organic chemistry from the University of Cologne, West Germany, in 1961 and in 1966 received the doctor's degree in the same field from the same university. In that year he became a member of the scientific staff of Bergbau-Forschung, GmbH, Essen, West Germany, where he is now head of the semitechnical coking test plant. Dr. Orywal's research interests include developing mathematical descriptions of the coking process, optimizing coking coal blends, and the development of new coke oven techniques.*

Hans-Dieter Beyer *graduated as Diplom-Physiker (graduated physicist) in physical chemistry after studying physics at the universities of Hamburg, Giessen, and Münster, West Germany. He received his doctor's degree in natural science from the Institute of Technology at Aachen, West Germany, in 1973. Dr. Beyer joined the staff of Bergbau-Forschung, GmbH, Essen, West Germany, in 1966 and now heads both the laboratory of coal technology and the semi-plant-scale models in chemical engineering. His research interests include certain chemical engineering problems in a new coal gasification process, the plastic behavior of coal at high pressure, and coke reactivity.*

6.1 INTRODUCTION

The heating of bituminous coal initiates complicated physical and chemical changes that lead to the splitting off of gases and condensable vapors. What remains at the end is a solid skeleton that consists mainly of carbon. With increasing temperature, the coal begins to soften and becomes more or less plastic. The individual coal particles fuse and stick together and form a porous coherent mass that expands, then solidifies. The final product is a solid porous structure—coke.

To investigate these complicated phenomena, a multiplicity of analytical processes has been developed. Investigations of the plastic properties of coal are of special interest because they are of paramount importance to the thermal refinement of coal, which even today accounts for the greatest consumption of coal.

6.2 METHODS FOR INVESTIGATING THE PLASTIC PROPERTIES OF COAL

The importance of coal's plastic behavior for many commercial processes is reflected in the large number of methods for measuring it. The object of all the investigations is to use physical measurement methods to obtain characteristic figures that permit classification of the coal and prediction of its behavior during combustion, gasification, or carbonization in commercial plants. Some measurement methods have been developed especially for one specific coal or for one specific commercial process. Therefore several measurement methods differ only slightly, whereas others are suitable only for specific applications.

6.2.1 Microscopic Investigations

The microscope is the only means for studying the behavior of macerals or individual particles in the plastic range during transition from coal to coke. For this purpose, samples withdrawn from a semitechnical coke oven after different periods of time were studied microscopically in polished sections.[1]

[1] H. Echterhoff and M.-Th. Mackowsky, *Glueckauf,* **96,** 618–626 (1960).

Changes in the macerals at the beginning of softening, up to the stage of resolidification, could be detected readily, as well as the temperature interval of the plastic range and the development of both anisotropic effects and the coke structure.

Another method[2] permits the study of coking processes in the laboratory under defined conditions, using a small cylindrical coking oven. This oven is heated only from the bottom and contains thermocouples spaced over the entire length of the cylinder for simultaneous temperature measurements. The sample is carbonized at a heating rate of 3K/min, and the heating up is terminated as soon as the coal at the heated surface has reached the semicoke stage. The cylindrical sample is then cut in the direction of its axis, and the progress of the carbonizing can be investigated microscopically and compared with the directly measured temperatures. This makes it possible to observe under the microscope changes in the coal as a function of rank and particle size.[3–5]

The influence of heating rate on the plastic behavior was investigated by coking granules.[4] Individual particles of coals of different rank were heated at rates of 0.5 K/min, 2–3 K/min, and about 300 K/min. To investigate the influence of packing density and particle size on the plastic properties, samples were coked in the Ruhr dilatometer (see Section 6.2.2.2) at different final temperatures, then studied under the microscope.[6]

These microscopic investigations were made on polished sections. The samples were evaluated not only qualitatively, as mentioned above, but also quantitatively: the change of the reflectivity of the macerals (without inertinite) on passing through the temperature interval of the plastic range was measured.[4,7,8]

[2] H. Ritter and G. Juranek, *Brennstoff-Chemie,* **41,** 170–176 (1960).

[3] H. Ritter and G. Juranek, *Brennstoff-Chemie,* **42,** 17–21 (1961).

[4] E. Wolff, Dissertation, University of Münster, 1963.

[5] H. Luther and B. Bussmann, *Brennstoff-Chemie,* **43,** 353–361 (1962).

[6] M.-Th. Mackowsky and E. Wolff, *Erdoel u. Kohle,* **18,** 621–625 (1965).

[7] F. Goodarzi and D. G. Murchison, *Fuel,* **56,** 89–96 (1977).

[8] M.-Th. Mackowsky, *4th Int. Conf. Coal Science,* Le Touquet, France, 1961.

The results obtained by microscopic investigations are a valuable supplement to technological and physicochemical investigation methods because they make it possible to better understand and judge the plastic behavior of coal.

6.2.2 Dilatometers with Different Heating Rates

The dilatometer test for coal measures changes in the volume and the fusing together of the plastic particles that occur during heating up and are important to the formation of coke. The dilatometer curve is used to characterize the suitability of coals for the coking process by characteristic figures. Furthermore, high-temperature dilatometers (operated at up to 1000°C) permit measurement of the change in volume during the transition from semicoke to coke.

Investigation of the softening and swelling of coal is made in the temperature range from 300–500°C and investigations on semicokes in the temperature range of 500–1000°C. Accordingly, the dilatometer sample is heated at a definite heating rate. About 3 K/min has been selected to characterize the coking properties with respect to the conventional coking process. The shortening of the coking time in modern high-performance ovens to about 11 hr[9] and the development of continuous coking processes[10] require considerably higher heating rates, which could be as high as 6–10 K/min or even as high as 100 K/min.[11] To adapt the laboratory methods to these new conditions, dilatometers with higher heating rates were developed.[12] If the instruments permit heating rates of more than 30 K/min, they are called shock dilatometers. The term "high-temperature shock dilatometer" is used if shock heating of the coal or semicoke sample to a temperature of 900°C or higher is possible.

6.2.2.1 *Audibert-Arnu Dilatometer*

Use of the dilatometer method to investigate coals was practiced first in France, where the volume change of the softening coal was related to the ability to form coke.[13] Following this, the currently used dilatometer principle was developed.[14] Finely ground coal is briquetted into a pencil, which is heated in a metal tube, and the change in length with temperature is measured. This method was further developed to a coal technological test method, which is still used as the Audibert-Arnu dilatometer[15,16] (Chapter 7).

In the "International Classification System for Bituminous Coal,"[17] the dilatometer test of Audibert and Arnu using coal pencils was chosen in addition to the Gray-King test as the final method for determining the "coking ability" (third digit in the code number). This method is used worldwide with only small modifications, which simplify the experiment but do not change the principle of the measurement.

6.2.2.2 Ruhr Dilatometer

The Ruhr dilatometer is used in the Federal Republic of Germany and in many other countries. This instrument is a further development of the Audibert-Arnu dilatometer and conforms to international standards. It is rather simple to use and is described briefly as a typical dilatometer method. The curves obtained with this instrument determine one characteristic value, the coking capacity *G*, which permits calculation in advance of the coke quality to be expected.[18,19]

To determine the progress of dilatation in the Ruhr dilatometer (Figs. 6.1 and 6.2), two pencils made of finely ground coal (<0.2 mm) are used simultaneously. They are heated under a load in a tube with a temperature increase of 3 K/min according to specifica-

[9] W. Rohde and K.-G. Beck, *Glueckauf-Forsch.-H.* **37**, 977–981 (1976).

[10] W. Reerink and W. Peters, *Brennstoff-Chemie,* **46**, 330–343 (1965).

[11] E. Ahland, J. Lehmann, and W. Peters, *Proc. Ironmaking Conf.,* **31**, 285–296 (1972).

[12] D. Habermehl, J. Lehmann, and W. Peters, *Glueckauf-Forsch.-H.,* **35**, 153–162 (1974).

[13] G. Charpy and J. Durand, *Compt. Rend.,* **171**, 1358 (1920).

[14] E. Audibert, *Rev. Ind. Miner.,* **6**, Part 1, 115–136 (1926).

[15] C. Arnu, *Rev. Ind. Miner.,* **13**, 155–173 (1933).

[16] C. Arnu, *Chim. Ind.,* **32**, 276–292 (1934).

[17] W. Mantel, *Brennstoff-Chemie,* **35**, 136–143 (1954).

[18] W. Simonis, G. Gnuschke, and K.-G. Beck, *Glueckauf-Forsch.-H.* **27**, 105–113 (1966).

[19] M.-Th. Mackowsky and W. Simonis, *Glueckauf-Forsch.-H.* **30**, 25–37 (1969).

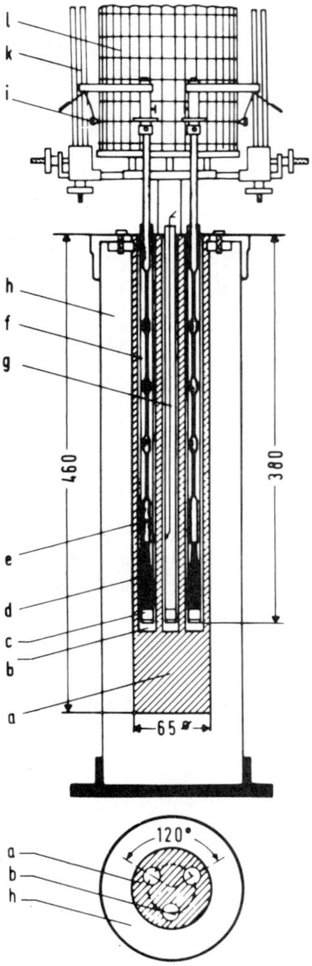

Figure 6.1 The Ruhr dilatometer: a = electrically heated brass block; b = hole for dilatometer tube; c = screw for closing the dilatometer tube; d = coal briquet; e = steel loading plunger; f = steel dilatometer tube; g = nickel/nickel-chromium thermocouple; h = insulation material; i = recording pen; k = guides for the recording pen; l = recording drum with recording sheet.

tions.[20,21] The axial changes of the lengths of the almost-cylindrical samples are transmitted to the plunger, which serves as the load and is connected to a device that records the changes in length on a drum. The heating rate is controlled by means of a nickel-nichrome thermocouple, which is located in a third opening of the dilatometer block. Figure 6.3 shows the dilatometer curves of different coals and their evaluation.

Contraction and dilatation are measured as a percentage of the length of the coal pencil used. The dilatation value is a measure of the "coking ability" and is used to determine the third digit in the code number of the International Classification System for Bituminous Coal (see Table 6.1).

Table 6.1 Numbers Characterizing Coking Ability[20]

Dilatation (%)	Gray-King Assay (Type)	Characterization (Third Digit)
No softening	A	0
Contraction only	B–D	1
0	E–G	2
0–50	G_1–G_4	3
50–140	G_5–G_8	4
> 140	> G_8	5

Caking coals soften during heating and the coal pencil collapses, and under the influence of the applied pressure of the recording plunger, it fills the empty space up to the wall of the tube. During this time the loaded plunger moves down. The maximum decrease in length, in percent, is called contraction. Theoretically a coal that softens well should have a contraction of about 32%, since in this case the total empty volume is filled by the coal that has become plastic.[22] Many coals do not show this theoretical value, since two opposing processes occur simultaneously: namely, the collapsing of the pencil and the swelling and degassing of the plastic mass. If the swelling begins before the empty space in the dilatometer tube is completely filled, the contraction is smaller. Many high-volatile coals show contraction values above 32%. The explanation for this observation must be found in secondary processes, for example, loss of mass during pyrolytic decomposition.[22] The extent of contraction often characterizes the coal and especially its softening behavior; however it does not characterize its coking ability.

The curve for some coals does not increase on reaching the lowest contraction value, and the behavior of these coals in the dilatometer

[20] Laborator Regulation 20/32/01, Verlag Glückauf GmbH, Essen, 1966.
[21] DIN 51739, Beuth-Vertrieb GmbH, Berlin W 15 and Cologne, May 1976.

[22] W. Lange, W. Radmacher, and H. Vierneisel, *Brennstoff-Chemie*, **42**, 312–319 (1961).

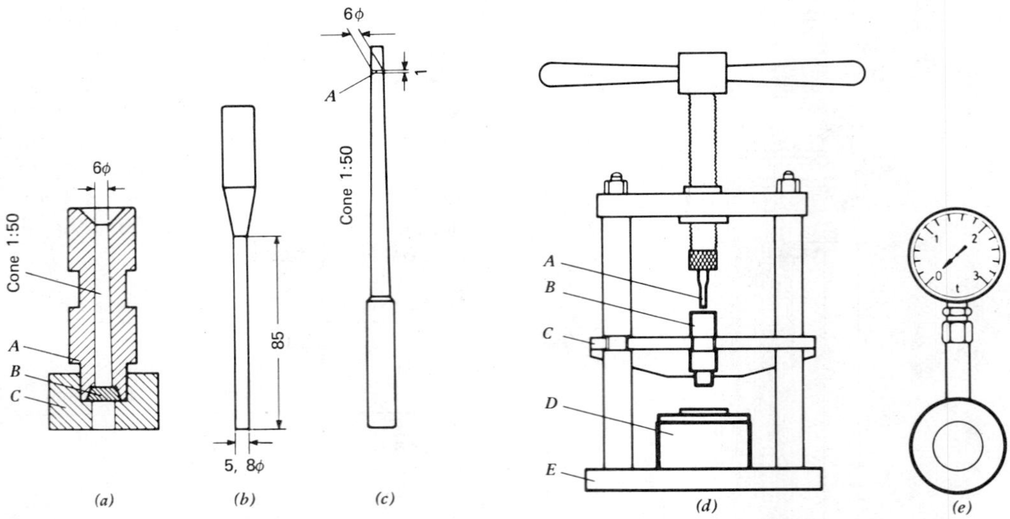

Figure 6.2 Supplementary equipment for the Ruhr dilatometer. (*a*) Briquetting device; *A* = cylinder with conical hole; *B* = sealing device; *C* = base. (*b*) Loading plunger. (*c*) Measuring device; *A* = reference marks. (*d*) Briquetting press; *A* = compression plunger; *B* = cylinder for compression plunger; *C* = swinging frame inserting compression cylinder; *D* = pressure measuring device; *E* = base plate. (*e*) Measuring device to indicate pressure.

is characterized as "only contraction" (examples 3 and 4, Fig. 6.3). In general the dilatometer curve has a true minimum and increases when the swelling of the plastic mass is higher than the contraction. This increase continues until the plastic mass is converted into solid semicoke (example 1, Fig. 6.3). From this temperature on, the curve is parallel to the abscissa and the dilatation process is completed. The distance of this final value from the zero line is called dilatation, measured in percent. This means that the coal has a "negative" dilatation when the zero line is not exceeded during the expansion process after the contraction (example 2, Fig. 6.3). The interval from the beginning of the contraction, when the dilatometer curve at the outset of softening leaves the zero line, until the resolidification of the plastic mass when the curve becomes horizontal, is known as the temperature interval of the plastic range. Its limits are the softening temperature and the resolidification temperature.

In addition to the values for the contraction, *K,* and the dilatation, *D,* both the softening and resolidification temperatures, *E* and *V,* are also characteristic values for the course of the dilatation of caking coals. They are used to calculate the dimensionless coking capacity *G,*[18,19] according to the following equation (Chapter 14):

$$G = \frac{E + V}{2} \frac{K + D}{VK + ED} \tag{6.1}$$

where *E* and *V* are expressed in degrees Celsius, *K* and *D* in percent.

In this process, the following three cases occur:

1. Contraction only: $G = 0$
2. $D < K$: $0 < G < 1$
3. $D \geqq K$: $G \geqq 1$

For high-quality coking coals, the *G* value is normally between 1.05 and 1.10.

Because coke quality can generally be predetermined with great accuracy for coals from one seam as well as for blends by means of the coking capacity *G,* it is unnecessary to obtain further data such as expansion and true dilatation from the dilatation test.[23]

Corresponding to the different softening and expansion properties of different coals in

[23] D. W. van Krevelen, H. A. G. Chermin, H. N. M. Dormans, and F. J. Huntjens, *Brennstoff-Chemie,* **39** (Special Issue), 77–83 (1958).

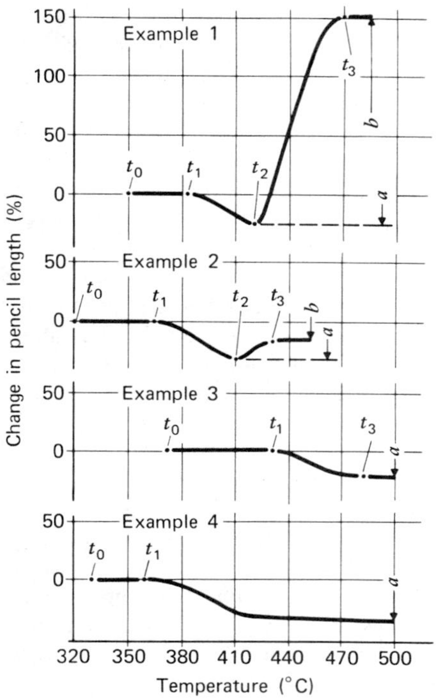

Figure 6.3 Dilatometer curves and their evaluation.[20]

the dilatometer, characteristic dilatation curves are obtained that can be quite different, as shown in Fig. 6.3.[20,22] An explanation of these different forms of the curves is given in Section 6.2.3.1.

6.2.2.3 *Dilatometers for Nonbriquetted Coal*

The Sheffield coking experiment also uses the dilatometer principle.[24] As in the Hoffmann dilatometer,[25] (see Chapter 7) which was introduced in the Saar district, bulk coal is used instead of pencils.

In the Sheffield dilatometer, the effective weight of the plunger on the coal can easily be changed by a counterweight. Since the diameter of the plunger is significantly smaller than the internal diameter of the dilatometer tube, the plunger, depending on its weight, may sink into the softening coal. This dilatometer thus fulfills the function of a penetrometer. Here, too, as in the Audibert-Arnu and Ruhr dilatometers, the curve for softening coals shows a sinking of the plunger to a minimum, which is also called contraction. However the Sheffield process is basically different. In the Audibert-Arnu dilatometer, the contraction occurs because the softening coal fills the empty space in the dilatometer tube. In contrast, in the Sheffield dilatometer the plunger sinks into the softening coal, which has completely filled the volume of the tube.

The Saar, or Hoffmann dilatometer,[25] is very similar to the apparatus described above. Here, too, the pulverized coal is poured into the unit and only slightly compressed. The diameter of the plunger is also significantly smaller than the diameter of the tube.

The curves obtained by this technique also differ significantly from those obtained in the Audibert-Arnu dilatometer. The different shape of the curves was used by Hoffmann to classify coals into four categories.[26]

The Hoffmann dilatometer is especially suited for the slightly coking coals found in the Saar district and in Lorraine. In addition, its greater diameter—samples of 23 mm in diameter were tested—permits the use of coarser coal blends so that the influence of particle size on the plastic behavior can be investigated too.[27]

6.2.2.4 *High-Temperature Dilatometer*

In principle, the high-temperature dilatometers are Audibert-Arnu dilatometers that can be used at temperatures of 1000°C or higher. They can be used to measure not only the above-mentioned changes in volume during softening and swelling of the coal, but also the changes in the dimensions of the sample in the postplastic range and especially the shrinkage when semicoke is converted to coke.

The technical conception used the experience with dilatometers available from the metallurgical and refractory industries.[28,29]

[24] R. G. Davies and R. A. Mott, *Fuel,* **12,** 294–303 (1933).

[25] H. Hoffmann, *Oel Kohle,* **40,** 531–551 (1944).

[26] H. Hoffmann and K. Hoehne, *Brennstoff-Chemie,* **35,** 202–211, 236–246, 269–275, 298–304 (1954).

[27] K. Hoehne, *Glueckauf,* **96,** 1203–1219 (1960).

[28] P. Chevenard, *Bull. Soc. Fr. Miner. Cristallogr.,* **76,** 165–188 (1953).

[29] P. Thormann and P. Buchmeyer, *Tonindustrie-Zeitung,* **91,** 218–220 (1967).

In all these methods,[30–33] the important factor is the contraction coefficient α or $\alpha.S$, which is defined as follows:

$$\alpha = \frac{1}{L_0}\frac{dL}{dT} \qquad (6.2)$$

in which L_0 is the initial length of the sample in meters and dL/dT is the change in length with increasing temperature (T in °C) which, as will be shown later, is important to both the strength and the size distribution of the coke.

6.2.2.5 *Shock Dilatometer*

Because of the low heating rate of only 3 K/min that is characteristic of the dilatometers described in Sections 6.2.2.1–6.2.2.4, they cannot be used to characterize briquetting blends that are used in continuous coking processes. At first a modified Ruhr dilatometer process capable of achieving higher heating rates was used.[34] It was heated to a predetermined temperature (e.g., 500 or 600°C), and this temperature remained constant during the entire experiment. This gave heating rates of up to 30 K/min which, however, did not remain constant during the total time of the experiment.

An increase in the heating rate generally leads to an increased dilatation and a shorter duration of plasticity. Thus with this modified dilatometer process, coals and coal char blends showed significant measurable dilatation values, although they did not show dilatation in the normal dilatometer.

The heating rate obtained is still significantly below the rates attained in modern continuous processes.[10,11,35–38] Furthermore blends that are used in hot briquetting (~ 30% caking coal, ~ 70% char)[39] show no dilatation. A valid determination of the behavior of bituminous coal at the high heating rates of the continuous processes is possible only if heating rates in the range of 100 K/min and higher can be reached. Such heating rates can be achieved only with the so-called shock dilatometer.[12,33]

This instrument uses the principle of the Audibert-Arnu dilatometer and permits heating rates between 1.2 and 700 K/min. A constant rate of heating and isothermal temperature operation can be used in succession at will. Since the samples have a diameter of 13 mm, coals of coarser size may also be studied. Investigation of coals poured without briquetting is also possible. The design of such a dilatometer is shown in Fig. 6.4.

The high heating rates are achieved with a resistance heating tube that simultaneously serves as the interior wall of the oven and is heated by currents that may be as high as 1500 A, causing very rapid heating. Temperature is controlled by a pid-controller with thyristors.

As shown in Fig. 6.5, the coal sample is placed in a tube that is closed at the bottom and made from high-temperature-resistant steel that is perforated to facilitate degassing. During the heating process the tube is located in a free-hanging quartz tube, which is flushed with inert gas to avoid oxidation of the sample during heating. The change in length is transmitted by a quartz plunger and transformed inductively by a displacement pick up that converts the changes in length into proportional electrical voltages. Basically the dilatation is measured as in the dilatometers described previously, except that the samples are generally heated to a final temperature of 550°C.

Figure 6.6 shows curves for the dilatation of a highly swelling coking coal as a function of the heating rate and the temperature of the oven. On one hand with the higher heating rates the dilatation is increased severalfold over the value of the conventional method with a heating rate of 3 K/min, and on the

[30] A.-F. Boyer, A. Ladam, and J.-L. Soulé, *Rev. Ind. Miner.*, **34,** 592–616 (1953).

[31] D. H. Gregory and A. E. Horton, *J. Inst. Fuel,* **43,** 389–396 (1970).

[32] F. Strunk, Dissertation, Clausthal, 1971.

[33] D. Habermehl, Dissertation, Aachen, 1972.

[34] F. Schönmuth, Dissertation, Clausthal, 1970.

[35] W. Peters, *Glueckauf,* **103,** 1273–1279 (1967).

[36] W. Peters, E. Ahland, and J. Langhoff, in *Technik und Entwicklung der Verkokung von Steinkohle (Technology and Development of Carbonization),* Information Meeting, Luxemburg (1970), pp. 215–235.

[37] E. Ahland, B. Beck, H.-J. Jagnow, J. Lehmann, and W. Peters, *12th Biennial Conf. International Briquetting Association,* Vancouver, 1971.

[38] E. Ahland, B. Bock, H.-J. Jagnow, J. Lehmann, and W. Peters, *Int. Symp. Advances Extractive Metallurgy Refining,* London, 1971.

[39] J. Lehmann, Dissertation, Aachen, 1970.

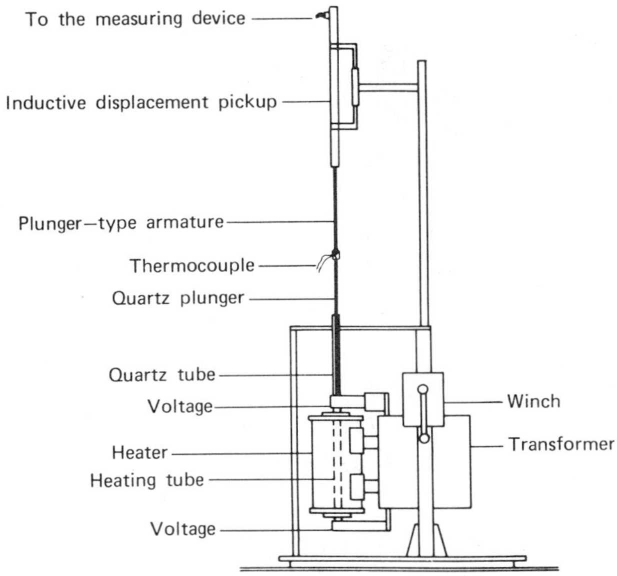

Figure 6.4 High-temperature shock dilatometer.[12,33]

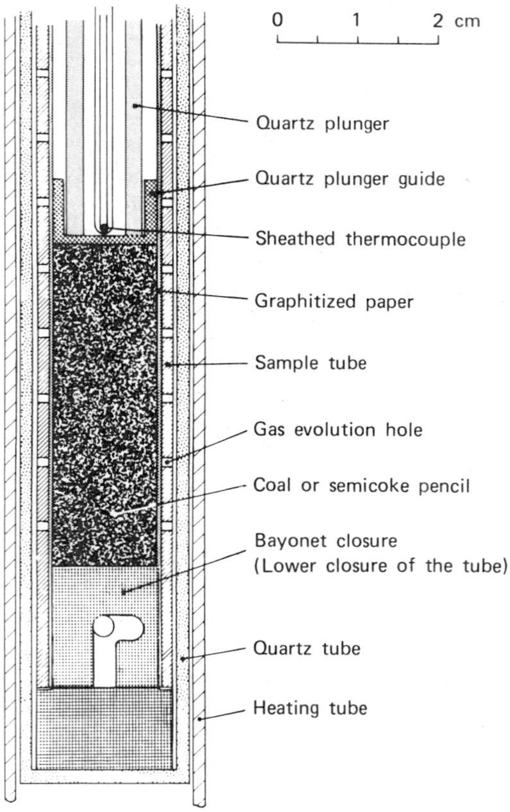

Figure 6.5 Measuring equipment in the high-temperature dilatometer oven.[12,33]

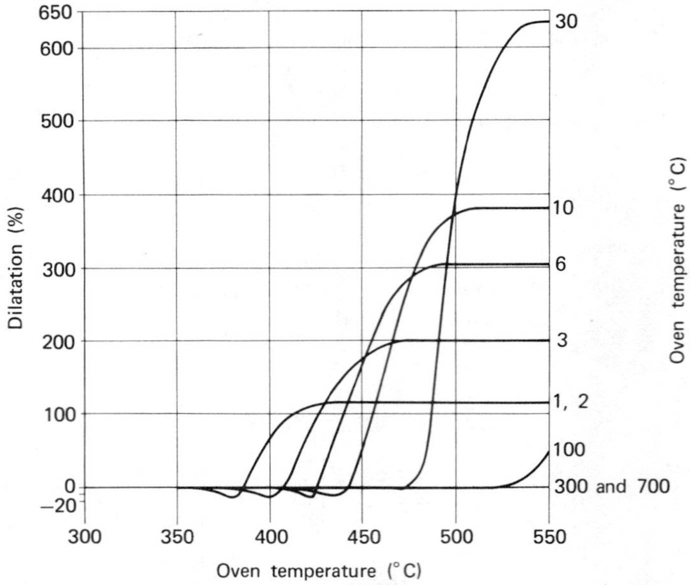

Figure 6.6 Dilatation of a good coking coal as a function of temperature and heating rate.[12,33]

other hand the total dilatation curve is shifted to higher temperatures. If the heating rate exceeds the value of 30 K/min, then the heating precedes the expansion, and the dilatation at a temperature of 550°C has not yet reached its final value (e.g., with a heating rate of 100 K/min) or it has not even begun (e.g., with an experiment with 300 and 700 K/min heating rates). Thus the temperature values characteristic of the progress of dilatation (softening temperature, resolidification temperature, and the interval of the plastic range) can no longer be determined.

For this reason it is advisable to show the change of dilatation with time as in Fig. 6.7. Obviously this requires that characteristics associated with time be introduced to describe the dilatation process. The softening temperature is replaced by the time at which softening begins, the resolidification temperature by the time of terminated resolidification, and the interval of the plastic range by the duration of plasticity.

The difference in the course of dilatation during slow and rapid heating is significant. At heating rates below 30 K/min, contraction occurs first, and the dilatation moves rather slowly to its final value. At heating rates greater than 30 K/min, contraction is completely absent.

Figure 6.8 shows the changes in the dilatation curves with heating rate for different coals. (The coal rank is indicated by the index, which shows the volatile matter content on a daf basis.) It can be seen clearly that the dilatation reaches its highest value when the heating rate exceeds 100 K/min. Coal rank can thus be characterized by the maximum dilatation in the shock dilatometer, which yields results parallel to those obtained with the Ruhr dilatometer, for example.

The high sensitivity of measurements made with the shock dilatometer permits it to characterize coals of different ranks or coal-char blends that do not show any effect in the Ruhr dilatometer.

6.2.2.6 *Influence of Heating Rate on Characteristic Dilatometer Values*

The dilatometer measurements in the shock dilatometer indicate two different ranges of heating rate: in the area of slow heating, with heating rates below 30 K/min, all characteristic dilatometer values depend to a great extent on the heating rate. In the area of rapid heating, with heating rates above 30 K/min, the corresponding values depend only slightly or not at all on the heating conditions.[12]

In dilatometer tests the "softening point" is the temperature, measured at the surface of the sample, at which the apparatus indi-

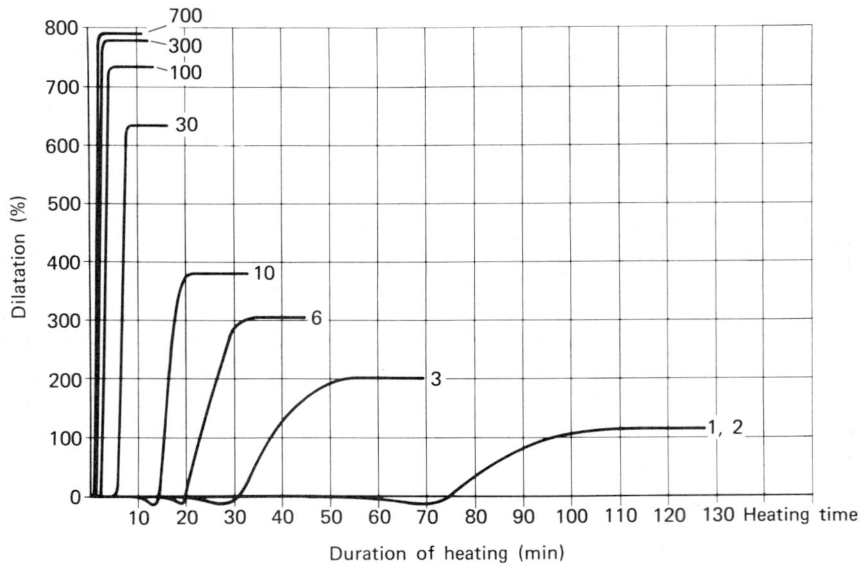

Figure 6.7 Dependence of dilatation on time and heating rate (K/min).[12,33]

cates contraction of the sample. Actually this condition becomes noticeable only when the coal grains in the axis of the cylinder become slippery because of the formation of oily constituents. One can safely assume that on the average, these conditions will appear on each coal particle at the same temperature. Because of the low thermal conductivity, a measurable effect will be shown only when this temperature is reached in the axis of the cylinder.

The "softening temperature" actually measured at the periphery is higher than the temperature in the center, and with increasing heating rate this deviation becomes greater. Therefore the measured softening temperature is moved to higher temperatures as the heating rate increases.

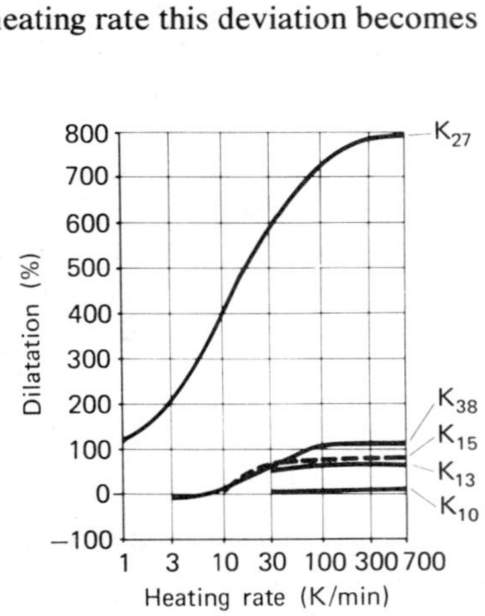

Figure 6.8 Dilatation of different coals as a function of heating rate.[12,33]

In a similar way, resolidification also becomes noticeable only when it has progressed to the axis of the cylinder. Again, it can be assumed that on the average, resolidification occurs in each coal particle at the same temperature. The "resolidification temperature" measured on the periphery is obviously also higher than the actual temperature in the middle of the sample.

The described effects change considerably when the heating rate is increased, because the individual coal particle remains in the plastic stage for an even shorter period. During slow heating at the beginning of softening in the axis of the cylinder, the coal on the periphery is still plastic. If the heating rate is increased, a limiting heating rate is reached at which the particles in the middle are softening at the same time that the periphery is already resolidified. At higher heating rates, the plastic zone travels in concentric hollow cylinders through the sample. Three different phases then appear simultaneously in the radial direction: near the axis is coal that has not yet softened, on the periphery the sample is already solidified semicoke, and in between is the plastic mass.

As far as the measuring device is concerned, the simultaneous existence of the different phases in the sample has several consequences. Although contraction in the measuring device can occur only if the true softening temperature has progressed to the axis of the cylinder, the dilatation may already have a measurable effect when the coal particles on the periphery begin to swell. For the case of the limiting heating rate produced by the apparatus, as described above, it means that when contraction starts, expansion is already beginning to occur on the periphery. Since the dilatation is usually greater than the contraction, contraction can not occur at sufficiently high heating rates, as shown in Fig. 6.9. Since thermal conductivity is limited, the duration of plasticity approaches the limiting value at very rapid heating rates. Then the duration of plasticity depends only on the radius of the sample (measuring device) and on the thermal conductivity of the sample, which changes very little.

The above-described phenomena can be neglected when using test equipment with a slow heating rate and small sample diameter, such as the Audibert-Arnu or the Ruhr dilatometers.

6.2.3 Plastometers, Penetrometers, and Other Testing Methods

The testing methods discussed in this section have been known for a long time and have been described sufficiently in many publications.[40] In the past several years, some instruments have been significantly improved and the influence of some parameters on the final result has been investigated in more detail.

6.2.3.1 *Plastometers*

Of the many plastometer types, the Gieseler plastometer[41] has evidently found worldwide application. The instrument was improved in certain details and is described in a new ASTM standard.[42,43] In the old instrument the stirrer was driven with a constant torque, which was obtained by a string with a weight attached to it. The new instrument has an electric motor of constant rpm with a shaft with a hysteresis brake, which controls the torque. The drum is directly connected to the shaft and the stirrer, and now it has 100 divisions instead of the previously used 360° dial. Each full rotation is counted photoelectrically. These improvements ensure uniform test operation, since it is no longer necessary to sometimes rewind the string. The system of rolls, string, and weight is used only to calibrate the instrument. Changes in the rotational speed with time are not automatically recorded, however, and this may be a disadvantage.

The influence of measuring conditions on the results obtained with the Gieseler plastometer has been widely investigated. It has been found that with increasing depth of the retort immersion in the metal bath, the maximum fluidity increases while the characteristic temperatures are not noticeably changed.[44] The stirrer arms should be renewed as soon as their diameter is smaller than 1.5 mm, since otherwise the value for the maximum fluidity shows erroneous increases. Too low a value, however, is obtained if the particle size is smaller than specified.[45]

Reproducibility of the results from different laboratories is still not quite satisfactory, perhaps because of differences in the construction of the individual instruments. Systematic investigation of North American coal showed[46] that the standard deviation (68.3% of the data within the limits $\pm\ \sigma$) for the softening and resolidification temperature is between 3.22 and 4.33 K, and for the plastic range it is 4.54 K.

Novel processes for the conversion of coal (e.g., the hot-briquetting processes or the gasification process) use much higher heating rates than are customary in the coke-oven process, to which testing conditions have been adapted. Therefore the Gieseler plas-

[40] *Chemistry of Coal Utilization,* 1st Supplementary Volume, H. H. Lowry, Ed., John Wiley and Sons, New York, 1963.
[41] K. Gieseler, *Glueckauf,* **70,** 178–183 (1934).
[42] ASTM Designation D 2639–74, 1974.
[43] U.S. Bur. Mines Bull. No. 638, *Methods of Analyzing and Testing Coal and Coke,* 1967.
[44] R. R. Allen, U.S. Bur. Mines Rep. Invest. No. 6559, 1964.
[45] T. Fukuyama, T. Miyazu, and S. Kimoto, *J. Fuel Soc. Jap.,* **51,** 628–638 (1972).
[46] W. R. K. Wu, U.S. Bur. Mines Rep. Invest. No. 7394, 1970.

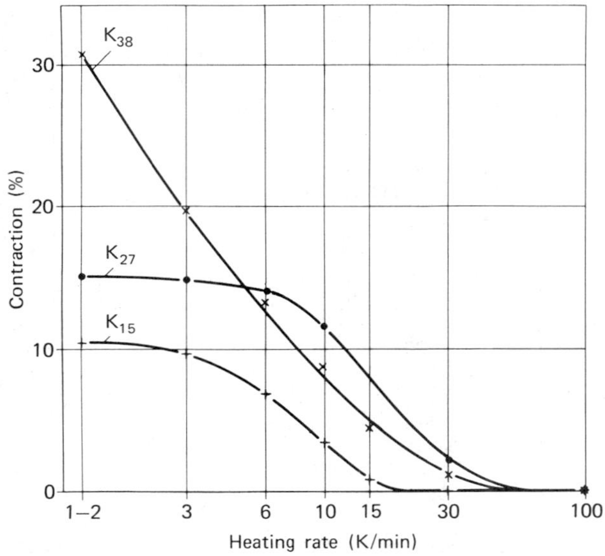

Figure 6.9 Contraction of different coals as a function of heating rate.[12,33]

tometer method was also used with higher heating rates. The retort with the sample was immersed in the heating bath, preheated to a constant temperature.[47,48]

With this method heating rates of about 100 K/min were attained. In this case the temperature of maximum plasticity is the temperature of the bath. An additional parameter was introduced: namely, the duration of plasticity, which is of great importance to the new processes. This parameter is also determined in the high-temperature shock dilatometer (see Section 6.2.2.5). The slope of the plastometer curve, the differentiated curve, has been proposed as another parameter for characterizing the plastic behavior of coal. To create conditions as they exist when, for example, heated coal is transported to the briquetting press, the influence of decreasing temperature on the plastic behavior after attainment of maximum fluidity has been investigated. The cooling was achieved by removing the retort from the heating bath for a short time.

The results show that at higher heating rates the softening temperature shifts to higher temperatures, the maximum fluidity is increased, the duration of plasticity is decreased, but the temperature of maximum fluidity is not affected. Cooling a sample after reaching maximum fluidity to a lower temperature that is still within the plastic range affects the maximum fluidity only slightly, but shortens the duration of effective plasticity.

The Brabender plastograph[40,49,50] is used in Germany and in Italy and has found wide application in industry to determine viscosity and grinding hardness. It has not been accepted for testing coal samples in other countries, although in comparison to the Gieseler plastometer it has several advantages: the test procedure is simpler, the result of the experiment is independent of the investigator, and the results have excellent reproducibility. Also, the curves obtained supply more information about the softening behavior of the coal than do those of the Gieseler plastometer, as illustrated by Fig. 6.10, which compares the curves obtained from three very different coal samples with the two

[47] A. Kijewska, *Koks, Smola, Gaz,* **18**(6), 164–170 (1973).

[48] A. Kijewska, *Koks, Smola, Gaz,* **18**(7), 191–195 (1973).

[49] H. Echterhoff, *Erdoel u. Kohle,* **8**, 294–298 (1955).

[50] H. Echterhoff, *Glueckauf,* **90**, 510–516 (1954).

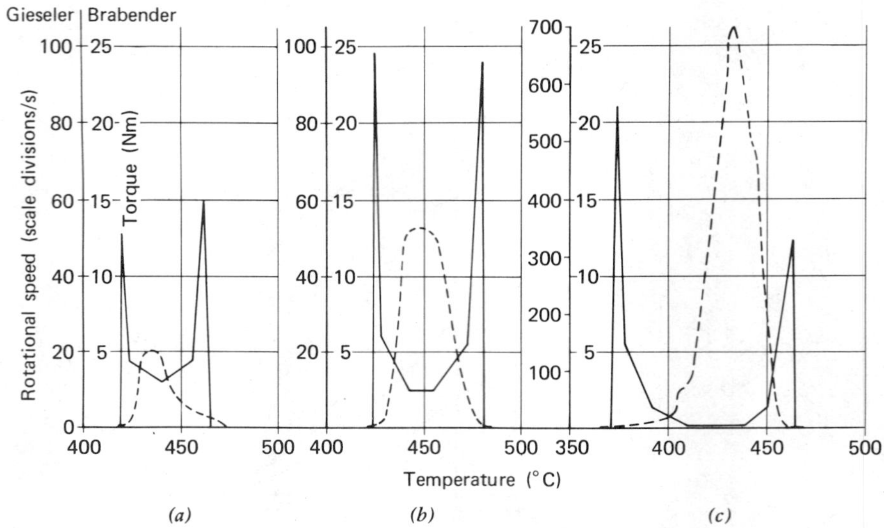

Figure 6.10 Comparison of curves made by the Brabender plastograph (solid curves) and by the Gieseler plastograph (dashes). (*a*) Coal from the Saar, 34.4% VM (maf); (*b*) coal from the Ruhr, 25.7% VM (maf); (*c*) North American coal, 41.7% VM (maf).

instruments. The diagrams show that the softening and resolidification points of both methods are in agreement and that the temperatures in the Brabender plastograph are better defined. The temperature of the maximum of the Gieseler curve coincides with the minimum of the Brabender curve. The Brabender curve supplies more information. The value of the first maximum characterizes the intensity of the beginning of softening, and the height of the second maximum represents a measure of the strengths of the formed semicoke.

Figure 6.11 presents typical Brabender plastograph curves for coals of different rank from the Ruhr district. For comparison, the Ruhr dilatometer curves are shown and the swelling indices given. The ability to form coke is qualitatively indicated by the dotted curve. However this curve does not represent the *G* values calculated by Equation 6.1.

Since the Brabender plastograph defines all phases of coke formation, it can be used for the preparation of the feed coal blends (e.g., for coking in slot ovens or for hot briquetting). It makes possible a determination of the proper ratio, for the coking proc-

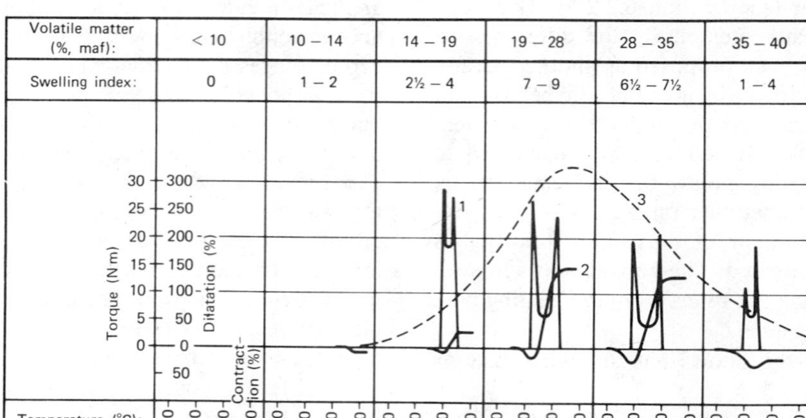

Figure 6.11 Torque curves of typical coals of different rank from the Ruhr district measured in the Brabender plastograph, compared with the course of dilatation and the swelling index: (1) plastic behavior, (2) course of dilatation, (3) coke-forming ability.

ess, of the dispersed and the dispersing phases in the plastic mass.[51,52] The relationships between the plastograph results and the expected coke strengths are also known.[51,53,54]

The following disadvantages of the Gieseler plastometer and the Brabender plastograph should be mentioned: (*a*) the stirrer arms in these units produce a complex system of shear stresses; (*b*) an expansion of the softened coal may occur in a part of the stirred container, which does not belong to the actual measuring location; (*c*) the stirrer arms produce grooves in the coal mass through which the blades of the stirrer may move with far less resistance; and (*d*) in the case of the Gieseler plastometer, a foam structure may develop. Because of the foam structure phenomenon, viscosities measured are often much lower than those actually present in the range of maximum fluidity.

For this reason a fully automatic rotation viscosimeter was constructed in which the shearing velocity gradient is almost constant.[55] It represents a concentric cylinder viscosimeter without end effects, in which the coal under test is contained in a 4-mm annulus between an outer rotating cylinder and a solid copper inner cylinder with milled surface. The surfaces of the two cylinders are kept at approximately equal temperatures, so that volatile products do not condense on them. The outer cylinder rotates with 2 rev/hr. The torque necessary to secure this speed of rotation is measured. A measurement requires 3 ml of coal ground—36 British Standard. This unit has not found wide application, although it was used successfully with Australian coals.

6.2.3.2 *Tests Using Unidirectional Heating*

The heat flow as it occurs in coke ovens is simulated by methods that supply heat only from one side. Among these measurement methods, the one developed by Sapozhnikov[56] is used in eastern European countries to characterize the plastic behavior of coals. It has recently undergone marked improvement.

The improved Sapozhnikov plastometer[57–59] now permits automatic recording of the conventional indices of this plastometer that is the thickness of the plastic layer, *y*, and the shrinkage, *x*, of the coal sample. In addition, it automatically measures the force needed for penetration of a needle into the plastic coal layer, which is proportional to the relative viscosity of the plastic mass.

Figure 6.12 is a schematic representation of the improved Sapozhnikov plastometer. A bottom-heated steel cylinder (1) is charged with 100 g of coal and heated at a rate of 3 K/min. The coal is weighted by a piston (2) with a pressure of about 1 bar. In the coal sample two channels for a thermocouple and the plastometer needle (12) are left open. The cylindrical plastometer needle with a diameter of 1.5 mm is introduced by a motor (8) into the channel, until it impinges on the plastic mass that has entered the channel. On further penetration, the needle meets the formed semicoke. As soon as the penetration force exceeds the value of 20 N (point *D* in Fig. 6.13*a*), the needle is retracted. After a predetermined time, the measuring process is repeated. During this procedure the force detector (11) measures the force that is needed to bring the needle into the coal mass, measures the length of the movement of the needle by means of a displacement pickup (14) and records the measurement. Movement of the lever (3), which transfers the load on the sample, is measured by a displacement pickup (6). This device is used to record dilatation and shrinkage. It also indicates the extreme values, to permit determination of the times at which the measurements should be made, as well as the direction of movement of the loading piston.

Figure 6.13 gives some typical plastometer curves for different coals. The upper diagrams *a–d* show the course of the contrac-

[51] G. Klepel, *Freiburg. Forsch.-H. A* **141,** 59–85 (1960).

[52] H. Echterhoff and M.-Th. Mackowsky, *Glueckauf,* **96,** 618–629 (1960).

[53] M. Engshuber, H. Telling, and E. Hanke, *Mitt. Inst. Energetik. (IFE), Leipzig,* No. 55, 300–307 (1963).

[54] J. Horacek, *Paliva,* **11,** 338–341 (1961) (in Czechoslavakian).

[55] J. N. Stephens, *Fuel,* **42,** 175–178 (1963).

[56] L. M. Sapozhnikov and L. P. Basilevitch, *Khim. Tverd. Topl.,* 3(2–3), 79–102 (1932).

[57] GOST 1186–62, *Coals, Determination of Plastometer Indices,* 1962.

[58] V. J. Donkov, *Coke Chem. USSR,* **3,** 44–47 (1967).

[59] V. J. Donkov et al., *Coke Chem. USSR,* **6,** 33–35 (1971).

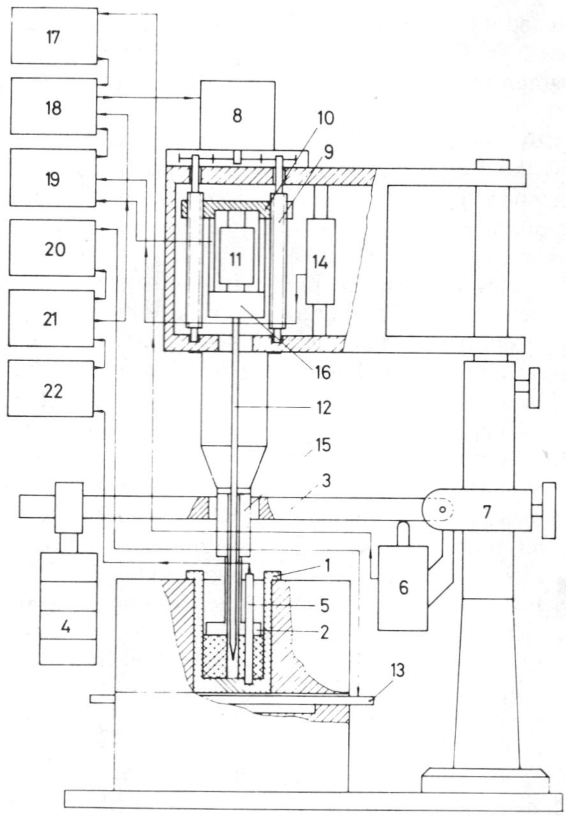

Figure 6.12 Schematic drawing of the improved Sapozhnikov plastometer.[58]

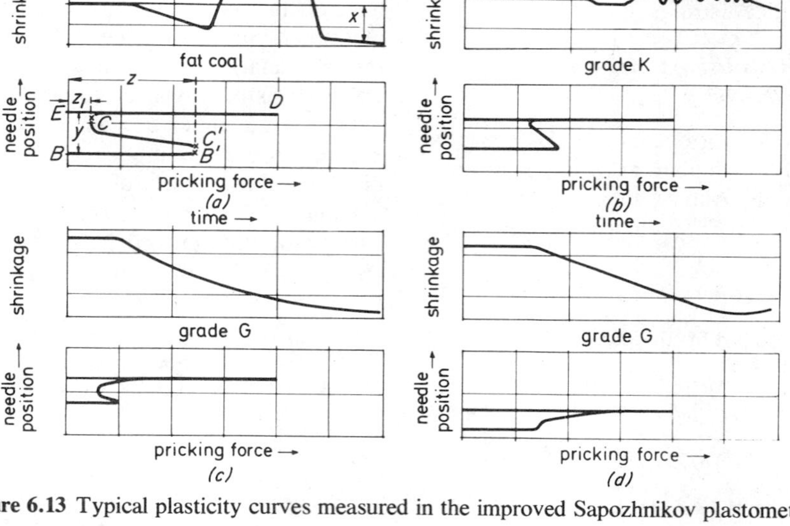

Figure 6.13 Typical plasticity curves measured in the improved Sapozhnikov plastometer.[58]

tion, x, and the lower curves represent the penetration force, which is measured at predetermined time intervals. When the needle enters the plastic mass, the penetration force increases until it reaches, for a short time, a constant value. The part of the curve designated B–B'–C' corresponds to the passage through the most dense and usually monolithic layer in the plastic mass. This is where the coal begins to soften. The part of the curve C'–C corresponds to the plastic zone, and part C–D of the curve is identified with the transition to semicoke. The minimum of the curve is proportional to the lowest viscosity.

It has been shown that the characteristic data obtained by the Sapozhnikov method are additive[51]; therefore these values can be used to determine the properties of coal blends.

When a coal sample is heated from only one side, the duration of the plastic state is proportional to the thickness of the plastic layer. Since heating from one side occurs in the conventional coke oven, these methods of measuring the thickness of the plastic layer have led to further investigations, especially with regard to the prediction of the caking and coking ability of coal by this method.

Accordingly, a modified Sapozhnikov method[60] was proposed that would determine the following parameters that characterize coking ability: (1) the thickness of the plastic layer, y', (2) the temperature at the beginning of the intensive thermal decomposition of the plastic mass, T_{dec}, and (3) the temperature drop inside the plastic layer, ΔT. The authors show,[60] using Soviet coals, that these values permit prediction of the coking ability of the coal and the strength of the coke produced. High-strength cokes were obtained from coals that have the following properties characterizing coking ability: $y' = 29$–34 mm; $T_{dec} > 410°C$, and $\Delta T = 125$–$135°K$.

The relationship between the thickness of the plastic layer and the strength properties of the coke was explored in a large laboratory oven, which was almost a section of a coke oven (width of the chamber: 400 mm, heating from two opposite walls).[61] Similar to the Sapozhnikov plastometer, the thickness of the plastic layer is measured with a needle. The average values of the thickness of the plastic layer measured in this oven are in good agreement with those of the Sapozhnikov plastometer. The best coke was obtained if an average thickness of the plastic layer of 14–17 mm was observed. These values differ significantly from those determined by other authors.[60] However it should be remembered that the thickness of the plastic layer alone is not sufficient to predict the coke properties.[60,62]

The determination of the thickness of the plastic layer with equipment similar to the Sapozhnikov apparatus was used for Indian[63] and Australian[64] coals, in addition to Soviet, European, and American[62] coals as a criterion for the composition of coal blends for the production of metallurgical coke.

Experimental data on expected coke strength were also obtained from another modified and partially improved Sapozhnikov method developed in Hungary. First the coal sample was compressed by different forces to simulate practical operation. Then the carbonized briquettes were tested for strength by dropping a 7-kg weight 5 times through a tube onto each coke briquette. The size consist of the fractured coke is then determined.[65]

6.2.3.3 *Other Methods*

The International Classification System for Bituminous Coal[17] for the characterization of coking ability requires use of the dilatometer or the Gray-King test, both of which are relatively time-consuming. Therefore quicker methods were sought. In Czechoslovakia a method for the determination of the "elasticity index" was developed.[66] This method is considered to be a measure of the ability of the plastic mass to form a more or less expandable elastic and strong skin under the

[60] M. D. Shapiro and L. S. Altermann, *Coke Chem. USSR*, **10**, 5–9 (1962).

[61] Yu. V. Pyzhov, *Coke Chem. USSR*, **7**, 17–19 (1963).

[62] M. J. Kovalik, D. E. Wolfson, and F. Fischler, U.S. Bur. Mines Rep. Invest. No. 6568, 1964.

[63] S. K. Sharma and N. N. Das Gupta, *J. Mines Met. Fuels*, **7**, 1–6 (1959).

[64] T. G. Callcott and J. A. McInnes, *Fuel*, **41**, 1, 15–34 (1962).

[65] M. Veltner, *Coke Chem. USSR*, **3**, 12–16 (1960).

[66] J. Kunc, *Brennstoff-Chemie*, **49**, 211–217 (1968).

action of the gases and vapors developed in the plastic mass. The "elasticity index" clearly depends on the dilatation of the coal under investigation. The crucible used for the free-swelling index test is being used for the determination; however it is covered with a sheet metal cover that has a 5-mm-diameter hole in the middle. It is weighted with a hollow cylinder having the same outside diameter. On coking at 850°C a bubble passes through the opening. The surface area of the silhouette of the cutoff bubble is measured by optical means and is called the "elasticity index."

The swelling index is one of the most frequently used methods of testing among those known so far. Therefore proposals have been made to automate this method.[67,68] For this purpose, the silhouette of the coke residue is no longer compared with a pattern, but is measured by means of optoelectric methods. Similarly, the heating is fully automated.

Attempts are being made to achieve better defined experimental conditions that improve reproducibility and lead to more uniform coke residues.[69] The modifications are concerned mainly with the heating equipment, the positioning of the crucible, and the sample preparation, since the current practice has considerable inaccuracies.

The range of particle sizes of coals of different rank was investigated autoradiographically to clarify whether after coking a complete mixing of the softened mass of the different coal types has still been achieved.[70] For this purpose different sized coal cubes were covered with radioactive rock dust ($CaCO_3$) and coked in a mixture with ground coal. The resulting coke was cut and the autoradiograms were taken on X-ray film. The results show that the particles of noncaking brown coals, poorly caking gas coals, and group K_2 coals tend to retain their shape and may function as centers of weakness in the coke. Thus these coals for use in carbonization blends must be crushed below 3 mm. Good caking gas and fat coals are able to penetrate and mingle with the surrounding charge and need not be crushed below 6 mm.

On the assumption that in the formation of coke the contact bridges formed between the softened coal mass and the coal particles are responsible for the electrical resistance, the electrical resistance of coke lumps has been measured and correlated with the thickness of the plastic layer, as determined in the Sapozhnikov plastometer.[71] The measurements confirmed this assumption; the electrical resistance decreases linearly with the thickness of the plastic layer.

The "inert binding capacity" (IBC) test[72] was developed in the Netherlands to obtain a better method of judging the coking properties of coals than indicated by the laboratory methods. The IBC test is similar to the Roga test (ISO 335) but is more closely related to actual operating conditions. It measures the inert binding capacity of coal, which is the amount, in grams, of an entirely noncoking coal that when added to 100 g of a softening coal, leads to a retort coke with sufficient abrasion strength. The limitation of this value was obtained from practical experience. The coking was done in a Jenkner retort[73] with a coal charge of about one kg.

G. E. Foxwell[74] devised a method for measuring the plastic stage of coal by measuring the flow resistance of a stream of gas flowing through the coal sample during continuous heating. This method was recently used in a modified form. In one case[75] the gas (nitrogen) entered from the bottom through a hollow coal cylinder of powdered coal (10 mm high, O.D. = 13 mm, I.D. = 4 mm). This form assures uniform heating of the sample. Other authors[76] used a somewhat modified Sapozhnikov plastometer for this purpose, so that the thickness of the plastic layer could be determined simultaneously. The Foxwell method has also been used to investigate the influence of high gas pressures

[67] Stamicarbon N.V., Holland, Brit. Pat. Spec. No. 998, 1961, p. 312.

[68] J. W. Igoe and E. A. Zawadzki, U.S. Pat. Off., No. 3 (1966), pp. 245, 308.

[69] R. K. Young and E. A. Zawadzki, *Coal Age*, 74–78 (1965).

[70] J. M. Glushchenko et al., *Coke Chem. USSR*, **6**, 9–12 (1971).

[71] E. M. Taits, N. K. Ermolaeva, and Yu. S. Nikolaev, *Coke Chem. USSR*, **2**, 19–21 (1970).

[72] H. A. W. de Vries, N. N. M. Dormans, and C. Bokhoven, *Brennstoff-Chemie*, **48**, 289–297 (1967).

[73] A. Jenkner, F. L. Kühlwein, and E. Hoffmann, *Glueckauf*, **70**, 473 (1934).

[74] G. E. Foxwell, *Fuel*, **3**, 122–128 (1924).

[75] J. N. Bogoslovsky, V. J. Kudrjasov, and G. N. Makarov, *Zavod. Lab.*, **29**, 198–199 (1963).

[76] G. K. Petrik and V. S. Nosochenko, *Coke Chem. USSR*, **7**, 4–7 (1968).

(up to 100 bar) on the plastic behavior of coals[77] (Section 6.3.4).

The method using gas permeability, which is a measure of the viscosity of the plastic mass, not only determines permeability itself but also permits determination of the temperature limits of the plastic range, the temperature of maximum softening, and the degree of oxidation of the coal. Investigations by different authors have confirmed that as the viscosity of the plastic mass decreases, the resistance offered by the plastic mass to the flow of gas increases, and the permeability decreases. A low permeability thus indicates a higher swelling index and a higher coking ability.[76] To obtain reproducible results with this method, it is necessary to use highly purified, oxygen-free gas, since small amounts of oxygen will oxidize the coal and significantly change its softening properties.

6.2.4 Comparison of Different Testing Methods

In addition to the content of volatile matter and/or the heating value, the International Classification System for Bituminous Coal[17] uses testing methods that in some form determine the softening properties of the coal. Thus the caking ability is characterized by the swelling index (ISO 501) or by the agglutinating coefficient according to Roga (ISO 335), and the coking ability by the dilatometer test (ISO 349) or by the Gray-King test (ISO 502).

The discussion of the different testing methods in the previous sections has shown that many other test methods have been developed, some of which are still in use. Of great significance among these methods are those for the determination of plastic behavior: the Gieseler plastometer, the Brabender plastograph, and the Sapozhnikov plastometer.

These methods of investigating the plastic behavior of coals were developed not so much to obtain information about the properties of the plastic mass itself, but rather to be able to predict the properties of the coke that will be produced. Since the softening of coal is an important prerequisite to the formation of metallurgical coke, it seems necessary to assume at first that the properties of the coke are closely correlated with the properties of the plastic mass. It has been shown, however, that such a close correlation does not exist and that the coke properties depend on a number of parameters connected with the raw material and the process. Therefore a variety of methods were developed to determine the properties of the solid product of the corresponding test. This permits more or less accurate conclusions of the properties of the coke produced in a commercial plant.

The different test methods can thus be compared from two points of view: (1) comparison of the test methods with respect to their ability to predict for a given coal the properties to be expected after coking on a commercial scale, and (2) comparison of the test methods with one another to answer the following questions: Which method supplies the most useful information with respect to the effort? and To what extent are the test results of the different methods interchangeable by virtue of correspondence to those of other methods?

Comparison of the different test methods for the determination of the plastic behavior on the basis of point 1 shows that according to the literature, there is no direct correlation between coke strength and the swelling index, the Roga agglutinating coefficient (ISO 335), or the dilatation.[53] In many cases the determination of the softening behavior in the Brabender plastograph is more suitable than the methods above. Coke strength increases with the length, L, of the curve drawn by the plastograph. This means that in general coke strengths are greater the higher the first and second maxima, the lower the minimum, and the wider the plastic range. The coke strength characterized by the Micum indices (ISO R556-1967) M_{40} and M_{10} of metallurgical coke can be predicted with an error of ±7–8% by functions of the form $M_{40} = f(\lg L)$ or $M_{10} = f[(\lg L)^{-1}]$.

The plastograph method in many cases is more suitable for coal blends than the other methods. It fails, however, if high-volatile coals (coals with high shrinkage) are used, and if such coals are part of the blend. It must also be considered that such factors as hardness and content of fissures, which determine coke quality, cannot be obtained from results of the Brabender plastograph.[53]

[77] C. Kröger and U. Benke, *Glueckauf-Forsch.-H.* **37,** 131–135 (1976).

Although there is no strict correlation between coke strength and the dilatation of the coal, it is still possible to obtain a rather reliable prediction of coke strength with the so-called *G* factor (Section 6.2.2.2), along with other parameters associated with the operating conditions and the raw material properties. The *G* value characterizes the total course of the dilatation, as discussed in Section 6.5.3. On account of this possibility to predict coke strength, the measuring of the course of dilatation is an important method of estimating the suitability of coals for the coking process. Current knowledge indicates that a comparison of the test methods on the basis of point 1 above will favor the dilatometer test, especially since this method is adaptable to the modern requirements for a higher heating rate.[12]

Nevertheless, the dilatometer and plastometer methods must be viewed critically. The dilatometer and plastometer curves, as well as the results of the IBC test, have some relationship to the strength of the coke. However they show no relation to the structure of the coal. For the formation of a coke of suitable quality, next to the amount and nature of the bitumen, the amount and nature of the remaining coal substance is a determining factor. The participation of the remaining coal substance is insufficiently visible in the measured curves. They more or less characterize only the melting parts of the coal (bitumen) and how they are influenced by the gas formation (dilatation) or by splitting and condensation reactions (plasticity).[78]

In comparing the different test methods according to point 2 above, it must be remembered that they generally measure different phenomena of coke formation, or the same phenomenon in a different manner. Strict correlation between the results obtained by the different test methods cannot be expected even if they describe the same phenomenon. The relationship between the reactions in the coal during coking and the external appearances as well as the influence of those reactions on the different testing methods are shown in Fig. 6.14, which is based on a publication by G. Klepel.[51] It makes evident the interrelationships of the different events and their influence on the testing methods. Comparison of the results obtained by different test methods and published in several papers[62,63,79] serves to confirm the conclusions above.

It is not possible to arrive at definite and quantitatively valid correlations for different coals and different test results in the sense that one method can be replaced by another. However qualitative correlations are recognizable. For example, the maximum fluidity measured in the Brabender plastograph is inversely proportional to the maximum fluidity according to Gieseler, which in turn increases with the thickness of the plastic layer *y* (Sapozhnikov). Similarly, the plastic range measured in the Brabender or Gieseler plastometer increases with a relatively large scatter with the thickness of the plastic layer. There is also a connection showing that the plastic range becomes larger as the coal becomes more fluid during heating. There is also a tendency toward an increase (which is strong at first and then weaker) in the swelling index with the plastic range, but the values are highly scattered.

There is no generally valid answer to the question of which test method is most useful with regard to practice. The selection of the test method depends on the purpose for which the coal will be used and on the process in which it will be used. It further depends on whether the coal properties will remain unchanged over long periods of time or whether there will be frequent and considerable fluctuations. For the manufacture of coke in slot ovens, the swelling index and the dilatometer test usually suffice, among the methods that describe the plastic behavior of the coal. In addition to the information obtained from the swelling index test, the dilatometer test supplies data about the softening and resolidification temperature of the coal and about the interval of the plastic range. It must be emphasized that definite conclusions on the results that can be expected in the production plant cannot be reached by means of one test method alone. For actual use of the coal, suitable test methods must be selected and compared to the results from the production plant.

[78] C. Kröger, *Brennstoff-Chemie,* **49,** 359–364 (1968).

[79] D. Totzev, M. Kurtschatov, G. Gruson, and H. Franke, *Izv. Inst. Obscy Neorganiceskoij, Sofia* **2,** 157–169 (1964).

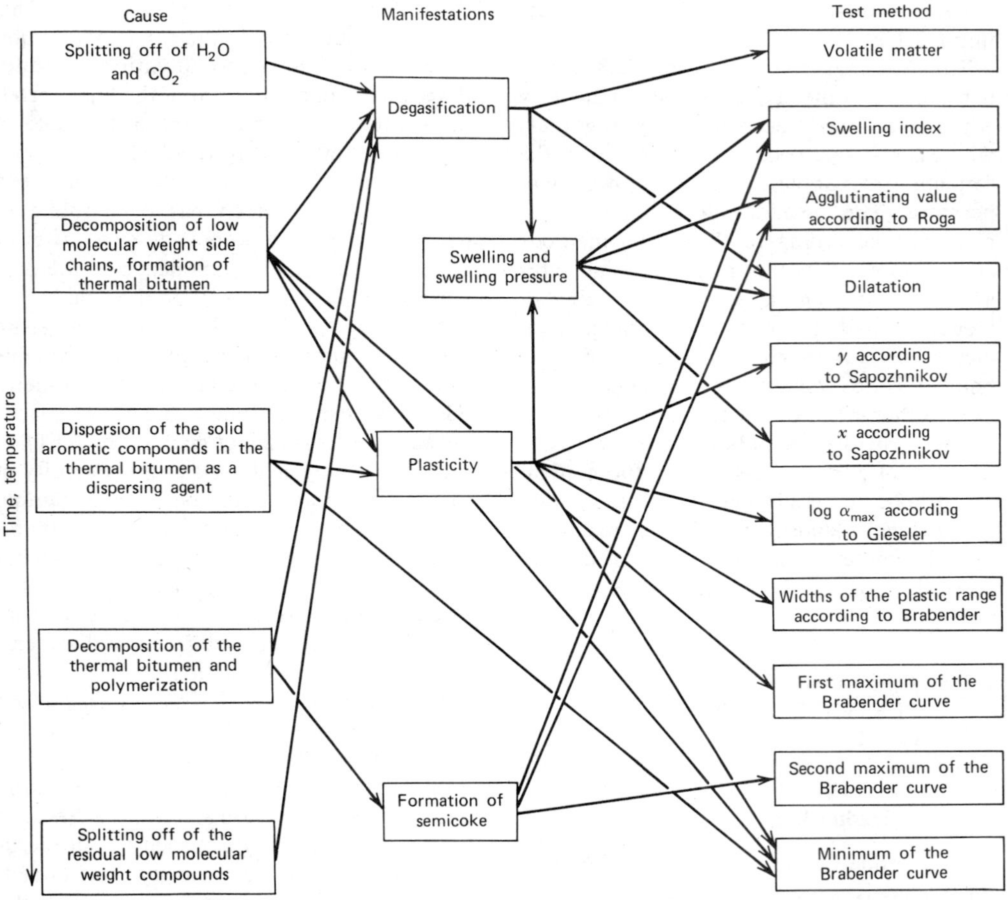

Figure 6.14 Relationships between causes and experimental events for the different methods of testing of coke formation according to Klepel.[51]

This is confirmed in the literature. For example, the maximum thickness of the plastic layer determined in the Sapozhnikov plastometer is not by itself sufficient to describe the coking properties of coals.[63] However there is a close relationship between the thickness of the plastic layer measured in the Sapozhnikov plastometer and in the large laboratory-scale oven. Therefore measurements in the larger unit are not necessary.[80]

According to investigations with Silesian coals, there is only a weak correlation between the dilatation and the swelling pressure, which does not permit calculation of the swelling pressure from dilatation, or vice versa. The dilatometer test is more sensitive, therefore it is possible to better differentiate between lower rank coals.[81]

Depending on the problem to be solved or to be considered in the area of coal use, different test methods are offered. The shrinkage of the coal charge during coking in the slot oven, for instance, is of great importance to practical operations. American authors find that the change in volume during coking correlates with the maximum fluidity determined in the Gieseler plastometer and that the contraction increases with increasing maximum fluidity.[82]

More detailed information about the reactions occurring in the dilatometer can be obtained by microscopic investigation. These

[80] Yu. V. Pyzhov, A. M. Scheikhet, and A. Ya. Obukhovskii, *Coke Chem. USSR,* **2,** 5–8 (1965).

[81] B. Roga and A. Rusin, *Pr. Glown. Inst. Gorn., Ser. B,* Komun. No. 312, Katowice, 1963.

[82] B. R. Kuchta and J. D. Clendenin, *Blast Furn. Steel Plant,* 648–653 (July 1961).

measuring methods were described in Section 6.2.1.[2,3]

Thus for coal from one seam (22.8% VM, maf) the dilatometer shows a softening only when coal particles are fusing together, whereas the microscope already shows the formation of bubbles, even at a lower temperature. The resolidification temperatures measured according to these two methods are the same; thus the microscope indicates a wider softening range. In the coking of blends of coal, the microscopic method indicates several softening and resolidification points for the individual components, whereas the other methods indicate average values.

The formation of bubbles and the fusing together of particles depend on the content of vitrinitic material. The compressed samples of coal blends give different results in the dilatometer, possibly because of the relative dilution of the vitrinite-rich component.[83]

6.3 INFLUENCES ON THE PLASTIC BEHAVIOR OF COAL

6.3.1 Heating Rate

Several properties associated with the softening of coals, such as the swelling, pressing, and gas permeability of the softened coal mass, depend on the viscosity of the plastic coal mass. If the viscosity is changed, obviously the properties associated with it will change as well.

An increase in the heating rate shifts the softening and resolidification points, therefore the point of maximum softening, to the range of higher temperatures. Since softening and resolidification temperatures do not increase to the same extent, there is usually a widening of the plastic range.[33,79,84] The viscosity of the plastic mass at the point of maximum softening decreases with temperature.[85] With very rapid temperature increases, an equilibrium between the temperature and the degree of decomposition of the organic coal material cannot be reached. This leads to a retardation of the polycondensation reactions of the decomposition residue, therefore to an enrichment of the liquid products in the plastic mass, so that the state of maximum softening is reached at a higher temperature. The plastic mass then has a lower viscosity, which has been found to decrease in inverse proportion to a power of the heating rate. The exponent and the proportionality factor depend on the coal.[84]

The decrease in the viscosity of the plastic mass with increasing heating rate explains the results obtained by the various test methods. In a loose coal sample the empty spaces between the particles of coal are better filled by the softening coal mass as the viscosity of the plastic mass decreases. In a sample through which a stream of gas flows, therefore, the pressure drop of the gas will increase as the softening increases and the viscosity decreases. This tendency is confirmed by the results published in the literature for the Foxwell method.[74]

All events connected with expansion and swelling pressure are determined by the viscosity and by the gas evolution. Thus, for instance, the dilatation increases significantly with the heating rate[12,33,79] (see Section 6.2.2.6). Swelling is decreased by all measures that increase the viscosity, thus impeding the filling of the empty spaces with the plastic coal mass and preventing the wetting of the particle surfaces (e.g., oxidation or increase of the fusinite content).[85] The maximum thickness of the plastic layer as determined in the Sapozhnikov plastometer increases with increasing heating rate.[62,63] However since endothermic reactions occur in the plastic stage, they create a heat sink that influences the thickness of the plastic layer. An increase in the heating rate leads to an increase in the heat flow, so that more coal becomes plastic per unit time and the plastic layer becomes thicker.[62]

6.3.2 Particle Size

The influence of particle size on the softening behavior has been explored in more detail in the dilatometer and the viscosimeter, according to Kushnirevich.[86–88] The dilatometer in-

[83] M.-Th. Mackowsky, *Erdoel Kohle-Erdgas-Petrochem.*, **15**, 441–451 (1962).
[84] N. S. Gryaznov, *Coke Chem. USSR*, **1**, 5–10 (1962).
[85] N. S. Gryaznov, *Coke Chem. USSR*, **8**, 6–9 (1960).
[86] N. R. Kushnirevich, *Tr. UKhINa*, No. 11, 66–75 (1960).

vestigations were performed according to ISO standards—for coal with a dilatation of about 200% and with maximum particle sizes of 0.075, 0.12, 0.3, and 0.6 mm—as well as with test samples cut from a coal lump. All samples, including the pencils, have the density of seam coal, which is between 1.2 and 1.3 g/cm³. Dilatation decreased with decreasing particle size, which means that the dilatation of the sample cut from coal was largest. The sample cut perpendicularly to the coal stratification had a greater dilatation than that cut parallel to the coal stratum. Contraction is caused by the empty spaces between the conical sample body and the wall of the dilatometer tube, as well as between the coal particles and possibly the pores of the coal, which on softening are more or less filled. The investigations have shown that the contraction increases with decreasing particle size. For the samples cut in one piece from the coal, the contraction was only about one-third that of the pencils. The softening temperature of the cut sample is higher than that of the pencils, for as soon as the grains in the pencils begin to soften, they may glide, but the whole body of the cut coal samples has to soften before it may fill the empty spaces. Here, however, degasification begins in the first softened portion before softening of the full sample, so that the contraction process is overlaid by expansion, which prevents the attainment of the maximum contraction. With decreasing particle size, this effect disappears.

Two Soviet coals were studied with the Kushnirevich instrument (36.5 and 33.3% VM, maf; Sapozhnikov plastometer index *y*, 8–9 mm and 11–12 mm, respectively). These coals were tested in five different size fractions in the range under 5 mm.[88] These tests indicate that with decreasing particle size the softening temperature increases and the resolidification temperature decreases slightly, which means that the plastic range is smaller. However the viscosity of the plastic mass increases with decreasing particle size. The particle size has an influence on the structure of the semicoke that is obtained. The coarser material (> 1 mm) leaves a semicoke with a greater proportion of pores than the fine particle coal. This change in the plastic properties of the coal is caused by different decomposition kinetics during pyrolysis.

As the particle size increases, the pressure within the particles caused by the pyrolysis reactions also increases, which in turn leads to condensation of the pyrolysis products. These events greatly affect the subsequent thermal changes in the coal substance, especially its plastic behavior.

6.3.3 The Nature of Coal and the Behavior of Blends

6.3.3.1 *The Influences of Petrographic Composition*

The chemical, physical, and technological properties of the coal depend on its rank and the composition of the macerals. The properties of the macerals generally change during coalification. The macerals themselves are not chemically uniform substances, even at a given rank. An investigation of the influence of the petrographic composition on the softening behavior of the coal must consider the possibility of an overlapping of the influences of rank and maceral composition.

The dependence of plastic properties on both rank (characterized by carbon content) and petrographic composition has been investigated with Australian coals. Measurements of the fluidity in the Gieseler plastometer, the dilatation, the swelling indices, and the Gray-King coke type were statistically evaluated.[89,90] More than 60 coals with ranks between 81 and 89% carbon (maf) were tested. The thermal decomposition was also studied by means of thermogravimetric methods.

The results show that some properties of these coking coals depend mainly on the rank and others on the vitrinite content, which is the main parameter in petrographic composition that influences the coking ability of the coal; still others depend in different measure on both these factors. Rank strongly influences the temperature of maximum fluidity and resolidification. These properties are directly related to the thermal decomposition,

---

[87] F. G. Kreulen-van Selms, *Brennstoff-Chemie,* **44,** 310 (1963).

[88] V. V. Gavrikov, E. D. Mozgovaya, and Z. G. Belan, *Coke Chem. USSR,* **1,** 7–11 (1967).

[89] P. L. Waters, *Fuel,* **41,** 1 (1962).

[90] H. R. Brown, W. R. Hesp, and P. L. Waters, *J. Inst. Fuel,* 130–138 (April 1964).

thus are determined more by the chemical structure of the coal as a whole than by the petrographic structure. The only properties of the coking coals investigated that depend almost entirely on the petrographic composition, not on the rank, were those caused by physical softening. These are the temperatures of the beginning motion in the Gieseler plastometer (at 0.1 dial division/min) or in the dilatometer. Thus the beginning of softening is inversely proportional to the vitrite content of the coal. The softening temperature was also independent of the temperature at the beginning of thermal decomposition. From this, and knowing that at heating rates of 3 K/min it is possible with the Gieseler plastometer to cover the softening range reversibly between 0 and 10 dial divisions/min, it can be concluded that the softening of the coal is similar to a physical melting process, such as the softening of pitch, and does not have the characteristic appearance of thermal decomposition.

Coal properties measured by the swelling index, the Gray-King coke type, the fluidity, and the dilatation are a result of the simultaneous action of physical and chemical phenomena and depend on the petrographic composition, as well as on the chemical composition (rank) of the coal. The swelling index and the Gray-King coke type are more sensitive to the vitrite content and to the content of volatile matter than are measurements of maximum fluidity and dilatation. For coals of the same rank, for instance, the swelling index and the fluidity of the softened mass depend strictly on the vitrite content. For coals of different rank, this is not valid since the caking properties of the vitrite change considerably with rank.[90] According to other authors, the maximum fluidity (Gieseler plastometer) correlates better with the content of microlithotype vitrite than with the total vitrinite.[91]

Similar results were obtained with coals from eastern Siberia.[92] It was found that coals of about the same rank and vitrinite content and the same thickness of the plastic layer may differ distinctly in the viscosity of the plastic mass or in their swelling ability, thus also in their caking and coking ability. The reason was found to be three different vitrinite types. Under the microscope they gave different appearances as a foamy, lumpy, or a defined cell structure, or a mixture of these structure types.

The behavior of different macerals and microlithotypes was elucidated by microscopic investigation[93] of a Bulgarian high-volatile coal (40.7% VM) that contained 80% vitrinite as the main component and had cutinite structural elements and small amounts of fusite, semifusite, and xylite. This work showed that the cutinite undergoes changes when it enters the plastic stage, whereas fusite, semifusite, and xylite remain unchanged, thus do not participate in the softening. The vitrinite base remains isotropic and homogeneous below 470°C and only above this temperature shows effects of anisotropy.

The Sapozhnikov plastometer was used on Indian coals of different rank to measure the maximum thickness of the plastic layer, y_{max}.[94] According to these investigations, y_{max} initially increases with carbon content, hydrogen content, and the heat of wetting, then passes through a maximum. A good coking coal with a carbon content of 87–89% (mmmf), hydrogen contents of 4.8–5.5% (mmmf), and heats of wetting of 4–6.5 cal/g (maf), has y_{max} values between 15 and 25 mm. It has been proposed to use y_{max} as an additional parameter for the classification of coals with respect to their coking ability.[94]

The qualitative relationships between the plastic properties of coal, the rank, and the petrographic composition are summarized in Table 6.2.

6.3.3.2 *Mixtures of Coals and Softening Additives*

The individual components of a coal blend soften quite independently of one another. Therefore it appears that the plastic range of coal blends can be broadened if plastic ranges

[91] A. R. Cameron and J. C. Botham, *Coal Sci.* **55,** 564–576 (1966).

[92] L. G. Olshanetskii, A. K. Vetrova, and K. S. Permitina, *Coke Chem. USSR,* **2,** 8–11 (1965).

[93] D. Rustschev, P. Dragostinov, W. Konstantinowa, and G. Schopov, *Brennstoff. Chemie* **46,** 141–145 (1965).

[94] R. Haque, S. K. Sharma, N. N. Das Gupta, and A. Lahiri, *Indian J. Technol.,* **1,** 322–327 (1963).

Table 6.2 Plastic Properties of Coal as a Function of Rank and Petrographic Composition (Vitrite Content) According to Brown et al.,[90] and Haque et al.,[94] Indicating Test Instrument[a]

Properties That Depend Primarily upon Rank (Characterized by C Content)	Properties That Depend on Rank and on Vitrite Content	Properties That Depend Primarily on Vitrite Content
Temperature of maximum fluidity (G)	Maximum fluidity (G)	Temperature at beginning of softening (G)
Solidification temperature (G)	Maximum dilatation (A)	Temperature at beginning of motion in dilatometer (A)
	Swelling index	
	Gray-King coke type	
	Maximum thickness of the plastic layer, (S)	

[a] G = Gieseler plastometer.
A = Audibert-Arnu dilatometer.
S = Sapozhnikov plastometer.

of the individual components have overlapping temperature intervals. In selecting coals for the preparation of coking coal blends, the parameters that influence the temperature level of the plastic range (e.g., the heating rate[95]) must be considered. The fluidity of the plastic mass of the blend results from the superposition of the fluidities of the individual components at the momentary temperatures. Thus the fluidity of a blend may be smaller than that of the individual components. One has to consider that the properties of the softened mass of the coal blend may be influenced to some degree by interactions between the softened or the softened and unsoftened components (e.g., by solution reactions). This possibility is also valid for mixtures of coal with pitch, asphalt, or other softening components, so that—depending on the kind and amount of additive—the independence of the softening behavior of the individual components can be influenced to a greater or lesser extent.

The possible behavior of coal blends just described has been made evident by investigations of mixtures of high-, medium-, and low-volatile coals in the Gieseler plastome-

[95] M. C. Chang, G. H. Denton, and J. L. Bayer, *Coal, Coke and Slag Ironmaking Conf. Proc.*, (1965), pp. 109–115.

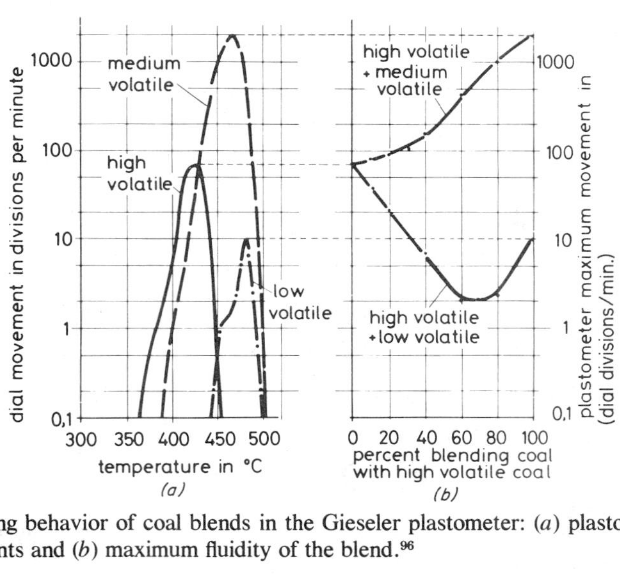

Figure 6.15 Softening behavior of coal blends in the Gieseler plastometer: (*a*) plastometer curves of the individual components and (*b*) maximum fluidity of the blend.[96]

ter.[96] Figure 6.15*a* shows the plastometer curves of the original coals, Fig. 6.15*b* the values of maximum fluidity for mixtures of the high-volatile and medium-volatile, as well as for different mixture ratios of the high-volatile and low-volatile coals. The shape of the curves in Fig. 6.15*b* depends to a great extent on the position of the plastic range of the individual components. At first during the continuously increasing temperature small amounts of low-volatile coal that soften at a higher temperature act as an inert material, which decreases the maximum fluidity. If higher temperatures are reached during the measurement, the high-volatile component is already coked and now acts as an inert component when the low-volatile components finally soften, so that in sum a lower maximum fluidity is obtained. Similarly, it is possible to explain the results of adding medium-volatile coal or other components to high-volatile coal. However the relations just described are at best qualitative, and in general the fluidity of a mixture cannot be reliably predicted on the basis of the fluidity of the individual coals.

The addition of asphalts or pitches lowers the softening temperature in the Gieseler plastometer and in the dilatometer, but the resolidification temperature remains practically unchanged. Asphalts or pitches also lead to an increase in the maximum fluidity in the Gieseler plastometer. The increase in the maximum fluidity is not always associated with a corresponding increase in the dilatation. Fluidity and the course of the dilatation of mixtures depend strongly on the values for the coals themselves; therefore the effect of additives is limited by the properties of the coal itself.[97,98] In the Sapozhnikov plastometer additions of up to 10% pitch to a medium-quality coking coal increased the thickness of the plastic layer by a factor of 4. Measurements with more than 15% pitch addition are not possible in the Sapozhnikov plastometer because the melting pitch then seeps out of the retort. Pitch additions of up to 10% gave metallurgical quality coke.

[96] W. G. Schulze and G. C. Soth, *Proc. Ironmaking Conf.*, (Chicago, 1960), Vol. 19, pp. 58–80.
[97] T. Miyagawa, S. Takahashi, and S. Ito, *J. Fuel Soc. Jap. (Nenryo Kyokai-shi)*, **51**, 172–179 (1972).
[98] V. Masek, *Brennstoff-Chemie*, **49**, 77–81 (1968).

6.3.3.3 *Mixtures of Coal and Noncoking Additives*

The softening properties of the coal become less evident if noncoking additives are used. This is because the inert material mixed with the softening coal may reduce fluidity by absorbing the primary decomposition products and the low molecular weight components that are present in the coal or are formed during pyrolysis, and act as plasticizers. Thus the plastic range is also decreased.[99]

This leads to the conclusion that the "depressing" effect of the inert material on the softening behavior is related to the specific surface of the inert material and increases as its particle size decreases. This has been confirmed by investigations with the Ruhr dilatometer and the Gieseler plastometer on mixtures of coal with coke or anthracite as inert components.[99,100] According to Pichler and Lauer,[99] the dilatation, D, or the maximum fluidity, α_{max}, in the Gieseler plastometer can be represented by Equations 6.3 and 6.4, respectively:

$$D_{\text{mixture}} = \phi_1 \cdot X \cdot D_{\text{pure coal}} \qquad (6.3)$$

$$\log \alpha_{\text{max, mixture}} = \phi_2 \log \alpha_{\text{max, pure coal}} \qquad (6.4)$$

where X is the percentage by volume of the coking coal in the mixture, and ϕ_1 and ϕ_2 the so-called inert factors, which describe the influence of the inert component on the softening coal. If the inert component has no influence on the softening behavior, ϕ_1 and ϕ_2 are 1. The inert factors generally depend on the percentage by volume of the inert component and its size consist. If, however, active carbon is used as the inert material, one finds a strong decrease of the inert factor with increasing content of the inert material, but no influence of the particle size of the inert material. The reason appears to be the magnitude of the specific surface. In active carbon the outer surface is negligible compared to the inner surface; therefore grinding has no effect on the plasticity. Other inves-

[99] H. Pichler and K. H. Lauer, *Brennstoff-Chemie*, **44**, 149–154 (1963).
[100] A. Kijewska, *Koks, Smola, Gaz,* **18**, 191–195 (1973).

tigations[101] have shown that removal of 2–3% of extractable bitumen destroys the coking ability completely. In the present case this is caused by adsorption. It increases with decreasing particle size, corresponding to the increase of the outer surface. When active carbon is used as an additive, only part of the inner surface participates, since the pores are filled with bitumen. However enough bitumen is removed so that the decrease in particle size has no effect.[99]

The events described also constitute one of the two important causes of the reduction in the ability to form coke with decreasing particle size of the coking coal: first, with decreasing particle size the gas pressure formed inside the particle by pyrolysis reactions is more readily reduced; second, the inert components of the coal, such as inertinite and mineral matter, offer an increasing surface that adsorbs the plasticizing material as the particle size decreases. As the coal size decreases and the active surface of the inert components increases, the negative influence on the ability to form coke becomes greater.[99]

The maximum thickness of the plastic layer as measured in the Sapozhnikov plastometer is also decreased by the addition of inert materials. Here the specific surface of the inert materials plays the decisive role. Thus the thickness of the plastic layer on addition of the same amount of inert material is decreased to a greater extent as the material becomes finer.[94,102]

The swelling index decreases with increasing addition of inert materials, at first slightly, then more strongly. Since this method uses high heating rates, the adsorption processes—which mean a transport of materials, thus depend on time—do not play as large a role.[99]

6.3.4 Pressure

The influence of gas pressure has been investigated in previous publications.[40] The problem gains significance as coal gasification processes operating at elevated pressures become important.

[101] H. Pichler and D. H. Lee, *Brennstoff-Chemie*, **42,** 241 (1961).

[102] J. Stefanescu, *Brennstoff-Chemie*, **50,** 354–358 (1969).

An increase in the external gas pressure during thermal decomposition leads to better softening of the coal so that its swelling and caking properties—therefore also the porosity of the coke produced—are increased. At pressures up to 50 bar, the volume of Soviet high-volatile bituminous coal and of mixtures of high-volatile and low-volatile coal increased by about 14% because of swelling.[103] The reasons for the better softening and caking of the coal are believed to be the increase in the boiling point of liquid tar-forming products and in their temporary accumulation in the coal charge during thermal decomposition.

The influence of gas pressure on coking results depends largely on the properties of the coal that is used. The increased gas pressure may increase strength, reduce it, or do neither. An increase in strength occurs when using slightly caking coals and coal blends (Sapozhnikov plastometer index $y <$ 10 mm) if the yield of primary tar is high ($>$ 5%), and when using highly caking coals ($y \approx 15$ mm), if the tar yield is small ($< 5\%$). The greatest changes in strength resulting from an increase in pressure were found in the range up to 15 bar.[103]

The caking properties of coal at elevated pressures were more recently investigated in a modified Foxwell apparatus.[77] The first experiment was conducted in quiescent gas. After cooling of the material the pressure drop of the fixed bed was determined. It was used as a measure of the fusion of the coal particles. Next the pressure drop was measured with gas flowing through the fixed bed of coal during heating up the sample. The caking ability measured according to this method in an atmosphere of nitrogen and hydrogen increases with pressure. As expected, even under pressure the caking ability depends on the coal rank and its degree of oxidation. In addition, it increases with increasing heating rate.

The thickness of the plastic layer in the Sapozhnikov plastometer also increases with pressure, giving improved caking properties.[62]

An effect similar to an increase in gas pressure is obtained by mechanically com-

[103] N. S. Grayznov and V. K. Petrov, *Coke Chem. USSR,* **1,** 27–32 (1960).

pressing the coal sample, to increase the pressure. For example, with Indian coals, an increase in pressure up to about 2 bar in the Sapozhnikov plastometer gave an increase in the thickness of the plastic layer of about 10.7 mm/bar. The shrinkage measured in this instrument also increases with pressure.[94]

6.3.5 Chemical Treatment and Additives

The caking properties of bituminous coal lead to problems in their use in coal gasification processes, especially in the processes using fluidized beds. Avoidance of such problems frequently requires the destruction of the excessive caking ability by suitable pretreatment of the coal. In other cases (e.g., in the preheating of coal—Section 6.3.6),[104] any decrease in caking ability is undesired. For these reasons, the effect of chemical treatment or of additives on the caking ability of coal has been frequently reported in the literature.

An important possibility for affecting the caking ability of coal is an oxidative treatment. The oxidation may be achieved in an oxidizing gas atmosphere, in solution, or by the addition of solid materials. The effect of oxidation with oxygen on the plastic behavior of coal is basically known, or has been confirmed by further measurements.[105] More recent investigations therefore emphasize a quantitative determination of this oxidizing effect.

An Audibert-Arnu dilatometer, with a heating rate of 2 K/min, was used to measure the decrease in dilatation due to oxidation. Before the tests the samples, with volatile matter contents between 25 and 32%, were stored up to 72 hr in air, at temperatures between 100 and 160°C. The results show that dilatation decreases exponentially with the oxidation time.[106]

Caking ability during oxidation of the coal is influenced mainly by the particle size of the coal, the gas pressure, the gas velocity, the oxygen concentration, the oxidation temperature, and the oxidation time. If investigations of the influence of oxidation on the caking properties do not separate these parameters very precisely, the effects of the different parameters are superimposed and the true correlations are obscured. To determine the causes of these associated events, detailed measurements on a high-volatile (41.1% VM) American coal were made using the Foxwell method, in which the foregoing parameters were changed systematically.[107,108] To characterize the caking properties of coal, the so-called agglutination number was introduced, which is the maximum pressure drop recorded when gas passes through the coal sample in the plastic range. Surprisingly, it was observed that at reaction temperatures above 200°C (up to 277°C), the oxidation time required to reduce the caking ability by the same amount is greater than at 200°C, although at the higher temperature a larger quantity of oxygen reacts with the coal. The authors believe that this occurs because there is a change with temperature not only in the oxygen reaction rate, but also in the relative rate of formation of oxidation products (CO, CO_2, H_2O, and O_2 combined with the coal). These effects can be expressed mathematically by means of an oxidation parameter (OP), which is also directly related to the caking ability, as measured by the maximum pressure drop.

The caking ability decreases with an increase in OP values. This means, according to the definition of the OP value, that the caking ability decreases if the formation of water and bound oxygen is at a maximum and the formation of the carbon-containing gases CO_2 and CO reaches a minimum. Using the OP value, the authors were able to express for any given particle size the decrease in caking ability independent of the mentioned oxidation conditions.

Treatment of coal in an oxidizing but very dilute solution of potassium permanganate (0.002–0.005%) increases the caking ability and the thickness of the plastic layer in the Sapozhnikov plastometer, independent of the type of coal. A more concentrated solution,

[104] C. Flockenhaus, W. Rohde, and D. Habermehl, *Proc. Ironmaking Conf.* (Atlantic City, N.J., 1974), Vol. 33, pp. 382–390.
[105] W. Devecchi, *Monatsbull., Schweiz. Ver. Gas Wasserfachmännern,* **46,** 233–243 (1966).
[106] J. B. Escudero Fernandez, *Bol. Inf. Inst. Nac. Carbon,* **12** (57), 51–63 (1963).
[107] A. Y. Kam, A. N. Hixson, and D. D. Perlmutter, *Chem. Eng. Sci.,* **31,** 821–834 (1976).
[108] A. Y. Kam, A. N. Hixson, and D. D. Perlmutter, *Ind. Eng. Chem., Process Des. Dev.,* **15,** 416–422 (1976).

however, decreases the caking ability.[109] A partially analogous result was obtained by determining the effect of the degree of oxidation on the swelling and the strength of the obtained semicoke. Combined with 1% oxygen, highly caking coals showed a definite improvement in the coking ability. If more oxygen is added, coke strength decreases rapidly. Slightly caking coal leads to a lower quality coke even with only slight oxidation.[110]

Destruction of the excessive caking ability of coal is desired in some processes. However sometimes (e.g., in the use of high-quality coking coal) one wishes to prevent it. Prevention has been attempted quite frequently by the use of oxidation inhibitors or so-called antioxidants, with varying success. If oxidation of the coal occurs through a chain reaction, antioxidants should interrupt the chain reaction. It has not been possible to demonstrate this process.[111]

When coking in a slot-type oven, the plastic layer progresses toward the inside and drives the moisture of the coal into the colder zones, where it condenses.[112] Therefore the moisture content of the coal in the interior of the oven increases and the temperature in the condensation zone remains constant at about 100°C. Investigations in the Sapozhnikov plastometer[113] have shown that for all coals this increase in moisture content leads to a decrease in the thickness of the plastic layer, and thus to a decrease in the coking ability. Up to a moisture content of about 8%, the decrease in coking ability is comparatively small, but becomes much larger at higher values. A similar result is obtained by treating the coal with steam. The steam was observed to react only with the outer coal layers. If the steam-treated material is further comminuted, the caking ability again increases. Accordingly, coals with a volatile matter content of 20–28% are more sensitive to steam than other coals with a higher volatile matter content. The coking ability of low-quality coal charges can be improved by decreasing the moisture content, whereas the coking ability of high-quality coking coals can be impaired by decreasing the moisture content. On the basis of rather extensive pilot-oven coking tests (250 kg), however, some authors conclude that for American high-quality coking coals, coking ability is not impaired by decreasing the moisture content.[114]

The efforts of the coking industry to produce coke economically require as much as possible the use of slightly caking coals, or the use of additives that improve coking ability and coke quality. The effect on the plastic behavior of coal of the addition of inerts has been investigated many times, and only a few such publications containing literature reviews can be mentioned.

Additives may have the following effects:

1. The fluidity of the plastic mass, therefore the caking ability, may be improved, reduced, or completely destroyed.
2. The softening range is shifted to another temperature range and either narrowed or widened.
3. The progress of degasification may be shifted in relation to its location to the plastic range; the degasification maximum can be lowered or increased, thus influencing the swelling properties of the coal.
4. The yields of tar, gas, and other by-products and their chemical composition may be changed.

The causes of these effects are not entirely clear. The following causes are discussed[115,116]:

1. Chemical reaction between coal and additives.
2. Catalysis.
3. Interaction of the additive with the surface of the plastic coal particle (wetting properties and adsorption are influenced).

[109] R. E. Leibowich, P. Ya. Goldberg, and G. G. Rabukhina, *Coke Chem. USSR,* **3,** 5–6 (1964).

[110] W. Duhameau, W. Fascotte, and M. Saussez, *Ann. Min. Belg.,* No. 7–8, 797–812 (1969).

[111] G. R. Yohe, *Fuel,* **44,** 135–146 (1965).

[112] B. Bogenschneider and W. Simonis, *Glueckauf-Forsch.-H.* **33,** 198–205 (1972).

[113] M. D. Shapiro, L. S. Altermann, S. R. Keitelgisser, and A. F. Konenko, *Coke Chem. USSR,* **1,** 4–6 (1967).

[114] H. E. Harris, Private communication, January 3, 1978.

[115] R. Agricola, *Brennstoff-Chemie,* **48,** 240–252 (1967).

[116] H. Barking and C. Eymann, *Brennstoff-Chemie,* **44,** 11–20 (1963).

4. Change in the thermal conductivity of the plastic mass.
5. Magnitude of the surface area of the additive if a particulate solid is being added.
6. Change in the gas permeability of the plastic layer.

In general, several of these causes are involved rather than one alone.

If a dilatometer is used to judge the effect of additives on the plastic behavior, then in many cases an exponential increase or decrease in the dilatation or the expansion (sum of contraction and dilatation) is observed with increased concentration of the additives.[117]

The literature indicates that inert coking additives (CaO, MgO, Fe_3O_4, SiO_2, Al_2O_3 coke breeze, etc.) only act to make the mixture leaner, whereas other materials (Fe_2O_3, H_3BO_3, MnO_2, $KMnO_4$, PbO_2, etc.) enter into reactions with the coking coal. A sharp limit is not indicated, since the term "inert in the coking process" cannot be defined precisely. Additions of up to 2% tar, pitches, phenols, ammonium carbonate, and calcium sulfate increase expansion; but higher concentrations decrease it. Boric acid causes the strongest decrease in expansion, and the slightest decrease is obtained when adding aluminum hydroxide or aluminum oxide and coke breeze. With many additives the strength of the coke goes through a maximum with increasing addition, for instance, with Fe_2O_3 or bentonite.

No connection was found between the kind and quantity of additive and the softening and resolidification temperatures, and thus the plastic range.

Soviet authors[118] explain the action of additives as follows: small additions of organic or inorganic materials cause, on heating, a change in the distribution of the hydrogen from the coal among the different liquid pyrolysis products (tarry materials and water). Associated with this effect is a change in the ratio of solid material to liquid material in plastic mass, which influences its caking properties. In addition, the surface properties of the coal particles are changed, thus their adsorptive capacity is also altered. This in turn influences the stability of the plastic mass and the caking ability. As a general rule, small additions of materials that reduce water and tar formation during pyrolysis improve the caking properties. Examples are the addition of β-naphthol, which improves caking ability, as opposed to the addition of ammonium sulfate, which decreases caking ability.

The action of boric and phosphoric acids on the coking behavior of the coal is explained by the esterification of the hydroxyl groups of the coal. There are indications of a mechanism that increases the softening temperature to such an extent that it reaches the unchanged resolidification temperature and destroys the softening and caking ability. The extent of this action apparently depends more on the nature of the radical of the acid than on the number of esterified hydroxyl groups.[119]

A reduction in the caking ability, which may be desired in combustion and gasification processes, can be achieved by treatment with alkalis (e.g., Na_2CO_3 or NaOH). The alkalis may be added in solid form or as a solution. In practice alkali solutions are of greater importance, since the uniform mixing of a powder with coal is difficult.[120]

The coking ability of coal apparently can also be influenced by radioactive radiation. A high-volatile coal (35% VM) was irradiated with 25×10^5 R and gave a dilatation of 14% in the dilatometer, compared to a dilatation of -6% for the nonirradiated coal.[121] Other results indicate that the yield of extract increases at first with irradiation, but as the dose is increased it again decreases.[122] Since, for example, it was found in the Gieseler plastometer that the maximum fluidity increases with the yield of extract,[123] the results obtained in the dilatometer can be explained. The cause of this action is probably the change in the molecular structure of the coal caused by the radiation.[122]

[117] H. R. Maybaum, *Brennstoff-Chemie,* **49,** 316–320 (1968).

[118] M. D. Shapiro and L. S. Altermann, *Coke Chem. USSR,* **9,** 10–15 (1964).

[119] A. C. Cunningham and W. F. Wyss, *Fuel,* **2,** 137–143 (1967).

[120] G. F. Crewe, U. Gat, and V. K. Dhir, *Fuel,* **54,** 20–23 (January 1975).

[121] A. Götte, West German Patent No. 1262 218 (27. 8. 1959) A.z. G 27 810 VI b/10a.

[122] M. M. Roy, *Fuel,* **42,** 125–130 (1963).

[123] J. Nadziakiewicz, *Fuel,* **37,** 361–363 (1958).

6.3.6 Thermal Pretreatment

As the use of preheated coal for the manufacture of blast furnace coke[124–126] (Chapter 14) grows in importance, the influence of thermal pretreatment on the plastic behavior of the coal used gains in interest. In discussions of the preheating of coking coal, it has been stated repeatedly that coking capacity may be impaired by such pretreatment. However improvement in coke quality has been observed,[104,127] and this is the main reason for the use of preheated coal. Both effects occur, but their results are not comparable; theories developed so far differ in their explanations.[104,128]

Investigations on a laboratory scale show that preheating to a temperature higher than about 170°C decreases the coking capacity of some Ruhr coals, even if the heating is done in an oxygen-free atmosphere (e.g., H_2 or N_2). With increasing temperature, the dilatation decreases only slightly at first. If one approaches the softening temperature, as determined in the dilatometer, a more abrupt decrease is observed. Preheating to 450°C for 30 min leads to complete disappearance of the dilatation. The plastic range of the coal generally changes very little up to the beginning of softening. However there are reports that the plastic range is widened.[129] On further increase of the pretreatment temperature, the plastic range becomes narrower, and it disappears entirely at 450°C. After this thermal pretreatment, the coal does not soften at all. Investigations in the Brabender plastograph have confirmed these results. Soviet coals also show the above-described influence of the treatment temperature.[130] No noticeable effect of the change in heating rate was observed in the range 2–30 K/min.[131]

All investigators agree that the oxygen content of the atmosphere used during heating has a significant effect: even slight oxidation causes great changes in the plastic properties of the coal[127,129,132] (Section 6.3.5).

The resulting requirement to conduct thermal pretreatment in an absolutely oxygen-free atmosphere is difficult to achieve in a commercial plant, especially for reasons of safety. One must expect an oxygen content of about 0.5–1.0% to be present in any case. The oxidation theory of bituminous coal,[133,134] however, indicates that the degree of oxidation at a given supply of oxygen depends on time, temperature, and the surfaces of both the coal particle and the transport pores. The surface can be decreased by using coarser coal feeds, but the coking process generally requires a given size distribution for technical reasons. A lower preheating temperature is desirable, but the addition of a process step such as preheating must be compensated by the use of sufficiently high preheating temperatures, to lead to a sufficient reduction in the coking time.

Since the oxidation of the coal also depends on time, the residence time of the individual coal particles in the preheat reactor should be as short as possible. This can be achieved in a flash dryer, which in this respect surpasses all other technical drying units. To avoid contamination of the environment, the preheating plant must be sealed from the atmosphere. It can be operated without difficulty with suction in the system, which is desirable from an operational point of view.

On the average for all coals used, systematic investigations in a commercial plant gave a slight decrease in the coking capacity G, due to preheating. This is evidenced by a decrease in the dilatation value.[135] An expla-

[124] J. Falque, *Tech. Mod.*, **64,** 2–4 (1972).

[125] R. H. Chambers and L. A. Watson in *Ironmaking Proc.*, Vol. 33, Atlantic City, N.J., 1974.

[126] W. Rohde and K.-G. Beck, *Glueckauf,* **109,** 348–354 (1973).

[127] D. Habermehl, G.-W. Lask, and H. Petak in *Informationstagung über Kokereitechnik der KEG (Information Meeting of KEG Coke Oven Technique),* Luxemburg, May 1975, pp. 227–237.

[128] R. Marcellini and D. G. Marting, in *Informationstagung über Kokereitechnik der KEG,* Luxemburg, May 1975, pp. 213–225.

[129] D. Breidenbach, *Erdoel Kohle, Erdgas, Petrochem.*, **17,** 276–278 (1964).

[130] Yu. B. Tyutyunnikov et al., *Coke Chem. USSR,* **2,** 3–7 (1972).

[131] I. N. Nikolaev and N. I. Kozlova, *Coke and Chem. USSR,* **4,** 7–9 (1964).

[132] H.-J. Schmidt-Holthausen, *Forschungsber. Landes Nordrhein-Westfalen,* No. 1859, Cologne, 1967.

[133] H. Sommers and W. Peters, *Chem.-Ing.-Techn.*, **26,** 441–453 (1954).

[134] W. Peters and H. Jüntgen, *Brennstoff-Chemie,* **46,** 56–62, 175–184 (1965).

[135] C. Meyer, H. Echterhoff, D. Habermehl F. Orywal, and K.-G. Beck, *Glueckauf-Forsch.-H.* **38,** 197–202, 231–236 (1977).

nation is not available at present, since a multiplicity of parameters must be considered, which may even compensate each other.

A slight oxidation of the coal particles during preheating is often assumed to be the cause of the decrease in coking capacity, since a similar effect is observed with well-defined oxidation experiments. The oxidation stages, however, cannot be detected analytically or microscopically. Others assume that purely thermal influences (e.g., polymerization or polycondensation reactions on the grain surface[136]) are responsible for the reduction in the dilatation value. However the analytical methods required to prove this are not sensitive enough.

For the reasons above, both effects—slight oxidation or chemical reaction on the surface—could appear in preheating on a commercial scale. Depending on the nature of the coal, one or the other influences will prevail.

The fundamental question that remains is whether the dilatation test conducted with preheated coals (i.e., after cooling) permits conclusions concerning the coking behavior that are comparable to test results with moist coals. In any case, at the time of the dilatation test the preheated coals have already undergone thermal pretreatment.

There are many unsolved questions that could lead to major future investigations. Regardless of the reduction in the coking capacity of a particular coal after preheating, and under otherwise equal conditions, however, a coke of equal or even better quality is sometimes produced, compared with that produced when moist coal is used, under the same heating parameters.[135] (Section 6.3.5).

Regardless of whether they are used moist or after preheating, coals with a high coking capacity *G* (Equation 6.1), lead to coke of about the same quality when carbonized at the same flue temperature. Under the same coking conditions, coals with a lower coking capacity produce cokes that are considerably more abradable if moist coal is used. In spite of the partial decrease in coking capacity, however, preheating yields less abradable coke.

Different opinions have been expressed about the influence of preheating temperature on coke quality.[127,128]

6.3.7 Mineral Matter Content

Section 6.3.5 described the effect of inorganic substances on the properties of the plastic mass. It is obvious that the mineral matter that accompanies the coal also influences its plastic behavior.

Systematic investigations[137–139] have shown that both swelling index and dilatation decrease remarkably with increasing mineral matter content. Lowering the resolidification temperature narrows the plastic range.

These dilatometer results were confirmed by tests in the Brabender plastograph. The reduction of the maximum fluidity shows, in addition, an increase in viscosity. Investigations with the Gieseler plastometer gave parallel results.

Changes in the plastic behavior lead to a decrease in pore volume and pore size distribution, which influences the reactivity of the coke.

In the stage of fissure formation, the size distribution of the particles of mineral matter has a significant influence. As centers of faults, the larger particles in particular have a considerable influence on coke quality.[140]

6.4 INTERPRETATION OF THE PLASTIC PROPERTIES OF COAL

Many methods have been used in attempts to explain the phenomenon of the plasticity of bituminous coals during heating at temperatures between 300 and 500°C, and to simulate it with model substances. Since the theories developed so far have been described repeatedly and extensively,[40,141–143] only one is discussed briefly here.

[136] K. Bunte, H. Brückner, and H. G. Simpson, *Fuel,* **12,** 222–232 (1933).

[137] H. Echterhoff, J. Langhoff, and W. Peters, *Brennstoff-Chemie,* **45,** 306–315 (1964).

[138] H. Echterhoff, J. Langhoff, and W. Peters, *Brennstoff-Chemie,* **46,** 110–117, 133–141 (1965).

[139] V. Stuchlick, *Glueckauf-Forsch.-H.* **31,** 204–210 (1970).

[140] J. Gibson and D. H. Gregory, *COMA Yearb.,* 137–161 (1975).

[141] D. W. van Krevelen, *Coal,* Amsterdam, 1961.

[142] R. Loison, P. Foch, and A. Boyer, *Le Coke,* Paris, 1970.

Basically, it can be shown that explanations of the plastic properties of coal have limitations resulting both from the test methods used and from the complex structure of coal. In general, the particular phenomenon can be explained only by making certain assumptions. Comparison of the different test methods (Section 6.2.4) shows that different limiting conditions often exist that do not permit any direct relationship between the individual measurements obtained. In the dilatometer test, for example, a condition such as heating rate may influence the plastic behavior (Sections 6.2.2.6 and 6.3.1). The same statement applies to a number of other factors.

Probably the most interesting theory of plastic behavior is the "metaplast theory,"[144,145] which permits quantitative description of experimental results.[141] It starts with the assumption that coke formation can be described by three consecutive reactions. Reaction I is a depolymerization reaction, in which an unstable intermediate phase, the so-called metaplast, is formed. This phase is responsible for the plastic behavior:

I. Coking coal $(P) \xrightarrow{k_1}$ metaplast (M)

Reaction II is a cracking process in which the tar evaporates and the nonaromatic groups are split off. This reaction is accompanied by recondensation processes and ends with the formation of the semicoke. It describes the resolidification of the plastic material. The primary gas formed during this process causes the swelling of the coal.

II. Metaplast $(M) \xrightarrow{k_2}$ semicoke (R) and primary gas (G_1)

Reaction III is a secondary degassing reaction in which, through the evolution of methane—and, at higher temperatures, through the evolution of hydrogen—the density of the semicoke is increased and coke is formed. During this reaction, a reduction in volume occurs.

III. Semicoke $(R) \xrightarrow{k_3}$ coke (S) and secondary gas (G_2)

These three reactions, which in a simplified form describe coke formation, are the basis for the following mathematical model.[141] It has been proved repeatedly that depolymerization and cracking reactions are first-order reactions[146,147]; therefore it is assumed that Reactions I and II are also first-order reactions. Although Reaction III is much more complicated, we begin by assuming that it is also first order. Thus the kinetics of the coke formation can be described by the following three equations:

$$-\frac{dP}{dt} = k_1 P \tag{6.5}$$

$$\frac{dM}{dt} = k_1 P - k_2 M \tag{6.6}$$

$$\frac{dG}{dt} = \frac{dG_1}{dt} + \frac{dG_2}{dt} = k_2 M + k_3 R \tag{6.7}$$

in which t represents time; k_1, k_2, and k_3 are the reaction rate constants of the different reactions; and P, M, G, and R are the masses of the reacting material P, the intermediate phase M, and the reaction products G and R.

The influence of Reaction III is negligible until Reaction II has proceeded to the point of maximum gas evolution.

Considering only the coke formation at lower temperatures, one may assume that below the temperature $T_{m,d}$, at which the degasification rate reaches its maximum, the following is valid:

$$\left(\frac{dG_2}{dT}\right)_{T<T_{m,d}} \approx 0 \tag{6.8}$$

or

$$(k_3)_{T<T_{m,d}} \approx 0 \tag{6.9}$$

[143] E. Stach and M.-Th. Mackowsky, in *Textbook of Coal Petrology,* 2nd ed., E. Stach, Ed., Gebrüder Borntraeger, Berlin and Stuttgart, 1975.

[144] D. Fitzgerald, *Trans. Faraday Soc.*, **52**, 362 (1956).

[145] H. A. G. Chermin and D. W. van Krevelen, *Fuel*, **36**, 85 (1957).

[146] P. K. Frolich and S. C. Fulton, *The Science of Petroleum,* Vol. 3, London, 1938.

[147] G. F. D'Alelio, *Fundamental Principles of Polymerization*, New York, 1952.

Equation 6.7 can then be expressed as:

$$\frac{dG}{dt} \approx \frac{dG_1}{dt} = k_2 M \qquad (6.10)$$

Since it is possible to show[145] that the reaction rate constants k_1 and k_2 for coking coals are almost identical, the following relation is valid:

$$k_1 \approx k_2 \approx k \qquad (6.11)$$

If the temperature is kept constant, Equation 6.5 has the following solution:

$$P = P_0 e^{-kt} \qquad (6.12)$$

and from Equations 6.6 and 6.12 one obtains:

$$\frac{dM}{dt} + Mk = P_0 k e^{-kt} \qquad (6.13)$$

Assuming as limit $M = 0$ at the time $t = 0$, then the solution of this differential equation is:

$$M = P_0 k t e^{-kt} \qquad (6.14)$$

Up to just below the temperature at which the degasification rate reaches a maximum, G is calculated as follows:

$$G \approx G_1 = P_0 - P - M \qquad (6.15)$$

Therefore the value G in first approximation can be calculated by the following equation:

$$G \approx P_0 [1 - (kt + 1)e^{-kt}] \qquad (6.16)$$

At a constant heating rate, if the temperature is given by the function $T = Vt + t_0$, the differential equations cannot be solved by simple analytical methods.

By replacing the linear temperature time function with the new function

$$e^{-\frac{2T_i}{T}} = bt \qquad (6.17)$$

an approximate solution can be found. This temperature-time function has a point of inflection at $T = T_i$. By arbitrarily selecting a value for T_i, the constant b can be determined.

The advantage of this function for the solution of the problem is that in the temperature range of $0.8 \leq T/T_i \leq 1.3$, this function can be approximated by a linear relationship. If, in addition, one uses instead of k a modified reaction velocity $\bar{k}$, which is given by the equation:

$$\bar{k} = \frac{k}{(E/2RT_i) + 1} \qquad (6.18)$$

then Equations 6.12, 6.14, and 6.16 can be used in their old form.[145] The activation energy, E, is described by the Arrhenius equation.

$$k = k_0 e^{-E/RT} \qquad (6.19)$$

At constant heating rates, the change in the masses of the products P, M, and G can be calculated as follows:

$$P = P_0 e^{-\bar{k}t} \qquad (6.20)$$

$$M = P_0 \bar{k} t e^{-\bar{k}t} \qquad (6.21)$$

$$G = P_0 [1 - (\bar{k}t + 1)e^{-\bar{k}t}] \qquad (6.22)$$

Chermin and van Krevelen[145] showed that this kinetic theory is in reasonable agreement with phenomena found experimentally during the heating of coking coal.

6.4.1 Degasification of Coal

To evaluate the reactions that occur during coking in relation to the softening behavior, the degasification behavior in the temperature range from the beginning of the heating to the final temperature is important. In 1928 Damm had already proposed that the process of gas evolution proceeds in three segments.[148] Subsequently this subdivision has been found useful for all coals.

In relation to the events that occur during coking, and considering the quantity of volatile matter, the following should be considered: the behavior of coal up to softening, conversions in the softening zone, and the resolidification of the softened coal and the completion of coke formation up to the end of the coking time.

The only process directly connected with plastic behavior is degasification from the beginning of softening to the resolidification

[148] P. Damm, *Glueckauf*, **64,** 1078 (1928).

of the plastic mass, that is the second segment and the beginning of the third segment. Changes in the reaction kinetics (e.g., an increase in the heating rate) change the quantity and composition of the gases obtained during the different periods. Thus there is an indirect relationship between predegasification (up to the beginning of the softening temperature) and postdegasification (during conversion from semicoke to coke) with respect to the plastic properties.

An accurate determination of the temperature limit of the three segments of degasification is affected by the different test methods for the plastic properties and by the conditions and investigating criteria, which do not always lead to the same temperature values (Sections 6.2.2.6 and 6.2.4). A convention is therefore necessary to determine the limits.[149]

The desorption of gases, especially methane, starts at about 100°C and ends at below 300°C. The process can be observed by means of mass spectrometry.[150,151]

Although at 250°C low-volatile saturated and unsaturated hydrocarbons begin to distill off as a result of initial thermal decomposition,[152] at temperatures up to about 330°C important changes in both the structure and the physical conditions cannot be observed.[1,83,153]

The following events can be responsible for causing thermal decomposition of the coal during coking: chemical reaction, the diffusion of gaseous reaction products from the coal or coke particle, and the input of heat.[154] The specific process that determines the rate in a given case depends on the reaction conditions, especially the particle size and the heating rate. In the conventional coking process, heating rates vary only within narrow limits around 3 K/min. But in the multistage continuous processes for the production of formcoke, much higher heating rates are used and their influence is of technical interest.

Investigations of the rapid degasification of bituminous coal led Peters[155] to a theory of the kinetics of rapid degasification, which showed that similar to drying, degasification is an evaporation process in the narrow sense of the word. The evaporating material, in this case called "primary bitumen," need not necessarily have the same composition while still in the coal. It is possible that the evaporation of the primary bitumen is preceded by a rapid primary decomposition during which a loose bond with the semicoke structure is broken. Figure 6.16, which gives the degasification curve and the corresponding temperature curves as a function of time, summarizes this theory. If a particle of coal is added to hot heat carriers while being intensely mixed, the outside of the particle is heated very rapidly. As long as there is no significant development of gas, heat transfer rates of more than 1000 kcal/(m^2)(hr)(K) can be assumed. If, for example, the temperature of the heat carrier gas is about 650°C, approximately 0.2 s is needed until the surface of a particle of high-volatile coal reaches a temperature of about 350°C.

The primary decomposition of the coal occurs with increasing temperature, during which a liquid material, the primary bitumen, is separated from a residue, the semicoke. As the outside surface of the particle reaches a temperature of 350°C, liquid primary bitumen is already present to a significant depth and now begins to evaporate; 350°C is assumed to be the narrow boiling region of the primary bitumen, which consists of a complex mixture of interconnected hydrocarbons. The evaporating distillation products now hinder the heat transfer considerably. The heat transfer rate decreases to about one-tenth of its previous value. Thus less heat is available for evaporation, which leads to a decreased evaporation rate until an equilibrium is reached between heat flow and gas transport. At a constant boiling temperature, the same amount of heat, per time unit, flows to the surface, from which a uniform quantity of bitumen evaporates. The inflowing heat supplies the heat of evaporation, disregarding

149 H. Echterhoff, *Erdoel Kohle*, **8**, 294–298 (1955).
150 J. Klein, Dissertation, Aachen, 1971.
151 K. H. van Heek, H. Jüntgen, and W. Peters, *Brennstoff-Chemie*, **48**, 163–173 (1967).
152 W. Eisenhut and W. Simonis, in *Thermische Technologie (Thermal Technology)*, Vol. 3 K. Winnacker and L. Küchler, Eds., Carl-Hauser-Verlag, Munich, (1971).
153 D. Rustschev, G. Shopov, W. Konstantinowa, P. Dragostinov, and D. Minkov, *Brennstoff-Chem.* **50**, 310–313 (1969).
154 K. H. van Heek and H. Jüntgen, *Chemie-Anlagen Verfahren*, No. 6, 39–42 (1969).
155 W. Peters, Habilîtation, Aachen, 1963.

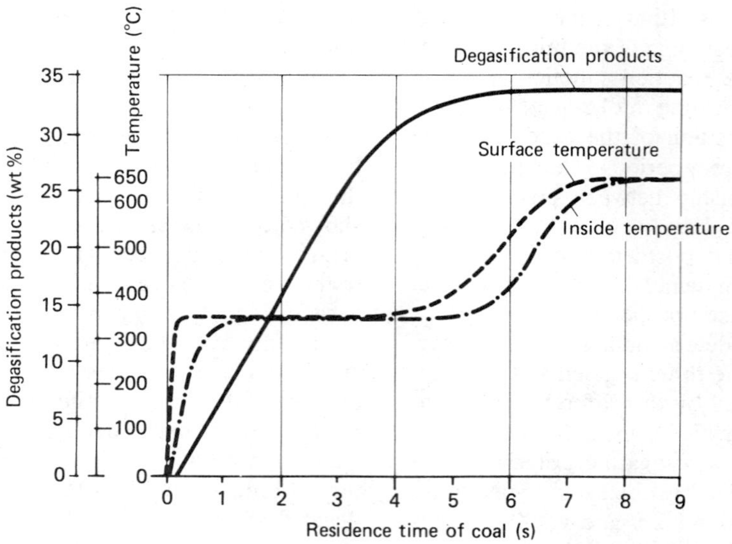

Figure 6.16 Course of degasification and temperature in the coal particle during rapid degasification (average particle size, 1.5 mm; heat carrier gas temperature, 650°C).[155]

the small amount necessary to increase the temperature of the inner portion of the particle.

When the surface of the particle reaches 350°C, the temperature inside the particle is considerably lower. For example, at a heat carrier gas temperature of 650°C, the temperature in the interior of the particle after 0.5 s is about 250°C. At constant surface temperature, the interior of the particle rapidly reaches the surface temperature. The limit of the zone of primary decomposition thus moves in concentric spheres to the inside. The gas bubbles formed during decomposition and the thermal expansion of the liquid primary bitumen move this to the outside and cause evaporation to occur on the outside for quite some time. Thus the zone in which the liquid primary bitumen "waits" for evaporation is broadened.

Eventually the evaporation front moves to the inside because insufficient primary bitumen is flowing to the outside. Initially this does not destroy the equilibrium. Because of the high heat conductivity of the coke outside the evaporation zone, the heat flow is not decreased in spite of the increased length of the path: thus the surface temperature of the coke remains practically constant. The large pores of the coke offer almost no resistance to the evaporation of the products. Only when the evaporation front has moved far to the inside does the equilibrium move slowly, and the evaporation curve decreases. Then the amount of distillable material decreases, and the quantity of product decreases considerably. In the final phase the outside zone of the particle is heated up. The semicoke splits off gaseous products, which become lower in molecular weight with increasing temperature. At last the temperature of the heat carrier is reached everywhere. Postdegassing reactions may take a long time to subside.

The influence of heating rate on degasification has been investigated within a wide range of heating rates between 10^{-2} and 10^6 K/min.[151,156–158] Measuring and analytical methods had to be developed to suit the heating rate. A detailed description is given in Chapter 12, which describes pyrolysis of bituminous coal.

6.4.2 The Softening Process

Microscopic observation shows that the visible softening of coal begins with a deformation of the individual particle that moves into the gap volume.[3–5] Large particles show formation of pores and flowing structures.

[156] P. Hanbaba, H. Jüntgen, and W. Peters, *Brennstoff-Chemie,* **49,** 368–376 (1968).
[157] V. Koch, H. Jüntgen, and W. Peters, *Brennstoff-Chemie,* **50,** 369–372 (1969).
[158] D. Schwandtner, Dissertation, Aachen, 1971.

They fuse with smaller particles. These show little or no indication of softening. High packing density improves the fusing of the grains immediately after the beginning of softening.[4,6]

With the beginning of softening, the so-called fusing pores begin to form, and at only a few kelvins higher the first degasification pores are formed in vitrinite. The temperature of the first formation of pores depends not only on the coal itself[1] but also on the particle size.[5] It correlates well with the reciprocal diameter. This correlation does not depend on temperature differences within the particle, but can be traced back to the influence of mass transport, with differences in the path length and pressure.[159] Diffusion is also of great importance in many reactions of coal and coke.[134]

The shape and reflectivity of the exinites remain unchanged at the beginning of softening, but immediately afterward their decomposition begins. The increasing softening is characterized by homogenization, which makes it impossible to distinguish between pores caused by fusion and by degasification. The viscosity of the plastic mass decreases and the gas formation increases. The disappearance of the particle boundaries indicates certain limitations of flow. However the particles do not flow together completely.[160]

Other authors[161] assume that the macerals of the exinite group soften only partially. The basis for this claim is probably the observation that exinite exhibits different behaviors when combined with vitrinite and when combined with inertinite.[162]

As the temperature increases further, the properties of the plastic mass change continuously. The viscosity increases, the pore volume reaches its maximum value,[163,164] and the reflectance increases.[1]

The development of areas of anisotropy and the size of such areas occurs in the plastic zone and is visible under the microscope.[4,8,165–168] Coal rank, particle size, and heating rate are of significant influence. The size of the areas of anisotropy is fixed when the plastic zone changes into the resolidified zone, and it does not change further. The intensity of the effects of anisotropy increases with increasing temperature.[1]

Special coking methods were developed on a laboratory scale to study the behavior of semifusinites[169–171] and micrinites[172] in the plastic zone. It was found that the limit between reactive and inert behavior of semifusinite depends on the rank of the vitrinite. Semifusinites, with a reflectance of 0.1–0.2% above that of the vitrinite, are inert. The micrinite does not change during transition through the plastic zone and must therefore be considered inert.

Several experimental facts led to the assumption that beginning softening is not connected with decomposition, but corresponds to the melting of a thermoplastic material (Section 6.3.3.1).[89,90,173] Investigations using the shock dilatometer[33] have shown that the increase in softening temperature with increasing heating rate, between 1.2 and 30 K/min, can be explained solely by heat conduction phenomena. At higher heating rates coals soften that are otherwise completely inert.

The influence of different parameters on the softening of coals has been described in Section 6.3 and is not covered here.

[159] H. Luther and S. Traustel, *Brennstoff-Chemie,* **44,** 65–67 (1963).

[160] G. Juranek, B. Bussmann, and G. Seifert, *Brennstoff-Chemie,* **44,** 97–104 (1963).

[161] P. G. Boiko et al., *Coke Chem. USSR,* No. 1, 8–10 (1974).

[162] M.-Th. Mackowsky, *Fortschr. Miner. (Prog. Mineral.),* **45,** 52–94 (1967).

[163] Y. Toda, *Fuel,* **52,** 36–40 (1973).

[164] V. F. Kiselera et al, *Coke Chem. USSR,* No. 5, 14–17 (1972).

[165] E. Wolff, *7th Int. Conf. Coal Science,* Prague, 1968, pp. 431–439.

[166] S. Pregermain and J. Deduit, in *Coal Science,* Am. Chem. Soc. Adv. Chem. Ser. No. 55 (1966), pp. 248–260.

[167] M. Iknatowicz et al., *Carbon (Oxford),* **4,** 41–50 (1966).

[168] G. H. Taylor, *Fuel,* **40,** 465–471 (1961).

[169] M.-Th. Mackowsky and E. Wolff, *C. R. 37, Congr. Int. Chim. Ind.,* Bruxelles (1967), pp. 576–579.

[170] H. R. Brown, G. H. Taylor, and A. C. Cook, *Fuel,* **43,** 43–54 (1964).

[171] G. H. Taylor, M.-Th. Mackowsky, and B. Alpern, *Fuel,* **46,** 431–440 (1967).

[172] E. Wolff, *16th Round Table Gespräch,* Essen, 1972.

[173] H. R. Brown and P. L. Waters, *Fuel,* **41,** 3–14 (1962).

6.4.3 The Swelling Process

The swelling of coal in the plastic zone is caused by the evolution of gas produced by pyrolysis. Such gases cannot escape rapidly enough because of the low permeability of the plastic mass. Therefore gas bubbles with very high internal pressures may form locally, swelling the viscous mass. The gas pressures decrease slowly through the formation of degasification pores.

This phenomenon can be observed in a single particle as well as in a loose mass of coal or a briquet. The main influence on the expansion of the individual particle, with simultaneous formation of degasification pores, is the ratio of gas evolution to diffusion within the particle.[3–5] Differences in the composition and size of the macerals also have an effect.

The influence on plastic behavior of the packing density and the heating rate was investigated by the so-called coking of granules and by coking coal in the Ruhr dilatometer.[4,6] Microscopic investigation has shown that formation of coke, especially with coals that coke badly under normal conditions, is influenced by these factors. High packing density improves the fusion of the particles immediately on softening. The size of the pores decreases, since expansion is made more difficult with increasing packing density.[174] High heating rates increase the formation of pores and increase the volume.[165,174]

The swelling process is strongly influenced by the test methods (Sections 6.2.2 and 6.2.3), the physical conditions of the experiment (Sections 6.3.1 and 6.3.4), the chemical composition and treatment (Sections 6.3.5 and 6.3.7), and the properties of the coal used (Sections 6.3.2 and 6.3.3).

6.4.4 The Resolidification Process

Exceeding the resolidification temperature irreversibly converts the softened coal into the solid state of semicoke. This process is strictly associated with thermal decomposition and, therefore depends more on the chemical constitution of the coal than on its petrographic structure.[90] The influence of rank is only indirect and is determined by the difference in chemical composition of the different types of coal.

The behavior of microlithotypes of homogeneous and heterogeneous composition and different rank was investigated under the microscope to observe the formation of the coke structure in the plastic zone.[175–177] According to these investigations, the formation of the detailed structure of coke begins before resolidification. It is indicated by the appearance of anisotropies, which shows, under the microscope, order in the molecular range.[1,2,83,167] It is surprising that the anisotropies increase in intensity only as the coking proceeds. Neither shape nor size of the areas of anisotropy is changed. These changes depend only on rank.[178] These observations were confirmed by studies with the electron microscope[166] and by X-ray structure analysis.[179]

Thus the structure of the coke is preformed when the resolidification temperature is exceeded. Further significant changes do not occur. Furthermore, the optical properties remain unchanged during further increases in temperature, if one disregards the small increase in reflectance and the previously described increase in the intensity of anisotropy. The latter is an expression of the increasing degree of order, which is accompanied by an increase in the density of the coke (Section 6.4.5).

6.4.5 Shrinkage after Resolidification

The product obtained during coking after resolidification is usually called semicoke. The properties of the coke are already evident; however a further temperature increase is required to complete the coke formation. In the course of coking, shrinkage of the semicoke occurs as a result of the pyrolysis reactions. This can be measured with suitable

[174] M.-Th. Mackowsky and E. Wolff, in *Coal Science*, Am. Chem. Soc. Adv. Chem. Ser. No. 55 (1966), pp. 527–548.

[175] I. V. Eremin, *Tr. IGI Minist. Ugol'noj Prom-St-*, **28**, 10–21 (1972).

[176] N. S. Gryaznov and L. V. Kopelivvich, *Koks Khim.*, **7**, 2–6 (1975).

[177] D. Hays, J. W. Patrick, and A. Walker, *Fuel*, **55**, 297–302 (1976).

[178] H. Marsh, J. M. Foster, G. Hermen, and M. Iley, *Fuel*, **52**, 243–261 (1973).

[179] K. A. Bogoyarlenskii and E. M. Soldatenko, *Coke Chem. USSR*, No. 6, 6–10 (1964).

methods on test samples. In the coke oven itself, shrinkage leads to the following events: formation of fissures, and horizontal as well as vertical contraction of the oven charge.

Brown coal and brown coal briquettes were used for the first detailed investigations of shrinkage, in which the correlation between degasification, shrinkage, and stability of the briquettes was determined.[180] The progress of shrinkage can be measured thermogravimetrically, that is, by measuring the mass and the density as a function of temperature. Results obtained with suitable dilatometers give the same values and have the advantage of greater accuracy.[181] The shrinkage is normally isotropic; in other words, cylindrical test samples shrink by the same percentage in the axial and radial directions. The shrinkage in length is thus an accurate measure of the total shrinkage in volume.[12] Only test samples with a significant gradient of density (e.g., briquettes made from coal blends and carbonized at a lower temperature) show any degree of anisotropy in shrinkage, since an increase in density leads to a decrease in the shrinkage coefficient.[182] Between the resolidification temperature and a temperature of about 1000°C, at which the shrinkage process has almost been completed, the linear shrinkage rate is about 10%. Coals with a high resolidification temperature therefore give lower shrinkage values than coals with a lower resolidification temperature. In addition, measurement of the shrinkage depends on the sample. Coal shows a greater shrinkage than semicoke made from the same coal at the resolidification temperature.[183]

The total shrinkage is independent of the heating rate; however the course of shrinkage depends to a large extent on time. The course of shrinkage is described by the shrinkage coefficient α, which is the first derivative of the length of the sample L, with respect to the temperature T, and refers to the lengths of the sample L_0 at the instant of resolidification (or in the case of semicokes, on the starting length):

$$\alpha = \frac{1}{L_0} \cdot \frac{dL}{dT} \qquad (6.23)$$

This coefficient has a first maximum in the vicinity of the resolidification temperature, a well-defined minimum at about 600°C, and a second maximum at around 700°C. Investigations of semicokes show only the second maximum. The first maximum is essentially defined by the relative positions of the main degasification and the resolidification temperature; consequently an increase in the heating rate leads to a slight decrease in the coefficient.[30] However the height of the second maximum is directly proportional to the heating rate[12] throughout the entire range of rates, from slow (1.2 K/min) to shock heating (100 K/min).

In the course of heating the oven charge, different coefficients of shrinkage lead to stresses, which can be calculated for a given temperature gradient[184,185] if the viscoelastic data (elasticity and Poisson modulus) and the flow properties (relaxation) are known. At points in the coke-oven chamber where the stresses are greater than the mechanical strength of the charge, fissures occur or are extended. These are most likely to be the areas of highest temperature (e.g., the areas next to the wall of the chamber), which explains the cauliflower shape of the coke lumps adjacent to the walls of the chamber. At first approximation, the distance between the fissures is directly proportional to the mechanical strength and indirectly proportional to the maximum shrinkage coefficient and temperature gradient. The second maximum is important to the formation of fissures in the interior of the charge.[185] Therefore the height of the shrinkage coefficient in the first maximum determines the lump size of the coke, and the height of the second maximum determines the strength of the coke—its resistance to degradation in shipping and screening.[31]

It is usually the higher volatile coals that

[180] O. Abel, Habilîtation, Clausthal, 1965.

[181] R. Tournant, R. Busso, and A. F. Boyer, *Rev. Gen. Therm.*, 3(29), 673–679 (1964).

[182] F. Strunk, *Glueckauf-Forsch.-H.* **36,** 91–97 (1975).

[183] G. Meimarakis and A. F. Boyer, *Rev. Gen. Therm.*, 3(31), 937–944 (1964).

[184] J. L. Soulé, *Bull. Soc. Fr. Mech.* **17,** 32–37 (1955).

[185] P. Chagnon and A. F. Boyer, *Rev. Gen. Therm.*, 3(50), 139–151 (1966).

tend to form fissures. To decrease this tendency, their plasticity is reduced by additives whose effect on the progress of shrinking varies[31,186]:

1. Medium-volatile coals in the coal blend decrease the first maximum. As shown in Fig. 6.17, the curve of shrinkage of the blend results from a graphical combination of the individual curves of shrinkage, with the left branch of the curve of the original coal almost removed.
2. Semicokes do not shrink in the vicinity of the temperature at which they were produced. Thus addition of a semicoke that has been produced at 500°C decreases the first maximum, while the remainder of the curve remains unchanged. Semianthracites show a similar relationship.
3. Coke breeze as an inert material gives an undesirable reduction of all sections of the curve of shrinkage, which means both maxima as well as the minimum. An optimum size consist of 0.1–0.5 mm is required. A similar action is shown by anthracite of very low volatile matter content.

Blends with optimum properties may often be obtained by a combination of lean additives.[187]

6.4.6 Conclusions

In the past 18 years, many individual phenomena of the plastic behavior of coals have been explained. Nevertheless, as pointed out in the first Supplementary Volume,[40] the mechanism by which certain coals soften, swell, and finally resolidify is not yet completely understood. The following is a simplified explanation of the phenomena that have been elucidated so far.

When certain coals are heated, gases are split off at first. When the softening temperature, which is characteristic for each of these coals, is exceeded, the coal begins to soften. This is most likely a purely physical process similar to the melting of plastic materials, especially since it has been shown to be reversible.[90]

On softening, the primary bitumen is first separated from the coke skeleton and liquefied.[155] The primary bitumen does not necessarily have to be present in the original coal in the same composition. It is possible that a very rapid primary decomposition occurs, which breaks a loose bond with the semicoke skeleton. On slow heating, volatile material is split off from the primary bitumen before its boiling point is reached. The residue, which has a higher boiling point, again decomposes into a volatile material and a higher boiling residue as the temperature is increased. Thus a chain of sequential reactions occurs before the boiling point of each of the decomposing materials is reached. The end of the last of these sequential reactions is the resolidification of the plastic mass, the formation of semicoke. For reasons described previously, this is an irreversible process.

The individual sequential reactions occurring during slow heating were investigated gravimetrically[188] as well as volumetrically.[189]

[186] P. Foch and G. Meimarakis, *Rev. Gen. Therm.*, **3**(31), 911–936 (1964).
[187] K. H. Flasche et al., *Informationstagung über Kokereitechnik der KEG,* Luxemburg, May 1975, Vol. 1, pp. 9–19.

[188] H. Luther, *Erdoel Kohle,* **13,** 947–955 (1960).
[189] K.-Ch. Traenckner, *Bergbau-Archiv,* **21**(3), 48–66 (1960).

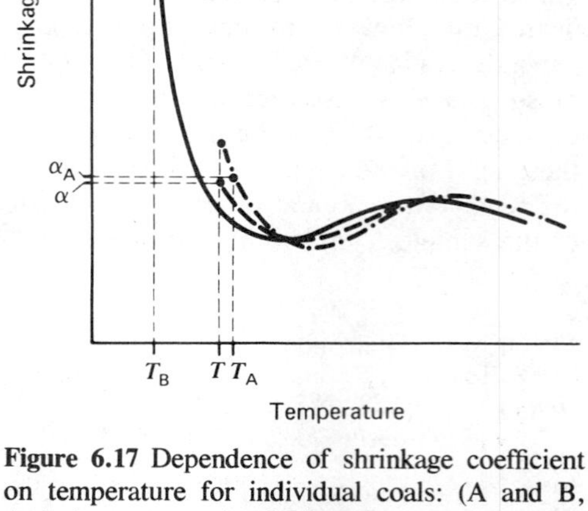

Figure 6.17 Dependence of shrinkage coefficient on temperature for individual coals: (A and B, dash-dot curve and solid curve, respectively) and their blend (dashed curve).[186]

It was shown that depending on the rank of the coal, the first of these sequential reactions in which tar is evolved starts more or less rapidly at about 320°C. This temperature can be considered as the boiling point of the primary bitumen.

With more rapid heating rates, these sequential reactions are run over and become indiscernible. The liquid primary bitumen is evaporated to a great extent before it has time to decompose. The portion that remains as a residue in the coke from the last sequential reaction during slow degasification is evaporated during rapid heating. This process explains the higher yield of pitch.

The formation of the structure of semicoke has already occurred in the plastic phase. It is indicated by anisotropies that do not change in form or magnitude during further heating. As a result of chemical decomposition reactions, a contraction of volume occurs.

The metaplast theory (which assumes an instable, intermediate phase that is responsible for the plastic behavior of coals) leads to a mathematical model that can be used to give a simple theoretical confirmation of some experimental results.

6.5 THE IMPORTANCE OF PLASTIC PROPERTIES IN THE FORMATION OF COKE

6.5.1 Formation of Coke

On charging cold coal into the coke oven, the wall temperature is suddenly decreased to about 600°C. This reduction in temperature is overcome rather rapidly. In the first step of the process the associated input of heat to the coal leads to the desorption of gases, later to the evaporation of water, and then to the start of decomposition. At a temperature of about 350°C, the coal begins to be plastic, and after a comparatively narrow temperature difference, reaches the point of resolidification. Semicoke is formed, which, on further increase in temperature, leads to high-temperature coke (Section 6.4).

The transfer of heat into the coal can be described with a modified Fourier differential equation, if instead of the heat conductivity, λ, a corresponding combined value λ^* is used, which includes conduction, convection, and radiation.[190] Describing the total charge of the coke oven presents special difficulties because instantaneous changes must be treated as an integral effect. However, it is possible to estimate the individual parameters,[191] thus to calculate the temperature distribution in the oven with satisfactory approximation.

Because heat flows at a right angle to the heated walls, two thin plastic zones are formed, and during the coking process they travel toward the middle. Ahead of them are evaporation and recondensation zones. The "impregnation" of coal that has not yet softened, by the recondensation of tar, prepares for conversion into the plastic state.[192] At this stage the formation of the fine structure of the coke begins (Section 6.4.2). On exceeding the resolidification point, the softened coal definitely reaches the solid stage. The structure of the coke at that point is already defined. Significant changes no longer occur, and pyrolytic decomposition leads finally to a denser coke.

Many studies have been made using individual coal particles, laboratory test methods, or semitechnical coke ovens (see the preceding sections) to investigate the processes occurring during coking. Systematic investigations on a commercial scale have also led to basic information[193–199] (see also Chapter 14).

On heating, gases, especially methane, are evolved at first. With further increases in temperature, chemical bonds are broken and liquid tarry products produced. The plastic mass is being formed. Thus begins the most essential phase of the coking process, during

[190] W. Rohde, W. Simonis, and W. Peters, *Brennstoff-Chemie,* **50,** 1–8 (1969).
[191] K. Hofherr and W. Simonis, *Glueckauf-Forsch.-H.* **32,** 35–49 (1971).
[192] R. Riedl, *Brennstoff-Chemie,* **48,** 52–55 (1967).
[193] K.-G. Beck, *Glueckauf,* **111,** 1125–1127 (1975).
[194] W. Weskamp, W. Dressler, and E. Schierholz, *Glueckauf,* **98,** 567–577 (1962).
[195] K.-G. Beck, W. Weskamp, and R. Beckmann, *Glueckauf,* **98,** 1559–1568 (1962).
[196] K.-G. Beck, R. Beckmann, W. Weskamp, and W. Dressler, *Glueckauf,* **100,** 839–852 (1964).
[197] W. Weskamp, *Glueckauf,* **103,** 215–225 (1967).
[198] W. Weskamp, *Glueckauf-Forsch.-H.* **29,** 295–303 (1968).
[199] W. Eisenhut, W. Dressler, and K.-G. Beck, *Brennstoff-Chemie,* **50,** 33–39 (1969).

which the important events of coke formation occur.

X-Ray structure analysis has been applied in a study of the fine structure of coke using changes at the molecular level. To explain the results, the investigators arrived at the following conclusion.[200] During the coking process, the carbon skeleton continuously becomes richer in thermally stable structural forms. This process occurs in four characteristic steps.

1. During the first step, between about 300 and 350°C, a partial breaking of macromolecular bonds occurs and leads to the evolution of low molecular weight products. The basic structure of the coal has not yet been changed.
2. In the second step, up to about 600°C, the coal substance is decomposed to a great extent and two-dimensional crystallization centers are formed, which slowly grow to ring polymers that are preferably oriented in these two dimensions.
3. This conversion continues in the third step, up to about 700°C. The carbon structures form larger units.
4. A large carbon system exists above 700°C, it is characterized by a high degree of order between the individual layers.

Microscopic methods (Section 6.4) and direct X-ray structural analysis lead to astonishing agreement in the descriptions of the processes occurring during the thermal conversion of coal into coke. There is still disagreement over the phase of the coking process in which fissures are formed, however, and the opinion[201] that the first breaks in the plastic mass are caused by overcoming the surface tension is being contradicted.[202] A network of such breakages would have to occur, with the result that during coking, only coke breeze would be produced, since the fissures so formed could not be repaired above the plastic range. More convincing is the theory that the evolution of coke lumps occurs during the formation of semicoke.

Theoretical considerations have led to the conclusion that the formation of fissures in the direction of heat transfer (i.e., at right angles to the heating wall) is based on the effective shrinkage and the corresponding temperature-time curve. Accordingly, it begins at about 530°C and terminates at about 620°C, since between 620 and 680°C the compression forces are larger than the forces caused by tensions. Furthermore the strengths of coke increase with temperature and the stresses decrease.

The formation of transverse fissures is thought to be caused by faults,[203] which are created by carbon formed by cracking and deposited on semicoke layers.

An attempt to describe the forces that are effective during the formation of fissures[204] led to a system of mathematical equations describing the distribution of bending and torsion moments and of shearing forces. Other authors[140] describe the fissures parallel to the wall as resulting from the incorporation of contraction of the semicoke and the deformation of pores under pressure at a temperature of about 550°C. Accordingly, the longitudinal fissures develop first at a right angle to the wall. They are mainly caused by stress concentrations in the vicinity of the nonsoftening components and become visible at about 650°C.

6.5.2 Knowledge of Coal's Plastic Properties and Conclusions Regarding the Production of Coke on a Commercial Scale

Knowledge of the plastic properties of coal and the possibility of influencing them lead directly to conclusions relating to the industrial production of coke. Basically coke and its properties are the result of the coal properties and the conditions used to convert coal into coke. Since there is no question that the plastic state and the properties of the plastic mass determine this technical process, all parameters that influence the plastic stage and the properties of the plastic mass are

[200] M. G. Sklyar, E. M. Soldatenko, and N. A. Valters, *Coke Chem. USSR,* No. 1, 20–24 (1973).

[201] G. P. Gordeev, V. P. Zaitsev, and A. G. Rembashevskii, *Coke Chem. USSR,* No. 1, 23–27 (1971).

[202] D. A. Muchnik, *Coke Chem. USSR,* No. 10, 14–18 (1972).

[203] V. P. Zaitsev, *Coke Chem. USSR,* No. 5, 16–20 (1973).

[204] A. P. Konyakhin, A. P. Fomin, and J. Z. Shatokha, *Coke Chem.,* No. 8, 16–19 (1973).

important. These include properties of the coal used, physical properties of the charge, and heat transfer conditions in the oven.

In the Federal Republic of Germany the possibilities for optimizing the coking of bituminous coal have been investigated for years. The basis for this study was the possibility of describing the coking of bituminous coal at high temperature when charged by gravity.[205] The work was extended by a mathematical description of coking with stamped charges.[206] Another facet of this work is the calculation of the gas pressure that occurs when the tar boundary is closed. This aspect is the reason for the mechanical pressure directed toward the coke-oven wall.[207]

A combination of these methods together with economic conditions has been applied with the aid of operations research to plants of different sizes.[208]

The properties of the coal are described by the coking capacity G (discussed in Section 6.2.2) and the content of volatile matter. In the range 0.95–1.10, the value G is additive, which means that the coking capacity of a blend can be calculated from the individual components according to their proportion in the blend.

To reach a maximum coke strength, each content of volatile matter is associated with an optimum coking capacity.[209] Thus the optimum composition of the coking coal blend can be calculated from any number of individual components.

The properties of a charge are characterized by the bulk density and the coefficient M_s, which is the deviation of the size distribution of the feed coal from that of the optimum size consist.[210] The authors assume that the charge to the oven should be as homogeneous as possible. This condition can be achieved by grinding. Below a certain particle size, the charge is likely to form bridges, which leads to great variations in the bulk density. In connection with theoretical considerations for piles of spheres, an optimum particle size distribution was obtained.

The above-mentioned M_s value now represents the sum of the positive or negative deviation of fractional portions in a given particle size distribution, in comparison to the optimum particle size distribution.

The optimum particle size distribution refers to the total charge. In view of the uniform distribution of the parameters characterizing the properties of the individual components, the components should be ground to such an extent that their absolute surfaces in the blend are equal.[209]

In the previously described model, the heat transfer conditions are expressed by the specific heating value K, which is an arithmetical connection between one-half the chamber width, the bulk density, and the rate of coking referred to one-half the chamber width.

A very good correlation with size consist and lump strength of the coke is given by the coking progress in Kelvins per minute.[1] It is calculated from the temperature difference between 350 and 550°C (the plastic stage of most coals is in this range) and the time needed to achieve this temperature increase. With increasing coking progress, the lump size of the coke decreases, as well as the lump strength. The coking progress in actual coke-oven operation, however, can be determined only by complicated methods.

Statistical evaluation of operating data has made it possible to obtain a mathematical description[205] of the yield of coking products. The author gives equations for the yield of coke, gas, light oil, crude tar, and ammonia. The content of volatile matter and the specific heating value K (Equation 6.26) were used as parameters. Another report indicates the possibility of calculating the sulfur content of the coke as a function of the sulfur content of the coal and the coking conditions.[211]

An important supplement to the process of optimizing the coking coal blend is the previously mentioned method for calculating the

[205] W. Simonis, *Glueckauf-Forsch.-H.* **29,** 103–119 (1968).

[206] E. Falk, W. Plate, and W. Simonis, *Glueckauf-Forsch.-H.* **33,** 234–239 (1972).

[207] E. Szurman and W. Simonis, *Glueckauf-Forsch.-H.* **34,** 68–74 (1973).

[208] W. Hermann, *Informationstagung über Kokereitechnik der KEG,* Luxemburg, May 1975, Vol. 1, pp. 35–46.

[209] W. Simonis, G. Gnuschke, and K.-G. Beck, *Glueckauf-Forsch.-H.* **27,** 181–192 (1966).

[210] W. Simonis and E. Rubrecht, *Glueckauf-Forsch.-H.* **26,** 301–308 (1965).

[211] H. Ernst, G. Gnuschke, K. Hofherr, and W. Simonis, *Glueckauf-Forsch.-H.* **30,** 197–202 (1969).

maximum gas pressure in the plastic zone when the two tar layers meet.[207] This gas pressure is the cause of the mechanical pressure toward the coke-oven wall during the so-called swelling of a coal (Chapter 14). One of the reasons for the importance of this procedure is that the calculated maximum gas pressure is a limiting condition in the optimization of coking coal blends.

When pressure-causing coals are used, these calculations give the measures necessary for the protection of the walls of a coke-oven battery: decrease of the bulk density, finer grinding of the feed coal, and increase of the coking time by a decrease in the flue temperatures.

In the U.S.S.R. another method was developed for describing the manufacture of coke, in which petrographic methods are used to characterize the coals and to prepare the optimum blend[212] (Chapter 3). The influence of the size consist is considered in relation to the preparation plant results, that is, to the distribution of the specific gravity fractions.[213] The thermal conditions are included by the form of the changes that occur in the plastic state.

Because of the great diversity of coals available in Japan, a process was developed in which a blend of as many as 20 components is compounded in such a manner as to obtain a predetermined coke quality.[214,215] The parameters used are the average mean reflectance and maximum fluidity according to Gieseler.

The feed coals are mixed in such a manner that an average mean reflectance of 1.2–1.3% is achieved. The limits for the maximum fluidity depend on the "working rate" of the coke-oven battery, a measure of the operating time:

$$\text{working rate} = \frac{\text{number of ovens pushed/24 hr}}{\text{number of ovens in the battery}} \times 100 \tag{6.24}$$

The lower limit of the Gieseler fluidity should be between about 60 and 200, and the upper limit between 1000 and 3000 dial divisions per minute.

Because of their behavior in the plastic stage, certain coking coal blends with a high content of volatile matter do not form coke suitable for use in blast furnaces if the coal is charged by gravity (top charging). In such cases, which are usually caused by the deposit of the coal, preheating[127] or an increase in the density of the oven charge (Chapter 14) by means of stamping[187] can lead to improved coke quality. Stamping at first leads to a decrease in the lump strength of the coke,[206] but it definitely improves the resistance to abrasion. Addition of inert components may decrease the formation of fissures, thereby improving lump size and strength (Section 6.5.4).

The influence of size consist on coke quality follows a course similar to that of the top-charging operation. The coal has to be ground more finely, however, to obtain optimum results.

In the stamping operation, the ash content has other effects in addition to its influence on fissure formation in the coke.[139] Because of the hydrophilic properties of the ash constituents, the statics of the stamped charge are reduced if the ash content is extremely low.

In practice this is not important because the preparation of coals with such low ash contents is not economic. Furthermore, the effect of ash depends on the composition of the mineral matter.

Another possibility for making more coals suitable for the coking operation is to use preheated coal, in which case the improvement in quality or the possible use of higher proportions of coal with a low coking capacity obviously depends on the preheating temperature[127,128,216,217] (Chapter 14).

The reduction in coking time[104] without diminution of quality results partly from better rheological properties of the charge. The nonuniform coking caused by recondensation[112] of the coal moisture is avoided; there-

[212] J. M. Glushchenko et al., *Coke Chem. USSR*, No. 8, 7–14 (1966).

[213] A. P. Fomin et al., *Coke Chem. USSR*, No. 4, 3–7 (1968).

[214] T. Miyazu et al., *Int. Eisen-Stahl-Kongress (Tron a. Steel Congress)*, Düsseldorf, 1974, Rep. No. 1.2.2.1.

[215] T. Miyazu et al., *J. Fuel Soc. Jap.* **49**, 736 (1970).

[216] G. Prudhon and R. Marcellini, *Informationstagung über Kokereitechnik der KEG*, Luxemburg, May 1975, Vol. 1, pp. 201–211.

[217] F. Orywal and D. Wagener, *ILAFA Coal Congr.*, Mexico City, 1976.

fore the charge can be pushed at a lower final temperature of the coke.[126] Other contributing factors include the temperature of the charged coal and avoidance of the evaporation of water.

6.5.3 The Plastic Properties of Coal and the Mechanical Properties of Coke

The mechanical properties of coke (Chapter 15) are usually tested by drum or tumbler tests, which are assumed to simulate the stresses in the blast furnace.[218] Standards for such tests have been developed in several countries.[219–222] The tests differ in the dimension of the tumbler drum, the number of rotations, the velocity at the circumference, the size and amount of feed, and the characteristic size consist of the mechanically stressed sample. In spite of these variations, the results among these tests are comparable if the same fractions are considered.[223,224] The tumbler test usually gives an indication of two criteria of the strength: the strength of the lump and the abradability.

Many proposals have been made to use other measured quantities to characterize coke strength, for example, using the grinding hardness to describe the strength of the crack-free matter[225] or the tensile strength.[226,227] The grinding hardness is the physical work required in the standardized grinding unit to decrease coke of a given size to another equally defined size consist.

Other authors describe the result of the tumbler test by the ratio of the average particle diameter before and after the mechanical stressing[228] or by the increase in the curves obtained from the changes in the average particle diameter with the increasing number of rotations of the tumbler.[140] Since the results of any tumbler test depend on the size of the coke, therefore do not even give an absolute indication of its mechanical strength,[223] it is not surprising that discussions on the characterization of coke strength take place all over the world, as evidenced by the examples cited.

A new strength index that includes the size of the coke requires further evaluation.[229] It is based on the assumption that a coke has a strength of 100% if none of its size fractions (not only the amount above 40 mm in the ISO Tumbler Test) are changed by mechanical stressing. The breakage quotients of all the size fractions are being used to describe the strength. These quotients follow the Rosin-Rammler-Bennett function and describe the size distribution before as well as after the mechanical stress. The M_{40} value describes the breakage quotient of the size class above 40 mm, which is only one point of the total function. Nevertheless tumbler test strengths are useful values for the coke-oven and blast-furnace operation. High tumbler strengths are the major objective of the processes described in Section 6.5.2 for optimizing coking coal blends and coke-oven plants, and in many places one can attempt to precalculate tumbler strengths from the properties of the coal. In some cases the operating conditions are assumed to be constant, and sometimes they enter the equations as variables. Among the latter cases is the mathematical description of high-temperature carbonization with top charging.[205] In the equation:

$$M_{40} = b + a'K + \alpha' M_s \qquad (6.25)$$

the properties of the coal are described by a', α', and b, where a' and α' are functions of the content of volatile matter and the coking capacity G; b depends only on the content of volatile matter. The coefficients can be calculated with equations obtained by a regression analysis of the process data from both semitechnical and full-scale operations.

[218] H. Kahlhofer and A. Send, *Stahl Eisen*, **76**, 141–144 (1956), Contribution to discussion, F. Wesemann, U. Graf, and R. Wartmann.
[219] DIN 51717.
[220] ASTM Designation D294-1964.
[221] AFNOR NF M 03-023-1962.
[222] Japanese Industrial Standard JIS K2151-1960.
[223] E. Szurman, F. Orywal, and K.-G. Beck: *Stahl Eisen*, **94**, 1096–1100 (1974).
[224] J. G. Walters, G. W. Birge, and D. E. Wolfson, U.S. Bur. Mines, Rep. Invest. No. 6482, 1969.
[225] Unpublished investigation of Bergbau Forschung GmbH, Essen.
[226] J. W. Patrick and A. E. Stacey, *Fuel*, **51**, 81–87, 206–210 (1972).
[227] J. W. Patrick, A. E. Stacey, and H. C. Wilkinson, *Fuel*, **51**, 174–179 (1972).
[228] J. Gibson, *Glueckauf*, **105**, 160–165 (1969).
[229] E. Szurman, *Stahl Eisen*, **97**, 1–6 (1977).

The specific heating value, K, includes bulk density, ρ (g/cm^3), chamber width, B (cm), and gross coking time, t_g (hr):

$$K = \frac{\rho \, (B/2)^2}{t_g} \tag{6.26}$$

The variable M_s is used to account for the effect of size distribution.

The range of variation of the data used for the regression analysis results in a limited validity, which is indicated by the following limits: (1) specific heating value K: 18–24 g/(cm)(hr), (2) volatile matter content (maf): 18–35 wt %, (3) coking capacity G: 0.95–1.10, (4) deviation of the sum M_s: $\leq$ 35%, (5) d_m (average particle diameter), according to Rosin-Rammler-Bennett $\geq$ 1 mm, and (6) inertinite content $\leq$ 20 vol %.

The coke abradability, M_{10}, can be calculated from the parameters K and M_s. The properties of the coal are indicated by the coefficients $m_i = f(\text{VM}, G)$.[205] The calculated M_{40} and M_{10} values correspond to the German standard DIN 51 712, which in 1967 was replaced by DIN 51 717. The new standard conforms to ISO-R 556.

For conversions, the following equations were given:

$$M_{40}(\text{DIN 51 717}) = 0.67 + 0.934\,[M_{40}(\text{DIN 51 712})] \tag{6.27}$$

$$M_{10}\,(\text{DIN 51 717}) = 2.75 + 0.766\,[M_{10}\,(\text{DIN 51 712})] \tag{6.28}$$

The system described has been used with good results for coals of different origin.[230]

Taking into consideration the coking capacity G and the content of volatile matter (VM also V), a statistical equation has been derived that leads, even for higher volatile coals, to satisfactory agreement between the calculated and the measured M_{40} values[231]:

$$M_{40} = 103.9 + 24.8 \times G - 1.196 \times V^5 \times 10^{-6} + \frac{2.57\,V^2}{T} - 88 \times \frac{V}{T} \tag{6.29}$$

where G is the coking capacity, V the content of volatile matter, and T the time needed to reach a temperature of 900°C in the center of the charge, with a chamber width of 18 in.

In connection with the characterization of the plastic properties of the coal by dilatation (G value), it has been shown that the components of a blend are not compatible if their plastic ranges overlap by less than 40 K. In practice this problem can be overcome by adding a third component that suitably overlaps the plastic ranges of the others.[232]

Another group of methods for the calculation of coke strength uses the results of petrographic analysis. In these methods the reflectance of the vitrinites is of great importance for describing the coking properties of the coal. In combination with the distribution of the maceral groups, reflectance is the basis for the different methods.

Fundamental principles[233] published in 1957 have been used by several groups of investigators, and changes have been made. The common basis of a number of methods[170,215,234] is to formulate the distribution of the ratio of reactive to inert components and a "strength index," which describes the coke-forming capacity of the "reactives," especially the vitrinites. There are, however, different opinions about which of the components of the inertinite group belong to the "inerts." A complete summary of these methods, containing material not previously published, appeared recently.[235]

Two processes based on strict statistical methods are difficult to use in practice. Nevertheless, they permit calculation of different desired properties using a large number of parameters.[236,237]

It has been demonstrated that the results of the tumbler test for the description of coke

[230] W. Kandler, E. Schatzl, and L. K. Bradacs, *Glueckauf,* **109,** 1204–1208 (1973).

[231] J. Gibson, *COMA Yearb.*, 182–201 (1972).

[232] J. C. A. Kaye and A. E. Horton, in *Informationstagung über Kokereitechnik der KEG,* Luxemburg, May 1975, Vol. 1, pp. 47–59.

[233] J. J. Ammossov, J. V. Eremin, S. F. Sukhenko, and L. S. Oshurkova, *Koks Khim.*, **12,** 9 (1957).

[234] N. Schapiro, R. J. Gray, and G. R. Eusner, *Blast Furn., Coke Oven Raw Mater. Comm., Proc.*, **20,** 89 (1961).

[235] M.-Th. Mackowsky, *J. Microsc.*, **109,** Pt. 1, 119–137 (1977).

[236] J. Lehmann and A.-G. Meyer, *Glueckauf-Forsch.-H.* **28,** 27–34 (1967).

[237] M. Gomez and K. Hazen, U.S. Bur. Mines, Rep. Invest. No. 7399, 1970.

strengths depend on size,[223] and on this basis attempts have been made to calculate in advance the size distribution of wharf coke. Using the same parameters that allow calculation in advance of the mechanical properties of the coke,[205] a characteristic number has been found that correlates with the average size as well as the percentage of lumps above 40 mm.[238]

6.5.4 The Influence of Nonmelting Additives

The influence of ash content (Section 6.3.7), nonmelting chemicals (Section 6.3.5), and inert components of the coal itself (Sections 6.5.1 and 6.5.3) has already been described.

The addition of inert or "semiinert" carbon-containing materials, also called lean components, serves to improve coke quality. These materials are especially useful if the basic coal leads to a well-fused coke containing many fissures or to a well-fused coke with few fissures but a weak structure.[239]

In the first case the contraction[31,142] is influenced, and the forces that cause fissures are distributed to a great number of contraction centers. This leads to an improved size distribution and lump strength (Section 6.4.5).

In the second case the decrease in the degasification rate in the plastic range together with a diminution of the number and size of the pores leads to an increase in structural strength. This effect is reinforced by the deposition of inert particles, which increases the strength of the walls of the cells. In coke testing this improvement in quality shows up as an improved abrasion resistance.

If one considers the effects on coke quality from these two points of view, it becomes evident that the different views on the mechanism of the addition of lean components[31,116,142,183,240–243] are only an apparent contradiction. Similarly, the effect of the size consist and the behavior during pyrolysis,[31,206,239,244] as well as the requirement of uniform distribution, can be explained without difficulty.

Of special importance is the addition of inert carbon materials during stamping operations. This compensates for the decrease in lump size and in lump strength, which is otherwise a primary effect of compaction of the charge. Furthermore there is a reversal of the effect of compaction, which means that the strength of the coke increases with increased compression because of the addition of coke dust.[245] This is apparently caused by better contact between the individual particles.

6.6 PLASTIC PROPERTIES AND AGGLOMERATION OF COAL DURING FLUIDIZATION

Caking properties of coals that are used in fluidized-bed processes for low-temperature carbonization, gasification, or combustion not infrequently cause difficulties. Such properties may lead to many process difficulties by formation of agglomerates, plugging of pipelines, formation of deposits, and slagging, and may have detrimental effects on control equipment. Fluidized beds make possible very rapid heating rates of the coal particles so that even coals that normally cake only slightly soften in the fluidized beds, and swell and fuse together if the softened particles are not kept apart. The addition of a caking coal in an amount equal to one-eighth of the total solid material in a fluidized bed may lead almost instantaneously to solidification of the fluidized bed.[246,247] In another area of coal use—namely, the continuous processes for the manufacture of coke from slightly caking coals or coal blends—it is important to heat the coal into the plastic range without destroying its caking properties.

Operating difficulties caused by excessive

[238] E. Szurman, F. Orywal, and K.-G. Beck, *Glueckauf-Forsch.-H.* **35,** 108–113 (1974).

[239] G. H. Taylor, *Glueckauf,* **98,** 164–172 (1962).

[240] H. Pichler and K.-H. Lauer, *Brennstoff-Chem.,* **44,** 149–154 (1963).

[241] K.-H. Lauer, *Brennstoff-Chemie,* **45,** 229 (1964).

[242] R. Agricola, *Brennstoff-Chemie,* **48,** 240–251 (1967).

[243] J. W. Patrick and A. E. Stacey, *Fuel,* **54,** 213–217, 256–264 (1975).

[244] L. Grainger, *COMA Yearb.,* 126–149 (1972).

[245] H. Lohrmann, *Glueckauf-Forsch.-H.* **26,** 233–240 (1965).

[246] K. Mc. G. Bowling and P. L. Waters, *Br. Chem. Eng.,* 7(2), 98–101 (1962).

[247] S. Werner, Dissertation, Aachen, 1972.

caking properties can be avoided by using two different procedures.[120,246] The first is to decrease the softening property of the coal by influencing its physical properties. This can be accomplished by selecting suitable raw materials and, if possible, noncaking coals, decreasing the caking properties by preoxidation, pretreating the coal with chemicals (e.g., boric acid, alkaline solutions), and partial pyrolysis. The second approach involves the selection of suitable process conditions: For including the coal with inert material (e.g., char) the process must be conducted in such a manner that there is relative motion between the coal and the inert material, possibly by simultaneous preoxidation using air as the fluidizing gas. In addition, the coal must be fed into the fluidized bed in a suitable manner.

The caking properties can be retained up to the plastic range by very rapid heating. Pretreatment of the coal can occur in a separate processing step, in which case a coal with changed physical properties is used for the main process, with the result that its caking properties are no longer detrimental. If the procedure for destroying the caking ability is immediately coupled with the process itself—for instance, in gasification and combustion—operating difficulties can be avoided only by using suitable process conditions.

When selecting the process for avoiding the agglomeration of caking coals used in the fluidized bed, bear in mind that separate pretreatment of the coal may have a negative effect with respect to the product yield, the throughput of the process, and the need for additional equipment, all of which may influence the process economics.[248]

6.6.1 The Influence of the Physical Properties of Coal

Pretreatment of coal to reduce its caking ability requires a test to judge the result that will be obtained. This can be done by means of the swelling index, with an agglomeration index test or with the so-called hydrogen test. The caking ability in general is sufficiently reduced if a swelling index of 1–2 is obtained; the lower value is considered to be more favorable.[249] The residue from the determination of the volatile matter can be used to determine the agglomeration index. The sample is said to be agglomerated if a weight of 500 g placed on top of the material does not pulverize it. Otherwise the sample is considered to be nonagglomerated. Depending on the conditions, both groups can be further subdivided.[43] In the hydrogen test the sample is heated in a quartz tube in a hydrogen stream to 600 or 900°C. The sample is considered noncaking if, after the test, it flows freely from the tube.[250]

Recent investigations concerned with the effect of oxidative treatment of coals with oxygen and the effect on their plastic properties were described in Section 6.3.5. Detailed investigations have been made of the important factors affecting the oxidative treatment in fluidized beds, especially the required residence time, the oxygen requirement, and the reaction temperature.[250] In discontinuously operating fluidized beds, it is possible to destroy the caking ability of coal, even of coal with a swelling index greater than 8, in the size fraction of 1.00—0.15 mm, by direct heating with inert gas or steam containing 0.2% oxygen at 430°C, and with a minimum residence time of 5 min. To destroy the caking ability means to reach a swelling index of the treated coal that is between 1 and 0.5. In continuously operating fluidized beds, more severe conditions are necessary. For the same coal at 430–440°C, about 2.3% of oxygen is required in the gas stream: this corresponds to an oxygen/coal ratio of about 0.025 m^3 of oxygen per kilogram of coal, at a residence time of 5 min. Finer sizes require a smaller oxygen/coal ratio. Lower temperatures require higher oxygen/coal ratios or longer residence times.

Using air as the fluidizing gas and an operating temperature significantly below the softening point, at 200°C a Pittsburgh Seam coal (36.6% VM) with a particle size between 0.105 and 0.074 mm requires a residence time of 30–40 min to destroy the caking ability. Smaller sizes require shorter residence times;

[248] J. Yerushalmi, *Chem. Eng. Prog.; Coal Process Technol.*, **3**, 156–165 (1977).

[249] S. J. Gasior et al., American Chemical Society, Fuels Div. Meeting, Washington, D.C., September 1971.

[250] A. J. Forney, R. F. Kenny, S. J. Gasior, and J. H. Field, U.S. Bur. Mines Rep. Invest. No. 6797, 1966.

however their fluidization is difficult. Treatment of this coal in a particle size range between 0.3 and 0.15 mm with inert gas having an oxygen content of 0.8–1.0%, at pressures up to about 22 bar and a temperature of 400°C, resulted in no significant reduction of the swelling index compared to treatment at atmospheric pressure. Pretreatment equipment for the destruction of the caking ability, the oxidation with oxygen, and the operating conditions are described elsewhere.[251]

Many of the proposed methods for destroying the caking ability cannot be used for economic reasons or because of process difficulties. For example, halogens and boron trifluoride can reduce or destroy the caking ability but are not suitable for gasification processes because of the cost of irretrievably lost reagents, the deterioration of the product, and other undesirable effects.[120] Possibly more successful would be the treatment of coal with dilute sodium hydroxide (NaOH) solutions. Sufficient reduction of the caking ability can be achieved with coals with a swelling index of 4.5 or 3.5, by absorption of 1–2 wt % NaOH (dry). If the NaOH is added to the coal as a powder, about 20 wt % is necessary to achieve the same result. The extent of the decrease in caking ability depends on the coal rank, the particle size distribution, and the mineral matter content. The reactions that reduce the caking ability occur at temperatures above 300°C. The actual mechanism and the effects on the reactivity of coal are not yet known.[120]

Another method for decreasing the caking ability is now being tested.[252] Caking coal is comminuted to below 20 mesh (0.841 mm) and is carried by a heated stream of gas consisting of oxygen and nitrogen under pressure through a tube 23 m long and about 12–13 mm I.D. Promising results were obtained with a highly caking coal (Pittsburgh Seam coal) at a temperature of 375°C, a pressure of about 8 bar, a gas velocity of about 3 m/s, and an oxygen/coal ratio of about 0.08 m³/kg.

In view of the use of bituminous coal in combustion processes, the swelling of individual particles has been investigated by optical methods.[253] These investigations indicate that the swelling involves two steps when heating occurs in air: after an expansion to about 2 or 3 times the original diameter of the particles, there follows a contraction to about 1.5 times the original value. The effect is independent of the coal rank or swelling index. Only anthracites with a volatile matter content below 5 wt % swell to a neligible degree or not at all.

6.6.2 The Influence of Processing Conditions

Fluidized-bed processes are especially suitable for the low-temperature carbonization, gasification, or combustion of pulverized caking coals, since under suitable conditions the fusion of the coal (agglomeration) can be avoided. In general, however, high fluidity of the softened coal and a wide plastic range are unfavorable for operations in fluidized beds. The tendency to agglomerate depends on the duration of the solidification reaction, that is on the time that the softened coal requires to solidify to semicoke under process conditions. It also depends on the fluidity and the actual stickiness of the molten coal mass itself. Of great significance in avoiding the formation of agglomerates is the mode of adding the solids. This must be done in such a manner that the coal particles are rapidly and uniformly distributed in the fluidized bed, so that they cannot stick together. Thus pneumatic feeding methods have been especially successful.[246,247,250] The number, arrangement, and orientation of the nozzles for feeding the coal are of great importance. Examples of different feeding methods are shown schematically in Fig. 6.18.[248,254–256]

When feeding from the top[254] (Fig. 6.18*a*), the coal passes a rotary valve into a horizontal pneumatic feeding line and drops onto the fluidized bed from the top. When feeding with a jet (Fig. 6.18*b*), the coal is blown in a

[251] S. J. Gasior, A. J. Forney, W. P. Haynes, and R. F. Kenny, *AIChE Symp. Ser.*, **72**(156), 117–126 (1976).

[252] G. E. Johnson et al., *Prepr. Am. Chem. Soc., Div. Fuel Chem.* **20**(3), (1975).

[253] R. H. Essenhigh and G. C. Yorke, *Fuel*, **44**, 177–189 (1965).

[254] H. G. Riedel, H. Tieroff, and M. Seidel, *Freiberger Forsch.-H.* A405, 7–20 (1967).

[255] B. Bock, Dissertation, Aachen, 1970.

[256] H. F. Feldmann, C. Y. Wen, W. H. Simons, and P. M. Yavorsky, *AIChE Symp. Ser.*, **69**(127), 74–76 (1972).

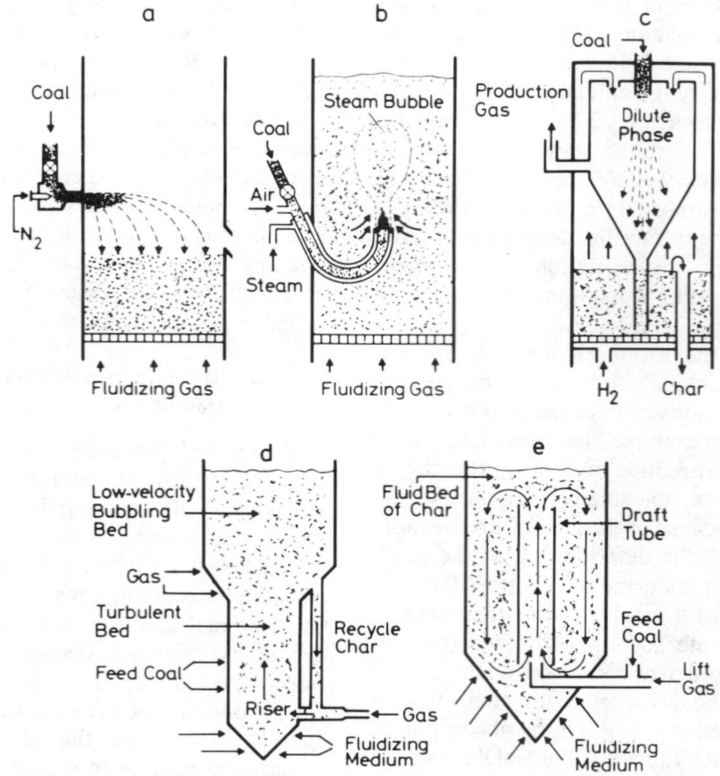

Figure 6.18 Possibilities of feeding a caking coal to a fluidized bed with avoidance of agglomeration: (*a*) feeding to the surface,[254] (*b*) pneumatic feeding with steam jacket,[247,255] (*c*) free-fall, dilute-phase feeding,[256] (*d*) pneumatic feeding to a turbulent fluidized bed,[66] and (*e*) pneumatic feeding with a draft tube.[66]

vertical direction from beneath. Air may be used as the feeding gas. The feeding tube may have a steam jacket, and the steam may leave at the end through a ring with nozzles. The high steam velocity causes a pressure drop at a right angle to the direction of the feed and helps in mixing. If the stream leaves the surface of the fluidized bed, it may entrain a significant quantity of solids.[255]

The use of screw feeders for feeding solids into fluidized beds has been unsatisfactory, at least for strongly caking coals. To decrease the risk of agglomeration, stirrers have also been used in the fluidized bed.[246]

In a new coal hydrogenation process, the hydrane process, a solids feeding method is used, which is based on an invention of R. W. Hiteshue.[257] Pulverized coal does not agglomerate if it flows cocurrently with hydrogen or a hydrogen-methane mixture in free-fall over a distance of only 1.5 m, and if the reactor walls are kept at a temperature above 725°C. Figure 6.18*c* is a schematic diagram of the free-fall dilute-phase solids feeding process. At first the coal drops cocurrently with the gas that has left the fluidized bed in the "dilute phase space," in which its caking ability is reduced to such an extent that it can enter to the fluidized bed by a diptube, without any tendency to agglomerate. The good heat transfer properties of the fluidized bed are used to maintain a constant reaction temperature.[256]

Rapid distribution of the coal can also be achieved by feeding into a turbulent fluidized bed,[248] as shown in Fig. 6.18*d*. In the upper part a bubbling fluidized bed is retained with a relatively low gas velocity, while the lower part is in the state of a turbulent fluidized bed. By proper sizing of this arrangement and without the addition of gas between the lower and upper fluidized beds, the state of a recirculating fast fluidized bed can be

[257] R. W. Hiteshue, *Proc. Synthetic Pipeline, Gas Symp.*, Pittsburgh, (1966), p. 13.

achieved, which also permits rapid distribution of the added coal. In Fig. 6.18*d* the coke recycle is outside the fluidized bed, but it is moved inside in the arrangement shown in Fig. 6.18*e*.[248] This is achieved by putting a tube on top of the coke feeder opening, through which the solid matter is transported upward. This method is difficult to use commercially because such inserts are hard to control, and in large fluidized beds a number of such tubes would be required. In addition, if only one tube is used, the feeding must be achieved through several nozzles. A further problem is the tendency of the fluidizing gas to flow preferentially on the periphery of the tube, so that inside the tube often only a moving bed with fixed extremities is attained.

In the process of preheating coal in the fluidized bed for use in the slot-type coke oven,[128] agglomeration is obviously avoided, since the temperature necessary to retain the caking properties must be below the softening point. Temperatures up to at least 300°C can be controlled in commercial operation without difficulty.[258]

It is not the destruction, but rather the retention of the slight caking ability during heating into the plastic range that is a fundamental problem in the development of continuous processes for the coking of slightly caking coals and coal blends. This retention can be achieved with the equipment that permits use of very high heating rates. In the U.S.S.R. a very promising route has been employed in which high heating rates of 130 K/s are achieved in vortex chambers, because of the high relative velocities between gas and solids.[259] The vortex chambers are horizontal, cylindrical spaces into which the gas enters in such a way that it is uniformly distributed along the circumference in a tangential direction. The solid material is fed in the center in the direction of the axis of the cylinder (e.g., by means of screw feeders) and is blown by the gas stream towards the end of the chamber, where it is withdrawn. To achieve better heat utilization, two-stage vortex chambers are recommended, in which the gas and solid material are carried in a countercurrent direction. The coal leaving stage I is dried and heated to about 200–260°C. In the second stage coal reaches 420–430°C, the final temperature. It can then be separated from the gas stream and introduced into the briquetting press. The flue gas from a preceding combustion chamber burning coke-oven gas with air is used as the heating and mixing gas. In the vortex chamber, solids concentrate in the zone near the wall, whereas the central zone is low in solids. Thus by suitable withdrawal of the gas and separation into solids-rich and solids-lean gas fractions, gas purification can be achieved more economically.

[258] W. Dietrich, M. Engshuber, and E. Fritsche, *Energietechnik*, **19**(4), 168–173 (1969).
[259] P. Z. Shubeko and V. P. Kozyrev, *Coke Chem. USSR*, No. 3, 20–24 (1972).

Using particle sizes up to 3 mm with gas velocities of 40–60 m/s and a free oxygen content in the gas of up to 12%, the plastic range was reached without significant loss of caking ability. To achieve this, a short residence time of the coal and a high relative velocity between gas and solid are needed. The oxygen content should not exceed the given value because of the danger of explosion.

In vortex chambers with diameters of 1.5–2.0 m and lengths of 2.0–2.5 m, coal throughputs of 50–100 tons/hr were achieved.

6.7 SUMMARY

Since the publication of the first Supplementary Volume, considerable progress has been achieved in describing and explaining the plastic behavior of coal during heating. Because of its commercial significance, the process of manufacturing coke was of primary importance in all these investigations. Although important mechanisms of the formation of coke are still unexplained, it has become possible to determine by means of statistical methods the quality and quantity of coking products; measurements of the plastic behavior, made by means of dilatometer tests and microscopic investigations, and their evaluation, have led to results of surprising accuracy.

Improvements in the instrumentation in the past few years have led on one hand to new test methods for the investigation of the plastic behavior, and on the other hand to improvements and simplifications of the existing methods. Interchangeability of test methods has not yet been achieved, since different phenomena are very often measured under conditions that are not comparable. In the past it was possible to observe the plastic behavior of coals only during coking in a laboratory or on a pilot-plant scale (ovens up to 500 kg); however the construction of full-scale testing plants allows the investigation of coke formation under commercial conditions.

The development of new coking processes, especially those with high heating rates, has in addition led to the development of new test methods or has necessitated the modification of existing methods to accommodate the new requirements. Correlation with the measurements obtained under the old conditions was possible in every case.

Although the number of investigations and the improvement of measuring methods have led to distinct progress in the study of the plastic properties of coal, many phenomena are not completely explained. Because of the complex nature of coal, sufficient information about its plastic behavior cannot yet be obtained.

CHAPTER 7

Coal Analyses, Trace Elements, and Mineral Matter

HAROLD J. GLUSKOTER, NEIL F. SHIMP, AND RODNEY R. RUCH

Harold J. Gluskoter *is a Senior Research Associate of Exxon Production Research Company, Houston, Texas. Prior to joining Exxon he was Head of the Coal Section, Illinois State Geological Survey. His personal research interests have been in the areas of mineral matter in coal, trace elements in coal, and inorganic geochemistry of coal. Currently Dr. Gluskoter is involved with a new program of research concerning geology as related to the exploration for and evaluation of coal.*

Neil F. Shimp *is affiliated with the Illinois State Geological Survey, Urbana, Illinois, as Principal Chemist and Head, Chemical Group. His recent research interests are analytical chemistry, trace element geochemistry, coal chemistry, coal analysis, use and disposal of coal solid wastes, and the development of national and international standard methods for the chemical analysis of coal. He was formerly Chairman of ASTM Subcommittee D05.21 on Methods of Coal Analysis. He is now Chairman of ASTM Subcommittee D05.29 on Major Elements in Coal Ash and Trace Elements in Coal, and Secretary of the main ASTM Committee D05 on Coal and Coke. Dr. Shimp is also Chairman of the Secretariat for Working Group 14 on Trace Elements in Coal, which is a part of Technical Committee 27 on Solid Mineral Fuels, International Standards Organization (ISO).*

Rodney R. Ruch *is affiliated with the Illinois State Geological Survey, Urbana, Illinois, as Chemist and Head, Analytical Chemistry Section. His current research interests include radiochemistry and trace element analysis with emphasis on coal and geologic materials. He is a member of ASTM Subcommittee D05.20 on Major Elements in Coal Ash and Trace Elements in Coal and Secretary of International Standards Organization (ISO) Working Group 14 on Trace Elements in Coal, which is part of Technical Committee 27 on Solid Mineral Fuels. Dr. Ruch is also a member of the National Research Council Panel on the Trace Element Geochemistry of Coal Resource Development Related to Health (PECH).*

7.1 CONVENTIONAL ANALYSIS OF COAL

7.1.1 Introduction

Renewed interest in the use of coal to generate electricity and produce liquid and gaseous fuels has stimulated development of new methods of analysis and has changed the meaning of "conventional coal analysis." Nevertheless, the inexpensive methods that have defined the determinations included in a conventional coal analysis will probably remain the most widely used coal standard methods for some time. These traditional methods will provide a basis for comparison of the cost, speed, precision, and accuracy of the new generation of instrumental methods now being applied to coal analysis.

Of the two groups of coal standard methods shown in Table 7.1, one was developed under the jurisdiction of the American Society for Testing and Materials (ASTM) Committee D-5 on Coal and Coke; the other was developed under the jurisdiction of the International Standards Organization (ISO), Technical Committee 27, Solid Mineral Fuels. The first group consists of national standards used in the United States and Canada. The ISO standards have been developed by ISO member nations for international trade, and, in many cases, are similar to the various national standards. These methods and the principles on which they are based have been extensively described and reviewed elsewhere.[1–8] They appear here as a convenient source for their most recent designations and to define for this text the determinations or procedures included in "conventional coal analysis." Nearly all are chemical methods for the analysis of whole coal; a few measurements are physical (e.g., plasticity and swelling tests for coking character), and one, ash fusibility, is performed on coal ash.

[1] *Chemistry of Coal Utilization,* H. H. Lowry; Ed., Supplementary Volume, John Wiley and Sons, 1963, pp. 202–231.

[2] D. E. Shipley *J. Inst. Fuel,* **37,** 510–514 (1964).

[3] J. G. King, *J. Inst. Fuel,* **33,** 550 (1960).

[4] O. W. Rees, Illinois State Geol. Survey Rep. Invest. No. 220, 1966; 55 pp.

[5] U.S. Bureau of Mines, Rep. Invest. No. 638, 1967, 85 pp.

[6] S. Ergun, R. A. Friedel, J. I. McCartney, and I. Wender, in *Encyclopedia of Industrial Chemistry,* Vol. 10, F. D. Snell and L. S. Ettre, Eds., Interscience, New York, 1970, pp. 209–262.

[7] American National Standards Institute, distributor for International Standards Organization Standard Methods (ISO) and recommendations (R).

[8] ASTM Standards, 1978, Part 26, pp. 187–458.

7.1.2 Collection of Coal Samples and Preparation for Analysis

During sampling and preparation, a representative portion of the coal to be studied is obtained. In this procedure a 50-g laboratory specimen may represent an entire 10,000-ton unit train. Overall requirements for collection of a gross sample are given in ASTM D2234 and ISO 1988. The sampling must be undertaken carefully because it is the basis for all subsequent analyses; however the procedure may vary because of the wide variety of coal-handling facilities. Some considerations pertinent to this process have been summarized by Savage (personal communication, K. E. Savage, Commercial Testing and Engineering Co., Chicago, 1977):

1. The "lot" of coal must first be defined [e.g., a single truck (20 tons), a barge (1500 tons), a unit train (10,000 tons), or a ship cargo (100,000 tons)].
2. The number of increments (e.g., the number of shovelfuls required to constitute the gross sample, which is usually 200–500 lb) must be established. For raw, uncleaned, or poorly cleaned coal, the minimum number of increments is 35. For thoroughly cleaned coal (i.e., maximum practical reduction of ash and sulfur), the minimum is 15. This minimum value, N_1, is increased by the lot weight according to the expression:

$$N_2 = N_1\sqrt{\frac{(\text{lot size, tons})}{1000}}$$

   where N_2 is the minimum number of increments for the entire lot. Experience indicates that the maximum size of a coal lot to be represented by one gross sample should not exceed 10,000 tons.
3. The precision used commercially (ASTM D2234) is based on one analytical determination falling within one-tenth of the true value 95 times out of 100. To reduce

Table 7.1 Selected ASTM and ISO Standard Methods for the Conventional Analyses of Coal

Coal Test	Designation and Title	
	ASTM Standards	ISO Standards
Sampling and preparation	D346*M* Collection and Preparation of Coke Samples for Laboratory Analysis D2013*M* Preparing Coal Samples for Analysis D2234*M* Collection of a Gross Sample for Analysis	ISO 1988*M* Sampling of Hard Coal ISO 2309*M* Sampling of Coke
Moisture	D1412*M* Equilibrium Moisture of Coal at 96 to 97 percent Relative Humidity and 30°C D2961*M* Total Moisture in Coal Reduced to No. 8 Top Sieve Size (limited-purpose method) D3173*M* Moisture in the Analysis Sample of Coal and Coke D3302*M* Total Moisture in Coal	ISO 331*M* Determination of Moisture in the Analysis Sample of Coal + (direct gravimetric method) ISO 348*M* Determination of Moisture in the Analysis Sample of Hard Coal (direct volumetric method) ISO 579*M* Determination of Total Moisture of Coke ISO 589*M* Determination of Total Moisture of Hard Coal ISO 687*M* Determination of Moisture in the Analysis Sample of Coke ISO 1015*M* Determination of Moisture Content of Brown Coals and Lingites (direct volumetric method) ISO 1018*M* Determination of Moisture Holding Capacity of Hard Coal
Volatile matter–fixed carbon	D3175*M* Volatile Matter in the Analysis Sample of Coal and Coke	ISO 562*M* Determination of Volatile Matter in Hard Coal and Coke
Ash	D3174*M* Ash in the Analysis Sample of Coal and Coke	ISO 1171*M* Determination of Ash in Solid Mineral Fuels
Proximate analysis	D3172*M* Proximate Analysis of Coal and Coke (gives scope, definition of the proximate analysis, and designation of applicable standards)	
Ultimate analysis	D3176*M* Ultimate Analysis of Coal and Coke (gives scope, definition of the ultimate analysis, designation of applicable standards, and calculations for reporting results on different moisture bases)	

Carbon and hydrogen	D3178*M* Carbon and Hydrogen in the Analysis Sample of Coal and Coke	ISO 609*M* Determination of Carbon and Hydrogen in Coal and Coke by the High-Temperature Combustion Method ISO 625*M* Determination of Carbon and Hydrogen in Coal and Coke by the Liebig Method
Nitrogen	D3179*M* Nitrogen in the Analysis Sample of Coal and Coke	ISO 332*M* Determination of Nitrogen in Coal by the Macro-Kjeldahl Method ISO 333*M* Determination of Nitrogen by the Semimicro Kjeldahl Method ISO 926*M* Determination of Nitrogen, Total Sulfur, Chlorine, and Phosphorus in Coke[a]
Oxygen	See D3176.	ISO 1994*M* Determination of Oxygen in Hard Coal[b]
Sulfur	D2492*M* Forms of Sulfur in Coal D3177*M* Total Sulfur in the Analysis Sample of Coal and Coke (includes three methods: Eschka, bomb washing, and high-temperature combustion	ISO 157*M* Determination of Forms of Sulfur in Hard Coal ISO 334*M* Determination of Total Sulfur in Coal and Coke (Eschka method) ISO 351*M* Determination of Total Sulfur in Coal and Coke (high-temperature combustion method)
Chlorine	D2361*M* Chlorine in Coal	ISO 352*M* Determination of Chlorine in Hard Coal (high-temperature combustion method) ISO 587*M* Determination of Chlorine in Coal and Coke Using Eschka Mixture
Carbon dioxide	D1756*M* Carbon Dioxide in Coal	ISO 925*M* Determination of Carbon Dioxide Content in Hard Coal (gravimetric method)
Calorific value	D2015*M* Gross Calorific Value of Solid Fuel by the Adiabatic Bomb Calorimeter D3286*M* Gross Calorific Value of Solid Fuel by the Isothermal-Jacket Bomb Calorimeter	ISO 1928*M* Determination of Gross Calorific Value of Solid Mineral Fuels by the Calorimetric Bomb Method and Calculation of Net Calorific Value[b]
Plasticity	D2639*M* Plastic Properties of Coal by the Constant-Torque Gieseler Plastometer	ISO 349*M* Audibert-Arnu Dilatometer Test for Hard Coal
Swelling and caking	D720*M* Free-Swelling Index of Coal	ISO 335*M* Determination of Caking Power of Hard Coal (Roga test)

Table 7.1 ***Continued***

Coal Test	Designation and Title: ASTM Standards	Designation and Title: ISO Standards
Swelling and caking	D720*M* Free-Swelling Index of Coal	ISO 501*M* Determination of Crucible Swelling Number of Coal
		ISO 502*M* Determination of Caking Power (Gray-King coke test)
Ash fusibility	D1857*M* Fusibility of Coal and Coke Ash	ISO 540*M* Determination of Fusibility of Coal and Coke Ash
Calculations	D3180*M* Calculating Coal and Coke Analyses from Ash Determined to Different Bases	ISO 1170*M* Calculation of Coal and Coke Analysis to Different Bases
Definitions and terms	D121*M* Definitions of Terms Relating to Coal and Coke	ISO 1213*M* Vocabulary of Terms Relating to Solid Mineral Fuels, Part II: Terms Relating to Coal Sampling and Analysis[b]
	D407*M* Definitions of Terms for Gross Calorific Value and Net Calorific Value of Solid and Liquid Fuels (discontinued 1976; replaced by D121)	ISO 1213*M* Vocabulary of Terms Relating to Solid Mineral Fuels, Part III: Terms Relating to Coke[b]

[a] Future status as ISO Standard uncertain.
[b] Currently being transformed from ISO Recommendation to ISO Standard.

Sources: 1978 Annual Book of ASTM Standards, Part 26 (pp. 187–458; 1976 Catalog, American National Standards Institute, pp. 109–146; International Standards Organization, Technical Committee 27, Solid Mineral Fuels, British Standards Institution, London.

this error by one-half, 4 times as many gross samples must be used.

4. The weight per increment varies according to the top size of the coal:

Pounds per Increment	Top Size (in.)
15	6
6	2
2	5/8

5. Increments must be systematically spaced. Stationary sampling employs a grid system, which may be a simple left front–middle center–right rear grid for samples from a railroad car or a surveyed grid system to take samples from a storage pile. Consideration must be given for variations in the depth of the storage pile.
6. Increments from a moving coal stream are often collected on a preset interval of time by a mechanical sampling device. The opening of the device must be 2 ½–3 times greater than the top size of the coal, and it should accommodate a full stream cut in both directions without disturbing the coal. For example, a mechanical cutter perpendicular to a falling stream of coal should move no faster than 18 in./s. The high rates of transfer (3000–5000 tons/hr) currently being designed for coal require the use of some form of mechanical or automatic samplers.

This method of gross sampling, based on whether the coal is raw or cleaned, is commonly used (see Keller[9] for a more detailed explanation). This method and the top size of the coal form the basis for ASTM and ISO standard sampling methods. A scheme for determining the precision of the various sampling steps as a function of coal preparation, coal top size as sampled, and ash content was tested on 100 low-volatile bituminous coals and lignites and later published in ASTM D2234.[10] From this work, which is summarized in Fig. 7.1, the following conclusions were drawn: (1) sampling variance (V_s) for the raw coals depends on the relative amounts of ash; (2) variance (V_t) is the sum of variances of sampling (V_s), field preparation (V_1), laboratory preparation (V_2), and analysis of ash (V_a), and increases with the ash content of raw coal; (3) the total variance (V_t) of sample preparation ($V_1 + V_2$) and analysis of ash (V_a) is nearly constant and is independent of ash content of washed coals; and (4) preparation variances are higher for raw coal and increase with ash content, especially for the raw coals.

Similar conclusions concerning sampling of run-of-mine and washed coals have been reported by Resler and Riedel, who found that total variance of sampling and analysis was 8–9 times greater for raw than for cleaned coals.[11] Errors were reduced by grinding the entire laboratory sample to < 0.2 mm or by separating 200-g portions after grinding the coal to < 1 mm and grinding these to < 0.2 mm.

Coal sampling from seaborne and rail consignments has been conducted in accordance with an ISO draft proposal.[12] All sampling (including belt, falling stream, grab, colliery, and power station sampling) was unbiased except for truck top sampling. Analyses of the coals for ash content showed that a satisfactory number of increments had been taken. Detailed test results and an outline of statistical principles for the tests were given.

Automatic sampling devices of various types have been described.[13] Automatic sampling of coking coal reportedly eliminates errors resulting from manual sampling and is faster and more precise.[14]

Preparation of coal samples for analysis (as described in ASTM D2013) includes the reduction and division of gross samples into individual portions for laboratory analysis. The gross sample is subjected to air drying, particle size reduction, mixing, and division in stages until a small analysis sample representative of the original gross sample is obtained. The minimum allowable weight of the sample at any stage of the gross sample

[9] G. E. Keller, *Trans. Am. Inst. Min. Metall. Pet. Eng.*, **232,** 218–226 (1965).

[10] S. J. Aresco and A. A. Orning, *Trans. Am. Inst. Min. Metall. Pet. Eng.*, **232,** 258–264 (1965).

[11] J. Resler and R. Riedl, *Sb. Vys. Sk. Chem. Technol. Praze, Technol. Paliv. D,* **24,** 145–167 (1972).

[12] A. Crawford and D. Smith, *J. Inst. Fuel,* 38, 263–273 (1965).

[13] T. E. Tibbets, S. F. Featherby, and R. I. Miller, *Can. Min. Met. Bull.*, **58,** 635–640 (1965).

[14] P. Morhauer, *Glueckauf,* **108,** 469–471 (1972).

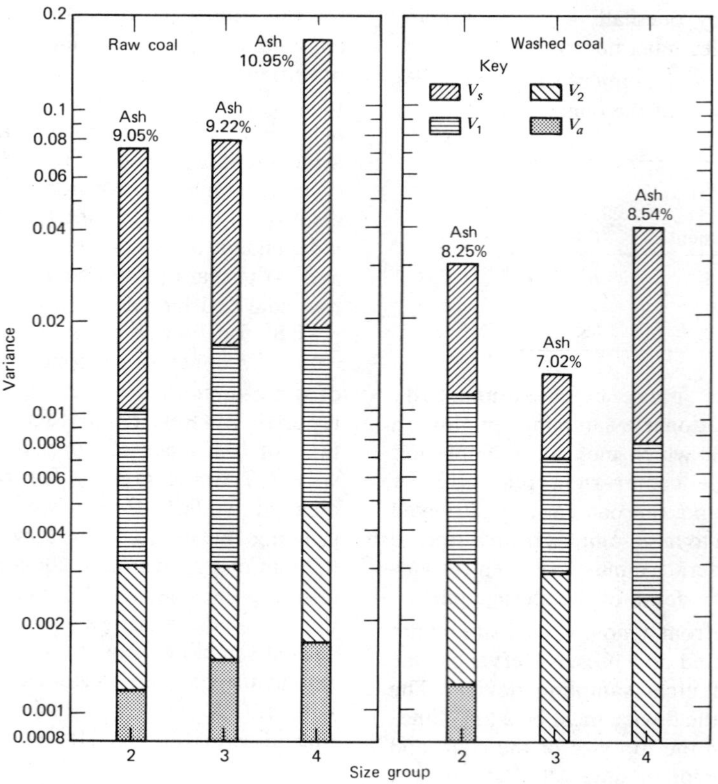

Figure 7.1 The dependence of coal sampling, sample preparation and analysis upon coal ash content. After Aresco and Orning.[10]

preparation, as prescribed by ASTM, is given in Table 7.2. Preparation of the laboratory analysis sample for moisture determination as specified by ASTM is shown in Fig. 7.2.

Results of studying analysis of coal microsamples (milligram amounts) indicate the possibility of obtaining acceptable results if a precision microbalance is used and if coal is ground to pass through a sieve finer than 200 mesh (74 μm).[15]

Storage of laboratory coal samples for subsequent analysis may be considered a part of proper sample handling. Long-term storage without change was achieved by placing the samples in a plastic bag containing dry ice, sealing them tightly in glass beakers, and storing them under vacuum.[16] Normally oxidation and deterioration of 60-mesh laboratory samples stored in air increase with decreasing particle size and decreasing rank of coal.

7.1.3 Moisture Analysis

The forms of moisture in coal and the relationship of inherent or natural bed moisture to classification of coal by rank on the moist-mineral-matter-free basis have been reviewed extensively by Krumin.[17] Krumin also reviewed the development of methods for the determination of equilibrium moisture in coal defined as the amount of moisture contained in coal at 30°C in a relative humidity of 96–97%. Test results were based on an equilibrium moisture method, similar but not identical to the method used in ASTM D1412, for which specifications were made regarding pressure (vacuum) of equilibrium chamber, saturated salt solution (K_2SO_4) for use in

[15] T. Fujimore and K. Ishikawa, *Fuel,* **51,** 120–124 (1972).

[16] G. Krafft and K. John, *Glueckauf,* **111,** 909–910 (1975).

[17] P. O. Krumin, Ohio State Univ. Eng. Exp. Sta. Bull. No. 195, 1963, 78 pp.

Table 7.2 Preparation of Laboratory Coal Sample

If moisture sample is required, the quantity of No. 4 or No. 8 sieve subsample should be increased by 500 g.

Crush to Pass at Least 95% Through Sieve	Minimum Weight for Division into Subsample (g)	
	Cleaned Coal	Uncleaned Coal
No. 4 (4.75 mm)	2000	4000
No. 8 (2.36 mm)	500	1000
No. 20 (850 μm)	250	500
No. 60 (250 μm) (100% through)	50	50

Source: Modified from ASTM Standards, 1978, Part 26, Designation D2013, p. 278.

attaining constant relative humidity, size of equilibration chamber, condition of coal to be tested (e.g., moisture content and particle size), equilibration time, and effect of hygroscopic salts in the coal.

Edwards[18] made a similar, 10-year investigation of methods for the determination of equilibrium (bed) moisture. Results included data from international testing programs, and he preferred a rapid method to the more lengthy ISO procedure.

No absolute method for determining the true moisture content of coal by heating has been devised because of the wide variance in the temperatures at which different forms of moisture (including surface, inherent, chemically combined moisture in the coal) and water of hydration in clay minerals are liberated. Abernethy et al. stressed the empirical nature of methods for determining the amount of moisture in coal.[19] In one method weight loss of coal was measured after heating to 106°C in a flow of dry oxygen-free nitrogen. Although two other methods (weight loss in air and direct measurement of moisture released in a nitrogen atmosphere), also at 106°C, yielded results within ASTM tolerances, those performed in a nitrogen atmosphere gave consistently higher results because heating in air sometimes oxidizes coal and offsets the moisture loss through addition of oxygen. Measuring indirect weight loss after heating in nitrogen at 130°C for 30 min with a flow rate of 90 oven-volume changes per hour was judged the most convenient method.

Moisture loss during pulverization is reportedly related to the kinds and amounts of banded ingredients in the coal.[20] Losses are least for vitrain and increase in order of vapor pressures for moisture (e.g., clarain < durain < fusain).

Roth has reviewed 10 methods for the measurement of moisture in solids, including intermittent methods (mechanical adaptation of conventional laboratory procedures such as multiple-sample moisture balance using a heat source, automatic Karl Fisher titration, and vapor pressure equalization) and continuous instrumental methods such as capacitance meter (dielectric constant), rf power absorption, microwave attenuation, and nuclear reflection.[21] Description of operating principles, specificity, speed, range, and applicable materials are included.

The significance of moisture in coal and methods for its determination have been extensively reviewed.[22] One review suggests that the best technique is to determine loss

[18] A. H. Edwards, *Fuel,* **47,** 29 (1968).

[19] R. F. Abernethy, E. G. Tarpley and R. A. Drogowski, U.S. Bur. Mines Rep. Invest. No. 6440, 1964, 12 pp.

[20] O. W. Rees, *Talanta,* 1027–1032 (1966).

[21] M. Roth, *Chem. Eng.* **73,** 83–88 (1966).

[22] H. G. Schaefer and H. Jansen, *Erodoel Kohle, Erdgas, Petrochem. Brennst.-Chem.,* **27,** 436–440 (1974).

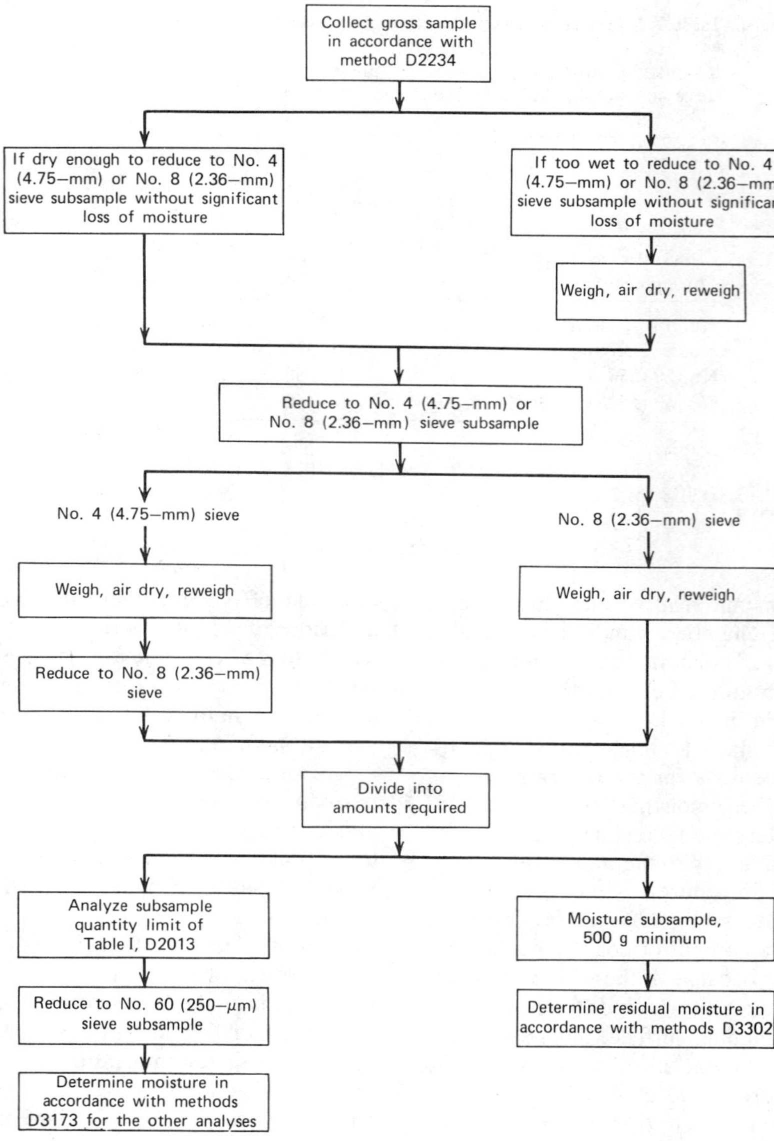

Figure 7.2 Gross sample preparation for moisture determination (nonreferee method). After ASTM Standards, 1977, Part 26, Designation D2013, p. 282.

during air drying (ASTM D3302) followed by codistillation of moisture with xylene.[23] Alternatively, moisture should be determined in an oven swept with dry nitrogen or another inert gas (ISO 589).

Two methods for the determination of the amount of moisture in coke have been investigated: (1) extraction of coke using anhydrous methanol and addition of CaH_2 from which the amount of released heat is measured; and (2) extraction of coke using anhydrous dioxane and measurement of the refractive index of the solution to determine its water content. Only the first method was found to be suitable.[24]

In determining moisture content in coke,

[23] H. D. Held, *VGB Kraftwerkstechnik*, **56**, 11D–16 (1976).

[24] J. Nerser and M. Kaloc, *Sb. Ved. Pr. Vys. Sk. Banskedstrawe, Roda Hutn.*, **19**, 129–140 (1973).

the temperature must be increased to at least 300°C.[25] Temperatures of 100–110°C are not high enough to remove all moisture.

It has been difficult to apply indirect weight loss procedures to the determination of moisture content in low-rank coals that have high moisture content and susceptibility to oxidation. Bull[26,27] reported that the British Standard (BS 1016, Part 3, 1957) minimum-free-space oven proved inadequate for determining the amount of moisture in analysis samples of brown coal. Small amounts of moisture were still released from samples that had been heated for 48 hr in a stream of nitrogen, and Bull observed no sharp cutoff point. He recommended removal of water from low-rank coal by distillation with toluene (a procedure similar to the one used in ISO 1015, in which a well-defined moisture cutoff is reached). Burns and Swaine[28] overcame these problems with brown coals by heating in silica dishes, doubling the nitrogen flow rate to 30 oven-atmosphere changes per hour, and extending the usual drying time of 1 hr for bituminous coals to 3 hr for brown coals. They achieved a well-defined cutoff point for moisture release without significant oxidation of the coal, and the results corresponded to results from tests in which the toluene distillation method was used. For this reason, ISO recommends use of toluene or xylene distillation methods for brown coals or direct measurement of the evolved moisture (ISO 331).[29]

It is clear that to achieve reproducible moisture values from low-rank coals, adherence to strict specifications is necessary. These specifications include fixed drying time rather than drying to constant weight, specified range of oven temperature, fixed rate of nitrogen flow, maximum oven loading, and stringent cooling procedures to exclude moisture.[30] Two methods that have been considered for the determination of moisture content in low-rank coals are measurement of weight loss caused by drying in a tube subject to a vacuum, which reportedly yields a sharp moisture cutoff,[30] and measurement of electrical capacitance,[31] which has been related to moisture content of brown coals containing up to 27.3% moisture (relative error 6.6%).

Continuous measurement of moisture in bituminous coal has been accomplished using a capacitance method in which coal moves between the plates of a capacitor.[32] The lower plate has a keel that extends into the stream. The dielectric constants, which are very different for dry coal and water, change in proportion to the moisture content of a particular coal. When a calibration curve is prepared for each coal tested, the stated accuracy for moisture content is ±1% absolute.

Continuous measurement of moisture in coal has also been accomplished by using a neutron moisture meter[33,34] to measure thermal or slow neutrons produced by irradiation of coal with a radioisotope (americium-beryllium) source. The thermal neutrons can be directly related to the hydrogen content of both bound and free moisture in coal. The free moisture in a single coal varies by only about 0.1% on a moisture-free basis, so suitable calibrations can be plotted for any single coal. For bituminous coals from the Pittsburgh Seam containing 4–10% moisture and 5–7% ash, values obtained over a 12-month period by the neutron moisture meter were within 0.2% of those obtained by ASTM standard methods.

A nuclear magnetic resonance (NMR) method for the continuous measurement of moisture in coal and coke is based on the NMR absorption spectra of hydrogen; this element produces a curve of intensity of sample absorption against the field strength.[35] The spectra are caused by combinations of hydrogen in water and hydrogen in the solid. The peak-to-peak line-width ratio of the NMR first derivative spectrum is related to inherent water plus free water (total) when a

[25] H. C. Wilkinson, *Fuel*, **44**, 191–198 (1965).
[26] F. A. Bull, *Fuel*, **42**, 203–210 (1963).
[27] F. A. Bull, *Fuel*, **44**, 161–163,165–168 (1965).
[28] M. S. Burns and D. J. Swaine, *Fuel*, **42**, 505–506 (1963).
[29] W. McMellan, *Fuel*, **44**, 308 (1965).
[30] D. G. Evans, *Fuel*, **44**, 92–94 (1965).
[31] A. Dumkowski, J. Kreska, and J. Golda, *Gorn, Odkrywkowe*, **17**, 102–106 (1975).
[32] D. A. Hall, G. F. Morris, and C. Scott, *Min. Min. Eng.*, **5** 30–40 (November 1969).
[33] R. F. Stewart and A. W. Hall, *Trans. Am. Inst. Min. Met. Pet. Eng.*, **238**, 269–272 (1967).
[34] A. W. Hall, J. L. Konchesky and R. F. Stewart, U.S. Bur. Mines Rep. Invest. No. 7807, 1973, 18 pp.
[35] W. R. Ladner and R. Wheatley, *J. Inst. Fuel*, **39**, 280–284 (1966).

sample of constant volume is used. A single calibration was found suitable for low-ash coals regardless of coal rank over a range of 3–30% moisture. Determination of moisture in high-ash coals (up to 50% ash) yielded results similar (within ±1% absolute) to moisture values determined by conventional weight loss methods. In coke, however, as amounts of ferromagnetic material increased, separate calibrations were necessary.

Another NMR study[36] described the application of a rapid-recovery, direct-current amplifier system for the determination of protons in coal and the conditions necessary for quantitative analysis. Rapid determinations were made for seven coals, and the results obtained agreed with chemical values.

Results of laboratory studies of microwave attenuation for continuous moisture measurements in coal warrant the further testing of this method in plant trials.[37] Although the calibration curve varies from one type of coal to another, a useful relationship between microwaves and the moisture content of coal has been found. Changes in characteristic microwave absorption in different coals are rank dependent and are probably associated with petrographic composition and pore structure.

Hampel has reviewed seven methods for the continuous measurement of moisture in coal: (1) electrical conductivity, (2) dielectric constant, (3) microwave attenuation, (4) neutron scattering, (5) nuclear magnetic resonance, (6) infrared, and (7) thermal conductivity.[38]

7.1.4 Volatile Matter Analysis

Volatile matter—as determined, for example, in ASTM D3175 or ISO 562—consists of the percentage of gaseous products, exclusive of moisture vapor, released during the heating of coal or coke under rigidly controlled conditions. The measured weight loss of the sample corrected for moisture establishes the content of volatile matter.

Factors affecting the determination of volatile matter, including coal temperature, rate of heating, size of furnace, and number and arrangement of crucibles for multiple determinations, were studied in relation to a revision of British Standard 1016, Parts 3 and 4.[39] Volatile matter results comparable to those obtained using the ISO method (ISO 562) for single samples were obtained from four determinations in a small furnace at 900°C.

The standard crucible method used for coal is unsuitable for coke because the temperature (850°C) at which the test is conducted is too low.[40] A thermogravimetric procedure in nitrogen at 1000°C for 30 min produced yields of volatile matter characteristic of various cokes.

Two detailed studies of the release of volatile matter in coal were reported by Chatterje.[41,42] The first of these studies deals with the relationship between the number of volatiles released by coals during rapid pyrolysis, the temperature to which they were subjected, and the ratio of hydrogen to carbon. The ratio measures the abundance of hydrogen on the carbon-linked structure of coal; hydrogen is known to be closely related to the amount of volatile material. Beginning at 200°C, volatile matter was released steadily up to 1000°C, at which point evolution was virtually complete. A simple linear relationship was found to exist between volatile matter yield at specified temperatures and the H/C ratio. Partially devolatilized coals yielded the calculated additional amount of volatile matter when heated to a higher temperature. The devolatilization process of a coal was related to the H/C ratio throughout the whole temperature range; coals having the same H/C value usually had the same basic structure and distribution of hydrogen in the structural unit.

In the second study[42] it was found that the quantity of volatiles released by coal during slow heating (beginning at room temperature) differed from the amounts released by fast heating (i.e., the instant application of final temperature). When the coal was heated slowly, the volatile matter yield was de-

[36] B. C. Gerstein and R. G. Pembleton, *Anal. Chem.*, **49,** 75–77 (1977).

[37] D. A. Hall, J. C. Sproson, and W. A. Gray, *J. Inst. Fuel,* **43,** 350–354 (1970).

[38] M. Hampel, *Glueckauf,* **110,** 129–133 (1974).

[39] D. E. Shipley, *Fuel,* **45,** 283–294 (1966).

[40] M. F. Kessler and L. Dockalova, *Freiberger Forschungsh. A,* **388,** 75–92 (1967).

[41] D. R. Chatterje, *J. Mines Met. Fuels,* **20,** 79–86 (1972).

[42] D. R. Chatterje, *J. Mines Met. Fuels,* **20,** 112–120 (1972).

creased and fixed carbon in the char was increased. The author speculated that a slow increase of temperature and pressure resulting from shearing or folding in a coal seam could cause an increase in the fixed carbon content with no consequent change in the H/C ratio. Thus the coal might show a higher H/C ratio than expected for the amount of volatile matter it contained. This type of slow heating of coal in place could explain certain differences encountered in the relationship of volatile matter to temperature developed for rapid heating, and it could explain improved coal quality. The study also treated effects of thermal metamorphism on the content of volatile matter in coal seams.

7.1.5 Ash Analysis

Coal ash is the inorganic residue remaining after high-temperature ignition of combustible material, as determined by prescribed methods (ASTM D3174 or ISO 1171). Coal ash is composed primarily of oxides and sulfates, and it should not be confused with mineral matter, which is composed of the unaltered inorganic minerals in coal. (See Section 7.3, "Mineral Matter in Coal," for further discussion.)

The retention of sulfur as $CaSO_4$ in laboratory coal ash (900°C) was studied[43] in relation to the amounts of calcium and sulfur in whole coal and the ashing temperature to which it was subjected. Evidence indicated that iron oxides formed during oxidation of pyrite contributed to the catalytic oxidation of sulfur dioxide (SO_2) to sulfur trioxide (SO_3), which reacts with calcium to form calcium sulfate ($CaSO_4$) during the ashing process. The amount of sulfur retained in power plant ash formed at temperatures exceeding 1000°C was found to be directly related to the concentration of unburned carbon.

A new method of ashing coal, low-temperature ashing (LTA) or plasma ashing, was first applied to the study of mineral matter in coal by Gluskoter.[44] When oxygen is passed through a radiofrequency field, a discharge occurs and forms an activated gas plasma which flows over the coal sample. Oxidation of the organic matter occurs at relatively low temperatures (~150°C) leaving the coal minerals (e.g., pyrite, calcite, clays) relatively unaltered. The low-temperature ash is much more useful than conventional coal ash for many purposes and can be used to correct various coal determinations to a mineral-matter-free basis.

A number of reports[45–47] have been published on continuous monitoring of the ash content of coal and coke using gamma rays from a radioactive ^{241}Am source. In this method the amount of forward scattering of gamma rays produced by a moving coal stream is proportional to the ash content, which may be automatically monitored by a digital analyzer. The method is very rapid for large tonnages of coal, and favorable comparisons of results with those from conventional ash determinations have been made.

Considerable work has also been reported on the use of X-rays for the continuous monitoring of ash in coal. Hudy and Deurbrock[48] evaluated a continuously recording X-ray ash meter that monitored the ash content of product coal from seven preparation plants. The apparatus was capable of preparing a suitable coal sample that was delivered to the X-ray unit for automatic analysis. The ash was then measured to within ±0.5% absolute at the 95% confidence level. This method is based on the measurement of reflected X-rays (backscatter). The absorption coefficient of the X-rays provides a linear correlation with ash content. Variation of the iron content in coal ash does not interfere with the ash determination when 1.79-Å X-rays are measured.

Similar studies using backscattered X-rays for the continuous determination of ash in coal have confirmed the capabilities of the method.[49–51] In addition, the forward scatter-

[43] O. W. Rees, N. F. Shimp, C. W. Beeler, J. K. Kuhn, and R. J. Helfinstine, Illinois State Geol. Survey Circ. No. 396, 1966, 10 pp.

[44] H. J. Gluskoter, *Fuel*, **44**, 285–291 (1965).

[45] F. W. Merhoff, Rep. No. EURISOTOP-52, NTIS, Springfield, Va. (1971) (in German).

[46] A. A. Rudanovskii, A. G. Vasil'ev, and A. M. Onischenko, *Izv. Vyssh. Uchebn. Zaved. Gorn. Zh.*, **16**, 1977–180 (1973) (in Russian).

[47] A. G. Lokshin, *Koks Khim.*, **4**, 11–12 (1973).

[48] J. Hudy Jr. and A. W. Deurbrouck, U.S. Bur. Mines Rep. Invest. No. 7101, 1968, 20 pp.

[49] C.G. Clayton, *Bull. Inf. Assoc. Tech. Energy Nucl., Suppl. 1972*, No. 93, 4-13.

[50] B. B. Ruzin, *Metall. Chem. Met.*, **1**, 162–166 (1973).

ing of beta rays from radioactive ^{90}Sr and ^{90}Y has been used to determine the ash content of coal.[52,53] Measurement of the beta rays is made with a scintillation detector that produces a counting rate directly proportional to ash content.

Energy dispersive X-ray fluorescence was employed to analyze coke samples for the major ash-forming elements contained in Al_2O_3, SiO_2, K_2O, CaO, and Fe_2O_3.[54] Sulfur and other minor and trace elements were also determined. Thus a coke sample was analyzed for all these elements simultaneously, and the total ash content and sulfur concentrations were calculated from this analysis. Similarly, the silica and alumina concentrations of whole coals have been used as a basis for the determination of the ash content of Belgian coals.[55] In this study equations for the calculation of ash were developed.

7.1.6 Proximate Analysis

Methods dealing with single determinations in the proximate analysis of coal are reviewed under their respective headings and are not repeated here. For reference to applicable determinations, see Proximate analysis, ASTM D3172, in Table 7.1.

Use of the thermogravimetric balance has been extended to cover determinations of moisture and ash as well as determinations of volatile matter in coal. Barbiroli[56] analyzed both coal and coke for these constituents by pyrolyzing the samples in oxygen and air. The weight loss at prescribed temperatures was measured by using a thermobalance.

Fixed carbon and percentages of moisture, volatile matter, and ash were measured in coal in less than 13 min using a thermogravimetric analyzer.[57] The system uses an automatic recording microbalance that has a sensitivity of 0.1 μg and a capacity of 5 g. Rapid cooling from 1000°C to room temperature can be attained in 15 min. For coal analysis, the sample is purged with nitrogen at 100°C to dry the sample and remove oxygen. The percentage of moisture (weight loss) is read at this point. The loss of volatile matter can be measured by heating the sample to 600°C at 160°C/min and heating at 600°C for 2 min; switching the purge gas to oxygen while heating to 900°C yields a residual weight equivalent to the ash content. Results from tests using this method have not yet been compared with results from tests using standard coal procedures.

7.1.7 Ultimate Analysis

Single determinations in the ultimate analysis of coal are reviewed under their appropriate headings if enough reports have been published to warrant inclusion. To avoid repetition of references, the papers concerning ultimate analyses that deal with multiple determinations are reviewed here. For reference to applicable determinations, see Ultimate analysis, ASTM D3176, in Table 7.1.

The carbon, hydrogen, and oxygen contents of coal have been determined on the basis of the content of carbon dioxide and water in the combustion products.[58] A graph for complete combustion of a fuel has been constructed on the basis of interrelationships among air and oxygen demands, theoretical weight of combustion products, amount of moisture, maximum and actual carbon dioxide content of dry combustion products, and the heat value for the fuel. The fuel composition was deduced from this graph, and its accuracy for control purposes and comparison of fuels of different types was determined.

Carbon and sulfur in a single coal sample have been determined by combustion in oxygen in an induction furnace.[59] Carbon was determined gravimetrically, and sulfur, titrimetrically.

Methods for the determination of nitrogen, carbon, hydrogen, oxygen, and sulfur in coal

[51] Anon., *Coal Age,* **81** (7), 203–204 (1976).
[52] K. S. Klemper and V. V. Parkhomenko, *Zavod. Lab.,* **40,** 421–423 (1974).
[53] M. Hampel, *Glueckauf,* **111,** 175–176 (1975).
[54] Ortec, Ortec TEFA Analysis Report of work at Monsanta Industrial Chemical Co., Columbia, Tenn., July 15, 1975, 15 pp.
[55] C. Block and R. Dams, *Bull. Soc. Chim. Belg.,* **83** (11–12), 457–462 (1974).
[56] G. Barbiroli, *Mikrochim. Acta,* **1,** 26–34 (1968).
[57] R. L. Fyans, *Instrum. News* (Perkin Elmer Corp.), **26:**2, 4–5 (1976).
[58] S. Kaspyzyk, *Arch. Procesow Spalania,* **3,** 59–71 (1972).
[59] L. L. Craddock, *W. Va. Acad. Sci. Proc.,* **45,** 272–282 (1973).

have been reported by Thuerauf.[60] The element nitrogen was determined using a thermal conductivity cell on the gases produced by combustion of coal in oxygen. Carbon dioxide, water, and sulfur dioxide were determined in the oxygen-combustion gases by infrared absorption. Oxygen evolved as CO was determined by infrared absorption on a separate sample heated in a nitrogen atmosphere.

At least one automated elemental analysis system that may be suitable for coal analysis is being manufactured.[61] Twenty three samples for simultaneous determination of carbon, hydrogen, and nitrogen can be accommodated; oxygen is determined on a separate combustion system. The manufacturer claims that this system will complete fully automatic pyrolization (1600°C), conversion, detection, and data processing in 4 hr. The system converts catalytically to N_2, CO_2, and H_2O, which are separated with a Poropak-O gas chromatographic column and detected through a dual thermal conductivity cell. Sulfur can be determined using a similar system that excludes the oxygen determination. Applications to analyses of coal have not been published.

7.1.8 Nitrogen Analysis

The semimicro Kjeldahl method (ISO 333) has become a widely used method for determining nitrogen in coal and coke,[62] although there is some question concerning the complete recovery of nitrogen in coke by this procedure.

The fusion of coal with lithium metal in a bomb to form lithium nitride, which is converted to ammonia in a phosphoric acid solution, has been reported to be an excellent reference method for nitrogen.[63]

Nitrogen determination by perchloric acid digestion of coal without catalysts is reportedly safe and much more rapid than the conventional Kjeldahl procedure.[64] Nitrogen was determined in the digest by the usual ammonia evolution and titration. Explosions during the digestion were prevented by additions of at least 15 ml of perchloric acid for every 0.2 g of sample. Extreme care should be taken whenever perchloric acid is used.

Nuclear reactions have been successfully applied to the determination of nitrogen in coal.[65] Coal has been irradiated with fast neutrons from a 14-MeV generator to produce ^{13}N, which decays via positron emission and has a 10-min half-life. Standard deviation for the nitrogen determinations was 0.07%. The instrumental neutron activation values were 6.8% higher (~0.1% N) on the average than those obtained using the standard Kjeldahl procedure. In the Kjeldahl procedure, loss of elemental nitrogen or nitrogen oxides or resistant heterocyclic nitrogen compounds can sometimes cause values to be low. The fate of nitrogen in the Kjeldahl method depends on its chemical form in the coal, the inorganic compounds added to catalyze the hydrolysis, and the amounts and types of compounds used to raise the boiling point of the mixture. Because results from instrumental neutron activation for nitrogen are independent of these factors, it has been suggested that the activation method is the more accurate of the two.[65]

7.1.9 Oxygen Analysis

Since 1960 many reports have been published concerning the direct determination of oxygen in coal; however the principal method used to estimate the oxygen content of coal is still to subtract from 100 the sum of the percentages of hydrogen, carbon, nitrogen, sulfur, moisture, and ash. All errors in these determinations and errors due to weight changes in the coal mineral matter are included in the oxygen-by-difference estimate. Unless a number of corrections and approximations similar to those described by Rees[4] are applied to such oxygen values, appreciable error in the oxygen value is possible. More accurate methods such as low-temperature ashing and fast-neutron activation procedures have only recently become available for the direct determination of organic oxy-

[60] W. Thürauf, *Erdoel Kohle, Erdgas, Petrochem. Brennst.-Chem.*, **27**, 135–138 (1974).

[61] Sanda, Inc., 4005 Gypsy Lane, Philadelphia, personal communication (1976).

[62] D. E. Shipley, *Fuel*, **37**, 510–514 (1964).

[63] V. W. Radmacher and A. Hoverath, *Glueckauf*, **96**, 1146 (1960).

[64] N. K. Prasad and S. S. Asthana, *Chem. Ind.*, **9**, 262–263 (1969).

[65] W. D. James, W. D. Ehman, E. E. Harmin and L. L. Chyi, *J. Radioanal. Chem.*, **32**, 195–205 (1976).

gen in coal. Moreover, recent investigations of coal conversion processes have emphasized the need for accurate oxygen data so that required amounts of steam or hydrogen can be calculated.

Because oxygen in coal includes both oxygen in organic matter and inorganic oxygen from mineral matter and moisture, demineralization or float-sink cleaning of coal is usually employed to remove inorganic oxygen. Using a modified Unterzaucher method similar to ISO methods R 1994, Oxygen in Coal, and R 602, Determination of Mineral Matter in Coal,[66] Abernethy and Gibson[67] analyzed both untreated and demineralized (HCl and HF extracted) coals for oxygen. Results of the direct determination of oxygen in untreated coals showed significant variations from oxygen values obtained by the difference method. The differences were primarily caused by SO_3 retained in the ash, sulfur in the pyritic form in the whole coal, and small quantities of inorganic oxygen. Much closer agreement was obtained when ash values used for calculation of oxygen by difference were corrected to a better approximation of mineral matter:

$$\text{corrected ash} = \text{ash}_{\text{determined}} - SO_{3\text{ash}} + 0.629\ S_{\text{pyritic}}$$

where pyritic sulfur in whole coal is added to the ash value, organic sulfur is added in the ultimate analysis as part of the coal substance, and CO_2 from carbonates is added to total carbon and oxygen values. Demineralized coals gave the best results for organic oxygen; values from untreated coals were usually higher because they included inorganic oxygen.

Other studies demonstrating the application of direct oxygen determinations to coal and coke have been published by Burns et al.,[68] Radmacher et al.,[69] Crawford et al.,[70] and Kirk et al.[71]

The use of gas chromatography to determine oxygen in coal[72] is considerably faster than the methods just described. The oxygen produced from coal pyrolysis in a vacuum was catalytically converted to carbon monoxide, and the total quantity of gas was measured. Gas chromatography was then used to measure the concentration of CO in the gas. Results were in good agreement for both whole and demineralized coals, and corresponded well with neutron activation results.

Kinson and Belcher[73] have developed a method employing reduction fusion using radiofrequency heating of coal for the determination of oxygen in coal. An induction furnace was used to heat an iron-tin bath in a graphite crucible supported in a bed of carbon black powder that was contained in a silica thimble. All were enclosed in a reaction tube through which argon carrier gas flowed. Pyrolysis products from a sealed coal sample (10 mg) dropped into the bath were purified, and oxygen was converted to CO_2, which was absorbed and weighed. Satisfactory results were obtained for organic oxygen in demineralized coal. For untreated coal, satisfactory results were obtained when a range of correction techniques for oxygen was applied to coals containing more than 5% ash. A standard deviation for the method of 0.1% absolute was attained.

The application of nuclear techniques to direct determination of oxygen and other constituents in coal is relatively new. Both oxygen and carbon have been determined by measuring the intensity and energy levels of gamma rays produced from the fast neutron irradiation of coal.[74] A fast-neutron generator was employed by Block and Dams[75] in the analysis of coal for total oxygen and silica, which was determined almost without interference by measuring the nitrogen nuclide from the reaction $^{16}O(n,p)^{16}N$.

Low-temperature ashing has been used

[66] ISO R 1994, American National Standards Institute, 1971.

[67] R. F. Abernethy and F. H. Gibson, U.S. Bur. Mines Rep. Invest. No. 6753, 1966, 21 pp.

[68] M. S. Burns, R. Macara, and D. J. Swaine, *Fuel,* **43,** 349–356 (1964).

[69] V. W. Radmacher and A. Hoverath, *Brennstoff-Chemie,* **40,** 97–104 (1959).

[70] A. Crawford, M. Glover and J. H. Wood, *Mikrochim. Acta,* **1,** 46–58 (1961).

[71] B. P. Kirk and H. C. Wilkinson, *Talanta,* **17,** 475–482 (1970).

[72] B. S. Ignasiak, B. N. Nandi and D. S. Montgomery, *Anal. Chem.,* **41,** 1676–1678 (1969).

[73] K. Kinson and C. B. Belcher, *Fuel,* **54,** 205–209 (1975).

[74] T. C. Martin, S. C. Mather, and I. L. Morgan, *Int. J. Appl. Radiat. Isot.,* **15,** 331-338 (1964).

[75] C. Block and R. Dams, *Anal. Chim. Acta,* **71,** 53–65 (1974).

with neutron activation analysis (NAA) of coal for the determination of organic oxygen and has been shown to be as effective in pretreatment for the determination of organic oxygen in coal as HF and HCl.[76,77] James et al.[65] used a 14-MeV neutron generator to induce the $^{16}O\,(n,p)\,^{16}N$ reaction in coal and low-temperature ash and counted the ^{16}N photopeaks. Organic oxygen was calculated from the following expression:

$$O_{organic} = O_{total} - O_{inorganic} \times \text{g LTA/g coal}$$

In a series of papers Volborth and coworkers studied the determination of oxygen in coal and coal ash by fast-neutron activation.[78–81] Discrepancies in comparing results for determined oxygen to those calculated by difference were found in coal. By determining oxygen content in coal samples as received and after they had been dried by heating to 105°C, Volborth inferred whether weight changes were from moisture only or included changes from oxidation of the coal or evolution of volatile gases. This is of considerable importance in studying easily oxidized, low-rank coals. Resulting moisture values in such cases were usually low, but the values for hydrogen calculated to the dry basis (see ASTM D3180, ISO 1170) were high, yielding low oxygen values when it was calculated by difference. For most of the coals analyzed, directly determined oxygen values (Table 7.3, column 3) were higher than those calculated by the conventional difference method (column 2). Oxygen-by-difference values calculated from the mean content of oxygen in coal ash (column 4) were in closer agreement with determined oxygen values. Closer agreement of determined and calculated oxygen values was also achieved when various formulas, such as the Parr or King-Maries-Crossley formulas,[4] were used to convert percentage ash (column 5) to percentage mineral matter for subsequent use in calculating oxygen by difference. It was recommended that the actual percentage of mineral matter, as determined in low-temperature plasma ash of coal, be used to achieve better stoichiometric summations in the ultimate analysis of coal. All the high-sulfur coals analyzed gave low summations when oxygen was determined separately, even though sulfur forms were accounted for by reporting them as other than organic. This was interpreted to mean that a part of the inorganic sulfur in some way competes for

[76] Y. Hasegawa, S. Ueda, Y. Maekawa and T. Sato, *Nenryo Kyokai-shi,* **52.** 917–925 (1973); *C.A.,* **80a,** 147573.
[77] C. E. Hamrin, Jr., P. S. Maa, L. I. Chyi and W. D. Ehrman, *Fuel,* **54,** 70–71 (1975).
[78] A. Volborth, G. E. Miller, C. K. Garner, and P. A. Jerabek, *Fuel,* in press.
[79] G. E. Miller and A. Volborth, *Proc. 4 Conf. Applied Small Accelerators,* Denton, Tex., 1977, in press.
[80] A. Volborth, G. E. Miller, C. K. Garner, and P. A. Jerabek, *Fuel,* **56,** 205–208 (1977).
[81] A. Volborth, G. E. Miller, C. K. Garner, and P. A. Jerabek, *Fuel,* **56,** 209–215 (1977).

Table 7.3 Oxygen Determined by Difference as Reported in Classical Coal Analysis Versus Oxygen Determined by Neutron Activation Analysis, and Oxygen Recalculated[70] (Assuming Ash Contained 46% O)

Coal No.	Oxygen as Difference, Reported by U.S. Bureau of Mines "As Received" (%)[a]	Oxygen Determined by NAA (%)	Oxygen as Difference, Calculated (%)[b]	Ash (%)
K-46566	36.6	42.34	44.34	7.70
K-46218	32.3	35.90	37.82	5.50
K-46217	31.7	36.15	37.89	6.20
K-46216	33.2	36.85	38.13	4.90
K-46565	38.8	43.28	44.19	5.40

[a] 100−(ash + C + H + N + S) = oxygen.
[b] 100−(ash × 0.46) + C + H + N + S = oxygen, where mean oxygen content of ash is 46%.

oxygen sites and perhaps has an organic attachment (see also Mössbauer spectroscopy, Section 7.1.11, "Forms of Sulfur").

7.1.10 Total Sulfur Analysis

Shimp et al.[82] reviewed methods for the determination of total sulfur in coal, including methods for Eschka fusion (ASTM D3177, ISO 334), bomb washing (ASTM D3177), high-temperature combustion (ASTM D3177, ISO 351), the peroxide bomb method, the induction furnace method, and X-ray fluorescence. Mohrhauer[83] compared sulfur results for the Eschka, lithium reduction, and two oxygen combustion methods and found them to be equally accurate.

An ion-exchange method,[84] in which coal samples were fused, dissolved in water, and passed through a sulfonated styrene–divinylbenzene column, was used for determination of total sulfur. The final solution was titrated with alkali.

Ratcliff and Cunningham determined total sulfur in coal using an oxygen flask method with thorin as a titration indicator.[85] The method proved unsatisfactory. Ahmed and Whalley[86,87] later used Arsenazo (III) successfully as an indicator. Precision of the method was similar to that of the Eschka procedure. The oxygen flask method is based on the combustion of coal in oxygen and adsorption of sulfur oxide gases in hydrogen peroxide. Excess barium perchlorate is added and then back-titrated with sulfuric acid using Arsenazo (III) as an indicator. Interfering metallic ions are removed by ion exchange prior to titration.

A modified oxygen flask procedure and a gas chromatograph equipped with a glass capillary and flame photometric detector were used to determine total sulfur with reasonable accuracy.[88] With the use of two oxygen flasks, sulfur determinations could be made in less than 5 min. A gas chromatographic method was also reported for determination of sulfur in coke.[89] Samples were burned in a stream of oxygen to produce sulfur oxides that were separated from other gases by adsorption on silica gel. Sulfur was determined with a thermal conductivity detector. Some difficulties were encountered because of retention of sulfur in the residues of high-ash coals.

The solution from oxygen bomb combustion of coal was titrated for sulfur with lead perchlorate in the presence of dioxane using a lead-selective ion electrode.[90] The method is said to be accurate and rapid, and the Btu determination can be made on the same sample.

Another titration procedure in which the Eschka melt was dissolved in sodium carbonate (Na_2CO_3) was used to determine sulfur as the sulfate.[91] Sulfate ions in a solution of HCl were titrated with barium nitrate [$Ba(NO_3)_2$] solution in the presence of an acetone–isopropyl alcohol solution of chlorophospho-azo III.

Oxidation of coal with potassium permanganate and peroxymono phosphoric acid was more effective when both reagents were used than when either was used alone.[92] Sulfur determinations were made on the digest, and the results were similar to those from the Eschka method.

The high-frequency induction furnace, which employs iodometric titration of SO_2, was rejected after thorough testing as a standard method for the determination of total sulfur in coal.[93] When an empirical constant is used, however, it can be successfully employed in quality control laboratories in which samples of the same coal are routinely analyzed. Combustion of the coal sample

[82] N. F. Shimp, J. K. Kuhn, and R. J. Helfinstine, *Energy Sources,* **3,** 93–109 (1977).
[83] P. Mohrhauer, *Gas Wasserfach,* **109**(21), 561–565 (1968).
[84] L. L. Okapkina, E. K. Kazakova, and N. D. Tyuryumina, *Izv. Nauchn.-Issled. Inst. Nefte-Uglekhim. Sin. Irkutsk. Univ.,* **12,** 95–98 (1970).
[85] D. B. Ratcliffe and A. T. Cunningham, *Fuel,* **47,** 89–92 (1968).
[86] S. M. Ahmed and B. J. P. Whalley, *Fuel,* **51,** 190–193 (1972).
[87] S. M. Ahmed and B. J. P. Whalley, *Fuel,* **53,** 61-62 (1973).
[88] L. J. Darlage, S. S. Block, and J. P. Weidner, *J. Chromatogr. Sci.,* **11,** 272–274 (1973).
[89] F. M. W. Olds, J. W. Patrick, and F. H. Shaw, *Analyst,* **92,** 54–56 (1967).
[90] J. E. Hicks, J. E. Fleenor, and H. R. Smith, *Anal. Chim. Acta,* **68,** 480–483 (1974).
[91] I. P. Avgushevich, *Khym. Tverd.,* **2,** 70–74 (1976).
[92] N. S. Rawat, *Chem Ind.,* **17,** 743–744 (1976).
[93] P. O. Krumin, Report to ASTM Committee D-5, 1959, 21 pp.

takes place in a tube furnace, within the field of an electric induction coil, and in a stream of purified oxygen. The combustion products are swept through an absorber containing an acid solution of potassium iodide (KI) and a starch indicator. Potassium iodate (KIO_3) is added to the absorbing solution as combustion proceeds, and when combustion is completed, the amount of iodate consumed can be measured by the sulfur content of the sample.

ASTM rejected this method of sulfur determination because SO_3 could not be measured in the combustion products by iodometric titration. Thus in the familiar KI-starch reactions

$$KIO_3 + 5KI + 6HCl \rightarrow 6KCl + 3H_2O + 3I_2$$
$$SO_2 + I_2 + 2H_2O \rightarrow H_2SO_4 + 2HI$$

only SO_2 reacts with liberated iodine; consequently the resulting sulfur analysis is in error (i.e., low) by the amount of SO_3 produced. Because combustion conditions vary for different coals, the amount of SO_2 produced varies, depending on the particular thermal equilibrium established for the reaction

$$2SO_2 + O_2 \rightarrow 2SO_3$$

Also, because chlorine is in coal,

$$RCl + O_2 \rightarrow Cl_2$$

and

$$Cl_2 + 2I^- \rightarrow I_2 + 2Cl^-$$

The oxidation by chlorine of iodide to iodine also results in a low value for sulfur.

Results of the ASTM study of this method gave sulfur recoveries ranging from 68 to 100% with a mean of 89.2%, compared with results obtained for the same coals using the standard Eschka method. During this study 15 coals whose total sulfur content varied from 1 to 22% were analyzed in four different laboratories.

A variation of the induction furnace, not yet tested by ASTM, may apply to a broad range of coals without using an empirical constant. It uses infrared absorption for the determination of sulfur and presumably includes both SO_2 and SO_3 in the analysis. Combustion of the sample, which may weigh 0.5 g or more, is aided by "accelerators." The products of combustion are carried by a heated delivery system into a "Luft" cell, where infrared absorption by sulfur oxides in the gas stream is measured. Analytic precision of ±3% of the sulfur content has been reported by one manufacturer. Such methods are well suited to automation and to display of digital or computer data.

X-Ray fluorescence has been used extensively for determining total sulfur in coal.[94–98] Ruch et al.[99] have recently described a method of preparing pressed coal samples that may be used to determine a variety of major, minor, and trace elements, including sulfur, in whole coal. The sulfur values they derived for more than 100 coals agreed closely with those determined by the ASTM standard Eschka procedure. The economy and speed of such X-ray fluorescence methods when used for multiple determinations (e.g., Al, Si, Ca, Mg, Fe, K, Ti, P, and S) in the same prepared coal sample are probably unsurpassed by any other method. The relative standard deviation for sulfur determined by the X-ray fluorescence method of Ruch et al. is 0.5%.

A neutron sulfur meter was developed for continuous monitoring of the sulfur content of coal from a preparation plant.[100] Neutrons from a radioactive source (^{252}Cf) produced thermal neutrons within the sample, which produced prompt gamma-ray emissions when captured by atoms in the coal. The gamma-ray energies characteristic of sulfur were measured by a sodium iodide crystal and associated electronic equipment. Several methods to eliminate interferences from other elements, moisture, and bulk density were evaluated. Results from the neutron sulfur

[94] M. Berman and S. Ergun, *Fuel,* **47,** 285–301 (1968).

[95] L. T. Kiss, *Anal. Chem.,* **38,** 1731–1735 (1966).

[96] T. R. Sweetman, K. Norrish, and R. A. Durie, C. S. I. R. O. Misc. Rep. No. 177, 1963.

[97] J. Frigge, *Brennstoff-Chemie,* **25,** 447–451 (1972).

[98] BCRA, Assoc. Coke Res. Rep. No. 71, 1972, 6 pp.

[99] R. R. Ruch, H. J. Gluskoter, and N. F. Shimp, Illinois State Geolo. Survey Environ. Geol. Note No. 72, 1974.

[100] A. W. Hall, J. W. Martin, R. F. Stewart, and A. M. Postox, U.S. Bur. Mines Rep. Invest. No. 8038, 1975, 12 pp.

meter for total sulfur in coal agreed closely with chemical analyses. The method was precise (±0.05% absolute), rapid, and suitable for on-line sulfur determinations.

7.1.11 Forms of Sulfur

Shimp et al.[82] have reviewed unpublished data from interlaboratory studies conducted by ASTM Committee D-5 on Coal and Coke on the determination of forms of sulfur in coal. The information given for sulfate, pyritic, and organic sulfur was selected from four studies,[101–104] which were the basis for specifications in the current ASTM Standard D2492. The same publication[82] also contains reviews of the various other methods available for determining forms of sulfur in coal, of difficulties that have been encountered in applying these methods to the determination of sulfur forms in coals containing nonpyritic sulfides (e.g., ZnS and FeS), and of the application of these methods to coal-related materials such as float-sink gravity fractions.

Since 1960 the accuracy of methods for determining varieties of sulfur in coal has been extensively investigated. Generally, these investigations were undertaken because it was known that very finely dispersed pyrite may be incompletely extracted with nitric acid in some coals.[105–109]

Fine grinding of coal (<200 mesh) tended to increase the amount of pyrite extracted by nitric acid (HNO_3), especially in high-rank British coals (>85% carbon content). Grinding to the usual < 60 mesh particle size has been successful, however, for U.S. coals.[101–104] Burns[110] also found that fine grinding was unnecessary for Australian coals when pyritic sulfur had been extracted from the coal residue instead of a fresh coal sample following the sulfate sulfur extraction.

A recent investigation[111] compared results of the ASTM oxidative method with results from a reductive method.[112,113] In the reductive method reduction of pyrite (FeS_2) to H_2S is accomplished with a strong reducing agent (e.g., lithium aluminum hydride). The H_2S is converted to sulfuric acid and titrated with a standard base. Sulfate sulfur was extracted by the standard HCl procedure, and organic sulfur was obtained by calculating the difference. These comparisons were made to determine whether extraction of pyrite was complete, whether organic matter was significantly affected by HNO_3 extraction, and whether organic sulfur can be obtained either by computation or by Eschka determination of residual sulfur following HNO_3 extraction of pyritic sulfur. The influence of coal particle sizes (< 60, < 200, and < 400 mesh) on the values determined by both methods was also studied.

Results for two coals, representing the nine coals studied in this investigation, are given in Table 7.4. The principal conclusions are summarized as follows:

1. Coal particle size did not significantly affect the amount of total sulfur determined by the Eschka method (Table 7.4, column 4).
2. X-Ray fluorescence values for total sulfur correlate closely with Eschka values (columns 4 and 5).
3. Pyritic sulfur values determined by the ASTM method were not significantly influenced by coal particle size (column 7).
4. Pyritic sulfur values obtained using the reduction method agreed with those from the ASTM procedure only when the coal was ground to <400 mesh (columns 7 and 8).
5. Agreement of values for total iron as determined by X-ray fluorescence with the

[101] P. O. Krumin, Report to ASTM Committee D-5 on Coal and Coke, 1958, 20 pp.
[102] P. O. Krumin, Report to ASTM Committee D-5 on Coal and Coke, 1959, 55 pp.
[103] P. O. Krumin, Report to ASTM Committee D-5 on Coal and Coke, 1960, 41 pp.
[104] P. O. Krumin, Report to ASTM Committee D-5 on Coal and Coke, 1961, 16 pp.
[105] H. R. Brown, M. S. Burns, R. A. Durie, and D. J. Swaine, *Fuel,* **43,** 409–413 (1964).
[106] A. H. Edwards, J. M. Jones and W. Newcombe, *Fuel,* **43,** 55–62 (1964).
[107] C. B. Belcher and T. G. Callcott, *Fuel,* **44,** 169 (1965).
[108] H. Mayland, *Fuel,* **45,** 97–98 (1966).
[109] R. G. James and M. I. Severn, *Fuel,* **46,** 476–478 (1967).
[110] M. S. Burns, *Fuel,* **49,** 126–133 (1970).
[111] J. K. Kuhn, L. B. Kohlenberger, and N. F. Shimp, Illinois State Geolo. Survey Environ. Geol. Note No. 66, 1973, 111 pp.
[112] W. Radmacher and P. Mohrhauer, *Glueckauf,* **89,** 503–511 (1953).
[113] J. W. Smith, N. B. Young, and D. L. Lawlor, *Anal. Chem.,* **36,** 619–622 (1964).

Table 7.4 Concentration of Iron and Varieties of Sulfur in Two Illinois Coals[111]

Sample and Size (mesh) (1)	Total Iron (%) ASTM (2)	Total Iron (%) X-Ray Fluorescence (3)	Total Sulfur (%) ASTM (4)	Total Sulfur (%) X-Ray Fluorescence (5)	Sulfur (%) SO_4, ASTM (6)	Pyritic ASTM (7)	Pyritic Reduction (8)	Calculated Organic ASTM (9)	Calculated Organic LAH (10)	Determined Organic ASTM (11)	Determined Organic Reduction (12)
C-18067											
< 60	4.28	3.80	7 .48	7.25	0.02	4.82	4.08	2.64	3.38	2.50	2.24
<200	4.26	4.28	7.53	7.45	0.02	4.76	4.44	2.75	3.07	2.48	2.21
<400	4.30	4.52	7.52	7.63	0.02	4.79	4.60	2.71	2.90	2.54	2.51
C-17167											
< 60	1.24	1.23	2.91	3.05	0.40	0.94	0.77	1.57	1.74	1.37	1.45
<200	1.29	1.31	2.99	3.09	0.39	0.92	0.88	1.75	1.68	1.26	1.25
<400	1.30	1.31	2.93	3.07	0.39	0.93	0.93	1.61	1.61	1.19	1.31

sum of the iron determined in the ASTM HCl and HNO_3 extracts was excellent (columns 2 and 3). Therefore all the pyritic iron (sulfur) must have been extracted by the HNO_3.

6. No pyritic iron remained in the residues of the nine coals studied after they had been subjected to either the ASTM oxidation method or the reduction method (on < 400-mesh coals only). Therefore values for organic sulfur calculated by subtracting the sum of sulfate sulfur and pyritic sulfur from total sulfur were deemed accurate (columns 9 and 10).

Investigations to improve the oxidative acid extraction procedures for pyritic sulfur are continuing. In one,[114] coal was ashed under standard conditions, then refluxed with HCl for 1 hr. Total iron was titrated with dichromate, and corrections were made on it for nonpyritic iron (sulfur) to yield a pyritic iron value. Comparisons with amounts of nitric acid extractable iron from whole coal were excellent. This method should be especially applicable where the whole coal matrix interferes with extraction of finely divided pyrite or where ash and pyrite are high in coal or coal-related materials.

Atomic absorption procedures for the determination of pyritic iron are being investigated by ASTM Committee D-5. Acid extracts prepared according to ASTM D2492 are analyzed for iron by atomic absorption methods instead of the ASTM-specified titration procedure. An independent comparison of iron results obtained by the two methods[115] showed that there was no statistical difference in the results when proper standardization of the atomic absorption procedure was made.

A method for determining forms of sulfur from the analysis of low-temperature coal ash for sulfur, which is equivalent to total inorganic sulfur, was described by Hamersma.[116] Sulfur determinations were made by several methods (Eschka, HNO_3 extraction, and X-ray fluorescence). Organic sulfur, which is reportedly volatilized by the low-temperature ashing process, was calculated by the difference between total sulfur and inorganic sulfur. Removal of carbonaceous material prior to acid extraction should promote a more quantitative recovery of the pyritic sulfur, especially in cases where pyrite is very finely divided. There is, however, some question concerning total loss of organic sulfur under all conditions and for all coals.

The determination of crystalline pyrite in coal by an X-ray diffraction method has been described by Schehl and Friedel.[117] Finely ground samples of coal of known pyrite concentration served as standards in a technique that uses digital computers to measure diffraction patterns and to calculate the percentage of pyrite in the coal sample. Accept-

[114] R. K. Young and E. A. Zawadzki, *Fuel,* **46,** 151–152 (1967).

[115] W. I. Galdfelter and D. W. Dickerhoof, *Fuel,* **55,** 360–361 (1976).

[116] J. W. Hamersma, presented to ASTM Committee D-5, New Orleans, La., May 19, 1977.

[117] R. R. Schehl and R. A. Friedel, NTIS P.B. Rep. No. 221867/5, 1973.

able precision within current ASTM tolerances was obtained for the coals studied, but interferences from reflections caused by high concentrations of kaolin in the coal were noted.

High-resolution X-ray spectroscopy was successfully employed to determine the chemical state of sulfur in coal.[118] Fluorescent X-rays from pressed coal powders, excited with X-rays from a chromium target tube, were analyzed using a germanium crystal. The sulfur K_α radiation increased with increasing oxidation number, allowing quantitative determinations for S^{2-} and S^{6+}.

X-Ray photoelectron spectroscopy (XPS) is used to determine the elemental composition of surfaces (at depths of 20–40 Å) irradiated by monochromatic X-rays. Binding energies of electrons ejected from the sample differ, and when they are accurately measured by XPS, valences or chemical forms of the elements present can be determined. When this method was applied to coal and coke,[119,120] it was found that iron sulfate peaks were caused by ferrous rather than ferric iron. Because of the apparent overlay of organic sulfur and sulfide peaks, the amount of pyritic sulfur could not be quantitatively estimated. During controlled surface oxidation of the pyrite to sulfate, a photopeak caused only by organic sulfur suggests a new method for determining organic sulfur.

Research on the nature and removal of organic sulfur compounds from coal has been extensive.[121–125] Unfortunately none of these investigations led to useful methods for determining varieties of organic sulfur in coal. In the past, when organic sulfur was determined, rather than calculated by difference as in ASTM D2492, the coal was first demineralized by methods similar to those described in the Standard Method ISO 602,[126] by Kuhn et al.[111] or by Smith and Batts.[127] In these methods sulfate and pyritic sulfur are first removed by acid dissolution, oxidation, or reduction reactions. The total sulfur determined in the coal residue is then considered to be organic sulfur. These methods were fairly reliable, although small losses of organic sulfur were unavoidable (Table 7.4, columns 9–12).

Hurley and White determined the amount of organic sulfur in coal[128] by employing X-ray fluorescence to determine the total sulfur from the K_β line intensity and then using a line ratio technique to find the proportion of sulfate sulfur (K'_β line intensity) in the total sulfur:

$$\%\ S_{SO_4} = \%\ \text{total S} \times \frac{SK'_\beta}{SK_\beta} \times M_1$$

where M_1 is the slope of a linear calibration of SK'_β/SK_β as a function of concentration. The peak shift of sulfur, K_β, was used to determine amounts of sulfide and organic sulfur:

$$\%\ S_{org} = \%\ \text{total S} \times \frac{\Delta SK_\beta - B}{M_2}$$

$$\% S_{pyr} = \%\ \text{total S} \times \left(100 - \frac{SK_\beta - B}{M_2}\right)$$

where the calibration constants M_2 and B are obtained from a linear calibration curve of the form $Y = MX + B$. Pyritic sulfur here is a calculated difference figure (i.e., the sum of the three forms of sulfur must equal 100%). Apparently the three sulfur values are not independently determined.

Sutherland,[129] using a microprobe for the direct determination of organic sulfur, analyzed the light gravity fraction of two coals (>80% organic sulfur). On one of the fractions organic sulfur was determined by wet chemistry and a polished section was analyzed 100 times for iron, and sulfur by using pyrite as a standard for direct comparison;

[118] Y. Gohshi and K. Yanagase, *Nenryo Kyokaishi,* **52,** 926–931 (1973).

[119] H. D. Schultz and W. G. Proctor, *Appl. Spectrosc.,* **27,** 347–351 (1973).

[120] D. C. Frost, W. R. Leeder, and R. L. Tapping, *Fuel,* **53,** 206–211 (1974).

[121] J. D. Brooks, *J. Inst. Fuel,* **29,** 82–85 (1956).

[122] N. P. Chopey, *Chem. Eng. News,* **79**(16), 86, 88 (1972).

[123] P. H. Given and J. R. Jones, *Fuel,* **45**(2), 151–158 (1966).

[124] L. Horton and R. B. Randall, *Fuel,* **45,** 151–158 (1948).

[125] R. K. Sinha and P. L. Walker, Jr., *Fuel,* **51,** 329–331 (1972).

[126] ISO 602, American National Standards Institute, 1974.

[127] J. W. Smith and B. D. Batts *Geochim. Cosmochim. Acta,* **38,** 121–133 (1974).

[128] R. G. Hurley and E. W. White, *Anal. Chem.,* **46,** 2234–2237 (1974).

[129] J. K. Sutherland, *Fuel,* **54,** 132 (1975).

low-iron regions (background) represented organic sulfur. A single factor was applied to the mean concentration of sulfur to correct it to the value for sulfur as determined by wet chemistry. The second coal fraction was identically analyzed, and the correction factor was applied to its mean value for sulfur content, which was obtained by direct comparison with pyrite. The result was the same as that obtained from chemical analysis of a pyrite-free separation of the coal. A direct determination of organic sulfur by the microprobe took 1–2 hr.

Another method of directly determining organic sulfur[130] is to subject a small (20–30 mg) sample of $<$ 200 mesh (74-μm) coal to LTA for 1–3 hr and to collect the sulfur oxides evolved in a cold trap. They are then absorbed in hydrogen peroxide (H_2O_2) and determined chromatographically using Dowex 10 resin. Preliminary results correlated closely with results from ASTM standard procedures. A relative standard deviation of about $\pm$19%, which is roughly equivalent to the ASTM method, was recorded for the organic sulfur. The procedure requires further evaluation to determine the effects of LTA on oxidation of very fine pyrite, small sample precision, temperature and LTA power ratings, sulfur losses from the system, and potential for conversion of organic sulfur to $CaSO_4$.

Using Mössbauer spectrometry, Smith et al.[131] have identified multiple-iron species in coal and charred residues without using a concentration step. Their results indicate that heat treatment of any kind, even at temperatures as low as 175°C, changes the nature of the iron species in coal. Furthermore, some kind of an association between the pyrite (FeS_2) in whole coal and the organic matrix was indicated. The pyrite also appeared to be altered when the coal was heated to 175°C.

In a coal-related study,[132] the forms of sulfur in hydrodesulfurized coal were determined by the application of ASTM D2492. When the pyrite (FeS_2) in coal reacts with hydrogen at a high temperature (400°C) and pressure, sulfides (FeS) and sulfites (SO_3) may be formed. These forms of sulfur are not determined as such by the ASTM method—they do not occur naturally in coal, and they become H_2S or SO_2 when treated with HCl. If organic sulfur were calculated in the product by difference [i.e., organic S = total S − (sulfate S + pyritic S)], the value would be high by the amount of sulfur evolved as a gas. This problem was solved by determining pyritic iron in the ASTM residue from the sulfate (HCl extraction) determination of both the whole coal (w–coal) and processed coals (p–coal) and calculating the sulfide sulfur (S_{FeS}):

$$\%\,S_{FeS} = \frac{32.06}{55.85}(\%\,Fe_{w\text{-}coal} - \%\,Fe_{p\text{-}coal})$$

which is in effect the amount of FeS_2 converted by the hydrogenation process to FeS and lost as H_2S from the reaction $FeS + 2HCl \rightarrow H_2S + FeCl_2$. Sulfite sulfur ($S_{SO_3}$) was then calculated from the percentage of total sulfur in whole coal, sulfate sulfur in the processed coal, and percentage of sulfur in the HCl-extracted residue of the processed coal (equivalent to organic plus pyritic sulfur):

$$\%\,S_{SO_3} = \%\,S_{w\text{-}coal} - \%\,S_{SO_4 p\text{-}coal} - \%\,S_{FeS} - \%\,S_{HCl\ residue}$$

It was further shown that S_{SO_3} was produced only from reduction of S_{SO_4} in the whole coal, not from pyrite. Thus

$$\%\,S_{SO_3} = \%\,S_{SO_4 w\text{-}coal} - \%\,S_{SO_4 p\text{-}coal}$$

since S_{SO_3} is lost as SO_2 on addition of HCl for the extraction of S_{SO_4}. Organic sulfur in the processed coal was then determined from the following expression:

$$\%\,S_{org} = \%\,S_{total\ p\text{-}coal} - \%\,S_{SO_4 w\text{-}coal} - \%\,S_{FeS} - \%\,S_{pyr\ p\text{-}coal}$$

The analytical scheme outlined above was a satisfactory method of analysis for both the raw and processed coals when the quantitative relationship between the weight of processed coal and the whole coal was known.

[130] B. Paris, (1977) Preprint, Meeting of American Chemical Society, Division of Fuel Chemistry Chicago, August 1977.

[131] G. V. Smith, J. H. Liu, M. Saporoschenco, and R. H. Shiley, *Fuel*, **57**, 41–45 (1978).

[132] W. L. Gladfelter and D. W. Dickerhoof, "Determination of Sulfur Forms in Hydro-desulphurized Coal," *Fuel*, **55**, 355–359 (1976).

7.1.12 Chlorine Analysis

Gluskoter and Rees[133] have emphasized the need for low-uniform chlorine content in the magnesium oxide used for fusion of coal by the ASTM Standard Eschka procedure (D2361). Their study included determinations of both water-soluble and total chlorine in 135 coals from the Illinois Basin. Although inorganic chlorine was not lost from the Eschka fusion mixture, which has a basic pH, it was unexpectedly evolved even at low temperatures (500°C) from coal containing pyrite. Wide variations of water-soluble chlorine expressed as a percentage of total chlorine were attributed to distribution of the coal pore size and possible evidence of organically combined chlorine; however a close relationship was found between the chlorine content of the coals and salinity of the groundwater.

A modification of the oxygen bomb combustion method (ASTM D2361) for the determination of chlorine[134] consisted of acidifying a solution of the adsorbed combustion products and titrating the chlorine potentiometrically. A potentiometric titration was also used for the determination of chlorine by the oxygen flask method.[135] Combustion products, including chlorine, were absorbed in sodium hydroxide (NaOH), and the chloride was measured using silver–silver chloride electrodes. Although there was no statistical difference in results obtained from potentiometric titration and the Eschka procedure, the latter was more precise.

Both sodium and chlorine were determined on the same 1-g coal samples by a neutron activation method.[136] For comparison, duplicates were analyzed for sodium and potassium by flame photometry; for total chlorine, the ASTM oxygen bomb combustion method was used. All values agreed closely. Chlorine in excess of stoichiometric sodium chloride (NaCl) was found in a number of samples in a form other than inorganic salt, and this chlorine was thought to be in organic combination.

The chlorine liberated during the determination of nitrogen in coal by the Kjeldahl method was absorbed in the alkaline hydrogen peroxide solution. It was then precipitated as silver chloride (AgCl), which was subsequently filtered, dried, and weighed. Nitrogen was determined in the usual manner in the remaining solution.[137]

7.1.13 Mineral Carbon Dioxide Analysis

Burns[138] conducted detailed studies of the British Standard manometric procedure for measuring the amount of mineral carbon dioxide in coal[139] and found consistently low results because adsorption of carbon dioxide by the coal-acid mixture caused a decrease in the volume of gas liberated and measured. A standardization procedure that overcame the difficulty was developed. The results of this investigation have been confirmed, and the modified manometric method was recommended.[140]

Three methods for determining mineral carbon dioxide in coal were investigated[141] using Scottish bituminous coals. It was concluded that Pringle's titrimetric method[142] was superior to either of the then-used British standard gravimetric or manometric methods (BS 1016, Part 6, 1958). The procedure involves the decomposition of carbonate minerals with hydrochloric acid and absorption of the evolved carbon dioxide in a mixture of benzylamine, ethanol, and dioxane. This mixture forms a stable salt of benzylcarbamic acid, which is then titrated with sodium methoxide. The method was said to be suitable for all concentrations of carbon dioxide. It is especially accurate for low concentrations, and it is much more rapid than other methods tested.

Knott and Belcher[143] reported a relatively simple and accurate semimicrogravimetric

[133] H. J. Gluskoter and O. W. Rees, Illinois State Geol. Survey Circ. No. 372, 1964, 23 pp.
[134] R. F. Abernethy, F. H. Gibson, and W. H. Frederic, U.S. Bur. Mines Rep. Invest. No. 6579, 1965, 34 pp.
[135] D. B. Ratcliffe and T. L. Young, *Fuel,* **47,** 185–192 (1968).
[136] H. J. Gluskoter and R. R. Ruch, *Fuel,* **50,** 65–76 (1971).
[137] J. N. Chakrabarti, *Fuel,* **53,** 108–109 (1974).
[138] M. S. Burns, *Fuel,* **41,** 239–248 (1962).
[139] British Standard 1016, Part 6, 1958.
[140] J. F. Cudmore, *Fuel,* **44,** 89–91 (1965).
[141] F. J. Allan, D. A. Lambie, M. C. McKissock, and T. C. Shaw, *Fuel,* **45,** 93–95 (1966).
[142] W. J. S. Pringle, *Fuel,* **42,** 63–68 (1963).
[143] A. C. Knott and C. B. Belcher, *Talanta,* **22,** 751–753 (1975).

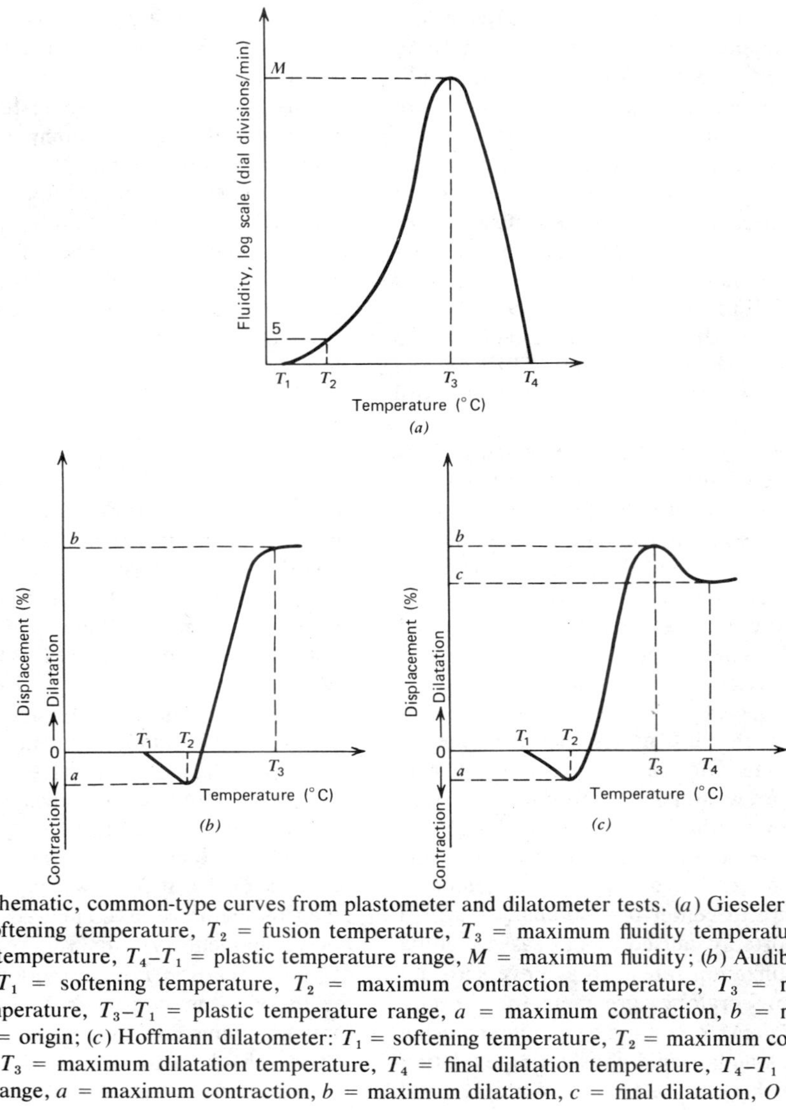

Figure 7.3 Schematic, common-type curves from plastometer and dilatometer tests. (*a*) Gieseler plastometer: T_1 = softening temperature, T_2 = fusion temperature, T_3 = maximum fluidity temperature, T_4 = solidification temperature, T_4–T_1 = plastic temperature range, M = maximum fluidity; (*b*) Audibert-Arnu dilatometer: T_1 = softening temperature, T_2 = maximum contraction temperature, T_3 = maximum dilatation temperature, T_3–T_1 = plastic temperature range, a = maximum contraction, b = maximum dilatation, O = origin; (*c*) Hoffmann dilatometer: T_1 = softening temperature, T_2 = maximum contraction temperature, T_3 = maximum dilatation temperature, T_4 = final dilatation temperature, T_4–T_1 = plastic temperature range, a = maximum contraction, b = maximum dilatation, c = final dilatation, O = origin. After Wu.[145]

method for determining mineral carbon dioxide in coal in which a purification train was used to remove water, halogens, and gaseous sulfur compounds.

7.1.14 Plasticity and Swelling Analyses

The fluid behavior or plasticity of coal in the temperature range 300–550°C is widely used to estimate coking characteristics and may be important for predicting coal properties with respect to coal conversion processes. Wu[145] has conducted precision studies for the most commonly used methods, in which the Gieseler plastometer (ASTM D2639), the Audibert-Arnu dilatometer (ISO 349) (Section 6.2.2.1), and the Hoffmann dilatometer[144] are used. Wu graphed the critical fluidity measurements in dial divisions per minute (plastometer) or percentage displacement (dilatometers) as a function of the temperatures at which they occur (Fig. 7.3). In general, the precision tests showed that (1) differences between means for the Gieseler maximum fluidity were 17% or less for a frequency

[144] H. Hoffman, "Die Bestimmung der Beldsamkiet von Steinkohlen nach Dilatometermethode," *Oel Kohle*, **40**, 531–550 (1944).

[145] W. R. K. Wu, U.S. Bur. Mines Rep. Invest. No. 7394, 1970.

distribution of 683 in 1000, or 33.7% or less for a frequency distribution of 955 in 1000; (2) differences between means for the Audibert maximum dilatation measurement were 5.2% or less and 10.4% or less for the same frequency distributions; (3) repeatabilities (one standard deviation) or characteristic temperatures for the Gieseler plastometer ranged from 3.22 to 4.33°C, for the Audibert-Arnu dilatometer from 2.64 to 2.75°C, and for the Hoffmann dilatometer 2.49 to 2.81°C; and (4) repeatabilities of plastic temperature ranges were 4.54, 3.57, and 3.80°C for the Gieseler, Audibert-Arnu, and Hoffmann tests, respectively.

Wu and Frederic[146] also reviewed factors affecting coal plasticity and conducted further studies, which indicate that test results from various instruments are not necessarily comparable. Included are comparisons of results of free-swelling and agglutinating tests with plasticity data from 84 coals obtained by the Gieseler plastometer and Audibert-Arnu and Hoffmann dilatometers.

Discrepancies have been noted in results obtained with the Gieseler coal plastometer. For example. Gieseler plasticity tests[147] performed on western United States coking coals showed that increased maximum fluidities were associated with greater retort immersion depths (distance from center of solder bath to center of coal charge), which were reportedly caused by increased melting and volatilization rates. To achieve satisfactory interlaboratory agreement for Gieseler plasticity data, tests must be conducted using the same standard and considering oxidation effects such as those from coal storage and transport, which rapidly lower coal plasticity.[148]

The swelling of coking properties of coal are measured by heating coal at a prescribed rate to 820°C and measuring the size or free-swelling index (FSI) of the coke button formed (ASTM D720). Like fluidity, the FSI is used as an indication of the coking quality of coal. Although gas and electric heating are both employed for the determination of FSI, the electric heating mode is preferred because the furnace provides better temperature control and convenience. Electric heating has been thoroughly tested[149]; results obtained in this way are comparable to those obtained by gas heating.

The ASTM method for the FSI of coal was modified by Abernethy and Snyder[150] so that small samples (0.3 g), especially coal macerals, could be used. The coke button areas were smaller than those obtained for the usual sample size (1.0 g), but they were directly related to the larger areas and yielded relative values.

7.1.15 Ash Fusibility

The softening or fusion temperature of coal ash is the physical characteristic most frequently used as an index of fireside fouling and slagging potential. The relationships of the composition of Illinois coal ashes and their softening temperatures (ASTM D1857, ISO 540) were investigated by Rees.[151] Three methods were used: (1) relation of acid-to-base ratio to ash-softening temperature (i.e., $SiO_2 + Al_2O_3/Fe_2O_3 + MgO + CaO + Na_2O + K_2O$); (2) relation of silica ratio to ash-softening temperature [i.e., $SiO_2 \times 100/(SiO_2 + Fe_2O_3 + MgO + CaO)$]; and (3) effect of bases in coal ash on lowering the ash-softening temperature as compared to that of a two-component system (SiO_2–Al_2O_3). None of these relationships provided a reliable means of estimating the softening temperatures of Illinois coal ashes.

The ash chemistries of various coal size and gravity fractions were studied in relation to ash fusion (softening) temperatures.[152] To develop methods or guidelines for predicting the fusibility of ash, Northern Great Plains lignites have been studied.[153] The investigation consisted of studying the effects on fusion temperature caused by adding a vari-

[146] W. R. K. Wu and W. H. Frederic, U.S. Bur. Mines Bull. No. 661, 1971.

[147] R. R. Allen, U.S. Bur. Mines Rep. Invest. No. 6559, 1964, 26 pp.

[148] M. S. Burns, T. Mryazu, and M. Shiboaka, *Fuel,* **51,** 164–165 (1972).

[149] R. K. Young and E. A. Zawadzki, *Coal Age,* **70**(11), 74–76 (1965).

[150] R. F. Abernethy and J. P. Snyder, U.S. Bur. Mines Rep. Invest. No. 7534, (1971), 17 pp.

[151] O. W. Rees, Illinois State Geol. Survey Circ. No. 365, 1964, 20 pp.

[152] R. W. Byers and T. E. Taylor, ASME Paper No. 75-WA CD-3, 1975, 12 pp.

[153] E. A. Sondreal and R. C. Ellman, U.S. Bur. Mines Rep. No. GFERC-RI-75-1, Pittsburgh, 1975, 127 pp.

ety of metal oxides, pyrite, clay, and dolomite to the ash. Results showed that sulfur oxides reduced the ash-softening temperature, and oxides of calcium and magnesium increased it. Sodium sulfate (Na_2SO_4) lowered or raised the hemispherical temperature, depending on whether fluxing or covolatilization with other fusible components prevailed. The objective of this investigation was to adjust coal ash fusion temperatures for boiler use or conversion processes.

A stepwise regression program was employed[153] to relate the ash composition of 1212 coals to ash fusion temperatures. The program derived an equation in the form:

$$y = \text{constant} + aX_1 + bX_2 + cX_3 + dX_4 + \cdots$$

where y is one of the ash fusion temperatures, and X_1, X_2, . . . are chemical properties such as percentage Fe_2O_3, percentage FeO, base-to-acid ratio, or percentage SiO_2. Fifty-one independent variables were used in solving for 14 dependent variables, including eight fusion temperatures. The derived equations predicted the ash fusion temperatures with an accuracy approaching or exceeding the accuracy of laboratory determinations when ash composition was expressed on a mole-percent, SO_3-free, normalized basis. The product of percentage SiO_2 and percentage FeO was the best single indicator of ash fusion temperatures under reducing conditions (i.e., as the product increased, ash fusion temperatures decreased). The product of percentage SiO_2 and percentage CaO was the best single variable for predicting oxidizing values (including negative correlation).

7.1.16 Comprehensive Coverage

A number of papers, books, and standards dealing with multiple analyses of coal have been published. For convenience and complete coverage, they are briefly reviewed here.

For methods of conventional coal analysis published prior to 1960, refer to the first Supplementary Volume of *Chemistry of Coal Utilization*.[1] A review by Shipley[2] of the methods and developments for British Standard 1016 included methods for moisture, volatile matter, ash, carbon and hydrogen, carbon dioxide, nitrogen, total sulfur, forms of sulfur, oxygen, and other analyses.

Rees[154] published an extensive review outlining many methods for the analysis of coal, including the ultimate, proximate, calorific, moisture, sulfur forms, mineral carbon, plasticity, swelling, and fusibility methods. Each discussion details the difficulties and limitations associated with the determination and gives the uses made of the determination. More important, means for overcoming difficulties are emphasized. Problems related to sampling, sample preparation, moisture loss and gain, oxidation, size consist, mineral matter calculation, and the empirical nature of coal analyses are discussed in some detail. Many of the procedures described are ASTM standards, although a number of foreign standards are included.

In 1967 the U.S. Bureau of Mines published their methods of analyzing and testing coal and coke.[5] This well-known, comprehensive set of methods covering virtually all phases of coal analysis includes determinations for ultimate and proximate analysis, sulfur forms, chlorine, carbon dioxide, calorific value, equilibrium moisture, and physical tests and calculations used for converting analyses to different bases. In addition, it includes methods for determining plastic, swelling, and agglutinating properties of coal as well as methods for determining the fusibility characteristics of coal ash and slag.

The *Encyclopedia of Industrial Chemistry*[6] contains extensive reviews on all phases of conventional coal analysis, including details on the standard specifications of tests taken primarily from ASTM, the U.S. Bureau of Mines, and British Standards.

Vacuum optical emission spectrometric techniques have been developed by Galletta[155] for the determination of total sulfur, organic, and pyritic sulfur in coal and coke. Samples as small as 400 mg were analyzed completely for ash, volatile matter, and the forms of sulfur present. Briquettes of coal blended with briquetting graphite and a lithium tetraborate internal standard were prepared and heated at 950°C in covered carbon

[154] O. W. Rees, Illinois State Geol. Survey Rep. Invest. No. 220, 1966, 55pp.

[155] F. A. Galletta, Unpublished method, Youngstown Sheet and Tube Co., Youngstown, Ohio, 1977.

crucibles for the determination of volatile matter. The volatile-free briquettes were then pulverized and rebriquetted for direct-reading vacuum spectrometer analysis of 0–16% sulfur and 0–50% ash by totaling the percentages of Fe_2O_3, SiO_2, Al_2O_3, CaO, MgO, TiO_2, and Na_2O. Spectrometer calibration was based on chemical and spectrometric analysis of ashed coals. The method was developed to analyze diamond drill core and sink-float samples of coal where resultant sample size precluded use of standard methods of coal chemical analysis. Galletta's technique shortens the analytical elapsed time for organic sulfur from 24 to 2 hr. A secondary check on the relation of pyritic sulfur content to pyritic iron content was also described. Spectrometric and chemical results demonstrated the agreement in Galletta's method.*

7.2 DETERMINATION OF MAJOR, MINOR, AND TRACE ELEMENTS IN COAL AND COAL ASH

7.2.1 Ash-Forming Elements

Except for oxygen and sulfur, elements that normally constitute the ash residues derived from coal combustion can arbitrarily be grouped as follows: *major elements*—elements in concentrations greater than 0.5% in the whole coal; these normally include aluminum, calcium, iron, and silicon; *minor elements*—those in the range of concentration of about 0.02–0.5% in the whole coal; these usually include potassium, magnesium, sodium, and titanium, and sometimes phosphorus, barium, strontium, boron, and others, depending on the geologic area; and *trace elements*—all other inorganic elements usually detected in coal at less than 0.02% (200 ppm) down to parts per billion and below. Most nonmetallic elements, although more volatile than metals, will leave some detectable residues in coal ash.

* After this chapter was written further coverage of conventional coal analysis, which concisely summarizes current ASTM and ISO methods, was published by W. J. Montgomery, in *Standard Laboratory Test Methods for Coal and Coke,* Vol. 1, C. Karr, Jr., Ed., Academic Press, New York, 1978, pp. 191–278.

The two sections that follow cite the majority of the references available on the analysis of trace elements in coal and coal ash without critically comparing them, offering the reader a comprehensive bibliography and some general guidelines for further study.

All the methods mentioned have some merit and some advantage. The specific method an individual laboratory might choose and adapt for a particular analysis problem would depend on the facilities and professional competence available and the sample demand.

The wet chemical or "classical" procedures, including many colorimetric and some electrochemical procedures, can be quite time-consuming. They are generally very accurate, however; they can be precise, and they do not require any special expensive equipment or facilities. These methods are less sensitive than some instrumental methods.

The newly developed "instrumental" methods (e.g., optical emission spectroscopy, atomic absorption, mass spectrometry, and nuclear methods) in contrast, can require a considerable initial expense to establish the facility and specially trained personnel to run and maintain the operation. Instrumental methods are more versatile than the classical methods. They have more extensive application, are faster and less tedious, and have better sensitivity in general. Instrumental techniques, in most cases, are not more accurate than classical procedures.

7.2.2 Determination of Major and Minor Ash-Forming Elements

Major and minor elements in coal, having concentrations easily detectable by most modern analytical techniques, can be determined by a number of acceptable procedures. A variety of approaches combining a number of specific procedures is frequently referenced in the literature. For example, the presently accepted ASTM procedure, D2795-69,[156] determines silicon, aluminum, iron, titanium, and phosphorus colorimetrically, calcium and magnesium chelatometrically, and sodium and potassium by flame photometry.

[156] ASTM Standards, 1975, Part 26, ASTM 2795-69 (reapproved 1974), p.339.

Other approaches[157–165] predominantly utilize colorimetry but also incorporate one or more of the following procedures: flame photometry, compleximetric titration, emission spectrography, and gravimetry. Durie et al.[166] employed a combination of wet chemical, atomic absorption, and X-ray emission procedures. Rhodes and Taylor[167] determined major elements in milligram amounts of coal dust by a combination of neutron activation analysis, alpha-particle scattering, and electron excitation techniques. An inherent advantage of using a variety of techniques is having a choice of methods for a particular element.

Conventional optical emission spectrography has been used successfully with minor variations to determine most major elements in coal ash.[168–176] In two studies[177,178] whole coal was utilized directly, without ashing first. Karacki and Corcoran[179] employed a high-temperature argon-plasma source and an echelle grating spectrometer to analyze coal ash that was dissolved after lithium metaborate fusion. Ikehata et al.[180] used a high-voltage, spark excitation technique.

Muter and Nice[181] discussed atomic absorption spectrophotometry (AAS) of major elements and some trace elements in coal ash after fusion with lithium tetraborate. A similar approach was described by Boar and Ingram.[182] Coutinho et al.[183] discussed the preparation of samples and standards for major element analysis by AAS. A procedure by Obermiller and Freedman[184] dissolves the coal ash with perchloric acid and hydrogen fluoride before it is analyzed by AAS.

Kuhn et al.[185] discussed the general application of X-ray fluorescence (XRF) to coal analysis for major elements, emphasizing the use of whole coal, rather than ash, as the sample. Kiss[95] discussed general matrix corrections and particle size considerations when analyzing brown coals. Sweatman et al.[96] and Copeman and Wolff[186] have developed XRF techniques for coal ash. Swanson et al.[187] reported on the analysis of major

[157] R. F. Abernethy, M. J. Peterson, and F. H. Gibson, U.S. Bur. Mines Rep. Invest. No. 7240, 1969.

[158] K. Archer, D. Flint, and J. Jordan, *Fuel,* **37,** 421–423 (1958).

[159] F. H. Gibson and W. H. Ode, U.S. Bur. Mines Rep. Invest. No. 6036, 1962, 23 pp.

[160] N. V. Novitskii and N. I. Ivanova, *Khim. Tverd. Tope.,* No. 3, 60–64 (1972).

[161] N.K. Prasad, *Chem. Ind.,* No. 2, 81–84 (1972).

[162] W. J. S. Pringle, *Fuel,* **36**(3), 257–276 (1957).

[163] G. I. Spielholtz and H. Diehl, *Talanta,* **13**(7), 991–1002 (1966).

[164] V. I. Volkhova, *Zavod. Lab.,* **29,** 804–805 (1963).

[165] J. Vorlicek and V. Leitnerova, *Tech. Zpr. Zelezorudm. Dolu Hrudkoven,* **2**(3), 38–41 (1963).

[166] R. A. Durie, H. N. S. Schafer, and D. J. Swaine, CSIRO Div. Coal Res. Misc. Rep. No. 178, 1963, p.13.

[167] John R. Rhodes and M. C. Taylor, NTIS PB Rep. No. 222491/3, 1972.

[168] C. H. Anderson and R. A. Hein, *Pittsburgh Conf. Anal. Chem. Applied Spectroscopy,* March 3–8, 1957.

[169] G. R. Angell and F. W. Bethell, BCURA Inf. Circ. No. 244, 1960.

[170] BSI, No. 1016, Part 14, 1963

[171] M. C. Clark and D. J. Swaine, *Fuel,* **42**(4), 315–318 (1963).

[172] A. I. Khrisanfova and G. N. Soboleva, *Tr. Inst. Goryuch. Iskop., Moscow,* **23**(3), 16–22 (1967).

[173] W. Radmacher and W. Schmitz, *Brennstoff-Chemie,* **38,** 225–230, 270–274, 308–312 (1957).

[174] A. K. Shubnikov, G. N. Soboleva, T. M. Bronovets, and A. I. Krisanfova, *Khim. Tekhnol. Topl. Masel,* No. 10, 1–5 (1957).

[175] R. L. Stenger, *Am. Lab.,* 19–22, 24–31 (July 1969).

[176] J. B. Zink, W. D. Washington, and M. J. Peterson, U.S. Bur. Mines Rep. Invest. No. 6985, 1967, 14 pp.

[177] M. F. Kessler and F. K. Voleska, *Paliva,* **37,** 158–160 (1957).

[178] M. V. Popov, V. A. Nikonova, and A. I. Krokhina, *Tr. Kolinin. Politekl. Inst.,* No. 13, 110–113 (1972).

[179] S. S. Karacki and F. L. Corcoran, Jr., *Appl. Spectrosc.,* **27**(1), 41–42 (1973).

[180] A. Ikehata, Y. Suzuki, S. Sayama, and T. Matsumae, *Jap. Anal.,* **14**(10), 874–880 (1965).

[181] R. B. Muter and L. L. Nice, in Am. Chem. Soc. Adv. Chem. Ser. No. 141, 1975, pp. 57–65.

[182] P. L. Boar and L. K. Ingram, *Analyst,* **95,** 124–130 (1970).

[183] C. A. Coutinho, H. T. Prates and C. M. Pereira, *Metal. ABM,* **31**(217), 817–823 (1975).

[184] E. L. Obermiller and R. W. Freedman, *Fuel,* **44**(3), 199–203 (1965).

[185] J. K. Kuhn, W. F. Harfst, and N. F. Shimp, in Am. Chem. Soc. Adv. Chem. Ser. No. 141, (1975), pp. 66–73.

[186] J. H. Copeman and W. T. E. von Wolff, *Gold Base Miner. S. Afr.,* **17**(1), 41 (1969).

[187] V. E. Swanson, J. H. Medlin, J. R. Hatch, S. L. Coleman, G. H. Wood, Jr., S. D. Woodruff, and R. T. Hildebrand, U.S. Dept. Interior, Geol. Survey, Open File Rep. No. 76-468, 1976, 503 pp.

elements in coal ash as determined by XRF; however magnesium and sodium were determined by atomic absorption spectrophotometry.

Other specific procedures reported for major element analysis include flame photometry,[188] fast-neutron activation analysis,[74,189] neutron activation analysis,[190] photon activation analysis,[191] colorimetric,[192,193] and chelatometric[194] procedures. A round-robin comparison study of various ash analyses utilizing colorimetric and spectrochemical methods was reported by Flint.[195]

Procedures mentioned in the literature for determining amounts of individual major elements include optical emission procedures for potassium,[196,197] sodium,[196] and titanium;[198] atomic absorption procedures for calcium,[199,200] potassium and sodium,[201] and magnesium[201,202]; colorimetric procedures for aluminum,[203] calcium and magnesium,[204] phosphorus,[205–220] and titanium[221]; flame emission procedures for calcium and magnesium,[222] and potassium and sodium[223–226] neutron activation analysis procedures for iron,[227] sodium,[228] silicon,[74,75,229–232] and aluminum[74,229–232]; chemical procedures for iron,[233–235] calcium

[188] A. Roynette and J. Baron, *Bull. Soc. Fr. Ceram.*, **24,** 3–14 (1954).

[189] M. M. Sattarov, Y. N. Talanin, and T. Khalikov, *Izv. Akad. Nauk Uzbek. SSR, Ser. Fiz-Matam. Nauk,* No. 2, 74–77 (1968).

[190] I. Obrusnik and B. Starkova, *Radioisotopy,* **14**(3), 437–447 (1973).

[191] W. H. Zoller, *Proc. 2nd Joint Conf. Sensing Environ. Pollutants,* 1973, pp. 185–189.

[192] N. N. Bannerjee and B. A. Colliss, *Fuel,* **34,** 571–583 (1955).

[193] N. V. Novitskii, I. A. Kuleshova, and N. I. Ivanova, *Khim. Tverd. Topl.*, No. 1, 62–67 (1968).

[194] A. C. Bhattacharyya and N. G. Banerjee, *J. Sci. Ind. Res. (India)*, **21C**(5), 140–144 (1962).

[195] D. Flint, *Fuel,* **43**(4), 321–330 (1964).

[196] C. H. Anderson and C. D. Beatty, *Anal. Chem.*, **26,** 1369–1371 (1954).

[197] A. F. Rekus, *Appl. Spectrosc.*, **12,** 141–142 (1958).

[198] F. Valeska and A. Havlova, *Sb. Pr. Hornischkeho Ustavu Csav,* No. 1, 182–190 (1959).

[199] P. L. Boar and J. V. Sullivan, *Fuel,* **46**(1), 47–52 (1967).

[200] S. Tardon and M. Bolcarkova, *Chem. Prum.*, **17**(10), 560–563 (1967).

[201] R. B. Muter and C. F. Cockrell, *Appl. Spectrosc.*, **23,** 493–496 (1969).

[202] S. Tardon and M. Balcarkova, *Chem. Listy,* **60**(3), 334 (1966).

[203] N. P. Fedorovskaya, I. M. Khaskina, and N. L. Surina, *Khim. Pererab. Topl. (Khim. Tekhnol.), Akad. Nauk SSSR, Inst. Goryuch. Iskop.*, 198 (1965).

[204] N. V. Novitskii and N. I. Ivanova, *Khim. Tverd. Topl.*, No. 4, 64 (1973).

[205] S. S. Asthana, *Chem. Ind.*, 190 (1967).

[206] R. Belcher, *Fuel,* **29,** 232 (1950).

[207] N. Biswas, S. N. Mukherjee, P. B. Dutta, and A. Lahiri, *J. Sci. Ind. Res., India,* **21D**(3), 101 (1962).

[208] British Standard 1016, Part 9, 1971.

[209] F. Ellington and W. N. Adams, *Fuel,* **30,** 272 (1951).

[210] T. Fujimori, T. Tokesue, and K. Ishikawa, *Nenryo Kyokai-shi,* **52**(2), 89 (1973).

[211] T. Fujimori, Y. Tsuru, and K. Ishikawa, *Nenryo Kyokai-shi,* **52**(2), 101 (1973).

[212] S. Heilpern, *Koks Smola Gaz.* No. 1, 2–6 (1961).

[213] B. P. Kirk and H. C. Wilkinson, *Talanta,* **19,** 80–82 (1972).

[214] B. P. Kirk and H. C. Wilkinson, *Fuel,* **43,** 105–109 (1964).

[215] N. V. Novitskii and N. I. Ivanova, *Khim. Tverd. Topl.*, No. 3, 133–135 (1971).

[216] N. K. Prasad, *Chem. Ind.*, No. 14, 444–445 (1968).

[217] R. P. Puri and S. P. Banerjee, *J. Sci. Ind. Res.* **India**), ***10B**, 86–88 (1951).*

[218] W. Radmacher, R. Walter, and W. Schmitz, *Brennstoff-Chemie,* **31,** 233 (1950).

[219] G. K. Shopov, L. A. Kostova, S. K. Stoyanov, D. D. Ruschev, E. E. Bekyarova, and I. S. Boshov, *Khim. Ind. (Sofia)*, **46**(7), 317–320 (1974).

[220] Lidia Wnekowska and S. Czubek, *Pr. Gl. Inst. Gorn., Katowice, Komun.*, 83 (1951).

[221] F. M. Kessler and L. Dockalova, *Zb. Pr. Vyzk., Ust.*, **A8,** 188–213 (1957).

[222] L. J. Edgcombe and D. R. Hewett, *Analyst,* **79,** 755–758 (1954).

[223] K. Dixon, *Analyst,* **83,** 362–371 (1958).

[224] P. J. Jackson and A. C. Smith, *J. Appl. Chem. (London)*, **6,** 547–559 (1956).

[225] W. Schuhknecht and H. Schinkel *Brennstoff-Chemie,* **38,** 275–277 (1957).

[226] W. Schuhknecht and H. Schinkel, *Brennstoff-Chemie,* **42,** 292–294 (1961).

[227] Yu. N. Pak and L. P. Starchik, *Koks Khim.*, No. 11, 10–11 (1974).

[228] H. J. Gluskoter and R. R. Ruch, *Fuel,* **50,** 65–67 (1971).

[229] R. Kurosawa, *Nippon Kogyo Kaishi,* **84,** 101–108 (1968).

[230] L. Loska and L. Gorski, *Radiochem. Radioanalyt. Lett.*, **10**(6), 315–322 (1972).

[231] T. C. Martin, J. T. Prud'homme, and I. L. Morgan, AEC Accession No. 33249, Rep. No. AED-CONF-65-155-10, Available from Gmelin, 1966, 16 pp.

[232] L. V. Navalikhin, V. A. Kireev, and Y. N. Talanin, *Khim. Tverd. Topl.*, **4,** 149–151 (1970).

[233] N. P. Fedorovskaya, I. M. Kazakova, and N. L. Surina, *Tr. Inst. Goryuch. Iskop. Min. Ugol. Prom. SSSR,* **23**(2), 14–19 (1967).

[234] P. J. Jackson, *J. Appl. Chem.*, **7,** 605–610 (1957).

[235] W. Radmacher and W. Schmitz, *Brennstoff-Chemie,* **40,** 161–164 (1959).

and magnesium,[236] potassium and sodium,[237–239] phosphorus,[240–244] and silicon[245,246]; electrochemical procedures for calcium,[247] magnesium,[247,248] and titanium.[248] Reviews and comparison of various techniques for the analysis of phosphorus[240,249–252] have been reported.

7.2.3 Determination of Trace Elements

Analysis of trace elements in coal has evolved from early semiquantitative emission spectrographic techniques and a variety of colorimetric procedures to the more recently developed and widely utilized instrumental approaches. Generally these newer methods are more sensitive for most elements and are faster and less tedious. Unless matrix effects are adequately resolved, however, the accuracy and precision of these methods may be unacceptable.

It is not possible, within the framework of this chapter, to discuss the merits of each analytical discipline. These topics are comprehensively covered in the annual reviews of *Analytical Chemistry.*

Since 1969 several excellent reviews of methodology for analysis of trace elements in coal have appeared,[253–257] including a book that had its origins in a 1973 American Chemical Society Fuel Division Symposium.[258] Extensive bibliographies concerned specifically with activation analysis[259] and atomic absorption analysis of coal[260] are available.

Several evaluative studies of methodology or "round robins" have been conducted on various trace elements in coal. A comparative study by the U.S. Bureau of Mines[261] concerning mercury concluded that double gold amalgamation techniques with atomic absorption constituted the best approach. An EPA multielement round-robin study[262] pointed out certain problems in trace element analysis. Further joint efforts with NBS and other laboratories resulted in a standard coal sample (SRM1632), which is certified for certain trace elements and has informational values for many others.[263,264]

Reviews or comparisons of techniques for specific trace elements have been reported

[236] R. Mineva, *Khim. Ind. (Sofia)*, **33**(3), 90–91 (1961).

[237] R. Belcher, A. J. Nutten, and H. Thomas, *Anal. Chim. Acta.* **11,** 120–127 (1954).

[238] J. M. Cardin Gonzolez and L. R. Pire, *Inst. Nac. Carbon (Oviedo, Spain), Bol. Inf.*, **7,** 227–233 (1958).

[239] D. Logie and J. E. Rayner, *J. Inst. Fuel,* **26,** 146–150 (1953).

[240] W. Radmacher and W. Schmitz, *Brennstoff-Chemie,* **39,** 274–279 (1958).

[241] B. P. Bhaduri, S. K. Ghose, and S. K. Majumdor, *Indian J. Technol.*, **2**(2), 62–64 (1964).

[242] A. C. Bhattacharyya, B. P. Bhaduri, and N. F. Banerjee, *Analyst,* **86,** 195–198 (1961).

[243] R. J. Cosstick and H. N. S. Schafer, *Fuel,* **38,** 277–281 (1959).

[244] S. S. Roy and S. Gupta, *J. Sci. Ind. Res. (India)*, **19B,** 495–496 (1960).

[245] H. Laske, *Wiss. Z. Martin-Luther Univ. Halle-Witlenberg, Math.-Nat. Reihe,* **11,** 873–878 (1962).

[246] P. S. Odnopozova and N. V. Novitskii, *Khim. Tverd. Topl.*, No. 1, 62–67 (1968).

[247] L. J. Edgcombe and G. R. E. C. Gregory, *Analyst,* **80,** 236–237 (1955).

[248] F. M. Kessler and L. Dockalova, *Prace ustavu pro vyzkum a Vyuziti Paliv 8,* Publ. No. 17–26 233 1957, pp. 214–233.

[249] K. Diering, *Koks Smola Gaz,* **10**(65), 202–206 (1965).

[250] F. H. Gibson and W. H. Ode, U.S. Bur. Mines Rep. Invest. No. 5743, 1961, 21 pp.

[251] L. Gonzolez, *Inst. Nac. Carbon (Oviedo, Spain), Bol. Inf.*, **11**(54), 58–71 (1962).

[252] A. K. Moitra, S. P. Banerjee, and N. N. Chatterjee, *J. Proc. Inst. Chem. (India)*, **29,** 203–206 (1957).

[253] P. Averitt, I. A. Breger, V. E. Swanson, P. Zubovic, and H. Gluskoter, U.S. Geol. Survey Prof. Paper No. 800-D, 1972 pp. 169–171.

[254] R. W. Freedman and A. G. Sharkey, Jr., *Ann. N.Y. Acad. Sci.*, **200,** 7–16 (1972).

[255] B. S. Ignasiak, T. M. Ignasiak, and N. Berkowitz, *Rev. Anal. Chem.*, **2**(3), 278–298 (1975).

[256] J. Konieczynski, *Koks Smola Gaz,* **14**(1), 13–18 (1969).

[257] U. S. Geological Survey Open File Rep. No. 76-481, 1976, 16 pp.

[258] *Trace Elements in Fuel,* S. P. Babu, Ed. Am. Chem. Soc. Adv. Chem. Ser. No. 141, 1975, 216 pp.

[259] G. J. Lutz, R. J. Boreni, R. S. Maddock, and J. Wing, U.S. Nat. Bur. Standards Tech. Note No. 467, 1972, 892 pp.

[260] I. Lang, O. Weisser, and V. Sychra, *Ropa Uhlie,* **17**(9), 499–509 (1975).

[261] M. D. Schlesinger and H. Schultz, U.S. Bur. Mines Rep. Invest. No. 7609, 1972, 11 pp.

[262] D. J. von Lehmden, R. H. Jungers and R. E. Lee, Jr., *Anal. Chem.*, **46**(2), 239–245 (1974).

[263] P. D. LaFleur, *Proc. Symp. Comp. Stud. Food. Environ. Contam.*, 1973, pp. 489–496.

[264] J. M. Ondov, W. H. Zoller, I. Olmez, N. K. Aras, G. E. Gordon, L. A. Rancitelli, K. Abel, R. H. Filby, K. R. Shaw, and R. C. Ragaini, *Anal. Chem.*, **47**(7), 1102–1109 (1975).

for arsenic,[265] gallium,[266] germanium,[267] and mercury.[261,268–271]

Most trace element references involve multielement studies. Since no single method applies to all elements, a multidiscipline approach to multielement analysis is more successful. Several laboratories have recently conducted surveys of trace elements in coals on which multidiscipline approaches had been used. The Oak Ridge National Laboratory[272] utilized instrumental neutron activation analysis, spark-source mass spectrometry, and atomic absorption spectrophotometry to study trace element mass balances in a coal-fired steam plant. In a similar study Kaakinen[273] determined 17 elements in the inlets and outlets of a pulverized coal-fired power plant by atomic absorption, X-ray fluorescence, wet chemical methods, and radiochemical analysis. The U.S. Geological Survey[187,274] surveyed some 799 U.S. coals for up to 44 elements by means of atomic absorption, optical emission, X-ray fluorescence, ion selective electrode, neutron activation, and colorimetric methods. The Illinois State Geological Survey[99,275] analyzed 170 U.S. coals from different coal-producing regions for up to 61 elements utilizing instrumental neutron activation analysis, radiochemical separations, atomic absorption, optical emission, X-ray fluorescence, and ion selective electrode procedures. A Soviet group[276] surveyed the Lvov-Volyn Basin for 24 elements using a combination of flame and spectrometric procedures. Pollock[277] has developed a multidisciplinary approach to coal analysis involving spark-source mass spectroscopy, atomic absorption (flame and flameless), colorimetry, and the ion selective electrode method.

There has been a sustained effort by many investigators to explore and develop a particular technique for the analysis of as many elements as possible. One must realize that any procedure when developed may not apply equally to all elements tested. The following discussion lists some of the available literature for various analytical disciplines.

In recent years many laboratories have attempted to develop a multielemental instrumental neutron activation analysis (INAA) scheme for coal.[190,278–289] High-resolution lithium-doped germanium [Ge(Li)] detectors have made it possible to determine up to 50 elements quantitatively in a single sample with varying degrees of accuracy and precision. This technique analyzes both whole

[265] Institute of Brewing Analysis Committee, *J. Inst. Brew.*, **50**, 462–463 (1954).

[266] N. I. Desyatskaya, *Issled. Ispol'z. i Standartizatsiya Uglei, Donetsk. Nauchn.—Issled. Ugol'n. Inst.*, 1965, pp. 125–131.

[267] M. Basinska and W. Rutkowski, *Chem. Anal. (Warsaw)*, **8**(3), 353–360 (1963).

[268] S. Chilov, *Talanta*, **22**(3), 205–232 (1975).

[269] J. V. O'Gorman, N. H. Suhr, and P. L. Walker, Jr., *Appl. Spectrosc.*, **26**, 44–48 (1972).

[270] M. D. Schlesinger and H. Schultz, U.S. Bur. Mines, Tech. Prog. Rep. No. 43, 1971, 4 pp.

[271] J. Weaver and D. J. Lehmden, *Abstr. 163rd National Meeting, American Chemical Society*, Boston, April 1972.

[272] N. E. Bolton, J. A. Carter, J. F. Emery, C. Feldman, W. Fulkerson, L. D. Hulett, and W. S. Lyon, in Am. Chem. Soc. Adv. Chem. Ser. No. 141, 1975, pp. 175–178.

[273] J. W. Kaakinen, *Environ. Sci. Technol.*, **9**(9), 862–869 (1975).

[274] V. E. Swanson and C. Huffman, Jr., U.S. Geol. Survey Circ. No. 735, 1976, 11 pp.

[275] H. J. Gluskoter, R. R. Ruch, W. G. Miller, R. A. Cahill, G. B. Dreher, and J. K. Kuhn, Illinois State Geol. Survey Circ. No. 499, 1977, 154 pp.

[276] V. S. Tikhonova, M. Yu. Fedushchak and S. B. Kazakov, *Geol. Geokhim. Goryuch. Iskop.*, No. 34, 104–107 (1973).

[277] E. N. Pollock, in Am. Chem. Soc. Adv. Chem. Serv. No. 141, 1975, pp. 23–34.

[278] K. H. Abel and L. A. Rancitelli, in Am. Chem. Soc. Adv. Chem. Ser. No. 141, 1975, pp. 118–138.

[279] S. Akcetin, E. Ayca, and J. Hoste, *Radiochem. Radioanal. Lett.*, **15**(1), 13–28 (1973).

[280] C. Block and R. Dams, *Anal. Chim. Acta*, **68**(1), 11–24 (1974).

[281] G. E. Gordon, *Proc. Conf.-Identification Measurement Environmental Pollutants*, B. Westley, (Ed., National Research Council of Canada, Ottawa, 1971), Paper No. 701668, p. 138.

[282] W. S. Lyon and J. F. Emery, *Int. J. Environ. Anal. Chem.*, **4**(2), 125–133 (1975).

[283] H. T. Millard, *Trans. Am. Nuclear Soc.*, **21**, 108–109 (1975).

[284] R. A. Nadkarni, *Radiochem. Radioanal. Lett.*, **21**(3–4), 161–176 (1975).

[285] L. A. Rancitelli, Rep. BNWL-1751 (Part 2), Battelle Northwest Laboratory 1973, pp. 56–57.

[286] R. R. Ruch, R. A. Cahill, J. K. Frost, L. R. Camp, and H. J. Gluskoter, *Trans. Am. Nuclear Soc.*, **21**, 107–108 (1975).

[287] D. W. Sheibley, in Am. Chem. Soc. Adv. Chem. Ser. No. 141, 1975, pp. 98–117.

[288] E. Steinnes and J. J. Rowe, *Anal. Chim. Acta*, **87**(2), 451–462 (1976).

[289] N. Tamura, *Radiochem. Radioanal. Lett.*, **18**(3), 135–142 (1974).

coal and coal ash accurately, thus allows checking for volatility losses.

Neutron activation analysis utilizing radiochemical separations (NAA-RC) is employed by investigators when the sensitivity for a particular element or group of elements is inherently low or when spectral interference for a given element in a specific matrix is too great to be adequately detected. This situation was more prevalent before the advent of Ge(Li) spectrometry when only low-resolution NaI(T1) detectors were available. Perez[290] utilized both NaI spectrometry and radiochemical separations to determine seven trace elements in coal ash. Frost et al.[291] have reviewed the various radiochemical procedures developed specifically for coal. An excellent group separation procedure for selenium, arsenic, zinc, cadmium, and mercury in whole coal has been reported by Orvini et al.[292]

Neutron activation analysis procedures (both INAA and NAA-RC) developed for specific elements include arsenic,[293] gold and rhenium,[294] cesium,[295,296] fluorine,[297] gallium,[298] germanium,[232] mercury,[299–302] rubidium,[295] selenium,[302,303] tantalum,[296] thorium,[304,305] and uranium.[304–307]

Although thermal (slow) neutrons derived from nuclear reactors are the most practical source of particles for nuclear excitation and generally provide the more useful reaction, that is, (n, γ), other excitation sources such as 14-MeV (fast) neutrons from commercially available accelerators or generators have also been applied to coal analysis. Block and Dams[308] and Sattarov et al.[189] used these "fast" reactions (n, p) and (n, α) to analyze multiple trace and major elements in coal.

Nuclear reactions such as (γ, n), accomplished by generating high-energy gamma rays or photons from the bombardment of particles with metal targets, may be applied to chemical analysis. Zoller[191] used this photon activation technique to determine up to 19 elements in coal. Utilizing a similar approach, Berzin et al.[309] determined several elements in coal including carbon, oxygen, and fluorine.

"Prompt gamma analysis," in which the instantaneous or "prompt" gamma ray emitted is detected when a nucleus traps a neutron, was evaluated by Rasmussen and Hukai[310] for possible application to coal. Gladney et al.[311] determined boron and cadmium in coal by a similar technique. Rasmussen[312] also investigated the utilization of neutrons from portable ^{252}Cm sources as an alternate approach to thermal neutron activation analysis. Other nuclear techniques in-

[290] A. M. Perez, *Ensayos Invest.*, **1**(1), 18–28 (1966).

[291] J. K. Frost, P. M. Santoliquido, L. R. Camp, and R. R. Ruch, in Am. Chem. Soc. Adv. Chem. Ser. No. 141, 1975, pp. 84–97.

[292] E. Orvini, T. E. Gills, and P. D. LaFleur, *Anal. Chem.*, **46,** 1294–1297 (1974).

[293] P. M. Santoliquido, *Radiochem. Radioanal. Lett.*, **15**(6), 373–377 (1973).

[294] G. S. Nikanorov, Yu. N. Samoilenko, and A. S. Chulyakov, *Yad. Metody Anal. Veshchestv.*, 42–46 (1971).

[295] A. A. Smales and L. Salmon, *Analyst,* **80,** 37–50 (1955).

[296] N. A. Solodov, Yu. P. Kostin, E. S. Meitov, V. I. Bel'kov, A. P. Zyryanov, and E. N. Pankov, *2nd Geol. Syr'evye Resur. Redk. Elem. SSSR, Tezisy Dokl., Vses. Soveshch.*, 1974, pp. 51–52.

[297] A. Ya. Botvina, V. V. Sokol'skii and N. R. Faizrakhmanova, *2nd Aktiv. Anal. Mater. Vses. Soveshch.*, (published 1971), pp. 189–191.

[298] P. M. Santoliquido and R. R. Ruch, *Radiochem. Radioanal. Lett.*, **12**(1), 71–76 (1972).

[299] E. J. Kennedy, R. R. Ruch, H. J. Gluskoter, and N. F. Shimp, *Proc. Am. Nuclear Soc. Top. Meet. Nucl. Methods Environ. Res.*, 1971, 205–215.

[300] H. L. Rook, T. E. Gills, and P. D. LaFleur, *Anal. Chem.*, **44,** 1114–1117 (1972).

[301] R. R. Ruch, H. J. Gluskoter, and E. J. Kennedy, Illinois State Geol. Survey Environ. Geol. Note No. 43, 1971, 15 pp.

[302] J. N. Weaver, *Anal. Chem.*, **45**(11), 1950 (1973).

[303] K. K. S. Pillay, C. C. Thomas, Jr., and J. W. Kaminski, *Nuclear Appl. Technol.*, **7,** 478–483 (1969).

[304] Zh. Ganzorig, L. Doshzeveg, O. Otgonsuren, G. Khuukhenkhuu, I. Chadraabol, and D. Chultem, *Radiokhimija,* **16**(2), 247–253 (1974).

[305] H. G. Meyer, *J. Radioanal. Chem.*, **7**(1), 67–79 (1971).

[306] D. C. Perricos and E. P. Belkas, *Talanta,* **16**(6), 745–748 (1969).

[307] J. N. Weaver, *Anal. Chem.*, **46,** 1292–1294 (1974).

[308] C. Block and R. Dams, *Environ. Sci. Technol.*, **9**(2), 146–150 (1976).

[309] A. K. Berzin, V. V. Sulin, V. I. Belov, G. Ch. Vitozhents, Yu. T. Martynov, V. G. Suslov, and S. I. Shornikov, *Proc. Symp. Radioisotope Instruments Industry Geophysics,* Warsaw, 1965, 34 pp.

[310] N. C. Rasmussen and Y. Hukai, *Trans. Am. Nuclear Soc.*, **10,** 29–30 (1967).

[311] E. S. Gladney, E. T. Jurney and D. B. Curtis, *Anal. Chem.*, **48**(14), 2139–2142 (1976).

[312] N. C. Rasmussen, *Trans. Am. Nuclear Soc.*, **14**(2), 931 (1971).

clude fission track[313] and autoradiographic[314] procedures for determining uranium.

Optical emission spectroscopy has been extensively applied to the quantitative and semiquantitative determination of multiple trace elements in coal ash[315–326] with a variety of excitation and optical parameters, internal standards, calibration standards, and procedures. Hegemann et al.[327] utilized a spark excitation technique to analyze whole coal for trace and major elements. Luedke and Holdt[328] employed a plasma arc approach for determining trace elements in pitch coke. Epstein and O'Haver[329] developed a flame emission spectrometric technique with repetitive scanning to determine six elements in coal ash.

Specific optical emission procedures have been developed for arsenic,[330,331] boron,[332–335] beryllium,[336] cadmium and zinc,[337] gallium,[338–341] germanium,[336,338,339,342–359] mercury,[360] mol-

[313] I. G. Berzina, D. P. Popenko, and Yu. S. Shimilevich, *Geokhimiya,* **8,** 1024–1027 (1969).

[314] E. Bujdoso, L. Medveczky, and A. Szalay, *Acta Phys. Acad. Sci. Hung.,* **8,** 195–201 (1957).

[315] R. F. Abernethy, M. J. Peterson, and F. H. Gibson, U.S. Bur. Mines Rep. Invest. No. 7281, 1969, 20 pp.

[316] I. Benko and G. Szadeczky-Kardoss, *Magy. Kem. Foly.,* **63,** 78–84 (1957).

[317] C. G. Carlsson and A. Larsson, *Jernkontorets Ann.,* **138,** 737–743 (1954).

[318] G. B. Dreher and J. A. Schleicher, in Am. Chem. Soc. Adv. Chem. Ser. No. 141, 1975, pp. 35–47.

[319] A. J. W. Headlee and R. G. Hunter, West Virginia Geol. Survey No. 13A, 1955, pp. 36–122.

[320] R. L. O'Neil and N. H. Suhr, *Appl. Spectrosc.,* **14,** 45–50 (1960).

[321] M. J. Peterson and J. B. Zink, U.S. Bur. Mines Rep. Invest. No. 6496, 1964, 15 pp.

[322] W. Radmacher and H. Hessling, *Z. Anal. Chem.,* **167,** 172–182 (1959).

[323] P. D. Rao, Alaska Univ. Mineral Industry Resources Lab. Rep. No. 15, 1968, 47 pp.

[324] B. I. Smirnov and L. P. Smirnova, *Geol. Geokhim. Goryuch., Iskop.,* **11,** 94–98 (1967).

[325] A. C. Smith, *J. Appl. Chem. (London),* **8,** 636–645 (1958).

[326] J. Szava and K. Szava Benocs, *Banyasz. Kut. Intez., Koslem.,* **11**(1–2), 339–349 (1967).

[327] F. Hegemann, K. Giesen, and H. Kostyra, *Ber. Dtsch. Ceram. Ges.,* **36,** 145–149 (1959).

[328] C. Luedke and G. Holdt, *Izv. Fiz. Inst. ANEB (At. Nauchnoeksp. Bazc), Bulg. Akad. Nauk,* **21,** 109–114 (1971).

[329] M. S. Epstein and T. C. O'Haver, *Spectrochim. Acta, B,* **30**(A), 135–146 (1975).

[330] R. H. Hall and H. L. Lovell, *Anal. Chem.,* **30,** 1665–1669 (1958).

[331] N. A. Kekin and V. E. Marincheva, *Zavod. Lab.,* **36,** 1061–1063 (1970).

[332] J. Konieczynski, *Pr. Gl. Inst. Gorn., Komun. Stalinogrod Ser. B,* 254, (1960).

[333] V. N. Kryukova, E. Shishlyannikova, and O. I. Koralis, *Geokhimiya,* No. 7, 1115–1119 (1975).

[334] J. Millet, *Congr. Groupe Avancé, Methodes Anal. Spectrogr. Prod. Met. 104,* Paris, 1957, pp. 229-235.

[335] S. Skalska and S. Held, *Chem. Anal. (Warsaw),* **1,** 294–300 (1956).

[336] O. A. Kulskaya and O. F. Vdovenko, *Khim. Fiz-Khim. i Spektraln. Metody Issled. Rsed Redkikk i Rasseyan. Elementov, Min. Geol. i Okhrany Nedr SSSR,* 1961, pp. 135–139.

[337] Y. Talmi, *Anal. Chem.,* **46**(8), 1005–1010 (1974).

[338] A. N. Bronshtein, T. I. Sendulskaya, and M. Ya. Shpirt, *Zavod. Lab.,* **26,** 973–974 (1960).

[339] A. Idzikowski and M. Sachanbinski, *Pr. Nauk Inst. Chem. Nieorg. Metal. Pierwiastkow Rzodkich Politech. Wroclaw,* No. 32, 145–152 (1975).

[340] M. L. Naser, A. B. Basily, and A. M. Raafat, *Indian J. Technol.,* **12**(8), 359–361 (1974).

[341] J. Widawska-Kusmierska, *Bull. Acad. Pol. Sci., Ser. Sci. Terre.,* **21**(1), 1–7 (1973).

[342] N. V. Arnautov, *Izv. Vost. Fil. Akad. Nauk, SSSR,* No. 1, 49–52 (1957).

[343] R. C. Corey, J. W. Myers, C. H. Schwartz, F. H. Gibson, and P. J. Colbassini, U.S. Bur. Mines, Bull. No. 575, 1959, 68 pp.

[344] W. J. Frederick, J. A. White, and H. E. Biber, *Anal. Chem.,* **26,** 1328–1330 (1954).

[345] N. C. Ganguly and D. P. Dutta, *J. Sci. Ind. Res. (India),* **15B,** 327–328 (1956).

[346] Z. Gregorowicz, *Chem. Anal. (Warsaw),* **3,** 777–782 (1958).

[347] J. E. Hawley, *Econ. Geol.,* **50,** 517–532 (1955).

[348] J. S. Machin and Juanita Witters, Illinois State Geol. Survey Circ. No. 216, 1956, 13 pp.

[349] M. Malinek, *Anal. Chim. Acta,* **19,** 502–504 (1958).

[350] S. Mantea, M. Petrescu, and V. Trita, *Acad. Rep. Pop. Rom., Stud. Cercet. Met.,* **4,** 537–545 (1959).

[351] M. Martin and M. Roca, *An. Quim.,* **67,** 585–590 (1971).

[352] G. A. Monnot, *Chim. Anal.,* **35**(11), 274–276 (1953).

[353] A. M. Perez, *Bol. Inst. Nac. Carbon,* **12,** 110–149 (1963).

[354] G. J. Pitt and M. F. Fletcher, *Spectrochim. Acta,* **7,** 214–218 (1955).

[355] E. A. Pometun, T. I. Chervina, and V. G. Tsygankova, *Izv. Akad. Nauk Tadzh. SSR, Otd. Geol.-Khim. Tekh. Nauk,* **2**(8), 19–26 (1962).

[356] K. P. Puzanova and R. I. Perkova, *Izv. Vyssh. Uchebn. Zaved., Gorn. Zh.,* No. 11, 5–10 (1966).

[357] J. A. Schleicher and W. W. Hambleton, Kansas State Geol. Survey Bull. No. 109, 1954, pp. 113-124.

[358] G. Szadeczky-Kardoss and I. Benko, *Acta Chim. Acad. Sci. Hung.,* **8,** 241–262 (1955).

[359] Misturu Ura, *Nippon Kagaku Zasshi,* **78,** 316–320 (1957).

[360] E. S. Peck, *Anal. Chim. Acta,* **80**(1), 75–83 (1975).

ybdenum,[339] manganese,[198] rubidium,[361] scandium,[362] selenium,[363] strontium,[361] vanadium,[339,364] and yttrium.[365]

Atomic absorption spectrophotometry (AAS) has been successfully applied to the multielemental analysis of coal by using acid-dissolved solutions of low-or high-temperature ash as samples. Utilizing the numerous analytical parameters and options available for AAS, several laboratories have developed multielemental schemes employing either "flame" or "flameless" excitation.[99,181,366–368] James[369] has reviewed the application of AAS to the analysis of coal ash.

Atomic absorption analysis procedures have been developed for individual elements, including arsenic,[370] beryllium,[371] cadmium,[337,372] chromium,[373] mercury,[374–386] lead,[387] strontium,[388] and zinc.[372]

Flame emission procedures have been developed for cesium,[389,390] and lithium and rubidium.[390]

X-Ray fluorescence analysis has recently been applied successfully to multielement analysis of coal by several groups[185,391,392] utilizing the various excitation and detection parameters available. High-resolution Si(Li) detectors, monochromatic and radioisotopic sources, and multichannel analyzers with computer capability have greatly increased the scope of XRF. Specific XRF procedures have been reported for boron,[333] germanium,[393] hafnium, thorium, and uranium,[394] mercury,[395] and selenium.[396]

Recently the application of proton bombardment for inducing X-ray emission (PIXE) has been highly successful. Simms et al.[397] and Walter et al.[398] have evaluated PIXE for multielemental analysis of coal. Shaboson et

[361] W. M. Tupper and D. H. Loring, *Geochim. Cosmochim. Acta,* **24,** 314–315 (1961).

[362] O. A. Kulskaya and O. F. Vdovenko, *Uk. Khim. Zh.,* **23,** 799–804 (1957).

[363] Y. Talmi and A. W. Andren, *Anal. Chem.,* **46**(14), 2122–2126 (1974).

[364] P. Buncok, *Ropa Uhlie,* **5**(5), 144–149 (1963).

[365] N. V. Arnautov and L. D. Shipilov, *Geol. Geofiz.,* **1,** 115–117 (1960).

[366] A. Attari, NTIS PB Rep. No. 223001/9, 1973, 39 pp.

[367] A. M. Hartstein, R. W. Freedman, and D. W. Platter, *Anal. Chem.,* **45,** 611–614 (1973).

[368] K. Mizuno, K. Yokogawa, N. Yasuda, and K. Hozumi, *Bunseki Kogaku,* **23**(9), 1010–1015 (1975).

[369] R. G. James, *BCURA Month. Bull.,* **33**(9), 204–207 (1969).

[370] G. I. Spielholtz, G. C. Toralballa, and R. J. Steinberg, *Mikrochim. Acta,* No. 6, 918–923 (1971).

[371] J. Owens and E. S. Gladney, *At. Absorpt. Newsl.,* **14**(4), 76–77 (1975).

[372] H. J. Gluskoter and P. C. Lindahl, *Science,* **181,** 264–266 (1973).

[373] K. Dittrich and G. Liesch, *Talanta,* **20**(7), 691–694 (1973).

[374] D. H. Anderson, J. H. Evans, J. J. Murphy, and W. W. White, *Anal. Chem.,* **43**(11), 1511–1512 (1971).

[375] B. W. Bailey and F. C. Lo, *J. AOAC,* **54,** 1447–1448 (1971).

[376] C. Feldman, *Anal. Chem.,* **46,** 1606–1609 (1974).

[377] H. Heinrichs, *Z. Anal. Chem.,* **273**(3), 197–201 (1975).

[378] C. Huffman, Jr., R. L. Rahill, V. E. Shaw. and D. R. Norton, U.S. Geol. Survey Prof. Paper No. 880-c, 1972, pp. 203–207.

[379] O. I. Joensuu, *Appl. Spectrosc.,* **25**(5), 526–528 (1971).

[380] G. W. Kalb, in Am. Chem. Soc. Adv. Chem. Ser. No. 141, 1975, pp. 154–174.

[381] G. W. Kalb, *Abstr. 163rd Natl. Meet. American Chemical Society,* Boston, April 1972.

[382] Fa-chun Lo and Brian Bush, *J. AOAC,* **56**(6), 1509–1510 (1973).

[383] J. Murphy, *At. Absorpt. Newsl.,* **14**(6) 151–152 (1975).

[384] T. C. Rains and O. Menis, *J. AOAC,* **55**(6), 1339–1344 (1972).

[385] D. Siemer and R. Woodriff, *Anal. Chem.,* **46**(4), 597–598 (1974).

[386] R. C. Streeter, *Abstr., 163rd Natl. Meet. American Chemical Society,* Boston, April, 1972.

[387] C. Block, *Anal. Chim. Acta,* **80**(2), 369–373 (1975).

[388] C. B. Belcher and K. A. Brooks, *Anal. Chim. Acta,* **29**(3), 202–205 (1963).

[389] J. Medek, *Paliva,* **36,** 370–374 (1956).

[390] V. F. Razdorozhnyi and A. F. Petrus, *Geokhim. Rudoobraz.,* No. 3, 42–44 (1974).

[391] M. Berman and S. Ergun, U.S. Bur. Mines Rep. Invest. No. 7124, 1968, 20 pp.

[392] C. J. Sparks, Jr., O. B. Cavin, L. A. Harris, and J. C. Ogle, in *Proc. Conf. Trace Substances Environmental Health,* Columbia, Mo., 1973, Paper No. 730613-4.

[393] W. J. Campbell and H. F. Carl, *Anal. Chem.,* **29,** 1009–1017 (1957).

[394] E. Donderer, *Neues Jahrb. Miner. Mh.,* No. 4, 167–175 (1972).

[395] R. L. Mykebust, M. M. Darr and K. F. J. Heinrick, Rep. No. COM-75-10686, U.S. Dept. Commerce, 1974, 21 pp.

[396] T. L. Belopolskaya and I. V. Serikov, *Khim. Tverd. Topl.,* No. 6, 73–79 (1969).

[397] P. C. Simms, F. A. Rickey, and K. A. Mueller, "Progress on Coal-Related Research at Purdue," Purdue Professors Rep., May 1976, pp. 9–12.

[398] R. L. Walter, R. D. Willis, W. F. Gutknecht, and J. M. Joyce, *Anal. Chem.,* **46**(7), 843–855 (1974).

al.[399] have evaluated an approach in which 40-MeV oxygen ions are used to induce fluorescene in coal ash.

Mass spectrometry (MS) is an excellent multielemental technique generally applicable to the analysis of coal. Carter et al.,[400] employing a highly accurate isotopic dilution technique, applied spark-source mass spectrometry (SSMS) and thermal emission mass spectrometry (TEMS) to the determination of some 19 elements in a study of trace element balance in a coal-fired steam plant. A group at the Pittsburgh Energy Research Center used SSMS to analyze more than 50 elements in coal dust.[401–403] Brown et al.[404] have reviewed various aspects of SSMS relevant to coal analysis. A specific procedure for beryllium has been reported.[405]

Many chemically based analytical procedures make use of one or more of approaches such as colorimetry, compleximetry, gravimetry, distillation, chromatography, fluorimetry, or electrochemistry to determine arsenic,[406–416] gold,[417] boron,[418–422] beryllium,[423,424] cadmium,[425] chromium,[426] copper,[426] fluorine,[427–430] gallium,[431–434] germanium,[344,434–456] mercury,[457–461] molyb-

[399] L. Shaboson, B. L. Cohen, G. H. Wedberg, and K. C. Chan, *J. Appl. Phys.*, **44**(10), 4749–4752 (1973).

[400] J. A. Carter, R. L. Walker, and J. R. Sites, in Am. Chem. Soc. Adv. Chem. Ser. No. 141, 1975, pp. 74–83.

[401] T. Kessler, A. G. Sharkey, Jr., and R. A. Friedel, U.S. Bur. Mines Tech. Prog. Rep. No. 42, 1971, 15 pp.

[402] T. Kessler, A. G. Sharkey, Jr., and R. A. Friedel, NTIS PB Rep. No. 214680/1, 1973, 13 pp.

[403] A. G. Sharkey, Jr., T. Kessler and R. A. Friedel, in Am. Chem. Soc. Adv. Chem. Ser. No. 141, 1975, pp. 48–56.

[404] R. Brown, M. L. Jacobs, and H. E. Taylor, *Am. Lab.*, **4**(11), 29–38 (1972).

[405] M. A. Phillips, *Environ. Lett.*, **5**(3), 183–188 (1973).

[406] R. F. Abernethy and F. H. Gibson, U.S. Bur. Mines Rep. Invest. No. 7184, 1968, 10 pp.

[407] R. G. Ault, *J. Inst. Brew.*, **67**(1), 14–28 (1961).

[408] R. G. Ault and A. G. R. Whitehouse, *J. Inst. Brew.*, **62**, 425–427 (1956).

[409] British Standards 1016, Part 10, 1960.

[410] A. Crawford, J. G. Palmer and J. H. Wood, *Mikrochim. Acta*, 277–294 (1958).

[411] S. R. Crook and S. Wald, *Fuel*, **39**, 313–322 (1960).

[412] L. J. Edgcombe and H. K. Gold, *Analyst*, **80**, 155–157 (1955).

[413] Z. Flum, *Paliva*, **37**, 33–36 (1957).

[414] E. Jackwerth and H. G. Kloppenburg, *Z. Anal. Chem.*, **186**, 428–436 (1962).

[415] F. H. Kunstmann and L. B. Bodenstein, *J. S. Afr. Inst. Min. Met.*, **62**, 234–244 (1961).

[416] A. D. Wilson and D. T. Lewis, *Analyst*, **88**, 510–515 (1963).

[417] N. L. Babenko and I. N. Chistyachenko, *Tr. Inst. Metall. Obogoshch., Alma-Ata*, No. 31, 53–55 (1968).

[418] F. J. Allan, F. S. Dundas, and D. A. Lambie, *J. Inst. Fuel*, **42**, 29–30 (1969).

[419] F. H. Kunstmann and J. F. Harris, *J. Chem. Met. Min. Soc. S. Afr.*, **55**, 12–15 (1954).

[420] B. F. Bohor and H. J. Gluskoter, *J. Sediment. Petrol.*, **43**(4), 945–957 (1973).

[421] L. V. Miesserova and N. P. Fedorovskaya, *Tr. Inst. Goryuch. Iskop. Min. Ugol, Prom. SSSR*, **25**(2), 3–9 (1969).

[422] A. E. Vasilevskaya, *Tr. Inst. Miner. Resur., Akad. Nauk Ukr. SSR*, No. 5, 22–31 (1971).

[423] R. F. Abernethy and E. A. Hattman, U.S. Bur. Mines Rep. Invest. No. 7452, 1970, 8 pp.

[424] I. V. Vasileva and V. A. Vekhov, *Met. Koksokhim.*, No. 32, 100-101 (1972).

[425] M. Weclewska, *Pr. Gl. Inst. Gorn., Komun.*, 242 (1960).

[426] N. P. Fedorovskaya and L. V. Miesserova, *Tr. Inst. Goryuch. Iskop., Moscow*, **23**(3), 3–7 (1967).

[427] R. F. Abernethy and F. H. Gibson, U.S. Bur. Mines Rep. Invest. No. 7054, 1967, 13 pp.

[428] P. Hermann, *Ziegelindustrie*, No. 4, 142–145 (1975).

[429] G. E. McGowan, *Fuel*, **39**, 245–252 (1960).

[430] Josephus Thomas, Jr.and H. J. Gluskoter, *Anal. Chem.*, **46**(9), 1321–1323 (1974).

[431] T. F. Borisova, G. F. Guren, L. N. Komissarova, and V. M. Shatskii, *Khim. Tverd. Topl.*, No. 4, 33–36 (1974).

[432] V. A. Nazarenko and S. Ya. Vinkovetskaya, *Zh. Anal. Khim.*, **13**, 327–331 (1958).

[433] M. K. Pakter, A. V. Pershin, L. A. Penchuk, S. S. Parifilova, L. G. Matyuschev, S. V. Postnikova, and R. A. Egorova, *Khim. Tverd. Topl.*, No. 1, 44–47 (1974).

[434] E. Urbanek, *Paliva*, **40**(3), 88–91 (1961).

[435] H. Almond, H. E. Crowe, and C. E. Thompson, U.S. Geol. Survey Bull. No. 1036-B, 1955, pp. 9–17.

[436] H. J. Cluley, *Analyst*, **76**, 523–530 (1951).

[437] L. A. Davidyuk, *Dopov. Akad. Nauk Ukr. RSR*, 401–405 (1962).

[438] M. Finkelsteinaite and J. Burskiene, *Liet. TSR Aukst. Mokyklu Mokslo Darb., Chem. Technol.*, No. 13, 47–52 (1971).

[439] M. Inagaki, *Nenryo Kyokai-shi*, **46**(485), 684–688 (1967).

[440] J. G. Iyer and N. Sundaram, *Indian Min. J.*, **3**(9), 182–183 (1955).

[441] N. D. Khizhnyak, L. S. Tsebrii, and Z. K. Lenkevick, *Ukr. Khimi Zh.*, **37**(4), 359–362 (1971).

[442] F. H. Kunstmann and E. F. E. Müller, *Analyst*, **84**, 324–325 (1959).

[443] M. A. Menkovskii, S. A. Gordon, and N. N. Nurminskii, *Zavod. Lab.*, **29**, 797–799 (1963).

denum,[443,445,462] manganese,[463] nickel,[464] lead,[425,465] rhenium,[466] scandium,[467–470] selenium,[396,471] uranium,[472–478] vanadium,[426,479] and zinc.[425,480–482]

7.2.4 Determination of Radioactivity

Various surveys using conventional counting procedures have been conducted for gross and specific radioactivity in coal and coal ash. Uranium and radium assay has been accomplished by measuring the emitted alpha radiation.[483,484] Gamma-ray counting has been used for estimating uranium, thorium, and potassium radioactivity in coal and lignite.[485] A combination of alpha-, beta-, and gamma-ray counting techniques[486] has been used to study coal dust inhalation. Other assays include the genesis and distribution of radioactivity in Soviet coals,[487] the comparison of thorium and uranium content before and after combustion in a coal-burning plant,[488] and radium content of flue ash.[489]

---

444 Hiroshi Nishida, *Bunseki Kagoku*, **5**, 389–392 (1956).

445 N. N. Nurminskii, *Khim. Tverd. Topl.*, No. 3, 128–130 (1971).

446 R. M. Saginashnili and V. I. Petrashen, *Zavod. Lab.*, **32**(6), 661–663 (1966).

447 G. E. Salikova and N. N. Sevryukov, *Zavod. Lab.*, **36**, 25 (1970).

448 M. L. Sazonov, T. E. Alymova, M. S. Selenkina, and A. A. Zhukhovitskii, *Khim. Tverd. Topl.*, No. 3, 64–70 (1968).

449 M. S. Selenkina, M. L. Sazonov, and A. A. Zhukhovitskii, *Tr. Vses. Nauchn.-Issled. Geologorazved. Neft. Inst.*, No. 112, 152–155 (1973).

450 D. P. Scherbov, R. N. Plotnikova, and I. N. Astaf'eva, *Zavod. Lab.*, **35**(5), 528–530 (1970).

451 N. I. Shuvalona, *Tr. Vses. Nauchn.-Issled. Geol. Inst.*, **117**, 73–74 (1964).

452 B. G. Simek, F. Coufalik, and A. Stadler, *Zpravg Ustavu Vedecky Vyzkum Uhli v Praze*, 1948, pp. 35–36.

453 P. Skrivan, *Sb. Vys. Sk. Chem. Techinol.-Pr., Miner.*, **6**, 391–394 (1962).

454 N. Wai and J. S. Wang, *Bull. Inst. Chem., Acad. Sin.*, **16**, 59–67 (1969).

455 Jy-Shik Wang, *Bull. Inst. Chem. Acad. Sin.* **12**, 34–39 (1966).

456 M. Weclewska and G. Popanda, *Chem. Anal. (Warsaw)*, **3**, 889–892 (1958).

457 N. Kh. Aidin'yan, *Byul. Nauchn. Tekhn. Inf. Min. Geol. Okhr. Nedr. SSSR*, **4**, 70–71 (1961).

458 M. K. Pakter, *Koks Khim.*, No. 7, 43–46 (1968).

459 A. E. Vasilevskaya and V. P. Shcherbakov, *Dopov. Akad. Nauk Uhr. RSR*, 1494–1496 (1963).

460 A. E. Vasilevskaya, V. P. Shcherbakov, and E. V. Karokozova, *Zh. Anal. Khim.*, **19**(10), 1200–1203 (1964).

461 A. E. Vasilevskaya, V. P. Shcherbakov, and V. I. Klimenchuk, *Zavod. Lab.*, **28**, 415 (1962).

462 E. Bekyarova, E. Khristova, and D. Ruschev, *God. Khim.-Tekhnol. Inst.*, **11**(1), 105–110 (1964).

463 F. M. Kessler and L. Dockalova, *Paliva*, **35**, 178–181 (1955).

464 N. P. Fedorovskaya and L. V. Miesserova, *Tr. Inst. Goryuch. Iskop., Moscow*, **23**, 3–7 (1967).

465 F. Cuttitta and J. J. Warr, Jr., *Geochim. Cosmochim. Acta*, **13**, 256–259 (1958).

466 V. V. Kuznetsova, *Zh. Anal. Khim.*, **16**(6), 736–737 (1961).

467 M. P. Belopolskii and N. P. Popov, *Zavod. Lab.*, **30**(12), 1441–1442 (1964).

468 G. Ya. Gornyi, *Ukr. Khim. Zh.*, **23**, 792–798 (1957).

469 B. Ya. Kaplan and I. V. Ol'shevskaya, *Zavod. Lab.*, **29**, 26–27 (1963).

470 N. A. Suvorovskaya and I. A. Shmarinova, *Tr. Inst. Obogashch. Tverd. Goryuch. Iskop.*, **1**(2), 127–131 (1972).

471 J. D. Sidelnikova, *Fiz.-Khim. Metody Anal. Knotr. Proizvod., Mater. Konf. Rab. Vuzov (Vyssh. Uchekn. Zaved.) Zavod. Lab. Yugo-Vostoka SSSR, 4th*, **3**, 54 (1971).

472 Gyula Almossy, *Magy. Tud. Akad. Kem. Tud. Osztolyanak Kozl.* **7**, 338–343 (1956).

473 N. S. Guttag and F. S. Grimaldi, U.S. Geol. Survey Bull. No. 1006, 1954, pp. 111–119.

474 M. Inagaki and U. Tokuga, *Tanken*, **10**, 97–107 (1959).

475 J. Korkisch, A. Farag, and F. Hecht, *Z. Anal. Chem.*, **161** 92–100 (1958).

476 B. C. Purkayastha and S. R. Saha, *Indian J. Appl. Chem.*, **25**(1), 20–23 (1962).

477 T. Schönfeld, M. El Garhy, C. Friedmann and J. Veselsky, *Mikrochim. Acta*, No. 5–6, 883–897 (1960).

478 C. Ujhelyi, *Mag. Kem. Foly.*, **61**(12), 437–442 (1955).

479 K. Sugawara, M. Tanaka, and A. Kozawa, *Bull. Chem. Soc. Jap.*, **28**, 492–494 (1955).

480 V. Strnad, *Sb. Pr. U.V.P.* (*Ustov Zyvk. Vyuziti Paliv*), No. 14, 56–61 (1969).

481 W. Thürauf and H. Assenmacher, *Brennstoff-Chemie*, **44**, 21–22 (1963).

482 C. Weaver, *Fuel*, **46**, 407–414 (1967).

483 R. J. Bayliss and H. M. Whaite, *Air Water Pollut.*, **10**, 813–819 (1966).

484 T. Heugi, *Schweiz, Mineral. Petrogr. Mitt.*, **46**, 377–387 (1966).

485 I. S. Safonov, *Elek. Sta.*, No. 3, 52–53 (1974).

486 I. P. Sklyarenko, P. A. Enyakova, F. E. Krigman, M. D. Krivitski, A. P. Kolesnik, and A. K. Tsapuk, *Gig. Tr. Prof. Zabol.*, **8**, 13–19 (1964).

487 A. A. Gipsh, G. G. Kapaturin, and Ya. E. Yudovich, *Izv. Vepsh. Uchebm. Zaved. Geol. Razved.*, **6**, 61–70 (1970).

488 H. Kirchner, E. Marz, and A. Schiffers, *Braunkohle*, **26**(11), 340–345 (1974).

489 W. Marquardt, R. Hoehle, and U. Schuh, *Z. Gesamte Hyg. Grenzgeh.*, No. 16, 188–191. (1970).

7.2.5 Volatility Losses

Any successful analytical procedure requires that the material being measured be representative of the original sample. Lack of sample homogeneity is a major problem that can be solved through adequate sampling and pretreatment. Volatility loss during pretreatment or ashing, however, can be a serious problem for some elements. The volatilization loss factor must be determined for every element considered when ashing or chemical dissolution is performed. For some analytical techniques (e.g., INAA and XRF), volatilization losses are no problem because whole coal may be used as the sample.

Although systematic effort has been reported in this area, more is needed. Vogel et al.[490] determined the volatility of 19 elements from 340 to 990°C. Studies by Schultz et al.,[491] Pollock,[277] and Guidoboni[492] have determined various trace element volatilization losses under different ashing and pretreatment conditions. Ruch et al.[99] investigated volatilization losses under low-temperature ashing conditions.

Studies of volatilization losses have been carried out for the following elements: arsenic,[415,493,494] beryllium,[495,496] fluorine,[497] gallium,[433,498–500] germanium,[356,393,499,501–508] mercury,[509,510] phosphorus,[511] titanium,[512] uranium,[513] and zinc.[482]

7.2.6 Distribution of Major, Minor, and Trace Elements in Coal

Some of the many discussions of the distribution of elements in coals are concerned with a single element, others with associated groups of elements, and still others with elements of a single coal seam or of coals in a certain part of the world. Several excellent review articles and bibliographies summarize much of the literature. The five listed below can direct the reader to more detailed descriptions of elemental distributions in coals.

1. A review by Bethell[514] very effectively summarizes the literature on the distribution of elements in coal published prior to 1962. The origin of the observed distributions is discussed, and more than 200 references are given. This article was also published in its entirety by the Institute of Fuel.[515]
2. Abernethy et al.[157] and Abernethy and Gibson[516] presented data on 10 major ele-

[490] G. J. Vogel, Argonne National Laboratories Rep. Ser. No. ANL/ES-CEN-1015, 1976, pp. 59–65.

[491] H. Schultz, E. A. Hattman, and W. B. Booher, in Am. Chem. Soc. Adv. Chem. Ser. No. 141, 1975, pp. 139–153.

[492] R. J. Guidoboni, *Anal. Chem.*, **45**(7), 1275–1277 (1973).

[493] K. Barton and J. Tampa, *Paliva Voda,* **28,** 234–235 (1948).

[494] H. Kirsch, S. Pollmann, and J. Ottemann, *Mitt. Vern. Grosskesselbesitzer,* **48**(1), 10–20 (1968).

[495] V. Jirele, L. Nechyba, and P. Pachner, *Cesk. Hyg.*, **11**(6), 329–339 (1966).

[496] I. V. Vasil'eva and V. A. Vekhov, *Khim. Tverd. Topl.*, No. 3, 131–132 (1971).

[497] R. Beising and H. Kirsch, *VGB Kraftwerkstechnik,* **54**(4), 268–286 (1974).

[498] B. I. Losev, Z. D. Nekrasova and I. V. Vasil'eva, *Khim. Tverd. Topl.*, No. 6, 108–111 (1972).

[499] M. Ryczek, *Prezeglad Gorn.*, No. 9, 420–426 (1959).

[500] I. Kh. Volodarskii, I. T. Grekhov, O. D. Malysheva, I. L. Tarnavskii, and M. Ya. Shpirt, *Khim. Tverd. Topl.*, No. 2, 94–97 (1974).

[501] V. I. Alekhina and D. M. Lisin, *Issled. Form Svyazi Germaniya Uglem Ego Povedeniya Pirolize Szhiganii,* 1972, pp. 36–59.

[502] N. V. Arnautov, *Tr. Inst. Geol. Geofiz., Akad. Nauk SSSR, Sibirsk. Otd.,* **32,** 46–50 (1965).

[503] B. I. Losev, A. N. Mel'nikova, R. N. Pitin, and I. L. Farberov, *Tr. In-ta Goryuch. Iskop. Akad. Nauk SSSR,* **13,** 164–166 (1960).

[504] A. Y. Meleshko, V. N. Morozov, A. S. Sambueva, N. A. Tatorinova, and S. A. Shipitayn, *Zh. Prikl. Khim. (Leningrad)*, **43**(5), 1027–1031 (1970).

[505] M. A. Menkovskii and A. N. Aleksandrova, *Zavod. Lab.*, **29,** 797–799 (1963).

[506] J. Veprek, *Sb. Pr. Vyzk. Chem. Vyuziti Uhli, Dehtu Ropy,* 2, 347–356 (1962).

[507] I. KL. Volodarskii, I. T. Grekhov, and M. Y. Shpirt, *Khim. Tverd. Topl.*, No. 1, 146–149 (1970).

[508] C. L. Waring and W. P. Tucker, *Anal. Chem.*, **26,** 1198–1219 (1954).

[509] C. F. Billings and W. R. Matson, *Science,* **176**(4040), 1232–1233 (1972).

[510] R. C. Diehl, E. A. Hattman, H. Schultz, and R. J. Haren, U.S. Bur. Mines Tech. Prog. Rep. No. 54, 1972, 9 pp.

[511] T. Fukimori, N. Ikebe, and K. Ishikawa, *Nenryo Kyokai-shi,* **52**(2), 108–112 (1973).

[512] F. H. Kunstmann and S. B. Gass, *J. S. Afr. Inst. Min. Met.*, **58,** 165–170 (1957).

[513] F. Cuttita and E. Brittin, U.S. Geol. Survey, U.S. Ato. Energy Comm. Rep. No. TEI-461, 1954.

[514] F. V. Bethell, *BCURA Month. Bull.*, **26,** 401–430 (1962).

[515] F. V. Bethell, *J. Inst. Fuel,* **36,** 478–492 (1963).

[516] R. F. Abernethy and F. H. Gibson, U.S. Bur. Mines Info. Circ. No. 8163, 1963, 69 pp.

ments and 34 minor elements in hundreds of coal samples of the United States. Nearly 400 references to published data are given, and separate chapters are devoted to germanium, gallium, and uranium in coal.

3. The excellent review of mineral matter and trace elements by Watt[517] has not been widely distributed. The title of Part 1 of the publication, "The Occurrence, Origin, Identity, Distribution, and Estimation of the Mineral Species in British Coals," is somewhat misleading. The publication is not limited to minerals in coal or to British coals, but contains an excellent review of the literature of the remainder of Europe and also of North America. Elemental and mineral distributions are discussed.
4. Swaine[518] reviewed the origin, mode of occurrence, and concentration of trace elements in coals in a symposium volume honoring U.S.S.R. Academician A. P. Vinogradov. The paper gives references to 97 articles about elemental distributions in coal.
5. Averitt et al.[519] compiled a bibliography of 175 selected references on trace elements in coal. It is intended to be a guide to readily available literature, most of which is applicable to the coals of the United States, and all the reports cited are in English.

Many of the articles referenced in the foregoing publications are concerned with the distribution of a single element or a group of elements. Of the many articles on the occurrence and distribution of germanium in coals that are available in the literature, a large number are published in Russian. Some of these have been translated, and articles by Ratynskiy et al.,[520] Kizilshteyn,[521] and Sofiev et al.[522] appear in a single volume entitled *Geochemistry of Germanium*, edited by Weber.[523]

The distribution of boron in coals has been researched intensively because of its possible use as an indicator of paleosalinity. Boron in coals of the U.S.S.R. has been studied by Timofeev et al.,[524] in Australia by Swaine,[525,526] in Czechoslovakia by Somasekar,[527] in Poland by Konieczynski,[528] in Scotland by Allan et al.,[529] and in the United States by Bohor and Gluskoter.[420]

The potential detrimental effects on the environment of elements associated with coal have prompted detailed studies of a number of elements in coal. Mercury has been studied by Karasik et al.,[530] O'Gorman et al.,[269] and Ruch et al.[301] An excellent review of the potential pollutants in fossil fuels, including coal and petroleum, prepared by Magee et al.,[531] gives special attention to mercury and sulfur but also considers many other elements.

Probably more articles have been published on sulfur than on any other single

[517] J. D. Watt, "The Physical and Chemical Behavior of the Mineral Matter in Coal Under the Conditions Met in Combustion Plant," Part I, "The Occurrence, Origin, Identity, Distribution and Estimation of the Mineral Species in British Coals," BCURA, Leatherhead, Surrey, England, 1968.

[518] D. J. Swaine, 1975, "Trace Elements in Coal," in *Recent Contributions to Geochemistry and Analytical Chemistry,* A. I. Tugarinov, Ed., Halsted Press, New York, 1975, pp. 539–550.

[519] Paul Averitt, J. R. Hatch, U. E. Swanson, I. A. Breger, S. L. Coleman, J. H. Medlin, Peter Zubovic, and H. J. Gluskoter, U.S. Geol. Survey, Open File Rep. No. 76-481, 1976, 16 pp.

[520] V. M. Ratyinskii, T. I. Sendul'skaia, and M. Ya. Shpirt *Dokl, Akad. Nauk SSSR,* **170**(5), 1177–1180 (1966).

[521] L. Ya. Kizil'shtein, *Dokl. Akad. Nauk SSSR, Earth Sci. Sect.,* **157**(4), 879–881 (1964).

[522] I. S. Sofiev et al., *Geochem. Int.,* No. 2, 297–300 (1964).

[523] Jon N. Weber, Ed., *Geochemistry of Germanium,* Benchmark Papers in Geology, Vol. 11, Dowden, Hutchinson and Ross, Inc., Stroudsburg, Pa., 1973.

[524] P. P. Timofeev, L. I. Bogolyuvoba, L. V. Miesserova, and N. P. Fedorovskaya, *Khim. Tverd. Topl.,* No. 5, 96–102 (1967).

[525] D. J. Swaine, *Aust. J. Sci.,* **25**(6) 265 (1962).

[526] D. J. Swaine, *Proc. 2nd Bowen Basin Symp.,* Geological Survey of Queensland, Rep. No. 62, 1970, pp. 41–48.

[527] B. Somasekar, *Trans. Inst. Min. Met. Ostrava, Min. Geol. Ser.,* 9–14 (1971).

[528] J. Konieczynski, *Pr. Gl. Inst. Gorn.,* No. 482, 3–26 (1970).

[529] F. J. Allan, F. S. Dundas, and D. A. Lambie, *J. Inst. Fuel,* **42**(336), 29–30 (January 1969).

[530] M. A. Karasik, A. G. Dvornikov, G. K. Talalaev, and K. K. Zemblevskii, *Koks. Khim.,* **9**, 14–16 (1967).

[531] E. M. Magee, H. J. Hall, and G. M. Varga, Jr., U.S. Environmental Protection Agency, Technol. Ser. EPA-R2-73-249, 1973, 293 pp.

element in coal. The monograph by Yurovskii,[532] originally published in Russian and now available in English, is an ambitious attempt at summarizing the occurrence, distribution, and genesis of sulfur in coal and the effects of sulfur on coal utilization. It also contains many references, especially to Russian literature published before 1960.

Many articles of broad interest are concerned with the distribution of trace elements in coals of a single basin, region, or country. Zubovic and co-workers at the U.S. Geological Survey have published a comprehensive and extremely valuable series of this type for North American coals of the Western Interior,[533] the Northern Great Plains,[534] the Eastern Interior,[535–537] and the Appalachians.[538] Significant elemental analyses on coals of the United States have been published by Abernethy and Gibson,[516] Abernethy et al.,[157] O'Gorman and Walker,[539] Swanson et al.,[187] and Gluskoter et al.[275]

A number of articles summarizing elemental distributions in Australian coals have been published by researchers at the Commonwealth Scientific Industrial Research Organization (CSIRO), including Durie,[540] Clark and Swaine,[541] Durie and Schafer,[542] Kemezys and Taylor,[543] and Swaine.[544]

Bulgarian coals have been analyzed for trace elements by Mincev and Eskanazi.[545] Similar reports have been published about Spanish coals by Perez,[546] and for Belgian coals by Block and Dams.[547]

The modes of occurrence of the trace elements in coals and the geochemical significance of the distribution and occurrences of the elements will be included in the more general discussions of modes of occurrence and geochemistry of mineral matter that follow in this chapter.

7.3 MINERAL MATTER IN COAL

In addition to the organic matter in coals, composed essentially of carbon, hydrogen, oxygen, nitrogen, and sulfur, all coals also contain inorganic constituents, which are generally grouped as "mineral matter" whether they are in distinct mineral phases or in organic combination. That is, all elements in coal except C, H, O, N, and S are relegated to an inorganic (mineral matter) status even if they are present as organic-metallics, chelates, or adsorbed species. Four of the five elements generally considered to be organic are also present in inorganic combination in coals. Carbon is present in carbonates of calcium, magnesium, and iron; hydrogen in free water and water of hydration; oxygen in oxides, water, sulfates, and silicates; and sulfur in sulfides and sulfates.

This mineral matter has been classified in different ways, depending on the interests of the person making the classification. Yancy and Geer,[548] in a volume on coal preparation,

[532] A. Z. Yurovskii, *Sulfur in Coals* (1960), Academy of Sciences of the USSR, Institute of Mineral Fuels, English translation, 1974, by Indian National Scientific Documentation Centre, New Delhi, for U.S. Department of Interior, 455 pp.

[533] Peter Zubovic, Nola B. Sheffey, and Taisia Stadnichenko, U.S. Geol. Survey Bull. No. 1117-D, D1-D33, 1967.

[534] Peter Zubovic, Taisia Stadnichenko, and Nola B. Sheffey, U.S. Geol. Survey Bull. No. 1117-A, 1961, 58 pp.

[535] Peter Zubovic, Taisia Stadnichenko, and Nola B. Sheffey, U.S. Geol. Survey Bull. No. 1117-B, 1964.

[536] Peter Zubovic, U.S. Geol. Survey Open File Rep., 1960.

[537] Peter Zubovic, in Am. Chem. Soc. Adv. Chem. Ser. No. 55, 1966, pp. 232–247.

[538] Peter Zubovic, Taisia Stadnichenko, and Nola B. Sheffey, U.S. Geol. Survey Bull. No. 1117-C, 1966.

[539] J. V. O'Gorman and P. L. Walker, Jr., U.S. Dept. Interior, Office of Coal Research, Research and Development Rep. No. 61, Interim Report No. 2, 1972, 184 pp.

[540] R. A. Durie, *Fuel,* **40,** 407–422 (1961).

[541] Marie C. Clark and D. J. Swaine, C. S. I. R. O. Tech. Commun. No. 45, 1962, pp. 1–20.

[542] R. A. Durie and H. N. S. Schafer, *Fuel,* **43,** 31–41 (1964).

[543] M. Kemezys and G. H. Taylor, *Symp. Inorganic Constituents of Fuel (Origin, Influence and Control)*, Inst. Fuel, Australian Sect., Melbourne, Paper No. 3, 1964.

[544] D. J. Swaine, *Sonderdr. Mitt. Naturforsch. Ges. Bern, Neue Folge,* **24,** 49–61 (1967).

[545] D. Mincev and Gr. Eskenazi, *God. SU. Geol.-Geogr. Fak.,* **64**(1), *Geologja,* 263–291 (1972).

[546] A. Martin Perez, *Ensayos Invest.* **1**(2), 10–19 (1966).

[547] Chantal Block and Richard Dams, *Environ. Sci. Technol.* **9**(2), 146–150 (1975).

[548] H. G. Yancy and M. R. Geer, "Properties of Coal and Impurities in Relation to Preparation," in *Coal Preparation,* J. W. Leonard and D. R. Mitchell, Eds., AIME, New York, Ch. 1, 1968, 55 pp.

referred to "segregated" and "disseminated" impurities. Watt[517] stated:

> For the practical user it is more useful to classify the mineral matter in coal as (*a*) "inherent mineral matter"—representing material that is too closely associated with the coal substance to be readily separated from it by methods available at present, or (*b*) "adventitious mineral matter"—representing the material that is less intimately associated with the coal and can be readily separated (pp. 2–3).

According to Marshall and Tompkins,[549] the term "inherent mineral matter" usually applies to the "inorganic" elements that formed an integral part of the plants from which the peat and then the coal were derived. The term "adventitious mineral matter" includes all the remaining inorganic constituents without regard for their time of genesis.

Researchers concerned with the origin of mineral matter in coal have generally used the terms "detrital," "syngenetic," and "epigenetic" to classify the inorganic constituents of coals.[550–553]

Most of the mineral phases in coal are listed in Table 7.5 arranged in columns representing syngenetic or epigenetic origin. Gluskoter[553] suggested that the minerals transported and deposited in the peat swamp by wind and water be called "allogenic" or "detrital," and that the remaining minerals, all of which formed in place (authigenic), be divided into those that formed contemporaneously with coal formation ("syngenetic") and those whose formation followed the initial stages of coalification ("epigenetic").

[549] C. E. Marshall and D. K. Tompkins, *Institute of Fuel (Aust. Membership), Symp. Inorganic Constituents of Fuel,* 1964, pp. 57–69.

[550] F. A. Williams and C. M. Cawley, "Impurities in Coal and Petroleum," *The Mechanism of Corrosion by Fuel Impurities,* H. R. Johnson and D. J. Littler, Eds., Butterworths, London, 1963, pp. 24–67.

[551] M.-Th. Mackowsky, "Mineral Matter in Coal," *Coal and Coal Bearing Strata,* D. G. Murchison and T. S. Westoll, Eds., Edinburgh and London, Oliver and Boyd, 1968, pp. 309–321.

[552] M.-Th. Mackowsky, "Minerals and Trace Elements Occurring in Coal," in *Coal Petrography,* E. Stach et al., Eds., Gebrüder Borntraeger, Berlin and Stutgart, 1975, pp. 121–132.

[553] H. J. Gluskoter, "Mineral Matter and Trace Elements in Coal," in *Trace Elements in Fuel,* S. P. Babu, Ed., Am. Chem. Soc. Adv. Chem. Ser. No. 141, 1975, pp. 1–21.

This classification, like all classifications of earth materials, is less than ideal, because some minerals can be categorized by subjective means only. Nevertheless a genetic classification can facilitate an understanding of the basic geochemical system and can apply to a variety of technologies concerning the utilization of coals.

General discussions of mineral matter occurrences and distributions in coals can be found in Williams and Cawley,[550] Guthörl,[554] Marshall and Tompkins,[549] Yancy and Geer,[548] Mackowsky,[551] Watt,[517] O'Gorman and Walker,[539] Mackowsky,[552] and Gluskoter.[553]

7.3.1 Separation of Mineral Matter from Coal

One of the principal impediments to the study of mineral matter in coal has been the difficulty of obtaining statistically valid (representative) samples of the mineral phases which are free of organic matter.

At first glance it appears that it would be simple to separate minerals from coal by specific gravity techniques. Coal, the organic portion, has a low specific gravity (in the range 1.2–1.4), whereas the minerals occurring in coal have specific gravities of 2–5 and greater. The finely disseminated nature of the minerals in coal precludes their complete separation, however. An incomplete separation based on specific gravity, which is the basis of modern coal cleaning or "washing," is effective in fractionating the sample into some portions that are relatively mineral rich and some that contain relatively pure coal. Specific gravity can be utilized to determine modes of occurrence of chemical elements and minerals in coal,[99,553,555] and this makes the procedure important economically. The technique does not provide an acceptable sample of minerals in coal, however.

High-temperature oxidation of the coal removes the organic matter but greatly alters the inorganic constituents. Treating the coal with a liquid oxidizing agent at low temperatures is a successful method of isolating the mechanically inseparable part of the mineral

[554] P. Guthörl, *Bergfreiheit,* **28,** 213–222, 256–265 (1963).

[555] B. B. Morgan, *BCURA Month. Bull.,* **28**(8), 325–334 (1964).

Table 7.5 Minerals in Coal, Classified as to Mode of Origin

	First Stage of Coalification: Syngenetic Formation, Synsedimentary–Early Diagenetic (intimately intergrown)		Second Stage of Coalification: Epigenetic Formation	
Mineral Group	Transported by Water or Wind	Newly Formed	Deposited in Fissures, Cleats, and Cavities (coarsely intergrown)	Transformation of Syngenetic Minerals (intimately intergrown)
Clay minerals	Kaolinite, illite, sericite, clay minerals with mixed-layer structure Montmorillonite, tonstein			Illite, chlorite
Carbonates		Siderite-ankerite concretions, dolomite, calcite, ankerite Siderite	Ankerite Calcite Dolomite Ankerite in fusite	
Sulfides		Pyrite concretions Melnikovite-pyrite Coarse pyrite (marcasite) Concretions of FeS_2-$CuFeS_2$-ZnS	Pyrite Marcasite Zinc sulfide (sphalerite) Lead sulfide (galena) Copper Sulfide (chalcopyrite) Pyrite in fusite	Pyrite from the transformation of syngenetic concretions of $FeCO_3$
Oxides		Hematite	Goethite, lepidocrocite ("needle-iron ore")	
Quartz	Quartz grains	Chalcedony and quartz from the weathering of feldspar and mica	Quartz	
Phosphates	Apatite	Phosphorite, apatite		
Heavy minerals and accessory minerals	Zircon, rutile, tourmaline, orthoclase, biotite		Chlorides, sulfates, and nitrates	

Source: Mackowsky,[552] p. 127.

matter. This technique was used in investigations by Nawalk and Friedel[556] and in studies of Australian coals by Ward.[557]

An electronic technique involving a radiofrequency-induced plasma has been adapted for the low-temperature oxidation of coals. In the device, termed a "low-temperature asher" or "plasma asher," oxygen is passed through a radiofrequency field. Activated oxygen then passes over the coal sample, and the organic matter is oxidized at a temperature usually less than 150°C.[44]

The effects of radiofrequency low-temperature ashing of the oxidizing gas stream on the minerals in coal are minor. Oxidation of mineral phases present in the coals has not been reported, and the only phase changes noted were those that would be expected at 150°C and 1 torr. Therefore the dominant mineral phases in coals, including the iron sulfides, kaolinite, illite, quartz, and calcite, are unaffected by the radiofrequency ashing.

Low-temperature ashing has been applied to a variety of investigations. Freeman and Fowkes[558] discussed low-temperature ashing of lignites from North Dakota. Augenstein and Sun[559] used low-temperature ashing to study the characteristics of anthracite refuse. Estep et al.[560] ashed coals before analyzing the minerals by infrared spectroscopy. Gluskoter[561] used low-temperature ashing to obtain clay minerals from coals, and Bohor and Gluskoter[420] analyzed boron content of clay minerals from coals obtained by low-temperature ashing to determine paleosalinities. Mineral matter in coals of Montana,[562] the Illinois Basin,[563–565] and a variety of other coals from the United States[539,566] has been prepared for study by the radiofrequency low-temperature ashing method. The technique has been used in Australia as a routine laboratory method for the quantitative determination of mineral matter content of coal.[567]

7.3.2 Content of Mineral Matter and Ash in Coal

It is difficult to determine, either qualitatively or quantitatively, the mineral matter content of a coal from the high-temperature ash. During high-temperature ashing, as designated by various standards (the ASTM standard temperature is 750°C), a series of reactions takes place involving the minerals in the coals. Of the four major mineral groups, only quartz is not significantly altered during high-temperature ashing.

The clay minerals in coal contain water that is bound within their lattices. Kaolinite contains 13.96%, illite 4.5%, and montmorillonite 5% bound water. In addition, the montmorillonite in the mixed-layer clays also contains interlayer or adsorbed water. All the water is lost during the high-temperature ashing.

During high-temperature ashing the FeS_2 minerals are oxidized to Fe_2O_3 and SO_2. Some of the SO_2 may remain in the ash in combination with calcium, but much is lost. If all the SO_2 were emitted during ashing, there would be a 33% loss in weight with respect to the weight of pyrite or of marcasite in the original sample.

The $CaCO_3$ is calcined to lime (CaO) during high-temperature ashing, and CO_2 is lost. This loss results in a 44% reduction in weight from the original calcite.

The stable mineral quartz (SiO_2) is the only major mineral found in coal that is inert during high-temperature ashing.

It has long been recognized that these changes in the mineral matter in coal occur during the ashing of coal. A number of for-

[556] A. J. Nawalk and R. A. Friedel, "Concentration and Minerals in Coal by Peroxidation of the Organic Matter," in *Proc. Sympo. Chemicals and Oil from Coals,* Central Fuel Research Institute, Dhnabad, India, 1969, pp. 526–530 (published 1972).

[557] C. R. Ward, *Fuel,* **53,** 220–221 (1974).

[558] P. G. Freeman and W. W. Fowkes, *Proc. N. Da. Acad. Sci.*, **24,** Part 2 (Papers), 1ı9–124 (1971).

[559] Dale A. Augenstein and S. C. Sun, *Trans. AIME, Soc. Min. Eng.* **256**(2), 161–166 (1974).

[560] P. A. Estep, J. J. Kovach and Clarence Karr, Jr., *Anal. Chem.*, **40**(2), 358–363 (1968).

[561] H. J. Gluskoter, *J. Sediment. Petrol.* **37**(1), 205–214 (1967).

[562] D. F. Wolfe, M. S. thesis, Montana College of Mineral Science and Technology, Butte, 1969, 64 pp.

[563] C. Prasada Rao and H. J. Gluskoter, Illinois State Geol. Survey Circ. No. 476, 1973, 56 pp.

[564] W. G. Miller, M.S. thesis, University of Illinois, Urbana, 1974, 145 pp.

[565] C. R. Ward, Illinois State Geol. Survey Circ. No. 498, 1977, 35 pp.

[566] J. V. O'Gorman and P. L. Walker, Jr., *Fuel,* **50,** 235–251 (1971).

[567] Frank Frazer and C. B. Belcher, *Fuel,* **52,** 41–46 (1973).

mulas have been devised to allow the calculation of the mineral matter content from various parameters determined during the analyses of the coal. Minimally, this involves a calculation based on total content of ash and sulfur. One of the first and simplest of these is the Parr formula, developed in 1909.[568]

A number of workers have attempted to refine this calculation by considering pyritic sulfur content, the loss of CO_2 from carbonate minerals, and the decomposition of chlorides during ashing. These modifications of the technique for calculating a "pure" mineral matter fraction of the coal are effectively summarized by Watt[517] and by Given.[569]

In a monograph on the problem of calculating a pure organic substance from the chemical analyses of coal, Given[569] clearly stated the problem, reviewed efforts to solve the problem over the past 70 years, and presented his suggestion for modifying the Parr formula. The formula he presented, which may also be found in Given et al.,[570] is:

$$MM = 1.13 \text{ ash} + 0.47 S_{pyr} + 0.5 \text{ Cl}$$

It was derived in essentially the same way as the Parr formula, but (*a*) Millot's value for the average water of decomposition of the clays is adopted, (*b*) only the pyritic sulfur is considered, and (*c*) the correction for chlorine is adapted from the KMC (King, Maries, and Crossley) formula (Ref. 659, p. 48).

7.3.3 Methods of Mineral Analysis

The short discussion that follows is concerned with the identification of minerals in coals, defining minerals in the strict sense as naturally occurring, inorganic substances with definite chemical composition and ordered atomic arrangement. It is not concerned with the analyses of chemical elements, which was discussed previously.

A number of instrumental analytical techniques have been applied to concentrates of minerals that were obtained from coals. The minerals may be in residues from peroxidation of coals or from radiofrequency low-temperature ashing, or they may have been mechanically concentrated from coal samples.

7.3.3.1 *X-Ray Diffraction Analysis*

The best developed and most widely used method for identifying mineral phases in coal is X-ray diffraction analysis. It has been used extensively by a number of workers, including Maruyama and Kobayashi,[571] Rekus and Haberkorn,[572] Gluskoter,[561] Wolf,[562] O'Gorman and Walker,[539] Paulson et al.,[573] Rao and Gluskoter,[563] Miller,[564] and Ward.[565] X-Ray diffraction is the most promising tool for the identification of minerals in coals because the equipment is readily available, and the techniques for identification have been extensively developed by mineralogists. These techniques may be applied to minerals from coal after the mineral matter has been separated from organic matter in the coal.

7.3.3.2 *Infrared Spectroscopy*

Infrared absorption bands from 650 to 200 cm^{-1} have been used at the Morgantown Energy Research Center to analyze, both qualitatively and quantitatively, minerals in low-temperature ash of coals. Resulting from these efforts are publications by Karr et al.,[574,575] Estep et al.,[560] and Kovach et al.[576] Estep et al.[560] reported an average error of 6.1% for the minerals determined (kaolinite, quartz, gypsum, calcite, and pyrite) and concluded that the method was satisfactory for

[568] S. W. Parr and W. F. Wheeler, Univ. Illinois Eng. Exp. Sta. Bull. No. 37, 1909, 67 pp.

[569] P. H. Given, U.S. Dept. Interior, Office of Coal Research Rep. No. SROCR-9, 1969, 40 pp.

[570] P. H. Given, D. C. Cronauer, William Spackman, H. L. Lovell, Alan Davis, and Bimal Biswas, *Fuel,* **54,** 40–49 (1975).

[571] T. Maruyama and K. Kobayashi, *J. Fuel Soc. Japan,* **44**(9), 640 (1965).

[572] A. F. Rekus and A. R. Haberkorn, *J. Inst. Fuel,* **39,** 474–477 (1966).

[573] L. E. Paulson, W. Berkering and W. W. Fowkes, *Fuel,* **51**(3), 224–227 (1972).

[574] Clarence Karr, Jr., Patricia A. Estep, and J. J. Kovach, *Chem. Ind. (London),* No. 9, 356–357 (1967).

[575] Clarence Karr, Jr., Patricia A. Estep, and J. J. Kovach, *Instrum. News,* **18**(2), 1, 5, 6 (1967).

[576] J. J. Kovach, *Symp. Recent Developments Fuel Analysis Techniques,* 157th Natl. Meet. American Chemical Society (Div. Fuel Chem.) Minneapolis, April 13–18, 1969.

multicomponent mixtures such as mineral matter in coal. O'Gorman and Walker[539] applied this method to an investigation of coals of North America. European workers including Fijal and Ratajczak,[577] Blum et al.,[578] and Petrova and Nikolov,[579] have also utilized infrared spectroscopy for the identification of minerals.

7.3.3.3 *Differential Thermal Analyses*

Warne[580] applied differential thermal analyses (DTA) in a nitrogen atmosphere to the identification of minerals in coals and reported the following detection limits: 0.5–1.0% for pyrite or marcasite; 1–2% for calcite, dolomite, and ankerite; 2.5–5% for siderite and kaolinite; 15–20% for montmorillonite and quartz; and as high as 30–35% for illite. Warne later refined the methods used to identify the silica minerals in coal by DTA[581] and reported the use of a carbon dioxide atmosphere for the identification of carbonate minerals.[582]

O'Gorman and Walker[539,583] used thermal methods to investigate minerals in coals of North America, including DTA, thermogravimetric (TG), and differential thermogravimetric (DTG) analyses. Combined thermal methods, termed "derivatography," have been used by Mukherjee et al.[584] to study Indian coals and by Lebedev et al.[585] to study coals in the U.S.S.R.

[577] J. Fijal and T. Ratajczak, *Przegl. Gorn.*, **25**, 529–533 (1969).

[578] I. Blum et al., 1969, *Freiberger Forschungsh.*, *A*, **473**, 73–84 (1969) (in German).

[579] R. Petrova and Z. Izv. Nikolov, *Geol. Inst. BAN, Ser. Neftena Vaglistna Geol.*, No. 19, 117–124 (1970) (in Bulgarian; Russian and German summaries).

[580] S. St. J. Warne, *Symp. Inorganic Constituents Fuel (Origin, Influence and Control)*, Paper No. 19, Inst. Fuel (Australian Sect.), Melbourne; also in *J. Inst. Fuel*, **38**(292), 207–217 (1964).

[581] S. St. J. Warne, *J. Inst. Fuel*, **43**, 240–242 (1970).

[582] S. St. J. Warne, *J. Inst. Fuel*, No. 142, 142–145 (1975).

[583] James V. O'Gorman and P. L. Walker, *Fuel*, **52**, 71–79 (1973).

[584] S. N. Mukherjee, A. K. Nag, and S. K. Majumdar, *J. Mines, Met. Fuels*, **20**(12), 363–367, 373 (1972).

[585] I. K. Lebedev et al., *Khim. Tverd. Topl.*, No. 5, 38–42 (1973).

7.3.3.4 *Electron Microscopy*

Although electron microscopy has become popular as a mineralogical research tool, until very recently it was rarely used to study coal and mineral matter in coal. Dutcher et al.[586] reported on what is probably the first investigation that used the electron probe to analyze mineral matter in coal.

Scanning electron microscopy with an energy dispersive X-ray analysis accessory system was used to identify some minerals as hosts for chemical elements in studies at the Illinois State Geological Survey.[99,372]

Recent studies on iron sulfide minerals in coals,[587] minerals in coals and residues from liquefaction processes,[588] and *in situ* investigation of minerals in coal[589] all used the scanning electron microscope (SEM) as the primary analytical tool. The ion microprobe mass analyzer (IMMA) is more sensitive than either the energy dispersive X-ray spectrometer or the wavelength dispersive X-ray spectrometer, both of which are used as accessories to an electron microscope. The ion microprobe was used in a study of minerals from coal by Ribbe.[590]

7.3.3.5 *Other Spectrometric Analyses*

Results of a study of ^{57}Fe Mössbauer spectra in coals have been published by Lefelhocz et al.[591,592] The study demonstrated the validity of the technique of Mössbauer spectroscopy applied to coal and suggested the presence of high-spin iron (II) in sixfold coordination in several of the samples studied.

[586] Russell R. Dutcher, E. W. White and W. Spackman, *Proc. 22nd Ironmaking Conf., Iron and Steel Div., AIME Metall. Soc., 1964, pp. 463–483.*

[587] H. J. Gluskoter, *Energy Sources*, **3**, 125–131 (1977).

[588] S. J. Russell, *Scanning Electron Microsc.*, **1**, 95–100 (1977).

[589] R. B. Finkelman, R. W. Stanton, and I. A. Breger, *Abstr., Geol. Soc. Am., Abstracts with Programs*, **8**(6), 865–866 (1976).

[590] P. H. Ribbe, *Abstr. Geol. Soc. Am., Abstracts with Programs*, **7**(7), 1244 (1975).

[591] J. F. Lefelhocz, R. A. Friedel, and T. P. Kohman, *Prepr., Am. Chem. Soc., Div. Fuel Chem.*, **11**(2), Print 2, 417–432 (1967).

[592] J. F. Lefelhocz, R. A. Friedel, and T. P. Kohman, *Geochim. Cosmochim. Acta*, **31**(12), 2261–2263 (1967).

Mass spectrometric investigations of isotopes in coal and coal minerals have also been very limited in scope. Rafter[593] published sulfur isotope data on 27 New Zealand coal samples but drew no conclusions from these data. Smith and Batts[127] determined the isotopic composition of sulfur in a number of Australian coals and concluded that it is possible to deduce the origin of the organically combined sulfur, the depth of penetration of seawater into underlying coal measures, and factors controlling reduction of sulfates to sulfides by biogenic residues from isotopic data on sulfur in coal.

7.3.3.6 *Petrographic Investigations of Minerals in Coal*

The coal petrographer uses the light microscope, usually illuminating the specimen with reflected light, to characterize the organic fraction (macerals) in coal. The science of coal petrography has become highly precise in the past few decades, but progress in the petrography of mineral matter in coal has not kept pace. Discussions of the use of petrographic techniques for determination of minerals in coal are found in Chapter 3 of this volume and in sections by Mackowsky in *Coal Petrology.*[552]

7.3.4 Minerals in Coals

A large number of distinct mineral phases have been reported in coals. Table 7.6, which shows most of those minerals, was constructed in part from similar tables published by Ode,[594] Williams and Cawley,[550] Francis,[595] Watt,[517] and Mackowsky.[551,552] Although lists of minerals in coal may contain as many as 50–60 minerals, most fall into one of four groups: aluminosilicates, sulfides, carbonates, and silicates (principally quartz). Several of these are shown in scanning electron photomicrographs in Fig. 7.4.

7.3.4.1 *Aluminosilicates—Clay Minerals*

Clay minerals are the most common inorganic constituents of coals and of the strata associated with coal seams.[552,553] Many different clay minerals have been reported within and associated with coals, but the most common clay minerals are kaolinite [$Al_4Si_4O_{10}(OH)_8$], illite [$KAl_2(AlSi_3O_{10})(OH)_2$], and mixed-layer illite-montmorillonite.

Rao and Gluskoter[563] investigated 65 coals from the Illinois Basin and reported a mean value of 52% for clay minerals in the total mineral matter. O'Gorman and Walker[539] also found that clay minerals were the dominant mineral group in most of the coals they studied. Ward investigated mineral matter content in 32 samples of the Springfield-Harrisburg (No. 5) Coal Member in the Illinois Basin and found that clay minerals dominated, constituting 37–73% of the mineral matter.[565] Aluminosilicates in coals have been studied by Kossovskaya et al.,[596] Murata,[597] and Dixon et al.[598]

Kaolinite-rich claystones are commonly found within and associated with coals in most of the coal basins of the world. They are generally called either "tonsteins" or "kaolin-tonstein."

Morphologically, tonsteins are classified as (1) crystal-tonstein, which is composed of large well-developed "books" and "worms"; (2) graupen tonstein, consisting of pellets or granules in a fine-grained matrix; (3) dense, fine-grained, generally massive claystone; and (4) pseudomorphs of kaolinite generally after feldspars. Recent descriptions of individual occurrences of kaolinite tonsteins have been written by Pietzner et al.,[599] May-

[593] T. A. Rafter, "Sulfur Isotope Measurements on New Zealand, Australian, and Pacific Islands Sediments," in *Biogeochemistry of Sulfur Isotopes*, M. L. Jensen, Ed., *Proc. Nat. Sci. Found. Symp. Yale Univ.*, April 1962, pp. 42–60.

[594] W. H. Ode, "Coal Analysis and Mineral Matter," in *Chemistry of Coal Utilization,* Supplementary Volume, H. H. Lowry, Ed., John Wiley and Sons, New York, 1963, pp. 202–231.

[595] Wilfrid Francis, *Coal,* Edward Arnold Ltd., London, 1961, 806 pp.

[596] A. G. Kossovskaya, V. D. Shutov, and V. A. Alexandrova, *C.R., 5th Int. Cong. Carboniferous Strat. Geol.*, Paris, 1963, pp. 519–529.

[597] Shigeo Murata, *Kyushu Inst. Tech., B (Sci. Tech.)*, No. 12, 79–86 (1962).

[598] K. Dixon, E. Skipsey, and J. T. Watts, *J. Inst. Fuel,* **43,** 124–128 (1970).

[599] H. Pietzner, M. Teichmüller and R. Teichmüller, *Fortsch. Geol. Rheinland. Westfalen,* **6,** 97–104 (1962).

Table 7.6 Minerals Identified with Coals

Mineral	Formula
Clay minerals	
Montmorillonite	$Al_2Si_4O_{10}(OH)_2 \cdot H_2O$
Illite-sericite	$KAl_2(AlSi_3O_{10})(OH)_2$
Kaolinite	$Al_4Si_4O_{10}(OH)_8$
Halloysite	$Al_4Si_4O_{10}(OH)_8$
Chlorite (prochlorite, penninite)	$Mg_5Al(AlSi_3O_{10})(OH)_8$
Mixed-layer clay minerals	
Sulfide minerals	
Pyrite	FeS_2
Marcasite	FeS_2
Sphalerite	ZnS
Galena	PbS
Chalcopyrite	$CuFeS_2$
Pyrrhotite	$Fe_{1-x}S$
Arsenopyrite	$FeAsS$
Millerite	NiS
Carbonate minerals	
Calcite	$CaCO_3$
Dolomite	$(Ca, Mg)CO_3$
Siderite	$FeCO_3$
Ankerite (Ferroan dolomite)	$(Ca, Fe, Mg)CO_3$
Witherite	$BaCO_3$
Sulfate minerals	
Barite	$BaSO_4$
Gypsum	$CaSO_4 \cdot 2H_2O$
Anhydrite	$CaSO_4$
Bassanite	$CaSO_4 \cdot 1/2H_2O$
Jarosite	$(Na, K)Fe_3(SO_4)_2(OH)_6$
Szomolnokite	$FeSO_4 \cdot H_2O$
Rozenite	$FeSO_4 \cdot 4H_2O$
Melanterite	$FeSO_4 \cdot 7H_2O$
Coquimbite	$Fe_2(SO)_3 \cdot 9H_2O$
Roemerite	$FeSO_4 \cdot Fe_2(SO_4)_3 \cdot 12H_2O$
Mirabilite	$Na_2SO_4 \cdot 10H_2O$
Kieserite	$MgSO_4 \cdot H_2O$
Sideronatrite	$2Na_2O \cdot Fe_2O_3 \cdot 4SO_3 \cdot 7H_2O$
Chloride minerals	
Halite	$NaCl$
Sylvite	KCl
Bischofite	$MgCl_2 \cdot 6H_2O$
Silicate minerals	
Quartz	SiO_2
Biotite	$K(Mg, Fe)_3(AlSi_3O_{10})(OH)_2$
Zircon	$ZrSiO_4$
Tourmaline	$Na(Mg, Fe)_3Al_6(BO_3)_3(Si_6O_{18})(OH)_4$
Garnet	$(Fe, Ca, Mg)_3(Al, Fe)_2(SiO_4)_3$
Kyanite	Al_2SiO_5
Staurolite	$Al_4FeSi_2O_{10}(OH)_2$
Epidote	$Ca_2(Al, Fe)_3Si_3O_{12}(OH)$
Albite	$NaAlSi_3O_8$
Sanidine	$KAlSi_3O_8$
Orthoclase	$KAlSi_3O_8$
Augite	$Ca(Mg, Fe, Al)(Al, Si)_2O_6$

Table 7.6 *Continued*

Mineral	Formula
Hornblende	$NaCa_2(Mg, Fe, Al)_5(SiAl)_8O_{22}(OH)_2$
Topaz	$Al_2SiO_4(OH, F)_2$
Oxide and hydroxide minerals	
Hematite	Fe_2O_3
Magnetite	Fe_3O_4
Rutile	TiO_2
Limonite	$FeO{\cdot}OH{\cdot}nH_2O$
Goethite	$FeO{\cdot}OH$
Lepidocrocite	$FeO{\cdot}OH$
Diaspore	$AlO{\cdot}OH$
Phosphate minerals	
Apatite (fluorapatite)	$Ca_5(PO_4)_3(F, Cl, OH)$

land,[600] Kralik,[601] Budkevich and Vaishel,[602] Kisch,[603] Spears,[604,605] and Stadler.[606] The literature on the genesis of tonsteins and their significance as stratigraphic marker horizons is too extensive to be reviewed adequately here. The modes of occurrence, distributions, or mineralogy of tonsteins have been reviewed by Eden et al.,[607] Pietzner and Werner,[608] Stoffler et al.,[609] Fotjik and Kralik,[610] von Schellendorf,[611] Barnsley et al.,[612] Kimpe,[613] Bouroz,[614] and Price and Duff.[615]

7.3.4.2 *Sulfide Minerals*

The iron disulfides pyrite and marcasite are the dominant sulfide minerals in coal; pyrite is the more abundant. Pyrite and marcasite are dimorphs, minerals with the same chemical compositions but different crystal forms. Pyrite is isometric, and marcasite is orthorhombic. These two minerals have been studied more than any of the other mineral phases in coals. They are readily observed and, to some degree, easily removed. They are especially interesting because they contribute significantly to the total content of sulfur, which can cause boiler tube fouling and corrosion, pollution by emission of SO_2 when the coal is burned, acid-mine drainage, and contamination of iron smelted from coal-derived coke.

Most articles about iron sulfides in coal are primarily concerned with its removal, either by physical (specific gravity) or chemical methods, but coal cleaning is not considered here. General descriptions of iron sulfide minerals are given by Watt,[517] Mackowsky,[552] and Gluskoter.[587] Sulfur forms in coals of many parts of the world have also been described. These include a series of papers about morphologies of iron sulfides in Soviet coals by Pyranishnikov,[616] Koporulin,[617] Zar-

[600] H. Mayland, *Geol. Mag.*, **101**(3), 285 (1964).
[601] J. Kralik, *Bergb. Arch.*, **25**(5), 57–63 (1964).
[602] M. Budkevich and Ia. Vaishel, *Bull. Acad. Polon. Sci., Ser. Sci. Geol. Geog.*, **11**, 11–14 (1963).
[603] H. J. Kisch, *Am. J. Sci.*, **264**, 386–397 (1966).
[604] D. A. Spears, *J. Sediment. Petrol.*, **40**(1), 386–394 (1970).
[605] D. A. Spears and C. M. Rice, *Sedimentology*, **20**, 281–294 (1973).
[606] G. Stadler, *Fortschr. Geol. Rheinland. Westfalen*, **18**, 79–100 (1971) (in German).
[607] R. A. Eden, R. W. Elliot, R. E. Elliot, and B. R. Young, *Geol. Mag.*, **100**, 47–58 (1963).
[608] H. Pietzner and H. Werner, *C. R. 5th Int. Congr. Carboniferous Strat. Geol.*, Paris, 1963, pp. 667–678.
[609] D. Stoffler, W. V. Engelhardt, and G. Muller, Mineral Institute University Tübingen, *Naturwissenschaften*, **50**(4), 124–125 (1963).
[610] Z. Fotjik and J. Kralik, *Sbornik (Ostrava)*, **10** (1–2), 35–45 (1964).
[611] G. B. von Schellendorf, *Neues Jahrb. Geol. Paleontol. Monatsh.*, No. 5, 299–305 (1964).
[612] G. B. Barnsley, J. M. Clowes, and W. Fowler, *Geol. Mag.*, **103**(6), 508–521 (1966).
[613] W. F. M. Kimpe, *Mededelingen Geol. Sticht.*, *Nieuwe Ser.*, No. 18, 31–38 (1967).
[614] A. Bouroz, *C.R. Int. Congr4. Carboniferous Stratigraphy Geology*, Sheffield, Vol. 2, 1967, pp. 539–546.
[615] N. B. Price and P. M. D. Duff, *Sedimentology*, **13**, 45–69 (1969).
[616] V. K. Pyranishnikou, *Tr. Inst. Goryuch. Iskop.*, **12**, 43–52 (1961) (in Russian).
[617] V. I. Koporulin, *Dokl. Akad. Nauk. SSSR. Earth Sci. Sect.*, in trans. **143**(1–6), 123–125 (1964) (published 1962).

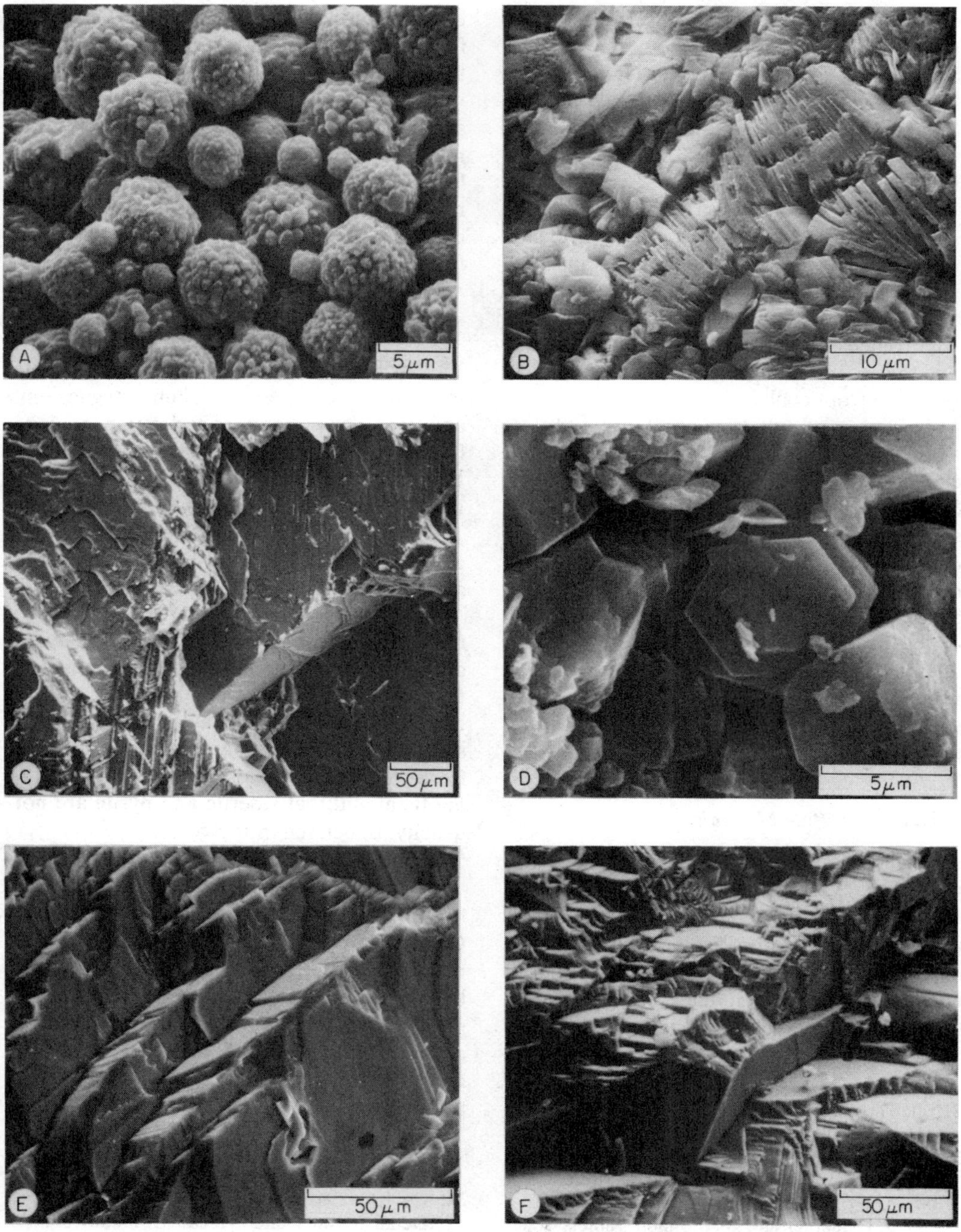

Figure 7.4 Scanning electron photomicrographs of minerals in coals from Illinois. The minerals were observed and photographed using equipment at the Center for Electron Microscopy, University of Illinois.

itskii,[618–620] and Kizilishteyn;[621] and two papers on iron sulfide minerals in Japanese

[618] P. V. Zaritskii, *Uchekn. Zap. Khar'kovsk. Gos. Univ. 125, Zap. Geo. Otd.*, No. 15, 5–14 (1962a); *C.A.*, **60,** 13021a.

[619] P. V. Zaritskii, in *Dokl. Akad. Nauk SSSR, Earth Sci. Sect.* **145** (1–6) 146–148 (1962b) (published 1964).

[620] P. V. Zaritskii, *Zap. Vses. Mineralog. Obshch.*, **92**(6), 727–733 (1963); *C.A.*, **61,** 447c.

[621] L. Ya. Kizil'shtein, *Khim. Tverd. Topl.*, **2,** 126–129 (March–April 1972) (in Russian).

coals by Nakayanagi.[622,623] Descriptions of iron sulfide minerals in coals of the United States have been published by Gray et al.[624] for a Pittsburgh Seam coal, by Gluskoter and Simon[625] for coals of Illinois, and by Khawaja[626] for the Springfield Coal Member (V) in Indiana.

The most comprehensive publication on sulfur in coal is the monograph by Yurovskii published in Russian in 1964 and translated into English in 1974.[532] The monograph has two major parts; the first deals with the occurrence and distribution of sulfur in coals, and the second is concerned with the removal of pyrite from coals.

A number of sulfide minerals, in addition to the iron disulfides, have been reported in coals (Table 7.6). Millerite from the Bulli Seam of New South Wales has been described by Lawrence et al.[627] Several publications have described the occurrence and distribution of relatively large quantities of sphalerite (ZnS) in coals of the midcontinent of the United States.[628–630] The sphalerite is an epigenetic mineral deposited along cleats and shears, and in clay dikes. Relatively high cadmium contents are associated with these sphalerites.[372]

7.3.4.3 *Sulfate Minerals*

The sulfate minerals identified in coals do not generally comprise a significant portion of the mineral matter in fresh, unoxidized coal samples. The iron disulfides oxidize rapidly after the coal is mined, however, and a number of hydrated sulfates have been reported in weathered coals and in coal refuse banks. The sulfates gypsum and barite are found in fresh coals. Most of the sulfates that form on weathering (oxidation) of pyrite are various hydrated states of ferrous and ferric sulfate.[625,631,632] Several of these iron sulfate minerals are soluble and contribute to problems of acid-mine drainage. They may also react with other minerals associated with the coals to produce calcium and sodium sulfates.[633]

7.3.4.4 *Carbonate Minerals*

The major cations found in the carbonate minerals in coals are calcium, magnesium, and iron. The rather pure end member calcite ($CaCO_3$) is dominant in coals of the Illinois Basin,[563,565] whereas siderite ($FeCO_3$) is dominant in some Australian coals.[543,634]

Both calcite and ankerite (a mixed crystal composed of Ca, Mg, and Fe carbonates) are abundant in British coals.[635]

The carbonate minerals may be related to the environments in which the coals were deposited and the geochemical conditions that existed during the first stages of coalification. Kemezys and Taylor[543] noted that siderite is abundant only in seams that have freshwater rather than marine rocks overlying them and that siderite and pyrite are not usually associated in the coals.

Carbonate petrifactions (fossil peat or "coal balls") are found within coal seams that had an apparent marine influence.[636,637] Calcite, dolomite, and pyrite commonly make up the major minerals in coal balls.[551,637] Paleobotanists prefer to study the purer carbonate coal balls because the degree of preservation of the plant fragments is much better in them than in the pyritized petrifactions.

622 Yasuo Nakayanagi, *Kozan Chishitsu,* **14**(68), 320–328 (1964b).

623 Yasuo Nakayanagi, *Kozan Chishitsu,* **14**(68), 320–328 (1964).

624 R. J. Gray, N. Schapiro, and G. Dale Coe, *Trans. AIME Soc. Min. Eng.,* **226** (1963).

625 H. J. Gluskoter and J. A. Simon, Illinois State Geol. Survey Circ. No. 432, 1968, p. 28.

626 I. U. Khawaja, Indiana Geol. Survey Spec. Rep. No. 9, 1975, p. 24.

627 L. J. Lawrence, S. St. J. Warne and M. Booker, *Aust. J. Sci.,* **23,** 87 (1960).

628 J. R. Hatch, H. J. Gluskoter, and P. C. Lindahl, *Econ. Geol.,* **71,** 613–624 (1976).

629 J. R. Hatch, M. J. Avcin, W. K. Wedge, and L. I. Brady, U.S. Geol. Survey, Open File Rep. No. 76-796, 1976, p. 26.

630 J. C. Cobb and S. J. Russell, *Geol. Soc. Am., Abstracts with Programs,* **8**(6), 816 (1976).

631 M. Kossenberg and A. C. Cook, *Min. Mag.,* **32,** 829–830 (1961).

632 Dennis Gruner and William C. Hood, *Trans. Ill. Acad. Sci.,* **64,**(2), 156–158 (1971).

633 Kameki Kinoshita, Honda Tsuneo, and Tenaka Nobuya, *Kyushu Kozan Gakkai-shi,* **34**(12), 377–380 (1966) (in English).

634 M. Smyth, *Fuel,* **45,** 221–231 (1966).

635 K. Dixon, E. Skipsey, and J. T. Watts, *J. Inst. Fuel,* **43,** 229–233 (1970).

636 N. S. Snigirevskaya, *Rev. Palaeobot. Palynol.,* **14,** 197–204 (1972) NCB.

637 T. L. Phillips, M. J. Avcin, and Dwain Berggren, Illinois State Geolog. Survey Educ. Ser. No. 11, p. 39, 1976.

Dixon et al.[635] described the distribution and composition of carbonate minerals in coals of the East Midlands coalfields in Great Britain. Several Soviet workers have studied carbonate concretions in coals of the Donbas Basin,[638,639] the Kansk-Achinsk Basin,[640] and the Bovov-Dol Basin.[641]

7.3.4.5 *Silica Minerals*

Quartz is the dominant form in which silica is found in coals, and it is ubiquitous. Mackowsky[552] distinguishes between clastic grains of quartz introduced by wind or water and authigenic quartz deposited from solutions. Quartz is also a major component of clay and siltstone partings in coal that are of detrital origin.

Kemezys and Taylor[543] reported the presence of well-crystallized quartz and chalcedony in Australian coals. Chalcedony was observed in cell fillings, veins, and lenses and may have been introduced by way of igneous intrusions. Unusual silicate spheres, thought to be space-derived dust, were found by Dolgov et al.[642] in peat specimens near the impact point of the Tunguska meteorite.

7.3.4.6 *Other Minerals*

A large number of minerals, in addition to those already discussed, have been reported to occur in coals (Tables 7.5 and 7.6). Not all have been positively identified, and often it is impossible to determine from the reports whether the mineral was intimately associated with the coal or was in the rock units making up the roof, floor, or a parting within the seam. Most of these other minerals are of limited significance in coal utilization, but a few are worth noting.

Authigenic apatite (carbonate fluorapatite) has been found in coals from widely separated areas of the world. Kemezys and Taylor[543] and Cook[643] reported fluorapatite from Australia, Gluskoter et al.[644] reported apatite petrifactions in Illinois, Ruch et al.[99] identified fluorapatite in Illinois coals, and Burger and Stadler[645] found apatite associated with tonsteins in Germany.

All sedimentary rock units, including coal, can host unusual detrital minerals and rocks. The larger of these are curiosities and are more likely to be described in the scientific literature. Zaritskii[646] noted such erratics in coals of the Donbas Basin; and Branson and Merritt[647] described an igneous cobble in a coal from Oklahoma and listed many such exotic occurrences in coals throughout the world.

Many recent publications concerning mineral matter in coals have attempted to characterize the minerals associated with coals in some specific region. O'Gorman and Walker[539] covered a large area in their study of mineral matter in coals of the United States. Other studies on coals of the United States have been limited to smaller portions of the country. Anthracites in eastern Pennsylvania have been studied by Spackman and Moses.[648] Lignites from the Great Plains were characterized for mineral matter content by Paulson et al.[649] Wolfe[562] reported on mineral matter in the Pawnee coal bed of Montana. Mineral matter in coals of the Illinois Basin have been studied by Hughes,[650] Rao and Gluskoter,[563] Miller,[564] and Ward.[565]

[638] P. V. Zaritskii, *Vestn. KHGU,* No. 2, 113–118 (1964a).

[639] P. V. Zaritskii, *Dokl. Akad. Nauk SSSR, Earth Sci. Sect.,* in trans. **157**(1–6), 143–146 (1964b).

[640] I. B. Volkova, *Lithol. Miner. Resourc.,* **6**(1), 24–30 (1971).

[641] Peecva-Sackova, *Pr., God. VMGI,* **13**(5), *Geologija,* 183–198 (1966–1967) (Summary in Russian and German, 1971).

[642] Yu. A. Dolgov, N. V. Vasil'yev, N. A. Shugurova, Yu. A. L'vov, Yu. G. Lavrent'yev, and Yu. A. Grishin, *Trans. Acad. Sci. USSR,* **200,** 212–214 (1971).

[643] A. C. Cook, *Aust. J. Sci.,* **25**(3), 94–95 (1962).

[644] H. J. Gluskoter, Louis Pierard, and Herman Pfefferkorn, *J. Sediment. Petrol.,* **40,** 1363–1366 (1970).

[645] K. Burger and G. Stadler, *Nordrhein-Westfalen,* No. 2125, 1971, 96pp.; No. 2126, 253 pp., Westdeutscher Verlag, Cologne, Opladen.

[646] P. V. Zaritskii, *Dokl. Akad. Nauk SSSR, Earth Sci. Sect.,* **213**(1–6), 156–157 (1975).

[647] Carl C. Branson and C. A. Merritt, *Okla. Geol. Notes,* **23**(10), 235–241 (1963).

[648] W. Spackman and R. G. Moses, *Proc. Anthracite Conf.* Bull. No. 75, Mineral Industries Exp. Sta., Pennsylvania State Univ., University Park, 1961.

[649] L. E. Paulson, W. Beckering, and W. W. Fowkes, *Fuel,* **71,** 224–227 (1972).

[650] Randall Edward Hughes, "Mineral Matter Associated with Illinois Coals," Ph.D. thesis, Department of Geology, University of Illinois, Urbana-Champaign, 1971.

Australian workers, including Kemezys and Taylor,[543] Marshall and Tompkins,[549] and Smyth,[651] have been actively investigating mineral distributions in coals of Australia. Snyman[652] related the major minerals in coals of South Africa to macerals in the coals and discussed the modes of origin of the minerals. Descriptions of mineral matter in coals of India have been published by Sen and Roy,[653] Sinha et al.,[654] and Moiz and Rao.[655] Maruyama and Kobayashi[656] have estimated the quantities of minerals in Japanese coals.

European coals have also been investigated for their mineral matter content in recent years. Guthörl[657] described minerals in Saarbrucken coal measures; Kovatsits and Gal[658] studied Hungarian coals; Sharlovskaya and Shugurov,[659] Bogdanova and Volkova,[660] and Sharnda and Moskvin[661] investigated mineral matter in coals of the U.S.S.R.

Dixon et al.[598,635,662] and Skipsey[663] have published a series of papers on the mineral constituents of the coals of the East Midlands. The first of the papers by Dixon et al. reviews the general chemistry and mineralogy of the coals, and the later papers deal with the clay minerals and the carbonate minerals specifically.

[651] M. Smyth, Tech. Commun. No. 48, Div. Coal Res. C. S. I. R. O., 1966, 35 pp.

[652] C. P. Snyman, Publ. Univ. Pretoria, Nuwe Reeks, No. 15, 1951, 59 p.

[653] P. Sen and A. N. Roy, *J. Min. Met. Fuels*, **10**, 7–14 (1962).

[654] A. K. Sinha, S. R. Chakraborty, and N. G. Banerjee, *J. Mines Met. Fuels*, **17**, 301–304 (1969).

[655] A. A. Moiz and N. Ramana Rao, *J. Mines Met. Fuels*, **24**(2), 63–65 (1976).

[656] T. Maruyama and K. Kobayashi, *J. Fuel Soc. Jap.*, **44**, 633–640 (1965) (in Japanese).

[657] P. Guthörl, *Bergfreiheit*, **28**(7), 256–265 (1963) (in German).

[658] M. Kovatsits and E. Gal, *Publ. Hung. Res. Inst. Min.*, No. 8–9, 271–281 (1965–1966).

[659] M. S. Sharlovskaya and V. F. Shugurov, *Teploenergetika*, 51–54 (1963) (in Russian).

[660] M. V. Bogdanova and I. B. Volkova, *Mater. Sovesch. Rabot. Lab. Geol. Organ., 19th*, Leningrad: Nedra, 1965, No. 8, pp. 82–88.

[661] I. I. Sharnda and V. I. Moskvin, *Lithol. Miner. Resour.*, **81**(1), 76–79 (1973).

[662] K. Dixon, E. Skipsey and J. T. Watts, *J. Inst. Fuel*, **37**(286), 485–493 (1964).

[663] E. Skipsey, *Univ. Nottingham Min. Dep. Mag.*, **18**, 57–65 (1966).

7.3.5 The Geochemistry of Mineral Matter in Coal

7.3.5.1 *Introduction*

The geochemical system "mineral matter in coal" is extremely complex because of the variety of physical and chemical conditions that existed within the environment in which the coal formed and because of the wide range of materials that were available in that environment. The system, which began as a coal-forming swamp, has been at low temperature and low pressure throughout its history; therefore metastable phases could have been present for long periods of time and may still be present. A large number of component phases have been involved and may have contributed to the resulting heterogeneous mixture now found in coals. The system has been open, and mobile components have been free to enter and exit. A further complication is that this system may have been active and therefore in the process of altering at any time since its genesis in the coal swamp (~ 300 million years for coals of the Pennsylvania System).

Some of the mineral matter is in organic combination in the coal, and interactions between organic and inorganic matter affect the phases produced. The system is complicated, but geochemical interpretations of the data collected on coals should not be impossible, because the minerals associated with coals (as with all other sedimentary rocks) are not randomly deposited. They are the necessary result of a definite set of physical, chemical, and biological conditions, all of which combined to provide an environment in which those minerals could be deposited or in which they could form. Attempts at describing the conditions that existed throughout the long history of coal formation are more likely to be correct, or at least consistent with all available data, if the modes of occurrence of the mineral matter are understood. Early classification of mineral matter in coals as (1) "inherent" and (2) "adventitious" or "extraneous" attempted to use genesis of the mineral matter as a basis for defining classes. Table 7.5 demonstrated a more sophisticated classification system used by Mackowsky,[552] featuring the terms "syngenetic" and "epigenetic" and "first stage of coalification" and "second stage of coalification." Table 7.7 attempts to describe the possible origins of mineral matter in coal by comparing two parallel genetic lines, one organic and the other inorganic. The diagram is intended to show that (1) mineral matter found in coals may be in either organic or inorganic combination, (2) mineral matter may have formed or been deposited in the coal swamp, or (3) it may have been formed, deposited, or otherwise adsorbed at any time since the time of the coal swamp, up to and including the geologic present.

The list of examples of mineral matter that formed at various stages of the development of the coal is certainly not all-inclusive. These minerals and elements are intended to serve only as examples of the types and kinds of associations that are present in coals and may reflect personal biases, given our experience with coals of the midcontinent of the United States.

Discussions of the geochemical significance of the many mineral phases found in coals are rare, perhaps because there are only a small number of quantitative mineral analyses on which to make such syntheses. It is also difficult to compare quantitative data on mineral content from different coals unless the exact method of sampling is known. Most coals contain clay partings, and many contain minerals in nodular form, both syngenetic and epigenetic. These partings, nodules, and other mineral forms may be collected with the coal sample or they may be excluded, greatly affecting the mineral matter analyses. A sample that represents the "run of the mine" would contain some portion of most of these rock units as well as portions of the strata underlying and overlying the coal. A more carefully collected "face sample" or "channel sample" would exclude roof and floor material and, depending on the purpose for which it was collected, would contain more or less of the mineral matter interbedded with the coal. This problem greatly complicates attempts to synthesize published data on mineral matter.

The more comprehensive discussions of the geochemistry of mineral phases in coals include a review by Watt,[517] the sections in *Coal Petrology* contributed by Mackowsky,[552] and the chapter entitled "Geochemistry of Coal-Bearing Strata" by Nicholls.[664]

7.3.5.2 *Organic and Inorganic Affinities of the Elements*

The geochemical aspects of trace elements have been discussed in the literature more widely than the geochemical aspects of mineral phases, primarily because of the more sophisticated level of analysis that has been attained for elements in coal and the subsequent increase in data for trace elements.

The modern investigations of trace elements in coals were pioneered by Goldschmidt, who developed quantitative chemical analysis by optical emission spectroscopy and applied it to coal ash. In these early works Goldschmidt[665] studied the chemical combinations of the trace elements in coals.

[664] G. D. Nicholls, "The Geochemistry of Coal-Bearing Strata," in *Coal and Coal-Bearing Strata*, D. Murchison and T. S. Westoll, Eds., American Elsevier Publishing Company, New York, 1968, pp. 266–307.

[665] V. M. Goldschmidt, *Ind. Eng. Chem.* **27**(9), 1100–1102 (1935).

Table 7.7 Parallel Genetic Lines in the Development of Coal

Organic	time	Inorganic
Coal swamp flora (C, H, O, N, S, Se)	↓	Detrital minerals (clay minerals, quartz, zircon, tourmaline)
Epigenetic (early)(first stage of coalification)(B, Ge)	↓	Syngenetic minerals (pyrite, calcite, ankerite, clay minerals)
Epigenetic (later)(second stage of coalification) (Ge, U, Br, Cl) →	Coal	← Epigenetic minerals (pyrite, calcite, kaolinite, sphalerite)

In addition to recognizing trace elements in inorganic combinations with the minerals in coal, he postulated the presence of metal-organic complexes and attributed the observed concentrations of vanadium, molybdenum, and nickel to the presence of such complexes in coal.

Goldschmidt[666] also introduced the concept of a geochemical classification of elements, in which elements are classified on the basis of their affinities and their tendencies to occur in minerals of a single group. The chalcophile elements commonly form sulfides, and lithophile elements commonly occur in silicates.

A large number of silicate, sulfide, and carbonate minerals have been identified in coal seams, and the elements composing them necessarily occur in coals in inorganic combination. Elements that occur in inorganic combination may also be present in organic combination, however, and the exact modes of combination for other elements that may be primarily in organic combination in coals are difficult to identify.

Nicholls[664] approached this problem by plotting the analytical data for the concentration of a single element in coal or in coal ash against the ash content of the coal. Diagrams depicting a number of such points for a single coal seam, or for a group of coal seams in a single geographic area, were interpreted to represent the degree of inorganic or organic affinity of the element. Nicholls[664] concluded:

> One element, boron, is largely, almost entirely, associated with the organic fraction in coals; some elements, such as barium, chromium, cobalt, lead, strontium, and vanadium are, in the majority of cases, associated with the inorganic fraction; and a third group including nickel, gallium, germanium, molybdenum and copper, may be associated with either or both fractions (p. 283).

He then subdivided the third group into nickel and copper, which are in inorganic combination when found in large concentrations, and gallium, germanium, and molybdenum, which are largely in organic combination when strongly concentrated.

A much more ambitious series of investigations of the organic-inorganic affinities of trace metals in coals was undertaken by Zubovic and co-workers at the U.S. Geological Survey.[535–537,667–669] In the more recent of these articles Zubovic[537,667] listed the following 15 elements in order of percentage organic affinity: germanium (87), beryllium (82), gallium (79), titanium (78), boron (77), vanadium (76), nickel (59), chromium (55), cobalt (53), molybdenum (40), copper (34), tin (27), lanthanum (3), and zinc (0).

Zubovic (Ref. 667, p. 50) then related the ranking of the elements in the table of organic affinity to the complexing ability of the metals with organic ligands and suggested that the metals having high organic affinities in coal are present as chelates.

Ruch et al.[99] and Gluskoter[553] published tables of organic affinities for 21 elements determined on four samples of Illinois coals that had been washed in the laboratory. The elements were listed in decreasing order of organic affinity, but numerical values were not given for the index.

Gluskoter et al.[275] calculated values for an organic affinity index for eight coals from the United States. The value of the index ranged from 0.08 to 2.02 for the elements determined; the higher values indicate a greater affinity of the element for the organic matter of the coal. Elements were divided into four groups: organic, intermediate-organic, intermediate-inorganic, and inorganic. The resulting list closely resembled Zubovic's list of organic affinities[667]—germanium, beryllium, boron, and antimony were classified as elements that have the greatest organic affinity, and zinc, cadmium, manganese, arsenic, molybdenum, and iron were classified as having the lowest organic affinities.

Mechanisms by which the elements are combined with the organic fraction have been postulated for many elements by Zubovic.[537] Manskaya and Drozdova[670] investigated the

[666] V. M. Goldschmidt, *Geochemistry,* Alex Muir, Ed., Clarendon Press, Oxford, 1954, 730 pp.

[667] Peter Zubovic, U.S. Environmental Protection Agency Technol. Ser. EPA-600/2-76-149, 1976.

[668] P. Zubovic, T. Stadnichenko, and N. B. Sheffey, U.S. Geol. Survey Prof. Paper No. 400-B, 1960, pp. 1384–1387.

[669] P. Zubovic, T. Stadnichenko, and N. B. Sheffey, U.S. Geol. Survey Prof. Paper No. 424-D, 1961, pp. D-345–D-348.

[670] S.M. Manskaya and T. V. Drozdova, *Geochemistry of Organic Substances,* translated and edited by C. Shapiro and I. A. Breger, Pergamon Press, New York, 1968, 345 pp.

manner of concentrations of uranium, germanium, and vanadium in coals. Eskenazy[671] postulated the presence of tungsten as organometallics in coals.

Manskaya et al.,[672] Burkser and Korniyenko,[673] and Bekayarova and Ruschev[674] have published articles in Russian on the form in which germanium is combined with the organic matter of coal. References to many more publications discussing the mode of occurrence of various elements in coal can be found in the review articles and bibliographies mentioned previously.

7.3.5.3 *Paleosalinities*

Researchers have attempted to determine the degree of marine influence (paleosalinity) on the coal-forming environment in several ways. The higher concentrations of sulfur in coal seams have been correlated with marine strata associated with coals in the U.S.S.R.,[532] in Australia,[543,634] and in the United States,[675,676] and in German brown coals.[677]

Mass spectrometric investigations of sulfur isotopes in coals have been very limited. Rafter[593] published sulfur isotope data on samples of New Zealand coals but did not draw from them any conclusions concerning paleosalinity. Smith and Batts[127] determined the isotopic composition of a number of Australian coals and concluded that from this type of data, one might deduce the origin of the organically combined sulfur as well as other paleosalinity parameters, such as the depth of penetration of seawater into the underlying coal measures.

Boron content of coals and other sedimentary rocks has been used as an indicator of the paleosalinity in which the sediments were deposited. The theory of boron paleosalinities and the difficulties inherent in making such interpretations have been described in relation to coal-bearing strata by Nicholls.[664] Boron contents in coals of the Bowen Basin and also in New South Wales have been correlated with marine influence by Swaine.[525,526] Timofeev et al.[524] found 3–6 times as much boron in the ash of coals from the Angara Basin (Siberia) that were deposited in a brackish-marine environment than in ash from a correlative coal that formed in a lacustrine-alluvial (freshwater) environment. Extremely high boron values in coals of New Zealand (values greater than 1% B_2O_3 in the ash) were attributed by Kear and Ross[678] to either enrichment by hot spring activity or seawater percolation.

Bohor and Gluskoter[420] separated the clay mineral illite from the remainder of the mineral matter of the Herrin (No. 6) Coal Member from the Illinois Basin and analyzed the illite for boron content. They concluded that the salinity of the coal-forming environment was no more than brackish. They also concluded that higher boron values correlated with more seaward parts of the coal-forming basin and that lower boron values correlated with fresher water distributary channels. Gluskoter et al.[275] noted that boron content was generally higher in coals of the midcontinental United States than in either the Appalachian Basin coals or in coals of the Rocky Mountain area, and concluded that this was probably due to higher paleosalinities, which affected the midcontinental coal swamps.

7.3.6 Utilization of Mineral Matter in Coal and Environmental Considerations

Much of the research on trace elements and minerals in coals that has been discussed previously was initiated when problems were encountered in utilizing the coals. The identification and removal of deleterious components have received much attention, and the recovery of potentially valuable coproducts and by-products of mining has also been considered. So much literature on the utilization of mineral matter exists that only the

[671] Gr. Eskenazy, *Chem. Geol.*, **19**, 153–159 (1977).

[672] S. M. Manskaya, T. V. Drozdova, and R. P. Kravtsova, *Geokhim. Akad. Nauk SSSR*, No. 2, 188–197 (1965).

[673] Ye. S. Burkser and T. G. Korniyenko, *Geochem. Int. 32*(6), 1022 (1965).

[674] E. E. Bekayarova and D. D. Rusrhev, *Fuel*, **50**(3), 272–279 (1971).

[675] H. J. Gluskoter and M. E. Hopkins, Illinois State Geol. Survey Guidebook Ser. No. 8, 1970, pp. 89–95.

[676] E. G. Williams and M. L. Keith, *Econ. Geol.*, **58**, 720–729 (1963).

[677] E. Stach, et al., Eds., *Coal Petrology*, Gebrüder Borntraeger, Berlin and Stuttgart, 1975, 428 pp.

[678] David Kear and J. B. Ross, *N.Z. J. Sci.* **4**, 360–380 (1961).

publications that review the literature are cited.

A study done at the West Virginia Coal Research Bureau on coal-associated minerals of the United States resulted in a seven-volume publication.[679] This study was not restricted to the minerals contained within the coals but also considered the rock units, such as shales and limestones, associated with the coals within the coal-bearing strata.

The two types of coal-associated waste products that have been most extensively investigated are fly ash and coal-cleaning refuse. Both of these waste products, as well as many other aspects of coal-associated wastes, are covered in a literature survey by Coalgate.[680] This publication lists, and in many cases annotates, several hundred references by waste categories. Trace elements in wastes from coal preparation plants was the topic of a literature review by Wewerka et al.,[681] which lists nearly 200 references and discusses the significant publications in great detail.

Summaries of the research that has been done on ashes and slags can be found in Coalgate[680] and Slonaker and Leonard.[682] A series of four international symposia has been held on the topic of ash utilization, and the proceedings have been published. These volumes comprise an extremely large resource of information. The proceedings of these conferences can be found in publications compiled by Faber et al.,[683–686] and were printed by either the U.S. Bureau of Mines or the U.S. Energy Research and Development Administration.

Many detrimental environmental effects of coal utilization have been recognized. Included among these are effects as disparate as acid-mine drainage, acrid fumes from burning refuse banks, and gaseous emissions from coal combustion. Many of these detrimental effects may be attributed to mineral matter associated with or contained in the coal. Recently there has been increasing concern about potential harm to the environment resulting from relatively new or projected uses of coal, principally the conversion of coal to gaseous and liquid products. The U.S. Environmental Protection Agency has held two symposia on "Environmental Aspects of Fuel Conversion Technology" and has published the proceedings, which were compiled by Ayer.[687,688] An extensive review of literature on the environmental and health aspects of disposal of solid wastes from coal conversion was recently prepared by Braustein.[689] Studies of coal liquefaction residues at Pennsylvania State University and at the Illinois State Geological Survey have resulted in the publications by Walker et al.[690] and by Russell.[588]

[679] O. B. Bucklen, C. F. Cockrell, B. A. Donahue, J. W. Leonard, C. R. McFadden, P. G. Meikle, L. C. Mih and H. E. Shafer, Jr., BCURA, Office of Coal Research, Coal Research and Development Rep. No. 8; 1965, in seven parts (seven volumes).

[680] J. L. Coalgate, Literature Survey, West Virginia Univ. Coal Res. Bur. Rep. No. 115, 1975, 111 pp.

[681] E. M. Wewerka, J. M. Williams, P. L. Wanek, and J. D. Olsen, U.S. Environmental Protection Agency EPA-600/7-76-007, U.S. Energy Research and Development Administration, ERDA LA-6600-MS, 1976, 61 pp.

[682] J. F. Slonaker and J. W. Leonard, West Virginia Univ. Coal Res. Bur. Rep. No. 87, 1973, 9 pp.

[683] J. H. Faber, J. P. Capp, and J. D. Spencer, U.S. Bur. Mines Info. Circ. No. 8348, 1967, 345 pp.

[684] J. H. Faber, N. H. Coates, and J. D. Spencer, U.S. Bur. Mines Info. Circ. No. 8488, 1967, 345 pp.

[685] J. H. Faber, W. E. Eckard, and J. D. Spencer, U.S. Bur. Mines Info. Circ. No. 864C, 1974, 317 pp.

[686] J. H. Faber, A. W. Babcock, and J. D. Spencer, U.S. ERDA, MERC/SP-7614, 1976, 688 pp.

[687] F. A. Ayer, compiler, *Symp. Proc.: Environmental Aspects of Fuel Conversion Technology,* St. Louis, Mo., May 1974, U.S. Environmental Protection Agency, Environmental Protection Technol. Ser. EPA-65012-74-118, 1974, 366 pp.

[688] F. A. Ayer, compiler, *Symp. Proc.: Environmental Aspects of Fuel Conversion Technology,* Vol. 2, Hollywood, Fla., December 1975; U.S. Environmental Protection Agency, Environmental Protection Technol. Ser. EPA-600/2-76-149, 1976, 404 pp.

[689] H. M. Braunstein, Ed., *An Information Assessment,* Oak Ridge National Laboratories, Oak Ridge, Tenn., 1977.

[690] P. L. Walker et al., College of Earth and Mineral Sciences, EPRI-336-1, PB-252 256/3WE 1975, (Available from NTIS, 111 pp.

CHAPTER **8**

Chemical Reactions and the Constitution of Coal

IRVING WENDER, LASZLO A. HEREDY, MARTIN B. NEUWORTH, AND IAN G. C. DRYDEN

Irving Wender *headed a group in Pittsburgh, Pennsylvania, that conducted research for about 25 years on the chemical reactions of coal and on the reactions of synthesis gas for the Bureau of Mines, the Energy Research and Development Administration, and the Department of Energy. The subjects mentioned remain his principal interests, and he has received a number of awards as a result of work in these fields. Dr. Wender was Director of the Pittsburgh Energy Technology Center, U.S. Department of Energy (DOE), from 1972 until 1978. He is presently Director of the Office of Advanced Research and Technology, Office of Fossil Energy, DOE.*

Laszlo A. Heredy *is Manager, Physical Chemistry Research, at the Energy Systems Group of Rockwell International Corporation, Canoga Park, California, where he is responsible for exploratory research in coal liquefaction, coal gasification, and electrochemical energy conversion. In addition to coal conversion Dr. Heredy's main research interest is in the chemical structure of coal, a subject he has investigated for a number of years in the laboratory of Consolidation Coal Company, Library, Pennsylvania, at the Carnegie Institute of Technology and in Hungary.*

Martin B. Neuworth *was involved with research and development on low-temperature pyrolysis of coal for the production of chemicals and fuels for 20 years with Consolidation Coal Company, Pittsburgh, Pennsylvania. Fundamental studies were carried out on the chemistry of pyrolysis and the relationship between coal structure and pyrolysis, using model compounds to highlight certain key structural features in coals. His work confirmed earlier ideas about the presence of aliphatic bridges linking aromatic structures in coal. In 1979, after serving as Deputy Director, Fossil Fuel Processing, in the Fossil Energy Division of the Department of Energy, concerned with research, development, and demonstration of processes for conversion of coal to liquid and gaseous fuels, Dr. Neuworth joined the Mitre Corporation as Chief Engineer, Energy Resources and Environment Division.*

Ian G. C. Dryden *has been involved in coal research since 1928, when he started to work for the British coal gas industry as a chemist and chemical engineer. In 1942 he joined BCURA and continued his research on coal until his retirement in 1969. He wrote the chapter "Chemistry of Coal" in the first Supplementary Volume of this work. He was also the principal contributor to "Coal" in the second edition of Kirk-Othmer's* **Encyclopedia of Chemical Technology** *(1965). He authored a UNIDO booklet* **Carbonization and Hydrogenation of Coal** *in 1973 and was General Editor of* **The Efficient Use of Energy** *published by IPC Science and Technology Press Ltd. 1979 Dr. Dryden has been Editor-in-Chief of the journal* **Fuel** *since 1975, he has published 70–80 papers in the scientific and technical press.*

8.1 INTRODUCTION

The title of this chapter needs some explanation, for most of the other chapters in this volume also deal with chemical reactions of coal: for example, gasification, liquefaction, combustion, and pyrolysis all fall into the category of "chemical reactions." But these reactions have some practical use and deserve larger and more detailed coverage than this chapter could afford. In addition, since they are usually carried out at elevated temperatures, they may reveal little about the constitution of coal. But it should be emphasized that an understanding of the constitution of coal must involve a comprehension of the various chemical reactions that take place during these "useful" reactions of coal. There is a tendency to separate studies on coal structure from investigations of reactions occurring during coal processing and conversion to liquids and gases, but the two areas are closely related.

This chapter covers at least four kinds of reactions. In the first category are reactions that are carried out at relatively low temperatures or with mild reagents; the hope here is that they will yield information on the essential structural features of coal. In such reactions the products should resemble—or in some logical way be related to—the original coal.

A second category consists of reactions that take place at somewhat higher temperatures or with powerful reagents but are specific reactions ordinarily used in organic chemistry to reveal structural characteristics; the catalytic dehydrogenation of coal falls into this class.

A third category includes both reactions that yield some relevant and/or practical chemistry of coal that is not treated separately in this volume, and those that are related to work discussed in other chapters; the chemical removal of sulfur from coal, related to work covered in Chapter 10, is one example.

The fourth class of reactions discussed here involves those that do not form the basis for a potentially useful process and do not seem to give useful information about the essential structural features of coal. These are included because an imaginative worker may well turn them to good use as a route to fuels or chemicals, or to the further elucidation of coal constitution. As an example, although it appears that the direct treatment of coal with fluorine gas is not too revealing, useful information can be obtained by suitable modification of this reaction.

And what about the phrases "constitution of coal" or "structure of coal"? Numerous articles discuss the structure of coal, but we seldom if ever read about the structure of petroleum. Petroleum is a liquid composed of hundreds of compounds, many identifiable; coal is an organic rock that varies tremendously in composition, contains many minerals, is very heterogeneous, and has relatively few entities that can be termed compounds. The term "constitution" of coal is more applicable than the often-used "structure" of coal; perhaps "the essential structural features" of coal would be a good term. In this chapter all three terms are used interchangeably, although "constitution" is to be preferred.

It is significant that the title of this chapter is a reversal of the similar chapter "Chemical Constitution and Reactions of Coal" by I. G. C. Dryden in the first Supplementary Volume.[1] Dryden's chapter starts with a discussion of the chemical constitution of coal, based largely on ultrafine capillary structure, X-ray diffraction, and spectrometric data. These approaches and their relation to the constitution of coal are covered in Chapter 4 of this volume. Statistical structural analysis, which flourished in that golden age of coal science that ended in the early 1960s, also formed part of Dryden's chapter; very little work has been done in this field since then.

Speaking of the golden age of coal science, those new in the field of coal chemistry (or coal science) will find that the distribution in time of the literature on the subject is rather peculiar, in that the latest important papers on many aspects of the subject were published about 1960 or earlier. It is indeed vital to appreciate that a number of classic papers published in the 1950s have never been superseded, not because they said the final word on the subject, but because the level of effort in most of the world's leading coal research laboratories dropped markedly in the 1960s. (Lowry's first Supplementary Volume appeared in 1963.) Hitherto prolific publishers of papers on coal chemistry pulled out of coal research altogether at this time. Research on coal at the Central Research Laboratory of the Dutch States Mines ground to a halt a few years after van Krevelen left in 1959. A flourishing group of coal researchers at the European Research Associates in Brussels was dissolved when the laboratory was closed about 1964. Work at the Coal Research Laboratory of the Carnegie Institute of Technology ended in the late 1950s. Dryden, Brown, and others at the British Coal Utilization Research Association made many contributions starting with their "little red book" of 1944, *Ultrafine Structure of Coals and Cokes*[2]; work on coal science at BCURA slowed down in the early 1960s and essentially ended in 1969 when BCURA was taken over by the National Coal Board. Research on coal also dropped off markedly in Australia, Japan, and India. Even work at the U.S. Bureau of Mines slowed appreciably.

During the 1950s, then, there were very active centers of coal research throughout the world. International conferences on coal science were held every two years in Europe, and at each a highly international group of experienced chemists met informally to discuss and debate recent results and ideas. These regular and frequent meetings of peo-

[1] *Chemistry of Coal Utilization*, H. H. Lowry, Ed., Supplementary Volume, John Wiley and Sons, New York, 1963.

[2] *Ultrafine Structure of Coals and Cokes*, London, 1944, 366 pp.

ple with related interests were to a large extent responsible for the great productivity of this period. The texts of the formal papers were duplicated for participants' use, and sets are still available in some libraries. The international meetings were held less frequently in the 1960s, and the last took place in 1968. The level of interest in coal science had declined to such an extent by then that it was felt that further meetings were no longer justified.

The OPEC oil embargo of 1973 showed the world that, as many people had pointed out for many years, our oil supplies were being depleted rapidly and were no longer cheap, and that the sources of supply were tenuous. New interest arose in all forms of energy, with coal and nuclear energy the best prospects for supplementing and replacing oil and gas. In 1975 the United States formed the Energy Research and Development Administration (ERDA), and this organization markedly raised the level of funding for coal technology. Funds for more basic coal work were also made available, but not at a level commensurate with those available for coal conversion technologies. Nevertheless ERDA in the United States and other laboratories in coal-producing countries around the world have given what is really a remarkable new impetus to coal research; as this is written in 1978, the seedlings are beginning to bear fruit. In the latter part of 1977 ERDA became part of the new Department of Energy (DOE) in the United States, and although the direction of coal research to be taken by this agency is not yet clear, it is hoped that support for work on coal science will, if anything, be increased. In the decade following the publication of this volume, there will very likely be a "quantum jump" in our knowledge of coal science and coal constitution.

8.1.1 The Heterogeneity of Coals

It is unfortunately true that much work dealing with the chemical reactions and constitution of coal is of dubious value because the investigators failed to take into account the heterogeneity of coal. Coals differ in rank and, although there is no single parameter of "rank," the term generally signifies major differences in the organic substance of coals, corresponding in a broad sense to the degree of maturity and, to a limited extent, to the age of the coal.

In this chapter "rank" generally refers to the percentage of carbon by weight, on a mineral-free basis, in the vitrinite of the given coal. In this sense then, coals differ in carbon content from 60 to 95%.

The reader can learn from Chapter 3 that coals are complex mixtures of materials known as macerals. The major macerals differ not only in reflectance and color in thin section, but also in chemical characteristics. It would seem to be of little value to discuss the chemical structure of whole coals; rather, the reactions and structures of the macerals in coal should be investigated. But separation of pure macerals is difficult and usually not attempted.

Vitrinite is the principal maceral in most coals; whole seam samples containing 70–90% of this maceral are common. For this reason, by far the greatest amount of information on coal macerals concerns vitrinite or vitrinite-rich samples, and there is validity in comparing vitrinites of various coals. However there are other macerals in coal, and their chemical structures differ significantly from that of vitrinite.

In addition, the properties of all the coal macerals show gradations reflecting the degree of metamorphism. Therefore, even for a single maceral, it is really meaningless to speak of a structure. There may be a set of structural features that characterize a particular maceral group, but the relative importance of the various features changes markedly with the degree of metamorphism.

Given and his co-workers[3] have pointed out further complexities in regard to coal chemistry. When chemical reactions are performed on coals, the mineral matter may serve as a catalyst or as an inhibitor. It is also likely that the reagent will solubilize part of the mineral matter. For example, oxidizing or reducing agents may remove pyrite, acidic reagents will leach ions from clays and dissolve carbonates, and alkaline reagents could disrupt clay structures and remove ions. Hence the mineral matter content and composition will change as a result of the reactions, and the relation of ash yield to mineral matter content will be different for starting

[3] Z. Abdel-Baset, P. H. Given, and R. F. Yarzab, *Fuel*, **57**, 95 (1978).

material and product. Therefore to investigate the effects of reactions on the organic matter by a comparison of ultimate analyses, it will be important to determine by a direct method the mineral matter contents of the original coal and of its reaction products. This assessment will be particularly vital when changes in oxygen content are in question.

Given has pointed out that this argument is simple and obvious, but that he knows of no single paper that has followed the recommendations. The quantitative conclusions about a number of reactions of coals in the literature may therefore be of questionable validity, since the authors have usually relied on maf analyses (treating all sulfur as organic).

The presence of mineral matter may have additional significance when a chemical reaction is used quantitatively to determine a functional group. For example, ion exchange with barium acetate may determine some of the acid groups in clays as well as carboxyl groups in lignites. For instance, the clay mineral montmorillonite is acetylated by acetic anhydride in pyridine to an appreciable extent (though not enough, fortunately, to have a significant effect on the determination of phenolic –OH in coals).

The presence of mineral matter may also interfere with the interpretation of infrared spectra of coals. If the mineral matter content is greater than about 10%, strong silicate bands near 1070, 1020, 570, and 525 cm^{-1} are likely to be seen, and –OH vibrations around 3300 cm^{-1} may be influenced to some extent by hydroxyl groups in silicates.

It is important that these thoughts about the heterogeneity of coal be kept in mind during the reading of the rest of this chapter. It is important, too, that the reader be aware that most of the work on coal constitution has been carried out on bituminous coals, all of which fall into the category, more or less, of higher rank coals. The bituminous coals include the caking coals, of course. It is not uncommon, therefore, to find papers that report work on "high-" and "low-" rank coals, the high-rank coal being a low-volatile bituminous coal and the low-rank coal being a high-volatile bituminous coal. The amount of work done on subbituminous coals and lignite is amazingly small, and little is known of their essential structural features.

8.2 CHEMICAL AND ELECTROCHEMICAL REDUCTION OF COAL

Coal can be converted into pyridine- and benzene-soluble products by chemical or electrochemical reduction under mild conditions. This conversion furnishes valuable clues in regard to the structure of coal and the mechanism of coal solubilization.

Both chemical and electrochemical reduction of aromatic hydrocarbons in amine solvents involve essentially two steps. Addition of a solvated electron to the substrate gives a radical anion which then, by addition of a proton, forms a free radical. Electron and proton additions are repeated until reduction to a stable hydrogenated structure is achieved.[4–11]

A study of the lithium-ethylenediamine reducing system with pure organic compounds showed its main effects to be the reduction of aromatic rings and the splitting of ethers

[4] H. W. Sternberg, R. E. Markby, I. Wender, and D. M. Mohilner, "Reduction of the Benzene Ring and of the Olefinic Double Bond by Electrolytically Generated Electrons," *J. Am. Chem. Soc.*, **91,** 4191–4194 (1969).

[5] H. W. Sternberg, R. E. Markby, I. Wender, and D. M. Mohilner, "Electrolytic Generation of Solvated Electrons and Reduction of the Benzene Ring in Ethanol Containing Hexamethylphosphoramide," *J. Am. Chem. Soc.*, **89,** 186–187 (1967).

[6] H. W. Sternberg, R. E. Markby, I. Wender, and D. M. Mohilner, "Electrochemical Reduction of Aromatic Hydrocarbons in Ethylenediamine," *J. Electrochem. Soc.*, **113,** 1060–1062 (1966).

[7] R. A. Benkeser, R. K. Agnihotri, M. L. Burrous, E. M. Kaiser, J. M. Mallan, and P. W. Ryan, "Highly Selective Lithium-Amine Reducing Systems. The Selective Reduction of Aromatic Compounds by Lithium in Mixed Amine Solvents," *J. Org. Chem.*, **29,** 1313–1316 (1964).

[8] H. W. Sternberg, R. E. Markby, and I. Wender, "Electrochemical Reduction of the Benzene Ring," *J. Electrochem. Soc.*, **110,** 425–429 (1963).

[9] A. P. Krapcho and A. A. Bothner-By, "Kinetics of the Metal-Ammonia-Alcohol Reduction of Benzene," *J. Am. Chem. Soc.*, **82,** 751 (1960).

[10] R. A. Benkeser, J. J. Hardra, R. F. Lambert, and P. W. Ryan, "Reduction of Organic Compounds by Lithium in Low Molecular Weight Amines. V. The Mechanism of Formation of Cyclohexanes. Utility of the Reducing Medium in Effecting Stereospecific Reductions," *J. Org. Chem.*, **24,** 854–856 (1959).

[11] L. Reggel, C. Zahn, I. Wender, and R. Raymond, *Reduction of Coal by Lithium-Ethylenediamine and Reaction of Model Compounds with Metal-Amine Systems,* U.S. Bur. Mines Bull. No. 615, 1965.

(Table 8.1). Depending on conditions, benzene rings are reduced to cyclohexane or to cyclohexene rings, sulfur is partially eliminated from heterocyclic compounds (Fig. 8.1), and some ether linkages are split. Other reactions include reduction of phenols with some cleavage of the carbon-oxygen bond, reduction of olefins, reduction of carbonyl group to alcohols, dehydrogenation of special types of cyclic dienes, and formation of certain nitrogen-containing heterocyclic compounds.[12–15] It is unlikely that carbon-carbon

[12] L. Reggel, R. A. Friedel, and I. Wender, "Lithium in Ethylenediamine: A New Reducing System of Organic Compounds," *J. Org. Chem.*, **22**, 891–894 (1957).

[13] L. Reggel, C. Zahn, I. Wender, and R. Raymond, U.S. Bur. Mines Bull. No. 615, 1965.

[14] L. Reggel, S. Friedman, and I. Wender, "The Lithium-Ethylenediamine Systems. II. Isomeriza-

Table 8.1 Reduction of Some Organic Compounds by Lithium-Ethylenediamine[12,13]

	Reactants			Products	
Compound	Amount (mole)	Li/Mole of Compound (moles)	Ethylenediamine (ml)	Compounds	Yield (%)
Benzene	0.30	4.8	375	Cyclohexane	51.0
				Cyclohexene	1.4
				Benzene	21.6
Anthracene	0.05	28.0	300	Dodecahydro- and tetradecahydroanthracenes	~95
Phenanthrene	0.05	14.4	200	Dodecahydro- and tetradecahydrophenanthrenes	~90
$\Delta^{9,10}$-Octalin	0.20	4.0	375	*trans*-Decalin	36.9
				cis-Decalin	Absent
				$\Delta^{9,10}$-Octalin	55.4
1-Heptene	0.32	1.0	300	Heptane	41.0
				2-Heptene	13.0
				1-Heptene	32.0
Phenol	0.30	14.0	400	Cyclohexanone	50.0
				Cyclohexanol	5.5
				Cyclohexane	Absent
				Phenol	21.3
2-Naphthol	0.10	5.5	200	5,6,7,8-Tetrahydronaphthol-2	44.3
				Tetralin	~1
Anisole	0.40	4.4	375	Phenol	53.9
				Δ^2-Cyclohexenone	6.5
				Cyclohexanone	1.6
				Cyclohexanol	1.1
				Cyclohexene	1.5
				Cyclohexane	1.6
Diphenyl ether	0.20	12.0	187	Phenol	83.0
				Benzene	1.0
				Cyclohexene	3.0
				Cyclohexane	~0.3
				Cyclohexanone	~0.1
Dibenzofuran	0.10	32.0	187	*o*-Cyclohexylphenol	~40
Di-*n*-hexyl ether	0.20	8.0	400	2- or 3-Hexene	0.5
				Hexanol[a]	8.4
				Di-*n*-hexyl ether	78.8
Benzyl alcohol	0.20	16.0	400	Hexahydrobenzyl alcohol	38.5

[a] Isomer not determined.

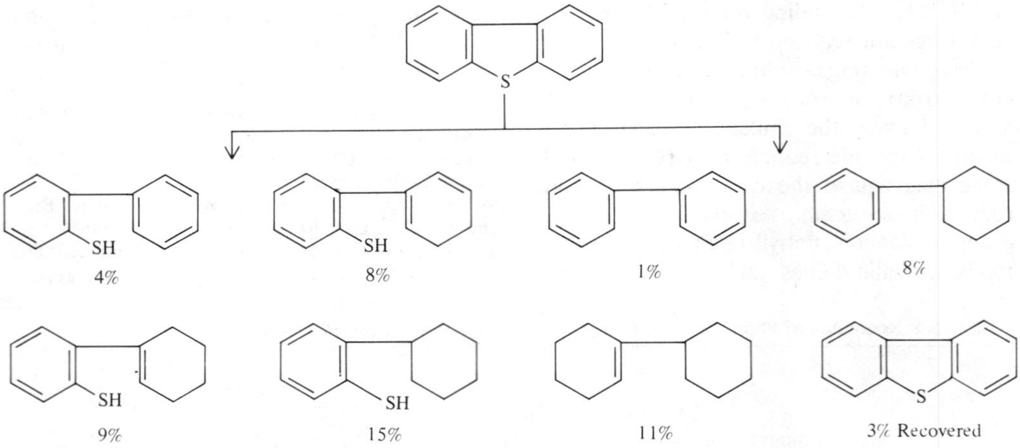

Figure 8.1 Products from lithium-ethylenediamine reduction of dibenzothiophene.[12,13]

bond cleavage occurs under these conditions, although certain very weak bonds of this type may indeed be broken.

A series of 13 coals—ranging in maf carbon content from 67.9 to 93.8%—and a graphite were treated at 100°C with lithium in ethylenediamine.[12,15–17] Some of the experimental results are shown in Table 8.2 and in Figs. 8.2 and 8.3.

Table 8.2 shows that hydrogen addition and pyridine extractability of the reduced coals increase with carbon content up to 90% carbon, then decrease as illustrated in Figs. 8.2 and 8.3. Hydrogen added to higher rank bituminous coals has a larger effect on increasing pyridine extractability than hydrogen added to lower rank bituminous coals. For example, addition of 21 hydrogen atoms per 100 carbon atoms to Bruceton (Pittsburgh) coal resulted in an increase in the amounts extracted by pyridine from 21 to 31%, and addition of 35 hydrogen atoms to Sewell coal raised the pyridine extractability from 3 to 91%. The sudden increases, between 84.9 and 88.7% carbon content, in both numbers of hydrogen atoms added and amounts of coal made soluble in pyridine, is

tion of Olefins and Dehydrogenation of Cyclic Dienes," *J. Org. Chem.*, **23**, 1136–1139 (1958).

[15] L. Reggel, R. Raymond, S. Friedman, R. A. Friedel, and I. Wender, "Reduction of Coal by Lithium-Ethylenediamine," *Fuel*, **37**, 126–129 (1958).

[16] L. Reggel, R. Raymond, W. A. Steiner, R. A. Friedel, and I. Wender, "Reduction of Coal by Lithium-Ethylenediamine. Studies on a Series of Vitrains," *Fuel*, **40**, 339–356 (1961).

[17] L. Reggel, I. Wender, and R. Raymond, "Reduction of Coal by Lithium-Ethylenediamine. A Re-evaluation of Previous Data," *Fuel*, **43**, 75–77 (1964).

Table 8.2 Addition of Hydrogen to Coals After Treatment with Lithium-Ethylenediamine and Amounts of Untreated and Treated Coals Extracted by Pyridine at Room Temperature[16,17]

Coal Bed and State	Rank	Carbon, (% maf)[a]	H Atoms[b]	Extractability (wt %) Untreated	Treated
Pittsburgh, Pa.	hvAb	82.5	21	21	31
Eagle, W. Va.	mvb	84.0	22	13	45
Pond Creek, Ky.	hvAb	84.5	32	13	38
Powellton, W. Va.	hvAb	84.9	24	21	37
Lower Branch, Va.	mvb	88.7	33	11	97
Sewell, W. Va.	mvb	89.4	35	3	91
Beckley, W. Va.	lvb	89.7	35	3	65
Pocahontas No. 3, W. Va.	lvb	89.8	48	4	63
Pocahontas No. 4, W. Va.	lvb	89.9	34	5	49

[a] Carbon content of untreated coal.

[b] Hydrogen added per 100 carbon atoms in starting material.

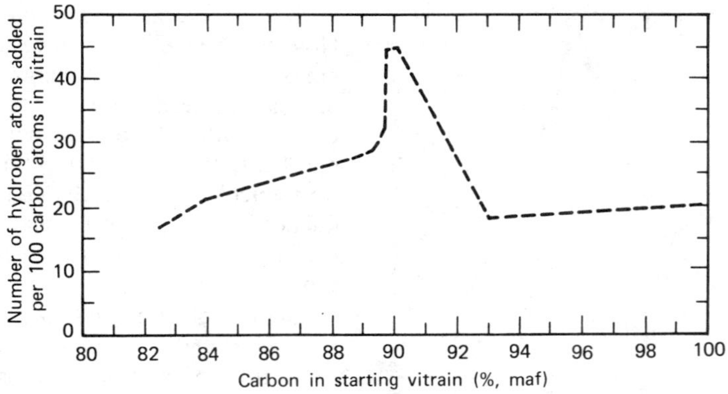

Figure 8.2 Addition of hydrogen to vitrains by treatment with lithium in ethylenediamine.[16]

interesting. The splitting of just a few ether linkages may help account for this phenomenon.

Lithium-ethylenediamine reduction results in a large decrease in sulfur content. The sulfur content of Bruceton hvAb coal was reduced from 1.13 to 0.11% and that of Pocahontas lvb coal from 0.58 to 0.12%.

Comparison of the infrared spectra of the original and the reduced coals showed increased intensities for the aliphatic carbon-hydrogen bands and decreased intensities for all the other bands, including the aromatic C—H bands at long wavelength. The reduced coals had a lower specific extinction coefficient and a lower free radical content.

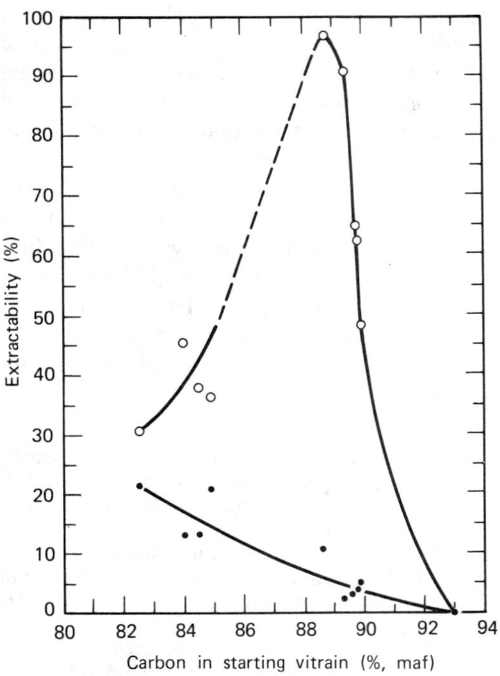

Figure 8.3 Amounts of vitrains extracted by pyridine at room temperature: ● = starting material, ○ = reduced material.[16]

X-Ray scattering intensities of seven vitrains and a graphite were obtained before and after treatment with lithium in ethylenediamine.[11,18,19] Extensive changes had occurred in the arrangement of carbon atoms, with the major changes occurring at low scattering angles. The findings suggest that conversion of planar aromatic layers to buckled hydrogenated rings had taken place.

Reduction of coals was also carried out with lithium in ethylenediamine at room temperature.[20] Less hydrogen is added, less sulfur is removed, and the reduced coals generally are less soluble in pyridine. On the other hand, the patterns of hydrogen uptake and amounts extracted by pyridine are the same at room temperature as at 100°C; that is, hydrogen uptake and amounts extracted by pyridine increase with coal rank up to a maximum of 89–91% carbon content, then decrease.

A series of six coals of increasing rank was reduced by lithium in ethylamine at its boiling point, 17°C.[21] The results are summarized in

[18] S. Ergun and I. Wender, "X-Ray Scattering Intensities of Anthraxylons Reduced with Lithium Ethylenediamine," *Fuel*, **37**, 503–506 (1958).

[19] S. Ergun and I. Wender, "X-Ray Scattering Intensities of Coals Treated with Lithium in Ethylenediamine," *J. Appl. Chem.*, **10**, 189–192 (1960).

[20] H. W. Sternberg, C. L. Delle Donne, L. Reggel, and I. Wender, "Reduction of Coal by Lithium-Ethylenediamine at Room Temperature," *Fuel*, **43**, 143–146 (1964).

[21] P. H. Given, V. Lupton, and M. E. Peover, "Studies of the Aromatic Systems in Coals by Chemical and Electrochemical Reduction Procedures," *Conf. Science Use Coal*, Institute of Fuel, 1958, Paper No. 3, pp. A38–42.

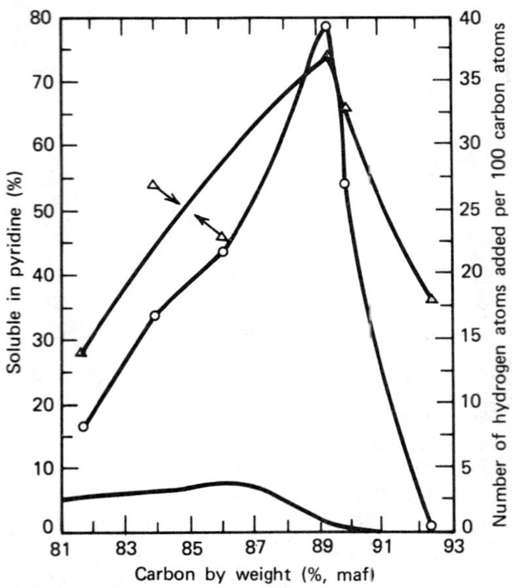

Figure 8.4 Reduction of coals with lithium in ethylamine at 17°C: △ = hydrogen atoms added per 100 carbon atoms, ○ = percentage of reduced coal soluble in cold pyridine; solid curve is percentage of original coal soluble in cold pyridine.[21]

Fig. 8.4. A direct comparison of the data obtained in this work with those obtained with lithium in ethylenediamine is not possible because different coals were used in these two sets of experiments. However the patterns of hydrogen uptake and amounts extracted by pyridine are the same as in the case of the lithium-ethylenediamine reduction. Infrared spectroscopic examination showed that absorption due to vibrations of hydroxyl groups increased in intensity in the reduced coals; the hydroxyl groups appeared to be mainly phenolic rather than alcoholic. The hydroxyl content of a high-rank (89.2%) coal determined by acetylation in pyridine was found to be 2.65% compared with 1.40% in the original coal. The amounts of this high-rank coal extracted by pyridine was 2.5% before and 75% after reduction. The molecular weights of the pyridine-soluble portions were 670 before and 1800–2000 after reduction. These molecular weights were determined in pyridine at 25°C from measurements of diffusion coefficients.

Addition of hydrogen has a marked effect on all subsequent pyrolysis; the reduced coal yields more tar and less coke than the untreated coal. There is evidence that the tars produced from coal on pyrolysis originate from the nonaromatic portions of the coal.[22] However Neavel[23] has suggested that tars do not originate from the nonaromatic structures in coal. Instead, hydroaromatic structures donate hydrogen to fragments formed on pyrolysis, thus minimizing repolymerization (to semicoke) and thereby enhancing tar production. Hence, all else equal, coals containing larger amounts of hydroaromatic hydrogen yield more tar.

Addition of hydrogen to high-rank coals, anthracite, and even graphite suggested the possibility of adding hydrogen to chars. Brooks and Silberman[24] and Reggel and his colleagues[11,17] explored this possibility and succeeded in adding substantial amounts of hydrogen to chars by treatment with lithium in ethylenediamine at 100°C. Brooks and Silberman prepared a series of chars by carbonizing an Australian coking coal (88.8% C) at various temperatures ranging from 300 to 900°C. Hydrogen addition increased from 38 hydrogens per 100 carbon atoms for 300°C chars to 48 for 550°C chars, reached a maximum of 51 for 600 and 650°C chars, then decreased to 19 for chars prepared at 900°C. Reggel and co-workers prepared a char at 550°C from a 90% C coking coal but were able to add only 19 hydrogen atoms per 100 carbon atoms to this char. This difference in hydrogen uptake may have been due to a difference in the starting coals and/or method of carbonization. Neither group reported whether or to what extent the reduced chars were soluble in pyridine.

Electrochemical reduction at room temperature of Pocahontas lvb coal (89.8% C) in ethylenediamine saturated with lithium chloride resulted in the addition of 53 hydrogens per 100 carbon atoms.[25] The electrochemically reduced coal was 78% soluble in pyridine and 30% soluble in benzene at room temperature.

[22] B. K. Mazumdar, S. Ganguly, P. K. Sanyal, and A. Lahiri, "Aliphatic Structures in Coal," in *Coal Science,* Am. Chem. Soc. Adv. Chem. Ser. No. 55, 1966, pp. 475–492.

[23] R. C. Neavel, private communication.

[24] J. D. Brooks and H. Silberman, "The Chemical Reduction of Some Cokes and Chars," *Fuel,* **41,** 67–79 (1962).

[25] H. W. Sternberg, C. L. Delle Donne, R. E. Markby, and I. Wender, "The Electrochemical Reduction of a Low Volatile Bituminous Coal—Nature of the Reduced Material," *Fuel,* **45,** 469–482 (1966).

Nuclear magnetic resonance (NMR) analysis and molecular weight determination of the benzene-soluble portion of the electrochemically reduced coal showed that this fraction contained a large amount (95%) of hydrogen on saturated carbon atoms and had a number average molecular weight of 800–900. Ultimate analyses and the amounts of the electrochemically reduced coal samples made soluble in pyridine showed that hydrogen added at the beginning of the reduction had a much greater effect on extractability than hydrogen added at a later stage. The removal of sulfur took place only after some hydrogen had been added to the coal. A similar pattern of sulfur removal was observed during the electrochemical reduction of dibenzothiophene.[26] In this case sulfur was removed only after a certain amount of hydrogen, corresponding to one hydrogen atom per four carbon atoms, had been added.

Electrochemical reduction of coal in dimethylformamide in the presence of phenol as a proton donor resulted in the addition of 14 hydrogens per 100 carbon atoms.[27]

8.3 CATALYTIC DEHYDROGENATION OF COAL

Dehydrogenation of coals and related materials has been carried out using palladium, ruthenium, and rhodium catalysts in high-boiling aromatic and heteroaromatic solvents.[28–31] The rank of coal affects the yield of hydrogen. For coals ranging from hvCb to anthracite, the hydrogen evolution decreases gradually with increasing rank of coal; lignites and subbituminous coals give less hydrogen than do low-rank bituminous coals. For bituminous coals, the hydrogen evolution (atoms of hydrogen evolved per 100 carbon atoms in the coal) approximates a linear function of the atomic H/C ratio and also of the atomic O/C ratio (Figs. 8.5 and 8.6). Lignites and subbituminous coals fall outside the lines defining the bituminous coals. These results support the view that low-rank bituminous coals contain large amounts of hydroaromatic structures that metamorphose to aromatic structures in higher rank bituminous coals.

Catalytic dehydrogenation of four macerals from an hvAb bituminous coal gave the expected results.[32] Vitrinite and exinite, which evolve 25 and 31 atoms of hydrogen per 100 carbon atoms, respectively, are evidently rich in hydroaromatic structures. Fusinite, which gave only 5 atoms of hydrogen per 100

[26] H. W. Sternberg, C. L. Delle Donne, and I. Wender, "Similarity Between the Electrochemical Elimination of Sulphur from Coal and from Dibenzothiophene," *Fuel,* **47,** 219–222 (1968).

[27] P. H. Given and M. E. Peover, "Polarographic and Electrolytic Reduction of the Aromatic Systems in Solvent Extracts of Coals," *Fuel,* **39,** 463–473 (1960).

[28] R. Raymond, I. Wender, and L. Reggel, "Catalytic Dehydrogenation of Coal," *Science,* **137,** 681–682 (1962).

[29] L. Reggel, I. Wender, and R. Raymond, "Catalytic Dehydrogenation of Coal. III. Hydrogen Evolution as a Function of Rank," *Fuel,* **47,** 373–389 (1968).

[30] L. Reggel, I. Wender, and R. Raymond, "Catalytic Dehydrogenation of Coal. Part 6. Yield of Hydrogen from Whole Coals," *Fuel,* **50,** 152–156 (1971).

[31] L. Reggel, I. Wender, and R. Raymond, "Catalytic Dehydrogenation of Coal. Part 7. The Effect of Some Catalyst Variables," *Fuel,* **52,** 162–163 (1973).

[32] L. Reggel, I. Wender, and R. Raymond, "Catalytic Dehydrogenation of Coal. Part 4. A Comparison of Exinite, Micrinite, and Fusinite with Vitrinite," *Fuel,* **49,** 281 (1970).

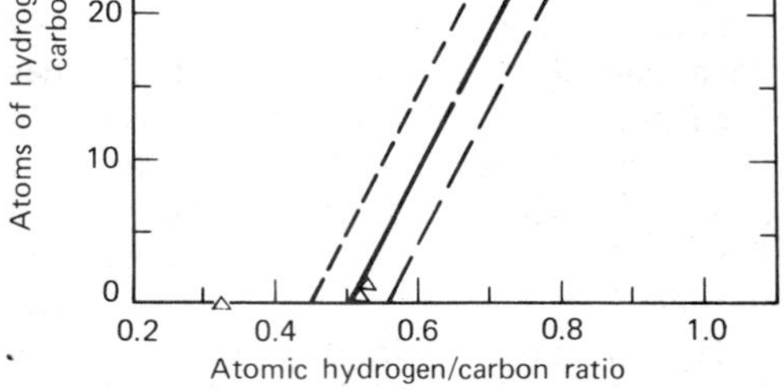

Figure 8.5 Dehydrogenation of vitrains with 1% palladium on calcium carbonate catalyst, phenanthridine vehicle, plotted as a function of the atomic H/C ratio: ○ = bituminous vitrains, ◎ = subbituminous vitrains, ● = lignite vitrains, △ = anthracite and semianthracite, □ = cannel coal. The solid line is the least squares straight line for the 24 bituminous vitrains. The dashed lines are drawn at a distance of twice the standard error of estimate.[29]

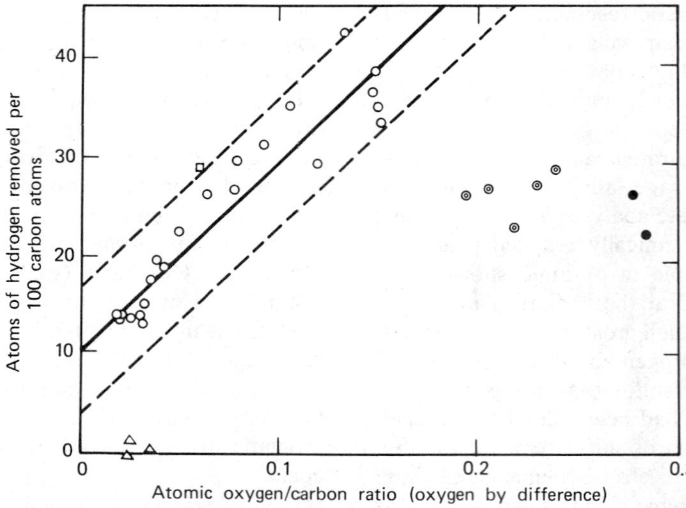

Figure 8.6 Dehydrogenation of vitrains with 1% palladium on calcium carbonate catalyst, phenanthridine vehicle, plotted as a function of the atomic O/C ratio (oxygen analysis by difference): ○ = bituminous vitrains, ◎ = subbituminous vitrains, ● = lignite vitrains, △ = anthracite and semianthracite, □ = cannel coal. The solid line is the least squares straight line for the 24 bituminous vitrains. The dashed lines are drawn at a distance of twice the standard error of estimate.[29]

carbon atoms, is thought to be mainly aromatic. Vitrinite and fusinite from an hvAb and an lvb coal showed the expected differences. For both coals the fusinite gave much less hydrogen than the vitrinite. Micrinite, which is similar to vitrinite in both ultimate and proximate analyses, gave only 18 atoms of hydrogen per 100 carbon atoms. Micrinite must therefore contain some type of hydrogen that is not hydroaromatic but does not greatly alter the ultimate or proximate composition from that of vitrinite. [Neavel[23] has pointed out, however, that micrinite (in the same sense of very fine—less than 1 μm—high-reflectance material) has never really been studied. Most "micrinite" studied has been fine-size semifusinite (also referred to as macrinite). This should not have the same analysis as vitrinite. If a sample does, it is likely to be contaminated with exinite or with vitrinite.]

The extent of dehydrogenation is relatively insensitive to prior mild oxidation of coal, but the yield of hydrogen is considerably reduced after oxidation at 150°C for 66 hr.[33] The yield of hydrogen is not greatly influenced by the catalyst support or the concentration of metal on the support. The particular metal used, however, has a large effect, palladium giving the highest yield of hydrogen.[31]

Hydrogen added to coal during chemical or electrochemical reduction can be quantitatively removed by catalytic dehydrogenation. Similarly, hydrogen added to catalytically dehydrogenated coal after reduction can again be quantitatively removed by catalytic dehydrogenation.[34] These results are summarized in Fig. 8.7, where the symbol Pd −27.5 H indicates that 27.5 hydrogen atoms per 100 carbon atoms had been removed during catalytic dehydrogenation, and Li +20.8 H indicates that 20.8 hydrogen atoms per 100 carbon atoms had been added during reduction with lithium in ethylenediamine. The symbols R-26, R-26A, and so on, are sample numbers. For example, the arrow labeled Pd −45.9 H indicates that 45.9 hydrogen atoms per 100 carbon atoms had been removed from sample R-26A during catalytic dehydrogenation. Similarly, the arrow labeled Li +36.0 H indicates that 36.0 hydrogen atoms had been added to sample R-41 during reduction with lithium in ethylenediamine.

The ability of hydrogen added to coal or to

[33] L. Reggel, I. Wender, and R. Raymond, "Catalytic Dehydrogenation of Coal. Part 5. Effect of Air Oxidation on Catalytic Dehydrogenation of Coal," *Fuel*, **49**, 287–288 (1970).

[34] L. Reggel, I. Wender, and R. Raymond, "Catalytic Dehydrogenation of Coal. II. Reversibility of the Dehydrogenation and Reduction of Coal," *Fuel*, **43**, 229–233 (1964).

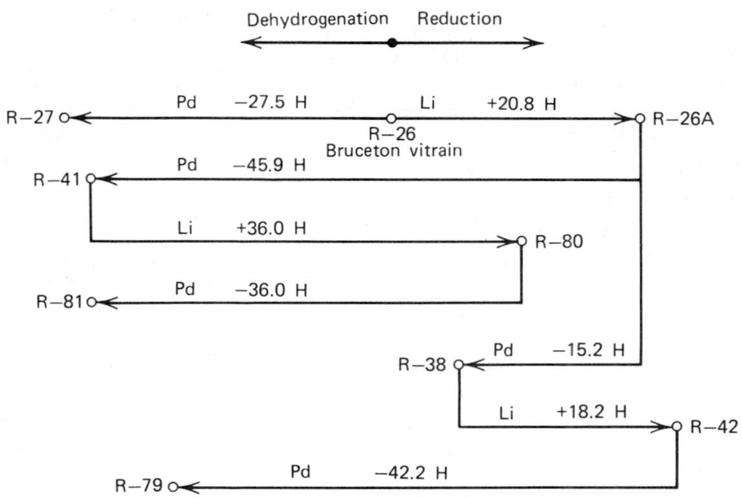

Figure 8.7 Reduction and catalytic dehydrogenation of Bruceton vitrain.[34]

catalytically-dehydrogenated coal to again be quantitatively removed by catalytic dehydrogenation as shown in Fig. 8.7 is strong evidence that coal has a hydroaromatic structure.

High-volatile bituminous coals may act as good hydrogen donors. Collins and his coworkers[35] found that this was the case and that Illinois No. 6 coal was even more effective as a hydrogen donor than tetralin. At 400°C the conversion of 1,1,2-triphenylethane and 1,1,2-triphenylethylene to diphenylmethane and toluene (the asterisk designates ^{14}C):

$$\left.\begin{array}{l} Ph_2\overset{*}{C}H\overset{*}{C}H_2Ph \\ \\ Ph_2\overset{*}{C}{=}\overset{*}{C}HPh \end{array}\right\} \rightarrow Ph_2\overset{*}{C}H_2 + \overset{*}{C}H_3Ph$$

was faster in the presence of vitrinite from Illinois No. 6 hvb coal than in the presence of tetralin. This may explain why this coal, when heated for a short time at 400°C, is converted to a pyridine-soluble material even in the absence of an external source of hydrogen.[36] A possible explanation may be that conversion of Illinois No. 6 coal to a pyridine-soluble product involves the cleavage of certain carbon-carbon and/or carbon-oxygen bonds and that the hydrogen required to stabilize the free radicals formed during thermolytic cleavage is furnished by the hydroaromatic structures present in the coal.

8.4 REDUCTIVE ALKYLATION OF COAL

Much work has been published on the formation, structure, and reactions of aromatic hydrocarbon anions.[37–39] These anions are readily formed When aromatic hydrocarbons, dissolved in a suitable solvent, are treated with alkali metals:

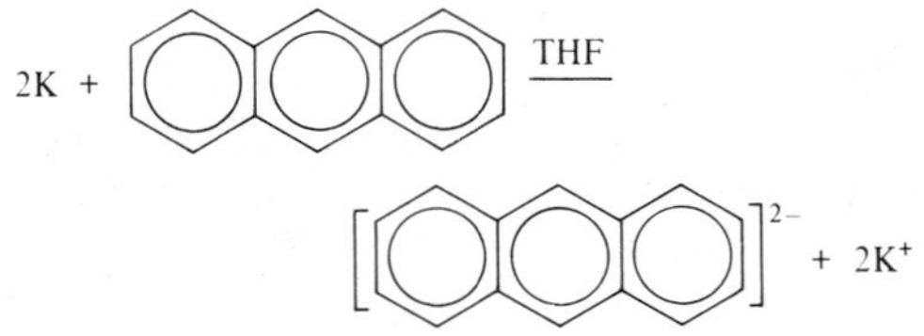

Aromatic hydrocarbon anions react with a wide variety of reagents such as alkyl hal-

[35] C. J. Collins, V. F. Raaen, B. M. Benjamin, and G. W. Kabalka, "Carbon-Carbon Cleavage During Asphaltene Formation," *Fuel,* **56,** 107 (1977).

[36] R. C. Neavel, "Coal Plasticity Mechanism: Inferences from Liquefaction Studies," *Proc. Coal Agglomeration Conversion Symp.,* West Virginia Geol. Econ. Survey, in cooperation with Coal Research Bureau, West Virginia University, Morgantown, May 5–6, 1975, West Virginia University, April 1976, pp. 120–133.

[37] E. DeBoer, in *Advances in Organometallic Chemistry,* F. G. A. Stone and R. West, Eds., Academic Press, New York, 1964.

[38] *Symp. Electron Affinities Aromatic Hydrocarbons and Chemistry Radical Ions,* in *Am. Chem. Soc., Div. Pet. Chem.,* **B:**2 (April 1968).

[39] N. L. Holy, "Reactions of the Radical Anions and Dianions of Aromatic Hydrocarbons," *Chem. Rev.,* **74,** 243–277 (1974).

ides,[38] carbon dioxide,[40] acetyl chloride,[41] aldehydes and ketones,[42] and compounds containing active hydrogen atoms such as alcohols and water.[43] For example, anthracene dianion reacts with methyl iodide[44] to give, among other alkylated products, 9,10-dimethyl-9,10-dihydroanthracene:

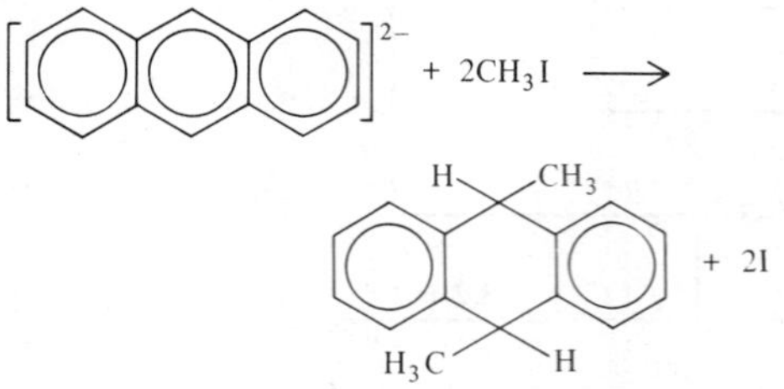

Some coals, when treated with potassium in THF in the presence of a small amount of naphthalene, are converted to coal anions.[45] An electron transfer agent such as naphthalene is required because neither potassium nor coal is soluble in THF. Naphthalene, being soluble in THF, readily reacts with potassium to form the radical anion ($C_{10}H_8^{\dot{-}}$). The latter is a strong electron donor and transfers its electron to the polycyclic aromatic hydrocarbons in coal as illustrated below.

$$K + C_{10}H_8 \rightarrow K^+ + C_{10}H_8^{\dot{-}}$$
$$K + C_{10}H_8^{\dot{-}} \rightarrow K^+ + C_{10}H_8^{2-}$$
$$\text{Coal} + nC_{10}H_8^{\dot{-}} \rightarrow \text{coal}^n + nC_{10}H_8$$
$$ArH \overset{e-}{\rightleftarrows} ArH^{\dot{-}} \overset{e-}{\rightleftarrows} ArH^{2-}$$

By this method, a large amount of coal can be converted to coal anion in the presence of a small amount of naphthalene. The number of charges associated with the coal anion was found to average 11.5 charges per 100 carbon atoms. When hexamethylphosphoramide (HMPA) is used as a solvent, an electron transfer agent is not required, since alkali metals are soluble in HMPA and can react directly with the coal.[46]

Alkylation of the coal anion by alkyl halide is analogous to that of anthracene anion and may be formulated as follows:

$$\text{Coal}^{n-} + nRX \rightarrow (nR) - \text{coal} + nX^-$$

where n^- is the number of negative charges associated with the coal anion. This number can be determined by titration and expressed as charges per 100 carbon atoms. These are the negative centers that are theoretically available for alkylation. The number of alkyl groups actually introduced is always less than the number of negative charges and the number of alkyl groups that can be added

[40] S. Wawzonek and D. Wearing, "Polarographic Studies in Acetonitrile and Dimethylformamide. IV. Stability of Anion-Free Radicals," *J. Am. Chem. Soc.*, **81**, 2067–2069 (1959).

[41] T. S. Cantrell and H. Shechter, "Acylation of Cyclooctatetraene Dianion and the Chemistry of Its Products," *J. Am. Chem. Soc.*, **89**, 5868 (1967).

[42] T. S. Cantrell and H. Shechter, "Reactions of Cyclooctatetraene Dianion with Aldehydes and Ketones and the Chemistry of its Products," *J. Am. Chem. Soc.*, **89**, 5877–5884 (1967).

[43] N. D. Scott, J. F. Walker, and V. L. Hansley, "Sodium Naphthalene. I. A New Method for the Preparation of Addition Compounds of Alkali Metals and Polycyclic Aromatic Hydrocarbons," *J. Am. Chem. Soc.*, **58**, 2442–2444 (1936).

[44] R. Gerdil and E. A. C. Lucken, "Action des Halogenures d'Alcoyles sur les Anions Anthracèniques. Quelques Nouveaux Deriveés Dihydroanthracèniques Substitués," *Helv. Chim. Acta*, **44**, 1966–1978 (1961).

[45] H. W. Sternberg, C. L. Delle Donne, P. Pantages, E. C. Moroni, and R. E. Markby, "Solubilization of an lvb Coal by Reductive Alkylation," *Fuel*, **50**, 432–442 (1971).

[46] H. W. Sternberg and C. L. Delle Donne, "Solvation and Reductive Alkylation of Coal via a 'Coal Anion' Intermediate," *Prepr., Am. Chem. Soc., Div. Fuel Chem.*, **12** (4), 13 (September 1968).

Table 8.3 Alkylation of Various Coals[47]

Coal	Carbon (% maf)	Charges Added per 100 Carbon Atoms	Alkyl Groups Added per 100 Carbon Atoms: First Alkylation	Alkyl Groups Added per 100 Carbon Atoms: Second Alkylation
Anthracite	96	11	0.6	—
Pocahontas lvb	90	12	9	—
Bruceton hvAb	82	26	11	16
Subbituminous	71	24	14	14

Table 8.4 Extractability of an Alkylated lvb Coal

Coal	Percent soluble	
	Hexane	Benzene
Original coal	Nil	0.5
Methylated coal	3	48
Ethylated coal	11	95
Butylated coal	17	93

vary with the type of coal,[47] as shown in Table 8.3.

Benzene and hexane solubility of the alkylated coal increases with the number and chain length of the alkyl groups introduced, as indicated in Table 8.4. Molecular weights of the benzene-soluble portions of two alkylated bituminous and one alkylated subbituminous coal are shown in Table 8.5.

These data indicate that molecular weight decreases with decreasing coal rank. This conclusion is justified with regard to lvb and hvAb coals, since in both cases the benzene-soluble fraction of the alkylated coal contains a large, therefore representative, portion (95 and 86%) of the coal. The benzene-soluble fraction of subbituminous coal, on the other hand, represents only 41% of the total coal. That 41% of subbituminous coal has a molecular weight as low as 700 is interesting because only 1% of the lvb coal has a molecular weight in the range of 900 based on gel permeation chromatography.[45]

These electron transfer agents are basic; they react with acidic hydrogen atoms such as those in phenols, in water, and in certain acidic hydrocarbons to yield aryloxides and carbanions.

$$ArOH + C_{10}H_8^{2-} \rightarrow ArO^- + C_{10}H_9^-$$
$$Ar_3CH + C_{10}H_8^{2-} \rightarrow Ar_3C^- + C_{10}H_9^-$$

Aromatic radical anions produced in the initial reactions reductively cleave ethers. The cleavage reactions of diaryl ethers proceed as follows[48]:

$$ArOAr + ArH^{\dot{-}} \rightleftharpoons ArOAr^{\dot{-}} + ArH$$
$$ArOAr^{\dot{-}} + ArH^{\dot{-}} \rightleftharpoons ArOAr^{2-} + ArH$$
$$ArOAr^{2-} \rightleftharpoons ArO^- + Ar^-$$

[47] H. W. Sternberg and C. L. Delle Donne, "Solubilization of Coals by Reductive Alkylation," *Fuel*, **53**, 172–175 (1974).

[48] C. G. Screttas, "Methylation of Aryl Ethers by Lithium Arenes," *Chem. Commun.*, 869–870 (1972).

A somewhat different mechanism has been proposed for aralkyl ethers[49]:

$$ArOR + ArH^{\dot{-}} \rightarrow ArOR^{\dot{-}} + ArH$$
$$ArOR^{\dot{-}} \rightarrow ArO^- + R\cdot$$
$$R\cdot + ArH^{\dot{-}} \rightarrow R^- + ArH$$

Arylbenzyl or dibenzyl ethers could cleave at either the radical anion or dianion stage to yield equivalent products. It is unlikely that highly reactive carbanions such as Ar^- or R^- would survive in the reaction medium until the addition of an alkyl halide. Rather, these strongly basic reagents would react to produce more stable carbanions, as shown by the reaction between an aryl carbanion Ar^- and a benzhydryl fragment:

$$Ar^- + Ar_2CH_2 \rightarrow ArH + Ar_2CH^-$$

Reductive cleavage of the labile carbon-carbon bond in polyarylalkanes can also occur.[39,50–54]

$$Ar_2CHCHAr_2 + 2ArH^{\dot{-}} \rightarrow 2Ar_2CH^- + 2ArH$$

Under the reaction conditions, the initial products obtained by electron transfer, proton abstraction, ether cleavage, elimination, and carbon-carbon bond cleavage reactions equilibrate as the coal matrix collapses to

[49] H. Gilman and J. W. Morton, "Organic Reactions" vol. VIII, Wiley, New York, 1954, Ch. 6.

[50] H. O. House and V. Kramar, *J. Org. Chem.*, **27**, 4146 (1962).

[51] J. J. Eisch, *J. Org. Chem.*, **28**, 707 (1963).

[52] H. Gilman and B. J. Gaj, *J. Org. Chem.*, **28**, 1725 (1963).

[53] A. Lagendijk and M. Szwarc, *J. Am. Chem. Soc.*, **93**, 5359 (1971).

[54] C. Elschenbroich, F. Gerson, and J. A. Reiss, *J. Am. Chem. Soc.*, **99**, 60 (1977).

Table 8.5 Number-Average Molecular Weights of the Benzene-Soluble Portion of the Ethylated Coals (alkyl-free basis)[47]

Ethylated Coal	Benzene-Soluble Portion (%)	Molecular Weight[a]
Pocahontas, lvb	95	2800
Bruceton, hvAb	86	2000
Subbituminous	41	700

[a] Molecular weights determined in benzene solution by vapor pressure osmometry.

yield a mixture of aromatic radical anions, aromatic dianions, stable carbanions, aryloxides, mercaptides, and nitrogen bases in accord with their relative stability. Fortunately few rearrangement reactions occur,[55] and the structures of the products should be indicative of structures in the original coal. When an alkyl iodide is added to the solution, the aromatic dianions, aryloxides, mercaptides, nitrogen bases, and stable carbanions experience alkylation:

$$ArH_2^- + 2RI \rightarrow ArHR_2 + 2I^-$$
$$ArO^- + RI \rightarrow ArOR + I^-$$
$$Ar_2CH^- + RI \rightarrow Ar_2CHR + I^-$$
$$ArS^- + RI \rightarrow ArSR + I^-$$
$$R_2N^- + 2R'I \rightarrow R_2NR_2'^+ + 2I^-$$

Naphthalene anion abstracts protons from phenols to give the corresponding phenoxide ions. Subsequent treatment of aromatic anions and phenoxide ions with ^{14}C-labeled ethyl iodide leads to an alkylated coal containing labeled ethyl groups attached to carbon $[ArH_2(C^*{}_2H_5)_2]$ and oxygen $[Ar—O—C^*{}_2H_5]$. When the ethylated coal is treated with naphthalene anion, only the alkyl groups attached to oxygen are split off. The number of $C^*{}_2H_5$ groups remaining after dealkylation represents the number of ethyl groups attached to carbon.[47] The difference between the number of $C^*{}_2H_5$ groups in the ethylated coal before and after treatment with naphthalene anion corresponds to the number of ethyl groups attached to oxygen.

By treating coal with an electron donor such as naphthalene anion in THF solution (i.e., in the absence of a proton donor), it is possible to transfer electrons to coal to form the corresponding coal anion. In the absence of a proton donor, the coal anion can be titrated to determine the amount of negative charges added.[45–47] (The coal anion, on treatment with water, liberates an equivalent amount of hydroxyl ion. The amount of hydroxyl ions formed can be readily determined by potentiometric titration with acid.) A comparison of the number of charges added to Pocahontas lvb coal and to anthracite (Table 8.3) shows that 12 electrons per 100 carbon atoms were added to the former and 11 electrons to the latter. Even though 11 electrons had been added to anthracite, less than one alkyl group was added during subsequent alkylation. In contrast, alkylation added nine alkyl groups to the Pocahontas coal.

A possible explanation for this difference is that the negative sites in the anthracite, in contrast to those in the Pocahontas coal, were not accessible to the alkylating agent due to steric hindrance.

It seems that solvated naphthalene anions are able to approach the periphery of the polycyclic rings in anthracite and to transfer electrons to the aromatic hydrocarbon structures. Apparently there is no steric hindrance to the spread of electrons across the condensed aromatic hydrocarbons, but approach of the alkylating agent to the negative sites is sterically hindered. In the lithium-ethylenediamine reduction, approach of ethylenediamine to the negative sites of the anthracite anion may also be sterically hindered. Since ethylenediamine serves as the proton donor, this could explain why less hydrogen is added to anthracite than might be predicted from its highly aromatic structure.

In this connection it is interesting to note that the sum of the amounts of hydrogen that can be added by reduction and removed by dehydrogenation is remarkably constant for a wide variety of bituminous coals, as may be seen from the data in Table 8.6. These data suggest that there may be a basic carbon skeleton structure common to most bituminous coals. This structure can take up hydrogen until the H/C ratio reaches a maximum of 1.0 or can give up hydrogen until the H/C ratio drops to a minimum of 0.5.

Sun and Burk[56] alkylated Illinois No. 6 coal with *n*-butyl iodide using a slight modification of Sternberg's procedure. About 90% of the coal was converted to products soluble in THF. The soluble coal alkylate was separated into 16 fractions by gel permeation chromatography. The number-average molecular weight of each fraction was determined by vapor pressure osmometry (Fig. 8.8). Sun and Burk found that about 55% of the butylated material had molecular weights of less than 12,000, and 90% had molecular

[55] E. Grovenstein and A. B. Cottingham, *J. Am. Chem. Soc.*, **99**, 1881 (1977).

[56] J. Y. Sun and E. H. Burk, "Coal Molecular Weight Distribution by GPC," in *The Fundamental Organic Chemistry of Coal; Proceedings of a Workshop Sponsored by the National Science Foundation,* Knoxville, 1975, p. 80, and private communications to L. Stock.

Table 8.6 Hydrogen Addition to and Hydrogen Evolution from Bituminous Coals[16,17,29]

Coal Bed	Rank	Carbon (%, maf)	Dehydrogenated Coal H/C[a]	Hydrogens per 100 Carbons Evolved	Untreated Coal H/C[b]	Hydrogens per 100 Carbons Added	Reduced Coal H/C[b]	Sum of H Added + H Removed
Rock Springs	hvCb	76.9	0.58	26	0.84	19	1.03	45
Bruceton	hvAb	82.5	0.50	30	0.80	21	1.01	51
Eagle	mvb	84.0	0.55	23	0.78	22	1.00	45
Pond Creek	hvAb	84.5	0.52	27	0.79	32	1.11	59
Powellton	hvAb	84.9	0.50	26	0.76	24	1.00	50
Lower Banner	mvb	88.7	0.55	18	0.73	33	1.06	51
Sewell	mvb	89.4	0.52	20	0.72	35	1.07	55
Beckley	lvb	89.7	0.50	14	0.64	35	0.99	49
Pocahontas No. 3	lvb	89.8	0.48	14	0.62	48	1.10	62
Pocahontas No. 4	lvb	89.9	0.49	12	0.61	34	0.95	46
Mean			0.52				1.03	51.3
Standard deviation			0.03				0.05	5.8

[a] H/C ratios of dehydrogenated coals were calculated by subtracting the hydrogens per 100 carbons evolved from the H/C ratios of the untreated coals.

[b] H/C ratios of untreated and reduced coals were determined experimentally.

weights of less than 20,000. Alemany and co-workers[57] examined the samples provided by Sun and Burk and found that the lower molecular weight fractions yielded well-resolved proton NMR spectra. These workers then studied the proton and carbon NMR spectra of the polybutylated coal. Their objectives were to determine whether the structures of the material depended on the molecular weight, and to determine the extent of alkylation on aromatic (sp^2) and aliphatic (sp^3) carbon atoms as well as on oxygen and nitrogen atoms. This information would provide insight concerning the most readily cleaved bonds and the relative significance of ether cleavage, reduction, alkylation, and reductive alkylation by ion or free radical processes during the butylation reaction.

[57] L. B. Alemany, S. R. King, and L. H. Stock, *Fuel*, **57**, 738 (1978).

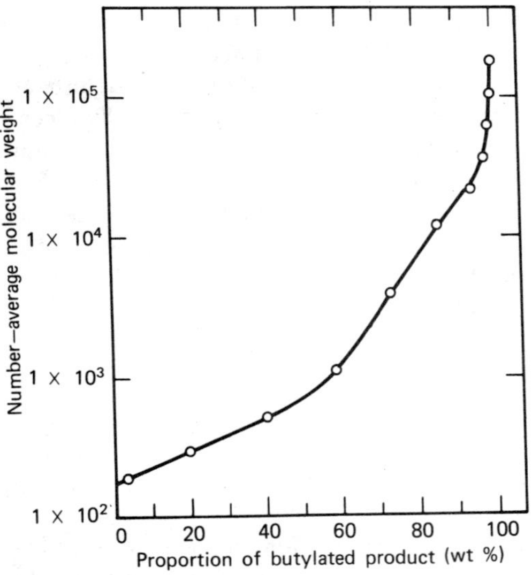

Figure 8.8 The number-average molecular weight distribution compared to the weight percent of the butylated product in the Sternberg alkylation of Illinois No. 6 coal as determined by Sun and Burk.[56]

The approximate molecular weight fractions for the butylated coal samples estimated from the results of Sun and Burk are presented in Table 8.7.

Stock and co-workers[57] showed that structure did indeed vary significantly with molecular weight. They also found that alkylation on aromatic carbons was an important reaction, probably occurring by way of a free radical rather than an ionic interaction.

When butyl iodide is added to the solution in THF, the aromatic anions, aryloxides, mercaptides, nitrogen bases, and stable carbanions undergo butylation.

In ethereal solvents such as THF, aromatic radical anions, not dianions, form preferentially. Aromatic radical anions are poor nucleophiles, but they are efficient electron transfer agents capable of converting alkyl halides to alkyl radicals[58]:

$$ArH^{\dot{-}} + BuI \rightarrow ArH + Bu\cdot + I^-$$

The concentration of alkyl radicals depends on the rate of electron transfer from the aromatic radical anion to the alkyl halide. Since a large halogen is better able to accom-

[58] K. A. Bilevich and O. Yu. Okhlobystin, *Russ. Chem. Rev.*, **37**, 954 (1968).

Table 8.7 Summary of Spectroscopic Observations[57]

	C-Butylation / O-Butylation		C-Butylation		
Fraction	C_γ	C_β	Quaternary sp^3	Tertiary sp^3	sp^2
1–4	—[a]	—[a]	—[a]	—[a]	—[a]
5–8	1.5	—[a]	1.4	1.0	2.7
9	1.4	—[a]	2.6	1.0	1.3
10	2.8	2.6	4.1	1.0	0.3
11	5.5	7.0	1.9	1.0	0.6
12	3.2	—[a]	0.7	1.0	0.4
13–16	—[b]		—[b]		

[a] Undetermined.
[b] No butylation.

modate the developing negative charge, the electron transfer reaction is greatest with iodides.[59] Indeed, the reactions of anion radicals with alkyl iodides usually convert 50–70% to free radicals.[60] Although radical intermediates obtained this way may dimerize or disproportionate to yield small molecules, they may also react with aromatic substrates:

$$Bu\cdot + ArH \rightarrow ArBuH\cdot$$

These radical intermediates can undergo rapid hydrogen atom transfer reactions to produce arylbutanes:

$$R\cdot + ArHBu\cdot \rightarrow RH + ArBu$$

As pointed out by Alemany and co-workers,[57] the presence of butyl radicals clearly enriches the chemistry of the Sternberg alkylation procedure.

Table 8.7 shows the NMR spectroscopic observations concerning the extent of C- and O-butylation, determined by examination of the C_γ and C_β signals, and the degree of C-butylation at various reaction sites, determined by examination of the C_β signals.

The spectroscopic results suggest that the lowest molecular weight fractions (Table 8.7), which represent about 2% of the product, have linear alkane fragments as the main structural feature. These substances appear to be closely related to the alkane hydrocarbons obtained by several groups in the investigation of solvent extracts of coal; they are liberated during the reduction and alkylation reactions that destroy the coal matrix.

The spectroscopic results also indicate that the structures of the molecules in the largest intermediate molecular weight fractions, which represent 46% of the product, are diverse. The extent of C-butylation compared to O-butylation ranges from 1.4 to 5.5. All the available evidence suggests that O-butylation has occurred exclusively on aryl oxygen atoms. The formation of the aryl butyl ethers is certainly expected. It is surprising, however, that there are no alkyl butyl ethers and that the ratio of O-butylation varies so greatly compared to C-butylation. The C-butylation reaction apparently has occurred to produce quaternary and tertiary aliphatic products as well as arylbutanes. These products are formed in quite different amounts in the various fractions.

The products with butyl groups bonded to quaternary and tertiary sp^3 carbon atoms presumably arise by alkylation of carbanions and aromatic dianions:

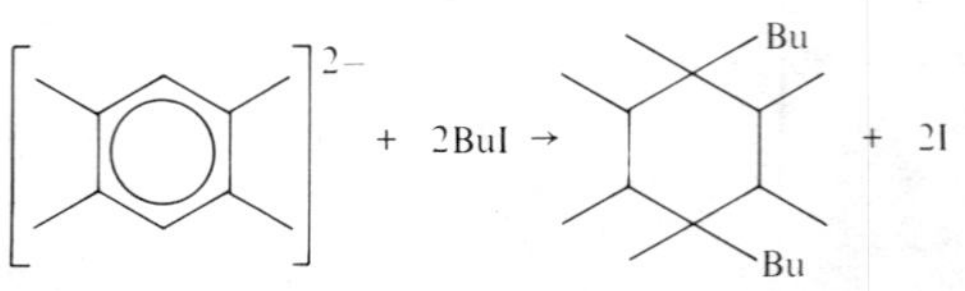

Free radical reactions between anion radicals and butyl radicals also yield compounds of this kind[61]:

[59] G. D. Sargent, J. N. Cron, and S. Bank, *J. Am. Chem. Soc.*, **88**, 5363 (1966).

[60] N. L. Holy, *Chem. Rev.*, **74**, 243 (1974); J. F. Garst, *Acc. Chem. Res.*, **4**, 400 (1971).

[61] G. D. Sargent and G. A. Lux, *J. Am. Chem. Soc.*, **90**,7160 (1968).

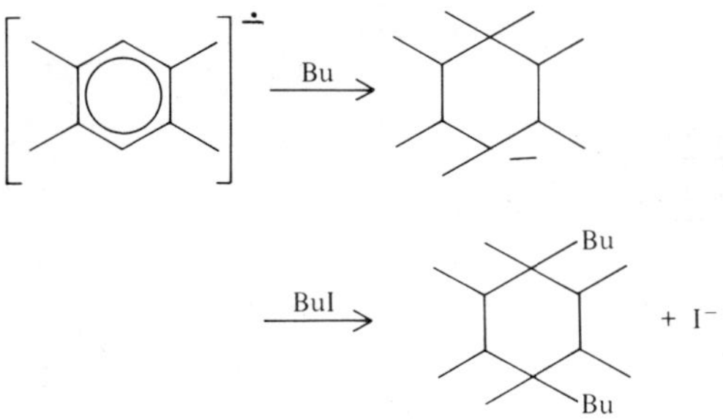

When butyl iodide is used in excess in these highly basic solutions, the secondary reactions of the initial products, the dibutyl compounds, yield tri- and tetrasubstituted compounds[62]:

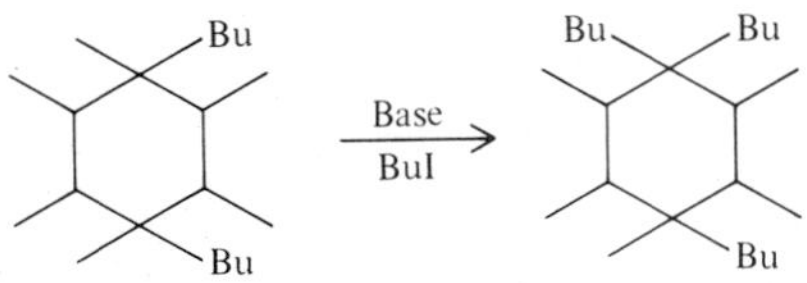

The spectroscopic results are in accord with the known chemistry of alkylation reactions and with a prominent role for reductive alkylation reactions as well as free radical alkylation.

Conversion of coal to a soluble product by reduction or reductive alkylation involves the addition of hydrogen, hydrogen and alkyl groups, or just alkyl groups to coal. Ether cleavage also takes place during reductive alkylation.[63] How important is ether cleavage in solubilizing coal compared to the other reactions mentioned above?

If an ether linkage connects two "basic units," as in Ar—O—R, cleavage will lower the molecular weight. If the oxygen is present in a ring, as in dibenzofuran, ether cleavage will not lower the molecular weight of the substrate. Wachowska and Pawlak, [63] investigating the reductive alkylation of coal, found that the amount of inert oxygen present in ring ethers (heterocyclic oxygen) varies with coal rank. In high-rank coals (89.9% C, maf), heterocyclic oxygen may be as low as 16% of the inert oxygen. In lower rank coals (82–85% C, maf), the percentage of heterocyclic oxygen varies from 73 to 93% of the inert oxygen. In view of these data, it may be reasonable to assume that at least half the inert oxygen is present as heterocyclic oxygen.

The molecular weights obtained by Sternberg and Delle Donne[47] (Table 8.5) indicate that a few ether linkages that were not in heterocyclic rings were split.

There is some experimental evidence[45] that ether cleavage is not sufficient to account for the increase in the amount extracted by benzene. Two experiments were carried out. In one experiment, the coal anion was protonated, in the other it was alkylated:

			Amounts (%) Extracted by:	
			Benzene	Pyridine
$\bar{R}CH_2R-\bar{O}$	$\xrightarrow{H_2O}$	$\bar{H}RCH_2R-OH$	3	13
$\bar{R}CH_2R-\bar{O}$	$\xrightarrow{C_2H_5I}$	$C_2H_5RCH_2R—OC_2H_5 + 2I^-$	95	95

In contrast to the alkylated coal anion, which was 95% soluble in both benzene and pyridine, the protonated coal anion was only 3% soluble in benzene and 13% soluble in pyridine. This is evidence that alkylation is primarily responsible for the conversion of coal to a benzene-soluble product.

Reductive alkylation converts coals with 89.9,[45] 82.5,[47] and 76.1%[56] C (maf) content into products very soluble in benzene and in pyridine (Table 8.8). In contrast, substantial conversion to a pyridine-soluble product by reduction by lithium-ethylenediamine is possible only with coals of high carbon content (89.6–88.7% C, maf).[16] This indicates that addition of alkyl groups is more effective than addition of hydrogen in converting coal to a soluble product. Conversion of Pocahontas 89.8% C (maf) coal, 4% of which is

[62] R. G. Harvey, D. F. Lindow, and C. N. Cortez, *J. Am. Chem. Soc.*, **94,** 5406 (1972).

[63] H. Wachowska and W. Pawlak, *Fuel,* **56,** 422 (1977).

Table 8.8 Amounts of Reductively Alkylated Coals Extracted at Room Temperature[47]

	Soluble (%) in			
Coal	Pyridine	Benzene	Hexane	Ethanol
Anthracite				
Original	0	0	—	—
Ethylated	0	0	—	—
Pocahontas, lvb				
Original	3	0·5	0	—
Ethylated	97	95	11	—
Bruceton, hvAb				
Original	25	0·5	0	0
Once-ethylated	81	74	22	15
Twice-ethylated	95	88	37	18
Dealkylated	91	68	10	25
Subbituminous				
Original	3	0·5	0	2
Once-ethylated	43	23	12	18
Twice-ethylated	73	41	30	35
Dealkylated	89	40	34	32

extracted by pyridine to a product that was 63% soluble in pyridine, required the addition of 45 hydrogen atoms per 100 carbon atoms.[16] Reductive alkylation, which converted this coal to a product 97% soluble in pyridine and 95% soluble in benzene, required the addition of only nine ethyl groups.[47]

With lower rank coals the effectiveness of alkyl versus hydrogen addition is even more pronounced. Addition of 21 hydrogen atoms to Bruceton 82.5% C (maf) coal increased the amount of that coal extracted by pyridine from 21 to 31%, and addition of ethyl groups increased the amount soluble in pyridine to 95% and the amount soluble in benzene from 0.5 to 88%.

There is yet another difference between the addition of hydrogen and alkyl groups that must be considered, namely, the effect on benzene solubility of the treated coal. Electrochemically reduced Pocahontas (89.8% C, maf) coal to which 53 hydrogens per 100 carbon atoms had been added was 78% soluble in pyridine and 30% soluble in benzene.[25] These data compare with a pyridine extractability of 95% and a benzene extractability of 48% for reductively methylated Pocahontas coal to which eight methyl groups had been added. The lower benzene extractability of the hydrogenated as compared to the alkylated coal may be due to the presence of phenolic hydroxyl groups in the hydrogenated coal, which could lower the amount of the hydrogenated coal soluble in benzene. In support of this view, note that dealkylation of the reductively methylated coal resulted in a decrease in benzene extractability from 48 to 12%. Dealkylation removed three methyl groups (per 100 carbon atoms) that were attached to oxygen and replaced these methyl groups with hydrogen. The remaining five *C*-methyl groups imparted 65% pyridine and 12% benzene extractability to the *C*-methylated coal. Decreases in amounts soluble in benzene due to dealkylation were also observed in the case of reductively ethylated Bruceton coal (Table 8.8), but the decrease there was less pronounced than with reductively methylated Pocahontas coal.

Although many studies suggest that coal is characterized by a macromolecular structure,[1,64] there is much uncertainty about the nature of the bonding between the basic units of coal structure. In part, this is caused by Blom's[65] conclusion that essentially all the oxygen in coals of 80–91% carbon content exists in hydroxyl or carbonyl groups; ether linkages and heterocyclic oxygen are supposed to be absent or to occur only occasionally.

[64] D. W. van Krevelen, *Coal,* Elsevier, New York, 1961, pp. 440–443.

[65] L. Blom, *Analytical Methods in Coal Chemistry,* Drukkery en Vitgeverij Joh. Luijk, C. V., Eindhoven, 1961. Also Ref. 64, pp. 160–175.

Fisher and Eisner[66] speculated about the removal of oxygen bridges by hydrogenolysis. Takegami and co-workers[67] concluded that formation of asphaltenes in coal hydrogenation might be due to cleavage of ether linkages. Ignasiak and Gawlak[68] investigated the possible relationship between the cleavage and removal of ether oxygen and molecular weight reductions. They used alkali metals as electron donors to form anionic species in THF or in liquid ammonia; carbon-oxygen bonds rather than carbon-carbon bonds are primarily attacked by these reactions.[39]

Ignasiak and Gawlak[68] treated an 89.1% C (maf) coal with potassium, THF, and naphthalene in the presence of methyl, butyl, octyl, and benzyl iodides. The number-average molecular weights (corrected for added alkyl groups) for all the pyridine extracts, regardless of the size of the introduced group, varied within a narrow range of 550–700. The number-average molecular weight of the same coal (corrected for introduced octyl groups) treated with sodium in liquid ammonia and octyl iodide was 590.

From the increase in hydroxyl function in the protonated coal complex, Ignasiak and Gawlak concluded that for each 4.6 ruptured covalent bonds, two might be ether linkages. If the cleavage of ether linkages contributed thus to lowering the number average molecular weight of the alkylated coal, the increase in number of hydroxyl groups should also be related to this number. The molecular weight calculated from the increase in number of acetylatable hydroxyl groups in the protonated coal concentrate after potassium-THF-naphthalene treatment was 760. This value may be too high if there is recondensation of free hydroxyl, particularly acidic hydroxyl groups in protonated coal.[68]

That the macromolecular network of vitrinite from a high-rank coal should be composed of relatively small units (number-average molecular weight about 670), interconnected by ether linkages, appears to be an unexpected preferential role for oxygen atoms. But these studies are compatible with results of studies on hydrogenolysis of coal in hydrogenating media[66] and aerial oxidation of bituminous coal.[69,70] About 20 wt % of these clusters have number-average molecular weights less than 500 (corrected for introduced groups) and the rest are characterized by relatively low uncorrected molecular weights (about 2000), apparently indicating that the published data[45] should be considered with some caution. But the coal used in this work was an mvb Cretaceous coal from British Columbia, Canada; it is possible that the response toward reductive alkylation of coals of the same rank but of different metamorphic paths could be entirely different. The Cretaceous bituminous coal used by Ignasiak and Gawlak is a much younger coal (60–130 million years) than the Carboniferous coals (300 million years) studied by Sternberg et al.[45]

When coal is treated with the potassium-THF-naphthalene system, some THF reacts with the coal. Hombach, however, has shown, using tests with ^{14}C-tagged THF, that only minimal amounts of this solvent or its dissociation products are incorporated into the coal.[71] Larsen, using ^{14}C-labeled naphthalene, has shown that significant amounts of naphthalene are incorporated into the coal and show up in the lower molecular weight nonpolar materials isolated from the reductive alkylation.[72] Ignasiak and Gawlak[68] found that reductive alkylation conducted with alkali metals in liquid ammonia eliminated these problems; it is possible, however, that some ammonia reacted with the coal.

The Sternberg reductive alkylation of bituminous coals gives products that are more soluble in pyridine than the coal used as the reactant. Sternberg believed this increase in extractability by pyridine was caused by:

1. Cleavage of ethers.
2. *C*-Alkylation of aromatic anions.
3. *O*-Alkylation of hydroxyl groups in phenolic structures.

Reductive methods are limited by the reduction potential of the electron carrier.

[66] C. H. Fisher and A. Eisner, *Ind. Eng. Chem.*, **29,** 1371 (1937).

[67] Y. Takegami, S. Kajiama, and C. Yokokawa, *Fuel,* **42,** 291 (1963).

[68] B. S. Ignasiak and M. Gawlak, *Fuel,* **56,** 216 (1977).

[69] B. S. Ignasiak, B. N. Nandi, and D. S. Montgomery, *Fuel,* **49,** 214 (1970).

[70] B. S. Ignasiak, D. M. Clugston, and D. S. Montgomery, *Fuel,* **51,** 76 (1972).

[71] H. P. Hombach, private communication.

[72] J. W. Larsen, private communication.

Hombach showed that the reduction potential in organic solvents could be raised by using solvated electron pairs in glycol ethers.[73] After quenching with protonating reagents, an increase in hydrogen contents of 20–30 hydrogens per 100 carbon atoms was observed. Anthracite can also be reduced in this way. Even with graphites of different origin, hydrogen increases of up to about 30 hydrogens per 100 carbon atoms were obtained.[74]

With coal, the hydrogen-rich derivatives exhibited no significant increase in weight or in oxygen content; evidently little or no incorporation of solvent occurs. Fractions of reduction products with H/C ratios as high as 1.44 were isolated.

Infrared spectroscopy showed that the reduced material had an increase in alkyl content due to sp^2-sp^3 hybridization. The number of methyl groups increased sharply with increasing hydrogen content, showing that one of the main reactions must be carbon-carbon bond cleavage. This was confirmed by special molecular weight measurement techniques.[75]

Therefore the main reactions that take place during reduction with free electrons, in addition to the three mentioned by Sternberg, are as follows:

4. Cleavage of C—C bonds in hydroaromatic moieties, for example: Ar-CH_2-AR′ → Ar-CH_3 + Ar′H
5. Cleavage of C—C bonds between aromatic units: Ar—Ar′ → ArH + Ar′H

The last reaction would be a reverse Scholl reaction,[76] and it has been shown that comparable conditions can give remarkable yields of cleavage products of polycyclic aromatic hydrocarbons.[77]

[73] H. P. Hombach and K. Niemann, Compendium 77/78, Suppl. Vol. *Erdoel Kohle, Erdgas, Petrochem. Breunstoff-Chem.*, 295 (1977).

[74] H. P. Hombach and K. Niemann, unpublished work.

[75] H. P. Hombach, *Erdoel Kohle, Erdgas, Petrochem. Breunstoff-Chem.*, in press.

[76] A. T. Balaban and C. O. Nenitzescu, in *Friedel-Crafts and Related Reactions*, Vol. 2, G. A. Olah, Ed., Interscience Publishers, New York, 1964.

[77] H. D. Stock, Thesis, University of Bochum, 1978.

8.4.1 On Heteroatoms in Coals

In many respects aromatic heterocyclics react similarly to their carbocyclic analogues. Oxygen, sulfur, and the—NH—group replace —CH═CH—, and —N═ replaces —CH═ in aromatic rings with some change in aromatic character. In general, the thermal stability of the heterocyclics is equal to that of their carbocyclic analogues. Acridine is less stable than anthracene, but carbazole, dibenzothiophene, and dibenzofuran are as stable as their iso-π-electron analogue phenanthrene.[78]

Aromatic heterocyclics are good electron acceptors; in some cases they are even better electron acceptors than their carbocyclic analogues. For example, the reduction potentials of pyridine and quinoline are less negative by 0.5V than the reduction potentials of benzene and naphthalene.[79] In other words, it is much easier to add electrons to pyridine or to quinoline than to benzene or naphthalene. The reduction potentials[80] of dibenzofuran (−2.46V) and dibenzothiophene (−2.43V) are almost identical with that of their carbocyclic analogue phenanthrene (−2.40V).

Alkylation of quinoline anion and naphthalene anion proceed by the same mechanism. In both cases the extent of alkylation is affected by the type of cation used.[81] Monoalkylated derivatives are produced in the presence of small-volume cations, whereas dialkylated derivatives are produced in the presence of large-volume cations. This difference has been attributed to the formation of solvent-separated ion pairs in the case of small-volume cations and to the formation of contact ion pairs in the presence of large-

[78] J. J. Madison and R. M. Roberts, "Pyrolysis of Aromatics and Related Heterocyclics," *Ind. Eng. Chem.*, **50,** 237–250 (1958).

[79] D. D. Whitehurst, M. Farcasiu, and T. O. Mitchell, "The Nature and Origin of Asphaltenes in Processed Coals," Annual report AF-252 prepared by Mobil Research and Development Corporation, Princeton, N.J., for Electric Power Research Institute, February 1976.

[80] R. Gerdil and E. A. C. Lucken, "A Polarographic and Spectroscopic Study of Dibenzothiophene and Some of Its Isologs," *J. Am. Chem. Soc.*, **88,** 733–737 (1966).

[81] W. H. Smith and A. J. Bard, "Electrochemical Reactions of Organic Compounds in Liquid Ammonia. III. Reductive Alkylation of Quinoline," *J. Am. Chem. Soc.*, **97,** 6491–6495 (1975).

volume cations.[82] Coal anion behaves like naphthalene or quinoline anion, in that alkylation of coal anion is more extensive in the presence of potassium cation than in the presence of lithium cation.[45]

The hydroaromatic heterocyclic 1,2,3,4-tetrahydroquinoline, like its carbocyclic analogue tetralin, is a powerful hydrogen donor solvent. Kölling and his co-workers[83,84] found that bituminous coal is converted to a soluble product by treatment with tetrahydroquinoline at temperatures as low as 300°C and that some quinoline is formed during the reaction. They also found that in tetrahydroquinoline at 390°C, phenyl benzyl ether and dibenzyl ether undergo quantitative hydrogenolysis to phenol and toluene, and to benzyl alcohol and toluene, respectively.

These examples show that hydroaromatic and aromatic heterocyclics resemble their carbocyclic analogues with regard to thermal stability, hydrogen donor, and electron acceptor properties.

8.4.2 On the Structure of Sulfur-Rich Rasa Lignite

The Rasa lignite of Istria, Croatia, Yugoslavia, is characterized by high organic sulfur content (total sulfur content, almost all organic, was 12.8% maf) and highly coking properties. Ignasiak and co-workers[85] applied the methods of reductive alkylation with sodium in liquid ammonia and reductive alkylation with potassium in THF to examine this unusual coal. They concluded that the highly thermoplastic properties of Rasa coal result chiefly from cleavage of C—S (C—O) bridges, low hydroxyl contents, and lack of sulfhydryl (—SH) groups. In contrast to normal bituminous coals, which are characterized by fairly wide distributions of the molecular weight of clusters formed on cleavage of ether bridges, about 80 wt % of Rasa lignite was found to be comprised of units of almost uniform molecular weight—about 500.

[82] P. W. Rabideau and R. G. Harvey, "Metal-Ammonia Reduction. IX. A Novel Metal Effect in the Reductive Methylation of Naphthalene," *Tetrahedron Lett.*, No. 48, 4139–4192 (1970).

[83] R. Brucker and G. Kölling, "Losen von Steinkohle in Tetrahydrochinolin," *Brennstoff-Chemie*, **46**(2) 41–43 (1965).

[84] D. Hausigk, G. Kölling, and F. Ziegler, "Losen von Steinkohle in Tetrahydrochinolin. II," *Brennstoff-Chemie,* **50,** 8–11 (1969).

[85] B. Ignasiak, J. F. Fryer, and P. Jadernik, *Fuel*, in press.

8.5 ACID-CATALYZED DEPOLYMERIZATION, ALKYLATION, AND ACYLATION OF COAL

Not only can coals be alkylated readily, they also can be used to alkylate aromatic compounds. With both types of reaction, soluble products are formed that can be analyzed with techniques not suitable for unreacted coal because they require soluble material. The two techniques give different and complementary products, which provide information about coal structure that is difficult to obtain by other means.

8.5.1 Depolymerization

Heredy and Neuworth[86] first realized that coal could be "depolymerized" by using it as an alkylating agent at temperatures as low as 100°C. The method was based on the assumption that coal contains aromatic structures linked by aliphatic bridges, such as methylene bridges, which are sufficiently reactive to participate in an acid-catalyzed transaryl-alkylation reaction. This general type of structure was proposed by Neuworth[87] and his co-workers to explain the chemistry of low-temperature pyrolysis of coal.

The acid-catalyzed reaction of coal with a pure reactive aromatic compound such as phenol would result in an aromatic interchange with phenol and a depolymerization of the coal; the overall reaction is illustrated in Fig. 8.9. One of the final products of such a depolymerization would be dihydroxydiphenylmethane.

The first step in this depolymerization reaction is undoubtedly the protonation of the aromatic moiety at the position of alkyl substitution. The next step probably depends on

[86] L. A. Heredy and M. B. Neuworth, *Fuel,* **41,** 221 (1962).

[87] E. A. Depp, C. M. Stevens, and M. B. Neuworth, *Fuel,* **35,** 437 (1956).

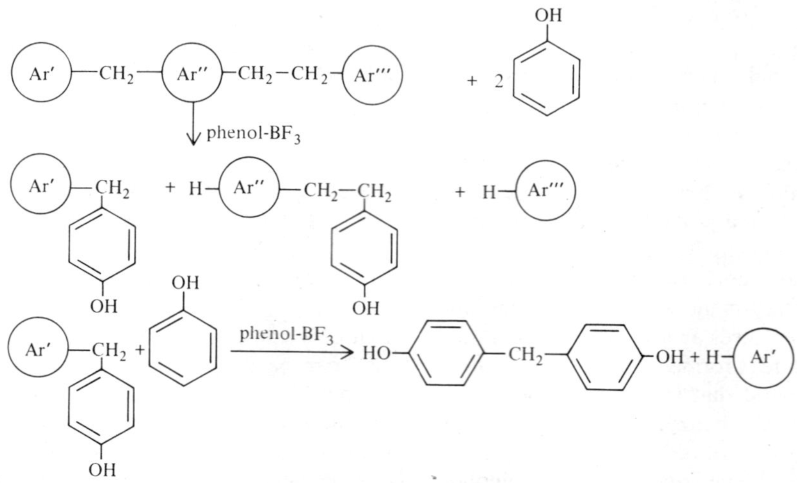

Figure 8.9 Coal depolymerization via transaralkylation with phenol-BF_3.[86]

the type of aromatic-aliphatic linkage that is attacked. The reaction may involve the formation of an alkyl carbenium ion. This mechanism could result in the conversion of primary alkyl carbenium ions to secondary carbenium ions; this problem was investigated by Heredy and co-workers,[88] who studied the transfer of the *n*-propyl group in the para-*n*-propylphenol/phenol (1-^{14}C) model system using boron trifluoride (BF_3) catalyst. It was shown that in a reaction period of 4 hr at 100°C, more than 80% of the *n*-propyl groups were transferred to the labeled phenol. Only 15% of the transferred propyl groups isomerized to isopropyl groups, indicating that in this particular system, the reaction does not proceed primarily by way of a carbenium ion mechanism. This latter reaction mechanism is more probable when a secondary or tertiary carbon is linked to the aromatic ring.

Another possible mechanism would involve a bimolecular substitution reaction. The interchange of benzyl groups in reactions of Type 2 [Ar—CH_2—Ph $\xrightarrow{BF_3}$ Ar(CH_2—Ph$)_2$ + Ar, where Ar is a substituted phenol] was shown to take place by a bimolecular substitution mechanism by Heredy.[89] An example for the bimolecular mechanism appears in Fig. 8.10.

The chemistry and selectivity of the depolymerization reaction were tested[86] on two model compounds containing a methylene bridge and an ethylene bridge, respectively, linking 2,6-xylenol and naphthalene. Reaction of the methylene bridge model with the phenol-BF_3 complex at 100°C for a reaction period of 4 hr resulted in 73 and 41% interchange of the 2,6-xylenol and naphthalene moieties, respectively. When the ethylene bridge compound was treated with phenol-BF_3 under the same conditions, 77% of the 2,6-xylenol was exchanged, and 2% of the naphthalene. It was concluded that the aromatic-aliphatic bridge linkage was more reactive in both model compounds on the 2,6-xylenol side than on the naphthalene side, and that the—CH_2—bridge was the more reactive of the aliphatic bridges. Complete interchange in the methylene bridge compound yields dihydroxydiphenylmethane, which was positively identified. This work established the potential utility of the aromatic interchange reaction under relatively mild conditions to depolymerize coal to soluble fragments, if similar structural types are found to exist in coal. Since phenol participates in the depolymerization process, labeled phenol (1-^{14}C) was used with coal to permit determination of the amount of phenol consumed and its impact on the analysis of the coal-derived depolymerization products.

Depolymerization of a high-volatile Pittsburgh Seam coal with phenol(1-^{14}C)-BF_3 at 100°C for 24 hr resulted in conversion of 73% of the coal to pyridine-soluble fragments.[86]

[88] L. A. Heredy, A. E. Kostyo, and M. B. Neuworth, *Fuel*, **42,** 182 (1963).
[89] L. A. Heredy, "Studies on the Chemistry and Mechanism of Transbenzylation," Doctoral dissertation, Carnegie Institute of Technology, Pittsburgh, 1962.

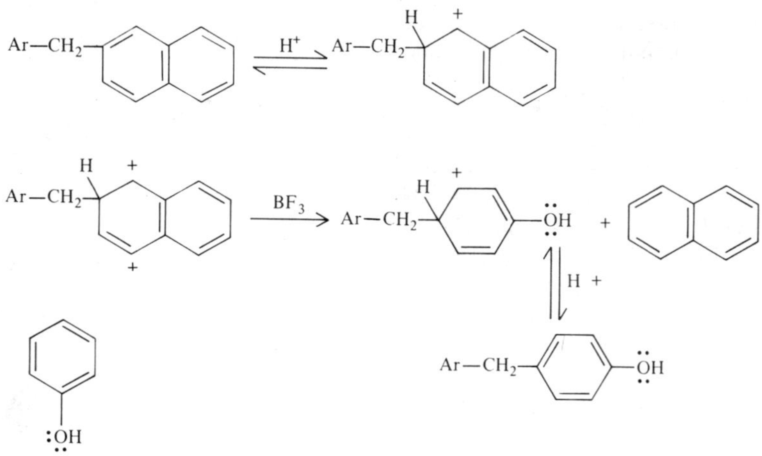

Figure 8.10 Illustration of the bimolecular reaction mechanism.[89]

The total phenol-soluble fraction was 62% of the coal, as compared to 20% in a blank experiment without BF_3. The extent of the depolymerization reaction provides strong evidence of the presence of aliphatic-linked aromatic structures that are sufficiently reactive to participate in an interchange with phenol-BF_3.

The work was extended to a series of coals of different rank[90] ranging in carbon content from 70 to 90%. The study included a lignite and high- and low-volatile bituminous coals. Petrographic analysis of the selected coals indicated that vitrinite was the principal petrographic component, representing 73–87% of the coals. Reaction time with phenol-BF_3 was reduced to 4 hr to minimize side reactions. The amount of phenol-soluble depolymerization products ranged from 9.8 to 75.2% of the coals (Table 8.9). The chemically combined phenol content of the soluble fractions ranged from 12.4 to 41.2%. These results indicate major differences in the number of reactive bridge structures in coals of different rank.

[90] L. A. Heredy, A. E. Kostyo, and M. B. Neuworth, *Fuel*, **44**, 125 (1965).

No systematic study has yet been reported regarding the relative reactivities of the aliphatic bridge structures that may be present in various coals. The following trends have been established, however, in model compound experiments.[86,88] Aliphatic bridge structures linked to nonactivated single benzene rings are the least reactive. The reactivity of bridge structures on naphthalene rings is still relatively low; those on a phenanthrene ring are more reactive. Experiments were done with the system consisting of retene (1-methyl,7-isopropylphenanthrene), phenol, and BF_3. It was found that the rate of isopropyl group transfer from phenanthrene to phenol was about as fast as it was from isopropylphenol to phenol (1-^{14}C). It was shown in other studies that alkyl group

Table 8.9 Depolymerization of Coals of Different Rank with Phenol-BF_3 at 100°C[90]

Coal Type	Carbon (%, dmmf)	Total Soluble Yield (%)[a]	Combined Phenol Content of Soluble Fraction (%)
Lignite	70.6	75.2	41.2
Subbituminous	76.7	23.4	32.8
hvAb	82.4	47.4	16.3
hvAb	85.1	28.8	12.4
hvAb	85.8	25.0	13.0
lvb	90.7	9.8	15.5

[a] Coal-derived part of phenol-soluble material.

or alkyl bridge transfer takes place faster from a phenolic ring than from a naphthalene ring.

A phenolic hydroxyl group has a strong activating effect, and bridge structures linked to single phenolic rings are easily broken during BF_3-catalyzed depolymerization. Of the aliphatic bridges, the methylene bridge is more reactive than bridges consisting of two or more carbon atoms.

It is probable that the breakage of aliphatic ether linkages contributes to coal depolymerization, particularly in the case of lignites and subbituminous coals. Although the reactivity of various ethers with phenol-BF_3 has not been investigated, several studies on the BF_3-catalyzed cleavage of various aliphatic ether linkages have been reported.[91,92] It was shown that the reaction of ethers (as in the case of alcohols and esters) with benzene gives alkylbenzenes. The relative ease of reaction varies considerably with the different ethers. Diisopropyl ether and dibenzyl ether react vigorously with benzene on saturation with BF_3. Isopropyl, phenyl, and benzyl ethyl ethers react violently. On the other hand, ethyl, isoamyl and *n*-amyl ethers did not react under ordinary conditions but required higher temperatures and pressures (150°C and 10–20 atm). No information was published on the reactivity of cyclic ethers.

The depolymerization reaction was modified by Ouchi and co-workers,[93] who substituted *p*-toluenesulfonic acid (PTSA) for BF_3 as the catalyst. The objective was to raise the reaction temperature to 185°C, thereby increasing the rate of the depolymerization and effecting a more complete solution of the coal. The yield of pyridine-soluble material increased to over 90% of the coal. Coals of different rank[94] were then treated with phenol and PTSA for reaction periods of 6–24 hr at 185°C. The pyridine-soluble extracts from coals with carbon contents of 70–86% amounted to 90–100% of the coal (Fig. 8.11). As rank increased to 89.6% C (maf), the pyridine-soluble fraction dropped to 30%,

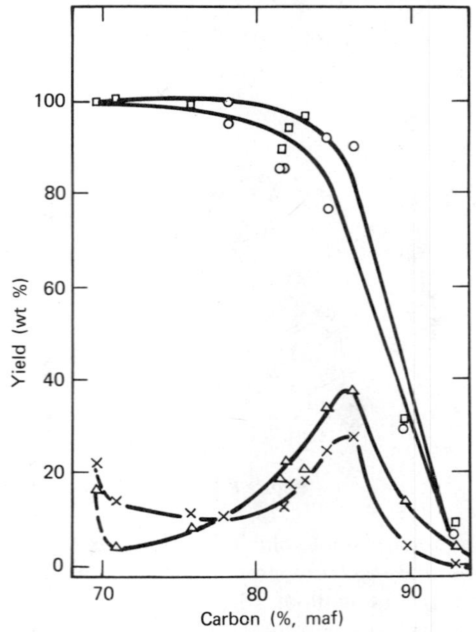

Figure 8.11 Yields of pyridine extracts from raw coals (×) blank tested coals (△), and depolymerized coals (□ = reacted for 24 hr, ○ = reacted for 6 hr).[94]

and finally to about 10% for a coal with 92.9% C (maf). Benzene-ethanol (7/3 vol/vol) was a more selective solvent for fractionating the depolymerization products. A linear relationship between the yield of soluble extract and carbon content of the starting coal was observed (Fig. 8.12).

A variety of other acids have been studied as depolymerization catalysts. Benzenesulfonic acid, PTSA, and concentrated sulfuric acid were the most effective.[95] Darlage and Bailey[96] investigated the effect of preoxidation, reaction temperature, and various solvents on depolymerization yields with a number of acid catalysts. In agreement with Ouchi, they found concentrated sulfuric acid a very effective catalyst at 180°C. Actually, the sulfuric acid reacts initially with the solvent (phenol) to produce *p*-hydroxybenzenesulfonic acid, which then serves as the depolymerization catalyst. It was found that BF_3 was an equally active catalyst at 150°C, but that its activity was markedly reduced at

[91] J. J. O'Connor and F. J. Sowa, *J. Am. Chem. Soc.*, **60**, 175 (1935).

[92] W. J. Monacelli and G. F. Hennion, *J. Am. Chem. Soc.*, **66**, 1722 (1941).

[93] K. Ouchi, K. Imuta, and Y. Yamashita, *Fuel*, **44**, 29 (1965).

[94] K. Ouchi, K. Imuta, and Y. Yamashita, *Fuel*, **44**, 205 (1965).

[95] K. Ouchi, K. Imuta, and Y. Yamashita, *Fuel*, **52**, 156 (1973).

[96] L. J. Darlage and M. E. Bailey, *Fuel*, **55**, 205 (1976).

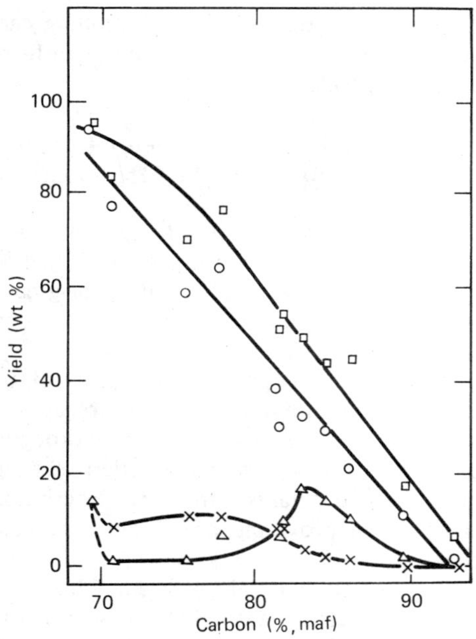

Figure 8.12 Yields of benzene-ethanol extracts from raw coals (×), blank tested coals (△), and reacted coals (□ = 24 hr, ○ = 6 hr).[94]

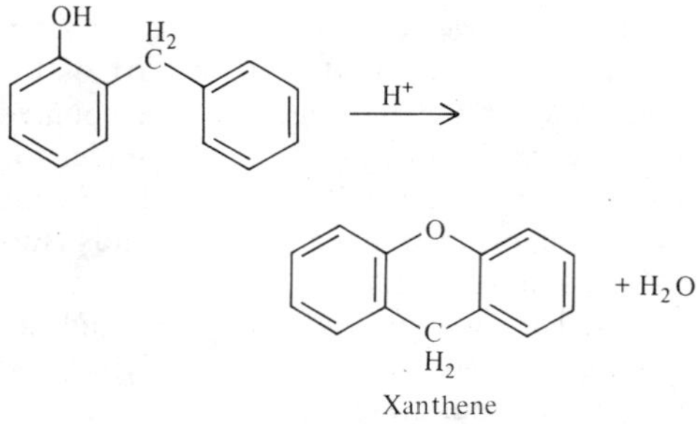

100°C. Metasubstituted phenols, such as *m*-cresol and resorcinol, were more effective solvents for depolymerization than phenol. Preoxidation of coal with dilute aqueous nitric acid considerably increased the yield of depolymerization products.[97] A review of the acid-catalyzed coal depolymerization method has been published by Larsen and Kuemmerle.[98]

Although the acid-catalyzed reaction of coal with phenol at 185°C results in a more rapid and complete solution of the coal, the higher temperature increases the number of side reactions, which complicates correlating the identity of the products with the original coal structure. A significant increase in organic sulfur was observed,[97] and this was attributed to the formation of sulfonate esters by reaction of hydroxyl groups in coal with the PTSA catalyst. Additional indication of secondary reactions was obtained from the work of Ouchi and Brooks.[99]

They showed that xanthene and 9-arylxanthenes, isolated from the higher temperature depolymerization experiments, were formed by cyclodehydration of hydroxylated diarylmethanes and triarylmethanes. The aryl group is derived from phenol, and only the aliphatic linking group can be derived from coal. It was concluded that the identification of 9-arylxanthenes was a definite indication of the existence of tertiary carbon atoms connecting aromatic ring structures. This may be an important experimental indication that at least a number of aromatic structures are connected in this manner to form a three-dimensional network.

Students of coal constitution can gain much insight into its structure once a large portion of the coal is available as soluble fragments produced under mild enough conditions to preclude any major undetermined structural alterations. It is important, however, to determine the extent of some of the side reactions that may take place during depolymerization. In general, depolymerization at 100°C should result in fewer side reactions than would occur at 185°C. One particular reaction, cyclodehydration of diarylmethanes or triarylmethanes, has been investigated in this regard by analyzing the benzene-soluble fraction from the depolymerization of a high-volatile bituminous coal with phenol-BF_3.[100] Hydrogen distribution analysis by NMR indicated 5.4% methylene bridge hydrogens. Dehydroxylation of the end product of the phenol interchange (dihydroxydiphenylmethane) should yield diphenylmethane. The identity of the recovered diphenylmethane was confirmed, and the amount corresponded to 80% of the methylene bridge hydrogens determined by proton NMR. This experiment does not ex-

[97] L. J. Darlage, J. P. Weidner, and S. S. Block, *Fuel*, **53**, 54 (1974).

[98] J. W. Larsen and E. W. Kuemmerle, *Fuel*, **55**, 162 (1976).

[99] K. Ouchi and J. D. Brooks, *Fuel*, **46**, 367 (1967).

[100] L. A. Heredy, A. E. Kostyo, and M. B. Neuworth, *Fuel*, **43**, 414 (1964).

clude the presence of xanthene and substituted xanthenes, but the major part of the methylene bridge hydrogens are accounted for. The work does establish that under the milder conditions used with phenol-BF_3, the methylene-bridged phenol compounds survive to a large extent.

In a more recent study, Franz[101] and coworkers applied ^{13}C NMR to depolymerized products to determine the distribution of carbon types in different structural positions. Successive depolymerization of a North Dakota lignite was carried out using Ouchi's method for reaction periods of 6 hr. Methanol extraction was used to separate the depolymerization products from the unconverted coal. It was estimated that 60% of the coal was soluble in methanol. Carbon-13 NMR analysis indicated the presence of diarylmethanes and triarylmethanes. Furthermore, the relatively severe depolymerization conditions probably resulted in cyclodehydration reactions with the conversion of these bis(hydroxyphenyl)methane derivatives to the corresponding xanthenes. In another study with model compounds, Franz[102] showed that methylene bridges could be formed from triarylmethanes during depolymerization both with PTSA and with phenol-BF_3 catalysts. The conversion of triaryl to diarylmethanes probably occurs by protonation and cleavage of an aryl group followed by hydride abstraction by the resulting carbenium ion from the starting material or from another hydride source:

$$Ar_3CH + H^+ \rightarrow Ar_2CH^+ + ArH$$
$$Ar_2CH^+ + Ar_3CH \rightarrow Ar_2CH_2 + Ar_3C^+$$

In view of these results, it is possible that some of the methylene bridges identified in coal depolymerization products originate from methane, not from methylene bridge structures in the coal.

Within the recognized limitations of its chemistry, depolymerization represents a uniquely valuable procedure for mild degradation of coal with the production of fragments whose detailed structural description is useful in providing an insight into coal constitution.

The molecular weight of the soluble fractions provides a measure of the size of the units linked by methylene bridges. In two depolymerization studies on coals of different rank, molecular weights from 300 to close to 700 were observed.[90,94] In the higher temperature study,[94] made at 180°C with PTSA catalyst, there was an indication of increasing molecular weight with increasing rank.

High-resolution proton NMR analysis gives detailed structural information about the soluble depolymerization products because it permits the classification of hydrogens into monocyclic and condensed aromatics and a number of groups of aliphatic hydrogens. The hydrogen distribution of the total soluble material from six coals, based on proton NMR, is summarized in Table 8.10. Depolymerization was performed at

[101] J. A. Franz, G. L. Tingey, J. A. Campbell, J. R. Morrey, D. M. Grant, and R. J. Pugmire, *Prepr. Am. Chem. Soc., Fuel Div.*, **20,** 12 (1974).

[102] J. A. Franz, J. R. Morrey, G. L. Tingey, W. E. Skiens, R. J. Pugmire, and D. M. Grant, *Fuel,* **56,** 366 (1977).

Table 8.10 Hydrogen Distribution of Depolymerization Products by NMR[86]

Coal Type	Carbon (%, dmmf)	Hydrogen Distribution in Total Soluble Material (%)				Phenol-Soluble Product (wt %, dmmf Coal)	
		Aromatic	Hydroxyl	α-Aliphatic[a]	β'-Aliphatic[b]	Total[c]	Net[d]
Lignite	70.6	30.7	4.6	40.0	24.7	75.2	73.0
Subbituminous	76.7	22.6	5.2	27.5	44.7	23.4	17.1
hvAb	82.4	19.6	4.7	36.9	38.8	47.1	29.1
hvAb	85.1	31.9	3.8	26.1	38.2	28.8	17.1
hvAb	85.8	24.0	3.5	24.3	48.2	25.0	18.5
lvb	90.7	31.0	3.9	25.2	39.9	9.8	8.7

[a] Hydrogens on carbon atoms directly attached to an aromatic ring.
[b] All other hydrogens on carbon atoms in chains attached to an aromatic ring.
[c] Coal-derived part of phenol-soluble material.
[d] Total soluble materials (coal-derived part) minus blank phenol-soluble material.

Table 8.11 Methylene Bridges and Associated Monomeric Units[103]

Coal Type	C (%)	Methylene Bridge Hydrogens (wt %, total hydrogens)	Average Molecular Weight of Soluble Fraction	Calculated Molecular Weight[a]	Phenol-Soluble Product (wt %, dmmf Coal)	
					Total	Net[b]
Lignite	70.6	11.6	328	346	75.2	73.0
hvBb	76.7	0.85	311	4,500	23.4	17.1
hvAb	82.4	3.5	689	1,100	47.1	29.1
hvAb	85.1	0.85	632	4,100	28.8	17.1
hvAb	85.8	0.89	570	4,100	25.0	18.5
lvb	90.7	0.57	344	7,600	9.8	8.7

[a] Calculated molecular weight of monomeric unit per CH_2 group, calculated as R—CH_2 (R is the organic coal structure per CH_2 group).
[b] Total soluble materials (coal-derived part) minus blank phenol-soluble material.

100°C with phenol-BF_3. The aromatic hydrogens range from 20 to 32% with no apparent relationship between aromatic hydrogen content and rank, as suggested by other workers. It should be emphasized that the values are derived from the soluble fractions and represent minimum levels, since the insoluble fractions show lower H/C ratios. Aliphatic hydrogens are the major type, ranging from 64 to 76% of the total hydrogen with relatively little relationship to coal rank. Aliphatic hydrogens include hydroaromatic types, since proton NMR cannot distinguish between these and the aliphatic portions of other substituted aromatic hydrocarbons. Catalytic dehydrogenation studies[104] indicate that a sizable fraction of the aliphatic hydrogen in coal is in hydroaromatic positions.

The methylene bridge hydrogen content of the soluble portions of the same six coals was also determined from the proton NMR spectra (Table 8.11). Methylene bridge hydrogens ranged from 0.57 wt % of the total hydrogens in the high-rank coal to 11.6 wt % in the lignite sample. The yield of soluble material appears to be proportional to the methylene bridge hydrogen content, indicating that cleavage of these linkages is a key to the depolymerization reaction. The average molecular weights of the soluble fractions, which were determined cryoscopically using sulfolane as the solvent, are also shown in Table 8.11. The molecular weights range from 300 to 700. The role of methylene bridges in depolymerization can be evaluated further by comparing the calculated value for the size of a monomeric unit per CH_2 group with the experimentally determined average molecular weight of the soluble products. In the case of the lignite, the experimental and calculated values are in good agreement, indicating that the methylene bridges are probably the sites for the depolymerization, and essentially all the bridges were cleaved. The very high yield of soluble products confirms this observation. Similarly, in the case of the bituminous coal with 82.4% C content, methylene bridges may have been responsible for the bulk of the depolymerization. The very large differences found in the other coals between the experimental and calculated values probably indicate that in these coals small monomeric units are connected to large polymeric clusters by CH_2 groups. Cleavage of the methylene bridge would yield a small soluble unit accompanied by a larger insoluble fragment.

The main features of coal structure derived from depolymerization studies to date are as follows:

1. Coals contain diaryl and possibly triarylmethane structures. Methylene linkages have been positively identified both by proton NMR and by chemical preparative separation.

2. The average molecular weight of the soluble depolymerized product ranges from 300 to 700. There is evidence that average molecular weight of the soluble portions increases with increasing rank.

[103] L. A. Heredy, A. E. Kostyo, and M. B. Neuworth, in *Coal Science*, Am. Chem. Soc. Adv. Chem. Ser. No. 55, (1966), p. 493.
[104] L. Reggel, I. Wender, and R. Raymond, *Fuel*, **43,** 229 (1964).

3. Aliphatic protons are the major type in the soluble depolymerization products. These include hydroaromatic hydrogens. More information is needed to define the ratio of hydroaromatic to aliphatic hydrogens.

8.5.2 Alkylation and Acylation of Coals

Alkylation of coal using various acid catalysts has been investigated. The basic Friedel-Crafts reaction was used in several investigations by Kröger and his co-workers; they used aluminum trichloride ($AlCl_3$) and alkyl chlorides[105] and $AlCl_3$ and olefins[106,107] for coal alkylation.

When high-vitrinite coals were treated with $AlCl_3$ and propyl chloride in carbon disulfide solvent,[105] a 15–25% weight increase was obtained. The pyridine extraction yield increased from 5 to 25% (before alkylation) to 40 to 68% for the isopropylated products.

In another investigation Kröger et al.[106] demonstrated that ethylene and propylene can be added to coal at 300–360°C and about 200 atm pressure in the absence of a Friedel-Crafts catalyst. In experiments with three different vitrinites at 300°C, the correlation between the amount of ethylene taken up and product extractability was determined. An ethylene takeup of 0.5 mol/100 g of coal corresponded to a weight increase of 25%; the pyridine extractability of the product was approximately 80%. The softening temperature of the coal was lowered by about 100°C by the ethylene addition. Structural investigations of the alkylated coals made by both methods (Friedel-Crafts alkylation or direct addition of olefins) showed a marked increase in the aromatic carbon content of the products. This was interpreted as a result of a reaction sequence including alkylation, cyclization, and aromatization. The last step of this sequence involves the breaking of carbon-carbon bridges and thereby increases the extractability of the products.

The addition of olefins to coal using molten sodium chloroaluminate catalyst was also studied by Kroger.[107] This melt contained mostly Na^+, $Al_2Cl_7^-$, and $AlCl_4^-$ ions.The reaction was carried out between 160 and 260°C, with the coal suspended in the melt. In a typical experiment, as the amount of reacted ethylene ranged from 0.4 to 1.35 mol/100 g coal, the ether-soluble product was 12–17% and the pyridine-soluble product 44–64% (the untreated coal gave 23% pyridine extract).

The sulfuric acid–catalyzed alkylation of several coals with isobutylene was investigated by Larsen and Kuemmerle.[108] The reaction proceeded immediately and rapidly at temperatures of 25–50°C with lignite, but long induction times were observed with bituminous coals. In the reaction with North Dakota lignite, a 73% weight increase was obtained; the pyridine solubility of the product was 45%. Bruceton (82.3% C) and Pocahontas (89.8% C) bituminous coals gave products with pyridine extractabilities of 43 and 30%, respectively. None of the alkylated coals showed any softening before decomposition. A possible explanation is that an acid-catalyzed polymerization reaction that decreases the extract yield competes with the alkylation. Interesting observations were made when the NMR spectra of the alkylated products were evaluated.[108] It was found that no aromatic protons could be detected in the spectra of the soluble portion of the alkylated product. At the same time, calculations based on the isobutylene uptake indicated that 80–90% of the aromatic protons must have remained unreacted during the reaction. It was concluded that surface alkylation must have taken place, with only the aromatic hydrogens at the coal surface undergoing substitution. When the alkylated coal is subsequently solubilized, the solution will contain microparticles whose surface is fully alkylated. The relaxation times of the protons inside the particles will result in very broad lines that will not be detectable in the spectra.

Friedel-Crafts types of alkylation of Pittsburgh Seam (hvb) coal using alcohols and olefins as alkylating agents and anhydrous hydrogen fluoride as the catalyst were investigated by Flores and co-workers.[109] Under

[105] C. Kroger and H. A. W. de Vries, *Liebigs Ann.*, **652,** 35 (1962).
[106] C. Kroger, H. B. Rabe, and B. Rabe, *Erdcel Kohle, Erdgas, Petrochem.* **16,** 21 (1963).
[107] C. Kroger, "Anlagerung von Athylen an Steinkohlen," *Forschungsber. Landes Nordhein-Westfalen,* No. 1488 (1965).

[108] J. W. Larsen, *Coal Liquefaction by Alkylation,* Final Rep. No. EPRI AF-42, 1976.
[109] R. A. Flores, M. A. Geigel, and F. R. Mayo, *Fuel,* in press.

conditions necessary to cause noticeable changes in the amount of coal that became soluble in THF and in pyridine (200°C for alcohols and 135°C for olefins), large quantities of by-product materials derived from the acid-catalyzed polymerization of the organic reagents were obtained. These by-products obscured the effects of alkylation on the amounts of coal that became extractable.

Acylation of four bituminous coals of different rank was investigated by Hodek and Kölling.[110] Acetyl chloride and chlorides of a number of higher carboxylic acids, through palmitic acid (C_{16}), were used as the acylating agents; the reaction was carried out in a carbon disulfide suspension using $AlCl_3$ as the catalyst. A coking coal (87.6% C, maf) and a dry steam coal (89.7% C, maf) showed particularly good reactivities, five acyl groups being added per 100 carbon atoms. A high-volatile bituminous coal (82.7% C, maf) and a semianthracite (91.0% C, maf) were much less reactive, with two to three acyl groups being added per 100 carbon atoms, depending on the chain length of the acyl group. The pyridine extractabilities of the acylated products obtained from the four bituminous coals are shown in Table 8.12. The data demonstrate clearly that this reaction is best applicable to bituminous coals of 87–89% carbon content. In the case of the dry steam coal, the chloroform extractability of the various acylated products was as high as their pyridine extractability. The acylated products obtained with the chlorides of lauric and palmitic acids ($C_{11}H_{23}COCl$ and $C_{15}H_{31}COCl$) were equally soluble in benzene, also, and the benzene extractability obtained with the other (shorter chain) acid chlorides was much lower.

The observation that coking and dry steam coals showed the best reactivity toward acyl chlorides was interpreted as an indication that coals of this rank contain the largest number and most easily accessible aromatic C—H groups. It was concluded, furthermore, that the chemical constitutions of the coking coal and dry steam coal must be quite similar. In view of the results of the acylation experiments, it is not likely that the formation of highly condensed aromatic ring systems begins to take place significantly during this phase of coalification (87.6–89.7% C).

[110] W. Hodek and G. Kölling, *Fuel,* **52,** 220 (1973).

8.6 OXIDATION OF COAL

Oxidation of coal has apparently aroused the curiosity of many coal researchers, judging by the large amount of effort expended on the topic.[111] Three principal goals of this research are expressed. One is to explain the occurrence of autoignition, the tendency of some coals to ignite spontaneously in storage piles. Air oxidation studies have been carried out at moderate temperature and atmospheric pressure, but little novel information that is useful in preventing autoignition has been produced since the appearance of the first Supplementary Volume.[111] A second area of interest has been destructive oxidation of coal as a method for gathering structural information. Development of new techniques for carrying out coal oxidation and for analyzing the products has reawakened interest in this topic. A third type of coal oxidation

[111] For a review of work prior to 1957 on oxidation of coal, see *Chemistry of Coal Utilization,* H. H. Lowry, Ed., Supplementary Volume, John Wiley and Sons, New York, 1963, pp. 272–289.

Table 8.12 Amount of Acylated Coal Soluble in Pyridine (wt %)[110]

Acylation with	High-Volatile Coal	Coking Coal (87.8% C)	Dry Steam Coal (89.7% C)	Semianthracite (91.0% C)
(Untreated)	18.5	27.2	3.5	2.0
C_2H_5COCl	14	32	—	5
$C_5H_{11}COCl$	18	32	27	—
$C_7H_{15}COCl$	22	63	85	14
$C_9H_{19}COCl$	—	62	84	17
$C_{11}H_{23}COCl$	28	85	85	—
$C_{15}H_{31}COCl$	20	85	85	13

work has to do with the conversion of coal to coal acids (e.g., aromatic polycarboxylic acids) or humic acids, intended for use as fertilizer. In spite of the large amount of research in this area, no commercial application of this work has yet been reported.

The term "humic acids" is used widely to describe the mixture of water- or base-extractable material formed when coal is oxidized by various oxidants, including air weathering. It is also used to refer to the base-soluble material obtained from decayed vegetation (humus). Many investigators equate these two substances, though there is no good evidence that they are similar. The following discussion considers only the "humic acids" formed by deliberate oxidation of coal, and these more properly would be called "coal acids."

8.6.1 Oxidation of Coal to Acids

Since the pioneering research on oxidation of coal to acids by air or oxygen in aqueous alkaline media was carried out at the Coal Research Laboratory of Carnegie Institute of Technology (now Carnegie-Mellon University),[112,113] an enormous amount of work has been devoted worldwide to the development of a commercial process based on this reaction. The products obtained from this oxidation that are of potential use include the crude coal acids, the humic acid fraction, the benzenepolycarboxylic acids (BPCA), individual BPCA—such as terephthalic acid—isolated from the mixture or formed by disproportionation reactions or partial decarboxylation, and oxalic acid. Because of the ready availability of suitable acids formed by oxidation of petroleum fractions at low cost, no commercial application has yet been made of coal oxidation, although natural humic acids from lignite have been tried as drilling-mud thinners, boiler scale inhibitors, and soil conditioners.[114–116]

Uses suggested for the coal acids have included preparation of resins by reaction with epoxides, polyols, and ethanolamine. These resins are reported to be water soluble, to have long storage life and good cure strength, and to be useful as foundry resins, binders for glass fibers, and removable size for yarns.[117] Even after the dramatic increase in world petroleum prices in 1973, coal was not considered competitive as a source of acids.[118]

Although oxidation by oxygen in aqueous alkaline media or alkaline oxygen oxidation, has been the principal method of converting coal to acids, some use has been made of nitric acid, electrolytic oxidation, permanganate, hydrogen peroxide, and chromic acid.

In general, the results from author to author agree qualitatively but differ quantitatively.

The products of oxidation, generally called coal acids or humic acids, are broken down into either water- or base-soluble and insoluble fractions. The following are some of the acids that have been separated from these mixtures and identified: oxalic acid; all the BPCAs; some alkyl-substituted BPCA; naphthalenecarboxylic acids[119–121]; phenanthrenetricarboxylic acids,[122] anthracenecarboxylic acids[120]; dimethylfluorenonecarboxylic acids[121]; pyridinecarboxylic acids[119]; quinonoid carboxylic acids[123–125]; protocatechuic acid[126]; and succinic,[120,126] glutaric,[120] and adipic acids.[120]

8.6.2 Alkaline Oxygen Oxidation

Conditions used by different workers for alkaline oxygen oxidation vary within a fairly

[112] N. W. Franke and M. W. Kiebler, *Chem. Ind.* **58,** 580 (1946).
[113] N. W. Franke, M. W. Kiebler, C. H. Ruof, T. R. Savich, and H. C. Howard, *Ind. Eng. Chem.*, **44,** 2784 (1952).
[114] V. A. Proskuryakov, T. A. Kukharenko, A. N. Chistyakov, and E. I. Rukin, *Khim. Tverd. Topl.*, No. 5, 42 (1971).
[115] G. Kapo and S. Calvert, *Ind. Eng. Chem., Process Des. Dev.*, **5,** 97 (1966).
[116] S. Tengler and D. Nanowska, *Chemik,* **26,** 65 (1973).
[117] W. L. Archer, R. S. Montgomery, K. B. Bozer, and J. B. Louch, *Ind. Eng. Chem.*, **52,** 849 (1966).
[118] A. E. Bearse, J. L. Cox, and M. Hillman, *Production of Chemicals by Oxidation of Coal,* Battelle Memorial Institute, Columbus, Ohio, 1975.
[119] T. A. Kukharenko, *Khim. Tverd. Topl.*, No. 1, 167 (1973).
[120] R. U. Lacal, *Combustible,* **18,** 163 (1958).
[121] Y. Kamiya, *Fuel,* **42,** 347 (1963).
[122] Y. Kamiya, *Kogyo Kagaku Zasshi,* **61,** 1593 (1958).
[123] Y. Kamiya, *Fuel,* **40,** 457 (1961).
[124] Z. A. Rumyantiseva, N. M. Karavaev, Z. I. Pevzner, and V. S. Polyanskaya, *Khim. Tverd. Topl.*, No. 5, 103 (1971).
[125] Z. A. Rumyantseva, B. B. Gartsman, Z. I. Pevzner, and N. M. Karavaev, *Khim. Tverd. Topl.*, No. 6, 56 (1974).

narrow range. Temperatures between 220 and 300°C, and pressures up to 1500 psi have been used. Sodium carbonate (Na_2CO_3), sodium hydroxide (NaOH), potassium carbonate (K_2CO_3), and potassium hydroxide (KOH) have all been used as alkalis, and sulfuric acid as the acidification reagent for neutralization of the salts of the acids. Most workers have used methyl ethyl ketone to extract the product acids from the acidified aqueous reaction product.

The Soviets have been the most prolific workers in investigating coal oxidation. In particular, they have studied the relation of type and quantity of product to rank and petrographic constituents of coal, and the effect of weathering on yield and ease of oxidation. Weathering has been found to increase the yield of organic acids (based on carbon content) and the ease of oxidation.[127–130] Coals of low rank are oxidized more readily than those of higher rank,[131] but the greatest yields of BPCA are obtained from the higher rank coals.[132,133] The use of sodium carbonate in place of sodium hydroxide is reported to avoid oxalic acid formation.[129]

Because weathered coal gave better results than unweathered coal in production of coal acids, an artificial weathering or preoxidation step was introduced, consisting of passing preheated air (120–180°C) over a fluidized bed of coal at 150–300°C.[134] Coal that was preoxidized and then subjected to alkaline-oxygen treatment showed increased yields and rate of formation of polycarboxylic acids compared to coal that was not preoxidized. By means of the two-stage oxidation, up to 90% yields of coal acids were formed.[135] Electron paramagnetic resonance measurements indicated that preoxidation created free radicals in unweathered coal, presumably making it more reactive in subsequent treatment, while it caused cross-linking in already weathered coal, thereby deactivating it. The oxidation that takes place during weathering and the alkaline oxygen reaction are believed to proceed by different mechanisms.[136]

Because of the large consumption of Na_2CO_3 or NaOH and H_2SO_4 and the large amount of dissolved matter in the wastewater effluent, the usual alkaline oxygen oxidation was deemed commerically impractical.[137,138] To overcome these deficiencies, two-thirds of the Na_2CO_3 was replaced by recycled Na_2SO_4 (formed during the neutralization step). As a result, consumption of Na_2CO_3 and H_2SO_4 was reduced to 1 and 0.96 kg/kg coal, respectively. Yields of coal acids and of BPCA were comparable to those obtained in oxidation using only Na_2CO_3. The two-stage oxidation has been incorporated with the use of mixed Na_2CO_3 and Na_2SO_4 to give improved yields of BPCA as well as decreased consumption of both base and acid.[139]

In a detailed study of the effect of reaction conditions on yields of acids from Polish coals, optimum operating conditions were found to be as follows: ratios of NaOH to O_2 to coal of 3 : 2. 5 : 1 by weight; 20% NaOH; initial O_2 pressure, 40 atm; 2.5 hr at 260°C. The yield of coal acids was 40%, based on maf coal, 75% of which were BPCA. Above 260°C, greater conversion to CO_2, was observed.[140–142]

Kamiya explored the potential commer-

[126] M. M. Roy, *Naturwissenschaften*, **43**, 497 (1956).

[127] T. A. Kukharenko, T. E. Vvedenskaya, E. A. Grigor'eva, and A. S. Savel'ev, *Izv. Akad. Nauk SSSR, Otd. Tekh. Nauk, Met. Topl.*, No. 4, 143 (1961).

[128] T. E. Vvedenskaya and E. I. Rukin, *Khim. Tverd. Topl.*, No. 6, 48 (1968).

[129] T. E. Vvedenskaya and E. I. Rukin, *Khim. Klassif. Iskop. Uglei, Akad. Nauk SSSR, Inst. Goryuch. Iskop.*, 69 (1966).

[130] D. T. Zabrammyi and N. I. Pobedonostseva, *Uzb. Khim. Zh.*, **14**(1), 60 (1970).

[131] T. E. Vvedenskaya, *Tr. Inst. Goryuch. Iskop., Akad. Nauk SSSR*, **21**, 3 (1963).

[132] T. A. Kukharenko, T. E. Vvendenskaya, and E. G. Grigor'eva, *Khim. Pererab. Topl. (Khim. Tekhnol.), Akad. Nauk SSSR, Inst. Goryuch, Iskop*, 55 (1965).

[133] E. A. Girgor'eva and T. A. Kukharenko, *Khim. Tverd. Topl.*, No. 5, 28, (1969).

[134] E. I. Rukin and T. E. Vvedenskaya, *Khim. Tverd. Topl.*, No. 2, 50 (1967).

[135] T. A. Kukharenko, T. E. Vvedenskaya, E. A. Grigor'eva, and E. I. Rukin, *Khim. Tverd. Topl.*, No. 1, 35 (1970).

[136] T. E. Vvedenskaya, *Tr. Inst. Goryuch. Iskop., Akad. Nauk SSSR*, **21**, 3 (1963).

[137] E. I. Rukin, *Khim. Tverd. Topl.*, No 3, 70 (1969).

[138] D. T. Zabramnyi and R. I Pobedonosteva, *Uzb. Khim. Zh.*, **15**(2), 82 (1971).

[139] E. I. Rukin, *Khim. Tverd. Topl.*, No. 5, 61 (1971).

[140] J. Bimer, D. Salbut, and A. Wielopolski, *Przem. Chem.*, **48**(8), 502 (1969).

cialization of the products from the oxidation of Japanese coals in a series of articles.[143–151] CuO was found to function as a catalyst for the oxidation, decreasing reaction time at 270°C from 120 min without catalyst to 10 min with catalyst, or allowing operation at lower temperatures (200°C) for the same time without changing product distribution. To prevent decomposition of product acids, an excess of base was necessary.[143,144] Although $Ca(OH)_2$ gives relatively little oxidation of coal when used as the base in the reaction at 260°C, respectable yields of coal acids are formed when NaCl, Na_2SO_4, Na_2CO_3, or NaOH is added. As much as two-thirds of the NaOH normally used can be replaced by $Ca(OH)_2$. The addition of neutral salts (e.g., Na_2SO_4) improves the yield of water-soluble acids.[145] In one series of reactions in which K_2CO_3 and Na_2CO_3 were used as the base, gases, including CO_2, were vented continuously during the oxidation to keep a favorable ratio of carbonate to bicarbonate.[146]

The yield of coal acids reached a maximum of 65% at 270°C, then decreased as the reaction proceeded.[147] This observation is consistent with the following suggested pathway: coal → H_2O-insoluble acids → H_2O-soluble acids → $H_2O + CO_2$.[148] Increasing the ratio of Na_2CO_3 to coal increased the reaction rate, as well as the yield of aromatic acids.[149] A maximum was reached at a ratio of approximately 8.7 g of base per gram of coal, which represents the amount of Na_2CO_3 needed to convert all the CO_2 to HCO_3^-. Yields dropped at higher concentrations of alkali and above 270°C. Efficient agitation was important.

Studies with model compounds indicate that at 270°C, under oxidizing conditions used for coal acid preparation, alkylbenzoic acids are oxidized to phthalic acids, and α- and β-naphthoic acids are converted to phthalic and 1,2,3- and 1,2,4-benzenetricarboxylic acids. Compounds having a hydroxyl substituent on the ring are readily oxidized further, salicylic acid going all the way to CO_2. Diphenic and anthraquinone-2-carboxylic acids are slowly oxidized to phthalic, as are anthraquinone and the hydrocarbons anthracene and phenanthrene. In general, oxidation rates are higher in Na_2CO_3 than in NaOH, but yields of phthalic acid are lower due to its lower resistance to further oxidation in Na_2CO_3 compared to NaOH. The stability of the benzenetetracarboxylic acids shows a reverse behavior, being more readily oxidized in NaOH than in Na_2CO_3. The stability of the model polycyclic compounds to oxidation is greater than that demonstrated by Japanese bituminous coal, implying that the coal contains few polycyclic systems that are not activated (e.g., by hydroxyl groups) and that the predominant aromatic systems are benzenoid and naphthalenic.

To improve the properties of the partially decarboxylated coal acids, a second oxidation with oxygen under pressure at 270°C was carried out.[150,151] This resulted in further loss of carboxyl groups and the apparent conversion of water-insoluble acids (e.g., iso- and terephthalic) to water-soluble acids (phthalic). The second oxidation increased the thermal stability of the product and made the product almost water-white. Similar results were obtained when coal acids were oxidized further with potassium permanganate.[152]

A low-temperature (60–100°C) kinetic study of coal oxidation using NaOH and oxygen in a flow reactor has been reported.[115] One of the most important factors affecting oxidation rate was surface area of the particle, or particle size. Increase of NaOH concentration, or of ionic strength (with NaCl or LiCl) up to 0.5*N* dramatically increased the oxidation rate, whereas increase above 0.5*N* just as dramatically decreased the rate. This phenomenon is ascribed to a reduction in rate of removal of product from the coal surface by a type of salting-out effect. Although cobalt salts failed to catalyze the

[141] J. Bimer and A. Wielopolski, *Koks Smola Gaz,* **14**(6), 159 (1969).

[142] J. Bimer and A. Wielopolski, *Koks Smola Gaz,* **14**(7), 221 (1969).

[143] Y. Kamiya, *Kogyo Kagaku Zasshi,* **59,** 197 (1956).

[144] Y. Kamiya, *Kogyo Kagaku Zasshi,* **62,** 106 (1959).

[145] Y. Kamiya, *Kogyo Kagaku Zasshi,* **61,** 337 (1958).

[146] Y. Kamiya, *Fuel,* **40,** 149 (1961).

[147] Y. Kamiya and Y. Nakano, *Kogyo Kagaku Zasshi,* **62,** 1571 (1959).

[148] Y. Kamiya, *Kogyo Kagaku Zasshi,* **63,** 752 (1960).

[149] Y. Kamiya, *Kogyo Kagaku Zasshi,* **61,** 197 (1958).

[150] Y. Kamiya, *Kogyo Kagaku Zasshi,* **62,** 942 (1959).

oxidation, both *p*-nitrotoluene and *o*-dinitrotoluene, compounds that undergo electron transfer reactions, did accelerate the reaction. A proposed mechanism involved in the base-catalyzed oxidation of coal is given below:

$$R—H + B: \rightarrow R^- + B{:}H^+ \quad (R8.1)$$

$$R^- + O_2 \rightarrow R\cdot + O_2^{\bar{}}\cdot \quad (R8.2)$$

$$R\cdot + O_2 \rightarrow R—O—O\cdot \quad (R8.3)$$

$$R—O—O\cdot + R^- \rightarrow R\cdot + R—O—O— \quad (R8.4)$$

$$R—O—O— + B{:}H^+ \rightarrow B: + R—O—O—H \quad (R8.5)$$

The rate-controlling steps are the electron transfer reactions shown in Reactions 8.2 and 8.4. By recycling a mixture of air and dilute NaOH through a fractured coal vein, it is speculated that an *in situ* process could be designed to make coal acids.

Oxidation of Central Asian coals by oxygen in an acid medium is reported to produce dibasic, bicyclic carboxylic acids containing *p*-quinoid structures.[153] When oxidized in basic solution, the same coals were found to yield, among other products, keto acids and a nitrogen-containing acid fraction that contained no pyridine or quinoline derivatives, though there were heterocyclic ring systems. Both glutamate and cysteine were identified.[154,155] The bituminous coals from Central Asia, especially Fan-Yagnob coal, are reported to oxidize readily at lower temperature (220°C) than other bituminous coals. Above this temperature degradation of the acids takes place.[156] This behavior is similar to that of lignite. In a comparison of the oxidation of Fan-Yagnob coal by oxygen (180°C, 20 atm) and by alkaline $KMnO_4$ (50°C), similar amounts of insoluble residue were formed. The $KMnO_4$ formed more CO_2 and oxalic acid, and lesser quantities of coal acids.[157]

Molecular hydrogen, believed to result from dehydrogenation of aliphatic and hydroaromatic components of coal, is reported among the gaseous products of alkaline-oxygen oxidation.[158]

Both molybdenum and vanadium catalysts were used to oxidize a low-rank coal with oxygen (75 kg/cm²) at 105°C.[159]

In acetic anhydride the rate of coal oxidation was slower than in water, and the acids formed were 2–3 times higher in molecular weight. Both cobalt nitrate and ammonium vanadate have been used similarly to produce humic acids.[160]

Oxidation of coal with air at 40 kg/cm² in acetic acid was studied as a function of temperature and acid concentration.[161] In dilute acetic acid the optimum yield of humic acids was obtained at 130°C. In concentrated acid (nearly 100%) only high molecular weight products insoluble in alkali were formed, whereas in dilute acid large amounts of low molecular weight organic acids were produced; 50% acetic acid gave high yields of humic acids.

8.6.3 Air-Ammonia Oxidation

The use of ammonia in place of alkali hydroxides or carbonates in the alkaline oxygen oxidation of coal results in humic acids containing nitrogen, much of it in the form of ammonium salts of coal acids, although heterocyclic nitrogen, as in isoindoles and amidines, has been observed.[162] No amino, nitro, or nitroso groups were found. The prinicipal purpose for carrying out this reaction has been to prepare fertilizers, and although one

[151] Y. Kamiya, *Kogyo Kagaku Zasshi,* **62,** 1563 (1959).

[152] T. Tsutsui, *Kogyo Kagaku Zasshi,* **63,** 756 (1960).

[153] Z. A. Rumyantseva, Z. I. Pevzner, B. B. Gartsman, N. M. Karavaev, and T. A. Kotova, *Khim. Tverd. Topl.,* No. 2, 81 (1968).

[154] N. M. Karaveav, Z. A. Rumyantseva, Z. I. Pevzner, and B. B. Gartsman, *Izv. Akad. Nauk Tadzh. SSR, Otd. Fgiz.-Tekh. Khim. Nauk,* No. 4, 36 (1966).

[155] Z. A. Rumyantseva, B. B. Gartsman, Z. I. Pevzner, N. M. Karavaev, and T. A. Kotova, *Khim. Tverd. Topl.,* No. 2, 130 (1971).

[156] R. V. Serebryakova, Z. A. Rumyantseva, and A. S. Fomina, *Dokl. Akad. Nauk Tadzh. SSR,* **11**(9), 35 (1968).

[157] R. V. Serebrakova, Z. A. Rumyantseva, and A. S. Fomina, *Khim. Tverd. Topl.,* No. 5, 59 (1971).

[158] Z. A. Rumyantseva, Z. I. Pevzner, B. B. Gartsman, and N. M. Karavaev, *Khim. Tverd. Topl.,* No. 6, 11 (1969).

[159] S. Nishida, K. Ogawa, and I. Miyashita, *Kogyo Kagaku Zasshi,* **70**(3), 337 (1967).

[160] S. Mishida, M. Ikariya, and H. Uetake, Japanese Pat. 70 10,723, April 17, 1970.

[161] S. Mishida, M. Yamane, and I. Miyashita,

report claims that in a coal humate containing 9.8% nitrogen, 50% of the nitrogen is in readily available form,[163] another report states that an ammoniated coal humate with 15% nitrogen did not release nitrogen or promote growth to any marked degree in greenhouse tests.[164]

The reaction between coal, usually lignite or brown coal, and air-ammonia mixtures generally has been carried out at 200–350°C at atmospheric pressure in a fluidized-bed reactor.[163–166] The reaction has also been carried out at 125°C under pressure (50 atm).[167]

8.6.4 Nitric Acid Oxidation

A method that has been extensively investigated for oxidizing coal to possibly usable products, such as humic acids or BPCAs, employs nitric acid. In spite of a large number of reports of the use of HNO_3, few generalities can be made about the work, and the reaction seems to have been explored in breadth, rather than in depth. There is agreement that nitration takes place, and the product formed is referred to as nitrohumic acids, although it contains many non-nitrated compounds, such as the BPCA. Nitric acid oxidation of lignitic coal is claimed to be the most suitable method for the production of humic acids.[168]

The nitrohumic acids from brown coal have been suggested as drilling-mud stabilizers[169] and as fertilizers. The cost of HNO_3 makes the oxidation economically unattractive. Overall yields of BPCA by HNO_3 oxidation are less than those produced by alkaline oxygen treatment, generally about 20% by weight of coal.[170]

A series of German patents describes a variety of technical modifications for conversion of coal to acids with HNO_3.[171–175] Yields of acids are improved either by preoxidizing the coal with air at 150–300°C or with H_2SO_4 at 180°C before HNO_3 treatment, or by bubbling air through the coal-HNO_3 solution.[176–180] In addition to the BPCA, products include picric acid, adipic acid, succinic acid, and glutaric acid.[181] Oxidation with HNO_3 tends to produce higher yields of mellitic acid than does air oxidation. The yield increases with carbon content up to coke and graphite, and this is suggested as an index of rank.[182,183] Coke and graphite can be used as sources of mellitic acid by HNO_3 oxidation.[184]

Several groups in Japan have been interested in HNO_3 oxidation of coal, for both commercial purposes and as a means of studying coal structure. In a series of studies,[185–190] one Japanese (Onoura) bituminous

Kogyo Kagaku Zasshi, **67**(12), 2121 (1964).
[162] A. Andrzejak and H. Wachowska, *Poznan. Tow. Przyj. Nauk, Pr. Kom. Mat-Przyr. Pr. Chem.,* **12**(6), 105 (1971).
[163] P. N. Mukherjee and A. Lahiri, Indian Pat. 62, 337, August 12, 1959.
[164] S. L. Chakrabartty, J. C. Wood, and N. Berkowitz, U.S. Bur. Mines Inf. Circ. No. 8376, 1968, p. 165.
[165] A. Lahiri, U.S. Bur. Mines Inf. Circ. No. 8376, 1968, p. 183.
[166] N. K. Sarma, T. V. Swininathan, T. V. Subramanian, and C. V. S. Ratnam, *Chem. Age India,* **18**(6), 405 (1967).
[167] D. T. Zabramnyi and N. I. Pobedonostseva, *Uzb. Khim. Zh.,* **15**(2), 82 (1971).
[168] D. T. Zabramnyi, S. Nasritdinov, K. A. Gorlenko, and N. I. Pobedonostseva, *Ugli Srednei Azii Puti Ikh Isopl'z,* 1968, p. 84.
[169] L. P. Proskurin, *Tr. Turkm. Fil. Vses. Neft. Nauchn.-Issled. Inst.,* No. 11 199, (1969).
[170] B. K. Mazumdar, A. Banerjee, J. M. Sanyal, S. Ganguly, A. K. Chatterjee, P. K. Sanyal, and A. Lahiri, *Proc. Symp. Chem. Oil Coal 1969,* (1972), pp. 45–65.
[171] A. Grosskinsky, B. Jüttner, and W. Thürauf, German Pat. 861,247, December 29, 1952.
[172] J. Ewers, O. Grosskinsky, and W. Thürauf, German Pat. 936,568, December 15, 1955.
[173] J. Ewers, O. Grosskinsky, and W. Thürauf, German Pat. 955,768, January 19, 1956.
[174] O. Grosskinsky, W. Klempt, and W. Thürauf, German Pat. 955,768, January 10, 1957.
[175] O. Grosskinsky and B. Jüttner, German Pat. 1,021,349, December 27, 1957.
[176] B. Jüttner, German Pat. 766,204, November 4, 1952.
[177] B. Jüttner, German Pat. 766,205, November 4, 1952.
[178] B. Jüttner, German Pat. 766,206, November 4, 1952.
[179] A. Benning, O. Grosskinsky, and B. Jüttner, German Pat. 879,103, June 11, 1953.
[180] O. Grosskinsky and B. Jüttner, German Pat. 954,239, December 13, 1956.
[181] A. Benning, O. Grosskinsky, B. Jüttner, and J. Karweil, German Pat. 939,574, February 23, 1956.
[182] B. Jüttner and H. Bertling, *Brennstoff-Chemie,* **46**, 277 (1965).
[183] B. Jöttner, *Brennstoff-Chemie,* **42**, 320 (1961).
[184] K. Handrick, D. Hausigk, and G. Kölling, *Brennstoff-Chemie,* **50**, 147 (1969).
[185] S. Kajiyama, C. Yokokawa, Y. Watanabe, and Y. Takegami, *Nenryo Kyokai-shi,* **38**, 86 (1959).
[186] C. Yokokawa, Y. Watanabe, S. Kajiyama, and

coal (81.5% C) was singled out for extensive study. The molecular weight of the humic acid obtained from the hydrogenolysis of this coal was estimated by cryoscopically measuring an aqueous solution of a potassium humate. An average structural unit of the coal was derived:

$$C_{24 \cdot 9}H_{22 \cdot 4}O_{1 \cdot 2}N_{0 \cdot 4}S_{0 \cdot 1}$$

In another study six Japanese coals of differing ranks were treated with boiling $1N$ HNO_3 for up to 72 hr. Low-rank coals oxidized fastest to give maximum yields of acids in a short time, followed by a gradual decrease in yield. Higher rank coals oxidized more slowly but gave higher yields of acids that had higher carbon contents and lower H/C ratios (more aromatic), and usually larger neutralization equivalents.[191] Similar results were obtained on three other coals.[192] There is evidence that in the initial stages of HNO_3 oxidation of coal, carbonyl and phenolic hydroxyl groups are formed, C—H bonds are decreased, and nitration takes place, whereas in the latter stages phenolic hydroxyl groups are destroyed and oxidative decomposition occurs.[193–197] A similar interpretation has been proposed for HNO_3 oxidation of pyridine extracts of Oyubari (Japanese) coals (86% C).[198] Nitric acid oxidation of brown coal is believed to occur in two stages: oxidation of functional groups followed by oxidative splitting of the molecules. The small amounts of CO_2 evolved in HNO_3 oxidation of coal below 100°C suggest that the principal change in forming nitrohumic acids is oxidative cleavage, particularly of ether linkages.[199]

Oxidation of Japanese coal with 12.7% HNO_3 gives humic acids more rapidly and more selectively than oxidation with oxygen at moderate pressure. This difference is attributed to selective cleavage of carbon-carbon bonds by HNO_3.[200]

Chinese (Zhalainuoer) brown coal was oxidized with 10–20% HNO_3 at 50–100°C to give above 95% of nitrated coal, 80% of which was soluble in alkali. The reaction proceeds in two stages as indicated by the rate of acid consumption; the first stage is 2.4 times faster than the second stage. As much as 50% of the nitrogen in the nitric acid used is recoverable as nitrogen oxides.[201,202]

In a kinetic study of the oxidation of anthracite, reaction with concentrated HNO_3 was carried out at 90–100°C for 24–1000 hr.[203,204] Measurement of the alkali-soluble portion of the acid-insoluble residue showed an increase with length of oxidation, reaching a steady state value after 192 hr of oxidation. This trend suggests that alkali-soluble products are converted to acid-soluble materials at the same rate as coal is converted to alkali-soluble material. Microscopic examination of the residue indicates that the action of the HNO_3 is not restricted to the surface of the particles.

Stepwise HNO_3 oxidation of some low-rank Indian coals led to the observation that the water-soluble acids that are formed are unreactive toward further oxidation by even fuming HNO_3. The acids, which are formed in 50–80% yield, contain a large number of

Y. Takegami, *Nenryo Kyokai-shi,* **39,** 590 (1960).
[187] Y. Takegami, C. Yokokawa, and S. Kajiyama, *Nenryo Kyokai-shi,* **40,** 620 (1961).
[188] Y. Takegami, S. Kajiyama, and C. Yokokawa, *Nenryo Kyokai-shi,* **40,** 702 (1961).
[189] S. Kajiyama, C. Yokokawa, and Y. Takegami, *Nenryo Kyokai-shi,* **40,** 761 (1961).
[190] C. Yokokawa, S. Kajiyama, and Y. Takegami, *Nenryo Kyokai-shi,* **42,** 176 (1963).
[191] S. Kajiyama, C. Yokokawa, Y. Watabe, and Y. Takegami, *Nenryo Kyokai-shi* **37,** 582 (1958).
[192] S. Nishida, N. Kato, and K. Ogawa, *Nenryo Kyokai-shi* **46,** 880 (1967).
[193] S. Nishida, M. Yamane, N. Kato, and I. Miyashita, *Kogyo Kagaku Zasshi,* **68,** 527 (1965).
[194] S. Nishida, M. Yamane, K. Ogawa, and I. Miyashita, *Nenryo Kyokai-shi,* **44,** 626 (1965).
[195] S. Nishida, N. Kato, K. Ogawa, and I. Miyashita, *Kogyo Kagaku Zasshi,* **69,** 2299 (1966).
[196] S. Nishida, N. Kato, and I. Miyashita, *Kogyo Kagaku Zasshi,* **69,** 2305 (1966).
[197] K. T. Moon and L. T. Hwa, *Chosun Kwahakwon Tongbo,* No. 3, 25 (1966).
[198] F. Yoshimura and Y. Hirayama, *Nenryo Kyokai-shi,* **51,** 700 (1972).
[199] M. M. Roy, *Research (London),* **10,** 77 (1957).
[200] S. Nishida, M. Ikariya, N. Kato, and I. Miyashita, *Nenryo Kyokai-shi* **44,** 797 (1965).
[201] K. Y. Ch'in and Y.-c. Chu, *Acta Foculio-Chim. Sin.,* **6,** 225 (1965).
[202] C.-p. Chu, *Acta Foculio-Chim. Sin.,* **6,** 235 (1965).
[203] M. A. Hammer, G. A. Brady, and J. W. Eckerd, U.S. Bur. Mines Rep. Invest. No. 5782, 1961.
[204] M. H. Kersten, J. W. Ramsey, and G. A. Brady, U.S. Bur. Mines Rep. Invest. No. 6535, 1964.

nitro groups. This makeup is in part confirmed by the formation of nitrophthalein derivatives by reaction with resorcinol.[162]

8.6.5 Oxidation with Potassium Permanganate

One of the first reagents used for determining the structure of coal was aqueous alkaline potassium permanganate. It was oxidation with this material that presented strong evidence for the presence of aromatic ring systems in coal and suggested the possibility of commercial application of oxidation to give humic acids and BPCA. One study considered alkaline $KMnO_4$ to be the most suitable oxidation method for production of acids to be used as chemical raw materials,[205] although this situation has not yet come to pass.

In addition to the BPCA and oxalic acid yielded in most coal oxidation processes, paper chromatography identified further products of bituminous coals oxidized with alkaline $KMnO_4$ as the aliphatic monocarboxylic acids from acetic through hexanoic.[206] Dibasic aliphatic acids up to C_{10}, as well as BPCA, were identified in the products from oxidation of Soviet brown coal.[207,208]

Structural information was inferred from alkaline $KMnO_4$ oxidation of four macerals,[209] as well as the original gassy coking coal, a carbon black, and graphite. All oxidations were carried to 94–100% of completion, but whereas oxidation of the whole coal took 30 hr, the graphite required 700 hr. Of the macerals, vitrinite required 40 hr (as did the carbon black), exinite took 50 hr, micrinite took 70 hr, and fusinite took 120 hr. Oxygen consumption was between 170 and 202 g of oxygen per 100 g of substrate, except for graphite, which used 238 g/100 g. Lowest acetic acid yields (0.3 and 0.4%) were from carbon black and graphite; highest (4.6%) was from micrinite. Aromatic acid yields varied from 34% from micrinite to 54% from the whole coal. Oxalic acid was formed in 0.7% yield from graphite, 5.7% from fusinite, 11.6% from micrinite, and 13.8% from the carbon black; the other produced 45–64%. A detailed analysis of the BPCA is given. In a similar comparison by coal rank, an increase in yield of BPCA and a corresponding decrease in oxalic acid formation was observed with increase in carbon content.[140] The same relationship was also observed for four Bulgarian coals. In addition, it was found that acetic and other aliphatic acid formation also decreased with increase in rank.[210] Reduction of the coals with sodium in liquid ammonia prior to oxidation with alkaline $KMnO_4$ increased their reactivity; the yield of volatile aromatic and aliphatic acids was much greater in the treated coals; the yield of oxalic acid was lower.[211]

The reactivity of brown coal toward alkaline $KMnO_4$ was found to be independent of the degree of natural oxidation, although the rate of oxidation is greater for weathered coal. The difference in reaction rate decreases with increasing oxidation temperature. The character and yield of products is the same from weathered and unoxidized coals, although oxalic acid content in the products is higher from naturally oxidized coal.[207,208]

Weathering of a bituminous coal (90% C) resulted in an increase in rate of oxidation by alkaline $KMnO_4$ as well as a higher yield of aromatic acids.[212] Comparison of oxalic acid yields for a number of weathered and unweathered coals showed a decrease in quantity with an increase in rank and an increase with degree of weathering.[213]

Alkaline $KMnO_4$ oxidation has been used for studying properties of hydrogenolysis products of a gas coal.[214] The oxidation product contained most of the BPCA, along with high molecular weight phenolic material.

[205] M. F. Kessler and A. Havlova, *Freiberger Forschungsh.*, **429,** 45 (1968).

[206] G. R. Yohe and J. M. Harris, *Fuel,* **40,** 289 (1961).

[207] V. I. Belikova, *Tr. Inst. Goryuch. Iskop., Moscow,* **24** (3), 3 (1969).

[208] T. A. Kukharenko and V. I. Belikova, *Khim. Tverd. Topl.,* 13 (1968).

[209] P. D. Salbut and A. Wielopolski, *Koks Smola Gaz,* **4,** 351 (1969).

[210] D. Totsev, G. Angelova, and N. Mateev, *Khim. Ind. (Sofia),* **44,** 387 (1972).

[211] D. Totsev, G. Angelova, W. Wielopolski, and D. Salbut, *Izv. Otd. Khim. Nauk, Bulg, Akad. Nauk,* **4,** 611 (1971).

[212] V. I. Belikova, *Khim. Klassif. Iskop. Uglei, Akad. Nauk SSSR, Inst. Goryuch. Iskop.,* 52 (1966); *Khim. Tverd. Topl.,* 160 (1969).

[213] T. A. Kukharenko and V. I. Belikova, *Dokl. Akad. Nauk SSSR,* **174,** 364 (1967).

[214] O. I. Pobedonostseva and D. T. Żabramnyi, *Uzb. Khim. Zh.,* **13**(1), 40 (1969); **12**(4), 43 (1968).

8.6.6 Oxidation Using Hydrogen Peroxide

The rate at which CO_2 was evolved while various coal powders were oxidized by a 20% solution of hydrogen peroxide (H_2O_2) at 20°C was suggested as a criterion for estimating spontaneous ignition behavior of powdered coal. Both the rate of formation of CO and CO_2 and their ratio in the gases produced from an isothermal H_2O_2 oxidation increase as the spontaneous ignition index of a series of powdered coals rises.[215]

Oxidation of coal by H_2O_2 in pyridine at 110°C has been suggested as an indicator of coal classification; hard coals produce high yields (> 30%) of insoluble residue with a low (< 5%) hydrogen content, brown coals give a low yield (< 20%) of residue with a higher (> 6%) hydrogen content, and oxidized hard coals form low yields of residue containing less than 4% hydrogen.[216]

An aqueous solution of coal acids obtained by alkaline oxygen oxidation of coal was subjected to further oxidation with refluxing H_2O_2. Large amounts of CO_2, but only small amounts of volatile acids (formic, acetic, propionic) were formed. The oxidation appears to be similar to the alkaline oxygen reaction in that once the rather stable coal acid structure is destroyed, only CO_2 and very low molecular weight acids are formed.[217]

At 70°C, H_2O_2 oxidation of a coal proceeded more vigorously and gave greater yields of acids than at 60°C, but the products were less stable and showed evidence of decomposition. Variations in humic acid yields suggest that the oxidation takes place in three stages: (1) formation of oxygenated groups and hydrolyzable bonds, (2) rupture of bonds and humic acid formation, and (3) destruction of the humic acids.[218] It should be noted that the reaction of coal and H_2O_2 is a catalytic reaction, with iron compounds in the coal functioning as catalysts for H_2O_2 decomposition and coal oxidation.[219,220]

A kinetic study of the H_2O_2 oxidation of a coal has been carried out. Decomposition of H_2O_2 and CO_2 evolution rates were independent of initial H_2O_2 concentration and directly related to temperature. Diffusion of oxygen to the reaction site is the rate-determining step.[221,222] The kinetics of coal oxidation with cumene hydroperoxide has been studied at 30–70°C in alcohol solution. Oxidation rate constants showed a direct dependence on content of phenolic groups in the coal.[223]

A new method of oxidatively degrading coal has been reported.[224,225] The products are aliphatic acids and alcohols, in dramatic contrast to other oxidative degradations (which produce benzenecarboxylic acids). In general, aliphatic systems are inert to this new method, whereas benzene rings are totally destroyed, leaving a carboxyl group as their only remnant. This method promises to provide information about the aliphatic components of coal structure that was heretofore unobtainable. To use the method, the reagent is made by mixing 10 ml of 30% aqueous H_2O_2 and 8 ml of trifluoroacetic acid (TFA) and adding 8 ml of 96% H_2SO_4 slowly with cooling. Then a sample of 0.8 g of −20 mesh coal is added. There is some rise in temperature. With lignites, a colorless solution forms on stirring for 1–3 hr. With bituminous coals, a colorless or near-colorless solution results on heating at reflux (85°C) with stirring for 1–4 hr.

Model compounds were added to the reagent over a period of 20 min in the belief that slow addition approaches the conditions of dissolving coals; temperatures were kept at 50–70°C.

An important *safety precaution* is to decompose excess peroxide with 10% platinum on asbestos at the conclusion of the reaction

[215] M. Lason, *Arch. Gorn.*, 55 (1959).

[216] O. B. Maksimov, *Tr. Dal'hevost. Fil. Sibirsk. Otd. Akad. Nauk SSSR, Ser. Khim.*, **6**, 60 (1962).

[217] R. S. Montgomery and K. B. Bozer, *Fuel*, **38**, 400 (1959).

[218] Z. A. Ryzhova, *Tr. Inst. Goryuch. Iskopaem., Akad. Nauk SSSR*, **12**, 15 (1961).

[219] A. Korta and M. Lason, *Arch. Gorn.*, 65 (1959).

[220] N. G. Titov, R. N. Smirnov, and V. A. Sukhodol'skaya, USSR Pat. 420,653, March 25, 1974.

[221] J. Kawecka, J. Klosinska-Drwalowa, and M. Lason, *Arch. Gorn.*, **6**(4), 347 (1961).

[222] J. Klosinska-Drwalowa, M. Lason, and W. Olpinski, *Arch. Gorn.* **7**(4), 451, (1962).

[223] V. I. Mozhaeva, G. V. Kharitonov, and Z. S. Dzhumagulova, *Tr. Frunze Politekh. Inst.*, No. 29, 67 (1958).

[224] N. C. Deno, B. A. Greigger, and S. G. Stroud, *Fuel*, **57** 455 (1978).

[225] N. Deno, B. A. Greigger, L. A. Messer, M. D. Meyer, and S. G. Stroud, *Tetrahedron Lett.*, 1703 (1977).

and wait until oxygen evolution ceases. A potassium iodide test for peroxide must be negative before the reaction mixture is concentrated. The products are identified by gas chromatography and NMR, either directly on the reaction mixture or on methyl esters of the reaction products.

Nuclear magnetic resonance is particularly suited for keeping an account of hydrogen bonded to carbon. In low-rank coals such as lignites 40–50% of the hydrogen bonded to carbon is preserved in the products. This percentage reflects the predominance of aliphatic carbons in such coals. In a high-rank coal such as Pittsburgh Seam, only 8–10% of the hydrogen bonded to carbon is preserved, which reflects the predominance of aromatic carbon in such coals.

The products observed from coals are shown in Table 8.13. The data indicate the wide differences in coal structure. The fact that the products from Illinois No. 6 coal consist of over 90% acetic, propionic, succinic, and glutaric acids suggests that this coal may be composed of relatively simple structures.

The origin of acetic, propionic, succinic, and glutaric acids as well as methanol is indicated by studies on model compounds.[224] An illustrative selection of model studies is shown in Table 8.14.

Acetic acid arises from arylmethyl groups. Arylmethyl models give 70% yields of acetic acid. The percentage of hydrogen that is present in coal as arylmethyl can be estimated by multiplying the percentage of hydrogen appearing as acetic acid (Table 8.13) by 10/7. This assumes that arylmethyl groups in coals will produce acetic acid in 70% yield. Similarly, the percentage of hydrogen present as arylethyl can be estimated by multiplying the percentage of hydrogen appearing as propionic acid by 10/7. This method is probably the best for determining the amounts of various arylalkyl groups in coals.

Succinic acid can arise from various Ar—CH_2CH_2—Ar structures such as diphenylethanes, acenaphthenes, and 9,10-dihydrophenanthrenes. It also arises from indane. It is a measure of —CH_2CH_2— units in coal. These units will produce succinic acid if attached to aryl rings. They will also produce succinic acid if attached to carbonyl groups and benzylic alcohols.

Glutaric acid can arise from 1,3-diarylpropanes and indane. It is a measure of —$CH_2CH_2CH_2$— units in coals.

Methanol arises from arylmethoxy units. The absence of higher alcohols indicates the absence of alkoxy groups higher than methoxy.

When the H_2SO_4 was omitted from the H_2O_2—CF_3COOH—H_2SO_4 reagent, large amounts of acetic acid were produced from certain model compounds in which CH_3 was absent.[224] It is not known how these hydrogen migrations occur or why H_2SO_4 is so effective in completely repressing such migrations. The same effect is seen in coals. In the absence of H_2SO_4, the coals still dissolve to give colorless solutions, but more acetic acid is formed with a corresponding decline in the yield of succinic acid.

The formation of small amounts of propionic and acetic acids from propyl aromatics[224] is not due to degradation of the dominant product, butyric acid. Butyric acid as well as 1- and 2-butanol and their ethers are stable to the H_2O_2—CF_3COOH—H_2SO_4 reagent.[226]

The failure to observe cyclohexene-1,2-dicarboxylic anhydride in the oxidation prod-

[226] N. Deno, E. J. Jedziniak, L. A. Messer, M. D. Meyer, S. G. Stroud, and E. S. Tomezsko, *Tetrahedron,* **33,** 2503 (1977).

Table 8.13 Yields of Hydrogen Appearing in Products from Coal plus H_2O_2-TFA-H_2SO_4[224]

Substrate	Carbon (wt %, dry basis)[a]	Organic Hydrogen in Coal[b] (wt %)	Ash (wt %, dry basis)	Moisture (wt %)	Yield of Hydrogen (wt %) Appearing in: Acetic Acid	Propionic Acid	Succinic Acid	Glutaric Acid	Methanol
Pittsburgh Seam Coal	79.6	5.2	6.1	1.5	0.9	0.4	4.4	0.5	0
Illinois No. 6 coal	70.8	4.3	8.4	17.7	6.1	0.7	13.4	2.2	0
Illinois No. 6 coal (Monterey)	69.7	4.4	10.8	12.8	3.2	—[c]	10.2	—[c]	0
Lignite, N. Dak.	65.3	2.8	8.4	35.7	4.2	—[c]	6.0	—[c]	16.2

[a] Wt % C in coal (as received) multiplied by the ratio of (wt of coal/wt of coal − wt of moisture).

[b] Wt % H (total) minus wt % H due to moisture.

[c] Not determined.

Table 8.14 Some Products from Oxidation of Alkylaromatics with H_2O_2–CF_3COOH–H_2SO_4[224]

Product	Substrate	Yield (%)
Acetic acid	Toluene	68–78
	o-Xylene	46
	p-Xylene	66
Propionic acid	Ethylbenzene	71
Butyric acid	Propylbenzene	73
	2-Propylphenol	71
	4-Propylphenol	70
	2-Propylthiophene	64
	2-Propylfuran	61
Succinic acid	1,2-Diphenylethane	73
	9,10-Dihydrophenanthrene	71
	Acenaphthene	64
	Indane	27
	4,7-Dimethylindane	42
Glutaric acid	1,3-Diphenylpropane	88
	Indane	27
	4,7-Dimethylindane	50
Butyrolactone	3-Phenyl-1-propanol	100
Cyclohexene-1,2-dicarboxylic acid	Tetralin	71
Malonic acid	Diphenylmethane	9
	9,10-Dihydroanthracene	36
Methanol	Anisole	90

ucts of coal mitigates against tetralin structures that are unsubstituted on the aliphatic ring. The absence of benzoic acid or any other aryl H mitigates against aryl carboxyl groups in at least the higher rank original coals. Such groups should have survived as some kind of substituted benzoic acid unless the aryl ring had been heavily alkylated or had been hydroxylated.

There are several clear paths for the future of this work. Foremost is to improve the yields from the 60–70% range to above 90%. Next is to pursue model studies on tetralins and indanes substituted on the aliphatic (alicyclic) ring. The minor products of oxidation need identification. Finally, after optimization, the technique can be applied to the cataloguing of the vast variety of coals.

8.6.7 Other Oxidants

A pyridine-extracted coal residue was oxidized with potassium dichromate ($K_2Cr_2O_7$) in acetic acid.[227,228] The reaction rate was a function of coal surface area, agitation, concentration of acid and of dichromate, and temperature. Increase in weight due to oxidation reached a maximum at a Cr(VI) consumption of approximately 5 mEq per gram of coal. Up to this point, the principal reactions were formation of quinones and ketones. Beyond this, oxidation to CO_2 and acids occurred. Dichromate oxidation has also been carried out in the presence of H_3BO_3 and H_3PO_4. The former acted as an inhibitor as a result of the formation of complexes with the coal.[229]

A series of coals of differing ranks was oxidized with KIO_3, which was found to be the best oxidant for peats and brown coals.[230,231]

A method for oxidizing bituminous coal was developed that involved bubbling chlorine through a suspension of coal at 240°C with MnO_2 in aqueous NaOH at reflux.[232]

[227] N. Fujiwara and F. Yoshimura, *Nenryo Kyokai-shi,* **51,** 335 (1972).
[228] N. Fujiwara and F. Yoshimura, *Nenryo Kyokai-shi,* **51,** 344 (1972).
[229] G. M. Mal'tseva and M. A. Sherova, *Sb. Tr. Aspir. Soiskatelei, Kirg. Univ., Estestv. Nauk,* No. 5, 33 (1971).
[230] G. Schlungbaum, W. Fischer, and L. Belau, *Bergbautechnik,* **17,** 129 (1967).
[231] G. Schlungbaum, W. Fischer, and L. Belau, *Bergbautechnik,* **19,** 599 (1969).
[232] B. Jüttner, *Brennstoff-Chemie,* **44,** 206 (1963).

This procedure was followed by a short air injection, filtration, acidification, and dechlorination. The method is claimed to have advantages over other oxidation methods, since no oxalic acid is formed and large yields of benzenetetracarboxylic and pentacarboxylic acids are formed; HNO_3 oxidation produces mostly benzenetricarboxylic acids. The preliminary oxidation step is considered essential for splitting the coal into small fragments.

Electrolytic oxidation of brown coal and hard coal has been carried out with cathodes of copper, iron, nickel, and platinum, and with NaOH, Na_2CO_3, and H_2SO_4 as electrolytes.[233–236] Sodium hydroxide gave the greatest current efficiency, and in solutions between 2 and 6*N*, its concentration had no effect on rate or products. Above 6*N*, the yield of oxalic acid increased with increasing concentration of NaOH. Increasing the excess anode voltage caused oxidation to CO_2. Relative oxidation rates without catalyst and with 1% (based on carbon) of Na_3VO_4, $CoCl_3$, $Ce(NO_3)_3$, KHF_2, $K_2Cr_2O_7$, and $KMnO_4$ were 28, 46, 58, 62, 65, and 100. Yields of BPCA were 60–70%. Preoxidation with HNO_3 or air improved current efficiency.

One of the more novel reagents suggested for oxidation of coal for structural studies is commercial household bleach (aqueous NaOCl, 0.8*M*). Use of this oxidant for coal was pioneered by Chakrabartty and co-workers,[237–241] who demonstrated that 3–5% of the carbon in a series of coals of varying rank (76–90% C, maf) could be oxidized to acetic acid and 1–2% to propionic acid. This observation was interpreted as placing lower limits on the number of propyl and butyl side chains, respectively, present in the original coals. In subsequent studies it was found that more than 50% of the carbon of each of a series of coals was oxidized to CO_2 by excess hypochlorite solution. The oxidations were carried out at 60°C until no more oxidant was consumed (i.e., for 2–3 hr). The pH of the reaction mixture, initially 12.6, had dropped to 8 by the end of the reaction. Coals with maf carbon contents above 83% required activation by nitration with nitronium tetrafluoroborate in acetonitrile prior to hypochlorite oxidation. In addition to the CO_2, 20–35% BPCA, 8–14% unidentified higher molecular weight (about 700) acids, and small amounts of low molecular weight aliphatic acids were identified. Since in hypochlorite oxidations carried out at high pH on a series of model compounds, only adamantane-type structures had been converted to CO_2 in high yield and several polycyclic aromatic hydrocarbons (naphthalene, anthracene, and phenanthrene) were unoxidized, Chakrabartty and co-workers concluded that between 55 and 60% of the carbon in coal is tetrahedral sp^3 hybridized, much of it in bridged tricycloalkyl or polyadamantane structures. The remainder was postulated to exist in aromatic structures consisting almost exclusively of noncondensed benzene rings.

This postulation stimulated studies by others whose results contradict the conclusions of Chakrabartty and co-workers. One objection was the failure of adamantane to react with tetralin under hydrogen donor conditions, which converted 60% of an Illinois No. 6 coal to cyclohexane-soluble liquids containing no more than traces of adamantane-type structures. These structures, according to the authors, should have survived the treatment.[242] It has also been noted that ^{13}C and ^1H NMR measurements of coal extracts,[238] aromaticity calculations based on infrared absorption,[243] C—C bond lengths determined by X-ray studies, and density correlations with structures are inconsistent with a largely

[233] B. G. Gaberman, L. N. Yushkova, and I. A. Shpakovskaya, *Khim. Tverd. Topl.*, No. 2, 31 (1967).

[234] B. G. Gaberman and I. A. Shpakovskaya, *Tr. Khim. Met. Inst., Akad. Nauk Kaz. SSR*, No. 2, 95 (1968).

[235] B. G. Gaberman and L. N. Yashkova, *Tr. Khim.-Met. Inst., Akad. Nauk Kaz. SSR*, No. 5, 3 (1969).

[236] B. G. Gaberman, E. A. Grigor'eva, I. A. Shpakcvskaya, M. G. Kipnis, and A. E. Marchenko, *Khim. Tverd. Topl.*, No. 4, 82 (1973).

[237] S. K. Chakrabartty and H. O. Kretschmer, *J. Chem. Soc., Perkin Trans.*, **II**, 222 (1974).

[238] S. K. Chakrabartty and H. O. Kretschmer, *Fuel*, **51**, 160 (1972).

[239] S. K. Chakrabartty and H. O. Kretschmer, *Fuel*, **53**, 132 (1974).

[240] S. K. Chakrabartty and N. Berkowitz, *Fuel*, **53**, 240 (1974).

[241] S. K. Chakrabartty and N. Berkowitz, *Fuel*, **55**, 362 (1976).

[242] T. Aczel, M. L. Gorbaty, P. S. Maa, and R. H. Schlosberg, *Fuel*, **54**, 295 (1975).

[243] H. L. Retcofsky and R. A. Friedel, *Fuel*, **55**, 363 (1976).

polyadamantane structure for bituminous coal.[244]

Even more pertinent to the controversy have been experiments on model compounds. It has been shown that both a naphthol-formaldehyde polymer and fluorescein are oxidized by NaOCl to give high yields—32 and 50%, respectively—of CO_2, although each contains but 10% of sp^3 carbon.[244] It was demonstrated that under the experimental conditions used on coal, 1- and 2-naphthols, 1-hydroxyanthracene, and 2-naphthoic and 2,3-naphthalenedicarboxylic acids[245,246] all gave BPCA on oxidation, and that even naphthalene[247] could be converted to phthalic acid at pH 8.5–9.

Experimental evidence indicates that the extent of oxidation of polycyclic aromatic species is very dependent on pH, hypochlorite concentration,[241,248] and even incident light. Thus the extensive conversion to CO_2 and the failure by Chakrabartty et al. to isolate appreciable amounts of polycyclic products are believed to be due to oxidation of aliphatic, alicyclic, and polycyclic aromatic components at pH <11, not to the presence of tricycloalkyl structures. A more conventional picture of coal structure is supported by hypochlorite oxidation studies of Illinois No. 6 coal, extracted with pyridine before oxidation. In this work Mayo and Kirshen[248] carefully oxidized finely divided coal at controlled pH and were able to isolate as a major product a fraction with an average molecular weight of 800, a neutral equivalent of 300–400, solubility in aqueous $NaHCO_3$ but not in water, and an average empirical formula of $C_{38}H_{31}NCl_{1.2}O_{21}$ (neglecting about 1% sulfur). Comparisons of the ^{13}C NMR spectrum of the coal initially extracted with that of the acids—together with carbon, hydrogen, and oxygen balances—indicated loss of about half the aromatic carbon during oxidation, mostly as CO_2 but partly by conversion to carboxyl groups. Aliphatic and ether-substituted carbon atoms were only slightly oxidized.

When larger particles (-60 mesh) of coal were oxidized, the products were recovered (insoluble) coal and CO_2, with only 3% of the carbon being converted to soluble acids. This is explained as the result of competition between insoluble coal and the dissolved intermediate acids for oxidant. When the coal is in the form of large particles, it offers fewer potential oxidation sites and the soluble fraction can compete more successfully for oxidant. The result is extensive oxidation of that portion already in solution at the expense of further oxidative solubilization of the insoluble particles.

It thus appears that careful control of reaction conditions during NaOCl oxidation can give rise to products resembling the conventional coal acids without excessive conversion to CO_2.

[244] G. Ghosh, A. Banerjee, and B. K. Mazumdar, *Fuel,* **54,** 294 (1975).
[245] R. G. Landolt, *Fuel,* **54,** 299 (1975).
[246] F. Mayo, *Fuel,* **54** 273 (1975).
[247] J. L. Angert, S. L. Gatton, M. T. Reilly, and R. G. Landolt, *Fuel,* **56,** 224 (1977).
[248] F. R. Mayo and N. A. Kirshen, *Fuel,* in press.

8.6.8 Oxidation of Coal with Alkaline Nitrobenzene

Alkaline nitrobenzene has been used as a mild and specific oxidant to study the structure of lignin because it is able to oxidatively cleave the lignin macromolecule into small, identifiable fragments. Because of the suspected structural similarity between lignin and lignite, a similar approach was tried on three Turkish tertiary lignites.[249,250] Heating the lignites with 2*N* aqueous NaOH and nitrobenzene to 200°C produced alkali-soluble fractions representing 40–90% of the lignite. The solubility varied with the geologic age of the lignite, the youngest having the greatest solubility. The soluble oxidation products, when examined by infrared, ultraviolet, and spot tests, showed evidence for the presence of phenolic aldehydes and ketones similar to those obtained from lignin. No polynuclear aromatic hydrocarbons were observed. Similar treatment of two bituminous coals[251] (80 and 82% C) resulted in 10% alkali-soluble material that had spectral properties similar to the soluble material obtained from lignite, implying that at least a small portion of the coal was ligninlike and probably was derived from lignin.

[249] A. Olcay, H. Evliya, A. F. Gaines, and J. Homer, *Fuel,* **52,** 20 (1973).
[250] B. Bozkurt, A. F. Gaines, A. Ogut, Y. Yurum, and A. Olcay, *Fuel,* **52,** 81 (1973).
[251] K. Bugur, A. F. Gaines, A. Gulec, S. Guyer, Y. Yurum, and A. Olcay, *Fuel,* **52,** 115 (1973).

8.6.9 Other Oxidations

By applying the techniques of gas chromatography, coupled with time-of-flight mass spectrometry, and high-resolution mass spectrometry to oxidation products of several coals, a number of polycyclic and heterocyclic ring systems not previously reported have been identified as methyl esters of the corresponding polycarboxylic acids.[252,253] The following principal ring systems were identified by oxidation with excess aqueous sodium dichromate ($Na_2Cr_2O_7$) in order of decreasing concentration: *from lignite:* benzene, xanthone, naphthalene, phthalan, fluorene, dibenzofuran, chroman, and pyridine; *from an Illinois hvb coal:* benzene, naphthalene, phenanthrene, dibenzofuran and benzonaphthofuran, xanthone, acridine, carbazole, fluorene, benzothiophene and dibenzothiophene; *from anthracite:* phenanthrene, fluorene, acridine, benzanthrone, and dibenzothiophene. Oxidation of lignite by H_2O_2 and acetic acid led to identification of methyl esters in the following approximate distribution: 36% benzenoid, 8% methylbenzenoid, 22% methoxybenzenoid, 16% furanoid, and 18% dibasic aliphatic. For bituminous coal the methoxy derivatives were half as abundant. A photochemical oxidation of the Illinois bituminous coal, using irradiation of a 10% aqueous HCl suspension of powdered coal by ultraviolet light from a high-pressure mercury lamp, resulted in identification of methyl esters of carboxylic acids of fluorene, xanthone, acridone, carbazole, pyridine, dibenzofuran, anthraquinone, phthalan, and quinoline, as well as the principal ring system, benzene, and also several aliphatic mono- and dicarboxylic acids. Lignite yielded mainly benzenecarboxylic acids and traces of pyridine tricarboxylic acids and xanthone carboxylic acids. Anthracite gave almost no observable products.

8.6.10 Decarboxylation of Coal Acids

Although most identification of individual acids formed by oxidation of coal has been by isolation or chromatographic separations of the acids or their esters, in some cases the coal acids have been decarboxylated and analytical techniques applied to the resultant neutral mixture. Mass spectrometry of the copper-quinoline decarboxylation product of alkali-oxygen oxidation of Pocahontas (89.8% C) coal has resulted in identification of the following nuclei[254–256]: naphthalene, acetophenone, indane, diphenyl, phenanthrene, fluorene, pyrene (or fluoranthene), terphenyl, phenylnaphthalene, and benzophenone (from diphenylmethane). Decarboxylation in the absence of quinoline yielded evidence for pyridine, phenylpyridine, quinoline, and phenanthridine. Copper salt decarboxylation of coal acids from HNO_3-air oxidation, followed by chromatographic separation, has led to identification of the following nuclei[257]: biphenyl, anthracene, phenanthrene, fluoranthene, 2,3-benzofluorene, 3,4- and 10,11-benzofluoranthene, 1,2-benzopyrene, 1,2-benzanthracene, xanthene, pyridine, quinoline, and diphenyl ether.

Partial decarboxylation has been applied to coal acids to simplify the BPCA fraction and improve the yields of the commercially useful phthalic acids. Thermal treatment in the presence of copper salts of the coal acids obtained from a Japanese bituminous coal (87% C) by alkaline oxygen oxidation resulted in overall conversion of 100 g of coal to 3.1 g of benzoic acid, 2.5 g of phthalic acid, 4.2 g of terephthalic acid, and 6.4 g of isophthalic acid.[121,146,147]

A second procedure for simplifying the coal-acid mixture utilizes the Henkel reaction, which converts most of the BPCA to terephthalic acid (as its salt) when the potassium salts are heated to 400°C in the presence of zinc or cadmium salts under CO_2 at 10 atm or more.[258]

Another method which has been used to convert the BPCA in coal acids to a simpler mixture—primarily benzoic, isophthalic, and terephthalic acids—is based on a decarbonylation reaction catalyzed by dicobalt octa-

[252] R. Hayatsu, R. G. Scott, L. P. Moore, and M. H. Studier, *Nature,* **257,** 378 (1975).

[253] R. Hayatsu, R. E. Winans, R. G. Scott, L. P. Moore, and M. H. Studier, *Prepr. Am. Chem. Soc. Fuel Div.,* **22**(5), 156 (1977).

[254] R. S. Montgomery, E. D. Holly, and R. S. Gohlke, *Fuel,* **35,** 60 (1956).

[255] R. S. Montgomery and E. D. Holly, *Fuel,* **36,** 63 (1957).

[256] R. S. Montgomery and E. D. Holly, *Fuel,* **36,** 493 (1957).

[257] F. Micheel, J. Bernsmann, G. Draeger, and J. P. Riebe, *Brennstoff-Chemie,* **45,** 360 (1964).

[258] B. Raecke, *Angew. Chem.,* **70,** 1 (1958).

carbonyl and its trialkylphosphine complexes.[259] By this procedure the BPCA fraction of a coal-acid mixture, composed largely of tri-, tetra-, and pentacarboxylic acids, was converted to a mixture containing only benzoic, isophthalic, and terephthalic acids in the ratio 4 : 10 : 3.

8.7 HALOGENATION

Earlier studies[260–262] on coal halogenation have not provided significant insights into the fundamentals of coal science. This section reports more recent work, which may suggest that halogenation studies can potentially make a contribution to a better understanding of coal chemistry and reactivity. Some of the reactions listed may stimulate future deduction of information on coal constitution.

8.7.1 Chlorination: Basic Studies

A series of papers by Oxtoby reported in depth on the controlled halogenation of coal, the course of the reaction, and the thermal behavior of the halogenated coals.[263–265] Chlorination at 20 and 60°C in aqueous and carbon tetrachloride (CCl_4) suspensions showed similarities in chlorine uptake, but differences in behavior during chlorination and differences in extractability of the products. The major difference in behavior in the two solvents was that dehydrogenation occurred to a much greater extent during chlorination in the aqueous suspension. The extractability of the coal chlorinated in the CCl_4 suspension was significantly greater than that of the product obtained from the aqueous chlorination. Coals chlorinated at 20°C were more soluble in acetone than the coals chlorinated at 60°C. The increased extractability following chlorination appeared to be due largely to a weakening of cohesive, secondary forces rather than to a rupture of primary chemical linkages. The similarities in the infrared spectra and the chemical composition of the untreated coal, the treated coal, and the extraction residue support the contention that halogenation does not substantially alter the coal structure. This conclusion is in agreement with earlier work by Pinchin at more severe conditions.[266] Although coal chlorination does not appear promising for the direct production of useful materials, it may be possible to use mildly halogenated coals as reactive matrices for the introduction of functional groups.

Nevertheless, in spite of the information above, changes must have occurred during chlorination. It appears that infrared spectra and chemical composition are not sensitive enough to be used as tools in comparing coal structures. Complete destruction of the coking properties of a strongly coking coal occurred after the uptake of small amounts of chlorine. As halogenation increases, thermal softening is progressively delayed, possibly because of cross-linking. Greater yields of char are obtained after chlorination in aqueous suspension at 20°C. The chars from the chlorinated coal have more disordered structures, lower apparent densities, and greater available surface areas than normal chars. The last paper in this series extends the earlier work to coals of different rank and their behavior on graphitization. Coals of higher rank show a progressive reduction in the extent of chlorination, and dehydrochlorination also becomes less pronounced. Cross-linking reactions probably occur in the early stages of thermal decomposition of the chlorinated coals, thus reducing the extent to which carbon layer planes can move toward parallelism. At temperatures below 2225°C any graphitization properties are severely inhibited by prior chlorination. At 2700°C, however, the chlorinated coals yield products that appear to be more graphitic than those from the original coals.

The chlorination of a lignite suspended in water or aqueous alkali was shown to result in oxidation and dehydration.[267] Analysis of

[259] S. Friedman, S. R. Harris, and I. Wender, *Ind. Eng. Chem. Prod. Res. Dev.* **9**, 347 (1970).
[260] *Chemistry of Coal Utilization*, Vol. 1, H. H. Lowry, Ed., John Wiley and Sons, New York, 1945, p. 337.
[261] *Chemistry of Coal Utilization*, H. H. Lowry, Ed., Supplementary Volume, John Wiley and Sons, New York, 1963, p. 270.
[262] V. A. Sapunov and V. S. Chupina, *Khim. Tverd. Topl.*, **1**, 18 (1975); *C.A.* **83**, 13000.
[263] J. C. Macrae and R. Oxtoby, *Fuel*, **44**, 395 (1965).
[264] J. C. Macrae and R. Oxtoby, *Fuel*, **44**, 409 (1965).
[265] R. Oxtoby, *Fuel*, **45**, 457 (1966).
[266] F. J. Pinchin, *Fuel*, **37**, 293 (1958).
[267] M. M. Roy, *Brennstoff-Chemie*, **46**, 407 (1965).

the products showed the presence of a variety of aromatic compounds, suggesting that the coal molecule contains structural units of the naphthalene, indene, thiopyrene, and anthracene types.

A study of the chlorination of coals of different ranks in the presence of water for periods of up to 20 hr showed a strong oxidizing effect as indicated by an increase in carboxylic acid groups and a lowering of the carbon and hydrogen contents.[268] The authors claimed these reactions occurred mainly on the side chains in the coal molecule, but some attack on aromatic rings also occurred.

The optimum temperature for chlorination in the dry state was shown to differ markedly for several types of bituminous coals.[269] In general, chlorine was added by substitution to flame (> 42% VM) and coking coals at 50–100°C, to lean (low-volatile) coals at up to 225°C and to anthracites at up to 250°C. The relative chlorination rates were highest at the following temperatures: flame through coking coals, 100–150°C; ortho coking coal (87–89% C), 200°C; lean coal, 225°C; anthracite, 250°C.

One report suggests that perhaps not all coals lose coking properties on chlorination. Several high-volatile coals chlorinated in a fluidized bed at room temperature for 5–180 min showed no significant change in coking properties.[270] The authors report, however, that gas coals (35–44% VM) showed a loss in swelling even after 5 min of chlorination. At these mild conditions chlorine substitution in aliphatic side chains appeared to be most likely to occur.

8.7.2 Other Halogenations

Several reports describe the use of irradiation or ultrasonics in attempts to increase the rate of halogenation or to utilize CCl_4 as a chlorinating agent. The chlorination of coal with CCl_4 using ^{60}Co as a source of radiation has been described.[271] The chlorination and fluorination of anthracite were also studied in the presence and absence of a ^{60}Co irradiator.[272] Gamma radiation did not have a significant effect on fluorination at 45°C and at 350°C as measured by differences in the weight, volatile matter, or ash content of the residue. At 350°C fluorination resulted in 90% conversion to volatile products with or without radiation. At 45°C a mild explosion occurred in the presence of the radiation, but the products formed appeared to be the same as those obtained in the absence of radiation. Using chlorine, there was no clear evidence of a rate increase in the presence of the ^{60}Co.

8.7.3 Miscellaneous Chlorination Studies

Yields of CCl_4 of 77–86% were obtained by the chlorination of coals for several days at temperatures up to 500°C in the presence of ferric chloride.[273]

The fixation of chlorine on anthracite and the resulting effect on electrical resistivity has also been reported.[274] In general, the resistivity decreased more on heat treatment in chlorine than in nitrogen; the differences at temperatures of 60°C and higher were especially significant. The decrease in resistivity observed after the thermal treatments is explained by a loss of functional groups on the peripheral aromatic carbon atoms.

8.7.4 Bromination

Chlorine and fluorine are too vigorous in their attack on coal to be useful reagents in elucidating coal structure, and iodine appears to be too mild a reagent. Bromine, however, appears to have possibilities.

The magnetic properties of chars and cokes prepared from a brominated lignite and a brominated coking coal were compared with those of chars and cokes prepared under the same conditions from the parent fuels.[275] The magnetic susceptibilities of the chars and cokes were governed by the mineral matter contained, but bromination of the fuels great-

[268] B. I. Losev, N. S. Vorob'eva and E. H. Zimakova, *Khim. Tekhnol., Topl. Masel,* **6**,26 (1961); *C.A.*, **55**, 26410 (1961).

[269] A. Wielopolski, J. Swierkat, and K. Kutkiewicz, *Freiburger Forschungsh.*, **A1968**, 429, 57; *C.A.*, **69** 79106 (1968).

[270] G. A. Obremskaya et al., *Khim. Tverd. Topl.*, **3**, 136 (1971); *C.A.*, **75** 65939 (1971).

[271] B. I. Losev and M. A. Troyanskaya, *Tr. Inst. Goryuch. Iskop., Moscow,* **25**(4), 147–150 (1960).

[272] R. Husack, G. A. Brady, and J. W. Eckerd, U.S. Bur. Mines, Rep. Invest. No. 6881, (1960), 17 pp.

[273] F. E. Kung, *Fuel,* **43** 225 (1964).

[274] W. O. Stacy, G. R. Imperial, and P. L. Walker, Jr., *Carbon (Oxford),* **4**(3), 343 (1966).

[275] A. Calimli et al., *Chim. Acta Turc.*, **2**, 42 (1974); *C.A.*, **82**, 33143 (1975).

ly reduced the magnetic susceptibility generated by subsequent pyrolysis. Electron spin resonance (ESR) measurements showed that relatively few unpaired electrons were formed when brominated coking coal was pyrolyzed, but bromination had a smaller effect on the number of unpaired electrons formed by pyrolysis of a lignite. These results agree well with the proposal that the pyrolysis of solid fuels consists of the disruption of a hydroaromatic system by the scission of aliphatic carbon-carbon bonds accompanied simultaneously by direct dehydrogenation. Simultaneous and subsequent condensation reactions result in the formation of a coke or char.

Changes in the hydrogen distribution in coals on bromination, on subsequent dehydrogenation, and on the carbonization of the brominated coals have been examined by NMR and chemical techniques.[276] The bromination of three coals (82.4, 85.3, and 89% C, dmmf) with bromine in glacial acetic acid added about six bromine atoms per 100 carbon atoms in each of the coals. The distribution of the bromine was dependent on the rank of the coal. In the lowest rank coal, four bromine atoms entered the nonaromatic portion and two joined the aromatic portion; the converse was true in the highest rank coal.

Cunningham and his colleagues[276] suggested reaction schemes for the bromination of the three coals. For the 85.3% C (dmmf) coal, the following was postulated:

Reaction	Atom Changes				
	Hydrogen			Bromine	
	CH_2	Ar	CH	Ar	Al
$3\frac{1}{2}\ ArH \xrightarrow{+Br} 3\frac{1}{2}\ ArBr$		−3½		+3½	
$5\frac{1}{2}\ {>}CH_2 \xrightarrow{+Br} 5\frac{1}{2}\ {>}CHBr$	−11		+5½		+5½
$2\frac{1}{2}\text{—CHBr—CH}{<} \xrightarrow{-HBr} 2\frac{1}{2}\text{—CH}=C{<}$		+2½	−5		−2½
Total	−11	−1	+½	+3½	+3
		−11½		+6½	

Bromination is accompanied by dehydrohalogenation caused by spontaneous elimination of hydrogen bromide. This is particularly evident in the low-rank coal. The NMR measurements showed that more than half the methylene hydrogen remained after bromination. To account for the observed changes in hydrogen distribution, it was postulated that much of the hydrogen removed was derived from tertiary aliphatic carbon-hydrogen groups. When heated, brominated coals lost hydrogen (by elimination of HBr) at a lower temperature than unbrominated coals, but by 500°C the ratio of hydrogen to carbon in the char was the same for both. At temperatures of 400–500°C virtually all the bromine in the coals was released as HBr. The facile elimination of hydrogen bromide is probably due to the readiness of hydroaromatic structures, especially when activated by adjoining aromatic systems, to aromatize.

Two other papers on bromination report on the properties and structure of the products obtained using brown coals in CCl_4 at 60°C[277,278] and on the course of the reaction. Bromination required 6 hr for completion, after which the coals contained 23–25% bromine. The authors suggest that breakdown of the high molecular weight coal structure proceeds by way of bromolysis of carbon atoms α to ether oxygen. This conclusion is based on the oxygen balance in the various oxygen-containing groups present in the raw coals and the bromination products after basic hydrolysis. It was also observed that pyrolysis of the brominated coal occurred at lower temperatures and was more uniform than that of the untreated coal.

8.7.5 Fluorination

Alkali fluorides were used to replace part of the chlorine in a chlorinated brown coal.[279]

[276] A. C. Cunningham, W. R. Ladner, R. W. Wheatley, and W. F. Wyss, *Fuel,* **45**,61 (1966).

[277] K. Kujawa-Buckat and A. Wielopolski, *Koks Smola Gaz,* **18** 114 (1973); *C.A.*, **79** 94601 (1973).

[278] K. Kujawa-Buckat and A. Wielopolski, *Koks Smola, Gaz,* **18** 136 (1973); *C.A.*, **79** 147988 (1973).

A coal containing 18% chlorine was shaken for 25 hr at 20°C with aqueous solutions of NH_4F, HF, KHF_2, and $NaHF_2$. Maximum substitution of fluorine for chlorine was in 10% $NaHF_2$ and 40% KHF_2; the fluorine contents of the coals were 7.3 and 4.2%, respectively.

The vacuum pyrolysis product of fluorinated bituminous coal has been suggested for use as an internal standard for unit mass assignment in mass spectrometry.[280] This material is especially advantageous in the high mass region, where it contributes a small peak at every mass number and does not mask the peaks of the sample under investigation. However because of the low intensity of fragment ions in the high-mass region of the fluorinated coal product, it is not recommended for use as a standard in high-resolution mass spectrometry.

An example of how halogenation studies can contribute to coal research is given in a paper by Huston and co-workers[281] on estimating aromaticity and structural features by way of controlled fluorination. The reaction was controlled by adding the initial fluorine at low pressures (50 torr) and gradually increasing the pressure as the remaining fluorinated coal became less reactive. The final product was a macromolecular solid, insoluble in dimethylformamide, which melted with decomposition at 380–400°C. The products obtained by pyrolysis under vacuum were almost completely volatile. The authors explain the lack of char formation by suggesting that (1) the products are more saturated than those from an untreated coal, therefore less likely to polymerize, and/or (2) the strong C—F bonds inhibit formation of free radicals. The spectral and gravimetric studies indicated that the bituminous coal investigated consisted of 70% aromatic carbon, 10% carbon in hydroaromatic structures, and 20% aliphatic carbon.

8.8 MISCELLANEOUS REACTIONS OF COAL

8.8.1 Hydrolysis

In organic chemistry the word "hydrolysis" generally means the acid- or base-catalyzed addition of water to such compounds as esters (to form acids and alcohols), amides, acetals, alkyl halides, nitriles, peptides, and a few others. It is doubtful, however, that compounds of these types exist in coal, although it is possible that small amounts of esters are present in low-rank coals.

In coal chemistry "hydrolysis" generally refers to the reactions of coal in the presence of alkali at elevated temperatures; acid is used in some cases. Dryden[282] pointed out that the products of the treatment of coal under these conditions are probably the result of hydrolysis, oxidation, and hydrogenation reactions; the overall reaction corresponded roughly to the addition of water to the coal.

Other reactions that could take place in the presence of alkali are[282]:

$$ROR' + H_2O \xrightarrow{NaOH} ROH + R'OH$$

$$RCH_2CH{=}CHR' + H_2O \xrightarrow{NaOH} RCH_3 + R'CH_2CHO$$

$$C + 2H_2O \xrightarrow{NaOH} CO_2 + 2H_2$$

It should be noted, however, that ethers are not acidic and do not usually react with bases. Still, they may react with bases at elevated temperatures, especially if they are benzyl ethers.

The expectation that the extent of hydrolysis decreases with increasing rank has been confirmed.[283]

Small molecules were observed in both alkaline and acid hydrolysis products. Simple phenols and fatty acids were identified in the liquid products from alkaline hydrolysis.[284] Other workers identified monosaccharides, mannitol, methylated sugars, and uric acid in

---

279 M. I. Savin, *Pisarzkevskaya, Tr. Dnepropets. Met. Inst.*, No. 44, 55 (1961), *C.A.*, **59**, 9697 (1963).

280 J. L. Huston, R. G. Scott, and M. H. Studier, *Org. Mass. Spectrum.* **11** 383 (1976).

281 J. L. Huston, R. G. Scott, and M. H. Studier, *Fuel*, **55**, 281 (1976).

282 I. G. C. Dryden, in *Chemistry of Coal Utilization*, H. H. Lowry, Ed., Supplementary Volume, John Wiley and Sons, New York, 1963, p. 289.

283 I. V. Eremin, *E A., Mater, Sovishch. Rab. Lab. Geol. Organ. 9th*, Kiev, 1965, p. 31; *C.A.*, **66**, 97312 (1967).

284 J. Nadziakiewicz and S. Heilpern, *Koks Smola Gaz*, **4** (5–6) 211 (1959); *C.A.*, **59** (1963).

the products of hydrolysis of a variety of coal substances in hydrochloric acid.[285]

More vigorous conditions of hydrolysis, undoubtedly accompanied by other chemical reactions, led to the conversion of 35–90% of Donetz and Kuznetsk coals to neutral or acid materials soluble in alkaline solutions.[286]

The hydrolysis of two bituminous coals in alkaline solution at temperatures of 250–350°C and pressures up to 200 atm was investigated with the objective of gaining knowledge of coal constitution.[287] Changes in composition were studied with the aid of H/C versus O/C diagrams. The conclusions of this work were that 5–10% of the total oxygen existed in the form of more easily split ether linkages. The hydrolysis reaction and the estimation of fresh hydroxyl groups introduced was complicated by degradation of the phenols in the alkaline solution.

Large increases in the extractability of coal were obtained by heating with ethanol and NaOH in autoclave experiments.[288] Although hydrolysis is a major reaction, other reactions, including hydrogenation, are also believed to contribute to the greatly improved extractability. Using equal parts of NaOH and coal with an excess of ethanol at 400°C, products were obtained whose extractability approached 70% in ethanol and 100% in pyridine in some cases. Blank tests with NaOH and ethanol showed that sodium ethoxide decomposed to sodium acetate and hydrogen with smaller quantities of ethylene. Corresponding tests in the presence of coal showed considerable uptake of hydrogen, and apparently some ethylene, by the coal. Elemental analysis showed that the hydrogen content increased from 6.8 to 7.5% in the soluble product. The yield of soluble product varied inversely with rank and directly with NaOH concentration and temperature. Maximum solubilization in ethanol was obtained at 400°C. At 450°C gas formation increased and product extractabilities decreased sharply. Infrared spectra showed an increase in OH absorption, apparently from the splitting of ether linkages. An increase in methyl and carbonyl groups is less certain, the former may possibly come from addition of ethylene and the latter by way of acylation with sodium acetate and/or hydrogenation of phenolic rings followed by dehydrogenation.

Hydrogen cyanide has been obtained by reacting coal with ammonia at 1250°C; yields as high as 0.7 ft³/ft³ of ammonia were obtained.[289] Coals with high-volatile contents gave the best yields.[290]

The reaction of amine radicals with coal was used to explain the data obtained by the irradiation of bituminous coal in the presence of ammonia.[292] No changes in the free radical concentrations in the irradiated coals were observed.

8.8.2 Sulfonation

The main reaction of concentrated sulfuric acid on coal is reported to be oxidation; considerable phenolic OH is formed.[291] Carboxylic acid groups are formed at higher acid concentrations and higher temperatures. Nevertheless, considerable sulfonation is also possible; a number of papers describe the production of sulfonic acid cation exchangers by the sulfonation of coals.[292,293]

8.9 FUNCTIONAL GROUPS

8.9.1 Phenols

The most important methods for determination of the phenolic content of coals were surveyed by Dryden.[294] Since that time, Giv-

285 E. Balabanova-Radonova and N. Matiev, *Khim. Ind. (Sofia),* No. 7, 297 (1970); *C.A.,* **74,** 12860 (1971).

286 V. I. Zabavin et al., *Iz. Akad. Nauk SSSR, Otd. Tekh. Nauk., Met. Topl. 1961,* No. 2 (1970); *C.A.,* **55** 22773 (1961).

287 J. N. Sharma and M. J. G. Wilson, *Fuel,* **40,** 331 (1961).

288 M. Makabe, Y. Hirano, and K. Ouchi, *Fuel,* **57,** 289 (1978).

289 G. E. Johnson, W. A. Decker, A. J. Forney, and J. H. Field. *Prepr. Am. Chem. Soc. Div. Fuel Chem.,* Part 2, **11**(2), 402 (1967).

290 A. Weinstein and P. L. Walker, Jr., *Fuel,* **40** 357 (1961).

291 M. S. Iyengar, S. N. Dutta, and A. Lahiri, *Proc. Symp. Nature Coal,* Jealgora, India, 1959, p. 230; *C.A.* **54** 23269 (1960).

292 D. R. Yunusova and A. T. Talzhiev, *Dokl. Akad. Nauk Uzb. SSR,* 25–27 (1968); *C.A.* **69** 88670 (1968).

293 B. I. Losev, A. N. Mel'nikova, and E. M. Zinchenko, *Khim. Tverd. Topl.,* 112 (1967); *C.A.,* **68,** 4700 (1968).

294 W. O. Stacy et al., *Carbon (Oxford),* **4**(3), 263 (1966).

en[295] has determined the phenolic contents of a large number of American coals using a modified base-catalyzed acetylation method employing ^{14}C-labeled acetic anhydride. The extent of acetylation was determined using a liquid scintillation counter to measure the radioactivity of the CO_2 produced on combustion of the sample. The possibility that the mineral matter of the coal could interfere in phenol determinations by acetylation was carefully investigated. The magnitude of errors that are likely to be induced by acetylation of clays or by weight losses of the mineral matter due to leaching of ions by the organic solvents used in the acetylation procedure was found to be small—less than 0.2% in most cases.

The phenolic hydroxyl contents were plotted against percentage of carbon for coals selected from the Interior, Eastern, and Rocky Mountain provinces of the United States. For this range of coals a linear relationship was found with a least squares correlation of 0.93. Thus the phenolic hydroxyl content may perhaps be considered to be a parameter defining coal rank in the same manner as is carbon content (Fig. 8.13). Over a wider range of coal rank, however, deviations from the linearity shown in Fig. 8.13 are apparent.[296]

A poor correlation was found when the same data were plotted as the percentage of hydroxyl oxygen of total oxygen versus carbon content. Although the phenolic hydroxyl content varies in a linear fashion with rank over a considerable range, it is not unexpected that the other forms of organic oxygen in coal must vary in a more random fashion with rank.

More attention has been drawn to the large role that the physical structure of coal plays in determining the accessibility of reagents to reactive sites. Exposure of coal to the vapors of polar solvents such as pyridine or acetone swells its microporous structure, thus exposing more phenolic hydroxyl groups for determination.[297] The effect was more pronounced for higher rank coals.

In a study of the nonaqueous titration in ethylenediamine[298] of phenolic groups, it was demonstrated that prior extraction of the coal

[295] Z. Abdel-Baset, P. H. Given, and R. F. Yarzab, *Fuel,* **57** 95 (1978).

[296] S. Friedman, M. L. Kaufman, W. A. Steiner, and I. Wender, *Fuel,* **40,** 33 (1961).

[297] B. Roga and M. Weclewska, *Brennstoff-Chemie,* **45,** 334 (1964).

[298] T. P. Maher and J. M. O'Shea, *Fuel,* **46,** 283 (1967).

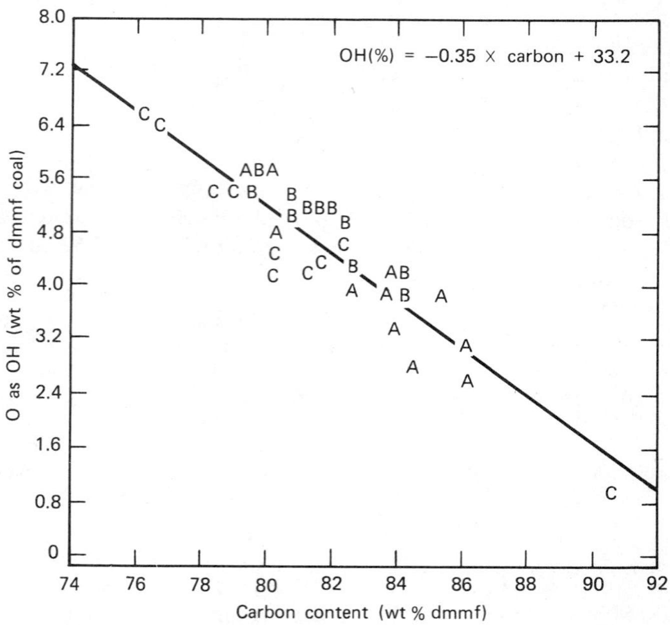

Figure 8.13 Phenolic hydroxyl content, dmmf coal (%, dmmf) related to carbon content: A = Eastern Province, B = Interior Province, C = Rocky Mountain Province.[295]

with aromatic, basic amine, or chlorinated hydrocarbon solvents increased the value of the sum of the phenolic contents of extract and residue to above that determined for the coal itself. The amount of coal extracted by each solvent was not related to the amount of additional phenol determined. The largest increase in the determined phenols was found using chloroform, which extracted only a minor amount of coal. It is interesting to note that the phenolic content determined in combination with chloroform extraction is in good agreement with that predicted by the linear relationship derived by Given and his co-workers[295] from acetylation results, but the determination by direct titration of coal without prior extraction was only 60% of the predicted value. Thus even though ethylenediamine is a powerful solvent for swelling coal by itself, it appears that when its use is combined with prior extraction with another solvent, phenols may be made even more accessible.

A modified method for determination of phenols by acetylation with acetic anhydride has been applied to lignites; the reaction is catalyzed by *p*-toluenesulfonic acid rather than by base, which is more commonly used.[299] Ethyl acetate was substituted for pyridine as solvent, thus eliminating the need to correct for the pyridine that invariably remains bound to the acetylated residues. Other advantages are claimed, including comparatively shorter reaction times (6–8 hr) and lower reaction temperatures (65–70°C). Thus the likelihood of interference due to undesirable side reactions sometimes found in the base-catalyzed reaction is diminished. Determinations of the hydroxyl contents of lignites made by acid-catalyzed acetylation were slightly greater than the values found by pyridine-catalyzed acetylation.

A novel method based on thermometric titration was introduced for determination of phenolic hydroxyl in coal.[300] Titration curves were obtained by measuring the temperature change when strong base was added to finely ground coal dispersed in a pyridine-acetone mixture. The end point was taken as the onset of the temperature rise caused by an indicator reaction, in this case, base-induced polymerization of acetone. The lapsed time for a complete titration is around several minutes. The method gives suitable results when applied to model compounds including monohydric, polyhydric, and sterically hindered phenols. With coals, a considerable variation was found according to the extent and manner of grinding. A standard method of grinding, which gave reproducible results, was adopted for all coals. The shapes of the titration curves for coal resemble those of sterically hindered phenols. Phenolic contents determined for coals of various rank are generally in accord with those values found by using acetylation (Fig. 8.13) and trimethylsilyl ether formation.[296] An exception is the somewhat higher than expected values for a high-rank (90.7% C) coal found by thermometric titration. Presumably this result reflects the diminished accessibility of the phenols in higher rank coals, which are generally less porous, to the reagent of greater steric bulk (acetic anhydride vs. sodium hydroxide). Since this method is based on titration of acidic groups, its use for determination of phenols is restricted to higher rank coals where contributions due to carboxyl groups are minimal.

8.9.2 Carboxylic Acids

The determination of carboxylic acids is largely confined to the analysis of low-rank coals, since the content of carboxylic acids rapidly declines with increasing rank. The carboxylic acid content has been related to the relative maturity of young coals.[301]

The carboxylic acid groups impart ion-exchange properties to low-rank coals. Appreciable amounts of inorganic ions are present as salts of these acids. The ion-exchange properties form the basis for methods of determining the acidity of low-rank coals. Schafer[302] has compared several related methods, including exchange with barium acetate and titration of the released acetic acid, exchange of sodium or barium chloride and titration, exchange with mixtures of $BaCl_2$ and $Ba(OH)_2$, and acid treatment of coal previously ex-

[299] C. Kroeger, K. Fuhr, and G. Darsow, *Erdoel Kohle,* **17,** 88 (1964); **18,** 368 (1965).
[300] G. A. Vaughan and J. J. Swithenbank, *Analyst (London)*, **95,** 890 (1970).
[301] A. Y. Huc and B. M. Durand, *Fuel,* **56,** 73 (1977).
[302] H. N. S. Schafer, *Fuel,* **49,** 197 (1970); **49,** 271 (1970).

changed with calcium acetate and determination of the released calcium. The decay of the 1700-cm^{-1} band due to carbonyl groups was a linear function of the extent of cation exchange until a limiting value was reached. Beyond this point exchange was due to phenolic groups. The extent of cation exchange is a function of pH. At pH 8.2–8.3 the exchange of carboxyl groups is complete; above this value phenolic groups exchange. Complications in assigning the amount of exchange due to these two different functional groups may arise if orthodiphenols, which exchange at pH less than 8, are present. However the pH must be maintained at 8.3 to ensure complete exchange of the carboxyl groups.

The determination of total acidity[302] is conveniently done with mixtures of $Ba(OH)_2$ and $BaCl_2$ at pH up to 13.1. Artifacts arising from absorption of $Ba(OH)_2$ on the exchanged coals were eliminated by careful washing techniques. The total acid group contents for 13 low-rank coals from Australia, Thailand, the United States, Yugoslavia, and Hungary range from 3.7 to 7.3 mEq/g.

The aqueous ion-exchange method[302] was compared with nonaqueous titration in ethylenediamine for five low-rank coals from various parts of the world.[303] The total acidities obtained by the nonaqueous titrations were from 0.1 to 2.8 mEq/g higher than the values obtained by the exchange methods. For three of the five coals investigated, the value for phenols, derived by difference, were in good agreement. The larger value for acids found by the nonaqueous titration may be related to the ability of ethylenediamine to swell coal and/or enhance the acidity of certain phenols. Thus some phenols that escape exchange with barium salts in water may be titratable in ethylenediamine. Similar discrepancies between titration and ion-exchange methods have been found with Rhenish lignites.[304]

Methods based on decarboxylation of lignite in refluxing acridine (344°C) with copper sulfate catalyst and those based on methylation with dimethyl sulfate followed by hydrolysis were less reliable.[304] The methods of ion exchange, decarboxylation, and potentiometric titration were in better accord when methylated and hydrolyzed lignite samples were used instead of raw lignite.

Other work on the ion-exchange method[305] and discussions on the structure of carboxyl groups in humic acids from lignites, which indicate that some water of hydration may be associated with these functions,[306,307] are cited.

8.9.3 Carbonyl Groups

Methods of carbonyl group determination were reviewed by Dryden.[294] More recently several methods for determination of carbonyl groups have been compared for their utility in a study of lignite.[308] The difficulty in carbonyl group analysis lies in the difficulty of knowing whether the reagents are truly specific for carbonyl, especially since both ketones and quinones may be present. Of six different determinations of total carbonyl content, the method based on sodium borohydride reduction followed by acetylation[309] and the one based on reaction with hydroxylamine were found to produce the largest values.[308] For example, the total carbonyl content of a Rhenish lignite of 30.7% oxygen was determined to be 7.5% by sodium borohydride reduction. These larger values were taken to be more reasonable than the rest, although no independent evidence was given that carbonyls and only carbonyls were determined. Methods based on reaction with phenylhydrazine or with Girard's reagent-T were not preferred, since they gave significantly lower values. The use of Girard's reagent-T for determination of the carbonyl content of vitrains in bituminous coals[310] has also been criticized,[311] since the method usually determines only the carbonyl compounds that are rendered water soluble by condensation with the reagent. It is unlikely that the

[303] T. P. Maher and H. N. S. Schafer, *Fuel,* **55,** 138 (1976).

[304] C. Kroeger, K. Fuhr, and G. Darsow, *Erdoel Kohle,* **18,** 368 (1965).

[305] G. S. De, S. Tumuluri, and K. C. Lahiri, *Fuel,* **49,**222 (1970).

[306] M. Falk and D. G. Smith, *Nature,* **200,** 569 (1963).

[307] D. Schwartz and L. Asfeld, *Nature,* **200,** 569 (1963).

[308] C. Kroeger, G. Darsow, and K. Fuhr, *Erdoel Kohle,* **18,** 701 (1965).

[309] S. W. Chaikin and W. B. Brown, *J. Am. Chem. Soc.,* **71,** 122 (1949).

[310] K. Littlewood and M. T. Monaghan, *Nature,* **192,** 128 (1961).

[311] W. F. Wyss, *Nature,* **193,** 870 (1962).

derivatives of high molecular weight carbonyl compounds would be completely extracted from the coal sample by water. In reply,[312] it has been pointed out that carbonyl compounds with molecular weights as large as 400 have been extracted with Girard's reagent from the pyridine-soluble portion of the vitrain. However the data for lignites[308] bear out the criticism, since less than half the total carbonyl as determined by sodium borohydride reduction was extracted from these lignites.

Quinones are usually distinguished from ketones by their relative ease of reduction.[294] Kröger and co-workers[308] found for Rhenish lignites that determination by reduction with titanium(III) chloride gave significantly larger values than reduction with zinc followed by acetylation. Determination of quinones by reduction with titanous chloride is based on the back-titration of excess reagent with ferric chloride. Thus the method rests on the untested assumption that only the quinone groups in the coal sample oxidize titanium(III). Depending on the method of determination, quinones accounted for a variable portion of the total of carbonyls, the difference being ascribed to ketones.[308]

It seems to be generally agreed that the quinone content of bituminous and higher rank coals is very small, although lignites may contain more substantial quantities.[313–315] The carbonyl content of coal probably has been overestimated.[68]

8.9.4 Ethers

A certain fraction of oxygen in coal is invariably found to be unreactive. This fraction is often attributed to ethereal oxygen. Dibenzofuran has been isolated and identified in an extract of a Pittsburgh Seam vitrinite.[316] This compound was also identified among the products of mild oxidation of Illinois No. 2 bituminous coal.[317] Related ethers that were also identified are as follows:

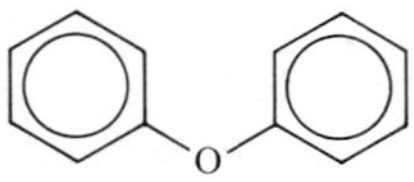
Dibenzofuran

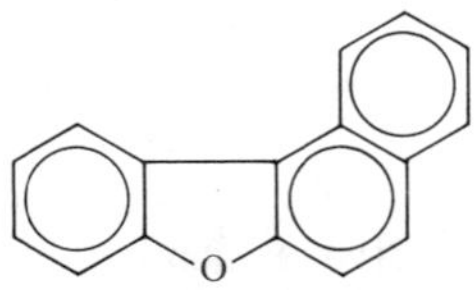
Benzonaphthofuran

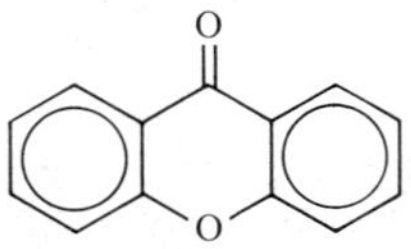
Xanthone

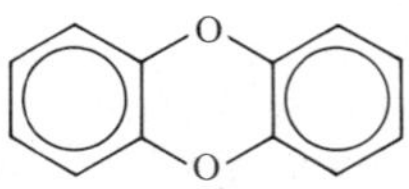
Dibenzo-*p*-dioxin

The amount of unreactive oxygen in coal is usually taken as the difference between the organic oxygen in coal and the sum of the oxygen values determined as phenolic, carboxyl, carbonyl, or methoxyl groups. All or part of this oxygen is likely present as ethers or in cyclic oxygen-containing compounds. Problems arise in obtaining an accurate determination of unreactive oxygen, since total oxygen is itself usually found by difference. Thus all errors accumulated in the ultimate analysis of the coal plus those of the determinations of reactive oxygen functional groups add to the uncertainty in the calculation of unreactive oxygen.

Earlier views held that the oxygen in coals of greater than 80% carbon content was contained largely in the hydroxyl and carbonyl form.[294] More recently the amount and importance of ether oxygen in coals has been reevaluated, primarily by use of alkali-metal reductions. Direct evidence has been presented that ether bonds are important bridges in the macromolecular structure of coal.[68,318] Protonation of a coal anion formed by treatment of the 1.35 float component of the coal with potassium and naphthalene in THF led to an increase in the number of hydroxyl groups available for acetylation from 0.7 to 2.9 wt %[68]: this result is in agreement with the increase determined in a similar manner on a Polish coal of about the same carbon content.[318] This increase is attributed to production of hydroxyl groups by the reductive

[312] K. Littlewood and M. T. Monaghan, *Nature,* **193,** 870 (1962).

[313] W. R. Ladner and A. E. Stacey, *Fuel,* **41,** 114 (1962).

[314] H. Tschamler and E. de Ruiter, *Brennstoff-Chemie,* **44,** 28 (1963).

[315] A. Halleux, S. Delavarenne, and H. Tschamler, *Nature,* **190,** 437 (1961).

[316] A. Halleux and H. de Greef, *Fuel,* **42,** 185 (1963).

[317] R. Hayatsu, R. G. Scott, L. P. Moore, and M. H. Studier, *Nature* **257,** 378 (1975).

[318] H. Wachowska and W. Pawlak, *Fuel,* **56,** 422 (1977).

cleavage of ethers. Cleavage of this number of ether bonds is said to contribute substantially to lowering the number average molecular weight of the reduced coal.

It was found that a fraction of oxygen in coal is unreactive even toward reduction with potassium naphthalenide.[318] After reduction of five Polish coals with oxygen contents ranging from 3.5 to 14.2 wt % (dry basis), the majority of the nonreactive oxygen was still found not to be subject to acetylation. This remaining unreactive oxygen was assigned to heterocyclic ethers, the largest amounts of which were found in low-rank coals. The basis for this interpretation of the results is contained in reports that aromatic ethers are cleaved by alkali-metal reductions under conditions similar to those used for these coals.[48,319] Cleavage apparently takes place by way of the dianions of the aryl ethers. Definitive data on the stability of heterocyclic ethers under reductive conditions is lacking, although the ESR spectrum of the radical anion of dibenzofuran has been obtained.[320] Thus heterocyclic oxygen could account for the remaining unreactive oxygen. However the absolute amount of heterocyclic oxygen in typical coals may still be in question, since the coals selected for study[318] seem to have an abnormally low amount of phenolic oxygen in comparison with other coals of similar carbon content.

Cleavage of ether groups has also been carried out by reduction with sodium in liquid ammonia.[308,315,321] Reduction of coal extracts[315] leads to reduction in the size of the 8.35-μm band in the infrared spectrum, attributed to the antisymmetrical stretching mode of C—O—C groups, in which at least one of the carbon atoms forms part of an aromatic ring. An increase in the 2.9-μm band due to —OH stretching accompanies reduction of the 8.35-μm band.

In contrast, the alkali-metal–liquid ammonia reduction of lignites is claimed to reveal no evidence for aryl ethers.[308] A small amount of methoxyl group oxygen was determined by reaction with HI. However when the method of Lahiri[322] was applied, the increment in phenolic groups was entirely balanced by a decrease in quinone groups. Thus reduction of quinones rather than ether cleavage is held responsible for the production of phenols. The low ether content for these lignites is surprising in view of the positive findings reported for higher rank coals.

In addition to the use of reducing systems in strongly basic solutions, cleavage of ether bonds has also been accomplished by refluxing coal in hydrogen iodide (130°C) for 8 hr.[322] Dialkyl and arylalkyl ethers are cleaved by this reaction, but diaryl ethers are known to be stable. Using this method, the ether contents of three coals of 79.4, 83.0, and 88.0% carbon (maf), respectively, were found to be 2.3, 2.2, and 1.6%.

From the results presently available, the general range of ether oxygen subject to cleavage in coal is 1–2.3%. There seems to be no discernible trend with rank. Although it is clear that the amount and importance of ether linkages in coal are of greater importance than previously thought, much more remains to be done to obtain an adequate knowledge of the relative amounts of the various types of ethers.

The importance of ethereal bonds to the structure of coal and coal solubilization is also discussed in Section 8.4, "Reductive Alkylation of Coal."

8.9.5 Nitrogen Compounds

Direct identification of acridone derivatives among the products derived by oxidizing Illinois No. 2 coal with aqueous $Na_2Cr_2O_7$ at 250°C was made by using gas chromatography and mass spectrometry.[256] Substituted pyridines, quinolines, and carbazoles were also identified in the products of a mild photochemical oxidation (7–10 days irradiation under a high-pressure mercury lamp of a 10% HCl aqueous suspension in the presence of air) of the same coal.[317] Heterocyclics such as carbazole and pyridine were found to be stable under the photochemical reaction conditions, but these ring systems were not isolated from among the products of the aqueous $Na_2Cr_2O_7$ oxidation, presumably because they are readily oxidized in this solution.

Nitrogen heterocyclics have also been found in the liquid products of coal hydro-

[319] D. H. Eargle, *J. Org. Chem.*, **28**, 1703 (1963).
[320] R. Gerdil and E. A. Lucken, *J. Am. Chem. Soc.*, **87**, 213 (1965).
[321] Y. Takegami, S. Kajiyama, and C. Yokkawa, *Fuel*, **42**, 291 (1963).
[322] J. N. Bhaunik, A. K. Mukherjee, P. N. Mu-

genation.[323] Both weakly acidic (pyrrole) and basic nitrogen heterocyclics have been found. The nitrogen bases interact with acidic phenols,[324] thus exerting a considerable influence on the physical properties of the coal-derived liquid. By analogy, it is considered likely that such effects are also important to the structure and properties of coal itself.

8.9.6 Overview

Reviewing the analytical techniques described in this section, it is fair to say that there has not been a "quantum jump" in analytical methodology employed for functional group analysis in coals since the first Supplementary Volume was published.[325] A persistent doubt remains about almost all the wet chemical methods. This doubt arises because of the possibility that reactions of analytical interest occur only at available surfaces of the coal. There may well be a certain portion of functional groups that escape detection because they are in an inaccessible region of the solid coal particle.

Several workers have addressed this question either by using strong solvents to swell the coal[297,298] or by analyzing ground coals of various mesh sizes.[300,326] In both the latter cases larger values were obtained for phenols on samples that passed through smaller size meshes. In one case[326] it is not clear whether this was due to the particle size of the sample analyzed or to segregation of a single sample of ground coal into size fractions of greater or lesser phenolic content. However difficulties arose in the analysis of the smallest particle size fractions. In one case it was difficult to wet the small particles because they accumulated electrical charge, and in the other case erratic results were obtained after extensive grinding, possibly because of oxidation. Both groups chose to use a standard grinding and sampling procedure, but the question of the degree of accessibility of the reagents was not entirely answered.

Research is obviously needed to develop instrumental techniques for functional group analysis on the whole coal. Only in this way, perhaps, can the problem of reagent accessibility encountered with the wet methods be overcome without using extreme methods to destroy the coal matrix, which would also vitiate the analysis.

8.10 SOLVENT EXTRACTION OF COAL

Solvent extraction is one of the oldest and most commonly used tools in coal structural studies. Depending on the rank of coal, on the conditions of the extraction, and on the solvent used, extract yields varying from less than 1% to more than 80% of the coal have been obtained. It has been shown that the chemical composition and many physicochemical characteristics of coal extracts are very similar to those of the parent coal, particularly in the case of high extract yields. Because of this, much effort has been made to relate the information derived from the study of the structure of coal extracts to the structure of the parent coal.

Dryden gave a comprehensive review of work on the solvent extraction of coal through the late 1950s in the first Supplementary Volume of this work.[327] Another excellent review on coal extraction is available in van Krevelen's book on coal chemistry.[328] According to van Krevelen, coal extraction methods can be classified into the following groups:

Nonspecific Extraction: Only a small amount (a few percent) of the coal is dissolved, preferably at low and moderate temperatures. Many common solvents, hydrocarbons, chlorinated hydrocarbons, and alcohols have been used at temperatures up to 150°C.

Specific Extraction: Between 15 and 40% of the parent coal is extracted, preferably at temperatures below 200°C. The most effective solvents have electron donor capacities and, in consequence, are nucleophilic. Pyri-

kherjee, and A. Lahiri, *Fuel*, **41**, 443 (1962).

[323] H. W. Sternberg, R. Raymond, and F. K. Schweighardt, *Science*, **188**, 49 (1975).

[324] S. R. Taylor, L. G. Galya, B. J. Brown, and N. C. Li, *Spectrosc. Lett.*, **9**, 733 (1976).

[325] *Chemistry of Coal Utilization*, Supplementary Volume, H. H. Lowry, Ed., John Wiley and Sons, New York, 1963.

[326] R. G. Ruberto and D. C. Cronauer, *Prepr., Am. Chem. Soc., Div. Pet. Chem.* August 1977, p. 264.

[327] Ref. 325, pp. 237–252.

[328] D. W. van Krevelen, *Coal, Typology-Chemistry-Physics-Constitution,* Elsevier Publishing Co., New York, 1961.

dine and ethylenediamine are typical examples.

Extractive Disintegration: This operation is carried out at rather high temperatures (> 200°C) and may lead to very high extract yields. Extraction with phenanthrene at its boiling point is a typical example.

Extractive Chemical Disintegration: The solvents used in this group are hydrogen donors (tetralin, hydrophenanthrenes); their chemical composition, as well as the composition of the extract, is significantly changed during the process. The extraction temperature is normally high, about 300–400°C.

Most of the recent work on "nonspecific extraction" has dealt with chloroform extraction in conjunction with studies on the characteristics of caking coals. A great amount of coal extraction work has been performed under operating conditions described for "specific extraction." Much less activity was apparent in the area of "extractive disintegration." Much work has been done on "extractive chemical disintegration" in conjunction with coal liquefaction studies. This branch of coal extraction, or hydroextraction, however, is treated in detail in Chapters 27 and 28 and is not discussed here.

8.10.1 Nonspecific Extraction

8.10.1.1 *Chloroform Extraction*

The chloroform-soluble coal extract and its behavior during the thermal softening of the coal have been associated with the caking properties of bituminous coals for some time. A systematic study of chloroform extracts of preheated caking coals has been performed by Dryden, Pankhurst, Brown, and their coworkers[329,330]; and the coking properties of the coal, expressed in its free-swelling index, have been correlated with extract yield.[331]

The function of chloroform extraction products in the coking process has been studied extensively by Brown and Waters.[332] Since the first part of this investigation includes much information regarding the mechanism of the formation and the properties of chloroform extracts, it is reviewed in some detail. The objective of this investigation was to examine the conditions of release of the chloroform-extractable material by relating extract yield to fluidity and weight loss for a wide range of coals heated at a standard rate of 3°C/min. The extracts obtained at different temperatures were characterized. Finally the origin of the chloroform-extractable material and its function in the coking process were discussed. Ten coals were tested. Eight were bituminous coals (83–88.5% C, maf); one lignite and one semianthracite completed the sample. Maximum extract yields for the bituminous coals were obtained at preheating temperatures of 410°C for high-volatile coals, and at 430–440°C for the low-volatile coals. The extract yield was generally 5–8%.

The chloroform extracts from the coking coals had fairly well-defined melting points, in the range of 100–205°C. the number-average molecular weights were about 500±70 and tended to be lower for the higher rank coals. The main conclusions of this work, with emphasis on chemical structural information, are as follows:

1. For a given coal, the composition and constitution of Extract I (recovered at preheating temperatures < 370°C) are closely similar to those of Extract II (i.e., extract released during coal decomposition, > 370°C), except for a slightly higher degree of aromaticity and oxygen content in the latter.
2. For a vitrinite of coking rank, an average chloroform extract molecule has about 36 carbon atoms arranged in seven fused rings; of these, five are aromatic. The remaining two naphthenic rings probably contain most of the aliphatic hydrogen, including methyl groups. No methylene bridges between aromatic rings could be detected.
3. As the coal rank increased over the range considered, there were only small changes in the extract structure. The fraction of aromatic carbon increased from 0.6 to 0.7, the oxygen content decreased, and there

[329] I. G. C. Dryden and K. S. Pankhurst, *Fuel,* **34,** 363 (1955).
[330] I. G. C. Dryden and W. K. Joy, *Fuel,* **40** 473 (1961).
[331] J. K. Brown, I. G. C. Dryden, D. H. Dunevein, W. K. Joy, and K. S. Pankhurst, *J. Inst. Fuel,* **31,** 259 (1958).
[332] H. R. Brown and P. L. Waters, *Fuel,* **45,** 17 (1966); **45,** 41 (1966).

was a slight decrease in the hydrogen and methyl group contents.

4. The authors concluded that most if not all chloroform-soluble material in coking coals released during preheating is present as such in the coal and is not a product of pyrolysis.

The kinetics of chloroform extraction of a bituminous coal (85.8% C, maf) and of preheated samples of it were investigated, and the corresponding apparent energies of activation for extraction were determined by Yoshii and Yoshimura.[333] The preheated coals were prepared in nitrogen-filled retorts at 300, 350, and 400°C. Chloroform extractions of the raw coal and the three preheated samples were performed between 20 and 60°C at 10°C intervals, and the extract yields were determined as a function of time. Since the magnitude of the activation energies was in the range expected for diffusion, it can be inferred from the results that the rate-determining step in chloroform extraction of preheated coals is the diffusion of the dissolved extract.

The initial rate of chloroform extraction from raw coal was relatively rapid; it may be inferred that material on the surface of coal particles will be dissolved quickly and that diffusion of the extract from inside the porous structure will be the rate-controlling factor in coal extraction.

[333] T. Yoshii and F. Yoshimura, *Fuel*, **48**, 229 (1969).
[334] T. Yoshii and F. Yoshimura, *Fuel*, **50**, 113 (1971).
[335] L. Lazarov, E. Zgurovska, and G. Angelova, *Fuel*, **50**, 338 (1971).

Further investigations have been performed by Yoshii and Yoshimura[334] on the nature of chloroform extraction of preheated coals. The properties of the extracts obtained (at 20 or 60°C) from the raw coal, and from the 300 and 350°C chars were very similar. Consequently it is inferred that for a given sample there is no significant change in the extracted components as the temperature and yield are increased. However a slight difference in the component characteristics could be observed at the two extraction temperatures in the case of the 400°C char.

It was of interest to note that the chloroform extraction yield from the raw coal was only 2.36% at 60°C, whereas the yield of the γ fraction (chloroform-soluble part of the pyridine extract) was about 13%, or roughly the same as the chloroform extract yield from the 400°C char. Apparently chloroform dissolves only the soluble material present on the surface of the coal particle.

Chloroform-soluble extract yields were determined by Lazarov and co-workers[335] using two bituminous coals that were pretreated by four different methods. The treatments involved four steps: (1) preheating the coal to the temperature of maximum plasticity, (2) treating the coal with sodium in liquid ammonia, (3) fractionating the pyridine extract of the coal, and (4) acetylating with acetic anhydride in pyridine. One of the coals (Gaskohle, G) had a carbon content (maf) of 82.2%; the other (Kokskohle, K) a carbon content (maf) of 88.0%. The extract yields obtained from the two coals after different pretreatments are shown in Table 8.15.

The authors suggested that in thermal pretreatment, as well as under the action of

Table 8.15 Chloroform-Soluble Extract Yields from Bituminous Coals[335]

	Yield (%)	
Extract	Coal G	Coal K
Chloroform extract from raw coal	0.8	0.9
Chloroform extract from preheated coal (G to 400°C, K to 450°C)	3.7	6.8
Chloroform extract from coal treated with sodium in liquid ammonia	3.2	11.2
Pyridine extract	23.4	10.8
Chloroform soluble fraction of the pyridine extract (γ fraction)		
Raw coal basis	6.9	5.9
Pyridine extract basis	29.5	54.8
Chloroform extract of acetylated coal	10.9	6.9
Chloroform extract of coal previously treated with pyridine under the conditions of acetylation (blank)	4.7	3.2

sodium in liquid ammonia, ether bond cleavage may have a very important role. On the other hand, the formation of chloroform-soluble substances by acetylation, as well as the formation of the γ fraction, may be explained by the dissociation of hydrogen-bonded structures.

8.10.1.2 *Benzene and Carbon Disulfide Extraction*

The benzene extraction of a bituminous coal was studied by Ouchi and Imuta[336] with the objective of separating the chemical constituents of the extract and isolating and identifying pure compounds. A low-sulfur Yubari coal (85.2% C, maf; 0.3% S) was used. The coal was extracted five times, each time with an equal weight of boiling benzene. The extract yield was 1.0 wt % of the dry coal.

The benzene solution of the extract was concentrated by distillation. After cooling the concentrated solution, a black crystalline material, identified as 4,11-dimethylpicene, precipitated (0.02% based on coal). The benzene was removed by vacuum distillation, and the residue was progressively extracted with petroleum ether, methanol, ethanol, and acetone. A product, soluble both in petroleum ether and in methanol, and representing 42% of the extract, was selected for chromatographic separation. The main groups of compounds identified were alkylbenzenes, alkylnaphthalenes, alkylphenanthrenes and anthracenes, and alkylchrysene and picene derivatives. In the latter fraction a dimethylchrysene was isolated and identified in crystalline form.

The results of a detailed analysis of the whole extract were published by the same workers.[337] Each solvent-separated fraction was further separated by liquid chromatography through an alumina column and finally by gas chromatography. The paraffins that were isolated probably came from exinite in the original coal; the most abundant in the normal paraffin series is C_{20}. There is a fairly large amount of diphenylene oxide (including benzodiphenylene oxide in the pyrene fraction) in the extract, suggesting that a considerable portion of inert oxygen in coal may be in the form of this inner ring type of oxygen.

Interesting observations were made regarding the relative abundance of condensed aromatic nuclei of different types. Phenanthrene was about 5–10 times more abundant than anthracene. Moreover, chrysene and picene derivatives were fairly abundant, but the pyrene concentration was low.

Spectral studies of the carbon disulfide extract of a vitrain-rich, high-volatile bituminous coal (82.6% C, maf) were reported by Retcofsky and Friedel.[338] These studies included the recording and quantitative evaluation of the ^{13}C NMR spectra of the extract; they represented the first application of this technique to a coal extract, permitting the direct measurement of the aromaticity of this material (see Chapter 4).

The carbon disulfide–soluble material, which represented approximately 4% the coal, was obtained by extracting the coal with pyridine followed by extraction of the pyridine-soluble material with carbon disulfide. (It is of interest to note that the extract yield was only 0.2% when coal was extracted directly with carbon disulfide.)

Both the carbon disulfide and the pyridine extracts were studied by ultraviolet, visible, and infrared spectrophotometry and by electron and proton magnetic resonance. Aromaticity values and other structural parameters were calculated for both the carbon disulfide and the pyridine extracts from the proton NMR data and the elemental constitution of the materials, using a slightly modified version of the method of Brown and Ladner.[339]

These investigations indicate that the aromaticity for the pyridine extract is higher than that for the carbon disulfide extract. In addition, the H/C ratio for the unsubstituted aromatic nucleus (H_{aru}/C) is found to be smaller for the pyridine extract, indicating the presence of larger ring systems. The value of 0.71 for the carbon disulfide extract indicates a ring size of two to three rings per mean structural unit. Carbon-13 NMR spectra of the carbon disulfide extract were recorded. The ^{13}C NMR method gives a direct determination of the aromaticity (f_a). An aromaticity of 0.6 ± 0.1 was obtained for the

[336] K. Ouchi and K. Imuta, *Fuel,* **42,** 25 (1963).
[337] K. Ouchi and K. Imuta, *Fuel,* **42,** 445 (1963).
[338] H. L. Retcofsky and R. A. Friedel, *Fuel,* **47,** 487 (1968).
[339] J. K. Brown and W. R. Ladner, *Fuel,* **39,** 87 (1960).

carbon disulfide extract, in good agreement with that deduced from the proton NMR spectrum.

8.10.2 Specific Extraction

8.10.2.1 *Pyridine Extraction*

Pyridine extraction of the vitrains from six Australian bituminous coals was investigated by Durie and co-workers.[340] The pyridine extracts, as well as the chloroform-soluble fraction of these extracts and of the shock-heated vitrains were studied by the combined application of high-resolution proton NMR and infrared spectroscopy. The H_{ar}/H_{al} ratios, which were determined by these measurements, were used to calibrate the infrared spectra of the extracts in the solid state, and the resulting calibrations were applied to obtain an estimate of the H_{ar}/H_{al} ratios of the parent vitrains.

The extraction yields, extract compositions, and some molecular weight data are shown in Table 8.16. With a few exceptions, the hydrogen distributions of the various extracts are remarkably similar. Approximately one-third of the protons are in aromatic positions, one-third in α positions, and the other third in other aliphatic positions.

[340] R. A. Durie, Y. Shewchyk, and S. Sternhell, *Fuel,* **45,** 99 (1966).

These investigations, which were performed to improve the infrared spectroscopic method for the determination of the H_{ar}/H_{al} ratio of vitrains, were not fully successful. The method was found to have a sizable margin of error, and any estimate for the H_{ar}/H_{al} ratio in a vitrain that is based on infrared spectra must be regarded with caution, even when the proton NMR spectra of extracts representing as much as 40% of the vitrain are used for the calibration of the infrared method.

The composition of pyridine extracts from a reduced and from an untreated vitrain of a bituminous coal was analyzed by high-resolution mass spectrometry by Kessler and co-workers.[341] The reduced material was prepared by treating the vitrain with lithium-ethylenediamine using the procedure of Reggel and co-workers.[342] The reduced vitrain contained 16 added hydrogen atoms per 100 carbon atoms. The mass spectra of the pyridine extracts were obtained in a high-resolution mass spectrometer where the inlet and source temperatures were 300°C. Approximately 30% of the extracts were volatilized in the spectrometer under these conditions. The following groups of compounds were identified.

[341] T. Kessler, R. Raymond, and A. G. Sharkey, Jr., *Fuel,* **48,** 179 (1969).

[342] L. Reggel, I. Wender, and R. Raymond, *Fuel,* **43,** 229 (1964).

Table 8.16 Yields and Elementary Analyses of Solvent Extracts From Vitrain Samples: All Data on Percent maf Coal Basis[340]

Carbon (%)[a]	Extract[b]	Yield (%)	Carbon (%)	Hydrogen (%)	Oxygen (%)[c]	Molecular Weight
81.7	PS	28.3				
	$CHCl_3$ extract of PS	5.7				525
82.8	PS	30.1	83.6	5.7	8.0	
	$CHCl_3$ extract of PS	11.3	84.8	6.6	5.7	644
84.3	PS	22.3	83.7	5.7	—	
	$CHCl_3$ extract of PS	4.5				
84.7	PS	31.4				
	$CHCl_3$ extract of PS	5.0				
87.1	PS	37.1	87.4	5.9	4.8	
	$CHCl_3$ extract of PS	15.7	88.4	6.8	2.8	
	$CHCl_3$ extract of shock-heated vitrain	7.5	84.3	6.4	4.4	
89.0	PS	37.9	88.3	5.6	4.4	
	$CHCl_3$ extract of PS	15.3	89.2	6.1	2.3	715
	$CHCl_3$ extract of shock-heated vitrain	8.4	85.7	6.2	3.9	
90.9	PS	0.6				

[a] Carbon content of vitrain.
[b] PS = pyridine soluble.
[c] Direct determination.

Oxygen Compounds. Phenols, benzofurans, acetophenones, naphthols, dibenzofurans, phenylphenols, and naphthylphenols have been identified. The concentration of phenols and naphthols was much greater in the extract from reduced vitrain than from untreated vitrain, indicating that rupture of ether linkages in the coal had occurred to form lower molecular weight phenolic compounds.

Sulfur Compounds. Thiophenes, thiophenols, benzothiophenes, dihydrobenzothiophenes, phenylthiophenes, dibenzothiophenes, dihydrodibenzothiophenes, acenaphthalenothiophenes, acenaphthenothiophenes, and benzonaphthothiophenes were identified. Of these groups of compounds, only the thiophenols, phenylthiophenes, and acenaphthenothiophenes were detected in the extract from reduced vitrain. Apparently all the other compound types were desulfurized during the reduction treatment. Sulfur values from ultimate analyses of the reduced and untreated vitrains were 0.14 and 0.95%, respectively. This reduction in sulfur content is in the same order as the decrease in sulfur compounds determined by mass spectrometry.

Hydrocarbons. Cyclohexadienes, acenaphthylenes, fluorenes, acenaphthenes, biphenyls, anthracenes, phenanthrenes, phenylnaphthalenes, benzofluoranthenes, and chrysenes were found. In addition, naphthalenes and di-, tetra-, hexa-, octa-, and decahydronaphthalenes were identified. A comparison of the total naphthalene and hydronaphthalene series was used to determine differences in the hydroaromatic content of the extracts. It was found that the concentration of naphthalenes was twice as great in the untreated as in the reduced vitrain. On the other hand, the amounts of dihydro-, tetrahydro-, and hexahydronaphthalenes were greater by factors of 4, 5, and 3, respectively, following reduction. The hydroaromatic compounds that were formed from naphthalene represented 29% of the reduced vitrain, compared with 7% for the untreated vitrain. The confirmation that large amounts of hydroaromatic compounds are formed during the lithium-ethylenediamine reduction of vitrain supports the conclusions of Reggel and co-workers[342] regarding the mechanism of that reaction.

Progressive extraction of a Pittsburgh Seam bituminous coal (83.6% C, maf) with pyridine was studied by Friedel and co-workers.[343] A small coal sample was extracted three consecutive times with fresh batches of pyridine, and the three extracts were recovered and analyzed separately by mass spectrometry. The pyridine extractions were carried out by mechanically shaking 1 g of coal with 2 ml of pyridine at room temperature. The extraction periods were 5 min, 30 min, and 17 hr, respectively. Low ionizing voltage mass spectrometry was used to determine, quantitatively, families of derivatives of various aromatic ring systems. The mass spectrometer had a heated inlet that usually was held at 290°C. The extract yields were 2.4, 4.6, and 12.3% over extraction periods of 5 min, 30 min, and 17 hr, respectively. One- and two-ring aromatics predominated in the extract obtained over a 5-min extraction period; the concentration of phenols was much greater in this extract than in the other two. The 30-min and 17-hr extracts contained approximately equal amounts of naphthalenes and three-, four-, and five-ring condensed compounds.

Investigation of chemical structural parameters of pyridine and carbon disulfide extracts of bituminous coals by means of proton and ^{13}C NMR spectrometry was reported by Retcofsky.[344] Six vitrains of high petrographic purity were selected for this study. Extractions with pyridine were carried out using the improved Soxhlet technique described by Raymond and co-workers.[345] Portions of the dried pyridine extracts were then extracted with carbon disulfide using essentially the same Soxhlet procedure. The results of this investigation—including extract yields, elemental analyses, NMR data, and calculated structural parameters—are summarized in Table 8.17 (opposite). In this table H^*_{ar}, H^*_{α}, and H^*_0 refer to the fraction of hydrogen atoms in aromatic (plus phenolic), in benzylic, and in other aliphatic positions, respectively: f_a is the carbon aromaticity (the ratio of aromatic to total carbon atoms); σ is the degree of aromatic ring substitution; and

[343] R. A. Friedel, J. L. Schultz, and A. G. Sharkey, Jr., *Fuel*, **47**, 403 (1968).
[344] H. L. Retcofsky, *Appl. Spectrosc.*, **31**, 116 (1977).
[345] R. Raymond, I. Wender, and L. Reggel, *Fuel*, **43**, 299 (1964).

Table 8.17 Elemental Composition, NMR Data, and Mean Structural Parameters for Coals and Coal Extracts[344]

Sample Identity			Analysis (%, maf)					Hydrogen Distribution			Structural Parameters		
Bed and Rank	Material	Extract Yield (%)	C	H	O^a	N	S	H^*_{ar}	H^*_{α}	H^*_o	f_a	σ	H_{aru}/C_{ar}
Adaville (subA)	Vitrain		76.3	5.2	16.5	1.3	0.7						
	Pyridine extract	6.7	77.5	6.4	14.0	1.7	0.4	0.20	0.30	0.44	0.64	0.59	0.76
	Carbon disulfide extract	1.5	83.4	9.4	6.3	0.4	0.5	0.15	0.31	0.52	0.44	0.57	1.0
Pittsburgh (hvAb)	Vitrain-rich coal		82.6	5.9	8.1	1.7	1.6						
	Pyridine extract	21.4	83.9	6.2	7.7	1.8	0.4	0.35	0.26	0.36	0.73	0.37	0.68
	Carbon disulfide extract	2.5	86.5	7.4	4.4	1.3	0.4	0.27	0.33	0.39	0.62	0.44	0.79
Powellton (hvAb)	Vitrain		85.1	5.3	7.1	1.5	1.0						
	Pyridine extract	20.9	83.0	5.8	8.5	2.3	0.4	0.27	0.29	0.39	0.72	0.47	0.59
	Carbon disulfide extract	5.1	86.9	7.2	4.6	0.9	0.4	0.29	0.35	0.34	0.61	0.45	0.86
Lower Banner (mvb)	Vitrain		86.1	5.5	5.6	1.8	1.0						
	Pyridine extract	19.3	87.1	6.0	4.6	1.8	0.5	0.32	0.28	0.37	0.73	0.37	0.57
	Carbon disulfide extract	5.7	88.1	6.9	3.1	1.3	0.6	0.29	0.36	0.33	0.64	0.44	0.76
Pocahontas No. 3 (lvb)	Vitrain-rich coal		90.0	4.4	1.1	0.7	3.8						
	Pyridine extract	2.8	88.4	5.2	4.2	1.8	0.4	0.55	0.31	0.13	0.85	0.27	0.63
	Carbon disulfide extract	0.8	89.9	5.8	2.8	1.1	0.4	0.53	0.36	0.11	0.81	0.29	0.71
Pocahontas No. 4 (lvb)	Vitrain		90.4	4.6	3.1	1.4	0.5						
	Pyridine extract	2.5	90.5	5.3	2.1	1.7	0.4	0.50	0.30	0.19	0.83	0.26	0.57
	Carbon disulfide extract	1.1	90.7	5.7	1.8	1.2	0.6	0.50	0.36	0.14	0.78	0.32	0.70

^{a} Oxygen by difference.

H_{aru}/C_{ar} is the H/C ratio of the hypothetical unsubstituted aromatic nuclei. For benzenoid structures H_{aru}/C_{ar} is 1.0, for naphthalene systems 0.8, for three-ring systems 0.72, and so on. With increasing rank, f_a increases from 0.44 to 0.81 for the carbon disulfide extracts and from 0.64 to 0.85 for the pyridine extracts. The fraction of the aromatic hydrogen increases from 0.15 to 0.53 for the carbon disulfide extracts and from 0.20 to 0.55 for the pyridine extracts. It should be emphasized that the f_a values for the carbon disulfide extracts were obtained directly from ^{13}C NMR spectra, whereas the f_a values for the pyridine extracts are estimates that were calculated from the proton NMR spectra using the aromaticity equation of Brown and Ladner.[339] In the latter case the presence of strong solvent resonances in the ^{13}C spectra prevented direct experimental determination of f_a values.

The H_{aru}/C_{ar} values for the extracts are shown as a function of the rank of the parent coals in Fig. 8.14. For the carbon disulfide extracts, the average number of aromatic rings per mean structural unit increases from one for the subbituminous coal to two to three condensed aromatic rings for the lvb coals. For the pyridine extracts the aromatic ring systems are much larger, varying from two to three rings for the lowest rank coal to five or more for the extracts from the lvb coals.

In the second part of this investigation two techniques were used to estimate the aromaticities of the parent coals: complementary NMR-infrared spectrometry and proton-enhanced nuclear induction spectroscopy. In the first method the proton NMR data for the solvent extracts were used to calibrate the intensities of the out-of-plane hydrogen bending vibrations on a "per aromatic hydrogen" basis. The f_a values obtained this way increased from 0.64 to 0.86 with increasing rank. The values obtained by proton-enhanced ^{13}C NMR were somewhat higher.

8.10.2.2 *Extraction with Various Electron Donor Solvents*

Solvent extraction of an Australian bituminous coal (Greta Seam, 82.4% C, maf) was investigated by Maher and O'Shea,[346] using a large variety of solvents. The principal objective of this investigation was the determination of phenolic hydroxyl groups in the extracts and the residues obtained with different solvents. Two extraction methods were used: digestion with the solvent at atmospheric pressure in a stirred extraction vessel, and Soxhlet extraction carried out at the boiling point of the solvent. The extract yields were 1–5% with alcohols, benzene, and chloroform; with pyridine, ethylenediamine, and dimethylformamide, 15–22% extract yields were obtained.

Acidic functional group analysis was carried out by the method of Brooks and Maher.[347] Most of the acidic groups were phenolic

[346] T. P. Maher, and J. M. O'Shea, *Fuel*, **46**, 283 (1967).

[347] J. D. Brooks and T. P. Maher, *Fuel*, **36**, 51 (1957).

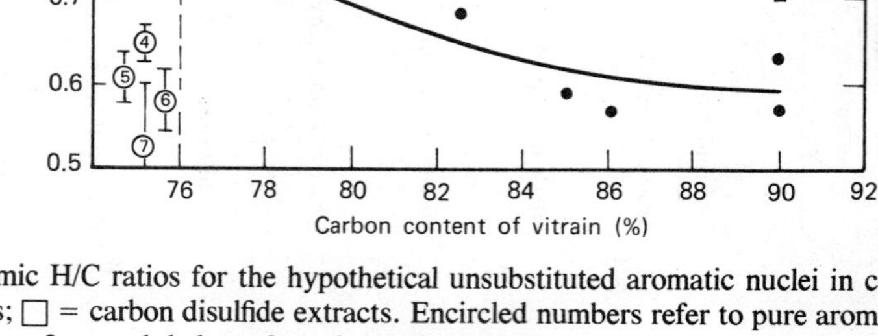

Figure 8.14 Atomic H/C ratios for the hypothetical unsubstituted aromatic nuclei in coal extracts: ● = pyridine extracts; □ = carbon disulfide extracts. Encircled numbers refer to pure aromatic hydrocarbons (e.g., 1 = benzene, 2 = naphthalene, 3 = phenanthrene).[344]

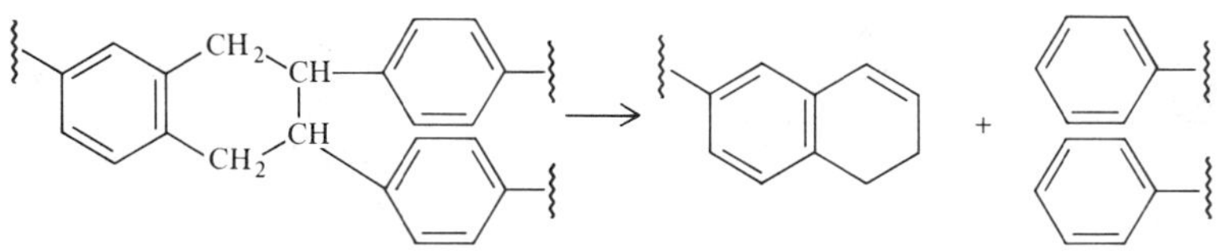

Figure 8.15 Possible reaction for bond-breaking and aromatization of a hydroaromatic structure during coal extraction with phenanthrene.[348]

hydroxyl groups. It is interesting that for every solvent except ethylenediamine, the sum of the phenolic group contents of the extracts and residues was higher than the expected value calculated from coal. Most probably, the explanation can be found in the increased accessibility of phenolic groups in the residue after the removal of the extract.

8.10.3 Extractive Disintegration

The extraction of three bituminous coals with phenanthrene at 350°C was investigated by Heredy and Fugassi[348] with the objective of determining whether any chemical interaction takes place between the coal and phenanthrene during the extraction. Two series of extractions were performed for this purpose: in one series the phenanthrene was labeled with tritium, in the other series it was labeled with ^{14}C. The extractions were made using a technique similar to that used by Orchin and co-workers.[349] Phenanthrene was refluxed in a flask, and the sample of the powdered coal (~ 6 g), wrapped in a piece of glass filter cloth, was placed for a period of 6 hr on a support directly under the reflux condenser so that it was in contact with the ascending vapor and the refluxing condensate of the solvent. The extract was separated into a benzene-soluble and a benzene-insoluble, phenanthrene-soluble fraction. Both extracts had low ash levels (0.1–0.15%) showing that effective separation of the extract from the mineral matter had taken place. Three high-volatile bituminous coals were studied (Pittsburgh Seam, 82–83% C, maf).

The extract yields were generally around 85%. It was shown that two chemical interactions took place between phenanthrene and the coal: hydrogen exchange and the incorporation of small amounts of phenanthrene into the coal. The first interaction took place to the extent of 10–15%, the second to the extent of 3–6%. It was shown by model compound experiments, using mixtures of tritiated phenanthrene with anthracene and with *p*-benzylphenol, respectively, that hydrogen exchange with these compounds took place only to the extent of 1% under identical experimental conditions. When coal was added to a mixture of tritiated phenanthrene and anthracene, the extent of exchange increased to 4%, indicating that coal promoted hydrogen exchange between the two condensed aromatic compounds. The infrared spectra of the coals and the extracts indicated that the relative intensity of the aromatic hydrogen absorption band at 3000–3050 cm^{-1} was greater in the phenanthrene-soluble extracts than in the original coals.

On the basis of the experimental results and observations summarized in the preceding paragraph, the formation of the coal extract was interpreted as a free radical chain reaction resulting in the breaking of certain bonds of coal and aromatization of some of the hydroaromatic structures. It appears that phenanthrene plays the role of a chain carrier in this process. An example of a reaction that might take place and would result in bond breaking and simultaneous aromatization of a hydroaromatic structure is shown in Fig. 8.15.

The high-temperature extraction of a high-volatile bituminous coal (78.0% C, maf) with tetralin, with naphthalene, and with dodecane was investigated by Neavel.[350] The experiments were conducted in small sealed stainless steel tubes. Normally 1.5 g of coal and 3.0 g of solvent were charged to the 8.0-ml reactor. Quick heating of the reactor was accomplished by agitating it in a preheated sand bath. Conversion to gases and extract

[348] L. A. Heredy and P. Fugassi, in *Coal Science*, Am. Chem. Soc. Adv. Chem. Ser. No. 55 (1966), p. 448.

[349] M. Orchin, C. Golumbic, J. E. Anderson, and H. H. Storch, U.S. Bur. Mines Bull. No. 505, 1951.

[350] R. C. Neavel, *Fuel*, **55**, 237 (1976).

was determined by weighing back the insoluble residue. The behavior of naphthalene and tetralin is very similar during the first 4 min of extraction when approximately 25% of the coal is solubilized. It is reasonable to assume that these benzene-soluble products are formed primarily by pyrolytic decomposition of macromolecular coal structures followed by the stabilization of free radicals through internal hydrogen disproportionation. Continued extraction in naphthalene (a nondonor solvent) decreased the yield of benzene-soluble material, probably due to the recombination into high molecular weight substances, further supporting the assumed free radical mechanism. The yield of benzene-soluble products (plus gases) increased continually with time in the tetralin-coal system, where after the first few minutes of contact time, hydrogen transfer from tetralin to the coal became the predominant process. The behavior of dodecane was similar to that of naphthalene.

8.10.4 Theoretical Studies of Solvent Extraction

A significant contribution to the fundamentals of solvent extraction was made in a paper by van Krevelen[351] in which two basic questions of coal-solvent interactions were examined. (1) Why does the potentially soluble fraction of coal decrease with increasing coalification, approaching zero at a carbon content of 92%? (2) Why have some solvents an optimum extracting effect at a specific degree of coalification? Explanations to both phenomena were sought on the basis of the theory of polymer chemistry, particularly the solubility of macromolecules.

The first problem was examined on the basis of Flory's theory[352] of trifunctional stepwise polycondensation. A characteristic feature of the three-dimensional polycondensation reaction is the occurrence of insoluble "gels" at a certain stage of the condensation. At the gel point the number-average molecular weight of the still soluble part reaches a maximum; beyond this point the average molecular weight of the still soluble part decreases. For the simplified theoretical case of two trifunctional monomers reacting with each other and the functional groups possessing the same reactivity, the distribution of the condensation products, w_x, as a function of the extent of reaction, p, is shown in Fig. 8.16. According to the theory the number average degree of polymerization before the gel point, $(\overline{X}_n)$ gel p, is $1/(1 - 1.5p)$. Since gelation occurs at $p = 0.5$, it follows that at the gel point $(\overline{X}_n)$ gel $p = 4$. This is the maximum value that $\overline{X}_n$ can reach. As the extent of polycondensation increases and the fraction of extractable material decreases, the value of $\overline{X}_n$ becomes smaller.

In applying the theory of three-dimensional polycondensation to coal genesis and coal

[351] D. W. van Krevelen, *Fuel*, **44**, 229 (1965).

[352] P. J. Flory, *Chem. Rev.*, **39**, 137 (1946); "Condensation Polymerization and Constitution of Condensation Polymers," *High Molecular Weight Organic Compounds*, R. E. Burk and O. Grummitt, Eds., Interscience, New York, 1949, pp. 211–282.

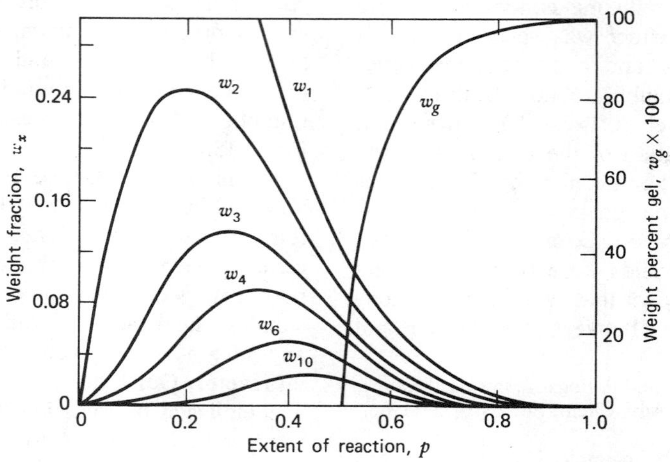

Figure 8.16 Weight fractions of various molecular species in a trifunctional polycondensation.[352]

extraction, one can assume that the "gel point" has been reached and exceeded at the end of the lignite stage. Using this analogy, the extent of reaction, p, reflects the rank of coal, and the fraction $(1 - w_g)$ in Fig. 8.16 represents the fraction of extractable matter in coal; this amount decreases continuously with increasing rank. The validity of this theory was tested against the experimental solvent extraction results obtained by Dormans and van Krevelen,[353] and by Wynne Jones and co-workers.[354] Strikingly good agreement with the theory was obtained for the extraction of a number of different coals. An additional interesting result of this work is the calculation of the average molecular weight of the "basic unit" in coals. This value was obtained from the measured molecular weight of the total extract by dividing it with $\overline{X}_n$, the number-average degree of polymerization. The results of these calculations indicate that the average "basic unit" has a molecular weight of about 400.

The second question dealing with the correlation between the optimum extracting effects of solvents and the rank of the coal was examined in terms of the so-called solubility parameter of coal. Calculated values of this parameter for coals of various rank and for a large number of solvents used in coal extraction have been derived from basic thermodynamic correlations dealing with the change of free energy when a polymeric substance either dissolves or swells in a solvent. The value of the solubility parameter, δ, of the solvents can be determined experimentally from the following correlation:

$$\delta = \left(\frac{\Delta H_{\text{vap}}}{V_M}\right)^{1/2}$$

where ΔH_{vap} is the molar heat of vaporization and V_M is the molar volume of the solvent.

The solubility parameter of coal as a function of rank is shown in Fig. 8.17. The solubility parameter shows a distinct minimum at a carbon content of 88.5%. It should be noted that the two curves in Fig. 8.17 enclose a range of δ values limited by a low and high value that were used for the fraction of aromaticity, f_a. Figure 8.17 also shows the δ values of a number of commonly used coal solvents. For good solubility, it is necessary that $(\delta_{\text{coal}}/\delta_{\text{solvent}})^2$ be less than or equal to 1. It is quite evident that substances such as benzene or chloroform are to be classed as poor solvents. Substances such as pyridine and ethylamine have δ values near those of high-rank coals; these solvents are, therefore, suitable for the extraction of high-rank coals. Ethanolamine, on the other hand, is more effective for the extraction of coals with a carbon content of 75–83%. This property is in good agreement with the high δ value of ethanolamine.

The δ values of pyridine and ethylamine approach sufficiently the δ values of high-rank coals to permit almost complete extractability. That this phenomenon is not observed is a consequence of increasing condensation and tridimensional network formation during the latter stages of coalification.

8.11 SMALL MOLECULES EXTRACTABLE FROM COAL

The extraction of trapped and/or occluded compounds in coal yields a source of hydrocarbons whose significance is being explored. Vahrman and his co-workers have presented a series of reports giving evidence to support the hypothesis that a quantity of simpler compounds can be separated from coal by physical methods, leaving a complex residue.[355–362]

Two groups of substances, found in both extracts and devolatilization tars (350–550°C)

[353] H. N. M. Dormans and D. W. van Krevelen, *Fuel,* **39,** 273 (1960).
[354] W. F. K. Wynne Jones, H. E. Blayden, and F. Shaw, *Brennstoff-Chemie,* **33,** 201 (1952).

[355] J. Roy and A. Lahiri, *Proc. Symp. Chem. Oils Coal,* Central Fuel Research Institute, Dhnabad, India (1972), p. 447.
[356] M. Vahrman, *Fuel,* **49,** 5 (1970).
[357] J. A. Spence and M. Vahrman, *Fuel,* **49,** 395 (1970).
[358] M. Rahman and M. Vahrman, *Fuel,* **50,** 318 (1971).
[359] T. J. Palmer and M. Vahrman, *Fuel,* **51,** 14 (1972).
[360] T. J. Palmer and M. Vahrman, *Fuel,* **51,** 22 (1972).
[361] M. Vahrman and R. H. Watts, *Fuel,* **51,** 130 (1972).
[362] M. Vahrman and R. H. Watts, *Fuel,* **51,** 235 (1972).

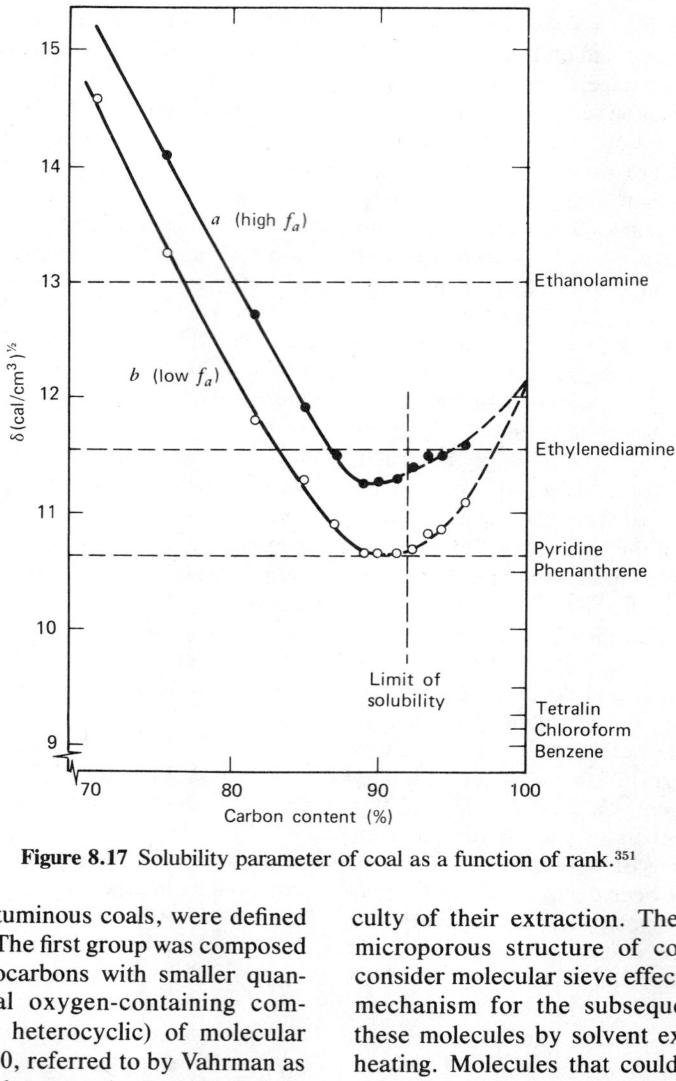

Figure 8.17 Solubility parameter of coal as a function of rank.[351]

derived from bituminous coals, were defined by Vahrman.[356] The first group was composed mainly of hydrocarbons with smaller quantities of neutral oxygen-containing compounds (mostly heterocyclic) of molecular weights up to 550, referred to by Vahrman as H-compounds; the second group, called O-compounds, was a series of alkylated hydroxyl-substituted aromatic compounds insoluble in light petroleum, having molecular weights up to about 1000.

In 1970 Vahrman found that after Soxhlet extraction was applied for 50 hr, small quantities of hydrocarbons could still be leached from bituminous coal. Soxhlet extraction of a British bituminous coal (81.4% C) with benzene-ethanol (7 : 3 by volume) for 250 hr gave extractable hydrocarbons. These findings and his later investigations were used to confirm the conclusion that low-temperature tar (350–550° C) consists largely of compounds preexisting in the microporous structure of coal, which might explain the difficulty of their extraction. The concept of a microporous structure of coal led him to consider molecular sieve effects as a possible mechanism for the subsequent release of these molecules by solvent extraction or by heating. Molecules that could easily fit into the micropores (e.g., straight-chain aliphatics) are retained to a greater degree than the branched and cyclic alkanes. His results showed that straight-chain alkanes became more predominant during later stages of extraction, whereas cyclic and branched aliphatic compounds were extracted in greater proportions during the earlier stages. Throughout Vahrman's studies molecules of larger critical diameter emerged earlier than those of smaller critical diameter; this was thought to be because the smaller molecules were trapped or occluded in the micropores.

Results on bituminous coals reported by Vahrman[360] showed that significant differences in the yields and compositions of tars could be correlated with the nature of the original coal. Molecular sieve effects, similar to those mentioned above, were noted in tar production. Branched chain and cycloaliphatic molecules emerged before straight-chain molecules. Low-boiling compounds not believed to be pyrolysis products evolved at temperatures much above their boiling points. These data may indicate that an activation energy for desorption or pore widening must be supplied to promote their evolution.

The biomarker compounds—pristane, phytane, farnesane, and other aliphatics—were identified in both tars and solvent extracts of coal.[356] In the early stages of extraction there was a preference for odd carbon number members to predominate in the alkanes, whereas a smooth odd-even distribution prevailed in the later stages. Vahrman speculated that the odd-prominent series may have originated from plant cuticles, and the smooth distribution may be associated with interior plant cells. An important conclusion to Vahrman's later work was that no accurate view of the structure of the main part of the coal, consisting mainly of complex O-compounds, could be obtained unless the smaller molecules, largely in micropores, were considered as separate entities. Taking into account adsorbed gases as well as liquids and solids, an orthohydrous coal contained a total of 11–12 wt % of the smaller molecules, about half of which were H-compounds, and a cannel coal contained 23 wt % of these substances, 60% of which were H-compounds. It would be useful to examine a broad range of coal-derived residues from complete extractions by using instrumental techniques available today and apply a mathematical analysis to the complete coal system, similar to that used by Ladner and Stacey.[363]

Extending such data to a number of coals may permit consideration of the classifications of coals in general. If one subtracts the contribution of the smaller molecules, the elementary compositions of the more complex parts of sub- and perhydrous coals, vitrinites, and exinites become very similar. Subtracting H-compounds only, not the extractable O-compounds (which are higher in hydrogen than the more complex coal macromolecules), results in an maf H % drop from 5.0 to 4.5% for orthohydrous coal and from 7.0 to 5.0% for the perhydrous coal.

Hayatsu et al.[364] carried out Soxhlet extractions of a Wyoming lignite (64.4% C, 5.3% H, 1.1% S, and 27.9% O, maf), a high-volatile Illinois bituminous coal (77.8% C, 5.4% H, 1.4% N, 2.1% S, and 13.3% O), and a Pennsylvania anthracite, Penn State PSOC No. 85 (91.3% C, 2.9% H, 0.6% N, 1.1% S, and 3.1% O) with benzene methanol (3 : 1) for about 48 hr. The lignite and bituminous coal extracts were separated on alumina by column chromatography, eluting with petroleum ether–benzene (9 : 1). The resultant chromatographic fractions and the total anthracite solvent extract were analyzed by gas chromatography and the effluent monitored by a mass spectrometer. From the three chromatograms they were able to identify or tentatively assign structures to 141 components. Table 8.18 summarizes the components identified by gas chromatography and mass spectrometry. The two fractions and one solvent extract represent 0.7, 1.4, and 0.04% of the lignite, bituminous, and anthracite coals (maf), respectively.

From gas chromatographic data of the solvent extracts, Hayatsu et al.[364] proposed a possible sequence of dehydrogenation-dealkylation reactions to demonstrate a coalification scheme by the existence of closely related compounds (Nos. 125, 133, 134, 120, and 108 of Table 8.18). Figure 8.18 illustrates the changes that may take place during the coalification process with loss of hydrogen and alkyl groups as the coal progresses from lignite to bituminous to anthracite. The relative abundance of components in this figure was obtained from three separate gas chromatograms of the coal extracts (lignite, bituminous, and anthracite). Because an internal reference was not used in the gas chromatography study, the different coal extracts should not be compared.

The concept that low molecular weight fractions of coal were retained in the micropores of a molecular sieve structure has been more recently supported by work of other

[363] W. R. Ladner and A. E. Stacey, *Fuel,* **40,** 295 (1961).

[364] R. Hayatsu, R. E. Winans, R. G. Scott, L. P. Moore, and M. H. Studier, *Fuel,* **58** 541 (1978).

Table 8.18 Compounds[a] Identified by Gas Chromatography, Mass Spectrometry, and High-Resolution Mass Spectrometry[364]

1. Straight-chain hexane
2. 2-Hexene
3. Dimethylbutane, methylcyclopentane
4. Cyclohexane
5. C_7-Alkene (B)
6. Benzene
7. Thiophene
8. C_7-Alkane (B), C_7-alkene (B)
9. C_7-Alkadiene (B) or C_7-alkyne (B)
10. Cyclohexene
11. C_7-Alkane (B)
12. Dimethylcyclopentane
13. 2- and 3-Methylhexanes
14. Heptene
15. 2,3-Dimethyl-2-pentene
16. Methylcyclohexane
17. Dimethylhexane
18. Heptyne
19. Trimethylpentane
20. Methylheptane
21. Methylheptene
22. Trimethylcyclopentane
23. 1-Methylcyclohexene
24. Toluene
25. Dimethylcyclohexane
26. Methylthiophene
27. C_9-Alkene (B)
28. Ethylcyclohexane
29. Trimethylcyclohexane
30. *n*-Propyl and/or isopropylcyclohexane
31. C_4-Alkylcyclopentane (%)
32. C_9-Alkane (B), C_9-alkene
33. C_9-Alkyne (?) and/or C_9-Alkadiene (?)
34. Ethylbenzene
35. Dimethylthiophene
36. *m*- and *p*-Xylene
37. *o*-Xylene
38. C_9-Alkene (B)
39. Tetramethylcyclohexane
40. C_{10}-Alkene (B)
41. C_4-Alkylcyclohexane
42. Diethylcyclohexane (?)
43. C_{10}-Alkane (B), C_{10}-alkene (B)
44. C_{10}-Alkene
45. Ethyloctane (?)
46. Trimethylthiophene
47. Propylbenzene
48. Methylethylbenzene
49. Trimethylbenzene
50. C_{11}-Alkene (B)
51. C_4-Alkylbenzene
52. C_{11}-Alkene (B) and C_4-Alkylbenzene
53. Tetramethylbenzene
54. 1-Methyl-4-isopropyl-3-cyclohexene (?)
55. Methylindane
56. C_{12}-Alkene (B)
57. Dimethylindane
58. C_5-Alkylbenzene
59. Tetralin
60. C_6-Alkylbenzene
61. C_{13}-Alkene, C_6-alkylbenzene
62. Naphthalene
63. C_{13}-Alkane (B), C_{14}-alkene (B)
64. 2-Methylnaphthalene
65. 1-Methylnaphthalene
66. C_3-Alkyldecalin
67. C_4-Alkyldecalin
68. Trimethylindane
69. Tetramethylindane and/or trimethyltetralin
70. C_9-Alkylcyclohexane
71. Biphenylene
72. 2-Ethylnaphthalene
73. 1-Ethylnaphthalene
74. Dimethylnaphthalene
75. Cadinane (4,10-dimethyl-7-isopropyldecalin)
76. Dihydrocadinene (T), C_9-alkylcyclohexane
77. Selinane and eremophilane (hydronaphthalenes)
78. Dihydroselinene (T) and/or dihydroeremophilene (T)
79. Dihydrocadinene (T)
80. C_5-Alkylindane
81. Methylcenaphthene, 2(?)-isopropylnaphthalene
82. Diphenylmethane
83. 3- or 4-Methylbiphenyl
84. C_{10}-Alkylbenzene
85. Tetramethylindan
86. $C_{15}H_{28}$-sesquiterpenoid hydrocarbon (?)
87. C_5-Alkyltetralin
88. Methylethylnaphthalene and/or trimethylnaphthalene
89. Trimethylnaphthalene
90. Fluorene
91. 1,6-Dimethyl-4-isopropyl-1,2-dihydronaphthalene (T)
92. Isobutylnaphthalene, Trimethylnaphthalene
93. 1-Methyl-4-isopropylnaphthalene
94. Eudalene (1-methyl-7-isopropyl-naphthalene)
95. C_5-Alkyltetralin (?)
96. 1-Methyl-2-propylnaphthalene (T)
97. Cadalene (1,6-dimethyl-4-isopropyl-naphthalene)
98. Tetramethylnaphthalene
99. 1,4-Dimethyl-6(?)-isopropyl-naphthalene, C_6-alkyltetralin (T)
100. C_6-Alkyltetralin (T)
101. 1,2,5,7-Tetramethylnaphthalene
102. Pristane
103. Methylfluorene
104. Pentamethylnaphthalene
105. Dibenzothiophene
106. Trimethyloctahydrophenanthrene
107. Methyltetrahydrophenanthrene (T)
108. Phenanthrene (Fig. 8.18)
109. $C_{14}H_{23}$ (*m/e* 191, base peak), $C_{20}H_{36}$ (M^+) tricyclic terpenoid (?)
110. Dimethyltetrahydrophenanthrene (T)
111. Ethyltetrahydrophenanthrene (T)
112. Anthracene
113. Naphthofuran
114. Methyldibenzothiophene
115. C_9-Alkyltetralin (?)
116. $C_{20}H_{32}$ (Abietadiene (?))
117. $C_{19}H_{30}$ (Tricyclic diterpenoid (?))
118. 2- and/or 3-Methylphenanthrene
119. 1- and/or 9-Methylphenanthrene
120. 1,7-Dimethylphenanthrene (Fig. 8.18)
121. Dimethyldibenzothiophene
122. C_{10}-Alkyltetralin (or C_{11}-alkylindane) (?)
123. Dehydroabietene (4,20-dimethyl-13-isopropyl-8*H*-phenanthrene)
124. Dimethylphenanthrene and/or dimethylanthracene
125. Dehydroabietane (Fig. 8.18)
126. $C_{20}H_{32}$(Tricyclic diterpenoid) (?)
127. Fluoranthene
128. C_{11}-Alkyltetralin (?)
129. Abietatetraene (T) (trimethylisopropyl-6*H*-phenanthrene)
130. 1,2,3,4-Tetrahydroretene (T)
131. Methylethylphenanthrene and/or trimethylphenanthrene
132. Pyrene
133. Simonellite (Fig. 8.18)
134. Retene (1-methyl-7-isopropylphenanthrene) (Fig. 8.18)
135. 1,2-Benzofluorene
136. 2,3-Benzofluorene
137. 3,4-Benzofluorene
138. Methylpropylphenanthrene and/or dimethylethylphenanthrene
139. Methylbenzofluorene
140. Tetramethylphenanthrene and/or tetramethylanthracene
141. Chrysene and/or triphenylene

[a] B = branched, T = identification tentative, ? = identification uncertain.

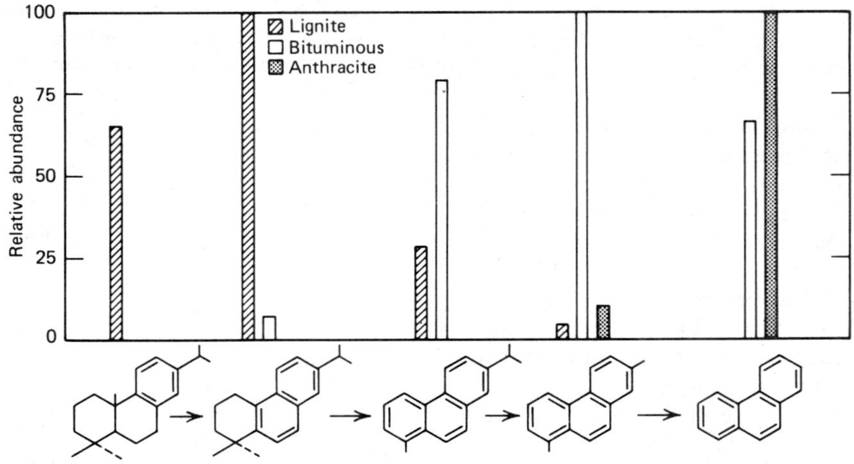

125 133 134 120 108

Figure 8.18 Transformation of diterpenoids in coals. Relative abundances in the solvent extracts of the three coals are normalized to the compound most abundant in the series shown. The numbers refer to compounds listed in Table 8.18.[364]

groups.[365–371] The relatively unfamiliar phenomenon of the extraction power of compressed gases (supercritical state) is being used to extract the occluded or trapped molecules from coal.[367,370–372] The method of supercritical gas extraction was suggested for coal by Wise in 1970.[373] The gas extraction technique is based on the ability of substances to vaporize in the presence of a compressed supercritical gas. Under favorable conditions the effect can be very large, and an increase in volatility of up to 10,000 times can be obtained.[372] It is claimed that this extraction technique has special advantages:

1. There is no strong chemical interaction at high temperature to decompose the coal.
2. Extraction takes place below the usual temperature of volatile evolution (i.e., 350°C; at this temperature some of the coal macromolecular structure or its occluded matter may undergo bond cleavage).
3. A compressed gas can contain more extractables per unit volume, and the amount can be continuously varied by altering the system pressure.
4. Pressure extraction at elevated temperatures helps to open up the micropore structure and increases the rate of diffusion of the solvent and solute molecules out of coal.

Blessing and Ross[370] have suggested more recently that whether an extraction medium is strictly supercritical (i.e., above the critical temperature) is of little consequence to the extraction. The changes in extraction yield they observed were attributed to temperature differences alone. The extraction efficiency they achieved, expressed as percentage of pyridine and/or tetrahydrofuran extractability, was attributed to the presence of a dense medium (solvent) at high temperatures.

Bartle et al.[365,367] have examined the supercritical gas extraction of a British bituminous coal (82.5% C). The extraction was

[365] K. D. Bartle, T. G. Martin, and D. F. Williams, *Chem. Ind. 313 (1975).*

[366] C. M. White, J. L. Shultz, and A. G. Sharkey, Jr., *Nature,* **262,** 620 (1977).

[367] K. D. Bartle, T. G. Martin, and D. F. Williams, *Fuel,* **54,** 226 (1975).

[368] M. J. McIntosh, *Fuel,* **55,** 59 (1976).

[369] R. E. Williams, R. Hayatsu, R. G. Scott, L. P. Moore, and M. H. Studier, "Examination and Comparison of Structure: Lignite, Bituminous and Anthracite Coal," Preprints, 1976 Coal Chemistry Workshop, Stanford Research Institute, Menlo Park, Calif.

[370] J. E. Blessing and D. S. Ross, *Prepr. Am. Chem. Soc., Div. Fuel Chem.,* **22,** 118 (1977).

[371] N. Gangoli and G. Thodos, *Ind. Eng. Chem.,* **16,** 206 (1977).

[372] J. C. Whitehead and D. F. Williams, *J. Inst. Fuel,* 182 (1975).

[373] W. S. Wise, *Chem. Ind.,* 950 (1970).

accomplished by a semicontinuous method, passing a stream of supercritical toluene at 350°C and 10 MPa (~ 1500 psia) through a bed of 1.6-mm coal. The gas was withdrawn through a sintered steel filter into a vessel at atmospheric pressure, where most of the extracted material precipitated. Any unprecipitated material was condensed and the toluene volatilized off when the pressure was reduced. The yield obtained was 17 wt %. Bartle and co-workers[367] used solvent fractionation of their coal extract to characterize the chemical compounds. A summary of the aliphatic components isolated as a petroleum ether column eluate of the supercritical gas extract of coal, representing 0.4% of the coal (maf), is given in Table 8.19. The relative abundance of these compounds agrees in general with Vahrman[362] and more recently with White and co-workers,[366] who found a smooth distribution of odd-even carbon atoms from a coal liquefaction product. They reported detailed compositional data on 85% of the coal extract (14.5% of the whole coal, maf). The chemical structural nature of the subfractions was determined by proton and ^{13}C NMR, low ionizing voltage mass spectrometry, and gas chromatographic analysis.

The macromolecular structure and substitution patterns of the extracts were similar to Vahrman's, corresponding to a generally open-chain polynuclear aromatic average structure with about one-third of the available aromatic sites carrying alkyl substituents. Bridging between aromatic structural units is thought to be through biphenyl-like bonds, methylene groups, and heteroatoms, mainly oxygen. It may be possible that a supercritical gas extraction produces in practice the limiting yields of small molecules postulated by Vahrman.

Bartle and co-workers[367] have concluded that the chemical nature of the supercritical gas extract is consistent with the view that the material has been separated from the coal in a structural state substantially unaltered from that in which it existed within the coal framework. They further explain that the nature of the isoprenoid and straight-chain alkanes, the absence of straight-chain alkenes, the presence of hydroaromatic structures, and the apparent open-chain structure of the aromatic skeleton suggest that most, if not all, of the coal extract (17%) is the product of coalification of plant chemicals and is not due to degradation during extraction. It is their[367] belief that the supercritical gas extraction process allows penetration of the coal structure and subsequent release of materials hitherto not recoverable without degradation. Their findings are in general agreement with Vahrman's molecular sieve theory of coal structure.

It would be most useful to examine products from both a supercritical gas extraction and a pyridine Soxhlet extraction of the same coal by instrumental and chromatographic

Table 8.19 Gas-Liquid Chromatographic Analysis of Petroleum-Ether Eluate of a Supercritical Gas Extract of Coal[367]

Component[a]	Percent of Eluate[b]
Octane	0.2
Nonane	0.6
Decane	0.2
C_{11}	0.5
Branched paraffins between C_{11} and C_{12}	5.5
C_{12}	0.6
2,6,10-Trimethylundecane	0.1
C_{13}	0.1
Farnesane	0.4
C_{14}	0.8
2,6,10-Trimethyldodecane	1.5
C_{15}	1.4
C_{16}	2.4
Norpristane	2.5
C_{17}	4.1
Pristane	8.4
C_{18}	4.5
Phytane	1.5
C_{19}	4.9
C_{20}	4.2
C_{21}	4.8
C_{22}	4.1
C_{23}	3.7
C_{24}	3.5
C_{25}	4.5
C_{26}	3.4
C_{27}	3.5
C_{28}	2.0
C_{29}	1.1
C_{30}–C_{31}	5.1
Total identified	80.1

[a] Straight chain unless otherwise noted.

[b] Total petroleum-ether eluate amounts to 0.4% of the maf coal.

techniques available today and to make direct quantitative comparisons, similar to the recent work of Kershaw.[374] Kershaw investigated the chemical nature of the aliphatic and aromatic hydrocarbons and the hydroxyl-containing compounds obtained from high-pressure hydrogenation, supercritical (toluene) gas extraction, and Soxhlet extraction of the same coal. His initial findings support the work of both Vahrman and Bartle.

Future investigations that would better detail the genesis of components extractable from coal should include studies of vascular plants that contributed to coal accumulation and are known from studies of petrified peat (coal balls). Plant materials in coal balls are direct evidence of swamp vegetation during specific geological time periods (e.g., the Pennsylvania period). An interesting summary of fossil plants and coalification has been presented by Phillips.[375] Material such as coal balls could be extracted and the components characterized; then these compounds could be traced through a proposed coalification scheme by an organized multidisciplinary approach of geologists, botanists, petrologists, and chemists. A more resolved picture of the constituents of coal could thus be achieved.

A more detailed study should also be made of the gasification and liquefaction of the residue left after supercritical gas extraction. It would be interesting to compare the coals that show a change in liquefaction behavior with those that do not.

8.12 REACTIONS OF COAL IN A HIGH-TEMPERATURE PLASMA

Coal can be pyrolyzed in the high-temperature gaseous plasma generated by an electric arc.[376–390] Pulverized coal, when passed through such a "plasma jet," is reacted by way of "plasma pyrolysis" and undergoes very rapid heating. Reaction times are only a few milliseconds and product quench rates are quite rapid. Under such conditions, the products consist principally of soot and gas; in most cases little or no tar is formed.

Owing to the short residence time of coal particles in a plasma jet, it is unlikely that thermodynamic equilibrium is attained. But thermodynamic data show that above about 1400°C, acetylene (C_2H_2) is the most stable hydrocarbon molecule, and it is the principal hydrocarbon formed from coal under these conditions. Since its free energy of formation is positive, high-temperature, acetylene-containing streams must be quenched rapidly to prevent decomposition to carbon and hydrogen.[386,391] At extreme arc temperatures (estimated to be >7000°K), C_2H_2 formed from coal in an argon plasma was decomposed to soot and H_2. Acetylene and ethylene have also been formed by very rapid high-temperature (1300°K) devolatilization in combustion

[374] J. R. Kershaw, *S. Afr. J. Chem.*, **30**, 205 (1977).

[375] T. L. Phillips, R. A. Petters, M. J. Avein, and P. F. Laughman, *Science*, **184**, 1367 (1974).

[376] P. R. Ammann, R. F. Baddour, M. M. Johnson, T. W. Mix, and R. S. Timmins, *Chem. Eng. Prog.*, **60**, 52 (1964).

[377] R. L. Bond, W. R. Ladner, and G. I. T. McConnell, in Am. Chem. Soc. Adv. Chem. Ser. No. 55 (1966), p. 650.

[378] R. L. Bond, W. R. Ladner, and G. I. T. McConnell, *Fuel*, **45**, 381 (1966).

[379] R. D. Graves, W. Kawa, and R. W. Hiteshue, *Ind. Eng. Chem. Process Des. Dev.*, **5**, 59 (1966).

[380] W. Kawa, R. D. Graves, and R. W. Hiteshue, U.S. Bur. Mines Rep. Invest. No. 6829 (1966).

[381] L. L. Anderson, G. R. Hill, E. H. McDonald, and M. J. McIntosh, *Chem. Eng. Prog. Symp. Ser.*, **64**, 81 (1968).

[382] P. R. Ammann, R. F. Baddour, and T. W. Mix, U.S. Pat. 3,384,467 (May 21, 1968).

[383] R. E. Gannon, V. J. Krukonis, and T. Schoenberg, *Ind. Eng. Chem. Prod. Res. Dev.*, **9**, 343 (1970).

[384] K. Littlewood, *Proc. Symp. Chemicals, Oils, Coal*, Central Fuel Research Institute, Dhnabad, India, 1969 (1972), pp. 517–525.

[385] R. Nicholson and K. Littlewood, *Nature*, **236**, 397 (1972).

[386] R. E. Gannon and V. Krukonis, *Arc-Coal Process Development*, Final Rep., Office of Coal Research, U.S. Dept. Interior, May 1966–April 1972.

[387] Y. Kawana, *Chem. Econ. Eng. Rev.* (Chemical Economy Research Institute, Japan), **4**, 13 (1972).

[388] V. J. Krukonis, R. E. Gannon, and M. Model, in Am. Chem. Soc. Adv. Chem. Ser. No. 131, L. G. Massey, Ed., 1974, p. 29.

[389] S. C. Chakravartty and D. Dutta, *Fuel*, **55**, 255 (1976).

[390] S. C. Chakravartty, D. Dutta, and A. Lahiri, *Fuel*, **55**, 43 (1976).

[391] D. M. L. Griffiths and H. A. Standing, in Am. Chem. Soc. Adv. Chem. Ser. No. 55 (1966), p. 666.

gases of a Utah hvBb coal in an entrained-flow reactor.[392]

Plasma pyrolysis of high-volatile bituminous coals in argon also forms H_2, CO, and small amounts of CH_4 and C_2H_4. Carbon dioxide, C_2H_6, C_3H_8, HCN, COS, CS_2, C_3H_4, and diacetylene have also been reported in the gaseous products. Under reaction conditions where the gas yields were at a maximum, CO production corresponded closely to the maximum theoretical yield based on the oxygen content of the coal; the amount of hydrogen appearing as H_2 or in hydrocarbons approximated the hydrogen content of the coal.[385]

Yields of C_2H_2 vary with the experimental apparatus and conditions used; for any particular conditions the C_2H_2 yield increases with increase in the coal volatile matter. Nicholson and Littlewood[385] found that for coals of 37 and 48% volatile matter, 36 and 46%, respectively, of the carbon in the coal was converted to C_2H_2. When the CO yields were also taken into account, the total yield of volatiles was up to 25% greater than the volatile matter. The authors hypothesize that C_2H_2 is formed principally by dehydrogenation of aliphatic and alicyclic components of the coal structure. Generally the proportion of nonaromatic material in coals is related to their volatile matter contents, imposing an upper limit on the yield of acetylene from a given coal. The maximum yields obtained from the four coals investigated by Nicholson and Littlewood were all very close to their limiting values. Possible means of increasing the limiting values are by adding more hydrogen to the coal structure by hydrogenation before pyrolysis or by pyrolyzing the coal in the presence of hydrogen. In an attempt to improve the acetylene yields, argon-hydrogen mixtures were used as the principal arc gas. A maximum acetylene yield of 74% by weight on a carbon basis was obtained from a coal with 36% volatile matter.

Others[386–388] also found that C_2H_2 formation increases in the presence of H_2 in the plasma. But this effect has also been explained in terms of hydrogen as H, CH, or C_2H reacting in a quenching step, to prevent the decomposition of acetylene to carbon black.[388]

Bond et al.[378] compared the reactions of various coals in Ar, Ar—H_2, N_2, and N_2—H_2 plasmas. In a nitrogen plasma a coal with 51% volatile matter gave, under their best conditions, a conversion of carbon to HCN of 40–56% and a 5–9% conversion to C_2H_2. The authors suggest that HCN (like C_2H_2) is derived mainly from the coal volatile matter. Thermodynamic data show that HCN and C_2H_2 are the expected products from the pyrolysis of coal at high temperatures in nitrogen.[391]

8.13 REACTIONS OF COAL IN A LOW-TEMPERATURE PLASMA

Coals can also be pyrolyzed in different types of electric discharges. These do not reach the high temperature that an electric arc does. In a microwave discharge coals of different ranks and types have been gasified in the presence or absence of argon.[393–399] The principal reaction in the gasification of coal in a microwave discharge is presumed to be rupture of the bonds in the coal structure caused by the bombardment of energetic species on the coal surface. Numerous reactive entities such as H-species, O-species, gaseous carbon, and hydrogenated carbon fragments (CH_x or C_xH_y) could be produced from coal in the discharge pyrolysis, and these in turn decompose or combine with each other to form the products—H_2, H_2O, CO, CO_2, and low molecular weight hydrocarbons (mainly C_2H_2). The extent of gasification increases with the volatile matter of coal. The presence of low pressures of gases

[392] R. L. Coates, C. L. Chen, and B. J. Pope, in Am. Chem. Soc. Adv. Chem. Ser. No. 131, L. G. Massey, Ed., 1974, p. 92.

[393] Y. C. Fu and B. D. Blaustein, *Chem. Ind. (London)*, 1257 (1967).

[394] Y. C. Fu and B. D. Blaustein, *Fuel*, **47**, 463 (1968).

[395] Y. C. Fu and B. D. Blaustein, *Ind. Eng. Chem. Process Des. Dev.*, **8**, 257 (1969).

[396] Y. C. Fu, *Chem. Ind. (London)*, 876 (1971).

[397] Y. C. Fu, B. D. Blaustein, and I. Wender, *Chem. Eng. Prog. Symp. Ser.*, **67**, 47 (1971).

[398] D. M. Bodily, S. C. L. Che, and W. H. Wiser, *Prepr., Am. Chem. Soc., Div. Fuel Chem.*, **18** (2), 221 (1973).

[399] S. C. L. Che, "Microwave Pyrolysis of Coals and Polynuclear Hydrocarbons," PhD. dissertation, University of Utah, 1974.

such as argon increased the rate of gasification of coal, and fast quenching of the primary products gave higher yields of C_2H_2 and other hydrocarbons (up to C_6). More than 30% of the carbon in an hvAb coal could be gasified by pyrolysis in an argon discharge.[395]

Reactions of coal in microwave discharges in H_2 and H_2O vapor[393,394] gave greater gasification and yielded more hydrocarbons than the discharge pyrolysis in the presence or absence of argon, indicating that H, OH, and O-species attack the coal surface and interact with the active species derived from the coal. In the H_2 discharge nearly 40% of the carbon in the hvAb coal could be converted to hydrocarbons, mainly C_2H_2. In a CO_2 discharge[397] the product species present are oxidized by the active O-species produced. The results are a higher yield of CO, a lower yield of hydrocarbons, and a greater extent of gasification.

Active nitrogen produced in the microwave discharge reacts with coal molecules[400] to yield gaseous products, namely: HCN, C_2H_2, small amounts of cyanogen and low molecular weight hydrocarbon gases, and nitrogen-containing compounds, in addition to the usual hydrogen and carbon oxides. More than 42% of carbon in an hvAb coal could be gasified. It has been suggested[401,402] that the C_2N_2/HCN ratio increases with rank, and that HCN is mainly formed from nonaromatic carbon, whereas C_2N_2 derives mainly from aromatic carbon. The formation of HCN is enhanced when coal is reacted in a N_2—H_2 discharge.

Coal has also been reacted with H-species produced in glow discharges[403–405] to yield CH_4, and C_2—C_4 hydrocarbons. As the sample is placed further downstream from the luminous zone of the discharge, the reaction rate decreases. It has been suggested that the C_2—C_4 hydrocarbons were produced from the reaction of H-species with nonaromatic "precursor" structures.[404] Other results on pyrolysis of polynuclear hydrocarbons in a microwave discharge,[398,399] however, have indicated that C_2—C_4 saturated hydrocarbons are produced from totally aromatic molecules.

If O_2 is passed through an electric discharge, chemically active oxygen species are generated. When these reactive oxygen species are passed over coal, they will oxidize the carbon and hydrogen in coal and the mineral matter can be recovered in a relatively unaltered state. The maximum temperature that the coal is exposed to by this procedure is around 200°C. This technique, called "low-temperature ashing," is discussed in Chapter 7.

Reactions of coal in both high- and low-temperature plasmas are interesting but give little or no new information concerning coal constitution.

8.14 ORGANIC SULFUR IN COAL

Until recently little interest was shown in determining the forms in which organic sulfur existed in coal, or in developing methods for its removal. What work was done was aimed at decreasing the sulfur content of coal for metallurgical coke production, much of it on the coke itself or in conjunction with the coking process. Public awareness of the possible consequences of air pollution increased interest in the chemistry of coal sulfur, and this interest grew quickly when the United States and other countries realized that a growing dependence on coal as an energy source was inevitable.

Sulfur is present in coal in three principal forms—as pyritic sulfur, as sulfate sulfur, and as organic sulfur. Inorganic sulfides are occasionally reported, and elemental sulfur is claimed to be present in small amounts.

Since there are few methods for determining the functionality of sulfur, especially in mixtures, and the few that are available either are ambiguous or give questionable results with coal, the nature and the bonding of the organic sulfur in coal are still unknown.

Extraction of coals with chloroform and

---

[400] Y. C. Fu, B. D. Blaustein, and A. G. Sharkey, Jr., *Fuel,* **51,** 308 (1972).
[401] S. Nishida and N. Berkowitz, *Fuel,* **52,** 262 (1973).
[402] S. Nishida and N. Berkowitz, *Fuel,* **52,** 267 (1973).
[403] M. Letort, A. F. Boyer, and P. Payen, *Bull. Soc. Chim. Fr.,* 1589 (1963).
[404] Y. Sanada and N. Berkowitz, *Fuel,* **48,** 375 (1969).
[405] K. Kobayashi and N. Berkowitz, *Fuel,* **50,** 254 (1971).

pyridine produced extracts with organic sulfur contents similar to those of the residues, indicating a uniform distribution in the coals analyzed.[406] Pyrolysis of coal (as in coking) produces a large number of sulfur-containing compounds, as does hydrogenation,[407] and many of these have been identified. This enumeration does not, however, lead to useful information concerning structures present in the parent coal, since both condensation and bond cleavage could be responsible for the observed species.[408]

Considerable work has been done on estimating the amount of —SH, —SR, and —S—S—groups in coal, utilizing the reaction with CH_3I to ultimately form $(CH_3)_3S^+I^-$. However carried out, the completeness of the reaction is not measurable, and so the values obtained may be a low estimate. Values determined for the total of the three functional groups by this procedure are generally low (5–15% of the total sulfur),[409–411] though for some high-sulfur coals it is reported to be 90%.[412] Similar low values (10–15% of total sulfur) for the three groups have been obtained by a method based on the catalytic decomposition of these groups to H_2S in the presence of Al_2O_3 at 450°C.[413] The same procedure resulted in values for thiosulfur groups, ranging from 76% in low-rank coals to 14% in anthracite.[414]

Treatment of coal with $AgNO_3$, followed by amperometric determination of excess Ag^+, led to a value of 0.4–4% of the organic sulfur for the thiol concentration of coals.[415] When a series of Assam (India) coals high in organic sulfur was oxidized by air or by HNO_3, between 84 and 100% of the organic sulfur was either eliminated or converted to sulfonic acids, implying that 0–16% of the organic sulfur in these coals was oxidation-resistant cyclic sulfur (Table 8.20).

The majority of the organic sulfur in coal is believed to be in the form of heterocyclic sulfur,[417] such as condensed thiophenes, benzothiophenes, dibenzothiophenes, and thioxanthenes; this type of sulfur cannot be determined by present analytical procedures. Some recent work on desulfurization by treatment with NaOH and with steam and oxygen indicates that up to 40% of the organic sulfur is amenable to removal by these

[406] See Ref. 64, p. 171.

[407] S. Akhtar, A. G. Sharkey, J. L. Shultz, and P. M. Yavorsky, *Prepr., Am. Chem. Soc., Fuel Div.*, **19**, 207 (1974).

[408] J. Aitken, T. Heeps, and W. Steedman, *Fuel*, **47**, 353 (1968).

[409] K. D. Gunderman, G. Hoffman, N. Schenck, and G. Fiedler, *Erdoel. Kohle, Erdgas, Petrochem. Brennst.-Chem.*, **25**, 58 (1972).

[410] G. Angelova, *Freiberger Forschungsh.*, **A141**, 97 (1960).

[411] G. Angelova and K. I. Syskov, *Izv. Akad. Nauk SSSR, Otd. Tekh. Nauk, Met. Topl.*, No. 5, 153 (1959).

[412] A. D. Baranskii, L. E. Rodionova, and N. B. Druzhinina, *Izv. Nauchn.-Issled. Inst. Neftl-Uglekhim. Sin. Irkutsk. Univ.*, **8**, 39 (1966) [Ref. *Zh. Khim.*, **12**, P30 (1958)].

[413] G. K. Angelova, *C.R. Acad. Bulg. Sci.*, **13**(5), 539 (1960).

[414] L. Wnekowska, *Pr. Gl. Inst. Gorn. Komun.*, No. 325 (1963).

[415] E. N. Prilezhaeva, N. P. Federovskaya, L. V. Miesserova, O. N. Domanina, and I. M. Khaskina, *Tr. Inst. Goryuch. Iskop. Akad. Nauk SSSR*, **21** 159–168 (1963).

[416] M. S. Iyengar, S. Guha, and M. L. Beri, *Fuel*, **39**, 235 (1960).

[417] M. M. Roy and S. K. Mitra, *Erdoel Kohle, Erdgas, Petrochem. Brennst-Chem.*, **25**(6), 343 (1972).

Table 8.20 Nature of Organic Sulfur (S_{org}) in Assam Coal[416]

Source of Sample	Total S_{org} in Original Coal (%, dmf)	S_{org} Eliminated During Oxidation as H_2S, SO_2, etc. (%)	S_{org} (as thiophenol) Oxidized to $—SO_3H$ (%, dmf)	S_{org} (as thiophene) Resistant to Oxidation (%, dmf)
Namdung	2.5	0.4	2.1	Nil
Laitrygnew	3.0	0.6	2.4	Nil
Margharita	3.1	0.9	2.2	Nil
Moulong	3.6	0.9	2.1	0.6
Mamlu Hills	5.2	2.2	2.4	0.5
Baragolai	6.0	2.7	2.3	1.0
Tikak	6.0	2.6	2.4	1.0

treatments, neither of which will remove sulfur from dibenzothiophene.[418,419]

8.15 DESULFURIZATION OF COAL

The initial interest in removal of sulfur from coal was to make the coal more suitable for metallurgical coke production, and much effort was put into pyrolytic methods of desulfurization. These include heating coal by itself or with reducing gases such as hydrogen or carbon monoxide and steam, with air or oxygen, air and steam, or with additives such as $AlCl_3$,[420] NH_4Cl, $KClO_4$,[421] or NaOH.

Extraction of pyritic sulfur from British coal for analytical purposes by boiling 2*N* HNO_3 is not always complete when 72-mesh coal is used. The degree of attack is related to rank, and the amount unattacked may be significant. When ground to pass a 300-mesh sieve, extraction is essentially complete for all coals.[422] Similar incomplete extraction of pyritic sulfur by 18–20% HNO_3 has been reported for Soviet coals.[423] Microscopic examination of coal treated with 5–20% HNO_3 at 20–130°C showed the presence of elementary sulfur formed from pyrite.[424] Attempts to extract pyrite from an hvAb coal with a variety of pyrite-dissolving aqueous reagents at temperatures not exceeding 100°C were relatively unsuccessful. Refluxing for 6 hr in a solution containing KNO_3 and KCN removed 38% of the pyritic sulfur. Solutions of KNO_3 with oxalic acid, $SnCl_2$ with HCl, and $Ca(OCl)_2$ were less effective.[425] The extraction of up to half of the organic sulfur from coal by *o*-chlorophenol, *p*-chlorophenol, *o*-cresol, and nitrobenzene has been claimed[426] on the basis of changes in total sulfur analyses on coal after treatment. These apparent decreases in sulfur values have been found to be due to dilution of the coal by adsorbed and reacted solvent.

8.15.1 Desulfurization by Pyrolytic Methods

Heating coal to coking temperatures results in evolution of gaseous components containing sulfur, such as H_2S and CH_3SH, and results in a decrease in the sulfur content of the residual char. Much of the sulfur lost is derived from the pyrite, but at higher temperature (1100°C) a decrease in organic sulfur also occurs.[427] Addition of $AlCl_3$ is reported to be the most effective way to remove sulfur during carbonization, but this approach is believed to be impractical because of corrosion problems. Simple pyrolysis, though less efficient, was considered the most practical desulfurization method.[427]

At about 500°C pyrite is known to dissociate into FeS and S, and some of the elemental sulfur becomes organically bound into the resulting coke. In an attempt to prevent the retention of this sulfur, an hvAb coal was carbonized at 600°C in the presence of reagents expected to produce volatile sulfur compounds from the elemental sulfur. Addition of 10% of ammonium chloride, sodium borohydride, *p*-terphenyl, or pyromellitic dianhydride produced chars containing 1.3–1.6% total sulfur, compared to 2.16% in the absence of additives.[425]

8.15.2 Desulfurization of Coal with Alkali

Treatment of bituminous coal with molten alkali (KOH/NaOH) at temperatures between 200 and 400°C, followed by aqueous washing, results in removal of all the pyritic and half the organic sulfur. The pyrite removal occurs at the lower temperature, with the extent of organic sulfur removal increasing with increasing temperature.[428] Molten Na_2CO_3, at

418 S. Friedman and R. P. Warzinski, *Trans. ASME, J. Eng. Power,* **99A,** 361 (1977).
419 S. Friedman, R. B. LaCount, and R. P. Warzinski, *Am. Chem. Soc. Symp. Ser.,* **64** 164 (1977).
420 A. F. Krasyukov, S. M. Slutskaya, and A. V. Tsin'ko, *Tr. Bashkirsk. Nauchn.-Issled. Inst. Pererab. Neft.,* No. 4, 76 (1960).
421 A. Z. Kulishenko and K. P. Medvedev, *Koks Khim.,* No. 7, 5 (1960).
422 A. H. Edwards, J. M. Jones, and W. Newcombe, *Fuel,* **43**(1), 55 (1964).
423 V. S. Kaminskii and S. Ya. Laites, *Khim. Tekhnol., Topl. Masel,* **4**(7), 27 (1959).
424 E. I. Schmuk, *Izv. Akad. Nauk SSSR, Otd. Tekh. Nauk,* No. 8, 126 (1957).
425 P. H. Given and J. R. Jones, *Fuel,* **45,** 151 (1966).
426 R. A. Meyers, British Pat. 1,298,175, November 29, 1972.
427 V. V. Kazmina, *Coke Chem.,* No. 6, 25 (1971).
428 P. X. Masciantonio, *Fuel,* **44**(4), 269 (1965); U.S. Pat. 3,166,483, January 19, 1965.

850–1100°C, followed by aqueous $NaHCO_3$ wash, removes much of the sulfur from coal.[429]

A process has been described[430] that removes 75% of the sulfur from 4% sulfur coal by heating with 20% aqueous Na_2CO_3 solution in a series of steps. The first step, at 250°C in the presence of air and hot flue gas, removes the water to form a "causticized" coal, which undoubtedly has been partially oxidized by the air. The second step, in the presence of steam, water gas, and hydrogen sulfide, is carried out at 450°C. The third step in the process allows the causticized char to be contacted with steam and hot flue gas for 30 min at 775°C. The final step involves a sodium stripping bed, where the char is heated to 900°C with superheated steam and flue gas.

When coal is heated with excess aqueous NaOH at 225 °C at the equilibrium pressure of the steam, followed by a dilute acid wash with H_2SO_4, HCl, or aqueous SO_2, extensive demineralization occurs to produce a coal containing as little as 1% residual inorganic material. This reduces the pyritic sulfur to 0.1% or less. Neutralization with CO_2 does not reduce the mineral content.[431] When the heating is carried out at 300°C, elimination of as much as 40% of the organic sulfur occurs.[418,419] In a variation of the process, $Ca(OH)_2$ is added with the caustic.[432] In one low-sulfur coal 70% removal of organic sulfur is claimed. Without the acid treatment, much of the sodium remains in the coal. The use of $Ca(OH)_2$ reduces sodium retention, probably by leaving calcium in its place. The process requires recycle of the sodium for economic practicality. There is evidence that it will not be possible to make the process economically viable.[433] The coal appears to lose none of its coking qualities at the lower temperature (225°C), but it becomes noncoking by 300°C treatment.[418]

[429] P. A. Lefrancois, K. Barclay, and J. P. Van Nook, U.S. Pat. 3,567,377, August 12, 1968.

[430] H. C. Messman, *Chem. Technol.*, No. 2, 114 (1971).

[431] L. Reggel, R. Raymond, I. Wender, and B. D. Blaustein, *Prepr., Am. Chem. Soc. Fuel Div.* **17**(1), 44 (1972); U.S. Pat. 3,993,455, November 23, 1976.

[432] E. P. Stambaugh, *Coal Age*, **80**(9), 72 (1975).

[433] R. R. Oder, L. Kulapaditharom, A. K. Lee, and E. L. Edkholm, *Am. Min. Congr. J.*, **33**(8), 42 (1977).

More than half the sulfur in a high-sulfur Assam (India) coal could be removed by air oxidation in the presence of Al_2O_3 or of Al_2O_3 plus ZnS.[434]

Pyritic sulfur has been removed from coal by bacterial action. Incubation of acidophilic *Ferrobacillus ferroxidans* with coal resulted in removal of 80% or more of the pyritic sulfur from coal in 3–4 days. The relative ease of desulfurization decreased with rank from bituminous to lignite, and increased with acidic pretreatment[435] and fineness of particle size. *Thiobacillus thioxidans* in the presence of an acidic medium was found to remove 52.5% of pyritic sulfur in 6 weeks from a Taiwanese coal.[436]

8.15.3 Desulfurization with Gases

Various attempts have been made to desulfurize coal by treatment with gases that would react with either pyrite or the organosulfur compounds to form volatile sulfur-containing products. Steam and air have been used by several investigators to remove pyritic sulfur.[437–439] Temperatures used have ranged from 350 to 550°C. Optimum sulfur removal was about 30%. Some organic sulfur was eliminated as H_2S at temperatures below 300°C.[440] Steam at 600°C removed as much as 61% of the total sulfur from a bituminous coal, including 25% of the organic sulfur in 15 min.[441] This process resulted in weight loss of 28% in the coal. Air was the most effective at 450°C, removing 38% of the total sulfur, including 20% of the organic sulfur; weight loss of the coal was 32%. Treatment with either nitrogen or hydrogen required

[434] M. S. Iyengar and J. L. Ghose, *Proc. Symp. Chem. Oil Coal,* Central Fuel Research Institute, Dhanbad, India, 1969, p. 531.

[435] M. P. Silverman, M. H. Rogoff, and I. Wender, *Fuel*, **42,** 113 (1963).

[436] C. M. C. Andnew and H.-h. Wang, *K'uang Yeh Chi Shu*, **11**(8), 8 (1973).

[437] A. Z. Yurovskii, M. M. Kifshits, and N. V. Milfort, *Koks Khim.*, No. 11, 14 (1938).

[438] I. Blum, I. Nistor, V. Cindea-Panaiteseu, and B. Ionescu, *Rev. Electrontech. Energet., Acad. Rep. Pop. Rom. Ser. B,* **8** (1), 113 (1963).

[439] I. Blum and V. Cindea, *Acad. Rep. Pop. Rom., Inst. Energet., Stud. Cercet. Energet.*, **11,** 325 (1961).

[440] V. Cindea-Muntean, *Acad. Rep. Pop. Rom. Stud. Cercet. Met.*, **8,** 331 (1963).

[441] S. S. Block, J. B. Sharp, and L. J. Darlage, *Fuel,* **54,** 113 (1975).

much higher temperature for effective desulfurization, with best results at 900°C. Hydrogen at 900°C removed up to 86% of the total sulfur; the weight loss of the coal was approximately 40%. Oxidative pretreatment for 10 min at 30°C with air, or with 25% HNO_3, increases the rate of sulfur removal by hydrogen, resulting in as much sulfur removal in 4 min as in 60 min without pretreatment. This accelerated removal is believed due to opening up of the pore structure of the coal. Steam at 600°C has been reported to remove 95% of pyritic sulfur and 60–70% of the organic sulfur for an overall sulfur reduction of 80–85%.[442] Air at 450°C removed more than 90% of the pyritic sulfur from some coals in as little as 10 min.[443] The efficiency of removal was felt to be controlled by the diffusion of oxygen into the coal and pyrite. Treatment of coal with steam and carbon monoxide alone, with carbon monoxide, or with nitrogen is less effective than air treatment for sulfur removal.[444]

Below 550°C hydrogen removes very little sulfur from coal.[445] At 600°C pyrite is completely converted by hydrogen to H_2S and FeS, which remains stable toward further desulfurization until 870°C. Removal of organic sulfur from Kentucky No. 9 coal by hydrogen[446] begins at 200°C and increases with increasing temperature, following a sigmoid curve until at 1000°C, 7% of the original organic sulfur still remains. The presence of H_2S in the gas stream hinders the desulfurization at higher temperatures because of equilibria between the H_2S and the desulfurized species.[446,447]

A nonisothermal kinetic study of coal hydrodesulfurization concludes that there are five successive kinetic processes involved.[448] In order of increasing temperature of observation, these are designated as Organic I, Organic II, pyrite, sulfide, and Organic III. This mechanism indicates that H_2S evolution is observed as a result of reaction of hydrogen with two distinct species of organic compounds at lower temperature, followed by a two-step hydrodesulfurization of pyrite, and finally the hydrogenolysis of a third type of organic sulfur compound.

Hydrogen treatment of coal containing an equal volume of CaO at 315–427°C resulted in elimination of substantially all the pyritic sulfur and 34% of the organic sulfur.[449] The use of the CaO considerably decreases the temperature at which hydrogenation of the FeS occurs. The CaS formed is regenerated by reaction with CO_2 and calcining, and the H_2S can, with difficulty, be recovered for sulfur production.

Manganese dioxide, in the form of an ore, has been used as an H_2S acceptor in hydrodesulfurization of coal.[450] The MnS is separated by air classification and reoxidized to MnO_2 for recycle. A bottom ash from a producer gas generator has been used as an H_2S acceptor during hydrogen treatment of a Kentucky coal at 870°C.[451]

8.15.4 Oxidative Removal of Sulfur from Coal

There have been numerous approaches to removal of sulfur, especially pyritic sulfur, from coal by use of oxidizing agents. An extensive study of oxidation of pyrite with dilute nitric acid has been carried out.[452] Between 20 and 90°C conversion of FeS_2 to $Fe_2(SO_4)_3$ increased with temperature and with HNO_3 concentration as expected. In the case of Donbas Soviet coking coals, almost all the pyritic sulfur is oxidized and 2–4% of the mineral matter removed by treatment with 6% HNO_3.

Coal has been treated with chlorine in

[442] H. H. von Hoessle and S. A. A. Quadri, *Pak. J. Sci. Ind. Res.*, **5**, 119 (1962).
[443] R. K. Sinha and P. L. Walker, Jr., *Fuel*, **51**, 125 (1972).
[444] R. K. Sinha and P. L. Walker, Jr., *Fuel*, **51**, 329 (1972).
[445] M. L. Vesetal and W. H. Johnston, *Prepr., Am. Chem. Soc., Fuel Div.*, **14**(1), 1 (1970).
[446] P. S. Maa, C. R. Lewis, and C. E. Hamrin, Jr., *Fuel*, **54**, 62 (1975).
[447] C. A. Gray, M. E. Sacks, and R. T. Eddinger, *Ind. Eng. Chem. Prod. Res. Dev.*, **9**, 357 (1970).
[448] A. L. Yergey, F. W. Lampe, M. L. Vestal, A. G. Day, G. J. Fergusson, W. H. Johnston, J. S. Snyderman, R. H. Essenhigh, and J. E. Hudson, *Ind. Eng. Chem., Process Des. Dev.* **13**, 233 (1974).
[449] B. Lee, and F. C. Schora, Jr., U.S. Pat. 3,640,016, February 8, 1972.
[450] J. D. Batchelor, G. P. Curran, and E. Gorin, U.S. Pat. 2,927,063, March 1, 1960.
[451] C. E. Hamrin, Jr., and P. S. Maa, *Fuel*, **54**, 288 (1975).
[452] A. Z. Yurovskii *Sulfur in Coals*, transl. for U.S. Dept. Interior by Indian National Sci. Documentation Center, New Delhi, 1974, pp. 355–385 (Russian edition publ. 1960).

aqueous suspension and with 3% aqueous H_2O_2 in attempts at desulfurization.[453] Both procedures remove pyritic sulfur. Chlorine treatment adds chlorine to the coal and has an unfavorable effect on coking properties. Hydrogen peroxide treatment does not change caking properties.

An Illinois bituminous coal has been treated with chlorine in suspension in 1,1,1-trichloroethane at 74°C for 1–4 hr, followed by hydrolysis with water of the filtered coal. Dechlorination was carried out at 340°C to produce a product that contained 2.6% chlorine but had its sulfur content reduced from 4.8 to 1.2%, a 70% reduction. Both pyritic and organic sulfur were reduced by 70–80%[454]

In the initial chlorination, which is very fast for the first half-hour with considerable decrease in rate thereafter, as much as 23% of chlorine is added to the coal. Hydrolysis removed about half the added chlorine, and the pyrolytic step removed all but 2–3% of the chlorine. Reactions proposed to explain the elimination of organic sulfur are as follows:

$$R{-}S{-}R' + Cl_2 \xrightarrow{H^+} RSCl + R'Cl$$

$$R{-}S{-}S{-}R' + Cl_2 \xrightarrow{H^+} RSCl + R'SCl$$

$$RSCl + Cl_2 \rightarrow RCl + SCl_2$$

$$RH + S_2Cl_2 \xrightarrow{slow} RS_2Cl + HCl$$

where R and R′ represent hydrocarbon groups. The resulting chlorinated organosulfur compounds can be hydrolyzed and oxidized in the presence of chlorine at a moderately high temperature to produce sulfates:

$$RSCl \xrightarrow{H_2O} RSO_2Cl \xrightarrow{H_2O,\ Cl_2} RCl + SO_4^{2-}$$

The sulfate formation occurs in the hydrolysis step.

An early attempt by Yurovskii[455] at removing pyritic sulfur by chemical reaction was based on the oxidation of FeS_2 by aqueous $Fe_2(SO_4)_3$ to give ultimately $FeSO_4$ and sulfur. This worker found that the reaction was retarded by mineral matter in the coal, especially by Al_2O_3, and by the iron sulfate itself, which apparently was deposited on the pyrite. This latter problem could be overcome by repeated washing with water. More recently[456,457] the reaction has been developed into a process that involves heating a coal-water slurry containing $Fe_2(SO_4)_3$ to approximately 100°C, holding at 130°C for extended periods of time—up to 23 hr—and including one or more changes of leach solution. The pyrite is oxidized to $FeSO_4$ and sulfur, and the oxidant, $Fe_2(SO_4)_3$, is reduced to $FeSO_4$. Elemental sulfur is removed by extraction with toluene or acetone, or by steam distillation following washing of the coal to remove the iron sulfates. The $FeSO_4$ is air oxidized to regenerate $Fe_2(SO_4)_3$, which is then recycled. The stoichiometry reported leads to the following overall reaction:

$$FeS_2 + 2.4O_2 \rightarrow 0.6FeSO_4 + 0.8S + 0.2Fe_2(SO_4)_3$$

The ferric sulfate indicated in this equation is in excess of the amount required for recycle. Pyrite removal by this process is around 90–95%, and fuel value loss is insignificant.

Oxidative desulfurization has been carried out using air and water under pressure.[418,419,458] At 150–200°C and at pressures between 300 and 1000 psi, essentially all the pyritic sulfur was removed from a series of coals; in addition, as much as 40% of the organic sulfur was eliminated. The sulfur in this case was converted to sulfuric acid. Fuel loss values are 10% or less. Similar sulfur removals from coal took place with air under pressure (up to 1000 psi) in the presence of such hydrocarbons as tetralin, decalin, and cyclohexane.[419] It is presumed that the hydrocarbons formed hydroperoxides, which served as oxygen donors. The reaction of oxygen at 300 psi on coal in aqueous slurry at 130°C has

[453] S. Mukai, Y. Araki, M. Konishi, and K. Otomura, *Nenyro Kyokai-shi,* **48,** 905 (1969).

[454] P. S. Ganguli, G. C. Hsu, G. R. Gavalas, and S. H. Kalfayan, *Prepr., Am. Chem. Soc., Fuel Div.* **21,** 118 (1976).

[455] Ref. 452, p. 385.

[456] R. A. Meyers, *Hydrocarbon Process.,* **54**(6), 93 (1975).

[457] J. W. Hamersma, M. L. Kraft, and R. A. Meyers, *Am. Chem. Soc. Symp. Ser.,* **64,** 143 (1977).

[458] J. Thomas and A. Warshaw, U.S. Pat. 3,-824,084, July 16, 1974.

been reported to oxidize the pyritic sulfur to sulfate. No loss of organic sulfur was reported.[459] When the reaction was carried out in the presence of 0.5–4.0*M* NH_4OH, up to 25% of the organic sulfur was removed with the pyritic sulfur.[460]

Nitrogen oxides and air react with coal at ambient pressure at 90°C followed by water and an NaOH wash[461] or under pressure (up to 1000 psi) at 150–200°C in the presence of an inert diluent (benzene) to effect removal of pyritic sulfur and some organic sulfur. Nitration of the coal is reported to occur.[418]

Hydrogen peroxide in H_2SO_4 has been used to remove pyritic sulfur from coal,[462] and in acetic acid to remove both pyrite and 50% of the organic sulfur.[463]

8.15.4.1 *Miscellaneous*

A patent has been issued[464] that combines chemical and magnetic methods for the removal of pyrite from coal. It is claimed that treatment of coal with iron pentacarbonyl vapor results in preferential decomposition of the carbonyl on the surfaces of pyrite and other minerals, increasing their magnetic susceptibilities. The treated coal is then passed through a magnetic separator, which removes much of the pyrite and mineral particles that have been made sufficiently magnetic.

8.15.4.2 *Hydrogenolysis*

Although hydrogenolytic liquefaction of coal has been studied for more than 50 years, only since about 1970 has the sulfur content of the liquid product been of major interest. It has been observed that liquefaction almost always results in elimination of a good part of the sulfur of coal. Since liquefaction usually involves separation of mineral matter by some physical separation technique, essentially all pyritic sulfur is removed. Liquefaction involves treatment of coal with hydrogen, with carbon monoxide and steam, or with hydrogen donor solvents at elevated pressures, and results in some hydrodesulfurization of organic sulfur. The most extensive desulfurizations reported produce a liquid product containing 0.2–0.3% sulfur and are carried out at 400–450°C in the presence of hydrodesulfurization catalysts such as cobalt molybdate.

8.16 ON THE ORIGIN OF SOME ORGANIC STRUCTURES IN COAL

The papers purporting to deal with the elucidation of coal constitution are not only very numerous, but they emanate from every country that has or has had significant coal reserves. As has been pointed out, coal is an organic rock containing a number of organic materials and mineral matter involving almost every element in the periodic table. In addition, coals found in different countries vary considerably from each other. Thus British coals are entirely bituminous coals, with low sulfur contents and, occasionally, comparatively high chlorine contents. Dutch coals are mostly bituminous. Although Germany has large deposits of brown coals, its scientists have devoted most of their study of coal constitution to work on bituminous coals.

The reader interested in coal constitution is advised to prepare by studying, at the least, the following three publications: Dryden's (Great Britain) critical chapter "Chemical Constitution and Reactions of Coal," which appeared in the first Supplementary Volume[465]; van Krevelen's (the Netherlands) comprehensive and readable book *Coal*[466]; and Given's excellent chapter "Towards an Understanding of the Chemical Structure of Coal," which appeared in the second edition of Francis' book *Coal*.[467] Given worked on

[459] J. C. Agarwal, R. A. Giberti, P. F. Irminger, L. F. Petrovic, and S. S. Sareen, *Min. Congr. J.*, **31**(3), 40 (1975).

[460] S. S. Sareen, *Am. Chem. Soc. Sym. Ser.*, **64**, 173 (1977).

[461] A. F. Diaz and E. D. Guth, U.S. Pat. 3,909,-211, September 30, 1975.

[462] E. B. Smith, *Prepr., Am. Chem. Soc., Fuel Div.*, **20**(2), 140 (1975).

[463] S. Friedman, unpublished work.

[464] J. K. Kindig and R. L. Turner, U.S. Pat. 3,938,966, February 17, 1976.

[465] I. G. C. Dryden, in *Chemistry of Coal Utilization*, Supplementary Volume, H. H. Lowry, Ed., John Wiley and Sons, New York, (1963), Ch. 6.

[466] D. W. van Krevelen, *Coal, Typology-Chemistry-Physics-Constitution*, Elsevier Publishing Company, New York (1961).

[467] P. H. Given, in *Coal, Its Formation and Composition*, W. Francis, Ed., Edward Arnold Ltd. London, 1961, Ch. 12.

English coals at the British Coal Utilization Research Association and then assumed a post at the Pennsylvania State University, where he has been involved in work on the whole range of American coals. However it is well to be aware that almost all the conclusions about coal constitution have been deduced from work on bituminous coals. Information on the constitution of subbituminous coals and of lignites is sparse.

It is unfortunate that a great many of the publications on coal structure only serve to confuse, dealing all too often with studies on poorly defined coal specimens and leading to conclusions about coal constitution that, on inspection, are not warranted.

The number of publications on this subject has grown enormously during the last few years, as funds for coal research have increased markedly. Many new researchers, on the upward slope of the learning curve, have initiated research on coal. One wonders whether critical discussion is possible with the present spate of publications—but the future is bright, as the newer researchers grapple with and overcome the obstacles to meaningful contributions leading to the elucidation of the constitution of coal.

Pérhaps the biggest block to progress in this field is that most scientists are accustomed to working with homogeneous specimens of good purity. There is no such thing as "pure" coal. Indeed, Given[467] has posed the question, "Has coal a structure?" Any specimen of coal, no matter how carefully chosen, differs from any other specimen, even one picked from the same seam. And a sample of coal from a high-sulfur, high-ash coal seam bears little resemblance to a lignite sample from North Dakota, a subbituminous coal from Montana, a coking coal from West Virginia, or a bituminous coal from the Ruhr Valley. And the ease of conversion of these coals to gases and to liquids reflects the differences not only of the organic coal arrangements, but of the particular kinds and amounts of inorganic constituents present.

It is also reasonable to suppose that the basic chemical structure of the various members of the metamorphic series differs, as it passes from peat, through lignite, subbituminous and bituminous coals, to anthracites. A single basic type of structure could not persist through so diverse a range of materials. But an understanding of the progenitors of coal furnishes evidence for the presence of certain common structures in coal.

Coal is derived from plants, which are composed of celluloses, lignins, flavonoids, lipids, proteins, tannins, resins, and so on. The average lignin content of the higher plants varies between 10 and 25%, and that of woody tissue can reach 35%.[468] The cellulose content varies from 20 to 55%; the concentration of the other classes of compounds is small.

Of all the plant constituents, carbohydrates are generally believed to be degraded by microorganisms most rapidly. The cellulose/lignin ratio in peats is probably considerably less than that in fresh plants, but there is little doubt that some polysaccharides persist. Raj[469] has pointed out that polysaccharides yield aromatic chars at relatively low temperatures. The sugar "monomer" presumably rearranges so that the sixth carbon atom replaces oxygen in the cyclic sugar structures; the product is then dehydrated to phenols:

$$C_6H_{12}O_6 \xrightarrow{-3H_2O} C_6H_3(OH)_3$$

Thus carbohydrates may well be precursors to phenolic structures in coals.

It appears likely, however, that lignin is the source of a major fraction of most peats, hence of a major fraction of the materials available for conversion into coal by metamorphic processes.[470] Higher plants contain a wide range of phenolic substances, and lignin is the most abundant substance in the higher plants; it also is insoluble in all solvents.

Bimer and co-workers[471] have pointed out

[468] F. E. Brauns and D. A. Brauns, *The Chemistry of Lignin,* Academic Press, New York and London, pp. 157–167, 577–615, 1960.

[469] S. Raj, "A Study of Some Chemical Structural Parameters of Coal," Ph.D. thesis, Pennsylvania State University, 1976.

[470] P. H. Given and C. R. Dickinson, in *Soil Biochemistry,* Vol. 3, E. A. Paul and A. D. McFarlen, Eds., Marcel Dekker, New York (1975), Ch. 4.

[471] J. Bimer, P. H. Given, and S. Raj, *Prepr., Am. Chem. Soc., Div. Fuel Chem.,* **22**(5), 169 (1977); *Am. Chem. Soc. Sym. Ser.,* **71,** 86 (1978).

that vitrinite, the principal maceral in most coals, represents coalified wood; hence lignin should be one of the important precursors to the vitrinites in coals. They accordingly set out to determine whether any chemical "fossils" related to lignin could be found in coals. Their experimental approach exploited a reaction, reductive degradation with sodium amalgam and hot water, that Burges et al.[472] applied to soil humic acids. Thin-layer chromatography of the ether-soluble part of the product (yield, ~ 20%) showed the presence of a number of phenolic acids and phenols, most of whose structures bore obvious relationships to known chemical and microbial degradation products of lignin:

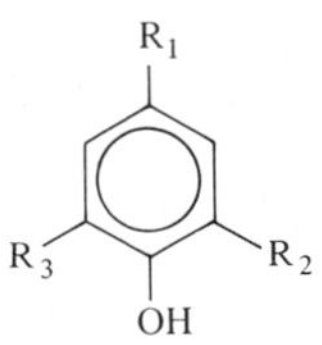

where R_1 could be —COOH, —CHO, —CH═CHCOOH, —CH_2COCOOH, and so on, and R_2 and R_3 could be —H, —OCH_3, or —OH. It is important to note that R_1 could be a side chain of one, two (rarely), or three carbon atoms, in various states of oxidation. Humic acids can be extracted from lignites but not from bituminous coals. However oxidation of bituminous coals with aqueous performic acid gives high yields (80–100% by weight) of materials that closely resemble humic acids.[473]

Raj[469] studied the constitution of humic acids obtained from peats and lignites (with 0.5*N* NaOH) and humic acids obtained from higher rank coals. Some 43 samples were studied, including coals of 78–87% C (dmmf). They were oxidized by performic acid (temperatures did not exceed 55°C), then treated (after purification) with 3% sodium amalgam according to the procedure of Burges et al.[472]

The yields of maf humic acids ranged from from 65 to 90% of the dmmf coal. It was found that the oxidation processes were confined largely to vitrinite materials. There was a marked increase in the acid yield with increasing rank (lignites up to hvAb coals). The phenols formed were converted to their trimethylsilyl ethers to make them more stable and more volatile for analysis by gas-liquid chromatography and mass spectrometry.

Although the chemistry of this procedure is not well known, it appears that many of the products identified must have been present in the starting substrates. The most striking result of the study was the ubiquity of cinnamic acid, $C_6H_5CH{=}CHCOOH$, found in 42 out of 43 materials analyzed; the one product that did not contain cinnamic acid did contain cinnamyl alcohol. In addition, substances closely related to cinnamic acid were found in some products, with the aliphatic side chain in a different state of oxidation (i.e., not as —CH═CHCOOH but as —CH_2COCH_2OH, etc.).

Cinnamic acid cannot be derived from lignin unless coalification includes mechanisms whereby a para-hydroxyl group can be removed, but this state of affairs is quite possible. Cinnamic acid commonly occurs in plants as a biosynthetic precursor to lignin, flavonoids, and phenylalanine, so it could have been present in peats.

Many of the products did contain *p*-hydroxycinnamic acid (*p*-coumaric acid). Again, similar compounds were found with the side chain in various stages of oxidation. Para-coumaric acid was found in about half the products from subbituminous and bituminous coals, but in only one of the lower rank materials. But *p*-hydroxybenzoic acid was found in almost all of the low-rank materials and in almost half of the products from the higher rank coals.

Vanillin (3-methoxy-4-hydroxybenzaldehyde), formed from 3-methoxy-4-hydroxycinnamic acid (ferulic acid) by microbial action, was found in all low-rank materials studied, in 12 of 15 samples from the Rocky Mountain Province, and in about half those from the other provinces. Ferulic acid itself, found in lignin, was present in 40% of the samples studied.

A few of the many structures of the compounds identified are shown on page 506.

[472] M. A. Burges, H. M. Hurst, and S. B. Walkden, *Geochim. Cosmochim. Acta,* **28,** 1547 (1964).

[473] P. H. Given, J. K. Brown, V. Lipton, and W. F. Wyss, *Proc. Institute Fuel Conference* "Science in the Use of Coal," Sheffield (1958), p. A-38.

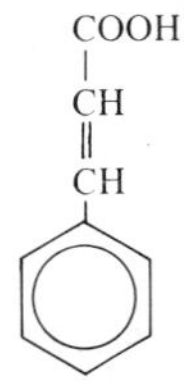

Cinnamic acid

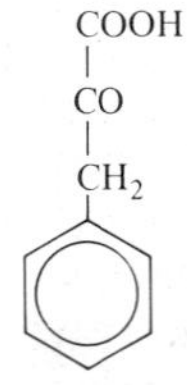

Phenyl, 3-pyruvic acid

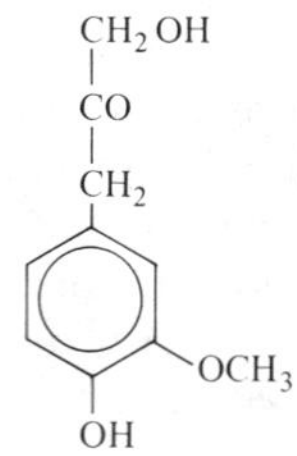

3-Methoxy, 2-hydroxyphenyl, 3-pyruvic alcohol

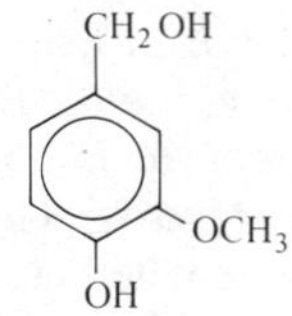

Vanillic alcohol

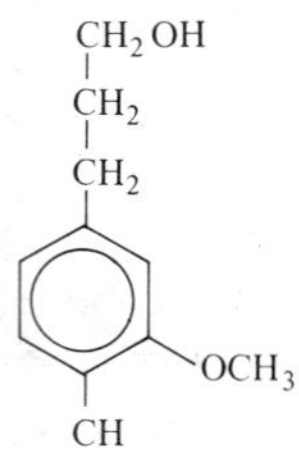

3-Methoxy, 4-hydroxyphenyl, 3-propanol

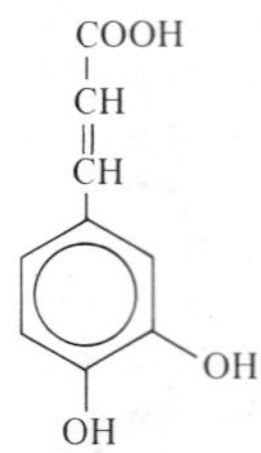

Caffeic acid

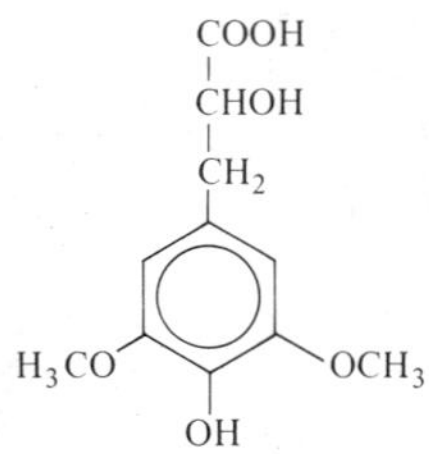

3,5-Dimethoxy, 4-hydroxyphenyl, 3-glyceric acid

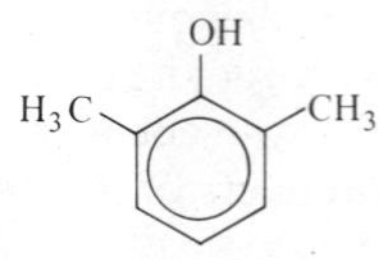

2,6-xylenol

The —COOH groups in compounds from higher rank coals were undoubtedly formed during the performic acid oxidation.

Interestingly, 2,6-xylenol was obtained from 22 of 43 samples studied. It appears to be a significant unit of many coals but no suggestion could be made about its origin.

In summary, it seems that a considerable proportion of the substances identified have clear associations with the common degradation products of the constituents of higher plants, particularly of lignin. The position isomers obtained are consistent with this conclusion, nor can methoxyl groups be introduced by oxidation. In addition the various C_3 side chains found in so many of the compounds are just what is found among the degradation products of lignin. There is little doubt that the lignin-related compounds identified are correctly described and that their carbon-oxygen skeletons existed as such in the original coals.

The presence of methoxyl groups in compounds identified from bituminous coals is at variance with the distribution of oxygen functional groups reported by Blom et al.,[474] who found methoxyl groups in lignites and in subbituminous coals but not in coals of higher rank. The internal surface area of coal available to reagents becomes less accessible with higher rank coals[475] and this could be a reason (among others) for the failure of Blom's use of the Zeisel method to detect methoxyl groups.

Given offered evidence that the structures identified in this study are very likely representative of significant features of the structures present in the vitrinitic features of coals. Some implications of this work are as follows:

1. Many benzene rings in vitrinites bear not only an —OH group but often one or two methoxyl groups as well; occasionally a second —OH group may be present instead of methoxyl. This would explain, among other things, the difficulty of sulfonating or nitrating vitrinites without accompanying oxidation.

2. As a corollary of item 1, methoxyl groups survive in vitrinites to a considerably higher level of coal rank than previously reported.

[474] L. Blom, L. Edelhausen, and D. W. van Krevelen, *Fuel,* **36,** 135 (1957).

[475] P. L. Walker, Jr., O. Cariaso, and R. L. Patel, *Fuel,* **47,** 331 (1968).

3. The structure of vitrinites, even of hvAb coals, appears to be similar, in part, to that of lignin, in being based to an important extent on phenylpropane monomer skeletons. However the nature of the connecting linkages, which must be readily broken by oxidation and/or reductive degradation, is obscure.
4. Either olefinic double bonds are abundant in vitrinite structures, contrary to previous ideas, or they are generated by the degradation processes used, in a manner difficult to envisage.

One would predict from the structures of the compounds identified that phenolic biphenyl derivatives should also be formed as degradation products; the phenol-BF_3 "depolymerization" work of Heredy and Neuworth discloses the presence of large numbers of such linkages in lignite. Lignite also has a three-dimensional structure, with numerous phenylpropane-substituted entities.[86,476] Indeed, some idea of the structure of lignite can be gained by examining a representative of the lignin molecule proposed by Ludwig and his co-workers (Fig. 8.19).

[476] I. Wender, *Catal. Rev.-Sci. Eng.*, **14,** 97 (1976).

[477] C. H. Ludwig, B. J. Nist, and J. L. McCarthy, *J. Am. Chem. Soc.*, **86,** 1196 (1964).

Bolker and Brenner,[478] using the Flory-Stockmayer theory[479] and molecular weights of lignin fractions, estimated the average length of primary chains in lignin to be 18 phenylpropane units. They found the average degree of crosslinking of intact lignin to be 0.277; that is, five out of every 18 phenylpropane units bear crosslinking benzyl-ether groups. Of the five, three are also etherified at their phenolic ends. This type of structure probably persists, to differing extents, throughout the various ranks of coal.

It is interesting that Sigleo[480] identified degraded lignin compounds in Triassic age silicified wood (~ 200 million years old) from the Petrified Forest National Park in Arizona. She found that the black carbonaceous samples of silicified wood contained up to 2.7% organic carbon. The remnants of the original wood polymers have evidently survived a 200-million-year history of diagenesis. Pyrolysis of this material yielded xylenes, trimethylbenzenes, benzofuran, methylbenzofurans, idene, phenol, cresols, xylenols, and 1-methylnaphthalene.

[478] H. I. Bolker and H. S. Brenner, *Science*, **170,** 173 (1970).

[479] P. J. Flory, *Principles of Polymer Chemistry*, Cornell University Press, Ithaca, N.Y., 1953, Ch. 9.

[480] A. C. Sigleo, *Science*, **200,** 1054 (1978).

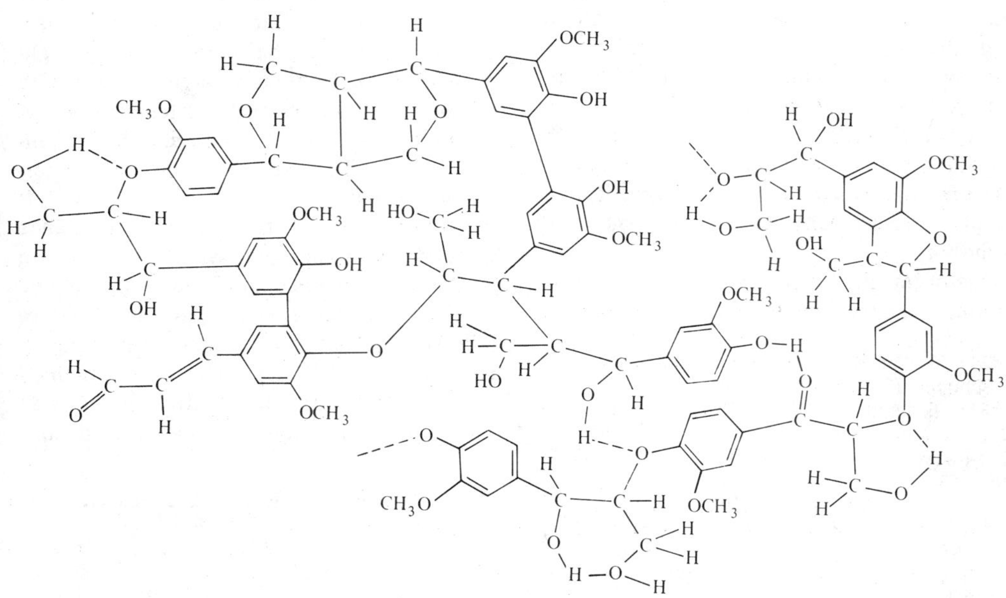

Figure 8.19 Representation of the lignin macromolecule with structural features suggested by NMR spectroscopy.[477]

8.17 THE CARBON AROMATICITY OF COAL

The carbon aromaticity of coal, generally designated as f_a and defined as

$$f_a = \frac{\text{aromatic carbon}}{\text{total carbon}}$$

has been and continues to be one of the most important, yet elusive structural parameters in coal research. The modern coal researcher, and indeed the modern organic chemist, must marvel at the ingenuity of the early workers who used such physical properties as the velocity of sound through a sample of coal[481] to estimate this very fundamental parameter. The early work on the aromaticity of coal (prior to 1961) has been reviewed by van Krevelen[466] and by Tschamler and de Ruiter[482]

The aromaticity of coal depends on the rank of the coal, its petrographic makeup, and probably the bed from which it came. When applied to a coal, the term "aromaticity" generally refers to the vitrinite or vitrain component. Although coals have generally been thought to be highly aromatic substances, this classical view of coal constitution was challenged as early as 1956[483] and as recently as 1974.[240] The main reason such controversy still persists is simply that at this writing there is no completely unambiguous method for determining the carbon aromaticity of a coal. Three more recent approaches (i.e., published after 1974) are the oxidation experiments of Chakrabartty and co-workers,[240] the reaction of coal with fluorine gas by Huston and his co-workers,[484] and the cross-polarization ^{13}C magnetic resonance measurements reported by a number of authors.[485–489]

Huston et al.[484] reported, on the basis of fluorination experiments, that the f_a value for a high-volatile Illinois coal was 0.69, in agreement with the classical view of coal structure. Fluorination of coal proceeds both by addition and by hydrogen replacement; thus these authors distinguished between the amounts of aromatic and nonaromatic carbons in the coal by determining both the total fluorination weight increase and that due to addition reactions by measuring the quantity of HF evolved. No serious disagreements with the interpretation of these results have yet appeared in the literature.

In any quantitative determination by chemical methods of the aromatic content of coal (or for that matter the amount of any functional group in coal), the possibility always exists that reactions are not as specific or as selective as proposed. The most recent approach to the problem of the carbon aromaticity of coal is a spectral one, which involves the use of newly developed NMR techniques. For coal extracts the aromaticity problem has been solved by conventional high-resolution ^{13}C NMR.[487] Typically ^{13}C spectra of whole coals show quite broad bands with little or no evidence for separate resonances due to aromatic and nonaromatic carbons. Dipolar line broadening by protons in the coals is the principal factor responsible for the broad lines. The NMR cross-polarization technique developed by Pines et al.[490] overcomes the problems of dipolar line broadening and has been used to estimate the aromaticity of a number of coals. The method utilizes high-power proton decoupling and leads to spectra consisting of two overlapping bands; one due to aromatic carbons, the other due to aliphatic ones. The spectra, being powder patterns, exhibit the effects of chemical shift anisotropy; that is, the aromatic resonance is asymmetric and somewhat broad because of the large chemical shift anisotropy of the aromatic carbons. Representative spectra (shown in Chapter 4) are much more highly resolved than ^{13}C spectra of coals obtained under ordinary broadline NMR conditions.[491] The cross-polarization spectra can be decon-

[481] D. W. van Krevelen, H. A. G. Chermin, and J. Schuyer, *Fuel*, **38**, 483 (1959).

[482] H. Tschamler and E. de Ruiter, in *Chemistry of Coal Utilization*, Supplementary Volume, H. H. Lowry, Ed., John Wiley and Sons, New York, 1963, Ch. 2.

[483] R. A. Friedel, *Nature*, **179**, 1237 (1957).

[484] J. L. Huston, R. G. Scott, and M. H. Studier, *Fuel*, **55**, 281 (1976).

[485] D. L. VanderHart and H. L. Retcofsky, *Fuel*, **55**, 202 (1976).

[486] H. L. Retcofsky and D. L. VanderHart, *Fuel*, **57** 421 (1978).

[487] H. L. Retcofsky, *Appl. Spectrosc.*, **31**, 116 (1977).

[488] V. J. Bartuska, G. E. Maciel, J. Schaefer, and E. O. Stejskal, *Fuel*, **56**, 354 (1977).

[489] A. Pines and D. E. Wemmer, Prepr. 1976 Coal Chemistry Workshop, Stanford Research Institute, Menlo Park, Calif., August 26–27, 1976, p. 229.

[490] A. Pines, M. G. Gibby, and J. S. Waugh, *J. Chem. Phys.*, **56**, 1776 (1972).

[491] H. L. Retcofsky and R. A. Friedel, *J. Phys. Chem.*, **77**, 68 (1973).

voluted to give the relative intensities of the aromatic and nonaromatic carbon resonances. This ratio corresponds to the ratio of aromatic to nonaromatic carbons in the sample.

A new dimension in resolution in cross-polarization NMR spectra of coals was achieved by spinning the sample rapidly (3 kHz) at the so-called magic angle with respect to the magnetic field.[488] Magic angle spinning reduces the anisotropic chemical shift contribution to the spectrum to zero, thus yielding isotropic chemical shifts similar to those observed in liquid samples. Representative spectra using this technique appear in Fig. 8.20. As shown in this figure, the magic angle spinning does indeed narrow the spectrum sufficiently, so that the aromatic and aliphatic carbon resonances are completely separated. Furthermore, the 3-kHz sidebands generated by the spinning (most clearly seen in the anthracite spectrum of Fig. 8.20*a*) do not interfere with signals in either region. Although there is some skewing, or asymmetry, of the aromatic carbon line of lignite under magic angle spinning conditions (Fig. 8.20*b*, lignite spectrum), no well-resolved structure is observed. The asymmetry is due to a dispersion of isotropic chemical shifts, reflecting the diversity of the chemical structure of this particular coal.

Although a number of problems have been posed that might invalidate the quantitative reliability of the cross-polarization NMR technique, most of these have been satisfactorily solved. The new NMR techniques offer the most reliable method for determining the carbon aromaticity of coals. More recent results for a number of coals, as well as petrographic components of one coal, are presented in Table 8.21. As can be deduced from the table, coals are highly aromatic materials, and their aromaticity increases more or less gradually with coal rank in agreement with the classical view of coal metamorphism. In the case of the macerals from an hvAb coal, the f_a values decrease in the order fusinite > micrinite ~ vitrinite > exinite.

8.18 THE MACROMOLECULAR STRUCTURE OF COAL

Coals of any rank up to anthracite are partially soluble in certain solvents. Depending on the character of both the coal and the solvent, the resulting solution is either largely molecular or largely colloidal. It has been shown by several authors that the infrared spectra of the extracted materials bear a striking resemblance to those of the original coal and also to those of the insoluble extraction residue. This similarity has been taken as an indication of a polymeric character. Studies on coal genesis and various methods of chemical coal degradation also suggest that coal has a macromolecular structure.[492] The next few paragraphs review this aspect of coal structure.

Some basic considerations regarding a polymeric structure of coal have been published

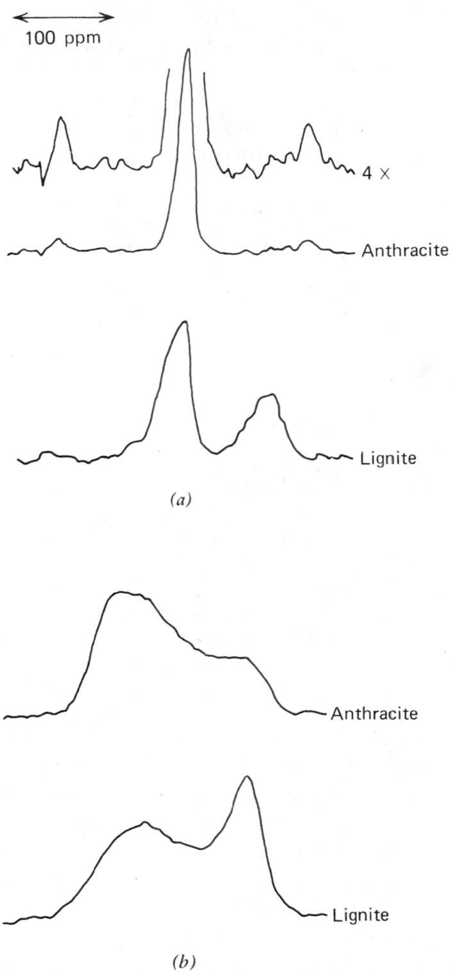

Figure 8.20 Cross-polarization ^{13}C NMR spectra of two types of coal obtained (*a*) with and (*b*) without 3 kHz mechanical spinning at the magic angle.[488]

[492] Ref. 328, Ch. 23.

Table 8.21 Aromaticities of Selected Coals and Petrographic Components as Determined by Cross-Polarization ^{13}C NMR[a]

Coals[b]	Carbon (%, maf)	Aromaticity, f_a
Unspecified anthracite[a]	?	0.99 0.94[d]
Dorrance anthracite	92.8	>0.96
Pocahontas No. 4 lvb	90.4	0.89
Lower Banner mvb	86.1	0.81
Powellton hvAb	85.1	0.80
Pittsburgh hvAb	82.6	0.76
Adaville subbituminous	76.3	0.76
Unspecified lignite[c]	?	0.72 0.65[d]
Wisconsin Peat	55.0	<0.5
Macerals from Hernshaw hvAb coal[e]		
Vitrinite	85.2	0.85
Micrinite	85.9	0.85
Exinite	86.2	0.66
Fusinite	91.5	0.93–0.96

[a] Values from VanderHart and Retcofsky,[486] except as noted.
[b] Samples are vitrains or vitrain-rich samples except as noted.
[c] From Bartuska et al.[488]
[d] Magic angle spinning technique utilized.
[e] See Retcofsky and VanderHart.[486]

by van Krevelen.[493] That study compared coal to a polymer that had been formed by trifunctional stepwise polycondensation. A characteristic feature of the three-dimensional polycondensation reaction is the occurrence of insoluble "gels" at a certain stage of the condensation. The number-average molecular weight of the still-soluble part reaches a maximum at the gel point: beyond this point the average molecular weight of the still-soluble part decreases.

In applying the theory of three-dimensional polycondensation to coal genesis, one can assume that the "gel point" has been reached and exceeded at the end of the lignite stage. Bituminous coal can then be represented as a three-dimensional polymeric network with a certain amount of soluble fraction of low degree of condensation entrapped in the interstices of the polymeric network. This theory predicts that the amount of extractable matter should be in the range of 15–35% for bituminous coals and that this amount should decrease with increasing rank. The average molecular weight of the "basic unit" of the soluble material was estimated to be about 400 and the average degree of condensation for this material 2–2.5, indicating a number-average molecular weight of 800–1000 for a typical pyridine extract. These theoretically predicted extract yields and average molecular weights are in good agreement with experimental results obtained by several investigators.[494,495]

The theoretical values estimated for the degree of condensation and the molecular weight of the structural unit of the extractable material are average values. The theory, therefore, allows for the presence of a significant amount of "monomers" or units of lower than average molecular weight. It has been shown by the work of Vahrman[362] and others that these units are indeed present.

Thus the application of the theory of three-dimensional polycondensation to coal genesis helps to explain the relation between three important components of the bituminous coal; namely, the basic tridimensional network, the (pyridine-) soluble extract with a number-average molecular weight of about 1000, and the smaller molecules that can be

[493] D. W. van Krevelen, *Fuel,* **44,** 229 (1965).

[494] H. N. M. Dormans and D. W. van Krevelen, *Fuel,* **39,** 273 (1960).

[495] W. F. K. Wynne-Jones, H. E. Blayden, and F. Shaw, *Brennstoff-Chemie,* **33,** 201 (1952).

recovered by vacuum distillation of benzene extraction with a molecular weight range of 100–300.

The polymeric theory of coal genesis and coal structure is also in satisfactory agreement with many of the physical properties of coals, such as poor solubility and swelling in certain types of solvent. There are, however, a number of problems that need to be discussed before we proceed further. The first problem involves the early stage of coalification or humification. The major coal-forming plant constituents—cellulose, and particularly lignin—are themselves polymers, and there is considerable disagreement in the literature[64] regarding the possibility and the extent of degradation of these polymers during humification. According to one important coalification theory, which was proposed by Enders,[496] the formation of humic acids from carbohydrates may take place during humification, when natural conditions become less favorable for the growth of microorganisms and changes occur in the metabolism process. Whereas the normal decomposition of carbohydrates results in carbon dioxide, the changed metabolism leads to the formation of methylglyoxal; this compound condenses rapidly, first to water-soluble quinonic compounds, then to humic substances. The Enders theory of humification and coalification supports van Krevelen's application of the theory of three-dimensional polycondensation to coal genesis. It should be kept in mind, however, that the extent of degradation of the original polymeric plant materials must have varied a great deal under different coalification conditions; therefore materials with a wide range of molecular sizes could have participated in the early phases of the coalification process.

This brings us to the second problem. Although it is clear that a major part of the coal substance consists of materials of very large molecular weight, coal is certainly not an ordinary polymer, composed of the same monomer or of repeating units. Perhaps coal is best described as a macromolecular material composed of an intimate mixture of interpolymers in which a large number of basic (or "monomeric") units are randomly distributed as building blocks in the macromolecular structure.

[496] C. Enders, *Angew. Chem.*, **56**, 281 (1943).

8.18.1 On the Chemical Structure of Coal

In the preceding section the conclusion was reached that coal has a macromolecular character. An analogy was drawn between the basic coal structure and that of a tridimensional polymeric network, with a sizable amount of extractable, lower molecular weight material being contained within the porous structure of the network. It was further concluded that the network is made of "basic units" having an average molecular weight of about 400.

This section discusses current knowledge regarding the chemical nature of the basic units and the linkages that connect them.

8.18.1.1 *Chemical Structure of the "Basic Units"*

Because of the uncertainties that have attended, especially through the 1960s, the structural interpretations of the results of X-ray, infrared, and other physical measurements on coals, and because it is fundamentally impossible to specify molecular structural models or formulas for coals by other than statistical parameters, van Krevelen and his colleagues developed methods of statistical structural analysis.[497] The first of these methods was based on using elemental analysis and specific volume data of vitrinite concentrates. It was assumed that the molecular volume can be regarded as the sum of mean atomic volumes and of correction factors relating to special structural features. A similar statistical method, utilizing the heat of combustion, was developed by both van Krevelen[497] and Dryden.[498] It should be pointed out that these statistical methods dealt with the structural features of the whole coal molecule, including both the basic units and the connecting linkages. Since the coal consists mainly of the basic units, however, the results of the statistical structural calculations are applicable to these basic (monomeric) units. The results obtained with these methods are summarized in Table 8.22.

An improved estimation of the distribution of carbon and hydrogen among different structural positions was obtained by supplementing the statistical method with the fol-

[497] Ref. 328, Ch. 24.
[498] I. G. C. Dryden, *Fuel*, **40**, 433 (1961).

Table 8.22 Structural Data of Basic Units in Vitrinites[497]

Carbon (%, maf)	Number of Carbon Atoms	Hypothetical Unsubstituted Aromatic Carbon Atoms	Aromaticity (f_a)	Ring Numbers per Basic Unit: Density Method	Graphical Density Method	Heat of Combustion Method
70.5	17	12	0.70	4.4	3.6	—[a]
75.5	17	13	0.78	4.3	3.8	—[a]
80.0	19	15	0.80	—[a]	—[a]	4.6
85.0	20	17	0.85	4.5	4.7	5.2
89.0	20	18	0.88	5.4	5.2	—[a]
92.5	20	18	0.96	6.4	6.1	—[a]
95.0	50	50	1.00	18.5	18.0	17.2

[a] No results available.

lowing data: (*a*) a chemical technique for the determination of methyl groups (Kuhn-Roth procedure),[499,500] (*b*) estimates of hydroaromatic carbon derived from dehydrogenation of coals with benzoquinone,[501] and (*c*) estimates of aliphatic-to-aromatic hydrogen ratios derived from infrared spectroscopic measurements.[502] High-resolution proton NMR spectra of coal extracts were used to improve the accuracy of the infrared measurements made on the corresponding coals. Average characteristics of vitrinite-rich coals of different rank considered in these studies, calculated by these methods, are shown in Table 8.23. The distribution of carbon and hydrogen among different structural positions calculated as above is shown in Table 8.24. The carbon and hydrogen distributions are plotted in Fig. 8.21.[503]

Structural parameters for pyridine extracts of bituminous coals have been calculated by Retcofsky,[344] and for depolymerization products obtained by the phenol-BF_3 method from coals of a wide range of rank by Heredy, Kostyo, and Neuworth.[90,103] These results are summarized in Table 8.25. The proton distributions were calculated from the high-resolution proton NMR spectra of soluble products representing 20–75% of the various coals. The values of carbon aromaticity, the degree of aromatic ring substitution, and the average number of aromatic rings per condensed cluster were calculated by the method introduced by Brown and Ladner.[339] The carbon aromaticities do not vary significantly with rank; they stay within the narrow range of 0.65–0.73. The average number of aromatic rings per condensed cluster increases from about two to five over the coalification range from lignite to medium volatile bituminous coal.

Typical polynuclear and condensed aromatic systems that have been identified, primarily in coal extracts, are biphenyl, naphthalene, triphenylene, and phenanthrene (Table 8.23). More detailed information is available from the spectrometric analysis of the pyridine extract of a Pittsburgh Seam coal (hvAb) as reported by Friedel and coworkers.[504] The extract, representing 19.3% of the coal, contained 10% benzene derivatives, 17% naphthalene and indane derivatives, 31% tricyclic condensed aromatics (acenaphthenes, fluorenes, phenanthrenes, anthracenes), 25% four-ring aromatics, and 17% five-ring aromatics. In contrast, investigations of Whitehurst and co-workers,[505] based on polarographic measurements, indicate that the degree of condensation of aromatic rings in bituminous coals may be smaller, closer to two on the average.

A considerable proportion of the potential hydrogen in the condensed aromatic units is

[499] R. Kuhn and R. Roth, *Ber. Dtsch. Chem. Ges.*, **663,** 1274 (1933).
[500] R. Bent, W. K. Joy, and W. R. Ladner, *Fuel,* **43,** 5 (1964).
[501] M. E. Peover, *J. Chem. Soc.*, 5020 (1960).
[502] I. G. C. Dryden, "Coal," in *Kirk-Othmer's Encyclopedia of Chemical Technology,* Vol. 5, John Wiley and Sons, New York, 1964, pp. 640–641.
[503] I. G. C. Dryden, *Fuel,* **41,** 301 (1962).
[504] R. A. Friedel, J. L. Shultz, and A. G. Sharkey, Jr., *Fuel,* **47,** 403 (1968).
[505] D. D. Whitehurst, M. Farcasiu, T. O. Mitchell, and J. J. Dickert, Jr., EPRI AF-480, Annual Report, July 1977.

Table 8.23 Average Characteristics of Coals Rich in Vitrinite[502]

Parameter	Vitrinite	Vitrinite	Vitrinite	Vitrinite	Vitrinite	Exinite[a]	Micrinite[a]
Carbon (%, dmmf)	82.5	85.0	87.5	90.0	92.5	~v	>v
Hydrogen (%, dmmf)	5.3	5.3	5.2	4.9	4.15	>v	<v
Oxygen (%, dmmf)	10.4	7.9	5.5	3.4	2.1	<v	$\gtrsim$v
Volatile matter (%, dmmf)	39	35.5	31	24.5	13.5	>v	<v
H/C (atomic)	0.805	0.77	0.73	0.66	0.54	>v	<v
CH_2 bridges	Absent						
Type of aromatic nuclei	Biphenyl, naphthalene, triphenylene, and phenanthrene may predominate						
Average number of rings in aromatic nuclei, (R_a)	>3	~3	~3.5	~4	~5	$\lesssim$v	$\gtrsim$v
Hydrogen substitution on periphery of aromatic rings (%)	50+	45+	40	30	10−		
Average stacking number of parallel aromatic clusters	1.3	1.35	1.5	1.8	2.15		
O as hydroxyl (% of coal)	6.0	5.0	3.5	2.0	1.0	<v	
O as carbonyl (% of coal)	~2.0	~1.6	~1.2	~0.9			
Carbon atoms per free radical	~8000	~6000	~4000	~3000	~1500	<v	>v

[a] An indication is given, where known, of the difference to be expected in the corresponding exinite and micrinite, that is, whether the parameter for these is greater (> v) or less (< v) than for the vitrinite.

Table 8.24 Carbon and Hydrogen Distribution in Vitrinite-Rich Coals (British School)[502]

Parameter	Vitrinite	Vitrinite	Vitrinite	Vitrinite	Vitrinite	Exinite[a]	Micrinite[a]
Carbon (%, dmmf)	82.5	85.0	87.5	90.0	92.5		
Aromatic hydrogen/non-aromatic hydrogen	0.23	0.26	0.35	0.54	1.27	<v	>v
Methyl hydrogen/total hydrogen	0.18	0.18	0.18	0.18	(0.19)		
Methyl hydrogen/nonaromatic hydrogen	0.24	0.24	0.25	0.285	(0.44)		
Methyl carbon/total carbon	0.048	0.046	0.044	0.040	(0.035)		
H/C (nonaromatic)	1.75	1.74	1.72	1.72	1.98		
H/C (aromatic)	0.24	0.25	0.28	0.30	0.34		
Mean aromatic hydrogen atoms per aromatic ring	1.1	1.2	1.25	1.3	1.4		
Fraction of carbon as aromatic (f_a)	0.665	0.68	0.705	0.76	0.88	<v	>v
Fraction of carbon as hydroaromatic rings	0.29	0.255	0.20	0.14	0.07	>v	
Fraction of carbon as aliphatic	0.045	0.065	0.095	0.100	0.05		
chains: (a + d)/100	1.00	1.00	1.00	1.00	1.00		
Atoms of carbon per 100 C atoms							
Aromatic	66.5	68	70.5	76	88	<v	>v
CH_3(side chains) (a)	4.8	4.6	4.4	4.0	3.5		
CH_2	15.7	14.2	12.3	9.2	4.5	>v	
CH	13.0	13.2	12.8	10.8	4.0		
Total nonaromatic[b] (b)	33.5[c]	32	29.5	24	12	>v	<v
Hydroaromatic (c)	29	25.5	20	14	7	>v	
(CH + CH_2) in side chains (d)	nil	1.9	5.1	6	1.5		
Total carbon	100	100	100	100	100		
Atoms of hydrogen per 100 C atoms							
Aromatic	16	17	19.5	23	29.7	<v	>v
CH_3 (side chains)	14.5	13.8	13.2	12.2	10.5		
CH_2	31.5	28.5	24.5	18.5	9	>v	
CH	13.0	13.2	12.8	10.8	4.0		
Total nonaromatic	59	55.5	50.5	41.5	23.5	>v	<v
Hydroxyl	5.5	4.5	3.0	1.5	0.8	<v	
Total hydrogen	80.5	77	73	66	54	>v	<v
CH and CH_2 in side chains (%)	0	7	20	30	18		

[a] An indication is given, where known, of the difference to be expected in the corresponding exinite and micrinite, that is, whether the parameter for these is greater (> v) or less (< v) than for the vitrinite.

[b] Nonaromatic = hydroaromatic, alicyclic, and side-chain aliphatic.

[c] a + c + d = b.

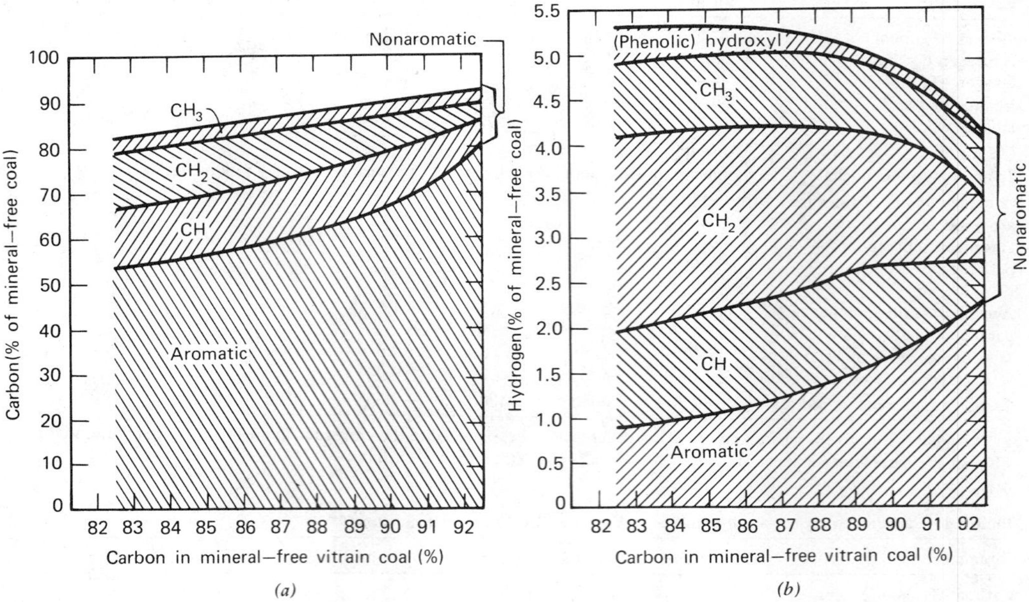

Figure 8.21 Distribution of carbon (*a*) and hydrogen (*b*) in vitrain coals. Distribution is based on the assumption that C_{CH_3}/C goes from 0.04 to 0.03 as the percentage of carbon in coal changes from 82.5 to 92.5.[503]

substituted. The degree of aromatic substitution, σ, varies from about 40–70% (Table 8.25). The substituting hydrocarbon groups are primarily hydroaromatic ring junctions and, to a varying but smaller extent, aliphatic groups. The aliphatic groups include methyl and other alkyl groups as well as methylene, ethylene, and other alkyl bridge structures.

It is remarkable that the degree of aromatic substitution is much higher for the depolymerization products than for the coal extracts. This trend is in agreement with the accepted mechanism of coal depolymerization. The reaction sites are at the activated positions of the aromatic units that are linked to a number of aliphatic or hydroaromatic structures (some of which are bridges): therefore these are the structures selectively dissolved during depolymerization.

Two sets of research data are available to estimate the relative proportions of hydroaromatic and aliphatic substitutions. British results[502] shown in Table 8.24, are based on estimating hydroaromatic carbons by the dehydrogenation of coal using benzoquinone.[501] These estimates indicate that in the bituminous coal range about 85% of the nonaromatic carbon is in hydroaromatic structures in a lower rank bituminous coal (82.5% C), and 15% of it is in aliphatic structures. This distribution gradually changes to about 60% in hydroaromatic structures and 40% in aliphatic structures in high-ranking bituminous coals (90.0% C). Another set of data, presented in this chapter, comes from the work of Reggel and co-workers,[506] who used catalytic dehydrogenation for the determination of hydroaromatic structures in coal. They found that in a lower rank bituminous coal (82.5% C) the amount of catalytically evolved hydrogen is 25–30 H atoms per 100 C atoms. This corresponds to the same number of carbon atoms (25–30/100 C) in hydroaromatic structures. This amount decreases with increasing rank and is about 15 H atoms per 100 C atoms for a coal with 89.8% C content. Since the aromatic carbon fraction in a lower rank bituminous coal is about 70% (Table 8.27), the dehydrogenation data indicate that 83 to 100% of the nonaromatic carbon is in hydroaromatic structures. The aromatic carbon fraction in the higher ranking coal is well over 80% (Table 8.21); therefore most of the nonaromatic carbon is in hydroaromatic structures in that coal also.

The conclusion can be drawn that both the benzoquinone and the catalytic dehydrogen-

[506] L. Reggel, I. Wender, and R. Raymond, *Fuel*, **43**, 75 (1964).

Table 8.25 Structural Parameters for Coal Extracts (CE) and Coal Depolymerization Products (CDP)

Structural Parameter	CDP[a]	Sum	CE[b]	CDP[a]	Sum	CDP[a]	Sum	CE[b]
Carbon content of coal or model (%, maf)	70.6		82.6	82.4		84.0		86.1
Carbon content of extract (%)	77.9		83.9	81.4		82.7		87.1
Yield of extract or depolymerization product (%)	75.2		21.4	47.4		30.4		19.3
Fraction of total hydrogen[c]								
As aromatic H, H^*_{ar}	0.30			0.20		0.25		
		0.35	0.38		0.25		0.29	0.35
As phenolic OH, H^*_β	0.05			0.05		0.04		
On nonaromatic carbon atoms that join two aromatic rings, $H^*_{\alpha 2}$	0.14			0.05		0.04		
		0.40	0.26		0.37		0.29	0.28
On nonaromatic carbon atoms next to aromatic rings, H^*_α	0.26			0.32		0.25		
On other nonaromatic carbon atoms, $H^*_{\beta+}$	0.25		0.36	0.38		0.42		0.37
Aromatic H/aliphatic H[d]	0.46		0.56	0.26		0.35		0.49
Carbon aromaticity, f_a[d]	0.72		0.73	0.65		0.67		0.73
Degree of aromatic ring substitution, σ	0.50		0.37	0.69		0.59		0.37
Average number of aromatic rings per condensed cluster	~2		3–4	3–4		4		5

[a] From Heredy et al.[90,103]
[b] From Retcofsky.[344]
[c] Separate resonances for H^2_α and H_α protons not observed; shape of resonances suggests only small number of H^2_α proton.
[d] Calculation assumes phenolic OH to be 0.03.

ation methods strongly indicate that most of the nonaromatic structures in bituminous coals are in condensed hydroaromatic rings. It is probable that some of the aromatic nuclei are linked together in the molecules through such hydroaromatic ring structures. Some of these rings can merge during coalification by loss of hydrogen with the aromatic units, thus increasing the size of the average aromatic cluster as the rank of the coal increases.

8.18.1.2 *Bridge Structures*

There is strong evidence for the existence of two bridge types which connect condensed aromatic structures in bituminous coals: aliphatic carbon structures such as methylene, ethylene, or longer paraffin chains, or cycloparaffin rings, and oxygen bridges. The sections that follow give a brief summary of more recent research results on the identification and characterization of such bridge structures.

Aliphatic Carbon Bridges. There is strong evidence from a variety of sources for the existence of aliphatic carbon bridge structures. Model compound studies performed to investigate the chemistry of coal depolymerization catalyzed by phenol-BF_3 showed conclusively that this reagent breaks some aromatic-aliphatic carbon-carbon bridges and that the reaction results in linking a phenol molecule to the aliphatic end of the bond that had been broken.[507]

No systematic study has yet been made of the relative reactivities of aliphatic bridge structures with phenol-BF_3. The following trends have been established, however, in model compound experiments.[86,88] Bridge structures linked to nonactivated single benzene rings are the least reactive. The reactivity of bridge structures on naphthalene rings is still relatively low, whereas those on a phenanthrene ring are more reactive. Phenolic OH has a strong activating effect, and bridge structures linked to single phenolic rings are easily broken during phenol-BF_3 catalyzed depolymerization. Of the aliphatic bridges, the single methylene bridge is more

[507] J. J. Eisch, *J. Org. Chem.*, **28**, 707 (1963).

reactive than bridges consisting of two or more carbon atoms.

Depolymerization experiments made with coals of various rank showed that lignite was by far the most reactive with phenol-BF_3. More than 75% of a lignite was depolymerized and converted to a phenol-soluble product with an average molecular weight of 330. The proton NMR spectra of the products indicated that most of the depolymerization took place at positions where single methylene bridges connected two aromatic rings. Phenol-BF_3 is less effective with bituminous coals. The yield of phenol-soluble depolymerization products varies from 30 to 50%. The NMR data indicate that the role of single methylene bridges in the depolymerization is much less pronounced in the case of bituminous coals than it is in lignite. Unfortunately, without further refinement it is difficult to identify positively some of the other aliphatic carbon bridge types by means of this reaction.

The breakage of some aliphatic carbon bridge types may contribute to coal solubilization in reductive alkylation when coal is treated with potassium, naphthalene, and an alkyl halide. It has been shown[502,508] in model compound experiments, for example, that an ethylene bridge connecting two naphthalene molecules will be cleaved at the center under these reaction conditions. In contrast, it is probable that few, if any, carbon-carbon linkages are broken in the lithium-ethylenediamine reduction of coal.

It is expected that more information will be obtained on the structure of aliphatic carbon bridges by means of a novel oxidative method[224] that has been more recently under investigation to study the aliphatic structures of coal. The method uses a mixture of 30% aqueous H_2O_2 and trifluoracetic acid, which selectively oxidizes the aromatic carbons of the coal while the aliphatic structures remain unchanged. For example, in the case of the model compound described in the preceding paragraph, 1,2[1,1′,-dinaphthyl]ethane, oxidation with this technique would give succinic acid ($HOOCCH_2CH_2COOH$), showing the destruction of the whole aromatic structure except for the two bridgehead carbons, which are oxidized to carboxyl groups. But further work is necessary to establish the usefulness of this method of oxidation in determining coal structure.

[508] A. Lagendijk and M. Szwarc, *J. Am. Chem. Soc.*, **93**, 5359 (1971).

Oxygen Bridges. Lazarov and Angelova[509] showed that approximately 50% of the oxygen in a coking coal (87.6% C, maf) was present in diaryl ether or arylalkyl ether linkages (2.1% of 4.1% oxygen in coal). In contrast, less than 4% of the oxygen in a high-volatile bituminous coal (82.2% C, maf) was found to be in similar structures (0.3% of 8.4% oxygen in coal). The determination was made by treating the coal with sodium in liquid ammonia. Diaryl ethers react easily with this reagent and are converted to phenols, but aliphatic ethers do not react under these conditions. Condensed heterocyclic compounds of the furan type also do not cleave when treated with this reagent. For example, dibenzofuran is reduced to dihydrodibenzofuran.

Similar results were obtained by Wachowska and Pawlak,[63] who determined the noncyclic aromatic or aromatic-aliphatic ether oxygen content of a series of coals of different rank by treatment with potassium in THF in the presence of naphthalene. It was found that 43% of the oxygen (2.35% of the 5.47% total oxygen) in a coking coal (87.8% C, maf) was cleavable ether. In the other coal samples the fraction of the cleavable ether oxygen was much less, ranging from 5 to 20% of the total oxygen content. In most coals 50–65% of the oxygen was in heterocyclic structures that are not cleaved by the potassium-THF-naphthalene treatment.

Ignasiak and Gawlak[68] converted a high-rank vitrinite concentrate (89.5% C, maf) to soluble products by reductive alkylation using potassium in THF in the presence of naphthalene with a variety of alkyl halides. They concluded that the macromolecular network of this vitrinite was composed of relatively small units (number average molecular weight 600–700) that are linked primarily by aromatic ether oxygens. The amount of oxygen ether linkages cleaved, calculated from the increase in phenolic hydroxyl content after the treatment, was 2.2 wt % of the vitrinite.

The distribution of organic oxygen be-

[509] L. Lazarov and G. Angelova, *Fuel*, **47**, 333 (1968).

tween oxygenated structures in a bituminous coal and in a subbituminous coal was determined by Cronauer and Ruberto.[510] Their data are summarized in Table 8.26.

This information reinforces previous data on the existence of aromatic ether bridges in high-ranking bituminous coals (about 2.8 wt % of maf coal). In a subbituminous coal approximately 1.0% aromatic or aromatic-aliphatic ether was identified (the ether was converted to phenol on hydrogenation). It is of special interest that approximately 4% oxygen in the subbituminous coal was present in unknown structural types. Together with the carbonyl oxygen, this latter type was eliminated as water during the hydrogenation process. It is possible that some of this unknown type of oxygen was present in aliphatic ether structures.

8.18.1.3 *Model Structures for Bituminous Coal*

One of the methods that can be used to simultaneously summarize and to illustrate the main known—and inferred—chemical structural features of coals is the construction of "coal model molecules." An important application of this approach was reported by van Krevelen,[511] who proposed formulas for the aromatic part of the structure of coals at different stages of coalification. Given,[512] Wiser,[513] and Gibson[514] have also proposed model molecule structures for high-volatile bituminous coals of approximately 82–83% C (maf) content. Despite significant differences among these proposed structures, several of their basic features are similar: they all contain relatively small condensed aromatic systems, consisting on the average of two to four condensed rings; phenanthrene- and fluorene-type condensed aromatic rings predominate; the nonaromatic part of the molecule consists mostly of cycloparaffinic and hydroaromatic ring systems; and there are few alkyl (mainly methyl) groups. Because of the great complexity of the coal structure, many features of these models are necessarily qualitative, and they need to be made more definite as new research data become available. Still, the derivation and construction of molecular model structures for coal are important because they help the coal research chemist to summarize and evaluate the consistency of experimental data from a structural viewpoint, and to identify key areas in which more research is needed.

Important related structural information is available from studies on molecular models of coal extracts and soluble products of coal hydrogenation or depolymerization. The structure of these products differs in various degrees from that of the parent coal, but

[510] D. C. Cronauer and R. G. Ruberto, EPRI Rep. AF-442 January 1977; *Am. Chem. Soc. Symp. Ser.*, **71**, 50 (1978).
[511] D. W. van Krevelen, *Coal*, Elsevier Publishing Company, New York (1961), pp. 433–480.
[512] P. H. Given, *Fuel*, **39**, 147 (1960).
[513] W. H. Wiser, "Some Chemical Aspects of the Conversion of Coal to Liquid and Gaseous Fuels," *Proc. EPRI Conf. Coal Catalysis*, Santa Monica, Calif., 1973.
[514] J. Gibson, "The Constitution of Coal and Its Relevance to Coal Conversion Processes," Robens Coal Science Lecture, London, BCURA, 1977; *J. Inst. Fuel*, 67 (1978).

Table 8.26 Analyses of Oxygen Functionality in Coals (wt %, maf coal basis)[510]

Coal	Burning Star (Bituminous)	Belle Ayr (Subbituminous)
Oxygen content as		
Hydroxyl (—OH)	2.4	5.6
Carboxyl (—COOH)	0.7	4.4
Carbonyl (—C=O)	0.4	1.0
Ether (—O—)	2.8	0.9
Total	6.3	11.9
Oxygen by difference		
Ash basis	5.9	16.2
Mineral matter basis	—	16.0

structural studies with such materials can be conducted with greater accuracy because their lower average molecular weights and their solubility allow the application of a number of high-precision analytical techniques. Two more recent structural studies of this type have been applied to the derivation of model molecule structures. Bartle and coworkers[367] studied the chemical nature of a supercritical gas (toluene) extract of a high-volatile bituminous coal. The extract was prepared at 350°C and it amounted to 17.0% of the coal. In the fractions with 500–600 molecular weight, the average molecular structure was represented by small condensed aromatic groups (naphthalene, dibenzofuran) linked by CH_2 and oxygen bridges (forming xanthene-type and dihydroanthracene-type structures). The nonaromatic structures were mostly C_1—C_3 side chains and a two-ring condensed hydroaromatic group. Farcasiu[515] reported a method for the fractionation and structural characterization of coal liquids produced by the solvent-refined coal (SRC) process. The proposed average structure of one of the major fractions ("polar aromatics") consists of a benzofuran ring that has a phenyl and a naphthyl group substituent. In another fraction the aromatic part of the proposed model molecule consists of a benzofuran, a pyridine, and a benzene ring. These rings are connected by means of a polycondensed cycloparaffinic network. No aliphatic side chains are indicated.

What follows is the presentation of a coal model molecule proposed by Heredy and Wender.[516] This model was constructed with two objectives: to incorporate into a coal model molecule new structural information that has become available in recent years, and to derive the model from the input data by means of a simple numerical analysis. The method applied by Whitehurst, Farcasui, and co-workers[517] was used as the basis for developing this analysis. It has the advantage that it can be easily modified to incorporate new or different input data or to evaluate the structural characteristics of other coals. Furthermore, it should be possible to improve the method by computerized operation. The investigation of Oka and co-workers[518] on computer-assisted molecular structure construction for coal-derived compounds exemplified very well the great potential in this area.

A high-volatile, vitrinite-rich bituminous coal of 83% C content was selected for the present study, since a large body of basic physical and chemical data is available for coals of this rank, and several structure-oriented and model studies have dealt with similar coals. The elemental composition and the general formula of the coal are shown in Table 8.27. The general formula was derived for a molecule containing 100 carbon atoms. Such a "molecular" size, corresponding to a "molecular weight" of about 1450, is, of course, completely arbitrary.

The aromaticity, or the fraction of aromatic carbon atoms (f_a), must be defined next. Dryden[502] used a value of $f_a = 0.66$, in structural studies for coals of 82.5% C (maf) content on the basis of measurements made with coals and coal extracts using infrared and high-resolution proton NMR spectroscopy. Work by Heredy and Neuworth,[90,103] based on depolymerization and high-resolution proton NMR on the depolymerized products, gave $f_a = 0.65$. Retcofsky[344] found $f_a = 0.73$ by studying coal extracts and using high-resolution proton NMR. The study of solid coal by Retcofsky by cross-polarization ^{13}C NMR gave $f_a = 0.76$. The average of these values ($f_a = 0.70$), was used in this calculation.

The aromatic structures included in the construction of the model molecule are shown in Fig. 8.22. They were selected on the basis of the following experimental evidence. Dryden[502] estimated that the average number of condensed rings in the aromatic structure of high-volatile bituminous coals with C = 82.5% (maf) was less than 3. Retcofsky[344] as well as Heredy and Neuworth[90,103] estimated that in the same type of coal the average number of condensed rings was 3–4. With regard to specific structures, it was shown by Kessler and co-workers[341] that a sizable fraction of the organic sulfur is in benzothiophene-type structures and that

[515] M. Farcasiu, *Fuel,* **56,** 9 (1977).
[516] L. A. Heredy and I. Wender, in preparation.
[517] D. D. Whitehurst, M. Farcasiu, T. O. Mitchell, and J. J. Dickert, Jr., in *The Nature and Origin of Asphaltenes in Processed Coals,* Annual Report 1976–1977, EPRI AF-480, Sect. 7.

[518] M. Oka, H. C. Chang, and G. R. Gavalas, *Fuel,* **56,** 3 (1977).

Table 8.27 Composition of High-Volatile Bituminous Coal Used in Model Structural Studies

Elemental Composition (dmmf)	Weight %	Atom %
Carbon	83.2	53.1
Hydrogen	5.5	42.0
Oxygen	7.7	3.7
Nitrogen	1.1	0.6
Sulfur	2.5	0.6
	100.0	100.0

General formula (100 C basis): $C_{100}H_{79}O_7NS$
Aromaticity, f_a = 0.70

much of the oxygen exists in benzofuran or dibenzofuran-type structures. Carbazole was identified as one of the nitrogen-containing aromatic structures in coal hydrogenation products by Sternberg and co-workers.[323] Among the simple condensed aromatic hydrocarbons, naphthalene and phenanthrene derivatives were found most frequently in coal extracts and in products obtained by mild hydrogenative degradation.[341] Four phenolic hydroxyl groups have been included per 100 C atoms on the basis of the work of Friedman et al.[296] The use of one aromatic ether linkage per 100 carbon atoms is based on the information discussed in the preceding section, "Oxygen Bridges."

The five constituent aromatic structures of the model molecule are interconnected by bridges. The minimum number of such bridges is five. One of these is the aromatic ether bridge mentioned in the previous paragraph; the other four are aliphatic hydrocarbon bridge structures.

On the basis of the input data given in Table 8.27 and Fig. 8.22, a formal numerical analysis was carried out to derive the general formula and the main characteristics of the aliphatic part of the coal model molecule. The details of this analysis are published elsewhere.[516] The essential points of this analysis are that a matrix is constructed, where the vertical columns correspond to a series of aromatic H contents (or degrees of aromatic substitution), and the horizontal lines correspond to different types of aliphatic substituents (single or multiple chains, single or condensed hydroaromatic or naphthenic rings). The structural characteristics of both the aromatic and the aliphatic parts of the model molecule can be expressed in terms of the number of positions that can accept substituents. Matching of the numbers of the aromatic and aliphatic structures in the matrix, which can accept the same number of substituents, identified the aliphatic structural configuration, as well as the percentage aromatic H content of the molecule. Therefore it is not necessary for this method to have independent experimental information available on the aromatic H content of the coal. If available, however, that information will improve the method by making the selection of the aromatic constituents of the molecule more accurate. In the case of the model molecule defined by the input data given in Table 8.27 and Fig. 8.22, the analysis gave an aromatic H content of 30%. Further-

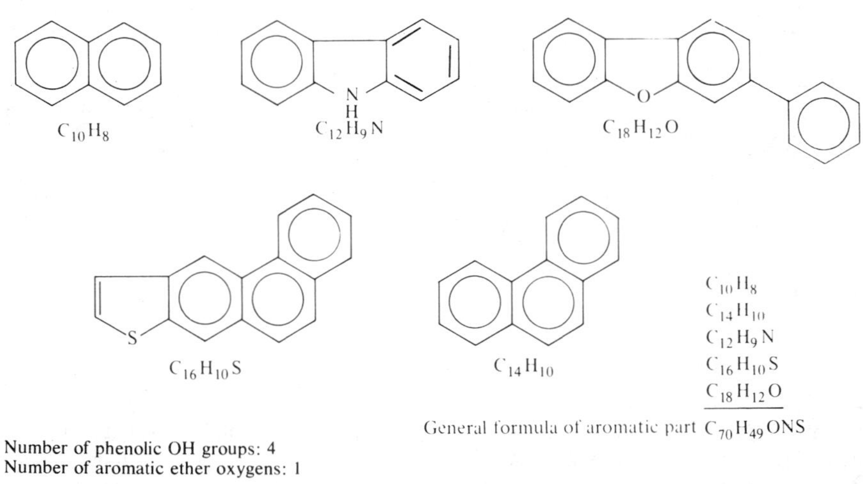

Figure 8.22 Aromatic structures considered (five molecules).

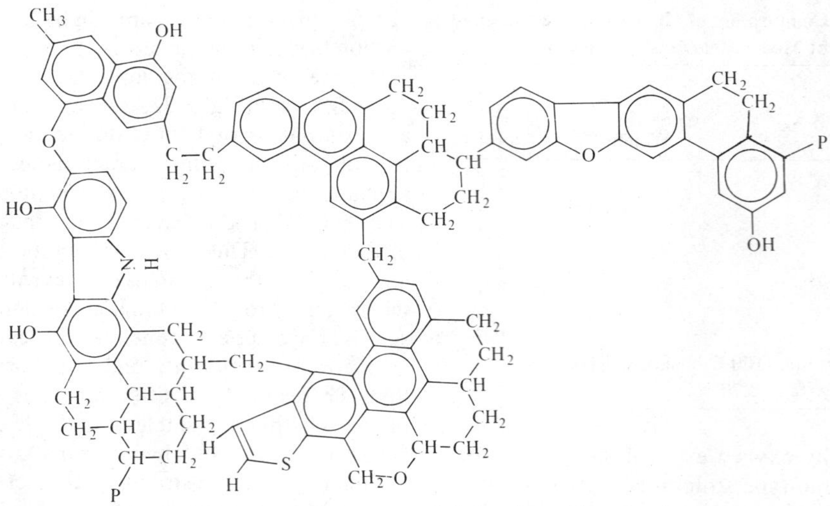

Figure 8.23 Proposed structure of the coal model molecule.

more, the analysis indicated that most of the aliphatic structures are composed of hydroaromatic or condensed hydroaromatic rings. On the basis of these considerations, hydroaromatic structures of these types were used to construct the nonaromatic part of the model molecule. The proposed structure of the coal model molecule is shown in Fig. 8.23.

The characteristics of the coal model molecule can now be examined. It is indicated (Fig. 8.22) that this "model molecule" is connected by two linkages to the rest of the coal structure and is thereby part of a larger macromolecular structure. The five constituent aromatic-hydroaromatic structures of the molecule are interconnected by one aromatic ether and four aliphatic bridges.

The distribution of carbon, hydrogen, and oxygen atoms among different structural positions in the model molecule is shown in Table 8.28. It should be emphasized that in addition to the general formula of the model molecule and the structure of the aromatic part, the input data defined only the distribution of the heteroatoms. In particular, no information concerning hydrogen distribution or concerning the distribution of carbon atoms among the different aliphatic positions was included in the input data. It is of great

Table 8.28 Characteristics of the Coal Model Molecule

		Aromatic		Aliphatic Ring				Other α-Aliphatic		Phenolic OH	
				Alpha		Beta					
Element	Total Atoms	No.	%	No.	%	No.	%	No.	%	No.	%
Hydrogen	79	23	29.1	21	26.6	20[a]	25.3[a]	11	13.4	4	5.1
Carbon	100	70	70.0	13	13.0	12	12.0	5	5.0	—	—

		Aromatic Ether		Heterocyclic			Phenolic OH	
Element	Total Atoms	No.	%	No.	%		No.	%
Oxygen	7	1	14.3	2	28.8	—	4	57.1
Nitrogen	1	—	—	1	100.0	—	—	—
Sulfur	1	—	—	1	100.0	—	—	—

[a] The protons attached to aliphatic carbon atoms in the β position to aromatic rings, but attached in the α position to oxygen atoms, are counted as β-aliphatic protons. Proton NMR analysis alone would count these protons as α-aliphatic protons, but it is assumed that they can be properly characterized by other methods.

Table 8.29 Comparison of Hydrogen-Type Distributions, Based on Coal of 82.5% C (maf)

Author and Reference	Methods Used	Fraction of Coal Represented (%)	Hydrogen Types (%)			
			Aromatic	Phenolic OH	α-Aliphatic	β-Aliphatic
Dryden [502]	Infrared extraction, high-resolution proton NMR	Whole coal	20.0	5.0	75.0	
Heredy et al.[90,103]	Depolymerization, high-resolution proton NMR	47.4	20.0	5.0	37.0	38.0
Retcofsky [344]	Extraction, high-resolution proton and ^{13}C NMR	21.4	35.0	3.0	26.0	36.0
Bartle et al.[367]	Extraction, high-resolution proton and ^{13}C NMR	17.0	24.4	4.8	40.2	30.8
Average of experimental data			24.8	4.5	35.1	35.6
Coal model molecule		Whole coal	29.1	5.1	40.5	25.3

interest, therefore, to compare the hydrogen structural type distribution in the coal model molecule with those determined experimentally in coal extracts and other coal products (Table 8.29). It is apparent that the values for the aromatic and α-aliphatic hydrogen contents agree within 5%. However, the average value of β-aliphatics of the extracts is about 10% higher than that of the coal model. This indicates an enrichment of aliphatic components in the extracts.

A large fraction of the hydrogen atoms (41 of the 79 in the model molecule) are in hydroaromatic structures. Of these, 24 hydrogen atoms would be expected to evolve as hydrogen gas under the dehydrogenation conditions used by Reggel and co-workers.[29] The experimentally obtained number for vitrain concentrates of 82.5–84.0% C content was 23–30 H atoms evolved per 100 carbon atoms. The expected effect of reduction with lithium-ethylenediamine of the model molecule was estimated by using the experimental data obtained with model compounds by Reggel et al.[11] It was estimated that the model molecule would take up 24 H atoms. The experimentally obtained number for a vitrain concentrate of 82.5% C content was the addition of 21 H atoms per 100 carbons.

The expected behavior of the model molecule in a depolymerization reaction catalyzed by phenol-BF_3 can be evaluated. The bonds on both sides of the CH_2 bridge would break because they are linked to OH-activated phenanthrene rings. Furthermore, the bond between the CH group of the hydroaromatic ring and the OH-activated aromatic system next to the thiophene ring would break. These interactions would release—and solubilize—the phenanthrothiophene-based fragment from the rest of the molecule. Additional bond breaking is expected to take place on both sides of the ethylene bridge; however the depolymerized products still will have high molecular weights, therefore probably will not be soluble. The yield of the soluble "depolymerization products" would be 27%. Heredy and Neuworth[90,103] obtained a net phenol-soluble depolymerized product yield of 29% with a coal of 82.4% C (dmmf) content.

The treatment of the model molecule with TFA—H_2O_2—H_2SO_4 would give acetic acid and succinic acid in a 3 to 8 ratio (on a hydrogen basis) as the principal low molecular weight products from the oxidation of the methyl group and the two —CH_2—CH_2— groups. The actual testing of high-volatile bituminous coals showed[216] that these two acid products predominate with an acetic acid to succinic acid ratio of about 1 to 3.

One can conclude that the aromaticity, the major oxygen functions, the hydrogen type distribution, and a number of important chemical characteristics of the model molecule are in good agreement with the experimentally observed characteristics of coals of this rank.

CHAPTER **9**

Transportation, Storage and Handling of Coal

TERRY L. THOMPSON AND FRED B. RAYMER

Terry L. Thompson *has been associated with Bechtel Corporation since 1964 in supervisory, sales, and managerial posts including Chief Slurry Engineer, Manager of the Slurry Systems Department, and Manager of Pipeline Division Engineering. He has been involved in the design of more than 10 long-distance slurry pipelines. He was responsible for preliminary design, development, and economic evaluation of the Black Mesa Coal Pipeline and start-up of the Bougainville Copper Concentrate Pipeline. Mr. Thompson has conducted numerous solids pipeline feasibility studies on materials ranging from woodchips to magnetite. He is recognized as an authority in solids pipelining and has written more than 20 technical articles on the subject.*

Fred B. Raymer *has been associated with Bechtel Corporation since 1962 and has held various supervisory posts, including Assistant Chief Mechanical Engineer, in the Mining and Metals Division. He has been for several years project engineer on the conceptual design of coal slurry pipeline terminal facilities, including preparation, dewatering, stockpiling, handling, and barge loadout facilities. Mr. Raymer has also been responsible for the design of materials handling systems on a large scale for coal power plants and nickel and aluminum processing plants.*

9.1 TRANSPORTATION

Coal is the most abundant energy source in the United States. Some estimates predict an increase from 650 million tons per year production in 1975 to approximately 1.1 billion tons in 1985.[1] All this coal must be transported, from a mile or less to 1000 miles or more. As of 1975, approximately 54% of U.S. bituminous coal was moved by all rail, 18% by river barge, 12% by truck, 6% on the Great Lakes, and the remainder by pipeline and conveyor belt or a combination of the other modes.[2] Figure 9.1 shows the major coal fields in the United States and the distance to major markets.[3]

9.1.1 Rail—Unit Train

In 1975 railroads transported approximately 405 million tons per year of coal, thus were the major long-distance transport alternative for coal.[2] The "unit train" is the most efficient method of moving coal by rail within the current state of the art. It should be noted, however, that about half the coal transported is in trainload lots, not strictly as unit trains. Much of the coal in the East is moved in 150–200 car trainload lots and will probably continue to be moved in this manner. The term "unit train" refers more to a management concept than to a technology. The U.S. Bureau of Mines gives the following definition[4]:

> The unit train is a management technique that permits efficient planning through long-term contractual commitment of producer and consumer and dedication of the equipment. Specifically, a unit train consists of a dedicated set of haulage equipment loaded at one origin, unloaded at one destination each trip, and moving in both directions on a predetermined schedule.

[1] "Project Independence Report," Federal Energy Administration, Washington, D.C., November 1974, Table II-21, p. 108.

[2] *Coal Traffic Annual, 1976 Edition,* National Coal Association, Washington, D.C., 1976, p. 1.

[3] *1975 Keystone Coal Manual,* McGraw-Hill Book Co., New York, 1975; *1974 Keystone Coal Manual,* McGraw-Hill Book Co., New York, 1974, p. 456.

[4] T. O. Glover, M. E. Hinkle and H. I. Riley, *Unit Train Transportation of Coal,* U.S. Bur. Mines Inf. Cir. No. 8444, 1970.

Unit trains came into existence in the early 1960s as a result of new competitive forces in the transportation industry including coal slurry pipelines, extrahigh-voltage (EHV) transmission, interruptible gas, and subsidized atomic energy.[5] Unit trains have been in use for roughly 15 years, and they now transport nearly half of all the coal tonnage handled by U.S. railroads. Whereas most unit trains contain approximately 100 cars, there are unit trains operating with as few as 20 cars. Unit trains are usually operated by a four-man crew that changes every 100–200 miles. The main reason for success of the unit train is radically increased use of rail cars. Labor productivity is also somewhat higher. It has been estimated[6] that the average non–unit train freight car in the United States spends 50% of its time in railroad yards, 40% in customers' yards or sidings, and only 10% of its time moving—either loaded or empty. Since unit trains spend virtually no time in railroad yards, the productivity is doubled at the outset!

9.1.1.1 *Capacity Relationships*

Based on the use of 100-ton cars and assuming loading and unloading times of 5 hr, Buck and Savage[7] have developed a nomograph to predict the number of unit trains needed to service a particular coal-fired power plant (Fig. 9.2). (See Sections 9.1.1.3 and 9.1.1.5 for discussion of car sizes and unloading rates.) In addition, they have prepared the more general equations. Use the following formula to calculate the time for a round trip from the mine to the power plant:

$$T = 2\left[\frac{D}{R} + t_1 C + L + \left(\frac{t_2 D}{500}\right)\right]$$

where T = total time per round trip (hr)
D = one-way distance (miles)
R = average speed in open country (mph)
C = number of cities between mine and plant

[5] R. D. Snouffer and V. D. Hanson, paper presented at the 26th Annual Meeting of the American Power Conference, Chicago, April 1964.

[6] J. P. Henderson, "Unit Trains," *Skillings Min. Rev.,* **1**(26), June 24, 1972.

[7] P. Buck and N. Savage, "Determine Unit-Train Requirements," *Power,* January 1974.

t_1 = extra time allowed for passing each city (hr)

L = time to load and unload the train (hr)

$D/500$ = number of federally required inspections

t_2 = time, to complete a federally required inspection (hr)

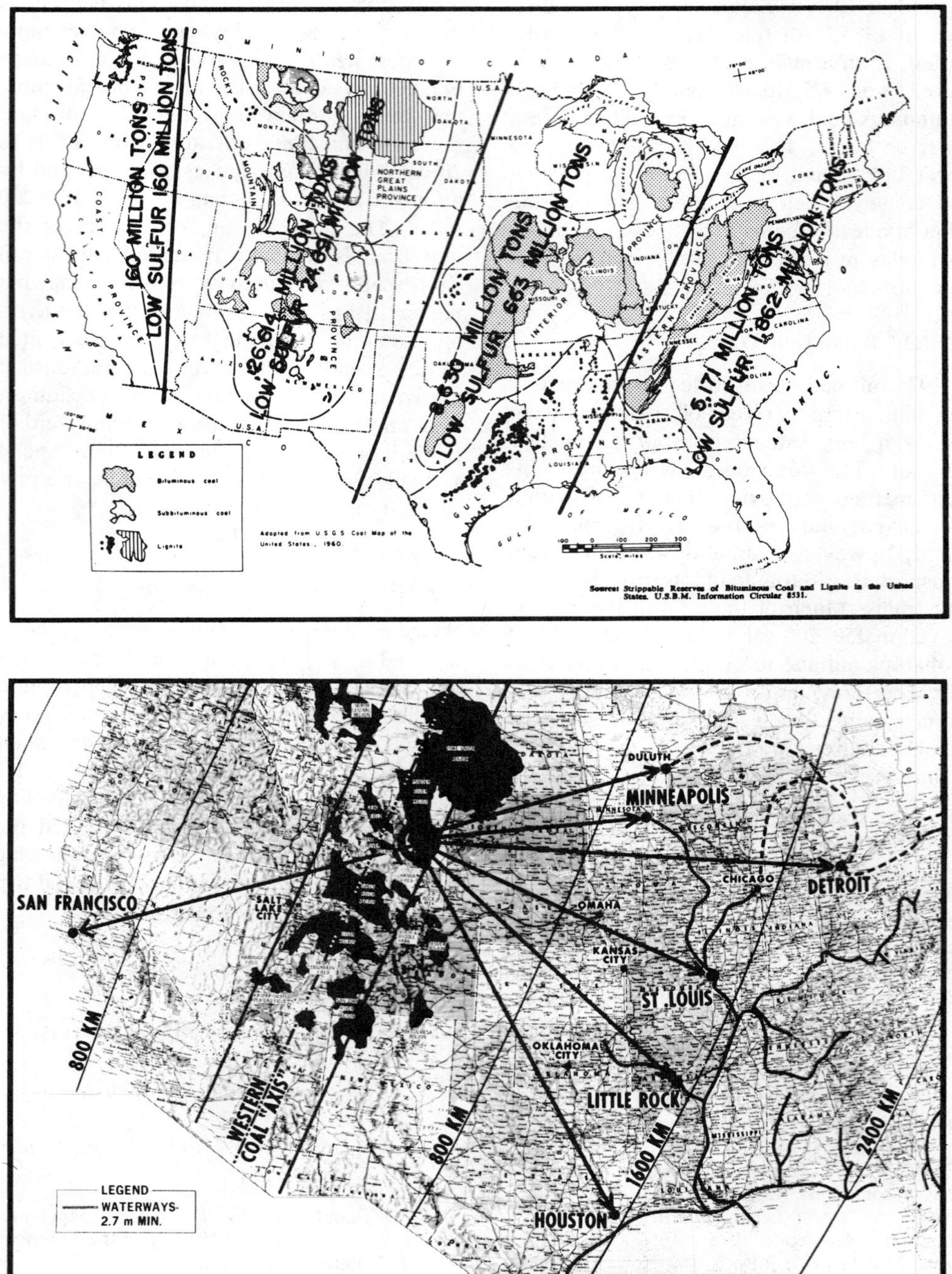

Figure 9.1 (*a*) Strippable coal reserves of the United States (lower 48) and (*b*) distances of Western coal fields to major markets.[3]

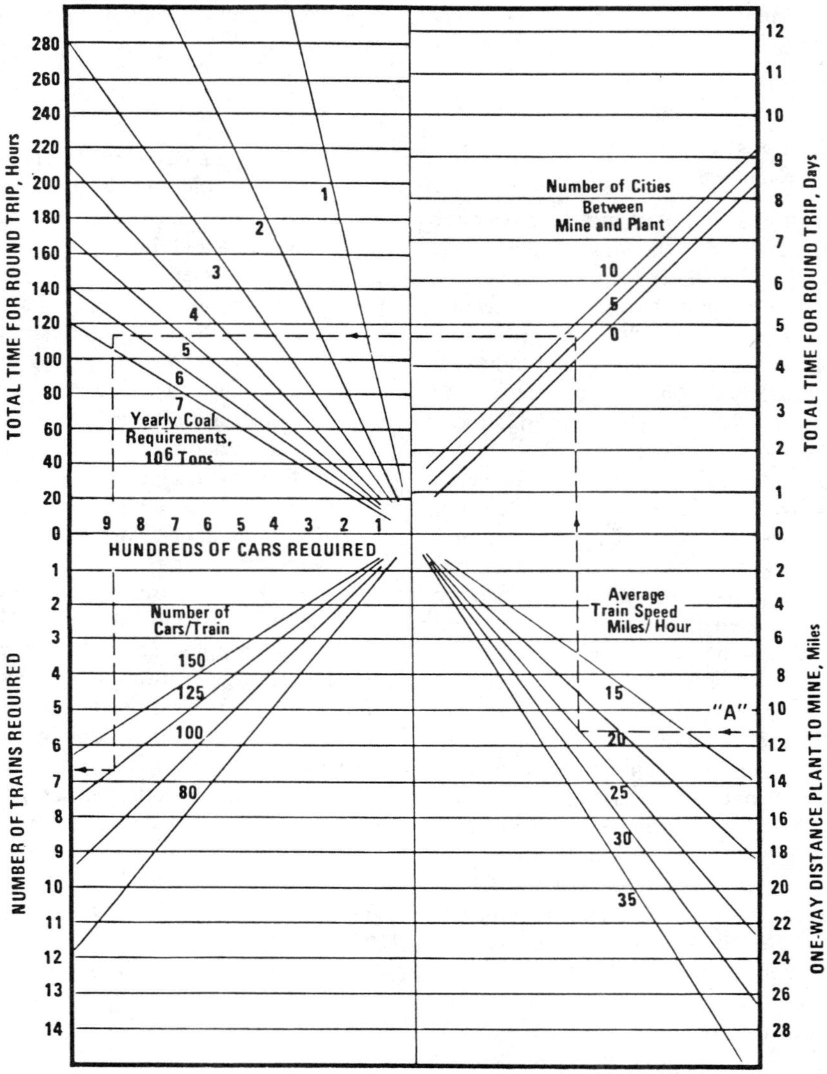

Figure 9.2 To determine unit train requirements rapidly with a nomograph, enter the nomograph at, for example, point *A*, the distance from plant to mine (1100 miles). Move horizontally to the left to the average train speed (25 mph), then proceed vertically to the number of cities between plant and mine (8). Go left to the yearly coal requirements (6.5 million tons), then down to the number of cars per train (125). The number of trains required to service this particular central station power plant (viz., seven) is shown at the left. Reprinted with permission from *Power*, January 1974.[7]

To calculate the number of trainloads required per year, use the ratio:

$$L = \frac{W}{SV}$$

where L = number of train loads per year
W = yearly power plant coal requirements (tons)
S = number of cars per train
v = capacity of each car (tons)

The number of trains theoretically required to service a power plant then is

$$N = \frac{LT}{8760}$$

This, of course, is the number of train sets required to be in service continually without allowing necessary spares for maintenance. To these "neat" calculations, an allowance for spares must be added. This number varies, but 10% extra cars and locomotives is a common figure in the industry.

9.1.1.2 *Motive Power Fuel Requirements*

The coventional source of motive power for unit trains is the diesel locomotive. Normally, a maximum of five or six locomotives is used with a 100–125 car unit train. Conventional locomotives are about 3000 hp per unit. Locomotives (except for electrified railways) use diesel fuel as the basic energy source. As of January 1, 1976, there were 27,823 locomotives in service owned by Class I railroads. Of these, 27,655 were diesel powered and only 167 were electric.[2] The literature reveals a wide variance in diesel fuel requirements. Weiss reports that " . . . in reasonably ideal conditions (minimum classification requirements, easy terrain, and efficient equipment) unit trains can operate at a rate of 0.5 gallons of fuel oil per 1000 net ton-miles. Under more realistic conditions, the fuel consumed may approach two gallons per 1000 net ton-miles."[8] This is compared to 4–6 gal/1000 net ton-miles for mixed train intercity freight. In a submission to the Interstate Power Commission for authority to construct and operate a 126-mile railroad between Douglas and Gillette, Wyoming,[9] Burlington Northern estimated a diesel fuel oil requirement of 1.8 gal/1000 net ton-miles. Caution should be used when comparing these figures with other transportation forms, since rail miles can differ significantly from truck, barge, or pipeline miles. Converting the 1.8 gal/1000 net ton-miles at 6 million Btu per barrel of fuel oil results in an approximate consumption of 250 Btu/ton-mile. In an environmental impact statement for a 1011-mile coal unit train,[10] the estimate was 2.2 gal/1000 net ton-miles. This is equivalent to 330 Btu/ton-mile.

[8] W. D. Weiss and R. H. Dunn, paper presented at Southwest Conference, American Institute of Mining, Metallurgical and Petroleum Engineers March 3, 1976.

[9] Burlington Northern Inc., submission to Interstate Commerce Commission, Finance Docket No. 27028, 1973.

[10] Final Environmental Impact Statement, Eastern Panhandle River Coal Basin, prepared by the U.S. Dept. Agriculture, Interstate Commerce Commission, and U.S. Dept. Interior, October 18, 1974, pp. VII-212, VII-213.

9.1.1.3 *Rolling Stock*

Typical coal cars carry 80–125 tons per car; 100 tons is "standard" for new equipment. Two basic types of car are used—the solid-bottom gondola and the hopper car. A rotary dump, as discussed in Section 9.1.1.5, can use either gondolas or hoppers, whereas the undertrack pit method of unloading requires hopper cars with bottom-opening doors. For rotary dumping, cars are provided with a swivel coupler at one end. The choice of type of car and unloading method involves evaluation of several tradeoffs. Solid-bottom gondola cars are the least expensive; they require minimal maintenance, and cars can be completely emptied with minimum coal hangup.[11] However they require an expensive rotary car dumper with sophisticated provisions for accurately stopping and spotting cars. The quick discharge bottom-dump car, on the other hand, can be unloaded more quickly, and the unloading can be done while the train is in motion over a relatively inexpensive track hopper. The bottom-dump gates make a hopper car more expensive and require more maintenance than a gondola. In addition, hopper cars may be more difficult to unload in freezing conditions. The life of a rail car varies from railroad to railroad, with a range of 10–20 years. Southern Railway reports ". . . a 14-year life is appropriate for both power and cars."[12]

9.1.1.4 *Maintenance*

The major reason for the success of unit trains is the drastically increased use of rolling stock. This in turn has resulted in new maintenance problems for the industry. Faster than "normal" wear of equipment has forced railroads to change traditional inspection, maintenance, and repair procedures. A survey of major railroads by *Railway Age* reported some areas demanding increased maintenance[12]:

[11] R. N. Bergstrom, "Latest Developments in Transporting Coal by Rail," Proc. ASCE, J. Power Div., **91**(P01), May 1965.

[12] G. Welty, "Unit Trains: Traffic by the Trainload," *Railway Age*, September 24, 1973.

1. Corrugation of rail.
2. Increased gauge maintenance.
3. Accelerated abrasive wear on rails.
4. Joints battered by heavier loads.
5. Harmonic effect of jointed rails (in empty cars the stress caused by harmonics may be greater than in fully loaded cars).

Weiss[8] reports

> Unit trains impose comparatively heavy and repetitive loading on the rails, which often necessitates use of heavier rails, thicker ballast sections, and more rigidly compacted subbase. For existing lines, this may entail either replacement of the rail or more frequent and intensive maintenance.

To combat these problems, the rail industry has generally gone to continuous welded rail to avoid harmonics associated with jointed rail. Furthermore, the use of thicker ballast sections with wider shoulders has provided a cushioning effect against the dynamic loading imposed by unit trains. Wheels are normally replaced from 150,000–300,000 miles, which is about 1½–3 years in normal unit train service.

9.1.1.5 *Loading and Unloading*

Glover et al.[4] explain the role of trackage in unit train loading:

> Three types of trackage are commonly used in unit train loading: (1) the loop or balloon track, (2) the single track above and below the loading station, and (3) parallel tracks that permit loading on both tracks. Since unit trains are generally over a mile in length, the loop must be at least this distance in circumference. A single track must be at least two miles long, one mile above and one mile below the loading point. Parallel tracks for a train of this length must stretch approximately one half mile above and one half mile below the loading point.

Loading systems are usually categorized by the type of equipment used to move empty cars past the loading station. The four systems in general use are locomotive, car haul, tripper conveyor, and hydraulic ram. The locomotive system moves the cars in one pass under a loading station. A car haul system utilizes parallel tracks moving two strings of cars in opposite directions under two loading stations. A tripper conveyor system loads two stationary parallel strings of cars by moving two-way loading chutes over them. In the hydraulic ram system, a locomotive initially positions the cars and then after the locomotive is removed, hydraulic pusher arms move cars successively into loading position.

In the first two systems noted, the train is loaded by moving it slowly under a flood feeder that fills each car individually while the train is in motion. Typical loading rates are 2500–4000 tons/hr. Thus a 100-car, 100-ton per car train could be loaded in 2½–4 hr. Weiss[8] reports that rates of 10,000–15,000 tons/hr have been achieved by Kaiser Steel in Utah. There trains pass through a tunnel directly below a silo and are loaded through two gates 5 ft in diameter.

There are two major unloading systems; one is the rotary dump, which will handle one or two gondolas or hoppers, and the other is the undertrack pit, which requires hopper cars (Fig. 9.3). In constructing the undertrack hoppers a 60° or greater slope of stainless clad plate or lining should be used.[5] Hoppers should be deep enough to hold more than a car of coal for each car length of pit. Normal unit train unloading rates are reported from 2500 to 3600 tons/hr. Since a typical positioning and dumping cycle time is 2 min, a 100-car train can be unloaded in less than 4 hr with a one car dumper and in approximately half the time with a tandem dumper. Shorter times can be achieved with bottom-dump cars if long below-track hoppers are provided so that the coal can be dumped while the train is moving. At the Salt River Project's Navajo Station, the hopper is sized such that six cars can be positioned and dumped simultaneously. The system is completely automatic, and unloading time for a 10,000-ton train is 15–20 min.[8]

9.1.1.6 *Windage Losses*

An important but often overlooked aspect of rail transport is loss of coal along the route by "windage" (i.e., coal blown from open-top cars by wind). This is both a direct economic loss of product and an environmental concern. Windage losses are a func-

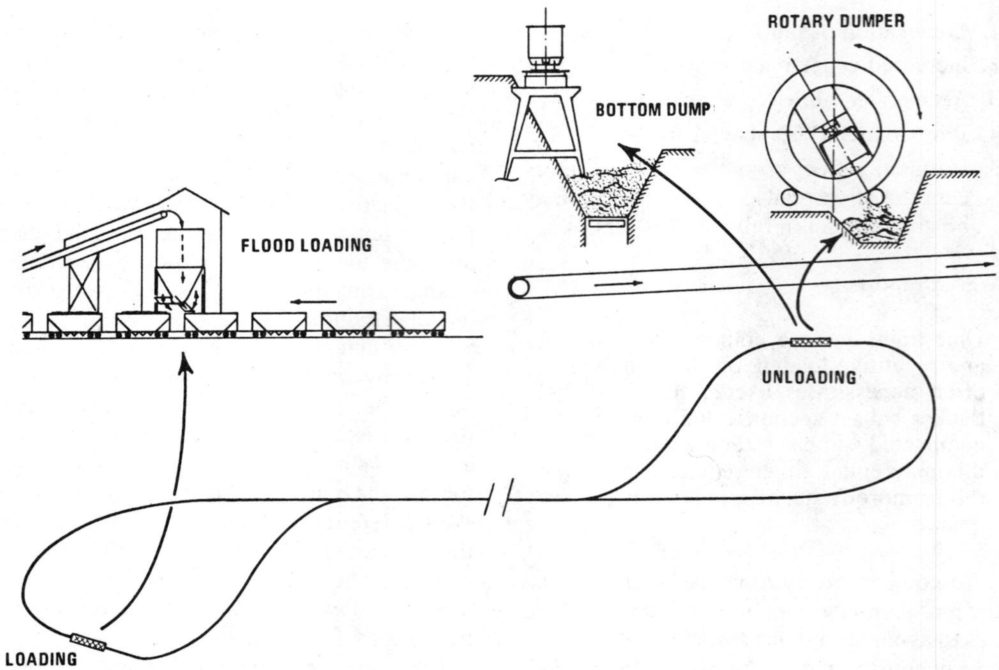

Figure 9.3 Schematic illustration of fast loading and unloading facilities.

tion of weather, amount of moisture, car design, percentage of car filled, train speed, and length of haul. For a 1000-mile unit train haul, it has been estimated that windage losses could be approximately 1% of the coal transported unless the cargo is wetted or oiled.[13] Other sources have indicated losses of up to 10% from the combination of windage and bottom losses. One unique method of combating this problem is a patented lid system for railroad cars. The cars, to be used by the Big Stone Power Plant in South Dakota, are 100-ton capacity, high-side, solid-bottom gondolas with "flip-top lids." The concept not only controls windblown loss, reduces dust, and protects air quality, but also keeps additional moisture from reaching the coal, thus reducing freezing problems.

Denton et al.[14] have reported encouraging results in applying a latex formulation to the exposed coal surface in rail cars. The latex binder solution produces a solid but flexible crust that consolidates the top 1½–2½ in. of the coal surface. Before this treatment losses on a particular run in Ohio were between 1 and 2%. After treatment losses were reduced by 50% or more. Additional experimentation may improve these results.

[13] "Energy and the Environment, Electric Power," prepared by the Council on Environmental Quality, August 1973, Appendix, p. 40.

[14] G. H. Denton, R. E. Hassel and B. E. Scott, "Minimizing in-Transit Windage Losses of OLGA Low Volatile Coal," paper presented to American Mining Congress, Cleveland, May 1972.

9.1.1.7 *Freezing in Cars*

A sometimes troublesome problem in the rail industry is the unloading of frozen coal cars. Moisture on the coal freezes both between coal particles and between the coal and the rail car. Often the center core of the coal is not frozen, but the outside layers along the top exposed surface, the size plates of the car, and the surfaces of the sloped hoppers are frozen.

When cars become frozen, rapid thawing is a necessity, especially in unit train operations. In some cases it is sufficient to thaw the cars only enough to permit unloading (i.e., skin thawing); at other times cars must be thawed more extensively so that the coal can pass through a specific size grating.

For skin thawing, it is only necessary to calculate the amount ot heat necessary to heat the steel, which is in contact with the frozen coal, to a temperature that will melt the ice:

$$Q = W(C_p)\Delta t$$

where Q = quantity of heat (Btu)

C_p = specific heat of steel [Btu/(1b)(°F)]

W = weight of steel in contact with frozen coal (lb)

Δt = required temperature rise for thawing. (°F)

If it is necessary to thaw to the complete depth of freeze, then to the Btu calculated for skin thawing must be added the heat to (1) bring ice to the freezing point, (2) change the state from ice at 32°F to water at 32°F, (3) bring the water up to the desired temperature above freezing, (4) bring the coal up to the desired temperature above freezing, and (5) change some of the water to steam. Mang[15] has shown typical calculations for thawing a 70-ton car with 12 inches of frozen coal thickness in which a 50°F temperature rise was desired. Total heat required was 2,917,000 Btu; only 420,000 Btu was required for skin thawing.

Numerous thawing techniques are used—from bonfires and oil-burning pans to sophisticated thawing sheds. For unit train operations, more rapid and efficient methods are required. Table 9.1 is a guide for estimating the approximate equipment needed for thawing coal cars at various rates.

Oil treatment of railroad car sides and bottoms is also used to minimize freezing and promote car unloading. In one 800-mile unit train shipment, spraying 1–1½ gal of oil per car was sufficient for about four round trips.

[15] G. Mang, "Coal Car Thawing Techniques," *Combustion*, April 1966.

9.1.2 Truck Transport

Trucks play an important role in the transportation of coal, especially in short-distance movements from the mine site to the next stop en route to the final destination or, in mine-mouth situations, directly to the power plant. A large portion are "off-highway" trucks, defined as vehicles that are too long, too wide, or too heavy for use on public highways. This discussion deals only with off-highway trucks for coal transport.

Up through the mid-1950s the most popular truck configuration included tandem rear-axel drive with a maximum size of 60 tons.

Table 9.1 Rail Car Thawing Equipment[15]

Thawing Rate (cars/8-hr day)	Car Stations	Thawing Equipment
2	1	2 Portable thawing tubes or 4 torches or 4 Radiarcs[a] or 2 thawing pits[a]
4	2	4 Portable tubes or 8 torches
	1	6 Radiarcs or 3 thawing pits
6	1	3 Pits or 8 Radiarcs or 1 set of 3 permanent tubes
10	2	6–8 Pits or 2 sets of permanent tubes
20	3	9–12 Pits
50	8	24–32 Pits[b]
100	12	36–48 Pits[b]
Infrared		
50	3	9–12 Pits plus 3 sets of sidecar heaters [32 panels per set, 60,000 Btu/(hr) (panel)][b]
160	6	18–24 Pits plus 6 sets of sidecar heaters[b]

[a] Windbreak used with pits and Radiarcs.
[b] Thawing shed and auxiliary unloading devices included.

Average size, however, was only 25–30 tons. During the next 20 years considerable progress was made in developing larger, more efficient coal haulage trucks. In 1976 trucks carrying more than 200 tons of payload were in use, and larger ones were being developed. (A prototype 350-ton truck was introduced in 1974.[16])

Basically, there are two major classifications of off-highway trucks: rear dump and bottom dump. A conventional rear-dump truck has the body mounted on the truck chassis and is dumped by means of an integrally mounted hydraulic hoist system. A bottom dump, by necessity, is a tractor trailer, since material cannot conveniently be dumped through the frame. In recent years bottom-dump tractor trailers have increased in use because of the low cost per ton of coal hauled.[17] However they are not well suited for hilly or rough-terrain operations because only a small portion of the vehicle weight is on the driving wheels. Payloads are generally higher for a tractor with the same horsepower as compared to an end-dump truck. This results in a lower horsepower-to-weight ratio; typically payloads 50–100% higher are carried in this manner.[16] Furthermore, with the payload moved from the back of the truck, both height and length restrictions on the box are eliminated. This is particularly valuable when transporting relatively light materials such as coal. Scrapers also are used for scraping coal and transporting it to the next transfer point. The scraper teeth normally gouge into the coal continuously, picking it up and conveying it into the scraper body while the scraper is moving at a good pace. Despite their higher initial and maintenance costs, they are used because of their excellent maneuverability and mobility in rough, hilly, or wet terrain.

9.1.2.1 *Capacity Relationships*

Rear-dump trucks have been tested in sizes up to 350 tons; however 200 tons is generally considered a "large" truck and 100 tons a "standard" off-highway truck. Burton[16] has summarized available off-highway trucks by manufacturer, model, transmission, and capacity; the reader is referred to Burton[16] for a detailed discussion of available trucks. Bottom-dump tractors are normally in the 35–180 ton range.

Calculation of truck fleet requirements is a function of the productive capability of an individual truck and the total tonnage required. The following general formulas, more fully described elsewhere,[16] can be used to calculate fleet requirements:

$$T = \frac{60(P)\,(W)}{C}$$

where T = tons per truck-hour (at 100% availability)
P = productivity factor
W = effective payload (tons/truck)
C = theoretical cycle time (min)

The number of operating trucks is calculated by dividing the hourly tonnage required by the tons per truck-hour. The actual operating fleet requirements may then be derived by dividing the number of operating trucks by the expected "utilization." Utilization refers to the truck's actual percentage of use, regardless of its availability. Utilization should average about 75% over the life of a fleet. Therefore:

$$F = \frac{R}{0.75T}$$

where F = number of trucks in fleet
R = hourly tonnage required

Theoretical cycle time (C) is composed of two parts: fixed time for loading and dumping, and variable time for travel. Obviously C is a function of distance, grade, altitude, road condition, truck performance characteristics, and loader characteristics, and it must be calculated for each individual set of conditions.

Productivity (P) reflects the fact of life that no operation can work continuously at 100% efficiency. Items that must be considered when calculating productivity are queuing, bunching, matching loader size with truck size, weather, and the human element. For study work a 50-min hour, which is equivalent to productivity of 83%, is commonly used.[16]

[16] A. K. Burton, "Off-Highway Trucks in the Mining Industry," *Min. Eng.*, December 1975 and January 1976.

[17] Anon., "Haulage Trucks," *Coal Age*, 80–87 (July 1974).

9.1.2.2 *Motive Power Fuel Requirements*

Trucks up to 85-ton capacity are generally powered by diesel engines with power shift transmissions and torque converters. From 85- to 100-ton capacity, the diesel power can be transmitted to the drive wheels by either mechanical or electrical means. At a capacity of 100 tons and above, the electric wheel drive dominates the market in the mining industry, and in the early 1970s was applied to the coal industry.[17] Basically the electric wheel concept uses a diesel engine that drives a generator to produce electricity, which in turn powers electric traction motors. The motors form the hubs in the truck wheels and transmit power directly to the wheels that surround them (Fig. 9.4). High horsepower thereby can be transmitted efficiently because differentials, U-joints, mechanical power transmission, clutches, and torque converters are not required.

Fuel requirements for representative dump trucks are listed in Table 9.2.

When fuel consumption is not known, it is usually calculated as follows, based on 0.42 lb of fuel/hp-Btu[16]:

$$F = \frac{0.42}{\gamma}(hp)(L)$$

where F = consumption of fuel per hour (gal)

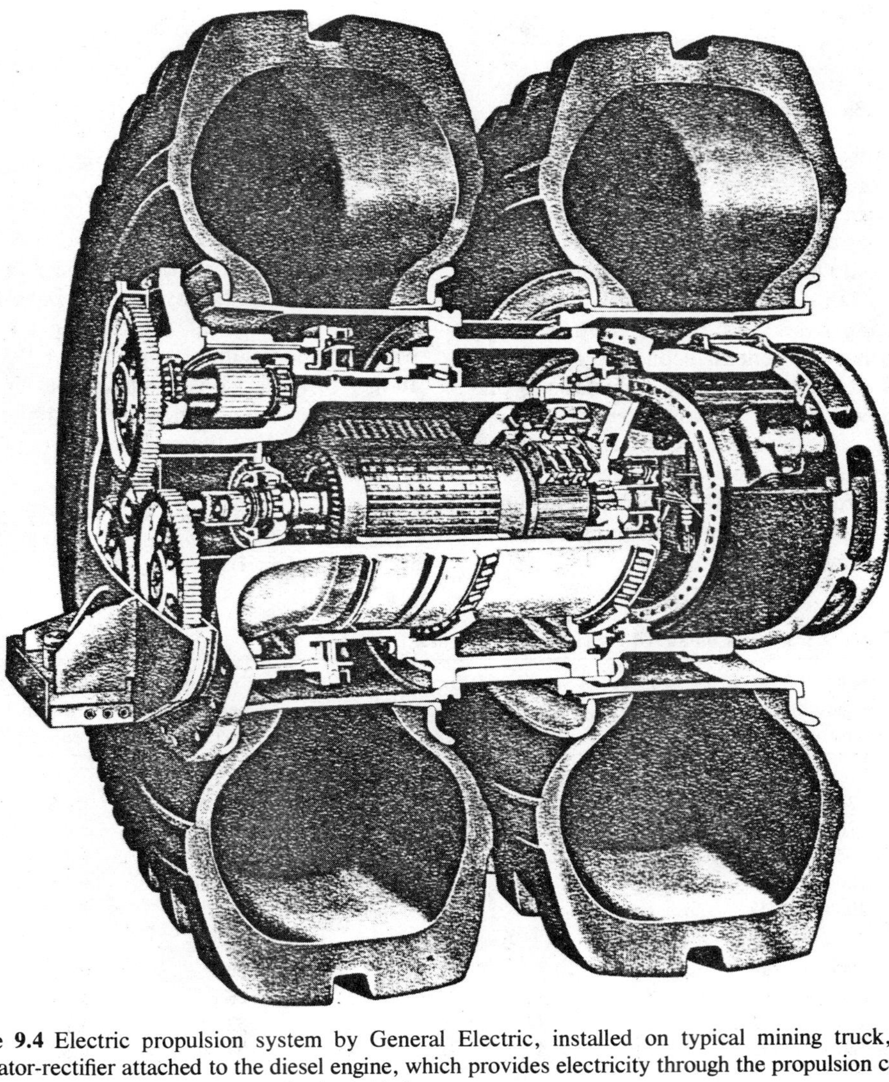

Figure 9.4 Electric propulsion system by General Electric, installed on typical mining truck, uses alternator-rectifier attached to the diesel engine, which provides electricity through the propulsion control system to two self-contained, motorized wheels.[7]

Table 9.2 Dump Truck Fuel Requirements

Truck	Fuel (gal/hr)	Gal/hr per Ton of Capacity
25 T Mach 25	6	0.24
35 T Cat 769B	8.2	0.23
50 T Cat 773	11.9	0.24
75 T Euclid R75	16	0.21
100 T Unit Rig M100	27	0.27
150 T Wabco 150B	36	0.24

Source: Derived from Burton.[16]

γ = density (for diesel, 0.9×8.3)
hp = gross horsepower
L = load factor

Load factor (L) can vary from 0.3 to 0.8 depending on the duty.

9.1.2.3 *Maintenance*

Operating costs for trucks include fuel, lubrication, tires, labor, and repairs. Repair costs can amount to 20–40% of total direct operating costs, depending on the size of the truck and whether it is rear dump or bottom dump. Table 9.3 gives typical 1974 truck operating costs. As shown, bottom-dump trucks have a general economic advantage, which explains why these vehicles are preferred where the terrain is suitable. Tires usually constitute the largest single identifiable expense in overall operating costs. Hence tire selection (like road maintenance) is of great importance in keeping operating costs in line.

9.1.3 Pipeline

Transport of material by pipeline is an efficient and proved method of conveyance. Primary application has been in the area of liquids and gases. In 1976 there were approximately 400,000 miles of pipeline in the United States. The same advantages of economy of scale, environmental protection, and simplicity of operation that apply to liquid and gas pipelines also apply to pipelines that carry solids. Solids pipelines are in operation carrying coal, iron concentrate, copper concentrate, limestone, phosphate, and waste materials. Virtually all these solids-carrying pipelines have used the slurry technique—the solids are ground to a fine size, mixed with a liquid vehicle such as water, and pumped in a conventional buried steel pipeline.

9.1.3.1 *Slurry Pipelines*

Concept and Background. Slurry pipelining is sometimes considered a new field; as long ago as 1891, however, a patent was issued for a method of pumping coal and water.[18] The landmark pipelines of 1957, the 10-in., 108-mile Consolidation Coal line and the 72-mile, 6-in. American Gilsonite line, however, marked the real beginning of long-

[18] E. J. Wasp, T. L. Thompson and P. E. Snoek, "The Era of Slurry Pipelines," *Chem. Tech.*, 552–562 (September 1971).

Table 9.3 Typical mid-1974 Truck Operating Costs[16]

Truck Weight (Tons)	Cost (as % of total) for: Fuel	Lubrication	Tires	Driver	Repairs	Typical Operating Cost ($/hr)	Cost per Hour per ton ($)
				Rear Dump			
35	13	5	15	31	36	21.00	0.60
50	15	5	15	23	42	27.80	0.56
75	18	6	17	20	39	31.60	0.42
100	24	8	22	18	28	36.00	0.39
200	17	6	34	8	35	81.00	0.41
				Bottom Dump			
60	14	5	29	31	21	20.60	0.34
100	15	5	34	24	22	26.80	0.27
120	17	6	34	20	23	32.60	0.27
180	17	6	41	12	24	54.70	0.30

distance slurry pipelines. The technology has been broadened and expanded greatly since the construction of these two lines. An important event was the start of commercial operation of the 273-mile, 18-in. Black Mesa coal pipeline in Arizona in 1970. This system provides all the fuel requirements for two 750-MW generating units in southern Nevada (about 5 million tons/year). The project is significant in that it was the first generating facility designed to use coal slurry directly from the outset of operations. Availability of the Black Mesa pipeline system has been on the order of 99%. Systems are being planned for moving 25 million tons or more of coal distances greater than 1000 miles. There are no significant technical problems in the design of such systems.

Thermal coal is well suited to pipeline transport because it must be ground until approximately 70–80% can pass through 200 mesh before it is blown into the boiler. Coal pipeline hydraulics require all material to pass through about 20 mesh. Therefore the slurry fineness requirements simply cause the required size reduction to be split—some at the head end of the pipeline and the remainder at the power plant. Typical pipeline coal slurry specifications are shown in Table 9.4.

Capacity relationships in a slurry pipeline are a function of material specific gravity, operating factor, slurry concentration, velocity of flow, and pipe diameter. For most long-distance commercial coal pipelines the range of these variables is fairly narrow, which allows some generalizations. Figure 9.5 shows pipe diameter as a function of annual tonnage for a coal pipeline with a 95% operating factor, 5.5 ft/s velocity, and 50% concentration of solids by weight. The required pipeline diameter can also be calculated as follows:

$$D = 0.01365\left(\frac{TPY}{WH\rho_s V}\right)^{1/2}$$

where D = required internal pipe diameter (in.)
TPY = annual requirement of bone-dry coal (short tons)
W = bone-dry solids (wt), (expressed as a fraction)
H = number of operating hours per year (usually 95–99% of 8760 hr)
V = bulk velocity (ft/s)
ρ_s = density of slurry

Annual coal requirements are often discussed on an "as-received" moist basis. The annual tonnage in the equation above is on a bone-dry basis. The reader is cautioned to verify the moisture level basis of the coal to be transported.

Feed Preparation. Feed preparation covers the physical and chemical processing necessary to give the slurry characteristics required for hydraulic transport and use. Preparation normally involves both size reduction (crushing and grinding) and addition of the liquid phase. Chemical treatment may also be part of slurry preparation for corrosion inhibition, thinning, and improving the characteristics of the final product.

In the case of coal transport, a particle size specifically suited for slurry transportation must be produced.[21] A balance has to be made between pumpability and dewatering characteristics. If sizing is too fine, pumpability may be good but the slurry may be

[19] E. J. Wasp, T. C. Aude and F. B. Raymer, "Terminal Facilities for Western Coal Slurry Pipelines," prepared for presentation at the 1976 Annual Meeting of the American Institute of Mining, Metallurgical and Petroleum Engineers, Las Vegas, Nev., February 22–26, 1976. (Not published; paper only.)

Table 9.4 Coal Slurry Specifications[19]

	Pipeline Concentration (%)	
Granulatory	Ohio	Black Mesa
+14 mesh	1– 3	0– 2
−100 mesh	35–40	35–45
−325 mesh	18–20	18–20

[20] T. C. Aude, T. L. Thompson and E. J. Wasp, "Economics of Slurry Pipelines," paper presented at Hydrotransport 3, Golden, Col., May 15–17, 1974. (Published by British Hydrodynamic Research Association Fluid Engineering, Cranfield, England, 1974.

[21] N. T. Cowper, T. L. Thompson, T. C. Aude, and E. J. Wasp, "Processing Steps: Keys to Successful Slurry Pipeline Systems," *Chem. Eng.*, 58–67 (February 1972).

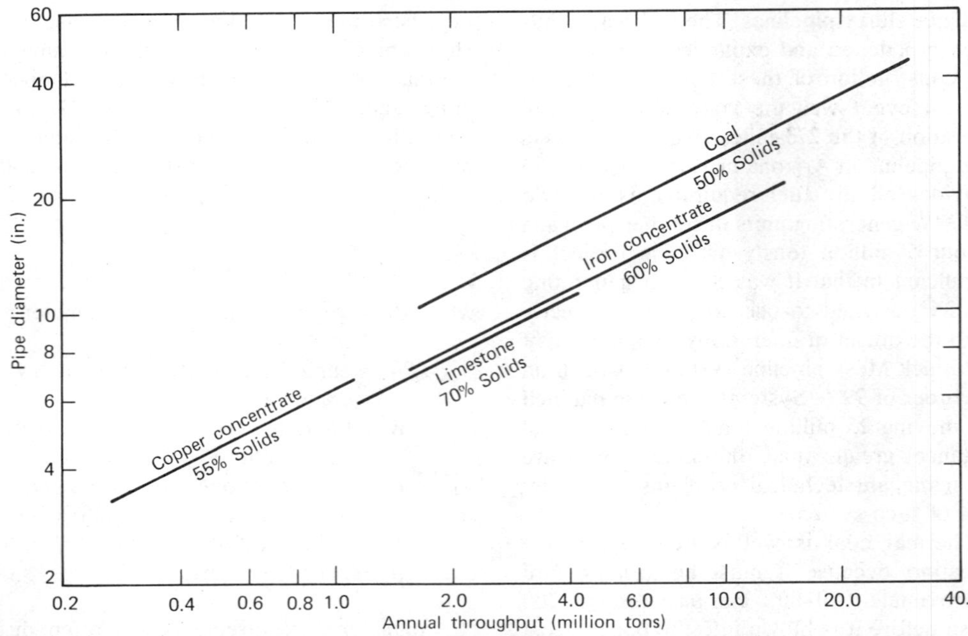

Figure 9.5 Estimating slurry pipeline diameter: 95% operating factor; velocity, 5.5 ft/s; solids concentrations, weight percent.[20]

difficult to dewater at the pipeline terminal. If the size is too coarse, the slurry is heterogeneous and must be pumped at higher flow rates to maintain a suspension. (The cost of pumping then goes up.) The choice of particle size for a slurry depends on (*a*) the overall cost of preparation, pumping, and dewatering of the slurry, and (*b*) operability of the slurry, including shutdown-startup characteristics and critical velocity.

Grinding normally involves conventional milling equipment. In the Black Mesa coal preparation plant rod mills (typical of those found in ore-dressing plants) are used for the final coal-grinding stage. Known milling technology of metal ore dressing can be applied to the grinding of coal.

In slurry preparation for pipelining, two variables become quite important (they may have only minor significance in other crushing processes). These are control of slurry density and control of product top size.

Control of slurry density is important primarily for the production of a consistent material, fitting the hydraulic design of the pipeline. Slurry concentration plays an important role in both the friction loss in the pipeline and the critical velocity of the slurry. The preparation step is usually performed at a solids concentration slightly higher than is required for the pipeline, with final dilution control provided downstream.

Top-size control is extremely important for any slurry transported over a long distance in a pipeline. Large quantities of coarse, fast-settling particles can cause pipeline plugs. To avoid this, safety screens are installed to prevent coarse particles from entering the pipeline. A discussion of grinding power requirements is given in Cowper et al.[21] The coal ultimately must be ground finer for thermal power use, and this must be considered carefully when comparing alternative transport modes (i.e., a credit should be taken).

Corrosion in a pipeline transporting a water-solids slurry is controlled by adding corrosion inhibitors. Laboratory-scale tests determine the most suitable type and dosage of inhibitor to be added to a slurry. Various inhibitors may be used, depending on the operative corrosion mechanism. For example, sodium dichromate will form a film on the pipe wall, protecting the pipe, and severe attack by dissolved oxygen can be controlled by adding an oxygen scavenger such as sodium sulfite. In other systems a high-slurry pH can control corrosion.[21]

As a final control step to ensure that only specification slurry is committed to the main pipeline, a "test loop" is normally included in the system. This is a piping section of the same diameter pipe used in the main slurry pipeline, and all slurry must pass through the section before entering the main pipeline. A test loop is useful in monitoring changes in slurry hydraulics. Pressure drop (usually by differential pressure cell) is measured in the loop along with flow, temperature, and density. Any increase above specified limits in the test loop pressure drop signifies a deleterious change in slurry properties and probably indicates that coarse, fast-settling particles are forming a bed in the loop. In such an event the off-specification slurry is either recycled or sent to a storage pond.

Pipeline and Pump Stations

Velocity. Pumping velocity is a critical variable in slurry pipelining. Figure 9.6 illustrates the general relationship between pressure drop and rate of flow; it also illustrates the concept of critical velocity.[22] The straight portion of the curve, between points *A* and

[22] E. J. Wasp, T. L. Thompson and T. C. Aude, "Process Basis and Economics of Pipeline Systems," *CIM Trans.* **73,** 335–339 (1970).

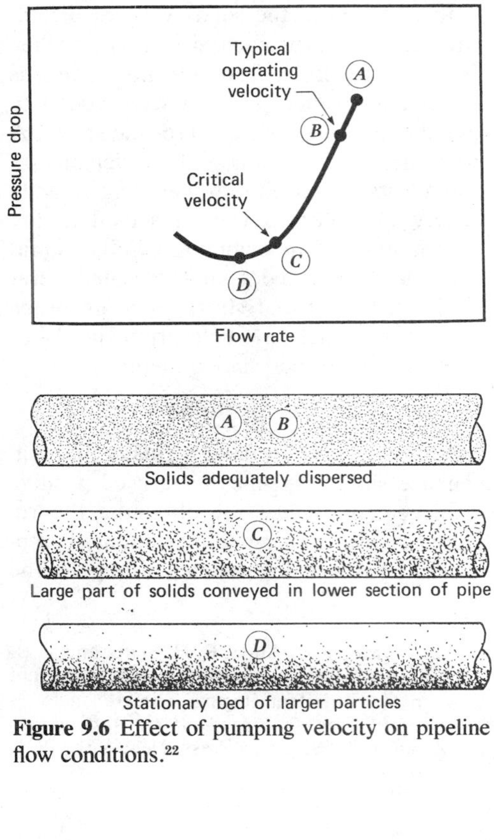

Figure 9.6 Effect of pumping velocity on pipeline flow conditions.[22]

C, encompasses the velocity range at which the solid material is fully suspended in the carrying liquid. However as the velocity is decreased from point *A,* the uniformity of the suspension gradually decreases. This decrease continues until at point *C,* part of the solids is being conveyed along the bottom of the pipe as a thickened layer. At point *D* a bed of solids has formed. This so-called critical velocity corresponds to the minimum point on the pressure drop–velocity curve.

Although all the material is being moved in suspension until a point close to *D* is reached, it is not desirable for the solid particles to be dragged along the bottom of the pipe, since this increases erosion of the bottom of the pipe, which is the area of unstable flow. Therefore the velocity at which the pipeline is operated must be somewhere above point *C.* Experience has shown that the practical operating velocity, point *B,* will vary between 4 and 7 ft/s.

Concentration. The pumping velocity and the critical velocity are dependent on the concentration of solids in the slurry. The viscosity of the slurry is directly related to the concentration of solids in the slurry, although this relationship varies from one type of material to another. (As used here, "viscosity" is the so-called limiting value at high rates of shear.) In general, the relationship between viscosity and concentration will be as shown in Fig. 9.7 for coal. The reader is referred to Thomas[23] for a good review article on the dependence of viscosity on concentration.

[23] D. G. Thomas, "Transport Characteristics of Suspension: VIII," *J. Colloid Sci.,* **20**(3), 267–277 (1965).

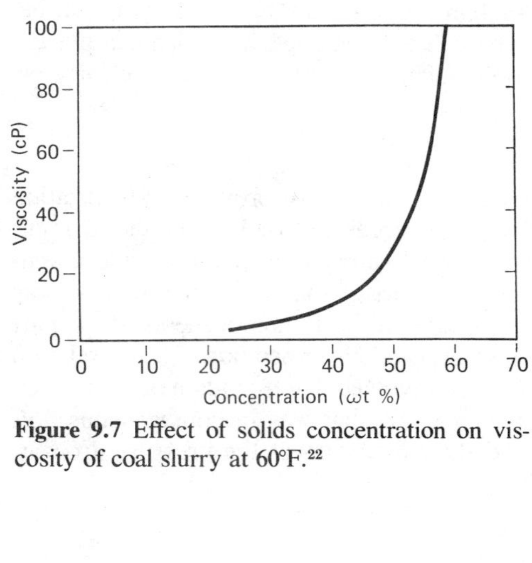

Figure 9.7 Effect of solids concentration on viscosity of coal slurry at 60°F.[22]

Experience with commercial slurry systems indicates that the most practical concentration of solids is in the general area where a small increase in solids concentration causes a large increase in viscosity (e.g., in Fig. 9.7 around 50 wt % solids). It is in this region, the so-called critical concentration range, that the slurry itself provides a maximum support for the coarse particles without creating an untenably high critical velocity. As implied by the viscosity curve, increasing the concentration of solids beyond the critical range results in sharp increases in pressure drop. More important, however, the critical velocity also increases sharply in this region. In terms of cost, this means that exceeding the critical concentration will require higher velocities with corresponding higher costs for pumping but with little real gain in increased system capacity.

During the preliminary task of selecting a transportation mode, there may not be enough information or sufficient time to carry out the experimental work necessary to establish the slurry rheology. For the purpose of a preliminary screening, however, a concentration of 40% solids by volume is an acceptable working figure, provided the concentration by weight does not exceed 65%.[22]

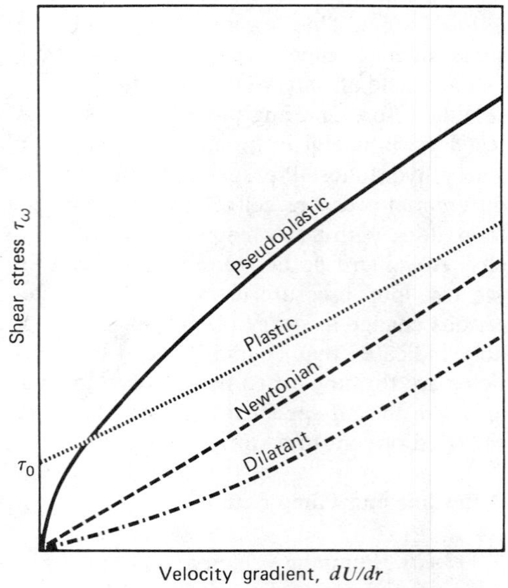

Figure 9.8 Fluid classification of slurries.[18] Reprinted with permission from *Chem. Tech.*, September 1971, pp. 552–562. Copyright by the American Chemical Society.

Rheology. Virtually all slurries exhibit non-Newtonian behavior. There are several characteristic shear stress–velocity gradient curves that can be representative of various slurries (Fig. 9.8). The most common is probably the Bingham plastic. With this type of behavior, the shear stress curve is shown as a straight line with an intercept on the shear stress axis. To evaluate any slurry pipeline system properly, the type of rheological behavior (i.e., pseudoplastic, Bingham plastic, Newtonian, or dilatant) must be clearly established.

Friction Losses. A previous investigation by Wasp successfully correlated the hydraulics of coal slurries that have a wide granulometry. Part of the slurry size consist was considered to be in homogeneous flow, part in heterogeneous or bed flow. The correlation method involved determination of the portion of the solids that was in homogeneous flow and the calculation of the pressure drop attributed to these solids. The remainder of the consist was then assumed to be in heterogeneous flow, and the pressure drop attributed to this portion of the solids was calculated. This was done for each size fraction; then the pressure drops from the homogeneous and heterogeneous portions were added to give the total pressure drop for the slurry. A discussion of this method and a demonstration of its versatility in handling a wide variety of materials was presented at the International Symposium for Solid-Liquid Flow in Pipes[24] and is not repeated here. Suffice it to say that slurry pressure drops may be accurately predicted by using laboratory-computer correlation methods.

Pump Stations. To date, virtually all long-distance slurry pipelines have used positive displacement pumps for obtaining the required pressures at pumping stations. Pumping stations are located according to the

[24] E. J. Wasp, T. C. Aude, R. H. Seiter and T. L. Thompson, "Hetero-Homogeneous Solid-Liquid Flow in the Turbulent Regime," in *Advances in Solid-Liquid Flow in Pipes and Its Application*, Iraj Zandi, Ed., Pergamon Press, New York, 1971.

ground profile and infrastructure access; however normal spacing is about 60–100 miles. Table 9.5 compares existing slurry pump capabilities. Pumping horsepower is calculated as follows:

$$hp = \frac{(\Delta P)\,(Q)}{1714e}$$

where hp = required horsepower
ΔP = required differential pressure (psi)
Q = Flow rate (g/min)
e = pump train efficiency (usually about 85% for positive displacement electric motor systems)

Total energy requirements for a slurry pipeline system include power for slurry preparation, pumping, and dewatering. As noted earlier, however, preparation energy is a credit to the power plant, hence should not be considered to be transport energy. Energy for pumping is dependent on the terrain traversed; however for initial comparison purposes, general energy consumption values shown in Table 9.6, which assume flat terrain, can be used. For long-distance (500-1000 mile) pipelines, Aude has estimated that transportation, including slurry preparation, pumping, and dewatering, requires the equivalent of 2½–3% of the energy transported.[20]

[25] T. L. Thompson, R. J. Frey, N. T. Cowper and E. J. Wasp, "Slurry Pumps—A Survey," paper presented at Hydrotransport 2, Coventry, England, September 1972. (Published by British Hydrodynamic Research Association, Fluid Engineering, Cranfield, England, 1973).

9.1.3.2 *Dewatering Terminal*

The principal aim in dewatering a slurry is to recover the solids with the minimum moisture content that still can be handled. Clarification of the liquid phase of the slurry is also required. In most cases centrifuges or filters are necessary.

Vacuum Filters. Continuous vacuum filters can be used for dewatering slurries. A typical vacuum filter incorporates a filter cloth attached to a moving frame. The frame moves through the thickened slurry, and the solids are caked onto the filter through the action of a vacuum. The frame continues its rotation to a point where the cake is removed by mechanical and pneumatic methods. The cake drops to a conveyor and the filtrate flows to a clariflocculator. The filter frame then moves back into the thickened slurry to repeat the cycle. For a practical guide on equipment selection and a better understanding of the subject of filtration, see Flood et al.[26]

Dewatering of slurry in Consolidation Coal Company's coal-slurry pipeline in Ohio was achieved by thickening and vacuum filtration, followed by thermal drying. (That power plant originally operated using rail-delivered coal.) See Wasp et al.[19] for a complete description of both this dewatering system and the Mohave system.

Centrifuges. Continuous centrifuges are often used for dewatering slurries. The solid bowl centrifuge consists of a rotating bowl

[26] J. E. Flood, H. F. Porter and F. W. Rennie, "Filtration Practices Today, Centrifugation Equipment," *Chem. Eng.*, 163–181 (June 20, 1966).

Table 9.5 Comparison of Existing Slurry Pump Capabilities[25]

Type	Maximum Pressures (psi)[a]	Maximum Flow (U.S. gal/min)[b]	Mechanical Efficiency	Maximum Particle Size	Comments
Plunger	3500–4000	920	85–90	8 mesh	Used for abrasive slurries
Piston	2500–3000	2,700	85–90	8 mesh	
Centrifugal	600–700	50,000	40–75	6 in.	Series installation, lower capital cost

[a] Maximum pressures refer to commercially viable slurry pipeline volumes.
[b] Positive displacement pumps are capable of these maximum flow rates only at pressures much less than the maximum pressures noted. For example, at 920 gal/min one manufacturer's 8-in. quintuplex plunger pump has a maximum pressure capability of 1000 psi.

Table 9.6 Pipeline Pumping Energy Requirements, Excluding Dewatering and Slurry Preparation[20]

Throughput (million tons/year)	Annual Btu × 10^9 Required per 100 miles Btu/ton-mile	Btu/ton-mile
5.0	135	270
10.0	190	190
20.0	270	135
30.0	330	110

and a rotating-screw conveyor section that revolves concentrically within the bowl. Slurry enters via a stationary feedpipe located within the hollow conveyor shaft. This feedpipe extends inward to a rotating chamber located within the screw conveyor. Here the slurry is accelerated and moves radially outward to the bowl. The liquid portion of the feed tends to form a cylinder of revolution within the bowl. Liquid depth is controlled by adjustable weir plates at the outboard end of the cylindrical section of the bowl. A rotating-screw conveyor keeps the solids moving up the conical slope. At the discharge ports the solids fall into a discharge hopper.

Recent tests have shown that with a screen bowl centrifuge, (Fig. 9.9) cake moisture can be reduced by 20–30% as compared with the cake moisture from a solid bowl centrifuge. A screen bowl centrifuge is divided axially into a conical section followed by a screen drum section. The drained solids in the conical section move by means of a rotating-screw conveyor to the screen section for final draining. The small quantity of solids passing through the screen is recovered in a thickener loop and recycled to the feed.

A slurry-dewatering facility using centrifuges is now in operation at the Mohave Generating Station, Clark County, Nevada. This system converts water-coal slurry into a coal cake containing about 20% total moisture. Slurry feed, drawn directly from an agitated-slurry storage tank, is heated and pumped to the centrifuges. From the centrifuge, the coal cake discharges directly to the pulverizer. The pulverized coal is then direct fired into the furnace. Centrifuge effluent contains a small percentage of very fine coal particles (−325 mesh). This coal is recovered in a thickening loop, where the fines are chemically flocculated and recovered in a thickener. The thickener underflow is directly fired in the boiler.

Water Requirements. Coal slurry pipelines usually use water as the vehicle, because of its wide availability. Actual requirements depend on the solids concentration pumped and the surface moisture in the coal delivered to the pipeline preparation plant. Typical water requirements are shown in Table 9.7.

Although the water requirements are sig-

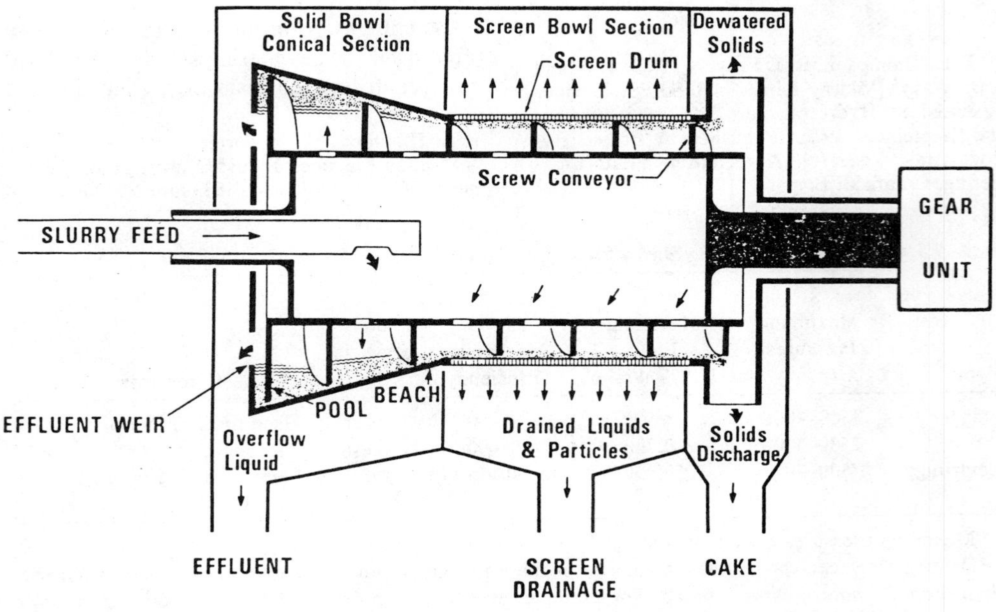

Figure 9.9 Screen bowl centrifuge.[21]

Table 9.7 Typical Water Requirements

Throughput (million tons/year)	Annual Water Requirements (acre-ft)
5.0	3,800
10.0	7,500
20.0	15,000
30.0	22,500

Source: Bechtel Corporation.

nificant, they normally are only about 15% of the required power plant makeup water. Hence the conventional practice is to use the slurry water as a part of the power plant makeup water.

Rights-of-Way. Rights-of-way for slurry pipelines are virtually identical to rights-of-way for oil and gas pipelines. Widths of 60–100 ft are adequate, and after the line has been buried, normal surface use of the property can continue. There is no technical problem in crossing rivers, roads, or railroads. There is a political problem in crossing railroads, however. Almost unanimously the railroads have refused to let the coal pipelines pass under them as they have allowed passage of lines for oil, petroleum products, gas, and water. Their objective is to prohibit competition in the movement of coal. Both project-specific legislation and eminent domain legislation (state and national) are under consideration.

9.1.3.3 *Capsule Pipelines*

Concept and Background. Capsule pipelining work began in 1958. Capsule pipelining is defined by Jensen[27] as follows:

> . . . a transportation system in which solid materials in the form of trains of cylinders or spheres are moved in a pipeline filled with a carrier liquid. The solid bodies have diameters slightly less than the pipe, and cylinders are 5 to 7 diameters long. The carrier liquid can be water, hydrocarbons or anything that can be pumped.

[27] E. J. Jensen, "Capsule Pipelining—The System and Its Potential," presented at Hydrotransport 3, Colorado School of Mines, Golden, Col., (May 1974). (Published by British Hydrodynamic Research Association, Fluid Engineering, Cranfield, England, 1974).

Capsules may be (*a*) rigid capsules comprised of cylinders or spheres cast from the material in the molten state, (*b*) semirigid capsules that consist of a flexible container filled with the solid to be transported, or (*c*) paste slugs made by mixing the comminuted solid with a liquid that is immiscible with the carrier liquid. In paste slugs the doughlike paste is formed into cylinders in extruders and inserted into the pipeline without a protective skin.

Much of the work on capsule pipelines has been done by the Research Council of Alberta underwritten by the government of Alberta.

State of the Art. As of 1976 there were no commercial capsule pipeline installations. When a complete system is developed, it will include capsule manufacture, capsule treatment before injection, injection machinery, pipeline hydrodynamics, pipeline pumps, booster stations and bypass systems, terminal retrieval systems, terminal liquid-solid separation and recovery, and instrumentation and control. To date, the bulk of the effort has involved pipeline hydrodynamics; however no general design equations covering all cylindrical capsules and pipelines exist. A model has been developed to predict the liquid velocity for a given capsule throughput and pressure gradient, but a general model for predicting both pressure gradient and liquid velocity for a given solids throughput does not exist.

Recently a breakthrough on a bench scale has been reported by Jensen for an in-line bypass and booster device.[27] The essential part of the bypass is a circular rotor with vanes. The off-line booster pump is connected to the housing of the bypass in such a fashion that the bypass assembly is pressure balanced.

Jensen summarizes the state of the art of capsule pipelines as follows[27]:

> Capsule pipelining is not a commercial reality and there appears to be a considerable reluctance on the part of private industry to venture into the implementation of the system and make it a part of the transportation network. It has not yet been possible to show convincingly that capsule pipelining is capable of solving specific problems more advantageously than existing systems. To do this requires

data from a demonstration line or full scale experimental line.

In looking forward to capsule pipeline applications, it is anticipated that initial applications will be for materials not amenable to slurry transport or materials with a high value per unit of volume whose inherent value can support a relatively high system transport cost. Coal does not appear to lend itself naturally to capsule pipelines.

9.1.4 Water Transport

"Transportation by water is the least expensive way to ship coal."[28] Water transport of coal includes barge and self-propelled vessels on the U.S. inland waterways system, the Great Lakes, and oceangoing transport. "The inland waterways system is a network of over 25,000 miles of navigable waterways. Coal is the largest commodity movement in the inland waterways market, accounting for 22% (1975) of the total tonnage flow."[28] About 117 million tons of coal and lignite was moved on the inland waterways in 1973.[29] Figure 9.10 is a map of the commercially navigable waterways of the United States.

9.1.4.1 *River Barge*

Size of barge tows is limited by the capability of the towboat, current velocity, width and depth of the channel, and lock size. Towboats have pushed up to 48 "jumbo" barges on the lower Mississippi River, where there are no locks (40 barges is more common, however). Fifteen jumbo barges (a 3 × 5 tow) is the maximum that can be towed for a single locking in a 1200-ft lock on the Ohio River.[31] The jumbo barge is 35 ft wide and 195 ft long; it has a 9-ft draft and a capacity of 1400–1500 short tons. It is well suited to locks on the Ohio, upper Mississippi, Illinois, and Tennessee rivers and is by far the most common type except in the upper Ohio Valley.[31] The largest tow is not always the most economical, however, since flexibility, turnaround time, size of dock facilities, and so on, must also be considered.

Almost all river coal is moved either in the jumbo barges or the smaller 175 × 26 ft barge, which is known as the "Standard" or "Pittsburgh Standard" coal barge. The Standard is well suited to the small locks on the Kanawha and Monongahela rivers and is the type used in the upper Ohio Valley.[32] The Standard carries about 900–1000 tons, compared to the jumbo's 1400–1500 tons.

Lock sizes are the major limitation in determining maximum tows. The inland waterways system contains approximately 114 locks. "The two most common sizes . . . are 110-ft by 600-ft and the more recent 110-ft by 1200-ft size. The average lock is designed to allow passage of vessels in 20 to 30 minutes."[33] Table 9.8 gives a summary of inland waterway locks.

Motive Power Fuel Requirements. "Tow size is presently at the maximum allowed by river and lock conditions, and towboats of the 10,000-horsepower size will handle the largest such tows."[35] It should be noted that on the inland waterways, "towboats" push rather than tow. Towboats generally range from 3000 to 10,000 hp.[32] They are usually

[28] "Using Waterways to Ship Coal," *1976 Keystone Coal Industry Manual,* McGraw-Hill Book Co., New York, 1976, p. 205.

[29] "Using Waterways to Ship Coal," *1976 Keystone Coal Industry Manual,* McGraw-Hill Book Co., New York, 1976, p. 212.

[30] "Waterways of the United States" by permission of The American Waterways, Inc., *1976 Keystone Coal Manual,* McGraw-Hill Book Co., New York, 1976, p. 217.

[31] "An Overview of the Potential Role of the Inland Waterways in an Integrated U.S. Transportation System," report prepared for U.S. Army Corps of Engineers and U.S. Dept. Transportation under Contract No. DACW 31-73-C-0082, October 1975, pp. 8–18.

[32] "River Transportation," *1971 Keystone Coal Industry Manual,* McGraw-Hill Book Co., New York, 1971, p. 170.

[33] "An Overview of the Potential Role of the Inland Waterways in an Integrated U.S. Transportation System," report prepared for U.S. Army Corps of Engineers and U.S. Dept. Transportation under Contract No. DACW 31-73-C-0082, October 1975, pp. 8–14.

[34] "Coal Transportation Capability of the Existing Rail and Barge Network, 1985 and Beyond," report prepared by Manalytics Inc., for Electric Power Research Institute, Palo Alto, Calif., EPRI EA-237, Project 437-1, September 1976, p. 53.

[35] "An Overview of the Potential Role of the Inland Waterways in an Integrated U.S. Transportation System," report prepared for U.S. Army Corps of Engineers and U.S. Dept. Transportation under Contract No. DACW 31-73-C-0082, October 1975, pp. 3–7.

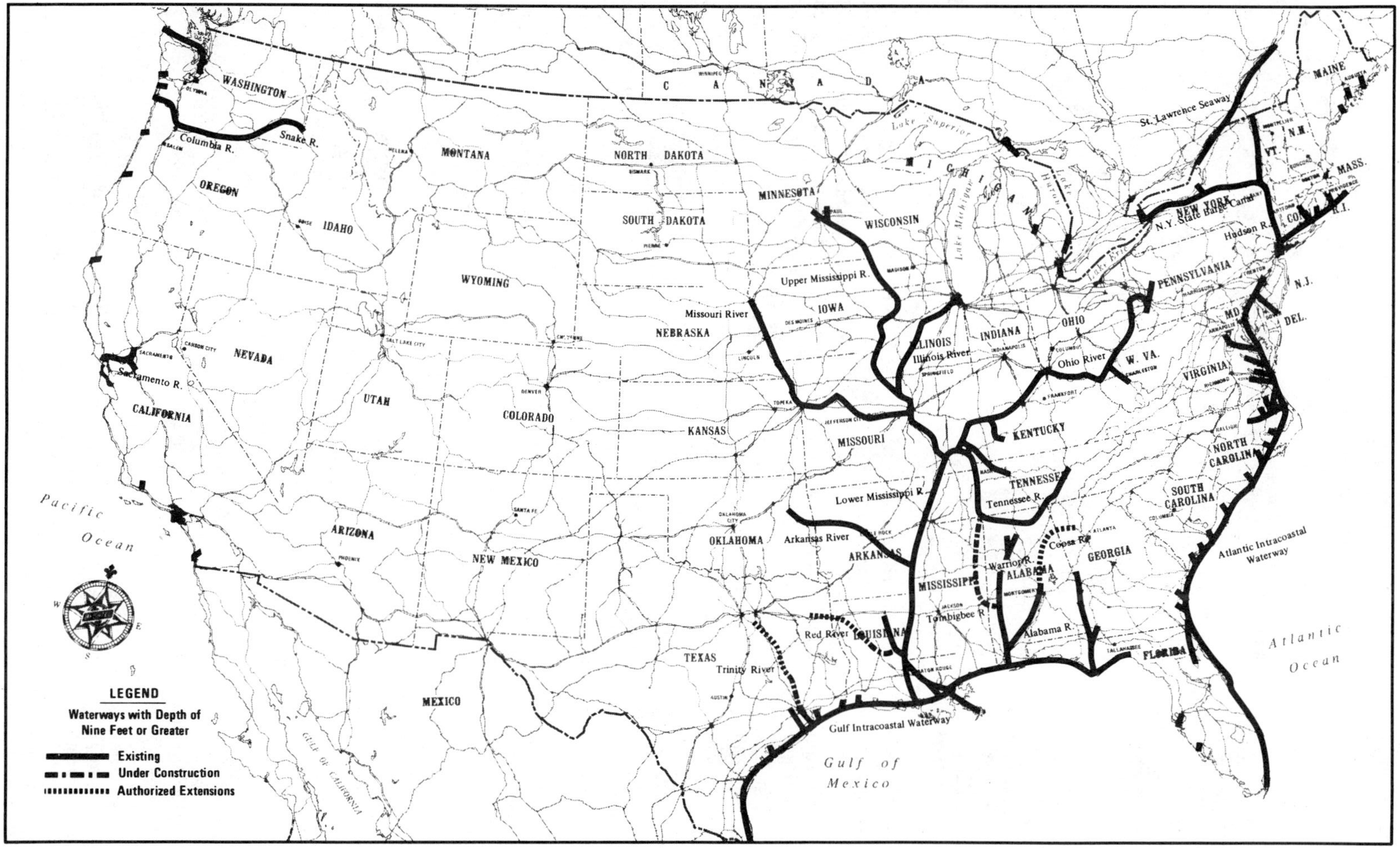

Figure 9.10 Commercially navigable waterways of the United States.[30]

Table 9.8 Summary of Inland Waterway Lock and Pool Characteristics[34]

Waterway	Distance (miles) Total	Distance (miles) Between Locks	Number of Locks Single Chamber	Number of Locks Multi-chamber	Number of Locks Total	Average Pool Size (miles)	Lock Size 600 × 110 ft	Lock Size Other
Upper Mississippi	858	669	25	4	29	24	25	8
Illinois River	327	247	8	—	8	35	7	1
Ohio River	981	957	4	17[a]	21	48	20	19
Allegheny River	72	55	8	—	8	8	—	8
Monongahela River	129	104	6	3	9	13	—	12
Kanawha River	91	52	—	3	3	26	—	6
Green River	108	54	2	—	2	54	—	2
Cumberland River	381	283	4	—	4	94	—	4
Tennessee River	652	580	6	3	8	83	6	6
Arkansas River	448	354	15	—	15	25	15	—
Gulf Intracoastal Waterway (West)	690	239	4	—	4	80	—	4
Intracoastal Waterway (East)	407	—	1	—	1	—	—	1
Port Allen–Morgan City Cut-off	64	25	2	—	2	25	—	2
Total	5208		85	30	114		73	73

[a] Includes one three-lock (McAlpine Lock).

propelled by two propellers driven by two to four diesel engines.[36] Speeds range from 3 to 12 mph, averaging about 6 mph.

Fuel requirements for inland waterway transport have been estimated by Rand Corporation, Oak Ridge National Laboratory, Carnegie-Mellon Institute for Research, the U.S. Department of Transportation, Peat Marwick Mitchell & Company, and the Federal Energy Administration. Their estimates range from 463 to 680 Btu/ton-mile with an average of 538 Btu/ton-mole.[37] Caution should be used when applying these figures, since the rivers wind and have a much longer actual transport mileage than the straight-line distance.

Loading and Unloading. Loading and unloading are very important because it is not unusual for the costs of these activities to equal or exceed transportation costs for short waterway transport.[38]

> River terminal capacity for the rapid transfer of large tonnages of western coal from unit trains to barges is presently (1974) limited. Two major river coal transfer and storage terminals are under development . . . ; a 20 million annual ton capacity terminal near Metropolis, Illinois, on the Ohio River and . . . a 28 million ton-per-year facility at a site near St. Louis.[39]

The facility at Metropolis will be able to load a tow of 15 jumbo barges with 22,500 tons of coal in 6 hr (i.e., 3750 tons/hr). Normal unloading rates for various types of unloaders were given by Grunthaner[40] (Table 9.9).

The continuous ladder unloader (Fig. 9.11) generally consists of two continuous bucket ladders suspended from a twin-girder cantilever. Grunthaner[40] explains its operation as follows:

Table 9.9 River Barge Unloading Rates[40]

Type of Unloading	Net Rate (tons/phr)
Stiff leg	200
Whirler	400
Straight, line Clamshell	900
Continuous ladder	1500–3000

[36] "An Overview of the Potential Role of the Inland Waterways in an Integrated U.S. Transportation System," report prepared for U.S. Army Corps of Engineers and U.S. Dept. Transportation under Contract No. DACW 31-73-C-0082, October 1975, pp. 8–19.

[37] "An Overview of the Potential Role of the Inland Waterways in an Integrated U.S. Transportation System," report prepared for U.S. Army Corps of Engineers and U.S. Dept. Transportation under Contract No. DACW 31-73-C-0082, October 1975, p. 5-4.

[38] "Using Waterways to Ship Coal," *1976 Keystone Coal Industry Manual,* McGraw-Hill Book Co., New York, 1976, p. 207.

[39] B. A. Ross and P. D. Martinka, "Western Coal Transportation—A Challenge," paper presented at American Institute of Mining, Metallurgical, and Petroleum Engineers Convention, Las Vegas, Nev., 1974, p. 9.

[40] G. L. Grunthaner, "River Transportation and the Public Utility," *Coal—Today and Tomorrow,* 13 (March 1964.).

[41] "Using Waterways to Ship Coal," *1976 Keystone Coal Industry Manual,* McGraw-Hill Book Co., New York, 1976, pp. 206, 207.

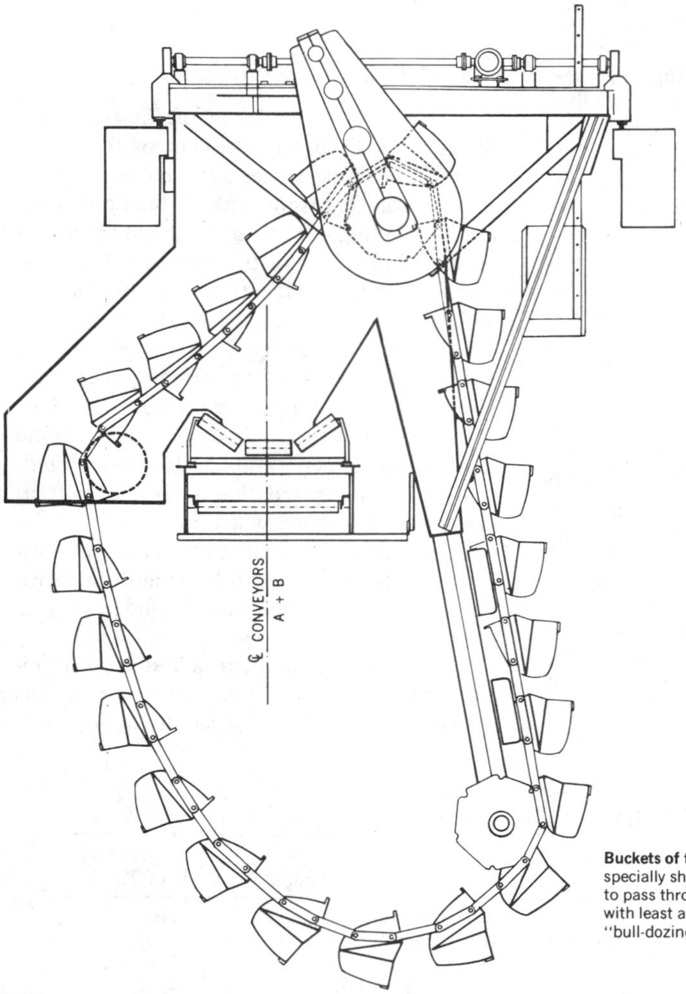

Buckets of the unloader are specially shaped so as to pass through coal with least amount of "bull-dozing."

Three passes under digging ladders of unloader efficiently empty barge: two with ladders together and one with ladders spread.

Figure 9.11 Barge unloading by continuous ladder unloader.[41]

This arrangement allows the digging ladders to be raised and lowered vertically. On the first pass, the bucket ladders are positioned together at the center of the cantilever and dig a trough through the center of the coal as the barge is being moved beneath the unloader. During the second pass, the ladders are spread to the sides of the hopper and dig to the barge inner bottom leaving a small window down the center to be removed in the clean-up pass.

Maintenance. Table 9.10 gives annual costs (1974) for typical inland waterway vessels. As shown, maintenance and repair costs for towboats can be 7% of total annual costs. Barge maintenance and repair costs are approximately 20% of total annual costs.[42]

[42] "An Overview of the Potential Role of the Inland Waterways in an Integrated U.S. Transportation System," report prepared for U.S. Army Corps of Engineers and U.S. Dept. Transportation under Contract No. DACW 31-73-C-0082, October 1975, pp. 8–25.

9.1.4.2 *Lake Vessel Transport*

Ever since the opening of the first Welland Canal in 1833 and the advent of the steam engine, the Great Lakes system has provided an important transport link for bulk commodities, including coal. The Great Lakes system is very important in the movement of coal because of the concentration of people and industry near its shoreline. Approximately 25 million people live in the Great Lakes megalopolis. Most of the Great Lakes' tonnage has been for the steel industry (i.e., iron ore, limestone, and coking coal). Some coal for power generation is also transported, and it is anticipated that with the shift to western coal, there will be a significant increase in thermal coal traffic. The two primary ports of entry will be Duluth-Superior in the United States and Thunder Bay in Canada.

The first transshipment and storage terminal on the Great Lakes for western coal was opened in June 1976. The facility in Superior,

Table 9.10 Annual Costs for Typical Inland Waterways Vessels[31]

	Average Annual Cost		Percentage of Total	
Cost Elements of a Towboat[a] (5000 hp/12-man crew)				
Towboat-related				
Fuels and lubricants	$264,000		32	
Maintenance and repairs	60,000		7	
Insurance—hull, protection and indemnity, property	48,000		6	
Depreciation	132,000		16	
Supplies	21,600		3	
Telephone and radar	1,200		1	
Subtotal		$526,800		65
Crew-related				
Wages and fringe benefits	$254,760		31	
Transportation	9,000		1	
Food and staples	19,200		2	
Miscellaneous	2,640		1	
Subtotal		$285,600		35
Total		$812,400		100
Cost Elements of a Barge (jumbo, covered)				
Depreciation		8,275		67
Repairs and maintenance		2,500		20
Insurance		1,650		13
Total		$ 12,425		100

[a] A. T. Kearney, based on field interviews, 1974 costs.

Wisconsin, is designed to handle Montana coal delivered by rail. The terminal is located at about the midpoint in the 1700-mile distance between the mine site and the power plant at St. Clair, Michigan. Transport is by a 770-ft self-unloader capable of moving about 43,000 tons of coal per trip (the largest coal cargo ever carried on the Great Lakes).[43] Round trips between Superior and St. Clair are made in 5½ days.(Vessels for this service that are under construction will be 1000 ft long and 105 ft wide; they will carry 62,000 tons of coal and will be able to discharge their cargo in 4 hr.)

The Superior terminal is designed to handle 12 million tons of coal annually and could be expanded to handle 20 million tons a year.[44] Coal delivered by 110-car unit trains is unloaded by automatic rotary car dumpers at a rate of 3500 tons/hr. The coal is then carried on an enclosed highline conveyor either to the stacker-reclaim facility or directly to the shiploader. The shiploader can operate at a rate of 11,000 tons/hr.

The unloading rate with the aforementioned 770-ft self-unloading coal vessel in 1976 was about 2000 tons/hr, allowing the vessel to be unloaded in less than 24 hr. Later addition of two radial stackers and a new conveyor system will increase the rate to 10,000 tons/hr and cut the unloading time to 5 hr.[45]

9.1.4.3 *Ocean Transport*

To date, ocean transport of coal has been primarily coking coal transported as a dry cargo. Conventional methods of loading and unloading, similar to those used in river and lake applications, are used. With the recent turmoil in the world petroleum picture, which resulted in crude oil prices many times their level of a few years ago, a new concept in world trade is developing. That concept is ocean transport of steam coal to the industrial nations of the world. Figure 9.12 shows world coal deposits near tidewater and distances for possible movements.

There are several possible schemes for transport of steam coal from the mine to overseas markets. Three of these schemes appear in Fig. 9.13. As shown, ocean transport can be coupled with pipeline or rail-land transport, and the coal can be loaded and unloaded in either dry or wet form. The technology for dry transport is well established; technology for slurry loading and transport is established for iron concentrate but not yet for coal. The Marconaflo process is illustrated in Fig. 9.14. Marcona Corporation presently operates 50,000-dead weight tons (DWT) oil-slurry carriers for the transportation of iron. Fraser and Connolly[47] describe the system as follows:

> The Marconaflo system involves one or more of the following process steps:
>
> 1. Loading into a storage tank, or in the hold of a ship.
> 2. Settling of the solids followed by decanting and/or draining off the excess liquid.
> 3. Repulping the settled solids by introducing high pressure liquid through rotating Marconajets projecting into the tank.
> 4. Collecting the repulped solids in appropriate sumps for pumping to storage and processing.
>
> The key to the system is the Marconajet, a rotating device which introduces a high pressure fluid (usually water) into the compacted solids in order to repulp them for easy pumping. The optimum fluid pressure will depend on many factors including the degree of compaction and specific gravity of the settled solids. For example, pressures in excess of 350 psi are used with fine iron ore concentrates (particle size: 65% −325 mesh); 200 psi appears to be satisfactory for solar salt (particle size: 95% +14 mesh).

This process has been applied to iron slurries,

[43] "Detroit Edison Dedicates Superior Coal Terminal," *Skillings Min. Rev.*, 18 (July 3, 1976).

[44] E. J. Jensen, "Capsule Pipelining—The System and Its Potential," presented at Hydrotransport 3, Colorado School of Mines, Golden, Col., (May 1974). (Published by British Hydrodynamic Research Association Fluid Engineering, Cranfield, England (1974) p. 19.

[45] "A New Coal Handling System," *Bechtel Briefs*, 4–5 (August 1976).

[46] E. J. Wasp and R. H. Derammelaere, "International Steam Coal: The New Energy Competitor," paper presented at Primer Seminario sobre la Utilizacion Integral del Carbon, Bogotá, Columbia, March 1974, unpublished.

[47] M. J. Fraser and L. P. Connolly, "Marconaflo and Its Use in New Mineral Developments," *Can. Min. Metal. Bull.*, March 1972.

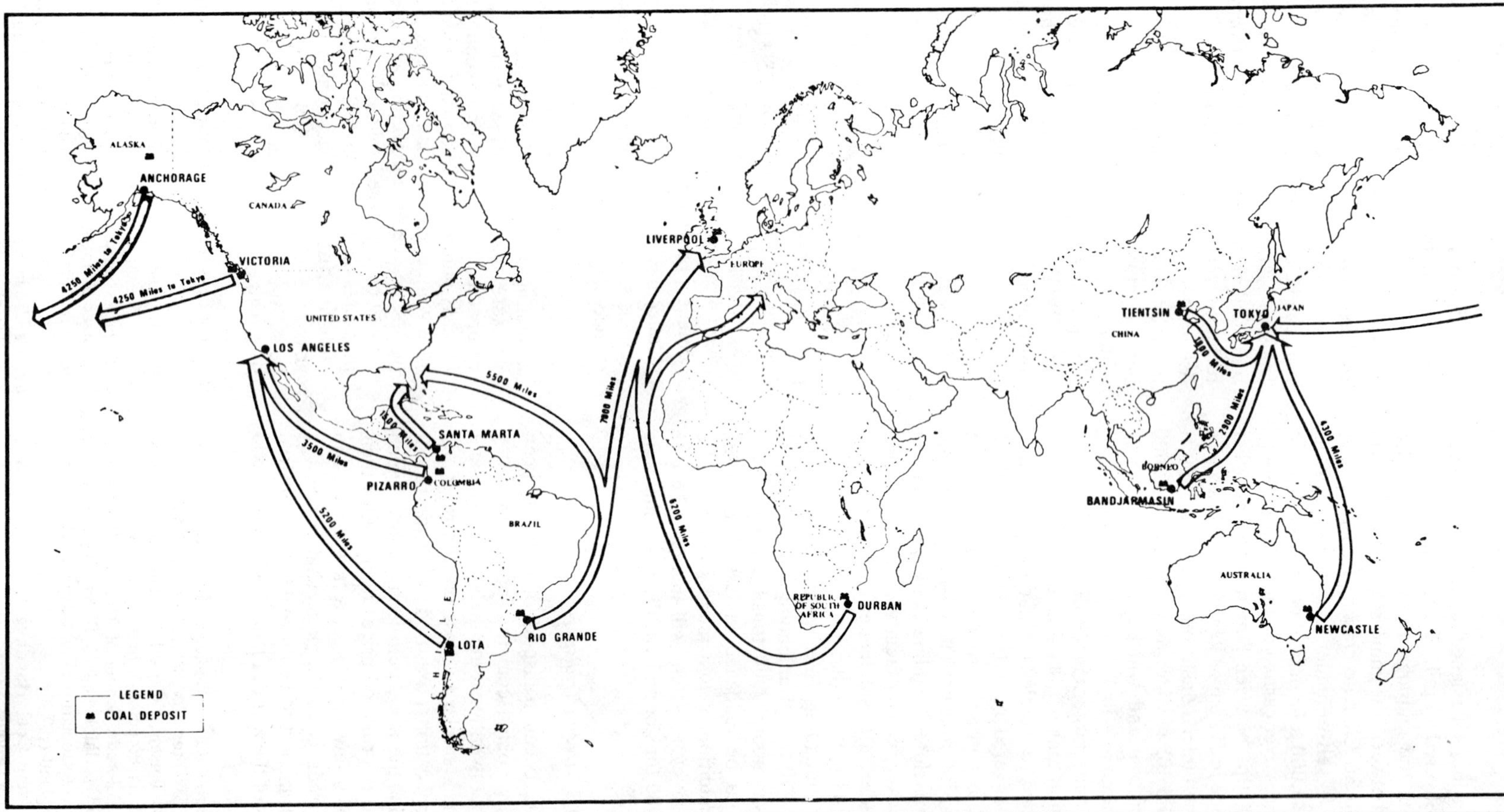

Figure 9.12 Coal deposits near tidewater and projected movements.[46]

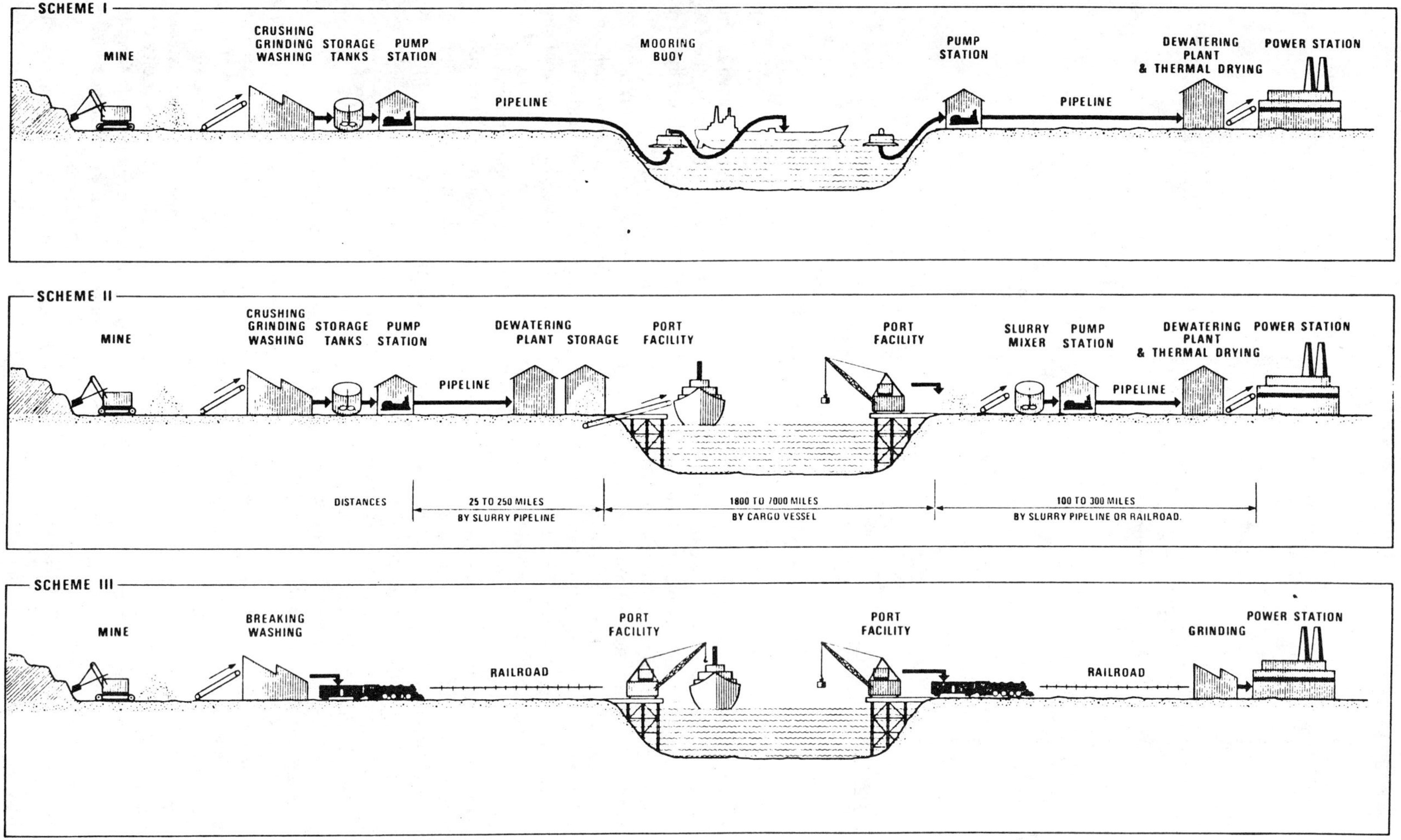

Figure 9.13 Concepts for overseas transportation of steam coal.[46]

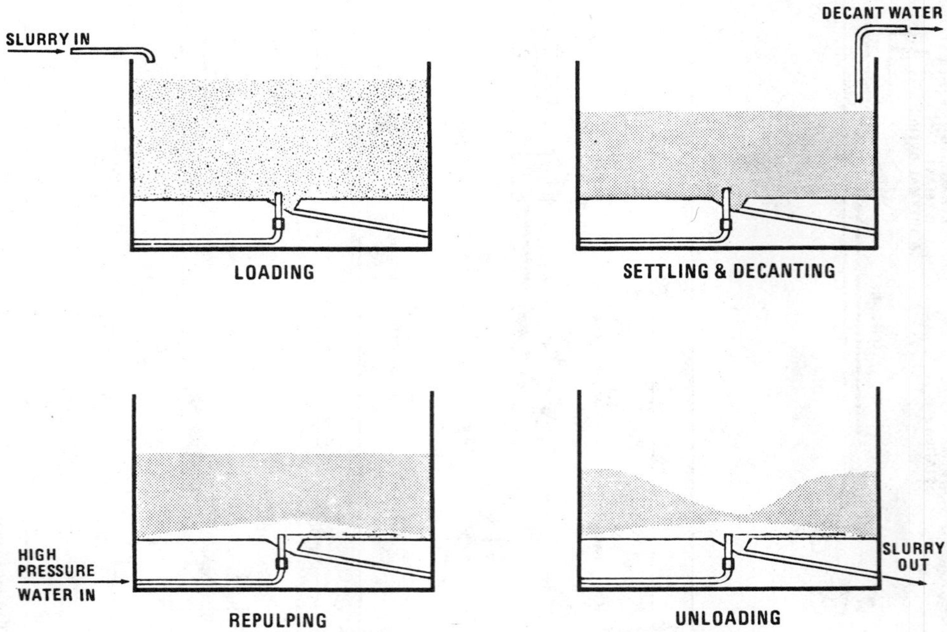

Figure 9.14 Basic elements of the Marconaflo system.[47]

which lend themselves to rapid gravity draining of the liquid phase, leaving the bulkhead with a firm mass of solids (around 90% solids by weight). Coal slurries, unfortunately, do not exhibit the same rapid draining characteristics, primarily because iron solids have a specific gravity of about 5.0, whereas the specific gravity of coal is 1.3–1.5. Transportation of the coal without adequate water removal could endanger the ship's stability if the center of gravity shifted and remained dislocated.[48] Furthermore, shipping excess water is uneconomical. Therefore additional work in this area must precede commercialization. The incentive, of course, is to avoid port size limitations by loading or unloading offshore (e.g., using a mono-mooring buoy) and taking advantage of the superior economics of larger vessels.

In Scheme II of Fig. 9.13 the pipeline product is dewatered at the coast and shipped in filter cake form. Conventional port facilities are used. In Scheme III the coal is only broken (e.g., 2 × 0 in.) and washed, and rail is used for overland transport. Here, of course, additional grinding is required at the power station to reduce the coal to the 150 mesh × 0 in. size required for burning in a pulverized coal boiler.

In terms of tonnage, coal is currently the third largest commodity of international commerce, exceeded only by oil and iron ore. Much of this coal tonnage is coking coal and is carried in relatively small ships (i.e., 60,000 DWT or less) using conventional methods of loading and unloading.[47] Supercolliers are also in use, and their numbers are increasing. A significant body of literature and experience in dry bulk cargo ocean transport exists and is not covered here.

9.1.5 EHV Transmission

Transmission by extra high voltage (EHV: 345 kV-A and above) transmission provides an alternative means to transport the energy derived from coal. Coal is delivered to the power plant and burned to produce steam, and ultimately electricity. This electricity can then be transported either in ac or dc form to the load centers. Owing to resistance losses in transmission, a larger power plant is required if long-distance EHV is used instead of transporting coal to a load center sited power plant.

[48] B. Fagerberg and A. Stavang, "Determination of Critical Moisture Contents in Ore Concentrates Carried in Cargo Vessels," paper presented at the 1st Int. Symp. Transport and Handling of Minerals, Vancouver, B.C., Canada, October 1971.

9.1.5.1 *Capacity Relationships*

As with other transport systems, EHV exhibits economy of scale. The 1970 National Power Survey[49] states:

> The cost per unit of energy transfer decreases with escalation of voltage levels even though total capital costs increase, because the capacity of transmission lines in stability-limited systems increases approximately as the square of voltage, while total cost increases at a lower rate. In addition, the line losses per unit of capacity are usually less with higher voltages. [For comparative costs, see Section 9.1.6.]

Voltage levels of 765 kV have been established for long-distance transmission that uses alternating current. The American Electric Power Company is an example of just such a long-distance operation, transmitting 765 kV over a 670-mile network, the first segment of which was energized in May 1969.[50] Higher voltages of 1100 kV, may be feasible in the future. High-voltage direct-current (HVDC) transmission of 400–600 kV is feasible and proved, and approximately a dozen major HVDC projects are in operation. Among these is the Northwest-Southwest Pacific Intertie Group, commissioned in 1970, which has a 1440-MW rating, is 846 miles long, and has a 400-kV voltage to ground.[51] HVDC has advantages over alternating current because of its much higher power density within a given right-of-way and the possibility of providing extremely rapid power control. However there are few installations, primarily because of the high cost of terminal equipment for direct current transmission. In addition, HVDC reliability to date has been lower than the United States requires for its bulk power supply.[51]

9.1.5.2 *Line Losses*

Transmission line I^2R losses vary both with the electrical load and with the energy transport distance. Figure 9.15 shows typical re-

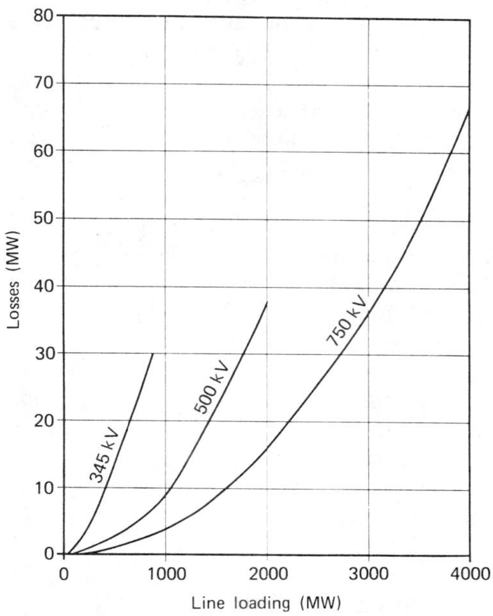

Figure 9.15 Resistance losses per 100 circuit miles of transmission.[49]

sistance losses per 100 circuit miles of transmission for 345, 500, and 765 kV, which are typical EHV line designs now in operation in the United States. Whereas the losses may be tolerable for distances of a few hundred miles or less, they become extremely significant for longer lines. For example, for a delivered load of 9000 MW using a 600-kV HVDC line over 1500 miles, a loss of 1400 MW (15.5%) is estimated.[52]

In selecting a power plant site, availability of water for cooling tower makeup is one of the main considerations. Therefore water availability economics may rule in locating a plant. A central point may be selected rather than a plant at the mine site or at the load center. Therefore EHV has application in spite of higher costs and losses. In addition, EHV has an advantage over other forms of energy transportation when connected to an integrated grid system. Reliability and firm power capacity are increased by such an interconnection.

9.1.6 Comparative Economics

In comparing coal transport alternatives, a major factor is cost—not only first cost but operating costs and the cost of inflation.

[49] "The 1970 National Power Survey," Part I, Federal Power Commission, p. I-13-7.

[50] G. S. Vassel, R. M. Maliszewski and N. B. Johnsen, "Experience with the AEP 765-kV System," *Proc. IEEE,* 1337–1344 (December 1, 1971).

[51] "HVDC Is No Longer Just an Exciting Concept, But a Practical Planning Option," *Electr. World,* July 1, 1974.

[52] "Coal Pipeline Beats Out Transmission Lines," *Electr. World,* **184**(12), 59 (December 15, 1975).

Unfortunately good general comparative costs on identical bases for all transport modes are not readily available. This is not difficult to understand, since transport costs are strongly influenced by terrain conditions, sunk cost elements (e.g., existing rail trackage with available capacity), and most important, inflation. The 1970s were witness to some of the most severe inflation in recent U.S. history. As a result, normal pricing and estimating practices were invalidated. At the same time, the implementation of environmental protection legislation began to have a significant impact on costs and on the time required to bring projects to fruition. Therefore the comparative data given here are from the most recent sources possible, to reflect these significant changes.

One recent study by Ebasco[52] compared long-distance coal energy transportation by rail, pipeline, rail-barge combination, and EHV transmission. This analysis was done for transport distances of 500 to 1500 miles and power plant sizes of 1600–9000 MW. As shown in Fig. 9.16, pipelines are most economic for long-distance, high-tonnage applications. Where some water transport is applicable, however, the rail-water combination is competitive. The cases analyzed assume that existing tracks would be utilized.

Another source of comparative costs is a recent study by the U.S. Department of the Interior.[53] Table 9.11 is adapted from this study for transport of 25 million tons per year, a distance of 1000 miles. As shown, rail and pipeline transport costs are virtually identical (before taking tariff escalation into account). Costs for EHV transmission are much higher than for either rail or pipeline transport costs.

Figure 9.17 compares rail, barge, and pipeline unit costs of transportation (annual costs of capital plus direct operating costs) as a function of distance and tonnage. In interpreting this chart, as well as the other cost comparisons in this section, the following generalizations should be considered:

1. Water transport is almost always the most inexpensive way to move coal. Since little coal is located adjacent to the nation's waterways, water transport usually involves another mode of transport to feed it. Combinations with rail, pipeline, and truck are all possible. See Table 9.12 for comparative water transport costs.
2. Truck transport is not competitive with rail or pipeline for high-tonnage, long-distance transport of coal (e.g., more than 50 miles and 1 million tons per year). Its primary application is in moving coal short distances from mine mouth to preparation plant or in transfer to a long-distance transport mode.
3. EHV transmission of coal (converted to electrical energy) is almost always more expensive than pipeline or barge and often more expensive than rail. Specific siting and system considerations for power plants, however, may dictate long-distance EHV transmission in spite of transport costs. Figure 9.18 shows EHV transport costs in mils per kilowatt-hour as a function of megawatts for a 200-mile transmission distance.
4. Railroads and slurry pipelines are usually the alternatives for high-tonnage, long-distance transport of coal. Costs must be compared for each individual coal movement; however, some generalities can be stated: (*a*) Railroad distances are normally longer than pipeline (often 30% or more) because of grade restrictions; therefore care must be taken when applying cents per ton-mile transportation rates. (*b*) Railroads often have very competitive initial tariffs when no new trackage has to be constructed. Where new trackage is required, pipelines are normally more economical. (*c*) Railroads have an advantage where there are numerous small delivery points. (*d*) Inflation effects are critical in an evaluation of pipeline versus rail. Pipeline transport tariffs are capital intensive; since about 70% is related to capital, only 30% is subject to inflation once a system has been installed. Railroads are the opposite, with 70% or more subject to inflation. Figure 9.19 shows the relative effects of inflation on pipeline and on rail tariffs, assuming that both start at about the same

[53] U.S. Bur. Mines, Hearings before Subcommittee on Energy Research, Development and Demonstration, 94th Congress, Vol. 4, No. 61, January 29, 1976, p. 406.

[54] E. J. Wasp, "Comparative Economics—Slurry Pipelines," lecture given at Hydrotransport 4, Banff, Alberta, Canada, May 1976. Unpublished.

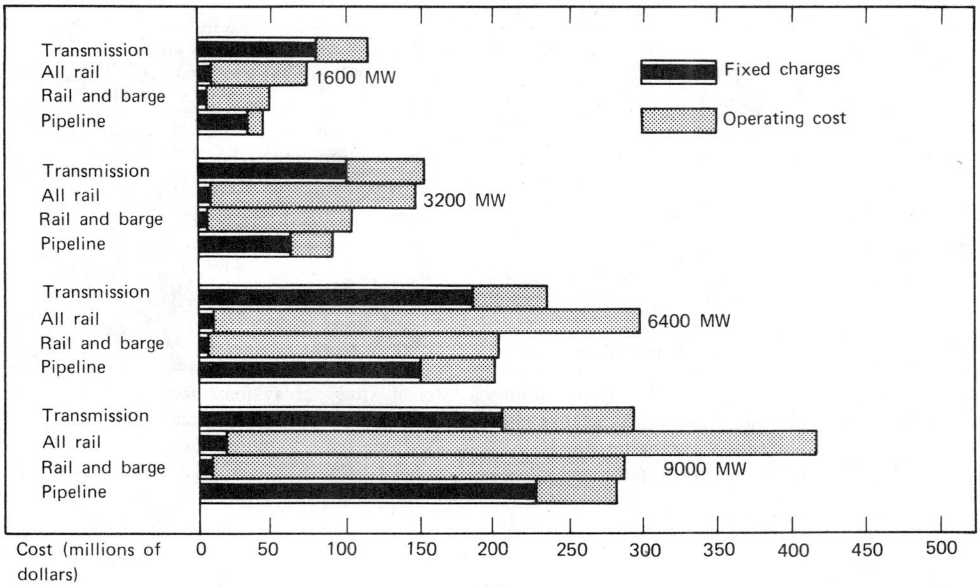

(a)

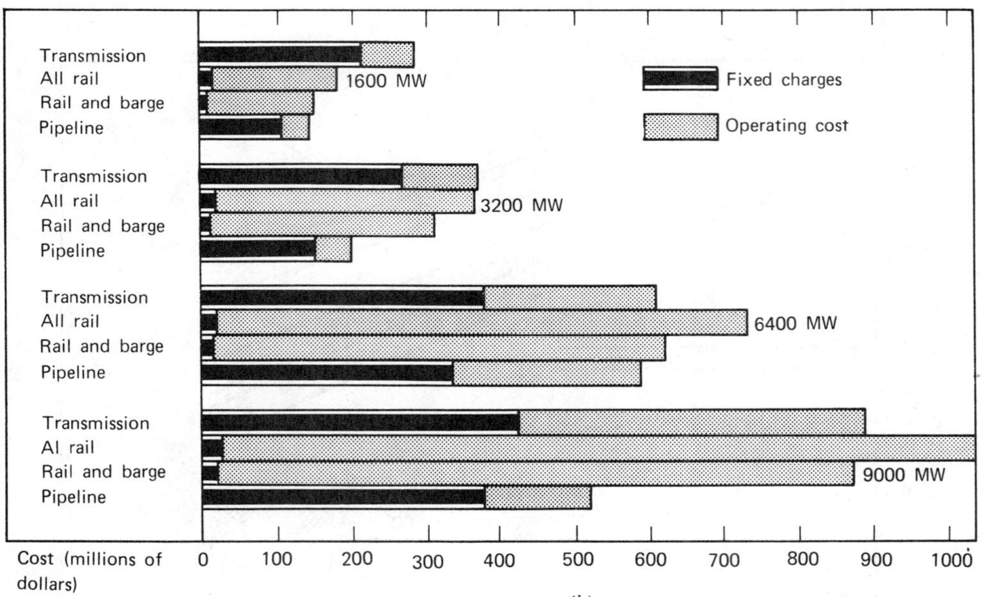

(b)

Figure 9.16 Comparative coal energy transport costs. (*a*) Annual levelized costs at 500 miles show dramatic effect of escalation of operating costs. (*b*) At 1500 miles pipeline increases advantage in levelized costs as power increases.[52] Reprinted from December 15, 1975, issue of *Electrical World* © copyright 1975, McGraw Hill, Inc. All rights reserved.

tariff level. (Since this figure was prepared the inflation rate has increased dramatically which favors pipeline systems even more.)

Cost comparisons under the best conditions are hazardous. The figures presented here should be considered as general screening devices only, to be supplemented by more detailed analysis. As costs inflate with time, the relative transport costs will change and could alter the comparisons presented. In

Table 9.11 Comparative Transport Costs[53]

Item	Delivered Cost[a] Including Coal ($ per million end-use Btu)		
	Unit Train	Slurry Pipeline	EHV
100% Equity	$6.23	6.18	8.20
75% Debt	4.56	4.43	5.74
	(Trillion Btu)		
Trillion Btu delivered as electricity	190	184	157

[a] Based on U.S. Department of Interior study of systems for providing coal-based energy to users 1000 miles southeasterly from eastern Wyoming coalfields. All cases assume that 25 million tons of coal is mined per year. Cost basis is early 1975.

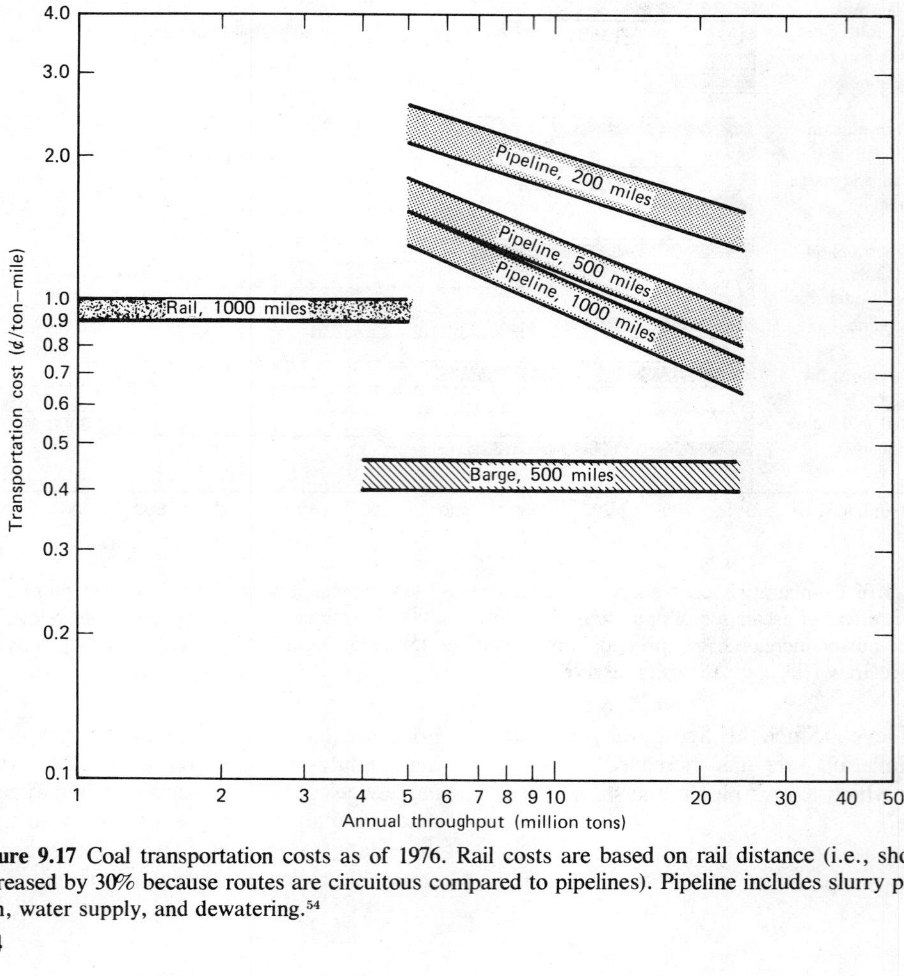

Figure 9.17 Coal transportation costs as of 1976. Rail costs are based on rail distance (i.e., should be increased by 30% because routes are circuitous compared to pipelines). Pipeline includes slurry preparation, water supply, and dewatering.[54]

Table 9.12 Transportation Economics[31]

Item	Barge	Rail	Truck
Transport cost (¢/ton-mile)[a]	0.1–1.1	Single/multiple car, 0.8–2.5; Unit train, 0.5–1.0 (Class I)	2.0–4.0 (Rail competitive)
Federal subsidies, fiscal year 1972 ($ million)[b]	291	(154.3)	325
Federal expenditures for right-of-way not covered by user charges ($ million)[b]	291	—	325
Federal subsidy to date (1974) ($ million)[b]	5700[c]	516[d]	?
Percent of revenues of			User charges Class I & II M.C.
Annual right-of-way costs[a]	—	21.0	5.9
Average annual operating expense[a]	77.8	72.4	91.5
Average annual depreciation[a]	7.5	6.9	4.1
Average annual federal taxes[a]	3.5	0.8	1.7
Average annual cash flow[a]	14.1	10.8	6.2
Average annual return on investment[a]	6.6	3.9	2.1
Average annual return on equity[a]	10.6	2.7	12.9
Average revenues as percent of assets[a]	52	34	205

[a] Kearney Report.
[b] U.S. Railway Association—preliminary plan.
[c] Association of American Railways—new construction and operation and maintenance.
[d] Aviation Advisory Service—considered repaid by tariff reduction to federal government.

addition, technological breakthroughs can be expected to alter these costs in the future.

9.2 STORAGE AND HANDLING

The increasing pressure to use coal as a source of energy to replace oil and gas in the production of electricity and in other industrial uses will inevitably increase the handling and storage of coal.

The modern power plant of 1000–3000 MW requires large piles of coal for both active and inactive (force majeure) storage. Handling these large quantities of coal in and out of stockpiles presents problems in the light of tighter environmental requirements, especially in and around populated areas.

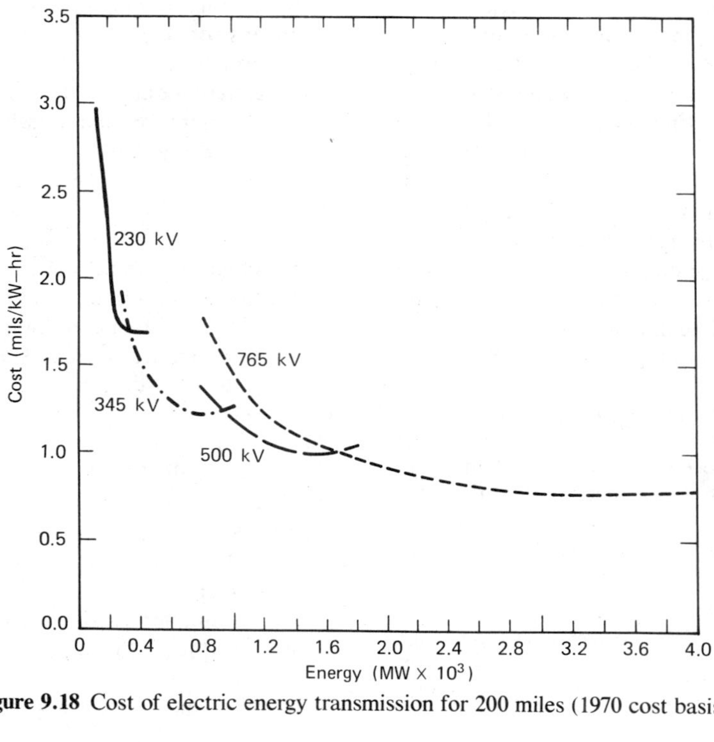

Figure 9.18 Cost of electric energy transmission for 200 miles (1970 cost basis).[49]

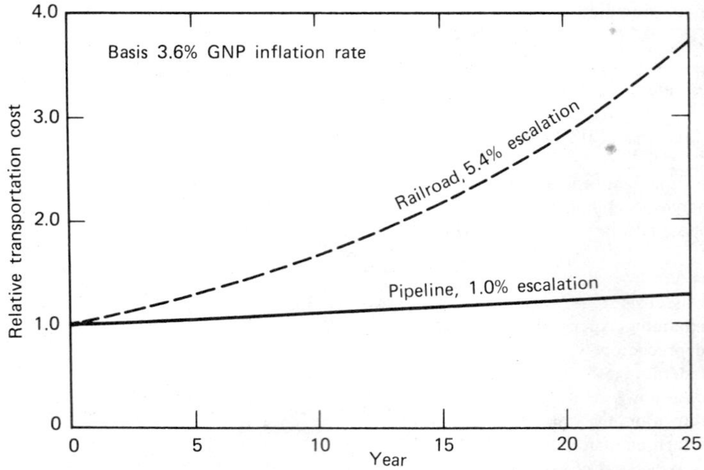

Figure 9.19 Effect on coal transport costs of inflation, based on 3.6% GNP inflation rate. *Source:* Bechtel Corporation.

The advent of coal slurry pipelines will involve the storage and handling of fine dewatered coal with controlled surface moisture. Such storage will be in large volumes, either in open piles or in new forms of covered storage.

9.2.1 Storage

The production of any raw material from mine to use point involves storage of that material at many points en route, and coal is no exception. With the future increase in the use of coal in utilities and industry, increasingly larger quantities will be moved by ship, barge, railroad, and other means. This necessitates large storage systems at transfer points between one mode of transport and another to allow for the different schedule patterns. Storage may be defined as space or material provided to ensure that future requirements will be met at a use point. Colijn[55] defines surge as follows:

> Space or material provided in a materials handling system to accommodate short-term fluctuations in materials handling rates between two adjacent components and which permits a system or groups of systems to operate at an average design handling rate.

In the case of an electric utility or a process industry using coal as its chief fuel, there are two important types of storage—namely, live and dead storage.

Live storage either is immediately within the reach of the principal means of reclaiming equipment or is subject to gravity flow.[55] An example in a coal-fired utility would be the active coal surge pile and the silos feeding the boilers.

Dead storage requires secondary or auxiliary means for reclamation, and, in a power plant, normally consists of a compacted pile with a capacity of about 60 days usage sealed against deterioration and weathering for environmental protection.

9.2.1.1 *Stockpiles*

Storage of large quantities of coal is generally accomplished in open stockpiles on the ground. Under certain conditions required to meet special environmental standards it is necessary to provide a roof over the stockpile.

In general, open-ground storage costs between \$5 and \$10 of capital per ton of storage capacity. Covered ground storage might cost between \$15 and \$20 per ton.[55] However considerably higher figures may be encountered for small installations.

Colijn[55] summarizes the basic criteria for any ground storage facilities as follows:

1. Optimize land usage by attaining highest

[55] H. Colijn, "Design Considerations for Storage and Reclaim Systems," in *Annual Seminars on Bulk Materials,* Vol. 1, University of Pittsburgh, 1976.

tons per acre values, consistent with storage and environmental requirements.

2. Maintain easy access to stored material.
3. Place the material in such a way that it can be reclaimed with the least effort.
4. Maximize the load-distance efficiency factor.
5. Minimize the overall cost in terms of dollars per ton handled, combining both operating and capital costs.
6. Maintain or improve the uniformity, integrity and quality of the stored material.

The characteristics of the land available for an open stockpile are important in the choice of the overall storage system. Size, shape, and load-bearing capacity determine the pile height and quantity that can be stored. In preparing ground for a stockpile, the topsoil must be removed and a stabilizing layer of crushed rock or similar material laid down. Adequate site drainage must be provided. Utility pipes and conduits should not run under piles. Long stockpiles should have a narrow side facing the prevailing winds to minimize dusting and oxidation.[56] The most common methods for the open stockpiling of coal are shown in Fig. 9.20. An indication of approximate storage volumes is given as a function of pile height, assuming a 40° angle of repose.[55]

A conical pile is laid down by a fixed stacker of appropriate height; a radial pile is formed by means of a pivoted radial stacker, which may have a boom with a radius of 250 ft or more; single- or double-windrow-type piles are formed by means of a traveling rail mounted single or double boom stacker. Single-windrow piles can also be formed by means of an overhead conveyor and tripper. The stockpiles just described are generally for live storage, and the amount reclaimable depends on the method of reclaiming.

The "ramped" and "block" piles in Fig. 9.20 are generally used for dead storage and require mobile equipment such as front-end loaders, bulldozers, and/or trucks, and scrapers to form these piles. A typical example is a force majeure coal pile for a utility or other large industrial user of coal as a primary fuel. Such piles may contain many thousands of tons and are generally surface sealed (e.g., by a chemical or bitumastic-type spray that dries to form a "skin") to prevent deterioration and environmental problems.

The following are guidelines for the capacity of coal stockpiles at various use points in the transportation train from mine to utility or industrial plant[55]:

1. *Shiploading or unloading:* 1½–2½ times maximum size of vessel anticipated, limited by port draft requirements (also a function of arrival times).
2. *Train loading or unloading:* 1½–2 times maximum train size—generally a unit train for coal is 10,000 tons.
3. *Barge loading:* 1½–2 times the weekly throughput.
4. *Barge unloading:* 2–3 operating shifts.
5. *Truck loading or unloading:* usually one day throughput.
6. *Overland conveyor transportation:* 1½–2 days supply.
7. *Power plant:* 2–3 months supply as force majeure or dead storage. Active or live storage will depend on the load swing characteristics of the power plant.

Figure 9.21 illustrates a typical large utility with surge and ground storage of coal. The coal may be brought in by barge if the facility is adjacent to a large river transportation system, or by unit train. Such unit trains will be unloaded at rates of 3000–4000 tons/hr, using bottom-dump pits, or rotary dumped with swivel couplings on the cars. Stacking out can be by stacker or stacker bucket wheel reclaimer. During stacking a portion of the feed may be diverted to the reclaim system and directed to the boiler feed bunkers.

In the arrangement pictured in Fig. 9.21 the live surge piles are on one side of the yard conveyor belt and may contain up to 100,000 tons of coal. On the other side of the yard, belt dozers form the dead pile of perhaps a million tons to serve as a reserve for emergency supply situations such as strikes. Such a dead pile is surface sealed.

In recent years there has been much pressure from environmental groups to limit the dusting from open coal stockpiles, especially

[56] W. F. Lauson, "Indoor-Outdoor Bulk Materials Storage Facilities," in *Annual Seminars on Bulk Materials*, Vol. 1, University of Pittsburgh, 1976.

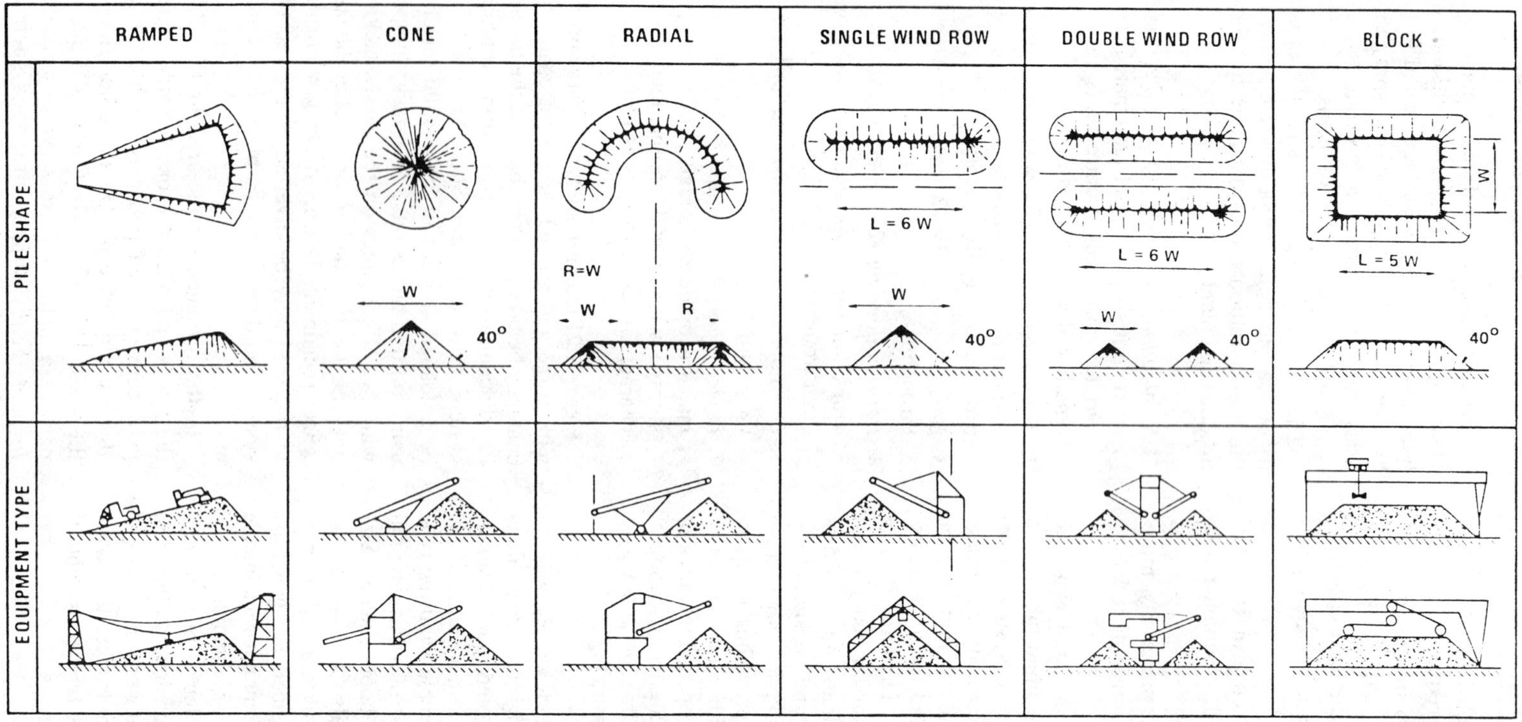

PILE HEIGHT (Feet)	25 50 75 100	25 50 75 100	25 50 75 100	25 50 75 100	25 50 75 100
BASE WIDTH (Feet)	60 120 180 240	60 120 180 240	60 120 180 240	2 x 60, 2 x 120, 2 x 180, 2 x 240	100 200 300 400
VOLUME (10^3 Cu. Yd.)	.85 6.8 23.5 55	5.8 48 164 385	10.5 85 294 685	21 170 588 1,370	46 370 1,250 2,950

Figure 9.20 Methods of stockpiling bulk material.[55]

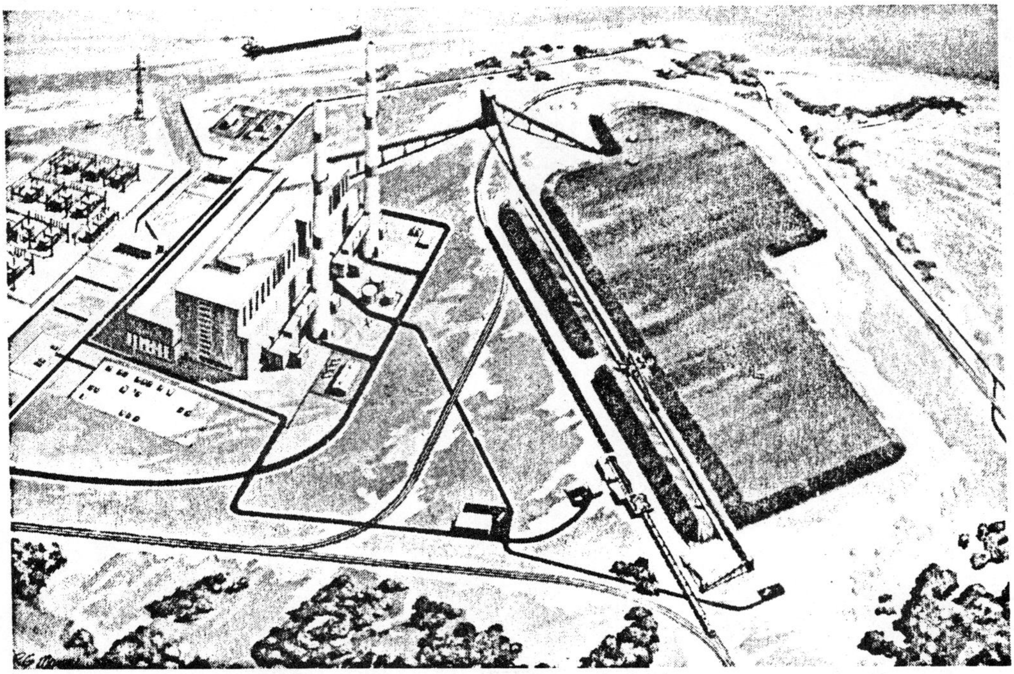

Figure 9.21 Surging and ground storage of coal for power generating station.[56]

those that are situated near populated areas. One solution has been to design large A-frame structures mounted over a V-shaped and lined slot in the ground (Fig. 9.22). Economic considerations obviously limit the span of such structures as well as the length. Coal is stacked out in such a structure generally by overhead conveyor and tripper mounted in the apex of the roof structure. Reclaim is by an automated rotary plow feeder worked in a slot formed in the lower part of the structure. A belt conveyor in the tunnel takes the reclaimed coal away.

9.2.1.2 *Bins and Silos*

As an alternative to open stockpiles on the ground and stockpiles within enclosures, coal is frequently stored in silos or bunkers. In general, this method of storage is more expensive than open-stockpile storage and the capacity is limited by the feasible size of such silos and bunkers. The large-diameter silo with a flat bottom will have multiple outlets to provide for a maximum of live storage by gravity flow.

Bunkers with a height-to-diameter ratio of 2 or 3 can generally be made self-cleaning by having a cone bottom of suitable angle to provide gravity flow through a single outlet. The flow of coal or any material from a bunker is discussed in detail in Section 9.2.2.4.

Silos and bunkers are constructed for coal storage in either concrete or steel and are generally round in cross section. Where single- or multiple-outlet cones are provided, a lining of stainless steel is a great advantage to ensure avoidance of corrosion, as well as hangup of the coal.

The construction of silos and bunkers requires careful design considerations of the subsurface conditions. Eccentric charging and eccentric withdrawing of the coal from the silo must be controlled to avoid dangerous offset loading on the foundations.

Silos and bunkers inherently offer a number of unique benefits. They control the dust and moisture content of the coal as well as runoff from rainwater, thereby making the storage unit environmentally more acceptable. Silos can also provide blending capabilities through the use of several units feeding a single reclaim conveyor system.[57]

[57] J. E. Sadler, "Coal Silo Design Considerations," *9th Annual Seminar on Bulk Materials, University of Pittsburgh, 1976.*

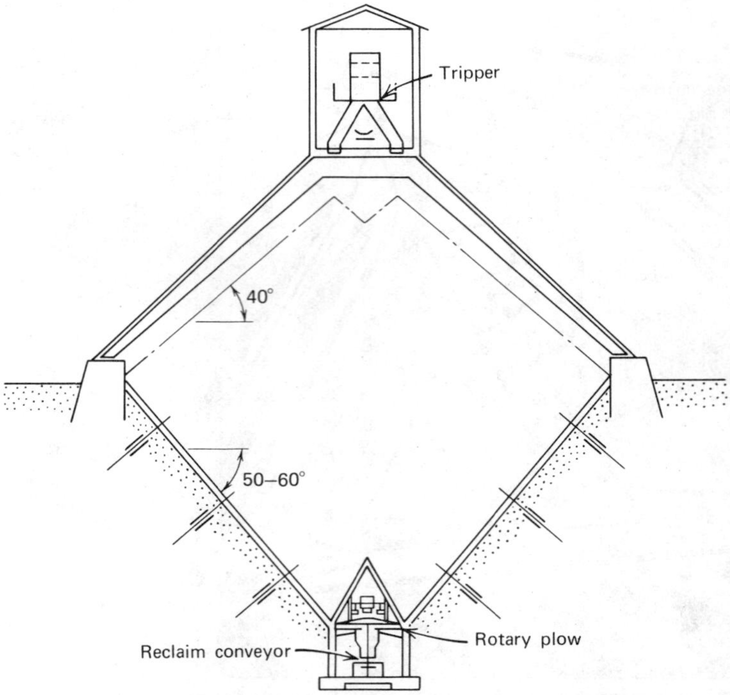

Figure 9.22 An A-frame structure designed to limit dusting from open coal stockpiles. *Source:* Bechtel Corporation.

Bins and Bunkers for Handling Bulk Materials—Practical Design and Techniques, the handbook by W. Reisner and M. V. Eisenhart Rothe, provides a good reference in this field.[58]

9.2.1.3 *Technical Considerations*

Designs for storing large quantities of coal in open or closed stockpiles or in silos or bunkers depend on the type of coal and its size consist. As an example, stockpiling of run-of-mine raw coal is often impractical because of size segregation, danger of self-ignition, and dusting. Precipitation and water used in dust suppression may aggravate the problem. Hence the stockpiling of raw coal should be avoided if possible, and where it must be used, the size of piles should be limited.

The ASTM categorizes coal by rank based on its stage of development in nature, from lignite to anthracite. Volatile matter, fixed carbon, moisture, and oxygen are all indicative of rank[59] (Chapters 1 and 3). These many varied properties of different coals present equally varied problems in storage, as discussed below.

Effects on Properties. The bulk of large stockpiles of coal en route from the mine to the final use point are of approximately 2 × 0 in. size consist. Except in the case of long-term storage (e.g., force majeure piles at power plants), the coal is laid down by automated stackers or trippers. The tendency is to segregate the various particle sizes by the gravity process of piling. It is essential that the mode and method of reclaiming allow for this, to pass on a reasonably uniform product to the next stage.

However much can be done in laying down a stockpile to minimize segregation. Coal should be laid down in layers about 0.5 m thick and compacted by suitable mobile

[58] Bergbau Betriebsblatt BB 23-015, "Empfehlungen fü das Auf - und Abhalden von Steinkohlen auf nicht ständig genutzten Lagern," Abteilung Normung und Prüfung des Steinkohlenbergbauvereins, Essen.

[59] H. Coligu and E. A. Vitunac, "Properties of Coal and Its Effect on Storage and Handling," *9th Annual Seminar on Bulk Materials,* University of Pittsburgh, 1976.

equipment. This will also greatly help to minimize the danger of spontaneous combustion.

Heating value and coking quality are reduced during storage. High-rank coals will lose about 1%/year in thermal value. Investigators have found that winter-stored coal lost 1.4% and summer-stored coal lost 2.1% of heating value during the first year.[59]

Spontaneous Combustion. A complete explanation for spontaneous combustion in coal is still available. However the various sequences for the occurrences of fines within piles have been rationalized to some extent.

In the lower rank coals the amount of carbon is low and the amount of oxygen high; the reverse is true of high-rank coals. Hydrogen remains a constant. Sulfur in coal is found as iron-free sulfur, organic sulfur, pyrite, and sulfate in varying percentages from less than 0.5% to more than 8%.[56]

Spontaneous combustion results from the oxidation of coal under conditions in which the rate of heat generated by complex chemical reactions exceeds the rate of heat dissipation.[59] Lignite and subbituminous coals require special handling to avoid spontaneous combustion. The lower the rank of coal, the easier it is to absorb oxygen. The lower rank coals are generally also higher in moisture, oxygen content, and volatile matter, the contributing elements to rapid oxidation.

According to Colijn and Vitunac,[59] the various stages and reactions leading to spontaneous combustion in a coal pile can be summarized as follows:

1. Coal starts to oxidize measurably in the range of 60–85°F, and the rate increases with an increase in temperature.
2. For each 15°F rise in temperature, the rate of oxidation doubles, causing further increases in temperature, until it reaches over 212°F and less than 280°F. Spontaneous combustion may occur within 3 days after the temperature reaches 170°F.
3. At 280°F carbon dioxide and water vapor are driven off.
4. Carbon dioxide liberation increases very rapidly as the temperature reaches 450°F, the temperature at which spontaneous combustion can occur.
5. At the 660°F level the coal ignites and burns vigorously.

It is known that the conversion of carbon to carbon dioxide generates about 94 kcal/mole (169 MBtu/lb-mole). If inorganic sulfur content is greater than 1%, its oxidation generates more than 200 kcal/mole (360 MBtu/lb-mole) and in addition, yields an intermediate pyrophoric compound.

It is suggested that the following combinations of reactions take place for the oxidation of pyrites. It is commonly believed that one possibility is:

or

$$2FeS_2 + 2H_2O + 7O_2 \rightarrow 2FeSO_4 + 2H_2SO_4 \quad (9.1)$$

$$FeS_2 + 3O_2 \rightarrow FeSO_4 + SO_2(\text{dry}) \quad (9.2a)$$

$$2SO_2 + O_2 + 2H_2O \rightarrow 2H_2SO_4\,(\text{wet}) \quad (9.2b)$$

Another possible explanation involves the reactions that follow:

$$FeS_2 + O_2 \rightarrow FeS + SO_2 \quad (9.3)$$

$$FeS + 2O_2 \rightarrow FeSO_4 \quad (9.4)$$

$$2SO_2 + O_2 + 2H_2O \rightarrow 2H_2SO_4 \quad (9.2b)$$

$$FeS + H_2SO_4 \rightarrow FeSO_4 + H_2S \quad (9.5)$$

$$2H_2S + SO_2 \rightarrow 3S + 2H_2O \quad (9.6)$$

The reaction represented by Equation 9.2b is reported to be 20 times faster than the formation of ferrous sulfate according to Equation 9.5. Equation 9.4 produces pyrophoric materials in a broad range of conditions. Equations 9.1 or 9.2a and 9.2b are exothermic reactions that as previously stated, can generate about 217 kcal/mole.

Fires start in storage piles for the same reasons that they start in silos. Prevention is possible by following practices developed from the experience of others and from research in the field that is intended to explain the mechanisms of self-ignition.

1. The storage area should be cleared of all combustibles having a low ignition tem-

perature, such as oil-soaked rags, wood, paper, and chemicals.

2. The area should be dry and on solid ground, free from cinders, dry-screened coal, or other porous materials, and sloped for easy drainage of any pockets of water.
3. Artificial heat from buried steam lines, sewer lines, hot water lines, or other sources should be avoided; likewise, do not store on hot, sunny days.
4. Because air enters the bottom of the pile and circulates upward, it must either circulate freely to dissipate heat of oxidation (to prevent a buildup of heat in fines) or be packed in layers of 1 ft or less to ensure the absence of air circulation.
5. Conical piles are susceptible to self-ignition because the larger pieces tend to roll to the outside, which provides a chimney effect for the fines in the center.
6. Vertical structural supports located in storage piles provide chimney effects.
7. The temperature of the pile should be taken periodically by temperature-measuring devices, located on 15-ft centers.
8. Stoker-sized coals may be treated with oil before storing to slow the absorption of oxygen and moisture. A thin layer of asphalt is sometimes used along the banks of a storage pile for the same reason.
9. Coal may be stored completely submersed in water indefinitely; no air, no oxidation.
10. Never spray a coal pile with plain water to keep dust from blowing: the subsequent drying encourages self-ignition.

Fires in silos and bunkers are best avoided by ensuring in the design of the structures that there are no dead storage areas.

Protection from Windage. The effects of wind on an open coal stockpile are obvious; hence the practical need for surface sealing. Tighter environmental standards now make sealing off of dead storage coal piles essential. Many chemical sealants have been and are being developed. They all work on the principle of spraying on a strong surface skin, which will not allow pickup of appreciable quantities of dust by the wind.

Such methods are somewhat impractical for active piles, and recourse must be had to maintaining the surface moisture of the coal on the outer faces of the pile at a level high enough to deter pickup by the wind.

When coal is handled in and out of stockpiles, the dusting resulting from transfer points on conveyors, and so on, is controlled by dust suppression. Essentially this consists of wetting airborne dust with a curtain of water or chemically treated water (surfactants or wetting agents). The effect is to agglomerate the micron-sized particles into larger and heavier ones to a degree that makes them settle out of the airstream.[59]

9.2.2 Handling

The handling of coal in and out of bulk storage piles, silos, and bins is a major part of any total system of utilizing coal. The size of coal-handling systems at transshipment terminals, electric utilities, and other industrial users has steadily increased and will probably continue to do so in the future. Massive items of equipment have been developed in the last 20 years, especially in West Germany, for stacking and reclaiming coal from stockpiles using the bucket wheel principle. Belt conveyors, various types of feeder, continuous weighing equipment, and so on, have also developed and increased in size to complement the giant reclaiming equipment. A total handling system design approach is normally used to integrate the various items of equipment and automate it so that a minimum of operating personnel is required.

Mobile handling equipment such as bulldozers, front-end loaders, compactors, and scrapers all play a part in reclaim and stockpiling operations. Most of this mobile equipment is similar to that used in the construction industry.

The final handling stage in the use of coal for firing a boiler usually involves a silo or bin system. Here in recent years the flow out of bins has been studied intensively in an effort to provide the continuous flowing system so vital to the user of coal.

9.2.2.1 *Stacker Reclaimers*

The equipment for stacking and reclaiming in coal stockyards has changed in basic concept over the last 20 years. Originally the process was discontinuous, using gantry bridges and grab buckets or dozers and front-end loaders. This concept resulted in low-capacity, high-equipment-cost installations. From this, the concept of handling coal or other bulk materials continuously with stacker bucket wheel reclaimers was developed.

The original machines were used as excavators in the brown coal fields of West Germany. In North America the first bucket wheel stacker reclaimer was installed in the Potomac Electric Power Plant at Chalk Point, Maryland, in 1963 (Fig. 9.23). Prior to this, the coal stockpiles in large power plants had been reclaimed by gravity systems onto a belt conveyor in a tunnel beneath the pile.

There were two significant features on this first U.S. machine:

1. *Counterweighted wheel boom.* The fixed counterweight of the original European design allowed the machine to overturn backward if the wheel rested on the ground because of a failure of the luffing system. By adding a movable boom counterweight located on the mast, it was possible to partially counterbalance the movable wheel boom. By means of this new design feature, the safety hazard was eliminated.
2. *Automation of operation.* Until 1963 bucket wheel stacker reclaimer operation in Europe had never been significantly automated. It was indeed venturesome in the first U.S. installation to break with long tradition and design for complete automation. With this new feature, the machine could be moved to the correct digging location on the coal pile and automatic operation begun. From this point on, the machine would automatically reclaim the coal pile, digging in the prescribed pattern as long as required by the power plant operator, who was at the remote main control panel. The predetermined digging pattern was to dig one slewing pass, advance the machine automatically, slew again, and so on. After a set number of slewing passes, the machine would travel backward, lower the wheel to the correct height for the next bench, and repeat the digging pattern. The automation operated satisfactorily, and most

Figure 9.23 First bucket wheel stacker reclaimer used in power plant layout in North America.[60]

subsequent rail-mounted machines used this labor-saving feature.

Probably the best measure of the success of this new storage method is that the owners chose similar reclaimers for their two later power stations. The only differences were a longer wheel boom and a higher tonnage capacity.[60]

The first bucket wheel stacker reclaimer at Chalk Point was rail mounted, and subsequent installations at other power plants have been the same. This type of installation lends itself to the formation of two parallel coal piles, which can be easily extended in the future if required.

Stacker reclaimers can also be crawler mounted, and in this mode they can cover "large-area" piles at transshipment terminals or force majeure piles at power plants. Movable "band wagon" conveyors are used in conjunction with these crawler-mounted machines to feed the reclaimed coal onto a fixed yard conveyor system.

The typical stacker reclaimer is set up to stack out coal as received from train or barge unloading systems and to feed all or part of it to the power plant boiler bins or to the stockpile. A high degree of automation is now involved in such installations with usually only one operator on the machine during stackout. Reclaim is normally automatic and controlled from the power plant control room.

Generally, a bulldozer is required to move the far side of the typical stockpile to within reach of the stacker reclaimer to provide full turnover of the coal to avoid deterioration, etc.

Stacking, Blending, Reclaiming, by R. H. Wohlbier,[60] contains tables detailing the various installations of rail-mounted and crawler-mounted stacker reclaimers in North America, most of which are handling coal at transshipment terminals or power plants.

9.2.2.2 *Conveyors*

The modern large coal-handling system, as found at a transshipment terminal or in a power plant or industrial plant, is dominated by the belt conveyor. It is an integral part of the stacking and reclaiming equipment, including bucket wheels, drum reclaimers, and so on. It provides the basic means of moving the coal from one place to another in the overall handling system. Bucket elevators have been all but eliminated by inclined belt conveyors, which carry coal smoothly from ground level to the top of power plant bunkers, shiploaders, and other such equipment.

In addition, there are now several installations of long-distance belt conveyors carrying coal over several miles from mine to loading points or power plants. Some of these conveyors use steel-cored belts and are made up of several sections of perhaps a mile or longer. A few use belts supported by steel cables; where one flight can be as long as 10 miles.

The design of belt conveyors for all bulk materials including coal has advanced tremendously in the last decade. Conveyor Engineering Manufacturers Association (CEMA) standards for engineering design of belt conveyors are universally acknowledged in North America and are covered in *Belt Conveyors for Bulk Materials*[61] and *Terms and Definitions of Conveyors.*[62]

A recent paper by Colijn, "General Design and Selection of Conveyor Systems,"[63] is a good reference for the fundamentals of selecting and sizing conveyors such as are found in coal-handling installations.

The belt conveyor also plays a large part in underground coal mines such as are operated in the eastern and midwestern United States. The belt conveyor is one of the principal methods of bringing the coal from the face to central collection points aboveground.

As a piece of equipment, the belt conveyor is extremely reliable and low in maintenance requirements. In recent years capacities have been increased greatly with the improvements in belting. Gregory, in his paper "Con-

[60] R. H. Wohlbier, Ed., *"Stacking, Blending, Reclaiming of Bulk Materials,"* Trans. Tech. Publications, Clausthal, West Germany, 1977.

[61] Conveyor Equipment Manufacturers Assoc., *Standards: Belt Conveyors for Bulk Materials,* 4th ed., Caxners Publishing Co., Boston, 1966.

[62] Conveyor Equipment Manufacturers Assoc., *Standards: Terms and Definitions of Conveyors,* 4th ed., Caxners Publishing Co., Boston, 1966.

[63] H. Colijn, "General Design and Selection of Conveyor Systems," *Annual Seminar on Bulk Materials,* Vol. 2, University of Pittsburgh, 1976.

veyor Belting in the 70s," makes the following statements about conveyor belting[64]:

> Not too many years ago the rubber industry had only cotton for a belt carcass and tree-grown rubber for the protective cover. Then came the development of monofilament rayon. This material had desirable elongation and elastic properties and was the first synthetic fiber to be used in place of cotton. But wet strength and other limitations restricted its use, and the conveyor belt industry turned to nylon.
>
> Initially, nylon was considered suitable only for the cross threads or filler threads of a belt carcass. Then technical advancements in heat setting and fabric weave created a major change in belt construction. As a result of these advances the industry was able to take advantage of the full tension strength of nylon cord. Another breakthrough in the use of synthetics was the tremendous increase in adhesion resulting from a chemical bond, rather than the limited physical bond available with cotton.
>
> The advantages of nylon had far-reaching effects. Manufacturers were able to design a 2-ply nylon belt that did the work of a 4- or 5-ply cotton or rayon belt. A 6-ply nylon belt now replaces a 12-ply cotton elevator belt. And nylon resists moisture, acid and mildew. It also has excellent resistance to impact and can be produced with controlled elongation and elastic properties that work to great advantage in reduced ply belt construction.
>
> Steel cable belts are now being used for high lift and long overland applications that require working tensions from 1000 to 5000 lb/in. of belt width.
>
> These belt carcass improvements have resulted in many significant advantages to the conveyor belt user.

1. Lower cost conveyor systems (on the basis of tension rating per inch of belt width, it is easy to find belts that cost 45% less than they did 20 years ago).
2. Longer center flights—fewer transfer points.
3. Higher lifts.
4. Excellent flexibility and ply adhesion (belts can operate in deep trough idlers, so tonnage is increased and spillage reduced).
5. Longer overland conveyor systems.
6. Movable conveyor systems that follow bucket wheel excavation.

A recent innovation in belt conveyor design is the catenary idler, which essentially consists of three to five rolls supported in a curve or catenary. It provides greater flexibility in loading and impact resistance than the conventional three-roll idler and is becoming more widely used in the coal industry especially for high capacity conveyors in mines. Because of the shape, there is less flexing and wear on the belt.

Other types of conveyors used in coal-handling systems include screw conveyors and flight conveyors. These units are generally confined to filling and discharge arrangements on silo or bunker installations.

9.2.2.3 *Feeders*

There are many types of feeders used in coal-handling systems. The most common are belt feeders, apron feeders, rotary-plow feeders, vibratory feeders, and reciprocating plate feeders.

Basically feeders are a link in the material handling system to provide a controlled flow from surge bins or hoppers onto belt conveyors or other equipment.

Belt Feeders. The belt feeder consists of a continuous rubber belt supported by closely spaced idlers and driven by a head pulley much the same as a conventional belt conveyor.[65] Length may vary from 6 to 50 ft, with widths as great as 84 in. The belt feeder is usually skirted for its entire length. As already mentioned, such a feeder is usually located in coal-handling systems under a surge bin with a slotted opening. Normal belt speeds are up to 100 ft/min.

In power plant silos located at the boiler, a unique type of belt feeder called a weigh feeder is used to feed coal to the pulverizers

[64] E. T. Gregory, "Conveyor Belting in the 70s," *Annual Seminars on Bulk Materials,* Vol. 2, University of Pittsburgh, 1976.

[65] H. Colijn and P. J. Carroll, "Feeders for Bins and Hoppers," *Br. Chem. Eng.* **15**(8), August 1970.

at controlled rates. Speed of the weigh feeder is variable, and the belt is supported on a weighing mechanism.

Apron Feeders. The apron feeder (Fig. 9.24) consists primarily of chain-linked heavy cast steel or manganese pans that run on rollers supported on rails. The drive is through a head sprocket and is usually limited to 50 ft/min. Capacities can run as high as 2000 tons/hr.[65] In coal-handling systems the feeders are usually found under rail-dump hoppers or truck-dump hoppers, where they are able to withstand the high impact of a load of coal falling on them.

Rotary Plow Feeders. This type of feeder was developed in West Germany for feeding lignite. In the United States it is becoming more common under coal storage piles. The plowing mechanism consists of a number of curved arms arranged to sweep the coal off a narrow shelf running the length of the storage pile. The plow is carried on a wheeled carriage that operates on tracks located over a belt conveyor in the reclaim tunnel. The plow traverses the length of the storage pile under automatic control.

This method of reclaim provides an alternative to aboveground methods such as bucket wheel reclaimers. It is uniquely suitable to reclaiming coal from storage piles enclosed under a A-frame roof (Figs. 9.22, 9.25).

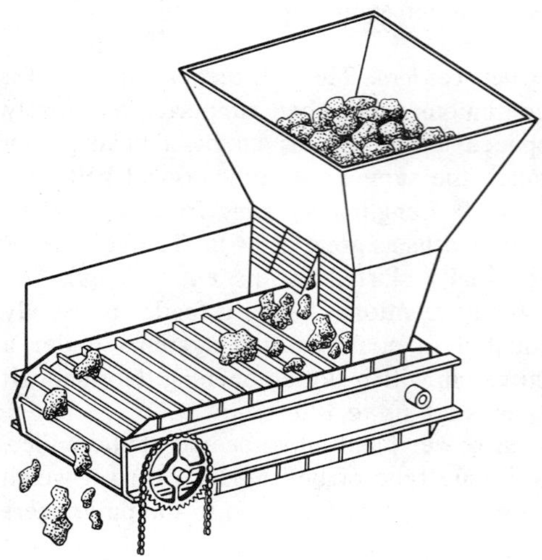

Figure 9.24 Apron feeder.[65] Reprinted from *British Chemical Engineering*, August 1970 issue, by permission of the editor of *Processing*.

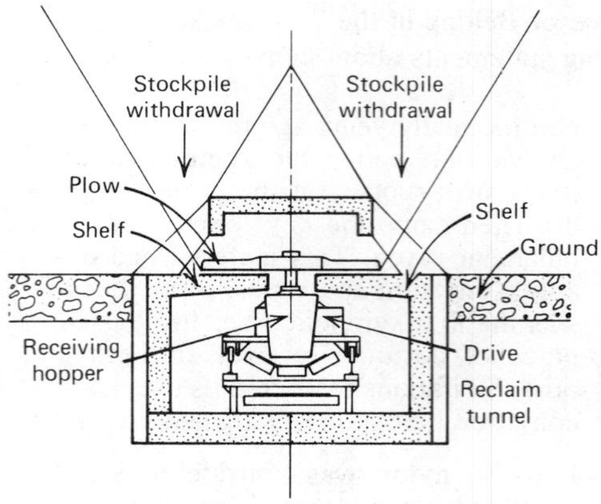

(*a*)

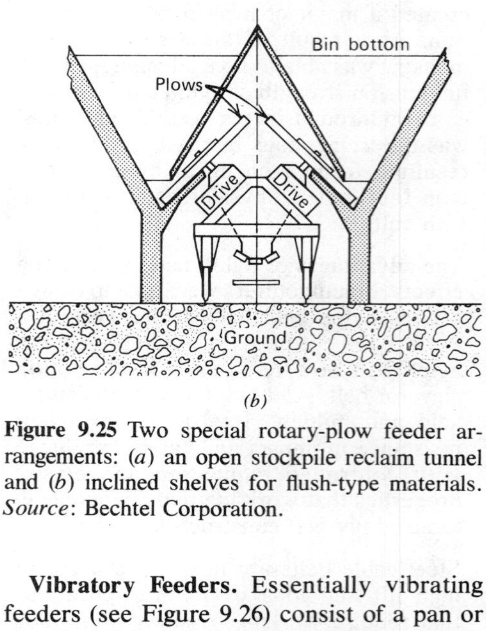

(*b*)

Figure 9.25 Two special rotary-plow feeder arrangements: (*a*) an open stockpile reclaim tunnel and (*b*) inclined shelves for flush-type materials. *Source*: Bechtel Corporation.

Vibratory Feeders. Essentially vibrating feeders (see Figure 9.26) consist of a pan or trough subjected to an applied vibratory force by means of eccentric weights on a rotating shaft or by means of an electromagnetic vibrator. Capacities of these units can run as high as 1600 tons/hr, and they can

Figure 9.26 Typical vibratory feeder.[65] Reprinted from *British Chemical Engineering*, August 1970 issue, by permission of the editor of *Processing*.

provide reasonably constant flow rates. In a coal-handling system they are often used under rail-dump pockets and silos. A comprehensive discussion of these feeders can be found in *Feeders for Bins and Hoppers* by Colijn and Carroll.[65]

9.2.2.4 *Mass Flow Bunkers (Bin Flow)*

An important link in the coal-handling system of a power plant or any industrial plant using coal to raise steam is the boiler feed silo or bunker. In a modern power plant burning coal, there are usually a number of round silos with conical bottoms arranged on each side of the boiler. These silos are usually sized to hold a surge capacity of 6–12 hr of the burn rate. The coal in the silos is usually ¼ × 0 in. to ¾ × 0 in. and feeds by way of weigh-type feeders to the pulverizers. Uninterrupted flow from these silos is essential.

To this end, much research has been done relative to the principles of bin flow, not only for coal but for all materials. From this, the concept of the mass-flow bins has evolved. The pioneer work in this area has been by Andrew Jenike and his associates.

According to Jenike in his article, "Why Bins Don't Flow"[66] the common source of trouble is "plug flow":

> The source of the difficulty lies in the fact that most bins now in use and on the drafting boards are designed for "plug flow," which occurs in all hoppers with sharp valleys as well as in all insufficiently steep conical hoppers. And a conical hopper whose slope is more than 20 degrees from the vertical is, invariably, insufficiently steep. In plug flow, a solid flows only in a vertical channel extending from the bin outlet, upward through the material (Fig. 9.27). The solid outside of the channel is stationary and flows into it by sloughing off the top surface.
>
> When an empty plug-flow bin is filled, the solid which first falls into the bin and rolls to the outside of the future channel will remain there until the bin is again completely emptied out. This is a serious drawback in the storage of solids which deteriorate with time; it also leads to obstructions to flow of solids which consolidate and gain strength when left under pressure, undisturbed, for extended periods of time. As the solid is drawn out

[66] A. Jenike, "Why Bins Don't Flow," *Mech. Eng.*, 40–43 (May 1964).

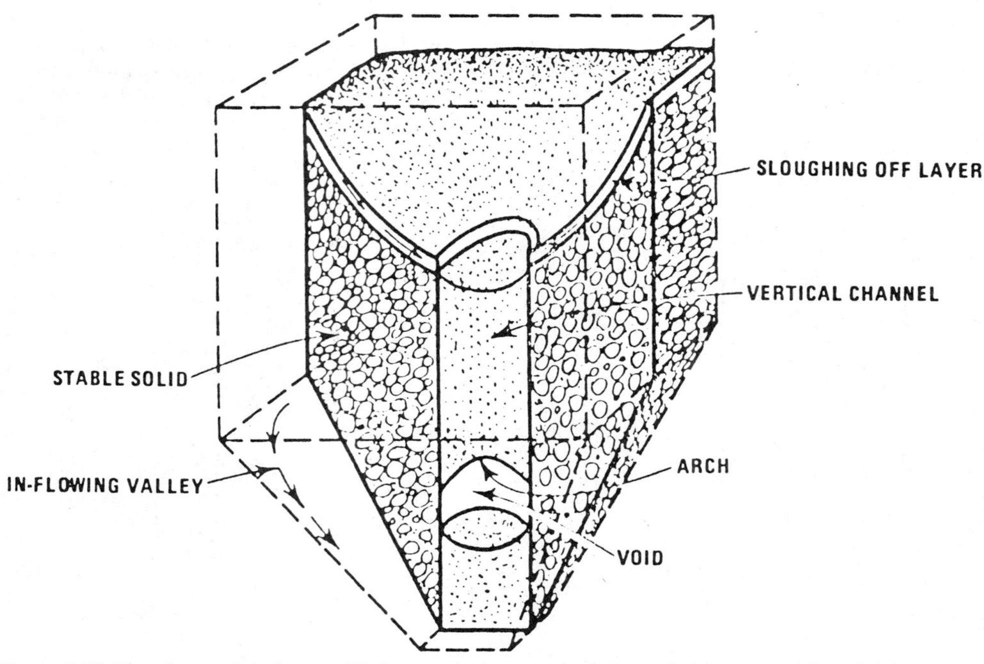

Figure 9.27 Plug flow, whereby a solid flows only in a vertical channel above the outlet of the hopper. It is fed by sloughing off the top surface.[66]

and the top surface drops to regions which have had time to consolidate at rest, sloughing off may cease; the channel empties out and all flow stops. This obstruction to flow is referred to as "channeling," "piping," or "rat-holing." It can reduce drastically the live capacity of a bin.

In research done at the University of Utah the concept of flowability or flow factor was developed. The lower the value of the flow factor of a hopper, for example, the better its flowability, and vice versa.

Figure 9.28 illustrates the flow patterns in plug-flow bins.

Jenike[66] further states:

> In a mass flow bin, the solid is in motion in substantially every point of the bin whenever any solid is drawn through the outlet. This flow pattern occurs when the walls of the hopper are sufficiently steep and smooth, the bin has no abrupt transitions, the hopper has no inflowing valleys, and the feeder or valve is active across the whole area of the outlet. An inflowing valley is of the type illustrated by the pyramidal hopper (Fig. 9.27); an outflowing valley by the chisel bin (Fig. 9.29). An outflowing valley retains little solid and does not hinder mass flow.

Mass flow has the following properties:

1. Channeling, hangups, surging, and flooding are absent.
2. Flow is uniform, and steady state flow can be approached closely.
3. Pressures throughout the mass and at the walls are relatively low, with resultant low consolidation and attrition.
4. Pressures are relatively uniform across any horizontal cross section of the hopper, causing uniform consolidation and uniform permeability.
5. There are no dead regions within the bin, hence there is a minimum of consolidation at rest.
6. A first-in, first-out flow pattern can be obtained, if desired. This is useful in the storage of solids which either deteriorate with time, or which segregate during charging into storage.
7. By circulating a mixture around a suitable mass flow bin, blending may be attained.

Typical shapes of mass flow bins are shown in Fig. 9.29. The chisel-shaped bin is simple and effective but has an outlet of length equal to the diameter of the bin. Some solids will also tend to stick in the valleys, though that can be prevented by rounding off the valleys with, say, a plastic filler. The bin (in Fig. 9.29*d*) has no valleys and a shorter outlet but requires more fabrication. The latter bin is recommended most frequently.

The hopper slope angles θ' are measured from the vertical, and are selected on the basis of the expected range of the angle of

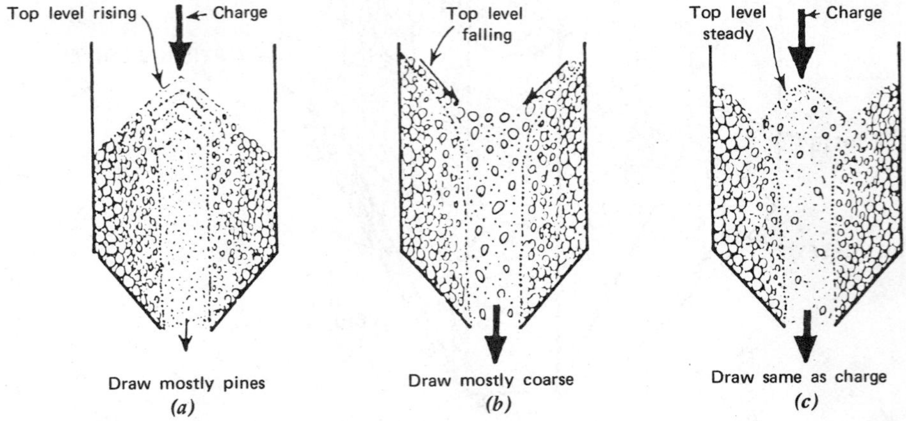

Figure 9.28 Flow patterns when (*a*) the charge coming in is greater than the draw out, (*b*) the draw is greater than the charge, and (*c*) draw and charge are equal.[66]

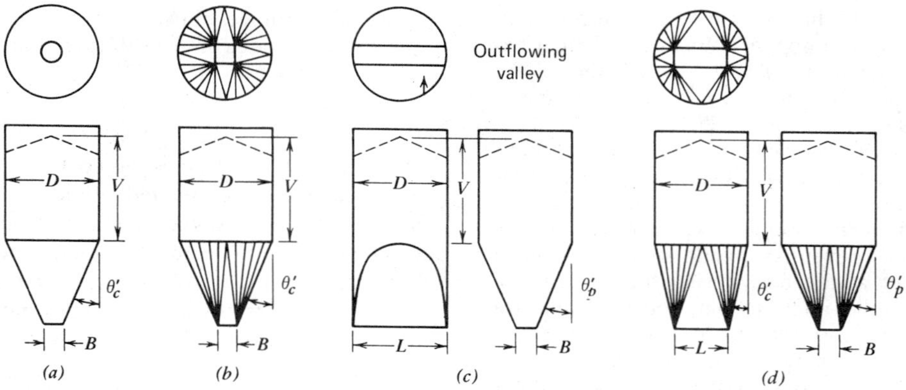

Figure 9.29 Four mass flow bin types. The chisel-shaped bin (*c*) is simple and effective but has a valley. Although harder to make, (*d*) has no valleys and is recommended most frequently.[66]

friction ϕ' between the solid and the material of the walls. The selected values of θ' should be such that the range of ϕ' lies within the solid block area in Fig. 9.30 for the sides of a slot outlet, top, and for the ends of a slot outlet, bottom, and for conical hoppers. A slot outlet should have a length-to-width ratio in excess of 3.

Suppose for instance, the range of the angle of friction ϕ' is from 23 to 30 degrees, then on the sides, θ'_p = 3 to 30 degrees, and at the ends, or in conical flow, θ'_c = 2 to 13 degrees. Suppose again that ϕ' = 32 to 36 degrees, then θ'_p = 0 to 28 degrees on the sides, and at the ends, or in conical flow, θ'_c = 0 to 3 degrees. Evidently, conical hoppers are not practical for the larger values of ϕ'.

It should be pointed out that the upper bound of the range in conical flow is particularly critical. If the conditions at a wall are such that point (θ'_c,ϕ) lies above that line, mass flow will not occur, even for quite free-flowing solids; the solid will either arch across the channel or plug flow will develop.

While this selection produces a hopper most conducive to the flow of a given solid, or group of solids, it does not guarantee flow. To do that, it is necessary to show that, under the consolidation which will prevail in the hopper, the solid (or solids) will not develop enough strength to form a stable dome across the hopper. The analysis of the flow no-flow criterion requires a knowledge of the flow properties of the solid.

The height V of the vertical part of a bin—more specifically, the vertical head of solid

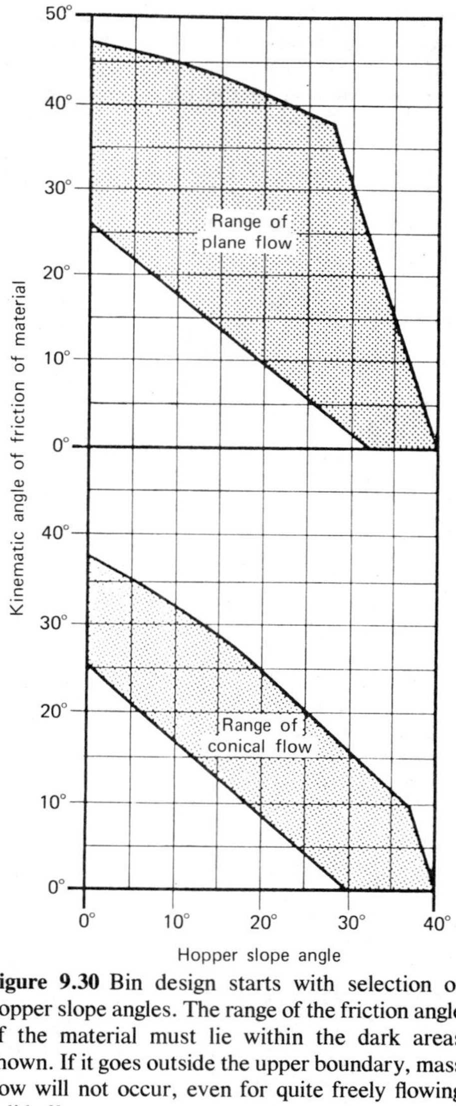

Figure 9.30 Bin design starts with selection of hopper slope angles. The range of the friction angle of the material must lie within the dark areas shown. If it goes outside the upper boundary, mass flow will not occur, even for quite freely flowing solids.[66]

above the hopper—governs the flow pattern in the hopper. When that height exceeds approximately one-quarter of the diameter of the bin, $D/4$, and the hopper is steep, say, $\theta'_p < 20$ degrees, $\theta'_c < 12$ degrees, the solid in the vertical part flows like a rigid plug. This enforces a first-in, first-out flow pattern in the bin and is particularly suitable when segregation is to be prevented. While segregation does occur at the top, during the charging of the solid into the bin, the fractions are completely remixed in the hopper. This type of mixing can be referred to as horizontal, or cross mixing. In practice, there is no significant segregation in the feed from such a bin as long as the level of the solid is not permitted to drop into the hopper. . . . If the hopper slopes are less steep than indicated, it is necessary to carry a greater minimum head of solid in the vertical part of the bin. The vertical part of a storage bin may be as high as desired, provided impact during charging is taken care of.

The advent of transporting coal by dewatered pipeline has prompted studies of mass flow in bins. If power plants and similar installations are to be able to handle pipeline-size coal with a controlled surface moisture content, the normal silo cone angles must be increased to around 78° to the horizontal. The actual slope is dependent on the shear strength characteristics of the coal. In addition, the lower part of the silo cone must be lined with stainless steel to ensure that corrosion and other problem factors do not obstruct the mass flow. Experimental results indicate good mass flow of dewatered coal under these conditions.

CHAPTER 10

Coal Preparation

ALBERT W. DEURBROUCK AND RICHARD E. HUCKO

Albert W. Deurbrouck *is Manager of the Coal Preparation Division, Pittsburgh Mining Technology Center, U.S. Department of Energy, Pittsburgh. It is his responsibility to plan and direct the coal preparation research program in all aspects of coal characterization, beneficiation, and environmental concerns.*

Richard E. Hucko *is Chief of the Physical Beneficiation Branch, Coal Preparation Division, Pittsburgh Mining Technology Center, U.S. Department of Energy, Pittsburgh. He has conducted investigations of the beneficiation of coal by selective flocculation and high-gradient magnetic separation. Mr. Hucko is in charge of the laboratory's research in physical coal cleaning, which includes dense-medium cyclone studies and coal-cleaning equipment performance evaluations.*

10.1 INTRODUCTION

The role of chemistry in coal preparation[1] has grown immensely over the last few years. Some of the most promising new processes and research studies for coal preparation are chemical or rely strongly on chemical treatment. Interest in them arose because of the urgent need to provide environmentally acceptable fuels. This need was a direct result of the Air Quality Act of 1963, which initiated a concerted effort by federal, state, and local governments for the preservation of the nation's air quality. The act called for an expanded federal research and development program and placed special emphasis on the problem of sulfur oxide emissions from the combustion of coal and oil in stationary plants. The Clean Air Act of 1970 led to the limitation of emissions of oxides of sulfur (SO_x) from new coal-fired sources and enforcement of state implementation plans for emissions from existing plants by 1975 with deferrals to 1977.

With this legislative mandate, an entirely new dimension in coal preparation research has emerged and, to an increasing degree, this research is based on the utilization of chemical methods to remove the various sulfur forms from the coal. Conventional coal preparation methods may be supplemented by new exciting processes and technologies that can play a very significant role in providing the required fuels for the last quarter of the twentieth century.

[1] Throughout this chapter the words "cleaning" and "beneficiation" are used as synonyms for "preparation." "Washing" connotes wet preparation.

10.2 WASHABILITY CHARACTERISTICS OF COAL

To determine the utilization potential of a coal, it is necessary to make a washability study of raw coal samples collected from the coal bed. This information is also needed to determine the preparation method and the equipment required to clean the coal to a selected level of ash and sulfur content.

A washability analysis is an evaluation of the physical properties of a coal that determine its amenability to improvements in quality by gravity separation methods. It includes stage crushing (in some studies) to show potential release of impurities, and specific gravity fractionation to show the quality and quantity of the cleaned product. A washability analysis is made by fractionating the coal sample at preselected, carefully controlled specific gravities. This is termed float-sink analysis or specific gravity separation. Heavy organic liquids or salt solutions are commonly used to obtain the desired specific gravities. Chemical analyses of the various specific gravity fractions of the coal are used to compile washability data, which indicate how well the coal can be prepared.

In 1972 the U.S. Bureau of Mines completed a 24-year survey[2] of the preparation characteristics of the coal beds of the Appalachian Region to determine their suitability for metallurgical use (Fig. 10.1). Washability data were reported on a county basis on coal beds from which samples could be collected, that contained significant reserves and were currently being mined. Classification standards of 1.25% sulfur and 8.0% ash were arbitrarily used to determine whether a coal was suitable or unsuitable for metallurgical coke.

In 1965 the National Air Pollution Control Administration (NAPCA) initiated a study of the sulfur reduction potential of the coals of the United States. The Illinois State Geological Survey, the Commercial Testing and Engineering Company, and the U.S. Bureau of Mines all became involved in this work through funding by NAPCA.

[2] A complete listing of the 62 Reports of Investigations generated during this survey can be found in J. E. Zeilinger, *Preparation Characteristics of Coal from the Ohio Coalfield,* U.S. Bur. Mines Rep. Invest. No. 7616 (1972), 72 pp.

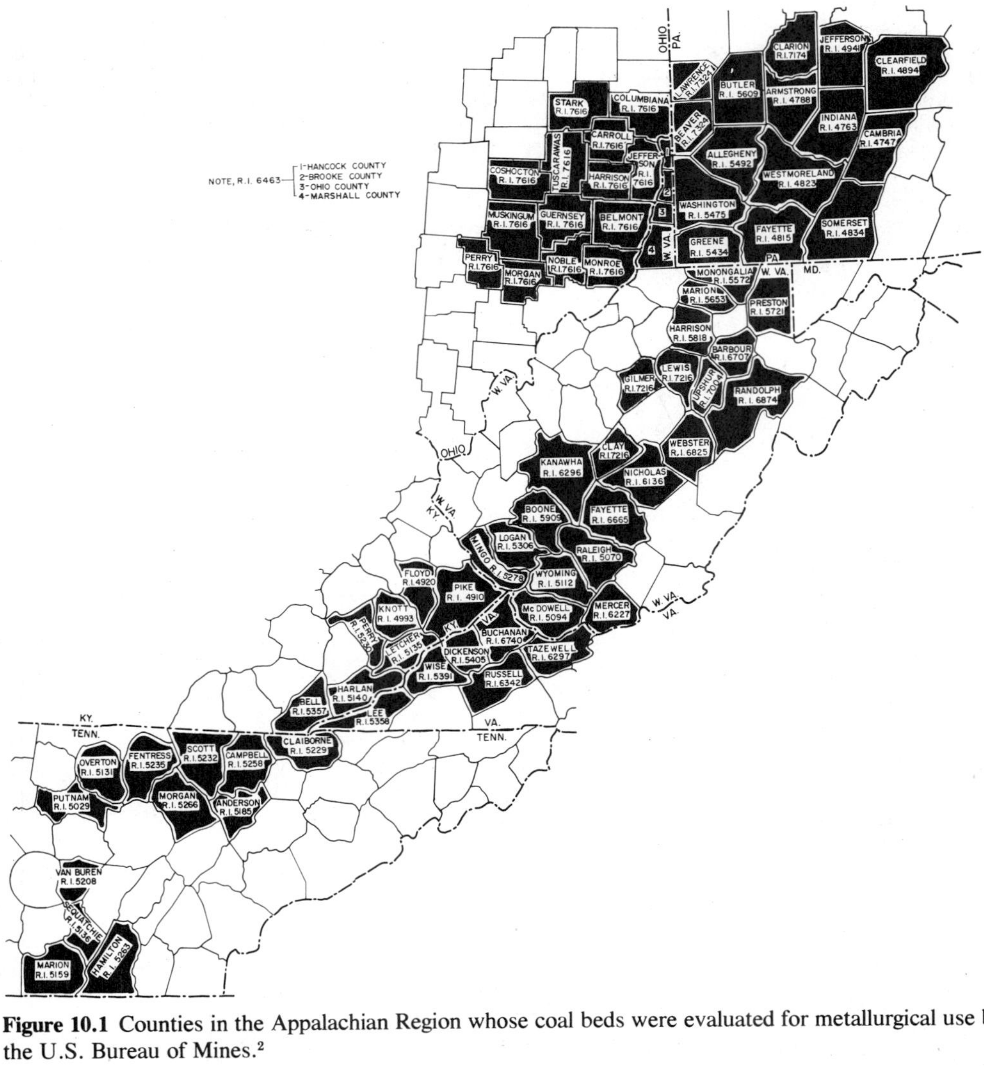

Figure 10.1 Counties in the Appalachian Region whose coal beds were evaluated for metallurgical use by the U.S. Bureau of Mines.[2]

The Illinois State Geological Survey published the results of a series of float-sink tests conducted on Illinois coal samples.[3,4] The sampling and testing data indicated that only a few coal beds could be prepared with a practical percentage of recovery and 1.5% or less total sulfur. The float coal fractions usually had less sulfur in the smaller coal sizes, but the differences were not great enough to make fine grinding a practical means of sulfur reduction for the coals tested.

Cavallaro et al.[5] recently reported an evaluation of 455 U.S. raw coal channel samples. These raw coals contained, on the average, 14.0% ash, 1.91% pyritic sulfur, and 3.02% total sulfur. Only 14% of the raw coal samples could meet the air quality standard for new-source SO_2 emissions of 1.2 lb of SO_2 per million Btu (Fig. 10.2), and even after nominal preparation only 24% of the samples could be upgarded to meet the SO_2 emission standard. However quite often very significant pyritic sulfur reductions are attainable at high coal recoveries with very little crushing. Some coals are especially suitable for desulfurization by physical means because their organic sulfur content is quite low and their pyritic sulfur is readily liberated by crushing.

The Mitre Corporation has also written a report[6] on the availabilities of coal by sulfur

[3] R. J. Helfinstine, N. F. Shimp, M. E. Hopkins and J. A. Simon, *Sulfur Reduction of Illinois Coals—Washability Studies,* Part 1, Illinois State Geol. Survey Cir. No. 462 (1971), 44 pp.

[4] R. J. Helfinstine, N. F. Shimp, M. E. Hopkins and J. A. Simon, *Sulfur Reduction of Illinois Coals—Washability Studies,* Part 2, Illinois State Geo. Survey Cir. No. 484 (1974), 32 pp.

[5] J. A. Cavallaro, M. T. Johnston and A. W. Deurbrouck, *Sulfur Reduction Potential of the Coals of the United States: A Revision of RI 7633,* U.S. Bur. Mines Rep. Invest. 8118 (1976), 323 pp.

[6] L. Hoffman, F. J. Lysy, J. P. Morris and K. E. Yeager, "Survey of Coal Availabilities by Sulfur Content," MTR-6086. Final report by the MITRE Corporation for the EPA, May 1972. Internal document.

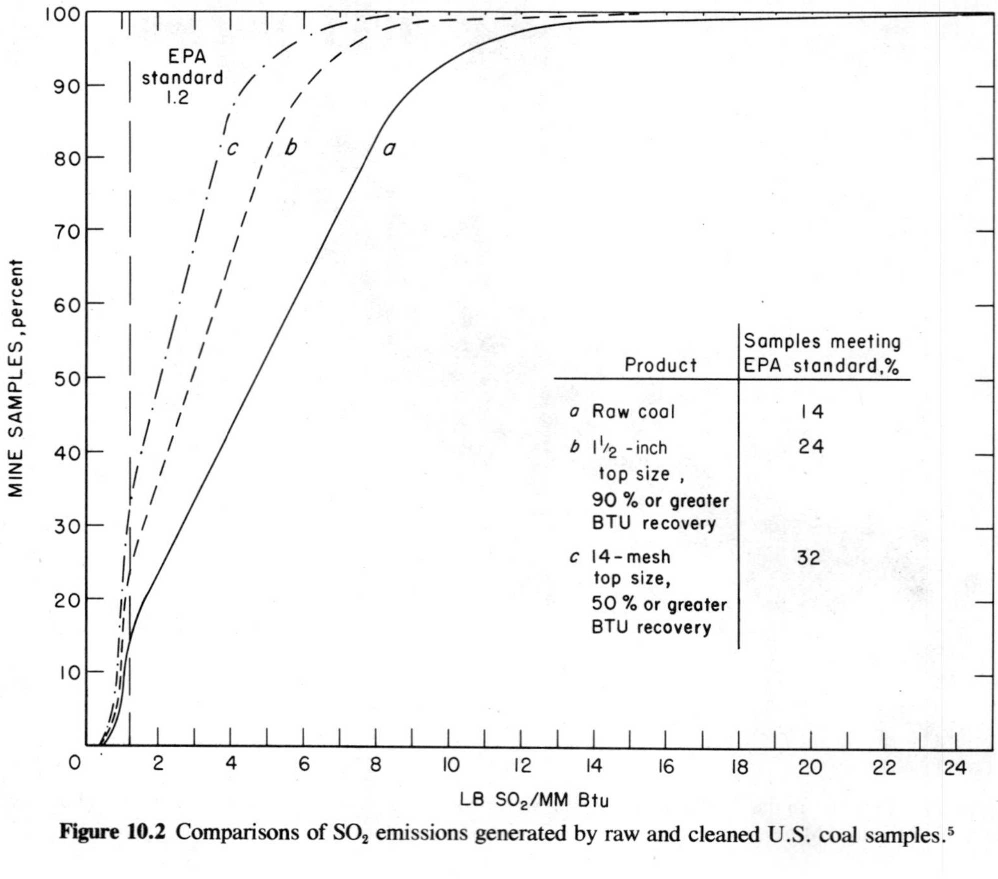

Figure 10.2 Comparisons of SO_2 emissions generated by raw and cleaned U.S. coal samples.[5]

content. This study examines the present and potential availability of commercial steam coal in terms of rank and sulfur content. Included are data on present and possible coal production growth rates in terms of coal source region, coal rank, sulfur content, present production levels, and considerations associated with new mining activities.

An extensive study on the reserve base of U.S. coals by sulfur contents has been published in two parts by the Bureau of Mines. Part 1 gives the coal reserve base for the eastern states,[7] and Part 2 gives the coal reserve base for the western states.[8] Table 10.1 lists the reserve bases of bituminous coals by sulfur level and state.

Sulfur is but one of the many elements found in coal; by virtue of its origin, coal contains nearly every naturally occurring element. Most of these are present in minor or trace amounts. Trace elements in coal are generally considered to be those present at a concentration of 0.1% (1000 ppm) or less[9] (Chapter 7).

[7] R. D. Thomson and H. F. York, *The Reserve Base of U.S. Coals by Sulfur Content.* Part 1. *The Eastern States,* U.S. Bur. Mines Inf. Circ. No. 8680 (1975), 537 pp.

[8] P. A. Hamilton, D. H. White, Jr., and T. K. Matson, *The Reserve Base of U.S. Coals by Sulfur Content.* Part 2, *The Western States,* U.S. Bur. Mines Inf. Circ. No. 8693 (1975), 322 pp.

[9] E. M. Magee, H. J. Hall, and G. M. Varga, Jr., *Potential Pollutants in Fossil Fuels,* Environmental Protection Technology Ser. EPA-R2-73-249, June 1973, 151 pp.

Table 10.1 Recoverable Reserve Base of Bituminous Coal and Lignite by Sulfur Level and State (million short tons)[7,8]

	Sulfur Level (%)				
State	<1.0	1.0–3.0	>3.0	Un-known	Total
Alabama	625	1,100	16	213	1,954
Alaska	11,459	184	0	0	11,643
Arizona	173	177	0	0	350
Arkansas	81	463	46	74	664
Colorado	7,476	786	47	6,547	14,856
Illinois	1,095	7,341	42,969	14,256	65,661
Indiana	549	3,306	5,262	1,504	10,621
Iowa	2	227	2,106	549	2.884
Kansas	0	309	696	383	1,388
Kentucky, east	6,558	3,322	299	2,729	12,908
Kentucky, west	0	564	9,244	2,816	12,624
Maryland	135	691	187	35	1,048
Michigan	5	85	21	7	118
Missouri	0	182	5,226	4,081	9,489
Montana	101,647	4,115	503	2,167	108,432
New Mexico	3,576	794	1	28	4,399
North Carolina	0	0	0	32	32
North Dakota	5,389	10,326	269	15	15,999
Ohio	134	6,441	12,634	1,872	21,081
Oklahoma	275	227	241	451	1,194
Oregon	2	0	0	0	2
Pennsylvania	1,037	16,731	3,800	2,299	23,867
South Dakota	103	288	36	1	428
Tennessee	205	533	157	88	983
Texas	660	1,885	284	444	3,273
Utah	1,969	1,547	49	478	4,043
Virginia	2,088	1,163	14	245	3,510
Washington	604	1,265	39	45	1,953
West Virginia	14,092	14,006	6,823	4,652	39,573
Wyoming	33,912	14,657	1,701	3,060	53,330
Total	193,851	92,715	92,670	49,071	428,307

Because of the vast tonnages of coal used, however, a concentration of only 1 ppm may result in the emission of hundreds of tons per year of a particular element. This problem, coupled with the ability of many trace elements to produce adverse effects (toxicity to plant and animal life, disposal hazards, catalyst poisoning, etc.), has led to numerous investigations to delineate the trace elements and their concentrations in various coals and in specific gravity fractions thereof. It has been recognized that reduction of mineral matter content as practiced in coal preparation affords the concurrent benefit of removing substantial quantities of trace elements from run-of-mine coal.

In 1935 Fuchs[10] analyzed the ashes of certain German brown coals for some 45 trace elements including lead, mercury, and cadmium. Later Headlee and Hunter[11] of the West Virginia Geological Survey published results of emission spectrographic analyses for 35 elements in ashes of West Virginia coals.

Abernethy et al.[12] recently reported spectrochemical analyses of minor elements in ashes from 827 U.S. coals used commercially. Coal samples were obtained from all the major coal-producing areas of the country and prepared for analysis by a low-temperature ashing technique. Twenty-two elements, including beryllium and lead, were detected in nearly all samples analyzed. Arsenic was found in 67% of coals from the Eastern Province, in 41% of coals from the Interior Province, and in only 16% of coals from the Western Province. Cadmium was detected in only 8.5% of all the coals analyzed. In addition to showing average trace element contents for coals from various broad geographical areas, these authors gave detailed analytical results according to state, county, and coal seam.

Many other investigators have contributed to the knowledge of trace elements in coal. (An extensive bibliography is presented in Ref. 9.) Of specific interest to coal preparation engineers, though, is the tendency of certain trace elements to concentrate in the inorganic (mineral matter) portion of the coal, and their potential removal by specific gravity fractionation. The most comprehensive study of this matter to date was reported by Ruch et al.[13] Four samples of Illinois coals were separated into six specific gravity fractions and analyzed for major, minor, and trace elements. Table 10.2 lists the trace elements in order of decreasing "organic" affinity. The sequence was determined by comparing ratios of the amount of an element in the lightest float fraction to the amount of that element in the heaviest sink fraction.

As shown in Table 10.2, certain elements are always in the group most closely associated with the clean coal and therefore have the greatest organic affinities. These include germanium, beryllium, and boron. The elements with the greatest inorganic affinity are mercury, zirconium, zinc, arsenic, and cadmium. Lead, manganese, and molybdenum are also near the bottom for three of the four coals. It is the latter group of elements whose reduction by coal preparation appears to be most feasible.

10.3 PROCESS DEVELOPMENTS

The period since the publication of the first Supplementary Volume has been one of great activity in the sphere of process development. The impetus for such development has stemmed from the need to minimize sulfur in steam coal (at maximum possible Btu recovery), and from the necessity for cleaning, sizing, and drying coal before subjecting it to conversion processes. The principal unit operations of coal preparation are discussed in this chapter; an overview of the applied processes is presented, along with a more thorough description of the significant processing developments that have occurred over the past several decades.

[10] W. Fuchs, "Rare Elements in Coal Ashes," *Ind. Eng. Chem.*, **27**(9), 1099–1100 (September 1935).

[11] A. J. W. Headlee and R. G. Hunter, "Elements in Coal Ash and Their Industrial Significance," *Ind. Eng. Chem.*, **45**(3), 548–551 (March 1953).

[12] R. F. Abernethy, M. J. Peterson, and F. H. Gibson, *Spectrochemical Analyses of Coal Ash for Trace Elements,* U.S. Bur. Mines Rep. Invest. No. 7281 (1969), 30 pp.

[13] R. R. Ruch, H. J. Gluskoter, and N. F. Shimp, *Occurrence and Distribution of Potentially Volatile Trace Elements in Coal,* Illinois State Geol. Survey EGN 72, (1974), 96 pp.

Table 10.2 Affinity of Trace Elements for Pure Coal and for Mineral Matter, as Determined from Float-Sink Data[13]

	Davis Coal	DeKoven Coal	Colchester (No. 2) Coal	Herrin (No. 6) Coal
Clean coal —lightest specific gravity fraction (elements in "organic combination")	B	Ge	Ge	Ge
	Ge	Ga	B	B
	Be	Be	P	Be
	Ti	Ti	Be	Sb
	Ga	Sb	Sb	V
	P	Co	Ti	Mo
	V	P	Co	Ga
	Cr	Ni	Se	P
	Sb	Cu	Ga	Se
	Se	Se	V	Ni
	Co	Cr	Ni	Cr
	Cu	Mn	Pb	Co
	Ni	Zn	Cu	Cu
	Mn	Zr	Hg	Ti
	Zr	V	Zr	Zr
	Mo	Cd	Cr	Pb
	Cd	Mo	Mn	Mn
Mineral matter—specific gravity greater than 1.60 (elements in "inorganic combination")	Hg	Pb	As	As
	Pb	Hg	Mo	Cd
	Zn	As	Cd	Zn
	As		Zn	Hg

10.3.1 Sampling

A general theory of sampling was presented by Visman.[14] He discussed methods whereby the precision of a given sampling procedure can be estimated in advance for a gross sample. Such a determination requires knowledge of two "sampling constants," A and B. These two factors are the "random variation" and "segregation" that are inherent in any consignment to be sampled. Visman also provided an equation for determining, in advance, the size and number of increments required to attain a preassigned precision of the gross sample that is to be collected. Visman's article triggered much further debate concerning sampling theory.[15–17]

Many practical aspects of bulk coal sampling systems are considered in various works.[18–22] Also, the latest ASTM standards applicable to coal sample collection and analysis are readily available in the literature.[23]

One facet of sampling that has changed

[14] J. Visman, "A General Sampling Theory," *Mater. Res. Stand.*, **9**(11), 8–13, 51–56, 62, 64 (November 1969).

[15] J. A. Duncan, "Comments on 'A General Theory of Sampling,' " *Mater. Res. Stand.* **11**(1), 25 (January 1971).

[16] J. Visman, "Further Discussion: 'A General Theory of Sampling,' " *Mater. Res. Stand.*, **11**(8), 32–37 (August 1971).

[17] J. Visman, "A General Theory of Sampling, Discussion 3," *J. Mater.*, 7(3), 345–350 (September 1972).

[18] C. D. Allman, "Automatic Coal Sampling System," *Trans. AIME, Soc. Min. Eng.*, **226,** 321–326 (September 1963).

[19] M. Bolek, "Some Possibilities in the Automation of Coal Preparation Plants," *5th Int. Coal Preparation Congr.*, Pittsburgh, October 3–7, 1966, Paper C1, pp. 121–130.

[20] D. W. Houston, "Automatic Sampling of High Capacity Material Streams," paper presented at the AIME Society of Mining Engineers Fall Meeting and Exhibit, Pittsburgh (September 19–21, 1973), 20 pp.

[21] T. E. Tibbetts, S. F. Featherby, and R. L. Miller, "Automatic (Mechanical) Coal Sampling," *Trans. Can. Inst. Min. Metall. Min. Soc. Nova Scotia*, **68,** 189–194 (March 1965).

[22] F. R. Huntington, "How to Select a Coal Sampling System," *Coal Min. Process.*, **9**(3), 42–45 (March 1972).

[23] ASTM Standards, 1972, Part 19, Gaseous Fuels; Coal and Coke, Committee D5 on Coal and Coke, Designation D2013-72, Standard Method of Preparing Coal Samples for Analysis, and Designation D2234-72, Standard Method of Collection of a Gross Sample of Coal.

substantially in recent times has involved the introduction of automatic continuous analyzers. These analyzers are covered in detail in Section 10.3.7, "Instrumentation and Automation."

10.3.2 Size Reduction

At one time coal preparation plant operators attempted to recover as much lump coal as possible, but today they must crush the coal before it is mechanically cleaned, so that impurities are liberated and a product of uniform size is obtained. In 1940 only 7.7% of the total bituminous coal production in the United States was mechanically crushed. This amount increased to 92.7% in 1974.[24]

Crushing and grinding installations have been designed and built to use increasingly larger machinery as it has become available, enabling larger throughputs to be obtained. Coal size reduction technology, however, is an area in which no dramatic changes have occurred in recent years.

The rotary breaker, an autogenous device, is now available with a hammermill incorporated near the discharge end. This ensures that all feed material passes through the peripheral discharge holes along the length of the rotating cylinder. Such a system minimizes the excessive breakage of soft coals, while helping to release carbonaceous material from the refuse.

The workhorse of coal size reduction is the roll crusher. Throughputs have been increased (2000 ton/hr of run-of-mine coal to 8-in. top size), and the units have been compactly engineered into two-stage crushing devices by setting one double-roll crusher on top of a second one.

Although the hammermill is the most commonly used coal pulverizer, the impact mill is increasing in popularity. Wet rod and ball mills, common to minerals processing, have been used or proposed for preparing coal slurry for pipeline transport. If future coal-cleaning techniques involve the deliberate production of finer sizes, these devices may also be used in coal preparation plants.

The roller mill, another variation of the ball–race mill concept, is being more widely applied as larger electrical power generation boilers are built. Also, air-swept ball mills are being considered for grinding harder coals or dewatered pipeline coal. The possibility of using ultrafine coal has generated renewed interest in attrition mills and jet mills.

The literature on coal crushing published over the past few decades has consisted basically of surveys.[25–27]

10.3.3 Sizing

The three principal reasons for sizing in commercial coal preparation practice are to separate raw coal into various sizes for marketing, to feed washing and dewatering units, and to recover the solids used to control the specific gravity in washing units.[28] Sizing of coal particles in preparation plants is done by screening or by classification. Both these operations may be performed wet or dry.

In screening, the most commonly used method for sizing coal, each particle in a mass of unconsolidated material is brought against an opening or aperture that allows it to pass through if small enough, or to be rejected if too large. Several novel screen configurations and modifications have evolved in recent years.

One example is the sieve bend,[28] which was developed by the Dutch State Mines in the early 1950s. This device is a rigidly mounted screen that operates on principles almost entirely different from those of common shaking or vibrating screens. It is a truly fixed screen having no moving parts, and if placed at an elevation lower than its source of feed material, it requires no power for its action. Benzon and Burgess[29] developed a rapping mechanism for use with sieve bends with very fine apertures for screening slurries

[24] L. W. Westerstrom and R. E. Harris, *Coal—Bituminous and Lignite,* U.S. Bur. Mines Minerals Yearb. Prepr. (1976), 66 pp.

[25] G. H. Roman, "Coal Crushers and Crushing," *Coal Age,* **72**(2), 108–122 (February 1967).

[26] K. Spink, "Comminution Methods and Machinery," *Min. Miner. Eng.,* **8**(2), 5, 7, 8, 11, 12, 15, 17, 19, 21, 23, 25 (February 1972).

[27] A. Ratcliffe, "Crushing and Grinding," *Chem. Eng.,* **79**(15) 62–75 (July 10, 1972).

[28] R. Q. Shotts, in *Coal Preparation,* J. W. Leonard and D. R. Mitchell, Eds., American Institute of Mining, Metallurgical, and Petroleum Engineers, New York, 1968, Ch. 8.

[29] W. Benzon and C. L. Burgess, "Development of Fine Aperture Sieve Bends for Coal Slurry Classification; and Application, in Conjunction with Classifying Cyclones, to Sulfur Reduction in Fine Coal," *5th Int. Coal Preparation Congr.,* Pittsburgh, October 3–7, 1966, Paper E2, pp. 297–311.

efficiently at 100 μm and at practical feed rates. In addition to performing sizing operations, sieve bends are used for dewatering. Since dewatering is generally considered the principal application of sieve bends, the reader is referred to Section 10.3.5, "Dewatering and Drying," for a more detailed discussion.

Another screen type that has gained wide acceptance in the coal preparation industry is the Vor-Siv, developed in Poland.[30] Like the sieve bend, the Vor-Siv is better known as a dewatering device and is described in Section 10.3.5.

A system that has been used in the United Kingdom to alleviate the problem of screening difficult run-of-mine coal at sizes between ⅜ and 1 in. is the loose-rod deck.[31] This type of deck was developed by the Inland Steel Corporation, in the United States, for the screening of difficult iron ores down to ¼ in. The screen consists of steel rods set at right angles to the direction of flow; the ends of each rod are positioned in oversize holes in spacer bars that lie in the flow direction. The rods are set at a suitable pitch to give the required equivalent screening aperture. The rotation of the rods tends to prevent the bridging and ultimate blinding of the apertures.

Electric heating of screens has been practiced for many years; however several new technological advances have been made.[32–34] Today the current is introduced into the screen by induction so that there is no physical connection between the transformer and the vibrating screen. This method eliminates heavy flexible cables, which were expensive to maintain because they crystallized under vibration and also dissipated heat. Transformer cores have also been redesigned so that energy loss due to heat has been reduced from approximately 10% to about 2%.

Replacement of the screen medium is a major operating cost, particularly when coal finer than ½ mm is being handled.[35] Plastic screen sections with great abrasion resistance have been developed.[36] Such screen material lasts 20–30 times longer than conventional screens in wet applications, and 10–20 times longer in dry usage.

Several excellent articles have been written on the application of screens as sizing devices in coal preparation plants.[37–39]

Although the greatest number of coal-sizing operations use screens, classifiers are used in some cases. Classification connotes size separation based on differing flow rates through a fluid. Dry classification uses air as the ambient fluid. Sizing by air classification involves higher fluid volumes and velocities than wet classification of equivalently sized particles.

Cyclones are often used as wet classifiers in preparation plants; they produce flotation feed by separating fine-size particles from slurries containing nominal ⅜-in. top size material.

For many years dry classification, or dedusting, has been a standard practice in Europe, where wet processing is usually applied to the coarser coal only. A new type of deduster has been developed and used extensively in Germany.[40] Dedusters employed in the past frequently clogged when the feed

[30] D. K. Cooper and J. C. Anderson, "Coal Preparation—1974," *Mining Congr. J.*, **61**(2), 86–90 (February 1975).

[31] M. P. Armstrong, A. Atkinson, D. E. Jenkinson, J. Roberts, and A. Turner, "The Dry Extraction of Small Coal and Fines from Moist Raw Coal," *6th Int. Coal Preparation Congr.*, Paris, March 26–28, 1973, Paper E4, 22 pp.

[32] R. Q. Shotts, in *Coal Preparation*, J. W. Leonard and D. R. Mitchell, Eds., American Institute of Mining, Metallurgical, and Petroleum Engineers, New York, 1968, Ch. 8.

[33] H. Farrimond, J. Smith, and C. Roberts, "Thermal Heating of Perforated Plates and the Effect on Screening Efficiency," *Mine Quarry*, **1**(4), 59, 61, 63, 65, 67 (August 1972).

[34] H. Farrimond, J. Smith, and C. Roberts, "Thermal Heating of Perforated Plates and the Effect on Screening Efficiency," *Mine Quarry*, **1**(5), 29, 31, 33, 35, 39, 41 (September 1972).

[35] F. X. Ferney, "Mechanical Coal Drying . . . Equipment and Methods," *Coal Age*, **70**(8), 63–68 (August 1965).

[36] Anon., "Modern Designs in Equipment for Today's Coal Preparation," *Coal Age*, **81**(1), 115, 116, 123–127, 139 (January 1976).

[37] Anon., "Screens," *Coal Min. Process.*, **10**(10), 40–51 (October 1973).

[38] G. E. Groseclose, "The Role of Vibrating Screens in Modern Coal Preparation Plants," *Coal Age*, **81**(1), 92–96 (January 1976).

[39] M. P. Armstrong, "Dry Fines Screening," *Colliery Guardian*, **223**(5), 162–164, 166–168, 170 (May 1975).

[40] B. Bogenschneider and K. H. Kubitza, "Preparation of Smalls Finer than 10 mm in the Coal Mining Industry of the Federal Republic of Germany," paper presented at the 1st Engineering Foundation Conference on Coal Preparation for Coal Conversion, Rindge, N. H. (August 10–15, 1975), 19 pp.

moisture increased sharply. This new air classifier, which can dedust relatively moist coal, operates with a closed air circuit that employs gravity combined with centrifugal action to separate the particles according to their size fractions. As shown in Fig. 10.3, the feed enters through a hollow shaft in the rotating center of the unit and is dispersed by a rotating distributor and centrifuged radially outward into the rising airstream, where the fines are carried upward and the coarse material continues to fall.

10.3.4 Beneficiation Processes

The essence of coal preparation is the separation of impurities from coal; consequently most process development work is concentrated in this area.

10.3.4.1 *Jigs*

Jigging continues to be the predominant washing technique throughout the world. A thorough explanation of the theory, construction, and operation of jigs was presented by Lovell.[41] Recent contributions to jigging theory have been made by Van Koppen,[42] Nagy and Krausz,[43] and Spetl et al.[44]

Jigging technology has advanced in recent years as competition by other processes has spurred research and development work. The Japanese invented the Tacub jig,[45] which embodies several modifications of the conventional Baum jig. The Tacub jig is pulsed with air from chambers directly below the jig screen bed. Since the air chambers are situated directly underneath the screen, the amount of liquid to be accelerated is minimal, and the distance to be covered by the liquid is about equal in all places. This increases the nominal working surface and gives uniformity of jig wave form.

A refinement of the Tacub jig is the Batac jig.[46] The Batac, derived from the names Baum and Tacub, incorporates the uniform air distribution system developed for the latter but has improved air valves and bed control mechanisms. Automation of the operation has made possible a more precise separation of coal from its associated impurities. Figure 10.4 illustrates a six-cell, three-compartment Batac jig. The heavy, dense material in the coal discharges through screen plate perforations and at the end of the compartments through shale ejectors. When treating coal sizes under ⅝ in., a feldspar bed may be used in certain compartments. A more thorough description of the Batac jig

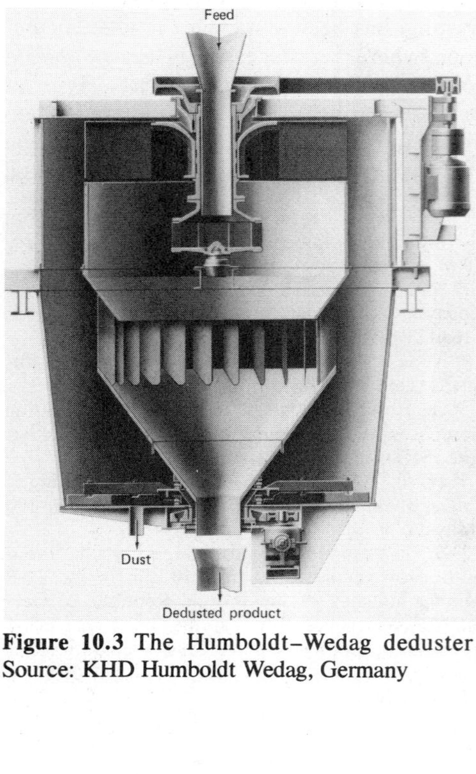

Figure 10.3 The Humboldt–Wedag deduster. Source: KHD Humboldt Wedag, Germany

[41] H. L. Lovell, in *Coal Preparation,* J. W. Leonard and D. R. Mitchell, Eds., American Institute of Mining, Metallurgical, and Petroleum Engineers, New York, 1968, Ch. 9, Part 2.

[42] C. W. J. Van Koppen, "A Contribution to the Fundamentals of the Jigging Process," *5th Int. Coal Preparation Congr.,* Pittsburgh, October 3–7, 1966, Paper B3, pp. 85–97.

[43] B. Nagy and S. Krausz, "New Optimization Principles of the Jigging Process for Fine-Grain Coal," *6th Int. Coal Preparation Congr.,* Paris, March 26–28, 1973, Paper 17E, 15 pp.

[44] F. Spetl, J. Puncmanova, and V. Prochazka, "Development and Results of a New Jig," *6th Int. Coal Preparation Congr.,* Paris, March 26–28, 1973, Paper 19E, 12 pp.

[45] T. Imaizumi, "Coal Preparation in Japan," *5th Int. Coal Preparation Congr.,* Pittsburgh, October 3–7, 1966, Paper H2, pp. 495–504.

[46] W. D. Hake, "Application of the Batac Jig for Processing Fine Coal," *Min. Congr. J.,* **62**(9), 52–55 (September 1976).

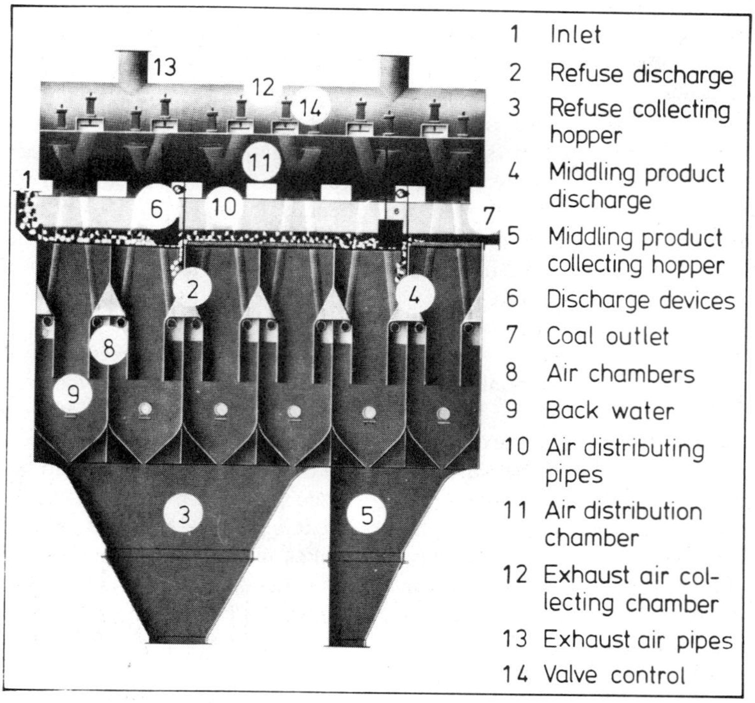

Figure 10.4 Batac jig cross section. Source: KHD Humboldt Wedag, Germany

along with typical performance data are presented by Zimmerman.[47]

10.3.4.2 *Dense-Medium Cyclonic Separators*

Although jigs currently wash more coal, dense-medium processes are gaining favor because of their ability to effect sharper separations over a wider specific gravity range. For example, preparation engineers found in one study that heavy-media cleaning yielded 2.7% greater recovery on a coarse coal fraction than did jigs.[48] Dense-medium processes have been used to clean coal for more than 50 years; thorough coverage of the history, types, and performance of dense-medium processes has been presented by Palowitch and Deurbrouck.[49]

The dense-medium cyclone, which was developed in 1945 in the Netherlands, was not installed in a U.S. coal-cleaning plant until 1961.[50] The growing importance of cyclones in coal preparation cannot be overemphasized. Bureau of Mines[51,52] performance data indicate that dense-medium cyclones effect a sharper separation between coal and impurity than can be obtained in

[47] R. E. Zimmerman, "Batac Jig—A New Improved Baum Type Jig for Cleaning Coarse and Fine Sizes of Coal," *Min. Congr. J.*, **60**(5), 43–49 (May 1974).

[48] J. L. Walker, Jr., "Future Trends in Preparation," *Coal Min. Process.*, **11**(1), 56, 57, 96 (January 1974).

[49] E. R. Palowitch and A. W. Deurbrouck, in *Coal Preparation*, J. W. Leonard and D. R. Mitchell, Eds., American Institute of Mining, Metallurgical, and Petroleum Engineers, New York, 1968, Ch. 9, Part 1.

[50] W. Benzon, "Fine Coal Cleaning with Heavy Medium Cyclones," *Min. Congr. J.*, **47**(12), 42–47 (December 1961).

[51] M. R. Geer, M. Sokaski, P. S. Jacobsen, and H. F. Yancey, *Performance of Dense-Medium Cyclone in Cleaning Fine Coal*, U.S. Bur. Mines Rep. Invest. No. 5732 (1961), 22 pp.

[52] A. W. Deurbrouck and J. Hudy, Jr., *Performance Characteristics of Coal-Washing Equipment: Dense-Medium Cyclones*, U.S. Bur. Mines Rep. Invest. No. 7673 (1972), 34 pp.

other types of cleaner that treat the same size range.

In a typical dense-medium cyclone (Fig. 10.5), a mixture of medium and raw coal enters tangentially near the top of the cylindrical section. Under the influence of centrifugal force, refuse moves along the wall of the cyclone and exits through the underflow orifice. Clean coal moves toward the longitudinal axis of the cyclone and passes through the vortex finder to the overflow chamber.

Dense-medium cyclones commonly clean coal in the ¾-in. × 35-mesh size range.[53] New installations, however, are extending the bottom size of feed, in some cases to zero. Section 10.3.8, "Plant circuitry," describes such an installation.

Japan's Swirl cyclone, illustrated in Fig. 10.6, is a modified coal-washing cyclone in which the apex is at the top (i.e., a conventional cyclone turned upside down).[54] An air column that is frequently unstable occurs inside conventional cyclones. An isolated air pipe, which is open to the atmosphere, stabilizes the air column in the Swirl cyclone. This cyclone modification is claimed to produce sharpness of separation and efficiencies comparable to conventional cyclones, and because of its larger apex diameter, the handling of larger sizes of coal (up to 3 in.) is also claimed. In addition, coarser magnetite can be used as media, resulting in lower magnetite losses.

The Soviets have recently developed three-product dense-medium cyclonic separators.[55] In this system the refuse from one cylindrical cyclonic unit becomes the direct feed to a second cyclone. This method may be used in place of two conventional cyclones in series. Cleaning in three-product cyclonic separators makes it possible to use one instead of two dense-medium recovery circuits, and also to eliminate the handling of a secondary feed for retreatment.

A novel cyclonic coal washer is the Vorsyl dense-medium separator.[56] This washing device (Fig. 10.7), was developed by the British National Coal Board. It is a cylindrical dense-medium separator designed for treating raw coal with a top size of about 1¼ in. The principal application to date has been in the production of coking coal in the United Kingdom. The Vorsyl commonly operates at a separating specific gravity lower than that normally achievable using Baum jigs.[57]

Another variation of cyclonic dense-medium separators is the Dynawhirlpool[58] (Fig. 10.8). Like the cyclone, the Dynawhirlpool separates coal and impurity with centrifugal force. It is essentially a cylinder that is inclined about 25° from the horizontal. The medium enters tangentially near the lower

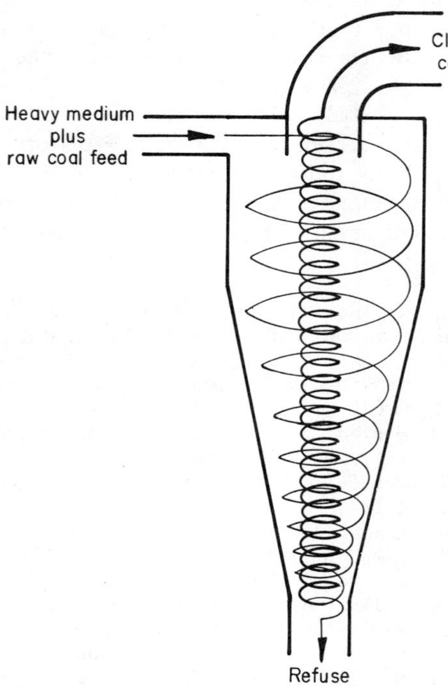

Figure 10.5 A dense-medium cyclone.

[53] G. L. Reynolds, "Use of Cyclones in Modern Coal Preparation Plants," *Min. Congr. J.*, **50**(6), 85–89 (June 1964).

[54] R. E. Zimmerman, "Japan's Swirl Cyclone, A New Approach to Dense-Media Cleaning," *World Coal*, **1**(3), 22–24 (May 1975).

[55] G. F. Efremov, L. S. Zarubin, M. B. Iofa, and V. I. Khaidakin, "Fine Coal Cleaning in Three-Product Dense-Medium Cyclones," *7th Int. Coal Preparation Congr.*, Sydney, Australia, May 23–28, 1976, Paper F1, 15 pp.

[56] E. Austin and W. Irwin, "The N.C.B. Vorsyl for Cleaning Small Coal," *Min. Miner. Eng.*, **6**(8), 48–54 (August 1970).

[57] Anon., "Mechanical Cleaning Spectrum Widens with Advent of New Equipment and Stricter Product Quality Demands," *Coal Age*, **81**(7), 193–216 (July 1976).

[58] M. Sokaski, M. R. Geer, and H. F. Yancey, in *Coal Preparation*, J. W. Leonard and D. R. Mitchell, Eds., American Institute of Mining, Metallurgical, and Petroleum Engineers, New York, 1968, Ch. 10, Part 1.

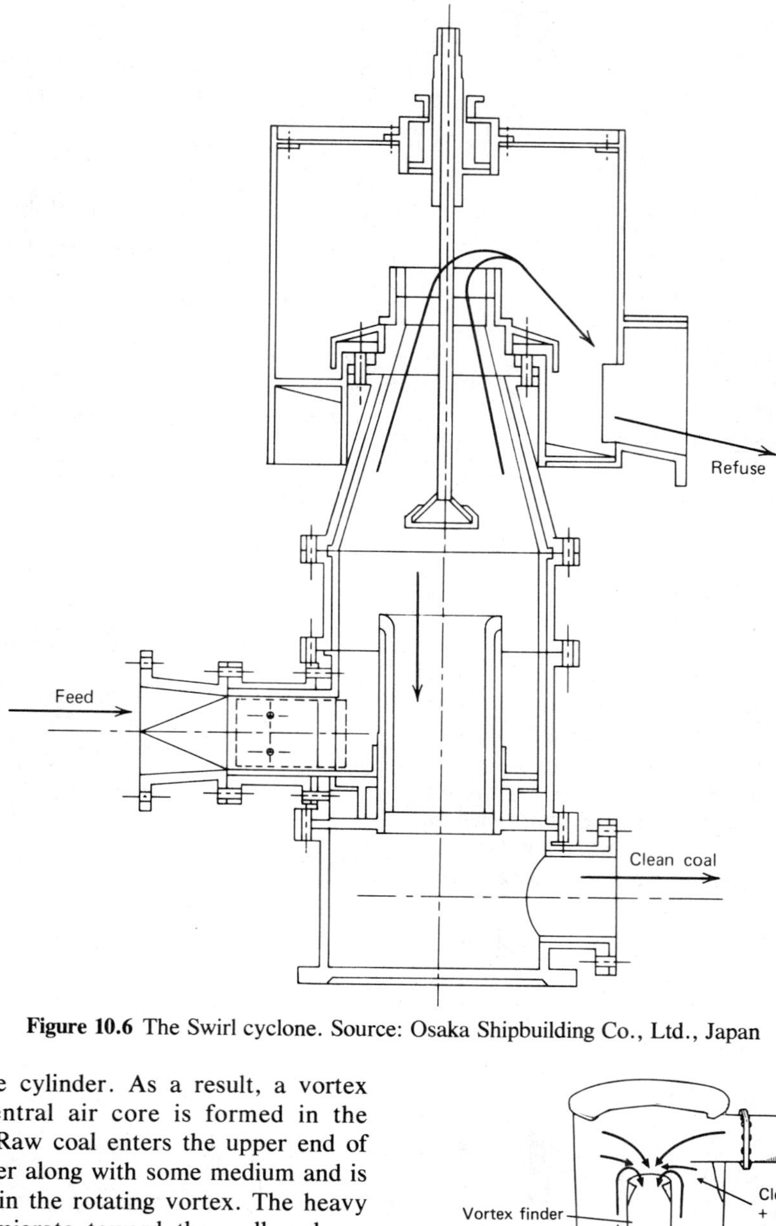

Figure 10.6 The Swirl cyclone. Source: Osaka Shipbuilding Co., Ltd., Japan

end of the cylinder. As a result, a vortex with a central air core is formed in the cylinder. Raw coal enters the upper end of the cylinder along with some medium and is entrained in the rotating vortex. The heavy particles migrate toward the wall and are discharged through a tangential outlet near the upper end of the cylinder. The washed coal, however, stays near the air core and is discharged through a central opening in the bottom end of the cylinder.

10.3.4.3 *Water-Only Cyclones*

Another cyclone gaining broad acceptance by the coal preparation industry is the water-only cyclone (Fig. 10.9). The water-only cyclone does not use an artificial density sus-

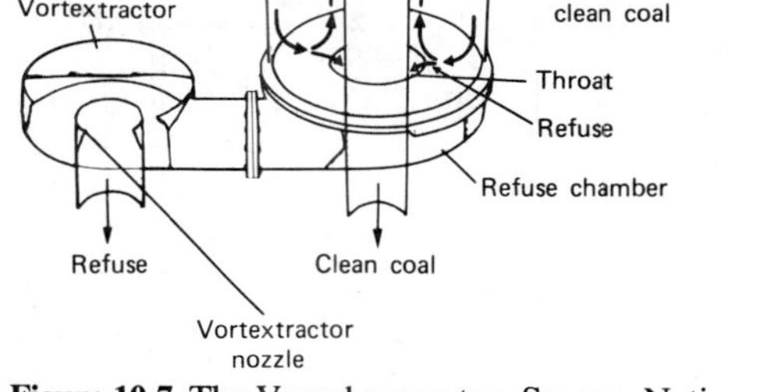

Figure 10.7 The Vorsyl separator. Source: National Coal Board, Great Britain

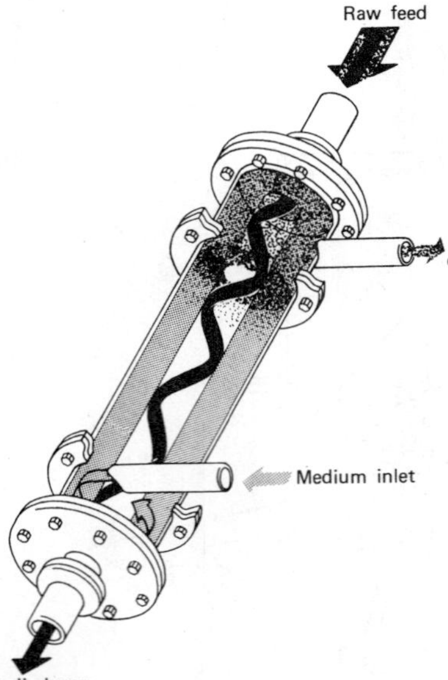

Figure 10.8 The Dynawhirlpool vessel. Source: James A. Redding Company

pension and is not regarded as a dense-medium cleaning device. Most investigators feel, however, that it does utilize an autogenous dense medium developed from the raw coal being cleaned.[58] The water-only cyclone differs from the conventional dense-medium cyclone in that the former has a much greater included angle in the cone—up to 120°—and a longer vortex finder.

The application of water-only cyclones in coal preparation circuits as roughers, followed by dense-medium cleaning, has proved to be a viable innovation. Although water-only cyclones do not achieve a separation as sharp as that of dense-medium cyclones, they provide clean coal products that are almost free of high-ash material in all but the finest sizes, as well as rejects that are almost devoid of slimes. However the rejects do contain large quantities of what would otherwise be clean coal. Nevertheless the water-only cyclone's simplicity, small space requirements, and economical operation have led to its adoption as a rougher device that decreases the load on downstream cleaning units. Many articles[59–64] are available concerning the theory and use of water-only cyclones in coal preparation plants.

[59] C. Krijgsman and J. N. J. Leeman, "The Application of Cyclone Washers in Coal Washeries," *4th Int. Coal Preparation Congr.*, Harrogate, Great Britain, May 28–June 1, 1962, Paper C1, pp. 143–154.

[60] J. Visman, "The Cleaning of Highly Friable Coals by Water Cyclones," *4th Int. Coal Preparation Congr.*, Harrogate, Great Britain, May 28–June 1, 1962, Paper C2, pp. 155–163.

[61] M. G. Akopov, "The Theory of Coal Washing Hydrocyclones," *4th Int. Coal Preparation Congr.*, Harrogate, Great Britain, May 28–June 1, 1962, Paper C3, pp. 165–172.

[62] W. E. Foreman and J. R. Lucas, "Current Status of Hydrocyclone Technology," *Min. Congr. J.*, **58**(12), 50–56 (December 1972).

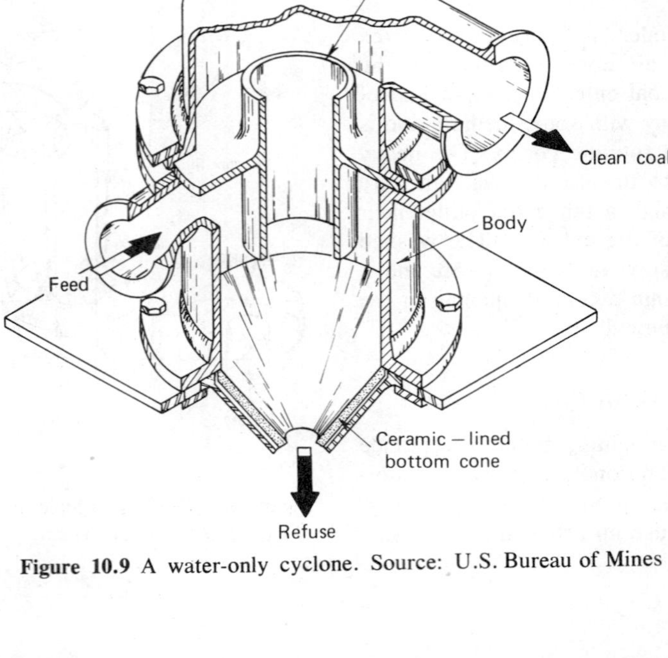

Figure 10.9 A water-only cyclone. Source: U.S. Bureau of Mines

10.3.4.4 *Concentrating Tables*

A common hydraulic separator is the concentrating table; it is one of the oldest cleaning devices and by far the most widely used in the United States for washing coals in the ⅜-in. × 0 size range.[65]

Although no recent technological breakthroughs have occurred in tabling practice, the desire for increased capacity per square foot of plant space has induced a change in table configuration. Twin-deck systems, which were introduced to the U.S. coal industry in 1957, were being supplanted by installations of four-deck stacks by 1966.[66]

Tables have also been studied in great detail to assess their potential for removing pyritic sulfur. Saltsman[67] concluded that free pyrite down to 64 μm in diameter was separated from the coal. His data also indicate that a relatively good removal efficiency is obtained on free pyrite down to 16 μm, whereas the −7-μm free pyrite distributes itself in roughly the same proportion as the water does.

10.3.4.5 *Froth Flotation*

Froth flotation is an integral part of the modern coal cleaning plant, for both steam coal and metallurgical grade coal. Application of this technology has expanded dramatically in recent times.

In froth flotation air is passed through a suspension of fine-size coal in water. Being hydrophobic naturally (or being chemically induced to be hydrophobic), the coal particles adhere to the air bubbles and form a froth atop the pulp. This froth is skimmed off, and the coal is thus separated from the refuse, which remains in the pulp.

Many factors affect coal flotation.[68] These include particle size, rank and state of oxidation of coal, pulp density, pH and other water characteristics, flotation reagents, flotation equipment, conditioning time, and temperature. The effects of these parameters on the kinetics of coal flotation have been the subject of several investigations. Burdon,[69] for example, conducted an exhaustive study to determine the influence of many of them on the rate of coal recovery by froth flotation. Bitter[70] examined in detail the specific aspect of pulp conditioning.

Since a great deal of the cost associated with coal flotation is for reagents, there has been considerable research in this area, particularly by the Soviets. Plaksin and Klassen[71] found, for instance, that coal flotation reagents are most effective when dispersed (as 3–5 μm droplets) into a water emulsion. At one Soviet coal preparation plant where kerosine (85%) and frothing agent (15%) were fed as emulsions, reagent consumption was reduced by one-third, with concurrent improvement in flotation results. Other reagent studies conducted by the Soviets have been reported at the International Coal Preparation Congresses.[72–74]

---

[63] E. J. O'Brien and K. J. Sharpeta, "Water-Only Cyclones: Their Functions and Performance," *Coal Age*, **81**(1), 110–114 (January 1976).

[64] J. W. Collins, "The Hydrocyclone—A Most Useful Preparation Tool," *Coal Min. Process.*, **7**(10), 53–55, 85 (October 1970).

[65] A. W. Deurbrouck and E. R. Palowitch, "Cleaning Fine Coal on Concentrating Tables," *4th Int. Coal Preparation Congr.*, Harrogate, Great Britain, May 28–June 1, 1962, Paper C5, pp. 181–192.

[66] A. W. Deurbrouck and E. R. Palowitch, in *Coal Preparation,* J. W. Leonard and D. R. Mitchell, Eds., American Institute of Mining, Metallurgical, and Petroleum Engineers, New York, 1968, Ch. 10, Part 2.

[67] R. D. Saltsman, "Removal of Fine Pyrite by Current Cleaning Methods—Tables," paper presented at the American Mining Congress Coal Convention, Pittsburgh (May 4–7, 1969), 13 pp.

[68] R. E. Zimmerman, in *Coal Preparation,* J. W. Leonard and D. R. Mitchell, Eds., American Institute of Mining, Metallurgical, and Petroleum Engineers, New York, 1968, Ch. 10, Part 3.

[69] R. G. Burdon, "Some Factors Influencing the Rate of Flotation of Coal," *4th Int. Coal Preparation Congr.*, Harrogate, Great Britain, May 28–June 1, 1962, Paper D6, pp. 273–285.

[70] J. H. Bitter, "Conditioning of the Pulp and Its Influence upon the Results of Flotation," *5th Int. Coal Preparation Congr.*, Pittsburgh, October 3–7, 1966, Paper A3, pp. 47–56.

[71] I. N. Plaksin and V. I. Klassen, "Scientific and Technical Progress in Coal Flotation," *5th Int. Coal Preparation Congr.*, Pittsburgh, October 3–7, 1966, Paper A4, pp. 47–56.

[72] I. N. Plaksin and V. I. Klassen, "Methods of Improving the Process of Coal Flotation," *4th Int. Coal Preparation Congr.*, Harrogate, Great Britain, May 28–June 1, 1962, Paper D5, pp. 263–271.

[73] N. A. Vlasova, I. M. Golovnin, M. G. Elyashevich, and G. V. Zhuvtyuk, "Main Trends of Coal Preparation Development by Flotation in the U.S.S.R.," *6th Int. Coal Preparation Congr.*, Paris, March 26–28, 1973, Paper 1E, 18 pp.

[74] L. A. Antipenko, V. M. Nazarenko, A. G. Tishchenko, and N. S. Vlasova, "Development of Coal Flotation Techniques and Technology," *7th Int. Coal Preparation Congr.*, Sydney, Australia, May 23–28, 1976, Paper J2, 14 pp.

Equipment commonly used for coal flotation in the United States is described by Zimmerman.[75] Soviet flotation machine technology has been outlined in several reports.[76–78]

10.3.4.6 *Oil Agglomeration*

Another technique applicable to fine-size coal preparation is oil agglomeration. Although the basic technology has been understood for many years (the process was first patented by Trent[79] in 1922), its commercial utilization has been hindered by lack of economic viability. With coal economics changing rapidly, however, it is logical to assume that oil agglomeration will grow in commercial importance, paralleling the development of froth flotation.

Basically oil agglomeration is the process of removing fine particles from suspension by selective wetting and agglomeration with oil. Many variations of the original Trent process have appeared over the years, but they differ only in the type of agitation employed to disperse the oil, the quantity and type of oil used, and the separation and treatment of the recovered coal agglomerates.

The Convertol process of oil agglomeration has been used in Europe for some time.[80] More recently, the Japanese installed a commercial size unit (10 tons/hr) of the Shell Pelletizing Separator.[81] As Table 10.3 shows, this technique is highly selective and quite capable of cleaning −200-mesh material. The renewed interest in oil agglomeration is reflected by ongoing research programs. The United States,[82] Canada,[83] and India,[84] are actively pursuing investigations in this area.

10.3.5 Dewatering and Drying

The continued drive to fully mechanized mining has resulted in greater quantities of fine-size coal, and increased underground dust suppression requirements have led to higher surface moisture contents in run-of-mine

[75] R. E. Zimmerman, in *Coal Preparation,* J. W. Leonard and D. R. Mitchell, Eds., American Institute of Mining, Metallurgical, and Petroleum Engineers, New York, 1968, Ch. 10, Part 3.

[76] I. N. Plaksin and V. I. Klassen, "Methods of Improving the Process of Coal Flotation," *4th Int. Coal Preparation Congr.,* Harrogate, Great Britain, May 28–June 1, 1962, Paper D5, pp. 263–271.

[77] N. S. Vlasova, I. M. Golovnin, M. G. Elyashevich, and G. V. Zhuvtyuk, "Main Trends of Coal Preparation Development by Flotation in the U.S.S.R.," *6th Int. Coal Preparation Congr.,* Paris, March 26–28, 1973, Paper 1E, 18 pp.

[78] L. A. Antipenko, V. M. Nazarenko, A. G. Tishchenko, and N. S. Vlasova, "Development of Coal Flotation Techniques and Technology," *7th Int. Coal Preparation Congr.,* Sydney, Australia, May 23–28, 1976, Paper J2, 14 pp.

[79] W. E. Trent, Process of Purifying Materials, U.S. Pat. 1,420,164, June 20, 1922.

[80] A. H. Brisse and W. L. McMorris, Jr., "Convertol Process," *Min. Eng.,* **10**(2), 258–261 (February 1958).

[81] E. Verschuur and G. R. Davis, "The Shell Pelletizing Separator: Key to a Novel Process for Dewatering and De-Ashing Slurries of Coal Fines," *7th Int. Coal Preparation Congr.,* Sydney, Australia, May 23–28, 1976, Paper H1, 9 pp.

[82] C. E. Capes, A. E. McIlhinney, R. E. McKeever, and L. Messer, "Application of Spherical Agglomeration to Coal Preparation," *7th Int. Coal Preparation Congr.,* Sydney, Australia, May 23–28, 1976, Paper H2, 16 pp.

[83] C. E. Capes, A. E. McIlhinney, and R. D. Coleman, "Beneficiation and Balling of Coal," *Trans. AIME, Soc. Min. Eng.,* **247**, 233–237 (September 1970).

[84] G. G. Sarkar, B. B. Konar, S. Sakha, and A. K. Sinha, "Demineralization of Coals by Oil-Agglomeration," *7th Int. Coal Preparation Congr.,* Sydney, Australia, May 23–28, 1976, Paper H3, 14 pp.

Table 10.3 Examples of Shell Pelletizing Separator Performance[81]

	United Kingdom Bituminous Coal, 80% < 75 μm		United Kingdom Bituminous Coal, 42% < 75 μm		Belgian Anthracite Coal, 70% < 75 μm		South African Bituminous Coal, 100% < 75 μm	
Item	Yield (%)	Ash (%)	Yield (%)	Ash (%)	Yield (%)	Ash (%)	Yield (%)	Ash (%)
Feed	100	40	100	58	100	20	100	16
Clean coal	59	4	49	18	81	3	89	8
Tailings	41	91	51	96	19	91	11	82

coal. Such factors have generally precluded the use of dry cleaning systems. After wet cleaning, however, coal contains significant moisture. This moisture is a contaminant that increases transportation costs per delivered Btu, can lead to freezing during transport, renders handling difficult, and decreases the effective Btu content of the coal. Thus the need to reduce the moisture content of washed coal is apparent.

10.3.5.1 *Screens*

Dewatering mechanically separates water from coal and is less expensive than drying, which connotes evaporative (thermal) separation. Among the most common dewatering devices are screens. Screening has, of course, been used as a dewatering technique in coal preparation ever since the introduction of wet cleaning methods. However, as mentioned in Section 10.3.3, several new screen configurations and modifications have appeared recently. One example is the Dutch State Mines sieve bend[85] (Fig. 10.10). It is now used throughout the world, particularly for the preliminary dewatering of coal before it moves to vibrating screens and centrifuges. Essentially it is a stationary wedge bar screen with the bars oriented at right angles and across the flow stream. The screening surface is usually curved in a 60° arc with a radius of 20–60 in., 30 in. being the most common. Screen widths up to 6 ft are normally used.[86] Because of the orientation of the sieve bend, only a portion of each slot is actually available for passage of coal particles. Thus the screen cannot be blinded, a particularly advantageous feature.

Another screen variation gaining wide acceptance in the coal preparation industry is the Vor-Siv[87,88] (Fig. 10.11). Developed by the Polish coal industry, the Vor-Siv is a stationary dewatering device designed to handle high volumes of solids in water slurry.

[85] M. R. Geer, A. W. Deurbrouck, and H. F. Yancey, *Fine-Screening of Coal: Testing of the Sieve Bend,* U.S. Bur. Mines Rep. Invest. No. 5698 (1960), 20 pp.

[86] D. C. Jones, "Mechanical Dewatering in Coal Preparation," *Coal Min. Process.,* **5**(10), 45–53 (October 1968).

[87] W. Jachna, "Vortex Dewatering Sieve-Operating Principles, Application, and Results," *5th Int. Coal Preparation Congr.,* Pittsburgh, October 3–7, 1966, Paper G4, pp. 435–445.

[88] D. C. Cooper, "New Vortex Sieve Works for Quarto Mining," *Coal Min. Process.,* **11**(9), 48–50 (September 1974).

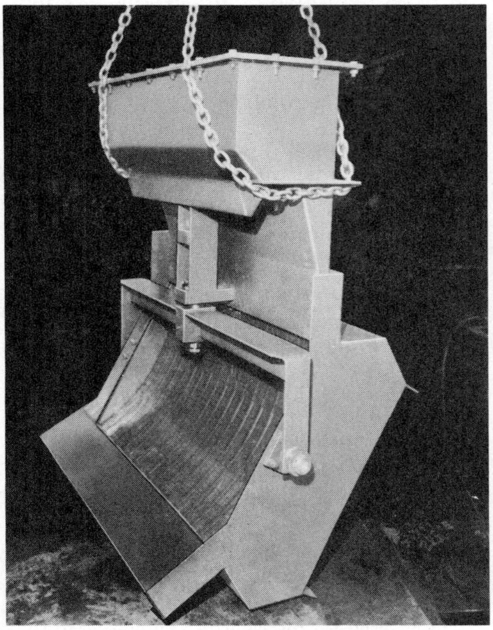

Figure 10.10 A sieve bend. Source: Heyl & Patterson, Inc.

Figure 10.11 The Vor-Siv. Source: National-Standard Company

The Vor-Siv combines certain characteristics of cyclones, sieve bends, and cross-flow screens. During operation solids in a water slurry are fed under pressure through the inlet nozzle, which channels the material into a circular raceway. After the slurry has rotated around the raceway 1–1½ times, it loses enough of its energy so that it drops down onto a conical basket, and the flow becomes a spiraling vortex. Because of centrifugally induced stratification, the water and very fine material circulate almost horizontally, whereas the solids rotate on a slightly steeper path toward the axis. By the time the slurry has reached the bottom quarter of the basket, almost all the water and fine material have filtered through the screen. This causes the coarse particles remaining on the screen to lose their rotational energy and tumble down the remaining quarter of the basket, where a final dewatering action takes place.

Conventional screens, when making a 28-mesh separation, can generally treat approximately 30–40 gal/min per square foot of wire surface, whereas the Vor-Siv commonly handles from 50 to 55, and up to 70 gal/min per square foot of screen area.[89]

A recent screening development designed to alleviate the blinding problem is the "sandwich screen" concept.[90] In this system one fine screen surface is placed on top of another, and they are held together with a single screen binding. When this new screen concept was combined with high-frequency–low-amplitude action, it was found that the blinding problem could be alleviated by never allowing the particles to become lodged in the free screen openings.

This screen may also be dressed with a slotted, natural rubber screening surface. This surface consists of slabs of rubber ¾ in. thick, with longitudinal rows of slotted openings that run perpendicular to the flow of the coal. The advantages of this surface are that it resists abrasion and, when used with the high-frequency–low-amplitude action, provides excellent dewatering characteristics without blinding.

Rose and Schreckengost[91] presented an excellent overview of conventional screen profiles as related to their dewatering characteristics.

10.3.5.2 *Cyclones*

In addition to functioning as classifying and beneficiating devices, cyclones are commonly used for dewatering.[92] They normally treat a pulp with a low percentage of solids and remove a portion of the water to produce a relatively thick suspension for further treatment. The feed enters the cylindrical section tangentially and spins around in the unit. During the spinning, centrifugal forces throw the solids to the outer wall, from which they slide down into the conical section and are discharged as a thicker slurry through the underflow orifice of the cone. The water and slimes in the center of the cylinder and cone are discharged through an overflow orifice at the top.

The feed to cyclones generally contains 5–20% solids and is pumped or flows under the influence of gravity into the unit at a pressure of 20–60 psi. The effectiveness with which the solids are removed depends on the pulp density, the size of the feed, and the ratio of overflow to underflow volumes, which is a function of the size of the nozzles.

10.3.5.3 *Centrifuges*

Centrifuges dewater coal by means of a spinning action. The water is separated from the coal to a degree that is dependent on the amount of centrifugal force developed by the spin, size of the particles, and retention time. There are two principal types available: solid bowl centrifuges, which dewater by means of centrifugal sedimentation, and screen bowl centrifuges, which dewater through centrifugal filtration. Although there are only two principal types, numerous variations exist.

Centrifuges are currently used to dewater material smaller than 1½ in. They are used

[89] R. H. Hooff, "The Vor-Siv Dewatering & Classifying Device," paper presented at the Annual Meeting of the American Institute of Mining, Metallurgical, and Petroleum Engineers, Las Vegas, Nev. (February 22–26, 1977), 16 pp.

[90] H. B. Charmbury, "Special Screen Cleans Up Water Clarification System," *Coal Min. Process.*, **13**(9), 54–55 (September 1976).

[91] C. G. Rose and E. D. Schreckengost, "Dewatering Characteristics of Screen Profiles," *Min. Congr. J.*, **53**(4), 100–105 (April 1967).

[92] D. C. Jones, "Mechanical Dewatering in Coal Preparation," *Coal Min. Process.*, **5**(10), 45–53 (October 1968).

in virtually every coal preparation plant utilizing wet cleaning. The application of centrifugal dewatering in coal preparation has been thoroughly covered in the literature.[93–97]

10.3.5.4 *Filters*

The final broad category of equipment applied to coal dewatering is the filter. Today the type most often used in coal preparation is the continuous vacuum disk filter.[98] Disk filters constitute about 90% of the total filters in use because they perform about as well as any other type; moreover they are considerably less expensive, per square foot of filtration area, both to purchase and to install, and they require less floor space.

This type of filter has a number of disk-shaped leaves that rotate on a shaft, through a tank of slurry. The vacuum inside the disk draws the water through the filter media; the solids form a cake on the leaf. In continuous operation the filter cake is lifted clear of the slurry tank and discharged by application of air pressure behind the filter media. For best performance, solids in the feed should be relatively coarse. Top size usually is 28 mesh, and a feed containing less than 25% −325-mesh fines is desirable. If there is an appreciable content of 10-mesh material in the feed, the filtering operation may be unsatisfactory because the vacuum will not pick up such coarse material. Also, if the feed contains an excessive amount of −325-mesh particles, the cake will not dewater sufficiently to discharge properly.

Although they cost more initially, drum-type filters find occasional use in coal preparation dewatering. They require less maintenance and are able to operate under varying load resulting from changes in feed rate. In addition, they are better able to handle high-ash feeds containing a large percentage of −325-mesh particles. This type of filter is made up of a cylindrical drum, which rotates slowly through a trough-shaped bath fed continuously with slurry. The outer surface of the cylinder is divided into shallow pans or sections running the length of the drum.

It has been demonstrated repeatedly that heating feed slurry will appreciably improve filter performance with respect to capacity, cake moisture, and effluent clarity.[99,100] To take advantage of this phenomenon, the Belgian Coal Institute developed the concept of steam filtration.[101] Their work showed that the application of superheated steam to fine-size coal cakes on vacuum filters produces a significant additional reduction of moisture content. This method basically involves covering a conventional vacuum filter with suitable hooding, then applying steam within the hood. As the steam is pulled through the filter cake by the vacuum system, it condenses and consequently releases latent heat, which raises the cake temperature. At the resultant higher cake temperature, the water viscosity is reduced; this greatly facilitates additional water drainage from the filter cake.

Results obtained in reported studies and in operational practice[102–104] show that a total moisture content reduction of 10–15% can be obtained with the use of steam on a typical 28-mesh × 0 flotation froth product. With a

[93] F. X. Ferney, "Mechanical Coal Drying . . . Equipment and Methods," *Coal Age,* **70**(8), 63–68 (August 1965).

[94] D. D. Dornenburg, J. Hurley, and D. L. Springston, "Comments on Coarse Coal Centrifuging," *Min. Congr. J.,* **52**(11), 65–69 (November 1966).

[95] R. L. Llewellyn, "Dewatering Flotation Slurries in Centrifuges at Stotesbury," *Coal Age,* **69**(4), 88–91 (April 1964).

[96] G. A. Mason and L. Armson, "Centrifuge Applications in Coal Preparation," *Steel Coal,* **186**(4936), 378–384 (February 22, 1963).

[97] F. X. Ferney, "Coarse Coal Centrifuging," *Min. Congr. J.,* **52**(9), 57–59 (September 1966).

[98] Anon., "Coal Preparation and Unit-Train Loading," *Coal Age,* **77**(7), 183–199 (July 1972).

[99] F. X. Ferney, "Mechanical Coal Drying . . . Equipment and Methods," *Coal Age,* **70**(8), 63–68 (August 1965).

[100] A. W. Deurbrouck, *Steam as a Coal Dewatering Aid During Vacuum Filtration: A Pilot Plant Study,* U.S. Bur. Mines Rep. Invest. No. 6718, (1966), 8 pp.

[101] G. Burton, "A New Process for Reducing the Moisture Content of Filter Cake," *4th Int. Coal Preparation Congr.,* Harrogate, Great Britain, May 28–June 1, 1962, Paper F2, pp. 363–369.

[102] K. Lemke, "Using Steam on Vacuum and Pressure Filters," *Min. Congr. J.,* **51**(3), 41–46 (March 1965).

[103] D. A. Dahlstrom and C. E. Silverblatt, "Reduction of Moisture Content by Filtration to Eliminate Thermal Drying of Smalls," *6th Int. Coal Preparation Congr.,* Paris, March 26–28, 1973, Paper 5E, 35 pp.

[104] R. von der Gathen, "Dewatering of Flotation Concentrate on a Steam Cap Filter," *Glueckauf,* **99**(25), 1374–1380 (December 4, 1963).

larger size of cleaned coal (¼ in. × 0), the total moisture content has been reduced from about 11% (without steam) to about 5% with steam applied. Steam filtration operation and capital costs have been determined to be strongly competitive with conventional thermal drying systems (discussed below).

The filter press is a form of pressure filter that usually consists of a series of rigid corrugated plates with intervening filter medium assembled in a framework so that the suspension to be filtered can be forced under pressure into the assembled press and the solids can collect as cake between the plates. These filters operate on a batch basis; however their operation has been substantially automated in recent years. Although many coal-producing countries employ pressure filters, the United States has been hesitant to adopt these labor-intensive devices.[105] At the time of writing, six pressure filter installations are on line or under construction in the United States. Radford and Smith[106] favor the mechanized filter press over other slime dewatering techniques.

10.3.5.5 *Thermal Dryers*

As indicated previously, thermal drying ordinarily involves moisture removal by convective heat transfer between hot combustion gases (or air) and coal; it adds substantially to the preparation cost of coal, but it is used when coal that has been partially dewatered by mechanical means still contains too much moisture for a specific market.

Table 10.4 provides a comparison of thermally dried production with both total and mechanically cleaned production for selected years between 1958 and 1975. Before 1967 the use of thermal dryers in coal preparation experienced a steady growth. The subsequent decline in usage was the result of stricter emission regulations that significantly increased the cost of thermal drying.

Table 10.5 indicates the trend in types of dryer used since 1958. Before that year the screen, vertical tray and cascade, multilouver, and suspension dryers dominated thermal drying hardware application. The fluidized-bed dryer, introduced to the coal industry in 1955, gained rapidly in favor after 1958. By 1967 it represented more than one-fourth of the total number of drying units in operation and was processing more than half the thermally dried production. In 1975 approximately half the dryers in operation were fluidized bed, and they were processing more than 70% of the thermally dried coal production.

Details of drying theory, practice, available hardware, and safety are presented in various publications.[108–113]

[105] J. C. Anderson, "Coal Waste Disposal to Eliminate Tailings Ponds," *Min. Congr. J.*, **61**(7), 42–45 (July 1975).

[106] B. C. Radford and N. P. Smith, "Slimes Treatment by Filter Press at Manton Colliery," *Mine Quarry*, **1**(6), 44, 45, 47, 49, 50, 52 (October 1972).

[107] This table is compiled from information appearing in various volumes of the U.S. Bureau of Mines Minerals Yearbook.

[108] J. W. Leonard and T. S. Spicer, in *Coal Preparation*, J. W. Leonard and D. R. Mitchell, Eds., American Institute of Mining, Metallurgical, and Petroleum Engineers, New York, 1968, Ch. 13.

Table 10.4 Thermal Drying of Bituminous Coal and Lignite in the United States[107]

Year	Total Production, 1000 tons	Mechanically Cleaned, 1000 tons	Thermally Dried, 1000 tons	Thermally dried, as Percent of: Total Production	Mechanically Cleaned
1958	410,446	259,035	31,542	7.7	12.2
1961	402,977	264,711	39,633	9.8	15.0
1964	486,998	310,203	58,702	12.1	18.9
1967	552,626	349,402	74,741	13.5	31.4
1970	602,932	323,452	64,165	10.6	19.8
1971	552,192	271,401	48,105	8.7	17.7
1972	595,386	292,829	53,235	8.9	18.2
1973	591,738	288,918	46,202	7.8	16.0
1974	603,406	265,150	36,045	6.0	13.6
1975	648,438	266,993	35,681	5.5	13.4

Table 10.5 Thermal Drying of Bituminous Coal and Lignite in the United States by Equipment Types[107]

Year	Fluidized Bed	Multilouver	Rotary	Screen	Carrier	Vertical Tray and Cascade	Suspension or Flash	Totals
				Number of Thermal Units				
1958	6	50	6	59	5	58	44	228
1967	67	49	5	35	—	35	42	233
1975	58	12	5	15	—	1	28	119
				Net Tons Dried (thousands)				
1958	3,337	9,416	405	7095	679	5775	4834	31,542
1967	38,102	15,090	1668	5994	—	4036	9853	74,741
1975	25,866	1,969	794	2798	—	70	4184	35,681
				Percent of Thermally Dried Coal				
1958	10.6	29.8	1.3	22.5	2.2	18.3	15.3	100.0
1967	51.0	20.2	2.2	8.0	—	5.4	13.2	100.0
1975	72.6	5.5	2.2	7.8	—	0.2	11.7	100.0

10.3.6 Water Clarification

Any device that dewaters coal simultaneously clarifies the associated water. Therefore the line between dewatering and clarification is somewhat nebulous. This issue is commonly resolved by a consideration of the principal flowstream emanating from a particular device. If that stream is coal, dewatering is said to occur; if the stream is water, clarification is taking place. We discuss next the pieces of equipment used to clarify water, thus helping close water circuits.

Almost all new coal preparation plants are constructed with some type of closed water circuits. Recirculation and treatment of wash waters is an integral part of preparation plant operation because of the need to decrease coal loss to the refuse, to prevent solids buildup in water recirculated to coal cleaning units, and to comply with water quality control regulations. State and local laws pertaining to stream pollution have become increasingly stringent during the last few years.

The primary system employed for water recovery in coal preparation plants today consists of a clarifying device, which produces effluent that is clarified sufficiently to be returned to the plant for reuse. The device most commonly used is the gravitational clarifier. This unit has the ability to produce overflow containing approximately 0.5% solids or less. A thorough discussion of the design and performance of conventional clarifiers and their application in modern coal preparation plants was presented by Dale and Dahlstrom.[114]

Attempts to reduce the structural space requirements of clarifiers have resulted in so-called lamella thickeners.[115] These are characterized by packs of sloping plates, arranged in parallel, installed in the settling area. This arrangement divides the settling area into many thin, sloping layers, within which the distance a particle must fall before contacting

[109] J. W. Leonard, "Improving Coal Values with Thermal Dryers," *Mechanization,* **27**(9), 35–38 (September 1963).

[110] R. E. Zimmerman, "Recent Progress in Thermal Drying Ultra-Fine Coal," *Min. Congr. J.,* **47**(9), 47–51 (September 1961).

[111] D. C. Jones, "Fluid-Bed Fine-Coal Drying," *Mechanization,* **23**(9), 69–71 (September 1959).

[112] H. A. Schrecengost and M. S. Childers, *Fire and Explosion Hazards in Fluidized-Bed Thermal Coal Dryers,* U.S. Bur. Mines Inf. Circ. No. 8258, (1965), 21 pp.

[113] D. C. Jones, "So You Are Considering Thermal Drying," *Coal Min. Process.,* **6**(10), 36–43, 74 (October 1969).

[114] L. A. Dale and D. A. Dahlstrom, "Design and Operation of Thickening Equipment for Closed-Water Circuits in Coal Preparation Plants," paper presented at the meeting of the American Institute of Mining, Metallurgical, and Petroleum Engineers, Chicago (February 14–18, 1965), 13 pp.

[115] W. E. Schlitter and W. Markl, "Cross-Flow Lamella Thickeners," *Min. Mag.,* **134**(4), 291, 293, 295, 297 (April 1976).

"bottom" is greatly reduced. Such thickeners are being employed at several coal preparation plants in the United Kingdom.

Space requirements can also be reduced by the Enviro-Clear thickener.[116] This unit was developed for the sugar beet industry and has been successfully applied in several American coal preparation plants with very encouraging results. Figure 10.12 is a schematic diagram of this bottom-fed thickener. The feed, which is generally mixed with a flocculant, enters the thickener at the bottom and strikes a deflector plate. It is directed horizontally into a sludge bed, which induces contact with previously formed agglomerates and generates additional coalescence of solids into relatively large flocs. The liquid then flows upward and is filtered by the sludge. This thickener has only two zones, a classified water zone and a compression zone. A sharp interface line between these zones is always present. In its first installation in the coal industry,[117] the Enviro-Clear thickener produced overflow clarities and underflow concentrations as good or better than free-settling thickeners at the same flocculant dosage, and handled at least 7 times the volume rate.

The deep settling cone,[118] which was developed in the United Kingdom, is essentially a sedimentation vessel of considerably smaller settling area than a conventional thickener. This device (Fig. 10.13) is cylindroconical. More than 60 such cones are presently operating on tailings and fine clay effluents in the United Kingdom.[119] The cones are usually 12.5–14 ft in diameter and about 25 ft deep and handle approximately 400 gal/min of dilute tailings direct from froth flotation cells.

Because of the excellent agglomerating properties of high molecular weight polyelectrolyte flocculants, clarified water is quickly produced at the top of the cone as the large flocs settle rapidly. The settling flocs are gently stirred to promote channels that allow

[116] N. P. Chironis, "New Clarifier/Thickener Boosts Output of Older Coal Preparation Plant," *Coal Age*, **81**(1), 140–145 (January 1976).

[117] A. A. Terchick, J. C. Anderson, and D. T. King, "The Application and Utilization of the Enviro-Clear Thickener in a U.S. Steel Coal Preparation Plant," paper presented at the AIME (Society of Mining Engineers) Fall Meeting, Acapulco, Mexico (September 23–25, 1974), 9 pp.

[118] J. Abbott, C. Dell, B. Denison, D. Knott, and N. W. Hill, "Coal Preparation Plant Effluent Disposal by Means of Deep Cone Thickeners," *6th Int. Coal Preparation Congr.*, Paris, March 26–28, 1973, Paper 20E, 18 pp.

[119] D. E. Jenkinson, "Coal Preparation Developments in the United Kingdom," *World Coal*, **2**(8), 37–40 (August 1976).

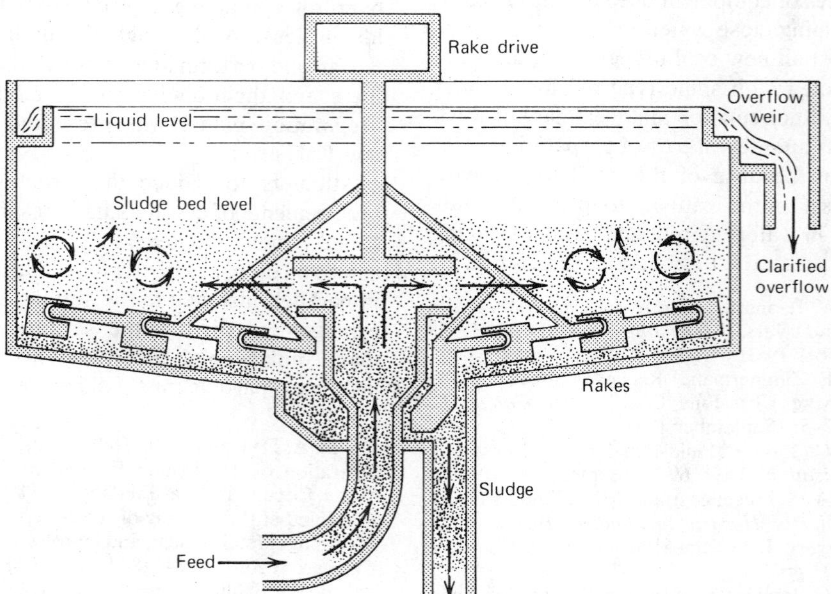

Figure 10.12 The Enviro-Clear thickener. Source: Enviro-Clear Division of Amstar Corporation

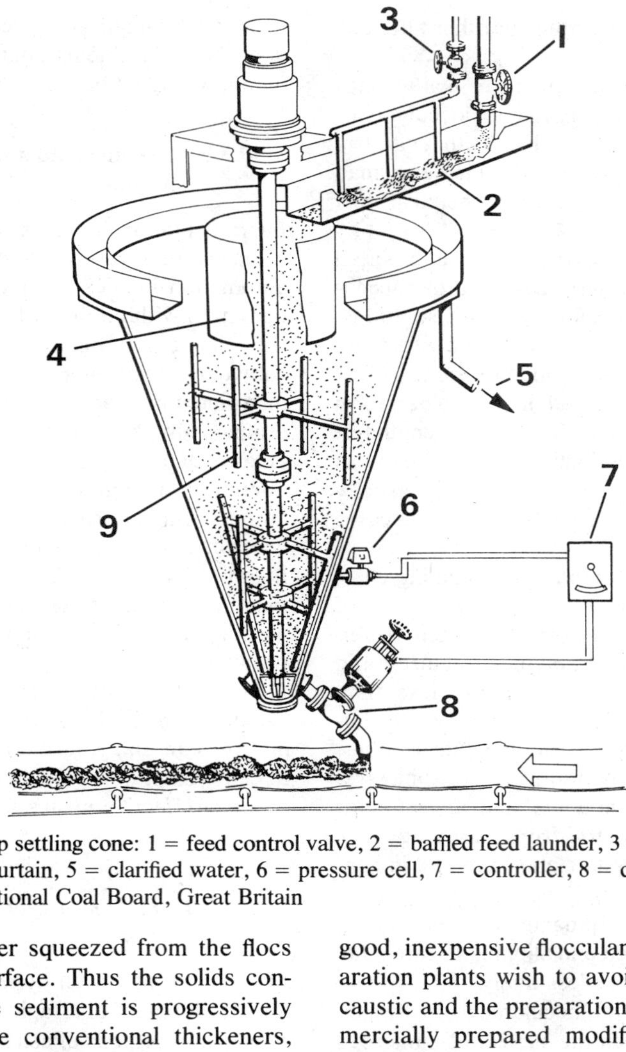

Figure 10.13 A deep settling cone: 1 = feed control valve, 2 = baffled feed launder, 3 = flocculant control valve, 4 = center curtain, 5 = clarified water, 6 = pressure cell, 7 = controller, 8 = discharge valve, 9 = stirrer. Source: National Coal Board, Great Britain

the expelled water squeezed from the flocs to rise to the surface. Thus the solids concentration of the sediment is progressively increased. Unlike conventional thickeners, deep settling cones for tailings are operated with a high solids load, which exerts pressure in the lower regions and assists in dewatering the floccules. A thickened discharge, having up to 65% solids, can be produced. Such a consistency enables the sludge to be transported on rubber-troughed conveyor belts.

Rarely do the nature of the refuse slimes and the need for clear water permit operation of a coal preparation plant clarifier without the addition of chemicals for flocculation. Many types of flocculant are employed. These include inorganic materials (e.g., alum, lime, iron salts, and sulfuric acid) as well as organics. In the organic group the natural or modified starches have long been employed. Raw starch is sparingly soluble and quite ineffective. When treated with caustic, however, starch is very soluble and usually a good, inexpensive flocculant. Yet many preparation plants wish to avoid the handling of caustic and the preparation methods. A commercially prepared modified starch called pregelatinized starch is readily soluble and, although more expensive, it is the type usually employed in preparation plants.

There are also many types of synthetic organic polymer on the market that are commonly used in clarification. These are high polymeric macromolecules that have long, threadlike molecules that cause flocculation by physically bridging the gap between particles. Their manufacture can be controlled to form longer or shorter chain compounds, the lower limits of molecular weight being determined by their effect on the speed of settling and the upper limit by the polymers gradually becoming insoluble in water. The use of these flocculants has spread rapidly and has resulted in considerable savings in the treatment of flotation tailings. They are also used as filter aids in the dewatering of flotation froth concentrates or other filterable solids.

Although these polymers are many times more expensive than inorganic flocculants or starch on a per pound basis, they are so powerful that they may be used in very small amounts. Thus the operating chemical cost can be less.

Because they are used in such small amounts, the polymers are added in concentrations ranging from 0.01 to 0.05% solutions. They usually form a better floc structure when used at these concentrations. Most polymers are purchased in a powder form and may be stocked for use at strengths of 0.5–1%. Stock solutions may be prepared as infrequently as twice per week, but usually they are made fresh every day, in some cases once per shift.

To remedy the troublesome handling characteristics of dry flocculants, a family of flocculants in liquid form has been developed.[120] These liquid polymers feature rapid solution preparation, commonly less than one-third the time required for dry flocculants. The liquid form also lends itself to use in automatic preparation and feeding systems. The liquid polymer and an activator are pumped directly from shipping containers, eliminating the labor required in handling bagged dry polymers. This provides safety and housekeeping advantages because no dust is generated and slipping hazards due to polymer spills are less likely.

The application of flocculants to coal washery waters has been the subject of numerous reports.[121–124] Because flocculation remains more an art than a science, washery water flocculation problems continue to be solved on an individual basis.

10.3.7 Instrumentation and Automation

Instrumentation involves the use of devices to measure given variables or conditions; automation implies making changes in plant operation without direct human intervention, the changes being based on indications shown by instruments. The application of instrumentation and automation is expanding dramatically in coal preparation today. Increasing labor costs have placed emphasis on operating with minimum manpower. Through automatic monitoring and remote centralized control, it is possible to reduce manpower, thereby eliminating the influence of the human factor and, at the same time, improving working conditions and the quality of the product.

Although quality controls were highly developed in many industries years ago, coal's progress in this direction has been slow. Recent years, however, have seen the emergence of effective means for monitoring coal quality parameters.

10.3.7.1 *Instruments for Determining Ash*

Much work has been done, for example, in the development of a rapid ash determinator. Most such methods of ash monitoring depend on the attenuation of electromagnetic radiation. The Cendrex continuous ash monitor,[125] developed by the Dutch State Mines, employs a beam of X rays from a cobalt X-ray tube. The beam is split into two equal parts by a rotary chopper, one part striking the coal sample and the other striking a reference of Plexiglas. Because of the difference in absorption coefficients of the coal and the Plexiglas, the intensities of the reflected beams generally are different. This difference is translated by means of a photocell into ash content values for the coal sample.

Another such device was developed by the

[120] G. F. McIntyre, "Polymers Optimize Plant Water Clarification," *Coal Min. Process.*, **12**(9), 61–63 (September 1975).

[121] G. Dru, "Study of Some Factors in the Flocculation of Coal Washery Tailings," *5th Int. Coal Preparation Congr.*, Pittsburgh, October 3–7, 1966, Paper D6, pp. 267–279.

[122] G. H. Matheson and J. M. W. Mackenzie, "Flocculation and Thickening Coal-Washery Refuse Pulps," *Coal Age*, **67**(12), 94–100 (December 1962).

[123] R. H. Oliver and K. C. Lindstrom, "The Application of Flocculants to the Clarification of Coal Washery Waters," *Mechanization*, **24**(3), 47–51 (March 1960).

[124] M. P. Corriveau, "Solving the Flocculation Problem," paper presented at the Annual Meeting of the American Institute of Mining, Metallurgical, and Petroleum Engineers, New York (February 14–18, 1960), 9 pp.

[125] H. Dijkstra and B. S. Sieswerda, "Apparatus for Continuous Determination of the Ash Content of Coal," *3rd Int. Coal Preparation Conf.*, Liège, Belgium, June 23–28, 1958, Paper F8, pp. 645–655.

National Coal Board and the Atomic Energy Research Establishment in Great Britain. The NCB/AERE ash meter[126] (Fig. 10.14) consists of a rotating presentation table above which are mounted a profiling system, a radioactive isotope source, and a radiation-measuring system. The coal is introduced by way of a feed chute (11) onto the turntable (5) near its center and, as the turntable rotates (19 rpm), a cam fixed under a horizontally mounted profile plate (2), just clear of the table top, pushes the coal radially outward across the turntable from the inner to the outer radius of the feed chute. Continuous introduction of coal into the created annulus forces the coal further across the turntable toward the peripheral sidewall of the profile plate. The coal is compressed vertically between the turntable and the profile plate and horizontally between the cam and the sidewall and emerges as a compacted band of coal with a smooth surface 7 in. wide by 1¾ in. deep for presentation to the radioactive isotope. More coal fed to the meter causes the original feed to be pushed out beyond an outer profile plate and discharged from the turntable onto a reject conveyor.

The radioactive source for the ash monitor consists of two disk-type low-energy γ-^{238}Pu isotopes of 7 mCi activity each, and having

[126] P. Cammack, G. Hannah, W. Irwin, H. McPherson, and G. F. Morris, "Design of an Automatic Small Coal Blending Plant," *6th Int. Coal Preparation Congr.*, Paris, March 26–28, 1973, Paper 23E, 14 pp.

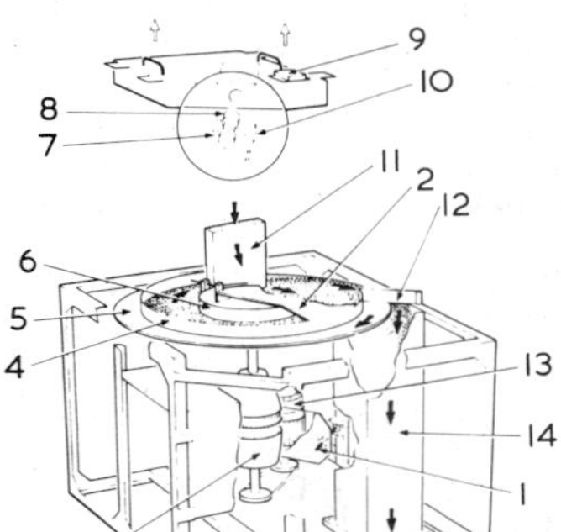

Figure 10.14 A phase 3A ash meter. 1 = belt tensioner with adjusting rod; 2 = profile plate and compression plate; 3 = gearbox; 4 = side wall; 5 = turntable; 6 = adjustable cam; 7 = steel block; 8 = detector; 9 = pre-amplifier; 10 = graphite block; 11 = feed chute; 12 = scraper; 13 = motor; 14 = discharge chute.

a half-life of 86 years. The radiation emitted bombards the surface of the coal, penetrating up to 1½ in. into the coal bed and being absorbed or backscattered, depending on the elements present in the sample. Elements of low atomic number (the combustible elements) backscatter well, whereas elements of a higher atomic number (elements in the ash) absorb the radiation. By measuring the backscattered radiation, a determination of the ash content of the coal sample can be made.

Direct-reading instruments have been designed for the rapid measurement of density as a means of estimating the ash content of coal. These are based on the known functional relationship between density and ash for coal from the same seam. The Constant Volume Bulk Density Meter[127] developed by the National Coal Board is one such instrument. In this device a constant volume of material is continuously weighed by a load cell, thereby providing a measure of bulk density.

Although these techniques have reached an advanced state of development, and continuous ash monitoring is used extensively in Europe, the United States does not have such a unit in operation. Also, as pointed out by Padberg and Becker,[128] these devices do not lend themselves well to laboratory application (as a replacement for the standard incineration method) because the accuracy is inadequate, and preparation of the apparatus for a new type of coal is very time-consuming.

10.3.7.2 *Instruments for Determining Moisture*

Moisture is another coal quality parameter for which rapid determinators have been developed. Although many successful processes exist for rapid moisture determination, only a few of these are suitable for coal preparation products. One method, devel-

[127] D. A. Hall, A. A. Hirst, H. MacPherson, and A. M. Wandless, "Automatic Measurement of Ash and Moisture in Coal: Methods and Their Application," *5th Int. Coal Preparation Congr.*, Pittsburgh, October 3–7, 1966, Paper F1, pp. 347–360.

[128] W. Padberg and R. Becker, "Experiences with the Cendrex Device for Determination of the Ash Content of Coal," *Glueckauf*, **101**(23), 1348–1351 (November 10, 1965).

oped by Stewart et al.[129] is based on thermalization of fast neutrons by the hydrogen in coal moisture. In this method, a 1-Ci americium-beryllium source and a boron-lined proportional tube are in the center of a large volume of coal. High-energy neutrons from the source penetrate the coal and are scattered and slowed by hydrogen atoms in it. An increase in moisture content increases the H_2 content and the density of slowed neutrons. Thus the electronic count rate is a direct measure of the moisture content.

Other moisture meters that are being used in coal preparation plants are described in the literature.[130,131]

10.3.7.3 *Instruments for Determining Sulfur*

At present the coal quality parameter of greatest concern is sulfur. Hall et al.[132,133] have developed a neutron meter for monitoring the sulfur content of coal streams. Neutrons from 80 μg of ^{252}Cf penetrate the coal, producing thermal neutrons that are captured by atoms of elements present in the coal. The excited atoms emit prompt gamma rays with energies characteristic of the capturing element. These gamma rays are detected with a 6 × 7 in. sodium iodide detector and processed electronically to produce a gamma-ray energy spectrum of the elements. Tests of this meter at a commercial coal preparation plant showed that sulfur in moving coal can be determined to a precision of 0.05% sulfur within 2 min. However no sulfur meter is being used as a standard piece of equipment in any coal preparation plant today.

10.3.7.4 *Automation*

Development of the instruments described in Sections 10.3.7.1–10.3.7.3 has spurred efforts to substantially automate the entire preparation process from preliminary storage to outloading. The subject of automation has been discussed in numerous papers and is too broad to be covered here. The reader may consult the literature[134–137] for a detailed description of automation as applied to all facets of operation of a modern coal preparation plant.

One objective of automation is to improve efficiency; today coal preparation engineers are continually striving toward this end by attempting to optimize flow scheme layout. The advent of modern computers has enabled designers to devise flowsheets for new plants by optimization techniques. A mathematical model is used to simulate plant operation, thereby pinpointing an optimum separation parameter (specific gravity, equipment type, etc.) on the basis of maximum clean coal yield subject to an ash or sulfur constraint. A number of papers[138–144] explain the approaches that have been employed in various countries.

[129] R. F. Stewart, A. W. Hall, and J. L. Konchesky, "Plant Tests of a Neutron Moisture Meter for Coal," *Min. Congr. J.*, **56**(11), 44–48 (November 1970).

[130] D. A. Hall, A. A. Hirst, H. MacPherson, and A. M. Wandless, "Automatic Measurement of Ash and Moisture in Coal: Methods and Their Application," *5th Int. Coal Preparation Congr.*, Pittsburgh, October 3–7, 1966, Paper F1, pp. 347–360.

[131] M. Hampel, "Automatic, Rapid Determination of Ash and Moisture in Small Coal in West Germany, and Its Use as a Preparation Process Regulator," *6th Int. Coal Preparation Congr.*, Paris, March 26–28, 1973, Paper 21E, 15 pp.

[132] A. W. Hall, J. W. Martin, R. F. Stewart, and A. M. Poston, *Precision Tests of a Neutron Sulfur Meter in Coal Preparation Plant*, U.S. Bur. Mines Rep. Invest. No. 8038 (1975), 12 pp.

[133] D. F. Summers, "Sulfur Meter Speeds Coal Blending," *Coal Age*, **82**(3), 82–85 (March 1977).

[134] W. M. Wallace and J. D. Menzies, "Towards Automation in Coal Preparation Plants," *4th Int. Coal Preparation Congr.*, Harrogate, Great Britain, May 28–June 1, 1962, Paper B4, pp. 119–129.

[135] M. Bolek, "Some Possibilities in the Automation of Coal Preparation Plants," *5th Int. Coal Preparation Congr.*, Pittsburgh, October 3–7, 1966, Paper C1, pp. 121–130.

[136] E. C. J. Cochet, B. L. J. Robert, and J. M. P. Maginot, "Automation and Control in the Coal Preparation Plants of the Collieries of the Lorraine Basin," *5th Int. Coal Preparation Congr.*, Pittsburgh, October 3–7, 1966, Paper C2, pp. 131–148.

[137] H. Paul, "Automation of the Feed to Raw-Coal Bunkers to Permit Storage of Unwashed Coal by Seam or Groups of Seams at the Bergmannsgluck-Westerholt Combined Mine," *5th Int. Coal Preparation Congr.*, Pittsburgh, October 3–7, 1966, Paper C3, pp. 149–164.

[138] P. N. Ivanov and M. N. Yampolsky, "An Economical and Mathematical Model of a Coal Preparation Flowsheet," *6th Int. Coal Preparation Congr.*, Paris, March 26–28, 1973, Paper 18E, 18 pp.

[139] M. K. Laverick, "Optimisation of Coal Washery Flowsheet Design and Operation With the Aid of a Computer," *6th Int. Coal Preparation Congr.*, Paris, March 26–28, 1973, Paper 22E, 16 pp.

In the same vein, many researchers have attempted to derive detailed mathematical models for several of the unit operations of coal preparation. Such models can be incorporated into larger ones defining the operation of an entire beneficiation process. Sztaba and Tajchman,[145] as well as Kubitza et al.,[146] have modeled coal flotation; Nowak and Aleksa[147] modeled the water-slurry circuit.

10.3.8 Plant Circuitry

As the role of coal in energy production becomes better defined and more important in a number of countries, specification coal is required. Ash, sulfur, and Btu contents may be specified alone or in some combination. Maximum recovery of coal may be desired even though the final coal product will not meet environmental limits without secondary cleanup such as flue gas desulfurization (see Chapter 22). Central preparation plants, where coal mined in one vicinity is processed in a common cleaning facility, will be designed to clean coals of differing washabilities. Such plants will have flowsheets that provide maximum flexibility, thus delivering final products of acceptable quality from a variety of coals by using different equipment configurations or just by varying the specific gravity of separation through medium control.

An interesting example of a new flowsheet configuration is the Homer City coal preparation plant built for the Pennsylvania Electric Company.[148] This plant yields three products; the first product is an intensively cleaned, high-Btu, low-ash and -sulfur product, suitable for firing in new units that must comply with an emission standard of 1.2 lb of SO_2 per million Btu. The second product is of intermediate sulfur and ash content, suitable for use in existing units outside critical air basins, where intermediate sulfur content coals are permitted to be burned, or where the burning units would be fitted with stack gas scrubbing processes. The third product is high-ash refuse.

The coals being processed at the Homer City plant provide a primary clean coal of less than 0.6% sulfur. Figure 10.15 is a simplified flow diagram of this plant. The raw coal is crushed to 1¼ in. top size and fed to the plant at the rate of 1200 tons/hr. This material is sized at ¼ in. The −¼-in. material is then deslimed at 2 mm. The ¼-in. × 2-mm material is fed to a primary dense-medium cyclone, which makes a separation at approximately 1.30 specific gravity. The overflow product is clean coal. The underflow product from this washing cyclone is then retreated in a secondary dense-medium cyclone at 1.8 specific gravity. The overflow product here is considered middlings coal; the underflow product is discarded as refuse. The −2-mm material from the desliming screen then goes to a classifying cyclone. The underflow from this cyclone, approximately 2-mm × 100-mesh material, is fed to a dense-medium cyclone also being operated at 1.30 specific gravity. The overflow from this cyclone is clean coal; the underflow goes

---

[140] D. Bartelt and W. Rosner, "Operational Use of Automation and Process Control in the West German Coal Mining Industry: Present Position and Experience Gained," *7th Int. Coal Preparation Congr.*, Sydney, Australia, May 23–28, 1976, Paper D1, 16 pp.

[141] T. C. Erasmus, "Predicting the Performance of a Coal Washer with the Aid of a Mathematical Model," *7th Int. Coal Preparation Congr.*, Sydney, Australia, May 23–28, 1976, Paper G1, 18 pp.

[142] L. H. E. Weyher, "Computer Assistance in the Analysis and Development of Coal Beneficiation Flowsheets," paper presented at the 1970 Coal Convention and Exposition, Cleveland, Ohio, (May 11–14), 31 pp.

[143] A. D. Walters and R. V. Ramani, "The Optimization of Coal Preparation Plant Yields Using Computer Programming," *7th Int. Coal Preparation Congr.*, Sydney, Australia, May 23–28, 1976, Paper G2, 21 pp.

[144] B. S. Gottfried, *Computer Simulation of Coal Preparation Plants,* Final Rep. U.S. Bur. of Mines Grant No. GO155030, November 1975, 241 pp.

[145] K. Sztaba and Z. Tajchman, "Regression Model of Coal Flotation Process," *7th Int. Coal Preparation Congr.*, Sydney, Australia, May 23–28, 1976, Paper J1, 10 pp.

[146] K. H. Kubitza, L. Leininger, F. Monostory, and T. Schieder, "Coal Preparation Research and Development in West Germany," *6th Int. Coal Preparation Congr.*, Paris, March 26–28, 1973, Paper 10E, 13 pp.

[147] Z. A. Nowak and H. Aleksa, "Mathematical Model of a Water and Slurry Circuit," *6th Int. Coal Preparation Congr.*, Paris, March 26–28, 1973, Paper 24E, 20 pp.

[148] J. F. McConnell and C. W. Statler, "Multi-Stream Coal Cleaning Strategy for Control of Sulfur," paper presented at the AIME Society of Mining Engineers Fall Meeting, Salt Lake City, Utah (September 10–12, 1975), 19 pp.

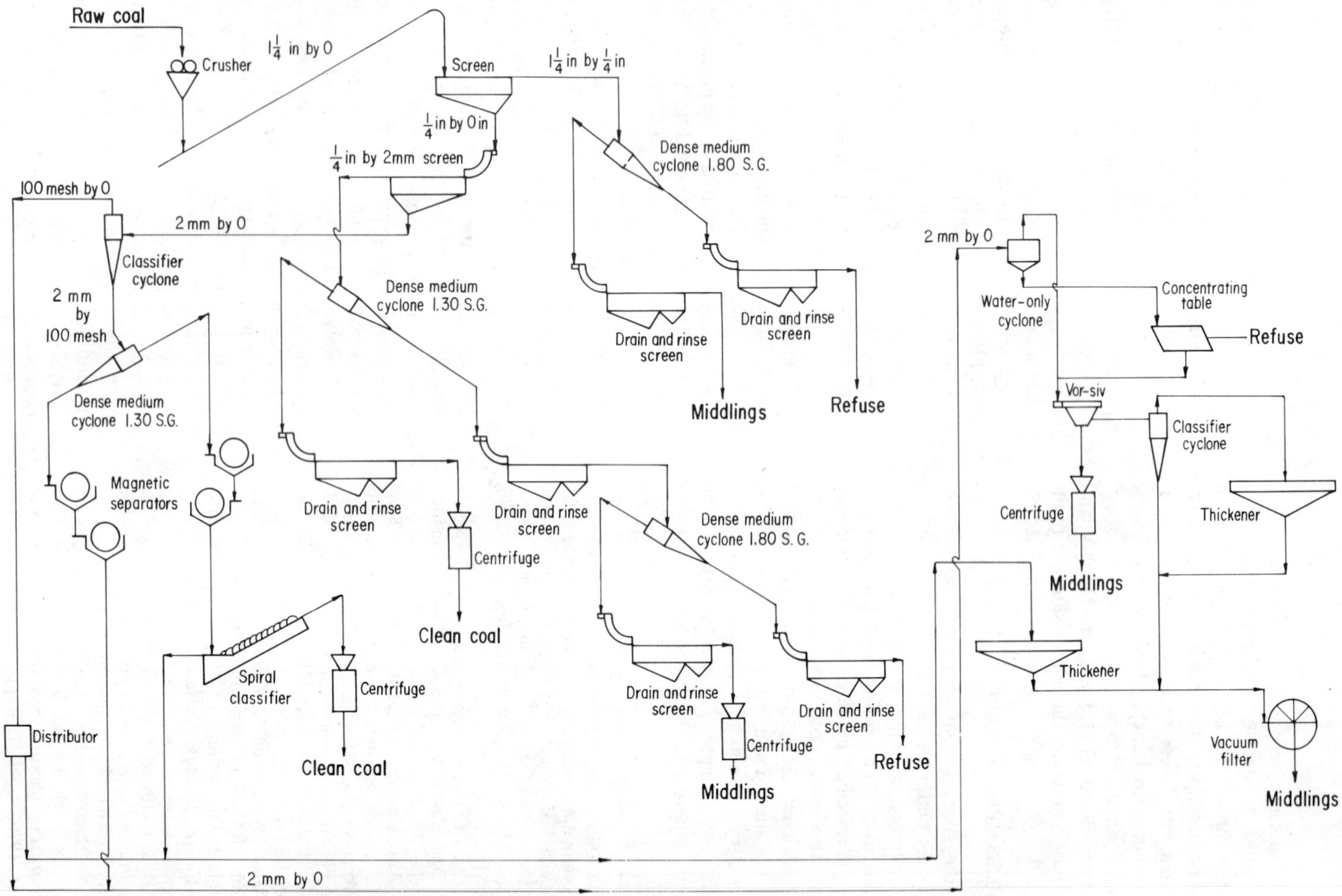

Figure 10.15 Simplified flow diagram of the Homer City plant.

to a scavenger circuit. The classifier cyclone overflow, 100 mesh × 0, also goes to the scavenger circuit. The 1¼ × ¼ in. material is fed to a dense-medium cyclone, washing coal at a 1.80 specific gravity. The overflow product here is considered middlings coal; the underflow product is refuse.

A water-only cyclone receives the total 2-mm × 0 scavenger circuit feed. The underflow from this cyclone is washed on a concentrating table; the heavies are rejected. The clean coal product reports with the overflow product from the water-only cyclone; both products are dewatered and combined with the final middlings.

Such a circuit relies on accurate control of the circulating medium to assure proper separating specific gravities. A flowsheet like the one described provides maximum flexibility, allowing the plant operator to change the specific gravity of separation quickly, hence to control the quality of the final product.

10.3.9 Miscellaneous Developments

Many miscellaneous improvements have appeared in coal preparation plants over the past several decades. The problems of corrosion and wear, for example, have been attacked through the use of plastic materials. Abrasion-resistant coal preparation equipment increases the life between repairs and promotes an increase in labor productivity. Application of plastics to flotation cells, bunker linings, vibratory feeders, and piping was discussed by Roberts.[149]

Plastic liners have an exceedingly low coefficient of friction; therefore materials with high slip characteristics are being utilized as liners for chutes and bins.[150] These have alleviated wear problems and eliminated blockages, and have the added advantages of being lightweight and easy to install.

Another material adopted for its abrasion resistance is cast basalt. Basalt is a volcanic rock that melts at approximately 1250°C and can be cast like iron. Cast basalt has been used in pipe, sumps, chutes, and froth flotation cells in the United Kingdom.[151] The problem of increasing wear resistance in preparation equipment has also been evaluated from a management viewpoint by Mason et al.[152] Various approaches taken to abate the abrasion problem in the preparation plants of the Soviet Union have been cited by Blagov.[153]

The inherent noisiness of many sizing and cleaning operations, coupled with the increased concern of regulatory agencies to lower noise levels in industrial plants, has brought about another application of plastics. Energy-absorbing polymer, which has the ability to absorb mechanical energy, has been applied directly to shaker screens, conveyor belt heads, jigs, and discharge chutes in several coal preparation plants, resulting in noise reductions of as much as 30 dBA.[154] The control of noise in washeries has also been studied by Spanaus and Allan[155] and Chironis.[156]

Aside from technological advancement, the subject of process developments encompasses the status of coal preparation throughout the world. A summary of coal preparation statistics for the major coal-producing countries has been prepared for each International Coal Preparation Congress. The latest such data available were compiled for the Seventh International Coal Preparation Congress held in Sydney, Australia. Table 10.6 shows the production of coal and the extent of preparation for many of the major coal producing countries of the world. Table 10.7

[149] J. Roberts, "Attacking Corrosion and Wear on the Coal Preparation Plant," *Mine Quarry,* **2**(4), 39, 43, 45–47, 49 (April 1973).

[150] Anon., "Plastic Liner for Chutes, Bins Shows Promising Results," *Coal Age,* **78**(8), 88–89 (August 1973).

[151] Anon., "Maintenance and Plant Design for Cleaning Efficiency," *Coal Age,* **81**(7), 212, 215, 216 (July 1976).

[152] G. A. Mason, R. S. Nicoll, and D. Sharples, "Coal Preparation: Aspects of Administration, Management and Control in the U.K.," *6th Int. Coal Preparation Congr.,* Paris, March 26–28, 1973, Paper 13E, 16 pp.

[153] I. S. Blagov, "Technical Progress in Coal Preparation in the U.S.S.R.," *6th Int. Coal Preparation Congr.,* Paris, March 26–28, 1973, Paper 15E, 12 pp.

[154] Anon., "Consolidation Coal Tests New Polymer to Reduce Plant Noise," *Coal Age,* **79**(12), 66–67 (December 1974).

[155] D. D. Spanaus and K. N. Allan, "Controlling Noise in Coal Preparation Plants," *Coal Min. Process.,* **11**(7), 38–40 (July 1974).

[156] N. P. Chironis, "Results of Noise Control Program at a New Coal Preparation Plant," *Coal Age,* **81**(1), 97–105 (January 1976).

[157] R. E. Zimmerman, "Preparation Trends," *World Coal,* **2**(8), 20–21 (August 1976).

Table 10.6 1974 World Coal Preparation Statistics (thousand tonnes)[157]

Coal	United States	West Germany	Poland	United Kingdom	India	Australia	South Africa	Rumania	France	Canada	Japan
Production of Salable Coal											
Lignite	14,041	126,000	39,826	—	3,040	27,278	—	17,500	1,604	3,485	91
Bituminous	533,357	88,747	161,647	99,650	83,118	56,972	58,440	7,130	16,140	12,541	19,403
Anthracite	6,003	6,129	—	1,178	—	—	—	—	6,313	5,076	—
Total Run-of-Mine Output											
Lignite	14,041	126,000	39,826	—	3,040	27,278	—	17,500	3,308	3,485	226
Bituminous	622,428	147,294	198,045	134,349	84,080	70,428	66,177	8,700	25,342	14,938	27,567
Anthracite	6,003	11,584	—	2,149	—	—	—	—	10,769	5,076	—
Input to Cleaning Plants											
Lignite	—	—	—	—	—	—	—	—	1,394	—	226
Bituminous	329,610	127,973	113,187	117,058	13,927	52,729	31,363	7,000	24,507	14,938	26,131
Anthracite	—	10,303	—	2,066	—	—	1,566	—	10,591	—	—
Output from Cleaning Plants											
Cleaned coal	240,539	70,727	72,217	82,931	9,803	37,967	24,196	3,700	18,968	11,070	14,934
Cleaned middlings	—	8,701	9,747	523	3,162	685	—	1,800	2,637	—	3,616
Discard	89,071	58,848	31,223	35,670	962	14,077	7,736	1,500	14,887	3,868	7,807
Number of Cleaning Plants											
In operation during 1974	387	50	87	202	15	55	60	4	31	9	22

Table 10.7 Percentages of Coal Cleaned by Various Methods in 1974[157]

Method	United States	West Germay	Poland	United Kingdom	India	Australia	South Africa	Rumania	France	Japan
Jigs	48.8	69.9	50.7	61.9	31.3	18.0	23.0	49.0	28.8	62.0
Tables	10.9	—	—	0.3	—	3.9	—	—	—	—
Classifiers	1.0	—	4.7	—	—	1.5	—	—	—	—
Launders	1.4	—	—	0.4	—	0.7	—	25.0	—	—
Hydrocyclones	—	0.1	—	0.3	2.2	9.2	—	—	—	—
Dense-medium (gravity separation)	30.9	22.4	38.4	22.0	43.9	15.6	62.5	14.0	53.3	10.0
Dense-medium (centrifugal separation)	—	0.5	—	6.1	21.9	38.8	12.0	—	5.7	16.0
Froth flotation	4.1	7.1	5.7	7.2	0.7	8.8	2.5	12.0	8.9	11.0
Pneumatic	2.9	—	0.5	0.3	—	—	—	—	2.1	—
Other miscellaneous	—	—	—	1.5	—	3.5	—	—	1.2	1.0

shows the preparation methods used in most of these countries. These tables are not complete because the information from some countries is not available.

Study of these data shows that jigs continue to clean more coal than other washers. Dense-medium cleaners are gaining in popularity, however, because of their ability to make sharper separations. In addition, because the feed to coal washing devices has decreased in size over the years (because of greater mechanization in mines for improved productivity and finer grinding in the preparation plant for impurity liberation), froth flotation plays an increasingly important role.

10.4 POTENTIAL AREAS FOR ADVANCES IN COAL PREPARATION

Recent emphasis on increased coal utilization has precipitated numerous coal-related investigations. Such research has led to the development of a great variety of new beneficiation schemes. Although they have not yet been applied commercially, a number of new processes have shown promise for future adoption by the coal preparation industry.

10.4.1 Processes for Sulfur Reduction

The importance of developing sulfur reduction technology to meet current and future stack emission regulations currently dominates coal preparation research. Consequently many newly developed processes have sulfur reduction as their principal goal. Five processes that fall into this category are briefly described below.

10.4.1.1 *Coal-Pyrite Flotation Process*

A two-stage froth flotation process was developed by the U.S. Bureau of Mines[158] to remove pyritic sulfur from fine-size coals. The process involves a first-stage standard coal flotation step to remove high-ash refuse and some of the coarser pyrite as tailings. The first-stage froth concentrate is then retreated in a second bank of flotation cells in the presence of a coal depressant and a xanthate flotation collector to selectively float the remaining pyrite.

The coal-pyrite flotation process has been tested successfully with a number of U.S. coals on both laboratory and pilot-plant scales. In one test more than 90% of the pyrite was removed, reducing the total sulfur content of the coal from 2.61 to 0.80%.

The process looks especially promising for removing pyritic sulfur from coals containing finely disseminated or unliberated pyrite that cannot be removed adequately by conventional single- or multiple-stage coal flotation.

10.4.1.2 *TRW Meyers Process*

Chemical leaching with aqueous ferric sulfate solutions at temperatures of 90–130°C is employed in the TRW Meyers process[159] to remove 90–95% of the pyritic sulfur contained in the coal matrix. The ferrous sulfate is regenerated at similar temperatures using air or oxygen, and elemental sulfur and iron sulfates are recovered as reaction products. The chemistry of the process is represented by the treating step:

$$FeS_2 + 4.6Fe_2(SO_4)_3 + 4.8\ H_2O \rightarrow 10.2FeSO_4 + H_2SO_4 + 4.8H_2SO_4 + 0.8S$$

and the solution regeneration:

$$2.4O_2 + 9.6FeSO_4 + 4.8H_2SO_4 \rightarrow 4.8Fe_2(SO_4)_3 + 4.8H_2O$$

10.4.1.3 *Hydrothermal Process*

In the hydrothermal process[160] developed at Battelle Memorial Institute, raw coal is crushed and mixed with a leaching fluid to form a slurry. The leachant used is an aqueous solution of sodium hydroxide and calcium hydroxide, with up to 10% NaOH and about 2% $Ca(OH)_2$. The exact proportions vary with coal characteristics. The slurry is pumped into pressure vessels, where it is held for up to 30 min at pressures between

[158] K. J. Miller, *Flotation of Pyrite from Coal*, U.S. Bur. Mines Rep. Invest. No. 7822 (1973), 15 pp.

[159] R. A. Meyers, "Desulfurize Coal Chemically," *Hydrocarbon Process.*, **54**(6), 93–95 (June 1975).

[160] E. P. Stambaugh, "Battelle Develops Leaching Process to Desulfurize Coal," *Coal Age*, **80**(9), 72–74 (August 1975).

350 and 2500 psi and temperatures between 225 and 350°C. The chemicals are then filtered off, regenerated, and recycled. The coal, after washing and drying, emerges low in sulfur and impregnated with alkali, which may capture additional sulfur during combustion. According to Battelle, this treatment is sufficient to remove nearly all inorganic sulfur. A unique feature of this process is that roughly half the organic sulfur is also removed.

10.4.1.4 *Magnex Process*

Developed by Hazen Research, Inc., the Magnex process[161] renders the mineral matter in coal (including pyritic sulfur) magnetic, thereby making possible its removal by conventional magnetic separation. In this process, which is an outgrowth of chemical vapor deposition technology, raw coal is treated with iron carbonyl vapor. This step puts a thin skin of magnetic material on the pyrite and other mineral matter, but does not affect the coal. The dry process has been applied to Lower Freeport Seam coal at temperatures of approximately 200°C. Up to 90% of the pyritic sulfur was removed and the ash was reduced from 25% in the feed to 9% in the clean coal.

10.4.1.5 *High-Gradient Magnetic Separation*

Coals are slightly diamagnetic; that is, when placed in a heterogeneous magnetic field, they tend to move toward the weaker regions of the field. Pyrite, on the other hand, is slightly paramagnetic and moves toward the stronger regions of the field. Even in recent years it was felt that the paramagnetism of pyrite was not sufficient to effect a separation from coal. Indeed, the majority of the early work in magnetic desulfurization of coal focused on enhancement of the magnetic susceptibility of pyrite.

Yurovskii and Remesnikov[162] achieved appreciable sulfur content reductions using steam-air treatment of some Soviet coals at temperatures ranging from 120 to 360°C for relatively short times, from 15 s to 5 min. Kester[163,164] thermally treated coal samples for 5–10 min at temperatures ranging from 175 to 280°C at various air/steam ratios. Raw, untreated coal also was magnetically separated for comparison. Kester's conclusion that superior coal quality and separation efficiency were obtained using untreated samples conflicts with the results of Yurovskii and Remesnikov. Ergun and Bean[165] studied the effect of storage, pulverization, heat treatment in a fluidized bed, and dielectric heating on the magnetic susceptibility of pyrite.

Recent developments in the field of high-gradient magnetic separation (HGMS[166]) have renewed interest in the possibility of removing sulfur from coal, without pretreatment, by way of magnetics. HGMS combines high magnetic field gradient, high field intensity, large working volume, and large surface area for particle collection in a design capable of attracting weakly magnetic materials from high-flow process streams at costs comparable with those of other extraction techniques. The magnetic field is generated by insulated, cylindrical electric coils that are cooled internally by water or another medium. These coils surround a canister that contains a matrix, such as steel wool, on which weakly magnetic particles are captured. The matrix serves to interrupt the overall magnetic field and set up minifields varying from very high to very low strength over a distance about equal to the filament diameter. Such magnet configurations appear to fill the requirements described above, and several research programs are seeking to demonstrate the feasibility of HGMS for removing pyritic sulfur from both wet and dry process streams.

[161] J. K. Kindig and R. L. Turner, "Dry Chemical Process to Magnetize Pyrite and Ash for Removal from Coal," paper presented at the AIME Society of Mining Engineers Fall Meeting and Exhibit, Denver (September 1–3, 1976), 19 pp.

[162] A. Z. Yurovskii and I. D. Remesnikov, "Thermomagnetic Method of Concentrating and Desulfurizing Coal," *Coke Chem.*, No. 12, 8–13 (1958).

[163] W. M. Kester, "Magnetic Demineralization of Pulverized Coal," *Min. Eng.*, **17**(5), 72–76 (May 1965).

[164] W. M. Kester, J. W. Leonard, and E. B. Wilson, "Reduction of Sulfur from Steam Coal by Magnetic Methods," *Min. Congr. J.*, **53**(7), 70–75 (July 1967).

[165] S. Ergun and E. H. Bean, *Magnetic Separation of Pyrite from Coal,* U.S. Bur. Mines Rep. Invest. No. 7181 (1968), 25 pp.

[166] N. R. Iammartino, "New Tasks for Magnetism," *Chem. Eng.*, **81**(1), 50–52 (January 7, 1974).

Liu and Lin[167] critically reviewed the recent developments and current status of the magnetic removal of sulfur as well as mineral matter from dry, wet, and liquefied coals.

10.4.2 Miscellaneous Processes

Processes have also been developed that attack problems other than sulfur reduction. For example, the Otisca process[168] is used for dense-medium washing of coal. The novel feature of this process is the heavy liquid employed: an organic compound that volatilizes readily. Therefore dense-medium separations can be effected without resorting to a pseudoheavy medium such as magnetite in water, and medium recovery can be achieved with minimal thermal input.

Developed by the Syracuse University Research Corporation, the chemical comminution process[169] is one in which chemical media induce nearly forceless breakage of coal. Liquid anhydrous ammonia and methanol have been found to be the most effective chemicals. The process involves the rapid migration of a chemical agent throughout the naturally occurring system of faults in the coal. The chemical disrupts the bonding forces acting across the surfaces that constitute these internal boundaries. No significant dissolution of the coal matrix has been observed, nor is there any apparent interaction between the noncoal constituents and the comminuting agent. With no mechanical crushing involved, the final size distribution of the treated coal is governed by the internal fault system in the coal, the comminuting agent, and the duration of exposure. Developers visualize the process to have application both in the preparation and the mining of coal.

A new coal-cleaning method developed by FMC Corporation employs a dry vibrating table.[170] This equipment is uncomplicated and is to be distinguished from wet tables and air tables, which use a fluid-separating medium. The dry table has a drive unit of the electromechanical exciter type used for virbrating feeders. In operation, raw coal is fed onto one side of the unit. Through a combination of density separation and size classification, the low-density coal particles migrate in a spiral path (Fig. 10.16); the denser rock, carbonaceous shale, and bone coal migrate in a smaller spiral and concentrate toward a backwall. The material discharges from the side of the table.

The Bureau of Mines has conducted investigations[171] into improving the efficiency of fine-size coal dewatering by vacuum filtration. The technique consists of washing the filter cake with hot water containing hydrocarbon-type surfactants before drying with air. The moisture content of the cake decreases rapidly up to a critical surfactant concentration, then remains essentially constant. In the laboratory the final moisture concentration of one vacuum-filtered coal was reduced to about 10% by optimizing the surfactant concentration and the quantity and temperature of wash water. This amounted to a moisture reduction of approximately 8 percentage points over that achievable on that sample by normal vacuum filtration.

The quantity of ultrafine-size particles processed in coal preparation facilities has risen steadily in recent years. Such "slimes" remain substantially unaffected by all conventional coal-washing methods and are simply discarded, resulting in the loss of a valuable resource, as well as water pollution and solid waste disposal problems. First demonstrated commercially for taconite ore beneficiation,[172] selective flocculation (i.e., the process of effecting a separation by pref-

[167] Y. A. Liu and C. J. Lin, "Assessment of Sulfur and Ash Removal from Coals by Magnetic Separation," *IEEE Trans. Magn.*, **MAG-12**(5), 538–550 (September 1976).

[168] D. V. Keller, Jr., C. D. Smith, and E. F. Burch, "Demonstration Plant Test Results of the Otisca Process Heavy Liquid Beneficiation of Coal," paper presented at the AIME Society of Mining Engineers Annual Meeting, Atlanta (March 6–10, 1977), 25 pp.

[169] P. H. Howard, A. Hanchett, and R. G. Aldrich, "Chemical Comminution for Cleaning Bituminous Coal," paper presented at Clean Fuels from Coal Symposium, Chicago (June 23–27, 1975), 18 pp.

[170] Anon., "Vibrating Dry Table Speeds Removal of High-Ash Material from Coal," *Coal Age*, **81**(6), 242 (June 1976).

[171] A. F. Baker and A. W. Deurbrouck, "Hot Surfactant Solution as Dewatering Aid During Filtration," *7th Int. Coal Preparation Congr.*, Sydney, Australia, May 23–28, 1976, Paper B2, 16 pp.

[172] J. W. Villar and G. A. Dawe, "The Tilden Mine—A New Processing Technique for Iron Ore," *Min. Congr. J.*, **61**(10), 40–48 (October 1975).

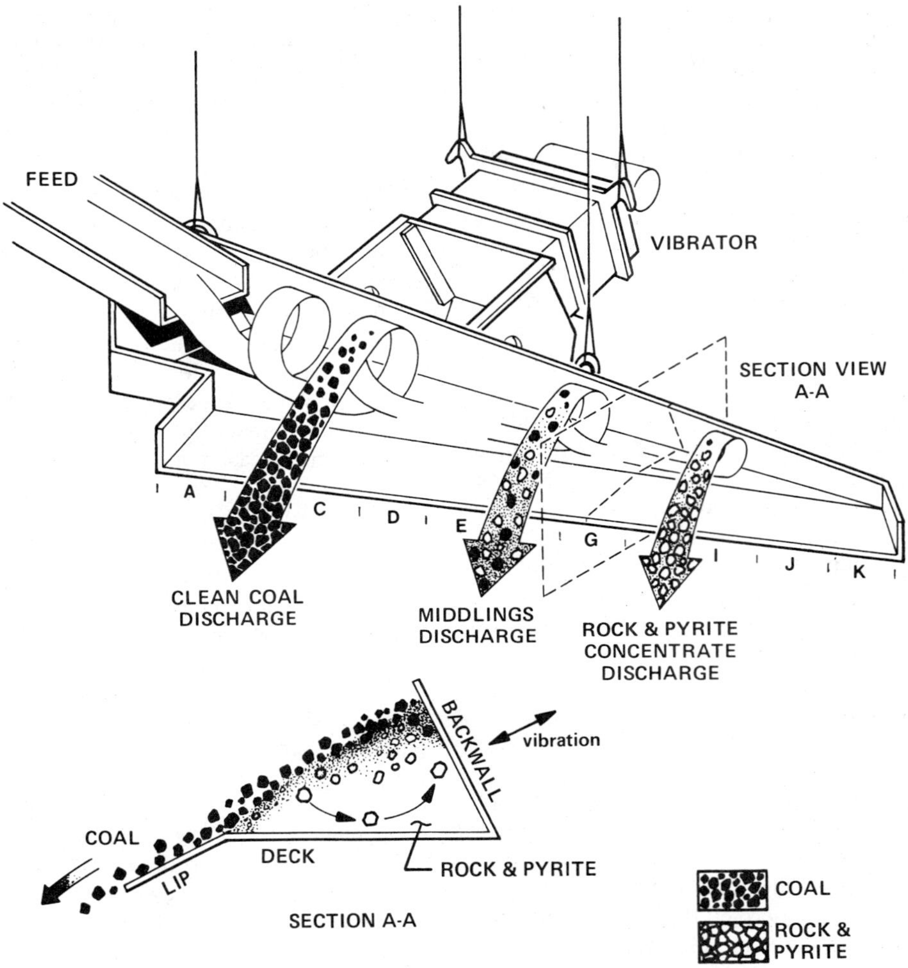

Figure 10.16 The FMC vibrating table. Source: FMC Corporation

erentially flocculating and sedimenting one component of an admixture) is being studied[173] as a possible means of recovering value from the ultrafine-size coal fraction. At present, costs are too high and benefits too low.

10.5 ECONOMICS OF COAL PREPARATION

Even though there is a definite cost associated with coal cleaning, there are also a number of factors that provide economic benefits. The cost components include those relating to cleaning plant capital, operation, and maintenance, and loss of Btu content in raw coal; the benefit factors result principally from a beneficiated product of increased heat content and lower ash and sulfur content per unit weight.

The cost associated with providing a ton of prepared coal at the plant site (assumed located at the mine) includes the sum of the costs of the following:

1. The amortization charge apportioned to each ton of prepared coal. This cost provides for discharging the capital liability and the cost of money associated with the construction of the coal preparation facility.

[173] Z. Blaschke, "Beneficiation of Coal Fines by Selective Flocculation," *7th Int. Coal Preparation Congr.*, Sydney, Australia, May 23–28, 1976, Paper F2, 13 pp.

2. The cleaning plant operation and maintenance cost apportioned to each ton of prepared coal. This cost covers, but is not necessarily limited to, operating wages and overhead and management burdens; cost of power, heat, water, and so on; cost of expendables and other supplies; repairs and replacements; energy required for dewatering and drying; water handling, treatment, and disposal; and refuse disposal. Some cost components scale inversely with coal yield, some are proportional to the refuse level, and some cost components scale to the throughput.
3. The value of raw coal that is used in providing a ton of prepared coal. Since impurities are removed in the cleaning process, it takes more than a ton of raw coal to provide one ton of prepared coal. The amount of raw coal used to produce a ton of prepared coal is simply the reciprocal of cleaning plant yield expressed as a decimal value.
4. The share of state and local taxes and insurance (on the preparation plant) allocated against each ton of prepared coal.

The "additional" cost to provide a ton of prepared coal can be estimated as the sum of the components listed in Table 10.8.

The readily identifiable economic benefit factors that are derived through burning prepared coal include the following:

1. Prepared coal has a higher heat content per ton than the raw coal from which it was produced. Therefore less prepared coal is required to provide a given heat content value.
2. Less prepared (higher heat content) coal need be shipped for a given overall heat content value. Therefore the cost of shipping is less.
3. Preparation reduces the mineral matter content of coal. Therefore the ash requiring disposal by a user (e.g., utility) is less.
4. Less prepared coal need be pulverized for a given heat content value; thus pulverizing cost is reduced.
5. There is evidence that savings in power plant maintenance costs can be obtained by using coal with reduced sulfur and ash content. Thus coal preparation can reduce maintenance costs.
6. For a given heat content value, fewer tons of prepared coal need be handled by the power plant. Therefore power plant units that are tonnage (utilization) limited will have an increased operational life.

The overall (nonenvironmental) user economics are necessarily related to the increase in F.O.B. mine selling price resulting from preparing coal compared to the economic benefits on the part of the user derived through the use of prepared coal. Economic benefits beyond those previously cited could also include those associated with meeting particulate and SO_x emission standards. In many instances a lower ash and lower sulfur content coal will permit the use of more economically attractive particulate removal and flue gas desulfurization elements, along with overall reduced costs.

The range of preparation conditions and cost components is such that specific economic generalizations are exceedingly difficult to make. Each situation, duly covering the raw coal characteristics, the degree or level of preparation, governing environmental restrictions, and so on, must be represented in estimating the overall cost of providing a ton of prepared coal. Nevertheless, several groups have endeavored to quantify coal beneficiation costs. For example, in a recent study funded by the U.S. Department

Table 10.8 Cost Factors per Ton of Prepared Coal

Amortization	$= \dfrac{\text{preparation plant yearly amortization cost}}{\text{tons of prepared coal per year}}$
Operation and maintenance	= based on identification of cost components, plant yield, current and historical cost values and escalation factors
Value of raw coal sent to refuse	$= \text{raw coal value per ton} \left(\dfrac{1}{\text{yield (decimal value)}} - 1 \right)$
Tax and insurance	$= \dfrac{\text{yearly tax and insurance burden}}{\text{tons of prepared coal per year}}$

Table 10.9 Plant Capital and Operating Costs (mid-1974 dollars) for Preparation of Eastern Bituminous Coal[174]

	Design Capacity (tons of clean coal per year)	Preparation Process	
		Level 3[a]	Level 4[b]
Plant A	3,000,000	$11,200,000	$25,200,000
Plant B	1,000,000	4,350,000	9,000,000
Preparation Cost		Cost ($/ton)	Cost ($/ton)
Operating costs[c]		0.30–0.60	0.80–1.60
Depreciation—return on investment[d]			
Plant A		0.93–1.33	2.10–3.01
Plant B		1.09–1.56	2.25–3.23
Total preparation costs			
Plant A		1.23–1.93	2.90–4.61
Plant B		1.39–2.16	3.05–4.83

[a] Removal of liberated mineral matter (80% recovery).

[b] Detailed, elaborate facility (70% recovery).

[c] Includes labor, power, and maintenance; no amortization or return on investment. Thermal drying accounts for about 25% of the capital costs and 30% of the operating costs.

[d] Depreciation is based on a 20-year plant life. The range shown is based on a 15–20% return on investment.

of Commerce,[174] the cost of coal preparation was determined to range from $1.23 to $4.83 per ton of clean coal at the mine (Table 10.9). Again, the cost of preparation reflects the complexity of the beneficiation process. Needham[175] analyzed the factors affecting the cost of coal preparation, and further economic studies of coal preparation were conducted by Kabiesz,[176] Cammack,[177] McNally et al.,[178] Quinlan and Venkatesan,[179] and the Hoffman-Muntner Corporation.[180]

[174] Commerce Technical Advisory Board, Report on Sulfur Oxide Control Technology, submitted to Rogers C. B. Morton, Secretary of Commerce, September 10, 1975, Commerce Technical Advisory Board, Washington, D.C.

[175] L. W. Needham, "Factors Affecting the Cost of Coal Preparation: An Analysis," *J. Inst. Fuel*, **38**(298), 475–480 (November 1965).

[176] A. M. Kabiesz, "Economic Effectiveness of Investments in the Preparation of Coarse Coal Sizes," *Pr. Gl. Inst. Gorn.* No. 215, 1–18 (1958).

[177] P. Cammack, "The Science and Economics of Coal Preparation," *Colliery Guardian Annual Review of the Coal Industry,* September 1973, pp. 84, 87–90.

[178] E. T. McNally, R. D. Woodhead, and J. L. Gamble, "Gaging the Costs in Coal-Preparation Plant Selection," *Coal Age,* **71**(8), 78–84 (August 1966).

[179] R. M. Quinlan and S. Venkatesan, "Economics Count in Coal Preparation Planning," *Coal Min. Process.*, **13**(9), 80, 85–87 (September 1976).

[180] Hoffman-Muntner Corporation, *Engineering/Economic Analyses of Coal Preparation with SO_2 Cleanup Processes for Keeping Higher Sulfur Coals in the Energy Market,* Final Report on U.S. Bur. Mines Contract No. JO155171, November 1976, 249 pp.

CHAPTER **11**

Briquetting

WALTER SCHINZEL

Walter Schinzel *worked on studies of coal, coke, and coke by-products in the Central Laboratory of Borsig-Kokswerke, Germany, in 1939 and 1940. At the Sudetendeutsche Treibstoff-Werke, Germany, he was involved in an effort to liquefy brown coal through gasification. Mr. Schinzel is Director of the Briquetting Institute of Bergbau Forschung GmbH, West Germany, and also directs research on the production of briquettes from coal for the Research Institute of the Steinkohlen-Bergbau-Gesellschaften.*

11.1 INTRODUCTION

The manufacture of fuel briquettes in countries that produce coal has been decreasing for several years. The briquetting of bituminous coal reached its peak in 1963; for brown coal the peak came during 1963–1964.[1,2] Table 11.1, which summarizes fuel briquette

[1] U.S. Geol. Survey, *Statistical Summary of the Mining Industry:* U.S. Bur. Mines, *International Coal Trade,* 1918–1958; *Chemistry of Coal Utili-*

manufacture in the most important of the world's briquette-producing countries that have provided reliable statistics, shows the decrease in production.

The increasing use of petroleum, gas, and electricity for heating purposes and the reduction in the number of individual heating plants are the main reasons for the reduction in the consumption of coal briquettes. During the last 20 years the intense competition, especially from petroleum products, as well as increasing demands by consumers for briquette quality, led to automation of the production of briquettes and a continuous improvement in quality.

Consolidation of briquette manufacture into fewer but larger plants and automation of several production steps improved the economics of the process. Development work improved the properties of the briquettes, and automated production reduced costs. Further important measures were the production of more ecologically acceptable briquettes, which develop little or no smoke, and the manufacture of briquettes for the production of formcoke.

11.2 BRIQUETTING OF BITUMINOUS AND ANTHRACITIC COALS WITH BINDERS

11.2.1 The Briquetting Process

The briquetting of bituminous coal with binders is a process that has been known for many years. There has been a growing trend toward employing only low-volatile coals in fuel briquettes for household use. In the German Federal Republic (West Germany) today only anthracitic coals are briquetted.[3] Coke-over tar pitch is still the most important binder. In manufacturing briquettes that smoke less or those used in producing formcoke, petroleum oil residues or water-soluble binders are being used increasingly.

The briquetting of bituminous coal with a binder requires several consecutive processing steps. The schematic flowsheet in Fig. 11.1 shows the following steps in a modern briquetting plant: comminution and drying of the coal (1–8), metering of the coal (9) and the binder (11), mixing of coal with binder (12), heating of the material in a steam-heated pug mill (13), evaporation of water contained in the material (14), and pressing in a double-roll press (15). The dust recovered in the electrostatic precipitator (19) is partly recycled to the briquetting process (20) and partly used for heating the coal dryer (5). Unsatisfactory (scrap) briquettes that leave the press are returned in metered amounts by way of a crusher (16) and recycle bunker (17) to the briquetting mixture in the mixer (12).

11.2.2 Drying and Comminution of Coal

Water occurs in the briquetting process as surface moisture of the coal, as steam that is added in the steam-heated pug mill for heating the coal-binder mixture, and as condensate from the steam.

Water has a great influence on the briquetting process, especially on the wetting and adhesion between coal and bituminous binder and on the strength of the resultant briquette. Systematic investigation[4] has shown that the moisture content of briquetting coal—before the addition of the bituminous binder—should be as low as possible, and should not exceed 4%. Figure 11.2 shows the effect of moisture content of the coal on the strength of the briquette, when using steam for the pug mill having a superheat temperature between 150 and 300°C and when using coke-over tar pitch as binder. Coal with a moisture content above 10% was successfully briquetted several times by adding surface-active materials such as tar or a similar bituminous binder.[5]

---

zation, Supplementary Volume, H. H. Lowry, Ed., John Wiley and Sons, New York, 1963, Ch. 16, p. 675.

[2] *Statistik der Kohlewirtschaft e.v., Steinkohlenbergbauverein,* 1959–1974 (West Germany), Personal information, 1977.

[3] H. Reintges, P. Schorn, E. Schröter and H. G. Willing, *Jahrbuch für Bergbau, Energie, Mineralöl und Chemie (Mining, Energy, Petroleum and Chemistry Yearbook),* Verlag Glückauf, Essen, 1974.

[4] W. Müschenborn and W. Schinzel, "Processes Occurring During the Briquetting of Coal," *Brennstoff-Chemie,* **46** (2), 52–56 (1965).

[5] G. Burton, *Ann. Min. Belg.,* No. 12, 1391–1400 (1967).

Table 11.1 Production of Briquettes in the Main Manufacturing Nations of the World (thousand tonnes)[2]

Country and Coal Type[a]	1959	1960	1961	1962	1963	1964	1965	1966
Belgium, Bit	1002	1079	1164	1602	2299	1433	1074	969
Czechoslovakia								
Bit								
Brn		436					791	
France, Bit	6561	6074	6083	6936	7994	6624	5809	5056
Germany, West								
Bit	4996	5566	5137	5939	6614	5505	4570	4006
Brn	16838	16774	17058	17302	17324	16747	13794	12788
Germany, East, Brn	54051	56047	57996	59727	60256	61504	60380	59426
Hungary, Bit						(including bituminous coal		
Japan, Bit	3300	3800	4100	4100	4100	4100	4400	4200
Netherlands								
Bit	1020	1139	1188	1250	1561	1355	1349	1222
Brn	64	62	74	71	63	66	45	43
Poland								
Bit		700					600	
Brn		300					300	
Rumania, Brn								
Spain, Bit	1300	1143	1119	1232	1184	1039	844	610
Great Britain, Bit	—	1435	1491	1573	1678	1351	944	964
United States, Bit	816	698	536	533	513	334	334	300[b]

[a] Bit = bituminous coal briquettes, Brn = brown coal briquettes.
[b] Estimate.

To achieve a coal with a constant moisture content, producers use different types of coal dryer such as cascade dryers, fluidized-bed dryers, and entrained-flow gas dryers, and they rely on the continuous measurement and control of the moisture content of the briquetting coal.[6] The entrained-flow dryers use the principle of pneumatic transport of particulate materials. Pneumatic drying almost always uses a hot gas stream that flows vertically upward. Coal is transported and thereby dried. Since settling velocity is dependent on the particle size of the coal and must be less than the velocity of the gas, entrained-flow dryers are economical when coal particles are smaller than 15 mm, or better, less than 10 mm.[7] However the fineness of the material can be selected.

At the end of the drying zone the coal is separated from the drying gas in cyclones. The dryer effluent is freed from dust in wet scrubbers or electrofilters. In modern plants the separated coal is screened and oversize material is continuously ground in a crusher and returned to the drying tube. Figure 11.3 is a schematic diagram of an entrained-flow dryer for briquetting coal, equipped with a wet scrubber for dust removal. Figure 11.1 shows the flow of materials during grinding and drying of the coal in an entrained gas dryer with dust removal by electrostatic precipitator. Operating results of the entrained gas dryer are summarized in Table 11.2.

Suspension or entrained-flow dryers are preferred because they require little floor space, have a low specific heat consumption, and require little labor. In addition, the short residence time of the coal in the drying tube permits control and, if necessary, rapid correction of the moisture content of the dried coal.

Studies of the comparative economics of suspension and drum dryers in the U.S.S.R. are described by Bolek.[8] The results are presented in Table 11.3, which shows investment cost and other data for the two dryer types. In addition to the use of easily regulated dryers with high output,[9] automatic control of the entire drying process has received great attention.

[6] I. Harveng, *Revue Univers. Mines*, **2**, 27–33 (1964).

[7] F. Baunack, *Gaswaerme*, No. 4, 107–110 (1959).

[8] M. Bolek, *Paliva*, **40** (7), 205–207 (1960).

[9] F. R. Zachar, *Min. Congr. J.*, **45** (1), 32–36, 49 (1959).

1967	1968	1969	1970	1971	1972	1973	1974
869	823	793	745	574	496	456	416
			—	—	—	—	—
			1356	1366	1343	1320	1307
4839	4549	4197	4301	3735	3351	3233	3397
3578	3693	3907	3725	2716	2427	2271	2249
11851	10669	10810	9804	7986	7000	6747	6560
56087	56389	56869	57078	55439	50801	50154	50061
briquettes)			1463	1308	1080	1060	1114
3900	3600	3200	3000	1700	1580	1200	1180
1090	1053	1051	886	585	465	251	—
29	21	—	—	—	—	—	—
			1496	1590	1675	1543	1618
			331	350	303	247	170
			—	—	97	169	—
300[b]	135	153	195	180	162	147	108
989	1040	1167	1201	1360	1252	1186	989
250[b]	200[b]	150[b]	120[b]	90[b]	80[b]	—	—

Leonard and Sandy[10] show better ways to control the whole drying process. Hilgraf[11] presented calculations for the drying of particulate solids in suspension dryers.

Drying and comminution (crushing the oversized coal) are usually combined in one process step in modern briquetting plants. The size consist of the bituminous coal used for briquetting has changed greatly in the last two decades. Finer and finer coals are being used; consequently large quantities of coal dust are being used in briquetting plants. High percentages of dust in the coal render the briquetting process difficult and usually require that more binders be added.

The increasing conversion to mechanized coal mining leads to a smaller size consist of the run-of-mine coal and a sizable increase in the dust content of that coal. Therefore for economic reasons fine coal particles must be used in the briquetting plant. Also in all grinding processes it is necessary to use grinding equipment that minimizes the production of additional dust.

Ponomarev[12] describes work with rapidly rotating hammer mills. Pietsch[13] published information on the development of grinding equipment, scientific results from testing breakage processes, and the decrease in size of individual particles, as well as on comminution kinetics. Favorable degrees of comminution, large throughout, and mild comminution with the production of little dust have been obtained with roll breakers.[14]

Table 11.4 shows the average size consist of briquetting coals from two German briquetting plants in 1975, comparing them to the average size consist in 1956. The percentage of coal particles smaller than 0.5 mm has doubled in the last 20 years for anthracitic briquetting coals. Solid binders are ground to a size below 1 mm and usually below 0.5 mm. Production of pitch dust is avoided as much as possible, since it leads to an increased bonding of binders.

11.2.3 Metering of Coal and Binder

To obtain briquettes with uniform mechanical properties and to retain uniformity of briquetting conditions, it is necessary to meter

[10] J. W. Leonard and E. J. Sandy, *Coal Age,* **67** (3), 94–98 (March 1962).

[11] P. Hilgraf, *Aufbereitungstechnik,* **6,** 228–294 (1976).

[12] I. V. Ponomarev, *Ugol,* **34** (4), 60 (1959).

[13] W. Pietsch, *Aufbereitungstechnik,* **7** (11), 655–665 (1966).

[14] J. Winterhalder, *Aufbereitungstechnik,* No. 4, 168–171 (1975).

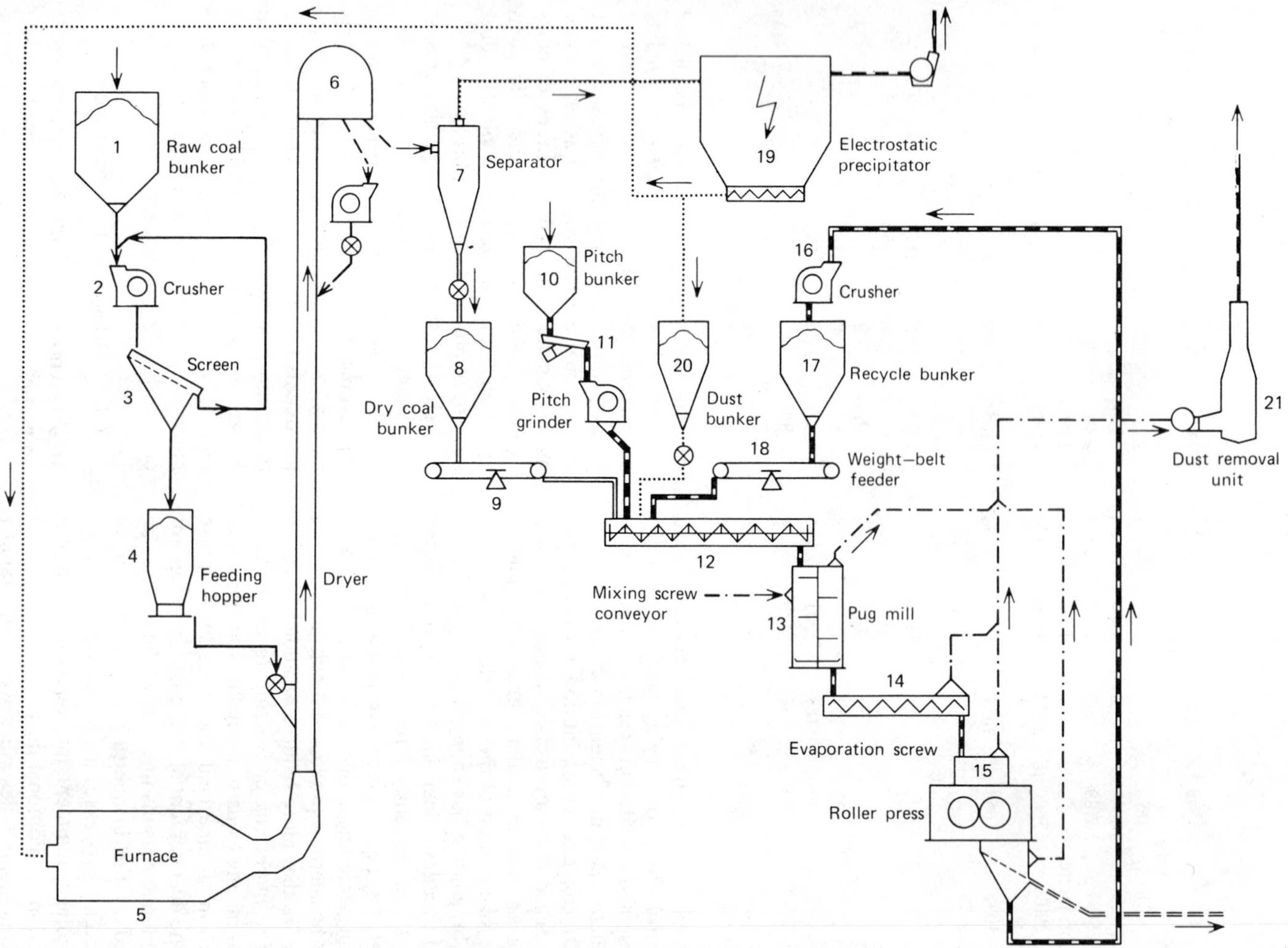

Figure 11.1 Coal briquetting with solid binder.

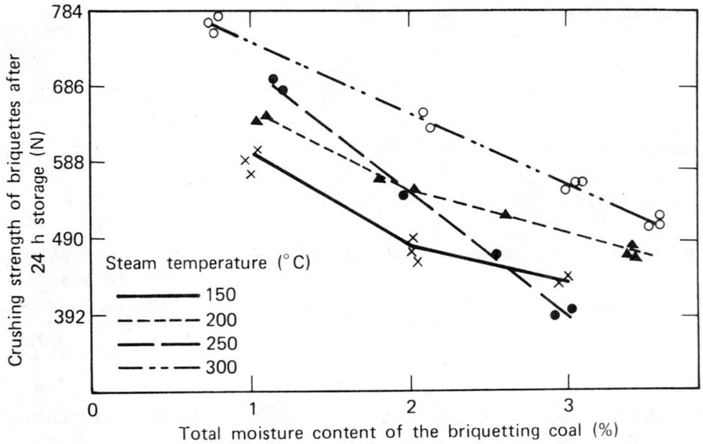

Figure 11.2 Influence of the moisture content of the briquetting coal on the briquette crushing strength.[4]

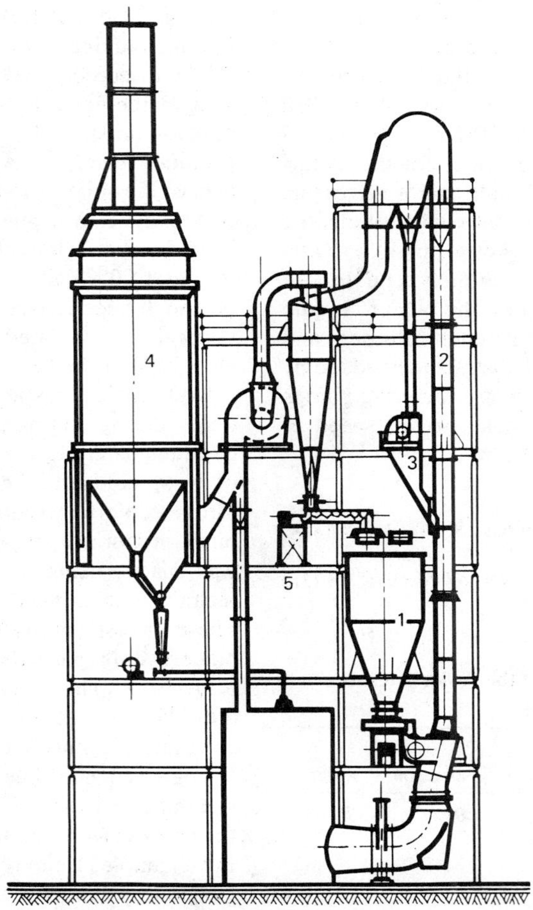

Figure 11.3 Flowsheet of a suspension dryer for briquetting coal: 1 = wet coal hopper, 2 = drying tube, 3 = oversize grinding, 4 = wet scrubber, 5 = dried coal.[7]

Table 11.2 Operating Results of an Entrained-Flow (suspension) Gas Dryer[7]

Variable	Value
Inner temperature	700°C
Exit temperature	110°C
Initial moisture content	8%
Final moisture content	0.5%
Specific heat consumption per kilogram of evaporated water	1106 kcal/kg 4631 kJ/kg
Electric power requirement per tonne of evaporated water	23 kW-hr/tonne 83 MJ/tonne
Method of dust removal	Electrostatic precipitator

accurately and uniformly the coal binder and auxiliary materials. Economic considerations also require reliable metering of the briquetting feed materials.

The problem of regulating and controlling the production of briquettes is the objective of many investigations. Lefèvre and Lusinski[15] found that one of the most important factors is a change in the ratio of coal to binder. They tested the suitability of metering equipment, which preferably is controlled remotely, using weigh feeders. The use of continuously operating scales for the metering of particulate feed materials is increasing in briquetting plants. Employment of electric weigh-belt feeders[16] makes it possible to meter the coal weight accurately and to use ratio control for metering the binder. Liquid binders, such as petroleum bitumen and water-soluble materials, are usually added by means of electrically driven metering pumps in which either the stroke or the speed of rotation is controlled.

[15] A. Lefèvre and J. Lusinski, *Revue Ind. Miner.* **42** (1), 66–76 (1960).
[16] R. Kohling and J. Reuter, *Glueckauf,* **98** (3), 129–141 (1962).

Table 11.3 Comparison of the Soviet Investment Costs of Two Dryer Types[8]

Kind of Investment	Suspension	Rotary
Space requirement (m^3)	9900	30,000
Investment for buildings (thousand rubles)	3245	5118
Investment for machinery (thousand rubles)	1168	3274
Electrical investment (kW)	697	973

11.2.4 Mixing and Heating

The step of mixing coal with the binder and heating the mixture determines, to a great extent, the quality of the final product. The proved double- or single-screw mixers are still used, as well as more modern intense mixing devices of different design. All mixers operate continuously.

The dried coal leaves the dry coal bunker at a temperature between 30 and 80°C, depending on size of the bunker and the transportation distance. After the metered addition of the solid or liquid binder, the mixture is conveyed to a pug mill heated by superheated steam, where its temperature is raised to about 20–25°C above the softening point of the binders. The proved horizontal or vertical steam-heated pug mill has changed very little in the last 20 years.

Systematic investigation[4] of processes that occur during mixing and heating of the briquetting material gave insight into the wetting of both cold and heated coal with pitch. Microscopic observation of the coal–binder pitch mixture at progressive heating stages was made, using a special grain-polishing technique to make the binder pitch visible. These investigations show that bituminous binders only partially wet the coal during briquetting. The individual coal particles in the mixture that feeds the presses are not completely connected by binder bridges.[4]

The residence time of the material in the steam-heated pug mill and its influence on the process were the subject of many studies. For example, Lefèvre and Lusinski[15] used a gamma-ray detector to determine the influence of the residence time on the quality of the briquettes. The level in the mixer is

Table 11.4 Size Consist of Anthracitic Briquetting Coals in Two German Briquetting Plants

	Percentage in Each Size Range		
Size Consist of Anthracitic Briquetting Coal (mm)	Plant A, 1975	Plant B, 1975	Average of Plants, A and B, 1956
>3.15	—	—	18.0
3.15–2.0	1.6	7.6	12.0
2.0–1.0	14.8	18.8	20.5
1.0–0.5	24.0	21.6	21.0
0.5–0.315	17.4	14.5	11.5
0.315–0.2	14.0	11.4	7.6
0.2–0.1	13.5	11.0	4.0
0.1–0.063	6.5	5.3	3.2
<0.063	8.2	9.8	2.2
	100.0	100.0	100.0
Particles below 0.5 mm (%)	59.6	52.0	28.5

controlled by means of this detector, and by means of an electronic unit regulated by the metering equipment. These investigations have also demonstrated the importance of uniform temperature of the briquetting feed before pressing.

11.2.5 Evaporation of Water (Tempering)

After leaving the steam-heated pug mill, the hot briquetting material is transported to the presses, usually by means of screw conveyors. These screw conveyors often have additional built-in shearing devices. In addition to transporting the material, they have two important functions: loosening the briquetting feed material during transport to permit evaporation of water, and cooling of the hot material before the actual pressing. A distribution device located above the presses supplements the function of the screw conveyors and also acts as a metering device for feeding the presses.

Insufficiently tempered material results in scrap briquettes, since the briquettes break after leaving the press. Properly controlled cooling of the material fed to the presses leads to the most suitable viscosity of the bituminous binder, and to an optimum bulk density of the free-flowing material. This is necessary to assure uniform and complete charging of the press molds.

11.2.6 Briquetting Presses

Double-roll presses and rotary table presses[17] are available for the briquetting of bituminous coal. Today, with few exceptions, only the double-roll presses are being used. The last Couffinhal rotary table press in West Germany was retired in 1975. Low output, lack of sales for large briquettes from bituminous coal, and the development of the double-roll presses, which today also permit the production of large briquettes, are responsible for the decline of the Couffinhal press.

Different processes for controlling the briquetting process[18] were derived from the development of methods for measuring pressing pressure[19] with double-roll presses. This work, as well as systematic investigation of events that occur during the briquetting process,[20] yielded valuable information that led to a continuous improvement of the double-roll press.

Novikov[21] made theoretical calculations of the changes in particle size during briquetting and found a connection between the relative decrease of the cross section of the column

[17] C. Berthelot, *Génie Civ.*, **112,** 221 (1938).
[18] R. Loison and J. Lusinski, CERCHAR, Publ. No. 1179, Paris, November 1961.
[19] E. K. Federov, *Gorn. Z.*, 64–65 (March 1963).
[20] E. Ertl, "Pressing Technology Problems," Dissertation, 1966.
[21] F. S. Novikov, *Ugol* No. 2, 50–52 (1963).

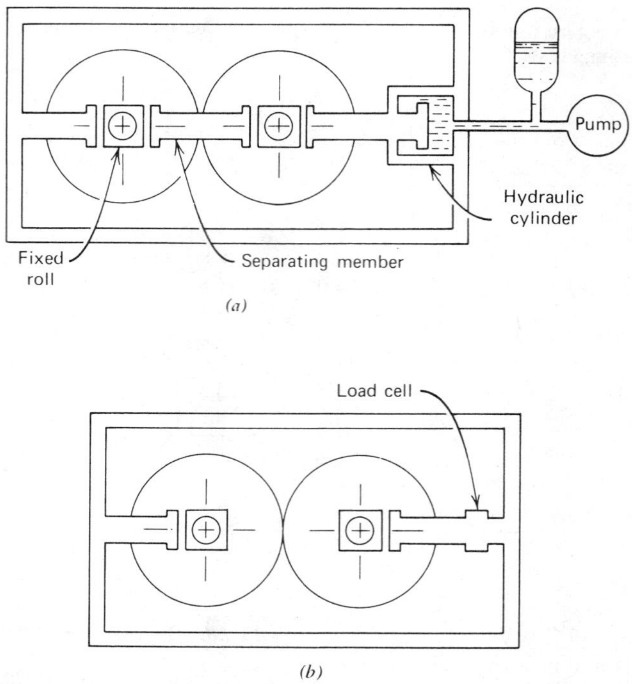

Figure 11.4 (*a*) A modern press with hydraulically loaded bearings compared with (*b*) a fixed bearing press.[24]

to be pressed and the pressure used during briquetting. Pietsch[22] used modeling functions to investigate the shape of different roll molds of briquetting machines in commercial use. The same author also studied basic problems of particle size increase in roll presses.[23]

The introduction of hydraulic systems was important to the development of the design of roll presses. A typical arrangement of a modern press with hydraulically loaded bearings is compared with a fixed-bearing press in Fig. 11.4. With the arrangement shown in Fig. 11.4*a* it is possible to preset an upper limit to the total briquetting load, and—provided sufficient material is fed to the press to ensure that the roll bearings are forced off the separating member—the briquetting load will remain constant at the preset value. If insufficient material is fed to the press, the briquetting load will be unknown but less than the preset value.

With fixed bearings (Fig. 11.4*b*), the total briquetting load varies with the amount of material fed to the press and is indicated by the load cell over the whole range of operation.

For free-flowing materials a simple charging funnel is used; it is equipped with a tongue for proportioning the feed opening of the free-feeding section (Fig. 11.5). If material of a low bulk density is involved, satisfactory mold filling cannot be achieved without some form of forced feeding, such as a charging screw (Fig. 11.6). Depending on the width of the rolls, more than one charging screw may be needed (Fig. 11.7).

Another step in the development of roll presses was automatic control of the feeding material to be pressed. Schematic arrangements of control units suitable for pressurized and fixed-roll presses are shown in Fig. 11.8. When overfilling occurs with the system

[22] W. Pietsch, *Aufbereitungstechnik,* **3,** 128–138 (1970).

[23] W. Pietsch, *Aufbereitungstechnik,* **3,** 120–127 (1976).

[24] T. G. Callcott, Australasian Institute of Mining and Metallurgy, Parkville, Australia.

[25] Köppern and Co., *Briquetting and Compacting with Rolling-Mills,* Information Pamphlet, Hattingen/Ruhr, West Germany, pp. 8–9.

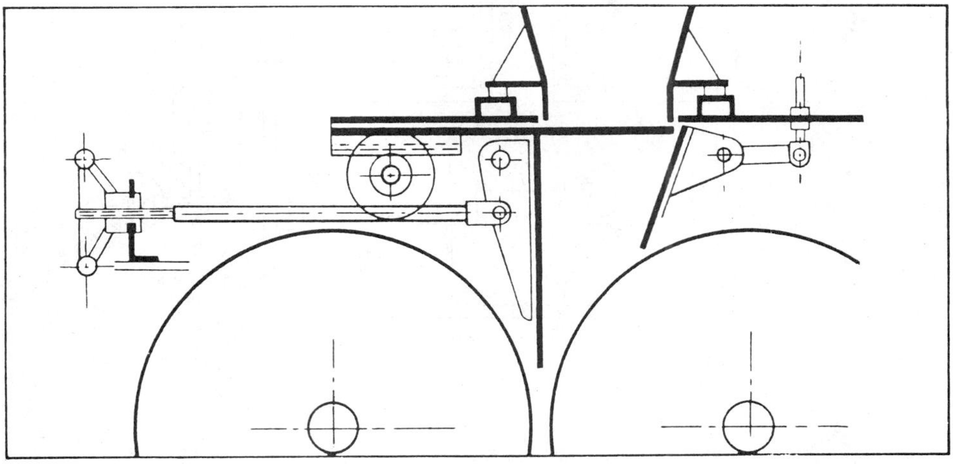

Figure 11.5 Charging funnel with slide valve control.[25]

shown in Fig. 11.8*a* the hydraulic system permits the free roll to yield. The displacement detector registers the amount of movement, and the controller passes a corresponding signal to the feed tongue actuators. With the fixed rolls (Fig. 11.8*b*) overfilling causes an increase in the load measured by the load cell, and the controller makes a corresponding reduction in feed tongue opening.

Normally double-roll presses have a roll diameter between 600 and 1500 mm. The compression ratio usually does not exceed 2.5 : 1. The briquetting pressure depends on the degree to which the molds are filled and the diameter of the rolls. As a rule, the compression time for the briquette form is less than 0.05 s. The surface speeds of rolls lie between 0.5 and 0.8 m/s.

Frequently, and especially in briquetting at elevated temperature, mold segments (Fig. 11.9) are used instead of the normally used mold rings. The use of form segments makes it easier to handle the thermal expansion of the forming gear.

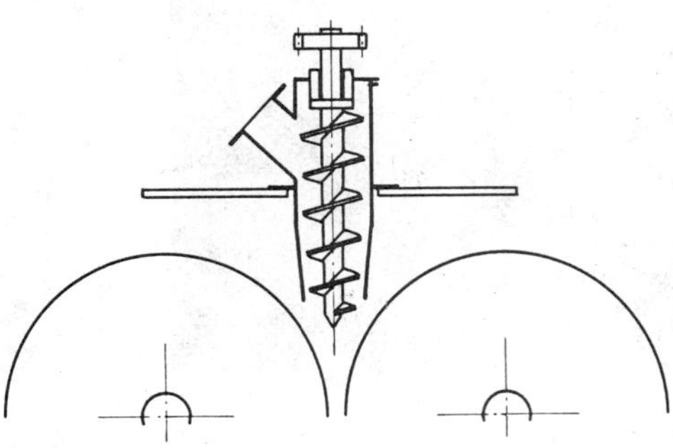

Figure 11.6 Charging screw.[25]

11.2.7 Properties of Briquettes

11.2.7.1 *Types and Sizes*

The shape and the size of bituminous coal briquettes depend essentially on their appli-

Figure 11.7 Double-roll press with two feed screws. Photograph courtesy of Koppern & Company, Hattingen, West Germany.

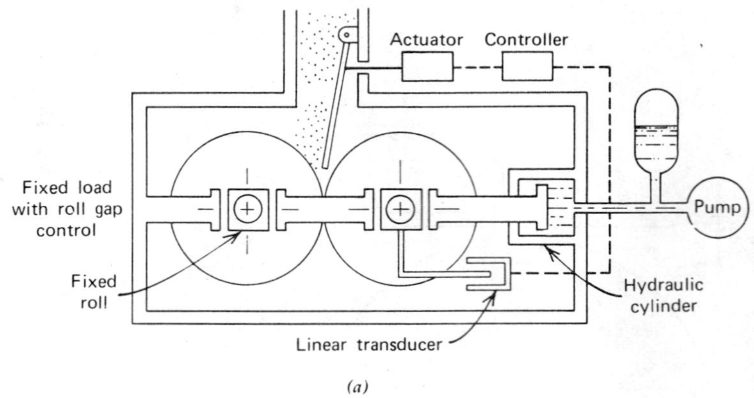

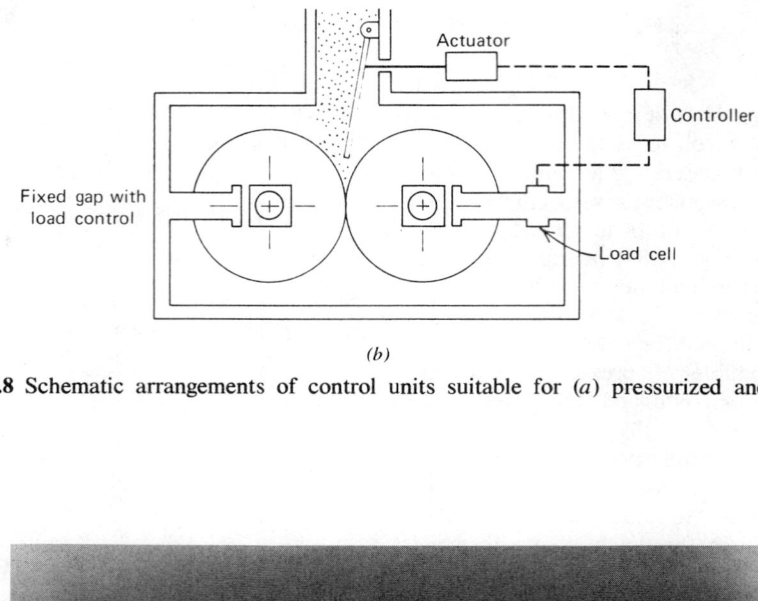

Figure 11.8 Schematic arrangements of control units suitable for (*a*) pressurized and (*b*) fixed-roll presses.[24]

Figure 11.9 Mold segment.

cation. Briquettes made on Couffinhal presses are usually cubes or bricks weighing between 0.5 and 2.0 kg. They are used only in industrial plants. Briquettes made on roll presses usually have the free form of eggs or pillows and, as a rule, weigh from 16 to 80 g. The surfaces of briquettes produced for household use often bear longitudinal indentations that identify the producer. The mold structure that impresses the grooves is supposed to give the material a higher density during briquetting. Household briquettes for closed stoves and cookers must be small enough to be loaded conveniently through the filling aperture, but not small enough to drop through the grate bars or to give high resistance to airflow, even in deep beds.

The shape of briquettes for industrial use can be of great importance, and interest in this subject over the past few years has led to a number of basic investigations. The most comprehensive study was made by Schultz and Abel[26] to determine the effect of the shape of briquettes on their properties. They also investigated flow problems using briquettes of different shapes in certain commercial applications, for instance, in the manufacture of formcoke or hot briquettes, which would replace the conventional lump coke in melting furnaces and cupolas or blast furnaces. The objective was to find briquette forms and shapes that have a high void volume and a small pressure drop. They investigated 85 different types of briquettes with volumes ranging from 50 to 300 cm³. It was found that only systematic changing of special forms in the direction of an increased proportion of voids produced more favorable shapes of briquettes. Results obtained with briquettes of conventional shape were used to improve the shape of industrial briquettes. Figure 11.10[27] shows some standard briquette shapes and sizes.

[26] H. J. Schultz and O. Abel, *Glueckauf,* **110** (13), S517–522 (1974).

11.2.7.2 *Mechanical and Burning Properties*

There are no internationally valid specifications or standard tests available for measuring the mechanical properties of bituminous coal briquettes. The properties currently

[27] E. Ahland, W. Goosens, W. Herman, W. Peters and F. Langhoff, *The Ancit- and the BFL-Processes—Two Hot Briquetting Processes for Formcoke Production,* Int. Iron and Steel Cong., May 27–30, 1974, Düsseldorf, Germany.

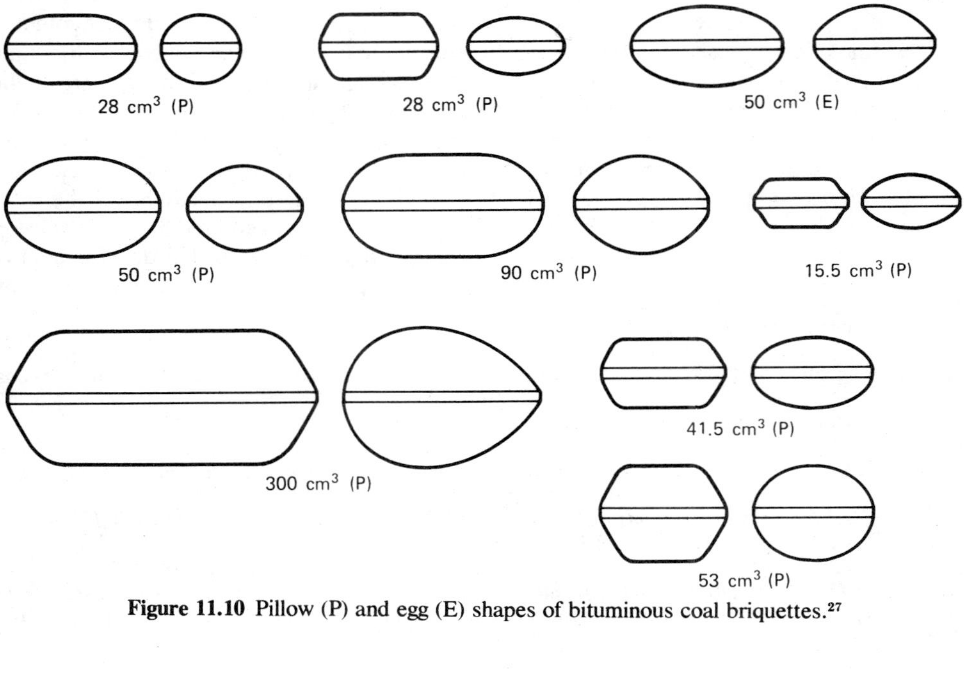

Figure 11.10 Pillow (P) and egg (E) shapes of bituminous coal briquettes.[27]

measured are the breaking strengths and the abrasion or attrition of the briquettes. In West Germany briquettes are being tested according to DIN 23081.[28] In many other countries the only standards available for judging briquette quality are commercial tests.

The attrition testing of briquettes by means of the so-called tumbler tests (see ASTM D441-45[29]) to some extent permits the comparison of briquettes that have the same weight but different shapes (breaking or crushing strengths of briquettes that vary in shape cannot be compared). This property is tested by placing the briquette between two parallel plates and loading one plate until the briquette breaks. Data from this testing method are expressed along a line or at one point—depending on briquette shape—in terms of newtons [N], a value that also depends on the shape of the briquette. The larger the surface of contact between the testing plate and the briquette, the larger the point compressive strength.

Several authors[18,30,31] have expressed dissatisfaction with existing testing methods and have proposed other tests of briquette strength. CERCHAR[32] has developed procedures to determine the scatter of the strengths of briquettes as it depends on fluctuating operating conditions, and to remove the causes of these fluctuations.

Competition from petroleum oils and the desire to produce ecologically acceptable, relatively smokeless household fuels from bituminous coal require that more and more attention be given to the burning properties of briquettes. Briquettes must have a high heating value; anthracite pitch briquettes average 7400–7700 kcal/kg[33] (30, 982–32,238 kJ/kg).

Furthermore, the briquettes must adequately retain their shape while burning in the furnace. Ledent and Marcourt[34] evaluated various briquettes for stability during burning. They tested their materials in a novel kiln with a rotary grate, and judgments were made objectively on the basis of the fuel content of the ash after the experimental firing. Standard test methods for determining the stability of briquettes during firing have not been developed.

Smoke evolution from briquettes and measurement of the noxious material that is vented into the atmosphere during combustion of the fuel are the subjects of new studies. In France, at Centre Technique des Industries de la Fonderie, a smoke density meter has been used that permits qualitative evaluation of the emissions during fuel combustion.[35] Also in France, the Bacharach pump measuring method[36] is being used on household furnaces. This test applies the Bacharach filter paper method,[37] and during combustion of coals with less than 15% volatile matter the Bacharach pump is actuated by 5 instead of 10 strokes, as in oil firing tests, to withdraw the products of combustion.

Several different measuring methods were also developed in West Germany. Baum and Brocke[38] described a testing method wherein 2.5 kg of fuel is burned, according to a schedule, in a furnace mounted on a scale, and a part stream of the products of combustion is analyzed. In accordance with a time schedule, a suction pump draws part of the products of combustion over a filter paper until the products of combustion are completely smokeless. The circular color spots caused by solids and aerosols on the filter paper are then evaluated by means of a comparative scale. A further development of this test procedure[39] permits determination

[28] DIN 23081, Guidelines for Factories Producing Bituminous Coal, June 1965, Beuths-Vertrieb GmbH, Berlin.

[29] ASTM Designation D441-45, Tumbler Test for Coal, Part 26.

[30] A. Dahme, *Rev. Ind. Mine.* **42** (1), 30–39 (January 1960).

[31] H. Rieschel, *Aufbereitungstechnik,* **16** (12), 647–651 (1975).

[32] R. Loison and J. Lusinchi, *Methods of Supervision of Briquette Manufacture,* CERCHAR, Pub. No. 1179, Paris, November 1961.

[33] *Ruhrkohlen-Handbuch (Ruhr Coal Manual),* Verlag Glückauf, GmbH Essen, West Germany, 1969, p. 56.

[34] P. Ledent and M. Marcourt, Institut National de l'Industrie Charbonnière *Bull. Tech.* No. 27, *Coal and Derivatives* 1963, p. 682–689.

[35] *Charbon et Chaufage,* Institut National Belgique de l'Industrie Charbonnière, Bull. Tech.-Mines No. 43, 1965, p. 13.

[36] Norme Française Homologuèe NFX 43-003, December 1966.

[37] ASTM Designation D2156-65, 1965.

[38] F. Baum and W. Brocke, *Proc. State Inst. Emission Soil Protection, NRW,* Vol. 17, Verlag Giradet, Essen, 1969, pp. 7–17.

of the weight of solids and aerosols in the off-gas of the combustion process. Liquid nitrogen cold traps followed by paper filters are used to separate quantitatively the emitted material from a sidestream of the combustion gases. The result is given in grams per kilogram of solid fuel tested.

The BSI presents a technique in BS 3841.[40] This comprehensive method is based on the use of electrostatic precipitators that collect the smoke particles; the index measured is the basis of smoke control legislation.

11.2.8 Binders

11.2.8.1 *Coke-Oven Tar Pitch*

Coke-oven tar pitch is still the most widely used binder in the production of bituminous coal briquettes. However it is being replaced to an increasing extent by petroleum-derived products.

The strict laws for the protection of the environment introduced in West Germany in 1976 permit combustion of pitch-bound briquettes only in continuously operating furnaces of the Universal type[41]; these are furnaces in which the flue gas undergoes a second combustion with secondary air.

In Germany the production of bituminous coal briquettes has been converted to use petroleum residues as binders. Since these materials develop little smoke, briquettes that burn without smoke are being produced. Soviet scientists[42] have investigated the most important properties of pitches as they influence the briquetting process. They found that pitches melting at intermediate and high temperatures have the same values for surface tension at different temperatures; however these pitches then have different viscosities and densities. They also found two stages at which a great change of viscosity occurs in the pitch. One is the transition from the solid into the plastic state; the other is the transition from the liquid state to that of high mobility.

Even today the composition of bituminous coal tar and tar pitches is not completely known. Lang[43] reports a total of 470 identified compounds. Dimitrovich et al.[44] have described a method for investigating the composition of pitches that employs a series of different solvents.

The coke-forming components of the pitch are of great importance for briquetting the currently preferred low-volatile coals, and especially for briquetting anthracitic coals. These coals do not cake or melt while burning in the furnace; thus the stability of the briquettes depends entirely on the skeleton of pitch coke formed in the briquette structure. It was found that the content of coke-forming components in the tar pitch increases[45] as the rate of pitch distillation is decreased.

In addition to the content of coke-forming material, the viscosity of the pitch strongly influences the briquetting process. In practice, it is measured by the softening point (expressed in °C). Usually the Kraemer-Sarnow method is used for the determination of the softening point, although this method is known to have some disadvantages.[46] Arend[47] has compared the suitability of pitch as binder with that of petroleum residues.

Soviet investigators[48] have reported studies on different pitches, in particular, their physicochemical properties. Additional valuable insights into the properties of pitches that are important to the briquetting process were derived from investigations and tests of pitches that are being used for the production of electrodes.[48]

X-Ray analyses[49] of electrode binder pitch-

[39] F. Baum, W. Brocke, und H. Brell, *Gesundheits-Ingenieur,* **10,** 295–306 (1969).

[40] BS 3841, British Standards Institution, 1972.

[41] W. Thieme, *Heizungs-Lueftung-Haustechnik,* **15,** (9), 305–308 (1964).

[42] T. J. Gogolewa and M. A. Stepanenko, *Koks-Chim.,* No. 4, 42–45 (1959).

[43] K. F. Lang, *Fortschr. Chem. Forsch.* **7** (1), 172–195 (1966).

[44] R. W. Dimitrovitch, R. M. Stickel and F. A. Smith, "Rapid Test Method for Determination of Benzene- and Quinoline-Insoluble Content of Pitches," presented at American Chemical Society 142nd Meeting, Division of Fuel Chemistry, Atlantic City, N.J., September 9–14, 1962.

[45] O. Scheffler, *Freiberger Forschungsh.,* **A355,** 47–54 (1965).

[46] K. Wiszniowski, *Koks Smola Gaz,* **7,** 150–152 (1962).

[47] A. G. Arend, *Colliery Guardian* **203** (5252), 724–725 (1961).

[48] J. P. Dobrovolskij and L. V. Kopeliovic, *Koks Chim.,* **4,** 23–25 (1966).

[49] S. S. Pollack and L. E. Alexander, "X-Ray Analysis of Electrode Binder Pitches and Their Cokes," American Chemical Society 135th Meeting, Gas and Fuel Chemistry Division, Boston, April 5–10, 1959.

es gave valuable information on the physical structure of the pitch. Dell[50] used pitches for the manufacture of carbon electrodes by means of their aromaticity—especially the C/H ratio. In addition to the C/H ratio of the pitch itself, the C/H ratio of the carbon disulfide–soluble portion of the pitch has been determined and is important in the manufacture of briquettes.

A comprehensive review of the scientific results obtained in studying the properties of pitches was prepared by Thomas.[51] When selecting coking conditions for coal, those that produce tar—thus pitch with suitable properties—should be chosen, among others. During the briquetting process the properties of the binder pitch are changed by heat, duration of the process, air, and steam. These factors have been shown by Müschenborn and Schinzel[4] to change the composition of the pitch in the thin layer present in the heated briquetting matrix. A number of components of the pitch were separated from the gaseous effluent of the steam-heated pug mill and later identified.

11.2.8.2 *Petroleum Residues*

For the petroleum residues used in the briquetting of coals, the collective name "bitumen" (asphalt) is being used. At first, only the materials developed for road building were used as binders. New insights and new requirements for binder quality have recently led to materials whose properties are significantly different from road-building bitumen.

The most important data used to characterize bitumens are the softening point, the penetration, the Conradson carbon value, and the plasticity range. These data are significant for the briquetting process since they indicate the thermoplastic behavior of the bitumen and its content of coke-forming components. These two properties are considerably different in bitumens and coal-tar pitches. Petroleum residues show a flatter solidification curve than tar pitches and are less sensitive to temperature influences. These disadvantages are especially apparent when the cooling briquettes move from the plastic to the solid state. Bitumen also has a lower content of coke-forming components than do coal-tar pitches, and this is a disadvantage when bitumen is used to bind briquettes of noncaking coals, especially of anthracites. Use of petroleum residues with an insufficient content of coke-forming components as indicated by a low Conradson carbon value produces briquettes that do not have sufficient stability during firing.

The objectives of many studies in briquette-producing countries were to overcome the disadvantages of bitumen in comparison with tar pitches as binders, and to develop suitable briquetting techniques for the use of petroleum residues in briquetting bituminous coal.

Zakar and Toth[52] reported on the briquetting of Hungarian coals with petroleum bitumen and indicated satisfactory results. According to Kraemer-Sarnow, similar satisfactory results were obtained in French briquetting experiments[53] with a bitumen that softened at 68°C and had a Conradson carbon value of 38%. The authors compared their favorable results with those of briquetting with coal-tar pitch as binder. Such comparative investigations were also made by Arend[47] and by a group of Soviet investigators.[54] These experiments, which used oxidized petroleum residues and bitumen originating from shale oil, also studied the influence of the addition of sodium humates on the briquetting mixture.

A group of Soviet investigators[55] studied the distribution of the binder film on the coal surface. Using radioactive marked atoms, the ability of the binder to wet the coal surface was quantitatively measured and determined for residues from petroleum and tar pitches. It was found that petroleum products distribute more readily on the coal surface than do pitches. Reduction of the maximum particle size from 6 to 3 mm increased the wetted surface of the coal by 50%.

[50] M. B. Dell, *Fuel,* **38,** 183–187 (1959).

[51] B. E. A. Thomas, paper presented at the meeting of the Midland Section of the Coke Oven Managers Association, Sheffield, 1966.

[52] P. Zakar and J. Toth, *Brennstoff-Chemie,* **39** (23–24) 373–379 (1958).

[53] M. Brot and M. Gaussorgeues, *Rev. Ind. Miner.* **42,** 95–102 (1960).

[54] L. V. Semenov and V. P. Davidov, *Ugol,* **37** (10), 37–38 (1962).

[55] T. F. Konolova, A. E. Smirnova and A. T. Elisevic, *Chim. Tverd. Topl.,* **10** (3), 135–139 (1976).

Table 11.5 Chemical and Physical Properties of Binders[57]

	Binder		
Testing Data	Coal-Tar Pitch	High-Vacuum Bitumen	Special Bitumen
Softening point (°C)[a]	82.0	91.5	101.0
Penetration (1/10 mm)			
At 25°C	1	8	1
At 40°C	5	20	7
Specific gravity (25°C/25°C)	1283	1059	1113
Benzene insoluble (wt %)	26.7	0.2	1.0
Conradson carbon value (wt %)	45.6	27.7	40.4
C/H atomic ratio	1.58	0.84	0.84

[a] Ring and ball.

The use of propane bitumen as a binder was also investigated.[56] Propane bitumen is the propane-insoluble portion of petroleum residues deasphalted with propane. It has a lower penetration value than bitumens obtained by distillation or by oxidation. An advantage of propane bitumen is its higher temperature sensitivity; a disadvantage is that in summer the briquettes stick together.

A special bitumen (SB)[57] is a modified high-vacuum bitumen and has been found to have advantages as an agglomerating binder in briquetting. Such material is similar in physical, chemical, and other properties to a conventional bitumen; however SB has a considerably higher Conradson carbon value.

Table 11.5 summarizes several chemical and physical properties of a coal-tar pitch, a high-vacuum bitumen, and a special bitumen.

The SB contains an addition of heavy cracked cycle oil, of steamcracker residue, and of furfurol extracts.[57]

The high Conradson carbon value of the binder SB leads to briquettes that have high stability during burning, even when anthracitic coals are used.

The new binders have been tested and new briquetting techniques developed at the Bergbau Forschung GmbH. The German briquetting plants have produced on a commercial scale bituminous coal briquettes that develop little smoke, using the SB developed by the German Shell AG and others of similar composition. These SBs are used as a liquid during the briquetting process and lead to anthracite briquettes that are stable during firing and produce considerably less smoke than briquettes made with coal-tar pitch.

[56] P. Zakar, *Bitumen, Teere, Asphalte, Peche Verw. Stoffe*, **11,** 438–443 (1967).

[57] L. Geldern and A. Grosser, *Erdoel Kohle, Erdgas, Petrochem., Brennst.-Chem.*, **26,** 148–151 (1973).

11.2.8.3 *Sulfite Liquor*

Sulfite liquor is a waste product obtained in great quantity when cellulose is produced from wood using the sulfite process. It has considerable sticking and adhesion properties, and thus fulfills one of the requirements of a satisfactory binder.

Carbohydrates are considered to be the components that supply adhesion. They are obtained when the wood is treated with sulfites of calcium, sodium, magnesium, and ammonia. Table 11.6 shows the properties of

Table 11.6 Composition of the Original Sulfite Liquor[58]

Components	Spruce Liquor (g/1)	Beech Liquor (g/1)
Evaporation residue	140–150	170–180
Organic portion	130–140	159–168
CaO	7.5–8.5	7.5–8.0
Free SO_2	1.0–1.8	1.5–1.6
Calcium lignosulfonate	85–95	90–105
Total sugar	38–45	60–70
Pentoses	9–12	48–56
Hexoses	29–33	12–14
Total volatile acids	4.0–4.5	15–16
Acetic acid	3.4–3.7	14–14.7
Formic acid	0.6–0.8	1.0–1.3
Methanol	0.4–0.6	0.5–0.7

Table 11.7 Properties of Concentrated Sulfite Liquors[58]

	Sulfite Liquor Binder Type			
Properties	Beech A	Beech B	Spruce A	Spruce B
Thickening (°Baumé)	20.0	31.0	32.5	31.0
Specific gravity at 15°C (g/cm^3)	1.16	1.27	1.29	1.27
Dry organic residue (wt %)	32.0	46.0	48.0	49.0
Ash in moist material (wt %)	2.0	5.0	6.0	4.0
Moisture (wt %)	66.0	49.0	46.0	47.0
Heating value				
kcal/kg	1000.0	1800.0	1900.0	1600.0
kJ/kg	4186.8	7536.2	7954.9	6698.9

original liquors obtained by treating spruce and beechwood with calcium sulfites. For use as a briquetting binder, the original sulfite liquors are concentrated to a solids content of 46–54%. Table 11.7 presents some properties of spruce and beechwood sulfite liquors after concentration by evaporation.

The average sulfur content of normal sulfite liquors with about 50% solids is about 3–5%. The solid material contained in sulfite liquor—the sulfite pitch—can be used as binder.

Both the sulfite liquors and the sulfite pitch remaining after water removal are water soluble. Briquettes made with sulfite liquor as binder can be made water resistant by heating them to 220–350°C. The conversion of the sulfite liquor into a water-resistant form in the briquette matrix is called hardening the briquette.[58]

There were attempts to make sulfite liquor less water soluble by the use of additives. Marcourt[59] reported a modified sulfite liquor, "Supracol C," produced by the l'Avébène Company of France. For hardening briquettes made with this binder, Marcourt recommends the sand-bed oven.[60]

Two other disadvantages besides water solubility hindered the commercial use of sulfite liquor as a binder. These are the low content of coke-forming components and the increased tendency for the briquetting equipment to break down when using this binder.

Newly developed briquetting technology[61] for using low-ash ammonium sulfite liquor, obtained by the treatment of wood with ammonium sulfite, made it possible to operate commercial briquetting plants without disruptions, even using anthracitic coals.[62] In this process the sulfite liquor is heated to about 30–50°C and added to the briquetting coal, which is heated to 100–120°C. In enclosed intense mixers, controlled amounts of water are continuously removed from the mixture, until the binder has reached its optimum adhesive properties. Since no heat is added from the outside during this concentration process, the undesired formation of deposits (skins) is completely avoided. A free-flowing material results, which will not stick in the molds during pressing.

Slightly sintering or lightly caking coals can be briquetted with calcium sulfite liquor or with magnesium or sodium sulfite liquor as binder. Briquettes that are stable during firing can be made from anthracitic coals only when using ammonium sulfite liquor. This low-ash binder produces a favorable coke skeleton in the briquette structure during combustion.[62] Figure 11.11 shows the appearance of different structures of cokes prepared with calcium and ammonium sulfite liquors when polished sections are viewed under the microscope. In West Germany an-

[58] F. Melms and K. Schwenzon, *Verwertungsgebiete für Sulfitablauge (Use Areas for Sulfite Liquor)* VEB Deutscher Verlag für Grundstoffindustrie, Leipzig, 1967.
[59] M. Marcourt, *Ann. Mines Belg., Liège,* **59,** 837–850 (1960).
[60] P. Ledent, *Ann. Mines Belg.,* No. 6, 782–802 (1964).

[61] German Pat. 1186825, 1671364.
[62] W. Schinzel, "Production of Smokeless Fuel by Briquetting Coal Using Sulfite Liquid as a Binder," *Erdoel Kohle,* **25** (2) 65–69 (1972).

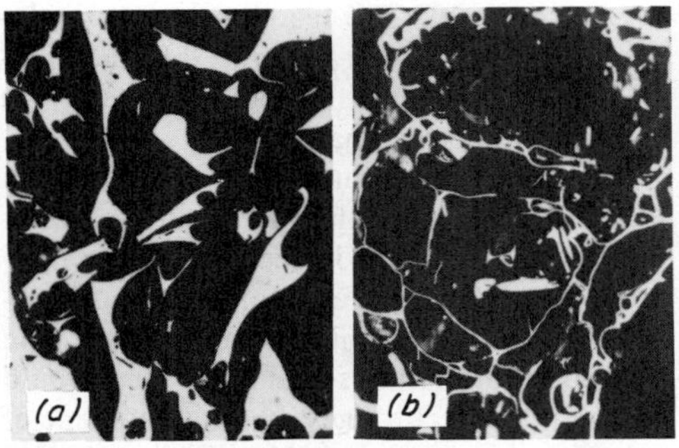

Figure 11.11 Polished sections of coke made with (*a*) calcium sulfite liquor and (*b*) ammonium sulfite liquor.[62]

thracitic coals with a binder of ammonium sulfite liquor are used in large-scale production of practically smokeless household fuel called "Extrazit."[63]

In this process dried, metered, briquetting coal of 2–0 mm size, at a temperature of 100–120°C, is mixed continuously in an intensive mixer with about 8–10% ammonium sulfite liquor (50% solids). Controlled suction removes the steam effluent and concentrates the binder to its optimum concentration for the briquetting process. The briquettes are made in roll presses and pass through a predryer into a heater, where they are heated to 230–250°C on an endless steel belt. From the conveyor oven the briquettes pass continuously through an immersion bath and to the loading dock. The immersion bath gives the briquettes a surface treatment that makes them water repellent and highly resistant to abrasion.

Table 11.8 summarizes the properties of 40-g Extrazit-Nussbriketts. The Extrazit briquettes burn practically without smoke and have great mechanical strength. The Extrazit production plant has a capacity of 50–60 tons/hr and is highly automated.

[63] W. Lawson and A. W. Sharpe, National Coal Board, Coal Res. Est. Rep. No. 33 R, August 1970.

11.2.8.4 *Starch Materials*

Because of the great adhesive power of starch, 1–3% starch can serve as binder for the production of coal briquettes of sufficient strength. Starch may be added to the coal as a solid or in a suspension.

The use of the dry binder makes it possible to use moist briquetting coals, since 8% water

Table 11.8 Properties of 40-g Extrazit Nussbriketts

Property	Raw	Moisture-free	Moisture- and Ash-free
Moisture (%)	1.0–2.0	—	—
Ash (%)	5.4–6.4	5.5–6.5	—
Volatile matter (%)	8.0–8.3	8.0–8.5	8.5–9.
Total sulfur (%)	1.1	1.1	—
Combustible sulfur (%)	0.8	0.9	0.9
Hydrogen (%)	2.9	3.0	3.1
Carbon (%)	83.1	86.1	89.7
Fixed carbon (%)	—	87.6	91.2
Oxygen (%)	4.0	4.1	4.3
Nitrogen (%)	1.7	1.7	1.8
Chlorine (%)	0.13	0.13	0.14
Carbon dioxide in carbonates (%)	1.4	1.4	—
Net heating value			
kcal/kg	7570–7420	7650–7570	8100
MJ/kg	31.72–31.09	32.05–31.72	33.94
Gross heating value (kcal/kg)	7720–7570	7800–7720	8250
[Heat of combustion] (kcal/kg)	32.35–31.72	32.68–32.35	34.57
Approximate compressive strength (kN)	1.47		
Approximate tumbler strength (%)	95		

Source: Briquette Plant Sophia-Jacoba, Hückelhoven, West Germany.

is necessary to produce the starch suspension.[64]

Because starch is very costly and lacks coke-forming components, it has not been practically applied as binder in fuel production. However it is important in the manufacture of charcoal briquettes, which are used in increasing amounts as barbeque briquettes.[64,65]

11.2.8.5 *Plastics*

Plastics have not been introduced into the commercial production of bituminous coal briquettes because of high manufacturing costs and difficulties in the briquetting process, and because briquettes having unsatisfactory strength properties were obtained.

Beglinger[64] used formaldehyde-urea as a binder in briquetting experiments; 4–6% of this binder produced briquettes that after drying at 200°C, had sufficient strength. The author stated, however, that the manufacturing costs were too high and the stability of the briquettes during burning unsatisfactory.

Bergbau Forschung, GmbH, Essen, made comprehensive investigations[66] of the possibility of using plastics for the production of bituminous coal briquettes. In these tests polymer preparations, plastic emulsions, thermoplasts, synthetic resins, and polyurethanes were tested as binders. To attain satisfactory briquette strength, 5–6% thermoplast (polystyrene, polyethylene) was needed as binder. Manufacturing costs were too high, and the briquettes were too weak during firing. Briquettes prepared with synthetic resins required a setting time of 110 min at 100–150°C, which is much too long. Anthracite briquettes made with 3% phenol-formaldehyde resins as binder were strong enough, but they performed unsatisfactorily in the fire.

Certain experiments have sought to use wet coal and avoid the drying process by adding polyisocyanates, which would react with the water to form polyurethanes in the briquetting material. However they have not been successful. The reaction to form polyurethanes was much too rapid, thus the binder could not be distributed successfully in the wet briquetting material.

The conclusion from these investigations is that only plastics that (*a*) require a maximum of 2% addition in the briquetting mixture, (*b*) permit easy adaptation of the available briquetting techniques, and (*c*) are able to produce bituminous coal briquettes, would be suitable. Accordingly, the low content of coke-forming components in all these plastics must be considered, as well as their high cost.

11.2.9 Progress in the Briquetting Process

With the general decrease in the production of bituminous coal briquettes in 1964, efforts were increased to make briquetting of bituminous coal more economical and to improve briquette properties.

In 1959, at the 5th Biennial Briquetting Conference in Glenwood Springs, Colorado, and the International Colloquium on Bituminous Coal Briquetting in Paris,[67] a great number of different investigations for the improvement of processing techniques and for decreasing costs were discussed. The reports treated questions of briquette structure, problems of several mechanical properties of the briquettes, and questions on automating control operations in briquetting plants. The interchange of experiences and information was continued in later congresses and discussions.

Soviet investigators[68] studied the influence of briquetting conditions on anthracite dust. In Germany one investigator tried to explain the individual events[69] that occur during the briquetting process. Accordingly, feeding of the binder, the method of operating the presses, and the influence of briquetting pressure were studied. A simplified briquetting technique has been described[70] that saves binder;

[64] E. Beglinger, paper presented at 7th Biennial Conf., International Briquetting Association, Jackson, Wyo., August 28–30, 1961, pp. 1–4.

[65] "Barbecuing with Lignite," *Coal Age*, **67** (1), 126–128 (January 1962).

[66] W. Schinzel, *Glueckauf-Forschungshefte*, **36**, 156–161 (1975).

[67] M. T. Mackowsky, R. Loison and J. Lusinchi, "Colloque sur l'Agglomeration des Fines de Charbon," *Rev. Ind. Miner.* **42** (1) 3–39 (January 1960).

[68] G. S. Zvenigorodskij and V. S. Kolomejew, *Ugol*, **12**, 36–40 (1959).

[69] F. Wulf, *Chem.-Ing.-Technik*, **32**, 198–201 (1960).

[70] E. J. Akselrol and A. G. Špachler, *Izv. VUZ, Gorn. Zh.*, **5** (9), 165–168 (1962).

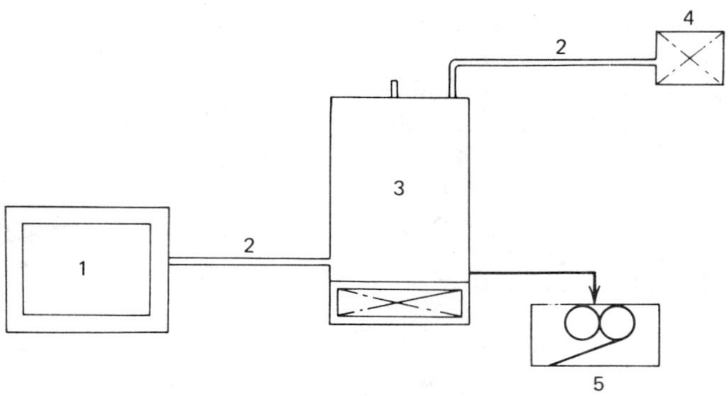

Figure 11.12 Flowsheet of briquetting using combined mixer-dryer with flue gas. 1 = mixing chamber, 2 = gas ducts, 3 = mixer-drier, 4 = fan, 5 = roll press.[70]

the drying and pug mill steam heating were combined into one stage, in a single unit.

Figure 11.12 is a schematic illustration of such a briquetting plant. It consists of the mixer-dryer (3) with drive, the fan (4), and the gas ducts. From the mixing chamber (1), flue gas at 300–500°C passes through the coal-binder mixture, suction being provided by the fan (4). After cooling, the material enters the roll presses (5) from the mixer-dryer (3). Advantages claimed for this process are less binder consumption, shorter treatment time, lower investment and operating cost, and smaller space requirements than in a conventional briquetting plant.

A briquetting process[71] developed by Bergbau Forschung GmbH, was rapidly introduced into a commercial briquetting plant. It solved the problem caused by the presence of large amounts of superfine coal and dust in the briquetting coal, and at the same time led to a reduction in the binder requirement.[72] In this process the dried coal is treated with 1–3.5% of an emulsion that consists of 5–40% oil, 10–40% sulfite liquor (50% solids), and 20–85% water before it is mixed with the bituminous binder.

Figure 11.13 is a schematic diagram of the coal pretreatment plant. The emulsion is metered and added continuously to the stream of coal. Agglomeration of small coal particles

[71] German Pat. 1185148.

[72] W. Müschenborn and W. Schinzel, *Glueckauf,* No. 7, 421–425 (1965).

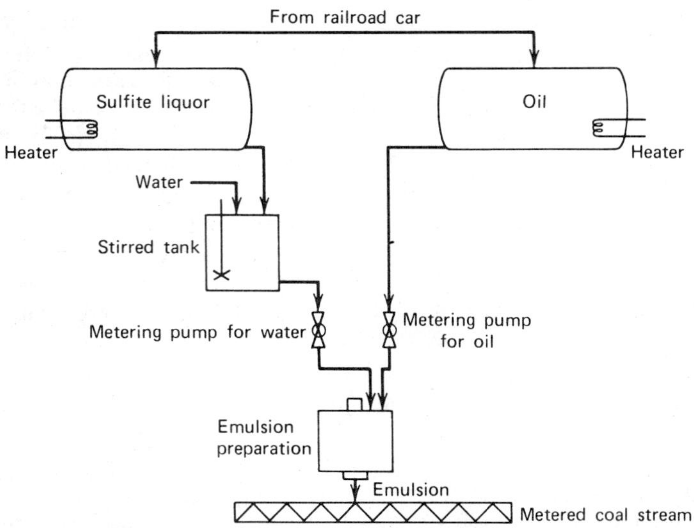

Figure 11.13 Flowsheet of coal pretreatment with an emulsion.[72]

and adhesion of coal dust to larger coal particles lead to an apparent increase in the size of the coal particles. Figure 11.14 shows the apparent increase in coal particle size of a dried briquetting coal after pretreatment with 2% emulsion. The presence of oil in the emulsion facilitates wetting of the coal surface by the binder pitch. With less binder, greater and more uniform briquette strength is achieved.

Because of the excellent binder distribution in bituminous coal briquettes that are made with emulsion, they burn less smokily than conventional briquettes. Figure 11.15 compares briquettes made with and without emulsion, from the same type of coal, using the same type and quantities of binders.

Additional advantages during the production of the briquettes include avoiding separation of the components in the bunker and decreasing the dust evolution during operation. Since 1962 almost all pitch-bound bituminous coal briquettes in West Germany have been produced by this process.

The possibility of improving the quality of the briquettes by using liquid pitch has been investigated.[73] It appears that a better distribution of the pitch in the material to be briquetted can be achieved.

The influence on briquette strength of preoxidizing the coal has also been studied. The oxidized coal may exert chemical binding forces that are the result of free radicals formed by condensation reactions and by the evolution of gases and water vapor.[74]

[73] J. Lusinski and M. Wals, CERCHAR Rep. No. 5, 301–305, 1965.

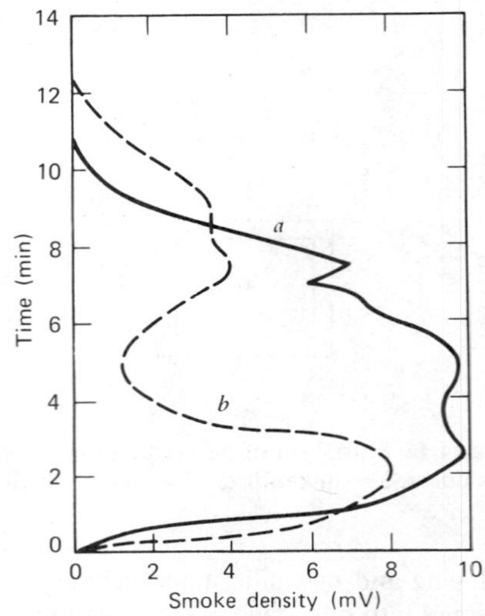

Figure 11.15 Smoke density of burning briquettes = curve *a*, untreated coal, curve *b*, coal pretreated with emulsion.[72]

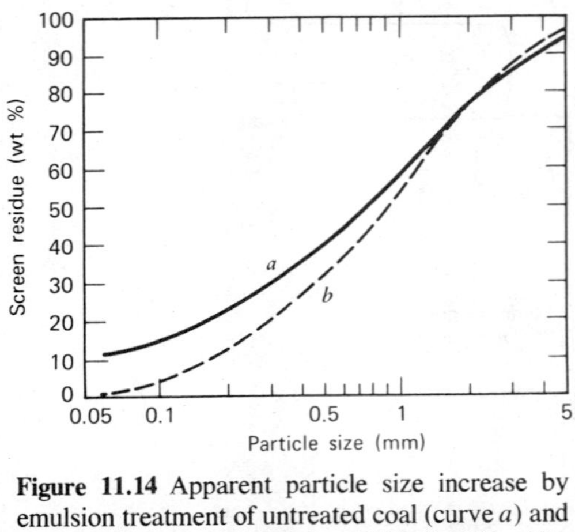

Figure 11.14 Apparent particle size increase by emulsion treatment of untreated coal (curve *a*) and coal treated with 2% emulsion (20% oil, 20% sulfite liquor, 60% water) (curve *b*).[72]

Testing of the briquetting behavior of modified bituminous coals[75] and precompression of the briquetting material[76] gave new insights into the briquetting behavior of bituminous coals and made it possible to improve the briquetting of materials that are difficult to agglomerate, for example, pulverized coke. Special attention was given to all measures that permit retaining constant conditions in the briquetting process.

Increased use of controlling and measuring instruments in briquetting plants has provided better control of the operating steps. Electromechanical weigh-belt feeders[77] were found especially suitable for metering coal, solid binders, recycling material, and auxiliary compounds.

Figure 11.16 shows a sample mixing unit for coal and recycled material employing weigh-belt feeders. Vibrating chutes that feed the weigh-belt feeder directly are used for

[74] C. Kröger and G. Brinkmann, *Brennstoff-Chemie*, **48** (7), 198–206 (1967).

[75] C. Kröger, I. Romey and G. Falkenhain, *Erdoel Kohle*, **24** (2), 99–103 (1971).

[76] W. Schinzel, *Glueckauf-Forschungshefte*, No. 35, 30–35 (1974).

[77] H. Schleputz, *Aufbereitungstechnik*, **1**, 11–15 (1976).

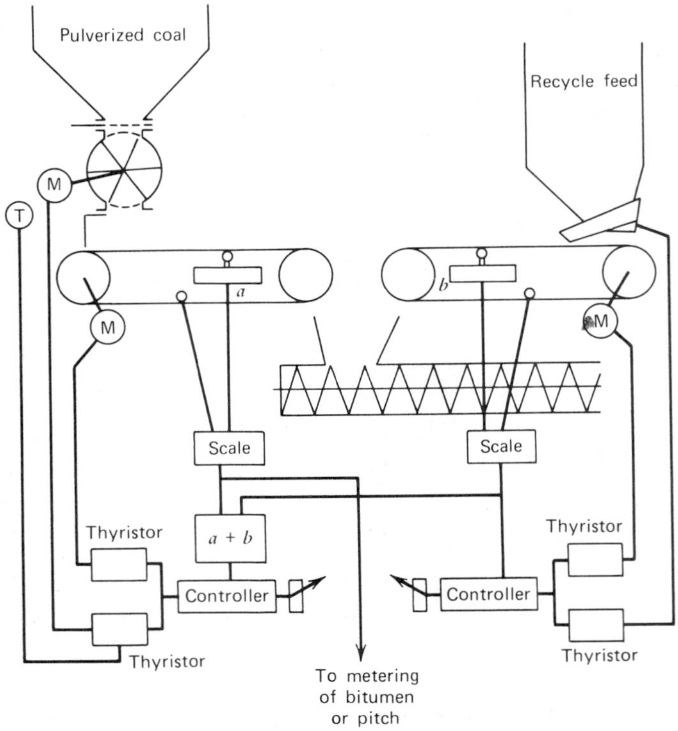

Figure 11.16 A mixing unit for briquetting coal and recycle feed using electromechanical weigh-belt feeders.[77]

discharging the recycling material. The controller adjusts the vibrating chute, as well as the velocity of the belt. By means of separate thyristor stages, the controller regulates the drive motors for the starwheel coal feeder and the belt feeder. The coal weigh-belt feeder supplies impulses to the metering of the binder.

Improved control and metering devices in the briquetting plants led to a decreased labor force and partial automation. Operating costs were thus lowered, and briquette quality was improved.

11.3 BRIQUETTING BROWN COAL AND BITUMINOUS COAL WITHOUT BINDER

11.3.1 Brown Coal

The production figures in Table 11.1 indicate that more than 80% of the world production of brown coal briquettes is concentrated in East and West Germany. During the last 20 years development and improvement studies on shaping brown coal occurred mainly in these two countries.

The briquetting of brown coal without binder involves the following steps: raw coal preparation, drying, dry coal treatment, and pressing.

11.3.1.1 *Raw Coal Preparation*

The object of the raw coal preparation step is to produce fine coal in the required size ranges of 6.3–0 mm, 3–0 mm, or even finer. Screens with a higher capacity than the normally used twin swinging screens have been introduced.[78] Today electrically heated resonance or vibrator screens are employed. Further progress was achieved by the use of classifiers instead of screens. The highest output in classifying coal was achieved with inclined-chamber classifiers,[79] in which wet

[78] E. Rammler, *Freiberger Forschungsh.* **A 147,** 22–43 (1960).

[79] A. Kubasch, J. Schmidt, et al., *Freiberger Forschungsh.* **A204,** 63–93 (1961).

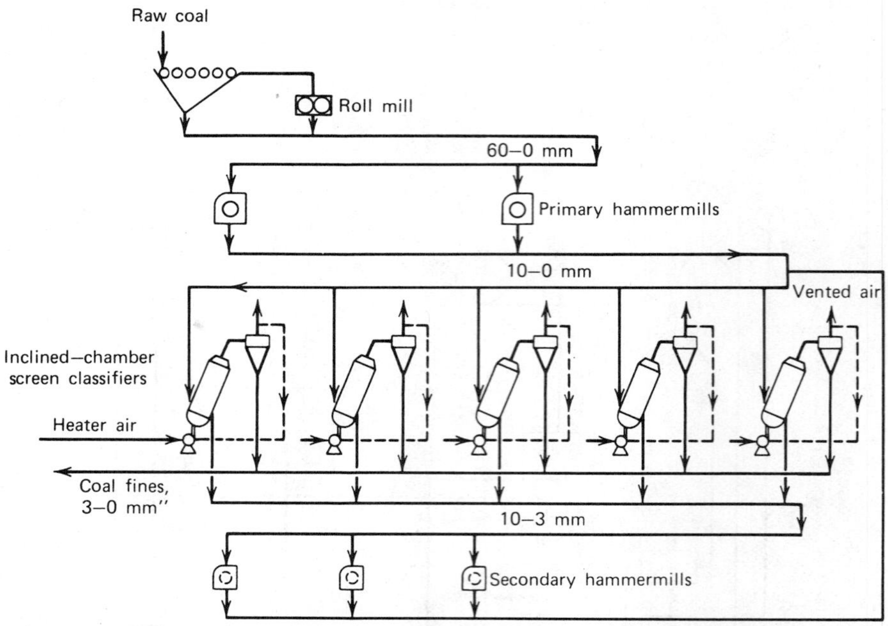

Figure 11.17 Preparation of raw brown coal with inclined-chamber screen classifiers.[80]

coal is so greatly comminuted that additional grinding of the dry coal is not necessary. Figure 11.17 is a schematic diagram of a raw brown coal preparation unit with inclined-chamber screen classifiers.

11.3.1.2 *Drying*

A distinct advance in coal drying was achieved by the development of tubular dryers with narrow drying tubes (75–83 mm I.D.). Previously tubes with a diameter of 94–104 mm were used. Grunewald and Müller[81] and other authors[82] showed that the hourly evaporation rate is considerably increased in the narrow drying tubes.

A Soviet scientist[83] calculated the relationship between the degree to which the tubular dryer is filled and the heating surfaces covered and determined the magnitude of the latter quantitatively.

Rammler and Weber[84] systematically reviewed concepts and parameters developed by different authors for the measurement, movement, and charging of the drying tubes, as well as other investigations important to the design and operation of large-scale dryers. They indicate that their results can be used for the design of large tubular dryers.

11.3.1.3 *Treatment of the Dry Coal*

In modern briquetting plants coal is comminuted while wet. Therefore the heart of the dry coal treating plant is the cooling of the dried coal. To increase the capacity of the plant and to improve briquette quality, improved cooling of the coal was studied intensively.

In 1957 the turbulent flow cooler developed by Kubasch[85] was introduced. Later the cooling chain conveyor[86] became available. It is used in addition to, or as a substitute for, the cooling drums in the briquetting plants.

The inclined-chamber classifier, the large tube dryer, and the fluidized-bed cooling chain conveyor are usually connected in one unit (Fig. 11.18) in which the brown coal is

[80] E. Rammler, *Freiberger Forschungsh.* **A393,** 193–212 (1966).

[81] H. Grunewald and W. Müller, *Freiberger Forschungsh.,* **A257,** 5–46 (1963).

[82] H. Grunewald and J. Hofmeister, *Bergbautechnik,* **15,** 129–137 (1965).

[83] S. P. Kuznecov, *Ugol,* **37,** 48–52 (1962).

[84] E. Rammler and M. Weber, *Freiberger Forschungsh.,* **A518,** 45–72 (1972).

[85] A. Kubasch, *Freiberger Forschungsh.,* **A209,** 47–90 (1962).

[86] J. Wilsdorf, *Bergbautechnik,* **14,** 462–475 (1964).

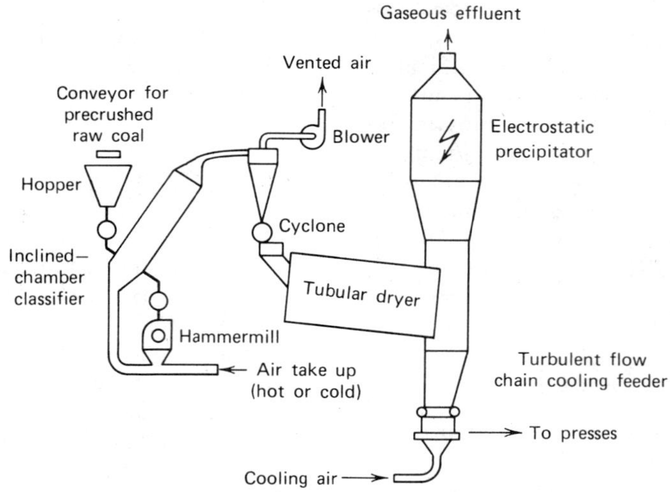

Figure 11.18 Inclined-chamber classifier, large tubular dryer, and turbulent flow chain cooling feeder combined in one unit.[80]

comminuted, classified, dried, and cooled. Such units can be centrally controlled, and automation appears to be possible.

11.3.1.4 *Pressing*

For the manufacture of briquettes from brown coal without binder, extrusion presses and ring roller presses are being used. The main characteristic of the extrusion press is the horizontal open shaping channel and the piston in it, which is alternately advanced and retracted. Extrusion presses are named for the linkage each type employs to transfer power from the drive mechanism (viz., reciprocating ram, toggle, and double-lever linkage presses). In practice an important characteristic is the number of pistons that work in one press. There are single-piston presses, double presses, twin presses, triple presses, double-twin presses, and other multiple-piston presses. Design and method of operation of these presses are described by Rammler and von Alberti in an extensive publication[87] on the utilization of brown coal.

For soft brown coals, the extrusion press is used, whereas the ring roller press may also be used for harder brown coals.[88] In modern briquetting plants the reciprocating ram press is preferred for the production of brown coal briquettes. Table 11.9 shows the characteristics of four types of reciprocating ram press.

For briquetting coals that are inelastic, rich in bitumen, and difficult to briquette, the ring roller press is more suitable. Rammler and von Alberti[87] describe the advantages and

[87] E. Rammler and H. J. von Alberti, *Technologie und Chemie der Braunkohlenverwertung (Technology and Chemistry of the Use of Brown Coal)*, VEB, Deutscher Verlag für Grundstoffindustrie, Leipzig, 1962.

[88] J. Klán and J. Němec, *Bergbautechnik*, **15**, 182–185 (1965).

Table 11.9 Characteristics Data for Reciprocating Ram Presses[80]

	Press Type			
Property	PSB-C	PSA	PSB-C	PSA
Maximum load (kN)	2500	2500	3000	3000
Largest width of channel (mm)	210	210	273	273
Stroke (mm)	220	220	220	220
Weight of press (tons)	50	100	65	130
Number of pistons	2	4	2	4
Width of press (mm)	2620	5600	3480	6720

disadvantages of these two press types. Rieschel[89] indicates that ring roller presses should be used only if the coals are so difficult to briquette that the extrusion press is not suitable. In the newer ring roller presses, briquetting pressures up to 294 N/mm² are achieved.[90] Feed material with 40–50% of the coal particles below 0.06 mm and with a water content of 8–9% can be briquetted. To improve productivity, it is important to decrease wear and tear of the forms, thus prolonging their lifespans.[91]

11.3.1.5 *Quality of Briquettes*

Coal with a normal size consist produces N-briquettes, and coal with a very fine size consist produces F-briquettes, as described below. A number of authors have studied the quality properties of brown coal briquettes. Special attention has been given to the accuracy of determining briquette strength and density.[92]

[89] H. Rieschel, *Braunkohle Waerme Energie,* **14,** 249–256 (1962).
[90] K. Nather, *Freiberger Forschungsh.,* **A158,** 27–54.
[91] W. D. Nolting, *Bergbautechnik,* **15,** 125–127 (1965).
[92] H. G. Schäfer and U. Wesemann, *Braunkohle,* **27** (4), 113–118 (1975).

Rammler[80] has described quality data for N- and F-briquettes (*Normalkorn* = standard grain, 0–6 mm, *Feinkorn* = fine grain, 0–2 mm) from two production plants (Table 11.10).

11.3.1.6 *Development Trends*

In all countries that produce brown coal, many investigations are under way to increase knowledge of the briquetting properties of such coal.

Because properties vary, even in raw coal from the same mine or area, the briquetting process must be adjusted.[93] The following differences are cited: strength, humic acid ratio, volume of capillaries, and others.

The influence of moisture content and briquetting pressure and temperature on the strength of the briquettes was tested, as well as the most favorable method for drying the coal. Accordingly, attempts were made to use previously gained knowledge to improve the processes occurring during briquetting without binder.[94] In particular, the develop-

[93] F. Haase, *Braunkohle, Waerme Energie,* **11,** 62–65 (1959).
[94] H. Borchert, *Braunkohle, Waerme Energie,* **11,** 62–65 (1959).

Table 11.10 Quality Data for Brown Coal Briquettes from Production Plants A and B[80]

Quality Designation	Plant	Hot-Compressive Strength N/mm²	Maximum Moisture Content (%)	Minimum Heating Value, Air-Dried kcal/kg	Minimum Heating Value, Air-Dried MJ/kg	Size Consist of Dry Brown Coal (mm)	Maximum Admissible Over-size (%)
N_1	A	7.1– 8.2					
	B	9.7–10.7					
N_2	A	8.2– 9.2			18.86	6.3–0	
				4500	17.60		
	B	9.7–11.7	19.5	4200			10
N_3	A	9.2					
	B	11.7					
F_0	A	9.2					
	B	11.7					
F_1	A	9.7	16.0	4800	20.10	4–0	
	B	12.2		4600	19.26		
F_2	A	10.2	16.0	4800	20.10	3.15–0	12
	B	13.3		4600	19.26		
F_3	A	—	—	—			
	B	14.3	15.5	4650	19.47	2–0	
F_4	A		—	—			
	B	17.3	11.0	4750	19.89	1–0	

ment of manufacturing briquettes from very fine coal and the attempts to produce carbonized briquettes and high-temperature coke from suitable green briquettes have led to intense investigations of the connection between the physical properties and the suitability of brown coal for the briquetting process.[95]

Thus the possibility of standardization[96] and automation of the briquetting process was indicated. Equipment for the uniform feeding of dryers, controlled by automatic temperature-indicating instruments, and the automatic control of the level in bunkers, are finding increased application in producing plants.[97] Bayer[98] describes these controlling and measuring devices in a new study on the status of automation in a briquetting plant.

11.3.1.7 *Briquetting of Hard Brown Coals*

There is a significant variation in the hardness of brown coals from different districts.[99] Table 11.11 compares the hardness of brown coal from German and Czech seams.

It is difficult to briquette the hard brown coals. The briquetting behavior of those from Czechoslovakia has been investigated repeatedly. Klán and Němec[99] proposed shock heating of hard brown coal to 800–1000°C for a short time. They concluded that even brown coals that have been considered unsuitable for briquetting may now be briquetted. Fine grinding of the coal is essential.

[95] J. Formánek, *Paliva,* **41** (9), 367–371 (1961).
[96] L. L. Gorinštejn and V. A. Zavjalov, *Meck. Ovtom, Proizv.,* **13,** 43–47 (1959).
[97] O. Rich, *Bergbautechnik,* **10** (2), 80–84 (1960).
[98] M. Bayer, *Braunkohle,* **25** (12), 365–368 (1973).
[99] J. Klán and J. Němec, *Paliva,* **42** (11), 330–333 (1962).

11.3.2 Bituminous Coals

A summary of the processes and the different presses used for briquetting bituminous coals without binder is given in Chapter 16 of the first Supplementary Volume.

Interest in briquetting bituminous coal without binder has significantly decreased in the last few years. Economic factors, the lack of progress in making an acceptable product under commercial conditions, the general decrease in the production of briquettes, as well as the shift toward producing smoke-free and hot briquettes by using higher pressures, have relegated conventional binderless briquetting to the background. Several studies, however, attempted to determine whether increased information about the details of binderless briquetting could lead to an improvement in the economics of such processes.

Iyencar[100] investigated the influence of the degree of coalification on briquette strength. He concluded that the strength of air-dried briquettes decreases in the following sequence: medium, high, and low degree of coalification.

Gregory[101] has pointed out that greater strengths can be obtained in bituminous briquettes without binder if, in addition to pressure, a shear stress is employed during briquetting. Each briquette is treated in sequence by two different compression plungers. Because of the high cost, these pilot-plant developments have not been commercialized.

Another study investigated the influence of

[100] M. S. Iyencar, T. A. Subramanian, A. Ghosal and A. Lahiri, "Binderless Briquetting of Coals," *J. Inst. Fuel,* **31** (208), 108–115 (1958).
[101] H. R. Gregory, *Iron Coal Trad. Rev.,* **180** (4791), 1091–1092 (1960).

Table 11.11 Hardness Number of Some Brown Coals According to Säuberlich[99]

Location	Moisture Content (%)	Hardness Number for 3-mm Diameter (N/mm²)
Cologne District, West Germany	60.0	43.2
Lusatia, East Germany	55.9	103.9
Saxony, East Germany	52.1	132.4
Czechoslovakia	27.5	721.8

oxygen in the air on briquettes obtained from bituminous coal.[102]

Many insights gained during binderless brown coal briquetting can be used advantageously to explain the events that occur during briquetting of bituminous coal without binder.[103]

11.4 CARBONIZATION AND OTHER THERMAL TREATMENTS OF BRIQUETTES

11.4.1 Fundamentals of the Carbonization of Briquettes

Possible feed materials for the original briquettes include coals of different properties, pretreated coals, low-temperature coke, high-temperature coke breeze, or mixtures of these. Various binders are used in manufacturing the briquettes. The binder may be obtained partly during the pretreatment of the coal or during the carbonization of the briquette.

The thermal posttreatment of the briquettes at different temperatures, with or without the influence of oxygen in the air, leads to briquettes that depending on the feed material, the method of treatment, and the treatment temperature, have properties more or less similar to those of coke. It is difficult to find a clear and orderly scheme for the different processes, since the many processes that have been developed overlap.

The importance of the development of carbonization of briquettes has grown[104] during recent years, as indicated by the increasing number of reports on the development of smokeless fuel briquettes, as well as the manufacture and experimental testing in almost all industrial countries of formcoke (see Chapters 13–15) in blast furnaces, cupolas, and electric furnaces.

The development of briquette carbonization and the processes developed up to 1958 for the thermal posttreatment of fuel briquettes is comprehensively described in Chapter 16 of the first Supplementary Volume.

In the last 20 years, nations—especially those lacking coking coal and those with strict laws on air quality—have shown great interest in further development of briquette carbonization.

As the content of volatile material in the feed coal increases, the production of briquettes often involves an increased number of processing steps.

If anthracite or low-volatile coals are used as feed materials, the comminuted coal is usually mixed with an additional binder such as pitch or sulfite liquor for the briquetting process. The briquettes undergo carbonization in one or several steps. The thermal posttreatment depends on the type of briquettes being treated and the properties desired in the final product. Heating rate and length and temperature of carbonization are the deciding factors in the carbonization of briquettes.[105] However the composition and properties of the raw briquettes have a basic influence on the processing steps and on the final product.[106,107]

Difficulties may result at high heating rates, since the bituminous coals undergo expansion or shrinking during heating. These occurrences may lead to cracking or bursting of the briquette. Tajc and Petrovskaja[108] investigated the influence of the heating rate on pitch-bound briquettes of different sizes from Soviet coals with different contents of volatile matter. They found that there are "critical temperature intervals" during the heating process, during which the rate of heating must be decreased. This reduces stresses in the briquette structure, avoids cracks to a great extent, and increases the final strength of the briquette.

If the binder pitch or bitumen softens, the briquettes may be deformed, or glued together as a result of the exuding of binder from

[102] M. S. Iyencar and R. S. Ghosk, *Fuel,* **39,** 113–117 (1960).

[103] A. G. Arend, *Colliery Guardian,* **203** (5246), 537–538 (1961).

[104] A. D. Fisher, *Metallurgical Coal for the Future,* Canadian Mining and Metallurgical CIM Bull. No. 65(721), May 1972, pp. 25–36.

[105] E. Ahland, J. Lehmann and W. Peters, "Production and Trials of Bergbau Forschung Formed Coke," 31st Ironmaking Conf., sponsored by Metallurgical Society of AIME, Chicago, April 1972. Published by AIME, New York: Vol. 31, 1973, pp. 285–296.

[106] D. Holze, *Bergbautechnik,* **18,** 270–271 (1968).

[107] G. Gruson, *Freiberger Forschungsh.,* **A443,** 9–19 (1968).

[108] E. M. Tajc and F. I. Petrovskaja, *Tr. IGI, SSSR,* **20,** 189–197 (1963).

the briquette surface. Such difficulties are avoided by an oxidative hardening pretreatment, by shock heating, or by keeping the briquettes quiescent during the temperature interval of softening of the binder.

If briquettes are made from coals with more than 14% volatile matter, the difficulties during carbonization increase. Such coals are usually totally or partly precarbonized. Then the char coke is briquetted alone, or blended with the input coal or even with another type of coal, using a binder.

Insofar as the procedure is economically attractive, tar is recovered for conversion to a suitable binder. Quite often, however, only gas and oil are obtained as by-products. The end products of briquette carbonization processes are smokeless fuels and formcoke with a great variety of properties.

The properties and the testing of formcoke as a substitute for conventionally produced coke are described in Chapters 14 and 15.

11.4.2 Carbonization of Brown Coal Briquettes

11.4.2.1 *Low-Temperature Carbonization of Briquettes*

Green briquettes with a low moisture content are used for the carbonization of brown coal briquettes at temperatures between 500 and 600°C. The moisture content should not exceed 12–15%.[87]

In the first processing step, the drying of the green briquettes, there should be no cracks when the briquettes are dried to a moisture content of 0%. The feed coal is usually comminuted to a size of less than 1 mm.

As a rule, the Lurgi Spülgasofen (flushing gas furnace) system is used for thermally treating the briquettes. It has an output of up to 500 tons of briquettes per day.

The construction and operation of a briquetting plant for the manufacture of carbonized briquettes from an Indian brown coal were described by Rieschel.[109] According to the author, the grinding and drying of the coal required special attention, and a series of measures necessary to overcome difficulties was described. The prepared fine coal was briquetted in double-twin extrusion presses, and the green briquettes were transported by conveyor belt to the carbonization plant. Originally the carbonization of brown coal briquettes had one objective: production of low-temperature tar as a by-product. However the emphasis has been switched to the production of high-temperature coke from brown coal (BHT coke).

11.4.2.2 *High-Temperature Coking of Briquettes*

The following processes are available[87] to carbonize brown coal briquettes for the manufacture of BHT coke:

1. *Two-stage processes:* carbonization of the coal (1): briquetting of the low-temperature coke and coking of the briquettes (2).
2. *One-and-one-half-stage processes:* carbonization of part of the coal (1): briquetting of the char in mixture with the part of the coal that has only been dried, and coking of the briquettes (1½).
3. *One-stage process with binder:* briquetting of the coal with binder and coking of these briquettes.
4. *One-stage processes without binder:* binderless briquetting of the coal and coking of the briquettes.

Historically, the oldest is the two-stage process in which the briquetting is conducted between the first stage (low-temperature carbonization) and the second stage (coking or postdegassing of the briquettes). The two-stage process has been developed especially in Germany, Yugoslavia,[110] and Hungary.[111]

In the two-stage process shown in Fig. 11.19, the coal is carbonized at low temperature, and the low-temperature coke is then wetted to a maximum moisture content of 10%. This is followed by briquetting, with pitch or heavy tar as binder, on roller presses. The briquetting mixture is frequently supplemented by caking coal or extraneous bitumen. The green briquettes are then coked in the second thermal stage at 500–1000°C. The first thermal step is preferably a flui-

[109] H. Rieschel, *Braunkohle, Waerme Energie*, **14**, 249–256 (1962).

[110] B. Popovic and M. Samec, *Symposium*, Portorŏz, Yugoslavia, 1960.

[111] B. Mory, *Freiberger Forschungsh.*, **A261**, 77–91 (1962).

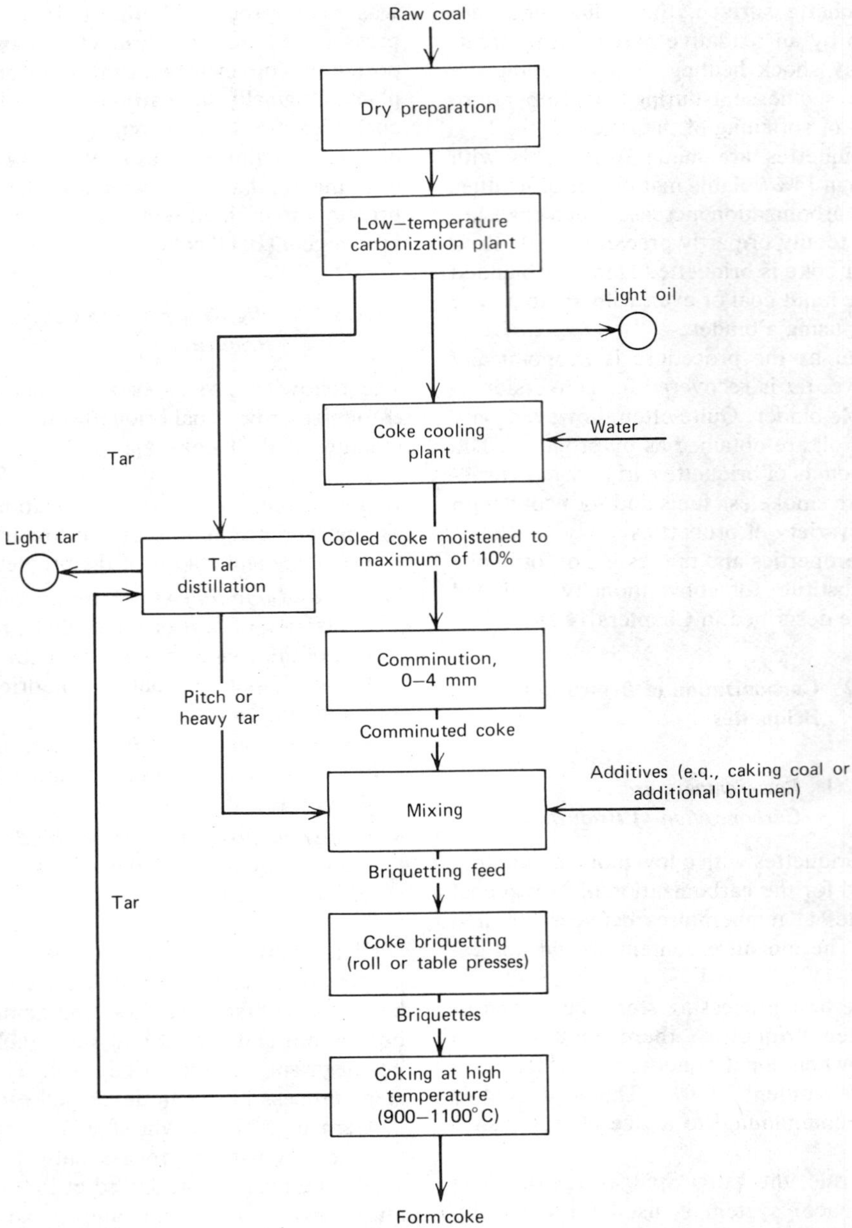

Figure 11.19 Flowsheet of the two-stage process for the manufacture of coke from brown coal briquettes.[87]

dized-bed, low-temperature carbonization step. Special attention is given to the cooling of the low-temperature coke as well as the BHT coke.[112]

The one-and-one-half-stage process for the manufacture of brown coal briquettes was developed in Rumania.[113] According to Fig. 11.20, part of the coal is carbonized at low temperature, then mixed with noncarbonized coal. To this mixture are added pitch and bituminous coal. Roll presses are used for

[112] H. Schädlich, *Freiberger Forschungsh.*, **A159**, 1–102 (1960); **A190**, 5–20 (1962).

[113] I. Barbu, I. Al. Milinovici, and N. Teodorescu, *Coscul Metalurgic* (Editora Technica), Bucharest, 1958.

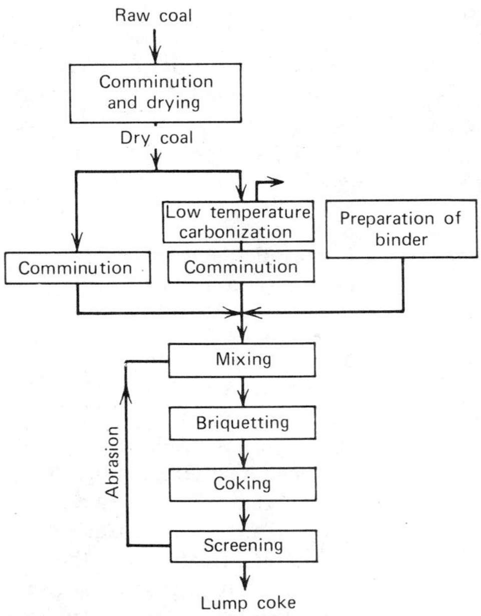

Figure 11.20 Flowsheet of the one and one-half-stage process for the manufacture of coke from brown coal briquettes.[87]

briquetting, and the green briquettes are coked at about 1000°C.

The one-stage process without binder has become more important commercially than other processes. It is suitable only for the readily briquetted soft brown coals. Commercial operation is conducted in the Lauchhammerwerk, East Germany. In principle, brown coal briquettes from superfine coal are produced without binder using special techniques. The briquettes are subsequently dried with flushing gas and then coked. Figure 11.21 is a schematic diagram of the process.

The briquetting coal is ground to a size of less than 1 mm with a minimum of 40% particles below a size of 0.25 mm. The briquettes made with this finely ground coal must be very strong. They are dried to a moisture content of less than 2% by means of flushing gases. The briquettes are coked at 1000–1100°C.[114]

Various methods have been described for

[114] G. Gerlach, *Bergbautechnik,* **9,** 7–16 (1959).

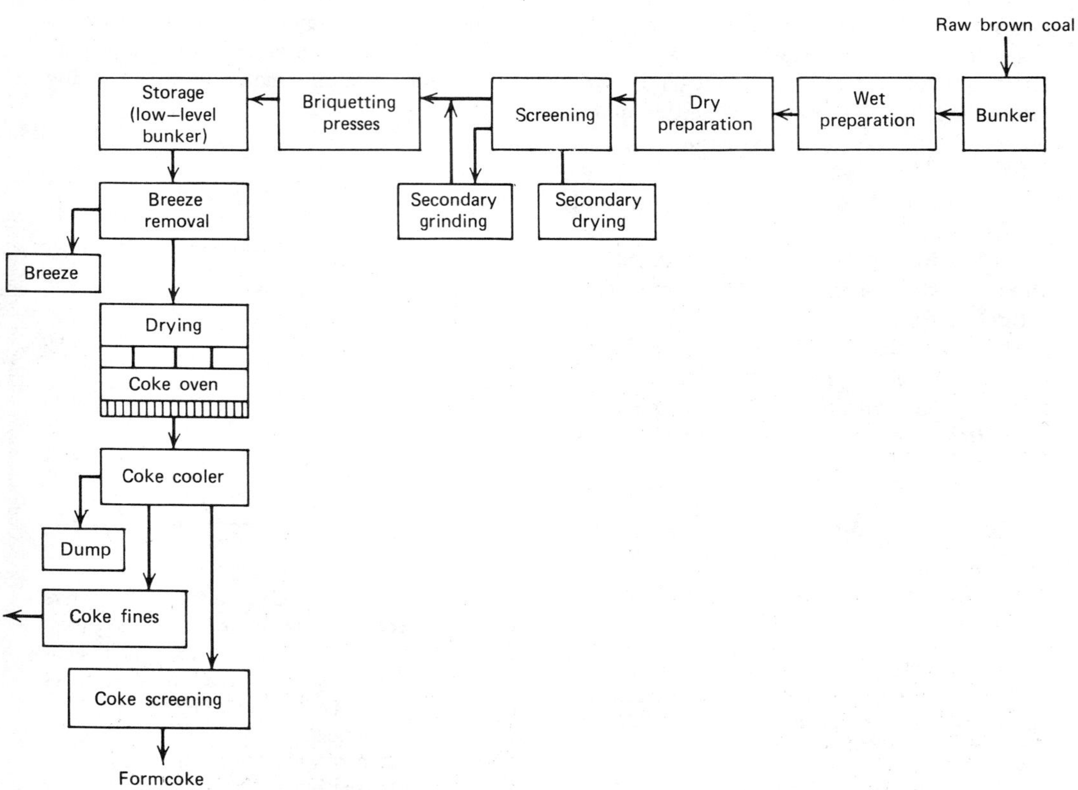

Figure 11.21 Flowsheet of the one-stage process for the manufacture of coke from brown coal briquettes.[87]

the carbonization of brown coal briquettes, mixed briquettes from brown coal and bituminous coal, or brown coal coke breeze briquettes. The Lurgi-Krupp method[115] uses ring roller presses for the manufacture of the briquettes. The carbonization of the green briquettes is done in Lurgi kilns. The process of Rammler-Bilkenroth[116] uses extrusion presses and a briquetting pressure of 117–127 N/mm² to manufacture briquettes with an average weight of 250 g. For carbonization, vertical refractory retorts, indirectly heated to 1000–1100°C, are used. A new development led to the construction of direct-fired shaft kilns in which three separate recycle gas streams are used to dry, coke, and finally cool the green briquettes. Pitch obtained from low-temperature brown coal tar is used as binder for coke breeze (< 0.75 mm) from brown coal, using roll presses.[117] The green briquettes are oxidized at 230°C with a flushing gas that contains 8% oxygen, then are carbonized at temperatures above 1000°C.

New publications[118] describe the carbonization of mixed briquettes consisting of brown coal and coke, and the manufacture of formcoke from brown coal according to a new process developed on a semicommercial scale. In this process low-temperature carbonization of the coke is followed by hot briquetting, cooling of the briquettes under pressure, and coking at temperatures up to 950°C.

In a cooperative effort of German and Soviet scientists,[119] the thermal decomposition of brown coals, of a high-volatile bituminous coal from East Germany, and of an anthracite from the Soviet Union, as well as mixtures of these, were investigated. These studies covered a wide range of factors and their influence on the strengths of the coke briquettes. The factors investigated were carbon content of the feed coal, coke content of the binder, binder content in the mixture, briquetting pressure, and final temperature during carbonization of the briquettes. It was found that the carbon content of the feed coal and the binder content of the mixture have great influence.

11.4.3 Carbonization and Other Thermal Treatments of Bituminous Coal Briquettes

The object of carbonizing bituminous coal briquettes is to convert smoke-producing briquettes into smokeless ones, to produce formcoke with properties similar to the conventionally produced high-temperature coke. These results are obtained in one-stage processes or in multistage processes. Carbonization consists of thermal treatment or thermal oxidizing treatment of the green briquettes.

The oxidative heating of the briquettes greatly influences their properties. Especially significant is the effect of oxygen on the strength and the smoke evolution of pitch-bound briquettes.

Figure 11.22 shows the effect on the strength of briquettes with 12% pitch as binder, as well as on the amount of binder converted into cokelike material, of oxidizing

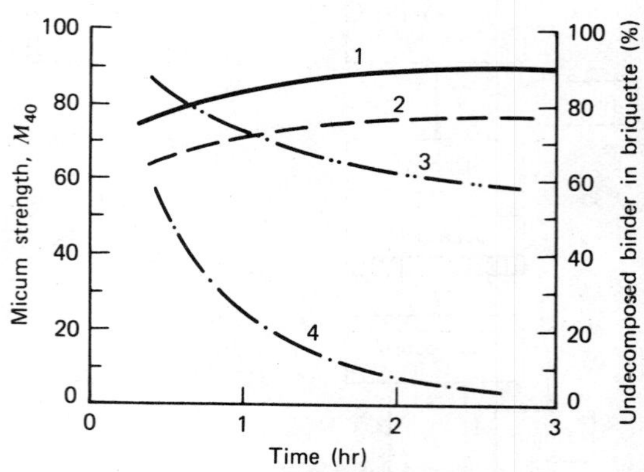

Figure 11.22 Effect of oxidative and nonoxidative heat treatment of briquettes on the mechanical strength and the degree to which binder is converted into a cokelike substance (binder content 12%): 1 = treatment with air at 220°C, M_{40}; 2 = oxygen-free treatment at 220°C, M_{40}; 3 = undecomposed binder, oxygen-free treatment at 220°C; 4 = undecomposed binder, treatment with air at 220°C.[115]

[115] H. Zielinski, *Formed Coke Manufacture,* Warsaw, Poland, 1972, Ch. 9 (Modern Coking Methods), 1967 "Silesia."
[116] E. Rammler and G. Bilkenroth, *Neue Huette* **7,** 513–519 (1962).
[117] A. Lissner and E. Rammler, *Freiberger Forschungsh.,* **A263,** 52–72 (1962).
[118] H. Krug, W. Naundorf, D. Trommer and D. Schonherr, *Neue Bergbautechn.,* **4** (4), 292–299 (1974); No. 8, 611–616.
[119] J. V. Birjukov, E. Klose, A. Seichter, W. Naundorf, and M. Hann, *Khim. Tverd. Topl.,* **10** (3), 108–114 (1976).

and nonoxidizing heating, at 220°C.[120] The oxidizing atmosphere causes profound chemical changes in the binder. The oxidation is an exothermic reaction, and the heat evolution is about 100 kcal/kg of briquettes.[121]

A number of older briquette carbonization processes such as the Carbocoal process, the Ocean Coal Company process, and the Suncole process, as well as the Carmaux process and others, are described by Zielinski.[115] These processes have not been developed further over the years, but their methods of thermal oxidation for treating bituminous coal briquettes have aided the development of newer, more economical processes that have to some extent replaced them.

The main emphasis of the development of the carbonization of briquettes during the last two decades was to further improve known processes and to develop new combinations of processes that are attuned to new standards.

11.4.3.1 *Anthracine Process*

The Anthracine process was developed in France.[122] Anthracite fines were briquetted with a pitch binder and treated with air-enriched flue gas at 350–380°C, thereby undergoing an oxidative thermal treatment. This hardening process initially used tunnel kilns, through which cars containing 2 tons of briquettes each passed continuously.[123] Today endless steel link conveyers are used. The kilns are 50 m long and 2.5 m wide, briquette output is 15 tons/hr.

The Rousseau plant, near Valenciennes, France, has a capacity of 500,000 tons of briquettes per year.[124] The main product is fuel briquettes that evolve little smoke.

11.4.3.2 *Coppée Process*

The Coppée-Houillères du Bassin du Nord et du Pas de Calais processes[125] uses higher volatile coals as feed material. Binders for the production of the briquettes are pitch and anthracene oil. In addition, an oxidizing material, preferably sulfur, is added to the briquette mixture.

The green briquettes undergo treatment similar to that of the Anthracine process. They are preoxidized in a belt kiln at 280°C and undergo the final thermal treatment at 380°C.

The plant in Zolder, Belgium, has an annual capacity of 400,000 tons.

11.4.3.3 *Synthracite Process*

The Synthracite process[122,126] has been developed in the Netherlands. The original process was identical with the Phurnacite process. Initially briquettes made with pitch as binder were carbonized in the Distico kiln.[115] Because of technical difficulties when using briquetting coal with more than 12% volatile matter, an oxidizing process was introduced in 1961. The briquettes are treated in an inclined kiln at 300–350°C, in a discontinuous operation, with recycled flue gas that is enriched with oxygen. The briquette layer in the kiln is about 1.5 m thick.

11.4.3.4 *Thermal Oxidizing Treatment of Briquettes in a Sand Bed*

The use of hot sand as a solid heat carrier for carbonizing coal briquettes has been known for quite some time. Two older processes, the Weber process and the process of the National Coal Board, have been described by Zielinski.[115] Several patents exist for the thermal treatment of briquettes with hot sand.[127]

The Institut National de l'Industrie Charbonnière (INICHAR) in Belgium has developed a new process. From 1964 to 1966 Belgian scientists described the development of and the theoretical basis for the thermal treatment of briquettes in a sand-bed kiln.[128–130]

[120] H. Zielinski, *Freiberger Forschungsh.*, **A119**, 61–71 (1958).
[121] M. Kaloč, *Hutn. Listy,* **18** (12), 847–850 (1963).
[122] K. Winnacker and L. Küchler, *Chemische Technologie,* Vol. 3, Hanser Verlag, Munich, pp. 45–46.
[123] F. Loison and P. Foch, *Rev. Ind. Miner.*, Note Technique, 1/1959, 6/1959, 2/1960, 1959.
[124] W. Finze, *Kurznachrichten Stbv.*, No. 59, Essen, West Germany, 1967, p 7.
[125] L. Meurise, *Ann. Min. Belg.*, No. 6, 735–753 (1964).
[126] West German Pat. 1170366; *Glueckauf,* **100,** 809 (1964).
[127] West German Pat. 848644, 1950.
[128] M. Marcourt, *Ann. Min. Belg.*, **63** 782–807 (1964).
[129] H. Dufour, *Ann. Min. Belg.*, **63** 865–882 (1964).
[130] P. Ledent et al. Bull. Tech. Houille, INICHAR, No. 32, 1966, pp. 775–786.

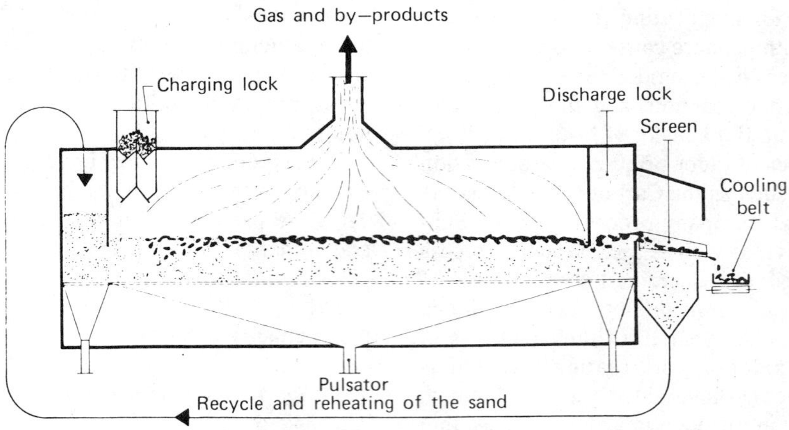

Figure 11.23 Schematic diagram of the thermal-oxidizing treatment of briquettes in a sand-bed kiln.[130]

The sand-bed oven developed in Belgium is a closed, long kiln (Fig. 11.23) with a bottom consisting of screen plates. These screens support a bed of sand that may be as thick as 50–60 cm. Intermittent blowing of steam, air, recycle gas, or inert gas is used to fluidize the bed. The briquettes in the kiln are heated by the hot sand and travel from the feed point to the extraction point; there they pass the screen and are separated from the sand. A process cycle usually comprises 4–5 s of sand fluidization, during which the briquettes float in the sand, then 80–90 s of settled bed. This method keeps the volume of fluidizing gas low. Heat carrier and briquettes travel concurrently through the kiln. The sand that has been separated from the briquettes at the kiln exit is recycled, reheated, and returned into the kiln.

The first experimental plant to approach commercial conditions was built in 1962 by Bergbau Forschung GmbH in West Germany.[131] Initially briquettes prepared with sulfite liquor as binder were hardened in the sand-bed kiln, at temperatures of 300–400°C. Later the main object of the development was the thermal oxidative treatment of pitch-bound briquettes to produce a smokeless fuel.

In 1963 and 1964, two production plants were put into operation with an output of 15 tons/hr at S.A. des Charbonnages de Wérister[132] and the S.A. des Charbonnages d'Hensies-Pommeroeul in Belgium. Other plants were built in Belgium and somewhat later in Cardiff, Great Britain.[133]

In Wérister coal briquettes made from low-volatile coal with 12–13% volatile matter and with pitch as binder were produced and converted into smokeless fuel. In three beds, each 1 m wide (Fig. 11.24), the briquettes travel through the 29-m-long kiln at Wérister. The beds have a slope of 2°. Sand enters the front of the kiln at 420°C and leaves at 250°C, thereby maintaining a mean oven temperature of about 320°C.

The thermal oxidizing treatment in the sand-bed kiln can be used[134] for briquettes from nearly all coals and with different binders.

Three heating methods are used: rapid degassing, progressive degassing, and progressive degassing with intermediate heat stages, in which a predetermined temperature is maintained for a predetermined time.

Table 11.12 shows the way properties change while briquettes are undergoing a treatment to make them smokeless.

According to Peters et al.,[131] if thermal oxidative treatment is applied as a means of smoke reduction, the best reaction temperature is 340°C. Higher temperatures applied to

[131] W. Peters, G. Schmeling and K. Kleisa, *Glueckauf-Forchungshefte*, No. 26, 67–78 (1965).
[132] M. Jamoulle, *Ann. Min. Belg.*, No. 6, 822–864 (1964).

[133] "Multiheat Plant at Cardiff," *Colliery Guardian*, **211** (5459), 724–726 (1965).
[134] P. Ledent, G. Burton and M. Marcourt, presented at the Information Meeting of EGKS, Luxemburg, April 1970, published in *Technology and Development of Carbonization of Bituminous Coals*, Central Office for Information and Documentation of European Associations, Luxemburg, pp. 191–214.

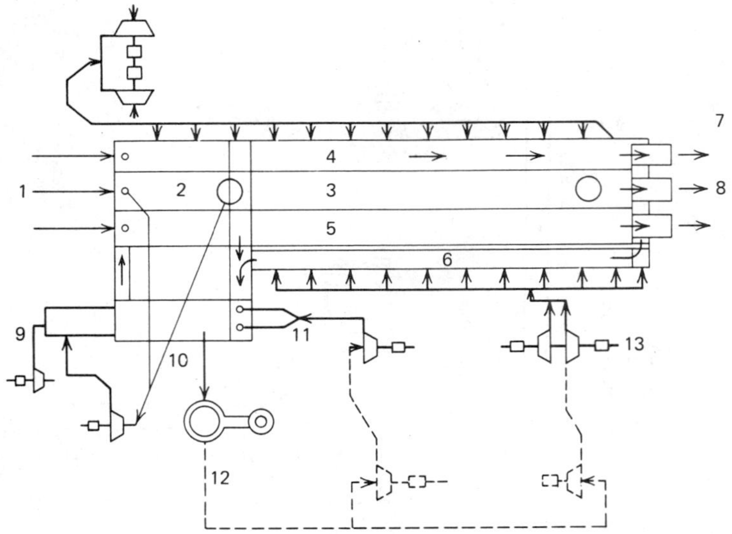

Figure 11.24 Schematic of the Wérister sand-bed kiln: 1 = green briquettes, 2 = preheating chamber, 3 = oxidation chamber, 4 = 2% slope, 5 = recycle gas exit, 6 = recycle channel for sand, 7 = sand separation screen, 8 = briquette with little smoke evolution, 9 = oil burner, 10 = reheating of sand, 11 = flue gas, 12 = dust cyclone, 13 = spare blower for kiln start up.[132]

pitch-bound briquettes lead to excessive evaporative losses of pitch. In addition, thermal oxidative treatment permits carbonization of coal briquettes at about 550°C. The carbonization of the briquettes is a more difficult process than the conversion to smokeless briquettes.[130]

Further development of the INICHAR sand-bed kiln process led to the INIEX process.

11.4.3.5 *INIEX Process*

In the Belgian INIEX process, pitch-bound briquettes are thermally treated in two stages.[134,135] In the first stage the green briquettes are carbonized in the sand-bed kiln at 550°C. The carbonized briquettes are then fed into a flushing gas kiln and are coked at about 1000°C.

The process was developed for use with briquettes made from anthracite and low-volatile coal with pitch binder. Later, for the carbonization, briquettes made out of mixtures of higher volatile coal with coke breeze were used. The investigations were made in an experimental plant with an output of 100 tons/day.

11.4.3.6 *BF Foundry Formcoke Process*

The process for manufacturing foundry formcoke developed by the Bergbau Forschung GmbH uses as feed material high-tempera-

[135] J. Langhoff, *Aufbereitungstechnik*, No. 4, 181–186 (1974).

Table 11.12 Properties of Briquettes Before and After Treatment in the Sand-Bed Kiln

Property	Green Briquette	Oxidized Briquette	Change (%)
Weight (g)	20.9	20.45	−2[a]
Moisture content (%)	1.6	0.7	—
Volatile matter (%)	16.2	14.5	−1.7[b]
Tumbler strength, M_{40} (%)	90.1	94.4	+5.0[a]

[a] Relative change.
[b] Absolute change.

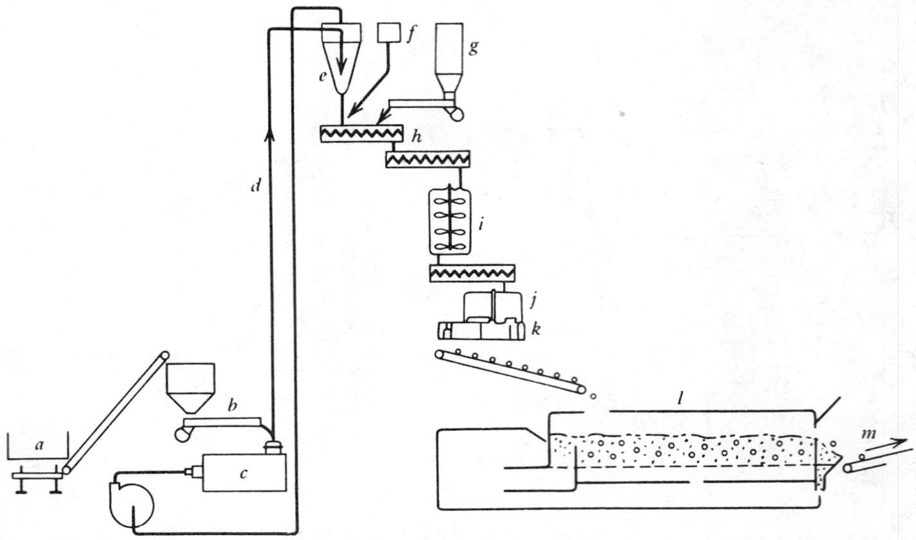

Figure 11.25 Schematic flowsheet of the BF process for the manufacture of foundry formcoke: a = coke breeze, b = metered feeder, c = combustion chamber, d = preheating, e = cyclone, f = emulsion, g = pitch, h = mixer, i = pug mill, j = distributing disk, k = briquetting press, l = sand-bed kiln, m = to loading rack.[136]

ture coke breeze particles smaller than 1 mm. The coke is briquetted with coal-tar pitch as binder and the green briquettes undergo thermal oxidative treatment in a sand-bed kiln like that in the INICHAR system.[136] Figure 11.25 is a schematic of a pilot plan for this process.

The dried breeze described above is metered and transported pneumatically with flue gas from a combustion chamber. Once it has reached the temperature suitable for briquetting, it is moved to a cyclone. After addition of emulsion[71] and pitch, the mixture is briquetted on roll or Couffinhal plunger presses (Fig. 11.25, k). The briquettes are oxidized in the sand-bed oven at 310°C, then cooled to 70°C before shipping.

It was found[136] that the 310°C treatment had to be applied for 2.5 hr to oxidize briquettes weighing 0.9 kg. As Fig. 11.26 shows, this temperature leads to a significant maximum in briquette strength.

Table 11.13 shows the properties of foundry formcoke. The formcoke produced in the BF-foundry process was made from green briquettes weighing about 0.38 and 0.9 kg. The coke breeze was bound with 10% tar pitch. Tests with these foundry formcokes in cupolas have led to higher outputs from those units.[137]

[136] W. Peters, J. Langhoff, W. Schinzel and G. Schmeling, *Glueckauf-Forschungshefte,* **28** (3), 101–109 (1967).

[137] A. Dahlmann et al., *Giesserei,* **54** (11), 297–304 (1967).

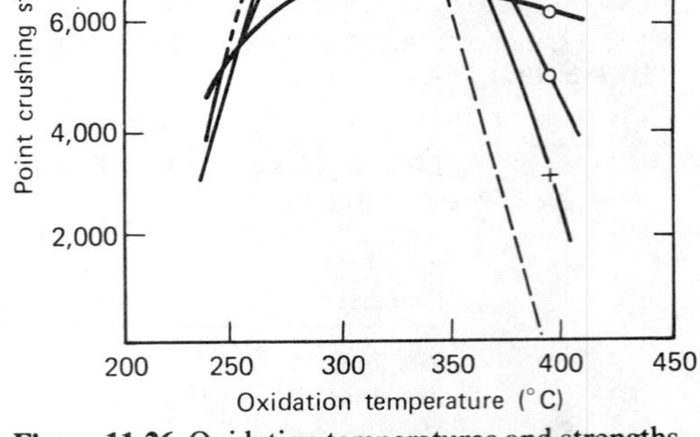

Figure 11.26 Oxidation temperatures and strengths of briquettes made by the BF foundry formcoke process. Oxidation times as follows: ○ = 1 hr, ● = 3 hr, + = 5 hr, Δ = 7 hr.[136]

Table 11.13 Properties of Thermally Oxidized Coke Breeze Briquettes[136]

	Coke Briquette Size	
Properties	380 g	880 g
Moisture (%)	0.8	0.8
Ash (%, mf)	7.4	6.8
Volatile matter (%, maf)	2.8	2.2
Sulfur (%)	0.82	0.88
True density (g/cm³)	1.89	1.82
Apparent density (g/cm³)	1.15	1.07
Porosity (%)	39.1	41.2
Reactivity (cm³/gs)	1.02	0.87
M_{10} Micum strength	19.3	22.6
Bulk density (kg/m³)	—	624.0

11.4.3.7 *IChPW Process*

Degassed coals that cake slightly or not at all are the feed material for the IChPW process (Institute of Chemical Coal Processing of Poland). It was developed in Poland to extend the range of coals used to make coke.

The process is divided into three stages[120,121]:

1. Carbonization of the coal at a temperature that is usually above 800°C, with production of low-temperature tar.
2. Comminution of the resultant coke or semicoke to a size below 4 or 3 mm, followed by briquetting. The heavy fraction of the coke's low-temperature tar is used as binder, or a foreign tar product that produces favorable hardening of the briquettes in the subsequent oxidizing thermal process may be used. Good results were also obtained with petroleum oil products as binder.[138]
3. Oxidation of the briquettes with oxygen

[138] H. Zielinski, *Koks Smola Gas*, 7 (5), 199–202 (1962).

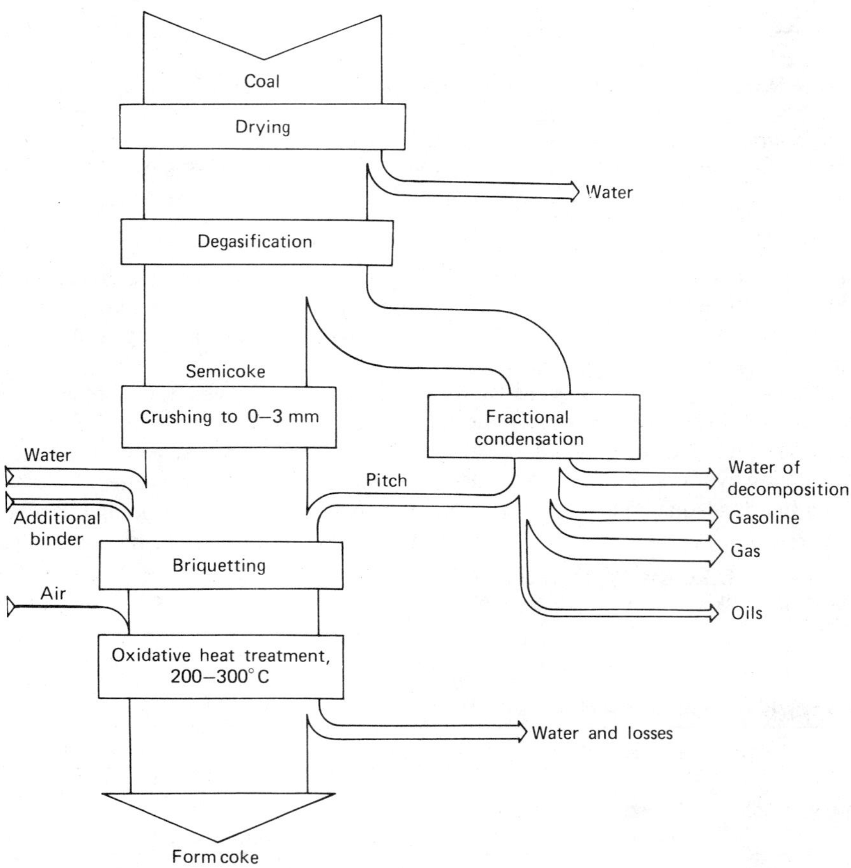

Figure 11.27 Sankey diagram of formcoke manufactured by the IChPW method.[121]

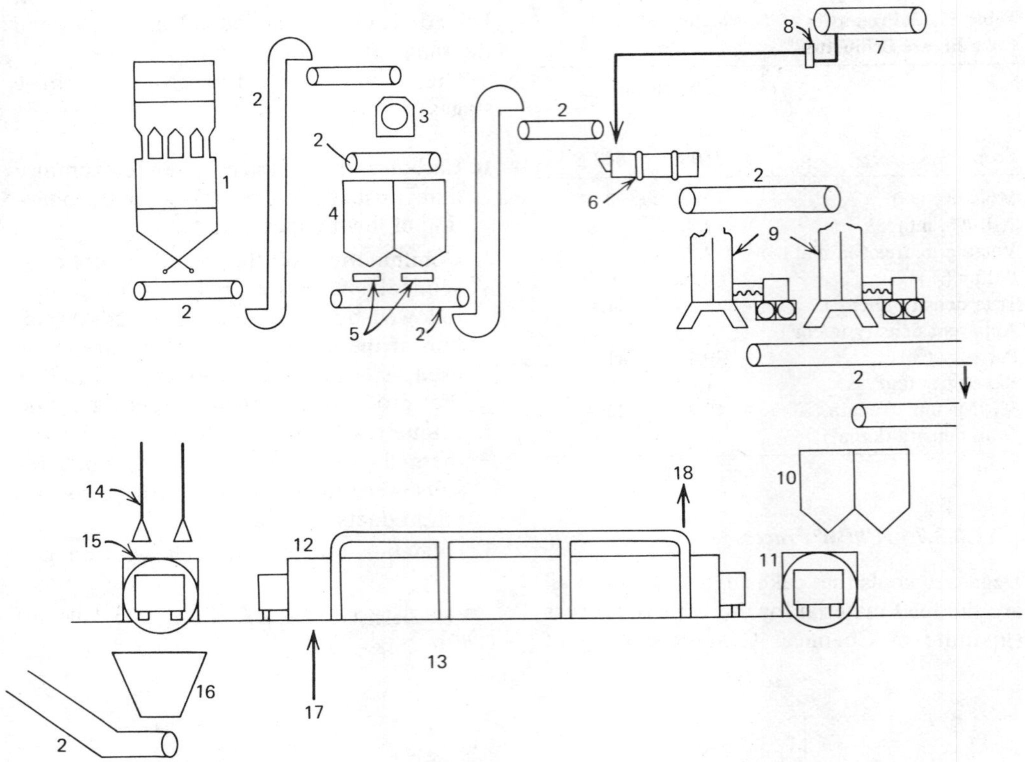

Figure 11.28 Flowsheet of IChPW coke-manufacturing process: 1 = shaft furnace, 2 = conveyors, 3 = hammermill, 4 = bunker for crushed semicoke, 5 = feeders, 6 = drum mixer, 7 = binder tank, 8 = proportioning pump, 9 = mixers and roll presses, 10 = charging equipment, 11 = shelf-type car, 12 = tunnel kiln, 13 = combustion chamber, 14 = water cooling, 15 = tippler, 16 = ground storage bin, 17 = air inlet, 18 = exhaust gases.[115]

containing gases at a temperature between 200 and 300°C or higher.

The most favorable method for degassing the coal in the first stage of the process was the subject of investigation.[139] Normally the degassing is conducted in a modified Lurgi shaft furnace; however degassing of the coal in a fluidized bed is also suitable.

[139] E. Cieckiewicz, H. Kornas, W. Medwecki, J. Rychly, and H. Zielinski, *Koks Smola Gas,* **9** (6), 203–208 (1967).

Figures 11.27 and 11.28 illustrate the IChPW process of formcoke production by a Sankey diagram and by a flowsheet, respectively.

Table 11.14 gives the properties of the semicoke that was used for the manufacture of the green briquettes according to Kaloč.[121]

Depending on its properties, 7–10% binder is needed. The fraction boiling above 300°C should be 60–66 vol %, having a Kraemer-Sarnow softening point of about 40°C and a Conradson carbon content of about 25–26%.[121]

Table 11.14 Analysis and Size Consist of Semicokes[121]

Properties of Semicoke					Size Consist (%)		
Volatile Matter (maf)	Ash (mf)	Carbon (maf)	Hydrogen (maf)	Sulfur (mf)	0–0.2 mm	0.2–0.3 mm	3–6 mm
5–7	9–10	92–94	1.5	0.5–0.6	25% (max)	50–60	Remainder

Table 11.15 Properties of IChPW Formcokes[121]

Property	Formcoke Brand		
	BW	BB	BO
Dimensions (mm)	76 × 73 × 47	100 × 145 × 55	93 × 135 × 55
Average weight of briquettes (g)	~ 225	~ 575	~ 510
Micum strength M_{40}	75–90	75–85	80–85
Attrition M_{10}	8–18	11–16	11–13
Bulk density (tons/m³)	0.6	0.6	0.5
Moisture content (%)	1.0–2.0	1.0–2.0	1.0–2.0
Volatile matter (%)	7–10	7–10	7–10
Ash content (%)	8–10	8–10	8–10
Heating value			
kcal/kg	~ 6,800	~ 6,800	~ 6,800
kJ/kg	28,500	28,500	28,500

On a commercial scale, various types of formcoke were produced using the process described. Coke PW is blast furnace coke, whereas the cokes BB and BO are suitable for use in foundries. Tentatively a type BY (size 140 × 112 × 75 mm; weight 800 g) was tested, too. Brand BM is mainly used as a smokeless fuel.

Data on the properties of the IChPW formcoke are summarized in Table 11.15.

11.4.3.8 *FMC Process*

The FMC process is a multistage process for the manufacture of coke briquettes from high-volatile coals. It was jointly developed by the Food Machinery and Chemical Corporation and the U.S. Steel Corporation.[140,141] The plant was built in Kemmerer, Wyoming, and has a daily capacity of 250 tons of coke briquettes.

In a treatise on the production of formcoke from coals in the United States, Holowaty[142] describes the FMC process. The comminuted coal is oxidized, carbonized at low temperature, and calcined. On cooling, the low-temperature tar or an extraneous binder is used for the briquetting. The briquettes pass through a posttreatment kiln (traveling-grate kiln) and are then heated in a shaft furnace with hot gases, to about 800°C. In the bottom part of the furnace, the briquettes are cooled to about 100°C.

Figure 11.29 is a schematic flowsheet of the FMC formcoke process, the properties of the Illinois No. 6 coal used as feed material are as follows:

Moisture (%)	Volatile Matter (%)	Formcoke (%)	Ash (%)	Sulfur (%)
11.0	30.4	53.5	5.1	0.53
—	34.2	60.1	5.7	0.65

The binder requirement is between 8 and 15%. Weakly caking coals, which are not suitable for the production of coke by the conventional process, can be converted by means of this process into a suitable blast furnace coke.[143–145] The capital costs of an FMC plant and of a conventional slot oven of the same capacity are reasonably close.[146]

[140] "New Opportunities for Western Coals," *Coal Age*, **65** (2), 26–28 (February 1960).

[141] D. E. Wolfson, *Congr. Int.*, Charleroi, 1966, A3.

[142] M. O. Holowaty, presented at the Meeting of the Technical Committee of the International Iron and Steel Institute, Brussels, March 28–29, 1976, published by International Iron and Steel Institute, Brussels.

[143] R. T. Joseph and E. Saller, *AIME Ironmaking Proc.* Vol. 30, Pittsburgh, April 1971.

[144] J. K. Holgate and P. H. Pinchbeck, *J. Iron Steel Inst.*, **211** Part 8, 547–573 (1973).

[145] T. J. Robert, paper presented at 13th Biennial Conf. Institute Briquetting and Agglomeration, Colorado Springs, August 1973, Institute for Briquetting and Agglomeration, Laramie, Wyo.

[146] R. F. Moran, R. T. Joseph, "Applicability of Coals to the FMC Coke Process," *Trans. AIME*, **260**, 29–41 (1976).

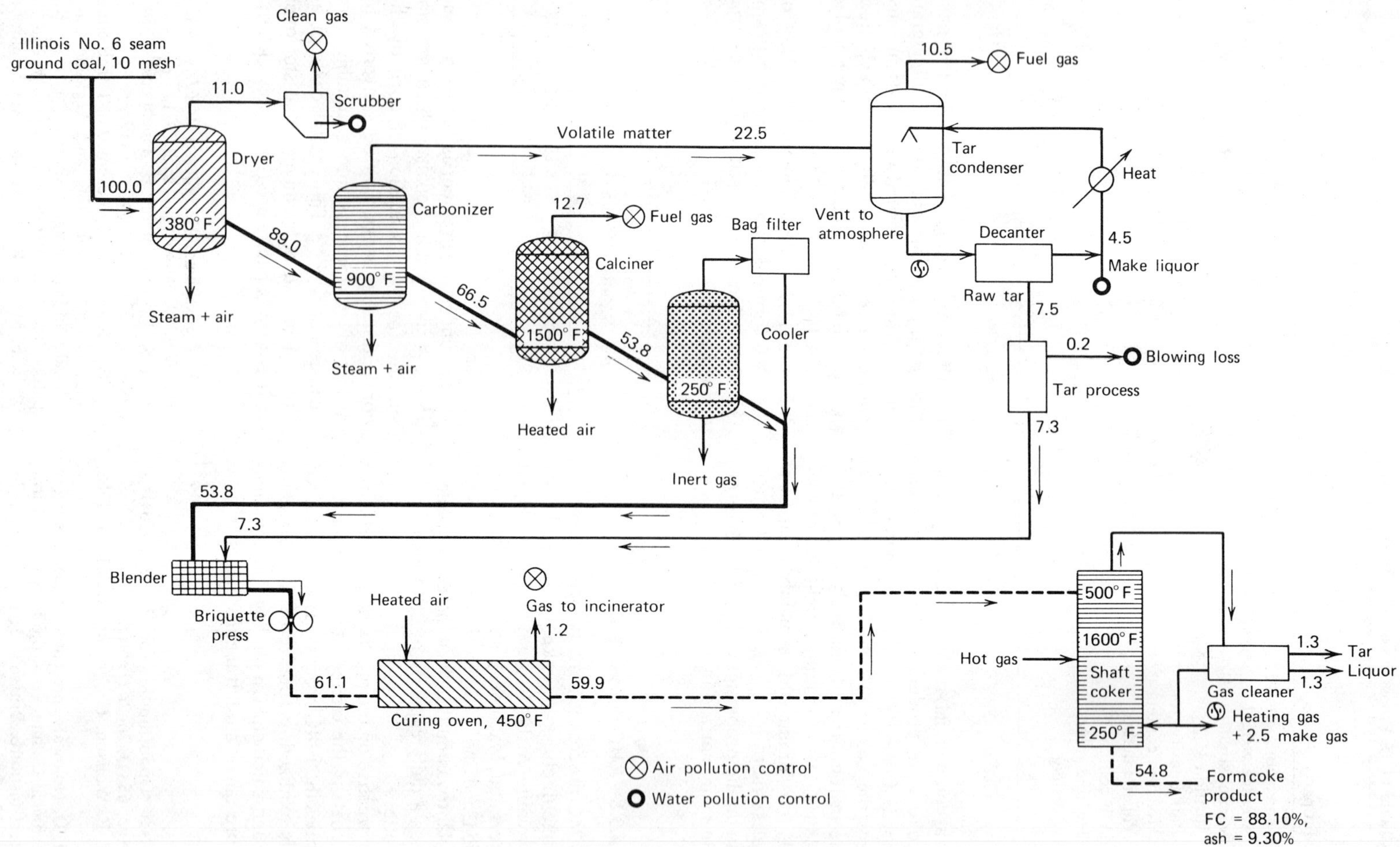

Figure 11.29 Flowsheet of the FMC formcoke process.[142]

11.4.3.9 *ICEM Process*

The ICEM process has been developed by the Institute of Energetics of the Rumanian Academy of Sciences.[115,147,148] This process for the manufacture of coke briquettes also starts with fine coal which is degassed at low temperature.

The green briquettes are prepared on roll presses using 9–10% low-temperature tar products or liquid petroleum residues as binder. Figure 11.30 is the technological flowsheet for the ICEM process. Of the particles to be briquetted, 80% are smaller than 2.5 mm and 55% are smaller than 1 mm. The green briquettes—weighing about 50 g—pass through a tunnel kiln on transport carts and are treated for about 4 hr at 247–270°C with hot gases containing 11–15% oxygen. A subsequent carbonization process at high temperature may reduce the content of volatile matter of the briquettes from 8–10% to less than 2%.

Table 11.16 summarizes the properties of various briquettes made by the ICEM process. The briquettes were made with different binders; some were oxidized and carbonized, and some were merely oxidized.

11.4.3.10 *Dr. C. Otto and Co. Process*

The Dr. C. Otto and Co. GmbH, West Germany, has developed a flushing gas retort for the continuous carbonization of briquettes.[149]

Flushing gas is heated in a combustion chamber and fed by special distributors to the center axis of the retort. The hot flushing gases pass in cross flow through the briquette

[147] J. Blum and J. Nistor, *Rev. Elektrotechn. Energ.* **6,** 151–162 (1961).
[148] J. Barbu, J. Jonescu and V. Czindea, *Metallurgy (Rumania),* **16,** 195–198 (1964).

[149] R. Meves, *Erdoel Kohle, Erdgas, Petrochem.,* **15** (7), 530–535 (1962).

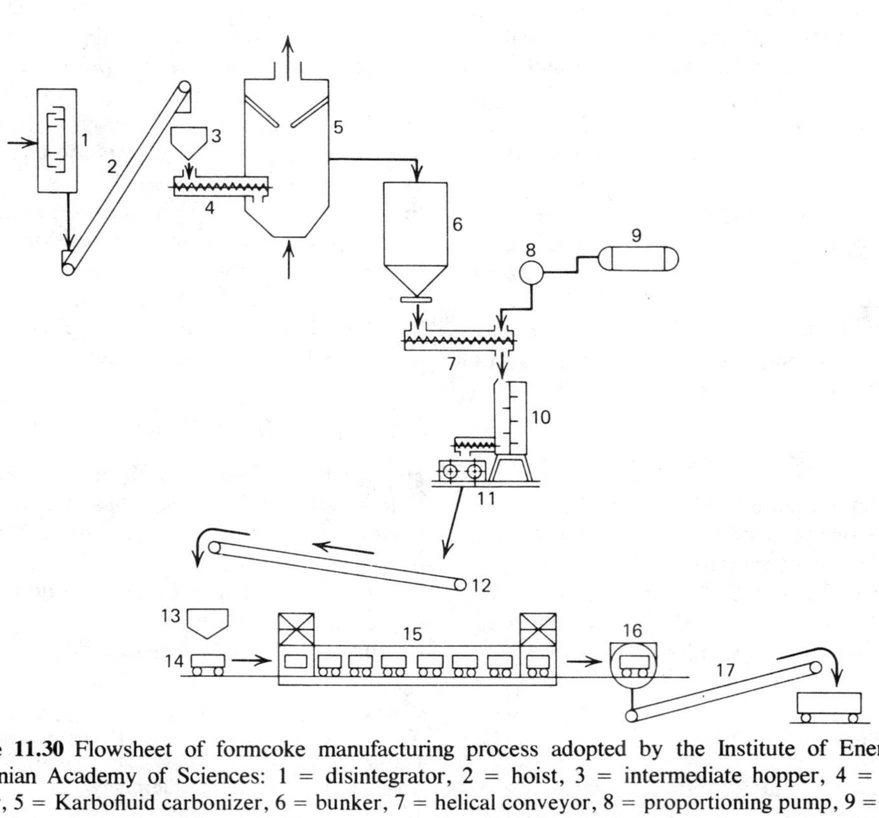

Figure 11.30 Flowsheet of formcoke manufacturing process adopted by the Institute of Energetics, Rumanian Academy of Sciences: 1 = disintegrator, 2 = hoist, 3 = intermediate hopper, 4 = helical feeder, 5 = Karbofluid carbonizer, 6 = bunker, 7 = helical conveyor, 8 = proportioning pump, 9 = binder tank, 10 = mixer, 11 = roll press, 12 = conveyor, 13 = charging equipment, 14 = car, 15 = tunnel kiln, 16 = discharging equipment, 17 = conveyor.[115]

Table 11.16 Properties of ICEM Process Briquettes[115]

Property	Oxidized Briquettes Containing Nonparaffin Binder	Briquettes with Paraffin Binder	
		Oxidized	Oxidized and Carbonized
Moisture (%)	3.11	3.39	2.62
Ash (%)	13.84	12.30	13.59
Volatile matter (%)	9.35	7.87	1.65
Calorific-value			
kcal/kg	6,776	6,877	6,747
kJ/kg	28,400	28,800	28,300
Ultimate analysis			
Carbon (%)	75.05	76.57	78.91
Hydrogen (%)	1.90	1.95	0.85
Sulfur (%)	2.15	2.05	2.12
Apparent density (g/cm³)	1.09	1.95	1.03
Micum indices			
M_{40}	85.2	84.4	79.8
M_{40}–M_{25}	4.40	6.0	6.9
M_{25}–M_{10}	1.0	1.2	3.1
M_{10}	8.4	8.4	10.2

in the retort and are withdrawn together with the evolved volatile material at the wall of the retort. Successful experiments with this process were conducted with various coals from developing countries.[115] In many cases mixtures were briquetted that consisted of the available noncaking coals, coke breeze, and added caking coals. As binder, 10–12% pitch was used. Addition of only 10% caking coal gave a formcoke with high attrition resistance.

A production plant containing two retorts, each with an output of 135 tons/day, is operating in Klipoortie, South Africa.[150]

11.4.3.11 *Schenk-Wenzel Process*

A further development of the Shawinigan traveling-grate process[151] by Schenk and Wenzel,[152] in cooperation with the German companies L. and C. Steinmüller, C. Still, and H. Koppers, led to a process that permits carbonization of briquettes on a traveling-grate kiln. Figure 11.31 is a schematic flowsheet of this traveling-grate kiln.

Brick-shaped briquettes prepared on a press with various binders are continuously passed through the traveling-grate furnace. The air needed for the combustion of the gas evolved during the coking process is added through channels at right angles to the briquette layer. During the combustion of the liberated gases, the heat necessary for carbonization is generated above the briquetting bed. The briquettes leaving the furnace are cooled with inert gas and water.

11.4.3.12 *HBNPC Process*

Houillères du Bassin du Nord et du Pas de Calais (HBNPC), France, has developed a one-stage process for the manufacture of formcoke.

After conventional binder briquetting of the coal mixture, the green briquettes are carbonized in a shaft furnace.[153,154]

[150] J. Langhoff, *Chemische Technologie,* Winnacker-Küchler, Vol. 3, Verlag C. Hanser, 1971, pp. 48–49.

[151] L. Gmelin and R. Durrer, *Metallurgie des Eisens (Iron Metallurgy),* Vol. 2, 4th ed., Verlag Chemie, 1968.

[152] W. Wenzel and F. Selig, *Klepzig Fachber.,* **76,** 71–76 (1968).

[153] M. Burteaux, F. Lesur, F. Pichon, M. Schneider and R. H. Busso, "Blast Furnace Test with Formed Coke," *Int. Iron Steel Congr.,* Verlag Stahleisen, Düsseldorf, Germany, Paper No 1, 1.2.2.6, 1974.

[154] L. Soubrier, "HBN process for the Production of Formcoke," *Technology and Development of Carbonization of Bituminous Coal,* Central Of-

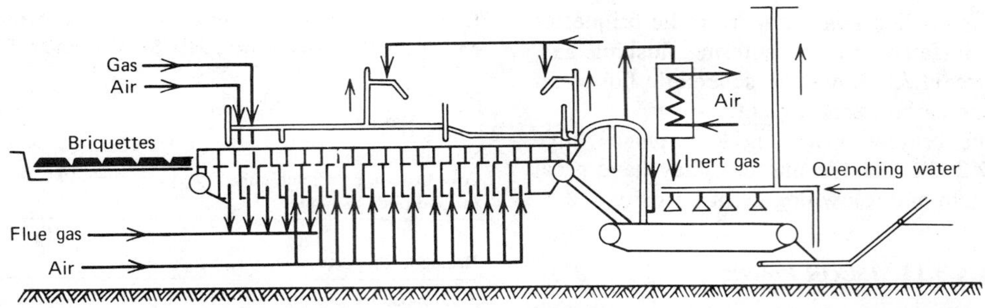

Figure 11.31 Carbonization of briquettes in a traveling-grate kiln according to Schenk and Wenzel.[150]

The briquetting coal usually consists of a mixture of two components: the so-called basic low-volatile coal (6–11% VM) and a binder coal. Basic coal with high caking properties must be oxidized or carbonized at a low temperature before briquetting takes place.[155] Depending on the properties of the basic coal, up to 10% binder coal is added. In addition, pitch or petroleum bitumen is needed as binder for the manufacture of the briquettes.

The shaft furnace is divided into several compartments, which are 3 m long, 0.5 m wide, and 6.5 m high. Countercurrent flow is used for the thermal treatment of the briquettes from the bottom to the top in three zones: the preheater, coking, and cooling zones. A temperature of 1000°C is achieved in the section of the retort in which the injected air reacts with the preheated gas.[156]

fice for Information and Documentation of European Associations, Luxemburg, 1975, pp. 241–252.

[155] W. Peters, "New Technologies for the Production of Blast Furnace Coke Under the Aspect of Enlarging the Coking Coal Basis," *10th Ann. Conf., International Iron and Steel Institute,* Osaka, Japan, October 13, 1976, published by International Iron and Steel Institute, Brussels.

[156] W. Wenzel, et al., *Aufbereitungstechnik,* No. 5, 233 (1976).

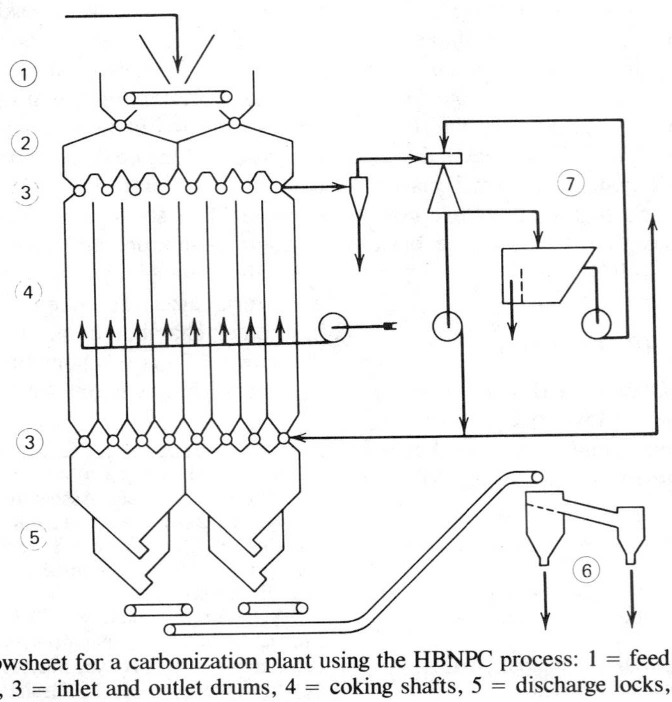

Figure 11.32 Flowsheet for a carbonization plant using the HBNPC process: 1 = feed of briquettes, 2 = charging bunker, 3 = inlet and outlet drums, 4 = coking shafts, 5 = discharge locks, 6 = separation of briquettes from coke breeze, 7 = gas purification.[155]

In the cooling zone heat from the briquettes is transferred to the entering flushing gas. Figure 11.32 shows the schematic flowsheet of one carbonization plant.

The briquettes leaving the furnace have only 2–3% volatile matter. They have great strength and a low degree of porosity.

11.4.3.13 *ISCOR Process*

The ISCOR (Iron and Steel Industrial Corporation, Ltd.) process developed in South Africa uses predominantly weakly caking coals that are briquetted with pitch and then coked in horizontal chamber ovens.[135,157] The coals usually have a swelling index of 1.5. The green briquettes partly agglomerate in the oven; therefore discharging causes no difficulties. Coke with satisfactory strength is obtained and has been successfully used in blast furnace tests.

11.4.3.14 *AUSCOKE Process*

The AUSCOKE process, developed by the Broken Hill Proprietary Company Ltd. in Australia,[158] is a multistage process. Compacts—which may be briquettes bound with a small percentage of binder—are individually heated in a fluidized bed so that they pass through the sticky plastic stage, then aggregate to form a feed that is suitable for conversion to a high-temperature coke. The pilot plant at Port Kembla (100 tonnes/day) used a shaft furnace to obtain the final temperature of 1050°C. The end product is a true coke, having undergone the normal plastic stage, semicoke, and high-temperature coking transformations. Short trial in the blast furnace was very satisfactory.[159]

11.4.3.15 *HBN (Oziole) Process*

In the HBN (Houillières du Bassin du Nord) process anthracite or low-volatile coal briquettes are carbonized and used as smokeless fuels.[160] The Fuvo-Erim flushing gas kiln is used for carbonization. The volatile material produced during the thermal treatment is burned with added air and supplies the heat of combustion.

Table 11.17 Compositions of the Briqueting Mixture and of the DKS Green Briquette Made from It[163]

Raw Material	% in Mixture
Coking coal	5–10
Noncaking coal	70–85
Coke breeze, petroleum coke	0–10
Tar and pitch as binder	10

Briquette Property	Value
Total moisture (%)	8–10
Ash (%, mf)	7–9
Volatile matter (%, mf)	22–29
Sulfur (%, mf)	0.4–0.6
Strength (N)	3500
Dimensions (mm)	53 × 53 × 35
Weight, moist (g/briquette)	65–66

11.4.3.16 *DKS Process*

The DKS process was developed in Japan. In the first step binders are used to form briquettes on roll presses. According to Sugasawa,[161] the green briquettes may be bound with a special "petroleum caking substance," as well as with tars and pitches[162]; about 10% binder is needed. Noncaking coals are used as feed material for the process. In many cases, mixtures of noncaking coals with additions of caking coal, coke breeze, or petroleum coke are preferred.

Table 11.17 shows the composition of the briquetting mixture and properties of green briquettes made from this mixture.[163] The green briquettes are coked in inclined-floor, horizontal chamber ovens for about 10 hr.

Figure 11.33 is a schematic of the process cooperatively developed by the Didier Engi-

[157] C.C. Grange, *Coal, Gold, Base Min. S. Afr.*, **18** (6), 67–71 (1971).

[158] T. G. Callcott, "The Auscoke Project," paper presented at Newcastle Conference, May–June 1972, p. 177.

[159] P. F. Snowden, *BHP Tech. Bull.*, **20** (1), 2-5 (April 1976), Melbourne, Australia.

[160] L. Soubrier, paper presented at Information Meeting of EGKS, Luxemburg, April 1970, published Central Office for Information and Documentation of European Associations, Luxemburg.

[161] K. Sugasawa, paper presented at 8th Regular Meeting of Committee on Technology, March 29–April 2, 1976 International Iron and Steel Institute, Brussels.

[162] S. Gomi, R. Takabashi, Y. Kiritani, T. Amagi and Y. Nishimura, paper presented at 9th World Petroleum Congr., Tokyo, May 1975.

[163] K. Sugasawa, K. T. Tasaka, Y. Sunami, K. Hino and C. Flockenhaus, *Stahl Eisen*, **93**, 24–30 (1973).

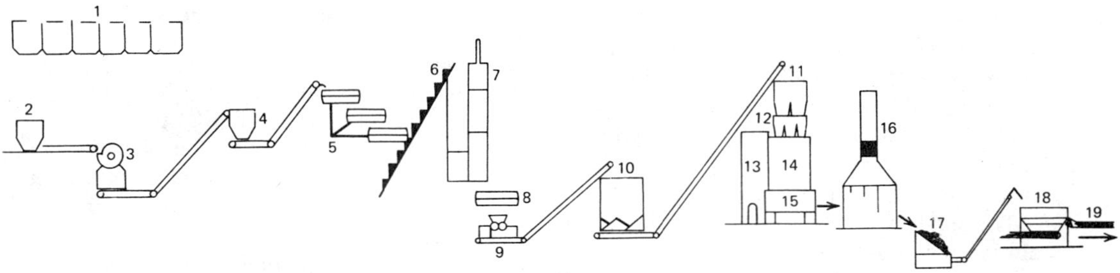

Figure 11.33 Flowsheet of the DKS process: 1 = coal hopper, 2 = binder hopper, 3 = crusher, 4 = metering hopper, 5 = mixer, 6 = chainlift, 7 = pug mill, 8 = feeder, 9 = briquetting press, 10 = storage hopper, 11 = feeding hopper, 12 = charging car, 13 = recuperator, 14 = coke oven, 15 = quench car, 16 = quench tower, 17 = slide, 18 = formcoke screen, 19 = to formcoke storage.[163]

neering GmbH, West Germany, the Keihan Rentan Kogyo Co. Ltd., and the Sumitomo Metal Industries, Ltd. (DKS).

The formcoke made from the green briquettes has been tested satisfactorily in large-scale experiments in blast furnaces.[164]

11.5 HOT BRIQUETTING

11.5.1 Fundamentals of Hot Briquetting

In hot briquetting the coal is shaped at a temperature in its softening range, generally between 400 and 500°C. If only one component is briquetted while hot, therefore 100% of the coal is shaped in the softened condition, this hot briquetting can be termed a binderless briquetting process.

In the two-component process, only one part of the coal—usually about 70% of the feed—is carbonized. The remaining 30% is mixed, either cold or preheated, with the hot, low-temperature coke and briquetted at 400–500°C in the softening zone of the added coal. This coal, which must have caking properties, provides the binding agent. It must soften in a manner that will facilitate incorporation of the grains of low-temperature coke into the briquette structure.

The caking coal is usually ground to a particle size of less than 10 mm. If moist, it may be thermally comminuted to the desired particle size by rapid heating.[165] The operating conditions and the composition of the feed materials determine the properties of the hot briquettes.[166]

While the feed coal is undergoing low-temperature carbonization, and the resulting coke is being mixed with the caking coal, gas, tar, and oil are liberated. The gas may be rich or lean, depending on the low-temperature carbonization process selected.[135] The tar usually has a high dust content, and its handling and recovery or conversion to gas and coke is difficult. For economic reasons it is desirable to recycle the tar into the process.

After leaving the briquetting press, the briquettes are slowly cooled to about 200°C and finally sprayed with water or passed through a water bath. The method by which the briquettes are cooled has a strong influence on their properties. If cooling is suitably controlled, the strength of the briquettes can be increased.[167]

Increased product quality is being demanded of the hot briquetting processes, especially if the hot briquettes are produced as formcoke for blast furnace use. Attempts are being made to increase the types of coal that may be used for the known processes, to use tar and gas as economically as possible, and

[164] N. Nakamura and H. Matsuoka, paper presented at 8th Regular Meeting of Committee on Technology, March 29–April 2, 1976, International Iron and Steel Institute, Brussels.

[165] W. Goosens and W. Herrmann, paper presented at 13th Biennial Conf. Institute for Briquetting and Agglomeration, Colorado Springs, August 1973, Institute for Briquetting and Agglomeration, Laramie, Wyo.

[166] G. V. Speranskaja, N. P. Kazakevich, L. A. Kleimenova and L. P. Malashenko, *Khim. Tverd. Topl.*, No. 2, 52–59 (1971).

[167] F. H. Beckmann, *Glueckauf*, **109**, 625–628 (1973).

to decrease the content of volatile matter in the final product. For the one-component hot briquetting process, higher volatile coals are preferable. In the two-component process, about 70% of the feed coal (to undergo low-temperature carbonization) may be coal at any degree of coalification (rank).[105]

The quality of the low-temperature coke always has a strong influence on the structural strength of the hot briquette. Only highly caking coals can be used as binding components. Their free-swelling index should be higher than 4.5, and they should have a volatile matter content of more than 20%.[135]

The interaction of the readily melting and the nonmelting petrographic components in the plastic range was studied,[168] as was the influence of the degree of coalification and the preheating rate on the temperature and the value of the maximum plasticity of the coal.[169] Semibituminous coals were found to be suitable as binder components if high temperatures are used. Valuable results were also obtained in investigations of the changes during thermal treatment in the preplastic, the actual plastic, and the postplastic zones of briquettes during carbonization.[170]

In the Soviet Union[171] the best conditions for the hot briquetting of mixtures of fine anthracite and high-volatile coal were studied in an attempt to widen the base of coals suitable for hot briquetting. Optimal results were achieved with 80% anthracite and 20% high-volatile coal ground to less than 0.5 mm. German[119] and Soviet[172] investigators tested the possibility of making formcoke with brown coal, sometimes combined with bituminous coal.

The best prospects for commercial use will be the hot briquetting processes that produce a plastic matrix, thus forming a uniformly porous structure in the manufactured formcoke.[173]

[168] *Coke Chem. USSSR,* No. 4, 18–23; No. 5, 17–20 (1974).
[169] M. Goedkopp, *Rev. Mine. Paris,* No. 1, 127–135 (1960).
[170] A. D. Dainton and W. G. Kaye, *J. Inst. Fuel,* **32,** 571–578 (1959).
[171] G. V. Efaskin, *Ugol Ukr.,* **6** (8), 47–48 (1962).
[172] A. K. Konovalov, *Tr. IGIAN, SSSR, Moskow,* **20,** 145–154 (1963).
[173] I. Barbu, I. Stefanescu and K. Panaitescu, *Coke Chem. SSSR, Moskow,* **6,** 20–24 (1974).

11.5.2 Sapozhnikov Process (IGI Process)

The well-known Sapozhnikov hot briquetting process, also known as the IGI process, was developed in the Soviet Union beginning in 1955 and continually improved over a long period of time. The flowsheet shown in Fig. 11.34 is now in use.[115,174,175]

The dried comminuted coal is heated to softening temperature (420–450°C) in a fluidized bed and transported by way of a cyclone into an intermediate bunker, where it remains for 1–10 min. The hot coal is then precompressed by means of an extruder and briquetted on roll presses at a pressure of only 0.5 N/mm². The briquettes thus produced are used as a smokeless fuel. For use as blast furnace coke, the hot briquettes are

[174] L. M. Sapozhnikov, *Tr. IGIAN SSSR, Moskow,* 1959.
[175] *Tr. IGIAN SSSR, Moskow,* **20** (1963).

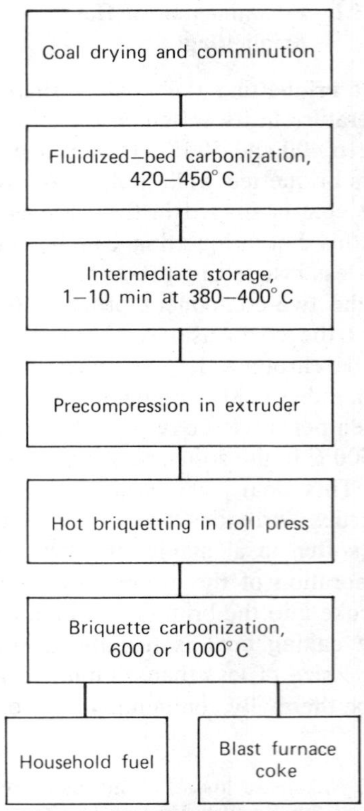

Figure 11.34 Flowsheet of the Sapozhnikov process.[176]

carbonized at 600 and 1000°C, respectively, in vertical or horizontal chamber ovens.

The characteristic phases of the process are as follows:[176]

1. Coal is heated to softening temperature in a few seconds to retain its optimal caking properties.
2. Softening temperature is maintained in the coal or briquette for at least 1 min.
3. Material is briquetted at a low pressure 0.2–1.0 N/mm².
4. Hot briquettes are carbonized at a heating rate of 1–3°C/min to avoid crack formation.

The selection of temperature and briquetting pressure depends on the properties of the feed coal and on the desired quality of the final product.

A pilot plant with an output of 5–10 tons/hr operates as shown in the schematic diagram in Fig. 11.35. The individual steps of this pilot plant differ from the flowsheet shown in Fig. 11.34. The hot briquettes produced from a coking coal 0–3 mm in size are heated at a slow rate in indirectly fired vertical retorts to 700°C.[177] In Charkow, U.S.S.R., a plant with an output of 400 tons/day has been built.[150]

[176] V. Stucklik, *Paliva,* **46,** (4), 141–143.

[177] W. I. Riva, H. P. Hudson, I. E. Landon and I. E. Walsch, *Can. Min. Metal. Bull.,* **57,** 52–56 (January 1964).

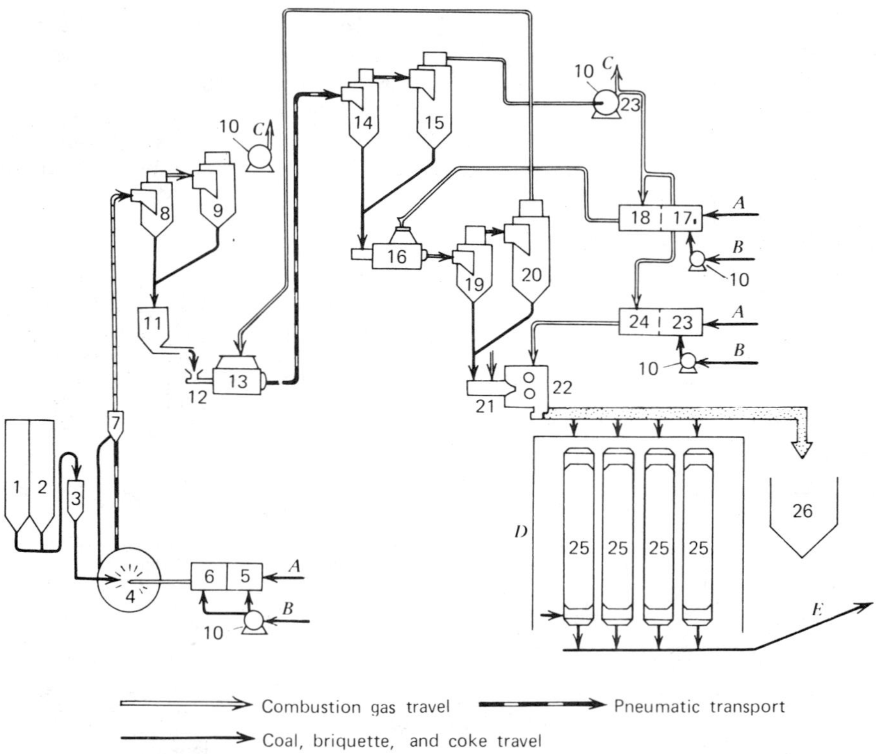

Figure 11.35 Flowsheet for Sapozhnikov process of manufacturing formcoke: 1, 2 = coal bunkers, 3 = intermediate hopper, 4 = mill with combustion gas blast, 5 = combustion chamber, 6 = mixing chamber, 7 = coarse grain separator, 8 = cyclone, 9 = multicyclone, 10 = blower, 11 = receiver, 12 = conveyor-type weigher, 13 = coal preheater, 14 = cyclone, 15 = multicyclone, 16 = preheater, 17 = combustion changer, 18 = mixing chamber, 19 = cyclone, 20 = multicyclone, 21 = worm feeder, 22 = briquetting press, 23 = combustion chamber, 24 = mixing chamber, 25 = briquette coking chambers, 26 = auxiliary bunker for uncoked briquettes, *A* = coke-oven gas, *B* = air, *C* = combustion gases, *D* = quenching water, *E* = dispatch.[115]

11.5.3 NCB Process

Development of a hot briquetting process in Great Britain was stimulated by the desire to produce high-volatile reactive smokeless fuels for open fires, and similar uses.[178]

In the National Coal Board (NCB) hot briquetting process[179] coal is carbonized at a low temperature and briquetted while still hot without binder or added caking coal; the process flowsheet is shown in Fig. 11.36. Coal with about 37% volatile matter is ground to a size smaller than 1.2 mm, dried, and carbonized in a fluidized bed at about 420°C. The residence time in the fluidized bed is roughly 35 min. The hot, low-temperature coke passes through heated conduits to the briquetting presses. The briquettes are cooled in the absence of air, then quenched with water. They weigh about 75 g, have a volatile matter content of about 20–23%, and are being used as smokeless household fuels.

Two commercial plants were built in Great Britain to exploit the NCB process. The Homefire Plant, with a capacity of about a million tons/year, uses reciprocating presses to produce dense hexagonal briquettes weighing 170 g.[180]

The Roomheat Plant, with a capacity of about 250,000 tons/year uses double-roll presses to produce a briquette weighing about 75 g.

[178] S. I. Evans, *Colliery Eng.*, **35,** 523 (1958).
[179] W. I. Jones and J. Ovens, *Fuel,* **35,** 404 (1962).

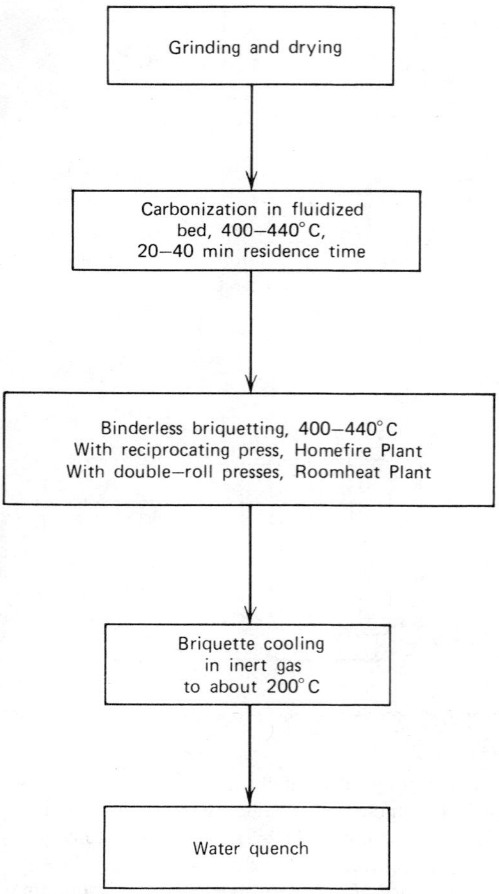

Figure 11.36 Flowsheet of NCB hot briquetting process. *Source:* D. G. Madley, Coal Research Establishment, NCB, Cheltenham.

11.5.4 BFL Process

The BFL hot briquetting process was developed by Bergbau-Forschung GmbH, Essen, in cooperation with the Lurgi Mineralöltechnik GmbH, Frankfurt, West Germany.[181,182] The process uses two-component hot briquetting. As shown in Fig. 11.37, hot, finely comminuted, low-temperature coke is mixed with caking coal, the second component. The ratio and the temperature of the low-temper-

[180] W. R. Watson, *Min. Eng.*, **127,** (93), 518–528 (1967–1968).
[181] W. Reerink and W. Peters, *Brennstoff-Chemie,* **46,** S330–343 (1965).
[182] W. Peters, *Glueckauf,* **103,** 1273–1279 (1967).
[183] E. Ahland and W. Peters, *Rev. Metall.*, **71,** 653–658 (September 1975).

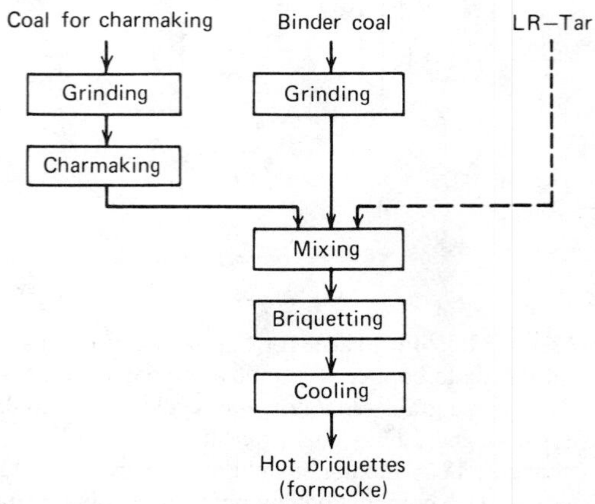

Figure 11.37 Scheme of BFL hot briquetting process.[183]

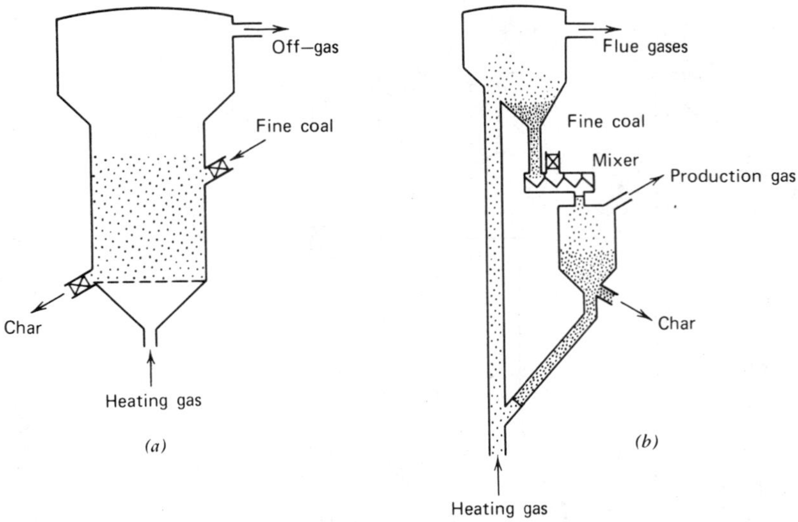

Figure 11.38 (*a*) Fluidized-bed process and (*b*) closed circuit of a heating gas carrier,[7] as used in the production of fine-grained char by the BFL process.[183]

ature coke must be selected so that the binder component shows the optimum degree of softening for hot briquetting. The plastic mixture is then briquetted on double-roll presses at 420–480°C. The hot briquettes are either cooled with recycle gas and quenched with water, or reheated to 600°C before cooling. If the coal used as binder does not cake strongly enough, another binder—for instance, tar from the carbonization process—can be added to the briquetting mixture to improve briquette strength.[184]

In manufacturing the low-temperature coke, two processes may be used, as shown in Fig. 11.38. In a simple process the coal is carbonized at about 750°C in a fluidized bed. The off-gas in this case is a mixture of fluidizing and carbonization gases, and has a low heating value.

If a rich gas is desired, the LR (Lurgi-Ruhrgas) process is used for the low-temperature carbonization. This process separates the heating gas from the gas obtained during carbonization. The pulverized feed coal is blended in a mixing chamber with 10 times its amount of recycled coke, heated to 750°C. The coal is degassed within seconds. The mixture then enters a second gas evolution vessel. Product gas is withdrawn, and the part of the coke not required for circulation is taken to the hot briquetting plant. This part of the coke can also be withdrawn from the collecting bin (Fig. 11.39). The coke in the recycle system is heated in an airlift to 750°C.

The BFL process may use almost all coals for low-temperature carbonization, unless their ash or sulfur content is too high. Satisfactory results were obtained with bituminous coals with a volatile matter content of 8–53%, as well as with a German brown coal.

Caking bituminous coals are suitable as the binder component. Whether a given coal has strong enough caking properties to serve the purpose or whether a binder such as tar must also be added can be determined with assurance in a hot briquetting laboratory experiment.[183,185]

After development work was done for about 10 years with small equipment, a demonstration plant with an output of 12.5 tons/hr was built at the Prosper coke-oven plant in West Germany (Fig. 11.39). In this plant,

[184] W. Peters, E. Ahland and J. Langhoff, paper presented at information meeting "Technology and Development of Bituminous Coal Coking," Luxemburg, April 23–24, 1970, published by Central Office for Information and Documentation of European Associations, Luxemburg.

[185] E. Ahland, J. Lehmann, W. Peters, W. Reerink and J. Langhoff, Instituto Latinoamericano del Fierro y et Acero-*Coal Cong.*, Mexico City, July 26–30, 1976.

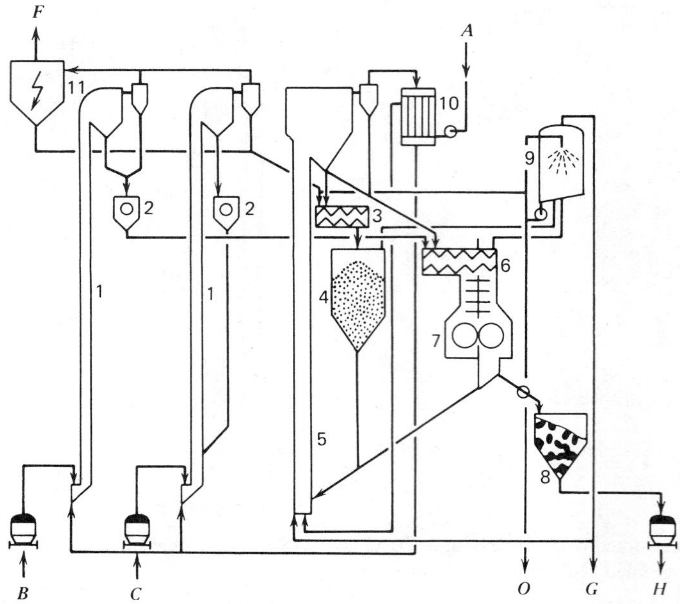

Figure 11.39 Flowsheet of the Prosper plant (BFL process): 1 = coal flash dryer, 2 = crusher, 3–5 = LR system (mixer, hopper, airlift), 6 = mixer, 7 = press, 8 = cooler, 9 = condensation, 10 = air preheater, 11 = electrostatic precipitator, *B* = binder coal, *C* = coal for char making, *O* = oil, *G* = coal gas, *H* = hot briquettes, *F* = flue gas, *A* = air.[155]

the LR process is used to manufacture the low-temperature coke.

A double-roll press with a diameter of 1.4 m is used to form the hot mixture into 20–300 cm^3 briquettes. The properties of BFL hot briquettes are compared with those of conventional coke in Table 11.18.

Products of the BFL process have been used successfully as smokeless fuel, and have substituted for conventional coke in cupolas and blast furnaces in large-scale experiments in Germany, Belgium, and England.[186]

The British Steel Corporation built a BFL hot briquetting plant in England with a production of 27 tons/hr.[187]

Table 11.18 Properties of BFL Hot Briquettes and Conventional Coke[155]

	BFL Hot Briquette		BF Coke III
Weight per piece (g)	101	48	—
Porosity (%)	39	49	50
Volatile matter (%)	10.1	7.1	1
Ash (%)	6.8	7.0	8.6
Sulfur (%)	0.91	0.83	0.86
Bulk density (kg/m^3)	561	508	437
Strength			
Micum (DIN 51717)			
M_{40} (%)	92.6	91.7	81.2
M_{10} (%)	6.7	7.8	6.5
IrSID			
I_{20} (%)	79.0		80
I_{10} (%)	20.8		18

11.5.5 Stamicarbon Process

The hot briquetting process[188,189] developed by the Staatsmijnen, the Netherlands, produces smokeless fuels. The principle of the process is shown in Fig. 11.40. Anthracite fines or weakly caking coal fines are heated to 550°C in a fluidized bed. Inert gas transports them into a cyclone, where they are mixed with caking coal (about 20% of the total). Sticking and plugging can be avoided if the caking coal component is added im-

[186] A. J. H. Metcalf, "Formed Coke," *COMA Yearb.*, 1974, p. 163.

[187] E. Rammler, *Stahl Eisen*, **95**, 433–442 (1975).

[188] British Pat. 787147, 1957.

[189] British Pat. 907107, 1962.

[190] G. Kardaun and D. Viets, *Gas Coke*, **25** (289), 249–253 (1963).

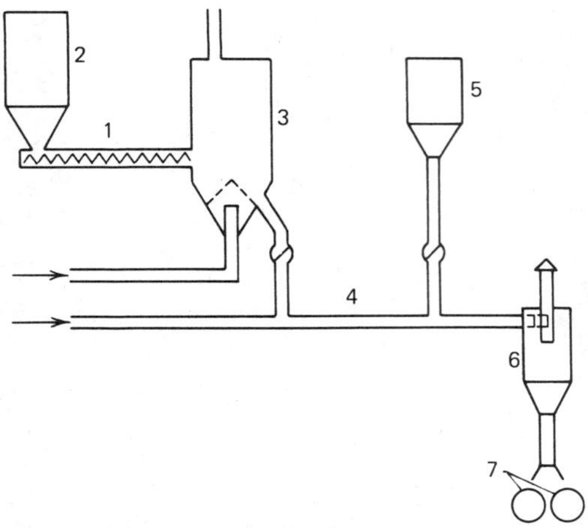

Figure 11.40 Diagram of Stamicarbon plant: 1 = screw feeder, 2 = anthracite bunker, 3 = fluidized-coal reactor, 4 = duct for hot anthracite transport in inert gas stream, 5 = coking coal bunker, 6 = cyclone, 7 = briquetting press.[190]

mediately ahead of the cyclone separator or in the cyclone itself; softening of the caking coal in the mixture then occurs only after passage through the cyclone. The hot mixture is briquetted at about 420°C on roll presses with a briquetting pressure of about 41 N/mm².

The process was developed by the Staatsmijnen to a large pilot-plant scale with an output of 2.5 tons/hr.

11.5.6 Ancit Process (EBV)

The Ancit process is a further development of the Stamicarbon process. In 1966 and 1967 the Eschweiler Bergwerks-Verein (EBV), West Germany, constructed a large experimental plant producing briquettes at 10–12 tons/hr. The plant was built with knowledge gained in the Stamicarbon process after patent rights covering that process were acquired.

In the meantime the Ancit process has been developed for large-scale production. Initially only production of smokeless fuels was attempted. Subsequent experiments have shown that the product can be used as a reduction medium in the blast furnace process.[191]

The Ancit hot briquetting process uses two components, an inert noncaking component such as low-volatile coal, anthracite, or coke fines, and a binder component consisting of caking coals.

The main feed coal (about 70%) is comminuted wet by being heated within tenths of a second to about 600°C. The off-gas from this reactor heats the binder coal (about 30% of the total feed) in another, smaller flow reactor.[192] Great care is taken to maintain the heat of the coking coal below its softening point, to avoid any possibility of sticking or plugging of the lines and cyclone that might occur in an equivalent stage of the Stamicarbon process. The two coals, comminuted at different temperatures, are mixed in a vertical, cylindrical mixer. The mixture reaches a temperature between about 460 and 520°C. At this temperature the binder coal is softened. The mixture is then briquetted on double-roll presses.

The gas produced has a heating value of 700–800 kcal/(Nm³) [2.9–3.4 MJ/(Nm³)]. In addition, the mixer yields small amounts of higher Btu gas with a heating value of about 6500 kcal/(Nm³) [27.2 MJ/(Nm³)].

The cooled briquettes (18 g, 36 × 35 × 19.5 mm) have a point compressive strength of about 850 N and contain 8% volatile matter. They are being used as smokeless fuel.

Two developments have improved the process still further and made it possible to use higher volatile coals for the Ancit process. Posttreatment of the briquettes in thermally insulated vessels immediately after they leave the press results in considerably increased briquette strength.[193] Such Ancit briquettes are suitable for use as blast furnace coke.[194] Low-temperature thermal pretreatment in a short suspension flow reactor can make higher volatile coals with a swelling index of up to 5–6 suitable as inert components in the Ancit process.

Figure 11.41 is a schematic flowsheet of the Ancit hot briquetting process showing posttreatment of the briquette. Ancit makes an excellent smokeless fuel, and formcoke made according to the Ancit process may replace conventional coke in blast furnaces.

[191] W. Goosens, *Glueckauf,* **109,** 521–524 (1973).

[192] W. Hermann, *Glueckauf,* **109,** 714–721 (1973).

[193] F. H. Beckman, *Glueckauf,* **109,** 625–628 (1973).

[194] W. Goosens, W. Zischkale, and R. Reiland, *Stahl Eisen,* **92,** 1039–1044 (1972).

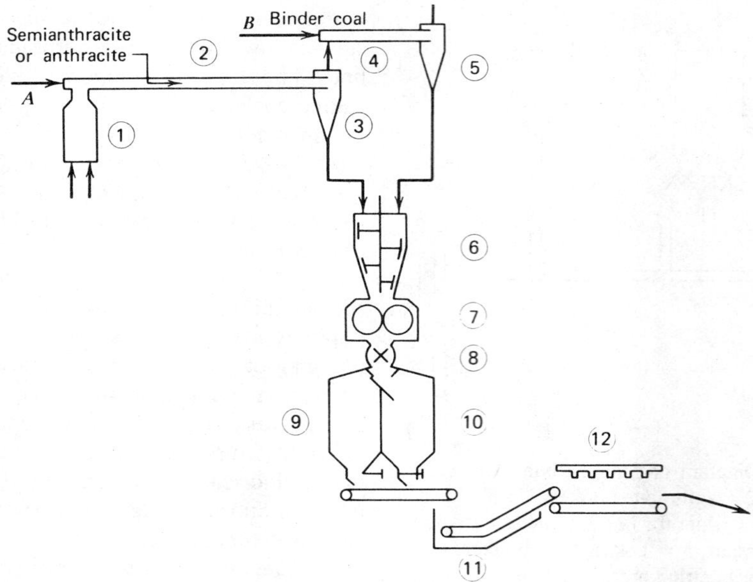

Figure 11.41 Flowsheet of the Ancit plant: 1 = combustion chamber, 2 = conduit (flue stream) with two anthracite inlets, 3 = anthracite cyclone, 4 = conduit with binder coal inlet, 5 = binder coal cyclone, 6 = mixer, 7 = double-roll press, 8 = discharger, 9–10 = hardening vessel, 11 = immersion cooling treatment, 12 = water sprays.[155]

11.5.7 CRIJ Process

The three-stage hot briquetting process developed by the Coal Research Institute of Japan (CRIJ) involves fluidized-bed preheating, hot briquetting, and high-temperature carbonization of the briquette.[163] The Coal Mining Research Center built an experimental plant in Japan with a briquette production capacity of 2 tons/hr.

Figure 11.42 is the flowsheet of the experimental plant. Using the "double-tower" method, the noncoking coal and the coking binder coal are heated in a high-temperature preheater and in a low-temperature preheater, respectively. Thus the binder coal is heated to a lower temperature than the noncaking coal, thereby avoiding agglomeration and plugging in the preheater. The temperatures are selected according to the softening behavior of the binder coal and the temperature needed for the hot briquetting of the mixture. The two components are mixed thoroughly in a paddle mixer before they enter the press. The hot briquettes are then carbonized in a circular furnace with heating gas. Table 11.19 shows the composition of two coal mixtures and the properties of the formcoke made from these mixtures.

11.5.8 NIPR Process

A small experimental plant producing briquettes at a rate of 2 tons/day was built by the National Institute for Pollution and Re-

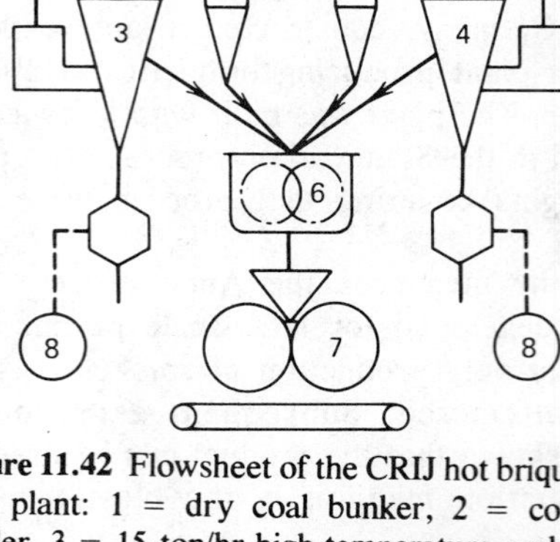

Figure 11.42 Flowsheet of the CRIJ hot briquetting test plant: 1 = dry coal bunker, 2 = constant feeder, 3 = 15-ton/hr high-temperature preheater, 4 = 10-ton/hr low-temperature preheater, 5 = cyclone, 6 = mixer, 7 = double-roll press, 8 = combustion chamber.[195]

Table 11.19 Properties of Formcoke[195]

	Mixture A	Mixture B
Blending Ratio (%)		
Caking coal (Australia)	40	40
Weakly caking coal (Japan)	10	10
Slightly caking coal (Japan)	—	30
Noncaking coal (USSR)	50	20
Properties		
Porosity (%)	39.1	39.9
Apparent density (g/cm³)	1.18	1.19
True density (g/cm³)	1.93	1.97
Drum index (DI^{30}_{15})	94.5	93.7

sources (NIPR) in Japan.[163] This process is similar to the CRIJ process described in Section 11.5.7 except that the hot briquettes are carbonized in a shaft furnace.

According to Yoshida,[195] the mixture of the feed coals has a volatile matter content of 25% and consists of a low-volatile coal and a coal with a higher volatile matter content. The hot briquettes leaving the press contain 17% volatile matter; after carbonization in the shaft furnace, they contain 2% volatiles and 13% ash.

11.6 PELLETIZING

11.6.1 Fundamentals of Pelletizing

The materials used for pelletizing fuels generally have a size consist below 0.1 mm; many of the particles are smaller than 0.060 mm. Addition of liquid on rotating disks or drums agglomerates the fuel to spherical shapes. The strength of the pellets depends on the water content of the agglomerated material. The strength is greatest when the optimum action of the capillary forces of the water is obtained.

Fuel pellets agglomerated with water do not have great strength; it amounts to about 0.49–0.98 N/mm². These pellet shapes cannot withstand major stresses during transport or when subjected to high loads. To increase the strength of the pellets, inorganic binders are often added, or sulfite liquor is added to the pelletizing feed.

The role of pelletizing is minor in comparison to that of coal briquetting. However the pelletizing of very fine coal and coke has recently gained increased attention[196–198] because of its economic attractiveness. Coal pellets can be used immediately after drying, or carbonized to improve their properties. Like hot briquetting, hot pelletizing can be used to agglomerate superfine fuels.

Developments in the area of pelletizing fine ores have been successfully applied to the pelletizing of fuels.

11.6.2 Cold Pelletizing

For some time coal sludges and other residues from the preparation plant have been commercially pelletized. Egan[199] reported the conversion of coal flotation residues to pellets in a Kentucky mine. Filter cakes are converted to pellets measuring 12–19 mm in diameter and containing less than 1.5% moisture. The author describes the plant equipment and experience gained during operation of the plant.

The principles of fuel pelletizing have changed very little over the last two decades. Nevertheless, some possibilities for further development have been investigated. A group of workers[200] studied pelletizing on flat surfaces by means of three-dimensional vibrations. Lucky and Spicer[201] summarized the work of investigators in the United States, Canada, England, and Germany on coal agglomeration. Capes et al.[202] describe the preparation and pelletizing of coal.

[195] Y. Yoshida, paper presented at 12th Biennial Conf. Institute for Briquetting and Agglomeration, Vancouver Canada, August 12–14, 1971, Institute for Briquetting and Agglomeration, Laramie, Wyo.

[196] J. R. Johanson, paper presented at 9th Biennial Conf. International Briquetting Association, Denver, 1965, pp. 17–29.

[197] H. Rausch and H. Seidel, *Neue Huette,* **11,** 647–657 (1966).

[198] B. A. Onusaitis and A. I. Petrovich, *Izv. Akad. Nauk SSSR, Otd. Met. Topl.,* **4,** 42–45 (1959).

[199] H. N. Egan, *Min. Eng.,* **19** (11), 73–75 (1967).

[200] S. M. Stoev and D. Watson, *Trans. Inst. Min. Met.,* **77** (736), 14–18 (1968).

[201] P. T. Lucky and T. S. Spicer, paper presented at 9th Biennial Conf. International Briquetting Association, Denver, 1965, pp. 61–75.

[202] C. E. Capes, A. E. McIlhinney, and R. D. Coleman, *Trans. AIME Soc. Min. Eng.,* **247** (3), 233–237 (1970).

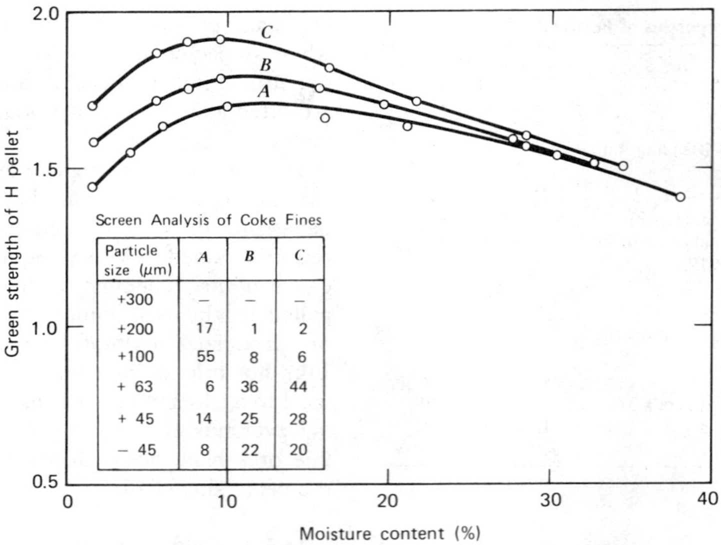

Figure 11.43 Moisture content and point strength of coke pellets without additives.[206]

Bethlehem Steel[203] operates an experimental plant for the recovery of fines from coking coal in which the filter cake is pelletized with additions of tar and sulfite liquor binders. The pellets are then dried in a fluidized bed.

Also in the United States the Rochester and Pittsburgh Coal Company, the Eastern Coal Corporation, and the Slab Fork Coal Company have pelletized fine coal on rotating plates and drums.[204]

Coke fines can also be satisfactorily pelletized. Agglomeration of coke breeze with tar as binder in West Germany has been described[205] and used in manufacturing fuel that develops little smoke.

Scientists at the Institute for Iron Metallurgy of the Technical University, Aachen, West Germany, developed a process that produces coke pellets suitable for sintering purposes from brown coal fines.[206] At their original size consist, these coke fines would have been useless for sintering.

Primary coke fines are converted to larger lumps by pelletizing. In the process, some of the properties that are important for sintering, such as reactivity and ignition properties, may change. The strength of coke pellets agglomerated with water only is small, and varies with the size consist of the coke fines used in the pelletizing, as shown in Fig. 11.43.

The coke pellets (diameter 1–3 mm) were produced on a disk rotating at 15 rpm and having a diameter of 1500 mm. The coke fines can also be agglomerated in a pelletizing drum.

Different additives have been tested in attempts to increase pellet strength. Table 11.20 shows several additives and the amounts in which they were added to the pelletizing mixture. The strength of pellets produced with additives increases during storage. An advantageous increase in pellet strength was obtained by adding lime. These

Table 11.20 Additives in Pelletizing of Coke Fines[206]

Additives	Amounts Used (wt %)
Filter ash	5, 10, 15, 20, 30, 40, 50
Cement	10
Gypsum	10
Bentonite	10
Iron ore fines	10, 20, 30, 40, 50
Iron sponge fines	6
Burnt lime	5, 10, 20, 30, 40, 50

[203] R. W. Utley, Bethlehem Steel, U. S. Pat. 3,655,350, April 11, 1972.

[204] W. Engelleitner, *Agglomeration—A*, 1975 Update, reprinted from *Miner. Process.*, 1975.

[205] H. Krug and W. Feiler, *Bergbautechnik*, **20** (9), 474–477 (1970).

[206] W. Wenzel, H. W. Gudenau, and S. A. Dwiwanto, *Aufbereitungstechnik*, No. 4, 194–197 (1975).

limed pellets were dried in a vessel in which the sintering plant off-gases were blown through the wet green pellets.

11.6.3 Carbonization of Pellets

The carbonization of coke pellets is attracting increased attention.[208] A Canadian[177] process involves a rotating drum 1.5 m in diameter and 2.4 m long, in which caking coal fines with a high content of superfine material are converted into pellets with the addition of about 2% bentonite. The mixture must have a moisture content of more than 18–23%. The pellets are dried in an 80°C stream of air, then carbonized in a shaft furnace. The best results were obtained when the mixture contained 60% caking coal. Larger amounts led to agglomeration of the pellets. To avoid this, about 10% coke breeze is frequently added to the mixture. The correct heating rate of the pellets has an important influence on the behavior of the charge in the oven.

Pelletizing of coal-ore mixtures, with an addition of lime if necessary, has been reported in Australia.[115] In this process the pellets are heated to 550–650°C in a "rilling furnace" and subsequently carbonized in a shaft furnace with sand as heat carrier.

A novel process for pelletizing caking coal fines and carbonizing the pellets was developed in West Germany.[207] The first step in this process is the conversion of caking coal fines into pellets. In the second stage the pellets are coated with hematite ore. The ore-covered pellets are then coked; during this process they soften and sinter together, forming a coherent coke body that fills the interior of the horizontal chamber oven. Under stress (e.g., when the mass is pushed from the oven into the quench car), the coke body breaks into the original pellet shapes. The hematite layer between the pellets destroys the caking properties in the surface layer of the pellets, thus permitting breakage to occur between the individual pellets. Coke of uniform size is obtained in the shape of uniform polyhedrons, called "polyhedron coke."

Addition of heavy oil to the coal increases the stability of the pellets in the oven. The amount of Fe_2O_3 is critical, and a definite level exists that cannot be exceeded if the coke is to retain its polyhedron shape. This shape is important, since it permits the coke to form a coherent body that can be pushed from the coke-oven chamber. In experi-

[207] W. Wenzel, H. W. Gudenau and I. Bernt, *Aufbereitungstechnik,* No. 5, 232–236 (1976).

[208] "Pelletized Coke," *Coal Age,* 77(10), 148–149 (October 1972).

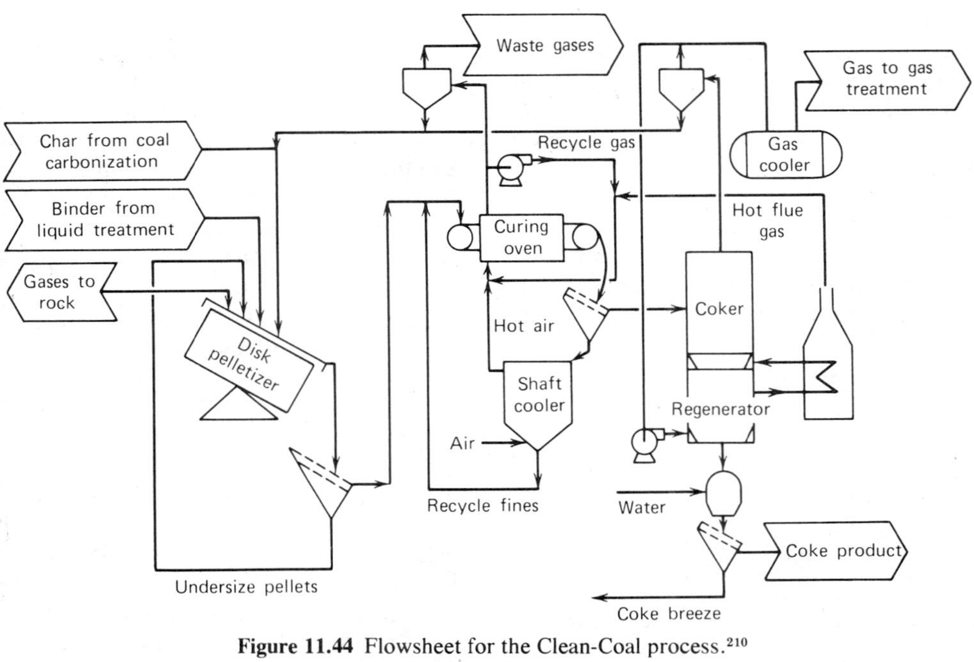

Figure 11.44 Flowsheet for the Clean-Coal process.[210]

Table 11.21 Properties of Formcokes and Metallurgical Coke[210]

Property	Metallurgical Coke	Clean-Coal Coke Pellets	FMC Formcoke
Tumbler strength[a] above 6.35 mm (%)	85	85	83
Apparent density (g/cm³)	0.94	1.0	0.90
True density (g/cm³)	1.90	1.93	1.69
Porosity (%)	50	48	47
Chemical analysis			
Volatile matter (%)	0.8	1.0	6.2
Fixed carbon (%)	92.2	91.5	87.1
Ash (%)	7.0	7.5	6.7
Sulfur (%)	0.7	0.5	0.75

[a] 9.97-kg (22-lb) sample, ASTM tumbler test.

ments, coals with 14–32% volatile matter were suitable as feed coal.

The United States Steel Corporation developed another pelletizing process called the Clean-Coal process.[209] It is a combination of the chemical conversion of the coal and the coking process (Fig. 11.44).

Low-temperature carbonization of the coal is combined with coal hydrogenation. The use of a hydrogen-rich recycle gas as fluidizing gas and of pressures above 165 atm in the fluidizing chamber permits reduction of the sulfur content of the feed coal from more than 2 to 0.5% in the coke product. The coke fines are then pelletized, using the binder oil obtained in the process. The pellets pass through a curing oven in which they are hardened with hot recycled coke fines. This is followed by calcination of the pellets in a shaft furnace at high temperature. Conditions necessary for obtaining coke pellets of high quality are sufficient fineness of the coke before shaping, addition of sufficient binder, and thorough mixing of coke fines and binder.[210]

Table 11.21 compares the properties of coke pellets with FMC formcoke and with metallurgical coke. Coke pellets for tests in blast furnaces will be produced in a 5-ton/hr plant.[211]

11.6.4 Hot Pelletizing

A hot pelletizing process shapes the coal in the temperature range of the softening of a caking coal, possibly in the presence of a binder, using pelletizing equipment. The advantages of hot pelletizing[210] include its simplicity and favorable economics.

A third advantage is the possibility of mixing in larger amounts of binder than can be done in briquetting. The binder component cracks thermally or cokes at the temperatures that are used during pellet shaping. Recently remarkable progress has been made in hot pelletizing, especially in the United States. The objective of the work is the production of formcoke for blast furnace use.

Consolidation Coal Company developed the Consol formcoke process (CCC process),[212] which is especially suited for converting the Eastern high-fluidity coals into formcokes for blast furnace use. Finely comminuted, preheated caking coal mixed with comminuted recycle coke (with the possible addition of recycle tar or coal extract) is fed into a heated pelletizing unit. Since the temperature is in the range of coal softening, spherical bodies are formed with a diameter of about 20–70 mm. These coke pellets are calcined in a subsequent flushing gas furnace.

New developments led to a desulfurization of the coke pellets during the calcining process by adding hydrogen under pressure.[213]

Consolidation Coal, together with four U.S. steel companies, built an experimental pelletizing plant in Library, Pennsylvania, with a pellet output of 30 tons/day. In the meantime, a 450-ton/day plant has been built in Sparrows Point, Maryland, to produce larger quantities of coke pellets for blast furnace tests.

[209] K. A. Schowalter and N. S. Boodman, *Chem. Eng. Prog.*, **70** (6), 76–82 (June 1974).

[210] R. W. Shoenberger, *Prepr., Am. Chem. Soc., Fuel Div.*, **20**(4), 46–69 (August 24–29, 1975).

[211] K. C. Krupinski, T. F. Johnson and R. J. Osterholm, *Prepr., Am. Chem. Soc., Fuel Div.*, **20**(4), 33–45 (August 24–29, 1975).

[212] Consolidation Coal, U. S. Pat. 507,375, 1963.

[213] E. H. Reichl, *Glueckauf*, **109**, 165–178 (1973).

CHAPTER 12

Fundamentals of Coal Pyrolysis and Hydropyrolysis

JACK B. HOWARD

Jack B. Howard *has been active in coal pyrolysis research since 1960–1961, when he conducted a series of coal carbonization experiments as a student in the Department of Mining and Metallurgical Engineering at the University of Kentucky. He subsequently studied rapid pyrolysis and associated ignition and combustion phenomena in pulverized coal flames as a graduate student in the Department of Fuel Technology at the Pennsylvania State University. Since 1965 Dr. Howard has been a faculty member of the Department of Chemical Engineering at Massachusetts Institute of Technology, where his research has included basic studies of coal pyrolysis and hydropyrolysis. The MIT work has been directed toward a better understanding of the complex chemical and physical phenomena governing the yields and distribution of products under conditions of potential commercial interest.*

12.1 INTRODUCTION

12.1.1 Previous Reviews

Considerable research on coal pyrolysis has been conducted over the years, but the results are widely dispersed in the scientific and technical literature. For example, the early work is reported predominantly in combustion and carbonization literature, whereas much of the more recent research, especially that concerned also with coal-hydrogen reactions, is discussed in coal gasification and coal liquefaction literature. Too little cross-fertilization has occurred.

Several past reviews are available. The best single source is probably Lowry's[1] compilation of review articles. In that work Howard[2] reviewed coal pyrolysis reactions (mostly information obtained by slow heating techniques), and several other authors reviewed subjects pertinent to pyrolysis. Subsequently Jones[3] reviewed a number of pyrolysis studies, most of which were done at low heating rates. Similarly, pyrolysis research at low heating rates was critically evaluated by Yellow,[4] and work at high heating rates was summarized by Badzioch.[5] Essenhigh and Howard[6] reviewed the pyrolysis literature bearing on combustion and explosion phenomena in coal dusts. With regard to pyrolysis in the presence of hydrogen, the Institute of Gas Technology[7] published a review of its extensive coal gasification research, including a brief recapitulation of some of the most important hydrogasification studies. Most recently, Anthony and Howard[8] reviewed the fundamental information on pyrolysis and hydrogasification with emphasis on reaction models of potential use in engineering work. The remaining critical assessments are contained in research papers referenced in the sections that follow.

Finally, it is important to mention that the two phenomena of pyrolysis in inert gas and pyrolysis in the presence of hydrogen are often treated separately, with their interrelationship, although acknowledged, remaining unexplored. Recent work included in the review of Anthony and Howard[8] addresses this area of omission, which has become an important focal point of many research and development efforts.

12.1.2 Scope of This Chapter

This chapter covers the fundamental coal pyrolysis literature published after the review of H. C. Howard.[2] The scope includes thermal decomposition of coal in the absence of added catalyst, both in inert gas (pyrolysis) and in hydrogen (hydropyrolysis). The primary emphasis is on the extent of conversion and product distribution (as affected by time, temperature, heating rate, pressure, hydrogen partial pressure, and particle size), on the interrelationship of pyrolysis and hydropyrolysis phenomena, and on the rapid reactions anticipated to be of most interest commercially. The presentation begins with a summary of background information, including the current technological basis for interest in fundamental pyrolysis work, coal composition and chemistry in relation to pyrolysis, and physical aspects of pyrolysis behavior. The actual review begins with a critical discussion of experimental methods, thereby recognizing that reported coal pyrolysis behavior is frequently biased by the equipment and technique used. Typical data are described, with attention paid to the method of acquisition and to possible influences of technique on interpretation. Specific results, correlations, and postulated mechanisms of coal pyrolysis are then analyzed.

[1] H. H. Lowry, Ed., *Chemistry of Coal Utilization*, Supplementary Volume, John Wiley and Sons, New York, 1963.

[2] H. C. Howard, "Pyrolytic Reactions of Coal," in *Chemistry of Coal Utilization*, Supplementary Volume, H. H. Lowry, Ed., John Wiley and Sons, New York, 1963, pp. 340–394.

[3] W. I. Jones, "The Thermal Decomposition of Coal," *J. Inst. Fuel*, **37**, 3 (1964).

[4] P. C. Yellow, "Kinetics of the Thermal Decomposition of Coal," BCURA Month. Bull. **20**(9), 285 (1965).

[5] S. Badzioch, "Thermal Decomposition," *BCURA Month. Bull.* **31**(4), 193 (1967).

[6] R. H. Essenhigh and J. B. Howard, *Combustion Phenomena in Coal Dusts and the Two-Component Hypothesis of Coal Constitution* (Penn State Studies No. 31, Administrative Committee on Research, Pennsylvania State University, University Park, 1971.

[7] E. J. Pyrcioch, H. L. Feldkirchner, C. L. Tsaros, J. L. Johnson, W. G. Blair, B. S. Lee, F. C. Schora, Jr., J. Huebler, and H. R. Linden, *Production of Pipeline Gas by Hydrogasification of Coal*, Res. Bull. No. 39, Vol. 1, Institute of Gas Technology, Chicago, December 1972.

[8] D. B. Anthony and J. B. Howard, "Coal Devolatilization and Hydrogasification," *Am. Inst. Chem. Eng. J.*, **22**, 625 (1976).

Since thermal decomposition is common to all coal conversion schemes, pyrolysis is treated alone before pursuing, in the section on hydropyrolysis, the apparently important effect of pyrolysis on coal reactivity to hydrogen.

Although there has been encouraging progress toward a basic understanding of the phenomena involved and efforts are underway to establish truly theoretical models for reactions involving coal, a heavy reliance on empiricism and semiempiricism is still necessary. The prevailing intense debate questioning the value of fundamental studies with small-scale equipment to pilot and demonstration plant studies must be recognized. The primary objective of all these efforts is to establish a data base suitable for design and scaleup of commercial plants. A genuine understanding of basic pyrolysis and hydropyrolysis phenomena will ease the difficult selection between alternatives and decrease the risk associated with building large-scale plants.

12.2 BACKGROUND

12.2.1 Technological

Coal represents a substantial fraction of the estimated recoverable fossil fuel resources of the world and may become an increasingly important source of hydrocarbon fuels and feedstocks. In the United States, where coal is estimated to be about 70% of the recoverable fossil fuel resources,[9] petroleum and natural gas resources are so low[10] that the development of alternatives is imperative. Recognition of the enormous immediate and long-range potential of coal has resulted in substantial efforts toward the development of new or improved technologies for coal gasification, liquefaction, and direct combustion. Pyrolysis of the coal occurs in all these approaches and in some cases dominates the process characteristics. Pyrolysis is also the central issue in metallurgical coke production, where improved technology is required because of decreasing reserves of good coking coals.

Present coal gasification efforts are directed toward low-Btu, intermediate-Btu, and high-Btu (pipeline-quality, essentially methane) gas. In a similar manner, coal liquefaction efforts include a range of product qualities. Choices among the various possible fuels from coal will be made locally and will be strongly influenced by required fuel characteristics and distribution considerations. However a relatively secure market for pipeline-quality gas will exist for many years, since most residential, many commercial, and some industrial users of natural gas cannot easily adapt to different qualities of gas or to other fuels, and a large amount of capital is already invested in gas transmission and distribution facilities. Therefore considerable attention has been focused on conversion of coal to methane.

The atomic hydrogen/carbon ratio of coals of potential commercial interest for conversion processes is typically 0.6–1.0, which is substantially less than the value for liquid and gaseous fuels commonly used (Table 12.1). Therefore the primary requirement of most coal conversion schemes is either to increase the hydrogen content or to reject carbon, thereby upgrading the hydrocarbon fraction. The latter effect can be accomplished directly by coal pyrolysis, which yields a hydrogen-rich volatile fraction and a carbon-rich solid residue (char and coke). The residue will react with hydrogen quite rapidly during or just after volatile release.

[9] P. Averitt, *Coal Resources of the United States, January 1, 1974,* U.S. Geol. Survey Bull. No. 1412, 1975, p. 100.

[10] B. M. Miller, H. L. Thomsen, G. L. Dolton, A. B. Coury, T. A. Hendricks, F. E. Lennartz, R. B. Powers, E. G. Sable, and K. L. Varnes, *Geological Estimates of Undiscovered Recoverable Oil and Gas Resources in the United States,* U.S. Geol. Survey, Circ. No. 725, 1975.

[11] H. M. Spiers, *Technical Data on Fuel,* 6th Ed., British National Committee, World Power Conference, London, 1962.

[12] P. H. Given, "How May Coals be Characterized for Practical Use?" in *Short Course on Coal Characteristics and Coal Conversion Processes,* Pennsylvania State University, University Park, May 19–23, 1975.

[13] Cameron Engineers, *Synthetic Fuels Data Handbook,* Cameron Engineers, Inc., Denver, Colo., 1975, pp. 265–266.

[14] R. H. Perry, C. H. Chilton, and S. D. Kirkpatrick, *Perry's Chemical Engineers Handbook,* 4th ed., McGraw-Hill Book Company, New York, 1963.

[15] H. R. Linden, "Conversion of Solid Fossil Fuels to High-Heating-Value Pipeline Gas," in *Hydrocarbons from Oil Shale, Oil Sands and Coal,* AIChE Chem. Eng. Prog. Symp. Ser., Vol. 61, No. 54, 1965, p. 76.

Table 12.1 Hydrogen-to-Carbon Ratio and Calorific Value of Selected Fossil and Process-Derived Fuels

Fuel Type	H/C, Atomic[a]	Gross Heating Value (kcal/g)[b]
Gaseous		
High-Btu		
Methane	4.0	13.3
Natural gas	3.5–4.0	11.7–11.9
Intermediate-Btu		
Hydrogen	∞	33.9
Coke oven gas	4.9	9.6
Water gas (H_2 - CO)	2.0	4.3
Low-Btu		
Producer gas (N_2-diluted; from bituminous coal)	1.2	1.2
Liquid		
Methanol	4.0	5.3
Gasoline	2.0–2.2	11.2–11.4
No. 2 fuel oil	1.7–1.9	10.7–11.0
No. 6 fuel oil	1.3–1.6	10.0–10.5
Crude shale oil	1.6	10.3–10.4
Bitumen (Athabasca tar sands)	1.4–1.5	9.8–10
Solid		
Kerogen (Green River oil shale)	1.5	10
Lignite	0.8	3.9–5.4
Subbituminous coal	0.8	5.5
Bituminous coal	0.5–0.9	6.7–8.8
Anthracite	0.3	8.4
Low-temperature coke	0.4	8.2
High-temperature coke	0.06	8.0

[a] Mineral-matter-free C and total organic H (i.e., including H bound with O, S, and N).

[b] Coals—moist, mineral-matter-free; cokes—dry, ash-free; 1 kcal/g = 1800 Btu/lb.

Source: Cokes—Ref. 11; coals—Ref. 12; shale oil, kerogen, and bitumen—Ref. 13; others—Refs. 14, 15.

This direct conversion contrasts markedly with one of the established technologies of complete gasification of the coal with steam and oxygen to produce primarily hydrogen and carbon monoxide at high temperatures. Methane, methanol, and other hydrocarbons can be catalytically synthesized from these gases after purification downstream of the gasifier. Process analyses indicate that catalytic methane formation is less efficient and more costly than direct methane formation by pyrolysis and reaction of the residue with hydrogen.[16,17]

The extent of direct methane formation in the various gasification processes now under development varies from negligible contributions to 95% of the final product gas.[18] The chronological record of developments reflects a strong emphasis on the enhancement of direct conversion. Similarly, in the coal liquefaction area, there is much interest in increased yields of valuable light oils produced directly by hydropyrolysis. In support of development and demonstration activities, accurate kinetic and yield data are required

[16] C. Y. Wen, C. T. Li, S. H. Tscheng, and W. S. O'Brien, "Comparison of Alternative Coal Gasification Processes," paper presented at the 65th Annual AIChE Meeting, New York, November 26–30, 1972.

[17] J. P. Henry and B. M. Louks, "An Economic Study of Pipeline Gas Production from Coal," *Chem. Tech.*, 237–247 (April 1971).

[18] H. C. Hottel and J. B. Howard, *New Energy Technology—Some Facts and Assessments,* M. I. T. Press, Cambridge, Mass., 1971.

for design and scaleup, and a mechanistic understanding of the controlling phenomena is essential for efficient progress toward full realization of the potential advantages.

12.2.2 Coal Composition

12.2.2.1 *Coal Petrography*

The organic material in coal is a heterogeneous mixture of "organic minerals" known as macerals. Coal petrographers have identified numerous types of maceral components by microscopic analysis with reflected and transmitted light.[19] The precise chemical nature of the macerals is not well established, but their botanical origins seem to be rather well understood.[20]

For ease of discussion, the various types of maceral component are often combined into three principal groups: vitrinite, exinite, and inertinite. With regard to physical and chemical properties, exinite is characterized by the highest hydrogen content, volatile matter content, and heating value: inertinite has the least. Inertinite has the highest density and the greatest degree of aromaticity, whereas exinite is the lowest in both these properties. Thus vitrinite, by far the most abundant of the three maceral groups as can be seen in Table 12.2, usually exhibits chem-

[19] W. Spackman, "The Nature of Coal and Coal Seams," in *Short Course in Coal Characteristics and Coal Conversion Processes,* Pennsylvania State University, University Park, May 19–23, 1975.

[20] D. W. Van Krevelen, *Coal,* Elsevier Publishing Company, Amsterdam, 1961.

[21] H. Gan, S. P. Nandi, and P. L. Walker, Jr., "Nature of the Porosity in American Coals," in *Short Course on Characterization and Characteristics of U.S. Coals for Practical Use,* Pennsylvania State University, University Park, October 25–28, 1971.

Table 12.2 Composition of Some American Coals[21]

			Ultimate Analysis (wt %, maf)					Vitrinite
		ASTM						Content
Number[a]	Location	Rank	C	H	N	S	O[b]	(vol %)[c]
85	Pennsylvania	Anthracite	91.2	3.8	0.60	1.1	3.1	97.5
80	Pennsylvania	Anthracite	90.8	2.6	0.75	0.64	5.1	96.3
130	West Virginia	lvb	90.5	4.3	1.1	0.55	3.4	72.7
127	Pennsylvania	lvb	89.5	5.0	1.0	0.83	3.6	77.7
135	Alabama	mvb	88.3	4.9	0.25	0.65	5.7	83.1
142	Oklahoma	mvb	88.2	4.5	0.20	1.0	6.4	—
4	Kentucky	hvAb	83.8	5.7	1.5	0.88	7.9	67.3
95	Washington	hvAb	81.6	6.0	1.0	1.7	9.4	89.6
105A	Indiana	hvBb	81.3	5.7	1.0	1.8	9.9	62.5
24	Illinois	hvCb	80.0	5.5	1.0	4.5	8.8	88.1
213	Kentucky	hvCb	78.8	5.6	1.7	4.2	9.4	73.4
22	Illinois	hvCb	78.7	5.8	1.5	2.8	11.0	88.3
26	Illinois	hvCb	77.2	5.6	1.1	7.4	8.3	88.5
197	Pennsylvania	hvBb	76.5	5.6	1.2	4.5	12.2	85.3
64	Missouri	hvBb	76.4	4.2	1.2	7.3	11.4	—
190	Illinois	hvCb	75.5	5.3	1.0	3.3	14.6	88.9
97	Wyoming	Subbituminous	75.0	5.5	0.58	1.3	17.5	86.7
138	Texas	Lignite	74.3	4.9	0.37	0.75	19.6	75.1
100	Wyoming	Subbituminous	72.0	5.2	0.81	0.55	21.2	86.0
141	Texas	Lignite	71.7	5.2	1.3	0.90	20.8	75.3
90	Montana	Lignite	71.5	4.8	0.83	0.50	22.2	60.9
89	North Dakota	Lignite	63.3	4.6	0.47	1.5	29.9	70.3

[a] PSOC numbers employed in Pennsylvania State University–Office of Coal Research (now ERDA) coal characterization study.

[b] Determined by difference.

[c] Mineral-matter-containing basis.

ical and physical properties between those of the other two.[22]

The different maceral groups exhibit markedly different behaviors under pyrolysis conditions,[2] and the same would presumably hold for hydropyrolysis. As might be expected from the order of volatile matter contents, the total yield of volatiles is usually in the order exinite > vitrinite > inertinite. There are also compositional differences among the products. For example, liquids from exinite have been characterized as primarily neutral oils, but those from vitrinite tend to be lighter and more phenolic.

The discussion of petrographic constituents points up the approximate character of available descriptions of pyrolysis and hydropyrolysis that treat coal as a single material. If extensive rate and yield data on the different maceral groups eventually become available, it should be possible to describe the behavior of different coals in terms of their petrographic compositions and kinetic parameters evaluated for each maceral group.

12.2.2.2 *Coal Chemistry*

The organic matter in coal consists primarily of carbon, hydrogen, oxygen, nitrogen, and sulfur, although trace quantities of other elements are also found. Typical elemental or so-called ultimate analyses of some American coals are given in Table 12.2, where organic and inorganic sulfur contents are lumped together in accordance with ASTM standards. In proximate analysis coal is divided into moisture, volatile matter, fixed carbon, and ash. Either volatile matter or fixed carbon on a moisture-free, ash-free (maf) basis, and the maf calorific value of the coal, permit classification of coals in the ASTM system[23] in decreasing rank as anthracitic, bituminous, or lignitic or brown. The rank of a coal is directly related to its carbon content, with a highly graphitic material representing the ultimate product of the coalification process. Unfortunately this information provides very little direct insight into chemical structure or petrographic composition.

There have been substantial efforts to elucidate the molecular structure of coal, but the task is exceedingly difficult because of the variety of coal types, the heterogeneity of a single coal, and the complexity of individual coal constituents. The results of numerous diverse measurements have been synthesized in attempts to develop a consistent picture (see Van Krevelen,[20] and Dryden[22]) but much work remains to be done. Given[24] presented the hypothetical structure shown in Fig. 12.1 for bituminous coal vitrinite possessing 82% carbon as one possible arrangement of the atoms consistent with most of the facts available at that time.

[22] I. G. C. Dryden, "Chemical Constitution and Reactions of Coal," in *Chemistry of Coal Utilization,* Supplementary Volume, H. H. Lowry, Ed., John Wiley and Sons, New York, 1963, pp. 232–295.

[23] ASTM 1974, Standards on Coal and Coke.

[24] P. H. Given, "The Distribution of Hydrogen in Coals and Its Relations to Coal Structure," *Fuel,* **39,** 147 (1960).

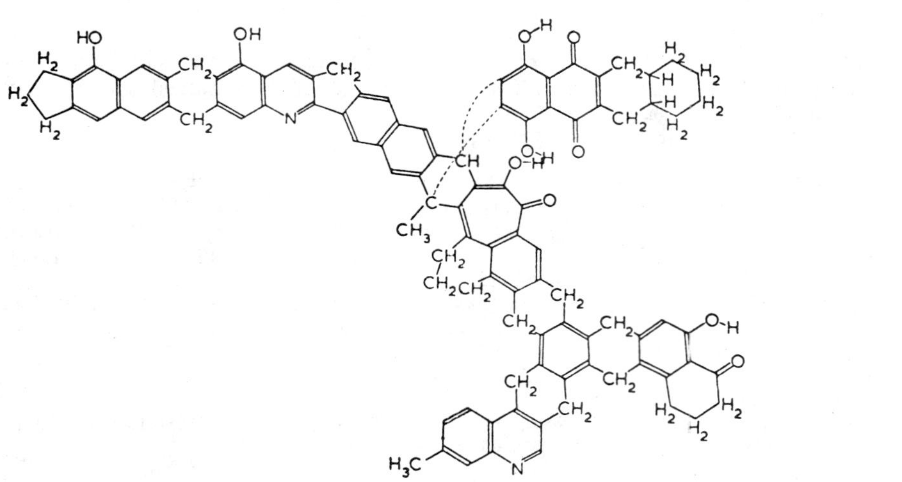

Figure 12.1 The Given model for molecular structure of coal (a vitrinite of 82% C (dmmf); $C_{102}H_{78}O_{10}N_2$).[24]

Although other structures have been proposed—some less complicated than Given's and clearly not in accord with recent findings,[25] and others including a larger variety of possible arrangements[26]—the Given model is generally accepted as a reasonable working hypothesis. This model is certainly suitable for the present discussion of pyrolysis behavior.

Immediately evident in the structure is the high degree of aromaticity involving about two-thirds of the carbon and about 20% of the hydrogen. Condensed aromatic clusters each comprised of one to three rings are in line with Hirsch's[27] X-ray scattering studies, which also showed that the number of rings per cluster increased with increasing rank of the coal, ultimately approaching a fully condensed graphitic structure. The linkages between clusters represent an area of considerable debate. The Given model is built around a 9,10-dihydroanthracene module. Dryden[22] recommended a dihydrophenanthrene module as being more compatible with the absence of methylene bridges, which was reported by Brown et al.[28] based on nuclear magnetic resonance (NMR) studies. In addition, other types of linkage such as short-chain aliphatics several carbons long, ethers, sulfides, and disulfides, are certainly possible. Much of the linkage structure, however, is hydroaromatic according to the results of Reggel et al.[29–31] Aliphatic side chains such as the methyl groups in the Given model are also distributed throughout the coal matrix. Oxygen appears predominantly in hydroxyl and carbonyl groups in all types of coal.[32] Carboxylic acid groups and ether linkages are common to lower rank coals only. The nitrogen in coal is shown in the Given model as a ring-substituted azine structure. Organic sulfur, although not shown, will also appear in heterocyclic aromatic rings as well as sulfide functional groups.

Blayden et al.[33] found that interatomic distances in coal are approximately 2–3 times greater in one dimension than in the other two. This result indicates a highly planar polymeric molecule in a layered structure with considerable internal pore volume and surface area. In fact, this layering phenomenon persists even in the intertwined and highly cross-linked structures of the less-ordered lower rank coals.[34] Internal surface areas of up to several hundred square meters per gram are typically found in all ranks of coal.[22] The porosity of various coals is reported by Sevenster[35] as ranging from 8 to 20%. Extensive data on the pore structure of numerous American coals are being generated at Pennsylvania State University.[12,21]

12.2.2.3 *Relationship to Pyrolysis Reactions*

Given[24] speculated that the course of pyrolysis anticipated from his model would consist of four steps: (1) a low-temperature (400–500°C) loss of hydroxyl groups, (2) dehydrogenation of some of the hydroaromatic structures, (3) scission of the molecule at the methylene bridges, and (4) rupture of the alicyclic rings. Wiser et al.[36] postulated a sequence of events initiated by the formation of two free radical species from a thermal

[25] W. Fuchs and A. G. Sandhoff, "Theory of Coal Pyrolysis," *Ind. Eng. Chem.*, **34**, 587 (1942).

[26] G. R. Hill and L. B. Lyon, "A New Chemical Structure for Coal," *Ind. Eng. Chem.* **54**, 36 (1962).

[27] P. B. Hirsch, "Conclusions from X-Ray Scattering Data on Vitrain Coals," *Proc. Conf. Science in the Use of Coal,* Institute of Fuel, London, 1958, p. A29.

[28] J. K. Brown, W. R. Ladner, and N. Sheppard, "Hydrogen Distribution in Coal-Like Materials by High-Resolution NMR Spectroscopy," *Fuel,* **39**, 79 (1960).

[29] L. Reggel, I. Wender, and R. Raymond, "Catalytic Dehydrogenation of Coal, II," *Fuel*, **47**, 373 (1968).

[30] L. Reggel, I. Wender, and R. Raymond, "Catalytic Dehydrogenation of Coal, III," *Fuel,* **49**, 281 (1970).

[31] L. Reggel, I. Wender, and R. Raymond, "Catalytic Dehydrogenation of Coal, VII," *Fuel,* **52**, 162 (1973).

[32] L. Blom, Thesis, University of Delft, 1960 (Results reproduced by Dryden, 1963.)

[33] H. E. Blayden, I. Gibson and J. L. Riley, *War-Time Bull.*, Institute of Fuel, London, February 1945, p. 117.

[34] P. B. Hirsch, "X-Ray Scattering from Coals," *Proc. R. Soc. (London)*, **A226**, 143 (1954).

[35] P. G. Sevenster, *J. S. Afr. Chem. Inst.*, **7**, 41 (1954).

[36] W. H. Wiser, G. R. Hill, and N. J. Kertamus, "Kinetic Study of the Pyrolysis of a High-Volatile Bituminous Coal," *Ind. Eng. Chem., Process Des. Dev.*, **6**, 133 (1967).

cracking of the linkages forming the aromatic clusters. These free radical species would then stabilize, either through a rearrangement of atoms within a fragment or by collision with another species. The resultant stabilized structure, depending on its vapor pressure, would either evolve as volatile material or remain as part of the residual char.

The amount of hydrogen in coal, if allocated appropriately to the carbon, would be sufficient to permit nearly complete volatilization, at least for medium- and low-rank coals. Unfortunately the structure of coal preferentially evolves hydrogen as chemical water (probably from the hydroxyl groups) and light, highly stable aliphatics (such as methane and ethane), depleting the remaining carbon of much needed hydrogen. Because of this inefficent use of intrinsic hydrogen, pyrolysis, even under the most favorable conditions, includes the formation of heavy tarry compounds and residual char. In the absence of external hydrogen, the aromatic clusters appear to be immune to internal rupture. In fact, prolonged heating at high temperatures results in only a slow dehydrogenation and further condensation of the clusters.

When the rate of reaction is considered, the free radical secondary reactions described by Wiser et al.[36] will be fast compared to the thermally invoked bond ruptures as radical formation steps. Given has already indicated the presence of a hierarchy of bond susceptibilities to scission within the original molecule. Intermediate char structures formed by free radical recombinations similarly will be subject to thermal bond rupture at high temperatures thus assuring a virtual continuum of rate-determining bond energies. This structural observation is the basis for the distributed activation energy model described later. Increased interest in coal research and the accessibility of powerful computers has prompted reexamination of more fundamental reaction mechanisms. In the past, however, a manageable set of fundamental reactions usually proved inadequate to describe experimental results, and a larger, more complete set resulted in too many degrees of freedom for useful modeling, since it was impossible to specify coal reaction rates a priori or to determine them independently.

12.2.3 Physical Factors

12.2.3.1 *Plastic Behavior*

A coal is characterized as plastic or nonplastic depending on whether it exhibits thermoplastic behavior, that is, softening and becoming deformable followed by resolidification to form a char, on being heated either through a certain temperature range or to a temperature above the softening point and there maintained for a certain period of time. Plastic coals are also referred to as caking coals because coal particles in the plastic state are viscous liquid masses capable of coalescence and, on resolidification, formation of a cake.

The macerals discussed above differ in physical behavior as well as in chemical composition. Neavel[37] states that plasticity is exhibited only by the macerals exinite (a minor constituent of typical coals) and vitrinite, of which only vitrinite has plastic capability in bituminous coals, except for a few very unusual coals. A plastic coal ground fine enough to effect substantial liberation of macerals may exhibit, under a given set of heating conditions, a wide range of individual particle behavior including no softening or swelling, moderate swelling with formation of approximately uniformly distributed macroscopic pores, and extreme cenosphere formation. In the present discussion the behavior attributed to a given plastic coal represents that of the dominant maceral, vitrinite.

The occurrence of the plastic state is associated with and kinetically similar in many ways to pyrolysis. The temperature limits and time duration of plasticity depend on heating rate and other conditions. At a constant heating rate of about 0.05 °C/s the plastic region is typically 420–500°C[38] with some variation among different coals. The plastic region moves to higher temperatures as heat-

[37] R. C. Neavel, "Coal Plasticity Mechanism: Inferences from Liquefaction Studies," *Symp. Plasticity and Agglomeration of Coal,* U.S. Energy Research and Development Administration, Morgantown, W. Va., May 5–6, 1975.

[38] D. W. Van Krevelen, F. J. Huntjens, and H. N. M. Dormans, "Chemical Structure and Properties of Coal, XVI, Plastic Behavior on Heating," *Fuel,* **35,** 462 (1956).

ing rate increases and is said to reach 2000°C in pulverized coal flames[39] where the heating rate is of the order 10^4 °C/s.

12.2.3.2 *Internal Structure*

The internal structure of pyrolyzing coal and that of the char product are important in the analysis of mass transfer associated with pyrolysis reactions. The porosity of the fine (molecular sieve) structure of the char from virtually any coal increases steadily as the temperature of its formation is increased, but the accessibility of these pores to penetrant molecules only increases below 500–600°C, then decreases dramatically in the range 600–1000°C.[20,40–43] Accessible surface area follows a similar trend for noncaking coals but is radically different for caking coals, showing a sharp minimum extending over the plastic region of the coal.[44]

The most striking macroscopic change occurring during pyrolysis of plastic coals is the formation and motion of gas bubbles within the liquid mass. Under a hot stage microscope the particles may be seen to soften, become rounded, and swell, while large bubbles of gas repeatedly break through the particle surface as pyrolysis continues.[45–48] The resulting char has a high large-scale porosity, presumably formed by entrapped gas bubbles.[46,49]

The development of large-scale porosity is carried to an extreme in the formation from plastic coal particles, heated at moderate to rapid rates in the absence of air, of swollen, essentially hollow, char particles called cenospheres.[50–58] Cenosphere structure varies considerably but is characterized by the presence of one to a few large cavities in a generally contiguous spherical shell and by a large degree of swelling, frequently as much as a forty fold increase in volume over that of the original coal.[54,55] Street et al.[58] noted an apparent absence of pores in the surfaces of cenospheres, suggesting that any large pores in the original coal do not survive passage through the plastic region.

12.2.3.3 *Transport Processes*

The growth and escape of gas-filled bubbles constitutes an important mode of volatiles

[39] R. F. Littlejohn, "Pulverized-Fuel Combustion: Swelling Under Rapid Heating," *J. Inst. Fuel,* **40,** 128 (1967).

[40] R. E. Franklin, "A Study of the Fine Structure of Carbonaceous Solids by Measurements of True and Apparent Densities, Part I, Coals," *Trans. Faraday Soc.,* **45,** 274 (1949).

[41] R. E. Franklin, "A Study of the fine Structure of Carbonaceous Solids by Measurements of True and Apparent Densities, Part II, Carbonized Coals," *Trans. Faraday Soc.,* **45,** 668 (1949).

[42] R. L. Bond and D. H. T. Spencer, *Industrial Carbon and Graphite,* Soc. Chem. Ind., London, 1958, p. 231.

[43] D. G. Evans and J. A. Hermann, "The Porosity of Brown-Coal Char." *Fuel,* **49,** 110 (1970).

[44] C. G. Cannon, M. Griffith, and W. Hirst, "The Carbonisation of Coals," in *Proc. Conf. Ultra-fine Structure Coals Cokes,* BCURA, London, 1944, p. 131.

[45] S. Ergun, H. J. O'Donnell, and B. C. Parks, "Microscopic Studies of Rate of Thermal Decomposition of Petrographic Components of Coal," *Fuel,* **38,** 205 (1959).

[46] M. F. Woods, G. M. Habberjam, K. Elsworth, and S. Bennett, "The Effect of Maceral Composition on the Binderless Briquetting of Hot Char.," *Fuel,* **46,** 193 (1967).

[47] W. Spackman and W. Berry, *Thermal Behavior of Coal Constituents,* motion picture produced at Pennsylvania State University, University Park, September 1968.

[48] N. Schapiro, personal communication, April 1975.

[49] H. F. Feldmann, K. Kiang, and P. M. Yavorsky, "Fluidization Properties of Coal Char.," *Prepr. Am. Chem. Soc., Div. Fuel Chem.,* **15**(3), 62 (1971).

[50] H. E. Newell and F. S. Sinnatt, "The Carbonisation of Coal in the Form of Fine Particles, I, The Production of Cenospheres," *Fuel,* **3,** 424 (1924).

[51] H. E. Newell and F. S. Sinnatt, "The Study of Cenospheres, II," *Fuel,* **5,** 335 (1926).

[52] H. E. Newell and F. S. Sinnatt, "The Study of Cenospheres, III," *Fuel,* **6,** 118 (1927).

[53] F. S. Sinnatt, A. McCulloch, and H. E. Newell, "The Carbonisation of Particles of Coal, Part 5. The Study of Cenospheres," *J. Soc. Chem. Ind.,* **46,** 331T (1927).

[54] F. S. Sinnatt, "The Formation and Structure of Cenospheres," *J. Soc. Chem. Ind.,* **47,** 152T (1928).

[55] F. S. Sinnatt, "Some Properties of Coal Dust and Pulverized Coal," *Fuel,* **8,** 362 (1929).

[56] D. M. Mason and F. C. Schora, Jr., "Coal and Char Transformation in Hydrogasification," in *Fuel Gasification,* Am. Chem. Soc., Adv. Chem. Ser. No. 69, 1967, p. 18.

[57] P. Lightman and P. J. Street, "Microscopical Examination of Heat Treated Pulverized Coal Particles," *Fuel,* **47,** 7 (1968).

[58] P. J. Street, R. P. Weight, and P. Lightman, "Further Investigation of Structural Changes Occurring in Pulverized Coal Particles During Rapid Heating," *Fuel,* **48,** 343 (1969).

transport in plastic coals. Thus when hydrodynamic and diffusional flows become greatly impeded by the essentially impenetrable pore structure developed during the period of plasticity, regions of high-pressure gas form within the particle and expand against the viscous and other forces to produce growing gas bubbles that may eventually burst through the particle surface as small jets. The above-noted decrease in the accessibility of the fine structure porosity of nonplastic coals also presumably leads to development of high-pressure regions, but these coals cannot flow and allow the incipient bubble to expand. Hence the pressure is expected to rise until the rate of hydrodynamic flow matches the rate of volatiles generation or the particle decrepitates. The latter phenomenon is in fact observed for nonplastic coals, particularly the sparking lignites.

Lewellen[59] has developed a bubble transport model for volatiles flow in plastic coals. The model describes the nucleation, growth, coalescence, and escape of volatiles-filled bubbles under the influence of viscous forces, pressure, and surface tension. The latter is relatively unimportant. Mass is added to and lost from the bubbles by volatiles formation and mass deposition accompanying secondary reactions at the bubble surfaces, respectively. A key feature of the model is the assumption, prompted by Fitzgerald's concept of fluid coal,[60–62] that viscosity of the decomposing mass is inversely proportional to the concentration therein of a plasticizing material formed by pyrolysis and released into the volatile (bubble) phase at a rate controlled by mass transfer. Thus fluidity is a transient property tuned to pyrolysis rate in a manner that accounts for experimentally observed characteristics of the softening and resolidification phenomena and permits realistic description of instantaneous bubble inventory and particle volume, periodic volatiles release from the particle, and other properties of the particle behavior. The model has been critically tested by comparing with experimental observations the predicted effects of particle size, pressure, and heating rate on volatiles yield and particle swelling. Although some questions remain, the predicted trends are in general agreement with experiments.

Other workers have also modeled volatiles transport in a coal particle in terms of bubbles in a melt. Thus Mills et al.[63] used equations for foams to describe the swelling of a coal particle. James and Mills[64] later built a pressure dependence into the model by use of an assumption that the velocity of bubbles through the melt is proportional to a pressure gradient along the particle radius. The model also included a number of other assumptions that the authors correctly identified as questionable, and further work is required.

The interest of Lewellen[59] and Mills et al.[63] in secondary reactions required a more sophisticated volatiles transport model than had been used previously when the main focus was on the relative importances of chemical kinetics and heat and mass transfer in the control of devolatilization rate. The previous workers usually resorted to highly simplified or even only implied concepts of particle structure and transport mechanism. Nevertheless the resulting empirical information, summarized below, is of much practical interest.

If pyrolysis of a coal particle were chemically controlled, the rate would be independent of particle size or structure. Badzioch and Hawksley[65] observed the same pyrolysis rate for 20- and 60-μm particles, and Howard and Essenhigh[66] found no particle size influence using conventional pulverized coal (50

[59] P. C. Lewellen, "Product Decomposition Effects in Coal Pyrolysis," M.S. Thesis, Dept. of Chemical Engineering, Massachusetts Institute of Technology, Cambridge, Mass., 1975.

[60] J. Bronowski, D. Fitzgerald, D. W. Gillings, and D. C. Rhys-Jones, "Plastic Properties of Coking Coals," *Nature*, **171**, 389 (1953).

[61] D. Fitzgerald, "Kinetic Study of Coal Carbonization in the Plastic Zone," *Fuel*, **35**, 178 (1956).

[62] D. Fitzgerald, "The Kinetics of Coal Carbonization in the Plastic State," *Trans. Faraday Soc.*, **52**, 362 (1956).

[63] A. F. Mills, R. K. James, and D. Antoniuk, "Analysis of Coal Particles Undergoing Rapid Pyrolysis," in *Future Energy Production Systems*, J. C. Denton and N. Afgan, Eds., Hemisphere Publishing Co., Washington, D.C., 1976.

[64] R. K. James and A. F. Mills, "Analysis of Coal Particle Pyrolysis," *Lett. Heat Mass Transfer*, **3**, 1 (1976).

[65] S. Badzioch and P. G. W. Hawksley, "Kinetics of Thermal Decomposition of Pulverized Coal Particles," *Ind. Eng. Chem., Process Des. Dev.*, **9**, 521 (1970).

[66] J. B. Howard and R. H. Essenhigh, "Pyrolysis of Coal Particles in Pulverized Fuel Flames," *Ind. Eng. Chem., Process Des. Dev.*, **6**, 74 (1967).

wt % < 35 μm; 80 wt % < 74 μm). Also, weight loss curves prepared by Wiser et al.[36] were unaltered for samples of 60–74 μm versus 246–417 μm. Nevertheless, under a given set of conditions, heat and/or mass transfer will become limiting at some critical particle size, the identification of which requires information on the pertinent heat and mass transfer mechanisms. Three limiting cases are discussed below with the net heat of the pyrolysis reactions being neglected because it appears[67–70] to be small enough to impose only second-order effects on heating under most conditions.

1. External heat transfer. If the dominant heat transfer resistance lies between the coal particle and its surroundings, the temperature of the particle is uniform during heatup and the rate of temperature rise decreases with increasing particle size. Pyrolysis rate during heatup under these conditions increases with decreasing particle size, but the amount of reaction occurring during heatup becomes negligible as heating rate is increased beyond some value. Thus for particles smaller than a certain size, pyrolysis rate is essentially chemically controlled and depends on the final temperature attained, not on heating rate or particle size. Calculations by Badzioch[5] indicate that the transition occurs at a particle diameter of approximately 100 μm, but the answer clearly depends on the assumed pyrolysis kinetics and mode of heat transfer to the particle.
2. Internal heat transfer. The external heat transfer coefficient can be high enough in some systems (e.g., fluidized beds or intensely mixed, entrained-flow reactors) to produce significant temperature gradients within the particle. In the extreme case of heat transfer rate being completely controlled within the particle, the transition between chemical and heat transfer control of the pyrolysis rate occurs at a particle size below which the center temperature closely approaches the surface value before significant reaction can occur. Peters[71] and Peters and Bertling[72] provided the following empirical equation to explain the apparently zero-order reaction observed for particles of 250–2000 μm diameter (nomenclature is at end of chapter):

$$\frac{dV}{dt} = \frac{0.03\,(T_a - 330)}{d^{0.26}} \tag{12.1}$$

This result was interpreted as representing the propagation of an evaporation front ($T \sim 330$°C) through the particle, with the rate of volatiles release being controlled by the overall temperature gradient. Koch et al.[73] used this result to predict, as a function of heating rate (presumably representing surface temperature under these conditions), the particle size at which devolatilization rate shifts from being controlled by a first-order reaction occurring uniformly throughout the particle to being controlled by heat transfer. The following selected values are representative of their predictions:

Heating Rate (°C/s)	Critical Diameter (μm)
100	2,000
1,000	500
10,000	200

Unpublished data from the Massachusetts Institute of Technology (MIT) indicate that the transition occurs at 100–1000 μm for a heating rate of 600°C/s.

[67] J. D. Davis and P. B. Place, "Thermal Reactions of Coal During Carbonization," *Ind. Eng. Chem.* **16**(6), 589 (1924).

[68] J. D. Davis and P. B. Place, "Thermal Reactions of Coal During Carbonization," *Fuel,* **3,** 434 (1924).

[69] J. D. Davis, "Calorimetric Apparatus for the Measurement of Reaction Heats at High Temperature," *Ind. Eng. Chem.,* **17**(7), 726 (1924).

[70] F. M. Gentry, *The Technology of Low Temperature Carbonization,* Williams and Wilkins, Baltimore, 1928, pp. 19–25.

[71] W. Peters, "Stoff- und Warme ubergang bei der Schnellentgasung feinkorniger Brennstoffe," *Chem. Ing. Tech.,* **32,** 178 (1960).

[72] W. Peters and H. Bertling, "Kinetics of the Rapid Degasification of Coals," *Fuel,* **44,** 317 (1965).

[73] V. Koch, H. Jüntgen, and W. Peters, "Nicht-Isotherme Reaktionskinetik der Kohlenpyrolyse, Teil III. Zum Reaktionsablauf bei hohen Aufheizgeschwindigkeiten," *Brennstoff-Chemie,* **50,** 369 (1969).

3. Mass transfer. Approaching the problem from a different point of view, Essenhigh[74] modeled the pyrolysis of larger coal particles (295–4760 μm) in terms of a shrinking spherical liquid core in a porous char matrix. Postulating that the rate is controlled by volatiles transport from the liquid surface through the char shell, he predicted the time required for devolatilization to be:

$$t_v = K_v d^2 \qquad (12.2)$$

where K_v is a constant inversely proportional to the char permeability and directly proportional to the volatile matter content of the coal. Essenhigh's single-particle data on 10 British coals in the stated size range confirmed this square-law behavior, and the mean value of K_v was found to be 0.9 s/mm^2. These findings are apparently contrary to the data of Peters discussed above, wherein pyrolysis time reportedly increases with particle diameter to the 0.26 power. However direct comparison is difficult because Essenhigh did not specify the temperature functionality of K_v which Peters found to be linear.

Experimental verification of Equation 12.2 does not directly distinguish between heat and mass transfer control. Since transient transport is usually described in terms of the Fourier group $\alpha t/r^2$, where α is mass or thermal diffusivity and r is radius, the expected time for either heat or mass transport is proportional to diameter squared.

Howard and Essenhigh[66] show that extrapolation of Equation 12.2 to particle diameters less than about 150 μm seriously underestimates pyrolysis times, implying that transport processes are not rate controlling there. However the absence of such control in small particles does not mean that mass transfer is unimportant with regard to the yield and composition of volatiles. As is discussed below, Anthony et al.[75] support the concept that chemical decomposition is rate controlling for fine particles, but yield is dependent on the outcome of competition between mass transfer and secondary reactions.

[74] R. H. Essenhigh, "The Influence of Coal Rank on the Burning Times of Single Captive Particles," *J. Eng. Power,* **85,** 183 (1963).

[75] D. B. Anthony, J. B. Howard, H. C. Hottel, and H. P. Meissner, "Rapid Devolatilization of Pulverized Coal," *15th Symp. (Int.) Combustion,* Combustion Institute, Pittsburgh, 1975, p. 1303.

12.3 EXPERIMENTAL METHODS

12.3.1 Experimental Conditions

A wide range of experimental conditions has been employed in fundamental research on coal pyrolysis. The conditions usually, but not always, reflect existing or projected commercial interest. From a review of proposed coal conversion processes,[18] the expected commercial conditions include coal types from lignite and subbituminous to caking and noncaking bituminous, temperatures of 500–1000°C, heating rates from 1 to 10^4 °C/s, particle sizes from several millimeters to 50 μm or less, pressures from 1 to 100 atm, and liquid or gaseous media that may be inert or reactive to pyrolyzing coal. The conditions projected for underground gasification are heating rates of the order 0.02–0.2°C/s and particle sizes ranging from a few centimeters to as large as meters. Given the complex phenomena of interests, unambiguous measurements are sometimes difficult to achieve. The equipment itself too often influences the results—for example, by introducing heat or mass transfer effects on conversion rates thought to be kinetically controlled. Precise definition of apparatus, procedure, and conditions is tantamount to meaningful comparison, interpretation, and application of the published results. Therefore Section 12.3.2 is devoted to a critical assessment of the experimental techniques that have been employed.

12.3.2 Experimental Techniques

The experimental techniques are grouped into two general classes: captive sample techniques in which the coal sample is stationary or fixed during the run (i.e., a batch experiment with respect to the solid phase), and continuous coal flow schemes in which the coal is continuously fed and withdrawn from the apparatus. The techniques and conditions employed in many of the studies are summarized in Table 12.3. Several other studies based on the heating of coal particles using shock tubes,[76] plasma jets,[77–79] xenon flash tubes,[77,80–82] and lasers[80,83–86] were conducted at such high heating rates as to virtually defy acquisition of kinetic data. Such work, discussed later, has typically been of an exploratory nature and confined to incomplete analysis of final decomposition products for varying exposure times and intensities.

[76] E. T. Woodburn, R. C. Everson, and A. R. M. Kirk, "Thermal Decomposition of Coal Dust in a Shock Tube," *Fuel,* **53,** 38 (1974).

[77] L. L. Anderson, G. R. Hill, E. H. McDonald, and M. J. McIntosh, "Flash Heating and Plasma Pyrolysis of Coal," in AIChE *Chem. Eng. Prog. Symp. Ser.,* Vol. 64, No. 85, 1968, p. 81.

[78] R. L. Bond, I. F. Galbraith, W. R. Ladner, and G. I. T. McConnell, "Production of Acetylene from Coal Using a Plasma Jet," *Nature,* **200,** 1313 (1963).

[79] R. D. Graves, W. Kawa, and R. W. Hiteshue, "Reactions of Coal in a Plasma Jet," *Ind. Eng. Chem., Process Des. Dev.,* **5**(1), 59 (1966).

[80] A. G. Sharkey, Jr., J. L. Shultz, and R. A. Friedel, "Gases from Flash and Laser Irradiation of Coal," *Nature,* **202,** 988 (1964).

[81] E. Rau and L. Seglin, "Heating of Coal with Light Pulses," *Fuel,* **43,** 147 (1964).

[82] A. F. Granger and W. R. Ladner, "The Flash Heating of Pulverized Coal," *Fuel,* **49,** 17 (1970).

[83] F. J. Vastola and A. J. Pirone, "Use of the Laser–Micropyrolysis–Mass Spectrometer in Studying the Pyrolysis of Coal," *Prepr. Am. Chem. Soc., Div. Pet. Chem.,* **11**(2), C-53 (1966).

[84] J. L. Shultz and A. G. Sharkey, Jr., "Gases from Laser Irradiation of Coal: Effect of Argon, Nitrogen, and Other Atmospheres," *Carbon (Oxford),* **5,** 57 (1967).

[85] W. K. Joy, W. R. Ladner, and E. Pritchard, "Laser Heating of Pulverized Coal in the Source of a Time-of-Flight Mass Spectrometer," *Fuel,* **49,** 26 (1970).

[86] H. Jüntgen and K. H. Van Keek, "Research in the Field of Pyrolysis at Bergbau-Forschung during the Last Fifteen Years," paper presented at the Meeting on Coal Fundamentals, Stoke Orchard, England, January 1977.

[87] C. Portal and H. Tan, "The Effect of Heating Rate on the Devolatilization of Coal," unpublished Joint 10.90 Rep. and S.B. thesis, Dept. of Chemical Engineering, Massachusetts Institute of Technology, Cambridge, Mass., 1974.

[88] D. Gray, J. G. Cogoli, and R. H. Essenhigh, "Problems in Pulverized Coal and Char Combustion," in *Coal Gasification,* Am. Chem. Soc. Adv. Chem. Ser. No.131. 1974, p. 72.

[89] H. Kobayashi, "Devolatilization of Pulverized Coal at High Temperatures," Ph.D. thesis, Dept. of Mechanical Engineering, Massachusetts Institute of Technology, Cambridge, Mass., 1976.

[90] J. H. Campbell, "Pyrolysis of Subbituminous Coal as It Relates to *In Situ* Gasification (Part 1, Gas Evolution; Part 2, Characterization of Liquid and Solid Products)," UCRL-52035, Lawrence Livermore Laboratory, University of California, Livermore, 1976.

[91] J. H. Campbell and D. R. Stephens, "Kinetic Studies of Gas Evolution During Pyrolysis of Subbituminous Coal," *Prepr. Am. Chem. Soc., Div. Fuel Chem.,* **21**(7), 94 (1976).

[92] R. W. Hiteshue, S. Friedman, and R. Madden, "Hydrogenation of Coal to Gaseous Hydrocarbons," U.S. Bur. Mines, Rep. Invest. No. 6027, 1962.

[93] R. W. Hiteshue, S. Friedman, and R. Madden, "Hydrogasification of Bituminous Coals, Lignite, Anthracite, and Char," U.S. Bur. Mines, Rep. Invest. No. 6125, 1962.

[94] R. W. Hiteshue, S. Friedman, and R. Madden, "Hydrogasification of High-Volatile A Bituminous Coal," U.S. Bur. Mines, Rep. Invest. 6376, 1964.

[95] H. L. Feldkirchner and H. R. Linden, "Reactivity of Coals in High-Pressure Gasification," *Ind. Eng. Chem., Process Des. Dev.,* **2,** 153 (1963).

[96] H. L. Feldkirchner and J. Huebler, "Reaction of Coal with Steam-Hydrogen Mixtures at High Temperatures and Pressures," *Ind. Eng. Chem., Process Des. Dev.,* **4,** 134 (1965).

[97] H. Jüntgen and K. H. Van Heek, "Gas Release from Coal as a Function of the Rate of Heating," *Fuel,* **47,** 103 (1968).

[98] K. H. Van Heek, H. Jüntgen, and W. Peters, "Fundamental Studies on Coal Gasification in the Utilization of Thermal Energy from Nuclear High-Temperature Reactors," *J. Inst. Fuel,* **46,** 223 (1973).

[99] M. L. Vestal, A. G. Day, J. S. Snyderman, G. J. Fergusson, F. W. Lampe, R. H. Essenhigh, and W. H. Johnston, "Kinetic Studies on the Pyrolysis, Desulfurization, and Gasification of Coals with Emphasis on the Non-Isothermal Kinetic Method," Rev. Final Rep., NAPCA, Phase I, Contract No. PH 86-68-65 (NTIS No. PB-185-882), Scientific Research Instruments Corp., Baltimore, 1969.

[100] A. L. Yergey, F. W. Lampe, M. L. Vestal, A. G. Day, G. J. Fergusson, W. H. Johnston, J. S. Snyderman, R. H. Essenhigh, and J. E. Hudson, "Nonisothermal Kinetic Studies of the Hydrodesulfurization of Coal," *Ind. Eng. Chem., Process Des. Dev.,* **13,** 233 (1974).

[101] F. Moseley and D. Paterson, "Rapid High-Temperature Hydrogenation of Coal Chars, Part 1," *J. Inst. Fuel,* **38,** 13 (1965).

[102] H. L. Feldkirchner and J. L. Johnson, "High-Pressure Thermobalance," *Rev. Sci. Instrum.,* **39,** 1227 (1968).

[103] J. L. Johnson, "Means of Relating Coal Characteristics to Chemical Engineering Data," *Seminar on Characterization and Characteristics of U.S. Coals for Practical Use,* Pennsylvania State University, University Park, October 25–28, 1971.

[104] N. Gardner, E. Samuels, and K. Wilks, "Catalyzed Hydrogasification of Coal Chars," in *Coal Gasification,* Am. Chem. Soc., Adv. Chem. Ser. No. 131, 1974, p. 217.

Table 12.3 Experimental Techniques and Conditions

Investigator	Technique	Residence Time(s)	Temperature (°C)	Heating Rate (°C/s)	Pressure (atm)	Ambient Gas	Particle Size (μm)
			Captive Sample Techniques				
ASTM Standard Proximate Analysis[23]	Crucible	420	950	15–20	1.0	Air (lid on)	≤250
Wiser et al.[36]	Crucible	300–72,000	400–500	15–20	1.0	N_2	250–420
Portal and Tan[87]	Crucible and basket	15–1200	550–1150	0.5–250	1.0	N_2	≤45–88
Gray et al.[88]	Crucible	420	950–1200	0.3–20	1.0	N_2	≤200
Kobayashi[89]	Crucible	1800–3600	270–1830	0.5–5	1.0	Ar	27–90
Campbell,[90] Campbell and Stephens[91]	Basket	Up to 18,000	110–1000	0.056	1.0	Ar, N_2, H_2, CO	1700–3600
Hiteshue et al.[92–94]	Fixed bed, gas swept	20–900	480–1200	10	18–400	H_2	250–600
Feldkirchner and Linden,[95] Feldkirchner and Huebler[96]	Fixed bed, gas swept	10–480	700–930	100–300 (seems high)	34–168	H_2, H_2O	840–1000
Jüntgen and Van Heek,[86,97] Van Heek et al.[98]	Fixed bed, gas swept	1000–10^6	400–1100	0.0001–0.5	1–70	He, H_2, H_2O	≤2000
Vestal et al.,[99] Yergey et al.[100]	Fixed bed, gas swept	600–60,000	≤1000	0.02–2	≤5	H_2, He	37–74
Moseley and Paterson[101]	Railway heater	15–165	815–950	25	18–95	H_2	150–300
Feldkirchner and Johnson,[102] Johnson[103]	Thermobalance	Several–7200	≤925	≤100	≤100	H_2, N_2, H_2O, etc.	420–850
Gardner et al.[104]	Thermobalance	25–3000	850–950	—	35–69	H_2	500–1000
Loison and Chauvin[105]	Elec. grid	0.7	≤1050	1500	Vacuum	—	50–80
Rau and Robertson[106]	Elec. grid	1–1.5	900–1200	600	1.0	—	250–420
Jüntgen and Van Heek[97]	Elec. grid	0.7	≤1000	≤Several–1000	Vacuum	—	50–60
Koch et al.[73]	Elec. grid	7	≤1500	167	Vacuum	—	75–630
Mentser et al.[107,110]	Elec. grid	0.05–0.15	400–1200	8250	Vacuum	—	44–53
Cheong et al.[112]	Elec. grid	<1–1800	300–1000	≤1000	Vacuum –3.4	—	90–360

Table 12.3 *Continued*

Investigator	Technique	Residence Time(s)	Temperature (°C)	Heating Rate (°C/s)	Pressure (atm)	Ambient Gas	Particle Size (μm)
Solomon[113]	Elec. grid	5–80	≤1000	600	Vacuum	—	44–370
Anthony et al.[75,108,109] Suuberg et al.[111]	Elec. grid	0.1–20	400–1100	100–15,000	0.001–100	H_2, He, N_2	53–1000
Blair et al.[114]	Elec. ribbon	0.02–20	350–1700	200–20,000	1.0	He	510–620
Squires et al.,[115] Graff et al.,[116] Dobner et al.[117]	Ring of coal, gas swept	1–6 (gas), 10–30 (solid)	600–1000	650	100	H_2	≤44
			Coal Flow Techniques				
Stone et al.[118]	Fluidized bed	10–2500	400–700	—	1.0	N_2	200–600
Peters,[71] Peters and Bertling[72]	Mechanically stirred bed	≤0.5–10	600–1100	50–200	1.0	N_2	1000–1500
Pitt[119]	Fluidized bed	10–6000	400–700	—	1.0	—	200–600
Kobayashi,[89] Kobayashi et al.[132]	Entrained flow	0.001–0.22	700–1800	10,000–200,000	1.0	Ar	38–45
Nsakala et al.[133]	Entrained flow	0.3	810	8000	1.0	N_2	50–180
Ubhayaker et al.[134]	Entrained flow	0.007–0.07	1500–2000	≤130,000	1.0	CO_2, H_2O, N_2	70% < 74
Belt et al.[140,143]	Entrained flow	≤1	815–1040	—	1–28	H_2, N_2	70% ≤ 74
Coates et al.[144]	Entrained flow	0.012–0.34	650–1370	—	1.0	H_2, H_2O, etc.	≤74
Moseley and Paterson[135]	Entrained flow	0.17–2.5	790–1000	1000+	50–520	H_2	100–150
Glenn et al.[136]	Entrained flow	2.4–10.4	920–970	—	70–84	H_2, CO, etc	≤44
Johnson[137]	Entrained flow	5–14	480–845	28 (or very fast)	18–52	H_2, He	75–90

Fallon and Steinberg[138]	Entrained flow	12–20	450–820	—	100–140	H_2	≤150
Shapatina et al.[139]	Free fall	0.45–14,400	≤550	—	45–490	N_2	150–200
Moseley and Paterson[142]	Free fall	A few	840–1000	—	45–490	H_2	100–150
Feldmann et al.[141]	Free fall	A few	650–900	—	35–205	H_2	150–300
Kobayashi[89]	Free fall	~1	760–1560	—	1.0	Ar	38–45
Jones et al.[120]	Fluidized bed	~2400	425–1095	1000+	1.0	N_2, H_2O	≤1000
Smith and Wailes[121]	Fluidized bed	—	—	—	1.0	N_2	—
Friedman[122]	Fluidized bed	1800–3600	300–650	—	1.0	H_2O	250–710
Birch et al.[123,124]	Fluidized bed	500–9500	500–950	—	21–42	H_2, etc	150–710
Hamrin et al.,[125] Maa et al.[126]	Fluidized bed	3600–26,600	600–870	0.06–0.25	1.0	H_2, H_2S, N_2	100–600
Eddinger et al.[127]	Entrained flow	0.008–0.04	625–1000	2500+	1.0	He	6150
Friedman et al.[128]	Entrained flow	0.004–2	400–1000	650–4300	1.0	He, N_2, H_2O	≤150, 74, and 5
Howard and Essenhigh[66]	Pulverized coal flame	0–0.8	200–1550	≤22,000	1.0	Air	80% ≤ 74
Sass[129]	Entrained flow	A few	540–650	10,000	1.0	—	25–80
Smith and Wailes[121]	Entrained flow	0.5	400–700	—	1.0	N_2	—
Badzioch,[5] Badzioch and Hawksley[65]	Entrained flow	0.03–0.11	400–1000	25,000–50,000	1.0	N_2	11–73
Kimber and Gray[130,131]	Entrained flow	0.012–0.11	780–2000	100,000–400,000	1.0	Ar	23–76

[105] R. Loison and F. Chauvin, "Pyrolyse Rapide du Charbon," *Chem. Ind. (Paris),* **91,** 269 (1964).

[106] E. Rau and J. A. Robertson, "The Use of the Microsample Strip Furnace in Coal Research," *Fuel,* **45,** 73 (1966).

[107] M. Mentser, H. J. O'Donnell, and S. Ergun, "Rapid Thermal Decomposition of Bistuminous Coals," *Prepr. Am. Chem. Soc., Div. Fuel Chem.,* **14**(5), 94 (1970).

[108] D. B. Anthony, J. B. Howard, H. P. Meissner, and H. C. Hottel, "Apparatus for Determining High Pressure Coal-Hydrogen Reaction Kinetics under Rapid Heating Conditions," *Rev. Sci. Instrum.,* **45,** 992 (1974).

[109] D. B. Anthony, J. B. Howard, H. C. Hottel, and H. P. Meissner, "Rapid Devolatilization and Hydrogasification of Bituminous Coal," *Fuel,* **55,** 121 (1976).

[110] M. Mentser, H. J. O'Donnell, S. Ergun, and R. A. Friedel, "Devolatilization of Coal by Rapid Heating," in *Coal Gasification,* Amer. Chem. Soc., Adv. Chem. Ser. No. 131, 1974, p. 1.

[111] E. M. Suuberg, W. A. Peters, and J. B. Howard, "Product Composition and Kinetics of Lignite Pyrolysis," *Ind. Eng. Chem., Process Des. Dev.,* **17,** 37 (1978).

[112] P. H. Cheong, M. Oka, and G. R. Gavalas, "Modeling and Experimental Studies of Coal Pyrolysis," NSF Workshop on Fundamental Organic Chemistry of Coal, Knoxville, Tenn., July 1975.

[113] P. R. Solomon, *The Evolution of Pollutants During the Rapid Devolatilization of Coal,* Rep. No. R77-952588-2 (NSF Grant No. AER 75-17247, monitored by Fossil Energy Div., ERDA), United Technologies Research Center, East Hartford, Conn., April 1977.

[114] D. W. Blair, J. O. L. Wendt, and W. Bartok, "Evolution of Nitrogen and Other Species during Controlled Pyrolysis of Coal," *16th Symp. (Int.) Combustion,* Combustion Institute, Pittsburgh, 1977, p. 475.

[115] A. M. Squires, R. A. Graff, and S. Dobner, "Flash Hydrogenation of a Bituminous Coal," *Science,* **189,** 793 (1975).

[116] R. A. Graff, S. Dobner, and A. M. Squires, "Flash Hydrogenation of Coal. 1. Experimental Methods and Preliminary Results," *Fuel,* **55,** 109 (1976).

[117] S. Dobner, R. A. Graff, and A. M. Squires, "Flash Hydrogenation of Coal. 2. Yield Structure for Illinois No. 6 Coal at 100 Atm," *Fuel,* **55,** 113 (1976).

[118] H. N. Stone, J. D. Batchelor, and H. F. Johnstone, "Low Temperature Carbonization Rates in a Fluidized Bed," *Ind. Eng. Chem.,* **46,** 274 (1954).

[119] G. J. Pitt, "The Kinetics of the Evolution of Volatile Products from Coal," *Fuel,* **41,** 267 (1962).

[120] J. F. Jones, M. R. Schmid, and R. T. Eddinger, "Fluidized Bed Pyrolysis of Coal," *Chem. Eng. Prog.,* **60**(6), 69 (1964).

[121] I. W. Smith and P. C. Wailes, "Oil from Coal by Flash Pyrolysis," in *Energy and Liquid Fuels,* Institute of Fuel, Royal Australian Chemical Institute, and Institution of Chemical Engineers, Adelaide, Australia, 1975, Paper No. 3.

12.3.2.1 *Captive Sample Techniques*

The best known example of the captive sample technique is the standard proximate analysis for volatile content.[23] A weighed sample

[122] L. D. Friedman, "Development of a Fluidized Bench-Scale Reactor for Kinetic Studies," *Prepr., Am. Chem. Soc., Div. Fuel Chem.,* **20**(3), 34 (1975).

[123] T. J. Birch, K. R. Hall, and R. W. Urie, "Gasification of Brown Coal with Hydrogen in a Continuous Fluidized-Bed Reactor," *J. Inst. Fuel,* **33,** 422 (1960).

[124] T. J. Birch, *Hydrogasification of Brown Coal,* Parts 1 and 2, Div. Chemical Engineering, CSIRO, Australia, Rep. No. CE/R25, 1969.

[125] C. E. Hamrin, Jr., P. S. Maa, and C. R. Lewis, *Measurement of Inhibition Isotherms of Kentucky Coal,* Project No. 4, Annu. Rep., Institute for Mining and Minerals Research, College of Engineering, University of Kentucky, Lexington, October 1973.

[126] P. S. Mas, C. R. Lewis, and C. E. Hamrin, Jr., Sulphur Transformations and Removal for Western Kentucky Coals," *Fuel,* **54,** 62 (1975).

[127] R. T. Eddinger, L. D. Friedman, and E. Rau, "Devolatilization of Coal in a Transport Reactor," *Fuel,* **45,** 245 (1966).

[128] L. D. Friedman, E. Rau, and R. T. Eddinger, "Maximizing Tar Yields in Transport Reactors," *Fuel,* **47,** 149 (1968).

[129] A. Sass, "The Garrett Research and Development Process for the Conversion of Coal into Liquid Fuels," paper presented at the 65th Annual American Institute of Chemical Engineers Meeting, New York, November 30, 1972.

[130] G. M. Kimber and M. D. Gray, "Rapid Devolatilization of Small Coal Particles," *Combust. Flame,* **11,** 360 (1967).

[131] G. M. Kimber and M. D. Gray, *Measurements of Thermal Decomposition of Low and High Rank Non-Swelling Coals at M.H.D. Temperatures,* BCURA Document No. MHD 32, January 1967.

[132] H. Kobayashi, J. B. Howard, and A. F. Sarofim, "Coal Devolatilization at High Temperatures," *16th Symp. (Int.) Combustion,* Combustion Institute, Pittsburgh, 1977, p. 411.

[133] N. Nsakala, R. H. Essenhigh, and P. L. Walker, Jr., "Characteristics of Chars Produced by Pyrolysis Following Rapid Heating of Pulverized Coal," *Prepr., Am. Chem. Soc., Div. Fuel Chem.,* **22**(1), 102 (1977).

[134] S. K. Ubhayaker, D. B. Stickler, C. W. von Rosenberg, Jr., and R. E. Gannon, "Rapid Devolatilization of Pulverized Coal in Hot Combustion Gases," *16th Symp. (Int.) Combustion,* Combustion Institute, Pittsburgh, 1977, p. 427.

[135] F. Moseley and D. Paterson, "Rapid High-Temperature Hydrogenation of Coal Chars, Part 2," *J. Inst. Fuel,* **38,** 378 (1965).

[136] R. A. Glenn, E. E. Donath, and R. J. Grace, "Gasification of Coal Under Conditions simulating Stage 2 of the BCR Two-Stage Super Pressure Gasifier," in *Fuel Gasification,* Am. Chem. Soc. Adv. Chem. Ser. No. 69, 1967, p. 253.

of ground coal in a crucible is suspended for a designated time in a vertical furnace at a specified temperature. The measured weight loss in excess of moisture is attributed to volatile release. The major disadvantage of such an apparatus for kinetic studies at constant temperature is the slow heating rate (estimated by Zielinski[145] and recently confirmed by Gray et al.[88] and Portal and Tan[87] to be approximately 15–20°C/s), during which most of the devolatilization occurs. Wiser et al.[36] employed a variation of this technique in studying coal pyrolysis rates. The initial reading was taken after 300 s when, for example, at a final temperature of 500°C, more than 60% of the final weight loss had already occurred. The problem is magnified at higher temperatures, since the rate of pyrolysis increases approximately exponentially with temperature. Kobayashi[89] employed a crucible technique for comparative measurements in a study based primarily on an entrained-flow technique described below. The crucible was used to measure asymptotic weight losses, not kinetics, for long residence times at temperatures up to 1830°C.

Pyrolysis under slow heating rates of interest in underground gasification was studied[90,91] using 50-g samples of coal particles in a basket positioned atop a bed of ceramic balls in a temperature-programmed furnace. Pyrolysis products were swept from the furnace by a constant flow of argon, which was preheated to the furnace temperature by passage through the ceramic bed. Time-resolved rates of evolution of hydrogen, carbon oxides, methane, ethane, ethylene, and propane were determined by mass spectrometry, while the temperature of the center of the sample, monitored by a thermocouple, was increased at a constant rate of 3.3°C/min. The study also included some analyses of liquid and solid products and comparative runs with other sweep gases. The extent to which the relatively large coal sample remains isothermal during heating, hence the meaning of the temperature measured, would appear to depend on the uniformity of the temperature and of the velocity of the preheated gas flowing through the sample, which in turn depends on the geometry of the sample and the closeness of its fit in the furnace passage. It is to be noted also that the observed time dependence (also temperature dependence, when heating rate is constant) of the flow rate of a given species in the carrier gas at a sampling station located some distance downstream of the sample, may not be the same as that of the rate of evolution from the sample if the holdup time in the flow system is not low enough to effectively eliminate axial dispersion of the species of interest. The occurrence of this problem as well as nonuniformity of sample temperature would lead to pseudokinetic parameters that depend in part on transport phenomena. The limited reported information does not permit quantitative assessment of the present apparatus with regard to these factors.

A semiflow technique called the hot rod reactor was employed for hydrogenation studies by the U.S. Bureau of Mines.[92–94] Several grams of coal particles in a fixed bed was placed in an electrically heated vertical tube and hydrogen was passed through the bed. The coal was heated to the final temperature in about 20 s and the resulting products were analyzed. Workers at the In-

[137] J. L. Johnson, "Gasification of Montana Lignite in Hydrogen and in Helium During Initial Reaction Stages," *Prepr., Am. Chem. Soc., Div. Fuel Chem.*, **20**(3), 61 (1975).

[138] P. Fallon and M. Steinberg, "Flash Hydropyrolysis of Coal," paper presented at the 173rd National Meeting, American Chemical Society, New Orleans, March 20–25, 1977.

[139] E. A. Shapatina, V. V. Kelyuzhnyi, and Z. F. Chukhanov, "Technological Utilization of Fuel for Energy, 1. Thermal Treatment of Fuels" (1960), reviewed by S. Badzioch, BCURA Month. Bull., **25,** 285 (1961).

[140] R. J. Belt, J. S. Wilson, and J. J. S. Sebastian, "Continuous Rapid Carbonization of Powdered Coal by Entrainment, and Response Surface Analysis of Data," *Fuel,* **50,** 381 (1971).

[141] H. F. Feldmann, W. H. Simons, J. A. Mima, and R. W. Hiteshue, "Reaction Model for Bituminous Coal Hydrogasification in a Dilute Phase," *Prepr., Am. Chem. Soc., Div. Fuel Chem.*, **14**(4), 1 (1970).

[142] F. Moseley and D. Paterson, "Rapid High-Temperature High-Pressure Hydrogenation of Bituminous Coal," *J. Inst. Fuel,* **40,** 523 (1967).

[143] R. J. Belt and M. M. Roder, "Low-Sulfur Fuel by Pressurized Entrainment Carbonization of Coal," *Prepr., Am. Chem. Soc., Div. Fuel Chem.*, **17**(2), 82 (1972).

[144] R. L. Coates, C. L. Chen, and B. J. Pope, "Coal Devolatilization in a Low Pressure, Low Residence Time Entrained Flow Reactor," in *Coal Gasification,* Am. Chem. Soc., Adv. Chem. Ser. No. 131, (1974), p. 92.

[145] E. Zielinski, "The Evolution of Volatile Matter from Pulverized Coal Particles," *Fuel,* **46,** 329 (1967).

stitute of Gas Technology (IGT) in Chicago, in the early 1960s, also used a semiflow reactor in which a few grams of coal char was charged batchwise into a vertical reactor and suspended on a screen above a bed of preheated alumina.[95,96] Various gas mixtures, primarily hydrogen and steam, were passed through the static charge and the effluent was analyzed. The initial char heating rates were allegedly several hundred degrees centigrade per second, and the first gas sample was taken after 20 s. It is interesting to note that one of the stated reasons for developing these techniques was an apparent need for rapid heating and the capability to analyze data for short residence times at high temperatures. Several other techniques described later were also developed in response to this need.

Workers at Bergbau-Forschung[86,97,98] have made extensive use of a gas-swept, fixed-bed reactor to study the pyrolysis, hydrogasification, and steam gasification of coal at low rates of heating (usually 1–10°C/min) and at pressures from 1 to 70 atm. Metered gas was passed at constant velocity through the bed, which was heated at a constant rate in an electric oven. The temperature history of the sample was monitored by a thermocouple placed in the bed. The product gas, swept away by the carrier gas, was purified and then analyzed by gas chromatography or mass spectrometry. The technique gave time-resolved emission rates of carbon oxides, water, methane, and certain other light hydrocarbon gases.

A technique similar to the foregoing was employed to study the evolution of hydrogen sulfide from coal pyrolyzed at constant heating rates in hydrogen or helium.[99,100] The coal samples were held in a quartz reaction tube inside a temperature-programmed furnace. The rate of evolution of hydrogen sulfide was determined as a function of time, hence of temperature, by continuous mass spectrometric analysis of the sweep gas.

Mosely and Paterson[101] used a novel method for heating and cooling coal and char powders. The particles were spread along a tubular reactor mounted horizontally above a railway, along which a series of burners were propelled, followed by jets for cooling. Heating and cooling rates of 25°C/s were achieved with a minimum of 15 s at temperature.

A thermobalance system was used by Johnson at IGT.[102,103] About 0.5 g of coal enclosed in a cylindrical wire mesh basket is lowered into a preheated reaction zone and the sample weight is continuously recorded. The maximum heating rate attainable is probably no more than a few hundred degrees per second, based on the work of Portal and Tan,[87] which included use of a similar wire mesh basket and measurement of bed temperature with a thermocouple and fast-speed recorder. Although the mesh basket is a substantial improvement over the crucible, the above-mentioned problem of slow heatup (and also cooling) still requires attention if a constant-temperature experiment is desired. A delay of 15 s or more before taking the first data prevents resolution of a regime of considerable reaction, as noted by both Mosely and Paterson and the Bureau of Mines workers. The observed initial rate is necessarily an average over this minimum residence time and the associated temperature-time history. Continuous weighing (e.g., Johnson) circumvents the difficulty of a minimum residence time, but the kinetic analysis is complicated because nearly all the pyrolysis and a substantial portion of the reaction with hydrogen occur while the temperature is rising, and this requires knowledge of temperature as a function of time and position within the sample. The thermobalance is suited primarily for studies of slower char gasification reactions (time scale of minutes) and has been so employed successfully at IGT and at Case Western Reserve.[104]

Several studies have involved devolatilization of approximately monolayer samples of coal particles on an electrically heated screen. Loison and Chauvin[105] achieved rapid heating to about 1050°C with this technique, and they measured yields of gas, liquid, and char; however they did not obtain rate data. Jüntgen and Van Heek[97] and Koch et al.[73] investigated the kinetics of evolution of several gaseous species with a similar technique combined with mass spectrometry. Apparently no attempt was made to weight or analyze the solid residue. Rau and Robertson,[106] on the other hand, heated their grid more slowly but held the sample at the final temperature for several seconds. They measured sample weight loss but did not analyze the volatile products.

Mentser et al.[107] pyrolyzed *in vacuo* 250-μg coal samples in microcylinders formed with 400-mesh stainless steel screen and determined total weight loss by microbalance weighings of the screen and sample. The screens were heated at a constant predetermined rate of 8250°C/s for different times, thereby generating data on the effect of final temperature on weight loss. The gases were analyzed by mass spectrometry in many of the tests. The data exhibited an unusual negative rate of weight loss for several coals over the temperature range 700–1000°C. One plausible explanation is that certain volatile species expelled from the coal at temperatures below 700°C cracked on the screen at higher temperatures and redeposited significant quantities of carbon. This effect might have been avoided by sweeping the products away with continuous pumping or by dilution with inert gases.

Anthony et al.[75,108,109] used approximately 10-mg samples and a dual heating circuit that permitted control of heating rate, from 100 to 15,000°C/s, and maintenance of a preselected final temperature. Weight loss was measured by weighing sample plus screen before and after the run, and the temperature-time history of the sample was obtained with a fast-response thermocouple and recorder. A weight loss versus time curve for a given set of conditions was obtained from a series of runs of different time duration. Time resolution of the weight-loss data was about 20 ms, adequate for kinetics studies, even for the most rapid heating rates. The heating element was enclosed in a vessel permitting both vacuum and high-pressure operation and hydrogenation as well as pyrolysis studies. The initial coal/gas ratio was small enough to ensure large dilution of the volatiles, to retard cracking and carbon deposition on the screens. The apparatus was recently modified to permit the collection and analysis of volatiles.[111,146] Complementary studies by Cheong et al.[112] and by Solomon[113] are also underway.

A disadvantage of the grid heating technique is the necessity for heavy electrical leads, which preclude continuous monitoring of conversion. Also, caution should be exercised in the use of such apparatus in the study of the composition of volatiles, some of which may be appreciably altered by catalytic cracking on the hot screens. Under certain conditions, carbonaceous deposits have been observed, indicating changes in some volatile species[75,109] recently found to be the tars.[111]

Blair et al.[114] pyrolyzed small samples of coal ranging from one to about 75 particles (0.04–3 mg) placed in a V-shaped, electrically heated ribbon of platinum or graphite. The technique permitted the study of controlled heating rates in the range 200–20,000°C/s and temperatures up to 1700°C. The ribbon was held in a chamber flushed with cold helium at 1 atm that carried the pyrolysis products through a capillary system to a measuring station. Total losses of weight and of nitrogen were determined by weighing and analyzing the solid residue, and yields of carbon oxides, methane, C_2 hydrocarbons, and nitrogenous species were determined by gas chromatography and by specific ion electrode analysis of an absorption solution. The technique did not permit closure of mass or nitrogen balances, since the analyzed species represented only 22–55% of the total weight evolved from the coal under the different conditions and about 20% of the evolved nitrogen.[114] Time-resolved rates of evolution of individual species from the coal particles were estimated from the measured signals by assuming the reactor to be well stirred and by calculating the dispersion of species in the capillary system. The measured signals occurred over a period of 10–20 s, while the actual evolution from the particles occurred on a much smaller time scale.

Use of a technique in which about 10 mg of coal particles are deposited in a thin ring on the inside surface of a stainless steel pipe through which hydrogen is passed was recently reported.[115–117] The pipe is electrically heated, and the volatiles are swept away for analysis of the lighter species, up to and including xylene. An objective of the design was to have means for determining the effect of vapor residence time on product distribution. The amount of carbon contained in the char, and presumably also in any material deposited in the hot system, was determined by measurement of carbon dioxide produced in a subsequent burnout with oxygen. The carbon contained in the species recovered from hydrogenation and oxidation was less than the carbon initially in the coal by an

[146] E. M. Suuberg, "Rapid Pyrolysis and Hydropyrolysis of Coal," Sc.D. thesis, Dept. of Chemical Engineering, Massachusetts Institute of Technology, Cambridge, Mass., 1977.

amount that decreased from 20% of the coal carbon at 600°C to nil above about 850°C. The lack of closure of the carbon balance at the lower temperatures was attributed to species less volatile than xylene that are trapped in tubing downstream of the reactor. The extent of cracking in the reactor tube is unknown, but it could be significant given (*a*) the large quantities of heavier hydrocarbons implied by the carbon balances and (*b*) previous measurements of cracking under similar conditions.[70,147,148] Finally, the acquisition of kinetic data, which was not an objective of the design, would appear to be difficult with this apparatus. Attempts to obtain such information from product gases would require accurate knowledge of residence time distributions and species production and loss in wall-catalyzed reactions. The latter knowledge would, of course, also be needed to infer the volatile product distribution and yield in the absence of the hot tube wall. In addition, since the solid phase cannot be monitored continuously, some means of rapidly heating and cooling the sample for run times of varying duration would be needed. Given the relatively large thermal inertia of the pipe, whose weight is more than 2000 times that of the coal, rapid quenching of the sample would be difficult.

12.3.2.2 *Coal Flow Techniques*

Fluidized Bed. Stone et al.[118] and Pitt[119] injected coal particles into preheated fluidized beds of sand at 400–700°C. The technique provided rapid heat transfer to the particles and sampling times were about 10–15 s. However nearly all the pyrolysis occurred prior to the first data point. Based on total volatiles released, Pitt's coal at the first data point was 62% devolatilized at 450°C and 89% at 650°C. Peters[71] and Peters and Bertling[72] studied much larger particles (1–1.5 mm) at higher temperatures (600–1100°C) and used improved sampling procedures that permitted data acquisition to start at a residence time of 1 s, essential for a process that was complete in 3–4 s. The technique involved the mixing of coal particles with heated sand, not in a fluidized bed but in a reaction vessel continuously stirred by mechanical paddles. Knowledge of particle temperature was imprecise, since the response of the center of a 1-mm coal particle will lag behind that of the surface by nearly 3 s. Laboratory studies with fluidized-bed pyrolyzers are reported by Jones et al.,[120] Smith and Wailes, [121] and Friedman.[122] Fluidized beds have also been used to study the hydrogenation of coal.[123,124]

Several researchers[125,126] have used a 22-mm diameter fluidized bed in gas chromatographic studies of the evolution of hydrogen sulfide during heatup under nitrogen and during a holding period at the run temperature in atmospheres of nitrogen, hydrogen, or mixtures of hydrogen and hydrogen sulfide. The solid residue was analyzed for weight loss, volatile matter, and ash, and pyritic, organic, sulfide, sulfate, and total sulfur. Fluidization of the coal with bottom ash from a producer gas generator was also studied[149] as a means for the reduction of hydrogen sulfide emissions and for the prevention of coal agglomeration, which is a problem with fluidized-bed pyrolysis of some bituminous coals.

Entrained Flow. Eddinger et al.[127] published results obtained by passing a gas-entrained stream of coal particles through a heated tube. The principal objective of this work was to achieve rapid heating conditions and simultaneously to obtain sufficient quantities of products for analysis and characterization. Friedman et al.[128] focused on the promotion of larger tar yields by reducing the extent of secondary reactions. Three entrained-flow reactors were employed to study several combinations of temperature and residence time. Howard and Essenhigh[66] reported data on rapid pyrolysis of coal particles as a part of a broader study of pulverized coal combustion. Samples were simultaneously collected and quenched with a water-cooled probe at points along the flow path through the pyrolysis zone of a vertical one-dimensional or plug-flow flame of pulverized coal (1.3 g/s) and air. Time-resolved records of total weight loss and weight loss by devolatilization were estimated from vol-

[147] C. T. Brooks, "Some Chemical Aspects of Processes Occurring in the Gas Recycle Hydrogenator," *Inst. Gas Eng. J. (England),* **7**, 492 (1966).

[148] H. W. Slotboom and J. M. L. Penninger, "The Role of the Reactor Wall in the Thermal Hydrocracking of Polyaromatic Compounds," *Ind. Eng. Chem., Process Des. Dev.,* **13**(3), 296 (1974).

atile and ash analyses of the collected particles and knowledge of the flow velocity. Particle temperature in this system typically rose from 200°C to about 1550°C with a maximum heating rate of approximately 22,-000°C/s.

Badzioch[5] and Badzioch and Hawksley[65] entrained finely ground and sized coal particles in a preheated laminar gas stream and collected the reacted particles in a water-cooled probe at designated distances downstream of the feed point. Heating rates of 25,000–50,000°C/s were reportedly attained for transit times of 30–110 ms and temperatures of 400–1000°C. Direct weight loss measurements were not possible, since the apparatus did not permit quantitative recovery of partially pyrolyzed particles stuck to the inside walls of the collector. However coal weight loss was estimated from an ash analysis of the residual material using ash as a tracer, and volatile matter remaining in the residue was determined by proximate analysis. Kimber and Gray[130,131] employed a similar apparatus but having higher temperature capability and extended the study to 2000°C for nonswelling coals, which presumably did not become sticky and cause the collection problem encountered with softening coals in the lower temperature study. These workers measured weight loss directly, thereby avoiding errors that would occur with the ash tracer technique because of ash loss at the higher temperatures.

A reactor of the Badzioch and Hawksley type was used at MIT[89,132] to obtain time-resolved data on the rapid pyrolysis of both a caking and a noncaking coal at temperatures up to 1800°C and heating rates in the range 10^4 to 2×10^5 °C/s. A steady feed of 2–3 mg/s of 38–45 μm diameter coal particles in a small laminar stream of cold argon carrier gas was injected axially into a larger laminar stream of plasma-preheated argon in a graphite muffle tube. The narrow particle stream and a portion of the surrounding annular stream were withdrawn at various distances from the feeder using a water-cooled probe, into which multiple water jets were directed to quench ($\sim 10^6$ °C/s) the particles rapidly and to prevent sticky coal particles from adhering to the inside walls of the collector. Direct weight loss measurements were achieved by filtering the residual particles from the flush water. Weight losses were also calculated using ash as a tracer and were found to deviate from the direct measurements in a manner indicative of ash vaporization at higher temperatures. In addition to ash content, the residual material was analyzed for volatile matter and elemental composition. The temperature-time history of the particles was calculated using a simplified model of the flow system, the measured temperature of the preheated gas, and particle velocities obtained with a laser Doppler anemometer sighted through radial observation parts of the furnace.

Another reactor of the Badzioch and Hawksley type was used in a coal pyrolysis study at Pennsylvania State University.[133] The recovery of char was incomplete, but the fractional weight loss was calculated from the measured ash content of the collected material using the ash as a tracer. These workers also measured densities and surface areas of the collected material.

Ubhayaker et al.[134] studied pyrolysis at temperatures up to 2000°C and heating rates up to 1.3×10^5 °C/s using a vertical entrained-flow cylindrical gasifier, in the top of which 0.5–15 g/s of pulverized coal in nitrogen carrier gas was rapidly mixed into hot combustion products from an oil burner operated with oxygen-enriched air. The mixture of char, pyrolysis products, and other hot gases was quenched with water jets near the bottom of the gasifier. Char was recovered from the quench water as a sludge, which was dried and subjected to ash analysis. The quenched gases were analyzed by chromatography. The yield of volatiles was determined by two methods, one based on the char analysis and the use of ash as a tracer, and the other based on the gas analysis and a calculation of the fraction of the total carbon remaining as a solid, assuming equilibrium to prevail at the measured gasifier temperature. The first method gave substantially larger yields, the difference apparently being a measure of the amount of soot formed by the cracking of volatiles in the hot gases. Such soot formation decreases the volatiles yield indicated by gas analysis but does not affect the yield indicated by char analysis when the soot is not collected with the char.

Moseley and Paterson[135] introduced a high-pressure apparatus in which finely divided particles of coal chars (100–150 μm) entrained in hydrogen were passed through a preheated reaction zone. Heating rates well over 1000°C/s and residence times as short as 0.17 s were possible. The unit could handle coal chars and dry lignite but failed to operate with fresh caking coal because of agglomeration. Glenn et al.[136] used an electrically heated, vertical continuous flow reactor into which coal particles (250 μm), slurried in water, were fed with various mixtures of synthesis gas. They reported that the addition of very fine silica particles (~3 μm) and a reduction in coal concentration in the slurry remedied the agglomeration problem. Johnson[137] studied lignite hydrogasification using a coiled tubular reactor that can be heated in stages to attain different heating rates. Fallon and Steinberg[138] used a vertical entrained-flow tubular reactor, through the top of which pulverized coal is fed into a preheated hydrogen stream and carried down to a cooling section, followed by a product recovery train.

Free-Fall Reactor. Coal particles are introduced and dispersed at the top of a large-diameter reactor and allowed to fall at terminal velocity through a preheated reaction zone. For example, Shapatina et al.,[139] as reported by Badzioch,[150] studied rapid pyrolysis of coal by dropping particles (150–200 μm) through a vertical furnace. Their work emphasized composition of volatile products instead of knetics, and most of their data were obtained at temperatures below 550°C. Feldmann et al.,[141] using a scheme suggested by Hiteshue et al.,[94] employed coal particles in the size range 150–300 μm and estimated a residence time of 0.33 s in a reaction zone 91 cm long. Mosely and Paterson[142] studied 100–150 μm particles and estimated a residence time exceeding 0.5 s for a 137-cm fall. Kobayashi[89] operated the entrained-flow reactor described above[132] under free-fall conditions to achieve longer residence times. By not employing the preheated argon stream, the injected coal particles were heated during their free fall to a collector located at the bottom of the stagnant muffle tube. An important feature of the free-fall technique is the ability to gasify coal particles under rapid heating conditions without agglomeration. The difficulty of estimating residence time for a particle undergoing significant density changes due to swelling and gasification is one of the most serious drawbacks to the use of this method in kinetics studies.

[149] C. E. Hamrin, Jr., and P. S. Maa, "Desulphurization and Deagglomeration of Western Kentucky Coal Using Bottom-Ash," *Fuel,* **54,** 288 (1975).

[150] S. Badzioch, "Rapid and Controlled Decomposition of Coal," *BCURA Month. Bull.* **25**(8), 285 (1961).

12.4 PYROLYSIS IN INERT GAS

12.4.1 Weight Loss Behavior

12.4.1.1 *General Characteristics*

Coal is the product of very slow decomposition of organic matter deposited in prehistoric times. The rate of decomposition can be increased by elevating the temperature. When the coal is heated at a more or less conventional rate, the decomposition becomes apparent at 350–400°C and the products consist of a carbon-rich residue and a hydrogen-rich volatile fraction, as mentioned earlier. The decomposition continues until a temperature typically around 950°C is reached, which if maintained for an extended time results in a residue of nearly pure carbon, with a structure approaching that of graphite. The accumulated volatiles are comprised of various gases and liquids, the relative proportions of which depend on the coal type and the manner of heating.

In the early 1960s investigators in Germany,[71,72] France,[105] England,[4] the Soviet Union,[150] and the United States[120] found that rapid heating techniques for coal permit substantially more volatiles than traditional slow heating methods. Furthermore, the liquid or tar fraction seemed most strongly influenced, resulting in considerable excitement over the prospects of developing a commercially viable coal conversion process based on pyrolysis. Much work has since focused on the kinetics, mechanism, and product distribution, to provide fundamental information of practical importance.

The effective volatile content of a coal must be carefully distinguished from volatile matter (VM), determined by proximate analysis, which is a convenient standard but a deceptive reference when the conditions of interest differ significantly from those of the proximate test. For example, it is often incorrect to associate VM with the potential yield of volatiles from pulverized coal particles widely dispersed in a gas. The volatiles escaping from particles in the packed bed employed in proximate analysis undergo secondary reactions including cracking and carbon deposition on solid surfaces, but the extent of such reactions within the bed is not well established.

Workers at the British Coal Utilisation Research Association (BCURA) studied the relationship (or lack thereof) between actual volatile yield and proximate volatile content of coal and char under a wide range of conditions. Gregory and Littlejohn[151] analyzed extensive literature data on retort carbonization of coals and found the yields to be less than VM. Badzoich and Hawksley[65] observed that the yield of volatiles from pulverized coal particles entrained and rapidly heated to 950°C in preheated inert gas exceeds the loss of proximate volatile matter found from the difference between the proximate volatile matter of coal and that of char. The results for several different coals were correlated by the empirical relation:

$$V^* = Q(1 - VM_c)VM \qquad (12.3)$$

where V^*, expressed as weight fraction of the original coal, is the ultimate yield of volatiles achieved by heating the coal for an infinite time ($t = \infty$), approximated by measurements at relatively long coal residence times; Q is the ratio of total weight loss to loss of proximate volatile matter; and VM_c is the fraction of the initial proximate volatile matter remaining in the char at $t = \infty$, obtained by extrapolation. Experimental values of Q varied from 1.3 to 1.8, depending on coal type, which led to values of V^*/VM greater than 1. Kimber and Gray[130,131] extended the measurements to 2000°C and found Q values from 1.3 to 1.95.

The observation of volatile yields exceeding proximate volatile matter is consistent with the results of other investigators who rapidly heated pulverized coal particles dispersed in gas or held in monolayers or thin beds.[75,105–107,114,120,121,127,132,134,146] As Table 12.4 indicates, the ratio V^*/VM for several American, British, and French coals usually falls within the range 0.75–1.36, although two of the higher temperature experiments indicate larger values. In a similar manner, both Peters[71] and Sass[129] found that flow schemes with rapid heating of coal particles suspended in gas gave significantly higher yields than the standard Fischer retort method. [The Fischer retort procedure involves slowly heating 20–100 g of coal to a designated temperature (400–600°C) and holding it there until evolution of volatiles ceases. The coal is heated in a special aluminum retort with an outlet tube leading to a condenser, where liquid products are collected.[152]]

An important observation from Table 12.4 is that not all coals give yields exceeding the proximate volatile content. Although the BCURA workers correctly note that some of their yields exceed by almost 100 % (value of Q in Equation 12.3) the *loss* of proximate volatile matter, none of the yields in the table except those of Kimber and Gray[130,131] and Ubhayaker et al.[134] exceeds the *total* proximate volatile matter by more than 36%. Residence times much longer than those (several seconds or less) associated with the data in the table would be required at temperatures less than 1200°C to complete the devolatilization of char, which is slow and accounts for about 5–8% of the original maf coal.[5,139,153,154] Therefore the experimental yields shown in Table 12.4 for pyrolysis at 1200°C or less are not truly V^* ($t = \infty$) as indicated, but only a close approximation. At the higher temperatures shown in the table (1800, 1900, and 2000°C), pyrolysis is extremely rapid as is discussed below, and the data are for essentially complete devolatilization. However the yields from one of the three high-temperature experiments[132] are within the range established by the lower temperature data. This raises the question of whether the difference between the excep-

[151] D. R. Gregory and R. F. Littlejohn, "A Survey of Numerical Data on the Thermal Decomposition of Coal," *BCURA Month. Bull.*, **29**(6), 173 (1965).

[152] F. Fischer, *The Conversions of Coal into Oils*, D. Van Nostrand Company, New York, 1925.

[153] J. B. Howard, "Mechanisms of Ignition and Combustion in Flames or Pulverized Bituminous Coal," Ph.D. thesis, Dept. of Fuel Technology, Pennsylvania State University, University Park, 1965.

[154] D. B. Anthony, "Rapid Devolatilization and Hydrogasification of Pulverized Coal," Sc.D. thesis, Dept. of Chemical Engineering, Massachusetts Institute of Technology, Cambridge, Mass., 1974.

Table 12.4 Comparison of Experimental Yields with Proximate Volatile Matter

Investigators, Coals and Conditions		Proximate VM	Experimental[a] V*	V*/VM
Loison and Chauvin[105]				
Maigre Oignies No. 2		8.4	8.4	1.00
Bergmannsgluck	Mf basis; heated to	18.1	23.6	1.30
Emma (fines)	1,050–1,100°C at	20.4	24.5	1.25
Lens-Lieven No. 10	1,500°C/s	24.4	26.7	1.09
Flénus de Bruay		31.0	39.6	1.18
Wendell III		33.9	34.5	1.02
Faulquemont		36.4	36.7	1.01
Rau and Robertson[106]				
Colver (mvb)	Mf basis, heated to	25.3	19	0.75
Kopperston No. 2 (hvAb)	950°C at 600°C/s	31.6	36	1.14
Federal (Pittsburgh hvAb)		37.7	49	1.30
Elkol (Wyoming subbituminous)		40.7	48	1.17
Orient No. 3 (hvb)		44.0	41	0.93
Kimber and Gray[131,130]				
NCB 902 30 μm	Maf basis; heated	38.3	71	1.85
(particle median 30 μm[b]	to 1,900°C at	38.3	72	1.88
diameter shown) 53 μm	10^5–10^6°C/s	40.0	69	1.73
Badzioch and Hawksley[65]				
NCB 203 (K)	Maf basis; heated	17.7	17.7	1.00
NCB 301a (J)	to 950°C at	22.7	20.6	0.91
NCB 301b (H)	25,000–50,000°C/s	25.2	26.0	1.03
NCB 602 (E)		34.3	44.2	1.29
NCB 401 (G)		34.4	35.1	1.02
NCB 601 (F)		35.3	45.1	1.28
NCB 802 (D)		36.1	47.0	1.30
NCB 902 (B)		36.4	40.0	1.10
NCB 802 (C)		37.8	42.8	1.13
NCB 902 (A)		37.9	42.9	1.13
Mentser et al.[110]				
Pocahontas 3	Mf basis; heated to	16.8	18.5	1.10
Lower Kittanning	1200°C at 8000°C/s	25.3	30.8	1.22
Pittsburgh		35.1	47.9	1.36
Rock Springs		37.7	42.4	1.12
Colchester		48.0	55.8	1.16
Anthony et al.[75]				
Pittsburgh No. 8	Maf basis; heated	46.2	53.7[c]	1.16
Montana lignite	to 1000°C at 650–10,000°C/s	46.2	41.1	0.89
Kobayashi et al.[132]				
Pittsburgh No. 8	Maf basis; heated	46.8	63	1.35
Montana lignite	to 1800°C at 2×10^5 °C/s	46.1	62	1.34

Table 12.4 *Continued*

Investigators, Coals and Conditions		Proximate *VM*	Experimental[a] V^*	V^*/VM
Suuberg[146]				
Montana lignite[d]	Maf basis; heated	46.1	50.0	1.08
Montana lignite[e]	to 1000°C at	44.3	44.7	1.01
Pittsburgh No. 8[e]	1000°C/s	44.6	52.2	1.17
Ubhayaker et al.[132]				
Pittsburgh Seam	Maf basis; heated to 2000°C at 1.3 × 10[5] °C/s	42.5	65	1.53

[a]Peak weight loss corresponding to V^* was achieved at the final (highest) temperature in all cases except Mentser et al., where the peak occurs at intermediate temperatures.

[b]Previous sample recycled.

[c]A range up to 56.2 given previously[8] is not used here, since the higher value reflects behavior at reduced pressure.

[d]From same master sample used by Kobayashi et al.[132]

[e]From same master sample used by Anthony et al.[75]

tionally high yields and those of the lower temperature experiments should be regarded as a measure of the more complete devolatilization or of some other characteristic of the difficult high-temperature techniques.

It can be seen from the data of Loison and Chauvin,[105] Rau and Robertson,[106] Badzoich and Hawksley,[65] and Mentser et al.,[110] each set of which covers a substantial range of coal ranks, that the tendency for V^* to exceed *VM* is strongest for the intermediate rank coals, which are in the bituminous range where the yield of tar also exhibits a maximum. As discussed later, volatile yields in excess of *VM* are often the consequence of reduced tar cracking and less carbon deposition, which are achieved by the reduction of contact between reactive tars and hot particle surfaces. The effect will clearly be more pronounced for coals that give large tar yields.

12.4.1.2 *Effect of Temperature*

Most of the available data on coal pyrolysis were obtained at temperatures below about 1000°C, although as indicated above, some measurements at higher temperatures have been reported. Figure 12.2 illustrates how pyrolysis yields depend on temperature in the range up to 1000°C. The data of each investigator have been normalized to a yield of 100% at 1000°C. This rather arbitrary basis is convenient because the yields observed in this temperature range approach a higher temperature asymptote, which is closely approximated by the yield observed at 1000°C. The curves on the plot require some explanation. Dryden's[155] correlation is based on the slow carbonization of a large number of American and British coals, here representing a coal having about 35–40% volatile matter (maf). The curve is said to be general in that it will accommodate any coal by simply sliding the abscissa to the right or left (~50°C to the left per 10% increase in *VM*). Shapatina et al.[139] dropped coal particles through a heated chamber (minimum residence time 0.45 s), then caught and soaked them in a heated trap for the duration of the run. Particles soaked for 14,400 s show a weight loss trend consistent with Dryden's correlation. The yields are less for shorter residence times (e.g., 30–60 s), and they are diminished further if only the heatup period of 0.45 s is considered. The additional weight loss observed between 0.45 and 30 s is roughly equivalent to the amount lost between 30 s and 14,400 s, which supports the existence of rapid initial decomposition followed by very slow degasification of char. Wiser et al.[36] determined values of the ultimate yield

[155] I. G. C. Dryden, "Chemistry of Coal and Its Relation to Coal Carbonisation," *J. Inst. Fuel*, **30**, 193 (1957).

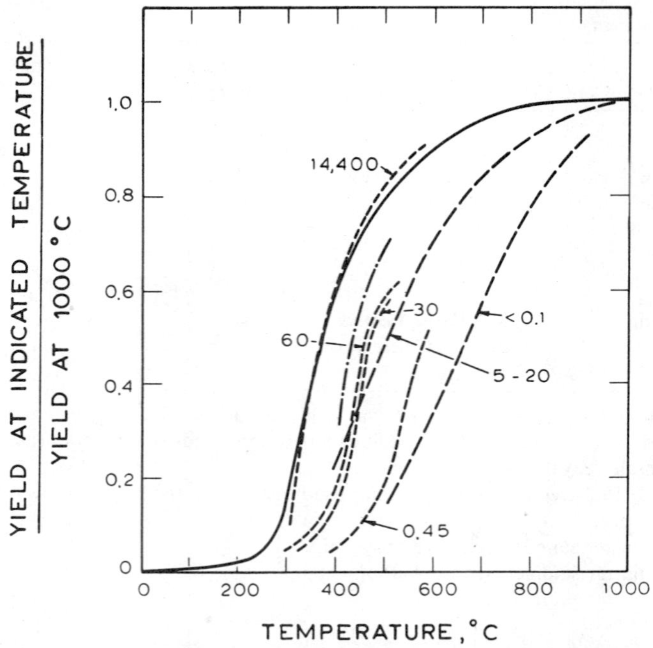

Figure 12.2 Effect of temperature on pyrolysis weight loss (yield) at different residence times (s, indicated on some curves). Curves: solid, Dryden[155]; short dashes, Shapatina et al. [139]; long dashes, Anthony et al.[75]; dots and dashes, Wiser et al.[36]

(V^*) by including the contributions of a final devolatilization stage identified as a constant rate period. The yields obtained by Anthony et al.[75] using bituminous coal at residence times of 5–20 s are in good agreement with those of Shapatina et al. at 30–60 s. The conversions obtained by Anthony et al. with rapid heating (10,000°C/s; conversion during heatup only) establish a trend similar to the yields of Shapatina et al. at 0.45s.

Since all data in Fig. 12.2 are normalized using the yields observed at 1000°C, the curves do not reflect the extent to which the largest yields exceed the proximate volatile content. Dryden and Wiser used experimental conditions comparable to those of proximate analysis and would not have been expected to see changes in total yield at high temperatures. Shapatina et al. calculated weight loss as the difference between the measured proximate volatile content of the fresh coal and that of the recovered char. This technique obviously cannot detect yields greater than proximate volatile matter, which very likely occurred in the dispersed-phase system employed. The yields of Anthony et al. at 1000°C and 5–20 s are about 20% larger than *VM*.

Although Fig. 12.2 supports the concept that the yield of pyrolysis products, for reasonably long experimental times, becomes independent of temperature above about 1000°C, recent data obtained at higher temperatures and presented in Fig. 12.3 suggest that additional devolatilization in relatively short times may be observed by heating the coal to temperatures well beyond 1000°C. This suggestion seems reasonable because the absolute time required for a given degree of completion of devolatilization is much less for the higher temperatures. For example, the heating of a coal from 1000 to 1800°C might produce in fractions of a second an additional weight loss that at 1000°C could be produced only in times too long to be available in laboratory experiments. Furthermore the slower rate of escape of volatiles from the particles at a lower temperature might present a larger opportunity for secondary cracking reactions and carbon deposition, which could eliminate the possiblity of ever attaining the weight loss achievable at a higher temperature. On the other hand, an equally plausible argument might be that an increased rate of cracking and carbon deposition at the higher temperatures might

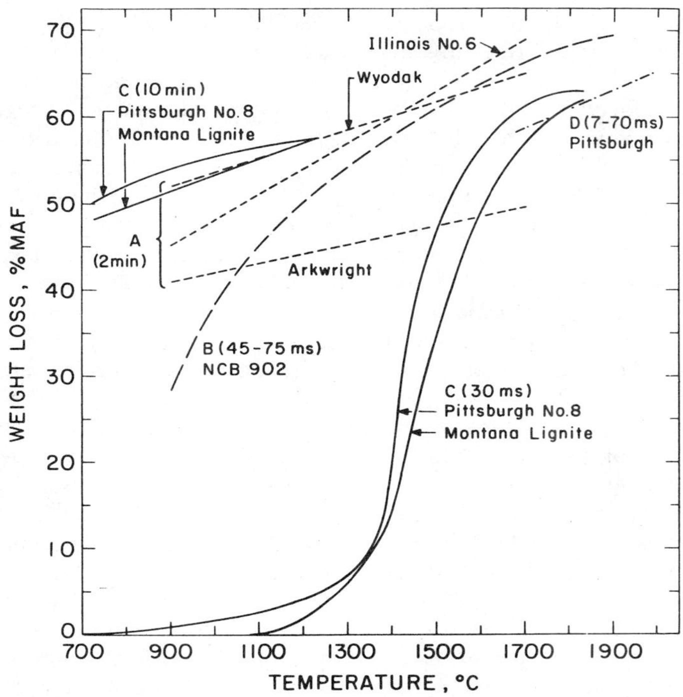

Figure 12.3 Effect of higher temperatures on pyrolysis weight loss at different residence times. *A*, Blair et al.[114]; *B*, Kimber and Gray[130]; *C*, Kobayashi[89] and Kobayashi et al.[132]; *D*, Ubhayaker et al.[134] Residence times [where available] and coal seams indicated on curves.

reduce the net weight loss to a value less than that achievable at lower temperatures, given enough time.

Comparison of the apparently asymptotic yields at the lower and higher temperatures is possible for two coals, data for which are included in Table 12.4. The Pittsburgh No. 8 coal for which a volatile yield of 53.7% at 1000°C was found by Anthony et al.[75] and a similar value of 52.2% was found by Suuberg,[146] in both cases using a thin layer of particles on a screen heater, gave a yield of 63% at 1800°C in the laminar flow reactor of Kobayashi et al.[132] (The master sample of Pittsburgh No. 8 coal used by Anthony et al. and by Suuberg appears to be very similar to the one used by Kobayashi et al., although the two samples were obtained about 2 years apart from the same mine.) Similarly, the volatile yields from the same master sample of Montana lignite were 50% at 1000°C on the screen heater[146] and 62% at 1800°C in the laminar flow reactor of Kobayashi et al. (As discussed by Suuberg, this sample of Montana lignite differs in several respects from the one used by Anthony et al. and by Suuberg in the main part of his study, although the two samples came from the same mine, about 2 years apart. Evidence of this difference is provided by the data of Suuberg in Table 12.4.) Although the high-temperature yields are within a band of possible extrapolations (not shown here) from the lower temperature data represented by the curves in Fig. 12.2, more work is needed to confirm the asymptotic behavior at the higher temperatures.

The solid phase resulting from the higher temperature experiments with bituminous coals includes both residues of coal particles and soot formed from volatiles. The reported volatile yields are based on separation techniques that include water scrubbing and filtration,[132] use of a cyclone,[130] and use of water scrubbing and sludge drying.[134] Small char fragments escaping the separator would lead to overestimation of volatile yield, and the collection of some of the soot with the char would cause an underestimation. The separation problem was studied by Kobayashi et al. using an ash tracer technique and by Ubhakayer et al. using calculations based on gas phase analyses, but the associated experimental errors deserve further investigation.

The effect of time on the variation of weight loss with temperature, which is illustrated in Fig. 12.2, can also be seen from the data of Kobayashi et al.[132] in Fig. 12.3. To achieve coal residence times longer than those available in the laminar flow reactor, the same reactor was operated in a free-fall mode by allowing the coal particles to fall at their terminal velocities into either a cooled or an uncooled collector at the bottom of the furnace. The data in Fig. 12.3 for a reaction time of 10 min represent a fall time on the order of 1 s and a soak period of about 10 min in an uncooled alumina collector. As in the case of the lower temperature data in Fig. 12.2, the increase of weight loss with temperature shown in Fig. 12.3 is much stronger for shorter reaction times. Thus as time increases, the effect of temperature on pyrolysis rate becomes less apparent, since the decomposition eventually approaches completion at any of the temperatures. Given enough time and the absence of secondary cracking and carbon deposition, the weight loss versus temperature curve would presumably be the horizontal line approximated in practice by the high temperature asymptote, which is defined above as the ultimate yield V^*. It is clear from Fig. 12.3 that V^* for a given coal unfortunately remains poorly established despite the availability of these high-temperature data.

12.4.1.3 *Effect of Heating Rate*

The fact that rapid heating techniques in coal pyrolysis give substantially larger yields of volatiles than are obtained by the slow heating of coal in conventional packed-bed carbonization retorts or standard volatile matter crucibles is sometimes attributed to an effect of the heating rate itself. This rationale is not proved by the data, since the techniques used to achieve faster heating, such as small particles well dispersed in a gas or thinly spread within a wire mesh heater, promote larger volatiles yields by avoiding some of the cracking reactions and carbon deposition associated with the conventional methods. In a critical test of the rationale, Anthony et al.[75] varied only the heating rate, in the range 650–10,000°C/s, while holding all other variables constant in a pyrolysis experiment on both a caking and a noncaking coal. The yield of volatiles did not change significantly (Table 12.4), thereby indicating that heating rate per se is not important with respect to volatiles yield under the conditions studied. Recent data of Suuberg[146] for the same two coals confirm this picture and extend it to include heating rates down to 270°C/s, and volatiles composition as well as yield (described later). Thus yield increases sometimes attributed to heating rate probably result primarily from the associated experimental conditions employed to achieve the faster heating. Other workers have reached essentially the same conclusion.[121]

The data of Kimber and Gray,[130,131] Ubhayaker et al.,[134] and to a smaller extent Kobayashi et al.[132] in Table 12.4 and Fig. 12.3 indicate, as discussed above, that increased volatiles yields may result from extensions of heating rate and final temperature to values above 100,000°C/s and to the range 1800–2000°C, respectively. Since heating rate and final temperature were not varied independently in these studies, the results do not resolve the separate effects of these two variables. However the same results in combination with the high-temperature but lower heating rate data of Blair et al.,[114] which are also shown in Fig. 12.3, support the view that temperature, not heating rate, is primarily responsible for the higher yields. Except for the Arkwright coal, which is not comparable to the other coals represented in Fig. 12.3, the weight losses observed by Blair et al. are generally as large or larger than those from the other three high-temperature studies. This set of data of Blair et al. was obtained by dropping coal particles onto a preheated graphite ribbon. Heating rate was not measured, but it seems to have been less than about 7000°C/s, which one calculates for the highest temperature run, the smallest particle size, and no limitation on heat transfer to the particle surface. The last condition presumably leads to a large overestimation, since the particle size was of the order 500–600 μm and only a small fraction of the particle surface actually contacted the heating element. Since the various high-temperature data in Fig. 12.3 represent heating rates from less than 7000°C/s to more than 200,-

000°C/s, the comparable weight losses suggest the absence of a significant effect of heating rate on this quantity.

Heating rate does have a clear effect on the temperature-time history of pyrolysis. Prior to the period covered by this review, the thermogravimetric measurements of Van Krevelen et al.[38] showed that as coal is heated, the temperature at which the rate of devolatilization assumes its maximum value increases with increasing heating rate. Furthermore the maximum rate of devolatilization was found to increase approximately linearly with heating rate. However the heating rates employed by Van Krevelen et al. were low and within a small range (i.e., 0.6–6.6°C/min). Consequently the temperature of the maximum devolatilization rate only increased from about 475°C to about 525°C, and the maximum rate only increased by a factor of about 10. More recently, much larger changes of this type were observed in association with the higher heating rates and higher temperature of interest in entrained-flow and fluidized-bed systems. For example, coal heated at rates up to about 22,000°C/s in a pulverized coal flame was found to attain temperatures well above 1000°C before the occurrence of the maximum devolatilization rate.[66]

Anthony et al.[75] used the electrical wire mesh heater described above to separate the time-resolved effects of heating rate and temperature. Different samples were heated at a given rate to a given temperature and held there for different times before breaking the heating circuit. Other samples were heated at the same rate but to lower temperatures, at which the circuit was immediately broken. Repetition of this procedure for different heating rates and different final or holding temperatures gave weight loss versus time data, of which Fig. 12.4 is representative. Elapsed time in the figure is the time at which the heating circuit was broken, but the measured weight loss includes that which occurred during the subsequent cooling period. Accordingly, the measurement of temperature-time history of each sample included the cooling period as well as heatup, and all this information was employed in the derivation of kinetic parameters described later.

The data in Fig. 12.4 are for Montana lignite heated at three different rates to 1000°C and at one rate to 700°C. Although increases of heating rate in the range 650–10,000°C/s for a final temperature of 1000°C do not affect the final weight loss, these changes clearly reduce the time required to achieve a given weight loss, hence

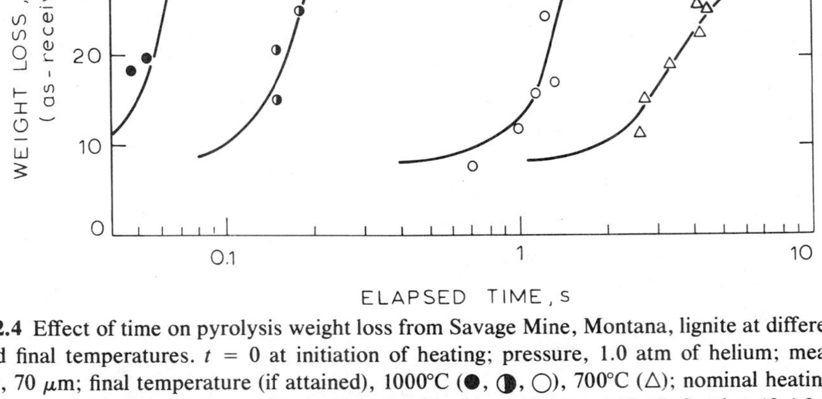

Figure 12.4 Effect of time on pyrolysis weight loss from Savage Mine, Montana, lignite at different heating rates and final temperatures. $t = 0$ at initiation of heating; pressure, 1.0 atm of helium; mean particle diameter, 70 μm; final temperature (if attained), 1000°C (●, ◑, ○), 700°C (△); nominal heating rate (°C/s), 10,000 (●), 3000 (◑), 650 (○), 180 (△). Curves: Equations 12.16 and 12.17 (Section 12.4.3.2) with $V^* = 40.6$ wt % of original coal, $E_0 = 48.72$ kcal/mole, $k_0 = 1.07 \times 10^{10}$ s^{-1}, and $\sigma = 9.38$ kcal/mole.[75]

increase the rate of weight loss. The curves drawn through the data in Fig. 12.4 represent the multiple parallel reaction model of coal pyrolysis, described later, with a single set of parameter values employed for all the heating rates and final temperatures. The data, as represented by the model and fitted parameters, are used to illustrate the high-temperature and high heating rate version of the same types of pyrolysis behavior (Figs. 12.5 and 12.6) described above for the lower temperatures and lower heating rates studied by Van Krevelen et al.[38]

Figure 12.5 shows the effect of heating rate on the temperature dependence of the extent of devolatilization. It can be seen, for example, that the temperature attained before devolatilization reaches 10% completion is 400°C at a heating rate of 1°C/s and 840°C at 10^5 °C/s. For 90% completion the coal reaches 860°C at 1°C/s and 1700°C at 10^5 °C/s. Figure 12.6 shows how the rate of devolatilization depends on heating rate and temperature. The maximum rate increases somewhat less than linearly with heating rate. As the heating rate is increased, not only does the temperature of the maximum rate shift to higher values, but also a given fraction of the devolatilization occurs over a wider temperature range. Thus as also noted above, the temperature range traversed by the coal in going from 10 to 90% completion of devolatilization increases from 400–840°C at 1°C/s to 860–1700°C at 10^5 °C/s. This behavior reflects the dependence of the amount of pyrolysis occurring at a given temperature on the time spent at that temperature, hence its dependence on heating rate. For the same reason, heating rate affects the cumulative amount of devolatilization occurring before a given temperature is attained, which in turn affects the instantaneous devolatilization rate at that temperature, since the rate is first order in the amount of pyrolyzable material remaining, as discussed later. These effects of heating rate on the temperature-time his-

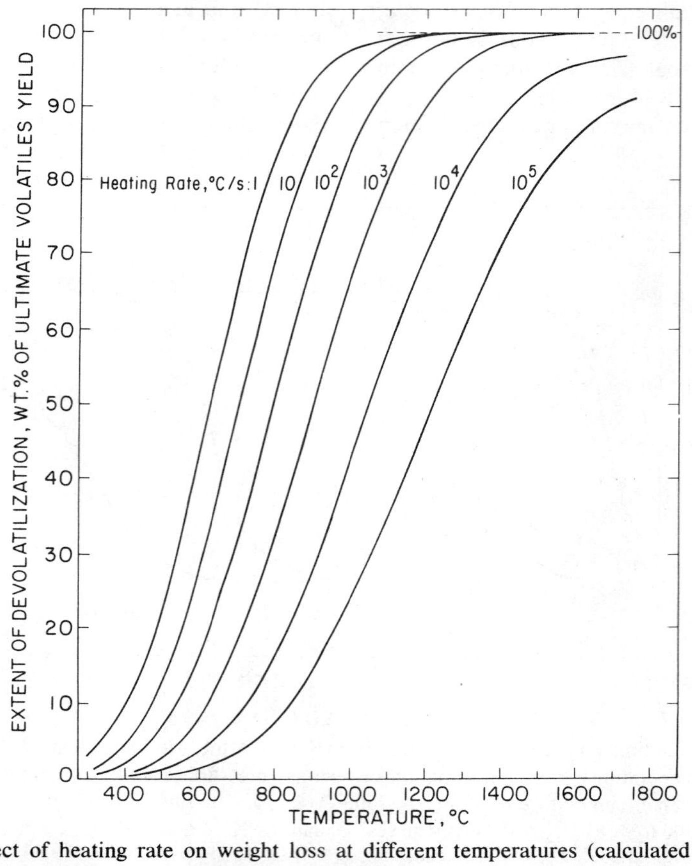

Figure 12.5 Effect of heating rate on weight loss at different temperatures (calculated with Equations 12.16 and 12.17 and parameters for Montana lignite from Fig. 12.4).

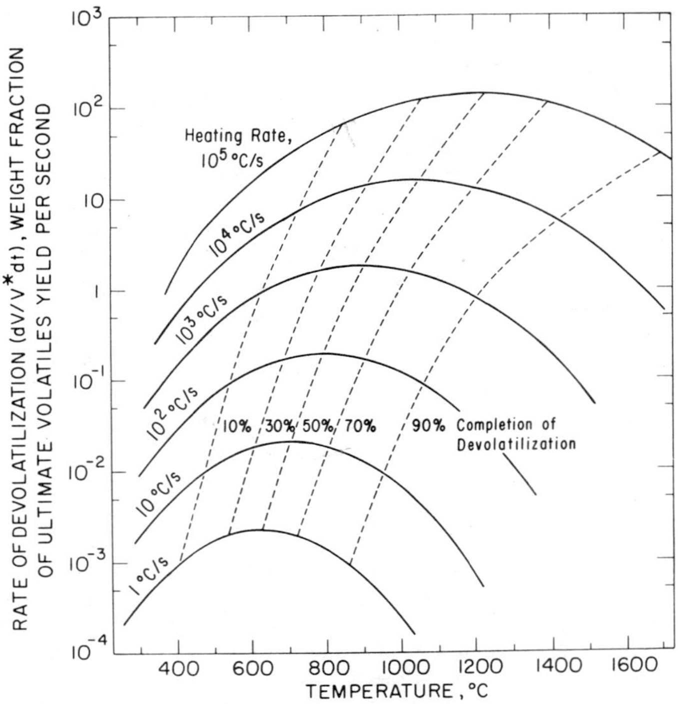

Figure 12.6 Effect of heating rate on rate of devolatilization at different temperatures (calculated with Equations 12.16 and 12.17 and parameters for Montana lignite from Fig. 12.4).

tory of pyrolysis can clearly influence the fate of the pyrolysis products in situations where secondary reactions are important. Furthermore, heating rate, through its effects on rate of volatiles evolution, would presumably affect the partial pressure of the ambient gas at or within a particle undergoing devolatilization. Such behavior has practical implications regarding the achievement of intimate contact between pyrolyzing coal particles and reactants in the ambient gas.

12.4.1.4 *Effect of Pressure*

It has been known for some time[2] that the pyrolysis of coal under reduced pressure results in larger weight losses, the main part of the increase being in the tar fraction. The behavior is generally attributed to secondary reactions, including cracking and carbon deposition, which can convert some of the tars to both heavier and lighter species (Section 12.4.2.4). These reactions are less favored at lower pressures owing to the smaller resistance to volatiles flow out of the coal particles.

The studies of low-pressure, essentially vacuum, pyrolysis were largely motivated on fundamental grounds. The technique offers a source of tars that, being less altered by secondary reactions, presumably bears a closer relationship to the original coal material. The recent growth of interest in pressurized coal conversion and combustion now presents practical reasons for extending the previous studies to considerably higher pressures. To this end, Anthony et al.[75] studied the effect of pressure in the range 0.001–100 atm on pyrolysis weight loss from two coals.

Data from Pittsburgh Seam bituminous coal heated at different rates to 1000°C are given in Fig. 12.7. The curves represent a correlation discussed in Section 12.4.3.3. A certain amount of carbon deposition from cracked volatiles occurred on the stainless steel screen sample holder, particularly at the lower pressures. Since weight loss was determined by weighing the screens before and after the run, the deposit reduced the observed weight loss by an error that was measured by weighing the screen without the coal residue. The corrected weight loss is represented by the dashed curve.

Weight loss is seen to decrease monotonically with increasing pressure, presumably

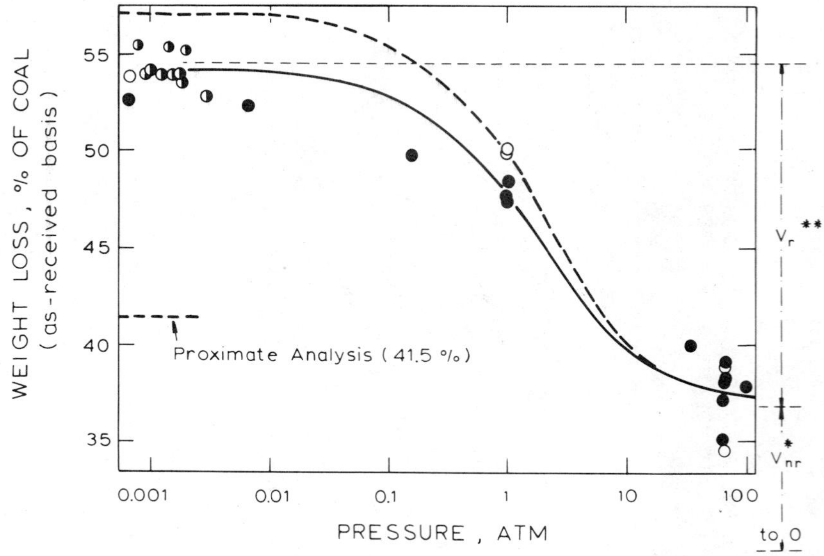

Figure 12.7 Effect of pressure on weight loss in the pyrolysis of Pittsburgh Seam bituminous (Ireland Mine) coal. Final (holding) temperature, 1000°C; experimental time, 5–20 s; mean particle diameter, 70 μm; atmosphere, helium; nominal heating rate (°C/s), 10,000 (○), 3000 (◑), 650–750 (●). Curves: solid, uncorrected for cracking deposition on sample holder; dashed, corrected for cracking deposition on sample holder. Curves from Equation 12.34 (Section 12.4.3.3) with V_r^{**} = 17.0 (solid curve) and 20.0 (dashed curve) % of original coal, V_r^* = 37.2% of original coal, and k_1/k_c = 0.56 atm^{-1}.[75]

reflecting an increasing extent of cracking and carbon deposition within the particles. The weight loss under atmospheric pressure is 50.0% of the as-received coal (ash = 12.2%; moisture = 1.65%), but the loss is 57.2% under vacuum and 37.2% at 100 atm. Furthermore, the trend suggests asymptotic behavior at the pressure extremes, but the scatter of the data precludes a firm location of the asymptotes. Such behavior would be expected theoretically. The pressure inside the particles during devolatilization probably becomes independent of the external pressure as the latter is reduced below a certain value, hence the low-pressure asymptote. The approach to a limiting yield at high pressures probably reflects the depletion of species that are susceptible to the secondary reactions. Accordingly, the weight loss achieved at the high-pressure extreme is the yield of nonreactive volatiles (V_{nr}^*), and the additional weight loss available by reducing the pressure to vacuum represents reactive volatiles (V_r^{**}).

The variation of the pressure effect with temperature is shown in Fig. 12.8 for the same coal as above.[154] Two points are obvious: the effect of pressure appears only above a certain temperature, here about 600°C; the high-pressure yield becomes roughly independent of temperature above the same temperature. In view of the foregoing interpretation of the pressure effect, these points imply that volatiles formed below a certain temperature are effectively inert to secondary reactions under the temperature, time, and other conditions prevailing in the particle, whereas most of the additional volatiles formed at the higher temperatures are reactive. Although this picture is quantitatively useful, it is clearly an approximation from a mechanistic point of view. Furthermore, the critical temperature here found to be 600°C is a function of both heating rate and time spent at the final temperature.

Similar experiments were performed with Montana lignite. The effect of pressure on weight loss was too small to be interesting in comparison with the foregoing behavior.

12.4.1.5 *Effect of Particle Size*

Although particle size has been varied in several studies, the results are often inconclusive because of failure to use the same temperature, or temperature-time history if time is small, for all particle sizes. Even small temperature differences associated with different particle sizes can cause larger changes in volatiles yield than would result from the particle size change at constant temperature. The problem is especially difficult because the effect of increasing particle size is a small to negligible decrease in volatiles yield, whereas this direction of particle size change usually results in lower temperatures, owing to the slower heatup of larger particles, and hence lower yields.

Anthony systematically varied particle size over the range 53–1000 μm in pyrolysis experiments with the wire mesh heater and an atmosphere of helium.[154] The effect of increasing the particle size of Pittsburgh Seam coal over this range was a decrease in weight loss of about 3 wt % (as received) of the original coal as shown in Fig. 12.9. This behavior apparently reflects an increased extent of secondary reactions and carbon deposition for larger particles, which is consistent with the view that larger particles offer more resistance to the escape of volatiles. In contrast to the bituminous coal, Montana lignite exhibited no significant effect of particle size within the stated range.

The rapid heating rate, high-temperature study of Ubhayaker et al.[134] involved the use of two size fractions, +200 mesh and −200 mesh, of pulverized Pittsburgh Seam bituminous coal. The mass mean diameters of the two fractions were 100 and 20 μm, respectively. The yield of volatiles from the fine fraction exceeded that from the coarse fraction by an amount that decreased with increasing residence time. When the particles were heated in a gas that was fed at a temperature of 1700°C, the difference in volatiles yield was 14 wt % (maf) of the coal at

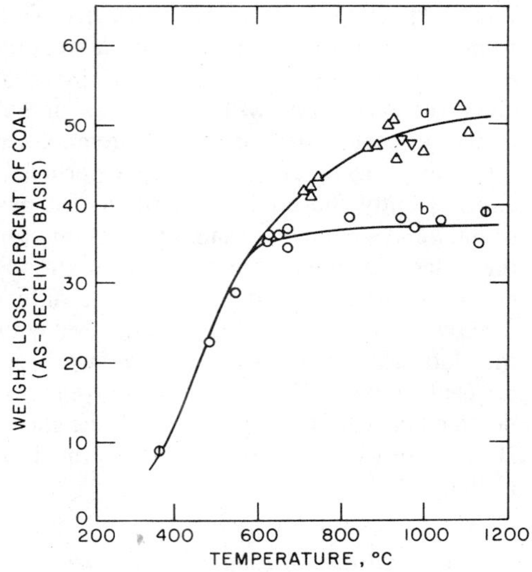

Figure 12.8 Effect of temperature on weight loss from Pittsburgh Seam No. 8 (Ireland Mine) bituminous coal heated in inert atmospheres at different pressures. Coal: VM = 39.8%, moisture = 1.6%; nominal heating rates, 65–750°C/s; mean particle diameter, 70 μm; residence times, 5–20 s. Heated in helium (▽) and nitrogen (△), 1 atm; helium (○), 69 atm; two-step heating, helium (⦶), 69 atm. Curves, Equation 12.53 with parameter values given in text; and a, P = 1 atm, P_{H_2} = 0; b, P = 69 atm, P_{H_2} = 0.[154]

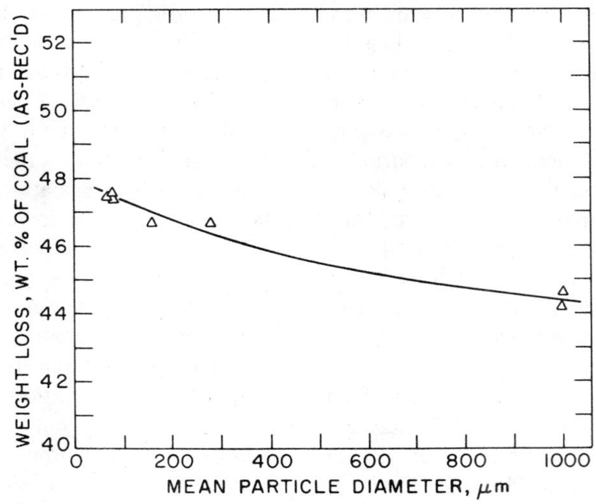

Figure 12.9 Effect of particle size on weight loss from Pittsburgh Seam bituminous coal heated in helium. Ash = 12.2%; moisture = 1.6%; 1000°C final (holding) temperature, 5–20 s residence time, 650–750°C/s nominal heating rate, 1 atm helium.[154]

7 ms, 7% at 25 ms, and 5% at 70 ms. This trend indicates the existence of a heating effect whereby the difference in yields at a given residence time within the range above is due at least in part to the requirement for longer heating times for the larger particles. Consequently the effect of particle size per se must have been less than 5%, as found for the longest residence time. In fact, extrapolation of this trend indicates that the difference would become negligible compared with the data scatter if residence time were increased to about 100 ms. The occurrence of no significant particle size effect in the range 20–100 μm would be consistent with the data in Fig. 12.9.

12.4.1.6 *Effect of Coal Type*

Only fragmentary information is available on the effect of coal type on weight loss in pyrolysis. According to the Dryden[155] correlation mentioned above (Fig. 12.2), the temperature at which a given fraction of the final weight loss is achieved decrease by about 50°C for each 10% increase in volatile matter content. Although this picture does not appear to have been tested critically in the recent literature, it is consistent with the data of Anthony.[154] Approximately the same relationship between the fractional attainment of the final (1000°C) weight loss and temperature was observed for a bituminous coal and a lignite having essentially the same volatile matter contents.

One of the most striking differences in the weight loss behavior of different coals is the potential for volatile yields to exceed the proximate volatile matter content when conditions are employed that avoid extensive tar cracking and carbon deposition on particles. This potential correlates with tar yield, as discussed above in connection with Table 12.4, being small to negligible for both low- and high-rank coals and strong for coals in the bituminous range. Another, but probably related, difference in behavior is revealed by the effects of pressure and particle size discussed above. Although weight loss from Pittsburgh Seam bituminous coal decreases with increases in either pressure or particle size, weight loss from Montana lignite is relatively insensitive to both these variables. Therefore, according to the interpretations given above, the extent of secondary reactions in the lignite seems to be relatively independent of changes in the resistance to volatiles escape from the particles. Unfortunately the recent weight loss data do not indicate whether the primary lignite volatiles are (1) extremely reactive (thereby permitting the completion of secondary reactions even for the lowest pressures and smallest particles studied) or (2) so inert to secondary reactions that almost no carbon deposition occurs, even for the highest pressures and largest particles studied.

Jones[3] discusses the concept that certain reactive aromatic units formed in pyrolysis can either take up hydrogen and be evolved as tar or recondense to form semicoke. Hydroxyl or other reactive oxygen groups can act to favor the latter route by consuming hydrogen that would otherwise promote tar evolution, or they can increase coke formation directly through cross-bonding. In view of the much larger oxygen content of lignite, such behavior would appear to be more prevalent for lignite than for bituminous coal. If the secondary reactions thus indicated are fast on the time scale of pyrolysis experiments, the extent of coke or semicoke formation associated with their occurrence would not change within the range of experimental conditions. Since the reactions of the oxygenated groups may indeed be fast, alternative 1 stated above seems reasonable.

Another measure of the difference between the weight loss behavior of Montana lignite and Pittsburgh Seam bituminous coal comes from a comparison of the augmentation in volatile yields observed by increasing the temperature or decreasing the pressure from a base case of 1000°C and 1 atm. Whereas for the bituminous coal the increase of volatiles yield effected by increasing the temperature to 1800°C[132] is approximately the same as that achieved by lowering the pressure to 0.001 atm,[75] the increase of yield from the lignite was relatively insignificant when the low pressure was employed but was comparable to that of the bituminous coal when the high temperature was used. In terms of the foregoing rationalization based on the supposed role of oxygenated groups, these observations support the view that the char-promoting reactions of the abundant oxygenated groups of lignite at the higher temperatures may not be as dominant over aromatic stabilization by hydrogen as they seem to be at 1000°C or less. Accordingly, pyrolysis of the lignite at the higher temperatures may avoid some of the char-forming secondary reactions that are relatively unaffected by physical augmentation of volatiles escape, such as reduction in pressure or particle size, at lower temperatures. However an alternative explanation that is also consistent with the data is that the effect of the higher temperatures on weight loss from the lignite reflects a larger extent of primary decomposition that is not observed at the lower temperatures because of insufficient residence times. Although this explanation cannot be ruled out for the lignite, it is in conflict with the bituminous data, since the high-temperature weight losses are not significantly larger than those available at lower temperatures through the use of reduced pressure.

12.4.2 Product Compositions

12.4.2.1 *General Characteristics*

Volatile product distributions reported by several authors are summarized in Table 12.5. Additional data are presented below in discussions of the effects of experimental conditions. Yields of tar (arbitrarily defined as including all liquids except water), gas, and chemically produced water, and the sum of these (denoted V^* as an approximation) are listed for several different coals and pyrolysis methods. The total yield of volatiles is compared with the proximate volatile content of the coal by the ratio V^*/VM, which exceeds unity in several cases. Tar yields are large, in the range 38–80% of the volatiles, except for the cases of Moscow brown coal, Wyoming subbituminous, Montana lignite, and the low-volatile Maigre Oignies coal, where tar is about 20% of the volatiles. It is clear from the two studies reporting Fischer assays that tar yields thus determined are much lower than the tar yields achievable in gas-swept stirred beds or in entrained-flow reactors.

Gas compositions associated with the data in Table 12.5 are given in Table 12.6. Methane is a major constituent in most cases (except for the Maigre Oignies coal and Montana lignite); carbon oxides, hydrogen, and light hydrocarbons account for the rest. Shapatina et al.[139] also reported some gas compositions, but interpretation is difficult because their experiments were not conducted at comparable time intervals. At 540°C with rapid decomposition, the gas analysis was 58.5% CO_2 ($+H_2S$), 22.9% CO, 2.4% H_2, 10.3% saturates, and 6.9% unsaturates. At higher temperatures the concentration of CO_2 decreased and that of CO and H_2 increased. Coal type is clearly a major factor in determining the composition of pyrolysis products, as discussed below.

12.4.2.2 *Effect of Temperature*

Volatiles Composition. Temperature is the most important variable affecting the composition of pyrolysis products from a coal. The role of temperature includes two primary effects, one on the decomposition of the coal and the other on the secondary reactions of volatiles. In the absence of secondary reactions, the yield of a given volatile component increases monotonically with increasing temperature, hence with the extent of the decomposition reactions producing that component. In the presence of substantial secondary reactions, an increase in temperature will enhance the yield of some species and retard the yield of others, reflecting of course species production and consumption, respectively, by the secondary reactions. The effects of temperature are clearly coupled with those of time, but the latter play a relatively secondary role if the rates are chemically controlled. The importance of time increases as heat or mass transfer limitations come into the picture.

Workers at Bergbau-Forschung[86] have made extensive measurements of the rate of gas evolution from coal as a function of temperature during heating at constant rates in the two captive sample apparatuses described in Table 12.3. Typical data obtained with the helium-swept fixed bed at 1 atm are presented in Fig. 12.10, using curves drawn through the data points. The clearly large variation, during heating, in the composition of the evolving gas reflects the predominance of different decomposition reactions in the different temperature intervals. The extent to which the observed compositions were affected by secondary reactions is not clear.

The formation of methane was interpreted as involving several mutually overlapping reactions, say two or four, proceeding in

Table 12.5 Distribution of Volatile Products of Coal Pyrolysis

		Coal		Volatile Products (wt% of dry coal)				
Investigator	Method	Type	VM[a]	Tar	Gas	Water[b]	V^*[c]	V^*/VM
Peters,[71]	Fischer assay, 600°C	Fürst Leopold	31.0	11.2	10.0	8.1	29.3	0.95
Peters and Bertling[72]	Stirred bed, 600°C		31.0	18.7	7.0	7.5	33.2	1.07
Shapatina et al.[139]	Slow heating, 0.08°C/s	Moscow brown coal	34.5	4.6	11.9	6.0	22.5	0.65
	Rapid heating to 500°C		34.5	4.0	8.1	9.6	21.7	0.63
Loison and Chauvin[105]	Thin layer of particles heated to 1050–1100°C at 1500°C/s	Maigre Oignies No. 2	8.4	1.8	5.1	1.5	8.4	1.00
		Bergmannsglück	18.1	14.7	5.9	3.0	23.6	1.30
		Emma (fines)	20.4	15.9	6.2	2.4	24.5	1.20
		Lens-Lievin No. 10	24.4	17.1	7.0	2.6	26.7	1.09
		Flénus de Bruay	31.0	25.2	9.6	4.4	39.2	1.18
		Wendel III	33.9	22.5	7.9	4.1	34.5	1.02
		Faulquemont	36.4	18.9	12.6	5.2	36.7	1.01
Friedman et al.[128]	Entrained flow, 600°C	West Virginia, mvb	31.6	18.7	1.8	6.5[d]	27.0	0.85
		Illinois, mvBb	34.6	19.6	19.7[e]	4.6[d]	43.9	1.27
		Pittsburgh, hvAb	37.7	31.5	22.0	8.0[d]	61.5	1.63
		Wyoming, subbituminous B	40.7	11.0	26.4	5.6[d]	43.0	1.06
Sass[129]	Fischer assay, 500°C	Western Kentucky, hvBb	41.0	16.3	5.0	9.4	30.7	0.75
	Flow tube, 530–650°C		41.0	33.0	6.6	1.7	41.3	1.01
Suuberg et al.,[111]	Thin layer of particles heated to 1000°C at 1000°C/s	Pittsburgh, hvAb	39.5	25.8	11.3	6.5	43.6	1.10
Suuberg[146]		Montana lignite	39.6	6.8	22.7	10.4	39.9	1.01

[a] Proximate volatile matter (wt %, mf).
[b] Chemically produced (i.e., not from coal moisture).
[c] Sum of tar, gas, and water, which only approximates the true V^*, since coal residence time is small.
[d] Estimated as weight difference between original dry coal and sum of tar, gas, and char yields.
[e] Reported value of 13.9% is said to be in error because of gas loss. Present value was estimated by difference, assuming chemically produced water to be the same as reported for 700°C.

Table 12.6 Composition of Gaseous Products of Coal Pyrolysis

Investigator and Conditions	Coal VM^a	Gas Composition (vol. %, mf) CO_2	CO	H_2	CH_4	Others
Loison and Chauvin[105]						
Thin layer of particles heated	8.4	2.2	10.1	66.2	8.6	12.9^b
to 1050–1100°C at 1500°C/s	18.1	1.6	5.0	67.2	21.4	4.8^b
	20.4	2.0	6.0	62.9	20.3	8.8^b
	24.4	3.8	9.6	64.2	21.0	1.4^b
	31.0	4.1	15.8	54.3	16.9	8.9^b
	33.9	2.4	10.0	52.8	21.8	13.0^b
	36.4	6.1	20.6	60.3	13.1	9.9^b
Peters,[71] *Peters and Bertling*[72]						
600°C Fluidized bed	31	12.0	12.3	14.7	50.5	10.5^b
600°C Fischer assay	31	10.3	8.7	22.4	54.2	3.7^b
Friedman et al.[128]						
Entrained flow, 600°C	31.6	9.2	3.6	19.4	24.7	38.0C_2's, 5.1C_3+
	34.6	11.7	0.0	23.0	40.7	17.1C_2's, 7.5C_3+
	37.7	9.5	9.5	17.1	42.9	4.3^b, 14.1C_2's, 2.6C_3+
	40.7	16.0	10.5	25.6	33.9	8.8C_2's, 5.2C_3+
Sass[129]						
Flow tube, 530–650°C	41	10.6	4.9	25	42.3	7.6^b, 5.6C_2H_6, 4.0C_3H_8
Suuberg,[146] *Suuberg et al.*[111]						
Thin layer of particles heated	39.5	3.2	9.9	57.8	17.8	3.4C_2H_4, 2.0C_2H_6, 6C_3+
to 1000°C at 1000°C/s	39.6	24.0	37.2	27.8	8.8	2.2C_2H_4

^{a} Proximate volatile matter (wt %, mf).
^{b} By difference; composition unknown.

parallel.[97] The release of hydrogen, which is spread over a relatively wide temperature range even at the low heating rates of 1 and 2°C/min, was interpreted as a combination of numerous overlapping first-order reactions with a statistical distribution of activation energies.[157] The rather minor ethane evolution in the range 80–300°C was said to represent desorption of gas formed during coalification and retained in the coal by adsorption. The second ethane peak appearing between 380 and 600°C was explained by a first-order decomposition reaction.[98] Pyrolysis water, oxides of carbon, and nitrogen are each formed by a set of reactions.[156] The release of hydrogen, carbon monoxide, and especially nitrogen is not finished at 1000°C, thus indicating that H, O, and N are strongly bound in the coke structure.

[156] J. Klein, "Untersuchungen zur Abspaltung von Wasserdampf, CO, CO_2, N_2 and H_2 bei der nicht-isothermen Steinkohlenpyrolyse unter inerter und oxydierender Atmosphäre," Thesis, Aachen, West Germany, 1971.

[157] R. Rennhack, "Zur Kinetik der Entgasung von Schwelkoks," *Brennstoff-Chemie,* **45,** 300 (1964).

Measurements similar to the foregoing were performed at Lawrence Livermore Laboratory.[90,91] The rate of evolution of product gases during the pyrolysis of a 50-g sample of coal at a constant heating rate (3.33°C/min) in an argon-swept fixed bed was determined by the measurement of species concentrations downstream of the bed in the metered sweep gas. The coal was Roland Seam (Wyodak Mine) subbituminous having the following composition (wt %, as received): moisture, 34.7; ash, 5.96; carbon, 43.6; hydrogen, 3.44; nitrogen, 0.73; sulfur, 0.48; oxygen, 11.1. The results (Fig. 12.11) show the temperature regimes within which the various species are evolved at the low heating rate studied. In view of the relatively large sample size and the associated opportunity for secondary reactions, the apparent pyrolysis behavior exhibited by these data is presumably affected by secondary reactions, but to an unknown extent.

The data in Fig. 12.11 are generally consistent with those in Fig. 12.10—for example, in reflecting the occurrence of numerous overlapping reactions—but some differences in detail are noteworthy. The emission rates

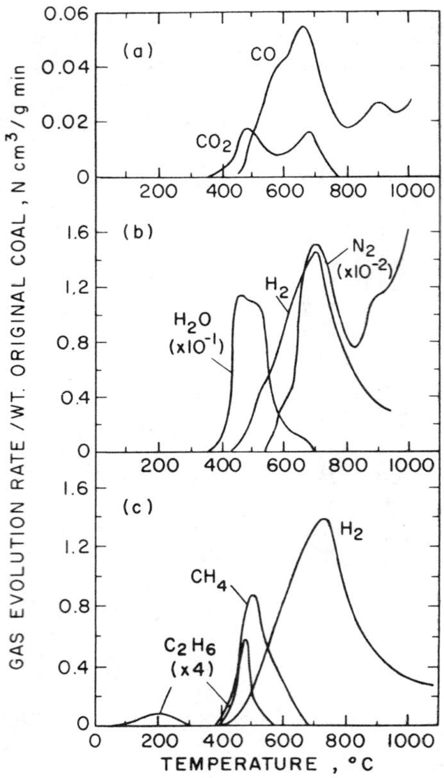

Figure 12.10 Variation of gaseous species evolution with temperature during coal pyrolysis at constant heating rate. (*a*) and (*b*), Data of Klein[156]: Gustav coal, VM = 29 wt % (maf), heating rate = 1°C/min. (*c*), Data of Jüntgen and Van Heek[97]: VM = 19.1 wt %; heating rate = 2°C/min.

and cumulative amounts of oxides of carbon are much larger in the case of the subbituminous coal (Fig. 12.11), as would be expected. In the case of carbon monoxide evolution, the change with increasing temperature beyond about 800°C is quite different for the two cases, being generally an increase in Fig. 12.10 and a clear decrease in Fig. 12.11. This difference supports the view that the processes forming carbon monoxide at the higher temperatures in the two studies differ in type as well as in rate.

The rates of evolution of aliphatic hydrocarbons ranging from C_5 to C_7, inclusive, and those of light aromatics are given on a relative basis in Fig. 12.11*d*. Although the absolute rates were not measured, these species probably constitute only a minor fraction of the total volatiles. These data indicate that the peak rates of formation of C_5—C_7 aliphatics, toluene, and xylene occur in about the same temperature range as do the peaks for ethane, ethylene, and propane (Fig. 12.11*c*), namely, about 500–550°C, which is somewhat lower than the temperatures characteristic of the methane peak. The benzene peak occurs between 650 and 700°C, or about 150°C past the peak of the alkylated aromatics. In the work of Shapatina et al.[139] mentioned above, the maximum concentration of hydrocarbons occurred at 500–700°C.

A recent study at MIT of pyrolysis product distributions[111,146] focused on the effects of temperature as well as those of other variables discussed below. The work was extended to conditions of rapid heating, and efforts were made to close a mass balance, thereby to account for all products or classes of products. To this end, the apparatus used by Anthony[154] to heat a thin layer of coal particles was modified to permit the collection and analysis of volatiles. To define the classes of products and the pyrolysis conditions, essential features of the technique are described below.

The coal sample was held and heated in a vessel by a folded strip of 325-mesh stainless steel screen positioned between two relatively massive brass electrodes. The heating circuit permits independent control of heating rate (10^2–10^4 °C/s), final temperature (150–1100°C), and solids residence time at the final temperature (up to 30 s). The temperature-time history of the coal was recorded by a thermocouple placed within the sample and connected to a fast-response recorder. Approximately 10–15 mg of powdered coal was spread in a layer 1–2 particles deep on a preweighed screen, which was reweighed and inserted between the brass electrodes. The sample temperature was raised at a desired rate to a desired final value, which was maintained until the circuit was broken. Sample cooling by convection and radiation then occurs rapidly, since the electrodes, the vessel, and its gaseous contents remain cold during the experiment, but not so rapidly as to avoid the occurrence of significant weight loss during cooling. The volatiles escape rapidly from the thin sample into the cold gas of the vessel, thereby reducing the opportunity for secondary reactions.

The yield of char, which remains on the screen, was determined gravimetrically. Products that condense at room temperature

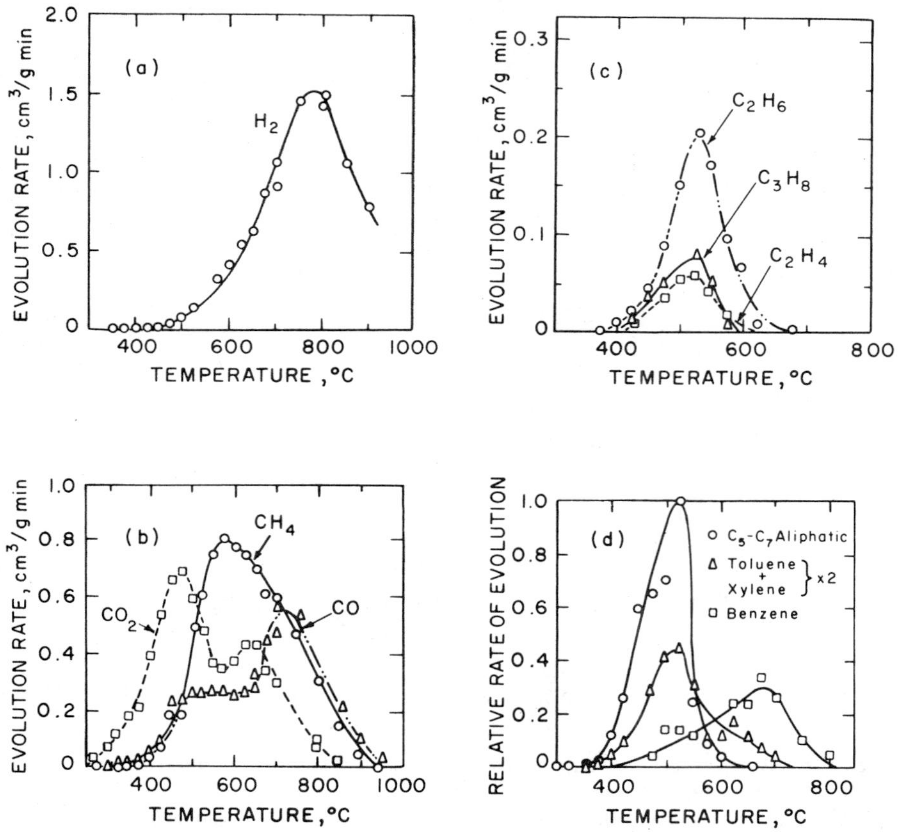

Figure 12.11 Variation of species evolution with temperature during coal pyrolysis at constant heating rate (3.3°C/min) for Roland Seam (Wyodak Mine) subbituminous coal. Rate basis for (*a*), (*b*), and (*c*): cubic centimeters at 25°C and 1 atm per gram of (dry) original coal per minute; rate basis for (*d*): relative to peak of C_5–C_7 curve. Points are original data. Curves from Equation 12.26 (Section 12.4.3.2) with parameters from Table 12.18 and τ = 0.[91]

(tars and oils, here defined as tars) were collected primarily on foil liners within the vessel and on a paper filter at the exit of the reactor. Any condensation on nonlined reactor surfaces was collected by washing with methylene chloride-soaked filter paper. The tar from all three collections was measured gravimetrically. Products in the vapor phase at room temperature were collected at the conclusion of a run by purging the reactor vapors through two lipophilic traps, the first operated at room temperature to collect light oils such as benzene, toluene, and xylene, and the second operated at −196°C to collect all the fixed gases except hydrogen, which was determined by direct vapor phase sampling. Products thus collected were analyzed by gas chromatography, and char and tar samples were subjected to elemental analysis.

Experimental assessment of the effect of the screen on the observed composition of volatiles included passivation of the surface with a vacuum-deposited layer of gold on some screens and copper on others, and variation of the number of layers of untreated screen through which the volatiles had to escape. None of these cases led to significant differences in the composition of products included in the gas chromatographic analysis. Therefore any error caused by the screen appears to be negligible for present purposes.

The effect of temperature on the distribution of pyrolysis volatiles obtained from a partially dried (as-received) Montana lignite is shown in Fig. 12.12. The composition of the lignite, along with that of a bituminous coal discussed later, is given in Table 12.7. Except for the data points carrying a double symbol, which are discussed below, the

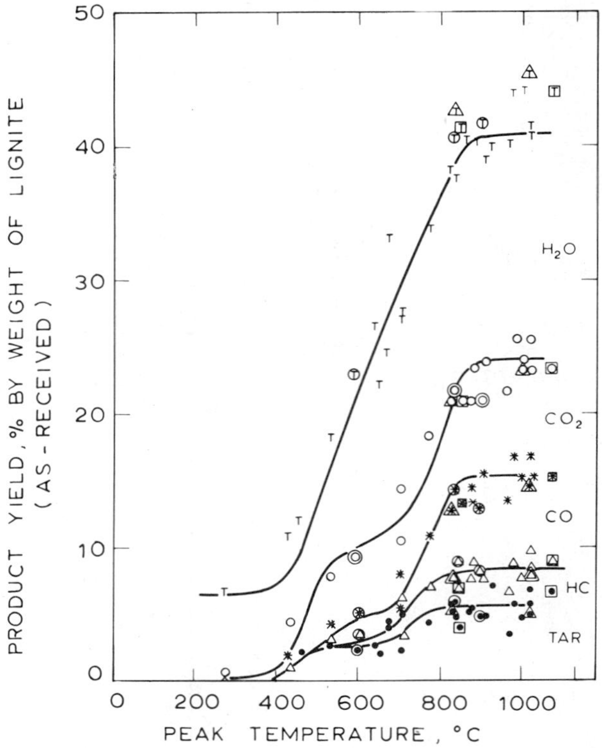

Figure 12.12 Pyrolysis product distributions from Montana lignite heated to different peak temperatures. ● = tar; △ = tar and other hydrocarbons (HC); * = tar, HC, and CO, ○ = tar, HC, CO, and CO_2; T = total (i.e., tar, HC, CO, CO_2, and H_2O). Pressure = 1 atm helium; particle diameter, 53–88 μm (average, 74 μm). Heating rates: 1000°C/s (single points); 7100–10,000°C/s (points in circles); 270–470°C/s (points in triangles); 1000°C, but two-step heating (points in squares). Curves, first-order model (Equation 12.27, Section 12.4.3.2), using parameters shown in Table 12.20 and τ = 0.[111]

product compositions shown in the figure were obtained when different samples of the lignite under 1 atm of helium were heated at approximately 1000°C/s to various peak temperatures indicated on the abscissa. The samples were cooled at roughly 200°C/s beginning immediately when the peak temperature was attained. The total yield of any product or group of products is given on the ordinate as a weight percent of the partially dried lignite. The curves were obtained by fitting a pyrolysis model to the data, as discussed in Section 12.4.3.2.

The lowest curve represents the yield of tar as defined above, which increases with increasing temperature to an asymptotic maximum of about 5.4% by weight of the lignite at temperatures above 750–800°C. The tar appeared to have two primary components. One was a heavy, tawny brown, liquidlike material that deposits as a film on surfaces. The other consisted of dark brown objects that give the impression of being solid pieces of lignite that were presumably broken off the parent material and carried away with the volatiles. These particles, some of which appeared to be agglomerates, were for the most part smaller than 10 μm across but sometimes as large as 25 μm. About 75% of the tar, including most of the tawny liquid and a fraction of the particles, appeared to dissolve in methylene chloride. Elemental analysis of the methylene chloride-soluble material gave an approximate empirical formula $CH_{1.5}O_{0.1}N_{0.01}$. The indicated oxygen content may be too high because of some oxidation in the presence of air.

The distance between the tar curve and the next one above it represents hydrogen and all hydrocarbons lighter than tar. The maximum yield of these species occurs at the higher temperatures and is only about 3.3% by weight of the lignite. The main components are methane (1.3%), ethylene (0.6%),

Table 12.7 Characteristics of Coals Studied by Suuberg[146]

Proximate Analysis, As Received			Ultimate Analysis, As Received			Petrographic Analysis, Mineral-Matter-Free Basis		
Variable	Bit.[a] (wt %)	Lig.[b] (wt %)	Variable	Bit.[a] (wt %)	Lig.[b] (wt %)	Component	Bit[a] (wt %)	Lig.[b] (wt %)
Moisture	1.4	6.8	Carbon	67.8	59.3	Vitrinites	84.5	69.7
VM	38.9	36.9	Hydrogen	4.8	3.8	Semifusinite	4.5	15.2
Fixed Carbon	48.4	46.4	Nitrogen	1.3	0.9	Fusinite	3.3	7.9
Ash	11.3	9.9	Sulfur	5.3	1.1	Micrinite	4.4	5.2
			Oxygen[c]	8.1	18.2	Exinite + resinite	3.3	2.0
	100.0	100.0	Moisture	1.4	6.8			
			Ash	11.3	9.9		100.0	100.0
				100.0	100.0			

[a] Pittsburgh Seam (No. 8) bituminous coal from the Ireland Mine of Consolidation Coal Company.
[b] Partially dried Montana lignite from the Savage Mine of Knife River Coal Mining Company.
[c] By difference.

and hydrogen (0.5%), with identified ethane, propylene, propane, benzene, and unidentified trace hydrocarbons making up the balance. The effect of temperature on the yields of methane, hydrogen, and ethylene is shown in Fig. 12.13. When the peak temperature is increased above 500°C, the methane and ethylene yields increase rapidly to small asymptotic values in the range 600–700°C. Further increase in temperature beyond 700°C effects a dramatic increase in the yield of both species, and a second asymptote is reached at about 850°C for ethylene and about 900°C for methane. The yield of tar also exhibits a similar two-step behavior; hydrogen production, on the other hand, appears to occur in one step at relatively high temperatures. The step behavior exhibited by these yield data is of course equivalent to the peaks exhibited by the rate data in Figs. 12.10 and 12.11.

The top curve in Fig. 12.12 represents the total yield of volatiles, and proceeding down-

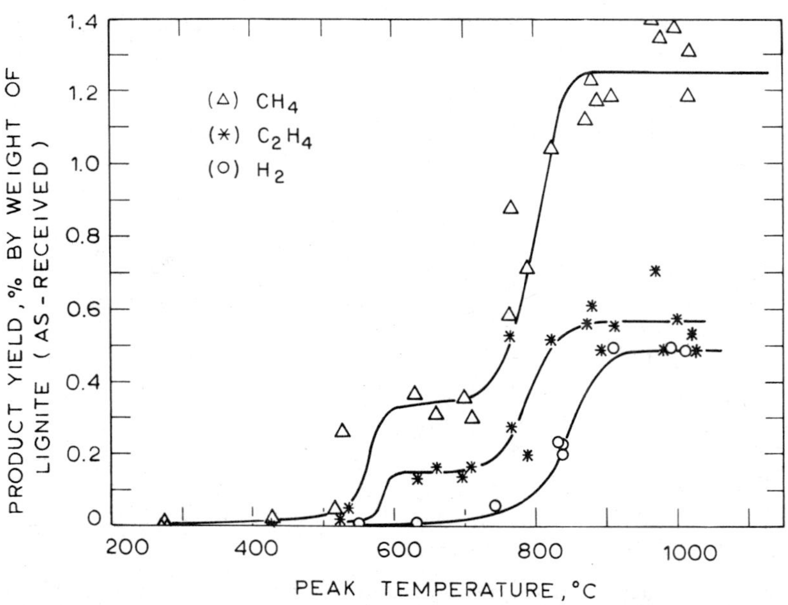

Figure 12.13 Yields of methane, ethylene, and hydrogen from Montana lignite pyrolysis to different peak temperatures. Same experimental conditions as in Fig. 12.12; curves, first-order model (Equation 12.27) using parameters shown in Table 12.20 and $\tau = 0$.[111]

ward, the first, second, and third regions between adjacent curves represent the yields of water (including the initial moisture), carbon dioxide, and carbon monoxide, respectively. The yields of these principal oxygenated species are shown in more detail in Fig. 12.14, where all three appear to approach high-temperature asymptotic yields of 16.5% for water, 8.4% for carbon dioxide, and 7.1% for carbon monoxide. The carbon oxides each exhibit also a lower-temperature asymptote.

Although most of the pyrolysis is complete for peak temperatures above 900–1000°C, there is yet a third step in the curves for the carbon oxides, which occurs at about 1100°C and therefore does not appear in Fig. 12.14. Since this temperature is the upper limit of the apparatus, investigation of the third step was accomplished by use of a longer residence time technique. The coal was heated at 1000°C/s to 1000°C and there held for 5–10 s rather than being immediately cooled as before. The resulting carbon monoxide yield exhibits a final asymptote of 9.4%; that of carbon dioxide is 9.5%. The yields of the other species were not changed by the additional residence time. Thus prolonged heating at 1000°C gave a total volatiles yield of 44.0% by weight of the lignite, which is close to the ASTM volatile matter plus moisture (43.7%).

Elemental balances were calculated for runs in which both volatile products and char were analyzed. For estimation purposes, trace hydrocarbons (total less than 1% by weight of the lignite) were assumed to be 90% carbon and 10% hydrogen by weight. The tar was assumed to have an empirical formula equivalent to that given above for the principal, methylene chloride-soluble, fraction. Typical results for four runs to different peak temperatures are presented in Table 12.8 along with total mass balances. Whereas the mass balances are excellent and the carbon and hydrogen balances are satisfactory, the oxygen balances are marginal. Since oxygen in char is determined by difference, uncertainties inherent in the other measurements are absorbed in the oxygen values.

Product heating value contents—that is, mass yield × net (lower) heating value on a mass basis—shown in Fig. 12.15 were estimated from the product yields in Table 12.8 and some additional tar and gas measurements. Heating values of the lignite and char

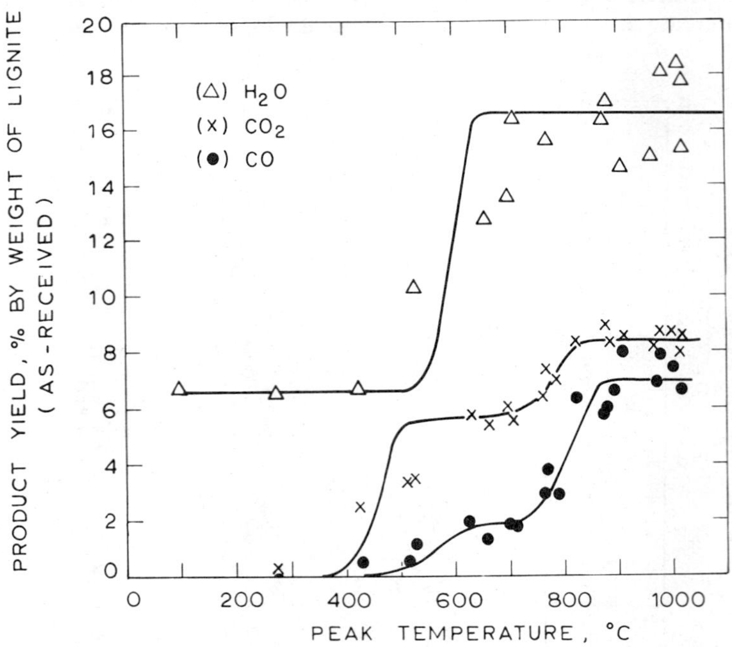

Figure 12.14 Yields of water, carbon monoxide, and carbon dioxide from Montana lignite pyrolysis to different peak temperatures. Same experimental conditions as in Fig. 12.12. Curves, first-order model (Equation 12.27) using parameters shown in Table 12.20, τ = 0, and lignite moisture (6.8%) added to the calculated H_2O.[111]

Table 12.8 Carbon, Oxygen, Hydrogen, and Total Mass Balances for Pyrolysis of Montana Lignite[111]

	Yield, wt % of Lignite (as-received)[a] in Runs to Four Different Peak Temperatures															
	430°C				710°C				910°C				1000°C			
Product	Total	C	H	O	Total	C	H	O	Total	C	H	O	Total	C	H	O
CO_2	2.5	0.68	—	1.8	5.5	1.5	—	4.0	8.6	2.3	—	6.3	8.7	2.4	—	6.3
CO	0.46	0.20	—	0.26	1.8	0.77	—	1.0	8.0	3.4	—	4.6	7.5	3.2	—	4.3
HC[b]	0.02	0.02	0.0	—	0.95	0.79	0.17	—	2.6	2.1	0.49	—	3.0	2.4	0.58	—
Tar	1.5	1.3	0.15	—	2.3	2.1	0.23	—	4.9	4.4	0.49	—	4.7	4.3	0.47	—
H_2O	6.6	—	0.73	5.9	16	—	1.8	14	15	—	1.7	13	19	—	2.1	17
H_2	0.0	—	0.0	—	0.01	—	0.01	—	0.50	—	0.50	—	0.50	—	0.50	—
Char	89.0	64.0	2.8	11.0	73.0	53.6	1.5	3.1	60.9	47.5	1.2	2.6	55.8	44.2	0.55	1.1
Total	100	66.2	3.68	19.0	99.9	58.8	3.71	22.1	101	59.7	4.38	26.5	99.2	56.5	4.20	28.7
Closure	100%	112%	81%	79%	100%	99%	82%	91%	101%	101%	97%	110%	99%	95%	93%	119%

[a] Lignite (as-received) is 59.3% C, 4.53% H, and 24.2% O, including H and O in lignite moisture (Table 12.7).
[b] Hydrocarbons other than tar.

were calculated from the elemental analyses using the correlation of Mott and Spooner[158] adjusted to a net heating value, as-received basis. The net heating value of the tar and that of hydrocarbons other than methane, ethylene, and tar were assumed to be 8.9 and 10.9 kcal/g, respectively. The heating values of the methane, ethylene, hydrogen, and carbon monoxide were, of course, known. The heating value contents of the gas and tar follow the two-step behavior associated with the appearance of many of the individual components. At the higher temperatures the

[158] R. A. Mott and C. E. Spooner, "The Calorific Value of Carbon in Coal: The Dulong Relationship," *Fuel,* **19,** 226, 242 (1940).

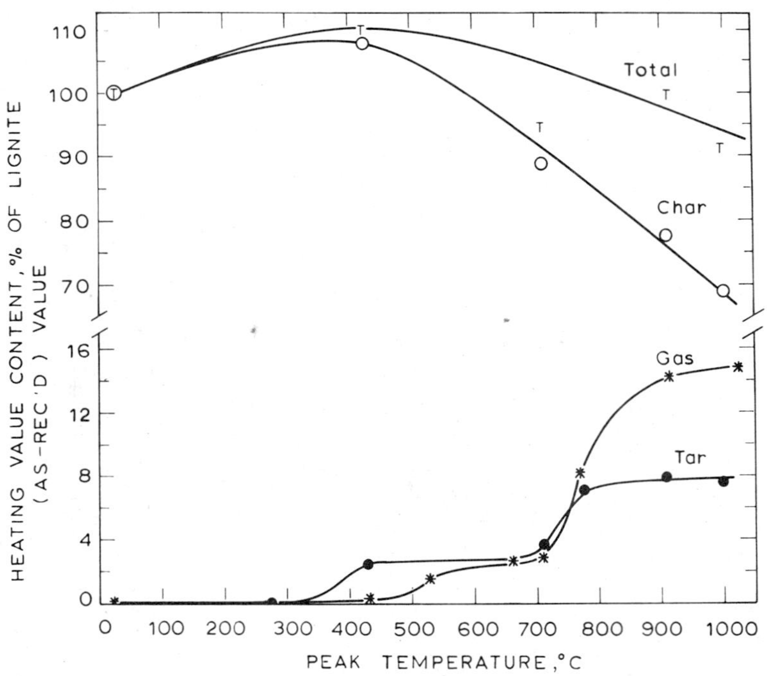

Figure 12.15 Distributions of heating value content of products from Montana lignite pyrolysis to different peak temperatures. Total (T), all products including char. Pressure, 1 atm, helium; heating rate, 1000°C/s. Basis: 100% on ordinate corresponds to 17.8 kcal/g (9900 Btu/lb) lignite (as received), the initial net (lower) heating value.[111]

gas accounts for a maximum of almost 15% of the heating value content initially in the lignite, and the tar for about 8%. At a temperature of about 1000°C the char retains almost 70% of the heating value on the same basis.

The total heating value content of all products appears to increase, then decrease, as the peak temperature increases. Pyrolysis appears to be endothermic over a range of lower to intermediate temperatures where the total heating value content of all the products is greater than that of the starting material. At higher temperatures, however, the total becomes less than the initial value; therefore pyrolysis to those temperatures may be thermally neutral or exothermic. The approximations used in this analysis preclude quantitative assessment of heats of pyrolysis.

Suuberg[146] also measured pyrolysis product yields from the Pittsburgh Seam bituminous coal described in Table 12.7, using the same apparatus and technique defined above. The overall distribution of products shown in Fig. 12.16 was obtained in a similar fashion to the lignite data in Fig. 12.12. Whereas the lignite yields in Fig. 12.12 are dominated by carbon oxides and water, the tar product (including all hydrocarbon liquids) is clearly dominant in the bituminous case. This result is of course in line with the data in Table 12.5.

The bituminous tar product appeared to be relatively homogeneous, in contrast to the

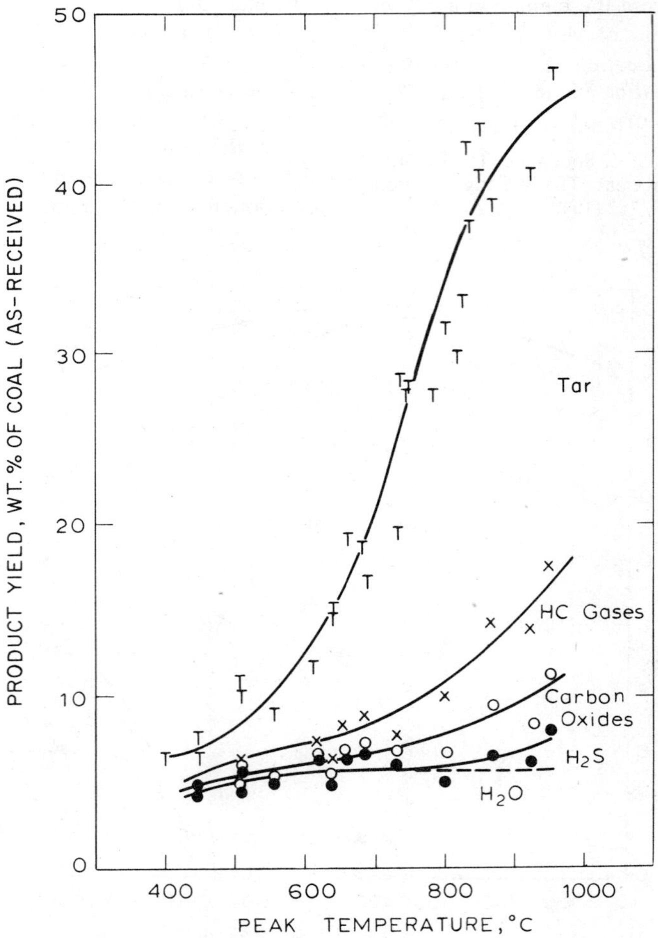

Figure 12.16 Pyrolysis product distributions from Pittsburgh Seam bituminous coal heated to different peak temperatures. ● = H_2O and H_2S; ○ = CO, CO_2, H_2O, and H_2S; x = hydrocarbon (HC) gases, CO, CO_2, H_2O, and H_2S; T = total (i.e., tar, including hydrocarbon liquids, HC gases, CO, CO_2, H_2O and H_2S. Pressure, 1 atm of helium; particle diameter, 53–88 μm (average, 74 μm); heating rate, 1000°C/s).[146]

lignite tar, and its color was a tawny yellow much like the lignite tar from high temperature runs. The material eventually became dark brown on standing in air, a well-known phenomenon presumably attributable primarily to oxidation, although the weight of oxygen involved was less than 1% of the tar. The elemental analysis of the tar was $CH_{1.1}O_{0.08}N_{0.01}$, which indicates a considerably lower H/C ratio than the lignite tar (for which H/C $\simeq$ 1.5). Nevertheless, in both cases the tars were significantly richer in hydrogen than the parent coals, but very similar to the parent coals in N/C ratio.

The yields of tar and water alone are presented in Fig. 12.17: the left-hand portion plots the yields as a function of peak temperature in the previous manner, and the data in the right-hand portion were obtained by providing a constant temperature holding period of 2–10 s at the peak temperature, rather than cooling the sample immediately on achieving the peak temperature. The yield at a particular holding temperature is of course higher than the yield at the corresponding peak temperature, except where the reaction (or reactions) has gone to completion even without the holding period.

The tar data do not show any two- or three-step behavior; the rather gradual effect of temperature is suggestive of a broad distribution of reactions, as discussed later. Pyrolytically formed water is seen to be primarily evolved at rather low temperatures (< 400°C). This behavior is in sharp contrast to that of the lignite, in which pyrolytic water is formed at temperatures higher by 200–300°C. The difference presumably reflects the difference in starting materials, mechanisms, or both.

The yields of gaseous products shown in Figs. 12.18–12.20 give no evidence of the two-step behavior seen during lignite pyrolysis. It is clear from Fig. 12.18 that methane evolution is virtually complete before hydrogen evolution begins. The methane is probably formed while there is still a significant amount of aliphatic hydrogen left in the coal, whereas much of the molecular hydrogen must arise from aromatization reactions. Although the temperature dependence indicated by the methane curve is suggestive of a wide spectrum of activation energies, the steep rise in the peak temperature interval 800–900°C is suggestive of one dominant mechanism. The gently sloping carbon dioxide curves in Fig. 12.19 also suggest a rather broad distribution of activation energies. The carbon monoxide curves are very reminiscent of methane in shape and temperature dependence, and the peak temperature interval of 800–900°C again seems to be important. The yields of four other hydrocarbon gases (propylene and propane combined)

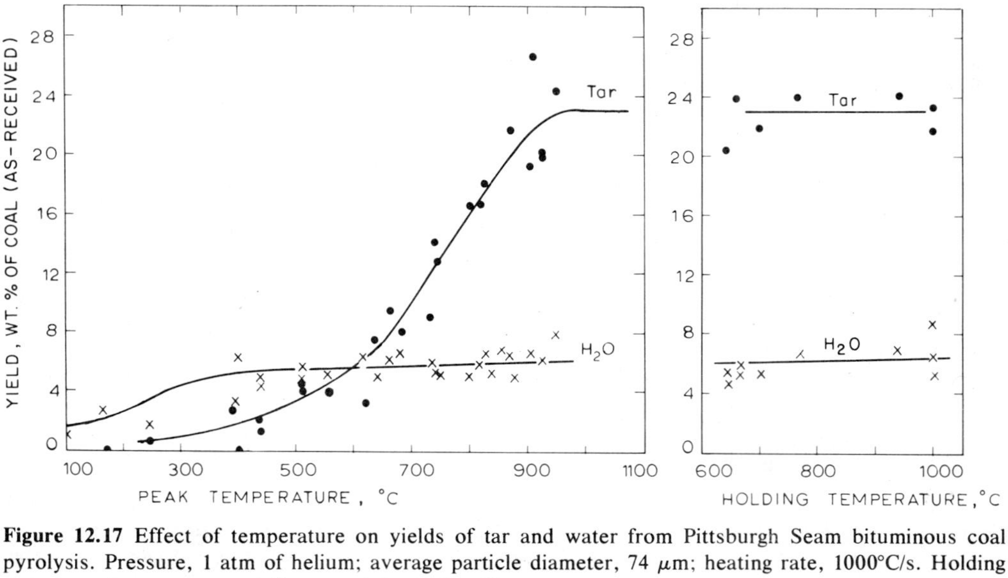

Figure 12.17 Effect of temperature on yields of tar and water from Pittsburgh Seam bituminous coal pyrolysis. Pressure, 1 atm of helium; average particle diameter, 74 μm; heating rate, 1000°C/s. Holding time at peak temperature, *left*: zero, *right*: 2–10 s.[146]

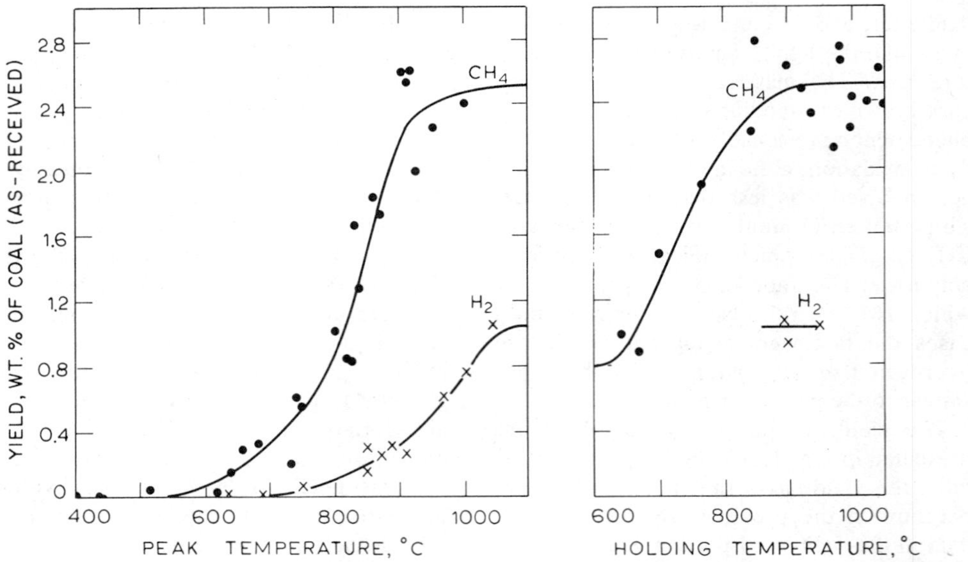

Figure 12.18 Effect of temperature on yields of methane and hydrogen from Pittsburgh Seam bituminous coal pyrolysis. Pressure, 1 atm of helium; average particle diameter, 74 μm; heating rate, 1000°C/s. Holding time at peak temperature, *left*: zero, *right*, 2–10 s.[146]

shown in Fig. 12.20 also exhibit a gradual increase with increasing temperature, except that ethylene yield exhibits a steep rise in the 800–900°C peak temperature interval and ethane yield would appear to be relatively independent of holding temperature.

Mass and element balances associated with the foregoing bituminous coal data and given in Table 12.9 were computed in a manner analogous to the balances in Table 12.8 for lignite. The balances typically fall in the range 100 ±10%, with the carbon balance

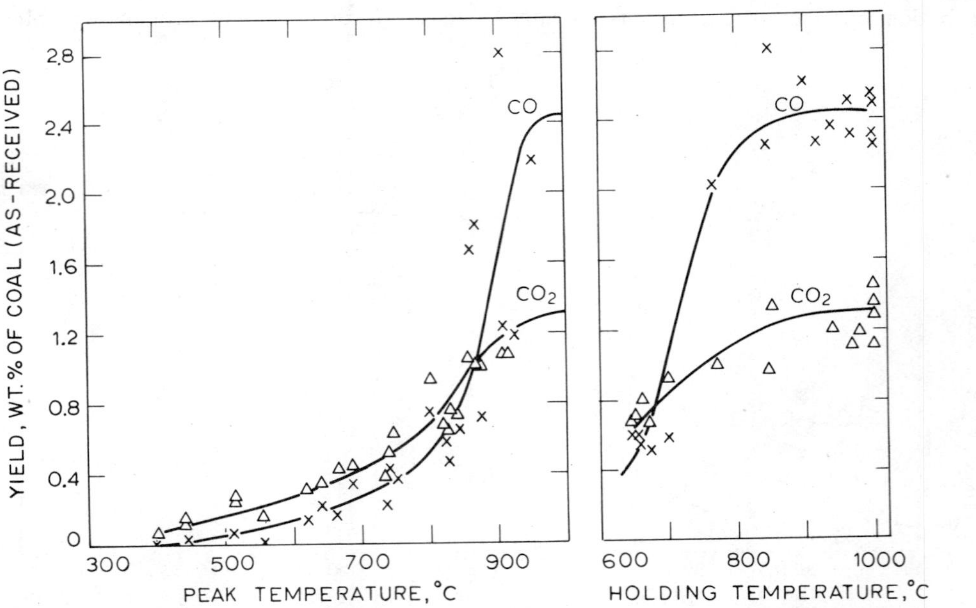

Figure 12.19 Effect of temperature on yields of carbon oxides from Pittsburgh Seam bituminous coal pyrolysis. Pressure, 1 atm of helium; average particle diameter, 74 μm; heating rate, 1000°C/s. Holding time at peak temperature, *left*: zero, *right*: 2–10 s.[146]

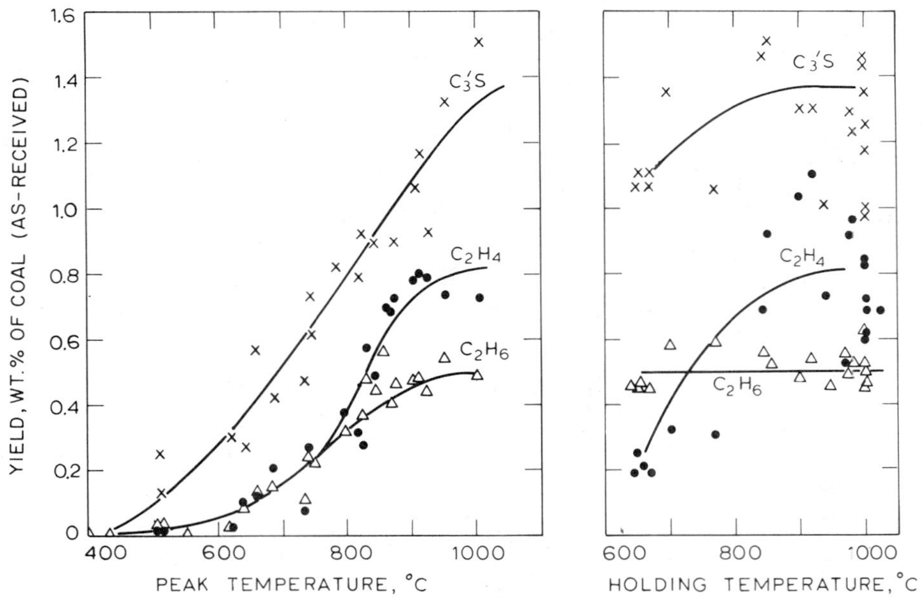

Figure 12.20 Effect of temperature on yields of various hydrocarbon gases from Pittsburgh Seam bituminous coal pyrolysis. Pressure, 1 atm of helium; average particle diameter, 74 μm; heating rate, 1000°C/s. Holding time at peak temperature, *left*: zero, *right*, 2–10 s.[146]

falling off at higher temperatures owing to the difficulty of collecting tar, which is 80 wt % carbon. It is striking that at most only about 10% of the carbon in the coal is found in the vapor phase products.

Char Composition. The composition of the char produced under different pyrolysis conditions can be an important source of global information on the volatiles. In addition, the study of char composition is a means for determining the partitioning of the sulfur and nitrogen of the coal between the solid and volatile products of pyrolysis. The latter information is especially pertinent to the selection and optimization of pollution control strategies.

The retention of elements in chars from the foregoing experiments of Campbell[90] and Suuberg et al.[111] are presented in Fig. 12.21. The data of Suuberg et al. are based on the measured total weight loss and measured compositions of the original lignite and the char. The numbers for the Campbell study were calculated here, using the reported elemental compositions and ash contents of the coal and char, and assuming that the weight of ash in the solid material remains constant. The percentage of each of the main original elements retained in the char is shown as a function of the final or peak temperature attained in pyrolysis experiments at constant heating rate. Since experimental runs in both studies were terminated when the peak temperature was reached, the residence time of the sample is a factor of 300 larger for the Campbell data than for those of Suuberg et al. because of the different heating rates. This difference probably accounts for the observation that the Campbell data exhibit generally lower retentions of the elements in the char.

Although more than 40% by weight of the lignite is volatilized at the higher temperatures (Fig. 12.12), only 22% of the carbon is volatilized (Fig. 12.21). Most of the volatile material comes from hydrogen and oxygen, which is consistent with the observed predominance of water among the volatile products (Fig. 12.14). Pyrolysis at the higher temperatures removes about 70% of the sulfur from the solid material, but the nitrogen is reduced by only about 25%. Consequently the sulfur content (percent by weight) in the char is lower than that of the lignite, but the reverse is true for nitrogen.

Compositions of bituminous coal chars from Suuberg's experiments are given in Fig. 12.22. The points represent pyrolysis at 1000°C/s to the indicated peak temperatures

Table 12.9 Carbon, Hydrogen, Oxygen, and Total Mass Balances for Bituminous Coal Pyrolysis[146]

Product	Average Yield, wt % of Bituminous Coal (as-received)[a]															
	Peak Temperatures 400–440°C				Peak Temperatures 620–690°C				Peak Temperatures 740–780°C				Holding Temperature 1000°C			
	Total	C	H	O	Total	C	H	O	Total	C	H	O	Total	C	H	O
CO_2	0.11	0.03	—	0.08	0.39	0.11	—	0.28	0.49	0.16	—	0.43	1.23	0.34	—	0.89
CO	0.0	0.	—	0.	0.21	0.09	—	0.12	0.40	0.17	—	0.23	2.42	1.04	—	1.38
H_2O	5.16	—	0.57	4.59	6.02	—	0.67	5.35	5.26	—	0.61	4.68	6.84	—	0.76	6.08
H_2	0.0	—	0.	—	0.0	—	0.0	—	0.06	—	0.06	—	1.01	—	1.01	—
CH_4	0.0	0.	0.	—	0.22	0.16	0.06	—	0.59	0.44	0.15	—	2.49	1.87	0.62	—
HC_{gas}[b]	0.0	0.	0.	—	0.90	0.69	0.21	0.0	1.56	1.32	0.24	—	3.95	3.39	0.56	—
Tar	1.18	0.94	0.09	0.10	9.45	7.50	0.69	0.78	15.0	11.9	1.09	1.23	25.4	20.2	1.85	2.09
Char	93.1	63.91	4.48	3.69	83.8	57.4	3.91	2.85	71.9	46.6	2.93	3.46	53.0	34.2	0.52	0.0
Total	99.6	64.9	5.14	8.46	101.0	66.0	5.54	9.38	95.4	60.6	5.08	10.0	97.3[c]	61.0	5.32	10.4
Closure	100%	95%	105%	90%	100%	97%	112%	99%	95%	89%	103%	106%	97%	90%	108%	110%

[a] Bituminous coal (as-received) is 67.8% C, 4.91% H, 9.45% O, including H and O in moisture (Table 12.7).
[b] Hydrocarbons other than CH_4 and tar.
[c] Includes 0.95% H_2S, not shown.

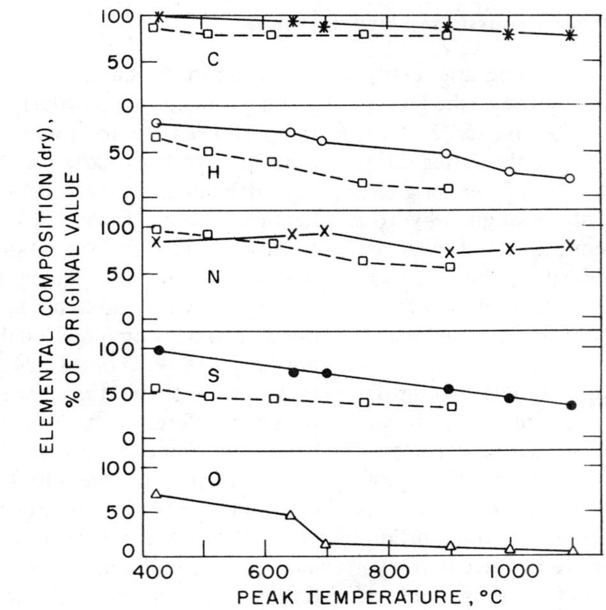

Figure 12.21 Elemental compositions of chars from lignite and subbituminous coal pyrolysis to different peak temperatures. Points connected by solid lines: Montana lignite (Savage Mine) heated at 1000°C/s in helium at 1 atm.[111] Points connected by dashed lines: Roland Seam (Wyodak Mine) subbituminous coal heated at 3.3°C/min in argon at 1 atm.[90]

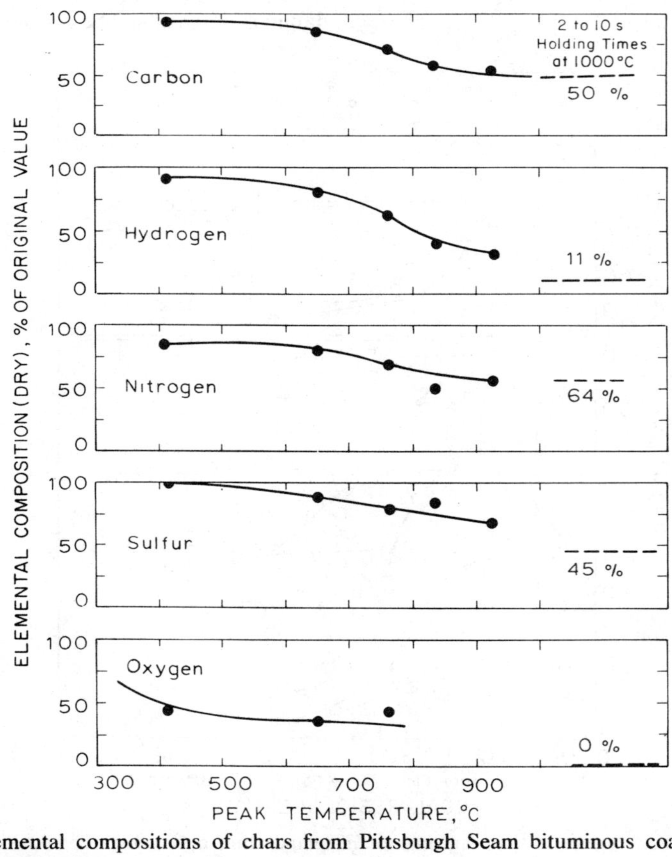

Figure 12.22 Elemental compositions of chars from Pittsburgh Seam bituminous coal pyrolysis. Coal heated at 1000°C/s in helium at 1 atm.[146]

followed immediately by cooling at roughly 200°C/s as described above; the dashed lines at the right represent the use of 2–10 s holding times at 1000°C. In the latter case volatilization of about 50% of the original carbon and about 90% of the original hydrogen occurs. To the extent that about twice as much carbon is volatilized in this case as compared with the lignite above, pyrolysis of the bituminous coal exhibits a more efficient utilization of hydrogen.

As in the case of lignite, almost two-thirds of the nitrogen remains behind in the char even at 1000°C. Significant sulfur volatilization occurs between the peak temperature of about 925°C and the 1000°C asymptote. It is in this temperature range that a large portion of H_2S evolution is believed to occur. Oxygen volatilization exhibits two steps, the first loss occurring at temperatures below 400°C, and corresponding to the evolution of pyrolytic water. The next stage must occur at temperatures greater than 770°C, presumably during the period of major CO evolution (800–900°C, Fig. 12.19). Virtually no oxygen remains in the char at 1000°C.

Kobayashi[89] measured the composition of chars produced in pyrolysis experiments at temperatures up to about 1800°C. Three types of pyrolysis experiment including dispersed laminar flow, free fall, and crucible heating of coal (Table 12.3) were employed to cover coal residence times from about 0.01 s to about 12 hr. The experiments were performed with Montana lignite and Pittsburgh Seam bituminous coal in argon at 1 atm. Figures 12.23 and 12.24 show the retention of each element in char as a function of time for different runs. The retentions were calculated from the original composition of the elements, the measured overall weight loss, and the composition of the elements in char. Since the temperature-time histories and other conditions were significantly different between the crucible runs and the laminar flow runs, the results of the crucible runs should not be interpreted as the extension of corresponding laminar flow runs at long residence times. The results of the free-fall experiment, plotted at 1 s residence time, may be compared with the corresponding laminar flow runs. The residence time for the

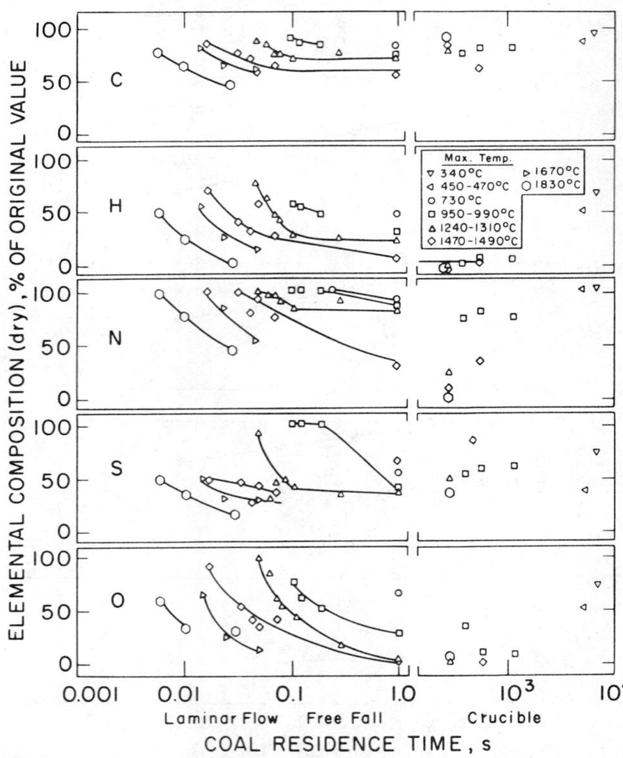

Figure 12.23 Effect of residence time and maximum temperature attained on composition of char from pyrolysis of Montana (Savage Mine) lignite.[89]

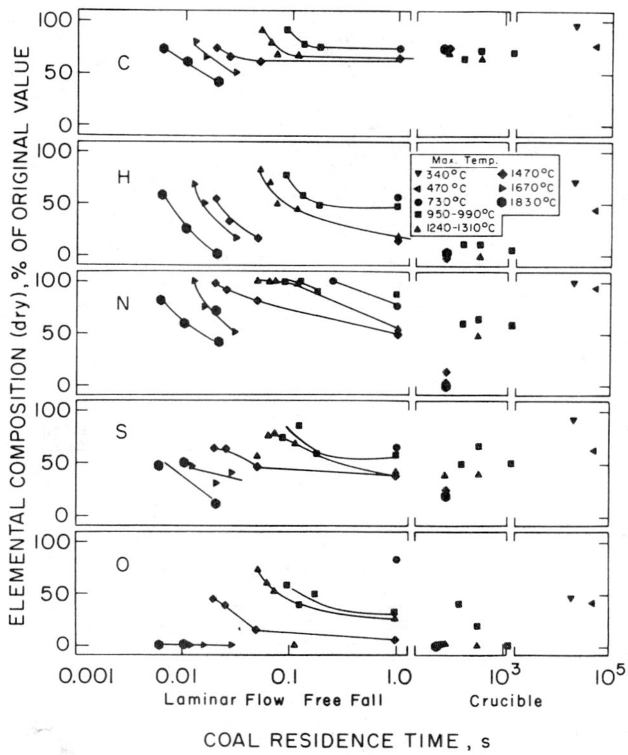

Figure 12.24 Effect of residence time and maximum temperature attained on composition of char from pyrolysis of Pittsburgh Seam No. 8 Bituminous (Ireland Mine) coal.[89]

free-fall runs is based on the particle velocities measured with a laser Doppler anemometer. Although only a limited number of samples were analyzed at each furnace temperature, the overall trends of carbon, hydrogen, nitrogen, and oxygen are clearly seen for the two coals. Both rates of volatilization and final losses of elements from the char increase with temperature within the laminar flow range. Perhaps the results for sulfur are relatively more scattered than those for the other elements because of some interactions of organic and inorganic sulfur.[89] Among the elements studied, oxygen and hydrogen are clearly the most readily removed. Essentially all the oxygen and hydrogen were lost in about 0.02 s at 1830°C, whereas about 50% of the nitrogen was retained in the char. The asymptotic values for carbon retention decrease with increasing temperature, as would be expected from the increased volatile yields with increasing temperature (Fig. 12.3).

The effect of temperature at the longer times provided by the crucible experiment is shown in Fig. 12.25, where the elemental retentions are plotted against the maximum temperature attained. The reaction times at the peak temperature ranged from about 5 min at 1480°C to 12 hr at 300°C. Weights of char (percentage of original coal, maf basis) are also shown, providing a reference curve with which preferential losses or retentions of the elements may be compared. It will be noted that weight loss in the crucible experiments reaches an asymptote at about 900–1000°C, which is considerably less than the value reached at higher temperatues when the coal is better dispersed and the volatiles are not confined with the hot solids (Fig. 12.3). As was the case in the laminar flow runs, hydrogen and oxygen are seen to be preferentially lost. Only about 10% of the original hydrogen and oxygen is retained at 900°C and at about 10 min of reaction time, which is consistent with the observation that little additional weight loss can occur beyond that point. Nitrogen exhibits the striking behavior of very little volatilization at the lower temperatures, consistent with Figs. 12.21 and 12.22, but there is essentially complete volatilization at temperatures above about 1600°C for the relatively long times here

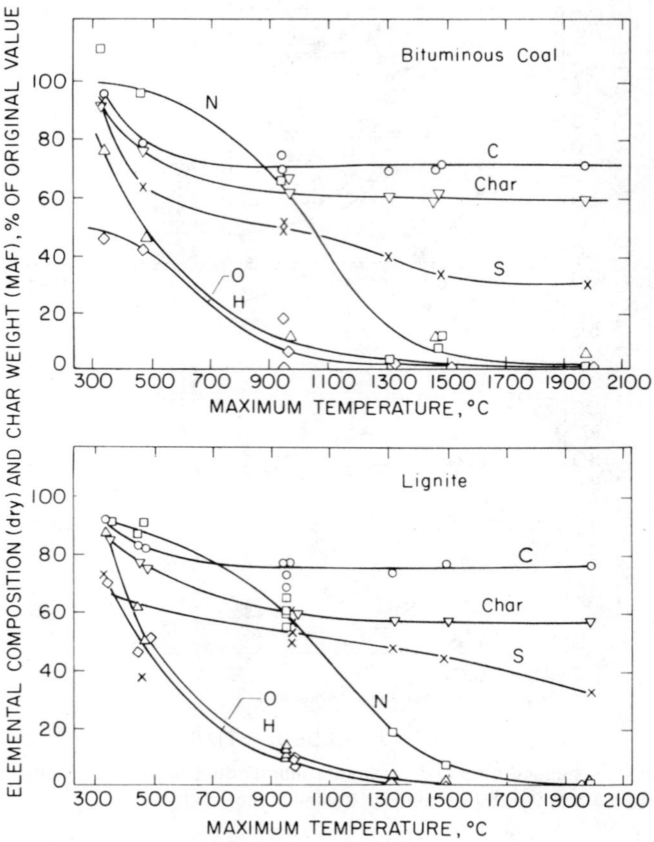

Figure 12.25 Variation of elemental composition and weight of char with maximum temperature attained in coal pyrolysis in a crucible.[89] (Data corrected for analytical error in Ref. 89.)

represented. For much shorter times such as those available in pulverized coal flames or slagging entrained-flow gasifiers (~ 1 or 2 s), nitrogen volatilization is expected to be limited by the time available even though the temperature may be as high as 1600°C[159]

12.4.2.3 *Effect of Heating Rate*

Peters[71] and Peters and Bertling[72] measured the yields of coal pyrolysis products produced under conditions of two very different heating rates. The slow heating technique was based on the use of a standard Fischer assay retort in which a fixed bed of coal was heated to 600°C at a rate of about 0.2°C/s. Heating rates of about 100–1000 times this value were achieved with the other apparatus, in which the coal was heated in a mechanically stirred bed of preheated coke or sand. Owing to the different rates of gas formation associated with the different rates of heating, the residence time of volatiles in the rapidly heated bed was much less than that in the Fischer retort. Hence the use of the two techniques constituted not a variation of heating rate alone but also a variation of the opportunity for secondary reactions. Although this important distinction was recognized by Peters and Bertling,[72] it is sometimes overlooked in references to their work.

The yield of tar from a given coal heated in the stirred bed first increases with increasing bed temperature, reflecting the effect of temperature on the tar-forming reactions, then decreases at higher temperatures, where tar cracking to form gases and char becomes dominant. The maximum tar yield was found at about 600°C, which is also typical of fluidized beds. The distribution of pyrolysis

[159] J. H. Pohl, "Fate of Coal Nitrogen," Sc.D. thesis, Dept. of Chemical Engineering, Massachusetts Institute of Technology, Cambridge, Mass. 1976.

products and some revealing characteristics of the hydrocarbon liquids (tar plus light oils) are summarized in Table 12.10. The combined yields of tar and light oils produced by rapid pyrolysis exceed, by about 12 wt % of the coal, the Fischer assay value. At the same time, the Fischer assay yields of coke and gas plus pyrolysis water are larger by about 10 and 2 wt % of the coal, respectively. This behavior supports the statement above that the conditions associated with rapid heating reduce the extent of secondary cracking reactions that convert some of the tar to gas and char. By virtue of the cracking, the Fischer assay tar contains a high fraction of light oil, whereas the tar from the rapid pyrolysis gives the highest yield of vacuum residue (pitch). Accordingly, the tar produced by rapid pyrolysis has a much higher molecular weight.

The foregoing observations pertain to the effect of volatiles residence time on the extent of secondary reactions. Similarly, in a study of pyrolysis in transport reactors, Eddinger et al.[127] and Friedman et al.[128] observed decomposition of tars and hydrocarbon gases to form char and hydrogen. They concluded that the achievement of a maximum yield of tar requires pyrolysis under conditions that avoid the loss of liquids to vapor-cracked carbon as opposed to the use of more severe thermal conditions that maximize the extent of devolatilization but with loss of liquids by cracking. In another entrained-flow pyrolysis study, Sass[129] observed deviations from Fischer assay yields that may appear in Table 12.5 to be larger than those observed by Peters[71] and Peters and Bertling.[72] However the entrained-flow data were obtained at 530–650°C, whereas the Fischer assay was performed at 500°C. This difference in temperature is not comparable to the common basis used in the other study. In the work of Shapatina et al.[139] with Moscow brown coal at 500°C (Table 12.5), rapid pyrolysis neither enhanced the volatile yield nor increased the tar fraction. This behavior implies the absence of significant secondary reaction effects for the stated coal and temperature.

Suuberg et al.,[111] using the electric wire mesh heater discussed above, measured volatiles compositions for different heating rates and also explored the effects of another type of variation of the temperature-time history of the coal. In contrast to the foregoing studies, these observations are for experimental conditions that essentially avoid secondary reactions except for those occurring inside the particles. Data points inside squares in Fig. 12.12 were obtained with the base conditions given above but modified as follows. Instead of being heated to the peak temperature in one step, the lignite was heated to an intermediate temperature, then cooled to room temperature as before. The resulting char was then heated to a higher temperature and again cooled. The figure shows cumulative product yields for both cycles. The intermediate temperatures for the points at 855 and 1070°C are 480 and 670°C, respectively. The yields of all products in both cases are not significantly different from those obtained when the lignite is heated directly to the final peak temperature. The observation is consistent with the previous conclusion from weight loss data[75,109] that the pyrolysis reactions occurring at higher temperatures are largely independent on those occurring at lower temperatures. Such be-

Table 12.10 Yields of Coal Pyrolysis Products at 600°C[72]

Product	Yields, wt % of Coal: Rapid Pyrolysis	Yields, wt % of Coal: Fischer Assay (slow)	Ratio of Rapid to Slow
Gas	7.25	8.44	0.86
Pyrolysis water	4.10	4.60	0.89
Tar and light oils (boiling as follows):	26.40	14.82	1.78
Fraction I, up to 140°C at 1 mm Hg	6.50	8.02	0.81
Fraction II, 140–230°C at 1 mm Hg	2.87	2.71	1.06
Residue at 230°C and 1 mm Hg	17.05	4.11	4.15
Coke	62.25	72.14	0.86

havior is indicative of multiple parallel independent reactions as opposed to competitive reactions.

The circled data points in Fig. 12.12 were obtained at heating rates of 7100–10,000°C/s, which is approximately 10 times higher than that of the base data. The points in triangles represent data taken at 270–470°C/s. No clear effect of heating rate is observed over the range here used. This behavior is expected for independent parallel, rather than competitive reactions.

The foregoing study of the effects of heating rate also included Pittsburgh Seam bituminous coal,[146] data for which are presented in Tables 12.7 and 12.11. In this case the coal was heated at different rates to a peak temperature in the range 850–1000°C and there held for 3–10 s. The exact peak temperature and the holding time within these ranges are insignificant, since the asymptotic yields are closely approached in all cases. The two-step heating process, which was based on a heating rate of 1000°C/s for each step, was similar to the one described above for the lignite except that a holding time of a few seconds at the peak temperature of each step was employed here. The difference between the sum of the yields of volatile products and the weight loss measured gravimetrically indicates mass balance closures to within 2.8–4.3 wt % of the original coal. A substantial part of the apparent mass loss was attributed to incomplete tar recovery.

The data in Table 12.11 show only small differences in yields between the different cases, which represent a change of heating rate from 350 to 15,000°C/s. The largest of these small differences are some yield reductions observed in going from the base case of 1000°C/s to the longer temperature-time history of the two-step process. Assuming the yield of hydrogen gas in the latter case to be at least 1.0 wt % of the original coal, the reduction in total yield of volatiles is approximately 2 wt % of the coal (1.8% according to the volatile product analyses, or 2.3% according to the weight loss measurement). This net effect is comprised primarily of reductions in the yields of tar (1%), carbon monoxide (0.4%), ethylene (0.5%), and hydrocarbon gases heavier than C_3's (0.4%),

Table 12.11 Effect of Heating Rate and Heating Path on Pyrolysis Product Yields from Pittsburgh Seam Bituminous Coal[146]

Coal composition, Table 12.7; average particle diameter, 74 μm; peak temperature, 850–1000°C; holding time at peak temperature, 2–10 s; atmosphere, helium; pressure, 1 atm.

	Yield, wt % of Coal (as-received)			
Product	350–450°C/s	1000°C/s (base case)	13,000–15,000°C/s	Two-Step Heating
CO	2.4	2.4	2.3	2.1
CO_2	1.6	1.2	1.7	1.6
H_2O^a	7.6	7.8	7.7	7.7
CH_4	2.2	2.5	2.4	2.5
C_2H_4	0.40	0.83	0.66	0.34
C_2H_6	0.59	0.51	0.59	0.56
C_3's	1.1	1.3	1.2	1.3
Other HC gas	1.1	1.3	1.4	0.86
Light HC liquid	2.3	2.4	2.7	2.4
Tar	22.4	23.0	23.0	22.0
H_2	—^{b}	1.0	—^{b}	—^{b}
Char	54.0	53.0	53.0	55.3
	95.7	97.2	96.7	96.7
Error (loss)	4.3	2.8	3.3	3.3
Number of runs	3	20	2	2

^{a} Includes coal moisture (1.4%); may include some H_2S.
^{b} Not measured.

which are partly offset by small increases in the yields of carbon dioxide (0.4%) and some other species. These effects, if significant, probably represent a small enhancement of secondary reactions in the two-step process.

Other evidence that heating rate per se has little if any effect on the cumulative yield of different volatile species is provided by Blair et al.[114] Indirect supporting evidence comes from the char analyses of Kobayashi,[89] given in Figs. 12.23–12.25. These data were used to prepare Fig. 12.26, which shows how the hydrogen and oxygen contents of the char, hence of the cumulative volatiles, vary with the extent of devolatilization. The figure includes data obtained under conditions of extremely different heating rates, namely, 0.5–5°C/s (crucible) and 10,000–200,000°C/s (laminar entrained flow), and for two very different coals. No obvious difference is seen between the data for the two different regimes of heating rate. Most of the scatter of the data was attributed to the values of oxygen content, which were determined as the difference between the weight of char and that of the sum of carbon, hydrogen, sulfur, nitrogen, and ash in char.[89] Thus the total hydrogen and oxygen content of the cumulative volatiles seems to be independent of heating rate in the range 0.5–200,000°C/s.

Jüntgen and Van Heek[97] studied the effects of heating rate on instantaneous rate of evolution of individual species during the heating of coal at constant rates of 10^{-2} to 10^{5} °C/min. Their work also includes the use of model compounds and both experimental and theoretical assessments of the effects of reaction parameters. The behavior of individual species is similar to that discussed above (Fig. 12.6) for the total volatiles. Thus both the temperature at which the rate of evolution achieves a maximum value and the width of the temperature interval over which a given percentage of the evolution occurs increase with increasing heating rate. In the case of ethane from a coal of 19.1% VM, the temperature of the peak evolution rate increases from about 410°C at 0.02°C/min to about 480 and 670°C at 2 and 10^{4} °C/min, respectively. The width of the temperature interval of ethane evolution for these three cases is about 140, 180, and 130°C in order of increas-

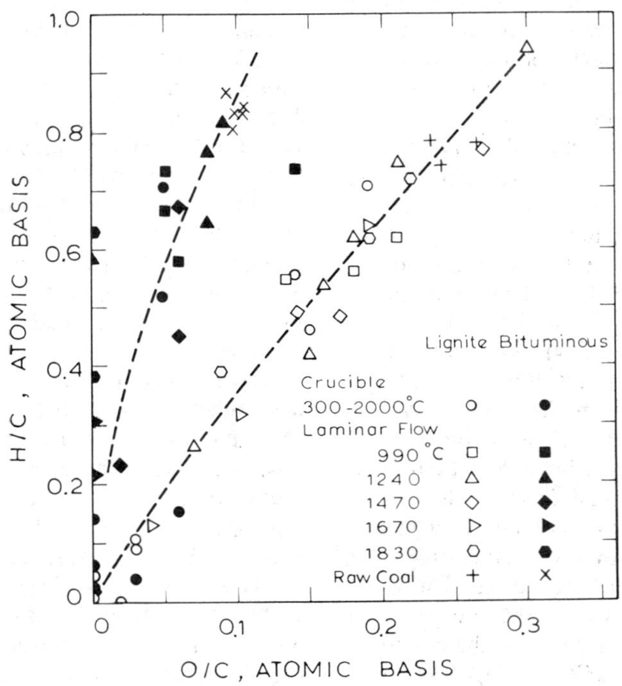

Figure 12.26 Hydrogen and oxygen contents relative to carbon content in original coals (Montana lignite and Pittsburgh Seam bituminous) and in chars from pyrolysis under different conditions. Heating rates: crucible, 0.5–5°C/s; laminar flow, 10^{4} – 2×10^{5} °C/s).[89] (Data corrected for analytical error in Ref. 89.)

ing heating rate. A demonstrated consequence of the widening of the temperature interval is an increasing extent of overlap of the evolution of the two species, as heating rate increases. These phenomena were shown theoretically to be consistent with the expected behavior of first-order reactions, from one to a large number of which being found to represent the evolution of individual species.

12.4.2.4 *Effect of Pressure*

Rennhack[157] studied the effect of pressure on the rate of evolution of hydrogen and of methane during the heating of a char that had been produced from a high-volatile bituminous coal at 580°C in a carbonization apparatus. According to the reported ash analysis, the weight of the char was about 90% of the weight of the raw coal, thus indicating only a small extent of devolatilization in the char preparation step. It appears from other data obtained at slow heating rates (Figs. 12.10 and 12.11) that little hydrogen would be expected to evolve during the char production at 580°C, but the same cannot be said for methane. Nevertheless the methane evolution that occurs subsequent to and at higher temperatures than the char production step is probably independent of that which occurred during char production, since (1) methane evolution behaves as multiple, independent, parallel reactions,[97,111] and (2) the yield of methane in the two-step heating experiment of Suuberg (Table 12.11) is essentially the same as that obtained in the one-step experiment. Therefore pressure effects observed by Rennhack for the char would be expected to pertain also to the coal.

Char in a nitrogen-swept bed was heated from 500 to 850°C at 5°C/min and, simultaneously, the rates of methane and hydrogen evolution were measured as a function of temperature. The experiment was performed at two pressures, 1 and 20 atm. The rate of methane evolution as a function of temperature increased with increasing pressure, while that of hydrogen decreased. These opposite effects were coupled such that the sum of the hydrogen evolved as methane and as hydrogen gas (i.e., $2CH_4 + H_2$) was independent of pressure throughout the range of temperatures studied. Rennhack interpreted these data to mean that hydrogen gas and methane represent products of two different reaction pathways involving hydrogen radicals, the formation of which is rate controlling. Since the stated characteristic of the combined hydrogen and methane evolution occurs over the entire range of temperatures studied, this interpretation implies that the hydrogen radicals are formed by a large number of reactions. Accordingly, Rennhack employed a combination of numerous overlapping first-order reactions with a statistical distribution of activation energies to correlate the data.

The coal gasification research at Bergbau-Forschung has included study of the influence of pressure on methane formation in pyrolysis.[86] In the pyrolysis of both a lignite and a high-volatile bituminous coal in helium, the rate of methane evolution during the heating of the sample increased with increasing pressure in the range studied (10–70 atm), but the effect was important only at temperatures above about 525°C for the lignite and about 550°C for the bituminous coal.

The work of Suuberg[146] included a study of the composition changes associated with the effect of pressure on weight loss. Data of the type shown in Figs. 12.12–12.14 and 12.16–12.20 for pyrolysis at 1 atm were also obtained under vacuum conditions (0.05 mm Hg). The results for Montana lignite heated under vacuum at 1000°C/s are presented in Fig. 12.27 and compared with the data obtained at atmospheric pressure. Up to a temperature of about 800°C, the total yield of volatiles and their composition are indistinguishable from those produced under 1 atm of helium at the same heating rate. Above 800°C, however, observable differences are noted in both total yield and product composition.

Experiments with the same lignite were also conducted at a helium pressure of 68 atm, and with Pittsburgh Seam bituminous coal at the same three pressures.[146] In these four cases the coal was heated at 1000°C/s to a peak temperature of 850–1070°C and there held for 2–10 s before being cooled. The yields at these peak temperatures are only slightly larger than would be obtained if the holding time were nil. Furthermore, the holding times and temperatures are all large enough so that the yields are essentially independent of time and temperature within the stated ranges. The results are given in

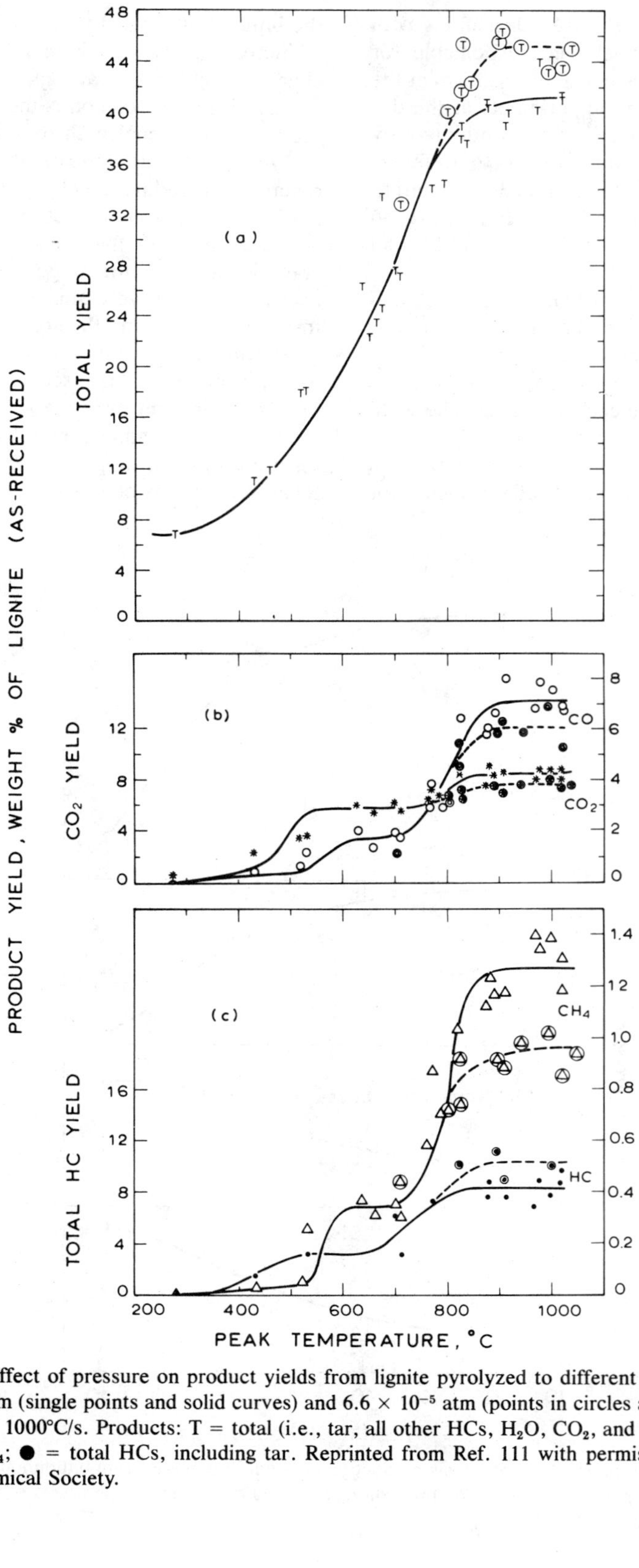

Figure 12.27 Effect of pressure on product yields from lignite pyrolyzed to different peak temperatures. Pressures: 1 atm (single points and solid curves) and 6.6×10^{-5} atm (points in circles and dashed curves). Heating rate = 1000°C/s. Products: T = total (i.e., tar, all other HCs, H_2O, CO_2, and CO); ○ = CO; * = CO_2; △ = CH_4; ● = total HCs, including tar. Reprinted from Ref. 111 with permission. Copyright by American Chemical Society.

Fig. 12.28 and Table 12.12. Difficulties with tar collection are believed responsible for much of the mass balance error shown at the bottom of the table. Experiments at the different pressures were also conducted by heating the coals at 1000°C/s to different peak temperatures, then immediately initiating cooling as described earlier. Data so obtained are presented in Figs. 12.29 and 12.30.

The main conclusion from Table 12.12 and Figs. 12.27–12.30 is that pyrolysis under higher pressures produces more char (i.e., less weight loss from the coal), less tar, more methane, and more carbon oxides. The yield of water decreases substantially with increasing pressure in the case of the lignite, but a somewhat smaller opposite effect occurs for the bituminous coal. By far the largest effect of increasing pressure is increased yields of char and light hydrocarbons at the expense of tar. This phenomenon is most striking for the bituminous coal with its relatively large tar yield. Thus it appears that secondary reactions including cracking and char formation play a role in determining product yields even under the present experimental conditions where the opportunity for such reactions is relatively small. In view of the thin layer of coal and the absence of particle-gas contact once volatiles have left a particle, the contribution of the secondary reactions in this experiment presumably occurs within the particles. A similar effect of secondary reactions on total weight loss from Pittsburgh Seam bituminous coal was observed by An-

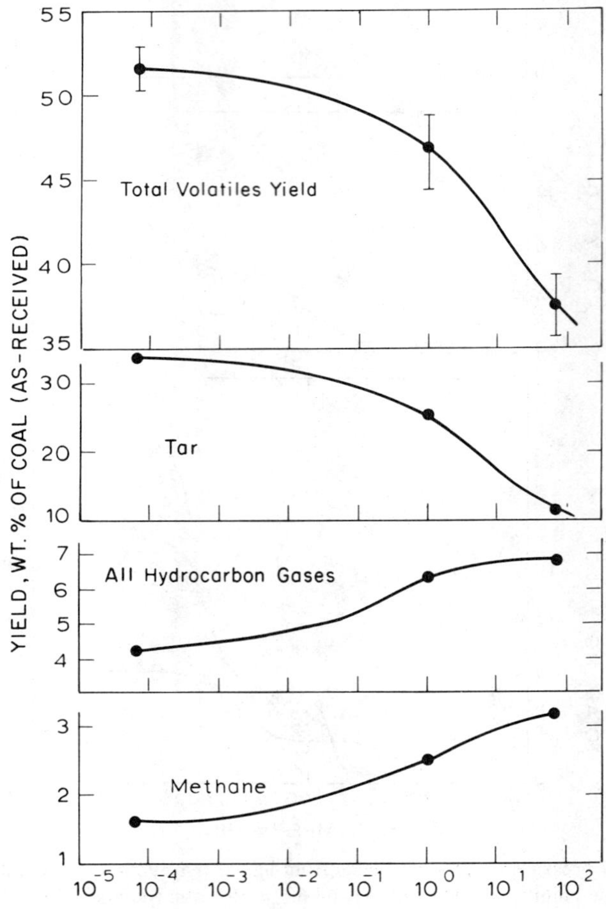

Figure 12.28 Effect of pressure on pyrolysis product yields from Pittsburgh Seam bituminous coal. Helium atmosphere, average particle diameter, 74 μm; heating rate, 1000°C/s; final temperature, ~1000°C; holding time at final temperature, 2–10 s).[146]

Table 12.12 Effect of Pressure on Pyrolysis Product Yields from Montana Lignite and Pittsburgh Seam Bituminous Coal Heated in Helium[146]

Coal compositions, Table 12.7; average particle diameter, 74 μm; heating rate, 1000°C/s; peak temperatures, lignite, vacuum and 1 atm, 900–1035°C, all other cases, 850–1070°C; holding times at peak temperature, lignite, vacuum and 1 atm, 0.0 s, all other cases, 2–10 s.

	Yield, wt % of Coal (as-received)					
	Lignite			Bituminous		
Product	Vacuum[a]	1 atm	69 atm	Vacuum[a]	1 atm	69 atm
CO	6.1	7.1	9.0	2.0	2.4	2.5
CO_2	7.6	8.4	10.6	1.4	1.2	1.7
H_2O[b]	17.7	16.5	13.4	6.8	7.8	9.5
H_2	—[c]	0.50	—[c]	0.75	1.0	—[c]
CH_4	0.94	1.3	2.5	1.6	2.5	3.2
C_2H_4	0.43	0.56	0.55	0.45	0.83	0.46
C_2H_6	0.21	0.18	0.17	0.44	0.51	0.89
$C_3H_6+C_3H_8$	0.46	0.37	0.38	0.71	1.3	0.71
Other HC gases	0.60	[0.47]	0.21	0.98	1.3	1.6
Light HC liquids	0.81		1.1	1.6	2.4	2.0
Tar	6.9	5.4	2.8	31.9	23.0	12.
Char	55.2	58.7	59.0	48.5	53.0	62.4
	97.0	99.5	99.7	97.1	97.2	97.0
Error (loss)	3.0	0.5	0.3	2.9	2.8	3.0

[a] 6.6×10^{-5} atm He.
[b] Includes coal moisture (lignite, 6.8%; bituminous, 1.4%); may include some H_2S.
[c] Not measured.

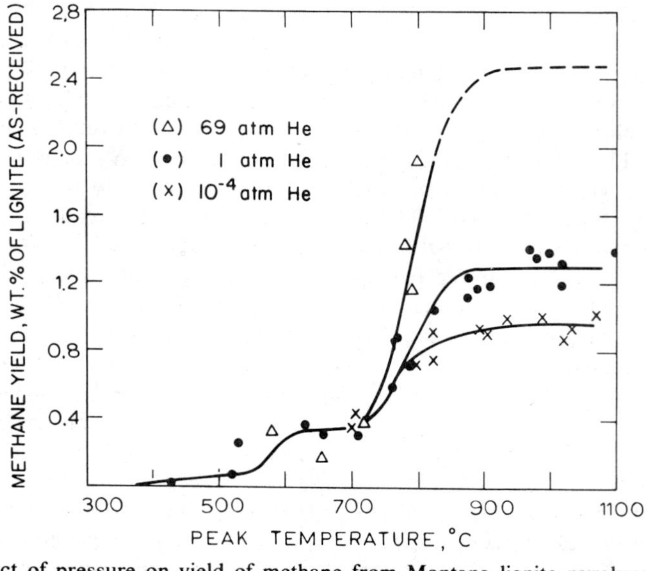

Figure 12.29 Effect of pressure on yield of methane from Montana lignite pyrolyzed to different peak temperatures. Helium atmosphere; heating rate, 1000°C/s; average particle diameter, 74 μm.[146]

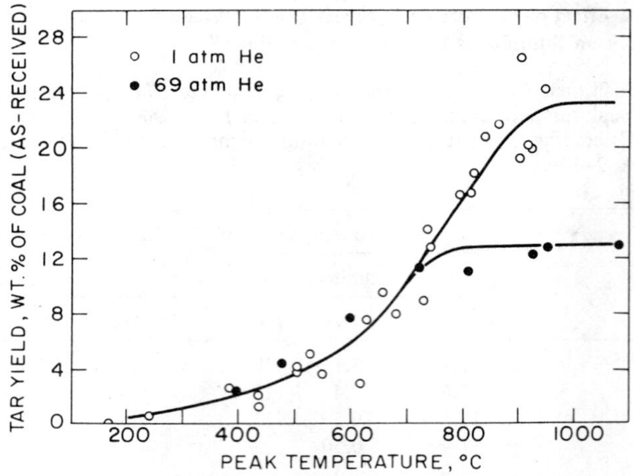

Figure 12.30 Effect of pressure on yield of tar from Pittsburgh Seam bituminous coal pyrolyzed to different peak temperatures. Helium atmosphere; heating rate, 1000°C/s; average particle diameter, 74 μm.[146]

thony et al.[75,109] Other workers (e.g., Howard[2]) also have observed increased yields of low-volatile species in vacuum pyrolysis of coal. Also Belt et al.[140] observed that the highest yields of char from their entrained-flow carbonizer were obtained at the highest inert gas pressures.

12.4.2.5 *Effect of Particle Size*

Suuberg[146] measured the effect of particle size on the yield of pyrolysis products from Pittsburgh Seam coal. The experimental conditions and results are summarized in Table 12.13. As particle diameter is increased in the range 74 μm to approximately 1 mm, a small increase in char yield (i.e., decrease in total volatiles yield) is observed, which agrees with Fig. 12.9, while the tar yield decreases and the yields of methane and oxides of carbon both increase. These changes are qualitatively similar to but much smaller than those discussed above for increasing pressure, and they undoubtedly reflect the enhancement of secondary reactions as particle size increases.

12.4.2.6 *Effect of Coal Type*

Workers at Bergbau-Forschung have included several different coals in their measurements of the evolution rate of gaseous species during the heating of coal in helium at constant rates of 1–2°C/min.[86] The cumulative yields obtained from four German coals are presented in Table 12.14 (coals 4–7) along with data on three American coals. The heating rates used with the American coals were 1000°C/s (coals 1 and 3) and 3.3°C/min (coal 2). In all cases in the table the peak temperature was approximately 1000°C.

The ranks of coal represented in Table 12.14 vary from lignite to anthracite. Proceeding in the direction of increasing rank (i.e., from coal 1 to coal 7), the oxygen content of the coal decreases by a factor of 10 and so does the yield of oxygen as water plus oxides of carbon. The corresponding decrease in the yields of both carbon dioxide and water is a factor of 16, whereas that of carbon monoxide is about 5. About half the oxygen evolved from the lignite appears as carbon dioxide. Although the hydrogen content of the coal decreases slightly with increasing rank, the tendency to form water in the low-rank coals results in the yield of hydrogen gas increasing significantly with increasing rank. The yield of methane goes through a maximum at intermediate ranks, although this point is not well established in the table. The yield of nitrogen gas from the four German coals decreases by a factor of almost 3 as the rank increases. Since the nitrogen content of these coals is almost identical, the nitrogen appears to be more strongly bound in the higher rank coals (assuming that the latter do not give larger yields of other nitrogenous species).

Table 12.13 Effect of Particle Size on Pyrolysis Yields from Pittsburgh Seam Bituminous Coal Heated in Helium[146]

Coal composition, Table 12.7; heating rate, 1000°C/s; peak temperature, 850–1070°C; holding time at peak temperature, 3–10 s; pressure, 1 atm.

	Yield, wt % of Coal (as-received)			
Product	53–88 μm (avg., 74 μm)	<300 μm[a]	300–830 μm	830–990 μm
CO	2.4	2.7	3.2	3.0
CO_2	1.2	1.1	1.2	1.3
H_2O^b	7.8	5.4	5.3	7.2
H_2	1.0	—[c]	—[c]	0.99
CH_4	2.5	2.9	3.0	3.2
C_2H_4	0.83	1.0	1.1	1.3
C_2H_6	0.51	0.50	0.55	0.63
C_3's	1.3	0.92	0.84	1.1
Other HC gases	1.3	1.4	1.1	1.2
Light HC liquids	2.4	2.5	2.6	2.7
Tar	23.0	24.2	21.3	18.4
Char	53.0	57.1	56.5	55.8
	97.2	99.7	96.7	96.8
Error (loss)	2.8	0.3	3.3	3.2
Number of runs	20	1	2	3

[a] 830–990 μm sample ground to pass 297-μm sieve.
[b] Includes coal moisture (1.4%); may include some H_2S.
[c] Not measured.

Although the sum of the gases shown in Table 12.14 (bottom line) represents a substantial part (70–75%) of the total volatile yield from low- and high-rank coals, the gases are a much smaller part of the total for the intermediate ranks where tar is a major product, especially in systems that avoid extensive secondary reactions. Thus the variation of tar formation in the primary reactions and the associated sensitivity to secondary reactions is an important effect of coal type.

With regard to the rate of species evolution during the heating of coal, Jüntgen and Van Heek and their associates obtained the type of data shown in Fig. 12.10 for several coals including the four German coals represented in Table 12.14 and a lignite (braunkohle).[89] With a few exceptions, the evolution rate versus temperature curves for the different coals are similar to those in Fig. 12.10 but displaced vertically in accordance with the differences in cumulative yield (hence in area under curve) discussed above. Two of the exceptions are discussed below.

The methane curve exhibits two overlapping stages of evolution, the first of which is dominant for the 19.1% VM coal in Fig. 12.10 *c*. In the case of the lignite, the peak rates of the two stages are approximately equal, and comparable to that of the lesser important second stage of a higher rank coal.

The second most important stage of carbon monoxide evolution is centered at about 580°C in Fig. 12.10 *a*. As rank decreases from coal 7 to coal 4 in Table 12.14, the contribution of this stage increases from being almost negligible to being clearly dominant.

12.4.3 Pyrolysis Models and Interpreted Behavior

12.4.3.1 *Single-Reaction Models*

The complex decomposition and transport phenomena involved in coal pyrolysis are not yet amenable to exact description. Many authors have approximated the overall process as a first-order decomposition occurring uniformly throughout the particle. Thus the rate of pyrolysis is expressed as:

Table 12.14 Effect of Coal Type on Yields of Gaseous Pyrolysis Products of Seven Coals

	Coal Number (see key in source note)						
	(1)	(2)	(3)	(4)	(5)	(6)	(7)
	Analyses (wt %) of coal						
C (maf)	71.2	73.5	77.7	83.6	87.5	89.3	91.0
H (maf)	4.6	5.8	5.5	5.1	4.6	4.3	4.0
N (maf)	1.1	1.2	1.5	1.6	1.7	1.6	1.6
S (maf)	1.3	0.81	6.1	0.95	0.67	1.2	0.97
O (diff.) (maf)	21.8	18.7	9.2	7.6	5.3	3.0	2.1
H_2O (as-received)	6.8	34.7	1.4	2.3	0.73	0.6	0.8
Ash (mf)	10.6	9.1	11.5	4.0	7.4	6.5	6.6
VM (maf)	44.3	—	44.6	39.5	28.9	19.1	10.0
Swelling index	0	—	—	3.5	7.5	9	0
	Products (wt % of coal, maf)						
CO	8.5	5.8	2.8	3.2	1.9	1.7	1.6
CO_2	10.1	11.3	1.4	0.81	0.75	0.71	0.63
H_2O	11.6	—	7.4	2.4	1.3	1.1	0.72
H_2	0.60	1.1	1.2	3.1	3.4	3.5	3.8
N_2	—	—	—	0.95	0.66	0.42	0.35
CH_4	1.6	4.8	2.9	4.2	—	—	—
C_2H_4	0.67	0.24	0.95	—	—	—	—
C_2H_6	0.24	0.89	0.58	—	—	—	—
C_3H_8	0.17	0.47	1.5[a]	0.29	0.26	0.12	0.01
	33.5	24.6	18.7	15.0	8.3	7.6	7.1

[a] Includes C_3H_6.

Source: Coal 1, Montana lignite, Savage Mine—Ref. 111; Coal 2, Roland Seam subbituminous, Wyodak Mine—Ref. 91; Coal 3, Pittsburgh Seam bituminous, Ireland Mine—Ref. 146; Coal 4, Fürst Leopold—Ref. 86; Coal 5, Gustav—Ref. 86; Coal 6, Dickebank—Ref. 86; Coal 7, Heinrich—Ref. 86.

$$\frac{dV}{dt} = k(V^* - V) \qquad (12.4)$$

where V is the cumulative amount of volatiles produced up to time t, expressed as weight fraction of the initial coal, k is the rate constant, and $V \to V^*$ as $t \to \infty$. Thus V^* represents the effective volatile content of the coal. The unknown parameters k and V^* have usually been the focus of kinetic studies.

The rate constant in Equation 12.4 is typically correlated with temperature by an Arrhenius expression:

$$k = k_0\, e^{-E/RT} \qquad (12.5)$$

where k_0 is the preexponential factor, E is the apparent activation energy, R is the gas constant, and T is the absolute temperature. A collection of experimental k values appears in Fig. 12.31, and the associated rate parameters and coal properties are summarized in Table 12.15. There is little agreement on the observed rates of pyrolysis, with several orders of magnitude discrepancy at a given temperature. Some, but by no means all, of the disagreement can be attributed to coal type even though the data are limited to high volatile coals. Differences in equipment and experimental procedures are undoubtedly also important.

[160] D. W. Van Krevelen, C. Van Heerden, and F. J. Huntjens, "Physiochemical Aspects of the Pyrolysis of Coal and Related Organic Compounds," *Fuel*, **30**, 253 (1951). (Numerical results modified by Jüntgen and Van Heek, 1970.)

[161] M. S. F. Boyer, *C. R. Assoc. Tech. l'Ind. Gaz France, Congr.*, 653 (1952) (as reported by Yellow, 1965).

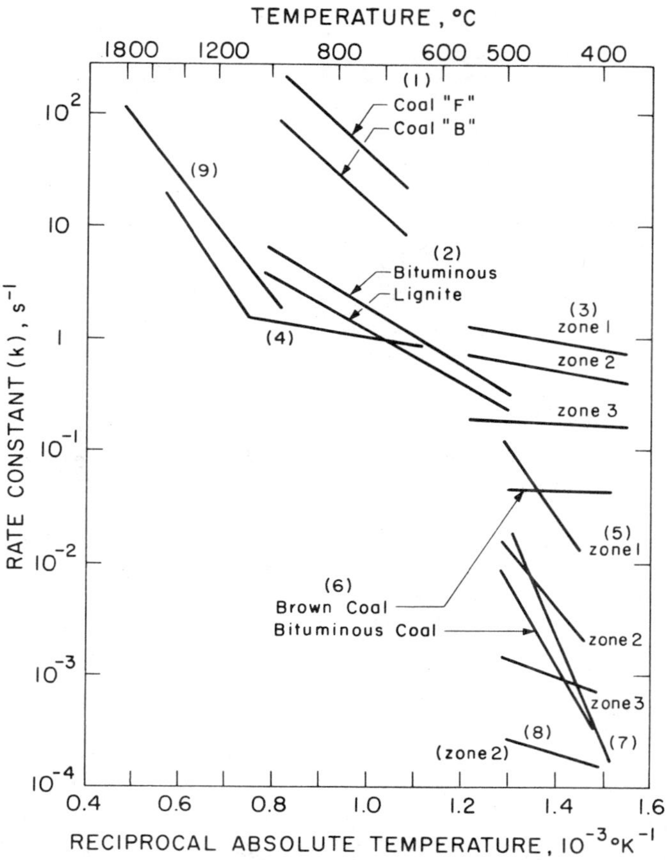

Figure 12.31 Comparison of simple first-order coal pyrolysis rate constants from different investigators. (1) Badzioch and Hawksley,[65] (2) Anthony et al.,[75] (3) Shapatina et al.,[139] (4) Howard and Essenhigh,[66] (5) Stone et al.,[118] (6) Van Krevelen et al.,[160] (7) Boyer,[161] (8) Wiser et al.,[36] (9) Kobayashi et al.[132]

Many authors have contended that a simple first-order model is inadequate. In one attempt to improve the utility of Equation 12.4 pyrolysis is described as a series of several first-order processes occurring in different time intervals. Thus Shapatina et al.[139] depict three stages (0–0.06, 0.06–0.1, and 0.1–180 s), and Stone et al.[118] use time intervals typically much longer than the preceding, and dependent on coal type and temperature. Another approach has been to use an nth-order expression:

$$\frac{dV}{dt} = k' (V^* - V)^n \qquad (12.6)$$

Thus Wiser et al.[36] found that $n = 2$ gave the best fit to their data over the first 3600 s of weight loss, whereas $n = 1$ was preferred for longer times (zone 2 of Fig. 12.31). Skylar et al.[162] found that values of n from 2 to 8 were required to fit nonisothermal pyrolysis data for different coals, which casts doubt on the validity of Equation 12.6. Pitt[119] successfully correlated his pyrolysis data with the empirical relation

$$\frac{V^* - V}{V^*} = A - B \log(t) \qquad (12.7)$$

where A and B are constants.

One of the most serious shortcomings of Equations 12.4–12.7 is that the apparently asymptotic yield of volatiles observed after some time at the final temperature, that is,

[162] M. G. Skylar, V. I. Shustikov, and I. V. Virozub, "Investigation of the Kinetics of Thermal Decomposition of Coals," *Intern. Chem. Eng.*, **9**, 595 (1969).

Table 12.15 Coal Properties and Pyrolysis Rate Parameters for Fig. 12.31

Investigators	Coal	VM (mf)	VM (maf)	E (kcal/mole)	k_0, (s^{-1})	Comments
Badzioch and Hawksley[65]	"B" NBC 902	35.2	36.4	17.8	1.14×10^5	
	"F" NBC 601	34.9	35.3	17.8	3.12×10^5	
Boyer[161]	St. Fontaine bituminous	—	36.0	45	1.00×10^{11}	Reported by Yellow (1963)
Howard and Essenhigh[66]	Pittsburgh Seam bituminous	35.9	37.4	27.7	4.92×10^4	$T > 1060°C$
				3.4	5.37	$T < 1060°C$
Shapatina et al.[139]	Moscow district brown coal	34.5	50.2	3.6	11.0	Zone 1 (0–0.06 s)
				3.8	6.63	Zone 2 (0.06–0.1 s)
				1.0	0.37	Zone 3 (0.1–180 s)
Stone et al.[118]	Pittsburgh Seam bituminous	40.7	42.2	27.3	5.41×10^6	Zone 1
				24.4	1.10×10^5	Zone 2
				7.0	0.131	Zone 3
Van Krevelen et al.[160]	Brown coal	—	51.0	0.7	0.07	Values corrected by Jüntgen and Van Heek[163]
	Low-rank bituminous	—	39.5	32.5	1×10^7	
Wiser et al.[36]	Utah bituminous	47.5	—	5.4	0.54	Zone 2 (> 3600 s)
Anthony et al.[75]	Pittsburgh Seam bituminous	40.5	46.2	11.8	706	
	Montana lignite	40.3	46.2	11.3	282	
Kobayashi et al.[132]	Pittsburgh Seam bituminous	41.6	46.8	25	6.6×10^4	
	Montana lignite	41.9	46.1	25	6.6×10^4	

the apparent value of V^*, is a function of final temperature; yet this is neither mechanistically consistent with the equations nor mathematically amenable to them. Also, as is reflected by the concept of time zones used in some of the correlations in Fig. 12.31, the relatively slow rate of weight loss observed after extended times at a given temperature requires a set of parameter values that differ markedly from those that fit the behavior at short times. Since coal pyrolysis clearly is not a single reaction but rather a multiplicity of overlapping decompositions concentrated in different time intervals for isothermal pyrolysis, or in different time and temperature intervals for the usual case of pyrolysis during heatup, any one set of parameter values for these equations cannot be expected to represent data accurately over a wide range of conditions.

The apparent activation energies of the data in Fig. 12.31 range from 0.7–45, but most are less than 28 kcal/mole. The preexponentials range from 0.07 to 10^{11} s^{-1} and are coupled with the apparent activation energy such that when E is large or small, so is k_0. Values of k_0 and E for the pyrolysis of a wide range of organic materials are listed in Table 12.16, along with the volatile products whose evolution is described by the paramters. Here again the parameters exhibit both wide ranges and a strong coupling (log k_0 exhibits significant correlation with E), but the values are substantially larger than those in Table 12.15.

It is instructive to compare data on coal pyrolysis with the behavior indicated when parameter values from Table 12.16 are used in Equations 12.4 and 12.5. Assuming that temperature increases at a constant rate $dT/dt = m$, these equations can be combined and integrated as follows:

$$\int_0^V \frac{dV}{V^* - V} = \int_0^T \frac{k_0}{m} \exp\left(\frac{-E}{RT}\right) dT \qquad (12.8)$$

Since $E/RT \gg 1$ is a good approximation for pyrolysis reactions, the solution to Equation 12.8 becomes[168]:

$$\frac{V^* - V}{V^*} = \exp\left(\frac{-k_0 RT^2}{mE} \exp\frac{-E}{RT}\right) \qquad (12.9)$$

where $m > 0$. This equation is plotted in Fig. 12.32 (dashed curves) for activation energies typical of the organic decomposition reactions and a typical preexponential factor of $k_0 = 1.67 \times 10^{13}$ s^{-1}. Also shown in the figure are data on the evolution of all volatiles (solid curve) and of methane (dotted curve) for lignite pyrolysis at constant heating rate.

The inadequacy of the single-reaction model, using what might be called reasonable parameters, in fitting the data on total yield of volatiles is evident in Fig. 12.32. However the different stages of methane evolution, which are representative of the general behavior of individual volatile species, are strikingly similar to the predicted curves. In fact, when families of such predicted curves comprising ranges of k_0 and E values are superimposed on data for various volatile species, sets of k_0 and E values can be found that closely match each stage of evolution of the different compounds. As discussed in Section 12.4.3.2, the best set of parameter values for a given evolution stage of a compound can be found by computerized curve fitting, but not without significant covariance of k_0 and E within the scatter of most data. Nevertheless, the parameters found are broadly consistent with those in Table 12.16. It is not surprising that coal, the chemical structure of which is far more complex than that of the materials in Table 12.16, decomposes thermally to produce numerous products, and these products exhibit different activation energies.

In contrast to the foregoing agreement, when the first-order, single-reaction model is used to describe the data on total volatiles evolution, the activation energy and preexponential factor are forced to be very low to fit approximately the temperature dependence that actually results from the occurrence of different reactions in different temperature intervals. The results are sometimes interpreted as reflecting transport limitations because the parameters are too low for organic decompositions. Thus Berkowitz[169] concluded from his pyrolysis study of Alberta subbituminous coals at 250–520°C that the rate is limited by mass transfer. After studies at higher temperatures, this conclusion was modified by postulating that movement of carbon lamellae is the rate-determining process.[170] Similarly Zielinski[145] collated pyrolysis

Table 12.16 Kinetic Parameters for Pyrolysis of Various Organic Materials[111]

Material Pyrolyzed	Product[a]	Exp. Temp. (°C)	Activation Energy (kcal/mole)	Preexponential Factor (s^{-1})	Reference
Ferulic acid ($C_{10}H_{10}O_4$)	CO_2	150–250	27.7	6.0×10^9	Jüntgen and Van Heek[97,163]
Perylene tetracarboxylic acid	CO_2	400–600	71.5	5.2×10^{17}	Jüntgen and Van Heek[97,163]
Anhydride ($C_{24}H_8O_6$)	CO	400–600	64.9	5.0×10^{16}	Jüntgen and Van Heek[97,163]
Protocatechuic acid ($C_7H_6O_4$)	$(H_2O)_1$	50–300	18.8	2.7×10^8	Jüntgen and Van Heek[97,163]
	$(H_2O)_2$	50–300	42.4	2.3×10^5	Jüntgen and Van Heek[97,163]
	CO_2	50–300	40.4	1.6×10^{15}	Jüntgen and Van Heek[97,163]
Naphthalene tetracarboxylic acid ($C_{14}H_8O_8$)	H_2O	100–250	33.5	1.2×10^{13}	Jüntgen and Van Heek[97,163]
Mellitic acid ($C_{12}H_6O_{12}$)	H_2O	230[b]	16.6	2.3×10^5	Jüntgen and Van Heek[163]
Tartaric acid ($C_4H_6O_6$)	H_2O	195[b]	42.9	6.7×10^{17}	Jüntgen and Van Heek[163]
Polystyrene $(C_8H_8)_n$	Overall	394[b]	77	8.3×10^{22}	Fuoss et al.[164]
	Overall	335–355[b]	58	9.0×10^{15}	Madorsky[165]
Teflon $(C_2F_4)_n$	Overall	575[b]	67–69	4.3×10^{14}	Fuoss et al.[164]
Polyethylene $(C_2H_4)_n$					
"Phase 1"	Overall	385–405	48	5.2×10^{11}	Madorsky[165]
"Phase 2"	Overall	385–405	71	8.7×10^{18}	Madorsky[165]
Hydrogenated polystyrene $(C_8H_{14})_n$	Overall	335–350	52	1.4×10^{14}	Madorsky[166]
Poly-*m*-methylstyrene $(C_9H_{10})_n$	Overall	333–353	59	7.2×10^{16}	Madorsky[166]
Poly-α-methylstyrene $(C_9H_{10})_n$	Overall	273–288	58	8.3×10^{18}	Madorsky[166]
Poly(methyl methacrylate) $(C_5H_8O_2)_n$					
Av molecular wt 150,000	Overall	240–270	33	3.6×10^9	Madorsky[166]
Av molecular wt 5,100,000	Overall	310–325	55	1.8×10^{16}	Madorsky[166]
Poly (methyl acrylate) $(C_4H_6O_2)_n$	Overall	285–300	37	1.4×10^{10}	Madorsky[166]
Cellulose $(C_6H_{10}O_5)_n$	Overall	250–1000	33.4	6.8×10^9	Lewellen et al.[167]
		250–350	35	3.3×10^{11}	Jüntgen and Van Heek,[163] Van Krevelen et al.[160]

[a] Denotes species whose evolution is described by the parameters given. Two different stages of water evolution are denoted by $(H_2O)_1$ and $(H_2O)_2$. "Overall" refers to all products combined.

[b] Temperature of maximum pyrolysis rate.

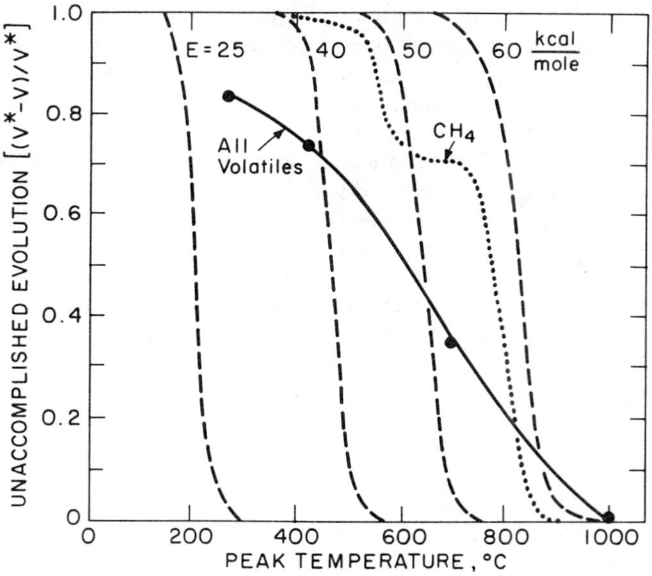

Figure 12.32 Comparison of lignite pyrolysis data with predictions from single-step, first-order pyrolysis model. ● = data on total yield of volatiles; dotted curve = data on methane evolution, from curve in Fig. 12.13; dashed curves = predictions from stated model.

data from several authors and concluded that heating rate is the controlling factor. In both cases these mechanisms were predicted on the basis of observations that perhaps are better explained by alternative theories. Furthermore, simple heat and mass transfer models are inadequate for dealing with the complexities of coal pyrolysis.

[163] H. Jüntgen and K. H. Van Heek, "Reaktionablaufe unter nichtisothermen Bedingungen," *Fortschritte der chemischen Forschung,* Vol. 13, Springer-Verlag, Berlin, 1970, pp. 601–699. Translated by Belov and Assoc., Denver, Col., APTIC-TR-0776.

[164] R. M. Fuoss, I. O. Salyer, and H. S. Wilson, "Evaluation of Rate Constants from Thermogravimetric Data," *J. Polym. Sci., A,* **2,** 3147 (1964).

[165] S. L. Madorsky, "Rates of Thermal Degradation of Polystyrene and Polyethylene in a Vacuum," *J. Polym. Sci.,* **9**(2), 133 (1952).

[166] S. L. Madorsky, "Rates and Activation Energies of Thermal Degradation of Styrene and Acrylate Polymers in a Vacuum," *J. Polym. Sci.,* **2**(5), 491 (1953).

[167] P. C. Lewellen, W. A. Peters, and J. B. Howard, "Kinetics of Cellulose Pyrolysis and Char Formation Mechanism," *16th Symp. (Int.) Combustion,* Combustion Institute, Pittsburgh, 1977, p. 147.

[168] K. H. Van Heek, H. Jüntgen, and W. Peters, "Kinetik nïcht-isotherm ablaufender Reaktionen am Beispiel," *Ber. Bunsenges. Phys. Chem.,* **71,** 113 (1967).

[169] N. Berkowitz, "Mechanism of Coal Pyrolysis, I," *Fuel,* **39,** 47 (1960).

[170] N. Berkowitz and W. Den Hertog, "Mechanisms of Coal Pyrolysis, V," *Fuel,* **41,** 507 (1962).

12.4.3.2 *Multiple-Reaction Models*

Models Based on Total Volatiles. The modeling of coal pyrolysis in terms of total volatiles yield is still important, even though the more powerful approach, introduced above and described in more detail below, is based on the individual compounds or classes of compounds that comprise the volatiles. Information on total volatiles, from weight loss measurements, has been more readily available than data on the evolution of individual species. Consequently much of the modeling work has been focused on total volatiles evolution or weight loss behavior, and valuable contributions have been made.

Wiser et al.[36] treated their data isothermally, integrating first- and second-order versions of Equation 12.6, then inserting an appropriate value of V^* depending on temperature. Similarly, Badzioch and Hawksley[65] correlated data with an expression derived by isothermal integration of Equation 12.4 and substitution of a temperature-dependent function for V^*. Thus neither group of investigators considered the difficult integration involved in nonisothermal cases where heatup is not instantaneous. Badzioch and Hawksley claimed that little decomposition occurred during their extremely rapid heatup, and Wiser et al. neglected the weight loss occurring before their sample reached the desired temperature (~ 300 s to reach temperature). The problem of how to include in kinetic studies the weight loss occurring during heatup has confounded many investigators attempting to evaluate coal decomposition on an isothermal basis.

Pitt's[119] recognition of this problem led him and later Rennhack[157] to adapt, as described below, Vand's[171] treatment of a large number of independent, parallel rate processes. The thermal decomposition of coal was assumed to consist of a large number of independent chemical reactions representing the rupture of various bonds within the coal molecule. Accordingly, marked differences in the strengths of chemical bonds throughout the coal molecule account for the occurrence of different reactions in different temperature intervals. Since the thermal decomposition of a single organic species is typically described as an irreversible reaction that is first order in the amount of unreacted material remaining, the rate of volatiles evolution originating from a particular reaction within the coal structure was described in a manner similar to Equation 12.4, with a subscript i used to denote one particular reaction:

$$\frac{dV_i}{dt} = k_i(V_i^* - V_i) \qquad (12.10)$$

If k_i is of the Arrhenius form (Equation 12.5), integration of Equation 12.10 for isothermal conditions to find the amount of volatiles yet to be released gives

$$V_i^* - V_i = V_i^* \exp\left(-kt \exp\frac{-E}{RT}\right)_i \qquad (12.11)$$

Values of k_{0i}, E_i, and V_i^* cannot be predicted a priori and must be estimated from experimental data, a problem that increases with the number of reactions postulated. The problem is simplified if it is assumed that the k_i's differ only in activation energy, that is, $k_{0i} = k_0$ for all i, and that the number of reactions is large enough to permit E to be expressed as a continuous distribution function $f(E)$, with $f(E)\,dE$ representing the fraction of the potential volatile loss V^* that has an activation energy between E and $E + dE$. Then V_i^* becomes a differential part of the total V^* and may be written

$$dV^* = V^* f(E)\, dE \qquad (12.12)$$

with

$$\int_0^\infty f(E)\, dE = 1 \qquad (12.13)$$

The total amount of volatile material yet unreleased is obtained by summing the contribution from each reaction or by integrating Equation 12.11 over all values of E using Equation 12.12. Thus

$$\frac{V^* - V}{V^*} = \int_0^\infty \exp\left[-k_0 t \exp\left(\frac{-E}{RT}\right)\right] f(E)\, dE \qquad (12.14)$$

By approximating the term

$$\exp[-k_0 t \exp(-E/RT)]$$

as a step function equal to zero for $E < E_t$ and unity for $E \geqslant E_t$, where $E_t = RT \ln(k_0 t)$ and $k_0 = 1.67 \times 10^{13}\ s^{-1}$, Pitt constructed an $f(E)$ curve to best fit his weight loss data. The resulting activation energy distribution is shown in Fig. 12.33 (other curves discussed later).

Pitt's results are not considered reliable in the initial stages of decomposition because he did not obtain data at solids residence times less than 10–15 s. Before the first sample was taken, devolatilization of the coal was more than 60% complete at 450°C and almost 90% complete at 650°C. Furthermore Pitt's determination of weight loss as the difference between the proximate volatile content of the coal and that of the char precluded the observation of yields larger than the standard volatile matter content.

Equations 12.11 and 12.14 are generalized

[171] V. Vand, "A Theory of the Irreversible Electrical Resistance Changes of Metallic Films Evaporated in Vacuum" *Proc. Phys. Soc. (London)*, **A55**, 222 (1943).

as follows to allow for nonisothermal processes:

$$V_i^* - V_i = V_i^* \exp\left(- \int_0^t k_i \, dt \right) \quad (12.15)$$

$$\frac{V^* - V}{V^*} = \int_0^\infty \exp\left(- \int_0^t k dt \right) f(E) \, dE \quad (12.16)$$

In addition, $f(E)$ can be approximated as a Gaussian distribution with mean activation energy E_0 and standard deviation σ. Thus

$$f(E) = \left[\sigma (2\pi)^{\frac{1}{2}} \right]^{-1} \exp \frac{-(E - E_0)^2}{2\sigma^2} \quad (12.17)$$

Assuming that $k_i = k_{0i} \exp(-E_i/RT)$ with k_{0i} being a constant having the same value (k_0) for all i, Equations 12.16 and 12.17 give

$$\frac{V^* - V}{V^*} = \frac{1}{\sigma(2\pi)^{1/2}} \int_0^\infty \exp \left[-k_0 \int_0^t \exp\left(\frac{-E}{RT}\right) dt - \frac{(E - E_0)^2}{2\sigma^2} \right] dE \quad (12.18)$$

Equation 12.18 permits correlation of coal decomposition data using four parameters (V^*, E_0, σ, k_0), which is only one more than is required in the simple single-step model represented by Equations 12.4 and 12.5 (V^*, E, k_0). This replacement of E with E_0 and σ, and the associated use of the somewhat more complicated equation, eliminate the temperature dependence of V^* and allow data on a given coal under different sets of experimental conditions to be correlated with one set of parameter values. In contrast, the simple model requires a different set of parameter values for each set of experimental conditions.[75,154] The curves fitted in Fig. 12.4 to data (points) on volatiles yields from Montana lignite at different heating rates and final temperatures illustrate the good correlations obtained with the improved model. The model can also accommodate a statistical distribution of k_0 or of some function of k_0. For example, Hanbaba et al.[173] and Jüntgen and Van Heek[163] used a Gaussian distribution for both the activation energy and the logarithm of the preexponential factor. In some cases, however, the associated introduction of yet another adjustable parameter (i.e., the single k_0 is replaced by a mean and standard deviation) is not necessary to fit the data (see Fig. 12.4). In many cases the additional adjustable parameter is not warranted because of large scatter in coal pyrolysis data.

The activation energy distribution found by Anthony et al.[75] for Montana lignite is compared in Fig. 12.33 with Pitt's results. Both the lignite and Pitt's Binley coal contained 46.2% (maf) volatile matter. Pitt used a fixed value of k_0 (1.67×10^{13} s^{-1}), whereas the value reported by Anthony et al. of 1.07×10^{10} s^{-1} was estimated directly from the lignite data. The higher value of k_0 tends to shift the distribution to higher activation energies. Thus refitting the lignite data with the statistical model with k_0 fixed at 1.67×10^{13} s^{-1} shifted E_0 from the previous value of 48.7 to 56.3 kcal/mole, while the width of the distribution remained almost unchanged (σ = 10.9 vs. 9.38 kcal/mole, previously). Use of the higher value of k_0 is justified, at least to some extent, on theoretical grounds. From transition state theory the frequency factor is calculated to have a value around 10^{13} s^{-1}. This is roughly equivalent to the preexponential factor (k_0) in the Arrhenius expression if the entropy of activation is close to zero.[174] Anthony et al.[109] recently found E_0 = 54.8 kcal/mole and σ = 17.2 kcal/mole with k_0 fixed at 1.67×10^{13} s^{-1} for a strongly caking Pittsburgh Seam coal also of 46.2% (maf) volatiles content.

Because Equation 12.18 is not widely used and its physical meaning is difficult to visualize, a few additional comments are warranted. Discussion is most simply based on isothermal reactions, even though these are physically difficult to achieve in short-duration, high-temperature runs. When the temperature is constant, the inner integral in Equation 12.18 is replaceable by $k_0 t \exp(-E/$

[172] H. A. G. Chermin and D. W. Van Krevelen, "Chemical Structure and Properties of Coal, XVII," *Fuel*, **36**, 85 (1957).

[173] P. Hanbaba, H. Jüntgen, and W. Peters, "Nicht-isotherme Reaktionskinetik der Kohlenpyrolyse, Teil II. Erweiterung der Theorie der Gasabspaltung und experimentelle Bestätigung an Steinkohlen," *Brennstoff-Chemie*, **49**, 368 (1968).

[174] S. W. Benson, *Thermochemical Kinetics*, John Wiley and Sons, New York, 1968, p. 75.

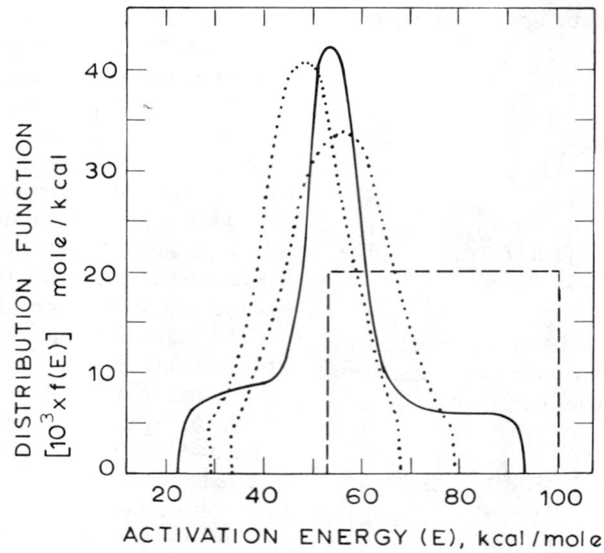

Figure 12.33 Comparison of activation energy distributions for coal pyrolysis.[8] Solid curves from Pitt,[119] dotted curves from Anthony et al.[75] *Left:* $k_0 = 1.07 \times 10^{10}$ s^{-1}, *right:* $k_0 = 1.67 \times 10^{13}$ s^{-1}. Dashes from Chermin and Van Krevelen.[172]

RT). Also, the lower limit on the left integral is conveniently altered from zero to $-\infty$ with negligible numerical effect. With the substitutions

$$\frac{E - E_0}{\sigma\sqrt{2}} \equiv x$$

$$k_0 t \exp\frac{-E_0}{RT} \equiv a$$

$$\sigma\frac{\sqrt{2}}{RT} \equiv b$$

Equation 12.18 becomes

$$\frac{V^* - V}{V^*} = \frac{1}{\sqrt{\pi}}\int_{-\infty}^{\infty} \exp[-a \exp(-bx) - x^2]\, dx = f(a, b) \qquad (12.19)$$

It may be shown that as $b \rightarrow 0$ (i.e., as E loses its spread around E_0), $f(a, b)$ becomes the term a itself; that is, the relation becomes that of a conventional first-order formulation:

$$\frac{V^* - V}{V^*} = \exp\left(-k_0 t \exp\frac{-E_0}{RT}\right) \qquad (12.20)$$

Values of $(V^* - V)/V^*$—the unaccomplished fraction of the total possible change in the reactants—obtained by numerical integration of Equation 12.18 are plotted versus $k_0 t \exp(-E_0/RT)$ for various values of $\sigma\sqrt{2}/RT$ in Fig. 12.34. The progress of all first-order reactions that can be characterized by a single activation energy used in the conventional first-order formulation is described by the single heavy curve. The ratio of the time to achieve 90% completion of the reaction to the time to achieve 60% completion is 2.5; and this ratio is not changed by temperature (only the time scale is changed). By contrast, for reactions of the type here discussed (parallel noninteracting reactions, Gaussian distribution of E around E_0, common preexponential term), the ratio of times to achieve 90% versus 60% completion of the reaction may vary widely with temperature. For example, if $\sigma = 8992$ cal/mole, and three temperatures of interest are 800, 1067, and 1600 K (making $b \equiv \sigma\sqrt{2}/RT = 8,\ 6$, and 4, respectively), the ratios of times to 90 and 60% completion are, from Fig. 12.34, 352, 93, and 22, respectively. In addition, of course, the absolute time involved is far less for the highest temperature. This effect can give the impression of an asymptote having been reached by a low-temperature decomposition when it is compared with one at high temperature (see Fig. 12.4). Thus the concept of a distribution of activation energies has introduced a highly useful single new degree of freedom, measured by $\sigma\sqrt{2}/RT$, into the conventional first-order process.

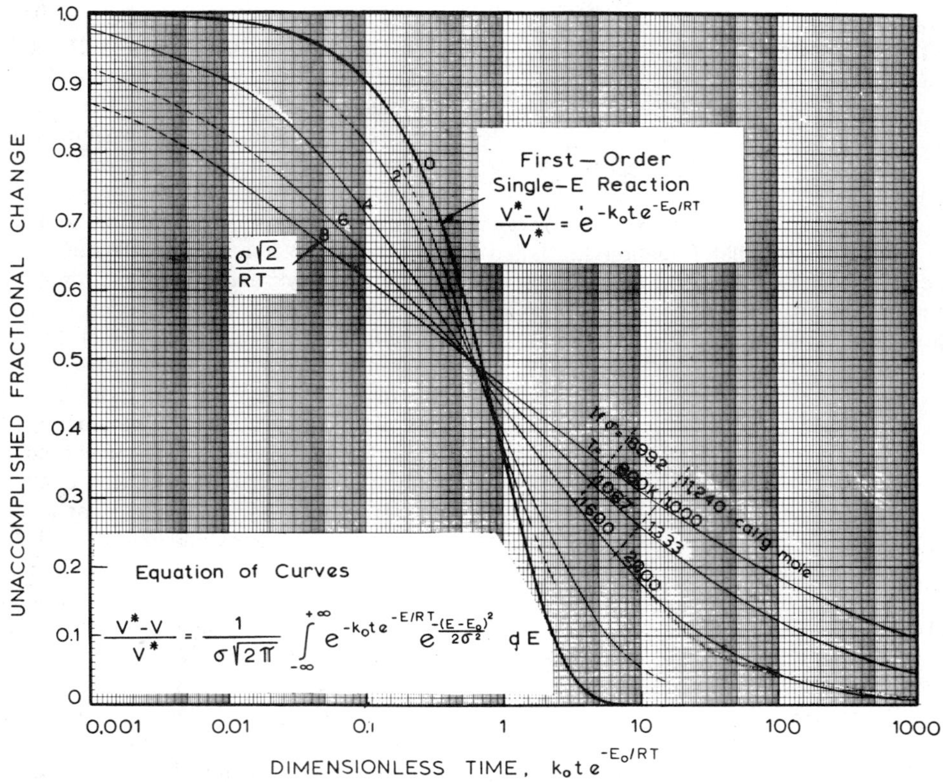

Figure 12.34 Properties of first-order reactions with Gaussian-distributed activation energy—compared with conventional first-order formulation.[75]

A valuable characteristic of the statistical distribution model is that it accounts for the low activation energies found with the single-step, first-order model. Data of Anthony et al.[75] give activation energies around 10 kcal/mole in the single-step correlation and around 50 kcal/mole in the multiple-step model as noted above. In a similar vein, Jüntgen and Van Heek[163] demonstrated theoretically that a set of overlapping, independent, parallel first-order reactions can be approximated by a single first-order expression having both lower activation energy and lower preexponential factor than any of the reactions in the set. To illustrate this phenomenon, weight loss versus time curves for four hypothetical independent first-order reactions having $k_0 = 2 \times 10^{10}\ s^{-1}$ and activation energies of 47, 50, 53, and 56 kcal/mole were added together and approximated successfully by a single first-order expression with $k_0 = 1.6 \times 10^6\ s^{-1}$ and $E = 30.3$ kcal/mole. This finding is clearly in accord with the experimental results of Anthony et al.

The foregoing analysis was limited to correlation of coal weight loss or volatiles evolution. Acquisition of other types of evidence during the devolatilization process has led to various alternative theories. Based on the melting or plastic behavior of some coals during heating, Van Krevelen[20] and Van Krevelen et al.[38] have pictured coal decomposition as a series of consecutive reactions [coal → metaplast (an intermediate product) → semicoke → coke] with primary volatiles formed in the resolidification of the metaplast and secondary gas evolved during the final coking stage. Recognizing the inadequacy of this simple sequence for explaining temperature-dependent yields, Chermin and Van Krevelen[172] (1957) theorized that secondary gas evolution from semicoke becomes progressively more difficult as the reaction proceeds. This behavior was represented by the conversion-dependent activation energy in the following rate equation:

$$\frac{dV}{dt} = k_0 (V^* - V) \exp \frac{-E_{max} - a(V^* - V)}{RT} \quad (12.21)$$

It can be shown that Equation 12.21 is a special case of Equation 12.14 where *f(E)* is a uniform distribution. Thus integration of Equation 12.12

$$\int_0^V dV^* = V^* f(E) \int_{E_{min}}^E dE \quad (12.22)$$

with

$$f(E) = (E_{max} - E_{min})^{-1}$$

and $$a = \frac{E_{max} - E_{min}}{V^*}$$

gives

$$E = E_{max} - a(V^* - V) \quad (12.23)$$

Chermin and Van Krevelen estimated values for the uniform distribution using data on the residual volatile content of chars. Their results are included in Fig. 12.33. The shape of the assumed uniform distribution is of course very different from that found by Pitt and Anthony et al. Furthermore, estimation of the distribution from data on residual volatile matter neglects yields in excess of proximate volatile matter, and predicted rates based on the results disagree badly with high-temperature data.

An alternative approach to modeling coal decomposition has been to assume the existence of competitive reactions by which the decomposing material may follow any of a number of reaction paths depending on the temperature-time pattern. Shapatina et al.[139] attribute to Chuckhanov a model in which coal is assumed to decompose through a combination of parallel and consecutive reactions each of different kinetic characteristics. Kobayashi[175] used a model comprised of two competing first-order reactions to correlate pyrolysis data published by Kimber and Gray[130] and Badzioch and Hawksley.[65] Recently the same model was used by Kobayashi et al.[132] to provide a better correlation of their data than is possible with the single-reaction model (Equation 12.4). Reidelbach and Summerfield[176] proposed a 10-step coal pyrolysis model and fitted a simplified five-step version of it to data of Badzioch and Hawksley[65] and Peters and Jüntgen.[73,173] The data were fitted by adjusting 15 rate and stoichiometric parameters, which of course guaranteed a good fit. In comparison, one of the main advantages of Equation 12.16 is the use of only four adjustable parameters.

Furthermore, Equation 12.16 is fundamentally a more realistic description of the coal pyrolysis process. Coal heated at low temperatures even for extended times will always devolatilize further when the temperature is raised. The foregoing competitive reaction schemes are inconsistent with this observation, since they predict that pyrolysis may be driven to completion even at low temperatures in times available in the laboratory. By contrast, Equation 12.16 predicts that the time required to achieve a certain degree of completion of the decomposition increases strongly with decreasing temperature and is so large at low temperatures as to be unattainable in practice. This effect can give the impression of an asymptote having been reached by a low-temperature decomposition when it is compared with one at high temperature as noted above, thereby explaining why the residual char from low-temperature experiments yields additional volatiles on being heated to a high temperature. Such additional volatiles are apparently formed by reactions having activation energies so large that they are immeasurably slow at the lower temperature. Application of the concept of a distribution of activation energies is thus seen to allow realistic description of multiple reactions without the need for a large number of parameters.

It would be premature to reject the competitive reaction schemes altogether, since competitive reaction behavior does indeed occur, at least in the form of secondary

[175] H. Kobayashi, "Rapid Decomposition Mechanism of Pulverized Coal Particle," M.S. thesis, Dept. of Aeronautics and Astronautics, Massachusetts Institute of Technology, Cambridge, Mass., 1972.

[176] H. Reidelbach and M. Summerfield, "Kinetic Model for Coal Pyrolysis Optimization," *Prepr., Am. Chem. Soc., Div. Fuel Chem.* **20**(1), 161 (1975).

reactions. The competitive schemes might be regarded as approximate global descriptions of the combined behavior of primary decomposition and secondary reactions. However each of these schemes will undoubtedly have to be modified by incorporating a distribution of activation energies in some manner. Furthermore, secondary reactions of volatiles are coupled with transport phenomena that cannot be properly addressed with the competitive schemes alone, or for that matter with Equation 12.18 alone. For a designer seeking a correlation for total volatiles yields, Equation 12.18 combined with a description of secondary reactions (see below) is presently the best recommendation if parameters for each volatile constituent are unavailable for the coal of interest.

Gavalas and co-workers[112,177,178] proposed a coal pyrolysis model based on considerations of chemical structure, chemical reactions, and transport phenomena. The model was said to be applicable within the range of primary pyrolysis (< 650°C). The chemical structure of coal was modeled as aromatic clusters having aliphatic side chains, bridges, and phenolic functional groups. Fifty-six thermal free radical reactions leading to the formation of tar, gases, and char were postulated, and their stoichiometry and rates were described by defining concentrations of the participating functional groups. To obtain activation energies and preexponential factors for the 56 reactions, literature values for the gas phase version of the reactions were modified by adjustable parameters to account for deviations, caused by cage and gel effects, of the condensed phase kinetics from predictions based on the gas phase behavior.

In the postulated transport model, the coal particle was partitioned by macropores and transitional pores into subunits in which tarry and gaseous products are transported by activated diffusion. A counting procedure was devised to predict the amount of aromatic fragments that escape from the subunits.

Appropriate initial conditions, structural data, and additional assumptions were employed to predict weight loss as a function of time and temperature. The results are qualitatively reasonable, but only limited quantitative comparisons with experimental data have been carried out. The approach is clearly ambitious given the lack of understanding of the chemical reactions and transport processes involved, but the work is an interesting attempt to place pyrolysis modeling on more fundamental grounds.

Models Based on Individual Volatile Constituents. An encouraging approach to modeling coal pyrolysis is to apply the first-order model to the evolution of single compounds or classes of compounds. As can be inferred from the data in Figs. 12.10–12.14 and 12.16–12.20, many products cannot be adequately described by one first-order process. However when the evolution of a species is dominated by only a few stages or peaks of reasonably simple shape on a plot of cumulative yield or rate of evolution against temperature, the kinetics can be described quite well by one, two, or three parallel reactions, depending on the observed behavior. For more complex patterns, a large number of reactions having distributions of parameters may be used for each stage of the evolution of a compound, and the number of stages can be chosen in accordance with the complexity of the behavior. The mechanistic implication is that a given compound in the simpler category may arise from a few different reactants or from a few different reaction pathways, whereas those in the more complex category may arise from a much larger number of reactants or pathways. Both approaches have been used by German workers,[97,156,163,173,179,180] and the first approach has been used in the United States.[91,99,100,111,146] Also, the first approach has been used to model the evolution of lumped hydrocarbon fractions, such as tar, in the absence of

[177] P. H.-K. Cheong, "A Modeling Study of Coal Pyrolysis," Ph.D. thesis, Dept. of Chemical Engineering, California Institute of Technology, Pasadena, 1977.

[178] M. Oka, H.-C. Chang, and G. R. Gavalas, "Computer-Assisted Molecular Structure Construction for Coal-Derived Compounds," *Fuel,* **56,** 3 (1976).

[179] P. Hanbaba, "Reaktionskinetische Untersuchungen zur Kohlenwasserstoffenbindung aus Steinkohlen bei niedrigen Aufheizgeschwindigkeiten," Thesis, Aachen, West Germany, 1967.

[180] H. Júntgen and K. H. Van Heek, "Fortschritte der Forschung auf dem Gebiet der Steinkohlenpyrolyse," *Brennstoff-Chemie,* **50,** 172 (1969). Translated by Belov and Assoc., Denver, Col., APTIC-TR-0779.

complete composition data.[111,146] These various studies are summarized below.

The use of multiple independent parallel first-order reactions to describe the evolution of a volatile product, that is, a compound or a class of compounds lumped together for convenience, proceeds with Equation 12.10 written for the contribution of reaction i or pathway i (hereafter denoted only reaction i) to the product. Assuming the rate constant to be $k_i = k_{oi} \exp(-E_i/RT)$, the contribution of reaction i to the rate of evolution of the product and to the amount of the product evolved up to time t is, for isothermal conditions,

$$\frac{dV_i}{dt} = V_i^* k_{oi} \exp\left(\frac{-E_i}{RT} - k_{oi}\, t \exp\frac{-E_i}{RT}\right) \qquad (21.24)$$

$$V_i = V_i^*\left[1 - \exp\left(-k_{oi}\, t \exp\frac{-E_i}{RT}\right)\right] \qquad (21.25)$$

Here V_i is amount of the product evolved from reaction i at time t, and V_i^* is the value of V_i at $t = \infty$. The total yield of the product, that is, the yield observed experimentally, is the sum of the contributions of all i. This summation need not be shown here.

If the coal is heated at a constant rate $dT/dt = m$ to a temperature T and there maintained for a holding time τ, the evolution rate and amount evolved become (Equation 12.9):

$$\frac{dV_i}{dt} = V_i^* k_{oi} \exp\left[\frac{-E_i}{RT} - k_{oi}\left(\frac{RT^2}{mE_i} + \tau\right) \exp\frac{-E_i}{RT}\right] \qquad (12.26)$$

$$V_i = V_i^*\left\{1 - \exp\left[-k_{oi}\left(\frac{RT^2}{mE_i} + \tau\right) \exp\frac{-E_i}{RT}\right]\right\} \qquad (12.27)$$

where $m > 0$. These equations are of course valid for all τ including the special case $\tau = 0$, where the coal is instantaneously quenched on reaching the peak temperature. To account for reaction occurring during the cooling at a linear rate $dT/dt = -m_c$, replace m with $mm_c/(m + m_c)$.

It is of interest to note that the predicted behavior occurring during heating at a constant rate is equivalent to that of an isothermal process of duration RT^2/mE at the peak temperature attained. Since $RT/E \ll 1$, the equivalent isothermal time at the peak temperature is much less than the time to heat up, which is of the order T/m. Furthermore, the extent to which most of the reaction occurs at or near the peak temperature increases with decreasing RT^2/mE.

In the calculation of evolution rate or yield of a product, application of Equations 12.24–12.27, or their equivalents for temperature-time histories not adequately approximated here, requires one equation and one set of parameters (k_{0i}, E_i, and V_i^*) for each reaction contributing to the product. Sets of parameters available in the literature are reviewed below.

The German workers[86,156] employed the nonisothermal rate equation to correlate data on the rates of evolution of water, carbon dioxide, carbon monoxide, and nitrogen obtained during the heating of four coals at a constant heating rate of 1°C/min. The data for one of the coals are presented in Fig. 12.10*a*, *b*. The evolution of each gas attributable to two peaks was described by two reactions, the fitted parameters for which are given in Table 12.17.

Gas evolution data obtained at a heating rate of 3.3°C/min were correlated in a manner similar to the foregoing by Campbell and Stephens[91] using one, two, or three peaks as indicated in Table 12.18. Excellent agreement was achieved between curves calculated with the parameters in Table 12.18 and the smooth curves drawn through the original data points in Fig. 12.11, but use of the parameters to predict yields under conditions of rapid heating fails badly. The same technique was used by Vestal et al.[99] and Yergey et al.[100] to correlate data on the evolution of hydrogen sulfide from coal during pyrolysis under conditions of constant low heating rate and atmospheres of inert gas and of hydrogen. The evolution was successfully modeled as five reactions, which appear to include both organic and inorganic processes.

In the German studies each of the major gaseous hydrocarbon species (C_1—C_5) was modeled as one, three, or four sets of reaction complexes.[163,173,179,180] Each reaction complex was described by a large number of

Table 12.17 Kinetic Parameters for the Evolution of Some Nonhydrocarbon Gases from Coals[86,156]

	Peak 1		Peak 2		
Coal[a]	E (kcal/mole)	k_0 (s^{-1})	E (kcal/mole)	k_0 (s^{-1})	V^*, Both Peaks (wt % of coal)
			H_2O		
1	33.5	3.3×10^6	34.0	5.8×10^5	0.72
2	32.0	1.3×10^6	33.5	1.7×10^5	1.10
3	31.4	1.1×10^6	33.1	1.5×10^5	1.31
4	31.0	1.0×10^6	33.0	1.4×10^5	2.43
			CO_2		
1	26.5	1.2×10^4	45.5	1.1×10^7	0.63
2	25.8	1.0×10^4	41.9	2.3×10^6	0.71
3	23.0	1.3×10^3	39.5	6.5×10^5	0.25
4	21.8	7.5×10^2	37.0	1.7×10^5	0.81
			CO		
1	27.0	3.3×10^3	41.0	2.0×10^6	1.59
2	26.5	1.5×10^3	40.2	1.7×10^6	1.70
3	25.5	1.3×10^3	38.5	3.8×10^5	1.94
4	22.0	1.7×10^2	33.0	3.3×10^4	3.15
			N_2		
1	50.4	4.2×10^7	72.5	1.3×10^7	0.35
2	46.0	7.5×10^6	67.0	8.3×10^6	0.42
3	42.6	1.5×10^6	64.5	2.5×10^6	0.66
4	39.0	3.3×10^5	59.5	1.0×10^6	0.95

[a] Analyses of coals given in Table 12.14; numbers as follows: Coal 1, Heinrich; Coal 2, Dickebank, Coal 3, Gustav; Coal 4, Fürst Leopold.

first-order parallel reactions having a Gaussian distribution of activation energies and a Gaussian distribution of the logarithms of the preexponential factors. Each distribution was characterized by a mean (E_0 and k_{00}) and a standard deviation (σ and $\Delta \log k_0$). The kinetic parameters derived from the data are listed in Table 12.19. Curves calculated with these parameters give excellent fit to the data, typical of which are those for methane and ethane in Fig. 12.10 *c*.

Suuberg et al.[111] used the first-order model and the measured time-temperature history of each experimental run to construct best fit curves for the lignite pyrolysis data in Figs. 12.12–12.14. Since the products from the lignite studied were dominated by a few species or classes of species, one objective of the work was to determine whether pyrolysis in such a case can be effectively modeled as only a few reactions representing the production of these key products. With the evolution of each product described by one, two, or three reactions, depending on the number of steps or evolution stages in the yield versus temperature data, the parameters listed in Table 12.20 were obtained. The curves shown in Figs. 12.12–12.14 were calculated using these parameters for a standardized case, representative of the experiments, in which coal is heated at 1000°C/s to the peak temperature, then cooled at 200°C/s back to room temperature. The curves fit the data well for most species, and modeling with only one or two reactions appears to be sufficient in all cases except the carbon oxides, which require a third reaction for data above about 1000°C (not shown here). Figure 12.12 shows how the various individual reactions cooperate to give a smooth total weight loss curve. This figure emphasizes that a large fraction of the weight loss from this lignite is due to the oxygenated species, as discussed above.

To compare the kinetic model based on individual products with the distributed ac-

Table 12.18 Kinetic Parameters for the Evolution of Hydrogen, Carbon Oxides, and Some Hydrocarbons from a Subbituminous Coal[91]

Roland Seam (Wyodak Mine) coal, analysis in Table 12.14

Gas	Peak	E (kcal/mole)	k_0, (s^{-1})	V^* (wt % of coal, mf)
H_2		22.3	20	1.02
CH_4	1	31.0	1.7×10^5	1.42
	2	31.0	2.8×10^4	1.56
	3	35.3	3.0×10^4	1.42
				4.40
C_2H_4		33.4	2.3×10^6	0.22
C_2H_6		33.4	1.7×10^6	0.81
C_3H_8		35.0	7.3×10^6	0.43
CO	1	18.0	55	1.58
	2	30.2	2.5×10^3	3.68
				5.26
CO_2	1	19.5	550	5.49
	2	23.0	230	4.75
				10.24
				22.38

tivation energy model based on total volatiles or weight loss, Suuberg et al.[111] derived (Fig. 12.35) the frequency distribution of the activation energies in Table 12.20. Data points in Fig. 12.35 represent the cumulative ultimate yields (V_i^*'s from Table 12.20) of all volatile components having activation energies less than or equal to those of the indicated points. Each point is labeled with the component whose ultimate yield is added to those of all other components shown on points to the left to give the indicated cumulative yield (e.g., the vertical distance between two adjacent points represents the ultimate yield of the

Table 12.19 Kinetic Parameters for the Evolution of Hydrocarbon Gases[163,173,179,180]

Gas	Coal[a]	E_0 (kcal/mole)	σ (kcal/mole)	$k_{00}/1.67$ (s^{-1})	$\Delta \log k_0$	V^* (wt % of coal)
CH_4	1	53.5	6.0	10^{12}	0.6	2.11
	1	58.0	6.0	10^{12}	0.6	0.84
	1	62.0	6.0	10^{12}	0.6	0.81
	1	69.0	6.0	10^{12}	0.6	0.49
C_2H_6	1	59.7	4.0	10^{15}	0.4	0.31
	1	56.0	5.0	10^{13}	0.5	0.39
	1	62.0	5.0	10^{13}	0.5	0.01
C_3H_6	1	60.0	4.0	10^{15}	0.4	0.26
C_3H_8	1	56.5	3.5	10^{14}	0.35	0.29
	2	58.3	3.0	10^{14}	0.3	0.26
	3	59.5	2.5	10^{14}	0.25	0.12
	4	62.0	3.0	10^{14}	0.3	0.01
cis-2-C_4H_8	1	60.0	4.0	10^{15}	0.4	0.03
n-C_4H_{10}	1	59.3	3.5	10^{15}	0.35	0.13

[a] Analyses of coals in Table 12.14; numbers as follows: Coal 1, Fürst Leopold; Coal 2, Gustav; Coal 3, Dickebank; Coal 4, Heinrich.

Table 12.20 Kinetic Parameters for Pyrolysis of Montana Lignite (Composition in Table 12.14[111])

Product	Stage	E (kcal/mole)	k_0 (s^{-1})	V^* (wt % of lignite, as received)
CO_2	1	36.2	2.1×10^{11}	5.70
	2	64.3	5.1×10^{13}	2.70
	3	42.0	5.5×10^{6}	1.09
				9.49
CO	1	44.4	1.8×10^{12}	1.77
	2	59.5	2.6×10^{12}	5.35
	3	58.4	5.9×10^{9}	2.26
				9.38
CH_4	1	51.6	1.6×10^{14}	0.34
	2	69.4	4.7×10^{14}	0.92
				1.26
C_2H_4	1	74.8	1.8×10^{20}	0.15
	2	60.4	7.1×10^{12}	0.41
				0.56
HC[a]		70.1	1.7×10^{16}	0.95
Tar	1	37.4	7.6×10^{11}	2.45
	2	75.3	2.0×10^{17}	2.93
				5.38
H_2O[b]		51.4	7.9×10^{13}	9.7
H_2		88.8	1.6×10^{18}	0.50
				37.2

[a] Hydrocarbons other than CH_4, C_2H_4, and tar.
[b] Does not include water from coal moisture.

component specified on the higher point). The slope of the smooth curve drawn through the data gives the frequency distribution curve $dV^*_{cum}/V^*_{tot}\ dE$, which is here normalized so that the area under a segment of the curve between two values of activation energy is the percentage of the total ultimate volatiles yield that is associated with activation energies in the specified interval.

Figure 12.35 also presents the two distributions (broken curves) from Fig. 12.33 obtained by Anthony et al.[75] for the same lignite used by Suuberg et al., but from the above-described analysis of weight loss data, which assumed a statistically large number of reactions having a Gaussian distribution of activation energies and all having the same preexponential factor. The left-hand curve was obtained when the preexponential factor ($k_0 = 1.07 \times 10^{10}\ s^{-1}$) was evaluated as one of the adjustable parameters, and the right-hand curve was obtained using a preferred fixed value of $k_0 (1.67 \times 10^{13}\ s^{-1})$. The mean activation energies (E_0) and standard deviations (σ) of the previous distributions are, in kcal/mole, (left) $E_0 = 48.7$ and $\sigma = 9.38$ and (right) $E_0 = 56.3$ and $\sigma = 10.9$. Statistical analysis shows that the product composition data are not significantly different from a Gaussian distribution with E_0 and σ being 53.3 and 11.5 kcal/mole, respectively. The similarity of the distribution derived from product compositions to the results based on weight loss is especially encouraging because k_0, which does influence E_0, was allowed to assume a different value for each reaction in the individual product model, whereas a single k_0 was used for all reactions in the model based on weight loss.

Although this rather simple, empirical model fits the lignite data rather well and most of the activation energy and preexponential values are broadly consistent with those shown in Table 12.16 for organic de-

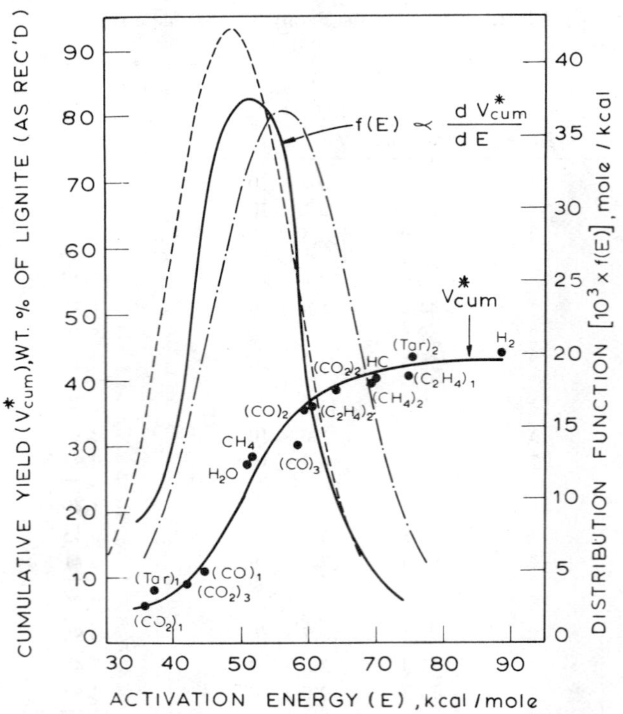

Figure 12.35 Distributions of activation energies of lignite pyrolysis reactions. ● = cumulative yields (from Table 12.20) of components indicated to left of and including a given point; solid curves = results based on individual products; broken curves = results (from Fig. 12.33), based on weight loss data and two different sets of kinetic parameters. Reprinted with permission from Ref. 111. Copyright by American Chemical Society.

compositions, caution must be exercised in its use. The fitted preexponential factors are closely correlated with the activation energies. Consequently an uncertainty of but a few kilocalories per mole in the activation energy of a particular step can lead to an uncertainty of factors of ten in the preexponential factor.

Application of a one-, two-, or three-step model for the bituminous coal data in Figs. 12.16–12.20 is not as straightforward as the lignite case above, since the bituminous coal data do not exhibit the well-defined stepwise behavior. Accordingly some preliminary modeling of the bituminous coal data was based on the use of a Gaussian distribution of activation energies for individual products, but a skewed distribution seemed to be more appropriate.[146] Therefore the method of Hanbaba et al.[173] reflected in Table 12.19 might be effective for these data.

Interpreted Behavior. The derivation of kinetic parameters for different stages of the evolution of individual products of coal pyrolysis not only constitutes an effective modeling technique, but it also offers the potential of helping to clarify the decomposition reactions responsible for each product. However progress toward a truly mechanistic understanding is now hampered by the large disagreements, evident in Tables 12.17–12.20, between the parameter values of different workers for the same species. Detailed assessment of these disagreements is probably not justified, but some broad observations can be made.

The activation energies and preexponential factors derived by use of only a few reactions are substantially lower in Tables 12.17 and 12.18 than in Table 12.20. The values in the first two tables are too low to be regarded as having fundamental significance with regard to the primary decompositions, nor is any direct fundamental significance attributed to the values in Table 12.20, although their orders of magnitude may seem relatively more meaningful. A possible reason for this difference rests with the contribution of secondary reactions. This speculation is supported by the observations, described earlier, that pyrolysis product yields from Pittsburgh Seam bituminous coal measured with the same technique employed for the lignite in Table 12.20 exhibit, in contrast to the lignite, much stronger effects of secondary reactions (i.e., pressure and particle size effects) and more gradual temperature dependences characteristic of smaller apparent activation energies and preexponential factors. Whereas Table 12.20 is based on data from a thin layer of lignite particles, the data underlying Tables 12.17 and 12.18 were obtained with packed beds of particles of higher rank coals. Therefore the products measured in the latter cases presumably had undergone more extensive secondary reactions, which would be expected to spread the stages of evolution over wider temperature intervals, thereby giving lower apparent activation energies and preexponentials. Although the data in Table 12.19 also came from application of the packed-bed technique to the same coals as those represented in Table 12.17, the use of parameter distributions has improved the resolution in a manner analogous to the activation energy and preexponential factor of weight loss models being substantially increased when a single reaction is replaced by multiple reactions with distributed parameters.

The choice of a model first order in product yet to be evolved in a given step does not necessarily imply that the true mechanism is a simple, first-order bond scission. The observed activation energies are too low to be representative of a set of independent bond cleavages (e.g., the bond strength of an average C—C single bond is approximately 80 kcal/mole). Suuberg[146] discussed the more likely possibility that the factor largely responsible for the observed behavior is a set of concerted or radical initiated reactions, or both. Radical reactions might be initiated, for example, by the cleavage within the coal structure of weak linkages such as methylene bridges, ether linkages, or heteroatom linkages. It has been known for some time[2] that free radicals are formed in coal pyrolysis at temperatures as low as 300°C, that substantial increases in the concentration of radicals occur at about 400°C, and that the presence of an added radical scavenger retards the thermal decomposition of the coal and inhibits the secondary polymerization of the tar phase.

The existence of free radical mechanisms in coal pyrolysis is not surprising, since pyrolysis of simpler hydrocarbons often proceeds by a radical chain mechanism. As an example, the pyrolysis of isobutene (IB) is believed to proceed by the mechanism shown below.[181]

[181] H. P. Leftin and A. Cortes, "High Severity Pyrolysis of Olefins; 1. Isobutene," *Ind. Eng. Chem., Process Des. Dev.*, **11**, 613 (1972).

	Activation Energy (kcal/mole)	Reaction
$CH_2{=}C(CH_3)_2$ (IB) $\xrightarrow{k_1}$ $CH_2{=}C(CH_3)\dot{C}H_2$ (IB·) + H·	8.25	(R12.1)
IB + H· $\xrightarrow{k_2}$ IB· + H_2	~5	(R12.2)
IB· $\xrightarrow{k_3}$ $CH_2{=}C{=}CH_2$ + $CH_3\cdot$	44	(R12.3)
$CH_3\cdot$ + IB $\xrightarrow{k_4}$ $CH_3{-}CH_2{-}\dot{C}(CH_3)_2$	6	(R12.4)
$CH_3{-}CH_2{-}\dot{C}(CH_3)_2$ $\xrightarrow{k_5}$ $CH_3{-}CH{=}C(CH_3)_2$ + H·	49	(R12.5)
IB· + H· $\xrightarrow{k_6}$ IB	0	(R12.6)

From a quasi–steady state analysis, the rate of disappearance of isobutene is

$$\frac{-d(\text{IB})}{dt} = 2\left(\frac{k_1 k_2 k_3}{k_6}\right)^{1/2}(\text{IB}) \qquad (12.28)$$

and the overall activation energy for the process is $(E_1 + E_2 + E_3 - E_6)/2 = 66$ kcal/mole. This prediction is in good agreement with the experimental value of 63 kcal/mole.

Penninger and co-workers[182,183] have studied the cracking and hydrocracking at about 500°C of various substances that are somewhat more reminiscent of coal structure than is isobutene. In a study of the cracking of the naphthenic ring in tetralin, they considered α ring opening, occurring by a radical process, and β ring opening, by a C—C bond scission:

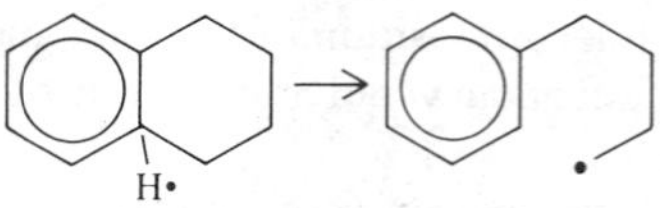

α Opening

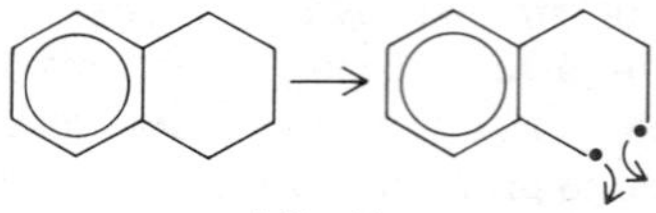

β Opening

The nature of the products suggested that the rate of the chain process ($E = 40$ kcal/mole) is much faster than that of the ordinary bond scission ($E = 69$ kcal/mole). Hydrogen atoms serve as the chain carriers in the reaction above, and the initiation step was hypothesized to be a scission of naphthenic C—H bonds. In similar work with tetrahydrophenanthrene, dihydrophenanthrene, and anthracene, Penninger and co-workers again found the rate of the radical pathways to be considerably faster than ring opening by bond scissions. These workers concluded that the overall sequence in the cracking of condensed polyaromatics is (*a*) hydrogen (radical) attack on outer aromatic rings, (*b*) successive naphthenic ring openings, and (*c*) side chain cracking (β scission).

[182] J. M. L. Penninger and H. W. Slotboom, "Kinetics and Mechanisms of Primary Cracking Reactions in High Pressure Hydrogenolysis of Tetralin and 'Indin'," *Rec. Trav. Chim. Pays-Bas,* **92,** 513 (1973).

[183] M. Van Boven, G. J. Roskam, and J. M. L. Penninger, *Int. J. Chem. Kinetics,* **7,** 351 (1975).

Suuberg[146] pointed out that the foregoing information on pure hydrocarbons supports the view that coal pyrolysis (and hydropyrolysis, discussed later), may be primarily a radical process, initiated by scission of weak bonds within the structure. Parameter values similar to those presented in Tables 12.19 and 12.20 were shown as follows to be consistent with this picture. Applying an analysis of the type presented above for isobutene pyrolysis, coal pyrolysis was tentatively assumed to involve the following reactions:

$$\text{radical-producing structures } (S_R) \xrightarrow{k_1} 2 \text{ radicals (R)} \qquad \text{(R 12.8)}$$

$$\text{radical (R)} + \text{structure } i\ (S_i) \xrightarrow{k_2} \text{intermediate } i\ (I_i) \qquad \text{(R 12.9)}$$

$$\text{intermediate } i\ (I_i) \xrightarrow{k_3} \text{product } i\ (P_i) + \text{radical (R)} \qquad \text{(R 12.10)}$$

$$2 \text{ radicals (R)} \xrightarrow{k_4} \text{stable structure (C)} \qquad \text{(R 12.11)}$$

The chemical natures of the radicals (R), structure i (S_i), structure (S_R), intermediate (I_i), and stable structure (C) were deliberately left vague to avoid reducing generality. The major feature of S_R is that it produces radicals that can interact directly, or via a chain reaction that is not rate limiting, with a particular structure i in the coal to yield a particular product i (e.g., CH_4). Applying the quasi–steady state assumption to species R and I_i, the rate of production of product i is found to be

$$\frac{dP_i}{dt} = k_2\left(\frac{2k_1}{k_4}\right)^{1/2}(S_R)^{1/2}(S_i) \qquad (12.29)$$

from which the overall activation energy is $(E_1/2 + E_2 - E_4/2)$.

The bond energies for cleavage of C—H bonds are of the order 100 kcal/mole, but the activation energy for formation of radicals from such a cleavage may be lower, depending on the nature of the compound (e.g., whether it can stabilize a radical by resonance). Data derived from pyrolysis experiments and compiled by Benson and O'Neal[184] indicate that the actual range may be from 80+ to 100+ kcal/mole. Similarly, cleavages of C—C bonds and C—O bonds are in the

[184] S. W. Benson and H. E. O'Neal, *Kinetic Data on Gas Phase Unimolecular Reactions,* U.S. National Bureau of Standards, NSRDS-NBS 21 (1970).

range 50–80 kcal/mole. Thus E_1 is roughly in the range 50–100 kcal/mole. Radical recombinations are extremely fast and may be considered to have zero activation energy ($E_4 = 0$). Reactions of type 2, that is, radical additions to stable species, typically have activation energies ranging from 10 to 20 kcal/mole.[174] Use of these values leads to overall activation energies in the range 35–70 kcal/mole. A similar estimation of the overall preexponential factor [i.e., $k_{0,2}\ (2k_{0,1}/k_{0,4})^{\frac{1}{2}}$] leads to values in the range 10^{10} to $10^{13}\ s^{-1}$. Although this analysis is clearly very rough, it is encouraging that the radical mechanism leads to a model first order in S_i, hence in product yet to be evolved, and that the predicted parameter values are similar to those derived from data using rather empirical models.

In comparison with the above, the work of Gavalas and co-workers[112,177,178] described earlier is a much more sophisticated consideration of radical reactions, involving detailed modeling with the aid of a computer. Because of the large variety of structures present in coal, use of a computing system with a statistical model of coal and a large library of possible reactions offers the potential of developing a truly detailed model of coal pyrolysis. Simulation results thus far have suggested that the product mix may be rather sensitive to the nature of the radical formation reactions. For example, the predicted total weight loss in pyrolysis at 500°C for 30 s drops by about 10% (from ~33 to ~23%) if the activation energy for the cleavage of ethylene bridges (ϕ—CH_2—CH_2—$\phi \rightarrow 2\phi$—$CH_2\cdot$) is raised by only 2 kcal/mole (from 48 to 50 kcal/mole). The strong sensitivity of predicted pyrolysis behavior to this parameter awaits further elucidation.

Various authors have discussed the order of appearance of the volatile species when coal is heated at a constant rate. In the previous review[2] the sequence after drying is given as chemical water, carbon dioxide, carbon monoxide, higher hydrocarbons (tars, etc.), ethane, methane, and finally hydrogen. As can be seen in Figs. 12.10–12.14, the order is not precise, and substantial overlapping occurs. Chukhanov[185] suggests that coal pyrolysis involves three stages: evolution of carbon oxides and water (450–550°C), formation of tar and hydrocarbon gases, and residue degasification.

The data of Suuberg et al.[111] in Figs. 12.12–12.14 reveal that there are apparently five principal phases of devolatilization in the case of Montana lignite. The first occurs at very low temperatures (~ 100°C) and is associated with moisture evolution. The second phase at 1000°C/s heating rate begins at about 450°C and is associated with a large initial evolution of carbon dioxide, probably from low activation energy decarboxylations. The loss of carboxyl groups as carbon dioxide at relatively low temperatures has been reported for lignites.[2] A small amount of hydrocarbon products is also evolved at this stage. The third phase involves evolution of water chemically formed in the range 500–700°C, where the only other significant product is carbon monoxide. It can be seen from Fig. 12.21 that carbon, nitrogen, and sulfur during this phase remain largely in the char, while the oxygen and hydrogen contents of the char sharply decrease. The fourth phase involves a final rapid evolution of carbon-containing species in the range 700–900°C: carbon oxides, tar, hydrogen, and hydrocarbon gases are all rapidly evolved, but little water is produced. The fifth phase is the high-temperature formation of carbon oxides.

To be compared with the foregoing, the data in Figs. 12.16–12.20 indicate that the pyrolysis of Pittsburgh Seam bituminous coal occurs in four somewhat less distinct stages. Again, surface moisture is driven off at low temperatures (< 100°C) followed by the liberation of pyrolytically formed water at peak temperatures below 400°C. Between 400 and 900°C the coal softens and the bulk of the hydrocarbons is evolved. Carbon monoxide and hydrogen are the principal products evolved above 900°C.

Suuberg[146] showed that his data on the two coals above agree with previous studies in drawing a correlation between carboxyl groups in raw coal and the amount of carbon dioxide evolved during pyrolysis at low temperatures, and between hydroxyl groups in raw coal and the amount of pyrolytically formed water. However the rather large difference in the temperature of water evolution for the two coals reflects either a substantial chemical difference between the hydroxyl groups, or a major difference in the pyrolytic processes. The lignite model structure of

[185] Z. F. Chukhanov, *Iz. Akad. Nauk SSSR, Otd. Tekh. Nauk,* No. 8, 7–22 (1954); *C.A.* **49,** 5808 (1955).

Wender[186] suggests that some of the hydroxyl groups in lignites may be present as alcoholic rather than phenolic structures, which are the only source of hydroxyls in higher rank coal. It is not clear whether such a difference is responsible for the observed behavior; however the narrow temperature interval for evolution of pyrolytically formed water from lignite may speak against a lignite structure that contains two very different types of hydroxyl groups.

Wolfs et al.[187] hypothesized that hydroxyl groups may play a key role in determining pyrolysis behavior. In the foregoing case of lignite pyrolysis, hydrocarbon formation is observed at temperatures between 500 and 600°C, and between 700 and 900°C, but not between 600 and 700°C (evident in Fig. 12.13). This plateau coincides with the principal part of the water formation step. During pyrolysis of bituminous coal, water formation occurs before the onset of the hydrocarbon formation phase, and no plateaus are observed. This behavior is consistent with the view that the hydroxyl groups consume hydrogen that could otherwise stabilize hydrocarbon radicals, and that these radicals repolymerize in the absence of hydrogen to form a solid cross-linked residue. Although the lignite initially has only a slightly lower hydrogen/carbon ratio than the bituminous coal (0.77 vs. 0.85), adjustment for the hydrogen removed during water formation increases the gap (0.55 vs. 0.73). Such a difference in H/C ratio may help to explain some of the large differences between the pyrolytic behaviors these coals, but other factors must also be important, since some bituminous coals with adjusted H/C ratios similar to that of the lignite show considerably more hydrocarbon formation than does the latter.

If the pyrolysis reactions that form water and hydrocarbons are in competition for hydrogen, perhaps the hydrocarbon yield could be increased by subjecting the coal to a favorable time-temperature history. For example, in the lignite results given in Table 12.20, tar-forming reactions have a higher activation energy than the water-forming reactions (75.3 vs. 51.4 kcal/mole), hence are favored by going to higher temperatures. According to Equation 12.27, very high rates of heating would be required to attain these higher temperatures without the water-forming reactions reaching completion first. At 1000°C/s the water-forming reactions are about 90% complete before the onset (defined at 1% completion) of the tar-forming reactions. This interpretation neglects the low-temperature, tar-forming reactions and assumes that the water- and tar-forming reactions are independent. At 6090°C/s the water-forming reactions are predicted to be 75% complete when the onset of tar formation occurs (at 700°C), and at 2.8×10^5 °C/s the water-forming reactions are only 35% complete at the onset of tar formation (at 800°C). Total volatiles yields significantly larger than those shown in Table 12.20 have in fact been observed when Montana lignite is heated at about 2×10^5 °C/s to about 1800°C (Fig. 12.3), but this result could simply be an effect of temperature on the extent of primary decomposition rather than a consequence of better hydrogen utilization.

12.4.3.3 *Secondary Reactions*

Mazumdar and Chatterjee[188] found the dimensions of the coke charge to be quite significant in determining volatiles yields, whereas heating rate was of secondary importance. Similar findings were reported by Gray et al.[88] and Anthony et al.[75] In the latter work sample dimensions as small as single particles 50 μm in diameter were studied. The effect of sample size is sometimes misconstrued as a heating rate effect because the time required to attain the final temperature usually increases with increasing sample size.

It is generally agreed that secondary reactions are responsible for these variations in yield, though Yellow[4] and Van Krevelen et al.[38] have suggested the possibility of fragments of a pyrolysis intermediate being physically carried away by rapidly escaping gases. There is also broad support for the picture discussed earlier that a major part of the secondary reactions is concerned with the

[186] I. Wender, "Catalytic Synthesis of Chemicals from Coal," *Prepr., Am. Chem. Soc., Div. Fuel Chem.* **20**(4), 16 (1975).

[187] P. M. J. Wolfs, D. W. Van Krevelen, and H. I. Waterman, "The Carbonization of Coal Models," *Fuel,* **39,** 25 (1960).

[188] B. K. Mazumdar and N. N. Chatterjee, "Mechanism of Coal Pyrolysis in Relation to Industrial Practice," *Fuel,* **52,** 11 (1973).

tar fraction of the volatiles. For example, the yields of char and light hydrocarbons increase at the expense of tar as pressure or particle size is increased, hence as secondary reactions are augmented. There is still disagreement over whether the primary volatiles are relatively small, free radical ring structures or large condensed ring structures (essentially identical to the tar ultimately evolved). Electron spin resonance (ESR) spectra[189] and measurements in the presence of a known free radical scavenger, nitric oxide,[190] support the existence of free radical processes, but the size of the primary species remains undetermined. In a hydrogen-deficient environment, reactive small species would polymerize to coke unless they could escape as tar at some intermediate stage of polymerization. The escaping tar can also undergo cracking reactions producing coke and gas. Mazumdar and Chatterjee[188] discounted cracking, since they observed no increase in gas yield as tar was lost in larger beds. However the data of Peters[71] and Peters and Bertling[72] (Table 12.5) do show an increase in gas yield under retort conditions, and gas formation clearly accompanies tar consumption during pyrolysis of thin layers of coal particles as discussed earlier. As Suuberg[146] pointed out, it is not clear whether the gas, which is dominated by methane, comes from the cracked tar molecule itself or whether the latter merely surrenders hydrogen, which promotes further reactions in the coal matrix. Griffiths and Mainhood[191] studied the cracking of freshly produced coal tar externally to the bed of coal and observed that the yields of light hydrocarbon liquids and gases increase as the extent of cracking increases. They also found activated carbon to be a more effective catalyst than silica or alumina for the cracking reaction, which raises the possibility that the coal char itself catalyzes cracking.

Regardless of the exact nature of the secondary reactions, their extent is influenced by the residence time and concentration of reactive species in contact with hot surfaces. Large packed beds and the accompanying slow rate of heating virtually ensure a minimum yield of volatiles by providing sufficient surface area and residence time for the secondary reactions. Similarly, as discussed above, pyrolysis under conditions of higher pressures and larger particle sizes increases the extent of secondary reactions by retarding the escape of reactive species from the particle, thereby increasing the internal concentration, and hence reaction rate, as well as residence time.

To develop an approximate model for the effect of secondary reactions on yields of volatiles, Anthony et al.[75] derived a selectivity expression describing assumed competition between diffusional escape and secondary reaction of reactive volatile species. The nomenclature in Table 12.21 was used to define different kinds of volatile formation or loss: V's without subscript refer to reactive volatiles V_r plus nonreactive volatiles V_{nr}. Priming the V refers to the amount of volatiles formed up to time t, absence of prime to the lesser amount that leaves the particle (i.e., the volatile loss). Double-asterisk superscript refers to total volatiles *formed* up to $t = \infty$, single-asterisk superscript to total volatiles *lost* up to $t = \infty$. For the hypothetical case of no solid deposition by secondary

[189] R. L. Bond, W. R. Ladner, and R. Wheatley, "ESR of Rapidly Heated Coals," *Fuel,* **47,** 213 (1968).
[190] W. Den Hertog and N. Berkowitz, "Mechanisms of Coal Pyrolysis, II," *Fuel,* **39,** 125 (1960).
[191] D. M. L. Griffiths and J. S. R. Mainhood, "The Cracking of Tar Vapor and Aromatic Compounds on Activated Carbon," *Fuel,* **46,** 167 (1967).

Table 12.21 Classes of Volatiles (g/g of original coal)

Cumulative Volatiles	Reactive	Nonreactive	Total
Formed up to time t	V'_r	V'_{nr}	V'
Lost from particle up to time t	V_r	V_{nr}	V
Formed up to $t = \infty$ (potential ultimate yield)	V_r^{**}	V_{nr}^{**}	V^{**}
Lost from particle up to $t = \infty$ (ultimate yield)	V_r^*	V_{nr}^*	V^*

reactions during coal pyrolysis in inert gas, $V = V'$ and $V^* = V^{**}$. In all cases $V_{nr} = V'_{nr}$ and $V^*_{nr} = V^{**}_{nr}$ by definition.

In an inert atmosphere, a material balance on the reactive volatile species present in a coal particle becomes:

$$\frac{dV'_r}{dt} - K_c(C - C_\infty) - k_iC = \frac{dC}{dt} \qquad (12.30)$$

where dV'_r/dt, the formation rate of reactive volatiles expressed as fractional mass per unit time, represents an irreversible decomposition reaction dependent on temperature and unaffected by the atmosphere contacting the decomposing material; C and C_∞ are vapor phase concentrations of reactive volatiles in the void volume of the coal particle and in the ambient gas, respectively, expressed as mass per unit mass of original coal; K_C is the overall mass transfer coefficient (s^{-1}), and k_1 is the overall rate constant (s^{-1}) of the deposition reaction, here assumed to be first order.

In Equation 12.30 a coal particle is assumed to possess a fixed internal void space of uniform composition and temperature. Reactive species released by thermal decomposition of the coal mass enter this designated volume and leave either by mass transfer into the ambient gas or by solid deposition. Since the mass of reactive species residing in the vapor phase within the particle is relatively negligible, a pseudo-steady state prevails ($dC/dt \approx 0$). If C_∞ is approximately zero, Equation 12.30 becomes

$$\frac{dV'_r}{dt} - K_cC - k_1C = 0 \qquad (12.31)$$

Since the loss of reactive volatiles from the particle is V_r, than $dV_r/dt = K_cC$ and, from Equation 12.31,

$$dV_r = \frac{dV'_r}{(1 + k_1/K_c)} \qquad (12.32)$$

Assuming that k_1/K_c is approximately independent of the temperature-time history, integration of Equation 12.32 from $t = 0$ to $t = \infty$ gives

$$V^*_r = \frac{V^{**}_r}{(1 + k_1/K_c)} \qquad (12.33)$$

where V^{**}_r, the integral of dV_r between the stated limits, is the potential ultimate yield of reactive volatiles corresponding to no mass deposition and infinitely long reaction time. Similarly V^*, the integral of dV_r from $t = 0$ to $t = \infty$, is the actual ultimate yield of reactive volatiles corresponding to infinitely long reaction time.

The mass transfer coefficient is assumed to be proportional to the diffusion coefficient of the volatiles. Thus $K_c = k_c/P$, where P is total pressure. Since V^* and V^*_{nr} are the ultimate yields of total volatiles and nonreactive volatiles (Table 12.21) and $V^* = V^*_r + V^*_{nr}$, use of Equation 12.33 leads to

$$V^* = V^*_{nr} + V^{**}_r (1 + k_1P/k_c)^{-1} \qquad (12.34)$$

This relation predicts the approach of V^* asymptotically to $(V^*_{nr} + V^{**}_r)$ and V^*_{nr} at low and high pressures, respectively. The variation with pressure of the observed yields of total volatiles indicated as weight loss in Fig. 12.7 is clearly consistent with this asymptotic behavior. Since the reaction times did not exceed about 20 s, the data in Fig. 12.7 are not truly ultimate ($t = \infty$) yields, although they appear to be close to the latter, based on the very slow rates of devolatilization for times exceeding a few seconds. If the data are taken as approximate ultimate yields, the values $V^*_{nr} = 37.2\%$ and $V^{**}_r = 17.0\%$ are obtained from the asymptotes. Using these values and an approximate value for k_1/k_c (= 0.56 atm^{-1}) in Equation 12.34, the predicted effect of pressure on volatile yield, represented by the curve in Fig. 12.7, is consistent with the data. The correction shown for deposition on the screen is significant at lower pressures (maximum 3% of sample weight) and leads to $V^{**}_r = 20.0\%$. According to the data in Table 12.12, this value is equivalent to about 60% of the vacuum tar yield. Since the reactive volatiles are known to be in the tar fraction as discussed above, an independent measure of V^{**}_r is the difference between the tar yield under vacuum and that under 69 atm of helium. The former is 31.9% from Table 12.12, the latter about 12% as indicated by the asymptotic value in Fig. 12.30, the difference being about 20% in good agreement with the value above from Fig. 12.7. This consistency lends support for the important role attributed to secondary reactions.

Although Equation 12.34 is consistent with the observed effects of pressure, the underlying description of the complex mass transfer and secondary reactions is highly simplified and the mechanistic details are concealed in the convenient ratio k_1/k_c. The escape of volatiles from the particle is not simple diffusion but a complex process involving a time-dependent pore structure and hydrodynamic flow during the period of rapid devolatilization. To account for this behavior, Lewellen[59] developed a considerably improved model of volatiles transport in caking coals, based on the concept that the transport occurs primarily through growth and escape of volatile-filled bubbles. The model, described earlier, predicts the variations described above with pressure and particle size and the effects of secondary reactions on volatiles yield, and offers a valuable approach for detailed study of these reactions.

A consequence of the secondary reactions occurring within a pyrolyzing coal particle is presumably a shift toward, though not necessarily close to, an equilibrium mixture of escaping volatiles. It is therefore interesting to compare observed product distributions with those that would be obtained if all the volatiles were in mutual equilibrium in the presence of char at the final temperature of the particles. To this end, Suuberg et al.[111] analyzed their data on lignite pyrolysis by assuming the char to be graphite and the mole fraction of the ambient helium adjacent to and within the particles to be negligible. Equilibrium relations for the reactions

$$CO + H_20 = CO_2 + H_2 \qquad \text{(R 12.12)}$$
$$C + H_20 = CO + H_2 \qquad \text{(R 12.13)}$$
$$C + 2H_2 = CH_4 \qquad \text{(R 12.14)}$$
$$CH_4 + H_20 = CO + 3H_2 \qquad \text{(R 12.15)}$$

in the presence of excess solid carbon and with an assumed total pressure of 1000 atm are shown as curves in Fig. 12.36, where the points are the data, obtained with a total pressure of 1 atm outside the particles. Two cases were considered, one including water from the lignite moisture in the volatiles (points) and the other including only the chemically formed water (not shown). The latter case covers the possibility that moisture is lost early enough to be absent from the reacting mixture. The complete or partial loss of other products could also be considered, but this was not done because moisture loss, which appears to offer the largest potential for an effect, was found to be of little significance in this rough analysis.

Figure 12.36 shows that the data, with some exceptions, tend toward less disagreement with the equilibrium values as temperature increases. When the total pressure of volatiles is assumed to be 1 atm, all the reactions are far from equilibrium except for Reaction 12.12 at the higher temperatures, where a rather close approach to equilibrium cannot be excluded within the large uncertainty of this calculation. Reaction 12.12 is the only one of the four that is independent of pressure. Improved agreement of the equilibrium values with the data is observed for the other reactions if the assumed total pressure of the volatiles is increased substantially, the values for Reactions 12.13–12.15 depending on pressure to the powers: -1, $+1$, and -2, respectively.

Volatiles pressures much larger than 1 atm within the pores of the particles are not inconsistent with calculated pressure drops associated with volatiles expulsion under the experimental conditions of rapid pyrolysis. The agreement in Fig. 12.36 between the equilibrium values and the data for an assumed pressure of 1000 atm is intriguing. It is not known to what extent such a high pressure properly reflects conditions developed within rapidly pyrolyzing lignite, but the authors noted that internal pressures of this order of magnitude *a* would be calculated if it is assumed that the formation of volatiles within the pores largely precedes the escape therefrom, and *b* would appear to be required to fracture the particles based on the strength of the material. Lignites rapidly heated are known to exhibit fracturing or "sparking," a phenomenon that could account for the particulate material found in the tar as noted above.

Although the foregoing equilibrium analysis is far from definitive, it is probably safe to conclude that the final mixture of volatiles is closer to equilibrium than are the primary volatiles; therefore the mass transport and secondary reactions within lignite particles do play some role in determining product distributions. Thus at temperatures above about 800°C, the enhanced mass transport condition provided by lowering the pressure from 69 atm to vacuum was found to lead to

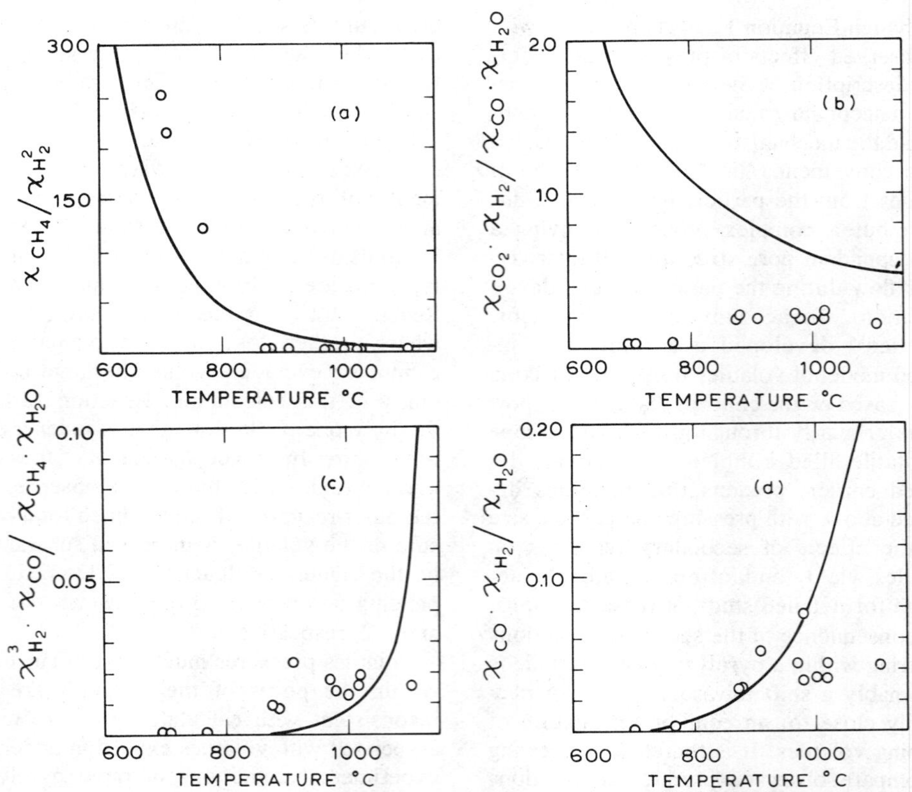

Figure 12.36 Comparison of gas composition data for Montana lignite heated to different peak temperatures with calculated equilibrium values under simplified conditions. χ_i = mole fraction of *i*, ○ = data; heating rate = 1000°C/s; mean particle diameter = 74 μm; curves = equilibrium values calculated for an assumed total pressure of volatiles of 1000 atm. Reprinted with permission from Ref. 111. Copyright by American Chemical Society.

somewhat higher yields of tars and correspondingly lower yields of light hydrocarbon and carbon oxide gases (Table 12.12). This behavior indicates that more volatiles cracking within particles occurs at 69 atm than under vacuum, but the difference affects not more than about 6% of the total lignite material. The small size of this effect would support the picture that the pressure inside lignite particles during rapid pyrolysis could indeed be high, therefore not strongly dependent on external pressure in the range 0–69 atm. If this were the case, the absolute amount of volatiles cracking within the pores could be much larger than the change observed in going from 69 atm to vacuum conditions. On the other hand, the yield of some components such as tar and methane is significantly affected by the changes in external pressure, which seems to suggest that the internal pressure cannot be extremely large compared with the imposed change of 69 atm.

In the case of bituminous coal, even modest pressure changes clearly affect the total yield of volatiles as well as the yields of individual species (Table 12.12). Although this behavior can be rationalized as described above in terms of the effect of pressure on diffusional escape of volatiles, the flow would appear to be hydrodynamic, based on the observed structure of residual particles as well as on calculated pressures as high as 100 atm in volatile bubbles within the molten particles during rapid pyrolysis.[59] Suuberg[146] suggests that this dichotomy may imply that both diffusive and hydrodynamic flow are important. According to his calculations, the vapor pressure of a large fraction of the coal tar may be no more than a few millimeters of mercury at the temperatures where cracking becomes important. Although the internal pressure of the molten coal particle may be high because of the rapid formation of gases, which certainly escape by pressure-driven flow, the escape of tar may involve evaporation and diffusion from the surfaces of the melt. The enhancement of tar transport by local shear may also occur.

The suggested role of mass transport and secondary reactions points up the empirical nature of pyrolysis models that do not explicitly account for these effects. The kinetic parameters obtained by application of the multiple parallel reaction model to a species originating in some combination of primary and secondary reactions are clearly overall properties. The derivation from available data of fundamental properties of the primary pyrolysis reactions must await a better understanding of mass transfer and secondary reactions in the particles. Nevertheless, the demonstrated success in correlating composition and yield data and the observation of significant correlations between coal structure and yields of certain pyrolysis products mean that the multiple parallel reaction picture together with the identification from the data of key "reactions" (e.g., stages in the evolution of a given product) is an effective representation of the decomposition kinetics as modified by the effects of mass transport and secondary reactions.

12.5 PYROLYSIS IN HYDROGEN ATMOSPHERE

12.5.1 Introduction

Terminology used below to describe coal pyrolysis in the presence of hydrogen gas is introduced by the following summary of the characteristic behavior. Hydrogen reacts much faster with the decomposing coal or with certain primary volatiles than with the residual char. Therefore the period of highest reactivity to hydrogen corresponds to the occurrence of pyrolysis and lasts only a few seconds or less at temperature approaching 800°C or higher.

The species that react with hydrogen during pyrolysis are not clearly identified and are referred to as rapid-rate carbon. Much evidence suggests that hydrogen interferes with the char-forming secondary reactions, presumably by hydrogenating reactive fragments sufficiently to stabilize them during their escape. Thus the rate of total product generation appears to be controlled by thermal decomposition of the coal, whereas competition among mass transfer, secondary reactions, and hydrogenation affects only the relative proportions of volatile and solid product formed. The transport of volatiles out of the coal particle coupled with the countercurrent diffusion of hydrogen is poorly understood, but modification of the foregoing pyrolysis models to include hydrogenation has been studied as described below.

Hydrogasification refers to the reaction of coal and its char with hydrogen-rich gas at 600–1000°C, with emphasis on methane production. Hydrocarbonization implies the use of lower temperatures (450–600°C) and longer coal residence times, with emphasis on liquid products. Hydropyrolysis, referring to pyrolysis in the presence of hydrogen-rich gas, includes thermal decomposition and the attendant hydrogenation, but does not include subsequent hydrogasification of char. Hydropyrolysis reactions are always included in coal hydrogasification, but processes based on hydropyrolysis use residence times that are too short for significant char gasification at the prevailing temperatures. Hydrocarbonization is in fact low-temperature hydropyrolysis. Thus hydropyrolysis includes the rapid-rate carbon phenomenon, which has received much study in the context of hydrogasification. The extraction of hydropyrolysis information from hydrogasification data requires some attention to the contributions of char gasification to the overall behavior.

12.5.2 Characteristics of Hydropyrolysis and Rapid-Rate Carbon

The residual fixed carbon in char reacts with hydrogen to produce methane, but the reported rates for this reaction are relatively slow, ranging from 0.167–0.833% of C/ks-atm hydrogen at 927°C.[95] For comparison, the rate of carbon conversion is more than 10^6 times faster in coal pyrolysis at 927°C than in this char reaction under 100 atm of hydrogen. Early work by Dent[192] showed

[192] F. J. Dent, "The Production of Gaseous Hydrocarbons by the Hydrogenation of Coal," *Gas J.*, **244**, 502 (1944).

that a substantial portion of the carbon in raw coal can be converted to methane more rapidly than can the carbon in char. Birch et al.[123,124] identified two distinct stages in their fluid-bed hydrogasification of raw brown coal, the second of which was comparable in rate to the hydrogenation of char. In the initial stage, described as a rapid hydrogenation of volatiles, the gas yield for residence times exceeding 360 s depended on temperature and hydrogen pressure but not residence time.

Feldkirchner and Linden[95] studied hydrogasification of two coals and a char by dropping a few grams of sample onto a preheated alumina bed through which gas mixtures containing hydrogen were passed. Their data (Fig. 12.37) also exhibit two distinct rate periods, with the first and more rapid stage occurring during heatup. Linear extrapolation from the slow-rate period to zero time indicates conversions roughly equivalent to proximate volatile contents, which supports the argument that the short-lived period of high reactivity is simply hydropyrolysis resulting in hydrogenation of volatile constituents.

Schroeder[193] and later Hiteshue et al.[92–94] were the first workers to report rapid-rate hydrogenation yields exceeding proximate volatiles content. They passed hydrogen through a heated bed of raw coal and measured weight loss from the coal as a function of time as illustrated by the data in Fig. 12.38. Certain amounts of time were required for heating the bed to temperature and for cooling at the end of the run. Zero residence time on the graph corresponds to a run in which the sample was quenched immediately after heatup (total time of 40 s spent at

[193] W. C. Schroeder (to Fossil Fuels, Inc.), Hydrogenation of Coal, U.S. Pat. 3,030,297 (April 17, 1962).

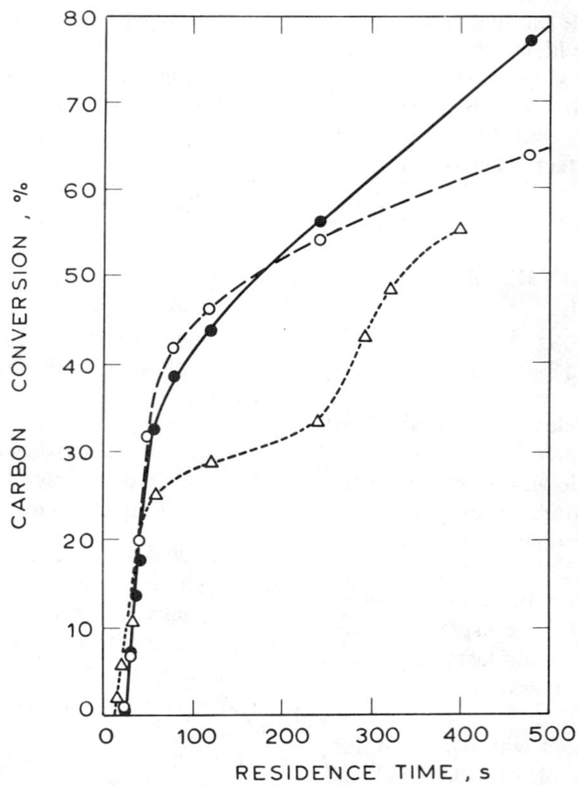

Figure 12.37 Effect of time on carbon conversion in the hydrogasification of coal and char. Final (holding) temperature, 927°C; hydrogen pressure, 100 atm; particle diameter, 841–1190 μm; ● = Pittsburgh Seam bituminous coal (VM = 35.5 wt % as received); ○ = North Dakota lignite (VM = 41.2 wt % as received); △ = bituminous coal char (VM = 17.3 wt % as received).[95]

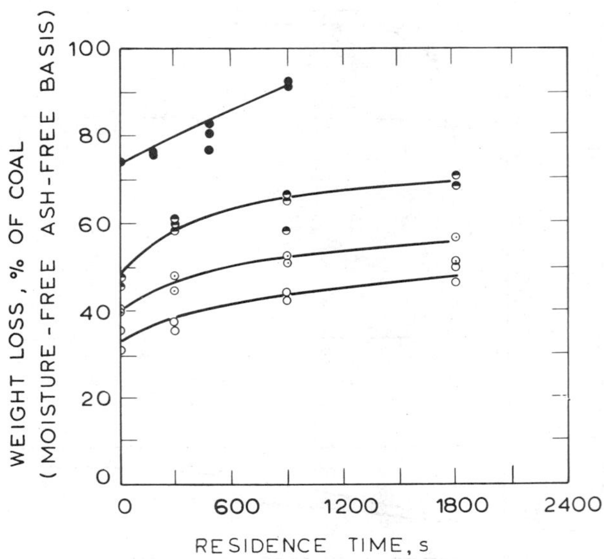

Figure 12.38 Effect of time on weight loss in the hydrogasification of Pittsburgh Seam bituminous coal (VM = 37.5 wt %maf). Final (holding) temperature, 800°C; hydrogen pressure: ○ = 18 atm; ⊙ = 35 atm; ◓ = 69 atm; ● = 409 atm and 1% molybdenum catalyst.[93,94]

temperatures above 300°C). The substantial initial conversions indicate that the early reactions are indeed rapid and that a significant portion of the coal's proximate fixed carbon participates in the rapid-rate period, which presumably involves hydropyrolysis.

Corroborative evidence was forthcoming in subsequent studies. Using a free-fall reactor at 900–925°C, Moseley and Paterson[142] obtained nearly complete conversion of raw coal to methane at 500 atm of hydrogen in a residence time of only a few seconds (data shown later). They also found that pretreating the coal by pyrolysis at 680°C in inert gas before contact with hydrogen substantially reduced reactivity to hydrogen, whereas pretreatment at 450°C had no effect on reactivity. In a similar manner, the thermobalance data of Johnson[103] presented in Fig. 12.39 show that the conversions obtained by gasification of raw coal at 927°C in 35 atm of hydrogen are much larger than those obtained when the coal is pyrolyzed in nitrogen at the same temperature and pressure, then subjected to the hydrogen treatment. Johnson's data also exhibit the early period of high reactivity of raw coal to hydrogen (curves *a* and *b*), where the yield is much larger than that obtained by pyrolysis alone (N_2, curve *c*), and show that the rate of char gasification with hydrogen in the slower second stage is approximately the same whether the coal was first heated in nitrogen or hydrogen. Data of Anthony[154] illustrated in Fig. 12.40 provide improved time resolation of the rapid first stage and show that a substantial part of the in-

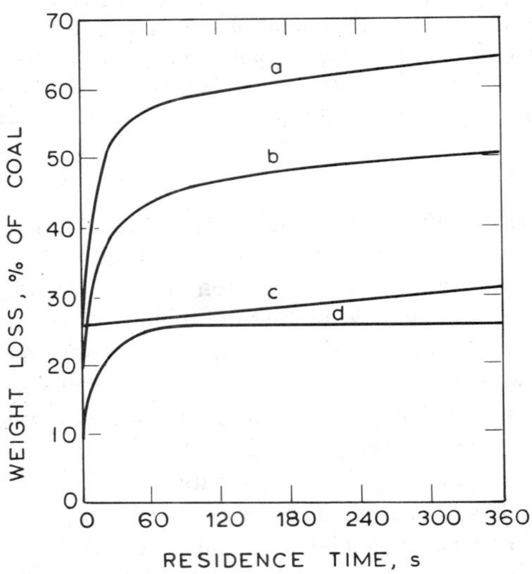

Figure 12.39 Effect of time on weight loss in the pyrolysis and hydrogasification of Pittsburgh Seam (Ireland Mine) coal and char. Final (holding) temperature, 927°C; (*a*) coal, 69 atm hydrogen; (*b*) coal, 35 atm hydrogen; (*c*) nitrogen-pretreated char, 35 atm hydrogen; (*d*) coal, 35 atm nitrogen.[103]

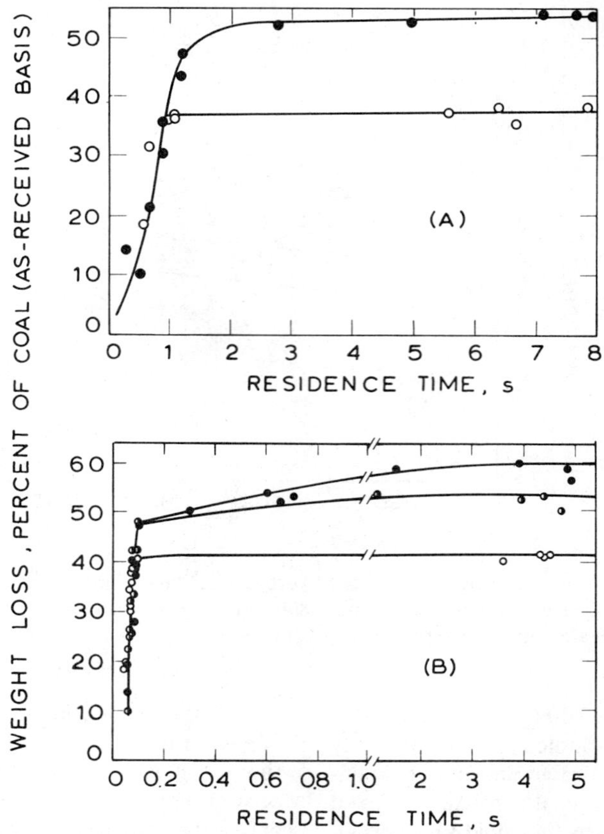

Figure 12.40 Effect of time on weight loss from coal heated in helium and hydrogen atmospheres. Mean particle diameter, 70 μm[109,154]. (A) Pittsburgh Seam bituminous coal (Ireland Mine): final (holding) temperature, 900°C; nominal heating rate, 750°C/s; pressure, 69 atm; ○ = helium; ● = hydrogen. (B) Montana lignite (Savage Mine): final (holding) temperature, 1000°C, nominal heating rate, 10,000°C/s; ○ = helium, 69 atm; ◑ = hydrogen, 69 atm; ● = hydrogen, 103 atm. Part (A) reproduced from Anthony et al.[109], by permission of publisher, IPC Business Press Ltd. ©.

crease in yield observed in hydrogen as compared to the pyrolysis yield observed in helium occurs as an extension of rapid pyrolysis, indicating that the rapid hydrogenation is coupled with thermal decomposition. The behavior is indicative of hydropyrolysis, the contribution of which to volatiles production is here seen to be substantial. By heating coal rapidly to 1000°C in 69 atm of hydrogen, yields 50% larger than the proximate volatile content were achieved in only 2 s with both a bituminous coal and a lignite. A large part of the increase in yield occurred during heatup, as did pyrolysis, even when the time for heatup was only 0.1 s (Fig. 12.40*b*). The data support the view that the rates of hydropyrolysis and pyrolysis are equal. With only limited information available, it appears that workers at the University of Utah have also achieved high yields (to 85 wt %) with a free-fall reactor (< 0.5 s residence time[194]). These yields were reportedly attained at substantially less severe temperatures (~ 500°C) with the aid of a catalyst.

Several of the studies discussed above have in common the heating of Pittsburgh Seam coal in hydrogen to temperatures in the range 800–927°C, thereby permitting comparison among different sets of time-resolved hydrogasification conversion data, which together cover residence times from

[194] W. H. Wiser, L. L. Anderson, S. A. Qader, and G. R. Hill, "Kinetic Relationship of Coal Hydrogenation, Pyrolysis and Dissolution," *J. Appl. Chem. Biotechnol.*, **21**, 82 (1971).

about 0.4–1800 s. Smooth curves that were drawn through the original data are collected in Fig. 12.41, where hydrogen pressure in atmospheres is indicated on the curves, and numbers in parentheses are cross-references to coal properties given in Table 12.22. Conversion is defined as weight loss excluding moisture, expressed as a percentage of the original coal on a dry ash-free basis, except for the data of Feldkirchner and Linden,[95] which are percentages of carbon conversion. In the present coordinate system, comprised of conversion on a rectilinear ordinate and residence time on a logarithmic abscissa, a rate that is first order in the amount of material yet to be converted will appear as a steep S curve. For example, the data represented by curves (1) were correlated by a nonisothermal model involving first-order kinetics (Fig. 12.40), and the upper portion of curve (4) also is consistent with first-order behavior. Furthermore, it should be noted that similar appearing slopes at widely different times actually reflect very different rates. Thus curve (3) and the upper portions of curves (2) indicate very slow rates and would appear as asymptotes on a rectilinear time abscissa. All the curves tend toward complete conversion, since carbon reacts with hydrogen, although very slowly.

The lower portions of all the curves in Fig. 12.41 were obtained under the nonisothermal conditions of heatup. The lateral displacements between curves are attributed almost entirely to different heating rates; that is, the apparent conversion rate at a given hydrogen pressure is dominated by the rate of temperature rise. For example, 50% conversion at 69 atm of hydrogen can be obtained in 1.2, 15, or 150 s (curves 1, 2, or 3) depending on heating rate. Furthermore, it appears that the conversions achieved in short times with rapid heating to 800°C or higher can be very large indeed if hydrogen pressure is sufficiently high. To emphasize the latter point, the data of Moseley and Paterson[142] are included in the figure (dashed lines, inset) even though these workers did not use Pittsburgh Seam coal. The range of residence times shown for their data was estimated by Anthony and Howard[8] based on calculated terminal velocities for the reported experimental

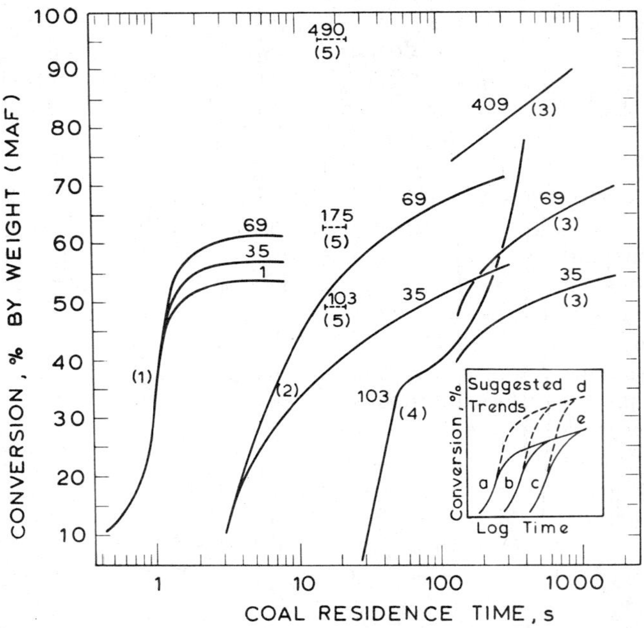

Figure 12.41 Effect of heating rate, hydrogen pressure, and time on coal conversion to gases in hydropyrolysis and hydrogasification.[8] Numbers in parentheses designate properties and investigators listed in Table 12.22; hydrogen pressure (atm) given as numbers on curves; for the 409-atm curve given for (3), 1% molybdenum catalyst was added. Final (holding) temperature: (1) 900°C, (2) 927°C, (3) 800°C, (4) 927°C, (5) 850–950°C. Coal type (Table 12.22): curves = Pittsburgh Seam bituminous; dashed lines = Warwickshire bituminous. Inset: idealized suggested trends.

Table 12.22 Properties of Coals Represented in Fig. 12.41

	Coal Number (see key in source note)				
	(1)	(2)[a]	(3)	(4)	(5)
	Ultimate Analysis (wt % of coal, maf)				
Carbon	77.7	79.8	82.8	83.0	78.4
Hydrogen	5.5	4.8	5.7	5.5	4.8
Oxygen	9.3	9.9	7.7	—	—
Nitrogen	1.5	1.5	1.7	—	—
Sulfur	6.1	4.0	2.1	1.6	—
	Proximate Analysis (wt % of coal)				
Volatile matter	39.8	28.4	33.9	33.5	34.5
Fixed carbon	46.3	60.7	56.5	56.9	54.0
Moisture	1.7	0	1.9	1.1	6.7
Ash	12.2	10.9	7.7	8.5	4.8

[a] Air-pretreated char.

Source: Coal 1, Pittsburgh Seam No. 8, Ireland Mine—Ref. 109; Coal 2, Pittsburgh Seam No. 8—Ref. 103; Coal 3, Pittsburgh Seam—Ref. 93, 94; Coal 4, Pittsburgh Seam, Montour No. 4 Mine—Ref. 95; Coal 5, Warwickshire Seam, National Coal Board No. 902—Ref. 142.

conditions. The coal, a noncaking high-volatile bituminous, appears to be less reactive than Pittsburgh Seam coal since, for example, the conversion at 15–22 s under 175 atm only slightly exceeds that of Anthony et al. for 69 atm and much smaller times. Nevertheless, the effect of increasing hydrogen pressure as indicated by the data of Moseley and Paterson is qualitatively similar to that observed by Anthony et al., and the former workers achieved more than 90% conversion in 15–22 s at 490 atm of hydrogen.

Although some of the behavior shown in Fig. 12.41 undoubtedly reflects effects of particle size, method of heating and contacting the coal with hydrogen, and other experimental conditions that vary among the different sets of data, the dominant effect on yield at a given final temperature is attributed to heating rate and hydrogen pressure while heating rate determines time scale. A conception of the trends is illustrated in the inset, where conversions achieved at different heating rates ($a > b > c$) under a given hydrogen pressure (d or e, $d > e$) eventually converge. This idealized view implies that heating rate does not significantly alter the path of secondary reactions or the reactivity of the residual char. Evidence supporting this hypothesis is limited, since heating rate is usually not varied without also changing apparatus, final temperature, or other conditions. The data of Anthony et al,[75,109] which include systematic variation of heating rate alone in the approximate range 100–10,000°C/s, are consistent with this view. The conditions have been extended recently to heating rates up to the order of 100,000°C/s and residence times as small as 50 ms in entrained-flow hydropyrolysis experiments at Cities Service[195,196] and Rocketdyne.[197–199] The observed conversions are in excellent agreement with the data of Anthony et al., which further strengthens the view that heating rate itself has little effect on the final overall conversion to volatiles. Also, since the Rocketdyne tests included coal feed rates

[195] B. H. Rosen, A. H. Pelofsky, and M. Greene (to Cities Service Company), Coal Hydrogenation to Produce Liquids, U.S. Pat. 3,960,700 (June 1, 1976).

[196] M. I. Greene, personal communication, February 1977.

[197] C. L. Oberg, *Partial Liquefaction of Coal by Direct Hydrogenation,* Month. Rep. FE-2044-17, prepared for ERDA Fossil Energy, Washington, D.C., by Rocketdyne Division of Rockwell International, Canoga Park, Calif. (January 1977).

[198] C. L. Oberg, A. Y. Falk, and G. A. Hood, *Partial Liquefaction of Coal by Direct Hydrogenation,* Ann. Rep. FE-2044-11, prepared for ERDA/Fossil Energy, Washington, D.C., by Rocketdyne Division of Rockwell International, Canoga Park, Calif. (January 1977).

[199] J. A. Gray and K. M. Sprouse, personal communication (March 1977).

up to 1 ton/hr, the agreement with the Anthony et al. data from a 10-mg batch reactor is noteworthy.

The idealized trends shown in the inset in Fig. 12.41 are for relatively short times (seconds). If time were longer, the curves would turn upward and eventually reach complete conversion of the carbon. Such behavior can be seen in Fig. 12.41 in the data of Feldkirchner and Linden (curve 4), and in data (not shown here) of Hiteshue et al.[94] for a temperature of 1200°C and time up to 1000 s or longer.

12.5.3 Weight Loss Behavior

Practical interest in the foregoing hydrogasification rate phenomena and the associated need for quantitative understanding of the underlying kinetics and mechanisms have led to the development of purely empirical correlations of the weight loss data as well as more sophisticated models based on proposed mechanisms with kinetics parameters evaluated from the data. The various approaches found in the literature and some additional data are discussed below.

12.5.3.1 *Empirical Correlations*

The principal correlative tool in the empirical correlations category is the simple model in which the rate of gasification of rapid-rate carbon is proportional to hydrogen partial pressure and the amount of rapid-rate carbon material remaining. Thus

$$\frac{dX}{dt} = k_4 P_{H_2} (X^* - X) \qquad (12.35)$$

where X is fractional conversion (see below) of rapid-rate carbon material, X^* is the total amount of this material originally in the coal, and k_4 is the rate constant, usually of the Arrhenius form. Equation 12.35 was used by Wen and Huebler,[200] Feldmann et al.,[141] Wiser et al.,[194] Jüntgen et al.,[201] and Anthony,[154] although Wen and Huebler replaced P_{H_2} with the difference between the actual hydrogen partial pressure and the value at equilibrium, on the grounds that the rate must be zero at equilibrium.

The present definition of X represents a necessarily broad standardization of four different nomenclatures used previously:

Moseley and Paterson[101]
Zahradnik and Glenn[202]
Jüntgen et al.[201]

$$X = \frac{\text{grams of carbon as methane}}{\text{grams of original carbon}}$$

Wen and Huebler[200]
Feldman et al.[141]

$$X = \frac{\text{grams of carbon gasified}}{\text{grams of original carbon}}$$

Wiser et al.[194]
Anthony[154]

$$X = \frac{\text{grams of coal gasified}}{\text{grams of original coal}}$$

Johnson[203]

$$X = \frac{\text{grams of fixed carbon gasified}}{\text{grams of original fixed carbon}}$$

(Moseley and Paterson also reported total carbon conversion and weight loss.) The definition that provides the best measure of conversion has not been established. Cross-correlations between weight loss, carbon conversion, and methane formation are provided below.

Values for k_4 as a function of temperature are shown in Fig. 12.42 for the various studies. Data points for Pittsburgh Seam bituminous coal at 69 atm hydrogen from Feldmann et al. and Anthony are plotted for comparison. Activation energies are low: 5–6 kcal/mole for bituminous coal. Anthony's data for lignite, the results of Jüntgen et al. for methane evolution (major hydrogenation product), and the data of Feldmann et al. at 207 atm exhibit lower rate constants. At low temper-

[200] C. Y. Wen and J. Huebler, "Kinetic Study of Coal Char Hydrogasification," *Ind. Eng. Chem., Process Des. Dev.*, **4**, 142 (1965).

[201] H. Jûntgen, K. H. Van Heek, and J. Klein, personal communication, July (1973).

[202] R. L. Zahradnik and R. A. Glenn, "Direct Methanation of Coal," *Fuel,* **50,** 77 (1971).

[203] J. L. Johnson, "Kinetics of Bituminous Coal Char Gasification with Gases Containing Steam and Hydrogen," in *Coal Gasification,* Am. Chem. Soc. Adv. Chem. Ser. No. 131, (1974), p. 145.

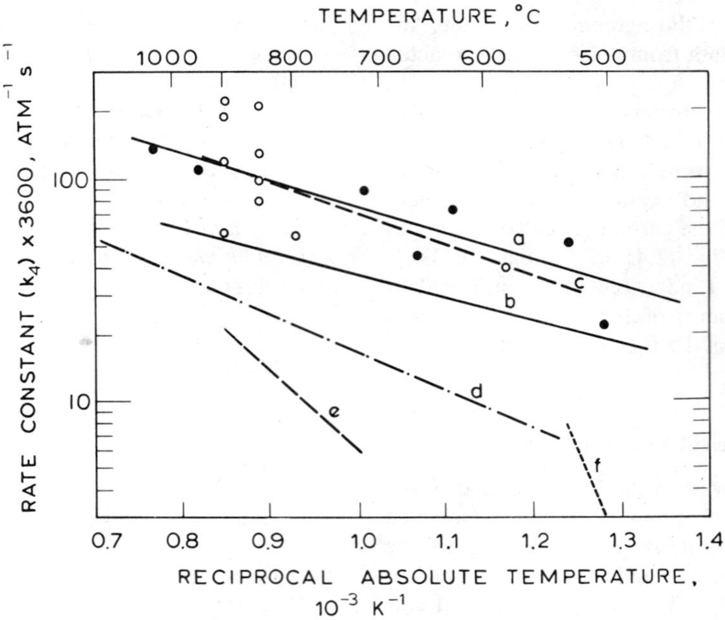

Figure 12.42 Comparison of hydrogasification rate constants from different investigators.[8] Data of Anthony[154], 69 atm of hydrogen: curve *a* and solid points, Pittsburgh Seam bituminous coal; curve *b*, Montana lignite. Data of Feldmann et al.,[205] 69 atm of hydrogen: curve *c*, and open points, Pittsburgh Seam bituminous coal. Data of Jüngten et al.,[201] 69 atm of hydrogen: curve *d*, CH_4 production only. Data of Feldmann et al.,[141] 205 atm of hydrogen: curve *e*. Data of Wiser et al.,[194] 137 atm of hydrogen: curve *f*.

atures Wiser et al.[194] report catalytic hydrogenation results with much higher activation energies. Data of Feldmann et al. at temperatures below 575°C are omitted here because of the mass deposition in the reactor at lower temperatures.[204]

The principal point from Fig. 12.42 is not that hydrogenation of coal has been successfully modeled, but rather that data obtained by widely varying experimental techniques are comparable in magnitude. In fact, the curves of Jüntgen et al. and Wiser et al. may not be entirely applicable in the discussion of rapid-rate conversion, since they were obtained at very low heating rates and involve a minimum time scale of 600 s or more where slow gasification of relatively old char is likely to be important. In any case, the value of the rate constant is poorly defined by this figure, partly because Equation 12.35 does not account for pyrolysis yields, thereby causing an apparent pressure effect on the rate constant that is particularly noticeable in the results of Feldmann et al.[205] at 69 and 207 atm. These workers attempted to correct for devolatilization at the higher pressure by use of an integration constant in Equation 12.35, representing the volatiles yield as instantaneous conversion at $t = 0$. The question of how to treat pyrolysis yields in the analysis of hydrogasification kinetics is discussed below.

The major shortcoming of Equation 12.35 is the need to know the amount of potential rapid-rate carbon (X^*) as a function of the pertinent parameters. Although the integrated form of the equation for constant temperature and hydrogen pressure is sometimes assumed to be

$$X = X^* [1 - \exp(-k_4 P_{H_2} t)] \qquad (12.36)$$

[204] H. F. Feldmann, personal communication, March 1976.

[205] H. F. Feldmann, C. Y. Wen, W. H. Simons, and P. M. Yavorsky, "Supplemental Pipeline Gas from Coal by the Hydrane Process," paper presented at the 71st National American Institute of Chemical Engineers Meeting, Dallas, Tex., February 20–23, 1972.

Wen and Huebler and Wiser et al. agree that X^* is a function of temperature, Feldmann et al. contend that X^* is dependent on hydrogen pressure and temperature, and Anthony found it to vary also with particle size.

12.5.3.2 *Active Species Model*

To explain the rapid decay of the initial reactivity of coal to hydrogen and the effects of temperature and pressure on hydrogasification yields, Moseley and Paterson[101] proposed that thermal decomposition of coal produces, in addition to volatiles, active sites that then form either methane by hydrogenation or relatively unreactive char by cross-linking. The resulting char was assumed to react very slowly with hydrogen to produce methane, and the primary methanation reaction was assumed to be rapid, resulting in no net consumption of active sites. The primary methanation reaction was expressed as:

$$\frac{dX}{dt} = k_4 P_{H_2} C^* \qquad (12.37)$$

where C^* is the concentration of active sites. By assuming that the rate of disappearance of active sites is controlled by a first-order thermal decomposition,

$$C^* = C_0^* \exp(-k_5 t) \qquad (12.38)$$

Equation 12.37 becomes

$$dX/dt = k_4 P_{H_2} C_0^* \exp(-k_5 t) \qquad (12.39)$$

Defining $X = 0$ at $t = 0$ and $k_3 = k_4 C_0^*/k_5$, integration of Equation 12.39 gives

$$X = k_3 P_{H_2} [1 - \exp(-k_5 t)] \qquad (12.40)$$

The destruction of active sites was assumed to be so fast that the exponential term in Equation 12.40 becomes negligible within fractions of a second. Thus the predicted rapid-rate yield was proportional to hydrogen pressure:

$$X = k_3 P_{H_2} \qquad (12.41)$$

Zahradnik and Glenn[202] used a somewhat similar model based on active species rather than sites. Equation 12.38 was modified to include disappearance of active species by methanation as well as by cross-linking. Thus

$$C^* = C_0^* \exp[-(k_4 P_{H_2} + k_5)t] \qquad (12.42)$$

and Equation 12.40 becomes

$$X = \frac{k_3 P_{H_2}}{1 + k_4 P_{H_2}/k_5} [1 - \exp\{-(k_4 P_{H_2} + k_5)t\}] \qquad (12.43)$$

Again, neglecting the exponential term after some minimum residence time, we get

$$X = \frac{k_3 P_{H_2}}{(1 + k_4 P_{H_2}/k_5)} \qquad (12.44)$$

Johnson,[203] on the other hand, viewed the active sites as catalyzing the reaction between fixed carbon and hydrogen and altered the form of Equation 12.37 as follows:

$$\frac{dX}{dt} = k_4 P_{H_2} (1 - X)^{2/3} \exp(-a'x^2) C^* \qquad (12.45)$$

Based on experimental data, Johnson found $a' = 0.97$ and concluded that

$$\int_0^X \frac{dX}{(1 - X)^{2/3}} \exp(-a'x^2) \simeq -\ln(1 - X) \qquad (12.46)$$

Therefore Equation 12.45 is approximately

$$\frac{dX}{dt} = k_4 P_{H_2} (1 - X) C^* \qquad (12.47)$$

Again, using C^* as given by Equation 12.38 and the approximation $\exp(-k_5 t) << 1$, integration of Equation 12.47 gives

$$-\ln(1 - X) = k_3 P_{H_2} \qquad (12.48)$$

Equation 12.48 was found to be inadequate at temperatures below 816° C, and a temperature-dependent relationship for C_0^* was proposed by Johnson[7]:

$$C_0^*(T) = C_0^*(T_0) \left[1 - \int_0^\infty f_s(E_s) \exp\left(-\int_0^t k_s\, dt\right) dE_s\right] \qquad (12.49)$$

or, assuming the ratio k_4/k_3 to be temperature independent,

$$-\ln(1 - X) = \left(\frac{k_4}{k_5}\right) C_0^*(T_0)P_{H_2}[1 - \int_0^\infty f_s(E_s) \exp\left(-\int_0^t k_s\,dt\right) dE_s] \quad (12.50)$$

where $f_s(E_s)$ is an assumed distribution of activation energies such that

$$\int_0^\infty f_s(E_s)\,dE_s = 1$$

and k_s is a rate constant of the Arrhenius form. The derivation of Equation 12.49 was based on the hypothesis that the active sites are formed by a very large (mathematically infinite) set of parallel reactions of differing activation energies described by $f_s(E_s)$. Furthermore, the rate of active sites formation was assumed to be much slower than their subsequent decomposition ($k_s << k_5$), in contrast to the models of Moseley and Paterson and Zahradnik and Glenn, where the formation of active sites or species is essentially instantaneous relative to their disappearance.

Johnson[137] later published data on methane and ethane yields from lignite and employed a kinetic model similar to that of Zahradnik and Glenn; that is, Equation 12.37 was substituted for Equation 12.47. Johnson, who assumed that the intermediate solid active species is destroyed in the methanation and ethanation step, arrived at a yield correlation identical in form to Equation 12.44. The temperature dependence of X, though, in contrast to the model of Zahradnik and Glenn, was assumed to arise from C_0^* rather than k_4/k_5, and Equation 12.49 was used to incorporate this phenomenon. Johnson drew heavily on detailed stoichiometric relationships, which are discussed below.

Moseley and Paterson[101] have also introduced evidence, although unaccounted for in their model, that the concentration of active sites is dependent on temperature but not pressure. Their data on the decay of methanation rate at different temperatures and hydrogen pressures are shown in Fig. 12.43.

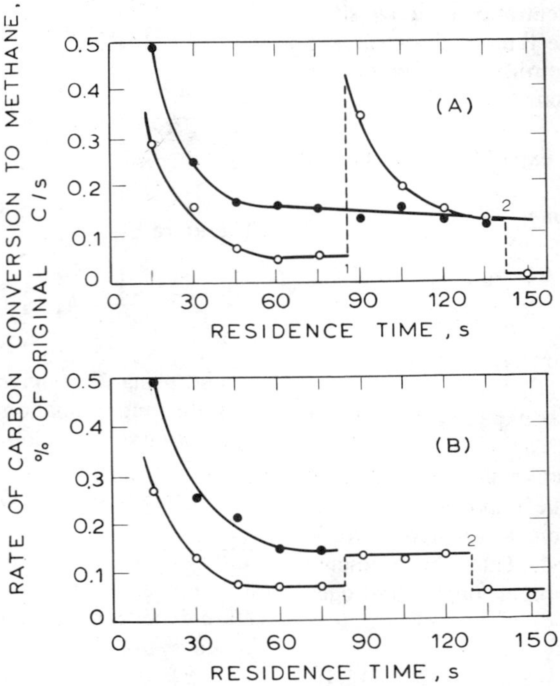

Figure 12.43 Effect of temperature and pressure on hydrogasification rate of Kingsbury char prepared at 700°C. (A) 96 atm of hydrogen: ● = 950°C; ○ = 815°C stepped to 915°C at (1) and back to 815°C at (2). (B) ● = 865°C, 96 atm hydrogen; ○ = 860°C, 42 atm hydrogen, stepped to 96 atm of hydrogen at (1) and back to 42 atm of hydrogen at (2).[101]

A step change in temperature from 815 to 915°C at 96 atm hydrogen resulted in a surge and subsequent rapid decay in rate (Fig. 12.43*a*), whereas a step change in hydrogen pressure from 42 to 96 atm at 860°C produced only a small increase in rate (Fig. 12.43*b*). This behavior is consistent with the picture that active species were produced by raising the temperature, whereas increasing the pressure only increased the rate of char gasification.

Zahradnik and Glenn[202] argue that the methane formation step is favored at high temperatures and support this position by showing a plot of $\ln(k_4C_0^*/k_5)$ versus $1/T$ to be a straight line with negative slope. The negative slope was taken as proof that the methane formation step is more highly activated than the thermal destruction of active sites. However the implied assumption that C_0^* is independent of temperature is inconsistent with the results of Johnson discussed above (Equation 12.49). Furthermore, values of C_0^* calculated from the constants reported by Zahradnik and Glenn correspond to an amount of carbon that when added to the amount released in pyrolysis, exceeds the total carbon in the coal. In revised versions of the model, the fractional available carbon is defined as unity.[206,207] Thus the issue of the proper method for treating the carbon evolved by thermal decomposition is raised again.

12.5.3.3 *Contribution of Volatiles*

Moseley and Paterson[142] did not distinguish between volatiles formed by pyrolysis and products of hydrogenation, based on the assumption that carbon in the volatiles is readily hydrogenated in the presence of hydrogen, and they suggested that their model (Equation 12.41) would be inaccurate at hydrogen pressures below 100 atm. As mentioned above, Feldmann et al.[141] introduced a constant in Equation 12.35 representing volatiles from pyrolysis. They reported the initial fractional conversions to be 0.22 at 725°C and 0.14 at 900°C, which seem inordinately low compared to the proximate volatile content of the coal (40.9%, maf). Furthermore, a decreasing volatiles yield with increasing temperature is the reverse of pyrolysis behavior. Zahradnik and Glenn[202] assumed that pyrolysis finishes before rapid hydrogasification begins and added a constant in Equation 12.44 to account for the contribution of pyrolysis products to the yield. However the values they obtained for this constant by fitting the modified Equation 12.44 to their hydrogasification data are small and essentially independent of temperature, unlike volatile yields. Zahradnik and Glenn based their assumption of extremely fast pyrolysis on the results of Badzioch.[5] Surprisingly, there are only a few such examples in the literature on hydrogasification in which reference is made to the large literature on thermal release of volatile matter from coal.

Recent research at MIT[109,154] has focused on the interaction of rapid hydrogasification and pyrolysis for a bituminous coal and a lignite at pressures, temperatures, and heating rates up to 100 atm, 1100°C, and 12,000°C/s, respectively. Volatiles yields (weight loss) from Pittsburgh Seam coal attained in 5–20 s at 1000°C and different pressures of hydrogen are shown as data points in Fig. 12.44, along with similar data for runs in helium discussed above (Fig. 12.7). Weight loss in hydrogen exhibits a peculiar minimum (48–50%; dashed curve discussed below) in the pressure range 1–20 atm, then rises with increasing pressure to about 60% of the original coal at 69 atm. As discussed above, weight loss in helium decreases with increasing pressure and is only 37.2% of the original coal at 69 atm, apparently reflecting increased char formation by secondary reactions of volatiles. These and other data were shown to be consistent with the picture that the rapid hydrogenation proceeds in two stages.[109] First, hydrogen can react with some of the volatiles to form light hydrocarbons sufficiently stable to avoid the secondary char-forming reactions. Thus volatiles yield is increased during pyrolysis (i.e., by hydropyrolysis) without affecting the rate of the primary decomposition. Second, hydrogen reacts with the fresh solid residue, as proposed by Moseley and Paterson,[101] at a

[206] R. L. Zahradnik and R. J. Grace, "Chemistry and Physics of Entrained Coal Gasification," paper presented at the 71st National American Institute of Chemical Engineers Meeting, Dallas, Tex. (February 20–23, 1972).

[207] R. L. Zahradnik and R. J. Grace, "Chemistry and Physics of Entrained Coal Gasification," in *Coal Gasification,* Am. Chem. Soc. Adv. Chem. Ser. No. 131 (1974), p. 126.

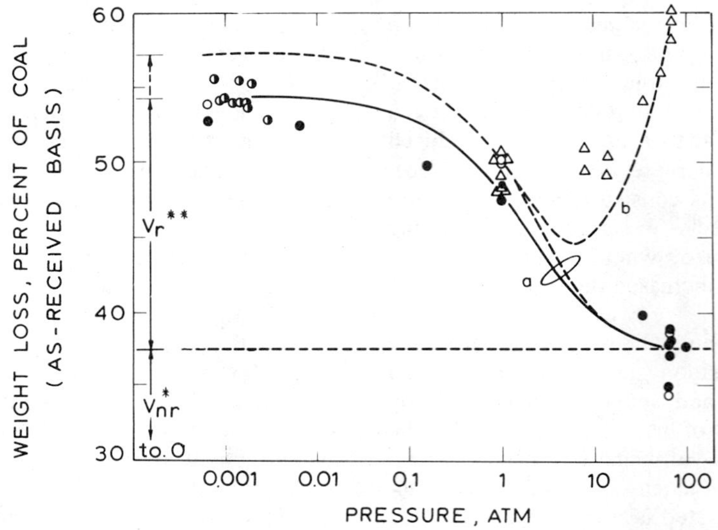

Figure 12.44 Effect of pressure on weight loss from Pittsburgh Seam (Ireland Mine) bituminous coal heated in hydrogen and helium atmospheres. Final (holding) temperature, 1000°C; residence time, 5–20 s; mean particle diameter, 70 μm; ○, helium, nominal heating rate = 10,000°C/s; ◑, helium, nominal heating rate = 3000°C/s; ● helium, nominal heating rate = 650–750°C/s, △, hydrogen, nominal heating rate = 750°C/s. Curve *a*, Equation 12.34; curve *b*, Equation 12.52: solid, V_r^{**} = 17.0% of original coal; dashed, V_r^{**} = 20.0% of original coal, corrected for cracking deposition on sample holder; V_{nr}^* = 37.2% of original coal. Reproduced from Ref. 109 by permission of the publisher, IPC Business Press Ltd. ©.

rate faster than hydrogasification of older char but slower than hydropyrolysis.

Volatiles yields were correlated as a function of total pressure and hydrogen partial pressure by modifying Equation 12.34 to account for hydrogenation as well as secondary reactions.[109] Thus the volatiles formed within a coal particle were assumed to include reactive and unreactive species as described in the derivation of Equation 12.34, and it was assumed that the reactive species may polymerize and/or crack and deposit carbon on hot coal surfaces, or escape either directly (if mass transfer is much faster than secondary reactions) or through the formation of stable species by reaction with hydrogen. A material balance on the reactive volatile species present in a coal particle may be written (see Equation 12.30):

$$\frac{dV_r'}{dt} - K_c(C - C_\infty) - k_1 C - k_2 C P_{H_2} = \frac{dC}{dt} \qquad (12.51)$$

where k_2 is the overall hydrogenation rate constant. Solution of this equation in a manner similar to Equations 12.30–12.34 gives an expression for the yield of reactive volatiles (V_r^*). The resulting correlation for total yield including unreactive volatiles (V_{nr}^*) and the contribution of heterogeneous hydrogenation (Equation 12.41) is

$$V^* = V_{nr}^* + V_r^{**}\left(1 + \frac{k_1}{k_c/P + k_2 P_{H_2}}\right)^{-1} + k_3 P_{H_2} \qquad (12.52)$$

The curves in Fig. 12.44 were obtained with Equation 12.52 using V_r^{**} = 20% (dashed) or 17% (solid), V_{nr}^* = 37.2%, k_1/k_c = 0.56 atm^{-1}, k_2/k_c = 0.02 atm^{-2}, and k_3 = 0.1% atm^{-1}. Although there are noticeable discrepancies, the model does predict the observed minimum yield in pure hydrogen. According to the model, the minimum arises from the opposing effects of solids deposition and homogeneous or heterogeneous hydrogenation (or both). According to the parameter values above, the dominant hydrogenation is homogeneous for the data in Fig. 12.44, but a more significant role for the heterogeneous reaction cannot be excluded. An overly simplistic view of particle structure and mass transfer phenomena is believed to be the principal source of error. Furthermore this approximate treatment of the heterogeneous reaction would require much caution in any application of Equation 12.52 for hydrogen pressures above those of the data on which k_3 is based (69 atm).

The different yield correlations discussed above are illustrated in Fig. 12.45, along with representative data of Moseley and Paterson[142] and Hireshue et al.[94] The model of Feldmann et al.[141] (Equation 12.36) is not presented here because it was intended for use only in the short period of rapid-rate behavior. The models presented apply to extended residence times for which their predicted conversions become approximately independent of time as discussed above. If Equation 12.36 were used under these conditions, the predicted conversion would be complete at all hydrogen pressures. The straight line predicted by Equation 12.48[203] has an intercept at zero hydrogen pressure, which is fixed at a weight loss equal to the proximate volatiles content in agreement with $X = 0$ in Johnson's nomenclature (see above). The model of Anthony et al.[109] is not extrapolated beyond the maximum hydrogen pressures covered in Fig. 12.44. It is interesting to note, however, that such an extrapolated curve would be shaped like the curve of Moseley and Paterson, reflecting use in Equation 12.52 of their heterogeneous reaction term, which becomes very significant at the higher pressures, and it would be roughly midway between the Moseley and Paterson curve and that of Johnson. The disagreement between the correlations can probably be attributed to differences in the coals and experimental conditions.

12.5.3.4 *Effect of Temperature*

Increasing temperature increases not only the rate of coal conversion to gases in rapid-rate hydrogasification, as shown in Figs. 12.42 and 12.43*a*, but also the extent of conversion achieved before the rate decays to that of very slow hydrogasification of residual char. Figure 12.46 illustrates the effect of temperature on weight losses attained in 5 to 20 s in hydrogen at 69 atm[109] and in 180 s and 3600 s in hydrogen at 35 atm.[7] The latter data for

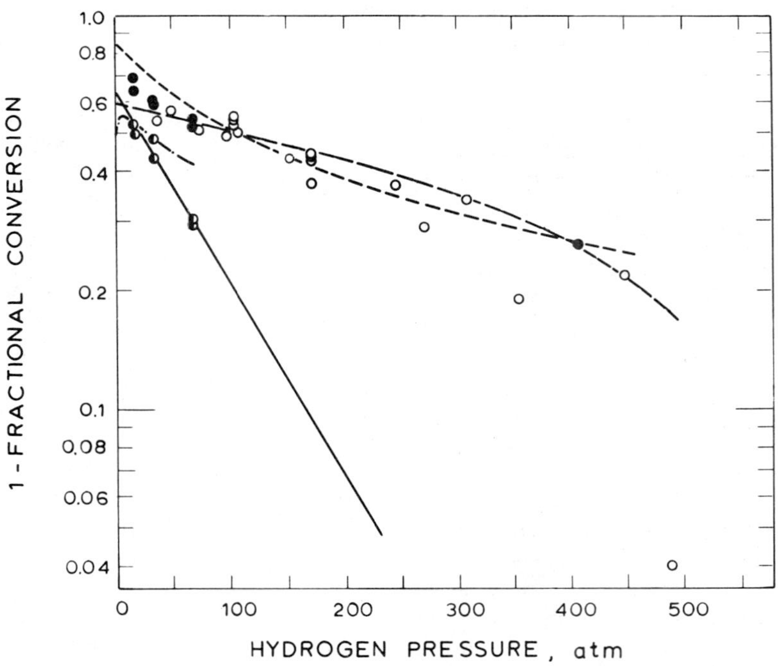

Figure 12.45 Comparison of effects of pressure on coal conversion in hydrogasification from different investigators, temperature, 800–1000°C.[8] Data points: ● and (◐), Hiteshue et al.[94], residence time 3–4 s (●), 1800 s (◐); ○, Moseley and Paterson[142], residence times, several seconds. *Curves:* long dashes, Moseley and Paterson[147]; short dashes, Zahradnik and Glenn[203]; alternating long and short dashes, Anthony et al.[109], data points in Fig. 12.44; solid, Johnson.[203]

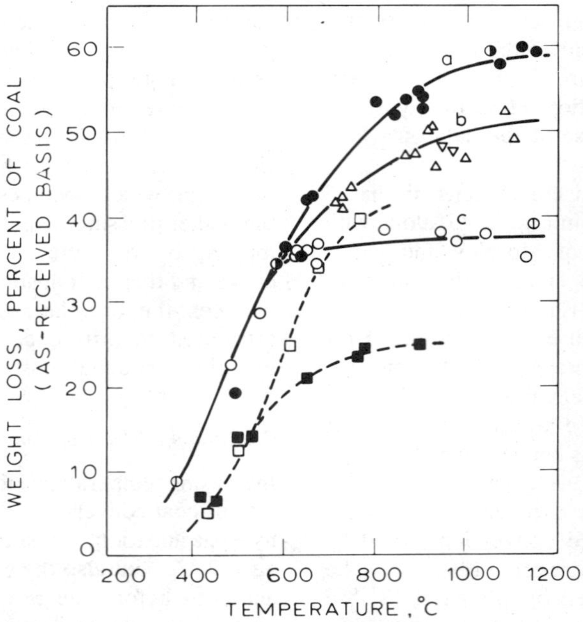

Figure 12.46 Effect of temperature on weight loss from Pittsburgh No. 8 Seam (Ireland Mine) bituminous coal and char heated in hydrogen and inert atmospheres at different pressures. Square points[7]: char (VM = 28.4%, moisture = 0); □, 35 atm of hydrogen; residence time, 180 s; ■ nitrogen; residence time, 3600 s. Circles and triangles[109]: coal (VM = 39.8%, moisture = 1.6%); nominal heating rates = 65–750°C/s; mean particle diameter, 70 μm; residence times, 5–20 s; ●, hydrogen, 69 atm; ▽, helium, 1 atm; △, nitrogen, 1 atm; ○, helium, 69 atm; ⦶, two-step heating, helium, 69 atm; ◑, two-step heating, hydrogen, 69 atm. Solid curves: Equation 12.53 with parameter values given in text and (*a*) $P = P_{H_2} = 69$ atm; (*b*) $P = 1$ atm, $P_{H_2} = 0$; (*c*) $P = 69$ atm, $P_{H_2} = 0$.

bituminous coal char were said to exhibit an effect of hydrogen at all temperatures, whereas Anthony et al. found no effect of hydrogen for bituminous coal at temperatures below about 600°C, and Moseley and Paterson, as noted above, observed no loss of reactivity to hydrogen in chars prepared at 450°C, but definitely observed a reactivity loss for chars prepared at 680°C and higher. An explanation for this apparent difference seems to be that Pyrcioch et al.[7] developed their temperature-dependency correlation from run times of 3600 s. In Fig. 12.46 the original data at residence times of 180 s are replotted, and these show much better agreement with the temperature dependency of rapid-rate yields found by Anthony et al.[109]

To compare the effect of temperature on hydrogasification yields with that on pyrolysis yields, other data of Anthony et al. on weight losses atttained also in 5–20 s, but in inert gas at 1 and 69 atm, are included in Fig. 12.46. Inspection of the complete set of data from Anthony et al. shows that yields increase equally under the different conditions up to about 600°C, above which the following effects are evident: (1) yields at 69 atm become much greater in hydrogen than in inert gas, (2) yields in inert gas become greater at 1 atm than at 69 atm, and (3) yields at 69 atm in inert gas are almost independent of temperature. Furthermore, with respect to item 1, a substantial part of the increase in yield achieved with hydrogen occurs at a rate comparable to that of rapid pyrolysis (Fig. 12.40). These data were shown to agree with the picture that the rate of total product generation under all these conditions is controlled by thermal decomposition, whereas competition among mass transfer, secondary reactions, and hydrogenation affects only the relative proportions of gaseous and solid products formed. Thus in terms of the discussions above, it appears that cases 1 and 2 reflect the roles of hydrogenation of reactive volatiles and increased rates of mass transfer, respectively, in the reduction of solid deposition by secondary reactions. In view of case 3, it appears that no more than minor amounts of unreactive volatiles are formed by raising the soaking temperature above about 600°C for the times involved here. Conversely, essentially all the reactive volatiles appear to be formed at temperatures above about 600°C, since neither increased mass transfer rate nor hydrogenation seems to augment the yield at lower temperatures.

Accordingly, the decomposition reactions forming reactive volatiles appear to have activation energies generally higher than those of nonreactive volatiles formation. Using the simplifying approximation that all the decomposition reactions forming reactive volatiles have activation energies exceeding E_1 and all those forming nonreactive volatiles have values less than E_1, Anthony et al.[109] showed that Equation 12.16 and the transient form of Equation 12.52 lead to the following expression for volatiles yield as a function of time, including the effects of hydrogenation and secondary reactions:

$$V = V_{nr}^{**}\left[1 - \left(\frac{V^{**}}{V_{nr}^{**}}\right)\int_0^{E_1} \exp\left(-\int_0^t k_0 e^{-E/RT}\,dt\right) f(E)\,dE\right] + \left\{V_r^{**}\left[1 + \frac{k_1}{k_c/P + k_2 P_{H_2}}\right]^{-1} + k_3 P_{H_2}\right\}\left[1 - \frac{V^{**}}{V_r^{**}}\int_{E_1}^{\infty} \exp\left(-\int_0^t k_0 e^{-E/RT}\,dt\right) f(E)\,dE\right] \tag{12.53}$$

In this model the time-dependent contribution to volatiles yield by heterogeneous hydrogenation during the short reaction times of interest here was assumed to be $k_3 P_{H_2} V'/V_r^{**}$, that is, the contribution is proportional to the fractional completion of reactive volatiles formation (V_r'/V_r^{**}). Accordingly, as can be seen in Equation 12.53, the reactive condensed phase constituents that participate in the heterogeneous hydrogenation are assumed to be formed by thermal decomposition reactions that are the same as, or at least kinetically identical to, the decompositions forming reactive volatiles.

Unlike Equation 12.52 and its limitation to the condition $t = \infty$, which is satisfied in only a few seconds at 1000°C, Equation 12.53 is applicable not only at such high temperature but also at lower temperatures, where the assumption $t = \infty$ is often not satisfied in available data owing to the associated slower reaction rates.

Equation 12.53 was fitted to all the data of Anthony et al. in Fig. 12.46 by using the experimental temperature-time histories, the above-stated values of k_1/k_c, k_2/k_c, k_3, V_{nr}^{*} (= V_{nr}^{**}), and V_r^{**} and $k_0 = 1.67 \times 10^{13}\ s^{-1}$. The distribution $f(E)$ was again assumed to be Gaussian (Equation 12.17). The curves shown in Fig. 12.46 were obtained with mean activation energy $E_0 = 54.8$ kcal/mole, standard deviation $\sigma = 17.2$ kcal/mole, and $E_1 = 61.4$ kcal/mole. In addition to correlating the hydrogasification data, these parameters constitute a reassessment of the pyrolysis kinetics in the light of the observed behavior in hydrogen.

The success of Equation 12.53 in correlating data obtained under a range of heating rates, final temperatures, inert gas pressures, and hydrogen pressures supports the underlying hypothesis that the rate of hydropyrolysis or rapid-rate hydrogasification at a given temperature or for a prescribed heating rate, is governed by the thermal decomposition of the coal, whereas mass transfer, secondary reactions, and hydrogenation affect the relative proportions of volatile and solid products formed. This picture agrees with recent data on product distributions,[146] discussed below, and it seems to be consistent with the role of pyrolysis in coal liquefaction described by Wiser[208] and Neavel.[37] Similar behavior has been described by Virk et al.[209] for simple aromatic hydrocarbons that decompose at the same rate in hydrogen as in inert atmospheres, although the product in hydrogen is predominantly methane, whereas substantial carbon is formed in inert atmospheres under otherwise similar conditions. The overall activation energies of these reactions were found to correlate with the calculated aromatic delocalization energies, and rate control was therefore attributed to destabilization of the aromatic structure. The implications of this mechanism to coal hy-

[208] W. H. Wiser, "A Kinetic Comparison of Coal Pyrolysis and Coal Dissolution," *Fuel,* **47,** 475 (1968).

[209] P. S. Virk, L. E. Chambers, and H. N. Woebcke, "Thermal Hydrogasification of Aromatic Compounds," in *Coal Gasification,* Am. Chem. Soc. Adv. Chem. Ser. No. 131, (1974), p. 237.

dropyrolysis reactions require further study. The fact that coal is not a pure aromatic undoubtedly has important consequences that must be considered.

According to the foregoing scheme, the primary effect of temperature on hydropyrolysis is simply the effect of temperature on pyrolysis. In contrast, as noted above, several authors have postulated a competition between gasification and decay of active sites or species as an explanation of temperature effects. However the data of Anthony et al.[109] as well as the collection of data in Fig. 12.41 show little influence of heating rate on yield. Pyrcioch et al.[7] also found little effect of heating rate on yields in hydrogasification. Furthermore, Anthony et al. reheated two samples of char from low-temperature runs at 69 atm to high temperatures at 69 atm and found that the cumulative weight loss in each case was essentially identical to that for raw coal heated directly to the same high temperature (Fig. 12.46). If, in fact, only a fixed concentration (C_0^*) of active sites were originally available as assumed in Equations 12.37–12.48, the extent of deactivation observed at a given temperature would increase with increasing residence time up to that temperature. Such behavior is not in line with the data above.

As noted above, Moseley and Paterson recognized that C_0^* would have to be a function of temperature, and Johnson replaced C_0^* in Equation 12.48 with an expression involving statistically distributed activation energies (Equation 12.49). To the extent that Anthony et al. used a statistical distribution of activation energies for pyrolysis and explained the temperature effects in terms of competitive reactions involving reactive species, their approach has an important similarity to that of Johnson, but there are also important differences. The model of Anthony et al. accounts for hydrogenation of reactive volatiles, whereas Johnson's model includes deactivation of active sites, for example, by cross-linking. Each phenomenon is omitted by the alternative model, although the Anthony et al. model does include solid formation from volatiles such as reactive tars, the effect of which is similar to that of cross-linking. Rate of conversion in both models depends on the formation rate of reactive species. Thus the change in conversion rate caused by a step change in temperature would occur instantaneously according to these models. The rate changes shown in Fig. 12.43*a* caused by a step increase from 815 to 915°C and a subsequent step back to 815°C appear to support this picture, but the time resolution of the data is not good enough to exclude a slower change. Furthermore, Blackwood and McCarthy[210] have shown that cycling the temperature of a devolatilized char produces repeated peaks in rate of methane formation from hydrogen. This phenomenon should be considered along with hydrogenation associated with further pyrolysis in analyzing the data in Fig. 12.43*a*, but its contribution indicated by the data of Blackwood and McCarthy is relatively small. It should be noted, however, that both the rate changes shown in Fig. 12.43*a* correspond to an apparent activation energy of 53–61 kcal/mole, which is consistent with the above-stated mean value of 54.8 kcal/mole found by Anthony et al. In a recent correlation of lignite data, Johnson[137] assumed a uniform distribution of activation energies, which was found to extend from 79.5 to 118.1 kcal/mole with a preexponential factor of 7×10^{20} s^{-1}. The use of k_0 values much larger than 10^{13} s^{-1} would imply that extremely large entropies of activation are associated with the solid phase reactions. In a recent use of the flat distribution of activation energies to describe the assumed first-order formation of active intermediates, Johnson[211] correlated hydropyrolysis data on lignite as well as subbituminous and bituminous coals and found the distribution to range from 40.8 to 62.9 kcal/mole with a preexponential factor of 1.97×10^{10} s^{-1}. These improved numbers are more in line with those found by Anthony et al.

12.5.3.5 *Effect of Particle Size*

Very little experimental evidence has been accumulated on the effect of particle size on hydrogasification or hydropyrolysis behavior. Although the available hydrogasification data were obtained using a wide range of particle sizes (Table 12.23), the effect of

[210] J. D. Blackwood and D. J. McCarthy, "The Mechanism of Hydrogenation of Coal to Methane," *Aust. J. Chem.*, **19**, 797 (1966).

[211] J. L. Johnson, "Kinetics of Initial Char Hydrogasification Stages," *Prepr., Am. Chem. Soc., Div. Fuel Chem.*, **22**(1), 17 (1977).

Table 12.23 Primary Particle Size Range Used in Hydrogasification and Hydropyrolysis Investigations

Investigators	Particle Diameter (μm)
Glenn et al.[136]	< 44
Belt and Roder[143]	70% < 74
Anthony et al.[75,109]	53–88[a]
Moseley and Paterson[142]	100–150
Feldmann et al.[141]	150–300
Wiser et al.[36]	246–417
Hiteshue et al.[94]	250–600
Johnson[203]	420–841

[a] Some runs performed with particles up to 1000-μm diameter in study of size effects.

particle size was usually not an object of study. Moseley and Paterson[142] reported a definite tendency for conversion to rise with smaller particles, but their experimental technique was unsuited to extensive studies of particle size. A particle size effect on conversion rate has been built into several models, but without direct experimental verification. Johnson (Equation 12.45) found rate to be proportional to $(1 - X)^{2/3}$, implying a shrinking core behavior. Pyrcioch et al.[7] rationalize this result by assuming that a coal or char particle consists of a large number of internally dispersed subelements that individually gasify with a shrinking core mechanism. Zahradnik and Glenn originally derived their model by assuming the methanation step (Equation 12.37) to be described as

$$\frac{dX}{dt} = k_3 P_{H_2} C^{*\,2/3} \qquad (12.54)$$

also a shrinking core postulate. In both cases the expression for yield was independent of particle size.

Anthony et al.[109] systematically varied particle size over the range 53–1000 μm in both hydropyrolysis and pyrolysis experiments. The effect of particle size on yield was found to be substantial for Pittsburgh Seam bituminous coal in hydrogen atmosphere but, as noted above, small for helium. The data are presented in Fig. 12.47 along with data of Moseley and Paterson.[142] At 69 atm of hydrogen and 1000°C, reduction of particle diameter from 1000 to 70 μm increases the weight loss attained in 5–20 s from 44 to 59% of the original coal. The observed trend indicates that still further reduction of particle size

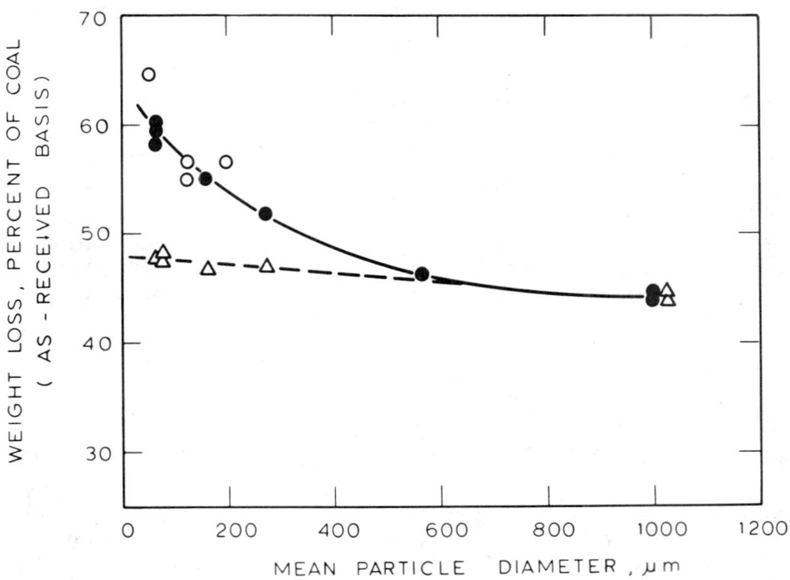

Figure 12.47 Effect of particle size on weight loss from bituminous coal heated in hydrogen and helium. ○, Moseley and Paterson[142], Warwickshire coal, 175 atm of hydrogen, 920°C final (holding) temperature, several second residence time; ● and △, Anthony et al.[109], Pittsburgh Seam coal, 1000°C final (holding) temperature, 5–20 s residence time, 650–750°C/s nominal heating rate, 69 atm of hydrogen (●), 1 atm of helium (△).

would result in larger increases in yield. In the case of Montana lignite, no significant effect of particle size was observed in either hydrogen or helium.

Since the yield from Pittsburgh Seam coal increases with increasing hydrogen pressure above about 10 atm (Fig. 12.44), the data in Fig. 12.47 can be interpreted as meaning that mass transfer of hydrogen to the interstices of the particle is constrained. Such behavior would be expected in view of the particle structure (see above) prevailing during pyrolysis when hydrogenation competes with solid deposition reactions. Anthony[154] cross-plotted the data from Figs. 12.44 and 12.47 to determine the effective hydrogen partial pressure as a function of particle size. According to this concept, the effective hydrogen partial pressure at 1000 atm and 1000°C was about 34% larger for a 70-μm particle than for a 1000-μm particle. Much more sophisticated mass transfer studies and experimental measurements on very fine particles are clearly needed.

12.5.4 Product Compositions

12.5.4.1 *General Characteristics*

Typical product distributions from rapid hydrogasification and hydropyrolysis (Table 12.24) exhibit the following trends. High temperatures and high hydrogen pressures give the highest yields (gases plus liquids), with methane the predominant fluid product. The production of heavier hydrocarbons is higher

[212] H. F. Feldmann, J. A. Mima, and P. M. Yavorsky, "Pressurized Hydrogasification of Raw Coal in a Dilute-Phase Reactor," in *Coal Gasification,* Am. Chem. Soc., Adv. Chem. Ser. No. 131 (1974), p. 108.

Table 12.24 Rapid Hydrogasification and Hydropyrolysis Product Distributions

Investigators and Residence Time Information	Temp. (°C)	H_2 Pressure (atm)	Yield (wt % of original coal, except as noted) CH_4	C_2H_6 + C_3H_8	CO	CO_2	H_2O	Oil + Tar	Char
Birch et al.[123] (includes slow gasification)	754	42	22.3	—	[15.7]		22.3	4.8	43.1
	850	42	45.0	—	[11.9]		29.3	0.9	29.8
	855	20	25.8	—	[18.6]		28.9	0.9	41.8
	951	42	61.1	—	[14.2]		28.4	0.4	16.4
Hiteshue et al.[94] (120 s)	800	35	8.2	1.8	—	—	—	10.0	59.6
	800	69	10.9	2.6	—	—	—	15.2	54.5
	1200	35	46.7	2.5	—	—	—	1.7	43.0
Moseley and Paterson[142] (several seconds)	900	106	35.7	—	6.8	—	—	—	50.0
Feldmann et al.[212] (several seconds)	850	69	14.5	0.9	1.5	1.0	4.1	3.0	64.6
	900	35	22.8	0.0	3.7	0.0	3.6	1.7	61.8
	900	76	27.2	0.3	3.5	0.4	3.2	0.5	66.3
Graff et al.[116,a] Vapor / Solids									
0.6 s / 30 s	817	100	27.7	7.3[b]	—	—	—	11.7	57
23 s / 30 s	814	100	44.7	0.5[b]	—	—	—	2.3	55
Fallon and Steinberg[138] (12–20 s)[a]	700–800	103–137	12–32	6–15[b]	[7–16]		—	11–26	11–64
Suuberg[146] (10–20 s)[c]									
Lignite	850–1000	69	11.4	1.7[d]	8.5	10.2	11.0	9.6	46.3
Bituminous	870–930	69	26.6	2.6[d]	—	1.5	—	19.8	46.0

[a] Yields expressed as carbon in product, percentage of carbon in coal.
[b] Ethane only.
[c] Yields expressed as weight percent of maf coal.
[d] Ethane only. All other hydrocarbon gases, excluding methane, were 5.2% for the lignite and 3.6% for the bituminous coal.

at lower temperatures even though the total yield is less. The yields at lower hydrogen pressures are lower in quality (i.e., hydrogen content) as well as in quantity. The data of Graff et al.[116] indicate that longer vapor residence times promote the conversion of oils and tars to methane at 100 atm of hydrogen. Benzene is a major constituent of the oils in the data of Graff et al. and Fallon and Steinberg.[138] In another study Growcock and MacKenzie[213] found the yields of methane and benzene from a lignite at 750°C and 100 atm of hydrogen to increase slightly ($\leq$ 10%) as a residence time of the vapor was increased from 0.3 to 9.7 s. In Suuberg's experiment, where the volatiles were rapidly quenched on being expelled from the particles into the cold ambient gas, the benzene was only 2.5% of the maf bituminous coal and about 1% of the maf lignite. Coal type is of major importance, probably accounting for much of the variation between the different sets of data, although differences in equipment and technique are also important. The coals employed in the three sets of more recent data were Illinois No. 6 (hvb) (Graff et al.), North Dakota lignite (Fallon and Steinberg), and Montana lignite and Pittsburgh Seam No. 8 bituminous (Suuberg). The present ability to predict product distributions is poor, and extrapolations to equipment of larger scale than that used to collect the data are risky.

Although it is difficult to predict the final destination of the many simultaneous and sequential reactions taking place, a potentially fruitful approach is becoming available. Gray et al.[214] recently presented correlations of coal carbon conversion with weight loss and coal hydrogen, nitrogen, and sulfur conversions. In many cases these correlations were independent of apparatus and conditions. Similarly Johnson,[137] as mentioned above, was able to infer stoichiometric relationships from correlations between methane and ethane yields and coal hydrogen. Although such correlations stop short of offering a full understanding of the detailed chemistry, they may approximate actual phenomena closely enough to be satisfactory for design purposes.

The varied definitions of conversion used in previous modeling attempts were introduced above; then ignored for purposes of simplicity. Objections to that simplification are partially mollified by the correlations of literature data in Figs. 12.48 and 12.49. A high degree of correlation between carbon conversion and weight loss is apparent in Fig. 12.48, and a somewhat poorer but nevertheless useful correlation between methane yield and carbon conversion is presented in Fig. 12.49. There is need for further work along these lines, especially on the development of stoichiometric relations initiated by Johnson.[137]

12.5.4.2 *Comparisons of Hydropyrolysis and Pyrolysis*

In the same way that Anthony[154] studied the contribution of hydropyrolysis to weight loss by heating coal in inert gas and in hydrogen under otherwise identical conditions, Suuberg[146] used the wire mesh heater technique to study the contribution of hydropyrolysis to product yields. Suuberg's pyrolysis results were discussed above, and some of his hydropyrolysis data appear in Table 12.24. He determined the total yield of volatiles both by a direct measurement of weight loss from the coal and by summing the yields of all volatile products. These data are in good quantitative agreement with those of Anthony, and show again that pyrolysis under hydrogen pressures exceeding 1–10 atm gives measurable increases in total yield for temperatures larger than a critical value that depends on the temperature-time history. Thus when heating rate is slow or a substantial holding time is provided at the highest temperature attained, the hydropyrolysis yields begin to diverge above the pyrolysis yields at a relatively low temperature (600°C in Fig. 12.46). Conversely, the temperature at which divergence begins is 700–800°C when the coal is heated at 1000°C/s to a peak temperature, then cooled at 200°C/s with zero holding time at the peak temperature. Thus the temperature required to achieve a given extent of hydropyrolysis is larger, the smaller the amount of time provided, which is behavior also exhibited by pyrolysis itself

[213] F. B. Growcock and D. R. MacKenzie, "Rapid Hydrogenation of a North Dakota Lignite," *Fuel,* **55,** 349 (1976).

[214] J. A. Gray, P. J. Donatelli, and P. M. Yavorsky, "Hydrogasification Kinetics of Bituminous Coal and Coal Char," *Prepr., Am. Chem. Soc., Div. Fuel Chem.,* **20**(4), 103 (1975).

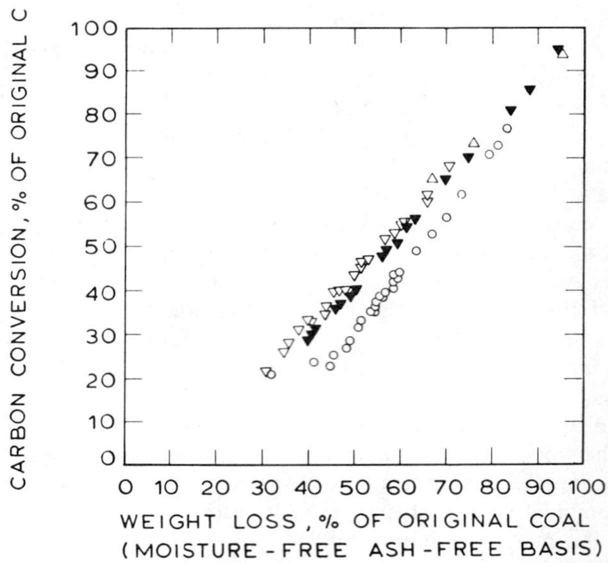

Figure 12.48 Relationship between weight loss and carbon conversion in coal hydrogasification and hydropyrolysis data of different investigators.[8] ○, Birch et al.[123]; △, Moseley and Paterson[142]; ▽ (800°C) and ▼ (1200°C), Hiteshue et al.[94]

(Figs. 12.5 and 12.6). Furthermore, the temperature range 700–800°C is identical to that over which increased inert gas pressure begins to show an effect on pyrolysis yield for the zero holding time case presented in Fig. 12.27, just as both hydropyrolysis and lower-pressure pyrolysis yields begin to diverge from higher-pressure pyrolysis yields at about 600°C (Fig. 12.46) when a few seconds is provided at the peak temperature. As discussed above, this behavior supports the view that hydropyrolysis reactions include stabilization of reactive volatiles that would otherwise be lost in cracking and carbon

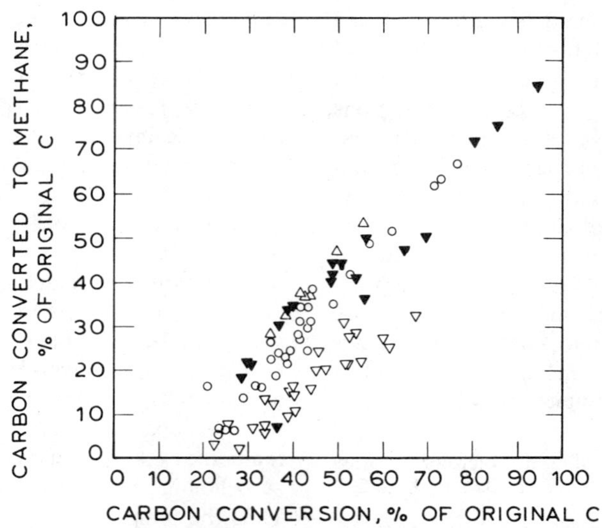

Figure 12.49 Relationship between methane yield and carbon conversion in coal hydrogasification and hydropyrolysis data of different investigators.[8] ○, Birch et al.[123]; △, Moseley and Paterson[142]; ▽ (800°C) and ▼ (1200°C), Hiteshue et al.[94]

deposition to an increasing extent as the inert gas pressure is increased. The products of this stabilization can escape the particle as volatiles.

Some insight into the nature of the volatiles stabilization phenomenon is provided by a comparison of tar yields achieved in pyrolysis and hydropyrolysis (Fig. 12.50). It is seen that the tar yields at higher temperatures in pyrolysis under helium are much larger at 1 atm than at 69 atm, which might be expected. The striking result is that tar yields produced under 69 atm of hydrogen are comparable to those under 69 atm of helium. [The indicated asymptote for tar yield under 69 atm of hydrogen or helium (12%) was established by relatively long residence time runs in which the coal was held for up to 20 s at temperatures between 850 and 1050°C.] Thus the hydrogen does not appear to stabilize tar as such, which lead Suuberg[146] to suggest the following possibilities for the yield-enhancing effect or hydrogen. The hydrogen may interfere with tar-forming reactions and stabilize tar precursors that would lead to carbon deposition in the presence of 69 atm of helium, or the hydrogen may serve to stabilize tar cracking (or hydrocracking) products. A third possibility is that the hydrogen physically interferes with the escape of tar and promotes its cracking in a manner analogous to that of the helium, with the yield enhancement being the result of unrelated (or indirectly related) chemical processes.

Summary comparisons of hydropyrolysis and pyrolysis product yields for the two coals studied by Suuberg are given in Tables 12.25 and 12.26. Table 12.25 indicates that hydropyrolysis of the lignite under 69 atm of hydrogen does give a small increase in tar yield over that obtained in pyrolysis at 1 atm. Also, the yield of light hydrocarbon liquids appears to be enhanced during hydropyrolysis, but the total yield of benzene almost never exceeds 1 wt %. The maximum total yield of tar plus hydrocarbon liquids obtained during 69 atm hydropyrolysis is approximately 10%.

In both tables it is clear that the main incremental yield due to hydropyrolysis is that of methane, with that of some other hydrocarbons also having significance. The reduced yields of carbon oxides are reasonable because reverse water gas shift and other water-forming reactions are thermodynamically favored. It is also interesting to note that as should be expected, the yield of

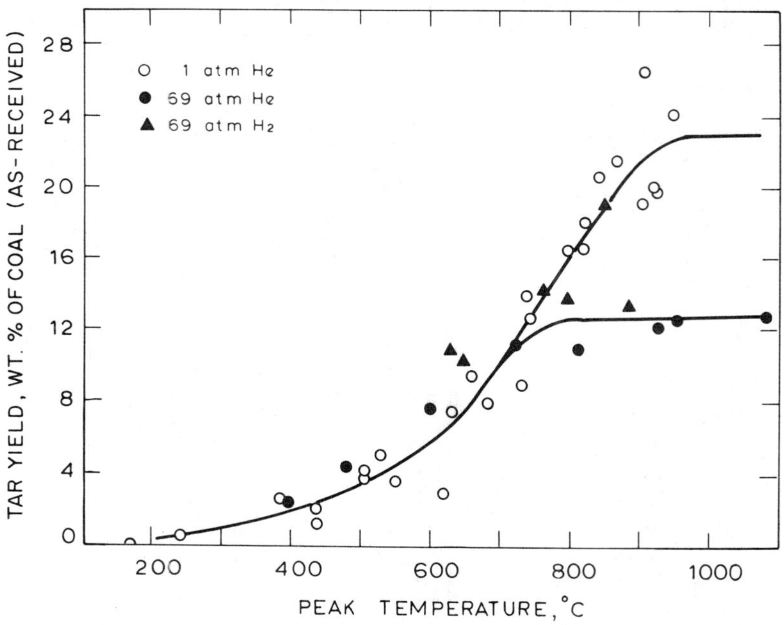

Figure 12.50 Comparison of tar yields from hydropyrolysis and pyrolysis of Pittsburgh Seam bituminous coal. Average particle diameter, 74 μm; heating rate, 1000°C/s Holding times at peak temperature: data points, zero; high-temperature asymptotes, up to 20 s (data points not shown).[146]

Table 12.25 Comparison of Yields from Pyrolysis and Hydropyrolysis of Montana Lignite[146]

Heating rate, 1000°C/s; average particle diameter, 74 μm; holding times and temperatures: pyrolysis under 1 atm He and 69 atm He, 3–10 s and approximately 1000°C, hydropyrolysis under 69 atm H_2, 10 s and 850–1000°C.

	Yield (wt % of lignite as-received)		
Product	1 atm He	69 atm He	69 atm H_2
CO	9.4	9.0	7.1
CO_2	9.5	10.6	8.5
H_20	16.5	12.9	16.0
CH_4	1.3	2.5	9.5
C_2H_4	0.6	0.6	0.2
C_2H_6	0.2	0.2	1.4
Other HC	0.8	1.7	4.1
Tar	5.4	~3	~8
Char	56.0	59.8	48.5

ethylene from hydropyrolysis is lower than that from pyrolysis, whereas the reverse is true for ethane.

The effect of temperature on methane yield is shown in Figs. 12.51 and 12.52. These data show an effect of hydrogen at temperatures as low as 600°C under the same temperature-time history for which the effect of hydrogen on weight loss is only apparent above 700–800°C. Comparison of the three sets of data in Fig. 12.51 shows that the enhanced methane yield observed during 600°C hydropyrolysis of the lignite is due to interaction of the lignite and external hydrogen, rather than being a manifestation of the autohydrogenation effect induced by the high external pressure alone, since the 69-atm helium curve does not diverge from the 1-atm helium curve until temperatures greater than 700°C are reached. In the same figure it is possible that hydropyrolysis exhibits the same two-step behavior as pyrolysis, but the hydropyrolysis curve is not well established because of the data scatter. Such behavior would seem reasonable, since it suggests that hydrogen participates mainly during the two principal hydrocarbon-forming steps.

In the case of the bituminous coal, it can be deduced from Fig. 12.52 and Table 12.26 that methane is the principal product formed during the isothermal period at the peak temperature. A total weight loss of about 41% of the coal (as received) occurs during heating under 69 atm of hydrogen to 900°C, where the methane yield shown in Fig. 12.52 is about 5%. During the isothermal period of 14–20 s, Table 12.26 indicates an additional weight loss of roughly 19% while the yield of methane has increased by roughly 18%. However the detailed weight loss data of Suuberg,[146] in agreement with those of Anthony in Fig. 12.40, show that most of the

Table 12.26 Comparison of Yields from Pyrolysis and Hydropyrolysis of Pittsburgh Seam Bituminous Coal[146]

Heating rate, 1000°C/s; average particle diameter, 74 μm; holding times and temperatures: pyrolysis under 1 atm He, 2–10 s and 850–1000°C; pyrolysis under 69 atm He, 2–10 s and 850–1070°C; hydropyrolysis under 69 atm H_2, 14–20 s and 870–930°C.

	Yield (wt % of coal, as-received)		
Product	1 atm He	69 atm He	69 atm H_2
CO	2.4	2.5	—
CO_2	1.2	1.7	1.3
H_2O	6.8	9.5	—
CH_4	2.5	3.2	23.2
C_2H_4	0.8	0.5	0.4
C_2H_6	0.5	0.9	2.3
$C_3H_6 + C_3H_8$	1.3	0.7	0.7
Other HC gases	1.3	1.6	2.0
Light HC liquids	2.4	2.0	5.3
Tar	23	12	12
Char	53.0	62.4	40.2

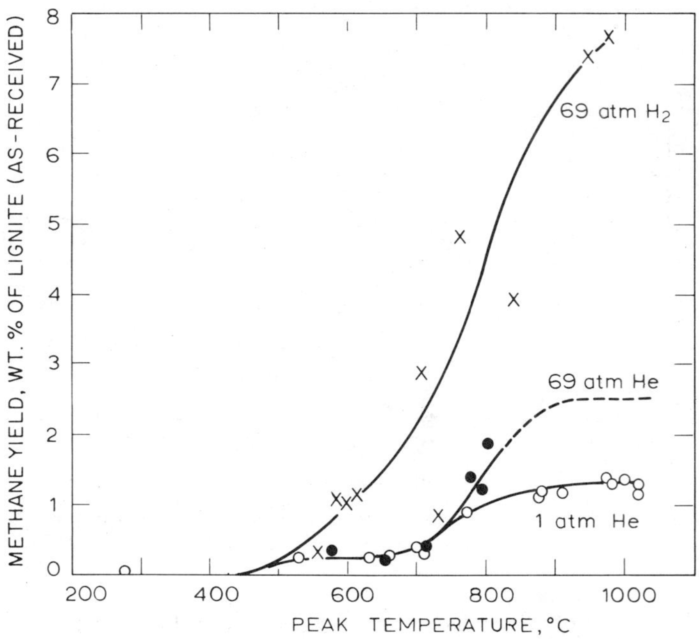

Figure 12.51 Comparison of methane yields from hydropyrolysis and pyrolysis of Montana lignite to different peak temperatures. Average particle diameter, 74 μm; heating rate, 1000°C/s; asymptote for 69 atm helium (dashed curve) established by runs with holding times of 2–10 s at the peak temperature.[146]

enhancement to weight loss occurring under 69 atm of hydrogen compared with 69 atm of helium occurs in less than 2 s for a 1000°C/s heatup to around 900°C. Since methane is the main constituent of the enhanced conversion, only a small part of the time spent in the isothermal period would appear to be sufficient to achieve most of the large methane yield shown in the table.

Interesting behavior of ethylene during hydropyrolysis, compared to pyrolysis, of the lignite can be seen in Fig. 12.53. Variation in inert gas pressure in pyrolysis seems to have little effect on ethylene yield, since the curves for 1 atm of helium and 69 atm of helium virtually coincide. Well-established two-step behavior is displayed by both sets of data. During hydropyrolysis it appears that ethylene is evolved during the first step only, in a quantity comparable to that evolved during the first step of pyrolysis, and that no further evolution occurs during the higher temperature hydrocarbon-formation stage (> 700 or 750°C). Based on this behavior and that of methane, the hydrogen apparently begins to manifest its greatest effect above 700°C. Data on evolution of hydrocarbon gases other than methane and ethylene

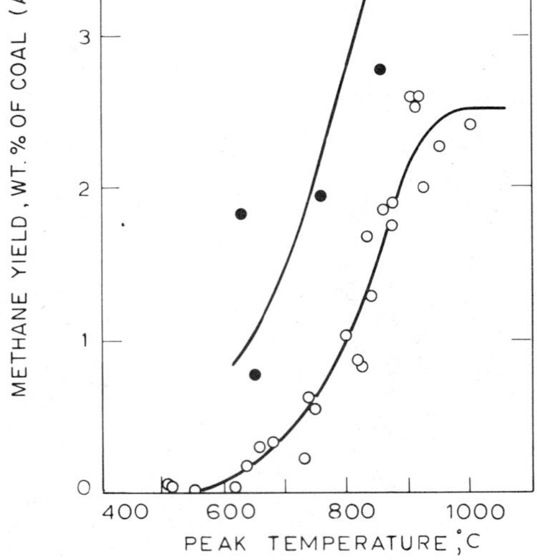

Figure 12.52 Comparison of methane yields from hydropyrolysis and pyrolysis of Pittsburgh Seam bituminous coal to different peak temperatures. Average particle diameter, 74 μm; heating rate, 1000°C/s.[146]

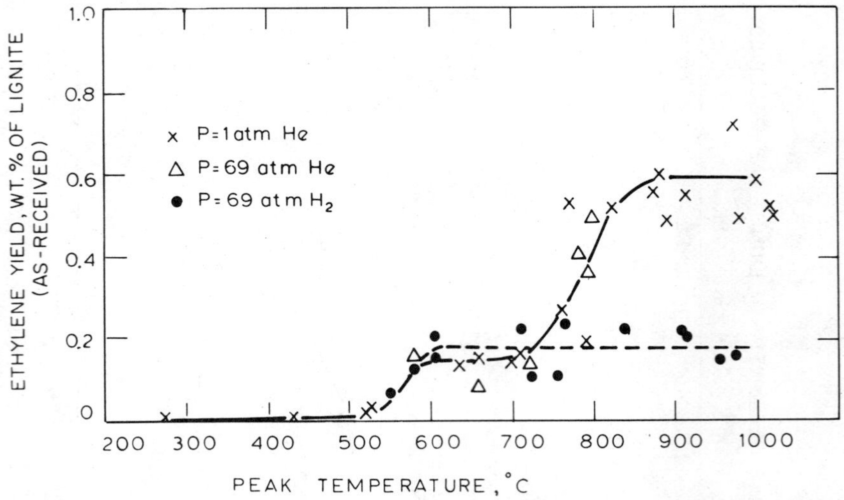

Figure 12.53 Comparison of ethylene yields from hydropyrolysis and pyrolysis of Montana lignite to different peak temperatures. Average particle diameter, 74 μm; heating rate, 1000°C/s.[146]

are shown in Fig. 12.54. Their behavior is analogous to that of methane, in that a low-temperature (< 600°C) effect of hydrogen is apparent.

The behavior of the oxygenated species in lignite hydropyrolysis is plotted in Fig. 12.55. The curves are from the lignite pyrolysis data presented earlier (points omitted). The agreement between the carbon oxide data for pyrolysis and hydropyrolysis is fair. The differences in the yields of carbon oxides from pyrolysis and hydropyrolysis as recorded in Table 12.25 appear to manifest themselves principally during the previously de-

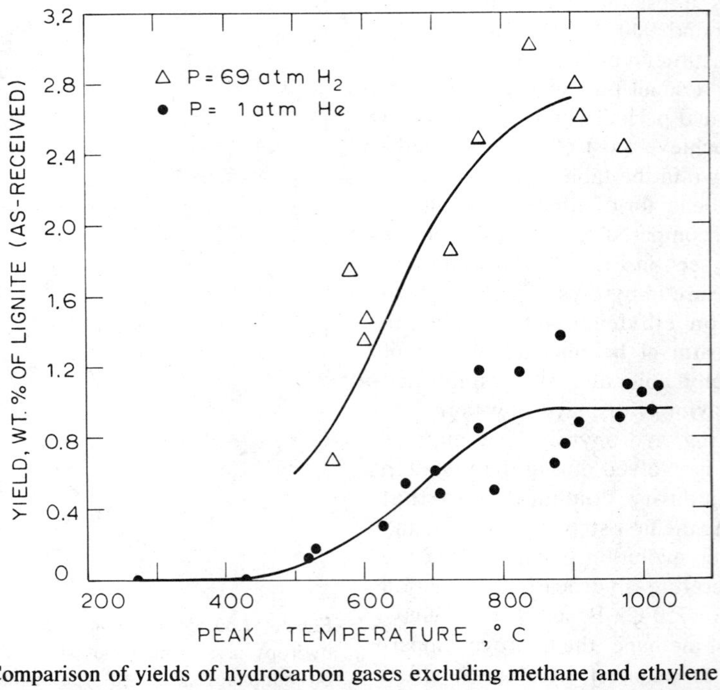

Figure 12.54 Comparison of yields of hydrocarbon gases excluding methane and ethylene from hydropyrolysis and pyrolysis of Montana lignite to different peak temperatures. Average particle diameter, 74 μm; heating rate, 1000°C/s.[146]

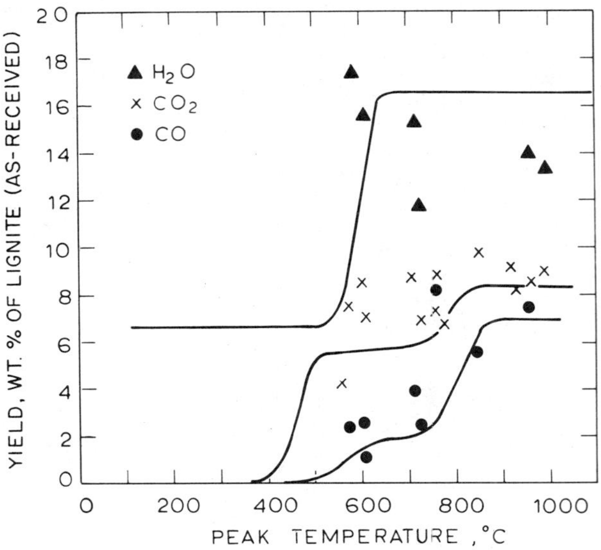

Figure 12.55 Comparison of yields of water, carbon dioxide, and carbon monoxide from hydropyrolysis and pyrolysis of Montana lignite to different peak temperatures. Data points, 69 atm of hydrogen; curves, from 1 atm of helium data (Fig. 12.14); heating rate, 1000°C/s; average particle diameter, 74 μm.[146]

scribed third, high-temperature evolution step, which is not shown.

Ultimate analyses of lignite chars produced by pyrolysis and hydropyrolysis are compared in Fig. 12.56. Carbon conversion is of course higher during hydropyrolysis than during pyrolysis. The retentions of hydrogen and oxygen (within the scatter of the latter) appear to be unaffected by the presence of hydrogen gas, implying that the mechanism of removal of these species is the same during hydropyrolysis as during pyrolysis. The volatilization of nitrogen is greatly enhanced by the presence of hydrogen; sulfur is curiously unaffected.

The effect of hydrogen on the hydropyrolysis behavior of lignite is addressed in Fig. 12.57. Data on total volatiles yield (total weight loss) and methane production are plotted as yield increments over that which could be obtained from ordinary pyrolysis at temperatures greater than about 850°C and holding times of 5 and 10 s. A value of zero on the ordinate implies a total weight loss of 44%, or a methane yield of about 1.3%. A simple functionality of pressure is clearly not implied, either for total weight loss or for methane yield. Also, the data point up how large a fraction of the incremental yield methane represents (though the actual fraction is not directly calculable, since there is no indication of what fraction of the hydrogen contained in the methane came from the coal itself). According to these data, higher temperatures are as effective as higher hydrogen pressures in increasing the yields. The total yield at 35 atm of hydrogen and 1100°C is virtually identical to that obtained at 100 atm and 850°C and not much lower than that at 100 atm and 1100°C. This behavior suggests a tradeoff of potential practical importance.

The evidence from the foregoing composition data, in agreement with the weight loss data of Anthony et al.,[109] suggests that the hydropyrolysis of coal proceeds at a rate comparable to that of pyrolysis. In view of the free radical nature of pyrolysis discussed earlier, hydropyrolysis would also appear to be a radical process whose rate is determined by the rate of initiation steps within the hydrocarbon itself. The chemistry of the overall process is changed by the probability of reactions of the type:

$$R\cdot + H_2 \rightarrow RH + H\cdot R \qquad (12.16)$$

being higher during hydropyrolysis than during pyrolysis. Hence a greater proportion of the chain carriers during hydropyrolysis will be hydrogen radicals rather than hydrocarbon radicals. This situation leads to a greater probability that species will be formed that

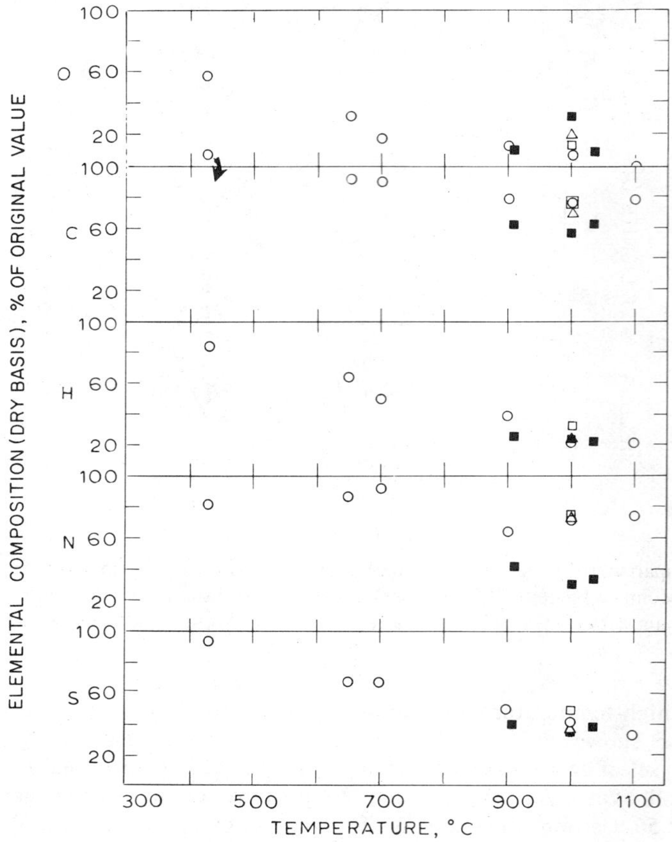

Figure 12.56 Comparison of elemental compositions of chars from hydropyrolysis and pyrolysis of Montana lignite. Average particle diameter, 74 μm; heating rate, 1000°C/s. Data points (using τ = holding time at peak temperature): ○, 1 atm of helium, τ = 0; △, 1 atm of helium, τ = 5–10 s; ■, 69 atm of hydrogen, τ = 2–30 s; □, 69 atm of helium, τ = 5–10 s.[146]

can escape the particle as volatiles, and a decreased probability of polymerization and char formation. Thus the distribution of products formed in hydropyrolysis differs from that of pyrolysis, the former having a higher ratio of light hydrocarbons to char. Mass transfer limitations to hydropyrolysis are indicated by the substantial effects of pressure and particle size in the case of Pittsburgh Seam bituminous coal, but such effects are minor in the case of Montana lignite. The latter observation does not exclude the possibility of mass transport limitations on a micro scale within the particle, nor can kinetic limitations be excluded in either case.

12.6 PYROLYSIS IN PLASMAS, SHOCK WAVES, AND FLASH RADIATION

Several additional studies that employed a range of rapid heating techniques and different atmospheres provide valuable insights into pyrolysis behavior. The studies tended to be exploratory and, for the most part, were focused on the primary products under the imposed conditions.

Banerjee et al.[215] investigated the materials formed by pyrolyzing coal in an electric furnace at 1150°C. A layer of powdered coal was spread uniformly along the bottom of an evacuated silica tube, which was introduced into a preheated furnace for 30 s. After withdrawal and cooling, a gas sample was obtained and analyzed by infrared spectros-

[215] N. N. Banerjee, G. S. Murty, H. S. Rao, and A. Lahiri, "Flash Pyrolysis of Coal," *Fuel,* **52,** 168 (1973).

copy for carbon oxides, light hydrocarbons, and hydrogen cyanide. Several runs were conducted under low pressures of nitrogen, argon, or ammonia, and the results were compared to those obtained under vacuum. The presence of inerts generally resulted in increased yields of unsaturated hydrocarbons, particularly ethylene, propylene, and acetylene, but the mechanism of this effect is not clear. The heating time of this experiment is probably long enough to permit significant reaction of the volatiles in the presence of hot surfaces.

Woodburn et al.[76] rapidly heated a dilute suspension of 40-μm-diameter bituminous coal particles in a shock tube. This method of heating exposes the ambient gas around the particles to an accurately calculated temperature-time history, including essentially a step increase to the experimental temperature and rapid cooling resulting from the expansion wave that follows the shock. Heating and subsequent cooling of a coal particle are basically by conduction through the gaseous boundary layer surrounding the particle. The wide separation of particles in the suspension allows the volatiles from a given particle to be diluted by the ambient gas without significant contact with other particles.

Volatile product distributions were obtained for the shock-heated coal particles under pressurized (4–37 atm) mixtures of argon, hydrogen, and iodine "catalyst." Calculated solids reaction times and temperatures ranged from 1.3 to 2.0 ms, and 475 to 875°C, respectively. Gaseous products up through the C_4 hydrocarbons were analyzed by chromatography, and yields but not the identity of benzene-extractable liquids were studied. Of most interest in the context of primary volatiles are the observations of increasing acetylene yield with increasing pyrolysis temperature, and of shifts to more methane at the expense of unsaturates (including acetylene) at higher hydrogen partial pressures. Acetylene is not favored thermodynamically at the relatively low temperatures of this study. Whether the observed acetylene stems from primary volatiles or from dehydrogenation of more saturated primary species is not clear. A similar question is whether the increase in methane yield with hydrogen pressure results from hydrogen stabilization of primary methyl radicals or from hydrogenation and cracking of more complex species.

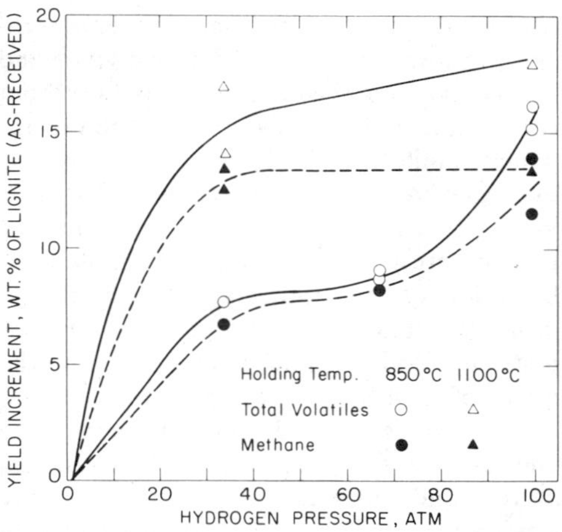

Figure 12.57 Effect of hydrogen pressure and temperature on incremental yields of methane and total volatiles in Montana lignite hydropyrolysis relative to pyrolysis. Average particle diameter, 74 μm; heating rate, 1000°C/s; holding time at peak temperature, 5–10 s.[146] Reference case (pyrolysis under 1 atm of helium at 850 or 1100°C): total volatiles = 44%; methane = 1.3%.

Volatiles generated by the pyrolysis of small coal particles in a plasma jet, formed by passing a high, low-voltage current through the gas-filled region between two concentric electrodes, were studied by Anderson et al.,[77] Bond et al.,[78] and Graves et al.[79] The coal was rapidly pyrolyzed as it passed in entrained flow through the plasma zone where temperatures, calculated by thermodynamic methods, were reported to be in the range 3000–15,000°C. The pyrolysis was carried out in argon, argon-hydrogen, argon-nitrogen, and argon-hydrogen-nitrogen plasmas. Gas chromatographic analysis of pyrolysis products showed in all cases that acetylene was the principal hydrocarbon product. Acetylene yield was decreased by quenching the reaction product mixture with a water spray immediately downstream of the plasma jet, and increased by the presence of hydrogen in the plasma mixture. The latter observation does not necessarily conflict with the opposite trend from the shock tube work of Woodburn et al., since acetylene is thermodynamically more stable than methane at the plasma temperature, whereas the opposite is true for the shock tube temperatures. Methane is known to react to form acetylene and hydrogen under plasma conditions,[77,79,216] but the reaction involves dehydrogenation, which should be slowed by the presence of hydrogen. A tenable explanation is that acetylene is not a primary pyrolysis product and that the presence of hydrogen promotes stabilization of free radicals among both the primary volatiles and the products of condensation and cracking reactions. Thus the effect of hydrogen on the pathways of secondary reactions could lead to more methane and less acetylene at the shock tube temperatures, and more acetylene (and probably less soot or other solid carbon) at the plasma temperatures. On the other hand, these data alone do not permit the exclusion of the possibility that some acetylene is formed as a primary pyrolysis product.

With regard to the observation of primary pyrolysis products, both the shock tube and the plasma technique provide rapid heating but have a disadvantage that arises from the indirect heating of the coal by conduction from hot gas surrounding the particles. This arrangement presents an excellent opportunity for the volatiles to react in the surrounding gas. Several radiative heating techniques have been used that eliminate the need for a gaseous heat transfer medium, yet permit the use of a gas when desired for other reasons, since most gases are virtually transparent to common forms of radiant energy. Rau and Eddinger[217] enclosed a 1-cm cube of coal in an evacuated flask and exposed the sample to the radiant flux from an arc image furnace for periods of up to 2 s. Pyrolysis products were identified by mass spectrometry and found to consist primarily of hydrogen and carbon oxides. The principal hydrocarbons were acetylene and methane, the relative yield of the latter being intermediate between the large yields observed in the low-temperature work of Woodward et al.[76] and the smaller yields reported from the high-temperature work of Anderson et al.[77] The data are consistent with the picture that acetylene formation is a secondary reaction. During the course of a run, the surface of the sample cube became white-hot, thus producing a high-temperature region conducive to acetylene formation. Lower temperatures in the remainder of the sample, however, might have permitted methyl or other radical formation and stabilization without cracking to acetylene to form the observed saturated products. This method of heating could be very useful in the study of primary volatiles if a dilute suspension of small coal particles were used instead of the solid cube to reduce the extent of cracking.

In a novel technique for promoting the yield of a valuable product instead of undesirable products of cracking, Kim[218] dispersed a pulverized high-volatile bituminous coal (44–88 μm particle diameter; analyses in Table 12.7) with finely divided CaO (-8 μm or -44 μm particle diameter) in hydrogen at 1 atm, and passed the suspension through a magnetically rotated arc. The residence time in the hot (several thousand degrees) zone was less than 10 ms. About 90% of the coal carbon was converted to calcium carbide (CaC_2), which remains stable during cooling to room temperature. Calcium carbide can of course be converted to acetylene by reaction with water in a subsequent step. When methane was used instead of coal and hydrogen, even a higher conversion of feed carbon to calcium carbide was obtained. The data indicate that free radicals formed from the coal in the presence of hydrogen at the very high temperatures can be captured by reaction with CaO, thereby avoiding soot formation, which otherwise would be extensive during cooling.

Several groups of workers have irradiated coal samples with high-intensity light pulses produced by capacitive discharge through a xenon flashtube. Extremely small discharge times (1 ms) may be obtained and energy fluxes capable of raising particle temperatures to 1000–1500°C within this time interval are easily realized.[80,81]

Sharkey et al. placed 150 mg of coal particles ground to pass a 74-μm screen into an approximately 15 cm long × 1 cm diameter tube, which was then evacuated. Rotation of

[216] H. W. Leutner and C. S. Stokes, "Producing Acetylene in a Plasma Jet," *Ind. Eng. Chem.*, **53**, 341 (1961).

[217] E. Rau and R. T. Eddinger, "Decomposition of Coal in an Arc-Image Furnace," *Fuel*, **43**, 246 (1964).

[218] C. S. Kim, "Formation of CaC_2 from CaO and 'Nascent' Carbon Species in a Rotating-Arc Reactor," Sc.D. thesis, Dept. of Chemical Engineering, Massachusetts Institute of Technology, Cambridge, Mass., 1977.

the tube prior to further treatment resulted in the deposition of about 11 mg of coal in a thin film on the tube walls; the remaining 140 mg presumably fell to the bottom of the tube. The tube was then placed horizontally into the center of a small helical flashtube and irradiated with the light produced by a 4000-J, 1-ms discharge. Although xenon radiates profusely in the ultraviolet, and such radiation is known to be photochemically catalytic, neither the tube material nor its transmissivity in the ultraviolet was reported. The products obtained were fractionated in a series of baths at different temperatures, analyzed by mass spectrometry, and compared to those observed under furnace and laser pyrolysis conditions. Hydrogen and carbon monoxide were the dominant products, and acetylene was the major hydrocarbon. When compared with data from both higher and lower temperature techniques, the acetylene and methane yields associated with the intermediate temperatures of flash pyrolysis are consistent with the trend of acetylene increasing and methane decreasing with rising temperature.

Anderson et al.[77] employed a similar but somewhat more refined technique. Coal particles of 10–40 μm diameter were deposited on the inner walls of evacuated Pyrex tubes and irradiated with a 5600-J pulse of 2–3 ms duration. Use of Pyrex, which is largely opaque to ultraviolet radiation, presumably reduced the extent of photolytic decomposition of volatiles. Nevertheless, carbon deposits found within the reactor indicate that some thermal or photochemical cracking did occur. The limited product characterization in this study consisted of the identification of two products, "Y" and "B," which coated the tube. Product Y was a yellowish, acetone-soluble, tarry substance formed in nonoxidizing atmospheres; it appeared to have a substantial content of saturated hydrocarbons. The acetone-insoluble B, on the other hand, was invariably formed under any pyrolysis conditions, and seemed to be a black carbonaceous material, somewhat aromatic in character. The infrared spectra of the two products, as well as that of the original coal, were compared and used to calculate the ratio of aromatic to aliphatic hydrogen present in each compound: Y was determined to be significantly less aromatic than either B or the parent coal.

The formation of the tarry substance Y is interesting because pyrolysis of coal under conditions such as these often appears to produce only gases or light liquids as volatile products.[78,80,81,84,85,217] These different apparent behaviors probably reflect different extents of tar conversion to lighter products and soot or char by cracking under the high-temperature conditions usually associated with these techniques.

Rau and Seglin[81] flash pyrolyzed coal samples with pulses between 800 and 3200 J, with corresponding flash durations of 0.2–0.8 ms. The samples consisted of approximately 10 mg of 5–10 μm diameter coal particles dusted onto the walls of an evacuated quartz tube. Following irradiation, the pressure inside the tube was measured and a gas sample was analyzed for hydrogen, carbon oxides, and light hydrocarbons. Several different coals were studied, and the results were compared on the bases of ASTM volatile content and flash energy. Some runs were conducted under low pressures (10–160 mm Hg) of hydrogen, nitrogen, or water vapor. Rapid quenching of gas phase reactions was attempted in some runs by suspending a thimble of liquid nitrogen or dry ice–acetone within the reactor prior to the irradiation. Only small increases in acetylene yield at the expense of saturates such as methane were observed with increases in flash intensity, and the acetylene yield actually decreased at higher hydrogen pressures. The latter trend is similar to the one observed in shock tube work and opposite to the plasma results, both discussed above. This implies that the effective temperature of the present flash technique is relatively low compared with that of the plasma systems, which is consistent with the stated characteristics of these experiments. The acetylene yields reported by Rau and Seglin are larger than those of the previous flashtube experiments. This result may be attributable to photolytic reactions induced by the high ultraviolet flux through the quartz reactor. A carbonaceous residue was found on the reactor walls, and even on the cooling thimble when present, indicating the occurrence of some gas phase reaction, which seems unlikely to have been thermal. Nitrogen, introduced into the reactor as an "energy absorber" for any free radicals produced, had no detectable effects on product distribution. Steam, however, produced results similar to those observed in the presence of hydrogen.

Granger and Ladner[82] investigated the flash pyrolysis of coal under a number of different conditions, including both inert and oxygen-containing atmospheres. Their technique involved dusting about 1 mg of finely ground coal (1–12 μm particle diameter) onto the walls of a silica tube approximately 50 mm long $\times$ 8 mm diameter. After evacuation and drying for 24 hr, the sample tube was pressurized to slightly above 1 atm with argon or an argon-oxygen mixture. Both the sample container and a cylindrical flashtube were then enclosed with a spherical reflective shell, and the flashtube fired with a 1500-J, 1-ms pulse to pyrolyze the sample. In most of the experiments the flashtube was surrounded prior to discharge by a red glass filter, which presumably reduced substantially the ultraviolet flux through the sample container. Product gases were analyzed by gas chromatography for light hydrocarbons, hydrogen, and carbon oxides. The solid residues were extracted with chloroform to determine tar yields by a light absorption technique applied to the resulting solutions. Use of the red glass filter resulted in an increase of light, saturated hydrocarbons and in a decrease of acetylene as compared with results obtained without the filter. Furthermore, carbon deposition on the reactor walls, which was observed in unfiltered runs, was eliminated by the presence of the filter. These observations demonstrate the occurrence of photochemical cracking reactions under high-intensity ultraviolet radiation, as mentioned above. Large quantities of tars were formed in these experiments. The authors indicate that sample temperatures were probably lower than those achieved by other flashtube workers; particle temperatures may have peaked at only 375–525°C. This factor might explain at least part of the increased yield of tar, since the extent of tar conversion to solids and lighter products would be reduced. The acetylene yield increased with increasing flash energy, but no further information was obtained on the nature of acetylene production.

Several groups have reported research on the pyrolysis of small quantities of coal by laser irradiation. Workers at the Bureau of Mines[80,84] focused the output of a 1.5–2 J ruby laser onto small solid blocks of coal, either *in vacuo* or under small pressures of argon, nitrogen, hydrogen, or helium, and analyzed the resulting gas mixture by mass spectrometry. Laser pulse duration in each experiment was on the order of 1–2 ms. The temperatures reached by the solid coal samples were not determined directly, but were estimated to be in the vicinity of several thousand degrees. Material not in the direct path of the beam naturally did not reach the maximum temperature, but the temperature distribution in the sample block was not estimated. As might be expected, acetylene was the primary hydrocarbon detected. Also observed were significant amounts of polyacetylenes, which are known pyrolysis products of acetylene, therefore indicative of secondary reactions among the evolved species. Increases in inert gas pressure over a series of runs led to increases in acetylene production. This observation lends further support to the hypothesis that acetylene is a secondary decomposition product. The increase was most pronounced under argon and nitrogen atmospheres. Acetylene production actually decreased under small pressures of helium. The authors attribute this anomaly to the high thermal conductivity of helium but do not explain why hydrogen—with an even higher thermal conductivity—appears to have no effect on acetylene yield, at least at the pressures reported. It seems doubtful that the high thermal conductivity of hydrogen and its capacity, if any, to somehow increase acetylene yield would exactly cancel under the reaction conditions in question, although this possibility cannot be excluded.

The most interesting work in the area of laser pyrolysis has been performed by Joy et al.[85] in a study specifically concerned with detection of primary free radical species formed during pyrolysis. These workers dusted a thin layer of 10-μm-diameter coal particles onto the end of a small quartz rod, which was inserted directly into the source chamber of a time-of-flight mass spectrometer. The sample was irradiated *in situ* for about 0.7 ms with an 0.8-J ruby laser pulse; the corresponding energy flux at the sample was as high as 800 kJ/m^2, although two lower flux levels were also utilized by optically spreading the laser beam with a system of lenses. Mass spectrometric analyses of the escaping volatiles were continuously performed for about 10 ms.

Irradiation of a sample at the lowest energy flux level produced a gas mixture rich in carbon monoxide, water, and saturated hydrocarbons, principally methane. Some higher hydrocarbons, including light aromatics, were also produced by pyrolysis under these conditions. Increasing the incident energy flux led to the production of hydrogen and of metal ions, primarily sodium. Of greater interest, significant quantities of acetylene and its polymers appeared under these conditions. Some evidence of methyl radical formation was observed at this energy level as well. Further increasing the energy flux to the maximum value indicated above led to the production of atomic hydrogen and yields of methyl radicals exceeding those of methane. Large yields of acetylene and polyacetylenes were also observed. Aromatic production tended to decrease from the level observed at lower radiation intensities. Some carbon vapor was detected, indicating the likelihood of thermal cracking reactions.

The coal pyrolysis runs were complemented by similar experiments involving the pyrolysis of certain high molecular weight model compounds in place of the coal particles. Solid 2-methyl anthracene and pyrene were ground with fine particles of carbon black, to increase total absorption of incident radiation during heating. The mixtures were then pyrolyzed under high-intensity irradiation. This treatment led primarily to the evaporation of the two materials. Some cracking and decomposition were observed, although the authors do not explain this observation in detail. Methyl radicals, however, were formed in large yield from the pyrolysis of 2-methyl anthracene. From the published spectra, it seems reasonable to conclude also that the pyrolysis of pyrene gave considerable yields of light products, but the authors did not identify them.

Clearly the work of Joy et al. constitutes a significant step in the study of primary volatiles. At some point between the low and intermediate energy levels, production of true primary volatiles might have been observed. At the medium and high levels, however, decomposition of the primary products—as evidenced by the appearance of acetylene and polyacetylenes—undoubtedly occurred. On the other hand, the lowest irradiation level used may have been inadequate to permit the escape of true primary volatiles. Greater attention to this energy region would have been invaluable, as would estimates of temperatures achieved at each of the three energy flux levels. Although the authors conclude that "except at very high temperatures it is sufficient to consider molecular entities only and to ignore radical species," the definition of "high" and the nature of the key molecular entities still need clarification.

NOMENCLATURE

a	constant (kcal/mole)
a'	dimensionless constant
A	dimensionless constant
B	dimensionless constant
C	vapor phase concentration of reactive volatiles within the coal particle, fraction of original coal mass
C_∞	vapor phase concentration of reactive volatiles in the ambient gas, fraction of original coal mass
C^*	dimensionless concentration of active sites
C_0^*	value of C^* at $t = 0$
d	particle diameter (mm)
E	activation energy for pyrolysis (kcal/mole)
E_0	mean of activation energy distribution (kcal/mole)
E_t	$RT \ln(k_0 t)$
E_1	maximum activation energy of reactions forming nonreactive volatiles, and minimum activation energy of reactions forming reactive volatiles (kcal/mole)
E_1–E_6	activation energies of free radical reactions (kcal/mole)
E_s	activation energy for active site formation (kcal/mole)
$f(E)$	activation energy distribution function for pyrolysis (mole/kcal)
$f_s(E_s)$	activation energy distribution function for active site formation (mole/kcal)
k	first-order rate constant for pyrolysis (s^{-1})
k'	nth-order rate constant for pyrolysis (g^{1-n}/s)
k_0	preexponential factor (s^{-1})
k_{oo}	mean of k_0 distribution (s^{-1})
k_1	overall rate constant for solid deposition reaction (s^{-1})

k_2 overall rate constant for hydrogenation of reactive volatiles (s^{-1} atm^{-1})

k_3 $k_4C_0^*/k_5$

k_4 rate constant for rapid-rate hydrogasification (atm^{-1} s^{-1})

k_5 rate constant for decay of active sites (s^{-1})

k_1–k_6 rate constants of free radical reactions (s^{-1} or cm^3/mole-s)

k_c pressure-independent overall mass transfer coefficient (atm/s)

k_s first-order rate constant for active site formation (s^{-1})

K_c overall mass transfer coefficient ($K_c = k_c/P$) (s^{-1})

K_v constant (s/m^2)

m constant heating rate (°C/s)

n reaction order

P pressure (atm)

P_{H_2} hydrogen pressure (atm)

Q ratio of total weight loss to loss of proximate volatile matter

R ideal gas constant [cal/(mole) (K)]

t time (s)

t_v time required for devolatilization (s)

T absolute temperature (K)

T_a ambient temperature (°C)

T_0 base temperature (K)

V total volatiles lost from particle up to time t (fraction of the original coal weight)

V_{nr} nonreactive volatiles lost from particle up to time t (fraction of the original coal weight)

V_r reactive volatiles lost from particle up to time t (fraction of the original coal weight)

V' total volatiles formed up to time t (fraction of the original coal weight)

V'_{nr} nonreactive volatiles formed up to time t (fraction of the original coal weight)

V'_r reactive volatiles formed up to time t (fraction of the original coal weight)

V^* total volatiles lost from particle up to $t = \infty$ (ultimate yield of total volatiles), approximated by measurements at long reaction times (fraction of the original coal weight)

V^*_{nr} nonreactive volatiles lost from particle up to $t = \infty$ (ultimate yield of nonreactive volatiles) (fraction of the original coal weight) (note: $V^*_{nr} = V^{**}_{nr}$)

V^*_r reactive volatiles lost from particle up to $t = \infty$ (ultimate yield of reactive volatiles)

V^{**}_r reactive volatiles formed up to $t = \infty$ (potential ultimate yield of reactive volatiles) (fraction of the original coal weight)

V^*_{cum} sum of V^*_i for all i with $E_i \leq E$ (fraction of the original coal weight)

V^*_{tot} sum of all V^*_i (fraction of the original coal weight)

V^{**} total volatiles formed up to $t = \infty$ (potential ultimate yield of total volatiles) (fraction of the original coal weight)

VM volatile content as measured by proximate analysis (fraction of original coal weight)

VM_c proximate volatile matter in char at $t = \infty$ (fraction of initial proximate volatile matter)

X conversion up to time t of original carbon material (carbon, coal, or fixed carbon) to products in rapid-rate hydrogasification or hydropyrolysis (fraction of the original weight of carbon material)

X^* total rapid-rate carbon material initially in coal (fraction of original weight of carbon material)

GREEK

α mass or thermal diffusivity (m^2/s)

$\Delta \log k_0$ standard deviation of $\log k_0$ distribution (dimensionless)

σ standard deviation of activation energy distribution (kcal/mole)

τ holding time at peak temperature (s)

χ mole fraction in a gas mixture

ADDITIONAL SUBSCRIPTS

i denotes one member of a large group of independent parallel pyrolysis reactions; or one compound

j denotes a single volatile species

max maximum value

min minimum value

CHAPTER 13

Low-Temperature Pyrolysis Technology

LEONARD SEGLIN AND SIDNEY A. BRESLER

Leonard Seglin *received his B.Ch.E. degree from New York University in 1938 and his A.M. in mathematics from Harvard University in 1939. Before joining Bechtel Corporation, he was Director of Process Research and Development for FMC Corporation during their development of the COED Process. Mr. Seglin is now President of Econergy Associates and an adjunct professor of chemical engineering at Manhattan College in Riverdale, New York. He holds 20 patents and is author or coauthor of 18 technical articles.*

Sidney A. Bresler, *President of Bresler and Associates, is a consulting chemical engineer. Mr. Bresler's principal interests are in chemical plant design, including coal utilization processes and methods of reforming the products of coal liquefaction and gasification into commercially usable substances.*

13.1 INTRODUCTION

Coal was once one of the few natural resources that could be used directly with little or no processing to change it from its natural form, but the march of civilization has altered that picture. Coal's direct use is disappearing and will be replaced in the future more and more by complex physical and chemical processing directed toward converting coal to more ecologically acceptable and/or more convenient forms.

Ecological motivation is now the driving force for removal of sulfur and particulates either from the coal before its use or from the coal combustion products (stack gas) after its use.

The rapid depletion of crude oil and natural gas, especially in the United States, and modern civilization with its emphasis on convenience, are now the main stimuli for converting the dirty black rock called coal to liquid fuels, clean fuel or pipeline gas, and more valuable chemicals.

Two of the methods for accomplishing some of these ends are reviewed in this chapter, namely, low-temperature carbonization and low-temperature hydrocarbonization processes. For these generic processes "low-temperature" means to temperatures less than 800°C.

The principal problems faced by these processes are how to do the following:

- Provide the necessary process heat.
- Handle the caking tendencies of the solids (coal).
- Separate the resulting three phases—gas, liquid (tar), and solids (char).
- Upgrade each of these phases to useful products.
- Most important, deal with toxic products and by-products.
- Accomplish all these tasks efficiently and economically.

Carbonization of coal is basically accomplished by heating it to a temperature at which the coal is broken down chemically and/or physically to simpler components, thereby converting the coal to a tar, and yielding gas, water, and a residual solid char. Low-temperature carbonization is typified by the following processes, which are discussed further in later sections:

- The FMC (Food Machinery Corporation) coke process.
- The Char Oil Energy Development (COED) process.
- The U.S. Steel Clean Coke process.
- The Occidental pyrolysis process (formerly the Garrett pyrolysis process).
- The Lurgi-Ruhrgas process.
- Work by Consolidation Coal.
- The work of the Commonwealth Scientific and Industrial Research Organization (CSIRO) of Australia.

The initial change that occurs in carbonization is the breakdown of the coal substance to simpler, distillable material. This breakdown is a function of the type of coal and of temperature, which, in turn, is directly dependent on the energy transferred to and into the solids. Therefore the basic controlling factor in these processes is either the rate of transfer of heat into the coal solids, or the rate of the decomposition reactions. The finer the grind of the coal, the more likely it is that the reaction rate will be controlling.

Hydrocarbonization of coal is accomplished in the same basic way as carbonization, with the added feature that hydrogen is present at a sufficient pressure to allow the mildly exothermic reaction of hydrogen with the organic (carbon) substrate in the solids, thereby enhancing yields of liquids and gases at the expense of solids (char). Hydrogen also stabilizes the reaction products by reacting with polymerizable species such as free radicals and olefins, which are the initial products of reaction. Hydrocarbonization is typified by the following processes, which are described and discussed further in later sections:

- The Coalcon process.
- The Schroeder process.
- The Cities Service short residence time (CS-SRT) hydropyrolysis process.

Since heterogeneous chemical reactions between species are involved in these processes, the basic controlling factor is either the rate of reaction of hydrogen with the organic substrate, or the rate of diffusion of hydrogen into—or of products out of—the coal-char substance. The finer the grind of the coal, the more likely it is that the reaction rate will be controlling. The use of catalysts would tend to shift toward diffusion being the controlling factor. Some processes have considered the use of catalysts.

Low-temperature carbonization is an old art. Developments and practice prior to 1962 are discussed in the first Supplementary Volume,[1] which reviewed some 41 different retort processes and 15 entrained or fluidized-bed processes. Since then, there have been essentially no new developments on retort systems. Practically all new developments have been based on entrained (transport) and fluidized-bed systems, plus significant activity in a new area—hydrocarbonization.

The retort processes ranged from batch to continuous, from vertical to horizontal, from complicated mechanical devices to relatively simple ones. In general, they all had one characteristic in common—their unsuitability for the construction of the huge-capacity units that may be required to achieve the economies of scale necessary to make synthetic fuels from coal a viable operation. Table 13.1 presents a summary profile of the

[1] *Chemistry of Coal Utilization,* Supplementary Volume, H. H. Lowry, Ed., John Wiley and Sons, New York, 1963, pp. 436–437.

Table 13.1 Summary Profile of Retort Processes for the Low-Temperature Carbonization of Coal

Type	Method of Heating	Number of Processes	Type of Coal
Vertical—batch	Indirect	8	Caking
Vertical—batch	Direct	3	Caking
Vertical—continuous	Indirect	7	Noncaking
Vertical—continuous	Partially indirect	3	Noncaking
Vertical—continuous	Direct	3	Noncaking
Horizontal—continuous	Indirect	5	Caking
Horizontal—continuous	Direct	3	Caking
Horizontal—conveyor, continuous		7	Caking
Horizontal—fixed, continuous		2	Caking
		41	

Source: H. H. Lowry, Ed., *Chemistry of Coal Utilization,* Supplementary Volume, John Wiley and Sons, New York, 1963, Ch. 10.

retort processes described in the first Supplementary Volume.

Hydrocarbonization is of more recent origin. The early investigations on it date back to the 1950s and 1960s,[2–13] and active process development appears to have been started by Union Carbide in 1960.[14] The trend in these developments has been toward continuous processing systems using either fluidized or entrained beds. For the same coal, agglomeration of solids is more severe in hydrocarbonization than in carbonization. Controlling the agglomeration has been one of the principal development tasks.

13.2 THERMODYNAMIC DATA

The thermodynamic properties of coal, coal carbonization, and coal hydrocarbonization systems are known only approximately. This is not surprising, considering the complex and diverse nature of the mixture that is coal.

An overall compilation of available data on coal carbonization was made by the Institute of Gas Technology (IGT).[15] The data presented below are based mainly on that study.

The sensible heat of the moisture- and ash-free (maf) coal solid above 77°F (25°C) is a function of temperature and the initial volatile matter of the coal, and is presented in Fig. 13.1, based on the IGT work.[15] The following data and correlations are used (see end of chapter for definition of terms used in equations):

- The evolution of volatile matter is approximated by the Gregory and Littlejohn correlation[16]:

$$W_v = 3.162 \times \frac{10^9}{t^{3.914}} \tag{13.1}$$

- Primary volatile matter is defined as that in excess of 10% of the maf coal. Its heat capacity is:

$$C_{v1} = 0.395 + 0.00081t \tag{13.2}$$

- Secondary volatile matter is defined as up to the first 10% of the maf coal evolved; if the total volatile matter is less than 10%, there are no primary volatiles. The heat capacity of the secondary volatiles is:

$$C_{v2} = 0.710 + 0.00061t \tag{13.3}$$

- The heat capacity of the fixed carbon is:

$$C_f = 0.165 + 0.00068t - 0.00000042t^2 \tag{13.4}$$

- Only the heat required for heating the solid is included, and additional sensible heat of the volatile matter above the temperature at which it is driven off is not included.

The sensible heat of the coal ash fraction is given in Fig. 13.2, based on:

$$C_a = 0.18 + 0.00014t \tag{13.5}$$

which is the average of the heat capacities of typical coal ash components.[15]

The heat of carbonization of coal is normally endothermic below 1200°F (649°C), corresponding to decomposition reactions, and is normally exothermic above 1200°F (649°C), corresponding to condensation reactions. The net heat of carbonization is exothermic by 25–68 Btu/lb (58–158 kJ/kg) maf coal.[15] The heat of carbonization cannot

[2] E. L. Clark, M. G. Pelipetz, H. H. Storch, S. Weller, and S. Schreiber, *Ind. Eng. Chem.*, **42**(5), 861–865 (1950).

[3] R. W. Hiteshue, R. B. Anderson, and M. D. Schlesinger, *Ind. Eng. Chem.*, **49**(12), 2008–2010 (1957).

[4] R. W. Hiteshue, R. B. Anderson, and S. Friedman, *Ind. Eng. Chem.*, **52**(7), 577–579 (1960).

[5] P. S. Lewis, H. H. Ginsberg, and R. W. Hiteshue, U.S. Bur. Mines Rep. Invest. No. 5908, 1961.

[6] R. W. Hiteshue, S. Friedman, and R. Madden, U.S. Bur. Mines Rep. Invest. No. 6027, 1962.

[7] R. W. Hiteshue, S. Friedman, and R. Madden, U.S. Bur. Mines Rep. Invest. No. 6125, 1962.

[8] K. C. Channabasappa and H. R. Linden, *Ind. Eng. Chem.*, **48**(5), 900–905 (1956).

[9] K. C. Channabasappa and H. R. Linden, *Ind. Eng. Chem.*, **50**(4), 637–644 (1958).

[10] S. Friedman, R. W. Hiteshue, and M. D. Schlesinger, U.S. Bur. Mines Rep. Invest. No. 6470, 1964.

[11] W. C. Schroeder, U.S. Pat. 3,030,297, April 17, 1962.

[12] W. C. Schroeder, U.S. Pat. 3,152,063, October 6, 1964.

[13] R. T. Eddinger, J. F. Jones, J. F. Start, and L. Seglin, Coal Science Conf., Prague, Czechoslovakia, June 1968.

[14] H. G. Davis, Union Carbide Corporation, private communication, August 1976.

[15] Institute of Gas Technology, *Preparation of a Coal Conversion Systems Technical Data Book*, for U.S. ERDA, Rep. No. FE-1730-21, 1976.

[16] D. R. Gregory and R. F. Littlejohn, *BCURA Month. Bull.*, **29** 173–180 (1965).

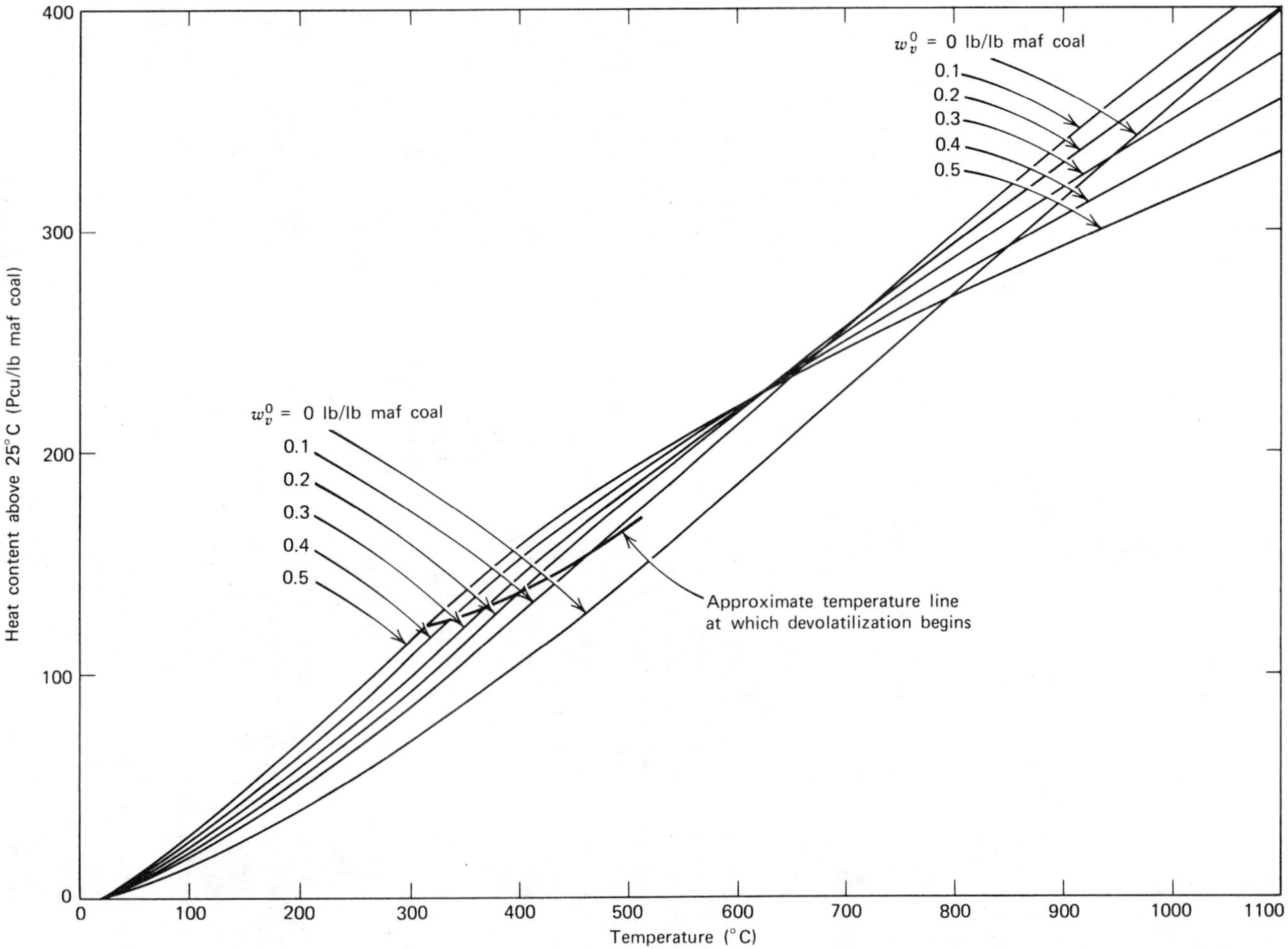

Figure 13.1 Heat content of solids from maf coal above 25°C for various amounts of initial volatile matter, (W_v^0). To convert Pcu/lb to Btu/lb, multiply scale by 1.8.

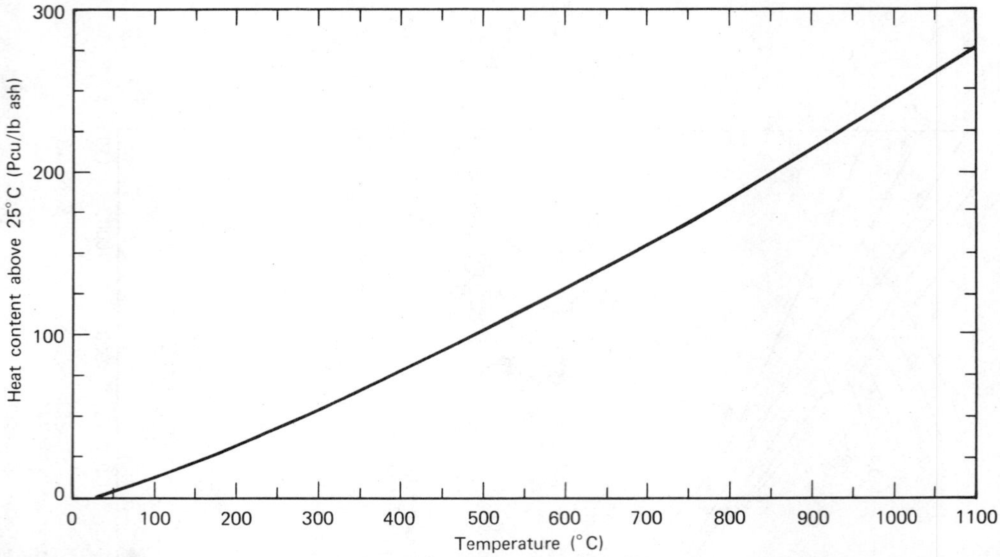

Figure 13.2 Heat content of typical ash in coal above 25°C. To convert Pcu/lb to Btu/lb, multiply scale by 1.8.

be included at this time because of the low quality of the available data and because there are indications that the heat of carbonization is small compared to the total sensible and latent heat effects.

The total energy required to heat the coal solids as received from 77°F (25°C) to some higher temperature *t* is then taken as the sum of the following:

- The heat content change of the maf coal from Fig. 13.1.
- The heat content change of the coal ash from Fig. 13.2.
- The latent heat of vaporization at, say, 212°F (100°C) plus the sensible heat change of the free water from 77 to 212°F (25–100°C). (The latent heat of vaporization of the tar is low and was not included.)

The effect of pressure on the total heat content of the system is probably negligible, and it can be neglected up to the top pressure envisioned for carbonization processes (150 psig or 1034 kPa).

Generalized data on the heat effects in the hydrocarbonization of coal are not available. A measured heat balance around a hydrocarbonization reactor is difficult in practice, mainly because of the problem of maintaining adiabatic conditions. In lieu of a direct heat balance, the soundest alternative would be to have a total material balance around the hydrocarbonizer, with all chemical components identified. With these data and the heats of combustion (or formation) of these components, it is possible to develop a heat balance, using well-established thermodynamic calculations.[17]

Unfortunately, such total material balances are not always available. Then one must resort to approximation methods, using correlations such as those given in Figs. 13.3 and 13.4, which are based on work by Holmes et al.[18]

The basic approach to predicting the heat balance for hydrocarbonization[19] is to add, taking into account signs, the total energy for carbonization and the heat liberated by the reaction of hydrogen. The energy for carbonization is derived, as previously described,

[17] O. A. Hougen and K. M. Watson, *Industrial Chemical Calculations,* 2nd ed., John Wiley and Sons, New York (1936), pp. 154–174.

[18] J. M. Holmes, H. D. Cochran, Jr., M. S. Edwards, D. S. Joy, and P. M. Lantz, *Hydrocarbonization Research, Phase 1 Report: Review and Evaluation of Hydrocarbonization Data,* Oak Ridge National Laboratory, ORNL-TM-4835, August 1975.

[19] L. Seglin, unpublished work.

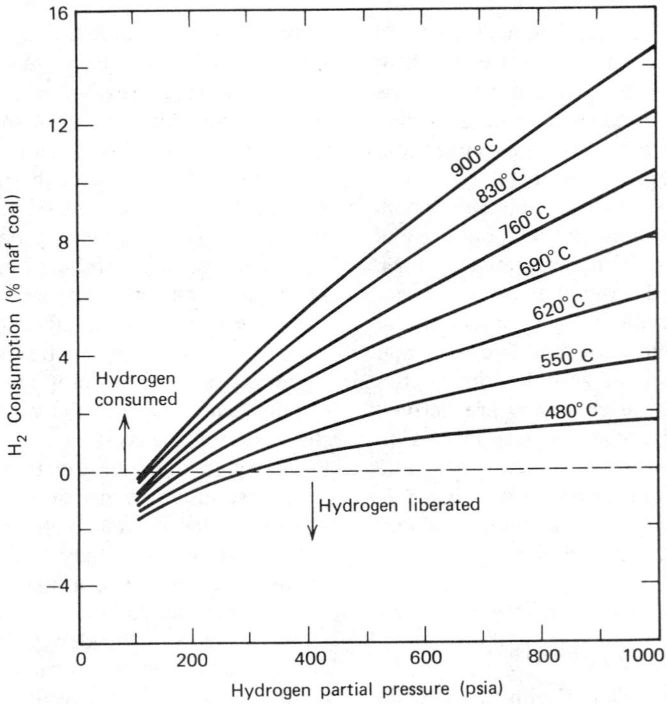

Figure 13.3 Hydrogen consumption for 10-min residence time.[18]

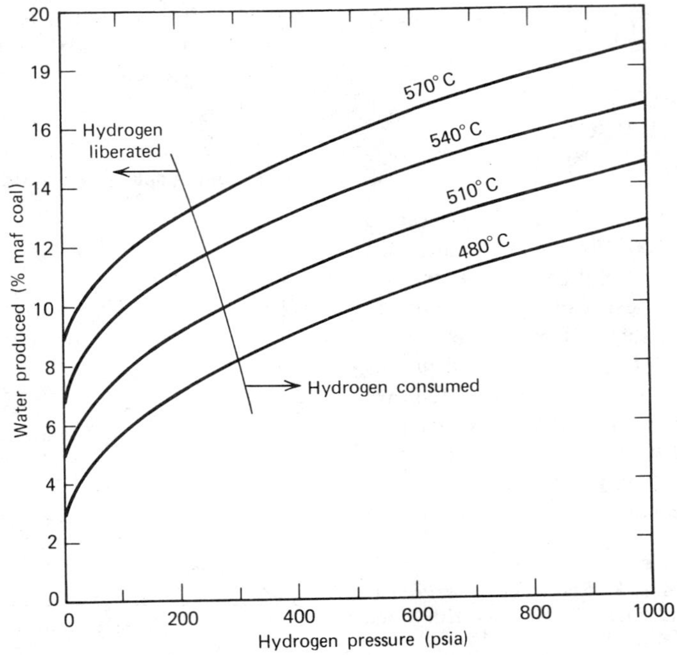

Figure 13.4 Water production in a fluidized-bed reactor.[18]

using Figs. 13.1 and 13.2. The heat liberated by hydrogen reaction is the sum of the heat of formation of water and the heat of the reaction with the organic (carbon) substrate to change the hydrocarbons (as by saturation, hydrocracking, etc.). The total hydrogen consumed is calculated from Fig. 13.3, which correlates consumption with temperature and pressure. The hydrogen consumed to form water is calculated from the difference between the total predicted water produced at the specified temperature and pressure and the water produced at zero hydrogen consumption. Both these quantities are derived from Fig. 13.4. The heat liberated at 25°C to form water is taken as −33.0 kcal/g per mole of H_2O [−138 kJ/(g)(mole)]; this figure is derived from the change in bond energies[20] for the following model reaction:

$$2H_2(g) + R_1{-}O{-}R_2(s) = R_1H(s) + H_2O(g) + R_2H(s)$$

The hydrogen that reacted with the organic (carbon) substrate is the difference between that predicted from Fig. 13.3 and that predicted from Fig. 13.4. Assuming that this hydrogen reacts in the same manner as observed in the hydrotreating of coal tars, the estimated heat liberated at 25°C by reaction of this hydrogen is −5990 cal/g per mole of H_2 [−25 kJ/(g)(mole)].[21]

13.3 ENGINEERING CALCULATIONS

The modern trend in the low-temperature carbonization of coal is toward continuous flow systems using fluidized or entrained solids reactors. These can be conducted in single or multiple stages. Heat is provided to the reactors by hot gases, generated *in situ* or externally, and/or by hot solids. The flows of solids and gases in such systems are predictable by heat balances around each reactor, using the method presented above.

A single-stage reactor system presents no unusual problems in defining the flows in and out and the energy requirements.

Multiple-stage reactor systems are a bit more complicated. One means for handling this is to set up simultaneous heat and material balances around each stage, and solve for the flows in and out of each reactor by trial and error. Another approach,[22] based on a graphical solution of these heat and material balances, is outlined next.

Figure 13.5 plots the effect of temperature on the heat content of the coal substance. Superimposed on this graph is a curve showing the temperature of the gases that provide the heat for the system and are either the fluidizing medium or the transport medium for the solids in each reactor of the system. There may be certain restraints on the system. First, the exit gases that convey the volatile yield of the coal (gases and condensable vapors) might have to be above some temperature, say the dew point of the condensables at the pressure of the system. Second, to avoid agglomeration of the solids, which might lead to bogging of a fluidized-bed reactor or to plugging of a transport reactor, the temperatures of the stages might be dictated by the melting point of the coal, especially in carbonization processes. However it might be possible to circumvent this problem by recirculation of dry solids; and especially in hydrocarbonization processes, one must also consider reaction rates and products of reactions.

[20] R. C. Reid and T. K. Sherwood, *The Properties of Gases and Liquids,* McGraw-Hill Book Company, New York, 1958, pp. 105–109.

[21] G. Gruber, FMC Corporation, private communication, December 1973.

[22] L. Seglin, unpublished work.

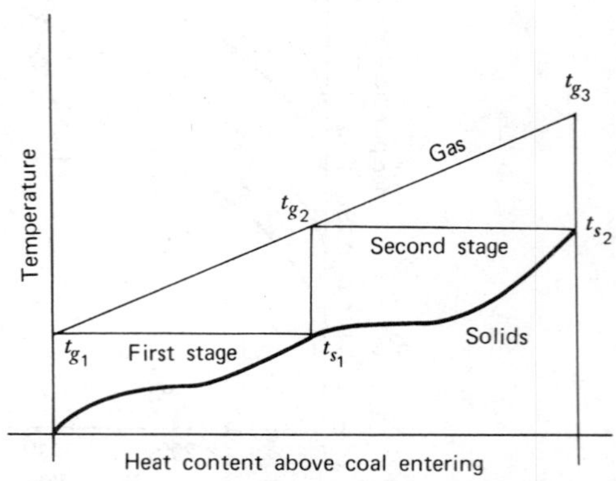

Figure 13.5 Two-stage coal carbonization with minimum heating gas inlet temperature.

Consider a two-stage reactor system as depicted in Fig. 13.5. Suppose the exit gases from the first stage must be at t_{g_1} to avoid condensation, and the solids in the second-stage reactor must be at t_{s_2} to obtain a reasonable reaction rate and/or volatile product mix. Assume further that the temperature of the exiting gases is the same as that of the exiting solids from each reactor stage. This is essentially true for fluidized-bed reactors, and almost true for transport reactors. Then, referring to Fig. 13.5, t_{s_1} (the temperature of the first-stage reactor) is equal to t_{g_1}, which is given, and t_{g_2} (the temperature of exit gases from the second stage) is equal to t_{s_2}, which is also given. The required gas entering temperature, t_{g_3}, is readily obtained by extending the line t_{g_1}–t_{g_2} to the heat content of the solids leaving the second stage reactor at t_{s_2}. This requirement is necessitated by the overall heat balance for the system. The ratio of hot gases to coal (solids) is simply calculated from overall heat balance for the system, which gives:

$$\frac{M_g}{M_r} = \frac{\text{heat content change of solids per unit of coal + heat content change of volatiles at } t_{g_1}}{(t_{g_3} - t_{g_1})\, C_g} \qquad (13.6)$$

These calculations can be simplified without undue loss of accuracy by neglecting the volatiles added to the gas in the second stage. If the addition of the volatiles is not neglected, the shape of the gas heat content line in Fig. 13.5 will be shifted, so that line t_{g_1}–t_{g_2} will have a lower slope than line t_{g_2}–t_{g_3}, reflecting the higher mass of gas in the first-stage reactor than in the second. The entering gas temperature, t_{g_3}, will be increased, and the calculated flow of hot gas entering will decrease.

The treatment of the two-stage coal carbonization system changes drastically when the temperature of the hot gas entering the second stage is higher than that calculated for the previous example. Increasing this temperature reduces the flow of hot gas to the second stage, thereby increasing the percentage utilization of heat from these gases. This decreases the gross energy requirements for carbonization, and decreases the size and cost of practically all the equipment handling these gases. There is an upper temperature limit of these gases, however, and this will be dictated by such factors as:

1. *Materials of construction:* Fluidized-bed gas distributors can be designed with relative ease for 1600°F (871°C) gases; above this temperature, significantly increasing costs are encountered. For all practical purposes, 2000°F (1093°C) may be considered to be the upper limit.
2. *Coal ash fusion temperature* (under reducing conditions): The lower this temperature, the lower will be the temperature of the hot gases made by partial combustion of char (as in the COED process discussed below) without encountering ash clinkering.
3. *Achievable temperatures:* Those above 2000°F (1093°C) can probably be achieved by use of special fluidized-bed designs developed in the Carbide-Battelle process.[23]

The two-stage reactor system with heating gases entering above the minimum temperature that was developed in the previous example can be handled in the following manner. The required quantity of heating gas is obtained by an overall heat balance around the system. The heat content–temperature plot of the solids and gases for this condition is shown in Fig. 13.6. Stepping off the stages, as in Fig. 13.5, shows that more heat is available from the inlet heating gases to heat solids from the first stage (entering at temperature t_{s_1}) to the temperature of the second stage, t_{s_2}. This surplus is indicated by E_1–E_1' in Fig. 13.6. The surplus heat can be used to heat a stream of solids, which circulates back to the first stage, where heat from these solids would be recovered. The gas temperature line t_{g_2}–t_{g_3} (in the second stage) will steepen because of the added heat transferred to the circulating solids. The gas temperature line t_{g_1}–t_{g_2} (in the first stage) will flatten because the heat transferred from the gas to the

[23] W. C. Corder, H. R. Batchelder, and W. M. Goldberger, *Proc. 5th Synthetic Pipeline Gas Symp.*, Chicago, October 29–31, 1973; Sponsored by AGA, and IGU, 109–124.

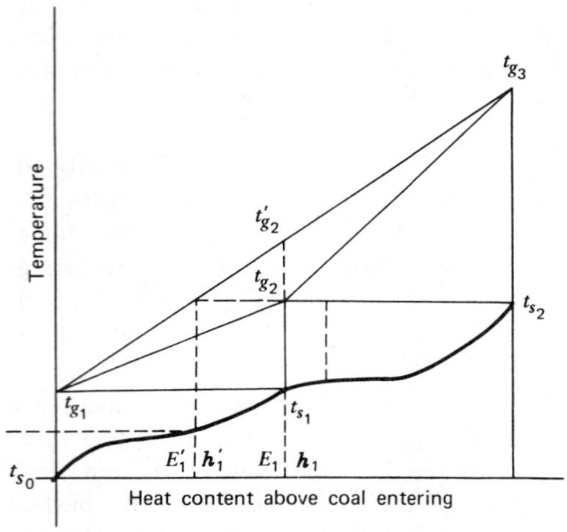

Figure 13.6 Two-stage coal carbonization with greater than minimum heating gas inlet temperature.

first stage is reduced by the amount of heat transferred from the circulating solids.

Assuming constant (average) specific heat for the gases and for the circulating solids, the heat content line for the gas will be straight. The actual temperature lines with solids-recycle to the first stage from the second are obtained by locating t_{g_2} as the intersection of the horizontal line through t_{s_2} and the vertical line through t_{g_1}. Having located t_{g_2}, the temperature lines t_{g_1}–t_{g_2} and t_{g_2}–t_{g_3} are constructed. The flow of solids is given by:

$$\frac{M_r}{M_g} = \frac{C_g(t'_{g_2} - t_{g_2})}{C_s(t_{s_2} - t_{s_1})} \tag{13.7}$$

It was found in the COED process that caking coals could not be handled in two stages without undue solids agglomeration, which caused bed bogging. Recourse to more than two stages was required to overcome this. The extension of this graphical method beyond two stages will be published later.

13.4 HEAT TRANSFER

The basic modes for transferring heat in coal carbonization and hydrocarbonization processes that might be encountered are as follows:

1. Hot gases to solids: (a) dense phase (moving-bed and fluidized-bed reactors) and (b) dilute phase (entrained reactors).
2. Hot solids to gas: (a) dense phase and (b) dilute phase.
3. Mixed phases (dense or dilute phases) to a wall.
4. Hot wall to mixed phases (dilute or dense phases).

These situations have been extensively explored both experimentally and theoretically. Unfortunately there is considerable disparity in the published results. A good review and analysis of this work was reported by Zenz and Othmer.[24] The following sections present recommended correlations and methods for each of the situations above.

13.4.1 Heat Transfer Between Fluids and Single Particles in Dilute Suspensions

A theoretical analysis of the transfer of heat from a sphere to a surrounding stagnant fluid yields Equation 13.8[25]:

$$Nu = \frac{h_c D_p}{k_m} = 2 \tag{13.8}$$

Johnstone et al.[26] analyzed the forced convective heat transfer between a sphere and a fluid. For Reynolds numbers greater than 500, they obtained

$$Nu = \frac{h_c D_p}{k_m} = 0.714\left(\frac{D_v v \rho_f}{\mu_f}\right)^{1/2}\left(\frac{C_p \mu_f}{k_f}\right)^{1/2} \tag{13.9}$$

Other investigators[27–31] obtained similar correlations, and these are compared in Fig.

[24] F. A. Zenz and D. F. Othmer, *Fluidization and Fluid-Particle Systems,* Reinhold Publishing Corp., New York, 1960, Ch. 13.
[25] J. Boussinesq, *Liouville J.*, **1,** 285 (1905).
[26] H. F. Johnstone, R. L. Pigford, and J. H. Chapin, *Trans. Am. Inst. Chem. Eng.*, **37,** 95 (1941).
[27] G. C. Williams, D.Sc. thesis, Massachusetts Institute of Technology, Cambridge, Mass., 1942.
[28] J. R. Carey, *Trans. ASME,* **75,** 483 (1953).
[29] H. Kramers, *Physica,* **12,** 61 (1946).
[30] D. N. Joukovski, *J. Tech. Phys. (USSR),* **10,** 999 (1940).
[31] Y. S. Tang, J. M. Duncan, and H. E. Schweyer, *NACA* TN 2867 (1953).

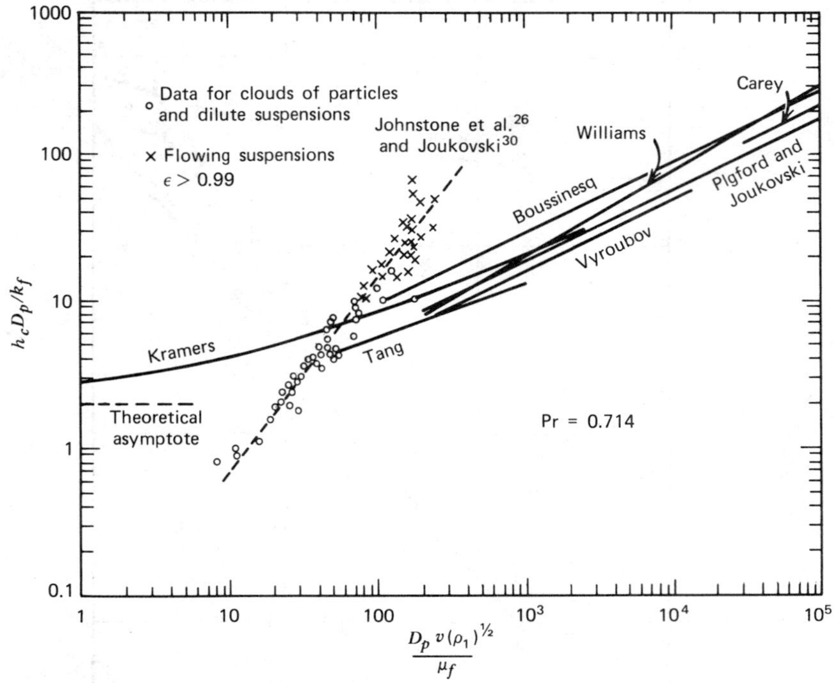

Figure 13.7 Forced convection heat transfer for spheres in air: Pr = 0.714; ○ = data for clouds of particles and dilute suspensions; X = flowing suspensions, $\epsilon > 0.99$.

13.7, which was taken from Zenz and Othmer.[24]

The transfer of heat to disperse suspensions was studied by Johnstone et al.[26] and Joukovski.[30] Their data are shown in Fig. 13.7, and indicate a distinct departure from the asymptote of $Nu = 2$ as Reynolds number approaches zero. According to Zenz and Othmer,[24] this effect is probably real and might be due to "an added degree of turbulence which would alter the heat transfer rate from that for a single sphere."

13.4.2 Heat Transfer Between Particles and Fluids in Fluidized Beds

Heat transfer coefficients between solids and gas are normally around 3–10 Btu/hr-ft²-°F (17–57 W/m²-K). However because of the large area of solids per cubic foot of fluidized bed, the volumetric heat transfer is very large indeed, resulting in a rapid approach of gas and solids temperatures. With reasonable gas distribution, thermal equilibrium is attained between the two phases within one inch (2.54 cm) above the distributor.[24]

Heat transfer coefficients between solid particles and fluidizing gases were investigated by Kettenring et al.,[32] Walton et al.,[33] Wamsley and Johanson,[34] Heertjes and McKibbins,[35] and others. Figure 13.8 compares the results of these investigators and indicates reasonable correlation of the data by Equation 13.10:

$$Nu = \frac{h_c D_p}{k_f} = 0.017 \left(\frac{D_p v \rho_f}{\mu_f}\right)^{1.21} \qquad (13.10)$$

Also noted in Fig. 13.8 is the Kramers correlation for heat transfer to a single sphere,[29] and the data of Johnstone et al.[25] and Joukovski[30] for dispersed suspensions. These indicate significantly higher coefficients for dispersed phases than for fluidized beds at Reynolds numbers less than 100.

[32] K. N. Kettenring, E. L. Manderfield, and J. M. Smith, *Chem. Eng. Prog.*, **46**, 139 (1950).
[33] J. S. Walton, R. L. Olsen, and O. Levenspiel, *Ind. Eng. Chem.*, **44**, 1474 (1952).
[34] W. W. Wamsley and L. N. Johanson, *Chem. Eng. Prog.*, **50**, 347 (1954).
[35] P. M. Heertjes and S. W. McKibbins, *Chem. Eng. Sci.*, **5**, 161 (1956).

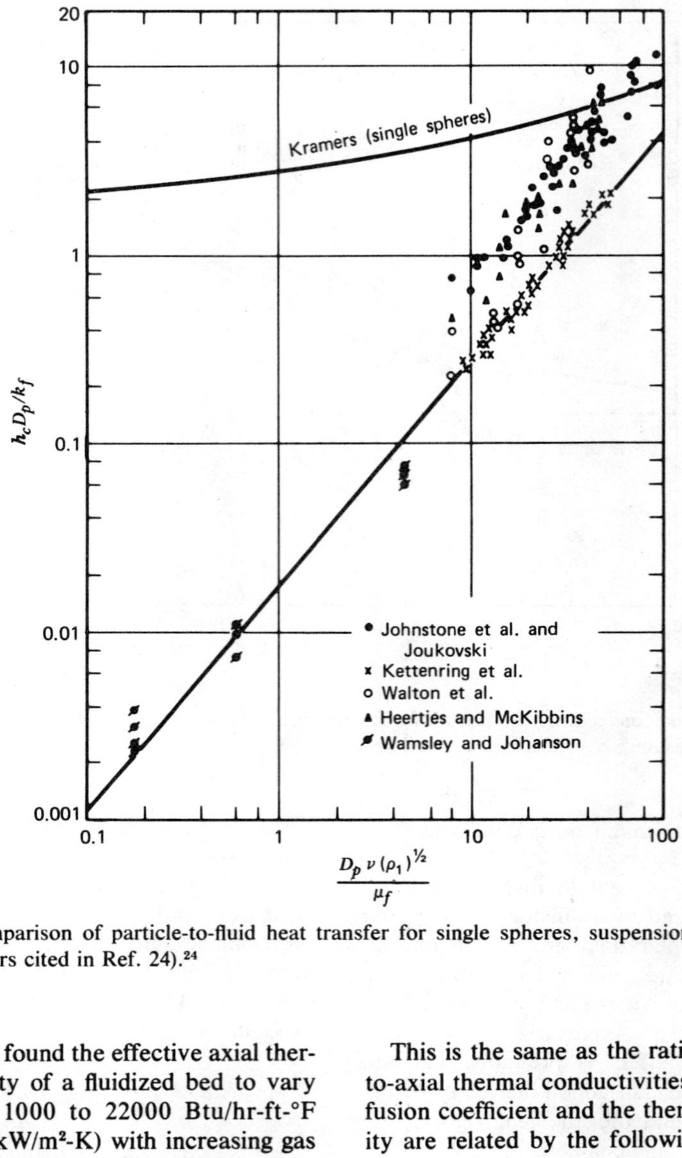

Figure 13.8 Comparison of particle-to-fluid heat transfer for single spheres, suspensions, and fluidized beds (investigators cited in Ref. 24).[24]

Shrikhande[36] found the effective axial thermal conductivity of a fluidized bed to vary linearly from 1000 to 22000 Btu/hr-ft-°F (5678–124,922 kW/m²-K) with increasing gas velocity. Radial coefficients appear to be considerably lower. Based on analyses by Kunii and Levenspiel[37] one can derive the following theoretical ratio of the radial and axial fluidized-bed diffusivities:

$$\frac{D_{sr}}{D_{sa}} = \frac{9}{16}\left(\frac{U_{mf}\delta}{\alpha(U_0 - U_{mf})\epsilon_{mf}}\right)^2 \frac{1}{1-\delta} \qquad (13.11)$$

[36] K. T. Shrikhande, *J. Sci. Ind. Res.*, **14E**, 457 (1955).

[37] D. Kunii and C. Levenspiel, *Fluidization Engineering*, John Wiley and Sons, New York, 1969, Ch. 5.

This is the same as the ratio of the radial-to-axial thermal conductivities, since the diffusion coefficient and the thermal conductivity are related by the following equations[37]:

$$D_{sa} = \frac{k_{sa}}{C_{ps}\bar{\rho}} \qquad (13.12)$$

and

$$D_{sr} = \frac{k_{sr}}{C_s\bar{\rho}} \qquad (13.13)$$

13.4.3 Heat Transfer Between a Wall and Fluidized Beds

The results have been published of numerous investigations on the heat transfer between the fluidized bed and a cooling or heating surface.

Table 13.2 repeats the summary by Zenz and Othmer[24] of the major studies in this field. According to Zenz and Othmer, no one has developed a fully satisfactory fundamental generalized correlation that explains all these reported data. One of the simplest approaches put forth by Zenz and Othmer[24] is given in Fig. 13.9, which correlates the Nusselt number with the Reynolds number and fluidized-bed voidage.

Figure 13.9 can serve as a rapid check or point of orientation in design predictions. Major deviations from this correlation are introduced by the effects of L/D, bed height, particle density, and the heat capacities of the solids and the fluidizing fluid.

Other investigators, utilizing data obtained from commercial or small fluidized units and employing an empirical approach, recommended the correlation shown in Fig. 13.10 for calculating the heat transfer to external surfaces, such as reactor walls, and the correlation shown in Figs. 13.11 and 13.12 for calculating the heat transfer to internal surfaces.

Many other correlations of heat transfer in fluidized beds have been presented in the literature, including studies of heat transfer to horizontal tubes,[64] vertical tubes,[65] and finned tubes.[66–68]

13.5 MASS TRANSFER

Mass transfer relationships in fluidized beds are, if anything, more complex and more poorly defined than are heat transfer relationships, although the phenomena are analogous. As in the preceding section, the material presented here should not be considered to be definitive. It is intended to provide some guidance to the reader and to suggest references in which additional material may be found.

13.5.1 Single Sphere to Fluid

Mass transfer coefficients from a sphere into a moving fluid may be calculated from Equation 13.14[69]:

[38] A. P. Agarwal and J. A. Storrow, *Chem. Ind. (London)*, 321 (1951).
[39] A. Baerg, J. Klassen, and P. E. Gishler, *Can. J. Res.*, **F28,** 287 (1950).
[40] R. N. Bartholomew and D. I. Katz, *Chem. Eng. Prog., Symp. Ser.*, **48,** 3 (1952).
[41] W. T. Brazelton, Ph.D. thesis, Northwestern University, Evanston, Ill., 1951.
[42] J. R. Campbell and F. Rumford, *J. Soc. Chem. Ind. (London)*, **69,** 373 (1950).
[43] W. M. Dow and M. Jakob, *Chem. Eng. Prog.*, **47** 637 (1951).
[45] A. Jacob and G. L. Osberg, *Can. J. Chem. Eng.*, **35,** 5 (1957).
[46] L. J. Jolley, *Fuel*, **28,** 114 (1949).
[47] R. Levelich and I. Caldas, Jr. *Am. Inst. Chem. Eng. J.*, **4,** 376 (1958).
[48] M. Leva, paper presented at General Discussion of Heat Transfer, London, September 11–13, 1951.
[49] M. Leva and M. Grummer, *Chem. Eng. Prog.*, **48,** 307 (1952).
[50] M. Leva, M. Weintraub, and M. Grummer, *Chem. Eng. Prog.*, **45,** 563 (1949).
[51] O. Levenspiel and J. S. Walton, *Proc. ASME Heat Transfer Fluid Mech. Inst.*, 136–146 (1949).
[52] H. S. Mickley and D. F. Fairbanks, *Am. Inst. Chem. Eng.*, **1,** 374 (1955).
[53] H. S. Mickley and C. A. Trilling, *Ind. Eng. Chem.*, **41,** 1135 (1949).
[54] R. W. Urie, M.S. thesis, Massachusetts Institute of Technology, Cambridge, Mass., 1948.
[55] A. P. Miller, Ph.D. thesis, University of Washington, Seattle, 1953.
[56] C. O. Miller and A. K. Logwinuk, *Ind. Eng. Chem.*, **43,** 1220 (1951).
[57] H. L. Olin and O. C. Dean, *Pet. Eng.*, **25,** C-23 (1953).
[58] R. D. Toomey and H. F. Johnstone, *Chem. Engr. Prog., Symp. Ser.*, **49,** 51 (1953).
[59] R. V. Trense, Ph.D. thesis, Northwestern University, Evanston, Ill., 1954.
[60] H. A. Vreedenberg, *J. Appl. Chem. (London)*, **2,** Suppl. 1, *A2*, 97 (1952).
[61] L. Wender and G. T. Cooper, *Am. Inst. Chem. Eng.*, **4,** 15 (1958).
[62] E. Wicke and F. Fetting, *Chem. Ing. Tech.*, **26,** 301 (1954).
[63] E. Wicke and K. Hedden, *Chem. Ing. Tech.*, **24,** 82 (1952).
[64] H. A. Vreedenberg, *Chem. Eng. Sci.*, **9,** 52 (1958).
[65] H. A. Vreedenberg, *Chem. Eng. Sci.*, **11,** 274 (1960).
[66] W. E. Genetti, R. A. Schwall, and E. S. Grimmett, *Am. Inst. Chem. Eng. Symp. Ser.*, **67**(116) (1971).
[67] W. J. Bartel, W. E. Genetti, and E. J. Grimmett, *Am. Inst. Chem. Eng. Symp. Ser.*, **67**(116) (1971).
[68] W. J. Bartel and W. E. Genetti, *Am. Inst. Chem. Eng. Symp. Ser.*, **69**(128) (1973).
[69] W. E. Ranz and W. R. Marshall, *Chem. Eng. Prog.*, **48,** 141 (1952); **48,** 176 (1952).

Table 13.2 Summary of Major Fluid Bed-to-Wall Heat Transfer Investigations[24]

Investigator	D_P (in.)	D_T (in.)	Fluid	Heat Transfer
Agarwal and Storrow[38]	0.000315–0.00394	1.5	Air	Wall to bed
Baerg *et al.*[39]	0.00462–0.0346	1.25–5.5	Air	Internal tube to bed
Bartholomew and Katz[40]	0.00333–0.00987	4.0	Air	Wall to bed
Brazelton[41]	0.00276–0.0432	1.96	Air	Wall to bed
Campbell and Rumford[42]	0.0095–0.0126	2.31	Air	Bed to coils
Dow and Jakob[43]	0.0008–0.0094	2, 3	Air	Wall to bed
Heerden *et al.*[44]	0.00295–0.025	3.35	Air, CH_4, CO_2, $H_2 + N_2$	Bed to wall Internal coil to bed
Jacob and Osberg[45]	0.0012–0.0132	3.0	Air, CO_2, He, H_2, N_2, Freon–12	Internal wire to bed
Jolley[46]	0.0014–0.063	6, 7	Air	Bed to internal surface
Lemlich and Caldas[47]	0.0019–0.027	2.0	Water	Wall to bed
Leva *et al.*[48–50]	0.00155–0.0179	2, 4	He, air CO_2, N_2	Wall to bed
Levenspiel *et al.*[51]	0.003–0.170	4.0	Air	Wall to bed
Mickley and Fairbanks[52]	0.00264–0.00324	4.0	He, air, CH_4, argon, Freon, NH_3	Internal tube to bed
Mickley and Trilling[53]	0.002–0.025	1–2.9	Air	Wall to bed Internal tube to bed
Miller,[55] Miller and Logwinuk[56]	0.0038–0.0098	2.0	Air, He, CO_2	Bed to axial tube
Olin and Dean[57]	0.0015–0.0062	4.0	Air	Internal tube to bed
Toomey and Johnstone[58]	0.00215–0.0334	4.73	Air	Bed to wall Internal tube to bed
Trense[59]	0.006–0.0437	3.92	Air	Internal tube to bed
Urie[54]	0.0016–0.0178	2.875	Air	Internal tube to bed
Vreedenberg[60]	0.0092–0.0236	1.35, 22.2	Air	Bed to internal tube
Wender and Cooper[61]	0.00173–0.0035	2–38	Air, synthesis gas	Bed to internal tubes
Wicke and Fetting[62]	0.0024–0.12	3.94	Air, CO_2, H_2	Internal tube to bed
Wicke and Hedden[63]	0.0114–0.174	3.94	Air	Internal tubes to bed

$$Sh = 2.0 + 0.6\, Sc^{1/3}\, Re^{1/2} \qquad (13.14)$$

Other investigators[70,71] have used the same form of this equation to correlate their data; their coefficients of the second term have ranged from 0.55 to 0.95, with an average of 0.6 taken over the entire sphere. Zenz and Othmer[24] developed the graphical representation (Fig. 13.13) of local values of this coefficient on the basis of investigations by Garner and Suckling.[71]

13.5.2 Mass Transfer in Fluidized Beds

Figure 13.14 presents a correlation that can be used to estimate mass transfer in beds of granular particles; Fig. 13.15 can be applied to beds of irregularly shaped particles.[72]

Mass transfer coefficients are difficult to measure in fluidized beds because of the rapid attainment of equilibrium between phases, and also because of the bubbling behavior in the bed, causing gas bypassing. The representative samples necessary for defining the driving forces, for example, are difficult to obtain.[37]

The foregoing deals with mass transfer across a gas film around shapes that essentially have an external surface only, no internal surface. Diffusion into and out of capillaries, which are created in coal during carbonization and hydrocarbonization, has not been systematically studied. Only cursory, totally empirical studies have been made on the mass transfer in the coal system during carbonization. The work of Anthony

[70] N. Froessling, *Gerlands Beit. Geophys.*, **52,** 170 (1938).

[71] F. H. Garner and R. D. Suckling, *Am. Inst. Chem. Eng. J.*, **4,** 114 (1958).

[72] J. C. Chu, J. Kalil, and W. A. Wetteroth, *Chem. Eng. Prog.*, **49**(3), 141 (1953).

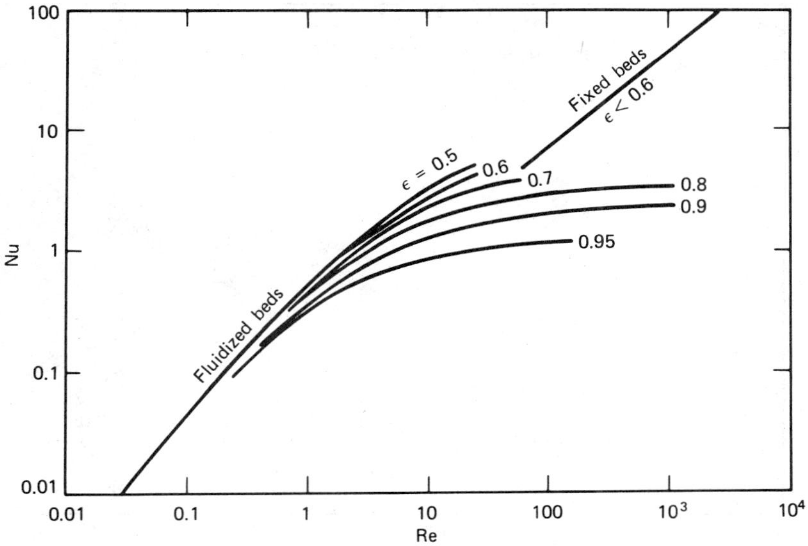

Figure 13.9 Heat transfer from a wall to fluidized beds.[24]

and Howard[73] has been an admirable step toward partially removing these faults. Qualitatively, the reactions that occur during carbonization or hydrocarbonization are the summation of a number of sequential reactions. These can be categorized as follows:

[73] D. B. Anthony and J. B. Howard, *Am. Inst. Chem. Eng. J.*, **22**(4), 625 (1976).

1. Transfer of energy into the coal-char body, as for carbonization; or the mass transfer of hydrogen into the coal-char, as for hydrocarbonization.

2. Reactions in the coal-char body: decomposition for carbonization or hydrogenation, plus decomposition for hydrocarbonization.

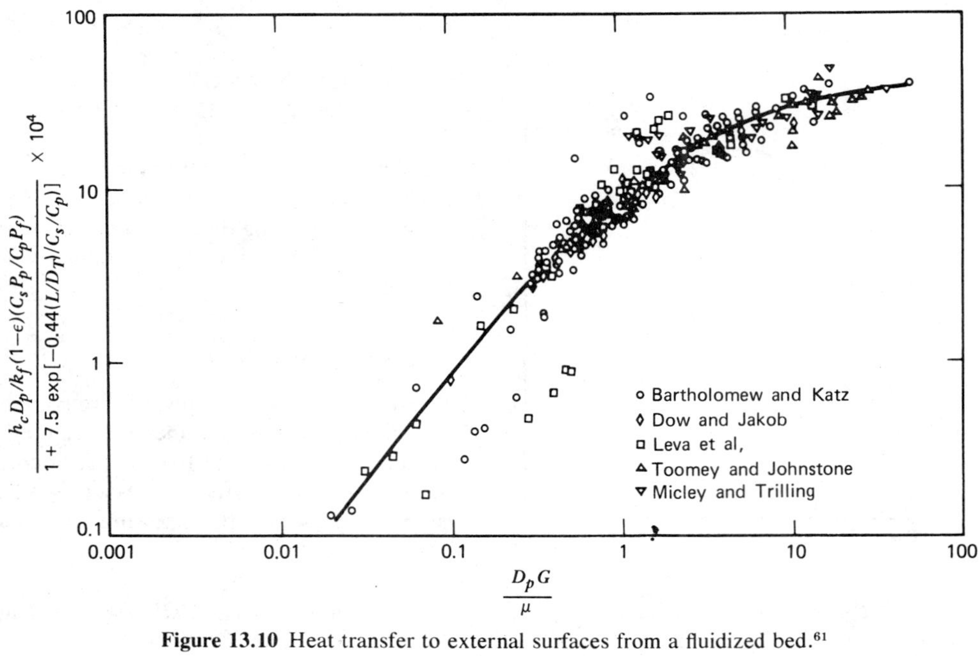

Figure 13.10 Heat transfer to external surfaces from a fluidized bed.[61]

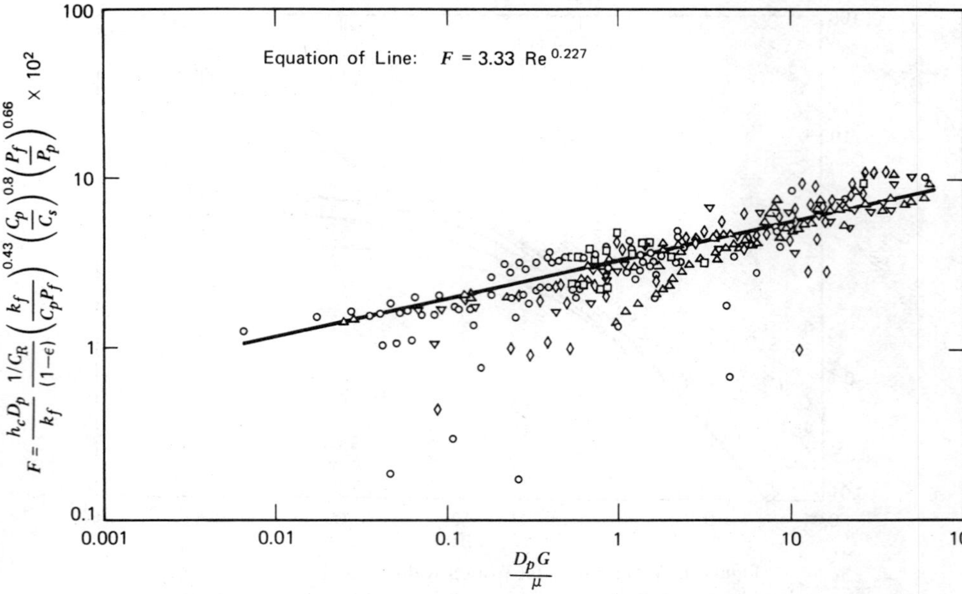

Figure 13.11 Heat transfer to internal surfaces of a fluidized bed.[61]

3. Mass transfer of products out of the coal-char body.

As the coal particle size increases, the mass transfer and/or the heat transfer rates decrease. As the particle size is increased (above zero), a particle size will be obtained above which these transfer rates could become slower than the chemical reaction rates. As a result, the overall process rate becomes controlled by mass and/or heat transfer. Conversely, for smaller particle sizes, these physical transfer rates could be faster than the chemical reaction rates, resulting in control of the overall process rate by the chemical reaction.

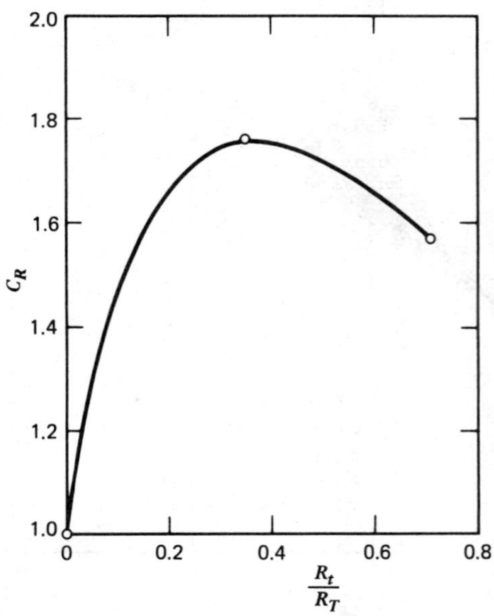

Figure 13.12 Corrections for nonaxial tube location.[61]

13.6 DYNAMICS OF PARTICULATE SYSTEMS

Details on particulate-fluid systems are thoroughly covered in various publications,[24,37,74] which the following discussion encapsulates. The reader is referred to the original works for details.

13.6.1 Fluidization

When a fluid is passed upward through a bed of solids, a pressure gradient is required to overcome the resulting friction. To increase the rate of flow, a greater pressure gradient is required. When the pressure drop (ΔP)

[74] G. G. Brown, et al., *Unit Operations*, John Wiley and Sons, New York, 1950.

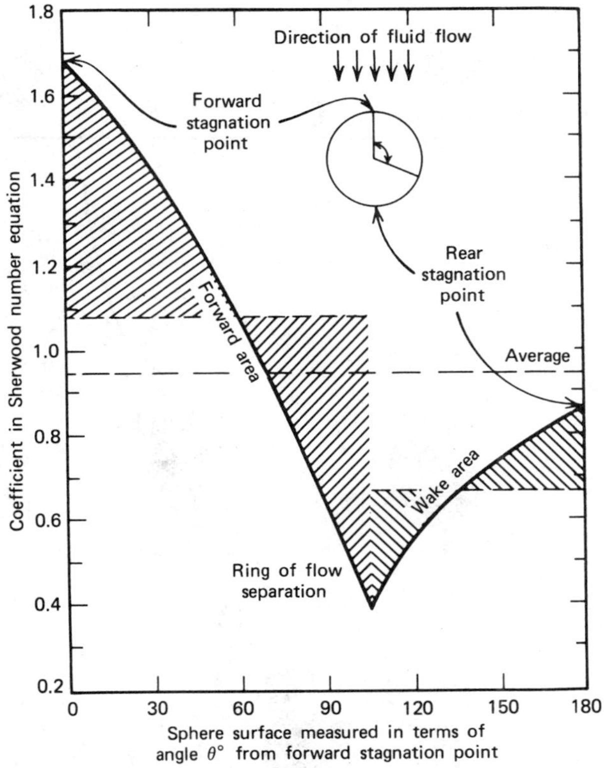

Figure 13.13 Distribution of mass transfer rates over a localized coefficient in Equation 13.14.[24]

approaches the weight of the bed over a unit cross-sectional area, the solids begin to move. This motion of the solids is created at superficial velocities far below the terminal free-settling velocity of the solid particles and constitutes the beginning of fluidization. The bed of solids expands, and the oscillation of each particle increases in velocity and frequency as the fluid velocity is increased further. Correspondingly, the porosity of the bed increases. This expansion continues with increasing fluid velocity until each particle behaves as an individual and is unhindered as a freely settling body by the action of any other solid particle.

A typical variation of ΔP with superficial velocity is shown in Fig. 13.16, where ΔP is plotted against the fluid velocity. The straight line from point A to point B corresponds to the increase of ΔP through the bed with increasing fluid velocity when there is no motion between the solid particles; this situation corresponds to fixed-bed operation. At point B the bed becomes unstable, and the particles begin to move in a rather restrictive sense. At point C the loosest arrangement of particles in contact is established, and any further increase in velocity of the fluid is accompanied by expansion of the bed and continuously greater agitation of the particles. Point C is known as the point of fluidization, and the corresponding fluid velocity is called the minimum fluidization velocity. The bed of solids begins to behave like a fluid: thus the name "fluidization." Expansion of the bed continues with increased velocities until the fluid velocity equals the terminal velocity of the particles. At this point the particles no longer flow in random motion, and the solids are entrained by the fluid.

The expansion of the bed during fluidization is indicated in Fig. 13.17, which plots the logarithm of the porosity of the bed against the logarithm of the velocity of the fluid. The horizontal portion up to point C corresponds to the porosity of the fixed bed. The straight line from point C to point E

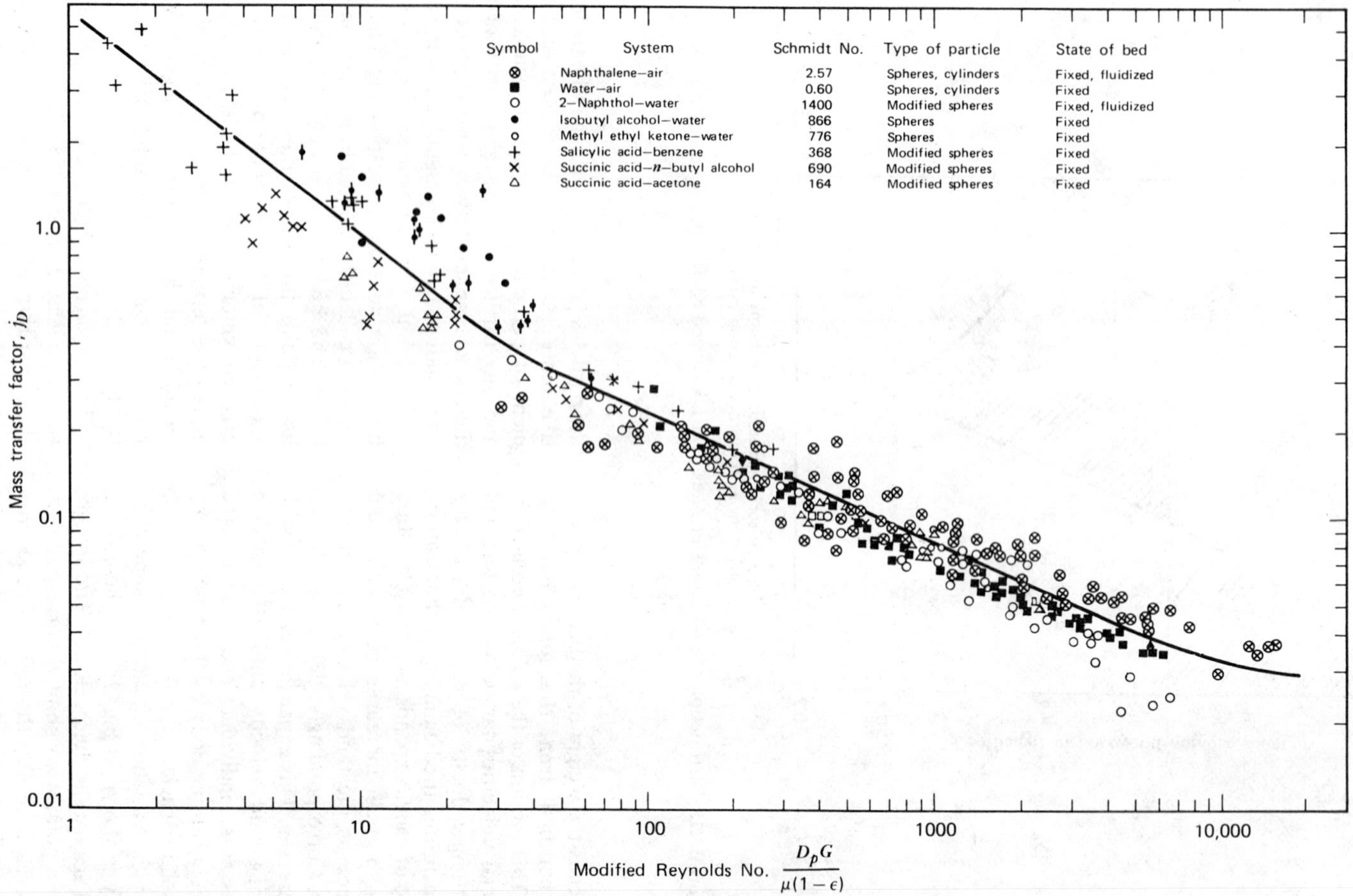

Figure 13.14 Mass transfer in fixed and fluidized granular beds.[72]

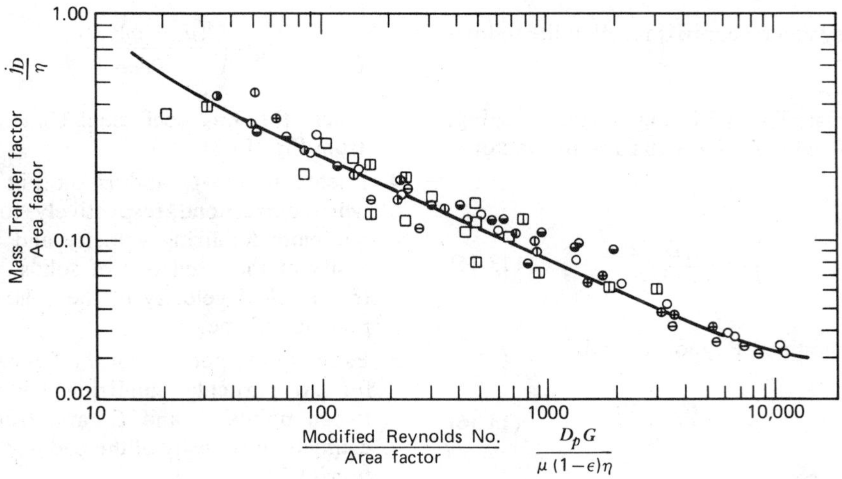

Particle	Area Factor	System
8–10 mesh flakes	0.80	2–Naphthol–water, fixed and fluid
14–18 mesh flakes	0.80	2–Napthol–water, fixed and fluid
½–in. Raschig rings	0.65	Water–air, fixed bed
1–in. Raschig rings	0.65	Water–air, fixed bed
2–in. Raschig rings	0.65	Water–air, fixed bed
¼–in. Berl saddles	0.65	Water–air, fixed bed
½–in. Berl saddles	0.65	Water–air, fixed bed
2–in. partition rings	0.65	Water–air, fixed bed

Figure 13.15 Mass transfer in fixed and fluidized beds containing tower packings or flakes; line through data point represents all the mass transfer data in fixed and fluidized beds.[72]

corresponds to the increasing porosity of the fluidized bed due to bed expansion with increasing velocity. Point *E* corresponds to the unhindered, free settling of the solids through the fluid. The fluid velocity at point *E* is equal to the free-settling velocity (terminal velocity) of the solids through the fluid, and the bed porosity is equal to one; that is, the volume of the particles is negligible compared with the voids.

Figures 13.16 and 13.17 can be obtained experimentally, but in the absence of such

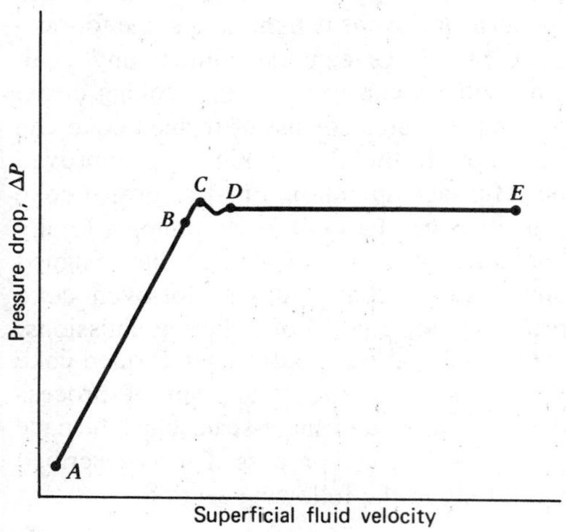

Figure 13.16 Fluidization of solids: *A*–*B* = fixed solids bed, *B* = bed becomes unstable, *C* = point of fluidization (minimum fluidization), *D* = complete fluidization, *E* = free settling.[74]

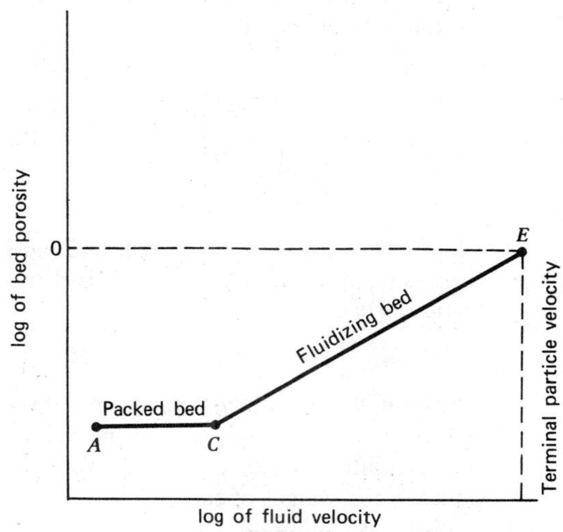

Figure 13.17 Effect of velocity (Reynolds number) on bed porosity.[74]

data, they can be approximated in the following way:

1. Calculate line AB in Fig. 13.16, using Fig. 13.18, which relates the friction factor f/F_f

$$\frac{f}{F_f} = \frac{2gD_p\Delta P_f}{Lv^2\rho F_f} \qquad (13.15)$$

to a modified Reynolds number

$$\mathrm{Re}' = \frac{D_p F_{\mathrm{Re}} v \rho}{\mu} \qquad (13.16)$$

where the factors F_f and F_{Re} are obtained from Figs. 13.19 and 13.20, respectively, which relate these factors to porosity (X) and sphericity (ψ). These properties are defined by

$$X = \frac{\text{volume of void space}}{\text{total volume of bed}} \qquad (13.17)$$

$$\psi = \frac{\text{surface area of sphere having a volume equal to that of the particle}}{\text{surface area of the particle}} \qquad (13.18)$$

It is necessary in these calculations to use a proper averaging method for the particle diameter. Equation 13.19 is recommended for this averaging:

$$D_p = \left[\frac{\sum \frac{m_i}{D_i}}{\sum \frac{m_i}{D_i^3}}\right]^{1/2} \qquad (13.19)$$

2. Fluidization will occur when the pressure drop is equal to the weight of the bed per unit area. Thus

$$\Delta P_f = L(1 - x)(\rho_p - \rho_f) \qquad (13.20)$$

The minimum fluidizing velocity (point C) will be approximately equal to the velocity calculated in Equation 13.15 for ΔP_f equal to the weight of the bed.

3. Estimate the terminal velocity of the particles from Equation 13.21:

$$v_t = \left(\frac{4(\rho_s - \rho_f)gD_p}{3\rho_f C_D}\right)^{1/2} \qquad (13.21)$$

where the drag coefficient C_D is derived from Fig. 13.21.

4. Locate points C and E on Fig. 13.17, which correspond, respectively, to (*a*) the minimum fluidizing velocity and the porosity of the fixed bed of solids, and (*b*) the terminal velocity of the solids and a porosity of one.
5. Estimate the porosity of the fluidized bed for any particular fluidizing velocity between points C and E, and from this, estimate the density of the bed from Equation 13.22:

$$\rho_b = (1 - x)\rho_s \qquad (13.22)$$

13.7 CARBONIZATION PROCESSES

13.7.1 Formed-Coke Processes

At last count some nine formed-coke processes had been reported in various advanced stages of development.[75,76] These are summarized in Table 13.3. About twice as many more have been tested on a smaller scale.

There are three basic reasons for this proliferation of formed-coke processes. First, the supply of good coking coal for feed in modern slot ovens is tight, and some formed-coke processes can use almost any coal, while others can use marginal coking coals. Second, the size consist of formed coke can be tailored, thereby making for improved blast furnace operation. Finally, formed-coke processes can be built in continuous, totally contained systems, hence can be designed more readily than today's slot-oven coke processes for control of polluting emissions.

One of the most advanced formed-coke processes, from the standpoint of process demonstration and large-scale blast furnace testing, is the FMC process. This is described more fully in the following section.

[75] N. R. Iammartino, *Chem. Eng.*, **83**(24), 30 (December 20, 1976).
[76] R. B. Jouvellier, *Chem. Eng.*, **84**(7), 5 (March 28, 1977).

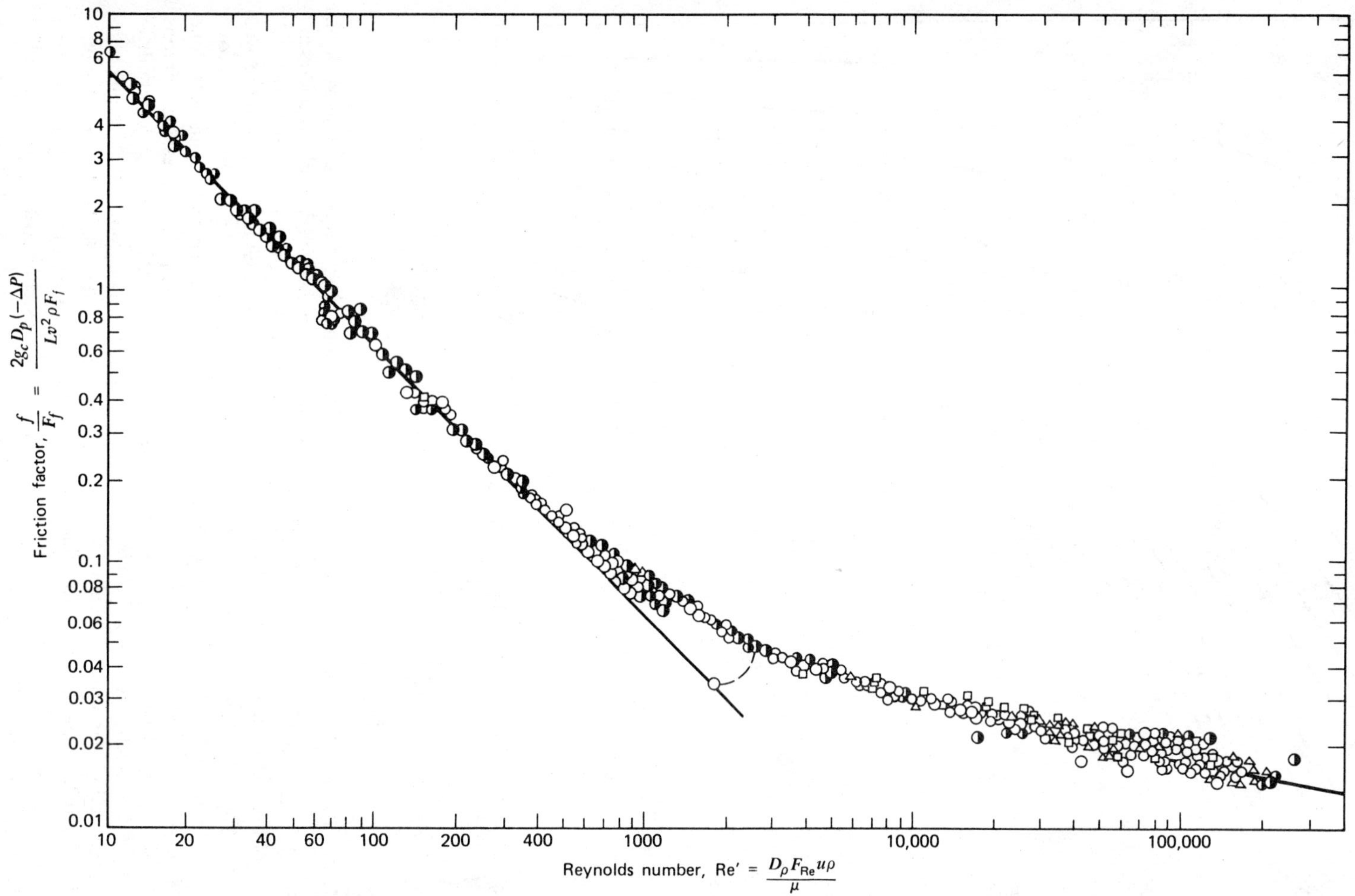

Figure 13.18 Experimental data on flow through porous beds plotted with friction factor (f/F_f) as a function of Reynolds number Re: □ = lead shot, ○ = glass rings, △ = Celite spheres, cylinders, Berl saddles, ◑ = glass spheres, nickel saddles, clay Raschig rings, Berl saddles.[74]

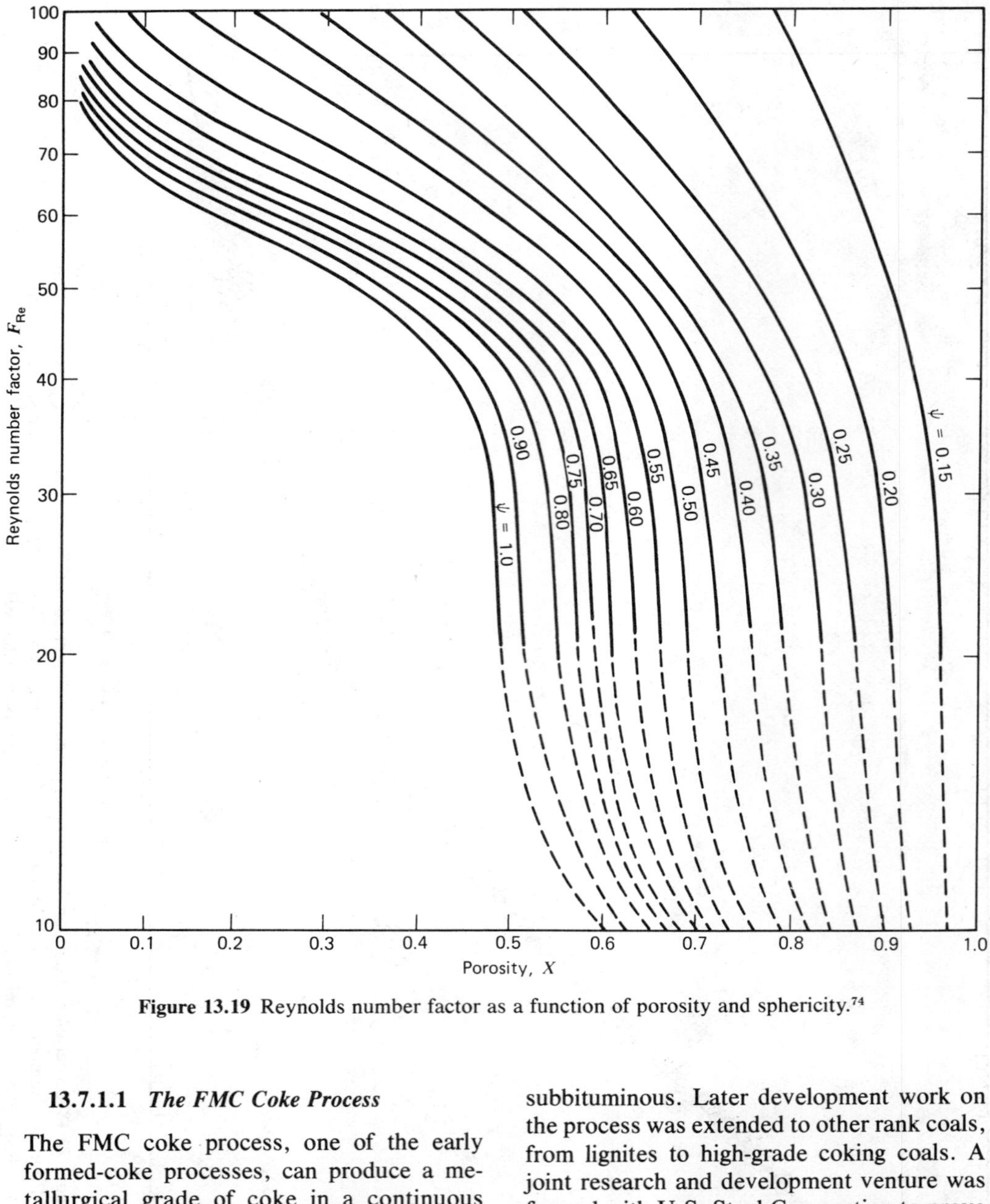

Figure 13.19 Reynolds number factor as a function of porosity and sphericity.[74]

13.7.1.1 *The FMC Coke Process*

The FMC coke process, one of the early formed-coke processes, can produce a metallurgical grade of coke in a continuous system from a wide range of coking and noncoking coals.[77,78] Initial incentive for development of the process was to provide coke for FMC's electric furnace phosphorus plant at Pocatello, Idaho, using locally available coals, which happened to be noncoking, subbituminous. Later development work on the process was extended to other rank coals, from lignites to high-grade coking coals. A joint research and development venture was formed with U.S. Steel Corporation to prove the process on a large scale. This was done in a demonstration plant processing 550 tons (500 tonnes) of coal per day. It was built in Kemmerer, Wyoming, in 1960. This plant has produced more than 300,000 tons of metallurgical coke. Part of this coke has been successfully tested in five different iron blast furnace trials, which are summarized in Table 13.4. The rest of the Kemmerer production has been routinely used in the Pocatello phosphorus plant. Operation of the furnaces with

[77] J. Work, *J. Met.*, **18**(5), 635 (1966).
[78] R. F. Moran, Preprints of papers, 168th National Meeting, American Chemical Society, Division of Fuel Chemistry, Altantic City, N.J., September 8–13, 1974.

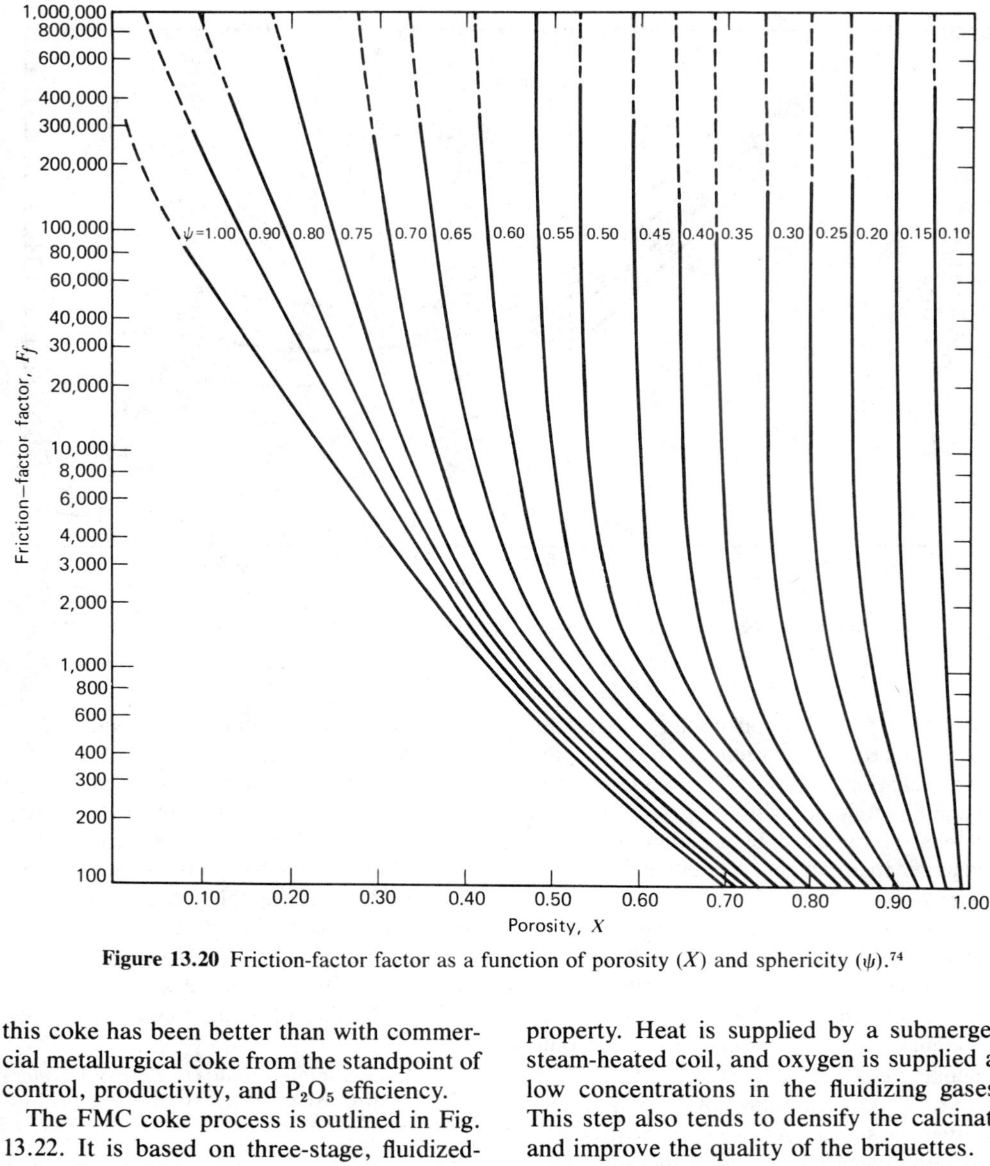

Figure 13.20 Friction-factor factor as a function of porosity (X) and sphericity (ψ).[74]

this coke has been better than with commercial metallurgical coke from the standpoint of control, productivity, and P_2O_5 efficiency.

The FMC coke process is outlined in Fig. 13.22. It is based on three-stage, fluidized-bed carbonization of coal to produce a solid calcinate (char) and tar, which are recombined, blended, and briquetted into a desired shape and size. The briquettes are cured by a controlled mild oxidation with air in a traveling-grate curing oven, thereby hardening the briquettes and reducing the volatiles to between 8 and 10%. The cured briquettes are coked in a shaft kiln to a uniform-sized and reactive coke, having about 2.5% volatiles.

The first fluidized stage is called the catalyzer. It operates at 300–500°F (149–260°C), depending on the coal. Here the coal is dried and mildly preoxidized to destroy its caking property. Heat is supplied by a submerged steam-heated coil, and oxygen is supplied at low concentrations in the fluidizing gases. This step also tends to densify the calcinate and improve the quality of the briquettes.

The coal from the catalyzer is transferred to the second stage—called the carbonizer, which operates at 900°F (482°C)—where it is fluidized by air. Tar and a low-Btu fuel gas are taken overhead, the tar recovered by condensing, and the gas used for plant fuel.

The char from the carbonizer contains about one-half the original coal volatile matter. This char is transferred to the third stage—called the calciner, which operates at about 1500°F (816°C)—where it is fluidized by additional air. The particles lose most of the remaining volatiles, and they shrink. This step is necessary for the formation of strong briquettes. The exiting calcinate is trans-

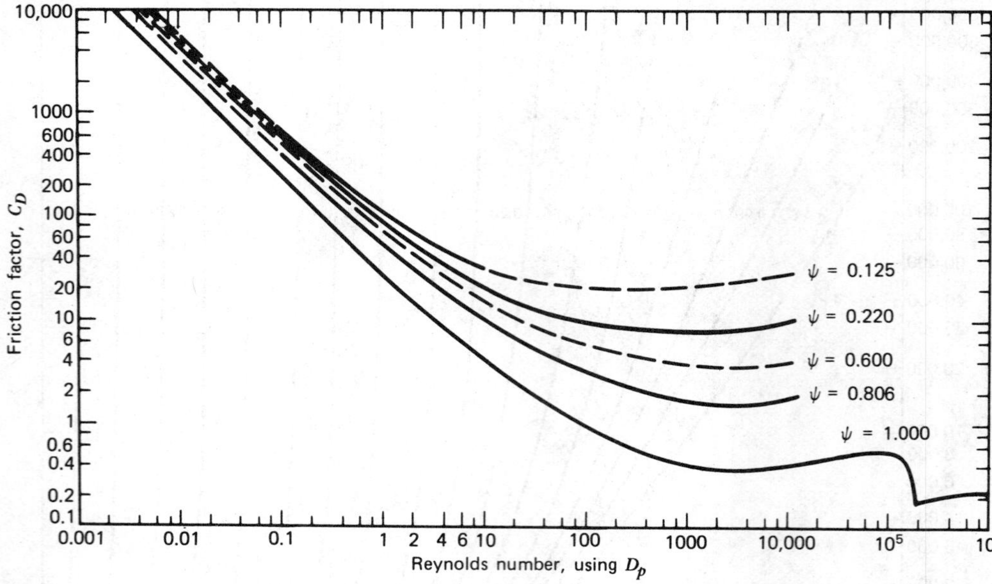

Figure 13.21 Drag coefficient versus Reynolds number for particles of different sphericities.[74]

ferred to a fluidized-bed cooler, where the solids temperature is reduced to allow safe downstream processing of the calcinate. The calcinate is fluidized with recycled inert gas, and heat is removed by a submerged heat exchanger that can generate either hot water or steam.

Economics for the process, derived from engineering designs based on the Kemmerer plant operation, are said to be competitive with the conventional slot-oven process. The economics are further improved where only noncoking coals are available.

An additional advantage of the FMC process over the slot-oven process is the potential ability to design more readily the means for control of polluting emissions. This feature is, of course, common to most continuous, totally contained processes such as this one. Batch-type systems like the slot-oven process produce difficult-to-control emissions when they are periodically opened and closed.

13.7.2 The COED Process

The COED process uses low-pressure, multistage, fluidized-bed carbonization of coal. The hot fluidizing gases flow countercurrently to the net forward flow of solids (coal-char), thereby heating the solids to successively higher temperatures. Heat and fluidizing gases are simultaneously generated for the process by the combustion of part of the residual char with oxygen in the last (hottest) fluidized stage. The process was developed by FMC under contract to the Office of Coal Research.[79–85] During this program, the process was successively carried through a 3-in. (7.62-cm) diameter bench-scale unit, an integrated continuous process development unit (PDU) using 1 ton (0.9 tonne) of coal per

[79] FMC Corporation, *Char Oil Energy Development,* O. C. R. Rep. No. 11, (Contract No. 14-01-0001-235); NTIS: PB-169 562/AS and 563/AS, March 1966.

[80] FMC Corporation, *Char Oil Energy Development,* O. C. R. Rep. No. 11—Supplement (Contract No. 14-01-001-235); NTIS: PB-173 916/AS and 917/AS, February 1967.

[81] FMC Corporation, *Char Oil Energy Development,* O. C. R. Rep. No. 56 Interim Rep. No. 1; NTIS: PB-234 166/AS, May 1970.

[82] FMC Corporation, *Char. Oil Energy Development,* OCR Rep. No. 56, Interim Rep. No. 2; NTIS: PB-234 167/AS, January 1971.

[83] FMC Corporation, *Char. Oil Energy Development,* OCR Rep. No. 56, Final Rep. NTIS: PB-234 168/AS, May 1972.

[84] FMC Corporation, *Char. Oil Energy Development,* OCR Rep. No. 73, Interim Rep. No. 1; NTIS: PB-234 020/AS, December 1972.

[85] FMC Corporation, *Char Oil Energy Development,* U.S. ERDA Rep. No. FE/1212/F—Final Rep., September 1975.

Table 13.3 Advanced Formed-Coke Processes

Process[a]	Coal[b]	Carbonization	Forming Method	Curing	Coking	Status
FMC	NC or C	Fluid bed, heat by internal combustion	Briquetting tar + calcinade	Low–0_2 gas	Moving bed	250–300 ton/day pilot; tested in large blast furnaces
Consol–BNR	WC	Rotary kiln	Pelletizing WC + char + binder	—	Lurgi designs	500-ton/day pilot
BFL	NC or C	Flash, heat by circulating char	Hot briquetting char + WC	—	No data given	300-ton/day pilot in startup; 650-ton/day under construction
EBV	NC + C	Flash of NC, preheat of C	Hot briquetting char + C	—	Yes, through retention	120–240 ton/day pilot
FCL	WC	Moving bed	Prepelletizing WC + binder	—	Included with carbonization	To ton/day pilot
DKS	NC + C	Moving bed	Prebriquetting NC + C + binder	—	Included with carbonization	130-ton/day pilot; 25-ton/day advanced pilot; tested in large furnace up to 50%
HBN		No data obtained				500-ton/day pilot; tested in large blast furnace
Broken Hill	MC or WC	Fluid bed of circulating hot sand	Prebriquetting	—	Moving bed	110-ton/day pilot; limited testing in blast furnaces
Sapozhnikov	C + NC	Flash (800°F)	Hot briquetting of plastic chars	—	Moving bed	90-ton/day pilot

[a] Processes: FMC = Food Machinery Corporation, Consol–BNR = Consortium of Consolidation Coal and Bethlehem Steel Company, National Steel Company, and Republic Steel Company, BFL = Bergbau-Forschung GmbH and Lurgi Mineralöltechnik GmbH, EBV = Eschweiler Bergwerke-Verein AG, FCL = Formed Coke Ltd. (England), DKS = Cooperative process of Didier Werke AG, Keikan, Reanten Industries, and Sumitomo Metal Industries, HBN = Houillères du Bassin du Nord, Broken Hill = Broken Hill Proprietary Company Ltd. (Australia).

[b] Coals: NC = noncoking, C = coking, WC = weakly coking, MC = mildly coking.

Table 13.4 Blast Furnace Trials Using FMC Coke[78]

Year	Trial	Size of Hearth (ft)	Tons Tested	Main Results
1962	U.S.Steel Corporation Experimental Furnace	4	2,000	Generally good but some problems indicated at high wind rates
1967	Armco-Inland	18½	9,000	Too dusty on handling; evidence of fines formation; low furnace productivity and efficiency
1970	Inland Steel	26½	270	Low dust formation; no operating troubles evident
1972	British Steel Corporation Cardiff, Wales	18	3,000	6 days at 100% FMC coke gave superior results compared to conventional coke
1973	Inland Steel Company, supported by Inland, U.S. Steel, Armco, Jones & Laughlin, and McLouth	26½	18,000	Good operation except for problems normally associated with sudden change of important furnace feed

day, and finally a 36-ton (33-tonne) per day pilot unit integrated with product oil-processing facilities.

The basic approach of this process is to minimize solids agglomeration while maintaining maximum inherent liquid yield from the coal by controlling the temperature in each of the successive carbonization reactors at a little below the fusion point of the coal (or char). After the solids are allowed to reside at this temperature for a period of time, the fusion point temperature is increased significantly. The solids can then be safely forwarded to the next reactor operating above the fusion temperature of the original feed coal.

As the caking tendency of the coal increases, the number of such carbonizing stages also increases. Thus it is possible with lignite or a subbituminous coal to use a minimum of two carbonizing stages; with a high-volatile bituminous B (such as Illinois No. 6 Seam), three stages; and with a high-volatile bituminous A (Pittsburgh Seam), possibly four stages (although completely successful demonstration was not achieved in the COED pilot project when feeding hvAb coal).

Figure 13.23 outlines the COED process using this staged temperature approach. Coal, superficially dried and milled to −10 mesh (2 mm), is fed to the first stage where it is heated typically to about 600°F (316°C) by hot, oxygen-free fluidizing gases entering at about 900°F (482°C). The balance of the free moisture, most of the inherent moisture in the coal plus some gas, and up to 10% of the tar yield, are driven off from the coal. The tar is recovered by condensation, and the noncondensables are reheated and recycled to stage one. The fusion temperature of the solids is increased, so that they can be transferred from the first stage to the second stage, which may operate at, say, 850°F (454°C). Heat is supplied to the second stage by hot gases from the third stage plus recycled char from one of the downstream stages. Most of the tar and some of the pyrolysis gases are distilled from the coal in the second stage. The overhead gases from this stage will have a substantial tar content. This precludes using these gases to supply the heat for the first stage, since considerable tar would be condensed from them at the lower first-stage operating temperature. Char from the second stage is forwarded to the third stage which may operate at, say, 1000°F (538°C). When operating with bituminous B or younger coals, the third-stage temperature is selected more to minimize vapor phase coking of the distilled tars than to avoid agglomeration of the solids. Heat is supplied to the third stage by hot gases plus hot recycled char from the fourth stage. The balance of the tar and much of the pyrolysis gases are distilled from char in the third stage.

The last stage of the process generates the necessary heat and fluidizing gases to pyrolyze the coal. Char from the previous stage is forwarded to the last stage, where air or a steam-oxygen mixture fluidizes the solids and provides the oxidant for combustion of char.

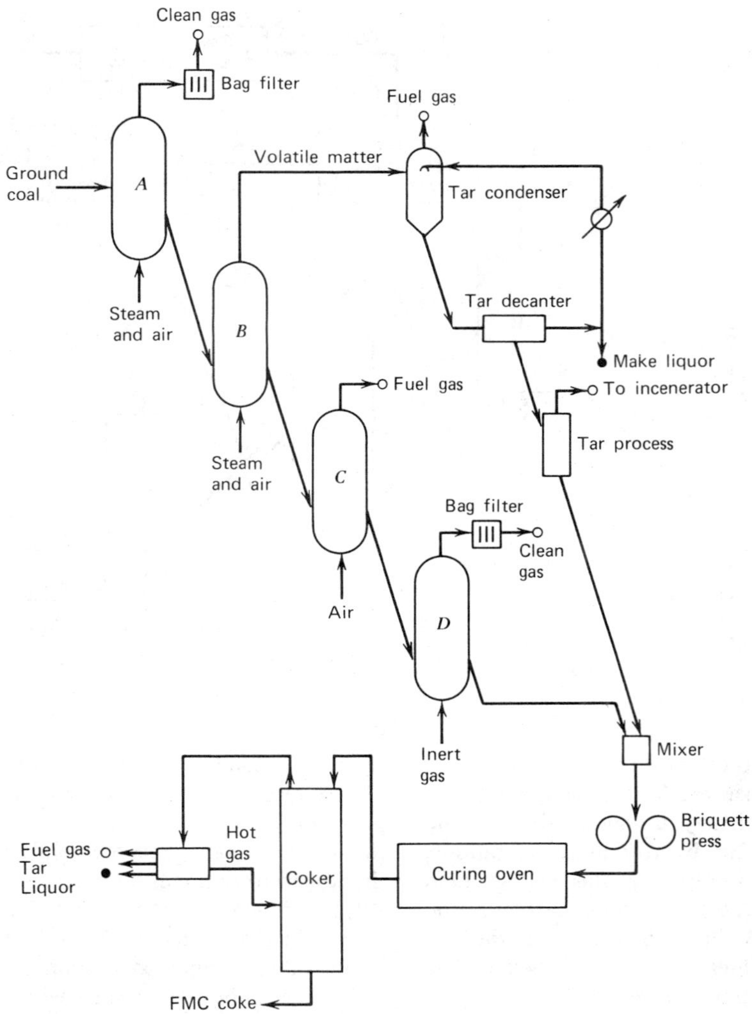

Figure 13.22 The FMC coke process: A = catalyzer, B = carbonizer, C = calciner, D = cooler, ○ = air pollution control, ● = water pollution control.[78]

Ideally the last stage is maintained at the highest temperature possible without ash clinkering. Temperatures as high as 1600°F (871°C) have been used. To avoid clinkering, it was found helpful to increase the gas velocity at the point where the air or steam-oxygen contacts the char. Only about 5% of the coal as char is required to supply the heat for the carbonization process. The rest of the char produced, which may represent upward of 60% of the dry coal, can be used in any number of ways, as suggested in a study by Battelle Memorial Institute.[86]

The overhead gases from the second stage are treated in one of two ways to recover the product tar. In the first system the gases are quenched by a recirculating stream of water in a venturi scrubber, with the resulting tar and water separated. A substantial amount of tar fog is formed in this quenching operation, and this is recovered by a second venturi scrubber followed possibly by an electrostatic precipitator. In the second approach the tar is removed from the pyrolysis gases by a staged oil scrubbing. A bottom, hot

[86] Battelle Memorial Institute, Final Report, *Study of the Identification and Assessment of Potential Markets for Chars from Coal Processing Systems*, OCR Rep. No. 44 (Contract No. 14-01-0001-1190); NTIS: PB-183 660/AS.), 1969

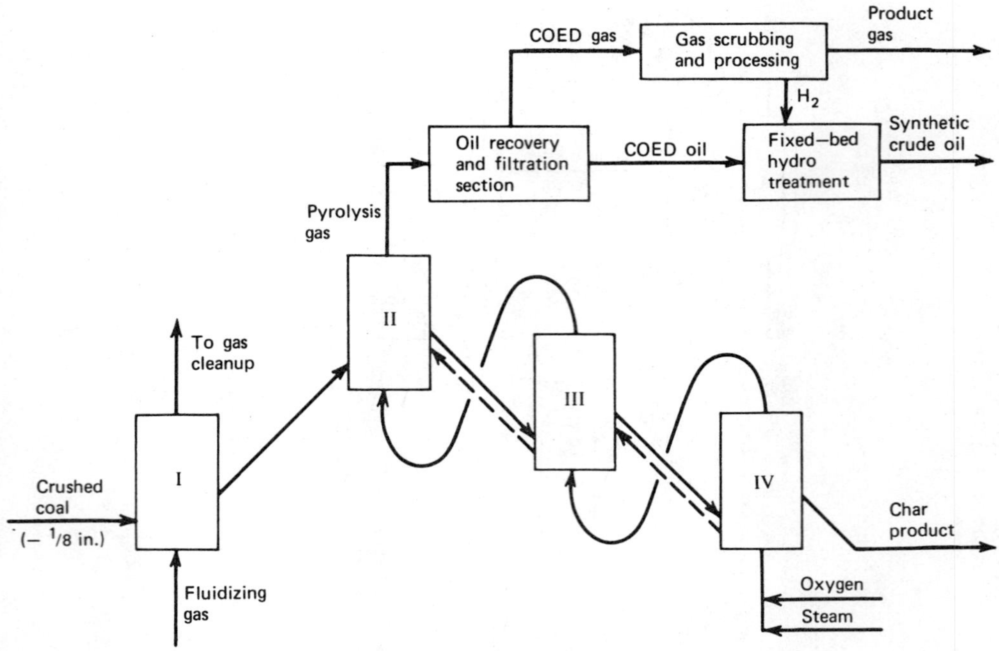

Figure 13.23 The COED coal pyrolysis.[18]

stage is used to knock down entrained solids and to condense the heavy ends of the tar. The resulting heavy tar-solids slurry could be recovered for hydrotreating operations, noted below. Alternately, this slurry could be pumped back to the third carbonization stage to crack these heavy ends, thereby attaining a lighter overall tar product. The lighter tar fraction, not recovered in the bottom section, is recovered in the lower temperature, top section. The absorber system potentially has the following advantages over the water quench system:

1. Ability to fractionate the tar, so that a light, more easily hydrotreated tar is recovered for hydrotreating.
2. Entrained solids from the second stage of carbonization can be substantially removed before condensation of the product tar, possibly eliminating the need for tar filtration.
3. Tar-fog formation is minimized, thereby eliminating the need for venturi scrubbing and/or electrostatic precipitators.

The recovered tar is filtered, using a precoated rotary pressurized filter, to remove entrained solids carried over by the gases. The filtered oil or the top oil from the absorber is forwarded to the hydrotreating step, in which the tar is reacted catalytically with hydrogen to remove the heteroatoms (O, N, S), to add hydrogen to the tar molecule, and to produce a distillable fuel oil or synthetic crude oil for refinery operation.

Typical pilot plant results with the COED process are given in Table 13.5.

In 1962, at the beginning of the COED program, it was envisioned that the net char produced could be used as a boiler fuel. Combustion tests on this char indicated its potential usefulness as such a fuel. The char contained about the same level of sulfur as the feed coal. The desulfurization of the char to ecologically acceptable levels was demonstrated using hydrogen plus an H_2S acceptor (dolomite), at about 1600°F (871°C). Separation of the sulfided dolomite and its regeneration proved to be more difficult than was economically tolerable.

Interest in the potential commercial use of char as a boiler fuel therefore waned, and use of this by-product was shifted to gasification processes, where it was employed as a feed. Ralph M. Parsons Company was retained by the U.S. Energy Research and

Table 13.5 COED Pilot-Plant Yields[18]

	Coal		
	Big Horn, Wyoming	Illinois No. 6 Seam, Peabody No. 10 Mine	Utah King
Pyrolysis			
Run number	P-12	P-22A	P-38
Yield (wt % of dry coal)			
Char	50.0	60.7	59.0
Oil	11.2	18.7	21.9
Liquor	11.6	5.8	
Gas	27.2	14.6	21.5
Hydrotreatment			
Run number	H-12	H-22	H-38
Yield (wt % of dry oil input)			
Syncrude	93.2		
Syncrude (C_6^+)		93.9	
Syncrude (C_5^+)			92.9
C_4–C_6			1.1
C_4 and C_5		0.3	
C_4^+	0.4		
C_1–C_3	4.0	0.8	2.3
Liquor	7.6	9.9	

Development Administration (ERDA) to evaluate the COED process in which this char was gasified to produce low-Btu fuel gas. Various gasification systems have been considered. One of the more attractive approaches is the COGAS process[87–89] (see Chapter 24). This system integrates the front end of the coal pyrolysis process of COED with a steam-carbon gasification of the surplus char. The endothermic heat for the steam-carbon reaction is supplied by the combustion of char fines with air, and the resulting combustion gases are used to heat a stream of char withdrawn from the gasifier and returned to the gasifier after reheating. The heat for the pyrolysis section is provided by the hot gasifier gases in much the same way as described previously, but the gasifier is substituted for the fourth stage of pyrolysis. This process is presently being considered by the Illinois Coal Gasification Group in a joint project with the U.S. Department of Energy.

The PDU used in the COED development was modified as shown in Fig. 13.24 and used to demonstrate the yields and operability of the Seacoke Process, which was being investigated by the Atlantic Richfield Company for the Office of Coal Research. The Seacoke Process is based on multistage, fluidized-bed pyrolysis of coal and residual oil together, rather than separately.[90]

The most important benefit of the Seacoke copyrolysis over separate pyrolysis is the higher hydrogen content of the derived oil, with attendant lower hydrogen yield in the gas. Additionally, the oil yield is slightly

[87] L. Seglin, L. D. Friedman, and M. E. Sacks, *Prepr., Am. Chem. Soc., Div. Fuel Chem.*, **19**(4), 31 (1974).

[88] H. R. Hoy and R. T. Eddinger, *Proc. 7th Synthetic Pipeline Gas Symp.*, Chicago, (October 1975), sponsored by AGA, ERDA, and IGU, Paper No. 85.

[89] R. Bloom, Jr., and R. T. Eddinger, *Proc. 6th Synthetic Pipeline Gas Symp.*, Chicago, (October 1974), sponsored by AGA, OCR, and IGU, 51–70.

[90] Atlantic Richfield Company, *Project Seacoke—Final Report,* Vols. I and II, OCR Rep. No. 29 (Contract No. 14-01-001-473); NTIS: PB-235 781, (February 1970).

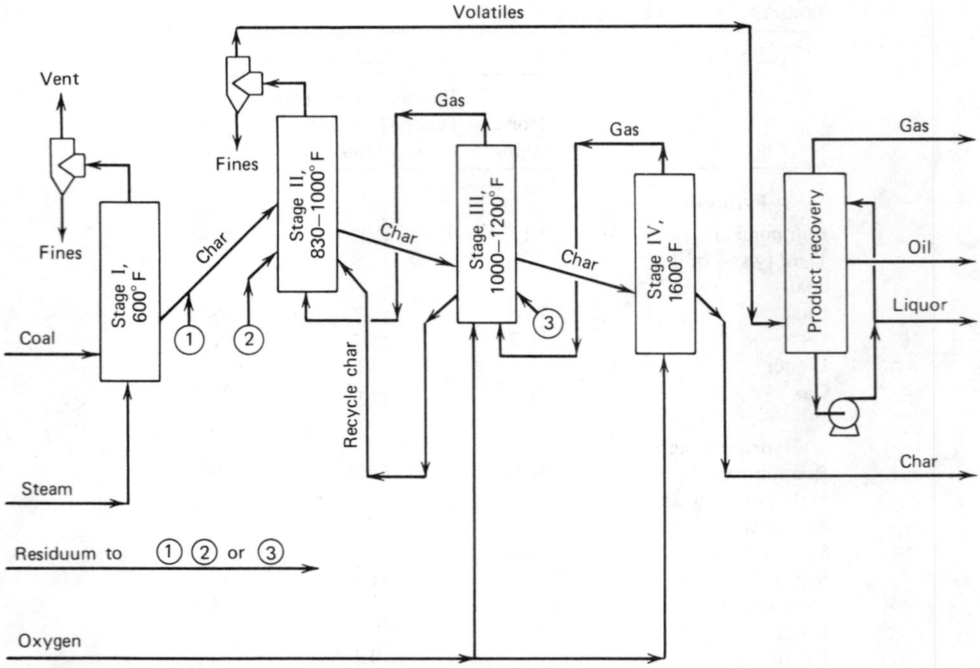

Figure 13.24 The Project Seacoke configuration in the COED process development unit.[90]

higher, and hydrocarbon gas yield correspondingly lower.

Two different coals and three residual oils were tested; the properties of these feeds are given in Tables 13.6 and 13.7. These tests confirmed the synergism observed earlier in the laboratory. This is shown in Table 13.8 for McKinley coal and short residuum in a 72 : 28 weight ratio. Similar results were obtained with Illinois No. 6 Seam coal, and with long residua.

13.7.3 The U.S. Steel Clean Coke Process

The Clean Coke process, based on concepts developed by U.S. Steel, is designed to employ both a carbonization and a hydrogenation step, operating in parallel, to produce metallurgical coke and also tars, oils, organic liquids, and gases.[91–94] The process is intended to be hydrogen- and energy-balanced.

Carbonization takes place in a two-stage vertical fluidized bed. Experimental PDU has been built and aspects of the process were tested under the sponsorship of ERDA. A simplified process flow diagram appears in Fig. 13.25.

After beneficiation, a portion of the coal is carbonized in a fluidized bed in the presence of a recycle stream of gas, rich in hydrogen and essentially free of sulfur. Most of the sulfur is removed from the coal. The char is pelletized, using a tar binder derived from the process. The pellets are cured and then calcined, producing hard, low-sulfur metallurgical coke and a hydrogen-rich gas.

The remaining portion of the coal is slurried with a process-derived carrier oil and then hydrogenated noncatalytically at 3000–4000 psia (20,680–27,580 kPa) and 900°F (482°C). Liquids and gases are removed from the residual solids and are processed into liquid fuels, chemical feedstocks, and oil fractions, which are recycled to the

[91] K. S. Schowalter and N. S. Boodman, *Chem. Eng. Prog.*, **70**(6) (1974).

[92] K. A. Schowalter and E. F. Petras, *Coal Proc. Tech.*, **2,** 109.

[93] T. F. Johnson, K. C. Krupinski, and R. J. Osterholm, American Chemical Society, Division of Fuel Chemistry Symp., Chicago (August 24–29, 1975).

[94] U.S.S. Engineers and Consultants, Inc., *Clean Coke Project,* OCR Annual Report, Washington, D.C., 1975.

Table 13.6 Analyses of Coals Used in Seacoke Pilot-Unit Runs[90]

	McKinley Coal: Nonagglomerating hvCb, New Mexico	Crown Coal: Mildly Caking, hvBb, Illinois No. 6 Seam
	Proximate Analysis (wt %)	
Moisture	10.81	10.86
Volatile matter	35.81	33.74
Fixed carbon	45.26	45.32
Ash	8.13	10.08
	Ultimate Analysis (wt %, dry)	
Carbon	70.82	68.60
Hydrogen	5.05	5.01
Nitrogen	1.28	1.29
Sulfur	0.58	3.79
Oxygen	13.83	10.50
Ash	8.44	10.81
	Analysis, Tyler Mesh (cum wt %)	
16	12.1	14.86
28	49.5	40.60
48	64.43	64.95
100	81.35	80.28
200	90.96	89.17
325	94.83	93.12
	Densities (lb/ft^3)	
Bulk	46.3	47.00
Pack	52.1	53.85
Particle Hg displacement	79.4	78.27
Heating value (Btu/lb, dry)	12,533	12,400

process. The anticipated product mix is shown in Table 13.9.

Initial bench-scale carbonization-desulfurization studies were conducted in a 3-in. (7.52-cm) batch fluidized-bed reactor, having a capacity of 800–1000 g of solids. The material studied was either Illinois No. 6 Seam coal, or a 50% mixture of this coal and recycled char. The carbonizing hydrogen was a mixture of hydrogen, methane, and carbon monoxide, and the superficial gas velocity was about 1.2 ft/s (0.37 m/s).

The results obtained in a typical run, using preoxidized coal and operating at 1365°F (741°C) and 80 psia (552 kPa, absolute) for a 30-min residence time, are shown in Table 13.10.

The effects of time, temperature, and pressure are shown in Table 13.11. [In the small experimental unit it took 70 min to heat the bed to 1150°F (621°C), and 100 min to heat the bed to 1365°F (741°C).] A comparison of the results in columns 1 and 2 indicates that most of the sulfur is removed during the warmup process. A comparison of the results in column 3 with the results in column 2—or more properly with the results in column 1—indicates that somewhat more sulfur is removed at the higher temperature. The results in columns 4, 5, and 6 indicate that more sulfur is removed as the operating pressure and the partial pressure of hydrogen are increased, even though the concentration of hydrogen in the recycle gas is decreased.

When a caking coal, such as Illinois No. 6 Seam coal, is heated to about 750°F (399°C), the solid particles of coal become plastic and stick together in an agglomerating mass, thereby bogging down the fluidized bed. Agglomeration of caking coals can be avoided by mild preoxidation at atmospheric pressure and 350°F (177°C) for 10 min. This treatment chemically alters the surface of the coal particles. Alternatively, bogging can be con-

Table 13.7 Petroleum Residua Used in Seacoke Pilot-Unit Runs[90]

	Long: Atmospheric Tower Bottoms	Intermediate: Vacuum Tower Bottoms	Short: Visbroken Vacuum-Flashed Bottoms
Percentage of crude	48.6%	23.7%	13.1%
Density at 60°	0.9422	0.9896	1.0375
API	18.6°	11.4°	4.8°, 8.646 lb/gal = 363.13 lb/bbl
Pour Point			160°
Flash Flash	345	590	
Ash (SO_4)	0.04	0.08	0.08
Ramsbottom/carbon	10% (9%)	17%	26.4
SUS viscosity at 210°F	129	2693	19,159
SUS viscosity at 250°F	79	891	3,610 (361 SF at 250°F; 86.3 SF at 300°F)
IBP[a] °F	455	772	820
1° Vol. %	509	832	868
3	561	880	915
5	596	898	947
10	647	924	1,007
	718	982	15% 1,039
30	762	~ 23% 998	
40	818		
50	863		
60	917		
70	996		
77	1030		
Rec. %	77	23	15
Res. %	23	77	85
Carbon	85.48	85.83	86.48
Hydrogen	11.74	10.81	9.14
Oxygen	0.51	0.71	0.85
Nitrogen	0.25	0.53	0.59
Sulfur	1.99	2.12	2.70
Ash	0.03		0.14
Gross heat of combustion	18,884 Btu/lb		17,872 Btu/lb
Metals			
Vanadium } Nickel			39 ppm

[a]ASTM distillation.

Table 13.8 Summary of Seacoke Process Tests in COED–PDU for McKinley Coal, Short Residuum[90]

Pyrolysis temperatures: stage 2 = 1000°F, stage 3 = 1200°F

	Yields (wt %)	
	Actual	Nonsynergism Prediction
Oil	34.1	32.3
Hydrocarbon gases	6.7	8.0
Char	42.4	44.4
$CO + CO_2$	18.9	20.0
Water	7.8	6.3
H content of oil	9.5	8.3

trolled by blending the coal with a noncaking char (1 : 1 ratio), which physically separates the softened coal particles in the plastic region.[93]

The results of a single test of unoxidized and preoxidized coal, at 1365°F (741°C) and 80 psia (552 kPa, absolute) with 30-min residence time, indicated that carbonization of unoxidized coal plus char results in somewhat better desulfurization, less char formation, and greater tar production. However these results were not verified by additional experimentation.[93]

Both methods of avoiding agglomeration worked well in the batch reactor, in which the charge was heated at about 10°F/min

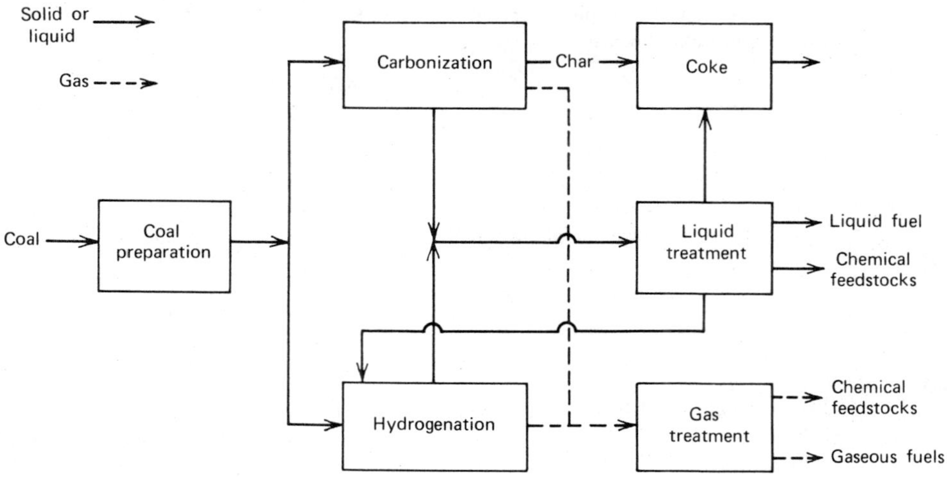

Figure 13.25 Simplified process flow diagram of the U.S. Steel Clean Coke process.[91]

(5.5°C/min). However difficulties arose when preoxidized coal was fed to a continuously operating bed, in which the heating rate was about 10°F/s (5.5°C/s). Sticky tars were generated at a faster rate than could be vaporized from the bed, and agglomeration resulted.

Work on this problem led to the development of a two-stage fluidized-bed system. Tests were continued in the batch reactor. The first stage, partial devolatilization, was operated at 840°F (449°C) and 125 psia (862 kPa, absolute). Minus 20-, plus 65-mesh Illinois coal, which had been preoxidized at 350°F (177°C), was added to the first-stage

Table 13.9 Properties of Raw and Beneficiated Coals and of Products of Clean Coke Process[92]

Percentage of raw coal:		Carbonizer Feed	Hydrogenation Feed	Refuse of Fuel	
Size:		42	42	16	
Component (wt %)	Raw Coal (Illinois No. 6)	1/8 in. to +100 mesh	−100 mesh		Anticipated Product Mix
Moisture	8–10	3.5	3.5		
Ash	15	5–6	14	43	
Sulfur	3.5–4	2	3	5–6	
Coke	—	—	—	—	38.4
Hydrogenation residue	—	—	—	—	11.5
Ammonia	—	—	—	—	0.6
Sulfur	—	—	—	—	0.9
Ethylene	—	—	—	—	6.3
Propylene	—	—	—	—	1.0
Phenol	—	—	—	—	1.3
Cresols	—	—	—	—	1.6
Xylenols	—	—	—	—	1.3
Benzene	—	—	—	—	5.0
Naphthalene	—	—	—	—	2.0
Other chemicals	—	—	—	—	0.3
Tar products	—	—	—	—	0.6
Process loss and fuels consumed	—	—	—	—	29.2

Table 13.10 Results of Typical Carbonization Test at 1365°F, 80 psia for 30 min[93] Clean Coke Process

	Amount in Charge (wt %)	
	Component	Total
Feed		
Coal (1.7% sulfur, 34% volatile matter)		100.0
Reducing gas		
Product		
Char (0.5% sulfur, 1.6% volatile matter)		66.4
Carbonization tar		
Chemical Oil (< 385 F)		
Phenols	1.5	
Benzenes	0.6	
Tar bases	0.2	
Naphthalenes	0.2	
Naphtheles		
Heterocyclics	0.3	
Middle oil (385–635°F)[a]	5.1	
Heavy oil (>635°F)[b]	6.0	
Subtotal		13.9
Gas (after H_2S removal)		
Hydrogen	5.8	
Methane	7.6	
Carbon monoxide	1.2	
Ethylene	trace	
Ethane	trace	
Propane	trace	
Carbon dioxide	trace	
Subtotal		14.6
Water		5.1

[a] Middle oil is used to slurry the coal feed to the hydrogenation unit.
[b] Heavy oil is used as a binder in the char pelletizer.

fluidized bed. When the entire charge had been added, the bed was maintained at these conditions for 20 min. A portion of the semichar from the first stage was fed to the second stage, which was operated at 1380°F (749°C) and 125 psia (862 kPa, absolute) with a retention time of 15 min. After the desired retention time was achieved, the remaining portion of the char was added to the second-stage fluidized bed, and the entire mass was held at 1380°F (749°C) for 15 min. It was found that the sulfur concentration had decreased from 2.28% in the preoxidized coal to 1.63% in the partially devolatilized coal, and to 0.33% in the carbonized material.[93]

An experimental PDU having a capacity of 500 lb/day (227 kg/day) has been built, and will be used for development of the Clean Coke process under the sponsorship of the Department of Energy. The PDU includes a 10-in. (25.4-cm) diameter carbonizer-desulfurizer, a 10-in. (25-cm) diameter coal hydrogenation unit, and a coal preoxidation unit.

First-stage carbonizer operation at 800°F (427°C) and either 80 or 160 psia (552 or 1103 kPa, absolute) resulted in the production of a semichar containing about 1.8% sulfur and 20% volatile matter. Second-stage carbonizer operation at 1300°F (704°C) or higher produced a char containing 0.6–0.7% sulfur and 2–3% volatile matter.[94] Operation at about 1400°F (760°C) and 150 psia (1034 kPa, absolute), with residence times of about 90 and 100 min, produced a char containing 0.4–0.5% sulfur and about 0.6% volatile matter.

Table 13.11 Effect of Time, Temperature, and Pressure on the Desulfurization of Preoxidized Coal[93] Clean Coke Process

	(1)	(2)	(3)	(4)	(5)	(6)
Temperature (°F)	1365	1365	1150	1365	1365	1365
Pressure (psia)	80	80	80	30	80	125
Time at temperature (min)[a]	0	30	30	30	30	30
Fluidizing gas composition (mole %)						
H_2	38	40	27	55	39	33
CH_4	52	52 }	73	34	52	60
CO	10	8 }		11	9	7
Sulfur content (wt %)						
Coal	1.76	1.76	1.76	1.74	1.74	1.74
Char	0.73	0.51	0.85	0.77	0.50	0.33

[a] The heating time to reach 1150°F was 70 min, and the time to reach 1365°F was 100 min.

Experimental work on PDUs for other sections of the plant is continuing.

13.7.4 The Occidental Pyrolysis Process

The Occidental pyrolysis process was formerly known as the Garrett pyrolysis process. The new name is the result of changes in corporate affiliations.

This process is designed to produce liquid and gaseous hydrocarbons and a char suitable for use as a power plant fuel. The process is based on rapid low-temperature carbonization of coal with short residence times, and it achieves higher liquid yields than any of the other carbonization processes.

Carbonization takes place in an entrained bed, in which minus 200-mesh coal is added to a stream of hot char. The heatup rate for the coal has been calculated as more than 500°F/s (278°C/s).[95] The operating pressure is 50 psig (344 kPa or less).[96] Pyrolysis occurs in fractions of a second. Because the residence time in the reactor is short (2 s less), secondary decomposition of the volatilized products is minimized, thereby maximizing the yield of liquids. The noncondensable gases are recycled through a compressor for use as the carrier gas for conveying the feed coal to the pyrolysis reactor.[97] The basic flow scheme is shown in Fig. 13.26.

A portion of the char collected in the cyclone downstream from the reactor is heated by direct heat exchange with char combustion gases. The contact time between the recycling char and the combustion gases is short, thereby minimizing the formation of CO (from CO_2 and carbon). This reduces heat losses to the combustion gases and significantly improves the process heat balance.[97] The reheated char is returned to the reactor and supplies the heat for carbonization.

Initial tests were conducted in a small semicontinuous pyrolysis reactor. The weight yields from a high-volatile eastern U.S. bituminous coal, reacted at 1075°F (579°C) to maximize liquid yields, were based on coal fed, 56.7% char, 35% tar, 6.6% gas, and 1.7% water. The pyrolysis gas is rich in hydrocarbons, and has a heating value of approximately 700 Btu/ft^3 (26 MJ/m^3) on a dry, H_2S-free basis.[95] The tar composition is shown in Table 13.12.[98]

The quantity of gas produced can be increased by raising the pyrolysis temperature and also by using a lower rank coal. Tests in a laboratory pyrolysis reactor produced the following gas, tar, and char yields from a typical Western subbituminous coal, as the temperature was increased from 1500 to 1700°F (816 to 927°C). In these tests the operating pressure was 2.5 psig (17 kPa), the coal was ground to minus 200 mesh, and the

[95] A. Sass, *Chem. Eng. Prog.*, **70**(1), 72 (1974).

[96] D. E. Adam, S. Sacks, and A. Sass, *Chem. Eng. Prog.*, **70**(6), 74, (1974).

[97] A. Sass, "Clean Energy from Coal" session, Am. Inst. Chem. Eng. 67th Annual Meeting, Washington, D.C., December 3, 1974.

[98] J. R. Longabach and F. Bauer, in *Fuels and Chemicals by Pyrolysis*, Am. Chem. Soc. Symp. Ser. No. 32, L. F. Albright and B. L. Crymes, Eds., 1976.

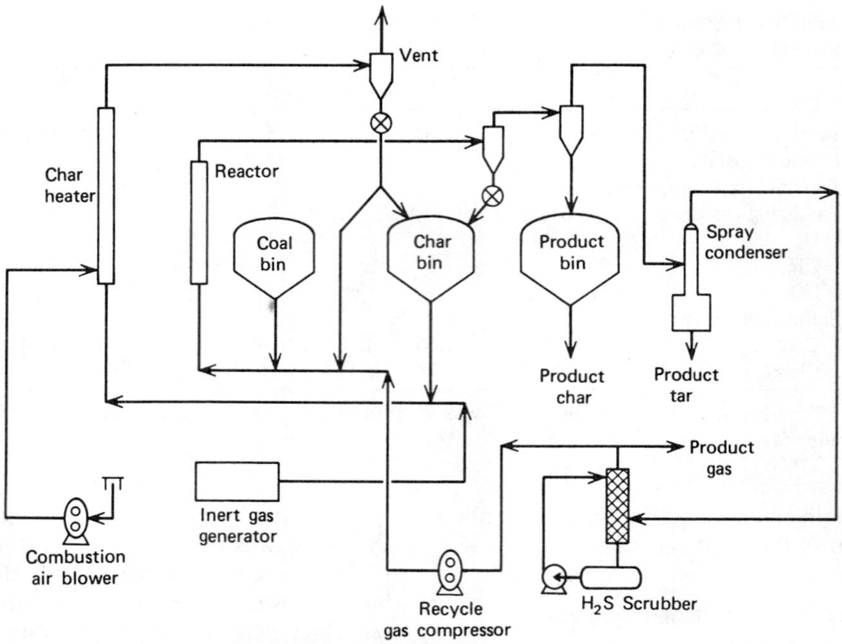

Figure 13.26 Coal conversion pilot plant.[97]

entraining gas for both coal and char was nitrogen.[99]

Product	Yield (wt %)
Dry gas	15–30
Tar	15–8
Char	55–40
Water	Balance

A typical gas produced at 1600°F (871°C) contained (in vol %), 35% H_2, 23% CO, 9% CO_2, 19% CH_4, and 14% C_2+ (dry, N_2-free and H_2S-free) and had a HHV of 646 Btu/scf (24 MJ/m³).[99]

A pilot plant was built in La Verne, California, in 1972. A flowsheet of this facility is given in Fig. 13.26. Since 1974 the unit has operated with coal over a wide range of operating conditions, and at rates up to 350 lb/hr (159 kg/hr) and char recycle at rates up to 2500 lb/hr (1134 kg/hr). Coal was pyrolyzed at a temperature of about 1100°F (593°C); recycle char usually was kept at 1200–1600°F (649–871°C). Results of the processing of a noncaking Western subbituminous coal are shown in Table 13.13.

A PDU having a coal capacity of 3 tons of coal per day and a design char circulation rate of 1000–1600 lb/h (454–726 kg/hr) has been built. Initial tests of the system have been satisfactory. Work is continuing, and subsystems are being constructed.[100]

A reactor design that will make it possible to pyrolyze caking coals without subjecting them to a preoxidizing step has been claimed by the developers. If finally proved, this design could be of considerable importance in the quest for means of overcoming the caking tendencies of bituminous coals. Initial pilot-plant tests of this design have been conducted satisfactorily.[97] More extensive pilot-plant tests are scheduled.

The marketability of the char that is produced, and the feasibility of using it as a power plant fuel, would be enhanced significantly if its sulfur content were reduced. Laboratory tests showed that if the pyrolysis char were subjected to a mildly acid aqueous leach to remove iron and calcium compounds

[99] H. G. McMath, R. E. Lumpkin, and A. Sass, "Production of Gas from Western Sub-Bituminous Coals by the Garrett Flash Pyrolysis Process," Am. Inst. Chem. Eng. 66th Annual Meeting, Philadelphia, November 1973.

[100] E. W. Knell et al., *Flash Pyrolysis Coal Liquefaction Process Development,* Quarterly Rep., July–September 1976, Occidental Research Corp. under U.S. ERDA Contract No. E(49-18)-2244.

Table 13.12 Tar from Occidental Flash Pyrolysis Process Operating at 1000°C[98]

Analysis (wt %)	
Moisture	0.0
Ash	0.0
Carbon	80.3
Hydrogen	7.0
Nitrogen	1.4
Sulfur	2.1
Oxygen	9.2
Other Properties	
Density (at 100°C)	1.04 g/ml
Viscosity (at 100°C)	127 cP
Heating value	15,342 Btu/lb
Boiling Range (% distilled)	
Initial boiling point	75° C
20%	75–200° C
25%	200–350° C
22.5%	350–520° C
Residue (32.5%)	>520° C
Distillate (75–350°C Fraction)	
Acids	25 vol %
Bases	—
Neutral oils	75 vol %

in the ash, the effectiveness of the hydrodesulfurization reaction would be improved markedly. The sulfur concentration of the resultant char is a function of the H_2S concentration in the H_2 used for hydrodesulfurization, the residence time of the leached char, the type of coal from which the char has been produced, and the reaction temperature. For example, hydrodesulfurization of a pyrolysis char produced from Kentucky bituminous coal, with a gas stream containing about 1 or 2% H_2S in the H_2, produced a char containing about 0.5% sulfur. Similar results were obtained from a char produced from Montana subbituminous coal, when the reaction time was less than 10 min with 2% H_2S, or 20 min with 1% H_2S. At longer reaction times the back-reaction of hydrogen sulfide and char caused the sulfur content of the latter to increase.[100] Additional tests of this process are to be made in a pilot plant.[97]

The tars that are produced by coal pyrolysis can be upgraded by hydrogenation. However these tars contain organometallic compounds, whose composition is a function of the coal from which the tar had been

Table 13.13 Pyrolysis Yields and Char Properties[97] Occidental Process

	Process Temperature 950°F	Process Temperature 1100°F
Pyrolysis Yields (wt %, maf)		
Dry gas	7.1	11.8
Water	9.0	11.9
Tar	13.5	16.3
Char	70.7	60.0

	Western Subbituminous Coal	Process Temperature of Pyrolytic Char 950°F	Process Temperature of Pyrolytic Char 1100°F	Process Temperature of Product Char 1400°F	Process Temperature of Product Char 1700 °F
Analysis of Coal and Char (wt %, maf)					
Carbon	73.05	78.01	82.54	88.96	94.00
Hydrogen	5.02	4.74	3.60	2.71	1.71
Nitrogen	1.21	1.38	1.56	1.50	1.18
Sulfur	0.65	0.53	0.58	0.68	0.57
Oxygen	20.08	15.84	11.74	6.15	2.54
Other Coal and Char Properties					
Atomic ratio (H/C)	0.82	0.65	0.52	0.36	0.22
Ash, typical (wt %)	6.61	9.0	10.3	13.8	14.5
Heating value (Btu/lb)	9400	11,430	11,610	11,940	12,310

produced. Organometallic compounds are potential hydrogenation catalyst poisons. This factor will have to be considered when selecting the proper catalytic hydrotreating process.[97]

The major feature of the Occidental Pyrolysis Process is its high liquid yields. Yields of tar as high as 35 wt % have been obtained. These are significantly higher than those obtained with the Coalcon hydrocarbonization process (29%) and the COED process (20–22%). Their high yield is attributed to the very short residence times. In addition, the tar yield is dependent on carbonization temperature, attaining a maximum at about 1070°F (577°C), as shown in Fig. 13.27.

13.7.5 The Lurgi-Ruhrgas Process

Lurgi-Ruhrgas AG has developed a system very similar to the Occidental process. Sommers[101] reports liquid yields of up to 180% of the Fischer assay, with yields of up to 30 wt % coal when processing high-volatile coals. Peters[102] explained these high yields by suggesting the possibility of simple distillation of the coal bitumen without the occurrence of decomposition, because of the short residence times. Caking coals can be used, but noncaking or lightly caking coals are preferred.

[101] H. Sommers, *VGB Kraftwerkstechnik,* **5,** 306 (May 1974).

[102] W. Peters, *Chem. Ing. Tech.,* **32**(3), 178 (1960).

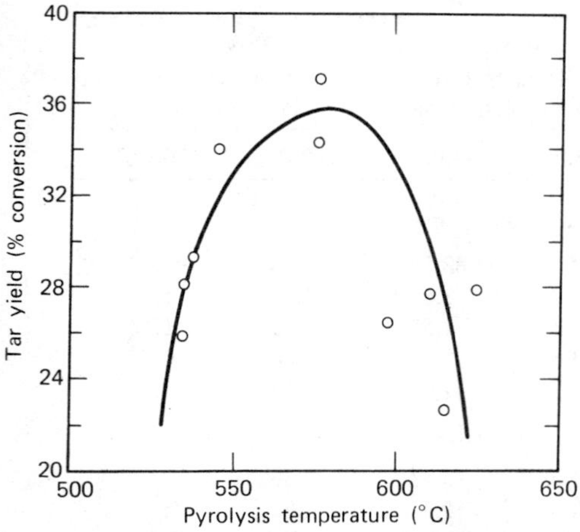

Figure 13.27 Tar yield versus temperature.[95]

Between 1949 and 1960 this process was tested over a total of 2100 operating hours in the Dorsten, Germany, pilot plant, which had a capacity of 10 tonnes of coal per hour. Continuous runs of up to 200 hr were conducted. Work was stopped because of increases in coal prices, decreases in petroleum product prices, and natural gas finds in Europe. Since 1960 these economic conditions have changed, and Lurgi has shown renewed interest in possible commercialization of the process. One approach is toward the integration of the process with electric power production: the coal would be carbonized in the Lurgi-Ruhrgas process to produce a significant yield of liquid products, which would displace an equivalent amount of crude oil; the by-product char and fuel gas would fire the utility boilers.

A flowsheet of the Lurgi-Ruhrgas process (Fig. 13.28) shows the process connected to a utility boiler. The coal is carbonized by heating it rapidly with a circulating stream of hot char in a mechanically agitated mixer. The mixer has two shafts with radially attached arms, which rotate in the same direction to prevent the collection of unwanted deposits of coal and char dust and tars. This mechanical system was replaced (at the Dorsten pilot plant) by a single high-pressure steam jet coal feeder. This new mixer led to a coarse char size consist, which created char circulation control problems and cyclone furnace combustion difficulties. These had not been remedied up to the publication date of the paper by Sommers[101] (May 1974).

After carbonization is complete, the char mix is split into two streams. One stream contains the product char, which consists of the finer size fraction of the total solids stream after carbonization. It can be forwarded hot as the fuel to cyclone furnaces, which provide heat to the utility boiler. The other stream is conveyed to a lift line and is pneumatically raised to the top of the line. At the bottom of the lift the char is picked up by air, which rapidly oxidizes a portion of the char, thereby releasing heat for reheating to a higher temperature the char leaving the carbonizer. The resulting combustion gases and elevated hot char are separated in a series of cyclone separators. The hot char is returned to the mixer to pyrolyze the feed coal. The combustion gases, containing about 6 vol % $CO + H_2$, are used for lift-line air

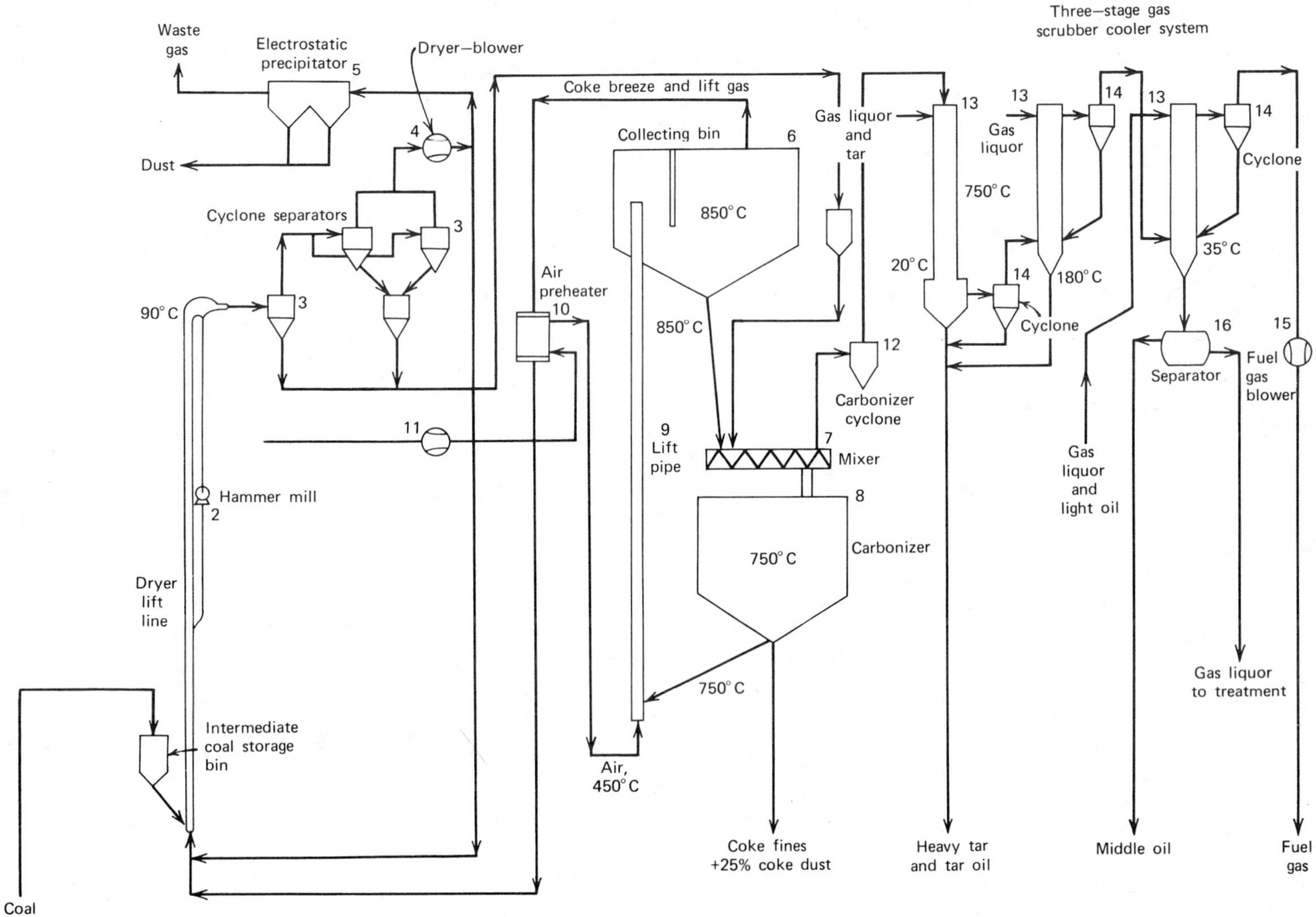

Figure 13.28 The Lurgi-Ruhrgas process.

preheating and coal drying. The CO and H_2 in these gases can be burned with air, introduced into the lift-line gases after removal of char.

Product gases and vapors driven off from the feed coal in the carbonizer are passed to cyclone separators to remove as much of the entrained solids as possible; then they are cooled and scrubbed in a three-stage, gas-liquid contact system. Tars, oils, and water are condensed and collected separately. The uncondensed portion is a fuel gas having a higher heating value of about 550–1090 Btu/scf (20,492–40,612 kJ/m^3; 4880–9730 kcal/m^3). The heating value increases with decreasing carbonization temperature (Table 13.14), and the quantity of fuel gas increases with increasing carbonization temperature.

Heat and material balance data obtained at the Dorsten pilot plant, operating at 10 tonnes/hr with coal from the Fürst Leopold–Baldur colliery in Dorsten, are given in Table 13.15. (Table 13.14 lists the properties of the products.)

Table 13.14 Properties of Products[101] Lurgi Ruhrgas Process

	At 920°C		At 590°C	
	Product Gas	Flue Gas	Product Gas	Flue Gas
Fuel Gas Less H_2S and Gasoline				
SO_2 (vol %)	—	0.1	—	0.1
CO_2 (vol %)	3.4	16.8	4.0	16.1
N_2 (vol %)	1.0	77.1	0.2	76.0
O_2 (vol %)	0.0	0.0	0.0	0.3
CO (vol %)	20.2	3.2	11.5	3.0
H_2 (vol %)	48.1	2.5	13.4	3.6
CH_4 (vol %)	23.9	0.3	50.5	0.9
C_2H_6 (vol %)	0.4	—	9.2	—
$C_2H_4 + C_2H_2$ (vol %)	2.8	—	3.2	—
$C_3 + C_4$ (vol %)	0.2	—	8.0 ($C_{3.1}H_{6.8}$)	—
Volume (m^3/ton of coal)	387	825	58	448
Density (g/m^3)	579	1347	942	1328
Heating value				
kJ/m^3	20,418	842	40,724	1236
Btu/scf	548	23	1,093	33
Coke				
Volatiles (%, maf)	1–2		7–8	
Ash (%, mf)	37.3		34.5	
Size range (mf)	20–50% (0.09mm)			
Sludge, Tar, and Oils				
Sludge (% mf)				
Coke	52		—	
Ash	31		—	
Tar	17		—	
Tar Softening point (°C)	~ 40		>100	
Dust Content (%)	~ 0.5		~3–10	
Middle oil	~ 50%			
Light oil	Naphthalene Chiefly benzol homologues		No naphthalene Gasoline	

Table 13.15 Heat Balance and Material Flow[101] Lurgi Ruhrgas Process

	990/920°C		650/590°C	
Products (% of coal)	Weight	Calorific Value	Weight	Calorific Value
Coke to boiler	54.0	48.5	63.4	59.7
Burnt coke	6.9	11.5	4.0	6.7
Dry sludge	6.1	6.2	—	—
Pitch and tar	3.8	6.1	14.2	20.8
Middle oil and naphthalene	0.8	1.6	1.7	3.3
Light oil	1.5	2.6	0.5	1.0
Gas without gasoline and H_2S	15.7	23.0	3.8	6.9
$H_2S + NH_3 + H_2O$	11.2	0.5	12.4	0.2
Balance	—	—	—	1.4[a]
	At 990°C		At 650°C	
Materials to Boiler	Weight	Calorific Value	Weight	Calorific Value
Flue gas (latent and sensible heat)	77.8	2.0[b] + 3.6	41.5	1.6[b] + 1.5
Steam (latent and sensible heat)	1.8	0.2	1.3	0.1
Coke (latent and sensible heat)	54.0	48.5 + 2.4	63.4	59.7 + 2.
Total	133.6	56.7	116.2	64.9
Flue gas heat from boiler	92.5	2.9	70	2.2
into coal during drying	93.5	0.4	93.5	0.4
Waste heat in vapors	99	1.2	76.5	1.1
Air preheating (from 60 to 450°C)	72.7	1.3	38.8	
Coal Efficiency (Summary)				
Fuel gas and hydrocarbons		33.3		32.0
Steam boiler (supply less requirement)		60.0[c]		62.7
Ratio of waste heat steam to losses		0/6.7		1.5/3.

[a] Not considered for heat balance.
[b] Of this amount, 0.5 or 0.3% is gasification heat.
[c] Including dry sludge.

13.7.6 Other Flash Pyrolysis Processes

Other flash pyrolysis processes have been proposed and studied. The National Coal Board has intensively studied high-speed carbonization in a dilute phase, through-wall heated transport reactor.[103] They indicated high liquid yields at short residence times. The liquid yields decreased as the residence times increased, and as the furnace (heat source) temperature increased above 1380°F (749°C). Their findings agree with those obtained in high-speed transport pyrolysis studies made in the early phases of the COED project.[79]

[103] National Coal Board (England), private communication (September, 1962).

13.7.7 Consolidation Coal Studies

Consolidation Coal studied the effect of sweep gas flow on the tar yields obtained by low-temperature carbonization of highly caking coals. To simulate large-scale operating conditions in a small-scale unit, they equipped their 6-in. (15.24-cm) diameter fluidized reactor with a stirrer. This device provided the necessary added mixing in the small reactor to assist in achieving rapid heat transfer and good temperature control, even when operating at low fluidizing velocities. In this way it is possible to use the same sweep gas rates in the small unit as in a large one, and still maintain the same residence times without having to use the same bed depths. If the same bed depth were used in the small unit, the resulting inordinately large height/diameter ratio in it would result in excessive slugging and unrealistic process simulation.[1] A schematic flowsheet of the unit is shown in Fig. 13.29.

The carbonizer was 6 in. (15.24 cm) in diameter and 20 in. (50.8 cm) deep in its lower bed section. The stirrer contained 13 three-blade propellers and rotated at 112 rpm. Baffles extended into the vessel between each two blades.

Tests were made on high-volatile bituminous Pittsburgh Seam coals from the Montour No. 10 and Arkwright mines in western Pennsylvania. The coals were ground to minus 28-mesh. The fluidizing gas was cleaned recycle gas. The carbonization bed was operated at 925°F (496°C) and 10 psig (69 kPa).

Comparison of the data obtained in the stirred bed with those from a 1.5-ton/hr (1361-kg/hr) pilot unit verified that the smaller unit gave the same tar yields when sweep gas rates were equal.

Results of this test are described below.

[104] R. T. Struck, P. J. Dudt, and E. Gorin, *Ind. Eng. Chem., Process Des. Dev.*, **6**(1), 85 (1967).

The sweep gas rate is defined as the number of cubic feet of gas entering the carbonizer bed (at reaction conditions) per pound of maf coal fed. The amount of preoxidation of the feed coal is defined as the weight percent of oxygen consumed, based on maf coal.

Tar yield decreased with decreasing sweep gas rate. It was found that the total tar yield for untreated coal decreased from 23–25 wt % maf coal at a sweep gas rate above 15 ft^3/lb (0.94 m^3/kg) maf coal, to about 19 wt % at 5 ft^3/lb (0.31 m^3/kg). During these tests the solids residence time was 45–120 min.

The amount of preoxidation required to avoid bogging of a fluidized carbonizer varies greatly with the feed coal. For the highly caking coals used in these experiments, approximately 4% oxygen addition is required for operation in an unstirred fluidized carbonizer.

Preoxidation of the coal had a detrimental effect on tar yields, as shown by a series of tests at constant sweep gas rate [10 ft^3/lb (0.62 m^3/kg) maf coal]. The tar yield decreased steadily from more than 22 wt % maf coal with no preoxidation to less than 14% with 6% preoxidation. Preoxidation was conducted at 725°F (385°C), and as much as 2% tar yield was recovered in this operation.

The loss in tar yield due to preoxidation is about one pound of tar per pound of oxygen

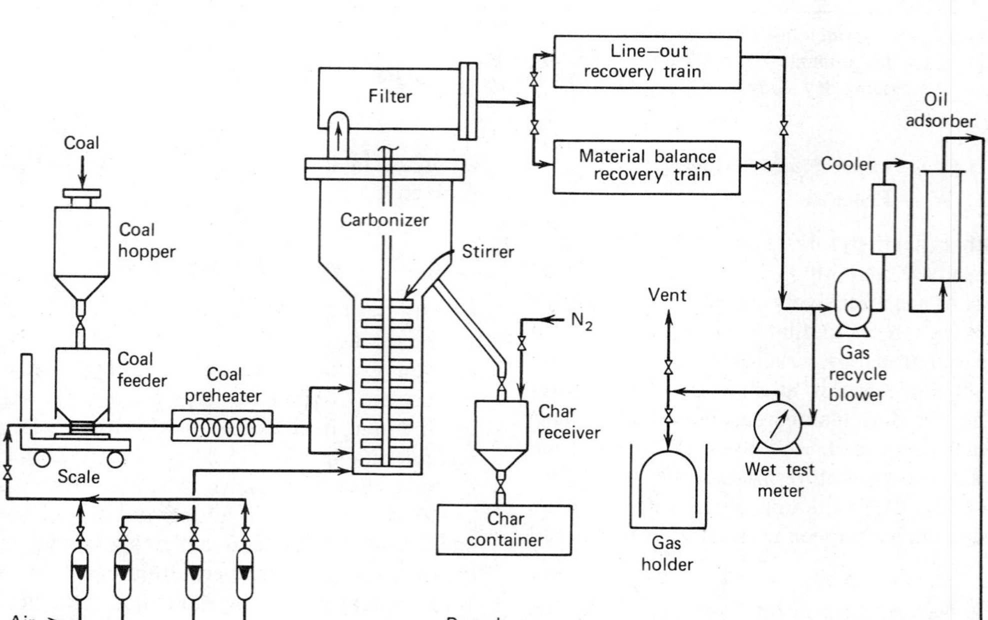

Figure 13.29 Stirred carbonizer unit.[104]

consumed. This is similar to the yield loss obtained in direct carbonization of untreated coal when oxygen is injected with the coal. The loss of tar is less when oxygen is injected below the coal feed point, indicating that oxygen preferentially reacts with char instead of with tar vapors when both are present.

Increasing the solids residence time from 45 to 120 min appeared to have no effect on tar yields at 925°F (496°C). Increasing the vapor residence time from 22 to 84 s in the stirred carbonizer appears to reduce tar yield by about 2 yield-percent. This phenomenon, however, may be confused by the decreased sweep gas flow associated with the increased vapor residence time.

Test work with this unit was discontinued a number of years ago.

The advantage of this process lies in its ability to handle nonpreoxidized, caking, high-volatile coals under nonagglomerating conditions, thereby achieving the inherently high liquid yields of these coals. The scaleup to commercial size of the stirrer used in this process does not appear to be feasible.

13.7.8 CSIRO Studies

Work by the Coal Research Section of the Commonwealth Scientific and Industrial Research Organization (CSIRO) was reported in the first Supplementary Volume.[1] CSIRO has studied fluidized carbonization in a pilot plant, with particular reference to the production of char for use in the coal mixes to coke ovens. Their reactor was 6 in. (15.24 cm) in diameter with a capacity of 50 lb (23 kg) of coal per hour. This coal was usually carbonized at 750–1110°F (399–599°C), using air or gas recycled from the process as the fluidizing medium. Figure 13.30 is a flowsheet of the plant.

When the coal was fluidized with air, the heat for carbonization was largely derived from combustion of a portion of the char product, although electrical heaters were employed when starting. When reduction of the

[105] K. McG. Bowling and P. L. Waters, *Br. Chem. Eng.* **13**(8); 1127 (1964).

[106] K. McG. Bowling and P. L. Waters, *Br. Chem. Eng.*, **7**(2), 98 (1962).

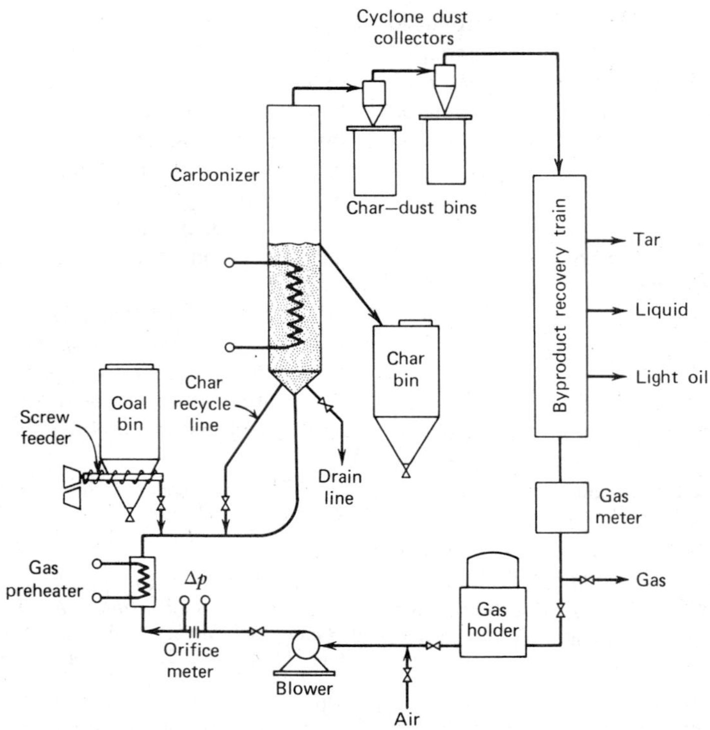

Figure 13.30 Flowsheet of the CSIRO pilot plant for the fluidized carbonization of coals.[105,106]

caking tendency was necessary, the coal was first oxidized at 300–350°C in a similar but separate bed. The tests showed that when air was used for fluidizing, the yields of char and tar from a medium-volatile, nonagglomerating coal were lower than with recycle gas. On the other hand, the gas yield was increased, although it possessed a low-heating value: 60–70 Btu/scf (2236–2608 kJ/m^3). During carbonization the coal particles swelled considerably, and a large proportion of the finer particles was carried over with the gases, so that efficient dust removal equipment was necessary.

Recent test work has concentrated on the development of a process that would maximize the yield of liquids.[107] Laboratory experiments were conducted using a very small flash pyrolysis reactor operating at atmospheric pressure and about 950°F (510°C). The coal feed was finely ground. It was mixed with sand, and the entire bed was fluidized by a stream of nitrogen.

Evolved vapors and gases were flushed away by a secondary stream of nitrogen introduced above the bed. The gas residence time was about 0.5 s. Under these conditions, the tar yield was 20 wt % (maf) for brown (lignite) coal feed, and 30 wt % for a subbituminous coal feed. When the tar produced from the subbituminous coal was submitted to catalytic hydrogenolysis, 63% was converted into a clear, volatile, easily ignited liquid and 27% was converted into gas and water.

13.8 HYDROPYROLYSIS PROCESSES

13.8.1 The Coalcon Process

The Coalcon process is the most technically advanced of the hydrocarbonization processes. It is based on work started in 1960[14] by Union Carbide Corporation, whose objective was to develop coal as a commercial source for chemicals. That objective was not attained because of the complex product distribution obtained by the hydrocarbonization of coal, and because of the adequacy—at that time—of inexpensive intermediates provided by the petrochemical industry. With the increases in prices for crude oil since 1973 and the rapid depletion of clean energy reserves in the United States, interest was renewed in this process. Union Carbide joined with Chemical Construction Corporation in 1973 to form a joint-venture company, Coalcon, to develop further this technology. For this purpose a contract between Coalcon and ERDA was entered into late in 1974. Detailed engineering and evaluation of this process, based on earlier experimental work, was completed in June 1977. It was decided then to defer construction of a demonstration plant until after more piloting of the process with Eastern caking coals. Some of the engineering evaluation results have been published.[108,109] Recently, Union Carbide took over 100% interest in Coalcon.

Figure 13.31 presents a flow diagram for the process. Using waste heat from boiler stack gases, 40 × 60 mesh coal is dried and preheated to about 620°F (327°C), then fed to the hydrocarbonization reactor via lock hoppers. The preheated coal is conveyed in dense phase by hydrogen into the hydrocarbonizer, which is fluidized with hydrogen entering at about 535°C.

One of the critical facets of this process is the feeding of raw coal in such a way as to avoid uncontrolled agglomeration in the reactor. The method chosen for this process involves mixing the raw feed coal with a large recycle stream of residual char as it is fed into the hydrocarbonizer. Either of two modes can be employed. One involves internal circulation of char; the other, external circulation.

The internal system (Fig. 13.32) was investigated by the British Gas Council,[110] which developed it to avoid excessive agglomeration of heavy oil when the oil is fed to a hydrogenator, employing a fluidized bed

[107] I. W. Smith and P. C. Wailes, *Proc. Inst. Chem. Eng., Symp. Energy and Liquid Fuelds*, Adelaide, Australia, 1975.

[108] Edward T. Coles, *Commercial Plant Process Evaluation Report*, Rep. U.S. ERDA, No. PER(A)-0, August 15, 1975, Contract No. E (49-18)-1736.

[109] Dravo Corp., *Conceptual Commercial Design and Commercial Feasibility Evaluation for Clean Boiler Fuel Facility*, Vol. 1, *Conceptual Commerical Design*, Vol. 2, *Selection and Evaluation*, Vol. 3, *Commercial Feasibility*, U.S. ERDA Rep. No. FE-1772-13, February 1976.

[110] A. G. Horsler, et al., *Chem. Eng. Prog.*, **65**(10) 59 (1964).

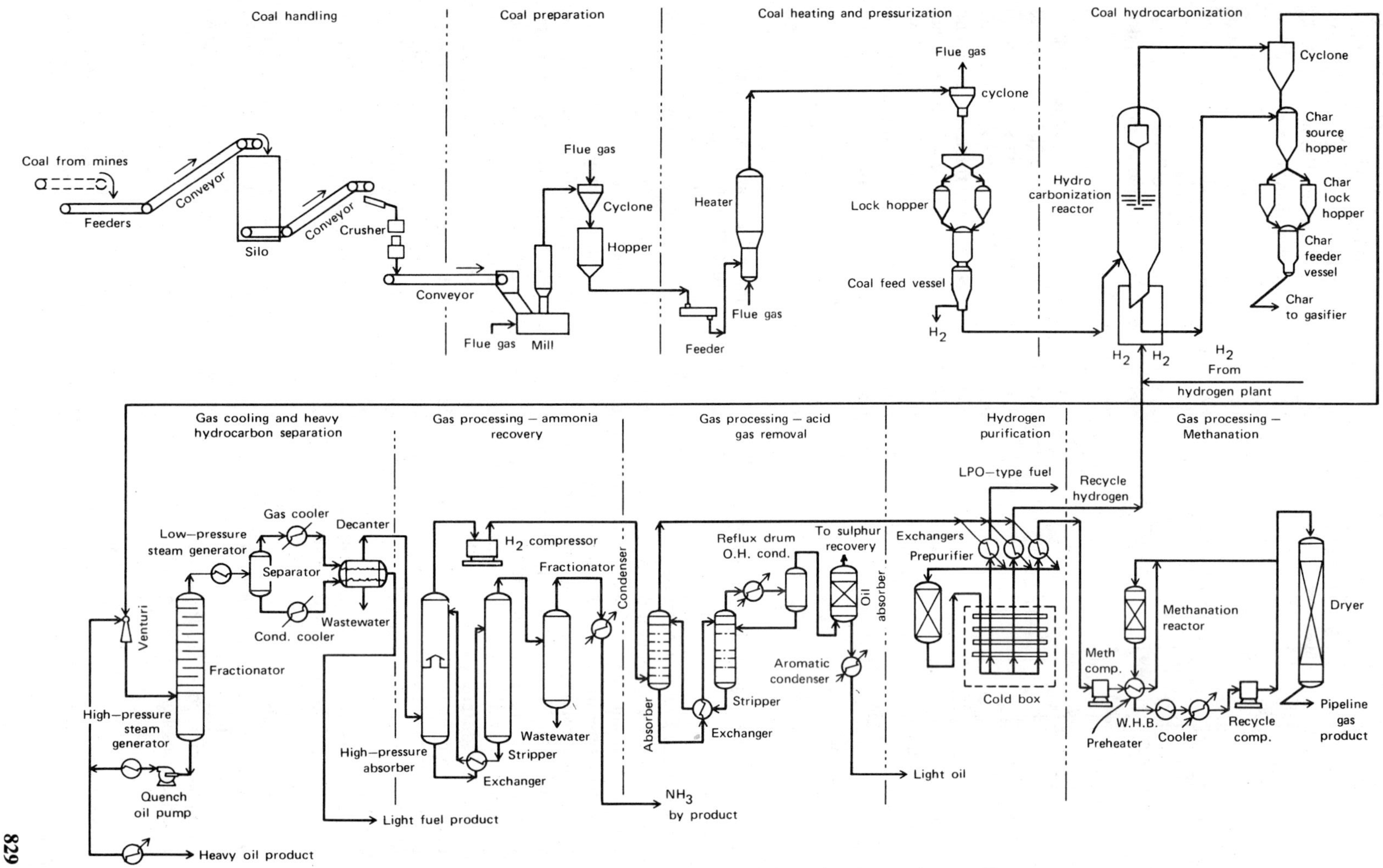

Figure 13.31 Coalcon's hydrocarbonization process flow diagram.[108]

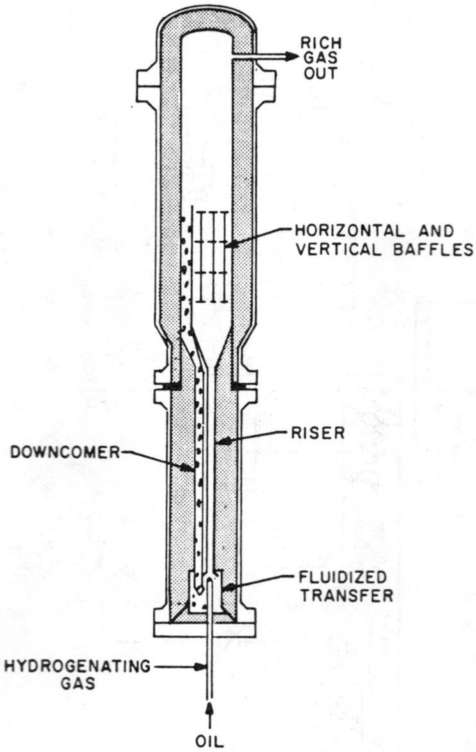

Figure 13.32 The fluidized-bed hydrogenator.[110]

of coke. Heavy oil could not be fed directly without undue agglomeration of the coke particles near the feed point. To overcome this, a draft tube was inserted in the fluidized bed. A lower solids-gas density exists in the draft tube than in the annular fluidized bed because of the higher gas velocity in the tube [~ 30 ft/s (9.1 m/s) vs. ~ 1 ft/s (0.3 m/s)]. This density differential causes a large flow of solids from the fluidized bed to the bottom of the draft tube, where the heavy oil was fed. The British Gas Council successfully operated the system with feedstock varying from light distillates to crude oil without significant agglomeration of the particles in the fluidized beds. Extension of this system to feeding coal appears to be obvious. Westinghouse Research Laboratories did so for their fluidized-bed coal gasification process, in which caking coals are fed directly into a devolatilizer-desulfurizer operating at 1300–1700°F (704–927°C). The recirculation of solids can be controlled by varying the gas velocity in the draft tube to give weight ratios of 100 parts of recycling solids to 1 part of coal feed.

This technique effectively prevents agglomeration of caking coal as it devolatilizes.[111,112] This concept was built into the Westinghouse coal gasification PDU at Waltz Mill, Pennsylvania. This unit has an inside diameter of 20 in. (50.8 cm) and operates at up to 1990°F (1088°C) and a maximum pressure of 300 psig (2070 kPa). The unit has gone through a successful 2-week test run and is considered to be operable.[113] A similar system has been demonstrated at the CSF pilot plant in Cresap, West Virginia.[114]

The external system (Fig. 13.33) is being considered for the Coalcon design. It operates on the same principle as the internal system: maintenance of a density difference in the external feed leg, where the circulating solids are mixed with the fresh coal feed and with the internal fluidized bed. The external system has certain advantages over the internal; namely, greater ease and access for maintenance, and possibly more positive control of gas flow into the risers.

Returning to the process depicted in Fig. 13.31, it is seen that this process is based on the noncatalytic hydrogenation of coal in a one-stage fluidized bed, at moderate temperatures [up to 1040°F (560°C)] and intermediate pressures [up to 1000 psia (6895 kPa, absolute)], with coal residence times up to 9 min. Hydrogen is used for fluidizing the reactor. Reactions between hydrogen and the coal are exothermic and liberate enough heat to provide for the system heat losses, and to provide heat for the coal and hydrogen fed to the hydrocarbonizer temperature. The amount of heat released by these reactions can be calculated if a complete chemical description of the inlet and exiting streams is known. Lacking such data, it is possible to approximate this heat balance, as noted in Section 13.2, "Thermodynamic Data."

One of the most important economic fac-

[111] D. L. Kairns, *Evaluation of the Fluidized Bed Combustion Process*, Vol. 1, EPA 650/2-73-048c, December 1973.

[112] W. C. Yang and D. L. Kairns, "Recirculating Fluidized Bed Reactor Data Utilizing a Two-Dimensional Cold Model," paper presented at the 75th National Meeting, American Institute of Chemical Engineers, Detroit, June 3–6, 1973.

[113] U.S. ERDA, publicity release, no date.

[114] *Pilot-Scale Development of the CSF Process*, OCR Rep. No. 39, Vol. IV, Book 3; NTIS: PB-234 131/AS, August 1971.

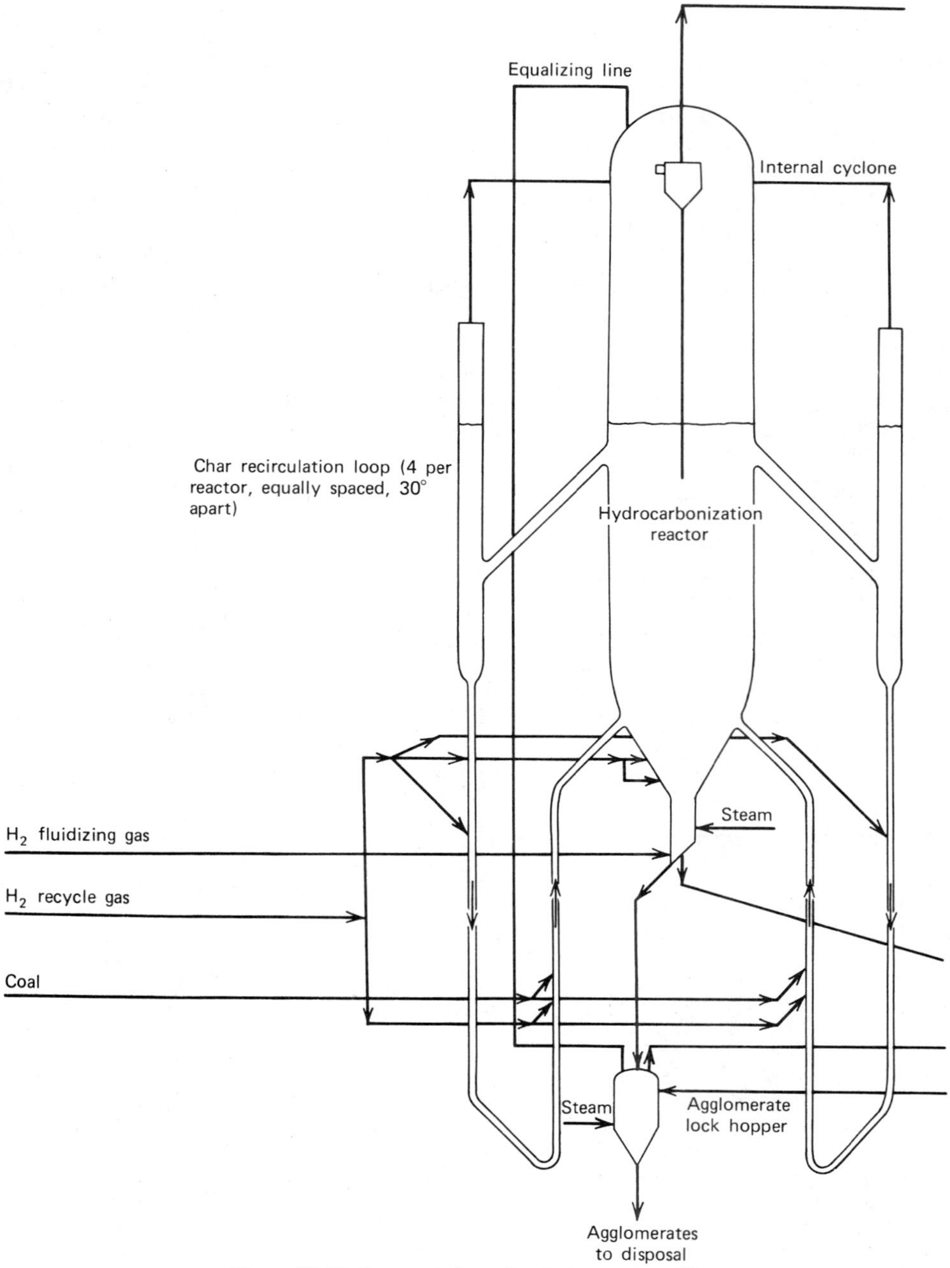

Figure 13.33 External char circulating system.[108]

tors in this process is the hydrogen consumption. Holmes et al.[18] correlated available data on subbituminous coal and lignite[115–118] by Equation 13.23:

[115] C. W. Albright and H. G. Davis, "A Process for the Hydrogenation of Low Rank Coals with High Yields of Phenolics," American Chemical Society Meeting, Division of Fuel Chemistry, Chicago (September 13–18, 1970.

[116] J. F. Jones and J. F. Start, *High Pressure Pyrolysis of Coal in a Hydrogen Atmosphere for Short Residence Times,* FMC Report No. PCR-319 under Contract No. 14-01-0001-235, Princeton, N.J., May 1963.

[117] E. L. Clark, *Ind. Eng. Chem.,* **42,** 861 (1950).

[118] T. J. Birch, K. R. Hall, and R. W. Urie, *J. Inst. Fuel,* **33,** 422 (1960).

$$H = 3.1089 \times 10^{-5}\, tP_{H_2} - 0.02066\, P_{H_2} + 0.39356\, P_{H_2}^{1/2} + 0.049920\theta - 5.4816 \quad (13.23)$$

Equation 13.23 is based on experimental data over the following ranges of temperature and pressure, and it is suggested that the equation not be extrapolated significantly beyond these ranges.

Temperature Range (°C)	Pressure Range (psia)	kPa
480–600	200–1000	1379–6895
700–760	150–600	1034–4137
840–860	20–600	138–4137
950–960	350–550	2413–3792

Extension of this correlation for hydrogen consumption to other coals, specifically the caking coals, would be of value in the future.

The effect of hydrogen partial pressure and temperature, for a 10-min residence time, on hydrogen consumption is shown in Fig. 13.3. The negative hydrogen consumptions noted in this figure correspond to the evolution of free hydrogen from the coal.

Albright and Davis[115] presented data on the Coalcon process, based on small-scale, fluidized-bed studies. The following equations correlate the yields of oil, gas, char, and water over a temperature range of 478–576°C and hydrogen pressure range 0–950 psia (0–6550 kPa). These correlations are shown graphically in Figs. 13.4 and 13.34–13.37.

$$Y_{oil} = 2.089\, t - 0.001906\, t^2 + 0.01683\, P_{H_2} - 559.179 \quad (13.24)$$

$$Y_{gas} = 1.7117 \times 10^{-5}\, tP_{H_2} + 10.8642 \quad (13.25)$$

$$Y_{char} = 253.032 - 8.0168\bar{t} - 0.8874\bar{P}_{H_2} \quad (13.26)$$

These equations correlate experimental data with Lake DeSmet, Wyoming, coal over a temperature range of 890–1070°F (477–577°C)

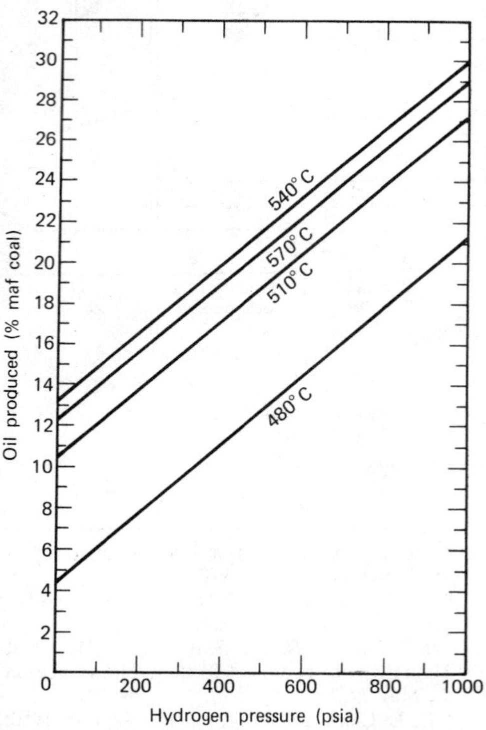

Figure 13.34 Oil yields as a function of pressure in a fluidized-bed reactor.[18]

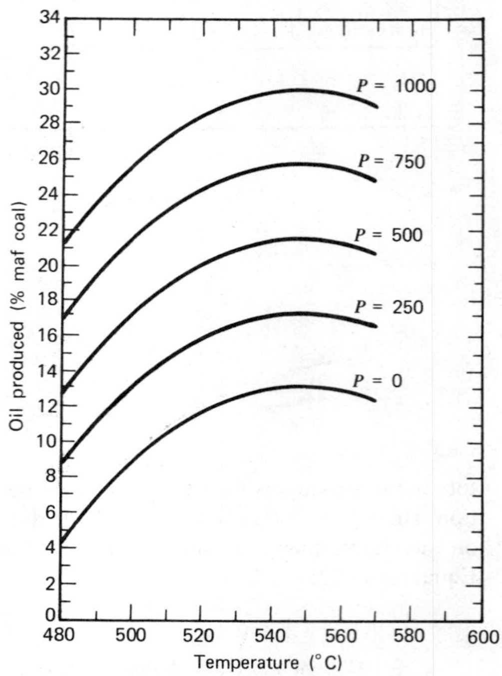

Figure 13.35 Oil yields as a function of temperature in a fluidized-bed reactor.[18]

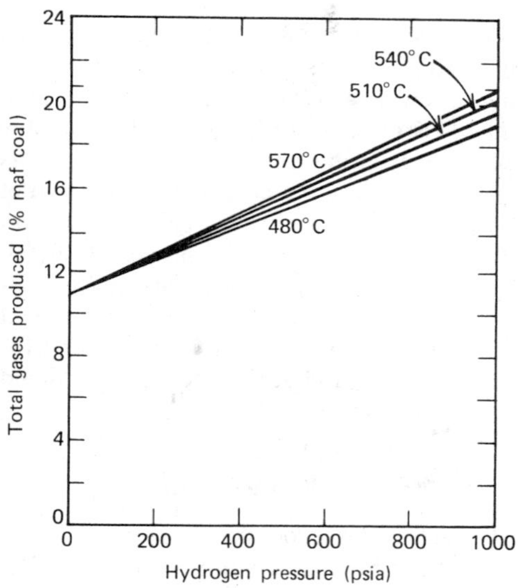

Figure 13.36 Total gases produced in a fluidized-bed reactor.[18]

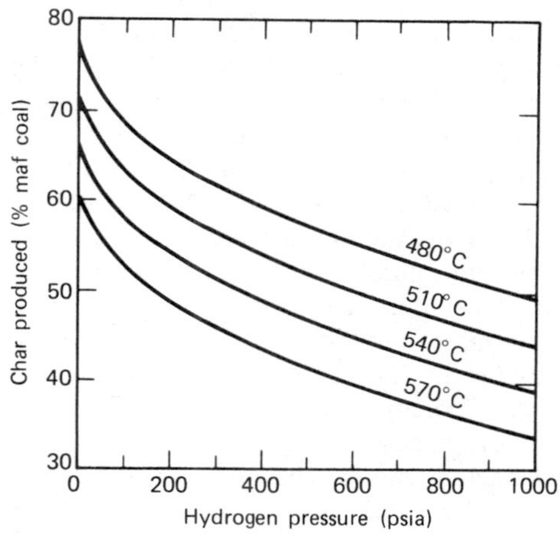

Figure 13.37 Amount of char produced in a fluidized-bed reactor.[18]

and a hydrogen pressure range of 0–950 psia (0–6550 kPa, absolute).

The hydrocarbonization process was also investigated in 1963, during the first phase of the COED project, to determine whether this approach should be considered in the ongoing investigation of that project.[116] The results were favorable and agreed reasonably well with those of Union Carbide. Work along these lines was discontinued, however, and COED continued toward development of low-pressure rather than high-pressure operation. The latter appeared to be well in hand by other investigators.

The hydrocarbonization investigations by the COED project included tests with subbituminous B and C coals (from Rock Springs and Kemmerer, Wyoming, respectively) and with a Montana lignite. Tests were made with hydrogen flowing through a bed of coal in a batch apparatus designed to provide rapid heatup and cooldown of the inlet and outlet gas stream. The range of conditions explored was as follows:

Pressure	100–5000 psig (689–34,474 kPa)
Temperature	700–1050°F (371–566°C)
Hydrogen rate	2.5–26 scf/hr per gram of coal [71–736 $m^3/(hr)(kg)$]
Solids retention time	10–120 s

Yields of oil and gas increased with increasing pressure and temperatures in the range explored. Hydrogen rate had no effect on yield. Retention time had no discernible effect on yield. Addition of a molybdenum catalyst showed no increase in yield. Oak Ridge National Laboratory[18] correlated the oil, gas, and char yields obtained in the COED project by the following equations:

$$\text{oil} = 4.7164t^{1/2} + 0.11173 P_{H_2}^{1/2} - 98.650 \qquad (13.27)$$

$$\text{gas} = 5.3079 \times 10^{-6} t P_{H_2} + 0.1145t - 43.1306 \qquad (13.28)$$

$$\text{char} = 314.185 - 10.3991t^{1/2} - 0.31658 P_{H_2}^{1/2} \qquad (13.29)$$

Maximum oil yields were not observed over the range investigated. The oil yields were lower than those observed by Union Carbide, and gas yields were approximately the same. However the latter were obtained by difference, since direct measurement was difficult because of the high dilution with the feed hydrogen. Thus they are subject to more error than the directly measured components.

Conceptual designs, assuming fluidized-bed reactors and the batch data obtained by COED, indicated that hydrocarbonization of coal was a potentially promising route for coal conversion. The process was basically the same as the Coalcon process.

The basic advantage of the hydrocarbonization process over carbonization lies in the increased yield of liquids and gases, as indicated in the comparative yields in Table 13.16. Increased yields result from the reaction of hydrogen with the coal substance. Additionally, the liquid products are claimed to be more fluid than those from carbonization. This improvement is said to result from lower molecular weight and lower heteroatom content. The major task to be overcome by the Coalcon process is demonstrating its ability to handle caking coals without excessive agglomeration. The Coalcon process has potentially significant advantages over the catalytic and noncatalytic hydrogenation processes, which use a coal-oil slurry as the reaction medium. It avoids the nasty problem involved in the separation of the unreacted solids from the produced oil (e.g., by filtration), as well as the need for recovery and recycling of catalyst. Since the system is dry, the task of separating solids from a viscous medium is eliminated. Additionally, the process has the capability of hydrogenating the heavy liquids formed in the hydrocarbonization of coal. These heavy liquids could be separated from the final product and recycled back to the hydrocarbonizer for hydrocracking to lower molecular weight oils with minimal loss to char production. Other possible advantages of the Coalcon Process are (*a*) its ability to liquefy the coal with a minimum of hydrogen consumption, similar to that claimed by the Schroeder process (below), because of the higher aromaticity of the product oils, and (*b*) its ability to operate at moderate pressures [~ 500 psia (3447 kPa, absolute)] and still achieve respectable liquid and gas yields.

13.8.2 The Schroeder Process

The Schroeder process involves the rapid reaction of hydrogen with pulverized coal at moderate temperatures and high pressures, with or without the benefit of a molybdenum catalyst. Typical conditions tested and the yields from a New Mexican coal (Table 13.17) are published in the patents.[119,120] These are

[119] W. C. Schroeder, Hydrogenation of Coal, U.S. Pat. 3,030,297, assignor to Fossil Fuels, Inc., April 17, 1962; assignee, Fossil Fuels, Inc.

Table 13.16 Comparison of Carbonization and Hydrocarbonization[18]

Yields, (wt. %, maf coal)	Carbonization in Nitrogen	Carbonization by Parry Process	Hydrocarbonization (1.6% H_2 reacted)
Char	67	67	49
Tar	12	12	25
Gas	9	8	14
Water	12	13	14
Light oil (< 260°C)	4	3	10
Phenols (< 230°C)	2	1.5	5.2
Phenol	0.5	0.5	1.9
Lake de Smet Coal Hydrocarbonization Operating Conditions and Yields			
Operating conditions			
Temperature (°C)		560	567
Hydrogen partial pressure			
psi		310	940
kPa		2137	6481
Residence time (min)		8.2	10.2
Yields (wt%, maf coal)			
Char		50.4	38.4
Tar		21.3	29.0
Water		13.2	19.2
Oil		16.0	16.2
Hydrogen		− 1.4	− 3.5
Unaccounted for		0.5	0.7
		100.0	100.0

Table 13.17 Analysis of New Mexico Coal[179]

	Analysis (wt %)	
Component	As Charged	Maf
Carbon	72.12	78.3
Hydrogen	5.52	6.0
Oxygen (by difference)	12.16	13.2
Nitrogen	1.25	1.36
Sulfur	1.05	1.14
Ash	6.38	—
Moisture	1.52	—
	100.00	100.00

presented in Table 13.18 and Fig. 13.38, which indicate that maximum liquid yield is obtained at gas residence time of about 20 s. Increasing the temperature favors gas formation over liquid formation, and favors benzene formation over substituted benzenes. Increasing the pressure increases the reaction rate but has little effect on the product distribution. Addition of 1% molybdenum on an maf basis increases coal conversion. No data were presented on the effect of this catalyst on the nature of the liquid products.

The process is claimed to give liquids rich in single-ring aromatics. Table 13.19 shows typical analyses of liquid products obtained at 1110 and 1470°F (599 and 799°C). At 1110°F (599°C), the total single-ring aromatics in the oil product equals 92.7 wt %, most of which were toluene and xylene. At 1470°F (799°C), the oil contained 83 wt % single-ring aromatics, which were less substituted than the product obtained at 1110°F (599°C). Apparently oil obtained over the range 600–1000°C contained essentially no paraffinic or saturated cyclic compounds. Because of the high aromaticity of the liquid products, low hydrogen consumptions are claimed for this process.

The suggested commercial process, based on the patent disclosures, is described in Fig. 13.39. It indicates a through-wall heated transport reactor in which finely ground coal (70% minus 200-mesh) is conveyed by an excess stream of hydrogen. The residence time is limited to less than 2 min (probably by design or by control of the hydrogen flow). The coal fed to the reactors is impregnated with ammonium molybdate solution, dried and reground before being pressurized into the system via lock hoppers. The catalyzed coal is fed into a stream of hydrogen under controlled conditions, and the mixture is conveyed into an externally heated transport reactor where the desired hydrogenation and conversion of the coal to gas and liquid products takes place. The reactor effluent is cycloned to separate most of the unreacted carbon, plus ash of the coal, from the gaseous products. The condensable liquids are then recovered by cooling, and the gases are separated and processed by conventional means.

This process has not been developed beyond the laboratory scale. Unique features claimed for the process include the following:

1. High selectivity to single-ring aromatic compounds.
2. Low hydrogen consumption because of the high aromaticity of the liquid products.
3. Large range in selectivity to liquid products, and the range can be controlled by temperature.
4. The process can be conducted in the presence or absence of a hydrogenation catalyst.
5. Recovery and separation of the products requires conventional technology only.

This process is the precursor to the Cities Service process (below), differing from it somewhat in the longer retention times and in the possible use of catalysts.

It should be possible to develop a conceptual engineering design for a commercial plant based on the data published. There is apparently some weakness in these data in that they do not reflect thermal effects within the hydrocarbonizer. These might be significant, and the ability to maintain an adequate temperature profile in an externally heated transport type reactor may indeed be difficult.

13.8.3 The Cities Service Short Residence Time (CS–SRT) Hydropyrolysis Process

The Cities Service process hydrogenates dry powdered coal noncatalytically.[121,122] The

[120] W. C. Schroeder, I. G. Stevenson, and T. G. Stephenson, Hydrogenation of Coal, U.S. Pat. 3,152,063, Assignor to Fossil Fuels, Inc., October 6, 1964; assignee, Fossil Fuels, Inc.

[121] B. H. Rosen, A. H. Pelofsky, and M. Greene,

Table 13.18 Rapid Hydrogenation of a New Mexico Coal to Liquids and Gases[119]

Tests 1–16: 1% molybdenum catalyst on maf coal basis; tests 17–18: no catalyst.

							Based on maf Coal			
Test No.	Temp. (°C)	Time at Temp. Above 300°C (min)	Time at Max. Temp. (min)	Pressure (psig)	Hydrogen Flow Rate (scf/hr)	Gas Residence Time (s)[a]	Fraction of Coal Converted (%)	Liquid Hydrocarbons Formed (%)	Hydrocarbon Gas Formed (%)	Nature of Oil
1	600	2.9	1.0	6000	20	32.2	67.6	15.1	32.6	Heavy
2	600	2.3	1.0	6000	20	32.2	66.1	16.2	26.7	Heavy
3	800	2.0	0	6000	20	26.2	73.1	9.9	37.2	Light
4	800	3.0	1.0	6000	20	26.2	67	13.8	41.2	Light
5	800	2.5	0	6000	50	10.21	83	23	40[a]	Very light
6	800	2.5	0	6000	100	5.2	90	39	40[a]	Very light
7	800	1.7	0	6000	100	5.2	88.4	48.9	31.8	Very light
8	800	1.8	0	6000	100	5.2	85.0	38.8	39.6	Very light
9	800	2.5	0	6000	228	2.3	91	31	35[a]	Very heavy
10	800	2.5	0	6000	228	2.3	97	39	42[a]	Very heavy
11	900	3.2	0	3000	20	12.0	78.3	14.8	40.7	Light
12	900	4.0	1.0	3000	20	12.0	83.7	17.4	46.3	Light
13	900	3.7	0	1000	20	4.0	67.0	21.2	21.3	Light
14	900	4.6	1.0	1000	20	4.0	69.9	23.8	24.1	Light
15	1000	3.8	0	500	20	3.7	66.8	11.6	18.7	Heavy
16	1000	5.2	1.0	500	20	3.7	73.0	27.0	29.0	Heavy
17	800	1.6	0	6000	100	5.2	73.4	31.4	38.9	Light
18	800	1.9	0	600	100	2.52	76.3	26.6	46.7	Light

[a] Estimated.

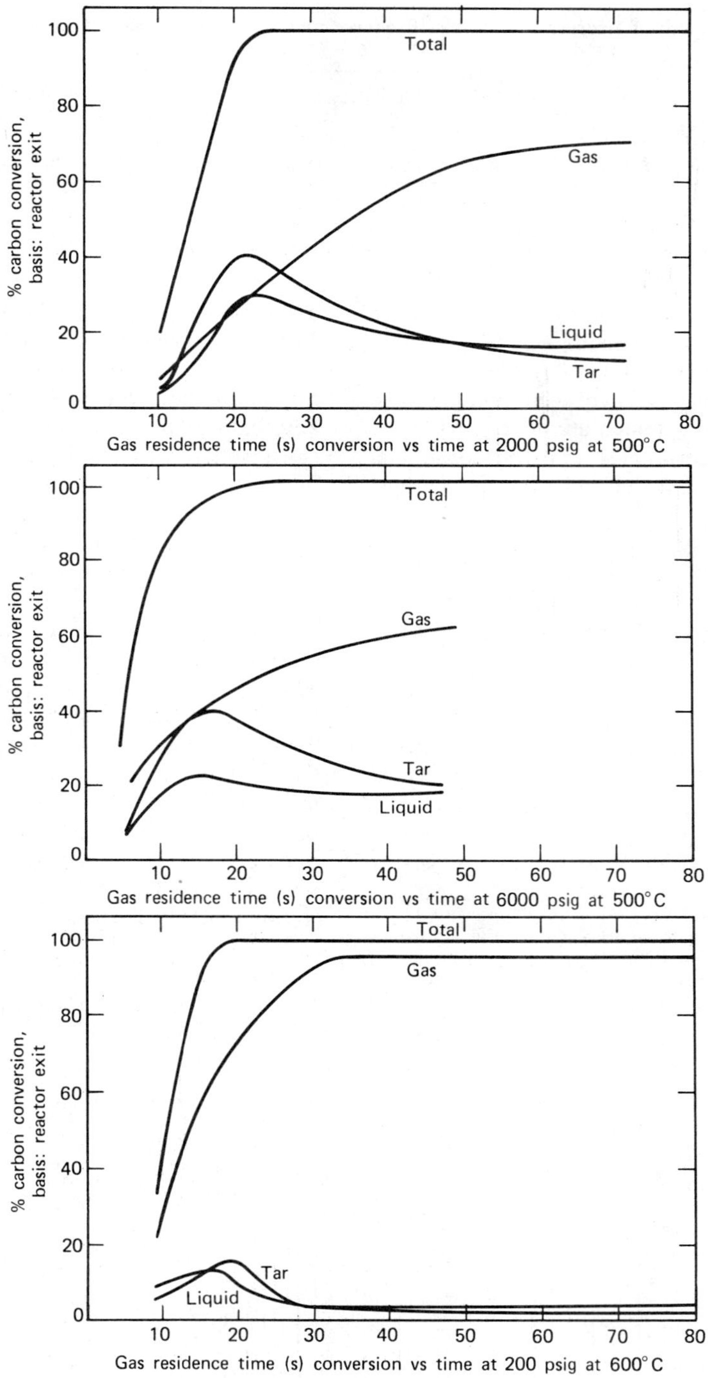

Figure 13.38 Effect of gas residence time, pressure, and temperature on carbon conversion and product distribution for the Schroeder process.[120]

Table 13.19 Analysis of Hydrocarbon Liquids Produced by the Hydrogenation of Coal[119] Schroeder Process

		Single-Ring Aromatics					Other	
Test No.	Temp. (°C)	Benzene	Toluene	Xylene	Others C_9–C_{10}	Total	Naphthalene	Unidentified
A	600	1.5	28.9	46.9	15.4	92.7	1.2	6.1
B	800	58.0	18.0	6.0	1.0	83.0	4.0	13.0

process is based on very short residence times (0.02–2 s), which include (*a*) rapid heatup to reaction temperature by mixing the coal with preheated hydrogen, and (*b*) rapid quench. A high-temperature heat recovery system is incorporated into the conceptual design of the rapid quenching step. A transport reactor is proposed, operating at 930–2730°F (499–1499°C) and from 300 to 3000 psig (2068–20,680 kPa). Potential commercial operation would probably be at about 1560°F (849°C) and 1500 psig (10,340 kPa), with a one-second residence time.

It is claimed that this process will produce directly—in one step—light aromatic liquids (essentially benzene), and high-Btu gas (CH_4, C_2H_6). In addition, heteroatoms in the coal are removed as H_2S, NH_3, H_2O, and CO_x. The residual char can be used to produce the hydrogen for the process. Short residence time, rapid heatup, and rapid quench are the key features of the process. These features can result in benzene-toluene-xylene (BTX) by-product yields that can dramatically lower the cost of producing commercial-scale quantities of pipeline-quality gas.

A continuous, bench-scale unit, processing 1–2 lb (450–910 g) of coal per hour has been in use since the summer of 1975 to study this process. Coal conversions of the order of 50–80 wt % maf coal have been attained with a light aromatics yield of around 12–16 wt % maf coal and a gas yield (C_1 + C_2 and CO_x) of about 34–68 wt %. Table 13.20 presents data on the performance of this process with various coals, ranging from lignite to bituminous, under varying conditions.

Additional data for North Dakota lignite[122] give the following indications:

1. Selectivity of carbon conversion to BTX goes through a maximum with increasing

Coal Hydrogenation to Produce Liquids, U.S. Pat. 3,960,700, June 1, 1976; assignee, Cities Service Company.

[122] A. H. Pelofsky, M. I. Greene, and C. J. LaDelfa, "Short Residence Time Coal Hydropyrolysis," American Chemical Society, Division of Fuel Chemistry, Annual Meeting, August 29–September 3, 1976.

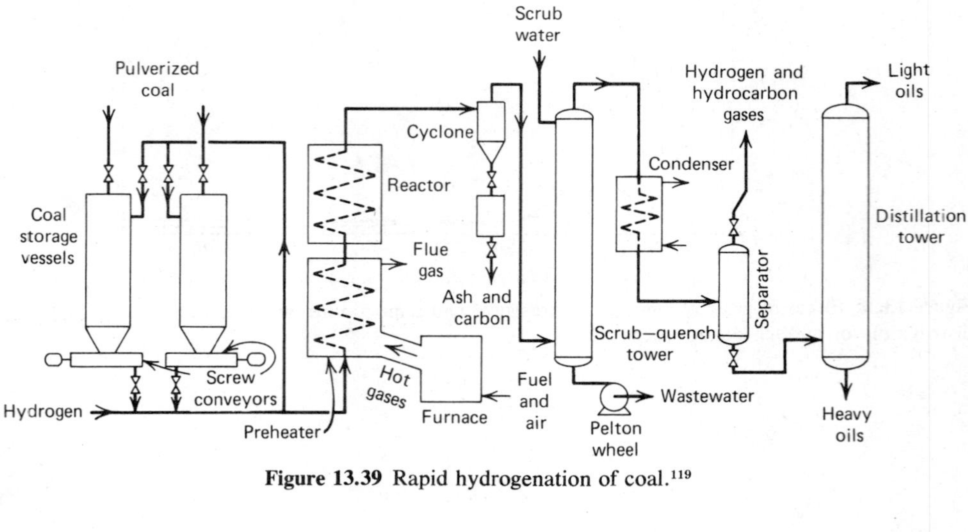

Figure 13.39 Rapid hydrogenation of coal.[119]

Table 13.20 Cities Service Research and Development Company CS–SRT Process: Typical Process Yields and Conditions[123]

	North Dakota Lignite	Wyoming Subbituminous	Western Kentucky hvAb	Mildly Air-Pretreated Western Kentucky hvAb
	Yields (wt. %, maf)			
$CH_4 + C_2H_6$	29.9	30.4	36	24.9
CO_x	11.9	10.0	1	8.3
BTX	7.5	9.0	—[a]	9.8
Light oils	0.1	1.7	—[a]	3.5
Tar	0.0	0.0	—[a]	Trace
Liquor + H_2S + NH_3	19.3	21.0	—[a]	14.4
Coal conversion (wt %, maf)	62.1	65.6	59	55.4
Carbon conversion (wt %)	50.0	54.2	52	45.7
H_2 consumption (scf/ton, maf)	24,270	24,500	—[a]	20,760
	Average Reactor Conditions			
Temperature (°F)	1600	1500	1450	1400
Pressure (psi)	1500	1500	1500	1500
Residence time (s)	< 2	< 2	< 2	< 0.5

[a] Not available.

residence time; above this optimum residence time, the BTX cracks to coke and hydrogen.

2. Carbon conversion increases with increasing residence time, increasing reaction temperature, and increasing heating rate (°C/s).

Figure 13.40, a block flow diagram of the system using the Cities Service CS-SRT process, is based on gasifying residual char plus lignite with steam and oxygen to provide the needed hydrogen. Total lignite to the system is assumed to be 22,044 tons maf/day (20,000 tonnes/day), and production is:

- *Methane:* 250 MM scf/day (7.1 MM m³/day).
- *BTX:* 6213 bbl/day (988 m³/day).
- *Ammonia:* 126 tons/day (113 tonnes/day).
- *Sulfur:* 188 tons/day (170 tonnes/day).

The overall thermal efficiency is calculated to be about 62% [based on a higher heating value for lignite of 11,220 Btu/lb (26,097 kJ/kg) (maf)], excluding the heating value of the by-product sulfur and ammonia.

[123] M. I. Greene, private communication, April 26, 1977.

An alternate system that could be considered is one based on steam reforming of part of the methane that is produced, using producer gas from air gasification of char to fire the steam reformer. The thermal efficiency of this system decreases to about 53%. However the capital investment would decrease substantially through elimination of the oxygen plant and its power requirement. This system has been evaluated, but results were unavailable at the time of writing.

13.9 CORROSION AND EROSION

The behavior of various materials of construction under the conditions existing in carbonization and hydrocarbonization processes has been reported by several investigators.[18,124] The principal problems that can exist in these processes are as follows:

1. Carburization and metal dusting.
2. H_2S attack in aqueous or gaseous environment.

[124] M. A. Howes, *Proc. 7th Synthetic Pipeline Gas Symp.*, Chicago, October 27–29, 1975 sponsored by AGA, ERDA, and IGU, Paper No. 31.

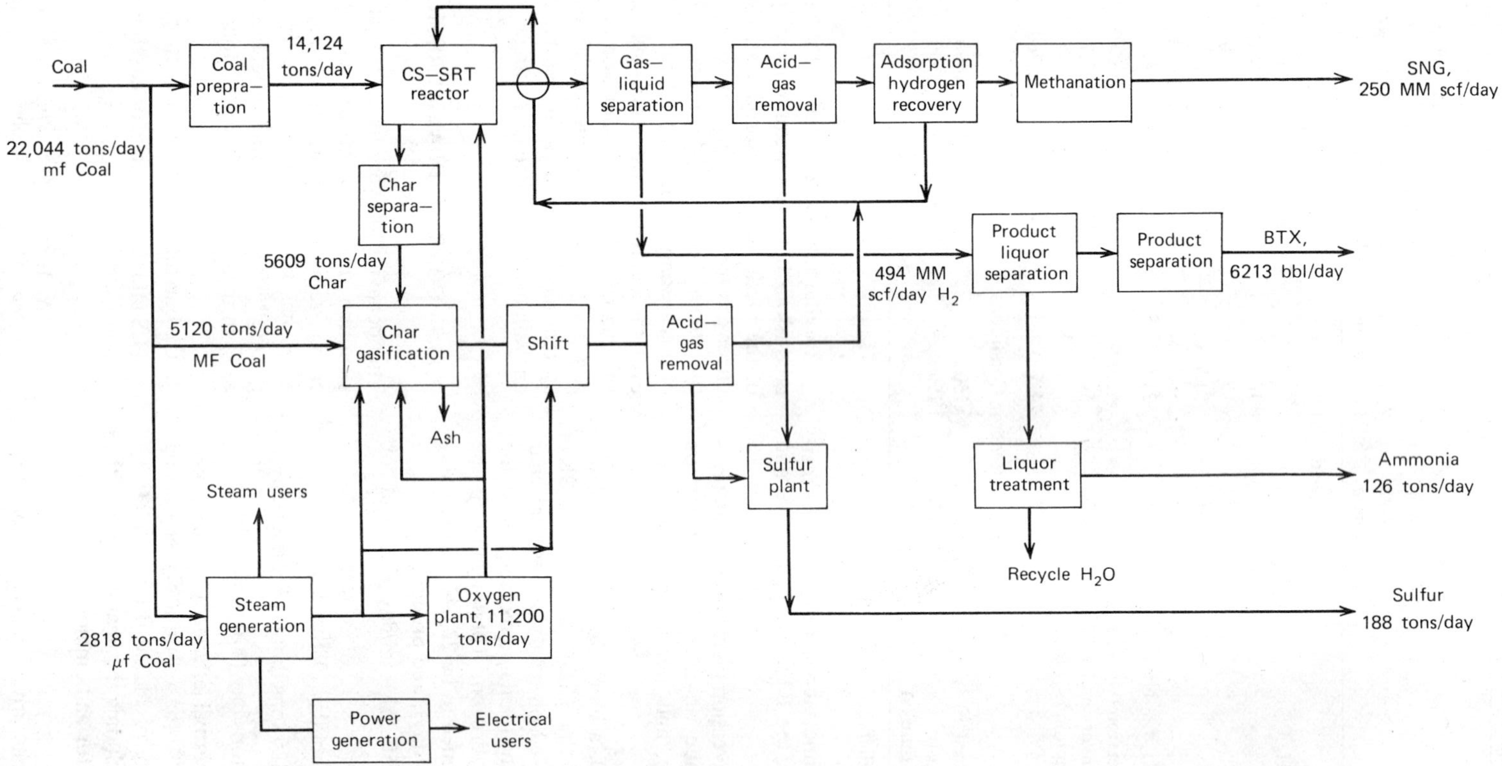

Figure 13.40 Conceptual scheme for Cities Service CS-SRT process: hydrogen production by way of char gasification.[123]

3. Formation of chromium oxide.
4. Grain boundary embrittlement.
5. Stress corrosion.
6. Hydrogen embrittlement.
7. Erosion.

Carburization is inhibited in some alloys by the presence of H_2S. The H_2S probably acts either by poisoning the active catalyst sites, which would have promoted deposition of carbon by cracking of carbon compounds, or by formation of a protective, tightly adhering scale. Erosion by char, for example, may remove the protective scale and expose the parent metal to further attack. Once carbon is deposited, there is a tendency for some metals in the parent alloy to diffuse out through this layer, leading to metal dusting. An iron-chromium-carbide complex has been identified. H_2S has inhibited this action in such alloys as Incoloy 800, Incoloy 600, and Alonized stainless 316. Some inhibition by H_2S was observed for stainless 316, but none with stainless 310.

H_2S aqueous attack, at temperatures up to 240°C and at H_2S concentrations in the gas up to 1.1 vol %, was moderate for all alloys tested in the Metals Properties Council project[124] except for carbon steel, cast iron, 410 stainless, Ni-Resist, Monel 400, and Al-bronze.

Table 13.21 Significant Test Results After 1000 hr of Tests[124]

0.1% H_2S		1.0% H_2S	
Acceptable	Unacceptable	Acceptable[a]	Unacceptable
At 1500°F			
800 Al	None	800	800 Cr
800 Cr		800Al	600
600		310Al	310 Cr
310 Al		309	601
310 Cr		314	302
601		446	304
309		793	316
304		50/50	
302		800	
		310	
314			
316			
446			
793			
50/50			
800			
310			
At 1800°F			
800 Al	800	800 Al	800
310 Al	800 Cr	310 Al	800 Cr
50/50	304	50/50	600
	600		310
	601		310 Cr
	310		601
	310 Cr		304
	302		302
	309		309
	314		314
	316		316
	446		446
	793		793

[a]Acceptable samples of corroded at a rate less than 29 mils/year.

Table 13.22 Alloys Being Studied in Phases I–IV[124] Metals Properties Council

Alloy Name	Standard Abbreviation	Used in Phase: I	III	IV	Used in Phase II Pilot Plants: HYGAS HT	HYGAS LT[a]	Consol HT	Consol LT	Synthane HT	Synthane LT	BI-GAS HT	BI-GAS LT[b]	Analysis Code[c]
Carbon steel A515			X			X		X		X		X	I
A515 aluminized						X				X			
Carbon-½Mo										X		X	N
2¼Cr-1Mo												X	N
Cast iron			X			X		X					I
Silicon iron						X							N
Ni-Resist (A439-D2B)			X			X		X				X	I
Ni-Resist (Cu) (A436-1B)	Ni-R (Cu)		X										N
21-6-9	21-6-9	X[e]			X		X		X				V
22-13-5	22-13-5		X		X		X	X	X				I
302	302	X[f]											I
304	304	X[f]	X		X	X	X	X	X	X		X	I
309	309	X		X	X		X		X		X		I
310	310	X		X	X		X				X		I
310 Aluminized	310 (Al)	X		X	X		X				X		
310 Chromized	310 (Cr)	X[f]					X						
312	312	X[e]											N
314	314	X			X						X		I
316	316	X[f]	X		X	X		X	X	X		X	I
321	321				X				X				V
329	329	X[e]	X			X		X		X		X	I
18Cr-18Ni-2Si	18-18-2		X			X		X					I
405	405											X	N
410	410		X			X				X			I
430	430		X		X	X			X	X			I
446	446	X		X	X		X						I
Crucible 26-1	26-1		X			X						X	I
18.2	18-2		X[g]										V
20Cb-3	20Cb-3		X					X				X	I
JS 700	JS 700		X[g]										N
Al Lud 6X	Al6X		X[g]										V
Al Lud 29-4	AL29-4	X[e]											V
Al Lud EX20	EX20	X[e]											V
RA333	RA 333	X[e]		X	X				X				V
Multimet N-155	N-155	X[e]											V
Crutemp 25	Cru-25	X[e]		X									V
RA 330	RA 330												N
HC 250	HC 250	X[e]											N
HL 40	HL 40	X[e]											N
HL 40-3Si	HL 40-3Si	X[e]											N
HD 45	HD 45	X[e]											N
CD4MCu	CD4MCu											X	N
Inco 800	800	X	X		X	X	X		X	X	X		I
Inco 800 aluminized	800 (Al)	X		X	X		X				X		
Inco 800 chromized	800 (Cr)	X[f]					X						
Inco 793 (811E)	793	X			X		X				X		I
Sani-Cro 32K	32K	X[e]											N
Wiscalloy 30-50-N	30-50-N	X[e]											N
Inco 600	600	X[f]			X	X			X	X			I
Inco 601	601	X		X	X				X				I
Inco 825	825		X		X		X	X	X			X	I
Inco 671 (50/50)	50/50	X		X	X		X				X		I
Hastelloy C-276	C-276		X										I

									Composition (wt %)	
C	Mn	Si	Fe	Ni	Cr	Co	Cu	W	MO	Other
0.22	0.62	0.2	Bal							
0.15[d]	0.45	<0.03[d]	Bal						0.5	
0.15[d]	0.45	<0.3[d]	Bal		2.25				1.0	
3.02	0.67	1.93	Bal	1.6					0.4	
1.1[d]	1.5		14.5	Bal	0.5		0.5		0.5	
2.80	0.91	2.81	Bal	19.8	3.1					
3.00	1.5[d]	2.80[d]	Bal	17.5[d]	3.5[d]		6.5			
0.03	8.7	0.62	Bal	7.2	20.6				0.2	0.27N
0.05	5.30	0.47	Bal	13.5	21.1				2.3	0.14 Cb, 0.18V
0.05	0.64	0.55	71.0	9.2	18.2					
0.05	1.45	0.54	70.0	9.1	18.8				0.34	
0.11	0.54	0.74	60.5	14.7	23.0					
0.06	1.71	0.68	52.2	20.2	25.0					
0.15[d]	2.0[d]	0.5[d]	Bal	9	30					
0.06	1.90	2.21	51.6	20.0	24.0					
0.05	1.65	0.43	65.3	13.9	17.1				2.5	
0.04	1.61	0.40	Bal	10.0	17.7		0.35		0.4	0.2Ti
0.05	0.50	0.38	Bal	4.3	27.1		0.14		1.4	
0.05	1.35	2.0	Bal	18.2	18.1		0.08			
0.08[d]	1.0[d]	1.0[d]	Bal		13					0.2Al
0.13	0.35	0.54	Bal	0.5	12.0				0.1	
0.06	0.36	0.62	Bal	0.4	17.3					
0.10	0.45	0.38	74.6	0.4	24.0					
0.02	0.27	0.43	Bal	0.3	25.6				1.0	0.42Ti
0.02	0.43	0.65	Bal		17.6				2.4	0.7Ti
0.03	0.56	0.50	Bal	33.0	19.4		3.1		2.1	Cb + Ta 0.91
0.03	1.7		Bal	25.0	21.0				4.5	0.3Cb
0.03	1.50	0.50	Bal	24.0	20.0				6.0	
0.01			Bal	0.1	29.6				3.9	0.01N
1.0	20		Bal	3.0	5.0					10Al
0.05	1.31	1.38	Bal	46.8	25.4		0.08	3.3	3.3	
0.10	1.34	0.60	Bal	20.9	19.8	19.5		2.8	3.0	Cb + Ta 1.08
0.06	1.68	0.32	Bal	25.2	24.5				0.3	
0.25[d]	2.0[d]	1.5[d]	Bal	35	18.5					
2.7	0.9	0.5	Bal	1.2[d]	28.0					
0.4	0.8	1.3	Bal	20	31					0.09N
0.4	0.8	2.3	Bal	20	31					0.09N
0.45	0.8	1.3	Bal	5	29					0.09N
0.04[d]	1.0[d]	1.0[d]	Bal	5	26		3		2	
0.03	0.80	0.33	47.1	30.8	20.6		0.1			0.32Al, 0.42Ti
0.03	0.72	0.40	43.2	32.2	21.4	0.02				1.76Al
0.8			Bal	31	21			3		0.35Ti, 0.35Al
0.55			Bal	47	27			4		
0.05	0.15	0.20	7.0	76.5	15.8	0.08				
0.04	0.24	0.19	15.9	59.5	22.5	0.03				1.2Al
0.04	0.63	0.27	29.7	41.9	21.6		1.9		3.1	0.75Ti
0.03	0.03	0.03	0.30	48.4	50.0	0.01				

Table 13.22 *Continued*

Alloy Name	Standard Abbreviation	Used in Phase: I	III	IV	Used in Phase II Pilot Plants: HYGAS HT	HYGAS LT[a]	Consol HT	Consol LT	Synthane HT	Synthane LT	BI-GAS HT	BI-GAS LT[b]	Analysis Code[c]
Hastelloy G	Hast G		X										I
Hastelloy X	Hast X				X		X		X		X		V
Crucible NI	NI	X[e]											V
Haynes 150	150	X[e]											V
Haynes 188	188	X[e]		X									V
Co-Cr-W-Nol	Co-Cr-WI	X[e]											N
Stellite 6B	6B	X[e]										X	V
VE 441	VE 441	X[e]											V
Titanium 50A	Ti-50A		X			X		X		X		X	I
Monel 400	M400		X			X				X		X	I
Al Bronze (CDA 954)	Al-Br		X			X		X					I
Al 1100	1100		X										I
Al 6061	6061		X										I

[a] Includes top of gasifier ~ 500–600°F.
[b] Proposed for BI-GAS plant.
[c] N = nominal, V = vendor analysis, I = IITRI analysis.
[d] Maximum.
[e] To be added.
[f] To be dropped.
[g] Tentative addition.

H_2S gaseous attack at elevated temperatures and up to 1% H_2S in the gas is summarized in Table 13.21, which lists the acceptable and unacceptable alloys observed after 1000 hr of exposure, under laboratory conditions, by the Metals Properties Council.[124] Compositions of the alloys are presented in Table 13.22. The maximum temperature encountered in low-temperature carbonization may be about 1500°F (816°C). At this temperature and at 0.1% H_2S, all the alloys listed were acceptable [corrosion rate less than 29 mils (0.74 mm) per year], and at 1% H_2S, about half the alloys were acceptable. At 1800°F (982°C), which might be encountered in the gasification section of a hydrocarbonization system, the only acceptable alloys were Alonized Incoloy 800, Alonized 310 stainless, and 50/50 chromium-nickel alloy.

Formation of chromium oxide ("green rot") can occur in intergranular areas under reducing conditions at 840–560°F (449–849°C).[18] This phenomenon can be prevented for austenitic stainless steels either by stabilization with titanium or niobium, or by annealing at 1100°C plus a rapid quench.

Grain boundary embrittlement can result from the carryover and deposition of trace metals such as lead and bismuth. This can occur at low ppm concentrations, resulting in grain decohesion and mechanical failure.

Stress corrosion can occur when an austenitic steel is exposed to a corrodant while under stress. Chloride ions, in locations below the water dew point of the gas, are one of the major corrodents to fear in this area.[18] Methods of avoiding water condensation during startup, operation, or shutdown are well known.

Hydrogen embrittlement results from the diffusion of hydrogen (atoms) into the metal. This is not a problem with the austenitic (300 series) stainless steels, but it causes difficulties with alloy and carbon steels.[18]

Erosion is a potential problem whenever solids are moved, and this problem is aggravated by high velocities or turbulence, changes in direction, and large pressure gradients. Careful engineering is required to

Composition (wt %)										
C	Mn	Si	Fe	Ni	Cr	Co	Cu	W	MO	Other
0.02	0.51	<0.01	5.7	Bal	16.0	2.2		3.9	15.4	0.06V
0.04	1.56	0.49	20.4	Bal	22.4	2.1	1.9	0.3	6.9	Cb + Ta 2.3
0.09	0.63	0.34	19.0	Bal	21.7	1.9		0.6	9.9	
		2.0		Bal	2					3Al
0.08	0.5	0.5	18.0	1.4	27.9	Bal			0.05	
0.08	0.72	0.34	1.8	22.5	22.7	Bal		14.3		0.05La, 0.22Al
2.5[d]					30	Bal		12		
1.0	1.4	0.47	1.9	2.7	29.1	Bal		4.9	1.2	
0.05			Bal						3.2	14.7Al, 0.01Zn
0.02			0.14							0.130, 0.14N, Bal Ti
0.14	1.06	0.14	1.9	63.6			33.2			
			3.3	2.0			Bal			10.38Al
	0.01	0.07	0.47				0.14			Bal Al
	0.05	0.6			0.2		0.3			0.01 Ti, 0.03Zn, 1.2Mg, Bal Al

avoid erosion. For example, changes in direction are many times handled expeditiously by a 90° tee, where the solids abrade against themselves rather than against the metal wall.[80]

NOMENCLATURE (IN CONSISTENT UNITS)

C_a heat capacity of coal ash (Btu/lb−°F)

C_D drag friction factor (Equation 13.21, Figure 13.21)

C_f heat capacity of fixed carbon (Btu/lb−°F)

C_g heat capacity of gas at constant pressure (Btu/lb−°F)

C_s heat capacity of solid particles (Btu/lb−°F)

C_R radial correction factor (Fig. 13.12)

c_{v1} heat capacity of primary volatile matter in coal (Btu/lb−°F)

c_{v2} heat capacity of secondary volatile matter in coal (Btu/lb−°F)

D molecular diffusion coefficient (ft²/s)

D_{sa} axial bed solids diffusion coefficient (ft²/s)

D_{sr} radial bed solids diffusion coefficient (ft²/s)

D_i indexed particle diameter (ft) (Equation 13.19)

D_p particle diameter (ft)

D_T reactor inside diameter (ft)

e base of natural logarithms

E energy (Btu)

f fluid friction factor = $2gD_p\,(\Delta P)/Lv^2\rho$

F_f friction factor factor for packed beds (Fig. 13.20)

F_{Re} Reynolds number factor for packed beds (Fig. 13.19)

g gravitational constant = 32 ft/s²

G mass velocity (lb/s−ft²)

G_m molal mass velocity (lb−mole/s−ft²)

h heat transfer coefficient between gas and heating or cooling surface (Btu/s−ft²)

h_c heat transfer coefficient between gas and solid particles (Btu/s−ft²)

H percentage of hydrogen consumed (wt

H$_2$ consumed/wt maf coal) × 100 (Equation 13.23)

j_D Colburn's mass transfer factor = $\frac{[K_g P_{bm}(Sc)^{2/3}]}{G^m}$

K_d mass transfer coefficient between gas and solid particles

K_G mass transfer coefficient between gas and solid particles (lb−mole/hr−ft²−atm)

k_f thermal conductivity of gas at film conditions [Btu/(hr−ft²−°F)/(ft)]

k_g thermal conductivity of gas [Btu/(hr−ft²−°F)/(ft)]

k_m mean thermal conductivity of gas across gas film

k_{sa} axial bed solids thermal conductivity [Btu/(hr-ft²−°F)/(ft)]

k_{sr} radial bed solids thermal conductivity [Btu/(hr−ft²−°F)/(ft)

L reactor height (ft)

L_H length of heat transfer surface (ft)

M_g gas flow (lb/s)

M_i mass fraction of ith index particle diameter (Equation 13.19)

M_r solids flow (lb/s)

N_u Nusselt number = $h_c D_p / k_g$

P_{bm} mean partial pressure of nondiffusing components in gas film (atm)

P_{H2} partial pressure of hydrogen, psia (Equations 13.23–13.29)

Pr Prandlt number = $C_g \mu / k_g$

Re Reynolds number = $D_p U_o \rho_g / \mu f$

Re$_{opt}$ Reynolds number at the point where the heat transfer coefficient is maximum

Re′ modified Reynolds number

R_t radial distance of tubes from axis of fluid bed (ft)

R_T radius of reactor (ft)

Sc Schmidt number = $\mu_f / \rho_g D$

Sh Sherwood number = $K_d D_p Y / D$

t temperature (°C)

t_g gas temperature (°F)

t_s solids temperature (°F)

U_0 superficial gas velocity (ft/s)

U_{mf} minimum fluidizing velocity (ft/s)

v velocity (ft/s)

V_{t2} terminal velocity of solid particles of diameter D_p (ft/se)

W_v mass of volatiles evolved per unit mass of original maf coal; this is less than or equal to the original volatile content, measured in same units

X porosity

Y logarithmic mean mole fraction of the nondiffusing gas

Y_{oil} yield of oil (% maf coal) (Equations 13.24 and 13.27)

Y_{gas} yield of gas (% maf coal) (Equations 13.25 and 13.28)

Y_{char} yield of char (% maf coal) (Equations 13.26 and 13.29)

α Ratio wake volume to bubble volume for gas bubble rising in a fluidized bed.

δ fraction of fluidized bed consisting of bubbles

ϵ void fraction of fluid bed

ϵ_{mf} void fraction of bed at minimum fluidizing

η area factor (Figure 13.15)

θ solids residence time, min (Equation 13.23)

ψ sphericity of a particle (Equation 13.18)

μ gas viscosity (lb/ft−s)

μ_b bed viscosity (lb/ft−s)

μ_f gas viscosity at film conditions (lb/ft−s)

ρ mean density of a gas-solid mixture (lb/ft³)

ρ_b density of bed (lb/ft³)

ρ_f density of fluid (lb/ft³)

ρ_g density of gas (lb/ft³)

ρ_p density of particles (lb/ft³)

ΔP_f pressure drop of fluid (lb/ft²)

CHAPTER **14**

High-Temperature Carbonization

WERNER EISENHUT

Werner Eisenhut *gained an engineering degree (Diplom Ingenieur) in the field of metallurgy from the Technical University of Clausthal, West Germany, in 1959 and received his doctor's degree in fuel chemistry from the same university in 1964. From 1964 to 1966 he was a technical consultant for gas works. He joined the scientific staff of Bergbau-Forschung, GmbH, Essen, where since 1971 he has been head of the Carbonization and Thermal Coal Processing Division. Dr. Eisenhut's main research interests are in heat engineering of coke ovens and environmental control of coking plants.*

14.1 PRESENT-DAY DESIGN OF HORIZONTAL CHAMBER COKE-OVEN PLANTS

14.1.1 Developments in Coke-Oven Construction

The pyrolytic decomposition of bituminous coal with final temperatures of about 900°C and above is called coking. The first attempts to coke bituminous coal (1584) date back to Duke Julius of Braunschweig.[1] It was reported that in the year 1585 in Sangershausen, at the initiative of Münzmeister D. Strumpfelt, coke was used in large-scale experiments for the reduction of ores, especially for copper schist. The results did not lead to commercial application.

The Thirty Years' War interrupted the experiments for the coking of bituminous coal, but they were resumed in England and developed there on a commercial scale. The commercial importance of coke as an effective reducing agent in ironmaking began with the year 1708, when A. Darby the Elder (1676–1717) succeeded in using coke from bituminous coal to produce suitable hot metal from iron ores. By the end of the eighteenth century, the English coking and blast furnace technology was transferred to the European continent (Chapter 1).

Initially bituminous coal was coked in piles that were used for the production of wood charcoal. Piles (also called kilns) appeared for the first time in 1657 in England and spread from there to the other European bituminous coal areas.

Around 1850 in Oberkirchen, in the Earldom of Schaumburg, the so-called Schaumburg kilns were used, which are half-open brick kilns from which circular mounds emitted odoriferous tar-containing distillation gases directly into the open. The next development was the closed "beehive" oven, which in its original form discharged the distillation and flue gases through a chimney at a greater height. It is interesting to note that until the end of the last century, England retained the old beehive oven. In the United States until 10 years ago, these ovens were used as spare units in case of large coke demands. In South Africa, South America, Australia, and the United States some beehive ovens are still in operation today.

The heat required for coking in the pile, Schaumburg, and beehive ovens is produced by partial combustion of the coal. This practice resulted in a substantial loss of material by combustion; therefore the coke yield in these ovens was at most 55% of the coal.

The further development of the pile led to the so-called flame oven, in which the coal was coked in chambers heated from the outside. The high heating value gases leaving the oven chamber were immediately burned in flues built into the walls of the ovens. Initially combustion air was introduced into the chamber; later the chamber was made airtight and the combustion air was led directly into the heating flues. In addition to a considerably higher yield of coke, a shorter coking time was also obtained. The flame ovens were developed in 1850 in Belgium and in the Saar District. Of the ovens built in the Saar District, the best known is the François Rexroth oven (around 1860); in Belgium the Evence-Coppée oven (1861) is the most famous. Because of favorable economics, the use of this type of oven spread rapidly throughout the world. The coke yield was about 75% in the ovens, which had a then-remarkable charge of 7 tons per oven and required 48 hr coking time. Such high yields, which of course depended on the kind of coal, lead one to believe that there was excellent exclusion of the air and tight oven walls. Even today's modern oven types do not exceed that yield to any great extent.

The first coke ovens that produced satisfactory blast furnace or foundry coke as the main product, and tar, ammonia, and later also benzene as by-products, were built around 1856. Their precursors were the "tar ovens" in which in 1764 in the Saar and 1761 in Scotland, for example, coal was coked only to obtain the tar. The oven was still heated with its own gas, but the by-products were recovered. Such ovens, which had an oven system that was only bottom-heated, were built in France especially by C. Knab, and by F. Carvés, who improved the Knab oven by also using the walls as heat transfer surfaces. Similar by-product recovery ovens

[1] F. M. Fess, *History of Coke Oven Technology,* Glückauf Verlag, Essen, 1957.

were built in Belgium. This new technology was not readily accepted, since the blast furnace people did not think that the coke was suitable for their process. This opinion was not always unfounded. The lack of acceptance on the European continent changed around 1880, after better coke-oven types were developed. In England and the United States the new technology was accepted just two decades later.

The development of modern coke-oven technology, which is very closely associated with the names A. Hüssener, C. Otto, G. Hoffman, F. J. Collin, F. Brunck, L. Semet, H. Koppers, and C. Still, was begun after 1880 in Germany and Belgium. The Otto-Hoffmann oven, which used regenerative heat recovery (regenerative oven with common longitudinal regenerators), was temporarily replaced by the waste heat oven. Later, in 1904, H. Koppers replaced longitudinal and group regenerators with single and cross regenerators, and reintroduced regenerative heat recovery. This arrangement of the regenerators became the prototype of all ovens, even the newest coke-oven systems. In 1907 Koppers successfully introduced the heating of the coke-oven walls with an outside source of gas (producer gas). In 1910 he and A. Wirtz (Friedrich-Wilhelm-Hütte, Mülheim-Ruhr) developed the so-called compound oven in which, according to their availability, rich gases (coke-oven gas, natural gas, mine gas, refinery waste gas, mixtures of coke-oven gas with LPG and air) or lean gases (blast furnace gas, mixtures of blast furnace gas with coke-oven gas, or blast furnace gas with LPG) could be used as the heating gas. This oven type has been used successfully for 50 years and is built even today. By heating with an outside source of gas, the coke-oven gas can be used for other purposes (steel mill, chemicals, town gas).

Since 1830 fire clay has been used as a raw material for the oven brick. After 1870 the use of "acid fireclay stones" slowly came into practice. Silica brick was introduced for building coke ovens around 1900; its use has been universal since 1914.

Coke is usually produced from fines of a medium-volatile coal or a blend of high- and low-volatile coals that, because of good coking and softening properties, produces an excellent coke. Coal quality is improved by use of finely ground coal rather than lump coal. To be able to use less suitable coals also, these are either mixed with highly caking coals, compressed by stamping, or preheated. The coking of a mixture of coals was first done in France in 1850, and later in Wales. The stamping process was developed in 1881 in Karwin (Mährisch-Ostrau) and is now used in Czechoslovakia, Poland, Lorraine, and Germany (Saar District and Sachsen). Preheating the feed coal was first done in 1935 in the Didier-Praetherm process and is now used in a few plants on a commercial scale in the United States, France, England, and West Germany.

The horizontal slot-type coke oven is of paramount importance for coking bituminous coal to produce high-quality blast furnace or foundry coke. Inclined or vertical chamber ovens with a small throughput of coal were once used in gasworks, but have been almost completely forgotten today and are not described here. For coke production in the horizontal slot-type oven, coal mixtures with a sufficiently high coke-forming ability are needed. Since the reserves of such coals are limited, it has become necessary for the coke-producing industry to improve the current process for the manufacture of coke or to develop new processes with a wider coal base.

Other reasons for further developing the current technology of coke making include the continuous increase in the cost of producing coke because of the increase in the cost of coal, as well as materials and wages, the cost of pollution control devices, and competition with the petrochemical industry, which led to reduced revenues for coke-oven by-products. For these reasons, great efforts have been made to improve the horizontal slot oven by new technologies (see Section 14.1.2).

Figure 14.1 is a schematic representation of a horizontal slot-type coke-oven plant (coke oven, auxiliary machinery, and buildings). The mixture of coal passes by means of a conveyor belt into the coking coal bunker (coal tower). From this it is charged into the charging car. It is then weighed on a track scale beneath the coal tower and is top charged into the group of connected slot

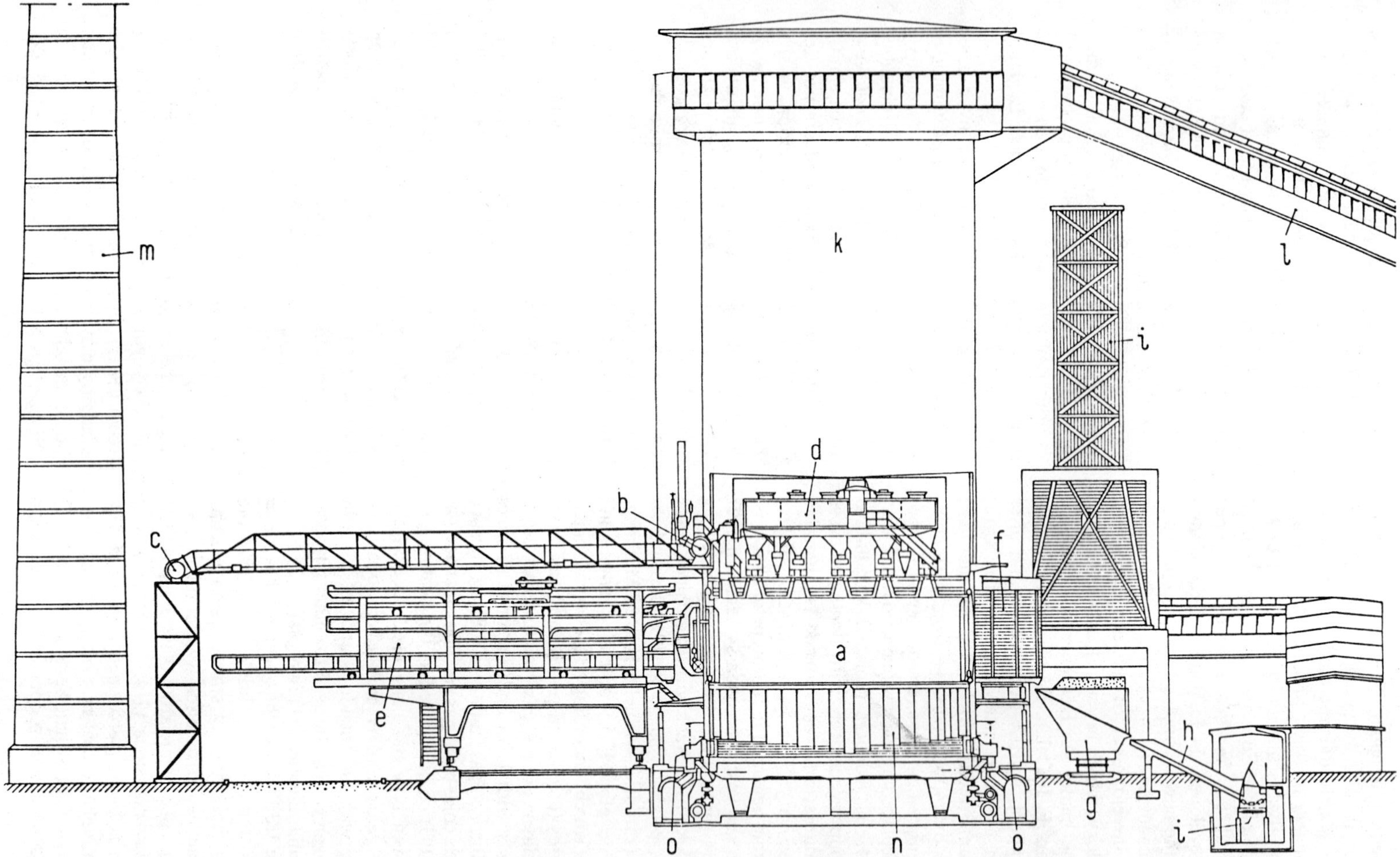

Figure 14.1 Schematic diagram of coke-oven operation: (*a*) coke oven, (*b*) raw gas collecting main, (*c*) raw gas suction main, (*d*) coke-oven charging car, (*e*) coke pushing machine, (*f*) coke guide car, (*g*) coke quench car, (*h*) coke wharf, (*i*) wharf conveyor belt to the coke-screening plant, (*j*) quench tower, (*k*) coal tower, (*l*) coal feed to tower, (*m*) chimney for off-gas, (*n*) regenerator, (*o*) inspection channel.

ovens. One battery consists of 20–80 or more ovens. After charging, the leveling bar is introduced through the leveling opening in the oven door to level the top of the charge. The leveling bar is a part of the coke-pushing machine. This operation produces an open space about 300 mm high (gas-collecting space) between the charge in the oven and the roof, through which the gases that are produced escape. In stamping operations the coal is compressed before charging by stamping; the stamped coal cake is inserted from the side into the oven chamber through the pusher side door. A leveling bar is not needed.

The oven walls are made up of a number of vertical flue heating ducts. The flue temperature (coking temperature) of the heating walls on both sides of the slot oven is usually between 1000 and 1400°C, which is below the upper temperature limit to which silica brick can be heated. The flue temperatures are measured by optical pyrometers through observation ports in the battery top. The ports can be closed. The heating gas required for coking is brought to the heating walls by heating conduits at the bottom of the regenerators. In this channel is also found the waste heat unit, through which the waste gas from the heating walls is fed to the stack flue after passing through the regenerators.

At the end of the coking cycle the oven doors are taken off and the glowing coke is pushed by the pushing machine from the narrow side of the slot oven (pusher side) through the door at the other side (coke side). It then passes through the coke guide car into the coke quench car. Following this, the doors are replaced and the oven chamber is refilled. The quench car travels to the quench tower, where the coke is quenched with water. The quenched coke is deposited on an inclined ramp (coke wharf) for further cooling. The coke travels by means of a conveyor belt to the coke-screening plant and then to the loading dock.

The gaseous products of coking pass through the gas-collecting space, which is located above the charge in the chamber, into a vertical ascension pipe (standpipe) and then into a collecting main, which is usually on the pusher side and extends the length of the oven battery. Many modern batteries have ascension pipes and collecting mains on both sides to control emissions. Each oven is connected to this collecting main and can be individually disconnected by valves. The raw gas from each oven is collected in the collecting main and the accumulated gas is conducted to the chemicals recovery portion of the plant.

The horizontal slot-type oven is the most modern and economical coke oven, because of its heat economy, its low operating personnel requirements, and its long life at high throughputs. The important elements of this construction are the vertical heating walls, which are subdivided into flues with the oven chamber in between, and the regenerators, which are arranged below the chamber. Up to 80 ovens are arranged into one battery.

The heat in the waste gas, which generally leaves the flues at a temperature above 1000°C, was previously used only in so-called waste heat boilers. Coke ovens using regenerators are more economical because the heat of the flue gas was recovered for use in the coking process. When using coke-oven gas, only the combustion air is heated in the regenerators. When using lean gas, that gas as well as combustion air is preheated. Coke-oven gas is not preheated, since there is a danger of carbon deposition from the cracking of higher hydrocarbons, which would result in plugging of the regenerators. At the present state of development the coke oven is usually built as a regenerative compound oven in which, as necessary, either its own distillation gas—coke-oven gas—or an outside source of gas (e.g., lean gas) can be used. A coke-oven plant is generally operated in conjunction with other industries (steel mills, chemical plants, gas plants) that use the product gas. Consequently an outside source of gas, either in the form of blast furnace gas or refinery waste gas, is available for heating the ovens, or else such gas is produced in the coke-oven plant itself, as producer gas, to make the coke-oven gas available for other purposes. The whole battery, as well as individual coke ovens, can be switched from one heating gas to the other in a very short time and without disturbing the operation. This compound operation is thus very simple and can accommodate the demand for coke-oven gas. Even today compound ovens are being built, especially in coke-oven plants connected with steel plants.

The upper part of the horizontal chamber oven is essentially composed of the coke-oven chambers with heated walls for indirect heat transfer to the coal as well as the oven roof, with standpipes for conducting the product gases. The oven chambers are narrow, long, and high. The heating walls consist of high-quality brick and are subidivided into a number of vertical flues through which the combustion gases flow. The number of flues for one wall depends on the length of the chamber and ranges from 22 to 36. Each heating flue is accessible from the top of the battery. The roof, which is about 1–1.5 m thick, contains, depending on the design, three to five charging holes per chamber through which the coal is charged into the oven. In ovens that use stamped charges, the oven roof has only one opening for removing and burning the gases that escape during charging.

To be able to push the coke charge more readily from the oven when degassing is concluded, the ovens are wider on the coke side. The increase in width amounts to 40–80 mm and depends on the kind of coal and the length of the oven. When stamping is used, the chamber walls are usually parallel.

The average width of the oven is 400–500 mm. Selection of the width of the chamber is determined by the kind of coking coal used (different coking rates) and the use of the coke (foundry coke consisting of large lumps, blast furnace coke). Usually an average chamber width of 450 mm is used for the production of blast furnace coke.

The height of the oven chamber is usually 4–7 m; the largest coke oven in the world has a chamber height of 7.65 m.[2] The useful length of the chamber is between 12 and 17 m; therefore the useful volume of one chamber is 20–53 m^3. The substructure of the battery consists of equipment for introducing gas and air into the flues that are arranged in the heating walls, into the regenerators, and into the ducts for bring combustion gases from the flues into the waste gas channels.

The regenerators are located below the oven chambers in the direction of the axis of the oven, that is, at right angles to the battery (cross-regenerators), in such a manner that each oven has its own pair of regenerators with the necessary reversing equipment. Their width and length are thus given by the measurements of the oven. The height of the lower part of the battery (regenerators) used to be about equal to the upper portion (oven chamber). In modern high-throughput coke ovens, which use shorter coking times, the ratio of useful height of the oven to the height of the regenerator space is about 0.5 : 1.[3] The volume of the regenerators depends on the size of the oven, the specific exchange efficiency of the regenerator material, and the desire for an economically optimal amount of heat recovery.[4]

Two groups of regenerators (one regenerator pair) are required for each oven. The combustion streams to be preheated (air, air and lean gas) pass through one, and waste heat is stored in the other. Groups are formed either by dividing the regenerator in the middle (''half-divided regenerator'') or into two ''pass-through'' regenerators, which in compound ovens are separated by longitudinal walls into spaces for air and lean gas (Fig. 14.2). In some types of oven the regenerators are subdivided into individual cells so that each pair or group of flues obtains predetermined amounts of air, lean gas, and flue gas. The subdivision usually depends on the system of flues selected. In all these regenerators the gases flow in a vertical direction.

When heating with rich gases, the regenerator has channels only for air and waste gas; when lean gas is used for heating the regenerator chambers, which in the former case were only fed with air, are now subdivided into equal-size channels for air and lean gas.

Because of the expansion behavior of the high-temperature brick, it is necessary to anchor the individual groups of ovens to avoid separation. Figure 14.3 shows, for similar construction elements, one of the many possibilities for anchoring. The coke-oven battery is stressed in the longitudinal axis by two reinforced concrete walls (*e*), which have positions that do not change. These walls are

[2] Y. Okada, T. Fujimura, and Y. Murakami, *Iron Steelmaker*, 4(5), 11–23 (1977).

[3] K. G. Beck, EGKS-Kommission Luxemburg, EUR 4520, d.f.e. 1970, pp. 283–297.

[4] D. Stalherm, C. Meyer, and D. Habermehl, *Glueckauf-Forschungsh.*, 37, 259–266 (1976).

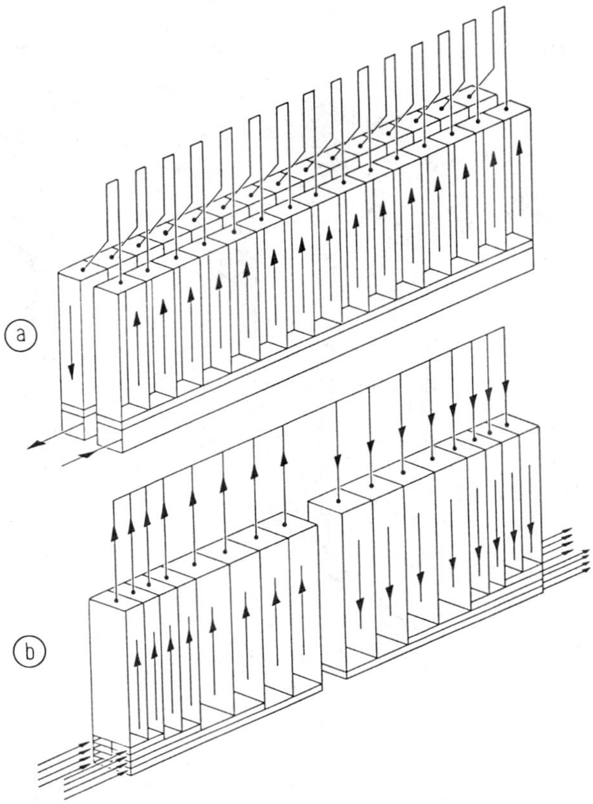

Figure 14.2 Regenerators for coke ovens: (*a*) "pass-through" type and (*b*) "half-divided" type. (Bergbau-Forschung GmbH.)

connected by the longitudinal anchoring (*c*), which is located above the oven roof. The number of lengthwise anchors depends on the width of the battery. On each narrow side (heads) the oven chamber is held together by wall protection plates with buck stays (*a*). Considerable force may act on the buck stays, so that to maintain their elasticity, they are equipped with springs of predetermined strength. At the foot—in the cold part—the buck stays are solidly anchored below the bottom of the oven; in the middle structure they are supported by the beam through springs (*d*) and obtain support on the top by the cross-tie rods (*b*), which are located in the roof of the oven, and by the head springs.

The building materials for the refractory part of the coke oven are constructed exclusively of silica coke-oven brick. For the parts that are not touched by coal, the "normal" (class 112) quality of brick is used; for the parts that are in contact with coal, the "consolidated" quality (class 117) is used; and for the bottom of the oven the "bottom stone" quality (class 117, higher cold compressive strength) is used.

New model coke ovens are operated with higher flue temperatures. Previously the regenerator walls were built of fireclay bricks; now, however, depending on the construction in the upper half, they are built entirely of silica material. The regenerators contain either soap-type bricks or conventional checker bricks. Several oven systems use both types of checker bricks to decrease the pressure drop in the regenerator. Depending on the coking temperature for which the coke oven is designed, these checker bricks consist either of silica or of fireclay materials. Just as in the case of the regenerator walls, bricks of both types are used simultaneously.

To decrease the surface heat losses of a coke oven, insulating brick is used as intermediate or surface insulation. Areas of es-

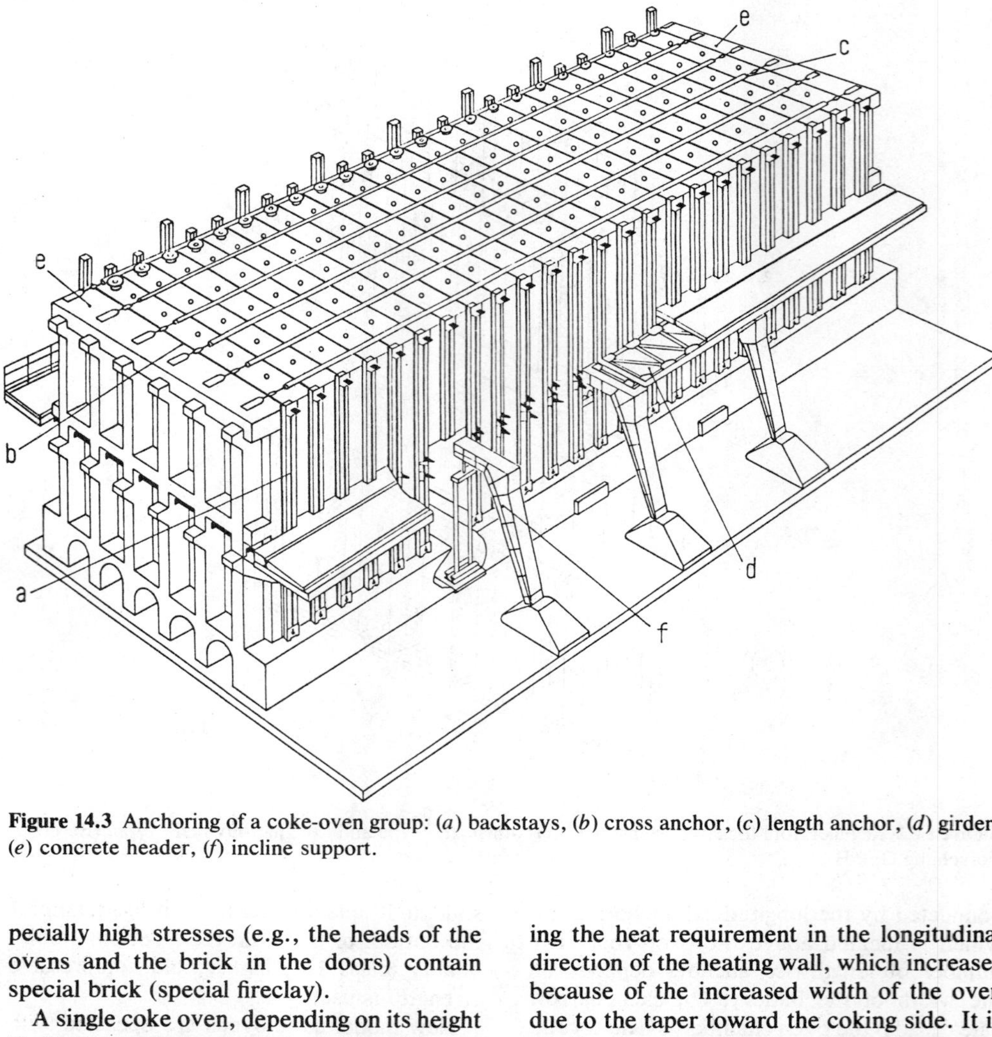

Figure 14.3 Anchoring of a coke-oven group: (*a*) backstays, (*b*) cross anchor, (*c*) length anchor, (*d*) girder, (*e*) concrete header, (*f*) incline support.

pecially high stresses (e.g., the heads of the ovens and the brick in the doors) contain special brick (special fireclay).

A single coke oven, depending on its height (3–7.65 m) and chamber length (12–17.25 m), contains 150–615 tonnes of refractory material. About 15–50 tonnes of steel is required per oven for steel and castings such as buck stays, oven doors and frames, receivers, standpipe waste gas receivers, and gas conduits.

The large heating wall surface must be heated in such a manner as to obtain a uniform coke with a uniform content of volatile matter. As a precondition for this, each place on the heating wall must obtain the amount of heat from the heating gases corresponding to the heat requirement at that point. By subdividing the heating wall into a large number of flues, the great quantities of gas flowing in a vertical direction along the whole heated wall can be controlled. Furthermore, the flues are necessary for adjusting the heat requirement in the longitudinal direction of the heating wall, which increases because of the increased width of the oven due to the taper toward the coking side. It is also important for the intensity of heat to be uniform in each flue in a vertical direction. Flames of lean gas that are long and nonluminous make it possible to heat uniformly without special precautions. Richer gas has a short luminous flame, so that heat distribution becomes nonuniform. To achieve a uniform temperature distribution along the heights of the flue with this richer gas, the following means are used: (1) distribution of gas nozzles in the individual flues at different hights, (2) admixture of inert gases (waste gases) to the products of combustion, and (3) addition of combustion air in stages. The heat content of the heating gases leaving the flues at a high temperature is recovered to a large extent in the regenerators and used to preheat the combustion feed streams.

A distinction is made between underjet

(bottom-burner) ovens and gun flue (top-heated) ovens, according to the method of adding rich gases to the flues. The underjet oven (Fig. 14.4) has an accessible room and a "nozzle basement" below the floor of the regenerators. Conduits from this basement lead to the nozzles, from which the heating gas is added in measured quantity to each flue by vertical rising channels in the separating walls of the regenerators. The waste gas channels are on the side of the battery.

The gun flue oven (Fig. 14.5) has conduits for the heating gas on the front side of the oven heads. Rich gases flow into this distribution channel, which is located between the oven chamber and the corresponding regenerator, and are distributed to the individual flues. The amount of gas required for each flue is controlled by calibrated ceramic nozzles at burner level. The waste gas channels are usually beneath the floor of the regenerators.

Depending on the flow of the heating and waste gases, a distinction is made between "semidivided" ovens, ovens with "twin-flue systems," and ovens with "flue system groups." In semidivided ovens one heating surface is exposed to the flame, and the waste gases flow through horizontal channels located above the flues in the other half. The twin-flue design consists of a section that is exposed to the flame and a section that is not. The two sections are separated only by a wall with an opening for the waste gases. This eliminates the horizontal channel. Half the flues—for example, the odd-numbered flues—are heated by the flame, and the other half is used for the waste gas. In the flue group oven, the heating walls are divided into flue groups. At both ends of the oven two flues make up a group, and between these groups there are four flues in a group. The flues of each group are connected at the top by openings for the waste gas in the flue separating walls. The groups of two operate like twin-flue groups; in the groups of four, heating of the inner and outer flues alternates with time.

An exception to those designs is an American system (Koppers-Becker oven) in which all flues on one heating wall are simultaneously heated and the waste gases led to the wall on the other side of the oven chamber. The different heating systems are connected with corresponding regenerator systems. To the semidivided oven belongs the "semidivided or cross-divided" regenerator; the twin-flue and flue group draft systems generally have "continuous" regenerators. One oven system connects the twin flues with the semidivided regenerator.

The direction of the flow of gases is usually reversed after 20 or 30 min; regenerators and heating wall halves exchange functions. Since the amount of gases flowing changes periodically, satisfactory control of the heating gas and waste gas flow is not easy to attain. The pressure in the gas distribution conduits, the gas pressure in the collecting main, and the draft in the waste gas channels is kept constant by controllers. To have the heating reflect the changing market for coke, quite often so-called intermittent heating is used, which means that the specific amount of gas (e.g., when changing from a coking time of 18 to 22 hr) remains constant, but the duration of heating is shortened by about 2 min. This eliminates regulating tasks, which are wage- and time-intensive.

The individual designs of coke ovens differ essentially with respect to the flue and regenerator systems and by the method of adding rich gases to the flues. The companies having the largest number of coke-oven installations in the world are Didier Engineering GmbH, Koppers Company, Inc. (U.S.), Krupp-Koppers GmbH, Dr. C. Otto, and C. Still. Their ovens are described in more detail in the following sections. Well-known companies that build coke ovens—for example, CEC (France), Simon-Carvés, Woodall-Duckham and Gibbons (England), and Wilputte (U.S.), have almost stopped building their own coke ovens and instead use the designs of the above-named companies, under license. New on the world market are the Japanese companies Nippon Kokan (NKK-Oven) and Nippon Steel (Nittetsue M-type-oven), which have oven systems similar to the four German coke-oven designs. In countries of the eastern bloc, most of the coke ovens are built by the companies Giprokoks (U.S.S.R.) and Koksoprojekt (Poland). These ovens also have basic characteristics of the oven systems described below.

Didier Engineering GmbH (Essen) builds the so-called control flow oven. It can be built with underjet or gun flue heating. The regenerators are subdivided into cells. Each cell is connected to a maximum of two flues

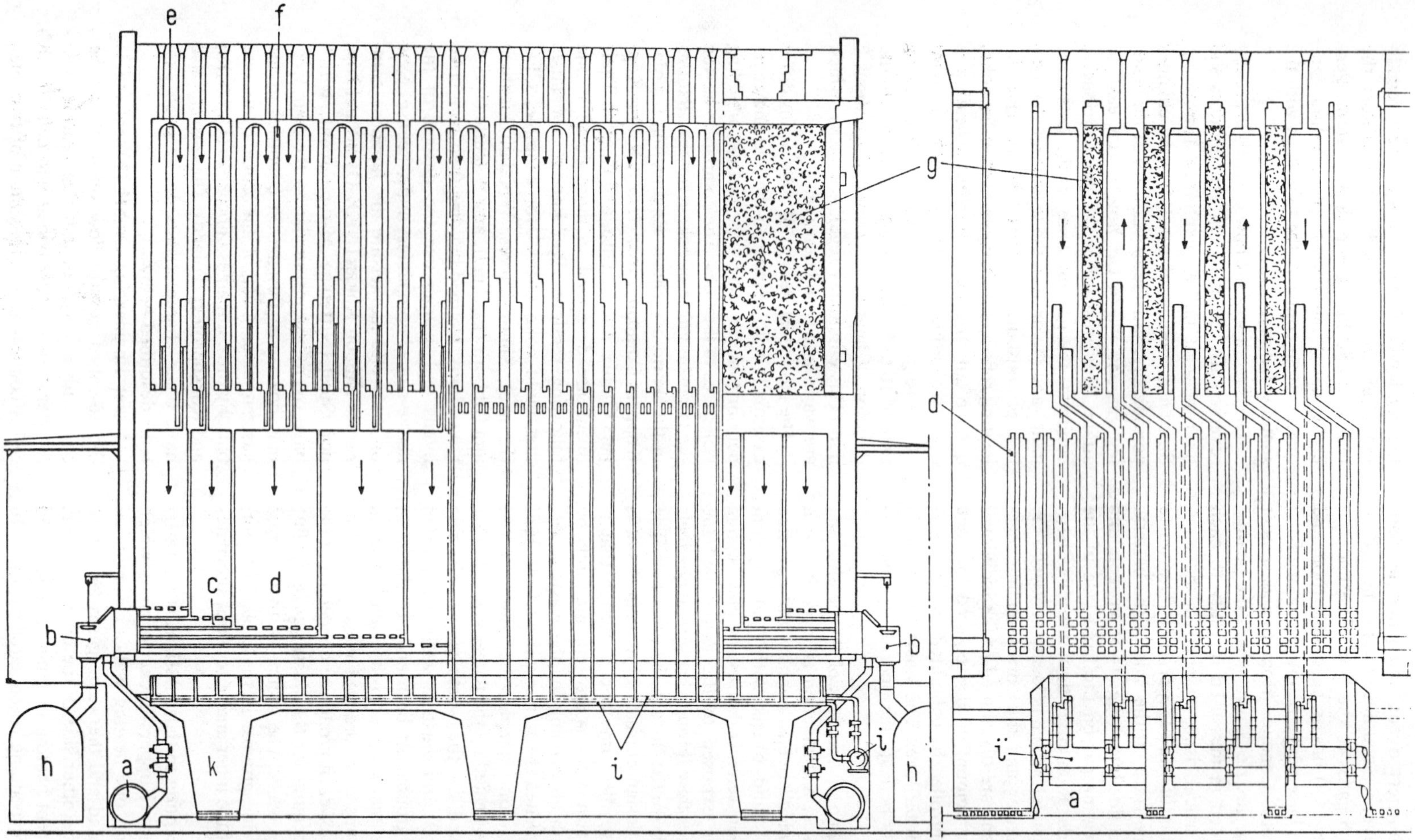

Figure 14.4 The group draft control flow oven (underjet design) of Didier Engineering GmbH, Essen: (*a*) low-Btu gas line, (*b*) off-gas valve with controller, (*c*) individual bottom channel, (*d*) regenerator cells, (*e*) double-flue group, (*f*) quadruple-flue group, (*g*) oven chamber, (*h*) off-gas chamber, (*i*) coke-oven gas line, (*j*) lines to the nozzles, (*k*) concrete substructure. (Bergbau-Forschung GmbH.)

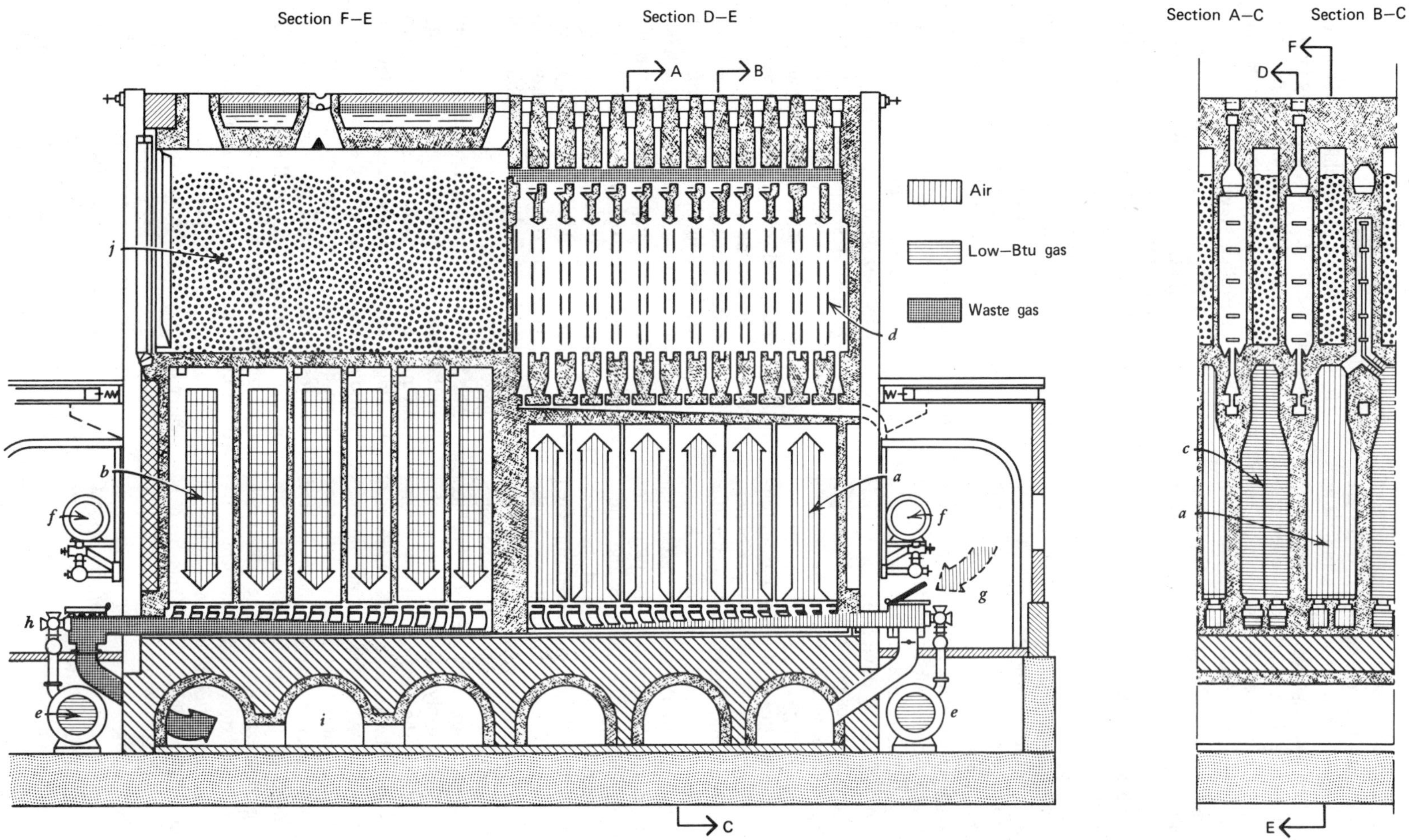

Figure 14.5 The compound coke oven of the Carl Still Company, Recklinghausen: (*a*) regenerator section carrying air; (*b*) regenerator section carrying waste gas; (*c*) regenerator section carrying lean gas; (*d*) stepwise heating; (*e*) lean gas line; (*f*) rich gas line; (*g*) air inlet; (*h*) waste gas exit; (*i*) waste gas channels; (*j*) oven chamber.

in one heating wall. Instead of a single bottom channel beneath the regenerators, there are as many single bottom channels arranged as there are regenerator cells. The combustion feed streams are individually controlled outside of the oven by control devices in the waste gas valve for each bottom channel. When built for low throughputs, the ovens are semidivided; for large throughputs they are built as flue group ovens. Figure 14.4 is a typical large-chamber flue group oven with underjet and compound oven construction.

Lean gas flows into the heating flue at the bottom and passes through a channel in the flue to a higher level, where it exits. Air also enters at the bottom and a second stream of air enters through a channel above the lean gas opening. When using rich gases, all these channels are used for air supply. Each of the four individual streams can be separately controlled by control devices in the off-gas valve. The nozzle for rich gases is kept very short to avoid formation of carbon deposits by thermal cracking.

The characteristics of the Kreisstrom (circular flow) coke oven of Krupp-Koppers GmbH (Essen) (Fig. 14.6) include a twin-flue system, differential heating, the use of Kreisstrom heating, and the semidivided cross-regenerator. Its connection to the twin flues is made possible by the shape of the brickwork between the regenerator and the heating wall. All flame-heated flues are connected to the regenerator half, which brings in the combustion feed streams. All waste gas flues are connected to the other half of the regenerator. When using rich gases for heating, the heat is uniformly distributed over the whole height of the heating wall by Kreisstrom heating. The dividing walls of the twin flues have openings at the bottom; through these openings part of the downward-flowing waste gas is sucked into the flame-heated flue because of the ejector action of the ascending combustion gases. There the combustion is slowed down and a uniform distribution of the heat of combustion is achieved over the whole height. The Kreisstrom is self-controlling for every operating condition. When using lean gas for heating, its effect is very small because of the special arrangement of the inflow of gas and air. By differential heating, the temperature of the gas-collecting space above the chamber charge can be influenced. Krupp-Koppers is building underjet as well as gun flue ovens. Figure 14.6 shows the top-heated type.

The main construction characteristics of the oven of the Dr. C. Otto Company GmbH (Bochum) include the use of the underjet principle, twin-flue separation in the heating wall, and controlled stagewise heating (Fig. 14.7). It is heated by a row of individual heating units that can be separately controlled. For this purpose the twin-flue system is connected to extended regenerators. The regenerators are divided into cells in such a manner that each twin flue has separate regenerator chambers for air, lean gas, and waste gas. Rich gases are added by underjets and distributed to the individual heating points; the amount of gas is controlled by nozzles in the nozzle conduits ahead of the rising channels. The flow of air and of lean gas can be controlled by the bottom channel as well as by nozzle plates, which are arranged at the bottom of the regenerator cells. The waste gas rates are also controlled by nozzle plates at the bottom of the regenerator cells. To obtain uniform heating in a vertical direction, the burners for rich gases are placed at different heights in the individual heating flues. The combination of continuous regenerators and independent twin-flue heating elements makes it possible to equip the oven with only one row of waste gas valves and one channel for collecting the waste gas. Combustion air is added from the opposite side. This gives the oven uniform flow conditions in the bottom channel independent of changes in regenerative heating.

The Carl Still Company (Recklinghausen) manufactures a semidivided oven, chiefly with gun flues (Fig. 14.5). It is characterized by staged combustion, which gives uniform heat distribution in a vertical direction, even in an 8-m-high oven chamber. The combustion feedstreams ascend in channels located in the separating walls of the heating wall; from these, depending on the height of the oven chamber, four to eight calibrated slits lead to both sides of the flues. When using lean gas for heating, the flow of gas alternates with the flow of air. These streams are always connected to the corresponding regenerator chamber. The combustion feedstreams that leave from opposite slits react with each other; thus the total combustion is separated into a number of individual combustion zones. When rich gas is used for heating,

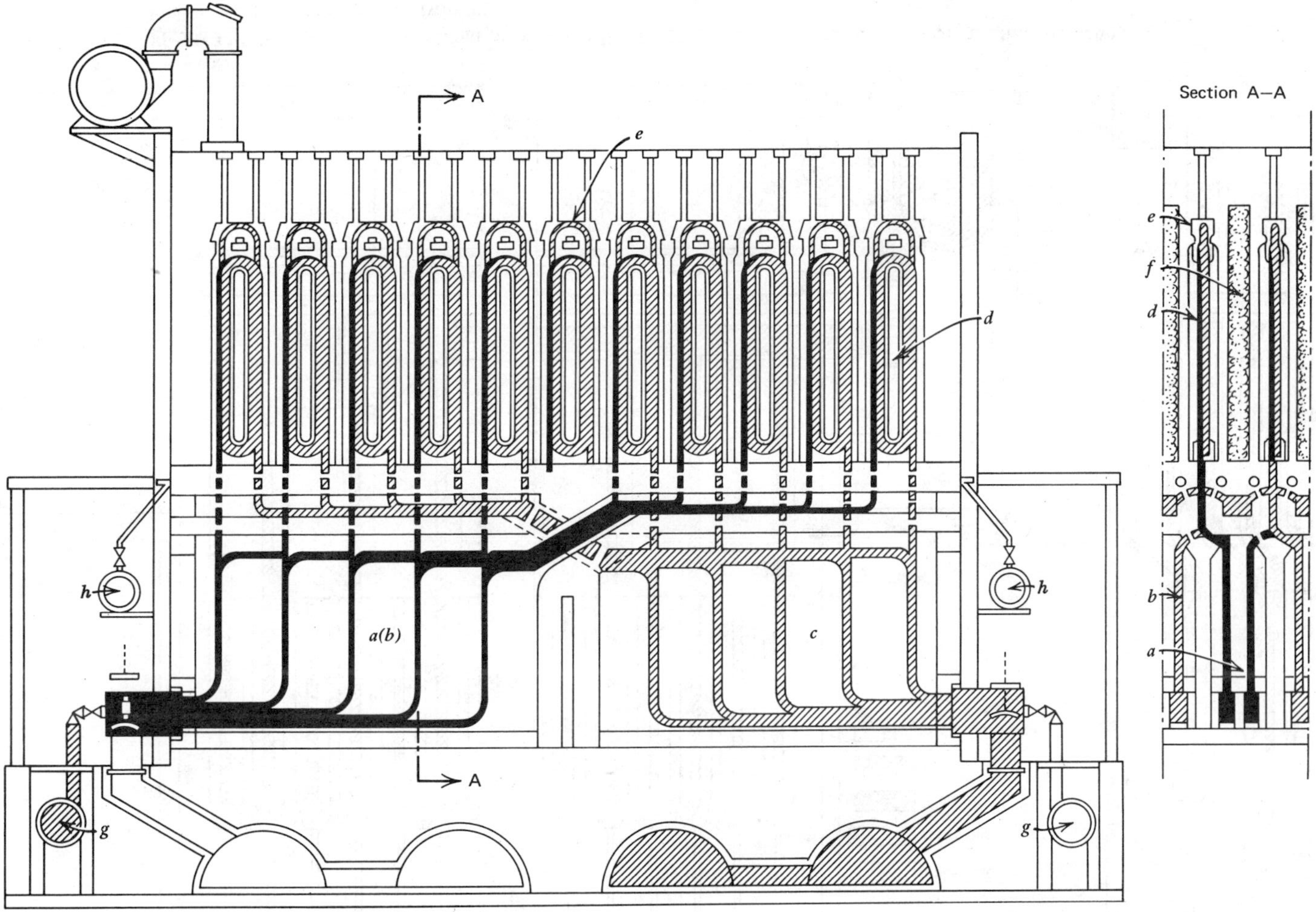

Figure 14.6 The Kreisstrom compound coke oven of Krupp-Koppers GmbH, Essen: *a*, regenerator section carrying air; *b*, regenerator section carrying gas; *c*, regenerator section carrying waste gas; *d*, twin flue for Kreisstrom heating; *e*, differential heating; *f*, oven chamber; *g*, lean gas line; *h*, rich gas line.

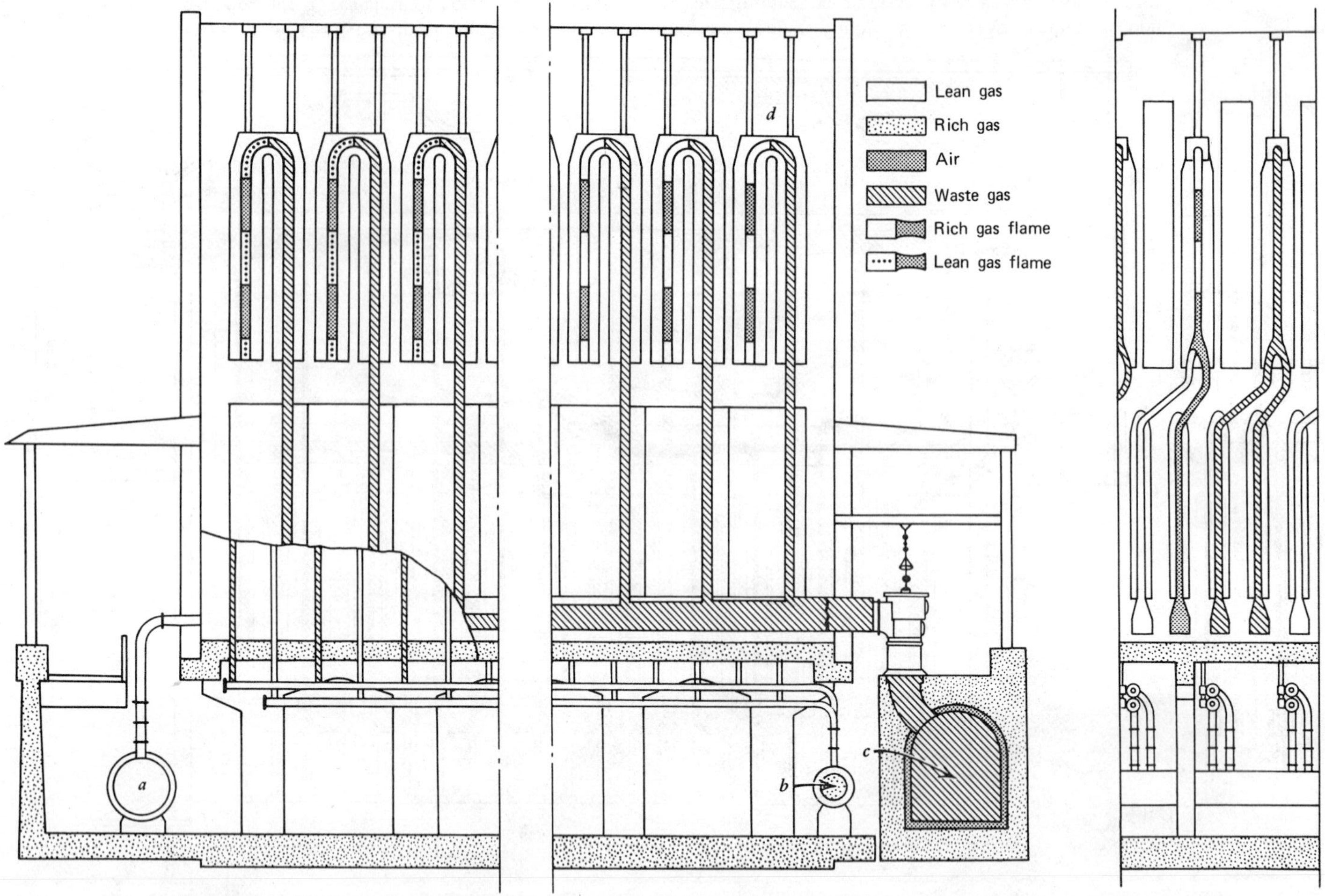

Figure 14.7 The twin-flue compound oven of the Dr. C. Otto Company, Bochum: *a*, lean gas line; *b*, rich gas line; *c*, waste gas channel; *d*, twin flue.

only combustion air is brought up in these channels. The rich gas flows through the heating flues and burns stepwise as it comes in contact with the air leaving the slits.

The Koppers-Becker oven (Koppers Company, Inc., Pittsburgh) is built as a gun flue oven or underjet system (Fig. 14.8). It is distinguished by its heating flue system. All flame-heated flues are on one side of the oven chamber and all waste gas flues are on the other side. The waste gases pass from the flame-heated wall by crossover flues in the roof of the oven into the other heated wall carrying the waste gas. In the direction of the oven group, two flame-heated walls are always followed by two walls passed by hot waste gas. The walls are connected to the regenerators, which follow each other in the axis of the oven group in sequence with the heating walls. Figure 14.8 shows the flow of gases, with rich gas heating used in a compound oven with bottom burners and a Kreisstrom principle.

The Nittetsue coke oven of Nippon Steel Engineering (Fig. 14.9) has twin flues in the heating walls and is built as a gun flue oven as well as an underjet oven if coke-oven gas is used for heating. When lean gas is used as fuel, the underjet design is employed. The continuous regenerator underneath the coke oven is subdivided into several independent regenerator cells. The two outside cells on the pusher and coke side are only half as wide as the regenerator cells located in between. The latter are again subdivided into two subcells. The number of regenerator cells corresponds to half the number of heating flues in one heating wall, plus one. The feeding of the cells is done with air and lean gas, through two bottom channels located beneath the regenerator. One bottom channel is always provided for air and one for lean gas. While air and lean gas are being heated in one row of regenerators, the waste gas (waste heat) flows downward in the neighboring regenerators. Each end cell and subcell of the regenerator is connected to one heating flue between two neighboring heating walls of the coking chamber, which is located on top of the regenerator. This arrangement of the regenerators permits easy control of the heating. Since the regenerator is divided into many subcells, the flow in the regenerator is uniform and permits excellent gas distribution.

To achieve uniform heating in a vertical direction in ovens with a height of more than 5 m, the combustion is distributed over three different heights.

14.1.2 Developments in Coke-Oven Machinery

Three influences have had worldwide importance in the further development of coke-oven machinery during the past decade. First, increasing public awareness of the environment and the resulting control laws have affected the design of all machines. Second, the necessity to increase productivity by the application of the high-volume oven and the use of shorter coking times has led to the building of sturdier machines and more powerful drives, and to considerable mechanization of time-consuming processes. Finally, the need to improve working conditions for the operating personnel has had to be considered.

The larry car originally served only to bring the feed coal from the coal tower to the individual oven chambers and to charge these chambers with coal. Considerable human labor was involved, and the job was made unpleasant by heat, dust, and gases. The strong evolution of gases during the charging process led to contamination of the atmosphere. Modern charging larries can be characterized in two main groups: (1) those with external suction of the charging gases, which means that dust removal equipment is located on the larry, or (2) those operating with a stationary dust removal station and having internal oven suction in which the charging gases are sucked through the standpipe into the collecting main by means of a steam ejector or high-pressure ammonia liquor.[5]

Another distinguishing feature of charging larries is the way that coal is fed from the charging funnels. This can be done by free fall or by means of mechanical charging devices. With mechanical charging, using a rotating disk, a starwheel feeder, or a horizontal screw feeder, it is possible to keep the coal feed rate constant during the charging period. This facilitates designing the suction capacity for controlling the charging gases.

[5] W. Eisenhut and H. Friedrich, *Glueckauf* **111**, 432–439 (1975).

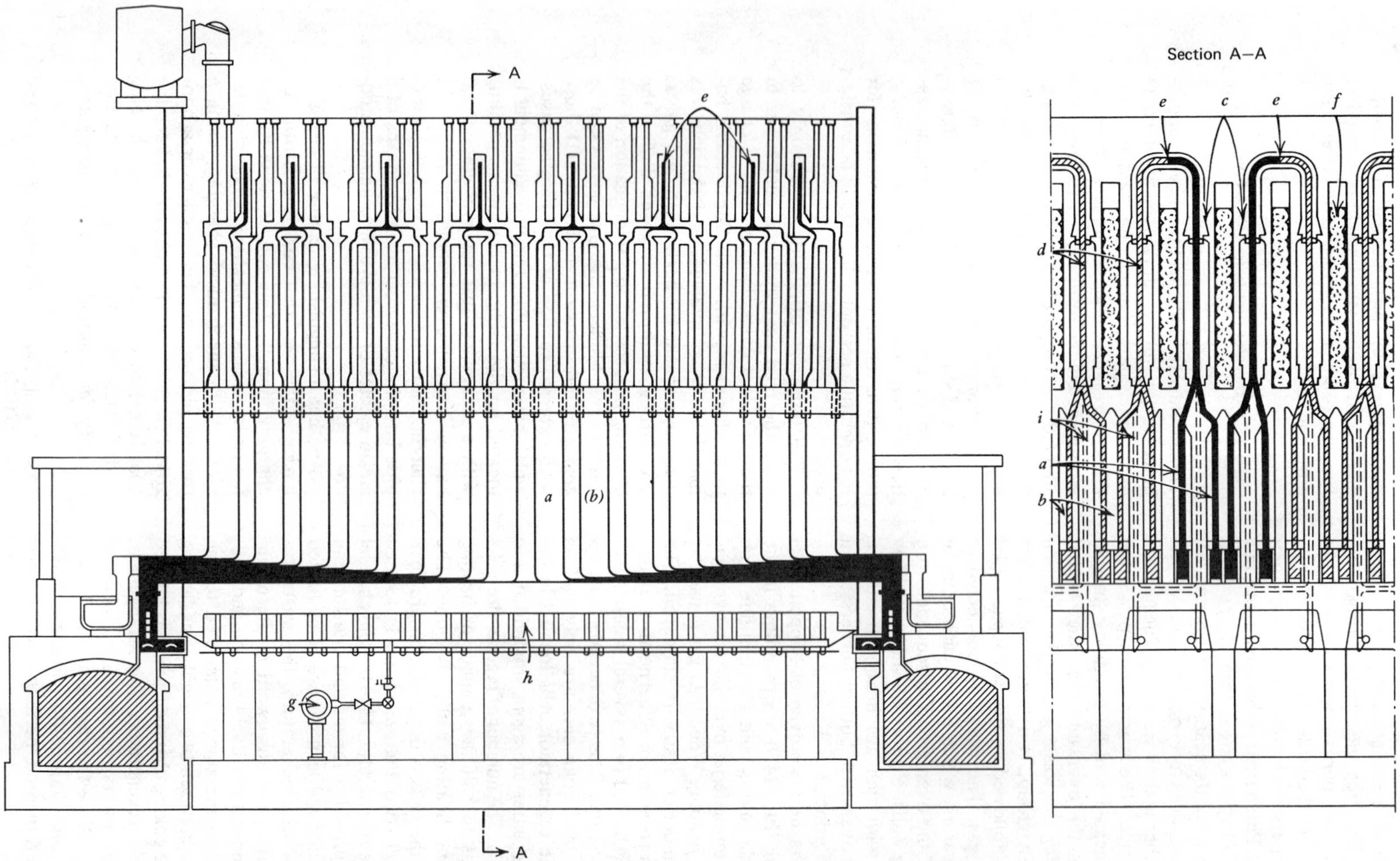

Figure 14.8 The Koppers-Becker compound coke oven with underjet burners from Koppers Company, Inc., Pittsburgh: *a*, regenerator for air; *b*, regenerator for waste gas; *c*, heated flues; *d*, flues carrying waste gas; *e*, transfer channel; *f*, oven chamber; *g*, coke-oven gas line; *h*, underjet burner equipment; *i*, channel for rising coke-oven gas.

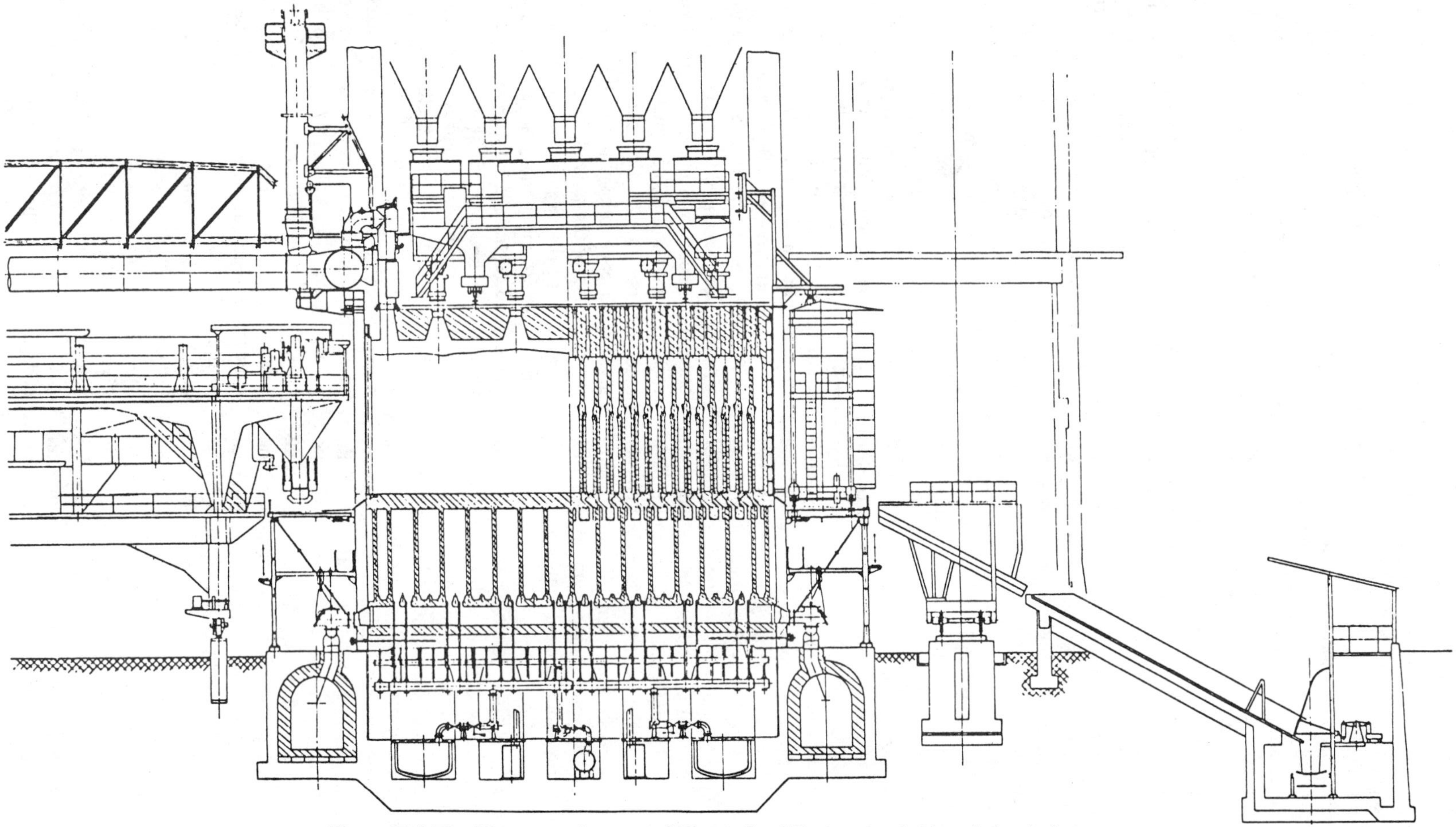

Figure 14.9 The Nittetsue coke oven of Nippon Steel Engineering (with underjet design).

In addition, the rate of charging the coal can be changed during the charging process. The worldwide trend to mechanical charging is intensified by the increasing use of finely ground feed coal, which does not always flow freely from the charging funnels of the charging larry, especially if the moisture content of the coal is high.

Today charging larries are built with a number of mechanical devices that make servicing easier. These larries work as "one-spot" machines, which means that they do not need to be moved during the charging process. The following operations are handled by mechanical devices: lifting and replacement of the magnetic charging cover; cleaning of the cover, the frame, and the standpipe and gooseneck elbow; pouring of a lime-fireclay mixture on the charging hole cover; vacuum cleaning of the oven top; and mechanical operation of the gates on the coal tower as well as the valves and lids on the standpipes. For the individual functions, hydraulic drives are usually used. This additional equipment increases the weight of the charging larry considerably. If one also considers the dust removal equipment with its water system, the air-conditioning unit for the operator cabin, DC electric equipment, and hydraulic equipment with storage vessels, the weight of a charging larry with large-volume ovens is more than 160 tons. Therefore consideration is being given to installing all the various mechanical units on an auxiliary car.

The coke-pushing machine has a leveling bar, which is used to smooth the surface of the coal charge in the oven, and a pusher bar to discharge the coke at the end of the coking cycle. Electric drive is used for the pushing machine; however it is always necessary to have a second power source (current from an independent source or air pressure) so that in the event of a power failure the leveling bar or pusher ram does not remain in the high-temperature oven, but can be removed from it. The pushing machine also has a device to unlatch and remove the oven doors and to close the oven doors and latch them. These operations may be activated pneumatically, hydraulically, or electromechanically. Machinery to move the doors is now designed in such a way that the oven doors can be removed or reset while the pusher machine is in position. In this case the oven door is automatically rotated by 90 degrees. In this position the doors are automatically cleaned by scrapers. At the same time, the door frames are automatically cleaned of adhering tar. The mechanical opening and cleaning of the leveling door and door frames has been installed on some new coke-oven batteries.

To avoid emissions through the leveling door opening, a so-called leveling shroud is used, which now comes supplied with a number of nozzles so that it can be pressurized by means of a blower. The leveling of the charged oven and the discharging of the next oven chamber is done without moving the machine: "one-spot position." At the same time, the coal that was withdrawn from the oven by the leveling bar and the coke that dropped down when removing the oven door are mechanically removed. In stamping operations the pushing machine is combined with the stamping machine. In the stamping mode the coal is compressed in several layers to a density of about 1 ton/m^3 by several reciprocating, hydraulically driven tampers.

The coke guide car bridges the operating platform, or bench, during pushing of the coke. It consists of a bottom chute covered with hard cast-iron plates and of a series of steel sections mounted on the sides spaced apart from each other. It can be moved in the direction of the axis of the oven to bring it as close to the oven chamber as possible. The coke dropping into the quench car is guided and restrained by a plate hanging in front of the opening of the guide car. The oven doors are inserted and removed on the coke side of the oven by a method similar to that used on the pusher side. The door removal machine and the coke guide car are sometimes constructed as one unit. However in large ovens with a chamber height of more than 6 m, it is preferable to build them separately. For cleaning the oven doors and frames, equipment similar to that on the pusher side is used. The coke guide car may also be built as a "one-spot" machine.

The reduction of emissions during coke pushing has been an aspect of coke-oven machinery technology that has received considerable attention during the past decade. Although a great number of constructive complex solutions have been suggested, they fall into two basic categories: shrouds with stationary or movable dust removal equipment, and container cars.[5,6]

The quench car travels from the coke oven to the quenching station, to the coke wharf, and to the next oven to be pushed. At the quenching tower the hot coke is doused with water and steamed out in the shortest possible time; then it is discharged on the wharf from which it is fed onto conveyor belts and transported to the screening station. The bottom and walls of the quench car are exposed to great temperature changes and stresses and are therefore designed and built for this special service. The quench car is usually moved by an electric locomotive that operates in a range of speeds. The locomotive has a compressor to produce compressed air for moving the discharge doors and the brakes on the quench car. Two kinds of cars are being used: the quench car with an inclined bottom and the car with a movable bottom. As the time between coke-oven discharges becomes shorter and shorter, the quench car may be the obstacle to further increases in the number of coke-oven discharges. New drives have therefore been developed. In one case a quench car with a compact two-axle drive unit was built. The high loads on this drive unit permit higher acceleration. In another type the quench car is moved by powerful winches. This type of quench car is well suited for automation.

Hand winches were originally used for the regular reversing of the heating units of coke ovens. These were later replaced by electrically driven reversing winches. Today the mechanical switch gear is sometimes replaced by units with a minimum of space requirement, which permits the use of hydraulically activated rods for moving gas valves, dampers, and other valves.

14.1.3 Environmental Control Facilities

The reduction of emissions from coke-oven plants is particularly difficult because emissions arise from several sources during the operation. These sources are separated from each other and can be brought together only with difficulty. In addition, the emissions are solid as well as gaseous or liquid. Usually two types of emission appear side by side, which makes removal especially difficult. The measures used to keep the air clean during operation of a coke-oven plant are described in the following section. (For information on cleaning of waste water, see Chapter 18, Section 18.5.)

The reduction of emissions during operation of a coke-oven plant presents great difficulties. Corresponding to the different processing steps during coke production, the emission sources in a coke oven can be divided into three groups: (1) dust emission resulting from the transport of coal and from coal preparation; (2) emissions of dust, gases, and tar-containing vapors occurring during coking (charging of the coke oven, the coking process itself, coke pushing, coke quenching, and heating of the coke ovens); and (3) dust emissions caused by transport and screening of coke.

So far it has not been possible to quantitatively determine the emissions that occur during the operation of coke ovens. Although data have not been lacking in the past, such data were estimated on the basis of a very few measurements. This has been particularly evident in the study of the Economic Commission for Europe,[7] which presents a wide range of data for individual coke-oven plants in several countries. This assessment problem is to be expected, since the determinable quantity of contaminating material depends on many factors, including the following: (1) the measuring and evaluating procedures, (2) the coal used and the method of grinding, (3) the charging process, which means top-charging or stamping operations, (4) the coking conditions, (5) the construction, age, and mechanical condition of the coke-oven plant, (6) the control equipment used during charging and discharging the coke oven, and (7) the gas used for heating the coke ovens, such as coke-oven gas, blast furnace gas, refinery gas, and LPG-air mixtures. In addition, the scale and type of emission control equipment installed in the coke-oven plant have a decisive effect on emissions. In the future, quantitative results

[6] J. Fiedler, *Bergbau*, **1,** 31–37 (1977).

[7] *Emissions from Coke Oven Plants,* ECE Document No. ST/ECE/COAL/26, UN Publications, New York, 1968.

will be obtained only when all the emissions can be measured.

In West Germany reliable measurements were made of dust emissions from the different operations in a coke-oven plant, with or without dust removal equipment (Table 14.1). Since there has been rapid development in dust removal technology, the maximum values are given for those emission sources having dust removal equipment. The current average value is below the given data. Quantitative data for the emission of gases from a coke-oven plant are not available at this time. Several research programs have been started in West Germany to obtain such data.

"Technical Guidelines for the Avoidance of Air Contamination"[8]—*TA Luft*—requires that new and additional coke-oven plants in West Germany operate using the following emission control measures:

1. Sulfur-free or partially desulfurized gas is to be used for heating coke ovens. The hourly average weight concentration of H_2S in this gas should not exceed 1.5 g/m^3 (62 gr/100 scf). That of other sulfur-containing compounds should not exceed 0.5 g/m^3 (20.7 gr/100 scf).

2. The gray value of the flue gas leaving the chimneys must be brighter than the No. 2 value of the Ringelmann scale. This value may be exceeded for a short time when changes are made in the heating cycle or when charging the coke ovens. The gray value must at that time, however, be brighter than No. 3.

3. The gases issuing from the coke oven during top charging must be completely injected into the raw gas, or must be treated before release into the atmosphere so that at least 90% of the solids is removed. During leveling, emission of charging gases must be prevented by maintaining a lower pressure in the gas-collecting main or by a tight seal of the leveling opening.

4. In coke ovens using stamping, a removal of 90% of the particulates is to be maintained if at all possible.

5. Measures to prevent emissions during the coking process itself are given in an operating procedure that emphasizes (*a*) avoiding leaks at all openings, (*b*) the importance of ensuring that only completely coked coke is pushed, and (*c*) avoiding the emission of unburned gases into the atmosphere.

6. The coke-oven charging machinery must have mechanical equipment for cleaning the sealing surfaces on oven doors and oven door frames, and this equipment must be properly maintained.

7. During coke pushing the waste gases must be collected and the dust removed. No particular process is specified, since a number of processes for eliminating emissions during coke pushing are still under development.[5,6]

8. Towers for quenching the coke must be of a shape, cross section, and height that results in a low emission velocity of the quenched vapors. Quench towers must be equipped with additional nozzles and inserts. No effluent may be used for quenching except the recycled water of the quench process itself. No organic substances are permitted in the quench water.

9. The surface piles of coking coal must be protected from the wind—for instance, by keeping the surface moisture at about 10% or by spraying on binders.

10. The long axis of coal and coke storage piles must be arranged in the direction of the prevailing wind. The height of dropping onto the pile and the number of reloading points must be at a minimum.

11. Roads inside the plant must be regularly cleaned of dust and breeze, and must be kept moist.

12. The waste gas from screening, crushing, and grinding plants for coke should have a maximum solids content of 75 mg/m^3.

[8] *Technical Guidelines for the Avoidance of Air Contamination (TA Luft),* Gem. Min. Bl. Ausg. A No. 24 (August 28, 1974).

Table 14.1 Dust Emissions from Coke-Oven Plants in West Germany[5]

	Emission (kg of dust/ton of coke)	
Operation	Without Dust removal	Maximum With Dust Removal
Charging	0.15	0.015
Pushing	0.4	0.040
Quenching	0.35	0.070
Coke treating	1.5	0.030

The "*TA Luft*" should make the freedom of the regulatory bodies as narrow as possible and necessary. In areas of strong pollution, however, the government, when granting permission to operate a plant, may require emissions to be below the limits given above or may require additional measures for the reduction of emissions.

Similar guidelines and instructions for preventing the deterioration of air quality have been promulgated worldwide in recent years, although there are differences depending on the country and the region.[9] Thus an industry might have to observe regulations of different organizations, such as local air pollution control authorities, state air pollution control boards, and the U.S. Environmental Protection Agency (EPA).[10] These authorities make the requirements increasingly more stringent. In addition, there are the regulations of the U.S. Occupational Safety and Health Administration (OSHA), which is interested in the conditions of the working environment. In 1977 the regulatory requirements were so comprehensive and time-consuming that each worker in a coke-oven plant required an environmental control supervisor.

In the United States in 1977 the threshold limit value of the benzene-soluble portion of polycyclic aromatic hydrocarbons was 0.2 mg/m^3 (standard volume) in the working environment.[11] Strict maintenance of these guidelines could make it impossible to operate most, if not all, coke ovens. In November 1974 a bureau of OSHA was formed to review coke-oven plant operations with the expectation of arriving at a more realistic standard.

The most important sources of emission in the horizontal chamber coke oven are the charging, pushing, and quenching operations, and leaking oven doors.[5,12] Therefore equipment that may reduce or prevent emissions in coke-oven operations should be mentioned briefly.

Figure 14.10 shows the points of emission during charging of the coke-oven chamber. Charging begins with opening of the slide gates of the coal hoppers of the charging larry. The coal fed into the chamber displaces the hot gas contained in it. The main constituent of this gas is nitrogen, and it also contains carbon dioxide, carbon monoxide, and small amounts of oxygen. This gas mixture escaping through the charging opening, the standpipe, and the levelling opening also contains coal dust (Fig. 14.10). When the moist coking coal touches the hot bottom of the oven and the hot walls of the chamber, evaporation of the water and coking of the

[9] C. E. Duncan, *AIME Ironmaking Proc.*, **34**, 238–241 (1975).

[10] E. H. Roe, *AIME Ironmaking Proc.*, **34**, 229–234 (1975).

[11] Threshold Limit Values for Chemical Substances in Workroom Air, adopted by ACGIH for 1975.

[12] N. Plaks, *AIME Ironmaking Proc.*, **33**, 109–118 (1974).

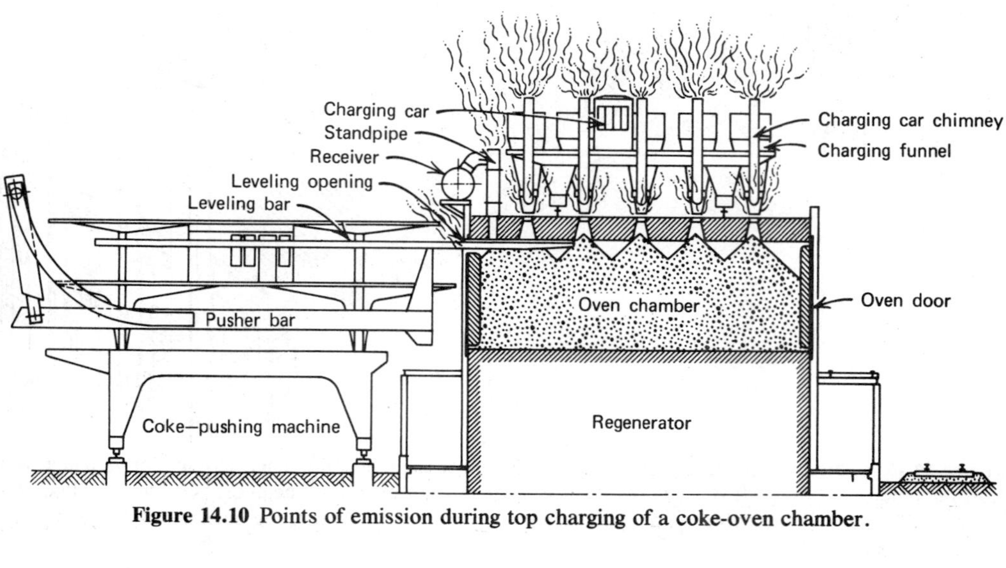

Figure 14.10 Points of emission during top charging of a coke-oven chamber.

coal begin immediately. The vapors and gases produced escape into the air. Tar mist gives the charging gases a yellow to brownish color. They may ignite and burn with a luminous flame.

When charging the coking coal, charging peaks occur and must be leveled by the leveler bar. In this way the gas-collecting space, a free space about 300 mm high, is formed between the coal charge and the oven roof. After the hoppers of the charging larry are emptied, and during the leveling, spillage around the sides of the charging holes is brushed into the ovens. The leveling bar is removed and the leveling opening closed by the leveling door. The charging hole covers are put in place, the standpipe lid is closed, and the oven chamber is connected to the collecting main by opening a valve. This action terminates the charging process.

In all processes for decreasing the formation of charging gas, the gas must first be sucked away. There are two main groups of processes: (1) injecting the charging gas into the raw gas, and (2) permitting the charging gas to escape to the outside air.

In some plants with top charging (by larry car), and with the stamping technique, the charging gases are sucked away by a conduit. The charging gases pass into the gas-collecting space of another coke oven and travel from there, together with the gases formed in that oven, into the collecting main by way of the standpipe.

Figure 14.11 illustrates this method of operation for coke ovens with top charging. During charging the oven chamber is connected to the collecting main, *A*, and the standpipe cover is closed. Before starting to feed the coal, it is necessary to remove the cover of the suction opening, *B*, of the oven being filled and to place the transfer pipe, *C*, in this opening. The end of the transfer pipe with the cover removal device, *D*, is now connected to the closed suction opening, *E*, of the coke oven that will receive the charging gases when its cover is removed, in turn. On both ovens the steam valves, *G*, which

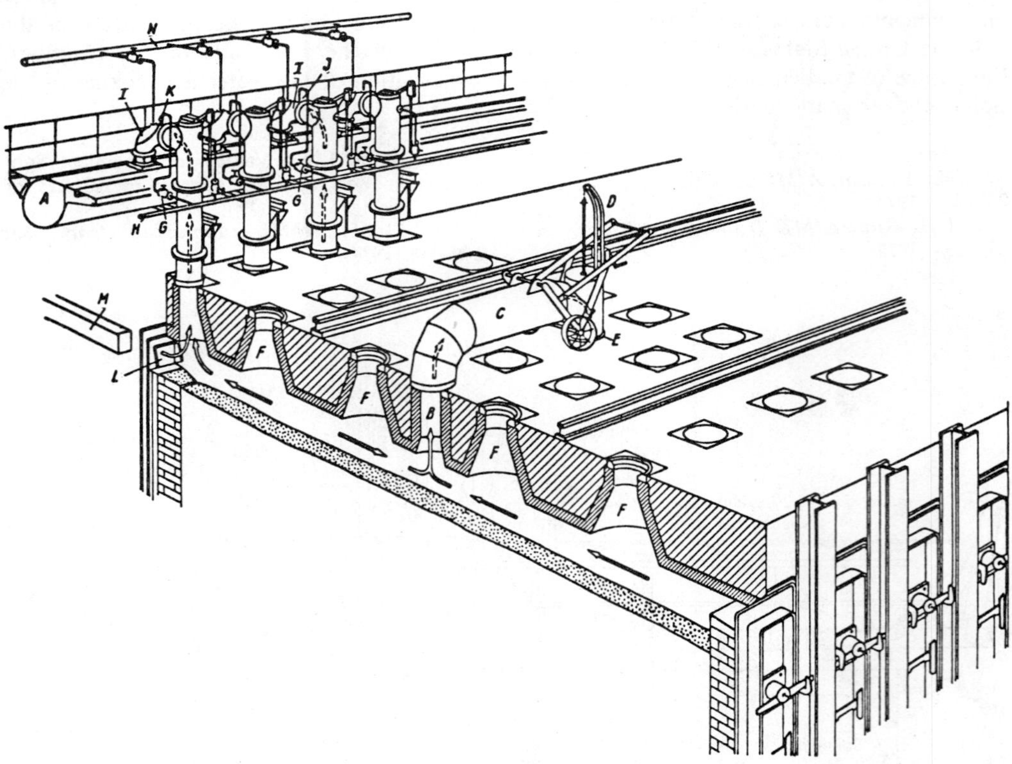

Figure 14.11 Suction of charging gases by a transfer pipe: (*A*) collecting main, (*B*) suction opening, (*C*) transfer pipe, (*D*) cover lift device, (*E*) suction opening, (*F*) charging openings, (*G*) steam valve, (*H*) steam supply line, (*I*) standpipe bend, (*J,K*) steam ejector nozzles, (*L*) leveling opening, (*M*) leveling bar, (*N*) spray for the receiver.

are open, are located in the connecting line between the main steamline, *H*, and the standpipe, *I*. The cover of the suction opening, *E*, is removed by means of a device inside the transfer pipe. Following this procedure, coal feeding from the hoppers of the charging larry into the charging openings, *F*, can be started. Steam entering through the steam jet nozzles, *J* and *K*, into the standpipe bend, *I*, causes suction whereby the charging gases from the coke-oven chamber travel to the collecting main, partially by the transfer pipe in the gas-collecting space of the receiving oven and partially by the standpipe of the oven to be charged. When charging is completed, the cover is removed by means of a lifting device and put on the suction opening of the next oven. Steam is shut off from the nozzle, *J*, and the transfer pipe is moved to the side. The suction opening *B* is closed by a cover, and steam supply to nozzle *K* is shut off.

The process of removing the charging gases by suction through a transfer pipe has many technical variations. These depend on the point from which the charging gases are evacuated from the coke oven, the selection of the coke oven to receive the charging gases, the method of producing the low pressure, the length of the transfer pipe, and the method of moving it.

At present, three processes are being used for the suction of charging gases through one or two standpipes. These processes are suitable for top-charged coke ovens only.

One of these processes uses a specially constructed free-fall charging car, which completely closes all charging openings to the outside during charging of the coal; therefore essentially all the gases issuing from the oven go by way of the standpipe into the collecting main. The flow of gas is accellerated by injecting steam into the standpipe. To keep the emissions during the charging process at a minimum, the charging larry must remain in place for removal of the charging hole covers, for charging, and for replacement of the charging hole covers.

To prevent the entrance of air, the unlocked leveling door is held closed during charging by a special device and is quickly opened only shortly before moving the leveler bar into the oven. Until such time, the flow of charging gases from the oven to the collecting main is maintained by uniform charging of the coal. Each of the five coal hoppers of the charging larry consists of a charging hopper and a funnel-shaped lower part, the transfer funnel, which can be hydraulically lowered. The feeding pipe of the charging funnel is freely inserted into the upper part of the transfer funnel. The transfer funnel is lowered onto the charging hole, so that its lower ring forms a tight seal against the outside air with the cone of the charging hole frame. The lower end of the transfer funnel can be closed by a butterfly valve, which is hydraulically operated and can be moved during charging. A capacitance-operated level-measuring device in the charging funnel automatically operates the butterfly valve so that coal is removed only to the point at which a sufficient quantity remains to act as a seal, but the lifting of the lower part is not hindered.

The oven chambers are charged according to the following sequence. The larry car travels to charging position to charge the oven. At first the charging hole covers are removed by means of a mechanical lifting device, one after another, and moved to the side. Immediately after that, the transfer funnels (drop sleeves) below the charging funnel are simultaneously hydraulically lowered and form a seal with the cones of the charging hole frame.

The automatically controlled charging process starts when all the transfer funnels (drop sleeves) are in place. Upon termination of the charging process, the charging hole covers are replaced, one after another, by means of the cover-lifting device. This step occurs immediately after lifting of the corresponding charging funnel. Charging can be done either automatically or manually, in which case the outside charging funnels begin to charge before the inner charging funnels.

The second process, known in England as "sequential charging," is used only in smaller plants with low throughput. In this process the charging hoppers are emptied one after another in a predetermined sequence. The best sequence with four charging holes is 1, 4, 2, 3, which makes it possible to remove the charging gas with sufficient suction through the standpipe into the raw gas.

The third process (Fig. 14.12), called "stage charging"[13] and used in the United States, is a further development of "sequen-

[13] F. M. Clark, *AIME Ironmaking Proc.*, **34**, 350–361 (1975).

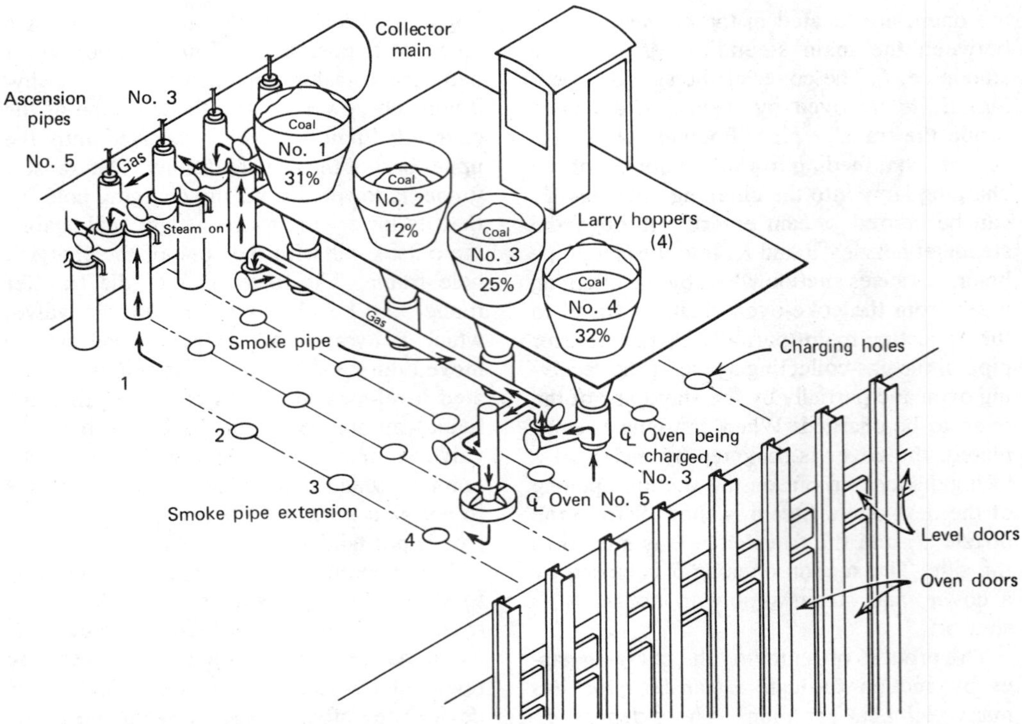

Figure 14.12 Stage charging diagram of smoke pipe and fifth boot jumper showing paths for gas exit.[13]

tial charging.'' In this case charging hoppers are emptied in the sequence 1, 4, 2, 3, with hoppers 1 and 4 emptied at the same time at the beginning. Hopper 2 is then emptied. Only when hopper 2 is completely empty is hopper 3 emptied. The quantity of coal charged in the individual charging hoppers is such that on emptying hopper 3, the leveler must be operated. On completion of the charging process, the charging hole covers are put in place one after another and the steam ejector shut off. The average charging time is 3.2 min.

In a different variation of the process, the charging gases are sucked away, cleaned in suitable equipment, and discharged into the atmosphere after scrubbing or combustion. This equipment may be separate or may be connected with the charging larry. In the combustion process, the charging gases are sucked away by means of a device, then burned. This process is used in a somewhat different form in top-charging and stamping operations.

In top charging the opening for the suction should be on the side of the oven roof opposite to the opening for the leveling bar. If an opening on the pusher side is used for suction of the charging gases, one has to be careful to avoid a short circuit with the outside air during leveling. Alternatively, a movable suction and combustion device connected to the top of the standpipe of the oven may be used for evacuation and combustion of the charging gases. The combustion car moves to the top of the suction opening and an extension piece is inserted. The combustion chamber is located on top of the extension piece, and above that is the chimney. A blower is used to compress air for sucking the charging gases into a nozzle arranged below the combustion chamber. During the charging process the oven chamber is not connected to the collecting main. To start combustion of the charging gases, which at first are difficult to ignite, raw gas is withdrawn from the collecting main. The blower is started when coal charging begins. This sucks in charging gases from the oven chamber as well as air that enters through the slots in the extension piece. The charging gases are ignited by the hot brickwork after mixing with air on top of the suction opening or in the combustion chamber.

A suction opening in the middle of the oven roof is used in stamping operations for evacuation of the charging gases. A compressed air or steam ejector or a blower sucks out the charging gases. The output of the ejector is adjusted so that tar mist at the door of the oven chamber and the leveling opening does not escape into the open. During charging the oven chamber is not connected to the collecting main and the ascension pipe lid is closed. The ejector is started at the beginning of feeding the stamped charge through the oven door. By this means charging gases from the oven chamber and additional air are sucked in through the extension piece. Either the charging gases are ignited by the hot brickwork or they must be ignited manually after mixing with air on top of the suction opening. After leaving the combustion chamber, the burned charging gas enters the atmosphere. Suction devices with a downstream wet scrubber have been developed for purifying the charging gases in top-charging operation, and these are among the accessories associated with the charging car.

As in any other dust removal process, the usefulness of the measures depends on two factors, namely, the possibility of attaining complete removal of the gases to be purified, and the efficiency of the dust removal equipment used. The suction and scrubbing devices used for purifying the charging gas are similar except for some design differences. Each of the four or five charging pipes for charging coal into the coke oven contains a suction shroud, which is lowered hydraulically or mechanically and pressed on the oven top before charging. Some of these shrouds have additional openings for air, which make possible the bringing in of air for the combustion of the charging gas. The charging gases are evacuated through a ring between the charging pipe of the car and the charging hole of the coke oven, and leave through the suction shroud either to a central conduit or to individual conduits leading to wet dust removal scrubbers of various designs (Fig. 14.13).

The suction capacity of the blowers, depending on the amount of charging gas, is about 100–200 m^3/min. Using a charging time of about 2 min, approximately 200–400 m^3 of gas-air mixture is sucked away and purified during the charging period.

The purification of charging gas in some processes is assisted by ignition in the suction shroud and raw gas lines. This especially reduces the tar content of the charging gases.

The water needed for the scrubbing process is pumped into a storage tank on the charging larry before each charging operation and, depending on the process, flows directly or by means of pumps to the scrubber. The sludge from the water scrubber is discharged to a settling plant after each charging process. Water consumption is about 0.8–1.5 l/m^3 of evacuated gas-air mixture; the total power requirement is between 30 and 60 kW.

Frequent attempts have been made to reduce the load on the wet scrubber by bringing part of the charging gas into the collecting main. The flow of gas is affected by injecting steam or high-pressure flushing water into the standpipe. In spite of this, wet scrubber charging cars are trouble-free only if preignition of the charging gases is assured. In Japan the efficiency of the wet scrubber charging car has been improved by sending the raw charging gases through a pipeline to a stationary dust removal plant.[14]

Effective equipment has been operated for many years to reduce emissions during the charging of coke ovens in coke oven plants. By comparison, it is problematic that emissions during the pushing of coke can be collected. A temporary measure tested in German coke-oven plants and applied in the United States is to spray the coke during pushing. The equipment for spraying is arranged around the coke guide units or on the upper rim of the quench car. The water sprayed on the coke evaporates immediately and rises in a cloud, together with coke dust, which it surrounds and wets. This measure permits deposition of much of the coke dust on the coke plant site.

In the early 1960s extensive experiments were made in a German coking plant with a suction shroud surrounding the coke guide car. These experiments did not lead to a technically or economically attractive solution. The emissions during the pushing of coke were removed from the coke quench car by means of a blower and sent through ducts to be purified in wet cyclones. Further development of these suction shrouds was

[14] Y. Kiritani, *Symp. Environmental Control Steel Industry,* Tokyo, February 18–21, 1974.

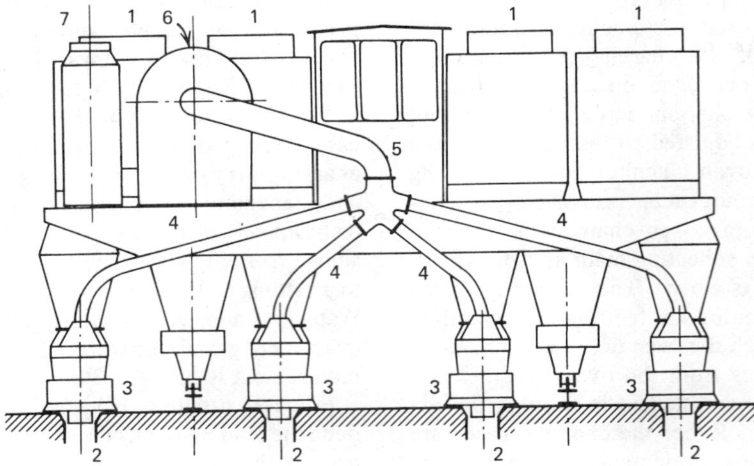

(a)

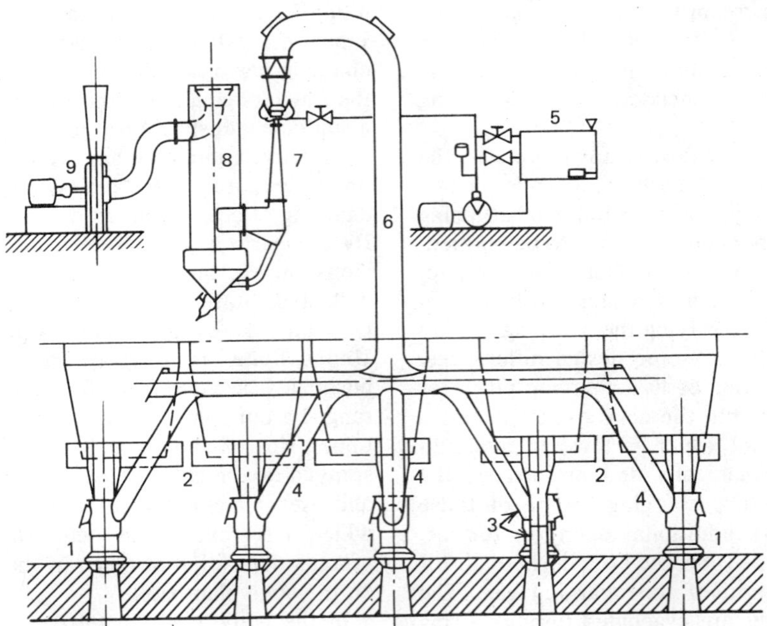

(b)

Figure 14.13 Charging car with equipment for the suction and purification of charging gases. (*a*): 1, charging funnel; 2, lowered charging tubes; 3, lowered telescope; 4, suction pipe for charging gas; 5, pipe for collecting charging gas; 6, wet scrubber; 7, water separator with vapor exit. (*b*): 1, telescope hoods; 2, rotating plate; 3, charging tube; 4, suction tube; 5, water level maintenance; 6, collecting tube; 7, venturi scrubber; 8, separator; 9, ventilator. (*c*): 1, impact scrubber; 2, centrifugal separator; 3, spray cooler; 4, coal-charging pipe; 5, telescope; 6, makeup water storage; 7, centrifugal water pump; 8, sludge container. (*d*): 1, telescope hood; 2, oil burner; 3, venturi cyclone; 3.1, venturi tubes; 3.2, vortex chamber; 4, excess pressure valves; 5, ventilator; 6, drop separator; 7, clean gas exit; 8, hand-operated fresh water valve; 9, water storage tank; 10, magnetic valve for spray water; 11, reducing valve; 12, magnetic valve for separator charging; 13, sludge valve for the collecting vessel; 14, sludge valve for the separator sump; 15, electronic control units; 16, sludge water outflow; 17, fresh water outflow.

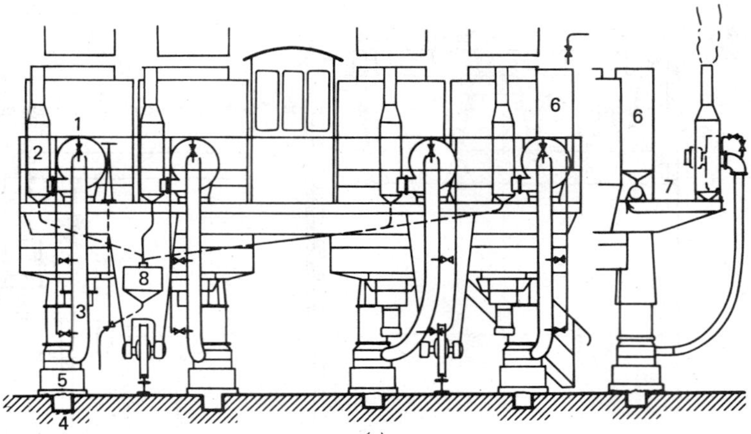

(c)

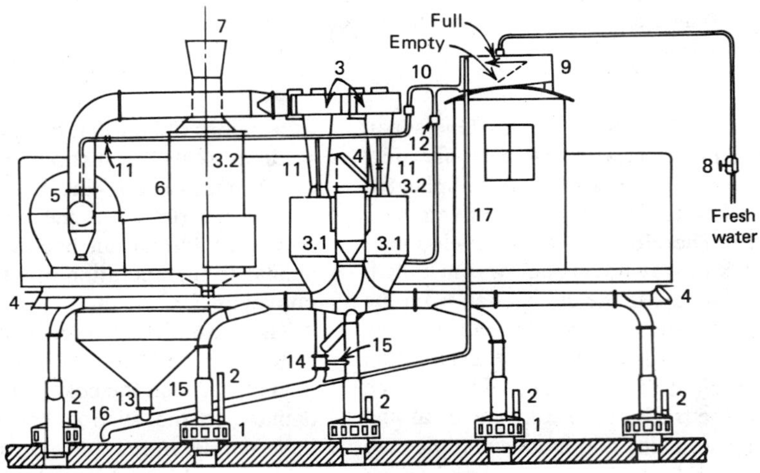

(d)

Figure 14.13 *Continued*

undertaken in Japan. Today all Japanese coke guide cars are equipped with a dust suction shroud, which is partially cantilevered over the track of the quench car or partly supported by a third rail, because of its size and weight. These dust shrouds are connected with a stationary dust removal plant by ducts. Each oven has a separate connection through which the mobile shroud is connected to the suction line.[14] These connecting units are always mechanical devices. Figure 14.14 gives a cross section of such a system. It shows the dust removal system (charging car and quench car dust removal) of a Japanese coke-oven plant.

In West Germany the principle of shroud suction by a mobile dust removal unit was put into operation in 1969, in the form of the "White Giant" installed on the Gneisenau coke-oven plant.[15,16] The mobile dust removal unit consists of eight steam-driven Rotovent dust removers (Fig. 14.15), a steam generating unit, and tanks for storing water, fuel oil, and sludge effluent. This equipment makes the car so heavy that the bench must be supported. Experience with this shroud, which contains heat- and corrosion-resistant material, has been excellent.

[15] W. Hűrter and A. D. Murphy, *COMA Yearb.*, 178–200 (1976).

[16] D. Breidenbach, *Glueckauf*, **109**, 219–223 (1973).

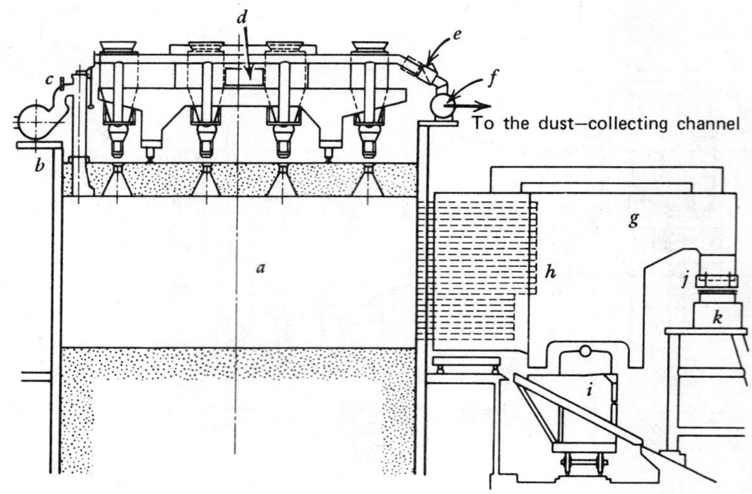

Figure 14.14 Dust removal equipment of a Japanese coke-oven plant (charging car and quench car dust removal): (*a*) oven chamber, (*b*) standpipe, (*c*) high-pressure flushing water nozzle, (*d*) charging car, (*e*) connection, (*f*) dust collecting line, (*g*) shroud, (*h*) coke guide unit, (*i*) quench car, (*j*) connection, (*k*) dust-collecting channel.[5]

Operating experience at the Gneisenau coke-oven plant showed that a mobile dust removal unit is technically too complicated and not suited for the rough coke-oven plant operations. Therefore in a later development a stationary dust removal plant was installed. The problem here is the connection from the shroud to the suction line. An entirely new concept was devised for the Minister Stein coke-oven plant. The shroud, which can be moved on the track of the coke guide car and a third rail, remains continuously connected to the suction line. The suction line is open on the top. This opening is closed by means of a rubber ribbon that because of the partial vacuum, seals the suction line. The connecting-shroud suction line is designed so that the rubber ribbon at the top of the shroud can be lifted and put back on the line again similar to the operation of a tripper on a belt conveyor.[15]

The shroud has three suction points arranged immediately above the door, the quench car, and the coke guide car. To avoid damage to the suction line by temporarily high temperatures, the shroud has a heat exchanger that reduces these temperatures. The stationary dust removal system uses

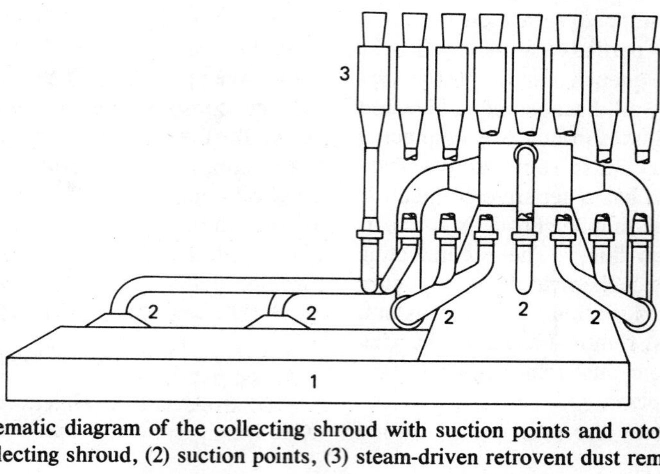

Figure 14.15 Schematic diagram of the collecting shroud with suction points and rotovent dust removal units: (1) dust-collecting shroud, (2) suction points, (3) steam-driven retrovent dust removers.[16]

current technology and is similar to the system used in the screening plant.

The Mitsubishi system was developed in Japan. The difference between this system and the Minister Stein system for controlling pushing emissions is the use of a heat exchanger for cooling the exhaust gases and the method of introducing the emissions into the exhaust duct.

Single-spot pushing was developed in West Germany. With this technique the coke is pushed into a container that does not move during pushing (Figs. 14.16 and 14.17). In that way the hood can be built in smaller dimensions and superimposed directly onto the container. This tight connection results in an effective capture of emissions. This technique is also practiced in operations in the United States.

A further possibility for decreasing emissions during the pushing of coke is to enclose the entire area of the coke guide car and

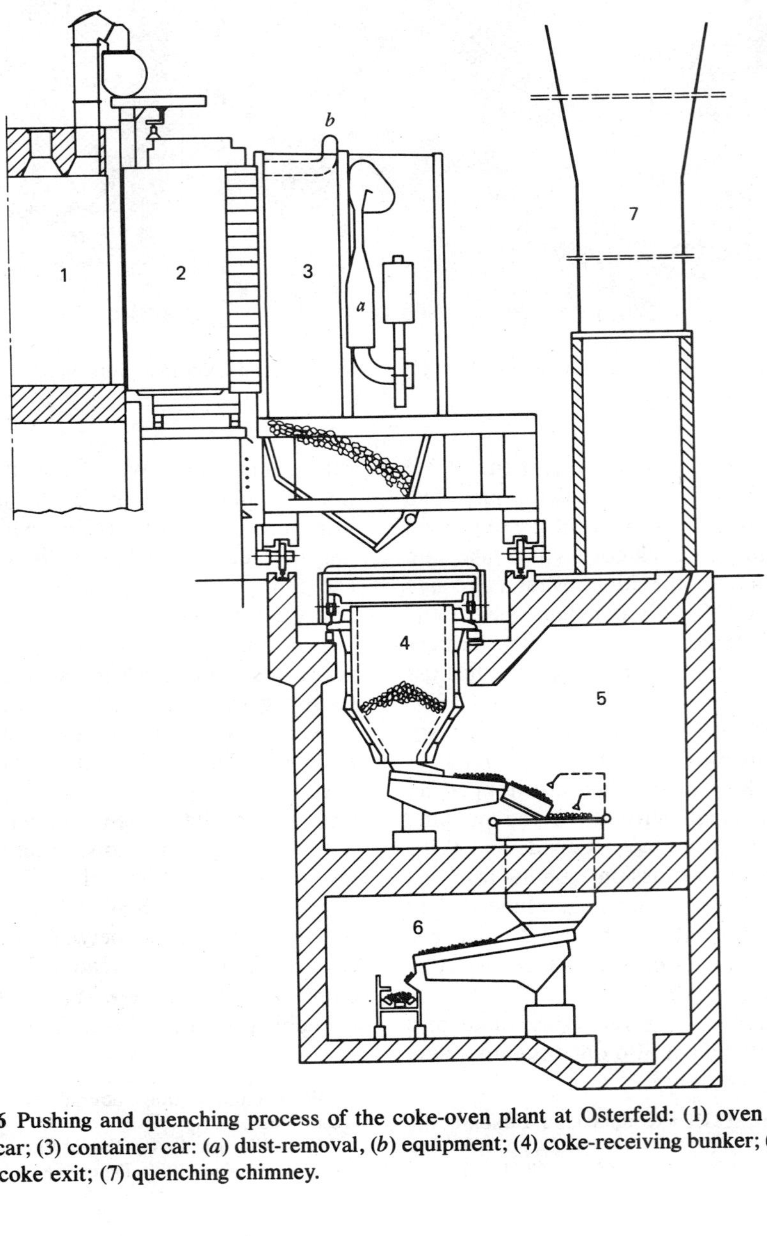

Figure 14.16 Pushing and quenching process of the coke-oven plant at Osterfeld: (1) oven chamber; (2) coke guide car; (3) container car: (*a*) dust-removal, (*b*) equipment; (4) coke-receiving bunker; (5) quenching station; (6) coke exit; (7) quenching chimney.

Figure 14.17 Coke-quenching vessel with dust removal hood. (Bergbau-Forschung GmbH.)

quench car track by a coke shed.[5,10,17,18] Emissions are removed through a duct. The suction and dust removal plant are at ground level. The quench track covers also take care of emissions from the doors of the coke side of the oven. The disadvantage associated with this technique is the excessive particulate fallout on the bench and on the hot car track.

When the doors on the pusher side are opened, emissions occur before pushing. In Japanese coking plants they are trapped by means of a small shroud on the pushing machine, which has a suction device. This unit also removes emissions coming from the leveling door during leveling. Modern dust removal equipment for the coke guide car and the oven doors on the pusher and coke side is shown in Fig. 14.18.

A new system for receiving coke and quenching it was put into operation in a new coke-oven plant in West Germany.[19,20] A stationary quench unit is fed by an enclosed transporting car (Fig. 14.16), which receives the coke during pushing without movement and transports it to the quench station. Before pushing, the dust removal plant of the transporting car is connected with the dust removal system. Both the air replaced while receiving the coke, and the flue gases, are sucked away, purified, and then released. Following this the receiving car is separated from the guide equipment and the car travels with the pushed coke and with the dust removal equipment in operation to an underground coke-quenching plant. There the cover of the receiving bunker is removed and a dusttight connection between the transport car and the bunker is made. The car is then emptied. From the low-level bunker of the quenching plant the hot coke is removed by

[17] V. Gobiet, EGKS-Kommission Luxemburg, EUR 5342, Part 1 (Ref. No. 13), 1975, pp. 149–165.
[18] Ano., *Mag. Met. Prod.*, **33**, 46–48 (1977).

[19] W. Brömmel, International Ironmaking Congress, Düsseldorf, 1974.
[20] C. Still, 6th Coke Oven Plant Technology Conf., Katowice, 1975, Paper No. XVIa.

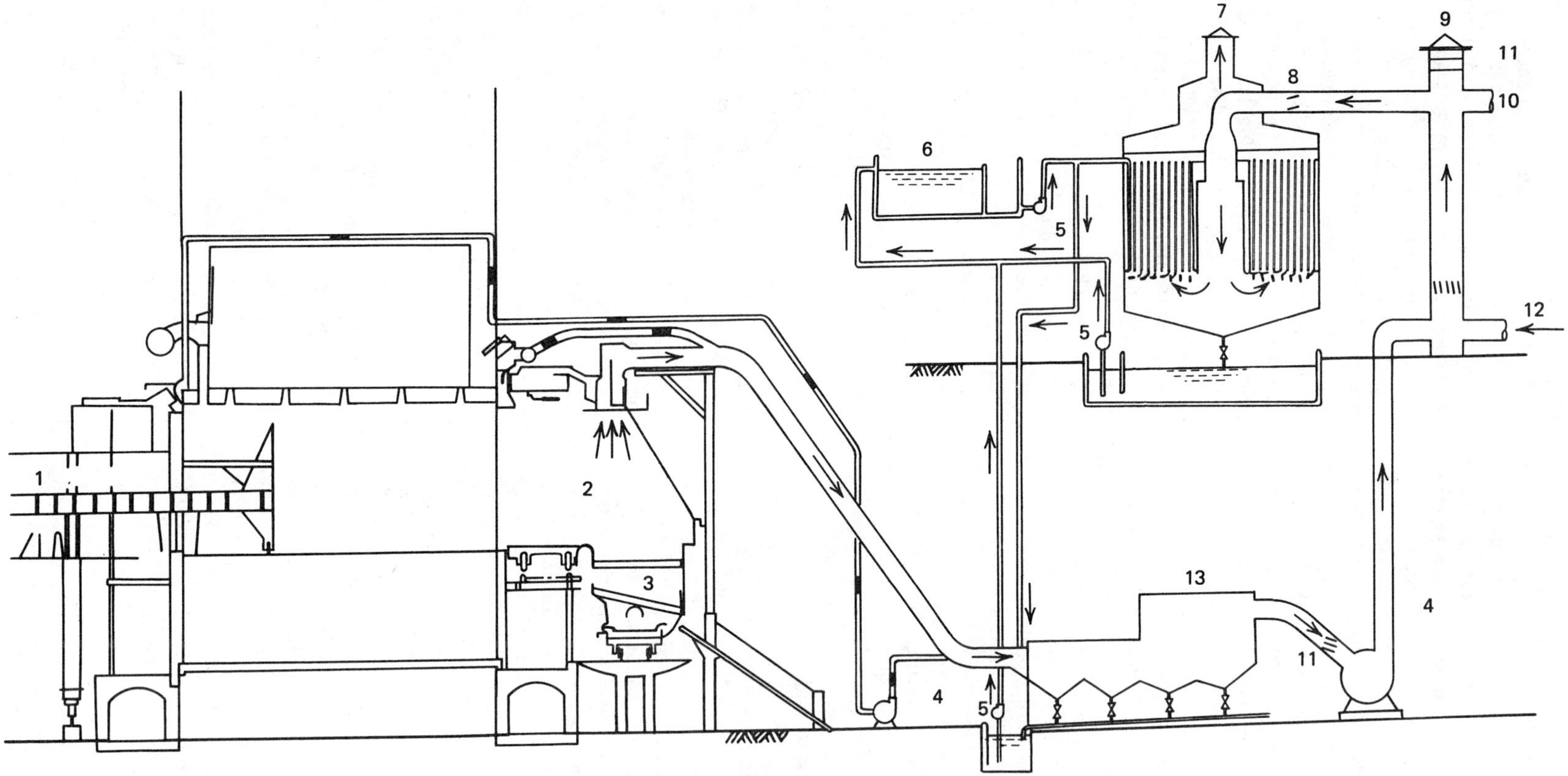

Figure 14.18 Dust removal system for the coke guide car and oven doors on the pusher and coke side: 1, coke-pushing machine; 2, coke guide car; 3, quench car; 4, blower; 5, pump; 6, water settling vessel; 7, electrostatic precipitator; 8, throttling valve; 9, mixing tube; 10, explosion valve; 11, throttling valve; 12, mixing device from charge car; 13, wet dust scrubber.[6]

means of a heat-resistant reciprocating-plate feeder and fed into a continuously operating quenching bunker. The quantity of spray water is controlled to satisfy the requirements of the throughput of the feed system. At present, the transporting car operates without difficulty. When the capacity of the dust removal system is increased, emissions will be almost completely collected. A similar plant was put into operation in the United States at about the same time. This concept can be used on new coke batteries in special situations only because of the elevations required for installation of the entire system.

A prototype of a mobile quench plant is also being tested in the United States.[21] Here a conventional quench car is equipped with a cover that is tightly connected to the coke guide equipment during pushing. Water nozzles in the cover are used to quench the coke. By controlling the water rate, a coke with low moisture content can be obtained. The dust produced during pushing and the steam formed during quenching is sucked into a cyclone dust remover by an exhauster. These units and all other auxiliary equipment are located on a second car.

In a German coke-oven plant a novel process for the emission-free discharging of coke and consecutive quenching of the hot coke has been tested in a pilot plant.[5] A prototype (Fig. 14.17) was completed in 1975 and put into operation for commercial tests. This technique employs new technology to push the coke—without creating any emission—into a container car that is stationary during pushing and can be closed with a cover. Quench water enters through the cover onto the coke and is withdrawn as steam at the bottom. This process has several advantages: (1) the coke is pushed approximately emission free and the water consumption for quenching can be reduced by approximately 50%, (2) no quench water is produced during quenching, (3) the moisture content of the coke can be controlled and reduced to approximately zero, and (4) the steam escaping from the grate at the bottom of the container is at a temperature between 800 and 300°C, which makes it possible to recover heat from the waste steam.

[21] R. S. Patton, speech at 67th Annual Meeting of the National Air Pollution Control Association, Denver, Colo., June 9–13, 1974.

In the usual method for quenching coke the hot coke in the quench car is driven underneath the quench tower and cooled by spraying with water (1–2 m^3 of water per ton of coke; depending on quenching method) squirted from nozzles. The steam produced during quenching passes up the chimney of the quench tower and into the open air. After about 60–180 s the water is turned off. The unevaporated water (50–60%) flows from the quench car into the settling basin; a small portion remains in the coke. The quench car remains a short time in the quench tower so that excess water may drain. Afterward the quench car travels to the ramp and is emptied.

The quench water is converted into steam by the hot coke, and because of the chimney action of the quench tower, the steam is subjected to a strong lifting action. Accordingly the openings at the bottom of the quench tower permit considerable quantities of air to be sucked in. The steam-air mixture carries coke particles with it and some of these particles escape with the steam and air into the atmosphere. The coarser particles drop mostly in the vicinity of the quench tower, the finer coke particles, depending on their size, are carried by the gaseous effluent greater or smaller distances. The quantity of air sucked into the tower depends on many factors—for instance, the quantity of steam; the height, shape, and cross section of the quench tower; and the weather conditions.

Dust emission is influenced to a great extent by the dimensions and shape of the quench tower, since the flow velocity of the effluent depends on these factors. The higher the quench tower, the greater the lift, and consequently the greater the velocity of the gaseous effluent. As a result, more dust is emitted. On the other hand, the quench tower should be tall enough to prevent environmental damage and abuse. In addition to quench towers of the usual design (Fig. 14.19*a*), others have been built to decrease the flow velocity of the effluent. Figure 14.19*b* shows a quench tower with uniform rectangular cross section. A decrease in the flow velocity of steam and air in the quench tower can be achieved by a funnel shape (Fig. 14.19*c*). Figure 14.19*d* shows a design in which the chimney is on the side of the quench cover.

A decrease in dust emissions can also be achieved by the use of grating, lattices, and

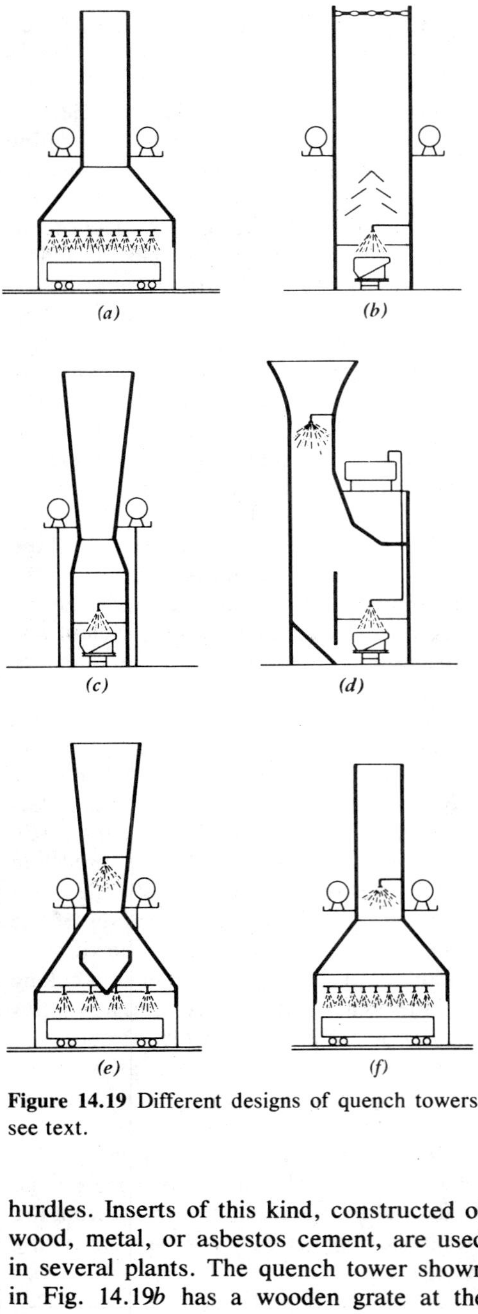

Figure 14.19 Different designs of quench towers: see text.

hurdles. Inserts of this kind, constructed of wood, metal, or asbestos cement, are used in several plants. The quench tower shown in Fig. 14.19*b* has a wooden grate at the outlet of the quench tower chimney. To improve dust removal in the lower third of this quench tower, roof-shaped baffle plates are also built in.

Dust emission during coke quenching can be effectively decreased by water spraying. In this process many dust particles contact water drops and fall to the bottom. Additional water nozzles are located in the chimney of the quench tower. Assuming that this additional water would result in too great an increase in the moisture content of the coke, a collecting gutter was built in above the main water nozzles that are used for quenching (Fig. 14.19*e*). In the quench tower shown in Fig. 14.19*d*, the spray water is also kept separate; additional water nozzles are found in the chimney beside the quench tower.

Since the greatest amount of dust is emitted during the first 20 s, additional spray water is used only during this period. The water added during this short time amounts to several hundred liters and has no influence on the moisture content of the coke. Therefore devices to separate this water are not necessary. Nozzles for spraying the effluent can also be installed in quench towers already in operation (Fig. 14.19*f*). Many investigations in coke-oven plants have shown that spraying water into the effluent reduces dust emissions by about two-thirds.

A further problem during operation of coke ovens is the elimination of emissions during the coking process.[5,6] Emissions occur shortly after charging. The sources include leaks in oven doors, charging hole covers, and standpipes. Different methods are used in attempts to avoid these emissions.

New pushing machines and coke guide cars are usually equipped with mechanical cleaning equipment for the oven chamber frame and door. Nevertheless it is frequently difficult to achieve complete sealing of the oven doors. Some Japanese coking plants have used suction devices on top of the door to catch the emissions. There is no expectation that the door seals in current use can be improved materially, and a better seal can be achieved only by careful and regular maintenance. The same is true for the seals on standpipe and the charging hole covers. In some coke-oven plants emissions from leaking doors are avoided by injection of steam beneath the door stops for about an hour after charging.

To avoid emissions from the charging holes, modern charging cars usually have a cover removal device with automatic rotation of the cover. The most efficient method for preventing the emission of gaseous and vapor effluents has been to pour a fireclay slurry, which sinters at about 850°C, on the charging hole closure.

Emissions from leaking standpipe closures are prevented by using standpipe covers immersed in water. The annular channels at the top of the standpipe, which contain the water and into which the cover rims fit, have continuous inflow and outflow of water to replace evaporated water. To facilitate maintenance work on the standpipes, mechanical cleaning equipment is used to remove graphite and other materials that reduce the cross section through which the gas passes.

14.2 COAL PREPARATION FOR COKE-OVEN USE

14.2.1 The Coal Base

Even today it is very difficult to accurately define the concept of "coking coal," since it depends not only on coal supply and demand, but also on the large differences that exist from country to country. In nearly all coke-oven plants in the world today, a high-quality coke is produced from a mixture of feed coals with very different coking properties and mineral impurities.

Volatile matter, dilatation, and caking properties usually suffice to define the suitability of a coal for coking. However they do not permit sufficiently accurate predictions of the properties of the coke to be produced. Nevertheless these three properties have been selected by the Coal Committee of the United Nations for Europe[22] as a basis for preparing an international classification system for bituminous coal.

The proportion of coal types and the three maceral groups—vitrinite, exinite, and inertinite—are very important for evaluating the suitability for the coking process. Therefore to evaluate the physical and technological properties of coal, it was necessary to develop an analysis according to coal type[23–26] and macerals[27] (Chapter 2). These results are used increasingly in connection with a device for determining the reflectance of coal.[28] Reference is made to an investigation for characterizing coking coal for high-temperature coking.[29] Furthermore American and German investigators[30] have tried to determine whether the reflectance of the vitrinites can be used as a measure of coal rank.

They analyzed the results of a large number of investigations and plotted the reflectance as a function of other parameters for characterizing the rank of the coal. For bituminous coal, an excellent correlation was found between volatile matter, carbon content, and atomic H/C ratio. Also, for the lower volatile anthracites, a satisfactory correlation between reflectance and hydrogen content was obtained. However for semianthracites, lignites, and so on, the correlation was less satisfactory. On the basis of these results, the authors proposed to correlate the limit values for the reflectance with the existing ASTM coal classification system and to use these limit values in the standard classification of other countries.

The primary method used for evaluating the coking ability of a coking coal is to describe the course of dilatation (Chapter 6). Using the course of dilatation and the reflectance of the vitrinites, a coking-coal classification system was recently proposed[31] in which the dilatation, measured in a Ruhr dilatometer, is shown as a function of the average maximum reflectance of the vitrinite. In addition to the volatile matter of a coking coal, the content of moisture, ash, and sulfur plays an important role in high-temperature carbonization, since the chemical composition of the coke has a significant influence on the blast furnace process.[32]

A quantitative description of the coking behavior of a coal and its influence on the strength of the coke can be derived from the specific heating parameter, *K*, for the coking

[22] *International Classification of Hard Coal Type,* prepared by Secretariat of the Economic Commission for Europe, Geneva, 1956.

[23] F. J. Huntjens et al., *Rev. Ind. Miner.*, 138–142 (1958).

[24] D. W. Van Krevelen et al., *Ann. Min. Bel.*, 1059–1063 (1957).

[25] K. Koetter, *Brennstoff-Chemie,* **41,** 263–272 (1960).

[26] W. Pickhardt et al., *Brennstoff-Chemie,* **46,** 44–48 (1965).

[27] *International Encyclopedia of Coal Petrology,* 2nd ed., Centre National de la Recherche Scientifique, Paris, 1963.

[28] J. T. McCartney et al., *Anal. Chem.*, **27,** 1320–1325 (1955).

[29] M. Th. Mackowsky and W. Simonis, *Glueckauf Forschungsh.*, **30,** 25–37 (1969).

[30] J. T. McCartney and M. Teichmüller, *Fuel,* **51,** 64–68 (1972).

[31] R. J. Marshall, *Fuel,* **55,** 346–348 (1976).

[32] J. H. Strassburger, *Blast-furnace Theory and Practice,* Vol. 2, Gordon and Breach Science Publisher, New York, 1969, pp. 576–579.

conditions and from the values for the size consist, volatile matter, and the dilatometer data (softening temperature, resolidification temperature, contraction, and dilatation).[33] The use of the parameter *G*, which measures coke-forming ability (see Chapter 6, Sections 6.2.2.2 and 6.4.2), for predicting coke strength has some limitations; however a retesting for accuracy on a number of coal mixtures from low- and high-volatile coals gave satisfactory results in many investigations.[34]

Coking coals have different particle sizes after preparation. Maximum strength of a coke depends on the homogeneous structure of the individual coal lumps, which in turn depends on the homogeneity of the coal use.[35–37] Through geometric and physical observations, the optimum size consist for coking coals has been found. Maximum strength and minimum abradability are obtained by this means. In addition, the quality of the coking products depends on the moisture and mineral matter content of the coking coal.[38,39]

Shrinkage as well as swelling of the oven charge occurs during high-temperature coking. These changes lead to considerable mechanical stress on the oven. Possible difficulties during pushing of the oven or damage to the wall are caused by high coking pressure and insufficient shrinkage of the coking coal. For smooth operation, therefore, the oven charge must shrink more strongly than it swells. The coking pressure is greatest with a moisture-free coal. It decreases with increasing moisture content, up to 8%, then decreases further only very slowly. In addition, the coking pressure is significiantly influenced by the bulk density, which depends on size consist and moisture content. The coking pressure increases with the bulk density. By finer grinding, which means reduction in the bulk density, and the addition of low-volatile materials such as anthracite or coke breeze, the coking pressure can be decreased. There has been no satisfactory correlation between swelling and shrinking on the one hand, and the composition of the coking coal (volatile matter, petrographic properties, and degassing behavior) on the other. A large number of results from small-scale ovens has been used in an attempt to mathematically formulate the correlation between swelling and shrinking of the oven charge, and the composition of the coking coal and the operation conditions.[40,41]

14.2.2 The Preparation of Coal

Since the reserves of the coals that are most suitable for the production of coke are being depleted, coal preparation for coking coal has recently gained in importance and for a modern coke-oven plant it is a necessity. In addition, the development of larger production units in coking and blast furnace technology has required better quality and uniformity of the feed coal.

The objects of mechanical coal treatment are to remove foreign matter and to obtain particle sizes suitable for the user. The float-and-sink test is used for judging the washability of a coking coal in the preparation plant.[42] This test furnishes information on the point at which the separation can be made without exceeding a predetermined average ash content and also shows the yield at that point (Fig. 14.20). An overview of the processes for coal preparation is given in Fig. 14.21; a sample flowsheet appears in Fig. 14.22[43,44] (see also Chapter 10). The different physical properties of the components to be separated are used for the separation into different types of material or kinds of coal. In dry preparation mechanical separating processes are usually used. For wet prepa-

[33] W. Simonis, G. Gnuschke, and K. G. Beck, *Glueckauf Forschungsh.*, **27,** 105–113 (1966).

[34] J. G. Walters et al. U.S. Bur. Mines, Rep. Invest. No. 6482, 1964; 1971.

[35] M. Th. Mackowsky, *Fortsch. Mineralog.*, **45,** 52–94 (1967).

[36] W. Simonis and W. Rubrecht, *Glueckauf Forschungsh.* **26,** 301–308 (1965).

[37] O. V. Verbitskya, Y. S. Putko, and K. J. Syskov, *Coke Chem. (USSR),* **1,** 15–18 (1961).

[38] K. G. Beck, R. Beckmann, W. Weskamp, and W. Dressler, *Glueckauf,* **100,** 839–852 (1964).

[39] H. Echterhoff, J. Langhoff, and W. Peters, *Brennstoff-Chemie,* **45,** 306–315 (1964).

[40] E. Szurman and W. Simonis, *Glueckauf Forschungsh.*, **34,** 68–74 (1973).

[41] E. Szurman, *Glueckauf Forschungsh.*, **37,** 50–54 (1976).

[42] Preliminary Standard DIN 23011, Guidelines for the Initial Operation and Control of Bituminous Coal Preparation Plants, 4th ed., 1954.

[43] *Bituminous Coal Preparation,* Part 1, Glückauf Verlag, Essen, 1966, p. 62.

[44] J. W. Leonhard and D. R. Mitchell, *Coal Preparation,* 3rd ed., New York, 1968.

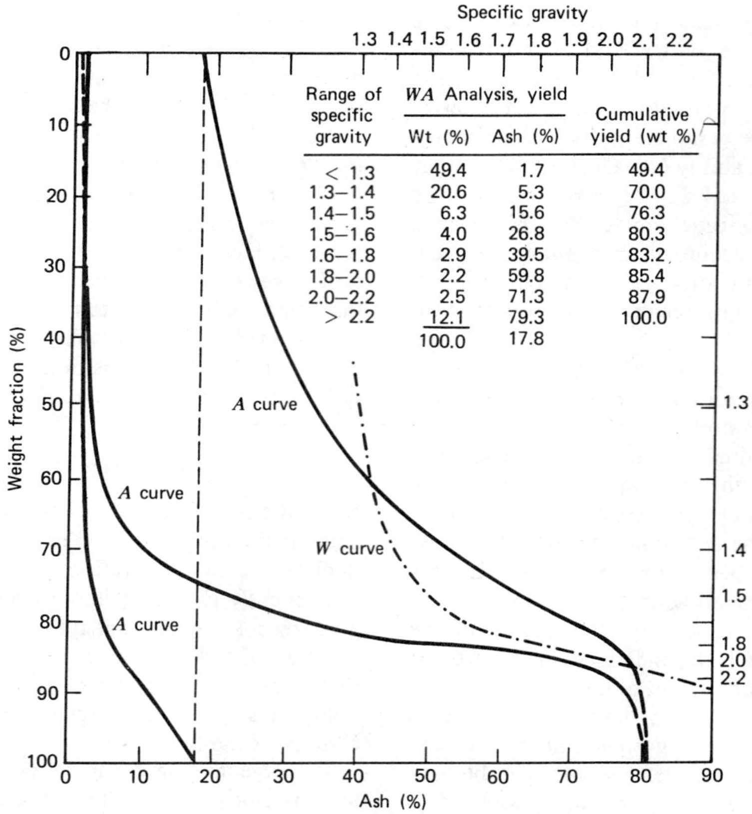

Range of specific gravity	*WA* Analysis, yield Wt (%)	Ash (%)	Cumulative yield (wt %)
< 1.3	49.4	1.7	49.4
1.3–1.4	20.6	5.3	70.0
1.4–1.5	6.3	15.6	76.3
1.5–1.6	4.0	26.8	80.3
1.6–1.8	2.9	39.5	83.2
1.8–2.0	2.2	59.8	85.4
2.0–2.2	2.5	71.3	87.9
> 2.2	12.1	79.3	100.0
	100.0	17.8	

Figure 14.20 Washability curve according to Henry Reinhardt. *W:* water curve; *A:* ash curve.

ration the differences in gravity are used in settling devices, including heavy-media separation processes. The most important process for wet preparation of fine coal is the flotation process.[45] It makes use of the various surface properties of the coal and is the only process for separating the finest particles below 0.75 mm, down to 5 μm. The flotation is usually carried out in several vessels in series, in which the flotation suspension is treated by means of flotation agents that must have collecting as well as foaming properties.

Classification is used to separate a mixture of different particle sizes into two or more size fractions. In practice, three separate methods for particle size classification have been developed. The first is screen separation,[46] in which a distinction is made between pre- and postclassification, depending on whether this process takes place before or after classifying. The second is air classifiers, which are used for dry separation of the finest particles from pulverized material. The third is hydroclassification, used as a process for wet separation of the finest particles.

Comminution disintegrates the coarse material, which can either undergo further preparation or can be used directly. One of the functions of the grinding process is to mix coking coals. Selectively acting breakers such as the drum screen breaker (Bradford breaker) and the swing screen breaker are used for the comminution of lump coal and refuse. In addition, impact mills and hammermills of all kinds are used. Measuring and mixing, which are especially important in the preparation of coking coal, occur at the end of the preparation sequence.

In the Longwy-Bürstlein process[47] (Fig.

[45] K. H. Kubitza et al., *Glueckauf Forschungsh.*, **30**, 53–65 (1969).

[46] W. Kluge, *Glueckauf Forschungsh.*, **27**, 163–180 (1966).

[47] C. Abramsky, *Glueckauf,* **91**, 714–727 (1955).

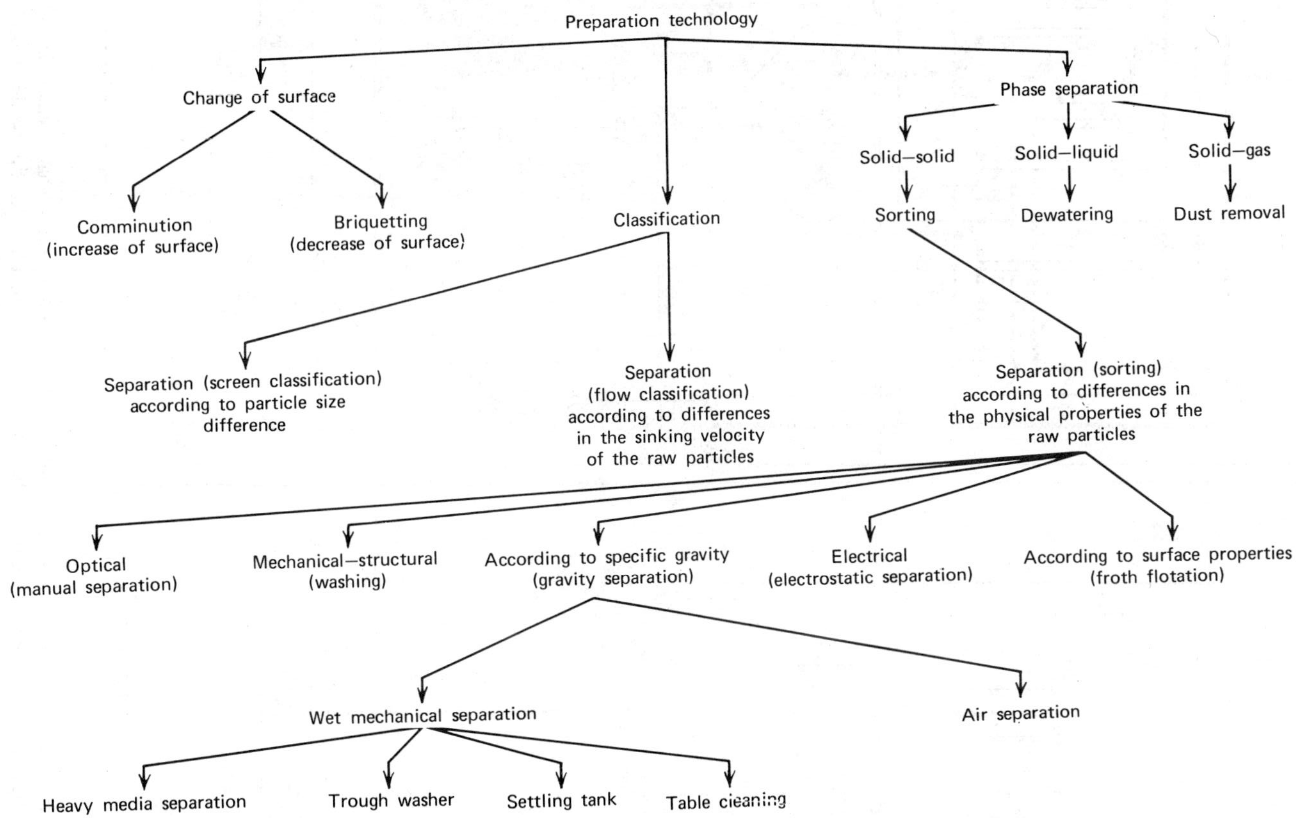

Figure 14.21 Overview of the areas of mining processing technology in bituminous coal preparation.[43]

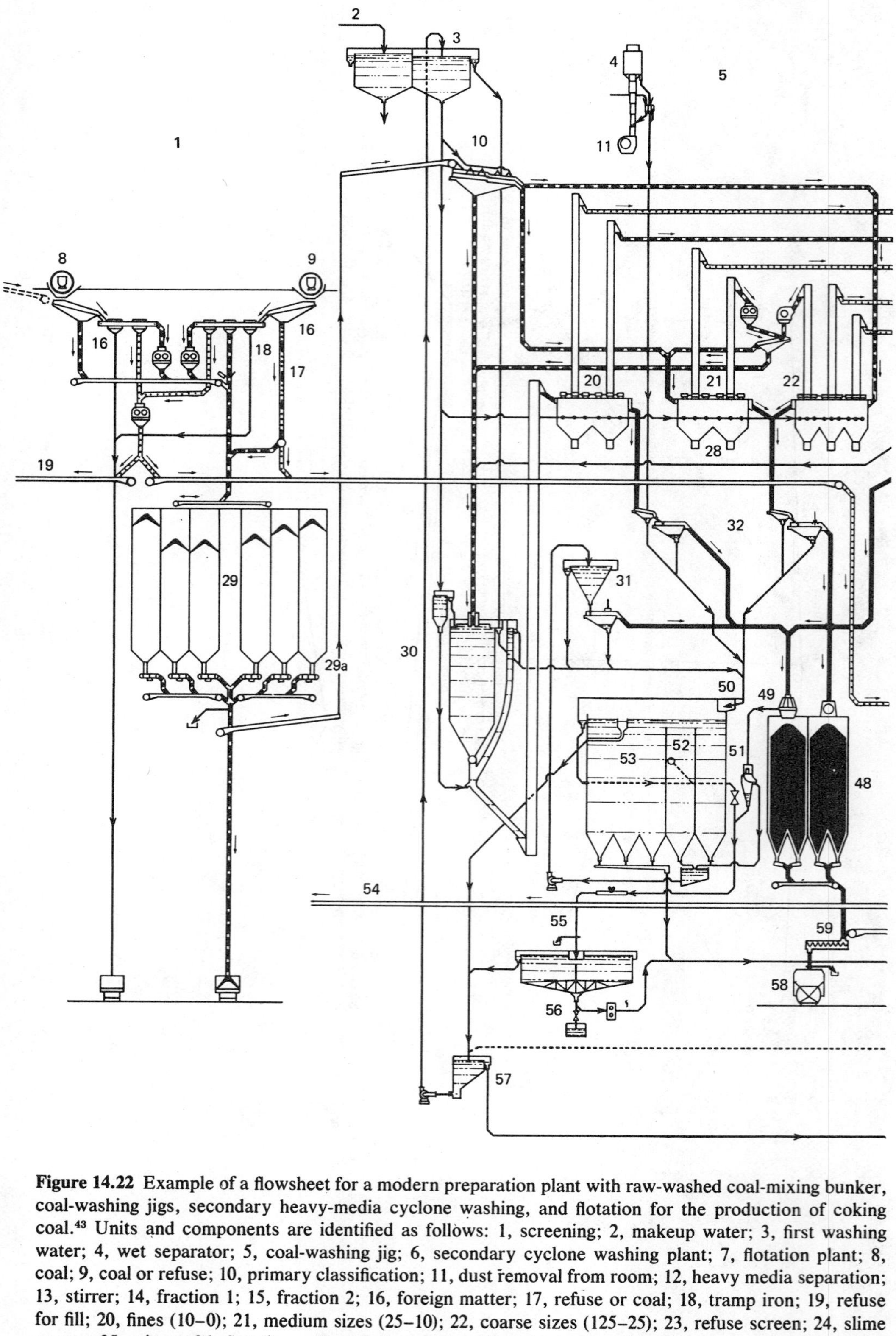

Figure 14.22 Example of a flowsheet for a modern preparation plant with raw-washed coal-mixing bunker, coal-washing jigs, secondary heavy-media cyclone washing, and flotation for the production of coking coal.[43] Units and components are identified as follows: 1, screening; 2, makeup water; 3, first washing water; 4, wet separator; 5, coal-washing jig; 6, secondary cyclone washing plant; 7, flotation plant; 8, coal; 9, coal or refuse; 10, primary classification; 11, dust removal from room; 12, heavy media separation; 13, stirrer; 14, fraction 1; 15, fraction 2; 16, foreign matter; 17, refuse or coal; 18, tramp iron; 19, refuse for fill; 20, fines (10–0); 21, medium sizes (25–10); 22, coarse sizes (125–25); 23, refuse screen; 24, slime screen; 25, stirrer; 26, flotation cells; 27, fine sludge; 28, coal-washing jig; 29, raw-washed coal mixing

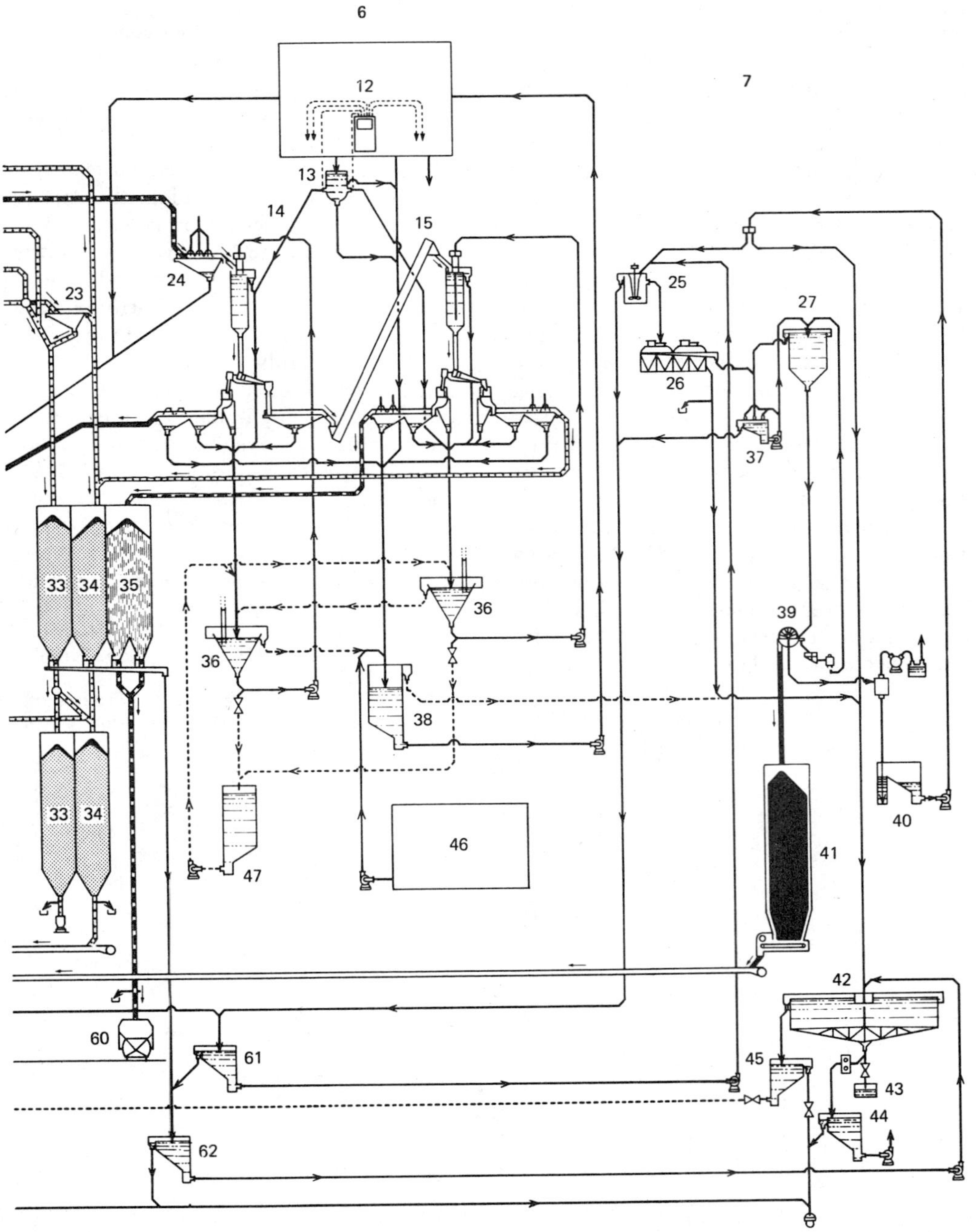

bunker; 29a, belt metering scale; 30, slime removal from fines; 31, coarse sludge; 32, dewatering screen; 33, backfilling refuse; 34, refuse to dump; 35, middle fraction; 36, slurry recycle vessels; 37, concentrate; 38, diluted heavy media; 39, filter; 40, filtered water; 41, sludge; 42, thickener for flotation refuse; 43, water drips; 44, to dump; 45, clarified water; 46, magnetite grinding; 47, slurry discharge; 48, ground lumps (<10); 49, centrifuged coal; 50, centrifuge; 51, screen; 52, preclarification; 53, clarification; 54, to dump; 55, thickener for wash water; 56, drip water; 57, clarified water; 58, coking coal; 59, mixer; 60, boiler fuel; 61, fine sludge; 62, drip water.

14.23), the poorly coking and noncoking constituents of the coal, which because of their hardness remain in larger particle sizes during grinding, are screened and separately ground, and screened again. The oversize is recycled. Conditions are such that the poorly softening or nonsoftening components are sufficiently comminuted and the strongly coking components are not unnecessarily decreased in size. This prevents the presence of coarse, poorly softening components in the mixture, which during the resolidification process during coking lead to stresses in the coke skeleton, an increase in the number of fractures, and a decrease in the lump size of the coke. Additional processing steps in the preparation of bituminous coal are drying and sulfur removal.

A report of the National Coal Board[48] on the current status of coal preparation describes froth flotation, disk and drum filters, filtration with assisting steam, and the use of centrifuges and filter presses. The dewatering of coal was also investigated[49] from both technical and economic points of view. This investigation is based on experience with experiments on a technical and semitechnical scale, and with commercial plants using centrifuges, filtration, and thermal drying as processing steps.

The removal of sulfur from coal is particularly important because the iron and steel industries, the largest consumers, require a low-sulfur coke. Several processes are available for desulfurizing coal.[50,51]

In the stamping process, another technical possibility for coal preparation, the density of the coal is increased by mechanical means. Stamping is used for high-volatile, poorly coking coals and has been practiced for decades in Silesia, Saxonia, Lorraine, and Czechoslovakia, and in the Saar District. By mixing with highly caking coals and by stamping, geologically younger coals can be converted into high-quality coking coal. High-volatile coal requires a noncoking additive if it is to be converted into blast furnace coke. In the past decade, low-volatile, highly caking coals have been scarce and expensive. Therefore, as substitutes, lower cost coals that are noncaking or only

[48] *Filtr. Sep. (NCB)*, **12**(4), 368–372 (1975).
[49] *Min. Congr. J.*, **60**(9), 116–121 (1974).

[50] U.S. Government Research Rep. No. PB 232083, 1974.
[51] "Sulfur Removal from Coal by Froth Flotation," *Proc. 6th Int. Coal Preparation Congr.* (1973), Paper No. 27E.

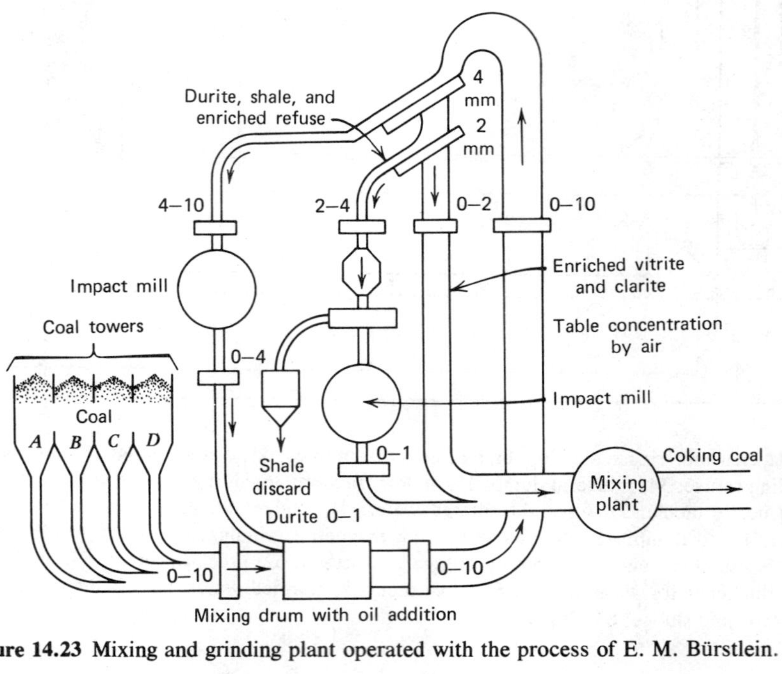

Figure 14.23 Mixing and grinding plant operated with the process of E. M. Bürstlein.

slightly caking, and different petroleum cokes, are being used. Another possible mixture is coke flour, which can be produced by drying and grinding coke breeze, a product of the coking process. Coke flour has a size consist of 75% below 0.09 mm.[52]

In top-charging operations the bulk density of the coal is about 700–780 kg/m³ (mf). By stamping, bulk densities of about 900–950 kg/m³ dry coal and above can be obtained. The higher densities increase the coking pressure to permissible extent, and this is desirable when nonswelling coals are used because the coal comes in closer contact with the heated walls of the oven, which improves heat transfer. The amount of stamping that is needed decreases with higher bulk density. It therefore has become necessary to grind coal to obtain particles of the following sizes: about 50% below 0.5mm, 30% between 0.5 and 2 mm, and 20% above 2 mm.

Lack of coking ability can be compensated by stamping only to a certain extent. Therefore admixtures to the coal are being used that largely supply the function of the caking bitumen of the coal. Such materials include bituminous coal-tar pitches and chemically similar petroleum-oil products.[52]

In top charging, and also in the stamping operation, the feed coal is located in the coal tower, which is on the side of the battery. In the stamping process coal from a separate storage bunker is fed to the movable stamping machine; then it slides into the actual stamping chamber. Special measures are required to prevent separation of the individual fractions when charging the stamping chamber. The greatest disadvantage of the stamping operation is that it is limited to oven heights below 4 m. Current development work is concerned with a new oven type said to be higher than 6 m.

14.2.3 The Blending of Coal

With the development in recent years of higher capacity ovens, and the increased demand for coking coals, it is not unusual for coal mixtures containing 10 or more individual components to be used. In Japan, Italy, and Sweden, where coking coal must be imported, 10–15 coals of various origins are often mixed into one feed mixture.[53] Such mixtures fulfill the coke-oven plant requirements for uniformity; volatile matter content; water, ash, and sulfur contents; and the *G* value (Section 14.3.3), all of which lead to improved coke quality.

The object of the mixing technology is to obtain a coking-coal mixture from individual components in such a manner that the coke produced has the highest possible strength. Investigations[54] have shown that the properties having an important influence on the quality of the coke are associated with the behavior of the charge in the plastic range (350–500°C) and in the postplastic shrinking zone (500–750°C). The behavior in the plastic zone can be determined by dilatometer measurements.

Results of the dilatation experiment (softening temperature, resolidification temperature, contraction, and dilatation) make it possible to derive a *G* value, which determines the coke-forming ability of the coking coal. Furthermore, the coke-forming ability, *G*, can be shown as a function of the content of volatile matter of the coal (Fig. 14.24). For all coal mixtures satisfying this correlation, the maximum coke strength M_{40} is obtained. With given properties of the feed coal and given coking conditions, the size consist of the coal should comply with the size consist $M_S = \pm 0$ if maximum coke strengths are to be obtained. Figure 14.25 shows how, from the conditions above, the M_{40} value can be calculated in advance (Section 14.3.3). With

[52] K. H. Flasche et al., in *Coke Oven Technology*, EGKS-Kommission Luxemburg, EUR 5342, (1975), pp. 9–20.

[53] M. Yoshinaga, M. Sanada, and M. Utsunomiya, *AIME Ironmaking Proc.* **35,** 256–275 (1976).

[54] J. Gibson, 39th Blast Furnace Conf., British Steel Corp., 1971.

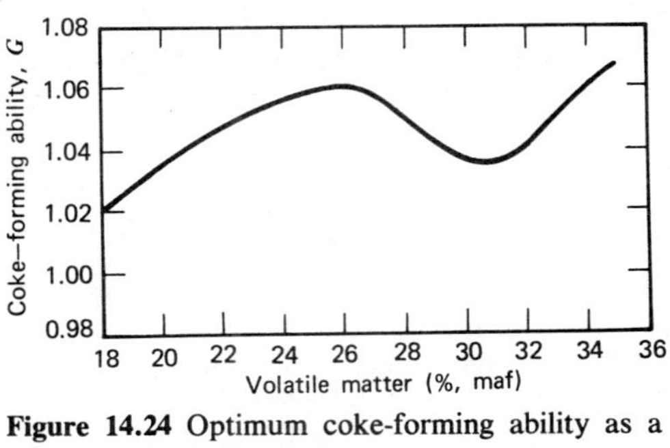

Figure 14.24 Optimum coke-forming ability as a function of volatile matter content.

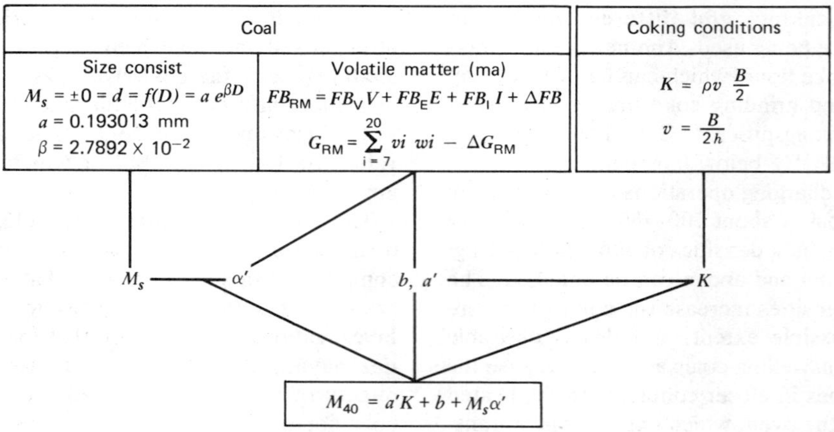

Figure 14.25 Determination of coke strength, M_{40}.

the help of this information, the optimum coking-coal mixture can be prepared from several components, to achieve the highest possible coke strength.

Other investigations[55] have shown that it is possible to determine with sufficient accuracy the coke quality for given coking conditions for different kinds of coal and coal mixtures. Caking ability and content of volatile matter can be used to calculate parameters for individual coals which, in turn, can be converted into mixture parameters that determine the mixing ratio for producing coke of a desired quality. Furthermore, on the basis of the mathematical description of high-temperature coking,[56] the relationships between raw material, coking conditions, yield, and product quality were used to develop an optimization model[57] from which an optimum coking-coal mixture can be determined. Thus not only the quality and quantity of the coking products, but also the economics of the process are obtained.

Also noted are investigations by T. Matsuo[58] on newer developments concerned with the use of poorly caking coals for the coking process. The author describes various possibilities for using slightly caking coals and discusses the effect of the feed coal mixture on the physical and mechanical properties of the coke.

14.2.4 The Briquetting of Coal

In West Germany in 1959, C. Still introduced the process of binderless prebriquetting of coal in a coke-oven plant in the Saar District. The briquetting presses were located beneath the discharge openings of the coal tower. All the charged coal passed through the presses and was briquetted. The briquetted product was carefully transported in the charging car to prevent breakage. Since no binder was used in briquetting, the strength of the briquettes was not very great. After pressing, only 75% of the coal was in the form of briquettes. This percentage was reduced during the various handling steps and reached 25% in the oven chamber. This gave a bulk density in the oven of 750 kg/m³ (mf), which corresponds to an increase in the bulk density of about 15%, compared to top charging without prebriquetting. The low bulk density results because for reasons of quality, this coal must be ground so that 100% is below 2 mm, and because it contains only 6–8% moisture.

The prebriquetting process has not been introduced into practice because this type of coal pretreatment gives results inferior to those of the stamping process. Not until the 1970s was prebriquetting investigated by Japanese coke-oven plant operators, and the process was modified to the extent that part of the feed coal was briquetted with binders.

[55] J. C. A. Kaye and A. E. Horton, EGKS-Kommission Luxemburg, EUR 5342, (1975), pp. 47–59.

[56] W. Simonis, *Glueckauf Forschungsh.* **29**, 103–119 (1968).

[57] W. Herrmann, EGKS-Kommission Luxemburg, EUR 5342, (1975), pp. 35–46.

[58] T. Matsuo, *SE Asia Iron Steel Inst. Q.*, **5**(2), 22–34 (1976).

In the meantime, this process has been applied in several Japanese coke-oven plants. It is estimated that about 35% of the Japanese coke output will be produced in this manner by 1980.

Two methods are being used for partial briquetting of the feed coal.[53,59] In the first method 20–40% of the feed mixture is separated and briquetted after grinding and addition of binder, but without drying the coal. Next the coal briquettes are mixed with the remainder of the feed coal and fed into the oven chambers by the charging car. The coke produced is homogeneously fused, just as if unbriquetted coal were used, and shows an increase in strength and a decrease in abradability. This process makes it possible, with the same quality of coal feed, to produce coke with improved strength, or if an increase in strength is not necessary, to replace highly caking components of the feed coal by weakly caking or noncaking coals.

If a substitution is made for the high-quality and expensive coking coal without the need of increasing the strength of the coke, the second method may be used. The noncaking portion of the feed coal may be mixed with a caking material and a binder, and separately briquetted. These briquettes are subsequently added to the unbriquetted main stream of the coal mixture before charging into the oven chamber (Fig. 14.26). The caking material is obtained from vacuum still oil residue. These oils contain more than 40% volatile components and have a high fluidity. The briquette usually weighs 34 g and measures 40 × 40 × 25 mm. Further data on the briquetting of bituminous coal can be found in Chapter 11.

[59] J. Kasperczyk, *Stahl Eisen*, **97**, 595–598 (1977).

14.2.5 Preheating Coal

The use of preheated coking coal for the production of coke has recently increased in importance throughout the world. Commercial plants have been in operation for some time in the United States, Great Britain, France, South Africa, and West Germany. Additional plants are planned or are under construction in Japan and Canada. Preheating was originally proposed in France and Germany to be able to use slightly coking or high-volatile coals for the production of coke. By the 1950s and increasingly in the 1960s, the concept was being advanced that drying and preheating of the coking coal should not be done in the coke oven but in a reactor that is more technically suitable and less complicated. The disadvantages of a two-stage process should be compensated by an increase in throughput, while retaining the quality of the coking products.

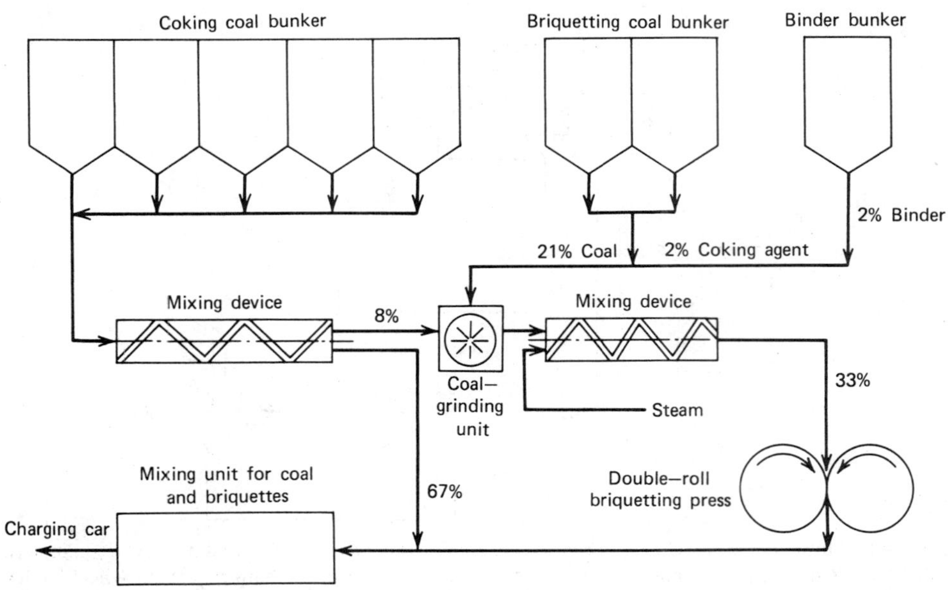

Figure 14.26 Partial briquetting of the feed coal with the addition of about 20% noncaking coal.

Practical investigations in existing plants that could use preheated coal showed that the expected results can be attained. Increases in throughput of about 50% were reached without changing the quality of the coke or of the coke-oven by-products. This means that with the same coking time as used in wet coal operation, an increase in the coke strength and a shift in the size distribution to larger lumps can be achieved.[60]

Comprehensive early investigations at the Marienau experimental station in France showed that thermal pretreatment of coking coal has special requirements and therefore cannot be considered to be a simple drying and heating process.[61] At the same time, special problems caused by the handling of preheated coal need to be solved. Three processes are now being used that differ, on one hand, in method of preheating and on the other hand in charging technique.

14.2.5.1 *The Coaltek Process*

The Coaltek process uses the grinder-dryer developed by Centre d'Etudes et Recherches des Charbonnages de France (CERCHAR). The main characteristic of the process is one-stage operation with simultaneous comminution of the coarse particles. This is combined with the "pipeline-charging" system, developed by Allied Chemical,[62] into one process. Figure 14.27 is a flowsheet of the Coaltek process. The wet coal coming from the feed bunker (*b*) moves cocurrently with steam heated to 570°C in a combustion chamber (*a*). Part of the sensible heat of the steam is used to vaporize surface moisture. After this predrying phase the suspension of hot vapor and coal particles enters—for final drying—the CERCHAR dryer, in which a hammermill (*c*) rotates in a fluidized bed. As soon as the particles are small enough, they are carried out by the gas. By controlling the velocity of the gas, the average size of the particles leaving the grinder-dryer can be controlled. Through a pneumatic transport line, the neutral vapors and the preheated coal flow to the dust removal plant, which consists of a series of cyclones (*e*). A portion of the inert gas and water vapor returns to the combustion chamber for temperature control. The excess inert gas containing a quantity of water equal to the moisture in the coal feed is vented after passing through a wet scrubber.

[60] C. Flockenhaus, W. Rohde, and D. Habermehl, *AIME Ironmaking Proc.*, **33**, 382–390 (1974).

[61] D. G. Marting and R. Marcellini, *Inf. Symp. Coke Oven Techniques*, Commission of the European Communities EUR 5342 Part 1, Luxembourg, 1975, 205–216.

[62] D. G. Marting and H. S. Auvil, UN Economic Commission Symp., Rome, 1973.

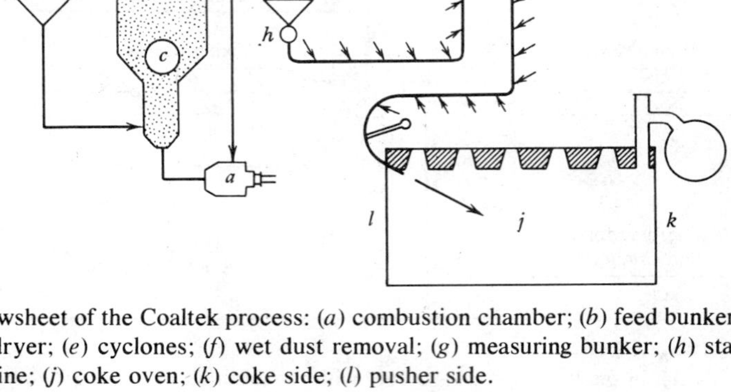

Figure 14.27 Flowsheet of the Coaltek process: (*a*) combustion chamber; (*b*) feed bunker; (*c*) hammermill; (*d*) CERCHAR dryer; (*e*) cyclones; (*f*) wet dust removal; (*g*) measuring bunker; (*h*) starwheel feeder; (*i*) pneumatic feed line; (*j*) coke oven; (*k*) coke side; (*l*) pusher side.

The feed system consists of a measuring bunker (*g*), which is charged with the preheated coal after adding oil (~ 0.5%). The preheated coal is discharged from the measuring bunker through a starwheel feeder and enters the pneumatic feed line (*i*), which contains several steam nozzles entering from the bottom and is connected to the coke oven (*j*). High-pressure steam is blown into the line through the steam nozzles, to keep the coal loose and to transport it. Immediately before entering the coke oven, the coal is separated from the steam in a bend in the feed line. Charging is accomplished with a connected manifold. The separated steam, which contains fine coal particles, passes into the gas space of a neighboring oven.[61,63,64]

14.2.5.2 *The Precarbon Process*

A different preheating and charging principle is used in the German Precarbon process, which employs a two-stage entrained dryer for thermal pretreatment (Fig. 14.28). The hot flue gases from the combustion chamber (*a*) mixed with recycled inert gas and water vapor for temperature control, first enter the second stage heater (*d*) where the preheating is begun, and then enter the first-stage or drying stage (*c*). This process fully utilizes all the advantages of the countercurrent principle—such as high efficiency and careful thermal treatment, which result from relatively low gas temperatures. The cyclones (*e*) are provided for separating the dried and preheated coal from the flue gases. The preheated coal from the measuring bunker (*g*) is charged into the coke oven by means of a chain conveyor and chutes connected to the charging openings. This technique, involving the gravity charging of preheated coal, leads to a higher bulk density than that obtained when top charging wet coals. The free-flowing properties of the preheated coal result in a more uniform charge. Because of better flow properties, leveling is not necessary.[60,65,66]

14.2.5.3 *The Simcar Process*

The Simcar process developed in England also uses a two-stage entrained dryer for preheating. Using available technology, the preheated coal is then charged into the coke oven with a modified hot-charging car. In contrast with the processes described in Sections 14.2.5.1 and 14.2.5.2, the charging gases containing the coal dust are filtered and purified by means of conventional equipment.[67]

[63] M. Whysall, *Iron Steel,* **45,** 25–33 (1972).
[64] R. F. Davis and V. W. Cekela, *AIME Ironmaking Proc.,* **34,** 339–349 (1975).

[65] K. G. Beck, *COMA Yearb.* 230–242 (1974).
[66] D. Habermehl, G. W. Lask, and H. Petak, *Inf. Symp. Coke Oven Techniques,* Commission of the European Communities, EUR, 5342 Part 1, Luxembourg, 1975, pp. 217–225.
[67] R. H. Chambers and L. A. Watson, *AIME Ironmaking Proc.* **33,** 233–248 (1974).

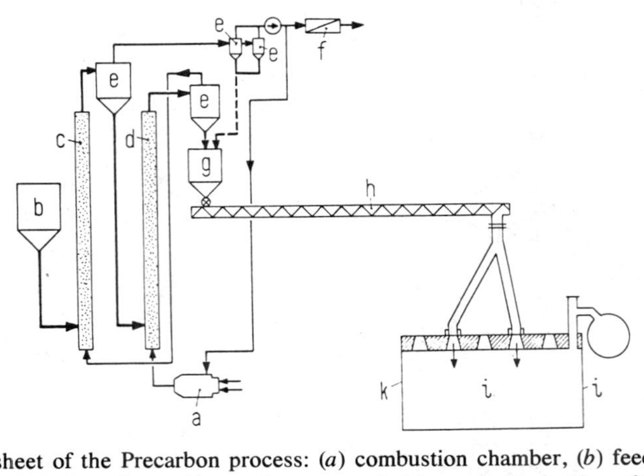

Figure 14.28 Flowsheet of the Precarbon process: (*a*) combustion chamber, (*b*) feed bunker, (*c*) drying stage, (*d*) heating stage, (*e*) cyclones, (*f*) wet dust removal, (*g*) measuring bunker, (*h*) chain conveyor, (*i*) coke oven, (*j*) coke side, (*k*) machine side.

14.3 COKING CONDITIONS IN COKE OVENS

14.3.1 The Influence of Coke-Oven Heating

The conversion of coal into coke requires a certain amount of heat, the quantity depending on the technical conditions of coking as well as on the properties of the coal used. The necessary heat is added to the process by the combustion of underfiring gas in the flues. The quantity of heat added is determined by the quantity and composition of the underfiring gas. If the quantity and heating value of the gas used for underfiring is known, as well as the amount of coal charged, the so-called underfiring requirement can be calculated. This is usually expressed in kilojoules per kilogram, and according to convention, the consumption is based on a feed coal containing 10% moisture.

The heat derived from burning the fuel gas is not used entirely for the conversion of coal into coke. Part of the heat is lost in the hot waste gases; another part is lost from the surface of the oven by convection or radiation. If these losses are taken into account, the heat of coking can be calculated and the thermal efficiency of the oven given. In modern coke ovens this is about 80%.

The heat of coking can be further divided into the specific heat of the coking products and the heat of reaction. Heat of reaction is the quantity of heat developed or required per kilogram of coal containing 10% moisture, to achieve coking under standard conditions (0°C, 1013 mbar) and with the coking products in the physical condition in which they leave the oven. The magnitude of this quantity of heat depends on the temperature at which coking occurs. An overview of the magnitude and composition of this heat is given in Fig. 14.29.

The heat of coking depends on the temperature.[68,69] Measurements of the heat of coking over a large temperature interval indicate an increase with temperature. Similarly, the heat content of the products of coking, the heat of reaction, and the waste gas losses increase with increasing temperature. Only the surface losses show a decreasing trend, since the heat losses from the surface increase considerably less than the heat brought into the oven by the increase in flue temperature.[70] The age of the oven also affects the heat consumption. On the average, the heat consumption increases by about 12.5 kJ/kg per year.

Process gas is usually used for underfiring, but other gases may be used. One distin-

[68] V. T. Chertok and R. E. Leibowich, *Coke Chem. USSR*, No. 1, 21–24 (1964).

[69] A. I. Voloshin, I. V. Virozub, and V. V. Kazmina, *Coke Chem. (USSR)*, **3**, 17–20 (1962).

[70] W. Weskamp, W. Dressler, and E. Schierholz, *Glueckauf*, **10**, 567–577 (1962).

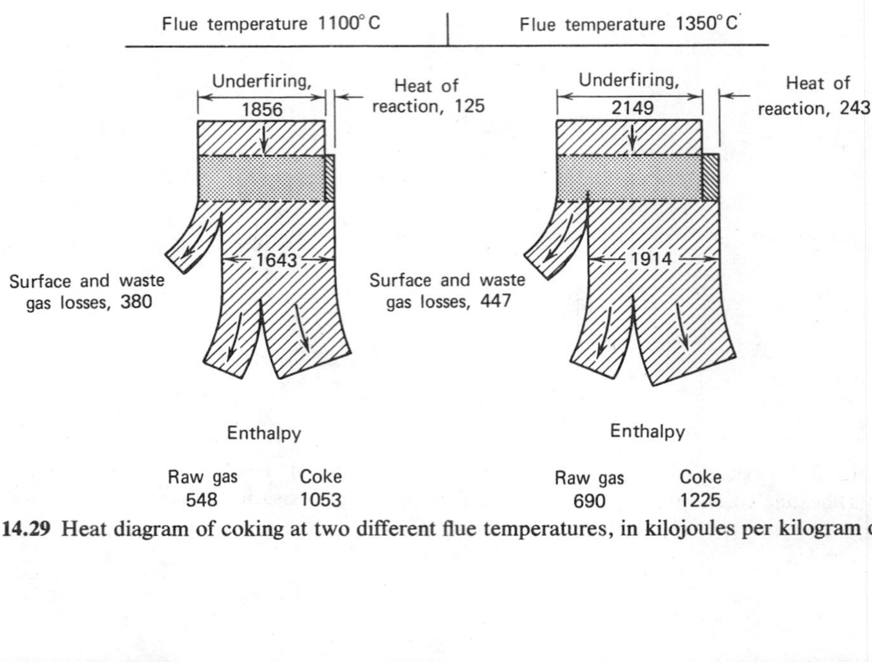

Figure 14.29 Heat diagram of coking at two different flue temperatures, in kilojoules per kilogram of coal.

guishes here between rich gases (coke-oven gas, producer gas, natural gas, residual synthesis gas, LPG-air mixtures) and lean gas (blast furnace gas or mixtures of blast furnace gas and coke-oven gas). Comparison of results with different gases used for underfiring shows that heat consumption is least with coke-oven gas. If one assigns a value of 100% as an index number for the heat consumption with coke-oven gas, the corresponding values for producer gas and for blast furnace gas would be 106% and around 110%, respectively. These increases in heat consumption are independent of the combustion temperature. Integrated steel plants of the future will probably see increased use of blast furnace gas for underfiring coke-oven batteries, to produce more surplus coke-oven gas for steel plant process heating needs.

An increased consumption of about 6–7% at 1200°C flue temperature is also observed when using natural gas or synthesis waste gas. The heat consumption increases linearly with temperature. It is possible to substitute an LPG-air mixture for the heating gas if the LPG consists of saturated hydrocarbons and if certain operating conditions are observed. Up to an addition of about 70% LPG, the heat consumption, compared with that for coke-oven gas, increases by about 12–17 kJ/kg of coking coal for each 10% of LPG. Above that, the additional heat consumption increases because of greater deposition of carbon, and at 100% LPG a value of 14% is reached. For a given percentage of LPG gas addition, the heat consumption increases with temperature.[71]

Under otherwise constant operating conditions, the flue temperature has a strong influence on coking time, thus on coal throughput. An increase in flue temperature of 250 K shortens the coking time by about 9 hr (Fig. 14.30). The flue temperature is usually between 1000 and 1400°C, which is below the upper temperature limit for heating the silica brick and the flues. However the operation of coke ovens in very high temperature zones causes different problems.[72] Increasing flue temperatures and decreasing coking time results in a reduction in coke strength or stability.

From the point of view of heating, the brick separating the flues from the actual reaction space causes considerable resistance to heat transfer. For this reason, investigations into the heat transfer properties of the heating wall were initiated (Fig. 14.31). There are two possibilities for improvement: increasing the heat conductivity of the brick material, and decreasing the thickness of the heating wall brick (stretcher brick).

Increasing the density of the silica material that is normally used results in improved thermal conductivity. The use of other materials has also been tested in the building of

[71] K. G. Beck, W. Weskamp, and E. Schierholz, *Progress Report of the State of Nordrhein-Westfalen No. 1807,* Westdeutscher Verlag Opladen.

[72] K. G. Beck and W. Weskamp, *Glueckauf* **107**(2), 43–51 (1971).

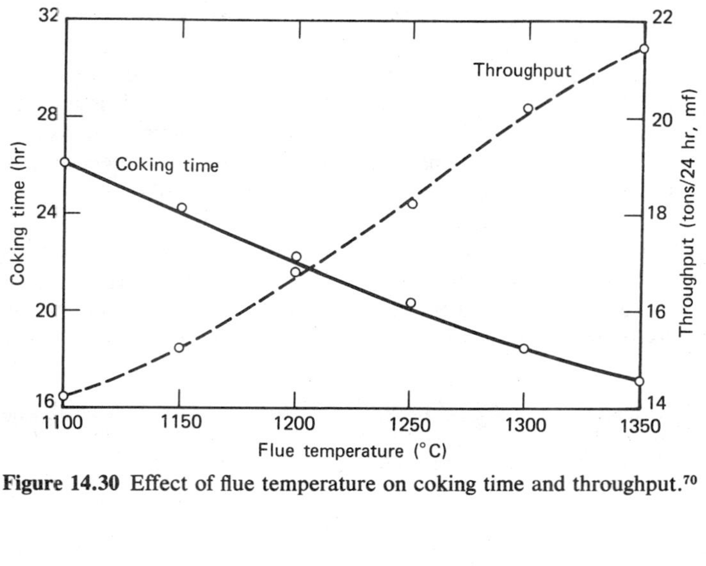

Figure 14.30 Effect of flue temperature on coking time and throughput.[70]

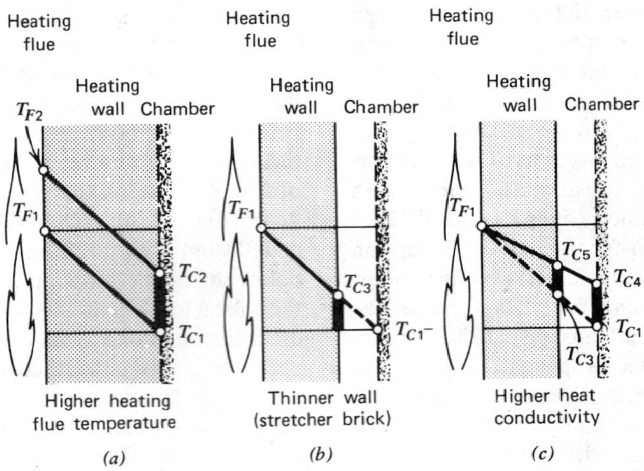

Figure 14.31 Possible ways of increasing specific throughput: (*a*) higher heating flue temperature, (*b*) thinner wall (stretcher brick), (*c*) higher heat conductivity. T_F: flue temperature; T_C: chamber temperature.

coke ovens.[73] A decrease in the thickness of the brick material from 110 to 70 mm resulted in a decrease in the coking time of 3–4 hr.[74]

Weskamp et al.[70] have investigated the quantity and composition of by-products, as they depend on flue temperature. Figure 14.32 summarizes their results.

14.3.2 The Influence of Coal Properties

In addition to the operating conditions of the coke oven, the properties of the coal used determine the results of the coking process in a decisive way. The duration of the coking process, as well as the heat consumption and the quantity and composition of the by-products, depends on the coal properties.

An important parameter in coking is the bulk density. For granular materials, the bulk density (mass of material per unit volume) is usually determined instead of the specific gravity. In top charging this parameter is determined from the weight of the charge, the dimensions of the oven, and the height of the charge. In stamping operations it is determined by calculating the weight of the stamped material. On the average, the bulk density is 700–780 kg/m³ of dry coal in top charging, and 900–1000 kg/m³ in stamping operations. Assuming a coal specific gravity of 1.3 g/cm³, it is evident that about 60% of the empty space contained solids in top charging, and 76% in stamping.

In normal top-charging operations in Western Europe, depending on the preparation, there is about 8–10% moisture in the coking coal; in the United States moisture is about 5–7%. The bulk density depends on the moisture content. It reaches a maximum for dry coal and decreases to a minimum, depending on the size consist of the coal, between 6 and 10% moisture, then increases again with further increases in moisture content (Fig. 14.33).[75,76]

As Fig. 14.33 indicates, the size consist of the coal determines not only the position of the minimum in the bulk density as a function of moisture content, but also its absolute magnitude. As a rule, fine grinding of the coal leads to a decrease in the bulk density. The formation of bridges between the coal particles is assumed to be the reason that no further filling of the empty space occurs.

It is possible to add small amounts (0.2–1%) of a suitable oil to the coal. This leads to a decrease in the bulk density when using dry coal and to an increase when using moist coal.[75] The additional oil makes it possible to achieve uniform bulk density even if

[73] F. Orywal, H. Echterhoff, and K. G. Beck, *Glueckauf Forschungsh.*, No. 5, 225–233 (1976).
[74] W. Rohde and K. G. Beck, *Glueckauf,* **17,** 977–981 (1976).
[75] W. S. Landers, M. Gomez, and D. J. Donaven, U.S. Bur. Mines Rep. Invest. No. 6502, 1964.
[76] K. G. Beck, W. Weskamp, and R. Beckmann, *Glueckauf,* **26,** 1559–1568 (1962).

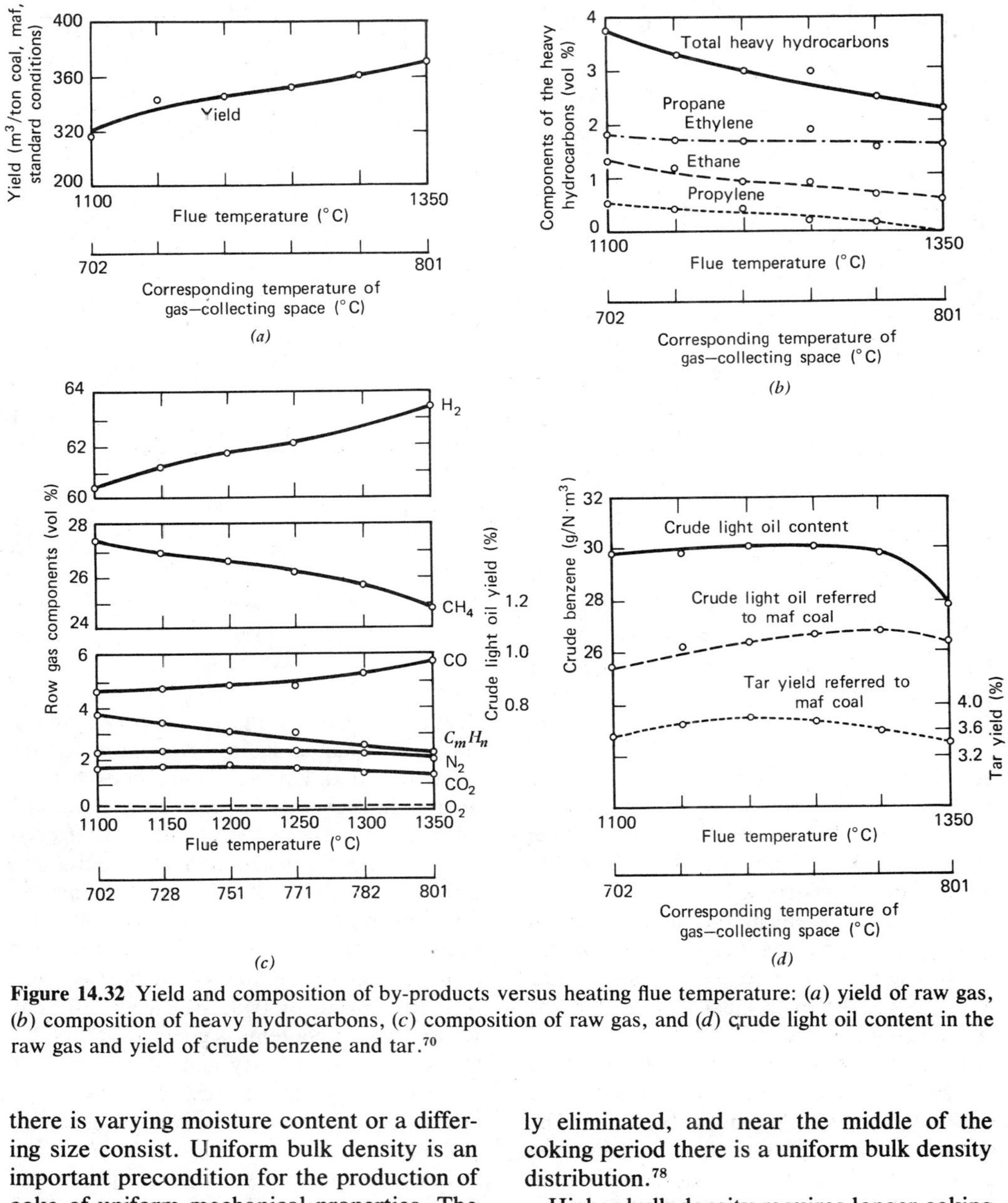

Figure 14.32 Yield and composition of by-products versus heating flue temperature: (*a*) yield of raw gas, (*b*) composition of heavy hydrocarbons, (*c*) composition of raw gas, and (*d*) crude light oil content in the raw gas and yield of crude benzene and tar.[70]

there is varying moisture content or a differing size consist. Uniform bulk density is an important precondition for the production of coke of uniform mechanical properties. The increase in the bulk density by the addition of oil often leads to a higher coal throughput.

Bulk density calculated from charge weight and charge volume is only an average value. As a result of using 3–5 charging holes, several different bulk density zones are present in the oven chamber.[77] The differences between zones may be greater than 100 kg/m³. During coking these differences are slowly eliminated, and near the middle of the coking period there is a uniform bulk density distribution.[78]

Higher bulk density requires longer coking times at the same flue temperature because of the greater quantity of coal to be coked. If the higher bulk density is caused by a higher moisture content, the heat consumption also increases.[76] Differences in the moisture content have only a slight influence on the yield and composition of the coal by-products.[76]

The effect of moisture content on shrink-

[77] J. G. Zubilin, *Coke Chem. (USSR)*, No. 6, 22–26 (1961).

[78] L. Hardt and W. Simonis, *Glueckauf Forschungsh.*, **5**, 245–252 (1966).

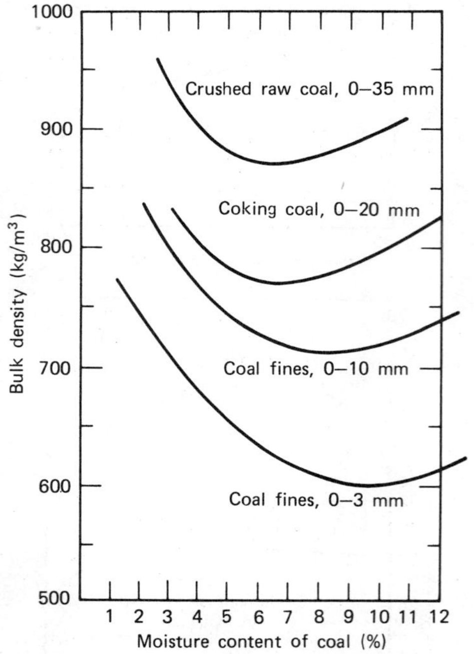

Figure 14.33 Influence of moisture content and particle size on bulk density.[75,76]

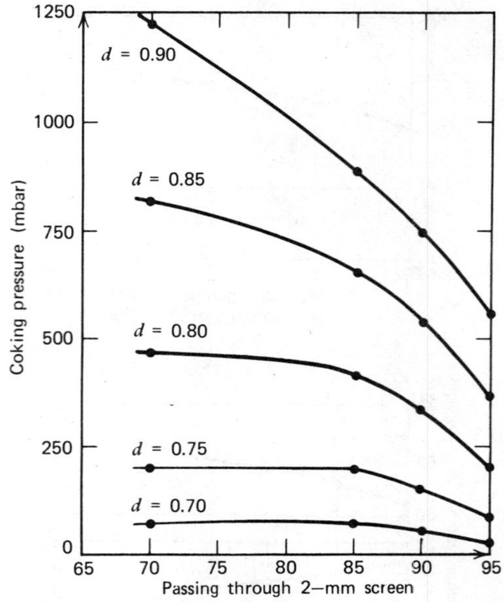

Figure 14.34 Coking pressure as a function of bulk density (d) and size consist.[80]

age of coking coal is opposite to its effect on bulk density and even at very low moisture contents results in swelling of the coke charge (decrease in bulk density), thus leading to difficult pushings.[76]

When using preheated coal, the lack of shrinkage becomes the factor that determines the coking time, since to assure easy pushing of the coke, the oven must remain standing after coking has ended.[79] For a swelling coal or coal mixture, the magnitude of the coking pressure is greatly affected by the bulk density and the size consist of the feed coal. The maximum coking pressure can be decreased considerably by a decrease in the bulk density and by fine grinding of the coal (Fig. 14.34).[80]

After the moisture content and the size consist, the raw material properties of the feed coal are the next factor in the coking process. One characteristic of the raw material is the volatile matter content, which can be traced to the different rank of coalification. A direct connection between coking time and volatile matter content does not seem to exist. However rank does have a great influence on the heat consumption.[81] With an increase in the volatile matter content, the heat consumption also increases; however at about 25–27% volatile matter (mf), the increase stops temporarily. At present there is no explanation for this behavior (Fig. 14.35). An overview of the heat balance in a coke oven for two coals with different volatile matter contents is given in Fig. 14.36. Obviously, the quantity and composition of the by-products change with changes in the volatile matter content. Details are given in Fig. 14.37.

To ensure the quality of the coke produced, the coking coal must have a certain swelling and shrinkage behavior. There are coals with such a high coking pressure that they can be used only in mixture with leaner additives. If coked by themselves, their excessive coking pressure may destroy the oven. For some coals the shrinkage often compensates for the swelling.

A direct relationship between the coking pressure and the raw material properties has

[79] W. Rohde and K. G. Beck, *Glueckauf,* **6,** 348–354 (1973).

[80] R. Loison, P. Foch, and A. Boyer, *Le Coke,* Dunod, Paris, 1970, pp. 406–419.

[81] W. Weskamp, German Coke Oven Committee Rep. No. 143, 1967.

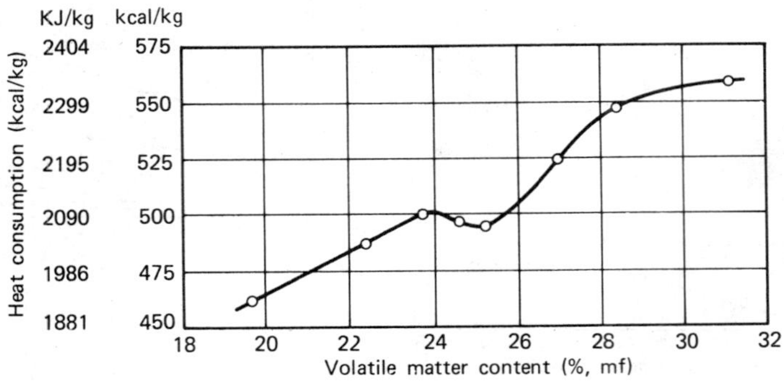

Figure 14.35 Heat consumption and volatile matter content of feed coal.

not yet been found. An attempt was recently made to calculate the maximum gas pressure in the plastic zone when the two zones meet in the center of the oven, since the gas pressure depends on the raw material properties and on the coking conditions.[82]

Coals with a volatile matter content between 17 and 25% often have a dangerous coking pressure. It is possible that coal mixtures may lead to coking pressures above that of the individual components. In this case additives such as anthracite or coke breeze must be used. For low-volatile coals, suitable mixtures are often achieved by adding high-volatile coals. In any case, careful preparation of the mixture is necessary.

The addition of coke breeze is also used to improve coke properties. The addition of coke breeze increases cohesion and simultaneously decreases fissuration in the coke so that a qualitatively better coke is produced.[83] It is possible to pretreat high-volatile or poorly coking coals to produce coke with the desired properties. One possibility, used mainly in some areas in Europe, is the mechanical compaction of the feed coal by

[82] E. Szurman, Dissertation, Technical University of Berlin, 1972.

[83] M. J. Kovalik, D. E. Wolfson, and L. Mafria, U.S. Bur. Mines Rep. Invest. No. 7718, 1973.

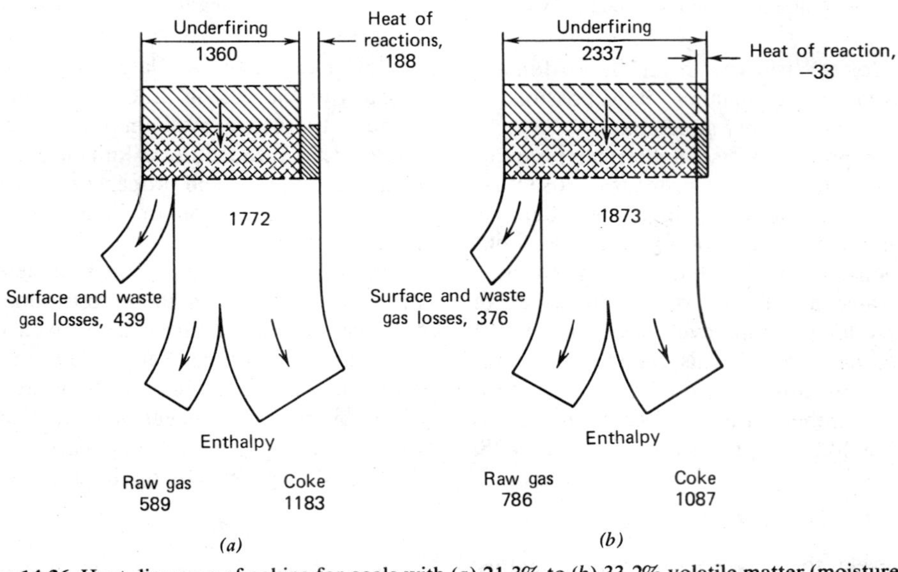

Figure 14.36 Heat diagrams of coking for coals with (*a*) 21.3% to (*b*) 33.2% volatile matter (moisture-free), in kilojoules per kilogram of coal.

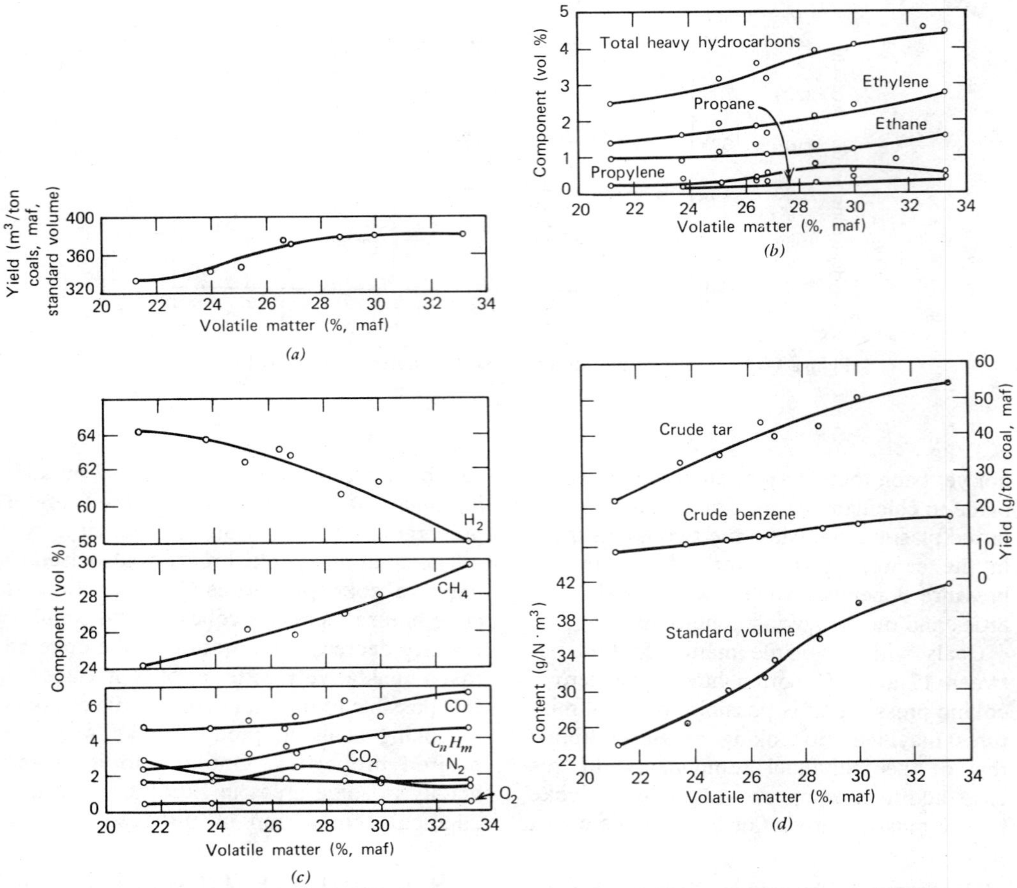

Figure 14.37 Yield and composition of by-products as a function of volatile matter content: (*a*) yield of raw gas versus volatile matter, (*b*) composition of the heavy hydrocarbons as a function of volatile matter content, (*c*) composition of the gases as a function of volatile matter content, (*d*) crude light oil content in raw gas and crude benzene and crude tar yields as functions of the volatile matter content.

stamping. With the high bulk density achieved in this manner, it is possible to obtain comparatively good coke.

Other processes use a thermal pretreatment of the coal by preheating (Section 14.2.5). In addition to the higher bulk density, preheating reduces coking time by about 60%, since the moisture does not need to be evaporated and the charge is already at an elevated temperature level. Measurements of the coking times of coals coked in the usual manner, in comparison with coals experimentally preheated in the Precarbon process (Section 14.2.5.2), are shown in Fig. 14.38.

14.3.3 Mathematical Models

At present there is no satisfactory mathematical model of the total coking process. The literature, however, reports a great number of attempts to calculate certain characteristics of the coking process. In many instances there are mathematical statistical models, since too little is known about the chemical and physical properties of the coal to permit preparation of kinetic dynamic models.

One of the most important parameters in the coking process is the coking time, and formulas for calculating it have been in existence for a long time. Reference 84 presents a summary of the formulas available at before 1950. By means of recent mathematical statistical methods, more complicated models for calculating the coking time have been derived.[85,86]

[84] P. Veit, *Brennstoff-Chemie,* 14–21 (1950).
[85] M. Gomez, J. G. Walters, and J. B. Gayle, U.S. Bur. Mines Rep. Invest. No. 7093, 1968.

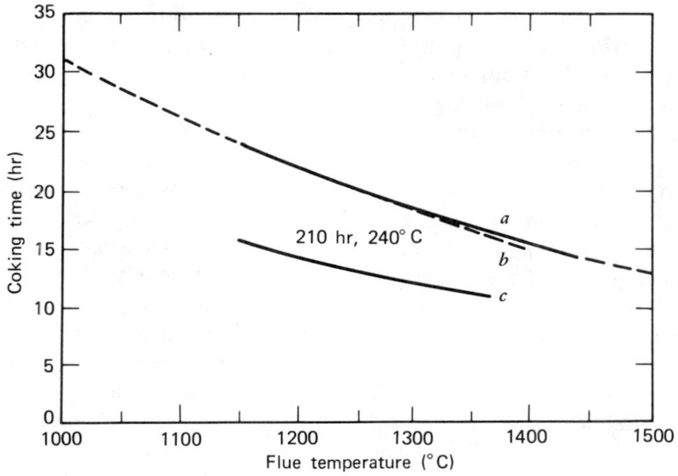

Figure 14.38 Coking time of moist and preheated coal as a function of flue temperature: *a*, moist charged coal (10% moisture content); *b*, test coke oven, 1969; *c*, preheated coal (200°C).

Reference 86 gives a formula for calculating the heat consumption of coke ovens.

On the basis of work by Baum[87] and Litterscheidt,[88] the changes in temperature regions in the coal charge were measured and used for further analysis. For this purpose a nonstationary temperature field was described by means of two coupled Fourier differential equations, which were numerically solved on an electronic data-processing unit.[89,90] This work makes it possible to determine the temperature field as it depends on the location and time. Figure 14.39 shows measured and calculated temperature fields.

[86] J. Kasperczyk, *Fuel,* **53,** 126–131 (1974).

[87] K. Baum, *Glueckauf,* **65,** 763–776,812–821, 850–860 (1929).

[88] W. Litterscheidt, *Glueckauf,* **70,** 77–84 (1934).

[89] W. Rohde, W. Simonis, and W. Peters, *Brennstoff-Chemie,* **50,** 1–8 (1969).

[90] J. Kasperczyk and W. Simonis, *Glueckauf Forschungsh.,* **31,** 149–154 (1970).

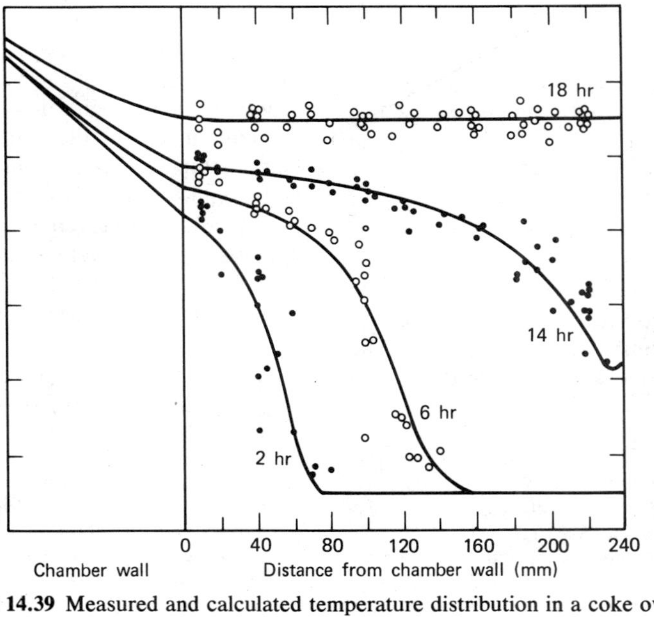

Figure 14.39 Measured and calculated temperature distribution in a coke oven.

By means of this model, it is possible, by selecting a suitable criterion defining the end of the coking period, to calculate coking time, heat consumption, and average heat flow in the wall (stretcher brick). Conclusions about the events occurring during coke formation appear to be possible.

Comprehensive statistical evaluation of coking tests in the U.S. Bureau of Mines, Tuscaloosa, Alabama, oven gave a series of formulas for calculating size consist, strength, and specific gravity of the coke produced.[91]

In the Steinkohlenbergbauverein in West Germany, a mathematical statistical model for coking was developed that makes it possible to calculate the lump strengths M_{40} as well as the material strengths M_{10} and the yield of by-products and coke.[56]

The parameters K, G, and M_s occupy central positions in this model. The K value, which is called the specific heatup value, is defined[92] as:

$$K = \frac{B^2 \cdot \rho}{\tau} \text{ (g/cm-hr)} \tag{14.1}$$

in which B is the half-width of the oven (cm), ρ is the bulk density (g/cm³), and τ is the coking time (hr).

The coke-forming ability G is derived from results of the Ruhr dilatometer test by Equation 14.2 (see Chapter 6):

$$G = \frac{E + V}{2} \frac{k + d}{kV + dE} \tag{14.2}$$

in which E is the softening temperature (°C), V is the resolidification temperature (°C), k is the contraction (%), and d is the dilatation (%). In the range from 0.95 to 1.10, the G value obtained has an additive behavior and can be used to calculate the properties of mixtures.[33,93]

Investigations based on spherical particle models have led[36] to the definition of an optimum size consist of the feed coal. The sum of the positive and negative deviations from this optimum curve is defined as the sum deviation, M_s. The model starts with the relation

$$M_{40} = a'K + b + \alpha' M_s \tag{14.3}$$

for calculating the lump strength M_{40}. The term a' describes the change in the M_{40} values as a function of the K values, and a' itself is a function of the content of volatile matter and the coke-forming ability G, as is α'. The parameter α' indicates the change in the lump strength as a function of the value of the sum deviation. The term b is an extrapolated coke strength for optimum size consist ($M_s = 0$), and $K = 0$. This specific material constant depends on the coal properties and is a function of the content of volatile matter.

Starting with the results of an analysis to determine the kind of coal and an analysis of macerals and the derived content of volatile matter of the maceral groups—vitrinite, exinite, and inertinite—it is possible to replace the parameter G for the coke-forming ability by the parameter G_{RM} (index R = reflectance analysis; index M = maceral group analysis), and to introduce a' and α' into the calculating formulas.[29] Use of these models with American coals showed that the determined and calculated values are in good agreement.[94]

Similarly, methods for calculating the strength of coke were developed[95,96] (Chapter 15), based on petrographic analysis. In this method the influence of different coking conditions is not considered.

Simonis[56] gives regression equations that permit calculation of the yield of coke, gas, raw tar, and raw light oil as these depend on the content of volatile matter, and the specific heatup value, K. Data for the quantity and composition of the raw gas produced during coking are given in Reference 97. The changes in these values with time are described as they depend on the K value and the content of volatile matter.

[91] M. Gomez, J. G. Walters, and J. B. Gayle, U.S. Bur. Mines Rep. Invest. No. 7024, 1967.
[92] K. G. Beck, H. Echterhoff, and W. Simonis, *Brennstoff-Chemie*, **46,** 48–51 (1965).
[93] W. Simonis, G. Gnuschke, and K. G. Beck, *Glueckauf Forschungsh.*, **27,** 181–192 (1966).
[94] J. G. Walters, C. Ortuglio, and D. E. Wolfson, U.S. Bur. Mines Rep. Invest. No. 7564, 1971.
[95] I. I. Ammossov, I. V. Eremin, S. F. Sukhenko, and L. S. Oskurkova, *Koks Khim.*, **12,** 9 (1957).
[96] N. Schapiro, R. J. Gray, and G. R. Eusner, *Proc. Blast Furnace, Coke Oven Raw Materials Committee*, **20,** 89 (1961).
[97] J. Kasperczyk and W. Kucharczyk, *Glueckauf Forschungsh.*, **33,** 107–115 (1972).

14.4 ANTICIPATED DEVELOPMENTS IN COKE-OVEN TECHNOLOGY BASED ON EXPERIMENTAL STUDIES

Future coking-coal requirements of the steel industry will increase worldwide. Although coke production is decreasing in some areas (United States, Western Europe), it will increase in Latin America, Australia, and in the Near and Far East. The increase in coke consumption will be smaller than the increase in steel production because the specific coke consumption per ton of pig iron will decrease further. The decrease in coke consumption was made possible by improvements in blast furnace technology, as well as by substitution of fuel oil for some of the coke. In West Germany the specific coke consumption per ton of pig iron decreased from about 830 kg in 1960 to about 500 kg in 1974 (Fig. 14.40). As a result of the energy crisis in recent years, the use of fuel oil has decreased slightly; therefore the specific coke consumption has had to increase. Theoretical calculation of the possible development of blast furnace technology indicates a lower limit for the specific coke consumption of about 350 kg/ton of hot metal.

Horizontal chamber coking will be predominant in the coke-oven technology of the 1980s. The productivity of this coking method will be increased, possibly up to 100%, with the development of new technologies (i.e., preheating, increasing chamber size, new charging processes, better heat exchange, process control). New processes for the production of formcoke will not become important on an industrial scale before 1985.

Requirements for the protection of the environment necessitate improvement of the existing processes, and the lack of coking-coal reserves furnishes an incentive for the quicker development of new processes.

In the development of high-temperature coking processes, the following factors will be of great importance: (1) improvement of the heat flow between the flue and the coking chamber (thinner wall bricks, use of different refractory materials such as magnesite and Starkite®, (a silicon carbide–containing material), (2) improvement of the regenerator inserts, (3) testing of large coke ovens, (4) improvement of the coking-coal preparation (mixing of coals), (5) broadening the number of coals suitable for coking (preheating, prebriquetting, stamping), (6) automation of the coke-oven operation, (7) dry quenching of the pushed coke, and (8) measures for the protection of the environment.

The object of the development of continuous processes for the manufacture of formcoke is to achieve, as far as possible, independence in the supply of coal.

Two groups of recently developed high-throughput silica coke ovens[74,99,100] were op-

[98] *Stahl Eisen,* **95**(7), 254 (1975).

[99] K. G. Beck, W. Eisenhut, and W. Rohde, *Int. Iron Steel Congr.* Vol. 1, Ref. 1.2.2.5, Düsselford, 1974.

[100] W. Rohde, D. Stalherm, and K. G. Beck, in *Coke Oven Plant Technology,* EGKS-Kommission Luxemburg, EUR 5342, Part 1, Ref. No. 7 (1975), pp. 75–94.

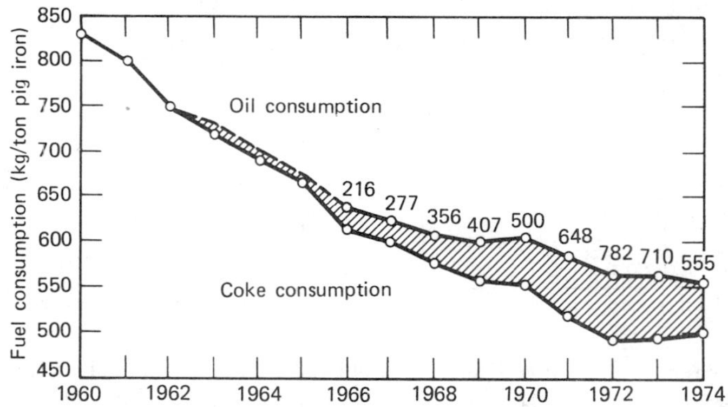

Figure 14.40 Coke and oil consumption of blast furnaces in West Germany for pig iron of all qualities, 1960–1974.[98]

erated at the Emil coke-oven test plant of the Bergbau-Forschung GmbH. To achieve an increased throughput, these ovens were built with oven wall (stretcher) bricks 85 and 70 mm thick, respectively. One group of ovens also has a new thin-wall regenerator brick. Both are equipped for programmed heating.

Reducing the thickness of the oven wall brick reduces the temperature gradient between the flue and chamber wall; therefore under otherwise equal conditions, the rate of heat transfer into the charge is increased. This change results in a new relationship between coking time and flue temperature.

Corresponding burner nozzle temperatures were obtained for 18, 16, 14, and 12 hr operating time in the new ovens equipped with a wall brick 70 mm thick. These values are shown graphically in Fig. 14.41. For comparison the values for ovens with 110-mm wall brick are shown also. These values agree with a modified Litterscheidt diagram.[101]

It can be seen that reducing the wall thickness from 110 to 70 mm while maintaining the same flue temperature leads to a reduction in operating time of 3–4 hr (Fig. 14.41), depending on the absolute magnitude of the flue temperature. For instance, at 1300°C flue temperature the operating time of ovens with 110-mm-thick walls is 18 hr, whereas in the new ovens it is 14 hr, a reduction of 22%. This result is essentially in agreement with previous calculations made in accordance with the heat transport model[102] for coking.

On the other hand, it is possible to obtain the same operating time with considerably lower flue temperatures than those in the ovens with 110-mm-thick walls. Combination of the two cases—reduction of the flue temperature and shortening of the coking time—is possible.

The foregoing results led to the construction of a battery of coke ovens containing 38 ovens with a wall thickness of 80 mm at the Prosper coke-oven plant in West Germany.

The higher heat flux required for coking necessitates improved heat exchange in the

[101] W. Litterscheidt, *Glueckauf,* **5**, 106–111 (1934).

[102] W. Rohde, *Glueckauf Forschungsh.*, **31**(.3), 149–154 (1970).

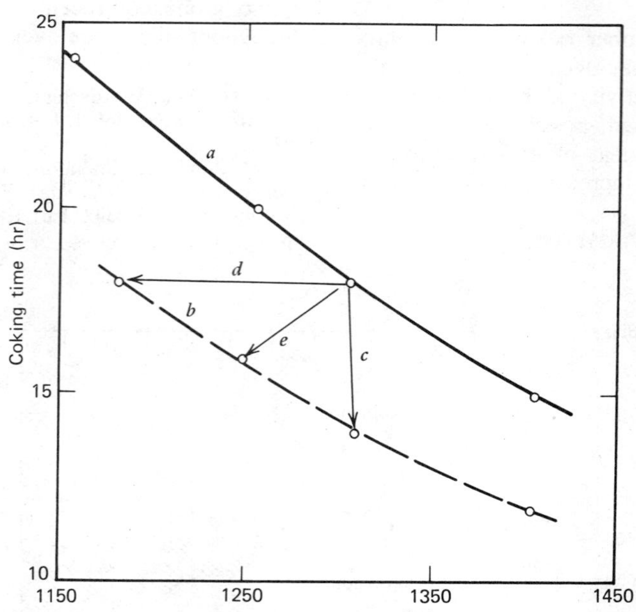

Figure 14.41 Coking time as a function of flue temperature at a chamber width of 450 mm: *a*, experimental oven with wall 110 mm thick (stretcher brick); *b*, high-throughput oven with wall 70 mm thick; *c*, short gross coking period at same flue temperature; *d*, lower flue temperature, same coking period; *e*, lower flue temperature, lower gross coking period.

Figure 14.42 Development stages of regenerator bricks.

regenerator.[4,100] An attempt is therefore made to attain a preheat temperature of the air that is as high as possible and a low exit temperature of the off-gases.

The optimum compromise for the coke oven must be found, taking into consideration (1) the highest possible heat transfer capability of the regenerator, which theoretically can be achieved by use of refractories with high specific heat exchange surface, (2) the operational possibilities for rapid reversal times, and (3) the necessity for structural stability and prevention of plugging of the channels.

At the experimental Emil coke-oven plant, three coke ovens were equipped with two types of regenerator bricks of different shapes (cassette and honeycomb stone),[4,100] which have a considerably higher specific heat exchange surface than the standard regenerator bricks. Figure 14.42 illustrates the stages of development of the regenerator bricks are shown.

The cross piece in the cassette brick, which has rectangular gas channels, is 6 mm thick. The connecting piece of the honeycomb brick, with its hexagonal channels, has an average thickness of barely 5 mm. By comparison, the conventional regenerator brick has a crosspiece thickness of 12 mm and the soap-type brick of 48 mm.

Table 14.2 shows the important data for heat transfer in the new regenerator bricks. For comparison, the values for a checker brick used in modern coke ovens, with channels of approximately rectangular cross section, are given. In the new regenerator bricks, the thickness of the crosspiece is halved. As a result, the specific heat surface exchange area, which is the quotient of the surface exchange area divided by specific volume, is more than doubled. The relative free cross section increases by 6–7% and the relative specific volume decreases by about 12–13%. Also of importance is the heat transfer surface area, which can be located in the regenerator space; it is almost twice as large in the new bricks compared with modern checker bricks. The larger heat transfer surface area achieved by the thinner crosspieces necessitates a shorter reversal time.

Heat measurements indicate that the new regenerator inserts in the shape of the cassette or the honeycomb bricks give satisfactory heat exchange for reversal times of up to 10 min. Comparatively high losses result from frequent reversals and the nonuniform heating of the pusher side and the coke side. What is desired is a better brick that works best at a reversal time of 10–15 min, and to minimize losses, heats the pusher and coke sides as uniformly as possible.

A further increase in the specific throughput of the oven can be achieved by using a

Table 14.2 Characteristic Data for Regenerator Bricks

Parameter	Cassette Brick	Honeycomb Brick	Checker Brick
Free cross section to total cross section (%)	43	44	37
Storage volume to total volume (%)	49	50	62
Surface exchange area to total volume (m^2/m^3)	129	120	69
Surface exchange area to storage volume (m^2/m^3)	265	238	112
Web thickness (mm)	6	<5	12

Table 14.3 Comparison of the Properties of Magnesite Brick and Silica Brick[73]

Parameter	Magnesite Brick	Silica Brick
SiO_2 (%)	≈1.5	≈96
Fe_2O_3 (%)	≈0.6	≈0.4
Al_2O_3 (%)	—	≈0.9
CaO (%)	≈3.3	≈2.2
MgO (%)	≈94.3	<0.1
C/S (%)	2.2	—
Alkali (%)	—	≈0.2
Volume weight (g/cm^3)	2.95	1.85
Apparent porosity (%)	16.5	19.4
Cold compressive strength (N/cm^2)	5490	5585
Refractoriness under load (K)		
t_b	1973	1933
t_e	1973	1943
Thermal conductivity (kJ/m-hr-K)		
At 600°C	20.5	5.0
At 1000°C	13.0	6.3
At 1400°C	8.8	7.3
Thermal expansion (%)		
At 600°C	0.95	1.2
At 1000°C	1.4	1.3
At 1400°C	2.0	1.25

refractory material with higher heat conductivity. Therefore many different kinds of bricks were tested for use under coke-oven conditions. It was found that a low-iron magnesite and the silicon carbide–containing material Starkite®[103] were especially suitable. Magnesite coke ovens have been built and operated in the semicommercial coke-oven laboratory of the Bergbau-Forschung.[73,104] In contrast to silica, magnesite bricks make it possible to reach a given wall temperature with a much lower flue temperature (e.g., 1250 vs. 1400°C) and to operate the oven with a shorter coking time.

Table 14.3 compares the technical data for a magnesite brick and for a silica brick. The bulk density of the magnesite is considerably higher. Shaping of the magnesite brick on hydraulic presses with very high specific pressures assures optimum compression of the mass, and the porosity is therefore smaller than with silica brick. Cold and hot properties of both materials are similar, so that one can count on satisfactory strength of the magnesite brick when used in the coke oven. Firing the brick at very high temperatures allows the formation of stable high-temperature ceramic phases, which permit volume stability under high thermal loads.

Magnesite brick has a considerably higher thermal conductivity than silica brick. Although the thermal conductivity of silica increases slightly with increasing temperature, the reverse is true of magnesite. In the temperature range of industrial interest for building coke ovens, on the average, about 50% higher thermal conductivity of the brickwork can be realized. The expansion when heating, however, is less favorable for magnesite. In contrast to the thermal expansion of silica, which when heated to 800°C expands by about 1.2%, then remains at a constant volume, the magnesite increases linearly and reversibly with about 0.15% expansion per 100 K temperature increase. This obviously must be taken into consideration in the design of operational coke ovens.

More than 2000 experiments were conducted in an experimental oven. The results led to the decision to build an experimental group of ovens on an industrial scale.

The British Carbonisation Research Association conducted experiments using a 350-kg oven with temperatures up to 1600°C.[105] The oven was heated with gas, and its walls were made of mullite. Three individual coals and three commercial coal mixtures were coked.

[103] W. Reerink, K. G. Beck, and H. Echterhoff, German DBP Pat. P2019078, (1974).

[104] F. Orywal, H. Echterhoff, and K. G. Beck, in *Coke Oven Plant Technology,* EGKS-Kommission Luxemburg, EUR 5342, Part 1, Ref. No. 8 (1975), pp. 95–107.

[105] V. J. Pater, *COMA Yearbo.,* 137–160 (1976).

The results showed that an increase in the wall temperature also increases the heat consumption per oven charge. At 1600°C wall temperature the throughput is 3 times as high as at a wall temperature of 1100°C. This observation is valid if coke is pushed when the temperature in the middle of the charge reaches 1050°C. The same throughput can be reached with preheated coal if it is coked at 1250–1300°C.

An increase in temperature leads to a considerable decrease in the average lump size and to a decrease in the Micum strength M_{40}. This decrease is caused mainly by the decrease in lump size. Between 1100 and 1600°C wall temperature the coke yield in the sizes from 80 to 20 mm increased from 23 to 58% (related to the quantity of coke in the chamber). The Micum value, M_{10}, for abradability became more favorable when coals of lower rank were used.

Reheating of the coke produced at 1100–1600°C had only minimal influence on its physical properties. Coking as well as reheating to 1600°C removes about 10% more sulfur from the oven charge than coking at 1100°C. The increase in coking temperature leads to a decreased heating value of the gas and decreases the quality of the tar.

Because of the disadvantages of using exceptionally high coking temperatures, this approach will not be practiced for quite some time, especially since most of the advantages can be gained by preheating the coal charge.

If the addition of heat in the horizontal chamber oven remains constant with time, too little heat is available at the beginning of the coking cycle and the quantity of heat near the end is too great. Thus heat is initially removed from the chamber walls and is later restored to the walls. As a consequence, a significant quantity of heat flows into the charge, which leads to an undesirable increase in the enthalpy of the coke. These temperature increases become more pronounced as the coking time decreases (Fig. 14.43). As a result there is an increased formation of graphite, a decrease in coke quality, and a larger heat consumption for coking. Programmed heating of the coke oven should supply the heat required at the proper time in the coking cycle.[106] Equipment suitable for industrial use in the coking process has the trademark CODECO® and is being tested at the new experimental Prosper coke-oven plant at Bergbau-Forschung.

The first experiments in this area were conducted from 1974 to 1975 at the experimental Emil coke oven plant, using high-throughput groups of ovens of the Otto and Still type. To decrease heat consumption, different heating programs were used. The optimum heating program consisted of five blocks changing with time (Fig. 14.44). At 14 hr coking time heating is turned off for a total of 5 hr. The heating itself occurs with only two different rates of feed gas.

[106] D. Stalherm and C. Meyer, *Int. Meeting Iron and Steel Making 1976*, Vol. 1a, Ref. 2.3, Brüssel-Düsseldorf, 1976.

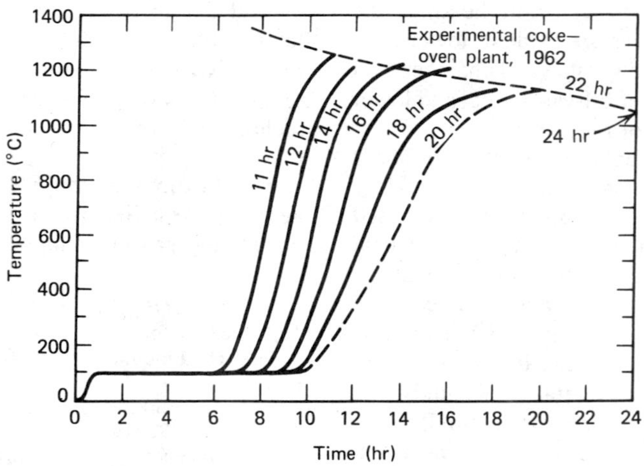

Figure 14.43 Variation in temperature in the middle of the chamber under different coking conditions: for 11-hr curve, coal moisture was 7%.

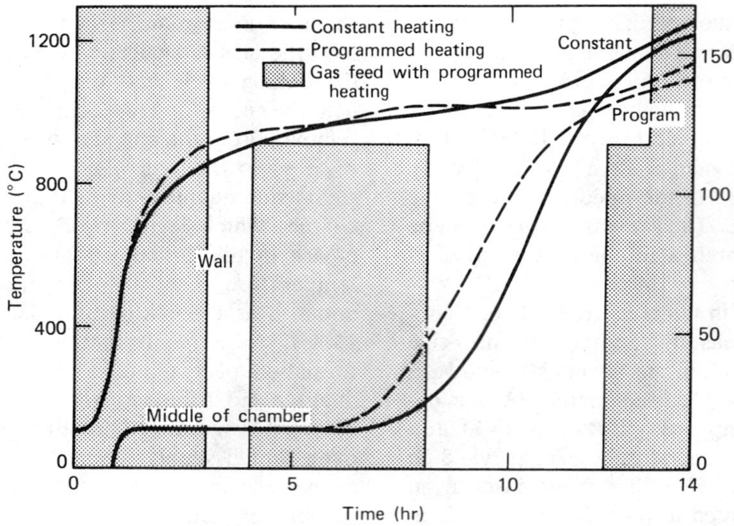

Figure 14.44 Variation in time and temperature with programmed and constant heating.

In checking the temperature throughout the charge, after 10 hr the total charge had already reached a uniform temperature of about 950°C. With constant heating, a uniform temperature of 1050°C is reached only after 12 hr. After 14 hr a final temperature of 1250°C is obtained. The programmed heating leads to a final temperature of less than 1100°C. The final temperature of the coke is decreased by 150–180 K, while maintaining the same satisfactory degree of coking.

In programmed heating the increase in temperature is more rapid at the beginning. Because of the long interruptions in heating, the heat content of the coke is considerably lower than in normal operation (Fig. 14.44). In comparison with constant heating, the following differences are evident: lower total gas requirement, lower flue and chamber wall temperatures, lower off-gas losses, and lower final temperature of the coke.

The lower final coke temperature of about 150–180 K gives a calculated energy savings of 12–15%, which corresponds to a decreased heat consumption of 185–220 kJ/kg of coal with 10% moisture. Reductions in waste gas and surface losses could not be determined quantitatively. The influence of the method of heating on the raw gas composition has not been examined in detail. This study will be done in the new experimental Prosper coke-oven plant. A further decrease in energy consumption appears possible.

An important goal in the future development of coke-oven technology is to increase productivity of the coke oven. The usual way to accomplish this aim is to increase the height and length of the oven chamber. The first oven chambers, 6 m high, were built in Germany in 1927.[107] These ovens were operated at low coking temperatures. The extensive use of 6-m-high oven chambers began in 1960 in Japan. Large coke ovens are now being built in many countries.[108,109] As important as the increase in the dimensions of the oven is the development of high-throughput systems,[110] which are characterized by an increase in heating temperatures, improvement in refractory material, and improved design of the oven. In the past flue temperatures of 1350°C were exceeded only rarely. Now, in high-throughput ovens, flue temperatures of 1450°C are being used. This technique decreases the coking time by about 4 hr.

Investigations at the Steinkohlenbergbauverein in West Germany have shown that the underfiring gas consumption increases with

[107] C. Still, *6th Coke Oven Technology Conf.*, Katowice, October 1975, Ref. 16.

[108] P. H. Pinchbeck, *COMA Yearb.* 101–113 (1973).

[109] B. G. Jenkins and G. A. Wade, *COMA Yearb.*, 161–192 (1973).

[110] D. Wagener, *COMA Yearb.* 243–264 (1974); D. Wagener, *Erdoel Kohle*, **27** (7), 372–378 (1974).

the volatile matter of the coal and also with increasing coking temperature.[72] For example, the underfiring gas requirement for a coal with 24% volatile matter increases from 2050 kJ/kg of coal at 1250°C to 2280 kJ/kg at 1350°C and to about 2385 kJ/kg at a flue temperature of 1400°C. An important factor contributing to the increase in underfiring gas is the higher sensible heat of the coke and gas. An additional influence can be attributed to more extensive cracking of the hydrocarbons in the gas. There is a direct relationship between the methane content of the gas and the flue and/or oven chamber temperature.[111] However an unequivocal relationship between cracking and oven height has not yet been found. An undesirable factor in the operation of high-throughput ovens is the so-called carbon formation. The deposits in the oven—especially at the roof—consist only partly of the so-called gas carbon. The magnitude of the deposits depends on the volatile matter and the size consist of the coal charge, on the coking temperature, and on the residence time of the gases in the oven. The temperature of the gas-collecting space does not seem to have the influence on the deposits that has been assigned to it in the past.[112] Finally, the operation of high-throughput ovens shows a decrease in the percentage of large lumps in the coke with increased operating temperature.[110,111] The same effect was found in an experiment of the British Carbonisation Research Association in a 350-kg oven.[105]

There is no consensus on optimum oven height. With greater oven heights, the specific investment cost increases because of thicker oven roofs, larger oven partitions, and the disproportionately greater weight of the steel parts. Furthermore, in ovens that are considerably higher than 7 m, operational difficulties can be expected (e.g., increasing gas pressures in the chamber). Increasing the length of the oven beyond 17 m makes pushing of the coke problematical.

In England 6.5 m is considered[113] to be the optimum oven height, but in the rest of the world opinions differ to a great extent (7–8 m and more).[107,110,114,115] Calculations for the optimum dimensions of a coke oven made by Giprokoks[116] led to the conclusion that with a chamber width of 410 mm, oven height of 7.4 m, and chamber width of 510 mm, an oven height of 6.8 m results in the lowest overall cost.

The economically recoverable reserves of high-quality coking coal in the world are limited.[117] Progress in mixing technology will therefore mitigate the results of the scarcity. Improved coking technology that also permits use of lower quality coking coals has led to a slightly decreased coke yield. The use of mixtures containing higher volatile coal has increased the ratio of coal charge to coke yield to a value of 1.4, which has been exceeded in some industrial countries. At present, for the countries of the European Economic Community (EEC) and the organization for Eastern Europe (COMECON), the ratio above averages 1.33; for the other large industrial countries it is 1.4. Within Europe it is expected that the value of 1.33 can be maintained, whereas in Asia and the United States a value of about 1.43–1.44 may be reached. In the United States and Japan the ratio will increase to a maximum of 1.47–1.48.[117] Details of coking-coal preparation were described in Section 14.2.

Among the technical developments (Section 14.2) that increase the kinds of coals suitable for coking is the stamping process, which has been described in Section 14.2.2.

In the course of improving the economics of and automation of the coke-oven operation, automatic control of the operating steps was introduced at several coke-oven plants.[114,118–121] There are two important rea-

[111] H. Guhr, *Glueckauf,* **106**,(8), 372–380 (1970).

[112] Y. Kyung-Suk, Diploma dissertation, Technical University of Clausthal, 1973.

[113] J. P. Graham, *COMA Yearb.* 29–37 (1976).

[114] E. F. Kümper, *Tech. Paper Krupp,* **35**(1), 17–24 (1977).

[115] E. Pries and W. Wackerbarth, *AIME Ironmaking Proc.,* **35**,(1), 17–24 (1977).

[116] V. P. Anokhin, E. M. Izraelit, L. M. Perfileva, and P. A. Toptygin, *Coke Chem. (USSR),* No. 11, 21–25 (1974).

[117] M. Rumberger, *Stahl Eisen,* **97**(15), 705–713 (1977).

[118] W. Eisenhut and H. Richter, 106th AIME Annual Meeting PTD/ISS, Rep. No. 4, Atlanta, 1977.

[119] F. Fautz and H. Lindenhorst, in *Coke Oven Plant Technology,* EGKS-Kommission Luxemburg, EUR 5342, Part 1, Ref. No. 9 (1975), pp. 109–119.

[120] T. Martens, in *Coke Oven Plant Technology,*

sons for introducing automation into the coke-oven plant: to relieve personnel of heavy and dirty manual labor, and to prolong the life of the mechanized and automated machines and equipment. Associated with the latter reason is an increased availability of machines and an assured production of coke. Advances in automation are absolutely necessary where prescribed by law. This is the case, for example, in some locking operations for machines that would otherwise require a high degree of attention and expertise from the machine operator. However the operation of coke-oven machines without personnel does not appear possible or desirable.

A precondition for automating the coke-oven operation, especially if one considers high coke ovens, is the mechanization of all processing steps that occur simultaneously or in sequence in the ovens or in the oven machines. For more extensive automation of the operating events, a series of other criteria is necessary. For the machines to move to defined positions, they must move according to predetermined lines. The driving motors of the machines must have a reduction ratio of about 100 : 1 to be able to accomplish this task. Such ratios can be obtained with DC motors with thyristor feeding, with three-phase motors with transductors or thyristor control, or with three-phase motors with a controllable eddy current brake.

It is also necessary in automated operation to supply information about the destination, the travel route, and the position of the machine. This information, which is called "place recognition," can be obtained at the desired position by noncontact switches and/or by a digital replication of the travel route of the machine. Strategically placed induction loops or ultrashort-wave transmitters and receiving equipment on the machines is usually used to transmit the information. The carrier frequency multiplex principle, or the time multiplex principle, or a combination of the two, may be used for transmission.

On the basis of the previously mentioned conditions, the automated coke-oven system ACOS® (Fig. 14.45) was developed, among others. This system makes operation and supervision possible, and assures reliable control and locking of the operating machines.[114,118]

In recent years dry quenching of coke has become important because it permits heat recovery. In dry quenching the sensible heat of the coke is recovered by quenching the hot coke with recycled inert gas, thereby cooling it from about 1000°C to about 200°C. The recovered heat may amount to 1050 kJ/kg of coke or 790 kJ/kg of charge coal.[122] The dry quenching units built so far utilize the heat to produce steam. As an alternative, it seems possible that there will be a combination of dry quenching with preheating of the coal, or that this heat will be used for heating in an area in the vicinity of the plant.

Dry quenching was introduced in the 1920s in Switzerland by the Sulzer Company and was used in several smaller coke-oven plants in gas works in Europe[123] and the United States. Since the 1960s the U.S.S.R. has apparently switched to dry quenching.[122,124] Other dry quenching units will be built in different countries. Two plants started operation in Japan in 1976.

Figure 14.46 shows a plan using the Giprokoks process. The coke is pushed into a special car and driven to the dry quenching plant. Here the coke container is lifted and discharged into the top of the cooling chamber. The upper part of the chamber is an intermediate bunker in which the level depends on the pushing cycle; the lower part is always completely filled. At this point the coke is cooled by inert gas. The cooled coke slides to conveyors. The heated gas passes through the dust separator to the waste heat boiler and from there returns to the cooling chamber.

Figure 14.47 is a schematic drawing of the coke dry quenching unit in the Homécourt plant (Sulzer process).[124,125] The two proc-

EGKS-Kommission, Luxemburg, EUR 5342, Part 1, Ref. No. 10 (1975), pp. 121–130.

[121] O. N. Pankrater, A. M. Shulga, et al., *Coke Chem. (USSR)*, No. 12, 42–44 (1973).

[122] J. P. Graham and J. E. Barker, *COMA Yearb.*, 187–207 (1974).

[123] W. Eisenhut and H. Friedrich, *Glueckauf*, **111**, 432–439 (1975).

[124] N. K. Kulakov and V. S. Kononenko, *Coke Chem. (USSR)*, No. 1, 19–26 (1975).

[125] M. Luchesi, *Circ. Cent. Doc. Sider.*, **26**,(10), 2211–2225 (1969).

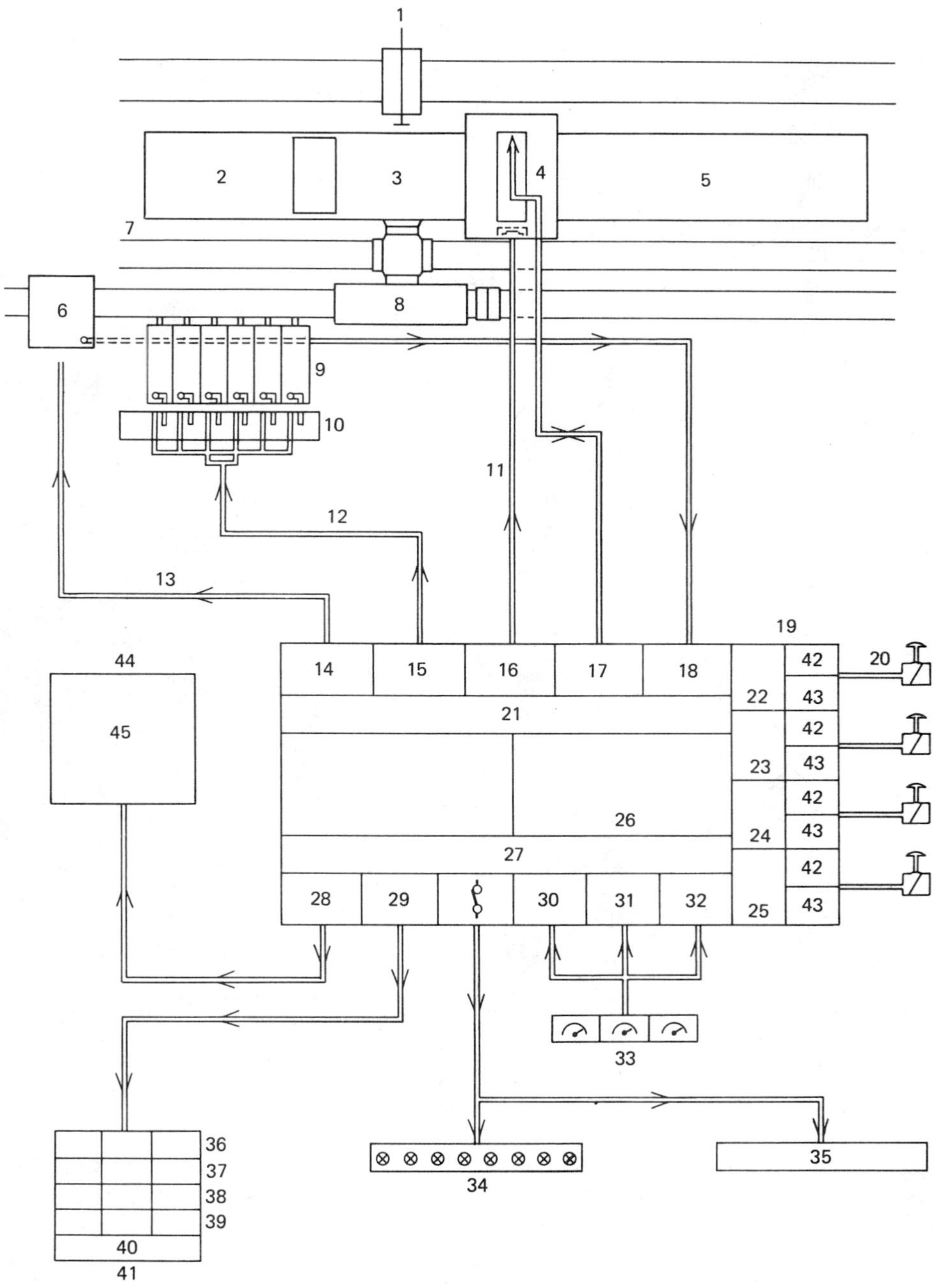

Figure 14.45 Diagram of an automated coke-oven operation: 1, coke-pushing machine (CPM); 2, battery 1; 3, coal-charging car (CCC); 4, coal tower; 5, battery 2; 6, coke quench tower; 7, coke guide car (CGC); 8, coke quench car (CQC); 9, coke wharf; 10, wharf conveyor; 11, steering for hydraulic system; 12, steering for ramp discharge; 13, steering for quench water; 14, quench tower; 15, wharf; 16, steering for coal tower; 17, scale coal tower; 18, quench tower; 19, data processing; 20, antenna; 21, input/output; 22, CQC; 23, CGC; 24, CCC; 25, CPM; 26, freely programmable control device; 27, input/output; 28, steering of printer; 29, oven number; 30, coking time; 31, quench time; 32, steaming time; 33, time check; 34, control lamps in control room; 35, plug for all ovens; 36, CPM; 37, CGC; 38, CCC; 39, CQC; 40, oven number; 41, position of machines; 42, transmitter; 43, receiver; 44, typewriter; 45, date, time, oven number (charged), oven number (pushed).[118]

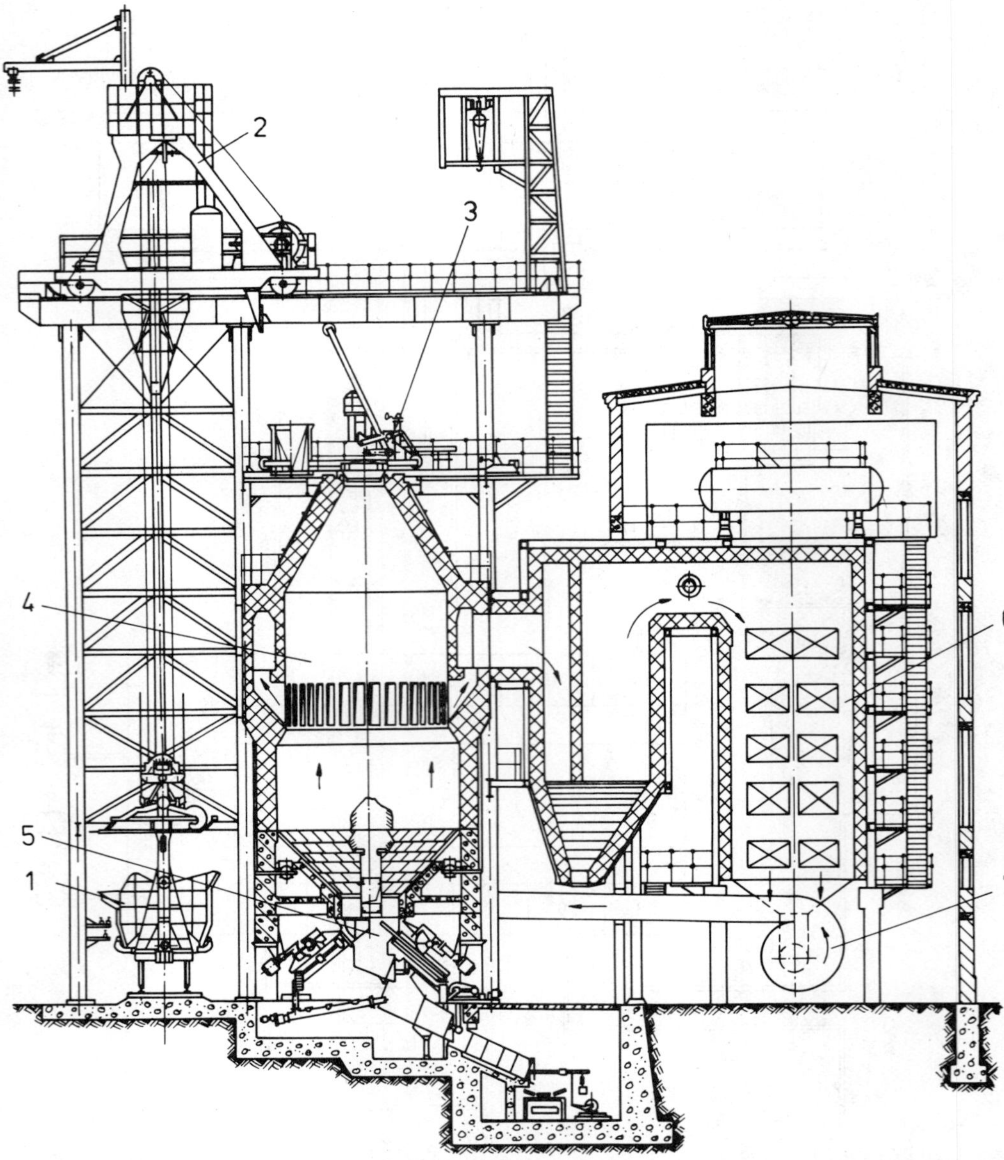

Figure 14.46 Schematic of the Giprokoks process coke dry quenching unit: (1) coke transport car, (2) elevator, (3) coke feed, (4) chamber, (5) coke discharge, (6) boiler, (7) blower, (8) cooling chamber.[127]

esses exhibit differences in construction and operating techniques.

Linsky[126] describes the utilization of recovered heat and the fuel savings that can be achieved by dry quenching of coke. Cost figures determined for the dry quenching processes of Giprokoks and Sulzer are given.

[126] B. Linsky, *J. Air Pollut. Control Assoc.*, **25**(9), 918–924 (1975).

Finally, the advantages and disadvantages of dry quenching in comparison with wet quenching are summarized.

Further development of conventional coking will be determined by measures for the protection of the environment (Section 14.1.3).

The manufacture of formcoke is described in Chapters 11 and 13. Its properties and use are treated in the next section.

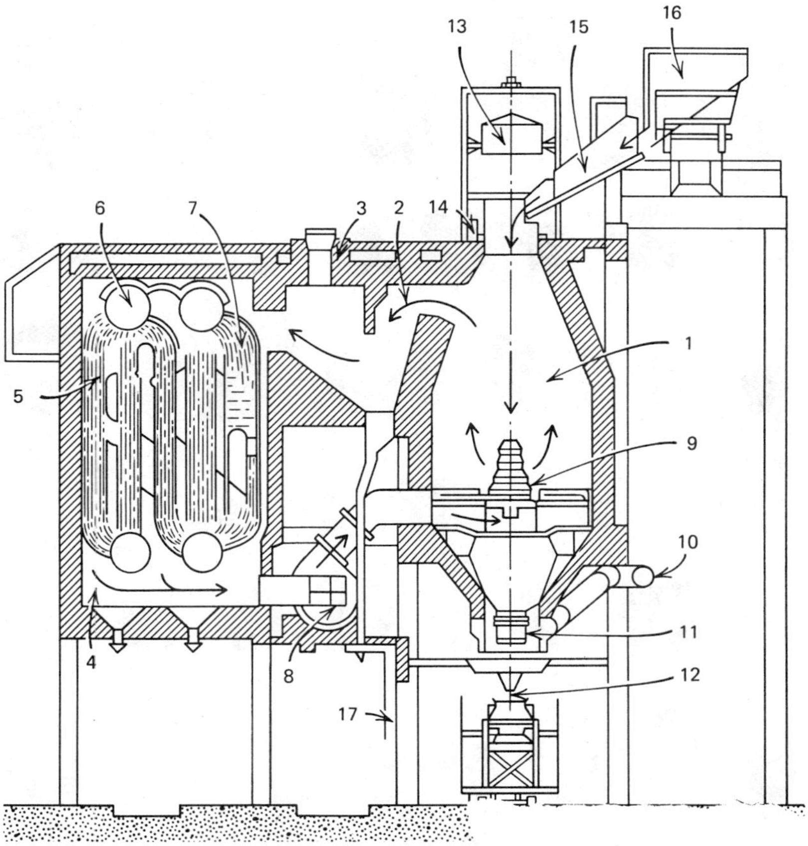

Figure 14.47 Schematic of the Sulzer process (Homécourt) coke dry quenching unit: 1, cooling chamber; 2, dust trap; 3, connecting duct; 4, waste heat boiler; 5, preheaters; 6, evaporators; 7, superheaters; 8, blower; 9, gas entry cone; 10, gas and dust injection channel; 11, cooled coke unloading conveyor; 12, compartmented transporter; 13, hood; 14, water seal; 15, inclined chute; 16, coke car; 17, special dust discharger.

14.5 FORMCOKE PROCESSES

The horizontal chamber coking process has been under continuous development for more than 100 years. The development of newer coking processes, especially a continuously operating formcoke process, did not really become significant until the decade from 1950 to 1960. In addition to considerations of economics and technology, the reason for such a development is the possibility of using a greater variety of coals for the production of blast furnace coke, foundry coke, and fuels for space heating. The best known processes for the manufacture of formcoke are listed in Chapter 11 and described in detail. This chapter gives several examples of the properties and testing of formcoke, as a replacement for oven coke (see also Chapter 15).

The use of formcoke in the blast furnace process is discussed first. The quality of a coke for use in the blast furnace is described by chemical composition (moisture, ash, sulfur, and volatile matter), and by physical properties (lump size, strength, and bulk density).

In addition to these material factors, flow properties are important. Figure 14.48[127] shows 85 different shapes of briquettes. Among these, only a few have desirable flow properties: namely, those that when charged, make point contact with each other. In addition to achieving a low pressure drop, the

[127] H. J. Schultz and O. Able, *Glueckauf,* **110**(13), 517–522 (1974).

Figure 14.48 Shape-dependent properties of briquettes.

coke and ore should have fitting properties, which means that the lump size ratio of ore to coke should be as close as possible to unity,[128] and the feed should not contain fines.[129] Practical blast furnace tests with formcoke were conducted in many countries with varying success.

In blast furnace tests with formcoke in Belgium,[130] a French formcoke produced according to the Carmonoix process was used (pitch-bound briquettes, coked in Koppers retorts), as well as a German formcoke produced according to the Bergbau-Forschung-Lurgi (BFL) process (hot-pressed briquettes from low-temperature coke with an addition of caking coal, and coked in a hot sand bed).

Although with the use of formcoke the pressure drop between the bustle pipe and top decreased and the top temperature and the temperature in the lower area of the furnace increased, high throughputs were obtained. However it must be pointed out that formcoke abraded in these experiments, and that the dust in the blast furnace gas led to deposits. Additional blast furnace experiments giving similar results are described in the literature.[131,132]

In two smaller blast furnaces in France oven coke was replaced by 33 and 44% of formcoke[133] produced in the HBNPC (Houillères du Bassin du Nord et du Pas de Calais) process. The blast furnaces operated properly. The pressure drop for the blast decreased somewhat, the coke consumption was decreased, and the dust content in the blast furnace gas increased slightly.

In West Germany Rheinstahl AG introduced an experimental program in 1970 with hot briquettes and with formcoke produced according to the BFL process.[134] The blast furnace had a hearth diameter of 6.8 m. It was operated with an oxygen-enriched blast, with controlled humidity, and with the addition of heavy oil. The production was between 1250 and 1350 tons of hot metal per day. The properties of the hot briquettes and of the formcoke in comparison with blast furnace coke are shown in Table 14.4.

The experiment with hot briquettes lasted 72 hr, and the furnace operated smoothly. The production remained unchanged at about 1350 tons/day. Although pressure of the burden in the furnace increased, hangup was not observed. Only the amount of dust in the blast furnace gas and the carbon content of the dust increased, which means that the hot briquettes were more abradable and resulted in a higher coke consumption. The experiments with formcoke showed similar results.

The first preliminary results with Ancit formcoke in the blast furnace were obtained by the Duisburger Kupferhütte[135] in 1970. Experiments with different shape briquettes were conducted in 1971 at the Burbacher Hütte, with furnace No. 3 having a hearth diameter of 5 m. The conversion from coke to Ancit was done in four stages, making a 25% substitution each time, without causing any problems.

In comparison to operation with oven coke, the pressure of the hot blast increased from 0.72 to 1.00 atm gauge pressure, the hot metal production was reduced by about 14%, and the quantity of blast furnace gas increased by about 4%. The dust content in the blast furnace gas was reduced from 12 to 7 kg/ton of hot metal.

Another series of blast furnace tests with the Ancit formcoke was conducted in 1973 in the Ilsede blast furnace plant of the Peine Salzgitter AG steelworks.[136] The furnace had a hearth diameter of 7.8 m. In the reference test oven coke of 20–80 mm size was used. When 50% of the oven coke was replaced by Ancit formcoke, the dust content of the blast furnace gas was greater and the blast pressure increased continuously, while at the same time the quantity of air blast decreased. This effect is caused by the regular shape of the formcoke, which, when charged, is more highly oriented, therefore causes a greater pressure drop than does the lump coke.

Pressure drop measurements indicated that in future blast furnace tests, coke and Ancit

[128] O. Able and H. J. Schultz, *Stahl Eisen*, **92**(1), 37–40 (1972).

[129] *Hutnik*, **36**(5), 214–221 (1969).

[130] A. Decker and A. Poos, *Stahl Eisen*, **88**(25), 1407–1414 (1968).

[131] *Circ. Cent. Doc. Sider.*, **28**(2), 449–467 (1971).

[132] INIEX Bull. Tech. Valoris, Utilis. Combust. No. 42, (1973).

[133] Rep. No. 1.2.2.6 for the International Iron and Steel Congress in Düsseldorf, 1974.

[134] O. Därmann, S. Henkel, and K. D. Haverkamp, *Stahl Eisen*, **90**(19), 1009–1012 (1970).

[135] W. Goossens, W. Zischkale, and R. Reiland, *Stahl Eisen*, **92**(21), 1039–1044 (1972).

[136] E. Teichert, V. N. Gupta, and W. Hermann, *Stahl Eisen*, **94**(26), 1330–1338 (1974).

Table 14.4 Properties of Hot Briquettes and Formcoke in Comparison with Oven Coke III[134]

Parameter	Oven Coke III	Hot Briquette	Formcoke
Size (mm)	40–100	48 × 61 × 35	47 × 60 × 35
Weight of one lump (g)	100–500	48	44
Ash (%, mf)	9.36	7.0	7.0
Sulfur (%, mf)	0.86	0.83	0.82
Porosity (%)	48–53	49.0	58.6
Breaking strength (kg)	—	185	260
Abradability (below 10 mm) (%)	7.5	7.8	7.1
Bulk density (tons/m³)	0.45–0.50	0.508	0.495

should be charged mixed instead of separately (Fig. 14.49). The final result of these tests showed that with Ancit briquettes of the shape used, about 25–33% of the conventional coke may be replaced in a blast furnace of 7.8-m hearth diameter without major problems.

In Japan in 1971 blast furnace No. 2 of the Kokura Works, with a hearth diameter of 8.4 m, was used for a test with formcoke.[137]

[137] K. Sugasawa et al., *Stahl Eisen,* **93**(1), 24–31 (1973).

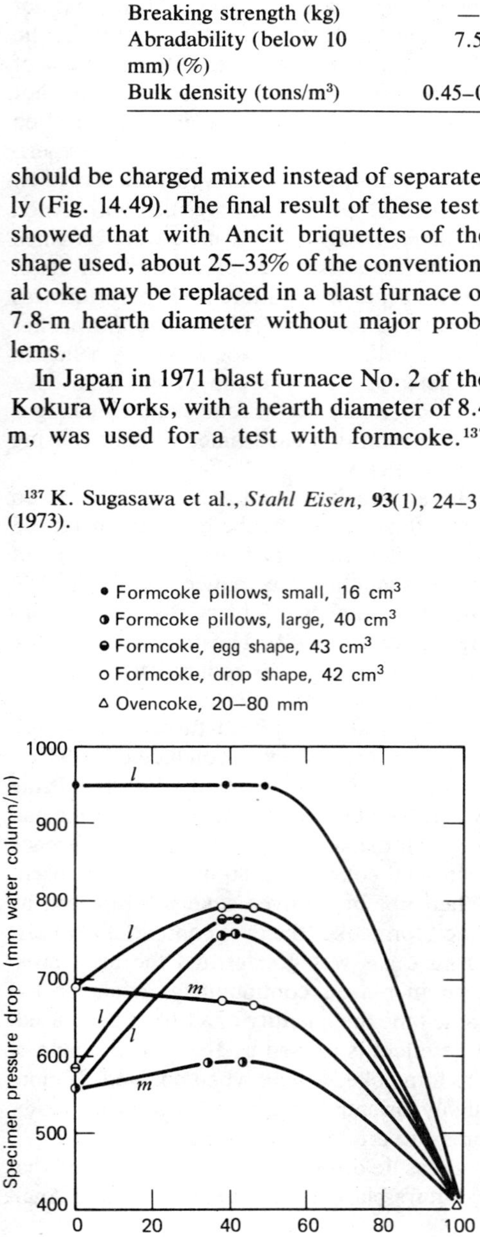

Figure 14.49 Pressure drop using oven coke–formcoke charges: *l* = charged in layers, *m* = charged mixed.

Table 14.5 compares the properties of the formcoke used with those of the oven coke used previously.

The operating results, when using formcoke in the blast furnace, showed that up to 50% of the blast furnace coke can be replaced by formcoke without damaging the stability of the operation. The blast furnace operated without difficulty during the experiment. There was no damage to the tuyeres, the burden did not hang up, and the blast rate and the weight of the burden did not have to be reduced because of nonuniform furnace conditions.

A very interesting experiment was conducted in Poland to determine the differences between the physical and chemical properties of formcoke and those of the standard oven coke.[138] In this experiment, performed in connection with the planned shutdown of a blast furnace with a useful volume of 350 m³, the furnace was charged before shutdown with a mixed burden consisting of a formcoke made by the Institute of Chemical Coal Processing (IChPW) process and an oven coke from the Bialy Kamién coke-oven plant.

The two types of coke exhibited similar behavior with respect to the following: toward the bosh of the blast furnace, the volatile matter, the hydrogen content, and the sulfur content decreased; the content of K_2O in the ash increased excepting the bosh, where it passes a minimum. The M_{40} values for both kinds of coke decreased in the same proportion between top and tuyere levels, whereas the M_{10} values for both increased to twice the original value.

[138] *Koks Smola Gaz,* **17** (7–8), 206–212; **17**(9), 258–264; **17**(11), 324–328 (1972).

Table 14.5 Comparison of the Quality of Formcoke and Usual Oven Coke in Japan[137]

Property	Formcoke	Oven Coke
Chemical composition (%)		
Moisture	0.6–2.8	2–4
Volatile matter	0.5–1.3	0.7–0.8
Ash	9.5–10.5	11.2–12.2
Sulfur	0.47–0.51	0.58–0.65
Size consist (%)		
50 mm	0	43–45
50–25 mm	90–97	48–49
25–15 mm	1–4	2–5
< 15 mm	1–5	1–5
Average lump size (mm)		
Length	47.5	25–75
Breadth	31.3	—
Mechanical strength (%)[a]		
DJ-15 (30 revolutions)	94–96	91.8–92.5
DJ-15 (150 revolutions)	83.5–87.0	79.0–82.0
Compressive strength (N/piece)	6600	6200[b]
High-temperature strength at 1273 K (%)	77–80	68–70
Composition of the ash (%)		
Al_2O_3	18.9	25.8
SiO_2	51.3	52.4
CaO	5.5	4.5
Ash behavior (K)		
Initial deformation	1528	1563
Hemisphere point	1618	1688
Porosity (%)	32.7	48.7
Apparent density (g/cm³)	1.31	1.00
Bulk density (g/cm³)	0.69	0.45
Moisture content after 24 hr in water (%)	8	17
Coke reactivity (%)	6.8–7.9	6–7

[a] Japanese standard.
[b] Particle size, this specimen, 30–40 mm.

A significant difference occurred in abradability. During its descent to the bottom, the proportion of formcoke in the size 60–40 mm decreased only from 97 to 88% and the shape of the briquette was essentially retained. The oven coke, however, showed a considerably greater change in shape. As a result, the average weight of a piece of formcoke in the blast area was 40 g, but only about 20 g for the oven coke. For this reason, the oven coke gave more favorable gas flow. The reactivity of the formcoke toward CO_2 was 8 times greater at the beginning than that of the oven coke. The difference decreased significantly when the cokes reached the tuyere area.

In summary, furnace operation became more difficult and irregular when formcoke was used. This behavior can be explained by the high abradability of the formcoke, which produced a large quantity of fines that impeded the burden, and by the greater weight of the lump and the low porosity of the formcoke, which resulted in slower combustion at the tuyeres.

In the years 1955–1965 blast furnace experiments were made by the Rumanian Coal Institute, with formcoke produced according to the ICEM process.[139] The experiments showed that the percentage of oven coke that can be replaced by formcoke in the blast furnace is only in the range of 30–50%. The upper limit was determined by the comparatively high sulfur content of the formcoke. In addition, it was found that the thermal and mechanical stresses were such that only a formcoke with a structure corresponding to that of a conventional blast furnace coke would be satisfactory. This similarity must extend to the microscopic structure, which means that the formcoke must pass through a plastic stage in which the molecular structure of the carbon can be rearranged.

In the U.S.S.R. formcoke produced according to the Sapožhnikov process was tested along with standard coke in the experimental plant at Kharkov, in a blast furnace having a volume of 700 m³. The percentage of formcoke was 25%, and later 50%, of the total coke charged. The results of the experiments[140] can be summarized as follows:

1. In comparison with normal oven coke, formcoke has a less ordered molecular structure, which can be explained by the kind of coal that was used in its manufacture.
2. Formcoke has pores of uniform size and thin pore walls. Because of this characteristic, the internal surface area is 1.5 times greater than that of normal coke.
3. The reactivity of formcoke is greater than that of normal coke because of the greater disorder of the material as well as the greater internal surface area.
4. Formcoke has high values for mechanical

[139] *J. Iron Steel Inst.*, **211**(10), 685–688 (1973).
[140] Ju. B. Tjutjunnikov et al., *Koks Khim.*, No. 1, 18–24 (1971).

strength; M_{40} averages 12–15% higher than for normal coke.

5. Unlike normal coke, formcoke retains its high mechanical strength down to the lower levels of the blast furnace.
6. When operating the blast furnace with 25% formcoke instead of the conventional coke, neither burning of the tuyeres nor collection of coke dust in the blast furnace was observed. Formcoke results in a burden with satisfactory gas permeability, which is indicated by the uniform operation of the furnace, reduction of the pressure drop, and the good discharge of liquid products from the blast furnace.

The British Steel Corporation[141] conducted blast furnace tests in England with two different kinds of formcoke. About 3000 tons of FMC formcoke and of BFL formcoke were used for periods of 5 days each in a blast furnace having 5.5-m hearth diameter. It became apparent that in the case of BFL formcoke, the normal operating values for blast pressure and the amount of blast could be maintained. The total carbon consumption was increased by about 3–10%, and carbon losses in the blast furnace gas were also increased. To summarize, optimum furnace operation was not achieved in the time available.

Optimum blast furnace operation was nearly achieved with 100% FMC formcoke. The results showed that the total specific fuel use was 2–5% below the values achieved with conventional blast furnace coke.

The first commercial-size blast furnace experiments using formcoke (FMC) were made in the United States in 1967 by the Armco Steel Corporation. In these experiments it was found that the strength of the formcoke had to be improved. The improved formcoke was tested in 1970 in a small experiment by Inland Steel Company and yielded satisfactory results. In further experiments in 1972 by the British Steel Corporation, with 3000 tonnes of FMC's Kemmerer formcoke, it was concluded that this formcoke could be used successfully as blast furnace fuel in the East Moors blast furnace No. 3.[141]

The most thorough experiment with FMC formcoke in the blast furnace was conducted in 1973 in blast furnace No. 5 of the Indiana Harbor plant of Inland Steel.[142] This furnace, having a diameter of 8 m, was operated at high throughput with oil blown in as fuel; it was equipped with numerous measuring instruments and was partly computer operated. To determine the standard blast furnace operation before and after the use of formcoke, an accurately controlled operation with conventional blast furnace coke was conducted. The formcoke test period was 33 days, during which the oven coke was replaced stepwise by the FMC formcoke. The furnace operated satisfactorily and produced quality hot metal. The burden remained permeable, and the blast velocity could be maintained without significant increase in the blast pressure.

Dust formation from the formcoke posed no problem either in the storage bunker or in the blast furnace gas. The formcoke briquettes retained their original shape and porosity while descending in the blast furnace. When a replacement proportion of 80% was reached, however, temperatures in the brickwork toward the top of the furnace and in the blast furnace gas increased, and the hot metal analyses and temperature varied in an unacceptable way. As the formcoke feed was stopped, temperatures returned to their normal level and the furnace again operated without difficulty. This test thus proved that formcoke can replace oven coke up to at least 50% in a medium furnace operated with high blast velocities.

The blast furnace tests with formcoke made so far do not permit an overall judgment of the feasibility of using formcoke in the blast furnace process, since the tests only lasted a few days and seldom used a formcoke percentage of more than 50%. This is not only because properties of the formcoke were not optimum in all cases, but also because the capacities of the experimental plants for producing formcoke were low.

Nevertheless it was shown that the formcoke, as now produced, is in a position to replace oven coke by up to about one-half, without creating difficulties in furnace operation. The greatest differences, compared to operation with blast furnace coke, lie in the increase in blast pressure because of the

[141] J. K. Holgate and P. H. Pinchbeck, *J. Iron Steel Inst.*, **211**(8), 547–566 (1973).

[142] F. W. Luerssen, 10th Annual Meeting and Conf. International Iron and Steel Institute, Osaka, Japan, 1976; P. K. Strangway and M. O. Holowaty, *Int. Symp. Agglomeration*, Atlanta, March 1977.

higher bulk density of the formcoke and in the increase in dust emission in the blast furnace gas due to the greater abradability of the formcoke. These disadvantages may be compensated for by the broader bases of coal for the production of formcoke.

The quality requirements for foundry coke are partly a result of the conditions in the cupola process and are associated with the required quality of the castings. During melting, the highest possible metal temperature should be reached, which is obtained by a correspondingly higher temperature in the combustion zone and sufficient contact time of the metal with the hot coke to assure carbon pickup. Of basic importance, therefore, is the low reactivity of the fuel, which reduces the proportion of the endothermic reaction $CO_2 + C = 2CO$.

Rapid melting in the cupola can be achieved only by a uniform gas flow, which is facilitated by coarse, uniform coke. A lump size above 100 mm is considered suitable.

Since coke undergoes mechanical stresses (e.g., when heavy scrap is used as feed), a high mechanical strength is required. As a lower limit, an M_{40} value of above 80% is generally required. Low abradability, characterized by the M_{10} value, not only improves gas permeation of the burden but also decreases the discharge of dust, which is detrimental to the environment. It is also important for the foundry coke to have a low ash content, because the total heat input increases and detrimental effects of the mineral matter on the quality of the casting are avoided. In addition, a low sulfur content is required. Sulfur deteriorates the castability of the metal and reduces the quality of the casting produced. Next to these physical and chemical properties, uniform quality over a long period of time is important.

The first publication on the manufacture and testing of coke briquettes for the cupola in Germany dates back to 1956.[143] Then it was reported that briquettes were prepared from coke breeze (86%), medium-volatile coal dust (6%), tar (6–6.5%), and pitch (1–1.5%) on roll presses in pillow and egg shapes and on Couffinhal presses in the shape of cubes and blocks. These briquettes were tested successfully in several foundries with cold-blast as well as hot-blast cupolas. Savings in coke charged of up to 30% and increases in melting throughputs of up to 35% were obtained.

Further tests with formcoke were made in 1967 in a cold-blast cupola of the Rheinstahl Hüttenwerke. This experiment showed that the high abradability of the coke briquettes investigated was a significant disadvantage. In addition to the problems caused by dust removal, the loss of fuel was much too high. The investigation showed on the other hand that the proportion of coke briquettes that actually reacts in the furnace is thermally very effective.

Because of the lack of sufficient quantities of foundry coke, foundries in Poland use coke briquettes to a great extent, together with the usual foundry coke. W. Kukielka[144] reported in 1966 on the influence of formcoke on the chemical composition of the castings. The influence of the percentage of coke briquettes in the charge on the content of carbon, sulfur, silicon, and manganese in the cast iron was observed for various ratios of foundry coke to coke briquettes. It was found that the carbon and sulfur content decreased with increasing proportion of coke briquettes in the charge, but silicon and manganese were not affected, or affected only slightly.

In addition to its use in the blast furnace and in foundries, formcoke is used as a fuel for small heating plants, for instance, as a household fuel. Coke for household use should definitely have a uniform size consist and high strength, since household furnaces are built to efficiently use only the uniform sizes. Smaller particles, breakage, and abrasion decrease efficiency because of increased carbon monoxide losses, and also favor the formation of clinker. Furthermore, a smokeless fuel is desired that considering the cost of the heat contained in it, should have small quantities of ballast material. Formcokes with these required properties have been produced in Germany by the BFL and Ancit processes,[135,145] and in Poland by the IChWP process.[146]

[143] W. Schreiber, paper presented at 72nd Session of the German Coke Oven Committee, March 9, 1956.

[144] W. Kukielka, *Przegl. Odlew.*, No. 2, 66–69 (1966).

[145] E. Ahland, J. Lehmann, and W. Peters, *AIME Ironmaking Proc.*, **31**, 285–296 (1972).

[146] E. Cieckiewicz et al., *Koks Smola Gaz*, **9**(6), 203–208 (1964).

CHAPTER 15

Solid Products of Pyrolysis

MICHAEL PERCH

Michael Perch *received his B.S. and M.S. degrees in chemical engineering from Carnegie-Mellon University. He joined Koppers Company in 1940 and has devoted his entire career to research in coal, coke, iron ore, and related fields. He is now Manager of the Engineering and Coke Research Section. At Koppers Mr. Perch assisted in development of the Koppers movable wall oven and carried on research that contributed significantly to the mechanism of carbonization pressure generation in coke ovens, behavior of various coals, and means of controlling pressure at safe levels in coke ovens. In connection with many new coke plants constructed by Koppers, he has made contributions to the utilization of safe blends that produced acceptable metallurgical cokes. He has expert knowledge of all the coalfields in the United States and in many foreign countries. He has supervised research projects in heat transfer in coke ovens, coke-oven design, production of active carbon from coal, production of formed coke from coal, iron ore pelletization and sintering, by-products from coal carbonization, removal of hydrogen sulfide from coke-oven gas, continuous casting of steel, and evaluation of coals for coal gasification. As chairman of two subcommittees and member of others in ASTM Committee D5 on Coal and Coke, he has contributed heavily to the standardization of many test procedures used in testing coal and coke.*

15.1 INTRODUCTION

More than 58 million tonnes of coke of all types was produced in the United States in 1973. This represented 15.2% of the world production. Only the U.S.S.R. produced more, about 21.3% of world production. Japan ranked third with 13.7%. Table 15.1[1] shows the distribution of coke production among the various countries. A grand total of 384 million tonnes was produced worldwide. About 85% of the coke produced in the United States was consumed in the blast furnace; the worldwide percentage was about 75%. A total of 55 million tonnes of coke was charged into blast furnaces in the United States, producing about 92 million tonnes of pig iron. Worldwide, about 271 million tonnes of coke was used in the production of about 493 million tonnes of iron. Although a predominant percentage of the coke production was consumed in the blast furnace, the remainder of the coke was used for such important applications as foundry coke, gas producers, industrial and residential heating, electric smelting, and iron ore sintering.

Since coke is used principally in the blast furnace, this chapter emphasizes the properties of coke that are relevant to this application, but other applications are also discussed where appropriate. Briefly, the chapter covers the chemical and physical properties of coke, the extent to which they are affected by the parent coal properties and coking conditions, the behavior of the coke in the blast furnace and cupola, and special carbonization products.

Regarding the evaluation of the properties of coke, many national standardization or-

[1] U.S. Bureau of Mines, Mineral Yearbook, (1973), pp. 433–434.

Table 15.1 Worldwide Production in 1973 of Coke of All Types[1]

Source: From U.S. Bureau of Mines, Minerals Yearbook, 1973

Producer	Million Tonnes	Percentage of Total
North America		
Canada	5.373	1.4
Mexico	1.935	0.5
United States	58.394	15.2
South America		
Argentina	0.360	0.1
Brazil	1.841	0.5
Chile	0.309	0.1
Colombia	0.590	0.2
Other	0.025	0.0
Europe		
Belgium	7.814	2.0
France	11.805	3.1
Italy	7.936	2.1
Netherlands	2.657	0.7
Spain	4.545	1.2
United Kingdom	18.000	4.7
West Germany	35.568	9.3
Other Western Europe	3.316	0.8
Czechoslovakia	11.123	2.9
East Germany	6.900	1.8
Poland	17.702	4.6
U.S.S.R.	81.700	21.3
Other Eastern Europe	3.499	0.9
Africa		
Republic of South Africa	3.703	1.0
Other	0.630	0.1
Asia		
India	9.146	2.4
Japan	52.714	13.7
Turkey	1.606	0.4
Other	0.611	0.2
People's Republic of China	26.054	6.8
North Korea	2.178	0.5
Oceania		
Australia	5.603	1.5
New Zealand	0.030	0.0
	383.667	100.0

ganizations have been active in promulgating reliable, tested methods for determining the quantitative values for the chemical and physical properties of coke. In the United States, the American Society for Testing and Materials (ASTM) Committee D5 on Coal and Coke has been very active in this area, and worldwide the International Organization for Standardization has been active through ISO Technical Committee 27. Working Group 8 of the technical committee dealt with the subject of coke up to 1968 and then was superseded by Committee SC3. There are about 24 ASTM methods[2] for coke analysis and testing, and about 26 ISO methods in the form of standards and recommendations.[3] Table 15.2 gives a listing of both designations as well as a correlation of the two. A number of years ago there was a wide diversity in the test methods for coke, as well as in the requirements or specifications for coke, but with the advent of the ISO there has been a general adoption, especially in Europe, of many of the ISO methods.

It is, of course, well recognized that unless a sample of coke for analysis or testing is taken in such a manner that it is representative of the total coke, the results of the analysis and testing are unreliable. Many theories of sampling have been proposed, but none has proved to be satisfactory in all respects. The International Standard[4] is probably the most satisfactory yet developed because it is based primarily on practical experience, including a substantial volume of experimental data collected in several countries. This method prescribes the number and minimum mass of increments, depending on the nominal upper coke size, for samples from a coke stream, wagon, ship, or stockpile that are intended for moisture determination and for physical tests. Table 15.3 summarizes these details.

The Japanese have thoroughly and systematically applied ISO sampling techniques and have advanced the technology of sampling in the field of mechanical sampling of coke from conveyor belts.[5–7] At least 16 plants in Japan now practice mechanical sampling and have presented experimental evidence that their procedures for mechanical sampling show no bias compared to manual sampling from the stopped belt, in determining total moisture, ash content, and physical properties of coke.

[2] ASTM Standards, 1975, Part 26.

[3] Current list of International Standards Organization Standards and Recommendations Relating To Coke, ISO/TC27/SC3, Document 28, 1976, British Standards Institution, London.

[4] International Standard 2309, Coke-Sampling, 1973.

[5] Outlines of Coke Sampling from Conveyor Belts in Japan, ISO/TC27/SC3, Document 33, 1976.

[6] Experiments on the Mechanical Sampling from Conveyor Belts, ISO/TC27/SC3, Document 36, 1976.

[7] One More Experimental Result on the Mechanical Sampling from Large Capacity Conveyor Belts, ISO/TC27/SC3 Document 49, 1976.

Table 15.2 Correlation of ASTM and ISO Test Procedures for Coke

ASTM Designation		ISO Designation	
D121	Definition of Terms Relating to Coal and Coke	R 1213	Vocabulary of Terms Relating to Solid Mineral Fuels, Part III, Terms Relating to Coke
D346	Sampling Coke for Analysis	ISO 2309	Coke. Sampling
D271	Laboratory Sampling and Analysis of Coal and Coke (discontinued 1975)	—	—
D346	Sampling Coke for Analysis	ISO 579	Coke, Determination of Total Moisture
D3173	Moisture in the Analysis Sample of Coal and Coke	ISO 687	Coke. Determination of Moisture in the Analysis Sample
D3175	Volatile Matter in Analysis Sample of Coal and Coke	ISO 562	Hard Coal and Coke. Determination of the Volatile Matter
D3174	Ash in the Analysis Sample of Coal and Coke	R 1171	Determination of Ash of Solid Mineral Fuels
D3177	Total Sulfur in the Analysis Sample of Coal and Coke	ISO 334	Coal and Coke. Determination of Total Sulfur. Eschka Method
		ISO 351	Coal and Coke. Determination of Total Sulfur. High-Temperature Combustion Method
D3176	Ultimate Analysis of Coal and Coke	—	—
D3178	Carbon and Hydrogen in the Analysis Sample of Coal and Coke	ISO 609	Coal and Coke. Determination of Carbon and Hydrogen by the High-Temperature Combustion Method
		ISO 625	Coal and Coke. Determination of Carbon and Hydrogen by the Liebig Method
D3179	Nitrogen in the Analysis Sample of Coal and Coke	ISO 333	Coal and Coke. Determination of Nitrogen. Semi-Micro Kjeldahl Method
D2795	Analysis of Coal and Coke Ash	—	—
D3180	Calculating Coal and Coke Analysis from As-Determined to Different Base	R 1170	Calculation of Coal and Coke Analyses to Different Bases
D2015	Gross Calorific Value of Solid Fuel by the Adiabatic Bomb Calorimeter	R 1928	Solid Mineral Fuels. Determination of Gross Calorific Value by the Calorimetric Bomb Method and Calculation of Net Calorific Value
D3286	Gross Calorific Value of Solid Fuel by the Isothermal-Jacket Bomb Calorimeter		
D293	Sieve Analysis of Coke	ISO 728	Coke. Size Analysis (Nominal Top Size Greater than 20 mm)
		ISO 2325	Size Analysis of Small Coke (Nominal Top Size 20 mm or less)
D167	Specific Gravity and Porosity of Lump Coke	ISO 1014	Coke. Determination of True Relative Density, Apparent Relative Density and Porosity
D292	Cubic Foot Weight of Coke	ISO 567	Coke. Determination of the Bulk Density in a Small Container
		ISO 1013	Coke. Determination of the Bulk Density in a Large Container
D3033	Drop-Shatter Test For Coke	R 616	Determination of the Shatter Indices of Coke
D3402	Tumbler Test for Coke	ISO 1881	Coke Not Greater than 60 mm Top Size. Determination of Mechanical Strength.
		R 556	Determination of the Micum Indices of Coke
D547	Index of Dustiness of Coal and Coke	—	—
D2677	Lightability of Barbeque Briquets	—	—
D1857	Fusibility of Coal Ash	ISO 540	Coal and Coke. Determination of Fusibility of Ash
D2361	Chlorine in Coal	ISO 352	Hard Coal and Coke. Determination of Chlorine. High-Temperature Combustion Method
		ISO 587	Coal and Coke. Determination of Chlorine Using Eschka Method
—	—	R 601	Determination of Arsenic in Coal and Coke
—	—	R 926	Determination of Nitrogen, Total Sulfur, Chlorine, and Phosphorus in Coke

15.2 CHEMICAL PROPERTIES

15.2.1 Proximate Analysis

15.2.1.1 *Moisture*

To control the operation of quenching, moisture content must be measured at the coke plant. Proper quenching obviates losses due to fires and excessive moisture additions. It is also important to determine coke moisture accurately and rapidly when the coke is weighed in a wet condition at the blast furnace, because the moisture value and the immediate compensation of coke weight are necessary for the proper thermal balance of the blast furnace. Coke moisture detracts from the carbon content of the coke burden, and its fluctuation in the burden leads to erratic operation of the furnace unless there is compensation for carbon shortfall. When reliable moisture values are not available, it

Table 15.3 ISO 2309 Standard for Coke Sampling

	Number of Increments for Coke in:			
	Stream	Wagons	Ships	Stockpile
For Moisture Sample (precision of ± 1% moisture)				
Coke debreezed	50	75	100	150
Breeze or coke <20 mm	20	—	—	—
For Physical Sample (precision of 1/10 of true mean size)				
Nominal upper size <20 mm	20	—	—	—
21–50 mm	50	75	100	150
51–100 mm	20	30	40	60
101–200 mm	10	15	20	30
Minimum Mass of Increment (kg)				
Size (mm)				
Nominal upper size <40			1	
41–80			2	
81–120			4	
>120			9	

is common practice to charge an excessive amount of coke (as much as 2–5%), to ensure that the furnace will not go cold. The recent trend of charging coke to the blast furnace on a volume basis rather than on a weight basis has meant that the moisture variation has a greater effect, thus increasing the importance of moisture measurement. Fischley[8] has found that a variation in the moisture content of coke of 4.0% can change the silicon content of the iron by 0.3%, the sulfur content by 0.04%. There is a 10–12 hr time lag between cause and effect.

Gee and Laslo[9] made a study in a plant of the mechanical sampling and measurement of coke moisture using 450-lb increments of coke, then making corrections to the coke weights to maintain steady silicon and sulfur contents in the iron as well as to maintain the iron temperature in the furnace. This study was continued with the installation of a special nuclear gauge integrated with a computer, which computed the coke moisture from the hydrogen density (measured with neutrons) and the bulk density (measured with gamma rays). Since the performance of this work, there has been a proliferation of studies and plant installations using neutron scattering[10–17] and nuclear magnetic resonance techniques.[18]

With respect to the quenching and storage of coke, the extent to which water is absorbed by the coke received little attention until Burden and Grey[19] conducted their in-

[8] J. Fischley, "Continuous Moisture Measurement and Control," *Int. Conf. Iron and Steel Making, Automation in Iron and Steelmaking*, Luxembourg, April 13–18, 1970, Centre National des Recherches Metallurgiques, Liège, Belgium.

[9] K. H. Gee and J. A. Laslo, "Study of Coke Moisture for Blast Furnace Control," *AIME Ironmaking Proc.*, **23**, 359–369 (1964).

[10] J. Dartnell, *J. Iron Steel Inst.*, **210**, 145–152 (March 1972).

[11] T. E. Reim, *Blast Furn. Steel Plant*, **57**, 465–471 (June 1969); T. E. Reim, R. F. Trenty, and J. F. Perko, *AIME Ironmaking Proc.*, **29**, 302–308 (1970); R. F. Trenty, J. F. Perko, and T. E. Reim, *AIME Ironmaking Proc.*, **31**, 265–273 (1972).

[12] J. M. van Langen, J. Maes, G. A. Flierman, and N. A. Hasenack, *J. Met.*, **20**(3) 68–74 (1968).

[13] J. Luckers, "Continuous Measurement of the Moisture Content of Coke By Means of a Neutron Probe," *CNRM (Belgium)*, **11**, 9–15 (1967).

[14] V. A. Smolyak, B. V. Shcherbitskii, N. T. Evseeva, and V. I. Vasilenko, *Steel USSR*, **4**(2), 94–96 (February 1974).

[15] E. C. Kovacs, J. O'Hanlon, and R. Hoffer, *AIME Ironmaking Proc.*, **29**, 259–264 (1970).

[16] M. S. Kudoyarov, I. P. Strokov, Yu. G. Parfenov, A. M. Nesterenko, and V. B. Korner, *Steel USSR*, **1**(9), 678–680 (September 1971).

[17] J. G. Crump, *Blast Fur. Steel Plant*, **52**, 325–327 (April 1964).

[18] W. R. Ladner, *Instrum. Pract.*, **18**(10), 1039–1043 (October 1964); W. R. Ladner, *J. Inst. Fuel*, **39**, 280–284 (July 1966).

[19] F. A. Burden and J. R. Grey, *Inst. Gas Eng. J.*, **3**(9), 533–552 (September 1963); F. A. Burden and J. R. Grey, *Gas World*, **156**(4084), 581–582 (November 24, 1962); F. A. Burden and J. R. Grey, *Steel Coal*, **186**(4941) 618–621 (March 29, 1963).

vestigation in which coke was totally immersed or exposed to the weather. In general, coke with a fine porous structure absorbed more water than other cokes. Smaller coke absorbed more water than large coke, reaching a maximum for breeze. The maximum moisture content attained was higher when the coke had an initial moisture content as a result of quenching than when it was initially dry. In stored coke the top 10–15 in. of the pile absorbed and lost water rapidly. The highest moisture content recorded for a storage pile after heavy rainfall was only just over 10%. The initial moisture content apparently was of greater importance than the rainfall in determining the moisture content of the large coke piles. It was concluded that coke of satisfactory moisture content could be recovered from the pile when attention was paid to rigid control during quenching, protecting recovery points in bad weather, and debreezing after recovery.

Ekholm and Kuchta[20] made a comprehensive study of the quenching process and reported the following findings:

1. Hot coke absorbs water more rapidly than cold coke.
2. Coke with small pores and thin walls absorbs much more water than coke with large pores and thick walls.
3. The coke moisture increases as the amount of quench water is increased.
4. The total moisture of a charge of coke being quenched increases as the bed depth increases, amounting to an increase of about 1.1% with each additional inch of coke depth.
5. Moisture increases as coke size decreases.
6. Uniformity of coke moisture is a function of the degree of uniformity with which the water was applied.
7. Variations in the depth of coke in the quench car cause serious variations in moisture content.
8. Coke moisture decreases as water temperature is increased.
9. Coke moisture is inversely related to the temperature of the coke when deposited on the wharf.

[20] L. G. Ekholm and B. R. Kuchta, *AIME Ironmaking Proc.*, **29**, 290–301 (1970).

15.2.1.2 *Volatile Matter*

The volatile matter content of coke has been criticized as a meaningless number because it has a small value that is difficult to determine with precision. Furthermore, it is usually determined at a temperature lower than that at which the coke was carbonized, therefore yielding an erroneous value for the volatile matter content. However the coke-oven expert does consider it a measure of the degree of coking. The volatile matter of coke, although obtained by accepted empirical methods, is sometimes unreliable when the test procedure is not carefully followed. When care is taken to follow the correct procedure, the results can be used to confirm the degree of coking, to check coke yields, and to provide a measure of relative reactivities.

A number of recent publications attempted to analyze the constituents of volatile matter in coke. They also suggested ways to improve the precision with which volatile matter content is determined. Queins and Schaefer[21] showed that volatile matter content determined under controlled conditions and plotted against the temperature at which the volatile matter was determined, exhibited a break at the carbonization temperature. Some of the sources of error in the determination were attributed to absorption of gases, oxidation, and surface reaction.[22] Nadziakiewicz and Heilpern[23] found in their test that moisture and some CO_2 were evolved up to 350°C and that H_2, CO, CO_2, and CH_4 were released above 700°C, with endothermic reactions occurring below 700°C and exothermic reactions above. One of the most comprehensive studies on this subject was conducted by Knauf,[24] who concluded that the volatile matter of coke comprised three groups of products: (1) products of secondary degasification, including H_2, CH_4, and CO products similar to primary gas evolution, (2)

[21] H. Queins and H. G. Schaefer, *Aachener Bl, Aufbereit. Verkoken-Briket.*, **16**(3–4), 107–172 (1966).

[22] I. Barbu, E. Falk, and G. Ursu, *Metallurgia (Bucharest)*, **19**(10), 591–594 (1967).

[23] J. Nadziakiewicz and S. Heilpern, *Koks Smola Gaz*, **12**(8), 202–208 (1967).

[24] G. Knauf, *Brennstoff-Chemie*, **47**, 69–76 (1966); G. Knauf and G. Cruson, *Freiberger Forschungsh. A*, **345**, 85–101 (1965); G. Knauf, *Freiberger Forschungsh. A*, **388**, 93–119 (1966).

products from decomposition of carbon surface complexes formed by oxidation resulting in the evolution of CO_2 and CO, and (3) occluded gases such as N_2 and CO_2, which were evolved below 300°C. The sum of the components evolved was significantly affected by oxidation of the coke and by the mineral matter present. These effects had to be taken into account if the volatile matter value was to be used to assess the completeness of carbonization of a coke. The amount of H_2 released above 700°C was a maximum for cokes of about 5% volatile matter.

15.2.1.3 *Ash*

Coke ash is derived from the mineral matter in the parent coal and is of little benefit to the coke user. In fact, it costs extra to remove it and dispose of it as slag in the blast furnace and cupola. Coke ash is a necessary evil that must be tolerated. In these metallurgical processes it reduces carbon input, lowers the iron temperature and, when excessive, increases the amount of slag. The ash content of coke varies from geographic location to location, and its amount and composition depend on the sources of the coals used in the production of the coke. Within economic limits, effort is made to minimize the amount of ash in the coke by selection and beneficiation of the precursor coals.

The ash content is under control of the coke-oven operator, who strives to keep it low and at a constant level by the suitable choice of coals and close control of the blending practice. As far as the blast furnace is concerned, it is necessary to measure the ash content of the coke being used, especially when there are changes in the amount of ash, so that process corrections can be made at the opportune time. Like moisture, ash detracts from the carbon content of the burden, and its variation can also lead to erratic operation of the furnace unless the carbon shortfall is adequately compensated.

It is therefore important to measure the ash content of coke. There are national standards as well as an international standard for the determination of ash (see Table 15.2). These methods are straightforward but are not without complication. For example, the ash content of coals and cokes with high calcium and sulfur contents can exhibit variable repeatability in the determination because of sulfation of the ash. Miyazu and Ishikawa[25] found that with such cokes the degree of fixation of the sulfur in the ash was closely related to the calcium content of the coke, not to the total sulfur content. Although sulfation of the ash is less likely with coke than with coal, Miyazu found that the organic sulfur in the coal, which is initially nonmineral, is fixed in the ash, thus resulting in high and variable ash contents for cokes made from high-sulfur and high-calcium coals. In the determination of the ash content of such coals, it is customary to heat the coal slowly in a well-ventilated furnace up to 500°C, so that the pyritic sulfur is oxidized and expelled before the calcite is decomposed and the SO_2 therefore is not picked up by the lime. This is not possible with coke, since there is no calcite present, and it is not possible to vaporize the carbon-sulfur complex by a slow heating rate up to 500°C. Miyazu and Ishikawa found, however, that incineration of the coke at 950°C (instead of 800°C) resulted in considerably less fixation of sulfur in the ash and better repeatability of the ash determination.

Brown et al.[26] found a close correlation between coke yield and the fixed carbon-plus-ash content of the proximate analysis. The formula they developed for coke yield (CY) at 900°C as a function of the percentage fixed carbon (FC) and the percentage of ash (A) was:

$$CY = 0.69(FC + A) + 25.4$$

at a standard error of ±2.0. A similar inverse relation was found between coke yield and volatile matter content. The ash content of the coke could then be calculated by dividing the ash content of the coal by the coke yield, as calculated above, with a standard error of ±1.2.

Two publications proposed formulas for estimating the mineral matter content of coke, useful for calculating the oxygen content of coke or the calorific value on a dry,

[25] T. Miyazu and K. Ishikawa, "Studies on Determination of Ash in Coal and Coke," *Symp. Science Technology Coal*, Ottawa, March 24–31, 1967, Canadian Institute of Mining and Metallurgy, Ottawa, (1967), pp. 132–148. Published by Mines Branch, Department of Energy, Mines and Resources, Ottawa.

[26] H. R. Brown, W. R. Hesp, and P. L. Waters, *J. Inst. Fuel*, **37**, 130–148 (April 1964).

mineral-matter-free basis. Barker and Mott[27] proposed a formula that was superseded by that of Leighton and Wald,[28] which was based on a statistical study of a large number of analyses. Leighton and Wald's formula, which appears to be superior, is:

$$\% \text{ MM} = \% \text{ ash} + 0.68 \text{ sulfur} - {}^{1}/_{7} \text{ iron} + 0.14$$

or if the iron content of the coke is unknown:

$$\% \text{ MM} = \% \text{ ash} + 0.51 \text{ sulfur} - 0.21$$

It is desirable, of course, to determine the ash content of coke in a short time and before the coke is used in the process, so that adjustments can be made in the burden. Including the time for sample preparation and that for ashing, a minimum of 2 hr is required. The British Carbonisation Research Association (BCRA)[29] determined that the Sheffield high-temperature method for the determination of carbon and hydrogen can be used to determine ash in coal or coke in 20–25 min. About 0.5 g of sample can be incinerated in a silica or alumina boat at 800°C in 15 min, in a 2 : 1 oxygen-nitrogen atmosphere, with results that agreed satisfactorily with the standard ash method. There was no evidence of sulfur fixation in the ash or of oxidation of the mineral matter. The British National Coal Board[30] developed the Stanley furnace for the rapid determination of ash, in which the samples of ash travel on a chain conveyor through a preheating zone, a hot zone (800–815°C), and a cooling zone. The time for the determination is about the same as for the conventional test, but a large number of samples can be processed in a normal working day. Although attempts have been made to measure ash in coal continuously using backscattered X rays or gamma rays, there has been little application to coke. Dumitrescu[31] determined the ash content of coke continuously in 30 min, but the coke samples had to be crushed before measurement. It is presumed that the coke suffers the same disadvantage as coal in that the intensity of backscatter absorption varies not only with the amount of ash but also with the composition of the ash, in particular, the iron and calcium contents.[32]

In metallurgical processes the composition of the ash of the coke is probably as important as the quantity of ash. In the blast furnace an ash containing large percentages of silica or alumina requires the use of more flux, as well as more heat, which means more coke. When there is a variation in the ash and its composition is unknown to the blast furnace operator, the result is a change in metal temperature and iron analysis. The operator then corrects the furnace by adding more coke, raising the blast temperature, or adding more flux.

The ash constituents also depend on the source of the coals used, the degree of beneficiation, and close control of the blending practice. These constituents in the coke ash are basically the same as those of the parent coal, since essentially all the ash reports to the coke on carbonization. The U.S. Bureau of Mines has published considerable ash composition data for most coals that are carbonized, as well as many others. Recent pertinent publications are those by Selvig and Gibson,[33] Abernethy et al.,[34] and Abernethy and Gibson.[35]

15.2.2 Ultimate Analysis

15.2.2.1 *Carbon and Hydrogen*

Coke is the only constituent charged into the blast furnace capable of providing the required thermal and reduction energy neces-

[27] J. E. Barker and R. A. Mott, *Fuel,* **39,** 363–365 (1960).

[28] L. H. Leighton and S. Wald, *Fuel,* **39,** 511–512 (1960).

[29] Anon., BCRA, Coke Res. Rep. No. 37, 1965, Wingerworth, Chesterfield, Derbyshire, England.

[30] H. C. Wilkinson, *COMA Yearb.,* 145–162, 1973, Waverly House, Wexborough, Yorkshire, England.

[31] T. Dumitrescu, *Metallurgia (Bucharest),* **22,** 226–230 (1970).

[32] Y. V. Stephanov and P. N. Gorelov, *Coke Chem. (USSR),* No. 10, 44–47 (1964).

[33] W. A. Selvig and F. H. Gibson, U.S. Bur. Mines Tech. Paper No. 679, 1945; W. A. Selvig and F. H. Gibson, U.S. Bur. Mines Bull. No. 567, 1956.

[34] R. F. Abernethy, M. J. Peterson, and F. H. Gibson, U.S. Bur. Mines Rep. Invest. No. 7240, 1969.

[35] R. F. Abernethy and F. H. Gibson, U.S. Bur. Mines Inf. Circ. No. 8163, 1963.

sary to produce iron; it also supports the burden column in such a way as to provide a permeable bed for the passage of the air-blast. The efficacy of coke for this task is measured by its carbon content, which is represented loosely by the fixed carbon content obtained in the proximate analysis, or carbon as determined by an ultimate analysis. All this carbon is not available for the reduction process, since a portion is needed to melt the ash of the coke, to eliminate its sulfur, and to provide the heat for the dissociation of the moisture content of the blast used for combustion of the coke.

A more practical measure of carbon for smelting was introduced by Flint[36] in the expression "effective carbon in coke." He defined this term as "the carbon remaining after setting aside the carbon required for (*a*) smelting the coke ash and sulfur and (*b*) smelting their associated fluxes"; in other words, "it is the carbon in coke available for performing outside work after the coke's own internal requirements have been satisfied." For this purpose he developed an empirical formula combining 21 burden and practice variables into a quantitative expression for calculating the effective carbon in coke. This formula of material and thermal balances has been very useful in planning and controlling ironmaking operations. A computer program for the formula was developed by Barnes and Lownie.[37] Further discussion of the formula and its usefulness are presented in a later section.

Although effective carbon content is perhaps the desired value for burden calculations, fixed carbon has been used; it is not satisfactory, however, because it can lead to high values or excessive thermal requirements. Ultimate carbon is now easier to determine, and determinations are more accurate with the new instrumental techniques. The ISO method in Table 15.2 (ISO 609) is essentially the Sheffield high-temperature method that was simplified by Radmacher and Hoverath.[38] A 60-g sample is burned at 1050°C in oxygen in a quartz tube that is unpacked except for the last third, which is packed with silver wool at 640°C to trap sulfur and halogens. The determination takes 25 min, and results have standard deviations of ±0.11% for carbon and ±0.05% for hydrogen. Ivanov et al.[39] have proposed the following formula for calculating the percentage carbon content of coke from the proximate analysis and sulfur content, determined on the dry basis:

$$\%\,C = 100 - (A + V + S + 0.55)$$

Where A, V, and S are the percentages of ash, volatile matter, and sulfur contents, respectively. It is claimed that the accuracy of the calculated value is about the same as that determined by actual analysis.

15.2.2.2 *Sulfur*

Most of the sulfur added to the blast furnace enters with the coke. Low sulfur content is, of course, desirable because high sulfur content results in irregular furnace operation, higher slag volume and higher sulfur content in the iron, or lower iron production. More sulfur entering with the burden means that more limestone must be added, which means that more coke must be used at the expense of ore, resulting in reduced iron production.

Given the growing scarcity of low-sulfur coals, sulfur is becoming an increasing problem in the operation of the blast furnaces around the world. Two recent publications[40] have pointed out the sources of low-sulfur coals in the United States. In some areas, such as England, coke sulfur content has reached seriously high levels and has become a major problem. As a result, it has become necessary in that country to make some sacrifices in the physical properties of the coke through the use of more poorly coking coals, to enable coke sulfur to be maintained at acceptable levels. Using the Flint formula mentioned in Section 15.2.2.1, it can be shown that for a blast furnace operating at a base condition, about 10 lb of coke per net

[36] R. V. Flint, *Blast Furn. Steel Plant,* **50,** 47–58, 74–76 (January 1962).

[37] T. M. Barnes and H. W. Lownie, Jr., *J. Met.,* **24,** 46–50 (February 1972).

[38] W. Radmacher and A. Hoverath, *Brennstoff-Chemie,* **41**(2), 52–54 (1960).

[39] E. B. Ivanov, V. I. Kal'chenko, and N. I. Kalach, *Metallurgia (USSR),* No. 4, 17 (1975).

[40] J. A. DeCarlo, E. T. Sheridan, and Z. E. Murphy, U.S. Bur. Mines Inf. Circ. No. 8312, 1966); F. E. Walker and F. E. Hartner, U.S. Bur. Mines Inf. Circ. No. 8301, 1966.

ton of hot metal is required for each 0.1% change in coke sulfur; where desulfurization must be accomplished by increasing the slag basicity, about 30 lb of additional coke is needed for each 0.1% change in coke sulfur; and where silica must be added to increase slag volume, about 70 lb of additional coke is needed for each 0.1% change in coke sulfur. This certainly demonstrates that the coke requirements increase substantially with high-sulfur coke when slag basicity must be practiced, or when slag volume must be increased by gravel addition. The sulfur content of the coke is therefore of utmost importance to the economics of iron making.

It has been conclusively proved that the sulfur in coke is derived from both the pyritic and organic sulfur compounds in the coal. The evidence shows that coke sulfur is derived about equally from both the organic and inorganic forms. In the coke oven both forms of sulfur begin to decompose at about 300°C, the pyrites yielding H_2S and ferrous sulfide (mostly as pyrrhotite) in the range of 400–500°C. About one-half the organic sulfur forms H_2S, and the remainder is partly volatilized and partly retained in the coke as a carbon-sulfur complex. The volatile H_2S and organic sulfur undergo further reaction during passage through the hot coke, yielding more carbon-sulfur complexes.[41]

[41] H. E. Blayden, *Coke Gas,* **20,** 26–29 (1958); Anon. BCRA Coke Res. Rep. No. 22, (1962), 31 pp.; Anon. BCRA Coke Res. Rep. No. 40, (1966), 43 pp.; H. E. Blayden and J. W. Patrick, *Fuel,* **49**(3), 257–270 (July 1970).

Pinchbeck et al.[42] published data on the distribution of sulfur in the coke in the oven, transversely across the oven as well as vertically. Rembashevskii et al.[43] also presented data for transverse sulfur distributions. Table 15.4 lists data from these two sources that exhibit retention of sulfur by the coke, in the form of a carbon-sulfur complex, as the volatile sulfur contained in the volatile carbonization products passes from the decomposition centers toward the heated walls and then vertically out of the oven. The cauliflower ends of the coke have higher sulfur contents than the interior portions of the coke, and the top of the coke charge has a higher sulfur content than the bottom of the charge.

The sulfurizing action of hydrogen sulfide on coke was demonstrated by Polansky et al.,[44] who passed H_2S at different concentrations over three cokes, at 800, 900, and 1000°C. The most reactive coke absorbed more than 3½ times and the least reactive coke absorbed about 2 times the total sulfur in the original coke. The amount of absorption was more significantly influenced by higher H_2S concentration than by reaction temperature.

[42] P. H. Pinchbeck, E. W. Nixon, and C. Riley, "The Role of Sulphur in Carbonizing and Ironmaking," *Congr. Int. Charleroi,* September 19–22, 1966, (1966), Paper No. C6. Published by European Coal and Steel Community, Luxembourg, 1967.

[43] A. G. Rembashevskii, A. M. Kisina, and V. V. Shipikin, *Coke Chem. (USSR),* No. 9, 23–25 (1960).

[44] T. S. Polansky, E. C. Knapp, and C. R. Kinney, *J. Inst. Fuel,* **34,** 245–246 (June 1961).

Table 15.4 Distribution of Sulfur in the Coke Within the Coke Oven[42,43]

	Sulfur Distribution in Half-Oven Widths (%)		
Coke Oven	Oven Wall	Midsection	Oven Center
Appleby-Frodingham	1.46	1.33	1.23
Brookhouse	1.83	1.76	1.70
Orgreave	1.48	1.44	1.36
Workington	1.57	1.43	1.25
Leningrad	0.64	0.61	0.59
Rutchenkovo	2.17	1.86	1.77
	Vertical Sulfur Distribution in 500-lb Oven		
Coal Blend	Top	Middle	Bottom
Orgreave	1.47	1.44	1.41

Pinchbeck[42] also reviewed the extent of sulfur retention and the formulas derived for the prediction of coke sulfur. For commercial cokes the well-known Lowry and the Blayden-Mott relationships are widely accepted and give virtually the same results. The following formula, newly proposed by the British Carbonisation Research Association, also gives comparable results for commercial cokes:

$$\% S \text{ (coke)} = 0.80\% S \text{ (coal)} + 0.064$$

Pinchbeck demonstrated that the scale of carbonization was an important factor; he showed that for one coal blend, the Gray-King assay (20 g) gave a sulfur retention in the coke of 73.5%, the Koppers test (80 g) gave a sulfur retention of 80.0%, and 500-lb pilot oven and full-scale oven tests yielded retentions of 82.0–82.5%.

Riley[45] studied the effect of carbonizing temperature in the range of 900–1600°C and proposed the formula:

$$R = 137 - 0.054t$$

where R = percentage of sulfur retention and t = final coke temperature (°C). Pinchbeck extended the temperature range to 2850°C and found that thermal elimination of sulfur progressively continues with increasing temperature, and is almost complete at 2500°C. The relationship is shown graphically in Fig. 15.1.

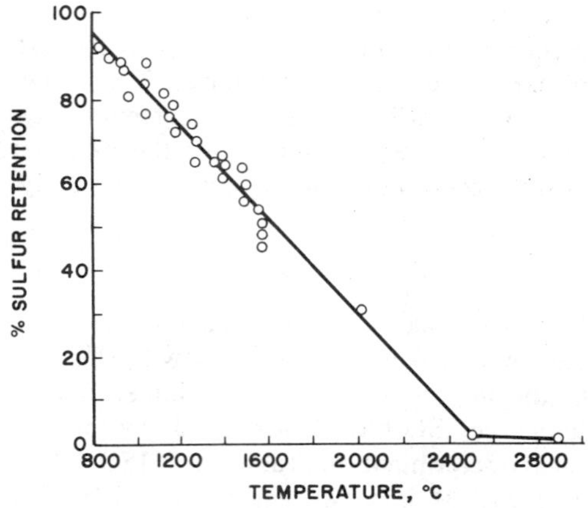

Figure 15.1 Effect of carbonizing temperature on sulfur retention in coke.[42]

Because of the importance of sulfur in the economy of ironmaking, efforts were made to reduce the sulfur in coke. In addition to the selection of lower sulfur coals, blending of low-sulfur carbon additives, and desulfurization of the coals by physical beneficiation or chemical treatment, attempts were made to desulfurize during and after carbonization. During carbonization desulfurizing gases such as hydrogen, steam, and ammonia have been tried, but these were found to be impractical because of the large volumes of gases required. Epure[46] added lime to the coal blend and afterward gave the coke a steam treatment under pressure, or quenched the coke to a high moisture content, then dried it under pressure. In both cases he claimed sulfur removals of up to 50%. No commercialization of either technique has been announced.

Probably the most active area of study has been the high-temperature treatment of coke. The work by Riley and by Pinchbeck in the United Kingdom was briefly described above. In addition, the BCRA[47] conducted pilot oven studies up to temperatures as high as 1600°C to study the effect of temperature on oven productivity and coke quality, in addition to the effect on sulfur. It was found that although coke sulfur levels could be significantly lowered, there was a considerable increase in productivity (2–3 times), but at a considerably increased rate of heat consumption (4 times). Similar research was conducted in the U.S.S.R. Kazmina[48] heated cokes to temperatures of 1300–1600°C. He found that 40–50% of the sulfidic sulfur is removed at 1100°C and 50–70% at 1500°C; he also found that about 10% of the organic sulfur was removed at 1100°C and 20–30% at 1600°C.

The use of higher carbonizing temperatures, up to 1425°C, would involve no unusual refractories in the coke oven, except perhaps the use of silica in the regenerators. This, according to the information given in Fig. 15.1, would amount to a sulfur retention

[45] H. L. Riley, *COMA Yearb.*, 268–280 (1962); Waverly House, Wexborough, Yorkshire, England.

[46] I. Epure, *Rev. Electrotech. Energ. (Rumania), Ser. B7*, 149–160 (1962); *Ser. B8*, 291–300 (1963).

[47] Anon., BCRA Carbonization Res. Rep. No. 12, 1974.

[48] V. V. Kazmina, *Coke Chem. (USSR)*, No. 6, 25–28 (1971).

of about 70%. Higher temperatures to attain higher sulfur removals would entail the use of expensive, exotic refractories. This, however, may not be necessary if current development of new technology for the external desulfurization of iron proceeds successfully.

15.2.3 Calorific Value

Not much has been published on calorific values for cokes since the review by Thibaut in the first Supplementary Volume.[49] Since then, the ISO has standardized the method for its determination (Table 15.2: ISO 1928). The ASTM has two methods, D2015 for determination by the adiabatic bomb calorimeter, and D3286 for determination by the isothermal-jacket bomb calorimeter. It has been recognized that it is more difficult to achieve accurate determination of the calorific value of coke than of coal, because coke is much less reactive. Special precautions are necessary to ignite the coke more readily and to burn it to completion. ISO documents of Working Group 8 of Technical Committee 27 (Nos. 188, 189, and 193) report these difficulties and describe a solution that is reproducible within 30 cal/g. Report No. 193 presented gross calorific values for seven foundry cokes that ranged from 7896 to 7969 cal/g on an maf basis; these cokes ranged in volatile matter contents from 0.7 to 1.5% and in ash contents from 6.4 to 8.5%. Spiers's latest book[50] lists gross calorific value of high-temperature coke calculated to the same basis as a range of 7200–8200 cal/g.

Brown[26] found no significant correlation between the heating value of coke and that of the parent coal, but did find a fairly high correlation between the calorific value (CV) of coke and its carbon content, which depended somewhat on the composition of the coal, but more on the carbonization temperature. He presented the formula:

$$CV = 141.7 \text{ carbon (maf)} + 538\ (\pm 237)$$

which had a correlation coefficient of 0.73.

[49] Ch. G. Thibaut, "Solid Products of Carbonization," in *Chemistry of Coal Utilization*, Supplementary Volume, H. H. Lowry, Ed., John Wiley and Sons, New York, 1963, Ch. 12, pp. 494–538.

[50] H. M. Spiers, *Technical Data on Fuel*, 6th ed., British National Committee, World Power Conference, London, (1962).

15.2.4 Other Properties

15.2.4.1 *Reactivity*

There have been numerous publications on reactivity since this subject was reviewed comprehensively by Thibaut[49] in the first Supplementary Volume. As in the past, these publications have reported on laboratory tests conducted at arbitrarily selected temperatures and using, in most cases, a CO_2 atmosphere. The significance of the results of such tests is moot in the light of Thibaut's review, further analysis by Thibaut in a later publication,[51] and another analysis by Marsh.[52] Marsh stated:

> As the coke falls through the stack of the blast furnace, its temperature, topographical structure, porosity, surface concentration of mineral matter, and its ambient gas composition with component diffusion coefficients all change progressively with interdependence. For example, the extent of diffusion control of rate of carbon gasification is a function of temperature, pore size distributions, gas compositions, and surface chemical activity. These important parameters change simultaneously. The term "coke reactivity" is thus a continuously changing, multidependent parameter and can have as many values as there are small pieces of coke in the blast furnace.

Marsh describes the complex chemistry of the gasification of coke by carbon dioxide in the presence of carbon monoxide and nitrogen in the blast furnace as follows:

> With a fixed coke condition, the reactivity is a function of particle size, temperature, and concentrations of nitrogen, carbon monoxide, and carbon dioxide. As the temperature of gasification changes, so do the parameters controlling the dependence of rate upon gas composition. As gasification proceeds and coke porosity develops, including macropore, transitional pores, and microporosity, then all the parameters relating rate to the above functions will change. Over and above these considerations there exist the influences of catalysts present as impurities in the

[51] C. G. Thibaut, *Centre Doc. Sider. Circ.*, No. 9, 1947–1965 (1963).

[52] H. Marsh, *J. Iron Steel Inst.*, **211,** 334–345 (May 1973).

coke. These influences are dependent upon the chemical composition of the catalyst, the shape and size of its physical form, and the position of the catalyst in terms of the porosity of the coke. By enhancing chemical rate, it can alter the overall balance between chemical and physical control of rate and bring about quite different chemical differences.

Although there have been many publications on the subject of reactivity, it is significant that there is no ASTM or ISO standard for its determination. However a group of experts under the sponsorship of the United Nations Economic Commission for Europe (ECE) drafted a method that was published and received some international attention.[53] This method is really that originally proposed by Koppers and Jenkner[54] with modifications by Dahme and Junker,[55] Hedden,[56] and Peters.[57] The test was said to give reproducible results and was able to distinguish between cokes made from different blends, temperatures of carbonization, and coking times, provided CO_2 gas flow and temperature of the test were controlled accurately. Reactivity was expressed in terms of a coefficient of reaction speed in cubic centimeters of CO_2 per gram per second for the reaction of carbon with the CO_2, producing 2CO.

Extensive testing with the modified test procedure has been conducted by the German Cokery Commission, and the results with 67 coke samples have been described by Peters[57] This test has found useful application in several countries. For example, the British Steel Corporation has adopted its use as a nonroutine test. Nixon and Brooks[58] have determined a significant linear relationship between this ECE reactivity test and the results of the microstrength test of Blayden et al.,[59] thus determining reactivity indirectly by performing the simpler microstrength test.

Among national standards, only the United Kingdom Critical Air Blast (CAB) test (BS 1016) and the U.S.S.R. test (GOST 10089) can be cited. The CAB test is really a combustibility test that was designed for testing domestic cokes. It is less suitable for testing blast furnace cokes because it simulates conditions under which coke is burnt in an open grate or in a closed stove, and combustion takes place at a relatively low temperature and with a limited supply of excess air.[60] The Soviet standard has been criticized by Kovalevskaya,[61] who contends that the specified testing of the same sample of coke at four successive temperatures (950, 1000, 1050, and 1100°C) is a serious failing of the test. Rate constants are difficult to calculate because the procedure does not specify the means for estimating the burnoff accurately at each temperature.

Another test that should be mentioned is the CORMA (Committee on Metallurgical Research) test in Norway,[62] which was really developed for the more reactive cokes used in electric smelting. In this test the flow of CO_2 to the coke bed is adjusted to give a definite composition of reaction gases, and the rate of flow, instead of the composition of reaction gases, is taken as a measure of reactivity. It is claimed, therefore, that the inhibiting effect of higher concentrations of CO on further conversion of CO_2 to CO is obviated by this procedure, thus giving more reliable values for highly reactive cokes.

The results of reactivity tests appear to be most useful in the characterization of cokes. Echterhoff et al.[63] reported the results of

[53] United Nations Economic Commission for Europe, *Method of Measuring Reactivity of Metallurgical Cokes,* 65, II, E-MIM22, ST-ECE-COAL-12, United Nations, New York, 1965.

[54] H. Koppers and A. Jenkner, *Arch. Eisenhuttenw.,* **5,** 543–547 (1932); H. Koppers and A. Jenkner, *Fuel,* **11,** 416–421 (1932).

[55] A. Dahme and H. J. Junker, *Brennstoff-Chemie,* **36,** 193–224 (1955).

[56] K. Hedden, *Brennstoff-Chemie,* **41,** 193–203 (1960).

[57] W. Peters, *Glueckauf,* **96,** 997–1006 (1960); *Brennstoff-Chemie,* **42,** 84–90, 323–328 (1961).

[58] E. W. Nixon and S. H. Brooks, "Effects of Chemical and Physical Properties of Coke on Blast Furnace Performance," *Proc. Conf. Coke Ironmaking,* London, December 10–11, 1969, Iron and Steel Inst., London, Publ. No. 127 (1970), pp. 105–112.

[59] H. E. Blayden, W. Noble, and H. I. Riley, *J. Iron Steel Inst.,* Advance Copy No. 2 (1937), 16 pp.; Advance Copy No. 3 (1939), 16 pp.

[60] T. Smith and H. C. Wilkinson, *J. Inst. Fuel,* **35,** 220–224 (May 1962).

[61] M. M. Kovalevskaya, V. I. Borsuk, V. G. Balanov, and D. D. Vorobev, *Coke Chem. (USSR),* No. 4, 14–20 (1969).

[62] Committee of Metallurgical Research, Royal Norwegian Council for Scientific and Industrial Research, Communication No. 18, 2nd rev. ed., Trondheim, Norway, 1964.

[63] H. Echterhoff, K. G. Beck, and W. Peters, *Proc. Blast Furn., Coke Oven, Raw Mat. Conf., AIME,* **20,** 403–418 (1961).

ECE reactivities for a number of cokes and found that reactivity was a function of the final coke temperature, not of the rate of carbonization. Reactivity decreased with increased coke temperature, and Echterhoff showed a parallel relationship between hydrogen content and final coke temperature. He concluded, therefore, that reactivity and hydrogen content are clearly related for the same feed coal.

Thompson et al.[64] applied a simple reactivity test to relate coke quality to coke-oven operation. Weight-percent loss of a 50-g, 18 × 40 mesh coke sample reacted for 2 hr at 1825°F (1000°C) was correlated with the maximum temperature of exposure and to coke-oven flue temperature. Reactivity was found to be an additive property for cokes made from coal blends, in proportion to the reactivities of the individual cokes. Reactivity of coke bore a fairly good relationship to the petrographic composition of the coal and was specifically related to the organic inert content of the coal that did not melt during the coking process. The reactivity test was applied to full-scale operation in three plants and was found to be an excellent tool for monitoring coke quality uniformity and quickly arriving at the variables in coke plant operation that led to nonuniformities in the coke.

Schapiro and Gray[65] made reflectance measurements of the polished surfaces of cokes and found a strong relationship between the structure of coke, as defined by pore volume, and the reactivity of the coke. Low-rank coals produce cokes having the highest pore volumes and the highest reactivities. As coal rank increases, the pore volume and the reactivity decrease to minimum values. High-rank coals produce cokes having intermediate pore volumes and reactivities.

15.2.4.2 *Combustibility*

Although reactivity is normally related to the reaction of carbon with carbon dioxide, the reaction of carbon with oxygen or air is referred to as oxyreactivity or combustibility. In general, these properties bear similar relationships to each other with different types of carbon. A combustibility test was described in Section 15.2.4.1, "Reactivity." This CAB test, useful for domestic cokes, has also been used by Smith and Wilkinson[60] to test a number of different types of coke, including blast furnace and foundry cokes. Figure 15.2 shows relationships among seven widely different carbonaceous materials as a function of mean size; the more reactive materials have lower CAB values. The same tool was used by Lee[66] to study the effect of the addition of activating materials to coke-oven charges on the coke reactivity. The addition of as little as 0.4–2.0% sodium carbonate or potassium carbonate reduced the CAB values significantly. The addition of increasing quantities of peat to the coal charge, up to 20%, also caused a small but distinct reduction in the CAB values.

Adair[67] gasified a blast furnace coke with oxygen to different degrees of burnoff, at temperatures of 820–1200°C, and studied the surface topography using a scanning electron microscope. The gasification process caused pitting of the previously smooth surfaces and deep fissuring. Pitting occurred at the lower gasification temperatures and fissuring at the higher temperatures or higher degrees of burnoff. It was concluded that drastic pitting and fissuring could reduce the abrasion resistance and strength of the coke passing down through the blast furnace.

Rebaudières and Guerin[68] studied the reactivity of metallurgical cokes to oxygen, with particular emphasis on the conditions under which the reactivity was determined solely by the chemical reaction rate. At 450°C, as the extent of combustion increased, the particle diameter below which the chemical reaction rate was the controlling factor was larger and larger. At 525°C, however, chemical reaction control did not occur even for the smallest particles tested, and at 600°C the reaction rate was entirely controlled by the rate of diffusion within the pores of the coke. Defining the absolute oxyreactivity as

[64] R. R. Thompson, A. F. Mantione, and R. P. Aikman, *Blast Fur. Steel Plant,* **59,** 161–177 (March 1971).

[65] N. Schapiro and R. J. Gray, *Blast Furn. Steel Plant,* 273–280 (April 1963); N. Schapiro and R. J. Gray, *Prepr., Am. Chem. Soc., Div. Fuel Chem.,* **6**(2) 256–267 (September 1962).

[66] G. W. Lee, *Coke Gas,* **23**(269), 398–405 (October 1961); **23**(270), 442–452, 469 (November 1961).

[67] R. R. Adair, *Fuel,* **51**(1), 57–63 (1972).

[68] P. Rebaudières and H. Guerin, *C.R. Acad. Sci. Paris,* **259,** 4639–4642 (1964).

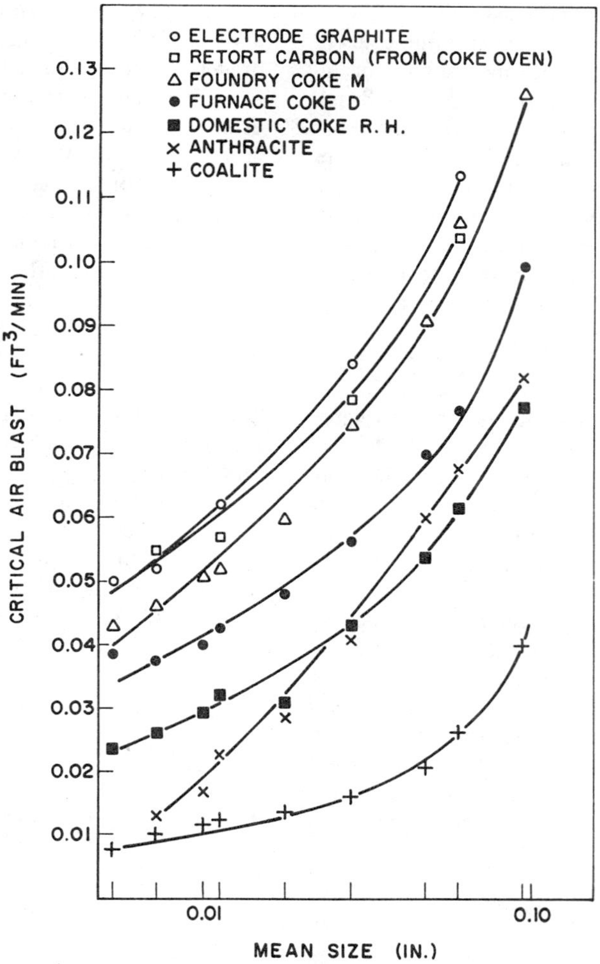

Figure 15.2 Relationship between CAB and mean size of coke.[60]

the reactivity per unit internal surface, it was found that this value was the same for all six cokes studied, indicating that the solid state structures were basically similar, although they appeared to have different textures and apparent reactivities. Wiesmann[69] made a similar but more comprehensive study. The surface area increased with gasification with air to a maximum at 20–40% combustion, then decreased; the greatest increase in surface area was found with cokes prepared at 800–1000°C. the rate of carbon gasification per unit surface area was high at first, then decreased to a constant value. This gasification rate per unit area at the maximum surface area decreased as the coking temperature increased, approximately following the reduction in hydrogen content in the temperature range 600–1000°C but occurring more rapidly between 1000 and 1500°C.

Another combustibility test involves the determination of the ignition temperature by the crossing-point method. Filonenko and Frolova[70] determined the ignition temperature of cokes prepared in the range of 600–1000°C and found the ignition temperature to increase with carbonization temperature, the increase being smaller with lower rank coals. The ignition temperature of cokes from blends could not be calculated additive-

[69] U. Wiesmann, *Brennstoff-Chemie*, **47**, 170–198 (1966).

[70] Yu. Ya. Filonenko and L. A. Frolova, *Khim. Tverd. Topl.*, **8**(5), 105–108 (1974); English Translation, *Solid Fuel Chem. (USSR)*, **8**(5), 84–87 (1974).

ly from the individual coals. Reduction in coke porosity by compression of the coals or by preheating caused an increase in the ignition temperature. Longanbach and Sass[71] determined ignition characteristics of chars from gasification and liquefaction processes and found that for these materials, the ignition temperatures were in the range of 400–500°C and were equal to or lower than those of the parent coals. Preoxidation of the chars raised the ignition temperatures.

15.2.4.3 *Electrical Resistivity*

Since electrical resistivity is closely related to reactivity and to the chemical nature of coke, the subject is treated here rather than in the section on physical properties. Resistivity is a measure of the disruption of the graphitic structure of the coke caused by pores, ash minerals, interfaces between coal inerts and coke, and foreign atoms such as hydrogen, oxygen, and sulfur. Electrical resistivity is one of the important properties for electrothermal or electrosmelting processes. High values of reactivity and resistivity are required. Normally coke is prepared for this purpose by using low-rank coals, low carbonization temperatures, or certain additives.

[71] J. R. Longanbach and A. Sass, *Fuel*, **54**(1), 29–33 (1975).

Gryaznov et al.[72] showed that resistivity and reactivity are conjugate properties, and presented the relationship of both properties to coal rank as shown in Fig. 15.3. Strakhov et al.[73] tested various coals at finish temperatures of 800–950°C and found that increasing the proportion of gas coals in the blend to the oven sharply increases the reactivity of the coke, and to a lesser degree the resistivity, which is also a function of the finishing temperature. To produce cokes of satisfactory strength, and satisfactory reactivity and resistivity properties, he recommended that coal charges contain 60–80% poorly caking coals and 10–15% strongly caking coals, and that they be carbonized to a temperature of 800–850°C.

An extensive fundamental study was conducted at the Australian Commonwealth Scientific and Industrial Research Organization (CSIRO) on electrical conduction in coals and cokes.[74] Measurements were made of the electrical resistivity of coal during carboni-

[72] N. S. Gryaznov, G. I. Nechaeva, and V. I. Sukhorukov, *Coke Chem. (USSR)*, No. 4, 15–19 (1973).

[73] V. M. Strakhov, V. S. Arkhipov, and V. G. Mizin, *Coke Chem. (USSR)*, No. 5, 32–35 (1966).

[74] *Coal Res. CSIRO*, No. 17, 2–6 (August 1962); No. 32, 2–5 (July 1967); P. L. Waters, *Proc. 5th Carbon Conf.*, Pennsylvania State University, University Park, June 19–23, 1961, Vol. 2, 1963, pp. 131–148.

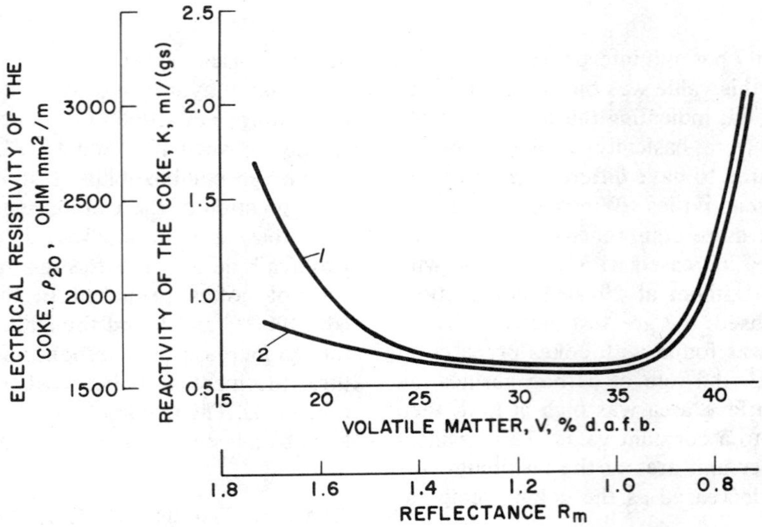

Figure 15.3 Variations in (1) electrical resistivity and (2) reactivity of the coke in relation to the rank of the coal carbonized.[72]

zation, and the results were interpreted in terms of the resultant changes in physical and chemical constitution. The resistivity of coke depends mostly on carbonizing temperature. The logarithm of resistivity is directly proportional to hydrogen content and linear shrinkage, and it is related to other properties in a complex way. It was explained that coal and coke behave as extrinsic semiconductors, where the main current conductors are π electrons from the basic polycyclic aromatic structures as well as "impurity" current carriers from free radicals. The progressive loss during carbonization of hydrogen atoms from the edges of the aromatic structures causes the graphitic crystallites to move closer together, thus improving current conduction from one crystallite to another.

The property of the electrical resistivity of powdered samples of coke was proposed by Galpern et al.[75] as a means of monitoring the degree of carbonization of coke in the ovens. They showed a linear relationship between coal composition and the logarithm of resistivity and found that there can be a twentyfold range of resistivity values in coke as a result of deviations in production conditions.

15.3 PHYSICAL PROPERTIES

15.3.1 X-Ray Diffraction

Lermusiaux[76] studied the structure of coke for a number of years, using the technique of prolonged X-ray exposure to produce the weaker diffraction rings. He found that metallurgical cokes had a basic structural unit consisting of two carbon hexagons separated by a distance of 3.4 Å. Macromolecules were formed by connection of these C_{12} units with C_2 radicals. At very small angles diffraction gave a mean radius for these macromolecules of 15 Å at 900°C and 20 Å at 1500°C. These C_{12} units were present in the coal in amounts that increased with the rank of the coal. The bitumen in the coal was responsible for fusion, swelling, and the formation of the C_2 linkages.

It has long been recognized that X-ray diffraction analysis can provide information on the structural changes that coals undergo during carbonization. In the U.S.S.R. Sklyar[77] carbonized four coals of different ranks, at temperatures from 300 to 1000°C, and determined the crystallite characteristics of the cokes by X-ray diffraction. The percentage of carbon in the crystallites increased with increasing temperature, the crystallite height (c-dimension) increased to a maximum at 500–600°C and then decreased, and the interlayer spacing decreased in this temperature range in an amount approximately equal to the increase in crystallite height. For the higher rank coals there was a second increase in crystallite height at 750–900°C. Sklyar proposed the following mechanism on the basis of this X-ray examination: (1) partial breakdown of the coal macromolecule up to 300–350°C, followed by extensive breakdown of the coal, up to a temperature of 600°C, with the formation of two-dimensional crystallization centers, and (2) mutual orientation of adjacent carbon layers, with initiation of crystallite growth up to 700°C and further orientation and extensive growth of crystallites above that temperature.

In the United States Schehl and Friedel[78] studied a high-grade metallurgical coke carbonized at different temperatures in the range of 500–1000°C. The calculated mean layer diameters by profile analysis correlated linearly with carbonization temperature and ranged from 6.5 Å for the 600°C coke to 14 Å for the 1000°C coke. Carbonization temperature had little influence on the stack height, which had an average of 14 ± 1 Å for the six cokes.

15.3.2 Optical Properties

Patrick and co-workers at the BCRA[79] studied the development of optical anisotropy in coke during its formation. Vitrains that were handpicked from a variety of coals having volatile matter contents of 10–37% and car-

[75] V. V. Galpern, S. I. Pinchuk, and A. G. Starovoit, *Coke Chem. (USSR)*, No. 8, 18–22 (1973).

[76] A. Lermusiaux, *Chim. Ind.*, **92**, 217–231 (1964).

[77] M. G. Sklyar, *Coke Chem. (USSR)*, No. 1, 20–24 (1973).

[78] R. R. Schehl and R. A. Friedel, U.S. Bur. Mines Rep. Invest. No. 7888, 1974.

[79] J. W. Patrick, M. J. Reynolds, and F. H. Shaw, *Fuel*, **52**(3), 198–204 (1973); Anon., BCRA, Coke Res. Rep. No. 75, 1972; Carbonisation Res. Reps. Nos. 13, 1975; and 25, 1976.

bonized at 370–1000°C were examined by polarized light microscopy, and point counts were made to determine the extent of isotropic and anisotropic areas, the latter broken down into fine grain, medium grain, coarse grain, and flow type. This study, which was conducted for several years, resulted in the following conclusions:

1. Caking coals generally produce fine-grain mosaic anisotropy, whereas the strongly caking coals form only medium-grain mosaic anisotropy, and weakly caking coals only isotropic.
2. With a decrease in coal rank, there is an increase in the coking temperature at which anisotropy is first detected, and there is a decrease in the extent and grain size of the mosaic anisotropy formed.
3. These results are explained as a function of coal plasticity and loss of volatile matter, and differences in the coal structures account for the variations found in optical anisotropy.

Similar studies were made by Goodarzi and Murchison[80] at the University of Newcastle-upon-Tyne, using reflected light microscopy on vitrinites from three coals—low-rank bituminous, coking, and anthracitic—carbonized slowly to various final temperatures (300–950°C). Maximum and minimum reflectances in air and oil were determined, as well as refractive and absorptive indices. With increasing temperatures, the reflectance of the low-rank and coking vitrinites increased slowly up to 400°C and rapidly at higher temperatures; for the anthracitic vitrinite the reflectance began to increase rapidly only above 500°C. In general, the absorptive index followed reflectance as temperature increased, but the refractive index reached a maximum value at 600°C, then decreased continuously at temperatures up to 950°C. All three vitrinites exhibited similar maximum and minimum reflectances at 950°C, demonstrating that under the carbonizing conditions imposed, even the low-rank vitrinite became highly anisotropic. Preoxidation had little effect on the reflectance of the anthracitic residues, but for the other two, reflectance increased up to 500°C, then decreased at the higher temperatures. The coking-coal vitrinite residues had good crystallite orientation with weak cross-linking; the other two residues exhibited strong cross-linking and poorer ordering.

Bayer and Denton[81] applied the technique of coke microscopy to practical coke plant operation. They established a basic relationship between the reflectance of coke cell walls and coal reflectance. Reflectance counts on the coke were used to calculate the proportion of different coals in the blend used. The efficiency of mixing of the coals, as well as their distribution and compatibility, could be determined. The presence of any oxidized coal in the blend could be detected. Coke reflectance was related to its reactivity, the final carbonization temperature, and carbonization pressure.

15.3.3 Pore Structure

Fundamental studies at the BCRA[82] have shown that the pore structure of coke is largely determined within the plastic range during carbonization and that the mechanism can be divided into four phases: (1) nucleation of the pore within the coal particle, (2) intraparticulate pore growth followed by swelling and fusion when interparticulate voids are completely filled, (3) postfusion pore growth to a maximum size, and (4) a decrease in pore size resulting in compaction of the structure near the resolidification temperature. Figure 15.4 presents a typical plot of the number of pores per centimeter traverse, the mean pore size, and the mean pore wall size versus the carbonization temperature. As can be seen from this figure, both the pore size and the pore wall size attain maximum values within the plastic zone. An increase in coal charge density results in a decrease in coke porosity, which is associated largely with the magnitude of the mean pore size, since pore wall size varies negligibly with the charge density. The pore structure of coke is not significantly influenced by the particle size distribution of the coal.

Toda et al.[83] made extensive studies of

[80] F. Goodarzi and D. G. Murchison, *Fuel,* **51**(4), 322–328 (1972); F. Goodarzi and D. G. Murchison, *Fuel,* **52**(3), 164–167 (1973).

[81] J. L. Bayer and G. H. Denton, *Blast Furn. Steel Plant,* **54**(12), 1133–1142 (1966).

[82] Anon., BCRA Carbonisation Res. Reps. Nos. 9, 1974, and 22, 1975.

[83] Y. Toda, *Fuel,* **52**(1), 36–40 (1973).

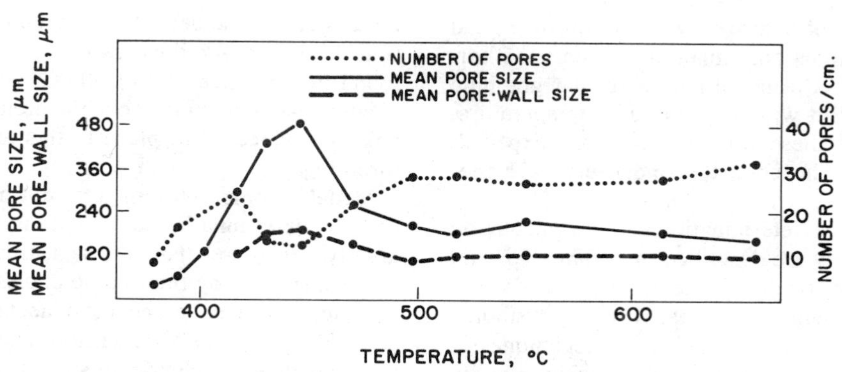

Figure 15.4 Variation of pore structure with carbonization temperature.[82]

changes in the macropore and micropore volumes of various coals during carbonization, with heat treatment up to 1200°C. Pore volumes calculated from the difference between reciprocals of densities measured in mercury and in n-hexane were almost equal to pore volumes measured by mercury penetration. It was concluded that micropores and transitional pores with entrance diameters larger than the diameter of the n-hexane molecule were not produced by heat treatment. The pore volumes for caking coals changed rapidly within the plastic range reaching a maximum at around 400°C; for noncaking coal no peaks were attained over the whole temperature range.

At the Institute for Chemical Processing of Coal in Poland (IChPW), Heilpern[84] observed the effects of a number of factors on the pore structure of coke. He found that micropore volume was reduced by an increase in the carbonization blend of the amount of coal with a viscous plastic stage, by an increase in the final carbonizing temperature, by an increase in coking time, and by a secondary thermal stabilization of the coke. An increase in coking time reduced the number of large micropores (4–7.5 μm) and significantly increased the number of small micropores (0.1–4 μm). The macropore volume (7.5–7500 μm) decreased as the Gieseler fluidity of the coal increased.

Filonenko[85] examined the changes that occur in the pore structure of coke as it descends through a blast furnace. Samples were taken at five levels from the throat to the tuyeres. The denseness of the pore structure decreased during the descent, with the proportion of larger pores ($>$ 0.5 mm) increasing and that of smaller pores ($<$ 0.1 mm) decreasing. The pore wall thickness decreased steadily, as did the structural strength of of coke. Laboratory tests on lumps of coke gasified with CO_2 for 20 min at 1000°C confirmed the results above, but also showed that with increasing depth from the surface of the lump, the porosity decreased and the pore wall thickness increased. These changes, however, extended only to a depth of about 15 mm. Coke of low porosity was consumed to a shallower depth than coke of high porosity.

15.3.4 True and Apparent Specific Gravity

The true and apparent specific gravities of coke are important properties, and generally are used in the calculation of porosity. The apparent specific gravity varies with the rank of coal carbonized, the bulk density of the coal charge in the oven, the carbonization temperature, and the coking time. The true specific gravity is mainly affected by the carbonization temperature and coking time, as well as the ultimate composition. The

[84] S. Heilpern, *Koks Smola Gaz,* **14**(7–8), 191–200 (1969); S. Heilpern and P. Wasilewski, *Koks Smola Gaz,* **16**(9), 230–236 (1971); **16**(10), 268–275 (1971).

[85] Yu. Ya. Filonenko, S. V. Vasilev, and L. A. Khasan, *Solid Fuel Chem. (USSR),* **8**(6) 43–45 (1974); Yu. Ya. Filonenko, V. P. Koloskova, V. M. Pronina, and E. S. Chebysheva, *Steel USSR,* **3**(9), 718–719 (1973); Yu. Ya. Filonenko, S. V. Vasilev, and P. G. Muravkov, *Coke Chem. (USSR),* **1,** 28–29 (1972); Yu. Ya. Filonenko, S. V. Vasilev, L. F. Bakhtarov, I. M. Dorofeev, and P. G. Muravkov, *Coke Chem. (USSR),* **8,** 24–26 (1972).

presence of hydrogen, oxygen, nitrogen, and sulfur atoms and minerals prevents the carbon atoms from forming a regular graphitic lattice. But with an increase in temperature, some of these foreign atoms are expelled, allowing the coke to condense to a higher density.

Both the determination and the calculation of porosity are described in ASTM D167 and ISO 1014 standards, and involve use of water displacement techniques. Kijewska[86] studied the use of various liquids and techniques in the determination of true specific gravity. Measurements were made using ethyl alcohol, water, methyl alcohol, gasoline, tetralin, benzene, and xylene as pycnometer liquids. The air from the cokes was eliminated by vacuum or boiling treatments. It was claimed that the best results were obtained by heating coke of less than 0.2 mm in methyl alcohol. The results obtained for two low-temperature and two high-temperature cokes are given in Table 15.5. Considerable differences were found with the use of the different liquids, and these differences were explained on the basis of the chemical and physical behavior of the cokes tested with respect to the liquids.

Kijewska also presented the relationship (Fig. 15.5) between true specific gravity and the volatile matter content of coke. A similar relationship was reported by Zielinski[87] between density and carbonization temperature (Fig. 15.6). A marked increase in density was found in the range of 500–800°C, which indicated that maximum consolidation of the coke substance takes place in this temperature range.

Medek[88] compared four different methods for the determination of apparent specific gravity. He found that the measurement of water displaced by the sample enclosed in a fine-mesh cage gave the most accurate results. Measurement of dimensions and weight of prismatic specimens gave similar, but less accurate values. Soaking the coke in boiling water, followed by cooling and weighing in water and measuring the water displaced by a wax-impregnated coke, gave consistently high values.

Brown et al.[26] reported that apparent specific gravity (ASG) and porosity were highly dependent on volatile matter (VM) and could be calculated from the following equation:

$$ASG = 1.327 - 0.013\,VM \text{ (air dried)}$$

with a standard error of $\pm$ 0.076.

15.3.5 Bulk Density and Permeability

Although the weight per cubic foot, or bulk density of coke, is important as a measure of the amount of coke being charged or as a measure of void volume, it is not determined directly to any great extent, especially in the United States, because of the difficulty of

[86] A. Kijewska, *Koks Smola Gas,* **5**(1), 3–9 (1960); ISO/TC/WG8 Document 136, Determination of True Specific Gravity of Coals and Low and High Temperature Distillation Cokes, 1960.

[87] Henryk Zielinski, *Present Methods of Coke Manufacture,* (1967). Translation from Polish by J. Bachrach, NTIS TT70-55025, 1972.

[88] J. Medek, *Freiberger Forschungsh. A,* **345,** 73–83 (1965).

Table 15.5 True Specific Gravities Obtained Using Various Liquids[86]

Liquid	Boiling or Vacuum	Low-Temperature Cokes 1	Low-Temperature Cokes 2	High-Temperature Cokes 1	High-Temperature Cokes 2
Water	Vacuum	1.503	1.632	1.796	1.790
Methyl alcohol	Vacuum	1.629	1.726	1.942	1.926
Methyl alcohol	Boiling	1.577	1.711	1.924	1.914
Ethyl alcohol (50%)	Vacuum	1.556	1.680	1.864	1.856
Ethyl alcohol (50%)	Boiling	1.536	1.667	1.853	1.842
Xylene	Vacuum	1.454	1.421	1.677	1.673
Xylene	Boiling	1.275	1.416	1.620	1.611
Benzene	Boiling	1.452	1.572	1.798	1.754
Tetralin	Boiling	1.402	1.506	1.797	1.753
Petroleum	Boiling	1.536	1.667	1.853	1.842

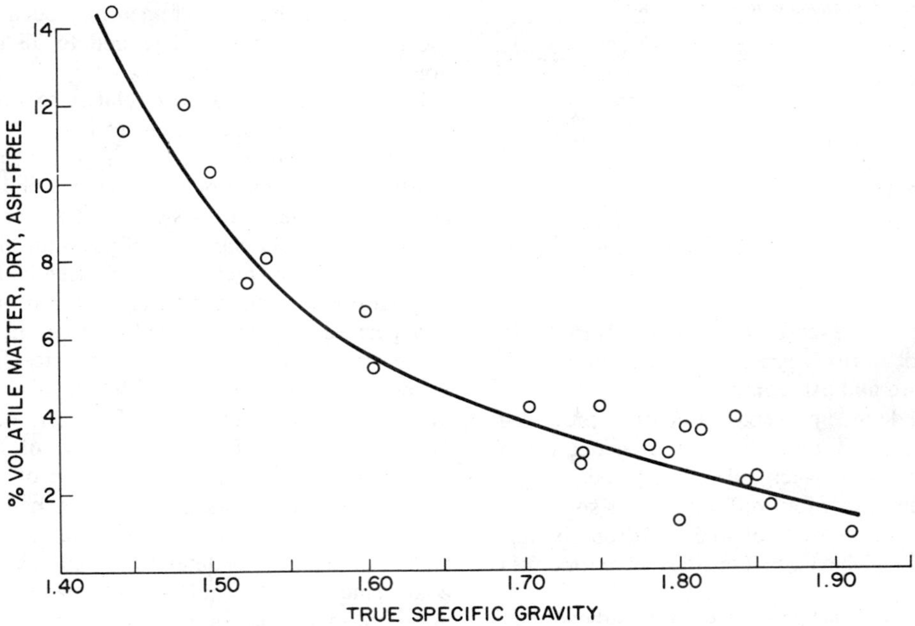

Figure 15.5 Relation between true specific gravity and volatile matter content of coke.[86]

measuring it. When coke is charged by volume, variations in coke bulk density can significantly affect furnace thermal load, but the effect on bed permeability may be even more important. Bulk density of coke is determined by the apparent specific gravity, size distribution, and moisture. Moisture has an additive effect, and the bulk density can be calculated to the dry basis if the moisture content is known. The measured bulk density, although it may represent the coke being charged, may change radically as the coke descends in the blast furnace. This change depends on the amount of size degradation resulting from mechanical stresses, and also on the amount of change in the apparent specific gravity due to the attack by reaction gases.

In the determination of the bulk density of coke, the results are influenced by the size distribution of the coke, and the size and shape of the container. Bose et al.[89] found that bulk density decreases with an increase in the average size of coke, and a higher value is obtained for bulk density with the larger container; he recommended that containers of at least 8 ft^3 capacity be used. The ASTM method specifies an 8-ft^3 box (D292), and the ISO has standardized two methods, one for a two hectoliter (7 ft^3) box, and the other for at least a 3-ton container such as a railway car or skip (ISO 567 and 1013, respectively).

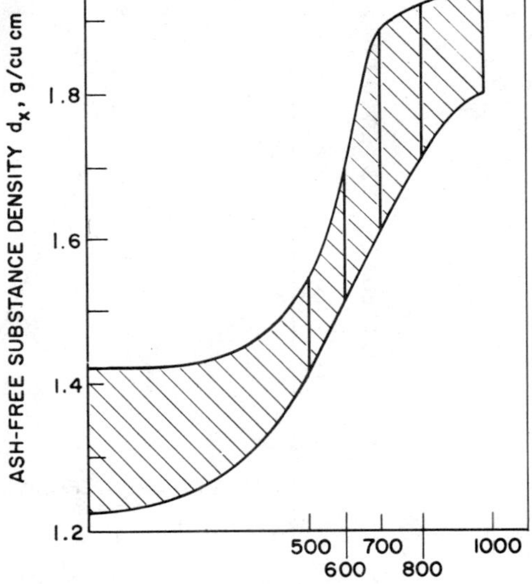

Figure 15.6 Relationship of coke density to temperature of carbonization.[87]

[89] R. N. Bose, M. V. P. Menon, K. Raja, and N. N. Das Gupta, Determination of the Bulk Density of Coke, ISO/TC27/WG8 Document 125, July 1960.

Table 15.6 Bulk Densities of Various Cokes[50]

Type of Coke	Density (lb/ft^3)[a] for: Run-of-oven	3/4 × 1½ in
Coke-oven coke	29–32	26–30
Horizontal retort coke	25–28	23–26
Vertical retort coke	23	21–22
Low-temperature coke	20	19

[a] Multiply by 16.018 to obtain kilograms per cubic meter.

Spiers[50] presented the data given in Table 15.6 for various types of coke with normal moisture and ash contents.

Clendenin[90] presented a chart, reproduced in Fig. 15.7, that shows a high degree of correlation between bulk density and average particle size of four different cokes subjected to various degrees of multiple-drop shattering. The highest bulk densities were for cokes that degraded the most, apparently because the small coke generated tended to fill the voids between the larger pieces.

Permeability, a property related to coke bulk density, has been found to be approximately proportional to the reciprocal of the fourth power of the void volume in a bed of coke.[91] It is therefore a significant factor in the attainment of a high flow of gases through the blast furnace stack, which is essential for smooth operation of the furnace. Good burden permeability is especially important in the case of the modern large-capacity furnaces, to ensure that high blast rates can be maintained and to ensure uniform gas flow distribution upward and even flow of the charge downward. In the upper part of the furnace permeability is largely determined by the iron portion of the burden, but in the bosh, permeability is determined only by the coke, since it is the only solid present. For good permeability in the large furnaces,

[90] J. D. Clendenin, COMA 3rd Carbonization Science Lecture, London, November 3, 1971. Waverly House, Wexborough, Yorkshire, England; in *Blast Furnace—Theory and Practice,* Julius H. Strassburger, ed., Gordon and Breach Science Publishers, New York, (1969), Ch. 8, 325–436.

[91] H. C. Wilkinson, *J. Inst. Fuel,* **37**, 480–484 (November 1964).

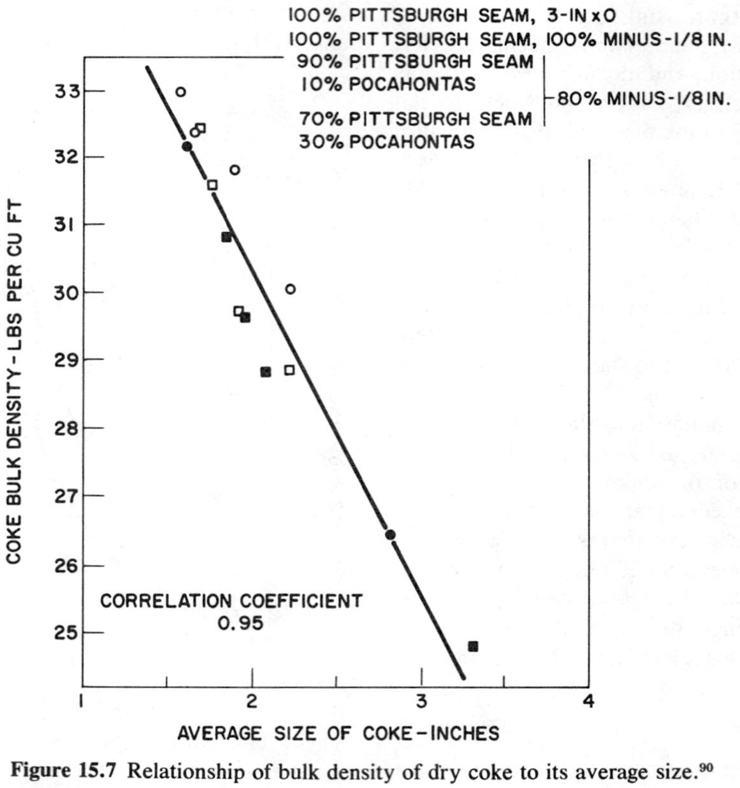

Figure 15.7 Relationship of bulk density of dry coke to its average size.[90]

Asai[92] found that coke strength is closely related to permeability. Increased coke strength improves furnace permeability, resulting in increased productivity.

A great number of papers have been published on the important subject of permeability, its measurement, and its effect on furnace operation. Permeability can be determined from measurements of airflow through a bed of coke and from pressure drop across the bed, or it can be calculated from the particle size distribution of the coke.[93] Figure 15.8 presents a correlation obtained by Gibson and Goleczka[94] at the British National Coal Board, using a number of blast furnace and foundry cokes having different size distributions. The flow resistance coefficient obtained from pressure drop measurements is plotted against the harmonic mean size (defined in Section 15.3.6) of the coke tested. The excellent correlation found shows that the flow resistance coefficient increases very rapidly when the harmonic mean size falls below 50 mm (2 in.); therefore small changes in blast furnace coke size below this size have a significant effect on permeability, and deterioration of the coke in the bosh area could be deleterious to good blast furnace operation.

Figure 15.9 presents Gibson and

[92] Hiromi Asai, "Coke Quality and Operation of Blast Furnaces in Yawata Iron and Steel Company," *Cong. Int. Charleroi,* (1966), Paper No. C_5. Published by European Coal and Steel Community, Luxembourg, 1967.

[93] A. S. Bruk and S. I. Pinchuk, *Coke Chem. (USSR),* No. 1, 11–14 (1970).

[94] J. Gibson and J. Goleczka, "Mechanical Properties and Bulk Permeability of Metallurgical Cokes," in *Coke Oven Techniques,* Symp. sponsored by Commission of the European Communities at Luxembourg, May 6–7, 1975. Published by the Commission, Luxembourg, April 1975. D. H. Gregory and J. Goleczka, "Permeability As a Measure of Coke Quality," *Symp. Assessment Physical Properties of Blast Furnace Coke,* London, October 17, 1972. Published by National Coal Board, Cheltenham, England.

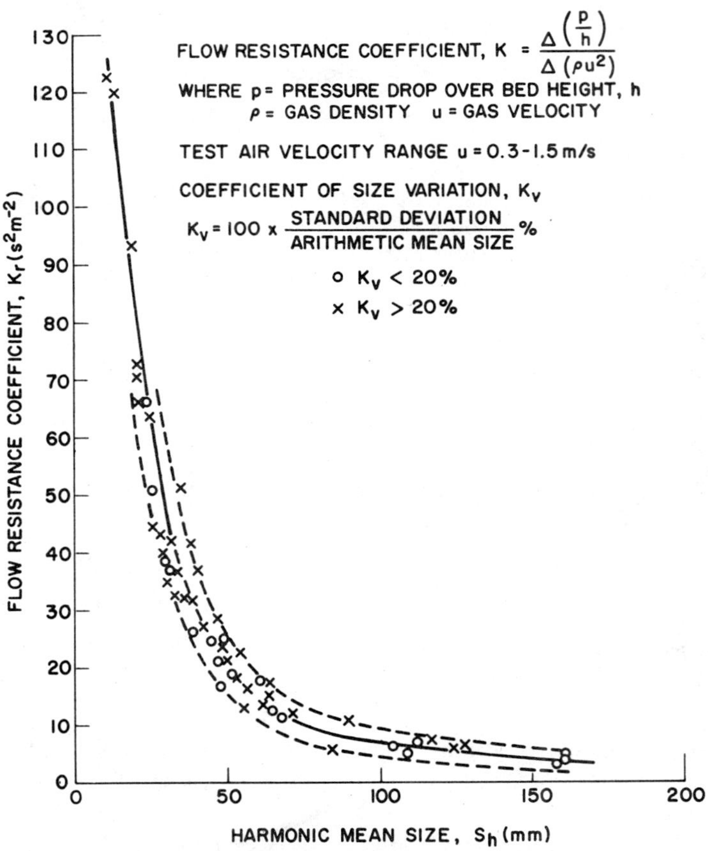

Figure 15.8 Effect of coke size and distribution on resistance of coke beds to air flow.[94]

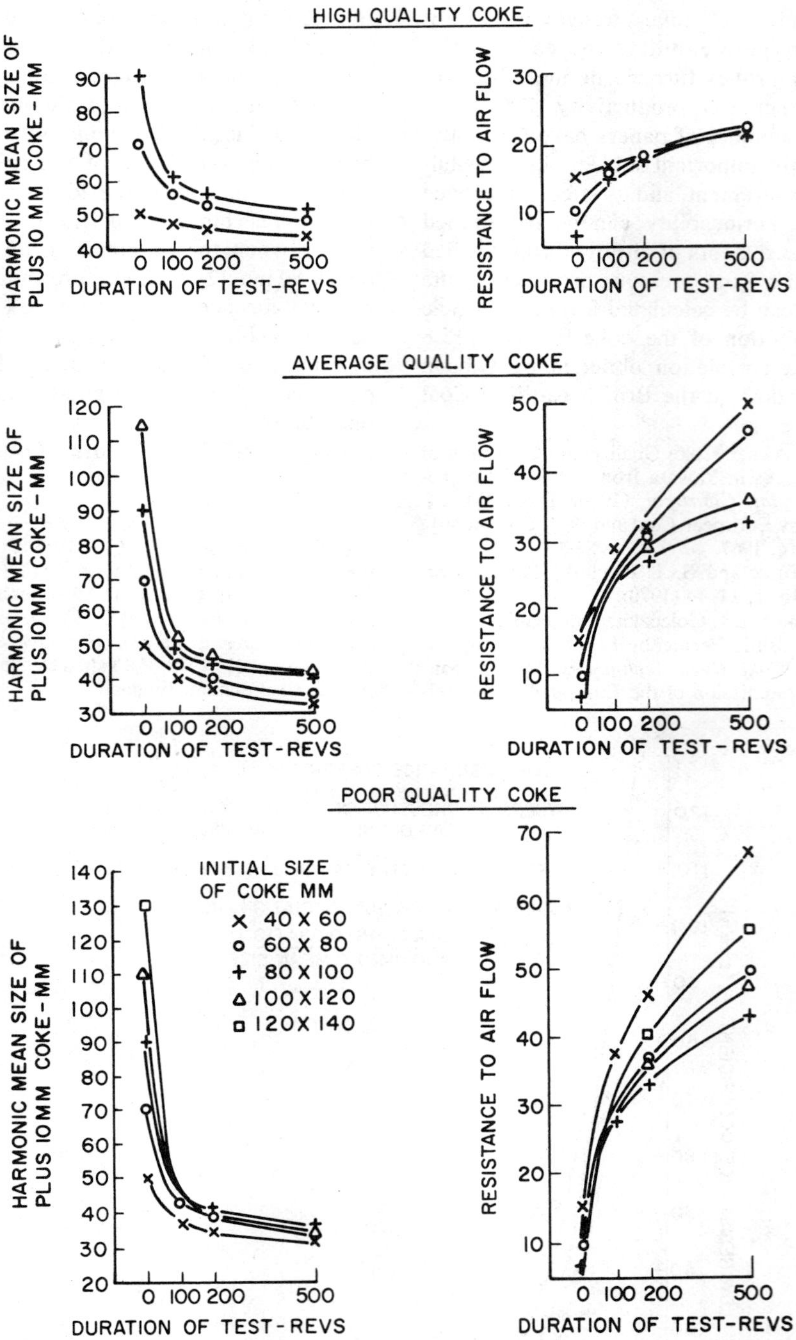

Figure 15.9 Harmonic mean size and resistance to airflow of plus +10 mm residues after degradation of closely sized fractions.[95]

Gregory's[95] determination of the harmonic mean sizes of three cokes of varying quality, which had been subjected to different degrees of degradation in the Micum tumbler. For

[95] J. Gibson and D. H. Gregory, *COMA Yearb.*, (1975), Waverly House, Wexborough, Yorkshire, England; L. Grainger, COMA 6th Carbonization Science Lecture, "Blast-Furnace Coke—Macro and Micro Considerations," London, November 6, 1974.

comparison, data on resistance to airflow are also plotted on the right-hand side. It should be observed that for the average-quality coke the resistance to airflow of the degradation products is about 1.5–2 times that of the products from high-quality coke. The products from the degradation of the poor-quality coke have air resistances 2–3 times that of the high-quality coke and about 1.35 times that of the average-quality coke.

In an effort to find practical ways of improving cupola operation by establishing appropriate size limits of coke for a cupola of given dimensions, as well as determining how much undersize coke can be tolerated in the charge, CSIRO made a study of the effect of blending cokes of various sizes on the percentage voidage and pressure drop of airflow upward through the bed of coke.[96] Figure 15.10 shows the relationships developed in the study. Voidage for blends of two size grades was always less than that of the individual components, and the pressure drop was always higher. When the difference in size was not great, the decrease in voidage, or the increase in pressure drop, was only slight. But when small coke was added to coarse coke, the voidage decreased markedly, and the pressure drop increased substantially to a minimum and a maximum, respectively, for a mixture containing 30–50% of the finer size.

15.3.6 Size Analysis

Size analysis of coke is important, as shown above, because of its effect on bulk density and permeability both in the blast furnace and in the cupola, and its effect on the rate of carbon gasification in these metallurgical units. It is also important in the world of commerce in describing or specifying the size of the coke. The determination of size analysis is described in practically all national standards as well as in two ISO standards (Table 15.2). The comprehensive discussion of size analysis presented by Thibaut in the first Supplementary Volume[49] is not repeated here. Suffice it to say that this is an important property of coke as far as the practical value of the fuel is concerned. Some of the important publications in recent years are treated here.

[96] Anon., *Coal Res. CSIRO*, No. 20, 2–5 (July 1963); P. L. Waters, CSIRO Div. Coal Res., Misc. Rep. No. 228, 1963.

The traditional procedure for size analysis of coke is the method of hand-placing coke on a predetermined set of sieves, which yields a group of several values depending on the number of sieves selected. This has the disadvantage of not determining a single value that can fully describe or characterize the sample, although many times a cumulative value on a certain sieve is selected at the discretion of the worker. Mean or average size is sometimes calculated from the values obtained. Thibaut described several ways of reducing size analysis to one or two values. Recently the British National Coal Board[94] introduced the term "harmonic mean size" for use with coke. This term was used in correlations in the previous section, and can be defined as follows:

$$S_h = \frac{100}{\sum w/s}$$

where w = percentage of an individual fraction in the sample, by weight, and s = arithmetic mean of the size limits of that fraction. The harmonic mean size, S_h, is a practical property because it is related to the surface area of the lumps, compared to the

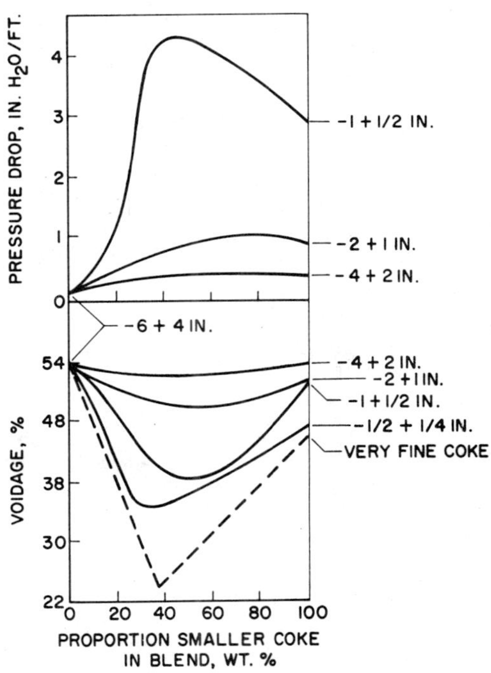

Figure 15.10 Voidage and pressure drop in beds with blends of large (−6 + 4 in.) coke and smaller sizes.[96]

normal mean size, which is related to their volume.

Turner and Wilkinson[97] studied what appears to be a useful method for determining the mean size of coke. It is based on the relationship of mean size of coke to the mean particle mass, by which the mean size is calculated from the mass and the number of individual particles comprising the sample. The correlation between these two factors is given in Fig. 15.11 and is described mathematically by the equation:

$$S_m = 17.59m^{0.303}$$

where S_m is the mean size (mm) and m is the mean particle mass (g). It is claimed that the results obtained from this method differ from observed values obtained by the traditional method by an average of only 1.3%. It was planned to apply this principle on a plant scale, for it appeared to be amenable to a mechanical system in which a known mass of coke particles can be converted into a single stream of individual particles, which can then be counted.

Semisalov et al.[98] studied the influence of coke lump shape on size indices and concluded that sizing of coke by sieve analysis yields an incomplete picture of coke size, since the separation is achieved only on the basis of cross-sectional dimensions. He developed a relationship between the length-to-breadth ratio and the cross-sectional dimensions, and found that as the absolute cross-sectional size increased, the length-to-breadth ratio decreased. Measurements of bulk density showed that bulk density increased as the length-to-breadth ratio decreased. Long lumps gave a lower bulk density than short lumps of the same cross-sectional dimensions. Thus the shape of the lumps affected the bulk density and also could affect the permeability.

[97] R. W. Turner and H. C. Wilkinson, *Ironmaking Steelmaking,* No. 1, 15–21 (January 1974).

[98] L. P. Semisalov, A. A. Lobov, D. M. Amstislavskii, Z. N. Vekselman, and A. V. Chebotarev, *Coke Chem. (USSR),* No. 9, 32–35 (1963).

15.3.7 Coke Strength

15.3.7.1 *Test Methods*

Coke is a brittle, porous, variable solid, and its strength is a difficult property to assess. A wide diversity of tests has been proposed and used, and all are empirical, reflecting little apparent attention given to the basic laws of breakage. One of the obstacles to the development of a test having a sound theoretical basis has been the paucity of fundamental knowledge of the mechanism of coke breakage, as well as correlation with blast furnace and cupola performance. As a result, the tests for coke strength adopted in various countries have been the product of custom rather than being based on scientific principles. In the United States the ASTM coke tumbler test has found favor for use with blast furnace coke, and the shatter test for foundry coke. Both tests are almost universally used in the United States by coke producers and users. In Europe the Micum and Irsid tests have gained the upper hand

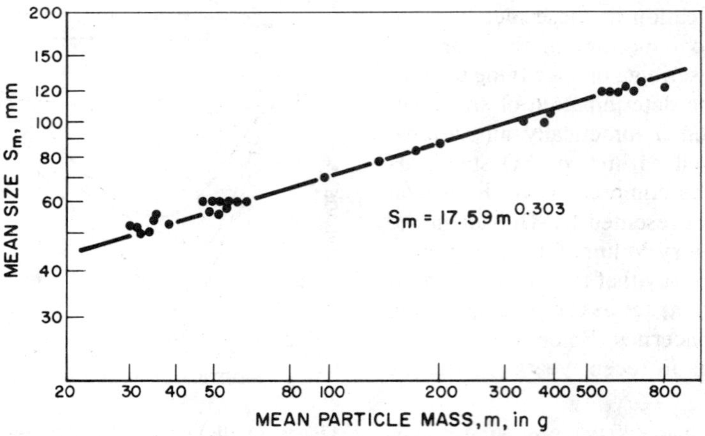

Figure 15.11 Relationship between observed mean size and mean particles mass.[97]

because of their standardization by the ISO. The United Kingdom previously used the B. S. (British Standard) abrasion and B. S. shatter tests, but these have been virtually abandoned in favor of the Micum and Irsid tests. Similarly, the U.S.S.R. has apparently dropped the Sundgren or barred drum in favor of the Micum and Irsid tests. Other countries in Europe and elsewhere have adopted the ISO methods. In Japan the Japanese Industrial Standard (JIS) drum was adopted, but both the ASTM tumbler and the Micum test are also used. Because these tests are the most popular, and publications present data on coke strength based on these tests, the treatment of the subject of coke strength in this chapter is limited in most cases to discussions based on their use. Special tests are described when appropriate.

The ASTM tumbler test, a standard since 1928 (D294), has remained virtually unchanged except for two recent developments. Because of the trend in the blast furnace industry toward the use of smaller coke, a need developed for a small coke tumbler test. A second procedure (D2490) was adopted in 1966, in which 2 × 1 in. coke, instead of 3 × 2 in. coke, is tested for 700 revolutions, instead of 1400 revolutions. This procedure found little use, probably because the values obtained were significantly different, thus led to confusion. As a result of this and because of a second development discussed below, D2490 was abandoned in 1975. The new development was the discovery that the use of 50% of 2 ½ × 2 in. plus 50% of 2 × 1 ½ in. sizes for the test sample, all other conditions remaining the same, yielded similar values for stability factor (total % on 1 in.) and slightly higher values for hardness factor (total % on ¼ in.) than did 3 × 2 in. coke in the established test. This new development has been incorporated into the new combined standard, D3402.

The Micum test and its companion, the Irsid test, are gaining worldwide attention because of the ISO standardization. These tests are now undergoing further changes within the ISO. Table 15.7 presents the essential details of the Micum, Irsid, and the

Table 15.7 Comparison of Standard Drum Test Methods

	ASTM	Micum[a]	Irsid	Latest ISO	JIS
Drum					
Diameter (mm)	910 (36 in.)	1000	1000	1000	1500
Length (mm)	455 (18 in.)	1000	1000	1000	1500
Thickness (mm)	6 (¼ in.)	5	5	5	—
Flights	2	4	4	4	6
Rotation (rpm)	25	25	25	25	15
Revolutions	1400	100	500	100/500[b]	30/150[c]
Sample					
Weight (kg)	10	50	50	50	10
Size (mm)	75×50 (3×2 in.)[d]	+60[e]	+20[e]	+20[e]	+50[e]
Moisture (%)	<1	<3	<3	<3	<3
Indices (mm)					
1	+25 (+1 in.)	+40	+20	+40	+15
2	+6.3 (+¼ in.)	−10	−10	+30	
3				+20	
4				−10	

[a] The half-Micum test is also used: one-half-length drum, 25-kg sample of +50-mm size yields nearly identical results.

[b] Round–hole sieves—no upper limit—in ISO test, sample proportioned according to size analysis of gross sample.

[c] Separate tests for 30 and 150 revolutions.

[d] Square-mesh sieves; sample can also be 50% of 2½ × 2 in. + 50% of 2 × 1½ in.

[e] After 100 revolutions sample is sieved to obtain Micum indices; fractions recombined, tumbled additional 400 revolutions, and sieved again for Irsid indices.

latest ISO combined version, together with the same information for the ASTM test, as well as the Japanese standard. The latest ISO method combines the Micum and Irsid tests and involves tumbling the coke for 100 revolutions in the Micum drum to determine the Micum indices, then recombining and tumbling the fractions an additional 400 revolutions to obtain the Irsid indices. This procedure would satisfy the coke manufacturer who prefers Micum values and the blast furnace man who desires Irsid indices. In addition, obtaining the size analysis of the coke sample at three different points (before the test, in between the two steps of the procedure, and at the end of the test) gives a better interpretation of coke breakage, particularly of the two facets of fissuring and abrasion.

It has been established by many,[94,99,100] in drum tests such as those listed in Table 15.7 that two breakage mechanisms occur in the drum: resistance to fracture and resistance to abrasion. Results after 100 revolutions in the Micum drum are biased because fracturing is still occurring, but at 500 revolutions bias is eliminated, and the true strength value of the coke is attained. The coke is characterized then, regardless of its previous history. As breakage proceeds, the distribution of sizes follows the Rosin-Rammler relationship. Breakage and abrasion occur at the same time, with breakage occurring predominantly at a low number of revolutions and abrasion predominantly after 500 revolutions. The transition from breakage to abrasion depends on the coke and can occur at any point between 300 and 500 revolutions. Figure 15.12, developed by Gregory and Goleczka,[94] demonstrates this bimodal breakage phenomenon for various sizes of coke and for various numbers of revolutions, up to 500, in the Micum drum. The rate of mechanical breakdown as a ratio of mean sizes after testing to before testing is very rapid for the first

[99] A. Milson, *COMA Yearb.*, 360–381 (1965), Waverly House, Wexborough, Yorkshire, England.
[100] Mechanical Degradation of Cokes and the Advantage of Testing at 500 Revolutions, ISO/TC27/SC3 Document 11, May 1974.

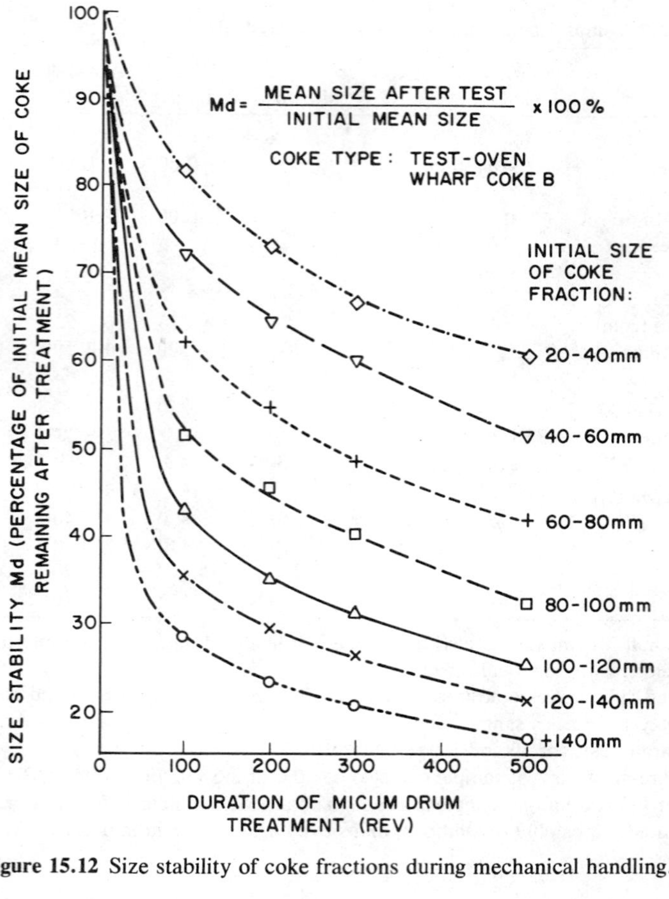

Figure 15.12 Size stability of coke fractions during mechanical handling.[94]

100–200 revolutions, when breakage is due to fissures and structural defects. After 200 revolutions the breakage slows down to a steady rate, at which point the abrasion phase is occurring.

Investigators have made comparisons of the various drum tests. The BCRA[101] made extended tests on four cokes in the ASTM, JIS, and half-Micum drums, and determined the effect of the degree of treatment on the relative reduction of the mean size of the coke. A linear relationship was found between the coke size-reduction ratio (size after to before crushing) and the logarithm of the number of drum revolutions for each drum. The reduction in coke size depended on the type of drum and the properties of the coke sample. The pattern of degradation varied from drum to drum. It was believed that in the JIS drum, the coke was degraded primarily by impact breakage, whereas in the ASTM tumbler a high proportion of the degradation resulted from rotational or abrasive reduction. The action in the half-Micum was at a point intermediate between these two. It should be pointed out that both breakage and abrasion can be useful measures of coke that are important to the operating behavior of the blast furnace.

Similar studies by the ISO Committee SC3 also compared the various drum tests.[102] ISO/TC27/SC3 document 13 concluded that the JIS drum is the most degradative and should be used to measure coke strength; however the Micum drum and the ASTM tumbler have more abrasion character, thus should be used to measure abradability. It found that the JIS test had the best repeatability and the ASTM tumbler had the best standard deviation between samples. ISO/TC27/SC3 document 45 discussed application of the Rosin-Rammler formula to the results of various coke strength tests and concluded that the destructive power of the JIS drum is greater than that of the Micum drum and the ASTM tumbler for an equivalent number of revolutions. ISO/TC27/SC3 document 39 confirmed these results, and calculated that the work expended on 1 kg of coke, in joules per revolution, was 25.37 for the JIS drum, 3.09 for the half-Micum drum, and 0.93 for the ASTM drum. ISO/TC27/SC3 document 53 presented a statistical evaluation of several test procedures using the Micum and JIS drums, and concluded that the Irsid test of 500 revolutions in the Micum drum was the most discriminating test, followed by the Micum and JIS (150 revolutions), ISO (100 revolutions), and finally JIS (30 revolutions).

Correlations have been reported in the literature for some of the drum tests, so that data from one drum can be converted to those of another drum. The U.S. Bureau of Mines made an extensive comparison of the ASTM and Micum coke test methods, and Walters et al.[103] reported the results. This study found that the Micum 40-mm index correlated to a high degree with the ASTM tumbler stability factor, but the standard deviation was too large to permit use of the equation that had been developed for predictive purposes. It was found that the Micum 20-mm index correlated with the stability factor with an acceptable standard deviation. Likewise, the Micum 10-mm index correlated favorably with the ASTM tumbler hardness factor. Table 15.8 presents some of the results obtained at the U.S. Bureau of Mines, and also gives equations, developed by the BCRA,[104] for predicting the Micum index from ASTM tests. Micum versus Irsid equations developed by the Institute of Chemical Utilization of Coal of Poland,[105] and Micum versus JIS drum equations published by Asai[92] are also included in Table 15.8. The M_{10} equation from the Bureau of Mines work was modified to give the customary values of passing through rather than retained on the 10-mm sieve. The M_{40} values calculated by the British formula are approximately 2 points higher than the Bureau of Mines values, but this may be due to the limited number of cokes included in the correlation. The table also presents equations from two

[101] Anon., BCRA Carbonisation Res. Rep. No. 5, June 1974.

[102] The Comparative and Semiquantitative Study of the Degradative Ability of Various Cohesion Tests, ISO/TC27/SC3 Document 13, May 1974; Comparison of Methods of Testing Coke Strength, ISO/TC27/SC3 Document 39, June 1976; Comparison of Testing Methods for Coke Strength Using R-Index and Coke Size Analysis, ISO/TC27/SC3 Document 45, (May 1976); Comparison of Micum, Irsid, and JIS on Coke, ISO/TC27/SC3 Document 53, June 1976.

[103] J. G. Walters, G. W. Birge, and D. E. Wolfosn, U.S. Bur. Mines Rep. Invest. No. 6482, 1964.

[104] Anon., BCRA Ann. Rep., 1975.

[105] M. Magdziarz, *Koks Smola Gaz,* **17**(1), 11–17 (1972).

Table 15.8 Correlation of Coke Strength Tests

Ref.	Equation[a]	Standard Deviation	Correlation Coefficient
103	$M_{40} = 25.89 + 0.884SF$	2.74	0.898
104	$M_{40} = 30.9 + 0.825SF$	—	0.954
103	$M_{20} = 75.68 + 0.264SF$	0.72	0.920
103	$M_{10} = 27.23 - 0.290HF$	0.75	0.683
104	$M_{10} = 33.7 - 0.376HF$	—	0.853
105	$M_{40} = 60.12 + 0.285I_{20}$	—	0.707
105	$M_{10} = 10.06 + 0.819I_{10}$	—	0.447
103	$M_{40} = 2.33\ S_{1.5} - 133.85$	2.36	0.926
106	$M_{40} = 1.893\ S_{1.5} - 97.6$	—	0.985
92	$D_{15}^{30} = 17.14\ (M_{40})^{0.39}$	—	—
92	$D_{15}^{30} = 103 - 1.54\ (M_{10})$	—	—

[a] Abbreviations:
M = Micum index on 40, 20, or through 10-mm sieves
I = Irsid index on 20 or through 10-mm sieves
SF = ASTM stability factor, total on 1-in. sieve
HF = ASTM hardness factor, total on ¼-in. sieve
S = ASTM or British shatter index on 1½ in. sieve
D_{15}^{30} = JIS index on 15-mm sieve after 30 revolutions

sources[103,106] relating M_{40} to 1½-in. shatter index by both the ASTM and the British standards. The results of these two equations are in rather close agreement.

Brown et al.[26] made a study of the interrelationships between various strength indices of coke and established correlation coefficients indicating the degree of dependence of one strength index on another. Table 15.9 presents these correlation indices. Good correlation was found between coke size (+3-in.) and the 2-in. and 1½-in. shatter indices, which evidently depended on the degree of fissuring resulting from carbonization. The stability factor and the ½-in. shatter index are partially related to shatter and hardness, implying that these result from the combined effects of shatter and abrasion. However the ½-in. shatter index appears to be more strongly related to the hardness factor than to the 2- or 1½-in. shatter index. Apparently specific gravity and shatter strength are significantly related.

[106] U.K. Report of Tests Used in the Assessment of Coke Strengths, ISO/TC27/WG8 Document 160, November 1960.

15.3.7.2 *Hot Tests*

Several investigators believed that coke should be tested at temperatures approximating those that it would be subjected to in the metallurgical processes. It was believed that such tests would provide more information on the behavior of the coke in the melting zone of the furnace or in the area of the furnace where coke properties are critical

Table 15.9 Linear Correlation Coefficients Between Coke Strength Indices[26]

	Coke Size + 3 in.	Shatter Index 2-in.	Shatter Index 1½-in.	Shatter Index ½-in.	Tumbler Test Stability Factor	Tumbler Test Hardness Factor
Shatter index						
2-in.	0.71	—	—	—	—	—
1½-in.	0.62	0.93	—	—	—	—
½-in.	0.28	0.40	0.60	—	—	—
Tumbler test						
Stability factor	0.28	0.43	0.57	0.65	—	—
Hardness factor	−0.31	0.49	0.34	0.84	0.47	—
Apparent specific gravity	0.32	0.63	0.66	0.48	0.51	−0.16

to furnace operation. At least 10 investigators have conducted high-temperature coke testing: two[107,108] made studies using a compression tester, one[109] used a ballistic apparatus, and the rest used a rotating drum. Birge et al.[110] used an ASTM tumbler, Echterhoff[111] used a Micum drum, Bradshaw and Wilkinson[112] used a miniature Micum, Das Gupta et al.[113] and Smolyaninova et al.[114] used special miniature steel drums, and Varga[115] and Wasa et al.[116] used silicon carbide tubes.

These investigators collectively arrived at the following conclusions:

1. Coke strength is poorer at high temperature than at ambient temperature (all experimenters).
2. Coke strength measured at room temperature is no indication of the strength at high temperature.[109,115]
3. High-temperature strength is decreased when the testing temperature is greater than that at which the coke was made.[111,112,116]
4. Long coking times improve the high-temperature strength of the coke.[115]
5. Good correlation was found between the testing apparatus using cold coke and conventional ASTM or Micum tumbler tests.[109–112,116]
6. Wet-quenched coke had a lower high-temperature strength because of internal thermal stresses, and annealing relieved the stresses.[111,114]
7. Cokes had maximum abradability in an oxidizing atmosphere and minimum abradability in a reducing atmosphere.[114]
8. Compressive strength was influenced by porosity—the higher the porosity, the lower the strength.[108]
9. Compressive strength was increased by a decrease in ash content or an increase in the melting point of the ash.[108]
10. The higher the cold coke Micum values, the smaller the reduction in high-temperature strength due to temperature.[108]

Although these testing methods and their results are informative, they merely represent research effort into a complex subject. None of the methods used has been applied as a routine procedure that can be used for process control, thus providing information relating hot strengths to furnace performance. Such tests are not practical because they are expensive and difficult to perform. Moreover, the methods used and the results can be criticized in that though the tests were made at high temperature, the cokes were not subjected to the gaseous and alkali attack that coke in the bosh area undergoes.

15.3.7.3 *Tensile Strength*

A unique procedure was developed by Patrick and Stacey of the BCRA for the measurement of the tensile strength of coke.[117] Cylindrical specimens, 1 cm in diameter and 8.5 mm long, are cut from cores drilled from

[107] M. O. Holowaty and C. M. Squarcy, *AIME Ironmaking Proc.*, **16**, 249–270 (1957); M. O. Holowaty and C. M. Squarcy, *J. Met.*, **9**, 577–581 (April 1957); P. J. Zorena, M. O. Holowaty, and C. M. Squarcy, *Blast Furn. Steel Plant*, **48**, 443–451 (May 1960).

[108] E. Mazanek, R. Kopec, and W. Garaczko, *Hutnik*, **29**(12), 451–457 (1962).

[109] Kenneth K. Humphreys, *Blast Furn. Steel Plant*, **51**, 1065–1069 (December 1963); U.S. Pat. 3,184,952, 1965; J. D. Clendenin, "Production of Specification Blast Furnace Coke in Coke Ovens," 3rd Carbonization Science Lecture, COMA and BCRA, November 3, 1971, Waverly House, Wexborough, Yorkshire, England; J. D. Clendenin, "Coal and Coke," in *Blast Furnace—Theory and Practice*, J. H. Strassburger, Ed., Gordon & Breach Science Publishers, New York (1969); Ch. 8.

[110] G. W. Birge, J. H. Lynch, Jr., and D. E. Wolfson, U.S. Bur. Mines Rep. Invest. No. 6814, 1966.

[111] H. Echterhoff, Bergbau-Forschung GmbH Rep. No. 118 of the German Coke Committee, February 3, 1961, published by Bergbau-Forschung GmbH, Essen, Germany.

[112] Anon., BCRA Coke Res. Rep. No. 38, March 1966, 20 pp.; K. Bradshaw and H. C. Wilkinson, *J. Inst. Fuel*, **42**, 112–117 (March 1969).

[113] N. N. Das Gupta, K. Raja, and P. K. Sarkar, *J. of Mines, Met. Fuels*, **12**, 235–238 (August 1964).

[114] N. M. Smolyaninova, A. A. Kaplin, and L. M. Vasileva, *Coke Chem. (USSR)*, No. 5, 24–27 (1961).

[115] John Varga, Jr., and H. W. Lownie, Jr., *Trans. A. Foundrymen's Soc.*, **64**, 217–220 (1956); John Varga, Jr., *AIME Ironmaking Proc.*, **19**, 605–617 (1960).

[116] A. Wasa, T. Nakai, and S. Sasaki, ISO/TC27/SC3 Document 12, 1974.

[117] J. W. Patrick and A. E. Stacey, Fuel, **51**(1), 81–87 (January 1972); No. 3, 174–179; 206–210 (July 1972); **52**(1), 27–31 (January 1973); **54**,(3), 213–217 (July 1975) **54**,(4), 256–264 (October 1975).

several full-length coke pieces and at several points along the length. These specimens are subjected to a diametrical compression test, and tensile strengths are calculated from applied load at breakage and specimen dimensions.

It was found that tensile strengths of industrial cokes varied widely from lump to lump and across each lump. The mean tensile strength of coke was found to be about 3.0–5.4 kN/m^2, but the maximum tensile strength of semicoke could be on the order of 7 kN/m^2. Therefore fissure formation was highly probable if the tensile stress developed within the semicoke exceeded 3.5 kN/m^2.

A very good correlation was established between the average tensile strength (S) and the Micum 10-mm index for a number of foundry cokes, which can be represented by the equation:

$$M_{10} = 35.6 - 18 \log S$$

However for blast furnace cokes, no statistically significant correlation was found between these values. Also, no correlation was found between either the M_{40} and M_{20} and the tensile strength for both foundry and blast furnace cokes. Since the M_{40} and M_{20} indices are actually measures of resistance to breakage due to fissures and flaws present in the coke, a relationship would not be expected. However the M_{10} index is a measure of resistance to abrasion; thus it would be expected to correlate with tensile strength. This was found in the case of foundry coke but not in the case of blast furnace coke. The reason for this was obscure, and the matter is still under study. Tensile strength was also found to correlate significantly with the total percentage porosity and the mean pore wall thickness.[118]

15.3.7.4 *Strengths of Conventional Cokes*

The trend in ironmaking over the past 25 years has been toward larger blast furnaces; the result has been more stringent demands on coke strength as well as quality. How well the coke producer has performed in providing a coke of higher strength and low ash and sulfur contents is demonstrated in Table 15.10. Not only have the cokes improved in these properties, but in some cases the ranges in the values have been reduced, which probably means more uniformity. For comparison, the cokes tested by the Commission Interprofessionelle des Essais des Charbons et des Cokes (Cominpecc) at 33 plants in continental Western Europe[122] averaged an M_{10} value of 6.9, with a range of 5.4–9, which corresponds to an ASTM stability factor of about 60 and a range of 50–67. The same European cokes ranged in ash content from 6.7 to 10.1%, and in sulfur content from 0.74 to 0.98%.

[118] Anon., BCRA Carbonisation Res. Rep. No. 18, September 1975.

[119] Anon., Coke Evaluation Report, American Iron and Steel Institute, New York, (1952).

[120] D. E. Wolfson, G. W. Birge, and J. G. Walters, *AIME Ironmaking Proc.*, **20**, 387–402 (1961).; U.S. Bur. Mines Rep. Invest. No. 6354, 1964.

[121] Various *AIME Ironmaking Proceedings*, 1966–1975.

[122] Ch.G. Thibaut, *Information Statistique sur la Qualité des Cokes en 1962*, Association Technique de la Sidérurgie Francaise, Sous-Commission des Cokes Cominpecc, CT/DW Services 541, No. 62–823, November 1962, 5 pp.

Table 15.10 Quality of Furnace Coke Produced in the United States in the Past 25 Years

Period	VM (%)	Ash (%)	Sulfur (%)	Stability Factor	Hardness Factor
1948–1949 (12 plants)[119]					
Average	0.9	11.5	0.80	34	65
Range	0.5–1.3	8.5–17.7	0.54–1.38	5–49	60–67
1959–1960 (13 plants)[120]					
Average	1.0	9.0	0.80	53	65
Range	0.6–1.4	7.5–10.7	0.61–1.11	43–57	60–68
1966–1975 (13 plants)[121]					
Average	0.9	8.2	0.73	56	68
Range	0.7–1.1	6.6–10.8	0.54–1.11	51–62	62–73

With the improvement in coke quality and strength over the years, together with adoption of the latest technology for preparing burdens, higher blast temperatures, fuel injection and so on, the productivity of the blast furnace has been significantly increased. Future improvement in coke properties is questionable because of a serious worldwide depletion of the better quality coals; any improvement that is achieved will depend on technology developments in the beneficiation of coals and in the manufacture of coke.

15.3.7.5 *Strength of Coke in the Blast Furnace*

A number of investigators have examined the properties of coke extracted from the tuyeres of the blast furnace, using the ASTM, Micum, and other tests. Simulation tests have also been made on cokes to determine the amount of treatment in the Micum or ASTM drum that is required to match the coke extracted. A discussion of the findings of these workers is deferred to a later, more appropriate section, which deals with the performance of coke in the blast furnace.

15.4 COKE PROPERTIES AND MANUFACTURE

15.4.1 Effect of Coal Properties

The properties of coke are influenced, of course, by the coal or coal blend from which the coke was made, as well as by the carbonizing conditions imposed. The rank of the coal, or the average rank of the blend of coals, profoundly affects the properties of the coke produced. One of the outstanding accomplishments of coal technologists a number of years ago was the development of standards for classifying coals. These have served as an aid to study coals and their cokes on a more rational, scientific basis. In the United States such a scheme has been standardized by ASTM (D388), and used by investigators to relate coal and coke properties to their relative rank in the system. Perch and Russell[123,124] pointed out that it had an important relation to various coking properties of coal. Internationally, the classification system developed by the United Nations Economic Commission for Europe[125] has received considerable attention in world trade and by researchers for its use in relating rank to coal and coke properties.

Coke strength and other properties of coal and coke show characteristic maxima when plotted against a parameter of ASTM rank, as shown in Fig. 15.13. These data were obtained at more or less constant conditions, but the data are or can be affected by differences in rank as well as in petrographic composition, mineral matter content, oxidation, pulverization, charge density, rate of heating, and other factors. Juranek[126] presented a similar correlation relating M_{40} values as a function of volatile matter for a large number of Ruhr coals.

Wolfson et al.[127] related coke properties to rank for cokes prepared in the U.S. Bureau of Mines–American Gas Association (BM–AGA) carbonization test, and compared the properties of a large number of BM–AGA cokes and industrial cokes; a high degree of correlation was obtained between the strength indices of the two kinds of coke. The Bureau of Mines has conducted BM–AGA tests since 1930, and more than 1200 coals have been evaluated. This storehouse[128] of information on coke properties represents a wealth of useful data on coking coals from all the principal coking coal regions of the United States. Brown et al.,[26] at CSIRO in Australia, used the BM–AGA car-

[123] C. C. Russell and M. Perch, *Proc. Blast Furn. Coke Oven Raw Mat.*, No. 2, 51–69 (1942).

[124] M. Perch and C. C. Russell, *Trans. AIME*, **184**, 205–214 (1949).

[125] United Nations Economic Commission for Europe, International Classification of Hard Coals by Type, E/ECE/247, E/ECE/Coal/110, UN Publication II. E. 4, 1956, 52 pp., United Nations, New York.

[126] G. Juranek, "Optimization of Coking Coal Blends Based on the Mathematical Model of Carbonization in Slot-Type Recovery Ovens," Seminar at Luxembourg, April 23–24, 1970, in *New Methods and Developments in the Field of Coke Production,* European Community for Coal and Steel Commission, Luxembourg, January 1971; pp. 301–315.

[127] D. E. Wolfson, G. W. Birge, and J. G. Walters, *Proc. Blast Furn. Coke Oven, Raw Mater. Conf. AIME,* **20,** 387–403 (1961); U.S. Bur. Mines Rep. Invest. No. 6354, 1964.

[128] C. Ortuglio, J. G. Walters, and W. L. Krobot, *Carbonization Yields, Analyses, and Physical Characteristics of Cokes from American Coals,* U.S. ERDA, PERC/B-75/1, July 1975.

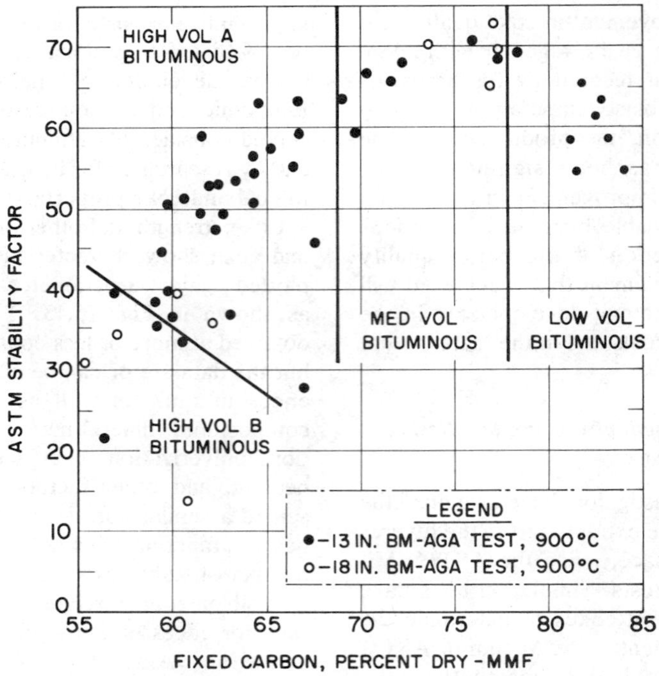

Figure 15.13 Relationship of relative coke strength to fixed carbon content (rank) of coals.[90,123]

bonization assay and presented a large number of correlations of coke size and coke shatter results with proximate and ultimate analyses, caking properties, and petrographic compositions of Australian coals.

It is common practice in blending coals destined for the coke oven to blend high-volatile with low-volatile and/or medium-volatile coals to improve the strength of the coke. The effect of adding varying amounts of low-volatile coal to a number of high-volatile coals of a broad range of coking propensities from various regions of the United States is shown in Fig. 15.14. It is of interest to note that improvement in the tumbler stability factor is effected in all cases; those with low stability factors are improved by low-volatile coal addition to a greater extent than are cokes from strongly coking high-volatile coals, and all curves seem to approach stability factors of 60–65 asymptotically.

Poorly coking or high-volatile coals of low fluidity are improved in those qualities making up a good metallurgical coke by the addition of medium-volatile coals. This was pointed out by Perch and Russell[124] in relation to a modification of the ASTM classification of coals by rank and was substantiated by Jackman et al.[129] and Landers et al.[130] Jackman blended six medium-volatile coals with coals from the No. 5 and No. 6 seams in Illinois and concluded that the medium-volatile coals, with their relatively high fluidity and strongly coking properties, may be used advantageously in blends with the Illinois coals to produce coke of good metallurgical quality. Landers made a similar study using Utah coals.

In 1961 a substantial advance was made in coal and coke technology when Schapiro et al.[131] applied the principles of coal petrography, using the reflected light method in a practical manner, to evaluate coals for coke making, specify optimum coking blends, and predict coke strength from petrographic composition (Chapter 3). It has been known for a long time that for a given set of carbonization conditions, the properties of the coke

[129] H. W. Jackman, R. L. Eissler, and R. J. Helfinstine, Illinois State Geol. Survey Circ. No. 278, (1959), 24 pp.

[130] W. S. Landers, M. Gomez, and C. C. Boley, U.S. Bur. Mines Rep. Invest. No. 6948, May 1967.

[131] N. Schapiro, R. J. Gray, and G. R. Eusner, *Proc. Blast Furn. Coke Oven Raw Mater. Comm. Conf., AIME*, **20**, 89–112 (1961); N. Schapiro and R. J. Gray, *J. Inst. Fuel*, **37**, 234–242 (1964).

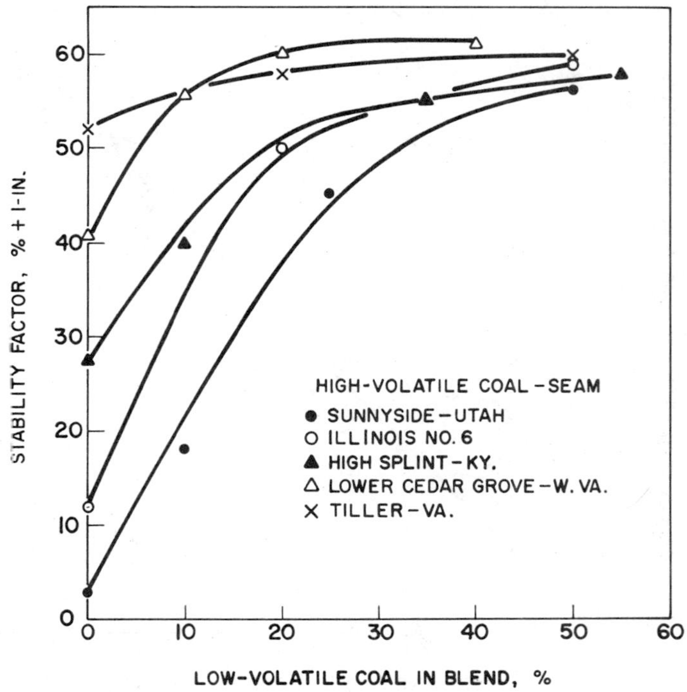

Figure 15.14 Effect on coke strength of additions of low-volatile coking coals to different high-volatile coals.[90]

produced depend on the rank and petrographic composition of the coal carbonized. The new development of Schapiro et al. established a technique for providing petrographic data in numerical terms and relating them to coke properties.

The method employs standard petrographic methods (ASTM D2798 and D2799) for determining the petrographic composition of coal and determining the mean reflectance of the vitrinite in oil, R_0. The macerals so determined are grouped into two classes—the reactives and the inert macerals. The reactives are defined as the macerals that melt on heating above approximately 425°C and consist of vitrinite, exinite, resinite, and one-third of the semifusinite and granular micrinite. The inert macerals do not fuse; they consist of fusinite, massive micrinite, and two-thirds of the semifusinite. The stability factor of the coke from a given coal or blend is determined by interpolating a series of empirical correlation curves relating petrographic composition to the stability factor. Predictions based on these curves are complicated by several factors, including the following: (1) coals have more than one type of petrographic constituent called vitrinite, and each vitrinite has a different optimum ratio of reactive to inert entities to produce the strongest coke, (2) the optimum amount of inerts is not always the average of the optimum amount of inerts for each rank of coal in the mixture, and (3) corrections must be applied for coals that are oxidized, have high ash contents, and are from periods other than the Carboniferous.

In spite of these exceptions, the method of Schapiro et al. appears to be one of the best available means for predicting the coking properties of coals and their blends. The accuracy of prediction is illustrated in Fig. 15.15, taken from Schapiro's publications. Although predictability appears to be good and many coke producers depend on coal petrography for guidance in the selection of carbonization blends, it is still necessary in important decisions to confirm the selected blends in a test oven or a full-scale oven. The use of petrography, however, reduces considerably the number of tests required.

This technique has received considerable worldwide attention and has been adopted as a routine analytical tool by nearly all the major steel and coal companies in the United States and in many foreign countries. Most

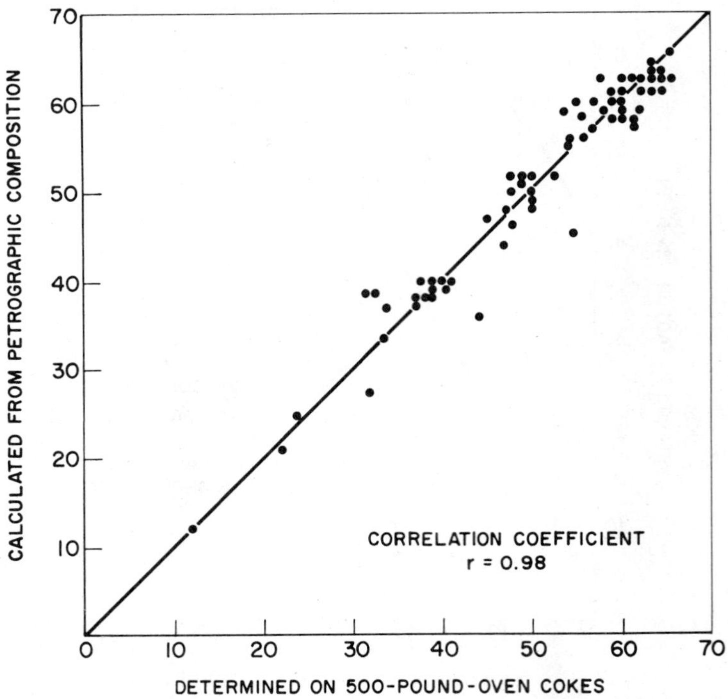

Figure 15.15 Relationship between the calculated stability factor and the stability factor determined on 500-lb oven cokes.[131]

workers have modified the empirical correlation curves originally presented, to suit their own test procedures and coals.

In Japan, which imports a large number of coals of a wide variety, it is necessary to practice judicious blending of these coals to obtain coke of uniformly high strength. An evaluation technique was developed, which is really a combination of the techniques coal petrography and Gieseler fluidity. This system makes use of the MOF diagram introduced by Miyazu and his colleagues Okuyama and Fukuyama[132] and shown in Fig. 15.16. Fluidity is plotted against the mean reflectance, and the coals are blended in such a way that the properties of the final blend fall within the cross-hatched area. To maintain the strength index D^{30}_{15} at the desired value of 92 or higher, the fluidity of the blend should be higher than 80 divisions/min at a coke-oven working rate of 150% (number of ovens pushed per day, divided by the total number of ovens). At a lower working rate (130%), the lower limit should be 200 divisions/min. The upper fluidity limit should be 3000 divisions/min for the low working rate and 1000 divisions/min for the higher working rate. The mean reflectance for both working rates should be between 1.2 and 1.3. However the Japanese have recently extended the reflectance range to above 1.3 by adding petroleum pitch to the coal blends, to offset the loss of fluidity, as described in Section 15.5.5.

Another procedure used to predict coke strength was developed at the Steinkohlenbergbauverein in Germany (Chapter 6). This method, developed by Simonis and Beck,[133] derives a *G* value, called caking capacity, which is based on Ruhr dilatometer results. The caking capacity, together with volatile matter content and due allowance for coal blend composition and carbonizing conditions, is used to predict the M_{40} coke

[132] T. Miyazu, Y. Okuyama, T. Fukuyama, N. Suzuki, and T. Mori, "The Evaluation and Design of Blends Using Many Kinds of Coal for Coke Making," *Int. Iron Steel Congr.*, Düsseldorf, Verein Deutscher Eisenhüttenleute, Düsseldorf.

[133] W. Simonis and K. G. Beck, *Glueckauf-Forschungsh.*, **26,** 201–207 (1965); J. G. Walters, C. Ortuglio, and D. E. Wolfson, U.S. Bur. Mines Rep. Invest. No. 7564, 1971.

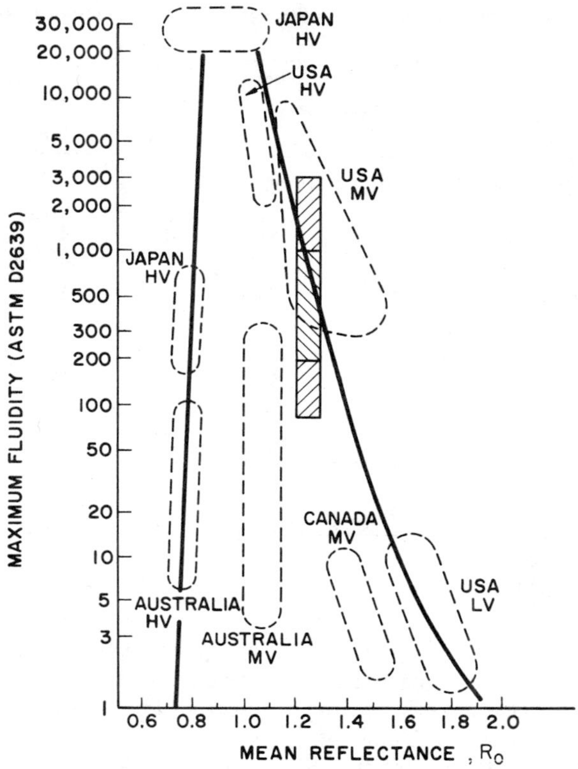

Figure 15.16 Relation between maximum fluidity and rank of coals (MOF diagram).[132]

strength. Gibson, at the British National Coal Board, made a thorough study[134] of this prediction method and found that the standard deviation between predicted and actual M_{40} values was 2.4 units for coals having a 20–30% volatile matter content, 3.3 units for those between 30 and 34%, and 13.3 units for higher volatile coals. He improved the prediction equation on the basis of 170 test oven charges, so that the M_{40} values could be predicted for all coals within standard deviation limits of 2.2 units. Prediction of M_{10} values by the Simonis method was disappointing, and the modification of the prediction equation to improve correlation was unsuccessful. Gibson also concluded that his dilatometric method can reduce the amount of laborious effort required in the test oven, but at this point it cannot supplant use of the test oven.

Many inert materials (nonfusible at coke-oven temperature) have been added to coals or coal blends for a variety of reasons—for example, to increase coke size or strength, to increase coke yield or add carbon value, and to reduce carbonization pressure. These inerts include coke breeze, anthracite fines, low-temperature coal char, petroleum coke, iron ore fines, lime, and bentonite. The inerts, called antifissurants or contraction modifiers, serve to reduce the shrinkage of the coke in the postplastic stage of coking, resulting in the decrease of fissuring. The type of inert, the amount added, and the particle size and size distribution, are important factors in determining the effect that the additive will have on the properties of the resultant coke. An excellent example of the effect of an inert on coke is shown by the addition of anthracite fines, as presented in Table 15.11. It can be concluded from these data that coarse anthracite fines substantially increase coke size (and shatter index) but decrease the tumbler test stability and hardness factors, and that fine anthracite fines increase the size of the coke to a moderate extent but increase the stability and hardness factors significantly. The first conclusion is the basis for foundry coke manufacture when large-

[134] J. Gibson, *COMA Yearb.*, 182–205 (1972), Waverly House, Wexborough, Yorkshire, England.

Table 15.11 Effect of Inert Additions on Size and Strength of Cokes Produced in a 12-in.-Wide Pilot Coke Oven[90]

Coking Blend	ASTM Stability Factor	ASTM Hardness Factor	Coke Size (cum %)			
			+4 in.	+3 in.	+2 in.	+½ in.
100% Pittsburgh Seam hvA coal	41	66	2	18	66	96
95% Pittsburgh Seam coal[a] 5% plus 30-mesh anthracite	35	58	15	48	80	—
95% Pittsburgh Seam coal[a] 5% minus 60-mesh anthracite	54	70	2	22	70	—
50% High-volatile[a] 40% Medium-volatile[a] 10% Low volatile[a]	58	69	2	19	74	98
50% High-volatile[a] 40% Medium-volatile[a] 5% Low-volatile[a] 5% Anthracite fines[b]	64	74	7	38	81	98

[a] 80% minus 1/8 in., 1–1.5% moisture.
[b] Nominal size, 20 × 200 mesh.

sized coke of high shatter strength is desired, and the second conclusion could benefit the production of blast furnace coke if the anthracite were sufficiently low in ash content.

Although addition of inerts to coals slated for the coke oven has been customary with foundry coke for many years, the practice has been limited in the United States with blast furnace coke, probably because of the lack of adequate sources of such materials, as well as the problem of economically reducing the particle to the size required. In the case of foundry coke, it is fortunate that anthracite fines of the proper size consist have been commercially available, and are used usually without further preparation. Adequate supplies of anthracite fines are becoming more difficult to acquire, and more and more foundry coke producers are installing coke breeze crushers to produce inerts of the required size from internal plant supplies. However on the basis of recent developments,[135,136] it appears that greater use will be made of inert addition in the near future for blast furnace coke manufacture. Larimer et al.[135] described experimental work and an operation in a plant in which coke breeze was crushed 85% through 20-mesh and recycled into the coal charge in amounts ranging from 1 to 3%. As a result, the coke breeze was upgraded to furnace coke without a significant effect on overall coke strength or coke size. A similar study by the Canadian Carbonization Research Association[136] concluded that coke stability could be maintained at the base level by adding 4–7% of minus 60-mesh coke breeze, depending on the coal blend and coking conditions. The more economical minus 35-mesh breeze also maintained stability at the base level, but the hardness factor tended to drop at the higher additions.

Experimental work has been conducted over the years on the effects of inert addition on coke properties. Foch and Meimarakis[137] added a combination of coke breeze and char to a coke-oven blend and found the effect shown in Fig. 15.17. The lone addition of coke breeze of minus-0.5-mm size improved the M_{40} index from less than 40 to greater than 60 with up to 12% addition; however, the simultaneous addition of both coke breeze and char up to 25% total inerts improved the M_{40} to better than 80. The effect on the M_{10} index was slight. The Bureau of Mines applied the same technology to U.S. coals and found similar effects with Utah

[135] T. M. Larimer, J. P. McGinness, and R. W. Shoenberger, *AIME Ironmaking Proc.*, **34**, 160–176 (1975).

[136] D. A. Reeve and H. N. Paulencu, *AIME Ironmaking Proc.*, **34**, 135–159 (1975).

[137] P. Foch and G. Meimarakis, *Reve Gen. Therm.*, **3**(31), 911–936 (1964).

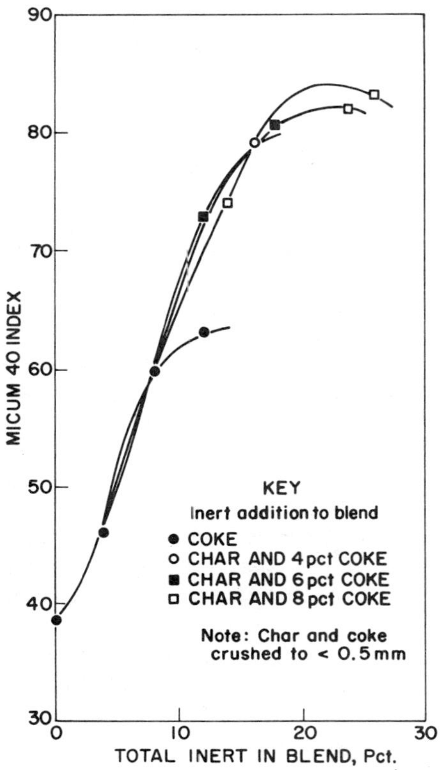

Figure 15.17 Effect of simultaneous additions of char and coke to a coke-oven blend.[137]

Lower Sunnyside coal as well as with Illinois and Kentucky coals.[138] Stability factors were improved from 13 to 43 with 6% additions of both breeze and char to Lower Sunnyside coal; a further addition of 10% low-volatile coal increased the stability to 51. Analogous improvements in the stability factor of an Illinois No. 6 coal were made with similar additions of breeze and char, with improvements from 21 to 44, and then to 47 with the 10% addition of low-volatile coal. The stability factor of Kentucky high-splint coal was improved in a similar manner, from 14 to 37 and then to 40 with the addition of low-volatile coal. According to Boley et al.[139] the use of char admixture alone also resulted in improvements in coke strength, but to a lesser extent, for both Lower Sunnyside coal from Utah and Allen coal from Colorado.

The synergistic effect of the combined inert addition was explained by Foch and Meimarakis as an effect resulting from char and coke breeze acting by different mechanisms on the properties of the resultant coke and the simultaneous use of both producing a desirable cumulative effect that is better than either one of the two antifissurants used alone.

In the studies at the U.S. Bureau of Mines[138] and at CERCHAR in France, the base coals used were marginally coking and of low fluidity. Fissuring is also a problem with high-volatile, highly fluid coals when low-volatile or medium-volatile blending coal is scarce. It was for this reason that CSIRO studied the effect of adding fluidized-bed chars made from high-volatile coals. It was found at CSIRO[140] that for high-volatile coals the addition of 10–15% finely ground char (50 × 70 mesh, or 35 × 50 mesh) improved the shatter index considerably and also significantly increased the abrasion hardness. For medium-fluidity coals the char was best prepared in the fluidized-bed apparatus at a temperature of 500°C, but for highly fluid coals the preparation temperature should be about 900°C.

15.4.2 Effect of Coking Conditions

In addition to the character of the coal charged to the coke oven, the carbonization conditions imposed on the coal charge influence the properties of the resultant coke. Carbonization conditions can range from the pulverization of the coal, the charge density, the rate of heating, and oven design, to special conditions such as preheating.

The preceding section demonstrated that the particle size of added inert materials has an important effect on the coke produced. It is equally important to pulverize the coals charged into coke ovens to a point where the inherent inert particles of the coal are reduced to an optimum degree, as well as to reduce the size of coal particles that attain only low fluidities. The importance of pul-

[138] M. J. Kovalik, D. E. Wolfson, F. Fischler, and L. Mafrica, U.S. Bur. Mines Rep. Invest. No. 7235 (1969), M. J. Kovalik, D. E. Wolfson, and L. Mafrica, U.S. Bur. Mines Rep. Invest. No. 7718 (1973).

[139] Charles C. Boley, M. M. Fegley, R. B. Porter, and D. C. Boteler, U.S. Bur. Mines Rep. Invest. No. 7599, 1972; C. C. Boley, M. M. Fegley, and R. B. Porter, U.S. Bur. Mines Rep. Invest. No. 7817, 1973.

[140] P. L. Waters and W. R. Hesp, *J. Inst. Fuel,* **34,** 323 (1961); W. R. Hesp and G. H. Taylor, *Coke Gas,* **24**(272), 5 (1962).

verization level is shown in Fig. 15.18, which gives examples of several coals having different coking abilities. Higher pulverization levels evidently tended to make a more homogeneous and intimate mixture of the reactive and inert components of the coals blended.

There are numerous other publications showing similar examples of the effect of pulverization. Among these is the work of Wenger and Neubaum,[141] who showed that the stability factor is influenced significantly by the degree of pulverization and is influenced to a lesser extent by the coal moisture content and oven flue temperature. Figure 15.19, from their publication, shows the combined effect of pulverization and moisture. Lines of isostability were drawn by contouring points obtained on the basis of pulverization and moisture. At a moisture level of 4% the stability factor increases about 10 points as the pulverization level is increased from 55 to 100% minus 1/8 in. The effect of moisture is less, but still significant, since an increase in moisture from 4 to 9% at an 80% pulverization level lowers the stability about 4 points. Figure 15.19 also shows that the hardness factor is influenced mostly by the moisture content and that pulverization has little effect. At 80% pulverization the hardness decreases about 5 points with a change in coal moisture from 4 to 8%.

The coal pulverization level and the moisture content of the coal charged into the oven were shown above to have an effect on coke tumbler indices. It is logical, therefore, that the coal bulk density in the oven should also be related to coke strength. This was demonstrated by Doherty et al.[142] for two typical coal blends, as shown in Fig. 15.20. Both the stability and hardness factors are directly related to the dry coal bulk density, a result that is achieved through the loss of moisture

[141] R. O. Wenger and V. A. Neubaum, *AIME Ironmaking Proc.*, **28**, 170–182 (1969).

[142] J. D. Doherty, E. J. Hodder, and I. N. Anthony, *Blast Furn. Steel Plant*, **50**, 248–260, 266 (March 1962).

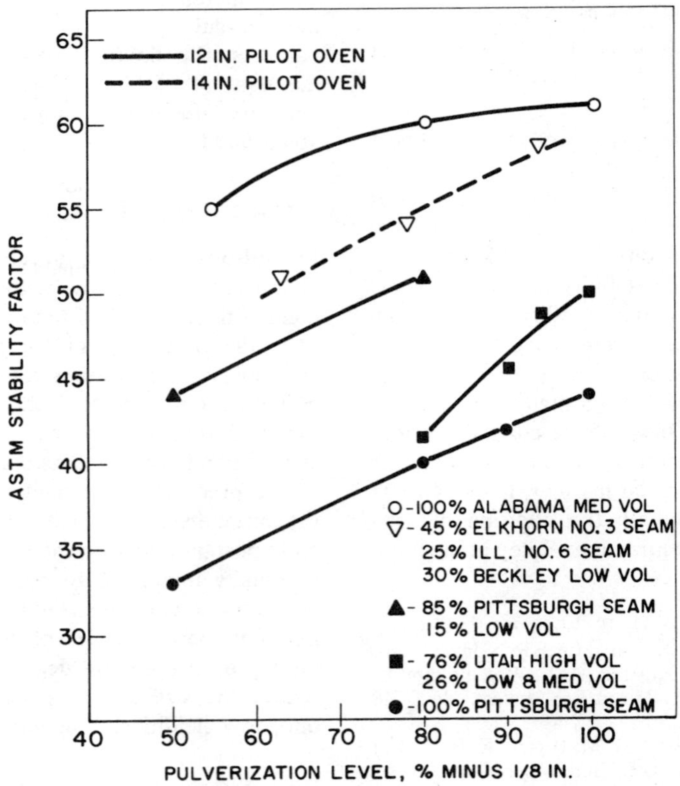

Figure 15.18 Effect of degree of pulverization of different coals and blends on the strength of the cokes produced.[90]

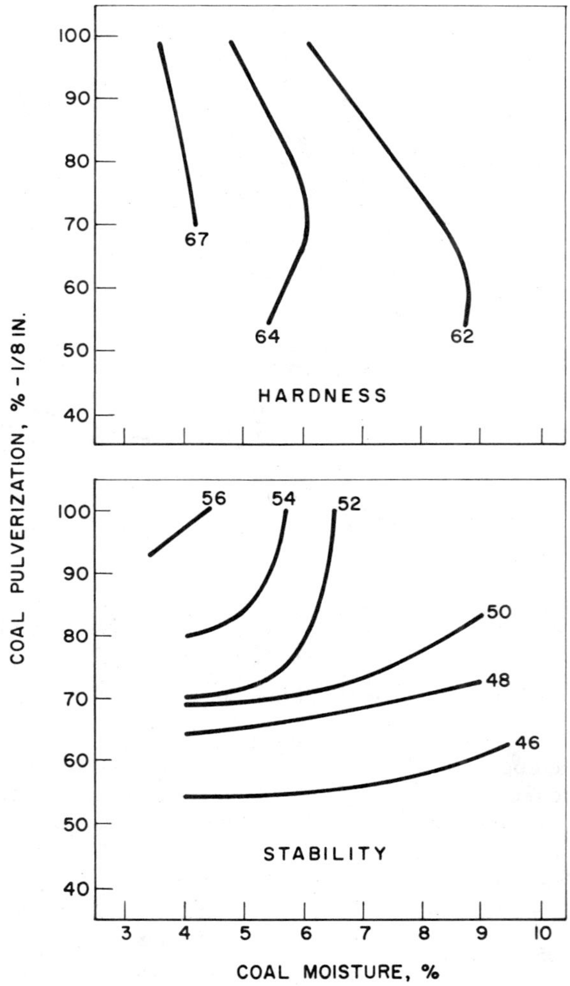

Figure 15.19 Effect of moisture content and pulverization level on coke tumbler indices.[141]

alone as discussed below. Since the relationship of these strength indices to wet coal bulk density was inconclusive, the authors stated that the plant operator would be wise to determine for each coal mixture the relation of dry coal bulk density to stability, to achieve the maximum stability that can be attained in the plant.

In recent years there have been particular efforts to increase the productivity of coke ovens by improved oven design and operating practice. The resultant effect on coke properties was also studied by various investigators. Table 15.12 presents a compilation of data obtained from three publications on the effect of coking rate on three different coals or blends tested in pilot ovens. The coking rate was varied over wide limits. In each case the coke size and the shatter index decreased significantly as the coking rate was increased. Likewise, the stability factors decreased and the hardness factors increased for the two U.S. coal blends as the rate of coking was increased. The reverse trend was found for the British coal; that is, the Micum values improved with an increase in coking rate. The British coal is apparently a poorly coking coal that without a blending coal additive, benefits from the faster heating rate. Additional work at the BCRA[47] on six other strongly coking coals exhibited the behavior of the U.S. coals, as shown in Table 15.12.

[143] G. W. Lee, *AIME Ironmaking Proc.*, **26**, 99–116 (1967).

[144] W. S. Landers, M. Gomez, and E. O. Wagner, U.S. Bur. Mines Rep. Invest. No. 6289, 1963.

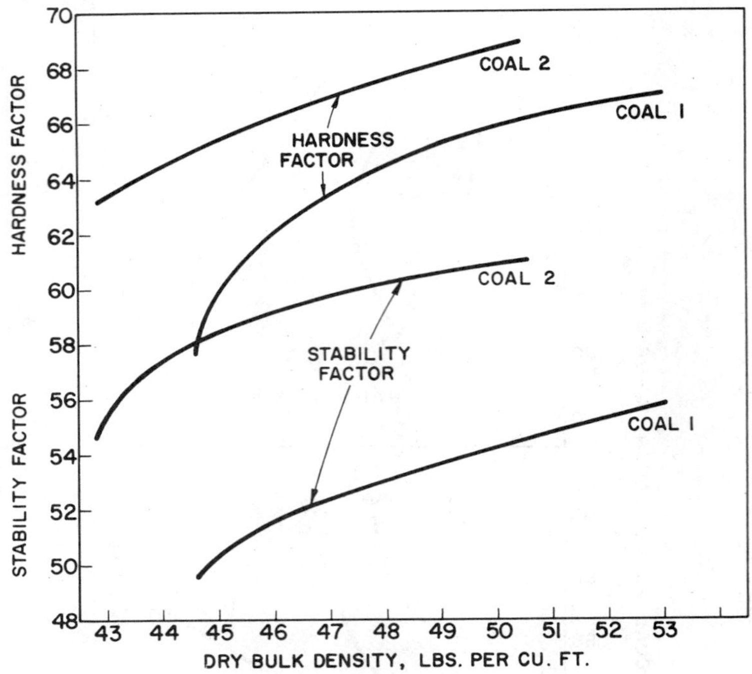

Figure 15.20 Effect of dry coal bulk density on coke tumbler test properties.[142]

It should be pointed out that although the increased coking rate reduces the size of the coke, the size becomes more uniform. For example, the 3- × 1-in. coke produced from the Eastern coal blend increased from 60.0 to 79.0% in the range of coking rates employed, and for the British coal the same size increased from 65.5 to 81.9%. Faster coking rate and the resultant "*in situ* crushing" is therefore an advantage to the iron makers who now crush coke to produce a closely sized burden for the furnace.

Since the first Supplementary Volume was published, two new technologies have been commercialized, and the effect of these on coke properties will be discussed. These new

Table 15.12 Effect of Coking Rate on Yield and Quality of Three Cokes[90,143,144]

	Eastern U.S. High-Rank Coal Blend[a]			Western U.S. Low-Rank Coal Blend[a]				British Medium-Rank Coal		
Coking Rate, in./hr	1.0	1.3	1.6	0.6	0.9	1.1	1.3	0.81	0.98	1.16
Coke size, (in.)										
+ 4	8.0	4.0	0.0	—	—	—	—	5.5	2.5	0.8
+ 3	33.0	22.0	11.0	—	—	—	—	26.5	18.7	11.0
+ 2	69.0	55.0	40.0	—	—	—	—	68.2	62.0	56.7
+ 1	93.0	91.5	90.0	—	—	—	—	92.0	93.1	92.9
Mean size (in.)	2.50	2.18	1.80	2.78	2.64	2.45	2.20	2.42	2.26	2.12
Shatter index, 1½-in.	—	—	—	88.0	86.0	82.5	78.0	74.5	74.6	72.6
Drum test										
Stability factor or M_{40}	57.0[b]	51.0[b]	45.0[b]	38.0[b]	38.0[b]	37.0[b]	33.0[b]	41.6[c]	43.2[c]	43.4[c]
Hardness factor or M_{10}	69.0	71.0	73.0	57.0	60.0	62.0	63.0	11.1	10.6	10.0
Bulk density (lb/ft^3)	—	—	—	28.0	29.0	29.1	28.2	24.7	24.5	24.9
Coke yield (%, mf)	—	—	—	68.8	70.3	70.7	70.9	69.0	68.7	68.9

[a] Values interpolated from curves.
[b] ASTM tumbler test.
[c] Micum test.

developments are the preheating of coal charges and the partial briquetting of the charge. A third development—formcoke—is in the demonstration plant stage and is discussed later in this chapter.

The effect of preheating on the coke properties has been described as a result of experimental studies by several investigators.[145–149] In general, these investigators have shown that the quality of coke from high-oxygen or marginally coking coals can be substantially improved by preheating the coal before carbonization.[146] In addition to reductions in the required coking times of 30–45%, preheating increased the 1 1/2-in. shatter index by about 3–6% and the M_{40} index by about 7–12%, and reduced the M_{10} index by 1.5–2.5%, depending on rank.[148] Cokes with stability factors of 50 or more usually could not be strengthened by preheating, but those with stability factors below 50 could be made substantially stronger by preheating. The low-volatile coal content of coal blends to the oven could be reduced substantially, and the loss of coke strength restored by preheating.[147]

At least three preheating systems have been commercialized and a number of plants constructed in the United States, Germany, and Great Britain. Publications[150–152] that describe these systems and plants include information on the properties of the cokes produced in comparison to the cokes produced from wet conventional charges. Some of the data from these publications appear in Table 15.13. In general, these commercial data confirm the conclusions made in the experimental studies and show that preheating improves the ASTM tumbler indices, and Micum indices, and the shatter indices, and that coke size remains essentially the same. However preheating tends to produce a higher degree of pulverization of the coal, which could contribute to increased coke strength.

The second technology that has recently been commercialized—partial briquetting of the coal charge to the coke oven—was advanced by developmental work in Japan. Although stamping of complete charges as a means of densification to improve coke from poorly coking coals has been practiced for many years, the earliest large-scale work on partial briquetting appears to have been done in Germany,[153] where poorly coking Saar coal, finely pulverized, was briquetted with about 8% water as the binder. Many of these briquettes broke up in handling, but the densification resulted in the production of a satisfactory coke from this coal, which was not possible by conventional means.

After a number of years of development, the Nippon Steel Corporation announced in 1975 that it had perfected its Briquette-Blend coking process, which was commercially installed in three plants. As a result, Nippon was able to decrease the amount of good coking coal in the charging blend from 70 to 50–55% and replace it, along with a tar or pitch binder, with a noncoking or weakly coking coal that could not be used otherwise, all this with no deterioration in the strength of the coke. If the coke strength is below that required, it can be improved by 2–4% if about 30% of the charge is in briquetted form. It was claimed this process extended the range of possible coals that could be used. In the evaluation scheme (MOF diagram) shown in Fig. 15.16, the limit of mean reflectance was extended from 1.2–1.3 to 0.95–1.3, and the Gieseler fluidity limit was

[145] F. W. Smith, G. W. Birge, D. E. Wolfson, and D. A. Reynolds, U.S. Bur. Mines Rep. Invest. No. 5418 (1958); D. E. Wolfson, U.S. Bur. Mines Rep. Invest. No. 7630, 1972.

[146] M. Perch and C. C. Russell, *Blast Furn. Steel Plant,* **47,** 591–597 (June 1959).

[147] H. W. Jackman and R. J. Helfinstine, Illinois State Geol. Survey Circs. Nos. 423 and 434, 1968; Nos. 449 and 453, 1970.

[148] J. P. Graham, *AIME Ironmaking Proc.,* **30,** 361–381 (1971).

[149] P. Foch, "Dry Charging and Preheating of Coking Blends: Development of the Processes," *Proc. Conf. Coke in Ironmaking,* London, December 10–11, 1969, publication No. 127, Iron and Steel Institute, London, pp. 47–52, 1970.

[150] D. G. Marting and R. F. Davis, Jr., *AIME Ironmaking Proc.,* **31,** 174–184 (1972); Charles E. McMorris, *AIME Ironmaking Proc.,* **34,** 330–338 (1975).

[151] D. A. Cooper, P. J. De Koker, J. Barnes, and H. J. van der Merwe, *AIME Ironmaking Proc.,* **35,** 450–462 (1976); W. Staniforth, *COMA Yearb.,* 208–231 (1974); Waverly House, Wexborough, Yorkshire, England.

[152] K. G. Beck, *COMA Yearb.,* 230–242 (1974); Waverly House, Wexborough, Yorkshire, England; C. Flockenhaus, W. Rohde, and D. Habermehl, *AIME Ironmaking Proc.,* **23,** 382–390 (1974).

[153] Karl Hoehne, *Glueckauf,* **96**(19), 1203–1219 (September 10, 1960); H. Hoffman and K. Hoehne, *Brennstoff-Chemie,* **40**(4); 129–136 (1959); **40**(7), 220–227 (1959).

Table 15.13 Comparison of Coke Properties Preheated Coal Versus Wet Coal in Commercial Installations[150–152]

System	Properties	Wet Charge	Preheated
Pipeline Charging[150]			
Allied-Ironton	Coke size, +3 in.	11.8	9.0
	Coke size, +2 in.	47.6	48.0
	Coke size, +1 in.	94.9	95.4
	Stability factor	51.8	55.2
	Hardness factor	62.1	64.5
Inland Steel	Stability factor	<60	>60[a]
Larry car charging[151]			
Brookhouse	Coke size	Same	Same
	Shatter index, 1½ in.	87.5	90.0
	M_{40}	68.0	74.0
	M_{10}	11.5	8.0
Iscor			
64% blend	M_{40}	60.4	68.4
	M_{10}	14.2	9.7
75% blend	M_{40}	58.0	67.6
	M_{10}	16.0	10.8
Conveyor charging[152]			
Emil plant	Coke size, >80 mm	65.0	61.2
	Coke size, 80 × 60 mm	19.5	21.3
	Coke size, 60 × 40 mm	11.7	13.5
	Coke size, 40 × 0 mm	3.8	4.0
	M_{40}	84.2	85.8
	M_{10}	6.0	6.8

[a] With reduced amount of medium-volatile coal in the blend.

extended from 80–1000 to 40–1000 divisions/mi.[154]

15.5 SPECIAL SOLID CARBONIZATION PRODUCTS

15.5.1 Coke Breeze

Coke breeze is not really a special product but is a by-product obtained from the coke oven, since it is not suitable because of its small size for use in the blast furnace or cupola. Coke breeze is sometimes collected in the form of separate fractions, but it usually consists of a mixture of screenings from the coke plant, at the blast furnace skips, quencher sump rakeout, sweepings, and so on. Depending on the coal and the operating conditions, breeze amounts to 4–8% of the coal. The uses for breeze are limited because of its fineness, its high moisture content, and its usually high ash content. Despite these limitations, it has found considerable use as a sintering fuel, in boiler firing, as inert material in foundry coke manufacture, for chemicals manufacture, for electric smelting, and in other applications.

The properties of breeze can vary widely depending on the source, the coal from which it is made, the carbonizing conditions, the screening process, and other factors. The breeze usually has higher volatile matter, ash, and sulfur contents than the lump coke from which it was separated. The breeze varies from plant to plant, depending on the bottom size of the coke desired for the furnace. Sieve analyses and proximate analyses of a number of coke breeze products are given in Table 15.14, as examples of the different properties.

[154] Taiyo Matsuo, South East Asia Iron and Steel Institute, "Recent Developments in Coking Processes Utilizing Low-Grade Coal," Manila Seminar, Philippines, (1976); *SEAISI Q.*, 22–34 (April 1976); T. Ikeshima, K. Sugasawa, and K. Akamatsu, "Developments in Metallurgical Coke Production and Applications of Those Developments to Latin American Coals," *Latin American Iron Steel Cong.* (ILAFA-16), Lima, Peru, September 1975, ILAFA, Santiago, Chile.

Table 15.14 Sieve Analyses and Proximate Analyses of Several Kinds of Coke Breeze

	Coke Breeze					
	Blast Furnace 1	Blast Furnace 2	Foundry 3	Foundry 4	Foundry 5	Beehive 6
	Sieve Analysis (%)					
On ½ in.	1.0	—	—	1.0	2.9	—
On ⅜ in.	8.0	—	—	11.0	16.7	—
On ¼ in.	15.0	26.7	2.0	23.0	19.7	4.6
On 6-mesh	24.0	13.3	23.0	15.0	20.5	20.0
Through 6-mesh	52.0	60.0	75.0	50.0	50.2	75.4
	Proximate Analysis (%, mf)					
Moisture	12.0	—	14.0	17.0	12.3	8.5
Volatile matter	1.2	2.2	—	—	1.7	10.5
Fixed carbon	88.2	90.4	—	—	91.1	68.0
Ash	10.6	7.4	—	—	7.2	21.5
Sulfur	1.12	0.60	—	—	0.58	1.04

The use of coke breeze as an inert in the manufacture of foundry coke or blast furnace coke was described previously. Its use in other applications is discussed in turn. In the sintering of iron ore fines, coke breeze and anthracite have been used predominantly, but chars, coals, and petroleum cokes have also been used. It is standard practice in sintering to provide sufficient heat energy for the sintering process by intimately mixing the proper amount of fuel with the raw charge. Particle size and volatile matter or reactivity of the fuel appear to be the most important properties. Bed permeability is very important. The presence of large fuel particles leads to local hot spots in the bed, as well as to a broad ignition zone. However a fuel of excessive fineness increases reactivity, decreases thermal efficiency, and adversely affects bed permeability. The fuel must therefore be properly sized for maximum production rate.

Myron et al.[155] made a systematic study of nine different fuels and found that the use of coke breeze yielded the highest production rates in sintering, and that the production rate tended to be at a maximum at 5% carbon; breeze also yielded the strongest sinter. Anthracite also gave good results, but high-volatile coal, lignite, petroleum coke, and fine anthracite gave poor results. Sieving out the 100-mesh fines from these materials improved the results. Dixon[156] found that the most suitable fuel must have low-volatile matter content, low reactivity, low ash content, and a top size of 3/16 in., with a minimum of dead fines. Bhrany and Pelczarski[157] studied the role of reactivity of fuel in the sintering process and concluded that the best sinter production rates are obtained when the fuel is relatively unreactive.

On the other hand, coke used as a reducing agent in electric smelting must be of high reactivity and resistivity. Generally breeze is mixed with larger sizes of coke to produce the proper reactivity and resistivity. Gasworks coke was ideal for this purpose in the past, but with the advent of natural gas this source has been eliminated. Coke made from straight high-volatile coal has been found suitable, although not in all respects. Deruelle[158] developed a continuous carbonization process conducted in rotary ovens, which produced a granulated coke low in volatile matter (4%), low in apparent density (0.63), high in porosity (64%), high in reac-

[155] T. L. Myron, M. R. Fagen, and R. L. Franklin, *AIME Blast Furn. Coke Oven, Raw Mat. Proc.*, **17,** 282–293 (1958).

[156] K. G. Dixon, *J. Iron Steel Inst.*, **200,** 120–125 (February 1962).

[157] U. N. Bhrany and E. A. Pelczarski, *Prepr., Am. Chem. Soc., Div. Fuel Chem.*, **6**(2), 157–168 (1962).

[158] J. Deruelle, "Manufacturing Process for Granular Coke for Electrometallurgy Using a Rotating Oven," *Inf. Symp., Coke Oven Techniques,* Commission of the European Communities, Luxembourg, May 6–7, 1975.

tivity (1 cm^3 CO_2/g at 1000°C), and high in resistivity [8000 Ω/(cm)(cm^2)], from high-volatile coals with a swelling index between 1 and 5. This synthetic coke has been used successfully by the electrometallurgical industry in France.

15.5.2 Semicoke or Char

Semicoke or char is not a conventional coke-oven product but is made in special equipment such as rotary retorts or fixed-bed, moving-bed, or fluidized-bed reactors, using low-temperature carbonization. A considerable amount of experimental work was done before 1960, especially in Great Britain, Germany, and France. An industry was developed for the production of domestic smokeless fuel as a result of the atmospheric pollution requirements in these countries. Semicoke has also been used in Europe in dust-fired power plants for many years. A comprehensive treatment of the subject of low-temperature carbonization was made by Wilson and Clendenin[159] in the first Supplementary Volume, and further discussion is not necessary. The U.S. Bureau of Mines published a collection of data from the low-temperature carbonization assays of some 400 North American coals.[160]

Since 1960, the Office of Coal Research in the U.S. Department of Interior (now in the Department of Energy), has sponsored a considerable amount of research on coal conversion. This research, dealing with the conversion of coal into liquid and gaseous forms of energy, produces a solid residue—char—which in a commercial venture would have to be used in the plant or sold in the marketplace to enable the process to be economically viable. As part of this research Battelle Memorial Institute assessed the potential markets for these chars and issued a report on the results of the study.[161] This report considered potential markets in electric power generation, sewage treatment, and in the manufacture of activated carbon, calcium carbide, blast furnace coke, foundry coke, and electric smelting reducing agents.

Table 15.15 shows the properties of chars produced by six experimental conversion processes using various coals, and Table 15.16 presents additional data on one of the chars (FMC) made from Illinois No. 6 coal. These chars are the residues from solvent treatment, pyrolysis, or partial gasification, and their properties depend not only on the rank of the coal, but also on the type and severity of the treatment. The ash contents of the chars range from moderately to considerably higher than those of the parent coals, and the sulfur contents are usually lower. Calorific values are usually lower for the chars than for the coals. Some of the values are inconsistently high on the ash-free basis because the analyses and calorific values were taken from two different tables and evidently pertain to different lots. All these chars show adsorptive properties that are approximately equal to those of commercial activated carbons for the purposes of waste water treatment and water purification. The uptake of amaranth red dye averaged about 1.2 mg/g for FMC and Consolidation chars, and 0.3 mg/g for the Institute of Gas Technology char.

15.5.3 Ferrocoke

After a considerable amount of work on ferrocoke, especially in the United States and Germany in the 1950s, interest in this special product died down considerably, judging from the amount of literature published since then. The plant at Everett, Massachusetts, of Eastern Gas and Fuel Associates, which conducted the most extensive manufacture and use of ferrocoke, operated successfully and economically for more than 2 years[162] and was shut down for other reasons. The Pokoj plant in Poland operated for at least a year,[163] but no reference to further operation of this plant can be found in the literature. In this plant about 7% Norwegian iron ore was added to the coking blend, resulting in an increase in M_{40} value from

[159] Philip J. Wilson, Jr., and J. D. Clendenin, "Low-Temperature Carbonization," in *Chemistry of Coal Utilization,* Supplementary Volume, H. H. Lowry, Ed., John Wiley and Sons, New York, (1963), Ch. 10.

[160] W. A. Selvig and W. H. Ode, U.S. Bur. Mines Bull. No. 571, 1957.

[161] H. W. Nelson, R. E. Schuler, M. J. Shilhan, and R. B. Engdahl, *Study of the Identification and Assessment of Potential Markets for Chars from Coal Processing Systems,* Office of Coal Research, Res. Dev. Rep. No. 44, 1969; NTIS PB-182966.

[162] G. Granger, M. J. Lach, and J. Mitchell, *AIME Proc. Blast Furn. Coke Oven Raw Mat. Conf.,* **18,** 152–161 (1959).

[163] W. Krause and Z. Kotas, *Koks Smola Gaz,* **3**(5), 174–180 (1958).

Table 15.15 Analyses and Properties of Various Chars for Office of Coal Research Coal Conversion Processes[161]

	Analysis (%, mf)				
	Volatile Matter	Fixed Carbon	Ash	Sulfur	Gross Heating Value (Btu/lb of dry char)
	FMC (COED)				
Pittsburgh-Federal	3.7	86.8	9.5	1.9	13,400
Illinois No. 6	3.5	76.4	20.1	3.1	11,870
McKinley	4.4	79.5	15.4	0.7	—
	Atlantic-Richfield (Seacoke)				
Illinois No. 6	4.2	76.9	17.7	3.1	11,630
McKinley	1.7	83.9	13.0	0.5	—
	Hydrocarbon Research, Inc. (H-Coal)				
Illinois No. 6	3–5	n.a.	44.0	n.a.	—
	Consolidation Coal Company				
Liquefaction Process					
Pittsburgh No. 8 (Ireland)					
80% depth of conversion	9.4	46.2	44.4	5.1	8,200
50% depth of conversion	12.9	62.8	24.3	4.6	11,200
Western high-volatile B					
80% depth of conversion	12.1	59.0	28.9	0.9	10,500
50% depth of conversion	14.6	71.4	14.0	0.9	12,500
Gasification Process—CO_2 Acceptor					
Pittsburgh No. 8 (Ireland)					
80% depth of conversion	1.0	56.4	42.6	1.0	8,300
50% depth of conversion	1.5	75.4	23.1	0.9	11,100
Western high-volatile B					
80% depth of conversion	1.5	69.7	28.8	0.2	10,300
50% depth of conversion	2.0	84.1	13.9	0.2	12,500
	Institute of Gas Technology				
Pittsburgh No. 8 (Ireland)	1.2	77.5	21.3	1.7	13,200

51.9 to 56.7 and in M_{10} value from 9 to 10.3. The use of this coke in the blast furnace reduced coke consumption about 7% in terms of fuel. In spite of favorable experience with the use of ferrocoke in the blast furnace, it has not become a widespread industrial practice. Some of the reasons that may be cited include the possibility of damage to the coke-oven walls, the extended coking time required, and higher sulfur retention in the coke. The economics for sintering and pelletization of iron ore may be more favorable in most locations.

A substantial amount of research was conducted by the BCRA[164] on the addition of blast furnace flue dust to coal. Two flue dusts of different composition and size consist were added to a high-volatile coal of about 7½ swelling index in amounts up to 30%. The addition of the flue dust increased the mean size, shatter index, and M_{40} index, but tended to reduce the abrasion resistance of the coke, confirming previous findings by others. The sulfur content of the coke was increased, but fine flue dust caused less retention of sulfur and had a more beneficial effect on the physical properties of the coke.

A considerable amount of research was conducted in recent years on the prereduction of iron ore by various techniques. Innes et al.[165] pelletized iron ore and coal or coke

[164] Anon., BCRA, Coke Res. Rep. Nos. 19 and 20, 1962.

[165] J. A. Innes, J. A. Gregory, and H. K. Worner, (Broken Hill Proprietary) Tech. Bull. No. 16, (April 1963), pp. 9–13; J. A. Innes, No. 17 (July 1963), pp. 28–34; J. A. Innes, *J. Met.*, **15**, 294–297 (April 1963).

Table 15.16 Analyses and Properties of FMC Char from Illinois No. 6 Coal[161]

Proximate analyses (% mf)	
Moisture	2.8
Volatile matter	3.7
Fixed carbon	72.5
Ash	21.0
Ultimate analyses (%, mf)	
Carbon	72.9
Hydrogen	1.2
Nitrogen	1.2
Sulfur	3.2
Oxygen	—
Ash	21.5
Gross heating value (Btu/lb)	11,595
Bulk density (lbs/ft³)	35.1
Hardgrove Grindability Index	114
Sieve analyses (cum wt %)	
on 8-mesh Tyler	2.1
16-mesh	27.0
28-mesh	59.7
48-mesh	81.3
100-mesh	93.7
200-mesh	98.0
325-mesh	100.0
Particle density (g/cm³)	0.964
Pore volume (cm³/g)	0.417
Surface area (m²/g)	61
Pore size (μm)	2.7

breeze with and without lime. Pellets of fine ore and coking coal were carbonized under appropriate conditions, yielding strong pellets in which the iron ore was partially reduced. Pellets produced from finely divided mixtures of iron ore and coke breeze were highly metallized (> 90%) and had high crushing strengths (200–400 lb on approximately 1/2-in.-diameter spherical pellets). By varying the amount of coke breeze, it was possible to control the amount of iron ore that was reduced and the percentage of residual carbon. A similar concept was commercialized in the Dwight–Lloyd McWane (D–LM) process in Mobile, Alabama, in which the ore concentrate, noncoking coal, and limestone were pelletized and carbonized on a sinter strand, and the reduced pellets provided the feed to an electric furnace.[166] Nandi and Iyengar[167] made ferrocoke briquettes from lignite-magnetite-lime mixtures. These briquettes had high compressive strengths (1000 psi) and a high metallization (64%) after carbonization at 900°C.

15.5.4 Formcoke

With the dwindling sources of prime coking coals in many parts of the world, and the more stringent requirements of the blast furnace, there has been greater incentive in recent years to develop a suitable coke from inferior coals. Besides the developments of the preheating of coal charges, the use of antifissurants, and the partial briquetting of coal charges, all of which make use of the coke oven, a development that should be mentioned is that of formcoke (Chapters 11, 13 and 14), which began in earnest just before 1960. Prior to this development considerable work on briquetting, particularly of noncoking coals, had already been done; a great deal of research and development is now being carried out, and the literature on the subject is voluminous. Emerging from this research are a number of processes that have attained demonstration or semicommercial status. Since comprehensive treatment of these formcoke processes is given in Chapters 11 and 14, this section is limited to the properties of formcokes, particularly those from the well-developed processes. Discussion of the behavior of formcoke in the blast furnace is included in Section 15.6.2.

The pertinent properties of formcokes produced by 14 advanced processes around the world, given in Table 15.17, are taken from a number of publications describing these processes.[87,168–173] The properties vary over

[166] T. E. Ban and B. W. Worthington, *J. Met.*, **12,** 937 (December 1960); T. E. Ban, B. W. Worthington, and C. D. Thompson, *Mod. Cast.*, **39,** 95–102 (March 1961).

[167] H. C. Nandi and M. S. Iyengar, *J. Sci. Ind. Res.*, **19A,** 84–87 (February 1960).

[168] E. Ahland, J. Lehmann, and W. Peters, *AIME Ironmaking Proc.*, **31,** 285–300 (1972); E. Ashland, W. Goossens, W. Hermann, W. Peters, and J. Langhoff, Int. Iron and Steel Congr. Düsseldorf, (May 27–30, 1974); Paper No. 1.2.2.7, Verein Deutscher Eisenhüttenleute, Düsseldorf; J. K. Holgate and P. H. Pinchbeck, *J. Iron Steel Inst.*, **211,** 547–566 (August 1973).

[169] R. W. Shoenberger, *Prepr., Am. Chem. Soc., Div. Fuel Chem.*, **20**(4), 46–69 (1975).

[170] K. Sugasawa, E. Hirao, T. Sakai, and D. Wagener, "DKS Formed Coke Process," Latin American Iron and Steel Congress (ILAFA), Mexico City, July 26–30, 1976, ILAFA, Santiago, Chile.

[171] M. Burteaux, M. Lescour, F. Lesur, M. Cordier, and M. Soubrier, *AIME Ironmaking Proc.*, **34,** 15–31 (1975); L. Soubrier, "The HBN Process for the Manufacture of Formed Coke," *Inf. Sym.*,

a broad range because of the wide variety of materials used and the various techniques employed to produce formcoke. The coals range from anthracite to subbituminous, and the processes include briquetting, pelletizing, and extrusion with binders of tar, pitch, coking coal, and a hydrogenate. The treatment prior to agglomeration can be drying, charring, and heating to within the plastic range of coal, and after agglomeration the coal can be oxidatively treated at low temperature to achieve carbonization at temperatures up to 1000°C. The shapes vary from pellets to briquettes to extrusions, and individual weights of average pieces range from about 15 to 100 g. In analysis, the volatile matter can range from 0.5 to 10%, ash content from 5 to 16%, and sulfur content from 0.5 to 2.8%. The range for apparent specific gravity is 0.9–1.3; porosity ranges from 30 to 60%, and bulk density from 500 to 740 kg/m^3. With respect to coke strength, the M_{40} values range from 70 to 95, M_{10} from 4 to 15, and I_{10} from 14 to 52. This wide range of properties is sure to affect blast furnace performance and in each case puts a burden on the blast furnace operator to determine the optimum procedure for operation of the furnace.

15.5.5 Coke from Pyrolysis Tars

Cokes have been produced from pyrolysis tars such as coal tar, or from other coal-derived bitumens, then used for electrode carbon, armature brushes, and other special applications. In addition, tars and/or pitches have been added to coal charges destined for the coke oven to improve the properties of the coke, especially when there is lack of fluidity, as well as to increase gas and coal-chemical yields. Tars and/or pitches are also used as binders in the briquetting of coal. Section 15.4.2 described the use of pitch in the partial briquetting of the coal charge, and Section 15.5.4 described the use of pitch binder in the production of formcoke.

Das Gupta et al.[174] added various kinds of pitch to coal charged to the coke oven and found that hard pitch had a more beneficial effect on the strength of the coke than did medium or soft pitches. The addition of hard pitch improved the coke produced from two weakly caking coals of 20% volatile matter (dmmf), yielding cokes with stability factors that met metallurgical requirements. However the same addition to a weakly caking, high-volatile coal did not improve the coke sufficiently for metallurgical use. With increasing pitch addition, maximum coke strength occurred between 10 and 25% addition, depending on the coal. A similar study in Bulgaria[175] showed that the addition of 3% pitch to a coal blend containing no low-volatile coal reduced the breeze yield by half, increased the M_{10} index by 5 points, but had no effect on the M_{40}. With another blend containing 10–15% low-volatile coal, similar improvements in breeze production and M_{10} values resulted, in addition to a considerably improved M_{40} index. Petroleum pitch addition was not as effective as coal-tar pitch addition. It appears that for the addition of coal-tar pitch to be satisfactory, the pitch must be highly aromatic, with a C/H atomic ratio of 0.5–0.6, and must have a softening point of about 200°C, as was found by Sugasawa et al.[176] Two petroleum pitches with these properties replaced 2–4% of the low-volatile coal in the coking blends without deterioration of coke strength.

Although Koppers Company has carbonized coal-tar pitch in coke ovens in the United States for years, the practice was discontinued a number of years ago. The carbonization of the pitch provided an expensive but desirable coke for the manufacture of high-purity electrodes for use by aluminum companies. Apparently this coke could not compete with petroleum coke. According to the literature, coke from carbon-

---

Coke Oven Techniques, Commission of the European Communities, Luxembourg, (May 6–7, 1975), pp. 229–240.

[172] J. W. Eckerd, R. E. McKeever, W. S. Sanner, and P. L. Woolf U.S. Bur. Mines, Rep. Invest. No. 6383, 1964; J. W. Eckerd, R. E. McKeever, P. I. Woolf, and W. S. Sanner, U.S. Bur. Mines Rep. Invest. No. 7157, 1968.

[173] P. L. Woolf, U.S. Bur. Mines, Rep. Invest. No. 6717, 1966.

[174] N. N. Das Gupta, K. Raja, B. N. Sharma, R. Hague, and A. Lahiri, *Gas Coke,* **24,** 342–347 (1962).

[175] P. Takacs and J. Korbuly, *Freiberger Forschungsh., A,* **405,** 33–42 (1967).

[176] K. Sugasawa, Y. Kiritani, and R. Takahaski, "New Ways of Producing Metallurgical Coke," Int. Iron Steel Congr., Düsseldorf, May 27–30, 1974, Paper No. 1.2.2.2, published by Verein Deutscher Eisenhüttenleute, Düsseldorf.

Table 15.17 Properties of Formcokes Produced by Various Processes

	BFL		FMC:	Ancit:	Consol-BNR:	DKS:	HBNPC:
Coal:	Ruhr	U.K.	Wyoming	Ruhr	Pennsylvania	Japan	France
Weight (g)	44–101	60	16	50–58	1¾ × ½ in. pellets	65	29.5
Volatile matter, (%, mf)	1–10	7	3.7–6.2	7.3–7.7	0.7	0.5–1.3	2–3
Ash (%, mf)	6.7–7.8	13	5.3–7.6	7.1–8.3	11.0	9.5–10.5	8.0
Sulfur (%)	0.85–0.91	0.98	0.72–0.75	0.85	0.9	0.47–0.51	—
Kg-cal/gm gross	7150–7460	—	—	—	7090	—	—
Apparent specific gravity	—	1.05	0.90	—	—	—	0.85–1.20
True specific gravity	—	1.63	1.69	—	—	—	—
Porosity (%)	39–59	35	47–53	25	—	32	30–50
Bulk density (kg/m³)	495–560	580	506–555	630–649	626	525	550–700
Micum 40	91.7–92.6	79.1	94.2[a]	95.4	—	94	—
Micum 10	6.7–7.8	12.9	4.1–7.0	4.4–6.2	—	5.6–7.5	7–9
Irsid 40	79.0[b]	31.8	76.5[a]	84.1[b]	—	—	—
Irsid 10	20.8	51.7	22.2	14.8	—	—	22–30

[a] Micum 30 or Irsid 30.
[b] Irsid 20.
[c] Micum 20.
[d] ASTM tumbler, 1400 revolutions, + 1 in. = 29–56%, + ¼ in. = 53–58%.
[e] ASTM tumbler, 700 revolutions, + ¼ in. = 85%.

ized coal-tar pitch is being produced in the U.S.S.R., Germany, and Japan. Some research work has been reported. Lapina et al.[177] described the thermochemical and structural changes that occur in coal-tar pitch of 70°C softening point when it is carbonized at various temperatures up to 800°C. Table 15.18 presents a compilation of these data. As the temperature is increased, the carbon content increases at the expense of asphaltenes and malthene constituents, until at 500°C a solid semicoke structure is developed and pitch is no longer thermoplastic. The softening point, the C/H ratio, and the volatile matter loss increase rapidly with temperature, especially in the range of 400–500°C.

A new development in the carbonization of coal-tar pitch is the delayed coking process, which has been used successfully by the petroleum refiner for the production of petroleum coke. At least two plants were built and operated in Japan,[178] using coal-tar soft pitch as feedstock and producing a pitch coke having volatile matter content of 7.5–9.5%, nil ash, and 0.28% sulfur, and a bulk density of 61.2 lb/ft³.

Another new development in the technology of producing electrode carbon is that of needle coke, which is gaining more and more prominence. This involves production of a material having a crystalline structure of high density, low resistivity, and low coefficient of thermal expansion, which is used for nuclear reactors and aerospace components, as

[177] N. A. Lapina, V. S. Ostrovskii, M. V. Averina, V. G. Nagornyi, and K. I. Syskov, *Coke Chem. (USSR)*, No. 12, 29–32 (1975).

[178] A. J. Gambro, D. Shedd, H. W. Wang, and T. Yoshida, *Chem. Eng. Prog.*, **66**(5) 75–79 (1969); Raul Remirez, *Chem. Eng.*, **76,** 74–76 (February 1969).

Table 15.18 Changes in Coal-Tar Pitch Properties on Carbonization[177]

Temp. (°C)	Loss of Volatile Matter (%)	Volatile Products (ml/g) H_2	CH_4	Malthenes (%)	Asphaltenes (%)	Carboids (%)	Softening Point (°C)	C/H Ratio	Microhardness (kg/mm²)
250	6.5	—	—	35	31	32	50	1.70	—
300	12.8	—	—	31	30	37	55	1.80	8
350	14.6	—	—	26	28	41	70	1.85	—
400	15.4	3.04	1.05	24	20	54	105	2.15	11
450	44.8	18.20	13.80	9	8	76	185	2.15	17
500	53.3	25.95	5.90	—	—	100	265	2.40	22
600	60.0	50.60	3.50	—	—	100	—	2.53	22
700	61.5	73.10	0.10	—	—	100	—	4.50	50
800	—	—	—	—	—	100	—	—	53

Carmonoix: France	IChPW: Poland	ICEM: Rumania	Clean Coke: Pennsylvania	USBM: Pennsylvania Anthracite	Auscoke: Australia	Sapozhnikov: U.S.S.R.	INIEX: Belgium
—	—	100	1¼ × ¾ in. pellets	—	50–87	—	20–26
2.2	8.5	—	1.0	1.0–1.4	0.5	—	3.9
9.5	7.9	12.5–15.8	7.5	9.8–12.3	—	8.8	9.0
0.74	0.48	1.8–2.8	0.5	0.7–0.8	—	1.45	—
—	6790	—	7430	6910–7190	—	—	—
1.25	—	—	1.00	1.28–1.33	1.04–1.25	—	1.18
—	1.66	—	1.93	—	—	1.70	—
38	—	—	48	—	—	50	34.5
—	—	—	512	640–740	—	—	—
—	83.3	70–78	—	—	—	89.9	93.6$_c$
—	14.6	7–10	—	—	—	6.0	5.7
—	—	—	—	—$_d$	85$_b$	—	74.7$_b$
—	—	—	—	—$_d$	14	—	23.6

well as in conventional graphite applications. It is produced predominantly from selected petroleum feedstocks by delayed coking techniques, but a patent by Shea[179] describes a way of producing it from coal-tar pitch. According to the patent, highly aromatic coal-tar pitch is a suitable feedstock, but only after removal of the insoluble matter (in quinoline) by centrifugation, or other means, followed by coking, taking care to eliminate high-temperature gradients and any agitation, especially when the charge is solidifying. An electrode made from such coke by conventional means can have an apparent density of 1.53, electrical resistivity of 26×10^{-5} Ω/in.^{3}, and a coefficient of thermal expansion of 18×10^{-7}/°C. A similar electrode made by the same procedure using untreated tar would have values for the properties above of 1.51, 53×10^{-5} Ω/in.^{3}, and 45×10^{-7}/°C, respectively.

15.6 EVALUATION AND SPECIFICATION OF COKE

15.6.1 Chemical and Physical Properties

Many publications have discussed the characteristics of good blast furnace coke. It is generally agreed that chemical impurities such as ash and sulfur should be as low as is economically feasible, and that coke strength should be as high as possible. The properties on which there is unanimity of opinion by users of coke have been gathered by Clendenin,[90] and the following is a digest of his list:

1. Physical and chemical properties, especially size and strength, and ash, sulfur, and moisture contents should be as uniform as possible.
2. The design of the furnace and the operating conditions determine the optimum properties of the coke required.
3. In addition to being uniform, the ash, sulfur, and moisture contents should be as low as possible, and the strength as high as possible, consistent with economic practice.
4. The size of the coke should be compatible with the sizes of the other burden materials, so that optimum gas permeability is obtained.
5. High production rates at low coke rates are attained with well-sized coke.
6. Larger furnaces with higher wind rates require stronger coke and are more sensitive to changes in coke quality.
7. Use of a strong coke results in maximum yields of furnace coke and minimum yields of breeze outside and within the furnace.
8. Weak coke degrades into small coke and breeze, outside and within the furnace,

[179] Frederick L. Shea, Jr., U.S. Pat. 2,775,549 (1956), British Pat. 847,840 (1960).

leading to poor gas permeability and poor furnace operation.

Numerical values for the chemical and physical properties describing the required coke have been derived from the experience of blast furnace operators over many years. Specifications for the desired properties are necessary to ensure agreement between the producer and the user; these agreed specifications can be the basis for bonus or penalty arrangements. The overall effect of using these specifications is consistency in coke quality, so important in metallurgical process performance.

The specifications for coke vary from nation to nation as well as within national regions. This variation is due to differences in the coal types available for coking, differences in the size and design of the blast furnace, and sometimes the whim of the operator. Several countries, such as the U.S.S.R. and Great Britain, have rigid standard specifications, but most countries do not have uniform national specifications. Each country and each operating plant is at the mercy of the kinds of coal available, with respect both to purity and to coking properties. This problem has become acute in many countries because of the lack of suitable coals or their depletion. As mentioned before, high ash content is not desirable because it requires more flux addition and more coke to melt the slag, thus adding to the cost of iron production. High sulfur content is undesirable for the same reason. Fortunately, however, many plants do not have to use coke that is highly variable in ash and sulfur contents. If the ash or sulfur content is high, it is consistently high, and the blast furnace burden can be adjusted accordingly.

Some of the chemical and physical specifications available in the literature for blast furnace cokes, according to country, have been assembled in Table 15.19; where definite specifications were not available, values

[180] J. P. Graham, "Specifications of the Properties of Coke for Ironmaking," *Proc. Iron and Steel Institute Conf. Coke Ironmaking*, London, December 10–11, 1968, Vol. 127, 1970, pp. 141–148. "Recent Developments in Metallurgical Coke Production," *Steel Times Annu. Rev.*, **203**(8), 739–748 (1975).

Table 15.19 Worldwide Specifications for Blast Furnace Cokes

	United States[a]		United Kingdom[180]	West Germany[180,183]
	Average	Range		
Moisture (%)	—	—	<3	<5
Volatile matter (%)	0.9	0.7–1.1	—	—
Ash (%)	8.2	6.6–10.8	<8	<9.8–10.2
Sulfur (%)	0.73	0.54–1.11	<0.6	<0.9–1.2
Size (mm)	>20 (¾ in.) or 20 × 51 (¾ × 2 in.)		20 × 63 (¾ × 2½ in.)	40 × 80
Shatter (1½-in. index)	88	—	>90	—
ASTM Tumbler				
Stability factor	56	51–62	—	—
Hardness factor	68	62–73	—	—
Micum				
M_{40} index[d]	—	—	>75	>84
				<6
M_{10} index[e]	—	—	<7	
Irsid I_{10} index	—	—	—	—
JIS D^{30}_{15} index	—	—	—	—

[a] No standard specifications (from Table 15.10,[121] coke produced 1966–1975 in United States).

[b] Depending on grade or class of coke.

[c] Maximum minus 25 mm = 3–7%.

[d] For Soviet cokes, high values for Donetz basin, low values for Eastern region.

[e] For Soviet cokes, low values for Donetz basin, high values for Eastern region.

for the general practice were used. General values are given for the United States because individual companies generally lay down specifications that can vary from plant to plant as well as from region to region. There are uniform specifications in Great Britain, the U.S.S.R., and Poland, where the industry is nationalized. In the U.S.S.R., for example, there is a very detailed set of standards (GOST 18686-73) in which coke size is specified by 11 grades and coke strength must meet three specifications. In addition, a recent publication[184] proposed new specifications for coke for the new generation of large blast furnaces in the U.S.S.R. This proposal recommended ash content of less than 9%, volatile matter below 1%, M_{25} values greater than 90, M_{10} values less than 6, coke size of 25 × 60 mm, with not more than 10% oversize and not more than 2 1/2% undersize, as well as low moisture content and low reactivity.

Table 15.20 was drawn up in a similar manner for the specifications for foundry coke. As with blast furnace coke, there are variations from country to country and from plant to plant. In general, the specifications for ash and sulfur contents in the coke are more stringent for foundry coke, and the cokes must be larger and stronger. Prime coking coals are generally selected in the manufacture of foundry cokes.

One other issue with respect to specifications should be mentioned. In 1960 the Coal Committee of the Economic Commission for Europe projected an international classification scheme by type and by size for high-

[181] P. Foch, *COMA Yearb.*, 304–323 (1974); Waverly House, Wexborough, Yorkshire, England.

[182] Anon., Reference Data (on coke quality standards), *Coke Chem. (USSR)*, No. 2, 52–53 (1962); GOST Standard 18686-73; S. G. Aronov, L. P. Semisalov, and I. M. Iazovskii, *Coke Chem. (USSR)*, No. 11, 13–18 (1974).

[183] G. Juranek and M. U. Otte, "The Production and Evaluation of Blast-Furnace Coke in West Germany," *Symp. Science and Technology of Coal*, Ottawa, Canada, March 29–31, 1967, pp. 121–126. Published by Mines Branch, Department of Energy, Mines, and Resources, Ottawa.

[184] Z. I. Nekrasov, *Koks Khim.*, No. (2), 8–10 (1976).

[185] J. D. Doherty and J. A. DeCarlo, *Blast Furn. Steel Plant*, **55**, 141–153 (February 1967).

France[180,181]	U.S.S.R.[b,180,182]	Poland[b,180]	Czechoslovakia[180]	Japan[180]	India[113]
—	<5–12	<6	—	3–4	—
—	<1.4–1.8	—	—	—	—
6.7–10.1	<10–12	<11.5–12.5	—	10–12	23–25
0.7–1.0	<1.79–2.00	—	0.6–0.8	0.6	—
40 × 80	40 × 80[c]	>40	>40	15 × 75	—
40 × 60	25 × 80[c]				
—	—	—	—	—	>85
—	—	—	—	—	>40
—	—	—	—	—	—
>80	I. >73–80	I. >63–69	>70–75	>75–80	>75
<8	II. >68–75	II. >52–63			
	III. >62–70	III >45–52			
	I. <8–9	I. <8–10	<7–8	—	<
	II. <9–10	II. <12			10–12
	III. <10–14	III. <13			
<20	—	—	—	—	—
—	—	—	—	>92	—

Table 15.20 Worldwide Specifications for Foundry Cokes

	United States[183]	United Kingdom[178]	France[179]	U.S.S.R.[180]	Czechoslovakia[178]
Moisture (%)	—	—	—	<4	—
Volatile matter (%)	<1.0	<1.0	1.0	—	—
Ash (%)	<7.0	<7.0	9–10	I. <12.5 II. <11.0 III. <12.0	—
Sulfur (%)	<0.6	<0.7	0.7	I. <0.6 II. <1.0 III. <1.4	0.6–0.8
Size (mm)	76 × 127 (3 × 5 in.) 102 × 152 (4 × 6 in.) 127 × 178 (5 × 7 in.) +102 (4 in.) +127 (5 in.) +152 (6 in.)	—[a]	60 × 90[b] 90 × 150[c]	I. >80 II. 60 × 80 III. 40 × 60	>90
Shatter (full-length pieces)	>90	—	—		—
3-in. index	>95	>90	—	—	—
2-in. index				—	
Micum					
M_{80} index	—	—	>65–70	—	—
M_{40} index	—	>80	—	I. >75	>85
M_{10} index	—	<10	—	II. >68 <12	<7
Porosity	45–50	—	45–52	<42	—

[a] Size 1/8–1/12 of cupola diameter.
[b] For 600-mm-diameter cupola.
[c] For 1000-mm-diameter cupola and larger.

temperature coke.[186] Each coke type is represented by a two-digit code number; the first digit is the group number, based on the purity parameter (volatile matter, ash, and sulfur content), and the second digit is the class number, based on the strength parameter. Coke size is classified by the lower and upper size limits; a 40 × 80 mm coke would have the code number 48. This classification system does not seem to be finding much use in international trade.

15.6.2 Behavior of Coke in the Blast Furnace

The behavior of coke in the blast furnace has been the subject of numerous publications, as has the correlation of coke properties with the performance of the coke in the furnace. Since the first Supplementary Volume, there has been a drastic increase in the size and productivity of the blast furnace, with corresponding reductions in coke rate. The most spectacular technological advances were made in Japan, where the average internal volume of 60 blast furnaces increased from a little over 800 m³ in 1960 to 2425 m³ (49 furnaces) in 1975, with some of the operating furnaces at or near 5000 m³. The daily average productivity was about 1.1 tonnes/m³ in 1960, and this was increased to about 1.87 tonnes/m³ in 1975. At the same time, the coke rate decreased from about 620 kg/tonne of hot metal in 1960 to 440 kg in 1975, but with an additional 52 kg of injected oil in 1975. Substantial but less impressive strides were made in other countries.

With a higher proportion of smaller, older furnaces, the U.S. coke rate has decreased in the same period of time from 750 to 605 kg/tonne of hot metal, with a corresponding increase in iron production. These improvements in furnace operation were accomplished by several factors, in addition to the use of larger furnaces, including coke improvement, higher top pressures, higher blast temperatures, oxygen enrichment, and agglomerated burdens. A detailed treatment of the practical operation of the modern blast furnace is given in a two-volume book edited by Strassburger.[187] The reader is referred to this work for information on all aspects of theory and practice of the blast furnace, including coke making, coke properties, and coke behavior in the furnace.

[186] I. F. Pakhalok, *Coke Chem. (USSR)*, No. 11, 29–30 (1961).

[187] *Blast Furnace—Theory and Practice*, Julius H. Strassburger, Ed., Gordon and Breach Science Publishers, New York, (1969).

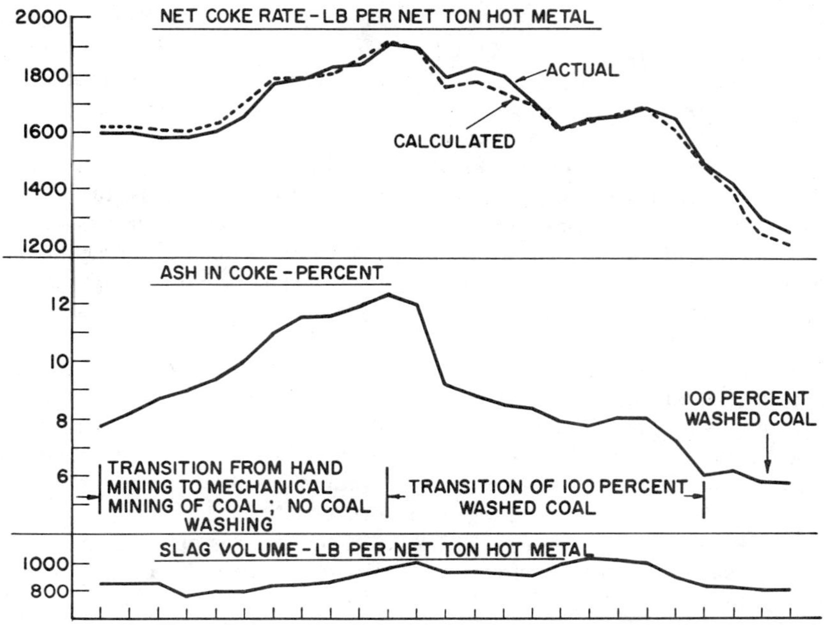

Figure 15.21 Coke rate chronological trend analysis for a practice with large changes in percentage of ash in coke.[36]

The Flint formula and its introduction of the term "effective carbon" has already been discussed (Section 15.2.2). This formula has been used successfully to predict coke rates in furnaces with various changes in operation, and Flint's publication[36] describes several of these changes. One example, the effect of coke ash, has been reproduced in Fig. 15.21, where the predicted versus actual coke rates compare extremely well for a variation in coke ash content from about 6 to 12%, over a period of 25 years of operation of several furnaces. During this period there was a gradual transition from hand mining of coal to mechanical mining, then to progressive installations of coal-washing facilities. Assuming all things to be equal, it appears that the variation in coke ash caused a variation in coke rate from about 1200 to 1900 lb/net ton of hot metal.

There have been other evaluations of the effect of coke ash on coke consumption. The exact relationship no doubt varies from furnace to furnace. Quigley and Sayles,[188] using Inland Steel's operating conditions, showed by use of the Flint formula that a 1% increase in coke ash content requires an 11% increase in flux and causes a 2.8% increase in coke rate. The effect of coke ash was demonstrated to them when an unsuspected increase in the coke ash of about 1.5% occurred suddenly, with the result that for 2 days more than half the iron was off-quality. In addition, Asai[92] reported a savings of 1.3% of the coke in Japanese furnaces for each percent decrease in ash in the coke. Craig[189] found a 1.6% increase in coke consumption in the furnace for each percent rise in coke ash content. With the high-ash Indian cokes, Ghosh and Roy[190] calculated that 1% more coke ash means about 2.5% more coke usage, and about 2.4% less production of hot metal.

There have been similar evaluations of the effect of sulfur on furnace performance. Section 15.2.2.2, using the Flint formula, showed that for a blast furnace operating at a base condition, about 5 kg of coke per net tonne of hot metal is required for each 0.1% change in coke sulfur content; where desulfurization must be achieved by increasing the slag basicity, about 15 kg of additional coke is needed for each 0.1% change in coke sulfur; and where silica must be added to increase slag volume, about 35 kg of additional coke is needed for each 0.1% change in coke sulfur. In an analysis of data from five British plants

[188] J. J. Quigley and N. Sayles, *AIME Ironmaking Proc.*, **33**, 157–166 (1974).

[189] J. Craig, *COMA Yearb.*, 114–125 (1973); Waverly House, Wexborough, Yorkshire, England.

[190] P. C. Ghosh and L. C. Roy, *J. Mines, Met. Fuels*, **23**, 114–115, 124 (February 1975).

by Ridgion and Whitehouse,[191] it was found that a decrease in the coke sulfur of 0.1% resulted in a decrease in the coking rate of 15–33 kg/tonne of hot metal, and a significant increase in metal production. Franke[192] evaluated data from four German furnaces and found that a 0.1% decrease in coke sulfur led to a decrease of about 22 kg coke/tonne of hot metal. Pinchbeck[42] found, as a result of three furnace campaigns in Great Britain, that a decrease of 0.1% sulfur in the coke resulted in a decrease of about 23 kg of coke/tonne of hot metal and an 8% increase in iron production. In addition, he reported the results of charging two different cokes having sulfur contents of 1.22 and 1.44% to the same blast furnace for consecutive 4-week periods. The higher sulfur coke required a 1% higher coke rate, and iron production was 11% lower. In contrast to the intensive efforts to charge coke with the lowest practical sulfur content, it was proposed by Agarwal and Elliott[193] to use less expensive, high-sulfur coke, and to externally desulfurize the hot metal after its discharge from the furnace. It was claimed that this concept may be economically justified.

It has long been recognized that the sizing of burden materials is very important for attaining good gas permeability in the stack, which results in greater furnace efficiency. Improvements were made when iron-bearing materials were agglomerated by sintering and pelletization and optimum sizes were charged. Some progress was also made in sizing of the coke. White[194] conducted plant tests and showed that as the percentage of coke in the range of 3/4 × 2 in. (20 × 51 mm) increased, there was an increase in the hot-metal production rate. In an experimental blast furnace Knepper and Sciulli,[195] using smaller, closely sized coke, were able to operate the furnace with a hot blast temperature 150°C higher than normal, thus reducing the coke rate by about 50 kg/tonne of hot metal. Kusuno et al.[196] and Zherebin et al.[197] reported the matching of coke, sinter, and ore sizes to improve permeability in a large blast furnace. This resulted in optimum operation of the furnace.

Other attempts were made to achieve savings of coke in the blast furnace. Ostrowski et al.[198] reported the use of ultrahigh blast temperatures in an experimental furnace and measured a decrease in coke rate of about 20 kg/tonne for every 100°C increase in blast temperature, and an increase of about 6% in iron production. Woolf[199] demonstrated in the same experimental furnace that prereduction of blast furnace burdens can result in large coke savings and greatly increased productivity, depending on the degree of prereduction.

Although it is believed that coke strength is a very important factor in the performance of the blast furnace, precise correlation has been difficult to achieve. Because there are so many variables affecting furnace behavior, it is necessary to apply statistical correlation methods. Normal operating practice is to permit as little variation as possible in the variables affecting the furnace, so that even with statistical methods it is difficult to obtain with good accuracy the degree of correlation necessary to relate coke strength to furnace performance. The Flint formula does not take into account any coke strength term.

In the first Supplementary Volume Thibaut[49] discussed the relationship of coke strength to furnace performance, and gave his proposed T-index, which combines the Micum or Irsid values with coke bulk density or apparent specific gravity. Although high correlation was claimed, no additional data have been found in the literature to support this relationship. On the other hand, Nixon and Brooks[58] had no success in correlating iron production or coke rate with the M_{40} index, although they found some relationship to the M_{10} index. In one case a one-point rise

[191] J. M. Ridgion and A. M. Whitehouse, *J. Iron Steel Inst.*, **184**, 249–257 (1956).

[192] W. Franke, *Neue Hütte*, **11**, 688 (1964).

[193] J. C. Agarwal and J. F. Elliott, *AIME Ironmaking Proc.*, **30**, 50–67 (1971).

[194] R. H. White, *Blast Furn. Steel Plant*, **51**, 968–973 (1963); R. H. White, *Blast Furn. Steel Plant*, **54**, 241–245 (1966).

[195] W. A. Knepper and C. M. Sciulli, *AIME Ironmaking Proc.*, **24**, 38–40 (1965).

[196] K. Kusuno, Y. Jyomoto, T. Kase, and T. Wada, *AIME Ironmaking Proc.*, **27**, 197–203 (1968).

[197] B. N. Zherebin, B. I. Ashpin, Zh. E. Steptsov, I. Z. Shatokha, and V. M. Dinel't, *Stal (USSR)*, No. 1, 6–13 (1972).

[198] E. J. Ostrowski, R. B. Schluter, and N. B. Melcher, *AIME Ironmaking Proc.*, **27**, 171–176 (1968).

[199] P. L. Woolf, *AIME Ironmaking Proc.*, **24**, 73–77 (1965).

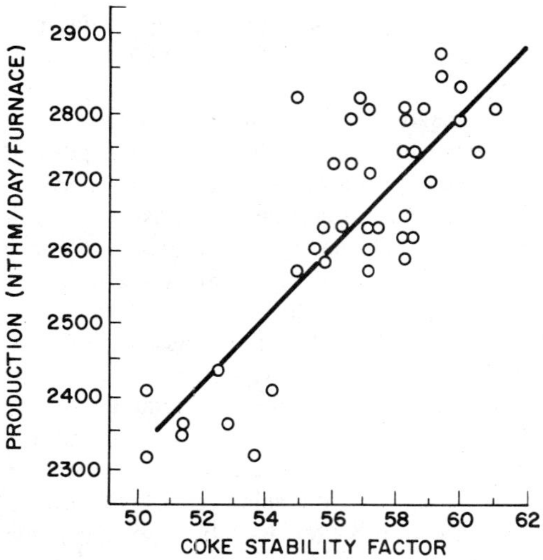

Figure 15.22 Effect of coke stability on production (NTHM = net tonnes of hot metal) for large furnaces.[188]

in the M_{10} contributed to a decrease in iron production of 10%, and in another case it increased the coke rate about 13 kg/tonne. Giedroyc and Matheson[200] related the M_{10} index to furnace permeability and to output in a blast furnace, but Simpson et al.[201] and Booley[202] found no correlation between Micum results and behavior in the furnace.

More success appears to be possible with correlating the ASTM tumbler stability factor and furnace behavior. Ostrowski[203] and George and Paulencu[204] presented correlations of coke rate with stability factor, but Quigley and Sayles[188] made regression analyses that provided quantitative relationships between stability factor and furnace performance. The relationship between iron production and coke stability obtained by Quigley and Sayles is illustrated in Fig. 15.22. The correlation coefficient for this relationship of 0.800 indicated that stability was positively associated with iron production, and that a change of about 64% in iron production is attributable to changes in stability, or that a 1% improvement in stability can lead to about a 2% increase in iron output. The statistical equation of the correlation is:

$$\text{production} = -40.131 + 47.29\,(\text{coke stability})$$

Figure 15.23 presents the relationship found between coke rate and stability factor. Quigley and Sayles[188] also conducted multivariate analysis of their data relating coke rate and iron output to pertinent furnace variables and reported the following predictive equations:

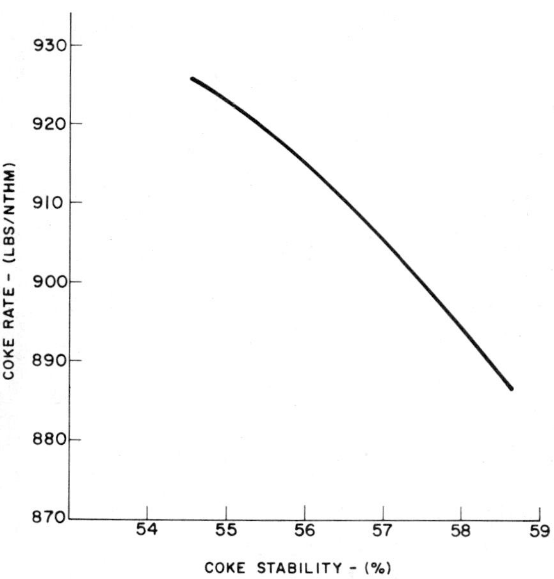

Figure 15.23 Effect of coke stability on coke rate.[188]

[200] V. Giedroyc and I. M. Matheson, "Influence of the Physical Quality of Coke on Blast Furnace Performance," *Proc. Iron and Steel Institute, Conf. Coke in Ironmaking,* London, (December 10–11, 1969), Vol. 127, 1970, pp. 123–130.

[201] J. G. Simpson, D. G. Williams, and E. Dodd, "Coke Properties of Importance to Blast Furnace Operation at Port Talbot Works," *Proc. Symp. Assessment of Physical Properties of Blast Furnace Coke,* London, (October 17, 1972), pp. 42–54. Published by National Coal Board, Stoke Orchard, Cheltenham, England.

[202] B. G. Booley, "Assessment of Coke Quality by Extended Abrasion Tests," *Proc. Symp. Assessment Physical Properties of Blast Furnace Coke,* London, (October 17, 1972), pp. 110–133. Published by National Coal Board, Stoke Orchard, Cheltenham, England.

[203] E. J. Ostrowski, *AIME Ironmaking Proc.,* **34,** 35–44 (1975); E. J. Ostrowski, *Ironmaking Steelmaking,* **2,** 34–36 (1975).

[204] D. W. R. George and H. N. Paulencu, *AIME Ironmaking Proc.,* **33,** 445–453 (1974).

Production (net tonnes/day) =
-564.05
$+12.85 \times$ coke stability
$-141.94 \times \frac{10^6}{(\text{slag production})^2}$ (net tonnes/day)
$+0.91 \times$ wind 10^{-2} (ft^3/min)
$+4.09 \times$ pellets (%)
$+5.32 \times$ coarse ore (%)
$+52.42 \times$ slag SiO_2 (%)
$-57.27 \times$ coke ash (%)
$+0.46 \times$ oil (net tonnes/day)

$R^2 = 0.913$

Coke (net tonnes/day) =
-18708.29
$+1072.06 \times$ stability
$-94.15 \times$ stability$^{3/2}$
$-147.64 \times \frac{10^6}{(\text{slag production})^2}$ (net tonnes/day)
$-0.75 \times$ oil (net tonnes/day)
$-214.51 \times$ average coke size (in.)
$+367.87 \times$ hot metal Mn (%)
$-337.25 \times$ (hot metal Mn)^2
$+0.35 \times$ wind 10^{-2} (ft^3/min)
$+711.92 \times$ slag sulfur (%)
$-14.54 \times$ slag CaO+ slag sulfur (%)
$-26.81 \times$ slag MgO (%)

$R^2 = 0.887$

A high degree of correlation was found. With iron production, most of the variation (84.4%) can be attributed to coke stability, slag production, and wind rate. For the coke rate equation, coke stability, slag production, and oil usage account for 78.2% of the variance.

Similarly, several correlations have been reported between the JIS coke strength and furnace operation, specifically gas permeability. Ikeda[205] found the relationship between gas permeability and JIS drum strength (D^{30}_{15}) shown in Fig. 15.24. Since the Japanese strive to maintain this strength index above 92, it is evident that good gas permeability, with resultant higher productivity and lower coke consumption, will be attained at the higher values of the index. Using multiple regression analysis, Ikeda also obtained the following equation relating coke size and strength and sinter variables to the same permeability index:

$$K(\times 10^{-6}) = -0.073D - 0.482\,D^{30}_{15} + 4.06X - 0.020SR - 0.015SI + 48.54$$

where K = permeability (ratio of pressure drop squared to blast volume)
D = mean coke size (mm)
D^{30}_{15} = JIS coke strength on 15 mm after 30 revolutions
X = ore/coke ratio
SR = sinter ratio in burden (%)
SI = sinter shatter strength (%)

[205] Tadashi Ikeda, "On the Coke for Use of Large Blast Furnaces," Latin American Iron and Steel Congress (ILAFA), Caracas, Venezuela (May 22–25, 1972). Published by ILAFA, Santiago, Chile.

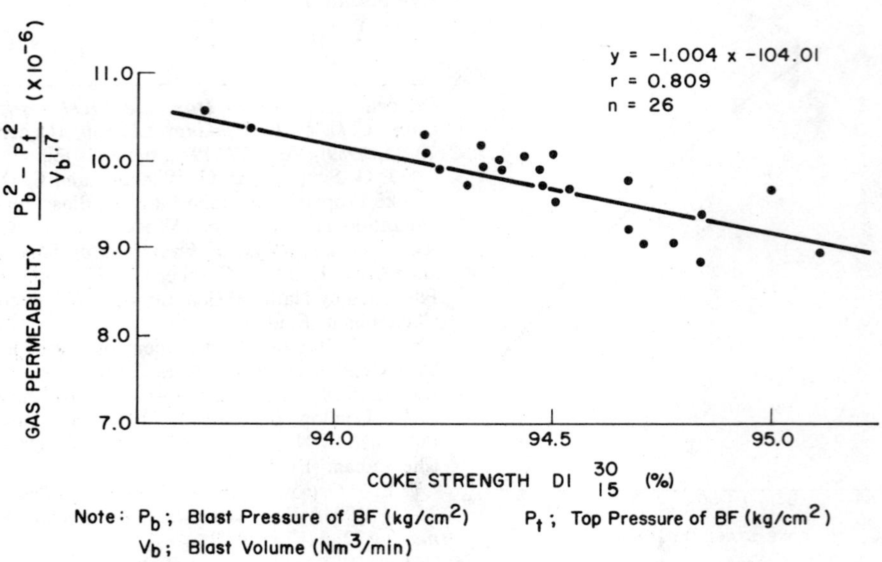

Figure 15.24 Relation between coke strength and gas permeability in the blast furnace.[205]

For 75 sets of data, the equation above had a correlation coefficient of 0.75. The equation shows that permeability is highly dependent on coke size and on strength and on the ore/coke ratio. The higher coefficient for the coke strength variable means that strength is a more important factor than coke size in governing permeability.

A new technique adopted in recent years by several investigators has been that of extracting coke from the tuyeres during short, temporary stoppage of the normal operation of the blast furnace, and determining the extent of coke degradation. The amount of coke degradation is then equated to the number of revolutions in the coke drum used for physical testing for equivalent degradation based on mean size. An ISO Committee SC3 document[206] contributed by Japan showed that coke deterioration in the stack of two blast furnaces was equivalent to 100 revolutions in the JIS drum or 240 revolutions in the ASTM tumbler. Similar studies in Great Britain[207] indicated coke degradation

[206] Observations on the Revolution Numbers of Drum for Coke Strength Test, ISO/TC27/SC3 Document 57 (1976).

[207] W. Hyslop and H. C. Wilkinson, "Consistent Iron," *Proc. Metals Society Technical Meeting*, London, April 24–25, (1974), pp. 38–51; Anon., BCRA, Res. Rep. No. 10 (December 1974).

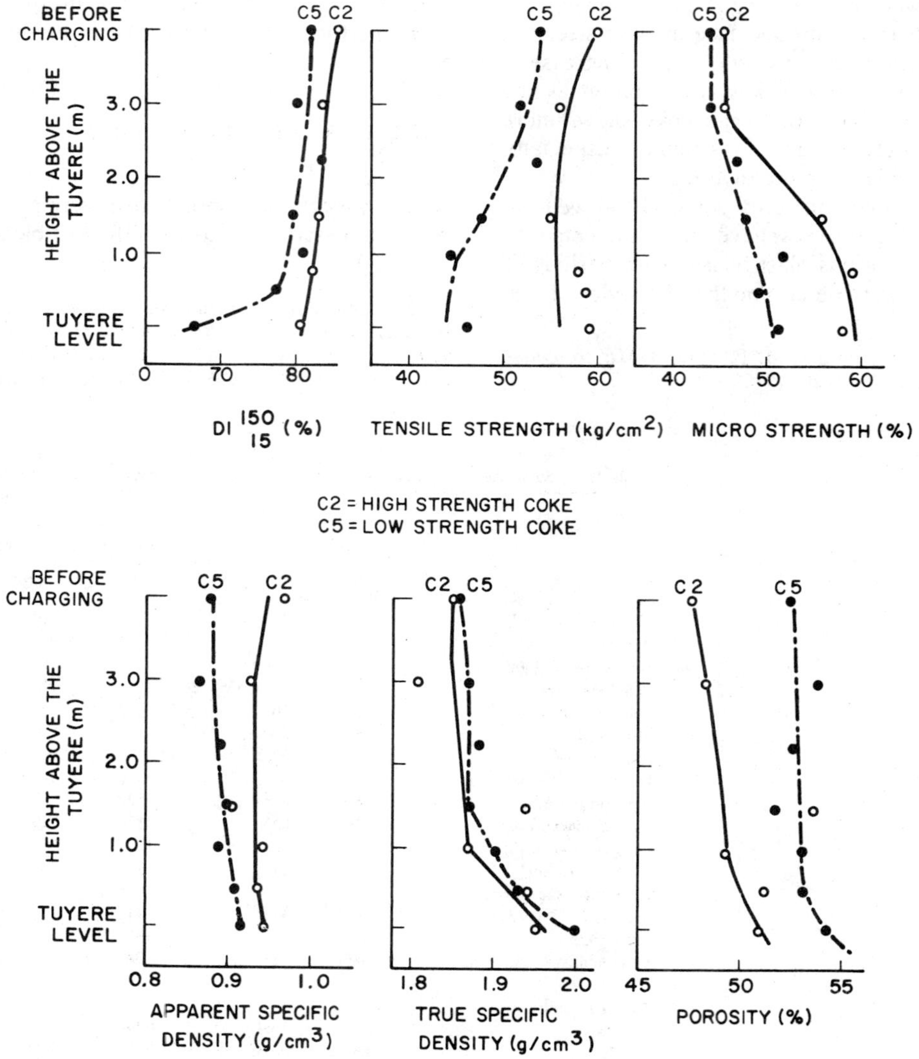

Figure 15.25 Changes of coke properties in the furnace.[208]

equivalent to about 90 revolutions in the Micum drum for a small furnace and about 390 revolutions for two large furnaces. It will be recalled for comparison that the standard coke-testing methods require 30 and 150 revolutions in the JIS drum, 1400 revolutions in the ASTM tumbler, 100 revolutions in the Micum, and 500 revolutions in the Irsid test. The degradation of both high- and low-strength coke as it proceeded down through an experimental blast furnace was measured, along with other properties, by Hatano and Fukuda.[208] Figure 15.25 shows the trends in JIS, tensile, and micro strengths, in apparent and true density, and in porosity properties of the two cokes at seven sampling points from the stockline down to the tuyeres. Samples were obtained soon after blowing out the furnace. The decrease in coke strength, except in micro strength, and the increase in porosity of both cokes are apparent as the coke descends; the weak coke shows more rapid deterioration in strength, especially near the level of the tuyere.

Formcokes made by some of the well-developed processes have been subjected to tests in various blast furnaces for periods of 3–30 days to determine their behavior. Table 15.21 was derived mostly from Schneider,[209] who summarized the pertinent results of these blast furnace tests. In each case, the results obtained by the use of conventional coke in the same furnace are included for comparison. Some of the results are for the total replacement of conventional coke, and some are for a replacement as low as 30%.

In general, the furnace performance with formcoke was encouraging, although tests were short. Because of the high bulk density and small size, the formcoke in all cases led to reduced permeability. In spite of this, iron production increased in some cases and coke rate decreased when formcoke was substituted for the normal coke. Dust loadings increased in most cases. The limited nature of the tests precludes definite conclusions as to the performance of formcoke in the blast furnace.

15.6.3 Behavior of Coke in the Cupola

Several aspects of foundry coke were previously discussed, such as the effect of blend-

[208] M. Hatano and M. Fukuda, *AIME Ironmaking Proc.*, **35**, 2–20 (1976).

[209] M. Schneider, "Utilization of Formed Coke in the Blast Furnace," *Inf. Symp. Coke Oven Techniques,* Commission of the European Communities, Luxembourg (May 6–7, 1975), pp. 251–262.

Table 15.21 Some Results of Industrial Blast Furnaces, Given with Hearth Diameters,

	FMC			
	British Steel, 5.5 m		British Steel, 5.5 m	
Formcoke (%)	0	100	0	100
Volatile matter (%)		3.7		7.0
Ash (%)		5.6		12.7
Bulk density (kg/dm³)		0.428		—
Pellet size (mm)		35 × 35 × 24		38.5
Pellet weight (g)		16		—
Micum 10		5.4		10–15
Irsid 10		22.2		—
Duration of tests (days)	—	6	—	5
Pig-iron production (tons/day)	805	873	752	587
Blast temperature (°C)	847	855	813	920
Dust (kg/ton of pig- iron)	12	13	12	24
Carbon content	37	21	39	48
Coke rate, (kg/ton of pig iron)				
gross	478	471	494	532
corrected	—	—	494	547
Permeability (reference = 100%)	100	92	100	69
Pig-iron quality				
% Si	1.05	1.04	0.86	0.76
% S	0.026	0.035	0.039	0.064
Slag residue	1.49	1.44	1.44	1.59

ing cokes of various sizes on the permeability, the tensile strength of foundry coke and its correlation with the Micum 10 index, the use of antifissurants in the manufacture of foundry coke, and the specifications in use worldwide for foundry coke. In the cupola, foundry coke not only provides the fuel for melting the iron charge, but also provides the carbon required for the composition of the cast iron. It therefore serves a thermal as well as a chemical capacity, a dual role that requires not only proper characteristics of the coke, but also the proper operation of the cupola.

Carter,[210] using a 24-in. experimental cupola, studied many of the possible variables affecting its operation, and verified the principles in large production cupolas. He studied the effects of coke composition, coke size, amount of coke, air volume, preparation of the bed and charge, bed height, size and type of scrap, control of slag, and other operating variables. This comprehensive study provided the principles of cupola operation and showed how cupola reactions could be controlled to produce consistent results.

An example of the findings of this study is given in Fig. 15.26, which shows the effect of average coke size on metal temperature and on metal chemistry. There is an optimum size of coke that leads to the maximum metal temperature; maximum temperature was obtained when the average coke size was 8–15% of the diameter of the cupola. Cokes smaller and larger than this optimum size tended to lower temperatures. Carbon pickup followed the same trend as the temperature curve. The sulfur content of the iron increased progressively as smaller sizes of coke were used. Silicon and manganese losses were lowest with intermediate coke sizes, and the amount of FeO in the slag was also lowest. The tests also showed that metal temperature reached a maximum at about 14% of the metal charge, but that more coke, with the appropriate amount of air, produced hotter iron, although the melt rate decreased. With more coke, the carbon and sulfur contents of the metal increased and the oxidation losses decreased.

The effect of the ash content of the coke on carbon pickup by the iron was discussed. With ash contents below 2%, a carbon content in the iron of nearly 4.80% was obtained. With an ash content of 15% in the coke, the greatest carbon pickup by the iron would be 2.70%. Sulfur pickup by the iron is also determined by the sulfur content of the coke. Iron sulfurs tend to be in the range of 0.120–0.160% if the coke sulfur is between 0.60 and 1.00%. With a coke sulfur of 0.40%,

[210] Sam F. Carter, *Foundry,* 88–95 (1960).

on FMC, BFL, HBNPC, and DKS Formcokes[209]

BFL				HBNPC				DKS	
Hattingen, 6.8 m				Longwy, 5.6 m		Denain, 7.0 m		Kokura, 8.4 m	
0	100	0	100	0	30	0	38	0	40
	7.1		1.7		2.1		2.1		0.9
	7.1		7.1		8.0		8.0		10.0
	0.51		0.495		0.7		0.7		0.69
	61 × 48 × 35		60 × 47 × 35		27 × 37 × 45		27 × 37 × 45		31 × 47 × 47
	48		44		29.5		29.5		44
	7.8		7.1		7.1		—		—
	—		—		15–25		15–25		—
—	3	—	7	—	30	—	24	—	16
1352	1348	1360	1265	809	781	1409	1450	2827	2847
1071	1048	864	865	935	923	990	940	1146	1150
23	49	36	59	14	21	25	29	8.7	7.7
—	—	—	—	16	22	—	—	45.7	45.3
595	636	571	586	615	602	540	540	499	497
595	601	571	582	615	601	540	527	499	503
100	81	100	80	100	90	100	97	100	90
—	—	—	—	0.43	0.41	0.65	0.80	0.69	0.64
—	—	—	—	0.046	0.051	0.027	0.027	0.039	0.032
—	—	—	—	1.30	1.30	1.12	1.15	1.19	1.17

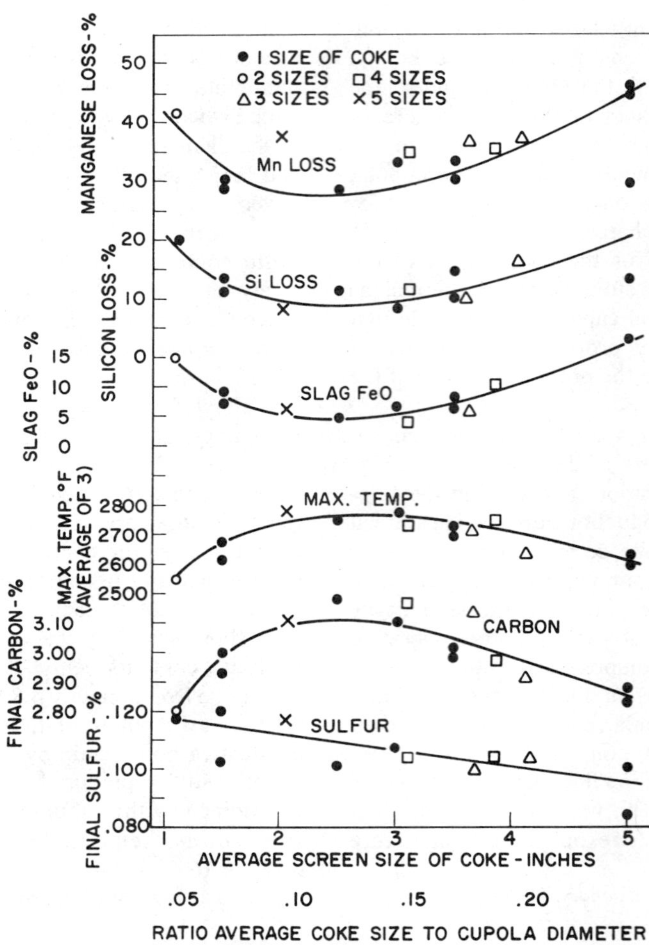

Figure 15.26 Effect of coke size on metal temperature and chemistry.[210]

the iron sulfur would be 0.080%, and a coke sulfur of 2% will result in a metal sulfur of 0.260%. The well-known principle that sulfur will keep out carbon, but carbon will not keep out sulfur, was verified. It was found that at 0.270% sulfur in the iron, carbon could not be dissolved above 3.00%. At 0.140% iron sulfur, the maximum carbon possible was 3.70%. With an iron sulfur of 0.030%, carbons near 4.20% in the iron are possible.

Leyshon,[211] in 1966, also discussed some of the factors that influence the efficient and economic operation of the cupola, as well as such new developments as the use of preheated air, the liningless water-cooled cupola, the use of oxygen, basic slag operation, the use of high-carbon coke, and supplementary fuels. The use of preheated air led to an improvement in the efficiency of operation. High-carbon irons of low sulfur content can be economically melted from high-steel scrap charges by applying basic slag operation. Injection of oxygen into the well improves thermal efficiency and reduces coke consumption. High-carbon coke also improves thermal efficiency of operation, but the benefits are not justified by the cost of such coke.

In 1974 Leyshon et al.[212] questioned the influence of coke size on melting performance of cupolas and presented the results of

[211] H. J. Leyshon, *Br. Foundryman,* **59,** 49–64 (February 1966).

[212] H. J. Leyshon, M. J. Shelby, and J. Briggs, *Foundry,* **102,** 52–54 (February 1974).

exhaustive tests conducted by the British Cast Iron Research Association. In separate tests in 30-, 60-, 72-, and 87-in. cupolas, the mean size of the cokes tested was increased from 3.4 to 5.9 in. Within this range, coke size had no significant effect on metal temperature and no appreciable effect on metal composition, although combustion ratio and stack gas temperature were increased. Only when the mean coke size was reduced to below 3.4 in. did size have an important effect on cupola behavior.

CHAPTER 16

Low-Temperature Tar

E. ARISTOFF, ROBERT W. RIEVE, AND HAROLD SHALIT

The authors of this chapter are researchers at ARCO Chemical Company, Division of Atlantic Richfield Company, and participated in Project Seacoke, Office of Coal Research Contract No. 14-01-0001-473. They have had broad experience in the development and testing of catalysts for processes involving the hydrogenation of coal, coal liquids, and heavy aromatics.

E. Aristoff *is a patent and technical information specialist.*

Robert W. Rieve *is Principal Engineer.*

Harold Shalit *is a Senior Research Associate.*

Low-temperature tar was reviewed by Clarence Karr, Jr., in the first Supplementary Volume of *Chemistry of Coal Utilization.* This chapter covers the subject from the early 1960s to 1975. It generally follows the format and definitions advanced by Karr, but some small increase in the upper temperature limit for production of low-temperature tar is used; for example, in the discussion of Project Seacoke, some of the temperatures in the final reactor are as high as 1200°F (650°C)—Karr had proposed a 500°C upper limit. We feel justified in this deviation, however, because the temperature of 1200°F in Seacoke occurred after the bulk of the volatiles had been removed, and it resulted from combustion of char to provide heat for earlier phases of the process.

The literature on low-temperature carbonization is not extensive. Indeed, application of this technique for the production of liquid hydrocarbons and other chemicals seems to be decreasing in importance with the growing emphasis on other more efficient procedures, such as hydrogenation and gasification. If this trend continues, serious doubt exists whether this subject will merit a separate chapter in succeeding editions of *Chemistry of Coal Utilization*.

16.1 PRODUCTION OF COAL TAR

16.1.1 Project Seacoke

Part of the discussion of production of tar in this chapter is devoted to Project Seacoke, which was sponsored by the U.S. Office of Coal Research from 1964 to 1969. It was carried out by investigators at the Atlantic Richfield Company and at FMC Corporation. In its final economic and design phases some work was subcontracted to the Blaw-Knox Company. The work is described completely in two reports to the Office of Coal Research[1,2] by A. E. Andersen, M. L. Becker, H. E. Jacobs, and P. J. White of Atlantic Richfield; J. F. Jones and M. R. Schmid of FMC; and R. J. Kendi and E. J. Vidt of Blaw-Knox.

Project Seacoke (Section 13.7.2) was an investigation of the cocarbonization of coal and petroleum residuum fractions in an effort to optimize the production of liquid tar fractions and coke. The tar fractions would be charge for a refinery, and char would serve as the fuel for a utility generating station. The objective was to obtain a synergism between the reactions of the petroleum and coal so that higher yields of the more valuable liquid fractions would be obtained when these reactions were carried out together. Initial laboratory work showed promise, and pilot-plant development was carried out in a modified char oil energy development (COED) process plant of FMC.

The principal conclusions of the experimental and pilot-plant work were that there appeared, indeed, to be some yield advantage for the cocarbonization over carbonization of the individual materials and that the products were richer in hydrogen than those of the separate operations. These conclusions are summarized in Figs. 16.1 and 16.2. In addition, a product from a pilot-plant cocarbonization of Crown coal and a short residuum was hydrogenated. Results (Table 16.1) indicated that the hydrogenated product was a satisfactory charge stock for a petroleum refinery.

On the basis of these experimental results, Blaw-Knox developed a design for a plant charging 19,500 tons of coal and 6500 tons of short residuum per day. The plant design incorporated some simplifications thought to be necessary. Under the conditions prevailing at the time of preparation of the report, the net product advantage was estimated at $105,000 per day, giving a daily net profit of $48,660 for a 15.5% return before taxes, based on a 15-year life. There appear to be a number of technical, logistic, and economic questions concerning the Seacoke approach. The yield advantages shown, though economically significant, might not be large enough to justify the added complexity of a plant operating in this manner. The economics of the scheme and the availability of petroleum residuum for the purpose need to be reviewed in the light of the current fuel situation.

16.1.2 Other Carbonization Studies

The FMC Corporation[3–5] has spent a large effort on a process for the development of a staged devolatilization of coal, the COED process (see Section 13.7.2). Their approach is particularly applicable to the agglomerating

[1] U.S. Government Rep. No. PB 235781, February 1970, 140 pp.

[2] U.S. Government Rep. No. PB 235782, February 1970, 174 pp.

[3] J. F. Jones, "Clean Fuels from Coal," paper presented at IGT Symp., Chicago, September 10–14, (1973), pp. 382–402.

[4] H. E. Jacobs, J. F. Jones, and R. T. Eddinger, *Prepr., Am. Chem. Soc., Div. Fuel Chem., Part 1*, **14**(4), 42–57 (1970).

[5] J. F. Jones, *Energy Sources*, **2**, 179–202 (1975).

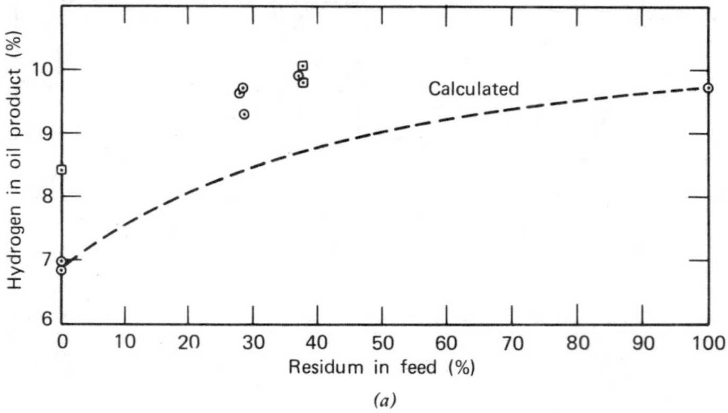

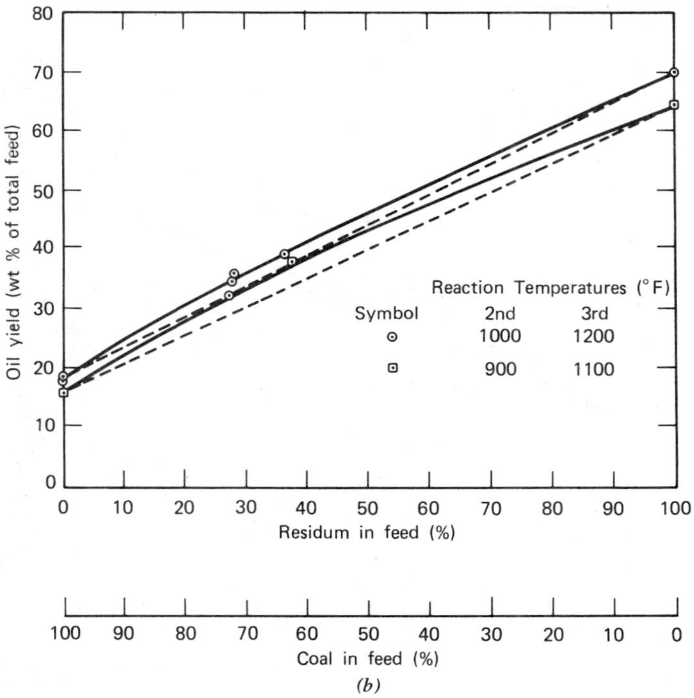

Figure 16.1 Project Seacoke: (*a*) hydrogen in oil product (%) and (*b*) oil yields for McKinley coal feedstocks (New Mexico hvCb), short residuum (4.8°API).[1]

coals that are difficult to carbonize under ordinary circumstances. In COED, after the coal is crushed and dried, it is heated to successively higher temperatures in a series of fluidized-bed reactors. The temperature of each bed is selected to be just lower "than the maximum temperature to which the coal can be heated without agglomerating and defluidizing the bed." Typical[3] is the use of four stages operating at 600, 850, 1000, and 1500°F. The number of stages and the temperature at which each is operated can be adjusted to suit the properties of the coal. Volatile products are released in the first stages and pass to a product recovery system. Char is burned in the last stage, the hot combustion gases being used to heat the earlier stages. Gas and oil are separated from the volatile products. The product oil is filtered, then subjected to a hydrotreating step in a fixed-bed catalytic reactor. This last refining step removes sulfur, nitrogen, and oxygen from the oil and produces a 25° API gravity synthetic crude. The gaseous product from COED can be used for fuel or converted to pipeline gas or hydrogen. Finally, residual

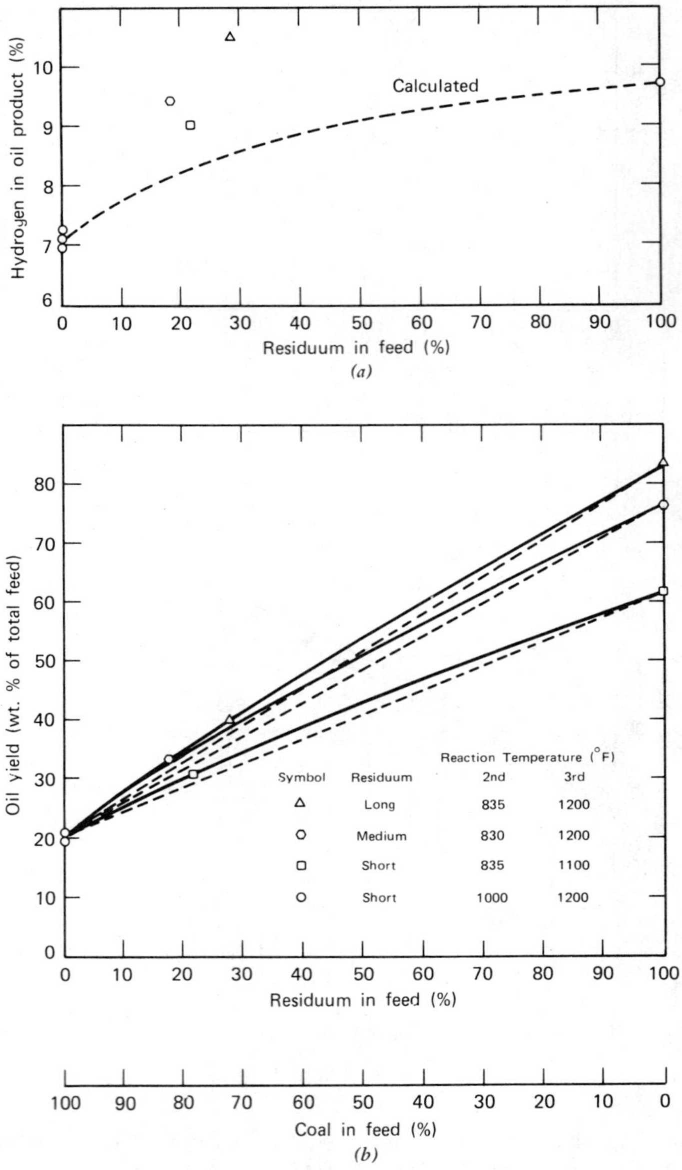

Figure 16.2 Project Seacoke: (*a*) hydrogen in oil product (%) and (*b*) oil yields for Illinois No. 6 Seam coal (hvBb); 4.8, 11.4, and 18.6°API for short, medium, and long residua, respectively.[1]

char from the final stage can be burned directly under a boiler, or it can be gasified by conventional technology to produce a fuel gas.

The COED process has been tested in a 100-lb/hr PDU (Fig. 16.3), and in a 36 ton/day pilot plant. Several different coals have been successfully processed through the pilot plant, including hvBb coal from Colorado and Illinois, a subbituminous coal from Wyoming, and a lignite from North Dakota. Typical operating conditions for an Illinois No. 6 coal include temperatures of 550, 810, 1050, and 1500°F in the successive stages. Some yield data are shown in Table 16.2. Jones states[5] that the commercial potential for COED is being studied by the Ralph M. Parsons Company, which developed a commercial plant design "sufficiently definitive so that a competent engineering contractor can complete the engineering and build a plant." In 1973 Chem Systems used the re-

Table 16.1 Hydrogenation of Carbonization Product[1]

Properties of Charge	
Gravity	−5.5° API
Oxygen	5.21 wt %
Nitrogen	0.86 wt %
Sulfur	2.80 wt %
Hydrogenation Results	
Hydrogen consumption	3400 scf/bbl
Product gravity	21.8° API
Oxygen removal	97.1%
Nitrogen removal	95.4%
Sulfur removal	98.8%
C_4 + liquid yield	115%

sults of COED hydrogenation study by the Atlantic Richfield Company to arrive at a value for the synthetic crude oil as a refinery feed. Jones thinks that the commercial potential for COED and similar processes yielding liquid and gaseous fuel from coal holds promise for the future. In his estimation, COED could be commercialized quickly if the economics were right, since it is based essentially on current mechanical equipment know-how.

Nyunt and Khin[6] studied the carbonization of Kalewa No. 2 coal of Burmese origin. The yields they obtained on heating this coal at various temperatures are given in Table 16.3. As can be seen, gas yields increase and coke yields decrease with increasing temperature. Tar increases, but liquor (on a water-free basis) remains essentially constant. The composition of the tar was determined by a combination of distillation, extraction, and solid-liquid chromatography. Distillation of the tar gave four fractions: a light oil (boiling to 205°C), middle oil (205–250°C), heavy oil (250–338°C), and a pitch. The yields of these fractions are shown in Table 16.4.

Extraction and chromatography of the light and middle oils resulted in the isolation and identification of the compounds having the greatest interest as chemical products. Nyunt and Khin back-calculated yields of these on the basis of the original coal.

For a number of years the Occidental Research Corporation has been investigating a carbonization process based on flash pyrolysis (see Section 13.7.4). The status of the development was described in a paper by S.

[6] M. H. Nyunt and U. A. Khin, *Union Burma J. Sci. Technol.*, 3, 15–31 (1970).

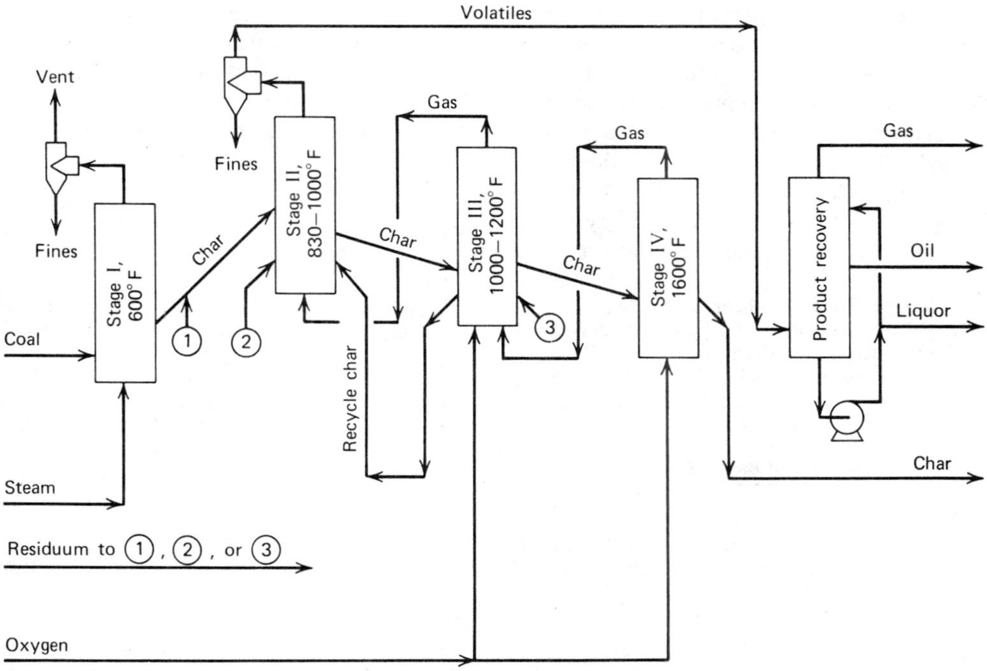

Figure 16.3 Project Seacoke: configuration in the COED process development unit.[3,5]

Table 16.2 Some Yield Data for Pyrolysis of Illinois No. 6 Seam Coal[3,5]

	Net Yields[a] from Coal Pyrolysis (wt %, mf)	Normalized to 100%
Char	60.68	59.12
Oil	20.13	19.61
Liquor	5.68	5.53
Gas	16.16	15.74
	102.65	100.00
	Net Process Yields	
Char	1177 lb/ton	
Oil	1.04 bbl/ton	
Liquor	7.1 gal/ton	
Gas[b]	8133 scf/ton	

[a] Process yields corrected for 10% reaction of steam with char.
[b] Gas free of N_2, H_2O, CO_2, and H_2S recovered from second-stage system.

C. Che et al.[7] The Occidental process subjects finely divided coal to very rapid heating; the rate is estimated at greater than 5500°C/s. The coal substance breaks down to gas, liquid, and char. The yields of liquids are high because the very short residence times prevent degradative reactions of the primary products. In practice, the rapid heating is accomplished by using circulating char as a solid heat carrier; the char is partly burned in air, raising its temperature, and it is then contacted with the incoming coal. In the process there is a net production of char. This work, presently supported by the U.S. Department of Energy, has progressed through a bench scale to a current three-ton/day pilot plant.

The liquid products from flash pyrolysis are heavy, viscous liquids, richer in hydrogen than the coals from which they are derived. Che et al. compared the products from a subbituminous and a bituminous coal (Table 16.5). It can be seen that the yield of liquid is considerably greater from bituminous coal than from subbituminous. On the other hand, oxygen elimination is greater with the lower rank coal, but sulfur elimination is less. The boiling ranges of these products are similar, with that from the subbituminous being slightly lower. The authors state that the higher liquid yield from the bituminous coal is offset by the lower sulfur and ash in the char from the subbituminous. Thus the two coals are of similar value as feedstocks.

[7] S. C. Che, G. E. Dolbear, and J. R. Longanbach, "Properties of Coal Liquids Produced by the Occidental Flash Pyrolysis Process," paper presented at the 1976 Coal Chemistry Workshop, sponsored by the Stanford Research Institute, Menlo Park, Calif., August 1976.

Table 16.3 Carbonization of Kalewa Coal[6]

	Yield (%, mf coal)		
	Test I	Test II	Test III
Carbonization temperature:	600°C	700°C	800°C
Coke	72.2	66.0	62.7
Tar	4.87	7.28	8.95
Liquor	8.1	9.15	8.94
Gas	9.83	17.7	19.5

Table 16.4 Distillation of Tar from Carbonization of Kalewa Coal[6]

	Yield (%)		
Carbonization temperature:	Test I 600°C	Test II 700°C	Test III 800°C
Light oil (boiling–205°C)			
Tar	27.2	14.6	11.85
Coal	1.2	0.97	0.96
Middle oil (205–250°C)			
Tar	22.0	21.7	22.4
Coal	0.97	1.43	1.81
Heavy oil (250–338°C)			
Tar	33.8	34.3	36.6
Coal	1.5	2.27	2.94
Pitch			
Tar	17.0	29.3	29.15
Coal	0.75	1.93	2.35

Table 16.5 Flash Pyrolysis of Coals[7]

	Wyoming Subbituminous	Kentucky No. 9 Bituminous
Coal		
Fischer Assay (%)	9.7	16.3
H/C, atomic ratio	0.82	0.89
Oxygen (%)	20.1	7.5
Sulfur (%)	0.65	3.5
Nitrogen (%)	1.2	1.7
Liquid product		
Yield (%)	21.0	33.0
H/C, atomic ratio	1.1	1.05
Oxygen (%)	14.1	7.9
Sulfur (%)	0.45	2.1
Nitrogen (%)	1.0	1.4

Che et al. also compared their product from bituminous coal with that made by COED from an Illinois No. 6 bituminous coal. The COED process yielded only 20.5% liquid, compared with the 33% reported above. The properties of the two products are roughly the same. They also show that yields from the slow pyrolysis of coal are even lower than those from COED when compared to the Fischer assay.

16.2 PROPERTIES, ANALYSIS, AND COMPOSITION OF LOW-TEMPERATURE TAR

Lang and Lacey[8] studied the distillate from low-temperature carbonization of an Alabama high-volatile bituminous coal. This material contained 28.5 vol % tar acids, 2.9 vol % tar bases, and 68.6 vol % neutral oil. Phenols can be isolated from the light distillate, which, in addition, can be catalytically reformed to aromatic hydrocarbons. The heavy distillate can be hydrogenated to a highly aromatic product.

Hsieh[9] showed that gel permeation chromatography (GPC) was applicable to the analysis of both the distillates and pitch of low-temperature tars. GPC was especially useful in analyzing the pitch where chemical extraction, fluorescent indicator adsorption, gas chromatography, and mass spectrometric methods were not applicable. The method, which avoids the hazard of thermal reactions inherent in the distillation of such high-boiling fractions, is discussed later.

Matthews[10,11] has shown that thin-layer chromatographic techniques can be used to identify coal tar and its fractions in various sewage discharges. In this method the oil (or its solutions) is spotted on a thin layer of adsorptive material, for example, alumina. The spots are then made to travel up the layer by developing them with various solvents such as petroleum ether. After drying, the chromatogram displays, under UV light, fluorescent patterns that are characteristic of various classes of oils. Matthews[11] published a table, reproduced here as Table 16.6, showing the characteristic patterns of various coal-tar fractions.

Sauerland and Zander[12,13] have investigated the application of gas chromatography to various coal-tar fractions. In particular, they discussed the utility of various stationary phases. For example, a column packed with poly-*m*-phenyl ethers (Varian GmbH, Darmstadt, West Germany) on Chromosorb G-AW (an acid-washed, flux-calcined diatomite; Supelco, Inc., Bellefonte, Pa.) was used successfully to analyze a raw tar in a temperature-programmed unit. Better separation of such aromatic hydrocarbons as phenanthrene and anthracene was obtained on a column comprising Bentone 34 plus Apiezon L on Gas-Chrom P (stationary phases and support available from Applied Science Laboratories, State College, Pa.). Even better for this purpose was a column made of a liquid crystal (No. 36317, Riedel de Haen AG, Seelze-Hannover, West Germany) on Chromosorb. They also used the fact[14] that triphenylphosphine oxides hydrogen bond to hydroquinones to construct a column of tris(3-methylphenyl)-phosphine oxide and Chromosorb R. This column separated the three isomeric cresols very well, even though the meta-para sepa-

[8] E. W. Lang and J. C. Lacey, Jr., *Ind. Eng. Chem.*, **52**, 137–140 (1960).

[9] B. C. B. Hsieh, Ph.D. dissertation, University of Utah, 1969, 118 pp. Available from University Microfilms, Ann Arbor, Mich., Order No. 69-18631.

[10] P. J. Matthews, *Water Pollut. Control (London)*, **67**, 584–591 (1968).

[11] P. J. Matthews, *J. Appl. Chem.*, **20**, 87–92 (1970).

[12] H. D. Sauerland, *Brennstoff-Chemie*, **44**, 37–43 (1963).

[13] H. D. Sauerland and M. Zander, *Erdoel Kohle*, **25**, 526–530 (1972).

[14] F. Ramirez and S. Dershowitz, *J. Org. Chem.*, **24**, 704–705 (1959).

Table 16.6 Comparison of Coal-Tar Products[11]

Band technique: activated alumina T–petroleum ether, 40–60°C, fluorescence at 350 nm.

Total R_f × 100	Color Sequence Color	Crude Tar R_f × 100 Color	Refined 32 EVT[a] R_f × 100 Color	Refined 34 EVT[a] R_f × 100 Color	Light Oil R_f × 100	Color
80	Pale blue	As total	Same as total	Same as 32	80	Pale blue
75	Purple	sequence	sequence but	EVT, but	75	Purple
70	Blue		note		64	Trace green
64	Green				60	Blue
60	Brown		extra		54	Purple
54	Blue		56 Green		48	Yellow
48	Yellow				0–48	Orange
34	Pale buff		34 Brown-yellow	34 Faint brown		Yellow
26	Pale yellow					
16	Brown		16 Brown-yellow	16 Faint		
0–10	Orange			brown-yellow		

[a] Equiviscous temperature.

ration is normally difficult. The tendency of the phosphine oxides to hydrogen bond was illustrated by the separation of biphenyl from indole, which occurred in the inverse order of their boiling points, indole being held back very much longer than biphenyl. Sauerland and Zander also demonstrated the applicability of a nitrogen selective flame ionization detector (made by Hewlett-Packard Co., Palo Alto, Calif.) for the analysis of nitrogen compounds in coal tar.

The composition of specific fractions of coal tar was discussed in a number of papers. Moustafa[15] showed by IR techniques that coal-tar pitch and asphaltic pitch differ principally in that the ratio of aromatic to aliphatic compounds is higher in the coal-tar material (2.11 vs. 0.116).

Bartle et al.[16] separated a neutral oil fraction from a fluidized-bed carbonization of Markham coal at 430°C. This material was then further fractionated by solvent extraction and urea adduction, to facilitate further analysis. The various fractions were examined by a number of methods, including proton magnetic resonance, low ionizing voltage mass spectrometry, gel chromatography with fluorescent indicator analysis, and cryoscopy. The distribution of hydrogen in the various fractions is shown in Table 16.7, together with elemental analyses, molecular weights, and hydrogen distributions. Analysis of these results indicates that this oil is about 30% paraffinic. A distribution of aromatic compounds determined by mass spectrometry is shown in Table 16.8.

Tesarik and Ghyczy[17] showed that certain capillary columns were suitable for the gas chromatography of the pyridine bases of coal-tar light oil. Their columns were coated with highly polar stationary phases such as polyethylene glycol 400, this material with 2% potassium hydroxide, Apiezon K mixed with Solvamin 20 (octadecylamine plus 20 moles of ethylene oxide), Amine 220 (1-ethanol-2-heptadecenyl-2-isoimidazole), and Reoplex 400. Pyridine, aniline, and quinoline derivatives were entirely separated on such media by the use of temperature programming.

Kusy[18] chromatographically analyzed the phenols in a brown coal tar. Using a column containing a nonpolar phase (Apiezon L) and a polar phase (sodium dodecylbenzenesulfonate), he isolated and identified the cresols, xylenols, 28 higher alkyl phenols, and 15 indanols and tetrahydronaphthols.

Vahrman and his co-workers have published a number of papers dealing with various compositional aspects of low-temperature coal tar. Vahrman found[19] that polycyclic aromatic hydrocarbons with one

[15] A. B. Moustafa, *J. Oil Colour Chem. Assoc.*, **57,** 44–47 (1974).

[16] K. D. Bartle, D. W. Jones, T. G. Martin, and W. S. Wise, *J. Appl. Chem.*, **20,** 197–202 (1970).

[17] K. Tesarik and S. Ghyczy, *J. Chromatogr.*, **91,** 723–731 (1974).

[18] V. Kusy, *Erdoel Kohle*, **23,** 575–580 (1970).

[19] M. Vahrman, *Fuel*, **49,** 5–16 (1970).

Anthracene Oil $R_f \times 100$	Color	Creosote 1 $R_f \times$	Color	Creosote 2 $R_f \times 100$	Color	Pitch $R_f \times 100$	Color
80	Pale blue	80	Pale blue	80	Pale blue	70	Mauve-blue
75	Purple	64	Trace green	70	Green	64	Green
70	Blue	60	Blue	64	Blue	60	Brown
64	Green	56	Trace green	56	Green	56	Green-brown
60	Brown	54	Very faint blue-purple	54	Blue purple	54	Blue
56	Green						
16–56	Purple	48	Pale yellow	48	Pale yellow	48	Bright yellow
16	Pale blue	34	Pale buff	0–34	Pale buff	16	Brown
0–16	Purple	0–34	Very pale buff			0–10	Orange

Table 16.7 Analyses of Neutral Oil and Its Seven Fractions[16]

	I	II	III	IV	V	VI	VII
Yield on oil (%):	100	38	61	91	9	29	9
Molecular weight:	201	212	155	173	210	235	219
	Elemental Analysis (%)						
Carbon	86.4	84.6	84.8	87.0	84.2	84.5	82.8
Hydrogen	10.2	13.5	9.1	9.9	14.5	13.4	14.0
Nitrogen	0	0.1	0.1	0.1	0	0	0
Sulfur	0.9	0	0.9	0.9	0.1	0.1	0
Chlorine	0.1	0.1	0.1	0.1	0	0.1	0
Oxygen (by difference)	2.4	1.7	5.0	2.0	1.2	1.9	3.2

	Hydrogen Distribution from PMR Spectra (%)							
Frequency (MHz):	220	60	60	60[a]	60	60	60	60
Compound[b]								
Aromatic	13.9	15	6	23	22	0	7	1
Olefinic	1.3	1	2	0	0	2	3	1
Hα	27.2	29	14	38	35	0	18	0
Hydroaromatic and cycloparaffinic	7.5	6	8	6	5	0	8	0
Hβ	31.1	31	45	19	22	71	38	68
Hγ	19.0	18	25	11	16	27	28	30
Compound		Fluorescent Indicator Analysis (%)						
Aromatics		64	33	94	74	5	37	6
Olefins		8	8	2	6	14	7	13
Paraffins		28	59	4	20	81	56	81

[a] 3% of hydrogen in this fraction was attributable to γ-butyrolactone.

[b] Hα = hydrogen of α-naphthalene type; Hβ = hydrogen of β-naphthalene type; Hγ = hydrogen of mesoanthracene type.

Table 16.8 Analysis of Neutral Oil by Low Ionizing Voltage Mass Spectrometrv[16]

	Mol. Wt of First Member		Number of Carbons in Alkyl Substituents								% of Fraction
			1	2	3	4	5	6	7	8	
Naphthalenes	128			2.4	6.4	6.3[a]	3.4[a]	2.4[a]	0.8[a]	0.3[a]	22.0
Acenaphthylenes Fluorenes	152 166 }			1.0	0.9	0.9	0.3				3.1
Acenaphthenes and biphenyls	154		0.8[b]	3.7[bb]	5.8[b]	5.7[b]	4.4[b]	1.9[b]	0.3[b]		22.6
Anthracenes and phenanthrenes	178	0.7	3.5	2.5	1.1	1.3[c]	0.6[c]	0.2[c]			9.9
Methylenephenanthrene and phenylnaphthalenes	190	0.3									0.3
Fluoranthenes and pyrenes	202	0.4	0.8[d]	0.8[d]	0.3[d]						2.3
Naphthacenes, chrysenes, benz[*a*]anthracenes, triphenylenes, benzo[*c*]phenanthrenes	228	0.1	0.2								0.3
Naphthobenzofurans	218	0.1	0.5	0.4							1.0
Carbazoles	167		1.4	1.5	0.8						3.7
Acridines	179	0.2									0.2
Tetrahydroacridine?	183	2.3									2.3
											67.7

[a] Isomeric with dibenzothiophen series.
[b] Isomeric with benzofuran series.
[c] Isomeric with naphthobenzothiophen series.
[d] Isomeric with dibenzofluorene series.

to six rings are present in coal tar; of these, alkylated benzenes and naphthalenes predominate. Vahrman and Rizvi[20] attempted to determine the presence of hydroaromatics in these materials. By a method of differential solubility in a series of acids of increasing strength, they separated an aromatic coal-tar fraction into a number of subfractions. The material, which was insoluble in a mixture of H_2SO_4 and P_2O_4, had properties that indicated the presence of hydroaromatic structures. Data for the fraction that reacted with this "activated" H_2SO_4 corresponded with those of a benzene ring having about $C_{10}H_{20}$ in alkyl side chains. The same authors[21] had previously worked out a column chromatographic method for the quantitative separation of total hydrocarbons from low-temperature tar. These hydrocarbons were almost free of heteroatoms such as oxygen.

In a further series of papers[19,22–25] Vahrman and associates studied the composition of low-temperature tars. They concluded that "much larger amounts of hydrocarbons than had previously been suspected exist in coal." These compounds exist in micropores and are liberated on heating up to 600°C, the aliphatics at less than 450°C. Palmer and Vahrman[25] reported obtaining tars from a number of coals by a standard Gray-King low-temperature carbonization at 600°C and showed that there were significant differences in the yields and compositions of the tars. Such differences could be correlated to the nature of the original coals.

These differences among ordinary humic coals were also observed by Karr and his coworkers at the Morgantown Energy Research Center (MERC).[26] After having investigated, in a series of papers,[27–29] the effect of changing carbonization conditions on tar composition, these workers undertook to determine the effect of coal rank on tar composition under constant carbonizer conditions. In a fixed bed they carbonized seven different American coals: two bituminous (hvA) coals from the same seam (Pittsburgh) but different counties (labeled BA and BC), one bituminous (hvB) coal from Utah (BS), one subbituminous coal from Colorado (SBE), one lignite from Texas (LS), and two cannel coals (C/C and CL) from West Virginia. The yield of commercially valuable products from each of these is shown in Table 16.9.

Karr et al. drew the following conclusions:

1. There was evidence of dependence of yield on rank within the humic coals.
2. The cannel coals have extrahigh yields of paraffins, olefins, and other nonaromatic neutral oil and light oil constituents.
3. The bituminous coals from different locations in the Pittsburgh Seam gave distinctly different results.
4. Total naphthalenes fell in the range of 4–11 lb/ton for all coals.
5. Highly alkylated quinolines made up the bulk of distillable tar bases from all seven coals.
6. There seemed to be a partial approach to thermodynamic equilibrium for the products from all coals; for example, *m*-xylene was the predominant xylene isomer; 2,4-xylenol was the most prevalent of the six xylenols, and there was more 2-methylnaphthalene than 1-methylnaphthalene.

Vahrman and Watts determined the composition of tars made in a short-path vacuum still.[30] Saprykin and Kulachkova[31] investigat-

[20] S. S. H. Rizvi and M. Vahrman, *Pak. J. Sci. Ind. Res.*, **14**, 535–537 (1971).

[21] S. S. H. Rizvi and M. Vahrman, *Pak. J. Sci. Ind. Res.*, **14**, 147–149 (1971).

[22] T. J. Palmer and M. Vahrman, *Fuel*, **51**, 14–21 (1972).

[23] J. A. Spence and M. Vahrman, *Fuel*, **49**, 395–408 (1970).

[24] M. Rahman and M. Vahrman, *Fuel*, **50**, 318–328 (1971).

[25] T. J. Palmer and M. Vahrman, *Fuel*, **51**, 22–26 (1972).

[26] J. R. Comberiati, C. Karr, Jr., E. E. Childers, J. O. Mapstone, Jr., and P. A. Estep, *J. Inst. Fuel*, **43**, 147–150 (1970).

[27] C. Karr, Jr., J. R. Comberiati, K. B. McCaskill, and P. A. Estep, *J. Appl. Chem.*, **16**, 22–27 (1966).

[28] C. Karr, Jr., J. R. Comberiati, K. B. McCaskill, and P. A. Estep, U.S. Bur. Mines Rep. Invest. No. 6839, (1966), 9 pp.

[29] C. Karr, Jr., J. R. Comberiati, P. A. Estep, and J. O. Mapstone, Jr., *J. Inst. Fuel*, **40**, 561–563 (1967).

[30] M. Vahrman and R. H. Watts, *Fuel*, **51**, 130–134 (1972).

[31] F. Ya. Saprykin and A. F. Kulachkova, *Khim. Tverd. Topl. (Moscow)*, 32–37 (1975).

Table 16.9 Yield of Commercially Valuable Products[26]

	Dry Coal[a] (lb/ton)						
	BA	BC	BS	SBE	LS	C/C	CL
Light oils	40.8	11.4	12.8	8.4	13.4	61.6	48.2
Aliphatics							
n-Paraffins	4.8	1.3	2.5	1.0	2.6	22.6	18.1
α-Olefins	2.7	0.9	1.8	0.6	2.2	20.7	12.9
Other aliphatics	10.6	3.0	1.5	1.6	3.0	9.3	7.9
Aromatics							
Benzene	1.6	1.0	0.8	0.4	1.4	2.2	2.9
Toluene	8.0	2.9	3.0	2.0	2.5	4.6	4.8
p-Xylene	1.8	0.3	0.6	0.4	0.2	0.3	0.2
m-Xylene	4.9	1.0	1.1	0.9	0.7	0.8	0.7
o-Xylene	1.9	0.4	0.6	0.5	0.3	0.8	0.2
Other aromatics	4.5	0.6	0.9	1.0	0.5	0.6	0.5
Tar	314	246	250	124	204	591	634
Resins	54.5	52.5	73.2	19.5	28.8	15.6	35.1
Resinoids	36.2	18.5	14.5	9.6	30.9	56.9	44.2
Tar bases	7.2	1.1	1.7	0.5	0.5	3.5	6.2
Other neutral oils	73.5	60.9	45.0	19.5	31.1	204	161
Tar acids	58.4	50.1	44.7	32.6	20.3	18.1	54.1
Phenol	0.8	1.0	0.8	1.1	1.7	—[b]	0.4
o-Cresol	1.9	1.8	1.2	1.0	1.3	0.4	1.0
p-Cresol	1.0	1.1	1.0	1.1	1.4	0.3	0.7
m-Cresol	1.7	2.0	1.4	1.7	1.5	0.3	0.6
Other low-boiling acids	4.6	5.5	2.7	1.6	2.8	4.1	4.4
High-boiling tar acids	48.4	38.7	37.6	26.1	11.6	13.0	47.0
High-quality neutral oil	84.7	63.2	70.9	42.5	92.6	293	333
n-Paraffins	9.8	3.9	11.6	7.4	8.3	46.8	42.8
Other paraffins	7.1	8.3	10.0	3.5	9.4	14.1	43.3
α-Olefins	3.2	2.7	8.0	4.8	10.9	23.5	37.9
Other olefins	3.2	3.7	7.0	3.8	13.2	69.4	48.4
Naphthalene	0.9	0.7	<0.1	0.2	0.4	0.3	0.1
Monomethylnaphthalenes	3.0	5.0	0.1	0.8	1.0	1.4	2.7
Dimethylnaphthalenes	3.5	2.4	4.1	1.1	1.3	3.4	4.2
Trimethylnaphthalenes	2.3	1.8	3.4	1.4	1.8	3.2	2.9
Tetramethylnaphthalenes	1.2	0.9	1.1	0.3	0.5	1.6	1.1
Pentamethylnaphthalenes	0.5	0.2	0.3	0.1	0.2	0.7	0.4
Biphenyls	0.3	0.6	0.4	0.5	0.5	0.5	1.2
Other aromatics	49.7	33.0	24.9	18.6	45.1	128	148

[a] BA, BC = Pittsburgh HvA bituminous coals (different counties)
BS = Utah hvB bituminous coal
SBE = Colorado subbituminous coal
LS = Texas lignite
C/C,CL = West Virginia cannel coals
[b] Not detected.

ed the composition of tars from the Lensk basin Sangar deposit coals.

The proposed new approach for the separation of coal-tar constituents[32–36] mentioned previously is the use of GPC, a technique first used in polymer chemistry for determining the molecular weight distribution of polymers and for fractionation of high molecular

[32] J. G. Hendrickson and J. C. Moore, *J. Polym. Sci., A-1*, **4**, 167–188 (1966).

[33] K. H. Altgelt, *J. Appl. Polym. Sci.*, **9**, 3389–3393 (1965).

[34] T. Edstrom and B. A. Petro, paper presented at the meeting of American Chemical Society, Chicago, September 1967.

[35] B. C. B. Hsieh, R. E. Wood, and L. L. Anderson, *Prepr., Am. Chem. Soc., Div. Fuel Chem.*, **13**(1), 44–60 (1969).

[36] R. B. Meiris, *Chem. Ind. (London)*, 642–643 (1973).

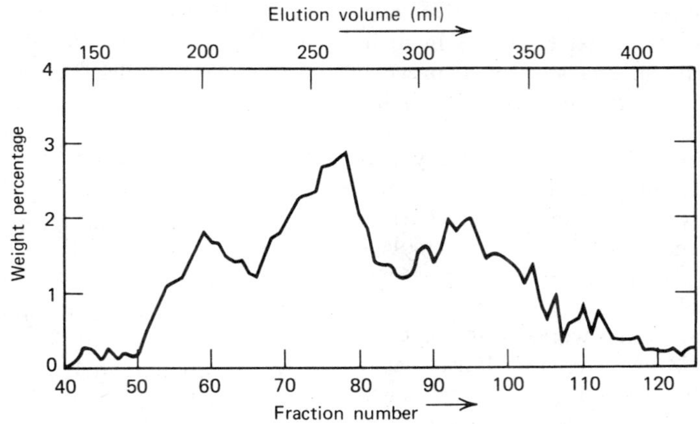

Figure 16.4 Elution curves (GPC) of low-temperature coal-tar distillates (150–350°C) from Spring Canyon coal.[35]

weight materials. The method consists of passing a solution of the coal-tar fraction down through a column packed with beads of a stationary gel phase. The tar material is eluted fractionally with the solvent, and the properties of the individual fractions are determined. Fractionation is accomplished according to molecular weight, the highest molecular weights passing through most rapidly; however chemical effects due to polar attraction of certain species on the column are superimposed on the size effect and must be taken into account.

Hsieh et al.[35] presented results on tars from three coals—Spring Canyon, Big Horn, and River King—carbonized at 550°C. Figure 16.4 illustrates the application of the method. The curve is clearly divided into three groups that can be identified by comparison with standard reference materials as predominantly aliphatic, predominantly aromatic, and (the last group) composed of polar compounds, respectively. As evidence for this last, the tar acids were extracted from the distillate, and both neutral oil and tar acid fractions were submitted to GPC. The results (Fig. 16.5) indicate that the first two groups occur in the neutral oil, whereas the third is in the acid fraction. GPC has been, and promises to be in the future, a procedure of great importance in characterizing such high molecular weight materials as coal tars.

There are, however, some caveats to be observed in the application of GPC to the analysis of low-temperature tars. One of the reviewers of this chapter[37] has pointed out

[37] D. S. Montgomery, Energy, Mines and Resources, Ottawa, Ont., Canada, private communication, 1977.

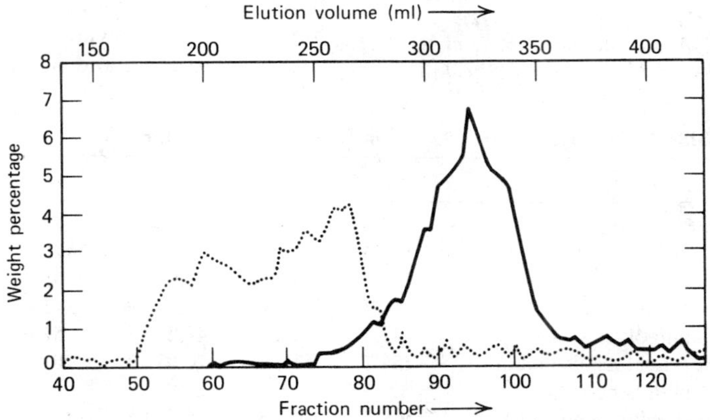

Figure 16.5 Elution curves (GPC) of neutral oil and tar acids from Spring Canyon coal-tar distillates (150–350°C).[35]

that these materials are quite reactive, and special precautions must be taken in their sampling, handling, and storage to prevent polymer formation and other changes, including those associated with heteroatoms (nitrogen, sulfur, oxygen). Montgomery also points out that polymer formation on the column itself is possible. In the case of GPC, it is also necessary to eliminate low molecular weight fractions of the polymer gel before use in analysis.

Sharkey and his co-workers at the U.S. Bureau of Mines in Pittsburgh and Denver[38–45] have used low-voltage mass spectrometry to determine the molecular weight distribution of various fractions of coal carbonization liquids. As an example of this work,[41] they determined the carbon number distributions for tar-acid fractions from five low-temperature tars. They found evidence for seven types of phenol in these tars: alkyl phenols, dihydric phenols or alkoxy phenol, indenols, indanols, naphthols, acenaphthenols and/or phenylphenols, and acenaphthylenols and/or fluorenols. The phenolic concentration at any carbon number (for each of the types) is a function of the carbon number and carbon/hydrogen ratio of the parent maf coal and is given by the following general equation:

$$Y = a + b_1X_1 + b_2X_2 + b_3X_3 \qquad (16.1)$$

where

Y = phenolic concentration at a given carbon number
X_1 = reciprocal of the carbon number
X_2 = square of the reciprocal of the carbon number
X_3 = carbon/hydrogen ratio of the parent maf coal

The values of the various constants for each of the phenol types, as determined by Sharkey and co-workers, are given in Table 16.10.

Shultz and Sharkey, collaborating with R. A. Brown of Esso Research and Engineering,[45] determined the mass spectrometric sensitivity data for a number of hydroaromatic compound types found in coal-tar liquids. The Pittsburgh group used high-resolution mass spectrometry to determine heteroatom species in coal carbonization products.[44]

16.3 PROCESSING AND CONVERSION OF TAR

The recent literature on low-temperature tar processing is devoted almost exclusively to hydrogenative techniques. The major objective has been to reduce the polynuclear aromatics content and to remove the heteroatoms (oxygen, sulfur, nitrogen). This work has been directed toward methods and catalysts that will require the least amount of hydrogen to produce distillable liquids of high fuel and chemical value.

The heteroatom and polynuclear aromatic

[38] A. G. Sharkey, Jr., J. L. Shultz, and R. A. Friedel, *Fuel,* **40,** 423–426 (1961).
[39] A. G. Sharkey, Jr., J. L. Shultz, and R. A. Friedel, *Fuel,* **41,** 359–371 (1962).
[40] A. G. Sharkey, Jr., J. L. Shultz, and R. A. Friedel, U.S. Bur. Mines Rep. Invest. No. 6318, 1963, 32 pp.
[41] M. Gomez, W. S. Landers, J. L. Shultz, and A. G. Sharkey, Jr., U.S. Bur. Mines Rep. Invest. No. 6586, 1965, 30 pp.
[42] A. G. Sharkey, Jr., J. L. Shultz, and R. A. Friedel, *Carbon (Oxford),* **4,** 365–374 (1966).
[43] R. A. Friedel and A. G. Sharkey, Jr., U.S. Bur. Mines Rep. Invest. No. 7122, 1968, 10 pp.
[44] J. L. Shultz, T. Kessler, R. A. Friedel, and A. G. Sharkey, Jr., *Fuel,* **51,** 242–246 (1972).
[45] J. L. Shultz, A. G. Sharkey, Jr., and R. A. Brown, *Anal. Chem.,* **44,** 1486–1487 (1972).

Table 16.10 Constants for Phenols in Low-Temperature Tars[41]

Phenol Type	a	b_1	b_2	b_3
Alkyl phenols	157.4	840.2	−7,484.3	−67.7
Dihydric or alkoxy phenols	110.5	839.5	−7,573.2	−29.1
Indenols	267.2	1642.9	−1,740.7	−56.5
Indanols	34.7	2825.6	−27,823.9	− 2.7
Naphthols	47.4	4332.5	−41,278.5	−52.6
Acenaphthenols and/or phenylphenols	256.2	535.3	−26,081.1	−101.0
Acenaphthylenols and/or fluorenols	308.7	−915.5	−16,279.5	−111.5

content of these tars makes them especially difficult to process. When they are heated to processing temperatures (usually 400–450°C), they undergo a variety of cracking and dehydrocondensation reactions leading to intractable solids. Thus they can readily foul furnace tubes and heat exchangers. The heteroatoms depress catalyst activity and may require several processing stages for their removal and subsequent conversion of the tar. To make matters worse, the tars also contain ash particles from the coal. Unless this ash is almost completely removed, it will plug up a fixed-bed catalyst. These ash particles are extremely small (a few micronmeters), so that conventional filtration is insufficient. When suitable tar feeds are prepared for processing by special filtrations and distillations, hydrogen pressures of about 135 atm or higher are required to sustain catalyst life; this, in turn, forces a higher than desired consumption of expensive hydrogen. These are the problems of tar processing that have received the most research effort.

16.3.1 Model Compound Studies

Workers in the field of model compound studies have investigated the hydrogenative conversion of model compounds to gain information on the products produced, the mechanisms, and the kinetics. Penninger and Slotboom[46] have studied the thermal hydrogenolysis of phenanthrene and anthracene at 80 atm of hydrogen and 475–495°C. They concluded that these polynuclear aromatics are hydrogenated and then cracked in a stepwise manner. The outer rings are hydrogenated as the first step, followed by opening of the saturated ring, then side-chain cracking. This cycle is repeated until, ultimately, benzene and light gases are formed. The same authors[47] have studied the thermal hydrogenolysis of 1,2,3,4-tetrahydronaphthalene and indane. Their data show the same general pattern.

Singh[48] studied the hydrogenation (nickel-tungsten-sulfide catalyst) and hydrocracking (clay; zeolite catalysts) of phenanthrene and anthracene. He also found that the hydrogenation reaction proceeds stepwise in an overall first-order reaction. Phenanthrene is hydrogenated at a much slower rate than is anthracene. Under hydrocracking conditions (400–550°C and 20–102 atm of hydrogen), the outer, hydrogenated rings were cracked. Even when the 9,10-dihydro derivatives were cracked, the pattern was the same. Singh postulates that at the high temperature required for cracking, hydrogens transfer from the center rings quite rapidly, so that only products from the cracking of tetrahydro and higher derivatives are seen.

Qader and co-workers[49–54] have studied the hydrogenation and hydrocracking of single polynuclear aromatics in a small stirred autoclave. This work measured the hydrogenation reaction using molybdenum oxide or sulfide, nickel-tungsten-sulfide, and tungsten disulfide catalysts. The hydrocracking studies included these hydrogenation catalysts mixed with acidic silica-alumina and zeolite or mordenite cracking catalysts. Data are also presented for dual-functional catalysts (hydrogenation and cracking components), which were prepared by impregnation.

It is concluded from these experiments that the overall reaction is first order with respect to the polynuclear aromatics and is best described by the dual-site mechanism proposed by Langmuir and Hinshelwood. Hydrogen pressures of 68–200 atm were needed to achieve reasonable reaction rates and to sustain the catalyst activity. In all cases the polynuclear aromatics were progressively hydrogenated, then hydrocracked in such a manner that only the outer, saturated rings underwent the cracking reaction.

[46] J. M. L. Penninger and H. W. Slotboom, *Erdoel Kohle*, **26**, 445–447 (1973).

[47] J. M. L. Penninger and H. W. Slotboom, *Rec. Trav. Chim. Pays-Bas*, **92**, 513–530, 1089–1094 (1973).

[48] S. Singh, Ph. D. dissertation, University of Utah, 1972, 184 pp. Available from University Microfilms, Ann Arbor, Mich., Order No. 72-22,-339.

[49] W. H. Wiser, S. Singh, S. A. Qader, and G. R. Hill, *Prepr., Am. Chem. Soc., Div. Fuel Chem.*, **13**(4), 300–312 (1969).

[50] S. A. Qader and G. R. Hill, *Prepr., Am. Chem. Soc., Div. Fuel Chem.*, **16**(2), 93–106 (1972).

[51] S. A. Qader, D. B. McOmber, and W. H. Wiser, *Prepr., Am. Chem. Soc., Div. Fuel Chem.*, **18**(4), 127–137 (1973).

[52] S. A. Qader, L. Chun Chen, and D. B. McOmber, *Prepr., Am. Chem. Soc., Div. Pet. Chem.*, **18**(1), 60–71 (1973).

[53] S. A. Qader, *J. Inst. Pet.*, **59**, 178–187 (1973).

[54] S. A. Qader and D. B. McOmber, *Am. Chem. Soc. Symp. Ser.*, **20**, 82–98 (1975).

One of the objectives of this work was to find a catalyst system or conditions such that the internal rings, which are hydrogenated first, could be selectively cracked to yield monocyclic and dicyclic aromatic products with a minimum of hydrogen consumption. No evidence for this selective process was observed.

A similar approach was studied under Project Seacoke.[55] In this work metal carbonyls (usually cobalt carbonyl) and a mixture of hydrogen and carbon monoxide were used for selective hydrogenation of various polynuclear aromatics. These catalyst systems are homogeneous (soluble catalyst) and introduce a limited amount of hydrogen into the polynuclear aromatic structures as compared to heterogeneous catalysts. The carbonyls are also attractive in that their activity is not affected by heteroatoms such as sulfur, oxygen, or nitrogen, nor does it remove these heteroatoms.

The Seacoke work showed that the carbonyls have a varying degree of hydrogenation selectivity, depending on the polynuclear aromatic substrate used, but in no case did they produce perhydro derivatives. Thus the carbonyls can be used to introduce a limited amount of hydrogen, even in the presence of high concentrations of heteroatoms. No data, however, are available to show that this type of hydrogenation selectivity can be coupled to a companion process to produce superior yields of desirable products.

Wu and Haynes[56] studied the hydrocracking of condensed ring aromatics over a nonacidic, chromia-alumina catalyst. This work was done in a small flow apparatus operating at 136–204 atm, 400–538°C, and 0.5–6.0 space velocity (weight of hydrocarbon per weight of catalyst per hour). The hydrocracking pattern followed the same general scheme as proposed by the other workers just discussed. Wu and Haynes do, however, show some evidence for the selective cracking of the internal ring (at the 9,10 position of phenanthrene). Their product, obtained at 482°C, 205 atm, and 1.0 space velocity, contained about 4 mole % of biphenyl and 1 mole % of 2-ethylbiphenyl. The yields from this suggested pathway are small, but it is of interest that it has been observed in a nonacidic catalyst system.

[55] U.S. Government Rep. No. PB 174926, December 1966, 52 pp.

[56] Wen-lung Wu and H. W. Haynes, Jr., *Prepr., Am. Chem. Soc., Div. Pet. Chem.*, **20**(2), 466–478 (1975).

16.3.2 Hydrotreating of Tars

The hydrotreating of tars and tar fractions can be a necessary step in processing to remove heteroatoms and/or to reduce the condensed aromatic system to give a liquid product.

Mirza et al.[57,58] studied the hydrotreating of a neutral oil fraction to produce a lamp or diesel oil. They found the optimum conditions over a nickel-tungsten-sulfide on alumina catalyst to be about 400°C, 100 atm, and 0.2 hr^{-1} space velocity. They were able to produce oils of 48–50 diesel index that were low in sulfur, nitrogen, and unsaturates.

Kluck and Riedel[59] hydrotreated a low-temperature tar before propane dewaxing. They found that the hydrotreatment facilitated the dewaxing but tended to lower the viscosity of the lubricating oil fraction. Thus the two effects must be optimized and will vary for a particular feedstock. Eddinger and Friedman[60,61] hydrogenated solid tars using iodine or hydrogen iodide as the catalyst and achieved more than 96 wt % yield of liquid product. The conditions ranged from 370 to 480°C, and 68 to 204 atm. The system is operable using pure hydrogen or a 50/50 mixture of hydrogen and carbon monoxide or hydrogen and nitrogen. The liquid product is not free of sulfur or nitrogen atoms and retains 0.3–0.5% of the iodine used. These compounds can be removed by further hydrotreatment over a cobalt-molybdate catalyst. Eqbal and Sarkar[62] used a similar system to hydrogenate a neutral middle oil (200–350°C boiling range). They compared iodine, iodoform, and a commercial cobalt

[57] A. Mirza, M. A. Masood, A. V. Ramaswamy, and R. Vaidyeswaran, *Brennstoff-Chemie,* **46,** 355 (1965).

[58] A. Mirza, M. A. Masood, M. M. Mallikarjunan, and R. Vaidyeswaran, *Brennstoff-Chemie,* **48,** 310–312 (1967).

[59] H. Kluck and G. Riedel, *Chem. Tech. (Berlin),* **18,** 362–365 (1966).

[60] R. T. Eddinger and L. D. Friedman, *Fuel,* **47,** 320–321 (1968).

[61] L. D. Friedman and R. T. Eddinger, U.S. Pat. 3,453,202, 1969.

[62] J. Eqbal and S. Sarkar, *Indian J. Technol.,* **9,** 313–314 (1971).

molybdate catalyst at 220–400°C in autoclave experiments using 5 wt % catalyst concentration and 100 atm of cold hydrogen pressure. Only the iodoform was active at 220°C. The products from iodoform and the cobalt-molybdate catalyst at 400°C had diesel indices of 36.4 and 38.4, respectively. These oils could be used as diesel fuels without further upgrading. (The iodoform catalyst was removed by a caustic wash.)

Berkebile et al.[63] hydrogenated a COED[3–5] low-temperature tar to produce a carbon black feedstock. The total tar, a > 270°C cut, and a > 315°C cut, were not acceptable feedstocks because of their high heteroatom contents (sulfur, oxygen, nitrogen). The total tar was hydrotreated (conditions and catalyst not given) to remove 58% of the oxygen, 16% of the nitrogen and 83% of the sulfur. The hydrogenated tar was fractionated to give 54 vol % of a >315°C bottoms, which gave an acceptable carbon black, but the yield of carbon black was lower than desired. A second hydrotreatment of the >315°C fraction gave an acceptable carbon at the desired yield. This final carbon black feedstock represented 35 vol % of the total COED tar. The >315°C fraction was hydrotreated to remove substantially all the sulfur, oxygen, and nitrogen, so that it could be further processed in a typical petroleum refinery. The hydrogenation of COED tars from three different coals was studied in some detail by Jacobs et al.[4] These tars were hydrogenated over a 3% nickel oxide and 15% molybdenum oxide on alumina catalyst (American Cyanamid's HDS-3A). The purpose of these studies was to produce a synthetic crude oil that could be processed in a petroleum refinery. The hydrogenations were conducted at 204 atm, 343–454°C, and 0.3–3.0 space velocity (weight of tar per weight of catalyst per hour). Hydrogen flow rate had no significant effect between 8000 and 12,000 scf/bbl. The removal of heteroatoms and gross hydrogen consumption was correlated by simple first-order kinetics. Hydrogenation severity needed to remove the heteroatoms varied with the coal source. The product oils were high in cycloparaffins and were deemed to be acceptable feeds for a conventional petroleum refinery. Jones[3] has also described a detailed scheme for the conversion of this synthetic crude oil into a full range of petroleum products.

16.3.3 Hydrocracking of Tars

Hydrocracking of the tars represents an approach for not only removing the heteroatoms and reducing the polynuclear aromatic content, but also producing lower molecular weight products that can be utilized directly as fuels or chemicals. The degree of hydrocracking can produce all the above or some compromise, along with heavy fuels or a synthetic crude oil.

Carpenter et al.[64] hydrocracked the low-temperature tar from North Dakota lignite and the creosote oil fraction from this tar (220–350°C boiling range). The work was done in a single-pass, fixed-bed reactor using both cobalt molybdate and zinc chromite catalysts. The conditions were 204 atm and 394–478°C. The cobalt molybdate catalyst removed most of the sulfur, nitrogen, and oxygen, while giving gasoline yields of 31–59 vol %. The gasolines were higher in aromatic and naphthenic hydrocarbons than the typical petroleum reformer feedstocks. Hydrocracking over the less active zinc chromite catalyst substantially increased the yields of phenolic compounds (less oxygen removal) of usable molecular weight over that available from the raw tar or creosote fraction. The hydrocracking over zinc chromite also improved the ease of caustic extraction of the phenolics.

Katsobashvili and Elbert[65] described a method for the hydrocracking of a "wide distillate" prepared from a low-temperature tar. The tar was distilled to remove materials boiling below 200°C, including phenolics, and a naphthalene fraction boiling up to a temperature of 360°C. The remainder, mixed with pitch distillate, constituted the feedstock. Hydrocracking was done at 500–550°C, 50 atm, and a hydrogen-to-feed ratio of 1000 liters of hydrogen per kilogram of feed. The <230 or <270°C fraction of the

[63] D. C. Berkebile, H. N. Hicks, and W. S. Green, *Ind. Eng. Chem., Prod. Res. Dev.*, **9**, 376–381 (1970).

[64] H. C. Carpenter, P. L. Cottingham, C. M. Frost, and W. W. Fowkes, U.S. Bur. Mines Rep. Invest. No. 6237, (1963), 13 pp.

[65] Ya. R. Katsobashvili and E. I. Elbert, *Koks Khim.*, No. 1, 43–47 (1966).

"hydrocrackate" was recycled along with benzene as a diluent. Under these conditions with the recycle ratio of "hydrocrackate" plus benzene to raw feed of 2 : 1 to 4 : 1, it was claimed that the yield of useful products from the tar could be as high as 84%. The useful products were given as naphtha, naphthalene, and phenolics. This was said to compare to a 37% yield from a nonrecycle process. No data were given on catalyst life in this process. The reported data showed a coke plus C_1 to C_4 gas yield of 11–30 wt %, but the coke yield cannot be estimated from these data. Since catalyst coking would be an important factor in this low-pressure operation, industrial application seems hopeless.

Qader and Hill[66] compared the hydrocracking of petroleum oils and low-temperature tar fractions (200–360°C boiling range). The tar fractions require a higher hydrogen pressure than the petroleum oils for successful hydrocracking. Both types of feed had to be hydrorefined (partially hydrogenated) before hydrocracking to ensure low coke yields and long catalyst life. The hydrorefining was done at 425°C, 100 atm, and 1.0 hr^{-1} space velocity over a cobalt molybdate catalyst. When the feeds were hydrocracked over a nickel-tungsten-sulfide on alumina catalyst at 500°C and 136 atm, they gave similar yields of naphtha: 77% for petroleum oil and 80% for tar oil. The tar oil consumed more hydrogen (1550 scf/bbl) than did the petroleum oil (1125 scf/bbl). The tar naphtha was higher in aromatics and had a research octane number of 90, compared to 78 for the petroleum-produced naphtha. Both feeds exhibited similar kinetic behavior with about the same activation energy (14–16 kg-cal/mole).

Hydrocracking in a static bed versus an ebulated bed was compared by Qader et al.[67] The ebulation was produced by a magnetically driven stirrer. The catalyst was nickel-tungsten-sulfide on alumina. An ebulated-bed technique is suitable for processing very heavy feeds, especially those containing ash (solids), since it provides for a continuous removal of small particles (i.e., ash and coke), which can physically or chemically foul a catalyst bed. The technique involves keeping the catalyst particles (of a definite size) in constant motion, so that smaller and less dense particles can be removed with the effluent. A typical low-temperature tar usually contains solid ash and coke particles, which must be removed. The data of Qader et al. showed that the static bed was somewhat better for removing heteroatoms and producing naphtha. Commercial ebulated-bed reactors (H-Oil process[68]) are now in use for the hydrotreating of heavy petroleum residua, and there is no doubt that this technology can be applied to tars and even coal. The somewhat better yields from a fixed bed, as compared to an ebulated bed, are offset by superior operability. Tar feeds containing solid ash particles will rapidly foul a fixed bed. The ash can be accommodated in the ebulated bed, then removed from the products and/or catalyst to achieve commercially acceptable stream times.

Flinn and Tarhan[69] have claimed the hydrocracking of tar over a cobalt molybdate on γ- or η-alumina catalyst with the recycle of the >300°C fraction to extinction. The conditions were 68–204 atm, contact times of 1–10 hr, and 345–400°C. The final product is an oil boiling below 300°C and rich in monocyclic and dicyclic aromatics.

Janardanarao et al.[70] hydrocracked a neutral oil fraction at 250–350°C over a nickel-tungsten-sulfide on alumina catalyst, a tin-nickel on silica-alumina catalyst, and a tin-nickel on activated carbon catalyst. The last two catalysts were not as active as the nickel-tungsten-sulfide catalyst. The best yields of low-boiling and kerosene fractions (40 and 34 vol %) were achieved at 450°C, 90 atm, and 0.5 space velocity (volume of oil per volume of catalyst per hour). All the catalysts showed an activity loss after a short time. The loss of catalyst activity in this and other "low-pressure" schemes is a very serious drawback to commercialization. Catalyst lifetimes of several months or more are required, and a hydrogen pressure of 135 atm

[66] S. A. Qader and G. R. Hill, *Ind. Eng. Chem., Process Res. Dev.* **8**, 462–469 (1969).

[67] S. A. Qader, W. H. Wiser, and G. R. Hill, *Prepr., Am. Chem. Soc., Div. Fuel Chem.*, **13**(4) 45–51 (1969).

[68] A. R. Johnson, R. G. Hippeli, R. H. Wolk, and G. Nongbri, *Prepr., Am. Chem. Soc., Div. Fuel Chem.*, **17**(2), 64–81 (1972)

[69] R. A. Flinn and M. O. Tarhan, U.S. Patent 3,503,872, (1970).

[70] M. Janardanarao, G. S. Salvapati, and R. Vaidyeswaran, *Erdoel Kohle*, **27**, 309–312 (1974).

or greater is the only proved method for achieving extended catalyst life.

16.3.4 Pyrolysis and Cracking of Tars

The thermal and catalytic cracking of tars and tar fractions have been investigated as ways of upgrading where the need for added hydrogen is not required. These methods were undertaken with the knowledge that hydrogen-deficient materials (e.g., coke) are necessarily formed in such reactions.

Berber, Rice, and Fortney[71,72] thermally cracked a >350°C tar fraction (40–50% of total tar) from Texas lignite at 790°C. The products included 15–25% coke, 20–40% cracked pitch, and 20–30% of oils plus gas. The 400°C end point oil was rich in aromatics and could be used as a feed for oxidative conversion to phthalic and maleic anhydrides. The oil residue (>400°C) showed good binding properties for carbon electrodes. The gas contained 10–15% of ethylene as well as hydrogen and methane. The coke was calcined and used with the >400°C oil residue to produce metallurgical electrodes totally from the lignite tar products. Sanner et al.[73] cracked the tar from high-splint, Harlan County, Kentucky, coal in a fluidized bed of coke at 600, 700, and 800°C. The yield of light oil passed through a maximum at 700°C. Waters[74] thermally cracked the tars from different carbonizers in a static bed apparatus at 500–1000°C, and 1–440 s. The main products were gases useful as fuels or for chemicals and coke that could be used for the manufacture of electrodes. Huettinger[75] studied the kinetics of the pyrolysis of four tars by a thermogravimetric technique. The calculated reaction orders and activation energies suggest that different rate-controlling steps are involved in different tars. X-Ray measurement on the carbonaceous residue made at varying times showed about the same structure regardless of the starting tar.

Ueda et al.[76] claimed a process for thermally treating tars at elevated pressures to produce a superior pitch for electrode manufacture. Tar was pyrolyzed at 320–470°C and 20 atm with internal recycle of condensable products. The noncondensable products were continuously withdrawn. The desired pitch was removed from the products, and the product oils further treated at a higher temperature (370–540°C) and a higher pressure (35 atm). The residence times were between 2 and 10 hr. The total yield of pitch was claimed to be at least 75% of the starting tar, with no difficulty due to coking of the reactor.

Janardanarao et al.[77] thermally cracked five neutral oil fractions with initial boiling points 210–223°C. The maximum yield of ethylene plus propylene (21.8 wt %) and naphtha (8.7% by volume to 210°C end point) occurred at 550°C. The highest yield of ethylene occurred at 700°C (19.6 wt %), and of propylene (11.5 wt %) at 550°C. The tar acids and lighter tar fractions gave less ethylene and propylene, and more coke. The same authors[78,79] have reported the catalytic cracking of neutral oils and tar acids over activated carbons and silica-alumina catalyst. Project Seacoke[55] contains data on the catalytic cracking of COED tars before and after hydrogenation using a cobalt carbonyl catalyst and a cobalt molybdate catalyst. No significant yield advantages were realized for the hydrotreatments, nor was the high coke production substantially reduced.

The Dow Chemical Company[80] has undertaken a project to evaluate the possibility of producing petrochemicals from coal tars using standard petrochemical processing. Under this project they have taken the COED[3–5] tar and distilled it into four fractions: naphtha (<177°C), middle distillate (177–343°C), heavy gas oil (343–454°C), and residue

[71] J. S. Berber, R. L. Rice, and D. R. Fortney, *Prepr., Am. Chem. Soc., Div. Fuel Chem.*, **11**(2), 366–377 (1967).

[72] R. L. Rice, D. R. Fortney, and J. S. Berber, *Ind. Eng. Chem., Prod. Res. Dev.*, **9**, 30–33 (1970).

[73] W. S. Sanner, B. W. Naugle, and D. E. Wolfson, U.S. Bur. Mines Rep. Invest. No. 7322, 1969, 19 pp.

[74] P. L. Waters, *Ind. Eng. Chem., Prod. Res. Dev.*, **9**, 194–202 (1970).

[75] K. J. Huettinger, *Erdoel Kohle*, **23**, 559–565 (1970).

[76] K. Ueda, J. Kimoto, and M. Moritake, U.S. Pats. 3,692,663, 1962: 3,835,024, 1974.

[77] M. Janardanarao, M. Ramacharyulu, and M. G. Krishna, *Brennstoff-Chemie*, **49** 10–14 (1968).

[78] M. Janardanarao, M. Ramacharyulu, and M. G. Krishna, *Brennstoff-Chemie*, **49**, 65–69 (1968).

[79] M. Janardanarao, M. Ramacharyulu, and M. G. Krishna, *Brennstoff-Chemie*, **49** 111–115 (1968).

[80] B. C. Peters, *Chemicals from Coal*, U.S. ERDA Contract No. E (49-18)-1534, Dow Chemical Co., Midland, Mich.

(454+°C). The distillate fractions were hydrotreated over a cobalt molybdate catalyst to remove the heteroatoms. The middle distillate and heavy gas oil fractions were then hydrocracked to reduce molecular weight. Next the straight-run naphtha and "hydrocrackate" naphtha were reformed to maximize aromatics. Fractions of "hydrocrackate" middle distillate and some portions of the naphthas were subsequently steam cracked to produce olefins (primarily ethylene). The yield of olefins was slightly less than expected from similar boiling range feeds derived from petroleum, but the aromatics yields were higher. These results suggest that a satisfactory product slate could be achieved. Work is continuing on these processes in an effort to optimize them for commercial evaluation.

16.4 UTILIZATION OF LOW-TEMPERATURE TARS

The complex nature of low-temperature tar ranks it in a class similar to a heavy crude oil or petroleum residuum. It is set apart from these materials by its higher content of aromatic carbons, lower hydrogen content, and higher heteroatom content. The utilization pattern has remained about the same over the years. Most of the tar is processed to liquid fuels, and selected portions are used to produce chemicals. As Section 16.3 shows, most of the recent process work on tar has been directed toward making it more nearly a substitute for petroleum crude oils (by hydrotreating), or to convert it directly to refined petroleumlike products (by hydrocracking). Portions of the tars are still used to produce phenols and phenolic chemicals, road tars, preservatives, and carbon binders that are somewhat unique to the tar as compared to petroleum. These uses, however, do not constitute a large enough market in themselves to support a major industry that is in keeping with available coal resources. The future of low-temperature carbonization, in our opinion, is not too bright. The future use of coal seems to be headed toward large-scale gasification and/or direct liquefaction. There could be some significant contribution to the overall energy picture by a process (or processes) such as the COED scheme (to produce tar and coke), but this trend is not yet clear.

CHAPTER 17

High-Temperature Coal Tar

DONALD McNEIL

Donald McNeil *was educated at Glasgow University (B.Sc., First Class Honors, 1935; Ph.D., 1938). He was awarded the Ramsey Fellowship in 1938 and after two years' postgraduate research at Manchester University, joined Research Department of Imperial Chemical Industries (Billingham Division). In 1946 he became Head of Speculative Organic Research Group. He was appointed Director of the British Coal Tar Research Association (CTRA) in 1948 and held that position until 1974, when CTRA merged with British Coke Research Association to form British Carbonization Research Association (BCRA). Since 1974 he has acted as Consultant to BCRA and a number of other chemical manufacturing and engineering firms. Dr. McNeil's publications include* Coal Carbonization Products *(Pergamon Press, 1966), "Chemical Structure and Physical Properties of Coal Tar Pitch," in* Bituminous Materials—Asphalts, Tars and Pitches *edited by Hoiberg (Interscience Publishers, 1966), and "Cresols" in Kirk-Othmer's* Encyclopedia of Chemical Technology, *as well as 25 scientific papers. He is a Fellow of the Royal Institute of Chemistry, member of the Institution of Chemical Engineers, and Fellow of the Chemical Society and the Society of Chemical Industry. Dr. McNeil was awarded the Order of the British Empire for services to technology in 1968.*

17.1 INTRODUCTION

High-temperature tar is one of the three liquid by-products associated with the manufacture of coke and town gas by the high-temperature pyrolysis of bituminous coal. Carbonization of coal is carried out either batchwise in coke ovens or continuously in vertical retorts. In coke ovens the primary aim is to make coke suitable for the reduction of iron ore in blast furnaces. The oven temperature, as measured in the flue, is around 1200–1350°C. The desired products in vertical retorts are town gas and a more reactive coke for domestic and industrial heating. The retort temperature generally varies between 1000 and 1250°C.

Carbonization involves a deep-seated decomposition and disproportionation of the coal substance, by which it is partly converted to smaller volatile molecules and partly to infusible, solid coke. The volatile products, in their passage through the hot coke, undergo a complex series of secondary reactions leading to stable products with molecular weights ranging from 2 (hydrogen) to several thousand. This mixture of altered coal decomposition products is cooled as it emerges from the oven or retort, to condense the water and most of the organic products as a tar. Some of the more volatile, normally liquid organic components are carried forward with the uncondensed gaseous products and are recovered as crude benzole by scrubbing the cooled gas with a suitable absorbing oil.

The three liquid products thus comprise aqueous liquor—a dilute solution of ammonia, hydrogen sulfide, and inorganic salts—crude tar, and crude benzole. The low-boiling fraction of the crude tar (~ 1% by weight) is similar in composition to the crude benzole and is usually blended with it for further processing.

The yields of these liquid carbonization products depend on the nature of the coal carbonized and the conditions of carbonization. On the average, the amount of crude tar produced at coke ovens is 4–5% by weight of the coal carbonized and the amount of crude benzole is about 1.5%. The corresponding yields of crude tar and crude benzole from continuous vertical gas-making retorts are 7.5 and 0.5%, respectively.

Crude high-temperature tar, as produced, is a dark brown to black, fairly mobile liquid, with a characteristic unpleasant smell, which has been likened to that of a solution of naphthalene in pyridine. It normally contains 3–5% water, suspended in it as small droplets, and few percent of finely divided carbonaceous matter to which it owes its black color.

17.2 PRODUCTION AND ECONOMICS

Crude tar, as recovered at carbonization plants, has little value except as a crude liquid fuel. Although some is used in this way, most of the crude tar produced is processed to yield a range of useful products and chemicals.

Historically, coal-tar distillation dates back to 1818, when Accum demonstrated that the tar produced as a by-product by the rapidly growing town gas industry in Great Britain could be distilled to yield a volatile oil, which could be used as a substitute for turpentine, and a residue that could be employed as an anticorrosion paint. Somewhat later other outlets for the bulk products of tar distillation developed particularly the use of the higher boiling oils, or creosote, as a wood preservative, and the distillation residue, pitch, for briquetting coal dust.

At the same time that the value of crude tar as the raw material for these bulk products was increasing, its value as a source of new, accessible, and versatile organic chemicals was growing. In 1848 Hofmann isolated benzene from coal-tar naphtha and prepared nitrobenzene and aniline from it. Even earlier, Runge had isolated phenol and aniline; Mansfield recovered toluene from neutral tar light oil in 1849; and Williamson isolated the cresols in 1854.

The ready availability of these and other aromatic chemicals provided by coal tar led to the development of dyestuff, pharmaceutical, and explosives industries based on them, and crude tar became valued mainly as a source of chemicals rather than bulk products.

It is perhaps worth recalling that up to the end of World War II, almost all the benzene, toluene, phenol, cresols, naphthalene, anthracene, and pyridine bases used industrially were derived from coal carbonization products. The explosive increase in the demand for these basic intermediates, brought about by the expansion of the plastics and man-made fibers industries, soon made the amounts available from the restricted supply of crude tar hopelessly inadequate and led to the development of the petrochemical industry. Today in the United States only 5.9% of the benzene, 1.3% of the toluene, and 2% of the phenol produced comes from coal carbonization. Percentages for Western Europe are slightly higher, with benzene from coke ovens accounting for 11.7% of total production, toluene from crude tar and crude benzole amounting to 4.5% of total output, and natural phenol contributing about 5% to the total.

Thus at present, as in the early days of the tar industry, the commercial value of crude high-temperature tar is mainly as a source of tar oils and pitch, not as a potential source of pure chemicals.

The amount and nature of the crude tar available for distillation depend on the activities of the steel and town gas industries and have been affected by changes in the technologies of these industries. A major change since the late 1950s has been the replacement of coal, first by petroleum oils, and more recently by natural gas, as the source of town gas. This has led to the virtual disappearance of vertical retort tar, and now almost all the high-temperature tar available comes from coke-oven installations. This change is illustrated by the statistics for tar production in Great Britain since 1955 (Table 17.1).

It will be noted not only that the total amount of crude tar has declined by more than two-thirds in the past 20 years, mainly because of the closure of gasworks retorts, but also that the output of coke-oven tar has fallen markedly. This is partly due to the recession in the U.K. steel industry since 1970, but the reduction in coke requirements by the steel industry is also responsible. The use of auxiliary fuels and other technological changes in blast furnace practice have reduced the amount of coke required to produce a ton of pig iron from about 2350 lb in 1955 to 1190 lb in 1975.[1] Another factor has been the replacement of low-temperature coke by other fuels for domestic and industrial heating. Statistics for crude tar production in the United States show a declining trend since 1970, indicating that any expansion in steel output was insufficient to compensate for the improved economy in coke use in blast furnaces. This is shown in Table 17.2.[2]

World output in 1974 of all types of coal tar is estimated at close to 20 million tons, of which about 15 million tons was of high-temperature origin. Production of high-temperature tar in the major producing countries is given in Table 17.3.

17.3 THE CHEMICAL NATURE OF HIGH-TEMPERATURE TAR

The chemical nature of high-temperature tar was admirably reviewed by J. F. Weiler in the first Supplementary Volume of *Chemistry of Coal Utilization*, and subsequent investigations have sharpened but have not materially altered the picture he outlined. This showed that high-temperature coal tar is an extremely complex mixture, the main components of which are cyclic aromatic hydrocarbons ranging from single-ring benzene and alkyl benzenes to compounds containing possibly as many as 20 or more rings. Heterocyclic compounds containing oxygen in a

[1] U.S. Bureau of Mines, *Mineral Industry Surveys, Coke and Coal Chemicals*, 1955 and 1975.
[2] D. L. Richards, *Chem. Mark. Rep.*, **209**(19), 3, 54 (1976).

Table 17.1 Coal-Tar Production (tonnes) in Great Britain 1955–1975

	1955	1960	1965	1970	1975
From coke ovens	1,101,294	1,187,215	1,076,515	935,469	699,883
From continuous vertical retorts	1,250,726	1,174,163	981,525	287,977	Nil
From low-temperature retorts	41,915	56,431	119,490	180,659	202,947
From other types[a]	687,388	449,057	328,342	54,017	20,646

[a] Horizontal, intermittent vertical, and inclined retorts, and carburetted water gas plants.

Source: Statistics derived from the Association of Tar Distillers, London.

Table 17.2 Coke-Oven Tar Produced and Distilled in the United States[2]

	Amount of Tar (thousands of U.S. gal)		
Year	Produced	Distilled	Burned
1960	694,000	608,600	85,100
1970	762,000	652,100	108,000
1972	747,181	592,507	119,030
1974	677,447	502,683	180,907
1975[a]	645,537	—	—

[a] Estimated.

five-membered ring, nitrogen as =N— in a six-membered ring or —NH— in a five-membered ring, or sulfur are present, and there is rarely more than one noncarbon atom in their ring systems. Small amounts of paraffinic hydrocarbons and partly saturated aromatic compounds also occur. The aromatic and heterocyclic structures occur in both unsubstituted and substituted forms, the main substituent groups being methyl, ethyl, or hydroxyl. Longer aliphatic side chains are encountered in small amounts in the lower boiling fractions, but as with hydroxyl, they tend to disappear in the higher boiling fractions. Molecules containing up to four rings are generally fully condensed, but in the more complex molecules, the degree of condensation is less complete and branched ring-chain structures, in which the majority of the rings are fused to not more than three other rings, predominate.

Continuous vertical retort tars differ from coke-oven tars in three major respects. Whereas coke-oven tars contain relatively small amounts of nonaromatic hydrocarbons, continuous vertical retort tars contain a relatively high proportion of normal straight-chain or slightly branched chain paraffins; the content of these decreases from about 20% in the lower boiling fractions to 5–10% of the higher boiling distillate oils.

The second difference is the content of phenolic components. Coke-oven tars contain an average of 3% of phenols—mainly phenol, cresols, xylenols, and naphthols—which are almost entirely confined to the fractions distilling up to 300°C. Vertical retort tars, on the other hand, contain 20–30% of hydroxy-substituted hydrocarbons, of which some 15% (on the tar) can be extracted from the oils distilling up to 350°C. These phenols in the tar from continuous vertical retorts are of complex composition and are mainly xylenols, methylethyl and diethyl phenols, dihydric phenols (e.g., catechol and resorcinol and their methyl and ethyl derivatives), hydroxy hydrindenes and their methyl and dimethyl derivatives, naphthols, and hydroxy-substituted acenaphthenes, anthracenes, and phenanthrenes.

Third, the aromatic hydrocarbon and heterocyclic structures in continuous vertical retort tars are predominantly substituted by one or more methyl groups, and the content of unsubstituted ring structures, which are the major components of coke-oven tars, is comparatively small.

The number of individual components in high-temperature tar can only be guessed at. Lang and Eigen,[3] in the latest compilation of components isolated from or identified in high-temperature tars, list 461 compounds. Since that publication, many more minor components have been identified by the new analytical methods that have become available.

The most rewarding of these newer methods is undoubtedly gas chromatography (GC), which has been developed and applied

Table 17.3 Worldwide Production of High-Temperature Tar

Country	Output (tonnes)	Ref.
U.S.S.R.	3,500,000	*a*
United States	3,000,000	*b*
Japan	2,000,000	*b*
West Germany	1,350,000	*b*
East Germany	1,000,000	*a*
Poland	750,000	*a*
United Kingdom	700,000	*b*
France	459,000	*b*
Italy	300,000	*b*
Benelux countries	390,000	*b*
Czechoslovakia	250,000	*a*
Spain	200,000	*b*

[a] Derived from coke output figures.
[b] Published statistics.

[3] K. F. Lang and L. Eigen, *Fortschr. Chem. Forsch.*, **8**(1), 91 (1967).

to the quantitative analysis of tars and tar distillate oils mainly by European workers.

17.3.1 The Application of Gas Chromatography to the Analysis of High-Temperature Tars and Tar Fractions

17.3.1.1 *The Method and Its Development*

Gas chromatography (GC) effects the separation of complex mixtures of compounds of similar chemical type by passing a small sample of the mixture, entrained in a carrier gas (the mobile phase), through a column containing a high-boiling liquid (the stationary phase), which is either absorbed on a suitable solid absorbent or present as a thin layer on the column wall. As the gas with the vaporized mixture passes through the column, the components of the mixture are repeatedly adsorbed by and desorbed from the stationary phase. The rates at which these mass transfer processes occur depend on the characteristics of the individual components, such as their relative volatilities, their solubilities and rates of diffusion in the stationary phase, and whether they interact in some way with the stationary phase. Because of differences in these properties, the components travel at different rates through the column and, if the column is long enough, emerge separately in the effluent stream. This stream is passed through a detection device that detects some difference in a property of the exit gas due to the presence of an organic contaminant and generates an electrical signal with a magnitude proportional to the concentration of the organic component in the gas stream. These signals are amplified and fed to a recorder to give a chromatogram consisting of a number of peaks, each peak representing the single components, or mixture of components, into which the sample has been separated.

To identify the peaks, it is necessary to calibrate the apparatus by injecting samples of pure components and chromatographing these under exactly the same conditions as for the sample. The retention time (i.e., the time elapsing from the injection of the sample to the appearance of the maximum peak height) under the closely controlled and standardized temperature, pressure, and flow conditions used, is characteristic of each component. In practice, since slight variations in conditions are difficult to avoid, the relative retentions (i.e., the ratio of the retention time of each component to the retention time of one particular component, or that of an added internal standard compound), are employed as the identification indices.

To obtain a quantitative measure of each component, the area under each peak is measured by planimeter, by multiplying the peak height by the peak width at half-peak height, by multiplying the peak height by the retention time, or by a built-in integrator. The peak area of each component is expressed as a percentage of the sum of all the peak areas, with a correction for the detector response to each component. These corrections are determined by prior chromatographing of different amounts of the pure components. This method—normalization or total summation—assumes that all the components are eluted. A more satisfactory method is to add to the bulk sample a known amount of a reference compound that ideally should be eluted separately from any of the components of the mixture. The column is first calibrated with individual components containing precisely known amounts of the reference compound, and factors giving the ratio of the detector response to the component and to the reference compound are established. These are used to correct the peak area ratios from the chromatogram of the mixture to obtain an estimate of the amount of each component in the mixture.

It is not within the scope of this chapter to give a history of the development of GC. Numerous textbooks explain the theory of the method, the development and the construction of the apparatus, and its use for different analyses. Its application to the qualitative and quantitative analysis of tars and tar fractions has been reviewed by Grant in 1964[4] and 1972,[5] by Petrowitz,[6] by Bricteux in 1967,[7] and by Bartle in 1972.[8] Only the

[4] D. W. Grant, *Ind. Chem.*, **40,** 240, 426 (1964).

[5] D. W. Grant, Coal Tar Research Association, Rep. No. 0511, 1972.

[6] H. J. Petrowitz, *Chem. Ztg.*, **88,** 235 (1964).

[7] J. Bricteux, *Ann. Mines Belg.*, No. 1, 45 (1967).

[8] K. D. Bartle, *Rev. Pure Appl. Chem.*, **22,** 79 (1972).

major developments concerning tar fraction analysis are considered here.

The gas chromatographic analysis of tars and tar fractions presents special problems. Tars and tar fractions contain large numbers of components of similar chemical types, many of which have almost identical boiling points. They contain relatively nonpolar (hydrocarbon) and strongly polar (phenolic and heterocyclic) components, and members of these two classes can have identical volatilities. The polar compounds are irreversibly absorbed on certain absorbents, giving rise to excessive losses and unsymmetrical peaks. They also contain small amounts of olefinic compounds, which tend to polymerize or oxidize on the column if the conditions are not properly selected. The major developments of the method as applied to tars and tar products have resulted from the efforts of various workers to overcome these difficulties.

The packed columns initially employed were of low efficiency, and only partial separation of mixtures of the complexity encountered in tar analysis could be expected. Packed capillary columns, which allowed an increase in efficiency of several orders, were introduced by Scott,[9] but their use was limited by the high pressures and low gas rates that had to be used because of the high pressure drops. The use of open capillary columns, in which the stationary phase was deposited as a thin film on the inner surface of the column, overcame this difficulty and increased separation efficiency from the 2000–5000 theoretical plates of packed columns to 50,000–250,000 theoretical plates. In use, however, uneven distribution of the stationary phase resulted in changes in the column characteristics and nonlinear absorption effects. This disadvantage in turn resulted in "tailing" of the peaks. These difficulties were solved by the use of porous layer open tubular (PLOT) columns, introduced by Halasz and Horvath[10] and developed by Grant.[11] In this type of column a thin layer of a solid support is fused to the inner wall of the column and serves to hold the stationary phase in an even and stable distribution.

These long capillary columns, usually arranged as closely packed spirals, required much smaller samples than the earlier packed columns and could be used only with detectors of much greater sensitivity than the catherometer detectors originally used. Detectors of the required sensitivity were developed. The most widely used detector now is the flame ionization detector (f.i.d.).[12] (Fig. 17.1*b*) This device measures changes in the electrical conductivity of the flame formed when the exit gas stream, mixed with hydrogen, is burned at a stainless steel jet. It is a further development of Grant's emissivity detector[13] and Scott's hydrogen flame detector,[9] which detected differences in the emissivity or temperature of the flame. Another detector that is sufficiently sensitive to use with high-efficiency capillary columns is the microargon detector, originally described by Lovelock[14] (Fig. 17.1*a*).

Even with the high separation efficiency made available by the use of capillary columns, however, separation of some of the components is not possible, and another strand in the developing fabric of GC analysis has been the search for selective stationary phases. Sauerland and Zander of Rütgerswerke AG, in Germany, have made important contributions to this aspect of GC. They pioneered the use of packed columns containing an absorbent impregnated with calcium or lithium chloride for the effective separation of aromatic hydrocarbons,[15] and of sugars such as dulcitol and methyl dulcitols as selective stationary phases for hydrocarbon and pyridine base mixtures.[16] In a later paper they describe the use of poly-*m*-phenyl ether and naphthyl methylene oligo-

[9] R. P. W. Scott, in *Gas Chromatography,* D. H. Desty, Ed., Butterworths Scientific Publications, London, (1958) p. 189.

[10] I. Halasz and C. Horvath, *Anal. Chem.,* **35,** 499 (1963).

[11] D. W. Grant, *J. Gas Chromatogr.,* **6,** 18 (1968).

[12] I. G. McWilliam and R. A. Dewer, in *Gas Chromatography,* D. H. Desty, Ed., Butterworths Scientific Publications, London, (1958), p. 36.

[13] D. W. Grant, in *Gas Chromatography,* D. H. Desty, Ed., Butterworths Scientific Publications, London, (1958), p. 165.

[14] J. E. Lovelock, *J. Chromatogr.,* **1,** 35 (1958).

[15] H. D. Sauerland and M. Zander, *Erdoel Kohle, Erdgas, Petrochem.,* **19,** 502 (1966).

[16] H. D. Sauerland, "Gas Chromatography of High Boiling Aromatic Hydrocarbons at a Relatively Low Temperature," paper presented at Colloquium on Gas Chromatography, Paris, November 21–22, 1960.

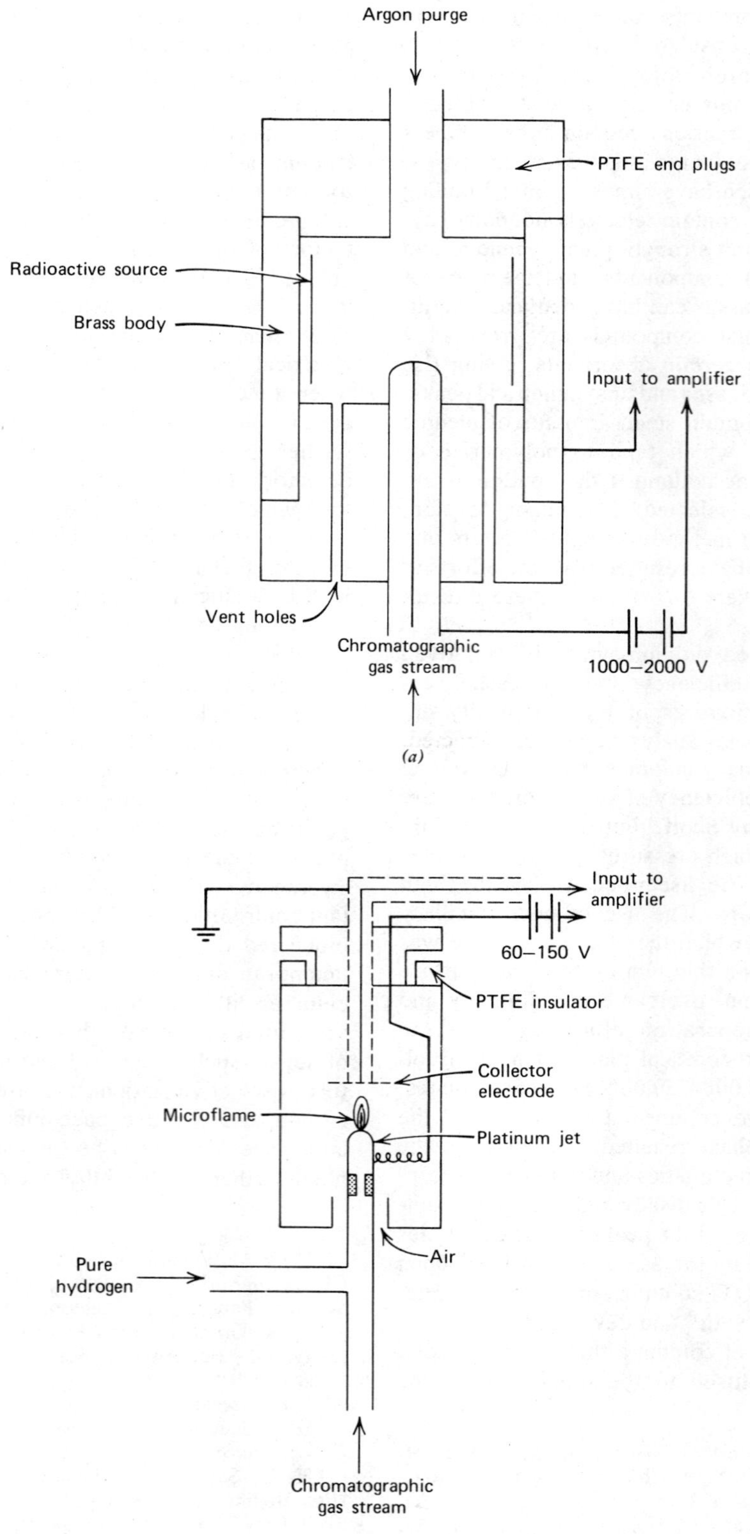

Figure 17.1 (*a*) Microargon detector and (*b*) flame ionization detector.[4]

mer to effect difficult separations in the polynuclear aromatic range, organoclay packing for the separation of anthracene and phenanthrene, and liquid crystals as selective stationary phases for the resolution of mixtures of isomers such as 2,6- and 2,7-dimethyl naphthalenes. They also claim that triaryl phosphine oxides used as stationary phases show high selectivity for phenols and enable the quantitative separation of meta- and para-cresol.[17] Grant[4] made use of organoclays such as Bentone 34 as a stationary phase for the separation of meta and para positional isomers and, in a later development, employed a packing of glass beads coated with a layer of an organoclay as the absorbent for the usual stationary phases to enable rapid quantitative analysis of difficult mixtures such as crude anthracene and cresylic acids. Dimethyl phthalate[18] and tricresyl phosphate[19] are other selective stationary phases recommended for the analysis of the lower boiling phenol mixtures: 4-amino benzaldehyde is advocated for the analysis of coal-tar bases.[20]

Another expedient is to convert the mixture to be analyzed into derivatives capable of easier separation. Grant[21] reacted mixtures of coal-tar phenols with hexamethyl disilazane, which converts them quantitatively to their trimethyl silyl ethers, and analyzed these on a 2-m packed column, with a silicone oil on Celite as the packing, using his emissivity detector. By comparing the chromatogram of the unreacted sample with that of the silylated sample, a mixture of phenol, all three cresols, all six xylenols, and the three ethyl phenols could be quantitatively analyzed.

If a mixture with a wide boiling range is subjected to GC at a fixed column temperature, the earlier peaks corresponding to the more volatile components tend to be compressed, and the later peaks become wide and shallow, thus difficult to measure for quantitative analysis. This problem was partly solved by repeating the analysis at a series of column temperatures, but a more satisfactory procedure is to program the temperature of the oven, in which the column and detector are housed, so that it increases at a steady and reproducible rate during the analysis.

The volatility of the stationary phase limits the quantitative use of GC to mixtures with components boiling below about 350–400°C at atmospheric pressure, and although very high boiling stationary phases have been used, such as silicone gum rubbers and high molecular weight polyethylene glycols, this limitation has not been entirely overcome.

The final limitation of GC lies in the identification of the peaks. For relatively simple mixtures, with components available in pure form for column calibration, there is no problem. But for tars and tar oils, numerous unidentified peaks are invariably present. There is the additional uncertainty over whether a peak corresponding to a pure component is due to that component alone, or to that component contaminated by other minor components with similar retention times. This problem has not been completely solved. The use of high-efficiency columns reduces the chance that major peaks are composite peaks, but it increases the number of unidentified peaks.

Ferrand[22] first demonstrated that the difference in the retention indices of hydrocarbons (i.e., the ratio of the retention time of a hydrocarbon to that of an *n*-paraffin or cycloparaffin of similar boiling point) on two stationary phases could be used to classify hydrocarbons by types. This information, together with the approximate boiling point as determined from the observation that the logarithm of the retention time of a hydrocarbon on a nonpolar stationary phase is linearly related to the boiling point, can be used to identify unknown peaks on a chromatogram. More recently Grant[23] described a multiple column system consisting of a

[17] M. Zander and H. D. Sauerland, *Erdoel Kohle,* **24,** 677 (1971).

[18] M. E. Neimark, *Coke Chem. (USSR),* No. 4, 40 (1967).

[19] J. Hrivnak and E. Beska, *J. Chromatogr.,* **54,** 277 (1971).

[20] H. O. Heinze, "Separation and Identification of Coal Tar Bases with the Help of Gas Chromatography and I.R. Spectroscopy," paper presented at Colloquium on High Temperature Chromatography, Essen, November 26–27, 1959.

[21] D. W. Grant, Coal Tar Research Association, Rep. No. 0254, 1960.

[22] R. Ferrand, *Separation Immediate et Chromatographie,* Groupe d'Avancement des Methodes Spectroscopiques, Paris, (1962). p. 132.

[23] D. W. Grant, Coal Tar Research Association, Rep. No. 0464, 1971.

high-resolution capillary column and two subsidiary columns so arranged that each peak on the main column yields two simultaneous chromatograms on the subsidiary columns. These contain different stationary phases, and if the retention times on them are plotted against each other, lines of different slope and location are obtained for different chemical types. Thus paraffins, alkyl benzenes, indanes, indenes, and coumarone homologues all lie on different straight lines, as do mono-ortho phenols, *m*- and *p*-substituted phenols, indanols, and so on. By plotting the ratios of the retention times on the subsidiary columns against each other and plotting the logarithm of the retention time on the main column against the retention times of compounds of known boiling point, both the boiling point and chemical type of the unknown component can be determined and its probable identity determined by means of the literature. This technique appears to be capable of considerable extension; the number of secondary columns need not be restricted to two, but so far its possibilities have not been exploited.

The other possibility, which also has not been exploited to nearly its full extent, is to couple a high-resolution GC column to a mass spectrometer so that the mass spectrum of each peak is recorded. In this development, lithium chloride supported on Chromosorb is a particularly useful column packing, combining good separation efficiency with freedom from vaporization or decomposition even at high column temperatures. Thus there is no contamination of the effluent gas with stationary phase vapors or decomposition products that might interfere with the identification of the mass spectra. The use of combined GC and infrared identification of the peaks is another technique with considerable possibilities, although it has limitations. To obtain sufficient material for spectroscopic analysis, either a number of repeat gas chromatograms must be carried out, or large samples must be analyzed, which can be done only on packed columns of relatively low separation efficiency. Heinze[20] used this method to identify the tar bases analyzed on a 2-m packed column.

The early GC apparatuses used by investigators in tar research laboratories were homemade. Nowadays, sophisticated commercial units are available that incorporate such features as automatic sample injection, sample splitting, temperature programming, dual columns with interchangeable detectors, peak integration, and readout. As the technique was developed, particularly suitable methods for the analyses required for the specification of tar products or for the control of tar-processing operations have evolved at each laboratory. Those used in the laboratories of the British Carbonization Research Association (BCRA) are listed in Table 17.4. Three sizes of GC columns are used and flame ionization detectors are standard. For the simpler analyses a 150 cm × 6 mm I.D. packed column (column 1) is used, for more difficult separations, a 275 cm × 6 mm I.D. packed column is used (column 2). For the determination of both major and minor components in a tar oil, a 25 m × 0.5 mm I.D. capillary column (column 3) is standard. Other conditions are listed in Table 17.4.

17.3.1.2 *Qualitative Analysis of High-Temperature Tars and Tar Fractions by GC*

Many of the references to the use of GC for tar analysis are concerned with the development of the technique rather than its application to carbonization products. Others are concerned with the identification of the major components of tars and tar fractions. Sauerland[16] obtained the chromatogram of a crude tar (Fig. 17.2*a*) on a 50-cm packed column employing silicone gum rubber as stationary phase and helium as carrier gas. Of the 17 major and about the same number of minor peaks, 23 were identified as indicated. A later paper[15] gave the chromatograms of similar tars on more efficient columns (Fig. 17.2*b* and 17.2*c*). Although the same mononuclear and polynuclear hydrocarbons and condensed heterocycles are shown to be the major components, the number of unidentified peaks increases with the efficiency of the separation. This is a general phenomenon. As the efficiency of separation applied to tars and tar fractions increases, the chemical makeup revealed is more complex and the number of individual components detected is greater. Marich and Lenkevich,[24] using a 50-

[24] L. I. Marich and Zh. K. Lenkevich, *Coke Chem. (USSR)*, No. (5), 35 (1972).

Table 17.4 BCRA Tar Analysis Methods

Analysis	Column[a]	Packing and Stationary Phase	Mobile Phase	Temperature or Programmed Rate	Internal Standard	Pretreatment of Sample
Determination of benzene, toluene, and xylenes in light oil	1	10% OV-101 (dimethylsilicone) on Chromosorb G[a]	N_2	80–100°C isothermal	None	None
Determination of naphthalene in crude tar, naphthalene oil or wash oil	1	5% OV-17 (phenyl methyl silicone) on Chromosorb G	N_2	130°C isothermal	Tetralin	Fractionate to remove high-boiling components
Determination of anthracene, phenanthrene, and carbazole in crude anthracene, anthracene oil, or crude tar	1	2.5% Liquid crystal[c] on Chromosorb G	N_2	200°C isothermal	None	None
Determination of phenol, cresols, and xylenols in crude phenol, cresols, or cresylic acids	2	5% Trixylyl phosphate on Chromosorb W[a]	N_2	75–125°C, 2°C/min	Anisole	Silylate with BSA[d]
Determination of pyridine and picolines in tar bases	2	5% Carbowax or polyethylene glycol on Chromosorb W	N_2	80°C isothermal	None	None
Characterization analysis of crude tar (simulative distillation)	2	5% OV-17 on Chromosorb G	N_2	120–240°C, 3°C/min	None	None
Analysis of total tar bases boiling to 240°C for major components or higher boiling bases for naphthylamines	2	5% Carbowax or polyethyene glycol 400 on Chromosorb W	N_2	80–180°C, 2°C/min	None	None
Quantitative analysis of light oil, naphtha, or neutral naphthalene oil for all components	3	OV-101	He	60–200°C, 2°C/min	None	None
Quantitative analysis of wash oil for all components	3	OV-101	He	100–200°C, 2°C/min	None	None
Estimation of acenaphthene and fluorene or diphenyl and diphenylene oxide in tars	3	OV-101	He	100–200°C, 2°C/min	None	Fractionate to remove higher and lower constituents
Quantitative analysis of anthracene oil for all components	3	OV-101	He	100–230°C, 2°C/min	None	None
Determination of all components in total or high-boiling tar acids	3	OV-101	He	80–250°C, 2°C/min	None	Silylate with BSA[d]
Quantitative analysis of creosote for all major components	3	Polyphenyl ether	He	150–230°C, 1°C/min	None	None

[a] Column 1 = 150 cm × 6 mm I.D. packed column; column 2 = 275 cm × 6 mm I.D. packed column; column 3 = 25 m × 0.5 mm I.D. capillary column.

[b] Chromosorbs G and W are diatomaceous earths specially prepared for GC analysis and supplied by Johns-Manville Company; they are usually used as a 60–85 mesh powder.

[c] The liquid crystal used is *N, N*-bis(*p*-methoxybenzylidene)-α,α'-bis-*p*-toluidine.

[d] Silylation can be carried out by reacting tar acids with hexamethyl disilazane or, better, with BSA (bistrimethyl silyl acetamide).

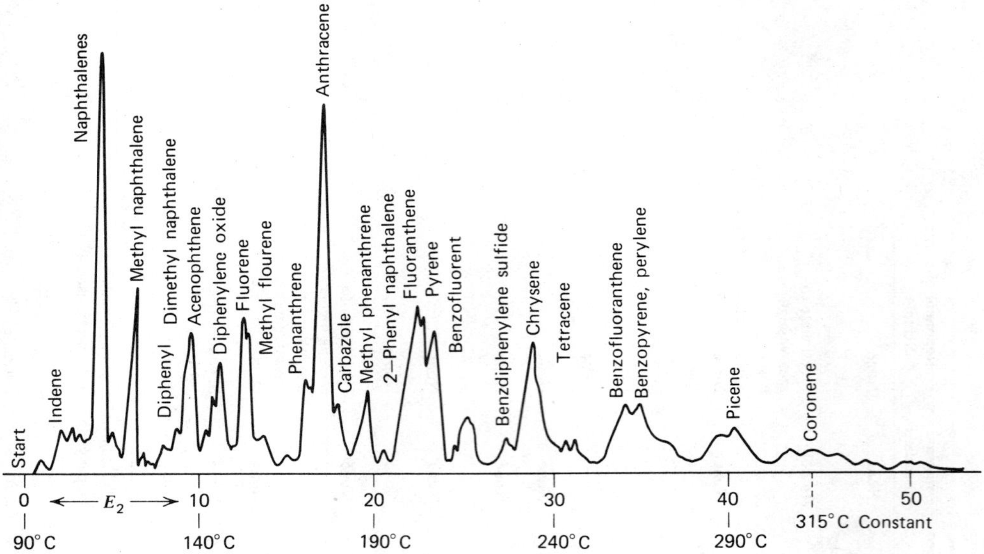
Start
Indene
Naphthalenes
Methyl naphthalene
Diphenyl
Dimethyl naphthalene
Acenophthene
Diphenylene oxide
Fluorene
Methyl flourene
Phenanthrene
Anthracene
Carbazole
Methyl phenanthrene
2–Phenyl naphthalene
Fluoranthene
Pyrene
Benzofluorent
Benzdiphenylene sulfide
Chrysene
Tetracene
Benzofluoranthene
Benzopyrene, perylene
Picene
Coronene
0
E_2
10
20
30
40
50
90°C
140°C
190°C
240°C
290°C
315°C Constant

(a)

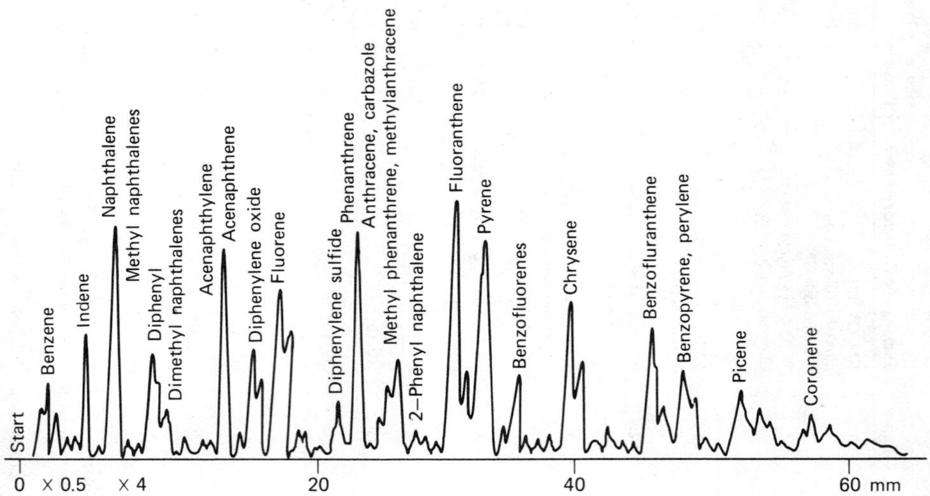
Start
Benzene
Indene
Naphthalene
Methyl naphthalenes
Diphenyl
Dimethyl naphthalenes
Acenaphthylene
Acenaphthene
Diphenylene oxide
Fluorene
Diphenylene sulfide
Phenanthrene
Anthracene, carbazole
Methyl phenanthrene, methylanthracene
2–Phenyl naphthalene
Fluoranthene
Pyrene
Benzofluorenes
Chrysene
Benzofluranthene
Benzopyrene, perylene
Picene
Coronene
0
× 0.5
× 4
20
40
60 mm

(b)

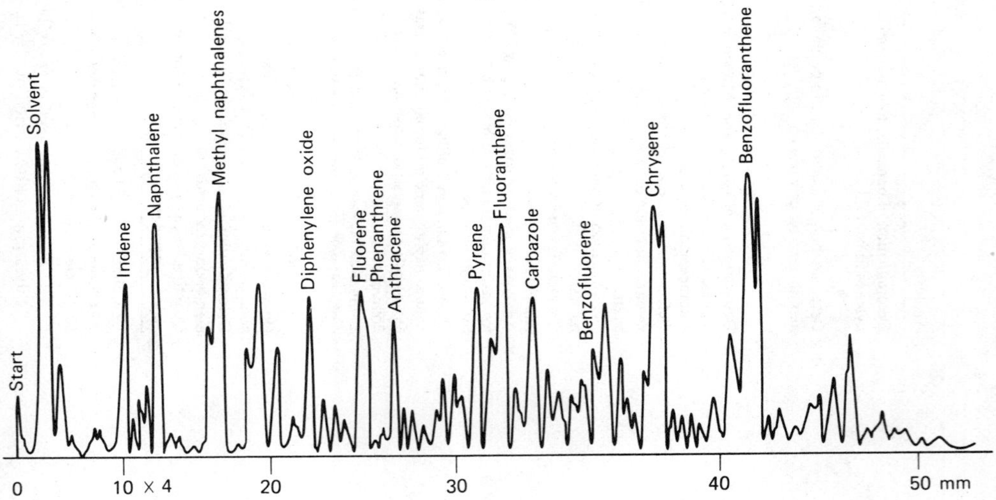
Start
Solvent
Indene
Naphthalene
Methyl naphthalenes
Diphenylene oxide
Fluorene
Phenanthrene
Anthracene
Pyrene
Fluoranthene
Carbazole
Benzofluorene
Chrysene
Benzofluoranthene
0
10 × 4
20
30
40
50 mm

(c)

m stainless steel capillary column with Apiezon L as stationary phase at 265°C, obtained a chromatogram for a coke-oven tar showing 242 peaks, of which 86 were identifiable.

Sauerland also showed the chromatograms obtained from various tar fractions on packed GC columns. The components identified in the fractions are those identified in the total tar. The primary information to be derived from the analyses of the fractions is the relative inefficiency of commercial tar distillation; thus naphthalene is present in light oil, naphthalene oil, wash oil (Solvay oil), and anthracene oil.

17.3.1.3 *Quantitative Analysis of Tars and Tar Fractions by GC*

The earliest attempt at the quantitative analysis of high-temperature tars is that of Du Pire, Grand'ry, and their co-workers at the Service de Recherches de la Société Carbochimique (SRSC) at Tertre. These workers analyzed four tars produced under different carbonization conditions by first subjecting them to a rough fractionation to obtain five distillate oils and a residue of medium-soft pitch. This residue was then further distilled to obtain pitch oils and a final residue with a softening point above 150°C (Kramer and Sarnow).[25] Each of the six distillate oils was treated to recover the tar acids and tar bases, and the neutral oils (and their acidic and basic counterparts) were submitted to GC analysis on 5-m packed columns with ground firebrick as absorbent, "Reoplex" (a sebacic polyester) as stationary phase for fractions distilling below 250°C, and silicone grease as stationary phase for the higher boiling fractions. Helium was the carrier gas, and catherometer detectors were employed. Column temperatures were 185°C for the two lowest fractions, 235°C for the next three fractions, and 295°C for the pitch oils. In all, 139 individual components or mixtures of components were separately eluted. Not all these were found in all tars; of the total, 37 peaks were not identified and 24 were tentatively identified on the evidence of their chemical types and their boiling points, as estimated from their relative retentions. The carbonization temperatures at which the tars were produced, as measured by the temperature in the oven at the top of the coal charge, were 800°C for tar G, 900°C for tar A, 1025°C for tar E, and 1100°C for tar M (Table 17.5).[26]

The same technique was used by Du Pire and Grand'ry[27] to analyze a road tar and the pitch and fluxing oils (light and heavy anthracene oils and drained anthracene oil) from which it was blended. Each material, without pretreatment to remove tar acids and bases, was chromatographed at three column temperatures, 185, 260, and 360°C. The anthracene-phenanthrene peak was resolved by infrared spectrometry. In all, 45 components or mixtures of components were definitely or provisionally identified and estimated, and the agreement between the tar composition as determined directly and as determined from the analyses of the pitch and fluxing oils was excellent.

The main components (>1.0% by weight of the tar by direct determination) were phenanthrene (5.2%), fluoranthene (5.1%), naphthalene (4.7%), chrysene (4.2%), pyrene (3.8%), an unidentified material with a boiling point of 398°C (3.3%), benzofluoranthene (2.0%), fluorene (1.5%), β-methyl naphthalene (1.5%), perylene and benzopyrenes (1.3%), carbazole (1.2%), methyl anthracenes and methyl phenanthrenes (1.0%), and unidentified components with boiling points, at atmospheric pressure, of about 430, 443, and 485°C. The sum of all the peaks amounted to 46.9% of the road tar.

Although the method used by Marich and Lenkevich[24] was unsuitable for quantitative

[25] *Standard Methods for Testing Tar and Its Products,* 6th ed., Serial No. Pt/2, 1967, p. 504.

[26] E. Grand'ry, "High Temperature Gas Chromatography of Coal Tar," paper presented at Colloquium on High Temperature Gas Chromatography, Essen, November 26–27, 1959.

[27] F. Du Pire and E. Grand'ry, *Goudron Routes,* No. 10, 5 (1959).

Figure 17.2 Gas chromatograms of crude tar: (*a*) 50-cm column, 20% silicone gum rubber on 30–60 mesh Chromosorb,[16] (*b*) 3-m column, 0.3% Apiezon L on 60–80 mesh glass beads; starting temperature, 80°C; flame ionization detector programmed at 4.8°C/min,[15] and (*c*) 4-m column packed with lithium chloride; starting temperature, 90°C; flame ionization detector programmed at 6.8°C/min.[15]

Table 17.5 Main Components of High-Temperature Tars as Reported by Grand'ry[26]

Component (wt %)	Tar G	Tar A	Tar E	Tar M	Component (wt %)	Tar G	Tar A	Tar E	Tar M
Toluene	0.29	0.15	0.08	0.03	2-Hydroxy diphenyl	0.12	—	—	—
m- and *p*-Xylene	0.57	0.17	0.13	0.06	2,3,?-Triphenylnaphthalene	0.45	—	—	—
o-Xylene	0.30	0.11	0.04	0.04					
Mesitylene	0.60	0.32	0.06	0.04	4-Hydroxy diphenyl	0.42	0.50	0.27	0.25
Ethyl toluene	0.42	0.22	—	0.03	2,5-Dimethyl-	—	0.18	—	—
Pseudocumene	0.70	0.31	0.14	0.08	diphenyl ?				
Hemimellitene	0.44	0.33	0.04	0.01	Unknown, b.p. ~ 290°C	—	0.15	—	—
Hydrindene	0.52	0.26	0.18	0.17					
Phenol	1.35	0.66	0.56	0.36	Fluorene	2.56	1.89	2.06	1.86
Aniline	0.16	0.02	—	0.01	1-Methyl diphenyl-	0.73	0.61	0.39	0.44
Unknown, b.p. ~ 190°C	0.39	—	—	—	ene oxide				
					2-Methyl diphenyl-	0.70	0.56	0.41	0.45
o-Cresol	0.84	0.19	0.26	0.31	ene oxide				
Durene	0.45	0.18	—	0.15	3-Methyl diphenyl-	0.09	—	0.15	0.12
Toluidines	0.15	0.06	—	—	ene oxide				
m- and *p*-Cresols	2.02	1.48	0.54	0.82	Dihydro anthracene	0.15	—	—	—
p-Tolunitrile ?	0.38	0.21	0.05	0.04					
Methyl indene	—	0.30	0.16	0.22	Unknown, b.p. ~ 314°C	0.37	—	—	—
m- and *p*-Xylenols	0.92	0.93	0.29	0.35					
Thionaphthene	—	0.08	0.15	0.12	2-Methyl fluorene	0.51	0.18	0.21	0.09
o-Xylenol	0.16	0.11	0.18	0.20	1-Methyl fluorene	0.19	0.53	0.20	0.27
Unknown, b.p. ~ 229°C	0.29	—	—	0.05	4-Hydroxy diphenyl	0.04	0.40	—	0.12
Pseudocumenol ?	0.28	0.18	0.05	0.09	Tetrahydro anthracene ?	0.76	0.44	—	—
Quinoline	0.27	0.41	0.35	0.56					
7-Hydroxy-Coumarone ?	0.25	0.10	0.03	0.02	1,5-Benzothionaphthene	0.81	0.15	0.64	0.28
β-Methyl naphthalene	2.39	2.62	1.92	1.73	Unknown, b.p. ~ 338°C	—	0.21	—	—
α-Methyl naphthalene	1.26	1.06	0.82	0.76	Phenanthrene and anthracene	2.87	3.75	5.45	5.07
Quinaldine	0.13	0.10	0.08	0.08	Unknown, b.p. ~ 342°C	0.26	—	—	—
Methyl quinoline	0.14	0.13	0.07	0.06	Acridine	0.17	0.15	0.15	0.11
Hemimellitenol	0.23	0.21	0.03	0.06	Phenanthridene	0.09	0.14	0.11	0.10
Indole	—	0.15	—	0.12	Carbazole	0.78	1.47	0.79	0.96
Diphenyl	0.42	0.43	0.33	0.50	Methyl phenanthrene	0.75	1.53	—	0.84
5-Hydroxy hydrindene	0.11	0.19	—	0.01	2-Hydroxy diphenylene	0.23	—	—	—
					oxide				
β-Ethyl naphthalene	0.41	—	—	0.19	Methyl anthracene	1.25	1.30	0.81	0.83
Unknown, b.p., ~ 261°C	0.13	—	—	—	Dimethyl phenanthrene ?	0.85	—	—	—
2,6- and 2,7-					Methyl phenanthrene	0.70	—	0.78	—
Dimethylnaphthalenes	1.04	0.88	0.55	0.58	Dimethyl anthracene?	1.0	0.74	0.27	0.16
1,7- Dimethylnaphthalene	—	—	0.19	0.14	Unknown, b.p. ~375°C	0.12	—	—	—
Unknown, b.p. ~264°C	0.13	—	—	—	Unknown, b.p. ~380°C	0.14	—	—	—
					Fluoranthene	1.46	1.78	1.96	1.77
1,6-Dimethylnaphthalene	1.10	0.93	0.37	0.21	*o*-(β-naphthyl) phenol	0.21	—	—	—
2,3-Dimethylnaphthalene	0.67	0.36	0.17	0.08	4-Phenanthrol	0.19	—	—	—
					Pyrene	1.41	1.09	0.97	1.08
Methyl indole	0.02	0.14	—	—	Unknown, b.p. ~398°C	—	—	0.13	—
1,2-Dimethylnaphthalene	0.24	0.28	0.14	0.03	Trimethyl phenanthrene	—	—	—	0.19
Unknown, b.p. ~273°C	—	0.17	—	—	Benzfluorene	1.86	0.89	0.49	0.60
					Methyl pyrene	0.46	0.34	—	0.13
Acenaphthene	1.28	1.47	1.86	1.73	Unknown, b.p. ~420°C	—	0.11	0.13	0.03
Trimethyl quinoline	0.21	0.09	0.04	0.02	Unknown, b.p. ~421°C	—	0.23	—	—
					Unknown, b.p. ~424°C	—	0.23	—	—
Diphenylene oxide	1.14	1.26	1.50	1.26	Benzofluoranthene	0.33	0.33	0.14	0.18
					Chrysene	0.79	0.70	0.49	0.51
					Perylene	0.38	0.26	0.11	—
					Hard pitch and tar	42.86	53.48	60.84	64.03

analysis, by splitting the tar into its various fractions and analyzing each fraction and the residual pitch at suitable temperatures, a method was developed for the quantitative assay of the 12 major components of crude coke oven tars. The values in Table 17.6 were obtained for the tars from three Soviet coking installations.

Gas chromatographic analysis was used by McNeil and Vaughan[28] to compare the composition of fresh wood preservation creosote with that left in the timber after long exposure. The technique used for examining the different zones of weathered wood was to pyrolyze a sliver of the timber in a heated platinum gauze basket situated at the column inlet, so that the combustion products, including the vaporized preservative, are carried through the column with the carrier gas. The technique was later modified by employing a PLOT column with a preheat column and injecting a small splinter of the timber, by means of a syringe, directly into the preheater column, which was maintained at about 100°C above the column temperature.[29] GC analysis is also used on a routine basis to monitor the composition of tar oil wood preservatives submitted for classification to the State Testing Station in Czechoslovakia.[30]

Grant and his co-workers have made more detailed analyses of the tar acids and tar bases extracted from tar distillate oils. In the case of the phenols, the acidic material extracted with 10% caustic soda from the total distillate oils, up to 300°C, was first fractionally distilled. After reaction with hexamethyl disilazane, each fraction was analyzed on 76-m capillary columns coated with silicone oil, at 100°C, with a microargon detector. Generally, between 40 and 50 individual peaks were found; however several of these could be attributed only to mixtures because of the similarity of the relative retentions of a number of alkyl phenols. The results are summarized in Table 17.7. The tar bases extracted from the distillate oils boiling up to 250°C were analyzed on a 275 cm × 6 mm I.D. column packed with 5% polyethylene glycol 400 on Chromosorb W at 80–180°C (2°C/min, rise), using an f.i.d. Table 17.8 summarizes the results.[31]

Table 17.6 Content of Major Components in Three Soviet Tars[22]

Component	Content (wt %) in Average Tar from:		
	Adveevka	Baglei	Zaporozhe
Naphthalene	11.56	10.52	9.83
β-Methyl naphthalene	1.24	0.82	0.82
α-Methyl naphthalene	0.62	0.35	0.44
Diphenyl	0.38	0.25	0.26
Acenaphthene	1.62	1.04	1.34
Diphenylene oxide	1.25	0.69	0.92
Fluorene	1.65	1.07	1.37
Phenanthrene	4.62	4.50	4.19
Anthracene	1.24	0.94	1.06
Pyrene	1.85	1.23	1.23
Chrysene	0.42	0.41	0.56
Fluoranthene	2.30	1.78	2.14

To complete this section on the GC analyses of tars and tar oils, the exhaustive investigation of the nature and amount of the minor neutral components in the lower boiling fractions of a coke-oven tar should be mentioned, both as an example of the possibilities of combined fractional distillation and GC analysis, and to demonstrate the extreme complexity of high-temperature tar composition. This study[32] involved only the lower boiling naphtha and naphthalene oil fractions, making up 13.5% of the coke-oven tar. Each of these oils was first fractionated on a 50-theoretical-plate column to yield twenty 10-ml fractions, each of which was analyzed by fast GC analysis on short packed columns and bulked according to the results of this preliminary characterization. The 13 bulked samples excluded those containing a high proportion of naphthalene, which were treated separately. These fractions were then subjected to GC analysis on (*a*) 6 m × 3.5 mm

[28] D. McNeil and G. A. Vaughan, in *Record of the 1964 Annual Convention of the British Wood Preserving Association,* Cambridge, 1964. Published by British Wood Preserving Association, London.

[29] D. W. Grant and G. A. Vaughan, in *Record of the 1967 Annual Convention of the British Wood Preserving Association,* Cambridge, 1967. Published by British Wood Preserving Association, London.

[30] F. Komona, *Holz-Zentralblatt (Stuttgart),* **125,** 1903 (1974).

[31] *The Coal Tar Data Book,* 2nd ed., Coal Tar Research Association, Gomersal, England, (1965), p. 14.

[32] D. W. Grant, Coal Tar Research Association, Rep. No. 0485, 1971.

Table 17.7 Composition of Phenols from High-Temperature Tar Oils[31]

	Coke-Oven Tars	Continuous Vertical Retort Tars
Total tar acids (wt % on dry tar)	1.84–7.61	14.51–15.83
Major components (wt % on tar acids)		
Phenol	16.2–22.9	8.6–8.9
o-Cresol	6.1–8.8	5.3–5.5
m-Cresol	11.0–13.6	6.0–6.4
p-Cresol	7.7–8.0	4.8–5.2
o-Ethyl phenol	0.4–0.7	0.9–1.3
2,5-Xylenol	2.8–3.4	2.1–2.7
2,4- and 3,5-Xylenol, 3-ethyl phenol	11.9–12.2	8.6–10.2
2-Isopropyl phenol, 4-ethyl phenol	0.6–2.7	1.3–2.3
2,6-Xylenol	1.4–3.0	2.2–2.3
2,3-Xylenol	0.9–1.2	1.2
3,4Xylenol, 3-methyl-6-ethyl phenol, 3-*iso*propyl phenol, 2-*n*-propyl phenol	1.7–1.9	2.1–2.2
2-Methyl-5-ethyl phenol, 4-methyl-2-ethyl phenol	0.7–1.0	1.2–1.4
3-Methyl-5-ethyl phenol, 2-methyl-4-ethyl phenol, 3-*n*-propyl phenol and catechol	2.7–3.6	4.0–4.5
2,3,5- , 2,4,5- , and 2,4,6-Trimethyl phenols, 2-methyl-3-ethyl phenol, and 4-*n*-propyl phenol	2.0–2.7	2.3–3.7
4-Indanol	0.3–0.4	0.8–1.6
5-Indanol	0.3–0.4	1.1
α-Naphthol	1.1–2.1	0.1–0.2
β-Naphthol	0.4–1.0	0.1–0.2
Minor components (wt % on total tar acids)[a]		
Other methyl ethyl phenols (2)	0.2–0.3	0.2–0.4
Other methyl isopropyl phenols (2)	0.1–0.4	0.3–0.4
Diethyl phenols (2)	0.1–0.2	0.3–0.4
4-*sec*-Butyl phenol	Trace-0.1	0.1–0.2
Methyl catechols (2)	0.1–0.3	0.6–0.9
Methyl resorcinols (2)	Trace-0.1	0.5–0.5
Methyl indanols (8)	0.8–1.4	3.3–4.4
Trimethyl phenols (2) and resorcinol	0.1	0.2–0.3
2,3,5,6-Trimethyl phenol	—	0.1–0.8
o-Phenyl phenol	Trace-0.1	0.1

[a] Numbers in parentheses indicate the numbers of such components.

I.D. columns packed with Celite 545 containing 5% trixylyl phosphate as stationary phase, using f.i.d. and nitrogen as mobile phase, and on (*b*) 60 m × 0.5 mm I.D. glass capillary columns with Reoplex 400 as stationary phase, nitrogen as carrier gas, and f.i.d. At the same time, the fractions were chromatographed on a large 9 m × 10 mm I.D. column packed with Celite containing 30% trixylyl phosphate as stationary phase and 9–10 mg fractions taken corresponding to the peaks obtained on chromatogram *b*. These fractions were examined by infrared spectroscopy to provide positive identification in some cases and to confirm the identity of peaks as predicted from the ratios of retention times on the two chromatograms.

The concentrated naphthalene fractions were subjected to zone refining, and the exudate from the pure naphthalene was fractionated. Each fraction was examined in the same way as the fractions that did not contain naphthalene.

Summarizing the results, in addition to the

Table 17.8 Analysis of Bases Extracted from Lower Boiling Tar Oils[31]

	Coke-Oven Tars	Continuous Vertical Retort Tars
Tar bases extracted from 0–250°C oils (wt % on dry tar)	1.85–2.77	2.04–2.13
Components (wt % on total bases in tar)		
Pyridine	1.0–3.7	1.6–1.8
2-Picoline	1.1–2.6	1.5–1.7
2,6-Lutidine	0.7–1.3	1.2–1.4
2-Ethyl pyridine	0.2	0.5–0.6
3-Picoline	0.8–1.0	1.0
4-Picoline	0.8–1.2	1.0
2,5-Lutidine	0.5–0.7	0.7–0.8
2,4-Lutidine	1.6–1.9	1.9–2.0
2,4,6-Collidine and 2,3-lutidine	1.5–2.5	3.1–3.5
Aniline and 3,5-lutidine	1.2–1.9	1.4–1.6
3,4-Lutidine	0.4–0.6	0.8–0.9
Toluidines	1.2–2.0	2.0–2.3
Quinoline	10.8–20.4	5.5–6.0
Isoquinoline	3.1–5.5	2.2–2.3
8-Methyl and 2-methyl quinolines	4.0–4.2	2.2–2.3
7-Methyl quinoline	Trace-1.0	0.6–0.7
6-Methyl quinoline	—	1.0
Unidentified components	1.4–3.0	8.4–9.0

known major components in this boiling range—benzene, toluene, and naphthalene—46 other components were identified and estimated. These included all the possible C_8 alkyl benzenes (xylenes and ethyl benzene), all the possible C_9 alkyl benzenes (ethyl toluenes, *n*-propyl benzenes, isopropyl benzenes, and trimethyl benzenes), nine C_{10} alkyl benzenes (tetramethyl benzenes and dimethyl-ethyl benzenes), thiophene and thionaphthene, indan and four methylindans, coumarone and four methylcoumarones, benzonitrile, the three tolunitriles, and the olefins—styrene, α-methyl styrene, *trans*-β-methyl styrene, and indene. Of these, the methyl styrenes, the six dimethyl-ethyl benzenes, and the 1-, 2-, and 5-methyl indans are not listed in Lang and Eigen's catalog of coke-oven tar components.

It will be noted that the components of this fraction of coke-oven tar consist mainly of (*a*) the three aromatic hydrocarbon structures—benzene, indan, and naphthalene—unsubstituted and substituted with methyl, ethyl, isopropyl, and *n*-propyl groups in every possible combination consistent with the boiling range, (*b*) oxygen- and sulfur-containing analogues of these hydrocarbons with a five-membered ring, the heteroatom replacing one of the carbons in that ring (such heterocyclic neutral oxygen and sulfur compounds also occur in methyl substituted form), (*c*) the parent hydrocarbons substituted by a nitrile group and methyl derivatives of these nitriles, and (*d*) olefinic compounds with one aromatic nucleus. There are also paraffins, representing 1.9% of the total oil sample; these paraffins appeared as a single peak at the beginning of the chromatogram and were not resolved.

Obviously, in this type of analysis, involving high-efficiency fractional distillation, high-resolution GC analysis, preparative GC, and identification of components by infrared or other means, we have a powerful tool for the elucidation of the fine structure of the volatile portion of high-temperature coal tar and if it were ever applied to the higher boiling fractions, in which the possibilities of isomers are much greater, the number of components identified in coal tars would multiply rapidly. However this time-consuming activity is not very attractive from the practical or commercial standpoint.

17.3.1.4 *GC Analysis Applied to the Chemical Structure of Coal-Tar Pitch*

The usefulness of GC analysis in investigating the chemical structure of the less volatile portion of coal tar is limited by the volatility and limited thermal stability of available useful stationary phases, by the absence of pure reference compounds, and by the tendencies, even at fairly high column temperatures, of the reference times of components to be long and of the peaks to be wide, shallow, and poorly defined. Nevertheless, some attempts have been made, with limited success.

Du Pire[33] chromatographed a pitch on a 2.5 m × 6 mm I.D. column packed with a refractory support, carrying 10% of high-vacuum silicone grease at two temperatures, 320 and 360°C. He used helium as the carrier gas and a sensitive tungsten wire catherometer detector. He obtained 18 peaks, which he claimed represented a separation of components with boiling points up to the equivalent of 720°C at atmospheric pressure. The identification of the peaks was very tentative, however, and the quantitative estimation was subject to considerable error: the total of the peaks from an estimated atmospheric pressure boiling point of 495–720°C amounted to only 9.5% by weight of the pitch.

The most successful attempt to employ GC to provide information on the chemical structure of pitch was made by Ferrand.[34] He employed a 7 m × 4 mm I.D. Pyrex column packed with crushed firebrick, carrying 5% of silicone gum rubber as stationary phase, helium as mobile phase, and f.i.d. The column had a silica inlet head that could be heated to 600°C. In this unit he analyzed the extract obtained from a coke-oven pitch by treating it with boiling xylene, as well as toluene extracts of coke-oven tars and a vacuum distillate of the pitch. This vacuum distillate—the so-called chrysene oil—was also studied by other methods to separate and identify individual components. These methods included high-efficiency, high-vacuum fractionation, treatment of the fractions by liquid-liquid chromatography on silica gel, extraction of the fractions with acid and alkali, treatment of the fractions with maleic anhydride, and fusion with sodium and sodamide. It was also fractionated on a preparative scale GC column and the fractions subjected to infrared, ultraviolet, and mass spectroscopy.

Pure polynuclear hydrocarbons chromatographed under the same conditions as the pitch extract gave retention times that when plotted as their logarithms against the number of carbon atoms, gave a narrow band with pericondensed compounds generally on the upper level. On the basis of this relationship, Ferrand gave each of his peaks a C_A number: that is, he attributed each peak to compounds with the same number of carbons or the equivalent. The magnitude of each C_A number fraction was estimated from the ratio of the relevant peak to that of the C_A 20 peak, which could be determined as a percentage of the "chrysene oil" covering the C_A 14–C_A 24 range. This amounted to 36% of the pitch. The major components assigned to each C_A range are those with this C_A value isolated from or identified in the "chrysene oil." The results are summarized in Table 17.9.

Lyjinsky and his colleagues[35] have described a study of the composition of the more complex neutral fraction of a "medicinal" tar, presumably a coke-oven tar topped to remove the more volatile components. The tar was pretreated to remove tar acids and bases by dissolving in cyclohexane, extracting with 90% aqueous methanol, and backwashing the methanol extract with fresh cyclohexane. The combined cyclohexane extracts were evaporated and the residue extracted with nitromethane. After removal of the solvent, the nitromethane extract was dissolved in benzene. The benzene solution was then chromatographed on a 2.42 m × 6 mm I.D. column packed with 60–80 mesh glass beads coated with silicone oil, using a microargon detector. Above the anthracene-phenanthrene range, the following polynuclear aromatics were identified and estimated to be present in the tar in the percentages, by weight, shown:

benzo[*a*]anthracene	0.696

[33] F. Du Pire, paper presented at Symposium on Analysis of Carcinogenic Air Pollutants, National Cancer Institute Monog. No. 9; U.S. Dept. of Health, Education and Welfare, (1962), p. 183.

[34] R. Ferrand, *Goudron Routes*, No. 3, 5 (1964).

[35] W. Lyjinsky, I. Domsky, G. Mason, H. Y. Ramahi, and T. Safavi, *Anal. Chem.*, **35**, 952 (1963).

Table 17.9 GC Examination of Coke-Oven Pitch After Ferrand[34]

Table gives weight percentages of the C_A14–C_A24 range, amounting to 36% of the pitch.

C_A Number	Result (wt %)	Probable Main Components
14	12.7	Anthracene and phenanthrene
15	3.3	Carbazole and its methyl derivatives, methyl phenanthrenes and methyl anthracenes, 4,5-dimethyl phenanthrene, 1- and 2-phenyl naphthalenes
16	21.5	Fluoranthene (11.6%) and pyrene (9.9%); peaks well separated
17	4.7	One of the most complex zones, with four separate peaks: present are 1,2- and 2,3-benzodiphenylene oxides, peri-1,9-benzoxanthene, methyl fluoranthenes and methyl pyrenes, benzofluorenes and methyl derivatives, 4,5-iminophenanthrene, 5,12-dihydrotetracene, benzo[*m,n,o*] fluoranthene, and partly hydrogenated derivatives of the chrysene type
18	16.5	Chrysene, triphenylene, tetraphene, and tetracene
19	5.2	Two peaks in this range: components include alkyl derivatives of 4-ring aromatics
20	16.3	Two peaks in this range: benzofluoranthenes, benzopyrenes, and perylene
21	4.4	A complex zone with a number of poorly defined peaks: components are benzopyrenes to picene
22	3.6	Picene and probably many other unidentified components
23	3.6	A number of poorly defined peaks that cannot be attributed to any compound
24	6.0	Coronene is the only component identified in this range; probably many unidentified components

benzo[*b*]chrysene	0.080
benzo[*i*]fluoranthene	0.045
benzo[*k*]fluoranthene	0.107
benzo[*g,h,i*]perylene	0.189
benzo[*a*]pyrene	0.176
benzo[*e*]pyrene	0.188

A technique that has not been used in the analysis of pitch but has been applied to the qualitative study of the polynuclear aromatic hydrocarbons (PAHs) in airborne particulate matter may prove useful in future studies of pitch composition. Lao et al.[36] coupled capillary or packed GC columns via a Markey-type Biemann-Wilson molecular separator[37] to a single-focusing mass spectrometer that is equipped with an electron impact source. They used this system to study the PAH fraction of airborne dust. The best results were obtained using a 3.6 m × 3 mm I.D. stainless steel column packed with Chromosorb W, carrying 6% of "Dexsil 300" (dicarbachlorododecaborane) as the stationary phase, and helium as the mobile phase, with an f.i.d. The column was programmed from 165 to 195°C at 2°/min. They claim that more than 70 PAHs with two to seven rings were separated and identified.

However the limitations of high-temperature GC have so far restricted its value for the analysis of pitch fractions, and other techniques have proved more rewarding.

17.3.2 Low-Ionization Mass Spectrometry in the Study of Coal-Tar Pitches

Mass spectrometry at ionizing voltages reduced to eliminate fragment ions, thus to produce a spectrum composed of molecular ions only, has been employed by Sharkey and his colleagues at the U.S. Bureau of Mines to study the chemical structures in coal extracts and coal pyrolysis products.

[36] R. C. Lao, R. S. Thomas, H. Oja, and L. Dubois, *Anal. Chem.*, **45**(6), 908 (1973).
[37] S. P. Markey, *Anal. Chem.*, **42**, 306 (1970).

With respect to the study of tar, its first use[38] was in the examination of close-cut phenol fractions from a continuous vertical retort tar. The results agreed well with those obtained from comparable samples by GC and chemical analyses, and they revealed for the first time the presence of hydroxy indenes.

The same technique was employed to examine three topped pitches from high-temperature tar.[39] The samples, A, B, and C, had softening points in the ranges 80–85, 100–110, and 140–150°C and represented approximately 70, 65, and 60% of the crude pitch residue, respectively. From the mass numbers of the molecular ions, 34 structural types and their alkyl derivatives were identified, and by making the assumption that the sensitivity (peak height per unit charge) was the same for structures having four or more condensed rings, the results were given semiquantitative expression, as shown in Table 17.10.

In a later paper[40] a more detailed analysis of sample A was undertaken by separating the polar compounds by extraction with 85% aqueous ethanol and partitioning the extract between petroleum ether and xylene. The material remaining in the aqueous ethanol was split into two fractions; 7.2% (on the pitch) was removed from the sides of the flask, where it had condensed during distillation, and 4% was obtained as a residue in the flask.

These fractions obtained from sample A were analyzed on a Consolidated Electrodynamics model 21-110B mass spectrometer at an inlet temperature of 300°C and low ionizing voltage. In all, 239 compounds corresponding to a minimum of 108 molecular types and their alkyl derivatives could be distinguished. They included 15 hydrocarbon types ranging from acenaphthene to benzo-

[38] A. G. Sharkey, Jr., J. L. Shultz, and R. A. Friedel, *Fuel*, **44**, 359 (1962).
[39] J. L. Shultz, R. A. Friedel, and A. G. Sharkey, Jr., *Fuel*, **44**, 55 (1965).
[40] J. L. Shultz, T. Kessler, R. A. Friedel, and A. G. Sharkey, Jr., *Fuel*, **51**, 242 (1972).

Table 17.10 Analysis of High-Temperature Pitches by Mass Spectrometry[39]

Structural Types, Including Alkyl Derivatives	General Formula C_nH_{2n-x} n	x	Sample (wt %) A	B	C
Indenes	9	10	0.3	—	—
Naphthalenes	10	12	0.1	0.2	0.2
Acenaphthenes, biphenyls	12	14	1.8	0.7	0.4
Acenaphthylenes, fluorenes	12	16	1.3	0.5	0.8
Anthracenes, phenanthrenes	14	18	7.5	2.8	1.1
Methylene phenanthrenes, phenyl naphthalenes	15	20	6.1	5.0	2.9
4-Ring aromatics (pericondensed)	16	22	10.2	9.5	4.0
4-Ring aromatics (catacondensed)	18	24	6.7	8.5	5.6
Methylene chrysenes, phenyl anthracenes, benzo-(*g,h,i*)-fluoranthenes	18	26	4.1	5.3	4.7
5-Ring aromatics (pericondensed)	20	28	8.1	11.9	10.6
5-Ring aromatics (catacondensed)	21	30	3.5	4.5	4.8
Benzo-(*g,h,i,*)-perylenes	22	32	3.9	5.6	6.0
6-Ring aromatics (pericondensed)	24	34	2.6	3.0	4.0
6-Ring aromatics (catacondensed)	26	36	0.5	0.7	0.9
Coronenes	24	36	0.5	0.4	0.5
Dibenzo-(*g,h,i*)-perylenes	26	38	0.8	0.4	0.9
Dibenzoperylenes	28	40	0.2	0.1	0.2
Oxygen compounds			2.3	4.8	3.6
Sulfur compounds			2.7		
Nitrogen compounds			7.7	7.5	6.8
Residue (involatile at 300°C/10^{-6} torr)			29.1	28.6	42.0

coronenes, 13 types made up of C and H atoms with one O atom corresponding to a range from phenylfuran to hydroxycoronenes, 8 types comprising carbon and hydrogen atoms with two oxygen atoms ranging from phenylbenzoquinone to naphthoanthraquinone, 11 types containing one nitrogen group in the molecule ranging from methyl isoquinoline to a structure corresponding to azabenzoperylene. There were 3 types with two nitrogen atoms per molecule corresponding to dimethyl azaindole, azabenzo-(*d,e,f*)-carbazole and other azacarbazoles, 10 types with one sulfur atom, 8 types with one oxygen and one nitrogen atom, and 3 types made up of carbon and hydrogen atoms with two oxygen and two nitrogen atoms in the molecule. These parent structures were accompanied by methyl, dimethyl (or ethyl), and trimethyl (or methylethyl, *n*-propyl, or isopropyl) derivatives, and it was noticeable that as the molecular complexity increased, the length of the side chains (or the number of methyl groups) decreased.

No attempt was made to express the results of this investigation quantitatively. The results are rightly claimed as the most extensive experimental evidence supporting the conclusion that the majority of the unidentified components in coal-tar pitch consists of higher molecular homologues of the known components. Throughout all the classes of hydrocarbons detected—and these make up the great bulk of the vaporized sample—the same skeletal ring structures are indicated. The series of components with molecular weights differing by 50 units indicates a linear or angular addition of one aromatic ring to the lower molecular weight nucleus, and the series differing by 24 mass units can be interpreted as ring closure to form a more highly condensed ring system.

17.3.3 Column Chromatography and Thin-Layer Chromatography in the Analysis of Coal-Tar Pitch

The application of GC and low-ionization mass spectrometry to the analysis of tars and tar distillate oils has introduced a quantitative dimension into our knowledge of the chemical structure of that part of coal tar having a molecular weight range of 50 to around 200, and gives a good qualitative picture of the nature of the components in the 200–300 molecular weight range. The remaining problems are to develop quantitative methods applicable to the 200–300 molecular weight range and to define the nature of the 20–25% higher molecular weight material represented by the petroleum ether–insoluble fraction of coal-tar pitch.

The first of these questions is assuming increasing importance in connection with industrial health requirements, since it is in the 200–300 molecular weight range that the PAHs known to have carcinogenic properties are likely to be found.

The original method for the separation of PAHs in particulate matter filtered from industrial atmospheres was by solid-liquid chromatography on silica gel columns[41–43]; but this method is time-consuming, requires relatively large amounts of sample, and can give erroneous results because of irreversible adsorption in the column. It has the advantage that up to 12 PAHs may be isolated, and for this reason is still recommended by the American Public Health Association as one of the tentative standard methods.

The main use for this method now is to effect a preliminary separation to provide fractions for more detailed separation by thin-layer chromatography (TLC). Liquid-liquid separation has also been used for the rough separation, and Matsushita et al.[44] have described the separation of 94 PAHs and the identification of 24 of them by liquid-liquid chromatographic separation followed by TLC analysis of the fractions, and identification by fluorescence spectra.

Benzo[*a*]pyrene (BaP) was the first PAH shown to have pronounced carcinogenic properties, and its content in airborne particulate matter, coal-tar pitches, and tar oils appears to be used—without, it may be said, sufficient justification—as an index of the carcinogenicity of such materials. The analysis of coal-tar fractions for PAHs has therefore been concentrated on the estimation of BaP. The most widely used separation meth-

[41] R. L. Cooper, *Analyst,* **79,** 573 (1954).
[42] P. Wedgewood and R. L. Cooper, *Analyst,* **78,** 170 (1953).
[43] V. C. Shore and M. Katz, *Anal. Chem.,* **28,** 1399 (1956).
[44] H. Matsushita et al., *Bunseki Kagaku,* **21,** 1471 (1972).

od is based on a preliminary separation of a BaP-containing fraction by column chromatography, usually on alumina, followed by TLC separation of the BaP concentrate on alumina,[45,46] acetylated cellulose,[47,48] or mixtures of these absorbents.[49,50]

A recommended procedure for the determination of BaP in atmospheric particulate matter or coal-tar fractions, including coal-tar pitch, is as follows.[51]

The material to be analyzed, which may be either a sample of pitch or tar heavy oil, or the particulate matter filtered from an industrial atmosphere, is exhaustively extracted with cyclohexane or other suitable solvent, and the extract is evaporated to about 100 ml under vacuum at 35–40°C. An aliquot, expected to contain 2–10 μg of BaP, based on previous analyses of similar materials, is taken. If the original solvent was not cyclohexane, the aliquot is carefully evaporated to dryness at a temperature not exceeding 40°C, and the residue is taken up in cyclohexane. The volume of the cyclohexane solution to be analyzed is adjusted to 50 μl by evaporation in a stream of cool air, and this sample is applied to 10 g of neutral chromatographic-grade alumina contained in a 30 cm × 10 mm glass column equipped with a bottom tap. Before use, the alumina, which should have a Brockmann activity of 1, is activated by heating at 105°C for one hour and cooled in a desiccator.

The column is eluted with pure benzene, and the movement of the BaP-containing zone is monitored by occasional illumination by UV light when it shows as a bright blue fluorescent band. The elute containing all this band is collected and evaporated by a gentle air current to 50 ml. If the sample is heavily contaminated, it is advisable to repeat this initial separation.

The cleaned-up BaP fraction is then streaked on Schleicher and Schull 21% acetylated cellulose thin-layer chromatography plates, while a current of cool air is directed on the 10 × 10 cm plate to evaporate the solvent. The plate is developed for 40 min in a mixture of ethanol-dichloromethane-water (20 : 10 : 1) and viewed in UV light. The bright blue fluorescent BaP zone is removed and Sohxlet-extracted with 15 ml pure benzene for 20 min, and the extraction apparatus is washed out with pure benzene. The combined benzene extract and washings are then carefully evaporated to under 5 ml and the residue is made up to exactly 5 ml with pure benzene. The BaP is estimated by UV spectroscopy, the purity of the sample checked by scanning the spectrum, and the concentration determined by the baseline technique.

17.3.4 High-Pressure Liquid Chromatography in the Analysis of Coal-Tar Pitch

The efficiency of column chromatography can be greatly increased by using a microparticulate absorbent, which requires that the developing and eluting solvent be pumped through the column under pressure. This technique—high-pressure liquid chromatography (HPLC)—is now being developed at a rapid rate. Since the separation is carried out in the liquid phase at ambient temperature, neither the volatility of the stationary phase nor the lack of volatility of the sample is a limitation; and since separation efficiencies comparable to those obtained in packed column GC are achieved and the characteristic UV absorption of PAHs makes their detection in trace amounts relatively simple, this method is particularly suitable for the analysis of pitch fractions.

The apparatus used consists of a stainless steel column up to 2 m long and usually between 2 and 4 mm I.D. This column is packed with silica or alumina of 5–20 μm particle diameter. A sample of 10–20 μl of the mixture to be analyzed is injected by microsyringe through a septum at the inlet to the column, and the chromatogram is developed with a single solvent (isocratic or equal power system) or by gradient elution, in which two solvents are employed. The rates of delivery of the two solvents are automatically varied so that the composition of the developing liquid varies regularly with time from pure solvent A to a predetermined mix-

[45] E. Sawicki, T. W. Stanley, J. D. Pfaff, and W. C. Elbert, *Chemist-Analyst,* **53,** 6 (1964).

[46] E. Sawicki, T. W. Stanley, W. C. Elbert, and J. D. Pfaff, *Anal. Chem.,* **36**(3), 497 (1964).

[47] G. M. Badger, J. K. Donnelly, and T. M. Spotiswood, *J. Chromatogr.,* **10,** 397 (1963).

[48] R. E. Schaad, *Microchem. J.,* **15,** 208 (1970).

[49] M. Koeler, H. Golder, and R. Schliesser, *Z. Anal. Chem.,* **206,** 430 (1964).

[50] D. F. Bender, *Environ. Sci. Technol.,* **2,** 204 (1968).

[51] BCRA Coke Res. Rep. No. 76, 1973.

ture of solvent A and solvent B. The eluate is continuously monitored by a variable wavelength ultraviolet detector. By repeating the analysis at different wavelengths, relative extinction coefficients at a number of wavelengths can be obtained for each component and used for identification and, in some cases, for the quantitative separation of composite peaks.

A recent report[52] describes the application of this method to the estimation of the PAHs in a filtered cyclohexane extract of a coke-oven electrode binder pitch. The apparatus used is shown in Fig. 17.3 and the construction of the column in Fig. 17.4. The column was 400 mm long × 4.4 mm I.D. and packed with 5-μm "Partisil" (a form of irregular silica particles).* Elution was by dry *n*-hexane containing 0.001% of dry methanol, and elution of all peaks was complete within 30 min.

To confirm the UV identifications and effect additional separation, the same sample was repeatedly chromatographed to obtain larger samples corresponding to each peak. These were subjected to thin-layer chromatography on plates coated with a 500-μm layer of Merck "Kieselgel G" containing 10% of Bentone 34, previously washed with a 50 : 50 water-methanol mixture. The plates were developed for 30 min in pure toluene, and the organic components from each fluorescent zone extracted and their UV spectra determined.

In almost all cases the TLC plates showed several separate zones, indicating that the HPLC separation was incomplete. However in many cases the peaks were found to consist of a major component contaminated by traces of other components. The extent of this contamination did not preclude the quantitative estimation of a number of the major PAHs present. The HPLC chromatogram is shown in Fig. 17.5. Table 17.11 gives the qualitative and, where possible, the quantitative interpretation.

The HPLC method described above does provide a sufficiently good separation of BaP from other PAHs for quantitative determination, and the following results give a comparison with the column chromatography–TLC method described earlier:

[52] D. W. Grant, BCRA Carbonisation Res. Rep. No. 28, 1976.

*Supplied by Whatman Laboratory Sales, Springfield Road, Maidstone, England.

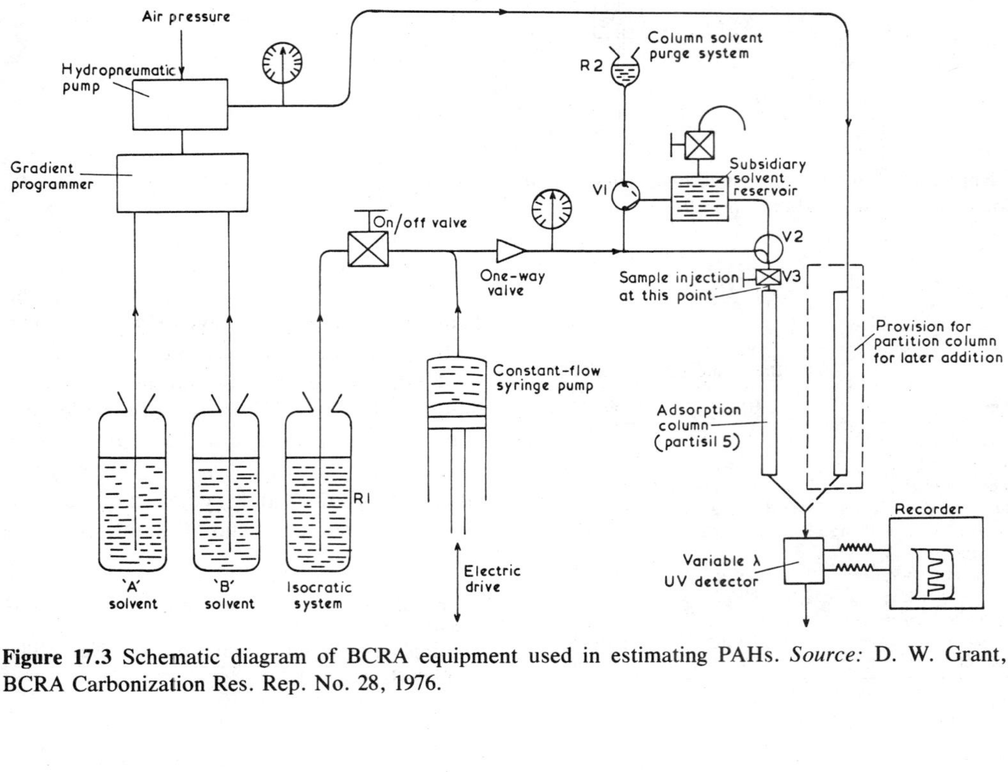

Figure 17.3 Schematic diagram of BCRA equipment used in estimating PAHs. *Source:* D. W. Grant, BCRA Carbonization Res. Rep. No. 28, 1976.

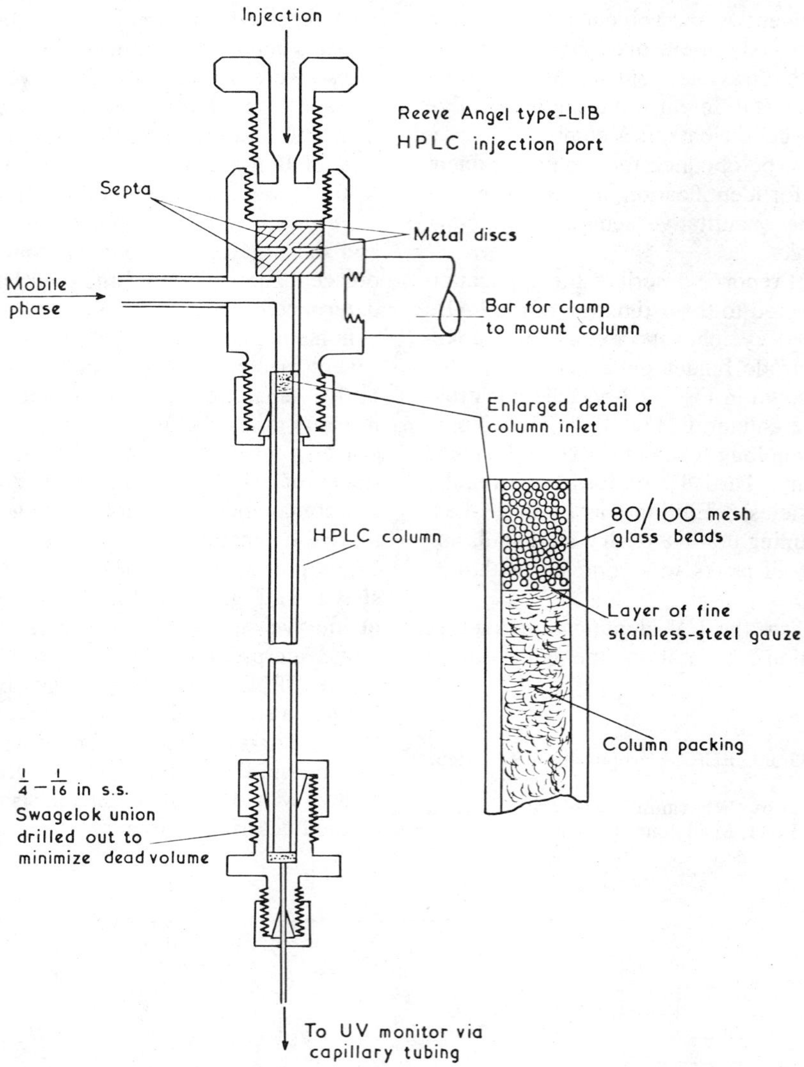

Figure 17.4 Construction of microparticulate columns used in apparatus shown in Fig. 17.3. *Source:* D. W. Grant, BCRA Carbonization Res. Rep. No. 28, 1976.

Sample	BaP Content Determined by: HPLC	Alumina-TLC
Coke-oven pitch	0.94 wt %	0.90 wt %
Coke-oven pitch	0.81 wt %	0.94 wt %
Coke-oven pitch	0.97 wt %	
Toluene-soluble fraction of	17.2 mg/l	18.8 mg/l
coke-oven pitch + 10 mg added BaP	18.9 mg/l	18.6 mg/l
Portions of high-volume	8.9 g	10.8 g
samplers after trapping	21 g	26.3 g
airborne particulates		28.0 g
Toluene extracts from	1.1 mg	1.56 mg
similar filters	4.4 mg	5.92 mg

As part of a collaborative study by the aluminum industry to develop an improved method for the characterization of coal-tar pitch volatiles, Bodin[53] has outlined an HPLC method that he claims is both rapid and accurate. It employs a 300-mm column packed with "Poracil," an 8–12 μl irregular silica absorbent* and cyclohexane as the developing solvent, and it determines the

[53] H. Bodin, *Chromatogr. Sci.*, **14**, 391, 204 (1968).

*Supplied by Waters Associated Ltd., New Mills, Stockport, England.

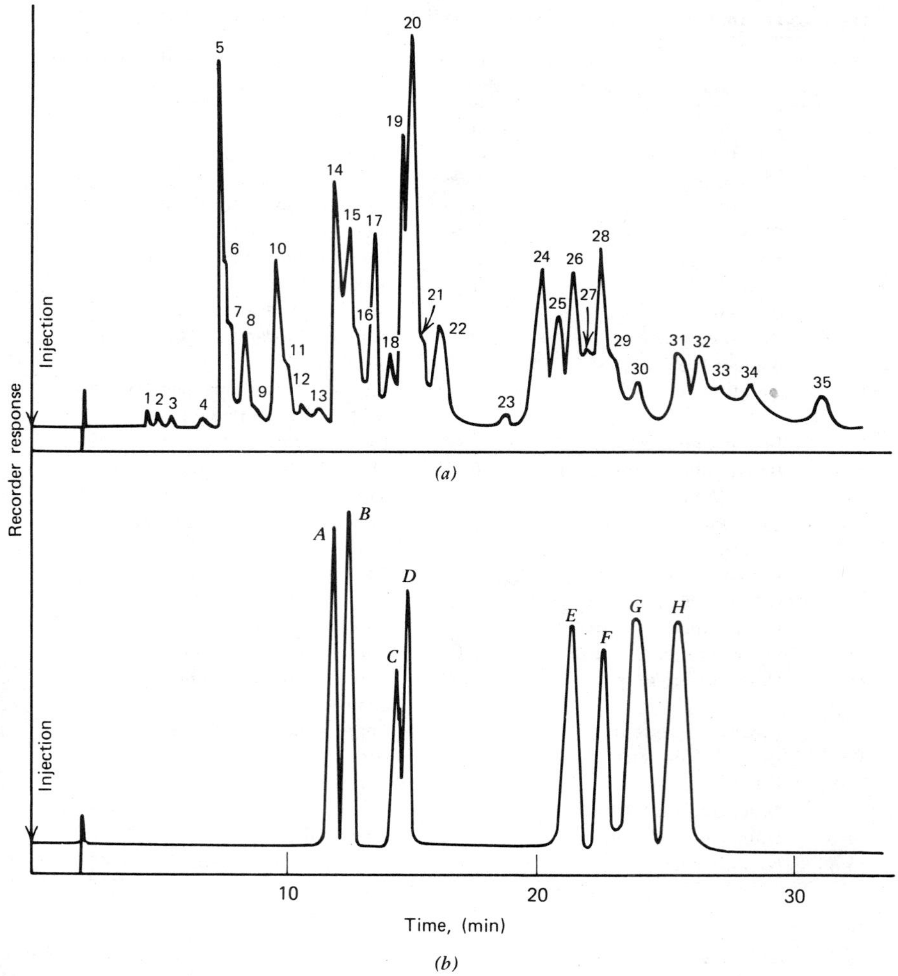

Figure 17.5 HPLC chromatograms of cyclohexane extract of (*a*) coke-oven pitch on Partisil and (*b*) a test mixture of the following nine components: *A* = benz[*a*]anthracene, *B* = benzo[*a*]pyrene, *C* = benzo[*k*]fluoranthene, *D* = benzo[*b*]fluoranthene, *E* = dibenz[*a,h*]anthracene, *F* = dibenzo[*a,i*]pyrene + dibenzo[*a,h*]pyrene, *G* = dibenzo[*a,e*]pyrene, *H* = dibenzo[*a,l*]pyrene. *Source:* D. W. Grant, BCRA Carbonization Res. Rep. No. 28, 1976.

BaP at a fixed wavelength of 383 nm. With this column Bodin found that benzo[*a*]pyrene and benzo[*k*]fluoranthene were separated sufficiently to give almost baseline peaks at moderate flow rates.

The immense advantage of the HPLC method is its rapidity. Bodin claims that a 5–10 min BaP determination is feasible with his method. Although the results tend to be slightly low compared with the established column and TLC method, there seems to be every possibility that improvement of the HPLC technique will result in its replacing the older method as the standard method of BaP estimation. It has the additional great advantage that the other major PAH components in the sample can be estimated at the same time. Since the same improvements in efficiency, selectivity, and precision that occurred during the development of GC can be expected as HPLC is developed, we can look forward to the provision of more detailed quantitative information about the chemical makeup of the more soluble fractions of coal-tar pitch through application of HPLC in the future.

Table 17.11 Analysis of PAHs in Coke-Oven Pitch by Combined HPLC–TLC–UV[52]

Peak No.	Major Components	Content (wt %)	Minor Components Shown by TLC
5	Pyrene	0.94	None
8	1-Methyl pyrene	—	None
10	Fluoranthene	0.96	3-Methyl pyrene
14	Benzo[*a*]anthracene	0.55	Benzo[*a*]pyrene (overlap)
	Chrysene	0.40	Benzo[*b*]fluorene
15	Benzo[*a*]pyrene	0.53	Benzo[*b*]fluorene (overlap), benzo[*c*]fluorene
17	Perylene	0.15	Benzo[*a*]pyrene (overlap)
	Benzo[*e*]pyrene	0.29	Benzo[*k*]fluoranthene (overlap)
	Anthanthrene	0.12	—
18	An alkyl benzo[*a*]pyrene ?	—	An alkyl benzo[*e*]pyrene and an unidentified component
19	Benzo[*k*]fluoranthene	0.40	Two unidentified trace components
20	Benzo[*j*]fluoranthene	0.29	Three unidentified trace components
	Benzo[*b*]fluoranthene	0.24	
22	3,4-*o*-Phenylene pyrene	—	Two unidentified components
	Coronene	—	
24	Benzo[*g,h,i*]perylene	0.31	Two minor components
	Dibenz[*a,j*]anthracene	0.04	
25	Unidentified components	—	Three components
26	Dibenz[*a,h*]anthracene	0.07	Three unidentified components
28	Dibenz[*a,i*]pyrene	0.07	Three minor unidentified components
	Dibenz[*a,h*]pyrene	0.05	Three minor unidentified components
30,31	Dibenz[*a,l*]pyrene	0.21	
	Dibenz[*a,e*]pyrene	0.92	Three unidentified trace components
32	Dibenz[*b,k*]perylene	—	
34	Dibenz[*e,k*]perylene	—	One unidentified component
35	Naphthacene	—	Two unidentified components

17.3.5 Statistical Structural Analysis in the Study of the Chemical Structure of Coal-Tar Pitch

The low volatility, restricted solubility, and extreme complexity of the petroleum ether–insoluble part of pitch, and the lack of pure reference compounds in the molecular weight range above 300, preclude application of the analytical techniques described so far to the so-called resin and C_2 fractions of pitch. It is unlikely that the individual components in these fractions will ever be separated and identified, and the best we can hope for is a general picture of their statistically average structure.

Statistical structural analysis, first applied to the structure of coal by Van Krevelen and his co-workers, depends on establishing by a study of the relevant data for pure compounds, relationships between chemical and physical parameters that are capable of measurement, and molecular structure.

The study made by Wood and Wilman,[54] referred to in the first Supplementary Volume (pp. 598 et seq.), is an example of the simplified use of this method. Using one of their examples, as follows: fraction 1B: fraction insoluble in methanol but soluble in benzene, amounting to 19.5% of a coke-oven pitch, which analyzed as C = 90.48%, H = 4.67%, total N = 1.62%, basic N = 0.53%, S = 1.26%, total O = 2.4%, phenolic O = 1.3%, ash, negligible; molecular weight = 910, the molecular formula could be written:

$$C_{68.6}\,H_{41.6}\,N_{0.27}\,N^*_{0.77}\,O_{0.67}\,O^*_{0.69}\,S_{0.35}$$

where N is nonbasic nitrogen, N* is basic nitrogen, O is nonhydroxylic oxygen, and O* is phenolic oxygen. In addition to the elemental analysis, the weight fraction of aro-

[54] L. J. Wood and W. G. Wilman, *Proc. Conf. Science in the Use of Coal*, Sheffield (1958), Paper No. C10.

matic CH groups was determined by infrared as 0.37.

Converting the molecular formula to the "equivalent hydrocarbon" by the "lift out and replace" procedure requires the following alterations:

1. For each basic N atom, the number of C atoms is increased by 1, the number of H atoms is increased by 1, and the number of aromatic H atoms is increased by 1.
2. For each nonbasic N atom, the number of C atoms is increased by 1 and the number of H atoms is increased by 1, but there is no change in the number of aromatic H atoms.
3. For each phenolic O, no change is made in the number of C or H atoms, but the aromatic H is increased by 1.
4. For each nonphenolic O atom, the number of C atoms is increased by 1 and the number of H atoms by 2, but there is no change in the number of aromatic hydrogens.
5. For each S atom, the total number of C atoms is increased by 1 and the total number of H atoms is increased by 2, but again no change is made to the number of aromatic H atoms.

Following these rules, the equivalent hydrocarbon formula for Wood and Wilman's fraction 1B is $C_{71.66}$ $H_{44.64}$, and H_{ar}, the number of aromatic hydrogens;—that is, the hydrogen atoms in CH groups in aromatic ring structures—is 26.84.

The number of nonaromatic hydrogens therefore amounts to 17.8 per average molecule. There is no way of assigning the nonaromatic hydrogens to particular groups, and assumptions must be made. If it is assumed that all the nonaromatic hydrogens are present as methyl groups, then there are 5.93 such groups per molecule and the unsubstituted equivalent aromatic structure has the formula:

$$C_{65.73}H_{26.84}$$

The number of rings per average molecule, R, is given for any aromatic hydrocarbon by $R = \frac{(C - H)}{2} + 1$, where C is the number of carbon atoms and H is the number of hydrogen atoms in the aromatic ring structure. Thus the equivalent hydrocarbon structure of fraction 1B is consistent with that of an aromatic structure of 20 rings containing six methyl groups.

If, on the other hand, it is assumed that the nonaromatic hydrogen is present as CH_2 groups in partly or completely saturated rings, the equivalent aromatic ring structure becomes $C_{62.66}H_{26.84}$ and there are 18.5 aromatic rings.

This approach is not very informative. For example, it cannot differentiate between such different structures as:

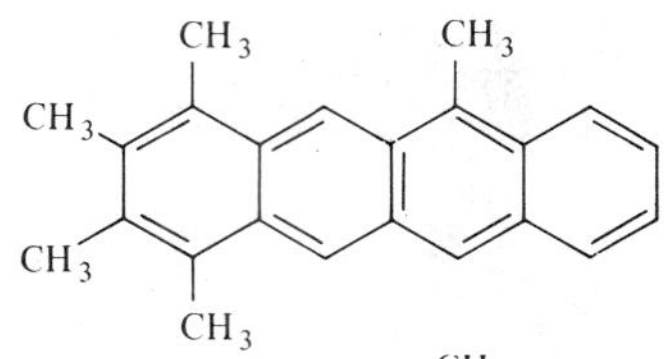

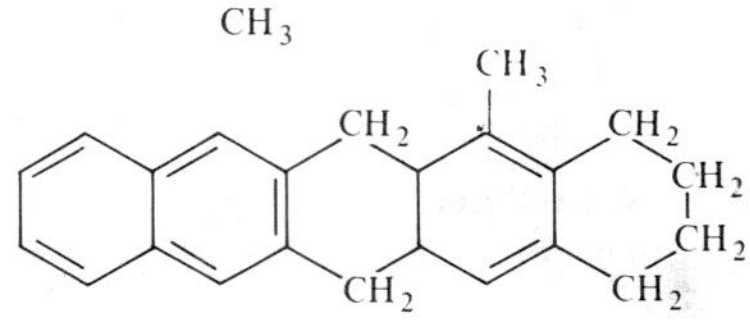

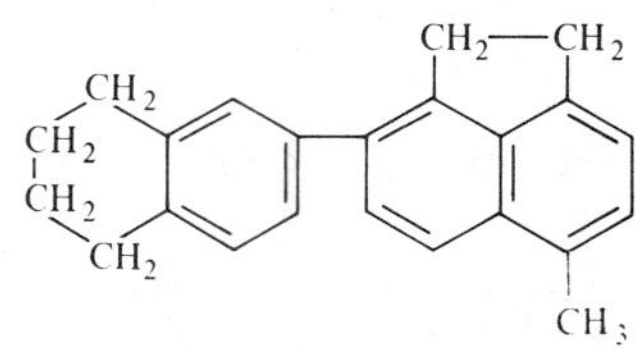

all of which (they are of illustrative significance only) have the molecular formula $C_{23}H_{22}$ and the same fraction, 0.318, of aromatic hydrogen.

Another approach, which has also been applied to coal structure, is based on the observation of Traube[55] that the molar volume of a liquid compound is equal to the sum of its individual atomic volumes, plus a certain volume due to end groups, less a certain amount of contraction due to structural factors. Rona,[56] as a result of a critical examination of this relationship for unsubstituted aromatic hydrocarbons, suggested the following expressions (V_m is the molar volume in cubic centimeters per gram-mole):

$$V_m = 6.6C + 3.45H + 31.2 \quad \text{for liquid hydrocarbons} \qquad (17.1)$$

[55] T. Traube, *Beri. Dtsch. Chem. Gess.*, **28,** 2722 (1895).

[56] V. Rona, *Fuel,* **39,** 187 (1960).

$$V_m = 6.0C + 3.45H + 25.0 \quad \text{for solid hydrocarbons} \quad (17.2)$$

Volkmann[57] used Rona's equations to relate the number of rings in a polynuclear aromatic to its molecular weight, density, and H/C atomic ratio, thus:

$$\text{molecular weight } M = dV_m = 12.01C + 1.008H \quad (17.3)$$

where d is the density at 20°C and V_m is the molar volume (cm³/g-mole) at this temperature; that is,

$$\frac{1}{C} = \frac{1}{M}\left(12.01 + 1.008\frac{H}{C}\right) \quad (17.4)$$

also

$$R = \frac{C - H}{2} + 1 \quad (17.5)$$

where R is the ring index or number of rings in the molecule, that is,

$$\frac{H}{C} = 1 - \frac{2}{C}(R - 1) \quad (17.6)$$

Substituting Equation 17.4 in 17.6 gives:

$$\frac{H}{C} = 1 - \frac{2}{M}\left(12.01 + 1.008\frac{H}{C}\right)(R - 1) \quad (17.7)$$

Solving for M gives:

$$M = \frac{(24.02 + 2.016\ H/C)}{1 - H/C}(R - 1) \quad (17.8)$$

and for R gives

$$R = \frac{(1 - H/C)M}{24.02 + 2.016\ H/C} + 1 \quad (17.9)$$

Taking the density as midway between that of the liquid and the crystalline solid (i.e., taking the mean of Rona's equations), we have:

$$C = \frac{V_m}{6.3} - 0.548H - 4.46 \quad (17.10)$$

$$H = \frac{V_m}{3.45} - 1.826C - 8.145 \quad (17.11)$$

[57] E. W. Volkmann, *Fuel*, **39**, 455 (1960).

From Equation 17.3 we write

$$C = \frac{M}{12.01} - 0.084H \quad (17.12)$$

$$H = \frac{M}{1.008} - 11.915\ C \quad (17.13)$$

Substituting Equation 17.13 in 17.10 and Equation 17.12 in 17.11 to obtain values for C and H in terms of V_m and M, and substituting these values in Equation 17.5: we obtain a value of R in terms of M and d:

$$R = \left(0.139 - \frac{0.186}{d}\right)M + 6.25 \quad (17.14)$$

Using Equations 17.8, 17.9, and 17.14, a diagram can be constructed (Figure 17.6) that relates molecular weight to the number of rings in the molecule, with density and H/C ratio as variable parameters, for unsubstituted PAHs.

Volkmann[57] points out the severe limitations on the use of this diagram to derive data for pitch fractions. The assumption that there are no aliphatic substituents (i.e., that $f_a = \frac{C_{ar}}{C} = 1$) is known to be untenable. No attempt is made to correct for the heteroatoms in the molecule, and the assumption that the density of the polynuclear aromatics in pitch lies between their liquid and solid densities may not be valid. The points for filtered pitch fractions lie between the lines for catacondensed and pericondensed molecules but closer to the lines for the former.

The value of f_a, the aromatic index of pitch, has been a matter of some controversy. Mazumdar et al.,[58] on the basis of dehydrogenation with sulfur at 200–360°C, which they say does not affect aromatic hydrogen, claimed that the ratio of hydroaromatic hydrogen to total hydrogen varied between 0.1 and 0.15. Their results have been challenged by Van Krevelen et al.,[59] who present evidence that unsubstituted aromatic hydrocarbons, such as anthracene, phenanthrene, chrysene, and pyrene, generate hydrogen

[58] B. K. Mazumdar, S. K. Chakrabartty, and A. Lahiri, *Fuel*, **38**, 115 (1959).
[59] D. W. Van Krevelen, M. L. Goedkoop, and P. H. G. Palman, *Fuel*, **38**, 256 (1959).

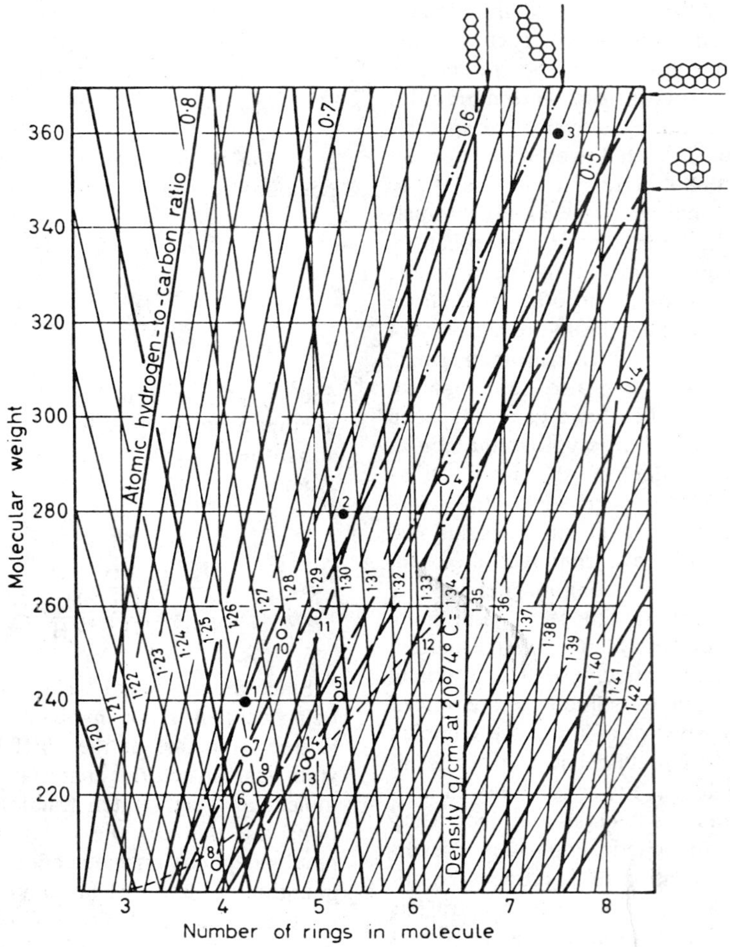

Figure 17.6 Structural analysis of various coal-tar pitches and pitch fractions.[57]

sulfide under the dehydrogenation conditions used.

Rao et al.,[60] in an attempt to resolve this point, studied the nuclear magnetic resonance (NMR) and vibrational spectra of the CS_2-soluble fraction of a medium-soft coke-oven pitch. This fraction had an atomic H/C ratio of 0.61. The absence of appreciable absorption in the infrared at 3300 cm^{-1} indicated the absence of OH or NH_2 in appreciable quantities. The spectra clearly showed that for this sample, f_a was not unity. The methyl content, determined quantitatively by the Kuhn-Roth procedure, indicated that methyl hydrogens comprised 9.45% of the total hydrogen. Making the assumption that all aliphatic C is present as CH_3 or CH_2 groups, their results gave the ratio $H_{CH_3} : H_{CH_2} : H_{ar}$ as 1 : 1.36 : 8.22 and the ratio of aliphatic to aromatic hydrogen as 1 : 16. The calculated ratio of aromatic hydrogen to aromatic carbon, $H_{ar} : C_{ar}$, was 1 : 1.97, implying a high degree of condensation in the ring system. They concluded that since about 72% of the nonaromatic hydrogen is bound to C atoms, this amount of hydrogen must be mainly accounted for by methyl groups attached directly to the ring, with the remainder present as methylene bridges and other small alkyl groups.

[60] H. S. Rao, G. S. Murty, and A. Lahiri, *Fuel*, **39**, (1960).

It is clear that proton magnetic resonance (PMR), which is capable of distinguishing between hydrogen atoms in different chemical environments and of estimating the amounts of different types of hydrogen in the molecule from the intensity of the signals, can be a considerable aid in structural analysis of coal-tar fractions. The absence of restrictions on volatility and molecular weight offers advantages over other spectroscopic methods. With lower resolution instruments (40–80 MHz), it is possible to estimate three hydrogen types—H_A, aromatic and phenolic hydrogen: H_α, hydrogen on saturated carbon atoms α to the aromatic ring; and H_β, hydrogen on other ring structures. In certain cases the hydrogen present as methylene bridges between aromatic rings, as in fluorene (H_F), can also be estimated. However such studies have generally been confined to the relatively low molecular weight materials, which are the only fractions that yield PMR spectra to which fine structure can be attributed at such resolutions. Bartle and Jones[61] have employed a system based on a superconducting magnet capable of giving spectra at 220 MHz, and have shown that the greatly enhanced resolution reveals new fine structure in the aromatic hydrogen of the more soluble, less complex fractions. The spectra of the more complex fractions are less precise because of the increasing complexity, lower solubility in non-proton-containing solvents, and increasing content of paramagnetic impurities. Nevertheless, the estimation of hydrogens of different types derived from such spectra has been employed by several workers as part of statistical structural analysis schemes for the elucidation of the structure of coal carbonization products. The method derived by Brown and Ladner,[62] applied in both original and modified forms to coal tars and pitches, is based on elemental analysis and PMR spectra from which the following parameters are estimated: H_{ar}/H, the ratio of aromatic to total hydrogen, assuming that 60% of the total oxygen is phenolic, H_α/H, and H_β/H. From these data, making the assumption that the atomic ratios of hydrogen to carbon in alpha and beta structures are X and Y, respectively, they calculate the following quantities:

$$f_a = \frac{C_{ar}}{C} = \frac{\dfrac{C}{H} - \dfrac{H_\alpha}{XH} - \dfrac{H_\beta}{YH}}{\dfrac{C}{H}} \tag{17.15}$$

σ, the fraction of available outer edge positions that are occupied by substituents:

$$\sigma = \frac{\dfrac{H_\alpha}{XH} + \dfrac{O}{H}}{\dfrac{H_\alpha}{XH} + \dfrac{O}{H} + \dfrac{H_{ar}}{H}} \tag{17.16}$$

and H_{aru}/C_{ar}, the atomic H/C ratio that the average aromatic skeleton would have if each substituent were replaced by a hydrogen atom:

$$\frac{H_{aru}}{C_{ar}} = \frac{\dfrac{H_\alpha}{XH} + \dfrac{O}{H} + \dfrac{H_{ar}}{H}}{\dfrac{C}{H} - \dfrac{H_\alpha}{XH} - \dfrac{H_\beta}{YH}} \tag{17.17}$$

Since the molecular weight of the material being studied is not involved, this approach is free from one of the chief limitations of statistical structural analysis of the higher molecular weight pitch fractions. Its limitations are that (*a*) the acceptability of the results depends on the accuracy of the PMR quantitative data, (*b*) no allowance is made for the presence of nitrogen and sulfur heterocycles, (*c*) the assumption regarding the H/C ratio in alpha and beta structures is arbitrary, (*d*) the assumption that completely or partly saturated rings are absent may not be justified, and (*e*) no information is derived on the size of the average molecule or the number of rings it contains.

It has, however, been used by several investigators. Brooks et al.[63,64] applied it to fractions of low-temperature tars. Friedel and Retcofsky[65] used this approach to determine the aromaticity of fractions of coke-oven pitch, which they found to average 0.94, and

[61] K. D. Bartle and D. W. Jones, *Fuel,* **48,** 21 (1969).

[62] J. K. Brown and W. R. Ladner, *Fuel,* **39,** 87 (1960).

[63] J. D. Brooks and J. R. Stevens, *Fuel,* **43,** 87 (1964).

[64] J. D. Brooks and J. W. Smith, *Fuel,* **43,** 125 (1964).

[65] R. A. Friedel and H. Retcofsky, *Proc. 5th Carbon Conf.,* Vol 2 (Pergamon Press, London), (1963), p. 149.

DeWalt and Morgan,[66] applying the original Brown and Ladner equations, found f_a for coke-oven pitch fractions to vary between 0.94 and 0.99, and calculated that between 6 and 12% of the available substitution sites were occupied. Oelert[67,68] combined this method with data derived from infrared and UV spectroscopy in a study of coal shock heating products and low-temperature tar fractions, and it has been applied to the structure of coal solvent extracts by de Ruiter and Tschamler.[69]

The statistical structural analysis scheme used by Chauvin and Deelder[70] to characterize the fractions of the "primary" tar, obtained by shock heating finely divided coal, was based on the elemental analysis of the fractions; the mean molecular weights, determined ebulliscopically in benzene; the ratios of aliphatic to aromatic hydrogen, determined from the infrared spectra of the fractions; the amount of methylene bridge hydrogen (H_f), determined by PMR; and the amount of hydroxylic hydrogen, determined by conventional chemical analysis. From these data, they derived the aromaticity index, f_a, of each fraction as follows:

$$\frac{H_{Al}}{H} + \frac{H_{Ar}}{H} + \frac{H_{OH}}{H} = 1 \qquad (17.18)$$

$$\frac{H_{Al}}{H} = \frac{H_{CH_3}}{H} + \frac{H_{CH_2}}{H} + \frac{H_{CH}}{H} \qquad (17.19)$$

where H, H_{Ar}, H_{Al}, H_{CH_3}, H_{CH_2} , H_{CH}, and H_{OH} represent the number of hydrogen atoms in the whole molecule (total hydrogen), in aromatic rings, in nonaromatic rings or alkyl groups, in CH_3 groups, in —CH_2—groups, in —CH or —CH=groups, and in OH groups, respectively. Assuming no quaternary carbon atoms, then:

$$\frac{C_{Al}}{H} = \frac{1}{3}\frac{H_{CH_3}}{H} + \frac{1}{2}\frac{H_{CH_2}}{H} + \frac{H_{CH}}{H} \qquad (17.20)$$

where C_{Al} is the number of nonaromatic carbon atoms. The aromaticity index is:

$$f_a = \frac{C_{Ar}}{C} = \frac{(C - C_{Al})}{C} = 1 - \frac{C_{Al}}{C}$$

that is,

$$f_a = 1 - \frac{C_{Al}}{H}\frac{H}{C} \qquad (17.21)$$

The ratio of aliphatic hydrogen to aliphatic carbon atoms is:

$$\frac{H_{Al}}{C_{Al}} = \frac{H_{Al}}{\frac{1}{3} H_{CH_3} + \frac{1}{2} H_{CH_2} + H_{CH}} \qquad (17.22)$$

It is not possible to measure CH_3 by infrared, and two limits are put on the value of H_{Al}/C_{Ar}. In the first case it is assumed that no olefinic structures such as indene are present, and a lower limit of H_{Al}/C_{Al} of 2 is used. The upper limit is fixed by assuming that no nonaromatic CH groups are present (i.e., that there are no branched-chain substituents). In this case, by combining Equations 17.19 and 17.20 we have:

$$\frac{C_{Al}}{H} = \frac{1H_{Al}}{3\ H} + \frac{1H_{CH_2}}{6\ H} \qquad (17.23)$$

which gives, for the lower limit of H_{Al}/C_{Al}, the expression:

$$\frac{H_{Al}}{C_{Al}} = \frac{6H_{Al}}{2H_{Al} + H_{CH_2}} \qquad (17.24)$$

On combining the limits for H_{Al}/C_{Al} in Equation 17.21, the limiting values of f_a are arrived at:

$$1 - \frac{1}{2}\frac{H_{Al}}{H}\frac{H}{C} \leqq f_a \leqq 1 - \frac{H}{C}\left(\frac{1}{3}\frac{H_{Al}}{H} + \frac{1}{6}\frac{H_{CH_2}}{H}\right)$$

R, the ring index, is related to f_a by the expression:

$$2\left(\frac{R-1}{C}\right) = 2 - f_a - \frac{H}{C} \qquad (17.25)$$

[66] C. W. De Walt, Jr., and M. S. Morgan, "Proton Magnetic Resonance Spectrometry in Characterization of Coal-Tar Pitches," in American Chemical Society *Division of Fuel Chemistry Symp. Tars, Pitches and Asphalts,* Atlantic City, N.J., September 9–14, 1962, *Preprints*, p. 33.

[67] H. H. Oelert, *Brennstoff-Chemie,* **48,** 362 (1967).

[68] H. H. Oelert, *Z. Anal. Chem.,* **231,** 81 (1967).

[69] E. de Ruiter and H. Tschamler, *Brennstoff-Chemie,* **45,** 15 (1964).

[70] R. Chauvin and R. Deelder, *Bull. Soc. Chim. Fr.,* No. 11, 3916 (1969).

when the H/C ratio is corrected for the heterocyclic atoms. In the case of the fractions studied by Chauvin and Deelder (see pp. 000–000), this was considered unnecessary and no correction was made. Substituting the limiting values for f_a in Equation 17.25 gives the following limiting values for R:

$$1 + \frac{C}{2}\left[1 - \frac{H}{C} + \frac{1}{6}\frac{H}{C}\left(2\frac{H_{Al}}{H} + \frac{H_{CH_2}}{H}\right)\right] \leqq R \leqq 1 + \frac{C}{2}\left[1 - \frac{H}{C} + \frac{1}{2}\frac{H_{Al}}{H}\frac{H}{C}\right]$$

UV spectroscopy is used to estimate the number of aromatic rings in the monomeric unit and in the calculation of the number of rings. (Chauvin and Deelder's analysis of their primary tar fractions and the results of hydrogenation of the tar led them to conclude that it has an oligomeric structure, the monomeric units being combined by means of oxygen atoms.)

To study the reactions occurring in exposure weathering of road tars, Bartle and Smith[71,72] developed a statistical structural analysis method based on PMR spectrometric studies of solvent fractions, first treated to remove any residual solvent and any paraffinic material. The molecular formulas were corrected by the "lift out and replace" procedure, and the numbers of H_α, H_β, and H_γ hydrogens were estimated from the PMR spectra (H_α representing hydrogen on saturated carbon atoms alpha to the aromatic ring; H_β, hydrogen atoms in alkyl substituent groups beta to the aromatic ring; and H_γ, hydrogens in terminal methyl groups three of four carbons long). If the assumption is made that the only alkyl groups present are methyl (*m*), ethyl (*e*), isopropyl (*i*), *n*-propyl (*n*), *n*-butyl (*b*), isobutyl (*s*), and *sec*-butyl (*c*), and letting the letters in parentheses stand for the numbers of such groups, the following equations are valid:

$$3m + 2e + 2n + i + 2b + 3s + c = H_\alpha \qquad (17.26)$$

$$3e + 2n + 6i + 4b + 5c = H_\beta \qquad (17.27)$$

$$3(n + b + 2s + c) = H \qquad (17.28)$$

[71] K. D. Bartle and J. A. S. Smith, *Fuel*, **44**, 109 (1965).

[72] K. D. Bartle and J. A. S. Smith, *Fuel*, **46**, 29 (1967).

Solution of these equations to give possible values for the number of alkyl groups per molecule is then possible if qualified assumptions are made about the relative distribution, based on GC or other analyses of those fractions amenable to such analyses, and the assumption is made that these relative distributions persist in the higher fractions.

An alkyl distribution parameter, P, may also be calculated:

$$P = \frac{\text{number of alkyl groups}}{H_{Ar} + \text{number of alkyl groups}} \qquad (17.29)$$

The number of ring carbon atoms, C_{Ar}, and the number of ring-joining carbon atoms, C_j, in the equivalent hydrocarbon structure may be calculated from the formulas:

$$C_{Ar} = C - C_{alkyl} \qquad (17.30)$$

$$C_j = C_{Ar} - \text{number of alkyl groups} - \text{number of fluorene-type } CH_2 \text{ groups} \qquad (17.31)$$

This method of structural analysis has given results for low molecular weight pitch and refined tar fractions that are quite consistent with the results of chromatographic analysis, but again the results from the higher molecular weight fractions of high-temperature tars and pitches have been inconclusive.

The introduction of high-resolution PMR has concentrated the work on statistical structural analysis of coal extracts, coal tars, and coal-tar pitches on schemes involving the distribution of hydrogen in different environments in the average molecule. The earlier method of attempting to allocate numbers to the different types of carbon atoms has not received much attention recently.

Binns[73] attempted to extend the earlier work of Wood and Wilman,[54] employing the same pitch fractions. He initially based his scheme on four types of carbon atom:

1. C_{Al}, the number of carbons in side chain CH_3, CH_2, CH, or C groups.
2. C_F, the number of cyclic CH_2's unsubstituted or substituted by an alkyl group.

[73] E. H. Binns, Coal Tar Research Association, Rep. No. 0286, 1962.

3. C_{Ar}, the number of aromatic CH groups or such groups substituted by an alkyl group.
4. C_j, the number of ring-joining carbon atoms.

The equations he set out to solve were:

$$C = C_{Al} + C_F + C_{Ar} + C_j \qquad (17.32)$$

where C is the total number of C atoms, and

$$H = 2C_{Al} + 2C_F + C_{Ar} \qquad (17.33)$$

$$C_{Ar} = H_{Ar} + C_{Al} \qquad (17.34)$$

$$V_m = V_1C_{Al} + V_2\,C_F + V_3C_{Ar} + V_4C_j + \frac{K}{M} \qquad (17.35)$$

where V_m is the molar volume (cm³/g-mole).

The second and third equations involve two assumptions that appear to be incompatible. In Equation 17.33 it is assumed that the average number of hydrogens associated with each alkyl carbon atom is two, whereas in Equation 17.34 it is assumed that all alkyl groups are methyl groups. Two sets of constants were derived for the molar volume, Equation 17.35. One was based on the full definition of C_j, and the other excluded carbon atoms at the juncture of nonfused rings.

The initial results of this restricted scheme were not successful, and it was concluded that a fifth type of carbon atom, joining nonfused aromatic rings, had to be included, requiring another equation. At the same time, a new molar volume equation was formulated based on a more extensive survey of the data for pure aromatic hydrocarbons and partly hydrogenated aromatic hydrocarbons. To obtain a full solution, attempts were made to develop a fifth equation involving molar refractivity, but a study of relevant data for pure compounds showed that if molar refraction was considered as an additive parameter, carbon atoms at the junction of saturated and unsaturated rings, as in fluorene, could not be grouped with ring-joining carbons in fully aromatic six-membered rings. Various assumptions were adopted to allow at least partial estimates of the different types of carbon atom in the average equivalent hydrocarbon molecule, but the results indicated either that the assumptions were incorrect and/or that the values of the coefficients derived for the molar volume and molar refractivity equations were not applicable to pitch fractions. The results, particularly those pertaining to the higher molecular weight fractions, gave only small, and in some cases negative, values for C_j.

17.3.6 The *n-d-M* Method

Karr and Comberiati[74] adopted the method developed by Hazelwood[75] for the determination of the total number of rings and the number of aromatic rings in highly aromatic hydrocarbons. This method is based on the observation that with an increasing proportion of alkyl groups, the index of refraction (n) and the density (d) of aromatic hydrocarbons and their oxygen analogues change in the direction of the values for the limiting paraffin or paraffin of infinite molecular weight (i.e., $d_4^{20} = 0.851$ and $n^{20} = 1.475$), and that these changes are approximately linear. If Δd is denoted by $d_4^{20} - 0.8510$ and Δn by $n_D^{20} - 1.4750$, it was empirically determined from an examination of data from a variety of aromatic petroleum fractions and alkyl naphthalenes and tetralins that if $M\Delta r_1$ is plotted against $M\Delta d$, a graph is obtained having isobaric lines for total ring content. In this plot M is the molecular weight and $\Delta r_1 = \Delta n - \Delta d/2$, which can be expressed as:

$$\Delta r_1 = n_D^{20} - \frac{d_4^{20}}{2} - 1.0495 \qquad (17.36)$$

Data for the slope of these total ring content (R_T) lines permit calculation of the relative values of the two factors and lead to a formula for calculating R_T:

$$R_T = 0.080M\,(\Delta d - 0.5\Delta n) + 1.0 \qquad (17.37)$$

Similarly, if $M\Delta d$ is plotted against $M\Delta d_1$, where $\Delta d_1 = \Delta d - \Delta n/2$, isobaric lines are obtained for hydrocarbons with different numbers of aromatic rings, leading to the

[74] C. Karr, Jr., and J. R. Comberiati, U. S. Bur. Mines Bull. No. 636, 1966.

[75] R. N. Hazelwood, *Anal. Chem.*, **26**, 1073 (1954).

formula for determining the aromatic ring content of the average molecule:

$$R_A = 0.040M\,(2.75\Delta n - \Delta d) + 0.58 \qquad (17.38)$$

A further correlation stems from the data presented by Martin and Sankin[76] relating to an additional 15 aromatic concentrates from petroleum, from which the following expression for the sum of the aromatic and naphthenic ring contents could be derived:

$$(3R_A + R_N) = M(0.0833)C_0 - 0.496H_0 \qquad (17.39)$$

where R_A and R_N are the aromatic and naphthenic ring contents, respectively, M is the mean molecular weight, C_0 is the percentage of carbon $\times$ 10^{-2} and H_0 is the percentage of hydrogen $\times$ 10^{-2}.

Karr and Comberiati point out that the Hazelwood *n-d-M* method is not directly applicable to high oxygen content samples such as coal-tar resins, since the substitution of an oxygen atom for a methylene group results in an increase in the refractive index and, particularly, in the density of a hydrocarbon. A modification of this method is therefore required before it can be applied to pitch fractions that are generally high in oxygen, particularly fractions derived from low-temperature or primary tars. From a study of the physical properties of more than 1200 oxygen compounds and their analogous hydrocarbons, these workers determined values for the decrease in refractive index and density in moving from an oxygen-containing compound to the hydrocarbon. The values differed for different types of oxygen-containing groups, but based on the known predominance of hydroxylic oxygen and the relatively minor contribution of aromatic ether oxygen (which was, in any case, largely canceled by the small contribution of carbonyl groups), values of 0.002 and 0.010 for the reduction in the refractive index and the density, respectively, per 1% oxygen in the molecule best represented pitch fractions. Thus in applying Hazelwood's equations to the statistical structural analysis of fractions of coal-tar pyrolysis products, the determined refractive index is reduced by subtracting 0.002 × wt % O. The determined density is reduced by subtracting 0.010 × wt % O.

When applied to oxygen compounds both alone and in blends, this modified method gives results that agree well with theory. The small discrepancies found are probably due as much to inherent limitations of the method as to errors introduced by the modification. In general, the values of R_T are slightly too large and those of R_A slightly too small.

To determine the arrangement of rings in a polynuclear hydrocarbon, Karr and Comberiati calculated the total number of rings per carbon atom R_T/C, and the atomic H/C ratio. They showed that there were seven well-defined groups ranging from entirely linear structures such as anthracene, phenanthrene, and picene, to highly globular structures such as coronene. This series is shown in Table 17.12.

Similar series for aromatic hydrocarbons with one or more methyl groups (or one ethyl group), and for aromatic hydrocarbons with one ring saturated, and with one methyl group on the saturated ring are listed by Karr and Comberiati. They point out that similar series can be drawn up for hydrocarbons with several rings saturated and with different degrees of substitution. There are limitations, however. Thus there is no way of distinguishing between a hydrocarbon of series 2 with one methyl substituent and a series 3 molecule with two methyl substituents. When the degree of alkylation increases beyond about four CH_3 groups or equivalent, only the extremes of the ring arrangement can be distinguished. From a knowledge of R_T, R_A, C, and H, however, it is possible to designate the permissible ring arrangements and the extent of alkylation, if this is not too great, for the equivalent hydrocarbon structure.

The *n-d-M* procedure outlined by Karr and Comberiati has been mainly applied to low-temperature tar resins. Its application to high-temperature pitch fractions would be of considerable interest.

17.3.7 The Molecular Weights of High-Temperature Pitch Fractions

It will be noted that most of the schemes adopted for the statistical structural analysis of tar fractions require an accurate value for

[76] C. C. Martin and A. Sankin, *Anal. Chem.*, **26**, 206 (1953).

Table 17.12 R_T/C and H/C in Terms of R_T for Polynuclear Aromatic Hydrocarbons[74]

Series	R_T/C in Terms of R_T	H/C in Terms of R_T	Representative Compounds
1	$\frac{R_T}{(4R_T + 2)}$	$\frac{(2R_T - 4)}{4R_T + 2)}$	Benzene, naphthalene, anthracene, phenanthrene, naphthacene, chrysene, triphenylene, picene
2	$\frac{R_T}{(4R_T + 1)}$	$\frac{(2R_T + 4)}{(4R_T + 1)}$	Fluorene, 1-*H*-benzonaphthene, 7-*H*-benzo[*d,e*]anthracene
3	$\frac{R_T}{4R_T}$	$\frac{(2R_T + 2)}{4R_T}$	Pyrene, fluoranthene, perylene, benzo[*a*]pyrene
4	$\frac{R_T}{(4R_T - 1)}$	$\frac{(2R_T + 2)}{(4R_T - 1)}$	4-*H*-Cyclopentano[*d,e,f*]phenanthrene, 6-*H*-benzo[*e,d*]pyrene, 4-*H*-benzo[*c,d*]fluoranthene
5	$R_T/(4R_T-2)$	$\frac{(2R_T + 2)}{(4R_T - 2)}$	1-*H*, 6-*H*-Benzo[*m,n,o*]acenaphthylene, 1-*H*, 5-*H*-dibenz[*c,d,m,n*]pyrene, 1-*H*, 6-H-dibenz[*k,m,n,o*]acenaphthylene
6	$R_T/(4R_T-2)$	$\frac{2R_T}{(4R_T - 2}$	Benzo[*g,h,i*]fluoranthene, benzo[*g,h,i*]perylene, dibenz[*c,d,j,k*]pyrene
7	$R_T/(4R_T-2)$	$\frac{(2R_T - 2)}{(4R_T - 4)}$	Coronene, benzo[*c*]coronene, naphtho(2,3-α)coronene

the molecular weight of the material being studied. The uncertainty in this determination, as evidenced by the widely differing values published by various workers for apparently similar fractions, is probably one of the factors that has led to the different ideas of the size and structure of high-temperature pitch fractions, based on the interpretation of statistical structural analysis results.

Although there is general agreement on the molecular weight range of the petroleum ether–soluble fractions, the correct values for the more complex fractions are still a matter of some controversy, and the values published appear to vary with the method of fractionation and the method of determination. Osmotic pressure measurements yielded a range of values between 213 and 2615 for a series of fractions of decreasing solubility prepared from a coke-oven pitch.[54] Greenhow and Smith[77] reported much smaller values, from cryoscopic determination, in phenanthrene, camphor, and particularly in catechol. Thus for a coke-oven resinoid fraction (material soluble in benzene but insoluble in petroleum ether), they obtained molecular weights in catechol of 160 and 183, whereas the values in phenanthrene (340 and 490) and in camphor (340 and 480) were much more consistent and were of similar magnitude to the values (383 and 482) quoted by Karr and Comberiati and obtained by microebulliometric determination in benzene.[74] These compare with values of 600–700 obtained by Wood and Phillips[78] by osmotic pressure measurement, 280 by Franck[79] by ebullioscopic determination in xylene, 815 by Green and Mukherji[80] by ebulliscopic measurement in chloroform, and 625 obtained by Binns[73] by ebulliscopic determination in pyridine.

There are several possible explanations for these discrepancies. The low figure quoted by Franck may be explained by the use of xylene as the extracting solvent, leading to the extraction of only 66% of the pitch instead of the normal 75–80% extracted by benzene. Also, since the methods used for molecular weight determination give number-

[77] E. J. Greenhow and J. W. Smith, *Fuel*, **40**, 69 (1961).

[78] L. J. Wood and G. Phillips, *J. Appl. Chem.*, **5**, 326 (1955).

[79] H. G. Franck, *Brennstoff-Chemie*, **36**, 12 (1955).

[80] S. J. Green and S. N. Mukherji, *J. Soc. Chem. Ind.*, **67**, 438 (1948).

average values, the lowering effect can be caused either by low molecular weight material left in the fraction because of incomplete separation from the more soluble material, or by solvent molecules that have not been completely removed. This effect is exaggerated and could account, at least partly, for some of the lower figures quoted. Incomplete solution in the solvent used for molecular weight determination tends to give high results, whereas any complex formation between solvent and solute leading to negative deviation from Raoult's law tends to give low results; the latter condition is thought to be the cause of the abnormally low values obtained cryoscopically in catechol.

Summarizing the current evidence, there appears to be general agreement that the petroleum ether–soluble fraction, which makes up 40–50% of a medium-soft coke-oven pitch, consists predominantly of three- to five-ring compounds with molecular weights in the range 175–300. Three groups of workers have shown the presence of relatively high molecular weight compounds in the benzene-soluble, petroleum ether–insoluble portion, which accounts for another 20–30% of the pitch. The conclusion is that molecular weights in this range increase from 300 to around 800. The molecular weight ranges of the benzene-insoluble fractions are less well defined. Greenhow and Smith[77] reported a value of 1240 for an acid-washed sample of the fraction insoluble in benzene but soluble in pyridine, by cryoscopic determination in phenanthrene. This agrees with the range of 1100–1800 for the same fraction (amounting to 10.6% of the pitch) reported by Wood and Wilman,[54] and the values of around 1000 reported by Bartle et al.[81] for the fraction insoluble in benzene but soluble in dioxan.

The inconsistent results of molecular weight determinations of high-temperature pitch fractions are not the only problem in applying statistical structural analysis to such materials. Other sources of imprecision include (*a*) the validity of assumptions that must be made about the nature and distribution of nonaromatic carbon and hydrogen and (*b*) the effect of the "lift out and replace" correction for heterocyclic atoms on the equations for additive parameters such as molar volume and molar refractivity. It would be wrong, however, to dismiss statistical structural analysis altogether. It has mainly been applied to low-temperature tar and pitch fractions and to coal extracts, and the findings by different schemes are in fair agreement here. Given more reliable values for molecular weight, improvements in the sensitivity and resolution of PMR and ^{13}C magnetic resonance measurements, more extensive data on group contributions to molar volumes, molar refractions, heats of combustion, and other additive properties, and a simultaneous application of the various approaches (i.e., that based on the modified *n-d-M* method, on hydrogen distribution as determined by PMR, and on carbon distribution by molar volume and other additive parameters), it should be possible to arrive at a more accurate picture of the average structure of the higher molecular weight end of high-temperature coal tar.

To date, however, this technique has made a minimal contribution. All that can be said is that none of the results are inconsistent and most are consistent with the view that the components of the benzene-insoluble fractions are of the same general type as those that have been shown to constitute the bulk of the distillate oils and the petroleum ether-soluble fractions of the pitch, and that they continue the progression to these same types with larger fused-ring systems.

17.3.8 The Determination of Functional Groups in Tars and Tar Fractions

17.3.8.1 *Oxygen-Containing Compounds*

The usual method for determining the presence of phenolic compounds in tars and tar distillate oils is to extract the sample, dissolved in a suitable solvent, with 10% caustic soda, backwash the extract to remove coextracted neutral oils and bases, neutralize the purified extract with mineral acid, and recover the precipitated phenols with ether or benzene. This procedure assumes that carboxylic acids are absent and that all the phenols are caustic soda extractable. The remainder of the oxygen, as determined by direct determination of total oxygen, is at-

[81] K. D. Bartle, J. A. S. Smith, and W. G. Wilman, *J. Appl. Chem.*, **19**, 283 (1969).

tributed to carbonyl oxygen in five-membered ring systems.

An elegant method for the determination of acidic hydrogen in organic materials was developed by Vaughan and Swithenbank,[82] based on enthalpimetric titration with acetone as an internal indicator. The sample is dissolved in acetone or, in the case of less soluble material, in a mixture of acetone and pyridine or nitrobenzene, and is titrated at an automatically controlled rate with normal potassium hydroxide in dry isopropanol. A sensitive thermistor measures the temperature of the solution during the titration and records this on a potentiometric recorder. The end point is shown by a sudden steep rise in temperature due to the exothermic dimerization of the acetone to diacetone alcohol, which occurs when a trace of excess alkali is present.

This method has been applied to tar oils and pitch fractions and to samples of coal, by grinding the sample with pyridine to which a small amount of acetone is added. The results agree well with potentiometric nonaqueous titration using *N*-butyl trimethyl ammonium hydroxide in dry methanol as titrant, but they are more repeatable. The method also gives a much better defined end point.

Greenhow and Sugowdz[83] have reported on the distribution of oxygen-containing functional groups in pitch and pitch fractions. Hydroxylic oxygen was determined by several methods: (*a*) nonaqueous titration with 0.2*N* sodium aminoethoxide in ethylene diamine with 2,6-xylenol added to sharpen the end point,[84] (*b*) mesylation with methane sulfonyl chloride in pyridine and estimation of hydroxyl content from the increase in the sulfur content of the sample, (*c*) methylation with diazomethane, and (*d*) the acetylation method of Blom et al.[85] Carbonyl oxygen was determined (*e*) by oxime formation with hydroxylamine hydrochloride and measuring the increase in nitrogen content, hydrolyzing and measuring the reduction in nitrogen content, and measuring the amount of ammonia evolved during hydrolysis of the oximes. Carboxylic oxygen was determined from the first inflexion point in nonaqueous titration (method *f*), and (*g*) by decarboxylation using Beroza's method.[86] Ester groups were determined (*h*) by the method used by Knoterus[87] for the saponification of blown bitumens, and 1,4-quininoid groupings were estimated from the increase in the nitrogen content following treatment of the sample with diazomethane on the assumption that these groups undergo addition of diazomethane to form pyrazoline rings:

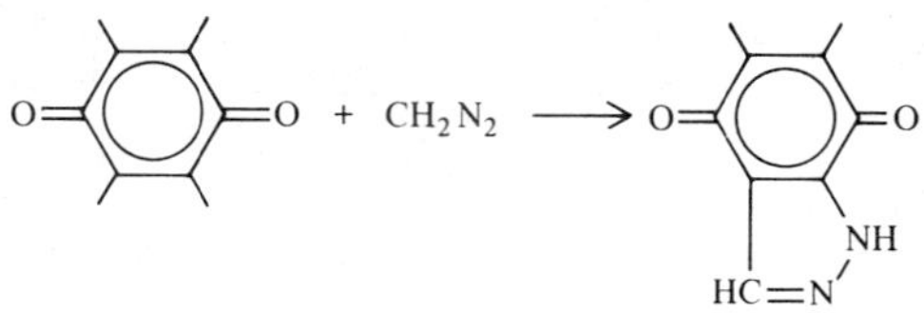

These methods were applied to the crystalloid (petroleum ether-soluble fraction), resinoid (petroleum ether-insoluble, benzene-soluble fraction), and C_2 (benzene-insoluble, pyridine-soluble fraction), and C_1 (pyridine-insoluble fraction) fractions of coke oven pitch. The results are summarized in Table 17.13.

Commenting on the results, the authors note the good agreement of the hydroxyl values determined by nonaqueous titration and mesylation of the crystalloids, resinoids, and C_2 fractions. They suggest that the low value for the hydroxyl content of the C_1 fraction by nonaqueous titration may be due to its relative insolubility. The results by acetylation of methylated fractions suggest that methylation may make functional groups more accessible to acetic anhydride. The suggestion is that several types of hydroxyl groups are present, which differ in their acidity or steric disposition.

Carboxylic oxygen in the resinoid and C_2 fractions, as determined by hydrolysis of methylated fractions, is higher than that determined by decarboxylation or by nonaqueous titration, although these methods are of poor repeatability. A possible explanation is that high carboxyl values and low hydroxyl

[82] G. A. Vaughan and J. J. Swithenbank, *Analyst*, **95,** 457 (1970).

[83] E. J. Greenhow and G. Sugowdz, "Distribution of Oxygen-Containing Functional Groups in Pitch from a High-Temperature Coal Tar," *Prepr., Am. Chem. Soc. Div. Gas. Fuel. Chem.*, **2,** 61–63 (1960).

[84] E. J. Greenhow and J. W. Smith, *Analyst,* **84,** 457 (1959).

[85] L. Blom, L. Edelhausen, and D. W. Van Krevelen, *Fuel,* **36,** 135 (1957).

[86] M. Beroza, *Anal. Chem.*, **25,** 177 (1953).

[87] J. Knoterus, *J. Inst. Pet.*, **42,** 355 (1956).

Table 17.13 Distribution of Oxygen in Pitch Fractions[83]

	Whole Pitch	Crystalloids	Resinoids	C_2	C_1
Pitch (wt %)	100	38.5	18.5	15.5	27.5
Total oxygen (wt %)	2.4	1.7	2.8	4.3	2.6
Oxygen as hydroxyl (wt %) by:					
Method *a*	—	0.9	1.9	2.0	0.55
Method *b*	2.10	0.95	1.63	2.5	1.2
Method *c*	0.39	0.36	0.83	0.57	0.57
Method *d*	0.96	0.46	0.52	0.59	0.64
Method *c*, then *d*	—	0.75	1.27	1.51	0.95
Method *d*, then *c*	—	0.46	0.52	0.59	0.64
Oxygen as carboxyl (wt %) by:					
Method *f*	—	<0.1	<0.1	1.1	0.5
Method *g*	0.23	0.19	<0.1	0.93	0.60
Methylation and hydrolysis	0.25	<0.1	0.58	1.47	0.26
Oxygen as carbonyl (wt %)	0.53	0.25	1.34	0.93	0.44
Oxygen as 1.4-quinone (wt %)	0.35	0.41	1.0	0.86	0.39
Oxygen as ester (wt %)	—	Nil	0.43	Nil	Nil

values are due to lactones of the fluorescein type, which give esters on methylation. The calculation of 1,4-quininoid groups from the nitrogen increase following methylation gives values lower than those for carbonyl oxygen (except for the crystalloids), but of the same order. This result could be interpreted to mean that much of the carbonyl is in the form of terminal ring 1,4-quininoid groups, although there is no other evidence to support this suggestion. Ester groups appear to be absent from all fractions except the resinoids. In this case, since the "ester-oxygen" content is approximately equal to the carboxylic oxygen, as measured by hydrolysis of methylated resinoids, methylation of lactone may have occurred.

If we take the mesylation results as the most reliable measure of hydroxylic oxygen, the decarboxylation figures as the best estimate of carboxylic or lactone oxygen, and the highest value of the carbonyl or 1,4-quinone oxygen as representing carbonyl or quinone oxygen, the oxygen balance is very good, as shown in Table 17.14.

The conclusion is that in the lower molecular weight pitch fractions, represented by the crystalloids and resinoids, the bulk of the oxygen is made up of hydroxylic and carbonyl oxygen, and that carboxy and ester groups are rare or absent. In the higher molecular weight range, although hydroxylic oxygen atoms are still most frequent, carboxylic oxygen is present in about the same amount as

Table 17.14 Oxygen Balance for High-Temperature Coal Tars[83]

	Components on Fraction or Pitch (wt %)								
	Crystalloids		Resinoids		C_2		C_1		Total on Pitch (wt %)
	Fraction	Pitch	Fraction	Pitch	Fraction	Pitch	Fraction	Pitch	
Hydroxyl oxygen	0.95	0.366	1.9	0.351	2.5	0.388	1.2	0.331	1.436
Carboxyl and lactone oxygen	0.19	0.073	<0.1	—	0.93	0.144	0.6	0.165	0.382
Carbonyl and quinone oxygen	0.41	0.158	1.34	0.248	0.93	0.144	0.44	0.121	0.671
Total oxygen by direct estimation	1.70	—	2.8	—	4.3	—	2.6		2.4
Totals of oxygen types	1.55	0.597	3.24	0.599	4.36	0.676	2.24	0.617	2.489

carbonyl oxygen. There is evidence to suggest that lactones occur particularly in the resinoid fraction and that 1,4-quinones are present in all fractions.

17.3.8.2 *Basic Nitrogen Groups in Pitch and Pitch Fractions*

Greenhow and Smith[88] have determined the basic nitrogen content in a series of pitches, including one derived from an Australian continuous vertical retort tar and one from an Australian coke-oven tar. As before, the pitches were solvent fractionated to give crystalloids, resinoids, and C_2 and C_1 fractions. The analytical method applied to these fractions consisted of adding an excess of an 0.1*N* solution of perchloric acid in acetic acid to the sample dissolved (or suspended) in a mixture of two parts of glacial acetic acid and one part of dioxan, and back-titrating potentiometrically with a standard base. Table 17.15 gives their results.

It can be seen that in the benzene-soluble fractions of both the continuous vertical retort pitch and the coke-oven pitch, slightly more than half the total nitrogen occurs as basic nitrogen, but in the benzene-insoluble fractions of the coke-oven pitch almost all the nitrogen is present in basic form. The low ratio of basic to total nitrogen in the C_1 fraction of the continuous vertical retort pitch is an anomaly and is probably due to analytical error resulting from the small size of the sample.

[88] E. J. Greenhow and J. W. Smith, *Aust. J. Appl. Sci.*, **11**(1), 169 (1965).

17.3.8.3 *Sulfur-Containing Groups in Pitch Fractions*

Karabon and Wiecek[89] provide data on the ratio of thiophenic to thioether sulfur (assuming that all the sulfur is made up of these two types) in pitch fractions. Their method of fractionation differed from that employed by Greenhow and Smith. The whole pitch of coke-oven origin was exhaustively extracted with benzene to yield soluble and insoluble fractions. The benzene-insoluble material was separated into an α_1 fraction (insoluble in benzene but soluble in quinoline) and an α_2 fraction (insoluble in quinoline); the benzene-soluble portion was split into a β fraction (soluble in benzene but insoluble in petroleum ether) and a γ fraction (soluble in petroleum ether). The analytical technique involved estimation of total sulfur by the normal method and of thioether sulfur by the procedure of Wnekowska,[90] which involves the selective destructive hydrogenation of thioether compounds over alumina, and the iodometric determination of the hydrogen sulfide generated. Their results are given in Table 17.16.

[89] B. Karabon and I. Wiecek, *Koks Smola Gaz*, **20**(9), 238 (1975).
[90] L. Wnekowska, *Prace G.I.G. Kommun.* No. 325, Katowice, 1963.

Table 17.15 Ratio of Basic Nitrogen to Total Nitrogen in High-Temperature Pitch Fractions[88]

	Pitch (wt %)	Total Nitrogen (wt %)	Basic Nitrogen (wt %)	Ratio of Basic to Total Nitrogen
CVR pitch[a]		1.6	0.99	0.619
Crystalloids	48.5	1.1	0.63	0.573
Resinoids	27.1	3.0	1.52	0.507
C_2	21.1	1.9	1.58	0.310
C_1	3.3	0.82	0.31	0.378
Coke-oven pitch[b]		1.3	0.71	0.546
Crystalloids	34.0	1.2	0.67	0.558
Resinoids	31.7	1.6	0.79	0.494
C_2	10.1	2.9	0.28	0.916
C_1	24.2	1.0	0.92	0.920

[a] Continuous vertical retort pitch; K and S softening point, 74.4°C.
[b] Coke-oven pitch K and S softening point, 71°C.

Table 17.16 Distribution of Thiophenic and Thioether Sulfur in Pitch Fractions[89]

Fraction	Pitch (wt %)	Total Sulfur (wt %)	Thioether Sulfur		Thiophenic Sulfur	
			Wt %	% Total S	Wt %	% Total S
γ	34.9	0.42	0.233	55.5	0.187	44.5
β	39.5	0.42	0.131	35.9	0.289	68.8
α_1	7.2	0.68	0.044	6.47	0.636	93.53
α_2	18.4	0.46	0.105	22.82	0.355	77.18

17.3.9 The Nature of the Insoluble Matter in High-Temperature Coal Tar

Crude high-temperature coal tar contains small suspended particles visible under the microscope. The nature of these particles has been a topic of some interest. One difficulty is the isolation of this intrinsically insoluble material. It is too small to filter out, it appears to be electrically uncharged, and it cannot be separated by electrophoretic or electrostatic precipitation. Efforts to concentrate it by ultracentrifuging hot tar or pitch have not been successful. The quinoline-insoluble fraction of pitch probably resembles the true tar-insoluble matter most closely, but it is probable that this fraction consists of the insoluble particles with adsorbed tar resins. Volkmann,[91] Weiler,[92] and other workers have shown that the X-ray diffraction pattern of the quinoline-insoluble fraction of coke-oven pitch resembles that of graphite but is of a less ordered structure.

Janek[93] has examined the toluene-insoluble fraction of a tar, and of the pitch derived from it, under the electron microscope. He concludes that this fraction is made up of particles of coke, fly ash, and resins. Most of the resin particles in the tar are about 100 Å in size, but are larger in the pitch. The coke dust particles are considerably larger and have dimensions around 500–1000 Å and in the pitch are probably bound to the resin particles.

It is generally acknowledged, although the effect has never been quantified and reported, that the extent of the free space at the top and the tightness of the oven have a great effect on the quinoline-insoluble content of a high-temperature coal tar. Light charges and/or older ovens lead to an increase in the tar quinoline-insoluble content compared to that obtained when the same coal blend is carbonized under the same time and temperature conditions but in a newer oven and/or with a heavier charge.

The insoluble carbonaceous particles in crude tar visible under the microscope are probably derived from two sources. The larger particles are due to carryover of coke dust and ash constituents; the smaller and much more numerous particles are "soot" derived from the vapor phase partial combustion of the tar polynuclear hydrocarbons in the free space at the top of the oven. The soot is, in fact, a form of carbon black, which it resembles in structure and properties. In tar or pitch, "soot-"derived particles almost certainly contain adsorbed tar resins. Although the suspended particulate matter makes up only a few percent of the tar, it is important in certain industrial uses of coal-tar pitch.

17.3.10 The Effect of Carbonization Conditions on the Yield and Nature of Coal Tar

The release of volatile matter from coal during carbonization is substantially complete at a temperature much lower than that attained in coke ovens or continuous vertical retorts, and the nature of the "primary" tar is very different from that of the commercial high-temperature tars. From assays and analyses of tars from low-temperature retorts, continuous vertical retorts, and coke ovens, the nature of the secondary reactions that occur as the "primary" tar passes through the hot coke bed can be determined. Low-temperature tars, continuous vertical retort tars, and coke-oven tars form a series in which the yield of tar decreases, the aromaticity of the tar increases, the content of phenols and

[91] E. W. Volkmann, *Fuel,* **38,** 445 (1959).
[92] J. Weiler, *Blast Furn. Steel Plant,* **55,** 238 (1967).
[93] M. Janek, *Brennstoff-Chemie,* **46,** 72 (1965).

paraffins decreases, and the ratio of substituted aromatics to unsubstituted aromatics and heterocycles falls.

The reactions accounting for these changes—cracking and cyclization of paraffins, dehydrogenation of cycloparaffins, dehydration of phenols, and dealkylation of aromatic hydrocarbons and heterocyclic ring compounds—are those that would be expected, on thermodynamic grounds, to occur at the prevailing temperatures in carbonization retorts.

However, the comparison of assays and analyses of commercial tars, even if such data were available, would not materially sharpen the details of this general picture for two reasons: (1) in commercial carbonization only a restricted range of coals is used, there are uncontrolled variations in the plant design and operation, and the range of conditions is deliberately kept to a minimum, and (2) the sole objective is to produce the highest yield of satisfactory metallurgical coke or town gas.

The results of laboratory and pilot-plant scale carbonizations do provide additional information, if it can be assumed that these results parallel those of large-scale operation, an assumption that is somewhat dubious. The most extensive report in this field is that of Walter et al.,[94] who have compiled the results obtained on carbonizing more than 500 U.S. coals by the Bureau of Mines–American Gas Association evaluation procedure.[95] It is claimed that this procedure permits close control of the process variables and produces yields of tar and light oils similar to those obtained commercially, in amounts sufficient to determine their general physical and chemical characteristics.

In this extensive project the selected coals were carbonized at temperatures in the range of 500–1000°C in the BM–AGA test retort. The collected tars were first dehydrated and their specific gravities and "free carbon" contents determined. This "free carbon" represents the residue left when the dry tar is exhaustively extracted with toluene in a Soxhlet apparatus and corresponds to the "C_2" and "C_1" fractions obtained by other methods of solvent separation. The dry tars were subjected to an assay procedure involving a flash distillation to yield total distillate oils and a residual pitch. A low-efficiency fractionation of the distillate oils split them into five fractions—up to 170, 170–200, 200–235, 235–270, and 270–350°C—and a residue. By cooling the 170–270° fractions and the 270–350°C fractions for one hour in an ice bath, and filtering and weighing the crystals produced, crude naphthalene and crude anthracene yields were determined.

Tar bases and tar acids were determined on the solid-free total distillate oils by sequential extraction with 1.5 volumes (in each case) of 20% sulfuric acid saturated with potassium sulfate and 10% aqueous caustic soda. To extract additional bases freed by the alkali extraction, sequential extraction was again performed with 20% sulfuric acid saturated with potassium sulfate. The total volume contraction on the first and final extractions was taken as the content of tar bases and that on the second extraction as the tar acid content.

Finally, the solid-, tar acid-, and tar base-free neutral oil was examined for its olefin, aromatic, and paraffin contents. This analysis was done by first shaking the neutral oil with 3 volumes of 80% sulfuric acid for 5 min, allowing it to stand for 20 min, and draining off the acid layer. The extracted oil was allowed to stand for another 30 min, and any additional acid separated before distilling to the 350°C end point. The difference in volume between the distillate and the original sample was taken as the olefin content. The olefin-free distillate was shaken for 5 min with 3 volumes of 98% sulfuric acid and allowed to stand for several hours before separating and measuring the unreacted paraffins.

Taking an arbitrary selection of the results from this report, in which the same coal was carbonized over the range 500–1000°C at 100°C intervals, and plotting the tar yields on an maf coal basis against volatile matter content on the same basis results in approximately linear curves at the highest and lowest carbonization temperatures. This is illustrated in Fig. 17.7. Intermediate carbonization temperatures give similar lines that are intermediate in slope between the two lines shown. The two anomalous coals are characterized by high moisture and high oxygen contents. It is clear that the effect of increas-

[94] J. G. Walters, C. Ortuglio, and J. Glaenzer, U.S. Bur. Mines Bull. No. 643, 1967.

[95] D. A. Reynolds and C. R. Holmes, U.S. Bur. Mines Tech. Paper No. 685, 1946.

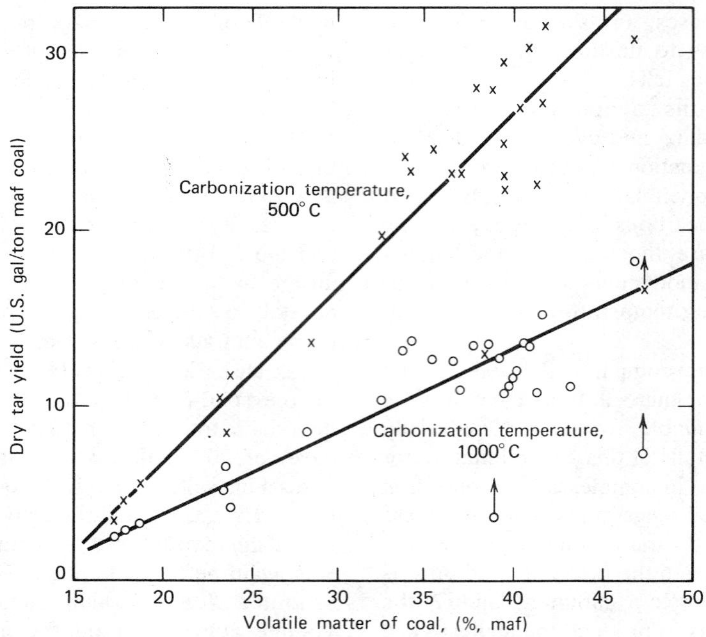

Figure 17.7 Dry tar yield versus volatile matter of coal.[94]

ing carbonization temperature is to reduce the tar yield; with high-volatile coals this reduction is greater than 100%. For example, a coal with an maf volatile matter of 35% yields about 22 U.S. gal/ton of tar at 500°C, falling to about 11 U.S. gal/ton at 1000°C. On the other hand, coals with volatile matter contents in the 20% range yield only about 6 gal/ton of tar at 500°C carbonization temperature and about 4 gal/ton at 1000°C.

Most of the tars examined by the Bureau of Mines–American Gas Association were in the 31–39% (proximate) volatile matter range, and considering the rather primitive method of assay, the tars obtained were reasonably uniform in properties. At the lowest carbonization temperature the specific gravities of the tars fell predominantly in the range 1.00 ±0.02, their pitch yields were between 35 and 45% of the dry tar, they did not deposit naphthalene or anthracene solids, and their tar acid contents ranged from 15 to 22% by volume on the dry tar. The variations within these limits are not readily attributable to any variation in the coal. There is a tendency for pitch yield and specific gravity to be lower with coals of higher volatile matter content, and for the tar acid content to be at a higher level with coals of higher oxygen content.

As the carbonization temperature increases, not only does the tar yield fall, but the properties of tar change markedly. The specific gravity rises to 1.20 ±0.02, the pitch yield increases to between 60 and 70% on the dry tar, and the tar acid content falls to 0.4–5.0% generally lying between 1.5 and 4%. Crude naphthalene and anthracene solids are not encountered below 800°C carbonization temperature. At 1000°C the yield of crude naphthalene is between 4 and 6% on the dry tar and that of crude anthracene solids between 1.0 and 2.5% on the dry tar.

The paraffin content of the solid-free neutral oils is generally in the 30–40% range at a carbonization temperature of 500°C, falling rapidly as the temperature increases, to 0.4–2.0%. The aromatic content shows the opposite trend, increasing from 45–50% to 85–90% by volume. The olefin content of the neutral oils varies within the range 6.5–15% at 500°C carbonization temperature, and is generally higher in the tars from coals of higher volatile matter content. It increases with increasing carbonization temperature to between 10 and 18% by volume.

Tar base content appears to depend on the nitrogen content of the coal and is not much affected by carbonization conditions. At the lowest carbonization temperature, the "free carbon" is generally below 1%, but increases with increasing carbonization temperature to as high as 8–9%. There appears to be no correlation between the "free carbon" and any other property of the tar or any of the reported parameters of the coal.

The tars derived from coals of lower volatile matter content have a higher specific gravity than those from more volatile coals carbonized at the same temperatures, and give a higher pitch yield, which is less affected by carbonization conditions. Thus coals with volatile matter contents in the range 16–20% yield tars at 500°C containing more than 50% pitch, and at 1000°C, tars with pitch contents in the range 70–75%. Low-temperature tars from such coals are more highly aromatic (75–80%) and contain less tar acids (6.5–8%) than low-temperature tars from the higher volatile coals, although the tars obtained at carbonization temperatures of 1000°C are generally similar to those obtained, at the same temperature, from coals of volatile matter contents of 33–38%.

Only a small number of the coals examined in this study had volatile matter contents above 40%, and the only abnormal features of the tars obtained from these coals—which may not be general characteristics—are the very high "free carbon" (21.16%) and lower aromaticity level of the neutral oil in the tars obtained at 1000°C carbonization temperature.

This study by the U.S. Bureau of Mines and the American Gas Association demonstrates quite clearly that the effect of relatively slight variations in commercially carbonized coals on the composition of the tar produced is negligible, and that any such effects would be outweighed by minor variations in carbonization conditions.

Some additional information on the effect of carbonization conditions can be derived from GC analysis of the tars produced at different carbonization temperatures, as reported by Grand'ry,[26] already discussed. Summarizing his results clearly shows the reduction in alkyl substitution of the aromatic components as the carbonization temperature is raised, as well as the reduction in phenolic content and the relative insensitivity of the tar base content to carbonization temperature:

	Carbonization Temperature (°C)			
Compounds Identified (%)	800	900	1025	1100
Monomethyl	13.59	11.47	8.0	7.29
Dimethyl	7.63	5.81	2.47	2.00
Trimethyl	2.96	1.47	0.36	0.49
Tetramethyl	0.45	0.18	Nil	0.16
Acidic	8.59	6.66	2.1	3.01
Basic	1.94	1.82	1.25	1.30

17.4 THE STRUCTURE OF TAR AND THE STRUCTURE OF COAL

The similarities in the appearance and properties of coal-tar pitch and the material extract from bituminous coal, after removal of the extracting solvent, have been remarked on by many investigators. Published values for molecular weights of coal extract fractions and pitch fractions are of the same order; infrared spectra and statistical structural analysis indicate a generally similar structure of condensed aromatic or partly hydrogenated aromatic ring clusters attached to each other by C—C bonds or through methylene or ether bridges.

The digestion of coal at elevated temperatures in solvents such as anthracene oil yields a colloidal solution that when filtered to remove the small quantity of undissolved coal and the bulk of the inorganic constituents, is similar in appearance to a coal-tar pitch. It differs in its properties, however, in that it displays markedly non-Newtonian flow properties. If, however, such a coal solution is filtered under pressure through a very fine filter to remove most of the colloidally dispersed coal substance, and the solvent is partly removed to leave a residue of medium pitch softening point, the residue shows a temperature-viscosity relationship and a flow pattern similar to those of a high-temperature pitch of the same softening point. It can be substituted for coal-tar pitch in many of its industrial applications.

These similarities between the portion of coal that is soluble in powerful solvents and coal-tar pitch have been advanced as evidence that the higher molecular weight components of coal tar represent fairly massive fragments of the original coal structure, which have survived the carbonization process substantially unchanged. Therefore it has been suggested that advances in our knowledge of the structure of the more complex fractions of coal tar can yield important information on the structure of coal.

Volkmann and Russell[96] go even further, pointing to the similarities in the properties of the less volatile part of the high-temperature coal-tar pitch and the vitrinites. They suggest that rather than assuming that the constituents of pitch are formed by condensation of small molecules or radicals formed by pyrolytic breakdown of the coal substance, it is more reasonable to postulate that the bulk of the constituents of pitch consists of coal substance carried out of the carbonization equipment, either by volatilization or mechanical entrainment in the gas stream, and altered only slightly in their passage. They envisage a two-step process in the course of carbonization, in which fragments of the vitrinite portion of the coal are first carried away from the bulk of the coal. On exposure to pyrolysis conditions they undergo some modifications, principally at the circumference of the molecule, which involve elimination of the oxygen-containing functional groups and of hydrogen atoms. This results in the conversion of aliphatic or alicyclic structures to aromatic structures of similar molecular weight.

However, there seems to be one insuperable obstacle to the view that the higher molecular weight components of high-temperature tar are fragments, altered or not, of the coal substance. Such fragments would be expected to make a large contribution to the benzene- (or toluene-) insoluble fraction of pitch. One would expect to find this fraction more prominent at the lower carbonization temperatures, under which further breakdown or alteration should be restricted. But the opposite is true. A comparison of the solvent analyses of pitches[97] shows that this fraction increases as the severity of carbonization increases. Thus a low-temperature pitch of 75.5°C ring and ball (R and B) softening point contained 7.2% benzene insolubles compared with 16.9% in a 78.9°C (R and B) softening point coke-oven pitch. Examination of the tars produced by carbonization of U.S. coals in the BM–AGA evaluation study clearly shows that the toluene-insoluble "free carbon" is formed by polymerization during carbonization. In tars produced at 500°C it rarely exceeds 1% of the tars, whereas in tars from the same coals produced at 1000°C, it averages 6.6% and can be as high as 20% of the tar.

It has been claimed that similarities in physical and chemical properties between high-temperature coal-tar pitches and coal macerals or coal extracts indicate that their components represent parts of the coal substance. But this argument can also be countered by noting that the pyrolysis of almost any carbonaceous substance yields a tarry by-product that on distillation, leaves a pitch-like residue. Itoh et al.[98] compared a coal-tar pitch with the pitch obtained from the tar produced as a by-product in the catalytic cracking of crude petroleum oil. The samples were vacuum fractionated and the fractions analyzed by elementary analysis, gel permeation, IR, UV, and NMR spectroscopy. It was concluded that the two materials were similar.

It may be, as Christyakov suggests,[99] that the two theories for the formation of polycyclic and heterocyclic compounds in pitch—that they result from the condensation of the low molecular pyrolytic decomposition products of the coal substance or that they represent straight structural units of the coal modified by pyrolysis—are complementary and not mutually exclusive. It seems clear, however, that the bulk of high-temperature tar and its more complex, higher molecular weight components are the result of secondary reactions during carbonization and that the chemical structure of high-temperature coal tar, or of the pitch derived from it, need not be related to the structure of the coal

[96] E. W. Volkmann and C. C. Russell, *Fuel,* **41,** 535 (1962).

[97] *The Coal Tar Data Book,* 2nd ed., Coal Tar Research Association, Gomersal, England, (1965), p. A2–1.

[98] H. Itoh et al., *J. Fuel Soc. Jap.,* **53**(12), 1021 (1974).

[99] A. M. Christyakov, *Coke Chem. USSR,* No. 6, 55 (1965).

from which it was derived. Any such relation, if it exists, can be accounted for only by the view that the breakdown of the coal substance on thermal decomposition, and the subsequent recombination of the "monomer" molecules and radicals, can build up structures similar to the original coal "polymer." Volkmann and Russell[96] hint at this possibility in the sentence "It may well be that the modification process of high-temperature pyrolysis conditions brings about, in a matter of seconds, the same reactions that nature applies over millions of years."

There is obviously much more justification for the view that "primary" tars (i.e., tars isolated from coal by heating under conditions that shield the initial material released from the coal from secondary reactions) should show some relation in chemical structure to the coal raw material.

Palmer and Vahrman[100] classify primary tars into three groups:

Class 1. Primary tars made at the lowest end of the decomposition range (up to 300°C) and obtained in relatively low yield (3–4% of the coal).

Class 2. Primary tars formed during the main decomposition of the coal, at 300–-500°C, and removed quickly from the hot zone so that secondary reactions are minimized. Such tars are obtained in relatively high yield (15–20% of the coal).

Class 3. Low-temperature tars that have undergone a greater or smaller degree of secondary cracking. This class includes commercial low-temperature tars, and at the lower end, Gray-King assay tars.

It is thus Class 1 and particularly Class 2 primary tars that should bear the closest relation to the structure of the original coal. The distillate, obtained in 3–4% yield from coal under molecular distillation conditions (i.e., Class 1 primary tar), has not been well characterized. These tars are often transparent, lacquerlike solids, which rapidly darken on exposure to air and light. Orning and Greifer[101] reported that the tar produced by high-vacuum, short-path pyrolysis of a sample of Pittsburgh Seam coal gave an infrared spectrum identical to that of anthraxylon, of which the original coal was largely composed. In a later paper Greifer,[102] on the basis of the infrared spectrum, molecular weight (270), molar refraction, and other properties, characterized this material as having a structure consisting of isolated aromatic rings fused into hydroaromatic configurations and containing aliphatic sidechains limited in size to methyl or ethyl. Juettner and Howard[103] heated a Pittsburgh coking coal under vacuum to just above 500°C. The vacuum tars they obtained were described as semisolid yellow or brownish materials that contained relatively small amounts of volatile matter. The volatiles contained about 30% of neutral ether-insoluble compounds, with an average molecular weight of 400. However the tars they obtained, like those obtained by Sun et al.,[104] belonged to Class 2, since the yields, which depended on the nature of the coal, amounted to as much as 21% of the coal.

Chauvin and Deelder[70] produced a primary tar from a Faulquemont coal of volatile matter (maf) 39.26% and consisting essentially of 78% vitrinite and 12% exinite. This coal was ground to between 80 and 160 μm and was subjected, entrained in nitrogen, to extremely rapid heating (>1000°C/s) for a constant time of less than one second. Heating was conducted in a tubular, electrically heated furnace, the configuration of which ensured that the coal particles did not come into contact with the furnace walls. The char was separated from the carrier gas and coal volatiles, and the latter were separated from the former by electrostatic precipitation and condensation. The tar yield increased with increasing furnace wall temperature (this was probably several hundred degrees higher than the temperature attained by the coal) up to a maximum of 17.1%, at a tube wall temperature of 850°C. Gas and water yields at this

[100] T. J. Palmer and M. Vahrman, *Fuel,* **51,** 22 (1972).

[101] A. A. Orning and B. Greifer, *Fuel,* **35,** 381 (1956).

[102] B. Greifer, *Fuel,* **37,** 345 (1958).

[103] B. Juettner and H. C. Howard, *Contrib. Carnegie Inst.,* **8,** 22 (1934).

[104] B. Sun, C. H. Ruof, and H. C. Howard, *Fuel,* **37,** 299 (1958).

temperature were low—on the order of 2% moisture and 4% gas. The gas contained carbon monoxide as the major constituent, with carbon dioxide, hydrogen, paraffins from methane to *n*-pentane, and olefins from ethylene to isobutene, and acetylene.

The tar produced in maximum yield was subjected to solvent fractionation with pyridine, benzene, and petroleum ether, in succession, and the fractions soluble in benzene and petroleum ether were then separated into acids, bases, and neutrals. The neutrals from these fractions were further separated by semipreparative GC into four fractions with boiling ranges equivalent to 0–200, 200–315, 315–370, and 370–425°C at atmospheric pressure. Each of these fractions was separated by chromatography on silica gel into paraffins and aromatics, plus other polar compounds.

The higher fractions insoluble in petroleum ether were subjected to the statistical structural analysis scheme previously outlined (in this chapter), with the results shown in Table 17.17.

In the low molecular weight fraction boiling up to 200°C, only small amounts of *n*-paraffins or *n*-olefins were found, but in the higher fractions, C_{13} + *n*-paraffins and *n*-olefins with terminal double bonds were the main components. The 200–315°C fraction was of interest in that several aliphatic hydrocarbons with isoprene groupings were found. Furthermore in the distribution of the C_{16}–C_{25} hydrocarbons in the higher boiling fractions, there was a higher content of components with an odd number of carbon atoms than with an even number. The same preponderance of odd numbers of carbon paraffins and olefins was also found by Vahrman and Watts[105] in their study of the hydrocarbons released from a weakly caking coal (NCB Rank No. 801: 81.4% C, maf; 5% H, maf) by heating in thin layers up to 400°C, and in the pyridine extracts of the Faulquemont coal.

The general structure of the benzene-soluble, petroleum ether-insoluble fractions of the primary tar from the Faulquemont coal shown in Table 17.17 agrees well with that derived by Karr and Comberiati[74] for similar fractions of low-temperature tar pitch from American coals. On the basis of their *n-d-M* statistical structural analysis, they calculate that the resin fraction from one sample of low-temperature pitch contained 5.4 rings, of which 3.8 were aromatic and 1.6 naphthenic, and that the average molecule contained six methyl or equivalent alkyl groupings. The corresponding figures for the benzene-soluble, petroleum ether–insoluble fractions of a low-temperature tar were as follows: total number of rings in the average molecule = 4.6, of which 3.8 are aromatic, and number of methyl groups or equivalent alkyl groups per average molecule = 7. These values correspond exactly with those determined by Kroeger et al.[106] for the resin fractions from a pitch obtained by distilling the tar from the vacuum pyrolysis of a micrinite-semifusinite and a vitrinite at 525°C.

Additional evidence for the structure of the low-temperature tar resins was deduced by Karr and Comberiati from UV and IR spectroscopy and by combined pyrolysis—GC analysis, leading to the overall conclusion that the low-temperature tar resins had a structure made up mostly of isolated aromatic rings and clusters of two fused rings, with saturated rings, including ether groups, completing the skeleton. The major substituent groups were phenolic hydroxyl and alkyl groups, which were predominantly methyls. Dehydrogenation suggested the presence of alcoholic hydroxyl groups on carbon atoms adjacent to aromatic rings, and partly saturated as well as wholly saturated aromatic rings that readily form three dimensional semigraphitic nuclei on dehydrogenation.

The tar and pitch resin structures were similar in general character to the benzene-insoluble, quinoline-soluble fractions from the quinoline extracts of bituminous and subbituminous coals, except that the latter contained a greater number of rings and the ring structure was more highly condensed. On this evidence, Karr and Comberiati concluded that low-temperature pitch resins probably consist of essentially unaltered coal molecules carried over with the tar vapors.

Chauvin and Deelder[70] are less dogmatic.

[105] M. Vahrman and R. H. Watts, *Fuel,* **51,** 131 (1972).

[106] C. Kroeger, H. G. Goebgen, and A. Klusman, *Brennstoff-Chemie,* **45**(6), 170 (1964).

Table 17.17 Statistical Structural Analysis[a] of Fractions of Tar from Shock-Heated Faulquemont Coal[70]

Fraction	Yield on Tar (%)	H_{Al}/H	H_{Al}/C_{Al}	f_a	R	R^*_{Ar}	C^*_{Ar}	m	R_{nAr}	Molecular Weight
Total tar	100	0.69 ±0.06	2.0–2.36	0.65–0.70	7.5–6.6	3.3 ±0.25	15 ±1	1.60–1.72	2.2–0.8	550 ±50
Insoluble in benzene	50	0.67 ±0.06	2.0–2.3	0.67–0.71	9.9–8.8	3.55 ±0.25	16 ±1	2.18–2.32	2.1–0.6	810 ±60
Soluble in benzene	50	0.75 ±0.06	2.0–2.34	0.63–0.68	5.9–5.3	3.15 ±0.25	14 ±1	1.17–1.27	2.2–1.3	380 ±30
Soluble in benzene, insoluble in petroleum ether	40	0.74 ±0.06	2.0–2.36	0.64–0.69	6.3–5.6	3.35 ±0.25	15 ±1	1.20–1.29	2.3–1.3	420 ±30
Soluble in petroleum ether	10	0.79 ±0.07	2.0–2.35	0.58–0.64	5.1–4.4	2.55 ±0.25	12 ±1	1.09–1.20	2.3–1.3	310 ±30

[a] H_{Al}/H = ratio of aliphatic H to total H in average molecule; H_{Al}/C_{Al} = ratio of aliphatic hydrogen to nonaromatic carbon a ms in average molecule; f_a = aromaticity index; R = ring index; R^*_{Ar} = number of aromatic rings in the monomeric unit; C^*_{Ar} = number of aromatic carbons in the monomeric unit; m = number of "monomer" units in the oligomer; R_{nAr} = number of nonaromatic rings in the average molecule.

They postulate that the pyrolysis of coal involves the following reactions:

1. Rupture of weak bonds between the monomer units, with the formation of free radicals.
2. Vaporization of a part of the monomer units, if the temperature is sufficiently high.
3. Condensation of part of the monomeric units into high molecular weight substances that remain in the semicoke.
4. Hydrogen transfer reactions, in which the monomer units are stabilized to molecules capable of later distillation.
5. Condensation of the vaporized monomer units in the gaseous phase by free radical reactions.
6. Stabilization of the vaporized monomer units in the gaseous phase by hydrogen transfer.
7. Secondary cracking at higher temperatures both in the solid and liquid phases.

They consider that their short-time, rapid-heating pyrolysis technique reduces the secondary cracking reactions to a minimum, favors the passage of monomer free radicals into the vapor phase, and reduces the importance of condensation and hydrogen transfer reactions. Thus the main components of their primary tar should represent the coal monomer units stabilized by the acquisition of a proton or by oligomerization. From the results of hydrogenation experiments, they postulate that the bridges between the monomer units in the oligomeric tar resins are oxygen atoms.

Accepting as valid the results of Chauvin and Deelder's, and Karr and Comberiati's statistical structural analysis of low-temperature tar and coal extract resins, and taking into account the more recent results of Vahrman and his colleagues, which are quite consistent, the picture of coal structure that emerges is of a three-dimensional polymeric structure with units consisting of condensed aromatic and hydroaromatic rings having methyl, hydroxyl, and possibly ethyl substituents on peripheral carbon atoms. These clusters, which probably contain from four to perhaps a dozen rings, are joined to each other by bridges that may be methylene groups, polymethylene or conjugated olefin chains, or ether linkages. This polymeric structure contains relatively small paraffinic, olefinic, and aromatic molecules adsorbed in the micropores, which can be released by solvent extraction, steaming, or mild pyrolysis.

That the origin of these hydrocarbons should be attributed to the desorption of material held in the micropore structure rather than to the breakdown of any part of the coal substance follows from the observation of Vahrman and Watts[105] that the complete evolution of volatile products, either by heating to 250–400°C, by steaming, or by solvent extraction at temperatures up to 100°C, takes a long time. This is also evident from the work of Loskutova and German,[107] who compared the composition of the neutral oil fractions of a primary tar obtained from a gas coal and the benzene-soluble part of the pyridine extracts of the same coal before and after preliminary heating to 320°C. They found that although the yields from the unheated coal were higher than from the heated coal, the paraffins, naphthenes, and aromatics of all the neutral oil fractions were very similar in composition.

These adsorbed hydrocarbons include straight-chain alkanes and branched-chain alkenes with terminal double bonds, as well as the lower polycyclic aromatics such as naphthalene, methyl and dimethyl naphthalenes, acenaphthene, and fluorene. They may be derived from the waxy resinous components of the coal, since in the alkanes and alkenes there is a preference for compounds with an odd number of carbon atoms, a characteristic found in certain crude petroleums,[108] and in the leaves and fruits of plants.[109]

Most of the decomposition of coal occurs in the temperature range of 450–500°C. The first primary reaction appears to be the depolymerization of the coal, releasing the monomer units as free radicals, and possibly giving rise to additional amounts of paraffins and olefins by breakdown of the structures between the monomer units. Some of the

[107] E. N. Loskutova and N. M. German, *Protsessy Term. Prevroshchaik Kamennykh Uglei,* (1969), p. 227.

[108] T. L. Martin, J. C. Winters, and J. A. Williams, *Nature,* **199,** 110 (1963).

[109] M. Vahrman, *Fuel,* **49,** 5 (1970).

released monomer units are stabilized by hydrogen transfer and can volatilize, to be condensed as primary tar. Others are stabilized by polymerization to more massive molecules that remain in the char. The secondary cracking, which accompanies even mild pyrolysis, is confined to some further splitting of bridging groups and dehydrogenation of hydroaromatic rings.

Solvent extraction at around 300°C leads to a similar, but less extensive decomposition to give monomer or oligomer units similar in general structure and showing similar spectral features to both the original coal and the primary tar fractions. Longer time and/or higher temperatures of carbonization are accompanied by progressively extensive secondary cracking reactions, resulting in the production of smaller, more highly aromatic molecules and, by the disruption of ether linkages, to phenolic compounds and the conversion of longer chain alkanes and alkenes to simpler paraffins and olefins. In commercial low-temperature tars, the original coal fragments have been degraded to a large extent. These tars contain only relatively small amounts of benzene-insoluble components, which may be products of polymerization rather than degradation. Their main constituents are paraffins, olefins, and alkyl- and hydroxy-substituted aromatics of relatively low molecular weight. The more drastic carbonization conditions in coke ovens lead to more extensive secondary reactions. The main secondary reactions are the dehydration of phenols, the dealkylation of substituted aromatics, the disproportionation of the free radicals produced by these reactions, and the formation of both low and high molecular weight aromatic structures by dehydrogenation. Accompanying reactions are the splitting of longer chain paraffins to gaseous paraffins and olefins, cyclization of some paraffins and olefins to naphthenes, and the aromatization of these and of naphthenes originally present to aromatics. It must be considered that most of the petroleum ether- and benzene-insoluble material in high-temperature tars arises as a result of these secondary reactions, and that any similarity to the original coal is fortuitous.

Karr[110] has pointed out that the distribution of phenolic isomers in low-temperature tars produced from subbituminous coals is frequently different from either the thermodynamic equilibrium distribution or that which might be expected from kinetic considerations, as calculated from their free valence numbers, assuming a free radical process for their formation.[111] He has suggested that the discrepancies between the observed and calculated isomer distributions may be explained by postulating that they are derived from a substituted 4-*n*-propyl-2-methoxy phenol, which can be considered to be a main building block of the lignin structure. This phenol therefore can be considered to be a frequent grouping in the coal polymer structure and in the monomer units derived from it. Waddington[112] has sought to extend this idea by considering the distribution of methyl indanols, trimethyl phenols, trimethyl benzenes, and dimethyl naphthalenes in different types of tar, and claims further support for this hypothesis. Thus the preponderance of 3,4,5- and 2,3,5-trimethyl phenol isomers and of 1,2,3- and 1,2,4-trimethyl benzenes in continuous vertical retort tars is used to support the hypothesis that they are formed by the following reactions:

[110] C. Karr, Jr., *J. Phys. Chem.*, **64**, 462 (1960).
[111] C. Karr, Jr., *Fuel*, **39**, 475 (1960).
[112] W. Waddington, paper presented at 4th Int. Conf. on Coal, Le Touquet, France, May–June 1961.

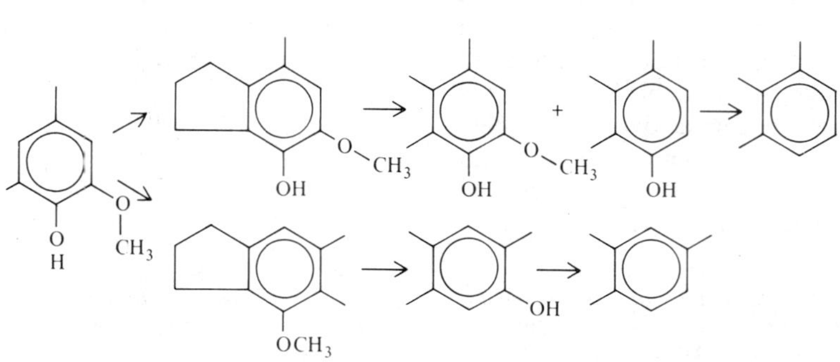

Evidence for the suggestions that there is a predominant structural unit in the coal polymer and in the "monomer" formed on its initial breakdown, and that the smaller molecules in tar are derived from this structure by the secondary pyrolysis reactions is, however, very tenuous and at present can be considered only as an interesting speculation.

17.5 TOXIC PROPERTIES OF HIGH-TEMPERATURE TARS AND TAR FRACTIONS

High-temperature coal tar, its fractions, and many of its components are toxic when ingested, inhaled, or absorbed through the skin. The most potent components are the more volatile mononuclear aromatic hydrocarbons, phenols, and tar bases, which, however, occur only in small amounts. The major polynuclear aromatic components are primarily skin and eye irritants and are tolerated by the human body when ingested in relatively large amounts. Naphthalene was once used therapeutically as an internal antiseptic and anthelminthic in daily oral doses of 0.1–0.5 g; the probable lethal oral dose is between 5 and 15 g. Anthracene is poorly absorbed from the gastrointestinal tract and mainly passes through the system unchanged. This is generally also true for the higher polynuclear hydrocarbons.[113] The lowest lethal dose for coal-tar creosote by oral intake for humans is given as 140 mg/kg body weight in the U.S. Department of Health, Education and Welfare 1974 Toxic Substances list.

In any case, the unpleasant smell and taste, and the immediate irritant effect of crude or refined coal tar, coal-tar creosote, or other tar oils are powerful deterrents to their oral consumption, and accidental poisoning by ingestion of these materials is extremely rare. The health hazard popularly associated with coal tar and its products is their carcinogenic activity. This hazard undoubtedly exists, but should be seen in perspective; it is by no means as serious as some reports would suggest, and it is constantly being reduced by greater precautions and better hygiene in the workplace. It is entirely an occupational hazard. The use of tar products in such outlets as road surfacing, preservation of telegraph and electric transmission poles, pitch-fiber pipes, and coal-tar enamels for corrosion protection of pipes for drinking water, constitutes no danger whatsoever to the general public.

The earliest recognition of cancer as an occupational hazard was by Potts in 1775, who noted the prevalence of scrotal cancer among London chimney sweeps and suggested that continual contact of the skin with soot was the cause. The chemical origin of this form of cancer was not universally accepted until 1922, however, when Passey[114] demonstrated that tumors could be induced on the skin of mice by painting with an ethereal soot extract. Skin cancer was also noted to be an occupational hazard of workers exposed to pitch dust in the coal briquetting industry,[115] and to workers exposed to contact with crude tar.[116]

Following these clinical observations, a number of workers attempted to induce tumors in experimental animals by the application of tar and pitch to their skins. In 1915 Yamagiwa and Ichikawa[117] were successful in producing tumors in the ears of rabbits by prolonged application of crude coal tar. Kennaway[118] showed that only the higher boiling fractions of coal tar were carcinogenic and that the fluorescence spectra of carcinogenic fractions was similar to that of benzo[*a*]anthracene. This led him to investigate the effect of pure polynuclear hydrocarbons, and he demonstrated that skin tumors could be induced by repeated topical application of benzo[*a*]anthracene to mice. This work was extended by Kennaway's colleague Cook and his co-workers, culminating in the isolation of a highly carcinogenic compound, benzo[*a*]pyrene, from coal-tar pitch.[119]

Since then many more PAHs have been synthesized and tested for their tumor-induc-

[113] N. V. Gerarde, *Toxicology and Biochemistry of Aromatic Hydrocarbons,* Elsevier Scientific Publishing Company, New York, (1960), p. 253.

[114] R. D. Passey, *Br. Med. J.,* **2,** 1112 (1922).

[115] A. Manouviriez, *Ann. Hyg. Publ. (Paris),* **45,** 459 (1876).

[116] R. Volkmann, *Beitrage zum Chirurgie,* Breitkopf and Hartel, Leipzig, 1875, p. 370.

[117] K. Yamagiwa and K. Ichikawa, *V. Jap. Path. Ger.,* **5,** 142 (1915).

[118] E. L. Kennaway, *J. Ind. Hyg.,* **5,** 462 (1924).

[119] J. H. Cook, C. L. Hewitt, and I. Hieger, *Nature,* **130,** 926 (1932).

ing effect. Some, such as dibenz[*a,i,*]pyrene and dibenz[*a,h*]anthracene, have been found to be even more potent carcinogens than BaP. However BaP appears to be regarded, for not very good reasons, as an index of the seriousness of the carcinogenic hazard involved in exposure to coal tar or petroleum oils.

As reported earlier in this chapter, the presence of BaP in high-temperature coal tar, coal-tar creosote, and coal-tar pitch has been demonstrated by numerous workers. The HPLC analysis of a pitch extract[52] also showed the presence of the other potent carcinogens, dibenz[*a,h*]anthracene, dibenz-[*a,e*]pyrene, dibenz[*a,h*]pyrene, dibenz[*a,i*]-pyrene, and dibenz[*a,l*]pyrene.

Although the values reported by Marek[120] (1.2–2.4% by weight in crude tars and 2.26–3.75% by weight in high-temperature pitches) and by other workers using the same analytical method for the BaP content of tars and pitches are believed to be high, there is no doubt that coal tar and the pitch derived from it contain dangerously high concentrations of PAHs, which have been shown to be potent carcinogens when applied repeatedly and over relatively lengthy periods to the skins of susceptible strains of experimental animals.

The recognition of cancerous lesions of the exposed skin as an occupational hazard to workers exposed to pitch dust undoubtedly supports the belief that these PAHs are the causative agents, although it has not yet been possible to demonstrate their carcinogenic action on animals, such as monkeys, more closely related to man.

Although high-temperature coal-tar creosote has been shown to contain BaP and to cause skin cancer when applied 3 times a week for several months in toluene solution to the shaved interscapular skin of a susceptible strain of mice,[121] reports of skin cancer outbreaks among workers in the timber creosoting industry are much less well documented. In any case the irritant effect of creosote on the skin is a sufficient incentive to workers at creosoting plants or those handling freshly creosoted timber to avoid such contamination by wearing protective clothing. Epitheliomas can result after prolonged exposure to creosote,[122] but there is no evidence that the large volume of creosote used throughout the world for the preservation of timber (150,000 tons/year in the United States alone) constitutes a hazard to the workers engaged in that industry or to the public. Similar large amounts of coal-tar pitch and tar oils are employed worldwide for the construction and maintenance of roads, and again no evidence has been adduced to link the use of road tar to any health hazard either among road-working personnel or the general population.

Repeated skin contact with molten coal-tar pitch is very unlikely, since the pitch used in industry, although a liquid, is hot enough to cause immediate, serious burns. As far as cancer of the skin is concerned, the sole hazard is long-term continual skin contact with finely divided pitch dust. This used to be a problem particularly at tar distilleries, where pitch was stored as a solid in bays. The breaking of the pitch by blasting and the manual loading of the broken pitch produced a considerable amount of dust. A similar hazard existed at the user end, particularly in the patent fuel industry, in which lump pitch was unloaded and ground up into a fine powder to mix with coal fines to make briquettes.

The Workman's Compensation Act of 1907 in England officially recognized that cutaneous epithelioma constituted an occupational hazard in these industries. From 1920 to 1945 pitch or tar was deemed responsible for 2229 of the 3753 notified cases of industrial skin cancer in Great Britain. During the decade 1946–1955, 2041 new cases were notified, and 1053 of these had a history of exposure to coal-tar pitch. Since then there have been fewer notifications from the tar distilling and patent fuel industries, and in most cases the contacts with tar or pitch could be traced to the period when little precaution was taken against this hazard. It is a characteristic of skin cancer due to contact with pitch that it frequently takes 30 or 40 years to develop. It is expected that the preventive measures that have been introduced since this occupational hazard was recognized will result in a reduc-

[120] V. Marek, *Coke Chem. (USSR),* No. 12, 30 (1964).

[121] W. E. Poel, *J. Nat. Cancer Inst.,* **18**(1), 41 (1957).

[122] H. I. Cookson, *Br. Med. J.,* **1,** 368 (1924).

tion in the occurrence of skin cancer due to pitch to a negligible figure.

There are three kinds of preventive measure. First of all, workers exposed to skin contact or inhalation of tar, creosote, or solid pitch are required to wear full protective clothing to avoid skin contact, particulate masks to avoid inhalation, and protective eyewear. They should also be furnished with barrier creams specially formulated to protect the skin from the irritant effects of PAHs. Second, a high standard of personal hygiene must be insisted on. Protective clothing should be laundered at frequent intervals, and operatives must shower before resuming their everyday attire.

But perhaps the most effective preventive has been the virtual elimination of pitch dust. Nowadays it is universal practice to store and deliver pitch to user plants as a liquid and to use it in this form. Where this is impossible because of the distance between the production plant and the user (e.g., in the export of pitch from Great Britain to Scandinavia), the pitch is produced in a dust-free solid form as "pencils," pastilles, or flakes. Almost all tar refineries that produce pitch for overseas export to aluminum smelters or briquetting plants have in recent years installed pitch-conditioning plants to produce their solid pitch in one of these easily handled forms.

Not all workers exposed to habitual contact with pitch dust will develop skin cancer, and not all skin lesions that susceptible workers may contract will be malignant. Hodgson and Whiteley[123] surveyed all personnel at a large Welsh coal-briquetting plant and compared the incidence of skin lesions of the 144 workers with a control group of 263 dermatological outpatients who had never had any contact with pitch. The lesions were classified as benign (papillomas), premalignant (squamous keratosis and pitch acanthomas, i.e., pitch warts), and malignant (squamous cell carcinoma and epithelioma). Benign proliferative papillomas occurred in 4.8% of the pitch workers in the age group 20–29 as opposed to 4.1% of the pitch workers in the same age group of the controls. In both groups papilloma incidence increased with age to 36.2% in the pitch workers aged 50–59 and to 29% in this age sector of the control group. Squamous keratosis was slightly more common in the pitch workers (12%) than in the controls (10%), as were acanthomas. But a 2-year follow-up revealed that the incidence of pitch warts among the pitch workers had increased to 15.7%, compared with a 2.7% incidence of acanthomas in the controls. Malignant squamous cell carcinoma affected 2.8% of the pitch workers compared with 0.4% of the control group.

All the pitch warts and carcinoma occurred in white-skinned workers, although 12.5% of the working population was Indian or Negro. These lesions occurred on the face around the eyes and nose, on the ears, and on the dorsum of the hands. In a few cases pitch workers developed carcinoma of the scrotum. Pitch warts are not themselves malignant, but if untreated they can develop into malignant carcinomas depending on both the length of exposure and the degree of skin contamination; their incidence was highest in men over 60 who had had 40–50 years of exposure, although this was not always the case. Some workers never had a wart after 50 years of work, whereas one worker developed a wart after only 3 years.

Pierre et al.[124] report on the incidence and treatment of skin lesions at a French coal-briquetting plant. For more than 8 years, they diagnosed and treated 24 cases of skin ailments in 10 members of the work force as a result of semiannual routine examination. These included nine papillomas, ten cases of prickle cell epithelioma, and four keratoacanthomas. Papillomas were treated selectively by electrocoagulation. This treatment need be employed only once, and there is never a risk of recurrence. Cases of keratoacanthoma usually heal spontaneously after excision, but if they do not disappear, electrocoagulation is employed. Malignant epithelioma, which is always obvious, must be treated by contact therapy only, usually a series of 10 consecutive visits. The total surface treatment is 5000 rads. This causes weeping of the skin, but after one month the lesion heals and the skin resumes its normal appearance. Skin pigmentation is altered in the treated area, but in all cases a cure is effected without local recurrence.

[123] G. A. Hodgson and H. J. Whiteley, *Br. J. Ind. Med.*, **27**, 160 (1970).

[124] F. Pierre et al., *Arch. Mala. Prof. Med. Trav. Secur. Soc.*, **T-26**, 475 (September 1965).

Thus to put the hazard of skin cancer due to tar and pitch in perspective, one can say that it is solely an occupational hazard confined to long-term contact of finely divided pitch with the skin. Susceptibility varies greatly from individual to individual, and only some 15–20% of the exposed workers are likely to be affected. It can be prevented, or at least greatly minimized, by the proper use of protective clothing, masks and goggles, and good personal hygiene. Periodic medical examination will detect it at an early stage, and treatment at this stage will effect a complete cure.

A more difficult problem is the risk of lung cancer resulting from inhalation of fumes from heated pitch or tar, which cannot be detected at an early and curable stage. How well proved and how serious is this hazard?

Sawicki et al.[125] have published figures for the PAH content of the atmosphere in urban and nonurban sites and sites adjacent to plants that used heated pitch or refined tar in their operations. The BaP content of nonurban air ranged from 0.01 to 1.9 μg/1000 m^3, and in urban air ranged from 0.1 to 61 μg/1000 m^3. This amounted to between 8 and 35% of the nine analyzed PAHs. The BaP content of the outdoor air close to pitch-fiber pipe-producing plants varied from 0.52 to 4.8 μg/1000 m^3, but this represented only 3–4% of the total PAHs because of the large amounts of phenanthrene and pyrene present.

However, the air close to the open or partly closed kettles used to heat pitch or refined tar in road-making or pitch-roofing operations showed a very different picture. Thus at a location 4 ft downwind from a partly closed kettle containing hot road tar, the atmosphere contained 6.13 g/1000 m^3 of PAHs (0.85 g of anthracene, 3.7 g of phenanthrene, 1 g of pyrene, 0.078 g of BaP, 0.024 g of perylene, 0.04 g of benzo[*g,h,i*]perylene, and 0.62 g of carbazole). This concentration fell to 855 μg/1000 m^3 at a site 10 ft downwind from the tar kettle, with a BaP concentration of 110 μg/1000 m^3. At a location downwind from the kettle holding heated pitch in a roofing operation, the total concentration of PAHs in the air amounted to 0.456 g/1000 m^3, of which 0.14 g/1000 m^3 was BaP.

In view of the high content of PAHs and, in particular, BaP in air contaminated with refined tar or pitch fumes, then, the question naturally arises whether workers exposed to these, and the general public, are at risk.

In an attempt to answer this question, Hammond et al.[126] decided to study the death rates, from different causes, of workers engaged in the roofing trade. The subjects were members of the United States Tile and Composition Roofers, Damp and Waterproof Workers Association. This union provides death benefits for both active and retired members, thus obtains copies of all death certificates. Through these records and the efforts of union officials, 96% of the 5939 active, probational, or retired members of the union in the 193 locals involved in pitch roofing and waterproofing were traced, and the deaths from various causes in this sample compared with those in the total male population of the United States. The death rates for the U.S. male population at various ages in 1960–1962 and 1963–1965 were used to calculate the number of deaths in the roofing worker sample under the null hypothesis that employment in that trade had no influence on the death rate. The results indicated that for total cancer deaths, deaths from cancer of the lung, larynx, and esophagus, and for cancer of the bladder, there was no statistically significant difference between the observed and expected deaths for the two age groups, 45–64 and 65–94. Thus actual lung cancer deaths totaled 43, compared with an expected total of 36.4, and bladder cancer deaths were 5, against an expected 4.8. Only in the case of deaths from skin cancer (5 observed vs. 1.9 expected) was there any indication of an occupational hazard, and in this case the numbers involved were considered too small to have any statistical significance.

The methodology of this epidemiological study can be criticized. It is not clear whether all members of the selected locals were actually exposed to pitch fumes; personnel involved in transport and supervisory duties may well have been included. No allowance was made for the smoking habits of the union

[125] E. Sawicki, F. T. Fox, W. C. Elbert, T. R. Hauser, and J. Meeker, *Am. Ind. Hyg. J.*, **23**, 482 (1962).

[126] E. G. Hammond, I. J. Selikoff, and P. J. Lawther, *Ann. N.Y. Acad. Sci.*, **271**, 116 (1976).

personnel. However despite these shortcomings, one would have expected that if a health hazard were involved in the regular inhalation of air contaminated with coal-tar pitch fumes, the results would have shown this quite clearly.

The results of this epidemiological study appear to contradict those of Mazumdar et al.[127] In this study the degree of exposure of three categories of coke-oven operatives to the fumes produced during the high-temperature carbonization of coal was determined by sampling and analyzing the atmosphere at various locations on the ovens and at various times during the carbonization cycle. These data were linked to the mortality data from the steelworkers' study conducted by the University of Pittsburgh's Department of Statistics,[128] which provided detailed information on the jobs held, length of time in each job, race, age, and cause of death, where applicable, of 1979 white and 2682 nonwhite coke-oven workers.

The information accumulated in this study is interpreted as indicating that coke-oven workers have a higher mortality due to cancer, particularly cancer of the lung, and that there is a strong relationship between the level of exposure to carbonization fumes and lung cancer mortality. The lowest exposure level of $\leqslant 199$ mg/(m^3)(month), which applies to operators such as benchmen, pusher car operators, and other workers operating at the oven sides or elsewhere, leads to lung cancer death rates similar to those of non-coke-oven workers For oven top men, who are exposed to a much higher level of air contamination, the age-adjusted rates of death from lung cancer range from 3 to 10 times those of comparison groups, depending on the degree of exposure. These results for oven top men refer to nonwhite workers. The white coke-oven workers overall experienced about the same mortality due to all cancers as the non-coke-oven worker controls, but a slightly higher lung cancer rate—11 per thousand versus 7 per thousand. The difference between lung cancer rates among white and nonwhite workers is probably due to the small number of white workers employed on the more hazardous oven top jobs.

This investigation by Mazumdar et al. confirms the earlier study by Redmond et al.,[128] whose data on the incidence of lung cancer in 4661 coke-oven workers indicated that workers employed exclusively for work on the top side had a probability of contracting lung cancer 6.87 times that of the population as a whole. For workers engaged both on top side and oven side with 5 or more years experience, the excess risk was 3.22 times, for special side oven workers (i.e., those with specialized duties, such as larry car drivers and pusher bar operators) with 5 or more years experience, it was 2.1 times, and for all men with less than 5 years experience, there was no significant difference.

A task force of specialists, organized by the Japanese Ministry of Labor and led by H. Sakabe,[129] has recently issued its report on a study of the incidence of lung cancer in coke-oven workers. It generally confirms the epidemiological studies of Mazumdar et al. and of Redmond et al. The basis of this investigation was a survey made in 1973 at the request of the Japanese Ministry of Labor into the incidence of lung cancer among current and retired coke-oven workers at 16 firms that had operated coke ovens over a long period. The task group studied the incidence of cancer among the retired workers of the 11 firms (4 iron and steel works, 4 city gas companies, and 3 coke manufacturing companies), which produced data on retired coke-oven operatives.

The 11 works manufactured blast furnace coke (oven temperature 1300°C), general-purpose coke (oven temperature about 1200°C), or foundry coke (oven temperature ~ 1000°C). A comparison of the incidence of all malignant neoplasia and lung cancer deaths among the 2178 retired workers and the population as a whole showed no significant difference. In the case of retired workers from the four iron and steel companies, the numbers contracting cancer were not significantly different from the general population, but the incidence of lung cancer was 2.37 times the expected number of 3.38. In

[127] S. Mazumdar, C. K. Redmond, W. Sollecito, and M. Sussman, *J. Air Pollut. Control Assoc.*, **25**(4), 382 (1975).

[128] C. K. Redmond, A. Ciocio, J. W. Lloyd, and H. W. Rush, *J. Occup. Med.*, **14**, 621 (1972).

[129] H. Sakabe, K. Tsuchiya, N. Takekura, S. Nomura, S. Koshi, K. Takemoto, M. Matsushita, and Y. Matsuo, *Ind. Health (Kawasaki)*, **13**, 57 (1975).

the case of the city gas companies the expected number of deaths from some form of cancer was 69.77 against an actual number of 51; there were 6 deaths from lung cancer, compared with an expected 7.36. Lung cancer tended to occur most frequently at the age of over 50 years, and always occurred at least 5 years after commencing employment. It frequently occurred 20–24 years after initial exposure.

Thus taking together the findings of these epidemiological studies on the occurrence of lung cancer in coke-oven workers, there appears to be strong evidence that long-term exposure to the high concentrations of coke-oven emissions (associated with work on the top of the ovens) leads to a statistically high incidence of lung cancer, but that the risk run by other coke-oven personnel, who are exposed to the lower concentrations at ground level or on the oven sides, is negligible.

It is unfortunate that Mazumdar et al. identified coke-oven emissions with "coal-tar pitch volatiles," and that results of these studies have been interpreted by some environmental protection authorities as evidence that exposure to the fumes generated when coal-tar pitch or refined coal tar is heated involves the risk of lung cancer.

Coke-oven emissions contain, in addition to pitch fumes, the whole of the tar in vapor or aerosol form, together with hydrogen sulfide, ammonia, the soot formed by vapor phase cracking, coke dust, and fly ash. There are grounds for the view that other components, possibly in combination with the PAHs in the "pitch fumes," are necessary factors in causing lung cancer to coke-oven top men. Laskin et al.[130] reported that rats that inhaled BaP and sulfur dioxide together were found to have changes in the bronchial mucous membrane and tumors of bronchogenic origin, but that no such changes occurred in rats that inhaled the two substances separately.

Although every effort should be made to avoid exposure of workers to the fumes generated in the handling of hot tar and pitch, evidence available from epidemiological studies and from practical experience suggests that inhalation of air heavily contaminated by coal-tar or pitch volatiles, even over prolonged periods, does not lead to the risk of lung cancer or any other form of cancer.

A relatively minor problem, though in both senses of the word an irritating problem, is experienced by workers exposed to pitch dust or to the fumes from hot creosote, refined tar, or pitch—namely, photosensitivity of the skin accompanied by acute erythema, which the operator call "smarts." In the study of workers at the South Wales briquetting works already mentioned,[123] 57.5% of the pitch workers claimed to have suffered from this erythema at some time. A similar condition is frequently experienced at creosoting plants and among men handling freshly creosoted telegraph poles. The condition affects only white-skinned workers and tends to be more frequent among fair people. Cold winds, particularly near the sea, increase the smarting discomfort, as do hot sunshine and sweating. Restriction of the skin area exposed and application of barrier creams to the exposed areas reduce the discomfort. According to sufferers, however, these are palliatives rather than cures, and the usual solution is for workers susceptible to this skin irritation to find other employment.

17.6 THE PHYSICAL PROPERTIES OF HIGH-TEMPERATURE TARS AND TAR FRACTIONS

Since the chemical structure of high-temperature tars varies with the carbonization conditions and the blend of coal carbonized, each sample of coal tar is unique. The properties, both chemical and physical, of an individual tar and of the oils and residual pitch derived from it will therefore differ to some degree from those of other samples of tar and their products. Therefore the study of the physical properties of tars and tar products has been concerned mainly with establishing correlations between readily determined chemical and physical characteristics and other properties, which can be used to predict the values of these other properties with an accuracy sufficient for such purposes as plant design.

[130] S. Laskin, M. Kuschner, and R. T. Drew, in *Inhalation Carcinogenesis,* M. G. Hanna et al., Eds., U.S. Atomic Energy Commission, Division of Technical Information, Washington, D.C., 1970, p. 321.

17.6.1 Density and Change of Density with Temperature

Mapstone[131] established the following equation for the variation of the specific gravity of coal tars with temperatures and their specific gravities at 60°F:

$$\frac{d(\mathrm{SG})}{dt°\mathrm{C}} = 0.001778 - 0.00098 \times (\mathrm{SG\ at\ 60°F}) \qquad (17.40)$$

[131] G. E. Mapstone, *J. Appl. Chem.*, **5,** 582 (1955).

and has used this expression to construct the nomographs shown in Figs. 17.8 and 17.9.

Gurevich et al.[133] studied the densities of a number of high-temperature tars and their distillation products, and confirmed that the density-temperature curves are linear apart from some deviation near the crystallizing temperature. They established correlations

[132] G. E. Mapstone, *S. Afr. Ind. Chem.*, **20,** 78 (1966).

[133] B. S. Gurevich, S. C. Sabirova, and V. M. Rednov, *Coke Chem. (USSR)*, No. 2, 33 (1973).

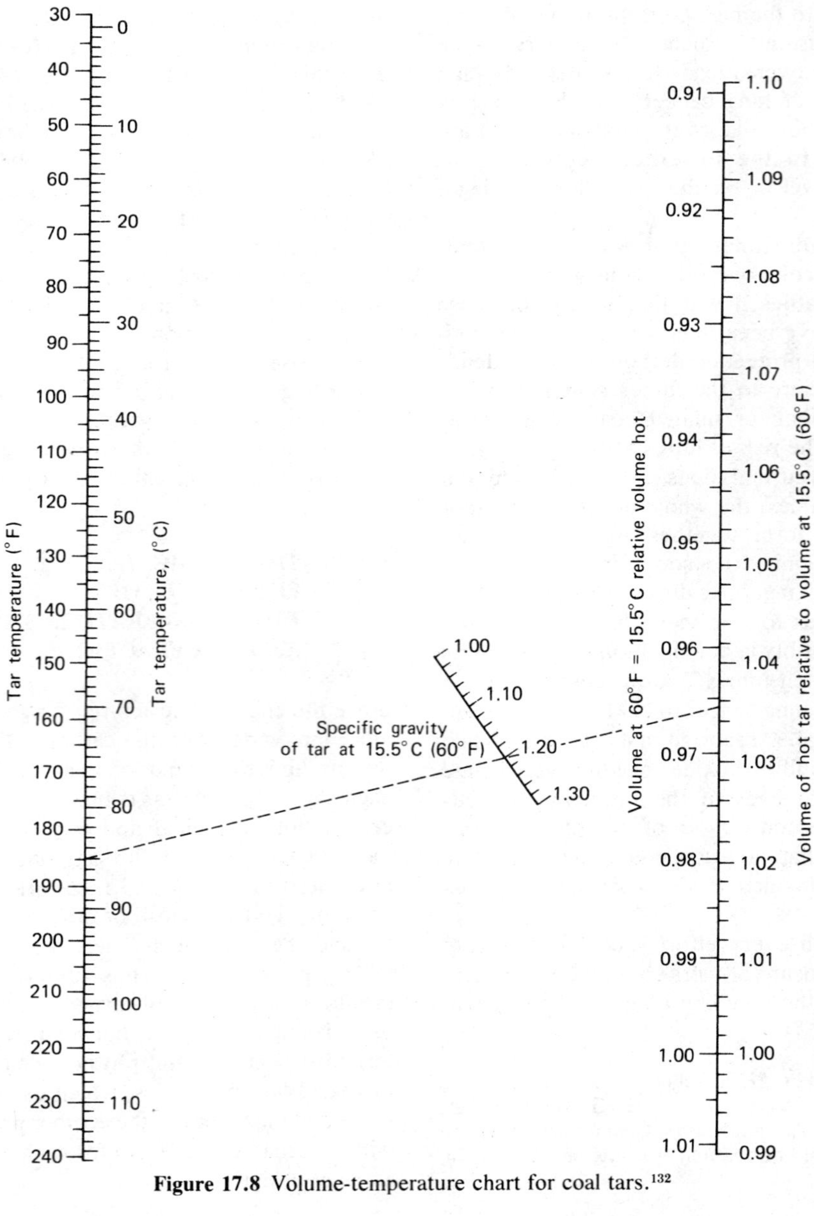

Figure 17.8 Volume-temperature chart for coal tars.[132]

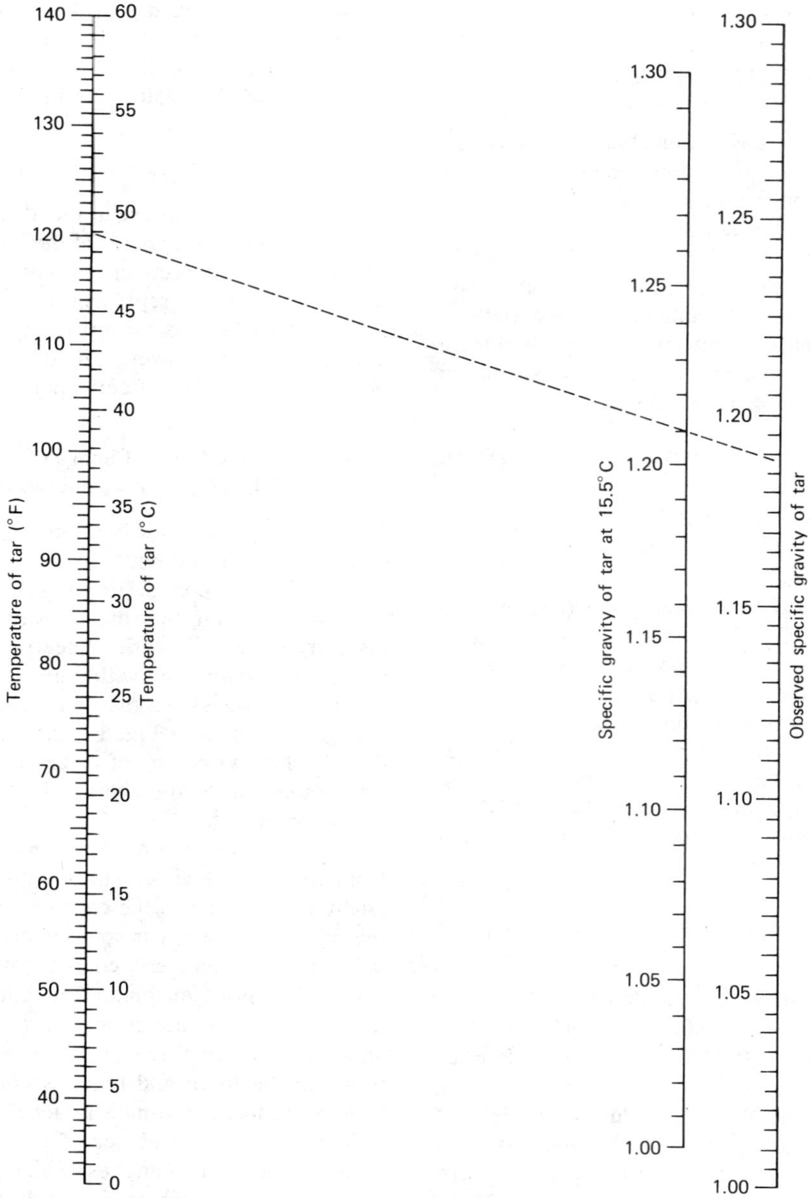

Figure 17.9 Nomograph for the specific gravity–temperature relationship for coal tars.[132]

between the density at 0°C (d_0) and the average molecular weight (M) and the average boiling point (t_b), defined as the mean of the temperatures in degrees centigrade at which 10, 20, . . . , 80, and 90% of the tar or fraction distills in a standard flask distillation. These correlations are:

$$d_0 = 1.877 \times 10^{-3}M + 0.808 \quad (17.41)$$

$$d_0 = 7.337 \times 10^{-4}t_b + 0.890 \quad (17.42)$$

The average relative temperature coefficient of density, β, given by the formula:

$$\beta = \frac{1}{d_0}\frac{d_0 - d_t}{t} \quad (17.43)$$

which has a negative value, is also correlated with M and t_b (°C) by the expressions:

$$\beta = \left(-10.5433 + \frac{0.0122}{t_b}\right)10^{-4} \quad (17.44)$$

and

$$\beta = (-12.6114 + 0.0346M)10^{-4} \quad (17.45)$$

According to Karabon et al.,[134] the average molecular weight and softening temperature of high-temperature pitches may be calculated from the results of a solvent analysis by the Broche and Nedelmann procedure.[135] This separates the pitch into material insoluble in benzene (α-component), material soluble in benzene but insoluble in petroleum ether (β-component), and material soluble in petroleum ether (γ-component):

$$t_s = 1.35\beta + 2.28\alpha \quad (17.46)$$

$$M = (0.049\alpha + 1.88)\beta + 4.9\alpha + 120 \quad (17.47)$$

where t_s is the softening point (Kramer and Sarnow) (°C), M is the average molecular weight, and α and β are the percentages by weight of the α and β fractions.

Briggs,[136] from a study of the densities of a range of British tar pitches and tar oils over the temperature ranges 30–100°C for tars, 20–140°C for tar oils, and 140–230°C for pitches, gives the following values for the constant b in the equation:

$$d_{t_1} = d_{t_2} - b(t_1 - t_2) \quad (17.48)$$

where d_{t_1} and d_{t_2} are the densities at temperatures t_1 and t_2, as 0.00068 ±0.00005 for tars and 0.00056 ±0.00005 for pitches. For tar oils, $b \times 10^5 = 162.7 - 86.3d_{20} \pm 8$, d_{20} being the density at 20°C. The values for b, derived from the data published by Gogoleva and Stepanenko[137] for coal-tar pitches, are 0.00054 for medium-soft pitch and 0.00048 for hard pitch. These authors state that the linear relationship between density and temperature remains valid at high temperatures; therefore the relationship between the densities of pitches of different softening points over the range 250–350°C can be represented as follows:

$$d_n^t = d_m^t + (t_n - t_m)0.0001 \quad (17.49)$$

where d_n^t and d_m^t represent the densities in grams per cubic centimeter of the harder and softer pitches, respectively (at any temperature, t) and t_n and t_m represent their softening points (°C). These constant density-temperature coefficients, however, hold only at temperatures above the softening point.

17.6.2 Viscosity and Change of Viscosity with Temperature

The viscosity of tar oil fractions is approximately constant and equal to 0.25 P at their average boiling points. From this information, and knowing that the logarithm of the viscosity of tar oils varies linearly with absolute temperature and with the average boiling point, Briggs[138] constructed the simple nomogram shown in Fig. 17.10, from which the absolute viscosity of a tar oil at any temperature up to the average boiling point can be predicted.

Viscosity is a more important property, from a practical viewpoint, in the case of coal-tar pitch than in the case of tar oils. At the relatively low shear stresses characteristic of most viscometers, coal-tar pitches display essentially Newtonian flow characteristics; that is, their rate of flow or deformation under an applied force or stress is proportional to the force and remains constant as long as the force continues to act. Viscoelastic behavior, with evidence of a yield point, is encountered when pitches of high viscosity are subjected to high loads, as discussed in Chapter 14 of the first Supplementary Volume.[139] But for most practical purposes, high-

[134] B. Karabon, I. Wiecek, and W. Solosski. *Pr. Nauk Inst. Technol. Nefty Wegla Politech. Wroclaw,* **11,** 197 (1972).

[135] H. Broche and H. Nedelmann, *Glueckauf,* **70,** 979 (1934).

[136] D. K. H. Briggs, Coal Tar Research Association, Rep. No. 0126, 1955.

[137] P. Ya, Gogoleva and M. A. Stepanenko, *Coke Chem. (USSR),* No. 4, 42 (1959).

[138] D. K. W. Briggs, Coal Tar Research Association, Rep. No. 0308, 1963.

[139] H. H. Lowry, Ed., *Chemistry of Coal Utilization,* Supplementary Volume, John Wiley and Sons, New York, 1963.

[140] *Standard Methods for Testing Tar and Its Products,* 6th ed., Serial RT 1/67, 1967, p. 154.

[141] *Standard Methods for Testing Tar and Its Products,* 6th ed., Serial No. Rt/3, 1967 p. 158.

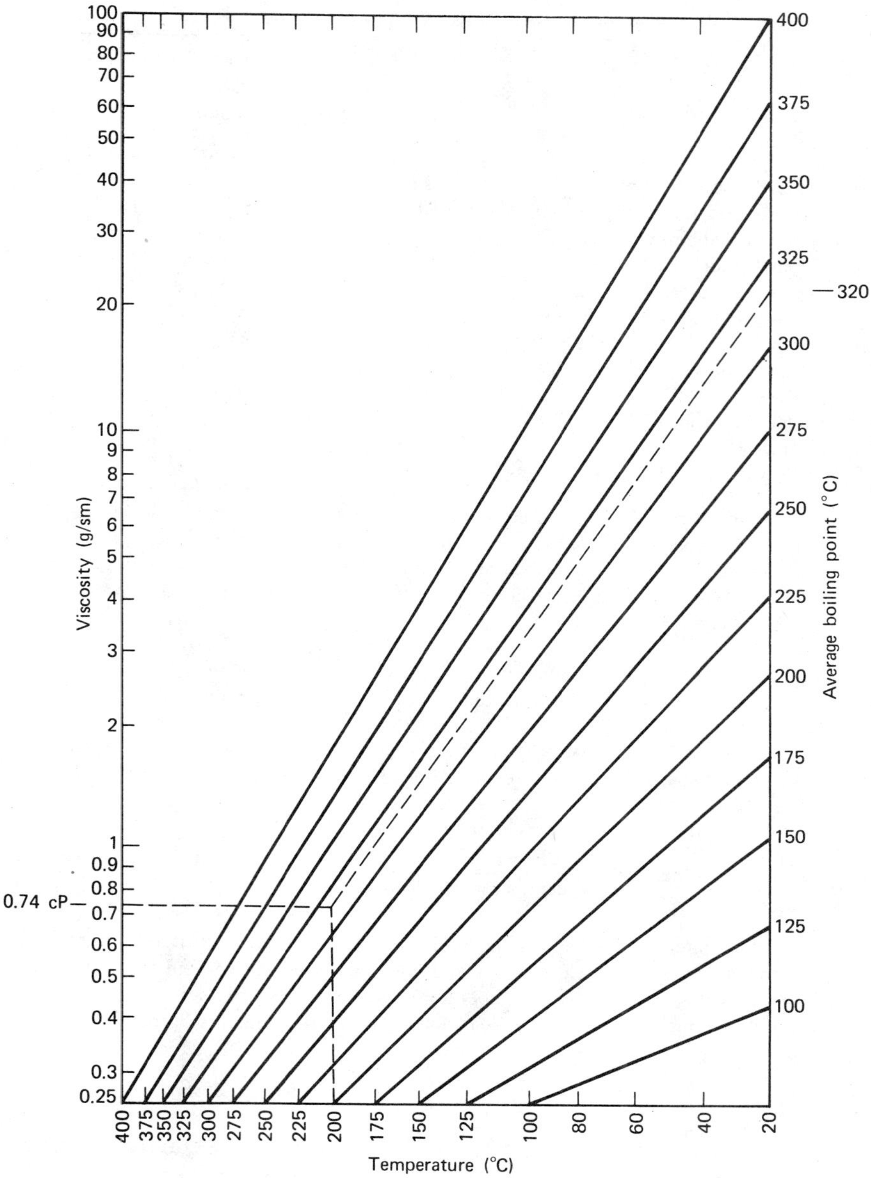

Figure 17.10 Nomograph for calculating the viscosity of tar oils from their average boiling points.[138]

temperature pitch may be regarded as Newtonian. This allows correlations between the viscosities as determined on different designs of viscometer and between various fixed points. For commercial purposes, pitches and fluxed pitches are characterized by their soft-

[142] *Standard Methods for Testing Tar and Its Products,* 6th ed., Serial No. Pt 3/3, 1967, p. 510.

[143] ASTM Standards, Part 15, 1961, Designation D5-61, Penetration of Bituminous Materials.

[144] ASTM Standards, Part 15, 1976. Designation D113-44, Test for Ductibility of Bituminous Materials.

[145] A. Fraas, *Asphalt Teer Stassenbautechnik,* **13,** 367 (1930).

[146] ASTM Standards, Part 15, 1977, Designation D2319-76, p. 604.

[147] ASTM Standards, Part 15, 1977, Designation D61-75, p. 114.

[148] R. N. T. Saal and G. Koens, *J. Inst. Pet. Tech.,* **19,** 1976 (1933).

Table 17.18 Viscosities of Fixed Points for Coal-Tar Pitches

Fixed Point	Corresponding Viscosity (Ns/m^2)	Difference (°C) Between Fixed Point and EVT	Ref.
Equiviscous temperature	25	—	141
R and B softening point (°C)[a]	800	−19	142
K and S softening point (°C)	5,500	−27	25
Penetration of 20 mm (i.e., penetration of 200) using 100-g weight at 25°C for 5 s[c]	2×10^4	−30	143
Penetration of 1 mm (i.e., penetration of 10) using 100-g weight at 25°C for 5 s	10^7	−51	
Ductility point (°C)[d]	10^7	−51	144
Fraas brittle point (°C)	4×10^7	−65	145
Transition to glassy state	10^{12}	−90	

[a] Two other softening point methods are employed, particularly in the United States. These are the cube-in-air softening point and the cube-in-water softening point.[146,147] According to information from Koppers Company, Inc., the relationship of these methods to the R and B softening point is approximately as follows:

cube-in-air softening point = R and B softening point + 4°C
cube-in-water softening point = R and B softening point + 10°C

[b] It should be emphasized that these relationships are only approximate, since as discussed later, the temperature coefficient of viscosity varies with the viscosity level.

[c] The relationship between viscosity and penetration[148] is:

$$\eta\ (Ns/m^2) = \frac{1.58 \times 10^9}{(\text{penetration})^{2.16}}$$

[d] When pitch is tested for ductility, the sample either suffers brittle fracture without elongation or elongates to the maximum distance without breaking. When tested at increasing temperatures, there is a particular point—the ductility point—at which the behavior changes from the first type to the second type.

ening points, their penetrations, or by other measurements involving viscosity. Table 17.18 lists such fixed points (i.e., measurements at which the pitch or fluxed pitch has a certain viscosity value). The approximate levels of these viscosities and the differences in temperature (°C) between the fixed points and the "equiviscous temperature" (EVT) are given. The EVT is the temperature (°C) at which the sample has a viscosity of 50s measured on the Standard Tar Viscometer[140] using a 10-mm cup. It is a parameter widely used to characterize refined tars and fluxed pitches in Great Britain and European countries.

Coal-tar pitches are characterized by a high viscosity-temperature coefficient compared with other thermoplastic materials. There is no simple relationship between the viscosity of a straight run or fluxed pitch and the temperature of measurement. Many arbitrary functions have been developed to provide an index of temperature susceptibility. The Evans and Pickard relationship[149]:

$$N = \frac{\log \eta_1 - \log \eta_2}{\log t_1 - \log t_2} \tag{17.49a}$$

[149] E. V. Evans and H. Pickard, *An Investigation into the Nature and Properties of Coal Tar*, South Metropolitan Gas Co. (London), Chemical Dept. Rep. 1931.

where N is the viscosity-temperature coefficient, η_1 and η_2 are the viscosities (poises) at t_1 and t_2, and temperature (°F) is linear only for products of 30–80°C EVT at temperatures below the EVT. It has proved useful in comparing road tars and soft pitches with bitumens and other materials in the same EVT range. The value of N for high-temperature refined tars and pitches varies from about 6.5 at 30°C EVT to 15 at 80°C EVT, whereas residual petroleum bitumens have values of 5–8 over this range.

This may seem, at first sight, to be inconsistent with the failure of the temperature interval between two fixed points (e.g., the EVT and the softening point) to vary with the softening point. If the viscosity, or more conveniently the logarithm of the viscosity, of a series of pitches of different softening points is plotted against temperature, a family of parallel curves is obtained, which can be made to coincide if displaced horizontally. This means that if pitches of different softening points are compared over the same viscosity range, their temperature susceptibilities are identical, but the curvature of the plots is such that as the temperature rises and the viscosity decreases, the relative change of viscosity corresponding to any temperature interval is less. By using a reduced temperature scale—that is, by plotting log viscosity (in poises) against the difference in the temperature at which the viscosity is measured and the R and B softening point—a single curve can be drawn (Fig. 17.11). From this, the viscosity of any pitch or fluxed pitch can be calculated from its R and B softening point.[150] The equation of the best curve through the points is:

$$\log \eta_t = -4.175 + \frac{711.8}{86.1 - t_s + t} \tag{17.50}$$

where η_t is the viscosity (poises) at a temperature of t °C and t_s is the R and B softening point (°C).

This relationship covers a temperature range of 180°C ($t - t_s$ from −40 to +140°C) and a viscosity range of 10^8. It cannot be used to predict accurately viscosity, since the softening point that represents the datum point is itself only an approximate viscosity measurement, and also because there are small differences in the temperature susceptibilities of different types of pitch. It does, however, provide a useful means for calculating viscosities of pitch at any temperature above the softening point, with an accuracy sufficient for most design purposes. The true

[150] E. H. Binns, Coal Tar Research Association, Rep. No. 0167, 1956.

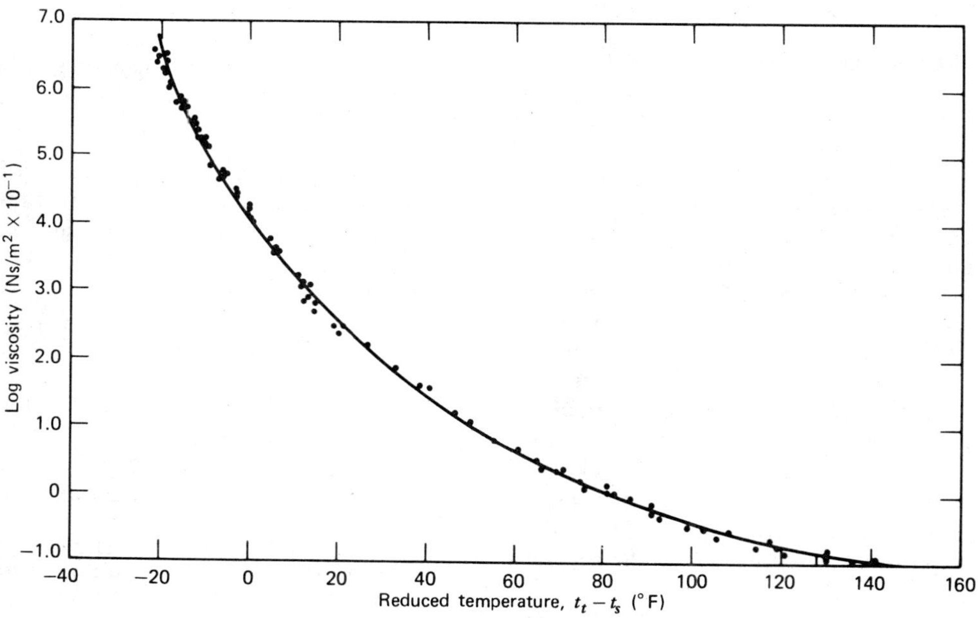

Figure 17.11 The general viscosity temperature curve for coal-tar pitches.[150]

value for any pitch at a given temperature will lie within the range +80 to −45% of the predicted value. This possible error is seen in its true perspective when it is realized that it corresponds to an error of 2.5°C in determining the softening point.

Another general equation for the temperature-viscosity function of coal-tar pitch has been proposed by Franck and Wegener[151] and is similar in form to that put forward by Walther for mineral lubricating oils. This takes the form:

$$\log \log (\nu + 0.8) = K + K_1 \log t \qquad (17.51)$$

where ν = kinematic viscosity (cS*t*)
t = temperature in Kelvins
K and K_1 = constants

The value of K depends on the softening point and nature of the pitch, whereas K_1, which can be regarded as an index of temperature susceptibility, can be calculated, according to Franck and Wegener, from the toluene-insoluble content (TI) of the pitch by the expression

$$K_1 = 5.8 - 0.03 \times \text{wt \% TI content} \qquad (17.52)$$

If the Kramer and Sarnow (K and S) softening point is taken as one fixed point, the viscosity (cS*t*) of the pitch at any other temperature, *t* °C, is given by the expression:

$$\log \log (\nu + 0.8) = 0.80 + 0.04 \log \text{TI} + (5.8 - 0.031\ \text{TI})\{\log(t_s + 273) - \log_t\} \qquad (17.53)$$

where t_s is the K and S softening point (°C).

This equation applies to a wide range of straight-run and heat-treated pitches, but not to fluxed pitches, since the value of K_1 depends not only on the TI content, but also on the nature of the fluxing oil. It is, however, limited in accuracy. Double-logarithm functions result in a broadening at the lower end of the scale and a compression at the upper end, thus are subject to error when extrapolated to higher values. Thus the Franck-Wegener equation is not satisfactory for calculating viscosities at temperatures below the EVT. Also, the viscosity unit in the general Franck–Wegener expression is the centistoke, (cS$_t$), and although it has been found in practice that the centipoise can be equally well used, with a slight adjustment in the value of K_1 (=5.7 − 0.03 TI), errors may be introduced when other viscosity units are employed. Schmidt[152] has published a viscosity correlation chart, relating EVT of road tars to efflux times in seconds on the standard tar viscometer (STV) at various temperatures. This is based on the following equation given by Franck and Wegener:

$$\log (\text{EVT} + 273) = \frac{\log \cdot \log \eta + 35.502}{14.4} \qquad (17.54)$$

In this case η is the viscosity in seconds at 30°C on the STV. Comparing the results obtained by using the Schmidt chart, by calculating from the general Franck–Wegener expression, and from the conversion chart in British Standard Specification No. 76, the following discrepancies occur:

	STV Viscosity (s), 10-mm cup at 30°C		
EVT of Tar (°C)	Franck–Wegener General Expression	Schmidt Chart	BS No. 76
35	124	140	120
40	322	550	315
45	903	2,600	850
50	2730	19,000	2500
55	7950	210,000	8000

Cornellison and Waterman,[153] as a result of examining materials, including petroleum fractions and bitumens, proposed an equation of the form:

$$\log \nu = \frac{A}{T^x} + B \qquad (17.55)$$

where ν is the kinematic viscosity, T is the absolute temperature, and x is a constant that depends for its value on the type of material. Wood and Downer[154] derived an equation of

[151] H. G. Franck and O. Wegener, *Brennstoff-Chemie,* **39,** 195 (1958).

[152] A. Schmidt, *Bautenschutz,* **8,** 48 (1959).

[153] J. Cornellison and H. C. Waterman, *J. Inst. Pet.,* **42,** 62 (1956).

[154] L. J. Wood and M. Downer, *J. Appl. Chem.,* **15,** 431 (1965).

similar form from the data relating to a range of road tars and coal-tar pitches varying in softening point from 54 to 104°C:

$$\log \eta = AT^{-5} + B \qquad (17.56)$$

where η is the absolute viscosity (cP). The value of A in this equation is closely related to the softening point and can be represented by the equation:

$$A = (0.358\, t_s + 3.74) \times 10^{12} \qquad (17.57)$$

when t_s is the R and B softening point (°C) using glycerol as the heating medium. Corresponding equations based on the R and B softening point using water as the heating medium, and on the EVT, are as follows:

$$A = (0.358\, t_s + 4.82) \times 10^{12} \qquad (17.58)$$

$$A = (0.358\, \text{EVT} - 2.00) \times 10^{12} \qquad (17.59)$$

Cornellison et al.[155] found a correlation between the two constants A and B for certain sets of similar substances, but no such relationship was evident for tar products. It was observed, however, that a relationship existed between the value of B and the constant K_2 in a modified Walther equation of the type:

$$\log \cdot \log \eta = K_1 - K_2 \log T \qquad (17.60)$$

where η is the absolute viscosity (cP), T is the absolute temperature, and K_2 is the slope constant or "temperature susceptibility constant" of the material. It thus appeared that the value of B in the Cornellison–Waterman equation for pitches was related to the TI content of the pitch in the same way as the value of K_1, in the Franck–Wegener equation. For straight-run pitches, the best expression, calculated by the method of least squares, was

$$B = 0.0206 \times \text{wt}\,\%\,\text{TI} - 0.378 \qquad (17.61)$$

There were not sufficient data to derive a similar relationship for fluxed pitches.

Using the Cornellison–Waterman equation, as applied to pitches by Wood and Downer, the viscosity of a pitch at any practical temperature can be calculated from two measurements in the range 10^4–10^6 cP, taking a value of 5.0 for x, with a precision similar to that which might be expected by direct measurement. By adopting the suggestion of Wayaku and Kikukawa,[156] and employing a reduced temperature measurement of (T – EVT) in kelvins, predictions of somewhat lesser accuracy can be made from one viscosity measurement. On the other hand, by employing the relationship between A and the softening point and B and the TI content, the viscosity of a pitch at any temperature can be calculated from a knowledge of these two parameters, without direct measurement and with an accuracy probably adequate for most engineering purposes.

17.6.3 Surface Tension

The surface tension of coal-tar distillate oils can be calculated from their densities and average boiling points by the equation[157]:

$$S_{20} = 93.8\, d_{20} - 0.0496\, T_b - 33.9 \qquad (17.62)$$

where S_{20} = surface tension (dynes/cm)
d^{20} = density (g/cm³ at 20°C)
T_b = average boiling point (°C)

Macleod[158] showed that the surface tension for a pure liquid was related to the fourth power of the difference between the liquid and vapor densities. In the case of tar fractions the vapor density is sufficiently small in comparison with the liquid density to be neglected, so that the expression

$$S = Cd^4 \qquad (17.63)$$

holds at any temperature up to the critical range; thus

$$S_t = \frac{S_{20} d_t^4}{d_{20}^4} \qquad (17.64)$$

The relationship between S_{20}, T_b, and d_{20} has

[155] J. Cornellison, H. C. Waterman, and J. A. Waterman, *Materialpruefung*, **2**, 265 (1960).

[156] M. Wayaku and H. Kikukawa, *Aromatics (Japan)*, **22**(1), 12 (1970).

[157] D. K. H. Briggs and K. E. Speak, *J. Appl. Chem.*, **16**, 137 (1966).

[158] D. B. Macleod, *Trans. Faraday Soc.*, **19**, 38 (1923).

been expressed in nomograph form (Fig. 17.12) by Mapstone.[159]

Volkmann et al.[160] concluded from their work on filtered and unfiltered tars that the surface tension of tars and pitches decreases linearly with temperature and that the value depends on the viscosity of the sample and the amount of suspended matter it contains. For a 60°C (R and B) softening point pitch, they gave a value of 43 dynes/cm at 25°C and 38 dynes/cm at 100°C.

Briggs[161] studied a wide range of coal-tar pitches using Adams's modification of Sugden's maximum bubble pressure method,[162] which is free from the limitations of the single tube apparatus employed earlier by Nellen-

[159] G. E. Mapstone, *Austr. Chem. Proc. Eng.*, **24**(10), 37 (1971).

[160] E. W. Volkmann, E. O. Rhodes, and L. T. Work, *Ind. Eng. Chem.*, **28**(6), 72 (1936).

[161] D. K. H. Briggs, *Fuel*, **43**, 439 (1964).

[162] N. K. Adams, *The Physics and Chemistry of Surfaces*, 3rd ed., Oxford University Press, London (1949), p. 375.

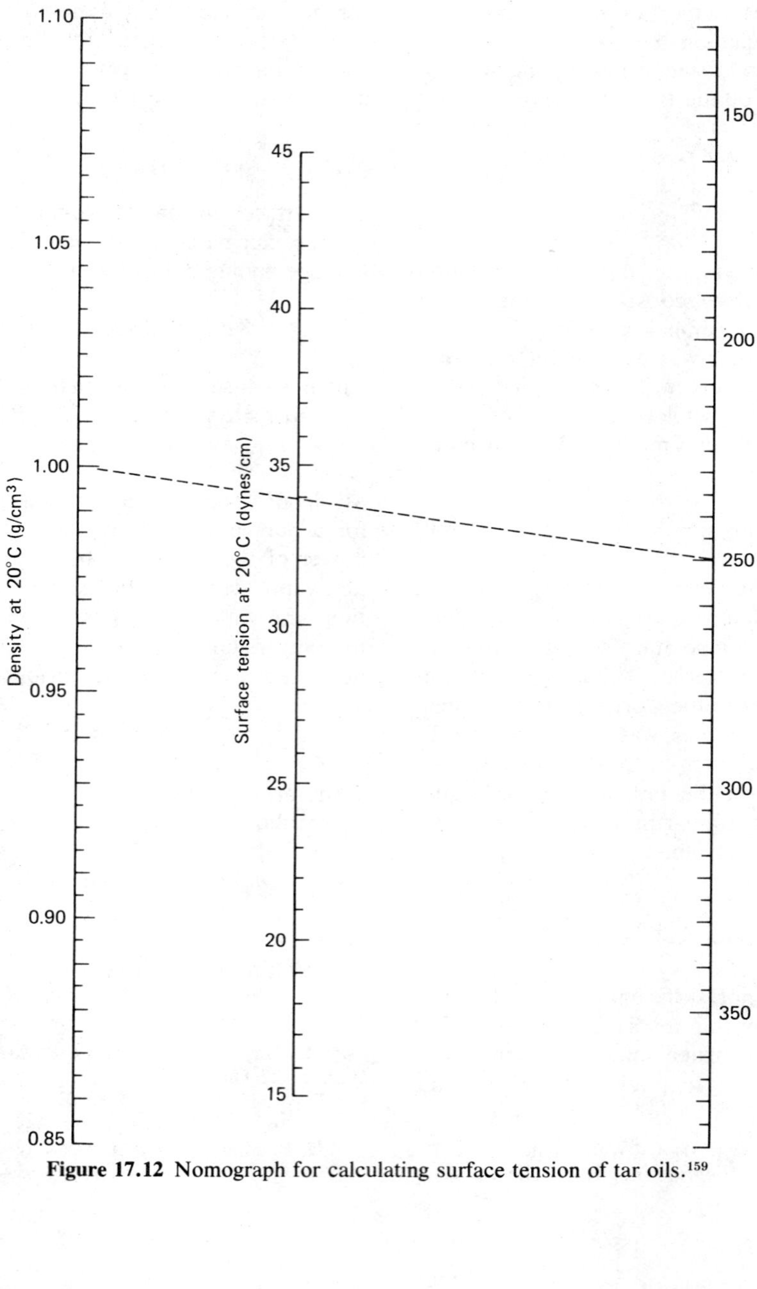

Figure 17.12 Nomograph for calculating surface tension of tar oils.[159]

steyn and Roodenburg.[163] He found that his results could be expressed by the simple equation:

$$S = 18.4 . d_t^4 (\text{dynes/cm}) \qquad (17.65)$$

The density at temperature t °C, d_t, can be calculated from d_{20} by the formula already given, $d_t = d_{20} - 0.00056(t - 20)$.

Higher values than quoted by Volkmann et al. or calculated by the Briggs formula are reported by Gogoleva and Stepanenko,[137] but the application of their differential weighing method to liquids of high viscosity is suspect.

17.6.4 Specific Heats and Heat Contents

It was found that the values obtained for the specific heats of 137 samples of coal-tar distillate oils could be correlated with the density at 20°C, the average boiling point (°C), and the tar acid content (the percentage by volume extracted by 10% aqueous caustic soda) by the expression[164]:

$$C_p = \frac{1}{d_{20}}(0.1758 + 0.2138 d_{20} + 8.61 \times 10^{-4} t) + (2.16 \times 10^{-3} - 5.28 \times 10^{-6} t_b)x \qquad (17.66)$$

where C_p is the liquid specific heat (cal/g) at temperature t °C, and x is the volume percent of tar acids.

A number of correlation equations have been proposed for the specific heats of tars and pitches. Hyman and Kay[165] give an expression relating the specific heats of tars and pitches to the specific gravities and the weight fractions of benzene-insoluble material:

$$C_p = \frac{1}{\text{SG}}(0.302 + 6.56 \times 10^{-4} t) - \text{BI}(0.321 + 6.94 t \times 10^{-4}) \qquad (17.67)$$

where SG is the specific gravity at 15.5/15.5°C, BI is the weight fraction of benzene insolubles, and C_p is the specific heat (cal/g) at temperature t °C.

Edler[166] prefers the following simpler formulas

$$C_p = \frac{1}{\text{SG}}(0.339 + 1.24 t \times 10^{-3}) \qquad (17.68)$$

for tars, tar oils, and pitches with softening points below 100°C, and

$$C_p = \frac{1}{\text{SG}}(0.381 + 8.1 \times 10^{-4} t) \qquad (17.69)$$

for hard pitches.

It is claimed that the correlation put forward by Briggs and Drake[167] is superior to others and predicts with an accuracy of ±5% the specific heats of tars and pitches, with densities ranging from 1.05 to 1.35 g/cm³, over the temperature range 30–210°C. This equation takes the form:

$$C_p = \frac{0.873}{d_{20}} - 0.413 + 9.3 t \times 10^{-4} \qquad (17.70)$$

where d_{20} is the density at 20°C.

17.6.5 Latent Heat of Vaporization

The expression given by Briggs and Popper[168] for the specific latent heat of tar oils at their atmospheric pressure boiling point has not been revised. It is:

$$L_s(\text{cal/g}) = d_{20}(116.1 - 0.143 T_b) \qquad (17.71)$$

where d_{20} is the density at 20°C and T_b is the average boiling point in kelvins.

17.6.6 Thermal Conductivity

The thermal conductivity (k) of commercial pitches is $(1.423 \pm 0.084) \times 10^{-5}$ W/m-K [$(34 \pm 2) \times 10^{-5}$ cal/cm-s-°C] over the temperature range 25–105°C, and it does not vary significantly with temperature. The conductivity of crude tars is slightly lower, at (1.34

[163] F. J. Nellensteyn and N. M. Roodenburg, *Kolloid Chem. Beih.*, **31**, 434 (1930).

[164] D. K. W. Briggs, Coal Tar Research Association, Rep. No. 0132, 1955.

[165] D. Hyman and W. B. Kay, *Ind. Eng. Chem.*, **41**, 1764 (1949).

[166] E. Edler, *Erdoel Kohle*, **6**, 543 (1953).

[167] D. K. H. Briggs and W. D. Drake, *Chem. Ind.*, 666, (1957).

[168] D. K. H. Briggs and F. Popper, *Trans. Inst. Chem. Engr. (London)*, **35**, 369 (1957).

$\pm 0.083) \times 10^{-5}$ W/m-K, whereas that for tar oils is similar regardless of the density or the source of tar, but falls slightly with increasing temperature, conforming to the expression:

$$k = (1.34 - 0.00084t \pm 0.084) \times 10^{-5} \text{W/m-K} \qquad (17.72)$$

where t is the temperature (°C).[169]

17.6.7 The Structure of Coal-Tar Pitch and Viscoelastic Properties

Although coal-tar pitch behaves as a supercooled liquid, it has long been accepted that the molecules of its individual components are associated or oriented in some way. The nature of this association—whether the molecules are in aggregates or dynamic swarms, and whether the associating forces are inductive forces between polar and polarizable molecules or steric entanglements—is still a matter for controversy.

Volkmann[91] used values for the activation energy of flow earlier derived by Raich[170] to calculate the magnitude of the associative forces in tars and pitches on the basis of the postulate of Glasstone et al.[171] that the energy of flow activation in an unassociated liquid should be between one-fourth and one-third of the energy of vaporization. They obtained a value of 28 kcal/g-mole at 20°C, reducing to 6 kcal/g-mole at 100°C for a refined tar. For a 50°C (R and B) softening point pitch, a value of 45 kcal/g-mole was derived at 25°C.

Shiraishi and Kobayashi[172] have investigated the diffuse X-ray scattering of pitch by using the Fourier transformation for the 002 band and Diamond's least-squares method for the 110 band. They calculated that the molecules are oriented in layers stacked parallel to each other, partly in groups of eight to nine layers, the average number of layers being 2.46. The distribution of layer size showed a peak at a diameter of 8.4–10 Å. The average layer size was 8.9 Å, corresponding to about 29 molecules per layer.

Shuvaev et al.[173] have studied the viscoelastic properties of solid high-temperature pitch cooled to below its glass transition point of approximately 10^{13} poises.

The natural vibrational frequency and compressive strength of coke-oven pitch specimens, in the form of rods of 25-mm diameter and 100-mm length, were determined. Young's modulus, E, was calculated by the formula:

$$E = \frac{16 L v^2 P}{D^2 g} \qquad (17.73)$$

where L is the length, D, the diameter, P, the weight of the specimen, and v, the natural vibrational frequency of the sample, all in consistent units. Values for v of a medium-soft pitch of 71.5°C (R and B) softening point (s.p.) were 5573 Hz; for four harder pitches made by air-blowing (s.p. 86°C), steaming (s.p. 91°C), and heat treatment (s.p. 88°C) the values were all between 5937 and 5967 Hz. A straight-run hard pitch gave a value of 5858 Hz. The values of E derived from these vibrational frequencies (in $N/m^2 \times 10^{-7}$) were 341 for the medium-soft pitch, 396 for the straight-run hard pitch, and 415 for the treated pitches. The compressive strengths (in N/m^2) were 118×10^{-3} for the medium-soft pitch, 46.8×10^{-3} for the straight-run hard pitch, 33.0×10^{-3} for the air-blown sample, 33.9×10^{-3} for the steamed sample, and 0.1 for the thermally treated pitch.

From these data, combined with the other parameters—Poisson's ratio (determined by plotting strain vs. compressive strength and found to be 0.33 for all the pitches tested) and the coefficient of volumetric expansion—Shuvaev et al. attempted to calculate values for d, the average diameter of the molecular associations, and W, the bonding energy between the molecular associations in the straight-run medium-soft pitch. They also derived values for the bulk elastic constant, K, the internal pressure, P_i, and the free

[169] D. K. H. Briggs and F. Popper, *Fuel*, **33**, 222 (1954).

[170] H. Raich, *Special Report on the Colloidal Structure of Bituminous Residuals*, Mellon Institute of Industrial Research, Pittsburgh (1952).

[171] S. Glasstone, K. J. Laidler, and H. Eyring, *Theory of Rate Processes*, McGraw-Hill Book Company, New York, 1941, p. 477.

[172] M. Shiraishi and K. Kobayashi, *Bull. Chem. Soc. Jap.*, **46**(8) 2575 (1973).

[173] E. A. Shuvaev, G. V. Dmitrieva, M. V. Vigutov, E. F. Chalykh, and S. V. Shulepov, *Coke Chem. (USSR)*, No. 9, 26 (1970).

energy of this material at room temperature, δ. The values they gave are as follows:

$$d = 22.3\ \text{Å}$$
$$K = 1.22\ \text{N/m}$$
$$W = 41.7\ \text{j/kmole}$$
$$P_i = 5.37 \times 10^{-4}\ \text{N/m}^2$$
$$\delta = 1.75\ \text{J/m}^2$$

The credibility of these values depends, however, on the doubtful validity of their simplifying assumptions, namely (1) that solid pitch is a disperse system with an isotropic structure, the disperse phase being the alpha fraction, and (2) that at low natural vibrational frequencies, where the length of the standing waves greatly exceeds the distances between adjacent associations, the disperse system can be replaced by the simplest continuous structure consisting entirely of spherical associations of molecules, each of identical mass and identical cross section, with identical bulk elastic moduli. Since the values they derived for K, δ, and W were quite small in absolute magnitude, they concluded that the molecular associations were only held together by van der Waals forces.

17.7 THE PROCESSING OF HIGH-TEMPERATURE COAL TAR

Since 1960 mergers and takeovers have considerably reduced the number of firms engaged in tar distillation. This has been partly the result of the reduction in the amount of crude tar available with the phasing out of coal carbonization at gasworks, which left some distillers without feedstock, and partly due to the realization that more efficient operation and, particularly, more careful planning of production to meet variations in the demand for tar products could be achieved with fewer, larger firms. In West Germany, for example, the merger in 1964 of the two major tar distillers, Gesellschaft für Teerverwertung and Rütgerswerke, effectively concentrated all tar processing into one organization. In Great Britain there were 33 tar distillers in 1960 if each Area Gas Board was regarded as a separate entity. Now three organizations, British Steel Corporation (Chemicals), Thos. Ness Ltd. (a subsidiary of the National Coal Board), and Croda Hydrocarbons Ltd., are responsible for processing coke-oven tar. Koppers Company Inc., United States Steel Corporation, and Allied Chemical Corporation treat more than 85% of the tar that is distilled in the United States.

This centralization of the industry has accelerated the trend toward the processing of tar in large continuous plants at strategically placed locations, and almost all the crude tar produced worldwide is now treated in such plants.

17.7.1 Primary Distillation

Although there are several designs of continuous tar stills in operation, they are all based on the same simple principles. The crude tar, after settling to remove as much entrained moisture as possible, is strained, doped with a small amount of aqueous alkali to neutralize ammonium salts and thus prevent corrosion, heated under a moderate back pressure to 150–180°C, and flashed into a dehydrator, where residual moisture and light oils are removed. The dehydrated tar is then heated to the temperature required to vaporize all the remaining distillate oils under the prevailing pressure and subjected to a second flash to separate the oil vapors from the residual pitch; the oil vapors are separated by fractional condensation into a series of fractions of increasing boiling range.

Different designs of continuous tar stills vary in such details as the extent to which the crude tar and dehydrated tar are heated by heat exchange, and whether the light oil from the dehydrator is recycled to the crude tar or to the dehydrator, injected directly into the fractionating column, or used as reflux in this unit. Other differences are in the design of the heating furnace, whether a single side stream column or a series of columns is used, and the pressure at which the fractionation is carried out. Tar stills can be placed in three general categories: single-pass atmospheric pressure plants, single-pass atmospheric and vacuum plants, and recycle atmospheric pressure plants. The following is a brief description of the salient features of the more common designs in current operation.

The Ab-der-Halden design, which is employed at French distilleries, is probably the simplest single-pass atmospheric pressure process.[174,175] After straining and alkali addition, the crude tar is heat exchanged with the distillation side streams and the hot pitch, passed through an economizer in the convection section of the furnace, and passed to the dehydrator, the temperature of which is maintained by circulating hot dehydrated tar through a coil in its base. The water and oil vapors go directly to the side stream fractionating column, while the dehydrated tar is pumped, under a pressure of 400–500 kN/m^2, through the main tube bank in the radiant section of the furnace. The tar is injected into the pitch flash chamber at a temperature of 350–360°C, together with superheated steam. The steam and oil vapors join the dehydrator vapors in the base of the fractionating column, from which are taken an overhead stream of light oil and water, the various oil fractions as side streams, and a heavy oil base product. The overhead stream is separated in a decanter and part of the light oil used as reflux to the fractionating column.

The Carl Still Company offers a similar design,[176] which differs only in the degree of heat exchange. Several descriptions have been published of the continuous tar stills used at Soviet tar plants.[177–179] These appear to be of the single-pass atmospheric pressure type, but they are noteworthy for employing a fractionating column with a bubble tray fractionating section and a baffled enrichment section, instead of the simple baffled cylindrical pitch flash chamber used in other designs. The heated dehydrated tar is injected into this atmospheric pressure column, superheated steam is added at the base, and residue pitch is taken off as the bottom product and anthracene oil as a side stream. The uncondensed vapors pass to the second bubble tray fractionating column, from which light oil and water are removed as overhead fractions, carbolic oil and naphthalene oil are taken as side streams, and the wash oil base product is partly recycled to the first column as reflux. The Soviet plants, which are designed by Giprokoks, the Kharkov-based state organization for design in the coal chemical industry, are also equipped with reboilers for both distillation columns.

The original Heinrich Koppers design,[180] which is still used at two plants in Great Britain, is also a single-pass atmospheric pressure design, but differs from other plants in employing separate bubble tray columns for each fraction. After filtering, doping, and dehydration, as for other designs, the tar is pumped through the main radiantly heated tube bank in the furnace, and is injected, at 360°C, into the first column, where pitch is separated from the oil vapors and steam is added to assist vaporization. These vapors then pass through four columns operated at atmospheric pressure and progressively lower temperature to give the various oil fractions as base products. The temperature at the top of each column is automatically controlled by passing back part of the bottoms from each column as reflux to the preceding column. The overheads from the final column consist of the light oil and steam; this is condensed, the two layers separated in a decanter, and part of the light oil product recirculated as reflux.

The simplest plant employing vacuum fractionation is that designed by Canadian Badger for Dominion Tar and Chemical Company of Hamilton, Ontario. In this plant the tar is dehydrated in the usual manner, and the dry tar is heated under pressure in the radiant section of the oil-fired tube heater and injected directly into a vacuum fractionating column, from which a light oil overhead stream, various side stream oil fractions, and pitch, as a base product, are taken.[181]

The modern Teerverwertung-Koppers design,[182] based on experience at the Rütgerswerke Meiderich refinery, is a single-pass, atmospheric and vacuum unit, which is illustrated in the flowsheet in Fig. 17.13. The crude tar is pumped through five heat exchangers in which it abstracts heat from all the oil products and the pitch, and is pumped into the dehydration tower. The light oil and

[174] *Mines*, **12**(3), 223 (1957).

[175] P. Le Roux, *Goudron Routes*, 12 (June 1958.).

[176] P. van Loyen and O. M. Stuhrmann, *Erdoel Kohle*, **13**, 758 (1960).

[177] P. Ya. Kuleshov et al., *Coke Chem. (USSR)*, No. 7, 38 (1960).

[178] A. M. Kaznacheev, *Coke Chem. (USSR)*, No. 3, 44 (1969).

[179] L. S. Elkind, *Coke Chem. (USSR)*, No. 6, 36 (1962).

[180] Anon., *Gas World*, 3, (April 1959).

[181] Anon., *Chem. Eng. Prog.*, **55**(6), 84 (1959).

[182] H. G. Franck and G. Collin, *Steinkohlenteer, Chemie, Technologie und Verwendung*, Springer Verlag, Berlin (1968), p. 35.

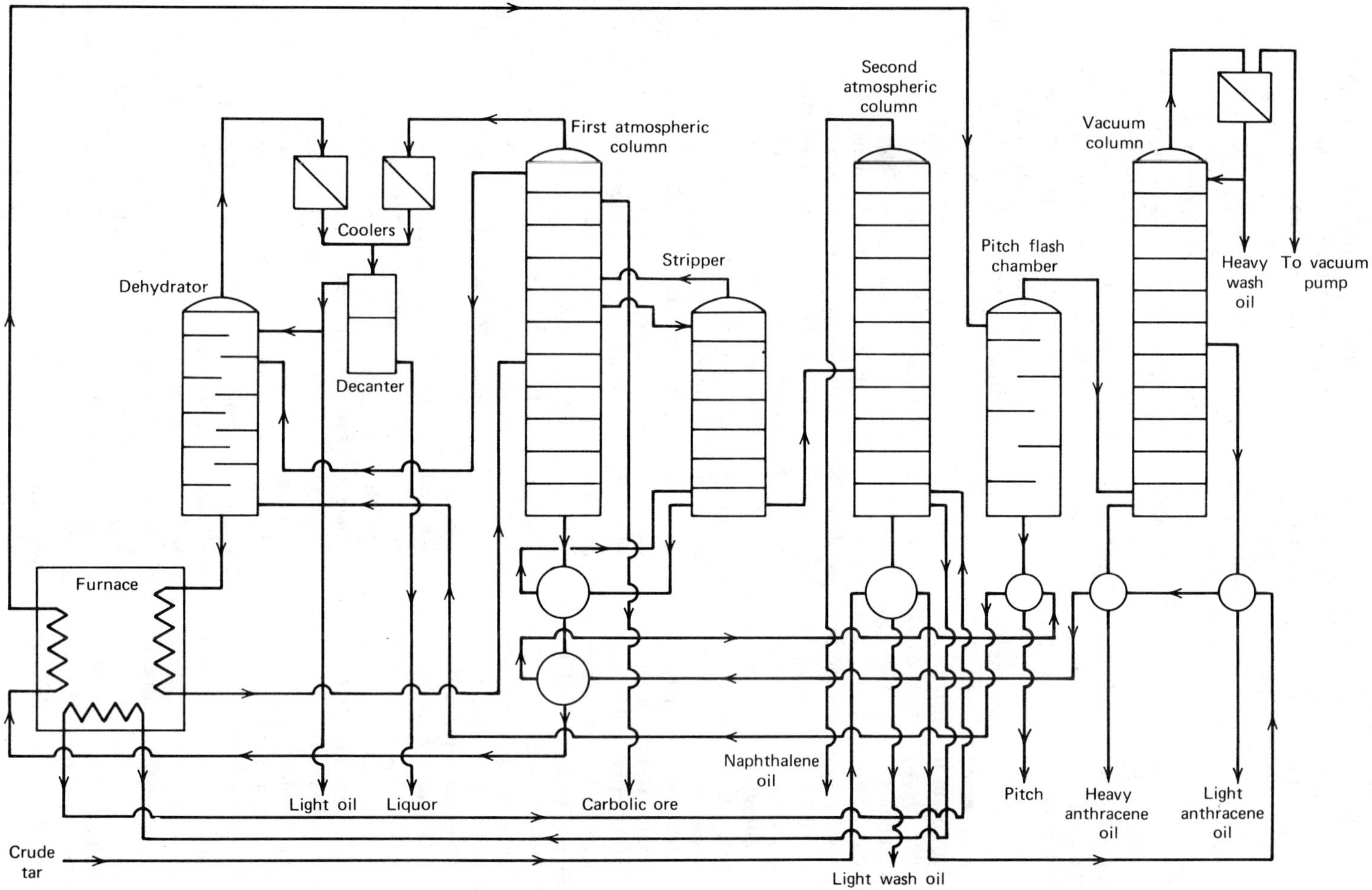

Figure 17.13 Teerverwertung-Koppers tar distillation plant.[182]

steam vapors from the top of the dehydrator are cooled, condensed, and separated, and some of the light oil passed back as reflux. The dehydrated tar is pumped through a heating coil in the tube furnace and injected into the atmospheric pressure fractionating stage, which consists of two fractionating columns and a splitting column. The products include residual light oils and water, taken as overheads from the first column and passed back to the dehydrator, carbolic oil as a side stream from the first column, naphthalene oil as the overhead from the second column, and wash oil as the base product from the second column.

The topped tar from the base of the first column is heated to about 300°C in a separate heating coil in the furnace and injected into the pitch flash chamber that, like the final fractionating column, is maintained under a vacuum of 100 mm Hg. The residual oil vapors separated from the pitch are divided into a heavy wash oil, and a light and heavy anthracene oil in the vacuum column.

The Rütgerswerke plant at Castrop-Rauxel[183] deserves special mention. One function of this plant is to produce closely fractionated concentrates of a number of polynuclear hydrocarbons, which are further separated and purified, to provide the main source of these chemicals in the noncommunist world. The crude tar is dehydrated by heat exchange with the dehydrator overheads and passage through a low-pressure steam heater. The dehydrated tar is heat exchanged with the carbolic oil stream and the pitch, and enters the middle of the atmospheric carbolic oil column at 250°C. This contains 40 bubble cap plates and operates at 16 : 1 reflux. Carbolic oil is taken off overhead, and a naphthalene oil side stream is upgraded in a 20-plate splitting column to give a fraction containing 84–88% naphthalene in an amount equivalent to 95% of the naphthalene in the crude tar. The topped tar is next treated in a 40-plate fractionating column and 20-plate splitting column to yield a concentrate of methyl naphthalenes, a light wash oil, and a fluorene-acenaphthane concentrate. The residual base tar from the bottom of this first vacuum unit passes, at 280°C, to the vacuum anthracene column operated at 70 mm Hg, in which the anthracene and heavy oils are separated from the pitch. The anthracene oil–heavy oil is further fractionated in a separate unit to give an anthracene-phenanthrene concentrate, a 95% carbazole fraction, and a heavy oil from which pyrene, fluoranthene, and chrysene can be isolated.

The special features of this plant, apart from the very efficient fractionation by the use of highly efficient columns operating at high reflux ratios, is the provision of a separate tube furnace to supply reboil to each of the main columns. It also includes heat exchangers in which the base product from the splitting columns is reboiled by heat from the bottoms product of the main columns.

One of the most recent plants installed in the United States is that at the Clairton refinery of United States Steel Corporation,[184] which is unique in employing a sequence of fractionation stages at increasing temperatures. Each stage comprises a helical coil heater, a vacuum flash drum, and a vacuum fractionating column. Dehydration and light oil removal is carried out at atmospheric pressure, and the dehydrated tar is pumped to the first stage to give phenolic oil as the overhead product. The bottoms pass to the second stage, which yields a close-cut naphthalene oil as the product. The third stage separates residual oils from the pitch residue and, in the fourth stage, the residual oils are separated into fractions of the boiling ranges required for making the various grades of creosote.

The Wilton design, supplied by Chemical and Thermal Engineering of Wilmslow, Cheshire, operates on the recycle principle. It is the most popular design in Great Britain and is also installed at tar refineries in South Africa, Australia, New Zealand, India, Spain, Argentina, and Korea. Its popularity stems from its combination of good thermal efficiency and ability to handle a wide range of throughputs. The latest version of the Wilton plant is illustrated in Fig. 17.14. It comprises a pipe still helical coil furnace, a dehydration column, two valve tray fractionating columns, and a pitch column that is divided into an upper stripping section and a bottom mix-

[183] H. G. Franck and G. Collin, *Steinkohlenteer, Chemie, Technologie und Verwendung,* Springer Verlag, Berlin (1968), p. 36.

[184] H. S. Turnbell, *Blast Furn. Steel Plant,* 411 (May 1962).

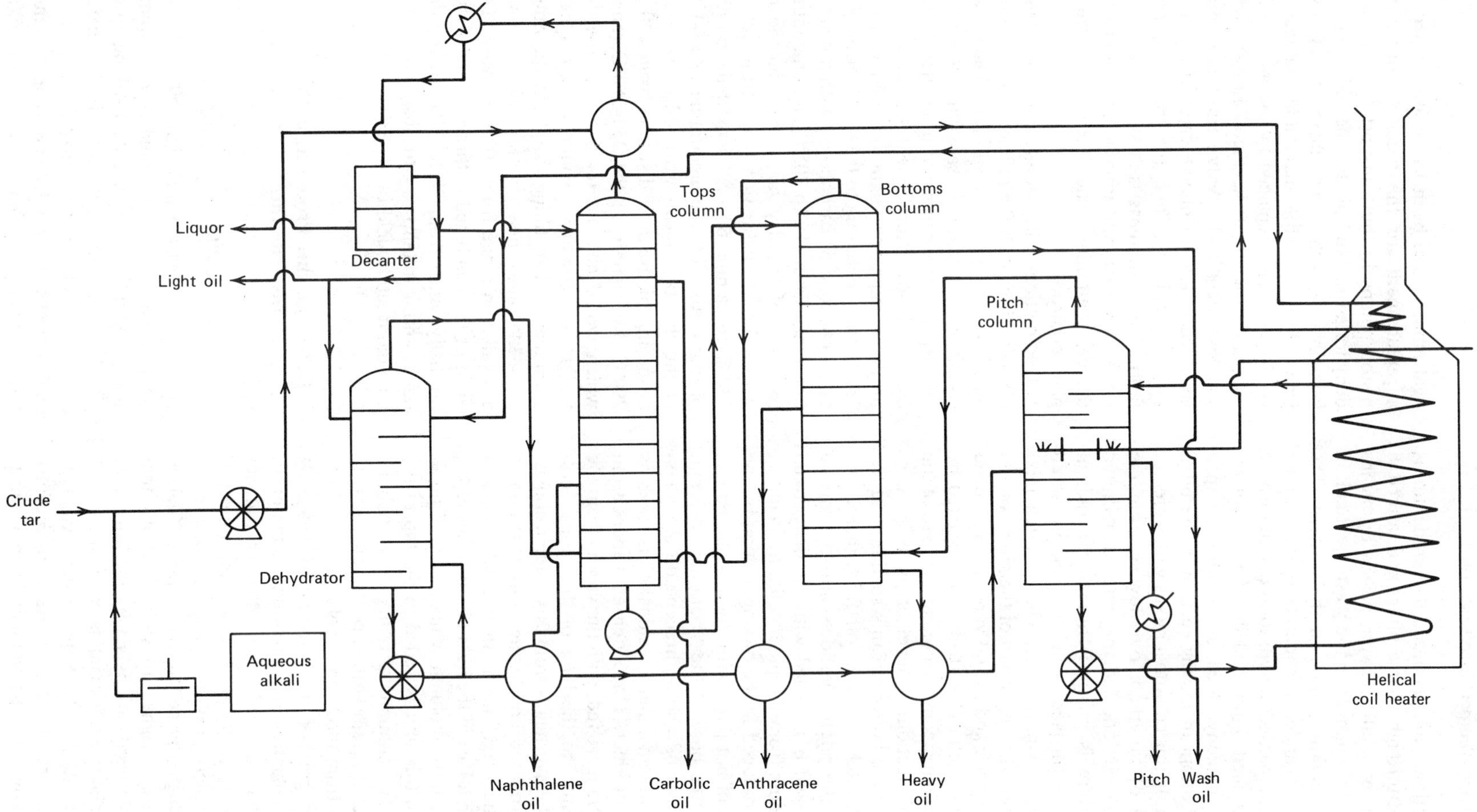

Figure 17.14 The Wilton tar distillation plant. Diagram courtesy of Chemical and Thermal Engineering Ltd., Wilmslow, Chesire, England.

ing section. In operation, the crude tar is pumped through coarse filters and, after the addition of alkali solution, passes by means of a heat exchanger to the waste heat coil in the convection section of the furnace. The heated tar, under a moderate pressure, then enters the baffled dehydration column where the entrained liquor and light oils vaporize, leaving the dehydrated tar, which is pumped to the bottom or mixing section of the pitch column. In this column it mixes with a greater volume of hot pitch overflowing from the stripping section, and this pitch-dehydrated tar mixture is pumped through the radiant coil in the furnace, which is back pressured to prevent vaporization. The tar-pitch mixture enters the upper part of the pitch column at 350–360°C, and the oil vapors are stripped from the pitch. This stripping is assisted by superheated steam injected at the base of the stripping section. A level of pitch is maintained at the bottom of this section, and the product pitch is drawn off from there. The rest of the pitch overflows into the mixing section and is recycled with the dehydrated tar. Vapors from the pitch column, including those produced in the mixing section by the contact of hot pitch with the dehydrated tar, pass to the base of the "bottoms" column. Here wash oil and anthracene oil are taken as side streams, and a residual oil is taken as a base product. Uncondensed vapors from this column are fed, with the vapors from the dehydrator, to the base of the "tops" column. The overheads from this column consist of light oil and liquor; these are first cooled in the heat exchanger, then condensed and separated. Part of the light oil furnishes reflux for the "tops" column, while the base product from this unit is added to the top of the "bottoms" column as reflux. The carbolic and naphthalene oils are taken as side streams from the "tops" column.

The degree of fractionation provided by different plants varies according to the product pattern desired. A tar distiller will not install more costly plants unless the improved performance provides products that command a higher price. Normally, tar distillation plants are designed to yield fractions that meet the specifications set by the purchaser of the plant; each plant is tailor-made and unique in its details. Some general changes in the nature of the plant items in recently installed units are worth mentioning. Centrifugal pumps have been successfully used as a replacement for the older reciprocating piston pumps in the transport of crude and dehydrated tar and pitch. The older type of box furnace with horizontal banks of cast-iron tubes in both radiant and convection sections is being replaced by the more compact and efficient helical coil heaters with stainless steel coils. Valve trays are being specified instead of bubble cap trays in fractionating columns, and the trays are now fabricated in stainless steel or special alloys rather than cast iron. Finally, a much greater degree of automation has been introduced in plant control.

The fractions taken at any individual refinery will differ with the efficiency of fractionation and the products desired. These two conditions are mutually satisfying. A typical primary product pattern at a distillery processing coke-oven tar to yield the full range of products might be: light oil boiling up to 180°C and amounting to 0.5–1.0% of the dry tar; a phenolic or carbolic oil with a distillation range from 180 to 205°C and amounting to 3–4% of the dry tar; a naphthalene oil with a 5–95% distillation range from 200 to 230°C containing 60–65% naphthalene (90% of that in the tar) and amounting to 10–12% of the dry tar; and a wash oil or benzole-absorbing oil fraction distilling from 240 to 290°C, but mainly in the 250–280°C range, which would account for 6–8% of the dry tar. Higher fractions would include a light anthracene oil distilling mainly in the range 250–330°C and a heavy anthracene oil having a distillation range from 300°C upward. The yield of the combined anthracene oils would normally range from 20 to 25% on the dry tar, resulting in (allowing for distillation loss) a medium pitch yield of 53–58%.

17.7.2 Further Processing of the Distillate Oils

17.7.2.1 *Light Oil*

The only processing that light oil might receive at the tar refinery would be a fractional distillation into crude benzole distilling up to 150°C, a naphtha distilling to 190°C, and a creosote residue. The benzole would normally be refined to produce benzene, toluene, and xylenes in combination with crude coke-oven benzole at the benzole refinery. The naphtha might be washed with alkali and acid to recover any tar acids and tar bases it contained, then treated with a small amount of concentrated sulfuric acid to remove sulfur compounds and olefins, and redistilled to give refined solvent naphtha, which is used as an industrial solvent.

17.7.2.2 *Carbolic Oil*

If a carbolic oil fraction is taken, the object is to recover the lower boiling phenols. Despite attempts to accomplish this by selective solvent extraction, the only method used in commercial practice is to contact the oil with a slight excess of 10% aqueous caustic soda, separate the crude phenate from the washed oil, and remove coextracted neutrals and bases from the phenate by steaming and/or washing with benzole. The purified phenate is then "sprung" with a suitable acid, and the crude tar acids are separated from the aqueous salt solution and fractionated.[185]

Carbon dioxide in the form of flue gas, or as produced in lime-coke kilns, is the acid usually used for decomposing the phenate. This is generally accomplished by passing the phenate solution down coke-packed towers countercurrently to the gas. This procedure has two disadvantages. One is that the crude tar acids contain 20–25% water and must be treated again with carbon dioxide to convert about half the residual sodium carbonate to sodium bicarbonate when the water content falls to around 10%. The other disadvantage is that on recausticizing the sodium carbonate–bicarbonate solution with quicklime, a calcium carbonate "lime mud" is produced, which is difficult to dispose of. Studies have been made of the economics of drying and recalcining this by-product to recover carbon dioxide and calcium oxide, but the energy costs of this conversion (the lime mud cannot be filtered or centrifuged to below about 40% moisture content) are excessive.

Sulfuric acid has been used for springing phenate and can be an attractive alternative if there is a market for sodium sulfate. Another alternative, which has been examined on the laboratory scale and appears to be attractive, is to partially electrolyze the sodium sulfate solution and employ the caustic soda–sodium sulfate mixture in the appropriate amount to extract more tar acids.[186]

Further processing of the crude tar acids consists of a preliminary distillation to separate water containing a small amount of phenol (this is recycled to adjust the caustic soda to the strength used), once-run tar acids, and a residue of phenolic pitch, which can be burned. The once-run tar acids are fractionated, either batchwise or continuously, to give phenol, orthocresol, and various grades of cresylic acid.[187]

17.7.2.3 *Coumarone-Indene Resins*

The naphtha fraction of coke-oven tar (i.e., the neutral fraction distilling in the range 150–200°C) contains indene. At some but not all coke-oven distilleries, a fraction rich in indene is made and polymerized. If this process is part of the refinery activities, the light oil fraction is first fractionally distilled to yield a crude benzole (distilling up to about 150°C), a solvent naphtha fraction, an indene fraction distilling in the range 170–185°C, and a residual heavy naphtha. After washing with 25–35% sulfuric acid to remove any bases, the indene concentrate is dried and heated with boron trifluoride–acetic acid complex as catalyst and neutralized with soda solution; the clear amber resin is recovered by removing unreacted naphtha by vacuum distillation.[188]

Coumarone-indene resin, as it is called, although it is almost wholly polyindene, is used as a tackifier in rubber compounding and in paints, and as an adhesive in the manufacture of flooring tiles.

17.7.2.4 *Pyridine Bases*

It was formerly universal practice to recover pyridine bases from light oils and carbolic oils. This was done by following the caustic soda wash with a water wash, then a further extraction with sufficient 25–35% sulfuric acid to combine with the bases present to yield the neutral base sulfates. The advent of

[185] D. McNeil, *Coal Carbonization Products*, Pergamon Press, Oxford (1966), pp. 88–97.

[186] BCRA, Carbonisation Rep. No. 27, 1976.
[187] Anon., *Ind. Chem.*, **32,** 477 (1956).
[188] R. C. Peter, *Ind. Chem.*, **31,** 141 (1955).

synthetic pyridine and picolines has made this process less attractive. At present, less than 20% of the coke-oven tar processed in Great Britain is treated for the recovery of tar bases.

After the crude tar base sulfate has been separated from the extracted oil, it is diluted with water and any resinous material that separates is skimmed off. The base sulfate solution is then boiled to expel any coextracted neutral material, cooled, and decomposed by the addition of alkali. The crude tar bases are separated, dried with solid caustic soda, or azeotropically with benzene, and fractionated to yield pyridine, α-picoline, and fractions containing β-picoline, γ-picoline, and lutidines.[189]

17.7.2.5 *Naphthalene Oil*

The recovery of naphthalene that is suitable for the production of phthalic anhydride or chemical intermediates may be carried out by fractional distillation or by crystallization and washing.[190] Fractionation is carried out on the primary naphthalene oil after removing tar acids. Fairly efficient bubble cap columns operating under vacuum, so that steam heating can be used, are usually employed. Distilled naphthalene is generally rather high in sulfur, since thionaphthene (b.p. 221°C) is not easy to separate from naphthalene (b.p. 218°C) and, for chemical purposes, distilled naphthalene must be desulfurized by treatment with concentrated sulfuric acid or metallic sodium, or by catalytic desulfurization.

Recovery of phthalic anhydride–grade naphthalene by crystallization is carried out industrially by first adjusting the melting point of the feedstock to about 55°C, by diluting with drained oil if necessary. The adjusted feedstock is allowed to crystallize in directly cooled, horizontal, stirred tanks to 30–35°C and the crystal-oil slurry is pumped to a centrifuge operating at 600–800 rpm, where it is spun dry, washed with warm water (sometimes containing up to 5% caustic soda or a small amount of a detergent), spun dry again, and removed. These operations are automatically timed. The washed naphthalene is melted, and the entrained water, which forms a lower layer, is removed. The drain oil from the centrifuge still contains 25–35% of naphthalene. It may be recycled to the primary distillation step with the crude tar, or added to the fractionating column as reflux at the appropriate point. Another method of recovering the naphthalene potential of the drain oil is to split it into two streams, distill one part to yield a "fronts" fraction, a naphthalene concentrate, and a "tails" fraction, and recycle the naphthalene concentrate with the other part of the drain oil to the primary naphthalene oil feedstock. The split of the drain oil necessary to yield a blended product of 55°C crystallizing point is determined by experiment before commencing the recovery.

Formerly the main outlet for coal-tar naphthalene was for oxidation to phthalic anhydride. Material produced from coke-oven naphthalene oil, either by distillation or crystallization, provided its crystallizing point is above 77.5°C, is satisfactory for this purpose. In the United States this is still a major outlet because the large tar-distilling firms have their own phthalic anhydride plants, but in Great Britain and Europe orthoxylene is the preferred feedstock for phthalic anhydride production. Therefore this use for naphthalene has shrunk to less than 10% of what it was in 1960. The main outlet now is for the manufacture of derivatives, such as α- and β-naphthols, and for this use a purer naphthalene is required.

Several novel crystallization methods have been developed for upgrading phthalic anhydride–grade or less pure forms to chemical-grade naphthalene. The simplest is the Pro-Abd refiner,[191] which makes use of the fact that when naphthalene is allowed to crystallize at a gradually reducing temperature, the initial crystals are of high purity. As crystallization proceeds, however, the crystals are covered by layers of solid that are of decreasing purity and lower melting point. The refiner is, in effect, a large box enclosing nests of coils through which either hot or cold water may be passed at closely controlled rates. In operation, it is filled with liquid naphthalene and the contents allowed to cool by circulating cooling water through the coils. When crystallization is complete,

[189] D. McNeil, *Coal Carbonization Products*, Pergamon Press, Oxford, (1966), pp. 109–112.

[190] D. McNeil, *Coal Carbonization Products*, Pergamon Press, Oxford, (1966), pp. 88–97.

[191] J. G. D. Molinari, *Ind. Chem.*, **37**, 323 (1961).

the cooling water is replaced by hot water and the contents slowly warmed to a final temperature just below the crystallizing point of the desired product. Oil formed by the melting of the less pure outer zones of the crystals drains from the base. When this drainage ceases, the rest of the material in the unit, consisting of the purified naphthalene, is melted and run out. This device is capable of improving a 73–76°C crystallizing point naphthalene to phthalic anhydride-grade, of upgrading phthalic anhydride-grade material to the 98% naphthalene required for β-naphthol production, or even to the 99% purity required for the production of H-acid.

A more sophisticated crystallization plant is used by one British firm to purify phthalic-grade naphthalene to very pure material. This is the Brodie crystallizer,[192] developed by Union Carbide Corporation (Australia) for separating ortho- and paradichlorobenzene. It comprises a series of jacketed horizontal tubes arranged in cascade fashion (Fig. 17.15). Each tube is equipped with a slow-moving axial screw conveyor. Tubes 1*a*, 1*b*, and 1*c*, which make up the refining section, are tubes of decreasing diameter. At the exit end of tube 1*a* there is a vertical tube equipped with a heating element at its base

[192] J. G. D. Molinari and B. V. Dodgson, *Chem. Eng.*, **287**(8), 460 (1974).

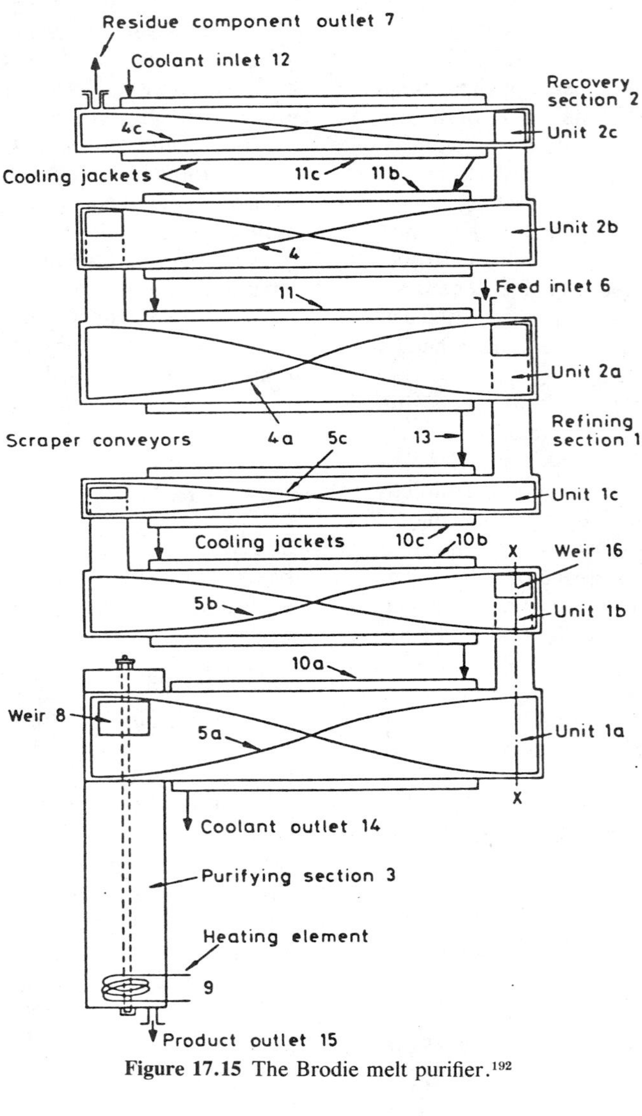

Figure 17.15 The Brodie melt purifier.[192]

that acts as the purifying section. Above the three tubes of the refining section are the three units, *2a*, *2b*, and *2c* of the recovery section, and these also decrease in size. The impure naphthalene feed is pumped to the lower tube of the recovery section and flows toward the cooler *2b* and *2a* tubes, depositing crystals when it reaches its crystallizing temperature. These crystals are carried by the screw conveyors toward the refining section. In this way a countercurrent movement between solid and liquid phases is established, and by the carefully controlled temperature gradient set up by the circulation of hot water through the cooling jackets, a concentration gradient is established in the liquid phase. Mass exchange between the phases occurs at each stage, with the liquid passing back tending to be enriched in the eutectic-forming impurities and the solid passing forward toward the purification section becoming enriched in naphthalene. Eventually, when equilibrium is attained, pure naphthalene crystals come into the purification section where they are melted, displacing pure liquid naphthalene back to serve as reflux. The pure product is taken off as a liquid from this section and an impure naphthalene from the residue component outlet 7. This can be recycled to the preliminary naphthalene recovery and upgrading stages.

Another type of multistage continuous crystallizer, which is claimed to be capable of upgrading coke-oven naphthalene oil of 65°C crystallization point (c.p.) directly to low-sulfur naphthalene of more than 80°C c.p. on the laboratory scale, is the rotating-helix crystallizer described by Betts et al.[193,194] The device is based on the apparatus devised by Schildknecht et al.[195–197] for accelerated zone refining. It consists of a column equipped with a cooling zone at the top and a heating zone at the base. A polished, closely fitting screw conveyor rotates in the column to force downward the crystals formed in the cooling zone. The helical conveyor, in effect, divides the column into separate compartments with little flow between them, so that as naphthalene crystals pass down the column and melt, a concentration gradient is set up in the liquid phase. Mass exchange between the solid and liquid phases leads to the higher melting component, in this case naphthalene, traveling toward the heating zone, and the impurities in the feed moving in the reverse direction toward the cooling zone. When equilibrium is established, the material at the base of the column is pure naphthalene and that at the top of the column is the multieutectic of naphthalene and the nonnaphthalene components in the feed. Fresh feed is added at the point in the column at which the temperature is close to the feedstock c.p. By carefully adjusting the feed rate, the takeoff rates of naphthalene and top product, and the rates of cooling and heating at the top and bottom of the column, respectively, this equilibrium can be maintained under continuous operation.

However this type of crystallizer has been operated only on the small pilot-plant scale to produce 5 kg/hr of pure naphthalene from coke-oven naphthalene oil. There are some doubts about the feasibility of its scaleup to commercial size, and this has not been attempted.

17.7.2.6 *Wash Oil*

The wash oil fraction is not further treated but is supplied to the coke ovens as benzole-absorbing oil or used in creosote blends.

17.7.2.7 *Anthracene Oil*

In Great Britain and Europe, though not in the United States, crude anthracene is isolated from coke-oven anthracene oils. The oil is allowed to crystallize in cooled stirred tanks to 30–35°C, and the crystals are separated by vacuum filtration or centrifuging. The initial product usually contains between 12 and 25% of anthracene and is upgraded by digesting with naphthalene drain oil, wash oil, or drained anthracene oil, and refiltering or recentrifuging. This brings the anthracene content to around 40%. Crude anthracene in this form is sold to the dyestuff manufactur-

[193] W. D. Betts, J. W. Freeman, and D. McNeil, *J. Appl. Chem.*, **17**, 180 (1968).

[194] W. D. Betts, J. W. Freeman, and D. McNeil, Coal Tar Research Association, Reps. Nos. 0483 and 0486, 1972.

[195] H. Schildknecht and H. Vetter, *Angew. Chem.*, **73**, 612 (1961).

[196] H. Schildknecht, K. Maas, and W. Krause, *Chem. Ing. Tech.*, **34**, 697 (1962).

[197] H. Schildknecht and J. E. Powers, *Chem. Zt. Chem. Appara.*, **90**, 135 (1966).

ers, who refine it to a purity suitable for conversion to anthraquinone by sequential crystallization from kerosine and high-boiling tar bases or acetophenone.

17.7.3 Coal-Tar Bulk Products

17.7.3.1 *Creosote*

The products of high-temperature tar refining include the "chemical" products: crude benzole, solvent and heavy naphthas, coumarone-indene resins, pyridine bases, phenol, cresols and cresylic acids, naphthalene, and crude anthracene; the residual oils: naphthalene drain oil, wash oil, strained anthracene oil, and heavy oil; and pitch. The "chemical" products account for about 10% of the tar, the remaining 85% (allowing for liquor and distillation losses) is composed of the tar oils and pitch, and it is the profitable sale of these bulk products that determines the economics of tar processing.

The oils are blended to give creosotes conforming to certain user specifications. The major outlet for coal-tar creosote is for the preservation of timber, a use that dates back to the 1850s. Although other fungicides have been increasingly used, particularly for the treatment of joinery timbers, coal-tar creosote, either alone or mixed with refined tar, is still by far the most widely used wood preservative. It is particularly effective in prolonging the life of timber artifacts that are used outdoors and are in contact with the ground, or are to be installed in seawater or freshwater. Examples are telegraph and electricity transmission poles, railway ties, marine piling, fencing, and farm buildings. To give these structures service lives of 50 years or more, they are impregnated under pressure with creosote so that the vulnerable sapwood cells are either filled with oil (by the Bethell or "full-cell" process generally used for marine piling and railway ties), or coated with a film of creosote (by the Rueping or Lowry "empty-cell" processes more generally used for poles, fencing, and farm buildings). Consumption of creosote for timber treatment in 1974 in the United States alone amounted to more than 150 million gal.

Other uses for creosotes, conforming to the appropriate specifications, are as horticultural washes to kill overwintering pests on fruit trees, for making disinfectant emulsions and sheep dips, and as a feedstock for carbon black manufacture. The last-named outlet has grown considerably since the rise in prices of crude petroleum oil at the end of 1973. Tar oils are also used to cut back coal-tar pitch to give refined tars for road binders and other uses.

17.7.3.2 *Coal-Tar Pitch*

Pitch is the major product of tar refining. Output in the United States in 1974 was 1,242,000 tons. Some 20 years ago the major outlets for pitch in Europe were (*a*) for producing creosote-pitch fuel, consisting of a blend of approximately 1 part of pitch to 1 part of tar oils and used mainly by the steel industry for firing open-hearth furnaces; (*b*) for road tar, a mixture of approximately 2 parts of pitch and 1 part of tar oils; and (*c*) as a binder for briquetting coal dust. The pattern in the United States differed in that road tar and briquetting were minor outlets for pitch, but roofing pitch, the adhesive in multilayer membrane roofs, was a major use. The use of pitch as a binder for the electrodes used in aluminum smelting, though of some importance, consumed only a minor volume, whereas other markets, such as for pipe-coating enamels, pitch-fiber pipes, and pitch-epoxy coatings, had not been developed.

The usage pattern is very different now. Electrode pitch is the major outlet. Briquetting, except for smokeless, precarbonized briquettes, has almost ceased. Roofing pitch, although still a good market for pitch in the United States and in some European countries (e.g., Holland), has lost considerable ground to petroleum bitumen. Road tar is now a minor use in Europe as well as in the United States. Table 17.19 shows the percentage of the total pitch potential used for various purposes in Great Britain in 1960 and 1975.

At the time that most of the continuous distillation plants now in operation were built, the main outlets for pitch included tar fuels, road tar, and briquetting pitch. Since these outlets could be best met by producing a base tar or a medium-soft pitch (80°C R and B s.p.), the distillation plants were designed to produce this grade of residue.

The aluminum companies, however, specify harder grades of pitch for their electrode

Table 17.19 Utilization of the Pitch Content of Tar in Great Britain in 1960 and 1975

Usage	Portion of Total Pitch Content (%) 1960	1975
Creosote-pitch fuel	40.5	14.8
Road binders	24.2	20.1
Coal briquetting	27.6	16.3
Binders for electrodes and/or graphite	4.4	34.2
Pipe-coating enamels	—	5.4
Pitch-fiber pipes	—	2.3
Miscellaneous	3.3	6.9
Total potential (thousands of tons)	1368.8	480.0

binders. For Soderberg paste electrodes, which are made up of around 30% pitch and 70% petroleum coke grist, the binder specification calls for a softening point of 95–100°C (R and B). In addition, it requires a quinoline-insoluble content of 8–12% and a toluene-insoluble content of more than 30%. The trend in aluminum production over the past decade has been away from paste electrodes to prebaked block electrodes, which require less pitch (14–18%) and, for most specifications, a still higher softening point binder in the range 100–110°C (R and B).

These specifications can be met by processing some coke-oven tars in some existing distillation plants, but it has been found more convenient, at least in Great Britain and Europe, to distill the crude tar, as the plants were designed to do, to a medium-soft residue. This material is hardened by soaking for several hours at a temperature of 385–400°C. This increases the toluene-insoluble content of the pitch and slightly increases its quinoline-insoluble content, which gives a measure of flexibility in that conditions can be changed to counter variations in the nature of the crude tar. Coke-oven tar distilleries are now usually equipped with plants for heat treating pitch in this way. Such plants may be batch or continuous. In some cases redundant pot stills arranged in cascade have been used. The more sophisticated pitch heat treatment plants consist of one or more steel or cast-iron vessels heated electrically and having close control of both the bulk liquid temperature and the inner wall temperature. The total capacity is sufficient to give a contact time of up to 10 hr. Where several vessels are employed, they are arranged in cascade, with the temperature increasing at each stage. Vapors from each vessel are passed into the preceding unit, those from the lowest temperature vessel being condensed and added to the heavy oil distillate.

Another change in tar distillation practice over the past few years has been the phasing out of pitch bays, from which the solid pitch had to be blasted for loading for export. The realization by tar distillers and their customers of the health hazard of pitch dust has led to the introduction of plants for producing solid pitch for export in the form of flakes, pastilles, or "pencils."

Pitch-shaping plants in a number of different designs are in operation. Cooled belt flakers are used at several plants. These consist of flexible steel belts about 1 m wide and up to 50 m long, with short rubber skirting at the edges. Molten pitch flows from a thermostatically controlled tank over a weir to give a flat, thin sheet on the moving belt, which is cooled from underneath by cold water sprays. Further cooling is achieved by sprays of cold water directly onto the ribbon of semisolid pitch from halfway along the belt to the end of its travel, where the now-solid pitch is broken into flakes by rotating tines. The pitch flakes are drained and transported to a storage silo by belt conveyor, during which time the surface moisture evaporates.

Direct water cooling is also used in the pitch pencilling plant developed by Rütgerswerke. In this plant pitch at about 50°C above its softening point is pumped through nozzles at a rate and pressure that give a series of almost horizontal jets. These nozzles are enclosed in annular pipes, through which cold water flows; the pipes extend horizontally for 25 m before bending in a quarter-circle. The cooling effect of the annular water stream solidifies the stream of pitch into a solid rod about 1–2 cm thick, and this breaks up into lengths as it is forced around the bend.

There are generally 6–10 of these annular pipes situated in parallel, and they deliver the solid pitch pencils to a rectangular cooling pond situated underneath the pipe assembly. A conveyor belt picks up the pencils from the pond and transports them to a chute. They fall from this chute onto another conveyor belt, which transports them to the storage silo. Surface drying of the pitch rods is accomplished by warm air jets.

Another type of pitch pencilling plant, designed by Chemical and Thermal Engineering, employs a vertical cooling tank. Liquid pitch at about 150°C held in a thermostatically controlled tank is circulated around a ring main by a centrifugal pump. A horizontal section of the main contains an extrusion manifold consisting of a number (usually 6–8) of 9–12 mm diameter nozzles set 20 cm apart. The circular streams of molten pitch from the nozzles fall into a deep rectangular tank containing water at 40–50°C, which causes the pitch to harden into solid rods. The rods break up into pieces about 6–10 cm. These are transported by a screw conveyor from the cooling tank to an inclined bucket elevator, which lifts them up through the water in the conveyor leg communicating with the cooling tank and delivers them by chute to the conveyor belt, which takes them to the silo. Heinrich Koppers of Essen also offers pitch-pencilling plants of generally similar design.

17.7.3.3 *Pitch Coke*

The manufacture of pitch coke provides a large tonnage outlet for high-temperature pitch in Germany, Japan, and the Soviet Union. Production in Germany and Japan is estimated to be 150,000 tonnes/year, and production in the USSR is believed to be on the same order. Pitch coke is used, either alone or mixed with other varieties of carbon, as the solid component of electrodes, carbon brushes, shaped carbon, graphite articles, and so on.

In Germany and Russia the main method used for the production of pitch coke is to carbonize a hard pitch in modified coke ovens, the construction and operation of which have been described by Sarkar[198] and Osthaus.[199] These ovens are faced with silica brick and are heated from the side. They are 10–13 m long, 2–3 m high, and 3.5–5 m wide, and take a charge of about 16 tons of pitch. The raw material is a hard pitch with a softening point (R and B) of 140–150°C. In Soviet plants this raw material is made by air-blowing a mixture of medium-soft coke-oven pitch and recycled coking oils to the desired softening point. The pitch is charged in a molten state to the ovens, over a period of 5 hr, and the charge is coked for 17–18 hr at a flue temperature of 1250–1300°C. Volatiles are condensed by direct water spraying, and the oils recovered are recycled to the feed preparation stage. Coke yield is 70%, oils amount to about 20% by weight of the pitch treated, and the gas yield (about 80% hydrogen) is 10%.

Many attempts have been made to produce pitch coke by delayed coking in the same way that needle coke is made from petroleum residues, but the only process that has been developed commercially is that designed by the Lummus Company, which is now in operation at two plants in Japan and one in Germany. A description of the first plant, operated by Ninnetsu Chemical Industries, has been given by Remirez.[200] Hot pitch from an adjacent tar distillation plant is preheated by heat exchange and pumped to the bottom of a divided fractionation tower, where it is further preheated by mixing with the vapors from one of the coking drums. The heavy fraction of the coking oils is condensed, and the mixture of preheated pitch and heavy oils passes to a tube furnace, where it is heated rapidly to a temperature just below the point of coke formation. It is then injected into one of the coking drums, which has been preheated by hot vapors that are circulated through it. These vapors come from the drum in which coking is taking place. The contact of the pitch at its incipient coking temperature with the hot walls of the drum initiates the exothermic coking reaction. This continues, without any additional heat, for about 20 hr, until all the pitch has been converted to green coke. At the end of the coking period the full drum is steamed and cooled, and the coke is discharged by high-pressure water lances. The empty drum is then steamed out, pressure tested, and preheated. It is ready for the next charge, by which time the second drum has reached the end of its coking period. Vapors from the fractionation tower are cooled, and the oil that condenses is recycled to the crude tar. Uncondensed gases are used as fuel in the process. The green coke is crushed and calcined in a separate kiln.

[198] S. Sarkar, *J. Mines, Met. Fuels,* **10,** 16 (1962).
[199] K. H. Osthaus, *Koppers-Mit.* No. 3 (1960), p. 115.
[200] R. Remirez, *Chem. Eng.,* **76,** 74 (1969).

The initial stages of pitch carbonization has been the subject of studies by several workers since Brooks and Taylor[201] first observed the formation of anisotropic spherical bodies in coal-tar pitch when heated in the range 400–500°C. These anisotropic spherulites, to which the term "mesophase" has been applied, can be observed under the electron microscope. When they have grown large enough, they can be seen as irridescent spheres when the pitch is viewed by transmitted polarized light with an optical microscope. Under favorable circumstances they grow until they coalesce into an oriented mosaic, which appears to be a necessary precursor to the formation of graphitic coke. Other carbonaceous substances such as petroleum bitumen and polyvinyl chloride show this mesophase formation on heating.

Marsh and his co-workers, adopting a technique involving the heating of specimens encapsulated in gold foil at pressures of 130–290 MN/m^2, have studied the formation of these liquid crystals during the carbonization of pure aromatic hydrocarbons,[202] mixtures of such hydrocarbons with heterocyclic compounds,[203] different types of coal-tar pitch and coals of different rank,[204] and unoxidized and oxidized vitrinite.[205] They found that both size and shape of the mesophase bodies formed when aromatic hydrocarbons are heated under pressure vary with the structure of the hydrocarbon, as does the rate of mesophase formation. This can be promoted by the addition of certain heterocyclic compounds such as phthalimide or phthalic anhydride, or inhibited by the addition of phenol.

More interesting with respect to pitch carbonization is the finding by both Marsh et al.[204] and Hüttinger[206] that the nature of the quinoline insolubles is critical to the growth of mesophase and, on further heating, to the formation of a graphitic pitch coke structure. Natural quinoline insolubles (i.e., the suspended carbonaceous matter present in normal coke-oven pitch) restricts mesophase formation by forming a boundary at the periphery of the mesophase spheres. Added carbon black has the same effect, but added graphite powder promotes mesophase growth and coalescence. The quinoline-insoluble matter, formed when pitch is heat treated (corresponding to Hüttinger's type 3 quinoline insolubles with a hydrogen content of 3% or more), favors the formation of mesophase and the production of a pitch coke that can be graphitized. This is not surprising, since the evidence is that mesophase consists of condensed polycyclic aromatic molecules with molecular weights in the range 1500–2000. These are believed to be formed by intramolecular dehydrogenation of the smaller aromatics, which make up the "crystalloid," "resinoid," and "C_2" fractions. The activation energy of 35–45 kcal/mole derived by Honda et al.[207] for mesophase formation is consistent with this theory. In the mesophase structure these condensed polynuclear aromatic molecules are arranged in a lamellar-oriented structure, with the lamellae curving at the edges to become approximately normal to the sphere surface.[201]

The observation that the quinoline-insoluble material normally found in coke-oven pitches retards the formation of graphitic coke has been reflected in patents claiming the delayed coking of pitch from which the quinoline-insoluble matter is first removed by pressure filtration[208] or centrifuging,[209] to yield a more readily graphitizable coke.

17.7.3.4 *Refined Tar*

The term "refined tar" originally referred to a tar from which the water and the lighter fractions had been removed by distillation, but it is now generally used to denote a blend of pitch or base tar and tar oils meeting a particular specification. The main outlet for this type of tar is for road making and maintenance. Road tars consist of pitch fluxed back with a mixture of a major proportion of anthracene oil (preferably strained anthra-

[201] J. D. Brooks and G. H. Taylor, *Nature,* **206,** 697 (1965).

[202] H. Marsh, J. M. Foster, G. Hermon, and M. Iley, *Fuel,* **52,** 234 (1973).

[203] H. Marsh, J. M. Foster, G. Hermon, M. Iley, and J. N. Melvin, *Fuel,* **52,** 243 (1973).

[204] H. Marsh, F. Dachille, M. Iley, P. L. Walker, Jr., and P. W. Whang, *Fuel,* **52,** 253 (1973).

[205] F. Goodarzi, G. Hermon, M. Iley, and H. Marsh, *Fuel,* **54,** 105 (1975).

[206] K. J. Hüttinger, *Erdoel Kohle,* **26,** 21 (1973).

[207] H. Honda et al., *Carbon (Oxford),* **8,** 181 (1970).

[208] Ninnetsu Kagaku Kobushiha K.K. and Nippon-Lummus K.K., British Pat. 1,365,493, 1974.

[209] Bergwerksverbank GmbH, British Pat. No. 1,462,166, 1975.

cene oil) and a minor proportion of wash oil or drained naphthalene oil. They are produced in a range of viscosities to suit the climatic conditions under which they are to be used. For road construction they are employed as binders for road bases and base courses, and for more lightly trafficked roads, as the binder in tar macadam wearing courses. In road maintenance they are used in surface dressing, a widely used, cheap method of restoring the riding quality of a road surface. First the road surface to be treated is swept free from dirt, and the profile is regulated by filling in any potholes. Hot tar is then sprayed on the road surface and immediately is covered with an evenly applied layer of single-size (6–18 mm) stone or slag chippings, which are rolled into the new road surface.

The use of tar as a road binder was never a major outlet in North America but accounted for a large tonnage in Great Britain and Europe. This market has declined drastically in recent years for a number of reasons. There is less crude tar available, for example, and the combined profits from selling the pitch component as electrode pitch and the oil component as wood preservative oil or carbon black feedstock are higher than that from the road tar, with the result that less road tar is being made. Road construction over the past few years in most industrial countries has been concentrated on high-speed, heavily trafficked motorways. These are built in concrete; or if they are of flexible construction, petroleum bitumen is used as the binder in the lower layers and either bitumen, lake asphalt, or pitch-bitumen for the hot-rolled asphalt or asphaltic concrete surfacing. This trend has coincided with a general reduction in the expenditure on both new road construction and road maintenance, because of the financial constraints that have affected most developed countries since 1973, so that the demand for road tar has fallen in line with the reduction in its potential supply. In Great Britain, for example, the use of tar products in road making and maintenance fell from 330,000 tons in 1960 to less than 100,000 tons in 1975.

There have been some interesting developments in the field of road binders. The addition of 20–25% of coal-tar pitch to the petroleum bitumen used as the binder for the hot-rolled asphalt surfacing of major highways has been shown to assist in maintaining long-term skid resistance. Pitch-bitumen is now the preferred binder for this type of road surface. A parallel development has been the use of tar-bitumen blends rather than road tar or cut-back bitumen for surface dressing. Such blends, containing 35–40% of coke-oven refined tar (or 40–50% of low-temperature refined tar), combine the low-temperature susceptibility of bitumen with the better wetting properties and resistance to softening, when in contact with petroleum oils, that is characteristic of refined tar. Special polymer-modified tars containing small amounts of PVC or synthetic rubbers have been developed for surface dressing areas where traffic stresses are high. Such modified tar binders have been successfully used to dress the surfaces of round-abouts, garage driveways, parking lots, and airfield runways.

Refined tars are used as extenders in epoxy and polyurethane resins to give surface coatings, which have developed a sizable market. These two-pack formulations harden into tough, abrasion-resistant, waterproof films when the two components, one containing the resin and the other containing the polyamine, polyamide, or isocyanate hardener, are mixed. They are widely used for coating underground storage tanks, structural steel work, pipelines, building work underground, water storage tanks, marine piling, and so on. They show excellent adhesion to prepared steel or concrete surfaces and are resistant to freshwater and seawater, fairly strong mineral acids, alkalis and salt solutions, detergents, and petroleum-derived oils.

Pinchbeck[210] has given a comprehensive account of the production, application, and properties of tar-epoxy paints. The tar extender in these systems should combine high aromaticity, low phenolic and basic nitrogen content, and a low level of quinoline-insoluble matter. Such properties can best be secured by blending the pitch produced by careful distillation of a centrifuged coke-oven tar with neutral high-boiling tar oils, to give the desired viscosity.

Additional uses for refined tars are as impregnants and binders for carbon-bonded refractories.

[210] P. H. Pinchbeck, *Aust. Oil Colour Chem. Assoc. Proc.*, **14**(6), 10; **14**(7), 6 (1977).

CHAPTER **18**

Coke-Oven Gas and Effluent Treatment

HERBERT A. GROSICK AND J. E. KOVACIC

Herbert A. Grosick, *a senior staff engineer in the Project and Technical Section of the Engineering and Construction Group of Koppers Company, Inc., has had 36 years of experience in the design and operation of coke ovens and coal chemical plants. Mr. Grosick served as Chief Project Engineer of his section for approximately 16 years, with responsibility for process design, plant layout, and specifications for coke and coal chemical plants throughout the United States and in many foreign countries.*

J. E. Kovacic *received his B.S. degree in chemical engineering from the University of Pittsburgh and has been employed in the Research Department of Koppers Company for the past 33 years. His entire professional career has been devoted to studies associated with carbonization of coal, by-product recovery, gas purification, and effluent treatment, including laboratory, pilot-plant and commercial-scale operations. Mr. Kovacic is Senior Research Scientist with Koppers Company.*

18.1 INTRODUCTION

The value of the coal chemicals recovered from the coke-making process has continued to decline since the first Supplementary Volume to *Chemistry of Coal Utilization* was published (1963). This trend has affected all the coke-producing nations, some to a greater extent than others. Statistics compiled by the U.S. Department of the Interior[1] reported that in the United States, the value of the recovered coal chemical products decreased from 37.5% in 1959 to 18.3% in 1975 compared to the value of the coal that is carbonized. The data tabulated in Table 18.1 reveal this trend, showing that between 1959 and 1975 the value of ammonia products declined to less than one-third of their value in 1959 and the value of surplus gas, tar, and light oil products declined to about one-half their 1959 values. In 1970 Reerink[2] reported a similar trend for the European nations. He stated that the values of the coal carbonization products changed since 1950 as follows:

[1] U.S. Bureau of Mines, *Coal and Coal Chemicals Survey, Minerals Yearbook* annual reviews 1959–1975.

[2] W. Reerink, *COMA Yearb.*, 349–372 (1970); Waverly House, Wexborough, Yorkshire, England.

1. The value of coke increased from 62 to 80%.
2. The value of coke-oven gas decreased from 20 to 13%.
3. The value of other products decreased from 18 to 7%.

He predicted that by 1980 the value of gas and other products will have decreased to approximately 15% of the total value of the carbonization products. These changes began with the start of the petrochemical industry and with the availability of large quantities of natural gas and petroleum distillate feedstocks. In 1976 there was some indication that the supplies of natural gas and petroleum reserves were being depleted, suggesting the possibility that the decline in the value of coal chemicals may be reversed in the near future.

The petrochemical industry has demonstrated that it is capable of producing the so-called coal chemicals at a higher purity level and at a more competitive cost than the coking industry can achieve. The petrochemical industry possesses the added advantage of being independent of steel production and, in turn, of the unsteady demand for the coke produced in the by-product recovery coke

Table 18.1 Percentage of Coal Costs Recovered as Coal Chemical Products, U.S. Production[1]

Coal Chemical Products	Coal Costs Recovered (%)								
	1959	1963	1967	1970	1971	1972	1973	1974	1975
Tar and its derivatives, used or sold	11.8	11.1	11.4	9.0	8.0	8.0	6.5	6.3	6.2
Surplus gas, used or sold	16.0	16.1	15.1	12.8	11.7	10.6	11.2	9.2	8.0
Ammonia	3.3	2.8	2.5	1.3	1.1	1.0	1.0	1.2	1.0
Light oil and its derivatives	6.3	5.0	4.4	4.3	3.8	3.2	2.3	3.5	3.1
Total coal chemicals	37.4	35.0	33.4	27.4	24.6	22.8	21.0	20.2	18.3
Value of coal carbonized ($/ton)	9.88	9.49	10.02	12.21	14.00	15.74	18.32	36.49	44.21

ovens. Regardless of all other considerations, the coal carbonization industry does not have the capacity to produce the volume of feedstocks required by the chemical industry. Haensel and Mechalis[3] and Ockerbloom[4] have published the U.S. benzene production figures for 1950–1972. These data are shown in Fig. 18.1. In 1972 the petrochemical industry in the United States produced 10 times the volume of benzene that was produced in the by-product recovery ovens. Haensel and Mechalis report similar statistics for the other, more common aromatic chemicals found in coke-oven gas. These two important factors—the greater production capacity of the petrochemical companies coupled with their higher product quality—have depressed the selling price of some coke plant products so severely that either the profit margin is small or, in some cases, the cost of recovery exceeds the value of the salable product.

18.1.1 General Comments

The major changes that have occurred in gas treatment facilities during the past 15–20 years were due either to economic conditions, as already discussed, or to the environmental restrictions that all producing nations have placed on the industry. Another factor becoming more prevalent in the literature is the increased concern for the health and safety of coke plant workers. Some countries have already established restrictions on working conditions, limits of exposure to specific chemicals, and so on. It can be expected that this trend will continue indefinitely. Thus far, most of these restrictions pertain to the coke-oven employees, but are also directed to the benzol and tar plant workers. The changes related to environmental and health restrictions are discussed at greater length in other sections of this chapter.

18.1.1.1 *U.S. Practice*

U.S. producers have begun to discontinue the recovery of gaseous constituents that are not profitable. It is expected that this practice will continue for as long as the economics dictate and for as long as it does not interfere with environmental regulations. For example, some plants wash the ammonia from the coke-oven gas, but the absorbing solution, which may or may not be combined with the waste ammonia liquor, is then distilled to recover the ammonia vapors, and these vapors in turn are incinerated. A few small plants do not recover the crude light oil, and a number of plants recover but do not process their crude light oil. Instead, it is sold to petrochemical firms for further processing. Much of the tar is processed to recover its more valuable derivatives. Some of the larger coke producers process their tar, but generally the crude tar is sold to the tar chemical processing companies. A large volume of the tar is used as fuel in the various steelmaking processes; this practice may take a sudden upswing if the more desirable fuels such as natural gas and heavy petroleum fractions become scarce or more expensive in the future. A very small volume of coke-oven gas is still purchased by the public utility firms for domestic use, but virtually all the surplus gas that is produced in the U.S. industry is consumed by the parent steel

[3] V. Haensel and N. Mechalis, *Prepr., Am. Chem. Soc., Div. Fuel Chem.*, **7**(1), 169–183 (1963).

[4] M. E. Ockerbloom, *Hydrocarbon Process.*, **51** 117–124 (November 1972).

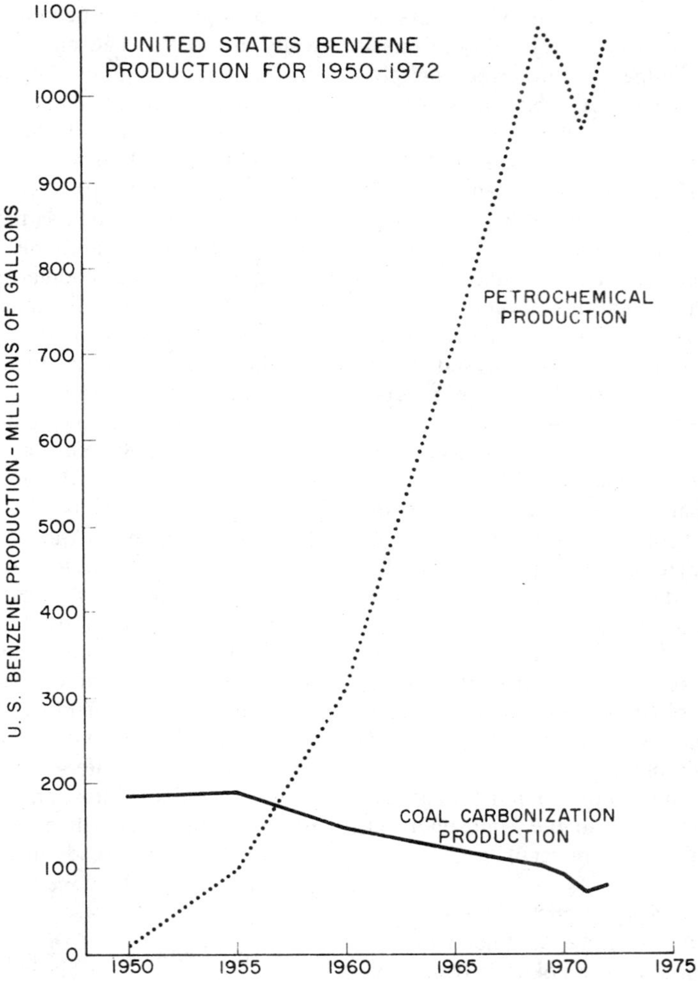

Figure 18.1

companies, either in their steel reheating furnaces or in the boilers to generate steam. Surplus gas is the volume of coke-oven gas that is available after fulfilling the requirements for heating the coke ovens.

One U.S. author[5] states that the steel industry's economy demands the lowest coke cost. He goes on to say that in the future nonrecovery ovens will be constructed instead of slot-type recovery ovens, An anonymous writer[6] reported that one U.S. firm would build 50 nonrecovery ovens to produce metallurgical coke. However it is inconceivable that the nonrecovery oven will ever make a significant inroad into coke production, especially in view of the concerted effort that is now directed toward the elimination of coke-oven emissions.

[5] J. W. Miller, *J. Met.*, **19**(6) 74–77 (June 1967).
[6] Anon.; *Chem. Week*, **99**(26), 21 (December 24, 1966).

18.1.1.2 *European Practice*

European operations have similarities to the American practice as well as differences. Generally, there is more similarity between Western European and American practice, and the differences from Eastern European practice are greater.

Western European Practice. The Western European nations have also felt the economic pressures on the recovery of coal chemicals. Consequently many operations do not recover ammonia and/or light oil as salable products. The increased availability of natural gas has become the preferred fuel for town gas use. Wagener,[7] who states that in West Germany " . . . coke-oven gas is in danger of being priced out of the market by the consumers' direct use of natural gas," also reports that the industry has been considering which segments of the by-product recovery system could be excluded without creating other difficulties. Some German plants have begun to incinerate the ammonia that is recovered from the gas, and some plants do not recover the light oil products. Many other authors have treated this subject in considerable detail, and they generally reach the same conclusions.[8–12]

A German patent[13] describes a method for the combustion of untreated coke-oven gas in furnaces. Ejectors are used to draw the raw gas from the collecting main through insulated pipes. The gas, heated to 500–700°C, is delivered at a velocity of 80–100 m/s through the burner orifice to the furnace. One obvious disadvantage of this proposal is that hydrogen sulfide (H_2S) and other sulfur compounds are converted to sulfur dioxide (SO_2) in the combustion gases. The discharge of SO_2 to the atmosphere is controlled by government regulations.

Eastern European Practice. The Eastern European producers have continued to follow the same practices of coke gas treatment that were common 25 years ago. These include the extensive production of ammonium sulfate as well as the refinement of light oil by the acid wash processes. More recently, however, many hydrorefining plants have been constructed to refine the light oil, and some authors report that the production of ammonia products is becoming less economically attractive. The coal carbonization industry in these nations supplies a major portion of their coal chemical needs. In the Soviet Union most of the naphthalene is derived from coal tar.

These countries also recover other coal chemical products such as tar acids and bases, tar chemicals, and the minor light oil constituents such as coumarone, indene, and styrene.[14–17] Eastern European practice differs because development of the petrochemical industry has been slower, the demand for coal chemical products, particularly by the agricultural industry, is heavy, and these countries, notably Russia, produce more coke than any other nation, hence greater volumes of coal chemicals. Even in these countries, however, the demand for coal chemicals has begun to decline. As other nations have learned, the flexibility and product quality of the petrochemical industry places the coal chemical producers at an unfavorable competitive disadvantage. The Eastern European manufacturers are experiencing restrictions with respect to pollution and to health and safety conditions for their workers similar to those that the other world manufacturers have encountered. Thus it is apparent that the trends in the industry are similar for all of the coke-producing nations.

18.1.2 Summary

Despite the competition from the petrochemical industry, the coke-producing nations continue to recover and sell huge volumes of coke-oven gas, ammonia, light oil, naphthalene, and other products. The monetary value of these products is impressive. For example, in 1975 U.S. producers sold surplus coke-oven gas and other recovered products valued at $560,927,000.[1] A partial listing of the materials produced and sold since 1959 is given in Table 18.2. Soviet authors Privalov and Zaichenko[15] reported in 1974 that the U.S.S.R. produced more than 6 million tons

[7] D. Wagener, *Blast Furn. Steel Plant,* **56,** 883–890 (1968).
[8] D. R. Jones, *COMA Yearb.,* 288–302 (1972); Waverly House, Wexborough, Yorkshire, England.
[9] F. Steer, *COMA Yearb.,* 168–181 (1972); Waverly House, Wexborough, Yorkshire, England.
[10] G. Nashan, *Glueckauf,* **108**(2), 67–72 (1972).
[11] W. A. M. Edwards, Benzole Producers Ltd., Inf. Circ. No. 1, 1967, 33 p. Watford, Hertfordshire, England.
[12] K. H. V. French, Benzole Producers Ltd., Res. Paper No. 2, 1963, 9 pp., Watford, Hertfordshire, England.
[13] Bergwerksverband, GmbH, German Pat. 2,-226,867, September 18, 1975.
[14] A. M. Akhtyrchenko et al., *Coke Chem. (USSR),* No 4, 47–51 (1974).
[15] V. E. Privalov and V. M. Zaichenko, *Coke Chem. (USSR),* No. 10, 24–27 (1974).
[16] M. S. Litvenenko et al., *Coke Chem. (USSR),* No. 11, 28–30 (1960).
[17] I. G. Revzin et al., *Coke Chem. (USSR),* No. 3, 27–29 (1970).

Table 18.2 Statistical Summary of U.S. Coal Chemicals Production: Multiply All Numbers by 1000[1]

Coal Chemicals: Production Sales and Value	1959	1963	1967	1971	1975
Coal carbonized (tons)	77,354	76,020	90,900	81,531	82,147
Crude tar					
Total production (gal)	653,728	671,876	780,334	679,377	645,537
Crude tar sales (gal)	334,717	290,698	354,674	334,076	284,841
Sales value	$42,633	$31,111	$33,686	$35,960	$98,898
Coke-oven gas					
Total production (ft³)	804,600	800,582	959,359	861,691	895,279
Surplus volume (ft³)	514,971	516,084	606,044	506,563	527,218
Sales value	$123,124	$116,721	$138,279	$134,362	$292,694
Ammonia (as sulfate equivalent)					
Total production (tons)	717	719	834	632	534
Ammonia sales (tons)	759	711	776	723	439
Sales value	$25,747	$20,617	$23,262	$12,167	$32,753
Light oil					
Total crude production (gal)	213,036	218,166	252,138	201,626	194,814
Crude light oil sales (gal)	19,515	32,610	94,514	91,423	99,005
Refined light oil sales (gal)	163,837	150,997	117,178	94,076	81,885
Crude sales value	$2,993	$3,633	$13,229	$8,654	$44,234
Refined sales value	$44,643	$32,180	$26,930	$18,338	$57,419
Total light oil sales value	$47,636	$35,813	$40,159	$26,992	$101,653
Total coal chemicals sales value	$257,952	$228,489	$263,534	$232,418	$560,927

of various coal chemical products per year, including 125 salable items. In 1972 the Soviet Union exported light oil, naphthalene, and various tar products valued at 14,830,000 rubles ($19,724,000).[18] It is apparent that even though the petrochemical companies will continue to manufacture the greater volume of these products, large quantities will be recovered from coke-oven gas. The industry continues to fund extensive research efforts to improve gas treatment facilities, in order to reduce operating costs, upgrade product quality, and develop new markets for the coal chemical products.

Some of the changes in gas processing that took place in the past few years are now being reversed for environmental or other reasons. The worldwide energy shortage and the subsequent steep rise in crude oil prices have resulted in a change in appraisal of the gas treatment process. In 1971 it was reported[19] that the British Steel Corporation was spending £ 1,600,000 in the Sheffield and Rotherham area to replace imported oils and natural gas with purified coke-oven gas. Rozentuller and Kulish[20] claim that it should be as economical to make ammonia from coke-oven gas as from natural gas. Kawat[21] reported that the German firm Linde AG developed a process for the low-temperature separation of coke-oven gas to produce ammonia synthesis gas or high-purity hydrogen. The author claimed that with this process it is cheaper to make ammonia from coke-oven gas than from natural gas. Patents have been awarded to a number of firms for the production of substitute natural gas (SNG) from coke-oven gas.[22–24] Colaianni[25] presented a brief description, with a flow diagram, of a new U.S. coke-oven gas treatment plant. One feature of this facility is the Phosam

[18] Anon., "Soviet Exports of Coal, Coke and Chemical Products," *Koks Khim.*, No. 2, 61 (1974).

[19] Anon., *Steel Times*, **199**(5), 418 (May 1971).

[20] M. M. Rozentuller and G. P. Kulish, *Coke Chem. (USSR)*, No. 7, 51–53 (1964).

[21] E. Kawat, *Trans. Chem. Eng.*, **43**, 294–301, 304 (1965).

[22] Harpener Bergbau, AG, German Pat. B.D. 1,244,329, 1967.

[23] Dr. C. Otto Co., German Pat. B.D. 1,245,920, 1967.

[24] H. Koppers, GmbH, British Pat. 1,080,129, 1967.

[25] L. J. Colaianni, *Chem. Eng.*, **83**(7), 82–83 (March 29, 1976).

process for the production of anhydrous ammonia. This new process, a development of the United States Steel Corporation that is generating much interest among U.S. coke plant operators, is described in Section 18.3.3.2.

18.1.3 Environmental Control

Perhaps the most significant shift in emphasis in coal chemicals processing in the past 20 years has been in the area of environmental control (see Chapter 31). This involves both air pollution and liquid waste disposal. Muder[26] in the first Supplementary Volume made only a passing reference to acid gas purification and said little about the disposal of ammoniacal liquor. Those familiar with the subject know that today these aspects of gas treatment plant operations are in some cases more important than the profitability of the recovered chemicals.

In the past, when coke-oven gas was used as town gas, it was necessary to remove the H_2S before the gas entered the distribution system, and many gas purification processes were developed for this purpose.[27] Once the utilities began to use natural gas, however, there was no longer a need to extract the H_2S except when the sulfur was detrimental to the steelmaking processes for which the coke-oven gas was used. Consequently interest in acid gas purification waned, and the practice was discontinued in many plants until strict government regulations were enacted to improve the air quality in industrial areas.[28] In some sections of the United States the H_2S concentration of coke-oven gas cannot exceed 50 gr/100 scf if the gas is used as a fuel. Privalov and Zaichenko[15] reported that the U.S.S.R. requires H_2S concentrations to be less than 20 gr/100 scf if the gas is to be burned as a fuel. It can be expected that future trends will be toward lower H_2S levels, in an effort to reduce the atmospheric levels of SO_2.

Similar changes have occurred in the treatment of the coke plant wastewaters. At one time it was necessary only to remove the tar acids and oils before discharging the plant effluent to natural water supplies, and only because these materials imparted a disagreeable odor or taste to the potable water. Today government regulations in many countries require extensive treatment of these waste streams before they are returned to the lakes and rivers, to improve the water quality. Conservationists and other concerned groups argue that coke plant wastes are detrimental to aquatic life and are eventually harmful to the population that consumes the fish and other products taken from these waters. Zitko and Tibbo[29] reported that a mass herring kill in coastal waters was attributed to the oily discharge from a coke plant.

The industry has spent vast sums of money for research and development on new processes for acid gas removal and wastewater treatment.[30] Large expenditures of funds have been made for the installation of environmental control equipment and it is projected that much more will be required in the years ahead. This is a subject of great interest to all operators today, and it is discussed more thoroughly in later sections of this chapter.

Compliance with environmental controls sometimes results in operational problems. One notable example is the regulation to reduce emissions from the coke-oven battery while the ovens are being charged (i.e., filled with coal: see Chapter 14). If emissions are to be reduced by venting the oven to the gas-collecting main and using steam ejectors in the standpipes, air may infiltrate the gas-collecting system, permitting the oxygen to take part in several undesirable reactions. For example, oxygen may react with hydrogen cyanide to form the thiocyanate, which is very corrosive. It may react with the ammonia (NH_3) and H_2S to form ammonium sulfate or with the H_2S and the iron from the gas piping to form elemental sulfur and iron oxide. These reactions contribute to corrosion in the gas-treating system and to the formation of solid deposits in the gas lines. Both situations are very troublesome and

[26] R. E. Muder, in *Chemistry of Coal Utilization,* Supplementary Volume, H. H. Lowry, Ed., John Wiley and Sons, New York, 1963, pp. 665–674.

[27] H. A. Gollmar, in *Chemistry of Coal Utilization,* Vol. 2, H. H. Lowry, Ed., John Wiley and Sons, New York, 1945, pp. 947–1007.

[28] J. L. Harrison, *Steel Times,* **202**(5), 367–384 (1974).

[29] V. Zitko and S. N. Tibbo, *Bull. Environ. Contaminants Toxicol.,* No. 6, 24–25 (1971).

[30] V. E. Marks, *Effluent Water Technol.,* **12,** 415–419 (1972).

costly to the plant operators. Massey and his co-authors[31] made an intensive study of these problems and provided recommendations for methods to avoid these adverse reactions.

Another problem becoming very important is that of health and safety. This does not imply that plant managers have ignored this subject in the past. On the contrary, the U.S. industry for many years has devoted considerable attention to the development and promotion of comprehensive safety programs in the coke plants. Awards and other signs of recognition were presented to those individuals or departments that achieved the highest safety records. The industry has also provided competent medical attention and the most modern medical facilities for employees injured in their work. The published literature[32–35] reveals that all the producing nations have taken action to protect their workers.

Despite this record, various government agencies in most nations have enacted new regulations to improve the health and safety conditions in the work place. In December 1970 the Congress of the United States passed the Occupational Safety and Health Act (OSHA).[36] This broad-based law covers all industrial areas and pertains to such matters as clean air, chemical toxicity, safe working conditions, and noise abatement. Similar laws have been enforced in other nations.[37,38] For the most part, the coke battery area has received more attention from government regulators than has the gas treatment plant, but there is concern also for the gas workers, especially the light oil and tar plant employees.

Industrial health specialists who have studied the effect of the coke plant working environment on the employees claim that the working conditions at the top of the coke-oven batteries are injurious to the workers' health.[39–42] Lloyd and Redmond made a statistical study of the health records of a large number of men employed in the coking industry since 1948. Their findings indicate that oven workers with 5 years experience or more tended to present cancer of the respiratory system, whereas nonoven workers, by-product plant workers, and others tended to show cancer of the digestive system and the pancreas. These studies indicated that certain aromatic compounds contained in tar are carcinogenic or otherwise harmful.[43,44]

Benzene has been known to be toxic for many years. Tauber[45] described the way that an employee reached an untimely death on exposure to a large light oil spill. Other authors[46–48] also reported on the toxicity of benzene and other light oil products, but there seems to be a controversy over the exact nature of benzene poisoning. A recent published report[49] declared that there was no link between benzene and leukemia. This statement was based on animal tests that did not result in leukemia when the animals were exposed to benzene vapors by inhalation. However the writer did state that some of these test animals had significant drops in lymphocyte and white cell counts. The research institution that performed these tests planned to continue the study.

Two British doctors have made statements that are contrary to the claims above.

[31] M. J. Massey, F. C. McMichael, and R. W. Dunlap, *Chem. Eng.*, **83**(18), 109–113 (August 30, 1976).

[32] B. J. Fuller, *J. Inst. Gas Eng., London,* **2**, 203–225 (1962).

[33] J. Fiedler, *Glueckauf,* **110**(117), 697–703 (1974).

[34] E. H. Mishin et al., *Coke Chem. (USSR),* No. 7, 28–30 (1974).

[35] D. G. Brinn, *Health and Safety Hazards at Coke Plants and Blast Furnaces,* British Steel Corp., Sheffield, England. [NTIS PB Rep. No. 240599, 1975, 13 p.]

[36] J. D. Hodgson, *Chem. Eng.*, **78**, 108–112 (June 28, 1971).

[37] G. Feldhaus, *Staub,* **34**(6), 160–163 (June 1974).

[38] F. H. Fearn, *COMA Yearb.*, 223–233 (1973); Waverly House, Wexborough, Yorkshire, England.

[39] J. W. Lloyd, *J. Occup. Med.*, **13**(2), 53–58 (February 1971).

[40] C. K. Redmond et al., *J. Occup. Med.*, **14**(8), 621–629 (August 1972).

[41] D. D. Walker et al., *Br. J. Ind. Med.*, **28**(4), 358–363 (1971).

[42] R. B. O'Conner, *Yearbook of the American Iron and Steel Institute* AISI, New York, 1970, pp. 155–160).

[43] V. B. Kapitulskii, *Hyg. Sanit.*, **21**, 408–412 (1966).

[44] S. Mazumdar, *J. Air Pollut. Control Assoc.*, **25**(4), 382–389 (April 1975).

[45] J. Tauber, *J. Occup. Med.*, **12**(3), 91–92 (1970).

[46] K. H. V. French, *Benzole Digest,* No. 2, 2–10 (1968).

[47] R. Doll et al., *Br. J. Ind. Med.*, **29**(4), 394–406 (1972).

[48] C. G. Hunter, *Ann. Occup. Hyg.*, **9**, 191–198 (October 1966).

[49] Anon., *Chem. Week,* **32**, 32 (May 26, 1976).

Hughes[50] referred to benzene as a silent insidious poisoner of the bone marrow, and Archibald[51] cited a reference that reported a death caused by acute leukemia after a worker's 5 1/2-month exposure to benzene vapors. A number of devices[52–55] have been developed to provide the plant safety or health officer with an instrument that will measure the amount of vapors or dust to which a worker has been exposed. From the statements found in the literature, one surmises that these devices are still in the developmental stage, but it can be expected that the plant operators will soon be acquainted with this equipment.

There are presently other aspects of the working environment that interest the occupational hygienist and safety officers. Work helmets, eye protective shields, respiratory masks, and safety shoes have become standard equipment for those employed in hazardous work areas. Now there are other working conditions that hygienists and safety officers are studying. Hughes[50] touches on some of these, including proper lighting, sufficient ventilation, and avoidance of excessive noise and vibration. He described the harmful effects to the workers attributed to each of these conditions. Thus the trend toward improving the working environment may well continue into the future.

18.2 GAS COOLING

Primary gas cooling and final gas cooling were not discussed in the first Supplementary Volume. Undoubtedly it was assumed that no major changes in the function or design of this equipment had taken place since the original volumes were published. This is no longer true, and a brief discussion of current designs and operation is now warranted.

[50] G. O. Hughes, *COMA Yearb.*, 107–125 (1972); Waverly House, Wexborough, Yorkshire, England.
[51] R. McL. Archibald, *COMA Yearb.*, 299–303 (1962); Waverly House, Wexborough, Yorkshire, England.
[52] A. N. A. Dicken, Institution of Chemical Engineers Symp. Emission Control, Paper No. 1, May 14, 1975.
[53] E. B. Hay, *Am. Ind. Hyg. Assoc. J.*, **25**, 386–391 (1964).
[54] R. J. Sherwood, *Am. Ind. Hyg. Assoc. J.*, **32**(12), 840–846 (1971).
[55] W. M. Smith, *J. Occup. Med.*, **13**(2), 69–74 (February 1971).

Richards and Taylor gave a detailed description of all phases of coke-oven gas cooling as of 1960.[56] Some of these data, as well as more recent developments, are presented here.

18.2.1 Primary Gas Cooling

The function of primary coolers has not changed: it is still to cool the raw coke-oven gas before it reaches the exhausters, to the lowest practical temperature that can be achieved without major operating problems or excessive maintenance. Direct coolers using recirculated liquor to cool the gas, and tubular indirect coolers are still in operation worldwide.

Standard designs for all primary coolers have had to be reevaluated in recent years because of the increased heat load per standard cubic foot of gas. This heat load[57] has in some cases increased by 25% or more because of higher coal moisture, faster coking, and higher coke temperatures.

18.2.1.1 *Direct Primary Coolers*

For many years all direct primary coolers consisted of towers packed with wooden hurdles. Naphthalene buildup causing high differential pressure across the tower was a constant problem requiring steaming and/or tar flushing. Although many packed coolers are still in service, essentially all units installed in the United States in recent years have been of the spray type.

Design details of spray-type primary coolers vary somewhat among coke plant designers but the end result is the same. Figure 18.2 is a flow diagram for a typical spray-type primary cooler. The gas is contacted with cold circulating liquor, and the circulating liquor is resprayed into the bottom stage, giving two countercurrent stages. The advantages of spray primary coolers are as follows:

1. Constant low differential pressure.
2. No naphthalene plugging.
3. Higher capacity—converting to the spray-

[56] R. O. Richards and A. Taylor, *COMA Yearb.*, 447–465 (1961); Waverly House, Wexborough, Yorkshire, England.
[57] H. Grosick, E. Helm, and J. Airgood, *Iron Steel Eng.*, 27–32 (May 1976).

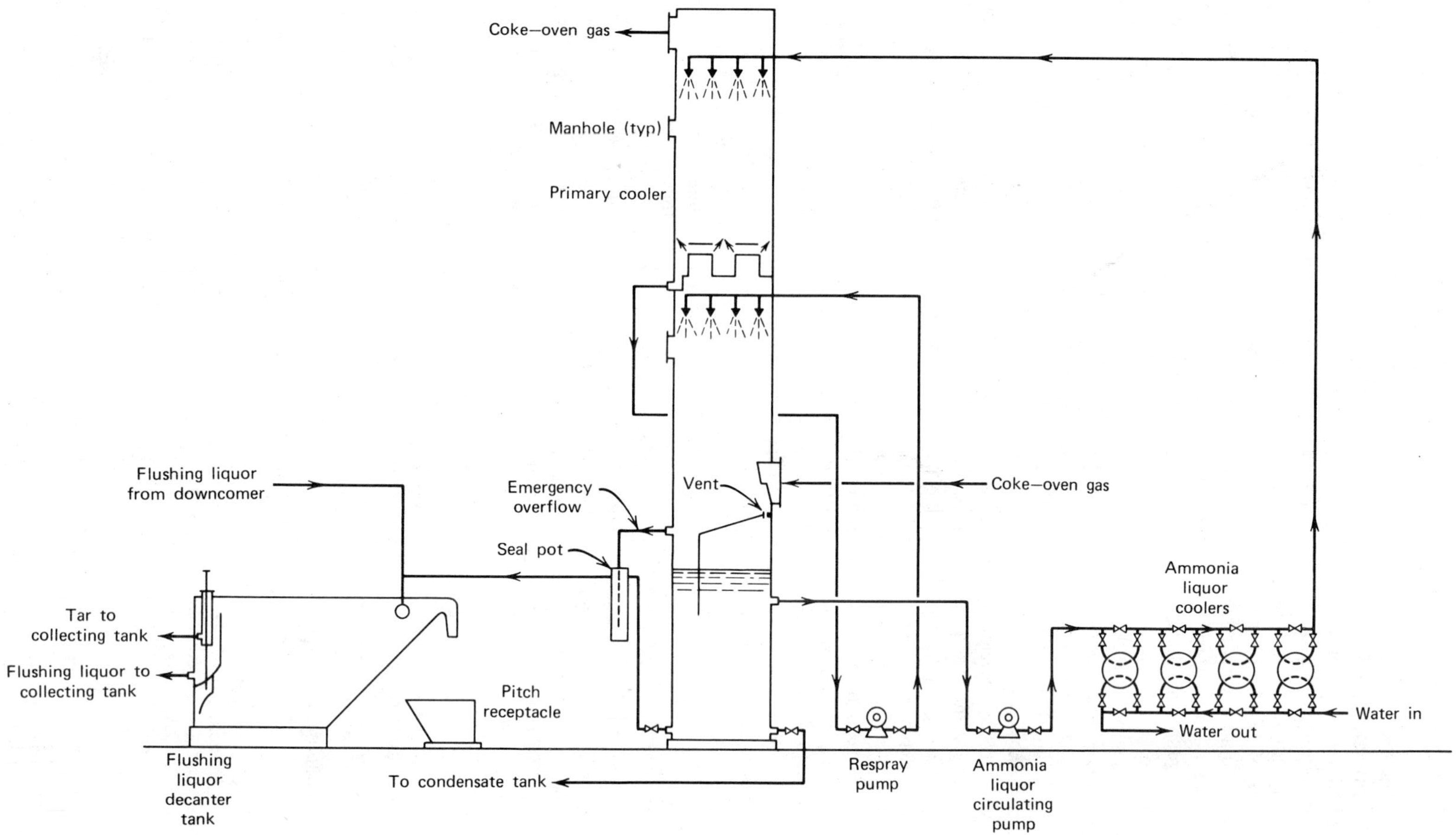

Figure 18.2 Spray-type primary cooler. Diagram courtesy of Koppers Company, Inc.

type unit doubles the gas-handling capacity for a unit of a given size.

One disadvantage is that the cooling efficiency of a spray cooler is not as good as that of a packed cooler. The normal approach of gas outlet temperature to liquor inlet temperature is somewhat higher for a spray cooler.

The type of heat exchanger employed to cool the circulating liquor has been evolved over the years. The standard practice was once to use pipe coils sprayed with water[58]; but radiator sections replaced the pipe coils, and they in turn were superseded by coolers of the shell and tube type. During the last 15 years spiral-type heat exchangers[59] have been installed in many North American and Western European plants. Spirals designed for this service are counterflow and have a high heat transfer coefficient. Because of the reduced surface requirement, spirals are generally less expensive than shell and tube exchangers. Very recently a few installations of plate-type exchangers[60–62] have been used to cool the circulating liquor. Overall performance of these units is still uncertain, but the very high heat transfer coefficient, small space requirement, and reduced cost make them appear attractive.

18.2.1.2 *Indirect Primary Coolers*

Indirect tubular primary coolers[58] have been in operation worldwide for many years. Although most new installations in North America are of the direct spray type, many indirect tubular coolers are still in service today. Essentially all indirect coolers in the United States and the U.S.S.R. employ vertical tubes (Fig. 18.3), whereas some Western European plants use horizontal tubes. The horizontal tube units bring the gas in at the top so that all the tubes are washed with tar; improved cooling and naphthalene removal are claimed. High capital cost has delayed their adoption in the United States.

Indirect tubular primary coolers have always had a problem with naphthalene. Steaming was the standard cleaning method for many years, but is expensive and time-consuming, and causes some atmospheric pollution. In recent years hot gas[63,64] has gained favor in the U.S.S.R. and North America as a cleaning agent. This method is accomplished by draining the water from the tubes and permitting a relatively small flow of hot coke-oven gas to heat the cooler, so that the naphthalene is melted and flows with the tar from the base of the cooler. Less than 1 hr is required per cooler.

A "tubeless" indirect primary cooler has been developed in the U.S.S.R.[65] This cooler consists of a series of horizontal water chambers and top gas inlets, similar to the older horizontal tube cooler. Good results are indicated, but costs appear to be high.

18.2.1.3 *Combination Indirect-Direct Cooler*

A combination cooling and washing system was proposed by Didier-Werke AG[66] in which the gas is first cooled indirectly to about 45°C. The condensate is then chilled and used as the cooling medium in a direct cooler, to cool the gas to 20°C. The process is claimed to be simpler and less expensive than other gas cooler-washers. A similar system without refrigeration has been installed in at least two U.S. plants. These units were quite expensive.

18.2.2 Final Gas Cooling

Final gas cooling received only brief mention[67] in the original volumes of *Chemistry of Coal Utilization* (1945). Final gas cooling

[58] P. J. Wilson, Jr., and J. H. Wells, in *Chemistry of Coal Utilization,* Vol. 2, H. H. Lowry, Ed., John Wiley and Sons, New York, 1945, p. 1375.

[59] A. M. Hargis, A. T. Beckman, and J. J. Loiacono, *Chem. Eng. Prog.,* **63,** 62–67 (July 1967).

[60] J. Marriott, *Chem. Eng.,* **78,** 127–132 (April 1971).

[61] C. T. Cowan, *Chem. Eng.,* **82,** 100–103 (June 1975).

[62] C. T. Cowan, *Chem. Eng.,* **82,** 102–104 (July 1975).

[63] M. Y. Finkel, *Coke Chem. (USSR),* No. 9, 35–38 (1969).

[64] M. Y. Finkel, V. V. Markov, V. P. Belonoshchenko, E. E. Kondrashov, V. V. Garmata, and A. N. Belitskii, *Coke Chem. (USSR),* No. 9, 34–35 (1972).

[65] M. D. Kuznetsov et al., *Coke Chem. (USSR),* No. 11, 45–47 (1973).

[66] AG Didier-Werke, West German Pat. 1,810,-601, 1970.

[67] A. R. Powell, in *Chemistry of Coal Utilization,* Vol. 2, H. H. Lowry, Ed., John Wiley and Sons, New York, 1945, p. 1235.

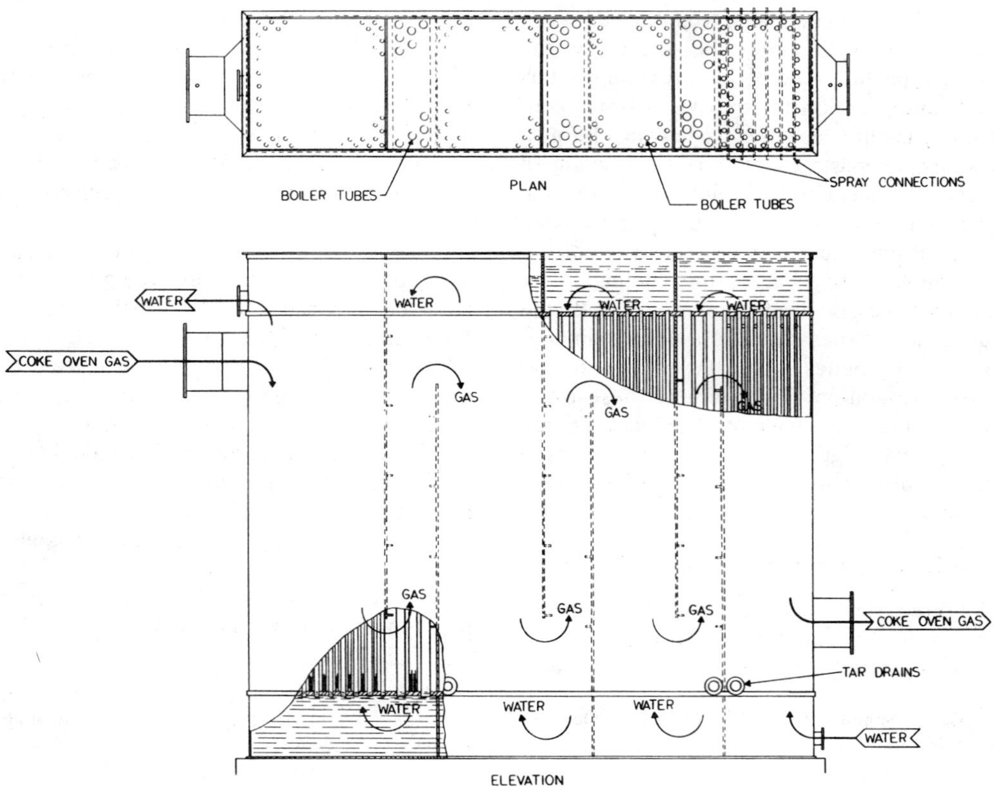

Figure 18.3 Primary cooler—indirect type. Diagram courtesy of Koppers Company, Inc.

units are required in all low-pressure coal chemical plants in which the semidirect process of ammonium sulfate production is employed. The coke-oven gas must be cooled before it is scrubbed in the benzol washers.

18.2.2.1 *Direct Water Final Cooler*

The direct water type of final cooler often consists of a wooden hurdle packed tower in which the gas is cooled directly by means of water recirculated over an open cooling tower. A fairly large quantity of naphthalene may be precipitated by this cooling process (particularly if the primary cooler outlet temperature is high) and a sump with skimmer is provided for removal of naphthalene from the surface of the water. The water also picks up traces of ammonia, some hydrogen sulfide, a sizable quantity of hydrogen cyanide (HCN), and smaller quantities of other contaminants from the gas.[68] The air leaving the cooling tower is therefore polluted, and this type of final cooler may be prohibited in the future. However many packed coolers are still in operation worldwide. In fact, a plane parallel packing has recently been developed in the U.S.S.R.[69]

Spray-Type Direct Water Final Cooler. As in the case of the primary cooler, the spray-type design doubles the capacity of a given size of final cooler. Coolers can be designed with or without respray, depending on the available water capacity. The greatest advantage of the spray-type cooler is that it does not plug with naphthalene, therefore does not require frequent steaming. Naphthalene is still handled in a sump, as in the case of the packed cooler. Many packed coolers in North America have been converted to the spray type to gain increased capacity.

Tar Bottom-Type Final Cooler. The han-

[68] L. N. Kutuzova, A. F. Kononenko, and Z. G. Sashevskaya, *Coke Chem. (USSR),* No. 7, 45–46 (1970).

[69] I. M. Khanin, O. V. Fedulov, V. A. Mizin, V. D. Barskii, A. P. Moroz, L. S. Maksimov, A. N. Tsikalovskii, and E. G. Friman, *Coke Chem., (USSR),* No. 1, 35–39 (1973).

dling of solid naphthalene from the sump can be avoided with either the packed or the spray-type final cooler by providing a tar-contacting section for the warm water containing naphthalene that is leaving the gas-cooling compartment. Crude tar is pumped from the collecting tank and returned to the decanter tank. Results have generally been good. Proper design is critical to avoid excessive tar carryover to the cooling tower and/or excessive water carryover with the tar. Schulte[70] has patented a design employing sloping plates in the tar section to improve separation. To minimize these problems, some plants have operated with a batch tar system. Figure 18.4 shows a spray-type final cooler system incorporating Schulte's design.

18.2.2.2 *Direct Recirculated Oil Type*

The direct recirculated oil type of final cooler, circulating a petroleum wash oil, apparently originated in Great Britain.[71] Originally this process was used to eliminate air pollution caused by the final cooler cooling tower and to eliminate naphthalene deposition in the ammonia washers of plants with indirect ammonia recovery equipment. It was successful in both cases. Various types of indirect cooler served to cool the recirculated oil.

Since 1960 several installations equipped with both packed and spray-type gas coolers have been constructed in North America. More units will probably be installed in the future, to eliminate the pollution caused by the open-type cooling tower. Some cooling efficiency is lost because of the need for indirect oil coolers, and the installation is more expensive, primarily because of the relatively poor heat transfer when cooling the wash oil. A larger oil circulation is also required because the specific heat of the wash oil is lower. Figure 18.5 is flow diagram of a typical unit. The methods for removing

[70] E. V. Schulte, U.S. Pat. 2,956,639, October 18, 1960.

[71] R. O. Richards and A. Taylor, *COMA Yearb.*, 458–464 (1961); Waverly House, Wexborough, Yorkshire, England.

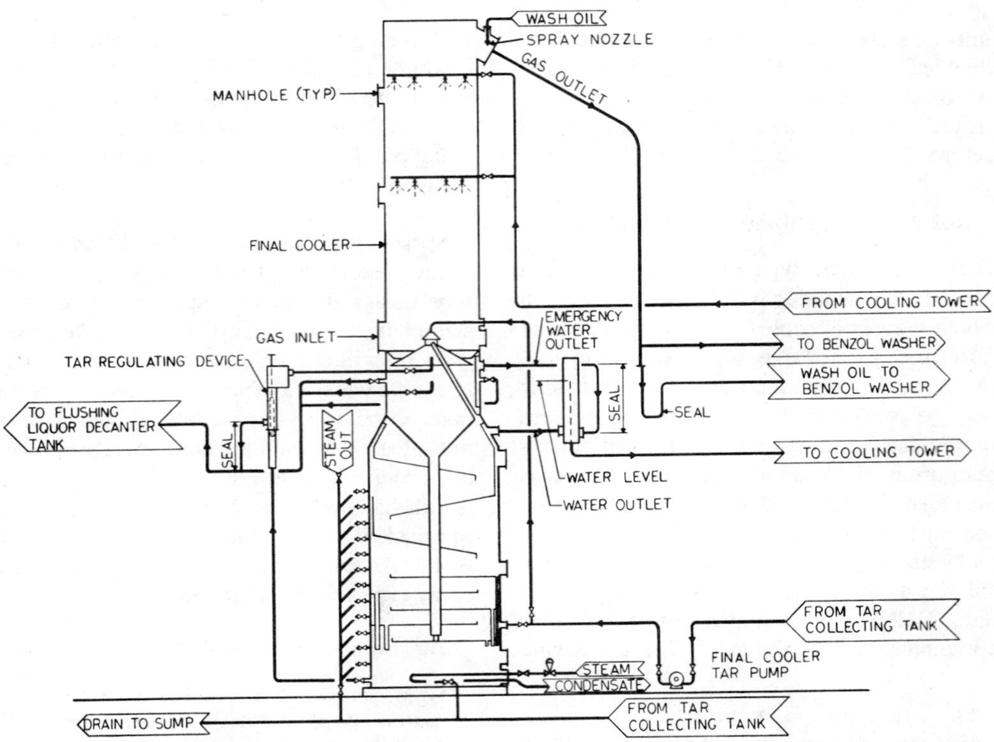

Figure 18.4 Tar bottom-type final cooler. Diagram courtesy of Koppers Company, Inc.

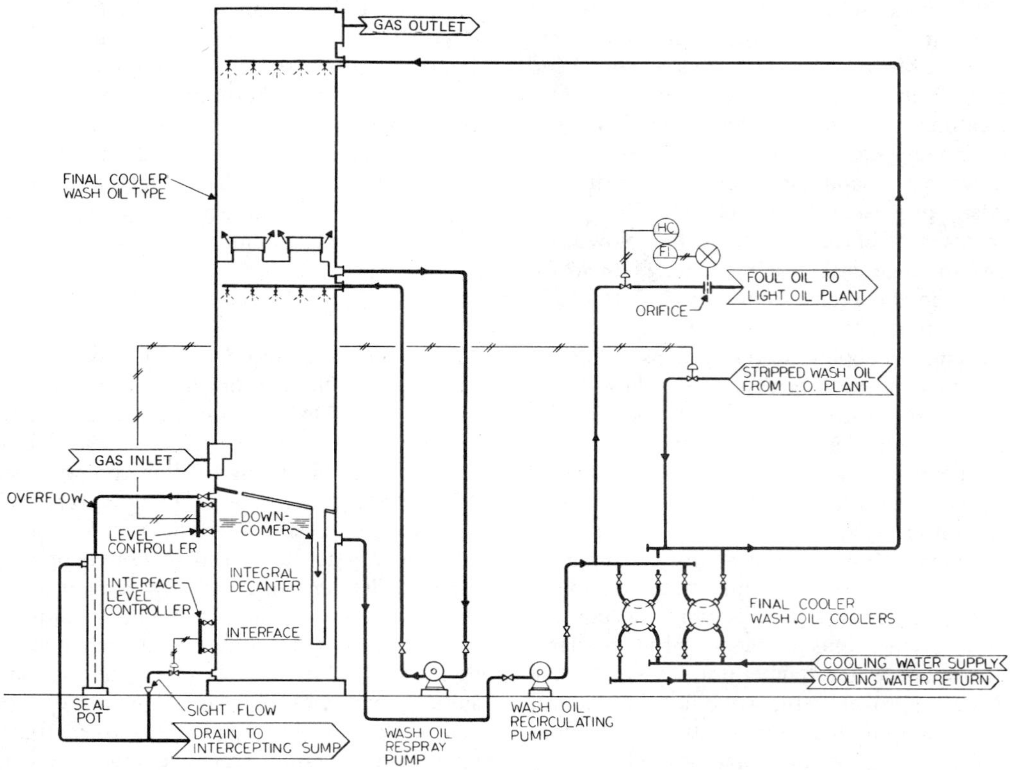

Figure 18.5 Wash oil spray-type final cooler with light oil plant. Diagram courtesy of Koppers Company, Inc.

naphthalene from the recirculated oil are discussed later in this chapter. It is obvious that the oil-type final cooler serves both as a gas cooler and as a naphthalene scrubber.

18.2.3 Closed-Loop Cooling Water Systems

Since the by-product coke plant was first put into operation it has been common practice to supply once-through cooling water for all services in which the water is not chemically contaminated. There has been a gradual shift in the last 15 years to closed-loop cooling water systems for these services. The reasons for this change include the following:

1. Fouling of heat exchangers at certain seasons due to very dirty water.
2. Extremely high summer water temperatures in rivers and lakes.
3. Scarcity of fresh water at some plant locations.
4. Thermal pollution regulations.
5. The possibility of leaks causing chemical pollution of the once-through cooling water.

The average coke plant does not use enough water to justify the use of huge natural draft towers, hence induced draft towers are normally installed for this service. Much work has been done by cooling tower manufacturers to develop various packing materials and arrangements.[72] Both counterflow and crossflow towers are available.

Another type of cooler, the wet surface air cooler,[73] is now being introduced in the United States. This cooler is a combination tubular cooler–cooling tower, in which air and water flow concurrently but countercurrent to the fluid to be cooled. The water is recirculated. Lower cost and less maintenance are claimed.

With either cooling towers or wet surface

[72] D. J. Tow, *Br. Chem. Eng.*, 191 (March 1960); 256–259 (April 1960).

[73] W. Kals, *Chem. Eng.*, 90–94 (July 26, 1975).

air coolers, water treatment is required to prevent scaling, corrosion, and biological growth. Much research[74–78] has recently been done in this area to develop less toxic water treatment chemicals, since regulations governing disposal of chromates in the blowdown are becoming much more stringent. Also, processes have been developed for removing[79] chromates from the blowdown and in some cases for reclaiming[80] them for reuse in the recirculating system. It is obvious that blowdown is necessary in all closed-loop cooling water systems to control the concentration of dissolved solids. Many coke plants solve the blowdown problem by disposing of it as quenching station makeup.

Closed-loop cooling water systems are certainly more expensive than once-through systems if an adequate supply of fresh water is available. However the ability to minimize the variation from minimum to maximum water temperature permits more constant operating conditions in the by-product plant. Maximum water temperature is set by the design approach to the wet bulb temperature, but winter minimum temperature can be controlled by fan operation and/or the bypassing of hot water to the cold well. In addition, recirculated treated water normally causes less fouling in the heat exchangers.

18.3 COAL CHEMICALS

Coal chemicals are the by-products of coal carbonization recovered during the purification of coke-oven gas. Historically these products have included tar, ammonia, light oil, and naphthalene. Sulfur should now be added to this list, since there are many sulfur recovery units in operation. Sulfur is recovered either as elemental sulfur in the Claus process and in the liquid oxidation processes, or as sulfuric acid in the contact acid process. Other chemicals recovered in limited quantities include hydrogen cyanide and the tar acids and bases. Generally these are recovered as salable products only in the Eastern European countries.

18.3.1 Light Oil

Light oil, also referred to as crude benzol, is derived from the carbonization of coal, at a rate of 3.0–3.5 gal/ton of coal. It is a mixture of aromatics, saturated and unsaturated hydrocarbons, and organic sulfur compounds. The relative proportions of the three types of hydrocarbon present in the light oil are related primarily to the carbonization temperatures. Oils derived from low-temperature processes contain a higher proportion of paraffins to aromatics, whereas at higher temperatures the opposite is true. The organic sulfur compounds vary as a function of the coal sulfur content. Light oil derived from the high-temperature process contains a great number of chemicals, but the aromatics account for 75% or more of the total volume. Generally the light oil is processed to upgrade the benzene, toluene, and xylene to acceptable levels of purity, and the remaining chemicals are considered to be impurities. However in countries where light oil is the chief or only source of some of these remaining coal chemicals, they are recovered as salable products. This practice is especially prevalent in U.S.S.R., where very large volumes of coke-oven gas are processed.

The separation of light oil from coke-oven gas may be accomplished by condensation, adsorption, or absorption. All three methods were developed years ago, and the physical principles involved in these processes are well documented. Today only an occasional description of a process improvement appears in the literature.

The condensation process usually is a combination of compression and refrigeration. Generally the process is used only when the gas is to be entered into long pressurized gas grids or the hydrogen of the gas is to be recovered for further processing. The process is very efficient but has the disadvantages of high capital investment and high energy

[74] P. Powley and P. Butler, *Process Eng.*, No. 4, 74–75 (April 1973).

[75] Z. A. Foroulis, Esso Research & Engineering, British Pat. 1302738, January 10, 1973.

[76] R. Nichols and R. Osborne, *COMA Yearb.*, 150–167 (1972); Waverly House, Wexborough, Yorkshire, England.

[77] V. A. Verbin, E. S. Troininkova, and P. S. Aizikov, *Coke Chem. (USSR)*, No. 2, 56–58 (1974).

[78] A. E. Stafford and A. J. McFarlane, *COMA Yearb.*, 163–178 (1975); Waverly House, Wexborough, Yorkshire, England.

[79] J. H. Haggenmacher and L. Lawlor, *Chem. Process.*, **40**(3), 166–167 (March 1977).

[80] Liquitech, Division of Thermotics, Inc., *Chem. Proc.*, **40**(3), 184 (March 1977).

requirements. Some authors[81,82] suggest that a combined process of refrigeration-compression with absorption would have the following advantages over absorption at atmospheric pressures: reduced equipment size, increased efficiencies, and smaller volumes of circulating absorbing oils. The Soviet authors claim that in the future it may be necessary to adopt such a process as the coking plants increase in size. The large-scale Soviet coke plants now are generating gas at a rate approaching 4.5×10^6 ft^3/hr. However these authors point out that the cost for compression and refrigeration is excessive when using present-day equipment.

In the adsorption process light oil in the vapor state is taken up by the pore structures of certain solids such as activated carbon, silica gel, and the zeolites. This process has been limited in use mostly to small gas plants. Coke-oven gas flows through fixed beds of active carbon until the carbon is unable to adsorb more light oil. The unit is then taken off stream and steam is back-purged through the bed to recover the light oil. The advantages of this process are simplicity of operation and light oil recovery efficiency equal to or better than for the absorption process. However a certain buildup of polymers and elemental sulfur takes place, which is not reversed by steaming. In addition, the carbon must eventually be removed from the process to be regenerated or discarded. Scott[81] presented a brief description and flow diagram for a plant in England that produces 20,000 gal of benzol daily.

Absorption continues to be the favored process for the separation of light oil from coke-oven gas. The physical properties of this process were described in detail in many literature sources documented by Glowacki[83] and Muder[84] in the original and supplementary editions of this work. The process is a closed-loop system that includes absorption towers, distillation columns, and sludge purifiers, plus the necessary pumps and heat exchangers. The gas is washed with a countercurrent stream of fresh absorbent oil to strip the light oil from the gas. The enriched oil is then pumped to the top of a steam distillation tower, where the ascending steam vapors distill the light oil from the descending oil. After the oil has cooled sufficiently, it returns to the benzol-absorbing towers to complete the cycle.

Two types of oil are commonly used as the absorbent: a creosote oil and a petroleum distillate. For many years European operators preferred the creosote oil because it was less expensive. The light oil is more soluble in the creosote oil, so that a smaller volume of oil needs to be circulated to effect good light oil recovery. In turn, this means reduced pumping and steam costs. However the creosote oil dissolves a sludge that forms in processing coke-oven gas. The viscosity of the creosote oil increases with the sludge content until the oil must be discarded. Operators in the United States have used almost exclusively a selected petroleum distillate, and this practice is becoming more common in Europe. This oil has a lower affinity for the light oil products and more oil must be circulated to separate the light oil from the gas, but the increased process costs are not excessive. The sludge that forms can be removed from the petroleum oil by distillation, thereby permitting long-term use of the absorbing oil.

The sludge formed in light oil absorption plants is the result of combination of polymerization and oxidation reactions. It is believed that polymerization occurs as the styrene, indene, and coumarone react under the influence of temperature to form resins. The presence of oxygen in the gas results in the formation of iron oxide, elemental sulfur, and thiocyanates by the reaction with the NH_3, H_2S, and HCN in the gas and the iron in the processing equipment. Claxton[85] reports that plants using vacuum stills experience very little sludge formation. The sludge is usually considered to be a waste product, but in some countries it is processed to recover the resins that are found there. Smolyakov[86] re-

[81] T. Scott, *Gas J.*, **33**, 300–307 (August 31, 1960).

[82] G. D. Kharlampovich et al., *Coke Chem. (USSR)*, No. 3, 23–28 (1973).

[83] W. L. Glowacki, in *Chemistry of Coal Utilization*, Vol 2, H. H. Lowry, Ed., John Wiley and Sons, New York, 1945, p. 1231.

[84] R. E. Muder, in *Chemistry of Coal Utilization*, Supplementary Volume, H. H. Lowry, Ed., John Wiley and Sons, New York, 1963, pp. 629–674.

[85] G. Claxton, *COMA Yearb.*, 322–336 (1962); Waverly House, Wexborough, Yorkshire, England.

[86] E. V. Smolyakov et al., *Coke Chem. (USSR)*, No. 6, 46–48 (1972).

ports that in Russia the sludge is the chief source of these resins and is valuable to the economy.

18.3.1.1 *Light Oil Refining*

Ordinarily, light oil is refined to bring the benzene, toluene, and xylenes to the levels of purity required by the users of the products. Light oil is refined either by acid washing or by catalytic hydrogenation. In either case the treated light oil must be subjected to crystallization procedures or to various distillation procedures to achieve an end product that meets the buyers' specifications.

The acid wash process is the older process, and in the early days of the industry it was the only refining process available. Concentrated sulfuric acid is mixed with the crude light oil and reacts with unsaturated hydrocarbons and the thiophenes and mercaptans to form an acid sludge, which is separated from the oil. The saturated hydrocarbons, straight-chain paraffins and naphthenes, and carbon disulfide are left untouched and must be removed by distillation and/or crystallization methods. The aromatic compounds are sulfonated to some degree, but this can be controlled by judicious manipulation of the process variables. Detailed descriptions of various procedures for acid washing can be found in the earlier editions of this work. Although catalytic hydrogenation has become the favored process, many plants still use the acid wash process. In some cases this choice is made because it is uneconomic to employ the hydrogenation process in a small plant or because it is desired to recover the unsaturated hydrocarbons as salable products.

Process development work has continued in an effort to improve the acid wash method. Claxton[87] described a method for reducing the thiophene content of benzene to less than 1 ppm in a two-stage treatment of benzene with liquefied sulfur trioxide (SO_3). A British patent[88] was issued for a process in which the addition of small quantities of dimethylamine to the acid wash and the application of an electric field to the mixture are claimed to enhance the reaction of the unsaturates with the acid. Gorback and Korobchanskii[89] described a unique liquid–vapor phase sulfuric acid process in which the addition of an unnamed unsaturated compound is used to produce benzene with less than 0.5 ppm thiophene. In this process crude light oil is washed with 90–93% sulfuric acid in a two-stage liquid phase unit. About 3–4.5% of additive is added, and the thiophene content is reduced to about 2 ppm. The treated light oil passes to a distillation column from which the benzene vapors enter the bottom of a vertical multistage vapor phase reactor. Additional additive is added (less than 2% wt/wt basis) with preheated sulfuric acid. The final product contains less than 0.5 ppm thiophene. No mention is made of the need for further processing to raise the crystallization temperature of the final product. The authors contend that the capital outlay for such a plant would be less than one-half that of a hydrorefining plant.

Disposal of acid sludge has always been a problem, especially today in view of the environmental control restrictions. A British firm[90] has developed a process for acid sludge treatment that is said to yield salable products and eliminate the disposal problem. The acid sludge is mixed with water and creosote to effect separation of a reclaimed acid layer and a solvent, in a three-stage countercurrent extraction process. The reclaimed acid is said to be clear, free of organic matter, and suitable for ammonium sulfate production. Part of the solvent is recovered from the process through distillation and sent to the crude light oil system. A residue is recovered from the still bottoms, which may be added to the crude tar. This process is said to cause severe corrosion and all equipment is designed for corrosive service.

The first commercial-scale catalytic hydrogenation plants for light oil refining appeared in about 1950. The German firm Scholven-Chemie AG adopted a process developed about 25 years earlier by Badische Analin and Soda Fabrik, AG. In this process light oil vapors are mixed with hydrogen and passed through beds of cobalt-molybdenum

[87] G. Claxton, *COMA Yearb.*, 160–172 (1963); Waverly House, Wexborough, Yorkshire, England.

[88] Howe-Baker Engineers, Incorporated, British Pat. 1,166,667, December 16, 1966.

[89] V. M. Gorback and V. I. Korobchanskii, *Coke Chem. (USSR)*, 28–31 (1975).

[90] J. G. D. Molinari and C. A. Roberts, *Gas World*, **166**, 74–78 (1967).

catalyst at approximately 350°C and 30–35 atm pressure. The sulfur compounds are converted to H_2S and paraffin gases, and the unsaturated hydrocarbons are hydrogenated to the saturated state, while the aromatics are left unreacted. The refined product contains 1 ppm or less of thiophene but has a low crystallization point because of the presence of the saturated hydrocarbons. This problem was resolved by extractive distillation, and the refined product from the combined unit meets the rigid specifications of the chemical industry. The high pressure equipment plus the extractive distillation unit are expensive. This process is not economically feasible except for large installations. A number of hydrorefining plants have been constructed in all the major coke-producing countries. Process descriptions and flow diagrams are found in several literature sources.[84,91–93]

The deficiencies inherent in the processes above prompted two American firms, the Bethlehem Steel Corporation and the Houdry Process and Chemical Company, to engage in a joint effort to develop a single-stage refining process. They succeeded in developing the Litol process,[94] a catalytic hydrogenation and dealkylation procedure capable of producing benzene with less than 0.5 ppm thiophene and a crystallization temperature of 5.4°C or higher.

The flow diagram for the Litol process is shown in Fig. 18.6. Crude light oil is preheated and mixed with hydrogen at pressures of 700–900 psi to prevent the polymerization of styrene and indene. The vapors are then preheated to about 600°C and pass through a two-stage catalytic reactor for further hydrogenation and finally dealkylation. The vapors from these reactors are quenched with cooled liquid product to reduce the temperature to 400–500°C before passing to a high-pressure flash drum, where the gases are separated from the liquid product. The H_2S-rich gas proceeds to the purification units before it is recycled to the process or enters the process fuel gas system. The liquid product flows first to a stabilizer to remove the trace gases that remained from the flash drum, then to clay treating towers to remove traces of unsaturated hydrocarbons. The refined benzene and toluene are recovered by straight distillation, and the residue from the toluene still bottoms is returned to the crude feed stream.

Typical chemical reactions are the following:

Hydrogenation

$$\underset{\text{styrene}}{C_6H_5CH\ \ CH_2} + H_2 \rightarrow \underset{\text{ethylbenzene}}{C_6H_5CH_2{-}CH_3} \tag{18.1}$$

$$\underset{\text{thiophene}}{2C_4H_4S} + 8H_2 \rightarrow \underset{\text{butane + hydrogen sulfide}}{2C_4H_{10} + 2H_2S} \tag{18.2}$$

$$\underset{\text{carbon disulfide}}{CS_2} + 4H_2 \rightarrow \underset{\text{methane + hydrogen sulfide}}{CH_4 + 2H_2S} \tag{18.3}$$

Dehydrogenation

$$\underset{\text{cyclohexadiene}}{C_6H_8} \rightarrow \underset{\text{benzene}}{C_6H_6} + H_2 \tag{18.4}$$

$$\underset{\text{cyclohexene}}{C_6H_{10}} \rightarrow \underset{\text{benzene}}{C_6H_6} + 2H_2 \tag{18.5}$$

Dealkylation

$$\underset{\text{toluene}}{C_6H_5CH_3} + H_2 \rightarrow \underset{\text{benzene + methane}}{C_6H_6 + CH_4} \tag{18.6}$$

$$\underset{\text{xylene}}{C_6H_4(CH_3)_2} + 2H_2 \rightarrow \underset{\text{benzene + methane}}{C_6H_6 + 2CH_4} \tag{18.7}$$

Undesired Reactions

$$\underset{\text{benzene}}{2C_6H_6} \rightleftarrows \underset{\text{biphenyl}}{C_6H_5{-}C_6H_5} + H_2 \tag{18.8}$$

$$\underset{\text{benzene}}{C_6H_6} + 3H_2 \rightarrow \underset{\text{cyclohexane}}{C_6H_{12}} \tag{18.9}$$

[91] R. Middleton and D. Bradley, *COMA Yearb.*, 174–194 (1960); Waverly House, Wexborough, Yorkshire, England.

[92] L. Wilsing, *COMA Yearb.*, 211–218 (1961); Waverly House, Wexborough, Yorkshire, England.

[93] I. I. Eru and A. A. Lange, *Coke Chem. (USSR)*, No. 8, 37–40 (1960).

[94] M. O. Tarhan and L. H. Windsor, *Chem. Eng. Prog.*, **62**(2), 67–72 (1966).

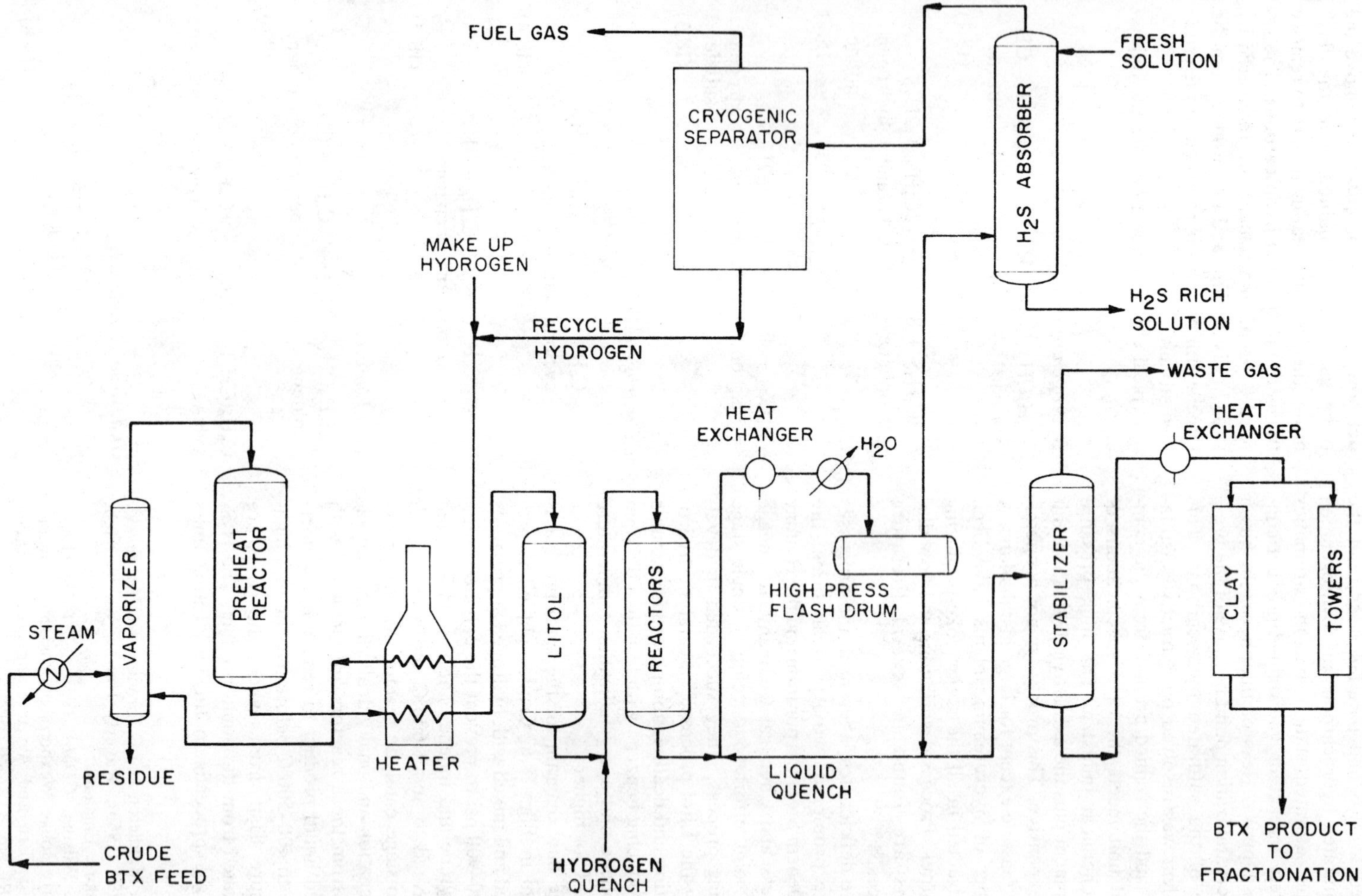

Figure 18.6 The Litol refining process.

As implied in these reactions, the process is capable of producing more benzene than was present originally at the expense of toluene and xylene, but the reactions can be controlled by adjusting the operating conditions of the reactors. An undesirable reaction is the conversion of benzene and its homologues to biphenyl and similar compounds. By recycling these materials back to the feedstream, an equilibrium is reached so that ultimately only a small amount of these undesired chemicals is produced in the process. The chief features that make this system attractive are the pretreatment reactors in which the resin formers, such as styrene and indene, are hydrogenated at high pressures to nonpolymerizable chemicals; and the Litol reactors, in which the high-boiling saturated and unsaturated hydrocarbons are converted to lower molecular weight compounds and benzene. The elimination of the polymerization products stops the fouling of catalyst beds, heat exchangers, and flow streams, and the hydrocracking and dealkylation reactions prevent the formation of the hydrocarbons, hexanes, and heptanes that are difficult to separate from benzene with ordinary distillation. Tarhan and Windsor[94] reported that at first the catalyst in the Litol reactors treated 460 gal of feedstock/lb of catalyst during a 340-day campaign. The catalyst was regenerated *in situ* with an atmosphere of nitrogen and oxygen and required 73 hr for regeneration. Thereafter it became normal practice to regenerate at the conclusion of operating cycles of 6–8 months, which could be accomplished in 45–50 hr.

The first Litol plant was installed in 1964 for the Bethlehem Steel Corporation. The plant has the capacity to refine more than 20 million gal of crude benzene-toluene-xylene (BTX) feed annually. The feedstock to this unit is first distilled in a foreruns distillation column to remove the low boilers such as carbon disulfide and cyclopentadiene. A second plant was constructed in England,[95] and the author reported that more than 90 million gal of crude coke-oven light oil was processed in 4 1/2 years of operation.[95] The article presented the following typical analyses to describe the feedstock and products that are processed in this plant.

Feed to the Vaporizer	Weight Basis (%)
Nonaromatics	1.5
Benzene	71.0
Thiophene	1.0
Toluene	19.0
Ethylbenzene	1.5
Xylenes	4.0
Styrene	0.5
Indene	0.1
Biphenyl	0.9
Naphthalene	0.2
Carbon disulfide	248 ppm

Overhead Product from Benzene Fractionation Tower	
Benzene crystallization temperature	5.46°C
Nonaromatics	0.11%
Toluene	<0.001%
Thiophene	<0.05 mg/kg
Total sulfur	0.5 mg/kg
Carbon disulfide	<0.1 ppm

Bottoms from Toluene Fractionation Tower (recycle back to feedstream)	Amount (%)
Nonaromatics	<0.1
Benzene	<0.1
Toluene	5.0
Xylenes	28.2
Ethylbenzene	21.3
Naphthalene	9.9
Biphenyls	31.3

The author claims that the plant consumes 14.2 ft^3 of hydrogen per gallon of feed treated and that processing costs are similar to those for acid washing.

Research and development continues in the search for improved methods of light oil refining. Undoubtedly some of these will achieve commercial status and some will remain laboratory curiosities. There has been considerable study of ways to improve the distillation procedures to separate the BTX oils from their impurities, particularly with respect to finding more efficient solvent systems for extractive distillation. Some of the solvent systems that have gained a measure of success are propylene carbonate,[96] *N*-

[95] T. Ward, *COMA Yearb.*, 208–222 (1973); Waverly House, Wexborough, Yorkshire, England.

[96] H. Koppers GmbH, British Pat. 1,139,710, January 25, 1969.

methylpyrrolidine and ethylene glycol,[97] sulfolane,[98] and *N*-formylmorphine.[99]

Studies have been made to discover suitable chemicals that will react with the impurities, especially the olefins and sulfur compounds, and permit the reaction products to be separated from the refined product, either by simple distillation or phase separation. It is claimed that the reaction of benzene and sulfuryl chloride yields benzene with a crystallization point of 5.5°C.[100] Also, benzene that is treated in a process that employs aluminum bromide, hydrogen bromide, and a methyl benzene having one to five methyl groups per molecule meets the rigid specifications for synthesis use.[101] In addition, gaseous boron trifluoride[102] is claimed to be an effective catalyst for thiophene removal.

The refining processes described above find application throughout the coal carbonization industry, according to the economic situation of each producer. For example, only three large U.S. plants use the medium-temperature hydrorefining process or the high-temperature hydrocracking-dealkylation process. Many of the American producers find it more economical to sell the crude light oil without further processing. West Germany resorts in large part to the medium-temperature hydrogenation process, and extensive research effort has been made in that country to improve the distillation methods required to bring the benzene and toluene fractions to the high standards that are needed. Great Britain still employs the acid wash treatment, coupled either with distillation procedures or benzene crystallization to yield high-purity product. Catalytic hydrogenation is used only in the largest plants. The Eastern European nations rely heavily on acid washing to recover the so-called impurities, mainly olefins, as salable products. Russia, however, which produces more than 20% of the total world's coke, has a number of hydrogenation plants in operation. The trend seems to be toward large centralized processing plants, to make use of the large-scale catalytic processes to produce a high-quality product. Many plants have been built with a processing capacity exceeding 50,000 tons/year. Japan finds it expedient to install the catalytic hydrogenation units in the newest plants, whereas India still resorts to acid washing in its many small-scale plants.

[97] E. Müller and G. Hohfeld, *Erdoel Kohle*, **24**(9), 573–578 (September 1971).

[98] I. S. Volfson, *Khim. Technol. Topl. Masel*, No. 2 (February 1963).

[99] E. Cinelli, S. Noe, and G. Paret, *Hydrocarbon Process.*, **51** 141–144 (April 1971).

[100] J. Dingle, *COMA Yearb.*, 320–335 (1966); Waverly House, Wexborough, Yorkshire, England.

[101] A. Schneider and E. J. Janoski, U.S. Pat. 3,148,226, September 1964.

[102] E. P. Babin, A. A. Kuz'menkov, and N. L. Zotova, *J. Appl. Chem. (USSR)*, **38**(6), 1237–1239 (June 1965).

18.3.2 Naphthalene Removal

The production of naphthalene during the carbonization of coal and the problems associated with its removal from coke-oven gas are well documented in the original volumes of *Chemistry of Coal Utilization*.[103] It has long been recognized that good naphthalene removal in the primary coolers greatly simplifies the residual cleanup in subsequent equipment. Gas outlet temperature is of paramount importance, but Lebedeva et al.[104] have shown that better contact with tar in the primary cooler can reduce residual naphthalene concentration to levels below the temperature equilibrium concentration. Some plants are taking advantage of this characteristic in both direct and indirect primary coolers.

Temperature equilibrium concentrations of naphthalene in coke oven gas have been reported in the literature, including the original volumes of *Chemistry of Coal Utilization*. Mikhailov et al.[105] disagreed with these data and showed somewhat higher concentrations above 15°C and somewhat lower concentrations below. Even if the data of Mikhailov et al. are correct, the problem is not greatly changed. It is still necessary to lower the naphthalene concentration to such

[103] A. R. Powell, in *Chemistry of Coal Utilization*, Vol. 2, H. H. Lowry, Ed., John Wiley and Sons, New York, 1945, pp. 1232–1242.

[104] G. N. Lebedeva, N. G. Bunakov, V. S. Patrikeev, and E. S. Predeina, *Coke Chem. (USSR)*, No. 6, 38–42 (1973).

[105] N. F. Mikhailov, V. M. Zaichenko, and L. P. Kolomiets, *Coke Chem. (USSR)*, No. 12, 26–28 (1969).

a level that it will not precipitate in the distribution systems.

The removal of naphthalene in all operations of the coal chemical plant was studied at the Avdeevka Coke and Chemical Works.[106] This study showed that 43% of the total naphthalene produced is removed in the primary coolers, further emphasizing the importance of adequate primary cooling. However in most conventional plants using recirculated water in direct final coolers, poor primary cooler operation is essentially counteracted because the naphthalene concentration leaving the final cooler is at the outlet temperature equilibrium and additional naphthalene is removed. However in the case of oil-type final coolers or naphthalene scrubbers, the naphthalene loading and process design are greatly influenced by the primary cooler outlet temperature.[107]

18.3.2.1 *Naphthalene Removal with Oil-Type Final Coolers*

In the case of oil-type final coolers, previously discussed in Section 18.2.2.2, it is essential that a small stream of oil be continuously removed from the recirculating system to maintain the desired naphthalene concentration. This bleed must be replaced by fresh oil or regenerated for reuse. If insufficient oil is removed from the system, the naphthalene concentration will increase so that less naphthalene is removed from the gas, and water-oil emulsions will become a serious problem. Naphthalene removal is important because essentially all the naphthalene leaving the primary cooler enters the oil-type final cooler, and it is imperative to keep the naphthalene loading to the benzol scrubbers within a reasonable range. We will consider the use of oil-type final coolers for plants recovering light oil and for plants without light oil recovery. In either case the use of oil-type final coolers is recommended only when ammonia is removed ahead of the final coolers.

[106] V. N. Ilyashenko, V. M. Chuishchev, and E. L. Volkov, *Coke Chem. (USSR),* No. 3, 44–46 (1972).

[107] R. O. Richards and A. Taylor, *COMA Yearb.,* 458–464 (1961); Waverly House, Wexborough, Yorkshire, England.

Oil-Type Final Cooler with Light Oil Plant. In some of the early oil-type final cooler installations in the United Kingdom,[107] with both semidirect and indirect ammonia recovery, a side stream of oil from the final cooler circulating system was taken directly to the benzolized oil stream, then to the wash oil still. Debenzolized oil was used as makeup to the final cooler oil system. Naphthalene removal was satisfactory only because the primary cooler outlet temperature was approximately 25°C and the total naphthalene to be removed in the wash oil still was relatively small. It was recognized by engineers in the United Kingdom that this system would not give satisfactory naphthalene removal if the primary cooler outlet temperature was in the range of 35–40°C, which is common under summer conditions in many parts of the world. Under these conditions the naphthalene to be removed in the wash oil still is increased by a factor of 2.2–6.6. At preset operating temperature and pressure, the only variables controlling naphthalene removal from the wash oil are steam rate and naphthalene concentration. Since the steam rate cannot be greatly increased, the naphthalene concentration rises in both benzolized and debenzolized oil. The naphthalene content of the debenzolized gas is then increased far beyond acceptable limits.

A new system[108,109] was developed in the mid-1950s to permit better naphthalene removal in the oil-type wash oil cooler and benzol scrubber system. In this system the oil bleed from the final cooler system is controlled at approximately 3–6% naphthalene, depending on inlet loading. This small bleed stream is contacted separately with the total wash oil still vapors, either in trays added to the top of the still or in a separate stripper. This intensive stripping of the small, naphthalene-rich stream permits more naphthalene removal at the normal steam rate, without adversely affecting the benzol scrubber operation. One plant of this type[110] has been operating successfully in the United States for approximately 15 years. Even with this design, primary cooler outlet temperature is important.

[108] J. Maloy, British Pat. 775,978, 1954.

[109] P. J. Wilson, Jr., U. S. Pat. 2,710,663, 1955.

[110] T. P. Dryja, *Blast Furn. Coke Ovens Raw Mater. Comm. (Proc.),* **21,** 463–467 (1962).

A separate naphthalene stripping column can also be used to remove the naphthalene from the bleed stream if the light oil plant is not large enough or if it is desired to return the naphthalene to the tar.[111] It is preferable that this stripper operate at a relatively high temperature (150°C) to minimize steam requirement. Still vapors may be returned to the light oil plant or simply passed through a vapor-to-oil heat exchanger and taken to the suction main ahead of the primary coolers. The stripped oil is cooled and recycled. One plant in North America uses this system and condenses the naphthalene vapors with the crude light oil vapors.

Oil-Type Final Cooler Without Light Oil Plant. One oil-type final cooler without a light oil plant has been in operation in the United States for several years. Naphthalene removal in this type of plant is much more critical than gas cooling. The two-stage spray unit provides cooling only at the bottom recirculating stage. The top stage serves as a second naphthalene removal stage. Naphthalene-rich oil is taken from the bottom stage to the stripping column, after which it is cooled and returned to the top stage. Since naphthalene removal must be very good, it is essential that a high percentage of the naphthalene be removed in the first stage. This is achieved by using a somewhat larger bleed stream to the still than would be required if a light oil plant were present for final cleanup.

The still vapors pass through a vapor-to-oil heat exchanger, then to the suction main. This vapor-to-oil heat exchanger is necessary to reduce the amount of vaporized wash oil carried into the flushing system. Too much oil can upset the tar system. Because of the high operating temperature (150°C), a wash oil purifier for sludge removal is mandatory. Adequate separation of the condensate from the naphthalene-rich oil is necessary in order to maintain the still temperature. A separate decanter is recommended. Naphthalene concentration in the rich oil is normally 2–3%, and less than 0.5% in the stripped oil.

[111] A. L. Kohl and F. C. Riesenfeld, *Gas Purification*, McGraw-Hill Book Company, New York, 1960, p. 529.

18.3.2.2 *Naphthalene Removal with Naphthalene Scrubbers*

Naphthalene removal with naphthalene scrubbers refers to a system in which an oil-type naphthalene scrubber is used to remove naphthalene before the ammonia is removed indirectly by means of water washing. The gas is contacted at the exhauster discharge temperature, and no cooling is involved. Because of both naphthalene loading and naphthalene partial pressure in the oil, good primary cooler operation is imperative. If benzol scrubbers follow the ammonia washers, it is necessary only to reduce the naphthalene concentration in the gas to the extent necessary to prevent naphthalene precipitation in the ammonia washers. If light oil is not recovered, much better naphthalene removal is necessary. These two cases are treated in later sections of this chapter. In either case condensation of water is undesirable in the naphthalene scrubber, and insulation of the gas piping and scrubber is recommended. A small oil heater may also be desired for winter operation.[112]

Naphthalene Scrubber with Light Oil Plant. When naphthalene scrubbers are operated in conjunction with a light oil plant, the rich oil bleed is handled as just discussed using additional trays on the wash oil still or a separate stripper. Because of the higher operating temperature, however, a two-stage scrubber is necessary for satisfactory naphthalene removal. Oil bleed to the additional trays on top of the wash oil still is critical, since too small a bleed will increase naphthalene concentrations in the scrubber, reducing removal efficiency, and too large a bleed will reduce still stripping efficiency with the same result.

This system of naphthalene removal will always achieve better naphthalene removal than a similar system without a light oil plant. However some plants do not include a light oil plant because of economic considerations.

Naphthalene Scrubber Without Light Oil Plant. The plant arrangement in which it is the naphthalene scrubber that must reduce

[112] R. O. Richards and A. Taylor, *COMA Yearb.*, 453 (1961); Waverly House, Wexborough, Yorkshire, England.

the naphthalene concentration to send-out gas levels presents the most difficult design problem. The high temperature level requires a relatively high oil bleed to the stripper to keep the naphthalene concentration in the rich oil to a low level. In addition, the stripping still must operate at high efficiency to reduce the concentration in the stripped oil to a very low level. Even under these conditions, multiple scrubbing stages are required to approach a permissible naphthalene concentration in the send-out gas. The flow diagram for this arrangement is similar to that described in the preceding section.

18.3.3 Ammonia

Ammonia has been an important constituent of raw coke-oven gas since the days when the first coke plant was operated. The production of NH_3 during carbonization and its recovery from the gas have been covered by many publications,[113–115] which discuss in great detail the production of ammonium sulfate, diammonium phosphate (DAP), and concentrated ammonia liquor. Recently processes for the recovery of ammonia as anhydrous ammonia and for its destruction by incineration have been developed. This section covers these developments as well as recent developments in Eastern Europe in the production of ammonium sulfate and DAP in addition to these new processes.

18.3.3.1 *Ammonium Sulfate and Diammonium Phosphate*

Very little work has been done in the United States or Western Europe to improve production of ammonium sulfate or DAP since the advent of spray-type ammonia absorbers and vacuum crystallizers, which were covered by the many references in the first Supplementary Volume.[114] However it is interesting to note that much work in this area has been done in Eastern Europe in recent years. Polish scientists[116] have extensively studied the production of DAP. Russian investigators[117–127] have published many reports of studies to improve the quality of ammonium sulfate as well as plant operations. These reports indicate that the recovery of pyridine in conjunction with ammonia recovery is still important in the U.S.S.R. Granulation[128,129] of ammonium sulfate to improve handling and storage properties has also been thoroughly investigated. One report[130] suggests granulation by fluidized-bed drying of alkaline depyridinized liquor. It seems apparent that coke-oven gas is a much more important source of ammonia in Eastern Europe than in Western Europe and the United States.

18.3.3.2 *Anhydrous Ammonia—United States Steel Phosam Process*

Anhydrous ammonia is the most concentrated and versatile form of fixed nitrogen, being a fertilizer ingredient and a chemical raw

[113] W. H. Hill, in *Chemistry of Coal Utilization*, Vol. 2, H. H. Lowry, Ed., John Wiley and Sons, New York, 1945, pp. 1008–1080.

[114] R. E. Muder, in *Chemistry of Coal Utilization*, Supplementary Volume, H. H. Lowry, Ed., John Wiley and Sons, New York, 1963, pp. 658–664.

[115] A. L. Kohl and F. C. Riesenfield, *Gas Purification*, McGraw-Hill Book Company, New York, 1960, pp. 315–336.

[116] F. Brytus and H. Gwiner, *Koks Smola Gaz*, No. 9, 264; No. 10, 293 (1972).

[117] M. Litvinenko et al., *Coke Chem. (USSR)*, No. 10, 21–26 (1964).

[118] D. Aivenko, *Coke Chem. (USSR)*, No. 9, 39–42 (1965).

[119] G. Lebedeva, L. A. Burmistrenko-Shashmurina, E. G. Kaplina, and O. G. Polanver, *Coke Chem. (USSR)*, No. 10, 37–42 (1965).

[120] M. Finkel, *Coke Chem. (USSR)*, No. 10, 31–34 (1964).

[121] M. Kartsynel et al., *Coke Chem. (USSR)*, No. 10, 47–51 (1969).

[122] R. Golubev et al., *Coke Chem. (USSR)*, No. 11, 29–31 (1972).

[123] M. Rabkin et al., *J. Appl. Chem. (USSR)*, **44**(5), 975–980 (May 1971).

[124] B. Litvinov et al., *Coke Chem. (USSR)*, No. 2, 35–36 (1974).

[125] M. Finkel et al., *Coke Chem. (USSR)*, No. 4, 30–33 (1966).

[126] G. Lebedeva et al., *Coke Chem. (USSR)*, No. 12, 33–37 (1970).

[127] Z. Ostrouskii et al., *Coke Chem. (USSR)*, No. 12, 44–45 (1974).

[128] L. Shashmurina et al., *Coke Chem. (USSR)*, No. 10, 33–35 (1968).

[129] K. Belov et al., *Coke Chem. (USSR)*, No. 6, 35–38 (1971).

[130] V. Privalov et al., *Coke Chem. (USSR)*, No. 12, 29–32 (1970).

material. The quantity of ammonia generated in coal carbonization is small and, until the 1960s, no process to recover anhydrous ammonia of commercial purity had been developed for coke plants. The development of such a process by the United States Steel Corporation, coupled with recent increases in the cost of energy and feedstocks for synthetic ammonia production, have made the production of anhydrous ammonia attractive in coke plants.

In the Phosam process[131–133] (Fig. 18.7) the ammonia is scrubbed from detarred coke-oven gas in a two-stage spray absorber countercurrently by means of an ammonium phosphate solution. The mole ratio of NH_3 to H_3PO_4 in the lean solution is less than 1.5 but more than 1.0. This solution is acidic enough to provide good ammonia removal (> 99%), while very little of the acid gases H_2S, HCN, and CO_2 is absorbed. The rich solution having a mole ratio of NH_3 to H_3PO_4 greater than 1.5 but less than 2.0 is then stripped of ammonia to regenerate the lean solution. This operation is carried out in a stripper operated at a pressure of 150–250 psia using live steam as the stripping medium. The stripped solution is then cooled and returned to the absorber for reuse. The stripper vapor, which contains 15–20% ammonia, is condensed and pumped to a fractionating column operated at a pressure of 200–250 psia, where it is rectified to refrigeration-grade anhydrous ammonia, which is produced as a liquid under pressure.

More than 12 Phosam plants are in operation worldwide.

18.3.3.3 *Ammonia Incineration*

In the mid-1950s the price of ammonium sulfate had fallen so low that even the cost of the acid could not be recovered. Actual losses of \$10–\$20/ton of sulfate produced were common in the United States. All coke plant managers were interested in removing the ammonia from the gas without producing ammonium sulfate. Schulte[134] proposed removing the ammonia by water washing, as in any indirect plant. Total ammonia would be distilled from this liquor and the crude liquor make; the combined vapors would then be concentrated by dephlegmation and incinerated. Wethly[135] proposed that only cooling

[131] R. D. Rice, U.S. Pat. 3,024,090, March 6, 1962.
[132] M. C. Fields and R. O. Rice, U.S. Pat. 3,-186,795, June 1, 1965.
[133] C. W. Hamilton and R. D. Rice, *AIME Ironmaking Proc.*, **34**, 622–637 (1975).

[134] E. V. Schulte, *Blast Furn. Coke Oven Raw Mater. Comm. (Proc.)*, **17**, 237–246 (1958).
[135] F. Wethly, U.S. Pat. 3,020,138, February 6, 1962.

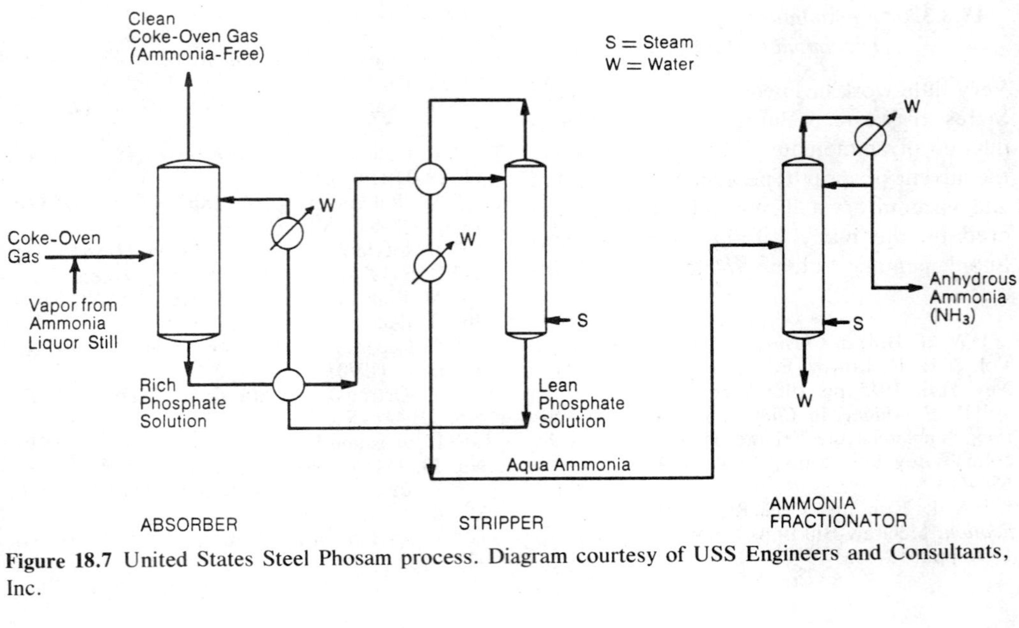

Figure 18.7 United States Steel Phosam process. Diagram courtesy of USS Engineers and Consultants, Inc.

and tar removal be performed prior to using coke-oven gas as fuel. The gas would always be kept at a temperature substantially above its dew point. He did not suggest that this gas could be used for oven underfiring, which would be a decided disadvantage. Schön[136] has patented a process in which ammonia is removed from the fuel gas only. The total ammonia is distilled and added to the surplus gas, which is heated to prevent condensation and is burned in the boilers. Underfiring the ovens with such a gas could present problems. Grulich[137] has patented a process for incinerating ammonia vapors in a strongly reducing atmosphere in which the combustion heat is used to generate steam. Many ammonia incineration or destruction plants have been installed, and several of these processes are described below. Despite the successful operation of these processes, their future use will be determined by pollution considerations and the economics of ammonia recovery.

The Koppers Incineration Process. Koppers Company, Inc., has installed several ammonia incineration plants essentially in accordance with Schulte's[134,138] process. One plant incinerates only the ammonia distilled from the crude ammonia liquor, and the others incinerate total ammonia (Fig. 18.8). In the plants handling total ammonia a naphthalene scrubber is provided ahead of the first ammonia scrubber, and cooling is provided in the bottom stage of this ammonia scrubber. The foul scrubber liquor is pumped from the base of this ammonia scrubber through a heat exchanger to the scrubber liquor still, where the ammonia is stripped with direct steam. The stripped liquor is cooled and recycled to the top of the second ammonia scrubber. The ammonia vapors from both the crude liquor still and the scrubber liquor still are dephlegmated and pass through a common line to the incinerator.

The incinerator is a single combustion chamber, with the products of combustion vented to the atmosphere by means of a lined stack. The addition of supplemental gas to the ammonia vapors is essential for good combustion. Contrary to some expectations, emissions of oxides of nitrogen (NO_x) have not been a problem when combustion temperature is properly controlled and reaction space is properly designed. Sulfur dioxide emissions are substantial, since it is impossible to absorb the ammonia in water without simultaneously absorbing 15–25% of the H_2S in the gas. In some parts of the country it is possible to meet local regulations by venting the combustion products to the oven stack, thereby greatly reducing the SO_2 concentration in the combined effluent. Another possibility is to use a deacidifier[139] ahead of the ammonia still to reduce the concentration of H_2S in the vapors to the ammonia incinerator.

A plant[140] using a process similar to the Koppers process has also been installed in Australia, and results have been very satisfactory.

The Carl Still Company Ammonia Destruction Process. The Carl Still Company (Firma Carl Still)[141–145] has done much work on ammonia destruction. Installations using techniques resulting from this work have been constructed in Europe and Japan. Some of these installations are similar to those using the Koppers process, previously described, in that the vapors are burned and the products of combustion are vented to the atmosphere. Other installations include a catalytic unit for ammonia and HCN cracking in a reducing atmosphere. The exit gases are then used as fuel with or without H_2S removal, or sulfur is recovered as elemental sulfur or sulfuric acid. These processes are de-

[136] Erich Schön, West German Pat. 1,671,346, August 24, 1972; U.S. Pat. 3,451,896, June 24, 1969.

[137] H. Grulich, West German Pat., 2,211,891, March 21, 1974.

[138] H. A. Grosick, *Blast Furn. Steel Plant*, **59**(4), 217–221 (April 1971).

[139] G. Wunderlich and H. Weber, British Pat. 1,318,731, May 31, 1973.

[140] R. Harrison and J. A. Canner, *COMA Yearb.*, 116–126 (1974); Waverly House, Wexborough, Yorkshire, England.

[141] K. Tippmer and J. Maloy, *COMA Yearb.*, 127–144 (1974); Waverly House, Wexborough, Yorkshire, England.

[142] A. Ostmann, Belgian Pat. 779,302, February 11, 1972.

[143] D. Laufhütte, *AIME Ironmaking Proc.*, **33**, 142–154 (1974).

[144] G. Coulat and B. F. Bell, *Gas World*, **160**, 346–353 (1965).

[145] F. Breitbach and G. W. Choulat, German Pat. 1,230,406, June 22, 1967.

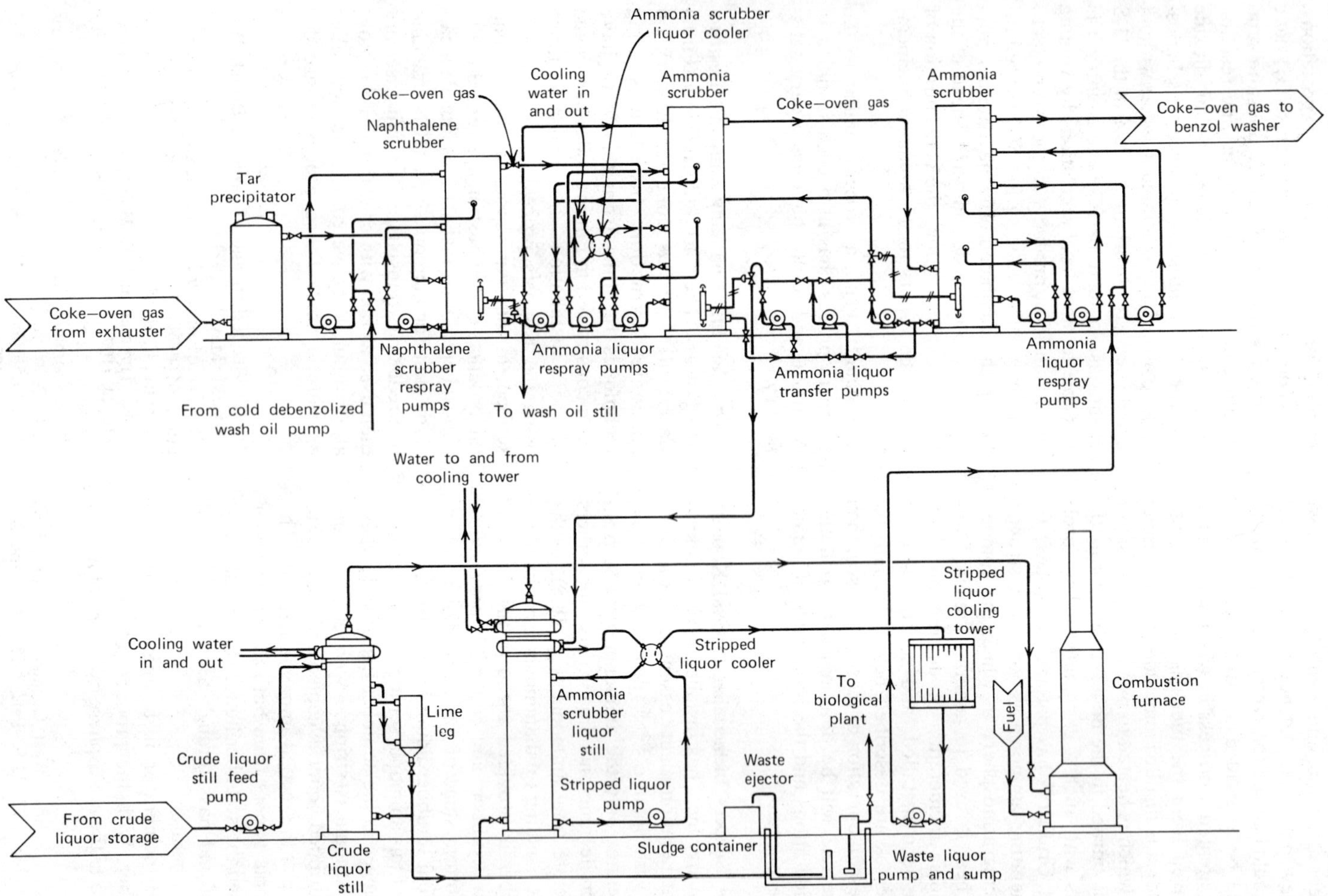

Figure 18.8 The Koppers ammonia destruction process. Diagram courtesy of Koppers Company, Inc.

scribed in more detail in connection with absorption-desorption.

Ammonia Incineration with Phosam. The Phosam system[146] for recovering the ammonia in coke-oven gas as anhydrous ammonia also is a very satisfactory process for use with an ammonia incinerator, in case the economics are not favorable for anhydrous ammonia production. The ammonia can be incinerated either as anhydrous ammonia or as the concentrated vapor from the stripping column. A Phosam plant designed solely for ammonia incineration would be less expensive, since the fractionator and auxiliaries for producing anhydrous ammonia could be deleted. The important advantage of using the Phosam system for ammonia incineration is that very little of the H_2S is removed from the coke-oven gas, and SO_2 pollution from the incinerator is almost negligible.

An ammonia removal and incineration process similar to the Phosam process has been patented in Japan by Hidemass and Jiro.[147] Heat recovery is included in this process.

18.4 GAS PURIFICATION

18.4.1 Cryogenic Separation Process

Coke-oven gas, which contains 50% or more hydrogen, has been an attractive source of hydrogen for ammonia synthesis and hydrogenation for many years.[148] The first coke-oven gas cryogenic separation plant was started by Linde in 1928, and many such plants have been installed worldwide since that time. Cryogenic separation is basically an old process and is not described in detail here. The state of the art as of 1960 was described by Funk.[149] He suggested the degree of gas pretreatment required before the gas is compressed and sent to the cold box. Problems are avoided by normal low-pressure removal of ammonia, light oil, and H_2S. Bose[150] described the problems of deposits and corrosion resulting when light oil and H_2S are removed after compression to 13–14 atm.

The cryogenic separation of coke-oven gas consumes a great deal of energy because of the relatively high pressures and low operating temperatures required. Ichihara[151,152] has been granted two U.S. patents for the cryogenic separation of coke-oven gas. Both heat and power savings are claimed. The newest coke-oven gas cryogenic separation plant for the production of anhydrous ammonia in the United States is the Keystone plant of United States Steel Corporation at Clairton, Pennsylvania. This plant introduced many innovations in the process of gas pretreatment ahead of the main gas separation plant (Fig. 18.9) and is described in detail.[153]

At the Clairton works all the gas from the ovens is cooled to approximately 106°F in primary coolers, after which it passes to the Keystone plant. In the gas processing section of the Keystone plant the gas is compressed by means of motor-driven axicompressors to approximately 35 psig. Water is added during compression to keep the gas cool, thereby minimizing gum and polymer formation and helping to condense and coalesce residual tar. The gases leave the compressors at 160°F and are cooled to 90°F in spray-type final coolers. Tar is injected into the spray water to dissolve condensed naphthalene. A sump is provided to separate the tar from the recirculated water, which is cooled indirectly and resprayed into the gas. The separated tar is sent to the tar plant for processing. Gases leaving the final cooler are treated for ammonia removal in a Phosam plant, as discussed in Section 18.3.3.2. The ammonia is recovered as anhydrous ammonia.

The gas from the Phosam plant proceeds to the main regenerators, which are towers filled with stones. As the gas passes through precooled stones, materials condense out and

[146] C. Hamilton and R. Rice, *AIME Ironmaking Proc.,* **34,** 622–641 (1975).

[147] T. Hidemass and I. Jiro, Japanese Pat. 7,-443,243, November 20, 1974.

[148] A. R. Powell, in *Chemistry of Coal Utilization,* Vol 2, H. H. Lowry, Ed., John Wiley and Sons, New York, 1945, pp. 1245–1247.

[149] H. Funk, *Blast Furn. Coke Oven Raw Mater., Comm. (Proc.),* **19,** 536–542 (1960).

[150] A. Bose, *COMA Yearb.,* 275–287 (1966); Waverly House, Wexborough, Yorkshire, England.

[151] K. Ichihara, U.S. Pat. 3,339,371, September 5, 1967.

[152] K. Ichihara, U.S. Pat. 3,340,698, September 12, 1967.

[153] U.S. Steel Corporation, private communication.

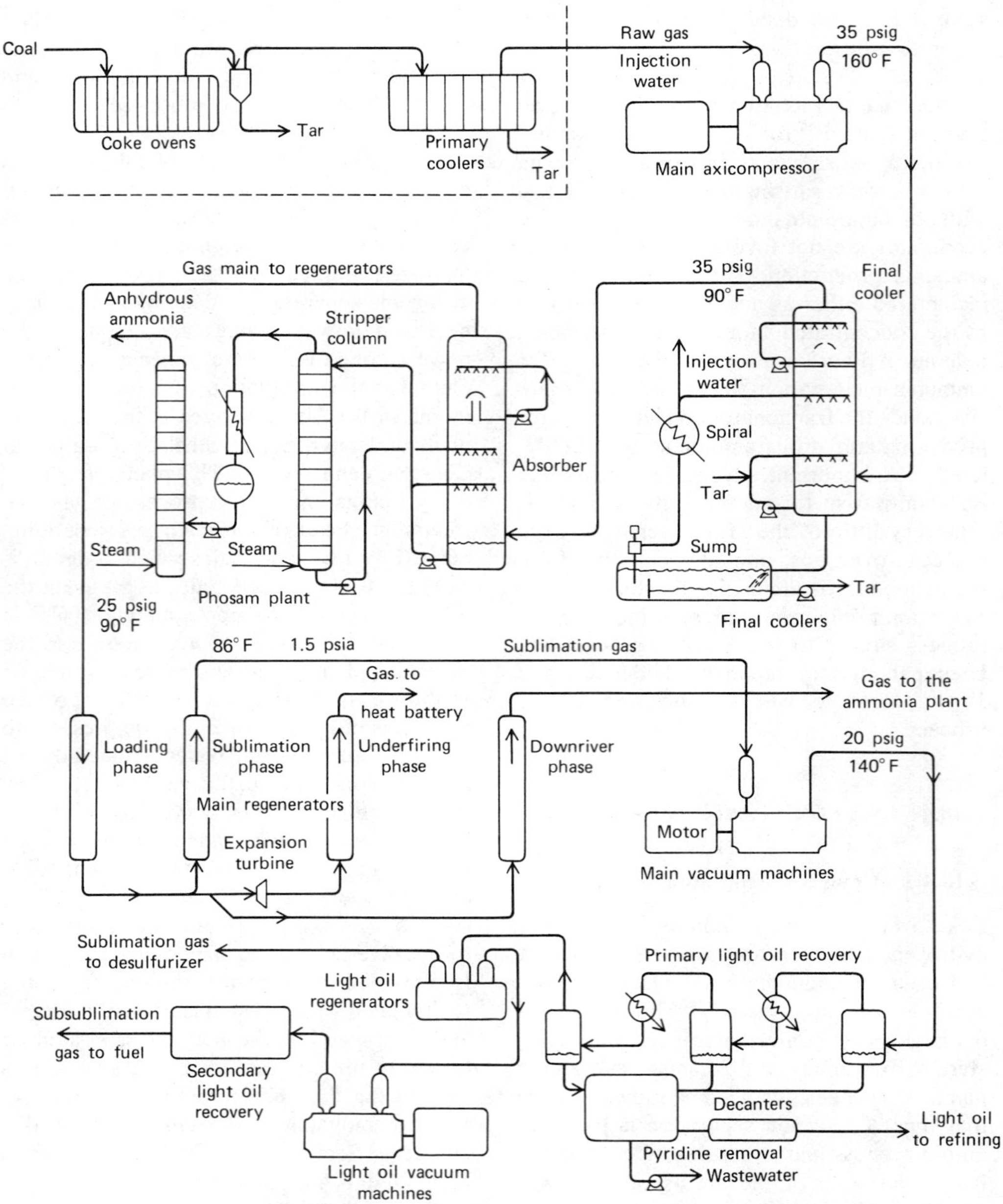

Figure 18.9 Gas-processing department of the Keystone plant, United States Steel Corporation, Clairton, Pennsylvania.

the gas is cleaned of light oils, sulfur, ethylene, and the remaining traces of tar by the time a temperature of −250°F is reached. The cold gas is then split into three parts. One part is passed up through a regenerator with clean stones, where the gas cools the stones and is heated to about room temperature before going to the synthetic ammonia plant. A second part of the gas is expanded in a power turbine for refrigeration, then is passed up through the stones and leaves the plant to be used as underfiring gas, after mixing with other gases.

The third portion of the gas under reduced pressure is passed up through the stones in the regenerators containing the condensed materials, to remove these materials from the stones. This gas, called sublimation gas, is compressed in vacuum machines to about 20 psig and is sent to heat exchangers where some of the light oils condense. The remaining light oils are removed from the gas in a second set of regenerators, which cool the gas to −80°F. This temperature is low enough to remove light oil but not sulfur compounds. The gas containing the sulfur compounds is then sent to a desulfurizing plant. The condensed materials in the light oil regenerators are again swept off under vacuum, compressed, and cooled for additional light oil recovery. The crude light oil is sent to the light oil plant for refining.

The clean gas stream from the main regenerators is sent to the ammonia plant, where all remaining components except hydrogen are removed. The hydrogen is then used to produce anhydrous ammonia by catalytic reaction with nitrogen from an air separation plant.

18.4.2 Hydrogen Sulfide Removal

The removal of hydrogen sulfide from coal gas had been important since coal gas was first used as town gas. At that time H_2S removal was required for domestic use of the gas. This was the situation when the first volumes of *Chemistry of Coal Utilization* were compiled. Volume 2[154] covered oxide purification, the Perrox process, the Thylox process, the Seaboard process, the vacuum carbonate process, Gluud's processes, the Feld process, the CAS process, the Katasulf process, the ammonia process, the alkacid process, Burkheiser's processes, the Ferrox process, the Manchester process, and the Staatsmijnen-Otto process, as well as various other processes for H_2S removal. The first supplement, published in 1963, covered gas purification only in the section on synthesis gas.[155] The only new process applicable to coke-oven gas is the Giammarco-Vetrocoke process. Since 1963 the control of air pollution has become so important that many new or revised processes have been developed to meet current regulations. These processes are described in the following sections. The major processes are summarized in Table 18.3.

18.4.2.1 *Absorption-Desorption Processes*

In absorption-desorption processes for H_2S removal the H_2S is absorbed in an alkaline solution, from which it is removed by steam stripping or by blowing with air. The original processes of this type included the ammonia process, the Seaboard process, and the vacuum carbonate process. The ammonia process and the vacuum carbonate process were recently improved and a new process of this type, the Sulfiban process, developed.

The Koppers Two-Stage Vacuum Carbonate Process. The Koppers original single-stage vacuum carbonate process[156] was designed to remove approximately 90% of the H_2S. The maximum removal that can be attained without excessive steam use is 93%. The two-stage process[157] was developed to achieve approximately 98% H_2S removal without excessive steam consumption. Figure 18.10 is a typical flow diagram of this process.

In the Koppers two-stage process, the foul gas enters the base of a two-stage absorber and is contacted in each stage countercurrently with an "actified" sodium carbonate solution (solution from which the H_2S has been stripped). The foul solution from the bottom stage absorber is pumped to the top stage of a two-stage vacuum actifier (steam stripper), and the solution from the top absorber is pumped to the bottom stage of the actifier. The actified solution from each actifier is then cooled and returned to its respective absorber stage. Stripping steam for both actifier stages is generated in the base of the bottom stage and passes through both stages in series. The solution circulation for the second-stage absorber is much less than that required for the bottom stage and is therefore stripped intensively in the bottom actifier, permitting excellent H_2S removal in the second absorber stage.

The stripped acid gases and steam leave

[154] H. A. Gollmar, in *Chemistry of Coal Utilization* Vol. 2, H. H. Lowry, Ed., John Wiley and Sons, New York, 1945, pp. 970–998.

[155] C. G. von Fredersdorff and M. A. Elliott, in *Chemistry of Coal Utilization,* Supplementary Volume, H. H. Lowry, Ed., John Wiley and Sons, New York, 1963, pp. 1017–1022.

[156] M. Massey and R. Dunlap, *J. Air Pollut. Control Assoc.*, **25**(10), 1019–1027 (October 1975).

[157] Anon., *33 Mag.*, **14**, 33–37 (October 1976).

Table 18.3 Summary of Major H_2S Removal Processes

Process	H_2S Removal Efficiency	Organic Sulfur Removal	Form of Recovered Sulfur	Regeneration Medium	Effluent Problems	Remarks
Absorption-desorption						
Vacuum carbonate						
Single stage	90–93%	None	Rich H_2S acid gas	Steam under vacuum	Small amount of circulating solution containing HCN and H_2S as condensates	Many single-stage plants in operation
Two-stage	98%	None	Rich H_2S acid gas			
Sulfiban	90–98%	Partial	Lean H_2S acid gas	Steam	Sludge from reclaimer, ammonia-contaminated condensate	Limited operating experience
Ammonia processes	90% normal, 98% maximum	None	Acid gas containing $H_2S + NH_3 + CO_2 + HCN$	Steam	Minimal	May require refrigeration
Oxidation type						
Stretford	99%+	None	Elemental sulfur	Air	Serious liquid effluent problem; effluent treatment required	Many plants in operation
Takahax	99%+	None	Ammonium sulfate or elemental sulfur	Air	Similar to Stretford except for Hirohax, which produces ammonium sulfate	In Japan only, at present
Fumaks-Rhodacs	90–99%	None	Elemental sulfur	Air	Similar to Stretford	In Japan only
Giammarco-Vetrocoke	<1 ppm	None	Elemental sulfur	Air	Arsenic contamination must be reclaimed	Sulfur is also contaminated with arsenic

Source: Author

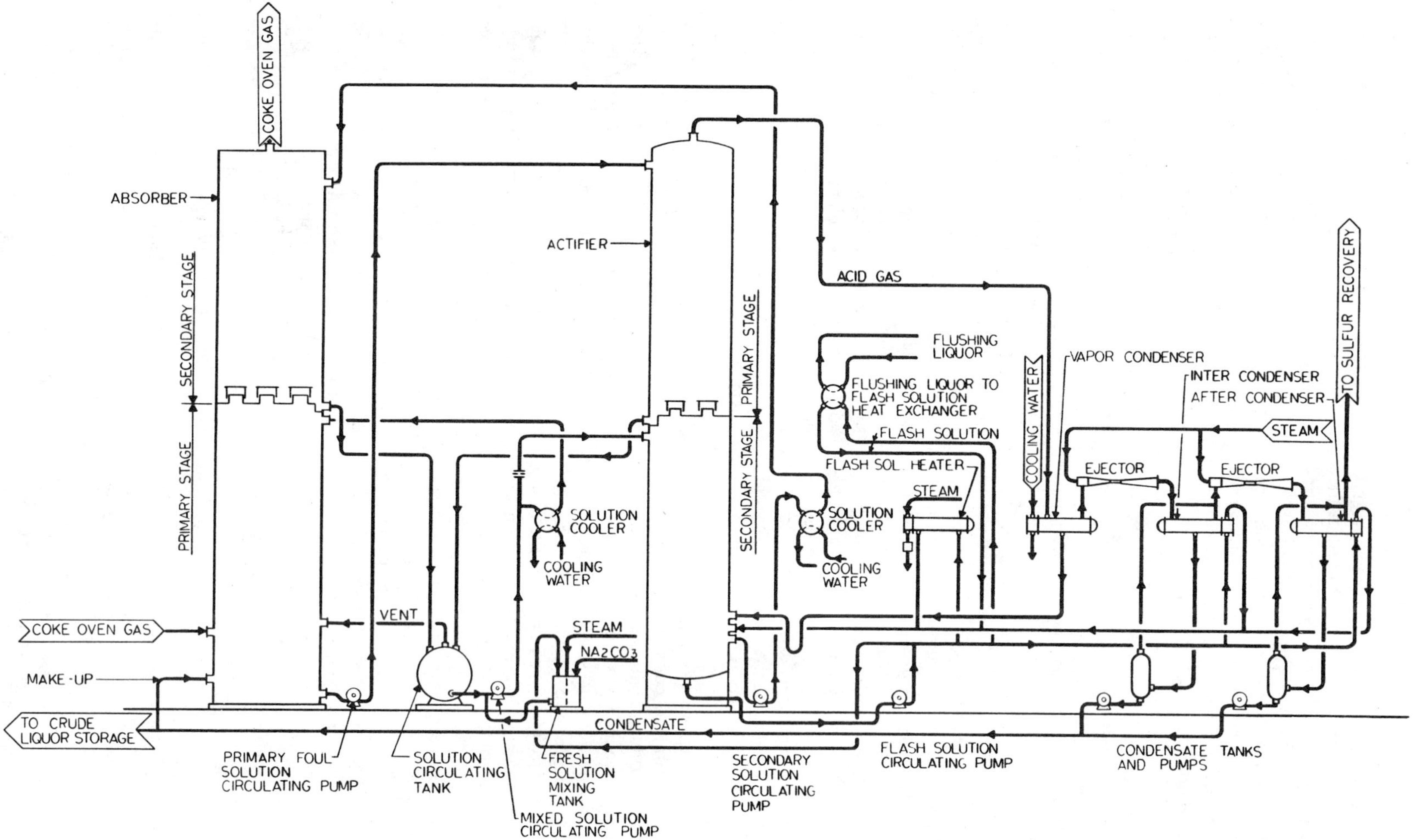

Figure 18.10 The Koppers two-stage vacuum carbonate process. Diagram courtesy of Koppers Company, Inc.

the top of the actifier under high vacuum and pass to vapor condensers, where most of the steam is condensed; the remaining vapors are pulled through the primary steam jet ejector, through an intercondenser, and into the secondary jet ejector, and are finally delivered to an aftercondenser. Subsequent treatment depends on the end product desired.

The Koppers heat recovery system, reclaiming heat from the steam jet ejectors and from flushing liquor heat exchangers, is provided with the two-stage vacuum carbonate system to minimize operating costs.

The Sulfiban Process. Monoethanolamine solutions have been used for many years to remove H_2S and CO_2 from natural gas. Such solutions were not used to purify coke-oven gas because of side reactions with other contaminants present in coke-oven gas. Recently Bethlehem Steel Corporation and Black, Sivals and Bryson adapted this process to coke-oven gas as a result of discovering additives that minimize side reactions and inhibit corrosion. This is the Sulfiban process.[158–162]

In the Sulfiban process (Fig. 18.11) the H_2S is removed from the coke-oven gas by countercurrent scrubbing with monoethanolamine solution in a packed or spray-type contactor, which is operated at slightly higher than atmospheric pressure. A liquid disengaging chamber is provided in the base of the contactor to minimize hydrocarbon contamination of the solution. The rich solution is pumped from the contactor through filters and a heat exchanger to the top of the stripping column, where it is actified by steam stripping. All stripping steam is generated indirectly using live steam. The lean solution is then cooled and recycled to the contactor. A side stream of the lean solution is continuously taken through a charcoal filter to remove degradation products. Another small stream is taken from the reboiler to a reclaiming system where it is concentrated by indirect steam. When solids build up in the reclaimer, the supply is shut off and the solution is concentrated to minimize chemical loss. The volume of residue discarded is very small. The acid gases and steam leave the stripper at 3–10 psig. The steam is condensed in a water-cooled condenser, and the acid gases leave for further processing to elemental sulfur or sulfuric acid. The condensate is returned as reflux to the stripper. A small amount of ammonia is released in the reclaimer; this accumulates in the reflux, and a small amount must be continuously discarded. This effluent can be mixed with the weak ammonia liquor that is produced.

The Sulfiban process is designed to easily reduce the H_2S concentration in the outlet gas to less than 10 gr/100 scf, and the developers claim that CO_2 removal can be limited to some extent. Several plants are under construction, and one plant has been started up, but no plant has yet been operated for an extended period.

Ammonia Processes for H_2S Removal. The use of ammoniacal solutions for H_2S removal is a very old approach.[163] It has always been appealing because it employs only the chemicals that are generated in the plant itself. The status of the process as of about 1958 was covered in great detail by Kohl and Riesenfeld.[164] It was not widely used for many years because of poor H_2S removal, high ammonia losses with the H_2S product, and extensive corrosion. Recently many improvements to this basic process have been developed, primarily in Western Europe, and it is used in many plants today.

In most processes employing ammonia the ammonia is scrubbed from the gas in an absorber using treated water or condensate, and the effluent ammonia washwater is introduced into the top of the H_2S absorber, which precedes the ammonia absorber in the gas stream. The foul liquor from the base of the H_2S absorber is pumped to a deacidifier, where the H_2S alone or with the total ammonia production leaves the system for fur-

[158] M. Massey and R. Dunlap, *J. Air Pollut. Control Assoc.*, **25**(10), 1019–1027 (October 1975).

[159] Anon., *33 Mag.*, **14**, 33–37 (October 1976).

[160] A. Singleton and G. Batterton, *AIME Ironmaking Proc.*, **34**, 604–621 (1975).

[161] O. A. Homberg, C. S. Sheldrake, and A. Singleton, U.S. Pat. 3,864,449, February 4, 1975.

[162] J. Williams and O. A. Homberg, *AIME Ironmaking Proc.*, **35**, 98–108 (1976).

[163] H. A. Gollmar, in *Chemistry of Coal Utilization,* Vol. 2, H. H. Lowry, Ed., John Wiley and Sons, New York, 1945, pp. 1000–1002.

[164] A. Kohl and F. Riesenfeld, *Gas Purification,* McGraw-Hill Book Company, New York, 1960.

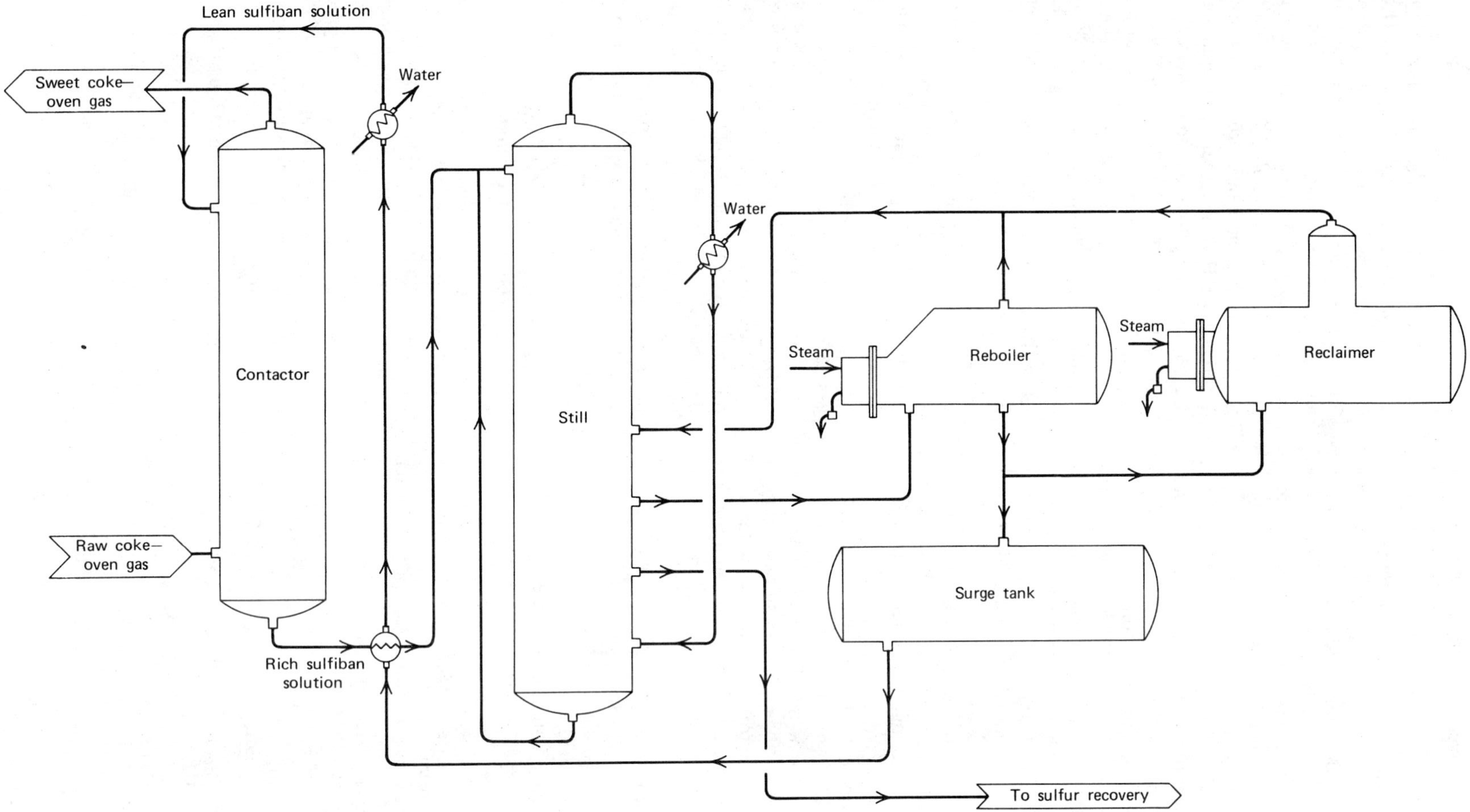

Figure 18.11 The Sulfiban system. Diagram courtesy of Applied Technology Corporation.

ther treatment. The ammonia is concentrated in the deacidifier, and the bottom liquor rich in ammonia is recycled to the H_2S absorber at a point somewhat below the top. An ammonia still is provided to strip the excess flushing liquor, and this ammonia vapor is piped to the deacidifier so that this ammonia can also be recycled.

Firma Carl Still[157,165–167] claims that the H_2S concentration in the purified gas can be reduced to 25 or even 10 gr/100 scf by careful control of the ammonia concentrations and temperatures. Figure 18.12 is a typical flow diagram. The many variations in plant design as well as in plant economics include operations to destroy ammonia by incineration or to produce ammonium sulfate (directly or indirectly) or anhydrous ammonia and elemental sulfur or sulfuric acid. All possible combinations are considered. Tippmer and Maloy[168] treat many of these processes in more detail.

Many patents and technical publications cover variations in the basic process and describe specific installations. Schmidt[169] described alterations made on an existing plant to convert it from ammonium sulfate and sulfuric acid production to ammonia combustion with sulfuric acid production. Wunderlich and Weber[170–172] and Wunderlich et al.[173] have been awarded many patents covering improvements to the basic process and the method of handling both the recovered ammonia and acid gases. Tippmer and Laufhütte[174] have patented a revised process including a separate ammonia concentrating still to further increase the concentration of ammonia in the liquor recycled to the H_2S absorber. Shibuya[175] has received a patent for a process using recycled ammonia, in which temperature, flow, and concentration are controlled to produce an H_2S product gas low in both ammonia and HCN. Radusch[176] has developed an ammonia process that uses a three-stage H_2S absorber with ammonia recycle. Normally total ammonia and acidic gases exit together at the top of the still.

Grosskinsky[177] patented a process to improve the recovery of H_2S using ammonia solutions by increasing the ammonia concentration in the raw gas. This is achieved by recirculating condensates to the gas-collecting main. Final H_2S removal is increased by scrubbing under pressure. Lebedeva et al.[178] studied the removal of H_2S from coke-oven gas by ammonia scrubbing in cases where the H_2S content is high. She recommended a two-stage process, with ammonia recycle leaving nearly all the ammonia in the purified gas. Approximately 2.5% of the ammonia is included with the acid gases. The efficiency of H_2S removal varies with temperature. Tests indicated 97% removal at 20–26°C, but only 85% removal at 28–30°C. Bussman[179] described a plant in which H_2S is removed from the gas with the aid of ammonia: the ammonia is catalytically destroyed, and sulfuric acid is produced from the products of combustion.

18.4.2.2 *H_2S Removal Processes Involving Oxidation*

In all H_2S removal processes involving oxidation, the H_2S is absorbed in an alkaline solution containing an oxidation catalyst in solution or in suspension. Processes using a suspension have always been beset with problems such as a contaminated sulfur product as well as plugging in various parts of the system. Only new processes using catalysts in solution are considered here.

[165] D. Laufhütte, *AIME Ironmaking Proc.*, **33**, 142–155 (1974).
[166] Firma Carl Still, British Pat. 1,348,937, March 27, 1974.
[167] M. Massey and R. Dunlap, *AIME Ironmaking Proc.*, **34**, 583–603 (1975).
[168] K. Tippmer and J. Maloy, *COMA Yearb.*, 127–144 (1974); Waverly House, Wexborough, Yorkshire, England.
[169] H. Schmidt, *Glueckauf*, **103**, 1170–1174 (1967).
[170] G. Wunderlich and H. Weber, West German Pat. 1,212,052, September 13, 1966.
[171] G. Wunderlich and H. Weber, British Pat. 1,318,731, May 31, 1973.
[172] G. Wunderlich and H. Weber U.S. Pat. 3,-819,816, June 25, 1974.
[173] G. Wunderlich et al., U.S. Pat. 3,822,337, January 24, 1972.
[174] K. Tippmer and D. Laufhütte, U.S. Pat. 3,-789,105, January 29, 1974.
[175] T. Shibuya et al., U.S. Pat. 3,880,617, April 29, 1975.
[176] P. Radusch, U.S. Pat. 3,556,721, January 19, 1971.
[177] O. Grosskinsky et al., U.S. Pat. 3,104,959, September 24, 1963.
[178] G. Lebedeva et al., *Coke Chem. (USSR)*, No. 3, 36–40 (1963).
[179] B. Bussman, *Stahl Eisen*, **94**(14), 627–631 (1974).

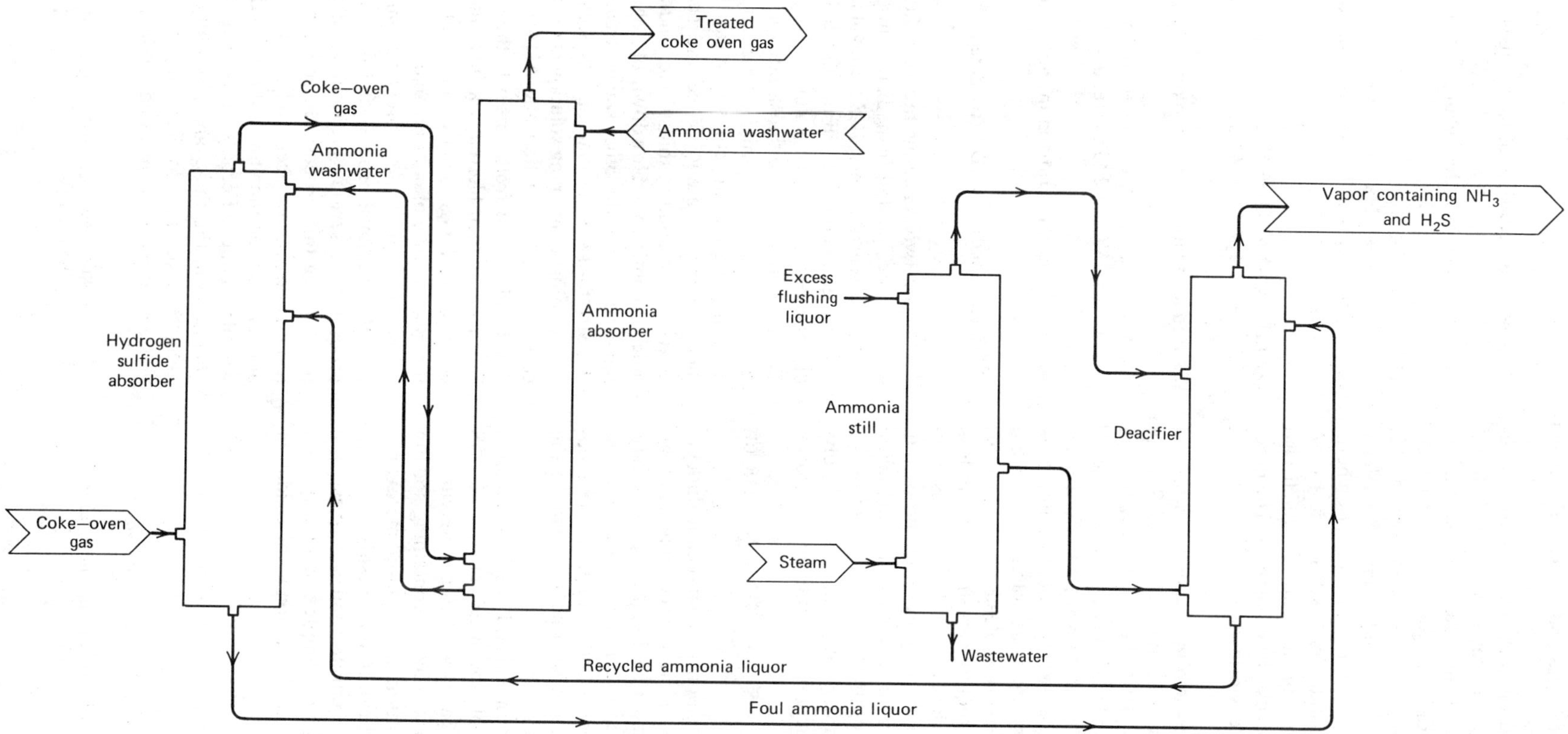

Figure 18.12 Coke-oven gas desulfurization with ammonia liquor. Diagram courtesy of Dravo/Still Corporation.

The Stretford Process. The Stretford process[180] was originally developed by the Northwest Gas Board and the Clayton Aniline Company of England for use in removing H_2S from town gas. It has been used for this purpose since 1965, and many plants have been built in Great Britain and elsewhere. Most of the installations were made by W. C. Holmes & Company, Ltd.,[181] which also made many process developments. The process is available worldwide through various licensees.

It is recommended that when the Stretford process is used for removal of H_2S from coke-oven gas, HCN be removed first from the gas stream. This is necessary because all the HCN will react with the Stretford solution to produce sodium thiocyanate, thus quickly building up undesirable salts in the system. An extensive bleed-off of the solution would be required, resulting in the loss of expensive chemicals. The HCN is normally removed by the polysulfide process in which the gas is contacted with ammonium or sodium polysulfide solution ahead of the Stretford absorber. With ammonium polysulfide, the reaction is essentially as follows:

$$HCN + NH_3 + S \rightarrow NH_4SCN \quad (18.10)$$

The concentration of ammonium thiocyanate builds up in the recirculating solution, and a portion must be discarded periodically. This portion represents one of the contaminated waste streams that required development of an effluent treatment system for the Stretford process.

After HCN removal and possibly ammonia removal, the coke-oven gas enters the H_2S absorber. The gas is washed countercurrently with Stretford solution containing sodium carbonate, sodium ammonium metavanadate, 2,7-anthraquinone disulfonic acid (ADA), sodium citrate, and a sequestering agent, to form a complex with any iron present. The overall reaction is the oxidation of H_2S to sulfur:

$$2H_2S + O_2 \rightarrow 2S + 2H_2O \quad (18.11)$$

[180] T. Nicklin and E. Brunner, *Inst. Gas Eng. J.*, **1**, 523–549 (1961).

[181] A. Moyes and J. Wilkinson, *Chem. Eng.*, No. 282, 84–90 (February 1974).

This reaction consists of four stages:

1. H_2S reacts with the alkali:

$$H_2S + Na_2CO_3 \rightarrow NaHS + NaHCO_3 \quad (18.12)$$

2. Sodium hydrosulfide reacts with metavanadate:

$$2NaHS + 4NaVO_3 + H_2O \rightarrow Na_2V_4O_9 + 4NaOH + 2S \quad (18.13)$$

3. Reduced vanadium is reoxidized by ADA:

$$Na_2V_4O_9 + 2NaOH + H_2O + 2ADA \rightarrow 4Na\ VO_3 + 2ADA\ (\text{reduced}) \quad (18.14)$$

4. The ADA is reoxidized in the oxidizer:

$$ADA\ (\text{reduced}) \xrightarrow{O_2} ADA\ (\text{oxidized}) \quad (18.15)$$

The first two reactions are very fast, and sulfur is precipitated immediately in the absorber. Because of this, H_2S removal is nearly complete, and nearly all the sulfide ion has disappeared before the solution is aerated. This tends to minimize thiosulfate formation, since it is now primarily a function of the oxygen content of the coke-oven gas. Nevertheless approximately 1% of the sulfur is converted to sodium thiosulfate, and a periodic or continuous blowdown of solution is required. This is the other waste stream from the Stretford system that must be treated.

The sulfur-containing solution is then treated in the oxidizer, where the ADA is reoxidized and a sulfur froth comes to the surface. This froth is then filtered, washed, and melted to produce high-quality sulfur suitable for the manufacture of sulfuric acid.

The disposal of the effluent streams from the HCN removal process and the Stretford process is a serious problem for all coke plants using this process. In addition, sulfur production was reduced and in some plants ammonia or ammonium sulfate production was also reduced because of the loss of these chemicals in the thiocyanate solution. Processes[182–184] are now available in which these waste streams are incinerated (under proprie-

[182] T. Nicklin and F. Farrington, German Pat. 2,355,839, June 20, 1974.

tary combustion conditions) to produce recyclable products for the process. This addition greatly increases capital cost but is essential to the continued use of the process.

More detailed information on the design and operation of the Stretford process is available in many publications[185–195] and patents. Figure 18.13 is a flow diagram of a typical Stretford plant.

The Takahax Process. The Takahax process is another oxidation-type H_2S removal process developed by the Tokyo Gas Company.[196–200] The original process was very similar to the Stretford process, except that a different catalyst, the ammonium or sodium salt of 1,4-naphthoquinone 2-sulfonic acid, was used. Hydrogen sulfide was adequately removed, but thiocyanate and thiosulfate built up in the system, requiring a relatively large blowdown and creating a disposal problem. The process has been revised to eliminate this problem, and four variations[201] of the basic process are now available. Many plants in Japan are now successfully using this process. Two of the variations, types A and B, use the ammonia in the coke-oven gas as the alkali; the other two variations, types C and D, use caustic soda or sodium carbonate as the alkali.

In the type A ammonia Takahax process, the absorber is located in the gas stream ahead of the saturator. The conditions in the absorber are controlled so that no colloidal sulfur is produced. The H_2S and HCN are oxidized to ammonium thiosulfate, ammonium thiocyanate, and ammonium sulfate in solution. The foul solution is then regenerated with air to reoxidize the catalyst from the hydroquinone state to the quinone state for reuse in the absorber. The air leaving the regenerator is washed with water or acid ammonium sulfate solution to remove and reclaim ammonia. To control the concentration of soluble salts in the recirculating solution, it is necessary to constantly remove a side stream from the recirculating system. This side stream is treated in a process developed by Nippon Steel Corporation.[202] The solution is pressurized to 70–90 atm, charged with compressed air, heated to 250–280°C, and sent to a reactor, where essentially all the salts are converted to ammonium sulfate in solution. This solution is concentrated in an evaporator and sent to the sulfate plant for crystallization. This combination of the ammonia Takahax process and the wet oxidation process is called the "Hirohax" process.

The type B system is similar to the type A system in that the same catalyst and alkali are used and the absorber is located ahead of the saturator. However process design and operation are altered to obtain maximum colloidal sulfur production. The recirculating solution is split into two streams: one goes directly to the regenerator and the other to a centrifuge or filter for sulfur removal. The filter underflow is again divided: one portion returns to the regenerator and the other goes to a waste treatment process. The waste stream containing thiocyanates and thiosul-

---

[183] F. Singleton et al., German Pat. 2,364,267, June 27, 1974.
[184] F. Singleton et al., U.S. Pat. 3,966,876, June 29, 1976.
[185] T. England, *COMA Yearb.*, 255–267 (1975); Waverly House, Wexborough, Yorkshire, England.
[186] M. Massey and R. Dunlap, *J. Air Pollut. Control Assoc.*, **25**(10), 1019–1027 (October 1975).
[187] T. Nicklin and B. Holland, *Gas World*, **158**, 273–278 (September 7, 1963).
[188] Anon., *33 Mag.*, **14**, 33–37 (October 1976).
[189] Anon., *Gas World; Gas Coke*, **158**, 827–831 (December 21–28, 1963).
[190] A. Moyes and S. Vasan, *Oil Gas J.*, **72**, 56–58 (September 2, 1974).
[191] C. Wood, *COMA Yearb.*, 81–100 (1973); Waverly House, Wexborough, Yorkshire, England.
[192] A. Moyes and J. Wilkinson, *Process Eng.*, 101–105 (September 1973).
[193] P. Elwood, *Chem. Eng.*, **71**, 128–130 (July 20, 1964).
[194] A. Moyes, J. Wilkinson and B. Mills, *COMA Yearb.*, 198–215 (1975); Waverly House, Wexborough, Yorkshire, England.
[195] J. Ludberg, *AIME Ironmaking Proc.*, **30**, 266–275 (1971).
[196] Iida Suetoshi et al., U.S. Pat. 3,459,495, August 5, 1969.
[197] N. Hasebe et al., *Kogyo Kagaku Zasshi*, **69**(8), 1419–1425 (August 1966).
[198] N. Hasebe, *Chem. Econ. Eng. Rev. (Japan)*, **2**(3), 27–32 (March 1970).
[199] C. Wood, *COMA Yearb.*, 91–94 (1975); Waverly House, Wexborough, Yorkshire, England.
[200] Anon., *J. Met.*, **24**, 6 (May 1972).

[201] Anon., *33 Mag.*, **14**, 33–37 (October 1976).
[202] M. Matumoto and T. Aono, U.S. Pat. 3,855,-390, December 17, 1974.

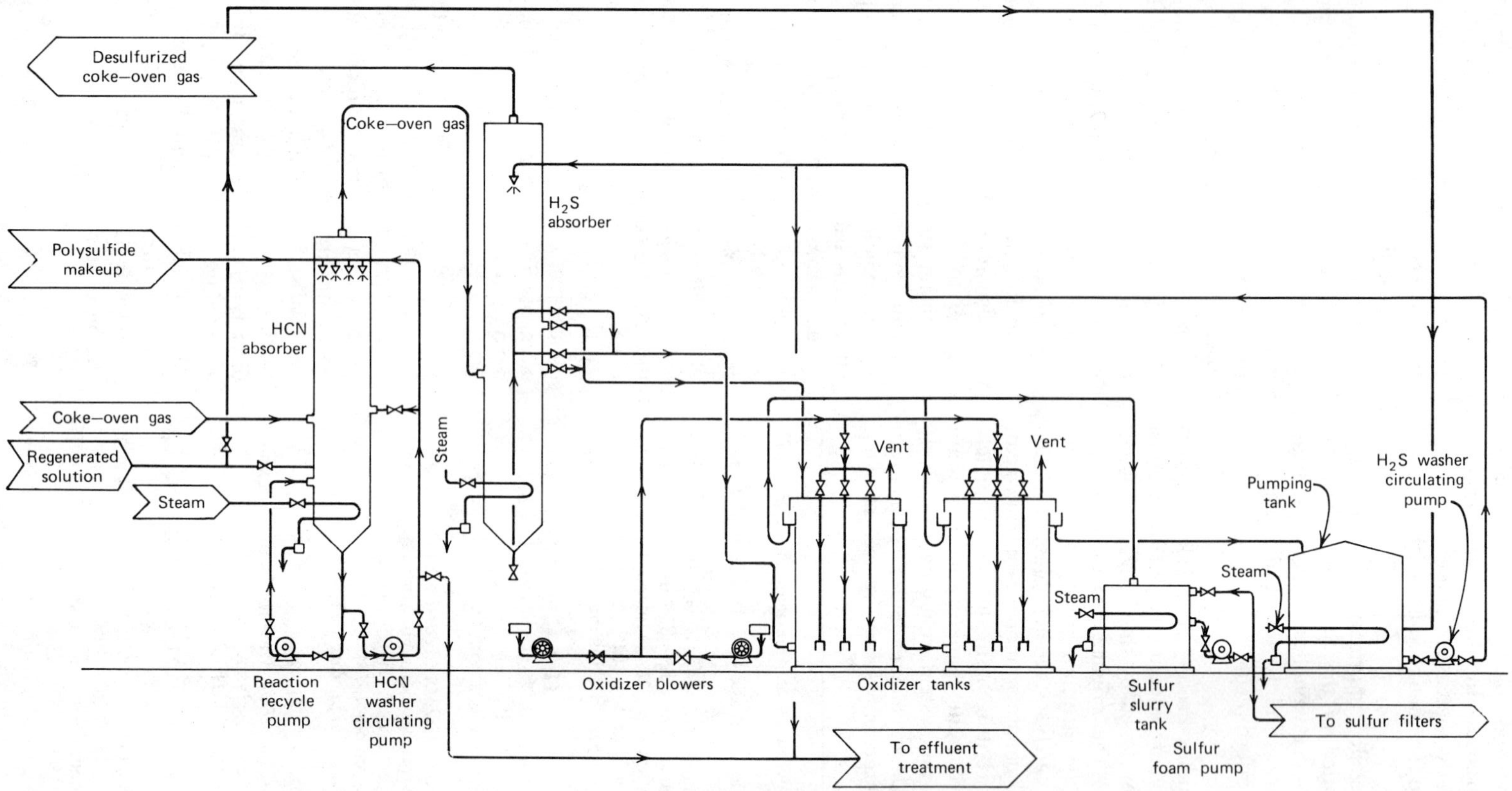

Figure 18.13 The Stretford desulfurization plant. Diagram courtesy of Woodall Duckham, Ltd.

fates is concentrated, then incinerated with the sulfur cake to produce SO_2, which is sent to a sulfuric acid plant.

In the type C and D systems the absorber is located in the gas stream after the ammonia recovery equipment, and sodium carbonate or caustic soda is used as the alkali. Both systems are operated for maximum sulfur formation. One advantage of these systems is that the air from the regenerator requires no scrubbing. In both type C and type D plants the sulfur is filtered from a side stream and autoclaved to molten elemental sulfur. In each process a portion of the filtrate is concentrated and burned in a submerged combustion furnace under reducing conditions.[203] The sodium thiocyanate and sodium thiosulfate are thermally decomposed to Na_2CO_3, H_2S, CO_2, and N_2. The carbonate solution is recycled to the recirculating system. The only significant difference between the type C and type D systems is the method of handling the off-gases from the submerged combustion furnace. In the type C system these gases are burned with the molten sulfur at the sulfuric acid plant, whereas in the type D plant they are scrubbed of H_2S by a side stream of the Takahax absorbing solution. The type D plant, therefore, recovers all the H_2S as sulfur that can be sold or used in an acid plant.

Many Takahax plants are in operation or are under construction in Japan. Removal of both H_2S and HCN is excellent. However the effluent treating processes are very new, and operating experience with these units is not extensive.

The Fumaks-Rhodacs Process. The Fumaks-Rhodacs process is also an oxidation process for H_2S removal with simultaneous HCN removal. The Fumaks[204–206] (H_2S removal) and Rhodacs (HCN removal) processes were developed separately by Osaka Gas Company, but since coke-oven gas contains both H_2S and HCN, the combined process is used in coke plants. The Fumaks-Rhodacs process is very similar to the Takahax process except that a different catalyst—picric acid—is used. The alkali can be either ammonia or sodium carbonate, but ammonia is preferred. Although removal of both H_2S and HCN is good, ammonium thiocyanate and ammonium thiosulfate build up in the recirculating solution, as in the Takahax process. An effluent disposal system similar to the Takahax or Stretford system is required. The sulfur cake does not autoclave well, and it is recommended that it be incinerated with the H_2S from the treated effluent. The SO_2 then goes to an acid plant.

Many plants have been built in Japan without effluent treatment. In some cases the effluent containing thiocyanates and thiosulfates was recycled to the coal stream to the ovens. This was not desirable, and one plant has added an effluent treatment plant of the type B Takahax process. Recently Ooka et al.[207] of Osaka Gas Company patented a process based on the Fumaks-Rhodacs process that also eliminates the effluent problem with a method similar to the Hirohax process. The Fumaks-Rhodacs process is marketed by Sumitomo Chemical Engineering Company, Ltd., in Tokyo.

The Giammarco-Vetrocoke Process. The Giammarco-Vetrocoke process[208–211] for H_2S removal from coke-oven and other gases was developed in Italy by G. Giammarco. Like the old Thylox[212] process, it uses arsenic-containing aqueous solutions to absorb the H_2S, and air for oxidation of the H_2S to elemental sulfur and for regeneration of the absorbing solution. However the process is a definite improvement on the Thylox proc-

[203] H. Tsuruta et al., U.S. Pat. 3,941,875, March 2, 1976.

[204] M. Maezawa and T. Kuragano, U.S. Pat. 3,099,535, July 30, 1963.

[205] M. Fukuzaki, *Chem. Econ. Eng. Rev.*, **4**(5) 42–46, 61 (May 1972).

[206] Anon., *Chem. Econ. Eng. Rev. (Japan)*, 41–42 (February 1972).

[207] I. Ooka et al., U.S. Pat. 3,953,577, April 27, 1976.

[208] G. Giammarco, Italian Pats. 537,564, 1955; 560,161, 1956; 565,320, 1957.

[209] S. Villa, paper presented at the Int. Conf. Control of Gaseous Sulfur Compound Emissions University of Salford, England, April 1973, in association with the Institution of Chemical Engineers.

[210] A. L. Kohl and F. C. Riesenfeld, *Gas Purification*, McGraw-Hill Book Company, New York, 1960, p. 303.

[211] R. Maddox and M. Burns, *Oil Gas J.*, **66**, 90–95 (June 3, 1968).

[212] H. A. Gollmar, in *Chemistry of Coal Utilization*, Vol. 2, H. H. Lowry, Ed., John Wiley and Sons, New York, 1945, pp. 973–981.

ess in that the concentration of H_2S can be reduced to less than 1 ppm, and side reactions are minimized.

The Giammarco-Vetrocoke process solution is composed of alkaline arsenite and arsenate, plus a small amount of hydroquinone that serves to catalyze the oxidation reaction. A minimum ratio of 3 moles of arsenate to 1 mole of arsenite must be maintained. The alkali can be either sodium carbonate or potassium carbonate. A solution of low alkalinity is normally used for treating gas for the removal of H_2S at normal temperatures. Usually the temperature is maintained at 40°C at the absorber top and at 40–70°C at the bottom. The solution composition can be adjusted so that H_2S and CO_2 are both removed simultaneously, or H_2S is removed selectively. Selective H_2S absorption is used for coke-oven gas. The reactions that occur during the different phases may be represented by the following equations:

Absorption

$$Na_2HAsO_3 + 3H_2S \rightarrow Na_2HAsS_3 + 3H_2O \quad (18.16)$$

Digestion

$$Na_2HAsS_3 + 3Na_2HAsO_4 \rightarrow 3Na_2HAsO_3S + Na_2HAsO_3 \quad (18.17)$$

Oxidation

$$Na_2HAsO_3S + 1/2O_2 \rightarrow Na_2HAsO_4 + S \quad (18.18)$$

The excellent H_2S removal attained by this process and the low thiosulfate formation are due to the low H_2S partial pressure of the monothioarsenate formed in the digestion reaction. In spite of this advantage, thiosulfate and thiocyanate build up in the system, and periodically a portion of the solution must be sent to a reclaiming system, since arsenic cannot be disposed of in the plant effluent. In the reclaiming system the solution is contacted with sulfuric acid at about 90°C for 5 hr. The arsenic is then precipitated as arsenic sulfide by addition of an alkaline sulfide. The arsenic sulfide is decanted, washed, and redissolved in the normal alkaline recirculating solution. The wastewater contains from 1 to 3 ppm total arsenic and can be discarded.

The Giammarco-Vetrocoke process can also be used for gas purification at high pressures and higher temperatures. In such cases the alkalinity must be high and an acidification step must be added before the oxidation step. The acidification uses CO_2 under a pressure of 4–6 atm. This step is required to assure sulfur precipitation in the oxidizer. This type of process is usually used for gases other than coke-oven gas.[213]

The advantages of this process are as follows:

1. Very good H_2S removal—to less than 1 ppm.
2. Small solution circulation.
3. Versatility.
4. Reduced formation of secondary products.
5. Low utility costs.
6. Little corrosion.

However there are also several disadvantages:

1. Limited capacity—maximum sulfur per day is 30–40 tons.
2. Toxicity of the arsenic solution.
3. Sulfur product not easily marketable without extensive purification.

18.4.2.3 *Other H_2S Removal Processes*

The preceding subsections have described the new H_2S removal processes being used on a large scale for coke-oven gas purification. However some older processes are still in use, and many new or improved processes have been patented or proposed.

For many years the standard method for the purification of town gas was the use of iron oxide boxes. This process is still widely used and has been improved in some plants by the incorporation of tower purifiers.[214] Honolulu Gas Company[215] has improved its system by using powdered iron and soda ash instead of iron oxide.

[213] E. Jenett, *Oil Gas J.*, **60**(18), 72–77 (April 1962).

[214] J. Burgess and J. Codman, *Gas World*, **151**, Coking Suppl., 17–28 (February 6, 1960).

[215] R. Fujita, *Gas World*, **162**, Gas Coke Suppl., 656–661 (November 27, 1965).

Mancke et al.[216] have patented a treatment process in which ammonia, H_2S, and HCN are removed by use of waste pickle liquor. In this process all the ammonia is recovered as ammonium sulfate, but only a portion of the gas is purified of H_2S and HCN. The H_2S leaves the system as iron sulfide. This process is interesting but would not meet present-day pollution control requirements.

Heinrich Koppers GmbH has been granted a patent[217] covering the use of carboxylic acid amides of morpholine as the absorption agent in the absorption-desorption process for acid gas removal. Pronounced selectivity for H_2S is claimed, as well as lower energy requirements for regeneration.

Hamblin[218] has patented a variation of the ammonia process for H_2S removal in which the ammonium hydrosulfide in the foul solution is catalytically oxidized to ammonium polysulfide, which is then treated to recover elemental sulfur. The ammonia is recycled and any excess is removed from the system. Gotza et al.[219] have patented another process in which the hydrogen sulfide and ammonia are absorbed in a solvent containing an arsenic, antimony, or tin compound as catalyst. An ammonium polythionate is formed in the presence of this catalyst. The polythionate is oxidized by air in the presence of hydroquinone or naphthoquinone to elemental sulfur.

Heinrich Koppers GmbH[220] has also been granted a patent for an oxidizing process for the removal of hydrogen sulfide, in which the alkali used is ammonia. The hydrogen sulfide is oxidized to elemental sulfur in the presence of a quinone-type catalyst. The principal feature of this process is the concentration of a bleed stream containing thiocyanate and thiosulfate. It is then recycled to the coal feed to the coke ovens. This process is very similar to ammonia Takahax or ammonia Fumaks.

The Konox[221] process for removing H_2S from gas streams is offered by Sankyo Process Services, Ltd., of Kawasaki, Japan. This process is a revised iron oxide process in liquid form. The main reaction is as follows:

$$4Na_2Fe^{VI}O_4 + 6H_2S \rightarrow 4NaFe^{III}O_2 + 4NaOH + 4H_2O + 6S \quad (18.19)$$

The regeneration reaction is as follows:

$$4NaFe^{III}O_2 + 4NaOH + 3O_2 \rightarrow 4Na_2Fe^{VI}O_4 + 2H_2O \quad (18.20)$$

The sulfur is reclaimed by filtration. Side reactions are claimed to be minimal.

Schmuk and Temmel[222,223] have patented a process for removing ammonia and hydrogen sulfide from coke-oven gas by treating it with a solution of ferrous sulfate. The ammonia reacts to form ammonium sulfate, and the ferrous hydroxide formed reacts with the hydrogen sulfide to precipitate ferrous sulfide. The ferrous sulfide is removed from the solution and treated with sulfuric acid solution to release the hydrogen sulfide and regenerate the ferrous sulfate for reuse. The hydrogen sulfide is then converted to elemental sulfur or sulfuric acid. No plants using this process have been installed. Similar processes using ferrous sulfate were developed many years ago by Feld et al.,[224] but side reactions prevented commercial success in every case.

The Diamox[225,226] process for H_2S removal from coke-oven gas is now offered by Mitsubishi. This adaptation of the ammonia process includes a naphthalene scrubber ahead of the H_2S absorber to remove naphthalene and control the temperature of the gas to the absorber. The H_2S is removed selectively in a deacidifier and sent to a Claus plant. The HCN concentration is controlled by distilling a side stream of the partially deacidified liquor and recycling the ammonia

[216] E. Mancke et al., U.S. Pat. 3,021,189, February 13, 1962).

[217] Heinrich Koppers GmbH, British Pat. 1,342,-836, January 3, 1974.

[218] R. Hamblin, U.S. Pat. 3,594,125, July 20, 1971.

[219] F. Götza et al., British Pat. 1,379,624, January 2, 1975.

[220] Heinrich Koppers GmbH, British Pat. 1,367,-820, September 25, 1974.

[221] T. Kasai, *Hydrocarbon Process.*, **54,** 93–95 (February 1975).

[222] J. Schmuk and F. Temmel, Canadian Pat. 726,618, January 25, 1966.

[223] Bethlehem Steel Corp., Australian Pat. 270,-228, October 13, 1966.

[224] W. H. Hill, in *Chemistry of Coal Utilization,* Vol. 2, H. H. Lowry, Ed., John Wiley and Sons, New York, 1945, pp. 1060–1061.

[225] Anon., *33 Mag.*, **14,** 25 (October 1976).

[226] W. Meuly and A. Seldner, U.S. Pat. 3,226,320, December 28, 1965.

and HCN to the inlet gas stream; 85–96% H_2S removal is claimed. Three plants using this process have been built in Japan.

Rhodia, Incorporated, offers the Cataban[227–229] process as the cheapest process for removing H_2S from gas streams. This process employs a one-step, continuous oxidation-reduction system. A complexed metal ion (preferably ferric) is kept in an aqueous solution by an organic chelating agent. The H_2S is oxidized to sulfur and the ferric ion is reduced to the ferrous ion. Injected air continuously regenerates the ferric ion needed for oxidation. Sulfur removal is claimed to be 99.9%. This process has not yet been tested on coke-oven gas, and side reactions with contaminants could cause problems. Also, it would not be desirable to introduce air into the gas stream for one-step operation.

18.4.3 H_2S Disposal

The absorption-desorption processes covered under Section 18.4.2.1 all recover the H_2S in the gaseous state. It was common practice for many years to burn the acid gases in a flare stack, from which the products of combustion entered the atmosphere. Regulations no longer permit this method of disposal; hence the H_2S must be converted to elemental sulfur or to sulfuric acid.

18.4.3.1 *The Modified Claus Process*

The chemistry of the Claus process[230] has been known for many years. The reactions are as follows:

$$H_2S + 1\tfrac{1}{2}\,O_2 \rightarrow H_2O + SO_2 \qquad (18.21)$$

$$2H_2S + SO_2 \rightarrow 2H_2O + 3S \qquad (18.22)$$

Combining Equations 18.21 and 18.22 gives

$$3H_2S + 1\tfrac{1}{2}\,O_2 \rightarrow 3H_2O + 3S \qquad (18.23)$$

The modified Claus process[231] is based on the same reactions: only the equipment and controls are different. The huge Claus kiln is replaced by a relatively small combustion chamber, followed by a waste heat boiler and a series of catalytic converters and heat exchangers. Figure 18.14 gives a schematic flow diagram of a typical modified Claus plant. In plants of this type the required amount of SO_2 is produced by burning the acid gas with a limited amount of air, so that one-third of the H_2S is converted to SO_2, giving an $H_2S : SO_2$ ratio of 2 : 1. Approximately 60–70% of the total sulfur yield is produced in the combustion chamber and is separated in the waste heat boiler. The gas stream is reheated and proceeds to a series of two or three catalytic converters operated at temperatures just above the sulfur dew point, where further reaction occurs in the presence of an activated alumina catalyst. At each interstage position condensers remove sulfur from the process stream and the gases are preheated before passing to the next converter. Overall sulfur yield varies from 90 to 97%.

Although plants of this type have been successfully processing H_2S-containing offgases from petroleum sweetening plants for many years, the application of the modified Claus plant to H_2S recovered from coke-oven gas is relatively recent. Operating problems[232,233] have been encountered when treating coke plant acid gases that had not been encountered in the plants treating petroleum plant acid gases. Extensive corrosion of equipment and catalyst bed plugging caused many plant failures. Incomplete combusion of HCN appeared to be the principal cause of both problems, and the Bethlehem Steel Corporation developed a unit to catalytically destroy the HCN, converting most of it to ammonia. Surprisingly, the combustion of the ammonia in the Claus plant was no problem.

Control of operating conditions in a modified Claus plant is critical. Norman[231] has studied the process in great detail and suggests many ways to achieve smoother operation. Although his recommendations are applicable to any Claus plant, they obviously

[227] W. Meuly and D. Ruff, *Paper Trade J.*, May 22, 1972.
[228] Anon., *Chem. Week*, **110**, 39 (April 5, 1972).
[229] J. Davis, *Chem. Eng.*, **79**, 66–68 (May 15, 1972).
[230] H. A. Gollmar, in *Chemistry of Coal Utilization*, Vol. 2, H. H. Lowry, Ed., John Wiley and Sons, New York, 1945, pp. 999–1000.
[231] W. Norman, *Oil Gas J.*, **74**, 55–60 (November 15, 1976).
[232] O. A. Homberg and A. Singleton, *J. Air Pollut. Control Assoc.*, **25**(4), 375–378 (April 1975).
[233] O. A. Homberg et al., U.S. Pat. 3,923,957, December 2, 1975.

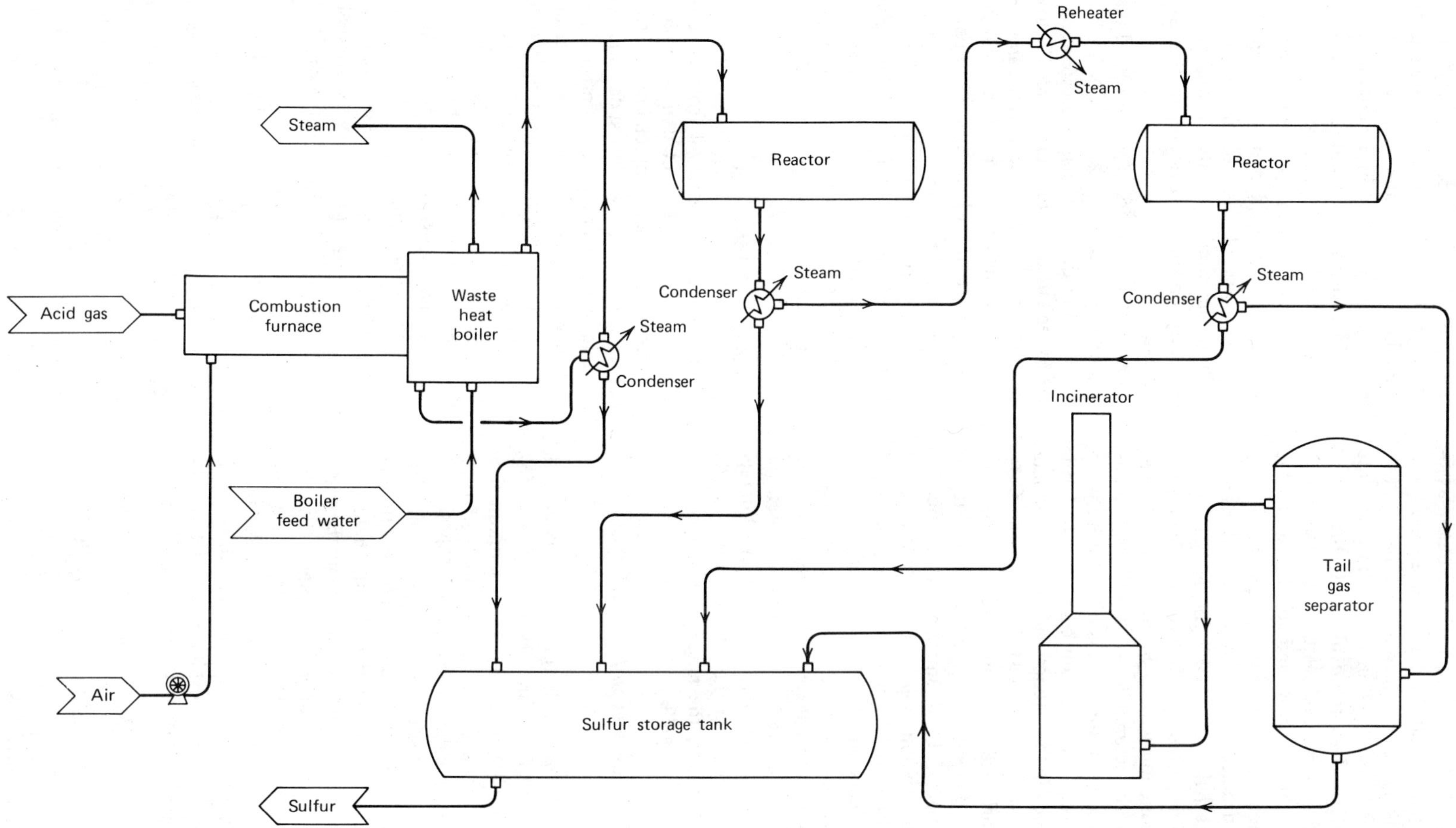

Figure 18.14 The modified Claus process.

are based on experience in petroleum-associated plants and do not cover all the problems likely to be encountered in handling acid gases produced in a vacuum carbonate or Sulfiban H_2S removal plant. Brodskii[234] studied this process with respect to conditions in the U.S.S.R. and concluded that the H_2S should be used to produce sulfur instead of sulfuric acid. The high purity of the sulfur product makes it valuable for many uses.

Modified Claus plants have also been used in conjunction with H_2S removal plants employing the ammonia process.[235,236] The gas to be treated in this case is different from the acid gases from vacuum carbonate or Sulfiban plants in that some or all of the ammonia is present, as well as H_2S, HCN, and CO_2. The Carl Still Company uses a variety of designs, depending on the final disposition of ammonia. If total ammonia is to be destroyed, the ammonia and HCN are cracked catalytically in a combustion chamber ahead of the Claus burner. The cracked gases are then concentrated before introduction into this burner. In case nearly all the ammonia is to be recovered as ammonium sulfate or anhydrous ammonia, catalytic cracking is still employed, but the concentration step is deleted because the amount of ammonia to be cracked is small.

18.4.3.2 *Contact Sulfuric Acid Production*

As an alternative to producing elemental sulfur by the Claus process, the H_2S-containing gases from an absorption-desorption H_2S removal process can be used as feed to a contact-type sulfuric acid plant. This is particularly desirable in plants where the ammonia is recovered as ammonium sulfate and there is a captive market for the acid produced. Even though an acid plant is much more expensive than a Claus plant, the possibility that an expensive tail gas cleanup process may be required for the Claus plant could eliminate the cost differential.

Many sulfuric acid plants have been built by the Carl Still Company[235,236] in conjunction with ammonia process–H_2S removal plants, primarily in Europe. These plants can even be operated satisfactorily when total ammonia is destroyed in the combustion furnace. A cooling step is required between the furnace and the catalytic reactor to permit production of concentrated sulfuric acid. Carl Still also builds acid plants in conjunction with recovery of ammonium sulfate or anhydrous ammonia. Plants of this type are currently under construction in the United States.

In the past builders of sulfuric acid plants required the ammonia and HCN to be removed from the acid gas before feeding the gas to the burner at the acid plant. Hydrogen cyanide removal was incorporated in several U.S. acid plants that burned the acid gases from the vacuum carbonate purification process. The Carl Still Company[237] has apparently solved this problem by control of combustion. Acid plant builders in the United States recently announced their willingness to accept acid gases with a relatively high concentration of HCN.

18.4.4 HCN Removal

Hydrogen cyanide removal from coke-oven gas is covered in great detail in the original[238] volumes and the first supplement[239] to *Chemistry of Coal Utilization*. Most developmental work in recent years has concentrated on the air and water pollution problems caused by HCN and the purification plant problems attributable to this gas. Simon Carves[240] patented a process to convert the ammonium thiocyanate produced by the polysulfide process (see discussion of the Stretford process) into ammonium sulfate, by treating it with sulfuric acid under pressure in an autoclave. A similar process was studied by Zaichenko et al.[241] Other processes for han-

[234] Yu Brodskii, *Coke Chem. (USSR)*, No. 7, 39–42 (1966).

[235] K. Tippmer and J. Maloy, *COMA Yearb.*, 139–140 (1974); Waverly House, Wexborough, Yorkshire, England.

[236] D. Laufhütte, *AIME Ironmaking Proc.*, **33**, 142–155 (1974).

[237] H. Schmidt, *Glueckauf*, **103**, 1170–1174 (1967).

[238] W. H. Hill, in *Chemistry of Coal Utilization*, Vol. 2, H. H. Lowry, Ed., John Wiley and Sons, New York, 1945, pp. 1083–1124.

[239] R. E. Muder, in *Chemistry of Coal Utilization*, Supplementary Volume, H. H. Lowry, Ed., John Wiley and Sons, New York, 1963, pp. 665–668.

[240] Simon-Carves Ltd., British Pat. 1,101,379, January 31, 1968.

[241] V. M. Zaichenko et al., *Coke Chem. (USSR)*, No. 7, 39–41 (1974).

dling the ammonium thiocyanate produced in conjunction with oxidation-type H_2S removal processes are covered in Section 18.4.2.2. Further items in this section consider only the HCN removal problems associated with the absorption-desorption H_2S removal processes.

18.4.4.1 *Catalytic Conversion*

The acid gases recovered in all absorption-desorption H_2S removal plants contain nearly all the HCN present in the inlet coke-oven gas. Operating problems have been encountered in several Claus plants in the United States when treating acid gases from absorption-desorption plants. Incomplete HCN combustion appeared to be the principal cause of the problem, which Homberg and Singleton[242] solved by catalytically converting the HCN to ammonia and carbon disulfide (CS_2), using an alumina catalyst at a relatively high temperature. The ammonia and CS_2 gave no trouble in the Claus plant.

Carl Still[235,236] solved the problem of incomplete HCN combustion by treating the acid gases from the ammonia process for H_2S removal. This was accomplished by catalytically cracking the ammonia and the HCN in a combustion chamber ahead of the Claus burner.

18.4.4.2 *HCN Removal by Water Washing*

Because HCN is much more soluble in water than is H_2S, water washing[239] can be used to remove HCN from acid gases. This is a very old process. If ammonia is being incinerated, the HCN-containing stream can be stripped along with the ammonia liquor, and the total dephlegmated vapor incinerated. The HCN can also be stripped and blended with the fuel gas for disposal. Good Claus plant operation was achieved when burning water-washed acid gases from a vacuum carbonate H_2S removal plant.

18.4.4.3 *Other Processes*

Koppers Company, Inc., has also had difficulties in achieving complete HCN combustion when burning acid gases from a vacuum carbonate plant in a Claus plant. However Koppers[243] believes that the difficulties in achieving complete HCN combustion have been overcome by the development of a more efficient acid gas burner for the Claus plant.

[242] O. A. Homberg and A. Singleton, *J. Air Pollut. Control Assoc.*, **25**(4), 375–378 (April 1975).

18.4.5 Tail Gas Cleanup

Modified Claus sulfur recovery plants have been in operation for many years in the petroleum industry, and many plants have recently been installed in coke plants. Sulfur recovery averages 90–97%, and it has been standard practice to incinerate the tail gases with the combustion products being vented to the atmosphere. The SO_2 emission from these stacks does not meet present-day air pollution regulations, and a variety of processes have been developed to treat Claus plant tail gases.

18.4.5.1 *The SNPA Process (Sulfreen)*

The Socíeté Nationale de Pétroles d'Aquitaine (SNPA) process[244,245] is really an extension of the Claus process and originally used activated carbon as a catalyst. Sulfur is deposited on the activated carbon and is recovered by hot gas circulation. Operation of the units is cyclic. The recovered sulfur is of good quality. More recently[246–248] a special activated alumina catalyst has replaced activated carbon as the catalyst. Efficiency of the process is limited, since COS and CS_2 are not removed to any great extent.

18.4.5.2 *The IFP Process*

The IFP process was developed by the Institut Français du Pétrole[249–250] specifically for treating Claus plant tail gases. It essentially

[243] J. A. Vietorisz, U.S. Pat. 3,732,070, May 8, 1973.

[244] J. Davis, *Chem. Eng.*, **79**, 66–68 (May 15, 1972).

[245] J. Hyne, *Oil Gas J.*, **70**, 64–78 (August 28, 1972).

[246] H. Krill and K. Storp, *Chem. Eng.*, **80**, 84–85 (July 23, 1973).

[247] Anon., *Hydrocarbon Process.*, **54**, 110 (April 1975).

[248] L. Cameron, *Oil Gas J.*, **72**, 110–118 (June 24, 1974).

[249] M. Hirai, R. Odello, and H. Shimamura, *Chem. Eng.*, **79**, 78–79 (April 17, 1972).

[250] Y. Barthel et al., *Hydrocarbon Process.*, **50**, 89–91 (May 1971).

involves the Claus reaction in solution with or without added catalyst. A high-boiling solvent of the polyglycol type is preferred; thus little cooling of the inlet gases is required. The catalyst is reported to be an alkali salt of a carboxylic acid. Removal efficiency is limited to 1500–2000 ppm H_2S, but the process is being altered to reach a removal efficiency of 300 ppm.

18.4.5.3 *The Beavon Process*

In the Beavon process,[251,252] which is capable of reducing total sulfur in the tail gas to very low levels, all sulfur compounds in the Claus plant tail gas (SO_2, S, COS, CS_2) are converted to H_2S. Before entering the reactor, the tail gas is mixed with the hot combustion products of fuel gas and with air to attain the reaction temperature and the correct reducing atmosphere. The gases are then passed over a cobalt molybdate catalyst, at which point all the sulfur compounds are converted to H_2S by hydrogenation and hydrolysis. Gas from the reactor is then cooled to approximately 100°F and treated in a Stretford process. Hydrogen sulfide concentration is reduced to less than 1 gr/100 scf.

18.4.5.4 *Shell Claus Off-Gas Treating (SCOT)*

The SCOT[253–255] process for treating Claus tail gas is another two-stage process in which the sulfur compounds are first converted to H_2S. This part of the process is very similar to the first stage of the Beavon process. The sulfur compounds in the tail gases are reduced to H_2S by a mixture of hydrogen and carbon monoxide in the presence of cobalt molybdate on alumina catalyst. The gases leaving the reactor are cooled by heat exchange and a direct water cooler before entering the H_2S removal stage. Condensate from the cooler is sent to a sour water stripper. The H_2S is removed from the cooled gases in an alkanol-amine absorption and regeneration plant. This plant is designed for high H_2S removal with reduced CO_2 removal. The concentrated H_2S-CO_2 stream is recycled to the feed gas stream for the Claus plant. Overall conversion of H_2S to elemental sulfur is excellent. The SCOT process is especially attractive in plants where the Claus plant is treating acid gases from an amine process–H_2S removal plant, since no additional amine regeneration facilities are required.

18.4.5.5 *The Wellman–Lord Process*

In the Wellman-Lord[244,245,256] process the Claus tail gas is first incinerated to convert all the sulfur compounds present to SO_2. Carbonyl sulfide and CS_2, as well as H_2S and any elemental sulfur in the gas, are converted to SO_2. The hot gases are cooled in a waste heat boiler, then quenched and fed to the SO_2 absorber. The SO_2 is absorbed in a solution of sodium sulfite by the following reversible reaction:

$$SO_2 + H_2O + Na_2SO_3 \rightleftarrows 2NaHSO_3 \qquad (18.24)$$

The bisulfite solution is then sent to an evaporator-crystallizer regeneration system where Reaction 18.24 is reversed. Sulfur dioxide and water vapor pass overhead to a condenser. A knockout drum separates condensed water for return to the absorbent dissolving tank and the concentrated stream of saturated SO_2 is piped back to the Claus plant feed or to other processing. Slurry is withdrawn from the evaporator, and the sulfite crystals are redissolved for recycling to the absorber. Overall sulfur recovery is excellent, and many Wellman-Lord plants have been installed in plants associated with petroleum production.

18.4.5.6 *Other Tail Gas Cleanup Processes*

In addition to the processes for treating Claus tail gases just described, many other less prominent processes have also been devel-

[251] Anon., *Hydrocarbon Process.*, **54,** 107 (April 1975).
[252] D. Beavon, *Chem. Eng.*, **78,** 71 (December 13, 1971).
[253] J. Naber, J. Wesselingh, and W. Groenendaal, *Chem. Eng. Prog.*, **69**(12), 29–34 (December 1973).
[254] Anon., *Hydrocarbon Process.*, **54,** 109 (April 1975).
[255] B. Goar, *Oil Gas J.*, **73,** 109–112 (August 18, 1975).
[256] Anon., *Hydrocarbon Process.*, **54,** 111 (April 1975).

oped: several of these are briefly outlined in this section.

The Cleanair[244,245] process offered by the Pitchard Company hydrolyzes COS and CS_2 while continuing the Claus reaction by cooling to virtually eliminate SO_2. The Stretford process is then used for final H_2S removal. This process operates most effectively when the Claus process leaves a slight excess of H_2S in the tail gas.

Rhodia, Incorporated, offers the Cataban[244,245] process for Claus tail gas treatment, but for good sulfur removal a hydrogenation step would be required ahead of this process.

In the Lucas[257] process, recently developed by Lurgi, the tail gas is first incinerated to convert all sulfur to SO_2. The partially cooled combustion products are then passed through a bed of hot coke, after which the SO_2 is removed from the gas stream by an alkali phosphate solution. This solution is regenerated to produce a concentrated SO_2 vapor, which is recycled to the Claus unit or otherwise treated.

Hohne[258] has patented a sulfur removal system that he claims can handle Claus tail gases. This system absorbs the sulfur produced by oxidation of H_2S on activated carbon. The sulfur is reclaimed and the activated carbon regenerated by treatment with hot inert gas. Efficiency of sulfur removal from Claus tail gas without preliminary treatment is questionable.

Koppers Company, Inc. has proposed a very simple tail gas disposal method for Claus plants installed in a coke plant. This method recommends conducting the Claus tail gas to the raw gas suction main ahead of the primary gas coolers. No serious problems are anticipated, but no such installation is yet in operation.

A. G. McKee & Company[259] is offering the citrate process for Claus tail gas cleanup. The total sulfur is incinerated to SO_2, which is absorbed in a water solution of sodium citrate and citric acid. Sulfur is precipitated by means of H_2S addition. Precipitated sulfur is heated and separated from solution by pressure decantation as liquid sulfur. Approximately 1% of the sulfur is converted to sodium sulfate, which is separated as Glauber's salt crystals.

18.5 EFFLUENT TREATMENT

18.5.1 Ammonia Liquor and Other Coal Chemical Plant Liquid Effluents

Ammonia liquor is a normal by-product of the carbonization of coal. Its handling and disposal have been problems since the beginning of the coking industry. Ammonia liquor production varies with coal moisture and chemical composition, and may range from 15 to 35 gal per net ton of coal carbonized. Extensive information on the production, analysis, treatment, and disposal of ammonia liquor is given in the original volumes of *Chemistry of Coal Utilization.*[260] The first Supplementary Volume[261] covers improved dephenolization processes and progress in effluent treatment. This chapter considers only the current practice, including recent developments.

Any coke plant effluent that has been in direct contact with the gas stream or any by-product stream must be contaminated to some extent. These streams include final cooler blowdown, light oil plant condensate water, purification plant condensate, and condensate from drip pots in the gas lines. These effluents are all less highly contaminated than the crude ammonia liquor, and it used to be common to dispose of these wastes in the plant sewer or to dispose of them as quenching makeup. With more stringent regulations on pollution, such disposal is no longer permissible, and new effluent treatment systems must handle increased effluent volumes, which in some plants may be nearly double the waste ammonia liquor.

[257] A. Doerges, K. Bratzler, and J. Schlauer, *Hydrocarbon Process.*, **55**, 110–111 (October 1976).

[258] R. Hohne, U.S. Pat. 3,634,028, January 11, 1972.

[259] F. Chalmers, *Hydrocarbon Process.*, **50**, 75–77 (April 1971).

[260] P. J. Wilson, Jr., and J. H. Wells, in *Chemistry of Coal Utilization,* Vol. 2, H. H. Lowry, Ed., John Wiley and Sons, New York, 1945 pp. 1371–1481.

[261] R. E. Muder, in *Chemistry of Coal Utilization,* Supplementary Volume, H. H. Lowry, Ed., John Wiley and Sons, New York, 1963, pp. 668–674.

18.5.1.1 *Ammonia Liquor Distillation*

Approximately 25% of the ammonia produced during high-temperature carbonization appears in the ammonia liquor make. It was formerly standard practice[260] to distill the ammonia liquor for recovery of both free and combined ammonia. At present, the liquor is distilled more as a preliminary step in meeting pollution requirements than as an ammonia recovery process. The following sections describe variations in the ammonia liquor distillation processes currently in use.

Conventional Ammonia Distillation with Lime or Caustic Soda. The distillation process[260] for ammonia liquor is basically the same as the process described in the original volumes except for design of equipment. Many modern plants operate stills constructed of carbon steel, whereas the stills were formerly of cast-iron construction. Modern stills can also be cleaned more easily.

Because of the high cost of caustic soda, lime has been the only alkali used for freeing the combined ammonia. However some plants in the United States are now using caustic soda to reduce capital cost and to eliminate still cleaning and the conventional lime sump. The overall economics are still in question.

The United States Steel CyAM process. The U.S. Environmental Protection Agency proposed effluent limitation guidelines[262] are so stringent that the United States Steel Corporation has developed the CyAM[263] process as an aqueous effluent pretreatment for feed to an activated sludge plant. The level of cyanide concentration in the effluent is one of the most difficult requirements to meet, and it was considered virtually impossible with conventional free and fixed (combined) ammonia stills. The CyAM process is basically an improvement of the conventional distillation process; technical details are still proprietary and have not been published. The process is claimed to provide improved acid gas removal, good ammonia removal, reduced steam consumption, and a fixed still that requires no cleaning if lime is used as the alkali. The process is in operation in two United States Steel Corporation plants treating total coke plant wastewater.

Air Stripping. Some investigators have proposed the use of air to strip the ammonia from the crude ammonia liquor. Barker and Thompson[264] cited work that was aimed at achieving this purpose. The ammonia-water system is described by the following equation:

$$NH_4^+ + OH^- \rightleftharpoons NH_4OH \rightleftharpoons NH_3 + H_2O$$

which signifies that an increase in hydroxyl ion (OH^-) shifts the equation to the right. It is claimed that at a pH level above 10, 85% of the ammonia can be liberated from the waste liquor by air blowing, but the volume of air required is very high (i.e., several hundred cubic feet per gallon of wastewater treated). It is necessary to heat the ammonia liquor to an elevated temperature because the solubility of ammonia decreases as the solution temperature increases.

Naso and Schroeder[265] reported that the Republic Steel Corporation installed an air stripping process at one of its coke plants. This unit is capable of stripping the ammonia from 75 gal of liquor per minute at 60°C, or 140 gal at 104°C. The flow diagram for this process appears in Fig. 18.15. The stripping tower is a packed column constructed of plastic reinforced by fiber glass. Ammonia liquor enters the top of the stripper after the pH is adjusted with caustic soda to about 11, while the stripping air enters at the base. The ammonia-enriched air passes to a sulfuric acid absorber, where the ammonia is converted to ammonium sulfate. The stripped liquor leaves the tower for further processing. The air from the acid absorber is recycled to the stripping tower to avoid air pollution and to avoid preheating the mass of circulating air.

[262] U.S. EPA, Development Document for Proposed Effluent Limitations Guidelines and New Source Performance Standards for the Steelmaking Segment of the Iron & Steel Manufacturing Point Source Category, EPA 440/1-73/024, February 1974.

[263] D. Glassman, *AIME Ironmaking Proc.*, **35**, 121–131 (1976).

[264] J. E. Barker and R. J. Thompson, EPA-R2-73-167, pp. 23–24 (April 1973).

[265] A. C. Naso and J. W. Schroeder, *AIME Ironmaking Proced.* **33,** 132 (1974).

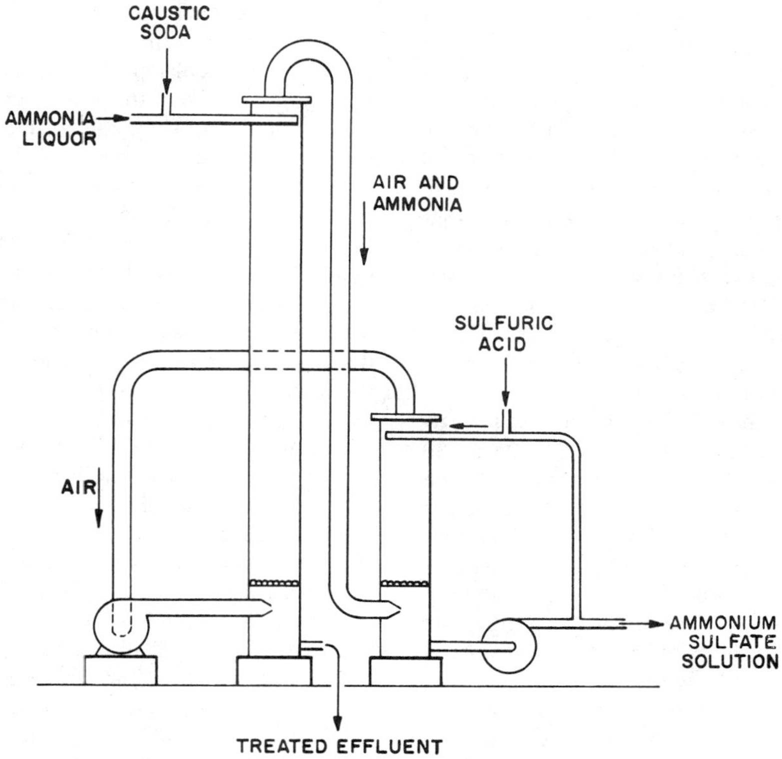

Figure 18.15 Scheme for ammonia removal in a packed column.

A patent was granted to a German firm[266] for a process in which a mixture of flue gas and coke-oven gas strips the ammonia from the waste liquor. The ammonia-rich gases are burned in furnaces in this process, and it may be more attractive to operators wishing to avoid the manufacture of ammonium sulfate.

A U.S. patent[267] was issued to the U.S. Environmental Protection Agency for a process in which the ammonia is stripped from wastewaters with air. The ammonia is then absorbed in water containing nitrifying organisms to oxidize it to NO_2 and NO_3, which are treated in a dentrifier to convert the NO_2 and NO_3 to nitrogen gas.

18.5.1.2 *Biological Treatment*

A coke plant generates about 50 gal of aqueous wastes per ton of coal carbonized, of which 50–60% is crude ammonia liquor. The remaining portion accumulates from light oil processing, surface water runoff, and other miscellaneous sources. This waste stream usually contains varying quantities of ammonia, phenols, chlorides, calcium, cyanides, thiocyanates, sulfides, thiosulfates, and possibly a considerable amount of suspended or dissolved solids. The concentrations of these components vary from plant to plant. They also vary within the same plant from one time to another, depending on the types of coal carbonized, the coal moisture, the carbonization rate, and the by-product plant operations. At one time it was common practice to dispose of this contaminated water by evaporation at the coke quench station or by discharge into the natural water courses. Neither practice is recommended anymore, first, because evaporation causes severe corrosion in the nearby plant area, and second because discharge to the natural waters causes excessive pollution. The contaminants from the coke plant wastes have three deleterious effects on the natural waters: (1) in municipal water works phenol reacts with chlorine to form chlorophenols,

[266] Stahlwerke Roechling-Burbach GmbH, German Pat. 1,642,416, September 1972.

[267] Environmental Protection Agency, U.S. Pat. 3,824,185, July 16, 1974.

which impart an objectionable taste and odor to the drinking water; (2) the contaminants are oxidized by the purifying organisms of the natural water and thereby consume the dissolved oxygen at a rate faster than the oxygen can be replenished; and (3) many of the contaminants are toxic to aquatic life.

It has been known for many years that sanitary sewage treatment plants are capable of purifying industrial wastes. Wilson and Wells[268] reported that the pioneering work on the purification of ammonia still wastes in bacteria beds was published in 1907, and a countless number of publications have appeared since that time on the biological treatment of coke plant wastes. Until 25 or 30 years ago, most biological treatment of coke plant waters was performed at municipal sewage plants, but since then there has been extensive research to develop processes that treat the coke plant water without dilution with sewage. As a result of this effort, the biological processes have gained widespread use throughout the world.

Biological treatment (Chapter 31) consists of a system in which a variety of bacterial strains is employed to ingest the undesirable components from a polluted effluent stream so that this stream can be discharged to the natural water courses. These bacterial strains are either aerobic (requiring oxygen for their metabolic process) or anaerobic (functioning in the absense of oxygen). The aerobic strains are employed in the coke works biological plants. In general, each strain of bacteria is capable of ingesting one type of substance from the plant's wastewater; thus a biological process must use several strains of bacteria to accomplish complete purification of a complex stream such as coke plant liquors. The metabolic actions of each bacterial strain may be inhibited by the other components of the system, and each strain has its own environmental requirements for maximum growth, such as temperature and pH range, nutrients, and oxygen levels. Thus, the biological process is complex, and any deviation from the ideal in feed composition and operating conditions can have serious consequences.

Abson and Todhunter[269] identified a number of bacteria employed in the biological purification of coke plant effluents and explained the nature of the inhibiting effects on these organisms. They named five strains of phenol-oxidizing bacteria, four strains capable of oxidizing thiosulfate, thiocyanate, and cyanide compounds, and two nitrifying bacteria that oxidize the ammonia. The thiobacilli oxidize the thiosulfates in preference to the cyanides and thiocyanates and do not function until most of the phenol has been destroyed. The nitrifiers act in two stages: the ammonia is first oxidized to the nitrite, then the nitrite is oxidized to nitrate. The nitrifying bacteria are solely dependent on inorganic matter to function, and organic substances such as phenol will either hinder or stop their activity.

The following discussion explains how the phenol-oxidizing bacteria are affected by conditions in the biological plant. The maximum growth rate for all five bacterial strains is achieved at the temperature range of 25–35°C, and temperatures much higher than 40°C are likely to kill the bacteria, whereas temperatures below 20°C greatly retard their growth rate. The optimum pH range is 6.5–8.0. The bacteria must also have a sufficient supply of oxygen and nutrients to maintain their maximum efficiency. Nutrients include nitrogen, phosphorus, magnesium, iron, manganese, chloride, and sulfate. Coke plant wastes usually contain enough nitrogen and chlorides, but it is often necessary to supply the other materials to the system.

Laboratory tests have shown that ammonia, sulfides, cyanides, and thiocyanates may inhibit the bio-oxidation rate. Ammonium ions in concentrations greater than 1700 ppm inhibit the reaction because of the excess, and concentrations of less than 300 ppm slow the reaction because of a deficiency of nitrogen in solution. Concentrations of sulfide ions greater than 25 ppm will cause a temporary or complete stoppage of phenol oxidation, but if the bacteria are allowed sufficient time to become acclimated to the sulfides, a concentration up to 100 ppm can be tolerated. The cyanides and thiocyanates

[268] P. J. Wilson and J. H. Wells, in *Chemistry of Coal Utilization,* Vol. 2, H. H. Lowry, Ed., John Wiley and Sons, New York, 1945, pp. 1425–1437.

[269] J. E. Abson and K. H. Todhunter, *COMA Yearb.,* 283–319 (1960); Waverly House, Wexborough, Yorkshire, England.

have little effect on the metabolic activity of the phenol bacteria if there is sufficient oxygen and if the organisms are properly conditioned. However the combination of sulfide and cyanide compounds may cause the biooxidation process to stop completely.

Various metallic ions may be toxic or may inhibit the phenol-oxidizing bacteria and so upset the biological process. Although 5 ppm of cupric ion will completely stop phenol oxidation, with proper conditioning the bacteria can tolerate concentrations up to 200 ppm at long retention times and reduce efficiency. Zinc has a similar effect; lead and chromium ions can be tolerated with no effect up to concentrations of 200 ppm. However chromate ion concentrations greater than 10 ppm can cause almost complete stoppage of phenol oxidation. The thiobacilli and nitrifying bacteria react to changes in feed composition and operating conditions in a similar manner.

There are two general types of biological waste treatment facility, the percolating or trickling filter and the activated sludge processes. The percolating filter process is the long-established method of biological treatment, especially for municipal treatment facilities. It consists of a static bed of porous media that supports the bacteria. The contaminated water is sprayed on top of the bed and the water is aerated as it trickles down through the porous bed. In addition to the bacteria, the bed contains a higher life form such as larvae, leeches, or worms, which consume the waste products from the biological process. Without this purifying element, the percolating bed would convert from an aerobic to an anaerobic process as the waste products accumulate, and so alter its characteristics. This process requires large volume dilutions of coke plant waters to avoid the toxic effect of its components on the larval life in the percolating beds. A variation of this process is the packed-tower filtering bed, which was developed in Europe.[270,271] Plastic packing with a very low bulk density is used in beds 20–25 ft high. The influent is sprayed on top of the bed, and the bacterial growth is supported on the plastic packing. Excess bacteriologic film sloughs off and is separated from the effluent in sedimentation tanks. The major advantage of this development is the large reduction in land space requirements because the process is built vertically rather than horizontally.

The Activated Sludge Process. The coking industry employs the activated sludge process almost exclusively for the biological processing of its wastes. This process uses large tanks in which the bacteria are kept in suspension by the action of large aerators and mechanical mixers. The process inflow is pretreated to remove tar and other extraneous matter. The pH and temperature are maintained at their proper levels, and nutrients are supplied as needed. The total volume of plant wastes and biomass (activated sludge) is kept in the aeration tanks until the phenols are reduced to an acceptable level. The outflow goes to large clarification tanks for separation of sludge from the effluent. The bacterial mass is kept active by discarding a portion of the sludge from the clarifiers. The bulk of the sludge is recycled to the aeration tanks. The treated effluent proceeds to the plant outflow streams or to final treating processes. With proper plant design and operation, this process is capable of reducing the phenols, cyanides, thiocyanates, thiosulfates, and ammonia to very low levels.

A large number of papers that pertain to this subject were published during the past 25–30 years. These articles give detailed descriptions of laboratory, pilot-plant, and full-scale plant tests and operations performed for the sole purpose of gaining information on the biological purification of coke plant aqueous wastes. Some of the most notable were published by the staff of the British Coal Utilisation Research Association (BCURA).[272–275] This group has conducted

[270] C. D. Furness, *Chem. Eng., London,* No. 282, 102–107 (February 1972).

[271] A. G. Ashmore, J. R. Catchpole, and R. L. Cooper, *COMA Yearb.,* 103–125 (1970); Waverly House, Wexborough, Yorkshire, England.

[272] A. G. Ashmore, J. R. Catchpole, and R. L. Cooper, *Water Res.,* **1,** 605–624 (1967).

[273] A. G. Ashmore, J. R. Catchpole, and R. L. Cooper, *Water Res.,* **2,** 555–562 (1968).

[274] A. G. Ashmore, J. R. Catchpole, and R. L. Cooper, *Water Res.,* **6,** 1459–1474 (1972).

[275] R. L. Cooper and J. R. Catchpole, paper presented at Information Symp. Coke Oven Techniques, EUR 5342, Park I, sponsored by the Commission of the European Communities, Luxembourg, May 1975, pp. 149–160.

extensive research studies on every aspect of the biological process as it applies to coke plant operations. Subjects of these studies include the effect of the concentrations of various contaminants found in coke plant streams, factors that hinder or accelerate the biological process, and the effect of design and operating parameters on the process. The authors contend that the waste streams vary considerably from plant to plant, and the only safe course of action to follow for plant design is to conduct a pilot-scale laboratory test for an uninterrupted period, for each plant's effluent. The test data will establish the dilution rate, the retention time, the sludge volume, and the oxygen requirements of the particular effluent stream. One can then determine the size of the aeration and clarification tanks, aerators, mixers, and pumps that will be required in the proposed process, and establish the capital outlay and operating costs for the plant. In addition, the laboratory tests will predict the effects of contaminant concentrations, such as the toxic and inhibitory actions, on the biological process.

These authors report that their studies clearly indicate the cause of many plant upsets that have occurred in operating biological plants. Some of their findings are cited below.

Sludge Condition. It was learned that to keep the bacterial mass active, a certain portion of the sludge that is separated in the clarifiers must be discarded. Otherwise the excessive sludge accumulates into large masses, and in time the inner portions of these masses become anaerobic. The result is deterioration of the process efficiency plus the evolution of odoriferous gases. Poorly conditioned sludge becomes unmanageable because of slow or incomplete settling characteristics or because the sludge becomes slimy and interferes with the operation of the aerators and mixers. If the sludge wastage rate is too high, the bacteria population decreases and again the process efficiency deteriorates. The authors state that the wastage rate for an unlimed ammonia liquor is ordinarily about 10%.

Effect of Contaminating Substances. The effects of ammonia, phenol, cyanides, and the thiosalts on one another in the bio-oxidation reactions were discussed earlier. It was learned that excessive calcium in the limed liquors has a significant effect on the process, which results from the deposition of solid calcium salts on pump impellers, aerators, and mixer blades. In addition, the flocculated sludge may become coated with the calcium deposits and become inactive.

It has been shown that some of the polyhydric phenols have an inhibiting effect on the process. For example, a laboratory experiment was conducted on a solution that contained 200 mg/l of ammonium thiocyanate. A minimum retention time of 9.3 hr was needed to reach complete oxidation of the thiocyanate salt, and 16.4 hr of minimum retention time resulted when 1000 mg/l of phenol was added to the solution. It is recalled that bio-oxidation of the thiocyanates will not occur until the phenol has been oxidized. However when 50 mg/l of oxidized catechol plus the 1000 mg/l of phenol was added to the solution, the minimum retention time increased to 27.5 hr. Pyrogallol, the trihydric phenol, inhibits the bio-oxidation of both thiocyanate and phenol. Other oxidized and/or polymerized substances of unknown composition affect the biological reactions. Additional research must be conducted before the full effect of many of the substances found in coke plant effluents can be known.

Effect of Bacterial Growth Promoters. The BCURA research group made a study to learn whether there were any growth stimulators that would be useful in coke plant biological processes. Many were discovered, including *p*-aminobenzoic acid, *N*- and isonicotinic acid, pyruvic acid, glycine, alanine, glucose, β-phenylalanine, and many others. This group of chemicals fitted into specific classes, such as the substituted benzoic acids (e.g., *p*-aminobenzoic acid), the pyridine carboxylic acids (e.g., the nicotinic acids), aliphatic amino acids (e.g., L-α-alanine), and substituted benzene compounds in which the growth-stimulating group is present as a side chain (e.g., β-phenylalanine). It is believed that these compounds work by exerting an influence on the bacterial enzyme system. The common denominator for this class of compounds appears to be the chemical breakdown to pyruvic acid, $CH_3COCOOH$.

Glucose was the growth accelerator selected for large-scale tests because of its low cost and ready availability. Laboratory tests confirmed that the growth factor concentration of 5 mg/l was sufficient to be effective. A test was conducted with this quantity of glucose added to the inflowing stream. For the particular test liquor, the growth factor effect was surprising. The minimum retention time decreased from 29.0 to 15.7 hr, and there was an improvement in clarity of the treated solution. When the glucose addition was stopped, the bio-oxidation rate decreased and accelerated again after glucose additions were resumed. It is not claimed that use of growth stimulators will offset poor plant design, but it was proved conclusively that the addition of small quantities of these compounds would be beneficial to the biological process. Patents were issued to BCURA and the British Steel Corporation for this process improvement.[276,277]

A large number of coke plants throughout the world are able to take advantage of the local municipal sewage treatment centers to treat their plant effluents.[268,278–280] This application is very effective primarily for the oxidation of phenols, but is susceptible to the inhibiting and toxic effects of the other contaminants contained in the coke plant waters. A large municipal sewage plant is required to provide the capacity for large-scale dilution to offset the effects of contaminants. Generally the municipal sewage department's treatment fees are high, but the effluent treatment facilities are located outside the plant property, and this attractive feature often offsets the high treatment fees.

Muller and Coventry[279] prepared a detailed report on a long-range test conducted by the Gary, Indiana, Sanitary District on the treatment of coke plant effluents from the local U.S. Steel works. The test period extended from December to June, during which time approximately 37.5 million gal of coke plant wastes was treated. The most difficult problem encountered in this test was the excessive calcium carbonate formation that resulted from mixing the carbonate-rich plant waters with the calcium-rich water of Lake Michigan. Additives to control scale formation were not effective. A high-flow, steady state condition was maintained for the month of May, during which time approximately 400,000 gal/day of plant wastes was treated. The results showed that the daily loading of 2290 lb of phenol, 525 lb of cyanide, and 16,366 lb of ammonia resulted in an effluent that contained 2.2 lb of phenol, 75 lb of cyanides, and 20,997 lb of ammonia after treatment of the combined municipal sewage and coke plant waters. The outflow from this sewage plant is chlorinated for final treatment before discharge into the river. The high residual ammonia that resulted from the coke plant waters increased the rate of chlorine usage to such a degree that the test was terminated. The added chlorine costs for treating 400,000 gal/day was $575/day.

It is now becoming common practice to construct an activated sludge biological plant at the plant site. Although space requirements are great, it is easier to control the process if coke plant wastes are treated separately. A large number of plants are now in operation throughout the world, raging in size from 100,000 to more than 1,000,000 gal/day of coke plant wastewaters.[281–289] Most of these plants were designed for biological oxidation of phenols and cyanides, but as the restrictions on plant outflows become more stringent, subsequent treatment facilities (tertiary treatment) are being added to the process for final treatment.

[276] British Coal Utilisation Research Association, British Pats. 1,210,492, 1970, 1,252,460, 1971.

[277] British Steel Corporation, British Pat. 1,281,-334, 1972.

[278] J. E. Barker and R. J. Thompson EPA-R2-73-167, April 1973, pp. 28–29.

[279] J. M. Muller and F. L. Coventry, *Blast Furn. Steel Plant,* **56**(5), 400–406 (May 1968).

[280] A. G. Ashmore, *COMA Yearb.*, 124 (1975); Waverly House, Wexborough, Yorkshire, England.

[281] M. Perch, Koppers Company, Inc., private communication.

[282] R. Nebolsine, *Iron Steel Eng.*, **34**, 141–145 (December 1957).

[283] P. D. Kostenbader and J. W. Flecksteiner, *J. Water Pollut. Control Fed.*, Part 1, **41** 199–207 (1969).

[284] J. E. Ludberg and G. D. Nicks, *Water Sewage Works*, **116**, *Ind. Wastes Suppl.* 10–13 (November 1969).

[285] S. A. Pearson, *COMA Yearb.*, 268–281 (1975); Waverly House, Wexborough, Yorkshire, England.

[286] D. E. Cunningworth, *Gas J.*, **303**, 438, 441–442, 445 (September 21, 1960).

[287] H. B. Hodgson, *COMA Yearb.*, 115–137 (1967); Waverly House, Wexborough, Yorkshire, England.

[288] R. Gross and A. Denne, *Stahl Eisen*, **88**, 280–285 (March 21, 1968).

[289] A. I. Shevchenko et al., *Coke Chem. (USSR)*, No. 5, 42–47 (1972).

Figure 18.16 is a flow diagram for a typical process designed for destruction of phenol and cyanides. The large tanks for storage of incoming wastes serve as separators for tar and suspended solids. The aerator and clarification tanks are constructed with a large ratio of surface area to height. These tanks are usually made from reinforced concrete to lessen the effect of corrosion and reduce heat losses from the circulating solutions. At the commissioning of the plant, the bacterial process is initiated by introducing a bacteria culture prepared from the treatment of a similar type of effluent. The feed waters are diluted so that the phenol and ammonia concentrations are low. As the bacteria population increases, the flows increase until the process is operating at the design rate.

The major problems common in the new plants are poor mixing and aeration, poor sludge characteristics, inconsistent chemistry of the feed stream, and foaming in the aeration tank. Foaming is caused by the generation of CO_2 gas and can be controlled by various defoaming techniques. The plant requires daily attention to the chemical analyses of inflow and outflow waters, to monitor operating efficiency. Careful attention must be paid to the flow rates, solution temperatures, and the pH of both the incoming water and the aeration tank solution.

Plant operators soon become experienced in plant control by observing the nature of the sludge, its settling properties, and the phenol concentration that remains in the treated waters. The bacteria employed in this process proved to be very hardy and are capable of reestablishing themselves to normal conditions despite some drastic upset in plant operations. The process is reliable and is the least costly method for reducing the phenol concentration of coke plant liquors to less than 1 ppm.

Final Effluent Treatment. Even though the process just described is able to achieve almost total phenol oxidation and usually achieves good oxidation of the thiocyanates and cyanides, there is an urgent need either to make vast improvements in the activated sludge process or to develop other treatment methods. If other treatment methods are adapted in time, it is most likely that these methods will be applied to the biologically treated effluent rather than to raw coke plant waters. Evidence from the latest research programs indicates that these added (tertiary) treatment schemes will be much more costly than the activated sludge process. Therefore it is logical to assume that the more costly processes will be applied to final treatment rather than to total treatment.

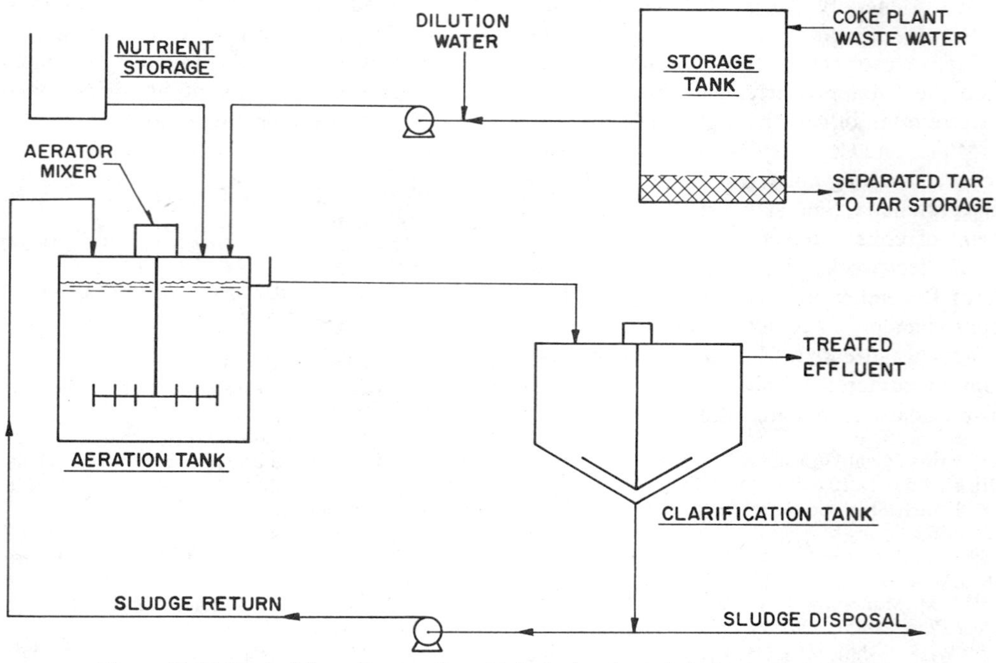

Figure 18.16 Typical flow diagram for a biological activated sludge treatment plant.

The biologically treated effluent resulting from current treatment methods does not meet the latest wastewater specifications that have been issued in many of the coke-producing nations. The objections are primarily related to the high ammonia levels, poor color, high turbidity of the treated waters, and high values for biological oxygen demand (BOD) or chemical oxygen demand (COD). The high ammonia level is lethal to the fish, and poor color and turbidity prevent the penetration of sufficient sunlight for normal plant growth. The BOD and COD values are empirical measures of the capacity of a contaminated water to absorb the dissolved oxygen from the receiving waters.

It was mentioned earlier that the bacterial population of an activated sludge process often contains ammonia-oxidizing bacteria. Since these nitrofers are autotropic, however, they are unable to function in a system that contains carbonaceous matter. Furthermore, these bacteria function at their optimum rate in a slightly alkaline system, whereas the phenol-oxidizing bacteria prefer a neutral system. Thus it is apparent that the activated sludge process ordinarily operates for total phenol destruction, but NH_3 oxidation is usually poor.

Turbidity is caused by suspended or colloidal matter that does not separate readily from the treated effluent. The reasons for poor color are not yet fully understood, but it is known that some of the polyhydric phenols and/or polymerized substances that pass untouched through the biological process develop some of the color. Also, it is believed that this material is lethal to fish life.

The BOD, COD, and other indices, such as permanganate value (PV) and total organic carbon (TOC) are measures of the residual polluting substances in the treated effluent. These include traces of the mono- and dihydric phenols, the polyhydric phenols, pyridine, and oils, plus the cyanides, thiocyanates, thiosulfates, and ammonia that leave the activated sludge process.

Multistage Activated Sludge Process. Barker and Thompson[290] prepared a detailed report of a year-long study of a three-stage biological process for treating coke plant wastewaters. A three-module pilot plant was constructed, with the first unit for oxidation of phenols and thiocyanate, followed by a nitrification unit for the aerobic oxidation of ammonia to nitrate, and the third unit for the anaerobic reduction of the nitrate to nitrogen gas (denitrification). Operating details of the first module are not discussed here, since it functions as a typical activated sludge unit.

The nitrification unit is similar to the first unit in that bacterial action oxidizes ammonia to nitrate in an activated sludge process. Two bacterial forms are employed: *Nitrosomonas,* which convert ammonia to nitrite, and *Nitrobacter,* which oxidize the nitrite to nitrate. Other nitrogen compounds such as cyanide, thiocyanate, and the organic nitrogen compounds are also oxidized in the process. Nitrification is accompanied by the release of hydroxyl ions, and since the nitrofers require a slightly alkaline environment, it is necessary to maintain close control of pH in this unit. Sodium carbonate was selected for this purpose, since it also provides the inorganic carbon that the bacteria need for their metabolic action. Although this module is very susceptible to the upsets in the first module, the test proved that with careful control, 75% or more of the ammonia can be converted in the process.

In the denitrification process, the organisms use the oxygen from the nitrates in their metabolic process only when oxygen is not available; therefore this process requires the addition of a reductant. For this test corn sugar molasses was selected as the reducing organic carbon. Operation of the unit was not troublesome, and it was learned that temperature and pH control were not as critical as had been expected. With the total unit at good steady operation, the denitrification unit converted 95% of the oxidized nitrogen to nitrogen gas.

The capital and operating costs were developed for a biological process designed to remove nearly all the phenol, ammonia, and thiocyanates, and to achieve good reduction in COD values. Bear in mind that the cost estimates were based on pilot-scale testing. The estimates were projected for a plant designed to produce 33,000 tons of coke per month. This facility would generate about 40,000 gal/day of excess ammonia liquor, which would contain approximately 6000

[290] J. E. Barker and R. J. Thompson, EPA-R2-73-167, April 1973, p. 178.

ppm COD and 4000 ppm ammonia. It was assumed that this liquor would be diluted to 80,000 gal/day before biological treatment in the carbonaceous oxidizing unit and further diluted to 25% (320,000 gal/day) before treatment in the nitrification and denitrification units. Anticipated results are 80% COD and 95% ammonia removal. In 1973 the authors estimated that the capital cost for the three-stage system would be $995,000, and the operating costs would be $235,000/year. This equivalent to 58¢/ton of coke produced. A three-stage process, similar to the above, was described by Bringmann and Kuhn.[291] These authors stated that domestic sewage was used as the hydrogen donor in the denitrification process.

Abson and Todhunter[269] described a three-stage process in which phenol is oxidized in an activated sludge unit in the first stage, followed by a trickling filter for oxidation of the thiosulfate, thiocyanates, and cyanide, and then by a second activated sludge unit for oxidation of ammonia to nitrate. The activated sludge units contained two aeration tanks in series operation for both phenol and ammonia oxidation. The average concentration of the inflow and outflow was as follows:

	Concentration (ppm)	
	Inflow	Outflow
Total ammonia	1950	74
Monohydric phenols	580	0.9
Total phenols	700	3.4
Thiocyanate and cyanide	120	0.3

There are many references to the operation of multistage activated sludge processes for the separate treatment of phenols (in the first stage), and the thiosulfate, thiocyanate, and cyanide (in the second stage). However Catchpole and Cooper[275] from the BCURA research laboratories state that a properly designed and operated single-stage process will reduce phenols and the cyanide and thiocyanates to the acceptable low levels.

Oxidation Methods. The municipal sewage treatment centers have used chlorination for the posttreatment of their biologically treated waters for many years. For a wastewater that contains low concentrations of phenol and little or no ammonia, the treatment costs are reasonable. This is not the case for coke plant liquors, which may contain well over 200 ppm of ammonia. Kostenbader and Flecksteiner[283] reported that attempts to remove cyanides by treatment of activated sludge effluent with chlorine, chlorine dioxide, and ozone were not economically feasible. Hepner et al.[292] tried to reduce the color of the biologically treated effluent from a commercial activated sludge plant. The effluent from this plant had a color that varied from 600 to 2900 Hazen units. The Commonwealth of Pennsylvania permits discharge of the effluent into its natural streams if the color index does not exceed 175 units. Hepner and his associates found that 19 lb of ozone was required per 1000 gal of effluent to reduce the color from 2200 to 175 units.[292]

Hall[293] reported that ozonization destroys the phenol and thiocyanates and reduces the permanganate value for biologically treated ammonia waste liquor. He did not mention the ammonia concentration in these effluents. He stated that ozonization is not an economical method for destroying cyanides.

Chemical and Physical Adsorption. Activated carbon and ion-exchange resins have been tested for use in the posttreatment of activated sludge effluent waters. Hepner et al.[292] reported that 75 lb of activated carbon was required to reduce the color of treated effluent from 2200 to 175 Hazen units/1000 gal, in a once-through process. Only one-half to one-third of this quantity was required for continuous operation with carbon regeneration. The authors concluded that the cost for this treatment would be high. Kostenbader and Flecksteiner[283] reported that activated carbon was not effective for the reduction of cyanides in treated wastewater.

Bernardin[294] reported that copper-impregnated activated carbon was capable of removing the cyanide. The reaction (in the

[291] G. Bringmann and R. Kuhn, *Gesundh. Ing.*, **86**(9), 276–279 (1965).

[292] R. D. Hepner, G. R. Tallon, and C. W. Fisher, paper presented before Eastern States Blast Furnace and Coke Oven Association, Pittsburgh, February 13, 1970.

[293] D. A. Hall, *COMA Yearb.*, 179–199 (1959); Waverly House, Wexborough, Yorkshire, England.

[294] F. E. Bernardin, *J. Water Pollut. Control Fed.*, **45**(2), 221–231 (1973).

presence of dissolved oxygen in solution) involves the oxidation of cyanide to cyanate. The copper then catalyzes the reaction of cyanate to ammonia by hydrolysis. The reaction conditions require that the effluent have a low concentration of suspended solids and metallic ions, especially iron. The pH value must be maintained at 6.5–8, and a sufficient level of oxygen must be provided. A laboratory test using coke plant wastes proved that the copper-impregnated carbon was able to reduce the cyanide content from 38 to 0.1 ppm. Iron, which is found in most coke plant wastes, inhibits the reaction. If the iron content is high, complete oxidation is not possible.

Cousins and Mindler[295] reported the results of their pilot-plant tests for ion-exchange treatment of the effluent from a plant-scale activated sludge process. The pilot plant consisted of two ion-exchange columns with a regeneration system. The first column is packed with a gel-type sulfonated polystyrene resin, and the second column is packed with a macroreticular sulfonated polystyrene. Typical feed to the ion-exchange columns contained 1200–1300 mg/l of ammonia, 2–15 mg/l of cyanide, and 30–250 mg/l of thiocyanate. The pilot tests showed a reduction in NH_3 to 40 or 50 mg/l with about 90% reduction in thiocyanates. In addition, the ion-exchange resins served to coagulate the suspended matter and collodial substances and to reduce the effluent color from dark brown to pale yellow. A by-product of the process is the production of ammonium sulfate by the reaction of absorbed ammonia with the regenerated sulfuric acid. To effect cyanide removal, the authors suggested a posttreatment process for the ion-exchange effluent. This entailed adjustment of the acidic effluent to a pH of 5 by passing it through a limestone column followed by addition of pickle liquor and hydrated lime slurry for final neutralization. A floc formed in the process, which readily settled to 10% by volume, and the supernatant phase was clear and almost water-white. The cyanide concentration was reduced to 0.2–0.4 mg/l. The authors went on the say that chlorination and ozonation of the treated effluent was not economically feasible.

[295] W. G. Cousins and A. B. Mindler, *J. Water Pollut. Control Fed.*, **44**(4), 607–618 (1972).

In 1972 the authors estimated that the cost for that year of a full-scale ion-exchange and tertiary treatment process to treat 400,000 gal of biologically treated wastes would require $700,000 in capital investmants plus $108 per day to operate the plant.

Miscellaneous Treatment Processes. For the removal of cyanides, several authors have suggested that the biologically treated effluent be coagulated by addition of ferric or ferrous salt solutions.[296–305] Patents have been issued for similar treatment processes. Ferric or ferrous salt solutions are reacted with cyanide to form the ferri-ferro cyanide complex. Generally the precipitate is difficult to separate from solution, but the degree of reaction is good. The finely divided precipitate has the added advantage of adsorbing the color-producing substances from the treated solution.

Patents[303,304] were issued to two Japanese firms for a process to reduce the cyanides and thiocyanates in the waste solution by electrolysis. The patent issued to the Seiichi Inoue firm claims that the CNS^- is electrolyzed to ammonia, oxalic acid, and sulfur, whereas the CN^- is converted to ammonia and oxalic acid. The Seibi Chemical Industry Company also proposes electrolysis as a means for reducing the thiocyanate formed in the liquid oxidation desulfurization process.

Hepner et al.[292] reported that the biologically treated effluent can be upgraded by soil

[296] A. G. Ashmore, *COMA Yearb.*, 118 (1975); Waverly House, Wexborough, Yorkshire, England.

[297] G. M. Dorogobid et al., *Coke Chem. (USSR)*, No. 9, 37–39 (1974).

[298] P. D. Kostenbader and J. W. Flecksteiner, *J. Water Pollut. Control Fed.*, Part 1, **41**, 199–207 (1969).

[299] W. G. Cousins and A. B. Mindler, *J. Water Pollut. Control Fed.*, **44**(4), 607–618 (1972).

[300] Bethlehem Steel Corporation, U.S. Pat. 3,-847,807, 1974.

[301] Fugi Electric Corporation, Japanese Pat. 74-70,456, 1974.

[302] Fugi Electric Corporation, Japanese Pat. 74-70,457, 1974.

[303] Seiichi Inoue Yamatogawa Dyeing Works, Ltd., Japanese Pat. 74-19,658, 1974.

[304] Seibi Chemical Industry Ltd., Japanese Pat. 74-90,677, 1974.

[305] P. J. Wilson and J. H. Wells, in *Chemistry of Coal Utilization*, Vol. 2, H. H. Lowry, Ed., John Wiley and Sons, New York, 1945, pp. 1457–1460.

percolation methods. Their test results showed almost complete removal of residual BOD, phenols, thiocyanate, cyanides, and color by soil percolation. Based on laboratory tests, it was estimated that one acre of ground would be capable of treating 35,000 gal/day of treated effluent. The prerequisite, of course, is land of good permeability. Land that was cared for with good vegetation growth and periodic plowing would be suitable. Disadvantages of this method might be an odor problem and possibly the need to impound untreated effluent during the winter season.

18.5.1.3 *Chemical-Physical Sorption Methods*

"Sorption" is a convenient expression that describes the absorption-adsorption action of a porous solid for use in the selective separation of specific components from the whole. A solid that has commonly been used for this purpose is activated carbon. Wilson and Wells[305] reported that activated carbon has been used to recover phenols from ammonia liquors since 1927. In the past the adsorbed phenol was recovered as a salable product from the carbon either by chemical extraction or by steaming. Ashmore[296] reported that a full-scale activated carbon plant was used in conjunction with ion-exchange resin for the total treatment of coke oven liquor. However the installation was abandoned because of high operating costs.

The manufacturers of activated carbon have continued their research studies to improve the processes for coke waste treatment. The Calgon Corporation of America and Bergbau-Forschung GmbH[306,307] have developed almost identical processes for the treatment of industrial wastes. Van Stone[306] suggested the flow diagram in Fig. 18.17 as a complete process to treat coke plant waste streams. The essential features of the process are pretreatment of the wastewaters to separate tars and other organics that may tend to clog the pore structure of the porous carbon. The filtered water is passed up through a bed of activated carbon for separation of the phenols. The wastewater then passes through a cyanide oxidation step to remove these salts. Ammonia and thiocyanates are not recovered in the process. It is reported that a process to treat 400,000 gal of wastewater per day would require only 6000 ft^2 of area, less than one-quarter of an acre. The author states that the cost of phenol removal can be approximated by using the ratio of 25 lb of phenol removed per 100 lb of carbon.

Crook et al.[308] and Zdybiewska et al.[309] reported good purification rates for coke-oven liquors with the use of ion-exchange resins. Crook and his co-workers reported that the anion Ambertile XAD was able to reduce the phenol concentration of a tested effluent from 6700 to <1 ppm. In this test the ion-exchange resin removed 87 g of phenol per liter of resin. Zdybiewska stated that a commercial anion resin reduced the phenol value of an effluent by 95–99% and also removed all of the chloride, sulfide, cyanide, and thiocyanate. Ashmore[296] reports that Soviet experimenters were successful in removing phenols with anion-exchange resins and removing pyridine with cation resins. Cousins and Mindler[295] reported laboratory tests for the removal of ammonia from biologically treated liquors and anion resins. Rashkevich and Dal[310] claimed that sulfonated coal could be used to successfully treat coke plant liquors. Reports state that phenol removal exceeds 99.5%; cyanide and sulfide removal were about 97 and 80%, respectively.

18.5.1.4 *Chemical Oxidation Methods*

Coke plant waters can be purified by oxidation of the contaminating substances. A number of oxidation reagents can be used, particularly chlorine, chlorine dioxide, ozone, and hydrogen peroxide. The Ohio River Valley Water Sanitation Commission conducted the first large-scale tests for the oxidation of dephenolized coke plant wastes.[311] The re-

[306] G. R. Van Stone, *Iron Steel Eng.*, **49**(4), 63–66 (1972).

[307] H. Jüntgen and J. Klein, *Prepr. Am. Chem. Soc. Div. Fuel Chem.*, **19**(5), 67–84 (1974).

[308] E. H. Crook, R. S. McDonnel, and J. T. McNulty, *Ind. Eng. Chem., Prod. Res. Dev.*, **14**(2), 113–118 (1975).

[309] M. Zdybiewska et al., *Koks Smola Gaz*, **11** 222–226 (1966).

[310] I. I. Rashkevich and V. I. Dal, *Ukr. Khim. Zh.*, **33**(2), 205–207 (1967).

[311] Ohio River Valley Water Sanitation Commission, *Phenol Wastes: Treatment by Chemical Oxidation*, Cincinnati, 1951.

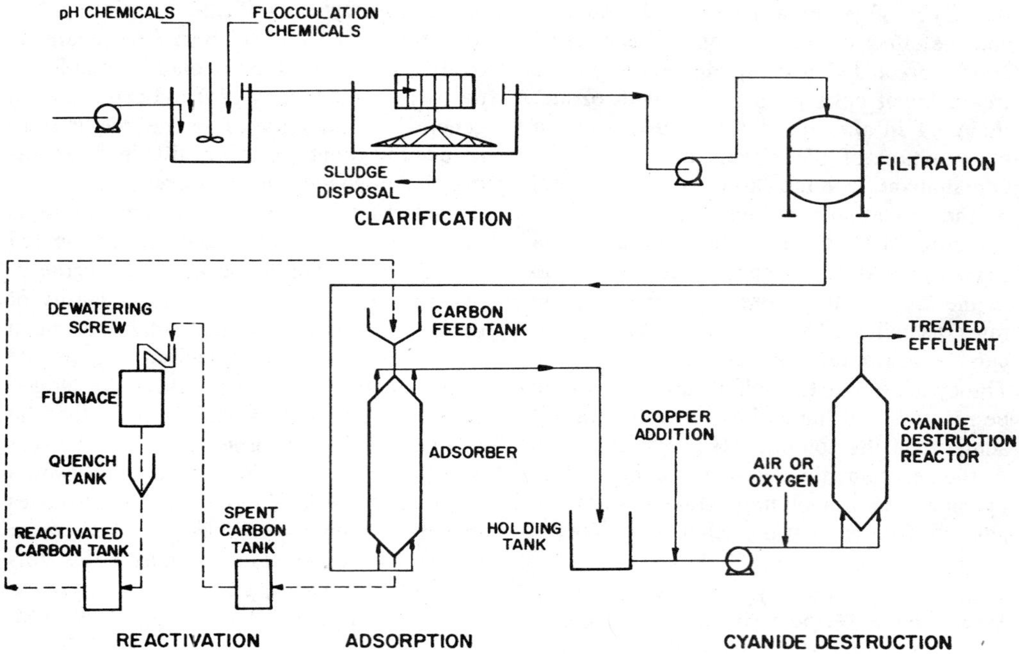

Figure 18.17 Flow diagram for activated carbon process.[306]

port contains reaction curves that show the effect of oxidizing dosage as a function of residual phenols in the treated waters. Although phenol is the major constituent treated by this method, one is not able to determine reagent requirements on the basis of the phenol concentration in the wastewaters. Other contaminants also consume reagents, especially ammonia, by reaction with chlorine gas. The authors made the following conclusions:

1. Chlorine treatment requires adjustment of both the temperature and pH of the wastewaters. Temperatures greater than 45°C may cause formation of chlorates, thus increasing the amount of chlorine needed. Adjustment of the pH to 7.0–10.0 is preferred for optimum reaction. Residual chlorine of several hundred parts per million is required to effect complete phenol removal, and it is necessary to provide posttreatment of the effluent with activated carbon to remove the excess chlorine. The chloride concentration of the treated effluent is increased in an amount equivalent to the chlorine dosage. Excess foaming may result during chlorination.
2. Chlorine dioxide and ozone treatment do not require temperature or pH adjustment, and excess reagents are not required to effect complete removal of phenol. Other contaminants (cyanides, thiocyanates, etc.) will also consume the oxidants. Ozonation may cause foaming problems, and chlorine dioxide treatment does contribute a small quantity of chlorides to the treated waters.
3. Oxidation treatment of coke plant wastes will be expensive, and treatment for complete removal of phenols would be prohibitive.

A number of papers have appeared in the literature on the use of ozone to purify phenolic wastewaters. Eisenhauer[312] and Adams et al.[313] published reports on the academic study of the ozonation process. These works examined the kinetics of phenol degradation,

[312] H. R. Eisenhauer, *Water Res.*, **5**, 467–472 (1971).

[313] C. E. Adams, W. W. Eckenfelder, and R. M. Stein, paper presented at lst Int. Symp. Ozone for Water and Wastewater Treatment, December 2–5, 1973. Published by International Ozone Institute, Waterbury, Conn., R. G. Rice and M. E. Browning, Eds.

the effect of temperature and solution pH, and the effect of catalysts on reaction rates. Hall[314] presented a complete report on the treatment of coke plant waters with ozone. Included in this laboratory study was an examination of the effects of the various contaminants common to coke plant waters on the ozonation reactions.

Figure 18.18 illustrates how the ammonia liquors from different coke plants respond to ozone treatment. These different reaction rates, as indicated by the shape of the curves, can be attributed to a number of factors. Thiocyanate and thiosulfate are the first to be reacted, and the optimum reaction rate is achieved if the solution pH is less than 8.0. In the reaction, thiocyanate is converted to cyanide, and ammonium sulfate and sulfuric acid are formed. On the other hand, phenol reacts rapidly only if the solution pH is adjusted to 10.0 or higher. The cyanides oxidize slowly until the phenol is completely destroyed. This study also indicated that in general the quantity of ozone required to reduce the phenol concentration increased as the phenol concentration decreased.

Table 18.4 lists the solution concentrations of the plant wastewaters that are represented in Fig. 18.18. This table lists the degree of oxidation that occurred and the quantity of ozone required. These data show that much more ozone would be required to treat the cyanide before this water could be discharged to the plant outflow. Also, it is noted that the reduction in the absorbed oxygen value (the permanganate value was used here) would not meet the specifications for treated wastes that have been imposed by the clean water regulatory bodies. Hall surmised that this was caused by the reaction products of ozonation and demonstrated that one such prod-

[314] D. A Hall, *COMA Yearb.*, 178–199 (1959); Waverly House, Wexborough, Yorkshire, England.

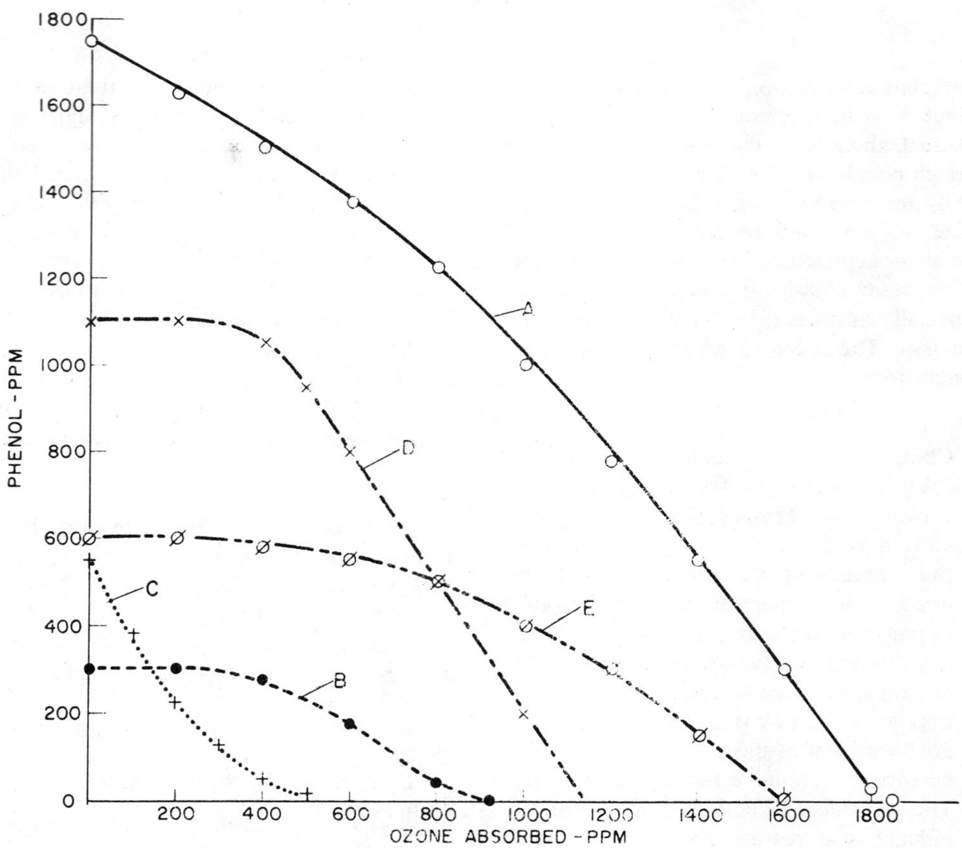

Figure 18.18 Ozonation of ammoniacal liquors: letters on curves designate coke plants listed in Table 18.4.

Table 18.4 Ozonation of Ammoniacal Liquors[314]

	Concentration (ppm)									
	Phenol		Thiocyanate		Cyanide		Oxygen Absorbed			
Plant	Before	After	Before	After	Before	After	Before	After	Ozone Absorbed	Ratio of Ozone to Phenol
A	1760	2.5	209	Nil	0.5	65	5750	1315	1870	1.1
B	300	8	209	Nil	65	72	1220	400	850	2.8
C	550	0.3	52	Nil	41	20	1610	230	700	1.3
D	1100	2.5	348	Nil	0.8	78	3710	900	1150	1.0
E	600	16	93	Nil	0.3	22	4040	1360	1600	2.6

Source: Modified from Ref. 314 Hall, D. A.

uct was oxalic acid. In the course of this laboratory study, it was learned that the wastewater was decolorized by ozone treatment. The liquor wastes first darken as ozonation commences, then become almost colorless as the phenol concentration is reduced to less than 1 ppm. Hall concluded that this program proved that ozone was both successful and efficient in destroying phenols and most of the other contaminants in coke plant waters. To achieve complete treatment of cyanides and oxygen-bearing substances, however, the cost of ozonation would be prohibitive. He predicted that ozonation would be much more attractive as a final cleanup process.

Eisenhauer[315] conducted laboratory experiments using hydrogen peroxide in combination with a ferrous salt to oxidize phenols. Optimum conditions were achieved when the reaction was carried out with 1 mole of ferrous salt and 3 moles of peroxide per mole of phenol. In this test a synthetic solution of 50 mg/l of phenol in distilled water was reduced to zero phenol in 5 min. Industrial wastes, however, contain a large concentration of substances that take up oxygen; thus the hydrogen peroxide requirements would be greater. In one experiment with a steel plant waste that contained 64 mg/l of phenol, 16 moles of peroxide reacted with only 35% of the phenols. The author concluded that in a process for industrial waste treatment where solution pH would be maintained at 3–5 and temperature maintained at 40–50°C, the treatment costs would approach $2.40/lb of phenol removed. This cost is about 2–3 times higher than that for biological treatment.

Ashmore[296] reported that the use of electrolytic oxidation for the treatment of phenolic wastes was studied in Poland, Russia, and the United Kingdom. Phenols, thiocyanate, and cyanide were successfully oxidized. The process is uneconomical and should be considered only for a final cleanup process. This author also cited Polish studies on the catalytic combustion of the total phenolic waters. It was reported that the phenols, HCN, H_2S, and NH_3 were almost totally oxidized.

18.5.1.5 *Soil Percolation*

Tallon[316] conducted a series of laboratory experiments to upgrade coke plant wastewaters by soil percolation. The wastewaters were sprayed onto percolation beds at a rate exceeding 3500 gal/(day)(acre). This resulted in more than 99% removal of the organic pollutants. The treatment began to fail after 3 months operation, apparently because of the accumulation of toxic salts. Tallon surmised that if the bed were an outdoor system, normal rainfall would leach the salts from the system. A percolation bed would then have an indefinite life, particularly if vegetation were maintained and the bed plowed occasionally to maintain good bed permeability. An example of test results was cited for a coke plant ammonia liquor. Soil percolation tests were conducted for 4 months at an irrigation rate of 5000 gal/(day)(acre). Results indicated phenol reduction in excess of 99%,

[315] H. R. Eisenhauer, *J. Water Pollut. Control Fed.*, **36**, 1116–1128 (1964).

[316] G. R. Tallon, Koppers Company, Inc., private communication.

and the color of the treated water changed from a dark brown to a pale yellow.

The main disadvantages of this process would be extensive space requirements, possible odor problems, and the suspension of operation during winter months if freezing temperatures prevailed.

18.5.1.6 *The Republic System*

The Republic Steel Corporation made extensive laboratory and semi-plant-scale studies to develop an overall process that will be capable of reducing all the contaminants in coke plant wastewaters.[317] The system is to be modular, to permit addition to or expansion of certain segments to tailor the process to a particular waste stream. Certain phases of this effort have been completed, and commercial-scale units are now under construction. Other segments of the process are still in the developmental stage.

The authors discussed the way in which their research studies were directed at the treatment of a particular plant's wastewaters. This coke plant has a production capacity of 6000 tons of coke per day and generates the following waste streams:

Stream	Flow (gal/min)	Contaminant Concentration (mg/l)			
		Ammonia	Phenol	Cyanides	Oils
Crude ammonia liquor	150	5000	2000	20	1,000
Barometric condenser water	1000	20	40	40	20
Intercepting sump waters	300	150	300	100	10,000

The barometric condenser water is the condensed vapor released by the crystallization and concentration of ammonium sulfate solution by vacuum evaporation. The intercepting sump water results from the various wastewater streams from the light oil plant.

Republic Steel concluded that all the current conventional methods of disposal were either inadequate or incomplete. These include (1) removal of NH_3 in a free and fixed NH_3 distillation system with the addition of lime slurry, (2) solvent extraction of phenols to produce sodium phenolate, (3) hot coke quenching with ammonia liquor and/or intercepting sump water, (4) recycling of the barometric condenser water, and (5) biological treatment. The goal was to treat the combined wastewater stream, to yield an effluent that contained no more than the following concentrations:

Ammonia	1 mg/l
Phenols	0.1 mg/l
Cyanides	0.05 mg/l
Oils	10 mg/l

A number of process modifications were made in the product recovery plant to reduce the burden for the proposed treatment facility. The barometric condenser was replaced by a surface condenser, which reduced the volume of the waste stream to 100 gal/min or less. The intercepting sump water contains neutral gums from the acid wash operation, which hinder proper clarification of this stream. It was also learned that removal of this contaminant reduced the ammonia concentration. The free and fixed ammonia stills with lime slurry treatment were discarded in favor of a system in which the crude liquor is adjusted with caustic soda, to a pH level of about 11, then is stripped with steam or air to reduce the NH_3 concentration to about 50 mg/l. The suspended solids content of the crude liquor was greatly reduced.

An activated carbon plant was designed for a Republic Steel coke plant after the existing plant was modified as above. The treatment plant was designed to treat 790,000 gal/day of combined plant wastewaters. A flow diagram is presented in Fig. 18.19. The

[317] A. C. Naso and J. W. Schroeder, *AIME Ironmaking Proc.*, **33**, 121–141 (1974).

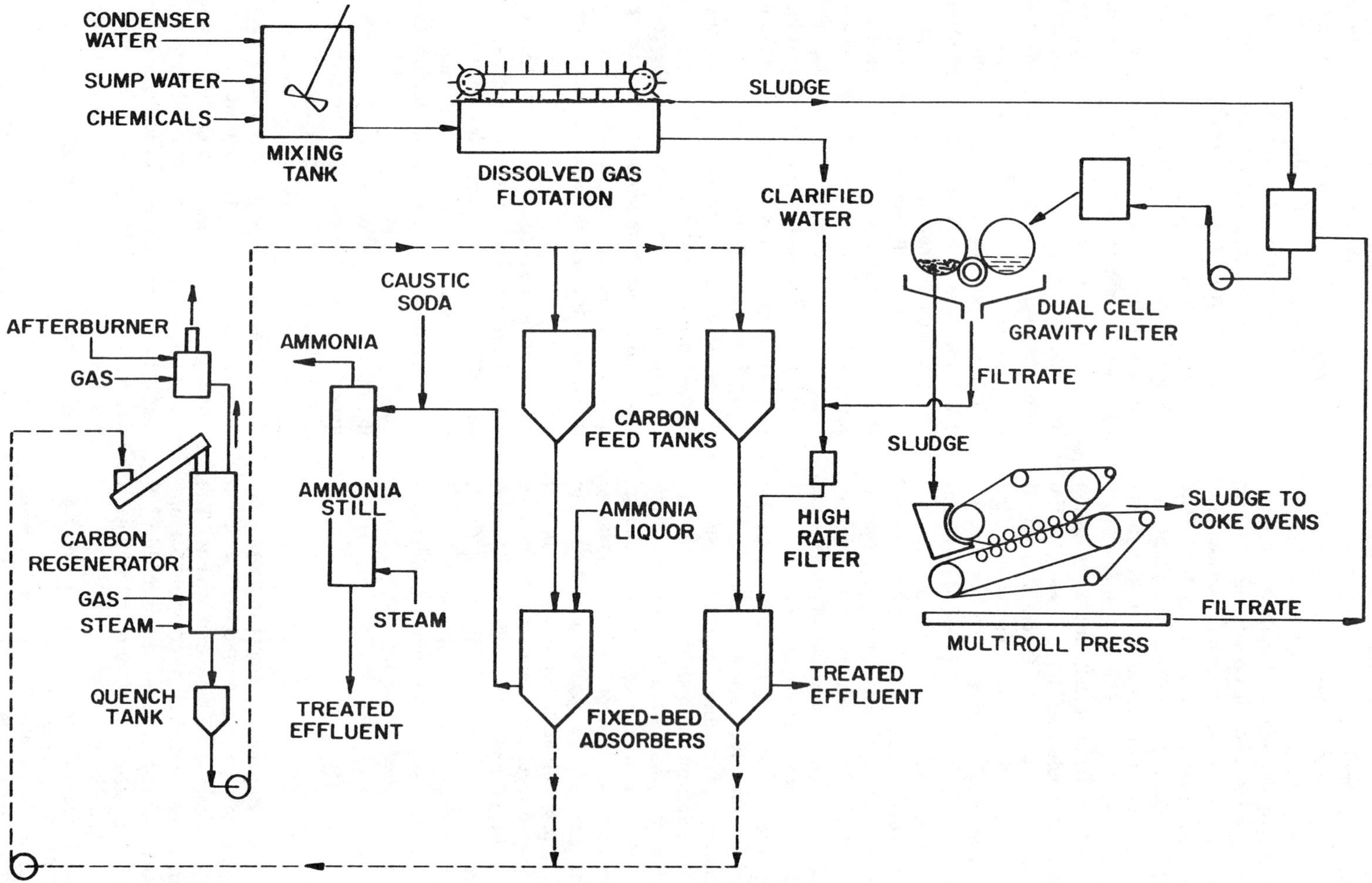

Figure 18.19 Schematic for an activated carbon coke plant water treatment plant.

crude ammonia liquor passes through a bed of activated carbon for phenol separation, then to an ammonia stripping still after caustic soda is added to adjust the pH. The intercepting sump waters are pretreated before undergoing activated carbon treatment to separate the oils. This is accomplished by adding waste pickle liquor and caustic soda to the wastewaters in a mixing tank. The reactions that occur are the precipitation of iron-cyanide complexes and ferrous hydroxide. The resultant mixture is pumped to a dissolved gas flotation cell for the separation of oils, suspended solids, and some of the cyanides. The clarified effluent is pumped through an activated carbon bed for final treatment. The sludge removed from the flotation cell is filtered and discarded by blending with the coal mixture to the ovens.

Schroeder and Naso[318] report that the estimated installation cost for this plant is about $8.5 million. The operating costs are estimated to be $4.95/1000 gal of wastewater treated, which is equivalent to $0.654/ton of coke. The authors made no claim that a plant based on this design would provide total treatment. In fact, they report that research is continuing on methods for the adequate removal of cyanides and thiocyanates. Laboratory tests on the use of ion-exchange resins seem to show promise. Ion-exchange resins appear to be capable of reducing the NH_3 concentration of the treated waters to the desired levels, but these processes have yet to be proved on a commercial scale. There was no reference to the BOD, COD, and TOC values or to the color of the treated effluent. It appears that much more information is needed before the process can be evaluated.

18.5.1.7 *Deep-Well Disposal*

The petroleum and chemical industries developed the technique of deep-well injection for the disposal of industrial wastes. The technical knowledge required prior to adopting this method has been documented.[319,320] A few U.S. steel firms have bored deep wells to inject both steel plant and coke plant wastes into underground formations.[321] This method can be used only in areas where there are proper subsurface formations: that is, a plant must be located above a limestone or sandstone formation covered by an impermeable strata that will seal the industrial wastewaters from the strata bearing the potable waters.

A number of states possessing such a subsurface structure have enacted laws that permit injection of wastewaters into deep wells under controlled conditions.[322] Certain geologic and hydrologic data must be acquired before the wastes can be injected. Primarily, it is necessary to obtain data on the permeability and porosity of the formation into which the wastes will be injected, plus the volume of the formation, the hydraulic pressure the strata will bear before fracturing, the long-term effect of waste storage in these formations, and the nature of the soil and rock structure above the storage area. The wells must be sealed with concrete and impervious liners to prevent the contamination of drinking water supplies.

It is necessary to remove tars and suspended matter from the wastes to prevent clogging of the pore structure of the rock strata, and some pretreatment may be required to avoid chemical reaction between the wastes and the limestone or sandstone layers.

The advantages of this waste disposal method are complete elimination of wastewaters, low initial and operating costs compared to aboveground treatment facilities, and small land requirements: in addition, this may be the only method available for disposing of certain toxic or strongly odoriferous materials. However there are the following disadvantages; the wells can be located only in areas with proper subsurface formations, the high-pressure injection may disturb these formations and create earthquake-type movement, and the wastewater may escape the underground storage area and contaminate other water sources. For example in the Denver, Colorado, area, high-pressure injection into wells 12,000 ft deep caused numerous earthquakes.[322] It was also found, also

[318] J. W. Schroeder and A. C. Naso, *Iron Steel Eng.*, **53,** 60–66 (December 1976).

[319] C. W. Querio and T. J. Powers, *J. Water Pollut. Control Fed.*, **34,** 136–144 (1962).

[320] G. W. Reid and L. E. Streebin, EPA-R2-72-001, December 1972, pp. 9–10.

[321] J. E. Barker and R. J. Thompson, EPA-R2-73-167, April 1973, p. 15.

[322] Anon., *Chem. Eng.*, **74**(2), 88–90 (1968).

that brine solution was forced to the surface in upper Michigan after injection into deep Canadian wells.[322]

Mukerji[323] reported that some coke plants in India that are conveniently located near the coal mines have pumped thier ammonia wastewaters into these mines. Precautions are taken to prevent seepage into potable water sources. The author did not elaborate on these precautions. He also stated that some deep wells have been bored for wastewater disposal.

[323] A. K. Mukerji, *COMA Yearb.*, 314 (1963); Waverly House, Wexborough, Yorkshire, England.

18.5.1.8 *Waste Treatment Economics*

There are virtually no data on the capital and operating costs for full-scale treatment facilities except in the case of the activated sludge biological process. Many authors projected estimated costs on the basis of data from laboratory or pilot-scale tests, but the data for these estimates may prove to be grossly in error. Ashmore,[296] in his paper on the review of waste treatment methods, compiled cost data from many sources and attempted to make cost comparisons for the treatment processes discussed in this chapter. Ashmore warned that since many of the reported data

Table 18.5 Operating Costs of Treatment Processes

Process	Data Year	Country	Cost (pence/1000 gal)[a]	Comments
Biological activated sludge	1968	Germany	104	
	1970	Germany	27–104	
	1970	United States	15–90	
	1968	Canada	105	Phenol removal
	1957	Great Britain	50	
	1971	Great Britain	170	Cost based on short-term
	1972	Great Britain	109	operation
Ozonation	1968	Canada	132	Phenol removal
	1971	Canada	23	Phenol removal, 300 mg/1
	1965	Great Britain	251	
	1965	Great Britain	40	Polishing of biologically treated coke-oven liquor
	1971	Great Britain	25	Removal of 30 mg/1 phenol
Electrolysis	1968	Canada	1230–2460	Phenol removal
	1969	Great Britain	1500	Removal of 15,000 mg/1 phenol
			180	Removal of 1500 mg/1 phenol
			70	Removal of 100 mg/1 phenol
	1971	U.S.S.R.	108	
Chemical oxidation				
Nitric acid	1951	Great Britain	40–220	
Hydrogen peroxide	1964	United States	1000	
Chlorine	1968	Canada	8	Incomplete data, phenol removal
Chlorine dioxide	1968	Canada	700	Phenol removal
Incineration	1973	Germany	103–167	NH_3 and H_2S oxidation only
	1970	Poland	772	All volatile constituents
Adsorption	1972	United States	72	Phenol removal
	1968	Canada	54	Phenol removal
	1973	Great Britain	45–63	Polishing of biologically treated effluent
Ion exchange	1972	United States	13–51	NH_3 removal after biological treatment
	1968	Canada	252	Phenol removal
Adsorption–ion exchange	1957	Great Britain	195	Full treatment of coke-oven liquor
Solvent extraction	1964	Germany	75	Phenol removal
Foam fractionation	1968	Canada	2	Phenol removal
Precipitation, HCHO	1968	Canada	112	Phenol removal

[a] These costs are for comparison only.

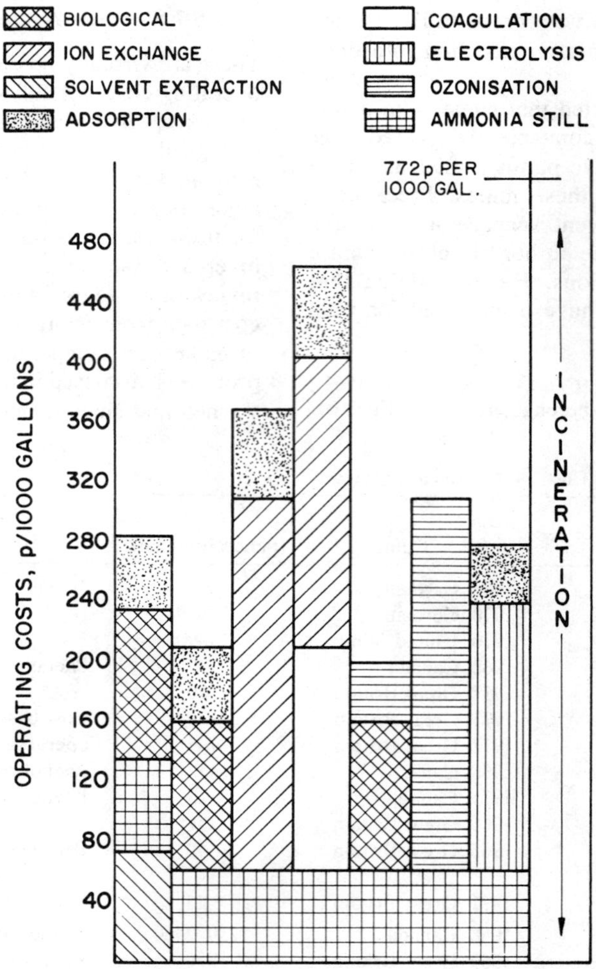

Figure 18.20 Estimated treatment costs (pence) for various combinations of processes.[296]

were accompanied by little detailed information, inaccuracies probably exist. Despite this, the compilation of data does serve as a means for comparison of the various treatment methods. Table 18.5 gives cost data for these methods, according to the country and year for which cost data were published. With these data, the author constructed a bar graph (Fig. 18.20), to illustrate the economics of a total treatment facility. It seems likely that such a facility will be a multistage process, since there is no single-stage process available today to provide the degree of treatment that must be achieved to meet plant outfall specifications. Based on these data, it appears that the most economical waste treatment plant will be a combined process of a biological activated sludge plant with ammonia still pretreatment to recover or incinerate the ammonia, followed by some type of posttreatment such as activated carbon adsorption or chemical oxidation by ozonation.

CHAPTER 19

Fundamentals of Coal Combustion

ROBERT H. ESSENHIGH

Robert H. Essenhigh *is E. G. Bailey Professor of Energy Conversion in the Department of Mechanical Engineering, The Ohio State University. Dr. Essenhigh engaged in coal combustion research in 1951, starting with studies on ignition and flame propagation in dust clouds in connection with coal dust explosions. He has also carried out research on coal/oil mixtures and combustion and pyrolysis of single captive particles of coal, particle groups in electric furnaces and particle clouds in flames. Experimental methods used have included single particles, drop (isothermal) tube, plug-flow furnace, and jet-stirred reactor. Additional studies have been on carbon combustion to explore more fundamental combustion mechanisms. Dr. Essenhigh's publications on coal and coal-related topics total about 50.*

19.1 INTRODUCTION

The combustion of coal in any system, whether a fixed bed, a fluid bed, or pulverized in suspension, is primarily a matter of combustion of carbon, with sequentially prior, or parallel combustion of volatiles. Thus the fundamentals of coal combustion involve the combustion or reaction kinetics of carbon and of the volatiles; the combustion of the volatiles tends to be dominated by their production during pyrolysis. The overall problem therefore breaks down into a general research program with four elements. The first is the classic simplification by which "pure" carbon studies are substituted for those on coal. The second is the study of pyrolysis kinetics (Chapter 12). The third is the study of volatiles combustion kinetics. The fourth is the study of coal itself in combustion. It must be recognized that coal is not graphite; but the results nonetheless must be interpreted in terms of the first three elements of the problem.

At the time of publication (1963) of the first Supplementary Volume,[1] progress on the general research program had been spotty. The status of the four research elements can be described as follows.

1. Substantial carbon studies had been reported,[2] and most investigators felt that the qualitative outlines of behavior were reasonably well understood. There was also good development of quantitative understanding based, essentially, on the Langmuir-Hinshelwood adsorption isotherm when the kinetics were not controlled by physical diffusion.
2. Understanding of pyrolysis kinetics was simplistic by comparison: there was speculation about complex kinetic schemes, but quantitative analysis still relied largely on a simple one-step, first-order reaction, in spite of evidence dating from the early 1900s that this approach was inadequate.
3. The study of volatiles combustion kinetics was virtually untouched.
4. The study of coal combustion ranged from one extreme to another across six subelements representing different experimental systems.
   (*a*) Combustion of individual coal particles or small groups of particles. Such studies date back decades, at least to Faraday and Lyell's pioneering investigations in 1844.[3]
   (*b*) Small laminar bunsen-type or flat flames with no backmix; these had contributed little of fundamental significance by 1960.
   (*c*) Backmix flames generated by swirled or unswirled jet flames; behavior tends to be dominated by heat transfer and combustion aerodynamics rather than by the combustion kinetics (although coupling exists between fluid flow and reaction where diffusion is significant). These flames have an experimental history of more than 100 years,[4] but significant research can, perhaps, only be dated from Sherman's experiments in 1931.
   (*d*) Explosion flames, with a history dating back to the 1800s.[5,6]
   (*e*) Coal combustion in fluid beds; there was little to report from such studies by 1960.
   (*f*) "Fixed" beds on grates, in vertical shaft gasifiers, *in situ* (underground), or in experimental combustion "pots." The literature in this area is substantial.

Overall, however, it can be said that such studies did more to define the problems than to solve them.

In the last 15 years there has been substantial advance in understanding in virtually all elements of the problem, except for the volatiles kinetics. The key has undoubtedly been greatly improved experimental design and definition, substantially aided by computer analysis of the results. The most particular developments have been in three areas: mechanisms of low-temperature carbon reactions, mechanisms of pyrolysis, and reaction behavior of coal, in contrast to carbon. However the extent to which we are now better fitted to predict combustion behavior in practical systems—which is the objective of such applied research—is still problematical, although it has been no deterrent to modelers and theoretical analysts.[7–25]

---

[1] *Chemistry of Coal Utilization,* Supplementary Volume, H. H. Lowry, Ed., John Wiley and Sons, New York, 1963.

[2] *Chemistry of Coal Utilization,* Supplementary Volume, H. H. Lowry, Ed., John Wiley and Sons, New York, 1963, Chs. 17, 18, 20.

[3] M. Faraday and C. Lyell, "Explosions in Coal Mines," *Phil. Mag.,* **26,** 16–35 (1845).

[4] R. H. Essenhigh, "A Short History of Pulverized Fuel Firing," *J. Inst. Fuel,* **34,** 16–26 (1961).

[5] R. H. Essenhigh, "Combustion Phenomena in Coal Dusts," *Colliery Eng.,* **38,** 534–530, (1961); **39,** 23–28, 65–72, 103–104 (1962).

[6] K. C. Brown and C. J. James, *Dust Explosions in Factories: A Review of the Literature,* SMRE Res. Rep. No. 207, HMSO, London, June 1962.

[7] T. F. Hurley, "Some Factors Affecting the Design of a Small Combustion Chamber for Pulverized Fuel," *J. Inst. Fuel,* **4,** 243–254 (1931).

[8] H. C. Hottel and I. M. Stewart, "Space Requirement for the Combustion of Pulverized Coal," *Ind. Eng. Chem.,* **32,** 719 (1940).

[9] R. H. Essenhigh, "Burning Times of Coal Par-

Most readers will be aware that there has been no lack of reviews on the subject (e.g., Refs. 26–64). Some of these[42–64] focused on

ticles in a p.f. Furnace." *Proc. 2nd Conf. on Pulverized Fuel,* Paper No. 2, Institute of Fuel, London, 1957, pp. B1–B7.

[10] R. H. Essenhigh, "Predicted Burning Times of Solid Particles in an Idealized Dust Flame," *J. Inst. Fuel,* **34,** 239–244 (1961).

[11] J. Csaba and A. D. Leggett, "Prediction of the Temperature Distribution along a Pulverized-Coal Flame," *J. Inst. Fuel,* **37,** 440–448 (1964).

[12] A. B. Hedley and M. E. Leesley, "Burning Characteristics of Pulverized Coal," *J. Inst. Fuel,* **38,** 492–500 (1965).

[13] D. Bhaduri, "A Theoretical Study of Coal Dust Flame," *Indian J. Technol,* **4,** 322–326 (1966).

[14] A. B. Hedley and E. W. Jackson, "A Simplified Mathematical Model of a Pulverized Coal Flame Showing the Effect of Recirculation on Combustion Rate," *J. Inst. Fuel,* **39,** 208–218 (1966).

[15] M. E. How, "The Influence of Turbulence on the Burning Rates of Fuel Suspensions," *J. Inst. Fuel,* **39,** 150–158 (1966).

[16] J. Szekely and N. G. Maroudas, "A Note on the Reaction Mechanism in the Underground Gasification of Coal," *Trans. Inst. Chem. Eng.,* **44,** T3–T6 (1966).

[17] J. Csaba and M. M. Baum, "Combustion of Single Pulverized Coal Particles, Part I." *J. Inst. Fuel,* **40,** 163–169 (1967).

[18] L. A. Klyachko, "Ignition of an Aggregate of Particles During Heterogeneous Reaction," *Pyrodynamics,* **6,** 29–38 (1968).

[19] F. E. Fendell, "The Burning of Spheres Gasified by Chemical Attack," *Combust. Sci. Technol.,* **1,** 13–24 (1969).

[20] M. M. Gibson and B. B. Morgan, "Mathematical Model of Combustion of Solid Particles in a Turbulent Stream with Recirculation," *J. Inst. Fuel,* **43,** 517–523 (1970).

[21] M. M. Baum and P. J. Street, "Predicting the Combustion Behaviour of Coal Particles," *Combust. Sci. Technol.,* **3,** 231–243 (1971).

[22] D. Bhaduri and S. Bandyopadhyay, "Combustion in Coal Dust Flames," *Combust. Flame,* **17,** 15–24 (1971).

[23] H. McKenzie, I. W. Smith, and G. a Donau Szpindler, "Calculation of Burnout of Polydisperse Suspensions of Low and High Reactivity Pulverized Fuels," *Proc. 4th Symp. Flames and Industry,* Paper No. 13, Institute of Fuel (1972), pp. 143–149.

[24] H. S. Caram and N. R. Amundson, "Diffusion and Reaction in a Stagnant Boundary Layer about a Carbon Particle," *Ind. Eng. Chem. Fundam.,* **16,** 171–181 (1977).

[25] J. T. Shaw, "Theoretical Work on Reaction Sequences in the Gasification of Coke by Carbon Dioxide and by Steam in Conditions Remote from Equilibrium," *Fuel,* **56,** 134–136 (1977).

[26] W. van Loon and H. H. Smeets, "Combustion of Carbon," *Fuel,* **29,** 119–121 (1950).

[27] R. H. Essenhigh, "The Mechanism of the Combustion of Carbon: A Theoretical Study," *Univ. Sheffield Fuel Soc. J.,* **6,** 15–24 (1955).

[28] H. E. Blayden, "Reactivity of Cokes and Chars: A Review," *Proc. Conf. Science in the Use of Coal,* Institute of Fuel (1958), pp. E1–E6.

[29] R. H. Essenhigh and M. G. Perry, "Combustion and Gasification: A Review," *Proc. Conf. Science in the Use of Coal,* Institute of Fuel (1968), pp. D1–D12.

[30] P. L. Walker, Jr., F. Rusinko, and L. G. Austin, "Gas Reactions of Carbon," *Adv. Catal.,* **11,** 133–221 (1959).

[31] R. H. Essenhigh and I. Fells, "Combustion of Liquid and Solid Aerosols," *Disc. Faraday Soc.* **30,** 208–221 (1960).

[32] M. W. Thring and R. H. Essenhigh, "Thermodynamics and Kinetics of Solid Combustion," in Ref. 2 (Ch. 17).

[33] R. A. Sherman and B. A. Landry, in Ref. 2 (Ch. 18).

[34] C. G. von Fredersdorff and M. A. Elliott, in Ref. 2 (Ch. 20).

[35] M. A. Field, D. W. Gill, B. B. Morgan, and P. G. W. Hawksley, "Combustion of Pulverized Coal," BCURA, Leatherhead, Surrey, England, 1967.

[36] M. F. R. Mulcahy and I. W. Smith, "Kinetics of Combustion of Pulverized Fuel: A Review of Theory and Experiment," *Rev. Pure Appl. Chem.,* **19,** 81–108 (1969).

[37] R. H. Essenhigh, "Dominant Mechanisms in the Combustion of Coal," ASME Paper No. 70-WA/Fu-2, 1970.

[38] D. Gray, J. G. Cogoli, and R. H. Essenhigh, "Problems in Pulverized Coal and Char Combustion," in *Coal Gasification,* Am. Chem. Soc. Adv. Chem. Ser., No. 131, (1974), pp. 72–91.

[39] I. W. Smith, "The Intrinsic Reactivity of Carbons to Oxygen," paper presented at Conf. Fundamentals of Carbon-Gas Reactions, sponsored by University of Newcastle, Institute of Fuel (Australia), Comb. Inst. (Australia), 1976.

[40] R. H. Essenhigh, "Combustion and Flame Propagation in Coal Systems: A Review," *Proc. 16th Symp. (Int.) Combustion,* Combustion Institute, (1977), pp. 353–372.

[41] R. H. Essenhigh, "Combustion of Coal," in *Coal Conversion Technology* (E. Stanley Lee and C. Y. Wen, Eds.), Addison-Wesley Publishing Co., Reading, Mass., 1979; Ch. 3.

[42] D. H. Bangham and J. G. Bennett, "The Chemistry of Gasification with Reference to Small Producers," *Fuel,* **19,** 95–101 (1940).

[43] R. F. Strickland-Constable, "Some Comments on the Work of Key on the Reactions between Coke and Carbon Dioxide and Between Coke and Steam," *J. Chim. Phys.,* **47,** 356–360 (1950).

[44] H. R. Batchelder, P. M. Busche, and W. P. Armstrong, "Kinetics of Coal Gasification, Proposed Mechanism of Gasification," *Ind. Eng. Chem.,* **45,** 1856–1878 (1953).

[45] F. A. Burden and T. D. T. Dore, "Measurement of the Reactivity of Coke," *Proc. Conf.*

specific aspects of the overall problem. Bangham and Burnett[42] reviewed the chemistry of gasifiers; Strickland-Constable,[43] some specific experiments of Key; Batchelder et al.,[44] the kinetics of coal gasification; Burden and Dore,[45] reactivity tests; Winter,[46] reactivity of oxide surfaces, with emphasis on the electronic aspects of the binding forces; Essenhigh,[4,5] large-scale dust explosions; Béer,[49] combustion of low-volatile pulverized coal; Dacey,[50] mobile adsorption; Lang and Magnier,[51] gasification of "pure" carbon at low pressures; Ergun and Mentzer,[52] the C/CO_2 reaction; Walker et al.,[53] catalysis of carbon reactions; Puri,[54] surface complexes; Turkdogan,[55] with Darken,[56] internal burning; Rosner with Allendorf,[57–60] O and H atom reactions; Eyring and Lin,[61] the activated complex; Wagner,[62] the oxygen activity concept; Soehngen,[63] the coal diesel engine data; Béer,[64] fluid bed combustion of coal. In particular, reviews have tended to emphasize (usually for good reason) either reactions of carbon with little or nothing on coal, or reactions of coal with relatively modest reference to the rich carbon literature. One objective of this chapter is to incorporate more of that carbon literature into the world of coal reactivity studies for its value in aiding understanding of coal reactions. Of all the reactions involved in coal combustion or gasification, the most important is still that of the char usually formed *in situ*. The char reacts at both external and internal surfaces at rates determined by physical and chemical processes that thanks to the carbon studies, are now quite well described qualitatively (Section 19.3). Quantitatively, however, there is still only partial understanding, and the extent of that understanding is the principal subject of this chapter.

Science in the Use of Coal, Paper No. 33, Institute of Fuel (1958), pp. E7–E13.

[46] E. R. S. Winter, "The Reactivity of Oxide Surfaces," *Adv. Catal.,* **10,** 196 (1958).

[47] J. M. Beer, "Combustion of Pulverized Coal," *Proc. 2nd Symp. Flames and Industry,* Institute of Fuel (London), 1962, pp. C1–C11.

[48] L. G. Austin and P. L. Walker, Jr., "Effect of Carbon Monoxide in Causing Nonuniform Gasification of Graphite by Carbon Dioxide," *Am. Inst. Chem. Eng. J.,* **9,** 303–306 (1963).

[49] J. M. Béer, "The Effect of Fineness and Recirculation on the Combustion of Low-volatile Pulverized Coal," *J. Inst. Fuel,* **67,** 286–313 (1964).

[50] J. P. Dacey, "The Interface Symposium—10," *Ind. Eng. Chem.,* **57,** 27–33 (1965).

[51] F. M. Lang and P. Magnier, "Action of Oxygen and Carbon Dioxide Above 100 Millibars on 'Pure' Carbon," in *Chemistry and Physics of Carbon,* Vol. 4, Marcel Dekker, New York, 1968, pp. 121–210.

[52] S. Ergun and M. Mentzer, "Reactions of Carbon with Carbon Dioxide and Steam," in *Chemistry and Physics of Carbon,* Vol. 1; Marcel Dekker, New York, 1965, pp. 203–263.

[53] P. L. Walker, Jr., M. Shelef, and R. T. Anderson, "Catalysis of Carbon Gasification," in *Chemistry and Physics of Carbon,* Vol. 4, Marcel Dekker, New York, 1968, pp. 287–383.

[54] B. P. Puri, "Surface Complexes on Carbons," in *Chemistry and Physics of Carbon,* Vol. 6, Marcel Dekker, New York, 1970, pp. 191–282.

[55] E. T. Turkdogan, "Fate of Oxidation of Carbon in Carbon Dioxide," *Proc. 3rd Int. Symp. High Temp. Tech.* Butterworths Scientific Publications, London, 1969, pp. 685–695.

[56] L. S. Darken and E. T. Turkdogan, "Adsorption and Heterogeneous Kinetics at Elevated Temperatures," in *Proc. Conf. Heterogeneous Kinetics at Elevated Temperatures,* Plenum Press, New York, 1970, pp. 25–99.

[57] D. E. Rosner, "Effects of the Stefan-Nusselt Flow on the Apparent Kinetics of Heterogeneous Chemical Reactions in Forced Convection Systems," *J. Heat Mass Transfer,* **9,** 1233–1253 (1966).

[58] D. E. Rosner, "Convective Diffusion Limitations on the Rates of Chemical Reactions at Solid Surfaces—Kinetic Implications," *Proc. 11th Symp. (Int.) Combustion,* Combustion Institute, 1968, pp. 181–194.

[59] D. E. Rosner and H. D. Allendorf, "Kinetics of the Attack of Refractory Materials by Dissociated Gases," in *Proc. Conf. Heterogeneous Kinetics at Elevated Temperatures,* Plenum Press, New York, 1970, pp. 231–250.

[60] D. E. Rosner, "High-Temperature Gas-Solid Reactions," *Annu. Rev. Mater. Sci.,* **2,** 573–606 (1972).

[61] H. Eyring and S. H. Lin, "The Activated Complex in some High Temperature Heterogeneous Reactions," in *Proc. Conf. Heterogeneous Kinetics at Elevated Temperatures,* Plenum Press, New York, 1970, pp. 1–23.

[62] C. Wagner, "The Concept of the Thermodynamic Activity of Atomic Species in Heterogeneous Kinetics," *Proc. Conf. Heterogeneous Kinetics at Elevated Temperatures,* Plenum Press, New York, 1970, pp. 101–111.

[63] E. E. Soehngen, "Development of Coal-Burning Diesel Engines in Germany," U.S. ERDA Rep. No. FE/WAPC/3387-1, 1976.

[64] J. M. Béer, "The Fluidized Combustion of Coal," *Proc. 16th Symp. (Int.) Combustion,* Combustion Institute, 1977, pp. 439–459.

19.2 MAGNITUDES

To set the combustion problem in quantitative perspective, order-of-magnitude values of some key practical and fundamental parameters are summarized in this section.

19.2.1 Magnitudes in Practical Systems

Coal is burned today in three ways: as lumps on a grate or in a shaft; crushed in a fluid bed; or pulverized, entrained, and burned in a dilute suspension. In principle, any particle size can be used in any of the three methods, but in practice there are engineering limitations[65] that establish preferred sizes for the three combustion methods. Size is also found to be the most significant "defining" parameter with respect to the dominant reaction mechanism and other thermal behavior (rate of heating, e.g., which can control volatiles yield and composition). Aspects of the following discussion of these points are summarized in Table 19.1.

19.2.1.1 *Particle Sizes*

Lump Combustion. Preferred sizes for lump combustion on a grate (or in a shaft gasifier), in the so-called fixed or broken bed, range from 5 to 50 mm. A narrow cut is preferred, but a size distribution with particles above and below the cited range is often accommodated. The combustible phase (col. 2 of Table 19.1) is dense, on the order of g/cm^3. Heating rate is slow (a few degrees per second or slower), which allows time for volatiles cracking and capture, both inside the pyrolyzing particle and on other particles. The combustion time ranges from minutes to hours, depending on particle size. Although the bed is constructed of independent lumps, it behaves as a coherent porous block, and it is usually described in physical models for subsequent mathematical formulation as a straight or tortuous channel, or as a porous solid (col. 7). The effective tube diameters are on the same order as the constituent particle sizes, and there is now substantial evidence (e.g., Eapen et al.[66]) that the combustion reactions (O_2 consumption) are diffusion controlled. However the gasification reactions (CO_2 and H_2O consumption) are not.[66] The tube description can also be applied to the channels of the graphite moderator of a nuclear power unit.

Fluid Beds. The fluid bed exists in two modes: ballasted and unballasted. In both modes the preferred size range is 1–5 mm, but again sizes above and below are frequently accommodated. In the ballasted mode the coal is injected into a bed of inert material (ash, limestone, shale, etc.). The combustible percentage is low—generally less than 1%—so that it is present in a dilute phase (order of grams per liter). The ballast acts as an attemperator, so that operating temperatures are generally under 1000°C; this is unusually low for combustors and can have some influence on the reaction mechanism.

In the unballasted mode (e.g., the Ignifluid design) there is no inert in the bed other than the natural ash contribution; this is continuously removed after agglomeration (Chapter 24). Temperatures run much higher, generally above 1000°C, and are usually adjusted to slag the ash of the particular coal being used. Combustible concentration is dense phase (order of grams per cubic centimeter).

In both modes heating rates are quite rapid (estimated at 10^3–10^4 °C/s), so that mass loss as volatiles could be anything up to twice the ASTM volatile matter percentage, because of the "Q-factor" effect (Chapter 12) described by Badzioch and Hawksley.[67] Particle sizes are still large enough for reaction to be controlled by boundary layer diffusion, and the burning times (col. 5), on the order of minutes, match prediction. In the ballasted bed reaction takes place only with oxygen; the temperatures of the bed are too low for the C/CO_2 reaction to contribute to solids gasification. Good theoretical models for char combustion have now been constructed; their principal theoretical problems are primarily

[65] R. H. Essenhigh, D. T. Pratt, H. E. Shull, and L. D. Smoot, "Current Status of Fluid Mechanics in Practical Heterogeneous Combustors," paper presented at Central States Society of Combustion Institutes Spring Meeting, Cleveland, 1977.

[66] T. Eapen, P. Blackadar, and R. H. Essenhigh, "Kinetics of Gasification in a Combustion Pot: A Comparison of Theory and Experiment," *Proc. 16th Symp. (Int.) Combustion,* Combustion Institute, 1977, pp. 515–522.

[67] S. Badzioch and P. G. W. Hawksley, "Kinetics of Decomposition of Pulverized Coal Particles," *Ind. Eng. Chem. Process Des. Dev.,* **9**, 521 (1970).

Table 19.1 Comparative Quantities in Coal Combustion Systems and Applications[a]

(1) Particle Description and Size	(2) System Descriptions and Applications: System Particulate Density (2a) Dense Phase (g/cm^3)	(2b) Dilute Phase (g/l)	(3) Characteristic Heating Rate (°C/s)	(4) Reaction Front (Flame) Propagation Speed (m/s)	(5) Characteristic Reaction Time (s) (5a) Volatiles	(5b) Char	(6) Fluid-Particle Mixing Description	(7) Reactive Element Description
Lump, <100 mm	1. "Fixed" or broken bed: including grate combustor, shaft gasifier; underground gasification 2. Graphite-moderated nuclear reactor cores		<1	10^{-5}–10^{-3}	$\sim 10^2$	$\sim 10^3$	Counterflow	Channel or tube (straight or tortuous); diffusion-controlled combustion
Crushed, <10 mm	Unballasted fluid bed (e.g., Ignifluid); operating temperature > 1000°C	Ballasted fluid bed; operating temperature <1000°C	10^3–10^4	Not relevant	10–50	100–500	Particle backmix: fluid plug flow	Sphere with diffusion-controlled combustion
Pulverized, < 0.1 mm		1. Pulverized-coal-fired furnaces and boilers 2. Compression-ignition engines (diesel) 3. Turbines (< 10 μm dia) 4. Explosion flames 5. Research reactors (plug-flow and stirred)	10^3–10^6 (10^7)	10^{-1}–10^3	<0.1	<1	Entrained particles and carrying fluid in parallel flow even with gross backmix flows	Sphere with mainly chemically controlled reaction

[a] The values cited are considered to be typical orders of magnitude for comparative purposes. Examples of exceptions can always be found.

related to bubble-exchange behavior, which is not relevant to this chapter (see Chapters 12 and 23). Theoretical models of volatiles combustion, however, do not exist (Section 19.5.6.2).

Pulverized Coal. In the pulverized size range 90–98% of most coal particles are below 100 μm, and reaction, it is now generally agreed, is dominated by the true chemical kinetics for both volatiles release and char burnout, which is rapid (< 1 s). Heating rates and flame propagation speeds are also high, and significantly greater than in the fixed-bed system. These also vary enormously between different combustion systems.

19.2.1.2 *Heating Rates*

The magnitudes and ranges in heating rates and flame propagation speeds have only indirect bearing on the principal subject matter of this chapter, but a brief summary is appropriate. From analyses discussed elsewhere[40] it is evident that heating rate is very much a function of method of heating and of mixing conditions; and heating rate, in turn, dominates the flame propagation speed. The fastest possible heating rates are achieved when cold particles are plunged into hot gases. The maximum value is still uncertain, but some recent unsteady state calculations relating to behavior in a stirred reactor[68] indicate that it cannot exceed about 10^6 °C/s for typical pulverized coal particle sizes. The calculations assumed an initial condition of uniform hot gas temperature up to the particle surface so that the results would also be relevant to heating in a shock tube. If a quasi-steady-state temperature profile is assumed, the heating rate for similar particle sizes cannot exceed about 10^5 °C/s. Heating rates at the upper value (10^6 °C/s) evidently occur in jet-stirred reactors,[69] in high-speed explosion flames,[40] and most probably in diesel combustion.[63] In radiation-stabilized laminar plane flames, with pyrolysis before ignition, heating rates are down to 10^3 °C/s.[70] With pyrolysis after ignition,[71–76] heating rates are about 10^4 °C/s.

The link to flame propagation speeds is now reasonably clear, with speeds ranging from 0.1 to 1000 m/s as heating rate rises from 10^3 to 10^6 °C/s.[40] Flame speed is determined primarily by the magnitude of an energy flux crossing a flame or reaction front, and the magnitude of this flux is controlled by the heat transfer mechanism, which also determines the heating rate. Order-of-magnitude heating rates and flame speeds can be estimated without reference to reactivity, but there is actually coupling between these three factors. This is implicit in the mathematical model of the pyrolysis-diffusion flame developed by Smoot et al.,[75] and it was explored explicitly by Cogoli et al.[77] in both experimental and theoretical evaluation. The coupling is demonstrated, but the details are still largely unscrambled.

19.2.1.3 *Particle Burning Times*

Burning times are integrated reaction rates. The biggest puzzle at present is the question

[68] P. M. Goldberg, "Stirred Reactor Study of Coal Dust Explosion Flames: Stirred Reactor Development," M.S. thesis, Dept. of Materials Science, Fuel Science Section, Pennsylvania State University, State College, 1978.

[69] P. M. Goldberg and R. H. Essenhigh, "Coal Combustion in a Jet-Mixed Stirred Reactor," *Proc. 17th Symp. (Int.) Combustion,* Combustion Institute, 1979, pp. 145–154.

[70] R. H. Essenhigh and J. Csaba, "The Thermal Radiation Theory for Plane Flame Propagation in Coal Dust Clouds," *Proc. 9th Symp. (Int.) Combustion,* Academic Press, New York, 1963, pp. 111–124.

[71] J. B. Howard and R. H. Essenhigh, "The Mechanism of Ignition of Pulverized Coal," *Combust. Flame,* **9,** 337–339 (1965).

[72] J. B. Howard and R. H. Essenhigh, "Combustion Mechanism in Pulverized Coal Flames," *Combust. Flame,* **10,** 92–93 (1966).

[73] J. B. Howard and R. H. Essenhigh, "Pyrolysis of Coal Particles in Pulverized Fuel Flames," *Ind. Eng. Chem. Process Des. Dev.,* **6,** 74–84 (1967).

[74] J. B. Howard and R. H. Essenhigh, "Mechanism of Solid-Particle Combustion with Simultaneous Gas-Phase Volatiles Combustion," *Proc. 11th Symp. (Int.) Combustion,* Combustion Institute, Pittsburgh, 1967, pp. 399–408.

[75] L. D. Smoot, M. D. Horton, and G. A. Williams, "Propagation of Laminar Pulverized Coal-Air Flames," *Proc. 16th Symp. (Int.) Combustion,* Combustion Institute, 1977, pp. 375–387.

[76] M. D. Horton, F. P. Goodson, and L. D. Smoot, "Characteristics of Flat, Laminar Coal-Dust Flames," *Combust. Flame,* **28,** 187–195 (1977).

[77] J. G. Cogoli and R. H. Essenhigh, "Computer Modelling of a Plane Flame Furnace Firing Low Volatile Fuels," *Combust. Sci. Technol.,* **16,** 177–185 (1977).

of burning times of both volatiles and char in the size range smaller than 100 μm. At the larger, diffusion-controlled sizes (> 100 μm) there is now good to excellent agreement between theory and experiment for the char burnout. This match is illustrated by Fig. 19.1. Experimental data generally obey the Nusselt "square law" variation of burning time with the square of the initial diameter. The equations, developed in Section 19.6, include the effects of oxygen vitiation and enrichment, volatiles loss, and swelling. Burning times (Fig. 19.1) range from fractions of a second to 10^3 s, as reported in Table 19.1 for the crushed and lump-sized particles.

For the volatiles combustion, in contrast, there is limited experimental information on particles larger than 100 μm to show that pyrolysis time, equated to burning time, also varies (empirically) with the square of the initial diameter,[80] but acceptable theories to predict this relationship are still lacking (Section 19.5.6.2). The substantial work on pyrolysis (Chapter 12) has not yet addressed this problem adequately.

For particles smaller than 100 μm the situation reverses. In this size range there is generally better theoretical understanding of pyrolysis than there is of the char combustion, although there is still a discrepancy to be resolved. For decades it was common to use a rule of thumb that the pyrolysis time was roughly one-tenth of the total burning time. Thus since burning times were generally given as about 1 s, pyrolysis times were about 0.1 s. This measurement was in general agreement, for example, with Howard and Essenhigh's data[73,74] and other flame sources. More recent investigations, however, using both the Badzioch and Hawksley isothermal furnace and other methods, including stirred reactors,[69] have shown that pyrolysis can be substantially complete in 10–20 ms at temperatures comparable to or lower than those found in boiler flames. This is an order of magnitude faster than previous expectations, and the results also appear to make good theoretical sense. It is evident, then, that pyrolysis can occur at a significantly faster rate than was envisaged a few years ago. This does not, however, invalidate the prior flame measurements of 0.1 s, which have yet to be reconciled with the developing theories and measurements showing faster pyrolysis. The difference may be related to different rates of heating.

Our present knowledge of char combustion burning times is summarized in Fig. 19.2 (see also Section 19.6). Flame measurements[40,63,73,82–89] on particles smaller than 100 μm clearly show that total combustion time is about 1 s, and this figure matches boiler design data. A characteristic of the sparse data is the indication that burning time may or may not be insensitive to particle size. In some drop-tube experiments[91–96] it was con-

[78] B. D. Katsnel'son and I. Ya. Marone, "Effect of Pressure and Oxygen Concentration on Ignition and Combustion of Fine Coal Particles," *Teploenergetika,* **11,** 11 (1964).

[79] G. P. Ivanova and V. Babii, "A Study of the Burnout Mechanism of Coal Particles," *Teploenergetika,* **13,** 54–59 (1966). Transl., *Therm. Eng.,* **13,** 70 (1966).

[80] R. H. Essenhigh and M. W. Thring, "The Measurement of Burning Times of Single Coal Particles," *Proc. Conf. Science in the Use of Coal,* Paper No. 29, Institute of Fuel, 1958, pp. D21–D27.

[81] H. M. Cassel, *Some Fundamental Aspects of Dust Flames,* U.S. Bur. Mines Rep. Invest. No. 6551, 1964.

[82] M. A. Nettleton and R. Stirling, "The Ignition of Clouds of Particles in Shock-Heated Oxygen," *Proc. R. Soc. London,* **A300,** 62–77 (1967).

[83] M. A. Nettleton and R. Stirling, "The Combustion of Clouds of Coal Particles in Shock-Heated Mixtures of Oxygen and Nitrogen," *Proc. R. Soc. London,* **A322,** 207–221 (1971).

[84] A. P. Ayling and I. W. Smith, "Measured Temperatures of Burning Pulverized-Fuel Particles, and the Nature of the Primary Reaction Product," *Combust. Flame,* **18,** 173–184 (1972).

[85] M. A. Nettleton and R. Stirling, "The Influence of Additives on the Burning of Clouds of Coal Particles in Shocked Gases," *Combust. Flame,* **22** 407–414 (1974).

[86] J. M. Béer and M. W. Thring, "Some Measurements of the Rate of Combustion of Pulverized Anthracite in a Controlled Mixing One Dimensional Furnace," *Proc. Anthracite Conf.,* Pennsylvania State Univ. Bull. Miner. Ind. Exp. Sta. No. 75, September (1961), pp. 25–35.

[87] J. G. Cogoli, D. Gray, and R. H. Essenhigh, "Flame Stabilization of Low Volatile Fuels," *Combust. Sci. Technol.,* **16,** 165–176 (1977).

[88] C. P. Fenimore and G. W. Jones, "Oxidation of Soot by Hydroxyl Radicals," *J. Phys. Chem.,* **71,** 593–597 (1967).

[89] C. T. Chamberlain and W. A. Gray, "Combustion of Coal in Oxygen," *Nature,* **216,** 1245 (1967).

[90] R. H. Essenhigh, "The Influence of Coal Rank on the Burning Times of Single Captive Particles," *J. Eng. Power,* **85,** 183–188 (1963).

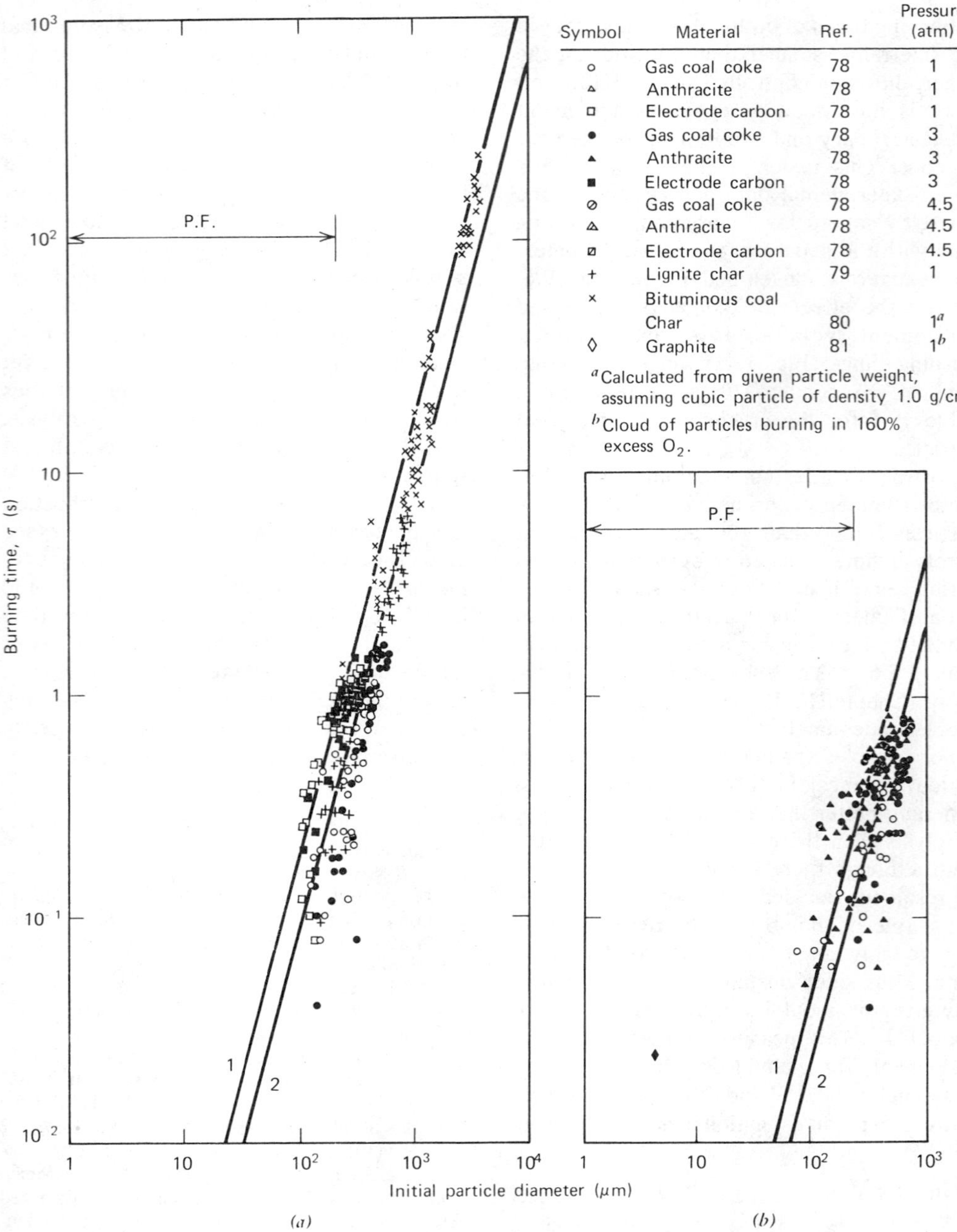

Symbol	Material	Ref.	Pressure (atm)
○	Gas coal coke	78	1
△	Anthracite	78	1
□	Electrode carbon	78	1
●	Gas coal coke	78	3
▲	Anthracite	78	3
■	Electrode carbon	78	3
⊘	Gas coal coke	78	4.5
⧄	Anthracite	78	4.5
⧄	Electrode carbon	78	4.5
+	Lignite char	79	1
×	Bituminous coal Char	80	1[a]
◊	Graphite	81	1[b]

[a] Calculated from given particle weight, assuming cubic particle of density 1.0 g/cm³.

[b] Cloud of particles burning in 160% excess O_2.

Figure 19.1 Comparison of experimental and theoretical particle burning times. (*a*) 21% v/v O_2, calculated from Equation 19.116 for 1500 K, with $K_c = 0$. (*b*) 98% v/v O_2, calculated from Equation 19.116 for 2000 K, with $K_c = 0$. Curves 1: Concentration $\sigma_A = 2.0$ g/cm³; curves 2: $\sigma_A = 1.0$ g/cm³.[36]

[91] M. D. Gray, G. M. Kimber, and D. E. Granger, "Internal Burning in Pulverized Coal," *Fuel,* **46,** 399–400 (1967).

[92] M. A. Nettleton, "Temperature Measurements on Burning Coal Particles in a Radiating Enclosure," *Combust. Flame,* **9,** 311–316 (1965).

[93] A. R. Ramsden and I. W. Smith, "Further Evidence for Internal Burning in Particles of Pulverized Coal," *Fuel,* **47,** 253–256 (1968).

[94] A. C. N. Tucker, "Internal Surface Burning in Anthracite p.f. Particles," *Fuel,* **48,** 105–119 (1969).

[95] M. A. Field, "Rate of Combustion of Size-Graded Fractions of Char from a Low-Rank Coal Between 1200°K and 2000°K," *Combust. Flame,* **13,** 237–252 (1969).

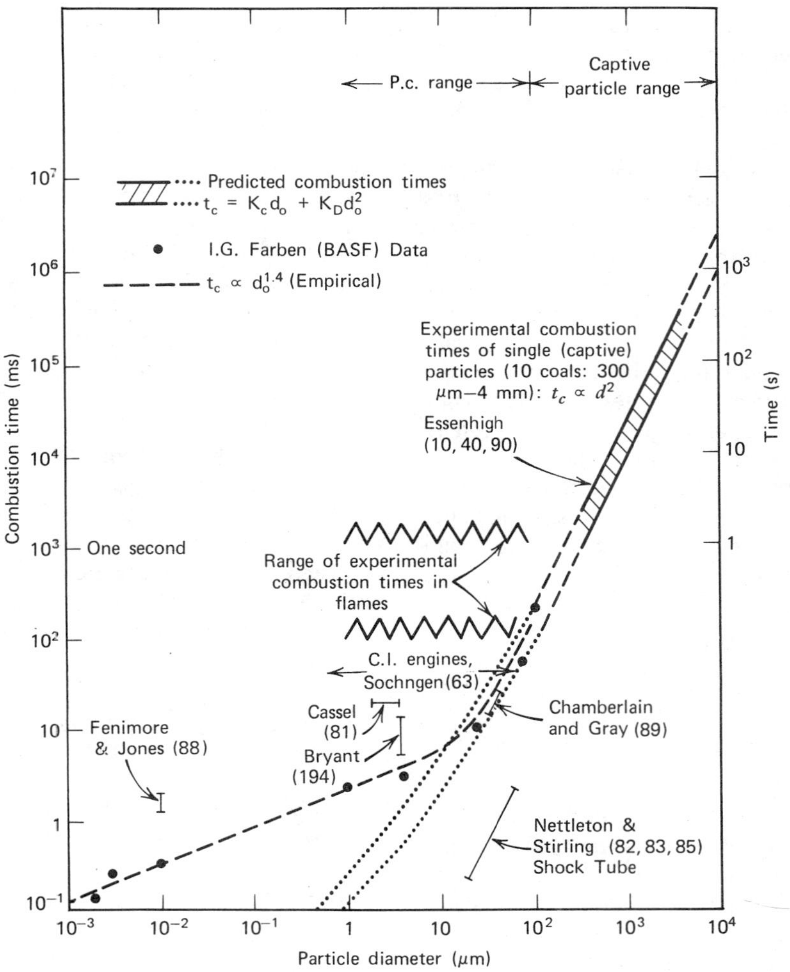

Figure 19.2 Variation of combustion time with particle diameter from various theoretical and experimental sources.

cluded that reaction took place mainly at internal surfaces, so that the particles burned at substantially constant diameter; and some recent flame experiments[87] have broadly substantiated this view, at least for coals and high-reactivity chars. As reactivity fell, it became increasingly sensitive to particle size; however this was also found to be the case for some high-reactivity coals in other experiments when the particles burned at constant density and diminishing diameter under some conditions.[97,98]

Burning times of about 1 s in flames under most usual conditions are therefore quite well documented, but there are data sources that indicate the possibility of much faster reaction under some circumstances. Explosion flames traveling at 300 ft/s (100 m/s) or more have flame lengths of 100–200 ft,[99] which yields burning times of 0.1 s. There may not have been complete combustion, however. Better documented are the diesel engine data summarized by Soehngen[63] (Fig. 19.2) from the reports of the German coal dust engine

[96] M. A. Field, "Measurements of the Effect of Rank on Combustion Rates of Pulverized Coal," *Combust. Flame,* **14,** 237–248 (1970).

[97] G. D. Sergeant and I. W. Smith, "Combustion Rate of Bituminous Coal Char in the Temperature Range 800 to 1700°K," *Fuel,* **52,** 52–57 (1973).

[98] R. J. Hamor, I. W. Smith, and R. J. Tyler, "Kinetics of Combustion of Pulverized Brown Coal Char Between 630 and 2200°K," *Combust. Flame,* **21,** 153–162 (1973).

[99] J. Grumer, "Recent Research Concerning Extinguishment of Coal Dust Explosions," *Proc. 15th Symp. (Int.) Combustion,* Combustion Institute, Pittsburgh, 1975, pp. 103–113.

projects running from 1916 to 1944. About 19 engines were investigated, operating between 160 and 1600 rpm and at pressures up to 11 atm. At 1600 rpm the total burning time available was 20 ms for combustion completed at a crank angle of 180°, and up to 96% combustion was reported even at the highest speeds. Similar combustion times are implied in the pulsating combustion reports of Traenckner[100] and Sommers.[101] Even shorter times are found in data reproduced in Fig. 19.2 attributed to the I. G. Farben laboratories, obtained during the coal dust engine program.[63] It may be noted that the burning times are comparable to the volatile release times at normal pressures; thus the two reactions are in competition. Since pressure adversely affects pyrolysis, it is possible that the combustion reaction under such conditions is predominantly heterogeneous.

19.2.2 Fundamental Magnitudes

19.2.2.1 *Physical Data: Densities, Internal Surface Areas, and Pore Structure*

Carbons and coals are porous solids of densities ranging from 1.1 g/cm^3 for low-rank coals to 2.33 g/cm^3 for high-density pyrolytic graphite. Natural graphite single crystals have about the theoretical maximum density of 2.26 g/cm^3.[102] Porosities (vol % or fractional free space) range from 0.2 to 90%. Internal surfaces for high-purity carbons range from 10^{-1} to 10^3 m^2/g for some chars, with a high value quoted by Brame and King[103] of 10^4 m^2/g for wood charcoal; high-density manufactured graphites are mostly in the range 0.1–10 m^2/g, with porous graphites rising to 100 m^2/g.

In coals, two surface areas are identified[104]: the lower value is for the macro or access pores, and the higher value is for the total. The higher values range from about 100 m^2/g (minimum values for middle-rank bituminous coals at about 30% VM) to about 500 m^2/g in both the higher rank (anthracitic) and lower rank (lignitic) coals,[104] as illustrated in Fig. 19.3. The minimum value at 30% VM (about 90% C) is a little misleading, since coal density is also a minimum at this point, implying that a substantial closed-pore volume exists that could become accessible on reaction. This internal surface is of crucial importance in combustion; it can be increased substantially by either pyrolysis or combustion, but in competition with reductions caused by heating (see Chapter 23 and the paragraphs on "Surface Area Changes with Heat Treatment" in Section 19.6.2.4).

The internal surfaces are measured by gas adsorption.[105] According to Mahajan and Walker,[105] CO_2 adsorption measurements at

[100] K. Traenckner, "Gas aus nicht-verkokungswürdingen Kohlen," *Gas Wasserfach,* **93,** 537–546 (1952).

[101] H. Sommers, "Experiences with Pulsating Tube Firing in an Experimental Installation," quoted in F. H. Reynst, *Pulsating Combustion,* Pergamon Press, New York, 1961, pp. 262–274.

[102] E. I. Shobert, *Carbon and Graphite,* Academic Press, New York, 1964.

[103] J. S. S. Brame and J. G. King, *Fuel,* 5th ed., Edward Arnold, London, 1955.

[104] H. Gan, S. P. Nandi, and P. L. Walker, Jr., "Nature of the Porosity in American Coals," *Fuel,* **51,** 272 (1972).

[105] O. P. Mahajan and P. L. Walker, Jr., "Porosity of Coals and Coal Products," in *Analytical Methods for Coal and Coal Products,* C. Karr, Ed., Vol. 1, Ch. 4. Academic Press, New York, 1978.

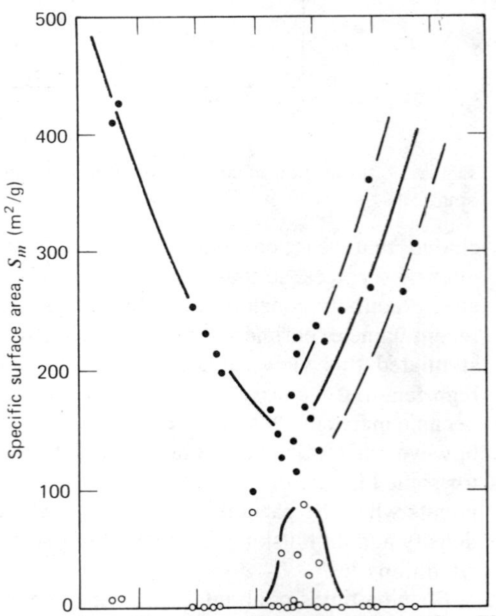

Figure 19.3 Influence of coal rank, designated by percentage of volatile matter, on internal surface of coals: ● = CO_2 surface area; ○ = N_2 surface area. Data source: Ref. 104.

0 and 20°C applied to the Polanyi-Dubinin equation yield total internal surfaces; but N_2 adsorption (using the BET equation) at 77 K gives only the macropore area. The latter value is presumed to be the area of the access pores. In coals the N_2 area varies as the inverse of the CO_2 area (Fig. 19.3), peaking at about 35% VM at about 100 m²/g, and declining to 10–20 m²/g into the anthracites and lignites. The implication is that the coals of lowest total internal surface have the largest access pores. In lignites, the access pores are initially small, but access is readily provided when pyrolysis opens up the structure. To a lesser extent, the same thing happens in the bituminous range. In the anthracites, however, the high internal surface is essentially inaccessible and remains so because the effects of pyrolysis are small.

The internal surface is provided by a network of internal pores of different diameter. The diameters other than those of the smallest pores can be estimated by mercury intrusion porosimetry.[105] This procedure yields graphs of cumulative pore volume as a function of (estimated) pore diameter. [The pressure required to force the mercury into the solid (up to 60,000 psi) may also open up some of the blind pore volume so that the pore structure and the internal surface measurements are not necessarily compatible. The effects are not considered to be large, however.] Carbons and coals vary widely in their pore size distribution structure; 20–80% of the pore volume can be in pores less than 10 μm; up to 33% of the pore volume can be in pores less than 500 Å. Pore diameters can be as small as a few angstroms, and a conventional classification has emerged[105] according to which pores having diameters greater than 200 Å are defined as macropores, those less than 20 Å are called micropores, and those between these values are termed transitional pores. Typically, up to 95% of the internal surface area is composed of micropores.

19.2.2.2 *Comparative Magnitudes: Mean Free Paths, Particle Sizes, and the Graphite Unit Cell*

The smallest pore sizes have the dimensions of molecules and are comparable to or less than the molecular mean free path of the penetrating gas up to quite high pressures. The mean free path depends on the molecular weight, but Equation 19.1, given by deBoer,[106] is a good approximation for most gases:

$$\lambda = 10^{-5}\left(\frac{T}{T_0}\right)\left(\frac{P_0}{P}\right)\text{(cm)} \qquad (19.1)$$

where T and P are the absolute temperature and pressure, respectively, and the subscript zero indicates standard temperature and pressure (STP).

Figure 19.4 is a plot of λ against the relative pressure P/P_0 for five different temperatures. An increase of temperature from ambient to 2500°C (2773 K) increases λ by a factor of 10. Figure 19.4 includes some other comparative quantities, as follows: the macro-, transitional, and micropore size ranges; the coal size ranges used in the fixed, fluid, and entrained combustors; the size range of soot particles; the range of visible radiation, and the CO_2 and H_2O bands. As shown, the radiation bands overlap the pulverized coal (p.c.) size range and exceed the soot particle range; this has relevance to radiative transfer and adsorption-emission characteristics.

The soot particle size range is roughly comparable with the macropore (access pore) size range. In the growth of soot particles, nucleation first forms crystallites of 20–30 Å. These grow to 100 Å by addition of carbon cracked on the surface. Although they may then agglomerate and cure to form particles, mostly of 500 Å, these still tend to be dense and unreactive, with little scope for developing micro- or transitional pores. Some porosity may develop in the agglomeration and curing, but this is likely to be low compared with manufactured carbons and coals. Soots and carbon blacks have little internal surface area. Such findings provide an independent basis for treating soot particles as impervious spheres as Fenimore and Jones did.[88]

Most important is the relation of mean free path to the pore size ranges, with pore diameters, as already noted, smaller in many cases than the mean free path. Thus there is a change from continuum to Knudsen diffusion, and this must be taken into account in any theoretical analyses to predict mass transfer into the particle interiors.

[106] J. H. deBoer, *The Dynamical Character of Adsorption*, Oxford University Press, New York, 1953.

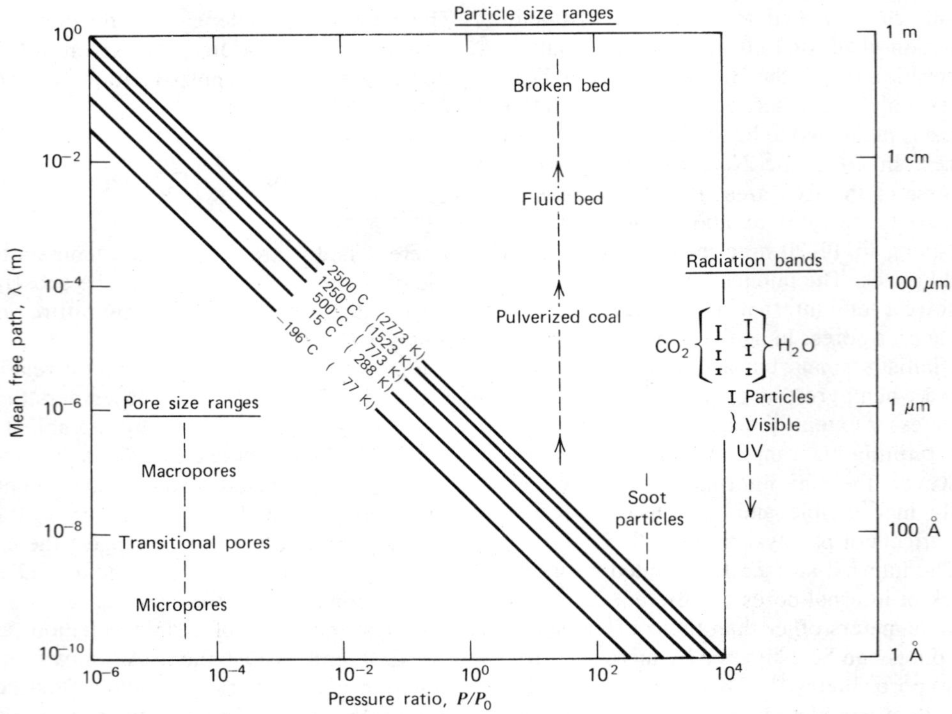

Figure 19.4 Variation of mean free path in gases with pressure ratio at different temperatures, indicating particle, radiation, and pore size ranges.

Another dimension of importance is the distance between the pores. If we reverse the rule of thumb for a bed of particles that the pore diameter is of the order of the particle size, the thickness of solid material between pores will likewise be of the order of the pore diameter, and this can be a few angstroms in some cases. The importance of this distance is twofold: first, it will affect the time before the pores have enlarged sufficiently during reaction for them to break into each other and coalesce (internal surface often goes through a maximum as the solids react away); second, it is an indication of how many atoms lying one below the other must be removed for burnout.

The interpore dimensions can be estimated from graphite crystal dimensions, and this can also be used to estimate surface areas. The graphite unit cell is illustrated with its dimensions in Fig. 19.5. The dimensions correspond to a volume of about 8.8 Å^3/atom or about 10^{23} atoms/g. If one gram was a thin wafer of 50 m^2 per side, it would be about 160 Å, or 80 atoms, thick. If the wafer is reacting on both sides, the important dimension is the half-thickness of 40 atoms. At 250 m^2/g (on two sides) the half-thickness would be about 8 atoms. At unit density, a wafer with 5×10^3 m^2/g would have a thickness of 2 Å. (This suggests that the cited internal surface of wood charcoal[103] is overestimated.) Extension of the estimations yields a maximum surface area of a wafer one atom thick of 2700 m^2/g (with some special case exceptions of no practical importance); a practical limit is in the region of 1000 m^2/g. An alternative calculation, assuming that the carbons are constructed of uniform spheres, yields a radius of about 160 Å for a particle of 100 m^2/g. Although the calculations are very rough, the answers are clearly compatible with each other and with porosimeter data.

19.2.2.3 *Chemical Data: Approximate Carbon Reaction Rates and Burning Times*

The rate of carbon atom removal can be roughly estimated either from surface adsorption times or from rates of molecular collision.

Adsorption Time Calculations. We assume that each carbon atom represents an adsorption site, and desorption of an adsorbed ox-

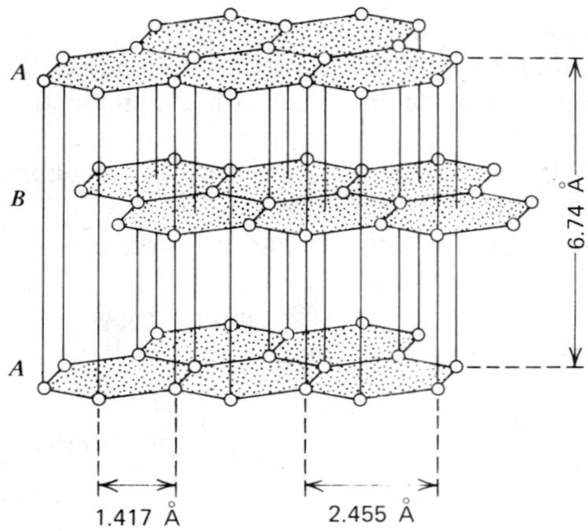

Figure 19.5 Crystal structure of graphite.[102]

ygen atom removes the underlying carbon atom to uncover a further adsorption site. If each new site is refilled immediately, implying a surplus of oxygen above the surface, the rate of carbon atom removal is controlled by the adsorption or residence time, τ, of the adsorbed oxygen molecules. This value can be used to estimate rates of carbon atom removal, or reaction times. deBoer[106] uses Frenkel's expression to calculate τ:

$$\tau = \tau_0 \exp \frac{Q}{RT} \tag{19.2}$$

where Q is the heat of adsorption, R is the gas constant, T is the absolute temperature, and τ_0 is a frequency factor. In one estimate of τ_0,[107] this factor was identified as the frequency of vibration of the constituent atoms of the solid for which Lindemann's equation could be used

$$\tau_0 = 4.75 \times 10^{-13} \left(\frac{MV^2}{3T_S}\right)^{1/2} \tag{19.3}$$

where M is the molecular weight of the solid, V is the molar volume (cm^3), and T_s is the absolute melting temperature. For graphite, deBoer[106] obtained a value for τ_0 of 5×10^{-14}. Thomas and Thomas[107] cited other, statistical mechanics studies that yielded a range of values for τ_0 of 10^{-12} to 10^{-16} and noted that the approximation of the values to the molecular oscillation time was in fact fortuitous. Taking τ_0 as $10^{-14 \pm 2}$, Equation 19.2 has been evaluated as illustrated in Fig. 19.6, which shows τ_0 as a function of Q for six different temperatures. The spread of $10^{\pm 2}$ in τ_0 creates some uncertainty, but it corresponds, alternatively, to an uncertainty in the value of Q

[107] J. M. Thomas and W. J. Thomas, *Introduction to the Principles of Heterogeneous Catalysis*, Academic Press, New York, 1967.

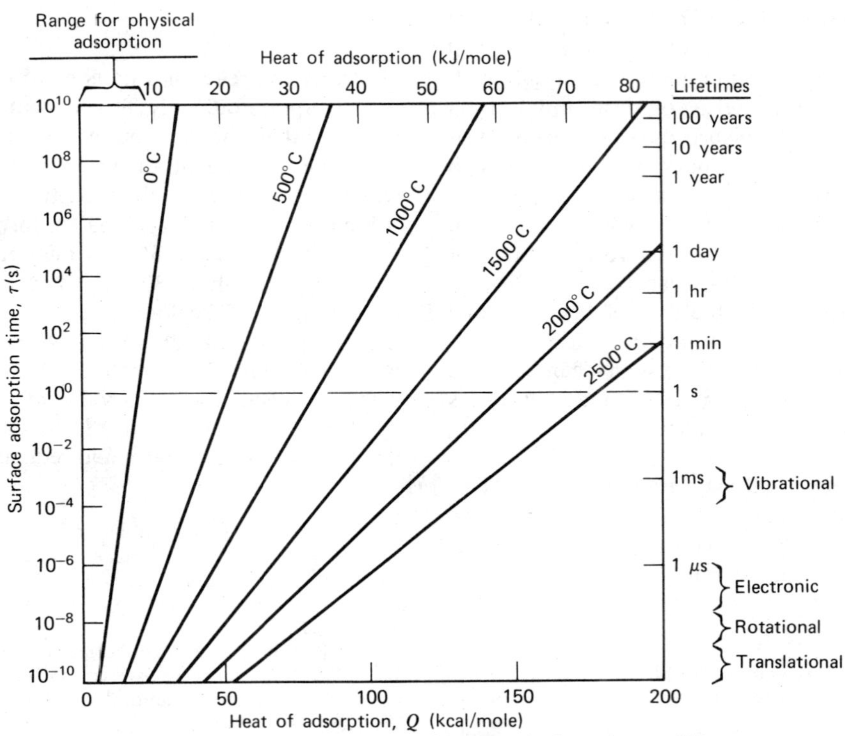

Figure 19.6 Variation of surface adsorption time with heat of adsorption at different temperatures.

of ±10 kcal/mole at 1000°C or of 20 kcal/mole at 2000°C (about 10 cal/°C), and this is smaller than the uncertainty in Q for oxygen adsorption on graphite.

Selection of a value for τ appropriate to carbon oxidation requires a value for Q, but there is considerable uncertainty in suggested values. Hayward and Trapnell[108] quote two values for Q for chemisorption of oxygen on carbon: a calculated value of 239 kcal/mole, which they consider to be too high, and an experimental value of 97 kcal/mole, which they consider to be too low. Since Q is a function of the extent of surface coverage, however, the low value is not impossible under some circumstances. At 100 ±10 kcal/mole at 1000°C we would have residence times of from one minute to one day, and residence times of centuries at 500°C, which would effectively preclude reaction below 1000°C. At 1500°C the times range from 10^{-1} to 10^{-3} s, and—assuming that each desorption removes a carbon atom—a wafer of 40 atoms half-thickness would be consumed in 0.04–4 s, and a wafer of 1 μm would react in 2.5–250 s. These values look much too high. The fault probably lies in the wrong choice of adsorption heat. The values cited are for oxygen adsorption under the assumption that the oxygen desorbs as O_2 (which it never does above −70°C). The relevant value here should be the heat of desorption to CO (which would also be more comparable with the activation energy of desorption). It is known that chemisorbed oxide films are stable at 500°C for hours or days, although reaction will take place at more active sites. A value of 75 kcal for Q would seem to be more realistic. It yields a reaction time of about 1 ms for 100-Å particles at 1500°C, a value in line with the data of Fenimore and Jones (Fig. 19.2); 10 ms for 10-μm particles at 2000°C, a value in line with Chamberlain's and Gray's data (Fig. 19.2); and about 0.1 s for a 1-μm particle at 1500°C.

Molecular Collision Rate Calculations. The preceding calculation is based on the assumption that the oxygen above the surface is in excess and all sites are always filled. If this is not the case, the reaction rate is reduced. Conditions under which this will occur can be estimated by comparing the time between impacts with the molecular residence time, τ.

If each atom of carbon is in a volume element of 8.7 Å, the side of the volume element is about 2 Å and its surface is 4–5 Å², which corresponds to about 10^{15} atoms/cm². The rate of surface impact by gas molecules $\dot{n}$ is given[107] by:

$$\dot{n} = 3 \times 10^{23}\left(\frac{T_0}{T}\right)^{1/2}\left(\frac{P}{P_0}\right) \text{ molecules/cm}^2\text{-s} \tag{19.4}$$

At STP this gives us 3×10^8 molecules/atom site–s, or 3.3×10^{-9} s between impacts (average) on each atom site. At other values of T and P this value is divided by $[(T_0/T)^{1/2}(P/P_0)]$. It is also assumed that all molecules adsorb. If only 20% are oxygen molecules, and only a fraction, $\exp(-E_{ads}/RT)$, have the necessary adsorption activation energy (E_{ads}), then, including a sticking (or orientation or entropy of adsorption) factor, η, the time interval between arrivals of oxygen molecules that adsorb becomes

$$\tau_{imp} = \left[6 \times 10^7 \eta \left(\frac{T_0}{T}\right)^{1/2}\left(\frac{P}{P_0}\right) \exp\frac{-E_{ads}}{RT}\right]^{\times 1} \tag{19.5}$$

Clearly the reaction rate is determined by the adsorption rate only if the impact interval is longer than the desorption time: $\tau_{imp} > \tau$. If the times are equal, the temperature T_{eq} at which this occurs is obtained from the resulting expression, which has the form $T_{eq} = (Q - E_{ads})M$; where M is a multiplier that has a value of about 30 for $\eta = 1$ and $P = P_0$. For $Q = 75$ kcal/mole at five values of E_{ads} ($E_{ads} = 0$, 10, 20, 30, and 40 kcal), T_{eq} takes the values 2150, 1850, 1530, 1214, and 890°C, respectively. For a value of $\eta = 10^{-2}$ (suggested by Laine et al.[109,110]), this overrides the square root term and reduces the T

[108] D. O. Hayward and B. M. W. Trapnell, *Chemisorption*, Butterworths Scientific Publications, London, 1964.

[109] N. R. Laine, F. J. Vastola, and P. L. Walker, Jr., "The Role of the Surface Complex in the Carbon-Oxygen Reaction." *Proc. 5th Carbon Conf., Vol. 2,* Pergamon Press, New York, 1963, pp. 211–217.

[110] N. R. Laine, F. J. Vastola, and P. L. Walker, Jr., "The Importance of Active Surface Area in the Carbon-Oxygen Reaction," *J. Phys. Chem.*, **67,** 2030–2034 (1963).

values to 1700, 1440, 1170, 900, and 650°C, respectively. There is now increasing evidence that E_{ads} has values in the region of 10 kcal/mole, which means that we should expect adsorption control only above 1400 or 1500°C, which is near peak flame temperatures, and desorption control below those temperatures. As discussed in more detail in the sections that follow, there are several adsorption sites with adsorption activation energies ranging from 5 to 15 kcal/mole.

In the calculations above it is assumed that there is rapid access of the oxygen to the particle surface. There are two problems here: holdup in the diffusion boundary layer, and holdup in pore diffusion. For the first, Equation 19.2 can be compared with an equation for the transit time, t_s, across a boundary layer of thickness δ[10]:

$$t_s = \frac{\delta^2}{2D_0}\left(\frac{T_0}{T}\right)^{1.75} \tag{19.6}$$

where D_0 is the STP diffusion coefficient for oxygen through nitrogen. This equation has been evaluated,[10] and t_s is compared with τ (from Equation 19.2) in Fig. 19.7, as both t_s and τ vary with temperature, for Q values of 50, 75, and 100 kcal/mole, and δ values of 1 cm, 1 mm, and 100 μm. In p.c. flames the interparticle distance is about 1 mm; thus most oxygen is already much closer to the particle surface than this. Since it is now accepted that boundary layer diffusion does not control reaction in small particle flames (<100 μm: but see the subsection "Comparative Magnitudes" in Section 19.6.3.2), the figure shows that Q must have the value of 75 kcal/mole or more, in agreement with the prior evaluation. If the effective boundary layer thickness is on the order of the particle diameters, however, Fig. 19.7 could indicate that diffusion could well dominate small particle combustion at about 2000°C.

19.3 QUALITATIVE BEHAVIOR IN COMBUSTION

This section describes the qualitative combustion behavior of carbons and coals as necessary background for modeling and quantitative analysis.

19.3.1 Graphite and Carbon Reactions

19.3.1.1 *Surface Reaction*

In early models of gas reactions with carbon it was common to assume that the surface is more or less uniform, at least as a statistical approximation. However this is far from

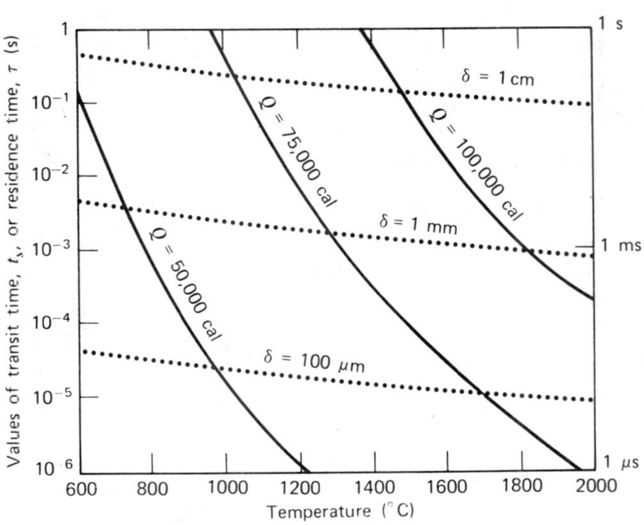

Figure 19.7 Variation with temperature of transit time (t_s) for an oxygen molecule diffusing to a solid surface through a film of thickness δ and residence time τ for an oxygen molecule adsorbed on a carbon surface with heat of adsorption Q: dotted curves, values of transit time for different δ; solid curves, values of residence time for different Q.[10]

being the case. As the result of mainly low-pressure (and fairly low-temperature) experiments (see Table 19.6, Section 19.6.1.1), the following qualitative picture has been developed.

Active Surface Reaction. The principal gases of concern are O_2, CO_2, H_2O, and H_2. All dissociate to chemisorb on carbon surfaces. The dominant consequence of the first three gas reactions is oxygen transfer. Adsorption is nonuniform, with preferential adsorption on the edge atoms of the graphite crystallites (Fig. 19.5). With very pure, highly graphitized materials many investigators regard the basal planes as virtually inactive at lower temperatures. In surface area measurements, it is then necessary to distinguish between active and inactive areas. The basal planes are included in measurements of the total surface area (TSA), but only a fraction of the TSA is specified as chemically active (ASA) because activity is mainly at the edges and there is little on the basal planes. When the low-temperature reactivity of the ASA is studied, only a fraction of that is found to be participating to any degree, with the fraction involved changing with temperature. On the basis of one interpretation,[109,110] the TSA can be divided into three regions:

1. The basal planes, where there is little or no reaction, and no chemisorption film to speak of.
2. The reacting fraction of the ASA (or RASA), where there is again little or no chemisorption film to speak of because reaction is so fast that the residence time of the oxygen atoms is negligible.
3. The unreacting fraction of the ASA (or UASA), which is covered by adsorbed oxide film; this is where reaction initially takes place.

A vacuum-cleaned, degassed sample, therefore, would adsorb oxygen for a finite time only on the UASA fraction of the surface at any given temperature, and the oxide film so formed would be essentially stable at the temperature of formation up to about 500°C. Reaction can still occur below 500°C, but it does so on the RASA. Above 500°C the oxide film on the UASA increasingly

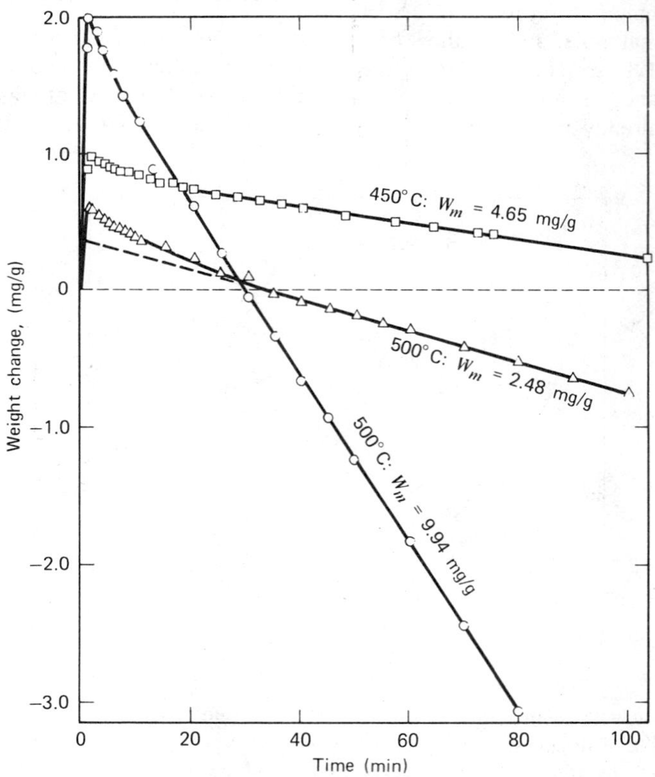

Figure 19.8 Weight changes observed during combustion in 20-torr O_2.[111]

participates in the reaction, typically with up to 95% removal at 1000°C in a vacuum.

This behavior is illustrated experimentally in Fig. 19.8, which shows the weight change of a turbostratic carbon black (Graphon) with time when reacting with a low oxygen pressure. The initial response is a weight increase followed by an almost linear fall. The initial rise in weight is due to the formation of the stable oxide film to provide coverage of the UASA of the ASA. The declining weight is due to reaction on the RASA. In Fig. 19.9 the graph of the total weight change for one of the runs has been broken into two components: the oxide film formation and combustion. This is typical of results of many other investigations. It clearly shows the weight increase due to surface oxide formation at the same time that the weight is declining because of combustion.

Quantitative investigation of behavior is complicated by changes in surface area as reaction (burnoff) proceeds. Laine et al.[109,110] found that the TSA of Graphon would about double with about a third burnoff; but the ASA increased nearly twenty fold. This accounts for the very different curves for weight loss at 500°C in Fig. 19.8 for two different levels of burnoff (W_m). Tucker and Mulcahy[111] investigated this effect by promoting burnoff in two ways: by combustion at constant temperature (which involved only the RASA), and by alternate adsorption (combustion) and oxide film desorption at low and high (975°C) temperatures, respectively—which also involved the UASA. The ASA was found to be increased more by the second procedure—high-temperature (stable) film desorption—than by the first. This finding has promoted the concept of the creation of "nascent" sites, which are considered to be more reactive. The increase in ASA was independent of combustion time or oxygen concentration. Because of these factors, there is no relationship between ASA and TSA (BET area). Likewise, there is no relationship between ASA and UASA. This was shown by Laine et al.,[109,110] who found that the fraction of ASA that was covered with film (UASA) depended on the tempera-

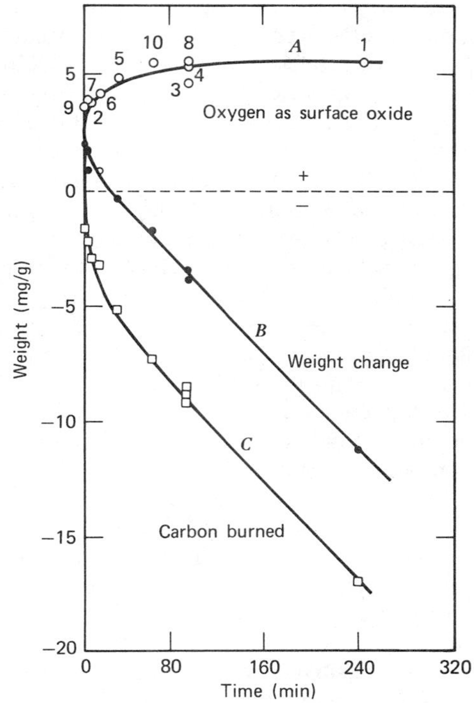

Figure 19.9 Progressive formation of surface oxide during combustion.[111]

ture of formation, with UASA decreasing with increasing temperature. This discovery also correlated with the CO_2 percentage in the film decomposition gases, which also declined with increasing film formation temperature. These conditions seem to imply that CO_2 is formed from the UASA film decomposition, and CO from the RASA reaction.

Bansall and associates[112,113] investigated the influence of temperature on the UASA fractional coverage and, from the results of their investigations, suggested the existence of several different types of adsorption sites. The UASA was determined by measuring O_2 adsorption, by means of oxygen loss and/or

[111] B. G. Tucker and M. F. R. Mulcahy, "Formation and Decomposition of Surface Oxide in Carbon Combustion," *Trans. Faraday Soc.*, **65**, 274–286 (1969).

[112] P. L. Walker, Jr., R. C. Bansal, and F. J. Vastola, "Studies on Ultra-Clean Carbon Surfaces. 1. Characterization of Surface Activity of Graphon by Room-Temperature Oxygen Chemisorption," in *The Structure and Chemistry of Solid Surfaces*, John Wiley and Sons, New York, 1969, pp. 81-1 to 81-16.

[113] R. C. Bansal, F. J. Vastola, and P. L. Walker, Jr., "Studies on Ultraclean Carbon Surfaces. II. Kinetics of Chemisorption of Oxygen on Graphon," *J. Colloid Inter. Sci.*, **32**, 187–194 (1970).

weight increase of the sample when the adsorption sites were saturated. The weight increase was checked by degassing at 1000°C to recover the adsorbed oxygen as CO (and CO_2).[114] The adsorption isotherms obtained were found to obey the Elovich equation (Section 19.5.3), and to consist of a series of straight-line sections. The different stages appeared only after a definite quantity of oxygen had been absorbed. The quantity was independent of temperature: the effect of increased temperature was to speed completion of each stage. This would seem to imply that adsorption is mobile, with the sites of lowest potential energy (highest energy of formation) always filling up first. From appropriate slopes determined from the Elovich isotherms, activation energies of adsorption were obtained, and Fig. 19.10 illustrates the variation of E_{ad} with quantity of oxygen absorbed. The values vary from roughly 5 to 15 kcal/mole. The step behavior is identified with the different adsorption sites.

A more positive identification was obtained when the experiments[115] were repeated with hydrogen adsorption. Figure 19.11 illustrates the results, showing parallel behavior to the oxygen adsorption, although there are only four steps compared with five in the case of oxygen. In both cases the different adsorption steps were identified with explicit inter-

[114] R. C. Bansal, F. J. Vastola, and P. L. Walker, Jr., "Studies on Ultraclean Carbon Surfaces. IV. Decomposition of Carbon-Oxygen Surface Complexes," *Carbon (Oxford)*, **8**, 443–448 (1970).

[115] R. C. Bansal, F. J. Vastola, and P. L. Walker, Jr., "Studies on Ultraclean Carbon Surfaces. III. Kinetics of Chemisorption of Hydrogen on Graphon," *Carbon (Oxford)*, **9**, 185–192 (1971).

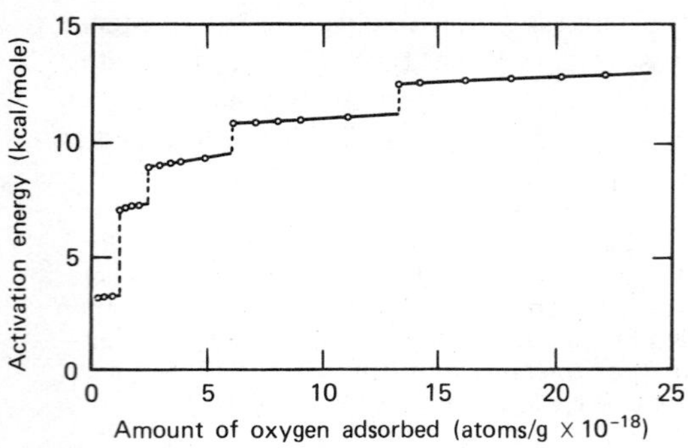

Figure 19.10 Change of activation energy for oxygen chemisorption on Graphon with amount adsorbed.[113]

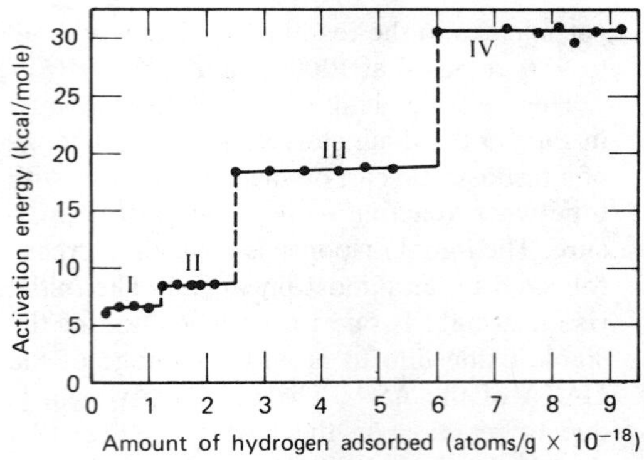

Figure 19.11 Change of activation energy for hydrogen chemisorption on Graphon with amount adsorbed.[115]

carbon distances, as follows: 1.42, 2.46, 2.84, 3.35, and 3.62 Å (Fig. 19.5). For hydrogen adsorption the activation energies were compared with the Sherman and Eyring calculations of 1932[116]: Table 19.2 shows the remarkable agreement. The Sherman and Eyring model showed that the activation energy had a minimum at a C—C distance somewhat greater than the H—H distance (H—H and O—O distances are 0.746 and 1.208 Å, respectively[102]). At small C—C distances the adsorbing molecules interfered with each other; at large distances they had to dissociate completely to be able to reach the C atoms on which they were adsorbing. Since the interatomic oxygen distance is greater than that for hydrogen, it is to be expected, as found experimentally (Fig. 19.10 and 19.11), that E_{ads} for oxygen adsorption is less than that for hydrogen.

Desorption was also found to occur in steps,[114] but the desorption activation ener-

[116] A. Sherman and H. J. Eyring, "Quantum Mechanics of Activated Adsorption," *J. Am. Chem. Soc.*, **54**, 2661 (1932).

Table 19.2 Comparison of Experimental and Theoretical Values of Activation Energy for Chemisorption of Hydrogen on Graphite[115]

Site Number	C–C Distance (Å)	E_{ads} (kcal/mole)	
		Experimental	Theoretical[a]
I	3.62	5.6	9
II	3.35	8.3	11
III	2.84	18.4	19
IV	2.46	30.4	28
V	1.42	—	50

[a] *Data source:* Sherman and Eyring.[116]

gies were not reported. There is a past history of such measurements,[117–122] but the only ones that are as detailed as those obtained by Bansal et al.[113–115] are measurements by Tucker and Mulcahy,[111] who established that the desorption activation energies varied with coverage. Their values were highest at zero coverage (in excess of 100 kcal), falling to 60–70 kcal/mole at full coverage.

Basal Plane Reaction and Catalysis. The discussion so far has focused on the edge sites. At the lower temperatures the basal plane atoms are not thought to participate to any great extent. Basal plane attack can occur at any temperature, however, if foci are provided by vacancies, dislocations, and other imperfections. Basal planes also seem to be the preferred location for attack when using catalysts. Catalysts are inert on perfect surfaces, but they aid attack at imperfections. The role of solid catalysts on carbon surfaces is to provide sites for dissociation of the reactive gas (but see Section 19.6.2.3), but only where the catalyst particle is in contact with an edge or an imperfection. The result is the formation of basal plane pits, which can also occur by reaction at imperfections without catalysts. The pits are generally hexagonal with parallel sides,[123–125] and growth of the pits can be observed optically. Activation energies for the $11\bar{2}0$ and $10\bar{1}0$ faces were measured by Thomas and Hughes,[123] and were found to be close, at 62 and 66 kcal/mole, respectively. Some pits found to spiral were thought to be on screw dislocations but are now considered to be the result of a spiral migration of a catalyst particle.[126,127] Photographs of catalyst particles show that they jump from place to place with almost continuous motion.

A wide range of elements has been found to catalyze the reaction. Noble metals have been widely investigated, but more mundane elements such as iron and calcium are also very effective, and this has practical consequences for coal combustion because these elements are present in the mineral matter (calcium is particularly important in lignites).

It appears that basal planes (Fig. 19.5) are otherwise attacked as actively as the edge sites by oxygen atoms.[128–136] Pitting is re-

[117] J. Gadsby, C. N. Hinshelwood, and K. W. Sykes, "The Kinetics of the Reactions of the Steam-Carbon System," *Proc. R. Soc. London,* **A187,** 129–151 (1946).

[118] J. Gadsby, F. J. Long, P. Sleightholm, and K. W. Sykes, "The Mechanism of the Carbon Dioxide-Carbon Reaction," *Proc. R. Soc. London,* **A103,** 357–376 (1948).

[119] E. A. Gulbransen and K. F. Andrew, "Reactions of Artificial Graphite: Kinetics of Oxidation of Artificial Graphite at Temperatures of 425 to 575°C and Pressures of 0.15 to 9.8 cm of Mercury of Oxygen," *Ind. Eng. Chem.,* **44,** 1034–1038 (1952).

[120] E. A. Gulbransen and K. F. Andrew, "Reactions of Artificial Graphite: Surface Oxide Formation and Surface Roughness Studies in Relation to Oxidation of Artificial Graphite at Temperatures of 25°C and Between 425 and 575°C," *Ind. Eng. Chem.,* **44,** 1039–1044 (1952).

[121] E. A. Gulbransen and K. F. Andrew, "Reactions of Artificial Graphite: Reaction of Carbon Dioxide with Pure Artificial Graphite at Temperatures of 500° to 900°C," *Ind. Eng. Chem.,* **44,** 1048–1051 (1952).

[122] J. F. Strange and P. L. Walker, Jr., "Carbon–Carbon Dioxide Reaction: Langmuir-Hinshelwood Kinetics at Intermediate Pressures," *Carbon (Oxford),* **14,** 345–350 (1976).

[123] J. M. Thomas and E. E. G. Hughes, "Localized Oxidation Rates on Graphite Surfaces by Optical Microscopy," *Carbon (Oxford),* **1,** 209–214 (1964).

[124] G. Hughes, J. M. Thomas, H. Marsh, and R. Reed, "Origin of Etch Pits on Graphite Surfaces," *Carbon (Oxford),* **1,**339–343 (1964).

[125] J. M. Thomas, "Microscopic Studies of Graphite Oxidation," in *Chemistry and Physics of Carbon,* Vol. 1, P. L. Walker, Jr., Ed., Marcel Dekker, New York, 1966, pp. 121–202.

[126] G. R. Hennig, "Surface Oxides on Graphite of Single Crystals," *Proc. 5th Carbon Conf.,* Vol. 1, Pergamon Press, New York, 1962, pp. 143–146.

[127] G. R. Hennig, "Electron Microscopy of Reactivity Changes Near Lattice Defects in Graphite," in *Chemistry and Physics of Carbon,* Vol. 2, Marcel Dekker, New York, 1966, pp. 1–49.

[128] J. Streznewski and I. Turkevich, "The Reaction of Carbon with Oxygen Atoms," *Proc. 3rd Symp. Carbon,* Pergamon Press, New York, 1959, pp. 273–278.

[129] J. D. Blackwood and F. K. McTaggart, "The Oxidation of Carbon with Atomic Oxygen," *Aust. J. Chem.,* **12,** 114–121 (1959).

[130] J. D. Blackwood and F. K. McTaggart, "Reactions of Carbon with Atomic Gases," *Aust. J. Chem.,* **12,** 533–542 (1959).

[131] C. E. Gleit, W. D. Holland, and R. C. Wrigley, "Reaction Kinetics of the Atomic Oxygen–Graphite System," *Nature,* **200,** 69 (1963).

[132] H. Marsh, E. O'Hair, R. Reed and W. F. K. Wynne-Jones, "Reaction of Atomic Oxygen with Carbon," *Nature,* **198,** 1195–1196 (1963).

[133] F. J. Vastola, P. L. Walker, Jr., and J. P. Wightman, "The Reaction Between Carbon and the Products of Hydrogen, Oxygen, and Water

ported to be as extensive as with catalysts. Reactivity also becomes much less sensitive to "intrinsic" carbon reactivity.

The overall picture is one of variable activation energy of both adsorption and desorption, which is hardly a surprise. The details in terms of the individual steps have been worked out mainly using Graphon. This is a channel black (Spheron 6), further purified at 2800°C to yield impurity levels at 150 ppm or less (about two-thirds of which is calcium[111]), with a surface composed almost entirely of basal planes of carbon crystallites.[137,138] The initial ASA was found to be only about 2.5% of the TSA (BET measurements), and about three-quarters of the ASA was the UASA—strictly, the fraction covered by stable oxide film that started to decompose above about 500°C.

Given these special circumstances, to what extent do the results also apply to other, impure carbons, and most particularly to coal chars? A definitive answer is not possible. It is suggested, however, that the catalysis results should be applicable at a microlevel. The overall behavior of a practical carbon or coal char is then something of a statistical average of the microcrystallite behavior where there is a range of activities on different crystallites (remembering that the sites on a given surface are occupied in adsorption successively from the highest activity to the lowest).

In practical circumstances, however, reaction at temperatures below 300°C (on chars, not coals) is essentially undetectable. It is readily measured on chars or cokes above 450°C,[45,139,140] but, in combustion, temperatures of interest generally range from 800–1600°C. At these temperatures it would appear that the relevant activation energies would be the limiting ones for the controlling mechanism of concern (Fig. 19.10).

In the case of adsorption, this would be the highest activation energy of the identified steps (Fig. 19.10) modified by catalytic effects. The effect of catalysts is generally to reduce the activation energy so that the values to be expected would be in the region of 10 kcal/mole (cf. estimates in Section 19.2.2).

In the case of desorption the Tucker and Mulcahy data[111] may be used as a guide. The (extrapolated) activation energy is highest at zero coverage (100 kcal), when adsorption would be controlling. At near full coverage (also by extrapolation), the value is closer to 60 kcal. Available data for impure carbons, discussed more fully below, are in the region of 40–45 kcal, which is a comparable figure if there is substantial reduction on account of catalytic effects. It would also be not unreasonable to expect the activation energy of the film decomposition to be lower if measured during the combustion reaction as compared with the value for the film decomposition alone. This would seem to be demonstrated by the data of Gulbransen and Andrew.[119–121]

19.3.1.2 *Diffusion Effects*

Diffusion can be important in two ways: as boundary layer diffusion and as internal pore diffusion. With particles that are small and dense, both diffusion effects can be ignored, as Fenimore and Jones[88] did for soot particles of about 100 Å that burned in about 1 ms. With particles larger than 100 μm or thereabouts, when boundary layer diffusion controls the reaction, internal reaction does not occur, even with the most porous carbons.

Microwave Discharges," *Carbon (Oxford)*, **1**, 1–11 (1964).

[134] M. Otterbein and L. Bonnetain, "1. Combustion du Graphite dans un Courant d'Oxygène Soumis à la Décharge Électrique. 2. Influence de la Température sur la Réaction du Graphite avec l'Oxygène Atomique," *Comp. Rend.*, **258**, 2563–2565 (1964); **259**, 791–793 (1964).

[135] H. Marsh, E. O'Hair, and W. F. K. Wynne-Jones, "Oxidation of Carbons and Graphites by Atomic Oxygen Kinetic Studies," *Trans. Faraday Soc.*, **61**, 274–284 (1965).

[136] R. C. Melucci and J. P. Wightman, "The Reaction of Carbon and Atomic Oxygen Produced in Microwave Discharge," *Carbon (Oxford)*, **4**, 467–472 (1966).

[137] E. A. Kmetko, "The Structure and Graphitization of Fine Coke Particles and Thermal Carbon Blacks," *Proc. 1st and 2nd Confs. Carbon*, University of Buffalo, Buffalo, N.Y., 1956, pp. 21–30.

[138] H. Akamutu and H. Kuroda, *Proc. 4th Conf. Carbon*, Pergamon Press, New York, 1961, pp. 355–369.

[139] H. W. Nelson, C. P. Brysch, and J. H. Lum, *Determination of the Minimum Temperature of Sustained Combustion of Solid Fuels*, Pennsylvania State College School of Mineral Industries Bull. No. 22, Miner. Ind. Exp. Sta., 1938.

[140] F. D. Moles and M. W. Thring, "Minimum Burning Rates in Coke Fuel Beds," *Proc. Conf. Science in the Use of Coal*, Paper No. 36, Institute of Fuel, 1958, pp. E26–29.

If the particle is spherical and the ambient gas is quiescent, calculation of reaction rates for large particles is accurate to within a few percent. In a flowing system, quite complex fluid mechanics aspects are involved whose value in prediction is still somewhat questionable: this topic is somewhat outside the scope of this chapter.

Internal Reaction. When internal reaction is important, the details, and particularly the quantitative details, depend very much on the particle size and the reactivity conditions. The qualitative pattern, however, is generally common, and the broad behavior has been understood for about 20 years. This pattern, the "three-zone" concept of porous solid reaction of Wicke,[141–143] is discussed in major detail by Walker et al.[30]

At low temperatures, when reaction is relatively slow, reactive gas can diffuse into the interior of a porous solid faster than it can be used by reaction. When diffusion and use come into balance, the reactive gas has penetrated into the center of the solid and is present in finite quantities throughout the solid. This is referred to as Zone I.

As the temperature rises, the rate of use exceeds the rate of diffusion, and the two come into balance again when all the in-diffusing gas is consumed in a reactive zone that does not penetrate to the center of the solid, leaving an unreactive core. This is Zone II.

At higher temperatures still, the reaction retreats to the superficial surface of the solid (Zone III). There is no significant penetration into the solid, and the reaction rate is controlled by the boundary layer diffusion rate.

It is also well established[30] that the activation energy in Zone II is half that in Zone I, and the apparent or effective reaction order can also differ from the true order (see Section 19.5.4).

This general picture is amplified particularly well by the experiments of Froberg[144]

[141] E. Wicke and M. Rossberg, "Primar und Sekundärreaktionen bei der Wasserdampfuergasung von Kohle," *Z. Elektrochem.*, **57**, 641–645 (1953).

[142] E. Wicke, "Contributions to the Combustion Mechanism of Carbon," *Proc. 5th Symp. (Int.) Combustion,* Reinhold Publishing Corp., New York, 1955, pp. 245–252.

[143] E. Wicke and G. Wurzbacher, "Konzentrationsprofile vor Einer im Sauerstoffstrom Verbrennenden Kohlenstoffoberfläche. I. Experimentelle Ergebnisse," *Int. J. Heat Mass Transfer,* **5**, 277–289 (1962).

[144] R. W. Froberg, "The Carbon-Oxygen Reaction: An Experimental Study of the Oxidation of Suspended Carbon Spheres," Ph.D. thesis, Dept. of Fuel Science, Pennsylvania State University, State College, June 1967.

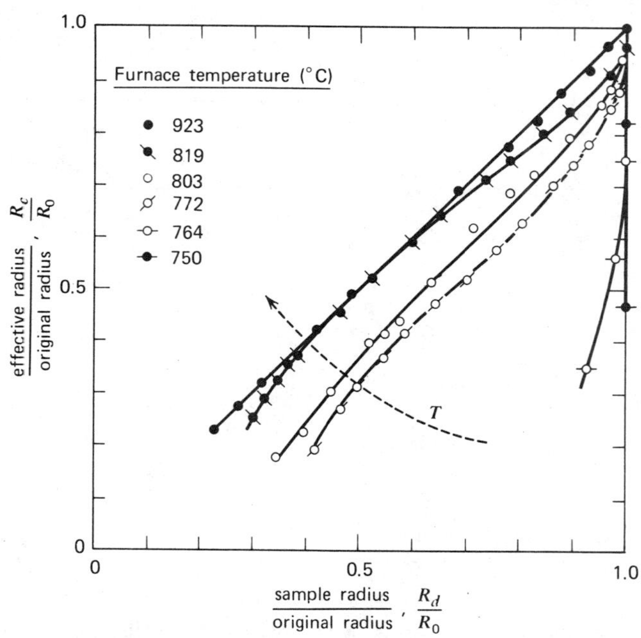

r air velocity of 1.0 cm/s (STP).[144,147]

and Kurylko.[145] Both used half-inch spheres of a particularly dense carbon (surface area 0.3 m^2/g) suspended from a balance in an electric tube furnace, à la Hottel et al.,[146] but with the difference that the particle diameters were measured simultaneously by time-lapse photography. As density changed, the mean density could then be calculated from the two simultaneous measurements. A more informative treatment of the data was developed, however, by calculating the effective radius (R_c) from the weight, assuming *no* change in density. If there was no change in density experimentally, then R_c had the same value as the experimentally measured radius, R_d. With internal reaction, the two radii differed. This is illustrated in Fig. 19.12. The normalized radii (R_c/R_0) and (R_d/Ro) are plotted against each other, where R_0 is the initial radius. There is an evident pattern. At the lowest temperatures (< 750°C for this carbon), reaction was in Zone I and was almost uniform throughout; and, as seen in Fig. 19.12, the calculated or nominal radius falls at constant actual radius. Zone III existed above 900°C, with R_c and R_d equal at all points and lying on the line of slope of 45°. Zone II lies between these two temperature ranges, with the extent of penetration falling with increasing temperature and indicated by the extent of deviation from the 45° line.

The actual penetration depth of the reaction in Zone II is difficult to measure because reaction is declining from the surface inward. This phenomenon is illustrated by density plots at successive times during reaction (Fig. 19.13). The reaction depth in such measurements is something of a value judgment; it is also a tedious measurement, not readily made. If the difference in radii (R_c - R_d) in Fig. 19.12 can be used as a measure of penetration depth, however, a pattern may be demonstrated. This is illustrated in Fig. 19.14, which shows the variation of the normalized radius difference with the normalized experimental radius. At the lowest temperature (835°C), there is a steady penetration into the particle. As temperature rises, the tendency is to develop a certain depth of penetration with a small reduction in outside radius, after which the penetration depth tends to remain fairly constant as the particle burns out. At higher temperature still, the initial penetration steadily withdraws, and the reaction zone is confined effectively to the surface with a reduction to about half the

[145] L. Kurylko and R. H. Essenhigh, "Steady and Unsteady Combustion of Carbon," *Proc. 14th Symp. (Int.) Combustion,* Combustion Institute, Pittsburgh, 1973, pp. 1375–1386.

[146] C. M. Tu, H. Davis, and H. C. Hottel, "Combustion Rate of Carbon: Combustion of Spheres in Flowing Gas Streams," *Ind. Eng. Chem.,* **26,** 749–757 (1934).

[147] R. H. Essenhigh and R. W. Froberg, "The Carbon-Oxygen Reaction: An Experimental Study of the Oxidation of Suspended Carbon Spheres," U.S. Office of Naval Research, Techn. Rep. No. F-567-2(u), 1967.

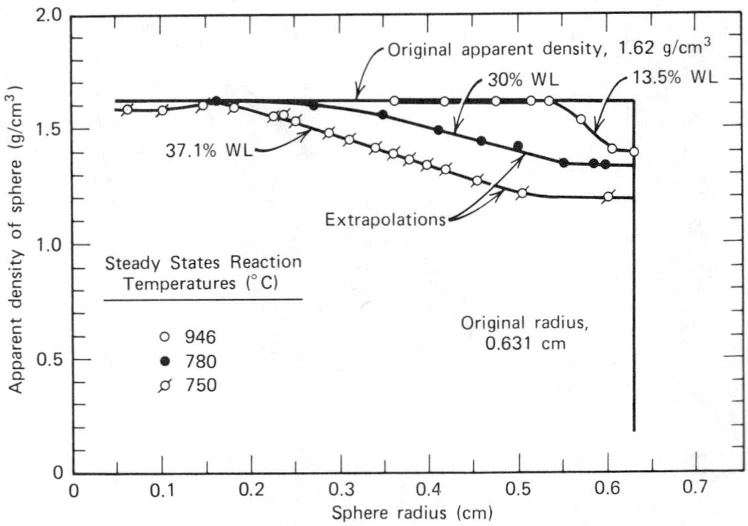

Figure 19.13 Apparent density of sphere versus radius: results of grinding experiment (WL = weight loss).[147]

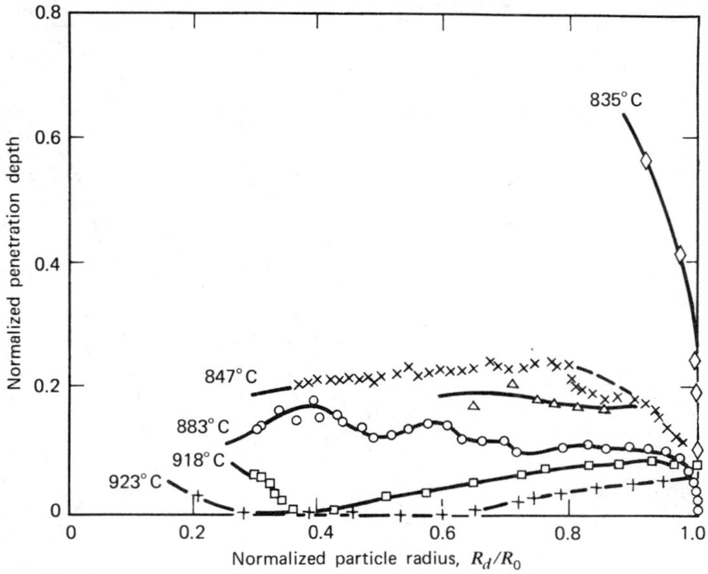

Figure 19.14 Variation of reaction penetration depth with particle radius (normalized) at five different temperatures.[144,147]

original radius, or a little less. Finally, and not included in Fig. 19.14, there is no penetration at all above about 900°C, and all points would lie along the x axis.

Internal Surface Changes. Plots such as Fig. 19.14, however, only tell half the story. As burnout proceeds, the internal surface area is also changing for the reasons outlined previously: the pores are expanding and increasing their area until they break through and coalesce, after which the internal area starts to decline. This phenomenon can be the reason for the contraction and readvance of the internal reaction zone shown in Fig. 19.14 by the lines at 918 and 923°C. If the total surface is measured up to 100% burnout, it goes through a maximum, as illustrated on a normalized basis in Fig. 19.15 (which is also shown to correlate closely with the reaction rate in this example).[142] If the area is corrected for burnout, given as area per unit mass, it behaves more regularly in this example, as illustrated in Fig. 19.16. The variation here is almost linear, but this may not be the most usual behavior. Dutta and et al.[148] have published data, discussed in more detail in Section 19.6.2.4, showing some curves with maxima and some with minima. They also advanced an empirical equation as a means of handling this complex behavior mathematically. Apparently, however, there is a need to extend such empirical knowledge of the trends by further experimental investigations.

The complexity of calculation has not deterred attempts to calculate. The results are bound to be artifacts of the assumptions to some extent, but with reasonably well-informed assumptions, the predicted trends are likely to be quite acceptably indicative of

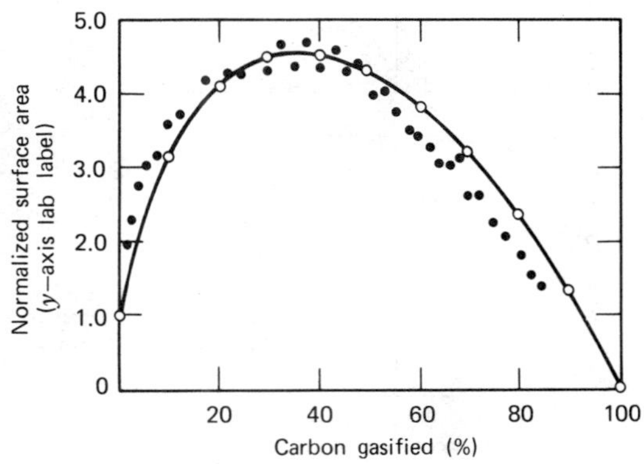

Figure 19.15 The whole carbon surface as determined by butane adsorption (○), and the gasification velocity measured at 1026°C (●), dependent on the degree of carbon consumption.[142]

[148] S. Dutta, C. Y. Wen, and R. J. Belt, "Reactivity of Coal and Char. 1. In Carbon Dioxide Atmosphere," *Ind. Eng. Chem. Process Des. Dev.*, **16**, 20–30 (1977).

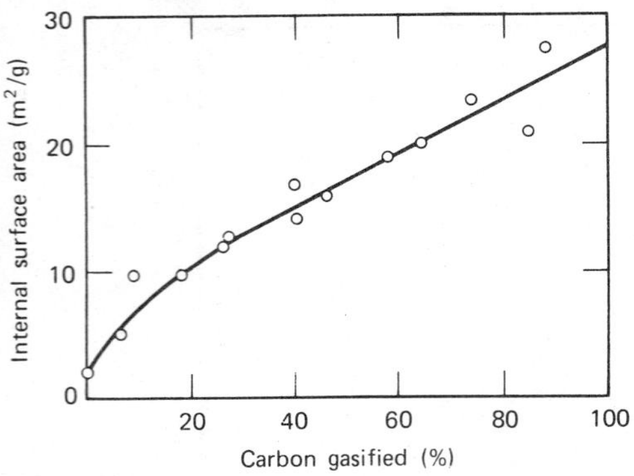

Figure 19.16 The surface per gram of carbon, dependent on the ratio of carbon gasified (by butane adsorption at 0°C; BET method).[142]

reality, even though there may be legitimate reservations about magnitudes. Such calculations are particularly valuable where measurement is difficult or impossible. Such is the case for gas concentrations inside the porous spheres discussed earlier, and to a lesser extent their concentrations outside.

Such calculations have been performed by Kurylko.[145] The pattern of behavior is illustrated in Figs. 19.17–19.19. Figure 19.17 shows reaction of a sphere in air, which produces CO and CO_2, with the CO_2 produced internally by gas phase combustion. Reaction is at low temperature (500°C), and it occurs almost uniformly throughout the sphere, with the oxygen concentration also

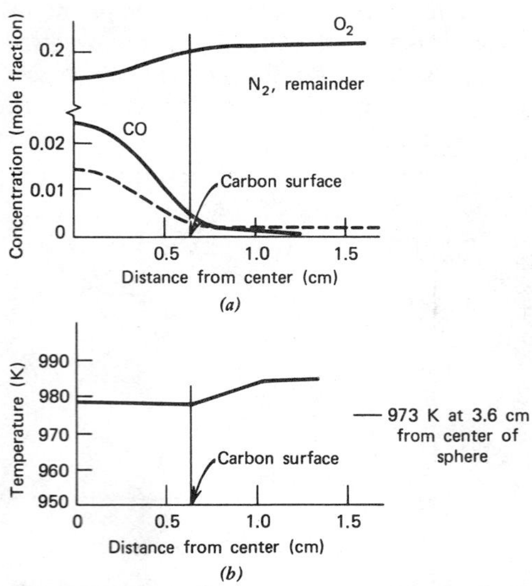

Figure 19.17 Calculated profiles of (*a*) gas concentration and (*b*) temperature; carbon sphere reacting in air at 973 K.[145]

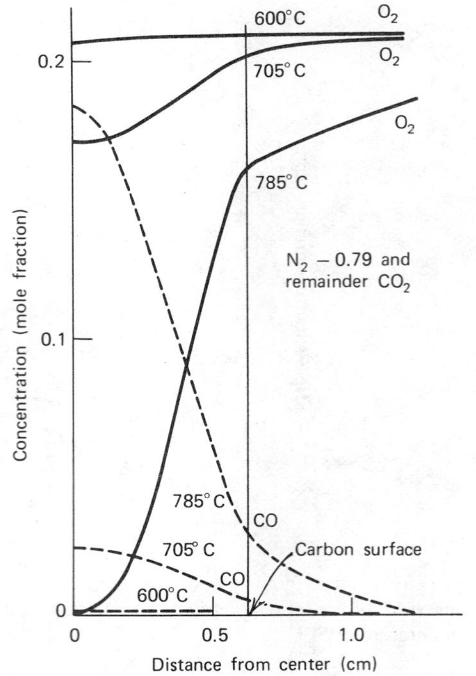

Figure 19.18 Calculated gas concentration profiles; carbon sphere reacting in air at different temperatures.[145]

almost uniform throughout the sphere. Diffusional escape of both CO and CO_2 is sufficiently fast that both are at very low concentrations in the boundary layer.

Figure 19.18 shows the effect of rising temperature. The oxygen concentration inside the sphere falls and the CO concentration rises. In this example the conditions at 785°C would correspond to transition from Zone I (total penetration) to Zone II (partial penetration). At higher temperatures, even at increased oxygen concentrations, the gas-solid reaction moves to near the surface and the CO reaction moves into the boundary layer. This effect is illustrated in Fig. 19.19, which also compares the effects of differences in particle porosity (factor of 10 difference in the gas-solid diffusion coefficient).

The most obvious characteristic in Fig. 19.19*a* is the sharp peak in the temperature profile, almost coincident with the peak in the CO_2 profile. The other significant characteristics are the insignificant penetration of oxygen into the solid and its low value at the solid surface, and the almost uniform CO concentration inside the solid. The effect of reduced gas diffusivities into the solid due to

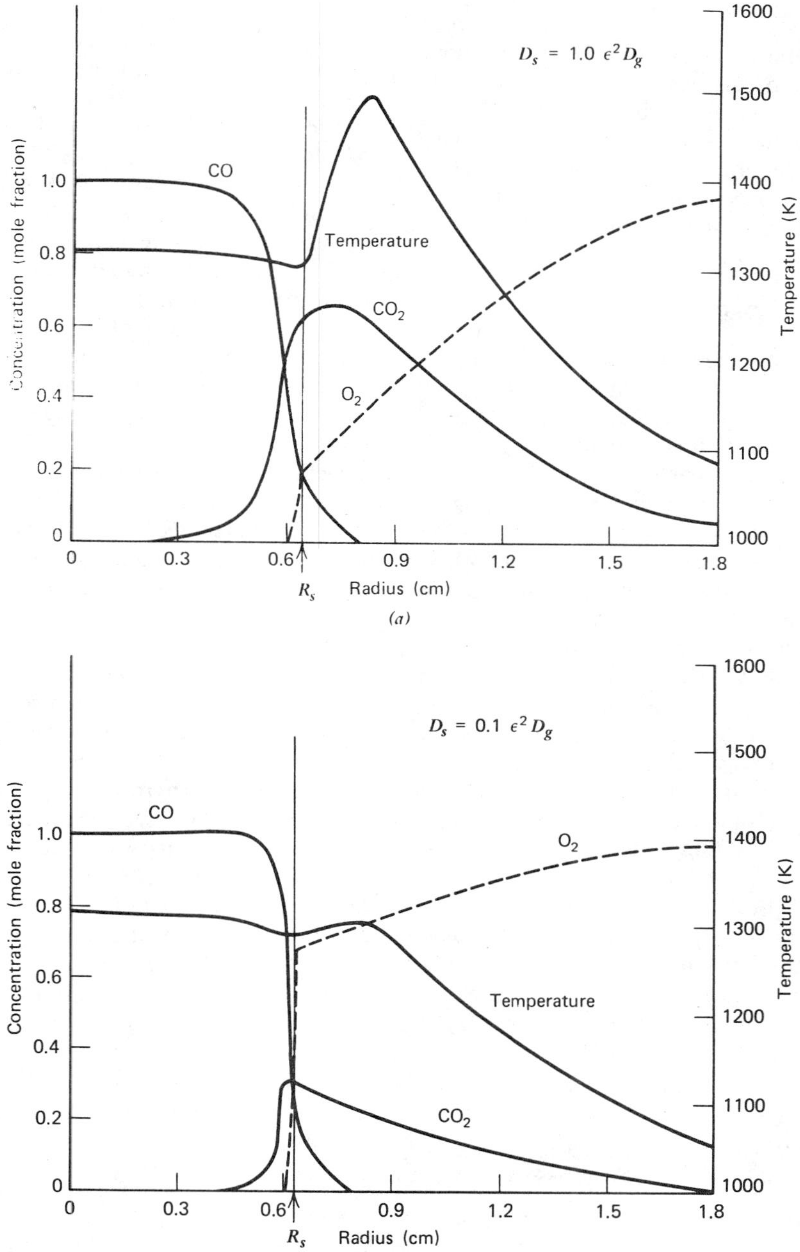

Figure 19.19 Calculated gas concentration and temperature profiles for a carbon sphere reacting in oxygen. (*a*) $D_s = \epsilon^2 D_g$: note the high-temperature excursion in the gas phase, partial penetration of CO_2 into the sphere, and low concentration of O_2 at the surface. (*b*) $D_s = 0.1\ \epsilon^2 D_g$: note the large concentration of oxygen at the carbon surface and the very shallow penetration of CO_2 into the sphere.[145]

lower porosity is shown in Fig. 19.19*b:* the boundary layer temperature peak is greatly reduced; the CO_2 peak is likewise reduced, and so also is its penetration depth into the solid; and the surface oxygen concentration is significantly greater.

Although the quantitative details are still uncertain and the couplings between mechanisms are highly complex, the qualitative behavior of the burning porous particle of carbon or char is nevertheless reasonably clear.

19.3.2 Coal Reactions

Coal generally pyrolyzes during combustion, splitting into volatiles and char. Combustion of the volatiles, as remarked in the introduction, is often determined by the rate of volatiles production. Char combustion is much slower; it can be described as graphite combustion with complications.

19.3.2.1 *Coal Structure*

Coal is not graphite. Its structure is discussed in detail in Chapter 8, but a brief summary of salient points is necessary here. Coal is generally represented as being composed of aromatic and hydroaromatic units, with functional groups at their periphery helping to form cross-links between units.[149] X-Ray analysis shows the units to consist of (X-ray) ordered and amorphous material arranged in layers or lamellae.[149–153] The layers are quite randomly ordered in the lower rank coals (below 90% C, maf), with 15–20 carbon atoms per layer (sometimes interpreted as two or three rings per layer). The somewhat random ordering of the layers, as depicted in Fig. 19.20, results in relatively large pores. This figure also illustrates the ordered layers cross-linked by the X-ray amorphous material.

As coal rank rises up to about 90% C (maf), the open structure contracts to some degree. The layer or lamella size increases slightly up to that point, but then increases rapidly, jumping from about 5 rings per layer at 90% C to about 30 at 95% C (maf). The structure converts to a much more ordered "anthracitic" structure, passing through a "liquid" transition structure at 90% C (Fig. 19.20).

The dimensions of this liquid structure are of interest. There are groups of layers about 8 Å across, and about 20 Å apart. These dimensions also relate to the size of the pores. This structure corresponds to about 30% (maf) volatile matter, and to the minimum in internal surface area illustrated in Fig. 19.3. Density is also at a minimum at this point, however, which suggests that there is substantial blind pore volume that may become available on pyrolysis during combustion.

Details of the cross-linked layer structure are still substantially speculative, but Fig. 19.21 illustrates one view,[40,41,105,150,152] showing the linked aromatic and hydroaromatic units, the cross-links being of various types that will break and/or re-form under different conditions. This illustration is intended to be indicative, not definitive, as discussed by Scheer,[155] but it does provide some possible insight into behavior on pyrolysis. As described by Walker[149] (see also Essenhigh[41]), the first stage of pyrolysis, beyond removal of water and CO_2, involves breaking of some cross-links, and removal of some of the functional edge groups as volatile matter (identified in some evaluations[156–158] as the Clark and Wheeler[159] "most-easily" evolved component). At higher temperatures the hydroaromatic groups lose hydrogen (the second volatile component of the two-component hypothesis of coal constitution[159]). They also form additional aromatic units that, with others, can then form small carbon crystallites or graphitize.

[149] P. L. Walker, Jr., "The Scientific Base of Coal Gasification," *Ind. Heat.*, **44**(11), 16–20 (1977).

[150] P. B. Hirsch, "X-Ray Scattering from Coals," *Proc. R. Soc. London*, **226A**, 143 (1955).

[151] R. Diamond, "X-Ray Studies of Some Carbonized Coals," Ph.D. thesis, University of Cambridge, Cambridge (1956).

[152] P. B. Hirsch, "Conclusions from X-Ray Scattering Data on Vitrain Coals," *Proc. Conf. Science in the Use of Coal* Paper No. 1, Institute of Fuel (London), 1958, pp. A29–A33.

[153] L. Cartz and P. B. Hirsch, "A Contribution to the Structure of Coals from X-Ray Diffraction Studies," *Proc. R. Soc. London*, **252A**, 557–602 (1960).

[154] L. B. Lyon and G. R. Hill, "New Chemical Structure of Coal—Why Not a Coal Refinery?" *Ind. Eng. Chem.*, **54**, 36 (1962).

[155] M. Scheer, in *Materials Problems and Research Opportunities in Coal Conversion,* Vol. 2, Ohio State University, 1974.

[156] R. H. Essenhigh, "Constitution, Rank Index, and Inflammability of Coals," *Fuel*, **34**, 497–500 (1955).

[157] J. L. Ayre and R. H. Essenhigh, "Carbon Ring Structure in Coals: A Theoretical Study," *Sheffield Univ. Fuel Soc. J.*, **8**, 44–59 (1957).

[158] R. H. Essenhigh and J. B. Howard, *Combustion Phenomena in Coals and the Two-Component Hypothesis of Coal Constitution,* Pennsylvania State Univer. Studies Monogr. No. 31, (1971).

[159] A. H. Clark and R. V. Wheeler, "The Volatile Constituents of Coal, Part III." *J. Chem. Soc.*, **103**, 1704 (1913).

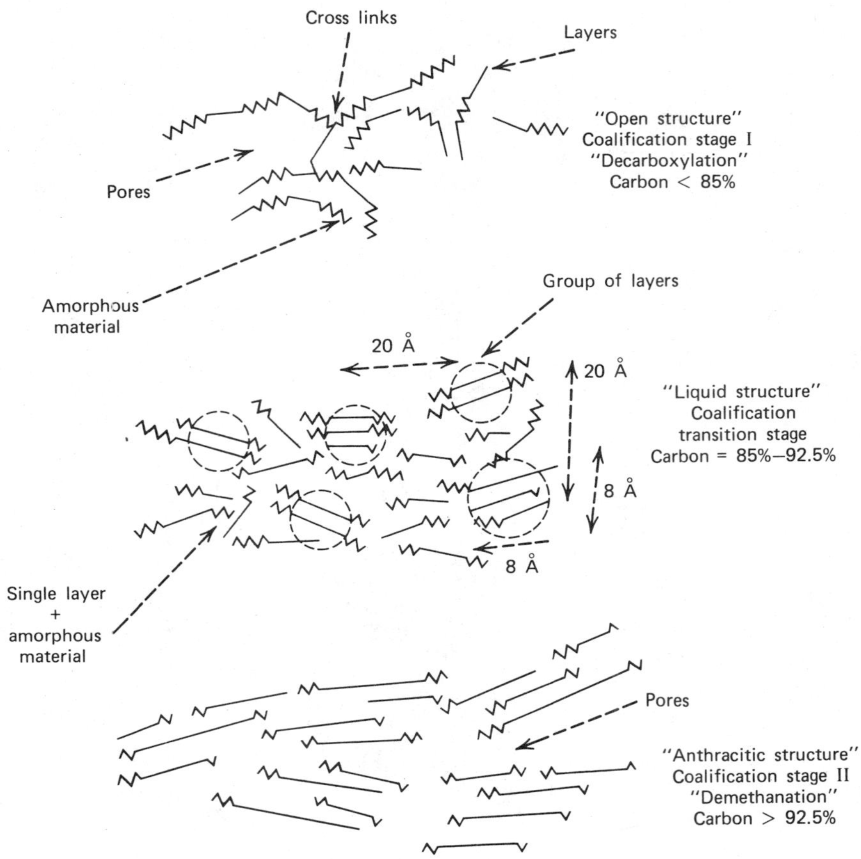

Figure 19.20 Schematic illustration of coal structure during coalification. After Hirsch[150,152] and Mahajan and Walker.[105] See also Essenhigh.[41]

19.3.2.2 *Volatiles Production: "True" VM Definition and the* Q *Factor*

As described earlier, the extent of cross-linking can affect the fluidity or mobility of pyrolyzing material, and this can affect the rate of escape of the volatiles. Light preoxidation enhances cross-linking and reduces mobility. Cracking of the volatiles can occur in the particle pores during escape or on the surface of other particles after escape. The standard ASTM VM test will yield reproducible results if followed exactly, but it corresponds to a standardized degree of volatiles cracking and capture. If the volatiles are removed with sufficient speed to prevent cracking and capture, the concept of "true" VM content emerges. If the concept is accepted, measurement of the "true" VM content of a coal would require satisfaction of the following criteria: a particle size small enough to prevent internal capture or cracking (probably $< 50\ \mu m$); densities of particle dispersions in gas low enough to prevent external carbon capture ($< 10^{-5}$ g/cm^3); "instantaneous" heating (exceeding at least 10^4 °C/s), to prevent reaction during heatup; and isothermal pyrolysis for "infinite" time at some temperature above which there is no further significant weight loss (excluding total dissociation and/or vaporization or sublimation of the residue): this may be 1000–1200°C. These conditions are approximately met in some types of experiment when particles are trapped in a heated wire mesh, or entrained and carried through an electrically heated furnace. When the conditions are not met, as in the ASTM VM test, anomalous VM values are obtained. In the case of the ASTM measurements, a proportional relationship has been found between weight loss in entrained particle experiments

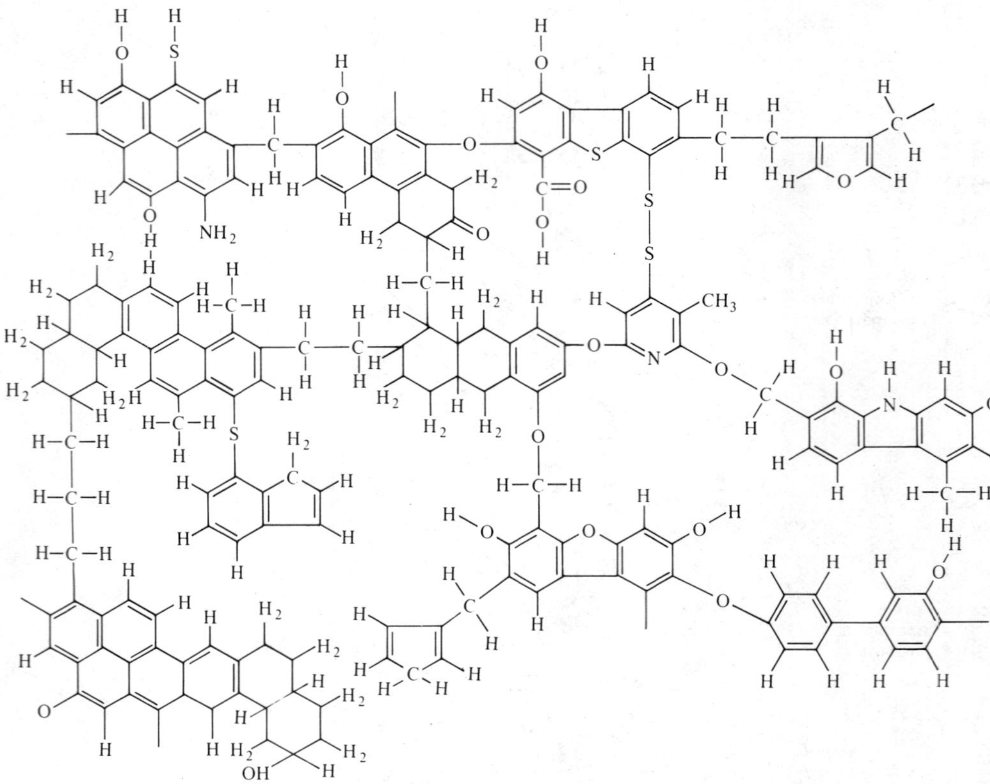

Figure 19.21 Postulated structure of coal.[154]

and ASTM-VM in the particle residues.[67,160] The ratio of the actual to apparent (from ASTM analysis) weight loss is known as the *Q* factor, with measured values in the region of 1.5–2. This ratio has been interpreted[38] as a factor related to the capture of VM by the char in the ASTM analysis procedure, and it provides a multiplier to obtain an estimate of the "true" VM value from the ASTM measurements. The capture factor is the reciprocal of the *Q* factor. Thus a *Q* factor of 2 is equal to a capture factor of 0.5. The capture factor is thought[38] to be made up of three effects: internal pore capture, external surface capture, and internal capture at such relatively slow heating rates that there would be time for a fundamental reordering or crystallization of the coal structure that binds material that would otherwise be lost at faster heating rates.

Conditions deviating from the "true" pyrolysis requirements as defined above will therefore generate a *Q* factor. There will also be a shift in the H/C ratio of the VM and of the char. Since carbon is removed from the volatiles by cracking, the volatiles formed in the true pyrolysis measurements will be richer in carbon than when formed under any other conditions. This is illustrated in Fig. 19.22 (Kobayashi data,[161] quoted by Ubhayaker[162]), which shows the variation of the H/C ratio in char for two coals measured at low and high density particulate conditions (entrained-flow and crucible measurements, respectively). Since the *average* temperature at which the volatiles are formed in the "true" VM measurements will also be equal

[160] N. Nsakala, R. H. Essenhigh, and P. L. Walker, Jr., "Studies on Coal Reactivity: Kinetics of Lignite Pyrolysis in Nitrogen at 808°C," *Combust. Sci. Technol.*, **16**, 153–163 (1977).

[161] H. Kobayashi, "Kinetics of Rapid Devolatilization of Pulverized Coal," Sc. D. thesis, Dept. Mechanical Engineering, Massachusetts Institute of Technology, Cambridge, 1976.

[162] S. K. Ubhayakar, *Pyrolysis of Coal*, AVCO-Everett Res. Lab. Inc. Rep. No. AMP 538, July 1977.

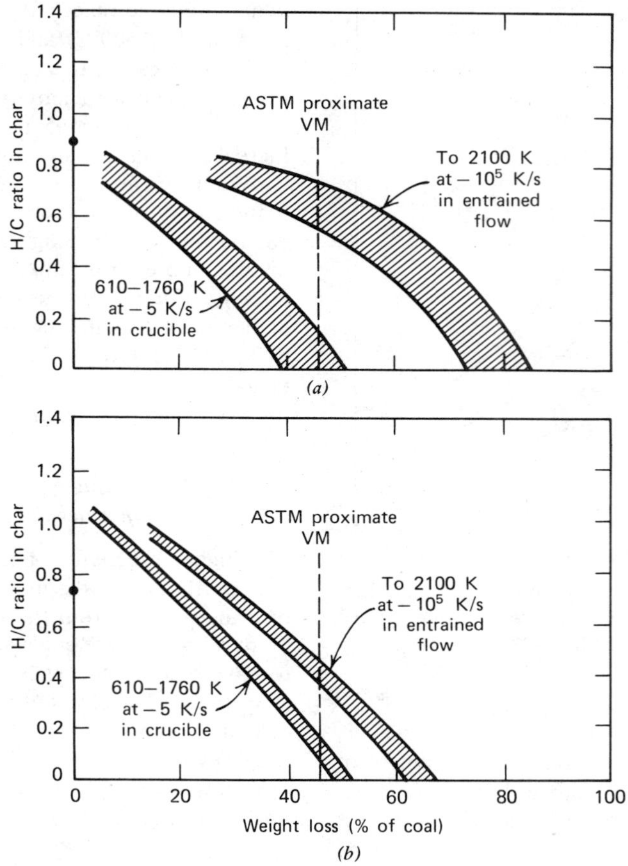

Figure 19.22 H/C ratios in chars from two coals: (*a*) Pittsburgh Seam bituminous, 58 μm and (*b*) Montana lignite, 48 μm.[161,162]

to or higher than it would be under any other conditions, secondary cracking, with or without carbon formation, can result in a lower mean molecular weight of the products, even though they may be richer in carbon. (The spread in molecular weights will be smaller.)

19.3.2.3 *Heat Treatment and Reactivity*

The structure, density, and porosity of the partially graphitized char, and therefore its reactivity, depend very much on the parent coal and on the heat treatment conditions. The rate of heating, the final heat treatment temperature (HTT: see, e.g., Refs. 163–167), and the time it is held at that temperature are particularly important. These variables affect the degree of alignment of the aromatic units as they form the carbon crystallites, which in turn affects the proportion of reactive edge sites; it also affects the pore structure that allows access to the interior. The larger and more regular the crystallites, the smaller the proportion of reactive edge atoms, and the smaller the access pore diameters, which depend largely on the degree of misalignment of the constituent units. Reactivity is, therefore, reduced by increased HTT and by longer treatment time, both of which increase crystallite size. The effect of HTT on char reactivity is illustrated in Fig. 19.23.

Crystallite size in the lower rank coals, below 90% C (maf), tends to be smaller on account of the smaller initial layer or lamella size. In the highest rank coals the lamella size is already large, the access pore volume is small, and the percentage of volatile matter that could open up pore volume is also small. This leads to lowest char reactivity. As carbon percentage drops, initial lamella size drops and pore size increases. The coals then

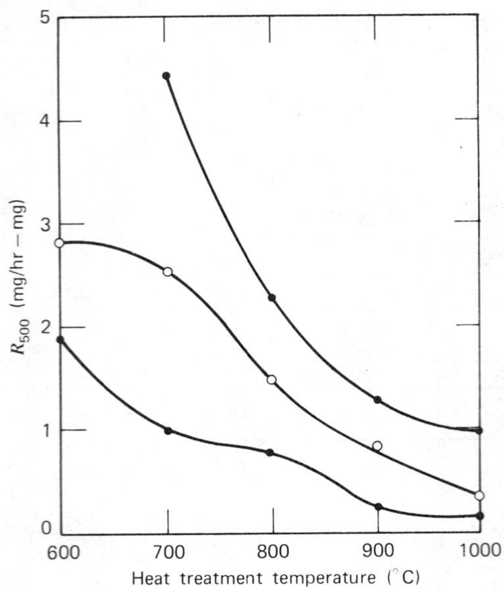

Figure 19.23 Effect of HTT on reactivity of chars in air at 500°C. Parent coals: ● = lignite, ○ and ■ = high-volatile bituminous.[164]

become increasingly thermoplastic, however, and the increased mobility aids alignment of the crystallites and helps to hold down reactivity. This is illustrated in Fig. 19.24,[164,165,166] which plots the reactivity of chars to oxygen (measured at 500°C) and to CO_2 (measured at 900°C) against the carbon percentage of the parent coals. Reactivity about doubles as carbon drops from 95 to 80%, but it then increases by nearly an order of magnitude as carbon drops further to 70%.

Reactivity of the thermoplastic coal chars can be improved by a light preoxidation. This creates more cross-linking, which reduces mobility during pyrolysis and also reduces alignment. Cross-linking is evidently assisted by —COOH groups, which are present in coals of less than 80% C. This is apparently important in lignite char formation.

19.3.2.4 *Catalytic Effects of Impurities in Coal*

The higher reactivity of the lignite chars is only partly due to the more open char structure that results from the large loss of mass on pyrolysis. These coals and chars also tend to have high impurities, particularly calcium (Fig. 19.25). The effect of removing or adding calcium is shown in Fig. 19.26: the influence is self-evident.

Variation in surface area is also a factor. The N_2, or access pore surface area also rises

[163] D. F. Williams and J. A. Pudifoot, "Chemical Reactivity of Carbonized Fuels," *Proc. Conf. Science in the Use of Coal,* Institute of Fuel, Paper No. 34, (1958), pp. E14–E21.

[164] R. G. Jenkins, S. P. Nandi, and P. L. Walker, Jr., "Reactivity of Heat-Treated Coals in Air at 500°C," *Fuel,* **52** 288–293 (1973).

[165] E. J. Hippo and P. L. Walker, Jr., "Reactivity of Heat-Treated Coals in Carbon Dioxide at 900°C," *Fuel,* **54,** 245–248 (1975).

[166] E. J. Hippo and P. L. Walker, Jr., *Effect of Cation Exchange on the Subsequent Reactivity of Lignite Chars to Steam,* Tech. Rep. No. 4 to ERDA, Contract No. E (49-18)-2030, 1977.

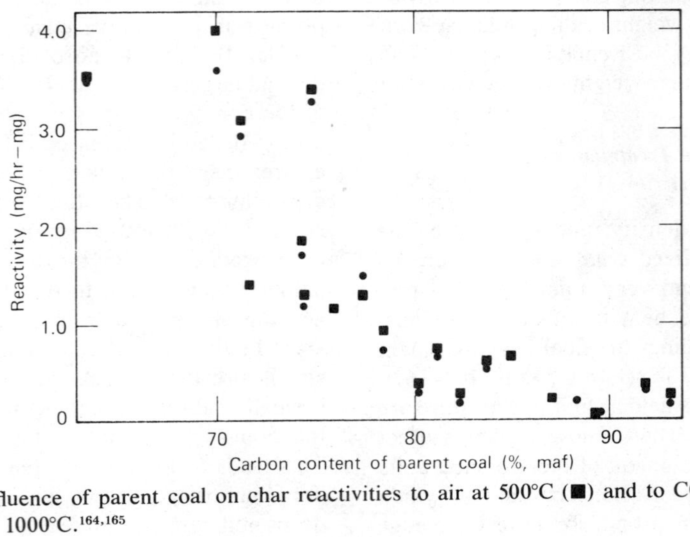

Figure 19.24 Influence of parent coal on char reactivities to air at 500°C (■) and to CO_2 at 900°C (●); chars formed at 1000°C.[164,165]

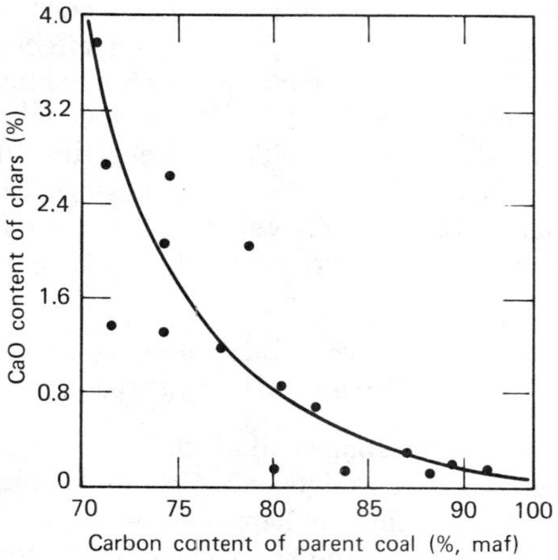

Figure 19.25 Correlation between CaO content of chars and carbon content of parent coals.[149]

dramatically as the percentage of carbon falls (Fig. 19.27). The resultant reactivity variations shown in Fig. 19.24 are therefore assumed to be largely due to the changes in Ca content and N_2 surface area. [Note that particularly with chars, N_2 may measure more than the access pore area and can include some of the micropore area.]

19.3.2.5 *Influence of Rate of Heating and Volatiles Cracking on Reactivity*

Rate of heating also affects reactivity. It can do so in two ways. As particles heat and pyrolyze, the volatile products can crack and, particularly with particles larger than 100 μm, cracking may occur inside the pores

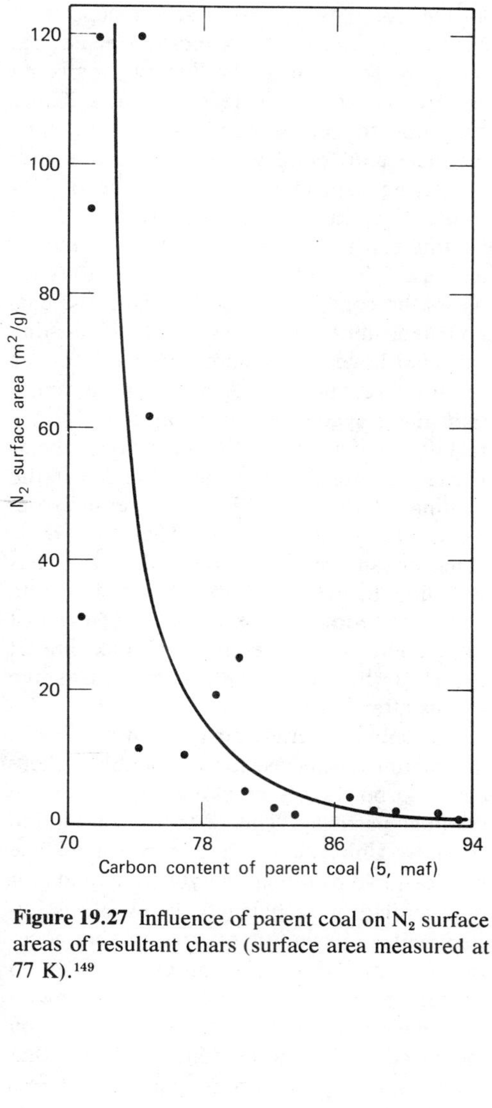

Figure 19.27 Influence of parent coal on N_2 surface areas of resultant chars (surface area measured at 77 K).[149]

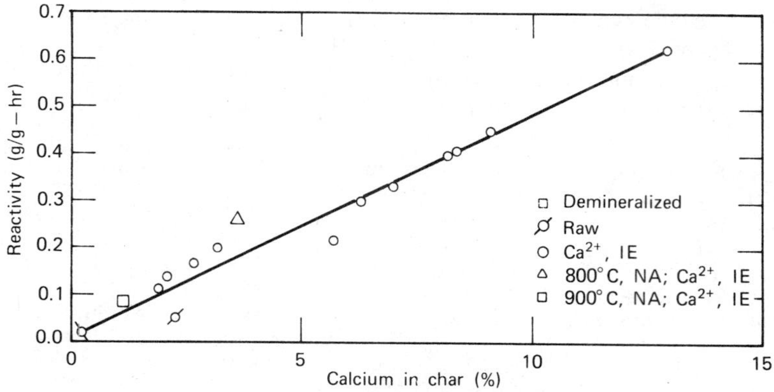

Figure 19.26 Influence of Ca^{2+} on reactivities of chars in steam at 650°C (chars formed from lignite at 800°C). Calcium variation was obtained by demineralization, to reduce it; and by ion exchange (IE) and by reaction following nitric acid (NA) treatment to increase it.[165,166]

as the volatiles escape. Carbon deposition can reduce reactivity (Kamishita et al., reported by Mahajan and Walker[105]). Kamishita studied the change in reactivity of a lignite char due to carbon deposited by cracking methane and found a steady decrease with increasing deposition. The cracking of VM inside pores can have the same effect. Rapid heating can reduce any such effect, since it increases the volatiles release rate, thus reduces the contact time for cracking and carbon deposition in the pores. Rapid heating can also have some more direct effects, as reported recently by Ashu.[168] Ashu found that the oxygen reactivity at 500°C (R_{500} reactivity) of a char that was previously heat-treated at 800°C was nearly doubled if the heating rate to the HTT was increased from 0.2°C/s to about 10^4 dgC/s. There was even some enhancement (~ 10%) when the char was first heated to 800°C at the slow rate and, after cooling, was reheated at the fast rate. There was some effect of soak time at the HTT, but reactivity became insensitive to this after about 30 min.

Not only may cracking affect internal pore deposition during escape of volatiles, there can also be further cracking on the surface of other particles. This is the interpretation given by Ubhayakar et al.[169] to an imbalance between two different analytical methods for determining mass balances in pyrolysis experiments. Missing mass—reasonably accurately identified as soot and cracked carbon on other particles—increased as a curve with declining slope as the particle concentration increased. Since the evidence indicates that such deposited carbon is significantly less reactive than the char substrate, the deposits can significiantly reduce reactivity in some cases, as observed by Kamishita. However this reduction may be of importance only in gasification reactions. In some current stirred-reactor experiments, Goldberg[69] has suggested that there is no such effect in combustion, even under fuel-rich conditions (equivalence ratio of 2). These combustion results are in line with the belief that the volatiles burn very rapidly, in which case there would be no volatiles available for cracking on the surface of other particles.

19.3.2.6 *Swelling and Cenosphere Formation During Pyrolysis*

As a further complication, the reactivity of the char is itself influenced by the ambient conditions. This can happen in two ways. As the particles pyrolyze, many swell to form blown-up "cenosphere" structures. The extent of swelling, like the structure of the swollen particle, is again a function of the rate of heating and the final temperature, but also of the ambient gas composition. These all precondition the char before the same factors help to control the reactivity on burnout as discussed above. Sinnatt and coworkers[170–173] are still usually quoted as the best source of information on swelling, with recent substantiation by Street et al.[174] (see also Littlejohn[175]).

The salient points[170–173] were that true cenosphere formation (an open-work porous structure of the char) occurred only under neutral or reducing conditions, not during oxidation, with maximum swelling occurring at 700°C. Overall swelling at higher temperatures was much less, particularly under oxidizing conditions.

Particle structures are also said to vary

[167] A. Linares, O. P. Mahajan, and P. L. Walker, Jr., "Reactivities of Heat-Treated Coals in Steam," *Prepr., Am. Chem. Soc., Div. Fuel Chem.*, **22**(1), 1–4 (1977).

[168] J. T. Ashu, N. Nsakala, O. P. Mahajan, and P. L. Walker, Jr., "Enhancement of Char Reactivity by Rapid Heating of Precursor Coal," *Fuel*, **57**, 250, 251 (1978).

[169] S. K. Ubhayakar, D. B. Stickler, C. W. v. Rosenberg, and R. E. Gannon, "Rapid Devolatilization of Coal in Hot Combustion Gases," *Proc. 16th Symp. (Int.) Combustion*, Combustion Institute, Pittsburgh, 1977, pp. 427–436.

[170] H. E. Newall and F. S. Sinatt, "The Carbonization of Coal in the Form of Fine Particles. I. The Production of Cenospheres," *Fuel*, **3**, 424 (1924).

[171] H. E. Newall and F. S. Sinatt, "The Carbonization of Particles of Coal in Air. II. The Study of Cenospheres," *Fuel*, **5**, 335 (1926).

[172] H. E. Newall and F. S. Sinatt, "The Carbonization of Particles of Coal. III. The Study of Cenospheres," *Fuel*, **6**, 118 (1927).

[173] A. MacCulloch, H. E. Newall, and F. S. Sinnatt, "The Carbonization of Particles of Coal: The Study of Cenospheres, Part 5," *J. Soc. Chem. Ind.*, **46**, 33IT (1927).

[174] P. J. Street, R. P. Weight, and P. Lightman, "Further Investigations of Structural Changes Occurring in Pulverized Coal Particles During Rapid Heating," *Fuel*, **48**, 343–365 (1969).

[175] R. M. Littlejohn, "Pulverized-Fuel Combustion: Swelling Under Rapid Heating," *J. Inst. Fuel*, **40**, 128 (1967).

considerably, from uniform porous particles at one extreme, to thin shelled, hollow spheres at the other. Street et al.[174] reported four different types: lacy spheres, balloon shaped, C-shaped, and solid. There was some dependence on the petrographic composition. The C-shaped and lacy spheres were considered most likely to contribute to unburned losses in flames. Alpern et al.[176] used electron microscopy to examine flame-formed particles and reported finding an intermediate semicoke stage before the coke cenospheres were formed; they also found evidence of swelling and *coking* of a "non-coking" coal.

If these extreme variations are really possible as a function of different firing conditions, and do not just occasionally happen, the variation in reactivity on that account could be substantial. But what is found by flame sampling may be more representative of what is not burning than of what is burning. There is scope here, however, for more detailed flame investigations.

Other experiments[174,175,177–181] seem to offer agreement that there is no correlation between standard swelling numbers and the degree of swelling of single particles or clouds in flames and furnaces. Essenhigh and Yorke[177] found a wide range in swelling factors of single particles (mostly of about 1 mm), but the average of many measurements was about the same (1.5) for nine of ten coals studied, the tenth coal being an anthracite that did not swell at all. The swelling factor was determined at the end of a period of initially greater swelling followed by contraction, as also reported by Shibaoka.[179] Lightman et al.[181] found that swelling decreased with particle size; Littlejohn[175] found it to be a function of time, temperature, and rate of heating; Street[174] substantiated Sinnatt's findings that it was also a function of ambient atmosphere, with swelling factors of 3.9 in N_2 reduced to 1.3 in air (close to Essenhigh and Yorke's value of 1.5). Shibaoka found greater expansion the more rapidly the particles were heated, which is evidently the reverse of the trend reported by Brookes.[178]

It is also evident that pyrolysis can directly affect reactivity. This was most clearly demonstrated by Johnson[182,183] in studies of hydrogenation (Chapter 23). As illustrated in Fig. 19.28, a comparison was made of weight loss as a function of time for coal heated first, in N_2 alone to determine the pyrolysis behavior (curve *A*); second, in N_2 followed by H_2 (curve *B*); and third, in H_2 at two different pressures (curves *C* and *D*).

The weight loss of the coal pyrolyzed in hydrogen (curve *C*) was about twice the loss of the coal that was first pyrolyzed in nitrogen before undergoing hydrogenation at the same pressure (curve *B*). The effect occurred in the first minute or so. It is attributed either (1) to higher chemical activity of the char as it is formed, sometimes described as "nascent site" formation, before the aromatic units have time to align and cross-link, or (2) to hydrostabilization of volatile radical fragments (tar-forming components) that would otherwise crack and deposit carbon in the presence of N_2. In the first case it is supposed that the coke is more reactive; in the second case less coke is formed than in the first instance. The second explanation is more in line with other phenomena, such as enhanced volatile yield (the *Q*-factor effect) on rapid heating of dilute suspensions.

19.3.3 Combustion of Coal

An overall picture of burning coal particles in a flame is presented in Sections 19.3.3.1–19.3.3.4.

[176] B. Alpern, P. Courbon, J. Plateau, and G. Tissandier, "Microscopic Examination of Samples Taken from a Pulverized-Fuel Flame," *J. Inst. Fuel,* **33,** 399–402 (1960).

[177] R. H. Essenhigh and G. C. Yorke, "Reaction Rates of Single Coal Particles: Influence of Swelling, Shape, and Other Factors," *Fuel,* **44,** 177–186 (1965).

[178] F. P. Brookes, "The Combustion of Single Captive Particles of Silkstone Coal," *Fuel,* **48,** 139–149 (1969).

[179] M. Shibaoka, "An Investigation of the Combustion Processes of Single Coal Particles," *J. Inst. Fuel,* **42,** 59–66 (1969).

[180] P. Lightman and P. J. Street, "Microscopical Examination of Heat Treated Pulverized Coal Particles," *Fuel,* **47,** 7–28 (1968).

[181] P. Lightman, P. J. Street, and R. P. Weight, "Swelling Test and Swelling Index," *J. Inst. Fuel,* **40,** 433 (1967).

[182] J. L. Johnson, "Kinetics of Bituminous Coal Char Gasification with Gases Containing Steam and Hydrogen," in *Coal Gasification,* Am. Chem. Soc. Adv. Chem. Ser. No. 131 (1974), pp. 145–178.

[183] J. L. Johnson, "Relationship Between the Gasification Reactivities of Coal Char and the Physical and Chemical Properties of Coal and Coal Char," *Prepr., Am. Chem. Soc., Div. Fuel Chem.,* **20**(4), 85–96 (1975).

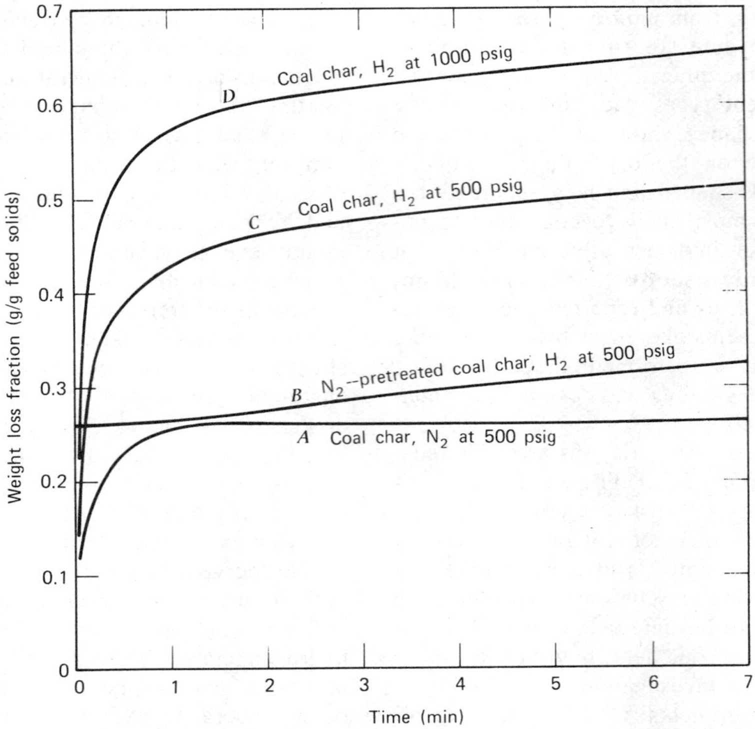

Figure 19.28 Influence of different methods of treatment on the variation of weight loss with time for a coal injected into N_2 (curve *A*) or H_2 (curves *B–D*) at 1700°F. Curve *B* shows the effect of prior treatment in N_2 (curve *A*) before injecting into H_2, compared with behavior without pretreatment (curve *C*). Curve *D* shows the effect of increased pressure.[182,183]

19.3.3.1 *Ignition*

As particles enter the flame, exposed to oxidizing conditions from the primary carrier air, they start to pyrolyze as they heat up, in competition with oxidation. With lump particles on a grate, the heating rate is so slow (Table 19.1) that pyrolysis sets in long before gas-solid reactions can start. With particle sizes outside the p.c. range, and sometimes inside that range, pyrolysis is so rapid that the particle surface is flushed clear of reactive gas.

In an oxidizing atmosphere, with sufficient heat, the volatiles can burn. As their evolution and combustion die down, reactive gas is able to penetrate to the char surface and begin reaction as described above. With crushed particles in a fluid bed,[184] which heat at about 10^3 °C/s, or with p.c. particles fired under conditions such that the heating rate is about the same as in the fluid bed,[40] pyrolysis also starts before combustion.

In a plug-flow p.c. flame, with particles heating at about 10^4 °C/s as they enter, oxidation of the (coal) surface is faster than pyrolysis[73] and the coal initially ignites heterogeneously, on the solid surface. There is reason to believe that this can happen in most p.c. flames. Pyrolysis normally starts at 300–400°C, but it is delayed under conditions of rapid heating.[185] Howard and Essenhigh[73] measured ignition temperatures in plug-flow p.c. flames of about 1000°C, with pyrolysis starting at 1200°C. These values can be compared with an estimated ignition temperature of 370°C—roughly the normal pyrolysis temperature—in Csaba's[186] very similar experi-

[184] M. M. Avedesian and J. F. Davidson, "Combustion of Carbon Particles in a Fluidised Bed," *Trans. Inst. Chem. Eng.*, **51**, 121–131 (1973).

[185] V. H. Jüntgen and K. H. van Heek, "Gas Release from Coal as a Function of Heating Rate," *Fuel*, **47**, 103–117 (1968).

[186] J. Csaba, "Flame Propagation in a Fully Dispersed Coal Dust Suspension," Ph.D. thesis, Dept. of Fuel Technology and Chemical Engineering, University of Sheffield (England), 1962.

ments. In some instances, however, diffusion flames are formed with the particles pyrolyzing downstream of the flame front (Section 19.3.3.3).

19.3.3.2 *Influence of Pyrolysis on Ignition*

In the experiments described above the reaction time for heterogeneous reaction before onset of pyrolysis was about 20 ms, a duration that allowed detection of mass loss. At still higher heating rates, probably reaching 10^6 °C/s, in a stirred reactor[69] operating at 1200°C, which approximates heating rates and temperatures in an explosion flame,[99] no prior heterogeneous reaction was detected. If ignition started heterogeneously, however, there would only be about a tenth of a millisecond before the heterogeneous reaction was quenched by the volatiles flush.

The volatiles flush does not necessarily quench the heterogeneous reaction in all cases, however. If the pyrolyzing particle behaves like a burning oil drop, there will be a volatiles flame around the particle at a standoff distance determined by the rate of volatiles escape. It is a matter of calculation[73] to show that there is a limiting rate of volatiles evolution below which the circumambient flame has regressed to the surface and there is a surplus of oxygen penetrating the flame region that is able to continue the heterogeneous reaction. The limiting rate depends on particle diameter, and in one set of experiments[73] an experimental estimate of 65μm compared quite well with a calculated value of 30μm.[73] It was also estimated from the experimental results that the continuing heterogeneous reaction was of minor account for particles down to about 15 μm, but significant for particles smaller than this size.

The heterogeneous ignition of the coal particles, before the volatiles appeared, suggested that ignitability was not conferred on the coal particles by the volatiles per se: there is an undoubted variation in both ignitability and explosibility between coals of different rank. One can conclude, however, that the volatiles are indices of some other parameter, such as coal particle reactivity.

Recent experiments[87] using coal chars have substantiated this general conclusion. With minimum VM in the chars, a wide range of reactivity was found, with two chars having as high a reactivity and flame speed as bituminous coal. High VM content is not necessary for rapid ignition. The determining factors seem rather to be internal surface and access to that surface. The chars investigated were found to have either high reactivity or low reactivity, with markedly different ignition times: about 0.1 and 1 s, respectively. Parent coal was not a factor.

Ashu[168] carried out comparative reactivity tests on the same chars, determining the R_{500} reactivity, and found that the results ordered the chars in the same way as did the furnace experiments, but the variation in reactivity was quite small. Neither was there any marked variation in the N_2 surface area. There seemed to be two factors of importance: (1) the percentage conversion of fixed carbon that occurred in the original gasification treatment, which evidently opened up the internal surface, and (2) the volume of pores wider than 130 Å, which seems to be a better index of interior accessibility than the N_2 surface area.

19.3.3.3 *Small (Diffusion) Flames*

The foregoing descriptions apply to flames of some size, with cross sections of 15 cm or more, contained in hot wall furnaces to maintain temperature for burnout. In such one-dimensional furnace systems with hot walls, flame speeds ranged from 50 to 150 cm/s. In many small flames,[75,76,187–197] however, the

[187] J. H. Burgoyne and V. D. Long, "Some Measurements of the Burning Velocity of Coal-in-Air Suspensions," *Proc. Conf. Science in the Use of Coal,* Paper No. 28, Institute of Fuel (London), 1958, pp. D16–D20.

[188] W. F. Marshall, H. B. Palmer, and D. J. Seery, "Particle Size Effects and Flame Propagation Rate Control in Laminar Coal Dust Flames," *J. Inst. Fuel,* **37**, 342–349 (1964).

[189] H. Hattori, "Flame Propagation in Pulverized Coal-Air Mixtures," *6th Symp. (Int.) Combustion,* Paper No. 74, Reinhold Publishing Corp., New York, 1957, pp. 590–595.

[190] H. M. Cassel, A. K. Das Gupta, and S. Guruswamy, "Factors Affecting Flame Propagation Through Dust Clouds," *Proc. 3rd Symp. (Int.) Combustion,* Williams and Wilkins Company, Baltimore, 1948, pp. 185–190.

[191] K. Saji, "Combustion Rates of Pulverized Coal in a Jet Stream," *Proc. 5th Symp. (Int.) on Combustion,* Reinhold Publishing Corp., New York, 1955, pp. 252–260.

[192] B. Ghosh, D. Basu, and K. K. Roy, "Studies of Pulverised Coal Flames," *6th Symp. (Int.) Com-*

thermal load on the tail of the flame is too high to maintain temperature for burnout. Thermal feedback to the incoming fuel is then so far reduced that the flame can be stabilized only at velocities of about 10 cm/s or slower. These have now been found to be diffusion flames[75] by a computer analysis of heating and reaction rates in the flame. The coal particles were unable to heat fast enough to pyrolyze at the leading edge of the flame. Pyrolysis occurred downstream, and the volatiles diffused upstream to form the diffusion flame. These flames have provided some additional information on pyrolysis but nothing on the char reactivity.

19.3.3.4 *Char Burnout*

Char burnout is critically influenced by porosity, with the result that the particles may burn in one of two extreme modes, or at a point somewhere in between. One extreme is reaction at the exterior surface, so that the particles shrink in diameter but at constant density. The other is reaction at all the internal surface of the particle, at constant diameter but decreasing density.

Experimental evidence supports both the extreme models and intermediate behavior under different conditions of reaction. In addition, particles larger than 100 μm in fluid beds or on grates have reactions that are modified by diffusional boundary layers; there is also evidence[89,95,96] that the boundary layer can influence particles as small as 10 μm at high enough temperatures (above 2000°C). This observation supports the conclusions of Section 19.2.2.

In p.c. flames Cogoli et al.[87] reported different behaviors of bituminous coal, high-reactivity char, low-reactivity char, and anthracite. The bituminous coal and high-reactivity char showed considerable insensitivity to particle size; the low-reactivity char showed considerable dependence on particle size; the anthracite was critically dependent on particle size.

The most comprehensive series of flame experiments in furnaces, however, has been carried out by Smith and co-workers (Smith,[198–200] Smith andTyler,[201,202] Ayling and Smith,[84] Sergeant and Smith,[97] Hamor, Smith, and Tyler[98]). Some of their experiments included comparisons with fixed-bed behavior (see Fig. 19.29). As Smith and Tyler[202] pointed out, the conclusions of that last paper in the series differ from those of the earlier ones. This would appear to be a joint consequence of a different method of mathematical analysis, improved experimental technique, and different fuels. Fuels used included anthracite, semianthracite, bituminous coal char, and brown coal char. The clearest cut results with the most sophisticated analysis were for the brown coal char. The particles burned with the reaction penetrating only partly into the solid (the Zone II described in Section 19.3.1.2), with the reaction rate determined jointly by the kinet-

bustion, Paper No. 75, Reinhold Publishing Corp., New York, 1957, pp. 595–602.

[193] H. M. Cassel, I. Liebman, and W. K. Mock, "Radiative Transfer in Dust Flames," *6th Symp. (Int.) Combustion,* Paper No. 76, Reinhold Publishing Corp., New York, 1957, pp. 602–605.

[194] J. T. Bryant, "The Combustion of Premixed Laminar Graphite Dust Flames at Atmospheric Pressure," *Combust. Sci. Technol.,* **2,** 389–399 (1971).

[195] J. T. Bryant, "Powdered Fuel Combustion: Mechanism and Particle Size," *Combust. Flame,* **20,** 138–139 (1973).

[196] T. A. Milne and J. E. Beachey, "The Microstructure of Pulverized Coal-Air Flames. I. Stabilization on Small Burners and Direct Sampling Techniques," *Combust. Sci. Technol.* **16,** 123–138 (1977).

[197] T. A. Milne and J. E. Beachey, "The Microstructure of Coal-Air-Flames. II. Gaseous Species, Particulate and Temperature Profiles," *Combust. Sci. Technol.* **16,** 139–152 (1977).

[198] I. W. Smith, *Studies of the Combustion Rate and Structure of Pulverized Fuels Burned in the C.15 Trials (1969) of the International Flame Research Foundation.* Part 1. *Measurement and Analysis of the Combustion Rate on Anthracite,* Div. Miner. Chem. CSIR Invest. Rep. No. 86, Chatswood, New South Wales, Australia, 1970.

[199] I. W. Smith, "Kinetics of Combustion of Size-Graded Pulverized Fuels in the Temperature Range 1200–2270°K," *Combust. Flame,* **17,** 303–314 (1971).

[200] I. W. Smith, "The Kinetics of Pulverised Semi-Anthracite Particles in the Temperature Range 1400–2200°K," *Combust. Flame,* **17,** 421–428 (1971).

[201] I. W. Smith and R. J. Tyler, "Internal Burning of Pulverized Semi-Anthracite: The Relation Between Particle Structure and Reactivity," *Fuel,* **51,** 312–321 (1972).

[202] I. W. Smith and R. J. Tyler, "The Reactivity of a Porous Brown Coal Char to Oxygen Between 630 and 1812°K," *Combust. Sci. Technol.,* **9,** 87–94 (1974).

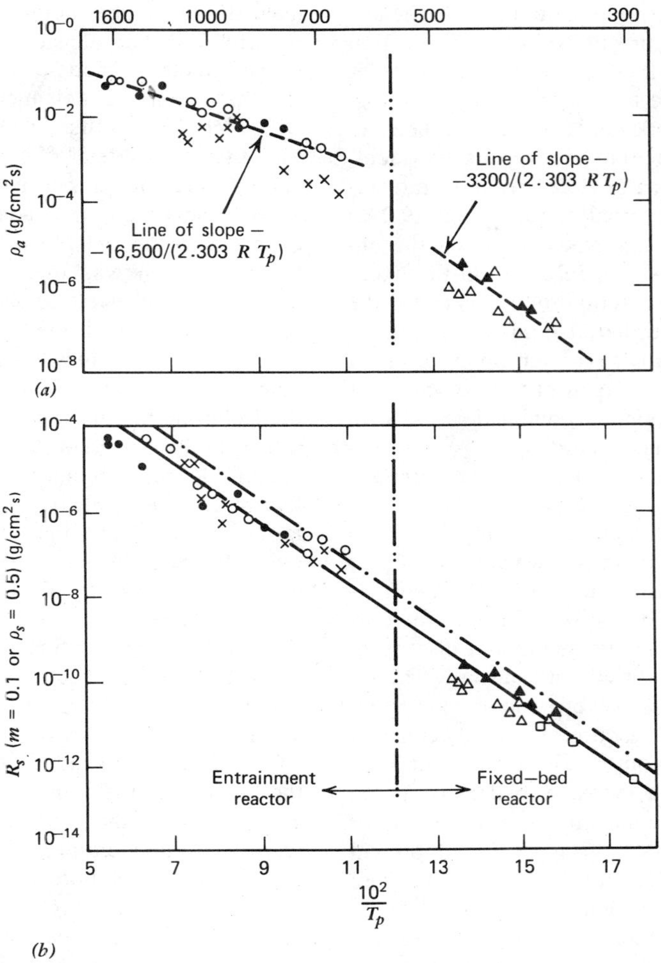

Figure 19.29 The dependence of rate on temperature. (*a*) Rate referred to superficial surface. (*b*) Rate referred to internal surface.

Initial Size of Particle (μm)	Entrainment Reactor	Fixed-Bed Reactor
89	●	▲
49	○	△
22	×	
1250	■	

$R_3 = 1.34 \exp[-32{,}600/(R\,Tp)]$ ($m = 0$); broken curve, ρ_s ($m = 0.5$).[202]

ics of diffusion into the pores and the chemical kinetic behavior at the pore walls.

For Zone II kinetics it can be shown (Section 19.5.4) that reaction order and activation energy as measured are apparent or effective values that differ from the true values. With some reservations, the authors[202] nevertheless concluded that the apparent reaction order for the brown coal char with respect to oxygen concentration was one-half, yielding a true order of zero (oxide film decomposition control). The apparent or effective activation energy was 16.5 kcal/mole, yielding a true activation energy for the zero-order reaction of 33 kcal/mole. This is a particularly significant result because of its excellent comparison with other data from a fixed-bed reactor in the temperature range below 900 K.

The remarkable agreement is illustrated in Fig. 19.29. The graphs show the variation of reaction rate with temperature on Arrhenius coordinates. Figure 19.29*a* is for the reaction rate referred to the superficial surface, yielding the effective activation energy. The discontinuity in slope between the fixed-bed and entrained (furnace) reactor results is evident. By comparison, the (true) reaction rate referred to the internal surface (Fig. 19.29*b*) also shows the agreement between the entrained and fixed-bed reactors. In the fixed-bed reactor the reaction fully penetrated the particles that burned at diminishing density (Zone I reaction). This is an opposite result to the finding of Eapen et al.[66] of constant density gasification in a combustion pot, with diameter declining accordingly.

The results of the earlier furnace work[84,97,98,198–201] are less definitive, but are nevertheless informative on many points. In all experiments there was a variation of particle density with burnoff that indicated partial penetration of the reaction into the particles, as in the experiments just discussed. Diameters also varied with burnoff. The bituminous char, however, was an exception: the data showed reaction at virtually constant density throughout.[97] There may, in fact, have been some increase in density at the start of the reaction, in line with an earlier finding by Field[95,96]; this is attributed to rapid burnoff of low-density, highly reactive fractions in the fuel sample.

In all other cases, however, the results were in agreement with the partial penetration model. Reaction was not necessarily accompanied by decreasing diameter, however. Diameter did decrease in tests on anthracite and semianthracite, but in experiments carried out by Hamor et al.[98] on brown coal char, the diameter was constant. An analysis by Sergeant and Smith[97] showed the constancy of diameter to be consistent with behavior of reaction of thin spherical shells with hollow centers.

In the prior work[84,97,98,198–201] the reaction order was not tested, but on the basis of Field's earlier work, was assumed to be unity except in the case of the brown coal char experiments,[98] when it was assumed to be half-order.

The results above are partly in agreement and partly at variance in some respects with both earlier and later work using mostly different experimental techniques. In earlier experiments both Field[95 96] and Street[174,196] used the drop-tube method. Anson et al.[203] reported reducing density in burnoff at constant diameter, a mode consistent again with the thin spherical shell model and possibly with Hamor's results, but at variance with the other fuels. Field[95,96] studied behavior of chars from a low-rank coal[95] and behavior of a set of seven different coals.[96] In the char experiments he established that boundary layer diffusion was quite significant as a factor affecting reaction, with the magnitude of the effect increasing with temperature. This result was somewhat in line with Hamor's experiments,[98] in which the ratio of the experimental to the predicted diffusional rate varied from 0.1 at 900°C to about 1 at 2000°C. In other findings concerning the effect of coal rank, Field reported that behaviors of swelling and nonswelling coals were quite different. In the nonswelling coals the diameter and density declined together; in the swelling coals the density first increased, then declined, as mentioned above. The rate coefficient of the reaction also varied with coal rank, changing by a factor of 8 between anthracite and the low-rank coal.

The most significant differences between the findings of Field and those of the Smith group were in the reaction order and the response to temperature. Field's reaction order was unity for both the preformed char and the flame-formed chars from the coals, compared with the most definitive value reported by Smith and Tyler[202] of one-half for the apparent order and zero for the true order. This difference is presumably due to the different materials and possibly the different methods of experiment. They need to be reconciled; it should be noted, however, that Smith and Tyler[202] were able to predict Field's results from their equations to within a factor of 2. This was possible even though Field's temperature variation did not fit a regular Arrhenius expression. The reaction rates instead were proportional to the temperature, a result shown by Gray and et al.[38] to be consistent with a full Langmuir with a low activation energy for the adsorption step (see Section 19.5).

In summary, it may be said that a pattern

[203] D. Anson, F. D. Moles, and P. J. Street, "Structure and Surface Area of Pulverized Coal During Combustion," *Combustion Flame*, **16**, 265–274 (1971).

of increased reactivity with decreasing coal rank has been established, in accordance with conventional assumption and expectation, but with some degree of recognition from Field's work of the difference between swelling and nonswelling coals. All the data support a contribution to the rate equations from the boundary layer diffusion, but in association with the kinetic components.

Recognition of this contribution by the boundary layer diffusion partly reverses a trend of the last 10–20 years to minimize or eliminate boundary layer diffusion as a factor in p.c. combustion. It would now appear that boundary layer diffusion can become dominant at the highest usual flame temperatures, and become controlling above 2000°C, in oxygen-boosted flames, for typical p.c. particle sizes. This fact could be significant in magnetohydrodynamic (MHD) applications. The reaction order and mode of reaction remain unsettled: there is clear experimental evidence for both first-order and zero-order (actual, with half-order effective) reaction. There is support for constant density burning and for constant diameter burning, and for modes in between. It is not yet possible to say under what conditions any given mode will prevail.

Other factors that are only now being considered include the effect of maceral type[204] and ash.[205,206] Nandi et al.[204] have found that combustibles in the ash are significantly increased by a high proportion of "inert" (i.e., low-reactivity) macerals. The ash itself is normally considered to be a secondary or even negligible quantity. In particular cases when it slagged on the particle surfaces, however, it led to uneconomic carbon loss.[205] Both factors (maceral type and ash) need more attention.

19.4 MECHANISMS

The objective of this section is to develop adequate descriptions of the reaction mechanisms, which are needed for mathematical formulations of the reaction rate equations in the sections that follow.

19.4.1 Char Combustion Mechanisms

19.4.1.1 *Overall Reactions*

The simple concept of a gas adsorbing at a solid surface to form an adsorbed layer that decomposes after a finite time to yield reaction products (Rhead and Wheeler[207–209]) has been conceptually elaborated in a number of different ways in the past.[107,108] There are, however, only two real alternatives for the initial adsorption step on carbon: (1) that the ambient molecules (O_2, CO_2, H_2O, H_2, and CO formed by reaction) adsorb as molecules, or (2) that these molecules dissociate as they adsorb. The experience of the last decade, however, seems to have led to the consensus that molecular chemisorption does not occur and that all chemisorption is dissociative, even if mobile.

All the pertinent gases will *physically* adsorb on carbon below certain temperatures. For H_2O, CO_2, and CO, the adsorption capacity of even activated carbon is very low. Oxygen and H_2 chemisorb very strongly at ambient temperatures. Physical adsorption of oxygen requires temperatures less than −70°C,[108] but there is evidence that parallel physical and chemical adsorption occurs on brown coal at ambient temperatures[210–217]

[204] B. N. Nandi, T. D. Brown, and G. K. Lee, "Inert Coal Macerals in Combustion," *Fuel,* **56,** 125–130 (1977).

[205] R. W. Borio, "A Laboratory Study of Anthracites Burned at a Spanish Generating Station," ASME Paper No. 74WA/Fu-2 1974.

[206] R. W. Bryers and Z. Kerekes, "An Examination of the Influence of Ash on the Combustion Characteristics of Anthracites," ASME Paper No 76-WA/CD-2, 1976.

[207] T. F. E. Rhead and R. V. Wheeler, "The Combustion of Carbon," *J. Chem. Soc.,* **101,** 846 (1912).

[208] T. F. E. Rhead and R. V. Wheeler, "The Mode of Combustion of Carbon," *J. Chem. Soc.,* **103,** 461 (1913).

[209] T. F. E. Rhead and R. V. Wheeler, "The Mode of Combustion of Carbon: The Effect of Drying the Oxygen," *J. Chem. Soc.,* **103,** 1210 (1913).

[210] L. D. Schmidt and J. L. Elder, "Atmospheric Oxidation of Coal at Moderate Temperatures: Rates of the Oxidation Reaction for Representative Coking Coals," *Ind. Eng. Chem.,* **32,** 249–256 (1940).

[211] P. G. Sevenster, "Diffusion of Gases Through Coal," *Fuel,* **38,** 403–418 (1959).

[212] P. G. Sevenster, "Studies on the Interaction of Oxygen with Coal in the Temperature Range 0–90°C, Part I," *Fuel,* **40,** 7–17 (1961).

[213] P. G. Sevenster, "Studies on the Interaction of Oxygen with Coal in the Temperature Range

(noting at the same time that coal is not carbon). A temperature for physical adsorption of H_2 does not seem to have been reported.

The evidence for the absence of molecular chemisorption is partly direct and partly inferential. Bonner and Turkevich,[218] Ergun,[219] and Kanter[220] all conclude from ^{14}C tracer studies, and Phillips et al.[221] from ^{13}C tracer studies, that CO is not chemisorbed to any important extent, although it is sufficient to be able to measure reaction rates.[222,223] From oxygen isotope studies[224–226] it is concluded, likewise, that neither O_2 nor CO_2 chemisorbs as molecules. It is also positively concluded that CO_2 is one of the primary products of reaction, accompanied by CO. As indicated earlier (Section 19.3.1.1), this may possibly be associated with the UASA film decomposition. Walker et al.[226] examined nine possible reactions for the formation of CO_2, and eliminated all but one. Mixtures of $^{16}O{-}^{16}O$, $^{18}O{-}^{18}O$, and $^{16}O{-}^{18}O$ were reacted with carbon in different proportions, and the calculated percentages of CO_2 produced by reaction using different oxygen isotope combinations were compared with experiment. Typically, for example,[226] oxygen with 2.6% $^{16}O{-}^{18}O$ was found to produce 50–52% of $C^{16}O^{18}O$, matching the percentages of the ^{16}O and ^{18}O percentages of the total original mixture. The percentage would have been 2.6% if the mixed-isotope CO_2 came only from the mixed-isotope O_2. Oxygen in the CO_2 molecules was found to exchange with time over carbon at 500°C (quite negligibly in the absence of carbon), but this change was slow and was responsible for only one or two percentage points of drift over a period of 90 min. It could not account for the results obtained in 1 or 2 min. Bonnetain et al.[227,228] found that the volumes of CO and CO_2 produced varied linearly with each other. These investigators also concluded that groups of molecules desorbed together, a finding that they suggested was possible, and likely, on account of electronic interaction in the graphite substrate. The dissociative chemisorption of H_2, on the other hand, would not seem to have been demonstrated directly, but it was implied from the agreement with dissociative adsorption isotherms.[229]

As a consequence of these findings, the sets of multiple equations examined by past reviewers (e.g., Walker et al.[30]) are reduced.

0–90°. Part II. Consideration of the Kinetics of the Process," *Fuel,* **40,** 18–32 (1961).

[214] D. L. Carpenter and D. G. Giddings, "The Initial Stages of the Oxidation of Coal with Molecular Oxygen. I. Effect of Time, Temperature and Coal Rank on Rate of Oxygen Consumption, *Fuel,* **43,** 247–266 (1964).

[215] D. L. Carpenter and D. G. Giddings, "The Initial Stages of the Oxidation of Coal with Molecular Oxygen. II. Order of Reaction," *Fuel,* **43,** 375–383 (1964).

[216] D. L. Carpenter and G. D. Sergeant, "The Initial Stages of the Oxidation of Coal with Molecular Oxygen. III. Effect of Particle Size on Rate of Oxygen Consumption," *Fuel,* **45,** 311–327 (1966).

[217] D. J. Allardice, "The Adsorption of Oxygen on Brown Coal Char," *Carbon (Oxford),* **4,** 255–262 (1966).

[218] F. Bonner and J. Turkevich, "Study of the Carbon Dioxide–Carbon Reaction Using ^{14}C as a Tracer," *J. Am. Chem. Soc.,* **73** 561–564 (1951).

[219] S. Ergun, "Kinetics of Reaction of Carbon Dioxide with Carbon," *J. Phys. Chem.,* **60,** 480–485 (1956).

[220] M. A. Kanter, "Diffusion of Carbon Atoms in Natural Graphite Crystals," *Phys. Rev.,* **107,** 655 (1957).

[221] R. Phillips, F. J. Vastola, and P. L. Walker, Jr., "Preliminary Study of the Exchange of Carbon Atoms Between Carbon Monoxide and Graphon," *Proc. 3rd Industrial Carbon Graphite Conf.* (1971), pp. 257–263.

[222] A. E. Reif, "A Study of the Reactions of Carbon-Monoxide with Coke," *J. Phys. Chem.,* **56,** 778–784 (1952).

[223] A. E. Reif, "The Mechanism of the Carbon Dioxide–Carbon Reaction," *J. Phys. Chem.,* **56,** 785–788 (1952).

[224] F. J. Vastola, P. J. Hart, and P. L. Walker, Jr., "A Study of Carbon-Oxygen Surface Complexes Using ^{18}O as a Tracer," *Carbon (Oxford),* **2,** 65–71 (1964).

[225] P. J. Hart, F. J. Vastola, and P. L. Walker, Jr., "Oxygen Chemisorption on Well Cleaned Carbon Surfaces," *Carbon (Oxford),* **5,** 363–371 (1967).

[226] P. L. Walker, Jr., F. J. Vastola, and P. J. Hart, "Oxygen-18 Tracer Studies on the Carbon-Oxygen Reaction," *Fundamentals of Gas-Surface Interactions,* Academic Press, New York, 1967, p. 307.

[227] L. Bonnetain, X. Duval, and M. Letort, "On the Role of Surface Oxides in the Graphite-Oxygen Reaction," *Proc. 4th Conf. Carbon* (1960), pp. 107–114.

[228] L. Bonnetain, "Desorption des Oxydes de Surface des Graphites Artificiels," *J. Chim. Phys.,* **58,** 34–45 (1961).

[229] F. J. Long and K. W. Sykes, "The Mechanism of the Steam-Carbon Reaction," *Proc. R. Soc. London,* **A193,** 377–399 (1948).

A further reduction in complexity is possible if it is assumed that there is no site specificity for different adsorbed gases. This would be a reasonable assumption from the principle of minimum hypotheses, but there is some experimental evidence to support it. There is no a priori reason to suppose that an oxide film formed by dissociative adsorption has any different structure whether the O atoms are supplied by O_2, CO_2, or H_2O. The film formed can thus be represented in each case as C(O); it is not necessary to write $C(O)_A$, $C(O)_B$, and so on, as some writers have done to represent different origins for the O atoms. The C(O) should decompose in the same way whatever its origin.

The net reaction steps for the adsorption and desorption processes in oxygen transfer are therefore considered to be:

$$2C_f - O_2 \xrightarrow{k_1} 2C(O) \qquad (19.7)$$

$$C_f + CO_2 \underset{k_{-2}}{\overset{k_2}{\rightleftharpoons}} C(O) + CO \qquad (19.8)$$

$$C_f + H_2O \underset{k_{-3}}{\overset{k_3}{\rightleftharpoons}} C(O) + H_2 \qquad (19.9)$$

$$2C_f + H_2 \underset{k_{-4}}{\overset{k_4}{\rightleftharpoons}} 2C(H) \qquad (19.10)$$

$$C(O) \xrightarrow{k_5} CO\ (+\text{ free site}) \qquad (19.11)$$

$$2C(O) \xrightarrow{k_6} CO_2 + C_f \qquad (19.12)$$

where C_f indicates a free site; C(O) represents chemisorbed oxygen atoms; and the k's are the velocity constants. The methane formation reaction is discussed separately under "Gasification" in Section 19.6.2.2

Equation 19.7 represents chemisorption of oxygen, with no film decomposition to recover free oxygen; in contrast, Equation 19.10 shows reversibly recoverable hydrogen gas. Equation 19.11 is shown to be irreversible on the basis of negligible CO adsorption. The reversibility of Equation 19.8 has been demonstrated directly; CO reacts with adsorbed oxygen to yield CO_2. Equation 19.12 represents direct primary formation of CO_2, where according to the results of Bonnetain et al.[227] Equations 19.11 and 19.12 are coupled in some way to preserve a fixed CO/CO_2 ratio during the course of a given experiment at fixed T and p_{O_2}. The reversibility of Equation 19.9 may be more arguable. It seems reasonable by analogy with the reversibility of Equation 19.8; it is included by Reif,[223] for example, but not by Walker et al.,[30] who show instead a reversible $C(H_2O)$ formation. Note that Equations 19.8 and 19.9, with 19.9 switched, constitute the water-gas shift reaction.

19.4.1.2 *Elementary Reactions*

The overall or net reaction schemes of Equations 19.7 to 19.12 are not unreasonable as initial summaries of experimental findings, but they do beg a certain number of questions, particularly regarding the details of the reverse reactions. In dissociative adsorption the gas may dissociate as it adsorbs, or before it adsorbs as an intermediate independent step.[135] Thus for CO_2 dissociating in adsorption we may write, in amplification of Equation 19.8:

$$\begin{aligned} C_f + CO_2 &\rightarrow C_f + O + CO \\ &\rightarrow C(O) + CO \qquad (19.8a) \end{aligned}$$

It is difficult to see, however, that the reverse reaction retraces the same intermediate path, although this is really implied by Equation 19.8 and is thereby implied by all authors who have proposed Equation 19.8. A more likely reverse path could be something like:

$$\begin{aligned} C(O) + CO &\rightarrow C(OOC) \\ &\rightarrow C_f + CO_2 \qquad (19.8b) \end{aligned}$$

which suggests that Equation 19.8 should really be written as two independent, irreversible reactions. If these speculations have any validity, the question of the meaning of the measured activation energies is also raised.

For Equation 19.9, dissociation into OH is the most likely, as suggested by Long and Sykes.[229] These authors also suggested that molecular hydrogen was formed on the surface that then rapidly desorbed, leaving behind an important fraction on the surface that retarded the reaction. In this instance we might construct a truly reversible sequence, as follows:

$$\begin{aligned} &C_f + H_2O \rightleftharpoons C_f + OH + H \rightleftharpoons \\ &C(OH)(H) \rightleftharpoons C(O) + 2H \rightleftharpoons C(O) + H_2 \qquad (19.9a) \end{aligned}$$

Again there arises the question of which activation energy would be measured.

Decomposition of the reaction into "elementary" steps can be carried further by conceptual or schematic representation of attachment of the molecules or atoms on the solid surface. Not forgetting that the process is really continuous,[116] critical stages in the sequence are generally recognized, with transition or activated complexes formed that could possibly (by assumption) revert to the prior stage. For oxygen adsorption, Fig. 19.30 is an extension of a scheme proposed by Marsh et al.,[135] showing the "stable" surface oxide formed as the result of two intermediate steps: an "activated complex" and a "surface intermediate." This mechanism is an elaboration of the Temkin-Pyzhev heterogeneous catalysis scheme described by Thomas and Thomas.[107] The schematic provides a basis for application of transition complex rate theory.[107]

Marsh and co-workers proposed that two intermediate steps were necessary to account for the observed CO and CO_2 production. The scheme of Fig. 19.30 may also account for the behavior described by Vastola et al.[224] who inferred formation of a reactive intermediate in the reaction, distinguishing it from the stable intermediate. The end of the sequence in Fig. 19.30 represents the final decomposition of the "stable" film. Such schemes are still essentially conceptual and speculative, awaiting mathematical definition and experimental test. An alternative approach is the quantum mechanics analysis examined, for example, by Grimley.[230] (See also Hwang.[231])

19.4.2 Volatiles Mechanisms

Combustion rates of volatiles may be controlled by gas phase kinetics in some cases, but they are often determined by the pyrolysis rate. Here we briefly consider both aspects.

19.4.2.1 *Pyrolysis Rates*

The pyrolysis of coal is examined in detail in Chapter 12. What follows here is a review of pyrolysis from the combustion point of view.

[230] T. B. Grimley, "Some Aspects of the Quantum Mechanics of Chemisorption." *Proc. Symp. Chemisorption,* July 1956, Butterworths Scientific Publications, London, pp. 17–26.

[231] S. T. Hwang, "Interaction Energy in Surface Diffusion," Am. Inst. Chem. Eng. Prepr. **29B**, Symp. Advances in Separation Processes, 62nd National Meeting, 1968.

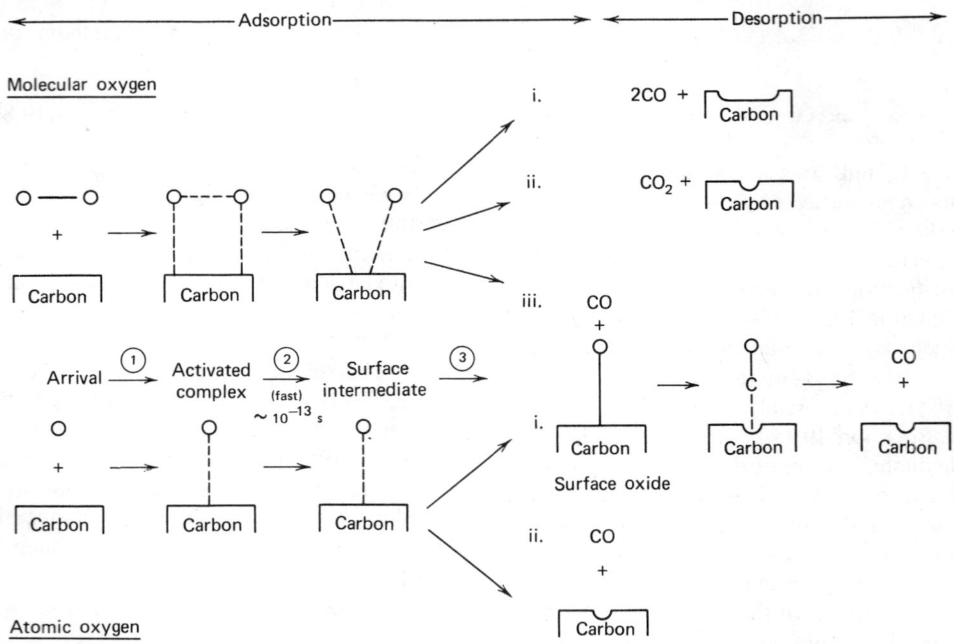

Figure 19.30 Schematic representation of sequence of events in molecular and atomic oxygen adsorption, with sequential oxide film decomposition. After Marsh et al.[135]

The basic coal model and essential behavior have been discussed earlier (Section 19.3.2 and Figs. 19.20 and 19.21). In summary, coal is an organic rock that decomposes on heating to form an immense range of products. The quantities of each product, and sometimes the existence of some of the products themselves, vary with operating conditions: pretreatment (if any), particle size, fuel bed size or particulate density, rate of heating, initial and final temperatures, exposure time at the final temperature, sweep gas rate (if any), and rate of cooling of the product gases. Part of the product variability is a consequence of secondary cracking of escaping volatiles and extent of cross-linking of the char being formed.

Theories of pyrolysis should take account of two stages of VM formation: the initial or primary formation, and the secondary cracking. In combustion, however, the evidence indicates that secondary cracking inside particles is important only for particles larger than about 1mm, which would include grate and possibly fluid bed firing, but should exclude p.c. firing; and cracking on other particles is to be expected only in very dense streams.[69]

Models of the primary VM formation have gone through several stages. The earliest approach was a form of "fundamental" analysis based on the assumption that coal was a pyrolyzable material that decomposed by a first-order, one-step reaction. The inadequacy of the assumption had been known for decades, but it did provide the start of a mathematical framework against which deviations could be measured. For certain purposes of evaluating or determining gross behavior, this simple approach still has some value, particularly when the pyrolysis is a secondary factor in a complex flame system.

In the more detailed and elaborate theories, the inductive development of the equations largely skimmed over attribution to mechanisms. To the extent that schemes were invoked, as an advance on simply writing down equations, these were essentially phenomenological, with general stages identified in such ways as: "formation of metaplast," "residue," primary and secondary tar, semicoke, coke, and gas. These stages exist; whether they are mechanistically important is still unknown. They do provide an appropriate basis for phenomenological theories, nevertheless, whose quantitative form is summarized later (Section 19.5.6).

The only model of pyrolysis known to exist that has a basis in coal constitution is a model of intermediate complexity. It is based on the model discussed in Section 19.3.2.1 where as already briefly indicated, the easily evolved or primary volatile matter is identified with the functional edge groups that connect the aromatic and hydroaromatic lamellae. The first stage of pyrolysis is conceived as the removal of this "easily evolved" component. It is not clear whether removal of this component, (designated as C_I), is a process of total volatilization, or whether the primary production splits C_I into solid (char), S_I, with volatiles V_I. My present view is that there is probably no primary char production from C_I. It may be assumed, nevertheless, that V_I will undergo secondary cracking under appropriate conditions. The aromatic and hydroaromatic lamellae are then identified as the second component, C_{II}. A principal product of C_{II} pyrolysis is H_2, which forms, typically, 50% of coke-oven or coal gas. The H_2 is derived mainly from the hydroaromatic components. Because H_2 is more firmly bound in these compounds, rapid evolution requires a temperature in excess of 700°C. The products from pyrolysis of C_{II} are then the solid char residue, S_{II}, and second-component volatiles, V_{II}. The two components can be expected to pyrolyze in parallel; the model also requires that C_I and C_{II} be fixed quantities, independent of particle size or heating rate. The model also leads one to expect the ratio of C_I to C_{II} to decrease with rising coal rank. This would have clearly evident influence on the variation of flammability with rank.

Quantitative application of the theory is discussed later. Application is hindered, at present, by lack of knowledge of the ratio of C_I to C_{II}.

19.4.2.2 *Combustion Mechanisms*

Almost nothing is known about the mechanisms of volatiles combustion. Direct investigation of the combustion of coal volatiles is confined to two studies,[232,233] neither of which

[232] D. Finch, "Ignition Temperatures of Volatiles Generated from Coal and Coal Chars," *Fuel,* **35,** 415 (1956).

developed any kinetic data. Only two kinetic studies of any solids volatiles (from cellulose) are known,[234,235] and the analysis used phenomenological kinetics with the assumption that the reaction rate was first order in both oxygen and fuel (volatiles). For the volatiles concentration it was necessary to use mass fractions. No mechanisms were invoked.

Field and et al.[35] devote a chapter to the problem, but again there is a lack of kinetic information beyond two approximations. The first approximation was the oil drop model used by Howard and Essenhigh,[73] but this assumes infinitely fast kinetics of the volatiles themselves. Furthermore, strictly speaking the particle temperature is not decoupled from the rate of heat supply in coal pyrolysis, as assumed in the oil drop analysis; they can be coupled, however, as shown by Waibel and Essenhigh[236] for combustion of thermoplastic polymers. The second approach was a nondiffuse model based on the common assumption that hydrocarbons react rapidly to form CO and H_2O, with CO burnup as the final, slow rate-determining step. A calculation using the stirred-reactor relation of Hottel et al.[237] for CO combustion yielded a time of 3 ms for the combustion of the volatiles at 1500°C. If an estimation of volatiles combustion times is necessary, this is probably the only basis that can be used at present. If the volatiles are written C_nH_m, then the first, very rapid reaction step is often written as follows[238]:

$$C_nH_m + O_2 \rightarrow nCO + \frac{m}{2}H_2 \qquad (19.13)$$

The subsequent reaction can then be treated as combustion of CO and H_2, using whatever kinetic sequence is appropriate to the problem (see, e.g., Smoot et al.[75]). There is evidence that secondary cracking is not a problem in combustion,[69] at least outside the particles. If this is found to be incorrect, the only approach available at present is again to use phenomenological kinetics with experimentally derived constants.

19.5 RATE EQUATIONS

19.5.1 General Assumptions on Carbon Reactions

Conversion of the carbon reaction mechanisms (Section 19.4) into mathematical form has mostly followed one of two paths. Up to 10 or 15 years ago, analysts of the carbon reactions invoked the Langmuir-Hinshelwood kinetics, but there has been increasing use of the Elovich equation. Transition state theory has also been used in a few cases.

The common assumption remains the Langmuir postulate of the existence of adsorption sites, but there are varying assumptions regarding the number and accessibility of the sites. The additional Langmuir assumptions of uniformly accessible sites on a uniform surface are most commonly invoked, although the real heterogeneity of the surface is well known. One method of dealing with heterogeneity is to assume that parts of the surface are locally homogeneous and to integrate over the surface, yielding standard equations.[108] Another is to assume or to demonstrate that sites of common potential energy are filled before the next suite of higher potential energy sites is involved. This approach entails dividing the surface area into conceptual regions of nominally uniform accessibility, with means of identifying the accessible fraction at any moment. Laine and co-workers[109,110] and subsequent investigators of the same group[112–115,221,224–226,239–242]

[233] G. Underwood and M. W. Thring, "Combustion of Coal Smoke," *Proc. Conf. Science in the Use of Coal,* Institute of Fuel (London), 1958, pp. D27–D32.

[234] B. K. Biswas and R. H. Essenhigh, "The Problem of Smoke Formation and Control," in *Air Pollution and Its Control,* Am. Inst. Chem. Eng. Symp. Ser. No. 126, Vol. 68 (1972), pp. 207–215.

[235] W. S. Shieh and R. H. Essenhigh, "Combustion of Computer Cards in a Test Incinerator: A Comparison of Theory and Experiment," *Proc. 5th Natl. Incin. Conf.,* ASME, 1972, pp. 120–134.

[236] R. T. Waibel and R. H. Essenhigh, "Combustion of Thermoplastic Polymer Particles in Various Oxygen Atmospheres: Comparison of Theory and Experiment," *Proc. 14th Symp. (Int.) Combustion,* Combustion Inst., 1973, pp. 1413–1420.

[237] H. C. Hottel, G. C. Williams, N. M. Nerheim, and G. R. Schneider, "Kinetic Studies in Stirred Reactors—Combustion of Carbon Monoxide and Propane," *Proc. 10th Symp. (Int.) Combustion,* Combustion Inst., Pittsburgh, 1965, pp. 111–121.

[238] R. B. Edelman, O. Fortune, and G. Weilerstein, *Proc. GM Symp. Emissions from Continuous Combustion Systems* (W. Cornelius and W. G. Agnew, Eds.) Plenum Press, New York, 1972, pp. 55–87.

[239] R. Phillips, F. J. Vastola, and P. L. Walker,

determined the various fractional regions experimentally, identifying them (Section 19.3.1) as TSA, ASA, and fractional coverage of the ASA, θ, by stable film. The area open to attack under the conditions of investigation was the unoccupied fraction of the ASA, which was (ASA) $(1 - \theta)$. This is also the reactive fraction (RASA) of the ASA. When the total (BET) surface area was used to calculate the rate constants, k, for a particular experiment at a given temperature, the value of k was found to vary drastically with time. It was effectively constant, however (apart from the initial surge explained by Fig. 19.8), when using the RASA. This, however, represented a level of detail not normally determined or determinable in many experiments, and most investigators still use equations applied to the whole surface, or just to the measurable superficial surface.

The kinetics can be modified, of course, by diffusion effects (Section 19.3.1.2). The common model adopted for pore diffusion is that described by Wheeler.[243] Boundary layer diffusion is mostly treated using the relevant Sherwood number, a function of Re and Sc.

The rate equations describe variation with time. In many cases the reaction reaches a steady state, or is assumed to reach that condition. In principle, the equations can be tested against experiment either in their time-dependent or in their steady state form. In the past, the latter has been more generally used but, with increasing sophistication in the necessary experimental methods, more use is being made of the time-dependent equations; this, however, is the only way the Elovich equations can be used. Since the time-dependent equations also make better use of available experimental data, it is to be expected that they will be more widely applied in the future as experimental conditions permit.

Jr., "The Effect of Oxygen Pressure and Carbon Burn-off on the Product Ratio of the Carbon-Oxygen Reaction," *Carbon (Oxford)*, **7**, 479–485 (1969).

[240] R. Phillips, F. J. Vastola, and P. L. Walker, Jr., "The Thermal Decomposition of Surface Oxides Formed on Graphon," *Carbon (Oxford)*, **8**, 197–203 (1970).

[241] R. Phillips, F. J. Vastola, and P. L. Walker, Jr., "Factors Affecting the Product Ratio of the Carbon-Oxygen Reaction. II. Reaction Temperature," *Carbon (Oxford)*, **8**, 205–210 (1970).

[242] D. L. Biederman, A. J. Miles, F. J. Vastola, and P. L. Walker, Jr., "Carbon–Carbon Dioxide Reaction: Kinetics at Low Pressures and Hydrogen Inhibition," *Carbon (Oxford)*, **14**, 351–356 (1976).

[243] A. Wheeler, "Reaction Rates and Selectivity in Catalyst Pores," in *Advances in Catalysis*, Vol. 3, Academic Press, New York, 1953, p. 249.

19.5.2 Langmuir-Hinshelwood (L-H) Kinetics

19.5.2.1 *General Rate Equation*

Reaction of free carbon sites, C_f, with gas, G (i.e., O_2, CO_2, H_2O, H_2) according to Equations 19.7 to 19.10 yields adsorbed atoms, X (i.e., O or H), with gaseous product G′ (i.e., CO, H_2, or nothing). The reactions are reversible in the case of Equations 19.8–19.10. The adsorbed gas film will also desorb to CO (and CO_2) when X is O (adsorbed oxygen). Any one of the reactions shown in Equations 19.7 to 19.10, with desorption, can be represented by

$$nC_f + G \underset{k_{-i}}{\overset{k_i}{\rightleftharpoons}} nC(X) + G' \qquad (19.14)$$

$$C(X) \xrightarrow{k_j} CX \qquad \text{for } X = O \qquad (19.15)$$

If site coverage is denoted by θ as a fraction of TSA, ASA, or RASA, or other appropriate designation, the rate of change of coverage, $d\theta/dt$, may be written:

$$\frac{d\theta}{dt} = r_i - r_{-i} - r_j \qquad (19.16)$$

where r is the rate of the forward reaction (subscript i), or reverse reaction (subscript $-i$), or film decomposition reaction (subscript j). In the forward reactions of Equations 19.7–19.10, all rates are proportional to the ambient gas concentration, p_G, and to the fraction of (active) surface not already covered $(1 - \theta)$. The reverse reaction of Equation 19.7, and the film decomposition reactions 19.11 and 19.12, are all proportional to the fractional surface that is covered, θ, and to the ambient gas concentration, $p_G{}'$, where this is relevant. In the cases of dissociative adsorption (forward reactions of Equations 19.7 and 10.10) and of associative desorption (reverse Equation 19.10), further factors are involved, such as the probability of the presence of unoccupied (or occupied) nearest neighbor sites. The expressions, as discussed, for example, by Hayward and Trapnell,[108] also depend on the different conditions of mobile or immobile adsorption, adsorption type, and coverage, but the simplest working approximations for the factors are $(1 - \theta)$ and θ, respectively.

Equation 19.16 may therefore be expanded into the general form representing the individual Equations 19.7–19.10:

$$\frac{d\theta}{dt} = k_i\, p_G(1 - \theta)^n - k_{-i}\, p_G{}'\theta^n - k_j\theta^m \qquad (19.17)$$

The parameters specifying the different reactions are set out in Table 19.3. The formation of CO_2 by Equation 19.12 is neglected as unimportant above 750°C. Hydrogen desorption is the only reaction likely to be associative, requiring $m = 2$, unless CO_2 formation by Equation 19.12 is included. It may be noted that Equations 19.8 and 19.9 are both linear in θ ($n = m = 1$). Therefore, when steam and CO_2 are the reactants (Equations 19.8 and 19.9), the equations are additive. (Even if steam is initially present alone, CO_2 is formed from CO by back-reaction of Equation 19.8.)

Examination of Equation 19.17 and Table 19.3 shows that either $n = m$ or, when $n = m$, then $k_{-i} = 0$ or $k_j = 0$. It is therefore convenient to write Equation 19.17 in the form

$$\frac{d\theta}{dt} = k_i\, p_G\left[(1 - \theta)^n - \left(\frac{\theta}{\theta_0}\right)^m\right] \qquad (19.18)$$

where:

$$\theta_0^m = \frac{k_i\, p_G}{k_{-i}\, p_G{}' + k_j} \qquad (19.19)$$

19.5.2.2 *Time-Dependent Solutions*

Time-dependent solutions are being used increasingly as criteria of mechanisms. Identification by this method requires care.

Desorption Only. All k's $= 0$ except k_5, yielding the following:

1. *For $m = 1$:* The general condition for film decomposition:

$$\theta = \theta_{in} \exp -\frac{t}{\tau} \qquad (19.20)$$

   where θ_{in} is the initial value of θ and τ is the characteristic time for desorption $= 1/k_5$.

2. *For $m = 2$:* The condition for associative desorption[108];

$$\left(\frac{1}{\theta}\right) = \left(\frac{1}{\theta_{in}}\right) - \frac{t}{\tau} \qquad (19.21)$$

Comparing the two equations, it will be seen that the variation of the fraction θ/θ_{in} with time is independent of the initial condition θ_{in}, in case 1 ($m = 1$), but not in case 2 ($m = 2$).

Adsorption Only. All $k_{-i} = 0$, and $k_j = 0$; hence $\theta_0 = \infty$. Again there are two solutions, for $n = 1$, and $n = 2$.

1. *$n = 1$:* This is nominally the case for CO_2 and H_2O adsorption only; it is often assumed to apply, however, to oxygen adsorption when the dissociation is either ignored or assumed not to occur. The conditions yield

$$\theta = \theta_m\left(1 - \exp\frac{-t}{\tau}\right) \qquad (19.22)$$

Table 19.3 Velocity Constants and Other Parameters for Rate Equation 19.17

Reaction	n	m	k_i	k_{-i}	k_j	G	G'
19.7, C/O_2	2	1	k_1	0	k_5	O_2	—
19.8, C/CO_2	1	1	k_2	k_{-2}	k_5	CO_2	CO
19.9, C/H_2O	1	1	k_3	k_{-3}	k_5	H_2O	H_2
19.10							
C/H_2	2	2	k_4	k_{-4}	0	H_2	—
$C/(H_2O + CO_2)$	1	1	k_2, k_3	k_{-2}, k_{-3}	k_5	H_2O, CO_2	H_2, CO

where θ_m is the maximum, or limiting value of θ, at $t = \infty$; $\tau = 1/k_2$ or $1/k_3$. Figure 19.31 is a typical plot for the C/CO_2 reaction using ambient CO_2 pressure as a measure of $1 - \theta$ so that the pressure declines exponentially.

2. $n = 2$: This is nominally the case for O_2 and H_2 adsorption only. The solution is

$$\theta = \theta_m \frac{t/\tau}{1 + t/\tau} \quad (19.23)$$

This equation is not known to have been used. Where θ/t data exist (or their equivalent is known, such as drop in ambient gas pressure or quantity of gas adsorbed), there is usually sufficient scatter for Equation 19.22 to be used as a good approximation to Equation 19.23. Both equations generate linear initial rises with time up to values of $t/\tau = 0.5$. Beyond that point the curves separate.

A log-linear plot may still confuse the issue, however. In Fig. 19.32 $\ln(1 - \theta/\theta_m)$

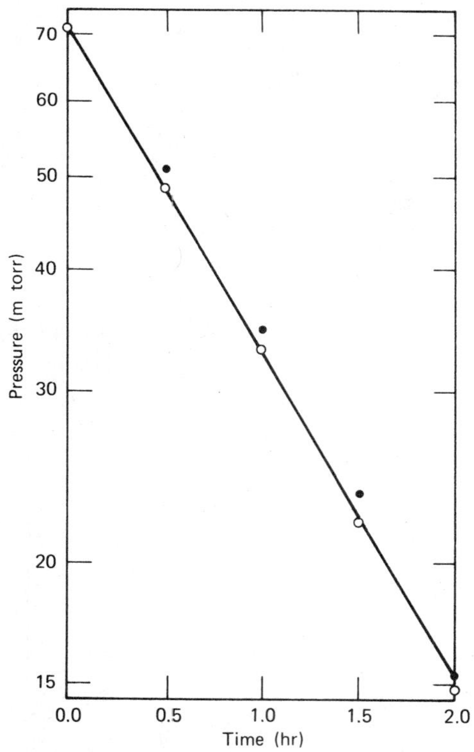

Figure 19.31 First-order plot for disappearance of CO_2 during SP-1 graphite–CO_2 reaction at 1100°C: CO_2 pressure was measured directly (○) and calculated (●) from measurements of CO pressure.[242]

plotted against t/τ yields the declining straight line of slope -1, which is what would be obtained from Equation 19.22. A plot of $\ln(1 + t/\tau)^{-1}$ against t/τ yields the (upper) curved line. If Equation 19.23 is governing, a plot of $\ln(1 - \theta/\theta_m)$ would follow the curve; but with a modicum of experimental scatter, the plot would appear to be linear with a displaced origin. With τ unknown a priori, the result could be interpreted as first order with some "initiation" behavior at the start of the reaction. Figure 19.33 illustrates two semilog plots for oxygen adsorption on carbon[244] at 500 and 200°C, showing two different modes of behavior. At 500°C the plot obeys Equation 19.22 (Fig. 19.34). This fact is no proof that the behavior represents dissociative adsorption. It merely shows that the results are consistent with it. The authors, however, attribute the curvature to the existence of two competing reactions.

Adsorption with Desorption. All $k_i \neq 0$. There are three solutions.

1. $n = m = 1$, $k_{-i} \neq 0$, $k_j \neq 0$. This applies to the CO_2 and H_2O reactions considered independently or simultaneously. The solution is

$$\theta = \frac{\theta_0}{1 + \theta_0}\left(1 - \exp -\frac{t}{\tau}\right) \quad (19.24)$$

where for the CO_2 reaction, $\tau = 1/k_2 p_{CO_2}$, and

$$\theta_0 = \frac{k_2 p_{CO_2}}{(k_5 + k_{-2} p_{CO})} \quad (19.25)$$

and for the H_2O reaction, $\tau = 1/k_3 p_{H_2O}$, and

$$\theta_0 = \frac{k_3 p_{H_2O}}{k_5 + k_{-3} p_{H_2}} \quad (19.26)$$

If both gases are reacting simultaneously, $\tau = 1/(k_2 p_{CO_2} + k_3 p_{H_2O})$ and

[244] F. J. Vastola and P. L. Walker, Jr., "The Reaction of Graphite 'Wear-Dust' with Carbon Dioxide and Oxygen at Low Pressures," *J. Chim. Phys.*, **58**, 20–24 (1961).

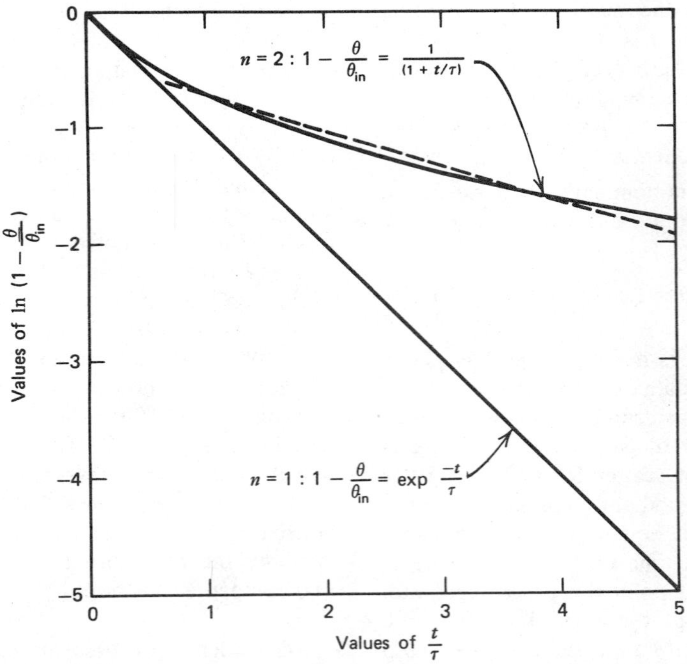

Figure 19.32 Growth of adsorption coverage, θ, with time (t). Comparison of equations for $n = 1$ and $n = 2$ (see text). Dashed curve for $n = 2$ is straight line approximated to solid.

$$\theta_0 = \frac{k_2 p_{CO_2} + k_3 p_{H_2O}}{k_5 + k_{-2} p_{CO} + k_{-3} p_{H_2}} \tag{19.27}$$

2. $n = 2$ *or* $m = 2$, $k_{-i} \neq 0$, $k_j = 0$. The value of n is for the hydrogen reaction with its reverse, but no independent desorption; the value of m is for the oxygen reaction with film decomposition to CO_2 but no back reaction. The solutions are of identical form:

$$\theta = \frac{\theta_0(1 - e^{-t/\tau})}{(1 + \theta_0)\left(1 - \frac{(1 - \theta_0)}{(1 + \theta_0)} e^{-t/\tau}\right)} \tag{19.28}$$

where for the hydrogen reaction, $\tau = \theta_0 / 2k_4 p_{H_2}$, and

$$\theta_0 = \left(\frac{k_4 p_{H_2}}{k_4}\right)^{1/2} \tag{19.29}$$

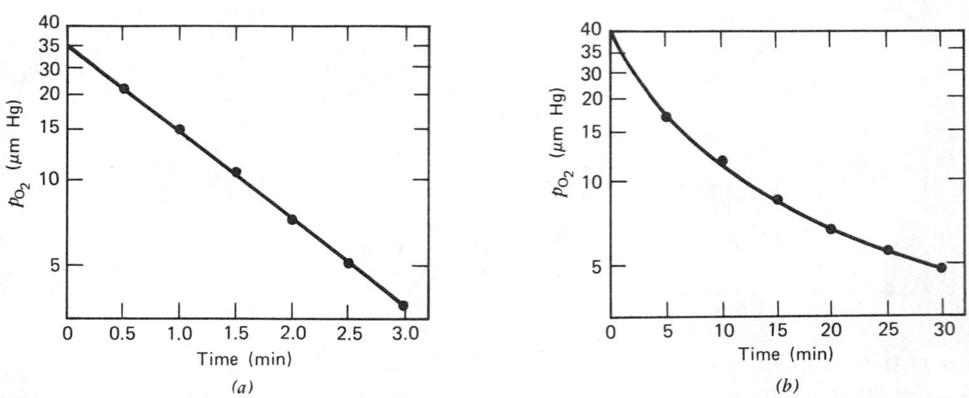

Figure 19.33 (*a*) First-order relationship of oxygen adsorption at reaction temperature of 500°C. (*b*) Deviation from first-order relationship of oxygen adsorption at reaction temperature of 200°C.[244]

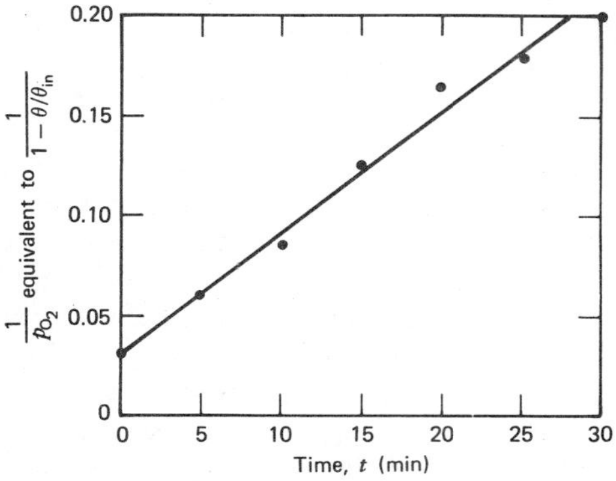

Figure 19.34 Variation of $(1 - \theta/\theta_{in})^{-1}$ with time, with P_{O_2} as estimate of $(1 - \theta/\theta_{in})$.[244] (Replot of Fig. 19.33b.)

and for the oxygen reaction, $\tau = \theta_0/2k_1 p_{O_2}$, and

$$\theta_0 = \left(\frac{k_1 p_{O_2}}{k_6}\right)^{1/2} \tag{19.30}$$

3. $n = 2$, $m = 1$, $k_{-i} = 0$, $k_j = 0$. This is the forward oxidation reaction with film decomposition to CO. The solution is

$$\theta = \frac{B(1 - e^{-t/\tau})}{1 - (B/A)e^{-t/\tau}} \tag{19.31}$$

where $\tau = 1/(1 + 4\theta_0)^{\frac{1}{2}} k_1 p_{O_2}$, and

$$A = \frac{[(1 + 4\theta_0)^{1/2} + 1]^2}{4\theta_0} \tag{19.32a}$$

$$B = \frac{[(1 + 4\theta_0)^{1/2} - 1]^2}{4\theta_0} \tag{19.32b}$$

$$\theta_0 = \frac{k_1 p_{O_2}}{k_5} \tag{19.32c}$$

Comment on the Solutions. There is a similarity in the basic (1 − exp) form of all but one of the solutions, with or without further modifying terms in time. All solutions also show a trend to a limit, given by several different expressions, as time goes to infinity. If the modifying time terms are small, which is possible if the preceding multipliers are small (e.g., B/A in Equation 19.31), then all the reactions will follow roughly similar rate behavior, and constants extracted from appropriate plots could make reasonable sense but yet be in error by misidentification.

Caution seems to be in order here.

19.5.2.3 *Steady State Conditions*

General Results. The L-H equations have been more commonly used and are better known in their steady state form, at θ = const, or $d\theta/dt = 0$. The required equations are directly obtainable from the relevant solutions. The interest is in the reaction rate at steady state. In all reactions but that for hydrogen adsorption this is given by

$$\text{rate} = k_5\theta_m \tag{19.33}$$

The following solutions are therefore obtained.

With only adsorption, the equilibrium is trivial, with $\theta = \theta_m$, where θ_m should be unity if appropriately defined. For all other cases there are only two solutions for θ_m in terms of θ_0 (Equations 19.25–19.27, 19.29, and 19.30).

1. For reaction of C with CO_2, H_2O, or H_2:

$$\theta_m = \frac{\theta_0}{(1 + \theta_0)} \tag{19.34}$$

2. For reaction with O_2:

$$\theta_m = \frac{2\theta_0}{(1 + 2\theta_0)}\left[1 + \frac{(1 + 4\theta_0)^{1/2}}{1 + 2\theta_0}\right] \tag{19.35}$$

Expanding these expressions by substitution for θ_0 yields mostly the familiar L-H equations, as follows.

For the C/CO_2 reaction we have

$$\text{rate} = k_5\theta_m = \frac{(k_2)p_{CO_2}}{1 + (k_2/k_5)p_{CO_2} + (k_{-2}/k_5)p_{CO_2}} \tag{19.36}$$

For the C/H_2O reaction we have

$$\text{rate} = \frac{(k_3)p_{H_2O}}{1 + (k_3/k_5)p_{H_2O} + (k_{-3}/k_5)p_{H_2}} \tag{19.37}$$

For the C/H_2O reaction in which CO_2, formed by reverse Equations 19.8, reacts by forward Equation 19.8, we have

$$\text{rate} = \frac{(k_2)p_{CO_2} + (k_3)p_{H_2O}}{1 + (k_2/k_5)p_{CO_2} + (k_3/k_5)p_{H_2O} + (k_{-2}/k_5)p_{CO} + (k_{-3}/k_5)p_{H_2}} \tag{19.38}$$

For the C/H_2 reaction we have

$$\theta_m = \frac{(k_4/k_{-4})^{1/2} p_{H_2}^{1/2}}{1 + (k_4/k_{-4})^{1/2} p_{H_2}^{1/2}} \tag{19.39}$$

The expression for the oxygen reaction is less familiar

$$\theta_m = \frac{k_5[(1 + 4k_1 p_{O_2}/k_5)^{1/2} - 1]^2}{(4k_1 p_{O_2}/k_5)} \tag{19.40}$$

The extreme values of large and small p_{O_2} are important. At large p_{O_2}, if $4k_1 p_{O_2}/k_5$ is large compared with unity, the numerator tends to this value, and the ratio of the numerator and denominator tend to unity, as expected. At small p_{O_2}, however, the square root term may be expanded to yield $(4k_1 p_{O_2}/k_5)^2$ in the numerator, which cancels with the term in the denominator to leave $\theta_m = (k_1/k_5) p_{O_2}$, which is first order in oxygen in spite of the assumption of dissociative adsorption.

H_2 Adsorption. Biederman et al.,[242] in one of the very few investigations of the C/H_2 reaction, reported that behavior followed an almost perfect dissociative Langmuir isotherm, obeying Equation 19.39. This finding contradicts to the result reported by Long and Sykes,[229] who concluded that the simple Langmuir equation was indeed obeyed but with the hydrogen concentration to the first power. However on evaluating their data[229] I found a better fit to a half-power expression. This is illustrated in Fig. 19.35, which is a reproduction of Long and Sykes's data plot, but with back-calculated lines using the half-power expression of Equation 19.39 in place of the original. The original, first-power lines are dotted in for comparison. At 750 and 850°C it is seen that there is very little to distinguish between the two expressions. There is a bigger difference at 800°C, with the half-power expression a somewhat better fit. Test of Equation 19.39 compared with a first-power expression is clearly very dependent on the experimental accuracy, and even then, in most cases there would probably be little to permit one to choose between the two expressions on the basis of fit to the curves or to the generated straight lines from plotting $(p_{H_2}^n/q)$ against $p_{H_2}^n$, with $n = 1$ or ½. Since theory predicts a half-power, not a first-power expression, however, the balance of evidence is in favor of Equation 19.39 (but see Section 19.5.3 on the Elovich equation).

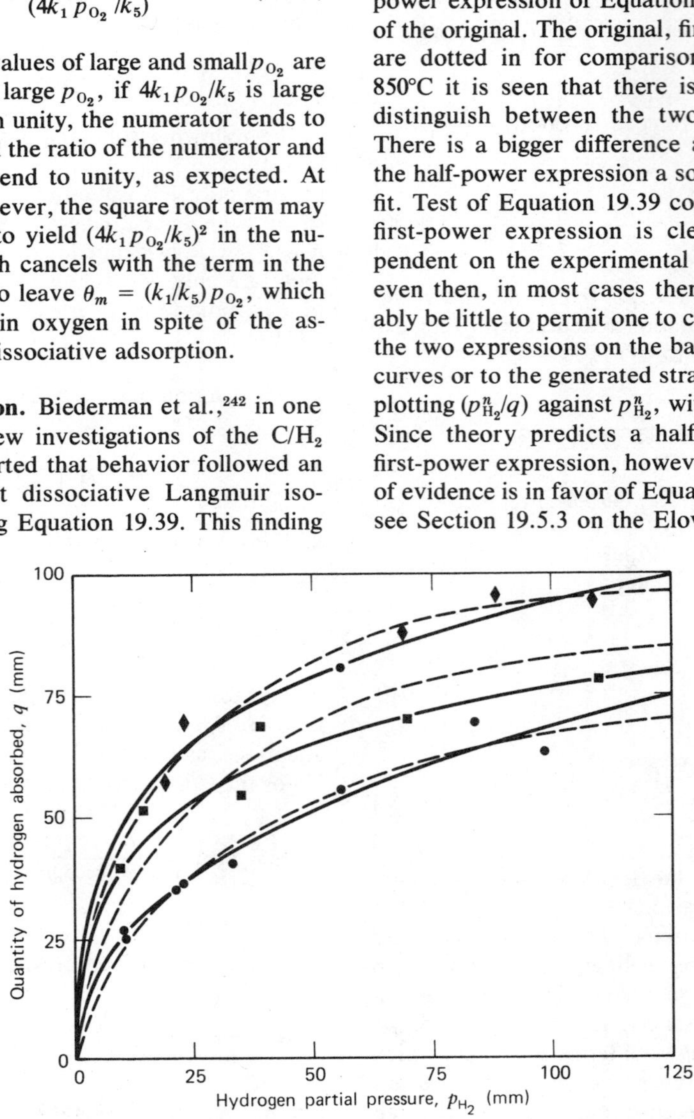

Figure 19.35 Variation of quantity of H_2 absorbed on carbon as a function of hydrogen partial pressure: ◆ = 850°C, ■ = 800°C, ● = 750°C. Solid curves: calculated $q = q_0[bp_{H_2}^{1/2}/(1 + bp_{H_2}^{1/2})]$; dashed curves: calculated $q = q_0[b'p_{H_2}/(1 + b'p_{H_2})]$.[229]

The C/CO_2 Reaction. This reaction has been much more extensively investigated (see Table 19.6). Many investigators have substantiated the L-H equation (Equation 19.36).[117,118,229,242,245–247] The presence of CO, produced by the reaction, has the effect of retarding it. The mechanism of retardation is the back reaction of Equation 19.8 [designated as 19.8(Reverse), or 19.8(R)]. An alternative mechanism (CO adsorption) is now discounted, since carbon monoxide does not adsorb with sufficient speed or to sufficient extent to be important (Section 19.4.1.1).

In some circumstances, Equations 19.8 and 19.8(R) react quickly compared to decomposition (Equation 19.11). Equation 19.8 is then close to equilibrium and, with k_5 proportionately small, Equation 19.25 can be written

$$\theta_0 = \frac{k_2 p_{CO_2}}{k_{-2} p_{CO}} = \frac{K_2 p_{CO_2}}{p_{CO}} \qquad (19.41)$$

where K_2 is the equilibrium constant for the CO and CO_2 in equilibrium with the surface. Equation 19.36 can then be written

$$\text{rate} = \frac{k_5}{1 + (p_{CO}/K_2 p_{CO_2}) + (k_5/k_2 p_{CO_2})} \qquad (19.42)$$

For the conditions under which this should be valid, the third term in the denominator would be small compared to the remainder. Ergun and Mentzer[52] report Ergun's investigations,[219] in which values of K_2 were obtained from measurements in a fluid bed and plotted as shown in Fig. 19.36. The line is reasonably straight over most of the plot, with a slope equivalent to a heat of reaction of about 23 kcal, for three different carbons. With coke, the rates were very close to those of the three carbons, with a slightly higher heat of reaction (27 kcal). Although slow, film decomposition did occur and the reaction rate yielding k_5 could be measured. The activation energies of the reaction of the three carbons in Fig. 19.36 were found to be the same (59 kcal), with three different frequency factors.

Grabke,[248,249] however, has suggested that the evaluation above should be regarded as only a first approximation. In the limit of true equilibrium with the surface, with no film decomposition, the forward reaction rate, r_2, of Equation 19.8 is exactly equal to the reverse, r_{-2}. If the decomposition rate, r_5, is negligible, Grabke showed that $r_5 = r_{-2}(k_5/k_{-2} p_{CO}) = r_2(k_5/k_{-2} p_{CO})$, so that, $(r_2/r_5 p_{CO}) = k_{-2}/k_5$, which should be a constant at constant temperature. Since $r_2/r_5 p_{CO} = k_2 p_{CO_2}/r_5 p_{CO}$, p_{CO_2}/p_{CO} should also be constant. Grabke showed that this was not the case in his experiments (which does not necessarily mean that it was not the case in Ergun's). Interpreting the results in terms of oxygen activity (a_O), citing Wagner,[62] k_2 was established as a function of a_O, taking the form $k_2(a_O) = k_2^0 a_O^{-m}$, with $m = 0.5$. The value of m increases with binding strength of the oxide on the surface: m was high on active charcoals but low (near zero) on natural graphite.

The C/H_2O Reaction. This reaction has also been quite extensively investigated (see Table 19.6) with results showing support for the steady state relation, Equation 19.37, paralleling Equation 19.36 for C/CO_2. Here also Ergun and Mentzer[52] examined the consequences of both Equations 19.8 and 19.9 being in equilibrium. For this to be the case we would have the ratio of the free to the filled sites proportional to the gaseous partial pressures, as follows:

$$\frac{1-\theta}{\theta} = \frac{k_{-2} p_{CO}}{k_2 p_{CO_2}} = \frac{k_{-3} p_{H_2}}{k_3 p_{H_2O}} \qquad (19.43)$$

which is also a form of the water-gas shift equilibrium equation. When these conditions are applied to Equation 19.37, it reduced to

[245] F. J. Long and K. W. Sykes, "The Effect of Specific Catalysts on the Reactions of the Steam-Carbon System," *Proc. R. Soc. London,* **A215,** 100–110 (1952); *J. Chim. Phys.,* **47,** 361 (1950).

[246] J. S. Binford and H. Eyring, "Kinetics of the Steam-Carbon Reaction," *J. Phys. Chem.,* **60,** 486–491 (1956).

[247] K. Hedden and A. Lowe, "Über die Vergasung vor Reaktorgraphit mit Kohlendioxid und Wasserdampf," *Carbon (Oxford),* **5,** 339–353 (1967).

[248] H. J. von Grabke, "Zur Kinetik der Reaktionen vor Graphit mit CO_2-CO- und H_2O-H_2- Gemishen," *Ber. Bunsenges.* **70,** 664–670 (1966).

[249] H. J. von Grabke, "Oxygen Transfer and Carbon Gasification in the Reaction of Different Carbons with CO_2," *Carbon (Oxford),* **10,** 587–599 (1972).

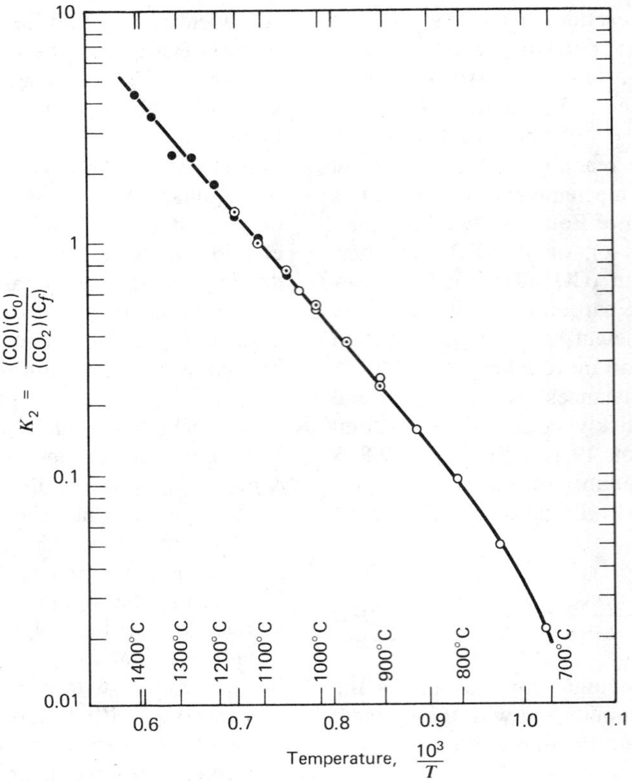

Figure 19.36 The equilibrium constant of the oxygen exchange reaction as a function of temperature (average heat of reaction between 800 and 1400°C is 23,000 cal/mole): ○ = activated carbon, ⊙ = activated graphite, ● = Ceylon graphite.[52]

$$\text{rate} = \frac{k_5}{1 + [p_{H_2}/K_3\, p_{H_2O}] + k_5/(k_2\, p_{CO_2} + k_3\, p_{H_2O})} \quad (19.44)$$

where, as in the case above, the third term in the denominator must be small. In reviewing their experimental results, evaluated in terms of the equation above, Ergun and Mentzer[52] showed that the shift equilibrium was only satisfied at 1000 and 1100°C at H_2O fractions reacted of 0.5 or 0.6. With good will on the part of the beholder (viewing a straight line through three data points), the Arrhenius plot for k_5 paralleled that for the C/CO_2 reaction with only a small displacement. This provided some substantiation for the assumption that all adsorption sites were common for reaction with both CO_2 and H_2O. They also showed that the surface equilibrium constants; $K_3 = [(p_{H_2}\theta)/p_{H_2O}]\,(1 - \theta)$ and $K_2 = (p_{CO}\theta/p_{CO_2})(1 - \theta)$, were very close in value. Direct calculation of the water-gas shift equilibrium, however, showed that equilibrium was only attained at quite high values of H_2 conversion at the temperatures investigated (1000, 1100, and 1200°C). This is of considerable importance in behavior of gasification of solid fuel beds where the shift reaction is assumed to equilibrate very rapidly at temperatures down to about 1000°C (below which temperature it is usually said to become very slow). In a study of shredded refuse, Rogers et al.[250] found the equilibrium to hold at about 1000°C, except near the reaction front in thin beds. Ergun and Mentzer's results, however, would suggest that the equilibration assumption should be used with care.

Retardation of Reactions. Equations 19.36–19.39 all show that resorption of the reaction product will retard the reaction by

[250] J. E. L. Rogers, A. F. Sarofim, J. B. Howard, G. C. Williams, and D. H. Fine, "Combustion Characteristics of Simulated and Shredded Refuse," *Proc. 15th Symp. (Int.) Combustion,* Combustion Institute, Pittsburgh, 1975, pp. 1137–1148.

occupation of active sites, as already discussed. An additional retardation effect is that of hydrogen on the C/CO_2 reaction. Hydrogen therefore inhibits three reactions: its own adsorption (Equation 19.10), the C/H_2O reaction (Equation 19.9) and the C/CO_2 reaction (Equation 19.8). Retardation of Equation 19.9 is due to reverse reaction. Retardation of Equation 19.8 is adsorption on vacant sites by Equation 19.10. At low CO_2 and CO pressures, Biederman et al.[242] conclude that the rate without hydrogen present (which they show is very nearly an impossible condition) is given by

$$\text{rate} = k_2 \, p_{CO_2}$$

The rate with hydrogen present, however, is given by

$$\text{rate} = \frac{k_1 \, p_{CO_2}}{1 + K_4^{1/2} \, p_{H_2}^{1/2}}$$

This suggests that the retardation by hydrogen may have been oversimplified when represented by Equation 19.37. Either the hydrogen term should be to the half-power or there should be an additional half-power term.

Symmetry would suggest that CO should retard the steam reaction if hydrogen retards the CO_2 reaction. This is not the case, although the subject has been discussed extensively in the literature. Mechanisms have been invoked involving the water-gas shift reaction and other factors. The simple answer, however, may be the most likely. The CO counterpart to the hydrogen adsorption reaction (Equation 19.10) exists but is not important. Without additional significant potential for filling vacant sites, CO can operate only through the reverse reaction of Equation 19.8. There is otherwise only an indirect effect noted by Gadsby et al.,[117] due to the water-gas shift reaction being accelerated over carbon. This can increase the hydrogen concentration, which is then able to participate in increased inhibition.

The C/O_2 Reaction. This reaction has been the most intensively investigated, and it may also be the least understood. It is important to note that according to Equation 19.40, the coverage, or rate, is first order with respect to oxygen concentration, p_O, at low p_{O_2}, in spite of dissociative adsorption. Further discussion is deferred (Section 19.6).

19.5.3 Elovich Kinetics

19.5.3.1 *Empirical Formulation*

The Elovich equation is usually expressed in terms of the quantity of gas adsorbed or desorbed (q). Textbook discussions[107,108] are generally confined to adsorption, but the equation has been successfully applied to desorption.[111] The empirical formulation is given in either the differential or the integral form. The differential form is

$$\frac{dq}{dt} = a e^{-\alpha q} \qquad (19.45)$$

where a and α are empirical constants. The relation to "fundamental" derivations is discussed below. The integrated form is

$$q = \frac{1}{\alpha}[\ln(t + t_0 \, e^{\alpha q_0}) - \ln t_0] \qquad (19.46)$$

where $t_0 = 1/a\alpha$ and q_0 is the quantity of adsorbate at $t = 0$: this can be q_0 in desorption or 0 in adsorption. This may be compared with the time-dependent solutions of the L-H equations, given in Section 19.5.2.2. A third form of the equation is obtained by substituting the solution, Equation 19.46, into the differential form, Equation 19.45. This yields

$$\frac{dq}{dt} = \frac{t_0}{t + t_0 \, e^{\alpha q_0}} \qquad (19.47)$$

The Elovich equation has been used in a number of carbon and coal investigations, including oxygen and hydrogen adsorption-desorption on carbon,[111,113–115] low-temperature oxygen adsorption on coal,[212,214–217] and volatiles production from coals.[251] In using Equation 19.46, it is quite common to disregard the t_0 term, and this practice is acceptable if t_0 or $t_0 \, e^{\alpha q_0}$ is small compared with t,

[251] E. T. Nelson, J. Worrell, and P. L. Walker, Jr., "Kinetics of Volatile Matter Release from Pennsylvania Anthracites," Paper No. 38 in *Coal Science,* Am. Chem. Soc. Adv. Chem. Ser. No. 55 (1966), pp. 602–620.

when a relevant plot is of q against $\ln t$ as illustrated in Fig. 19.37. The validity of this procedure has nevertheless been questioned. Without an independent method of estimating t_0, the fitting procedure that requires adjustment of t_0 to produce a straight line is regarded by some as too arbitrary and empirical.[252] It may be noted that Tucker and Mulcahy[111] disregard t_0 in the time differential form, Equation 19.47 (i.e., $dq/dt \propto b/t$). Their integral form is correspondingly simpler, but must be in error at low t, since with $t_0 = 0$, $q \to \infty$ as $t \to 0$.

19.5.3.2 *Comparison with Other Isotherms*

The L-H isotherms of Section 19.5.2, and others such as those of Freundlich and Temkin,[107,108] are steady state expressions for equilibrium between adsorption and desorption. The Elovich equation is an expression for the rate of one of the elementary steps, r_i or r_{-i} in Equation 19.16. It is not a steady state expression. Hayward and Trapnell[108] give general expressions for the reaction velocities, r, in terms of the accommodation coefficient, σ, the fractional coverage, θ, and

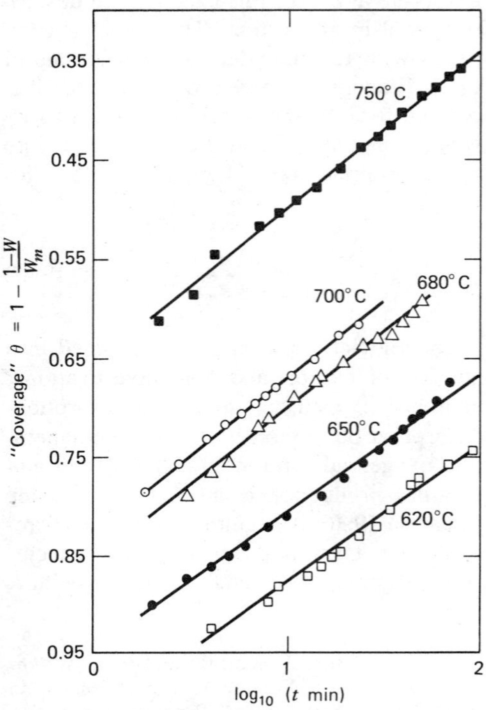

Figure 19.37 Elovich plots for desorption of surface oxide.[111]

the activation energy, E. For the Langmuir assumptions, σ and E are assumed to be constants. In general, however, they may be functions of θ, thus

$$r_i = \frac{\sigma(\theta) p_i f_i(\theta) \exp - \dfrac{E_i(\theta)}{RT}}{(2\pi mkT)^{1/2}} \tag{19.48}$$

$$r_{-i} = K f_{-i}(\theta) \exp \frac{-E_{-i}(\theta)}{RT} \tag{19.49}$$

where $f_i(\theta)$ and $f_{-i}(\theta)$ are functions of coverage [$= (1 - \theta)^n$ and θ^n, respectively, for the L-H kinetics]. At equilibrium, when $r_i = r_{-i}$, a general adsorption isotherm yields $p_i \propto \exp(-Q/RT)$, where Q, the heat of adsorption, is the difference $E_i - E_{-i}$. It is well known that the Temkin isotherm is obtained from the Langmuir isotherm by inserting the condition that Q declines linearly with coverage, and by judicious neglect of the ratio f_i/f_{-i}. This is true for either a uniform or a nonuniform surface.

The Elovich equation may be shown to be related to the Temkin isotherm in the following way. If $Q = E_i - E_{-i}$, then $Q = Q_0(1 - b\theta)$ is a linear fall in Q with θ that can reflect a similar linear fall in E_i or E_{-i} or both. Writing $E = E_0(1 - \alpha'\theta)$, and assuming for the moment that θ is proportional to q, the quantity adsorbed, we obtain

$$\begin{aligned} r &= \frac{dq}{dt} \\ &= \text{const} f(q) \exp \frac{-E_0(1 - \alpha q)}{RT} \end{aligned} \tag{19.50}$$

which is of the form of Equation 19.45 with the additional term $f(q)$ if σ is a constant. Where $f(q) = (1 - q)$, as in the simplest condition discussed above, it is generally supposed that when q is not close to unity, $(1 - q)$ has a small variation compared with the exponential term. Thus the Elovich equation is obtained from the elementary analysis for a uniform surface. If the surface is nonuniform, an integral is obtained that can be integrated if it is assumed that there is a first subset of surface elements of high E_0 that have low coverage, whence $f(q)$ is near unity; and a second subset, of low E_0, that have high coverage, whence $f(q)$ is near zero. Integration over all sites then yields the same relation, Equation 19.45.[108]

This derivation provides a basis for evaluating results of Elovich applications. If the Elovich equation is found to apply, the slopes of the lines obtained at different temperatures can be interpreted as due to variable activation energies (e.g., Tucker and Mulcahy[111]). This is the conventional interpretation. The derivation, however, requires that E change quite drastically with coverage for $f(q)$ to be effectively constant, even on a small proportion of a uniform surface. It is not required that the surface be nonuniform. When E was progressively measured, however, by Bansall et al.,[113,115] the values at the different steps were found to be almost constant (Figs. 19.10 and 19.11). This is saying that α in Equation 19.50 is almost zero, and the principal variation of dq/dt should be due to $f(q)$, which is no more than a Langmuir condition unless σ varied significantly with coverage.

19.5.3.3 *Other Derivations*

The Derivation of Boudart. In one view the Elovich equation is totally empirical, and a search for a theoretical derivation is held to be fruitless. Low[252] has provided several such comments, citing Boudart[253] in particular to the effect that the logarithmic expression may represent a number of different processes, both chemical and physical. More recently, however, Boudart[254] has described an analysis to show that a relationship for a nonuniform surface can be established between the Freundlich-Temkin (or Frumkin-Temkin, also known as the Slygin-Frumkin[107]) and Elovich isotherms. A general rate equation is obtained for a two-step, reversible reaction sequence that also incorporates a surface nonuniformity parameter (γ).

The three isotherms then appear as special cases of the general rate expression. The Freundlich and (Frumkin-) Temkin isotherms are obtained for reversible one-step adsorption (such as hydrogen adsorption) with finite and zero values, respectively, of the surface nonuniformity parameter. The Elovich equation is obtained as an extension of the (Frumkin-)Temkin isotherm for which $\gamma = 0$, but with a two-step reaction and with the adsorption step rate determining. This development for *uniform* surfaces is in line with the Bansall findings.[113,115]

The three isotherms then appear as special cases of the general rate expression. The Freundlich and (Frumkin-Temkin) isotherms are obtained for reversible one-step adsorption (such as hydrogen adsorption) with finite and zero values, respectively, of the surface nonuniformity parameter. The Elovich equation is obtained as an extension of the (Frumkin-Temkin) isotherm for which $\gamma = 0$, but with a two-step reaction and with the adsorption step rate determining. This development for *uniform* surfaces is in line with the Bansall findings.[113,115]

The Derivation of Taylor and Thon. An alternative approach was suggested by Taylor and Thon,[255] who assumed that there was increasing coverage with coverage per unit molecule adsorbed; that is, $d\theta/dq$ was a function of θ. Taylor and Thon use number of sites, N, in place of θ, but if it is assumed that the total number of sites, N_0, remains invariable, then the fraction filled, N/N_0, may be equated to θ. Adapting Taylor and Thon's equations and distribution assumption, we write:

$$\frac{d\theta}{dq} = \alpha(1 - \theta) \tag{19.51}$$

where q, as before, is the quantity of fluid adsorbed. [*Note.* Taylor and Thon[255] actually assume $d\theta/dq = -\alpha\theta$. This has to be wrong, since it yields $\theta = \exp(-\alpha q)$, which means that θ (or N, the filled number of sites) must *decline* with increasing q. Moreover, it gives $-d\theta/dt = (\alpha a)\theta^2$, which integrates to a time-dependent function with $t_0 = 0$, which is impossible.]

Integrating yields

$$\theta = 1 - e^{-\alpha q} \tag{19.52}$$

[252] M. J. D. Low, "Kinetics of Chemisorption of Gases on Solids," *Chem. Rev.*, **60**, 267–312 (1960).

[253] M. Boudart, *Kinetics of Chemical Processes*, Prentice-Hall, Englewood Cliffs, N.J., 1968.

[254] M. Boudart, "Heterogeneous Catalysis," in *Physical Chemistry*, Vol. 7, H. Eyring, B. Henderson, and W. Jost, Eds., Academic Press, New York, 1975, Ch. 7.

[255] H. A. Taylor and N. Thon, "Kinetics of Chemsorption," *J. Am. Chem. Soc.*, **74**, 4169–4173 (1952).

This shows that the assumption used in Section 19.5.3.2, that q is approximately proportional to θ, is only true for small q.

If Equation 19.48 expresses the rate of molecular adsorption, this is dq/dt. If we then have σ and E_i both constant, and $f_i(\theta) = 1 - \theta$, as for the Langmuir isotherm, then

$$\frac{dq}{dt} = a(1 - \theta) \tag{19.53}$$

$$= ae^{-\alpha q} \tag{19.53a}$$

from Equations 19.52 and 19.45.

We also find

$$\frac{dq}{dt}\frac{d\theta}{dq} = \frac{d\theta}{dt} = a\alpha(1 - \theta)^2 \tag{19.54}$$

which is Equation 19.18 with $m = 0$ and $n = 2$. Ritchie[256] recently established the same result by making the empirical assumption that $d\theta/dt$ was proportional to $(1 - \theta)^n$, then showing that the best-fit value of n to selected experimental results (including those of Bansall et al.[115]) was $n = 2$. This means that the Elovich equation can be represented equally well as an L-H isotherm with increasing coverage per adsorbed molecule as θ increases. Difficulties in differentiating between alternate assumptions is partly a consequence of the equations being developed explicitly or implicitly in terms of coverage, θ, when what is measured is the quantity adsorbed, q, which may or may not be proportional to θ.

It may be assumed that on a nonuniform surface, the equations above then apply to each uniform section. The value of this approach is that it is the only one consistent with Bansall's experimental findings that E is apparently constant for a given subset of adsorption sites, not highly variable. In effect, it reconciles the L-H equations developed in terms of coverage, θ, with isotherms as measured in terms of adsorbed or desorbed quantities, q.

One other interpretation is one of those suggested by Boudart, that the behavior modeled by the Elovich equation is physical, not chemical. This is well illustrated by Fig. 19.38, showing the variation of gas loss with time from anthracite. The two curves represent chemical release and diffusional loss. The close agreement between the two up to 95% is quite clear.

19.5.4 Internal Diffusion Kinetics

19.5.4.1 *Models*

General Description and Parameter Definition. The general behavior when reaction occurs inside the solid is described in Section 19.3.1.2. The physical model is of a porous solid with reactant diffusing into the pores, reacting as it penetrates. Mathematical for-

[256] A. G. Ritchie, "Alternative to the Elovich Equation for the Kinetics of Adsorption of Gases on Solids," *J. Chem. Soc. Faraday I*, **10**, 1650–1653 (1977).

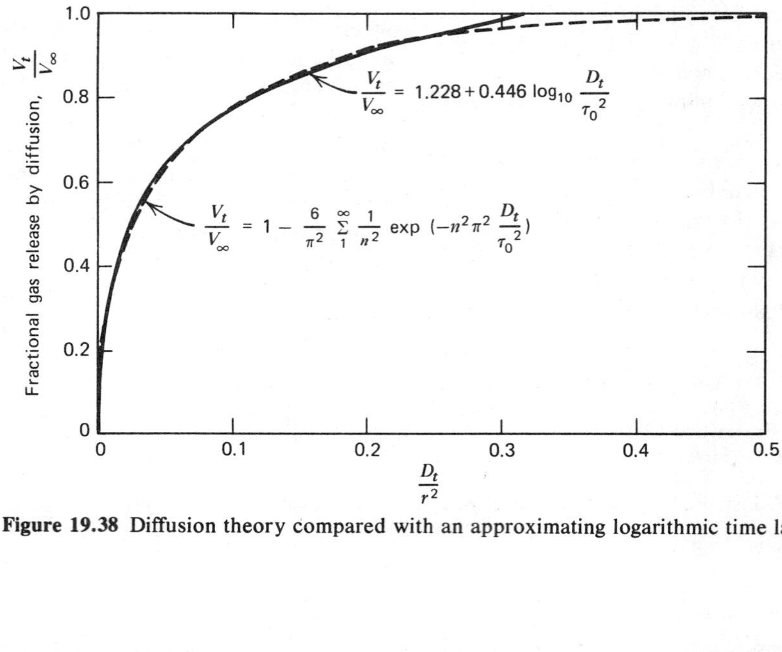

Figure 19.38 Diffusion theory compared with an approximating logarithmic time law.[251]

mulations of the model were initiated by Thiele[257] for reaction inside porous catalysts. Enlargements of the catalysis model were reviewed by Wheeler[243] and Weisz and Prater.[258] Applications of the model to the carbon reactions were extensively reviewed by Walker et al.[30] There are also brief to extensive descriptions in numerous texts such as those by Bird et al.,[259] Levenspiel,[260] Aris,[261] and Thomas and Thomas (Ch. 4 of Ref. 107). There is also a useful treatment by Eyring and Lin.[61]

There are several possible ways of formulating the problem. The first, originating with Thiele[257] is to represent the porous body as a set of uniform, parallel, cylindrical pores penetrating into the body in a direction perpendicular to the external surface and parallel to the direction of diffusion inside the body. Variations on this include tortuous pores, nonuniform pores (variable radius along their length), uniform or nonuniform pores with a distribution of mean radii, noncylindrical pores, and intersecting pores. The study of reversible adsorption often shows hysteresis loops in plots of quantity adsorbed against relative pressure and, on the basis of the five different types of hysteresis loop identified (types A to E), five sets of capillary shapes have been proposed to account for the observed behavior. These are well described by Thomas and Thomas.[107]

Coal is recognized as having chamber and aperture types of pores. These will change shape and size on swelling and reaction, but the original structure can affect some aspects of reaction behavior. Specifically, carbon from cracked gases can deposit in the apertures, and the reduced access to the internal surface can be quite out of proportion to the quantity of carbon deposited.

[257] E. W. Thiele, "Relation Between Catalytic Activity and Size of Particles," *Ind. Eng. Chem.*, **31**, 916–920 (1939).

[258] H. L. Weisz and C. D. Prater, "Interpretation of Measurements in Experimental Catalysis," in *Advances in Catalysis*, Vol. 6, Academic Press, New York, 1954, pp. 143–196.

[259] R. B. Bird, W. E. Stewart, and E. N. Lightfoot, *Transport Phenomena*, John Wiley and Sons, New York, 1960.

[260] O. Levenspiel, *Chemical Reaction Engineering*, John Wiley and Sons, New York, 1962.

[261] R. Aris, *Introduction to the Analysis of Chemical Reactors*, Prentice-Hall, Englewood Cliffs, N.J., 1965.

With such complexity of pore shape, size, direction, and other characteristics, one other approach to analysis is to represent the pore behavior on more of a phenomenological basis. A thin section or a thin shell is characterized by assigning a local value of porosity, $\epsilon = (1 - \rho/\sigma)$; true density, σ; solid density, ρ; surface area per unit volume, S_v; surface area per unit mass, S_m; and effective diffusion coefficient, D_e. Relationships exist between these assigned values and the dimensions of some equivalent pore. We have

$$\rho S_m = S_v = \frac{2\epsilon}{r} \qquad (19.55)$$

where r is the pore radius, and

$$D_e = \frac{D_T\,\epsilon}{\tau} \qquad (19.56)$$

where τ is a tortuosity factor, and D_T is the molecular or Knudsen diffusion coefficient (respectively D_M and D_K), or a sum of their reciprocals

$$\frac{1}{D_e} = \frac{1}{D_M} + \frac{1}{D_K} \qquad (19.57)$$

The diffusion coefficients are given by, respectively, $D_M = \frac{1}{3}\lambda\bar{u}$, and $D_K = \frac{2}{3}r\bar{u}$, where the mean Maxwellian velocity $\bar{u} = (8RT/\pi M)^{\frac{1}{2}}$.

General Equation for Diffusive Flow of Fluid Through a Porous Body Accompanied by Reaction. For a change of concentration, p_j, of some reactant, Eyring and Lin[61] give the following equation:

$$\partial p_j / \partial t = \nabla \cdot (D_e \nabla p_j) - \nabla \cdot (p_j \mathbf{v}) + (R_j)_v \qquad (19.58)$$

where $\mathbf{v}$ is the velocity of the fluid center of mass and $(R_j)_v$ is the reaction rate per unit volume for reactive component j. This is the conservation equation for a thin element or section in which there is reaction, with reactant supplied both by diffusion and by bulk or convective (forced) flow.

Convective or forced flow is due to a pressure difference that is created either by a sufficient pressure gradient in the ambient fluid, which is rare, or by fluid generation or absorption inside the porous body that cannot be immediately rectified by convective flow. A diffusion coefficient for forced flow can be written[61] as $D_p = r^2 Y_T R T/8\eta)$, where Y_T is the total molar concentration of all molecules and η is the fluid viscosity. Evaluation yields the estimate of about 0.1 cm^2/s for D_p at STP, for gases of 10^{-4} cP in pores of 1 μm. For pores larger than 100 μm, forced flow generally is so rapid that significant excess pressures would not develop, and forced flow is likely to dominate the fluid transfer. In small pores bulk diffusion can compete with forced flow, and some internal pressures can develop.

Significant volume generation is not likely to occur, however, in the carbon or char combustion; thus the flow term can normally be ignored. There are two usual sources of volume generation or reduction.

1. Pyrolysis, and also the consequences of significant change in volume on reaction, as in the case of one O_2 molecule creating two CO molecules. The flow term is likely to be dominant in the case of coal pyrolysis, but this is obviously not a factor when pyrolysis has finished. As pyrolysis terminates, however, back diffusion of oxygen against the flow could occur, allowing simultaneous pyrolysis and oxidation. This possibility needs to be explored.
2. The second source of volume change, due to reaction, can be incorporated in the basic differential Equation 19.58. Thiele[257] considered this problem, showing that there was an additional term in the equation: $\nu - 1$, which represents the increased number of moles. By assuming that Y_T (total number of moles) was constant, Thiele obtained an analytical solution. The validity of the assumption is discussed by Wheeler[243] and by Thomas and Thomas.[107] It is considered good enough to give a qualitative picture of the behavior. Alternatively, the effects of the volume change are sometimes incorporated in the common diffusion coefficient.

Simplified Rate Equation for Diffusion with Reaction. Equation 19.58 is usually applied under constraints that simplify it considerably. The usual constraints are steady state (zero time derivative), no forced flow, and either linear, cylindrical, or spherical symmetry. It is convenient to transform the volumetric reaction rate term, $(R_j)_v$, into the product of the volume-specific surface, S_v, and the surface reaction rate term, $(R_j)_s$, given by Equation 19.33. Then R_j can be written as the product of an effective velocity constant, $\bar{k}$, and a concentration function, $f(p)$, where $\bar{k} = k_5$ and $f(p) = \theta_m$ if Equations 19.34–19.40 are used. It is also convenient and common to use a simple power approximation p_j^n for $f(p_j)$; thus incorporating Equation 19.33 we have

$$R_j = k_5\theta_m = \bar{k}f(p_j) = \bar{k}p_j^n \qquad (19.59)$$

in place of the more complete but more cumbersome equations. This also defines $\bar{k}$. Dropping the general subscript j, Equation 19.58 becomes

$$d\left(a^s \frac{dp}{da}\right) = \lambda^2 a^s f(p)\, da \qquad (19.60)$$

where a is an appropriate linear or radial dimension, s is a shape factor (= 0 for a plane, 1 for a cylinder, and 2 for a sphere), and λ is given by

$$\lambda^2 = \frac{S_v \bar{k}_s}{D_e} \quad \text{for a porous body} \qquad (19.61a)$$

$$\lambda^2 = \frac{2\bar{k}_s}{D_T} \quad \text{for a cylindrical pore} \qquad (19.61b)$$

Equation 19.60 can also be written in the alternative form:

$$d\left(a^2 \frac{dp}{da}\right)^2 = 2\lambda^2 a^{2s} f(p)\, dp \qquad (19.62)$$

19.5.4.2 *Solutions*

The information primarily required from integration of either Equation 19.60 or 19.62 is the value of the slope, dp/da at the solid surface, at $a = a_s$, since this is proportional to the flow of reactant into the solid and therefore to the total reaction rate. This requires only one integration of either equation. The second integration may be required to determine necessary coefficients set by boundary conditions.

There are two sets of boundary conditions in each case. As described in Section

19.3.1.2, the indiffusing gas decreases in concentration as it reacts in the pore. The extent of reaction depends on the rate of diffusion and on the pore reactivity. If reactivity is low with respect to the rate of diffusion, the reacting gas will penetrate to the center of the reacting solid, and its concentration at the center will be finite. If reactivity is high, on the other hand, the reactant concentration will fall to zero some distance from the center and there will be an unreacted core at any given time.

In this section we focus on the behavior of a plane solid of half-thickness, a_s, and on a sphere of radius a, with distance, a, measured hereafter from the point or plane of symmetry. If reaction occurs throughout the solid, there can be a finite value of $p = p_i$ (internal value) at $a = 0$. If reaction penetrates only to some distance "a_0" from the center, $p = 0$ at $a = a_0$. These two conditions define the "Zone I" and "Zone II" behaviors, respectively (Section 19.3.1.2).

The two different sets of boundary conditions usually generate two sets of solutions. Different solutions are also generated by the different shape factors ($s = 0$ for the plane, and $s = 2$ for the sphere), and the different reactivity behavior—governed by reaction order. Solutions given here are restricted to the plane and sphere, and for the most part to zero- and first-order reactions.

Plane: Shape Factor $s = 0$. For this condition, and using Equation 19.59, Equation 19.62 integrates to

$$\frac{dp}{da} = \sqrt{2}(\lambda)\left[\frac{p^{(n+1)} - p_i^{(n+1)}}{n+1}\right]^{1/2} \qquad (19.63)$$

If p_i can be taken as zero, for the Zone II condition, Equation 19.63 can be used to calculate the reaction rate by taking the value of dp/da at $p = p_s$, where p_s is the gas concentration at the solid surface. If this condition of $p_i = 0$ is not met, the equation must be integrated further. Simple solutions only exist for limited values of n. The solutions of greatest interest to us here are for zero- and first-order reactions.

1. *Zero-order reaction: $n = 0$.* For Zone I, with $p = p_i$ at $a = 0$, Equation 19.63 integrates to

$$\frac{p}{p_s} = 1 - \frac{\lambda^2}{2p_s}(a_s^2 - a^2) \qquad (19.64)$$

and for Zone II, with $p = 0$ at $a = a_0$, we have

$$\frac{p}{p_s} = \frac{\lambda^2}{2p_s}(a - a_0)^2 \qquad (19.65)$$

2. *First-order reaction: $n = 1$.* For Zone I, with $p_i = p$ at $a = 0$, we obtain

$$\frac{p}{p_s} = \frac{\cosh \lambda a}{\cosh \lambda a_s} \qquad (19.66)$$

and for Zone II, with $p = 0$ at $a = a_0$

$$\frac{p}{p_s} = \frac{\sinh \lambda a - \sinh \lambda a_0}{\sinh \lambda a_s - \sinh \lambda a_0} \qquad (19.67)$$

noting that the product (λa_s) is often written as ϕ, the Thiele modulus.[257]

3. *For other reaction orders.* A general solution is not known to exist for such cases. Intermediate reaction orders can be handled only by inserting an appropriate function for $f(p)$ in Equation 19.62, such as Equation 19.59, and integrating once to give Equation 19.63. Such methods are based, however, on the implicit assumption that the adsorbed film coverage is small and fixed, or large with decomposition independent of coverage. Thomas and Thomas[107] show that if coverage is θ, the equations obtained previously apply, but with a term $(1 - \theta)$ appearing in the constant λ. This assumes, however, that θ is constant along the pore, which obviously is unlikely. Eyring and Lin[61] showed that if the variation of θ is important, the Langmuir equation can be used instead for $f(p)$. Writing $f(p) = (k_1 k_2 p / \bar{k}) / (k_1 p + k_2)$, integration yields

$$\frac{dp}{da} = \lambda\left(\frac{2k_2}{\bar{k}}\right)^{1/2}\left[(p - p_i) - \left(\frac{k_2}{k_1}\right)\frac{\ln(k_1 p + k_2)}{(k_1 p_i + k_2)}\right] \qquad (19.68)$$

This can be used if $p_i = 0$, which is valid for Zone II.

Sphere: Shape Factor $s = 2$. With $s = 2$, no general first integration is possible using Equation 19.59. Equation 19.60, however, becomes

$$\frac{d^2p}{da^2} + \frac{2dp}{ada} = \lambda^2 p^n \qquad (19.69)$$

This is a variant of the astronomical Emden[262] or Lane-Emden[263] equation of order n, and exact solutions are known for $n = 0$, 1, and 5. Only the first two concern us here. A general series solution in terms of n and a was developed by Airey (see Refs. 262, 263), and tables were computed corresponding to values of n from 0 to 6. The variation lies in the sign of λ^2, which is positive in this formulation and negative in the astronomical form. The solutions to the astronomical equations can be used, however, with $i\lambda$ altering solutions in functions to solutions in hyperbolic functions.

As before, we consider only the zero-order and first-order cases.

1. *Zero-order reaction: n = 0.* For Zone I we obtain

$$\frac{p}{p_s} = 1 - \frac{\lambda^2 a_s^2}{6p_s}\left(1 - \frac{a^2}{a_s^2}\right) \qquad (19.70)$$

with $p = p_i$ at $a = 0$.
For Zone II we obtain

$$\frac{p}{p_s} = 1 - \frac{\lambda^2}{6}(a_s^2 - a^2) - K\frac{1}{a} - \frac{1}{a_s} \qquad (19.71)$$

where

$$K = \frac{1 - (\lambda^2/6p_s)\,(a_s^2 - a_0^2)}{(1/a_0) - (1/a_s)} \qquad (19.71a)$$

2. *First-order reaction: n = 1.* A general solution can be written

$$\frac{p}{p_s} = \left(\frac{a_s}{a}\right)\left(\frac{\sinh \lambda a}{\sinh \lambda a_s}\right)\left[\frac{1 - \dfrac{\cosh \lambda_a \sinh \lambda a_0}{\sinh \lambda_a \cosh \lambda a_0}}{1 - \dfrac{\cosh \lambda a_s \sinh \lambda a_0}{\sinh \lambda a_s \cosh \lambda a_0}}\right] \qquad (19.72)$$

When a_0 becomes zero, this expression, which is valid for Zone II, reduces to the more commonly used expression valid in Zone I:

$$\frac{p}{p_s} = \frac{a_s \sinh \lambda a}{a \sinh \lambda a_s} \qquad (19.73)$$

A number of sources use as a boundary condition $dp/da = 0$ at $a = 0$ (center), or $a = a_s$ (surface), or $a = a_0$ (penetration depth). Differentiation of the solutions above shows that the gradient is not necessarily zero at the locations given.

19.5.4.3 *Rate Equations*

Effectiveness Factor and Thiele Modulus. The local, internal rate of reaction is given by Equation 19.33. The diffusion equations given above provide the basis for summing the local values to give the total internal reaction rate. For both practical use and for theoretical evaluation it is then convenient to express the total internal reaction rate in terms of the total rate per unit of external or superficial surface, R_s.

The overall maximum possible rate is obtained when all internal surface is involved at the maximum possible rate. For this to occur, the internal gas concentration must be uniform, at the surface value, p_s. If the volume of the reacting solid is V_p with an external surface S_p, the internal surface is $V_p S_v$, and the internal reaction rate using Equation 19.59 is $V_p S_v \bar{k}f(p)$. Thus the maximum specific reaction rate ($R_{s,m}$) is given by

$$R_{s,m} = \frac{V_p}{S_p} S_v\, kf(p) \qquad (19.74)$$

The actual specific reaction rate is obtained, as described above, from the value of the rate of diffusion of reactive gas into the solid. For unit area of surface this gives

$$R_s = D_e \left(\frac{dp}{da}\right)_{(a\,=\,a_s)} \qquad (19.75)$$

The ratio of the actual to the maximum reaction rate is the reaction efficiency or effectiveness factor, η; thus

$$\eta = \frac{R_s}{R_{s,m}} \qquad (19.76)$$

[262] H. T. Davis, *Introduction to Nonlinear Differential and Integral Equations,* Dover, New York, 1962.
[263] C. G. Lambe and C. J. Tranter, *Differential Equations,* Row, Peterson & Company, New York, 1961.

Evaluating Equation 19.76 for the plane and the sphere, *for first-order reaction in Zone I,* by differentiating Equations 19.66 and 19.73 and substituting appropriately, we obtain the commonly quoted expressions

$$\eta = \frac{\tanh \lambda a_s}{\lambda a_s} \quad \text{for the plane} \quad (19.77)$$

$$\eta = \frac{3}{\lambda a_s \tanh \lambda a_s}\left(1 - \frac{\tanh \lambda a_s}{\lambda a_s}\right) \quad \text{for the sphere} \quad (19.78)$$

noting that these expressions are all in terms of the Thiele modulus, λa_s or ϕ.

For other reaction orders and different zones, other expressions are derived. The most useful general expression may be obtained by considering application of Equation 19.63 to Equation 19.76, with the special condition that $p = 0$, and with $p = p_s$. The result is

$$\eta = \frac{1}{(\lambda a_s) p_s^{(n-1)/2} \left(\frac{n+1}{2}\right)^{1/2}} \quad (19.79)$$

This defines a modified Thiele modulus, $\phi = 1/\eta$, for reaction order n. For $n = 1$ it reduces to $\eta = 1/\lambda a_s$, which is an approximation of Equation 19.77 for λa_s greater than about 2 when tanh λa_s is close to unity. Petersen[264] therefore suggested that a general Thiele modulus could be written

[264] E. E. Petersen, *Chemical Reaction Analysis,* Prentice-Hall, Englewood Cliffs, N.J., 1965.

$$\phi_g = \frac{1}{S}\lambda a_s p_s^{(n-1)/2}\left(\frac{n+1}{2}\right)^{1/2} \quad (19.80)$$

Curves of η against ϕ_g using the general modulus were found to fall on something of a common curve, reproduced in Fig. 19.39. The values for zero-order reaction deviate from the more common curve at values of ϕ_g near unity. For $\eta = 1$, the curve is two straight lines with values $\eta = 1$ for $\phi_g < 1$, and $\eta = 1/\phi_g$, as given by Equation 19.79, for $\phi > 1$. In general, however, the value $\phi = 1$ is a good guide as a dividing line between Zone I and Zone II reactions.

Clearly, the two zones do not change discontinuously from one to the other, except in the zero-order case, so the existence of a transition region between zones should be borne in mind. The assumption of uniform film coverage (unless Equation 19.68 is used) should also be remembered. In general, reaction order can be expected to change from zero to unity with increasing penetration depth of the reaction, except in the case of the very slowest reactions, where there is zero-order reaction throughout. Some authors, therefore, give more stringent criteria for Zone I reactions. These have been discussed by Walker et al.,[30] who prefer to take the value $\phi_g < 0.3$ as defining Zone I. This gives $\phi^2\eta$ approximately equal to 0.1, as another form of the criterion.

[265] R. Aris, "Transition Between Regimes in Gas-Solid Reactions," *Ind. Eng. Chem. Fundam.,* **6,** 315–318 (1967).

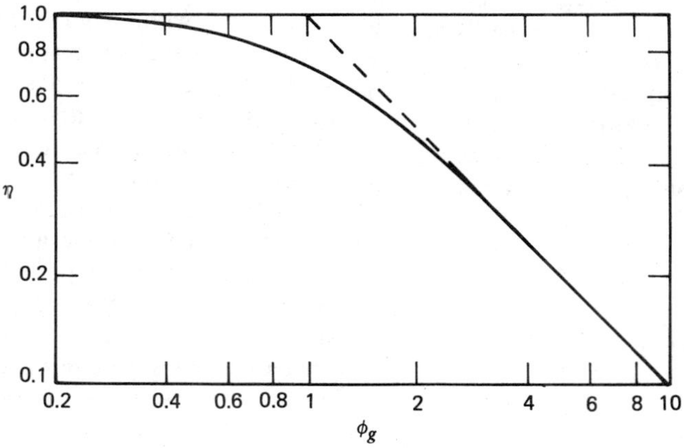

Figure 19.39 Mean effectiveness factor for all shapes.[265]

Rate Equation. Combining Equations 19.74a and 19.76 and using 19.59, a general rate equation may be written:

$$R_s = \frac{a_s}{s}(S_v\bar{k})p^n\eta \qquad (19.81)$$

$$= \frac{2D_e}{a_s\, s\,(n+1)}(\phi^2\eta)p_s \qquad (19.81a)$$

1. *For Zone I.* When η is approximately 1, Equation 19.81 reduces to

$$R_s = \frac{a_s}{s}(S_v\bar{k})p^n \qquad (19.82)$$

and the reaction order is the true order in the solid.

2. *For Zone II.* The value of η is approximately $1/\phi_g$, and Equation 19.81 reduces to

$$R_s = \frac{a_s}{s}(n+1)^{-1/2}(2S_vD_e\bar{k}_s)^{1/2}p_s^{(n+1)/2} \qquad (19.83)$$

This is the well-known result yielding a modified reaction order, $(n+1)/2$, and a temperature coefficient that is the square root of the true value. For a true reaction order of zero, the effective order becomes one-half; and for a true activation energy of E, the apparent value is $(E/2) - (3RT/8)$, which is approximately $E/2$ for large E. The correction term derives from the dependence of D_e on T.

19.5.5 Kinetics with Boundary Layer Diffusion

19.5.5.1 *Total Rate Equations*

The equations in the foregoing section describe behavior due to internal reaction, with removal of internal mass. This is the behavior that changes density. The surface may also regress, however, and the total reaction rate, R_T, must include the effect of surface regression. The total rate equation for reaction at unit surface may be written

$$R_T = \psi\bar{k}f(p_s) + \frac{a_s}{s}(S_v\bar{k})f(p_s)\eta \qquad (19.84)$$

$$= \bar{k}f(p_s)\psi\,(1 + a_sS_v\eta/s\psi) \qquad (19.84a)$$

where ψ in a roughness factor for the external surface. Clearly, for Zone II conditions with small penetration, ψ can be replaced by ψ_{eff}, given by the term in parentheses. For Zone I conditions we could expect the second RHS term in Equation 19.84 to be dominant, allowing us to neglect the first RHS term.

Reaction will always include boundary layer diffusion, which may or may not be dominant. It may be included quite generally by writing k_D for a boundary layer mass transfer coefficient, or velocity constant, whence

$$R_T = k_D(p_O - p_s) \qquad (19.85)$$

where p_0 is the mainstream gas concentration. Equations 19.84 and 19.85 must be solved simultaneously; Caram and Amundson[24] describe some very detailed solutions.

19.5.5.2 *Velocity Effects*

Inclusion of the diffusion velocity constant introduces the effects of ambient flow. These are secondary to the main purpose of this chapter and are not discussed in detail. The circumstances of interest are quite limited, and are largely determined by the particle size (see Table 19.1).

In fluid beds the particles are mostly smaller than a few millimeters; and in entrained beds they are smaller than about 100 μm. Treating the particles as spheres, k_D in Equation 19.85 is commonly incorporated in the Sherwood number: $\text{Sh} = 2a_sk_D/D_M$. The Sherwood number can be represented as a function of the Reynolds (and Schmidt) number(s), but for 100-μm particles or smaller, the Reynolds numbers are always so small at any realistic relative velocity that the Sherwood number can be taken at its limiting value of 2 with insignificant error; the difficulties in developing appreciable velocities between particles and gas in a cloud are not always fully appreciated. The Schmidt number may be affected by turbulence, however, if the scale is small and the intensity is high.

At the particle densities of interest in entrained clouds, cooperative or interactive mechanisms between particles can reduce or eliminate the normal effects of elutriation, and the cloud tends to move as a unit, including the surrounding gas.[266] The treatments often used to estimate the acceleration

[266] R. G. Boothroyd, *Flowing Gas-Solids Suspensions,* Chapman and Hall, London, 1971.

behavior of single particles in a cloud are usually inapplicable on that account.

In the fluid bed, where the particles are an order of magnitude larger, the relative velocities are still so low that the limiting Sherwood number of 2 is close to what has been measured. Measured values, in fact, have been below the normal limit of 2, with average reported[184] values of 1.4 attributed to the effects of blockage by surrounding particles.

In a fixed bed, as for lumps on a grate, or in a shaft gasifier, for example, the model changes from a spherical particle to a set of intersecting tortuous pores (like the pores in a porous particle on a larger scale). Here, also, correlations with Reynolds and Schmidt numbers for mass transfer in a porous bed are available and have been successfully used.[259] A recent investigation[65] suggests

$$k_D = \text{const}\left(\frac{V}{a_s}\right)^{0.7} \qquad (19.86)$$

where V is the flow velocity in the system. The more difficult problem lies in determining the best value for the constant, but this is not of concern here.

The problems of fluid mechanics in heterogeneous combustion have been discussed elsewhere.[65]

19.5.5.3 *Combined Kinetics*

The diffusional and chemical kinetics are usually combined by eliminating the local surface gas concentration, p_s, between Equations 19.84 and 19.85. This requires selection of an appropriate function for $f(p_s)$. It also involves knowing η (i.e., the zone in which the reaction occurs). At either of the two extremes—first, of being so far into Zone I that η is nearly unity and we can write $\psi = \psi_{\text{eff}}$, and second, of being barely inside Zone II so that η is small and we may disregard the second RHS term of Equation 19.84—Equation 19.84 reduces in either case to

$$R_T = \psi_i \bar{k} f(p_s)$$

where ψ_i is either ψ or ψ_{eff}. For the function $f(p_s)$, we can use either Equation 19.59 or the Langmuir isotherm. Using the former, we obtain

$$R_T^{1/n} + \frac{\psi\bar{k}}{k_D}R_T - \psi\bar{k}p_O = 0 \qquad (19.87)$$

In the particular case when $n = 0.5$ gives the best empirical fit, the first term is R_T^2. Equation 19.87 is then a quadratic yielding two solutions, one of which would be negative, however.

A quadratic is also obtained if the Langmuir isotherm is used for $f(p_s)$, which should be somewhat more accurate than $f(p_s) = p^n$, and this quadratic has two positive roots. If $\psi\bar{k}f(p_s) = (k_1k_5p_s)/(k_1p_s + k_5)$ is used, and p_s is eliminated in Equation 19.85, the following is obtained[267]:

$$R_T^2 - \left(k_D p_O + k_5 + \frac{k_D k_5}{k_1}\right)R_T + k_D k_5 p_O = 0 \qquad (19.88)$$

where k_1 and k_5 are the velocity constants for O_2 adsorption and C(O) decomposition (Equations 19.7 and 19.11). By successively taking k_D, k_5, and k_1 to be large, three important approximations of Equation 19.88 are obtained.

1. k_D *large*. This corresponds to very rapid boundary layer diffusion and Equation 19.88 reduces to the Langmuir isotherm with p_O in place of p_s, thus

$$R_T = \frac{k_1k_5p_O}{k_1p_O + k_5} \qquad (19.89)$$

2. k_5 *large*. This corresponds to very rapid film decomposition. Equation 19.88 reduces to

$$R = \frac{p_O}{1/k_D + 1/k_1} \qquad (19.90)$$

   This is sometimes referred to as the "resistance" equation on account of the summation of the reciprocal velocity constants. Resistance is not an analog, however, since Equation 19.88 indicates where reciprocal velocity constants are not additive.

3. k_1 *large*. This corresponds to very rapid adsorption. The residue of the quadratic factorizes as follows[267]:

[267] R. H. Essenhigh, R. Froberg, and J. B. Howard, "Combustion Behavior of Small Particles," *Ind. Eng. Chem.*, **57**, 32–43 (1965).

$$(R_T - k_D p_O)(R_T - k_5) = 0 \qquad (19.91)$$

This equation is interpreted to mean that reaction either is chemically controlled or is diffusionally controlled. There is no transition from one to the other. The rate curves are continuous, but the slopes of the rate curves are discontinuous. If the point is accepted, many apparent examples of such discontinuous transition can be found. One such example is illustrated in Fig. 19.40. In spite of the very low oxygen concentration, Gulbransen and Jansson[268] positively identify the low-slope section of the curve as due to diffusion. The slope of the chemical section of the curve corresponds to an activation energy of 39 kcal, which the authors identify, without substantiating reason, as due to mobile adsorption. On the basis of values given in the subsection "Molecular Collision Rate Calculations" (Section 19.2.2.3), however, the value is more consistent with that for film decomposition (possibly with reaction just in Zone II). That interpretation is consistent with the discontinuous slope according to Equation 19.91.

[268] F. A. Gulbransen and S. A. Jansson, "Vaporization Chemistry in the Oxidation of Carbon, Silicon, Chromium, Molybdenum and Niobium," *Heterogeneous Kinetics at Elevated Temperature*, Plenum Press, New York, 1970, pp. 181–207.

The possibility of a discontinuous transition from diffusion to chemical control suggests that there is need to distinguish between reaction "control" and reaction "dominance." In the first case the reaction rate could presumably be calculated to within an acceptable margin of error based on only one reaction step being of importance. In the second case such accuracy of calculation would be possible in the limit when other reaction steps were indeed of low importance; but in general one reaction step would only dominate the reaction rate, not control it.

4. *For both k_1 and k_5 very large*. Reaction at the surface is very rapid and, if p_s in Equation 19.85 is very small, we have the diffusion-limited condition (diffusion control of the reaction rate)

$$R_T = k_D p_O \qquad (19.92)$$

Under these conditions the reaction is in Zone III.

19.5.6 Volatiles Kinetics

The general mechanisms of pyrolysis and volatiles combustion have been described in Section 19.4.2. The rate of volatiles combustion may be determined, in the limits, either by the rate of volatiles release when com-

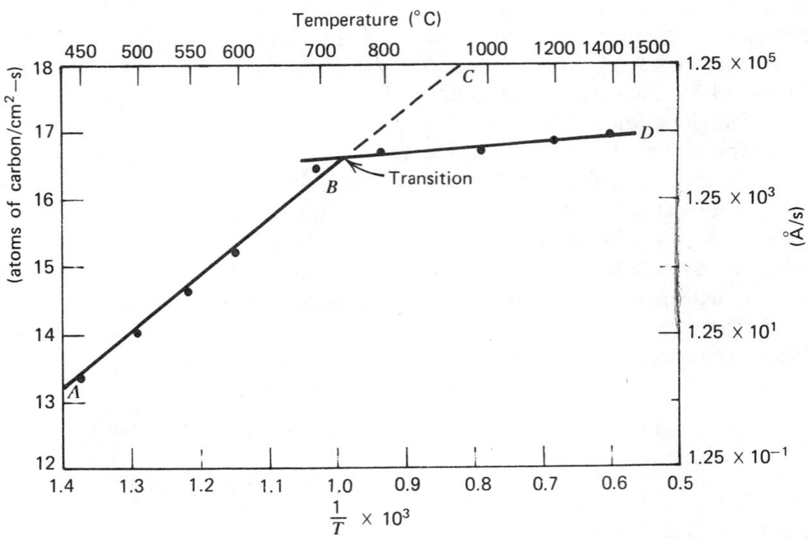

Figure 19.40 Oxidation of graphite, 38 torr O_2, 6.2-cm² area: *ABC* = chemical control, *BD* = gas diffusion control.[268]

bustion is very rapid, or by the rate of volatiles combustion when pyrolysis is very rapid. The first problem, of release, must be considered briefly here, but it is discussed in much greater depth in Chapter 12. Very little is known about the second problem.

19.5.6.1 *Pyrolysis Kinetics*

When the rate of combustion of volatiles is determined by their rate of release, the rate of release itself can be controlled, in the limit, by one of two factors. With small particles, this will be the chemical rate of thermal degradation of the coal structure to form volatiles (the pyrolysis mechanism). With large particles the rate of escape of the volatiles through the porous char matrix becomes the controlling factor. The dividing line between "small" and "large" in this context may be taken as approximately 100 μ. This is only a rough measure, however. As outlined in Section 19.3.2.1, one working model of coal constitution is consistent with a two-stage model of pyrolysis. There is now evidence[160] that the first stage is very rapid by comparison with the second stage with first-stage pyrolysis possibly still influenced by particle size down to 60 μm, but second-stage pyrolysis is evidently uninfluenced by particle size up to about 200 μm and possibly larger. The limiting particle size should also be a function of the pyrolysis temperature and ambient gas composition, since these influence the extent of swelling, cross-linking, fluidity, and char porosity.

Large Particles. The problem of large-particle pyrolysis seems to have been given little theoretical attention. In principle, it can be assumed that Equation 19.58 should apply. Actual application, however, presents considerable difficulties because of lack of information. Several optional assumptions are possible, but their consequences have not yet been explored.

One problem is evaluation of the term $(R_j)_v$ in Equation 19.58. As one option, it can be assumed that pyrolysis may take place very rapidly throughout the particle, generating pressure, and that escape is then determined by forced flow through the porous matrix. This behavior would be pressure regulated, with pressure varying throughout the particle and declining throughout with time. Then we would assume that the volatile material already exists in the particle before Equation 19.58 is applied, and $(R_j)_v$ for such conditions is zero.

A second possibility is that further pyrolysis occurs at any one location only as the volatiles escape (pressure-controlled pyrolysis). Assumptions are then required for the $(R_j)_v$ term.

In a third assumption,[90] a shrinking-drop behavior inside the porous matrix was postulated. This is essentially a variant of the first option, but with the assumption that the volatiles form, in effect, a "liquid" drop, of radius a, inside the char. Darcy's permeability equation was applied to the rate of volatiles transfer from the drop surface to the outside. At the "liquid drop" surface the flow rate is given by

$$v_a = -\frac{r^2 P_0}{a^2 \eta}\frac{dc}{dr} \qquad (19.93)$$

where P_0 is the assumed permeability of the char, η is the viscosity of the flowing volatiles, and dc/dr is the volatiles concentration gradient. Assuming a concentration of "pure" volatiles of c_a at the drop surface (assumed to be a constant), integration yielded

$$v_a = \frac{c_a(P_0/\eta)}{a(1 - a/a_s)} \qquad (19.94)$$

Since v_a was the mass rate of change (per unit area) of liquid to vapor, it was proportional to the rate of shrinkage of the drop, da/dt. Using this condition and integrating again gave

$$t_v = K_v d_s^2 \qquad (19.95)$$

where K_v is the pyrolysis constant $= (\sigma\eta/P_0 8 c_a)(V/100)$, σ is the "drop" density, and V is the volatile matter content of the coal. Such a square law relationship has been obtained experimentally for pyrolysis of large particles,[80,90] with values for K_v of about 100 s/cm^2.

A fourth model depends solely on heat transfer. The temperature of particles being heated from ambient follows a (1 − exp) curve with time. The characteristic time for the heating also has the form of Equation 19.95. The value of the constant is about 250 s/cm², which is of the same order as K_v. Measurement of the pyrolysis time is not from the start of heating, however; it begins when the particles have already reached a substantial temperature.

Small Particles. Particles smaller than 100 μm have had much more attention from theorists. The most significant models of pyrolysis have been classified recently in a table by Ubhayakar[162] (Table 19.4). In essence, five basic schemes are listed, with variations on each. The first scheme, with $n = 1$, is the

[269] E. A. Shapatina, V. V. Kolynzhni, and Z. F. Chuklanov, *Dokl. Akad. Nauk SSSR,* **72** (1950).

[270] D. W. Van Krevelen, C. van Heerden, and F. J. Huntjens, "Physico-chemical Aspects of the Pyrolysis of Coal and Related Organic Compounds," *Fuel,* **30,** 253 (1951).

[271] H. N. Stone, J. D. Bachelor, and H. F. Johnstone, "Low Temperature Carbonization Rates in a Fluid Bed," *Ind. Eng. Chem.,* **46,** 274 (1954).

[272] D. B. Anthony, J. B. Howard, H. C. Hottel, and H. P. Meissner, "Rapid Devolatilization of Pulverized Coal," *Proc. 15th Symp. (Int.) Combustion,* Combustion Institute, Pittsburgh, 1975, pp. 1303–1317.

[273] W. H. Wiser, G. R. Hill, and N. J. Kertamus, "Kinetic Study of the Pyrolysis of a High-Volatile Bituminous Coal," *Ind. Eng. Chem. Process Des. Dev.,* **6,** 133 (1967).

[274] M. G. Skylar, V. I. Shustikov, and I. V. Virozub, "Investigation of the Kinetics of Thermal Decomposition of Coal," *Int. Chem. Eng.,* **9,** 595 (1969).

[275] G. J. Pitt, "The Kinetics of the Evolution of Volatile Products from Coal," *Fuel,* **41,** 267 (1962).

[276] D. D. Stickler, R. E. Gannon, and H. Kobayashi, "Rapid Devolatilization Modelling of Coal," paper presented at Eastern Section of Combustion Institute, Fall 1974.

[277] D. Fitzgerald and D. W. Van Krevelen, "Chemical Structure and Properties of Coal. XXI–The Kinetics of Coal Carbonization," *Fuel,* **38,** 17 (1959).

[278] J. L. Johnson, "Kinetics of Coal Gasification in Hydrogen During Initial Reaction Stages," *Joint U.S.–U.S.S.R. Symp. Coal Gasification/Liquefaction,* Moscow, October (1976), p. 42.

[279] H. Reidelback and M. Summerfield, "Kinetic Model for Coal Pyrolysis Optimization," *Prepr., Am. Chem. Soc., Div. Fuel Chem.,* **20**(1), 161 (1975).

Table 19.4 Coal Pyrolysis Models[162]

1. Single Reaction (at any instant of time)
Coal → Solid Residue (*S*) + Volatiles (*V*)

nth-Order Arrhenius reaction

$$\frac{dV}{dt} = k(V_\infty - V)^n$$

$$k = A \exp\frac{-E}{RT}$$

V = volatile released at time t

V_∞ = total volatile released at infinite time

References

$n = 1$ Shapatina et al.,[269] Van Krevelen et al.,[270] Stone et al.,[271] Howard and Essenhigh,[73] Jüntgen and van Heek,[185] Badzioch and Hawksley,[67] Anthony et al.[272] Kobayashi[161]

$n = 1$–2 Wiser et al[273]

$n = 2$–8 Skylar et al.[274]

Non-Arrhenius reaction

$$\frac{dV}{dt} = BV_\infty/t$$

B = constant

Reference: Pitt[275]

[a] This scheme is intended to describe the primary pyrolysis only. Secondary cracking requires additional considerations. Only this scheme has independent support from considerations of coal constitution.

Table 19.4 Coal Pyrolysis Models[162] (continued)

2. Multiple Parallel Reactions (at any instant of time)

Two first-order Arrhenius reactions[a]

$$\text{coal}\begin{cases} C_1 \xrightarrow{k_1} [S_1] + V_1 \\ C_2 \xrightarrow{k_2} S_2 + V_2 \end{cases}$$

$$\frac{dC_1}{dt} = -k_1C_1$$

$$\frac{dC_2}{dt} = -k_2C_2$$

$$k_1 = A_1\exp - \frac{E_1}{RT}$$

$$k_2 = A_2\exp - \frac{E_2}{RT}$$

C_1 = concentration of component 1

C_2 = concentration of component 2

Reference: Nsakala et al.[160]

Multiple first-order reactions with statistical distribution of activation energies

$$\text{coal} \quad \sum_{i=1}^{n} C_i \xrightarrow{k_i} \sum_{i=1}^{n} \quad (S_i + V_i) \rightarrow S + V$$

$$\frac{dV_i}{dt} = k_i(V_{i,\infty} - V_i)$$

$$k_i = A \exp - \frac{E_i}{RT}$$

$$E_i \rightarrow E_0, \sigma, f(E)$$

where $\int_0^\infty f(E)\, dE = 1$

V_i = amount of volatile released of ith component at time t

$V_{i,\infty}$ = amount of volatile released at infinite time

References: Pitt,[275] Anthony et al.[272]

3. Multiple Competing Reactions (at any instant of time)

Two first-order reactions

$$\text{coal (mass } m) \begin{cases} \xrightarrow{k_1} (1-\alpha_1)S_1 + \alpha_1V_1 \\ \xrightarrow{k_2} (1-\alpha_2)S_2 + \alpha_2V_2 \end{cases}$$

$$m = -(\alpha_1k_1 + \alpha_2 k_2)m$$

$$k_{1,2} = A_{1,2}\exp - \frac{E_{1,2}}{RT}$$

$\alpha_{1,2}$ = stoichiometric constants

References: Kobayashi,[161] Stickler et al.[276]

Multiple first-order reactions

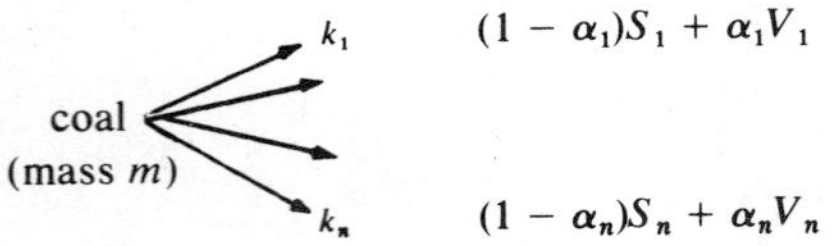

$$\dot{m} = -\sum_{i=1}^{n} \alpha_i k_i m$$

Reference: Discussed by Kobayashi[161] but not explored

4. Complex Schemes

Multiple consecutive parallel first-order reactions

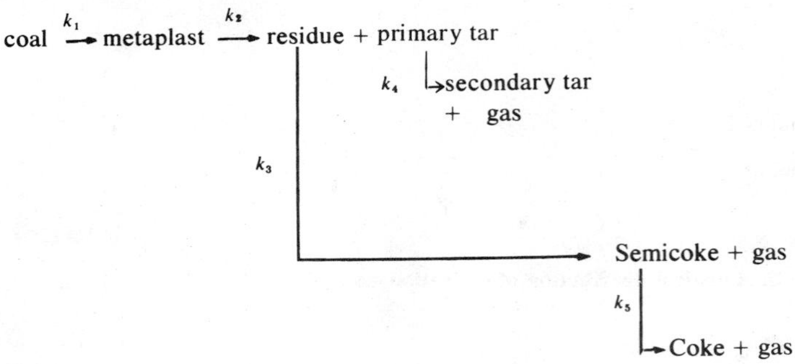

$$k_i = A_i \exp -\frac{E_i}{RT} \qquad i = 1, 5$$

Reference: Fitzgerald and Van Krevelen[277]

Parallel competing first-order reactions

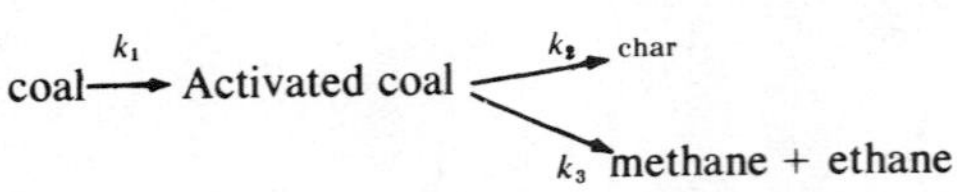

$$k_{1-3} = A_{1-3} \exp \frac{-E_{1-3}}{RT}$$

Reference: Johnson [278]

5. Schemes Involving Secondary Char-Forming Reactions

Consecutive competing char-forming reactions

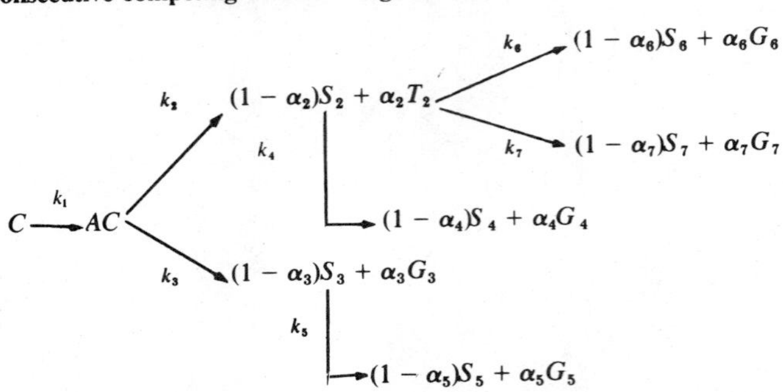

Table 19.4 Coal Pyrolysis Models[162] (continued)

where C = coal

S = solid residue

T = tar

G = gas

AC = activated coal

$$k_i = A_i \exp \frac{-E_i}{RT} \qquad i = 1, 7$$

α_i = stoichiometric constants

Reference: Reidelbach and Summerfield[279]

Parallel Competing Char-Forming Reactions

$$\text{coal} \rightarrow \sum_{i=1}^{n} C_i \xrightarrow{k_i} \sum_{i=1}^{n} (S_i + V_i) \rightarrow S + V \begin{cases} \xrightarrow{k_e} S + V \\ \xrightarrow{k_c} [S + S_c] + V_c \end{cases}$$

$$\text{where } k_i = A \exp - \frac{E_i}{RT};\ E_i \rightarrow E_0, \sigma, f(E), \text{ where } \int_0^\infty f(E)\, dE = 1$$

k_e = rate of escape of volatiles

k_c = rate of cracking of volatiles to solids (S_c) and cracked volatiles (V_c)

Reference: Anthony et al.[272]

simple one-step, first-order reaction that has been known for decades to be an oversimplification, although it still has value in complex reaction schemes. Complexity grows in all the following schemes. A detailed evaluation and comparison is inappropriate here, but one is given by Ubhayakar.[162] It is appropriate to note here only that the simpler schemes (1, 2, and 3 in Table 19.4) are not designed to show the effects of secondary reactions such as cracking of volatiles, since they were constructed for experimental conditions in which secondary reactions were considered unimportant.

Of all the schemes proposed, only that for two first-order Arrhenius reactions (Table 19.4) has any independent basis in coal constitution, as discussed in Section 19.4.2.

19.5.6.2 *Combustion Kinetics*

There is very little to be said about volatiles combustion kinetics because so little is known (Section 19.4.2.2). The only direct investigations or analyses of volatiles kinetics from any source (Biswas and Essenhigh[234]) resorted to phenomenological kinetics. First-order reactions in fuel and in oxygen were assumed for an overall second-order reaction. For lack of knowledge of molecular weights, it was necessary to use mass fractional concentrations in place of the more usual molar volume or mole fractions. Use of mass fractions has the interesting consequence that all preexponential factors then have dimensions of reciprocal time, whatever the reaction order.[280] The equation used was

$$\frac{dC_v}{dt} = -k_o C_v C_{ox} \exp - \frac{E}{RT} \qquad (19.96)$$

where C is the mass fraction concentration, with subscripts v for volatiles and ox for oxygen. The other symbols have their usual significance. Activation energies in the two sets of experiments were 16.5 and 20 kcal/mole (70 and 84 kJ/mole), respectively, with mass basis frequency factors of about 10^6 and 10^5 s^{-1}, respectively.

[280] R. H. Essenhigh, "An Introduction to Stirred Reactor Theory Applied to Design of Combustion Chambers," in *Combustion Technology,* H. B. Palmer and J. M. Béer, Eds., Academic Press, New York, 1974, Ch. 14.

The alternative is to use the indirect approach of assuming some initial one-step irreversible behavior to CO and H_2, as described in Section 19.4.2.2, followed by conventional CO and H_2 kinetics. This was the approach used by Smoot et al.,[75] who adopted the expression of Edelman et al.[238] for the initial irreversible one-step reaction for C_nH_m going to CO and H_2:

$$\text{rate}_{C_nH_m} = -\frac{dC_{CnHm}}{dt} = \frac{5.52 \times 10^8}{p^{0.825}} C_{C_nH_m}^{1/2} C_{ox} \exp\left(-\frac{12{,}400}{T}\right) \quad (19.97)$$

The overall rate of combustion of the CO and H_2 formed is then governed by the usual rate equations for the individual constituents, for example:

$$R_j = k_j \Pi_i (\rho\sigma_i)^{\alpha'_{ij}} \quad (19.98)$$

$$R_{-j} = k_{-j} \Pi_i (\rho\sigma_i)^{\alpha''_{ij}} \quad (19.98a)$$

where ρ is the gas mixture density, σ_i is the component i concentration, and α_{ij} is the stoichiometric coefficient of the ith component in the ith reaction.

19.6 EXPERIMENTAL RESULTS

In this section attention is mostly focused on the more "applied" results of carbon, char, and coal investigations, with particular attention given to the reaction rate data and kinetic constants derived therefrom. The results are partly discussed in the context of common experimental methods such as the shock tube, or common objectives such as study of ignition; but most emphasis is on common phenomena or aspects of behavior without regard to the experimental sources of the data. A major concern is evaluation and interpretation of rate data, principally as a function of temperature. Some acquaintance with or awareness of the experimental methods used is necessary nonetheless.

19.6.1 Experimental Methods and Data Compilation

Experimental methods include techniques of experiment and techniques of measurement. Details of the methods are not described; they are available in the cited sources. Additional qualitative background can be found elsewhere.[281–294] What follows is a summary of methods, for reference, with a brief critical evaluation of some aspects of measurements and reported data. Few, if any, of the exper-

[281] W. E. J. Broom and M. W. Travers, "Reaction Between Carbon and Certain Gases," *Proc. R. Soc. London,* **A135** 512–537 (1932).

[282] J. R. Bowring and H. G. Crone, "La Combustion du Carbone," *Proc. Symp.,* Nancy (1949), p. 85.

[283] J. A. Graham, A. R. G. Brown, A. R. Hall, and W. Watt, "The Rates of Reaction of Carbon and Graphite Materials with Combustion Gases at High Temperatures," *Conf. Industrial Carbon and Graphite,* Soc. Chem. Ind., September 1957.

[284] F. Boulangier, X. Duval, and M. Letort, "The Kinetics of the Reactions of Carbon Filaments with Carbon Dioxide and Water Vapor at High Temperatures and Low Pressures," *Proc. 3rd Conf. Carbon,* Pergamon Press, New York, 1957, pp. 257–264.

[285] F. M. Lang, P. Magnier, and S. May, "Etude de l'Oxydation des Carbones par l'Air et l'Anhydride Carbonique en Fonction de leur Texture et de leur Etat de Pureté," *Proc. 5th Carbon Conf.,* Vol. 1, Pergamon Press, New York, 1963, pp. 171–196.

[286] D. W. L. Griffiths, W. J. Thomas, and P. L. Walker, Jr., "Effect of Oxidation on the Surface Heterogeneity of Some Graphitized Carbons," *Carbon (Oxford),* **1,** 515–523 (1964).

[287] J. G. H. de Graaf, "Über den Mechanisms der Verbrennung von Festem Kohlenstoff," *P.W.K. (VDI),* **17,** 227–231 (1965).

[288] D. Kisch, "Über den Temperaturverlauf vor im Sauerstoffstrom Brennenden Graphitoberflächen," *Ber. Bunsenges.,* **71,** 60–64 (1967).

[289] M. Shelef and P. L. Walker, Jr., "Transient Phenomena in the Gasification of Graphite by High-Purity Carbon Dioxide," *Carbon (Oxford),* **5,** 93–105 (1967).

[290] S. W. Spokes and S. W. Benson, "Oxidation of a Thin Film of a Carbonaceous Char at Pressures Below 10^{-4} Tons," in *Fundamentals of Gas-Surface Interactions,* Academic Press, New York, 1967, pp. 318.

[291] H. Guérin, T. Siemieniewska, Y. Grillet, and M. François, "Influence de la Chimisorption d'Oxygéne sur la Détermination de l'Oxyreactivité des Combustibles Solides. I. Etude d'un semi-coke de lignite préparé à 550°C," *Carbon (Oxford),* **8,** 727–740 (1970).

[292] D. Dollimore and A. Turner, "Kinetics of Oxidation of α and β Resin Extracts from Coal Tar Pitch," *Trans. Faraday Soc.,* **66,**2655–2660 (1970).

[293] M. Kito, M. Ishimaru, S. Kawahara, T. Sakai, and S. Sugiyama, "Combustion of Carbon Particles Generated from Liquid Hydrocarbon Fuels," *Combust. Flame,* **17,** 391–397 (1971).

[294] C. L. Wagoner and E. C. Winegartner, "Further Development of the Burning Profile," ASME Paper No. 72 WA/Fu-1, 1972.

imental methods have no deficiencies. More often than not it has been possible in the past to show that data are open to two or more equally acceptable interpretations, and the conclusions drawn are therefore ambiguous. Attempts to eliminate ambiguities have rarely been successful, so that virtually all data and conclusions are open to criticism. To make the discussion useful, however, most data and their interpretations are taken at face value. This is an acceptable approach, but a healthy skepticism toward categorical assertions of interpretations is warranted.

19.6.1.1 *Techniques of Experiment and Data Tabulation*

Table 19.5 classifies 25 techniques into seven groups. The listing includes selected references to work in which kinetic data were obtained, and the data are listed in Table 19.6. These data and their significance are reviewed in Section 19.6.2.

The obvious characteristic of the listing in Table 19.5 is that only two methods account for about half the investigations. One method is the use of reactant gas flow, through or over fixed beds, and the other is the use of the thermobalance; these approaches presumably are considered to be either the easiest or the most informative techniques. Most of the other methods were designed to solve specific problems or to obtain data under particular conditions. Examples are the use of jets as an alternative to low pressure in eliminating the diffusional boundary layer; or the use of flames without backmix to determine the difference in reaction behavior between particles in isolation and in groups.

19.6.1.2 *Techniques of Measurement*

Measurement techniques refer to quantitative methods of following the progress of the reaction to determine kinetic constants, or for other evaluation purposes. Only three methods exist for obtaining rate data, although they can profitably be used simultaneously. The methods are (1) measurement of gas composition, (2) the variation of weight with time, or (3) determination of the surface regression rate. These are all differential methods. There is also the determination of the total reaction time, but this is an integral method and rate data are only obtained inferentially, not directly. The measurements must also be made in conjunction with proper determination of the boundary conditions, such as reacting gas concentration at the inlet, inlet flow velocity, and temperature. The type of measurement that can be made is also constrained by the experimental design; such factors as particle size, flow field, and geometry provide additional constraints.

Gas Analysis. In this method either the drop in the reactant gas concentration or the rise in the products concentrations is measured as a function of time. The procedure can be used for carbon reactions, pyrolysis reactions, VM combustion reactions, and coal reactions; in the last two cases, however, the nature of the products usually creates problems of interpretation. When the reactants are solids, the rate of reaction is generally calculated in terms of unit area of superficial surface. If the rate is to be assigned to the real (internal) surface, necessary surface area measurements must be made simultaneously. No continuous method of doing this is known; it has to be done by a series of batch measurements. Penetration depths of the reaction must also be known for a full description plus (local) density variations with time if a full interpretation over time is to be developed or tested. Gas analysis measurements by themselves can be informative, particularly if the potential for internal reaction has been eliminated by using very thin samples. The procedure averages the reaction over the whole sample, however, which can be misleading if there is significant variation in rate over the sample surface.

Weight Loss. This technique is used almost as often as gas analysis. It is applicable only to solids, and it is necessary to be able to hold the sample captive or to be able to determine weight loss in successive batches or by some tracer method.

Continuous weighing of single captive particles has been carried out down to about a millimeter but is impracticable much below that. If smaller particles are to be weighed captive, they must be in a small basket or sample holder. This raises questions about possible interference with the reaction by boundary layer effects.

Table 19.5 Classification of Experimental Methods Used in Obtaining Kinetic Data

Method	Investigators	Ref.
	REACTANT GAS FLOW	
Flow tube	Hawtin et al. (1964)	297
	Hawtin and Murdock (1964)	298
	Hawtin and Gibson (1966)	295, 296
	Heuchamps and Duval (1966)	299
	Johnstone et al. (1952)	300
Fixed beds		
Flow over bed		
	Binford and Eyring (1956)	246
	Blackwood and McTaggart (1959)	129, 130
	Bonnetain et al. (1960)	227
	Carpenter and Sergeant (1966)	216
	King and Wise (1963)	301
	Marsh et al. (1965)	135
	Mosely and Paterson (1965)	302
	Rosner and Allendorf (1965)	303
	Rosner and Allendorf (1965, 1968)	304, 305
	Streznewski and Turkevich (1959)	128
Flow through bed:	steady state	
	Blackwood (1959, 1962)	306, 307
	Blackwood et al. (1967)	308
	Blackwood and Ingeme (1960)	309
	Blackwood and McGrory (1958)	310
	Carpenter and Giddings (1964)	214 I
	Carpenter and Giddings (1964)	215 II
	Eapen et al. (1977)	66
	Feistel et al. (1977)	311
	Gadsby et al. (1948) (1946)	117, 118
	Hedden and Lowe (1967)	247
	Van Heek et al. (1973)	312
	Hottel et al. (1966)	313
	Kayembe and Pulsifier (1976)	314
	Lambert (1936)	315
	Long and Sykes (1948)	229
	Sevenster (1961)	212, 213
	Thring and Price (1954)	316
	Tyler and Smith (1975)	317
	Zielke and Gorin (1955,1957)	318, 319
Flow through bed:	unsteady state (crossing point)	
	Sebastian and Mayers (1937)	320
	Sherman et al. (1941)	321
Electrically heated fibers or rods	Meyer (1938)	322
	Nagle and Strickland-Constable (1962)	323
	Okada and Ikegawa (1953)	324
Single particles (captive)	Blyholder and Eyring (1957,1959)	325, 326
	Gleit et al. (1965)	131
	Lewis et al. (1964–1965)	327, 328
	Tyler et al. (1976)	329

Table 19.5 Classification of Experimental Methods Used in Obtaining Kinetic Data *(continued)*

Method	Investigators	Ref.
	MOVING BED	
Fluid bed	Birch et al. (1960)	330
	Ergun (1955,1956)	219, 331
	Fuchs and Yavorsky (1975)	332
	Johnson (1974)	182
	Lewis et al. (1949)	333
Rotary drum	Schmidt and Elder (1940)	210
	ENTRAINED PARTICLE FLOW	
Drop (isothermal) tube	Field (1969,1970)	95, 96
	Gray et al. (1974)	38
Plug-flow furnace (downflow)	Béer and Thring (1961)	86
	Béer et al. (1964)	334
	Essenhigh (1970)	37
	Howard and Essenhigh (1967)	73, 74
	Leesley and Siddall (1972)	335
	Leesley and Hedley (1972)	336
Up-flow furnace	Ayling and Smith (1972)	84
	Hamor et al. (1973)	98
	Smith (1970,1971)	198–200
	Smith and Tyler (1972,1974)	201, 202
	Sergeant and Smith (1973)	97
Standing flame (mostly soot)	Fenimore and Jones (1967)	88
	Lee et al. (1962)	337
	Magnussen (1971)	338
	Tesner and Tsybulevskii (1967)	339, 340
Dispersed (free) particles	Nagy and Surincik (1966)	341
	THERMOBALANCE	
Single particles	Allardice and Walker (1970)	342
	Austin and Walker (1963)	48
	Blakely and Overholser (1965)	343
	Essenhigh et al. (1965)	267
	Froberg (1967)	144
	Gregg and Tyson (1965)	344
	Gulbransen and Andrew (1952)	119–121
	Gulbransen et al. (1963, 1964, 1965)	355–358
	Heintz and Parker (1964)	359
	Kurylko and Essenhigh (1973)	145
	Overholser and Blakely (1965)	360
	Parker and Hottel (1936)	361
	Petersen et al. (1955)	345, 346
	Popov (1959)	347
	Rakszawski et al. (1963)	348
	Rakszawski and Parker (1964)	349
	Rellick et al. (1975)	350

Table 19.5 Classification of Experimental Methods Used in Obtaining Kinetic Data (continued)

Method	Investigators	Ref.
	Rodriguez-Reinoso et al. (1974)	351
	Strange and Walker (1976)	122
	Tu et al. (1934)	146
	Turkdogan et al. (1968)	352
	Turkdogan and Vinters (1970)	353
	Walker et al. (1953)	354
Multiple particles	Allardice (1966)	217
	Dutta and Wen (1977)	148, 362
	Gardner et al. (1974)	363
	Hippo and Walker (1975)	165
	l'Homme and Boudart (1968)	364
	Jenkins et al. (1973)	164
	Johnson (1974,1975)	182, 183
	Tomita et al. (1977)	365, 366
	Tucker and Mulcahy (1969)	111
	Turkdogan and Vinters (1968)	353, 367
	STATIC (NONFLOW)	
Batch reactor	Bansal et al. (1970)	113, 114
	Bansal et al. (1971)	115
	Biederman et al. (1976)	242
	Gadsby et al. (1948)	118
	Hart et al. (1967)	225
	Laine et al. (1963)I,II	109, 110
	Long and Sykes (1948, 1950–1952)	229, 245
	Lussow et al. (1967)	368
	Ubhayakar and Williams (1976)	369
Diffusion cell	Davis and Hottel (1934)	370
	Yang and Steinberg (1977)	371, 372
	"AERODYNAMIC"	
Shock tube	Lightman and Street (1973)	180
	Nettleton and Stirling (1967, 1971, 1974)	82, 83, 85
	Park and Appleton (1973)	373
Rotating disk	Acharya and Olander (1973)	374
Jet	Day et al. (1958)	375
	Evans and Emmons (1977)	376
	Ladd and Walsh (1966)	377
	Matsui et al. (1975)	378
	Walls and Strickland-Constable (1964) (1965)	379, 380
Stirred reactor	Goldberg and Essenhigh (1978)	69
Explosion gallery	No kinetic data yet reported from investigations	
	OPTICAL	
	Thomas and Hughes (1964)	123

[295] P. Hawtin and J. A. Gibson, "The Effect of Diffusion and Bulk Gas Flow on the Thermal Oxidation of Porous Carbons. II. Diffusional Effects in Graphite at High Temperatures," *Carbon (Oxford),* **4,** 489–500 (1966).

[296] P. Hawtin and J. A. Gibson, "The Effect of Diffusion and Bulk Gas Flow on the Thermal Oxidation of Porous Carbons. III. Ungraphitised Carbons," *Carbon (Oxford),* **4,** 501–508 (1966).

[297] P. Hawtin, J. A. Gibson, R. Murdoch, and J. E. Lewis, "The Effect of Diffusion and Bulk Gas Flow on the Thermal Oxidation of Nuclear Graphite. I. Temperatures Below 500°C," *Carbon (Oxford),* **2,** 299–309 (1964).

[298] P. Hawtin and R. Murdock, "The Role of in-Pore Mass Transport Resistance in the Reaction of Porous Solids with Gases. The Air Oxidation of Large Tubes of Graphite," *Chem. Eng. Sci.,* **19,** 819–834 (1964).

[299] C. Heuchamps and X. Duval, "Effet des Catalyseurs sur les Caracteristiques Cinétiques de la Combustion du Graphite," *Carbon (Oxford),* **4,** 243–253 (1966).

[300] H. F. Johnstone, C. Y. Chen, and D. S. Scott, "Kinetics of the Steam-Carbon Reaction in Porous Graphite Tubes," *Ind. Eng. Chem.,* **44,** 1564–1569 (1952).

[301] A. B. King and H. Wise, "Reaction Kinetics of Hydrogen Atoms with Carbon Films," *J. Phys. Chem.,* **67,** 1163 (1963).

[302] F. Moseley and D. Paterson, "The Rapid High-Temperature Hydrogenation of Coal Chars. I. Hydrogen Pressures up to 100 Atmospheres," *J. Inst. Fuel,* **38,** 13–23 (1965).

[303] D. E. Rosner and H. D. Allendorf, "High Temperature Oxidation of Carbon by Atomic Oxygen," *Carbon (Oxford),* **3,** 153–156 (1965).

[304] D. E. Rosner and H. D. Allendorf, "High-Temperature Kinetics of Graphite Oxidation by Dissociated Oxygen," *AIAA J.,* **3,** 1522–1523 (1965).

[305] D. E. Rosner and H. D. Allendorf, "Comparative Studies of the Attack of Pyrolytic and Isotrpic Graphite by Atomic and Molecular Oxygen at High Temperatures," *AIAA J.,* **6,** 650–654 (1968).

[306] J. D. Blackwood, "The Reaction of Carbon with Hydrogen at High Pressure," *Aust. J. Chem.,* **12,** 14–28 (1959).

[307] J. D. Blackwood, "The Kinetics of the System Carbon-Hydrogen-Methane," *Aust. J. Chem.,* **15,** 397–408 (1962).

[308] J. D. Blackwood, B. D. Cullis, and D. J. McCarthy, "Reactivity in the System Carbon-Hydrogen-Methane," *Aust. J. Chem.,* **20,** 1561–1570 (1967).

[309] J. D. Blackwood and A. J. Ingeme, "The Reaction of Carbon with Carbon Dioxide at High Pressure," *Aust. J. Chem.,* **13,** 194–209 (1960).

[310] J. D. Blackwood and F. McGrory, "The Carbon-Steam Reaction at High Pressure," *Aust. J. Chem.,* **11,** 16–33 (1958).

[311] P. P. Feistel, K. H. van Heek, and H. Jüntgen, "Gasification of a German Bituminous Coal with H_2O, H_2, and H_2O-H_2 Mixtures," *Prepr., Am. Chem. Soc. Div. Fuel Chem.,* **22**(1), 53–58 (1977).

[312] K. H. van Heek, H. Jüntgen, and W. Peters, "Fundamental Studies on Coal Gasification in the Utilization of Thermal Energy from Nuclear High-Temperature Reactors," *J. Inst. Fuel,* **46,** 249–258 (1973).

[313] H. C. Hottel, G. C. Williams, and P. C. Wu, "The Reaction of Coke with Carbon Dioxide," *Prepr., Am. Chem. Soc., Div. Fuel Chem.,* **10**(3), 58–72 (1966).

[314] N. Kayembe and A. H. Pulsifer, "Kinetics and Catalysis of the Reaction of Coal Char and Steam," *Fuel,* **55,** 213–216 (1976).

[315] J. D. Lambert, "The Oxidation of Carbon. Parts I and II," *Trans. Faraday Soc.,* **32,** 452–462, 1584–1591 (1936).

[316] M. W. Thring and P. H. Price, "La Relation entre les Facteurs Physiques et Chimiques dans la Détermination de la Vitesse de Réaction Anhydride Carbonique–Carbone dans des Lits Épais de Combustibles," *Proc. Conf. Int. Gaséification Intégrale de la Houill Extraite,* Liege, 1954.

[317] R. J. Tyler and I. W. Smith, "Reactivity of Petroleum Coke to Carbon Dioxide between 1030 and 1180°K," *Fuel,* **54,** 99–104 (1975).

[318] C. W. Zielke and E. Gorin, "Kinetics of Carbon Gasification," *Ind. Eng. Chem.,* **47,** 820–825 (1955).

[319] C. W. Zielke and E. Gorin, "Kinetics of Carbon Gasification," *Ind. Eng. Chem.,* **49,** 396–403 (1957).

[320] J. J. S. Sebastian and M. A. Mayers, "Coke Reactivity: Determination by a Modified Ignition Point Method," *Ind. Eng. Chem.,* **29,** 1118–1124 (1937).

[321] R. A. Sherman, J. M. Pilcher, and H. N. Ostborg, *A Laboratory Test for the Ignitibility of Coal,* ASTM Bull. No. 112, (1941), pp. 23–34.

[322] L. Meyer, "The Surface Reaction of Graphite with Oxygen, Carbon Dioxide, and Water Vapour, at Low Pressures," *Trans. Faraday Soc.,* **34,** 1056–1061 (1938).

[323] J. Nagle and R. F. Strickland-Constable, "Oxidation of Carbon Between 1000–2000°C," *Proc. 5th Conf. Carbon (1962),* Vol. 1, Pergamon Press, New York, 1963, pp. 154–164.

[324] J. Okada and T. Ikegawa, "Combustion Rate of Artificial Graphites from 700°C to 2000°C in Air," *J. Appl. Phys.,* **24,** 1249–1250 (1953).

[325] G. Blyholder and H. Eyring, "Kinetics of Graphite Oxidation," *J. Phys. Chem.,* **61,** 682–688 (1957).

[326] G. Blyholder and H. Eyring, "Kinetics of Graphite Oxidation. II," *J. Phys. Chem.,* **63,** 1004–1008 (1959).

[327] J. B. Lewis, P. Connor, and R. Murdoch, "The Order of Reaction for the Oxidation of Nuclear Graphite in Dry Oxygen-Nitrogen Mixtures," *Carbon (Oxford),* **2,** 311–314 (1964).

[328] J. B. Lewis, J. Dix, and R. Murdoch, "The Order of Reaction for the Oxidation of Nuclear Graphite in Moist Oxygen-Nitrogen Mixtures," *Carbon (Oxford),* **3,** 321–322 (1965).

[329] R. J. Tyler, H. J. Wouterlood, and M. F. R. Mulcahy, "Kinetics of the Graphite-Oxygen Reaction near 1000°K," *Carbon (Oxford),* **14,** 271–278 (1976).

[330] T. J. Birch, K. R. Hall, and R. W. Urie,

"Gasification of Brown Coal with Hydrogen in a Continuous Fluidized-Bed Reactor," *J. Inst. Fuel,* **33,** 422–435 (1960).

[331] S. Ergun, "Flow Experiments in Studying Kinetics: Role of Diffusion and Influence of Flow Pattern in Gasification Reactions," *Ind. Eng. Chem.,* **47,** 2075–2080 (1955).

[332] W. Fuchs and P. M. Yavorsky, "Gasification of HYDRANE Char in Reactions with Carbon Dioxide and Steam," *Prepr., Am. Chem. Soc., Div. Fuel Chem.,* **20**(3), 115–133 (1975).

[333] W. K. Lewis, E. R. Gilliland, and G. T. McBride, "Gasification of Carbon by Carbon Dioxide in Fluidized Powder Bed," *Ind. Eng. Chem.,* **41,** 1213–1226 (1949).

[334] J. M. Béer, K. B. Lee, C. Marsden, and M. W. Thring, "Flammes de Charbon Pulvérisé à Mélange Controlé," *Cinqième Journées de la Combustion et de la Conversion et de l'Energie,* Paris, May 19–23, 1964, pp. 141–150.

[335] M. E. Leesley and R. G. Siddall, "The Combustion Rate of a Pulverized Anthracite Dust Cloud of Known Initial Particle Size Distribution," *J. Inst. Fuel,* **45,** 169–173 (1972).

[336] M. E. Leesley and A. B. Hedley, "The Effect of Particle Size Distribution on the Combustion Rate of a Pulverized Anthracite Dust Cloud," *J. Inst. Fuel,* **45,** 224–230 (1972).

[337] K. B. Lee, M. W. Thring, and J. M. Béer, "On the Rate of Combustion of Soot in a Laminar Soot Flame," *Combust. Flame,* **6,** 137–145 (1962).

[338] B. F. Magnussen, "The Rate of Combustion of Soot in Turbulent Flames," *Proc. 13th Symp. (Int.) Combustion,* Combustion Institute, Pittsburgh, 1971, pp. 869–877.

[339] P. A. Tesner and A. M. Tsybulevskii, "Kinetics of Dispersed Carbon Gasification in Diffusion Flames of Hydrocarbons," *Combust. Flame,* **11,** 227–233 (1967).

[340] A. M. Tsybulevskii and P. A. Tesner, "Gasification of Dispersed Carbon in Hydrocarbon Diffusion Flames," *Combust. Explo. Shock Waves* (transl.), **3,** 64–68 (1967).

[341] J. Nagy and D. J. Surincik, *Thermal Phenomena During Ignition of a Heated Dust Dispersion,* U.S. Bur. Mines Rep. Invest. No. 6811, 1966.

[342] D. J. Allardice and P. L. Walker, Jr., "The Effect of Substitutional Boron on the Kinetics of the Carbon-Oxygen Reaction," *Carbon (Oxford),* **8,** 375–385 (1970).

[343] J. P. Blakely and L. G. Overholser, "Oxidation of ATJ Graphite by Low Concentrations of Water Vapor and Carbon Dioxide in Helium," *Carbon (Oxford),* **3,** 269–275 (1965).

[344] S. J. Gregg and P. F. S. Tyson, "The Kinetics of Oxidation of Carbon and Graphite by Oxygen at 500°–600°," *Carbon (Oxford),* **3,** 39–42 (1965).

[345] E. E. Petersen, P. L. Walker, Jr., and C. C. Wright, "Reaction of Artificial Graphite with Carbon Dioxide: Surface Area Development Within Artificial Graphite Rods Reacted with Carbon Dioxide from 900°–1300°C," *Ind. Eng. Chem.,* **47,** 1629–1634 (1955).

[346] E. E. Petersen and C. C. Wright, "Reaction of Artificial Graphite Rods with Carbon Dioxide from 900°C–1300°C," *Ind. Eng. Chem.,* **47,** 1624–1629 (1955).

[347] V. A. Popov, *Development of the Combustion Process of Moving Particles in Nonstationary Hydrodynamic and Thermal Conditions,* Akademia Nauk U.S.S.R. Tech. Sci. Section Bull. No. 12, (1958), pp. 90–95. Translations: John Hopkins University, IG 230–1132.

[348] J. F. Rakszawski, F. Rusinko, and P. L. Walker, Jr., "Catalysis of the Carbon–Carbon Dioxide Reaction by Iron," *Proc. 5th Carbon Conf.,* Vol. 2, Pergamon Press, New York, 1963, pp. 243–250.

[349] J. F. Rakszawski and W. E. Parker, "The Effect of Group IIIA–VIA Elements and their Oxides on Graphite Oxidation," *Carbon (Oxford),* **2,** 53–63 (1964).

[350] G. S. Rellick, F. Rodriguez-Reinoso, P. A. Thrower, and P. L. Walker, Jr., "Transient Rates in the Reaction of CO_2 with Highly Oriented Pyrolitic Graphite," *Carbon (Oxford),* **13,** 81–82 (1975).

[351] F. Rodriguez-Reinoso, P. A. Thrower, and P. L. Walker, Jr., "Kinetic Studies of the Oxidation of Highly Oriented Pyrolytic Graphite," *Carbon (Oxford),* **12,** 63–70 (1974).

[352] E. T. Turkdogan, V. Koump, J. V. Vinters, and T. F. Perzak, "Rate of Oxidation of Graphite in Carbon Dioxide," *Carbon (Oxford),* **6,** 467–484 (1968).

[353] E. T. Turkdogan and J. V. Vinters, "Kinetics of Oxidation of Graphite and Charcoal in Carbon Dioxide," *Carbon (Oxford),* **7,** 101–117 (1969).

[354] P. L. Walker, Jr., R. J. Foresti, and C. C. Wright, "Surface Area Studies on Carbon–Carbon Dioxide Reaction," *Ind. Eng. Chem.,* **45,** 1703–1709 (1953).

[355] E. A. Gulbransen, K. F. Andrew, and F. A. Brassart, "The Oxidation of Graphite at Temperatures of 600°C to 1500°C and at Pressures of 2 to 76 torrs of Oxygen," *J. Electrochem. Soc.,* **110,** 476–483 (1963).

[356] E. A. Gulbransen, K. F. Andrew, and F. A. Brassart, "Ablation of Graphite in Oxygen and Air at 1000 to 1400°C Under Flow Conditions," *Carbon (Oxford),* **1,** 413–424 (1964).

[357] E. A. Gulbransen, K. F. Andrew, and F. A. Brassart, "Oxidation of Graphite, Molybdenum, and Tungsten at 1000–1600°C," *Prog. Astronaut. Aeronaut.,* **13,** 227–250 (1964).

[358] E. A. Gulbransen, K. F. Andrew, and F. A. Brassart, "Reaction of Graphite with Carbon Dioxide at 1000–1600°C Under Flow Conditions," *Carbon (Oxford),* **2,** 421–429 (1965).

[359] E. A. Heintz and W. E. Parker, "Catalytic Effect of Major Impurities on Graphite Oxidation," *Carbon (Oxford),* **4,** 473–482 (1966).

[360] L. G. Overholser and J. P. Blakely, "Oxidation of Graphite by Low Concentrations of Water Vapor and Carbon Dioxide in Helium," *Carbon (Oxford),* **2,** 385–394 (1965).

[361] A. S. Parker and H. C. Hottel, "Combustion Rate of Carbon: Study of Gas-Film Structure by Microsampling," *Ind. Eng. Chem.,* **28,** 1334–1341 (1936).

[362] S. Dutta and C. Y. Wen, "Reactivity of Coal and Char. 2. In Oxygen-Nitrogen Atmosphere," *Ind. Eng. Chem. Process Des. Des.,* **16,** 31–36 (1977).

[363] N. Gardner, E. Samuels, and K. Wilks, "Catalyzed Hydrogasification of Coal Chars," in *Coal Gasification,* Am. Chem. Soc. Adv. Chem. Ser. No. 131 (1974), pp. 217–236.

[364] G. A. l'Homme, and M. Boudart, "The Ignition of Carbon and Its Catalysis by Platinum," *Proc. 11th Symp. (Int.) Combustion,* Combustion Institute, Pittsburgh, 1967, pp. 197–202.

[365] A. Tomita, O. P. Mahajan, and P. L. Walker, Jr., "Catalysis of Char Gasification by Minerals," *Prepr., Am. Chem. Soc., Div. Fuel Chem.,* **22**(1), 4–6 (1977).

[366] A. Tomita, O. P. Mahajan, and P. L. Walker, Jr., "Reactivity of Heat-Treated Coals in Hydrogen," *Fuel,* **56,** 137–143 (1977).

[367] E. T. Turkdogan and J. V. Vinters, "Effect of Carbon Monoxide on the Rate of Oxidation of Charcoal, Graphite and Coke in Carbon Dioxide," *Carbon (Oxford),* **8,** 39–53 (1970).

[368] R. O. Lussow, F. J. Vastola, and P. L. Walker, Jr., "Kinetics of Oxygen Interaction with Graphon Between 450° and 675°C," *Carbon (Oxford),* **5,** 591–602 (1967).

[369] S. K. Ubhayakar and F. A. Williams, "Burning and Extinction of a Laser-Ignited Carbon Particle in Quiescent Mixtures of Oxygen and Nitrogen," *J. Electrochem. Soc.,* **123,** 747–756 (1976).

[370] H. Davis and H. C. Hottel, *Ind. Eng. Chem.,* **26,** 889 (1934).

[371] R. T. Yang and M. Steinberg, "The Reactivity of Coal Char with CO_2 at 1100–1600°C," *Prepr., Am. Chem. Soc., Div. Fuel Chem.,* **22**(1), 12–16 (1977).

[372] R. T. Yang and M. Steinberg, "A Diffusion Cell Method for Studying Heterogeneous Kinetics in the Chemical Reaction/Diffusion Controlled Region Kinetics of $C + CO_2 \rightarrow 2CO$ at 1200–1600°C," *Ind. Eng. Chem. Fundam.,* **16,** 235–242 (1977).

[373] C. Park and J. P. Appleton, "Shock-Tube Measurements of Soot Oxidation Rates," *Combust. Flame,* **20,** 369–379 (1973).

[374] T. R. Acharya and D. R. Olander, "The Rate of Oxidation of the Basal and Prismatic Surfaces of Pyrolytic Graphite in the Transition Regime Between Chemical and Diffusional Control," *Carbon (Oxford),* **11,** 7–18 (1973).

[375] R. J. Day, P. L. Walker, Jr., and C. C. Wright, "Carbon-Oxygen Reaction at High Temperatures and High Gas Flow Rates," *Conf. Industrial Carbon and Graphite* (Soc. Chem. Ind., September 1957), *Industrial Carbon and Graphite* (1958), pp. 348–370.

[376] D. D. Evans and H. W. Emmons, "Combustion of Wood Charcoal," *Fire Res.,* **1,** 57–66 (1977).

[377] I. R. Ladd and P. N. Walsh, "Complex Temperature Dependence of the Oxidation of Pyrolytic Graphite by CO_2," *Carbon (Oxford),* **4,** 539–541 (1966).

[378] D. Matsui, A. Koyama, and K. Uehara, "Fluid-Mechanical Effects on the Combustion Rate of Solid Carbon," *Combust. Flame,* **25,** 57–66 (1975).

[379] J. R. Walls and R. F. Strickland-Constable, "Oxidation of Carbon Between 1000–2400°C," *Carbon (Oxford),* **1,** 333–338 (1964).

[380] J. R. Walls and R. F. Strickland-Constable,

Direct weighing of clouds of particles is obviously impossible. The problem occurs when particles are being fed in batches or continuously into electric tube furnaces or into flames. The particles are sampled at different locations in the furnace or flame and weight loss is determined by ashing. On the assumption that the ash is unchanged by heating, it can be used as a tracer and the weight loss is determined by comparing the ratio of combustible to incombustible material in the successive samples. This method is most commonly used for coal studies, although it has been used for char combustion studies. (It must be remembered that coal ash is derived from mineral matter on heating.[382] The tracer procedure using ash is still valid in principle, nonetheless.)

Surface Regression Rate. This measurement has been most commonly used for shrinking particles and for regression of segments eroded by jets. If the measurements can be made, it is one of the most direct methods available. It can also be one of the most misleading if reaction is taking place in part internally, with the regression rate of the superficial surface not clearly related to the total burning rate. Such partial internal burning can occur at low temperatures in quiescent velocity fields (e.g., Froberg[144]). With high-velocity jets there is not believed to be any significant internal reaction and, if there is, and regression is steady state, any internal reaction can probably be ignored because it will average out if the penetration depth does not change.

Combined Methods. Sometimes the various drawbacks of the different methods taken individually can be partially eliminated if the methods are combined. Froberg[144] and others used simultaneous weighing and photography to determine changes in internal density, for example.

"The Reaction between CO_2 and Graphite at Temperatures Between 1500 and 2100°C," *Carbon (Oxford),* **3,** 350 (1965).

[381] F. C. Cowlard and J. C. Lewis, "Vitreous Carbon—A New Form of Carbon," *J. Mater. Sci.,* **2,** 507–512 (1967).

[382] P. H. Given and R. F. Yarzab, *Problems and Solutions in the Use of Coal Analysis,* Tech. Rep. No. 1 for DOE (ERDA), prepared under Contract No. E(49-18)-390 by Pennsylvania State University, November 1975.

Table 19.6 Activation Energies and Reaction Orders for the Gas Reactions of Carbon

Investigator	Ref.	Reaction	Carbon Type	Pressure	Temp. Range	Activation Energy (kcal/mole)	Reaction Order: Experimental	Reaction Order: Assumed	Evaluation Equation	Experimental Method
LOW PRESSURE (< 1 ATM)										
Reaction with Oxygen										
Bansal et al. (1970)	113	O_2 chemisorption	Graphon	100 mtorr	78–100°C	3.1–12.4 in 5 steps	1	—	Elovich	Static (batch)
Blyholder and Eyring (1957, 1959)	325, 326	O_2	Graphite (spectroscopic grade, cylinders)	1–100 μm	600–1300° C	80 (thick samples), 42 (thin samples)	0.5 (thick samples)	—	Pore diffusion	Fixed single particles in gas flow
Bonnetain et al. (1960)	227	O_2	4 Artificial graphites	10^{-4}–10^{-1} mm Hg	400–700°C	6–8.5 (for CO/CO_2 variation)	—	—	Evaluated CO/CO ratio	Reactant gas flow over sample
Gregg and Tyson (1965)	344	O_2	Carbon and graphite	7.6 mm Hg	500–600°C	44 (most graphitized); 44–67 (function of HTT)	—	—	Rate ∝ weight with simple Arrhenius	Thermobalance (single particle)
Gulbransen and Andrew (1952) I	119–121	O_2	Artificial graphite	0.15–9.8 cm Hg	425–575°C	36.7	1 (interp. of results)	—	Simple	Thermobalance (single particle)
Gulbransen et al. (1963, 1964)	355, 356	O_2	AGKSP graphite	2–76 torr	600–1500° C	39 ($T <$ 700°C), 3.6 ($T >$ 500°C)	1 ($T <$ 700°C); 0.32 ($T >$ 800°C)	—	Arrhenius d (wt)/dt and simple Arrhenius	Thermobalance (single particle)
Gulbransen et al. (1964)	357	O_2 and air	AGKSP graphite	2–19 torr	1000–1400° C	39	—	—	Simple Arrhenius	Thermobalance (single particle)
Hart et al. (1967)	225	O_2 chemisorption	Graphon	10 mtorr	25–400°C	7.4 for single site	—	1	Dissociative adsorption	Static (batch)
Laine et al. (1963) I	109	O_2 adsorption	Graphon	30–500 μm Hg	300–675°C	35–40 for O_2 depletion; 43–48 for C gasification	1 in p_{O_2}, 1 in ASA	—	First-order ASA with Arrhenius	Static (batch)

Laine et al. (1963) II	110	O_2 adsorption	Graphon	40 μm Hg	575–675°C	36 for O_2 depletion; 44 for C gasification	1 in p_{O_2}, 1 in ASA	—	First-order ASA with Arrhenius	Static (batch)
Lussow et al. (1967)	368	O_2	Graphon	0.5 and 700 torr	300–625°C	29 for O_2 consumption	—	—	First order in P_{O_2} and ASA with Arrhenius	Static (batch)
Rodriguez-Reinoso et al. (1974)	351	O_2	2 Pyrolytic graphites	20 torr	700–800°C	47–53 (edge atom reaction)	0.5 and 0.6		Arrhenius	Single particle or thermobalance
Thomas and Hughes (1964)	123	O_2	Graphite (Ticonderoga)	1 cm Hg	800–875°C	66 and 62 for <10$\bar{1}$0> and <11$\bar{2}$0> directions	—	—	Arrhenius	Optical microscopy of etch pits
Tucker and Mulcahy (1969)	111	O_2	Graphon	1.2–125 torr	500–750°C	43 ±2	0.5		Arrhenius	Thermobalance
Tyler et al. (1976)	329	O_2	AGKSP graphite	15–250 torr	900–1000° K	64	0.34 at 15 torr; → 0.19 at 245 torr		Arrhenius	Flow over single cylindrical particles
Reaction with Carbon Dioxide										
Biederman et al. (1976)	242	CO_2 (with H_2)	Natural graphite	<108 mtorr	960–1120°C	103.5	1 (CO_2); 0.5 (H_2)	—	Langmuir-Hinshelwood	Static (batch)
Gadsby et al. (1948)	118	CO_2	Coconut char	10–760 mm	700–830°C	68 for CO_2 adsorption	1	—	Langmuir-Hinshelwood	Static (batch) and flow through multiple-particle bed
Gulbransen and Andrew (1952) I	119, 121	CO_2	Artificial graphite	7.6 cm Hg	500–900°C	—	—	—	Simple Arrhenius	Thermobalance (single cylinders)
Gulbransen et al. (1965)	358	CO_2	AGKSP graphite (cylinders)	19 torr	1000–1600° C	88 ± 5 (T <1350°C)	—	—	Simple Arrhenius	Thermobalance (single cylinders)
Ladd and Walsh (1966)	377	CO_2	Pyrolitic graphite	400 torr	1500–3200 K	74.5 for CO_2 adsorption	1 for CO_2 adsorption	—	Nagle and Strickland-Constable	High-velocity jet

Table 19.6 (continued)

Investigator	Ref.	Reaction	Carbon Type	Pressure	Temp. Range	Activation Energy (kcal/mole)	Reaction Order: Experimental	Reaction Order: Assumed	Evaluation Equation	Experimental Method
Reaction with Carbon Dioxide (continued)										
Meyer (1938)	322	CO_2	Graphite	$< 5 \times 10^{-2}$ mm Hg	2000–2700 K	90	0	—	Arrhenius	Electrically heated filament
Rellick et al. (1975)	350	CO_2	Pyrolitic graphite	50 torr	1050–1100° C	81	—	—	—	Thermobalance (single particles)
Strange and Walker (1976)	122	CO_2	Natural graphite (SPI)	15–150 torr	900–1007° C	99 for CO_2 adsorption	1	—	Langmuir-Hinshelwood	Thermobalance (single particles)
Reaction with Water										
Binford and Eyring (1956)	246	H_2O	Graphite (spectroscopic grade, cylinders)	1–100 μm	900–1300° C	60.3 ($T <$ 1100°C); 43 ($T >$ 1200°C)	0 ($T <$ 1100°C); 1 ($T >$ 1200°C		Basic Langmuir-Hinshelwood type	Reactant gas flow over sample
Gadsby et al. (1946, 1948)	117, 118	H_2O	Coconut char, coal char	10–700 mm	700–800°C	Not evaluated	Appt. fract. order		Langmuir-Hinshelwood	Flow through multiple particle bed
Hedden and Lowe (1967)	247	H_2O in He (with H_2)	Reactor graphite	0.2–0.8 mm Hg for H_2	940–1030°C	68 in adsorption	1 $(H_2)O$; 1 (H_2)		Langmuir-Hinshelwood	Flow through multiple particle bed
Long and Sykes (1948, 1950)	229, 245	H_2O (H_2)	Char	10–700 mm	680–800°C	62 for adsorption, 20 for desorption	1 (H_2O), 1 (H_2)		Langmuir-Hinshelwood	Static (batch); and Flow through multiple particle bed
Reaction with Hydrogen										
Bansal et al. (1971)	115	H_2 adsorption	Graphon	0.02–0.47 torr	80–600°C	5.7–30.4 in 4 steps	1	—	Elovich	Static (batch)

NORMAL PRESSURE

Reaction with Air and Oxygen

Acharya and Olander (1973)	374	O_2		Pyrolitic graphite	1 atm (O_2, 0.05–1%)	1100–2000 K	34.1 (basal plane); 56.5 (prism plane) for $T < 1400$ K	0 (basal plane), 0.25–0.6 (prism plane)		Nagle and Strickland-Constable	Rotating disk
Ayling and Smith (1972)	84	Air		Coal (pulverized) (semianthracite)	1 atm	1400–2200 K	18.8	—	1	Semiempirical with simple Arrhenius	Upflow furnace tube
Béer et al. (1964)	334	Air		Anthracite	1 atm	1500–1750 K	18.9	—	1	Simple Arrhenius	Plug-flow furnace
Cowlard and Lewis (1967)	381	Air		Vitreous carbon	1 atm	700–850°C	32	—	—	(Unstated)	(Unstated)
Day et al. (1958)	375	O_2		5 Rod carbons (manufactured)	1 atm	1500–2000 K	5.3, dropping to 2.3 with increased velocity	1	—	Simple Arrhenius	High-velocity jet
Dutta and Wen (1977) II	148, 362	O_2/N_2		6 Gasification chars	1 atm	400–1100° C	31		1	First-order, semiempirical with Arrhenius	Thermobalance (multiple particle)
Eapen et al. (1977)	66	Air		2 Cokes, 1 coal	1 atm	950–1550°C	50 for C/CO_2 reaction		1	Arrhenius	Flow through, fixed bed (combustion pot)
Essenhigh (1970) [Béer (1961) data]	37	Air		Anthracite	1 atm	1350–1750 K	22 and 44	0	—	Arrhenius	Plug flow furnace
Essenhigh et al. (1965) [Hottell (1934) data]	267	Air		Carbon	1 atm	900–1600° C	40 ($T <$ 1100°C); 3.4 ($T >$ 1250°C)		O ($T <$ 1100°C); 1 ($T >$ 1250°C)	Modified Langmuir	Thermobalance, (single sphere)
Evans and Emmons (1977)	376	Air		Wood charcoal	1 atm	750–1100° C	18		1	Simple Arrhenius	Jet
Field (1969)	95	O_2/N_2		Anthracite char (pulverized)	1 atm	1200–2000 K	30→10 for increased T	1		Empirical	Drop (isotherm) tube

Table 19.6 (continued)

Investigator	Ref.	Reaction	Carbon Type	Pressure	Temp. Range	Activation Energy (kcal/mole)	Reaction Order: Experimental	Reaction Order: Assumed	Evaluation Equation	Experimental Method
Reaction with Air and Oxygen (continued)										
Field (1970)	96	Air	11 p.c, chars	1 atm	1200–2000 K	Linear rise of rate constant with T	1		(Incomplete evaluation)	Drop (isotherm) tube
Gray et al. (1974) [Field (1969) data]	38	Air	Coal chars	1 atm	1200–2000 K	37 (n = 0); 6 (n = 1)		0 1	Langmuir-Hinshelwood	Drop tube
Hamor et al. (1973)	98	Air	Brown coal (pulverized)	1 atm	630–2200 K	16.2 (dia > 25 μm) (T) > 760K)	0.5		Empirical nth-order with simple Arrhenius	Upflow furnace tube
Hawtin and Murdoch (1964)	298	Air	Graphite (pile grade A)	1 atm	400–550°C	23.5 as measured 45.4 (corr. for diffusion)	0.6		(Detailed flow analysis)	Flow tube
Hawtin et al. (1964)	297	Air	Graphite (pile grade A)	1 atm	400–500°C	44.5 (corr. for diffusion)	0.6		(Detailed flow analysis)	Flow tube
Hawtin and Gibson (1966) I	295	Air	Graphite (pile grade A)	1 atm	550–675°C	25 (appt) 48.6 (corr. for diffusion)	0.8 (appt); 0.6 (true)		(Detailed flow analysis)	Flow tube
Hawtin and Gibson (1966) II	296	Air	Baked carbon	1 atm	300–420°C	55.2 (falling with T)	0.78		(Detailed flow analysis)	Flow tube
Howard and Essenhigh (1966, 1967)	73, 74	Air	Coal	1 atm	500–1500° C	30–65		1	Simple Arrhenius	Plug-flow furnace
Jenkins et al. (1973)	164	Air	21 Heat-treated coal chars	1 atm	500°C	—	—	1	Rate ∝ weight with simple Arrhenius	Thermobalance

Lambert (1936)	315	O_2/N_2	Diamond, graphite, anthracite char, coconut char	1 atm	500–800 K	44, 49, 31,29 (respectively)		1	Arrhenius	Flow through multiple particle bed
Lee et al. (1962)	337	Air	Soot	1 atm	1300–1700° C	39.3	1		First-order Arrhenius	Flame (soot)
Leesley and Hedley (1972)	335, 336	Air	Anthracite (13 size cuts)	1 atm	1200–1600° C	20	0.87	1	Computer fit. on surface reaction	Plug-flow flame
Lewis et al. (1964)	327	O_2/N_2 (dry)	Graphite (pile grade A)	1 atm	575–655°C	Not reported	0.54–0.6		Empirical *n*th order	Flow over single cylinder
Lewis et al. (1965)	328	O_2/N_2 (moist)	Graphite (pile grade A)	1 atm	575–655°C	Not reported	0.65		Empirical *n*th order	Flow over single cylinder
Magnussen (1971) [Lee et al. (1962) data]	338	Air	Soot	1 atm	1300–1700° C	39 for adsorption		1 (O_2)	Langmuir	Flame (soot)
Froberg (1967)	144	Air, O_2	Graphitized carbon (L-56 HH)	1 atm	600–1100° C	45 (Zone I), 22.5 (Zone II)	0 0.5		Arrhenius	Thermobalance (single sphere)
Kurylko and Essenhigh (1973)	145	Air, O_2	Graphitized carbon (L-56 HH)	1 atm	600–1100° C	48.1 (Zone I), 25.5 (Zone II)	0 0.5		Arrhenius	Thermobalance (single sphere)
Matsui et al. (1975)	378	Air	Carbon	1 atm	1000–1800 K	47 and 53	—	1	Fluid mechanics analysis	Jet flow
Nagle and Strickland Constable (1963)	323	O_2 or O_2/N_2	1 Carbon, 2 graphites	up to 0.6 atm	1000–2000° C	30 ($T <$ 1400°C)	1 (O_2 adsorption)		Nagle and Strickland-Constable	Electrically heated rods
Okada and Ikegawa (1953)	324	Air	Artificial graphite	1 atm	700–2000° C	12.7 ($T <$ 800°C)	0		Arrhenius	Electrically heated rods
Parker and Hottel (1936)	361	Air	Brush carbon	1 atm	1000–1400 K	44		1	"Resistance"	Thermobalance (single cylinder)
Popov (1959)	347	Air	Carbon	1 atm	1000–1400° C	19.4		1	"Resistance"	Thermobalance (single sphere)

Table 19.6 (continued)

Investigator	Ref.	Reaction	Carbon Type	Pressure	Temp. Range	Activation Energy (kcal/mole)	Reaction Order: Experimental	Reaction Order: Assumed	Evaluation Equation	Experimental Method
Reaction with Air and Oxygen (continued)										
Sebastian and Mayers (1937)	320	Air	15 Cokes	1 atm	200–600°C	10–38 function of coal type	—	1	Crossing point analysis	Crossing point (ignition)
Sergeant and Smith (1973)	97	Air	3 Coal (bituminous) chars, pulverized	1 atm	800–1700 K	27		1	Empirical *n*th order with Arrhenius	Drop (isotherm) tube
Smith (1970)	198	Air	Anthracite (pulverized)	1 atm	1400–2200 K	18.8		1	Empirical *n*th order with Arrhenius	Upflow furnace tube
Smith (1971) I	199	Air	Anthracite, coke, char (pulverized)	1 atm	1200–2270 K	16–18 (for 3 fuels)		1	First order with Arrhenius	Upflow furnace tube
Smith (1971) II	200	Air	Semianthracite (pulverized)	1 atm	1400–2200 K	19		1	First order with Arrhenius	Upflow furnace tube
Smith and Tyler (1972)	201	Air	Semianthracite (pulverized)	1 atm	1400–2200 K	20		1	First order with Arrhenius	Upflow furnace tube
Smith and Tyler (1974)	202	Air	Brown coal char (pulverized)	1 atm	630–1812 K	32.6		0	Empirical *n*th order with Arrhenius	Upflow furnace tube
Tu et al. (1934)	146	Air	Graphite (brush carbon)	1 atm	900–1700 K	40 ($T < 1000°$ K)		1	"Resistance"	Thermobalance (single sphere)
Ubhayakar, and Williams (1976)	369	O_2 and N_2	Electrode carbon	1 atm	2000–3500 K	18			Empirical *n*th order with Arrhenius	Laser ignition of free particles
Walls and Strickland Constable (1964, 1965)	379, 380	O_2	2 Pyrographites	1 atm	900–2400° C	30 for O_2 adsorption	—	—	Nagle and Strickland-Constable	Electrically heated rods with jet impingement

Reaction with Carbon Dioxide

Austin and Walker (1963)	48	CO_2	Graphite (cylinders)	1 atm	900–1300° C	54 (appt. 110.7 at low *T*)	Not given; implied *n* = 1 at surface; *n* = 1 effective		Langmuir-Hinshelwood	Thermobalance (single particle) cylinder
Blakely and Overholser (1965)	343	CO_2 in He	ATJ graphite	1 atm	775–925°C	60	0.6–0.7		Langmuir-Hinshelwood	Thermobalance (single sphere)
Dutta et al. (1977) (I)	148, 362	CO_2	6 Gasification chars	1 atm	840–1100° C	2.5 for stage 1; 59 for stage 2	1		Semiempirical rate ∝ $p^n_{CO_2}$ × char weight	Thermobalance (multiple particles)
Eapen et al. (1977)	66	CO_2 (see data for C/air)	2 Cokes, 1 coal	1 atm	450–1550°C	50 for coke 35 for coal		1	Arrhenius	Fixed bed (combustion pot)
Ergun (1955, 1956)	331, 219	CO_2	Carbon and 2 graphites	1 atm	700–1400° C	59 (all 3 samples)		1	Modified Langmuir-Hinshelwood	Fluid bed
Hedden and Lowe (1967)	247	CO_2 (CO)	Nuclear graphite	0.1–1 atm	980–1120°C	71 in adsorption; 50.5 in desorption	1, CO_2; 0.5, CO		Langmuir-Hinshelwood (modified)	Flow through multiple particle bed
Hippo and Walker (1975)	165	CO_2	16 Heat-treated chars	1 atm	900°C	—		1	Rate ∝ weight with simple Arrhenius	Thermobalance (multiple particle)
Hottel et al. (1966)	313	CO_2	Coke	1 atm	1500–1900° F	Various values	see text		Langmuir-Hinshelwood; Semechkova and Frank-Kamenetskii; Temkin	Flow through fixed bed of particles
Lewis et al. (1949)	333	CO_2	Anthracite, retort coke	1 atm		86 kBtu/lb mole (coke); 59 kBtu/lb mole (anthracite)			Langmuir-Hinshelwood	Fluid bed
Overholser and Blakely (1965)	360	CO_2 in He	Graphite (Speer, moderator 2)	1 atm	875–1025°C	55	0.7		Langmuir-Hinshelwood	Thermobalance (single sphere)

Table 19.6 (continued)

Investigator	Ref.	Reaction	Carbon Type	Pressure	Temp. Range	Activation Energy (kcal/mole)	Reaction Order: Experimental	Reaction Order: Assumed	Evaluation Equation	Experimental Method
Reaction with Carbon Dioxide (continued)										
Petersen et al. (1955)	345, 346	CO_2	Graphite	1 atm	900–1300° C	38–48			Arrhenius	Thermobalance (single cylinder)
Tesner and Tsybulevskii (1967)	339, 340	CO_2	Soot	1 atm	1800–2200 K	75	1		Arrhenius	Flame (soot)
Thring and Price (1954)	316	CO_2	Coke	1 atm	1400–1650° C	23.3		1	"Resistance" with Arrhenius	Flow through multiple particle bed
Tyler and Smith (1975)	317	CO_2	Graphite, petroleum, coke, electrode carbon	~ 1 atm	750–950°C	Graphite, 70; petroleum, coke, 48–57; electrode carbon, 45–57	0.6		Empirical *n*th order with Arrhenius	Flow through multiple particle bed
Walker et al. (1953)	354	CO_2	Baked carbon, artificial graphite	1 atm	900–1200° C	47, 48 ($T <$ 1200°C); 26 ($T >$ 1200°C)		1	Arrhenius	Thermobalance (single cylinder)
Yang and Steinberg (1977)	371, 372	CO_2	Graphite (pile grade)	1 atm	1200–1600° C	—	1		Diffusion analysis	Diffusion cell
Reaction with Water										
Blakely and Overholser (1965)	343	H_2O in He	ATJ graphite	1 atm	775–925°C	50	0.6–0.7		Langmuir-Hinshelwood	Thermobalance (single sphere)
Johnstone et al. (1952)	300	H_2O	Graphite	1 atm	850–950°C	Variable with burnoff; see text	1, H_2O; 1, H_2		Langmuir-Hinshelwood	Flow tube
Overholser and Blakely (1965)	360	H_2O in He	Graphite (Speer, moderator 2)	1 atm	825–1025°C	$60 \rightarrow 40$, drops with increased T	0.72		Langmuir-Hinshelwood	Thermobalance (single sphere)

Oxygen (Shock Tube)										
Lightman and Street (1968)	180	Air	8 Coals	1.5–20 atm	500–3000° C				Microspical evaluation	Shock tube
Nettleton and Stirling (1967)	82	O_2	Coal and graphite	1.5–30 atm	700–1600 K				Ignition studies	Shock tube
Nettleton and Stirling (1971)	83	O_2/N_2	Coal	1.5–50 atm	1000–3000 K	13.5 (57 kJ)		1	Simple Arrhenius	Shock tube
Nettleton and Stirling (1974)	85	O_2, air	Coal with additives	Not given	1100–1500 K					Shock tube
Park and Appleton (1973)	373	O_2	Soot	0.05–13 atm	1700–4000 K	34 ($T < 2000$ K)		(Complex)	Nagle and Strickland-Constable	Shock tube
Hydrogen, Water, and Carbon Dioxide										
Birch et al. (1960)	330	H_2	Brown coal	300–600 psi	500–950°C					Fluid bed
Blackwood (1959)	306	H_2	Graphite, char, charcoal, coal char	1–40 atm	650–870°C	30	1		Rate $= k\, p_{H_2}$	Flow through multiple-particle bed
Blackwood (1962)	307	H_2 and CH_4 (mixture)	Coconut shell char	1–40 atm	650–870°C	Parameter values cited			Rate $= \dfrac{k_1 p_{H_2}^2 - k_2 p_{CH_4}}{1 + k_3 p_{H_2} + k_4 p_{H_2}^2 + k_3 p_{CH_4}}$	Flow through multiple particle bed
Blackwood et al. (1967)	308	H_2 and CH_4 (mixture)	Coconut char and brown coal chars	up to 25 atm	700–950°C	7 for H chemisorption			Composite	Flow through multiple particle bed
Blackwood et al. (1967)	308	H_2	6 Coal chars	20 atm	750–950°C	30		1	Rate $= k^* p_{H_2}$; k^* = composite	Flow through multiple particle bed
Blackwood and Ingeme (1960)	309	CO_2	Coconut shell char	up to 40 atm	790–870°C	Parameter values cited			Rate $= \dfrac{k_1 p_{CO_2} + k_5 p_{CO_2}^2}{1 + k_2 p_{CO} + k_3 p_{CO_2}}$	Flow through multiple particle bed

Table 19.6 (continued)

Investigator	Ref.	Reaction	Carbon Type	Pressure	Temp. Range	Activation Energy (kcal/mole)	Reaction Order		Evaluation Equation	Experimental Method
							Experi-mental	Assumed		
Hydrogen, Water, and Carbon Dioxide (continued)										
Blackwood and McGrory (1958)	310	H_2O	Coconut charcoal	1–50 atm	750–830°C	Parameter values cited			$\text{Rate} = \dfrac{k_1 p_{H_2O} + k_4 p_{H_2} p_{H_2O} + k_5 p^2_{H_2O}}{1 + k_2 p_{H_2} + k_3 p_{H_2O}}$	Flow through multiple particle bed
Feistel et al. (1977)	311	H_2O	Coal (39% VM)	1–70 atm	600–1100° C	—	—		$\dfrac{dx}{dt} = k(1-x)$; $k = \dfrac{ap_{H_2O}}{1 + bp_{H_2O}}$	Flow through multiple particle bed
Feistal et al. (1977)	311	H_2	Coal (39% VM)	1–70 atm	600–1100° C	—	—		$\dfrac{dx}{dt} = k(1-x)$ $k = \dfrac{ap_{H_2}}{1 + bp_{H_2}}$	Flow through multiple particle bed
Fuchs and Yavorsky (1975)	332	CO_2 and H_2O	Coal char (from Illinois No. 6)	up to 70 atm	700–900°C	56	0		Arrhenius for C(O) decomposition	Fluid bed
Gardner et al. (1974)	363	H_2	Coal chars (+ catalysts)	0–1000 psi	0–1000°C	$E = E_0 + \alpha x$ $E_0 = 29.3$	3/2		$\dfrac{dx}{dt} = k\, p^n_{H_2}$ $(1-x) \exp \dfrac{-E}{RT}$	Thermobalance (multiple particles)
Johnson (1974)	182	H_2 and H_2O (mixture)	Coal char	1–70 atm	1500–2000° F		$M(X_r) a\, p_{H_2}$		Integral correlation function $(M(X_R)$	Thermobalance multiple particles
Johnson (1975)	183	H_2 and H_2O (mixture)	Coal char	35 atm	1400–1700° F		—		Integral correlation function	Thermobalance (multiple particles)
Kayembe and Pulsifer (1976)	314	H_2O	Coal char	1 atm	600–850°C	60.5, (uncatalyzed) 34.4, (catalyzed)		1	First order in carbon residue	Flow through multiple-particle bed
Mosely and Paterson (1965)	302	H_2	Coal char	1–100 atm	600–1000° C			1	2-Term for $\dfrac{d(CH_4)}{dt}$	Flow over thin bed in horizontal tube

Tomita et al. (1977)	365, 366	H_2	18 Coal chars	2.7 MPa (26 atm)	980°C	14–52 (60–220 kJ)	1		Normalized mass rate loss	—
Turkdogan et al. (1968)	352	CO_2	Graphite	1–40 atm	900–1100° C	73 (low P) → 49 at 30 atm			$\frac{dw}{dt} = kw$; w = weight	Thermobalance (single sphere)
Turkdogan and Vinters (1969)	353	CO_2	Graphite, charcoal	0.03–40 atm	700–1300° C	68 (both samples)	0.5		(Oxygen activity)	Thermobalance (multiple particles)
Turkdogan and Vinters (1970)	367	CO_2	Coconut char, graphite, coke	0.03–20 atm	700–1400° C	60 for CO_2 dissociation			(Oxygen activity)	Thermobalance (multiple particles)
Turkdogan and Vinters (1970)	367	CO_2	Coconut char, graphite, coke	0.03–20 atm	700–1400° C	69 for CO adsorption	—		(Oxygen activity)	Thermobalance (multiple particles)
Turkdogan and Vinters (1969)	353	CO_2	Electrode graphite, coconut charcoal	0.03–40 atm	700–1300° C	68	0.5		(Oxygen activity)	Thermobalance (multiple particles)
Turkdogan and Vinters (1970)	367	CO_2 (with CO)	Coconut charcoal, electrode graphite, metallurgical coke		700–1400° C	60.7 (all samples)	0.5 (No CO) 1.0 (with CO)		(Oxygen activity)	Thermobalance (multiple particles)
Van Heek et al. 1973)	312	H_2O	6 coals; 3 chars	1–70 atm	400–1100° C	25–40 for samples tested		1	Arrhenius for first order, one step	Flow through multiple-particle bed
Zielke and Gorin (1955)	318	H_2	Disco char	1–40 atm	1500–1700° F	15–50 (rising with fraction gasified)			$\frac{d(CH_4)}{dt} = ap^2_{H_2}(1 + bp_{H_2})$ (semiempirical)	Tube reactor
Zielke and Gorin (1957)	319	H_2	Disco char	1–30 atm	1500–1700° F	40–75 (rising with fraction gasified)				Tube reactor
SPECIAL REACTIONS										
Oxide Film Decomposition										
Bansal et al. (1970)	114	C(O) → CO (film decomposition)	Graphon	10^{-8} torr	100–950°C	—	—	—	Elovich	Static (batch)

Table 19.6 (continued)

Investigator	Ref.	Reaction	Carbon Type	Pressure	Temp. Range	Activation Energy (kcal/mole)	Reaction Order: Experimental	Reaction Order: Assumed	Evaluation Equation	Experimental Method
Oxide Film Deposition (continued)										
Gadsby et al. (1948)	118	C(O) → CO(film decomposition)	Coconut char	10–760 mm	700–830°C	38		0	Langmuir-Hinshelwood	Static (batch); and flow through multiple-particle bed
Gulbransen and Andrew (1952) II	120	C(O) → CO (film decomposition)	Artificial graphite	10 mm Hg	790–850°C	40		0	Simple Arrhenius	Thermobalance (single particles)
Strange and Walker (1976)	122	C(O) → CO (film decomposition)	Natural graphite (SPI)	15–150torr	900–1007° C	87	0		Langmuir-Hinshelwood	Thermobalance (single particles)
Tucker and Mulcahy (1969)	111	C(O) → CO (film decomposition)	Graphon (2 samples)	10^{-5} torr	600–750°C	Sample: 70 → 43; and 57 → 47; decline with increasing burnoff			Elovich	Thermobalance (multiple particles)
Radical Reactions										
Fenimore and Jones (1967)	88	Products of combustion (OH)	Soot	1 atm	1530–1890 K	32	1 (for OH)		Arrhenius	Flame (soot)
Gleit et al. (1963)	131	O atom	Graphite (spectroscopic grade)	(P_{O_2} = 10^{-4} to 0.3 atm) 1.3 mm Hg	215–300°C	6.5	—	—	Arrhenius	Flow (HF discharge in O_2)
King and Wise (1963)	301	H-atom	Carbon films (evaporated)	10^{-6} mm	365–500 K	7.1, 9.2 (2 separate reactors)	1 at low H; 0.5 at higher H		Arrhenius	Flow reactor
Blackwood and McTaggart (1959)	130	O-atom	Natural graphite, wood char, diamond	0.1 mm Hg	up to 200°C	—	—	—	—	Sample in flow tube (HR discharge in O_2)

Marsh et al. (1965)	135	O-atom	3 Amorphous carbons; graphite	0–12 torr	14–350°C	10 ($T <$ 200°C) O ($T >$ 200°C)	1.2		Arrhenius	Flow (HF discharge in O_2)
Rosner and Allendorf (1965)	303, 304	O-atom in Ar	Commercial graphite	1 torr	1100–2000 K	6.7 (O); 31 (O_2); ($T <$ 1300 K)	1 (O); 0.56 (O_2)		(Oxygen probability)	Flow (HF discharge in O_2)
Rosner and Allendorf (1968)	305	O-atom	Pyrolytic and isotropic graphites	1 torr	1100–2000 K	6.7 (O-atom) for both graphites	—	—	(Oxygen probability)	Flow (HF discharge in O_2)
Streznewski and Turkevich (1959)	128	O-atom	Carbon-condensed films	Not given	20–100°C	0	1		—	Flow (HF discharge in O_2)
Low Temperature Adsorption										
Allardice (1966)	217	O_2 (adsorption)	Brown coal char	up to 1 atm	25–200°C	11.1 ± 1.5	O		Elovich	Thermobalance (multiple particles)
Carpenter and Giddings (1964) I	214	O_2 adsorption (air)	6 Coals	1 atm	75–115°C	7–19 function of coal and degree of reaction	—	—	Elovich	Flow through multiple-particle bed
Carpenter and Giddings (1964) II	215	O_2 adsorption (air)	6 Coals	1 atm	95°C	—	0.5 (mostly)		Elovich	Flow through multiple-particle bed
Carpenter and Sergeant (1966)	216	O_2 adsorption (air)	3 Coals	1 atm	300–400°C	2–6 function of coal particle size	—		Elovich	Flow through multiple-particle bed
Schmidt and Elder (1940)	210	Air	8 Coals	1 atm	30–99°C	11	0.61		Arrhenius	Rotary drum

Table 19.6 (continued)

Investigator	Ref.	Reaction	Carbon Type	Pressure	Temp. Range	Activation Energy (kcal/mole)	Reaction Order		Evaluation Equation	Experimental Method
							Experi-mental	Assumed		
Low Temperature Adsorption (continued)										
Sevenster (1961)	212, 213	Air (adsorption)	5 Coals	1 atm	30–90°C	65	—		Elovich	Flow through single disks of coal
Catalysis (also see Table 19.)										
Allardice and Walker (1970)	342	O_2	Natural graphite with boron	2.5–200 torr	550–650°C	45.8–52.9 (variable with B concentration)	~0.6		Simple Arrhenius	Thermobalance (single particles)
Heintz and Parker (1966)	359	O_2	UCP-1-200 graphite with transition metals		600–700	48.8, no catalyst; 15–125, with catalyst		1	First order with simple Arrhenius	Thermobalance (single particles
Heuchamps and Duval (1966)	299	Air	Graphite with catalysts	—	500–700°C	40–60 (function of catalyst)			—	Flow tube
Kurylko and Essenhigh (1973)	145	Air, O_2	Graphitized C (L56HH) with Pt	1 atm	600–1100° C	48.1 (Zone I); 25.5 (with Pt)	0		Arrhenius	Thermobalance (single particles)
Long and Sykes (1950)	245	H_2O	Coconut char with Fe, Na, Ca, Al	—	750°C	—	—	—	—	Static (batch)

l'Homme and Boudart (1967)	364	O_2/Ar	Spheron 6 with Pt	1 atm, 0.1–1 atm	up to 600°C	25.4 (no Pt) 39.8 (with Pt)	1 → 0.5 (T increasing) 0 → 0.5 (T increasing)		Arrhenius	Thermobalance (multiple particles)
Rakszawski and Parker (1964)	349	Air; O_2	Graphite UCP 1-200; with elements and oxides of Groups III A –VI A	1 atm	700–800°C	54 in Zone I in air; 63 in O_2	1 in Zone I		Arrhenius	Thermobalance (single particles)
Rakszawski et al. (1963)	348	CO_2	Graphite (SP-1) with Fe	1 atm	900–1200° C	60	—	—	Arrhenius	Thermobalance (single particles)

Total Reaction Time. This technique has been used almost exclusively for captive coal particles[78–81,90,177–179,203,383] and for batches of small particles of coals and chars in a crucible[384] or injected into furnaces.[385–387] It gives an integrated value of behavior, in contrast to the differential values or rates of the other methods. To use this method for testing theories, it is necessary to assume a rate mechanism or mechanisms and to integrate the relevant equations. Agreement provides good support for the assumed mechanism, but there is usually a lack of uniqueness, since in principle there may always be other mechanisms whose integral would have the same result. Direct rate measurements are always preferable.

19.6.1.3 *Diagnostics for Data Evaluation*

The data tabulations of Table 19.6 contain values of kinetic constants such as activation energy and reaction order. These are evaluated and compared, together with relevant frequency factors, in the following section. However comparisons of the kinetic constants must be made with some care: the numbers do not necessarily mean what they appear to say. The source of the problem is essentially one of correct identification of reaction mechanisms, sometimes compounded by problems of experimental technique and accuracy. When a mechanism is identified, the appropriate equation is used to calculate the reported kinetic constants. Agreement with a particular equation is then often taken as proof of correct identification of the assumed mechanism. However the argument may be circular. Where there is only a reasonably plausible mechanism, the procedure is acceptable; but where the reaction process is complex, there may be several equations that show equally good agreement with the experimental data. Some diagnostic procedure is then needed to reduce the number of possible mechanisms for consideration. Careful experimental design can sometimes be used to reduce the number of mechanisms that may be involved. Even with good experimental design, however, it is still impossible in many cases to reduce the complexity of the experimental process to the point that the possible rate equations are unambiguously testable. Identification of mechanisms, then, is likewise ambiguous, and the final conclusion regarding a mechanism and its associated kinetic constants more usually represents the investigator's best value judgment rather than unarguable certainty. Some of the sources of ambiguity are reviewed briefly here as background to the data evaluations following.

Aspects of Experimental Design. Of all factors interfering with accurate determination of identified kinetic mechanisms, external and internal diffusion seem to be the most important. The most common means of reducing or eliminating the effects of boundary layer diffusion has been the use of low pressures, an approach originating with Langmuir; but extrapolation of the results to higher pressures is a procedure of arguable usefulness unless it is known that total pressure is not a factor in the fundamental kinetics. Furthermore, in many low-pressure experiments the samples were heated but the reacting gases were cold and the data are therefore suspect. The errors due to such temperature differences were explored by Blyholder and Eyring[325,326] and were shown to be significant. An alternative to low pressure is the use of high-velocity jets to sweep away the boundary layer mechanically; but this procedure has been criticized as one leading to mechanical erosion rather than reaction, and it does not eliminate the problem of the cold gas and hot sample. The diffusion cell[370–372] and the rotating disk[374] are alternative approaches in which the

[383] G. Beeston and R. H. Essenhigh, "Kinetics of Coal Combustion: The Influence of Oxygen Concentration on the Burning-Out Times of Single Particles," *J. Phys. Chem.*, **67**, 1349–1355 (1963).

[384] J. A. Dulhunty and P. L. Harrison, "A Laboratory Method for Measuring Some Combustion Properties of Solid Fuels," *Fuel*, **31**, 220–225 (1952).

[385] H. K. Griffin, J. P. Adams, and D. F. Smith, "Rate of Burning of Individual Particles of Solid Fuel," *Ind. Eng. Chem.*, **21**, 808–815 (1929).

[386] A. L. Godbert, "The Combustion of Pulverized Fuel," *Fuel*, **9**, 57–75 (1930).

[387] M. A. Nettleton, "Burning Rates of Devolatilized Coal Particles," *Ind. Eng. Chem. Fundam.*, **6**, 20–25 (1967).

boundary layer effects can, in principle, be designed in, in a calculable, thus controllable way.

When internal diffusion is involved, it is necessary to work either at such low temperature that reaction is all in Zone I (Section 19.3.1.2), or to use such thin samples that there can be no internal reaction. The reaction mechanism identified at low temperature, however, may not be the dominant mechanism at higher temperatures—indeed, this almost invariably will not be the case—and the problem of extrapolation remains. The thin-sample approach for high temperatures is feasible, but not always practicable. Blyholder and Eyring[325,326] found it was possible to rub carbon samples on ceramic rods to produce marks estimated at 2 μm thick, and then to follow the reaction by gas analysis, but the analysis technique had to be correspondingly sensitive. The method obviously has limited application.

Another severe problem is the temperature difference that develops between the gas and the reacting sample as the result of reaction. One solution is to operate at low reactive gas concentrations to prevent too much heat release; but then the test of the theoretical predictions is limited to low gas concentrations. At high reactant gas concentrations where there is significant heat release, sample temperature measurement is necessary, but is not always practicable or properly performed. In a fixed bed, for example, a sheathed thermocouple is incapable of distinguishing between possibly small but potentially significant differences between the solid sample and the reacting gas. This problem, as discussed below, is very likely the source of considerable error in past identifications of reaction order and therefore of mechanisms.

An alternative method of reducing the boundary layer effect is to use very small particles; a figure commonly quoted is <100 μm. However these small particles then present significant problems for measuring the extent or determining the existence of internal reaction. This is compounded by the problem of the temperature difference; Field,[95,96] for example, cites differences of as much as 300 K at 2000 K between particles and gas in a furnace, based on gray body emission from the particles to the surrounding walls. On that account, increasing use is being made of two-color pyrometry[82–85,387] to determine particle temperatures. In spite of known errors of sheathed or bare thermocouples in flames, however, these are still being used to a surprising extent. A healthy skepticism toward reported temperatures is often in order. A further problem associated with flames is the uncertainty attached to the history of material sampled from the furnace. It was to eliminate this problem that the plug-flow furnace[37,73,74,86,334–336] was developed. It may also have been eliminated in the upflow furnace,[84,97,98,198–202] although here there is uncertainty over the possibility of some backmix flow in a flame-stabilizing cone at the bottom of the furnace.

A further problem that is particularly acute with coals is the possibility of reaction during heatup. The shock tube appears to provide an answer to this. It is also a means of studying the effect of pressure under reasonably controlled conditions. A dual problem of the method that is currently in the process of being solved is that of particle dispersion and of determining the particle concentration and consumption. The method is limited, however, in the possible range of concentrations and reaction times (generally to milliseconds).

Diagnostic Factors in Rate Equations. When experimental design is not able to reduce behavior to a single dominant or controlling mechanism, mechanism identification must depend on diagnostic elements in the rate equations. The three most useful parameters are usually represented as the temperature coefficient of the reaction, usually characterized by the activation energy, E; the reaction order; and the effect of velocity on the reaction rate. Some results are unambiguous, but they can be misleading.

The Temperature Coefficient of the Reaction. This is usually characterized by the activation energy, E. If boundary layer diffusion is important, the temperature coefficient is governed by T^n, where n is about 0.75 (depending on the gases involved). This is often approximated by an apparent or effective activation energy, E, given by $E = nRT$. The effective E would therefore take the values 1–3 kcal/mole in the temperature range 1000–2000 K. Any activation energy significantly over this range of values can be attributed with considerable confidence to nondiffusional behavior. It does not mean, however, that boundary layer diffusion is not important; only that it is not controlling. In this instance it could be appropriate to make the distinction between controlling and dominant. Activation energies within the range, however, are not necessarily indicative of boundary layer dominance. In some conditions the reaction rate peaks with temperature (see Section 19.6.2.6) and the apparent activation energy becomes negative. Measurements over a small temperature range in this region can be ambiguous. In some instances, also, E may be very small anyway (see Section 19.2 and Table 19.6).

The Order of Reaction. Measurements have shown that reaction order of the carbon and char reactions with respect to oxygen can vary from zero to one (with a problematical second order suggested,[338,388] for the experiments of Lee et al.[337]). The theoretical predictions are given in Section 19.5. A zero-order reaction is reasonably interpreted as a desorption-dominated, or controlled reaction. A first-order reaction can be due either to adsorption control (or dominance) or to boundary layer diffusion; to make the differentiation, recourse in this instance must be had to the temperature coefficient or some other diagnostic point. A half-order reaction or something close to it has even more interpretations. It is predicted explicitly for internal pore diffusion control of an intrinsic zero-order reaction (Equation 19.83) in Zone II, and also for dissociative adsorption as in Equation 19.39. It is a possible value, however, for a Freundlich isotherm; and it is also a good approximation to the middle range of a Langmuir isotherm. The converse is true as well: Blyholder and Eyring[325,326] showed that some of their data fitted a half-power expression in oxygen concentration (which may be the correct expression to use), but the data were found by this writer to fit a Langmuir isotherm equally well.

[388] K. B. Lee, Ph.D. thesis, Dept. of Fuel Technology and Chemical Engineering, University of Sheffield, England, 1961.

The Influence of Velocity. It is a common assumption that a chemically controlled reaction is insensitive to or independent of ambient velocity, and that conversely, velocity dependence is indicative of boundary layer diffusion. This is usually, but not always, the case. As discussed previously,[32] in the special case of reaction in a channel the diagnostic characteristics can reverse. On account of dilution along the channel, the overall mean reaction rate can be influenced by velocity in the chemically dominant region; and on account of limiting boundary layer development, the overall mean reaction rate can become insensitive to velocity in the diffusion region.

In summary, diagnostic behavior intended to identify mechanisms is not always as unambiguous as it is often suggested to be, and conclusions from such experiments should be accepted with care.

Equation Approximations as Sources of Error. The implied assumption in the foregoing is that the equations used in the kinetic expressions are accurate or "correct." There are, however, two extreme methods of treating raw data to determine the rate coefficients. One extreme is to assume a set of "accurate," "elementary" reaction steps or mechanisms that may be combined into a composite equation, such as the Langmuir-Hinshelwood expression, in one way or another. The composition may be strictly a priori analytical, or it may be partially a priori analytical and combined with fitting to inductively modified equations to eliminate misfits between data and equations.

The other extreme is to fit the raw rate data to the simplest possible nth-order, semiempirical relation, usually of the form

$$\text{rate} = kp^n c^m \tag{19.99}$$

where "rate" is the rate of loss of mass, or rate of regression of a surface, or some other; n is the "operational" reaction order with respect to the partial pressure (p) of the reactant gas; and m is the "operational" reaction order with respect to the solid concentration, c, where c may be identified in several different ways, the two most common being the surface area of the sample and weight of the sample. The expression is then often normalized to yield a "specific" rate by dividing by the surface area in the first case or by the sample weight in the second.

The fundamental and semiempirical rate expressions approach each other if the complexity of the reaction being studied can be reduced by appropriate experimental design, as discussed above. This would be the case, for example, if the only reaction being studied were the adsorption process alone or the desorption process alone. In general, the two analytical approaches converge if, as discussed earlier, pressures and temperatures are low, and samples are thin, since these conditions tend to eliminate diffusional complications and to allow only one reaction to predominate. The evaluation problem, therefore, can be alternatively expressed as a matter of judging how good an approximation the equation used may be to the "true" fundamental mechanistic equation (which latter, indeed, may not be known).

Where the convergence between a fundamental equation and the empirical or approximate operational equation is not good, or is unknown, it is then a good question what the empirical order index, n, or rate coefficient, k, really represent, and whether inclusion of such coefficients in data compilations is justified.

Experimental Characterization as a Source of Error. An obvious yet subtle source of error is to be found in incorrect or incomplete characterization of the reacting sample. This point can be illustrated with two examples.

The first concerns the importance of knowing the sample temperature, which has been emphasized several times. It was illustrated particularly well by Froberg.[144] The experiments involved 1/2-in. spheres suspended in a tube furnace. Using the *furnace* temperature (which is often employed), the rate data taken in air and in oxygen showed a considerable influence of oxygen partial pressure; using the *sample* temperature, however, no such effect was found, as illustrated in Fig. 19.41. All air and oxygen data lie on the same line, implying a zero-order reaction of the carbon with respect to the oxygen concentration.

This sort of result immediately makes many kinetic results suspect, particularly with respect to reports of the oxygen reaction

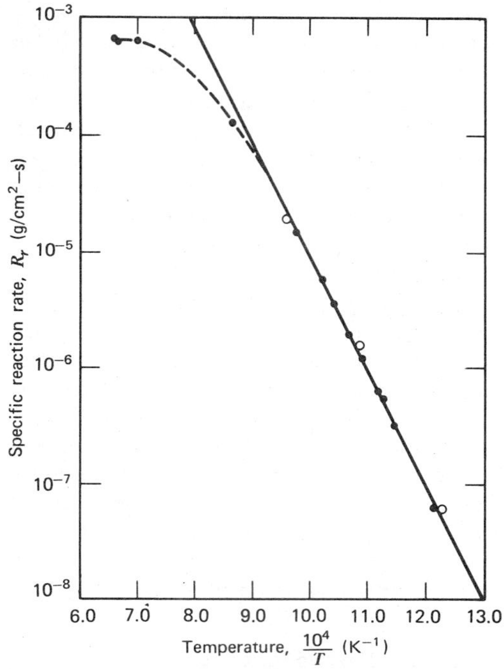

Figure 19.41 Arrhenius plot for combustion of ½-in. carbon spheres in air (○) and in oxygen (●). (Zone I reaction over straight-line portion. Insensitivity to oxygen concentration implies zero-order reaction.[144,147])

order. In particular, if the comment made above is recalled—that small temperature differences between sample and gas are often unmeasurable or undetectable, as in a fixed bed with temperature measured by a sheathed thermocouple—it is more than likely that many reaction orders are in error on account of such undetermined temperature differences.

The second example, the effect of internal surface on reaction rate, was described briefly in Section 19.3.1.2. The developing surface obviously presents a major problem of evaluation of reaction order, since as with the temperature problem discussed above, changes of reaction rate can be wrongly attributed to one parameter rather than to another.

In particular, as the result of changing internal surface and pore structure with burnoff, changes in reaction rate can be incorrectly identified with a changing reaction order. Froberg[144] found that this could be taken into account by changing the reactant gas concentration during reaction in steps up and down. This technique was brilliantly exploited by Tyler et al.[329] to establish the "true" reaction order of a high-purity graphite (type AGKSP, with 2.2 ppm impurity). The method of experiment was to switch either the total pressure or the oxygen concentration from a "high" to a "low" value and back again. Independent experiments established that reaction was uniform throughout the sample (Zone I).

Figure 19.42 illustrates the typical response obtained, converted into specific reaction rate as a function of burnoff. The upper diagram shows the variation of reaction order. In the example cited, the reaction order was determined as 0.27. This included a correction for a difference in temperature between the two oxygen levels. Without the temperature correction, the apparent order was found to be 0.37, the effect being of the same type as that found earlier by Froberg.[144]

19.6.2 Reation Rate Measurements

This evaluation of the rate data is carried out in six parts. Many of the low-pressure data have already been discussed in previous sections. The principal focus is on the kinetic constants (including the preexponential factors that are discussed in a comparative manner) and on the high-pressure behavior. Attention is also given to the negative temperature coefficient behavior.

19.6.2.1 *Activation Energies and Reaction Orders*

Reference to Table 19.6 shows that taking the data at their face value, with the exception of hydrogen adsorption, no very clear picture emerges. The oxygen activation energies cover a particularly wide range.

Hydrogen. With only one investigation cited, there is nothing to add to the discussion in Section 19.3.1 of Fig. 19.11 and Table 19.2. It is appropriate to reemphasize the high reactivity of carbon with hydrogen via chemisorption, even at normal temperatures and pressures.

Water. The activation energies lie mostly between 60 and 70 kcal/mole, with first-order reaction with respect to the water concentration implying that the process is adsorption. Blakely and Overholser[343,360] report nonunity reaction orders and cite values down to 0.5 obtained by other investigators. It is possible that this simply represents an apparent order that approximates the L-H equation in transition from first to zero order (see also "The C/H_2O Reaction" in Section 19.5.2.3).

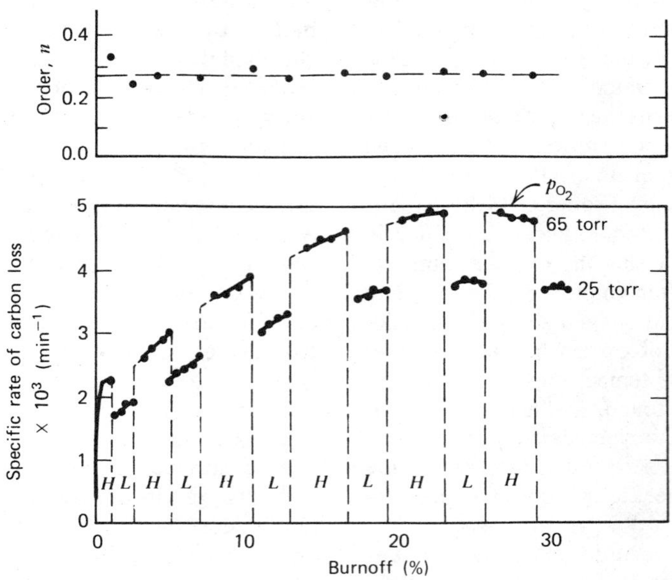

Figure 19.42 Specific reaction rate (grams of carbon per gram of carbon remaining per minute) and reaction order, n, at 993 K as function of burnoff, using pressure step method (see text).[329]

Carbon Dioxide. This reaction shows a wider spread in the activation energies: from 50–100 kcal/mole. There is mostly agreement that the reaction is first order with respect to the gas, implying adsorption. The Blakely[343] and Overholser[360] papers report lower orders than unity and, again, it is not clear whether this is not just an approximation to the L-H equation in a transition region.

Oxide Film, C(O). The few direct measurements of film decomposition[111,114,118,120,122] yield values ranging from 40 to 80 kcal/mole, with Tucker and Mulcahy[111] showing a linear drop with increasing burnoff. This range is interestingly close to the values for the H_2O and CO_2 reactions and also, as will be seen below, to the values for the O_2 reaction. This may be coincidence, but if there had been an error of order identification in the H_2O and CO_2 reactions, the activation energy values would be identified with desorption rather than with adsorption.

Oxygen. This is the most complex and confusing of all the reactions. Table 19.6 shows reaction orders ranging from zero to 1; and activation energies ranging from 3 to 80.

In evaluating these numbers it is informative to divide the activation energies into groups corresponding to first order, fractional order and zero or very low order, taking the reaction orders at their face value in the first instance. For zero-order reaction the activation energies range mostly from 40 to 80 kcal/mole. For fractional(mostly half-) order, the activation energies range from roughly 20 to 50 kcal/mole. For first order the energies range from about 3 to 30 kcal/mole.

If the values of Bansall et al.[113] for the activation energy of adsorption are considered to be the best documented, Fig. 19.10 shows a stepwise increase in E with coverage from 3 to 12 kcal/mole. Table 19.6 also shows values obtained by other investigators in roughly the same range (up to 15 kcal/mole). Of the remaining activation energy values outside this low range, it may be noted that many are tabulated under this group on the basis only of the *assumption* that the reaction was first order, and therefore was an adsorption-limited step.

In some cases where there appeared to be experimental support for a first-order reaction, later analysis showed that the conclusion was not well substantiated. Béer and Thring,[86] for example, reported a first-order reaction, but this was based on a method of interpolation that assumed the validity of linear interpolation between experimental points (the same was true of Lee et al.[337]). The result of an alternative method of interpolation that makes no such assumptions is illustrated in Fig. 19.43. This can be interpreted in two ways: either the measurements were too erratic to be of any value at all, or the reaction was essentially insensitive to oxygen concentration, implying that the order was close to zero. When the rate data are plotted on an Arrhenius plot, Fig. 19.44 is obtained, showing activation energies for the different runs of either 22 or 44 kcal/mole.[37] The excellent order here suggests that the measurements were not erratic; the results also indicate both Zone I and Zone II reactions.

The data of Gulbransen and Andrew[120] were reinterpreted by Blyholder and Eyring.[325] Gulbransen and Andrew originally claimed support for the first-order reaction by a plot of rate against oxygen concentration that did not go through the origin (as also did Lee et al.[337]). However Blyholder showed that the Gulbransen data fitted a half-order plot, allowing alternative interpretations of either dissociative adsorption or pore diffusion–controlled internal reaction.

Of other apparent first-order data, the sequence of papers by Smith and associates[84,97,98,198–202] is particularly striking. These are mainly concerned with furnace experiments with some fixed-bed experiments. It is interesting and possibly significant that the proposed reaction order changed with time, in the sequence of papers, from first to zero. The papers by Hamor and Smith[98] and by Smith and Tyler[201,202] are particularly important in that respect. They both show more careful evaluation of the reaction order, partly representing the superior technique developed by the longer experience.

One other paper of interest is that by Field,[95] whose data, by reevaluation,[38] are clearly first-order dependent at the higher temperatures, but possibly not at the lower. Field reported that no simple Arrhenius expression would fit his data, on coal char, which were close to linear in their variation

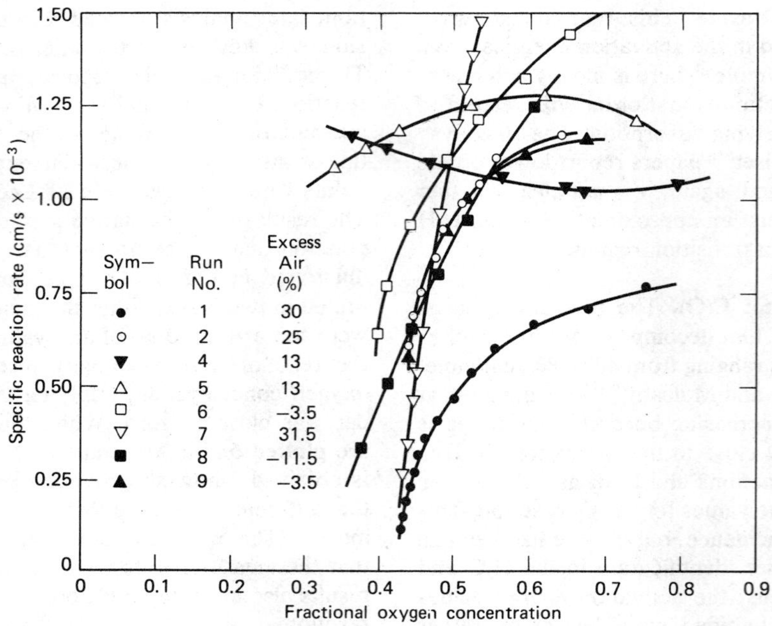

Figure 19.43 Variation of reaction rate with fractional oxygen concentration (temperature variable through runs); anthracite flame in plug-flow furnace.[37,389]

with temperature. Later data[96] on swelling and nonswelling bituminous coals showed the same behavior. Reanalysis of the earlier char data,[38] however, showed that they were well fitted by a simple Langmuir equation, with an adsorption activation energy of 6 kcal/mole and a desorption activation energy of 37 kcal/mole.

The present tentative conclusion, therefore, is that there is experimental support,

[389] J. M. Beér, "Investigation into the Combustion Mechanism of Pulverized Fuel Flames with Special Respect to the Mixing of Flames with Secondary Air," Ph.D. thesis, Dept. of Fuel Technology and Chemical Engineering, University of Sheffield, England, 1960.

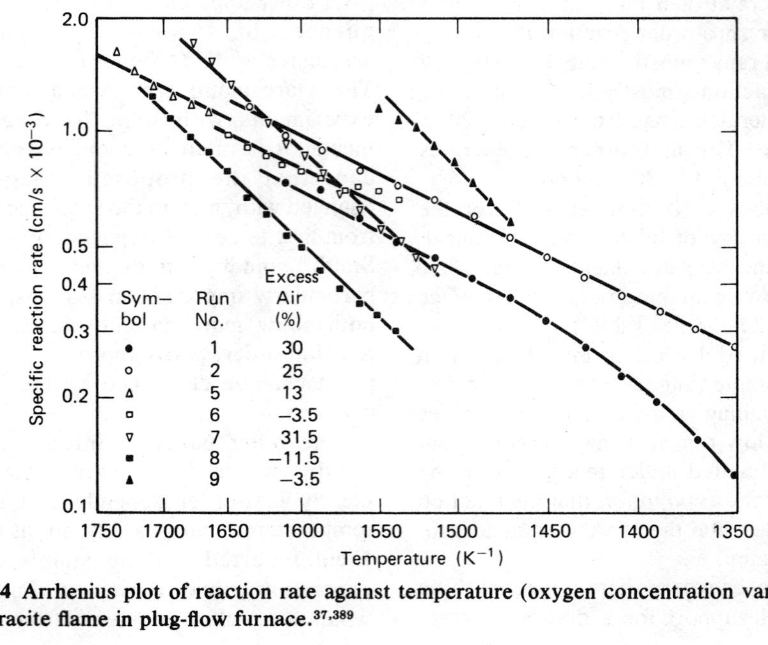

Figure 19.44 Arrhenius plot of reaction rate against temperature (oxygen concentration variable through runs); anthracite flame in plug-flow furnace.[37,389]

from several sources, for adsorption activation energies in the range from 3–15 kcal/mole, accompanied by first-order reaction. True desorption activation energies would then lie in the range roughly 40–80 kcal/mole, with true reaction orders of zero. Half-order reactions are then obtained under conditions of pore diffusion control, with true reaction orders of zero and apparent orders of 0.5, or in that region. This evaluation is also supported by data discussed below on measurements of internal surface and particle density during reaction (Section 19.6.2.4).

The outstanding question that remains is why true desorption activation energies vary so widely. The tentative answer seems to be that a combination of degree of graphitization and catalysis by impurities (see Section 19.6.2.3) is most probably responsible. The purest material generally shows the highest activation energies.

The effects of heat treatment temperature and graphitization have been discussed earlier (Section 19.3.2 and Fig. 19.23), but there are additional factors to be considered. Although Bansall et al.[114] did not determine the activation energies for film decomposition, they did show that film decomposition also occurred in steps, paralleling the stepwise adsorption (Section 19.3.1). The decomposition steps were identified over the temperature range 300–900°C, the range over which many rate measurements have been carried out. When reaction of very pure material is examined under carefully controlled conditions, separation of steps is clearly possible. When reaction of manufactured graphite is carried out at a fixed temperature, with simultaneous adsorption and desorption, overlapping of the steps is to be expected. The activation energy measured over a range of temperatures would then be a composite of the different steps, and the makeup of the composite could very well change with temperature thus leading to different activation energies for the same sample measured under different reaction conditions. At the higher temperatures (possibly above 1000°C) this can be further complicated if there is increasing graphitization, in competition with reaction, tending to change the composite mix of activated steps still further. It is therefore a problem to know just what the activation energy really represents other than being an empirical index governing the temperature coefficients of reaction whose value is very dependent on reactivity conditions. Extrapolation to other conditions then becomes arguable (though it will be done for want of anything better). In short, elucidation of the factors governing the temperature coefficient of the oxidation reactions of carbon (and possibly the others as well) is still a challenge. The behavior is not simple.

19.6.2.2 *Reaction Under Pressure*

Reaction under pressure can be divided into combustion (reaction with O_2) and gasification (reaction with CO_2, H_2, and H_2O). Gasification under pressure has been studied for obvious reasons; combustion under pressure is often involved as the first stage of gasification, but it has not had the same degree of attention on that account. A potentially important practical application of pressurized combustion is in diesel engines and possibly gas turbines, and some work has been carried out in those connections.[63,390–392] Rate data for combustion under pressure, however, are only available from shock tube experiments.

A general result that seems to be common to both combustion and gasification under pressure, however, is that no reaction rates appear to be significantly influenced directly by pressure. This is particularly well illustrated by Fig. 19.45 for carbon oxidation in the shock tube at 10 atm. The shock tube rate data, recorded as g/(cm^2)(s)^1, are sandwiched between low-pressure and normal pressure results, with fitting to the semiempirical Nagle and Strickland-Constable expression.

Pressure does, or can, affect the reaction in other ways. Rates of heat transfer, and therefore rates of heating, can be affected.[391] In the gasification reactions the most significant result is the favorable thermodynamic shift toward methane formation. This is so unfavorable at normal pressures that meth-

[390] H. P. Hazard and F. D. Buckley, "Experimental Combustion of Pulverized Coal at Atmospheric and Elevated Pressures," *Trans. ASME,* **70,** 729–737 (1948).

[391] T. T. Omori and A. A. Orning, "Effect of Pressure on the Combustion of Pulverized Coal," *Trans. ASME,* **72,** 591–597 (1950).

[392] W. Wentzel, "The Ignition and Combustion Process in the Coal Dust Engine," *Fuel,* **11** 177–196, 222–228 (1932).

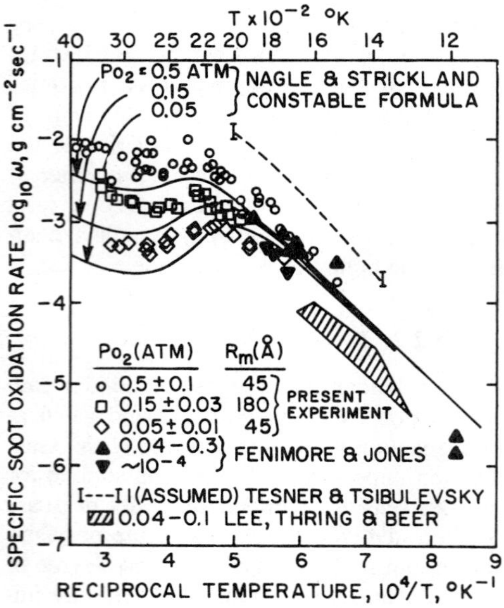

Figure 19.45 Specific soot oxidation rate measurements versus temperature as a function of oxygen partial pressure measured in a shock tube. Comparisons are with normal pressure and low-pressure data.[373]

ane production from hydrogasification is then negligible.

Combustion. Early work on pressurized combustion was in connection with the diesel,[63,392] with somewhat later work on gas turbine combustion.[390,391] Pressures were relatively low: up to 11 atm in the diesel experiments and only up to 5 atm in the gas turbine experiments.

No rate data were obtained from these experiments, but they did yield some interesting magnitudes. In the diesel experiments the maximum rpm was 1200, so that the residence time was only 20 ms. Most data were obtained at much lower rotation rates, with reaction times mostly in the range 0.1–0.2 s. Some particles were quite large, up to 125 μm, with reaction times quite comparable to normal pressure data, as indicated in Fig. 19.2. Ignition delay times were also relatively long: maximum delays were about 50 ms, with most values in the range 10–30 ms. These can be compared with Goldberg's recent data in a stirred reactor[69] of residence times of 10–20 ms and ignition times estimated at about 1 ms.

In the gas turbine and related pressurized combustion experiments[390,391] pressures were lower. Qualitative results were essentially the same. Hazard and Buckley[390] reported that the combustion reactions were "not measurably affected" by the pressure and concluded that design could be based on normal pressure rate data. Omori and Orning[391] argued that the pressure reduced the rate of heat transfer and reduced the rate of heating so that in their equipment, which was not designed for complete combustion, the reaction efficiency dropped as pressure increased. This should not be a problem in a combustor designed for complete combustion, however.

Only in the shock tube experiments[82,83,85,180,373] have rate data been obtained, as listed in Table 19.6. Nettleton and Stirling[82,83,85] obtained an activation energy of 13.5 kcal/mole for an assumed first-order reaction. In the limiting condition of low p_{O_2}, Park and Appleton[373] require a first-order reaction with an activation energy of 30 kcal/mole. The Nagle and Strickland-Constable expression used, however, is essentially semiemperical, and Nagle did not demonstrate directly the first-order rate at, essentially, low oxygen partial pressures or temperatures. What is evident in Fig. 19.45 is that the data are clearly separated by low oxygen partial pressure at the highest temperatures, where the slopes of the curves are flattening out or going negative; but where the points are showing the greatest slope, they are becoming insensitive to oxygen partial pressure, supportive of zero-order reaction in this region.

Pressures and temperatures in the shock tube are much higher than in the engine experiments, with values up to 30 atm and to 4000 K. Reaction times are correspondingly reduced; Nettleton and Stirling reported ignition times of a few hundred microseconds and reaction times of 0.1–2 ms.

Gasification

General Behavior. In gasification, carbon reacting with CO_2, even under pressure, only forms CO. The mechanism is Equation 19.8, but reaction under pressure can be enhanced by significant CO chemisorption according to Blackwood and Ingeme[309] by the reaction $C(CO) + CO_2 \rightarrow 2CO + C(O)$. The modified rate expression is given below (Equation 19.100).

With hydrogen supplied directly, or with

hydrogen formed first from steam, reaction with carbon forms methane. Mosely and Paterson[302] describe the methane formation as a two-stage process: an initial rapid reaction is followed by a slower production. This phenomenon is seen in Fig. 19.28. One interpretation is provided by Tomita et al.,[365,366] who identified an initial "induction" stage that was diffusion controlled, followed by a slow, chemically controlled stage. Blackwood[306,307] reported that gasification with hydrogen was very much dependent on there being oxygen present in the char.

In gasification with water, oxygen is present either in the char, or the reactant gas, or both. The four gases involved in the water-gas shift reaction are therefore present, and this reaction might affect the local concentrations, thus the rate of conversion or the rate of methane formation, which can be significantly influenced by the presence of catalysts.

Kayembe and Pulsifer[314] found that catalysts such as potassium, sodium and lithium carbonates, potassium and sodium chlorides, and copper oxide were effective; calcium and iron oxides were ineffective, in contrast to their behavior in catalyzing the oxidation reaction. With the catalysts, the water-gas shift reaction was in equilibrium; without the catalysts it was not. Blackwood et al.[308] also investigated the effect of heat treatment temperature: the effect on the gasification rate was described as an exponential function of the reciprocal of the preparation temperature. In other work these workers[393] also found a heating rate effect (cf. Ashu et al.[168] with respect to char oxidation).

In a mixture of hydrogen and steam, Wen et al.[394] found that the reactions proceeded simultaneously, but apparently indepently of each other, without being detectably affected by the consequences of concentration changes due to the water-gas shift reaction.

In the hydrogasification of coal, pyrolysis is an additional, competing reaction. Johnson[182,183] describes the process in three stages: devolatilization, rapid methane formation, and slow methane formation.

Rate Data. The kinetic constants listed in Table 19.6 show values of activation energies for both the CO_2 and H_2O reactions that, taken at face value, are roughly of the same magnitude as those for normal and low-pressure conditions. The detailed rate expressions, however, are considerably more complex than those developed in Section 19.5.

There are essentially two reasons for this. First, the only hydrogen reaction considered earlier was the reversible, dissociative chemisorption: at pressure it is also necessary to consider the processes leading to methane formation. Second, CO chemisorption evidently plays a much more significant role at high pressure than at normal pressure or below. Blackwood and Ingeme[309] derived and tested the following rate equation based on the assumption of reaction with chemisorbed CO:

$$\text{rate} = \frac{k_1 p_{CO_2} + k_5 p_{CO_2}^2}{1 + k_2 p_{CO} + k_3 p_{CO_2}} \qquad (19.100)$$

Compared with the standard L-H expression (Equation 19.36), the difference is the addition of the $p_{CO_2}^2$ term. It can also be compared with the Semechkova and Frank-Kamenetskii[395] equation, which has the same form as Equation 19.100 but with $(-p_{CO}^2)$ in place of $(+p_{CO_2}^2)$.

For reaction with hydrogen and steam, a number of different expressions have been proposed. The one best known for hydrogen is probably that of Zielke and Gorin[318,319]:

$$\text{rate} = \frac{k_1 p_{H_2}^2}{1 + k_2 p_{H_2}} \qquad (19.101)$$

Interestingly, at high hydrogen concentrations, this equation reduces to a first-order expression. With methane in the reactant gas to inhibit the reaction, Blackwood[307] obtained the variant:

$$\text{rate} = \frac{k_1 p_{H_2}^2 - k_2 p_{CH_4}}{1 + k_3 p_{H_2} + k_4 p_{H_2} + k_4 p_{H_2}^2 + k_5 p_{CH_4}} \qquad (19.102)$$

[393] J. D. Blackwood, D. J. McCarthy, and B. D. Cullis, "The Carbon-Hydrogen Reaction with Cokes and Chars," *Aust. J. Chem.*, **20**, 2525–2528 (1967).

[394] C. Y. Wen, O. C. Abraham, and A. T. Talwalker, "A Kinetic Study of the Reaction of Coal Chars with Hydrogen-Steam Mixtures," in *Fuel Gasification,* Am. Chem. Soc. Adv. Chem. Ser. No. 69 (1967), pp. 253–271.

[395] A. F. Semechkova and D. A. Frank-Kamenetskii, *Acta Physiochim. (USSR),* **12,** 879 (1940).

In developing their equations, both investigators proposed a sequence of elementary reactions in which first one, then two, then three, hydrogen atoms were attached to a single carbon edge atom.

For reaction with steam, Feistel et al.[311] found that a simple Langmuir expression was satisfactory (i.e., Equation 19.37 with the retarding hydrogen term deleted); or for reaction with hydrogen and steam, the retarding hydrogen term was retained (but see below). Blackwood and McGrory[310] found a more elaborate expression to be necessary:

$$\text{rate} = \frac{k_1 p_{H_2O} + k_4 p_{H_2} p_{H_2O} + k_5 p_{H_2O}^2}{1 + k_2 p_{H_2} + k_3 p_{H_2O}} \tag{19.103}$$

This equation shows product and squared terms added to the L-H expression. It is based on essentially the same assumptions regarding the methane formation from hydrogen; what is added is a term for oxide film [C(O)] formation and decomposition.

In the face of this complexity many investigators have reverted, as in the case of oxidation, to the semiempirical nth- and mth-order approximation of Equation 19.99. The carbon concentration is commonly designated by the weight (w) remaining, with $m = 1$, so that the rate $= dw/dt$, with $c = w$. This was used by Wen et al.[394] and by Van Heek et al.[312] both of whom found that the rate for the carbon-steam reaction was insensitive to steam concentration (i.e., in Equation 19.99, $n = 0$). This follows, of course, in Equation 19.37 or Equation 19.103 when the steam concentration is high and the retarding terms are low. Under these limiting conditions, the C(O) film decomposition must be the rate-limiting step and this, indeed, was the conclusion in separate experiments on carbon-steam by Fuchs and Yavorsky.[332]

For the C/H_2 reaction, the semiempirical evaluation of Wen et al.[394] yielded a first-order reaction (i.e., $n = 1$ in Equation 19.99). This result is also obtained at high values of hydrogen concentration according to Equation 19.101. Feistel et al.[311] obtained similar results working over a wider range of steam and hydrogen concentrations. Expressing the carbon concentration in terms of carbon conversion, X, a rate equation was written that is essentially a variant of Equation 19.99:

$$\frac{dX}{dt} = \bar{k}(1 - X) \tag{19.104}$$

The "constant" $\bar{k}$ in this instance, however, was developed in terms of compound expressions of the elementary rate constants and gas partial pressures. For reaction with steam alone, $\bar{k}$ was found[311] to take the form of the Langmuir equation. For reaction with hydrogen alone, $\bar{k}$ took the form of the RHS of Equation 19.101. For a mixture of steam and hydrogen, $\bar{k}$ took the form of the L-H equation (Equation 19.37), with the hydrogen partial pressure not appearing in the numerator.

The most elaborate expressions, however, have been developed by Johnson,[182,183] involving an integral correlation function that is an integrated form of a more complex version of Equation 19.104. Gardner et al.[363] used a somewhat similar expression in a study of catalysts in the hydrogasification reactions. The details of the derivations and results are out of place here. It will suffice to say that Johnson found it necessary to include an expression for the pretreatment temperature of the char. He was then able to use the expression to collate and correlate data taken under a variety of different conditions to show good agreement between predicted and experimental values in a variety of different reactors. Several different parameters were used, such as reaction rate constants, methane formation rates, and carbon oxide formation rates.

19.6.2.3 *Catalysis*

Much or most of what is known about catalysis of carbon reactions was extensively reviewed in 1968 by Walker et al.[53] with emphasis on very pure carbons. In 1977 Hippo and Walker[166] reported that there had been little addition to the literature of catalytic studies of very pure carbons since 1968. It may be concluded, therefore, that the 1968 review[53] remains the present definitive review article on the subject. With a caveat to be mentioned below deriving from a paper by Allardice and Walker,[342] the conclusions of that review seem to be the best present position. Key elements are summarized here. Readers are referred to the review for detailed substantiation.

The phenomenological behavior of catalysts is described in the literature in a number of different ways, but most conveniently by examining the variation of rate data or kinetic constants with the level and type of impurity. Some such data are listed in Table 19.7. These data can be compared with the behavior of "pure" carbons with "no" impurities. To determine baseline uncatalyzed data, it is possible only to make measurements over a range of impurity concentrations and to extrapolate to zero concentration on the assumption that the trend is monotonic. On the basis of relative unanimity for values of activation energies for highly purified carbons, Walker and co-workers then quote "baseline" values for the O_2 and CO_2 reactions as follows: for C/O_2, $E = 61 \pm 3$ kcal/mole (references cited by Walker:[125,142,285,349,401]) and for C/CO_2, $E = 87 \pm 3$ kcal/mole (references cited by Walker:[142,285,400,402–404]). No values are given for the C/H_2O reaction. The values may be compared with those in Table 19.6, and with the evaluation of those data in Section 19.6.2.1.

Two models have been proposed to explain catalytic behavior. The first is the oxygen transfer model. The catalyst enhances dissociation of the gas being adsorbed (O_2, CO_2, or H_2O), and the oxygen atoms so formed are then substantially more reactive than the molecular form, as shown in independent reactions using O atoms.

The second model is the electron transfer mechanism that originated with Long and Sykes.[245] It requires electronic binding between the adsorbed oxygen atoms and the carbon atoms, which can also be the case with the oxygen transfer model. It is then assumed, however, that the presence of a catalyst modifies the electronic binding in such a way that the C=O bond is strengthened and any C—C bonds are weakened. The model requires the ability to accommodate electrons in unfilled energy bands, and it is supposed that the catalyst action induces a preferred distribution of π electrons whose mobility allows catalysts to influence events some distance from active sites. Since transition metals have unfilled *d* bands, they have been given particular attention as catalysts, as also have the lanthanide series with unfilled *f* bands.[359]

The first model emphasizes the catalyst-gas interaction; the second emphasizes the catalyst-carbon interaction. Comparing the two, Walker et al.[53] concluded that the oxygen transfer mechanism was the more probable. In a later paper, however, Allardice and Walker[342] seemed to reverse that conclusion. Boron was used as the catalyst, and the findings showed that the reduction of activation energy qualitatively paralleled the reduction in the Fermi levels in the boronated carbon. They conclude that "this correlation supports clearly, for the first time, the Long and Sykes concept that electron transfer between the graphite and an impurity can be important in affecting the rate of gasification of the graphite by an oxidizing gas."

There are still problems, however. The impurity affecting the electronic configuration is presumed to allow a change in both adsorption and desorption activation energies, and the best evidence is that the first is involved in the CO_2 reaction and the second in the O_2 reaction. Walker et al.[53] state that the oxygen transfer model tacitly assumes that chemisorption is the rate-limiting step. This appears to be the case at first sight, and it would rule out catalysis of a zero-order oxygen reaction. The evidence indicates, however, that catalysts have little or no effect on the edge atom reactions, and that they act to enhance the basal plane reactions taking

[396] G. R. Hennig, "Catalytic Oxidation of Graphite," *J. Inorg. Nucl. Chem.*, **24**, 1129 (1962).

[397] H. Amariglio, Thesis, Docteur en Science, University of Nancy, France, 1960.

[398] E. A. Heintz and W. E. Parker, cited by Walker, Shelef, and Anderson (Ref. 53) as private communication.

[399] Y. Kawana, "Reactivity of Coke. III. Effects of Some Metallic Additions on the Surface Area of Cokes from Humic Acid and on the Absolute Reaction Rates of the Coke–Carbon Dioxide System," *Bull. Chem. Soc. Jap.*, **27**, 574 (1954).

[400] K. Hedden, H. H. Kopper, and V. Schulze, "Volatile Inhibitors and Catalysts for the Reaction of Carbon with Carbon Dioxide," *Z. Phys. Chem.*, **22**, 23 (1959).

[401] C. Heuchamps, Thesis, Ingénieur-Docteur, University of Nancy, France, 1960.

[402] J. F. Strange, Ph.D. thesis, Pennsylvania State University, State College (1964).

[403] M. Rossberg, "Experimental Results Concerning The Primary Reactions in the Combustion of Carbon," *Z. Elektrochem.*, **60**, 952 (1956).

[404] A. F. Armington, Ph.D. thesis. Dept. of Fuel Technology Pennsylvania State University, State College, 1960.

Table 19.7 Effect of Catalysts on Activation Energy E, of Carbon Gasification[53]

Carbon Characteristics	Reaction	Temp Range (°C)	Kind and Amount of Impurity in Pure Sample	E for pure sample (kcal/mole)	Mode of Addition (or removal) of Impurity	Kind and Amount of Impurity in Catalyzed Sample	E of Catalyzed Reaction (kcal/mole)	Ref.
Pechiney EDF	C + air	580–640	<1 ppm total	64	Chlorination, 2700°C; heating, 3500°C	V, 38; Fe, 6; Na, 0.5; Mn, 0.2 ppm	60	285
Pechiney spectroscopic	C + air	580–640	Na, 0.5; V, <0.1 ppm	64	Heating, 3500°C	Na, 3; V, <0.1 ppm	58	285
Natural Madagascar agglomerated	C + air	560–620	Na, 0.1; Mn, <0.5 × 10^{-3}; Ta, <10^{-4} ppm	64	Heating, 3500°C	Na, 9; Mn, 0.46; Ta, 0.1 ppm	59	285
Natural Ticonderoga single crystals, prismatic planes	C + 0_2	550–700	?	46	Deposition on surface	Colloidal Fe	10	396
Nuclear graphite	C + air	<600	~1 ppm total	62	Impregnation from solutions	~100 ppm of Ba, Cu, Mn, or Na	40 ± 3	397
United ultrapurity graphite	C + O_2	700–800	<60 ppm ash	63	Added as powders	0.1 mole % of Groups IIIA–VIA	29–52	349
United ultrapurity graphite	C + air	600–800	<60 ppm ash	49	Added as powders	3*d*, 4*d*, 5*d* transition metals	—	398
Pechiney EDF	C + CO_2	In Zone I	<1 ppm total	87 ± 3	Chlorination, 2700°C: heating 3500°C	V, 38; Fe, 6; Na, 0.5; Mn, 0.2ppm	87 ± 3	285

Coke made from humic acid	C + CO_2	700–900	2.5% ash	95	Admixing	Large amounts of Na, Fe, or Al	Lowered by 25–40	399
A whole series of carbons including very pure	C + CO_2	~1000	Varies	81–92.5	Added to gas	1–3% $POCl_3$	59.0–47.5	400
Charcoal	C + CO_2	700–900	400 ppm, mainly B and Si	74	Extraction by HCl and HF	3.5% ash, mainly Na_2O, K_2O, Fe_2O_3, Al_2O_3, SiO_2	68.0	245
Charcoal	Adsorption step	700–900	400 ppm, mainly B and Si	68.4	Extraction by HCl and HF	3.5% ash, mainly Na_2O, K_2O, Fe_2O_3, Al_2O_3, SiO_2	58.8	245
Charcoal	Desorption step	700–900	400 ppm, mainly B and Si	66.0	Extraction by HCl and HF	3.5% ash, mainly Na_2O, K_2O, Fe_2O_3, Al_2O_3, SiO_2	38.0	245
Charcoal	C + H_2O, desorption step	700–900	400 ppm, mainly B and Si	83 ±3	Extraction by HCl and HF	3.5% ash, mainly Na_2O, K_2O, Fe_2O_3, Al_2O_3, SiO_2	55 ± 7	245
High-purity sugar C	C + CO_2, desorption step	600–900	0.07% ash	61.2	Impregnation from solutions	7% Fe	22.8	395
High-purity sugar C	C + CO_2, desorption step	600–900	0.07% ash	61.2	Impregnation from solutions	5% Al	55	395
High-purity graphite (SPI)	C + O_2, dry and moist	550–650	<1 ppm B	53	Added as powder and heat treated	B.6 × 10^{-6} to 10^{-2} B/C atom ratios in 6 steps	53 to 45 falling as B/C ratio increases	342
Graphitized carbon	C + air,	600–1100	<1 ppm ash	48	Added as salt, reduced with H_2	Pt	25.5	145

place at imperfections (in accordance with what would be expected from O atoms).

There is, indeed, some evidence that catalysis can be regarded as two parallel reactions. The oxygen transfer model is clearly consistent with a change in adsorption activation energy. If the rate control is desorption, then there will again be a difference between catalyzed and uncatalyzed reactions, since there will be significantly different extent of involvement with the basal plane, and this could result in an apparent difference in activation energy. The electron transfer mechanism, on the other hand, does not account for the apparent preferential attack on the basal planes in the presence of catalysts.

Taking into account the additional arguments by Walker et al.,[53] and in spite of the later findings of Allardice and Walker,[342] it nevertheless seems that the best present explanation for the influence of solid catalysts remains that of enhancing dissociative adsorption.

19.6.2.4 *Internal Surface Effects*

The significance of internal surface area is mentioned earlier in several contexts. The variation that can be found is illustrated in Figs. 19.15 and 19.16 (with discussion in Section 19.3.1.2). The importance of the variation lies in the effect that surface area changes can have on reactivity that, if identified as due to some other cause, can lead to errors in values of activation energies and other kinetic constants. The point is made more explicitly by Equation 19.84, as a general equation for the total reaction rate of a sample, combined with Equation 19.79 for the effectiveness factor (η). Both depend on the internal surface (S_v): η is related through the parameter λ given by Equation 19.61. If S_v is a function of conditions such as burnoff, it is necessary to establish that if rate data are obtained, for example, at different temperatures to determine activation energies, the rates must be measured at the same values of internal surface and, even then, the distribution of surface area between pores of different diameter may change for different histories of burnoff. There is no known report of this having been done. Incomplete pore diffusion, with reaction confined to the outer layers, is a particular source of developing anisotrophy of porosity and pore distributions.[405–410]

Figure 19.42 illustrates an alternative approach in which the operating conditions are switched back and forth. This method is not always feasible, however, and direct measurements of surface area are ultimately required, particularly for analyzing entrained-flow reactor experiments. The surface area required would be the ASA, and this may depend on the previous history of the carbon.

Surface Area Measurements. The extent of internal reaction is governed in part by the internal surface area and in part by the accessibility of that internal surface. Accessibility is governed by the pore structure, as mentioned briefly in Section 19.5.4.1. The most significant parameter is probably the distribution of pore radii, but their form—whether they are tortuous, multiple intersecting, of variable diameter, and so forth—can be almost as important. The measurements that establish internal pore structure therefore usually require some degree of interpretation.

The three most important measurements are true density, usually determined by helium displacement; internal surface, by fluid adsorption; and internal volume, by mercury intrusion porosimetry. Methods of measurement and interpretation have been reviewed recently by Mahajan and Walker[105] (see also Thomas and Thomas[107] and Chapter 4 in this

[405] P. L. Walker, Jr., F. Rusinko, Jr., and E. Raats, "Anisotropy of Effective Gas Diffusion Coefficients in Extruded Carbon Rods," *Nature*, **176**, 1167–1168 (1955).

[406] P. L. Walker, Jr., F. Rusinko, Jr., and E. Raats, "Changes in Macropore Distributions in Carbon Rods upon Gasification with Carbon Dioxide," *J. Phys. Chem.*, **59**, 245–249 (1955).

[407] P. L. Walker, Jr., and E. Raats, "Changes in Physical Properties of Graphitized Carbon Rods upon Gasification with Carbon Dioxide," *J. Phys. Chem.*, **60** 364–369 (1956).

[408] V. M. Weekman and R. L. Gormig, "Influence of Volume Change on Gas-Phase Reactions in Porous Catalysts," *J. Catal.*, **4**, 268 (1965).

[409] W. J. Thomas, "Effect of Oxidation on the Pore Structure of Some Graphitized Carbon Blacks," *Carbon (Oxford)*, **3**, 435–443 (1966).

[410] R. H. Tien and E. T. Turkdogan, "Incomplete Pore Diffusion Effect on Internal Burning of Carbon," *Carbon (Oxford)*, **8**, 607–621 (1970).

volume). It will suffice here to mention some salient points.

Fluids used in internal surface measurements have mostly been gases, with low-temperature N_2 adsorption and normal temperature CO_2 adsorption evidently the most commonly used gases at present. In 1949 Franklin[411] described results with several liquids such as methanol, water, *n*-hexane, and benzene. Some liquids show anomalous behavior because of induced swelling effects, however, and the more recent work has focused on gases, although there is revived interest in the use of water.[412] Gases have included, for example N_2 at 78 K, Kr at 195 K, CO_2 at 195 K, Xe at 273 K, and CO_2 at 298 K (Walker and Kini,[413] Marsh et al.[414–416] Razouk et al.[417]). The method of experiment requires measuring the quantity of fluid adsorbed in the interior of the solid. The surface area is then estimated by means of equations based on different adsorption assumptions. The BET equation was originally used,[418] but the most commonly used today is the Dubinin-Polanyi equation (Dubinin[419]). Comparisons of the results from the two equations have been carried out by a number of investigators (e.g., Walker and Patel,[420] Medek[421]).

Different gases at different temperatures yield different values of surface area. The differences are interpreted to mean that the gases penetrate to different degrees into the fine pore structure of the solids. Those investigators who use CO_2 adsorption at 298 K and N_2 adsorption at 78 K identify the reported areas as the total internal area and the macropore area, respectively (Fig. 19.3).

Pore volume distributions are mostly measured by the mercury porosimeter, with some recent use of transmission electron microscopy.[422] Figure 19.46 illustrates some results for both unburned and burned particles. In this example, little pore volume is associated with pores from 1 μm down, although there may be substantial surface area involved.

Surface Area Changes with Heat Treatment. Both porosity and surface area can change with reaction or heat treatment.[423–427] Considerable attention has been given to the effects of carbonization of coal, and also those of heat treatment of chars. With coal there is, initially, a small to considerable increase in surface area with rising temperature, followed by a decrease. The extent of

[411] R. E. Franklin, "A Study of the Fine Structure of Carbonaceous Solids by Measurements of True and Apparent Densities. Part I. Coals. Part II. Carbonized Coals," *Trans. Faraday Soc.*, **45**, 274–286, 668–682 (1949).

[412] O. P. Mahajan and P. L. Walker, Jr., "Water Adsorption on Coals," *Fuel*, **51**, 308 (1971).

[413] P. L. Walker and K. A. Kini, "Measurement of the Ultrafine Surface Area of Coals," *Fuel*, **44**, 453–459 (1965).

[414] H. Marsh, "The Determination of Surface Areas of Coals: Some Physicochemical Consideration," *Fuel*, **44**, 253–268 (1965).

[415] H. Marsh and T. Siemieniewska, "The Surface Areas of Coals as Evaluated from the Adsorption Isotherms of Carbon Dioxide Using the Dubinin-Polanyi Equation," *Fuel*, **44**, 355–367 (1965).

[416] H. Marsh and W. F. K. Wynne-Jones, "The Surface Properties of Carbon. I. The Effect of Activated Diffusion in the Determination of Surface Area," *Carbon (Oxford)*, **1**, 269–279 (1964).

[417] R. I. Razouk, F. Z. Saleeb, and A. M. Youssef, "The Effect of Thermal Treatment on the Surface Area of Coal," *Carbon (Oxford)*, **6**, 325–331 (1966).

[418] C. Pierce, "Computation of Pore Size from Physical Adsorption Data," *J. Phys. Chem.*, **57**, 149–151 (1953).

[419] M. M. Dubinin, "Porous Structure and Adsorption Properties of Active Carbons," in *Chemistry and Physics of Carbon* Vol. 2, Marcel Dekker, New York, 1966, pp. 51–120.

[420] P. L. Walker, Jr., and R. L. Patel, "Surface Areas of Coals from Carbon Dioxide Adsorption at 298°K," *Fuel*, **49**, 91–94 (1970).

[421] J. Medek, "Possibility of Micropore Analysis of Coal and Coke from the Carbon Dioxide Isotherm," *Fuel*, **56**, 131–133 (1977).

[422] L. A. Harris and C. S. Yust, "Transmission Electron Microscope Observations of Porosity in Coal," *Fuel*, **55**, 233–236 (1976).

[423] S. P. Nandi, V. Ramdass, and P. L. Walker, Jr., "Changes in the Ultrafine Structure of Anthracite upon Heat Treatment," *Carbon (Oxford)*, **2**, 199–210 (1964).

[424] S. P. Nandi and P. L. Walker, Jr., "Diffusion of Argon from Coals of Different Rank," Paper No. 23 in *Coal Science*, Am. Chem. Soc. Adv. Chem. Series No. 55, 1966.

[425] P. Chiche, S. Durif, and S. Pregermain, "Development of the Internal Surface Area of Coals During Carbonization," *Fuel*, **44**, 5–28 (1965).

[426] Y. Toda, "A Study by Density Measurement of Changes in Pore Structures of Coals with Heat Treatment. Part I. Macropore Structure. Part 2. Micropore Structure," *Fuel*, **52**, 36–40, 99–104 (1973).

[427] J. Berger, T. Siemieniewska, and K. Tomkow, "Development of Porosity in Brown-Coal Chars on Activation with Carbon Dioxide," *Fuel*, **55**, 9–15 (1976).

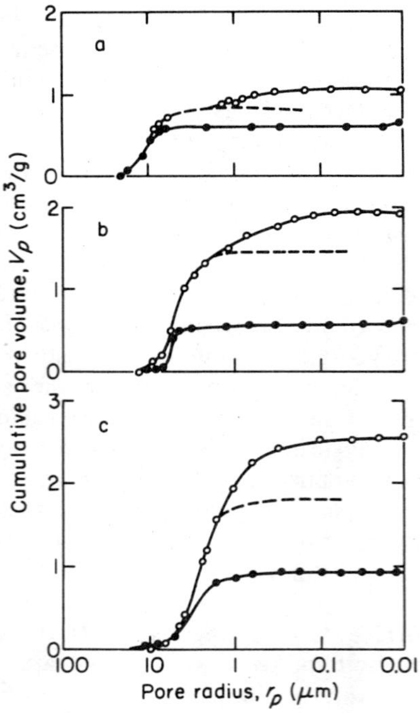

Figure 19.46 Pore volume distributions for raw (●) and burned (○) materials in terms of initial diameter (d_0) and fluidizing velocity (U). (*a*) d_0 = 78 μm, U = 0.38; (*b*) d_0 = 49 μm, U = 0.55; (*c*) d_0 = 22 μm, U = 0.81. Dashed line denotes assumed limit of interparticle voidage.[201]

the increase or decrease is a function of rank. For semianthracite, Nandi et al.[423] found that the initial increase was quite slight, but the surface area dropped sharply above an HTT of 900°C. Diffusion of gases such as argon in and out of coal (Nandi and Walker[424]) was activated; the activation energy increased with HTT, which indicated that the micropore structure was closing. In other investigations by Chiche et al.[425] and by Berger et al.,[427] bituminous coals showed a very marked increase in surface area up to about 750°C.[425] Brown coals were found to have well-developed macropores.[427] Coking and noncaking coals showed different behavior above 750°C: with the coking coals the porosity steadily fell to zero as the temperature rose; with the noncaking coals the porosity changed, but a substantial fraction survived in sealed pores.

The time required for development of pore structure is also important, particularly if the results are to be related to behavior of small (p.c.) particles during rapid combustion or gasification. Nsakala et al.[428] recently examined the development of surface area of a lignite at 800°C following rapid heating (8000°C/s). Both CO_2 and N_2 areas increased, but there was little change in the first 0.2 s, although there was substantial mass (volatiles) loss in that time and this was correspondingly the principal period of density change. Since the N_2 area corresponds to increase in access pore volume, it is surprising but potentially most significant to find that it opened up after 0.2 s and only after considerable pyrolysis has occurred.

Surface Area Changes with Reaction. In the context of this chapter, the surface area changes during reaction are the more important. They are needed to extract rate data from raw measurements, for analysis and determination of rate constants. They are also needed for prediction of particle histories throughout reaction, either for comparison with such data as measured in plug-flow furnace experiments, or for design predictions.

In prediction of lifetime histories, there are two possible approaches. The first is the detailed mechanistic modeling from "first principles" of a reacting pore that enlarges during reaction, with length extension and coalescence with other pores. This requires appropriate knowledge of the initial pore shape, structure, length, and kinetic response to the reactive gas.

The second method is the "assigned parameter" modeling, which is more phenomenological. This method utilizes assumptions about the existence of porosities and surface areas. It is therefore more general, since it is not limited to a particular pore structure model. It does require, however, development of empirical relationships between the internal surface changes and burnoff, and these must be obtained by experiment.

The First-Principles or Mechanistic Approach. As an example of this approach,

[428] N. Nsakala, P. L. Walker, Jr., and R. H. Essenhigh, "Characteristics of Chars Produced from Lignites by Pyrolysis at 808°C Following Rapid Heating," *Fuel,* **57,** 605–611 (1978).

Finson et al.[429] consider reaction of a single parallel-sided pore as a base for overall relations. The internal surface and porosity are then given by the first and second moments, with respect to radius, of the distribution function of the pore radii. The two integrals are taken from a minimum value, being the smallest pore that can be penetrated by a gas molecule, to a maximum that does exist but does not need to be closely specified. The integrals can be evaluated if the pore size distribution functions can be specified. Using data from Stacy and Walker[430] and from Berger et al.,[427] an inverse r^3 function was proposed. This stated the initial conditions.

As the reaction proceeds, the pore distribution function changes, with the pores growing. A full model must take into account pore opening, coalescence, and related behavior.

Turkdogan et al.[431] emphasize the particular role that isolated pores can play as reaction gets under way. As much as 50% of the initial pore volume can be isolated in some cases. As reaction starts, anything from a small to a large fraction of the isolated pores can become accessible as reaction breaks through the pore walls. At the same time, the existing pore structure is modified, primarily by widening the pores, and this in turn can significantly influence the rates of diffusion into the particles by altering the internal diffusivities. Equation 19.56 relates the normal molecular diffusion coefficient (D_T) to the internal diffusivity (D_e). Typically, the ratio D_e/D_T is around 0.1, but this can change significantly with burnoff. Some data for coke and graphite are illustrated in Fig. 19.47. The initial ratios are much lower than 0.1 to start with, but they have doubled or tripled after only 10% burnoff.

Since the increased diffusivity allows more rapid access to the particle interior, other

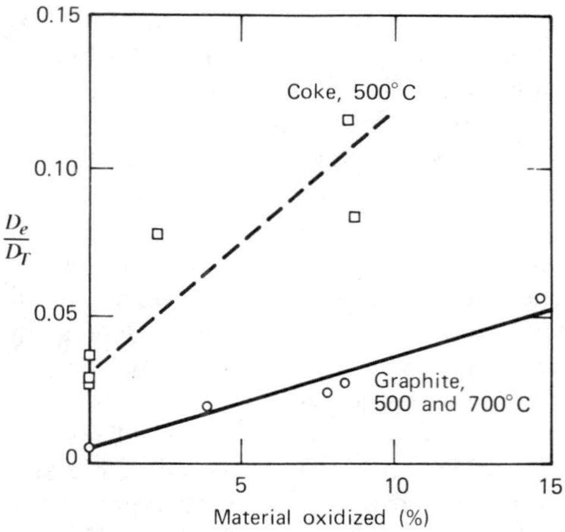

Figure 19.47 Effect of oxidation on the D_e/D_T for coke and graphite.[431]

things being equal, the particle is more likely to burn in Zone I with penetration of the reaction to the center. If burnoff is accompanied by increased surface area, however, which is more generally the case, more gas will react closer to the surface, less will be available to react internally, and the reaction will tend to move into Zone II. The result of these competing factors is that the penetration depth of the reaction can become unstable, possibly accounting, for example, for the somewhat erratic wanderings illustrated in Fig. 19.14.

The problem is compounded in the case of coal by the joint effects of pyrolysis and swelling. Pyrolysis initially retards penetration into the interior; swelling can enhance or retard it depending on the extent to which thermoplastic behavior may close pores or thermosetting behavior may help to keep them open.

If reaction is in Zone II, reaction will be fastest at the pore mouth, and the pore will change shape as well as grow. In the Finson model,[429] calculation has been confined to the kinetic limit with the proposition that mass removal is proportional to the pore area. The result is that all pores grow at the same rate.

The model has been used to predict reaction rates for comparison with experimental data obtained by Smith.[198–200] Finson et al.[429] point out that the Smith data are reported at unknown levels of burnoff; their calculations are therefore compared with the experimen-

[429] M. L. Finson, P. F. Lewis, G. A. Piper, G. A. Simmons, and K. L. Wray, *Modeling of Coal Gasification for Fuel Cell Utilization,* Final Rep. to ERDA, Div. Conservation Research and Technology, Contract No. E (04-3)1254, 1977.

[430] W. D. Stacy and P. L. Walker, Jr., *Structure and Properties of Various Coal Chars,* Rep. Coal Res. Sect., College of Earth and Mineral Sciences, Pennsylvania State University, September 1972.

[431] E. T. Turkdogan, K. G. Olsson, and J. V. Vinters, "Pore Characteristics of Carbons," *Carbon (Oxford),* **8,** 545–564 (1970).

tal data at 0, 50, and 100% burnoff, for different particle sizes, and the experimental results are indeed bracketed by the predictions after appropriate selection of pore length as an adjustable parameter. The calculations so far completed ignored pore coalescence; consequently, the surface area was overestimated, and so also was the reaction rate, near completion of the reaction. With fitting by the pore length, agreement with experiment could, otherwise, be obtained to within a factor of 2.

Such calculations are not new, of course. Petersen,[432,433] for example, used both a single-pore model and an intersecting-pore model, and compared his predictions with data on reactions of cylinders carried out by Petersen and Wright.[346] He assumed a uniform pore diameter, however. There are also other model calculations in the literature, some on very complex pore assumptions[261,434–463] (see

[432] E. E. Petersen, "Reaction of Artificial Graphite with Carbon Dioxide: Uniform Gasification of Artificial Graphite with Carbon Dioxide at 1100°C," *Ind. Eng. Chem.*, **47**, 1630–1634 (1955).

[433] E. E. Petersen, "Reaction of Porous Solids," *Am. Inst. Chem. Eng. J.*, **3**, 443–448 (1957).

[434] R. Aris, "On Shape Factors for Irregular Particles," *Chem. Eng. Sci.*, **6**, 262–268 (1957).

[435] J. O. Mingle and J. M. Smith, "Effectiveness Factors for Porous Catalysts," *Am. Inst. Chem. Eng. J.*, **7**, 243–249 (1961).

[436] J. M. Ausman, and C. C. Watson, "Mass Transfer in a Catalyst Pellet During Regeneration," *Chem. Eng. Sci.*, **17**, 323–329 (1962).

[437] J. J. Carberry, "Mass Diffusion and Isothermal Catalytic Selectivity," *Chem. Eng. Sci.*, **17**, 675–681 (1962).

[438] N. Wakao and J. M. Smith, "Diffusion in Catalysts Pellets," *Chem. Eng. Sci.*, **17**, 825–834 (1962).

[439] P. B. Weisz and J. S. Hicks, "The Behaviour of Porous Catalyst Particles in View of Internal Mass and Heat Diffusion Effects," *Chem. Eng. Sci.*, **17**, 265–275 (1962).

[440] K. B. Bischoff, "Accuracy of the Pseudo-Steady State Approximation for Moving Boundary Diffusion Problems," *Chem. Eng. Sci.*, **18**, 711–713 (1963).

[441] N. Wakao and J. M. Smith, "Diffusion and Reaction in Porous Catalysts," *Ind. Eng. Chem. Fundam.*, **3**, 123–127 (1964).

[442] G. W. Roberts and C. N. Satterfield, "Effectiveness Factor for Porous Catalysts," *Ind. Eng. Chem. Fundam.*, **4**, 288–293 (1965).

[443] J. Shen and J. M. Smith, "Diffusional Effects in Gas-Solid Reactions," *Ind. Eng. Chem. Fundam.*, **4**, 293–301 (1965).

[444] K. B. Bischoff, "An Extension of the General Criterion for Importance of Pore Diffusion with Chemical Reactions," *Chem. Eng. Sci.*, **22**, 525–530 (1967).

[445] M. Ishida and C. Y. Wen, "Comparison of Kinetic and Diffusional Models for Solid-Gas Reactions," *Am. Inst. Chem. Eng. J.*, **14**, 311–317 (1968).

[446] M. Ishida and C. Y. Wen, "Effectiveness Factors and Instability in Solid-Gas Reactions," *Chem. Eng. Sci.*, **23**, 125–137 (1968).

[447] C. Y. Wen, "Noncatalytic Heterogeneous Solid Fluid Reaction Models," *Ind. Eng. Chem.*, **60**, 34–54 (1968).

[448] R. S. Cunningham and C. J. Geankoplis, "Effects of Different Structures of Porous Solids on Diffusion of Gases in the Transition Region," *Ind. Eng. Chem. Fundam.*, **7**, 535–542 (1968).

[449] C. McGreavy and J. M. Thornton, "Prediction of the Effectiveness Factor and Selectivity for Highly Exothermic Reactions," *Chem. Eng. Sci.*, **25**, 303–309 (1970).

[450] J. Szekely and J. W. Evans, "A Structural Model for Gas-Solid Reactions with a Moving Boundary, I.," *Chem. Eng. Sci.*, **25**, 1091–1107 (1970).

[451] A. Calvelo and P. E. Cunningham, "Criterion of Applicability of the Moving Boundary Model," *J. Catal.*, **17**, 143–150 (1970).

[452] B. N. Mehta and R. Aris, "Communications on the Theory of Diffusion and Reaction. VII. The Isothermal *p*th-Order Reaction," *Chem. Eng. Sci.*, **26**, 1699–1712 (1971).

[453] J. Szekely and J. W. Evans, "A Structural Model for Gas-Solid Reactions with a Moving Boundary. II. The Effect of Grain Size, Porosity, and Temperature on the Reaction of Porous Pellets," *Chem. Eng. Sci.*, **26**, 1901–1913 (1971).

[454] C. Y. Wen and L. Y. Wei, "Simultaneous Nonisothermal Noncatalytic Solid-Gas Reactions," *Am. Inst. Chem. Eng. J.*, **17**, 272–280 (1971).

[455] M. Ishida and C. Y. Wen, "Comparison of Zone-Reaction Model and Unreacted-Core Shrinking Model in Solid-Gas Reactions. 1. Isothermal Analysis," *Chem. Eng. Sci.*, **26**, 1031–1041 (1971).

[456] M. Ishida, C. Y. Wen, and T. Shirai, "Comparison of Zone-Reaction Model and Unreacted-Core Shrinking Model in Solid-Gas Reactions. II. Nonisothermal Analysis," *Chem. Eng. Sci.*, **26**, 1043–1048 (1971).

[457] H. Y. Sohn and J. Szekely, "A Structural Model for Gas-Solid Reactions with a Moving Boundary. III. A General Dimensionless Representation of the Irreversible Reaction Between a Porous Solid and a Reactant Gas," *Chem. Eng. Sci.*, **27**, 763–778 (1972); **28**, 1169 (1973).

[458] R. Abed and R. G. Rinker, "Reaction with Mole Changes in Porous Catalysts in the Molecular, Transition, and Knudsen Regimes," *Am. Inst. Chem. Eng. J.*, **19**, 618–624 (1973).

[459] M. M. Lih and K. Lin, "Simultaneous Heat and Mass Transfer with and Without Reactant Exhaustion in Catalysts Particles for Reaction with Volume Change," *Am. Inst. Chem. Eng. J.*, **19**, 832–838 (1973).

[460] K. Hashimoto and P. L. Silveston, "Gasification. Part 1. Isothermal, Kinetic Control Model for a Solid with a Pore Size Distribution," *Am. Inst. Chem. Eng. J.*, **19**, 259–268 (1973).

also Section 19.5.4), but almost all are unverified by comparison with experiment. Good model validations are still awaited. Many of the models are extensions of the Yagi and Kunii[464] "shrinking core model," which is not applicable to porous carbons.

The "Assigned Parameter" Approach. In this modeling method, the significant parameters whose values are unknown a priori are identified and measured in appropriate experiments to determine their behavior. In many cases reasonable assumptions can be made regarding values that bracket actual behavior. Development of experimental values may also indicate regularities or valid simplifications that a priori (or mechanistic) analyses do not easily reveal.

In the particular example of the carbon reactions, adoption of the phenomenological approach requires assigning local parameter values of porosity, density, surface area, and internal diffusion coefficient. Some experimental information on all these parameters is now available that may in time provide a basis for prediction of behavior of unknown materials.

Most information is available on the variation of surface area with burnoff for a given sample. This is commonly represented in one of two ways: as total surface area (absolute, or relative, or normalized) variation with burnoff, or as surface area per unit volume or per gram. Figure 19.15 shows the first: total (normalized) surface area variation; Fig. 19.16 shows the second.

Further data are available from a number of other sources, and it would seem that there are many variations in behavior to be found. Blake et al.[465] show curves of area per unit initial weight against burnoff that have the same characteristic form as shown by Fig. 19.15. A relationship not given by the authors, but calculable from their data the area per gram at burnoff, against burnoff, has the shape of Fig. 19.16. The initial curved portion is controlled by an initial development attributed by Tomkow et al.[466,467] to an early period of breaking into blind micropores. If this occurrence is abrupt, there can be a sharp increase in area per gram with very little burnoff. The rising curve can then flatten quite abruptly.

The surface area profile shapes illustrated in Figs. 19.15 and 19.16 have been reported by Kawahata and Walker,[468] Wicke,[142] and Blake et al.[465] In contrast, Jolley and Stantan,[469] Smith and Tyler,[202] and Tomkow et al.[466,467] report curves of surface area *per gram* that have the form of Fig. 19.15; that is, they show peak areas per gram at some optimum burnoff. Taken at face value, these curves indicate a significant difference in behavior compared with the first-quoted sets of data, where area per gram increased monotonically with burnoff. There is some question, however, whether there are indeed two different sets of results. It is not clear in the case of the last-quoted data sets[466,467] whether the weight used as reference is the weight at burnoff or the initial weight. If it is the latter, the curves belong to the family of Fig. 19.15, from which curves similar to Fig. 19.16 may be constructed.

[461] K. Hashimoto and P. L. Silveston, "Gasification. Part II. Extension to Diffusion Control," *Am. Inst. Chem. Eng. J.*, **19**, 168–277 (1973).

[462] J. Szekely and M. Propster, "A Structural Model for Gas Solid Reactions with a Moving Boundary. VI. The Effect of Grain Size Distribution on the Conversion of Porous Solids," *Chem. Eng. Sci.*, **30**, 1049–1055 (1975).

[463] J. Szekely, C. I. Lin, and H. Y. Sohn, "A Structural Model for Gas-Solid Reactions with a Moving Boundary. V. An Experimental Study of the Reduction of Porous Nickel Oxide Pellets with Hydrogen," *Chem. Eng. Sci.*, **28**, 1975 (1973).

[464] S. Yagi and D. Kunii, "Studies on Combustion of Carbon Particles in Flames and Fluidized Beds," *Proc. 5th Symp. (Int.) Combustion,* Reinhold Publishing Corp., New York, 1955, pp. 231–244.

[465] J. H. Blake, G. P. Bopp, J. F. Jones, M. G. Miller, and W. Tambo, "Aspects of the Reactivity of Porous Carbons with Carbon Dioxide," *Fuel,* **46,** 115–125 (1967).

[466] K. Tomkow, A. Jankowska, F. Czechowski and T. Siemieniewska, "Activation of Brown Coal Chars with Oxygen," *Fuel,* **56,** 101–106 (1977).

[467] K. Tomkow, T. Siemieniewska, F. Czechowski, and A. Jankowska, "Formation of Porous Structures in Activated Brown-Coal Chars Using O_2, CO_2, and H_2O as Activating Agents," *Fuel,* **56,** 121–124 (1977).

[468] M. Kawahata and P. L. Walker, Jr., "Mode of Porosity Development in Activated Anthracite," *Proc. 5th Carbon Conf. 1961,* Vol. 11, Pergamon Press, New York, 1963, pp. 251–259.

[469] L. J. Jolley and J. E. Stantan, "Reactivity to Steam and Air of Coke Made from Coals of Different Caking Properties and Rank," *Proc. Conf. Science in the Use of Coal,* Paper No. 35, Institute of Fuel (London), 1958, pp. E22–E26.

The position is clearer with respect to data obtained by Dutta et al.[148] whose typical results are illustrated in Fig. 19.48. The authors also report that some of the curves go through a minimum, not a maximum. The relative pore surface parameter, α, is defined by the authors as the ratio of pore surface per unit weight at burnoff to initial pore surface per unit (initial) weight. Their experimental data were also fitted to an empirical equation

$$\alpha = 1 \pm 100 x_c^{\nu\beta} e^{-\beta x_c} \qquad (19.105)$$

where β is a char characterization parameter and ν is a multiplier for β that equals the conversion, x_c, at which the surface area curve reaches a maximum or a minimum. The positive sign is used for the first case and the negative for the second.

These curves show two significant characteristics. The first is the optimizing, noted. The second is the slow rise in area with burnoff to 0.2 conversion, in marked contrast to the more rapid rise shown in Fig. 19.16. As discussed above, the more rapid rise is

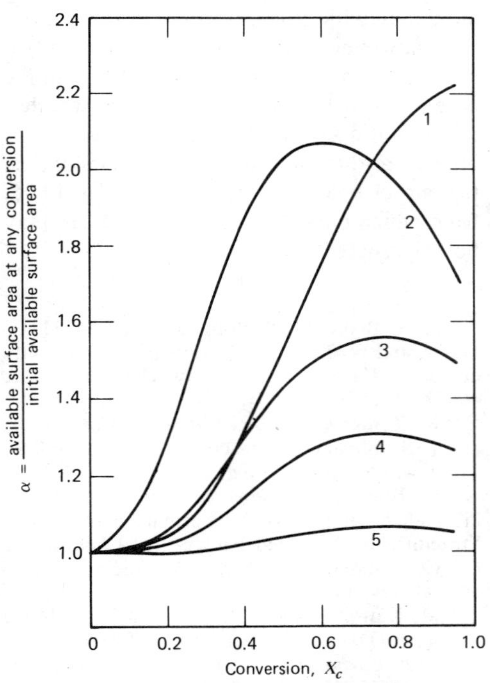

Figure 19.48 Relative available pore surface area versus conversion: 1, Hydrane Char No. 150; 2, Pittsburgh HVAb coal; 3, IGT Char No. HT155; 4, Illinois coal No. 6 and Synthane Char No. 122; 5, Hydrane Char No. 49.[148]

attributed to the opening up of blind pores. The difference in behavior does not seem to be due to different reactant gases. Wicke,[142] Smith and Tyler,[202] and Jolley and Stantan[469] used air or oxygen, with significantly different trends. Kawahata and Walker,[468] Blake et al.,[465] and Dutta et al.[148] used CO_2. Tomkow et al.[466,467] used O_2, CO_2, and H_2O, with comparable results in their experiments. Field,[95,96] burning chars in air, obtained a different result again, with relative surface area dropping monotonically, either linearly (for chars from swelling coals), or by a two-thirds power law (for chars from nonswelling coals).

There is good reason to suspect that the variation is partly a function of the rate of reaction; this is a part of the anisotropy development mentioned above. The nature of the reacting gas is then involved, but in combination with the temperature. Dutta and Wen[362] reporting on reaction of the same coals and chars, but in oxygen, found that reactivity was much less dependent on the parent coal than was the case for the CO_2 reactivities, and it depended much more on the pore development due to reaction. The pore modification in O_2 was much more drastic than in CO_2.

An investigator's response to this observation must be a little uncertain at this time. The more fundamental pore model calculations could be of value here if used to compare models, to determine the sensitivity of the models to different assumptions, and to indicate what dominant parameters might exist that would also result in the sort of surface area variations found experimentally. The present alternative is to use the empirical equation,[362] Equation 19.105, with arbitrarily adjusted values of the coefficients. With only two adjustable coefficients this is quite a promising approach. It is also hoped that as more experimental data become available, the dependence of the empirical coefficients on char type and operating conditions will become evident.

The problem may still be a little simpler than appears at present, however. The basis for this conclusion is Fig. 19.49. The curve represents the variation of burnoff with normalized time of a number of chars in oxygen (air). Similar results were obtained for reaction in CO_2, H_2O, and H_2. The temperatures and pressures were 405°C at 1 atm for air,

900°C at 1 atm for CO_2, 910°C at 0.022 atm for H_2O, and 980°C at 27.2 atm for H_2. Chars from 16 parent coals were used, and measurements were made on submillimeter particles in a thin bed carried on a thermogravimetric balance. In all cases the variation of burnoff with time was found to have the same sigmoid shape.

Furthermore, inspection showed that if the curves were normalized, on the time axis, against the time for 50% burnoff (being already normalized for weight loss), the data for a given reactant gas tended to form a single composite, as illustrated in Fig. 19.49. Even more pertinent, the composites for the different gases also were close to a common composite. The comparisons were tested using a cubic equation to determine an empirical fit of the individual curves to the respective composites, and of the respective composites to a common composite. A fitting parameter, which is essentially a correlation coefficient, was calculated to determine the closeness of fit to the respective common composites, represented by the empirical cubic equation, of the common composites. For the data on air, CO_2, H_2O, and H_2, the fits were 96, 99.2, 99.1, and 96.6%, respectively. The fit to a common composite for all data using all reactant gases, for 50 char-gas combinations, was 98.2%.

Some perspective on this commonality can be obtained from the following considerations. In the first place, it may be noted that the experimental conditions were chosen to eliminate problems of both boundary layer and internal pore diffusion. It may be initially accepted for purposes of analysis that reaction was essentially uniform throughout the particles in all cases. The commonality of behavior would also suggest that the internal surface development was probably very similar in all cases. We assume that reaction is slow enough for surface regression to be only nominal. For a particle of mass m and density σ reacting at constant radius, we have $m/m_0 = \sigma/\sigma_0$, where the subscript zero denotes the initial values. From Equation 19.84 we have a relation for R_T with the first term on the RHS zero to denote no change in radius; and the effectiveness, η, is unity for uniform reaction throughout the particle. Inserting $s = 3$ for a sphere, we can write Equation 19.84 with replacement of m for σ as

$$R_T = \frac{dm/dt}{4\pi a_0^2} = \frac{a_0}{3} S_v \bar{k} f(p_s) \qquad (19.106)$$

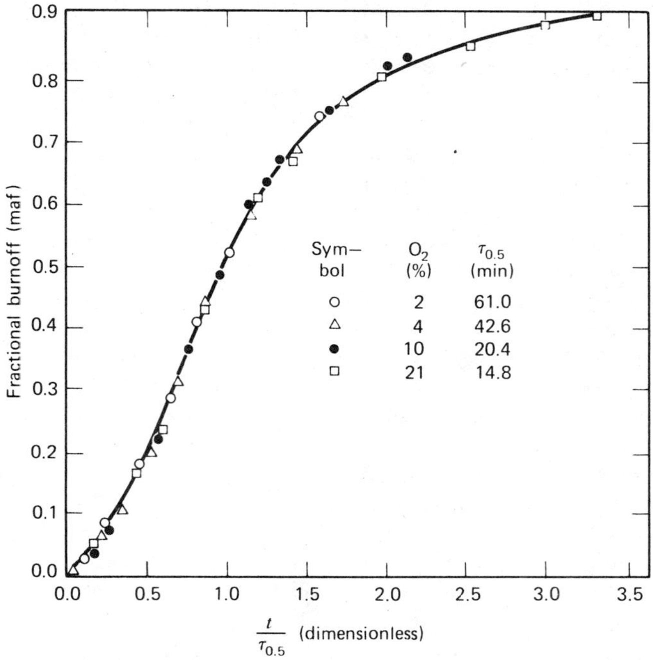

Figure 19.49 Variation of fractional burnoff with normalized time ($t/\tau_{0.5}$) for char particles burning in air at 450°C, showing unification of data by normalization.[470]

Writing S_v as a function of burnoff, or mass remaining (m/m_0), $g(1 - m/m_0)S_v^0$, and rearranging, we obtain

$$\frac{d(m/m_0)}{m/m_0} g\left(1 - \frac{m}{m_0}\right) = \frac{dt}{\tau} \quad (19.107)$$

where the characteristic time τ is given by

$$\frac{1}{\tau} = S_v^0 \bar{k} f(p_s) \quad (19.108)$$

If the surface area development function, g, is common for all reactions, we have an integrable equation with separated variables that will have a common function as a solution. Calling this $G(1 - m/m_0)$, the solution can be written

$$G\left(1 - \frac{m}{m_0}\right) = \frac{t}{\tau} \quad (19.109)$$

with the integration constant $G(0) = 0$. If $t = \tau_{0.5}$, for $m/m_0 = 0.5$, as written by Walker et al.,[470] $G_{0.5}$ is a constant for all reactions, whence $\tau_{0.5} = G_{0.5}\tau$. This justifies the use by Walker et al.[470] of $\tau_{0.5}$ as a normalizing parameter for the time, provided the internal surface area development with burnoff (the function g) is the same for all the reactions in the composite curve. It is also notable that the characteristic time parameter, τ, is a function of the initial surface area per unit volume only, and of the overall kinetics of the reaction. Writing $\bar{k}f(p_s)$ as $k_{\text{eff}}\, p_s^n$, to introduce, as before, an operational reaction order, n, I have determined the value of n for the few oxygen reaction data provided.[470] The best value was about two-thirds, indicating one of two possibilities: (1) only an approach to full penetration of the particles with a true zero-order internal reaction (it would be 0.5 in Zone II with true zero order); or (2) reaction in the transition between true first and true zero order, possibly with some modification from partial penetration.

Further use can be made of these data, or of further data from similar experiments. The procedure will be briefly sketched out here. The empirical cubic equation fitting the curve of Fig. 19.49 can be differentiated to give an equation in g as a function of reduced time. It is in fact a quadratic, from which some idea of the nature of the function g could be obtained. Alternatively, the function obtained could be refitted by Equation 19.105. A third alternative would be to integrate Equation 19.107, using Equation 19.105 for the function yielding g, and with the coefficients appropriately adjusted to suit. In this way it may be possible to build up a more coherent data bank on the variation of surface area with burn-off for different reactions.

19.6.2.5 *Comparative Rates*

If reaction rates and the governing equations are accurately known, the rates can be described by the equations and the three kinetic coefficients: the activation energy, the reaction order, and the frequency or preexponential factor. Just as the first two kinetic constants in practice are often global or operational values, combining a number of different factors, so also the preexponential factor is (more often than not) a compound, most usually incorporating an unknown surface area. Because there is, also, an effective coupling between activation energy and preexponential factor, it is usually not too informative to compare values directly. In this section they are compared indirectly by graphic comparison of reaction rates on Arrhenius plots.

Figures 19.50 and 19.51 provide two comparisons for the oxidation reaction.

Reaction with O_2. The graphs of Fig. 19.50, from the well-known 1969 review by Mulcahy and Smith,[36] and Fig. 19.51, from a new (1976) review by Smith,[39,471] are thought to be the most authoritative rate evaluations in existence. The evaluation methods are different, and they provide an important contrast. The data from the primary investiga-

[470] P. L. Walker, Jr., O. P. Mahajan, and R. Yarzab, "Unification of Coal Char Gasification Reactions," *Prepr., Am. Chem. Soc., Div. Fuel Chem.*, **22**(1), 7–11 (1977).

[471] I. W. Smith, "The Intrinsic Reactivity of Carbons to Oxygen," *Fuel*, **57**, 409–414 (1978).

[472] R. A. Sherman, "Burning Characteristics of Pulverized Coals and the Radiation from their Flames," *Proc. 3rd Int. Conf. Bituminous Coal*, Vol. 2; 1931; p. 510; *Trans. ASME*, **56**, 401 (1934).

[473] S. N. Mironov, "Experimental'noe issledovanie Protsessa Szhiganiya Antratsitovoi Pyli," *Teploenergetika*, No. 1; 78–80 (January 1965) (English transl.).

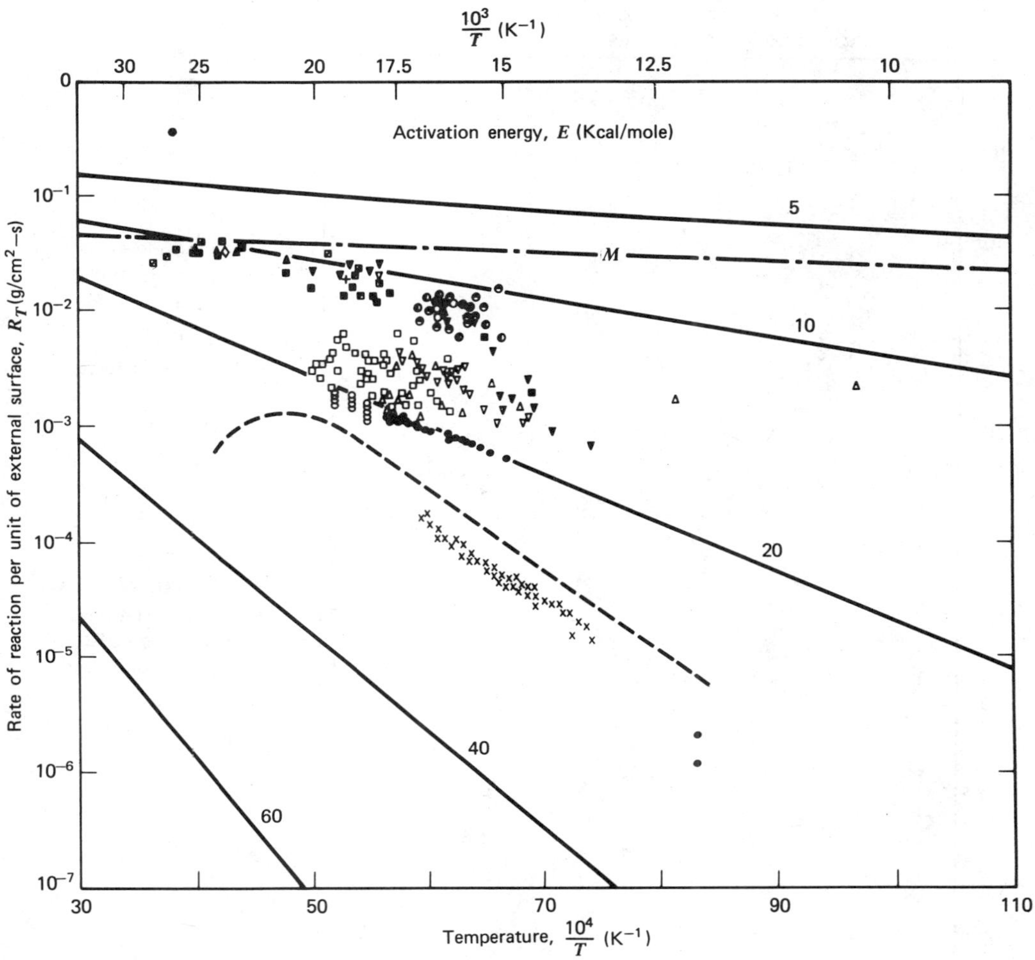

Ref	Symbol	Material
86	●	Anthracite
472	▼	Bituminous coal (low volatile)
472	●●	Bituminous coal (high volatile)
73	△	Bituminous coal (high volatile)
473	□	Anthracite
337	×	Soot
35	■ ▣	Charcoal
88	⊙	Soot
86	○	Anthracite
35	▼	Charcoal
35	▽	Anthracite
35	▪	Bituminous coal char
35	+	High temperature coke
35	▣	Charcoal
35	▲	Charcoal
35	◇	Anthracite
380	---	Bulk graphite

Figure 19.50 Burning rates of particles (calculated to p_{O_2} = 0.1 atm), with M, mass transfer (40-μm particle).[36]

tions had to be converted to a common basis for comparison, particularly of oxygen concentration.

In Fig. 19.50 the implicit assumption was that reaction occurred only on the outside of the particles; but the data for Fig. 19.51 were corrected for internal reaction, as described below. For Fig. 19.50 it was also assumed explicitly that reaction was first order, with a correction to 0.1 atm O_2. For Fig. 19.51 different reaction orders were used or adopted, all lying between 0.3 and 1, using experimentally derived values for the rate-order index, n, wherever possible, and with correction to 101 kPa (1 atm). Figure 19.50 has, additionally, a line estimating a diffusion limit (for a 40-μm particle), this being the maximum possible rate that can be achieved if the

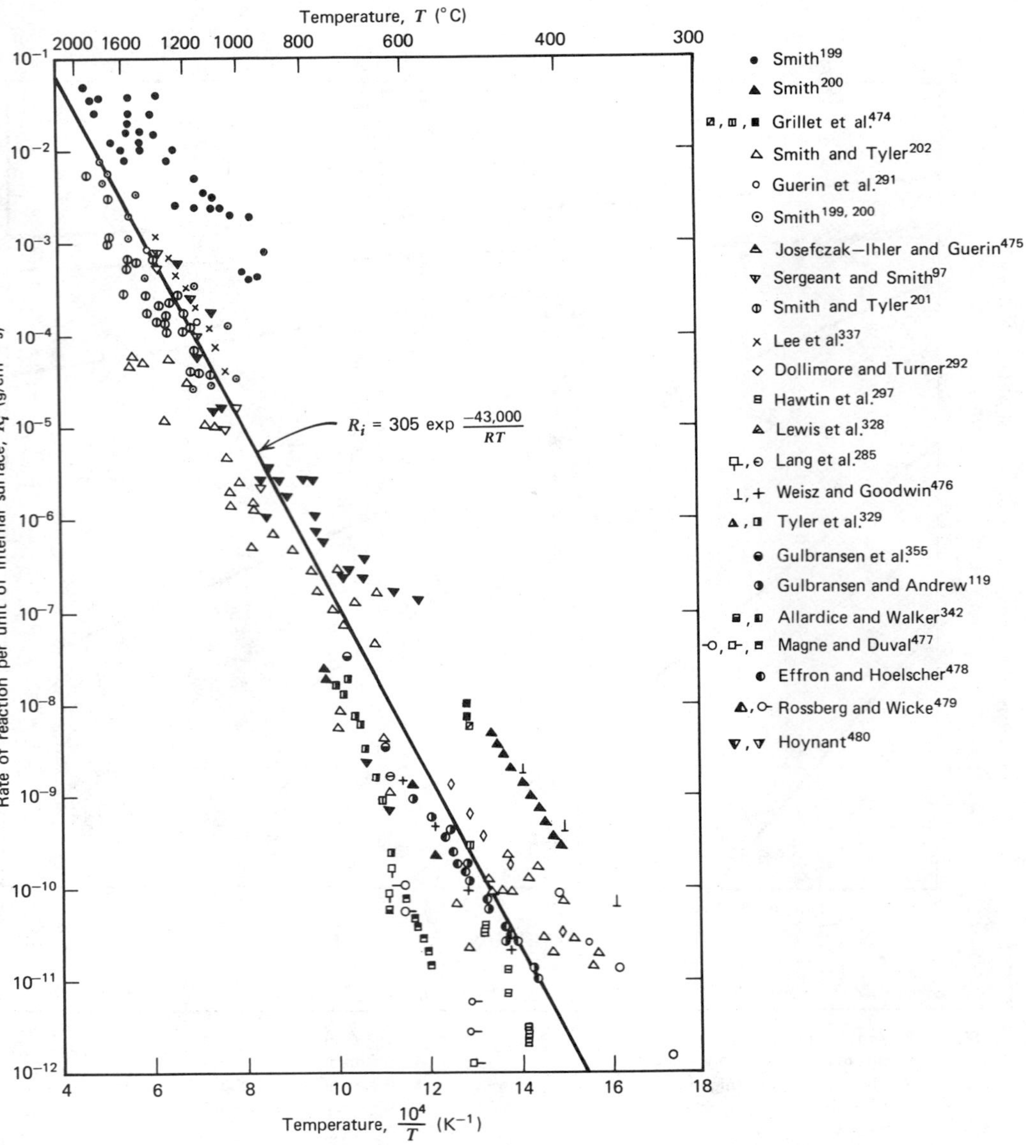

Figure 19.51 Burning rate data of carbons in air per unit area of internal surface as a function of temperature.[39]

[474] Y. Grillet, P. Rebaudiéres, and H. Guérin, "Sur la réactivité des carbones Pregraphitiques. IV. Comparaison des Oxyréactivités de carbones Pregraphitiques d'Origines Diverses," *Bull. Soc. Chim. Fr.*, **7**(429), 2423 (1976).

[475] M. Jozefczak-Ihler and H. Guérin, "Sur la Réactivité des carbones Pregraphitiques. III. Examen de Quelques Anthricites," *Bull. Soc. Chim. Fr.*, **6**(334), 2018 (1966).

[476] P. B. Weisz and R. B. Goodwin, "Combustion of Carbonaceous Deposits Within Porous Catalyst Particles. I. Diffusion-Controlled Kinetics," *J. Catal.*, **2**, 397–404 (1963); **6**, 227 (1966).

[477] P. Magne and X. Duval, "Cinétique d'Oxydation du Graphite dans N_2O; Comparaison avec l'Oxydation dans l'oxygène," *Bull. Soc. Chim. (Fr.,) 5(262), 1585 (1971).*

[478] E. Effron and H. E. Hoelscher, "Graphite Oxidation at Low Temperature." *Am. Inst. Chem. Eng. J.*, **10**, 388 (1964).

[479] M. Rossberg and E. Wicke, "Transportvorgange und Oberflächenreaktionen bei der Verbrennung graphitischen Kohlenstoffs," *Chem. Ing. Tech.*, **28**, 181 (1956).

[480] G. Hoynant, Thesis, Ingéniéur-Docteur, University of Nancy, France, 1950.

rate is controlled by boundary layer diffusion.

The correction for internal reaction used by Smith[39] utilized Equations 19.84 and 19.85 combined, assuming either no surface regression, or that the surface regression contribution was rolled into the equation constants as in Equation 19.84a. In addition, the general function $f(p_s)$ was written in the power approximation form of Equation 19.74a, as p_s^n. Equation 19.85 was used to eliminate p_s in favor of the mainstream value, p_0. The effectiveness factor, η, was calculated for each case and was found to be unity for half the data used in the study. Where it was not unity, the calculated value was used. Putting the equations together thus established a relation for the effective rate coefficient, $\bar{k}$, in terms of measured rate data, R_T (the total reaction rate per unit area of the external surface); the mainstream oxygen concentration, p_0; and other measured or calculable parameters. The dimensions of $\bar{k}$, however, depend on the reaction order, so that this parameter is not the best on which to base comparisons. A standard concentration, $\bar{p}_i$, was therefore chosen (101 kPa) to convert $\bar{k}$ to an "intrinsic" reaction rate, $\bar{R}_i$, giving

$$\bar{R}_i = \bar{k}\,\bar{p}_i^n = R_T\left(\frac{\bar{p}_i}{p_0}\right)^n \frac{3}{a_s\,S_v\,(1 - R_T/k_D\,p_0)^n\eta} \tag{19.110}$$

Figure 19.51 is more convincing taken at face value than Fig. 19.50, but mainly because there are more data points, and the temperature range covered is greater (300–2000°C in Fig. 19.51 vs. 1000–2500°C in Fig. 19.50). The added data are mainly at the lower end of the temperature scale. The more convincing appearance is a little misleading, however. The alternative treatment has not diminished the spread in the data at a given temperature. In both cases the spread in rate, R, is 2–2.5 orders of magnitude for temperatures above 1000°C. For temperatures below that there is obviously no basis for comparison; the spread increases, however, as commented on by Smith,[39] to a maximum of 4 orders (at 500°C).

The best fit activation energies and their range can be said to be comparable, with a slightly higher average in the later evaluation. In Fig. 19.50 the values seem to be lying between 20 and 40 kcal/mole, with no common average assigned. In Fig. 19.51, the average is 43 kcal/mole (179 kJ/mole), but the spread for individual data sets is stated[39] to be 30–70 kcal/mole. The reason for this increase over the older evaluation seems to have been that the more recent work was better planned experimentally, with the result that the data were generally less affected by extraneous diffusion behavior, which cuts the activation energy in two. Many of the new data, particularly at the lower temperatures, were obtained from much purer carbons that by reducing catalytic effects also allowed the activation energy to rise. Smith[39] examined this point explicitly and found that for particularly pure carbons, the common Arrhenius plot had a best-fit activation energy of 60 kcal/mole (250 kJ/mole); this can be compared with the value of 61 ± 3 kcal/mole cited by Walker[53] as the best "baseline" value for pure carbon (Section 19.6.2.3).

The spread in activation energy values still has to be accounted for, and the conclusions of the previous discussion on this, in Section 19.6.2.1, still seems to hold. The present results are consistent with a range of true activation energies of 40–80 kcal/mole, with the value increasing with sample purity, and with Zone II diffusion reducing these values to the range 20–40 kcal/mole.

The spread in activation energies still does not explain the spread in the rate data of 2–4 orders of magnitude, however. Again, Smith[39] considered this, and concluded that it was a function of degree of graphitization. This idea is discussed further in examination of the negative temperature coefficient effects (Section 19.6.2.6). Smith examined, in particular, data from materials in which there was little chance of internal reaction. These included diamond, pyrolytic graphite, vitreous carbon, and soot. The majority of the data fell clearly *above* the average line of Fig. 19.51, implying *higher* reaction rate per unit area of (true) surface than for the less ordered material.

At present there seem to be only two possible explanations for this. If the intrinsic rate is the maximum possible rate when all other rate-limiting factors have been excluded, then the lower rates of the less organized carbons, where there is a chance of diffusional limitations, must mean that the diffusional limitations inside the carbon sample have not been fully accounted for. A diffusional limitation can take one of two forms: a strict limitation as discussed in Section 19.5.4, or else an unrecognized one that results in an overestimate of the surface area involved in reaction. At the lowest temperatures, the difference between BET and ASA area is recognized, as discussed in Section 19.3.1. The difference between them can be as much as 2 orders of magnitude. It is possible that the rates are being divided by far too large an area.

At the higher temperatures it appears that the difference between BET and ASA area is closer. It is possible that the effective area is still being overestimated, however, if the measured area includes the micropores, since in fast reactions the micropores may not participate and may not contribute to reaction area to the extent normally assumed. To some degree this would seem to be at variance with some of the findings mentioned in the preceding section on surface area (Section 19.6.2.4), where there are reports that the surface area opens up slowly or rapidly by reaction in the micropores. However provision of access to blind micropores does not necessarily mean that much subsequent reaction takes place in them. Here again there is a potential factor of several orders of magnitude by which the surface area may have been overestimated, and that would lead to corresponding underestimation of the true rate per unit area.

A comparison of the "superficial" and "intrinsic" rates in Fig. 19.50 and 19.51 is as expected. Superimposing the mean line of Fig. 19.51 on Fig. 19.50 places nearly all the data of Fig. 19.50 just above the line. There is a factor of 10 difference in the standard oxygen concentrations and, if the reaction is first order, the imposed line will be shifted down by a further order of magnitude, or a half order of magnitude for a half-order reaction; for a zero-order reaction the line will not be shifted at all. Thus dividing by the internal surface, the "true" rates, as expected, are lower than the rates calculated for the superficial surface, by amounts that differ by up to 2 orders of magnitude, depending on the reaction order assumed. In particular, the rates agree quite well with soot data discussed below (Section 19.6.3.2), suggesting that the spread in Fig. 19.50 may be largely due to variations in internal surface. The two graphs provide some basis for estimating true specific reaction rates if the superficial rate is known but the corresponding internal surface is not.

Finson et al.[429] suggest an alternative explanation for the spread, which could be due to change of reactive gas pressure. This can be seen in Fig. 19.55, discussed in more detail below. The assumption was that the reaction rate was zero order at the lower temperatures, and the bounding straight line was the limiting case of reaction control by film decomposition. This bounding line is almost identical with Smith's mean line, and the activation energy for the limiting line is very close to Smith's value: 39 kcal/mole compared with 43 kcal/mole. However a simple *n*th-order correction for the differences due to oxygen concentration will not contract the plots to a single line, particularly at the higher temperatures.

Reaction with CO_2 and H_2O. Few attempts have been made to develop comparable Arrhenius plots for the CO_2 and H_2O reactions. Figure 19.52 is one of the few and more recent. This plot represents the limiting term of an L-H equation identified as the desorption step in the reaction of carbon with CO_2 and H_2O. This is, therefore, for the oxide film decomposition where the film was formed during the reaction of carbon with the two stated gases.

Comparison of these data with the oxygen reaction data of Figs. 19.50 and 19.51 show that they lie up to 5 orders of magnitude below the oxygen data, except at the very highest temperatures.

It is difficult to know what to make of these values. If the reaction is indeed oxide film decomposition, then earlier discussions, particularly of Ergun's data ("The C/H_2O Reaction" in Section 19.5.2.3, and Fig. 19.36), would indicate that the nature of the film is the same whether it was formed by oxygen transfer from CO_2 or from H_2O. If the film is formed by oxygen transfer from O_2, it would be most reasonable to expect the film to be the same again. In that event, however, the data should match the bounding line on Fig. 19.50 for (presumed) film decom-

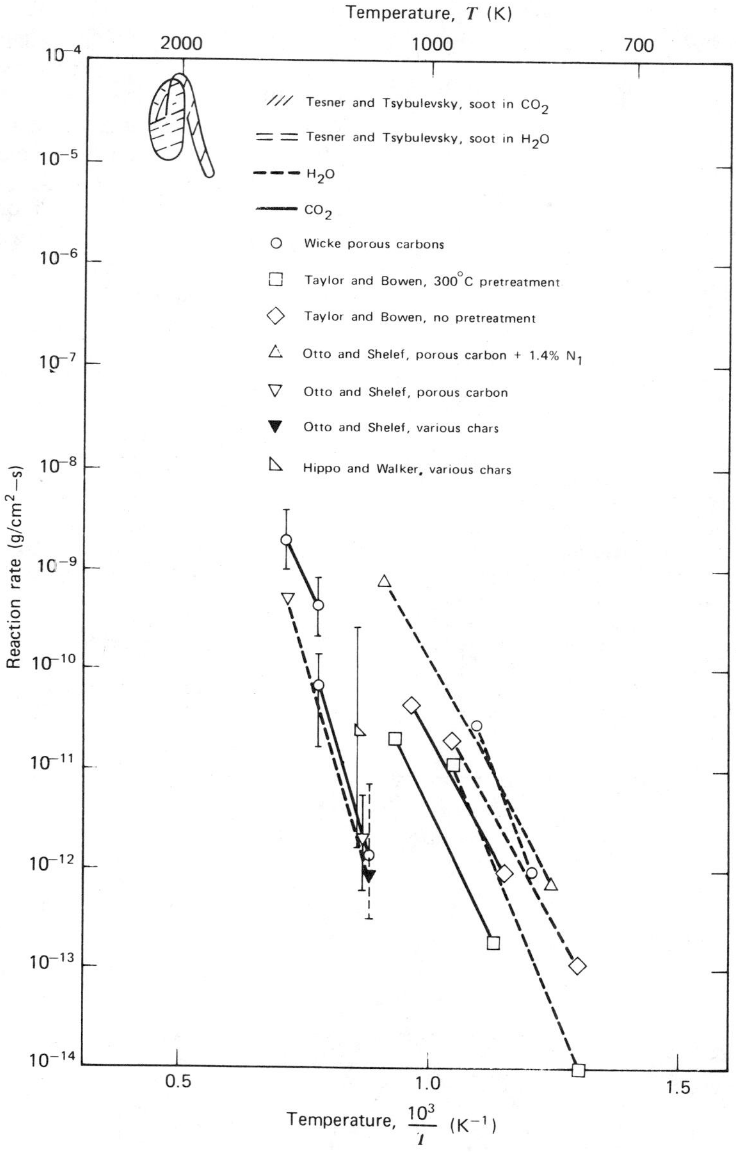

Figure 19.52 Desorption rates in the presence of CO_2 and H_2O.[429]

position. This is clearly not the case. Therefore the following conclusions can be drawn.

1. If the conditions for film decomposition have been correctly identified, the film formed in oxygen is, indeed, different from that formed in CO_2 or H_2O.

2. If the films formed in the three gases are the same, then one or other of the two plots, Fig. 19.51 or 19.52, is wrongly identified as representing film decomposition.

3. If the identification is correct and the films are the same, there must be a substantial error in some calculation parameter used to determine the reaction rates, and the erroneous parameter is most probably the surface area.

In summary, the status of the evaluation of the rates of the carbon-oxygen reaction is not exactly satisfactory, but it appears that progress is being made. The same cannot yet be said for the C/CO_2 and C/H_2O reactions.

19.6.2.6 *Negative Temperature Coefficient*

The rate of reaction generally increases with temperature. Figure 19.53 illustrates a situation, however, where reaction rate develops a negative temperature coefficient. This behavior was first observed by Griffin et al.[385] and by Smith and Gudmundsen[482] in wet air, and it has subsequently been reported in a number of other investigations.[59,303–305,322,323,371–373,377,379,380,483–489]

Figure 19.53 illustrates both some specific results and also the typical general pattern reported by others. The curve of reaction rate with temperature exhibits two extrema, respectively a maximum and a minimum as illustrated by the dashed (calculated) line. Different investigators have then reported results covering part or all of the overall variation. Such a region of negative temperature coefficient in the rate plot has been reported, usually with at least one of the extrema, under the following wide-ranging

[481] G. Blyholder, J. S. Binford, and H. Eyring, "A Kinetic Theory for the Oxidation of Carbonized Filaments," *J. Phys. Chem.*, **62**, 263–267 (1958).

[482] D. F. Smith and A. Gundmundsen, "Mechanism of Combustion of Individual Particles of Solid Fuels," *Ind. Eng. Chem.*, **23**, 277–285 (1931).

[483] R. F. Strickland-Constable, "The Interaction of Oxygen and Carbon Filaments at High Temperatures," *Trans. Faraday Soc.*, **40**, 333 (1944).

[484] A. Euken, "Der Reaktions mechanismus der Kohlenstoff verbrennung bei geringen Drucken," *Z. Angew. Chem.*, **43**, 986 (1930).

[485] L. Meyer, "Der Mechanismus der Primärreaktion zwischen sauerstoff und graphit (Kinetik der Kohlverbrennung)," *Z. Phys. Chem.*, **17(B)**, 385 (1932).

[486] V. Sihvonen, "The Influence of Keto- and Ketene Groups, Adsorbed Molecules and Ions on the Mechanism of Carbon Oxidation," *Trans. Faraday Soc.*, **34**, 1062 (1938).

[487] X. Duval, "Réaction de l'Oxygène avec des Filaments de Carbone, À Hautes Témperatures et sous Très Basses Pressions," *J. Chim. Phys.*, **47**, 339 (1950).

[488] X. Duval, "Les Reactions du Graphite avec l'Oxygène," *J. Chim. Phys.*, **58**, 3–11 (1961).

[489] W. G. Bradshaw, R. M. Bonesteel, and J. N. Ong, *Carbon (Oxford)*, **3**, 347 (1965).

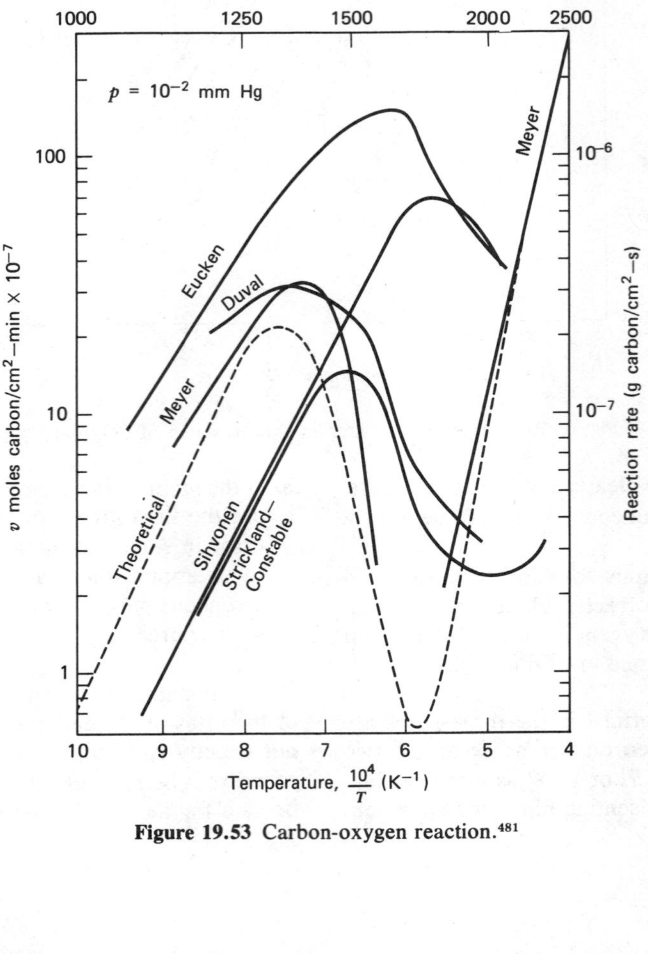

Figure 19.53 Carbon-oxygen reaction.[481]

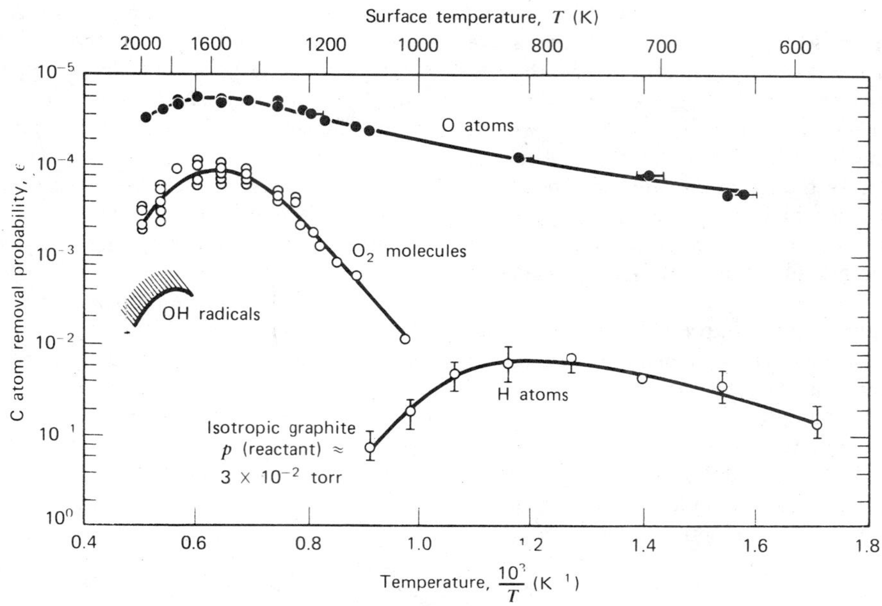

Figure 19.54 Temperature dependence of the attack of isotropic graphite by O, O_2, H, and OH.[59]

conditions: reaction of carbon with O and H atoms, OH radicals, and O_2 molecules (Fig. 19.54) at low pressure and high temperature[59]; reaction with O_2 at high velocity at high temperature and near-normal pressure;[323,379,380] reaction with CO_2 under diffusion-dominated conditions at high temperature and normal pressure[371,372]; and reaction with O_2 at high pressure in a shock tube.[373] In addition there are several reports[390,392] of reaction rate becoming almost independent of pressure at high temperature, which may represent onset of peaking (or may not); and similar reports, at low temperature, in reaction with O atoms.[128,132] Similar peaking and double-extremum behavior have also been found in a number of other reactions[59] such as oxygen with tungsten, rhenium, silicon carbide, boron, and boron nitride, and the reaction of chlorine with molybdenum.

An acceptable explanation of this behavior is still not forthcoming, although equations have been constructed with mechanistic bases that describe most of the observed results[323,379,380,481] for the carbon reactions.

A mechanism common to all these reactions is unlikely. It could only mean that they all reacted by similar reactions in essentials, or else the reaction rate was determined only by the gas properties and was independent of the properties of the reacting solid. For the oxygen-tungsten reaction, however, this reaction pattern appears to be due to the competitive formation of a number of different oxides (see Walsh et al.[490]), which is evidently not the case for graphite, particularly in the case of H atom and H_2 molecule reactions.

The simplest explanation for the peaking behavior requires a very low activation energy of adsorption,[27] coupled with adsorption control of the reaction. The reaction is then governed by Equation 19.48, which has two temperature terms. The $T^{1/2}$ term is usually neglected in comparison with the Arrhenius group, but it becomes important at low E_{ads} values. If there are any other temperature-sensitive terms in the rate equation, they may be incorporated by an empirical generalization by writing T^n. The expression $(e^{-E/RT}/T^n)$ is found to have a maximum at a temperature given by $T_{opt} = E/nR$. For $n = ½$ and a value of T of 1500 K at the peak rate, the activation energy required is 1500 cal/mole. This value does look improbably low, but not impossibly so if n is greater than ½. For $T = 2000$ K and $n = 1$, for example, $E = 4$ kcal/mole, which is in the region of

[490] P. N. Walsh, J. M. Quets, and R. A. Graff, "Kinetics of Oxygen-Tungsten Reaction at High Temperatures," *J. Chem. Phys.*, **46** 1144–1153 (1967).

the values found at low temperature as described in Section 19.3.1, and is compatible with a few values obtainable from other investigations at high temperatures. This explanation can conceivably account, however, only for the maximum in the curve; it cannot account for the minimum and the subsequent increase at a higher activation energy. About the only possibility would be pickup of the reaction by utilization of the CO_2 combustion product.

Annamalai and Durbetaki[491,492] have suggested an alternative explanation based on boundary layer combustion of the CO to CO_2, with an initial trend to final CO production at an intermediate temperature and a trend to CO_2 production at the higher temperatures. Intuition would suggest the opposite trends. The predictions are in fact contrary to experiment, as illustrated, for example, in Fig. 19.67 (below), and contrary to other calculations, and the explanation also fails in the low-pressure cases where there is no boundary layer.

A more recent explanation is provided by Finson et al.,[429] who compared prediction with a number of experiments, as illustrated in Fig. 19.55. The rate equation used is essentially of the L-H form. The factor in the numerator was the usual product of the adsorption and film decomposition velocity constants. The factor in the denominator, however, was the sum of the adsorption and film decomposition velocity constants, but it also included in the sum a velocity constant for oxygen desorption without change. As discussed earlier (Section 19.4.1.2), however, there is no evidence of recovery of adsorbed gaseous oxygen unchanged. The equation form may be correct, nevertheless, even if the original basis given is open to doubt. Attempting to provide an alternative basis for the equation should be a worthwhile path to pursue.

The most widely adopted explanation is the two-site hypothesis of Blyholder et al.,[481] used mostly in the modified form developed by Nagle and Strickland-Constable.[323] In the original derivation a fraction (f), of the total sites was assumed to be a Type 1, and the rest, a fraction $(1 - f)$, was a Type 2 of different reactivity. The reaction rates at each set of sites were then assumed to be governed by a simple Langmuir isotherm (Equation 19.89), but with different rate constants, and with the following additional aspects. Film decomposition on a Type 1 site was assumed to uncover a further Type 1 site. Additionally, Type 1 sites were converted to Type 2 by "annealing" at a first-order rate; and Type 2 sites were converted to Type 1 at a rate proportional to the number of Type 2 sites and to "the rate of Type 1 oxides leaving the surface." (The rationale for the assumption quoted is obscure.) At any given temperature the fraction of sites, f, was assumed constant, and simplifications were obtained by assuming that the Type 1 site reactions were first order and Type 2 site reactions were zero order.

Nagle and Strickland-Constable[323] adopted and modified the analysis, notably by allowing "annealing" in only one direction (from 2 to 1) and dropping the final assumption. In addition, the reaction rate at Type 1 sites was retained in the Langmuir form, and reaction at Type 2 sites was assumed to be first order, not zero. The outcome was an equation for the reaction rate that took a somewhat similar form by either derivation:

$$R_T = \frac{k_1 k_5 p}{k_1 p + k_5} f + k_2 p(1 - f) \qquad (19.111)$$

where $f = 1/(1 + k_T/k_2 p)$; k_1 and k_2 are the velocity constants for adsorption on the Type 1 and Type 2 sites, respectively: k_5 is the desorption constant (for the Type 1 sites); and k_T is the velocity constant for Type 2 to Type 1 thermal "annealing." In some formulations p is small and k_5 drops out.[377]

The equation has been successfully used to correlate data from quite an array of different sources[373,374,377]; it has also been used for comparisons of other data that do not necessarily show the peak. In those comparisons corrections to other oxygen concentrations are sometimes made on the basis of Equation 19.111, thereby adopting the reaction order assumptions as correct (which is open to question). The wide applicability of the equation is impressive at its face value. It should be recognized, however, that it

[491] K. Annamalai and P. Durbetaki, "Combustion Behavior of Small Char/Graphite Particles," Spring Meeting, Central States Sect. Combustion Institute, Columbus, Ohio, 1976.

[492] K. Annamalai and P. Durbetaki, "A Theory on Transition of Ignition Phase of Coal Particles," *Combust. Flame*, **29**, 193–208 (1977).

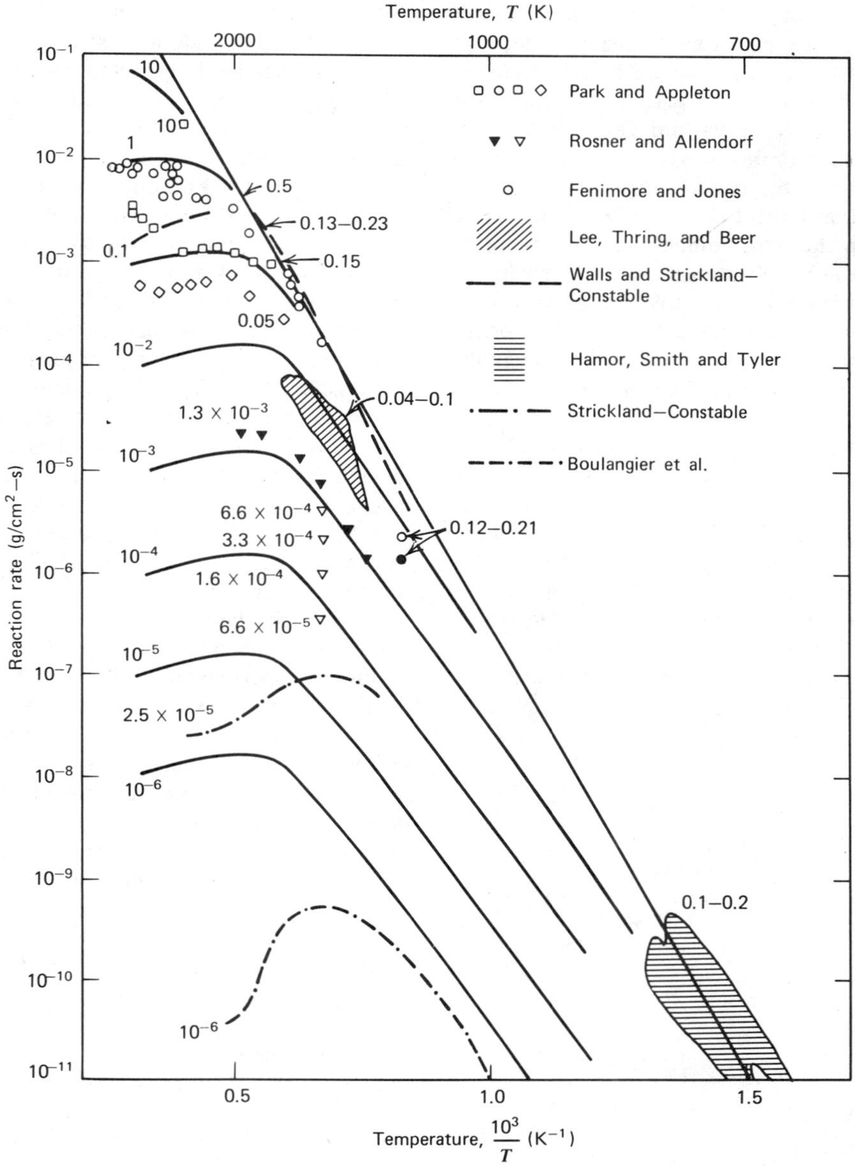

Figure 19.55 Adsorption-desorption kinetic rates for carbon and oxygen. The appropriate pressures (atm) are indicated.[429]

contains eight adjustable constants in its most commonly used form, and six when k_5 is eliminated. Ability to fit to a curve with only two extrema and six to eight coefficients is, perhaps, not too surprising. There is also information to suggest that the basic model requires review.

In the original model[481] it was assumed that there were two different types of *edge* site, and Blyholder and co-workers proposed a fairly precise identification with "hydrocarbon or hydrogen disturbed sites" (regarding hydrogen mainly in the light of an impurity). If the impurity concept is valid, then the peaking would not occur with very pure material, which leaves open the question whether the samples used by different investigators were pure or contaminated (even at the ppm level). The need for impurities is not essential, however, since there is plenty of evidence, discussed in Section 19.3.1, for the existence of several different sites, even in very pure material. The evidence also indicates, however, that the sites operate in sequence, not in parallel (Section 19.3.1), a fact that undermines the two-site concept, at least so far as edge sites are concerned. It could, however, apply to edge and basal plane sites. Acharya and Olander[374] measured reaction rates of basal and edge (prism) planes using the rotating-disk method. They showed, as expected, a significant difference between the two, finding also that the reaction of the basal planes was zero order (more in accordance with Blyholder et al. than with Nagle and Strickland-Constable). [The prism plane rates were fitted using the Langmuir isotherm only (Equation 19.89), which assumed, in effect, nondissociative adsorption (cf. Equation 19.29 for dissociative adsorption).]

The two-site concept is not supported, however, by results of O atom reactions. Since these evidently attack basal and edge atoms impartially,[59] use of O atoms should not show the peaking effect, but Fig. 19.56 shows that it does. This figure also shows that differences between the two carbons used, which have significantly different reactivities to molecular oxygen, are almost eliminated when using atomic oxygen. This phenomenon can be interpreted as the result either of eliminating a significant adsorption "resistance" or of increasing the surface area available to attack, with the rider that the total (BET) surfaces would have to be comparable. In short, the O atom behavior does not seem to support the two-site theory.

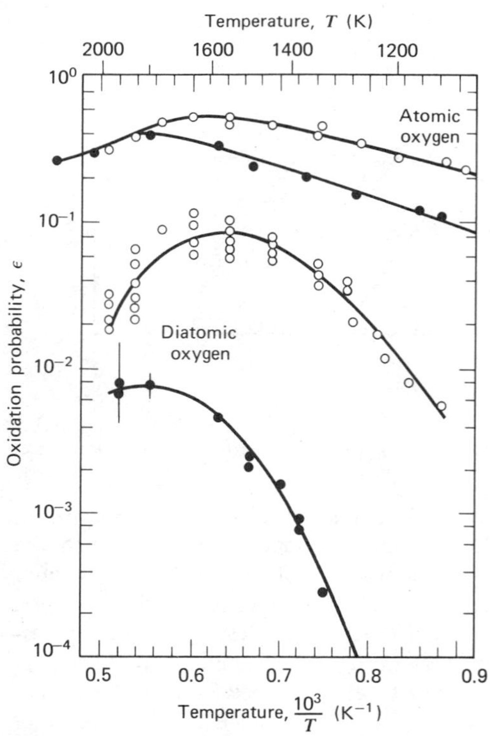

Figure 19.56 Arrhenius diagram displaying true oxidation probabilities for pyrolytic graphite (●) and for isotropic graphite (○) exposed to atomic and diatomic oxygen at 3×10^{-2} torr.[59]

There are also arguable points regarding the CO_2 behavior. Figure 19.57 parallels Fig. 19.53, but with the difference that the extrema occur at appreciably higher temperatures, near 2000 K in Fig. 19.57 compared with about 1500 K in Fig. 19.53. The existence of the peak in CO_2/C reactions has recently been confirmed by Yang and Steinberg[371,372] in the diffusion cell, also at 1500 K, but it is notably absent in other data plots for the carbon-oxygen reaction (Fig. 19.67, below). It is undetermined in Fig. 19.68 (for the C/CO_2 reaction) because of an uninvestigated temperature interval from about 1600 to 1950 K due to experimental difficulties.[493] A peak in this region, however, is not inconsistent with the data obtained. Another effect shown in Fig. 19.68 is the increased rate above 2000 K that might be looked at as paralleling the final rise in Fig. 19.57 in spite of the very different experimental conditions. This rise was also observed by Walls and Strickland-Constable.[380] Ladd and Walsh[377] suggest that reaction is turning into sublimation in their case. Walls and Strickland-Constable ascribe it to reaction with oxygen obtained by CO_2 dissociation. Golovina and Khaustovich[493] describe some sort of breakup, also reported previously by Martin and Meyer[494]: "a very fine glowing dust was given off by the burning surface. It amounted to about 20% of the carbon consumed," and the rate values above 2000°C were corrected accordingly. It was not observed during the C/O_2 reaction, nor was it inhibited by imposition of any boundary layer so that high-temperature reaction rates with CO_2 exceeded those with

[493] E. S. Golovina and G. P. Khaustovich, "The Interaction of Carbon with Carbon Dioxide and Oxygen at Temperatures up to 3000°K," *8th Symp. (Int.) Combustion,* Paper No. 84, Williams and Wilkins Company, Baltimore, 1962, pp. 784–792.

[494] H. Martin and L. Meyer, "The Mechanism of the Primary Reaction Between Carbon Dioxide or Oxygen and Graphite," *Z. Elektrochem.,* **41,** 136–146 (1935).

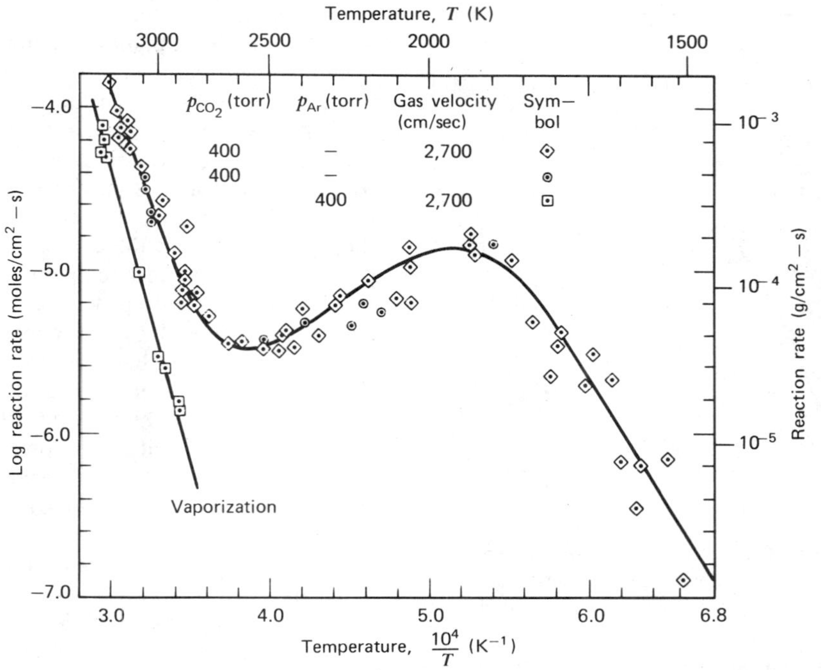

Figure 19.57 Temperature dependence of the rates of graphite-CO_2 reaction and of graphite vaporization. Temperature scale based on an assumed emittance of 0.82. Gas velocities calculated at the nozzle exit.[377]

O_2 (cf. Figs. 19.67 and 19.68 at temperatures above 2000°C). It does not seem that the rate increase above 2000°C can possibly be due to reaction of some Type 2 "annealed" sites, and even if the two-site hypothesis is acceptable for the C/O_2 reaction, it seems that the mechanism for the C/CO_2 reaction is different in spite of the phenomenological similarity in behavior. The effect of unrepressed gas phase dissociation may be the best explanation.[380]

It may also be noted here that assumptions about reaction orders are also open to question. The point has already been made that the reaction order of the basal planes appears to be zero, not first, and that the order of other reactions is something other than unity. This would change the structure of the equations somewhat and would change, likewise, the numerical values obtained for the constants in fitting, which thereby complicates interpretation.

At present, therefore, the two-site hypothesis to explain the extrema in the curves seems to be open to question. It is possible, nevertheless, that the explanation for the observed behavior can, indeed, relate to some form of annealing or recrystallization, by heat treatment, particularly above 2000 K.

Some of the effects of heat treatment have already been discussed in Section 19.3.2.3 (Fig. 19.23) for coal chars. Gregg and Tyson[344] also reported earlier heat treatment experiments on carbons and graphites up to much higher temperatures. Carbon made from compressed petroleum coke and pitch was heat treated in argon for 30 min at temperatures ranging from 1000 to 3000°C (~ 1300–3300 K). This procedure is known to change the degree of turbostraticity of the structure (the degree to which basal plane layers are randomly oriented with respect to each other; see Kaye[495]). This occurred with the carbons under treatment, which were then investigated for change of reactivity in pure dry oxygen using a vacuum microbalance. Reactivities were measured in the range 500–550°C to obtain data for Arrhenius plots, and Fig. 19.58 illustrates the variation of activation energy (measured at 500–550°C) with preparation or heat treatment tempera-

[495] G. Kaye, "Structural Changes in Heat Treated Carbon Blacks," *Carbon (Oxford)*, **2**, 413–419 (1965).

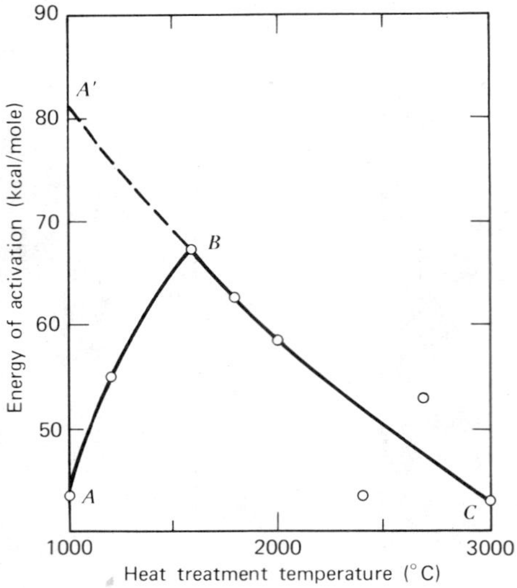

Figure 19.58 Oxidation of carbon in oxygen at 7.6 cm Hg pressure, showing the relationship between the experimental energy of activation and HTT.[344]

ture. Taking these data at face value, an increase of activation energy corresponds to potential for reducing the reaction rate. In Fig. 19.59 the reaction rates have been extrapolated to the HTTs from the rate data determined at 500–550°C. If the extrapolation is permissible, the result is a clear peaking of

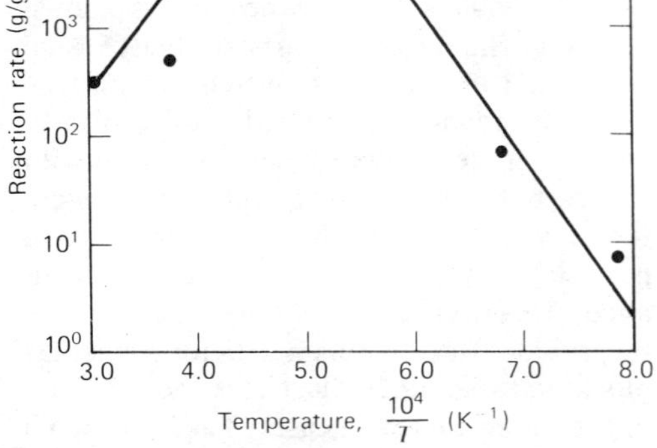

Figure 19.59 Variation of (log) reaction rate with (reciprocal) HTT for heat-treated carbon. Data obtained by extrapolation from rate measurements at 500–550°C to the HTTs of the carbons (cf. Fig. 19.58).

the rate data, due to the changes of activation energy and preexponential factor resulting from the heat treatment.

The extrapolation is, in fact, questionable for the reason given by the authors[344] for the peaking in the activation energy. It was attributed to a change from Zone II to Zone I behavior with rising temperature. The authors concluded that the carbons had lowest porosity at the highest temperatures, therefore the reaction fully penetrated the sample and was in Zone I (see Section 19.3.1.2). Thus the activation energies were the true values, which are then seen to decline with rising temperature along line *BC* in Fig. 19.58. Heat treatment at the lower temperatures, however, resulted in sufficient porosity for the reaction to move into Zone II (partial penetration), and resulted in line *AB*. The activation energies recorded were assumed to be only half the true values, so the true activation energy for condition *A* should be twice the recorded value, or 88 kcal/mole, which, at point *A′*, is on the continuation line of *CB*, extrapolated backward. (The authors do not explain why other points on line *AB* should not also be doubled.) The points clearly established by Gregg and Tyson remain unarguable, nevertheless. Heat treatment temperature significantly affects the reactivity. This was reconfirmed by Jenkins et al.[164]; *rate* of heating is also important[168,496,497] (Section 19.3.2.3).

There can also be other problems of interpretation in some instances on account of behavior of some manufactured carbons described by Kaye[495] and by Froberg.[144] Where carbons are conventionally manufactured out of binder and filler, the two constituents can retain different reactivities in spite of high HTTs (up to 3200°C in Froberg's case) under some experimental conditions (generally reaction at the lower temperatures). The partially graphitized binder would react out, allowing the filler particles to drop off, largely unreacted, or apparently so. Kaye[495] used more highly graphitized filler particles ("Basilblak"), and with these it was found that the

[496] S. Mizushima, "Rate of Graphitization of Carbon," *Proc. 5th Carbon Conf.*, Pergamon Press, London, 1963, pp. 439–447.

[497] P. L. Walker, Jr., and Y. Sasaki, "On the Growth of Crystals in Carbon Films," *Carbon (Oxford)*, **4**, 536–538 (1966).

external particle surface was mostly basal plane. The structure below the surface consisted of well-ordered planes, until the center was approached; there the planes became more disoriented with respect to each other. Where the surface basal planes butted against each other, edge reaction could develop, and this penetrated radially inward to react out the more disoriented particle centers, creating particles "with dense peripheries around spongy centers."[495]

Applying all the foregoing results to carbons, chars, and the like presents a further difficuly. If the materials are heat treated beforehand, the HTT is known. If, however, the material is being heat treated, in effect, as it reacts, there is competition between the gas reactions and the heat treatment reactions. The latter are likely to be rather slow, but not enough is known about the rates to be able to predict behavior. The competition may exist, not between reactions of two different kinds of site, but between the gas reaction and the graphitization reaction. This would have about the same overall result as would competition between sites of different reactivity. There are, conceptually, two curves for variation of reaction with temperature for the high- and low-reactivity materials, appropriately displaced from one another. During reaction, material will track up the curve for the higher reactivity with increasing temperature until graphitization is sufficiently fast to be competitive with the gas reactions. The reactivity plot can then move to the lower reactivity curve by a path exhibiing a lower slope (apparent activation energy), which in some instances could appear as a negative activation energy. It would be pointless to elaborate this model at this time. Appropriate experiments are now needed that use carbons heat treated at very high temperature ($> 3000°C$) to determine whether this heat treatment would eliminate the extrema in the rate curves. If it does, support for a graphitization mechanism would be strengthened.

In summary, the mechanism for the negative temperature coefficient behavior is still unknown. The original two-site theory of Blyholder et al.[481] is not disproved, but neither is it well substantiated. The alternative graphitization theory is qualitatively reasonable, but it is untested. At the same time, the idea of the reaction being governed by a low activation energy of adsorption at high temperatures cannot yet be ruled out.

19.6.3 Aspects of Flame Behavior: Particle Ignition, Extinction, and Combustion

The behavior of great numbers of particles in flames is often somewhat different from the behavior of the individual particles, or even from the behavior of small groups of particles in experiments such as the drop tube. There are interactive or "cooperative" influences in particle packings or dense dispersions that may retard or accelerate ignition and combustion. The phenomenon of ignition itself takes on additional significance; the same can be true of extinction. Other phenomena such as flame propagation become evident.

Full descriptions of complete flame systems are out of place in this chapter. They are dealt with generally or specifically in other chapters. The systems of concern include fixed and fluid beds, and stationary and traveling (explosion) flames. (See Chapters 20 and 24 for details; see also Essenhigh.[41]) It will suffice here to note that complete descriptions include the combined influences of reaction kinetics (as discussed in this chapter), heat transfer, and fluid mechanics. Mechanisms of flame propagation are integral components of complete flame theories, although they can also be discussed somewhat in isolation.[41] However, the fluid mechanics aspects present particular complications that are still being unraveled.[65] A good description of the present status of the fluid mechanics aspects of particle combustion in entrained flow is to be found in the recent book by Pratt and Smoot.[498] From the point of view of determining the particulate interactions in flames—the "cooperative" influences—an effective approach is to design out the fluid mechanics interactions as far as possible. In principle this can be achieved either by the plug-flow or nonbackmix furnace or, for entrained flow combustors with backmix, by the perfectly stirred reactor. In fixed or fluid bed combustors it is impossible to design out the fluid mechanics, but it is easier in those systems to take it into account by empirical

[498] L. D. Smoot and D. T. Pratt, Eds., *Pulverized Coal: Combustion and Gasification,* Plenum Press, New York, 1979.

correlations[65] (the so-called experimental modeling). In reality, also, mixing in the stirred reactor is never perfect, and this should be taken into account.

With the fluid mechanics aspects designed out or otherwise accounted for, we are left with interactions between reaction and heat transfer. These are particularly important for ignition and extinction, although only marginally so for particle interactive effects on combustion.

19.6.3.1 *Ignition and Extinction*

Ignition and extinction are of major importance in a number of different practical situations. Somewhat perversely, ignition studies are of greatest relevance to unwanted flames (explosion flames and spontaneous combustion); and extinction studies are of greatest relevance to boiler and furnace flames. One wants to prevent explosion flames from starting and prevent wanted flames from stopping. There has also been a study of extinction of single particles, evidently from the point of view of developing kinetic data. Most investigations have been of ignition, however, and most have involved single particles and clouds,[499–502] but have included coal on a grate[503] and ignition of graphite.[504]

The appropriate basis for analysis of a reacting gas-solid system is the Semenov thermal explosion theory for batch reactions, with the Vulis extension[505] to heterogeneous combustion and to continuous flow backmix reactors. The practical situations of interest in ignition are: spontaneous combustion underground in coal mines (particularly in the goaves), in coal mine spoil heaps, and in stocked coal; ignition of dust clouds leading to explosive propagation underground or in coal grinding and conveying equipment; and ignition in p.c. flames. It is also of interest in the experimental one-dimensional flames. Extinction is concerned, as noted, with flameout in boilers or furnaces.

Thermal Explosion Analysis. A detailed development of the equations of thermal explosion theory is out of place here. These relations are reviewed elsewhere,[280] with particular emphasis on the continuous flow reactors. It will suffice here to summarize specific results.

Steady State Combustion. For the continuous flow, backmix reactor we define the reaction efficiency, η, as $\eta = 1 - c_R/c_R^o$, where c_R^o and c_R are the inlet and outlet concentrations of reactant, for a one-step, nth-order reaction between reactant and oxidant, c_{ox}; the reaction efficiency is related to the temperature, T, by the relation[280]

$$\frac{\eta}{(1-\eta)^r(1-\phi\eta)^{n-r}} = (c_R^o)^{r-1}(c_{ox}^o)^{n-r}\tau_S(\rho^{n-1}k_0)e^{-E/RT} \qquad (19.112)$$

where η is also a dimensionless expression for heat generation, ϕ is the equivalence ratio, r is the reaction order with respect to the reactant, τ_S is the residence time in the reactor, and k_0 and E are the kinetic constants of the reaction.

Equation 19.112 has mostly been applied to gas combustion in stirred reactors[280] and to oil combustion in gas turbines (see review by Odgers[506]). It was also used by Biswas and Essenhigh[234] to determine the kinetic constants for volatiles combustion as described in Section 19.5.6.2, taking $r = 1$ and $n = 2$. It could be applied to coal in flames with sufficiently intense backmix, and has been to a simplified flame-ball model of coal

[499] S. K. Ubhayakar, "Burning Characteristics of a Spherical Particle Reacting with Ambient Oxidizing Gas at Its Surface," *Combust. Flame,* **26,** 23–34 (1976).

[500] B. Ghosh and A. A. Orning, "Igniting Pulverized Coal," *Ind. Eng. Chem.,* **47,** 117–127 (1955).

[501] J. O. Cutress, T. J. Peirce, and A. C. N. Tucker, "On the Ignition of Pulverized Anthracite," *Combust. Flame,* **8,** 289–298 (1964).

[502] C. S. Finney and T. S. Spicer, "The Influence of Particle Size and Volatile Content on Ignitibility of Pulverized Bituminous Coal," *Pyrodynamics,* **1,** 231–249 (1964).

[503] P. O. Rosin and H. R. Fehling, "The Ignition of Coal on a Grate," *J. Inst. Fuel,* **17,** 102 (1938).

[504] P. R. C. Goard and M. F. R. Mulcahy, "A Study of the Ignition of Graphite," *Carbon (Oxford),* **5,** 137–153 (1967).

[505] L. A. Vulis, *Thermal Regimes of Combustion,* McGraw-Hill Book Company, New York, 1961.

[506] J. Odgers, "Current Theories of Combustion Within Gas Turbine Chambers," *Proc. 15th Symp. (Int.) Combustion,* Combustion Institute, Pittsburgh, 1975, pp. 1321–1338.

combustion in a boiler.[507] It is implied in effect in the Béer and Lee investigations[508] of pulverized anthracite combustion in a furnace fired with a swirled burner. It has recently been applied to pyrolysis of coal in a stirred reactor, assuming a first-order pyrolysis reaction ($n = r = 1$).[69]

The stirred or backmix reactor is only one of three idealized conditions of general interest. The other two, as noted, are the batch reactor—such as the coal pile—and the plug-flow flame. The first is the unsteady state system originally analyzed by Semenov and others, using the thermal explosion theory. The second was first analyzed by Vulis as an extension of perfectly stirred reactor (PSR) theory. Recently[280] it has been shown that all three situations can be related. A theoretical framework therefore exists for analysis of such systems, and first steps have been taken to apply it to the plug-flow flame and to batch samples of dust in an electric tube furnace.

The equation form also applies to single particles. Vulis[505] has shown this to be true for the equation for the specific reaction rate at the surface of a single sphere of carbon, if reaction is first order following diffusion through a boundary layer. The equations for the burning sphere are therefore an analogue of the stirred reactor.

Ignition. The equations leading to Equation 19.112 are constructed from considerations of heat generation and heat loss. When the generation and loss are in balance, the reactor is operating stably in steady state. An ignition criterion is provided, however, when the curves for heat generation and loss are tangent to each other, as described in the relevant literature sources.[280,505] This condition is known as the "tangency criterion."

The equations of the thermal explosion analysis do not always permit the use of the tangency criterion, however, to define ignition. Vulis[505] has explored this particularly carefully and has demonstrated that in the $\eta - T$ space there are two regions that he designates "critical" and "noncritical." The tangency criterion and the temperature jump occur only in the critical region. Outside that region the slopes of the heat loss curves are always less than the slopes of the heat generation curves, and there is never more than one intersection for a stable condition. Ignition then has to be defined by some other arbitrary criterion. In the case of a plug-flow flame it can be shown that the ignition cell where the tangency criterion can be satisfied in the middle of the flame is also at the point of inflection.[509] When the tangency criterion cannot be satisfied and all cells in the flame lie in the noncritical region, probably the best ignition criterion is again the point of inflection.

Ignition of Particles. Ignition of particles has several different aspects. Original interest focused on underground mine fires, and this precipitated studies of self-heating at low temperature as oxygen was drawn through fixed beds (e.g., Winmill[510]). Some studies derived from that work were carried out with the sample in a furnace being steadily heated as the oxygen was drawn through. The "crossing-point"[320,321] method of determining ignition temperatures was developed from those investigations. Similar methods have also been developed as described below. Later, interest arose in ignition on grates.[503]

In the studies of self-heating at low temperatures, initial interest largely focused on the gases produced (Winmill[510]). The more recent investigations[211–217] have included the determination of the evolved gases, but principal interest was on the low-temperature adsorption of oxygen with some related studies of diffusion of gases through powders. Behavior at temperatures up to 200°C was examined, with most experiments in the lower range (< 100°C). One interesting finding,

[507] R. T. Waibel and R. H. Essenhigh, "Requirements for the Characterization of Coal Chars for Combustion," Seminar on Characterization and Characteristics of U.S. Coals for Practical Use, Pennsylvania State University, State College, October 1971.

[508] J. M. Béer and K. B. Lee, "The Effect of Residence Time Distribution on the Performance and Efficiency of Combustion," *Proc. 10th Symp. (Int.) Combustion*, Combustion Institute, Pittsburgh, 1965, pp. 1187–1202.

[509] R. H. Essenhigh, "Application of the Perfectly Stirred Reactor (PSR) Theory to Analysis of One-Dimensional Flames," Paper No. WSCI-66-28 presented at Western States Sect., Combustion Institute Meeting, 1966.

[510] T. F. Winmill, "The Absorption of Oxygen by Coal, Parts I and II," *Trans. Inst. Min. Eng.*, **46** 563–578 (1913–1914); **48,** 503–544 (1914–1915).

by Allardice,[217] was that the adsorption occurred in two stages. The first was a physical, reversible adsorption, unlike adsorption on carbon except below −70°C. The second stage was an irreversible chemisorption.

Carpenter and Sergeant also reported a two-stage adsorption, with the first stage physical and the second stage chemical.[216] In both investigations, and in earlier one by Sevenster,[211–213] it was found that the chemisorption was best described by the Elovich equation (Section 19.5.3), with very low activation energies as reported in Table 19.6. Carpenter and Giddings[215] also determined the reaction order and found it to be 0.5. This was in fair agreement with earlier experiments by Schmidt and Elder[209] in which coal was rotated in a drum. Schmidt and Elder also found that the oxidation rate was an inverse function of the friability. As the friability decreased the oxidation rate increased; this increase in oxidation rate could be related to the degree of graphitization and cross-linking developed in the coals.

These low-temperature studies have little directly to do with ignition, but there is a significant indirect connection. In studies of relative flammability as measured by the quantity of inert dust required to prevent ignition, Godbert and Wheeler[511] found that the relative flammability correlated very well with a relative reactivity, this last being the number of cubic centimeters of oxygen adsorbed in an hour by 100 g of coal at 100°C. This evident connection deserves further exploration.

More pertinent to ignition are the crossing-point,[320,321,512] critical air blast,[513] and other tests.[6,140,514] All involve blowing air or CO_2 through a bed of particles and measuring some parameter—for instance, the minimum air blast required to sustain reaction,[512] the burning temperature at some specified air rate, or the crossing-point temperature. Some methods involve small-scale bench experiments; others determine the scale effect through use of combustion pot, laboratory-scale studies, or equipment of larger size. Interest in such devices has largely vanished in the last 20 years, but with the renewed attention to coal use of all types, particularly in fixed-bed gasifiers, interest in the methods of testing may also revive.

Blayden[28] and Burden and Dore[45] have provided useful overviews of this subject that should be consulted for comment on the relation of such tests to practical applications. It will be found that discussion of the relationships is largely if not entirely qualitative. There is an absence of sufficient theoretical analysis to permit the most effective use of the reactivity test results. Sebastian and Mayers[320] initiated a theoretical analysis of the crossing-point method that involved determination of kinetic constants, and Sherman et al.[321] elaborated this work, relating the activation energy to the temperatures at which the rates of temperature rise were 15–75°C/min. More recently, Chauvin[512] reported further results using the same method of analysis. Hashimoto[515] has also described an analysis of such low-temperature data, using the data of Scott for testing his analysis.

It is notable, however, that the analyses are essentially empirical, depending on somewhat artificial or operational (i.e., experimental) definitions of "ignition" temperature. No one apparently has yet applied the Semenov thermal explosion analysis to these systems.

More has been made, theoretically, of the behavior of single particles, although the available data are sparse. Apart from the original theoretical analysis of the single particle by Vulis,[505] which was only tested qualitatively, the first analysis of the single particle in ignition, with experimental comparison, seems to have been that of Cassel and Liebman[516,517] in their studies of mag-

[511] A. L. Godbert and R. V. Wheeler, Safety in Mines Res. Board Paper No. 56, HMSO (London), 1929.

[512] R. Chauvin, "Measure de la Réactivité des Combustibles Solides par la Méthode du Point de Croisement," *Rev. Gen. Therm.*, 1–19 (February 1964).

[513] H. E. Blayden, W. Noble, and H. L. Riley, "A Simple Laboratory Method for Assessment of the Combustible Nature of Coke," *Gas. J.*, 201 (January 24, 1934).

[514] A. L. Godbert, "Laboratory Methods of Determining the Inflammability of Coal Dusts," Safety in Mines Res. Board Paper No. 31, HMSO, London, 1931; SMRE Res. Rep. No. 58, HMSO, London, 1952.

[515] K. Hashimoto, "A Theory on the Mechanism of an Outbreak of Spontaneous Combustion of Coal, II." *Mem. Fac. Eng., Hokkaido Univ.*, **11**(3) (No. 49), 269–291 (1962).

[516] H. M. Cassel, and I. Liebman, "The Cooperative Mechanism in the Ignition of Dust Dispersions," *Combust. Flame*, **3**, 467–475 (1959).

[517] H. M. Cassel and I. Liebman, "Combustion

nesium ignition. They mention in a footnote that some coal particles from the Pittsburgh Seam were also examined experimentally with unstated but similar results. What is of interest is the method of analysis, which has been applied here to some experimental results of Bandyopadhyay and Bhaduri[518,519] that were only partially analyzed.

The analysis developed by Cassel and Liebman[516] followed the Semenov principles of constructing the equations for heat generation and heat loss for a particle. For unit area of the surface, the heat generation was taken, in effect, to be given by Equation 19.84 multiplied by the heat of reaction, and with the following simplifications: first-order reaction, with no internal reaction, and unity roughness factor. With no significant reaction preceding ignition, boundary layer diffusion was assumed to be negligible as a factor influencing ignition; the surface oxygen concentration would be the ambient value. Only conduction heat loss was considered, with this inversely proportional to the particle radius. Equating the heat generation and loss, and also the gradients of the two equations for the ignition condition, an expression in T and r (particle radius) was obtained that included the Arrhenius exponential. The Arrhenius exponential was simplified by approximating it to a power expression in T, T^{α}; it was also assumed that the ambient gas temperature (T_{amb}) and the ignition temperatures (T_{ign})were sufficiently close that $\ln(T_{ign}/T_{amb})$could be replaced by the first term in the expansion. By rearrangement, it was then found that $T_{ign}/T_{amb} = \alpha/(\alpha - 1)$, and

$$T_{amb}^{(\alpha - 2)} r = \text{const} \qquad (19.113)$$

Cassel and Liebman[516] obtained the necessary experimental data but did not actually test the equation (19.113) they derived. Both their data for magnesium and Bandyopadhyay and Bhaduri's data for coal do, in fact, obey the equation, determined by plotting log (T_{amb}) against log(r). Good straight lines were obtained in both cases, with slopes of value $\alpha - 2$. The values of α were 11.2 for coal and 14.9 for magnesium, so that the ratios T_{ign}/T_{amb} were 1.22 and 1.16, respectively. The activation energies correspond to the ranges 20–25 and 30–35 kcal/mole, respectively. Figure 19.60 illustrates the experimental and calculated results: both T_{amb} and T_{ign} are plotted against particle size, where the ambient temperatures are the experimental values, with the line corresponding to Equation 19.113 drawn through, and the ignition temperatures calculated from the ambient temperatures using the α relationship derived.

These values, surprisingly, appear to be the only extant values of ambient and ignition temperatures for coal, other than four values obtained by Tretyakov quoted by Khitrin[520] and by Bandyopadhyay and Bhaduri.[518] Both the experimental and theoretical results show clearly that smaller particle sizes have higher ignition temperatures, and therefore then require higher ambient temperatures to achieve ignition, with a difference of 150–200°C. The reason for the inverse relationship is clear from the theory: the heat loss increases as the particle size drops, and the surface temperature must rise for the heat release rate to match the increased heat loss rate.

Bandyopadhyay and Bhaduri also calculated the temperature-time relationship for some lignite and anthracite particles as shown in Fig. 19.61. The general shape of the curves is characteristic for ambient temperatures near the critical value for ignition[521,522]: there is an initial rapid rise to a plateau, followed by slow reaction on the plateau during which the temperature rises

of Magnesium Particles, Parts I and II," *Combust. Flame,* **6,** 153–156 (1962); **7,** 79–81 (1963).

[518] S. Bandyopadhyay and D. Bhaduri, "Effect of Particle Size and Distribution in the Combustion of Coal Dust," *J. Inst. Fuel,* **44,** 377–380 (1971).

[519] S. Bandyopadhyay and D. Bhaduri, "Prediction of Ignition Temperature of a Single Coal Particle," *Combust. Flame,* **18,** 411–415 (1972).

[520] L. N. Khitrin, M. B. Ravitch, and L. L. Kotova, "The Characteristics of the Combustion of the Coke Residue of Solid Fuels," *Proc. 8th Symp. (Int.) Combustion,* Paper No. 85, William and Wilkins Company, Baltimore, 1962, pp. 793–800.

[521] P. H. Thomas, "Some Approximations in the Theory of (*a*) Self-Heating and Thermal Explosion," *Trans. Faraday Soc.,* **56,** 833 (1960); (*b*) "Effect of Reactant Consumption on the Induction Period and Critical Condition for a Thermal Explosion," *Proc. R. Soc. London,* **262A,** 192 (1961).

[522] P. H. Thomas and P. C. Bowes, "Thermal Ignition in a Slab with One Face at a Constant High Temperature," *Trans. Faraday Soc.,* **57,** 2007 (1961).

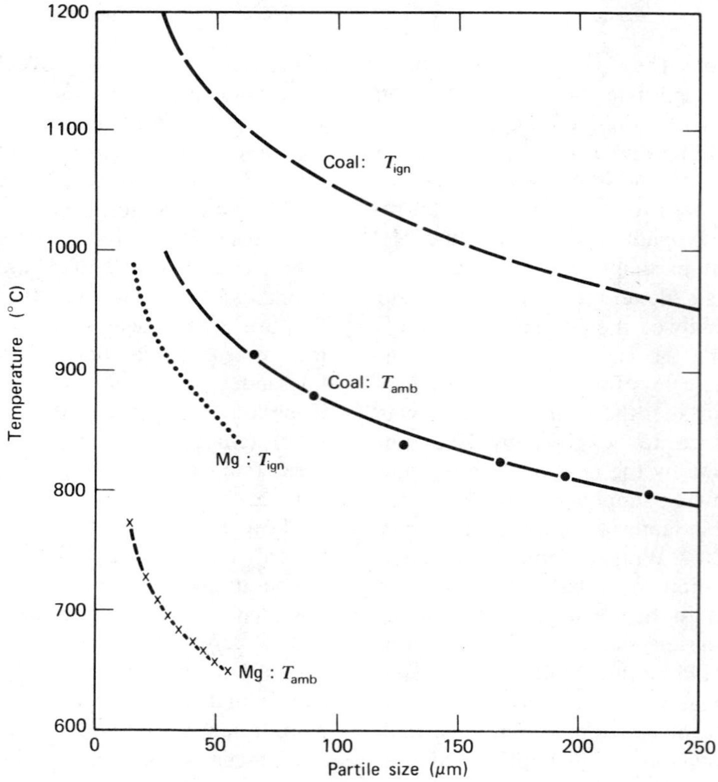

Figure 19.60 Variation of ambient (gas) and ignition temperatures for single particles of bituminous coal (●)[518] and magnesium (×).[516] Values of T_{amb} are experimental; values of T_{ign} are calculated from T_{amb}.

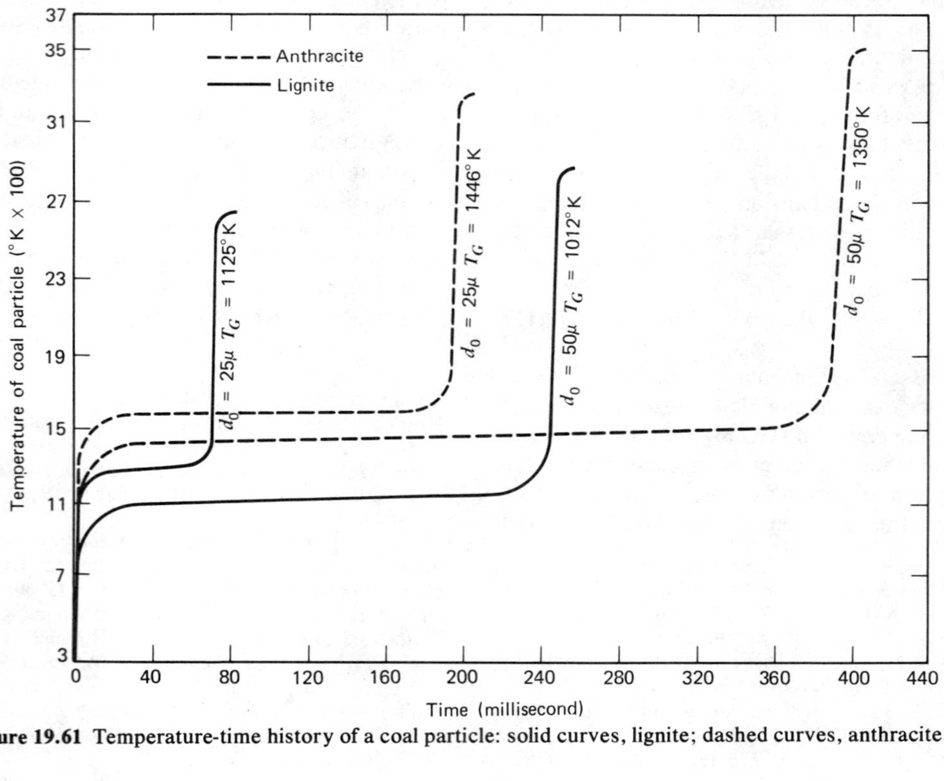

Figure 19.61 Temperature-time history of a coal particle: solid curves, lignite; dashed curves, anthracite.[518]

very slowly to ignition. Stevenson et al.[523,524] hav obtained similar plots for coal. At ignition the temperature jumps (the Semenov jump) to the final burnout temperature. The time between the start and finish of the plateau is the ignition delay or induction period. The temperature of the plateau increases with decreasing particle size, but the induction period decreases with decreasing particle size. Stevenson et al.[524] also reported that under some conditions their calculations predicted behavior outside the critical region, and rise to burnout temperature was obtained without the Semenov temperature jump (for a general analysis of noncriticality, see Vulis[505]).

Annamalai and Durbetaki[492] reviewed the heterogeneous ignition analysis and introduced, for comparison, a further analysis of pyrolyzing particles which were assumed to be igniting by the Faraday mechanism. This is a homogeneous mechanism in which ignition is assumed to occur in the volatiles. The background to this is discussed in more detail in connection with "Ignition in Flames," Section 19.6.3, but it derives from an experimental finding[73] that coal particles can ignite heterogeneously. Annamalai and Durbetaki concluded that the ignition temperature when the particles ignited heterogeneously was an inverse function of radius and oxygen concentration, as previously established by Cassel and Liebman,[516] noting that the dependence on oxygen concentration is a function of the reaction order assumptions made initially. If the reaction order is zero, all conclusions depending on this parameter are voided. The question of homogeneous ignition was addressed also by using the thermal criterion of ignition, although the appropriateness of this criterion for gases might be considered questionable. The conclusion was that particles smaller than 350 μm may not ignite in the gas phase. However, this conclusion would seem to be contrary to the findings of Orning,[525] who showed that the ignition of the volatiles depends very much on the temperature of the ambient gases.

Ignition and ignition delay were also part of the shock tube measurements described earlier (Section 19.6.2.2;[82,83,85,179,373]). At the temperatures, pressures, and heating rates involved (up to 4000 K, 30 atm, and 10^6 °C/s, respectively) both the ignition times (induction lag) and burning times were much shorter than in the single-particle experiments, these being about 1 ms for ignition and 1–3 ms for combustion. Nettleton and Stirling used the Frank-Kamenetskii[526] approach to the Semenov analysis to determine ignition temperatures. Ignition in the shock tube was found to be at about 1800 K, which is substantially above the values in Fig. 19.60, obtained at normal pressure and at relatively low heating rates. Lightman and Street,[180] contrarily, reported that ignition temperature in comparative experiments they carried out on the drop tube and the shock tube were the same; clearly more work on this is needed. Nettleton and Stirling also found the ignition temperatures to vary with oxygen concentration; they assumed first-order reaction in their analysis. In the analysis, simplifications due to Merzhanov[527] were used.

The shock tube studies went beyond single particles. Multiple particles or small clouds are necessary. Although the investigators using the shock tube[82,83,85,180,373] did not comment on the point, the presence of other particles can modify the results. This fact has been given surprisingly little attention, however, in spite of the widespread use of batch cloud ignition devices in safety testing to determine the explosion hazard of dusts. Typical methods of experiment involved projection of a sample of dust through a furnace, or over some ignition source such as a heated coil or a spark.

The methods date back 100 years or more.[514] In the earliest experiments the temperature of the ignition source was measured and reported as the "relative ignition temperature." In the course of time, the word "relative" was dropped, and many values of

[523] G. R. Thomas, A. J. Stevenson, and D. G. Evans, "Ignition of Coal Particles Without Temperature Jump," *Combust. Flame,* **21,** 133–136 (1973).

[524] A. J. Stevenson, G. P. Thomas, and D. G. Evans, "Modelling the Ignition of Brown-Coal Particles," *Fuel,* **52,** 281–287 (1973).

[525] A. A. Orning, "The Mechanism of the Combustion of Pulverized Fuel," *Proc. Conf. Pulverized Fuel, Vol. 1,* Institute of Fuel, London, 1947, pp. 45–64.

[526] D. A. Frank-Kamenetskii, *Diffusion and Heat Transfer in Chemical Kinetics,* 2nd ed., Plenum Press, New York, 1969.

[527] A. G. Merzhanov, "Thermal Explosion and Ignition as a Method for Formal Kinetic Studies of Exothermic Reactions in the Condensed Phases," *Combust. Flame,* **11,** 201–211 (1967).

ignition temperatures are available in data books that are, in fact, misleading, being ignition source temperatures only. The temperature of the ignition source at ignition varies inversely with its size.[528] The temperature may also be a function of the sample size, and particle size in the dust sample. The sample size may or may not affect concentration; most experiments are ambiguous on account of other variable factors such as dispersion velocity, which can be affected by sample size, dispersion pressure, and so forth.

The clearest illustration of the effect of particle interaction—the "cooperative" effect—is given by Cassel and Liebman,[516] though for magnesium, unfortunately, not for coal. They showed clearly that the ignition temperature is reduced when many particles are present, and the reduction increases with particle concentration. Even more interesting, the effect is greatest with the smallest particles, and diminishes with increasing particle size. The result is that the ignition temperature of the smallest particles is higher than that of the largest at low cloud concentrations, in accordance with Equation 19.113; but at the highest concentrations, the order is reversed: the smaller the particle size, the lower the ignition temperature.

Little theoretical attention has otherwise been given to analysis of such batch cloud devices. Only Nagy and Surincek[341] have attempted such an analysis, for the Godbert-Greenwald[529] test furnace, which is an electrical tube furnace, erected vertically, through which the sample of dust is blown. The thermal explosion analysis was used, but only results on cornstarch are presently available; there is nothing on coal.

Ignition in Flames. Ignition in flames is a problem of a different dimension. It includes the problem of defining ignition, but with an added problem of rate of heat supply. In the case of jet flames, compared with plug-flow flames, the heat supply is dominated by recirculation or other backmix flows. This can supply 60–95% of the heat required for ignition[47]; the problem of practical interest is then less one of ignition and more one of extinction, or the conditions under which stable flames are maintained. The extinction problem is discussed in the subsection that follows; the jet flame behavior is otherwise part of the aerodynamic problem that is outside the scope of this chapter (see Chapter 20). The remaining plug-flow problem has two different theoretical aspects; there is also a question regarding the actual mechanism of ignition of coal particles.

Plug-Flow Flames. As outlined elsewhere,[280] the plug-flow flame can be treated in principle, following Vulis,[505] as a sequence of perfectly stirred reactors feeding successively into each other, with the possibility then of identifying an "ignition" cell in the sequence that provides a basis for defining ignition in the middle of the flame[509] without resorting to arbitrary definitions.

Only three flame investigations applying this approach are known.[507,509,530] The earliest[509] was a general examination of the equations in dimensionless form and without any thermal feedback, so that the equations simulated only the true thermal explosion system. The results showed the existence of both a critical and a noncritical region that depended on concentration. In the critical region the Semenov jump existed; in the noncritical region it did not. In the critical region, however, it was possible to show that the ignition temperature was at the point of inflection of the temperature curve. For the noncritical region (the lower concentrations) the criterion of ignition therefore adopted was the point of inflection.

In the second investigation[530] a convective backmix flow was included, and in the third[77] a radiative feedback was included. Of these two, the first was simulating a gas flame and the second was simulating a coal flame. In both, the parameter values were so chosen that the predicted values of flame temperature, flame speed, and so on, were realistic values. In both cases, unexpectedly, only a noncritical region was found to exist in the

[528] R. H. Essenhigh, "Dust Explosions in Factories: Ignition Testing and Design of a New Inflammator," SMRE Rep. No. 188, HMSO, London, 1960.

[529] A. L. Godbert and H. P. Greenwald, U.S. Bur. Mines Bull. No. 389, 1935.

[530] M. Kuwata and R. H. Essenhigh, "Comparative Influence of Turbulence and Convection in Combustor Performance," *Proc. 2nd AGA/IGT Conf. Nat. Gas. J. Res. Technol.* (1972), p. IV.3.1.

range of realistic values of predictable parameters. In essence this meant that the first cell was always the "ignition" cell, even though a point of inflection in the temperature curve existed both theoretically and experimentally some distance downstream. Computation was iterative and required selection of some initial value of temperature for the first cell. If the value chosen was high enough, the temperature rose to a steady state value on successive iterations. If the value chosen was too low, the downstream temperatures were too low to feed back the necessary heat to the first cell, and the cell temperatures steadily dropped to a value marginally above ambient.

The method of analysis requires exploration over a wider range of parameters, but a tentative conclusion from the results above is that the application of thermal analysis to any backmix or thermal feedback system requires a different criterion of ignition.

Effect of Experimental Conditions on Ignition Temperature and Time. In the absence of an acceptable theoretical definition of ignition in flames using the thermal explosion criterion, as discussed above, it is necessary to use some arbitrary assumption. The easiest approach is to postulate an "ignition temperature" (T_{ign}) and determine this experimentally if possible. This technique was the basis of the original Nusselt radiation analysis[531] for a plane flame for infinite ignition distance. Essenhigh and Csaba[70] extended that analysis to the condition of finite ignition distance, representing the reality of the plug-flow flames, particularly with the water-cooled tube bank as a boundary.[73] The Essenhigh and Csaba paper was principally concerned with flame speed, but it included a basis for calculating ignition time for an arbitrarily chosen ignition temperature.

Field et al.[35] evaluated the given equations,[70] describing ignition time as a function of input velocity, and their graphs are reproduced in Fig. 19.62. The graphs show, first, that a "fundamental" input velocity exists at which the ignition time is infinite; this "fundamental" velocity corresponds to a "fundamental" flame speed.[70] Ignition time then drops rapidly with reduced input velocity for a given concentration, with a finite ignition time, t_i, at zero input velocity: this would correspond to a true "induction lag."

To date, only one experimental substantiation of this predicted behavior is known.[87] The data were obtained for a fixed fuel supply rate with varying input velocity so that the input concentration was varying, in contrast to Fig. 19.62, where the input concentration was held constant. The results are illustrated in Fig. 19.63, which shows ignition time at around 0.1–0.2 s, and this is in the region of the values predicted in Fig. 19.62.

The predictions assume a fixed ignition temperature, and to the extent that the equations are tested by the experimental results in Fig. 19.63, it can be concluded that the ignition temperature was indeed substantially constant. Taking the ignition temperature as the gas temperature at the flame front, the available experimental data do further support the relative insensitivity of ignition temperature to such conditions as concentration and possibly fuel type.[77] In other calculations,[77] taking ignition as the 1% combustion efficiency point, the ignition temperature was again found to be insensitive to conditions. Ignition time, however, was found experimentally to be very sensitive to fuel reactivity. Chars with little volatile matter were used, and Cogoli et al.[87] reported that the ignition time for high-reactivity chars was around 0.1 s, as in Fig. 19.63; but the ignition time for low-reactivity chars, with very similar ignition temperatures, concentrations, and particle sizes, was as high as 1 s. This cannot yet be accounted for properly in the theoretical analyses[77,532,533] and further investigation is required.

Mechanism of Ignition. The mechanism of ignition in the case of char particles is self-evident: the primary reaction is one of reaction of oxygen with the solid particle, although there may be production of CO that does not burn in the particle boundary layer

[531] W. Nusselt, "The Process of Combustion in Powdered-Coal Firing," *V.D.I.*, **68**, 124 (1924).

[532] H. Krier and J. L. Krazinski, "Theory of Coal-Dust/Air Flame Propagation," Paper No. 74-1111, presented at AIAA/SAE 10th Propulsion Conf., San Diego, Calif., October (1974).

[533] J. L. Krazinski, R. O. Buckius, and M. Krier, "Modelling Coal-Dust Air Flames with Radiative Transport," paper presented at Spring Technical Meeting, Central State Sect., Combustion Institute, Cleveland, 1977.

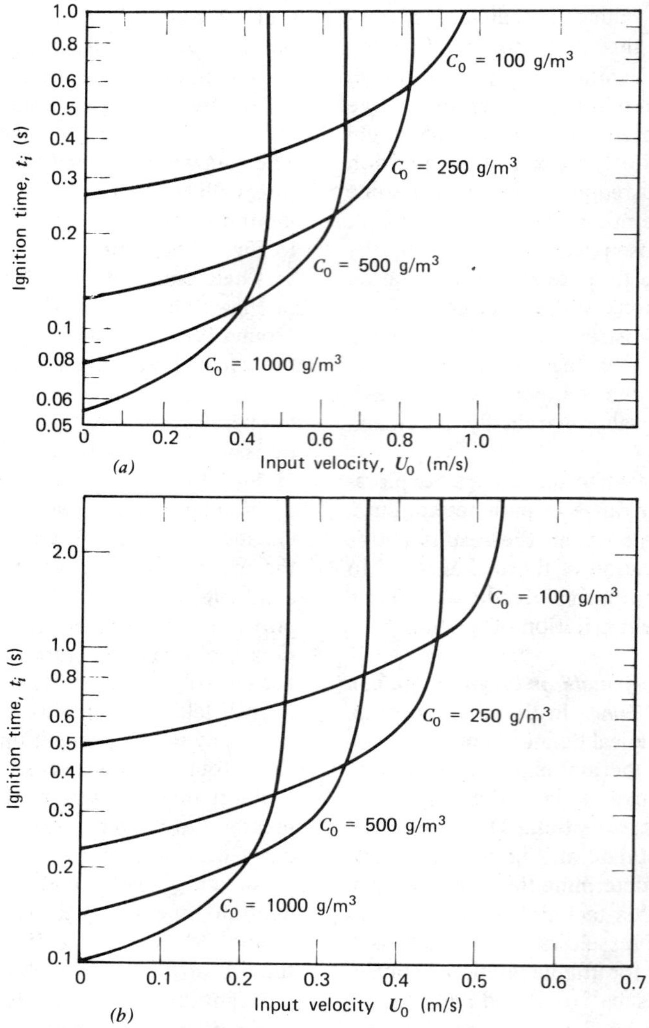

Figure 19.62 Theoretical variation of ignition time with input velocity, coal concentration, and ignition temperature, T_i', in plug flow: particle diameter = 30 μm, inlet temperature = 350 K, I_f = 100 Kcal/m² – s. (*a*) T_i' = 650 K and (*b*) T_i' = 900 K.[35]

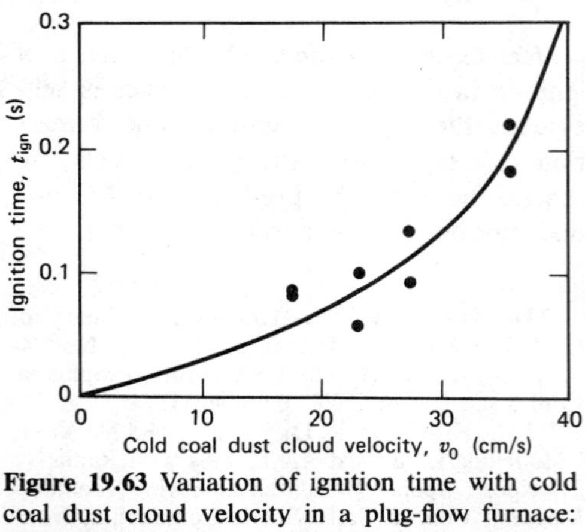

Figure 19.63 Variation of ignition time with cold coal dust cloud velocity in a plug-flow furnace: equation of curve: $t_{ign} = 0.105 \ln (1 - v_0/s_0)$.[87]

and burns in the space between. In the case of coal, however, there is scope for pyrolysis before or after ignition, and this can affect the mechanism of ignition. These alternatives have been mentioned cursorily in earlier sections.

Faraday and Lyell[3] first proposed that burning coal particles might pyrolyze as a flame front approached, so that the flame was mainly homogeneous. A number of variants have been suggested since then, such as (1) following prepyrolysis the flame may be almost exclusively homogeneous, hardly involving the flame-formed char, (2) prior ignition of the volatiles is required for the char to ignite (the "touchpaper" theory), (3) the ignition and combustion of the char is independent of the prior ignition and combustion of the volatiles, or (4) the coal particles can ignite without pyrolyzing; this is known as whole coal or heterogeneous ignition. That heterogeneous ignition might be possible was first proposed by Wheeler[534] 68 years ago to explain some results of gas analyses taken in explosion flames.

Prior release and ignition of the volatiles occurs with large particles (about millimeter size or above) when heating rates are relatively slow (order °C/s). Volatiles can be seen to emerge and to ignite, while the charring coal particle can be seen to be unignited.[80,90] When char ignition occurred after the generation, ignition, and combustion of the pyrolysis products, Orning[525] reported that "ignition of the solid residue generally occurred at one point and then moved as an ignition wave over the surface" (see also Thomas et al.[535]); ignition of the volatiles was not required for the char to ignite after completion of pyrolysis.[525] Essenhigh[536] indeed noted that on experiments on single particles in the particle size range 300–4000 μm, only two particles out of 1300 showed their volatiles igniting when the particle size was smaller than 776 μm. The cutoff particle size would be a function of gas temperature, however.

The evidence for heterogeneous ignition is limited. From experiments using a modified Hartmann explosion chamber, Carpenter[537] drew the inference that the results did not favor the volatiles ignition hypothesis. His argument was that a sudden change in the explosibility characteristics of the coals used would be expected as the VM content changed, since a coal of low VM could not generate sufficient volatiles for propagating a gaseous flame, and no such change was identified by the author. Some small change of explosibility characteristics at about 20% VM can be discerned, however, in the results reported.

More direct evidence was obtained by Howard and Essenhigh.[73] The experimental system was a laminar, plug-flow flame in a vertical refractory furnace, with a water-cooled tube bank as the well-defined inlet boundary. Two stable flame fronts were observed, 2–3 cm apart. The upstream flame front coincided with the plane of the appearance of CO_2, disappearance of O_2, and disappearance of solid mass, measured by ash tracing. The plane of VM loss was coincident with the second (lower) flame front. Between the flame fronts the VM content remained constant in spite of the recorded combustion (O_2 and CO_2 changes). Spatial resolution of the sampling locations was estimated as millimeters, and sample repeatability was good. The conclusion was that reaction without change in VM content was possible only if there was whole coal (heterogeneous) ignition and combustion.

These results and conclusions have been criticized on several grounds, including the following: the Q factor effect (Section 19.3.2.2) was not taken into account,[538] the char samples were unrepresentative,[538] there was differential heatup of the particles,[538] measurements were imprecise, with poor spatial resolution in the flames,[196] the flames were unstable,[196] and there was a loss of fixed carbon between the two flame fronts.[535,539] The assertions lack factual basis, however, and it is my opinion that the criticisms lack merit. The reader can make a personal evaluation by referring to the original papers.

Thomas et al.[535] attempted a direct demonstration of heterogeneous ignition, examining behavior photographically. They used 200–1000 μm captive particles of brown coal in cool surroundings, heated by air at 500°C flowing over the particles at 12 ft/s. At such relative velocities the major portion of the volatiles would be swept away from the particle as it was generated. The photographs showed that the behavior described earlier by Orning[525] occurred: namely, the spread of an ignition wave over the particle, with pro-

[534] R. V. Wheeler, in *2nd Report to Explosions in Mines Committee,* HMSO (London), 1912.

[535] G. R. Thomas, I. J. Harris, and D. G. Evans, "The Ignition of Pulverized Brown Coal," *Combust. Flame,* **12,** 391–393 (1968).

[536] R. H. Essenhigh, "The Burning Times of Single Coal Particles," Ph.D. thesis, Dept. of Fuel Technology and Chemical Engineering, University of Sheffield, England, 1959.

[537] D. L. Carpenter, "The Inflammability of Coal Dusts," *Proc. New South Wales Inst. Eng.,* **69,** 36–50 (1954); "The Explosibility Characteristics of Coal Dust Clouds Using Electric Spark Ignition," *Combust. Flame,* **1,** 63–93 (1957).

[538] S. Badzioch, in Field et al., Ref. 35.

[539] G. M. Kimber and M. D. Gray, *Combust. Flame,* **11,** 362 (1967).

duction and ignition of volatiles a short time later. The experiment is ambiguous, however. It demonstrated ignition of volatiles only after ignition of the solid particle; there was no indication of whether the particle was already pyrolyzing significantly before solid ignition ensued, so that pyrolysis ignition could have occurred first but for the artificial conditions of the gas stream sweeping the volatiles away.

Heterogeneous ignition can be explained by delay in pyrolysis. As shown by Jüntgen et al.,[185] this is what happens under conditions of very rapid heating. Under normal heating conditions, at about 1°C/s, the onset of pyrolysis, taken as the point of 1% loss of VM, is in the range 350–400°C for most bituminous coals. At a 10^4 °C/s rate of heating, the 1% loss occurs at around 1200°C. It is an effect not of temperature but of time.

In summary, particles heating at low rates will generate volatiles that may well ignite (but not necessarily) before pyrolysis is complete. As heating rate increases, the temperature required to generate a significant quantity of volatiles also increases. At high enough heating rates, the particles may well be hot enough for them to ignite heterogeneously before pyrolysis, with pyrolysis occurring somewhat later. The volatiles production in the case of the larger particles may be able to screen the solid surface from oxygen attack and extinguish the heterogeneous reaction. In the case of smaller particles this shielding may not occur, and reaction will proceed, as described in summary in Section 19.3.2.

Extinction. There is very little information on extinction. Only two studies have been identified, one on an idealized flame[507] and the other on single particles.[369] Both used the thermal explosion theory as the basis for analysis; the first was only theoretical; the second included experimental verification.

Waibel and Essenhigh[507] examined an idealized flame configuration in cold-wall surroundings (a boiler) to determine the characteristics likely to be important in flameout. The context of the problem was that low-reactivity chars were expected to be produced from coal gasification and liquefaction plants; for the plant to be economically viable, the char would have to be disposed of in boilers to raise steam. The question at issue was the potential for inadvertent flameout, which carries a risk of subsequent explosion if chars, assumed to be of low reactivity, are used in bituminous coal boilers. One conclusion was that the conditions for extinction could indeed become very sensitive to small changes in reactivity; and there might be little warning by way, for example, of a significant fall in flame temperature, as the extinction conditions were approached.

A partial resolution of the safety problem was obtained from a different source when it was found that since some chars had reactivities as high as that of bituminous coal,[87] there would be no barrier to their use in bituminous coal boilers. Other chars were found to have reactivities greater than that of anthracite, which would allow use of U- and W-flame boilers (see Chapter 20) for their combustion, although these are usually more expensive to construct, and less efficient. Tests of an FMC-COED char in a U-flame boiler substantiated the expectation of satisfactory operation.

A study of single-particle extinction (and combustion) was carried out by Ubhayakar and Williams.[369] Particle sizes ranged from 50 to 200 μm. Combustion, in O_2/N_2 mixtures, was initiated by a laser that carried the temperature to 2000–3500 K, as measured by two-color pyrometry. With the combustion in cold gas and in cold surroundings, the particles extinguished before burnout, a behavior predicted and demonstrated qualitatively by Vulis[505] among others. The Ubhayakar and Williams experiments are believed to be the first quantitative substantiation of the extinction theory.

Analysis focused on extinction, using the thermal explosion theory, and employed the semiempirical nth-order equation (Equations 19.74 and 19.74a) to represent the kinetics. Only two mechanism models were considered: the occurrence of and the absence of O_2 dissociation in the gas phase. Dissociative adsorption was not considered. The kinetic constants listed in Table 19.6 were obtained by curve fitting; the reaction order (lying between 0.5 and 1), in particular, does not have a high reliability on that account, but the experimental method, of course, is not suited to accurate determination of kinetic constants. An aspect of interest is that the particle sizes and temperatures were evidently great enough for boundary layer diffusion to be important in the analysis, as Fig. 19.50 indicates might be the case.

At lower temperatures results such as these might have application to particle extinction near the water wall in a boiler.

19.6.3.2 *Burning Time Measurements*

Because of the difficulty of making rate measurements, particularly on small particles, many early experiments were confined to measurement of total burning times. Evaluation of the results required assumption of rate equations and their integration. The early investigations assumed simplifications such as boundary layer control of the reaction (e.g., Nusselt[531]), or impermeable particles. Here these approximations are applied to somewhat more general equations to demonstrate the context of the simplifications.

Equations for Burning Time. Expressions for the total reaction rate, R_T, are developed in Section 19.5.5.3 in somewhat simplified form, with the more general expressions given in the previous subsections. For a spherical particle of radius a_s, mass m, and density σ, $m = (4/3)\pi a_S^3 \sigma$. The total specific reaction rate can be written

$$-R_T = \frac{a_S}{3}\frac{d\sigma}{dt} + \sigma\frac{da_S}{dt} \qquad (19.114)$$

where R_T is given generally by combining Equation 19.84 with 19.85. The result is an equation in σ, a_s, and t that is inexact. Limiting solutions are quite easily obtained: the conditions are either that reaction occurs only at the superficial surface ($d\sigma/dt = 0$, implying boundary layer control or impermeable solids), or that reaction is in Zone I with uniform reaction throughout the particle ($da_s/dt = 0$). Both these possibilities are explored here.

Intermediate solutions require other approaches. One is to develop a totally mechanistic model for the pore burnout to develop the necessary relation between σ and a_s. The other is to devise some rule that will provide a relationship between σ and a_s on a phenomenological basis. These alternatives are considered briefly later.

No Internal Reaction ($\dot{\sigma} = 0$). Depending on the extent of the simplifications used, R_T is a simple or an involved function of a_s. With simultaneous diffusion, adsorption, and desorption, the quadratic of Equation 19.88 is obtained. The three simplifications of Equations 19.89–19.91, however, provide four simple approximations. The first is that R_T is a constant (k') with respect to a_s (but is a function of T and p_O). The same result is obtained from the second solution to Equation 19.91. The first solution to Equation 19.91 yields $R_T \propto 1/a_s$. Equation 19.90 yields the sum of the two as a "general" approximate differential equation that integrates to

$$t = K_C(d_0 - d) + K_D(d_0^2 - d^2) \qquad (19.115)$$

where d_0 is the initial diameter, and K_C and K_D are the "chemical" and "diffusional" burning constants, respectively, given by Equations 19.119 and 19.120, discussed below. At zero diameter, Equation 19.115 yields[10] the equation for the burning time, t_b

$$t_b = K_C\, d_0 + K_D\, d_0^2 \qquad (19.116)$$

If K_C is very small, the equation reduces to the Nusselt "square law" relation.[531]

No External Reaction ($\dot{a}_s = 0$). If we assume at the limit that there is uniform reaction throughout the particle, R_T is given by Equation 19.114 with the second RHS term zero, and by Equation 19.84 with the first RHS term zero, $\eta = 1$, and $s = 3$ (for a sphere). This yields: $\dot{\sigma} =$ const. Integration yields a linear decline of density with time, giving an expression for the burning time, at $\sigma = 0$:

$$t_b = K_C' \qquad (19.117)$$

where the structure of K_C' is given in the following paragraphs.

The structure of the three burning constants, K_D, K_C, and K_C', depends on several factors.

1. K_D. Two models exist for the diffusion burning constant, the single-film and the double-film assumptions. In the single-film model oxygen is transferred across the film boundary to react at the solid surface to produce either CO or CO_2. In the double-film model oxygen diffuses across an outer film, to react with CO at the interface of an inner film to produce CO_2, the CO_2, as the oxygen carrier, then diffuses to the particle surface to react back to CO. For a sphere in an infinite, quiescent gas, the diffusional velocity constant for mass transfer can be written: $k_D = D/a_s$ (with parameters appropriately defined to balance units), if the reactant concentration is low. For oxygen in air, the low reactant concentration requirement is satisfied for concentrations less than 21%. Above that limit, convective effects are involved because there is a net mass transfer to the solid surface (in the case of CO_2 formation). The variation of reaction rate with reactant concentration changes from a linear to a logarithmic form. This can be represented in Equation 19.85 either as a modification to k_D, or as a rearrangement of the equation itself. When Equation 19.85 is used as a definition of k_D, the effect of the convective flow is to modify k_D. The velocity constant, k_D, is also influenced by ambient velocity. This is taken into account by incorporating k_D in a Sherwood number.

Combining all effects, the burning constant can be written

$$K_D = \frac{\sigma_C^0}{4NM_C} \cdot \text{Sh}\, DL_0 \ln(1 - p_O) \qquad (19.118)$$

where σ_C^0 is the initial (or constant) density of the particle; N either is the number of films or takes the value 2 if O_2 reacts only to CO; M_C is the (carbon or char) molecular weight; Sh is the Sherwood number; D is the diffusion coefficient; L_0 is the total molar concentration (moles/cm³), given approximately by ρ/M_0, where ρ is the ambient gas density and M_0 is the molecular weight of the ambient gas; and p_O is the ambient oxygen mole fraction.

2. K_C, K_C'. Expressions for these constants vary with reaction order, and they vary in similar ways.

For a first-order reaction we have (neglecting the convective flow and ambient velocity effects):

$$K_C = \frac{\sigma_C^0}{M_C L_0 k_1 p_O} \qquad (19.119)$$

for reaction of a nonporous solid

$$K_C' = \frac{K_C}{S_V} + 2/3\, K_D\, d_0^2 \qquad (19.119a)$$

for uniform reaction of a porous solid

where S_V is the internal surface area per unit volume of the reacting solid.

For zero-order reaction the expressions reduce to

$$K_C = \frac{\sigma_C^0}{M_C L_0 k_5}$$

for reaction of a nonporous solid (19.120)

$$K_C' = \frac{\sigma_C^0}{M_C L_0 S_v k_5}$$

for uniform reaction of a porous solid (19.120a)

As can be seen, the last two expressions differ only by the multiplier of the internal surface per unit volume, S_v.

Comparisons with Experiment: Large Particles. The equations above provide the theoretical framework for some results already described, as well as others, to be described.

Captive Particles. Figure 19.1 illustrates agreement between theory and experiment for particles larger than 100 μm burning under diffusion dominance or control, with the theoretical lines calculated from Equation 19.116 with K_C small. Most results cited are for precarbonized coal chars.

Essenhigh et al.[80,90,177,383] investigated the behavior of coal particles, measuring both the pyrolysis combustion times and the combustion time of the chars formed during the pyrolysis of the coal, and included in the analysis the effects of pyrolysis and swelling. The corrected values of K were given by[90]

$$K_r = \frac{K_D(C_f/100)}{f} \qquad (19.121)$$

where C_f is the fixed carbon, correcting for change of density on VM loss; and f is a swelling factor measured experimentally, and found unexpectedly to have approximately the same value (1.5) under the conditions of experiment for all the coals tested except the anthracite.[177] Agreement between theory and experiment was thereby greatly improved over the agreement shown in Fig. 19.1. The effect of oxygen concentration was also investigated in greater detail. Figure 19.64 illustrates the agreement obtained[383] between experimental values of burning time and values calculated from Equation 19.116 with K_C

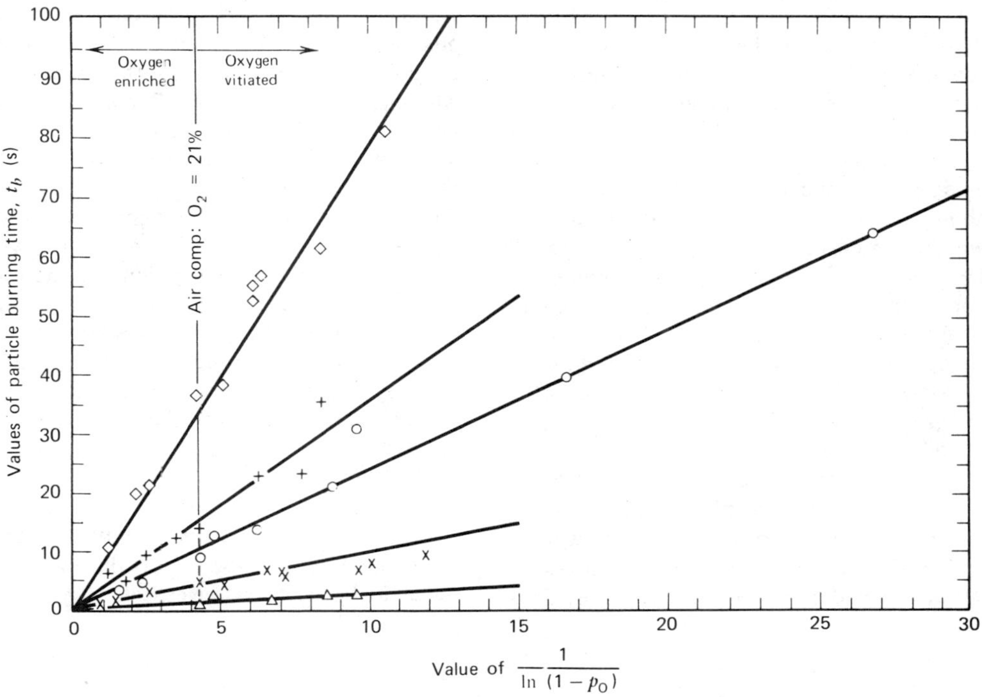

Figure 19.64 Linear plot of burning time, t_b, against the reciprocal of the oxygen partial pressure function $\ln(1 - p_O)$; d_0 values (μm): ◇, 1870; +, 1300; ○, 928; ×, 649; △, 388.[383]

small and K_D given by Equation 19.118. The best-fit lines on the plot were found to be in almost exact agreement with a priori calculation.

Alternative behavior for particles burned under conditions similar to those of Essenhigh et al [80,177] has been suggested by Brookes,[178] and by Anson et al.[203] Anson photographed captive coal particles and substantiated the earlier reports of Essenhigh and Yorke[177] and of Shibaoka[179] that there was considerable and variable swelling during pyrolysis. (Previous investigations, e.g., Newall and Sinatt,[170] had determined only the end results of such studies.) Anson used a lamp as a radiation source to heat the particles very rapidly, with sudden onset, and described three types of structure developed. There was an overall degree of swelling before start of combustion of only 10% compared with Essenhigh and Yorke's 50%. Anson also reported burnout at constant diameter, but without numerical documentation. This is possible for particles that blow up into thin spherical shells with hollow centers, but not for any particles of the size used that retain any degree of internal cokelike structure of reasonably uniform density if burnout is controlled by diffusion. Brookes[178] reported somewhat similar behavior, with a similar conclusion that reaction was at somewhat constant diameter, since "the particle retained its size and shape to a large extent." This conclusion was not supported by the reported data, however: the shells that were supposed to be reacting at almost constant diameter were found to have burning times that followed the diameter square law, with diameters being reduced to 50 or 60% of the initial (swollen) value. Moreover, the photographs used the light from the burning particles, which included the light from the circumambient flame, whereas Essenhigh and Yorke used backlighting to eliminate the interference from the flame, which otherwise gave spuriously large diameters. The behavior was analyzed theoretically by Brookes using a thick-shell, hollow-center model, but the hollow center contained only 5–15% of the particle volume, which meant that the real behavior was very similar to the behavior analyzed above for the regressing surface model with the surface regressing to zero diameter.

Swelling Effects. Figure 19.65 illustrates the typical swelling behavior found by Essenhigh and Yorke,[177] plotting d^2 against time for a single particle, in accordance with Equation 19.115 for K_C assumed small. It will be noted that the last measurements made correspond to diameter reductions of about 50%, or volume reductions of 85–90%, which also correspond to the values quoted by Brookes.[178] Combustion did continue beyond that point, but measurement on the film strip was difficult. Shibaoka[179] also measured changes in diameter with time and showed, in some records, the relative volume going to zero at burnout.

Shibaoka[179] also reported different degrees of swelling from different macerals (photographed on a hot-stage microscope), as might be expected. He also found greater swelling at higher heating rates, in contrast to Brookes, who reported the opposite. This variance could be due to the different methods of heating: Shibaoka essentially used conduction from a heated thermocouple, whereas Brookes employed convection-radiation, using Essenhigh and Thring's technique of electrically heated flat wire spirals. Shibaoka also reported greater swelling with finer particles, but—to prevent combustion before reaching operating temperature with the finer particles—the ambient gas was diluted with nitrogen. Since it is well known that the degree of swelling is influenced by the ambient gas composition[170] (greater swelling occurs in inert or reducing atmospheres), conclusions regarding greater swelling with smaller particle sizes do not necessarily follow.

Ash Effects. Aside from speculations that there might be catalytic effects that were too subtle to be detected, investigators[80,90,178,179,203]

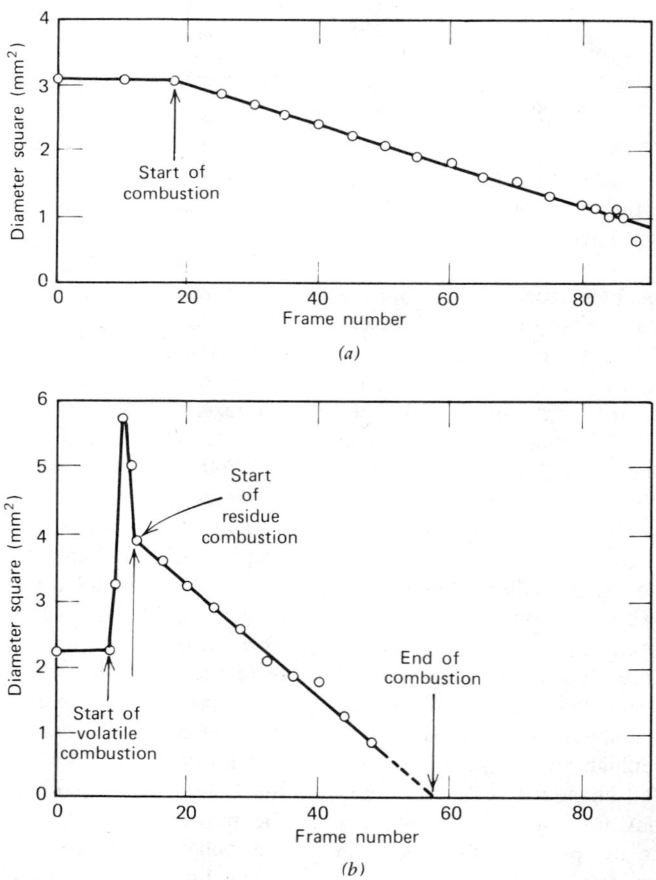

Figure 19.65 Combustion of single coal particles indicating difference between nonswelling and swelling coals. (*a*) Anthracite (Stanylld) and (*b*) bituminous (Cowpen) particles.[177]

agreed that ash had no discernible effect on the combustion rate. The ash diffusion control model with a shrinking, impermeable core,[464] which has been widely adopted in some sections of the literature (e.g., Levenspiel[260]), is not supported.

Reactions in Fluid Beds. Combustion and gasification of char particles in a fluid bed have also been investigated.[52,184,540,541] In the first effective analysis of such behavior,[184] an expression for the variation of diameter with time was obtained having the form

$$t = f(U, U_0, X)\left(1 - \frac{d^3}{d_0^3}\right) + K_D\, d_0^2\left(1 - \frac{d^2}{d_0^2}\right) \qquad (19.122)$$

where the function $f(U, U_0, X)$ is a function of the bed and bubble properties, such as the fluidizing and minimum fluidizing velocities, U and U_0, and other parameters. At burnout, the equation reduces to an equation of the form of Equation 19.116, with a first RHS term given by $f(U, U_0, X)$, which is a constant for given fluidizing conditions. The precise function is given in the original papers[184]; it is irrelevant in this context and also of little meaning without a derivation, which is out of place here. It will suffice to note that the function $f(U, U_0, X)$ is also proportional to the size of the particulate charge injected into the fluid bed, m_0. Experimental measurements of burning times were then found to agree very well with Equation 19.122 for $t = t_b$ at $d = 0$, as illustrated in Fig. 19.66. Figure 19.66*a* shows the variation of burning time with d_0^2, with an extrapolated intercept corresponding to the "constant" $f(U, U_0, X)$. Figure 19.66*b* illustrates the variation of burning time with the charge size, m_0. Evaluation of the numerical values of the slopes and intercepts of the figures also showed good agreement with prediction.[184]

There are, nevertheless, some arguable aspects of these data analyses. For the fluid bed analysis, Avedesian and Davidson[184] assumed a double-film model that is kinetically unacceptable at the temperatures involved; the temperatures were too low for significant CO_2 reaction, and the double film (see below) is known to require a much higher temperature.[493] It has also been challenged on other grounds.[542] Agreement on the value of K_D was good nonetheless, with a value determined from the experimental results for the Sherwood number, Sh, of 1.4. This reduced value of Sh below the usual limit of 2 was attributed to blockage by adjacent bed particles and was in fair agreement with independent estimates cited by Avedesian and Davidson. In Equation 19.118, which was used by Avedesian and Davidson in a first-order form (not the logarithmic form), the number of films, N, therefore should be 1, not 2. Numerical agreement between prediction and experiment is then possible only if either Sh is twice the estimated value (2.8), or CO reaction in the boundary layer is delayed, to burn in the bed generally, when N is again 2. Since the bed can release CO under conditions close to the limit of stable operation,[543] the latter explanation is the more likely.

Double-Film Combustion. A double film can be developed, it is believed, under appropriate conditions. The effect of a possible double film is illustrated in Fig. 19.67, which is for combustion of 15-mm spheres of electrode carbon. Chemical reaction control causes a rapid initial rate of rise of the reaction rate with temperature, with a flattening close to 1000°C, as first observed by Tu et al.[146] due to imposition of a single diffusion film (or to a combined diffusion film and adsorption with a low activation energy[267]). The data of Tu et al. define this initial behavior in very much greater detail. Above 1200°C, however, the reaction rate increases again (not reported by Tu), jumping to a second diffusion limit, attributed[493] to CO formation. The two thin lines on the plot

[540] E. K. Campbell and J. F. Davidson, "The Combustion of Coal in Fluidized Beds," *Proc. Fluidized Combustion Conf.*, Institute of Fuel, (1975) pp. A.2.1–A.2.9.

[541] J. T. Stewart and E. K. Diehl, "Fluidized-Bed Coal Gasification Process and Equipment Development," *Proc. 3rd Int. Conf. Fluidized Bed Combustion* (Hueston Woods, Ohio), Environmental Protection Technology Ser. EPA-650/2-73-053, December 1973, pp. III.1.1–III.1.18.

[542] P. Basu, J. Broughton, and D. F. Elliott, "Combustion of Single Coal Particles in Fluidized Beds," *Proc. Fluidized Combustion Conf.*, Institute of Fuel (1975), pp. A3.1–A3.10.

[543] D. J. Skinner, *Fluidized Combustion of Coal*, National Coal Board (London), (1970).

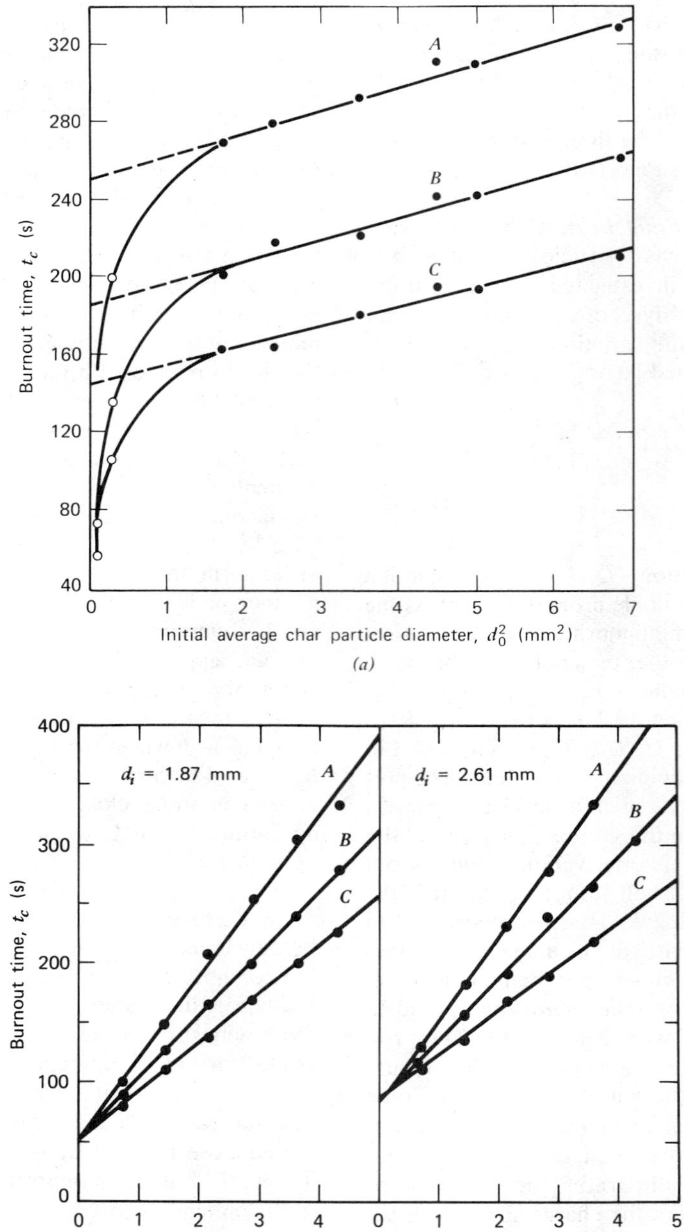

Figure 19.66 Particle burnout time in a fluid bed as a function of (*a*) initial average char particle diameter (m = 3.6 g of carbon), open points not included in least squares calculation), and (*b*) initial charge of char. See Equation 19.122. Values as follows: d_0 = 0.39 mm, U_0 = 46 mm/s, ρ_c = 720 kg of carbon/m³. Curves A, U = 214 mm/s; curves B, U = 300 mm/s, curves C, U = 383 mm/s.[184]

represent the calculated limits for diffusion, with single-film production of CO_2 (lower line) and double-film production of CO (upper line). Strictly, it is not possible to distinguish just from these results between a single film with only CO production and net CO production from a double film. Logic would suggest the latter explanation. If the rate is diffusion limited with CO_2 production at about 1000°C (for this particle size), the CO/O_2 reaction in the boundary layer is fast enough to consume all the CO. If CO escapes

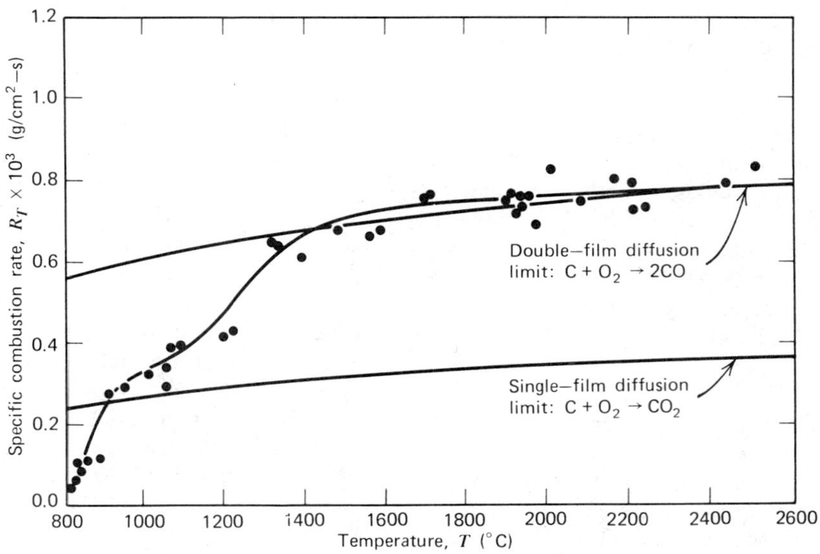

Figure 19.67 Specific combustion rate R_T versus temperature for reaction of carbon in air.[493]

from the boundary layer in the temperature range 1200–1400°C, thus allowing additional oxygen to reach the surface, the CO/O_2 reaction must be slowing up at the higher temperatures unless a double film is developing. The greater likelihood is that CO_2 close to the surface is participating, with a full double film established at 1500°C. The conclusion is not clear-cut, however, since further experiments by the same investigators[493] on the C/CO_2 reaction showed that to match the upper limit rate required 100% CO_2 (Fig. 19.68) when the CO_2 at the boundary interface in the presumed double-film condition would be 21%. There is an ambiguity here to be resolved. The CO_2 data also

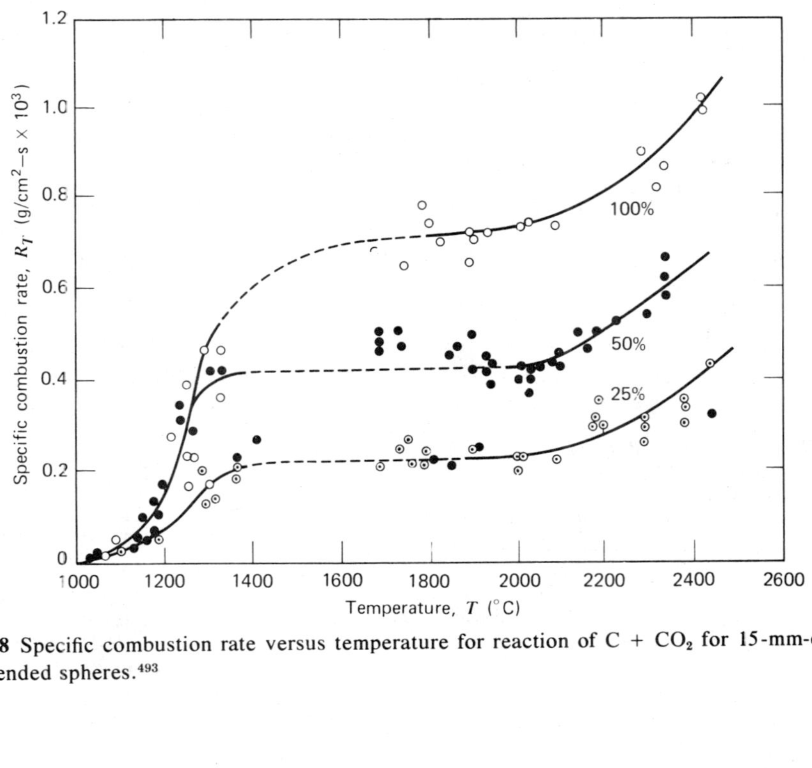

Figure 19.68 Specific combustion rate versus temperature for reaction of C + CO_2 for 15-mm-diameter single suspended spheres.[493]

show an unexpected rate increase above 2000°C that is also unexplained (but cf. Fig. 19.57).

Comparisons with Experiment: Small Particles. Information on the behavior of small (<100μm) particles is both more and less complete than that available for larger particles. Substantially more information exists on burning rates of the smaller particles, but there is substantially less on burning times. Figure 19.2 summarizes most of what is known.

Early Work. Early attempts to measure burning times were made by dropping small groups of particles into heated furnaces,[385,386] and some approximate measurements in terms of seconds or fractions of a second were obtained. Godbert's estimates of less than 5 ms[386] are on the short side; probably they represent combustion time of the volatiles only, and possibly a quenched time on account of the experimental method (batch-fed, entrained dilute flow through an electrically heated furnace). Times were mostly measured by streaks on a photographic plate, and the investigators were universally cautious in reporting the results as final burnout, since they were concerned with dependence on the sensitivity of the film (q.v., Brookes[178]).

Measurements in Flames. [73–76, 84, 86, 87, 98, 187–189, 191–196, 198–202, 334–336, 544–550] When small flames are used,[75, 76, 188, 196] it is now known that combustion is mainly of the volatiles, with char not being completely burned. In larger flames it is necessary to use such quantities of material (kg/hr) that there is always a substantial particle size distribution that may affect results, even when the burning time is somewhat insensitive to particle size (the cost of sieving out narrow cuts is usually prohibitive). There is also difficulty in determining the end point of reaction, which can be in error by 10 or 20% or more, and the results are ambiguous in any case unless nonbackmix or plug-flow flames are used. Data are therefore sparse, and the best available furnace data are summarized in Fig. 19.2 by the band between 100 ms and 1 s, for particles between 1 and 100 μm.

Engine Measurements. As summarized in Section 19.2.1, Soehngen[63] has described compression-ignition engine measurements 1–1.5 orders of magnitude shorter than the flame data (for mid-90% combustion); these are measurements under pressure, although this condition is not believed to influence the burning time. It could also be that the final burnout of the last few percent could significantly increase the time, since the final residue has been heat treated, in effect, and may be less reactive. Equally, this could be offset by a probable faster heating time which, according to Ashu's results (Section 19.3.2.5), would increase the reactivity. Figure 19.2 illustrates some data showing still faster burning times (I. G. Farben data¸`). These are also presumed to have been obtained in engines, although the original raw data source now seems to be irrecoverable¸`; my instinct is to view such data with considerable reserve, given the absence of the raw data source, and the small particle sizes cited (down to 20 or 30 Å). Fenimore and Jones,´´ however, have provided fully documented data yielding burning times (of 1–4 ms) of 100-Å particles at normal pressure, as indicated in Fig. 19.2. The I. G. Farben data are thus not particularly out of place.

[544] J. A. Browning, T. L. Tyler, and W. G. Krall, "Effect of Particle Size on Combustion of Uniform Suspensions," *Ind. Eng. Chem.*, **49,** 142–147 (1957).

[545] S. Yagi and D. Kunii, "Studies on Combustion and Gasification of Pulverized Coal in a Model Cyclone Generator," *6th Symp. (Int.) Combustion,* Paper No. 73, Reinhold Publishing Corp., New York, 1957, pp. 584–590.

[546] H. von Effenberger, "Untersuchungen der Kohlenstaubflamme an einem Modell," *BWK,* **12,** 351–355 (1960).

[547] R. H. Essenhigh, "Bemerkungen zu Forchungsarbeiten über Kohlenstaubflammen," *BWK,* **12,** 356–360 (1960).

[548] J. M. Beer, and R. H. Essenhigh, "Control of Reaction Rate in Dust Flames," *Nature,* **187,** 1106–1107 (1960).

[549] A. S. Ippolitov, "Skorost rasprostraneniya Fronta Plameni v Turbulentnom Pylevozdushnom Fakele," *Inzh. Fiz. Zh.,* **4N** 3–8 (October 10, 1961); (English transl.) *J. Eng. Phys.,* **10** (1961).

[550] V. Wróblewska, J. Zelkowksi, and S. Wójcicki, "The Optimum Conditions for the Combustion of Low-Calorific Coal Dust in Swirl Burners," *Proc. 16th Symp. (Int.) Combustion,* Combustion Institute, 1977, pp. 401–410.

Soot Combustion. Fenimore and Jones obtained their data[88] from combustion of soot, some of it burned in a secondary burner almost immediately after formation, and some of it collected and burned later after drying. Other soot combustion experiments are those by Lee et al.,[337] and by Tesner and Tsybulevskii.[339,340] The experiments of Fenimore and Jones are the best documented and probably the best designed. The other experiments used diffusion flames, and the local gas concentration can be disputed.

Lee's group reported a first-order reaction with respect to oxygen at constant temperature (obtained by a linear interpolation), but the lines obtained showed a substantial intercept on the oxygen axis[337] and, as Lee noted[388] and Magnussen[338] later rediscovered, the data better fitted a second-order reaction than a first. In the wide range of reaction orders reported in the literature, all but this value lie between 0 and 1.

Fenimore and Jones[88] eliminated such ambiguity by experimental design, creating the soot on one burner and burning it, premixed, on a second burner. Particle sizes, found to be remarkably uniform, were initially about 200 Å, and fell to about 100 Å (less than 15% of the original mass) by reaction in the distance of measurement. Soot mass flows were measured and, assuming spheres of uniform initial size with reaction only at the exterior, the relative change in diameter (d/d_0) is given by the cube root of the soot weight percent; the effective d/d_0 was found to decay linearly with time, in accordance with Equation 19.115 with K_D small, as illustrated in Figure 19.69. Figure 19.69 is important because it provides the only extant data for the case of nondiffusional reaction at a known external surface. The burning time is then given by $K_C d_0$, where K_C is the slope of the lines in Fig. 19.69 (appropriately corrected for units). Fenimore and Jones reported values of burning time, t_b, from the graph slopes (in fact as the reciprocal), although the reaction was not taken to burnout.

Interpreting their results kinetically, Fenimore and Jones[88] determined an activation energy for the reaction in the region of 22 kcal/mole, but also reported that the oxygen concentration was either appreciable or near zero. The reaction rates were fairly close, however, in spite of the difference in oxygen concentration. Other results in dry gas (at lower temperatures) were more oxygen dependent than those in wet gas, and significantly lower than temperature extrapolation of the wet gas data would predict. The authors concluded that reaction in the wet gas was significantly determined by OH and showed, by estimation, that there was a rough correlation between the reactivity parameter and the OH concentration. If the mechanism of Equation 19.9a is accepted, it is clear that OH can be involved when water is present. The difference in viewpoints is between OH obtained by dissociation in the ambient gases and by dissociation on the carbon surface. It is not clear why it was necessary to appeal to OH from the flame gases rather than using H_2O. The contribution of CO_2, which was also present in substantial quantities, was also ignored.

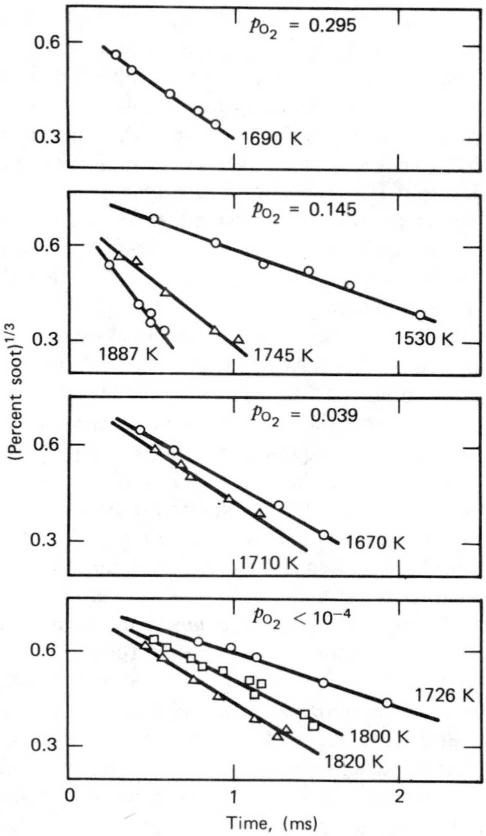

Figure 19.69 Cube root of the soot content versus time. Soot is expressed as a percentage of the carbon fed in the ethylene; time is inferred from the known gas velocity. Cube root of soot content is proportional to diameter. Test of Equation 19.115 with K_D small.[88]

Tesner and Tsybulevskii,[339,340] however, reported that with no oxygen the soot reaction in their experiments was proportional to the CO_2 concentration, and apparently independent of H_2O. They reasoned that this H_2O independence could be due to inhibition by CO and H_2. Adsorption of either CO or H_2, however (with CO adsorption of small account according to the discussion of Section 19.4), should affect both the C + CO_2 and C + H_2O reactions equally if there is no site specificity for adsorption. The back reactions of Equations 19.8 and 19.9 can occur, at rates in proportion to the CO and H_2 concentrations and the relevant kinetic constants. There is also coupling of the gas concentrations through the water-gas shift reaction, which is evidently accelerated over a carbon surface and could well be in equilibrium. This could then control the net importance of one gas over another in the overall reaction, depending on the gas concentrations and temperatures. Noting that Tesner's CO_2 and H_2O concentrations were somewhat lower than Fenimore's, with higher temperatures (1800–2200 K vs. 1500–1900 K), it appears that the discrepancy must remain unresolved at present, but the differences in concentrations and temperatures are strongly suggestive of overall control by the shift reaction. There can also be a difference in the nature of the soot. Fenimore provided good evidence that there was no internal reaction in his experiments. Tesner, however, indicated some internal surface and internal reaction, at least up to 2000 K. The activation energies reported are also indicative of different reactions, with 22 kcal/mole found by Fenimore, and 75 kcal/mole found by Tesner, which is certainly in the range of values to be expected for C/CO_2 reactions.

Tesner also observed the negative temperature coefficient behavior, discussed in Section 19.6.2.6, as illustrated in Fig. 19.70.

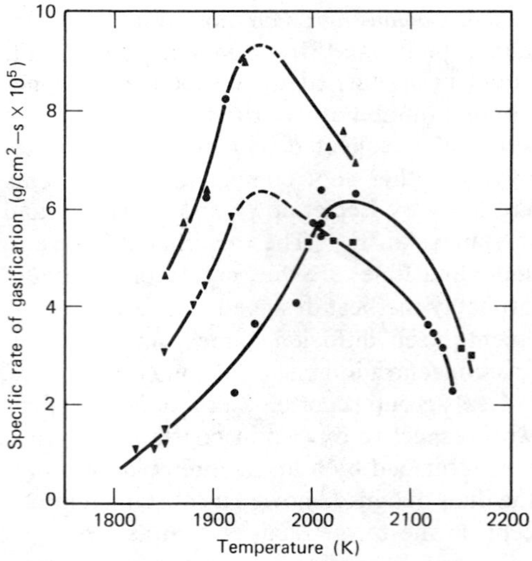

Figure 19.70 Relation between specific rate of gasification and temperature for soot particles obtained by flame cracking of fuel. Values of P_{CO_2} as follows: curve 1, 0.09 atm; curve 2, 0.21 atm; curve 3, 0.33 atm.[339]

Comparative Magnitudes. The data of Fenimore, Tesner, and others can be used to provide some comparative values of reactivity. Data from Fenimore and Tesner can be compared at the point of temperature overlap, 1850 K. At this temperature Fenimore reported rates of reaction of 10^{-3} g/(cm²)(s). Tesner reported (3 ± 1) × 10^{-5} g/(cm²)(s). They differ by more than 1.5 orders of magnitude, suggesting that there is probably considerable difference in degree of graphitization if Fenimore's reactions in combustion products can be compared with Tesner's in CO_2 and H_2O mixtures. Surface areas are comparable, being within a factor of 2 or better. Activation energies are, respectively, 22 and 75 kcal/mole. [Tesner and Tsybulevskii[339,340] also calculated E using kinetic theory following a previous paper and obtained completely different values and trends. They assumed, however, that the accommodation coefficient (Equation 19.48) was unity and the validity of the assumption is not evident.] The preexponential factors, combined with any gas concentration function involved, are then about 2, and 2 × 10^4, respectively. (i.e., correcting for the exp $-E/RT$ term).

The figures above provide interesting comparisons with large-particle behavior. From Equation 19.115 it can be shown that the reaction rate in grams per square centimeter-second is given by $\sigma_C^0/2K_C$, if there is no significant diffusion term; and by $\sigma_C^0/4K_D d$ if diffusion is controlling. The chemical term is independent of diameter and, with σ_C^0 between 1 and 2, $1/K_C$ is roughly the rate of reaction. Extrapolating Fenimore's data to 1000°C (1273 K) gives rates of about 10^{-4} g/(cm²)(s), which yields values of K_C of about 10^4. Values of K_D, however, are only in the region of 10^3, which means that the chemical term in Equation 19.115 should be evident at diameters of 1 mm to 1 cm (roughly the experimental range of large-particle measurements).[78–80,90]

This calculation is not borne out by experiment, however. If K_C was important in any large-particle burning time experiments, it could show up as an intercept by plotting Equation 19.116 in the form (t_b/d_0) against d_0. Figure 19.71 illustrates the typical plot obtained, with the line evidently passing through the origin. A "zero" intercept requires a value for K_C that is less than 1 (possibly much less), so that the possible specific reaction rate must be greater than or much greater than 1 g/cm^2-s; this is to make the chemical rate very large compared with the diffusion rate, which then varies roughly from 10^{-3} to 10^{-1} g/cm^2-s over the range 1 cm to 100 μm. It may be noted that the reaction rates in Figs. 19.67 and 19.68, for 15-mm spheres, are in the range 10^{-4} to 10^{-3} g/cm^2-s, which values bracket the correct range for diffusion at 1000°C. If the "intrinsic" reactivities inside the spheres are of the much lower values found by Fenimore and Tesner, then the difference between the intrinsic and nominal burning rates can be ascribed to a difference between the nominal (external) and actual (internal) surface area, with only about one order of magnitude required between the difference in the two areas for the reaction rates to match (cf. Section 19.6.2.5 conclusions comparing Figs. 19.50 and 19.51).

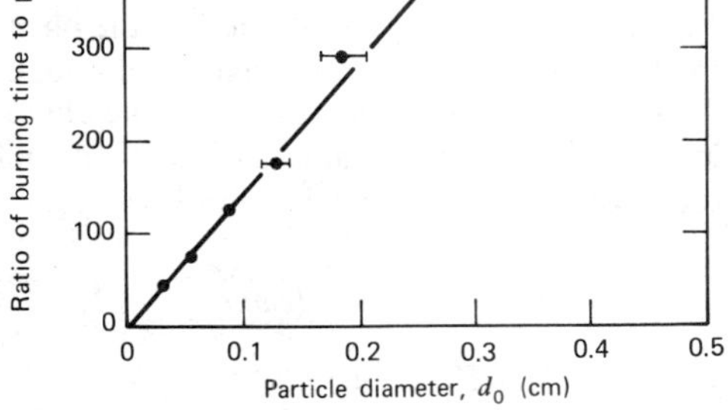

Figure 19.71 Variation of ratio of burning time to particle diameter with diameter, evaluating Equation 19.116 for Garw coal (Cwm Tillery), 28% VM.[536]

The discrepancy between the real values of K_C of 10^4, determined from Fenimore and Jones data, and the estimated values of less than 1 required for K_C to be small in Fig. 19.71 is startling. The only viable explanation at present would seem to be that the term $K_C\, d_0$ in Equation 19.116 cannot exist. It derives from Equation 19.90, which is the limit approximation for fast desorption with first-order adsorption following boundary layer diffusion. Concerning the two other limit approximations: if there is boundary layer diffusion, the approximation of Equation 19.89 cannot apply; if adsorption is fast, however, the approximate Equation 19.91 would alternatively apply. According to Equation 19.91, reaction is *either* diffusion controlled, in which case Equation 19.116 would apply with K_C zero; *or* desorption is controlling, in which case K_D is zero. The burning time is then given either by $K_C\, d_0$, with K_C given by Equation 19.120; or by $K_C{}'$ (Equation 19.117), with $K_C{}'$ given by Equation 19.120a. In either case, there should then be an abrupt change of slope in the double-log (t_b vs. d_0) graph, as there appears indeed to be in Fig. 19.2, with a change in slope in the first case from 2 to 1 as d_0 declines, and, in the second, a change from 2 to 0. Best evidence is that the second case is the more likely, at least for bituminous coals and similarly reactive solids, with the first case possible for anthracite.

The above argument may break down, however, at high oxygen concentrations. Chamberlain et al.[551] carried out experiments with coal in oxygen at high temperature (measured by line reversal at about 3000 K), burning at rates of about 6 g/s (50 lb/hr) after grading into seven size cuts from about 10 to 70 μm. They reported burning times (for 80% mass loss) as a function of diameter. Interpretation of the results is a little ambiguous. The results as reported graphically[89] show burning time increasing linearly with diameter with some scatter, and with a finite intercept on the t_b axis at zero diameter. The existence of this intercept can be interpreted

[551] C. T. Chamberlain, W. A. Gray, and J. L. Smith, "A Burner for the High-Intensity Combustion of Coal in Oxygen," *J. Inst. Fuel,* **39,** 120–125 (1966).

as the combination of internal reaction with surface regression, describable in principle by the solution to Equation 19.114. The intercept corresponds to integration of the $\dot{\sigma}$ term, and the slope corresponds to integration of the $\dot{a}_s$ term. The slope and intercept should therefore relate approximately to the terms K_C and K_C', given by Equations 19.119 and 19.119a. Since the reciprocals yield approximately the rates of burning, they are of interest for comparison with the values given above. They are approximately 0.5 and 200 g/cm²-s, respectively, which are much higher values than those obtained by Fenimore and Tesner. Such values are possible if reaction is internal, in which case there is an internal surface area component that can account for the difference (noting also the significantly higher temperature and reactant gas concentration).

The authors pointed out, however, that a fair correlation was also obtained by plotting the burning time against d_0^2. The implication, of course, is that the boundary layer has some influence in spite of the small particle size, but it is nonetheless possible on account of the high temperature and oxygen concentrations. It also implies a non-zero-order reaction, and the relevant equation would either be Equation 19.116 or Equation 19.119a. Testing Equation 19.116, a plot of (t_b/d_0) against d_0 generated a pronounced curve. A plot of t_b against d_0^3, however (for Equation 19.119a), generated the lines illustrated in Fig. 19.72. The two lines represent two different oxygen/fuel ratios (2.11 and 1.51). The intercept corresponds, as before, to K_C/S_v (and is dependent on the oxygen/fuel ratio), but the slope is now 2/3 K_D and is *insensitive to or independent of the oxygen/fuel ratio.* The value of the slope is about 500, which is close to the calculated value of about 300, from Equation 19.118 for combustion at 3000 K in pure oxygen. It is concluded from this agreement that boundary layer diffusion is operative at this temperature, but is neither controlling nor dominant, since the reaction data are also consistent with there being partial to full penetration of the particle. The actual values of the burning time are of interest, ranging from 10 to 40 ms.

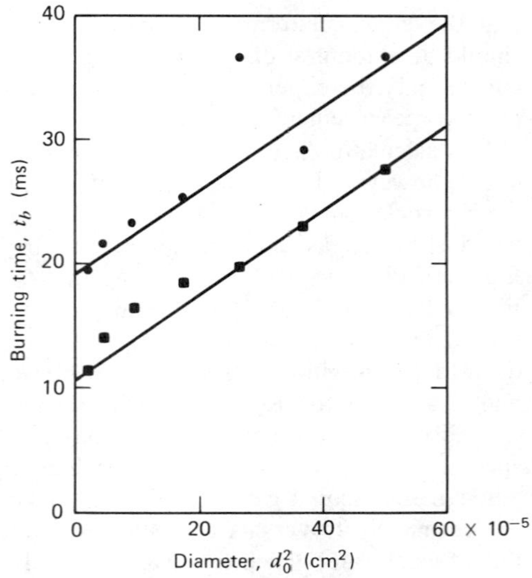

Figure 19.72 Variation of burning time, with diameter squared for coal particles (38% VM) in oxygen at two oxygen/fuel ratios. O_2/fuel = 1.51 (●) and O_2/fuel = 2.11 (■).[89]

Oxygen Depletion in Flames. Bryant[194,195] obtained burning times for graphite in a propane-oxygen-nitrogen flame, with brightness temperatures ranging from 1600 to 1900 K. Oxygen percentages ranged from 50 to 100%. The particles were not too well sized, with some particles up to 20 μm but most in the range 2–5 μm. The burning times in pure oxygen were 4–8 ms, indicating roughly comparable values to those of Chamberlain and Gray[89] (they are about half their intercept values). The principal objective in the experiments was to examine the effect of oxygen depletion along a flame. This topic is discussed briefly in the first Supplementary Volume.[32] The burning time is obviously extended if reaction is non-zero-order on account of the oxygen depletion. For boundary layer diffusion control, Nusselt[531] was the first to show that the burning time could be written $F_D K_D d_0^2$, where F_D is multiplier for the burning time that is a function of excess air alone. This was later extended to include the chemical term, and the combined chemical and diffusional terms, to yield the following[10]:

$$t_b = F_C K_c d_0 + F_D K_D d_0^2 \qquad (19.123)$$

where F_C and F_D are given by Equations 33 and 34 of Thring and Essenhigh.[32] The multipliers F_C and F_D would also attach to K_C and K_D, respectively, in Equation 19.119a. Bryant[194] adopted Equation 19.123 for test against his experiments, but he assumed that the squared term was small and wrote $K_C d_0$ in a modified form as (K_C''/p_0) (see Equation 19.119), where the diameter is incorporated in K_C''. It then follows that $K_C'' = t_b p_O/F$. Values of K_C'' were then calculated for the 18 experiments reported and found to be essentially constant, with a value of 2.6 ±0.2 ms-atm. It was concluded from this that reaction was indeed first order (although the range in p_O was so relatively small that there might not have been much difference if the correct form had been $p_O^{1/2}$). Bryant does not give a temperature coefficient for the constant K_C'', only remarking, on the basis of a citation from Essenhigh,[10] that the temperature dependence is expected to be quite small. The data reviewed in this chapter indicate a value of less than 10 kcal/mole for adsorption reaction control. For the temperature range involved in Bryant's experiments, the reaction rate would change by ±20% for $E = 10$ kcal/mole, but 250% for $E = 40$ kcal/mole. For 5-μm particles, K_C is about 10, so that the reaction rate is about 0.1 g/cm²-s, which is of the order of the values obtained by Chamberlain and Gray. This is not unexpected in view of the similar burning times, but it could be inferred that Bryant's particles were also burning internally, and that Chamberlain and Gray's particles were also burning with a very small temperature coefficient.

Bryant evaluated the multiplier F_C and plotted $t_b p_O$ against the oxygen/fuel equivalence ratio (R) for comparison with prediction using the F_C multiplier. This comparison between prediction and experiment is illustrated in Fig. 19.73, which is the only known test of the equation describing the effect of oxygen consumption in a flame.

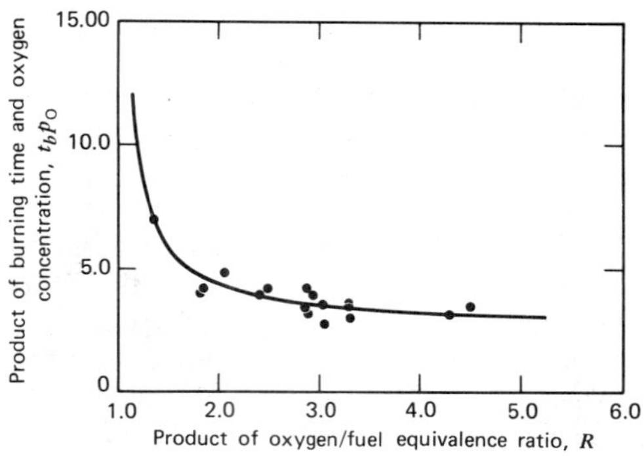

Figure 19.73 The dependence of product of particle burning time and initial oxygen concentration on oxygen/fuel equivalence ratio: solid points indicate experimental determination; curve, 2.60 E_c (calculated); fuel, graphite particles, 2–5 μm.[194]

19.7 CONCLUDING REMARKS

Any review of an actively developing subject must be essentially open-ended. There are always more views to be expressed, and more topics that could be included. Topics of potentially direct relevance to coal combustion fundamentals are the fate of the fuel nitrogen and the associated NO_x reduction, and the fate of the sulfur, or the sulfur/carbon interaction. Some work has been done on these topics,[552–559] but they are more appropriately discussed in the context of the phenomenological behavior of NO_x and SO_x and their control, about which there is considerably more information (Chapters 22 and 31). The same is true of the problem of the

[552] D. G. Madley and R. F. Strickland-Constable, "The Kinetics of the Oxidation of Charcoal with Nitrous Oxide," *Trans. Faraday Soc.*, **49**, 1312–1324 (1953).

[553] J. O. L. Wendt and O. E. Schulze, "Effect of Diffusion-Reaction Interactions on Fuel Nitrogen Conversion During Coal Char Combustion," preprint of paper presented at Eastern Sect. Meeting, Combustion Institute, Fall 1974.

[554] R. A. Vogt and N. M. Laurendeau, "NO_x Formation from Coal Nitrogen: A Review and Model," paper presented at Central States Sect. Combustion Institute Meeting, Spring 1976.

[555] W. O. Stacey, F. J. Vastola, and P. L. Walker, Jr., "Interaction of Sulphur Dioxide with Active Carbon," *Carbon (Oxford)*, **6**, 917–923 (1968).

[556] R. K. Sinha and P. L. Walker, Jr., "Removal of Sulphur from Coal by Air Oxidation at 350–450°C," *Fuel*, **51**, 125–129 (1972).

[557] R. K. Sinha and P. L. Walker, Jr., "Desulphurization of Coals and Chars by Treatment in Various Atmospheres Between 400 and 600°C," *Fuel*, **51**, 329–331 (1972).

[558] A. L. Yergey, F. W. Lampe, M. L. Vestal, A. G. Day, G. T. Fergusson, W. H. Johnston, J. S. Snyderman, R. H. Essenhigh, and J. E. Hudson, "Nonisothermal Kinetics Studies of the Hydrodesulfurization of Coal," *Ind. Eng. Chem. Proces. Des. Dev.*, **13**, 233–240 (1974).

[559] R. T. Yang, C. R. Krishna, and M. Steinberg, "Fluidized-Bed Coal Combustion with Lime Additives; The Phenomenon of Peaking of Sulfur Retention at a Certain Temperature," *Ind. Eng. Chem. Fundam.*, **16**, 465–467 (1977).

inorganics[560] and of such additional matters as decomposition of gases,[561] or disproportionation[562] over carbon surfaces, or electrical conductivity of the products of combustion in relation to use of coal in MHD.[563]

The content of this chapter may also be reviewed in the context of the comments in the Introduction, which describe the four elements of interest in the fundamental reactions of coal: pure carbon studies, pyrolysis kinetics, volatiles kinetics, and whole coal behavior in reaction. The Introduction also summarized the standing of current work: thus the first topic, as is clearly evident, has advanced significantly; the second, substantially (see Chapter 12); the fourth, usefully; but the third—the study of the coal volatiles combustion kinetics—not at all. That this should be so is almost unbelievable.

The balance between topics from the coal point of view is evidently somewhat uneven, reflecting as always, the profile of interest of the past. Much of the critical, ultra pure carbon work would never have been carried out but for the nuclear interests. For the future, with the applied interest turning more to coal, there will clearly be more developments in the study of total coal behavior. Some of the more fundamental pure carbon aspects should not, however, be dismissed on that account.

Over the whole range of topics, some of the questions of interest or significance that have emerged include the following.

1. An acceptable explanation of the negative temperature coefficient and/or the extremum behavior of the curve of reaction rate with temperature in many experiments is still required. If this is a matter of carbon crystallization, it could also be of major significance to the burnout of flame-formed char.
2. There is evidence that the oxide film [written as C (O)] formed from the CO_2 and steam-oxygen transfer reaction is the same, so that the decomposition behavior and kinetic constants are also the same. It should be a good assumption that it is also the same when formed by the oxygen reaction; but the evidence is not good and tends in some instances to indicate the opposite. This contradiction needs to be resolved.
3. Understanding of pyrolysis in the context of phenomenological kinetics is improving rapidly. What is still lacking, however, is an adequate working model of coal constitution that will enable us to translate results from one coal to another. Without such a model there can still be nasty surprises. This has bearing as well on the whole coal behavior in reaction.
4. The lack of knowledge of the volatiles combustion kinetics in any form needs no further comment.
5. There are several questions relating to whole coal behavior, some of which, as remarked, will depend on an understanding of coal constitution. These relate in the first place to a range of physical properties such as initial porosity, density, and surface area, and about which there is already considerable knowledge as a function of coal rank. What is still largely absent is knowledge of the subsequent behavior of these parameters on heating and/or reaction. This is partly a matter of the need for very extensive data gathering; but again it points up the need for an adequate model of coal pyrolysis behavior that can be used to codify and generalize available experimental data. Without that, individual measurement and testing of nearly every coal will be necessary.
6. A more focused problem is represented by the somewhat erratic knowledge concerning combustion of coal and char particles. As discussed in Section 19.6, it would appear that the particles may burn only at the surface, with diameters

[560] H. R. Brown and D. J. Swaine, "Inorganic Constituents of Australian Coals. Part 1. Nature and Mode of Occurrence," *J. Inst. Fuel*, **37**, 422–440 (1964).

[561] A. A. Orning and E. Sterling, "Oxygen Transfer Between Carbon Dioxide and Carbon Monoxide in the Presence of Carbon," *J. Phys. Chem.*, **58**, 1044–1047 (1954).

[562] S. A. Pursley, R. A. Matula, and O. W. Witzell, "The Kinetics of Carbon Dioxide and Carbon Formation from Carbon Monoxide," *J. Phys. Chem.*, **70**, 3768–3770 (1966).

[563] W. A. Gray and J. I. Smith, "Conductance of Coal Combustion Products," *Nature*, **212**, 393–394 (1966).

shrinking at constant density, or the opposite, with density dropping at constant diameter. It seems possible at present that the same coal or char may behave in the two extreme ways under experimental conditions that are not—apparently—significantly different. Clearly, there must be some significant differences, and these must be identified. The suspicion is that the behavior may be of the "catastrophe" variety,[564] with the possibility of flip-flop behavior when there are changes in values of apparently insignificant boundary conditions.

7. Another important question involves the influence of the ash. Its potential catalytic behavior may possibly require its detailed analysis, and knowledge of its transformation during reaction, if detailed models are to be of any value. This is a wide open problem.
8. At the theoretical level there is a question regarding the level of detail that may be needed in equations used to evaluate data, particularly when coals and chars are being investigated. Increasingly, experimentalists are using the "*n*th-order" semiempirical approximation equation (Equation 19.74a), coupled with "rate of weight loss is proportional to weight." Sophistications like TSA, ASA, UASA, and the like, are overlooked completely, and it is not a matter of ignorance. At the same time, modelers are using increasingly complex models and testing them against empirically evaluated data. Are the complex models necessary?
9. There is the question of handling all these results from the practical and operational point of view. The currently favored method is detailed computer modeling. This is a valid method of collating and generalizing the results, and it is essential in one form or another. It is obviously critical in effective modeling for the modeler to work closely with experimentalists who are developing and evaluating the raw data and testing the fundamental equations developed in the basic analyses. It also seems to me that more attention should be given in modeling to the question of uniqueness: that is, demonstrating necessity as well as sufficiency (the mathematician's "if and only if"). This approach is difficult, given the nature of models, which are "iceberg" constructs, with so much of their critical content in the computer that is inaccessible and unevaluatable.

   Without doubt, the use of modeling will expand. The target should remain the production of models from "first principles." At the same time, however, continued effort on development and testing of the unique basic principles is of equal importance.
10. The influence of coal rank remains the dominant factor, nevertheless. At present, the effect of rank is treated essentially empirically, with reliance on individual testing of nearly all coals. Broad trends exist as general guides, but they still lack precision. The views summarized in Section 19.3.2 provide a possible basis for a systematic treatment to resolve the problem. Interestingly, Kessler's[565] recent maceral analyses suggest more regularity and correspondence to whole coal analyses than has been assumed in the past; this could prove to be a valuable and significant simplification. To break away from the present empiricism however, it will be necessary to establish an adequate model of coal constitution that can provide acceptable predictive correlations. This may well be the most important problem to be solved in combustion and gasification research.

Acknowledgments

I have great pleasure in acknowledging the following valuable contributions to this chapter. Dr. Ian W. Smith, of CSIRO, Australia, very kindly provided the complete text of a paper "The Intrinsic Reactivity of Carbons to Oxygen," previously given only in brief at an Institute of Fuel (Australian Membership) Conference, from which Fig. 19.51 was taken, along with permission to use that figure,

[564] E. C. Zeeman, *Catastrophe Theory,* Addison-Wesley Publishing Company, Reading, Mass., 1977.

[565] M. F. Kessler, "Interpretation of the Chemical Composition of Bituminous Coal Macerals," *Fuel,* **52,** 191–197 (1973).

now published in *Fuel* (see Ref. 471). Dr. Phillip L. Walker, Jr., of the Pennsylvania State University, granted permission to use material from a paper given at a PSU course on Advances in Combustion Technology in July 1977; this includes Figs. 19.23–19.28 and related text in Section 19.3.2. Dr. Harry Marsh, of the University of Newcastle, England, reviewed Sections 19.2, 19.3, and 19.4 in an early draft and made valuable and much-appreciated comments and suggestions for some critical revisions.

CHAPTER 20

Combustion Process Technology

F. J. CEELY AND ERNEST L. DAMAN

F. J. Ceely *has spent his entire professional career of 38 years with Foster Wheeler Energy Corporation in the study of fuel properties and in the design and testing of fuel-burning apparatus. His particular interests have been in the development of equipment for burning low-volatile coals and cokes in suspension and in the laboratory tests of such fuels to determine their ignition characteristics. Mr. Ceely has been a long-term member of National Fire Prevention Association committees on Pulverized Fuel Systems and Prevention of Boiler-Furnace Explosions.*

Ernest L. Daman *is Chairman of the Board, Foster Wheeler Development Corporation, and Vice President–Research, Foster Wheeler Corporation. Mr. Daman's principal technological interests have been in heat transfer, fluid flow, and combustion, with particular emphasis recently on the development of fluidized-bed combustion systems.*

20.1 FUNDAMENTALS

The practical uses of coal in various types of combustion equipment ranging from domestic to public utility service involve fundamental problems that apply to the whole range of applications. The research and development efforts addressed to these problems in combustion technology normally fall into one of two categories: fuel-oriented studies and equipment-oriented studies. Much of the work reported in both categories has been characterized by ready applicability to important practical questions of design or operation. It is just this type of development that is emphasized here.

The overall objectives of combustion equipment designers have been to achieve continuous economical operation, free of slagging or fouling, while minimizing the discharge of sulfur, nitrogen oxides, and particulates to the surroundings. Equally challenging, as subsequent sections of this chapter illustrate, has been the need to design for a much wider range of coals and coal-derived fuels, for equipment both smaller and larger than previously used, and for more exacting operating conditions.

In direct response to these needs, extensive work has been performed in measuring fuel properties and judging their significance as indices or guides to combustion, slagging, and other aspects of performance. Some of these efforts have been motivated by the inadequacies of conventional coal and coal ash analyses in predicting fuel behavior in full-scale equipment; others, by the lack of accurate data or the failure to correlate available data with actual performance. Among the more important measurements and techniques studied in the period under consideration are tests for fuel reactivity or combustibility,[1–4] examination of coal structure and extent of graphitization by electron scanning and X rays,[5,6] proliferation of thermal analysis procedures typified by the "burning profile,"[7,8] and combustion of small samples in special laboratory furnaces.[9,10]

Two major system developments that promise to utilize coal in novel ways are well

[1] G. G. Thurlow, "Combustion and Energy Conversion," *BCURA Month. Bull.*, Review No. 238, **28**(9) (1964).

[2] C. S. Finney and T. S. Spicer, ASME Paper No. 62-WA-195, 1962.

[3] F. J. Ceely and R. I. Wheater, *Trans. ASME,* **79,** 1177 (1957).

[4] R. Chauvin, *Rev. Gen. Therm.,* No. 26, 159 (February 1964).

[5] W. Beckering, H. L. Haight, and W. W. Fowkes, U.S. Bur. Mines Inf. Circ. No. 8471, 1970.

[6] R. W. Borio, ASME Paper No. 74-WA/Fu-2, 1974.

[7] C. L. Wagoner and A. F. Duzy, ASME Paper No. 67-WA/Fu-4, 1967.

[8] C. L. Wagoner and E. C. Winegartner, *Trans. ASME,* **95** *J. Eng. Power* (Ser. A, No. 2), 119 (1973).

[9] J. M. Beer, M. W. Thring, and R. H. Essenhigh, *Combust. Flame,* **3,** 557 (1959).

[10] J. B. Howard and R. H. Essenhigh, *Proc. 11th Symp. Int. Combustion,* Combustion Institute, Pittsburgh, 1967, p. 399.

under way: fluidized-bed combustion and magnetohydrodynamics (MHD). Fluidized-bed combustion is being studied intensively for the entire size range of coal-burning equipment. In this system of burning approximately ¼ × 0 in. crushed coal is fed to a bed consisting of inert particles of sand or ash, or a chemically reactive material such as limestone or dolomite, with particle sizes close to those of the incoming coal feed. In most cases the heat transfer surface is submerged in the bed and the system is operated with a bed temperature of approximately 1600°F. In beds containing limestone or dolomite the sulfur dioxide released during combustion reacts to form calcium sulfate. This material can be discarded, or it can be regenerated to calcium oxide and reused. The low bed temperature inhibits the formation of nitrogen oxides and also prevents the coal ash from reaching fusion temperature, thereby avoiding serious slagging, ash deposition, and ash corrosion. Because of the last phenomenon it should be possible to use higher thermodynamic cycle fluid temperatures than are practical in pulverized-coal firing, thus improving cycle heat rates.

MHD is a longer range possibility, intended to use coal for large-scale power generation at higher efficiencies than those achieved thus far by other means. In this system coal is burned at pressures up to 10 atm with very high heat release and with the addition of alkali metal oxide to obtain the electrical conductivity of the combustion gas or plasma needed to generate electrical energy in the subsequent stages. Chemical reactions of the gases with the alkali seed materials are used in the MHD system to control the emission of oxides of nitrogen and sulfur.

The engineering community can also anticipate, in the near future, the large-scale production of new fuels from solvent-refining and coal gasification plants. Several such substances have been made available recently in test quantities for initial evaluation of their handling, pulverizing, and burning characteristics.[11–13]

[11] R. W. Borio, *Solvent Refined Coal Evaluation: Pulverization, Storage, and Combustion,* final report to EPRI, Report No. KDL-75-F-15, April 1976.

[12] S. M. Barrick and S. J. Vecci, *Investigating Storage, Handling, and Combustion Characteristics of Solvent-Refined Coal* (Tasks A and C), final report to EPRI, Rep. No. 5185, March 1976.

In addition to understanding the processes associated with the combustion of individual particles of fuel, a better knowledge is required of the gross phenomena that occur in equipment burning pulverized coal in suspension or coarse coal in beds. An example of the type of information needed is a thorough comprehension of the transfer of heat by radiation and convection in pulverized-coal furnaces. This subject has been studied recently by several investigators.[14,15]

These and other developments related to technological applications of coal combustion are further discussed in the following sections of this chapter and elsewhere in this volume.

20.1.1 Fixed-Bed and Suspension Firing

Fixed-bed firing, as exemplified by mechanical stokers, and suspension firing using pulverized coal are the major methods of burning coal. As with all well-established processes, the level of fundamental research and development work is minimal. Perhaps the most active fundamental work in support of these traditional coal-firing methods has been in studies of coal properties by laboratory procedures. This work involves the continuing effort to devise laboratory tests that can predict fuel behavior during combustion and can serve as guides in equipment design and fuel selection. The ultimate purpose of these efforts is to avoid such problems as poor ignition or burnout of the fuel, troublesome ash accumulations, and corrosion of pressure parts or their structural supports by ash constituents.

Of the two principal methods of combustion in use, fixed-bed firing has received the least attention in the recent past. Some of the reasons for this neglect are discussed in Section 20.2.1, "Stoker Firing." For those approaching the subject of fixed-bed firing

[13] W. Downs, F. L. Jones, and W. L. Sage, *Investigating Storage, Handling, and Combustion Characteristics of Solvent-Refined Coal* (Task C), final report to EPRI, Rep. No. 5186, March 1976.

[14] I. McC. Stewart and T. F. Wall, "Temperature and Heat Transfer in Pulverized Coal Furnaces," paper presented at 1st Australasian Conf. Heat and Mass Transfer, Monash University, Melbourne, May 23–25, 1973.

[15] A. Lowe, T. F. Wall, and I. McC. Stewart, *Proc. 15th Symp. (Int.) Combustion,* Combustion Institute, Pittsburgh, 1974, pp. 1261–1270.

for the first time, an extensive discussion of its fundamental nature is given by Gumz.[16] In addition, Section 20.2.1.1, "Classification of Equipment," discusses stoker types according to bed geometry and gives some characteristics of commercial equipment (Table 20.2). Areas needing prompt attention to support the presently anticipated expansion of stoker application are the proper uses of overfire air and fly carbon reinjection and the application of automatic controls to these functions. It is hoped that work recently undertaken by a committee of major American manufacturers will provide answers to these and other questions, at least for the spreader stoker (the most frequently used stoker type).[17]

A promising effort toward relating the performance of various coals on a stoker to one or more of their readily determined properties was reported in 1959 by Zawadzki and du Breuil.[18] Their apparatus (the BCR 7-lb furnace) had a small crossfeed stoker (with about 1 ft^2 of grate area) and was essentially a simplification of that used by prior investigators. Zawadzki and du Breuil managed to isolate some of the properties identifying coals suitable for crossfeed burning, but not accurately enough to be decisive in borderline cases, as noted by Sherman and Landry.[19] The same apparatus was used later, however, by Benedict and Berry in a broader study reported in 1966.[20] Of particular interest to us is their attempt to relate petrographic characteristics to combustion performance in the test furnace. The early result of this attempt is shown in Fig. 20.1, which relates grate burning rate (for the particular, small crossfeed stoker considered) to free-swelling index, ash fusion temperature, coal size, and mean reflectance of the coal vitrinite. In general, as might be expected, coals requiring more grate area for good bed conditions and burnout were those of higher swelling index and reflectance and of lower ash fusion temperature. Benedict and Berry recognized the tentative nature of their results and anticipated continuing the study but never did.

The porosity of coal and char structures, their surface areas, and their ability to adsorb gaseous compounds are likely to be related to their reactivity to oxygen in either fixed-bed or suspension firing. This is particularly true under ignition conditions and with low-volatile coals or other comparatively inert fuels. Hence further effort should be made to extend work on the micro- and macroporosity of coals and to relate these characteristics to the combustibility (or at least the ignitibility) of the same fuels.[21–24]

In the various studies of coal combustibility conducted over the past two decades, low-volatile fuel flames have received most of the attention, because their combustion is more susceptible to the effects of the input variables and less subject to the overlapping of a volatile flame and the heterogeneous burning at the solid surface.[25] In addition to emphasizing this use of low-volatile fuels to minimize the effects of flame complexity, Beér and co-workers have produced an apparatus and a technique that greatly reduce the complexity of the flow patterns existing in most other test furnaces, namely, the "one-dimensional flame with controlled mixing."[9] Closely related procedures and equipment have been used by Csaba[26] and by Essenhigh and his co-workers at Pennsylvania State University (the "plane flame furnace").[10] It is largely because of the use of this technique that we now have not only a better understanding of particle ignition in a pulverized-fuel cloud, but also extremely helpful comparative data on the relative com-

[16] W. Gumz, *Kurzes Handbuch der Brennstoff- und Feuerungstechnik,* 3rd ed., Springer-Verlag, Berlin, 1962.

[17] "A Testing Program to Update Equipment Specifications and Design Criteria for Stoker Fired Boilers," Inter-agency agreement number IAG-D7-E681 (EPA) and U.S. Dept. of Energy number EF-77-C-01-2609 (not completed at time of publication).

[18] E. A. Zawadzki and F. du Breuil, *Mech. Eng.,* **81**(12), 104 (1959) (ASME–AIME Fuels Conf. Paper No. 59-Fu-3).

[19] R. A. Sherman and B. A. Landry, in *Chemistry of Coal Utilization,* Supplementary Volume, H. H. Lowry, Ed., John Wiley & Sons, New York, 1963, Ch. 18.

[20] L. G. Benedict and W. F. Berry in Am. Chem. Soc. Adv. Chem. Ser. No. 55, 1966, p. 577.

[21] W. O. Stacey and P. L. Walker, Jr., Report to Office of Coal Research, Contract No. 14-01-0001-390, August 1968.

[22] P. L. Walker, Jr., and K. A. Kim, *Fuel,* **44,** 453 (1965).

[23] H. Gan, S. P. Nandi, and P. L. Walker, Jr., *Fuel.,* **51,** 272 (1972).

[24] R. G. Jenkins, S. P. Nandi, and P. L. Walker, Jr., *Fuel,* **52,** 288 (1973).

[25] J. M. Beér, *J. Inst. Fuel,* **37**(282), 286 (July 1964).

[26] J. Csaba, Ph.D. thesis, University of Sheffield, England 1963.

bustibility of some important coke and char residues.[27]

The newer ideas concerning particle ignition were summarized in two letters,[28,29] by Howard and Essenhigh and later elaborated on in a paper[10] by the same authors. The paper contrasts the behavior of particles undergoing a high flux of pyrolysis products in one case, and a much lower flux in the other. Generally, a high flux of volatile material was thought to shield the particle surface and to burn as a gaseous flame in the space surrounding the particle. If the flux was small (as with low-volatile fuels or at the end of a devolatilization period), it was thought that heterogeneous and gaseous burning might proceed simultaneously. In a series of tests in a plug-flow furnace of the one-dimensional flame type patterned after Csaba's,[26] Howard and Essenhigh showed that for the prevailing commercial conditions flame ignition will most often start as heterogeneous burning on the surface of the finer particles, regardless of their volatiles con-

[27] R. H. Essenhigh and J. G. Cogoli, *Prepr., Am. Chem. Soc., Div. Fuel Chem.*, **20**(3), 134 (1975).
[28] J. B. Howard and R. H. Essenhigh, *Combust. Flame*, **9**, 337 (1965).
[29] J. B. Howard and R. H. Essenhigh, *Combust. Flame*, **10**, 92 (1966).

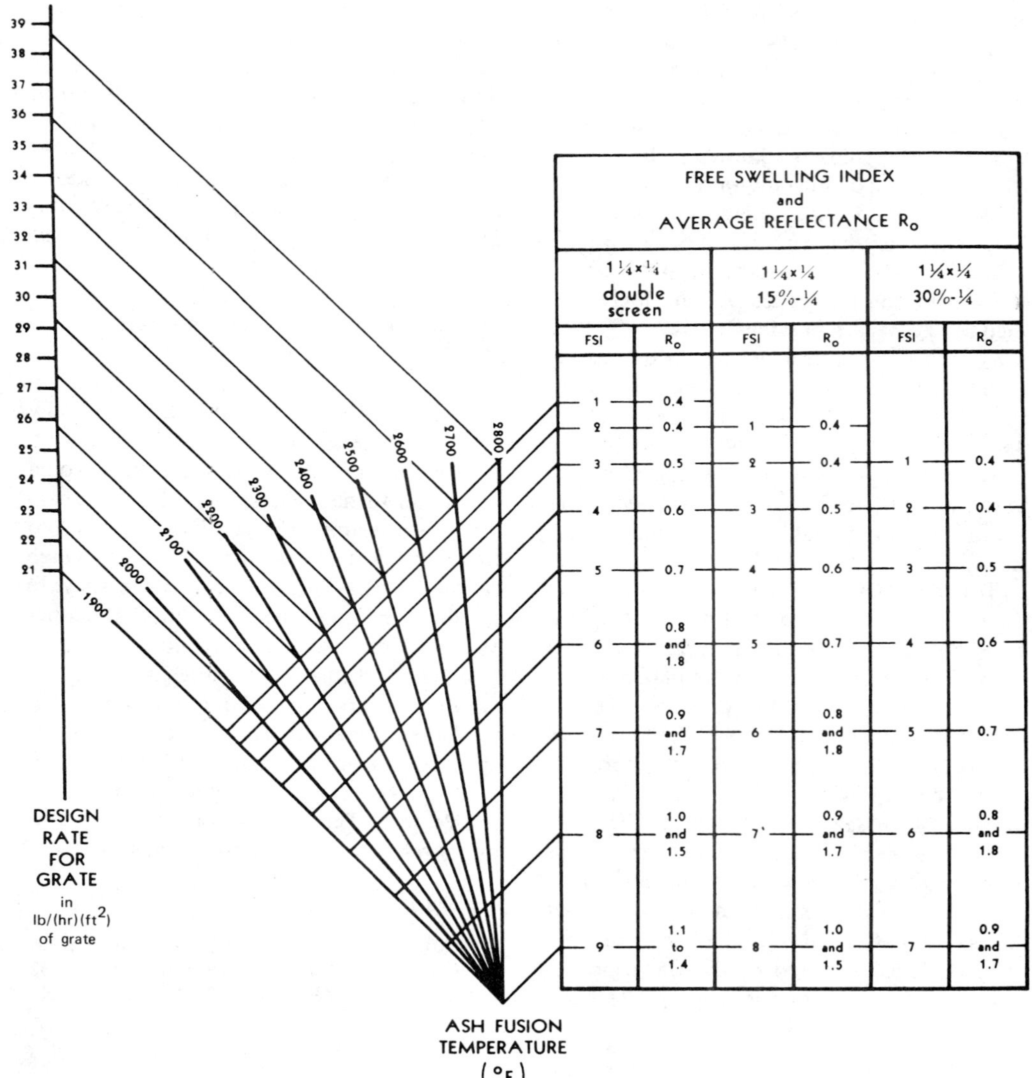

Figure 20.1 Relationship between reflectance, free-swelling index, screen size, ash fusion temperature, and grate loading.[20]

tent. Specifically, they found that the surface flux of volatiles decreases as particle size decreases. For particles above 65 μm in diameter, the flux is too high for oxygen to reach the surface; for particles smaller than 65 μm, gas phase and heterogeneous burning proceed together; and for particles about 15 μm in diameter or smaller, the initial burning is apparently completely heterogeneous.

These experiments of Howard and Essenhigh were run with preignition heating rates of about 10^4 °C/s and maximum furnace temperatures of about 1550°C—both of which are comparable to conditions in large pulverized-coal-fired furnaces. Also, in such furnaces most of the coal fired is smaller than 65 μm, and perhaps 20–30% is smaller than 15 μm. It may be concluded that these fine particles are the first to ignite and that they do so in a surface-burning mode.

The greater role now known to be played by heterogeneous burning in establishing ignition and stabilizing flames increases the importance of the careful study of surface characteristics, pore structure, and adsorption/desorption properties of solid fuels typified by the papers previously cited.[21–24] This topic is discussed at some length by Essenhigh and Cogoli.[27] They also indicated how the plane flame furnace may be used to assess the relative combustibility of a fuel by comparing its performance in the test furnace with that of better-known fuels burned in the same apparatus under comparable conditions.

The parameters of particular interest in such a comparison are the furnace temperature profile and the distance of the flame front from the point where the particles first receive furnace radiation. Such data are given by Essenhigh and Cogoli[27] for two types of coal and several potentially important chars, one of which was derived from an Illinois No. 6 seam coal by the COED process. The tests indicated that the char was more reactive than a Pennsylvania anthracite tested at the same time, even though the char had less volatile matter and a higher reactivity index (as measured by the Carnegie Research Laboratories crossing-point test).[3,30]

This result, which Essenhigh and Cogoli attributed to the porosity and surface structure of the char, is confirmed by full-scale firing tests conducted on a commercial anthracite-fired steam generator (similar to that shown in Fig. 20.33*e*) using 500 tons of COED char containing 4–6% volatiles derived from a Utah A seam coal.[31] These tests, conducted at several loads without ignitors or supporting fuel and for test runs lasting up to 18 hr, showed better combustion results bustible loss, than were experienced with anthracite in the same steam generator. The char combustibility appeared to be inferior to that of bituminous coals, but further testing might show modified, horizontal, turbulent burner firing to be satisfactory. In any event, commercial steam generators can definitely be designed for the COED chars and most, if not all, of the other residues resulting from gasification processes. This is an extremely important point, since the economy of many of the proposed gasifier projects depends on the use of the residues in steam generators fired without such supporting fuels as oil, gas, or higher volatile coal.

More difficult to judge is the utility to designers of the investigations reported by McCann et al.[32] and Demeter et al.,[33] in which low-volatile (5–15%) chars and a Pocahontas coal were burned in a horizontally fired water-cooled furnace. Part of their stated objective was to see how the chars would behave in furnaces unsuitable for the chars they were testing. The results indicated that either supporting fuel or a relatively high degree of char preheat was necessary to stabilize ignition of the chars, but that neither expedient was necessary for the Pocahontas coal. The fact that the burning profiles of the Pocahontas coal and the 5% volatile char were the same, whereas their combustibility appeared to be different, was taken by the authors as evidence of a defect in the burning profile test. This conclusion may be valid, but surely not much can be said in favor of

[30] R. A. Sherman, J. M. Pilcher, and H. N. Ostborg, ASTM Bull. No. 112, 1941, pp. 23–34.

[31] FMC Corporation, *The Combustion Performance of COED Char,* ERDA Contract No. 14-32-0001-1212, R & D Rep. No. 73—Interim Rep. No. 4, 1975.

[32] C. R. McCann et al., *Prepr., Am. Chem. Soc., Div. Fuel Chem.,* **16** (1971).

[33] J. J. Demeter, C. R. McCann, and D. Bienstock, ASME Paper No. 73-WA/Fu-2, 1973.

testing poor flames in inappropriate apparatus and judging combustibility by how much supplemental fuel is required to achieve good burning.

Other substances considered to be potentially important fuels for industrial furnaces are coal-derived, deashed solids produced by solvent refining. Process descriptions and the properties of various products of solvent refining are given in Chapters 28 and 29 of with char, as judged by ignitibility and com- this book. The handling, pulverization, and combustion characteristics of small quantities of solvent-refined coal (SRC) were tested several years ago,[11–13] and again recently—but with much larger quantities available. The older, small-scale tests revealed sticking in hoppers, feeders, and transport lines; softening and plugging in pulverizers; and heat-induced deposits on burner parts. Despite these handling problems, the SRC, when fired in suspension, burned readily.

A more recent paper by McCann et al.[34] confirmed that SRC can be burned efficiently in horizontal turbulent firing similar to that usually used for bituminous coal applications. To avoid burner fouling by softened SRC, McCann et al. had to use water cooling for the burner parts contacted by the fuel and modify the burner tip to shield the SRC from furnace radiation until it had left the fuel nozzle. Apparently, to utilize this fuel properly, equipment and operating procedures will have to be used that carefully control its temperature and that of the parts contacted by it prior to discharge into the furnace.

The largest scale burn test of SRC to date is the utilization of 3000 tons in June 1977 at the Georgia Power Company's Plant Mitchell. As in previous successful SRC firing tests, the burner fuel nozzles were water cooled. To suppress NO_x formation and to inhibit heating of burner parts, the burners utilized were of the dual-register type, which provides improved control of the turbulent mixing pattern between fuel and air. Conventional ball-and-race pulverizers were employed, and cold ambient air was used in the milling system. As reported by Harrison and

[34] C. R. McCann, J. J. Demeter, and D. Bienstock, *Trans. ASME, J. Eng. Power,* **99** (Ser. A, No. 3), 305 (July 1977).

An interesting result observed thus far in the SRC tests reported by McCann et al.[34] McRanie,[35] 18 days of testing revealed no major problems in utilizing this fuel. and in the recent large-scale runs at Plant Mitchell is the apparent production of lower quantities of NO_x than were observed when high-volatile bituminous coals were burned in the same equipment. This situation is particularly strange because the N_2 content of the SRC per 10^6 Btu is higher than that of the coals. Most likely, in these early burning tests, the major variables affecting NO_x formation, such as excess air and flame turbulence, were not optimized for either fuel. Thus a comparison of NO_x formation is premature.

In addition to the refined product, the SRC process leaves behind a high-ash residue that, in the operations conducted thus far, has contained as much as 60% ash. Despite this high content of inert material, the residue has a heating value of about 5000 Btu/lb and has burning properties—as determined by laboratory tests—similar to low-volatile bone coals or refuse from coal-cleaning operations. The residue can therefore be burned in the type of furnace long used for coals of low heating value and poor ignitibility. (See "Single U-Flame and Double U-Flame," in Section 20.2.2.2, and Fig. 20.36.)

Among the most important recent developments in laboratory-scale evaluation of solid fuels are the attempts to use thermal analysis techniques to characterize their combustion or pyrolysis properties. Thermal analysis has been used for a long time to determine the response of substances to temperature change.[36–38] Following the introduction in the 1950s of simple, inexpensive apparatus for making the desired measurements, interest in the technique broadened rapidly. For purposes of quality control or rapid evaluation of a fuel's suitability for a given use,

[35] W. B. Harrison and R. D. McRanie, *Proc. Am. Power Conf.,* **40,** 350 (1978).

[36] H. C. Howard, in *Chemistry of Coal Utilization,* Supplementary Volume, H. H. Lowry, Ed., John Wiley and Sons, New York, 1963, Ch. 9.

[37] R. C. MacKenzie, Ed., *Differential Thermal Analysis,* Vol. 1, *Fundamental Aspects,* Academic Press, New York, 1970.

[38] W. J. Smothers and Yao Chiang, *Handbook of Differential Thermal Analysis,* Chemical Publishing Company, New York, 1966.

these new instruments have been used to determine proximate analysis, heating value, and free-swelling index.[39,40] The application of thermal analysis for prejudging combustibility is the subject of several interesting papers leading to the use of the now common term "burning profile" and the descriptive phrase "combustion fingerprint."[7,8]

Whatever the object of the thermal analysis procedure may be, the usual technique is to continuously measure a physical or chemical change in a sample in an environment where temperature is increased at a uniform rate. The measured variable may be the change in weight, temperature, or dimension of the sample; the flow of heat into or out of the sample; or some other quantity.

If the change in weight of a sample is measured and plotted as a function of furnace temperature, the resultant curve is called a thermogravimetric analysis (TGA) thermogram. The usual instrumentation, however, yields a continuous plot of time rate of weight change versus furnace temperature—called a derivative thermogravimetric analysis (dTGA) thermogram or, more simply and aptly, a burning profile.

Figure 20.2 shows burning profiles for a number of solid fuels of various types to illustrate the nature of the information given by this test. In using such profiles to design fuel-firing systems, the profiles of well-known fuels are compared with those of new or untried fuels. Thus valuable judgments can be made concerning events occurring throughout the course of the combustion process, such as the onset and completion of burning and the maximum rate of combustion of the sample. These judgments contrast with the comparatively meager (though very useful) results obtainable from tests for ignition temperature, flammability index, extinction temperature, and the like.

In addition to providing ignition and burnout information as outlined above, the study of burning profiles has identified a number of fuels showing false ignition at low temperatures. This phenomenon is illustrated by burning profiles such as those shown in Fig. 20.3. These fuels show a small peak at some relatively low temperature, followed by near extinction and subsequent reignition. This behavior is thought to result from the presence in mixed form of several fuel species of different reactivities. Obviously, a designer could be disastrously deceived by false ignition in the absence of a burning profile for the fuel.

It must not be concluded that the use of burning profiles has gone uncriticized. As already mentioned, their ability to distinguish between two fuels of apparently different combustibility has been questioned by De-

[39] R. W. Bryers, B. K. Biswas, and T. E. Taylor, "Fuel Evaluation Using Differential Thermal Techniques," paper presented at 68th Annual Meeting of American Institute of Chemical Engineers, Los Angeles, 1975.

[40] R. L. Hassel, *Coal Min. Process.*, 74 (May 1976).

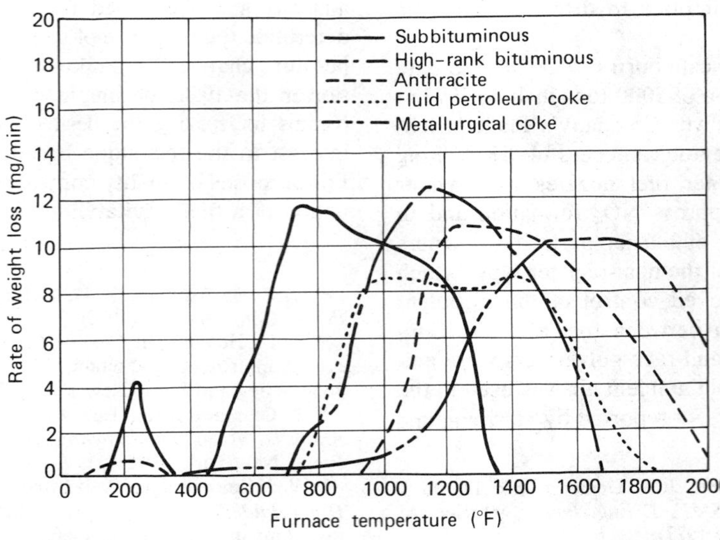

Figure 20.2 Burning profiles for various solid fuels. Modified from Refs. 7 and 8.

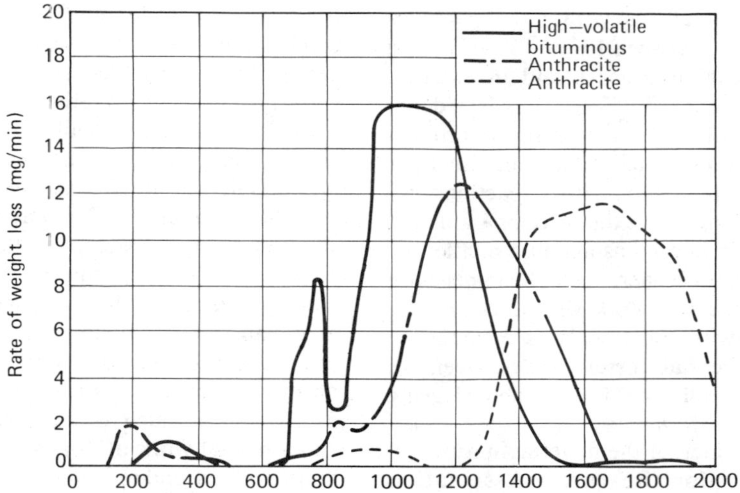

Figure 20.3 Burning profiles for various solid fuels showing false ignition.[7]

meter et al.[33] Others have pointed out that the heating rates in the burning profile test are very low (≃ 15°C/min) compared with heating rates in industrial equipment of 10^3–10^5 °C/s. Obviously, the way in which such high heating rates affect physical size or shape, surface conditions, or amount and type of volatile matter emitted from different fuels is not accounted for in burning profile tests. The effect of heating rate on devolatilization is particularly important, and reference should be made to Gray et al.[41] for a recent study of this relationship. Undoubtedly more experience and data are needed to apply thermal analysis with full confidence, but it is hardly contestable that it represents a marked advance in the tools available to estimate combustion characteristics.

The sizing of a pulverized-coal-fired furnace for a given fuel and heat duty is perhaps the most important technical and economic problem facing the designer. Work reported by Stewart, Wall, and Lowe[14,15] in Australia has contributed toward its solution by advancing an understanding of the radiative and convective heat transfer processes in large pulverized-coal-fired steam generators. In their work measurements have been carried out in the furnaces of two corner-fired steam generators: a 20-MW unit and a 300-MW unit. Results so far have provided means for a more rational approach to furnace design calculations, rather than the almost totally empirical approach formerly used. Some major findings of this work relate to the better understanding of the amount of heat absorbed by the cooler layer of gas along the furnace walls. In addition, the role played by particulate radiation has been quantified and a beginning made in accounting for its contribution to radiative heat transfer as affected by ash quantity and ash particle size.

Fuel ash composition and properties strongly influence the design of coal-burning apparatus. Evaluation of the fouling and slagging characteristics of coals is especially important. Correlations of slagging potential as a function of various ash characteristics are widely used as guides in prejudging troublesome fuels. Many of these correlations are strongly empirical but have proved to be useful nonetheless. Attempts are being made constantly to develop better methods of analysis, such as LTA, the low-temperature ashing method described by Gluskoter[42] (Chapter 7). With this technique a sample is placed in an evacuated chamber and subjected to an oxygen-bearing plasma (an ionized gas containing free radicals and other reactive species) that can oxidize organic material at low temperatures. Plasma ashing does not depend on heat; the oxidation energy comes from the electrical energy of the plasma. The only heat reaching the sample is that produced by the exothermic oxidation reactions them-

[41] D. Gray, J. G. Cogoli, and R. H. Essenhigh, in Am. Chem. Soc. Adv. Chem. Ser. No. 131, 1974, p. 72.

[42] H. J. Gluskoter, *Fuel—J. Fuel Sci.,* **44** (July 1965).

selves. Temperatures within the chamber typically range between 100 and 125°C during the entire ashing process and in extreme cases do not exceed 250°C. Thus less dehydration and little or no loss occur through volatilization of metallic elements such as sodium and potassium. Also reactions between the sample and the test apparatus are minimized. In normal ashing in a muffle furnace at 950°C or higher, such phenomena are a serious concern. Work with low-temperature ashes, with some coals, is expected to yield more accurate correlations between ash analysis and full-scale fouling and slagging combustion performance.

Although much of the material in a British Coal Utilisation Research Association (BCURA) review series[43] does not readily fit in with the discussions of this chapter, their subjects and ours are closely related. Their material is strongly recommended for study, as is a recent review by Essenhigh[44] dealing with combustion and flame propagation. In addition, both these references offer extensive bibliographies.

20.1.2 Combustion in Fluidized Beds

Much of the early work to develop a theoretical model for the behavior of a fluidized-bed combustor has been an attempt to describe the process in terms of the model of the two-phase reactor proposed by Orcutt et al.[45] One phase of this model consists of almost particle-free bubbles; the other is a dense phase of gas and particles with a void fraction of 0.5. Matched to this model is the two-stage combustion process of a coal particle (i.e., volatiles evolution and burning followed by char burning). The interrelationship is obvious, since the mixing of gas-air-solids in the bed determines the rate at which combustion can proceed. Furthermore, the heat transfer rate to cooling surfaces within the bed is controlled by the fluid dynamics, combustion rate, and particle size within the bed, with the last well treated in Gelperin and Einstein[46] and Botterill.[47]

Cold-model tests of fluidized beds with the cooling tube surface in the bed have shown that bubble size is affected by tube spacing and that frequency of bubble formation is affected by the spacing of the distributor plate from the tube surface.[48] Other cold-model work has shown that coalescence and/or expansion of bubbles does not occur to any marked extent in systems where tubes occupy the bed. It is thus apparent that we cannot apply much of the experience and experimental work in "open" fluidized beds that involved understanding the fluid mechanisms of particle and gas behavior.

The fundamental understanding of the combustion process in a fluidized bed is largely based on theory and hypothesis. Major differences in the combustor from processes previously studied concern the relatively low combustion temperature (750–900°C), the particle size (0.3–10 mm), the low concentration of coal particles in the bed (1–5%), the chemical reactions with the bed material, and the rapid physical circulation of the bed mass.

It has been assumed that rates of devolatilization of the coal particle can be based on the findings of Field et al.,[43] in that the rate is temperature dependent and is affected by rank and swelling index. Using these theories, a 100-μm particle may take 5–10 ms to reach 95% devolatilization. A 1-mm particle, however, would require 0.5–1 s to devolatilize to the same level—a time closer to the residence time of the bubbles in a fluidized bed.

The mechanics of combustion of volatiles in a fluidized bed have not been experimentally treated. Reports from small pilot plants have implied that volatile combustion occurs in the freeboard above the bed. Rate data on the burning of volatiles reported by Biswas

[43] M. A. Field et al., "Combustion of Pulverized Fuel," *BCURA Month. Bull.*, Parts 1–8, Review Nos. 255, 257, 259–264, **30**(9, 12), 301, 455 (August–September 1966, November–December 1966), **31**(3–8), 81, 193, 241, 285, 371, 429 (February–March 1967, April 1967, May 1967, June 1967, July 1967, August 1967).

[44] R. H. Essenhigh, in *Proc. 16th Symp. (Int.) Combustion,* Combustion Institute, Boston, 1976.

[45] J. C. Orcutt, J. F. Davidson, and R. L. Pigford, Chem. Eng. Prog., Symp. Ser. No. 38, Vol. 58, 1962, p. 1.

[46] N. I. Gelperin and V. G. Einstein, in *Fluidization,* J. F. Davidson and D. Harrison, Eds., Academic Press, New York, 1971, Ch. 10.

[47] J. S. M. Botterill, *Fluid-Bed Heat Transfer,* Academic Press, New York, 1975.

[48] J. A. Bazan, *Proc. 4th Int. Conf. Fluidized-Bed Combustion,* MITRE Corp., McLean, Va., 1976, pp. 271–283.

and Essenhigh[49] may be relevant to the fluidized-bed combustion case.

Combustion of char in the fluidized bed is thought to follow a stepwise pattern with (1) diffusion of oxygen from the bed toward the particle surface and (2) adsorption and reaction at the surface or in the pores, with the following main reactions likely:

$$C + \tfrac{1}{2} O_2 \rightarrow CO \qquad (20.1)$$

$$C + O_2 \rightarrow CO_2 \qquad (20.2)$$

as well as a simultaneous, homogeneous reaction:

$$CO + \tfrac{1}{2} O_2 \rightarrow CO_2 \qquad (20.3)$$

From past work it is known that temperature, coal rank, particle diameter, and—we must infer—bed dynamics dictate whether and where these reactions take place. The relative rate of these processes has been studied by Field et al.,[43] Wicke,[50] Field,[51] Smith,[52] and Gray et al.[41] On the bases of the temperature range of fluidized-bed combustors and the particle size, it is likely that diffusion controls the combustion process.[53] For smaller char particles ($< 100\ \mu m$) reaction rates are less than those set by diffusional limits.[52]

Avedesian and Davidson[53] developed a mechanistic model for the combustion of char particles in a fluidized bed and tested the model experimentally for burnout time and gas composition. Campbell and Davidson[54] extended the development of this combustion model to treat a continuous operating bed. The theory successfully predicts maximum combustion rates. The experiments yielded a definite CO content in the bed, which indicates that the surface temperature of the char particles is not high enough for the Boudouard reaction

$$2CO \rightleftharpoons CO_2 + C \qquad (20.4)$$

to occur rapidly.

Basu et al.[55] conducted experiments to determine reaction mechanisms of carbon in fluidized beds. When $POCl_3$ is used to inhibit the homogeneous oxidation of CO to CO_2, the data indicate that the $CO \rightarrow CO_2$ reaction proceeds around the particle at the test temperature of 1560°F. The authors concluded that the Boudouard reaction does not occur at all.

Associated with fluidized-bed combustion work is the interest in removing SO_2 generated during combustion by reacting it with the bed material. It is known that lime is most reactive to capture SO_2 at temperatures of 1400–1800°F, yielding

$$CaO + SO_2 + \tfrac{1}{2} O_2 \rightarrow CaSO_4 \qquad (20.5)$$

Hence limestone and dolomites are the preferred bed materials when sulfur removal is desirable.

Regeneration or reuse of the bed material for sulfur capture is the subject of intensive work. Regeneration reactions are carried out in reducing atmospheres at temperatures up to 2100°F. The most desirable reactions are:

$$CaSO_4 + CO \rightarrow CaO + SO_2 + CO_2 \qquad (20.6)$$

and

$$CaSO_4 + H_2 \rightarrow CaO + SO_2 + H_20 \qquad (20.7)$$

The concentration of SO_2 in the off-gas will be approximately 5 mol%. At this level of concentration various processes are available to reduce the SO_2 to elemental sulfur. Work in this field has been reported most recently by Hubble et al.,[56] Montagna et al.,[57] Ruth,[58] and Snyder et al.[59]

[49] B. K. Biswas and R. H. Essenhigh, Chem. Eng. Prog., Symp. Ser. No. 68, Vol. 126, 1972, p. 207.

[50] E. Wicke, *Z. Elektrochem.*, **60**, 774 (1956).

[51] M. A. Field, *Combust. Flame,* **13**, 237 (1969).

[52] I. W. Smith, *Combust. Flame,* **17**, 303 (1971).

[53] M. M. Avedesian and J. F. Davidson, *Trans. Inst. Chem. Eng.*, **51**(2), 121 (1973).

[54] E. K. Campbell and J. F. Davidson, "The Combustion of Coal in Fluidized Beds," paper presented at Fluidized Combustion Conf., London, September 1975.

[55] P. Basu, J. Broughton, and D. E. Elliot, "Combustion of Single Coal Particles in Fluidized Beds," paper presented at Fluidized Combustion Conf., London, September 1975.

[56] B. R. Hubble et al., *Proc. 4th Int. Conf. Fluidized-Bed Combustion,* MITRE Corp., McLean, Va., 1976, p. 367.

[57] J. C. Montagna et al., paper presented at Fluidized Combustion Conf., London, September 1975, p. 393.

[58] L. A. Ruth, paper presented at Fluidized Combustion Conf., London, September 1975, p. 425.

[59] R. B. Snyder et al., paper presented at Fluidized Combustion Conf., London, September 1975, p. 439.

20.1.3 Ambient Pressure Fluidized-Bed Combustion

The concept of fluidized-bed combustion at temperatures between 1400 and 1700°F with the heat transfer surface submerged in the fluidized medium was pursued in the United Kingdom by the Central Electricity Generating Board (CEGB) in 1962.[60] In 1963 BCURA proceeded with the development of a fluidized-bed, coal-fired package boiler.[61]

The National Coal Board (NCB) began work in 1965 to develop fluidized-bed combustion for utility steam generator applications.[62] In the United States work started in 1965[63] with the objective of package boiler development. The work undertaken by Pope, Evans and Robbins (PER) was supported by the Office of Coal Research (OCR) and the U.S. Environmental Protection Agency (EPA) and later by the U.S. Energy Research and Development Administration (ERDA) now a part of the Department of Energy (DOE).

Starting in 1970, interest in fluidized combustion heightened because the potential for SO_2 and NO_x emission reduction was recognized. Emission of SO_2 could be reduced by the use of limestone or dolomite bed material; NO_x emission could be reduced in part by the relatively low bed temperature at which combustion occurs.

Table 20.1 lists test units used for investigations of combustion and pollution control. Experiments performed with these units have resulted in the establishment of operating parameters for optimum performance and economy of operation and provide the basis of design for commercial-size units now under construction.

[60] J. M. Botterill and D. E. Elliott, *Engineering*, 146 (1964).

[61] M. H. Barker, A. G. Roberts, and S. J. Wright, paper presented at Steam Plant Convention, Institution Mechanical Engineers, London, 1969.

[62] J. McLaren and D. F. Williams, *J. Inst. Fuel*, **42**(343), 303 (1969).

[63] S. Ehrlich et al., in Am. Inst. Chem. Eng. Symp. Ser. No. 126, Vol. 68, 1972, p. 231.

Table 20.1 Summary of Experimental Fluidized-Bed Combustion Units[64]

Rig	Bed Size	Construction	Operating Conditions		
			Temperature (°F)	Superficial Velocity (ft/s)	Particle Size
Consol	4 in. dia	Refractory-lined	1700–1900	2–3	−16 + 28 mesh
Exxon Research (pressurized)	3 in. dia	Refractory-lined	1500–1800	2.6	450, 930 μm
BCURA	12 in. dia	Refractory-lined	1380–1800	7.2–10.6	30 mesh
BCURA	27 in. dia	Refractory-lined	1300–1830	7–14	−¼ + 30 mesh
NCB-CRE	8 in. dia	Refractory-lined	1300–1470	1–3	−10 mesh
NCB-CRE	12 in. dia	Refractory-lined	1560	3	−1/16 mesh
NCB-CRE	3 ft²	Refractory-lined	1300–1560	2–4	−10 mesh
NCB-BCURA (pressurized)	48 × 24 in.	Refractory-lined	1470	2	−1/16 mesh
USBM	18 in. dia	Water jacket	1700	—	0.08-0–0.55 in.
CSIRO	8 in.	Refractory-lined	1800	2	−1/16 mesh
Esso Petroleum Ltd.	2 × 4 ft	Refractory-lined	1600	4–6	−1/8 mesh
Battelle Memorial Institute	20 in.	Refractory-lined	1900–2000	6–8	−10 mesh
Pope, Evans and Robbins	2 × 4 ft	Waterwall	1550	6	—
Exxon Research (pressurized)	12 in. dia	Refractory-lined, steam-cooled bundle	1500–1800	5–10	−8 + 25 mesh
U.S. Bureau of Mines, Morgantown, W. Va., 2 units	18 in. dia	Refractory-lined, cooling surface in bed	1600	3–6	2000–5000 μm
Argonne National Laboratories	6 in. dia	Refractory-lined	1600	3.5–7.4	Limestone, 490–680 μm
Foster Wheeler Energy Corporation	18 in. dia	Refractory-lined, cooling surface in bed	1600	3–10	2000–5000 μm
Sheffield University	12 × 12 in.	Refractory-lined	1380–1630	2–4	—
Combustion Power Company (refuse-fired)	9.5 ft dia	Insulated steel vessel	1270–1450	5–7	16-mesh sand
University of West Virginia (refuse-fired)	15 in.	Refractory-lined	—	3–8	−8 mesh
Pennsylvania State University (gas-fired)	16 in. dia	Insulated steel drum, water-cooled	1800–2350	6–10	1-in.-thick bed of alumina

Information in several important areas continues to be developed in the pilot plants listed. The first area relates to the combustion efficiency as affected by bed dynamics, coal rank, and excess air. Results from work by the CEGB on this subject are shown in Fig. 20.4. Essentially there is little CO at bed temperatures above 1300°F. The work reported by Hoy and Stanton[65] shows that mechanical versus pneumatic coal feeding, particularly feed distribution, has a measurable effect on combustion efficiency. On the basis of this work it is presently accepted practice to provide a coal feeder for each 10 ft² of bed cross-sectional area. Excess air as low as 2% has proved to be feasible[66] with good combustion efficiency. Bed temperatures below 1300°F with 5–15% excess air give poor performance.[67–69]

[64] R. W. Bryers and W. E. Kramer, "The Status of Technology of Fluidized Beds for the Combustion of Solid Fuels," paper presented at 1976 TAPPI Engineering Conf., Houston, Tex., October 4–7, 1976.

[65] H. R. Hoy and J. E. Stanton, "Fluidized Combustion Under Pressure," paper presented at Joint Meeting, Chemical Institute of Canada and American Chemical Society, Toronto, May 1970.

[66] D. G. Skinner, *The Fluidised Combustion of Coal,* National Coal Board, Research and Development Dep., London, 1970.

[67] R. Bailey, M. J. Cooke, and D. F. Williams, CRE Tech. Memo. No. 361, August 1968.

[68] D. J. Keating and S. J. Wright, BCURA Inf. Circ. No. 302, 1966.

[69] A. A. Orning and C. R. McCann, "A Study of Fluidized Bed Combustion," paper presented at 97th Annual Meeting, American Institute of Mining, Metallurgical and Petroleum Engineers, New York, February 1968.

A second important area of information development is the optimum size and the effect of size distribution of coal feed on performance. Here the problem is one of elutriation of unburned carbon. It would appear that for normal combustion applications, coal size should be ¼ × 0 in. Depending somewhat on crusher and coal type, as well as on gas velocity and freeboard above the bed, as much as 15–20% of the carbon can be elutriated from the bed. Recovery of this material by conventional cyclones is possible. The recovered material can be recycled to the main combustion bed or to a separate fluidized bed, called a carbon burnup cell, which operates at lower superficial velocities and higher temperatures than the main beds. The latter concept was proposed and developed by PER.[70] With this technique, acceptable carbon utilization is achieved.

A third important area of information development is the matter of heat transfer in the bed. The recognized variables affecting heat transfer in cooling tubes in a fluidized bed are bed particle size, bed temperature, superficial velocity, and tube spacing. Extensive investigations were conducted by Wright et al.,[71] McLaren and Williams,[62] Rice and Coates,[72] Skinner,[66] Gelperin and Einstein,[46]

[70] E. B. Robison et al., PER report to EPA, Contract No. CPA 70-10, February 1972.

[71] S. J. Wright, R. Hickman, and H. C. Ketley, *Br. Chem. Eng.,* **15**(12), 12 (1970).

[72] R. L. Rice and N. H. Coates, *Proc. 3rd Int. Conf. Fluidized-Bed Combustion,* Vol. 1, 1972, p. 117.

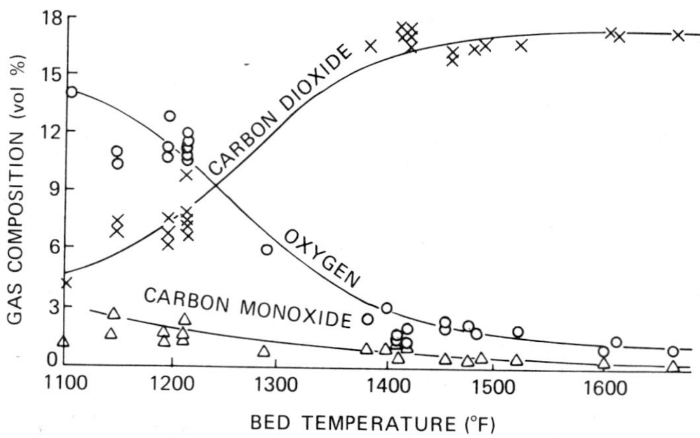

Figure 20.4 Effect of bed temperature on exhaust gas composition for combustion of char in CEGB combustor.[64]

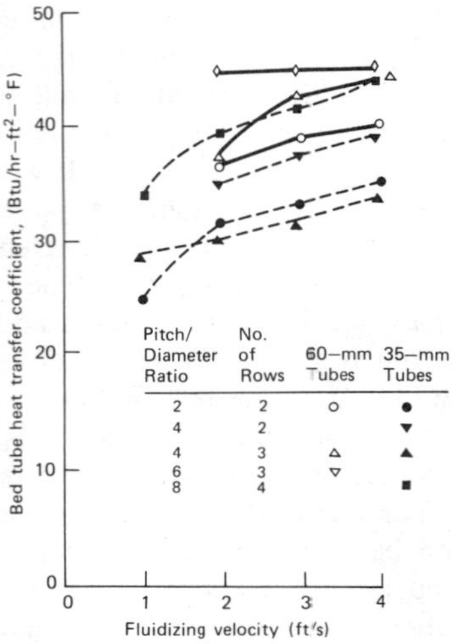

Figure 20.5 Variation in bed tube heat transfer coefficient with fluidizing velocity.[66]

and CSIRO.[73] Elliott and Virr[74] conducted experiments with extended surface tubes in shallow beds. Results of work by Skinner are shown in Fig. 20.5. Heat transfer coefficients from 25 to 100 Btu/hr-ft^2-°F have been reported depending on the above-mentioned variables.

A fourth important area of information development is the effect of bed material size on both fluidization characteristics and sulfur capture in those cases where limestone or dolomite are used. Figure 20.6 plots gas velocity versus particle size, giving the boundaries of terminal velocity, where material elutriates, to minimum fluidizing velocity. From this plot the critical nature of bed material size distribution is apparent. Experimenters also report a great difference in the decrepitation behavior of various stones in fluidized beds. Therefore it is not possible at this time to predict the behavior of raw limestones or dolomite stones from their mechanical or chemical analysis regarding equilibrium size distribution. The same is true about sulfur capture of different stones. Figure 20.7 shows results by PER of SO_2 adsorption for various Ca/S mole ratios for two different stones. Figure 20.8 shows a composite SO_2 adsorption behavior derived from other investigators. All experimenters have shown that optimum sulfur capture temperature is relatively independent of stone type and occurs between 1500 and 1600°F.

Depending on stone SO_2 capture efficiency and mechanical properties, it is necessary to feed fresh stone and remove spent stone continuously. Figure 20.9 gives some indication of bed material size distribution as fed and after 50 hr of operation. In addition, most investigators feel that it will be necessary to circulate the bed material through a classification loop to remove oversized agglomerates and tramp material that would tend to settle to the bottom of the bed and adversely affect fluidization characteristics.

Emission of NO_x from fluidized-bed combustors has been reported by several investigators. As may be expected, the emission is relatively low and appears to be almost

[73] Commonwealth Scientific and Industrial Research Organization, *Coal Res. CSIRO,* **34,** 2 (1968).

[74] D. E. Elliott and M. J. Virr, *Proc. 3rd Int. Conf. Fluidized-Bed Combustion,* Vol. 2, 1972, p. 1.

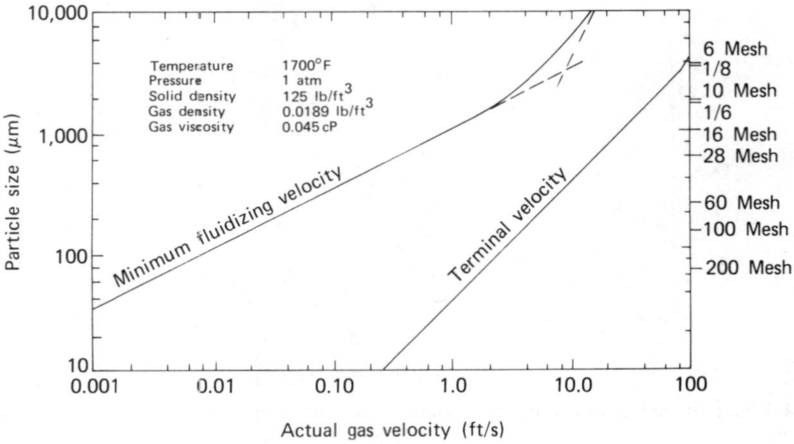

Figure 20.6 Fluidizing velocity and terminal velocity for 1-atm system.[64]

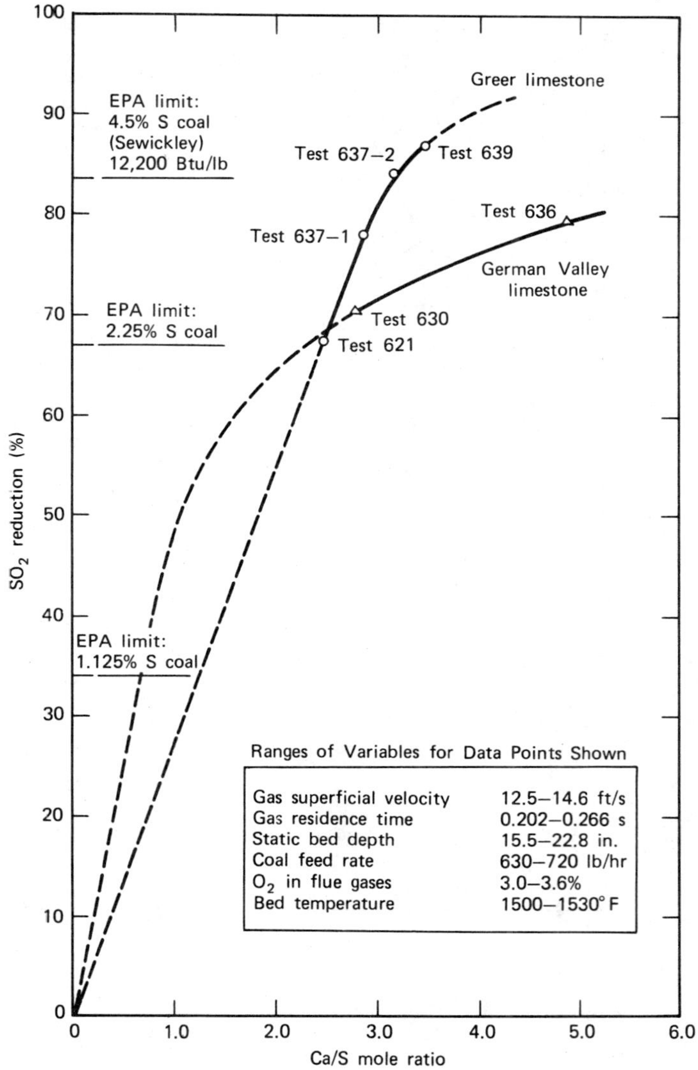

Figure 20.7 Comparison of performance of Greer and German Valley limestones. (Data uncorrected for differences in operating variables.)[75]

entirely dependent on fuel-bound nitrogen rather than on fixation of air-nitrogen. Figure 20.10 gives some normalized results of NO_x emission reported by various investigators.

A fifth important area of information development involves the investigations to determine whether ash fouling and corrosion present any problems in fluidized-bed combustion operation. To date, no significant corrosion and no fouling have been encountered on surfaces of materials operating at normal steam generator metal temperature conditions. This result is not surprising, since

[75] J. E. Mesko, *Multicell Fluidized-Bed Boiler Design, Construction and Test Program,* Monthly Progress Status Rep. No. 39 to ERDA, Contract No. E(49-18)-1237, December 1975, p. 31.

[76] S. J. Wright, in *Proc. 3rd Int. Conf. Fluidized-Bed Combustion,* Vol. 1, 1972, p. 135.

[77] D. H. Archer et al., *Evaluation of the Fluidized-Bed Combustion Process,* prepared for EPA, Contract No. CPA 70-9, November 1971.

[78] G. M. Goblisch and E. A. Fletcher, "Nitric Oxide Emissions from the Fluidized Combustion of Coke," paper presented at the Combustion Institute Central States Sect. Spring Meeting, April 1976.

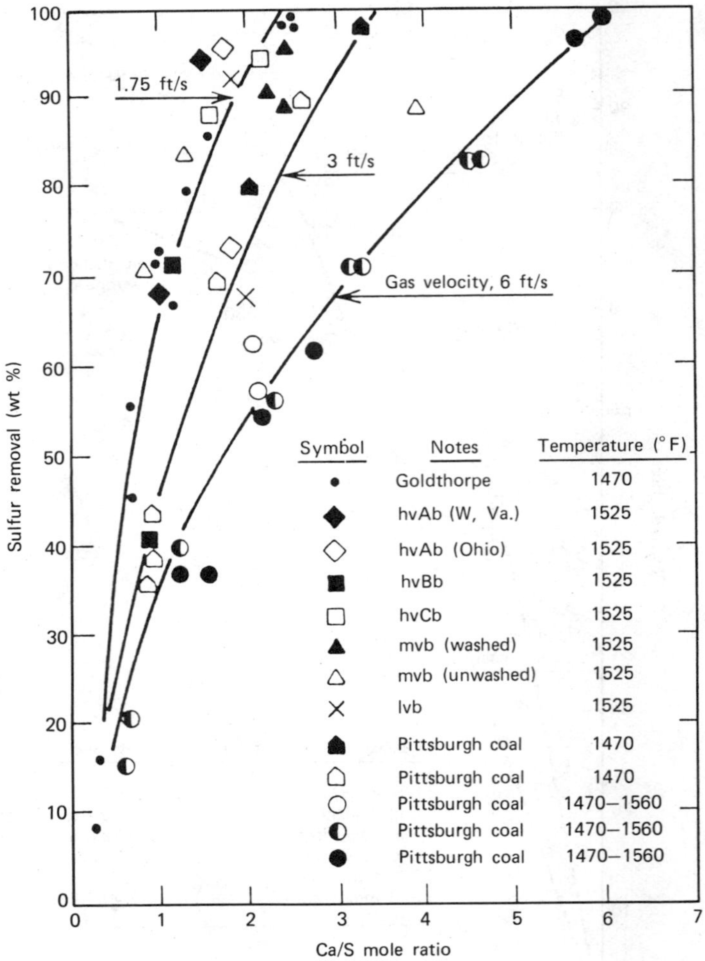

Figure 20.8 Effect of calcium sulfur mole ratio on sulfur removal.[76]

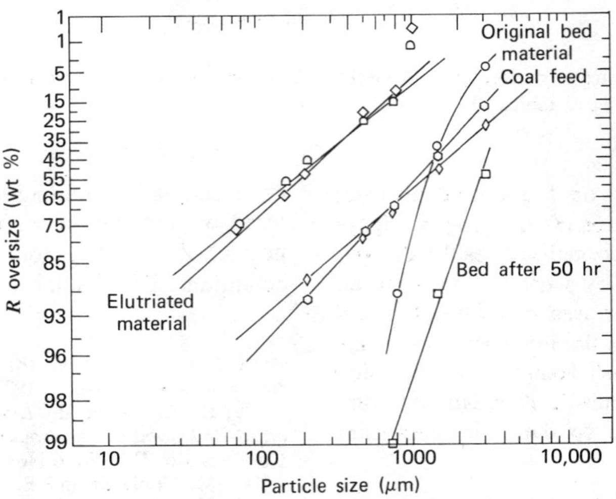

Figure 20.9 Size distribution data reported by BCURA.[77]

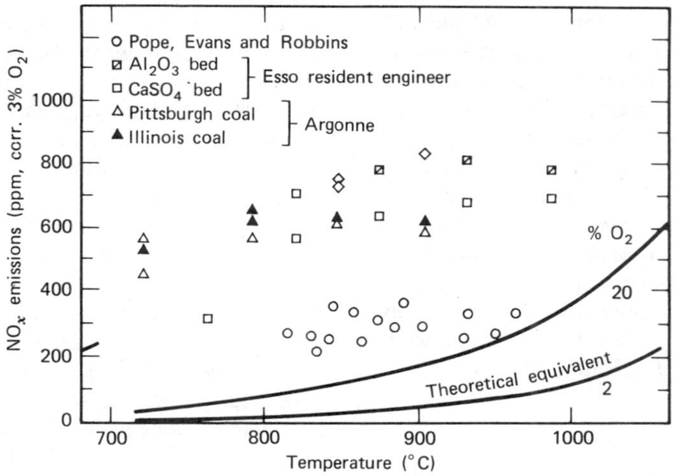

Figure 20.10 Comparison of NO_x emissions from fluidized-coal combustion.[78]

the combustion temperature is considerably below the ash fusion temperature of coals. Also, vaporization of alkali sulfates, silicates, and carbonates at these temperatures is so low that subsequent condensation on cooler metal surfaces is generally insignificant. Work in this area has been summarized by Westinghouse.[79]

20.1.4 Pressurized Fluidized-Bed Combustion

The interest in pressurized fluidized-bed combustion is based on the recognition that as combustion pressure is increased, the size of the combustor is decreased, as is the heat transfer surface requirement, thus reducing investment cost. Also, at combustion pressures above 4–6 atm, a gas turbine can be driven by the combustion gases to compress the combustion air and generate electric power. The bed-cooling surface can be steam cooled, resulting in a combined steam-gas turbine cycle, or the bed-cooling surface can be a closed, gas-cooled loop—helium-air-CO_2. The desired result is to obtain improved power generation efficiencies.

This approach was introduced in the United Kingdom in 1968 when BCURA built the first pressurized combustor.[65] Work in the United Kingdom up to 1970 is described by Skinner.[66]

As the work progressed, two important facts became apparent: (1) the Ca/S mole ratio required for sulfur capture was less than had been found in ambient pressure combustors, and (2) the emission of NO_x was even less than that found in ambient pressure systems. Results of work by BCURA and NCB were reported by Hoy and Roberts in 1972.[80]

Pressurized combustion work in the United States was reported by Jonke et al.,[81,82] Hammons and Skopp,[83,84] Hoke et al.,[85] and Vogel et al.[86] Bed sulfur capture by dolomite in pressurized combustors has been found to be superior to that by limestone because with limestone ($CaCO_3$) the calcination reaction does not proceed at temperature levels of 1500–1700°F and several atmospheres pressure. With dolomite ($CaCO_3 \cdot MgCO_3$), how-

[79] D. E. Thomas et al., *Energy Conversion Alternatives Study—ECAS—Westinghouse Phase I,* Final Report, Vol. 2, *Materials Considerations,* Contract No. NAS 3-19407, NASA CR-134941, February 12, 1976.

[80] H. R. Hoy and A. G. Roberts, in Am. Inst. Chem. Eng. Symp. Ser. No. 126, Vol. 68, 1972, p. 225.

[81] A. A. Jonke et al., ANL Annual Report to EPA, ANL/ES-CEN-1004, July 1970–June 1971.

[82] A. A. Jonke et al., in Am. Inst. Chem. Eng. Symp. Ser. No. 126, Vol. 68, 1972, p. 241.

[83] G. A. Hammons and A. Skopp, ASME Paper No. 71-WA/APC-3, 1971.

[84] G. A. Hammons and A. Skopp, Am. Inst. Chem. Eng. Symp. Ser. No. 126, Vol. 68, 1972, p. 252.

[85] R. C. Hoke, H. Shaw, and A. Skopp, in *Proc. 3rd Int. Conf. Fluidized-Bed Combustion,* Vol. 1, 1972, p. 93.

[86] G. J. Vogel et al., *Proc. 3rd Int. Conf. Fluidized-Bed Combustion,* Vol. 1, 1972, p. 35.

ever, endothermic decomposition and half-calcination occur at temperatures several hundred degrees lower than calcination for $CaCO_3$, even at higher pressures. Results of sulfur capture tests are shown in Fig. 20.11.

Additionally, results of tests show low rates of NO_x emission even at very high excess air, as indicated in Fig. 20.12. The values of NO_x emission are generally lower in pressurized systems than in ambient pressure systems, although in both cases these values fall well below the EPA allowable emission rate of 0.7 lb/10^6 Btu.

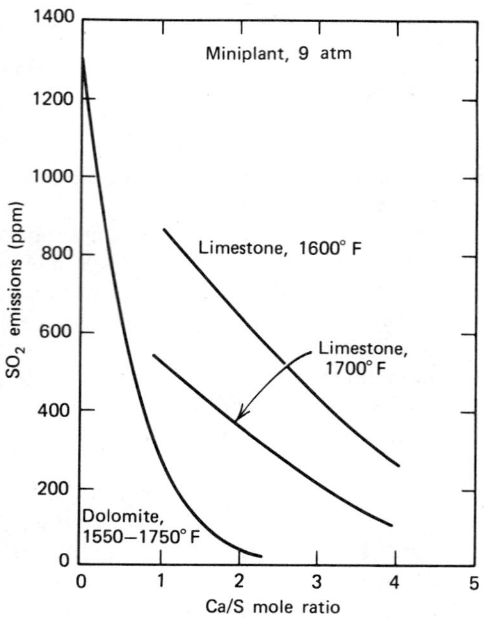

Figure 20.11 Effect of calcium/sulfur mole ratio on SO_2 emissions.[87]

Results of the pilot work have confirmed that heat release per unit area of bed surface can be increased in a pressurized system in proportion to absolute pressure without excessive carryover of solids from the bed. The same is true for volumetric heat release, although this is practically limited by the volume required for heat transfer surface to control bed temperature. (The heat transfer does not increase to any measurable extent with pressure over the range of practical application, i.e., up to 10 atm.) The relationship between bed heat release, coal size, and superficial velocity is shown in Fig. 20.13.

Problems concerning deposits and corrosion in the beds have been addressed for pressurized combustors for conventional tube materials and tube metal temperature conditions associated with steam plants. No discouraging results have been obtained. Work is progressing to determine resistance to corrosion of higher alloys operating at temperatures in excess of 1100°F for use as heat transfer surface associated with air-cooled, closed, or closed-open cycle systems, as well as with alloys used in gas turbines that expand the combustion gases in combined cycles.

The work in progress on corrosion, erosion, and deposits is referenced in papers by

[87] R. C. Hoke, in *Proc. 4th Int. Conf. Fluidized-Bed Combustion,* MITRE Corp., McLean, Va., 1976, pp. 565–567.

[88] R. C. Hoke, M. S. Nutkis, and D. D. Kinzler, in *Proc. Fluidized Bed Combustion Technology Workshop*, CONF-770447-P-2, Vol. 2, April 13–15, 1977, p. 157.

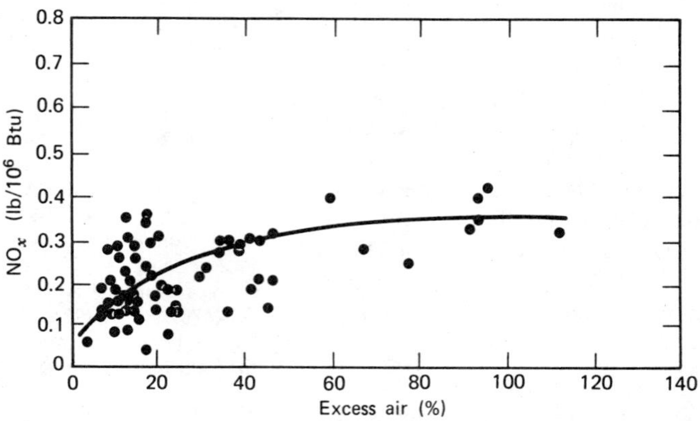

Figure 20.12 Effect of excess air on NO_x emissions.[88]

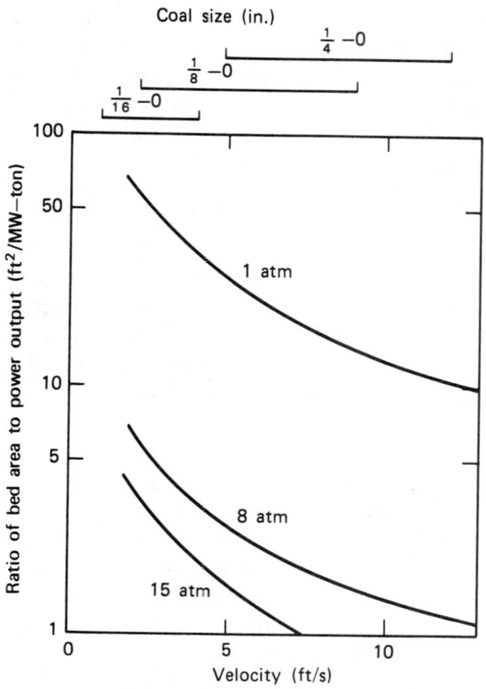

Figure 20.13 Ratio of area of bed to power output versus fluidizing velocity in fluidized bed combustion.[89]

Highley,[90] Wigton,[91] and Westinghouse.[79] Early results of turbine blade erosion-corrosion indicate little likelihood of coal ash erosion problems, principally because the ash, which is formed at temperatures below its fusion point, is fluffy and free of fused silicate materials. On the other hand, levels of sulfur and alkali carried over, although low, are higher than those allowable for the commonly used gas turbine alloy materials. This problem is being attacked by the development of high-efficiency particulate removal equipment, since it appears that the alkali material is condensed on particulate surfaces. Work in this area is discussed by Wade.[92]

[89] H. B. Locke et al., in *Proc. Fluidized-Bed Combustion Technology Workshop*, CONF-770447-P-2, Vol. 2, April 13–15, 1977, pp. 69–90.
[90] J. Highley, *Proc. 4th Int. Conf. Fluidized-Bed Combustion*, MITRE Corp., McLean, Va., 1976, pp. 199–210.
[91] H. F. Wigton, *Proc. 4th Int. Conf. Fluidized-Bed Combustion*, MITRE Corp., McLean, Va., 1976, pp. 305–321.
[92] G. L. Wade, *Proc. 4th Int. Conf. Fluidized-Bed Combustion*, MITRE Corp. McLean, Va., 1976, pp. 287–304.

20.1.5 Combustion of Pulverized Coal Under Pressure

Attempts to use the products of combustion from pulverized coal under pressure to supply a gas turbine have been aborted principally because of corrosion and erosion problems. Work since the late 1960s has been connected with the development of MHD combustors. Peculiar to this development is the need to produce a plasma with electrical conductivities of 5–10 mho/m. This requirement essentially means very high combustion temperatures (> 2600 K), as well as the addition of alkali metal salts to the working fluid (MHD) plasma. Furthermore, the plasma must be free of nonuniformities or flow instabilities. These requirements pose a severe problem on materials of construction, since a melt of high-alkali ash known to be very aggressive is produced.

Fundamental work on MHD development is extensively reported in the literature.[93] This chapter deals only with the aspects of MHD relating to the coal combustion discussed in Section 20.3, "System Application Development."

20.2 APPLICATION OF PRINCIPLES

Although the use of coal during the past 20 years in large installations has remained high in the face of competition from oil and gas firing, some markets for smaller coal-burning equipment have all but disappeared. Among these lost markets are applications of coal stokers to residential, small commercial, locomotive, and marine requirements. However the various types of apparatus and systems appear to have survived without exception, are still in widespread use, and will be available for application in any general return to coal firing in the future. The lost markets are a part of a general decline in stoker firing that also extends to a marked degree to the larger stokers used in industrial plants. As might be expected during such a period of almost complete neglect by potential users, little research or development has

[93] W. D. Jackson, M. Petrick, and J. E. Klepeis, *Trans. ASME, J. Eng. Power*, **92**, 217 (1970).

been reported. Although this is especially true for large stokers, lack of interest in smaller equipment and even in fundamental laboratory studies of burning in beds is also apparent.

Despite this recent history of abandonment of stokers, it is conceivable that they can now make a comeback, particularly in the larger sizes, under present and projected conditions of energy supply and demand. A good discussion of the factors involved in the increased future use of stokers is contained in a 1975 Battelle Energy Program Report.[94] Although the report deals with industrial-sized units, much of its general discussion is applicable to smaller installations. Increased cost and uncertain future availability of oil and gas were found to be the principal factors promoting renewed interest in stokers. The available data showed that stoker coal of acceptable quality could be supplied in the quantities required to meet any likely degree of expanded use. The main obstacle to coal firing was found to be environmental protection laws, particularly those limiting stack emission of sulfur compounds. It was also thought that the possible use of coal-based alternate fuels (such as those resulting from coal gasification or solvent refining) and the use of fluidized-bed combustors would potentially restrict the increased use of stokers. Since these coal-based alternative fuels and systems would not be available commercially for some 10 years or more, the report concluded that stokers could maintain a market through that period if stack gas emission requirements were relaxed in locations where reasonable primary ambient standards could be met. The authors anticipated that such relaxation may occur as a result of national fuel policy and because of the difficulty of developing successful stack gas cleanup equipment in the small sizes necessary.

In the field of pulverized-coal firing, on the other hand, the situation is quite different. In addition to retaining and improving older equipment types, manufacturers have developed new pulverizers and burners while greatly increasing the capacities of both old and new designs to meet recent demands of the electric utility industry. During the period of this review the average size of new units on order for U.S. utilities—by far the largest consumers of the coal burned—has increased from about 200 to 600 MW. However a better appreciation of the astounding size extrapolation achieved is produced by noting that coal-fired 1300-MW units have already been operating successfully for several years. This growth was not a continuous, uniform expansion; rather, it took place in big increments, and the whole process was challenged at every stage by a general deterioration of the coal and ash characteristics most troublesome to equipment designers. This situation has demanded a high degree of design innovation by steam generator manufacturers, particularly in response to the changing requirements among the fuels to be fired. Especially noteworthy is the increased use of lignite and subbituminous coals, the use of poor grades of anthracite, and the use and testing of such fuels as coal-water slurry, coal-oil slurry, coal chars, and SRC. Of 67 utility steam generators on order and scheduled for startup in the next 7 years, 61 will burn various ranks of coal and only 6 are designed for oil or gas firing. Of the 61 units intended for coal, 42 are designed for lignites or subbituminous coals,[95] some of which have very poor slagging and/or fouling characteristics.[96,97]

Industrial installations also represent a reviving market for pulverized-coal-fired steam generators, usually in sizes considered large for stokers (> 250,000 lb/hr of steam) or for fuel not well suited to stokers, such as fine coal, highly caking coal, or coal with a very low ash fusion temperature.

20.2.1 Stoker Firing

20.2.1.1 *Classification of Equipment*

Before discussing individual stoker applications and new developments, let us classify the various types of apparatus employed. Since classifications have been made on the

[94] R. D. Giammar, J. F. Foster, and R. B. Engdahl, *Stokers for Industrial Boilers: Assessment of Technical, Economic, and Environmental Factors,* Battelle Memorial Institute Energy Program Report, 1975.

[95] J. Graham, *Electr. World,* **186**(10), 43 (November 15, 1976).

[96] G. F. Moore and R. F. Ehrler, *Trans. ASME,* **98**(Ser. A, No. 1), 97 (January 1976).

[97] G. H. Gronhovd, R. J. Wagner, and A. J. Wittmaier, *Proc. Am. Power Conf.,* **28** (1966).

basis of mechanical design in the previous volumes of this series,[19,98] in various equipment-oriented books,[99,100] and more recently in other publications,[94,101,102] such material is not emphasized. A more fundamental classification based on the way the fuel and combustion air flow through the stoker is of greater use in analyzing the principles of operation and such operating characteristics as susceptibility to smoking, the influence of fuel properties, and allowable rates of load change. The earliest attempts to classify the flow geometry of fuel beds of different types with the object of understanding their underlying principles were made by Nicholls and Eilers[103,104] and later refined by Landry and others.[105,106]

In all stokers the fresh coal, when initially subjected to heat, gives off moisture and volatile matter. The solid residue, the evolved moisture and volatile, and the combustion air react in different ways, depending mostly on the basic flow configuration. Three relatively pure patterns of feeding coal and air have been developed and are identifiable both singly and in combination in available commercial equipment. These patterns are called overfeed, underfeed, and crossfeed. Because there are many examples of commercial equipment with more than one pattern of air and fuel feed, disagreement is sometimes found in the literature over which type predominates in a given stoker application.[98] However this is a minor matter and will not invalidate an analysis of characteristics made on the basis of flow geometry.

In the "overfeed pattern" (Fig. 20.14*a*) the flow of coal and air is countercurrent. In applications of this type the fuel is fed onto the top of the bed and flows down as it is consumed, while the air flows up through successive layers of ash, incandescent coke, and fresh coal. The newly charged coal is

[98] M. A. Mayers, in *Chemistry of Coal Utilization,* H. H. Lowry, Ed., John Wiley and Sons, New York, 1945, Vol. 2, Ch. 33.

[99] Babcock and Wilcox Company, *Steam: Its Generation and Use,* New York, 1975.

[100] A. J. Johnson, Ed., *Fuels and Combustion Handbook,* McGraw-Hill Book Company, New York, 1951.

[101] *Power,* **118**(3), S-1 (March 1974).

[102] T. E. Ctvrtnicek, S. J. Rusek, and C. W. Sandy, *Evaluation of Low-Sulfur Western Coal Characteristics, Utilization, and Combustion Experience,* EPA 650/2-75-046, May 1975.

[103] P. Nicholls and M. G. Eilers, *Trans. ASME,* **56,** 321 (1934).

[104] P. Nicholls, *Inst. Mining Met., Trans./Coal Div.,* **119,** 183 (1936).

[105] B. A. Landry, in *Encyclopedia of Chemical Technology,* R. E. Kirk and D. F. Othmer, Eds., Vol. 6, Interscience Publishing Co., New York, 1951, p. 903.

[106] E. P. Carman, E. G. Graf, and R. C. Corey, *Combustion of Solid Fuels in Thin Beds,* U.S. Bur. Mines Bull. No. 563, 1957.

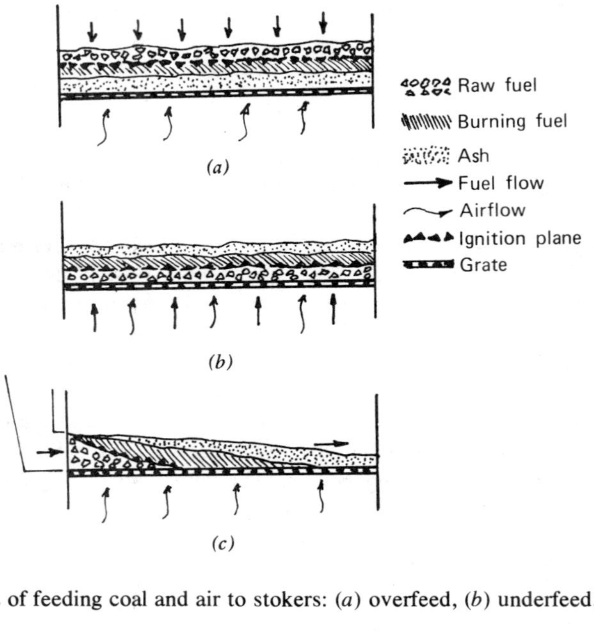

Figure 20.14 Patterns of feeding coal and air to stokers: (*a*) overfeed, (*b*) underfeed, and (*c*) crossfeed.

heated by contact with the burning coal beneath it and by the counterflow of combustion gases. The combustion products, as well as some volatile combustible material and particulate matter, rise into the furnace space above the grate without passing through a hot layer of partly burned coal. Generally, overfire air must be supplied above the fuel bed to help burn the combustible gases and solids. As combustion proceeds in the bed, the burned-out material builds up immediately adjacent to the grate and is periodically removed by dumping, by shaking or vibrating the grate, or in some stokers by continuous grate travel. Major examples of stokers with predominantly overfeed firing are the spreader stoker and some overfeed inclined-grate stokers, both front- and side-fed. Of course, most hand-fired grates are almost purely overfeed devices with the notable exception of the downdraft heaters discussed later.

The "underfeed pattern" appears in Fig. 20.14*b*. The flow of coal and air is parallel and usually upward. The evolved volatile matter, moisture, and combustion air pass through a layer of burning fuel; therefore this type of fuel bed produces less smoke during feeding and low-load operation than is encountered in overfeed firing. Nevertheless many underfeed stokers, particularly the on-off type, are arranged for overfire air admission during all or part of the off period. Single- and multiple-retort stokers are principally the underfeed type combined with crossfeed action, depending on the extent to which some of the coal flows at right angles to the flow of air.

The "crossfeed pattern" (Fig. 20.14*c*) is a widely applied method of air and fuel feed. In this pattern the fuel normally moves horizontally, with the air moving upward at right angles to the fuel. This firing pattern is embodied in front-hopper-fed stokers equipped with chain grates, traveling grates, and vibrating, reciprocating, or oscillating grates. Pure crossfeed would occur if the angle between the ignition plane and the grate were 90°. In practice, however, the incoming fuel is ignited on top and the ignition travel and airflow bear the same relationship in this first stage of burning as prevails in underfeed firing. The subsequent combustion of the incandescent fuel beyond the ignition plane takes place in the crossfeed mode. Since the makeup of the fuel bed and the rates of burning vary along the path of coal travel, the air supply to the bed in crossfeed stokers is frequently zoned into a number of crosswise compartments arranged along the grate from the coal feed end to the ash discharge.

For a more detailed discussion on fuel bed geometry, reference should be made to the sources cited above, particularly Landry.[105] However even this brief analysis will provide much of the material necessary to interpret the summary of stoker operating characteristics given in Table 20.2. Three of the four methods described in Table 20.2 are illustrated in Figs. 20.15–20.17. We have not attempted completeness, but have limited the tabulation to the main features of equipment most likely to find widespread use in the future.

20.2.1.2 *Coal-Fired Heaters*

As previously noted, except for its continued use with older equipment in existing installations, coal has almost disappeared in residential and small commercial service applications. In recent years in the United States, for boilers and hot-water heaters up to several million Btu/hr duty, not one new installation in a thousand was served by a stoker.[108] This proportion has held for some years and is largely a result of the lower overall cost, greater cleanliness, and increased automation achievable in firing oil and gas. A quantitative idea of the extent of the decline of small stokers in these markets lies in the astonishing figures quoted by Landry and Sherman indicating sales of 176,000 such units in 1941 and 172,000 in 1946. Even between 1954 and 1958 small stoker sales averaged 11,600 units per year.[19]

From applications made in recent years and the lack of new designs, it is clear that if coal is applied to an increasing degree for residential and other small uses, it will likely be fired by screw-operated underfeed stokers or in some form of downdraft heater, perhaps arranged for continuous charging and automatic control. Although there has been no significant research or development of residential stokers since the 1950s, there has been further refinement of the so-called downdraft heater for domestic coal firing. The improvements in such heaters have been part of a long-term search for smokeless combustion of high-volatile coals in compact apparatus. In the United States, the United Kingdom, and Germany, heaters have been developed that achieve good efficiency and

Table 20.2 Characteristics of Various Bed-Burning Methods

Description	Fuel Bed Type	Feed Location	Feed Equipment	Principal Characteristics
Spreader-fed stoker (Fig. 20.15)				
Traveling grate Oscillating grate Vibrating grate Dumping grate	Overfeed-crossfeed (combined with suspension burning)[a]	Seeded from above—more or less uniformly over grate area	Mechanical or pneumatic thrower	Burns almost any properly sized bituminous, subbituminous, or lignite coal.[b] Allowable range of ash fusion temperature and swelling index is greater than for front gravity feed on the same types of grate, possibly because of preoxidation in suspension.
				Can be operated over load range down to 15% of maximum on most coals. Operates with thin fuel bed, thus responds quickly to load changes. Since much of the coal burns in suspension, enough particulate matter is carried out of the furnace to require a dust collector (or in some instances an electrostatic precipitator). Fly ash reinjection is usually practiced, in which case combustibles loss is optimized against the costs of the refiring system and increased operating expenses, such as power, system maintenance, and erosion of boiler parts. Overfire airjets are usually used to eliminate low-load smoking and reduce combustibles loss. Spreader stoker maximum heat release rates per unit of grate area are usually 50–60% higher than can be achieved with other types of stoker. Can be made in a wide range of sizes from less than 10,000 to 400,000 lb/hr steam.

Table 20.2 Continued

Description	Fuel Bed Type	Feed Location	Feed Equipment	Principal Characteristics
Single retort (Figs. 20.16 and 20.23)	Underfeed-crossfeed	Coal pushed in under coal bed	Reciprocating ram or screw feeder	Burns all ranks of coal from anthracite to lignite. Designs imparting considerable agitation to the coal can handle many Eastern caking coals. Although ash-softening temperatures as low as 2100°F can be handled at reduced grate liberations, these stokers operate best on coals with ash-softening temperatures above 2400°F. Can be operated over a very wide load range from banked to maximum rating.[c] Operates with a sizable fuel bed, therefore responds slowly to load demands. Operates with very low particulate carryover. Although relatively smokeless in operation even at low loads, overfire air is frequently provided to prevent smoke formation following fast load changes. Made in a range of sizes from residential use to about 40,000 lb/hr steam.
Multiple retort	Underfeed-crossfeed	Coal pushed in under coal bed	Reciprocating plate feeder or rams	With inclined grate and rear ash discharge, burns narrower range of coals than single-retort side discharge stoker. Best coals to use are those that cake but break up when agitated. Free-burning coals drift toward rear of stoker and highly caking coals impede airflow. High-volatile coals (>30%) tend to smoke. Can be operated over a wide load range, but most suitable for relatively constant loads—high or low. Low particulate carryover. Can be made in very large sizes—300,000 lb/hr steam or more. Not often used because of cost and performance limitations.

Front-hopper fed (Fig. 20.17)			
Traveling grate Chain grate Vibrating grate Oscillating grate Reciprocating grate	Crossfeed	Deposited across front of grate by gravity[d]	Burns all ranks of coal except strongly caking coals (because fuel bed is not agitated). If free-swelling index tends to get too high, double-screened coal 1½ × ⅜ in. is recommended.[e] Ash-softening temperature should be above 2200°F. Can be operated over a wide load range. Load changes can be accommodated faster than with underfeed stokers, but not as fast as with the spreader. Low to moderate particulate carryover from fuel bed. Dust collector required in some instances. Amount and size consist of feed should be uniform across width of stoker. Vibrating grates are usually water cooled (on both front-fed and spreader stokers), so require relatively little maintenance. Can be made in a wide range of sizes (15,000–300,000 lb/hr steam).

[a] From 25–40% may be burned in suspension; greater amounts have been claimed for certain ranks of coal and favorable feed sizing.

[b] With all stokers, best feed sizing depends on fuel being fired, furnace size, stoker-grate release rate, frequency and extent of load changes, pollution control requirements, and other lesser factors. Refer to more detailed discussions[99,100,107] and to literature and operating instructions issued by major manufacturers.

[c] See page 1343.

[d] Frequently assisted mechanically by reciprocating plate feeder or similar device.

[e] Coals of fairly strong caking characteristics have been burned on vibrating, oscillating, and reciprocating grates.

Figure 20.15 Spreader-fed stoker with traveling grate.

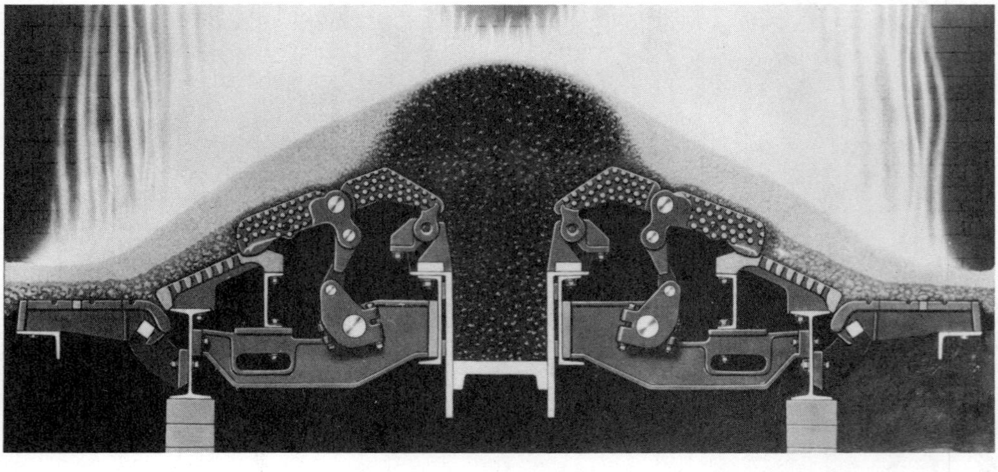

Figure 20.16 Single-retort stoker with side ash discharge and undulating grates.

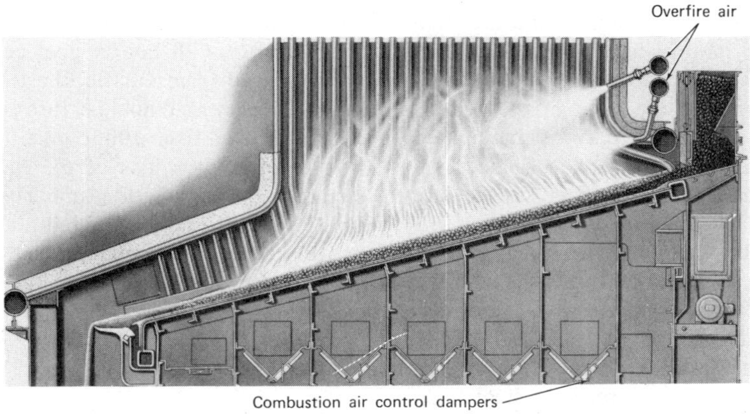

Figure 20.17 Front-hopper–fed, water-cooled, vibrating-grate stoker.

smokeless combustion even with intermittent charging and relatively simple airflow control. Some of this development occurring before 1950 has been described by Landry and Sherman[109] and Fellows.[110] Smokeless heaters described 20 years later than the units detailed in these two references have shown only small differences in configuration and air admission pattern.[111,112] In fact, the basic fuel and air-feeding patterns were published in 1934.[103]

The Rayburn Smoke-Eater,[112] a fan draft unit developed by the National Coal Board of Scotland, is said to have achieved acceptance when burning bituminous coals in strictly controlled smokeless zones in the United Kingdom. The New Zealand Coal Research Association imported several of these heaters for testing on indigenous coals and found that they complemented nicely the Juno heater,[111] a natural draft heater better suited to freer burning subbituminous coals.

The Rayburn Smoke-Eater is noteworthy as being typical of the more successful downdraft heaters marketed to date. Figure 20.18 shows that the draft through the fuel bed is essentially downward from the upper free surface of the bed to the bottom rear of the heater, where the gases issue into the combustion chamber and are mixed with secondary air. This type of bed burning is a combination of the underfeed and crossfeed patterns illustrated in Fig. 20.14. A heater of this type is started by igniting a layer of lump coal on the grate by any available means of kindling. The depth of this layer is then increased, rather slowly at first. After a good fire and draft have been established, however, the final charging can be quite rapid without smoke being formed. The primary air passing downward and across the coal bed burns the coal within and above the starting layer, and any smoke formed passes

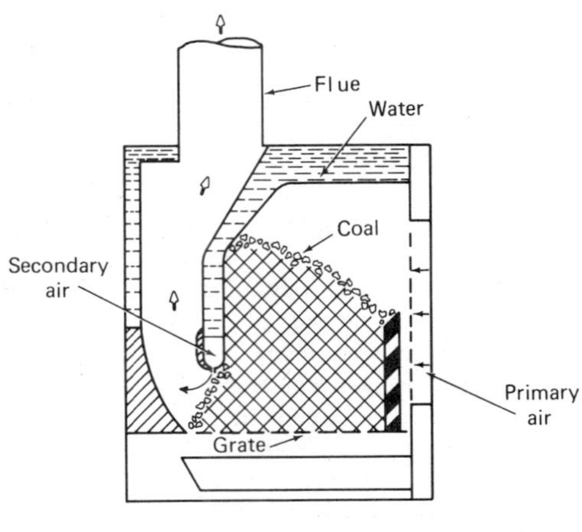

Figure 20.18 The Rayburn Smoke-Eater.[112]

[107] M. O. Funk, *Proc. Am. Power Conf.* **25,** 505 (1963).

[108] Confidential statistics of American Boiler Manufacturers Association and Hydronics Institute, 1976.

[109] B. A. Landry and R. A. Sherman, *Trans. ASME,* **72,** 9 (1950).

[110] J. R. Fellows, Univ. Illinois Eng. Exp. Sta. Circ. Ser. No. 58, 1948, pp. 38–58.

[111] "Down Draught Space Heaters," *Heat. Air Cond. Refrig. Vent.,* November 1968.

[112] D. V. Shannon and D. Welsh, *Fact. Manage.,* **4**(2), 36 (1973) (Auckland, New Zealand).

through the hottest portion of the fuel and burns, partly in this region and finally by mixing with the secondary air in the combustion chamber at the back of the heater. From this description it may be anticipated that the pattern of airflow and its distribution between primary and secondary sources are key elements in the proper functioning of this apparatus. The shape of the coal-charging chamber and the downdraft of air cause the coal to subside in the chamber as it is burned, thus providing the gravity feed so highly desirable in these simple devices. As the coal approaches the grate, it receives heat from the combustion zone, and gaseous constituents begin to distill off and are carried through the layer of charred coal and into the combustion chamber. The charred coal, having now descended into the grate area, burns in a crossflow of air, and the resultant gases also pass into the combustion chamber.

These installations operate in an on-off control mode, effectively producing two levels of output, depending on whether the fan is operating. The fan may be operated on a timed cycle, or it may be thermostatically controlled to maintain a given room or fluid temperature. With the more reactive coals, reestablishing ignition after a shutdown of a day or more is accomplished by merely turning the fan on again.

Some of the test results recorded by the New Zealand Coal Research Association are summarized in Table 20.3 and shown in Figs. 20.19 and 20.20. The quantities tabulated are for tests run with hourly coal charging and are averaged from curves like those in Fig. 20.19. The average flue gas loss is calculated and subtracted from 100 to give the combustion efficiency. A loss of 6% has been assumed to cover radiation, unburned combustibles, and other unaccounted-for items; this figure is subtracted, and the result is reported as "thermal efficiency." Total effective heat

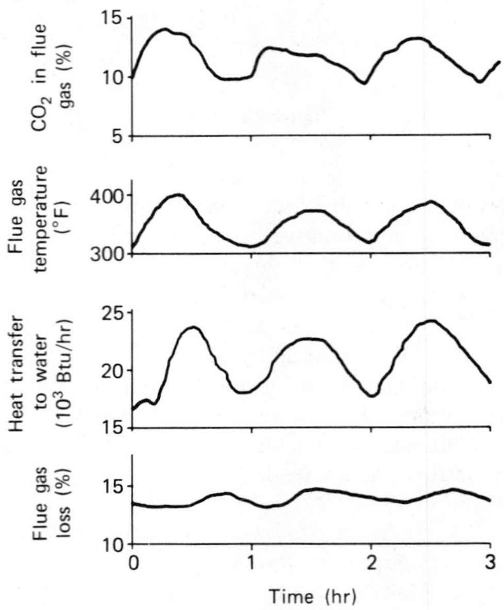

Figure 20.19 Tests of the Rayburn Smoke-Eater with Alpine coal (hourly charging).[112]

Table 20.3 Results of Rayburn Smoke-eater Tests with Hourly Coal Charging[112]

	Coal Type		
Data	Strongman Bituminous	Alpine Bituminous	Smith's Subbituminous
Proximate analysis (%)			
Moisture	7.0	11.2	15.8
Ash	4.2	4.5	2.8
Volatile matter	38.0	36.9	37.1
Fixed carbon	50.8	47.4	44.3
Calorific value (Btu/lb)	13,030	11,830	10,820
Swelling number	3	1	0
Rate of feed (lb/hr)	3.25	3.75	4.00
Average gas temperature (°F)	342	347	437
Average CO_2 content (%)	10.8	11.4	14.5
Combustion efficiency (%)	86.2	86.2	84.7
Thermal efficiency (%)	80.2	80.2	78.7
Total effective heat (Btu/hr)	34,000	35,600	34,000
Heat transfer to water (Btu/hr)	20,700	20,500	27,900
Bonnet heat (Btu/hr)	13,300	15,100	6,100

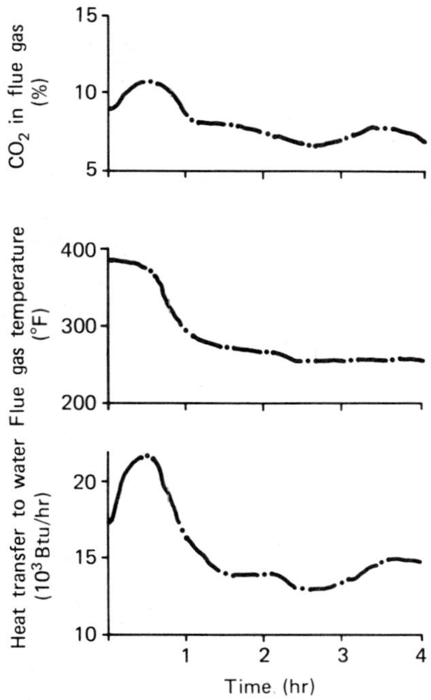

Figure 20.20 Tests of the Rayburn Smoke-Eater with Strongman coal (4-hr charging).[112]

is the hourly heat fired multiplied by the thermal efficiency. Bonnet heat is the total effective heat minus the measured heat added to the water. Considering the intermittent nature of the fuel-charging and airflow cycles, these results are quite good. Coals of low sulfur content and moderate swelling characteristics are most suitable. In particular, the swelling number should probably be no greater than 1½ for the overnight operating period, but it may be higher during normal daytime operation, where small disturbances incidental to charging and other activities tend to prevent bridging.

Research has also continued on the downjet burning principle for use as either a combustor or a coal gasifier. Downjet combustion is a fixed-bed burning technique intended to achieve efficient, high rates of burning, flexibility in control, smokeless operation, gravity feed, and simple air admission. A good discussion of the principles involved, the theoretical goals, and the work done up to 1957 can be found in the first Supplementary Volume.[19] There downjet combustion is more or less written off as succumbing to difficulties of clinker removal, irregular feeding of fuel, blockage of airflow, blowing fuel from the bed, and other largely fuel-dependent difficulties.[113]

An exception to the general lack of success is the application to the burning of sod peat, reported by Miller.[114] The sod peat used is sun dried to 30–40% moisture content and is a completely noncaking fuel. The design of furnace and air jet is essentially that devised by Thring (Fig. 20.21).

Another satisfactory application of the downjet configuration is in the combustion of brown coal briquettes, as reported by Laing.[116] Apparently, some hundreds of furnaces with capacities between 100,000 and several million Btu/hr have been operated in Australia, burning Victoria brown coal. Furthermore, Evans[117] has reported favorable results in the range of domestic heat outputs, whereas previous investigators had been least hopeful of success in this area.

The present state of development is illustrated by an interesting paper by Palmer[118] and a subsequent exchange of comments between Laing[119] and Palmer.[120] In addition

[113] P. M. Frazier and R. B. Engdahl, Battelle Memorial Institute Rep. No. 274 to Bituminous Coal Research, Inc., December 1949.

[114] H. M. S. Miller, *Int. Peat Symp.*, Dublin, Sect. D2, July 1954.

[115] M. W. Thring, *Trans. Faraday Soc.*, **42,** 366 (1945).

[116] K. W. Laing, "The Downjet Combustion of Briquettes," paper presented at Institute of Fuel (Australian Membership) Conf., 1968.

[117] D. G. Evans, "Downjet Combustion of Coal at Low Burning Rates," paper presented at Institute of Fuel (Australian Membership) Conf., 1966.

[118] E. R. Palmer, *J. Inst. Fuel,* **44**(370), 653 (December 1971).

[119] K. W. Laing, *J. Inst. Fuel,* **45**(375), 280 (May 1972).

[120] E. R. Palmer, *J. Inst. Fuel,* **45**(378), 481 (August–September 1972).

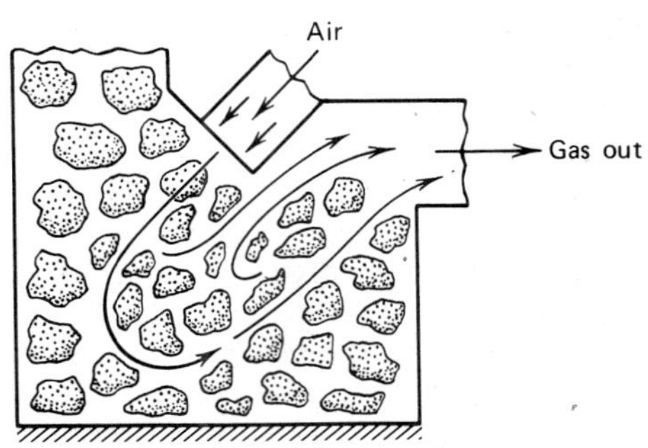

Figure 20.21 Principle of down-jet firing as proposed by Thring.[115]

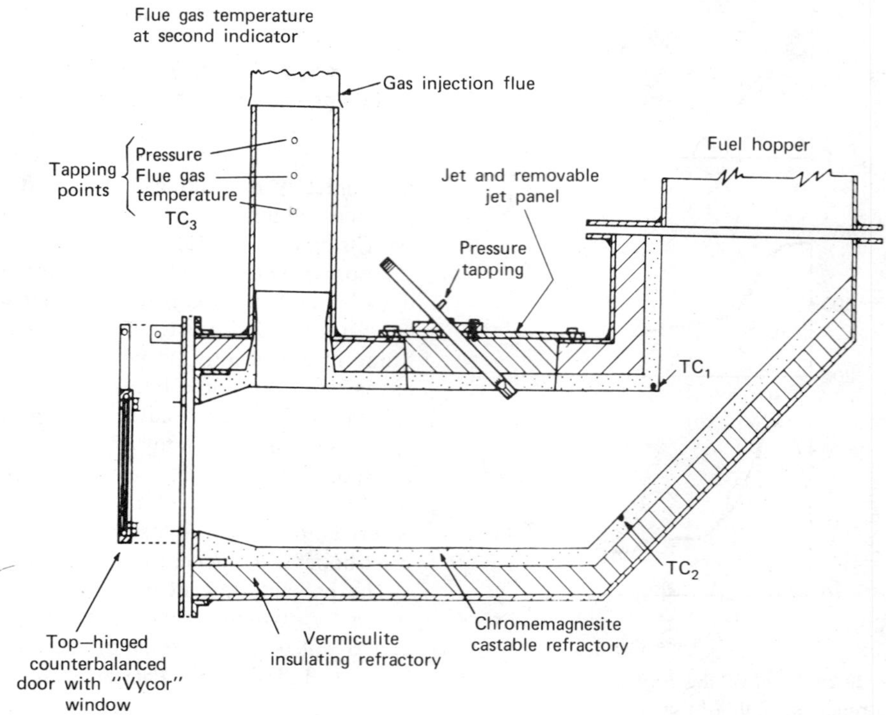

Figure 20.22 Experimental down-jet furnace.

to burning tests in the 9 × 9 × 24 in. (40,000 Btu/hr) combustion chamber shown in Fig. 20.22, Palmer used a cold-water flow model to study jet mechanics and particularly the entrainment of gases in the free space above the bed, with a view to better mixing and completion of combustion in a smaller volume. Essentially, he moved the jet away from the fuel bed and determined a curved combustion chamber configuration, which largely eliminated secondary turbulence and to a stable and effective mixing pattern. Although Palmer has made some progress toward a fundamental treatment of the subject, it is apparent that correlation of performance with laboratory values is not altogether reliable because of the sensitivity of designs to a number of coal and coal ash properties. It seems that coal suitability is determined satisfactorily only by burning tests under the proposed conditions of unit size and operating regimen.

This discussion illustrates the continuing efforts to develop downdraft heaters and downjet combustors, whose potential for wider application will be enhanced by a further understanding of their operation, a better knowledge of suitable coal properties, and improved methods of coal charging and ash removal. It must be emphasized, however, that screw-fed underfeed stokers with manual removal of ash or clinker comprise the greater proportion of coal-firing applications to dwellings and small commercial enterprises. This situation is likely to continue well into the future.

20.2.1.3 *Small Stokers*

Although the application of small stokers to institutional and commercial installations in capacities up to about 30,000 lb/hr steam will surely increase in the next few years, a review of available literature and correspondence with equipment manufacturers reveals that no fundamental changes in stoker designs have been made since the 1950s. Various refinements of existing types have been achieved, but it is probable that new installations in the sizes considered here will be of the types already found most suitable for these classes of service. The types most often used are the single-retort underfeed design and a crossfeed stoker with a vibrating or oscillating grate like that applied to the series of Coal Pak boilers[19,121] developed in the mid-1950s by Bituminous Coal Research, Inc. Spreader stokers using dump, vibrating, or even traveling grates may be used in the upper portion of the capacity range.

Underfeed stokers are usually fed by a screw feeder or a reciprocating ram or pusher. In the smaller sizes the screw feeder predominates, and there are numerous manufacturers making successful equipment of this type. All this equipment dates from the period preceding this review and has been well described, with details of its operating characteristics and the range of coals it can handle. Although the underfeed stoker is suitable for all ranks of coal from anthracite to lignite, it does work less favorably, and at reduced ratings, with coal that has a low ash fusion temperature (< 2400°F) and produces a strong clinker. There are many recorded attempts to circumvent these limitations by providing grate mechanisms affording bed agitation and by using several methods that produce a degree of preoxidation of the coal before ignition,[16,19] thus achieving the same conditioning effect that seems to result from the suspension phase of spreader stoker firing. Equipment designs affording preconditioning by oxidation and/or partial devolatilization are also described by L'Hermitte and Lemaire in a paper on small automatic boilers[122] and by Feldhoff[123] in a paper on the current status of stoker development.

A method of assessing the clinkering potential of a coal in bed burning is described by Dumoutet and Desseine.[124] In their work coal was fed into a laboratory pot furnace, and measurements were taken leading to determinations of apparent specific volume, permeability, and mechanical strength of the resultant clinker. Analysis of a series of such tests led Dumoutet and Desseine to conclude that the usual tests, such as ash fusion temperature and free-swelling index, often misguide testers in their efforts to prejudge clinkering behavior and may be replaced advantageously by a standardized test on a grate in a laboratory combustion pot like theirs.

The single-retort underfeed stoker is particularly suited for steady load operation and may be the only satisfactory device providing automatic operation on load cycles that include long banking periods or extended periods of very light loads. Many residential and commercial applications where the duty is primarily space heating have load cycles of this type. The fundamental characteristic that makes the single-retort stoker suitable for this type of operation is its ability to maintain ignition over a relatively small part of the total grate area, then to automatically expand the active burning area as the controls call for more fuel and air. As shown in Fig. 20.23*a*, at light loads the reduced feed of fuel and air permits the bed to burn out over the discharge grates at the sides, yet maintain active burning over the retort in the center. It is infeasible with other stoker types to reduce the heat release to a low banking rate over a major portion of the fuel bed area without losing ignition temperature.

Generally, the feed rates of small underfeed stokers are set so that the stoker runs about 75% or more of the time while satisfying normal demand. In the "off" periods some smoke may be produced by continued devolatilization of coal without sufficient air supply. This smoke can be reduced by an overfire air system that may start as the stoker motor stops or after an adjustable time delay. In some units the overfire air supply has been controlled by electronic smoke density detection cells.

Other stoker types particularly suitable for the range of firing considered here (roughly 10^6 to 40×10^6 Btu/hr) are the front-fed and spreader-fed vibrating-grate designs. The stoker used in the Coal Pak series of package boilers and hot-water heaters, previously mentioned, is the front-hopper-fed type. These boilers are now marketed in sizes up to 15,000 lb/hr steam (around 20×10^6 Btu/hr), although this is probably only a temporary commercial plateau. Other package boilers, which are being offered in sizes up to about 25,000 lb/hr steam, employ spreader-fed vibrating grates. Many of these are of the two-drum baffleless type shown in Fig.

[121] P. O. Kock, unpublished ASME paper No. 57-A-229, 1957.

[122] L'Hermitte and Lemaire, "Combustion of Low or High Volatile Coal in Automatic (Central Heating) Boilers," *J. Combust. Convers. Energ.*, 597 (1964).

[123] K. Feldhoff, *Tech. Mitt. (Essen)*, 63 (May 1970).

[124] P. Dumoutet and J. Desseine, *J. Combust. Convers. Energ.*, 419 (May 1964).

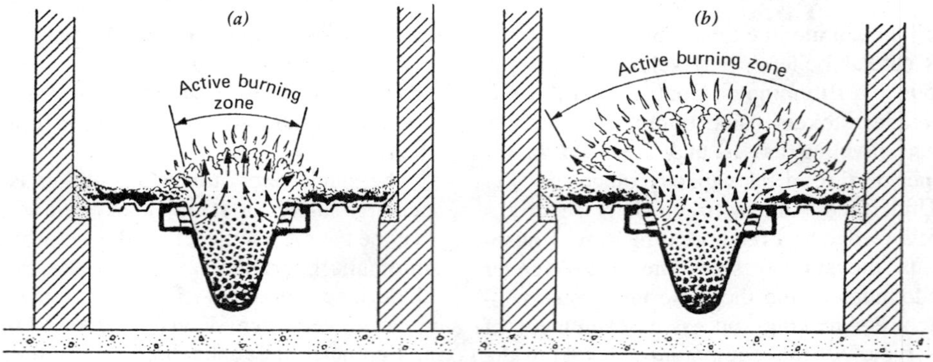

Figure 20.23 Single-retort stoker showing active burning zone at (*a*) partial load and (*b*) full load.

20.24—a good arrangement for achieving low draft loss and erosion-free operation. Well-designed boilers of this type are provided with an overfire air system that handles as much as 20% of the required combustion air, depending on the fuel specified and the contemplated operating cycles. This air is introduced over the grate, usually through the rear wall, at sufficient pressure (25–30 in. H_2O) to penetrate the furnace volume and create the desired turbulence and mixing. Usually fly carbon reinjection in such small units is limited to that collected in the boiler hoppers, because any further reduction in combustible loss gained by reinjecting from more collecting points is offset by the additional cost and operating problems.

The vibrating grate oscillates intermittently for a period of a few seconds at intervals of several minutes. The oscillations are large, around ⅛–3/16 in., and of low frequency (e.g., 400–600 cycles/min). Their action lifts and throws the coal bed toward the ash discharge end, producing a fuel bed of very uniform thickness like that produced by a fair degree of bed poking or stirring. As a consequence, the hopper-fed vibrating-grate stoker has been used for almost as wide a range of coals as the spreader stoker. Coals with ash content up to 20%, volatile as low as 16%, heating value down to 9000 Btu/lb, and swelling index up to 9 have been fired. When coal containing a high proportion of fines is fired, the undergrate air may be throttled automatically while the grates are vibrated to minimize particulate carryover.

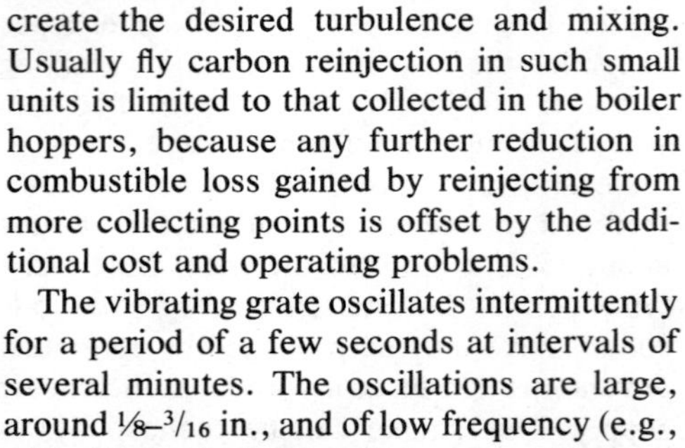

Figure 20.24 Stoker-fired package boiler.

20.2.1.4 *Large Stokers*

The whole range of industrial applications of stoker firing, comprising individual units of about 20,000–300,000 lb/hr steam (or more) is now completely dominated by spreader stokers with the various applicable types of grate, such as stationary or dumping, traveling, and reciprocating, vibrating, or oscillating. This observation is substantiated by Fig. 20.25, which shows the spreader grate area as a percentage of the total grate area sold in the United States between 1950 and 1975.[125] If properly converted to a heat release basis (no heat release figures were available), the percentages would be even higher, since spreader stokers normally operate at grate releases above those used with underfeed or front-hopper-fed stokers.

The advantages of spreader stoker firing are so attractive that they are likely to assure its continued popularity in the absence of startling innovations in other equipment. Besides the higher achievable grate release rates, which permit more economical boiler designs, its wide range of suitable coal properties and its ability to follow rapid load changes are potent arguments for its use. Nevertheless, costly allowances must be made in spreader stoker applications for ac-

[125] Confidential statistics of American Boiler Manufacturers Association.

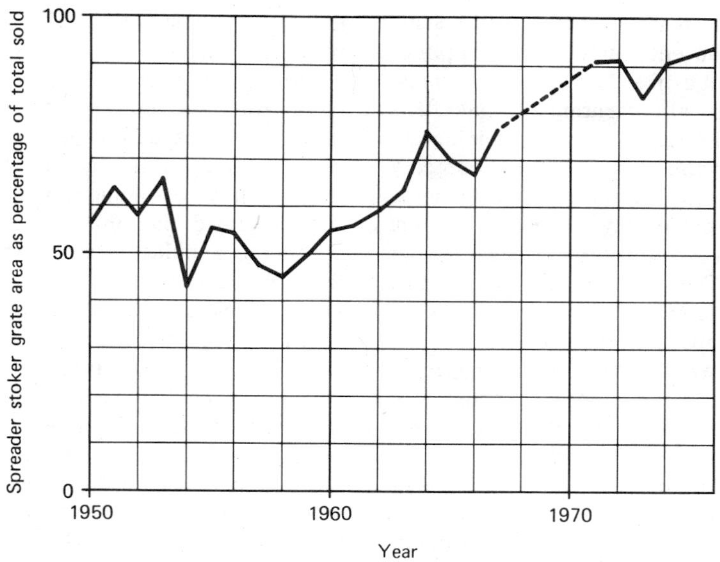

Figure 20.25 Spreader stoker grate area as percentage of total grate area sold in the United States, 1950–1975.[125]

knowledged difficulties with smoke at low loads and particulate carryover at high loads. Furthermore, the full range of stoker equipment types described in older publications is still available for application, with underfeed stokers getting some preference in the smaller sizes and front-gravity-fed traveling or oscillating grates maintaining a market in the mid- to larger sizes.

In view of its dominant position, it is almost certain that future research, development, and testing of stokers will be mainly devoted to spreader stoker firing. This conclusion is borne out by a proposal recently accepted by the U.S. ERDA (subsequently DOE) from a group of major stoker manufacturers to develop design and operating parameters for industrial stoker applications.[17] The proposal recommends testing only spreader stokers until a later stage of testing, which might then include another type of stoker, such as a front-fed chain or traveling-grate unit. The intent is to test five installations representing a variety of furnace and stoker designs and sizes to determine response to changes in coal analysis and sizing, degree of fly ash reinjection, and overfire air admission and to determine emitted pollutants. The frankly stated positon of the major manufacturers indicates that they have insufficient information on the foregoing items to accurately design for the wide range of coal and ash properties now confronting them and to meet the laws and regulations governing stoker-fired boiler operation.

Prior to this recently accepted test program, no doubt inspired by the 1973 oil shortage, the efforts of the Spreader Stoker Research Committee (1948–1951) represent the last major stoker development work undertaken.[94] One of the principal efforts of this committee was to study the effect of overfire air (or steam) jets on reducing combustible loss and smoke and fly ash emission.[126] The committee also investigated the effects of fly ash reinjection on combustible loss and particulate emission.[127] Some of the data published,[127] particularly the astonishing reduction in carbon loss, have long been regarded with incredulity. In fact, very few data that are really dependable, accurate, and widely applicable were published on the variables mentioned. The lack of such information is most apparent to anyone who has tried to compare and evaluate competitive stoker bids within the past 20 years.

The combustible loss incurred in stoker firing is higher in general than that experienced with the same fuel burned in pulver-

[126] W. C. Holton and R. B. Engdahl, *Trans. ASME,* **74,** 207 (1952).

[127] C. H. Morrow, W. C. Holton, and H. L. Wagner, *Trans. ASME,* **75,** 1363 (October 1953).

ized form in a steam generator of the same capacity. The recent escalation in fuel prices has made combustible loss one of the most important items in comparing alternate fuels or modes of firing. Therefore obtaining accurate data on this subject will undoubtedly be a prime object of future investigations. The many variables involved in predicting combustible losses form the greatest barrier to progress. These factors can be grouped into three categories: fuel properties, design features, and operating conditions.

Among the most important fuel properties are coal rank, ash content, maximum feed size, feed size distribution, ash fusion temperature, and free-swelling index.

The standard laboratory tests for some of these properties correlate imperfectly with actual performance, and investigators are actively attempting to devise improvements or uncover new relationships. The efforts of Dumoutet and Desseine[124] on a clinker test that might include the permeability of the clinker, and of Benedict and Berry,[20] who tried to correlate grate release rate in a laboratory furnace with the petrographic character of the coal, have already been mentioned.

Design features of importance in determining combustible loss are heat release per unit of grate area, amount and location of carbon reinjection, amount and location of overfire air, effectiveness of air seals, grate type and design, and furnace size and shape.

These and similar factors, dependent on the skill and experience of individual designers in the past, now may get the type of industry-wide attention necessary to meet present-day fuel supply conditions and environmental protection requirements.

Among operating conditions demanding skill and care to achieve optimum combustible loss, the foremost probably is maintenance of the firing equipment and combustion controls. In addition, fuel deliveries should be monitored to assure proper fuel characteristics, particularly size consist, and coal should be stored and reclaimed in such a manner as to avoid size segregation. These and other factors relating to good design and operation are covered in a recent updating of industrial boiler practice.[128]

In 1963 the stoker manufacturers presented a set of carbon loss charts, averaged from the data in their files.[107] These charts illustrate the influence of many of the above-mentioned fuel and design variables. The carbon losses given are for bituminous coal only and are based on the assumptions that effective overfire air jets are provided and that the furnace volume is adequate. Finally, it is assumed that the coal sizing will be within the manufacturer's recommended limits, and curves are given for the various stoker types to show the sizing required. The resultant carbon loss charts (Figs. 20.26–20.29) for the most common stoker types are said to represent consistently achievable values. However poor operation and maintenance, improper coal sizing, and small furnaces (e.g., those on older units or retrofit stokers) will result in losses much higher than those shown by these curves.

Figure 20.27 shows the favorable influence of fly ash reinjection on the combustible loss with spreader stoker firing. Unfortunately, the greater the extent of collection and reinjection of refuse (ash plus carbon) leaving the furnace, the greater the dust loading passing through the boiler, heat recovery, and dust collecting equipment. Of course, the amount leaving the dust collectors to the stack is also increased. Reinjection from a high-efficiency collector can increase recirculating load and stack emission manyfold; therefore full reinjection is rarely considered.

In the absence of accurate test data concerning the relationship among combustible loss, stack emission, and recirculating load, we can refer to a paper by Mullen[129] giving helpful methods for approximating the desired relationships by calculation. Mullen starts by presenting test data showing the percentage of the fired fuel ash (A_F) that is discharged as bottom ash (A_G). Figure 20.30 for spreader stokers and Figure 20.31 for front-fed traveling grates are reproduced directly from that paper. These curves show A_G/A_F at high grate loadings to range from 0.60 to 0.75 for spreader stokers of the continuous ash discharge (CAD) type and to be higher and less variable for front-fed traveling-grate stokers. The variation is attributable to differences in fuel sizing, fuel rank, and, most important, to the grate heat release rate. Then, with the help of assumptions that are approximate but adequate for illustration,

[128] *Power*, **121**(1), S1 (February 1977).

[129] J. F. Mullen, ASME Paper No. 64-WA-Fu-4, 1964.

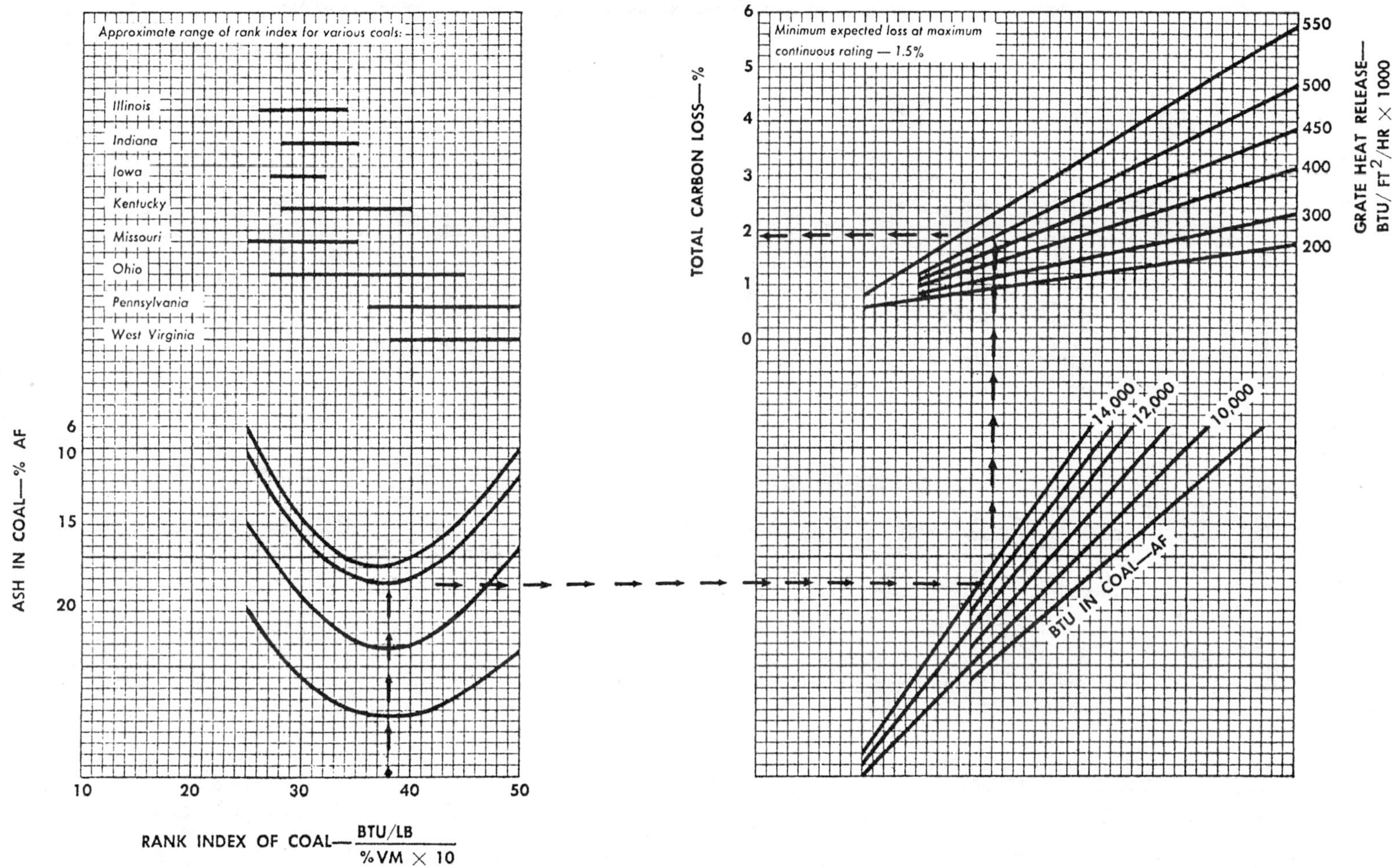

Figure .0.26 Total carbon loss in overfeed stokers, including ash in pit, fly carbon, and siftings.[107]

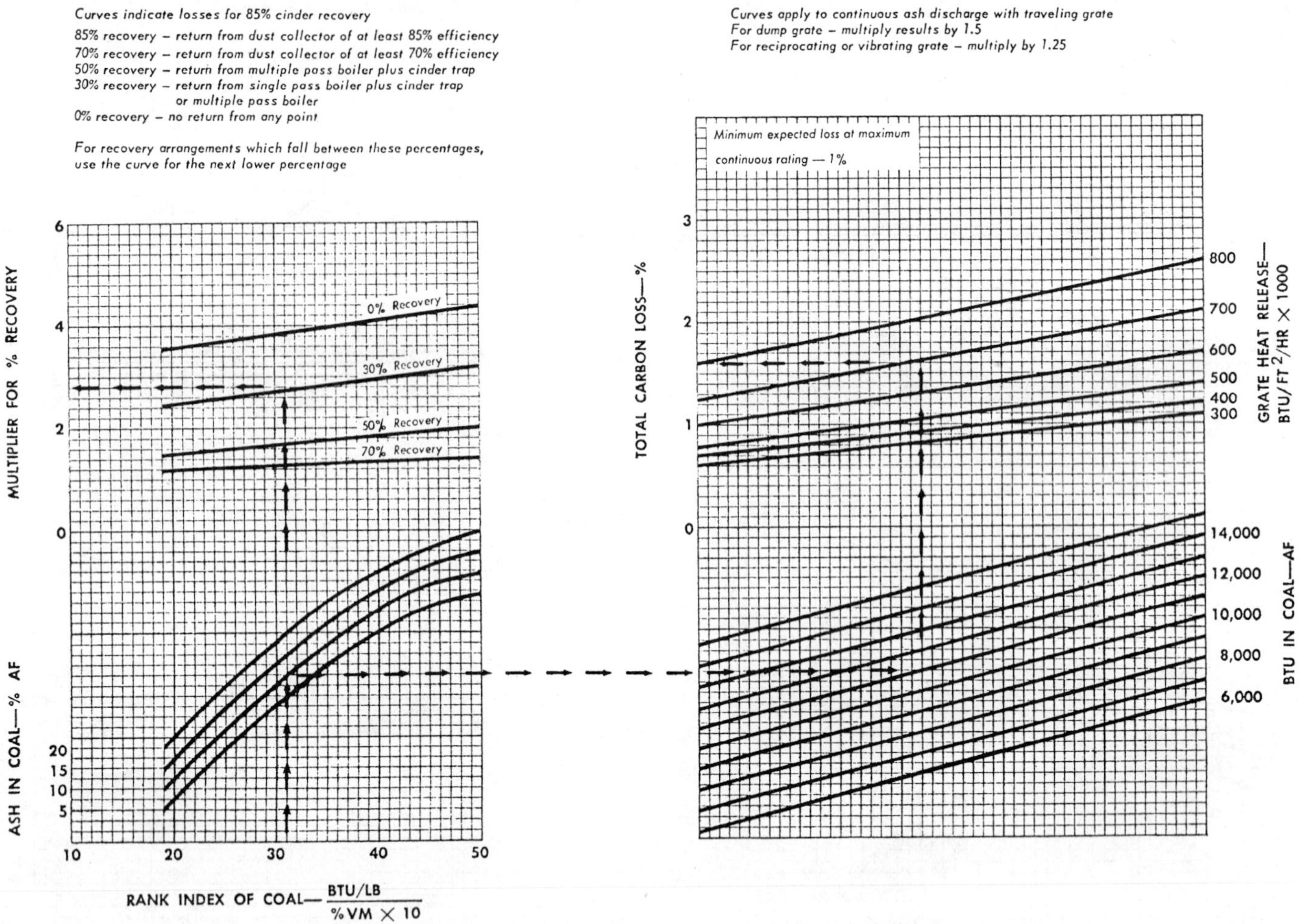

Figure 20.27 Total carbon loss in spreader stokers, including ash in pit, fly carbon, and siftings.[107]

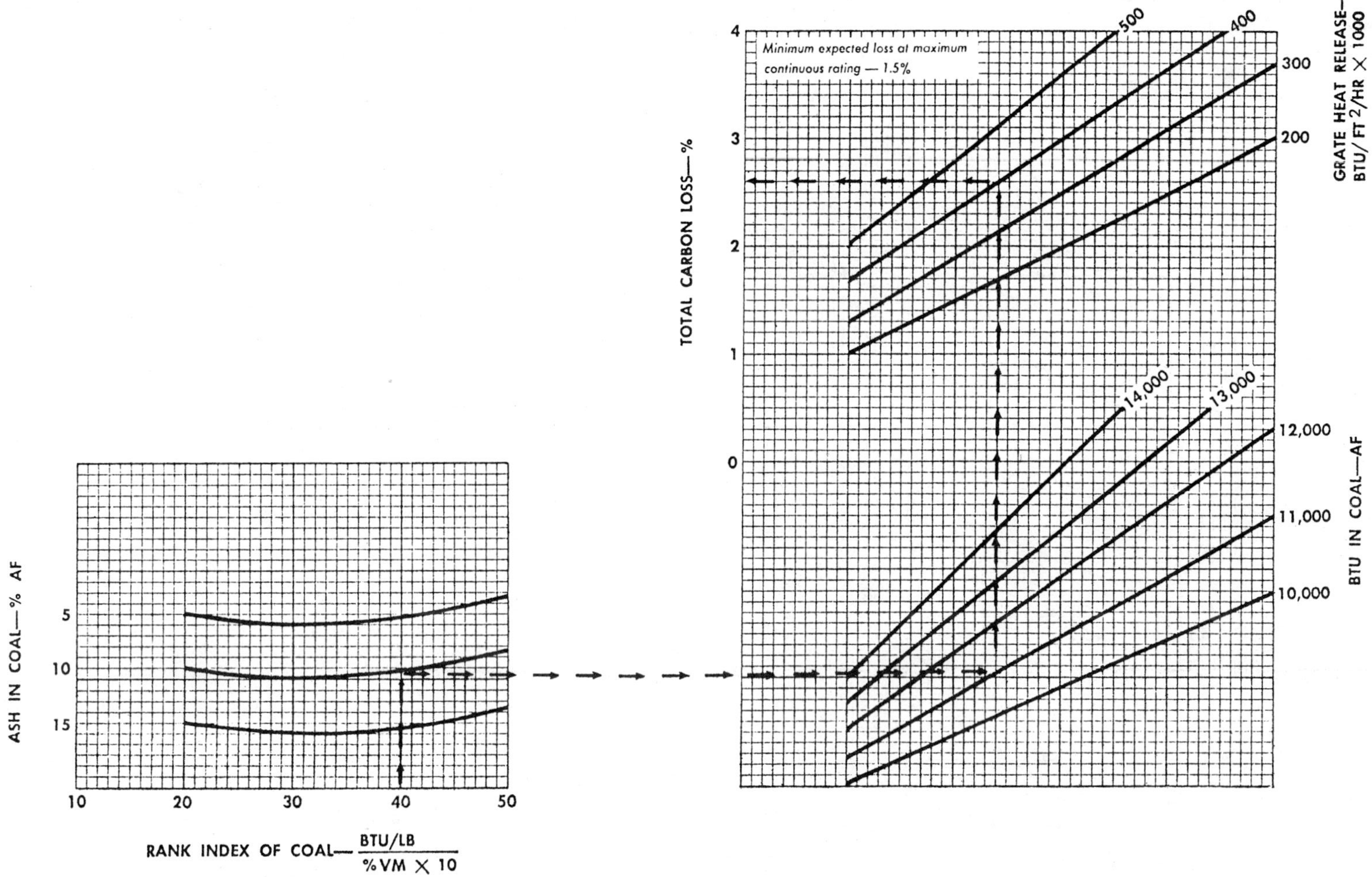

Figure 20.28 Total carbon loss in underfeed stokers with side ash discharge, including ash in pit, fly carbon, and siftings.[107]

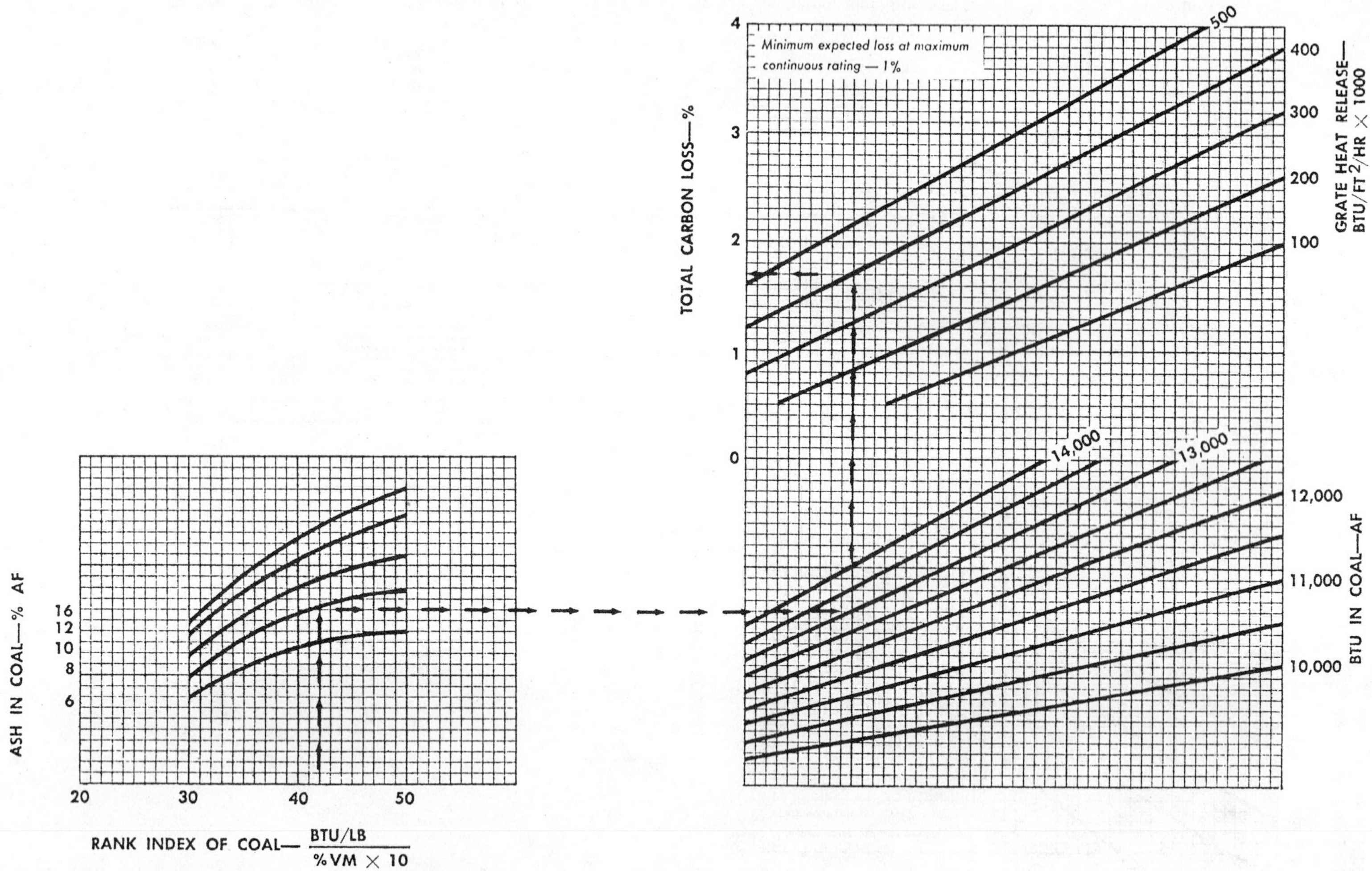

Figure 20.29 Total carbon loss in underfeed stokers with rear ash discharge, including ash in pit, fly carbon, and siftings.[107]

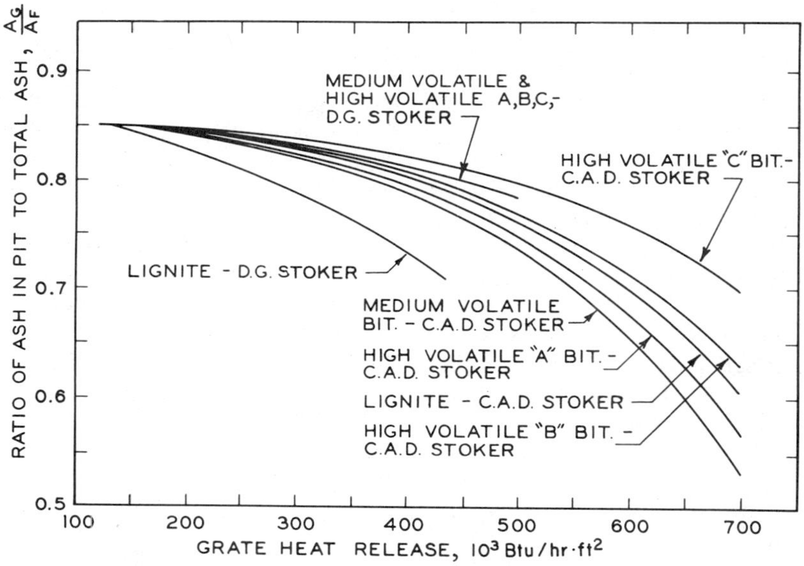

Figure 20.30 Ratio of ash in pit to total ash ($A\hat{G}/A\hat{F}$) versus heat release for coal-fired, low-set spreader dump grate, and continuous discharge stokers with partial reinjection.[129]

Mullen calculates several interesting examples showing why full reinjection of fly ash is not practiced.

In the first calculation using only ash quantities (assuming complete combustion in the furnace), Mullen[129] shows a tenfold increase in recirculating load and stack emission when going from boiler and economizer hopper reinjection to full reinjection from a dust collector. Mullen proceeds to a second calculation in which specific combustible contents are assumed in the boiler, economizer, dust collector, and stack refuse—once again comparing stack emission and recirculating load for partial and total reinjection. These increase thirteenfold for a reduction in carbon loss of about 1% under the assumed conditions of combustible content. Actual

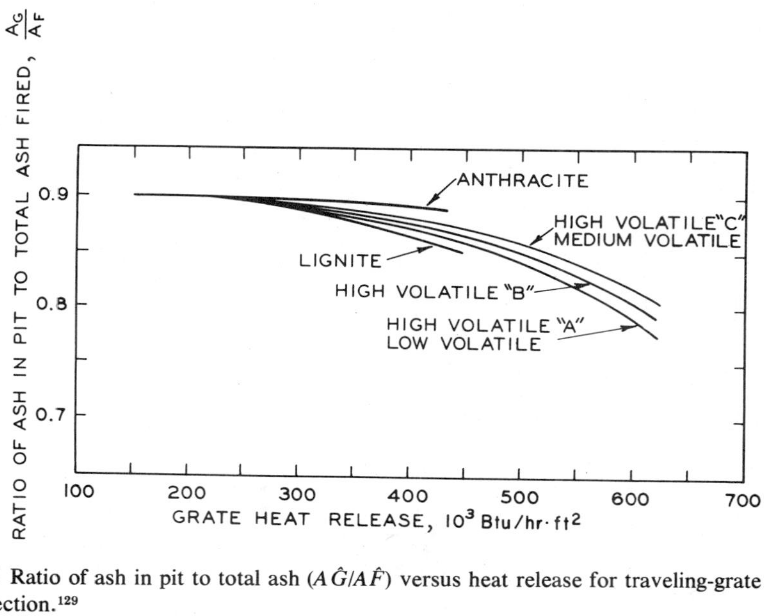

Figure 20.31 Ratio of ash in pit to total ash ($A\hat{G}/A\hat{F}$) versus heat release for traveling-grate stokers with partial reinjection.[129]

test data do not give nearly as great an increase in recirculating load or as small a reduction in carbon loss as obtained by Mullen's calculations. The calculations reflect the following assumptions:

1. The percentage of entering refuse collected in hoppers and dust collectors will not increase as the recirculating load increases. Experience shows that it frequently does increase.
2. The combustible content of the refuse at various extraction points will not decrease as the recirculating load increases. Experience shows that it does decrease.

The principal conclusions of this paper are, therefore, exaggerated, but they are by no means invalidated. Mullen's methods of calculation are clear and simple and provide a way of performing rough checks on the reasonableness and compatibility of claims regarding combustible loss and ash reinjection and of checking bad test data.

Figure 20.30 shows that a spreader stoker will emit from the furnace between 25 and 40% of the fuel ash fired at a normal full-load rating of 600,000 Btu/hr-ft². This compares to about 12–18% from a front-fed traveling-grate stoker at 500,000 Btu/hr-ft².

Data published by an industry committee[130] on 54 emission measurements for spreader stokers showed the total emission in grains per standard cubic foot to fall into two distinct groups reflecting the effect of firing with and without reinjection of ash from boiler and dust collector hoppers. The group representative of reinjection (25 measurements) fell between 2.6 and 5.2 gr/scf, with a mean of 3.71. The lower group (17 measurements), representative of no reinjection, averaged 0.364 gr/scf. These values give an idea of the range encountered for a wide variation of undefined fuel and operating conditions. Further testing is much needed, with accurate control of the variables previously mentioned.

Figure 20.32, from the same committee publication, gives an idea of particle size distribution in the emissions from several types of stokers. The committee recommended the data for general reference only, not for equipment specifications, because so few measurements were available and the oper-

[130] Industrial Gas Cleaning Institute–American Boiler Manufacturers Association, *Criteria for the Application of Dust Collectors to Coal-Fired Boilers: The Results of an IGCI–ABMA Joint Technical Committee Survey.*

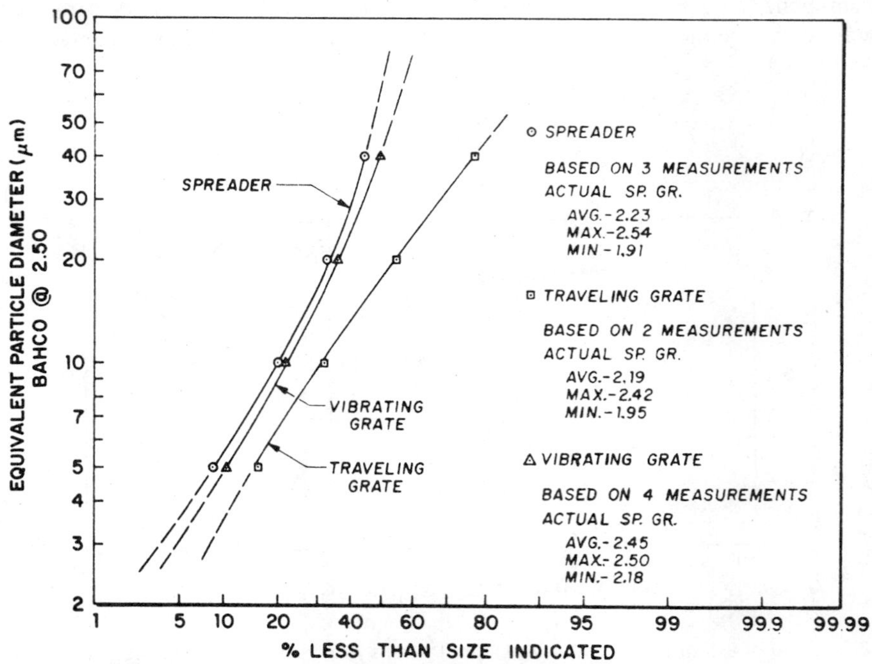

Figure 20.32 Ash particle size distribution for stoker-fired boiler.[130]

ating conditions for most of them were unknown.

20.2.2 Pulverized-Coal Firing

Most coal-fired industrial steam generators with a capacity greater than about 250,000 lb/hr steam, and virtually all coal-fired electric utility units, are fired by pulverized coal rather than by stokers. At present, the largest single utility unit produces about 9,500,000 lb/hr steam (1300 MW), and several companies are operating units of this size.[131,132] Experience with these and other very large units has not revealed any specifically size-related difficulties, and further size increases can be made when justified by economics and load conditions.

The capacity of pulverized-coal firing to accommodate readily all the historical demands for increased size is one of the strongest motivations for its almost universal adoption. However the most powerful incentive for its use over the entire range of sizes has probably been its astonishing freedom from restriction by the extremes of coal properties.

Compared with other types of firing, such as stokers or the use of granular coal in cyclone furnaces, the selection of pulverized coal as a firing method is almost unaffected by coal characteristics. Of course a particular installation will be restricted in the kinds of fuel it can burn, but in general some form of pulverized-coal system is suitable, with proper design provisions, for virtually any coal mined in the world. In addition to natural coals, the cokes and chars derived from them in various gasification processes and the solid resulting from solvent refining offer no technically insurmountable problems to firing in pulverized form.

If pulverized-coal firing is defined as the burning of finely ground coal in suspension, the major applications either already accomplished or envisioned for the near future fall into two categories:

1. The use of natural coals, coal residues, SRC, or partly dewatered slurries of these fuels—all pulverized and blown into suspension with transport air in the conventional manner.
2. The use of pulverized coal that has been slurried in water, oil, methanol, or some other combustible medium and blown into suspension in the form of a finely divided spray. (Slurry firing is discussed in Section 20.2.2.4.)

20.2.2.1 *Classification of Equipment*

Pulverized-coal-fired units may be classified broadly into two types according to whether their furnaces are designed for removal of ash in dry or molten form. The dry ash furnaces are frequently referred to as "dry bottom" furnaces, and the molten ash furnaces are normally called "slag tap" or "wet bottom" furnaces.

Dry ash furnaces are the more common of the two and in the past 20 years have steadily increased their proportion of the installed capacity. They are now almost the only type sold in the United States, even for the coals of high ash content and low to moderate fusion temperature formerly thought to be most suitable for wet bottom firing. Considering all aspects of design and operation, it is apparently economical in most instances to provide enough furnace volume and cooling surface so that all the ash fired may be withdrawn in dry form, with no objectionably large accumulations occurring in the furnace or convection passes.

The operation of dry ash units is essentially simpler, more flexible with regard to load range and fuel properties, and more reliable than that of slag tap units. Their main disadvantages are that they are usually bigger (thus more costly) and that 80–90% of the ash fired must be removed from boiler and precipitator hoppers in the form of very fine dust.

The intent in developing slag tap units was to avoid this problem of dust disposal as much as possible, by getting the ash into a heavier, granular form and retaining much more of it in the furnace than is possible in a dry ash unit. In wet bottom units the streams of molten ash flowing from the furnace are chilled by cold water that breaks them into the desired coarse product. Ash retentions of more than 80% have been achieved with some designs. At the same

[131] C. M. Davis and R. F. Hering, ASME Paper No. 72-WA/Pwr-11, 1972.

[132] G. W. Bouton and W. A. Hansen, *Proc. Am. Power Conf.*, **30,** 383 (1968).

time higher volumetric and area heat release rates are used in an effort to reduce equipment cost. More extended discussions of wet and dry bottom firing, considering in detail the advantages and disadvantages of each, are given in works by Gumz,[16] Schröder,[133] Ledinegg,[134] and Dolezal.[135] The extensive use of molten ash firing in Central Europe, as reflected in these references, results from the properties of the coals available there and from other local social and economic pressures.

Cyclone furnace firing is a special form of wet bottom operation not generally classified as suspension burning, since a large portion of the fuel is burned on the surface of a moving slag layer. In cyclone firing one or more small combustors, in which burning is essentially completed, are mounted directly on a wall of the main furnace, the combination thus forming a two-stage combustion chamber. Although most American cyclone furnace units are fired with coal crushed to about ¼-in. top size, foreign (particularly German) practice uses partially pulverized coal, sometimes as coarse as 25% through a 200-mesh sieve. With suitable coals, good designs of this type show a retention in the furnace of 80–90% of the coal ash. Other advantages and disadvantages of cyclone firing are treated in Section 20.2.3, "Cyclone Furnace Firing," and in the references previously cited.[16,99,133–135]

20.2.2.2 *Dry Bottom Firing*

In dry bottom firing the ideal furnace is so designed that the flames clear the walls, at most lightly brushing them; the temperatures developed near the walls produce little or no ash melting; and the products of combustion are cooled sufficiently before leaving the furnace to prevent troublesome ash adherence in the convection banks. This ideal is rarely achieved, however, and most actual units experience some slagging and/or fouling of heat transfer surfaces. The most frequently used furnace and burner configurations are shown in Fig. 20.33. These arrangements cover firing systems suitable for all ranks of coal from peat to metaanthracite[136] and for the lowest coal qualities as measured by high ash or moisture content, low heating value, low ash fusion temperature, and high potential for slagging or fouling.

Horizontal and Opposed Horizontal. These furnace types (Fig. 20.33*a, b*) are usually fired by circular turbulent burners spaced uniformly across the width of the furnace in either the front or rear wall (or both). In such arrangements each burner has its own relatively independent flame envelope and ignition point, and the firing system may be designed so that an individual burner can be placed in service, adjusted, or removed from service independently of the other burners. Some small units are fired by burners in one or both *side* walls, but this is rare. Figure 20.34 shows a 900-MW steam generator arranged for opposed-horizontal firing with turbulent burners in the front and rear furnace walls. This unit is served by eight ball-tube mills firing high-volatile bituminous coal, with each mill feeding coal to four burners in a horizontal row spread uniformly across the width of the furnace. The furnace is 80 ft wide and 46 ft deep. The furnace heat release rate including flat projected finishing superheater surface is 75,500 Btu/hr-ft^2.[137] Other references to recently installed horizontally fired units have already been given.[131,132]

Tangential and Opposed Inclined. Particularly in tangential (or corner) firing, but to a certain extent also in opposed-inclined (turbofurnace) firing (Fig. 20.33*c, d*), the burner turbulence ranges from nil to moderate, according to the designer's intent, and is replaced by the overall furnace turbulence—that is, the furnace is the burner. In such furnaces there is effectively a single overall flame envelope and, if desired, an extended zone of slow turbulent mixing. These characteristics give the designer a degree of control over the peak flame temperature that can favor low levels of NO_x formation, although care must be taken to avoid flame instability at partial loads when the individual burners tend to assume their own identity. It has also

[133] K. Schröder, *Grosse Dampfkraftwerke,* Vol. 3, Part A, Springer-Verlag, Berlin, 1966, pp. 51–59.

[134] M. Ledinegg, *Dampferzeugung,* 2nd ed., Springer-Verlag, Berlin, 1966.

[135] R. Dolezal, *Large Boiler Furnaces,* Elsevier Publishing Company, Amsterdam–New York, 1967.

[136] ASTM Standards, Part 26, Designation D388, Classification of Coals by Rank, 1976.

[137] *The 1976 Electric Utility Generation Planbook,* McGraw-Hill Book Company, New York, 1976, pp. 25–34.

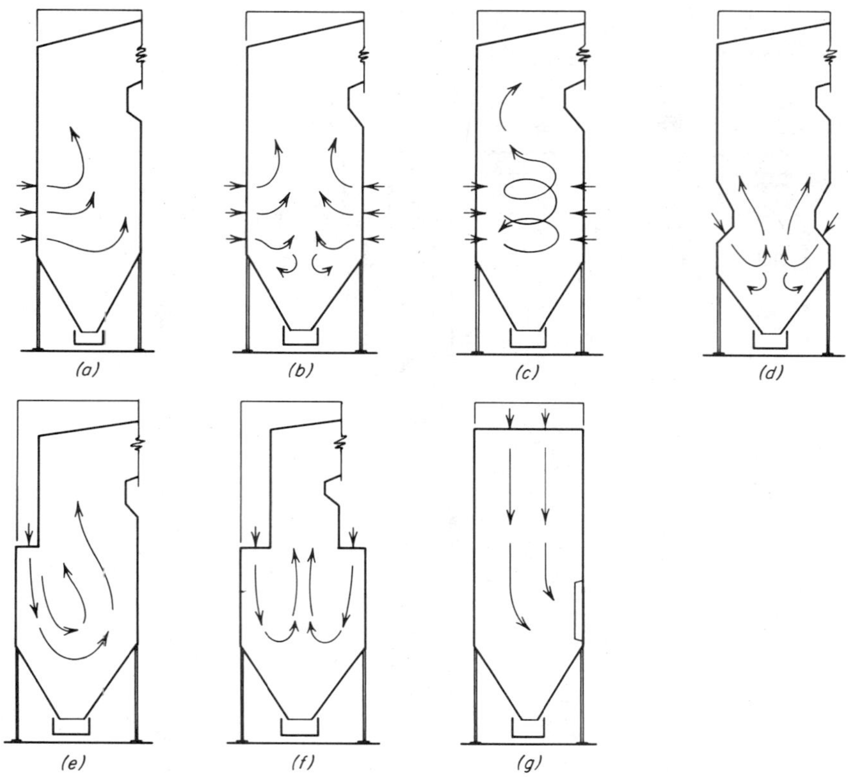

Figure 20.33 Dry bottom furnace and burner configurations: (*a*) horizontal (front or rear), (*b*) opposed horizontal, (*c*) tangential, (*d*) opposed inclined, (*e*) single U-flame, (*f*) double U-flame, and (*g*) vertical.

been argued that such firing arrangements are less vulnerable to unequal proportioning of air and fuel to the burners, but it is doubtful that this tolerance can be relied on any more than in any other firing arrangement without incurring objectionable imbalance of superheater or reheater element temperatures or the occurrence of local or generalized tube wastage. Figure 20.35 illustrates a recently built public utility unit of the tangential firing type. Of the many published descriptions of units of this type, the one given by Hanzalek and Boudon[138] discussing the units at Keystone Station is particularly informative. Their paper goes into unusual detail concerning the arrangement of the coal nozzles, warmup guns, igniters, and flame scanners. It also shows a line diagram of the pulverizer fuel and air flow system. Additional units of this type are described, with interesting information relating to subbituminous coals, by Hamilton et al.[139]

Single U-Flame and Double U-Flame. U-Flame firing (Fig. 20.33*e, f*) is used for firing such fuels as anthracite and coke that are hard to ignite and that burn slowly. The fuel, with its transport air and a portion of the necessary combustion air, is fired vertically downward through a burner arch located about halfway up the furnace. The remaining air required for combustion is admitted through the vertical walls below the arches and joins the fuel on the downward portion of its path. Radiation from the rising portion of the coal flames and from the burners in the opposite arch (in double U-flame units) helps to assure stable ignition over a good load range. Figure 20.36 shows a 350-MW unit that operates on anthracite fines containing 15% surface moisture, 32% ash, and 6%

[138] F. J. Hanzalek and P. E. Boudon, *Proc. Am. Power Conf.*, **26,** 409 (1964).

[139] T. B. Hamilton, R. B. Knust, and R. P. Hensel, *Proc. Am. Power Conf.*, **30,** 394 (1968).

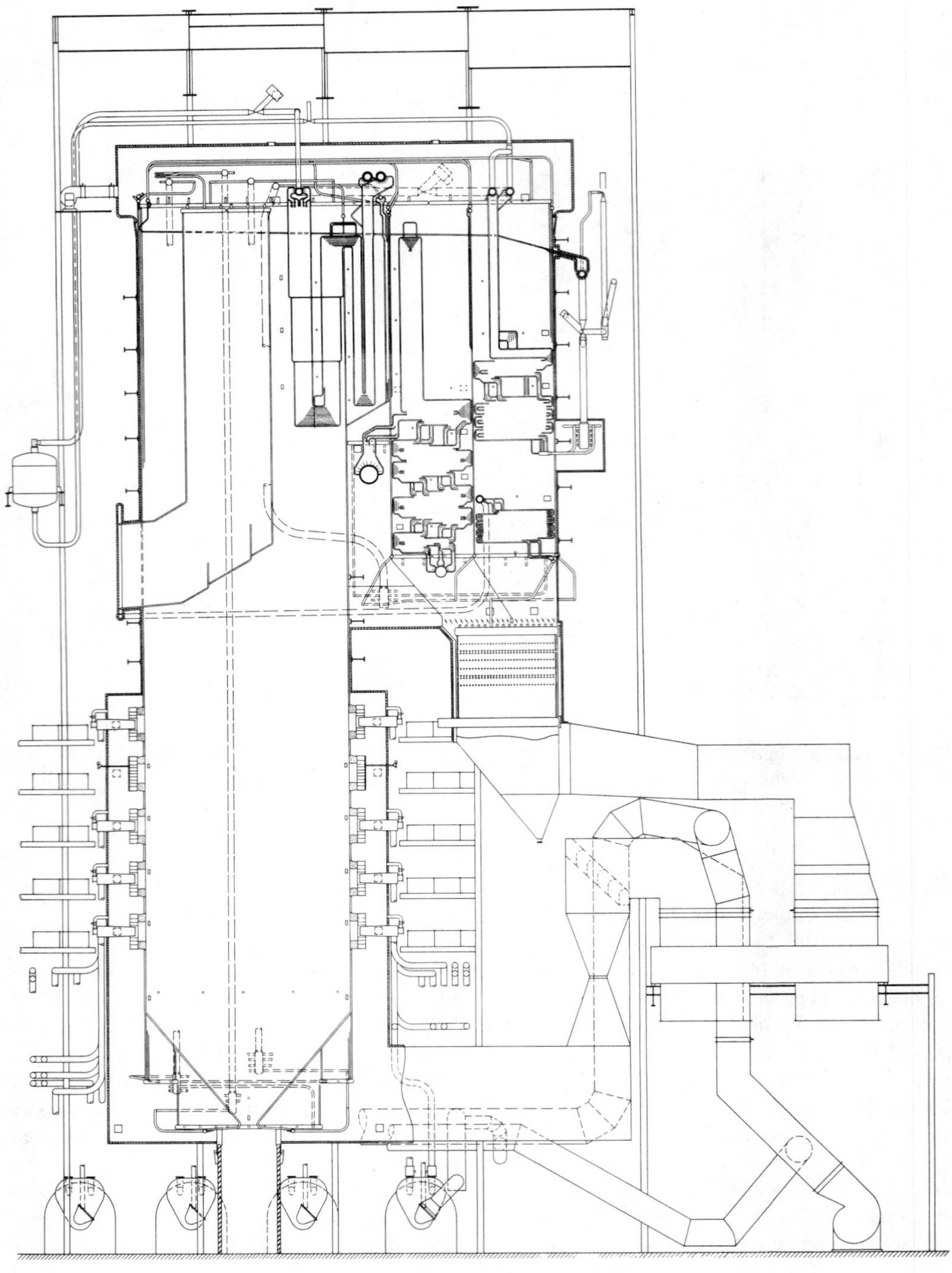

Figure 20.34 Steam generator arranged for opposed-horizontal firing.

volatile matter. This coal is fired alone, with no supporting fuel from about one-quarter load to full load. At maximum output, with all mills in service, the coal fineness is 85–90% through a 200-mesh sieve.

In addition to coal fineness, other important variables in this type of firing include (1) furnace volume and proportioning of volume between lower and upper parts, which must conform to the unusually difficult ignition

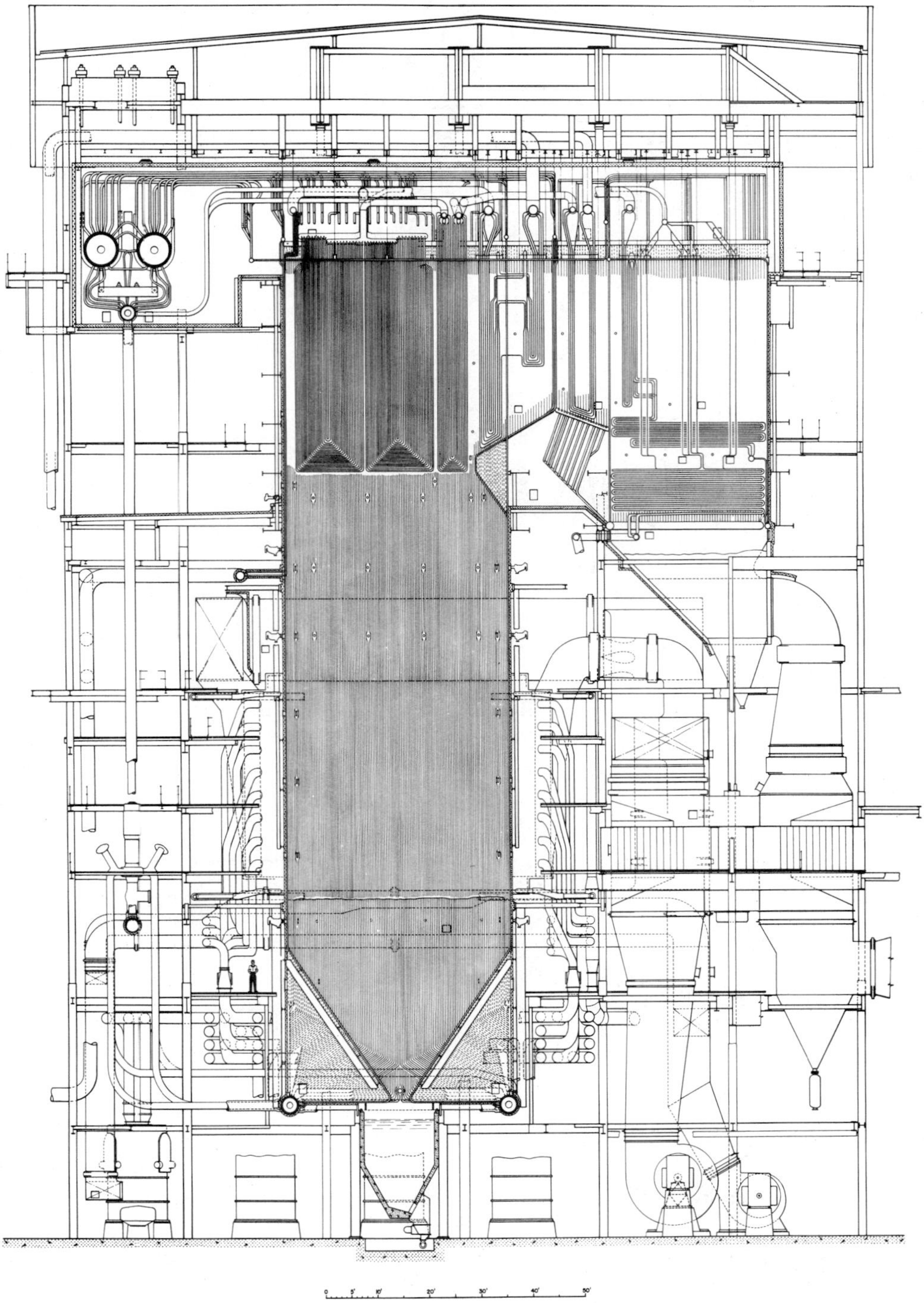

Figure 20.35 Steam generator arranged for tangential firing.

to the natural flow dynamics of the air and fuel streams before and after ignition; (3) amount, velocity, and location of burner arch air and lower wall air; and (4) amount and location of refractory-covered furnace wall surface.

Experience has shown that relatively small and burnout characteristics of anthracite and

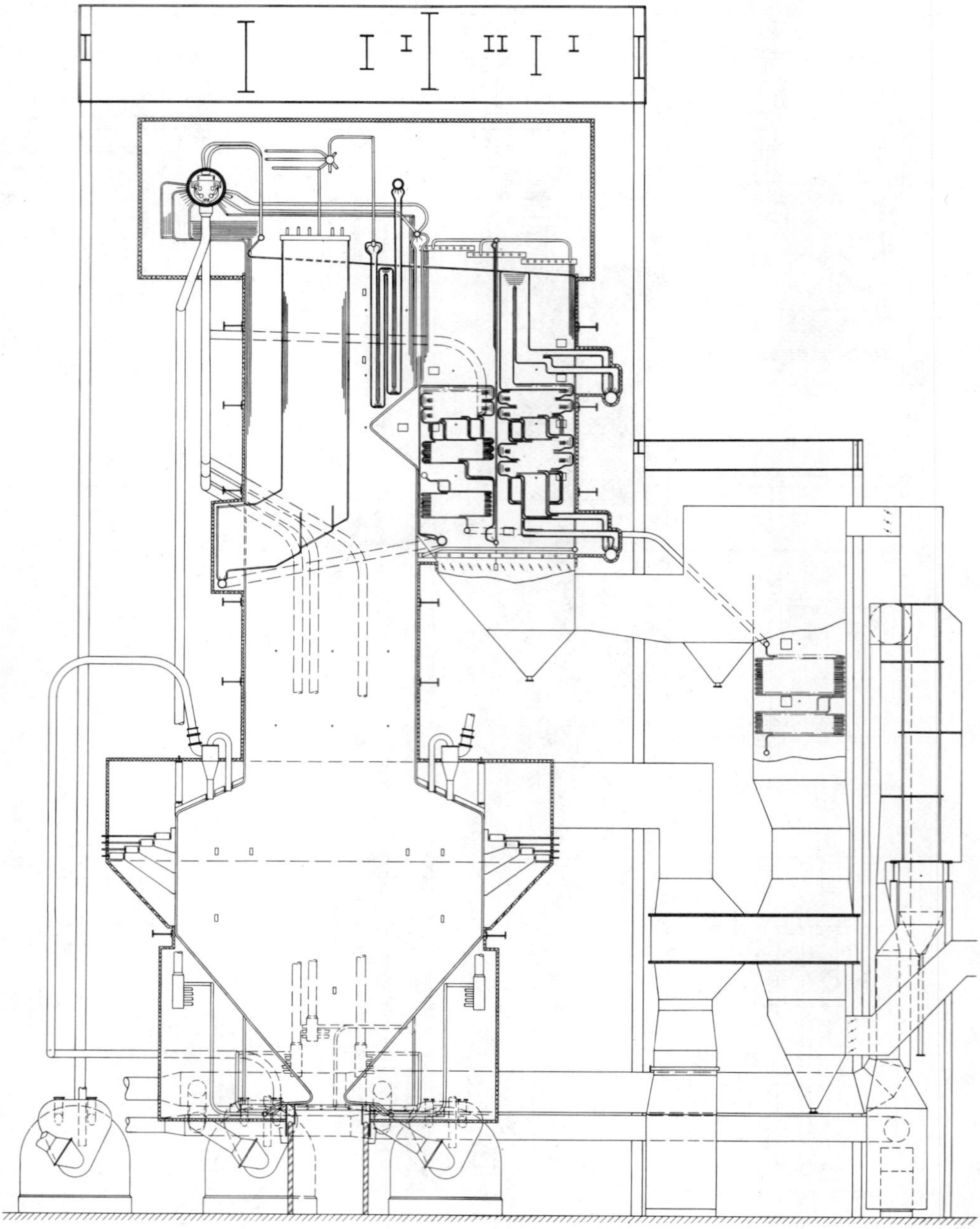

Figure 20.36 Steam generator arranged for double U-flame firing.

coke; (2) furnace shape, which must conform differences in the foregoing design factors can markedly affect the operating results. Cuttress and Pierce,[140] Chauvin et al.,[141] Loison et al.,[142] and Salier[143] strongly emphasize the distribution of air between the various points of admission; but our experience indicates that the resultant changes in overall furnace flow pattern are probably more important than the recorded shifts in quantity of air from one place to another. Wilwertz

[140] J. O. Cutress and T. J. Pierce, *J. Inst. Fuel*, **38**, 54 (February 1965).

[141] B. Chauvin, G. Tissandier, and P. Duquesne, *Bull. Inf. Cent. Electr.*, No. 35, 5 (1962).

[142] R. Loison, P. Duquesne, and G. Tissandier, *Rev. Gen. Therm.*, 705 1963).

[143] N. Salier, *Assoc. Ing. Elect. Inst. Electrotech. Montefiore (AIM) Bull.*, **75**, 488 (July 1962).

and Genart[144] give a good idea of the results achievable with U-flame firing of low-grade coal cleaning-plant rejects (containing more than 50% ash) and emphasize the need for high fineness of pulverization.

Other interesting aspects of low-volatile fuel firing and the special fuel properties and characteristics encountered are given by Ceely and Wheater,[3] Borio,[6] and Bryers and Kerekes.[145] These papers discuss some methods of assessing fuel combustibility by laboratory testing and relate some of the results to actual experience.

Vertical or Roof. The vertical or roof arrangement (Fig. 20.33*g*) is not frequently used, but it has been applied to units below about 150 MW. Above this size an uneconomically large furnace cross section would be needed for the required burners to be placed in the roof. Although downward firing is inherently stable and may result in a unit occupying less floor space than other designs, such units have drawbacks that usually rule them out. The most important disadvantages are a tendency toward burner overheating, damage from slag falls, and the difficulty of purging combustibles from the top of the furnace.

Since in all the dry bottom furnace configurations shown in Fig. 20.33, 80% or more of the ash fired leaves the furnace in the form of very fine dust, electrostatic precipitators are almost always installed. Many precipitator applications are extremely difficult because of the unfavorable physical and chemical properties of the fly ash, as discussed in Chapter 22. One of the principal sources of difficulty is the extreme fineness of the fly ash, as indicated in Fig. 20.37, which gives the results of 69 determinations of particle size distribution on dry bottom units, plotted to show the most probable distribution and the 50, 80, and 90% confidence intervals. Dust collector inlet concentrations for the same tests are plotted in Fig. 20.38. Unfortunately there is considerable scatter, since the dust loadings are presented as a function of percentage of ash in coal rather than some basis like pounds of ash per million Btu, which probably would have produced a better correlation.

[144] C. Wilwertz and J. C. Genart, ASME Paper No. 63-WA-129, 1963.

[145] R. W. Bryers and Z. Kerekes, *Trans. ASME, J. Eng. Power,* **99** (Ser. A, No. 3), 348 (July 1977).

20.2.2.3 *Wet Bottom Firing*

Early wet bottom furnaces, described in the 1920s, were simply open, single-stage furnaces. The high temperature necessary for ash melting and retention was produced by locating burners close together and near the furnace bottom, covering the floor and nearby surfaces with refractory to inhibit heat absorption, and using high coal fineness and highly preheated air. The furnace type used was usually one of those shown for dry ash firing in Fig. 20.33*a–d,* modified by replacing the hopper bottom with a flat floor with one or more slag discharge holes. Units of this simple type, which are satisfactory for fa-

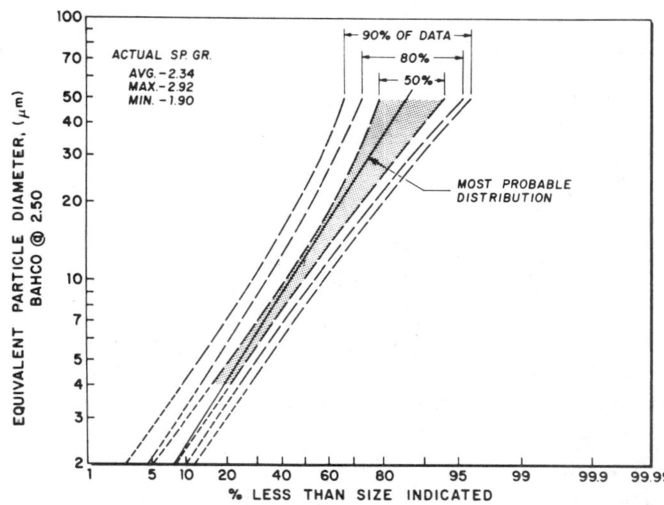

Figure 20.37 Ash particle size distribution for pulverized-coal-fired boilers.[130]

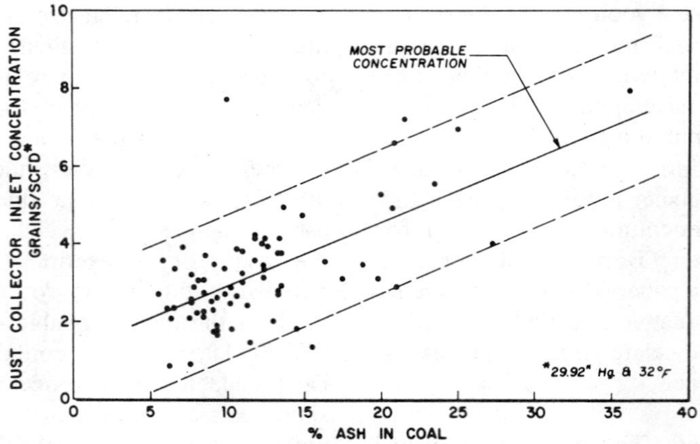

Figure 20.38 Ash concentration for pulverized-coal-fired boilers.[130]

vorable fuels and for a limited turndown range, are sold periodically even today.

For coal ash that is difficult to melt, however, and when a load range of 4 : 1 (or greater) is essential, more sophisticated designs of the closed type have been developed, a number of which are shown in Fig. 20.39. All these later furnaces are essentially two-stage combustion chambers in which the two parts fulfill the dual requirements of sustaining a temperature above the ash flow temperature throughout the first stage, including the discharge openings, and cooling the gases and their entrained ash below the point of troublesome adherence to convection surfaces in boiler, superheater, and reheater.

The ingenuity displayed in these designs has been directed primarily toward increasing the retention of ash and extending the load range over which successful slag tapping can be accomplished. The percentage of ash retained (without reinjection) for various open and closed designs is shown roughly in Fig. 20.40. Values in the upper part of the given

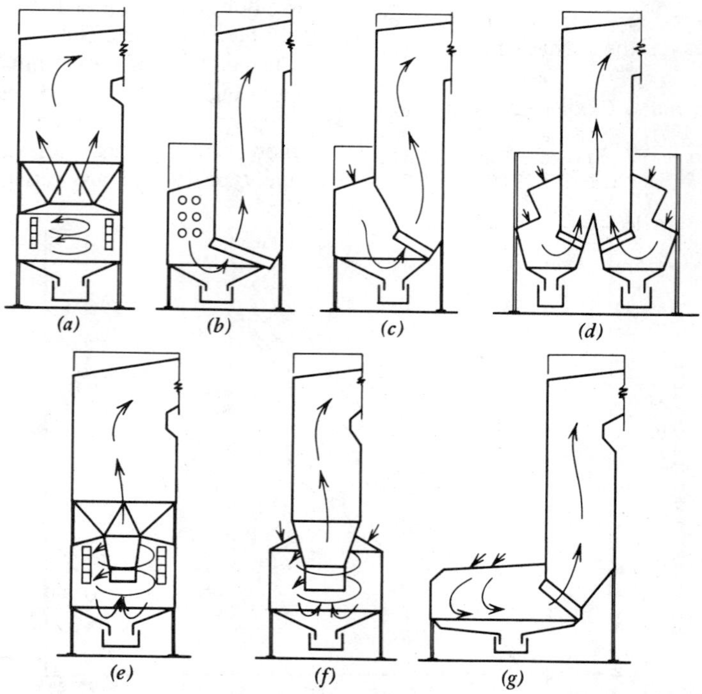

Figure 20.39 Furnace and burner configuration for two-stage slag tap firing.

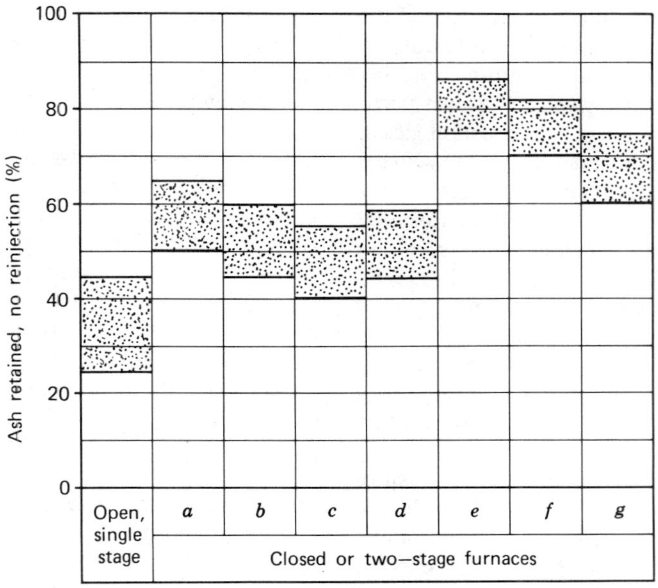

Figure 20.40 Percentage of ash retained for various open and closed wet bottom firing designs.

ranges are achievable only with favorable fuels and the best design and operating conditions. Low coal fineness, low air preheat, poor furnace proportions, and other deficiencies frequently result in retentions below the lower limits shown.

Improvements in the design of wet bottom units are continuously being made in the areas where they are frequently applied, particularly Germany and the Soviet Union. The improvements usually relate to better retention of ash or better tapping at partial loads. A good example of the latter is the compartmented furnace (Teilkammer) design described by Otting.[146] In this variation three division walls divide the melt chamber into four compartments, each fired by its own pulverizer (Fig. 20.41). By selecting mills for a coal fineness of 75–80% through 200 mesh, a continuous load of about 20% of unit rating can be carried with no tapping problems.

In favorable circumstances—that is, given fuel properties and design provisions that have led to trouble-free melting—the ash caught in the dust collector (or even that from adjacent boilers) may be reinjected into the furnace to further decrease the net production of fly ash and to reduce the combustible loss. The operation of a reinjection system involves additional equipment, auxiliary power, and maintenance costs and frequently results in intolerable particulate emission and erosion of boiler tubes, fans, and dust collectors. Many such installations have been run for only a short time, then abandoned as uneconomical.

Another problem associated with reinjecting fly ash is frequently decisive in determining the success or failure of a reinjection system, namely, the difficulty of achieving a

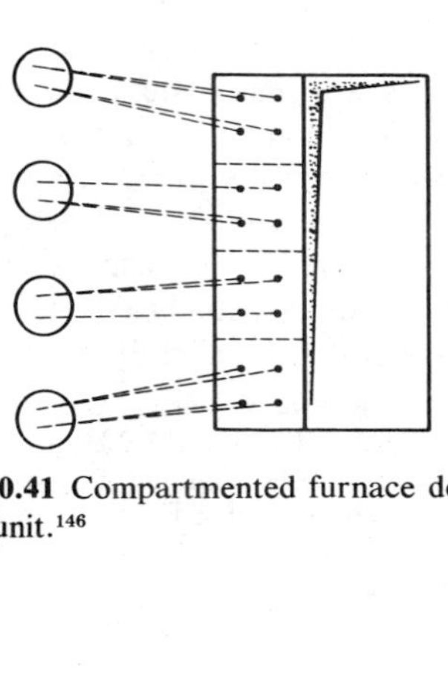

Figure 20.41 Compartmented furnace design, wet bottom unit.[146]

[146] H. F. Otting, *VGB-Kraftwerkstechnik,* **53**(1), 25 (1973).

high degree of retention in the furnace of the refired dust. Retention is particularly difficult in units arranged for dry bottom firing because of the ease with which the dust particles are reentrained in the combustion gases. Dolezal[135] describes small slagging cyclones that, when mounted on the walls of dry bottom furnaces, can be used to fuse the reinjected ash. These meltdown cyclones can be fired with the recirculated dust alone if its carbon content is above 50%: otherwise some coal can be fired with the dust to furnish the required heat of fusion. An interesting solution to this problem—although applied to a wet bottom furnace—has been discussed by Afanasev et al.[147] Their paper describes a system of reinjecting the collected fly ash of a 300-MW anthracite-fired unit in small, oil-fired cyclones mounted on the walls of the main furnace. Since the combustibles content of the anthracite refuse is quite high, this sytem is also designed to burn out its high carbon content.

The unit in question is a semiopen furnace design similar to that shown in Fig. 20.39*a*. The ash-melting cyclones are fired by oil at a rate of 0.25 lb of oil per pound of ash, amounting to about 8% of the fuel consumption of the boiler. The hot gas discharging from the cyclones to the main furnace is directed toward the incoming anthracite to aid ignition.

It is difficult to decide what type of slag tap design is necessary for a specific coal or whether the coal is suitable at all for wet bottom firing. A provocative series of articles has been presented by Geissler[148] giving a number of numerical coefficients (read from charts) that can be used to estimate the suitability of a given coal for various systems of slag tap operation. Although this method does not directly include all the pertinent parameters (in particular coal fineness and excess air) and has been criticized on other grounds, it is worth studying for the insights it gives into this complex problem.

Wet bottom firing is purely an expedient aimed at easing the probem of ash disposal by getting nearly all the ash in an easily handled form in essentially one location. It may also result in a slightly higher boiler efficiency with fuels containing less than 20% ash because of the lower combustible loss incurred and the lower excess air that may be tolerated. With higher ash contents, the losses caused by fusion and sensible heat of the slag generally outweigh any gains and a net decrease in efficiency results. Other disadvantages—compared with dry ash firing—that have led to its almost complete decline in the United States are less flexibility of fuel selection, higher incidence of fouling and external corrosion of pressure parts, higher levels of NO_x formation, and lower average steam generator availability.

[147] B. P. Afanasev et al., *Teploenergetika*, **18**(9), 25 (1971) (Transl. UDC 621.181.003.1).

[148] T. Geissler, "Die Eignung In- und Ausländischer Brennstoffe für den Betrieb von Schmelzfeuerungen" [transl. of Parts I and II, *BCURA Month. Bull.*, **27**(3) and **27**(4) (1963)].

20.2.2.4 *Slurry Firing*

The firing of furnaces with slurries of fine coal mixed with water or some other liquid has been proposed repeatedly by several generations of engineers. The teachings and claims of an 1879 patent issued to Munsell and Smith[149] cover the preparation and potential uses of essentially all the slurry fuels subsequently considered. Pipelining of liquid and gaseous fuels had, of course, been practiced for many years before 1879, and the first U.S. solids pipeline patent was issued in 1891. In recent years combustion trials in a number of test and permanent installations have proved the general feasibility of burning slurry fuels.

In most of the work conducted since the early 1940s, the solid-liquid mixtures have been merely stabilized suspensions of finely ground coal in water or oil, or a slurry made of all three components together. In contrast, many of the proposals and experimental efforts before 1940 consisted of treating pulverized coal with hot oil or tar, sometimes under pressure, so that the suspended solid in the resultant slurry was not coal but a coal-derived residue. Also, the liquid medium was altered by dissolving or reacting with some of the coal substance, as related, for example, by Brownlie.[150,151]

A notable but by no means unique example of this type of technology recently reported is the Methacoal process,[152] in which hot coal

[149] H. M. Munsell and H. R. Smith, U.S. Pat. 219,181, 1879.

[150] D. Brownlie, *Ind. Eng. Chem.*, **28**, 629 (1936).

[151] D. Brownlie, *Ind. Eng. Chem.*, **28**, 839 (1936).

[152] R. M. Jimeson, "Methacoal: Synthetic Energy Symbiotic with Water Deficient Coal Resource Regions," paper presented at American Institute of Chemical Engineers, 3rd Natl. Conf. Complete Water Re-Use, Cincinnati, Ohio, June 1976.

is mixed with methanol under slight pressure and with moderate agitation. It is claimed that the fine coal feed is further comminuted by this treatment; that mineral constituents are separated and can be readily removed; and that the coal residue is lighter, more porous, and much more reactive than the original coal.

In all cases, whether forming simple mechanical mixtures or altering the constituents by physicochemical means, the principal goal has been to achieve a fuel that substitutes readily available coal for scarcer oil and is easier to transport, store, and use than coal. The proven technology for transporting solids economically over long distances by pipeline, borrowed in part from the mineral industries, has certainly been another strong attraction and has assured even the simple coal-water slurry of continuing attention. The overall cost reduction offered by pipeline delivery of coal to utilities and other large users promises to outweigh the problems and loss of thermal efficiency arising from the use of coal-water slurry. A good general review of U.S. slurry pipeline projects and a useful discussion of pipeline operating problems and cures are given in a paper by Kiefner that also contains an extensive bibliography.[153] The present status of mechanical and thermal dewatering equipment is reviewed well by Shen.[154]

Since coal-water slurries present difficult problems in dewatering and can be used directly from the pipeline in only a few special types of boiler-furnace system, oil or alcohol slurries would promise wider applicability. Experience has shown that coal-oil slurries can be used in some existing oil-fired steam generators without requiring more than nominal changes to the steam generator itself, although extensive overall plant changes are necessary for slurry preparation, pollution control, and so on.

Coal-Water Slurry. Several major public utilities have used coal-water slurries in full-scale commercial equipment for long periods. These first installations have not been entirely trouble free, but they have confirmed the feasibility of commercial operation with coal-water slurry and have defined areas of difficulty needing future improvement.

In the late 1950s at Eastlake Station, coal was delivered from the now famous 108-mile-long Cadiz-Eastlake pipeline and burned in one of the plant's steam generators. Details of the slurry preparation plant, pipeline, and dewatering facilities initially installed at Eastlake are given in a paper by Hanson and Regan,[155] and the dewatering operations are the subject of a paper by Dauber and Gill.[156] These initial facilities combined mechanical and thermal drying, with the 50% water pipeline slurry subjected successively to three stages of dewatering: (1) thickeners, the underflow of which goes to vacuum filters; (2) vacuum filters, the filter cake of which is conveyed to flash driers; and (3) flash dryers, the dried product (containing 5% moisture) of which is fed to bunkers for subsequent pulverizing in a conventional firing system.

Since thermal dewatering involves a very high energy loss and is usually associated with dust, fires, and reliability problems, the Eastlake installation was used to develop a mechanical system of dewatering. The resultant centrifuge-pulverizer system installed on one of the unit pulverizers is described by Jonakin and Caracristi[157] and Halvorsen.[158] They found that the centrifuges could dewater the high-rank bituminous coal to a total moisture content of 18–19% (with the equilibrium or inherent moisture about 3%). Troubles were experienced in feeding the centrifuge cake to the pulverizer, but the pulverizer readily dried and pulverized the cake once it was received.

Essentially the same type of centrifuge-pulverizer system has been used on the two 790-MW units operating at Mohave Station since 1971. Design features of the Mohave plant have been given by Coffey et al.,[159] and

[153] J. F. Kiefner, "Review of Slurry System Projects in the U.S.," paper presented at 1st Int. Technical Conf. Slurry Transportation, Columbus, Ohio, February 1976.

[154] S. S. C. Shen, "Dewatering Equipment for Coal Slurry Pipelines," paper presented at 2nd Int. Technical Conf. Slurry Transportation, Las Vegas, March 1977.

[155] V. D. Hanson and T. J. Regan, "The Coal Pipeline," paper presented at American Mining Cong., Pittsburgh, 1958.

[156] C. Dauber and N. Gill, *J. Pipeline Div., ASCE,* October 1959.

[157] J. Jonakin and V. Z. Caracristi, *Coal Util.,* February 1963.

[158] W. J. Halvorsen, "Slurry Preparation, Separation, and Utilization," paper presented at 1st Int. Technical Conf. Slurry Transportation, Columbus, Ohio, February 1976.

[159] R. C. Coffey, H. G. Lyons, and A. C. Oakes, *Proc. Am. Power Conf.,* **31,** 392 (1969).

operating and maintenance results on the steam generators and dewatering equipment are discussed by Dina[160] and Halloran.[161] Dina reports the following major changes from the original coal-slurry system to improve operation and reliability:

1. The centrate was originally thickened in clariflocculators and the underflow reinjected into the centrifuges, as in the Eastlake trials. The clariflocculator underflow is now injected into the furnace through spray nozzles at the top burner elevation. Some slagging has resulted from this practice.

2. In each steam generator about 40×10^6 Btu/hr of natural gas is burned in the mill air ducts to keep the mill exit temperature up to 175°F. Prior to this change, with low mill outlet temperatures, plugging and subsequent fires were experienced between mills and furnace.

3. Centrifuge wear parts were eventually coated with high-alumina ceramic and hardened stainless steel to achieve a wear life of 10,000 hr.

4. A slurry heating system, using intermediate-pressure turbine exhaust extraction steam, was installed to preheat the slurry to a maximum of 190°F. The current slurry temperature is limited to 140°F, however, because of mechanical considerations in the centrifuges. According to Dina, the use of turbine extraction steam incurs a cycle thermal loss of 1.8%; but by reducing the moisture content of the pulverizer feed, one mill can be taken out of service for maintenance without dropping unit load.

Figure 20.42 is a full-load slurry flow diagram for one of the Mohave units showing the dry coal concentration at various points in the system for a centrifuge slurry temperature of 140°F. More details on Mohave are given in the Dina and Halloran papers previously cited.

As noted in the foregoing, certain areas of coal-water slurry usage need further development in spite of the successful burning experience accumulated at Eastlake and Mohave. Possibly the most important areas requiring development, from the utilization point of view, are those involving long-term storage and reclaiming of both slurry and centrifuge cake, dewatering by thermal and mechanical means, and the direct burning of thickener underflow. More detailed discussions of these subjects are given in the papers already referenced and by Aude and Gandhi.[162] These authors point out the impracticality of connecting too closely the transportation and utilization steps, as was done in the Black Mesa–Mohave system.

The alternative of direct firing of coal-water slurry was demonstrated over a period of several weeks at Werner Station in 1961. The boiler selected for the trial firing was a cyclone furnace–fired unit (Section 20.2.3, "Cyclone Furnace Firing") with a steam capacity of 475,000 lb/hr. This unit, originally designed for Pittsburgh Seam coal from the Fairmont, West Virginia, district, is equipped with two 9-ft cyclones. The tests are well described in a paper by Kelcec et al.[163] and in one given a few years later by Duzy[164] on firing high-moisture fuels. The results at Werner Station showed that a suitably designed cyclone furnace can be fired directly with slurries of high-volatile coal, with solids concentrations of 62%. Such slurries, when fired with about 25% excess air, will provide a calculated adiabatic cyclone temperature above 3400°F, which is the criterion proposed by Duzy[164] for successful cyclone operation. The turndown ratio obtainable was surprisingly good. Tapping was reportedly comparable to that obtained with the original design coal, which had shown a lower limit of approximately one-half load. Presumably there was no increase in combustibles loss, since this is not specifically mentioned. The lower

[160] M. L. Dina, "Operating Experiences at the 1580-MW Coal Slurry-Fired Mohave Generating Station," paper presented at 1st Int. Technical Conf. Slurry Transportation, Columbus, Ohio, February 1976.

[161] J. J. Halloran, "Coal Slurry Dewatering Equipment Maintenance Developments: Techniques and Costs," paper presented at Slurry Transport Association Annual Meeting, Houston, Tex., CECO Publ. No. TIS-5039, August 1976.

[162] I. C. Aude and R. L. Gandhi, "Research and Development for Slurry Pipeline System Design," paper presented at 2nd Int. Technical Conf. Slurry Transportation, Las Vegas, March 1977.

[163] G. Kelcec, P. P. Olivadoti, and A. F. Duzy, *Coal Util.*, March 1963.

[164] A. F. Duzy, U.S. Bur. Mines Inf. Circ. No. 8516, 1968, p. 57.

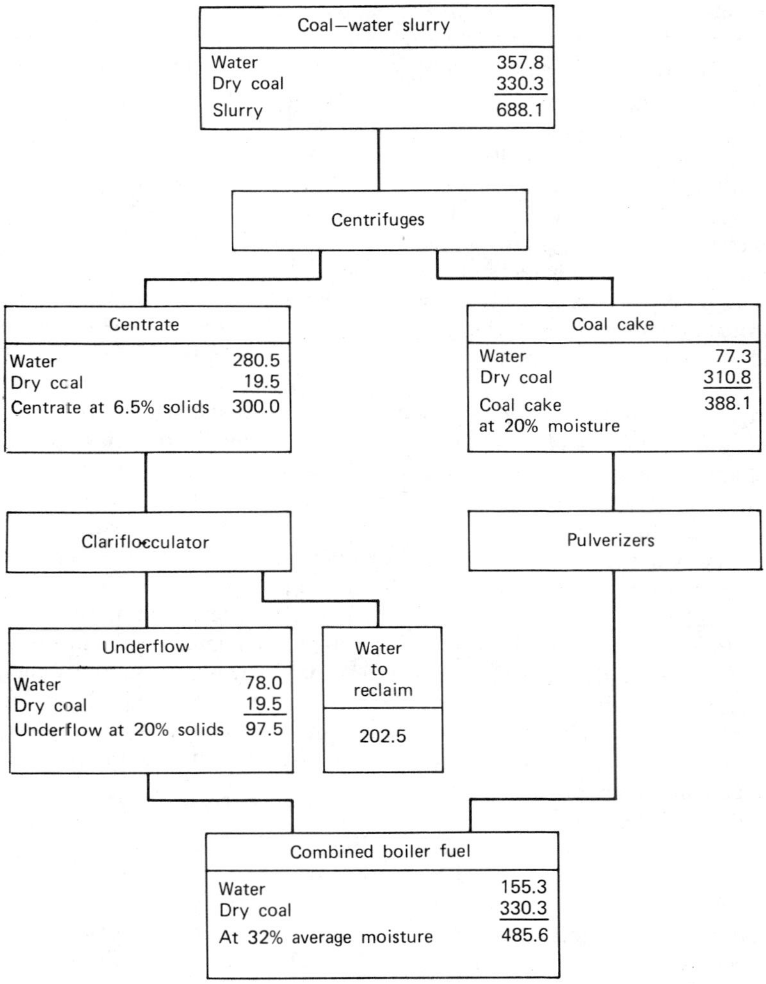

Figure 20.42 Full-load slurry flow diagram, Mohave generating station: all quantities in tons per hour; slurry temperature 140°F to centrifuges.

efficiency, amounting to about 4.5%, was caused primarily by the water in the fuel and the slightly increased exit gas temperature.

In thinking about the favorable results of the Werner Station tests, it is logical to wonder whether other boiler designs, in particular designs such as those illustrated in Fig. 20.39, would perform equally well. Some indication that this is so is given in the work reported by Schwarz and Merten.[165,166] In the first of these papers, Schwarz[165] gives the results of a number of theoretical calculations (flame temperatures, gas radiation, etc.) and concludes that depending on the coal properties in each case, a number of slag tap boilers could be converted to direct slurry firing. In the second paper Schwarz and Merten[166] report favorable results with direct firing in a small dry bottom boiler at Kellermann Station.

[165] O. Schwarz, *Brennstoff-Wärme-Kraft,* **16**(6) (1964).

[166] O. Schwarz and H. Merten, *Brennstoff-Wärme-Kraft,* **18**(10), 474 (October 1966).

Coal-Oil Slurry. Slurries of coal in oil, or of any reactive hydrocarbon solid in any combustible medium, present a less severe burning problem than do the coal-water slurries previously considered. Nevertheless it is not likely that any of these fuels, should they become available, will be suitable for use in most existing oil-fired equipment designed for good grades of oil, especially at or near the maximum steam outputs originally realized. The closest approach to success will be achieved with those large, conservatively sized, industrial or utility units installed initially for use with heavy residual fuels. Successful conversion of oil-fired boilers to coal-oil slurry firing will require design provisions suitable for the quantity and characteristics of the ash being fired, the gas temperatures prevailing throughout the unit, and the gas velocities and spacing of tubes in the convection banks.

In spite of these requirements, there is no doubt that new units can be readily designed for coal-oil slurries and that some existing units with minor changes in the steam generator itself can produce reasonably high outputs when fired with these fuels.

In the older literature the report by Barkley et al.,[167] with its inclusion of the major results of Idris-Jones in Great Britain, is a good confirmation of the ease of atomizing and burning a suspension prepared by dispersing in heavy fuel oil 40% by weight coal ground to about 98 or 99% through a 230-mesh sieve. Tests were made in three different plants, and good results—as measured by CO_2 in flue gases, absence of smoke, and general absence of ash problems—were obtained on all but the smallest boiler. Barkley's paper concluded that pumps, heaters, and other essential oil-firing system equipment could be readily selected or devised for this fuel and that combustion was good.

As a direct result of coal-water slurry investigations in Germany, a study was made by Meier zu Köcker and Hüning[168–170] of three-component suspensions of coal, water, and oil. Their study reports deal with pumping and stability,[168] theoretical analyses of burning characteristics,[169] and laboratory testing of the analytical conclusions.[170] According to the authors, the purpose of adding oil to the coal-water slurry was to upgrade the mixture and make it much more widely applicable than a simple coal-water slurry. The amount of oil generally approximated that required to vaporize the water present.

Additional work on coal-oil combustion is reported by Brown and Lee.[171] These authors give results of laboratory-scale (100–150 lb/hr) burning of a Western Canada lignite and a very nonreactive bituminous coal reject. The contrast in results between the two fuels is startling, with very low burnouts observed on the bituminous coal. Especially strange is the report that increasing the fineness of the bituminous coal from 70% −200 mesh to 94% −200 mesh had no marked effect on burnout.

Interest in the use of coal-oil fuels as a means of quickly reducing oil consumption is now the subject of several tests funded in part, and essentially promoted, by ERDA (now DOE). These activities range from the establishment of a combustion test facility at the Pittsburgh Energy Research Center to the field testing of coal-in-oil fuels in a range of industrial and utility steam generators. The program is outlined in a paper by Foster,[172] and results of a portion of it—consisting of a 1000-hr test on a 100-hp firetube boiler—are given by Demeter et al.[173]

20.2.3 Cyclone Furnace Firing

Horizontal cyclone-furnace firing, as illustrated in Fig. 20.43, is a form of two-stage, wet bottom design and is therefore related to some of the furnaces described in Section 20.2.2.3, "Wet Bottom Firing." Cyclone furnace firing carries the furnace staging concept further than do the other designs shown in Fig. 20.39 by more effectively confining and intensifying the ash-melting process, so that a much smaller area (and amount) of slag must be kept in the molten state until discharged.

Cyclone furnace units generally conform to one of the basic arrangements shown in Fig. 20.43. In the screened furnace type, now largely superseded by open furnaces, the cyclone gases discharge first into a small primary chamber and then through a slag screen into the main gas-cooling furnace.

[167] J. F. Barkley, A. B. Hersberger, and L. R. Burdick, *Trans. ASME,* **66,** 185 (1944).

[168] H. Meier zu Kocker and R. Huning, *Brennstoff-Chemie,* **47** (1966).

[169] H. Meier zu Kocker and R. Huning, *Chemie-Ingenieur-Technik,* **39**(1), 7 (1967).

[170] H. Meier zu Kocker and R. Huning, *Brennstoff-Chemie,* **48** (1967).

[171] T. D. Brown and G. K. Lee, Canadian Combustion Research Laboratory Divisional Rep. No. FRC-72/95-CCRL, 1972.

[172] C. B. Foster, ASME Paper No. 77-WA/Fu-4, 1977.

[173] J. J. Demeter, et al., ASME Paper No. 77-WA/Fu-5, 1977.

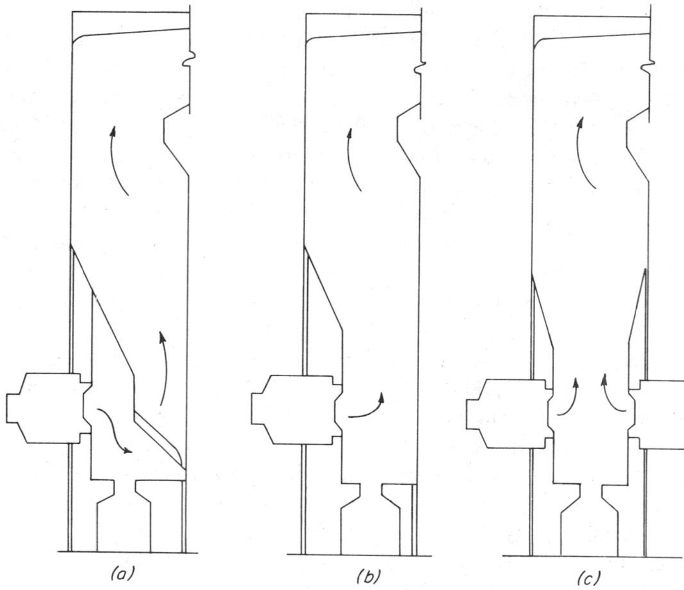

Figure 20.43 Horizontal cyclone furnace arrangements: (*a*) screened furnace, (*b*) and (*c*) one-wall and opposed open furnaces.

The two open furnace arrangements shown in the figure have evolved for larger units. One-wall firing is used where sufficient space is available on either the front or the rear wall for the necessary cyclones. For the largest units this arrangement becomes uneconomically wide, so cyclones are located in both front and rear walls. An informative, though uncritical, discussion of cyclone furnace firing systems, principles of operation, suitability of fuels, and operating results is given in a book published by the Babcock & Wilcox Company.[99]

The development of the horizontal cyclone furnace took place rapidly in the United States in the mid-1940s,[174,175] although the concept of cyclone or vortex-type firing is an old one. A number of earlier U.S. patents showed cyclone-type cylindrical combustors arranged both horizontally and vertically.[176–178] The best known of these was the Burg combustor[176] which had at least one trial in Europe.[179] The literature, however, shows no indication that any real attempt was made in the United States before 1944 to develop any of the cyclone (or near-cyclone) firing concepts, except the work done in 1926–1927 by Cushing and Moore[180] at Huntley Station on the so-called well-type furnace.

The various objectives of horizontal cyclone furnace firing are the same as those for other two-stage, wet bottom furnace designs. All objectives have been achieved in different degrees and combinations, depending on the specific design and its particular fuel and operating conditions. This subject is discussed best in the works previously cited,[16,133–135] where balanced attention is paid to both benefits and shortcomings.

The fuel characteristics of greatest interest in cyclone firing are the fusibility and viscosity of the ash lining the walls of the cyclone. In addition, it has been suggested recently by Holyoak[181] that the wetting character of the slag with respect to the new and partially burned fuel on the slag surface is also important. Holyoak reports a high carryover of unburned carbon when test firing subbituminous coal in cyclones originally designed

[174] A. E. Grunert, L. Skog, and L. S. Wilcoxson, *Trans. ASME,* **69,** 613 1947.

[175] G. W. Kessler, *Proc. Am. Power Conf.,* **16,** 78 (1954).

[176] E. Burg, U.S. Pat. 1,618,808, 1927.

[177] C. G. Hawley, U.S. Pat. 1,591,679, 1926.

[178] J. M. Schutz, U.S. Pat. 1,657,698, 1928.

[179] M. Schimpf, Glueckauf, **65,** 295 (1929).

[180] H. M. Cushing and R. P. Moore, *Trans. ASME,* 1927–1928.

[181] R. H. Holyoak, ASME Paper No. 73-WA/FW-4, 1973.

for coal of higher rank. The carryover was reduced by adding 15–25% of the high-volatile C bituminous coal for which the cyclones were designed. Wetting properties of coal slags are also treated, with other slag properties, in a paper by Raask,[182] who found that slag *did not* wet carbon, graphite, coal, or coke. Possibly Holyoak's unburned combustible resulted from some property other than the wetting or nonwetting character of its slag or merely from too low a cyclone temperature and pressure drop for the low-rank coal he was testing. This possibility is also suggested by the successful applications reported by others of cyclone firing on a range of high-moisture, low heating value fuels, such as subbituminous coal, lignite, coal-water slurry, and wood bark.

Firing of low-rank coals in cyclones is particularly important now because of the increasing use of such coals and the prospect of partially avoiding the inordinate preparation costs associated with firing high-moisture, low heating value fuels in pulverized form. (See remarks at the end of this section concerning NO_x formation in cyclones.) The utilization of a number of high-moisture fuels in cyclone furnaces is the subject of a paper by Duzy,[164] which presents in general terms the fuel properties necessary for satisfactory cyclone operation over a reasonable load range. Essentially the ash (either as-fired or together with a flux) should be of low to moderate fusion temperature, and the moisture content per unit of heating value should be limited to that producing a calculated adiabatic cyclone temperature above 3400°F. Fuels that cannot produce such a temperature can be assisted by the supplementary firing of oil, gas, or other rich fuel, or by predrying in one of the "conditioning" systems described in Duzy's paper.[164]

The higher moisture, low-Btu lignites (moisture $> 36\%$ and as-fired Btu < 7000 Btu/lb), were found to require the type of predrying system described by Dolezal[135] and illustrated in Fig. 20.44. In this system the vapor-laden air from the crusher is separated from the lignite in a mechanical collector and bypasses the cyclone furnace. Further particulars on the testing of this type of cyclone firing system and the effect of such variables as moisture content, excess air, and supplementary oil firing on adiabatic gas temperature are given by Rusanowsky.[183]

[182] E. Raask, *Trans. ASME,* **88** (Ser. A, No. 1), 40 (1966).

[183] N. P. Rusanowsky, *Proc. Am. Power Conf.,* **29,** 475 (1967).

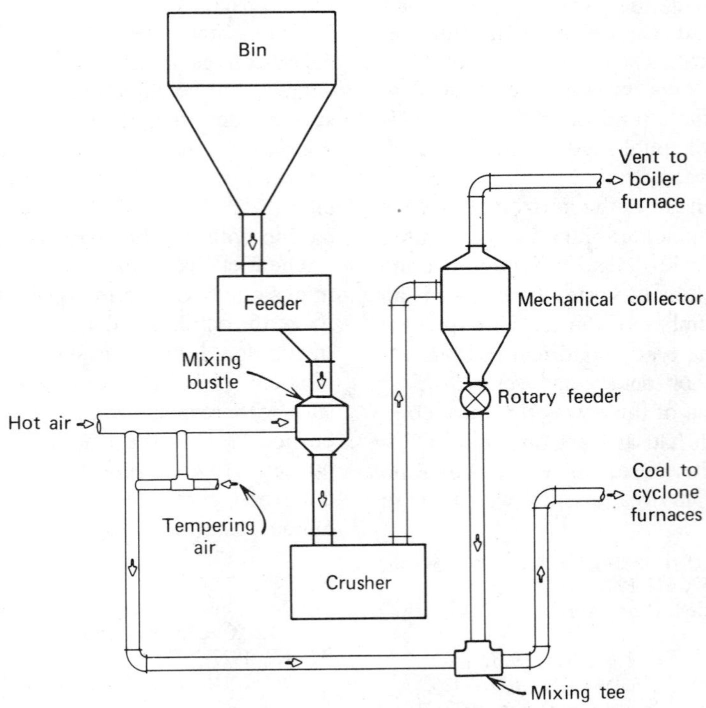

Figure 20.44 Direct-fired predrying and cyclone bypass system.[164]

A later paper by Duzy and Rudd[184] gives some operating results on the first lignite-fired cyclone installation in the United States to use the predrying and moisture-venting system illustrated in Fig. 20.44. They report that firing has been acceptable in all respects, with lignite moistures as high as 40%. Ash carryover from the cyclones has been about 25% of the ash fired.

As previously stated, the fusibility and viscosity of the fuel ash are important factors in judging cyclone suitability and in selecting such design parameters as air preheat, cyclone pressure drop, and coal feed size. Fusibility and viscosity of *lignite-type* ash (defined as ash having more CaO plus MgO than Fe_2O_3) are discussed in another paper by Duzy.[185] In that discussion he reminds us that previous work, such as that by Sage and McIlroy,[186] did not cover a wide range of ash composition and, in particular, failed to include ash with low silica ratios and a high percentage of basic components. Duzy finds that the previous correlations of fusibility-viscosity with ash composition need modification for fuels with lignite-type ash, and he presents charts showing a new correlation based on what he calls the dolomite ratio:

$$\frac{(CaO + MgO)100}{CaO + MgO + Fe_2O_3 + Na_2O + K_2O} \tag{20.8}$$

A full discussion of coal ash viscosity-temperature-composition relationships is given in Chapter 21 (Section 21.3.1).

As previously mentioned, wet bottom firing produces higher levels of NO_x than does dry ash firing, This is the case to an even greater extent with cyclone furnaces because of their elevated operating temperatures. The latest Standards Support and Environmental Impact Statements (October 1976) presented to EPA indicate that measured No_x emissions from cyclone furnaces reach values 2–3 times as high as those from an average dry ash unit. With lignite firing, though, and perhaps with other fuels producing reduced cyclone temperatures, the cyclone NO_x emissions may be about the same as those resulting from uncontrolled dry ash firing of the same fuels. However the two most common methods of NO_x reduction with pulverized coal firing—off stoichiometric operation and reduction of combustion turbulence—appear to be infeasible for cyclone firing.[187] Consequently the use of cyclones in future applications does not appear likely until the NO_x control problems are solved and existing cyclone units may be derated or decommissioned.

The amount of particulate emission from cyclone furnace firing is less than that from firing pulverized coal in suspension, but the sizes emitted are much finer. This size difference is evident when comparing the cyclone furnace firing data shown in Fig. 20.45 with the pulverized-coal firing data shown in Fig. 20.37. An idea of the amount emitted from each system can be obtained by comparing Fig. 20.46 with Fig. 20.38.

20.3 SYSTEM APPLICATION DEVELOPMENT

New application developments in coal combustion have been undertaken in the last 15 years, with the principal objectives of conserving fuel, controlling environmental discharge from combustors, and achieving good economics in investment and operating costs. None of these recent developments has fully reached commercial status. However limited commercial application has occurred in some cases, and a great deal of system-oriented development work has been accomplished and is continuing. This section deals with some systems using the products of coal combustion to provide the energy source directly. It should be recognized, however, that there is another important system now under development that gasifies the coal, cleans the gasified product of sulfur and particulates, and then uses the clean gas to fuel an energy converter, such as a boiler, a gas turbine, or a combination of the two. One preferred process, known as an integrated coal gasification–gas turbine–steam turbine cycle, is shown in Fig. 20.47.

[184] A. F. Duzy and A. H. Rudd, *Proc. Am. Power Conf.*, **33,** 554 (1971).

[185] A. F. Duzy, ASME Paper No. 65-WA/Fu-7, 1965.

[186] W. L. Sage and J. B. McIlroy, *Trans. ASME,* **82,** 145 (1960).

[187] W. Bartok et al., *Systematic Field Study of NO_x Emission Control Methods for Utility Boilers,* Office of Air Programs, Environmental Protection Agency, GRU.4GNOS.71, December 1971.

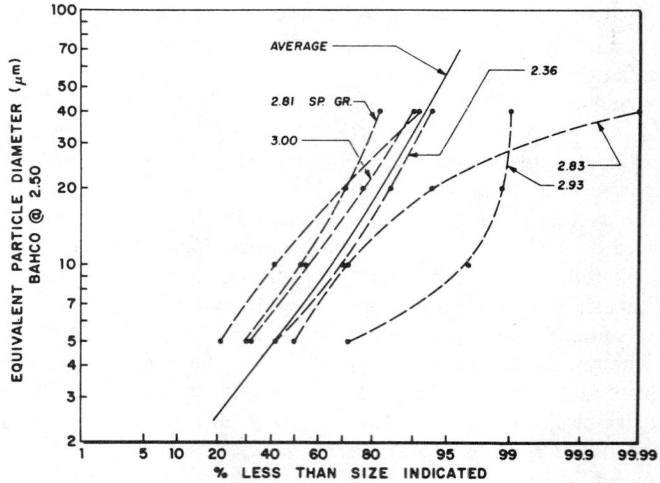

Figure 20.45 Ash particle size distribution for cyclone-fired boilers.[130]

20.3.1 MHD Combustor Development

MHD has long been recognized as a method of electric power generation with potentially very high thermal efficiencies. Various types of cycle configurations have been proposed, including open and closed cycles, all featuring the direct conversion of thermal energy to electrical energy in an MHD duct. Reviews of MHD systems are given in the literature.[93]

In open-cycle systems coal will have to be burned with preheated oxidant under pressures up to 10 atm. Alkali seed material is added to the combustion gas before it expands through the MHD channel. Many cycle configurations, including closed cycles, have been devised using liquid metal coolants. The open direct-fired cycle shown in Fig. 20.48 is the more popular one. Work on coal combustor development for this cycle is summarized and discussed extensively by Henry et al.[189]

Typical of the performance of these com-

[188] AVCO Everett Research Laboratory, "MHD Energy Conversion: A Presentation to the Electric Research Council Task Force on MHD," May 7, 1970.

[189] J. M. Henry, R. C. Kurtzrock, and D. Bienstock, *Proc. NSF–OCR Engineering Workshop MHD Materials,* November 1974, p. 259.

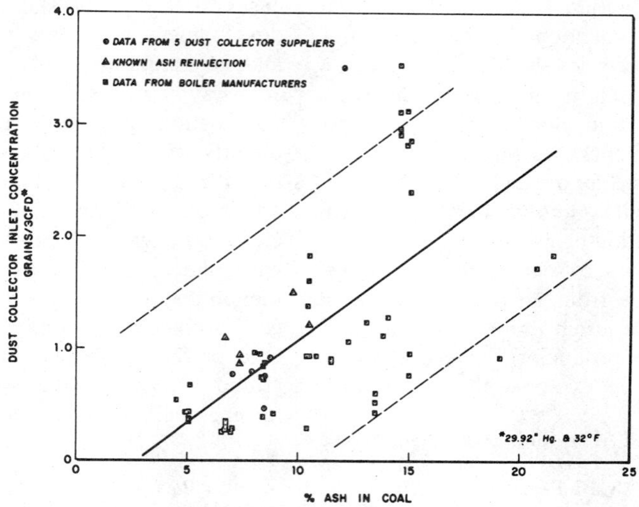

Figure 20.46 Dust collector inlet concentrations for cyclone-fired boilers.[130]

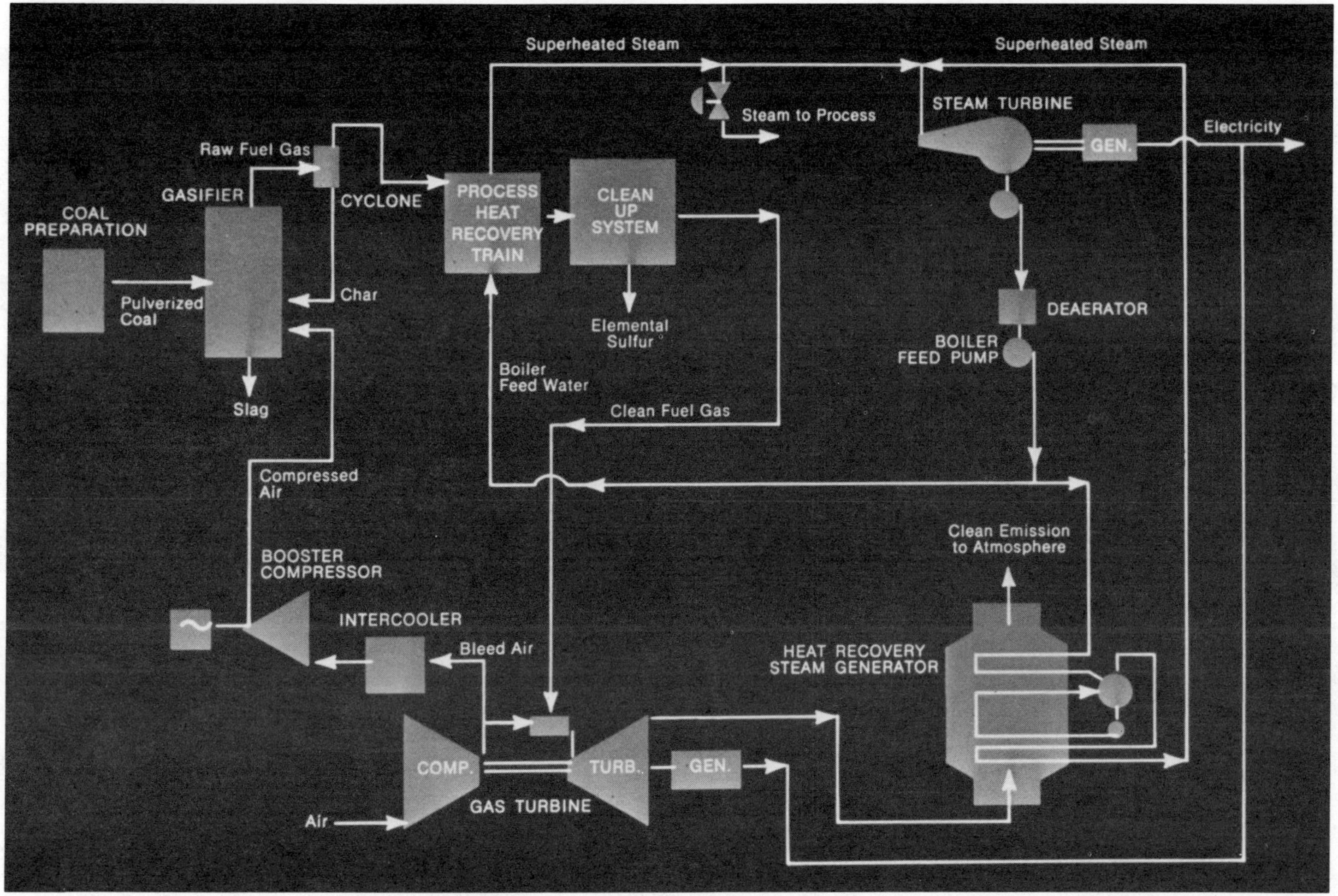

Figure 20.47 Integrated coaı gasification–gas turbine–steam turbine cycle.

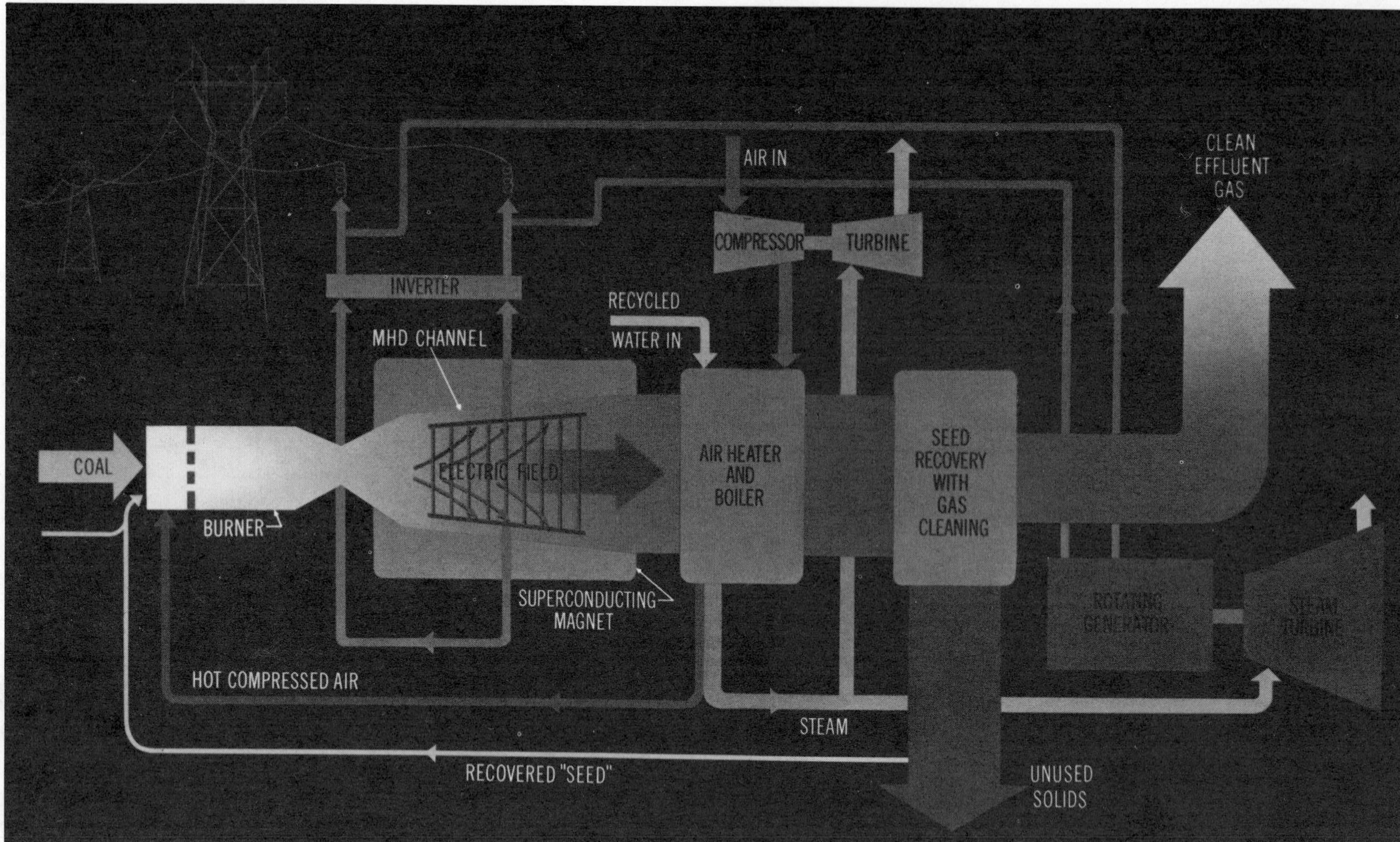

Figure 20.48 MHD power cycles.[188]

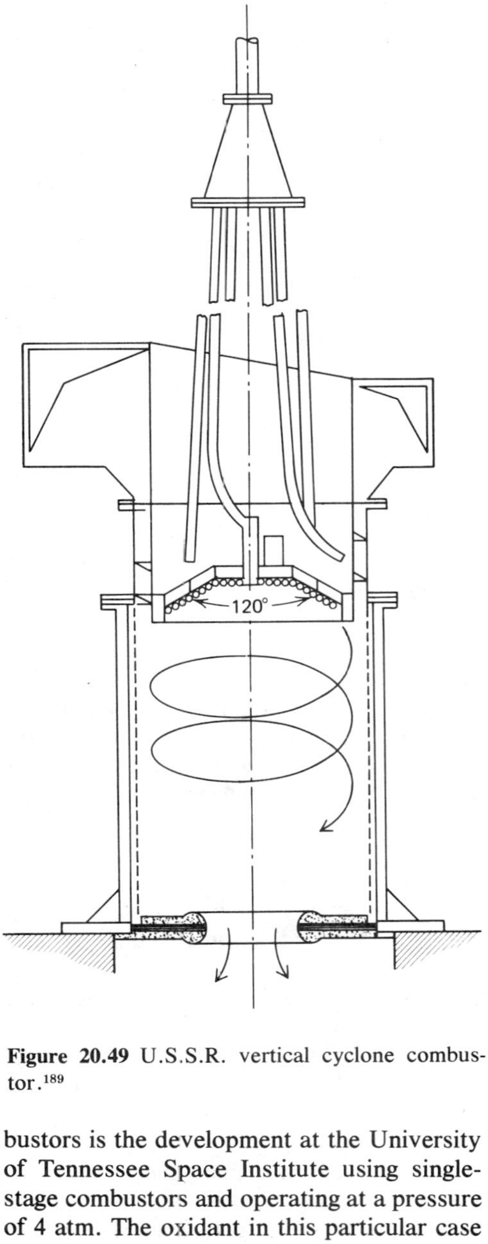

Figure 20.49 U.S.S.R. vertical cyclone combustor.[189]

bustors is the development at the University of Tennessee Space Institute using single-stage combustors and operating at a pressure of 4 atm. The oxidant in this particular case is pure oxygen, the combustor is water cooled, and the volumetric heat release rate is 320 MW/m^3. Plasma conductivities as high as 10 mho/m have been achieved.[190]

Others [191,192] have worked on cyclone combustors in an effort to separate coal ash as slag prior to injection of seed material (Figs. 20.49 and 20.50). Bienstock[192] showed that with 95% ash rejection in the combustor, 99% of the seed can be recovered from the seed–fly ash mixture by simple extraction Bureau of Mines (USBM) combustor is shown in Table 20.4. Another approach to achieving high ash-slag rejection is the two-stage combustor proposed by Way in 1974,[193] shown in Fig. 20.51. The same approach can be used by running the first stage fuel rich.[194] with hot water. Performance of the U.S.

The USBM has constructed a three-stage gasifier combustor with a heat input of 5 MW to operate at 8 atm. This combustor is shown in Fig. 20.52. The first stage gasifies raw coal and recycled char with air and recycled gas at approximately 2000 K. Slag is rejected at the bottom. The second stage is a fluidized bed of char, using the gases from the first stage as a fluidizing medium and operating at 1400 K. Raw coal is added to this stage and devolatilized. The off-gases are cleaned of particulates in cyclones, and the gas is finally combusted in the third stage to produce the desired plasma.

Thus coal combustion systems for MHD plasma production are of two basic types:

1. Single-stage devices that are simple but reject little ash or slag and, because of the high temperature in the presence of ash, vaporize a considerable portion of the ash; these disadvantages make seed recovery inefficient, therefore increase operating costs.
2. Multistage combustors that separate most of the ash as slag but are complex; this type of system alleviates material problems and, most important, makes possible highly efficient seed recovery.

20.3.2 Ambient Pressure Fluidized-Bed Combustion Systems

Application of ambient pressure fluidized-bed combustion is being developed to take the

[190] R. V. Shanklin et al., paper presented at 13th Symp. Engineering Aspects of MHD, Stanford University, 1973.

[191] S. A. Tager et al. *Proc. 5th Int. Conf. MHD Power Generation*, Vol. 1, 1971, p. 471.

[192] D. Bienstock et al., paper presented at 13th Symp. Engineering Aspects of MHD, Stanford University, 1973.

[193] S. Way, paper presented at AIAA 6th Biennial Gas Dynamics Symp., Northwestern University, AIAA Paper No. 65-633, 1967.

[194] H. R. Hoy, A. G. Roberts, and D. M. Wilkins, in *Open Cycle MHD Power Generation*, J. B. Heywood and G. J. Womack, Eds., Pergamon Press, New York, 1969, pp. 262–308.

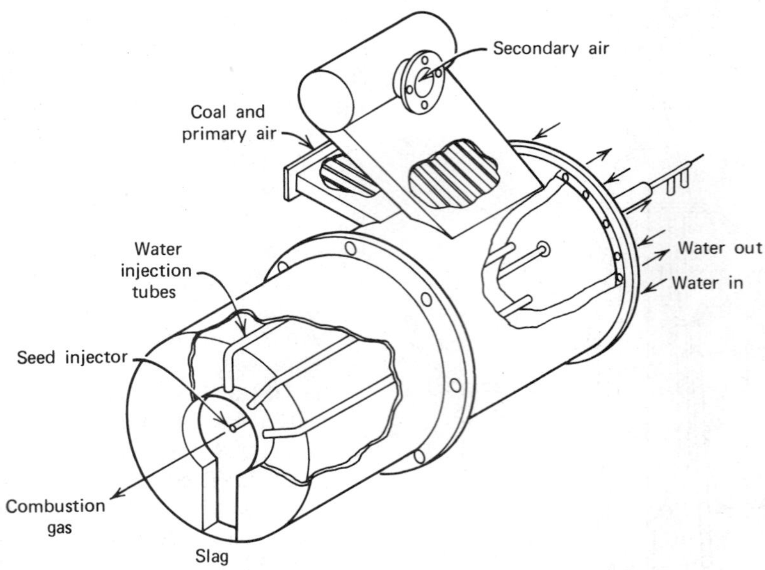

Figure 20.50 USBM horizontal cyclone combustor.[189]

place of pulverized-coal and stoker-fired systems to generate steam. Because of the nature of the combustion process, it is possible to control sulfur and nitrogen oxide emissions, to burn a wide variety of coal without encountering problems of slagging and corrosion, to reduce the amount of heating surface required, and to eliminate a large radiant furnace cavity.

A schematic diagram (Fig. 20.53) shows the application of fluidized-bed combustion to steam generation. Figure 20.54 shows a conceptual design of a fluidized-bed combustion steam generator for an electric utility generation system. The concept depicted in Fig. 20.53 is one in which individual and separate fluidized beds are stacked vertically. Each cell, except the carbon burnup cell at the bottom of the unit, has the heating surface submerged in the bed. The expanded (fluidized condition) bed depth in each cell is 48 in. Above the bed is a freeboard space, then the convection surface. The gas leaving each cell is put through a cyclone separator whose rejects comprise the feed for the carbon burnup cell. On leaving the cyclones, the combustion gases are ducted to an air preheater (not shown). Each cell has its own air supply—controlled by inlet dampers—and its own coal and limestone feed system.

Table 20.4 Operating Data on USBM Horizontal Cyclone Combustor[189]

Fuel	High-volatile bituminous coal
Heat input (fuel)	0.76 MW
Air preheat temperature	1000 K
Oxygen content of "air"	33 vol %
Operating pressure	1 atm
Gas residence time	30 ms
Plasma temperature	2600 K
Carbon combustion efficiency	99.97%
Slag rejection	75–80%

The unit shown in Fig. 20.54 is designed for 150 MW at 3500 psi at 1000°F and reheat at 1050°F. For a larger capacity unit, width is increased. In this arrangement an overbed coal feed system and a common convection pass are used.

Fluidized-bed mechanics and basic combustion rate heat transfer and chemical parameters have been reasonably well established. System development work is now underway to determine scaleup parameters and to investigate the more practical aspects of coal feeding, limestone feeding, bed material classification, materials of construction, compatibility, and control system development.

The first practical size experimental system was constructed in Scotland by Babcock Combustion Systems Ltd. (BCSL) in 1975. Early results indicate satisfactory operation and confirmation of smaller pilot-plant data.

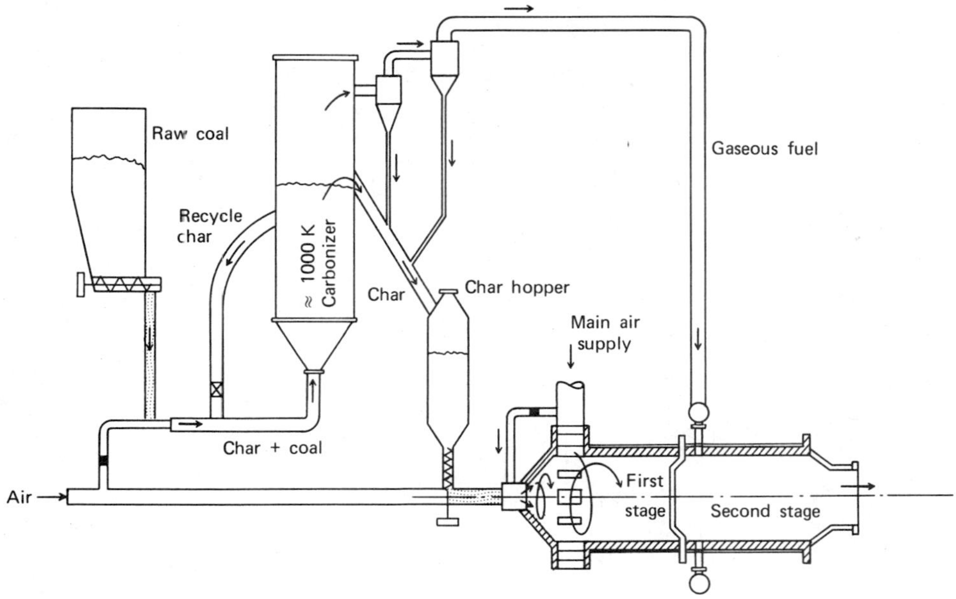

Figure 20.51 Westinghouse stoichiometrically staged cyclones combustor.[194]

Figure 20.55 shows the schematic of this unit. A 10 × 10 ft fluidized bed was fitted into an existing stoker unit, the bed taking the place of the stoker. Progress on this work was reported by McKenzie.[195]

In the United States under sponsorship of ERDA (now DOE), a 300,000-lb/hr steam unit was built, with operation commencing in 1976. The unit (Fig. 20.56, 20.57) is installed in the Rivesville, West Virginia, plant of the Monongahela Power Company. Table 20.5 presents the design conditions for the 300,000-lb/hr unit. As shown in Fig. 20.56, this unit consists of four cells of individual fluidized beds in a horizontal arrangement.

Additionally, a number of industrial fluidized-bed combustion units, again under ERDA (subsequently DOE) sponsorship, are currently in construction. A conceptual drawing of one of these units appears in Fig. 20.58.

In a summary paper, Gibson[197] reports industrial fluidized-bed developments in England. Among these are units being developed for burning of colliery wastes as well as shell-type steam generator units being converted from oil to coal firing by the introduction of a fluidized-bed combustor in the steam generator setting. It is important to point out that at this stage in the development many hardware and operational problems are being identified that alter the configuration of the fluidized-bed units. Major problems to be resolved deal with the need to develop better fuel feed systems. As shown in previous sections, fuel feeding every 10 ft^2 of bed area has proved necessary on the basis of pilot-plant experience. Applying this technique to large units creates a severe problem. Feeding methods that can reduce the number of feed points per square foot of bed are being sought. Also important is the development of feed systems that can handle a wide range of coal sizes, such as those resulting from a crushing operation, without the need for product screening. Such a feed system has not yet been developed.

Operationally, the load range of individual fluidized beds is severely limited by the constraints of bed temperature and fluidizing velocity range. Several approaches to managing this problem are being investigated. The principal approach is modularization. A possible alternative approach yet to be applied to large units is localized fluidization,

[195] E. C. McKenzie, *Heat. Air Cond. J.*, March 1977.

[196] R. L. Gamble and F. R. Warshany, "Commercial Development of Atmospheric Fluidized Bed Utility Steam Generators," paper presented at 1975 Joint Power Generation Conf., Portland, September 28–October 2, 1975.

[197] J. Gibson, "Present Status of Fluidized-Bed Combustion," Paper No. 62, presented at 3rd Int. Conf. Coal Research, 1976.

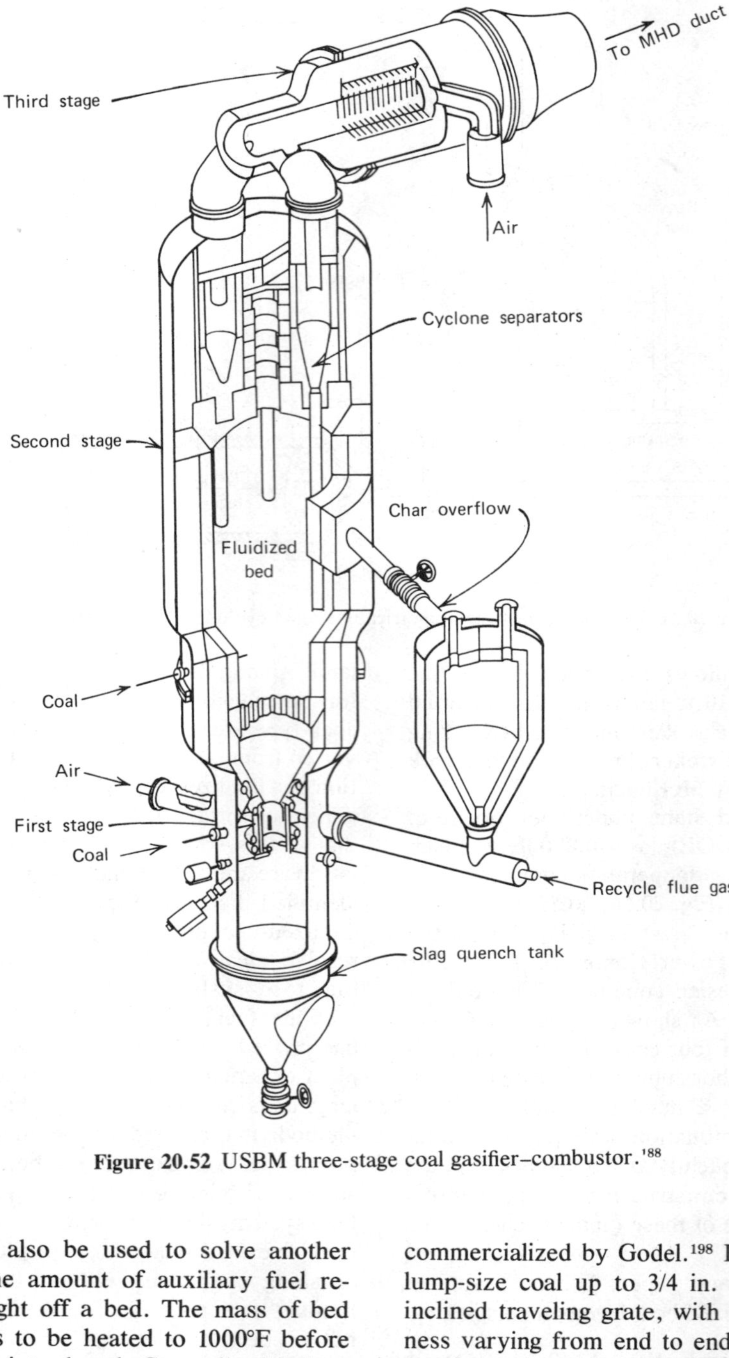

Figure 20.52 USBM three-stage coal gasifier–combustor.[188]

which may also be used to solve another problem: the amount of auxiliary fuel required to light off a bed. The mass of bed material has to be heated to 1000°F before coal can be introduced. Generally this heat requirement means that igniters of very large capacity must be used, with consequent economic penalties in distillate or gaseous fuels as well as in first cost.

Another form of fluidized-bed combustion is the ''Ignifluid'' system introduced and commercialized by Godel.[198] In this process lump-size coal up to 3/4 in. is fed into an inclined traveling grate, with the bed thickness varying from end to end. Substoichiometric (60%) air is blown through the grate; the airflow is compartmental, to allow adjustment for bed thickness and superficial velocities of about 15 ft/s. The combustible gas generated on the grate is then burned

[198] A. A. Godel, *Rev. Gen. Therm.* April 1966.

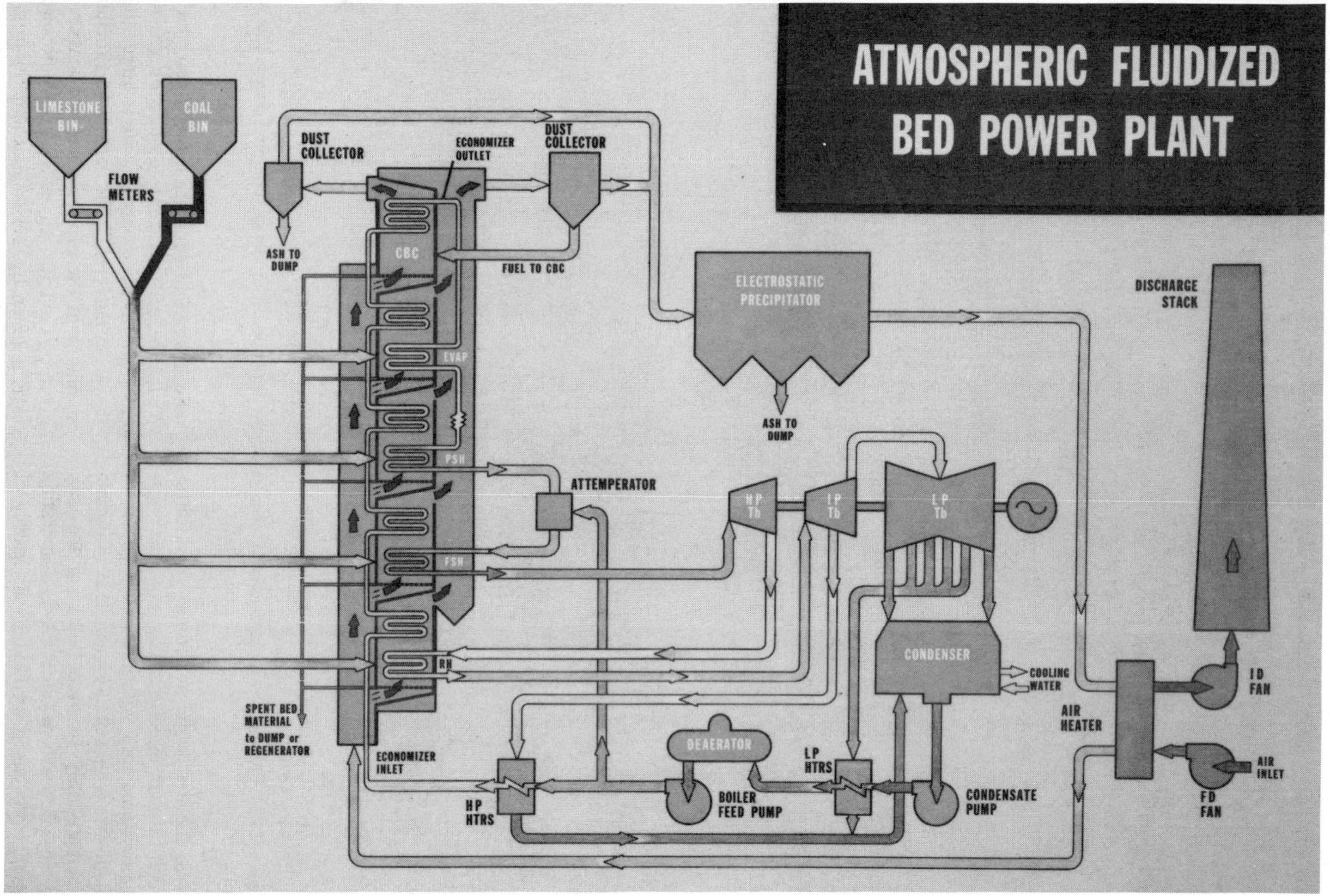
ATMOSPHERIC FLUIDIZED BED POWER PLANT
LIMESTONE BIN
COAL BIN
FLOW METERS
DUST COLLECTOR
ASH TO DUMP
ECONOMIZER OUTLET
DUST COLLECTOR
CBC
FUEL TO CBC
EVAP
PSH
FSH
RH
ATTEMPERATOR
ELECTROSTATIC PRECIPITATOR
ASH TO DUMP
DISCHARGE STACK
HP Tb
IP Tb
LP Tb
CONDENSER
COOLING WATER
ID FAN
AIR HEATER
AIR INLET
FD FAN
SPENT BED MATERIAL to DUMP or REGENERATOR
ECONOMIZER INLET
HP HTRS
DEAERATOR
BOILER FEED PUMP
LP HTRS
CONDENSATE PUMP

Figure 20.53

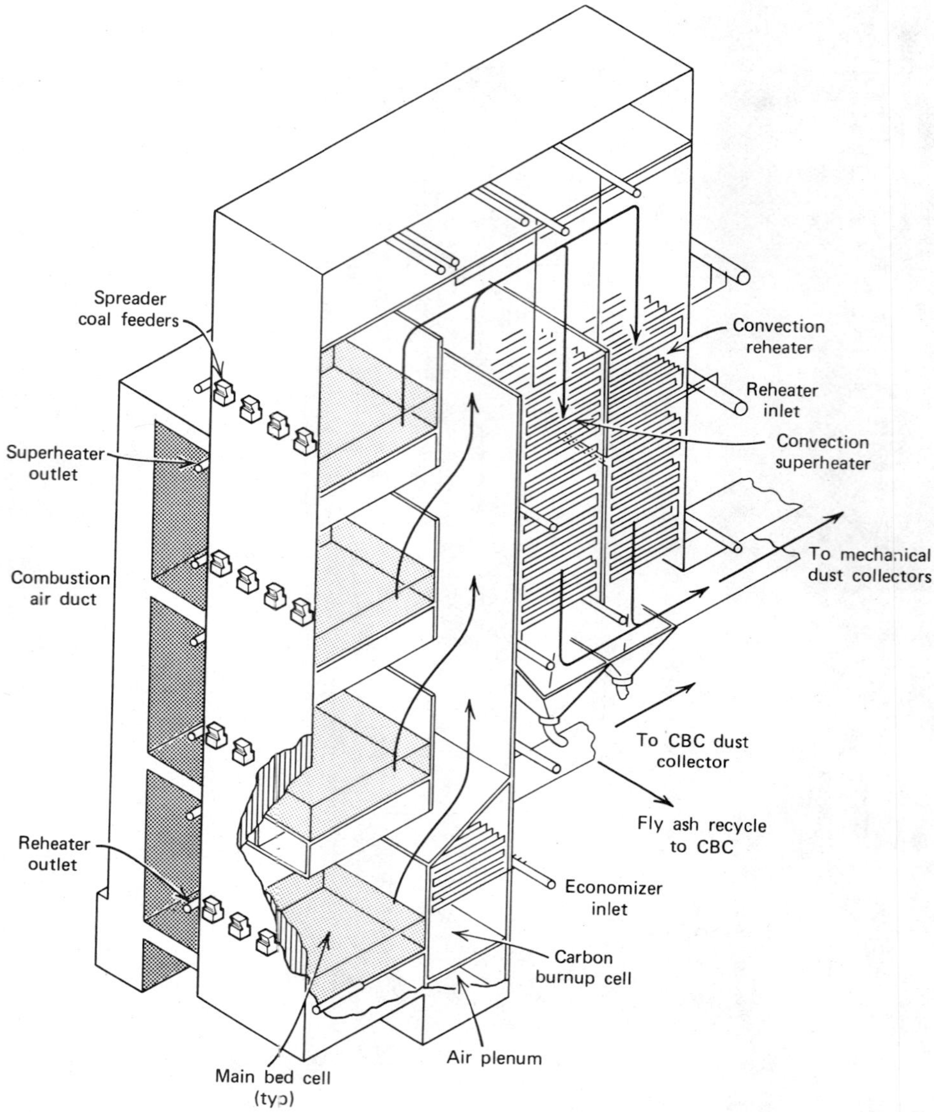

Figure 20.54 Fluidized-bed combustion steam generator for modern utility generation system, 150-MW unit.

above the bed while additional air is introduced through injectors. The interesting discovery connected with the development of this device is that sticky or molten ash tends to adhere only to other ash, not to coal. Recent efforts to develop ash-agglomerating gasifiers are based on Godel's discovery of this phenomenon.

Application of Ignifluid boilers seems particularly suitable for anthracite coals, although bituminous coals are being fired in several installations. Advantages claimed are reduced investment, ability to burn low-volatile coals, and generation of fly ash that is easier to separate from combustion gas because the average particle size is larger than that from pulverized-coal firing. An excellent summary of the principle and application is given by Godel;[198] an analysis of combustion tests is given by Yerushalmi and Squires.[199] Figure 20.59 gives typical side and end views of an Ignifluid boiler.

[199] J. Yerushalmi and A. M. Squires, *Notes on the Ignifluid Boiler and the Godel Phenomenon*, City College, City University of New York, 1973.

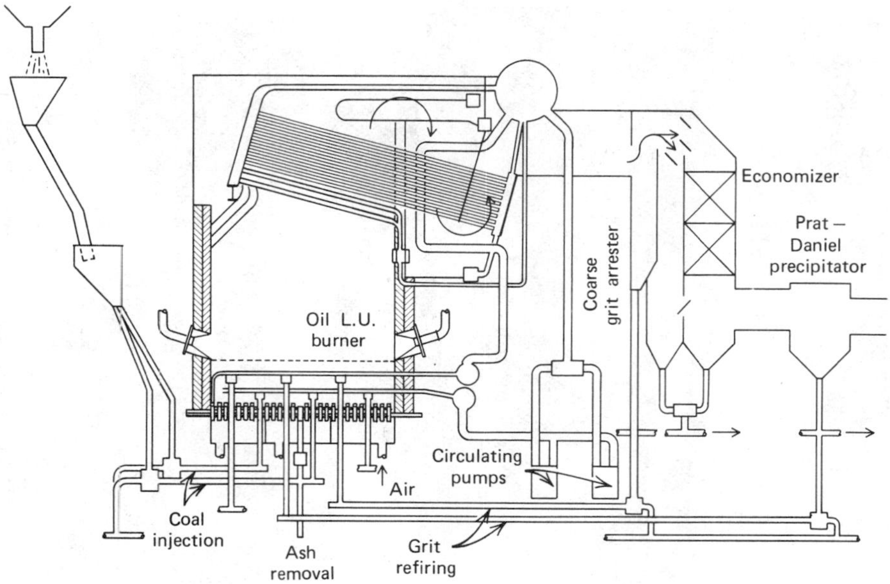

Figure 20.55 BCSL fluidized-combustion water tube boiler.[89]

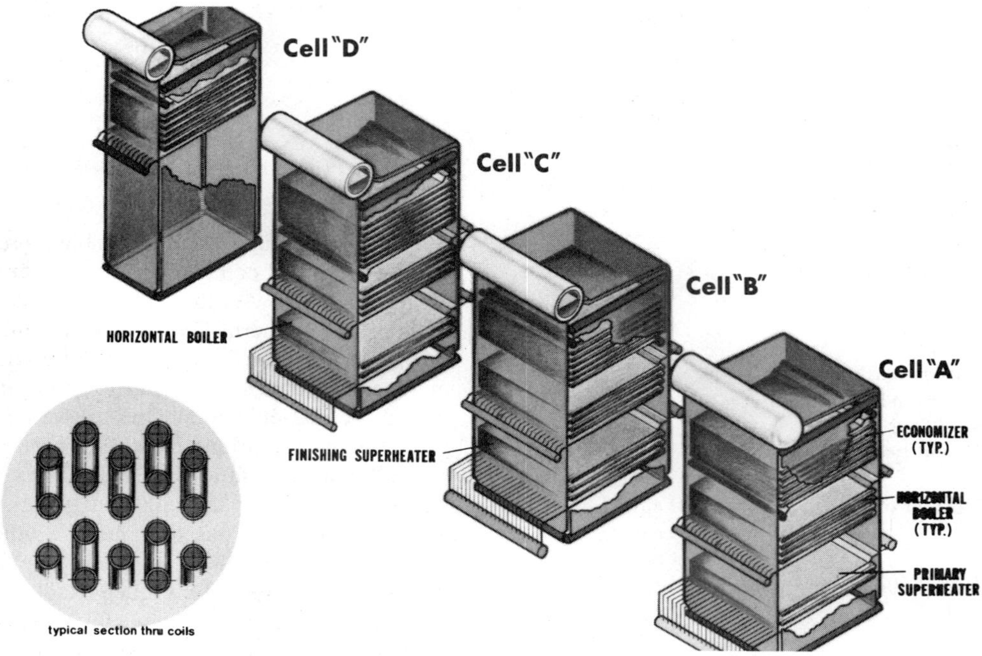

Figure 20.56 Cell arrangement of Rivesville, West Virginia, 30-MW demonstration unit.[196]

Figure 20.57 The 30-MW demonstration unit at Rivesville under construction.[196]

Table 20.5 Design Conditions for Rivesville 300,000-lb/hr Steam Generator

Steam flow	300,000 lb/hr
Steam pressure	1350 psig
Steam temperature	925°F
Bed temperature	1550°F
Carbon burnup cell temperature	2000°F
Excess air	18%
Carbon burnup cell excess air	25%
Limestone feed size	⅛ × 0 in.
Coal feed size	¼ × 0 in.
Circuit type	Forced circulation
Number of cells	4 (including 1 carbon burnup cell)
Overall dimensions of erected unit	12 ft wide × 38 ft long × 25 ft high

20.3.3 Pressurized Fluidized-Bed Combustion Systems

The same advantages listed for ambient pressure fluidized-bed combustion apply to pressurized systems, with the added prospect of attaining higher thermal efficiencies by introducing gas turbines in the cycle. Two basic approaches are being pursued to attain improved thermal efficiency. One uses a combined steam–gas turbine cycle; the heat extracted from the fluidized bed serves to generate steam, and the combustion gas is expanded through a gas turbine. The other uses an inert gas or air in a closed cycle for bed heat extraction. To reduce stack gas temperature, however, steam still must be generated downstream of the combustion gas

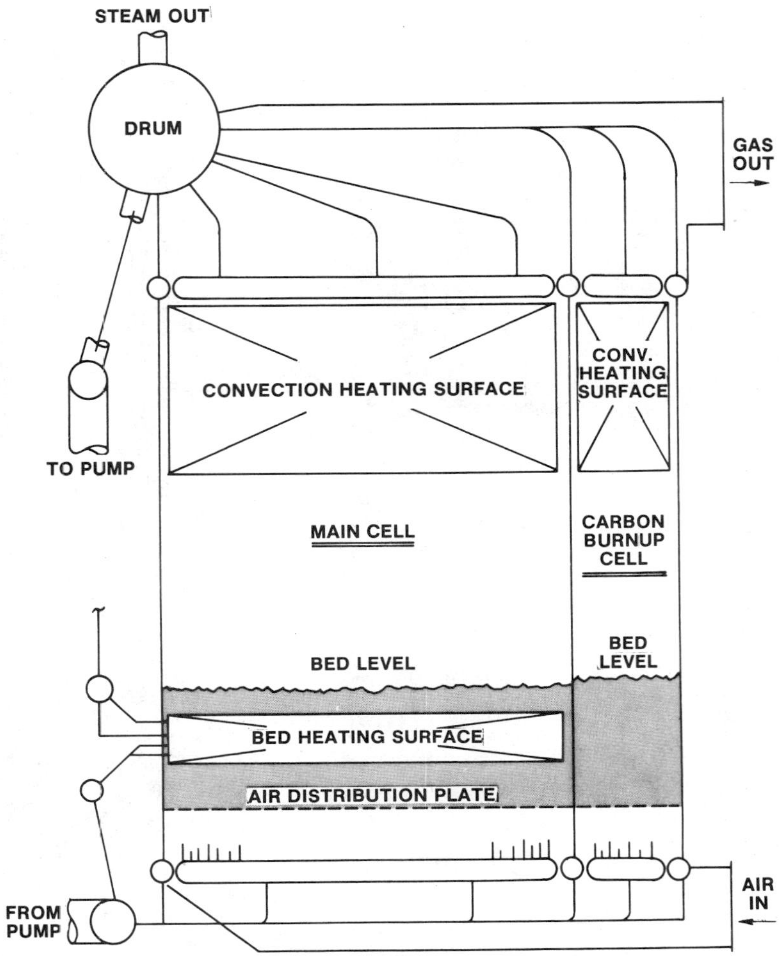

Figure 20.58 Design for a fluidized-bed steam generator

turbine as well as in the interstage coolers of the closed-cycle turbine.

Another approach, which is not particularly efficient thermally but is lower in capital cost, is to run the fluidized bed as a gas turbine combustor, controlling the bed temperature by excess air.

Figures 20.60–20.62 schematically illustrate the configuration of the cycle variations discussed. Figure 20.63 indicates the generating efficiencies obtainable by these cycle configurations.

Another combined-cycle system under development is the alkali-metal-vapor topping cycle. Here very high thermal efficiencies (~45%) are predicted with the potassium turbine operating at 1540°F. A schematic diagram of such a cycle is shown in Fig. 20.64. The combustor system shown is a pressurized fluidized-bed system, although an ambient bed could be used in its place at a modest reduction in efficiency.

The question of gas turbine protection from corrosion is not yet resolved. Work is in progress to approach this problem by removing particulate matter ahead of the gas turbine, thereby reducing the concentration of sulfur and alkali to a level compatible with turbine blade alloys. This, in effect, means particulate removal at temperatures of 1600–1750°F and pressures of 4–10 atm, with an efficiency of 99.9+%.

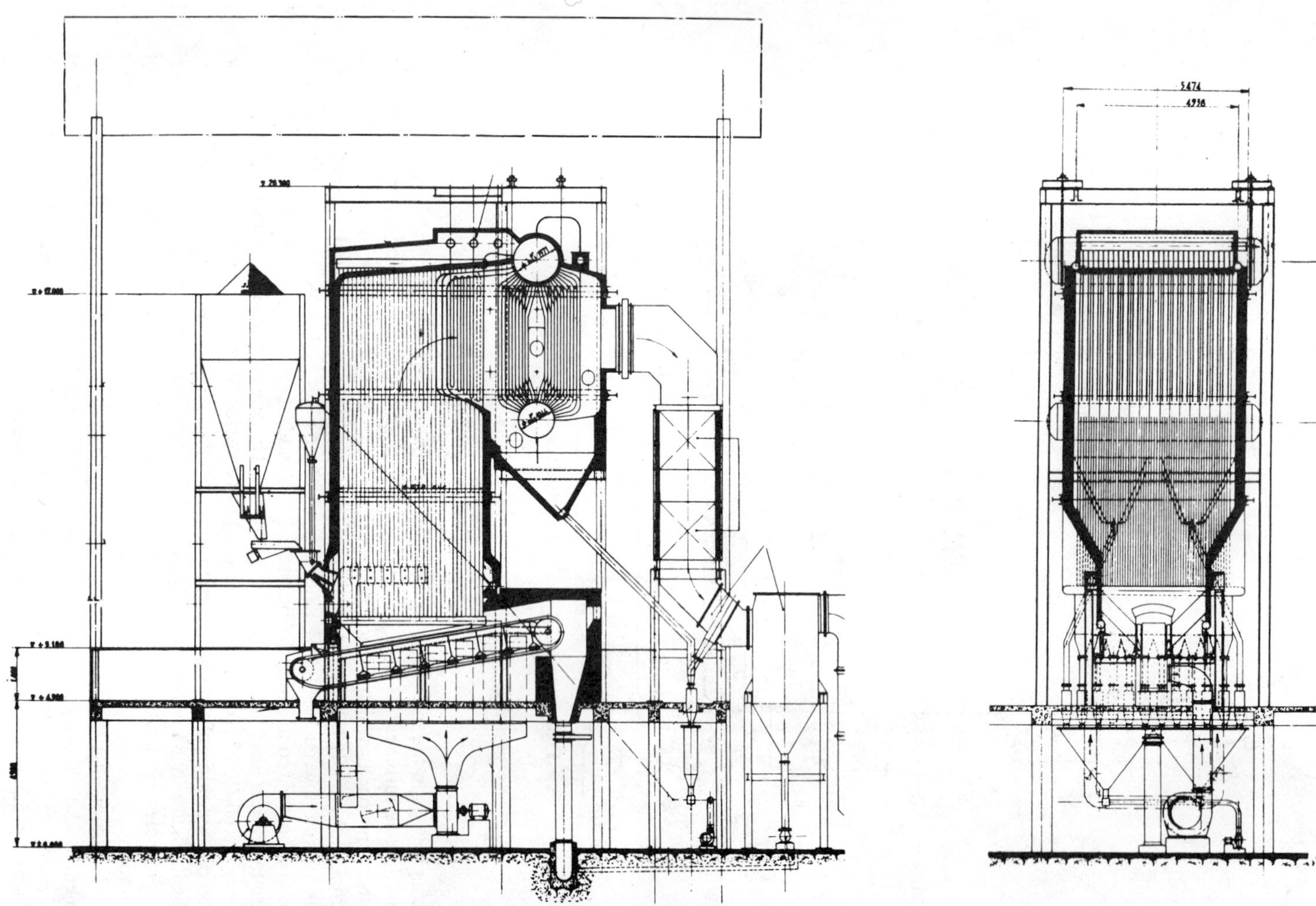

Figure 20.59 A 50–55 ton/hr boiler burning unwashed fine anthracite coal, fitted with an ignified combustion system built by Société Française des Constructions Babcock and Wilcox.[198]

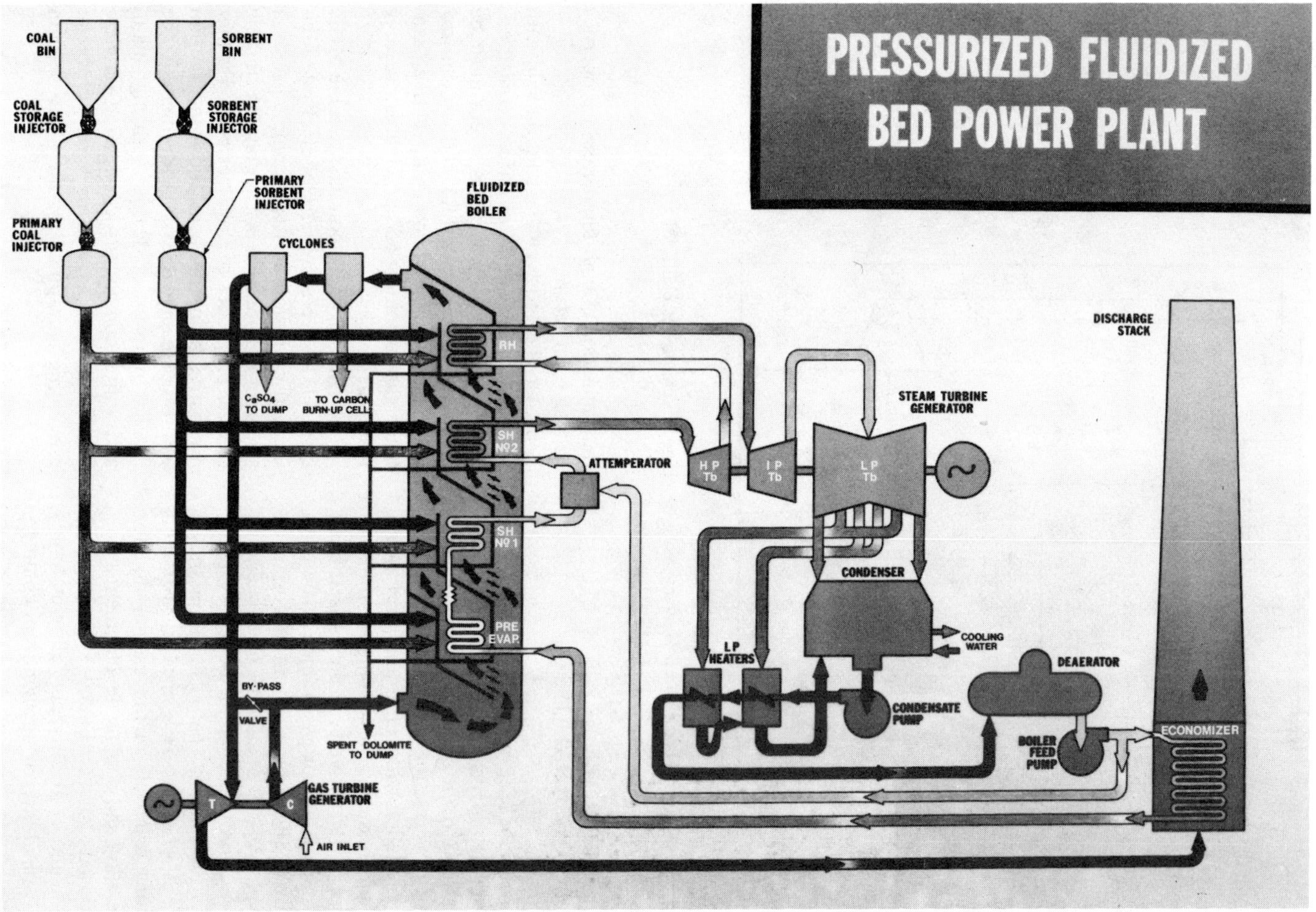

Figure 20.60 Coal-fired combined-cycle power plant.[200]

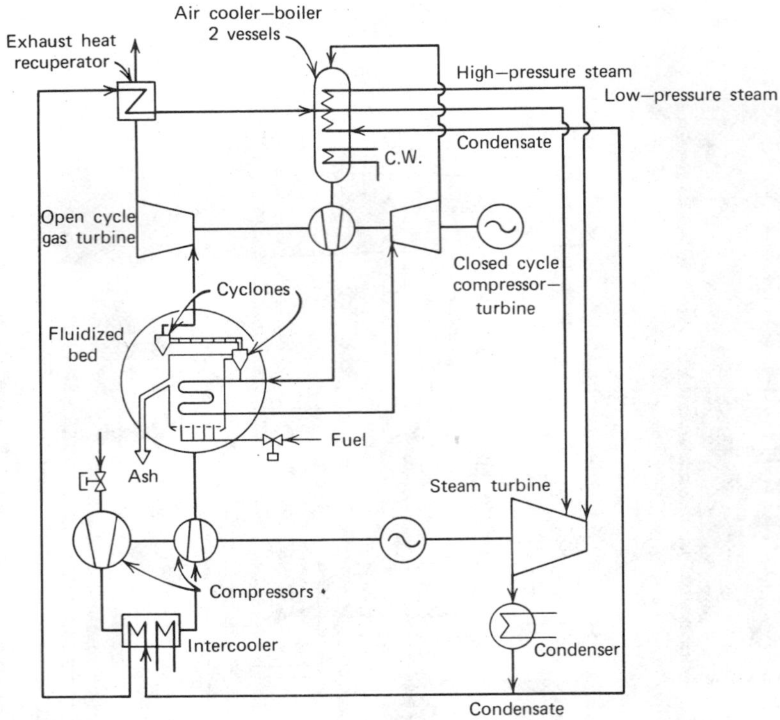

Figure 20.61 Gas-air-steam combined-cycle power plant.[200]

20.4 MATERIALS CONSIDERATIONS

One important area of development in new coal combustion systems concerns materials of construction. In pressurized fluidized-bed combined cycles, the gas containing particulates and laden with sulfur and alkali needs to be cleaned to a level not customarily achieved. Alternatively, a new gas turbine alloy development or gas turbine blade cooling would alleviate a potential corrosion problem, the latter at a sacrifice to generating efficiency. In other combined-cycle systems, such as the one for coal gasification shown in Fig. 20.47, there is a need for work on refractory materials in the gasifier and for turbine alloy and cooling developments.

In closed-cycle combined systems using either helium or alkali vapor, alloy materials that have had little or no significant commercial application need to be applied. This application calls for the investigation of fabrication processes and their effect on the properties of these materials, as well as the determination of high-temperature properties in long-term service.

By far the most challenging and difficult materials problem is in direct coal-fired, open-cycle MHD generators. Here must be faced the requirement for refractories exposed to very high temperatures with an ash-laden gas containing additional alkali seeding material.

Downstream of the combustor in the MHD channel, a severe problem of electrode fouling and corrosion exists, accompanied by a stringent requirement of maintaining suitable electric properties of the electrodes. Further downstream in the heat recovery section where a steam cycle is used, there is the familiar problem of dealing with high-alkali and sulfur compounds condensing on heat transfer tubing, as well as a new problem: preventing entrapment of the alkali seed in

[200] General Electric, proposal to Office of Coal Research, Vol. 2, RFP No. 14-32-0001-1726, Proposal No. CFES-P-75-1, January 31, 1975, pp. 2-31, 2-32, 2-41, 2-137.

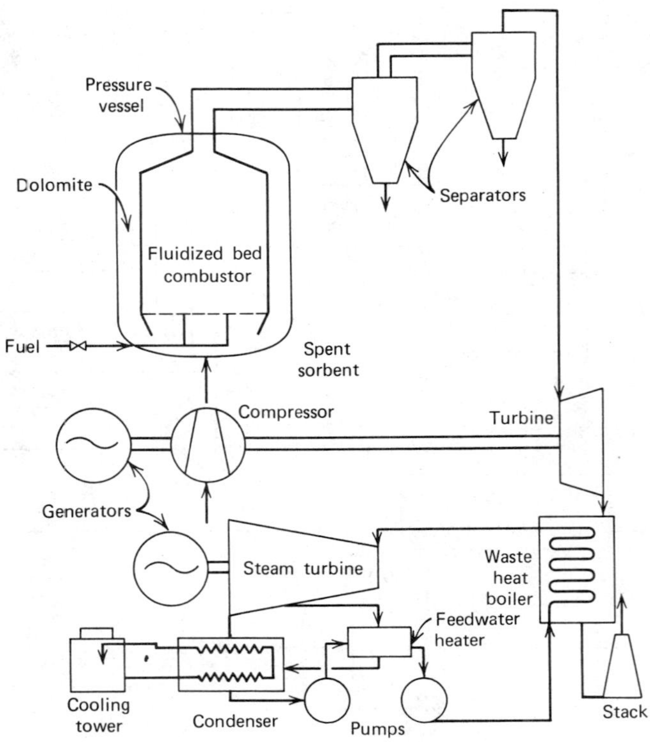

Figure 20.62 Exhaust heat boiler, combined steam–gas cycle power plant.[200]

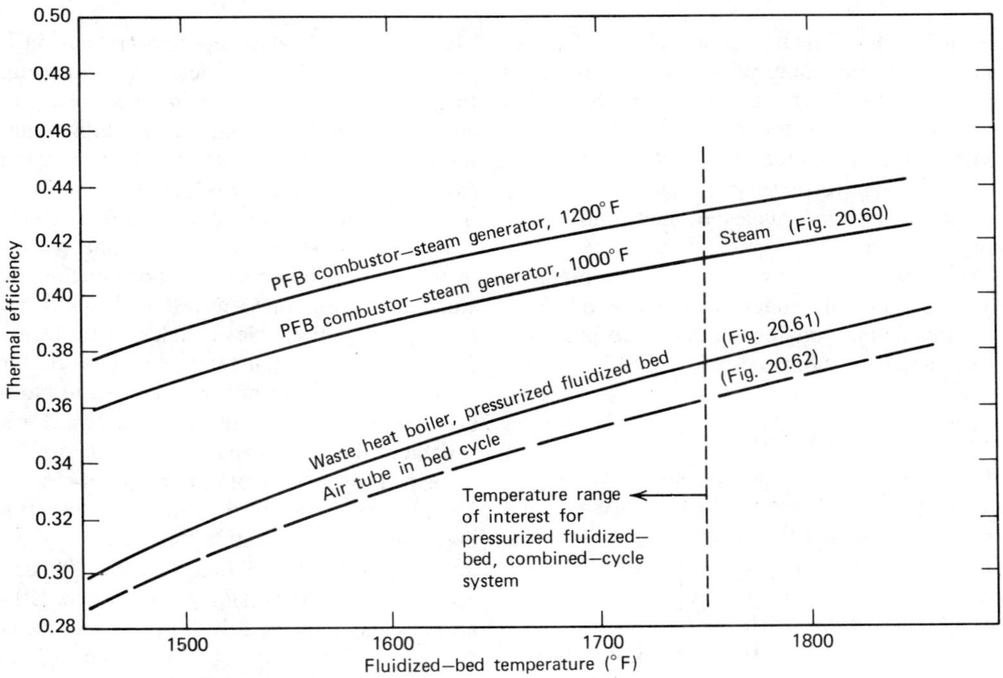

Figure 20.63 Comparison of alternate cycle efficiency versus fluidized bed combustor temperature.[200]

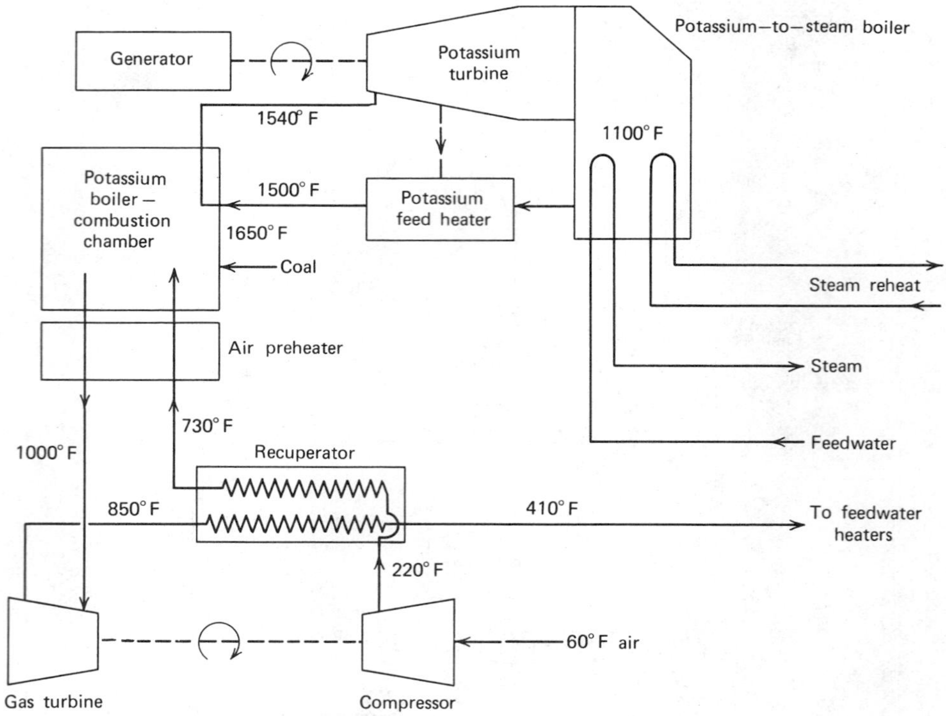

Figure 20.64 Alkali-metal-vapor topping cycle. Schematic diagram courtesy of Oak Ridge National Laboratory.

coal ash (slag). This new problem is a serious one because economy of operation depends on the ability to recover and reuse the seed material. If air is the oxidant, very high-temperature recuperative air heaters are required, using refractory compositions exposed to alternate cycles of high and low temperatures.

Much work is being done in these areas. This section only touches on some of the developments, allowing the reader to pursue more details in the listed references.

20.4.1 Materials Development

Fluidized-bed combustion is claimed to have advantages over other combustion processes insofar as it avoids the formation of vitreous ash compounds and precludes corrosion and deposits on heat transfer tubing caused by operation at temperatures below the ash fusion point. In various experimental rigs these advantages have been confirmed. A series of NCB tests in a 12-in. ambient pressure combustor is reported by Westinghouse.[79] The same work also reports on tests in a pressurized rig 4 × 2 ft bed at pressures up to 5 atm. Materials used in these tests included conventional heat transfer tubing materials used by the boiler industry—ranging from carbon steels (SA-210) to chromium-molybdenum steels (T-22) to high-alloy steels (austenitic Type 304-316, etc.). The results of these tests, however, are not conclusive, since the testing time was relatively short: no more than 1000 hr. Nevertheless, heat transfer tubes operating at temperature levels for which they are commonly employed in modern combustion systems generally perform satisfactorily in fluidized-bed combustors.

Tests at higher metal temperatures—up to 1500°F—were run by the National Research Development Corporation (NRDC).[79] Preliminary indications are that Incoloy 800 can give acceptable corrosion performance. Other work has indicated that, in fluidized beds at metal temperatures of 1500°F and higher, cladding of the Incoloy 800 with Inconel alloy 671 may provide long-term corrosion protection.[79]

The higher temperatures become of interest in closed-cycle systems using helium, air, or potassium in the closed loop. Some recent work has shown that Haynes 188 may be a promising alloy for use at temperatures up to 1600°F. This alloy is of interest because of its compatibility with liquid potassium and sodium.

In pressurized fluidized-bed combined cycles the products of combustion from the bed are expanded through a gas turbine at temperatures up to 1750°F. Carryover of particulates, together with alkali and sulfur compounds from the combustor, will depend largely on bed operating characteristics. Present turbine alloys cannot tolerate alkali sulfur compounds above 1500°F, even when the concentrations are in the ppm range. Other than the development of new alloys for gas turbines—a goal that is being pursued intensively by manufacturers—the approach to protect the turbine is to introduce a highly efficient particulate separator. It is presumed that the alkalis and sulfur corrodents are attached to ash particles, thus can be removed with the ash. The ash itself is unlikely to cause an erosion problem, since it is formed at temperatures below that of ash fusion in the bed and appears as a fluffy, soft substance.

Other forms of combined cycle, such as coal gasifiers with gas and steam turbines, exhibit different materials problems. Reducing atmospheres are prevalent in gasifiers and downstream heat recovery equipment, and materials must be selected with that in mind. No new material development appears to be required except in the case of slagging gasifiers, where it is expected that the refractory throats exposed to liquid slag may show the need for development work.

Basic, however, to these cycles are the materials development problems associated with the gas turbine. Gas turbines in this form of combined cycle must be operated at gas inlet temperatures above 2200°F to achieve reasonable economics.[79]

Substantial work is required to develop materials suitable for this service. The approach being used incorporates techniques for blade cooling, alloy development, and ceramic blade and stator development.[79]

Of the system developments discussed, the open-cycle MHD presents the most severe and most speculative materials problems. The subject is so complex yet so basic to the success of the MHD program that generalizing or abstracting here could be grossly misleading. The reader is referred to the ECAS study[79] for an extensive discussion.

CHAPTER 21

Coal Ash—Its Effect on Combustion Systems

WILLIAM T. REID

William T. Reid *was educated as a chemical engineer but has spent his entire professional life in fuels technology, first at the Pittsburgh Station of the U.S. Bureau of Mines, and then at Battelle Memorial Institute, Columbus, Ohio, until his retirement in 1972. He has specialized in the properties of coal ash and coal ash slags and is a recognized international authority on problems of slagging and fouling in large, pulverized-coal-fired boiler furnaces. Mr. Reid also has done extensive research into the causes of metal wastage attributable to the mineral matter in coal. Since his retirement, he has served as a consultant on fuels utilization problems to Battelle, the Electric Power Research Institute, and many utilities, government agencies, and industrial concerns.*

21.1 MINERAL MATTER IN COAL

The inorganic matter in coal continues to be a major source of problems in using this most abundant of our fossil energy resources. Whether in solvent coal refining, where filtering an intermediate product to remove ash is a required and troublesome step, or in combined cycles, where even traces of inorganic matter can foul open-cycle gas turbine blades and nozzles, the noncombustible material in coal leads to many problems. But it is in the more conventional use patterns, for example, burning pulverized coal to generate steam—currently our largest single end use of coal—that difficulties with mineral matter are most apparent. Large central station coal-burning power plants are still limited in performance, reliability, and efficiency by slagging and fouling problems and by external corrosion of heat-receiving surfaces. Relatively little coal is burned today on grates, where clinkering is troublesome, but the growing interest in gasification in fixed beds will again bring attention to sintering of coal ash at moderate to high temperatures. Slagging gasifiers are basically similar to slag tap, pulverized-coal-fired boiler furnaces (except for the reducing rather than oxidizing conditions). Operation of such gasifiers will emphasize the behavior of coal ash slags in the range where they behave as Newtonian fluids or as pseudoplastic solids.

Many advances in our understanding of the behavior of the mineral matter in coal have made in the past two decades, but much more information is needed before designers and operators will be able to use coal for energy conversion systems with maximum effectiveness. This chapter points out significant advances in our understanding of the behavior of the mineral matter in coal in energy conversion systems and identifies problem areas in which coal ash will be particularly troublesome in the next few years.

The characterization of ash constituents to identify the minerals and to determine the quantity of mineral matter in coal is a subject of growing importance. Mineral matter in coal is a major cause of problems when coal is burned or gasified, and it causes myriad problems with filtration, catalysis, and abrasion when coal is converted, as in liquefaction and solvent coal refining. Improved laboratory procedures to characterize mineral matter are now being developed that promise to be extremely helpful as the uses of coal are expanded during the next few decades. Such techniques include maceral analyses, optical and electron microscopic techniques, and X-ray diffraction, infrared spectroscopy, and spectrochemical methods of examining low-temperature ash (LTA)[1] (Chapters 4 and 7). Through the use of these techniques, combined with computerized systems such as CESEMI (computer evaluation of scanning electron microprobe images), analytical procedures are being developed that are contributing substantially to understanding the role of inorganic matter in the use of coal.

21.1.1 Control of Ash Fusibility by Coal Preparation

Coal preparation by gravity separation methods (Chapter 10) was not the general practice for coals used for steam generation. In the past, utilities buying large quantities of coal on long-term contracts established the prices they paid on a Btu basis, without much concern about coal characteristics or ash properties if the coal came from geographical areas that produced coal of generally known characteristics. Industrial and commercial users of coal were even less critical, partly at least because they were unable to define their coal needs other than by heating value. The steel industry, with its rigid specifications for coke, placed the greatest demands on coal preparation plants. Thus for many years the principal emphasis of coal preparation has been reduction in ash content and control of size for coals going into boilers.

The situation is different today. In particular, limitations on the emission of sulfur oxides have led to emphasis in coal preparation on lowering the sulfur content of boiler fuels. To minimize transportation charges when shipping low-sulfur, low-rank western coals long distances to eastern markets, it is essential to reduce both the moisture and ash contents. Ash quality has also become more critical. Greatly increasing capital costs for large coal fired boiler furnaces led a few years ago to more compact units, where higher heat release rates tend to cause increased slagging and fouling. More recently, the trend has been to more expensive boilers capable of operating with fewer unscheduled outages. Research is now in progress to identify the minerals in coal that lead to increased operational problems in large steam generators and to determine the fate of these minerals when coal is washed.

21.1.1.1 *Pyrites*

Iron pyrites, FeS_2, causes problems because it usually provides about half the sulfur in most coals, melts at a moderate temperature (1171°C), dissociates readily to release sulfur and FeS (1193°C), and when oxidized, is a significant source of Fe_2O_3 to lower the viscosity of coal ash slags, thereby to increase slagging and fouling. The high density of FeS_2 (5 g/cm^3) compared with the low density of coal (1.2–1.7 g/cm^3) explains why gravimetric separation of pyrites is possible. Since pyrites occurs as discrete particles, even down to submicrometer levels, the major problem lies in freeing the particles of pyrites from the coal substance to permit gravity separation. The ability to free pyrites is shown by washability measurements where the coal is stage ground, then float-sink tested at different specific gravities, typically 1.30, 1.40, and 1.60, and occasionally at 1.90. Data on washability for sulfur reduction are available on many coals in the United States for the sizes of crushed coal commonly used to enhance liberation of pyrites.[2]

[1] P. L. Walker, Jr., W. Spackman, P. H. Given, E. W. White, A. Davis, and R. G. Jenkins, *Characterization of Mineral Matter in Coals and Coal Liquefaction Residues,* EPRI Research Project No. 366-1, December 1975.

[2] J. A. Cavallaro, M. T. Johnston, and A. W. Deurbrouck, *Sulfur Reduction Potential of the Coals of the United States,* U.S. Bur. Mines, Rep. Invest. No. 8118, 1976, 323 pp.

Table 21.1 shows the average reduction in pyritic sulfur content of U.S. coals crushed to pass 3/8-in. and 14-mesh sieves and separated at 1.60 specific gravity. It is evident that finer crushing liberates more pyrites, so that sulfur reduction by coal washing is more effective as the particle size becomes smaller. Less stringent preparation conditions would show appreciably less pyrites removal; but the relationships are complex, and the behavior of each coal is unique. The wide variability in the manner in which pyritic sulfur can occur in a single mine is shown in Fig. 21.1 for 37 samples of coal from the Pittsburgh bed.[4]

Removal of pyrites has four principal effects on furnace performance: (1) decreasing the likelihood, in pulverized-coal-fired boiler furnaces, of the sulfidation of heat-receiving surfaces that can result from impingement and the capture on wall tubes of particles of pyrites; (2) lowering the Fe_2O_3 content of the coal ash, thereby increasing the viscosity of any molten slag formed within the furnace; (3) lessening the possibility that pyrites particles will be captured on the hearth of slag tap furnaces or of slagging gasifiers, to accumulate as a pool of molten iron sulfides with consequent failure of the furnace bottom; and (4) lowering the sulfur level in the flue gas, thus decreasing the possibility of external corrosion of superheaters and reheaters.

[3] U.S. Bureau of Mines, Rep. Invest. No. 8118, 1976.

[4] Harold J. Rose and Richard A. Glenn, "Coal Cleaning in Relation to Sulfur Reduction in Steam Coals," ASME Paper 58-A-197, 1958.

Table 21.1 Reduction in Pyritic Sulfur of U.S. Coals by Float-Sink Separation[3]

Float-sink separation at 1.60 specific gravity.

Region	Average Pyrites Content (%)		
	Raw Coal	Washed Coal 3/8 in.	Washed Coal 14 mesh
Northern Appalachian	2.01	0.90	0.70
Southern Appalachian	0.37	0.23	0.20
Alabama	0.69	0.48	0.39
Eastern Midwest	2.29	1.12	0.98
Western Midwest	3.58	1.90	1.58
Western	0.23	0.15	0.13

Sulfidation of wall tubes rarely occurs in boiler furnaces operated with normal excess air, but becomes a possible source of trouble with staged combustion for NO_x control. It is not yet possible to estimate the extent of sulfidation in boiler furnaces of the future. For gasifiers, sulfidation will be a serious problem.

Ash fusibility in some coals is substantially affected by the removal of pyrites through coal preparation. During combustion any pyrites in the coal is oxidized to FeO or Fe_2O_3, which is then able in combination with the other fluxes (CaO, MgO, Na_2O, and K_2O) to lower the viscosity of the silicate melt. Since these fluxes are fairly comparable in their ability to decrease slag viscosity, any lowering in Fe_2O_3 content can significantly increase slag viscosity.

Of 29 coals analyzed by the U.S. Bureau of Mines,[5] 18 had almost twice the Fe_2O_3 content that would have resulted from complete oxidation of the pyrites in the coal. In other words, roughly half the Fe_2O_3 in the ash of these coals would have come from pyrites and half from other iron-containing minerals in the coal. Hence significant lowering of pyrites in these coals by a preparation plant would substantially decrease the Fe_2O_3 in the ash, thereby raising the viscosity of any resulting slag.

It is important to note that pyrites in coal is *not* the sole source of the iron that appears in the ash of many coals. Thus even complete removal of pyrites by coal washing can still leave appreciable amounts of Fe_2O_3 to serve as a flux with many coal ashes.

Slag tap furnaces are less popular today than they were in the 1930s and 1940s, but with the growing use of Western coals with low-fusion ash, wet bottom furnaces may return. Slagging gasifiers are also attractive, so that future combustion systems converting the coal ash to molten slag appear likely. Pyrites will pose serious problems with such combustors, through capture by the molten slag and attack of the hearth bottom by accumulated molten iron sulfides. Design modifications, such as improved cooling methods for the bottom, could minimize these problems.

[5] R. F. Abernethy, M. J. Peterson, and F. H. Gibson, *Major Ash Constituents in U.S. Coals,* U.S. Bur. Mines Rep. Invest. No. 7240, 1969.

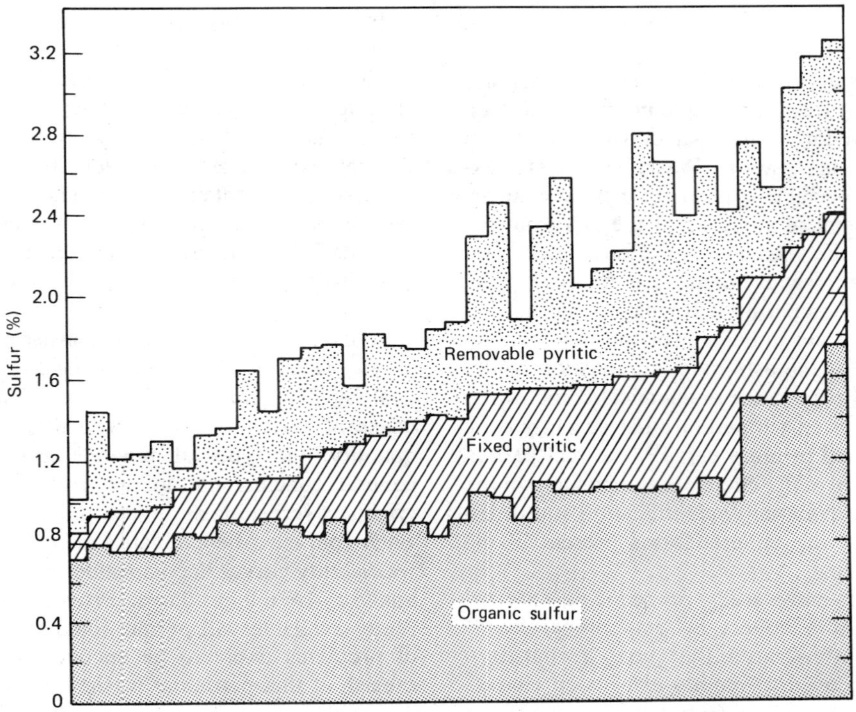

Figure 21.1 Variations in total sulfur, sulfur forms, and removable pyrite at ⅜-in. 48-mesh, of face samples of Pittsburgh bed coal from a single mine in Greene County, Pa.[4]

Coal preparation, by removing pyrites, can have a highly beneficial effect both on ash fusibility and in decreasing operational problems associated with sulfide sulfur. Liberation of pyrites by finer grinding, froth flotation, and coal beneficiation techniques other than the conventional gravity separation processes may be necessary to reduce the pyrites content to the desired low level.

21.1.1.2 *Mineral Matter*

Coal preparation can modify the chemical composition of ash by selective removal of some mineral species. The *inherent* ash in coal is changed only slightly by coal washing, and then only by the presence in the reject of some of the coal substance that may be particularly rich in an inorganic constituent. It is the *extraneous* mineral matter in coal that is mainly removed by coal preparation and that has the greatest effect on ash fusibility. Partings of shale, for example, often occur in discrete pieces large enough to be readily separated from the coal by conventional gravity separation methods. Since these shale partings generally have a high fusion temperature—they consist mainly of silica and alumina—their removal lowers the fusion temperature of the ash left in the coal product. Calcite and dolomite likewise may be present in coal in discrete particles large enough to be liberated when the coal is crushed prior to washing, hence may be removed. Unlike shale, calcite and dolomite are effective fluxes for coal ash, and their removal raises the fusion temperature of the ash remaining in the coal product, except for unusual coals where the calcite content is very high. Mineral matter rich in iron compounds other than pyrites or marcasite, or containing substantial amounts of sodium or potassium, may be removed by coal preparation, thereby raising ash fusibility. Few data are available here; each coal must be considered individually. A research investigation supported by the Electric Power Research Institute at Battelle, which began in 1976, is aimed at relating the occurrence of mineral forms in coal to the fouling and slagging behavior of that coal in large utility boiler furnaces. Pennsylvania State University is conducting related research for the U.S. Energy Research and Development Administration.

It is important to note that washability curves by themselves give no information on the behavior of the ash of washed coal in combustion systems. Washability tests, except in special cases, do not include comparative chemical analyses of the ash from the raw and the washed coals. Cone fusion temperature measurements are of little help, since the temperature at which cones deform is not a good indication of the flow characteristics of the ash under combustion conditions. The influence of chemical composition on ash fusion temperature has recently been determined in more detail.[6] The real parameters involve the chemical composition of the ash and, eventually, identification and distribution of the mineral forms present in the washed coal.

An extensive study on reducing external corrosion in boilers by coal preparation[7,8] also demonstrated that coal preparation could be used to modify ash fusion characteristics. Based on 60-ton lots of coal from six different coal seams, the alkalis Na_2O and K_2O tended to be in higher concentration in the larger particles (plus 1/4-in.) and in lower concentrations in the smaller sizes (minus 48-mesh). This was also the case with CaO and MgO, but with a large variation among different coals. Preparation consisting only of screening, therefore, could result in a substantial difference in ash characteristics with some coals. Float-sink separation showed no strong trends, some coals tending to concentrate MgO at higher specific gravity, whereas others showed less MgO in the heavier fractions. Lime and Na_2O and K_2O likewise differed widely enough under float-sink separation that no general trends were apparent. This study mainly demonstrated that the complex way in which alkali and alkaline earth minerals occur in coal makes it impossible to predict changes in concentration of these constituents when coals are washed at different specific gravities. Nevertheless, adjustment of the calcium and magnesium content of the ash of corrosive bituminous coals has been found to have a marked effect on reducing wastage of superheaters and reheaters.

Coal preparation, for size consist and by gravity separation, influences the use of the coal by changing its ash composition. Because of variability in the size distribution, quantity, and types of minerals present, no general conclusions can be drawn on the effects of screening or of float-sink treatment. Each coal must be evaluated individually by washability curves accompanied by chemical analyses of the products. Procedures currently available will permit some prediction of problems likely to be encountered with the ash of that particular washed coal.[9]

21.1.2 Behavior of Mineral Matter when Heated

The mineral matter in coal undergoes many modifications as the coal is heated. The modifications may be quite minor, such as changes that occur during carbonization, or they can be extensive—for instance, in pulverized-coal firing when temperatures may easily exceed 1650°C and the flow patterns in the furnace can cause segregation of ash particles. In magnetohydrodynamic (MHD) systems, where the temperature may approach 2800°C, radical changes can be expected. As a further complication, at temperatures above about 800°C interactions can occur as some mineral species begin to melt. Such particles may then react with other particles of different chemical composition, often yielding a product quite unlike the reactants. The heterogeneous appearance of fly ash under a microscope demonstrates the wide variation in the response of coal mineral matter to high temperatures.

[6] E. C. Winegartner and B. T. Rhodes, "An Empirical Study of the Relation of Chemical Properties to Ash Fusion Temperatures," ASME, Paper 74-WA/CD-1, 1974; *Trans. ASME, J. Eng. Power, Ser. A*, **97**, 395–406 (July 1975).

[7] E. B. Wilson, J. W. Leonard, R. P. Hensel, R. W. Borio, and R. C. Ulmer, "Reducing the Corrosive Properties of Utility Coals Through Modification of Current Coal Cleaning Practices," paper presented at ASME–IEEE Joint Power Generation Conf., Detroit, September 25–27, 1967.

[8] R. W. Borio, *The Control of High-Temperature Fireside Corrosion in Utility Coal-Fired Boilers* U.S. Dept. of Interior, Office of Coal Research, R & D Rep. No. 41, 1969.

[9] R. W. Bryers and T. E. Taylor, "An Examination of the Relationship Between Ash Chemistry and Ash Fusion Temperatures in Various Coal Size and Gravity Fractions Using Polynomial Regression Analysis," ASME Paper 75-WA/CD-3, 1975.

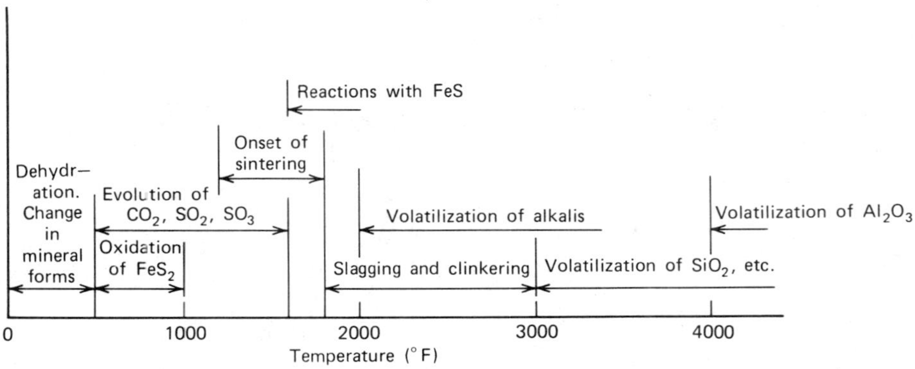

Figure 21.2 Effect of heating on minerals in coal.[10]

Figure 21.2 illustrates the progressive changes that can occur with coal minerals as the temperature is raised. Loss of water of hydration takes place first. Next, CO_2 and SO_x are evolved and pyrites is oxidized. Sintering begins at about 650°C, and liquid phases are formed that lead to melting and reactions between particles to produce clinker and slag above about 1000°C. Volatilization of alkalis (Na_2O and K_2O) becomes appreciable as the temperature rises above 1100°C. Silica begins to volatilize above 1650°C under oxidizing conditions, and at a significant rate above 2100°C.

Under reducing conditions, as in the presence of carbon and hydrogen in fixed-bed gasifiers, and in the temperature range 1750–1950°C, the amount of silica volatilized as silicon monoxide can be calculated from equilibrium thermodynamic data.[11]

Figure 21.3 shows the vaporization rate of small particles of quartz under conditions similar to those in an MHD combustor, with less loss of silica as the system became less strongly reducing.[12] The sudden increase in vaporization rate above 2150°C is unexplained.

Figure 21.4 shows the calculated loss by evaporation of the constituents in a typical coal ash containing 23% Fe_2O_3, 5% CaO + MgO, and 2% Na_2O + K_2O, assuming a neutral atmosphere and no interactions between the oxides.[13] Alkalis are not included in this figure because they have such high vapor pressure at temperatures above 1100°C. It is evident that the only solids remaining above 2200°C would be CaO and Al_2O_3, and above 2700°C only Al_2O_3.

Few data are available on the transformations that actually occur in mineral matter as the coal is burned. This system of transformations is complex, involving many mineral

[10] W. T. Reid, *External Corrosion and Deposits: Boilers and Gas Turbines,* Elsevier Publishing Company, New York, 1971, p. 55.

[11] E. Raask and D. M. Wilkins, "Volatilization of Silica in Gasification and Combustion Processes," *J. Inst. Fuel,* **38,** 255–262 (June 1965).

[12] British Coal Utilisation Research Association, *Annual Report,* 1966, p. 73.

[13] F. Hals, L. Keefe, and A. El Bindari, "Studies of Seed Recovery and Corrosion in Coal Fired MHD Power Plants," ASME Paper No. 64 WA/ENER-5, 1964.

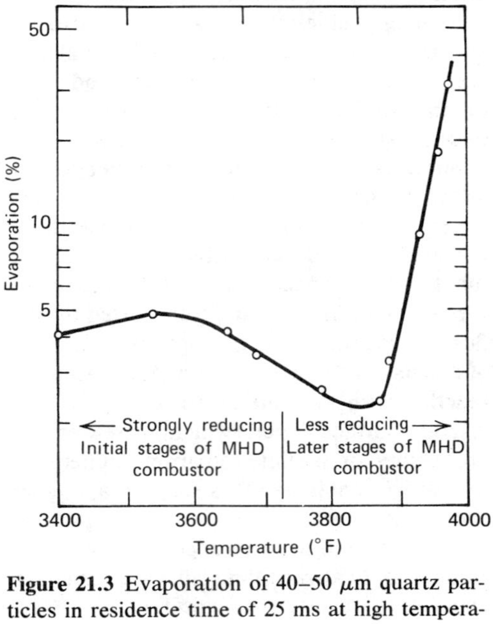

Figure 21.3 Evaporation of 40–50 μm quartz particles in residence time of 25 ms at high temperatures.[12]

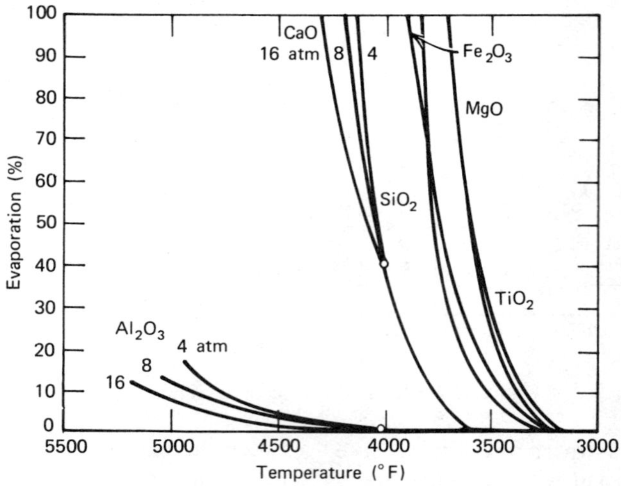

Figure 21.4 Calculated evaporation of mineral constituents in a typical coal ash at high temperatures.[13]

forms heated at varying rates under conditions that may shift from strongly reducing to oxidizing in a few milliseconds. With the growing interest in gasifiers and in new combustion systems such as fluidized beds, more interest is being developed in the form in which inorganic matter is present in coal, how it reacts physically and chemically during combustion, and the way it affects combustor performance, such as in fouling or the emission of particulates.

21.1.3 Importance of Alkalis

Sodium and potassium are major contributors to fouling problems in large boiler furnaces as a result of the ease with which the alkalis are volatilized from coal minerals and from organic alkali compounds in coal. The alkalis, once volatilized, then condense on cool boiler surfaces and on particles of suspended fly ash as the flue gas moves through the furnace circuit. There is general agreement that these condensed alkali-rich surface films on wall tubes and superheater elements facilitate the capture of particles being transported by the flue gas. Furthermore, the high concentration of alkalis condensed on ash particles leads to reaction with the particle to produce a thin viscous surface layer, which increases the probability of particle-to-particle adherence following a collision. This surface layer also encourages sintering because of the thermal gradient, resulting in stronger deposits that adhere to heat-receiving surfaces and are exposed to furnace temperatures.

The mechanisms by which alkali-bonded deposits form are complex.[14] There is abundant evidence that increased alkali content in coal leads to increased fouling, with condensation and surface concentration as the most logical explanation.

Figure 21.5 shows the vapor pressure of the alkali oxides increasing by an order of magnitude with a temperature rise of about 110°C.[15] Potassium has a much higher vapor pressure than sodium, roughly 10 times greater, over the range of temperatures in boiler furnaces. Although most attention is usually given to sodium, potassium could lead to worse fouling problems as a result of volatilization during combustion. Moreover, although U.S. coals on the average contain equal quantities of potassium and sodium on a molar basis, deposits formed at a moderate temperature are likely to contain more potassium than sodium. This has not been borne out in commercial practice, where equal amounts of potassium and sodium in a fuel resulted in selective deposition of sodium on a controlled-temperature superheater surface.[8] Volatilization of alkalis is still poorly understood.

The sintering test to predict the fouling tendency of fly ash from different coals discussed in the first Supplementary Volume[16]

[14] F. G. Ely and D. H. Barnhart, in *Chemistry of Coal Utilization,* Supplementary Volume, H. H. Lowry, Ed., John Wiley and Sons, New York, 1963, p. 838.

[15] L. Brewer, *Chem. Rev.,* **52,** 1 (1953).

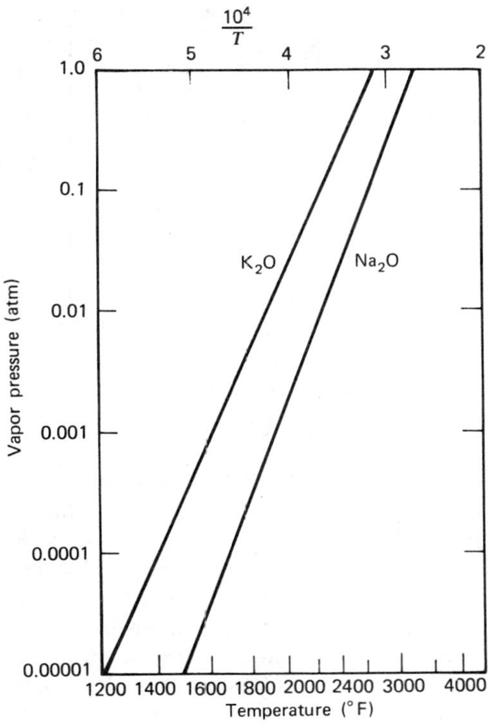

Figure 21.5 Vapor pressure of alkali oxides.[15]

showed the close relationship between sodium content and fouling tendency. Since then, a correlation has been developed between the total ash composition and the fouling tendency, with sodium having the greatest effect. A distinction has also been made between water-soluble sodium in ash prepared by a low-temperature ashing technique and the sodium in ash from the conventional ASTM proximate analysis; the LTA tended to form stronger deposits at the same sodium level, suggesting some unknown modification in mineral structure at the higher ashing temperature.

Two fouling indexes related mainly to sodium have been proposed to predict the extent of fouling of convective heat-receiving surfaces.[17,18] Both apply to Eastern coals, rather than to the lignites in which the CaO + MgO content of the ash may be greater than the Fe_2O_3 content. The indexes are:

$$Rf = \frac{\text{base}}{\text{acid}} \times Na_2O \text{ (ASTM ash)} \qquad (21.1)$$

$$Rf' = \frac{\text{base}}{\text{acid}} \times \text{water-soluble } Na_2O \text{ (LTA ash)} \qquad (21.2)$$

where "base" is the sum of Fe_2O_3, CaO, MgO, Na_2O and K_2O, and "acid" is the sum of SiO_2, Al_2O_3, and TiO_2, all in percentage units. Sodium is determined conventionally on the ASTM ash, and on the water-soluble portion of LTA ash.

Table 21.2 shows how these fouling indexes relate broadly to chemical composition and more specifically to sodium content. Increasing the base/acid ratio leads to a more fusible ash, hence as this ratio is larger, the tendency is greater for inducing sintering. The dominant factor appears to be sodium.

These measures of fouling are based largely on experience with many large pulverized-coal-fired steam generators, and relate mainly to superheaters and reheaters where heat transfer is by convection. The relationships are not precise; they serve mainly as a guide to performance when a choice must be made among different coals with unknown fouling tendency.

These indexes are related to sodium. It is probable that potassium acts similarly, but no data are available on its relative fouling tendency.

Other fouling indexes have been proposed for lignitic ash where CaO + MgO is greater than Fe_2O_3, showing low fouling when the

Table 21.2 Fouling Indexes of Eastern-Type Coals[18]

Fouling Tendency	Rf	Rf'
Low	<0.2	<0.1
Medium	0.2–0.5	0.1–0.25
High	0.5–1.0	0.25–0.7
Severe	>1.0	>0.7

$$Rf = \frac{\text{base}}{\text{acid}} \times Na_2O \text{ (ASTM ash)}$$

$$Rf' = \frac{\text{base}}{\text{acid}} \times Na_2O \text{ (water-soluble, LTA ash)}$$

[16] F. G. Ely and D. H. Barnhart, in *Chemistry of Coal Utilization,* Supplementary Volume, H. H. Lowry, Ed., John Wiley and Sons, New York, 1963, p. 842.

[17] R. C. Attig and A. F. Duzy, "Coal Ash Deposition Studies and Application to Boiler Design," *Proc. Am. Power Conf.,* Vol. 31, 1969, pp. 290–300, American Power Conference, Chicago.

[18] E. C. Winegartner, *Coal Fouling and Slagging Parameters,* Research Dept., ASME, 1974, 34 pp.

Na_2O content of the ash is less than 2%, and severe fouling when it exceeds 8%.[19] Such correlations, based on measurements in a laboratory combustor, are approximations at best and must be used with caution. Nevertheless, they show the pronounced effect of sodium in influencing fouling.

A detailed study[20] of fouling when burning Australian brown coal in a laboratory combustor at the rate of 23 kg/hr produced enough data to develop, by a multiple linear regression analysis, a correlation between the composition of the coal ash and the accumulation of deposits on simulated superheater tubes. This showed that SiO_2, Fe_2O_3, and Na_2O are consistently significant, and that MgO and CaO are of varying significance. The final expression:

$$Y = 0.030\,[SiO_2] + 0.092\,[Fe_2O_3] + 0.061\,[CaO] + 0.264\,[MgO] + 0.423\,[Na_2O] - 10.6 \qquad (21.3)$$

where Y is the weight of the deposit and the chemical symbols are the percentage values, shows that the coefficient for sodium dominates. Garner points out, however, that other factors such as the sulfur oxides in the flue gas may be important because they affect the formation of calcium and sodium sulfates, which, in turn, influences deposit accumulation.[20] Nevertheless, sodium is probably the most important factor in the composition of coal leading to massive deposits.

A double salt, $MgSO_4 \cdot Na_2SO_4$, with a melting point of only 680°C, has been considered as a possible contributor to fouling by the formation of a low melting point liquid phase.[21] Garner[20] notes that this mixed salt has been found in deposits from Morwell brown coal.

Problems with alkalis increase as the chlorine content of coal increases, not because of the chlorine itself, but because it serves to indicate easily volatilized alkalis in the coal mineral matter. Figure 21.6 shows the effect of chlorine on the deposition rates of the alkali sulfates in a pulverized-coal-fired boiler burning English East Midlands coals with similar sulfur contents.[22] It is evident that no appreciable transport of the alkalis occurred until the chlorine content exceeded about 0.3%. Few U.S. coals contain this much chlorine, yet serious fouling occurs with many coals essentially free of chlorine. Hence the volatilization of alkalis leading to fouling can occur by routes other than as chlorides, probably through mineral and organic compounds of the alkalis not yet completely understood.

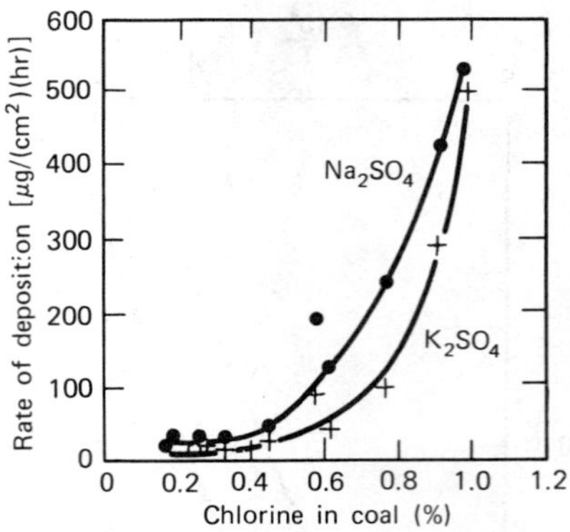

Figure 21.6 Effect of chlorine in coal on deposition of alkali sulfates in pulverized-coal-fired boilers.[22]

The alkalis are also responsible for external corrosion through formation of the alkali iron trisulfates, $Na_3Fe(SO_4)_3$, and the potassium equivalent. Because these trisulfates melt below 600°C, depending on the availability of SO_3, they produce a liquid phase that causes rapid metal wastage. Although full agreement has not been reached on the exact mechanism by which the trisulfates are formed, it is generally accepted that deposition of Na_2SO_4 or K_2SO_4 on the metal surface is the first step. Since this is a condensation phenomenon, even small amounts of alkalis in the coal can eventually lead to the accumulation of enough alkali sulfates to cause trouble. The alkali level in coal has not been found to be a limiting parameter in forming the trisulfates. Hence although high alkali content of coal may lead sooner to deposits of appreciable alkali sulfates on heat-receiving surfaces, as the first step in external

[19] G. H. Gronhovd, A. E. Harak, and P. H. Tufte, *Lignite Symposium,* U.S. Bur. Mines Inf. Circ. No. 8471, 1969.

[20] L. J. Garner, *J. Inst. Fuel,* **40,** 107 (1967).

[21] C. H. Anderson and E. K. Diehl, ASME Paper No. 55-A-200, 1955.

[22] E. Raask, "Reactions of Coal Impurities During Combustion and Deposition of Ash Constituents on Cooled Surfaces," in *The Mechanism of Corrosion by Fuel Impurities,* H. R. Johnson and D. J. Littler, Eds., Butterworths Scientific Publications, London, 1963, pp. 145–154.

corrosion, even low alkali levels can be objectionable over long periods.

21.1.4 Utilization of Ash

Essentially all the mineral matter in the coal being burned in large central station power plants is recovered at some stage and leads to a disposal problem. Bottom ash and the slag from slag tap and cyclone furnaces have found little use and are generally put in a landfill. Attempts were made many years ago to recover the iron content of coal ash slags as part of the burden in a blast furnace, but the practice was not widely adopted.

Most of the inorganic matter in coal ends up as fly ash, with a composition essentially the same as the bulk composition of the ash in the coal from which it comes. The principal difference is that the fly ash consists of particles that have been heated to temperatures ranging from about 1400 to 1700°C. As a result, some chemical reactions took place when two dissimilar ash particles adhered after a collision: some of the individual particles—such as pyrites, feldspar, and mica—may melt, whereas others—such as calcite, dolomite, and kaolin—may be calcined; at least part of the alkalis volatilized in the flame front are likely to condense on particles as the flue gases cool passing through the boiler circuit; and sulfur oxides, mainly SO_3, are partially absorbed on the particles as the temperature falls below about 500°C. Indeed, a major source of problems in disposal is this heterogeneous mixture of spherical, dark, iron-containing fused solids; thin-walled, hollow, glassy spheres known as cenospheres; and unfused ash particles, together with varying but usually small amounts of unburned carbon. Alternative uses of fly ash have been investigated thoroughly,[23] but dumping in a landfill is still practiced widely.

Commercial uses of fly ash (Chapter 30) mainly include concrete and concrete products, cement manufacture, lightweight aggregates, soil stabilization, asphalt paving mixes, and ceramic products.[24] End use depends on the size consist of the fly ash, its chemical composition, and the availability of the material. The pozzolanic properties of fly ash are important in many uses, particularly in concrete, soil stabilization, and ceramic products. Unfortunately, pozzolanic activity—the ability to react chemically with hydrated lime to improve cementitious properties—is difficult to assess. Except that finely divided fly ash appears to be more pozzolanic than coarse material, no good yardstick is available to choose between different fly ashes.

The major market for fly ash has been for use in concrete. No single characteristic of fly ash is important here, but the fly ash should have a surface area (Blaine fineness) of 2800–3,000 cm^2/g and a size consist of 88% through a 325-mesh sieve (44 μm). Generally high silica and high alumina content is desirable. Fly ash addition to concrete results in lower water requirement, higher ultimate strength, improved workability, lower ultimate permeability, and less heat generation during curing.

In cement manufacture, fly ash added to the raw batch before burning can provide an inexpensive source of silica and alumina, thereby partly replacing shale or other argillaceous substances that mixed with limestone or other calcareous materials, form cement clinker when heated to a high temperature. The dominant factors here are cost and the relative availability of fly ash compared with shale or a similar substance.

Specifications have been developed for fly ash used for cement manufacture.[25] In general, depending on the pozzolan class, the fly ash should contain a minimum of 70% SiO_2 plus Al_2O_3 plus Fe_2O_3 and not more than 5% MgO or 4% SO_3; it should have a moisture content less than 3%; it should not lose more than 10% on ignition, and it should contain less than 1.5% available alkalis as Na_2O.[26]

Ash from lignite, often containing 25–30% CaO, may be pozzolanic without addition of lime. Such lignite ash can develop greater strength when hydrated than ash from bituminous coal.

[23] J. P. Capp and J. D. Spencer, *Fly Ash Utilization*, U.S. Bur. Mines, Inf. Circ. No. 8483, 1970.

[24] M. J. Snyder and H. W. Nelson, *A Critical Review of Technical Information on the Utilization of Fly Ash*, Rep. to Prime Movers Comm., Edison Electric Institute Res. Project No. 8, from Battelle Memorial Institute, July 13, 1962, 106 pp. Available from EEI, Washington, D.C.

[25] "Tentative Specifications for Fly Ash—For Use in Portland Cement Concrete," ASTM Paper No. C618-72, Part 10, 1973, pp. 354–358.

[26] O. E. Manz, *Lignite Fly Ash Utilization*, U.S. Bur. Mines Inf. Circ. No. 8471, 1970, pp. 138–149.

The trends in use of ash seem to have been moving up on a tonnage basis, but down as a percentage of total ash and slag available. Thus total ash production rose from 25.2 million tons in 1966 to 33.4 million tons in 1969, but the total percentage used dropped from 17.5% in 1968 to 15.9% in 1969.[27] At the beginning of the 1970 decade, total ash production was growing at a more rapid rate than use.

Lightweight aggregates can be made by sintering some fly ashes. Such aggregates have not yet been widely used.

Soil stabilization with fly ash for road beds, airport runways, parking areas, and foundations has been practiced fairly extensively. In many cases lime is added to the fly ash to improve strength and resistance to weathering. No correlations have been developed between the chemical composition of fly ash and its suitability for soil stabilization, but the carbon content should be less than 10%, and at least 75% of the fly ash should pass a 325-mesh sieve.

Asphalt paving mixes can incorporate fly ash to serve as a replacement for other mineral fillers. Fly ash has no unusual characteristics that give it an advantage over other additives in bitumastic mixes, its use depending mainly on economic factors.

Ceramic products, where fly ash is mixed with clay, have not provided much of a market for fly ash. Economic factors again do not favor fly ash except in unusual circumstances.

A great deal of attention is currently being paid to "modified" fly ash, the slurry of fly ash and reacted limestone from wet scrubbers that is used for flue gas desulfurization (FGD).[28] This method of emission control is being projected for use in plants totaling 90,000 MW generating capacity by 1980,[29] although the public utilities are accepting FGD methods reluctantly because of economic and operational problems. The amount of modified fly ash produced annually will be very large if this goal is reached. Typically a 1000-MW power plant is expected to produce as much as 730,000 tons/year of the ash-limestone slurry on a dry basis, and about twice this much wet as discharged from the scrubber. This huge bulk of material will pose serious problems of disposal. Regeneration of the reacted limestone, with a sulfite/sulfate mole ratio of about 9, is being proposed. Fly ash would be rejected if a practical regeneration system were devised, but the practicality of such a system is still in doubt.

Until about the end of the century, as coal continues to provide most of our energy for generating electricity, problems with fly ash disposal will cause much concern. More effort to find uses for fly ash is certainly warranted. It also appears that slagging gasifiers may be widely used. Most of the mineral matter in their coal will be converted to molten slag, apparently suitable mainly for landfill. Useful end products from such slag may be developed; but for the present no market is foreseen for coal ash slag, and disposal will become increasingly difficult as the tonnage increases.

21.1.5 Recovery of Useful Products from Ash

Although the prospects for using fly ash and bottom ash as they come from boiler furnaces have been thoroughly studied, little research effort has been expended on the potential for recovering useful materials by chemical processing of ash residues.

One early attempt was to recover germanium, based at least in part on the finding that germanium present in coal was concentrated in the fly ash. Coal originally containing only 2.4–9.0 ppm germanium produced fly ash with 15–60 ppm germanium, or an enrichment factor ranging from 3.3–18.7.[30] The highest concentration was found in fly ash from a chain-grate-fired boiler furnace with 530 ppm germanium, and from a cyclone furnace with 290 ppm germanium. But because the occurrence of germanium in coal varies and selective mining is not economi-

[27] *Ash Utilization,* EEI-NCA-BM Symp., Pittsburgh, 1967, U.S. Bur. Mines, Inf. Circ. No. 8488, 1970, 351 pp.

[28] *Electr. World,* **178,** 40 (November 15, 1972). [Based on an unidentified report by L. Z. Condry, W. F. Lawrence, and R. Muter, Coal Research Bureau of West Virginia.]

[29] *Status of Stack Gas Technology for SO_2 Control,* EPRI Rep. No. 209, Part II, 1975, 109 pp.

[30] R. C. Corey et al., *Occurrence and Determination of Germanium in Coal Ash from Powerplants,* U.S. Bur. Mines Bull. No. 575, 1959, 68 pp.

cally feasible, together with the trend in the semiconductor industry to develop devices based on silicon rather than on germanium, no commercial germanium industry using coal ash residues has developed.

Zinc in the form of sphalerite has been found in Illinois coals in concentrations greater than 5000 ppm.[31] Separation of sphalerite from the coal by washing is practical and has been suggested as an economically attractive process. Recovery from ash apparently has not been evaluated.

Selenium in fly ash occurs in a concentration much greater than that in the original coal, with most of the selenium condensing on the smaller particles of fly ash that may escape an electrostatic precipitator.[32] This has posed problems in growing vegetation on fly ash landfills, but since the finer sizes of fly ash contain about 1100 times more selenium than the bottom slag, it also suggests a commercial source of this element if, as in the past, other selenium sources cannot meet the commercial demand.

In general, trace elements in coal occur in about the same concentration as in the earth's crust.[33] Only three elements—cadmium, boron, and selenium—are enriched, and then only by about one order of magnitude. This suggests that coal ash may not offer much promise as a source of useful products except that as shown earlier, volatile substances may be concentrated appreciably in fly ash. Or, as in the case of sphalerite, coal preparation might yield a commercially useful product. There seems to be little possibility that the inorganic matter in coal will find a market as a source of useful raw materials in the near future.

Among the attempts to derive useful products from coal ash have been efforts to produce alumina. By using a lime-soda sinter process, alumina was produced from fly ash not essentially different than that produced from a clay as the raw material. Costs were excessive compared with bauxite. Other alumina-based by-products from coal ash have been found to be equally unattractive economically.[34]

Mineral wool has been blown from ash usually fluxed with limestone to achieve adequate fluidity. Also, bloating of coal ash has been successful in producing a lightweight product suitable as thermal insulation for buildings. Best results have been reported when the Fe_2O_3 + CaO + MgO in the ash is not more than 8% of the SiO_2 + Al_2O_3, or more than 13% of the SiO_2.[35] This represents an ash of high fusion temperature, typically with a viscosity at 1425°C of more than 10,000 poises.

21.2 SULFUR IN COAL DURING COMBUSTION

Although the sulfur in coal contributes insignificantly to the thermal energy released when coal is burned, it leads to very serious problems during combustion and gasification. In combustion systems, sulfur is a major contributor to external corrosion, both at high and at low temperatures, through the formation of aggressive liquid films leading to rapid metal wastage. At temperatures over about 830°C the alkali sulfates are molten; at 540°C the alkali iron trisulfates form liquid films if adequate SO_3 is present; and at 160°C the dew point is reached with normal excess air, and SO_3 condenses to form sulfuric acid. Hence from wall tubes to air heaters, sulfur compounds pose corrosion problems in steam generators where conditions overall are oxidizing.

In addition to these sulfate problems, serious problems occur under reducing conditions when sulfides are present. Essentially all available alloys intended for high-temperature service are unsuitable when brought into contact with sulfides, both chromium and nickel being susceptible to attack along grain boundaries. Hence in gasifiers where reducing conditions must be accepted, the problem of finding suitable construction materials is particularly acute.

[31] J. R. Hatch, H. J. Gluskoter, and P. C. Lindahl, "Sphalerite in Coals from the Illinois Basin," *Econ. Geol.*, **71**, 613–624 (1976).

[32] W. H. Gutenmann et al., "Selenium in Fly Ash," *Science*, **191**, 966–967 (March 5, 1976).

[33] R. R. Ruch, H. J. Gluskoter, and N. F. Shimp, *Occurrence and Distribution of Potentially Volatile Trace Elements in Coal*, Environ. Geol. Notes, No. 72, Illinois Geol. Survey, August 1974, 96 pp.

[34] H. T. Stirling, "Beneficiation of Fly Ash," U.S. Bur. Mines, Inf. Circ. No. 8488, 1970, pp. 300–309.

[35] Tae Yoon Song, "Study on the Utilization of Coal-Refuse," *J. Korean Inst. Min.*, **4**(1), 21–18 (March 1967).

Furthermore, alkali sulfates lead to increased deposition of solids on the heat-receiving surfaces of steam generators, such as wall tubes, superheaters, and reheaters, by the formation of a condensed surface film to which solid ash particles may adhere. These sulfates may originate in the gas stream or result from sulfation of condensed films of alkalis such as the oxides or hydroxides. There is also a problem in combustion turbines with alkali sulfates.

Sulfur, in short, is a major offender in essentially all uses of coal or coal-derived fuels.

21.2.1 Influence of Forms of Sulfur

21.2.1.1 *Pyrites*

Of the three forms in which sulfur occurs in coal (pyritic, organic, and sulfate), it is the pyritic sulfur that causes the worst problems in furnace operation. Pyrite occurs in essentially all coals, in particles ranging in size from a few micrometers to massive nodules. Because pyrites has a density of 5, it can be removed from coal by conventional coal-washing processes if the particles are freed from the coal by crushing. Unfortunately, in some coals, pyrites occurs in particles too small to be liberated, so that coal beneficiation may extract only some of the pyritic sulfur. Thought is being given to removing pyrites from pulverized coal between the mill and the burner, but no promising approaches have been developed. Also, several schemes are under development to increase the magnetic susceptibility of particles of pyrites for separation in a high-gradient magnetic field.

Marcasite, also FeS_2, has a crystal structure different from that of pyrites and a somewhat lower density (4.87); but in general it behaves like pyrites during combustion or gasification.

Much attention has been given to pyrites because of its significant influence on slagging and corrosion. Thermal decomposition occurs in at least three distinct ways: in the absence of air, in air, and under reducing conditions.[36] In the absence of air, the reaction occurs in two steps:

[36] J. D. Watt, "The Physical and Chemical Behavior of the Mineral Matter in Coal, Part II," British Coal Utilisation Research Association, August 1969.

$$FeS_2 \rightarrow FeS + \tfrac{1}{2}S_2 \qquad (21.4)$$

$$FeS \rightarrow Fe + \tfrac{1}{2}S_2 \qquad (21.5)$$

but these precise compositions are unlikely and the sulfur may polymerize.

Discrete particles of pyrites may not be fully dissociated if they pass too quickly through a flame front. Hence under laboratory conditions 100-μm particles of pyrites heated for 0.5 s to furnace temperatures, in an atmosphere containing 4% oxygen, lost sulfur as shown in Table 21.3. The conversion of FeS_2 (53.4% sulfur) to FeS (36.4% sulfur) evidently occurs in less than 0.5 s at 1100°C. At higher temperatures further loss of sulfur occurs, and the second reaction predominates. The calculated pressure of sulfur over FeS_2 at 1100°C is more than 1000 atm, so that gaseous sulfur will diffuse rapidly through a hot particle of pyrites. For FeS, the equilibrium pressure of sulfur is only about 10^{-6} atm at 1100°C, so that further loss of sulfur should occur much more slowly than with FeS_2. Nevertheless, Table 21.3 indicates that appreciable dissociation of FeS does occur in 0.5 s at temperatures above 1200°C.

In air, this thermal dissociation of iron sulfides is complicated by oxidation of the sulfur to SO_2 and of the iron to FeO and eventually to Fe_2O_3. Under reducing conditions, sulfur evolved from FeS_2 could form sulfides if it migrated through ash or slag deposits to tube surfaces at temperatures below, say, 600°C, as on superheater elements.

Figure 21.7 is a phase diagram of the system Fe-S, which is particularly important because a liquid phase is present at temperatures down to 985°C, with about 33% sulfur.

[37] W. D. Holstead and E. Raask, "The Behavior of Sulphur and Chlorine Compounds in Pulverized-Coal-Fired Boilers," *J. Inst. Fuel,* **42,** 344–349 (September 1969).

Table 21.3 Residual Sulfur in 100-μm Particles of Pyrites Heated for 0.5 s in 4% Oxygen[37]

Temperature (°C)	Sulfur in Particle (%)	Total Sulfur Lost (%)
Untreated	53.1	O
1100	28.5	46.3
1200	22.7	57.2
1300	17.8	66.5
1400	15.1	71.6
1500	10.1	81.0

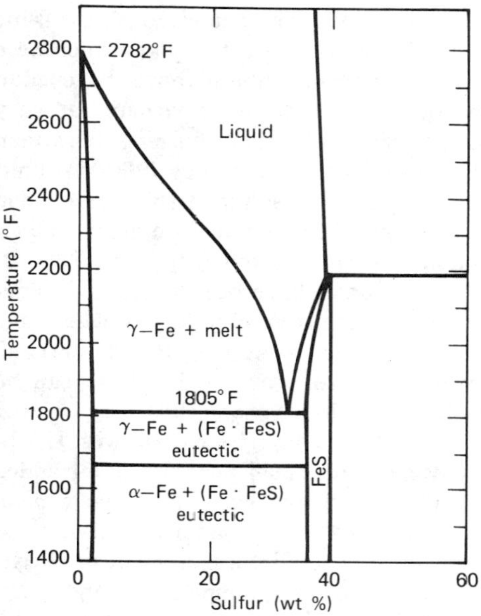

Figure 21.7 Phase diagram of the system Fe-S.[38]

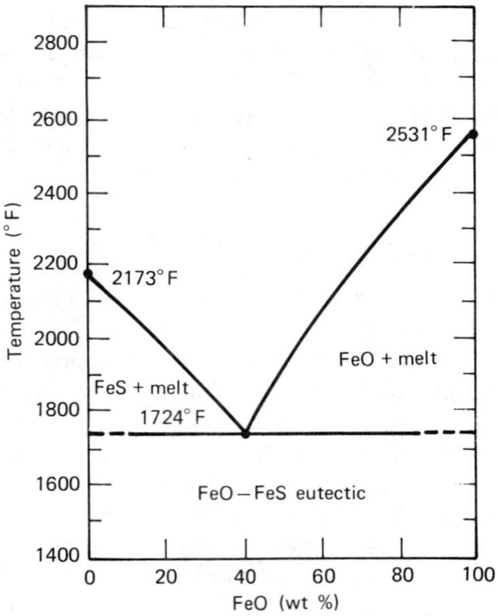

Figure 21.8 Phase diagram of the system FeO-FeS.[38]

Figure 21.8 shows the system with oxygen. Here there is a eutectic at 940°C with 40% FeO.

Although much of the work on pyrites was done many years ago, fireside problems associated with pyrites in coal continue to be troublesome.[39] Slagging and fireside fouling became more acute as volumetric heat release rates increased, and even in recent years problems continue with slag formation on furnace walls near burners. In some cases such deposits have contained as much as 85% iron reported as Fe_2O_3, leading to the conclusion that pyrites is responsible. In recent work Bryers used thermal gravimetric analysis (TGA) to demonstrate that high ash levels may increase the "ignition" temperature of pyrites but not the reaction rate, which proceeds as a simple first-order reaction. Oxidation of the pyrites takes place in two stages, from FeS_2 to FeS and then to some form of iron oxide. The first stage has about the same reaction rate as the combustion of coal, but the second stage is much slower, as was shown earlier in the discussion of the thermal dissociation of FeS. Figure 21.9 shows the TGA curves.

Recent trends in air pollution control reflect much attention devoted to removal of pyrites from coal. This will aid in burning the

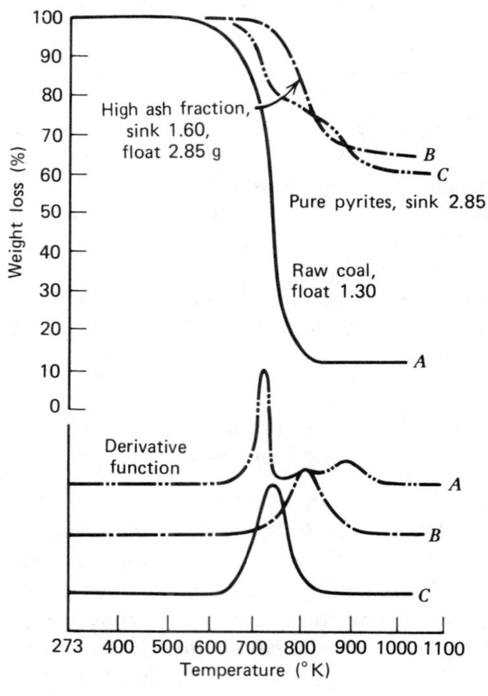

Figure 21.9 Comparison of thermal gravimetric analyses of raw coal and pyrites.[39]

[38] R. C. Corey, B. J. Cross, and W. T. Reid, "External Corrosion of Furnace-Wall Tubes. II. Significance of Sulphate Deposits and Sulphur Trioxide in Corrosion Mechanism," *Trans. ASME*, **67**, 289–302 (1945).

[39] R. W. Bryers, "The Physical and Chemical Characteristics of Pyrites and Their Influence on Fireside Problems in Steam Generators," ASME Paper No. 75-WA/CD-2, 1975.

coal with less trouble from slagging and fouling, but the residual pyrites in cleaned coal will probably continue to cause problems. No data are available to indicate how low the pyrites content of coal must be to eliminate this cause of operational difficulties.

21.2.1.2 *Organic Sulfur*

Little attention has been paid to organic sulfur in coal as a source of trouble in steam generation. Although it usually represents roughly half the total sulfur in coal, organic sulfur converts in a complicated fashion mainly to SO_2, with a very small fraction oxidizing further in the flame to SO_3. Sulfur dioxide by itself has little effect in a boiler furnace, but further catalytic oxidation of SO_2 to SO_3 and the subsequent formation of sulfates or sulfuric acid do lead to serious problems. These include the formation of alkali iron trisulfates beneath slag deposits, where the required high levels of SO_3 may be reached. Also heterogeneous catalysis on fly ash particles and on active boiler surfaces raises the SO_3 level in flue gas, thus raising both the dew point and the tendency for corrosion in air heaters. It is unlikely that coal treatment processes now being developed will decrease sulfur levels sufficiently to eliminate these problems with SO_3.

21.2.1.3 *Sulfate Sulfur*

The quantity of sulfate sulfur in coal usually does not exceed a few percent of the total sulfur; most of the sulfur is pyritic and organic. Weathering of coal in stockpiles or in outcrops is largely responsible for sulfate sulfur. No specific problems with coal ash have been attributed to sulfates in the coal, and the amount present can generally be neglected.

21.2.2 Sulfur-Oxygen Reactions During Combustion

Pyritic and organic sulfur in coal react under combustion conditions in a complex fashion. This is partly because of the nature of diffusion flames in pulverized-coal-fired boiler furnaces, where there is a central zone that is fuel rich surrounded by the additional air needed to complete combustion. The scale and intensity of turbulence along with the reactivity of the fuel then establish the flame pattern. Very little is actually known of these factors, but engineering advances have led to the design of successful burners for very large furnaces, typically burning more than 400 tons of coal per hour in a 1000-MW unit.

Participation of sulfur in such flames had to be deduced from studies on much simpler and more easily controlled systems. Typically these studies have been based on examination of premixed H_2S flames where the visible flame is very thin (e.g., 0.5 mm) and the reaction zone above the flame can be probed optically or physically.

Other studies have been done with H_2-O_2-N_2 flames with small percentages of added SO_2.[40] Such flames have the same general characteristics as the sulfur-free flames, so that a great deal of information already exists on the composition, temperature, and kinetics of the burning gases. Also the spectra attributed to the added SO_2 stand out clearly because radiation from this system normally is weak except for strong hydroxyl bands in the ultraviolet and the infrared. The reaction zone is also thin, in this case 0.1–1 mm. As a result of such studies, addition of SO_2 was found to show S_2 bands, SO bands, and an SH band, from which it was concluded that for up to 0.25% added SO_2, the amount of S_2 never exceeds 10% of the sulfur. With 4% added SO_2, half the sulfur may be present as S_2. The amount of SO_2 remaining in such flames depends on the H_2/C_2 ratio; SH_2 appears only in fuel-rich flames, and S_2 becomes important only for large amounts of added sulfur.

More recent work done with thin, premixed H_2S flames has led to further understanding of the complex reactions of sulfur-containing flames, with particular emphasis, from the practical standpoint, on the flame-induced formation of SO_3 in boiler furnace flue gas.[41] Mass spectrometric attempts to determine species concentrations of samples collected from the H_2S flame with a micro-

[40] A. Demerdache and T. M. Sugden, "Some Reactions of Sulphur Compounds in Flames," in *The Mechanism of Corrosion by Fuel Impurities*, H. R. Johnson and D. J. Littler, Eds., Butterworths Scientific Publications, London, 1963, pp. 216–227.

[41] A. Levy and E. L. Merryman, "The Microstructure of Hydrogen Sulphide Flames," *Combust. Flame*, **9**, 229–240 (September 1965).

probe were not reliable, but in combination with a quantitative colorimetric wet analysis procedure using barium chloranilate, data were obtained on SO, SO_2, and SO_3 at various positions before, in, and following the visible flame zone.

Figure 21.10 shows these compositional profiles for a typical H_2S-O_2-N_2 flame. The normalized SO_2 profile results from unexplained variances in stoichiometric levels shown by the mass spectrometer.

Figure 21.11 illustrates how the SO profile was deduced by subtracting these normalized SO_2 values from the combined SO + SO_2 data, shown by the wet analysis for total sulfur as SO_4^{2-}. The relative position of the curves is difficult to determine, but the error is probably not significant. This figure also shows the SO_3 profile, based on wet analysis of condensed SO_3 caught in a cold trap.

This SO_3 profile demonstrates an important point in these studies—showing that about 65 ppm SO_3 forms before combustion reactions have proceeded far enough to produce a visible flame. This SO_3 subsequently disappears early in the flame reaction, but beyond the flame front, SO_3 again is formed, the maximum concentration appearing about one flame thickness beyond the visible region. Thereafter, although sufficient SO_2 and O_2 are present and the temperature has not changed significantly, no additional SO_3 appears. Figure 21.12 shows this clearly as a rate profile.

The early formation of SO_3 prior to a visible flame is attributed to the following reaction:

$$SO + O_2 + M \rightarrow SO_3 + M \quad (21.6)$$

Disappearance of SO_3 early in the flame could result from:

$$SO_3 + H_2 \rightleftarrows SO_2 + H_2O \quad (21.7)$$

Table 21.4 lists 19 reactions that could be involved in H_2S flames. The first step, reactions 1–6, indicates that heat transferred into

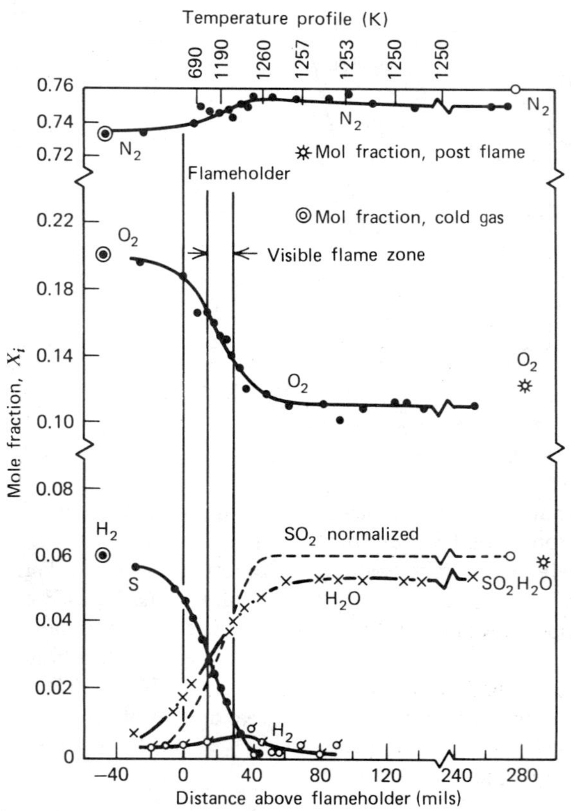

Figure 21.10 Flame composition profiles: initial compositions 6.06% O_2, 73.5% N_2; total flow rate, 40.3 cm^3/s. * = calculated mole fraction, postflame region, ◉ = cold gas mole fraction.[41]

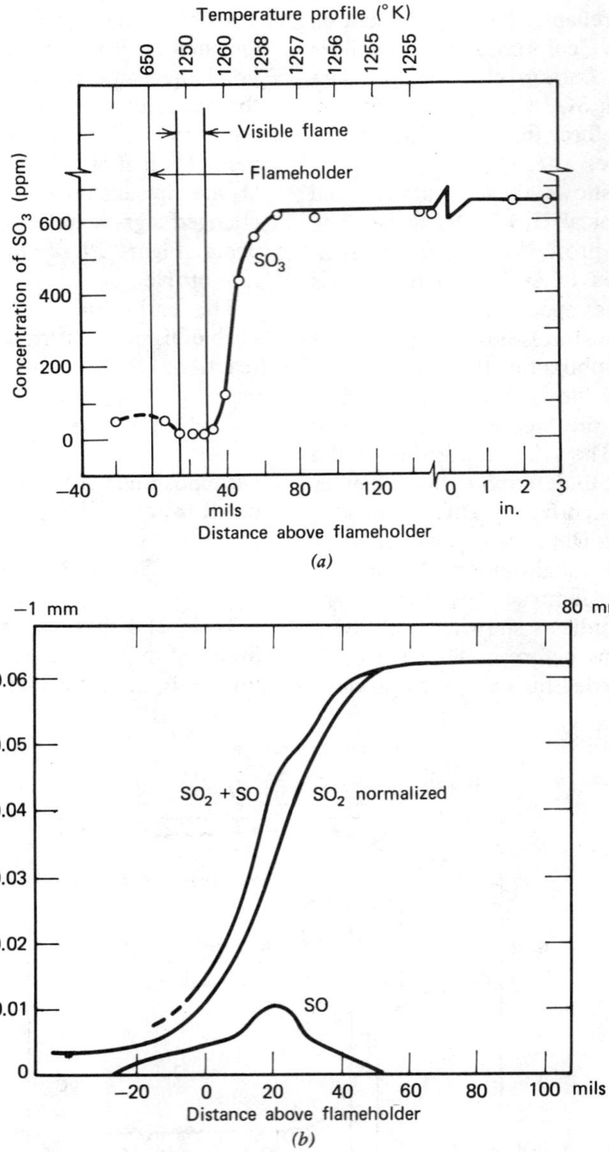

Figure 21.11 Flame profiles for (*a*) SO_3 and (*b*) SO_2 and SO.[41]

the incoming gas stream causes some dissociation of H_2S. Reactions 2 and 3 then form H_2 and SO, respectively, as the first reaction products. Water vapor and SO_2 (water vapor probably first) then appear. Reactions 17 and 18 finally yield SO_3.

Other work on the formation of SO_3 in flames has been done by spraying sulfur-doped liquid fuel into the combustion products obtained by burning a fuel that is similar but sulfur free.[42]

[42] A. B. Hedley, "A Kinetic Study of Sulphur Trioxide Formation in a Pilot Scale Furnace," in *The Mechanism of Corrosion by Fuel Impurities*, H. R. Johnson and D. J. Littler, Eds., Butterworths Scientific Publications, London, 1963, pp. 204–215.

Figure 21.13 shows that in such cases the initial yield of SO_3 exceeded the maximum theoretical yield calculated from thermodynamic considerations. But as combustion proceeded, the amount of SO_3 present gradually decreased until the theoretical level was reached. If the SO_3 had been formed by reaction between SO_2 and O_2, the yield of SO_3 would have been at the thermodynamic

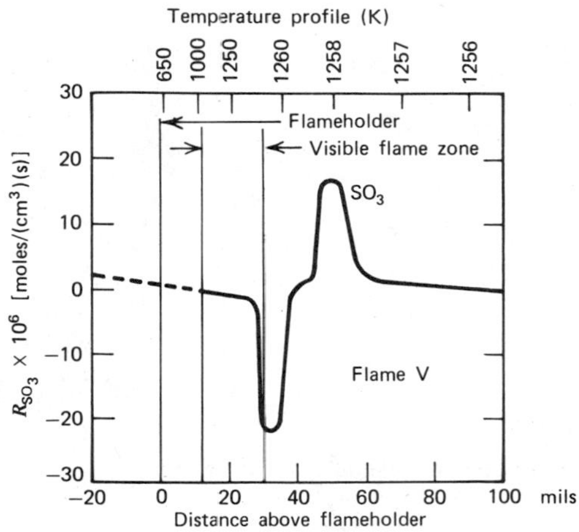

Figure 21.12 SO_2 rate flame profile.[41]

equilibrium value. That this was not the case indicates the true course of the reactions as:

$$SO_2 + O \rightarrow SO^*_3 \quad (21.8)$$

$$SO^*_3 \rightarrow SO_3 + h\nu \quad (21.9)$$

$$\text{or } SO^*_3 + M \rightarrow SO_3 + M \quad (21.10)$$

Flame-produced SO_3 therefore depends on the availability of oxygen atoms, which in turn must come from other flame reactions. Accordingly, because the SO_2 level with sulfur-bearing fuels is invariably greater than the oxygen atom concentration, the amount of SO_3 formed in flames depends on factors that provide oxygen atoms.

21.2.3 Catalytic Formation of Sulfur Trioxide

Oxidation of SO_2 to SO_3 to produce sulfuric acid is practiced widely with commercial catalysts. The same effects are obtained in

Table 21.4 Reaction Mechanism for Hydrogen Sulfide Flames[41]

Reaction	Reaction Number
$H_2S \rightarrow SH + H - 91$ kcal	1
$H + H_2S \rightleftharpoons SH + H_2 + 19$ kcal	2
$SH + O_2 \rightleftharpoons SO + OH + 16$ kcal	3
$OH + H_2S \rightleftharpoons H_2O + SH + 22$ kcal	4
$SO + O_2 \rightleftharpoons SO_2 + O + 18$ kcal	5
$O + H_2S \rightleftharpoons OH + SH + 19$ kcal	6
$SO + O \rightarrow SO_2^* \rightarrow SO_2 + h\nu$	7
$SO + SO \rightleftharpoons S_2O_2 + 49$ kcal	8
$SH + SH \rightleftharpoons H_2 + S_2 + 26$ kcal	9
$4S_2 \rightleftharpoons S_8$	10
$SH + OH \rightleftharpoons H_2O + S + 27$ kcal	11
$S + O \rightleftharpoons SO$	12
$S + S \rightleftharpoons S_2$	13
$H + O_2 \rightleftharpoons OH + O - 16$ kcal	14
$O + H_2 \rightleftharpoons OH + H - 2$ kcal	15
$OH + H_2 \rightleftharpoons H_2O + H + 15$ kcal	16
$SO_2 + O + M \rightleftharpoons SO_3 + M + 82$ kcal	17
$SO + O_2 + M \rightleftharpoons SO_3 + M + 93$ kcal	18
$SO_3 + H_2 \rightleftharpoons SO_2 + H_2O + 38$ kcal	19

boiler furnaces when hot flue gases containing SO_2 and oxygen from excess air pass over convection tube banks with catalytically active surfaces of iron oxides. Fly ash particles also can be catalytic. Early efforts to identify the source of SO_3 in flue gas pointed out the importance of heterogeneous catalysis.[43] Although for many years it was argued whether SO_3 in flue gas originated in the flame or was catalytically produced later in a boiler furnace, it is now generally conceded

[43] F. G. Ely and D. H. Barnhart, "Coal Ash—Its Effect on Boiler Availability," in *Chemistry of Coal Utilization,* Supplementary Volume, H. H. Lowry, Ed., John Wiley and Sons, New York, 1963, pp. 855–863.

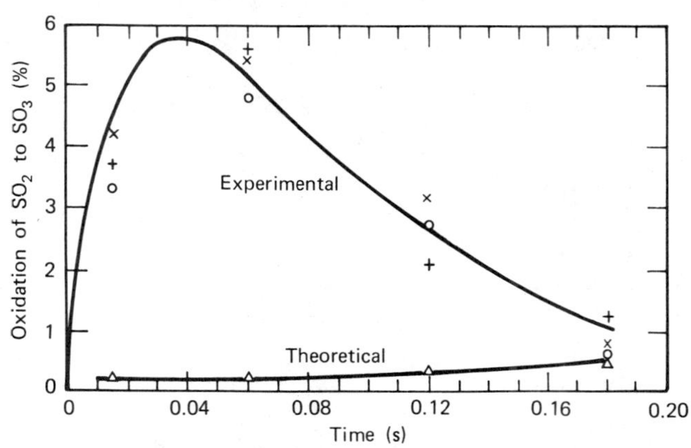

Figure 21.13 Formation of SO_3 in a one-dimensional flame.[42]

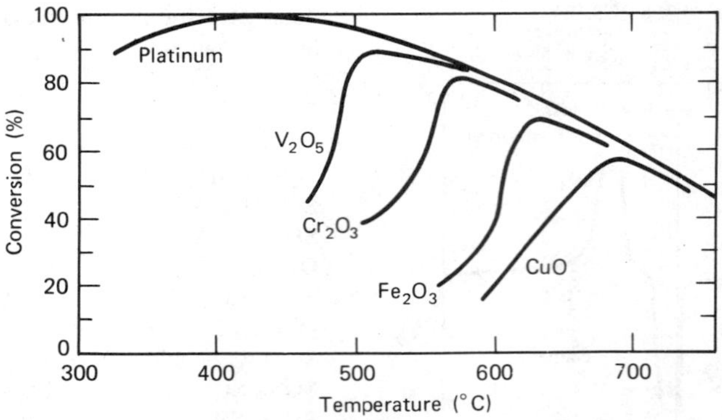

Figure 21.14 Conversion of SO_2 to SO_3 in the presence of several catalysts.[45]

that both actions occur and that catalysis accounts for the high concentrations of SO_3 (>250 ppm) that lead to catastrophic metal wastage through the formation of the alkali iron trisulfates.

Early work on the catalytic formation of SO_3, based on tests in boiler furnaces by the Boiler Availability Committee in England, was not conclusive.[44] It showed that both steel and refractory surfaces led to increased SO_3 and that gas temperature was a significant factor. But overall, problems in conducting tests in full-scale working boilers limited the interpretation of the results.

A better understanding of the formation of SO_3 has come from laboratory experimentation. Catalytic activity depends on thermodynamic equilibria as well as on the influence of the catalyst on reaction kinetics. Hence each catalyst has a specific temperature at which it is most effective in converting SO_2 to SO_3.

Figure 21.14 shows the relative activity of several metal oxides and platinum, with the oxides about equivalent to platinum at higher temperatures. For example, above 625°C, Fe_2O_3 is capable of converting about 70% of the SO_2 to SO_3, and Cr_2O_3 at 580°C can convert 80%. Since the surface temperature of superheaters and reheaters is commonly 590–650°C, it is evident that their oxidized surfaces should be effective catalysts for SO_3 formation.

Figure 21.15 demonstrates how the gradual

[44] Boiler Availability Committee, Final Report of the Barking Panel, London, April 1962. [This paper was widely distributed in England during 1950s and 1960s; major technical libraries there should have it.]

[45] A. Levy and E. L. Merryman, "Sulfur Chemistry and Its Role in Corrosion and Deposits," ASME Paper No. 62-WA-124, 1962, 9 pp.

[46] R. E. Barrett, "High-Temperature Corrosion Studies in an Oil-Fired Laboratory Combustor," *Trans. ASME, J. Eng. Power, Ser. A*, **89,** 288–296 (1967).

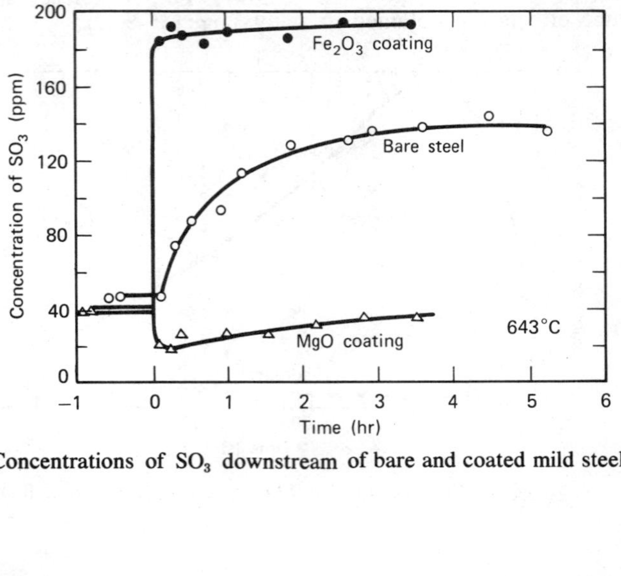

Figure 21.15 Concentrations of SO_3 downstream of bare and coated mild steel specimens.[46]

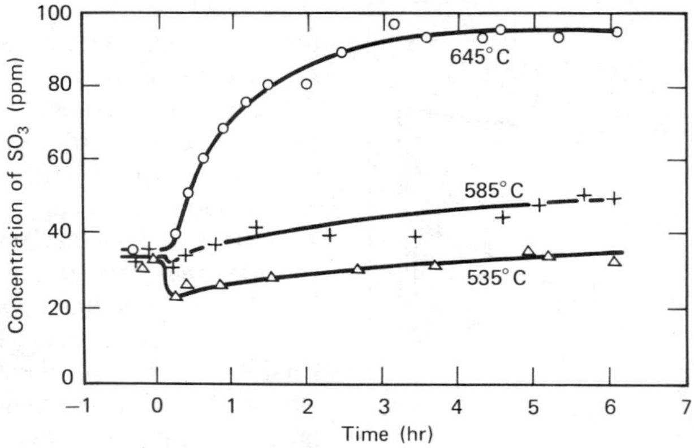

Figure 21.16 Concentrations of SO_3 downstream of F-22 surfaces at various temperatures.[46]

development of an oxidized surface layer on mild steel increased the SO_3 concentration from 40 to 140 ppm in 5 hr. This experiment was made under closely controlled conditions in a laboratory furnace, with small tubular specimens simulating a tube bank. When the tubes were coated before the test with Fe_2O_3, the SO_3 level rose at once from 40 to 190 ppm. When the surfaces were coated with MgO, there was an initial drop in SO_3 followed by a very gradual gain in SO_3.

Figure 21.16 shows how temperature influences the rate of formation of a catalytically active surface and the increase in SO_3 level on initially bare specimens of F-22. Since this alloy contains 2¼% Cr, a somewhat longer time was required to produce an active surface. And, in agreement with Fig. 21.14, temperature is important: the most rapid increase in SO_3 occurs above 600°C, where Fe_2O_3 is undoubtedly present and Cr_2O_3 is forming.

Figure 21.17 also illustrates this temperature effect on surfaces precoated with Fe_2O_3. No gain in SO_3 above the flame-produced level occurs at 480°C, but the SO_3 increased to 120 ppm at 650°C.

Figure 21.18 shows similar results. Here a bare polished disk of mild steel was exposed to a furnace atmosphere containing SO_2, with a gas sample for analysis drawn from the edge of the tilted disk. Again, the polished surface gradually became oxidized and the SO_3 level increased; when the disk was precoated with Fe_2O_3, the increase in SO_3 occurred immediately, and when the disk was precoated with MgO as a protective layer, SO_3 increased slowly.[47]

Fly ash suspended in a flue gas stream or

[47] R. E. Barrett, J. D. Hummell, and W. T. Reid, "Formation of SO_3 in a Noncatalytic Combustor," *Trans. ASME, J. Eng. Power, Ser. A*, **88**, 165–172 (1966).

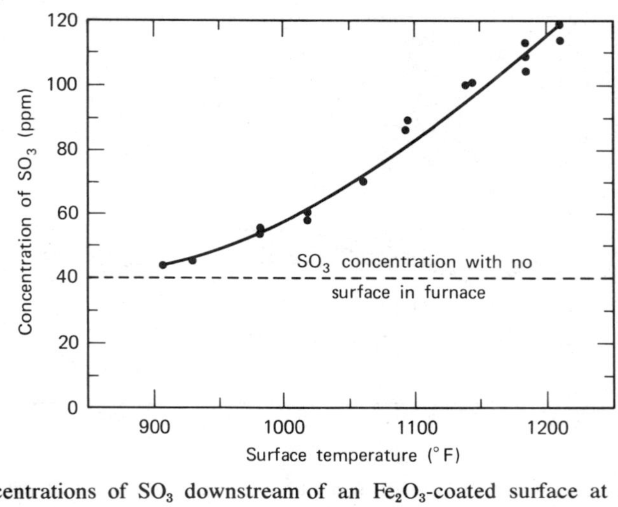

Figure 21.17 Concentrations of SO_3 downstream of an Fe_2O_3-coated surface at several temperatures.[46]

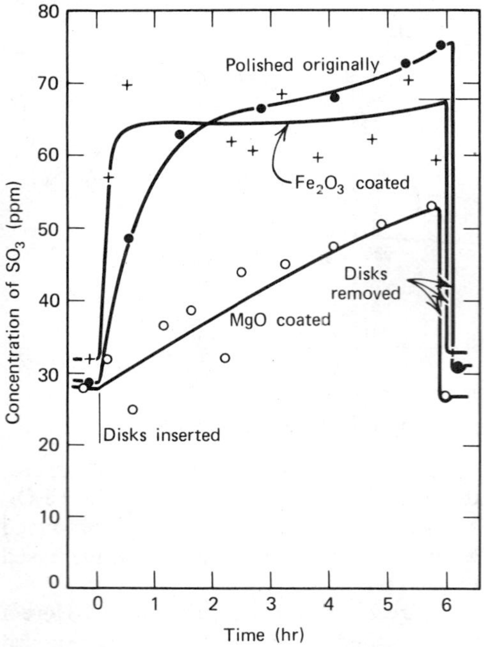

Figure 21.18 Concentrations of SO_3 in flue gas after passing over catalytic surface: fuel, natural gas containing 5.5 wt% sulfur; air level, 113–115% of stoichiometric; disk temperature, 1200–1250°F.[47]

deposited on heat-receiving surfaces is generally considered to be catalytically active toward SO_2. But the wide variation in composition and size consist of fly ash has made it difficult to decide which characteristics are most important. The catalytic activity of three fly ashes was compared under controlled laboratory conditions in the same combustor and under conditions similar to those used for metallic specimens.[48] Here the tubular test elements were precoated with fly ash or mixtures of kaolin and Fe_2O_3.

Figure 21.19 compares the behavior of these fly ashes with that of Fe_2O_3-coated specimens. The fly ashes were similar in composition except that the Fe_2O_3 contents were 16.8 and 17.0%, respectively, for the Astoria and Port Jefferson samples, and 30.4% for the Joppa fly ash. Although this difference was expected to show increased catalytic action with the Joppa fly ash, it is evident that all three samples behaved similarly and that there was no pronounced catalysis.

Figure 21.20, for similar specimens coated with a mixture of Fe_2O_3 and kaolin in different ratios and with Fe_2O_3 alone, shows that physically diluting Fe_2O_3 with kaolin had little effect on catalytic activity; the three specimens behaved similarly.

It was concluded from this work that any catalytic activity of fly ash cannot be attributed to its Fe_2O_3 content, so that the way in which Fe_2O_3 is chemically combined with other fly ash constituents is probably important. As a check on this conclusion, further tests were made with mixtures of Fe_2O_3 and kaolin, which were ball milled and then fired at 1095°C for 16 hr before being ground and applied to the specimen tubes. These showed little catalytic activity compared with that of Fig. 21.20, indicating that a reaction between

[48] R. E. Barrett, "Alkali Iron Trisulfate Formation Within Deposits in an Oil-Fired Laboratory Combustor," *Trans. ASME, J. Eng. Power, Ser. A*, **91**, 173–181 (1969).

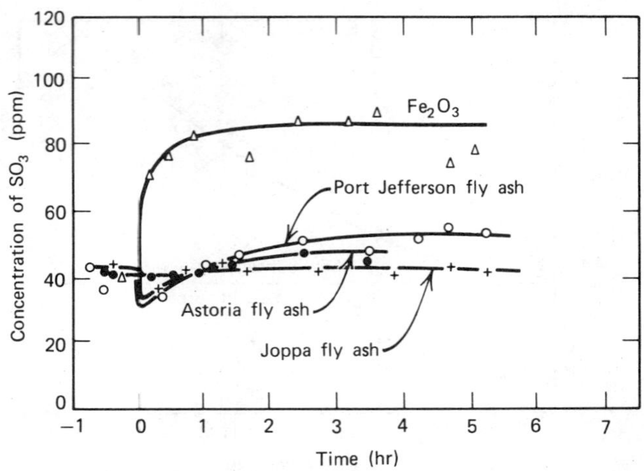

Figure 21.19 Concentrations of SO_3 downstream of specimens coated with fly ash and with Fe_2O_3.[48]

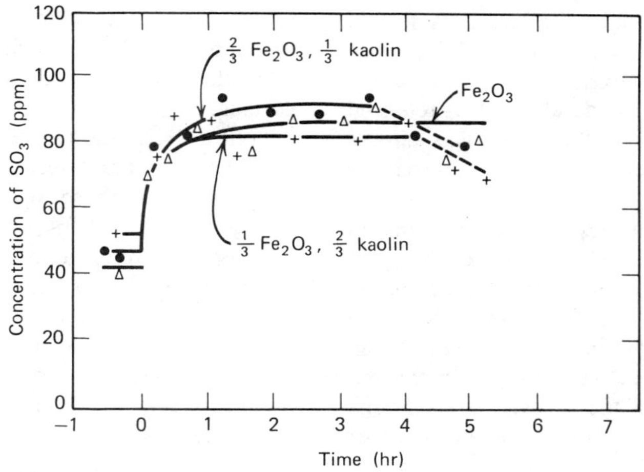

Figure 21.20 Concentrations of SO_3 downstream of specimens coated with Fe_2O_3 and with Fe_2O_3-kaolin mixtures.[48]

the Fe_2O_3 and the kaolin had eliminated the catalytic activity of the Fe_2O_3.

As will be discussed later, higher levels of SO_3 than can be supplied by flame reactions alone are required to form the extremely corrosive alkali iron trisulfates. These higher SO_3 concentrations must result from catalysis, but the amount of SO_3 in the bulk flue gas stream seldom exceeds 3% of the SO_2 present and is more often near 1%, or roughly in the range of 25–75 ppm SO_3. Because the trisulfates form only beneath slag deposits where gas velocities are very low, or where stagnant conditions probably exist, the catalytic oxidation of SO_2 to SO_3 under such conditions was investigated by withdrawing samples through a microprobe from a mixture of nitrogen (74.8%), oxygen (19.4%), and sulfur dioxide (5.8%) flowing at 36 cm/s over an active surface of Fe_2O_3 at 593°C.[49]

Figure 21.21 shows the steep gradient in SO_3 concentration between the bulk gas stream (5 mm from the surface) and the Fe_2O_3 surface itself. It is evident that the decreased velocity within the boundary layer provides

[49] H. H. Krause, A. Levy, and W. T. Reid, "Sulfur Oxide Reactions: Radioactive Sulfur and Microprobe Studies of Corrosion and Deposits," *Trans. ASME, J. Eng. Power, Ser. A,* **90,** 38–44 (January 1968).

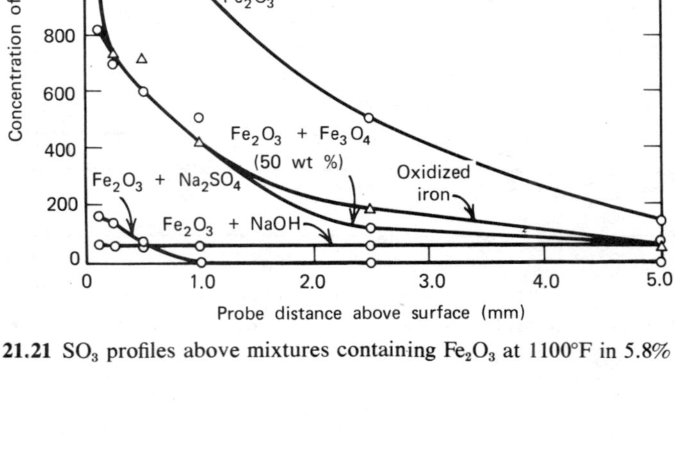

Figure 21.21 SO_3 profiles above mixtures containing Fe_2O_3 at 1100°F in 5.8% SO_2.[49]

additional time to oxidize a larger fraction of the SO_2. Hence concentrations of SO_3 close to equilibrium may exist at the surface, which are sufficient to provide all the SO_3 required to produce the trisulfates. As was shown earlier, an oxidized iron surface approaches Fe_2O_3 in catalytic activity. Other substances (e.g., NaOH or Na_2SO_4) appear not to inhibit the catalyst, but at 1100°F (593°C) they appear to remove free SO_3 from the gas phase as rapidly as it is formed. It is evident that an approach to equilibrium levels of SO_3 can be attained over active catalysts in the immediate vicinity of the surface, even though bulk flue gas compositions show much lower SO_3 levels.

Because SO_3 is such a significant contributer to corrosion and deposits in boiler furnaces, and to a high dew point and metal loss in air heaters, minimizing its formation by deactivating catalytic surfaces would eliminate many problems. Additives have been proposed to control SO_3 and, as indicated in Fig 21.15, a coating of MgO on a polished steel surface was effective for a few hours in preventing the steel specimen from oxidizing to form a catalytically active surface, probably because the coating reacted with whatever SO_3 was present to produce $MgSO_4$. An even better way would be to eliminate any catalytic action by "poisoning" the surface of tubes or of fly ash. If any substance is effective as a catalyst poison, less of it would be needed to prevent the formation of SO_3 than the substantial amounts of additives such as MgO required to neutralize SO_3 once formed on an active surface.

Figure 21.22 shows the results of an experimental program searching for such catalyst poisons. In these tests when a synthetic flue gas containing 3.6% O_2 and 0.25% SO_2 was passed over a mixture of Na_2SO_4, K_2SO_4, and Fe_2O_3 in a mole ratio of 1.5 : 1.5 : 1.0 at 593°C, a gain in weight indicated the oxidation of SO_2 to SO_3, which then combined with the "standard corrosion mix" to form alkali iron trisulfates. Thus weight gain here was a measure of catalytic activity. In some tests fly ash was substituted for Fe_2O_3 in the mixture, and as the figure indicates, a weight gain of about 8% occurred over 100 hr with such a mixture. Furthermore, as the test proceeded, the previously loose mixture of solids became bonded, showing the presence of a liquid phase attributed to the trisulfates.

Addition of Sb_2O_3 to this corrosion mix completely prevented SO_3 formation. Even with only 0.5% added Sb_2O_3, the catalytic properties of the corrosion mix were destroyed, with an insignificant weight gain over 100 hr. Other tests based on passing the

[50] W. Nelson and B. S. Lisle, "A Laboratory Evaluation of Catalyst Poisons for Reducing High-Temperature Gas-Side Corrosion and Ash Bonding in Coal-Fired Boilers," *J. Inst. Fuel,* **37**(284), 378–385 (September 1964).

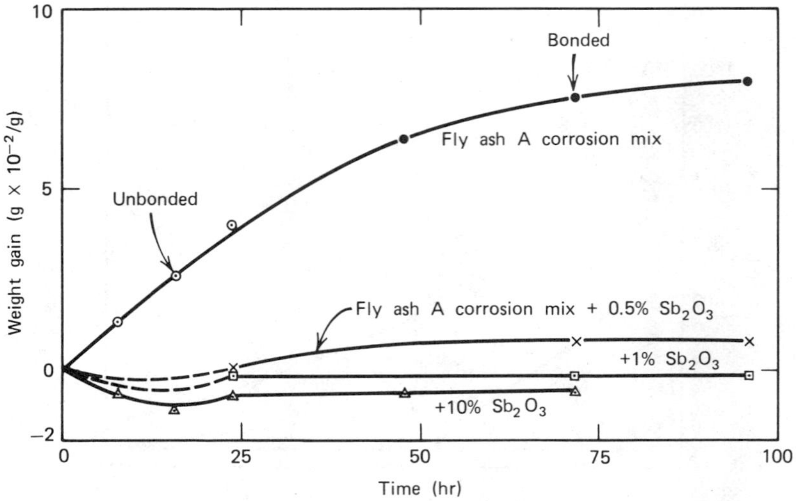

Figure 21.22 Effect of antimony oxide (Sb_2O_3) on corrosion reactions involving SO_3: ● = bonded; ○ = unbonded.[50]

synthetic flue gas over various mixtures in a boat at 593°C and then analyzing the gas showed little formation of SO_3 when Sb_2O_3 was present in the mixture compared with when it was absent.

These tests demonstrated conclusively that Sb_2O_3 can eliminate catalytic oxidation of SO_2 under laboratory conditions. No other substance was equally effective, although some 85 compounds were examined.

The toxicity of antimony has prevented its use to control SO_3 formation in boiler furnaces. For the present, it appears that chemical poisoning of catalytic surfaces is not a practical answer to the problem of SO_3 formation in boiler furnaces.

21.2.4 Interaction of SO_3 with Solid Ash Particles

As ash is released during the burning of a particle of coal, it is exposed to variations in the atmosphere, which at first is reducing when the carbonaceous matter predominates and later is oxidizing as combustion is completed. The ash particles also are exposed to SO_2 and the low levels of SO_3 produced in the flame. Since sulfates are unstable at furnace temperatures, few reactions between SO_3 and ash particles occur until the temperature has dropped appreciably. But with ash accumulated on heat-receiving surfaces, the temperature can be low enough that deposits on wall tubes and superheaters can begin to react chemically with SO_3. As the temperature of the flue gas drops still further, equilibrium begins to favor the formation of additional SO_3, which then can condense as a thin film on the surface of fly ash particles still suspended in the flue gas. This film has a profound effect on the electrical resistivity of the fly ash, hence on the collection efficiency of electrostatic precipitators.

Tests made in a laboratory apparatus with a high temperature gradient, in which ash was exposed to a synthetic flue gas, showed that maximum SO_3 pickup occurred at 560°C; but at this temperature the trisulfates evidently were formed as one step in external corrosion.[51]

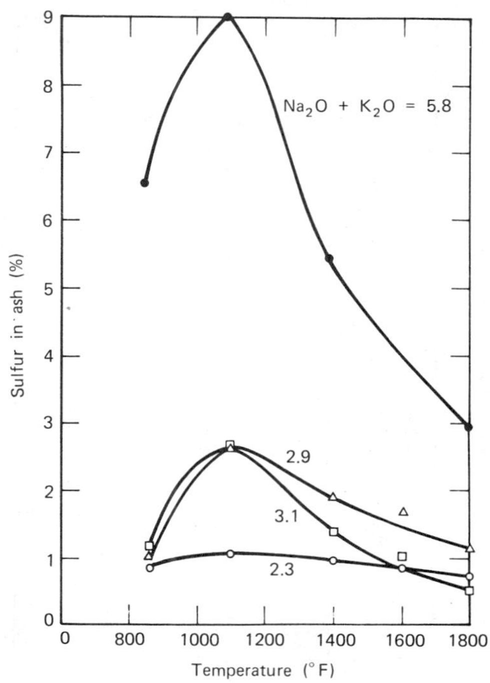

Figure 21.23 Sulfur in ash after exposure to synthetic flue gas at various temperatures. Synthetic flue gas contained 2.5% SO_2, 10% O_2, 15% CO_2, and remainder N.[51]

Figure 21.23 shows that the amount of sulfur reacting with the ash was a maximum at about 1050°F (560°C), but more important, the extent of the increase depended largely on the alkali content of the ash, as would be expected if the SO_3 had reacted with the ash to form $Na_3Fe(SO_4)_3$ and $K_3Fe(SO_4)_3$. Figure 21.24 shows too that as the equilibrium level of SO_3 increased in the flue gas, the amount of sulfur picked up by the ash was also greater.

Coal ash high in CaO and MgO will also react with SO_3 in flue gas when the temperature drops below about 1000°C, as the flue gas passes through the boiler. This has been observed particularly with Australian brown coals,[52] but the same action can occur with high-CaO Western coals. In the case of the brown coals containing from 20.7% CaO + MgO for a sample from the lower part of the Yallourn Seam, to 71.3% CaO + MgO for a

[51] M. Weintraub, S. Goldberg, and A. A. Orning, "A Study of Sulfur Reactions in Furnace Deposits," *Trans. ASME, J. Eng. Power, Ser. A*, **83**, 444–450 (October 1961).

[52] K. Grant and J. H. Weymouth, "The Nature of Inorganic Deposits Formed During the Use of Victorian Brown Coal in Large Industrial Boilers," *J. Inst. Fuel*, **35** (255), 154–160 (1962).

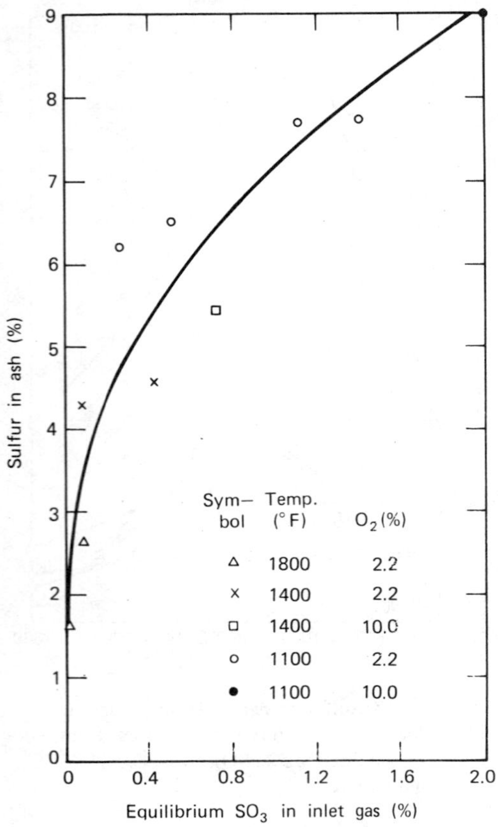

Figure 21.24 Correlation of sulfur in ash and calculated equilibrium content of SO_3 in flue gas.[51]

similar sample from the Morwell Seam, the reaction with SO_3 in furnace deposits produced appreciable amounts of thenardite (Na_2SO_4) and anhydrite ($CaSO_4$). Laboratory combustor studies confirmed that both materials formed rapidly. When these brown coals are burned for steam generation, the principal bonding agent on screen and superheater tubes is anhydrite; but in later sections, as the temperature is lower, thenardite becomes more prominent. The same general action was observed in early attempts to control SO_2 emission by injecting pulverized limestone into a pulverized-coal-fired furnace. One of the serious operational problems, massive accumulations of $CaSO_4$, was solved in part by coordinating soot blower operation and increased frequency of blowing.

Although the reaction of SO_3 and ash constituents at higher temperatures has the important effect of causing problems with deposits and metal wastage, the interaction between SO_3 and ash at lower temperatures in affecting electrical resistivity of fly ash is probably more significant.

Electrostatic precipitators are sensitive to the electrical resistivity of the ash, with best performance at resistivities not exceeding 10^{10} Ω-cm. Although resistivity depends on several factors, the dominant one appears to be the presence of a conducting film on the surface of fly ash particles, which in turn depends mainly on the SO_3 content of the flue gas. The effect of such a surface film on electrical resistivity is shown in Fig. 21.25: the resistivity was measured on a sample of fly ash as it entered a precipitator and on the same sample after degassing in a vacuum at 360°C. Below 200°C it is evident that the removal of a surface film by heating *in vacuo* radically changed the electrical conductance of the ash particles. It is also worth noting that above 200°C, the resistivity of this fly ash is not affected by surface conditions, and

[53] H. G. T. Busby and K. Darby, "Efficiency of Electrostatic Precipitators as Affected by the Properties and Combustion of Coal," *J. Inst. Fuel,* **36**(268), 184–197 (May 1963).

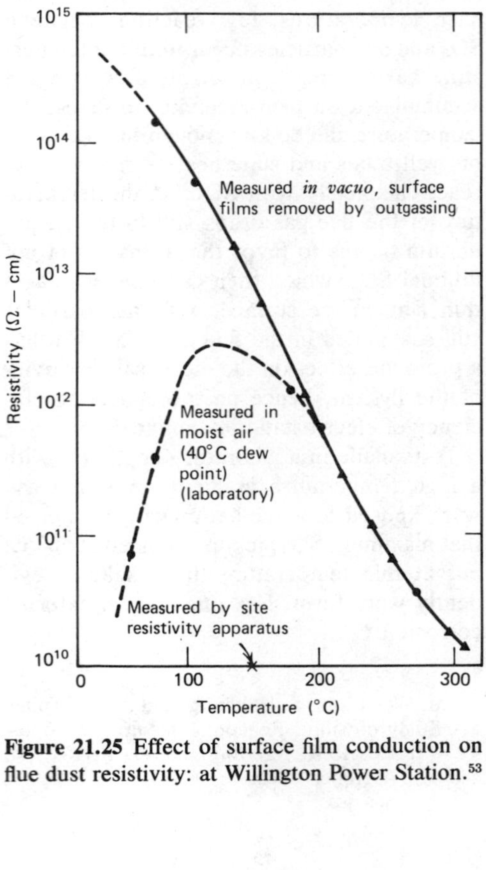

Figure 21.25 Effect of surface film conduction on flue dust resistivity: at Willington Power Station.[53]

the resistivity continues to fall as the temperature increases.

Under some conditions SO_3 in flue gas may not affect electrical resistivity of fly ash. This occurs mainly with low-rank Western coals with ash that is high in CaO and MgO. According to studies by the U.S. Bureau of Mines in a laboratory combustor burning 75 lb of coal per hour, fly ash resistivity was found to be related to (CaO + MgO)/(Na_2O + SO_3).[54]

Table 21.5 shows how resistivity is related to ash composition. The correlations here are good. Neglecting Na_2O has minimal effect, and the role of Na_2O is not clear, although it is possible that an alkali sulfate on the surface of fly ash particles may be partially equivalent to a condensed film of H_2SO_4 in decreasing fly ash resistivity. This suggests in turn that adding a sodium compound plus SO_3 would have more effect on fly ash resistivity than adding SO_3 alone, as is now practiced occasionally in conditioning fly ash.

Analysis of flue gas for SO_3 has been troublesome because many wet procedures involving sorption of the gas by a liquid reagent allow some oxidation of SO_2 to SO_3 during sample collection, particularly if even traces of copper are present. A technique based on physical separation of SO_3 from SO_2 by condensation at a temperature between 60 and 90°C eliminates this problem.[55] At these temperatures only a negligible amount of sulfuric acid can exist in the vapor phase, and since the dew point for water vapor generally falls below 50°C, no water is condensed and no SO_2 is retained.

The equipment required is simple: a heated sampling line to prevent condensation of SO_3 during sample collection; a glass spiral condenser and a sintered glass disk maintained at about 80°C; and a burette for titrating the acidity of the condensed sample flushed from the spiral condenser and the disk. Repeatability is good, gas-sampling rates may be high, and the equipment is simple and easy to operate. Less than 15 min is required for a determination.

[54] S. J. Selle, P. H. Tufte, and G. H. Gronhovd, "A Study of the Electrical Resistivity of Fly Ashes from Low-Sulfur Western Coals Using Various Methods," paper presented at Air Pollution Control Assoc., 65th Annual Meeting, Miami Beach, Fla., June 18–22, 1972.

[55] H. Goksoyr and K. Ross, "The Determination of Sulphur Trioxide in Flue Gases," *J. Inst. Fuel,* **35**(255), 177–179 (April 1962).

21.2.5 Formation and Behavior of Sulfates

It has been known since the early 1940s that the formation of alkali iron trisulfates—$Na_3Fe(SO_4)_3$ and $K_3Fe(SO_4)_3$—could lead to catastrophic external corrosion. Two possible mechanisms have been suggested. In one, the Fe_2O_3 existing as an oxide on the metal reacts with Na_2SO_4 and SO_3, thereby removing the protective film and leading to further oxidation of the metal. The other assumes that Fe_2O_3 in the overlying ash deposit reacts to form the trisulfate, which then attacks the metal directly. Both paths probably occur. Both also assume the presence of Na_2SO_4 or K_2SO_4 and of SO_3 at a moderately high level.

21.2.5.1 *Alkali Sulfates*

The exact way in which sulfates form in boiler furnaces is poorly understood. It is generally assumed that alkalis are volatilized in the flame front, possibly convert almost immediately to the oxides, react quickly with SO_2 or O_2 or with SO_3 to form the sulfates, and condense on fly ash particles and on heat-receiving surfaces. Probes inserted into boiler furnaces often collect a white film as a precursor to ash and slag deposits; this film

Table 21.5 Effect of Composition on Electrical Resistivity of Fly Ashes[54]

Fly Ash Composition (%)			Ratio of	
CaO + MgO	Na_2O	SO_3	CaO + MgO to Na_2O + SO_3	Electrical Resistivity at 150°C ohm-cm
22.2	14.0	19.3	0.7	$< 3 \times 10^8$
26.6	7.6	11.6	1.4	3×10^8–1×10^9
24.7	4.5	5.1	2.6	1×10^9–5×10^9
39.3	3.3	5.9	4.3	5×10^9–1×10^{10}
27.1	1.3	4.0	5.1	$> 1 \times 10^{10}$

generally is rich in alkalis and is probably the best indicator of the presence of sulfates in the flue gas or of condensable alkalis reacting rapidly with sulfur oxides. Because of the relatively low melting points of Na_2SO_4 (882°C) and of K_2SO_4 (1075°C), and because the minimum melting point of the system Na_2SO_2-K_2SO_4 occurs at about 833°C, liquid films of the alkali sulfates can be present even on cooled surfaces where there is a steep temperature gradient. At temperatures above 800–1000°C, as on hangers and uncooled metal parts, even Na_2SO_4 is not deposited.[56] At more moderate temperatures, molten sulfates can serve as a liquid phase to increase the likelihood of strongly bonded deposits. More important, such liquid films can provide an electrolyte to encourage rapid metal wastage by electrochemical processes.

21.2.5.2 *Pyrosulfates*

The earliest identification of the alkali iron trisulfates indicated the possibility that small amounts of the pyrosulfates might also be present, but the estimated level in the "enamel" deposits found in corrosion areas probably did not exceed 2%. The pyrosulfates melt at moderately low temperatures, about 400°C for $Na_2S_2O_7$ and 300°C for $K_2S_2O_7$; but these salts dissociate rapidly in atmospheres not containing SO_3. At 538°C about 50,000 ppm (5%) SO_3 is necessary to convert Na_2SO_4 to $Na_2S_2O_7$, and about 8000 ppm (0.8%) is required to form $K_2S_2O_7$.

Despite requirements of SO_3 levels well above those that could be reached by complete conversion of SO_2 to SO_3 in boiler furnace atmospheres, the pyrosulfates have been considered, particularly in England, as likely causes of corrosion. Work by Crossley of the Central Electricity Generating Board (CEGB),[57] for example, indicated that pyrosulfates in superheater deposits were largely responsible for fireside corrosion. This conclusion was based in part on laboratory studies where mixtures of $K_2S_2O_7$ and Fe_3O_4, heated to temperatures up to 538°C, formed iron sulfates. At higher temperatures, up to 593°C, iron oxide instead of iron sulfate was present, leading Crossley to expect that a new type of corrosion would be likely when superheater metal temperature reached 593°C, as for a steam temperature of 538–566°C. Failure to identify $K_2S_2O_7$ in areas of corrosion at these temperatures can be explained by the insensitivity of X-ray diffraction analysis, by thermal dissociation, and by the high chemical activity of the pyrosulfates, converting them to other forms. Because analyses of the inner layers of wall tube and superheater deposits showed that the ratio of SO_3 to $Na_2O{\cdot}K_2O$ ranged from 1.7 to 2.5, with an average of 2.07, and since this ratio is 2 for the pyrosulfates, Crossley concluded that the first stage in the development of a potentially viscous deposit is the formation of acid sulfates approximating the pyrosulfates.

This position has not been supported by others, who point to the high levels of SO_3 necessary to maintain pyrosulfates at temperatures above about 425°C. The complexity of sulfur chemistry under boiler furnace conditions still provides considerable opportunity for further research.

Figure 21.26 summarizes more recent efforts to provide basic information on this system. It shows that Na_2SO_4 does not react with SO_3 at partial pressures less than 2500 ppm to form a liquid phase, which is about the maximum concentration of SO_3 attainable burning a 4% sulfur fuel. With $K_2S_2O_7$, heated at a constant rate in a fixed concentration of SO_3, the weight loss becomes rapid up to about 410°C, followed by complete conversion of the melt to solid K_2SO_4. Liquid K_2SO_4-$K_2S_2O_7$ melts can be maintained at temperatures as low as 408°C with as little as 200 ppm SO_3. In support of Crossley's contentions, a liquid phase containing a high percentage of $K_2S_2O_7$ can exist between 407°C and 495°C if potassium is the only alkali metal and if equilibrium is reached between SO_2 and SO_3 for flue gas containing 2500 ppm SO_2.[58]

21.2.5.3 *Alkali Iron Trisulfates*

Little additional information has been developed on the trisulfates since they were first implicated in catastrophic metal wastage in

[56] P. J. Jackson, "Researches in the CEGB into Fireside Fouling and Corrosion of Boiler Plant," *J. Inst. Fuel,* **40,** 335–341 (August 1967).

[57] H. E. Crossley, "External Fouling and Corrosion of Boiler Plant: A Commentary," *J. Inst. Fuel,* **40,** 342–347 (August 1967).

[58] A. W. Coats, D. J. A. Dear, and D. Penfold, "Phase Studies on the Systems Na_2SO_4-SO_3, K_2SO_4-SO_3, and Na_2SO_4-K_2SO_4-SO_3," *J. Inst. Fuel,* **41,** 129–132 (March 1968).

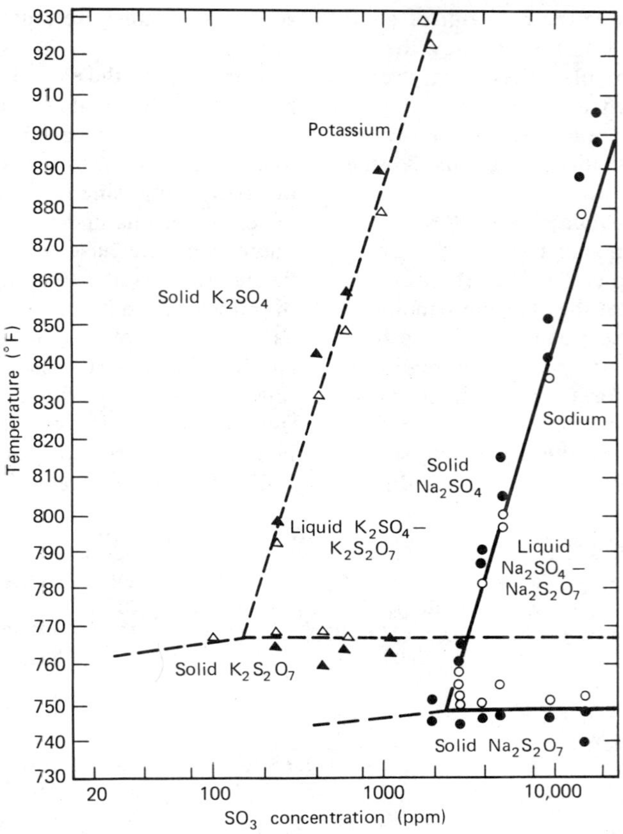

Figure 21.26 Melting points in systems Na_2SO_4- SO_3 and K_2SO_4-SO_3: ●, ▲ = solid phase, ○, △ = liquid phase.[58]

the 1940s. Most of the subsequent work has been on modes of formation of the trisulfates, availability of the SO_3 required for their formation and maintenance, and methods of detection. A phase diagram of the system K_2SO_4-$Fe_2(SO_4)_3$ is now available, as shown in Fig. 21.27. It indicates two allotropic forms of $K_3Fe(SO_4)_3$ as well as another complex sulfate, $KFe(SO_4)_2$. It is now known that in this system mixtures of K_2SO_4 and 12–23 mole % $Fe_2(SO_4)_3$ will separate solid trisulfate when cooled below 703°C, but some liquid phase will persist until the temperature is down to 627°C. Likewise, mixtures containing less than 12% $Fe_2(SO_4)_3$ will freeze below 627°C but will retain some molten phase at temperatures approaching 1090°C. If the system contains more than 25% $Fe_2(SO_4)_3$, it will freeze below 693°C. Ideal mixtures such as these do not exist on boiler tubes, but the

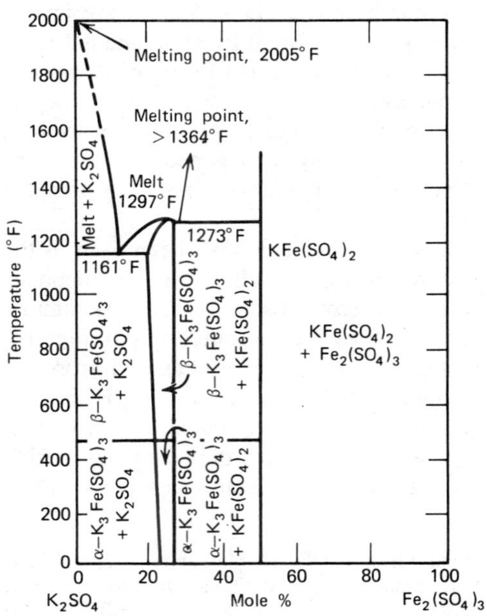

Figure 21.27 Phase diagram of the system K_2SO_4-$Fe_2(SO_4)_2$.[59]

[59] A. Rahmel and W. Jager, "Investigation of the System Potassium Sulfate–Ferric Sulfate," *Ztg. Anorg. Allg. Chem.*, **303**, 90–95 (1960).

basic factors affecting the behavior of complex mixtures of sulfates are now better understood. In particular, they suggest conditions under which a liquid phase can be present, and these are of great importance in assessing the possibility of serious fireside corrosion.

The problem of availability of SO_3 at concentrations high enough to form the trisulfates was discussed earlier in Section 21.2.3. It is known now that the stagnant conditions beneath layers of slag can provide the time and the conditions for sufficiently high SO_3 concentrations to form and to maintain the trisulfates at temperatures up to about 595°C.

Possible sulfate-producing reactions include the following[60]:

Reaction	ΔF (kcal)	
$Fe_2O_3 + 3SO_2 + {}^3/_2O_2 \rightleftharpoons Fe_2(SO_4)_3$	−29	(21.11)
$Fe_2O_3 + 3SO_3 \rightleftharpoons Fe_2(SO_4)_3$	−18	(21.12)
$2Fe_3O_4 + 9SO_2 + 3O_2 \rightleftharpoons 3Fe_2(SO_4)_3$	−111	(21.13)
$2Fe_3O_4 + 3SO_2 + 2O_2 \rightleftharpoons 2Fe_2O_3 + Fe_2(SO_4)_3$	+ 7	(21.14)
$2Fe_3O_4 + 3SO_3 + \frac{1}{2}O_2 \rightleftharpoons 2Fe_2O_3 + Fe_2(SO_4)_3$	−43	(21.15)
$2Fe_3O_4 + 2SO_2 + O_2 \rightleftharpoons 2Fe_2O_3 + 2FeSO_4$	−24	(21.16)
$FeO + SO_3 \rightleftharpoons FeSO_4$	−17	(21.17)
$FeO + SO_2 + \frac{1}{2}O_2 \rightleftharpoons FeSO_4$	−20	(21.18)
$Fe_3O_4 + 3SO_3 \rightleftharpoons 3FeSO_4 + \frac{1}{2}O_2$	−11	(21.19)
$Fe_3O_4 + SO_3 \rightleftharpoons FeSO_4 + Fe_2O_3$	−12	(21.20)

Of these, Reactions 21.11 and 21.12 are probably minor, but 21.13 and 21.16 are important, as demonstrated by experimental work in the laboratory. But like many thermodynamic analyses in which kinetic data are lacking or unreliable, such information is most useful only as a general guide to the behavior of components in such a complex system.

Several techniques have been used in the past to detect the trisulfates in boiler deposits or in laboratory tests. Massive accumulations, such as the "enamel" found in corrosion areas on wall tubes, pose no problems because of appearance, the low pH in solution, and identification by x-ray diffraction. Minuscule amounts of the trisulfates, which might form in a few hours as a precursor to serious problems, are more difficult to identify.

A procedure developed by Battelle[61] for identifying individual crystals of $Na_3Fe(SO_4)_3$ and $K_3Fe(SO_4)_3$ is based on measuring the index of refraction of crystals found in deposits by immersing them in standard index-of-refraction oils and viewing them under a microscope. Because the trisulfates exist as hexagonal crystals, two values of the index of refraction can be determined, one for each axis. Thus by rotating the microscope stage, and by trial-and-error matching the index of refraction of the oil with the specimen, it was found that the indices of refraction for $Na_3Fe(SO_4)_3$ were 1.590 and 1.628, and for $K_3Fe(SO_4)_3$, they were 1.584 and 1.622. Furthermore, because these trisulfates usually occur mixed, it is possible to determine the relative proportions by the same method.

Figure 21.28 gives the index of refraction at the two positions for mixtures of $Na_3Fe(SO_4)_3$ and $K_3Fe(SO_4)_3$, since the index of refraction of a mixture forming a solid solution changes linearly with composition. Using a typical example, the figure illustrates that a mixture of two-thirds of the potassium compound and one-third of the sodium compound follows this relationship.

This technique has been applied successfully to identifying the trisulfates in samples prepared in the laboratory under simulated field conditions, with exposures as short as

[60] A. Levy and E. L. Merryman, "Interactions in Sulphur Oxide-Iron Oxide Systems," *Trans. ASME, J. Eng. Power, Ser. A*, **89**(2), 297–303 (April 1967).

[61] Battelle Memorial Institute, *Formation of Alkali Iron Sulfates and Other Compounds Causing Corrosion in Boilers and Gas Turbines*, 2nd Summary Rep. to ASME Research Committee on Corrosion and Deposits, June 30, 1967, pp. 17–18. [Internal report; issued by ASME to the committee.]

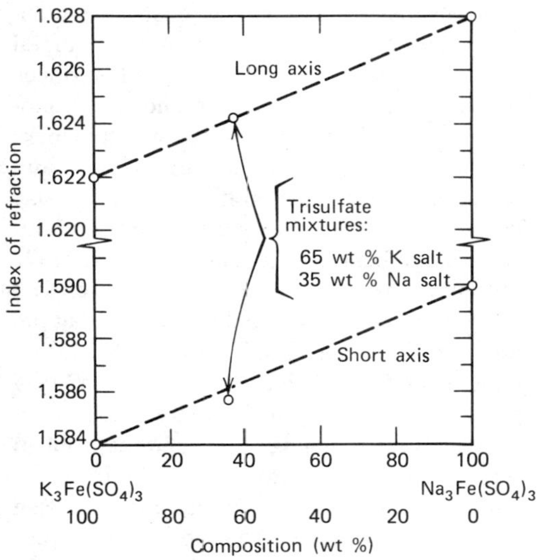

Figure 21.28 Refractive index of alkali iron trisulfates.[61]

65 hr. It is much more sensitive than X-ray diffraction methods. This optical technique offers a powerful tool for tracing the course of corrosion reactions and shedding light on the reaction mechanism. It should also serve effectively for examining deposits on corrosion probes long before any metal loss can be detected and before enough trisulfates are present for detection by pH or X-ray techniques.

Other useful techniques for studying the formation of the trisulfates and for investigating sulfur reactions involve ^{35}S. This isotope emits low-energy beta radiation (0.167 meV) and has a half-life of 86.7 days. Safety hazards are minimal, and a statistically significant radioactive count can be obtained over reasonable time periods. Because ^{35}S behaves chemically exactly like the sulfur in fuels, and because extremely small quantities of this isotope can be detected, even when the ratio of ordinary sulfur to ^{35}S is as high as 10^{10}, the isotope is highly useful for comparing such factors as the relative reactivity of SO_2 and SO_3, or the influence of various surfaces on the formation of the trisulfates.

21.2.6 Sulfidation Reactions

Since most coal is burned under oxidizing conditions, sulfur is present as oxides in the products of combustion, and as sulfates in deposits. Under reducing conditions (e.g., in gasifiers), sulfur appears as H_2S or other sulfides, and as sulfides in deposits. Compared with oxidizing conditions, relatively little information is available on sulfide formation. Attention is being given today to sulfide chemistry for several reasons; among them is the speculation that control of NO_x by "biased" or "two-stage" combustion may result in reducing conditions on heat-receiving surfaces, or alternating oxidizing and reducing atmospheres under conditions that are not yet definable. Furthermore, in gasifiers, suitable construction materials to withstand sulfide sulfur pose very great problems.

Most of the attention paid to sulfides in energy systems has been confined to gas turbines, because inlet temperatures have increased steadily and problems are growing with sulfates on high-temperature, highly stressed turbine components. Although liquid fuel rather than coal is burned in combustion turbines, problems with sulfidation are common to both fuels.

Early attempts to induce corrosion in laboratory experiments where metal coupons were immersed in molten Na_2SO_4 were only partly successful unless carbon was also present, in which case rapid attack occurred.[62] The reactions involved occur in two steps:

$$Na_2SO_4 + 3R \rightleftarrows Na_2O + 3\,RO + S \tag{21.21}$$

$$M + S \rightleftarrows MS \tag{21.22}$$

where R is any reducing agent and M is a metal forming a sulfide, and:

$$Na_2SO_4 + 3MS \rightleftarrows 4S + 3MO + Na_2O \tag{21.23}$$

$$4M + 4S \rightleftarrows 4MS \tag{21.24}$$

Reaction 21.21 triggers 21.22 and Reaction 21.23 triggers 21.24. Later it was decided that sulfidation can proceed without the need for carbon or an external reducing agent, since the metal itself may be a sufficient reducing agent. Most workers in the field seem to agree that "condensed" Na_2SO_4 is necessary, at least in gas turbines.

[62] J. Stringer, *Hot Corrosion in Gas Turbines,* Metals and Ceramics Information Center, Battelle Memorial Institute, Metals and Ceramics Information Center No. 72-08, June 1972, 52 pp.

The metallurgical aspects are confusing. For example, nickel reacts with sulfur compounds in a complex fashion, according to the following possible reactions:

$Ni + \frac{1}{2}O_2 \rightleftharpoons NiO$	(21.25)
$NiS + 2O_2 \rightleftharpoons NiSO_4$	(21.26)
$3Ni + S_2 \rightleftharpoons Ni_3S_2$	(21.27)
$Ni_3S_2 + {}^1/_2S_2 \rightleftharpoons 3NiS$	(21.28)
$2Ni + S_2 \rightleftharpoons 2NiS$	(21.29)
$Ni_3S_2 + {}^3/_2O_2 \rightleftharpoons 3NiO + S_2$	(21.30)
$2NiS + O_2 \rightleftharpoons 2NiO + S_2$	(21.31)
$2NiSO_4 \rightleftharpoons 2NiO + S_2 + 3O_2$	(21.32)
$4NiO + Ni_3S_2 \rightleftharpoons 7Ni + S_2 + 2O_2$	(21.33)
$2NiO + 7NiS \rightleftharpoons 3Ni_3S_2 + \frac{1}{2}S_2 + O_2$	(21.34)
$3NiSO_4 + NiS \rightleftharpoons 4NiO + 2S_2 + 4O_2$	(21.35)

Pyrites in coal is a common cause of trouble when particles of FeS_2 impinge on heat-receiving surfaces such as wall tubes in pulverized-coal-fired boiler furnaces. If FeS_2 manages to reach such surfaces before converting to FeS, severe metal wastage will occur. Particles of FeS or FeS_2 captured in the pool of molten slag in slag tap furnaces can cause serious operational problems because of the low melting point of FeS (1190°C) and the formation of the FeO-FeS eutectic melting at 940°C.

Detection of sulfides in deposits or even in grain boundaries is relatively easy, using a technique which depends on the dissociation of sodium azide.[63]

Sulfidation also may become more of a problem when low excess air is tried in coal-fired systems. Although moderately low excess air has been demonstrated with coal firing, the low levels of excess air possible with fuel oil, with as little as 0.2% O_2 in the flue gas, have not yet been attained with coal.

21.3 COAL ASH SLAG

Slag tap furnaces for steam generation were widely installed in central station power plants during the 1930s and 1940s because high heat release rates could be tolerated and low ash fusion coals at low cost were plentiful. These furnaces suffered from metal wastage of wall tubes. Fouling of surfaces between the slagging section and the economizer was troublesome because somewhere in the furnace or in the superheater and reheater sections, the molten droplets of slag carried by the flue gas had to freeze without adhering to heat transfer surfaces. Since 1972 the annual reviews of plant design prepared by *Power*[64] have listed only two new slag tap units in 1972, one in 1973, two in 1974, and none in 1975 and 1976. Hence of the 247 new plants contracted over these 5 years, only 5, or 2%, were slagging. Part of the problem results from NO_x emission, which is enhanced by the higher temperatures in slag tap units. Part results from operational problems. Nevertheless, with the increasing availability of low ash fusion coals, the utilities may again buy slag tap furnaces if the NO_x problem can be solved. For gasification, slagging operation appears highly attractive, both in fixed-bed and entrained-flow systems, so that interest in the properties of coal ash slag can be expected to increase.

21.3.1 Viscosity-Temperature-Composition Relationships

The basic parameters relating viscosity to slag composition, and rate of change of viscosity with temperature, date back to the 1930s and the extensive measurements of slag flow characteristics made at that time by the Bureau of Mines. These have been adequately reported.[65] A detailed listing of the parameters relevant to slag behavior in large boiler furnaces also is available.[18] An excellent review of more recent work in England on slag viscosity,[66] and an account of measurements made in that country of the separating of a solid phase in molten slag[67] gen-

[63] W. A. Szymanski, "Method for Rapid Detection of Sulfur Embrittlement of Nickel," *Mater. Prot.*, **1**, 100 (1962).

[64] Annual Plant Design Reports, *Power*, November 1972–1975.

[65] R. C. Corey, *Measurement and Significance of the Flow Properties of Coal-Ash Slag*, U.S. Bur. Mines Bull. No. 618, 1964, 64 pp.

[66] J. D. Watt and F. Fereday, "The Flow Properties of Slags Formed from the Ashes of British Coals. Part 1. Viscosity of Homogeneous Liquid Slags in Relation to Slag Composition," *J. Inst. Fuel*, **42**, 99–103 (March 1969).

[67] J. D. Watt, "The Flow Properties of Slags Formed from the Ashes of British Coals. Part 2. The Crystallizing Behavior of the Slags," *J. Inst. Fuel*, **42**, 131–134 (April 1969).

erally confirm the findings of the Bureau of Mines, with more precise treatment of the experimental data. These studies emphasize the "silica percentage" concept of relating viscosity to composition, as well as the temperature of critical viscosity (T_{cv}) for the transition of a slag on cooling from a Newtonian fluid to a pseudoplastic solid.

Other treatments of the influence of composition on the viscosity of coal ash slags have relied on the ratio of "bases" to "acids," or of Fe_2O_3 + CaO + MgO + Na_2O + K_2O to SiO_2 + Al_2O_3 + TiO_2, as is commonly done in evaluating the properties of metallurgical slags. In pyrochemical terms, an "acid" is a metal oxide capable of reacting with CaO or similar oxides. A "base" is a metal oxide such as CaO that can react with an acid, typically SiO_2. Such terms are poorly defined but offer one method of reducing complex systems with many compositional parameters to a more manageable size.

Metallurgical slags have been evaluated by the base/acid concept for a very long time, principally because high-temperature systems, such as open hearths or converters, used a basic refractory lining such as dolomite, or an acidic lining such as silica. In coal-burning systems, no such limitations occur. Nevertheless the base/acid ratio has been demonstrated as a useful measure of slag viscosity. It is a simple procedure to apply, if the curves of temperature for a specified viscosity as a function of the base/acid ratio are available.[68] No mathematical development of these relationships has been provided. But since the base/acid ratio and the silica percentage are closely related (because both involve the same general parameters), it is possible to convert from one to the other. For practical purposes, the following relationships apply:

$$\text{silica percentage} = 92.33 - (66.67)\,(\text{base/acid ratio}) \quad (21.36)$$

$$\text{base/acid ratio} = 1.385 - (0.015)\,(\text{silica percentage}) \quad (21.37)$$

This makes possible the conversion from one system to the other for those who prefer either the better accuracy and predictability based on the silica percentage or the simpler relationship of the base/acid ratio.

Figure 21.29 illustrates the relationship between the base/acid ratio of coal ash slags and the temperature where the viscosity is 250 poises, the maximum viscosity suitable for tapping slags from cyclone furnaces.[68] Similar plots are available for 50 and 100 poises. This relationship is stated for 20 ferric percentage, in other words, when 20% of the total equivalent Fe_2O_3 in the slag actually is

[68] W. L. Sage and J. B. McIlroy, "Relationship of Coal-Ash Viscosity to Chemical Composition," *Trans. ASME, J. Eng. Power, Ser. A,* **62,** 145–155 (1960).

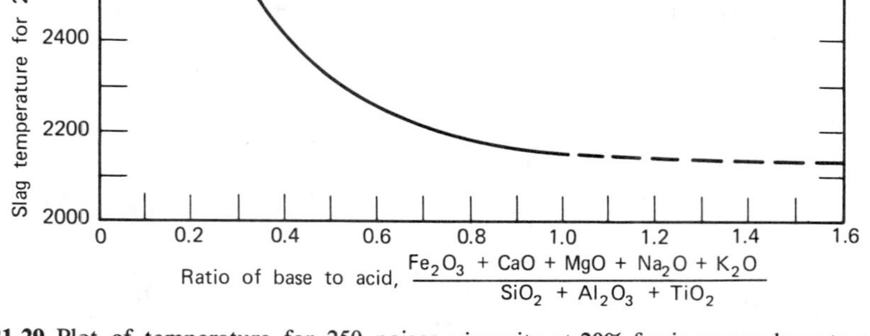

Figure 21.29 Plot of temperature for 250 poises viscosity at 20% ferric versus base-to-acid ratio.[68]

present as Fe_2O_3. A change in the state of oxidation of coal ash slags does not substantially affect true Newtonian viscosity, but it does have a pronounced effect on T_{cv}. Also, since only the equivalent Fe_2O_3 in the slag responds to oxidizing or reducing conditions, the change in T_{cv} is greater with high-iron slags than with low-iron slags.

A recent study[69] emphasizes further empirical correlations of viscosity-composition-temperature relationships by a stepwise regression analysis.

Table 21.6 shows that such techniques can indeed predict the viscosity of general types of coal ash slags within fair limits. In this example, the Eastern-type ash was from 36 Kentucky coal samples, and the Western-type ash from 13 Canadian coals. Arbitrarily, "Eastern-type" ash is defined as that in which Fe_2O_3 exceeds the sum, in terms of percentages, of CaO + MgO, and "Western-type" ash contains more CaO + MgO than Fe_2O_3, with the sum of the three constituents exceeding 20% of the total ash analysis. No simple equation has come yet from this study, each ash category requiring the generation of eight expressions corresponding to eight preselected viscosity levels.

Regression analysis also has been used on a group of 1250 ash analyses of both Eastern and Western coals to indicate how ash composition affects cone fusion determinations.[6] The best single indicator of ash fusion by the ASTM method under reducing conditions is $SiO_2 : FeO$, with FeO as the second best single variable. Under oxidizing conditions, $SiO_2 : CaO$ is most applicable. Other relationships are complex. This study mainly demonstrated that it is possible to analyze ash fusion data with modern computer techniques to obtain a better understanding of the actions taking place in a cone fusion determination.

The most extensive study in recent years of slag composition and viscosity was conducted by Watt and Fereday in England.[66] Based on viscosity measurements of 113 coal-ash mixtures over the range of 1–10,000 poises at temperatures up to 1800°C, equations were developed allowing more precise estimates of viscosity than have thus far been available. The relationship is complicated, but the confidence limits based on instrumental errors, analytical errors, and irregular behavior of slag were higher at ±0.23 than those of the experimental determinations themselves, at ±0.15.

Using this correlation the viscosity of a coal ash slag, fully melted with no solid phase present, can be represented by:

$$m = 0.00835SiO_2 + 0.00601Al_2O_3 - 0.109 \quad (21.38)$$

$$c = 0.0415SiO_2 + 0.0192Al_2O_3 + 0.0276Fe_2O_3 + 0.0160CaO - 3.92 \quad (21.39)$$

$$\log \text{viscosity} = \frac{10^7 m}{(t - 150)^2} + c \quad (21.40)$$

where t is in degrees Celsius, and the analyses are expressed on the basis of:

$$SiO_2 + Al_2O_3 + Fe_2O_3 + CaO + MgO = 100$$

with Fe_2O_3 being the total equivalent Fe_2O_3, given by analysis of the slag for forms of iron. Statistical evaluation shows that no significant error results from excluding Na_2O,

[69] S. M. Barrick and G. F. Moore, "Empirical Correlation of Coal Ash Viscosity with Ash Chemical Composition," ASME Paper No. 76-WA/FU-3, 1976. [Published as a pamphlet by Fuels Division, at Winter annual meeting of ASME, 1976; does not appear in any ASME transaction.]

Table 21.6 Measured and Calculated Temperatures (°C) for Specified Viscosities of Coal Ash Slags Under Reducing Conditions[18]

Viscosity (poises)	Eastern-Type Ash		Western-Type Ash	
	Measured	Calculated	Measured	Calculated
50	1488	1416	1274	1266
250	1302	1257	1157	1208
1,000	1196	1178	1146	1184
10,000	1116	1099	1143	1171

K_2O, TiO_2, and S from this relationship. For convenience, MgO is not included in the log viscosity equation because it is necessary mathematically to express only four of the oxides in calculating c.

A somewhat simpler relationship called the "S^2 correlation" also comes from England[70]:

$$\text{log viscosity} = 4.468\left(\frac{S}{100}\right)^2 + 1.265\frac{10^4}{T} - 7.44 \qquad (21.41)$$

where S is the silica ratio 100 $SiO_2/(SiO_2 + Fe_2O_3 + CaO + MgO)$, and T is temperature (K).

The simplest correlation was that initially derived when the silica percentage concept was originated.[71] It can be expressed by:

$$\log(\mu) = (0.05784)(SiO_2) - 1.8452 \qquad (21.42)$$

where μ is the viscosity at 1425°C and SiO_2 is the silica percentage when $SiO_2 + Fe_2O_3 + CaO + MgO = 100$.

Table 21.7 compares the accuracy of these methods of relating viscosity to composition. It must be emphasized that these procedures are based on slags where all chemical reactions have taken place, the temperature is above that where any solid phase is present, and the slag behaves as a Newtonian liquid. Generally, slags held at temperatures above about 1450°C for at least an hour meet these requirements, but very refractory coal ashes with cone fusion temperatures too high to be measurable may not. Under ideal conditions it is scarcely necessary to measure viscosity of coal ash slags in a viscometer; the relationships described above should suffice for most coal ashes.

The temperature of critical viscosity T_{cv} is less easily predicted. Since it depends on the separation of a solid phase in the molten slag, a process that may require several hours for viscous melts, T_{cv} may go undetected in a routine viscosity measurement until so much solid has accumulated that the slag freezes. Yet only trace amounts of solids may lead to thixotropic slags that do not have the flow characteristics of Newtonian slags.

Although the relationship is poor, the ASTM cone fusion temperature can be a general guide to T_{cv}. Figure 21.30 compares the measured T_{cv} of a group of coal ash slags with the cone ash-softening temperature plus 110°C (200F) added to improve the correlation. Yet the variance sometimes exceeds 165°C (300F).

Watt[67] has made a detailed analysis of the behavior of slags near or below T_{cv}, including additional parameters involving the liquidus (the highest temperature at which liquid and solid phase can coexist in equilibrium), the solidus (the lowest temperature at which liquid and solid can coexist in equilibrium), and the temperature at which the viscosity curve, on reheating, rejoins the curve corresponding to the Newtonian condition. The last parameter is largely time dependent. Attempts to relate the crystallization properties of a slag to the temperature at which the ash fused to a hemispherical shape, using a procedure similar to the ASTM cone fusion test, were unsuccessful. This led Watt to conclude that no useful information on deviations from Newtonian behavior could be derived from fusion behavior.

[70] H. R. Hoy, A. G. Roberts, and D. M. Wilkins, "Behavior of Mineral Matter in Slagging Gasification Processes," *J. Inst. Gas Eng.*, **5**, 444–469 (June 1965) (Communication No. 672).

[71] W. T. Reid and P. Cohen, "The Flow Characteristics of Coal-Ash Slags in the Solidification Range," *Trans. ASME, Furnace Performance Factors*, **66**, 83–97 (1944).

Table 21.7 Comparative Reliability of Methods of Predicting Viscosity of Coal Ash Slags from Chemical Composition[66]

TEMP	Log Viscosity (poises), Calculated—Observed					
	Mean			Confidence Limits (0.05 level)		
(°C)	Reid and Cohen	Hoy et al.	Watt and Fereday	Reid and Cohen	Hoy et al.	Watt and Fereday
1200	−1.42	—	−0.024	± 0.52	—	± 0.29
1300	−0.031	0.182	−0.009	± 0.42	±0.31	± 0.24
1500	0.032	0.023	0.006	± 0.28	±0.25	± 0.23

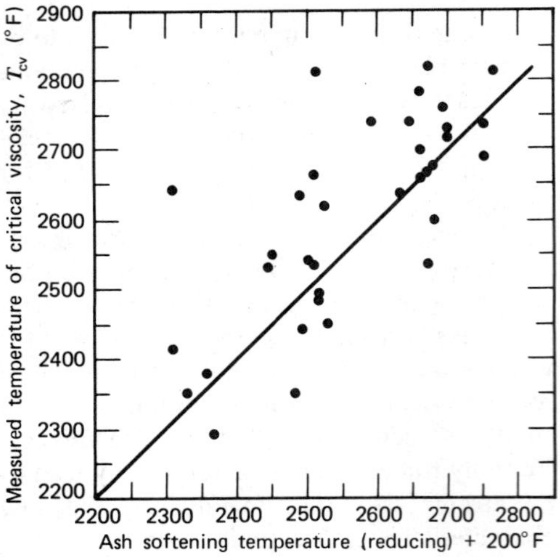

Figure 21.30 Comparison of measured T_{cv} with value calculated at 20 ferric percentage.[68]

Based on viscosity measurements made on 113 coal ash slags, Watt arrived at an expression showing a definite relationship between composition and T_{cv}:

$$T_{cv} = -1470\left(\frac{SiO_2}{Al_2O_3}\right) + 360\left(\frac{SiO_2}{Al_2O_3}\right)^2 - 14.7(Fe_2O_3 + CaO + MgO) + 0.15(Fe_2O_3 + CaO + MgO)^2 + 2290 \qquad (21.43)$$

where T_{cv} is in degrees Celsius, and the composition of the slag is recalculated so that $SiO_2 + Al_2O_3 + Fe_2O_3 + CaO + MgO = 100$. In this case the standard deviation of the residuals was 65°C, but the equation is expected to estimate T_{cv} with a standard deviation of about 54°C.

Although this thorough examination of the temperature at which a slag deviates from Newtonian behavior indicates an ability to relate T_{cv} to composition, it must be concluded on practical grounds that its usefulness is limited. The trend today is to assume that T_{cv} occurs approximately 110°C above the ASTM cone-softening temperature.

21.3.2 Physical Properties

21.3.2.1 *Density*

Few data are available on the physical properties of coal-ash slags. Hoy has provided a good summary.[70]

Table 21.8 lists practical values for the density of coal ash slags. Generally, slag density increases as the Fe_2O_3 content becomes greater, but the exact relationship is probably complex. These are approximations at best, but they permit some estimates of the quantity of slag on a hearth or after tapping into a water stream as slag "frit."

Table 21.8 Density of Coal Ash Slags[70]

Material	Density (kg/m³)
Liquid slag, 1400°C	2200–2400
Nonporous solid slag	2600–2700
Solid slag layers in plant	1900–2600
Slag frit	1000–1400
Porous pumice slag	560–640
(Quartz)	2600
(Limestone)	2500

21.3.2.2 *Surface Tension*

The ability of coal ash slags to become attached to surfaces such as wall tubes or superheaters by a wetting action is a complex function of the gas phase and the solid phase as well as the surface tension of the slag droplet. The wetting characteristics of slags have been investigated by Raask,[72] who measured the contact angle between slag droplets and a graphite surface at a temperature up to 1500°C in an atmosphere of dry N_2. Although these data are of limited usefulness with regard to fouling, they do have a bearing on combustion phenomena, since slag is formed on a particle of burning coal. It was concluded that a layer of sintered ash can interfere with combustion, but if the ash melts and forms a series of isolated nonwetting droplets, the effect is probably negligible.

Interesting scanning electron micrographs have been made of the ash released on the surface of coal burning in a type of fluidized bed.[73] Typical individual droplets are shown with a diameter of about 0.1 μm, apparently formed from the inherent ash in the coal. Eventually these droplets coalesce and react with extraneous ash to produce massive slag forms.

[72] E. Raask, "Slag-Coal Interface Phenomena," *Trans. ASME, J. Eng. Power, Ser. A*, **88**, 40–44 (1966) (ASME Paper No. 65-WA/FU-1, 1965.)

[73] J. Yerushalmi, M. Kolodney, R. A. Graff, and A. M. Squires, "Agglomeration of Ash in Fluidized Beds Gasifying Coal: The Godel Phenomena," *Science*, **187**, 646–648 (February 21, 1975).

Measurements of the surface tension of molten coal ash slags were made by Reid in the 1940s using the maximum bubble pressure technique, but the results were never published because correlations with composition could not be found. The surface tension of typical coal ash slags at 1450°C was measured at 350–400 dynes/cm.

Hoy,[70] using an expression developed for commercial glasses, suggests that the surface tension of coal ash slags may behave similarly, according to:

$$\text{surface tension}_{1400°C} = 3.24SiO_2 + 5.85Al_2O_3 + 4.4Fe_2O_3 + 4.92CaO + 5.49MgO + 1.12Na_2O - 0.75K_2O \quad (21.44)$$

On this basis, slag formed from coal from the Pittsburgh No. 8 bed in Ohio would have a surface tension at 1400°C of about 410 dynes/cm, in good agreement with the maximum bubble pressure measurements.

Few attempts have yet been made to relate surface characteristics of coal ash slags to problems with slag fouling in large boiler furnaces. This would appear to be a fruitful field for research.

21.3.2.3 *Thermal Conductivity and Emissivity*

Because coal ash accumulated on heat-receiving surfaces can interfere so seriously with heat transfer, many investigators have concerned themselves with measurements of the thermal conductance and the emissivity of ash and slag deposits. Such measurements are difficult because of the irreversible changes occurring in ash as the temperature increases and sintering and melting take place. Most measurements have been made at temperatures lower than 1200°C, hence are more applicable to superheater deposits than to wall slag.

Figure 21.31 shows the general relationship between temperature and thermal conductance for a variety of deposits. The most significant factor is the change in conductance when sintering occurs, the conductance of fused particles being some 4–5 times greater than that of the particulates. Also, once fused, conductance increases substantially with an increase in temperature. The meas-

[74] M. F. R. Mulcahy, J. Boow, and P. R. C. Goard, "Fireside Deposits and Their Effect on Heat Transfer in a Pulverized-Fuel-Fired Boiler. Part I," *J. Inst. Fuel,* **39**(308), 385–394 (September 1966).

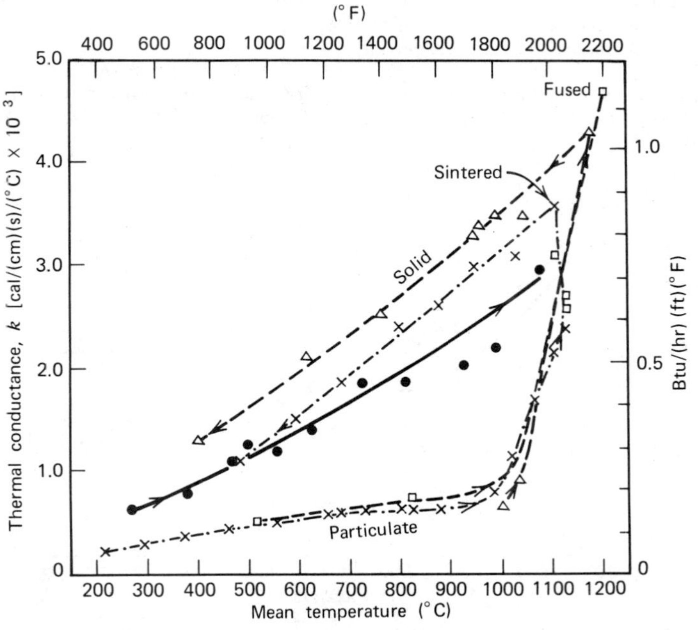

Figure 21.31 Effective thermal conductance of several samples of combustion chamber deposits as a function of mean temperature: △, ×, □ = samples *F, G, H*; ■ = solid disk.[74]

ured value for a fused deposit at 1200°C (2200°F) of about 5×10^{-3} cal/(cm)(s)(°C), compares with about 100×10^{-3} cal/(cm)(s)(°C) for steel, or approximately one-twentieth the conductance.

Others[75] measuring the coefficient of thermal conductivity have demonstrated with Nazarovo coal of unspecified ash composition that the coefficient increases almost linearly with temperature, from roughly 0.07 kcal/(m)(hr)(K) at 550 K to 0.17 at 1000 K. The Australian brown coals investigated by Mulcahy were somewhat higher. As might be expected, the coefficient of thermal conductivity depends on the density of the deposit, increasing steeply as the density is greater.

Emissivity of coal ash depends on the physical and the chemical characteristics of the surface. For example, high-Fe_2O_3 slags in an oxidizing environment are likely to have a surface of hematite, whereas low-Fe_2O_3, high-CaO slags may have a surface that is essentially a glass with no solid phase present. Hence the measurement of emissivity can be troublesome. Figure 21.32 shows that emissivity decreases with temperature and that coal ash does not behave strictly as a "gray" body.[76] For all practical purposes, the emissivity can be considered to be about 0.5 at furnace temperatures. A similar investigation in the Soviet Union[77] showed the emissivity to range between about 0.7 and 0.9 at 500 K to between 0.65 and 0.75 at 950 K for a wide variety of Donetsk coals.

Since both emissivity and the logarithm of thermal conductance vary linearly with the logarithm of particle size, the relationship between them is also linear. Figure 21.33 demonstrates that this is indeed so at 500°C. This relationship apparently does not depend on composition, thermal history, or the nature of the particles. It cannot properly be extrapolated beyond the limits shown. Within these limits, however, the following relation presumably holds:

$$\text{total emittance} = 0.51 \log k + 2.44 \qquad (21.45)$$

where k is expressed in cal/(cm)(s)(°C)and the temperature is 500°C.

21.3.2.4 *Heat Capacity*

Establishing a heat balance on large boiler plants calls for an allowance for the thermal energy carried off by slag, bottom ash, and fly ash. Since slag and bottom ash usually leave the boiler furnace at temperatures approaching the furnace temperature, the total amount of heat lost can be significant. For slag tap furnaces, where half the ash in coal

[75] A. A. Abryutin and E. S. Karasina, "Thermal Conductivity and Thermal Resistance of Ash Deposits in Boiler Furnaces," *Teploenergetika*, **17**(12), 36–39 (1970).

[76] P. J. Street and C. S. Twamley, "Fuel Particle Emissivities," *J. Inst. Fuel*, **44**(367), 477–478 (August–September 1971).

[77] A. A. Abryutin and E. S. Karasina, "Emissivity Factor and Absorptivity of Ash Deposits in Boiler Furnaces," *Teploenergetika*, **17**(10), 43–46 (1970).

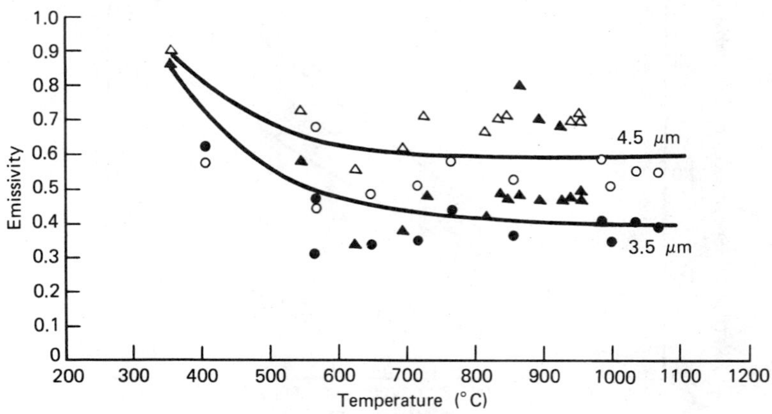

Figure 21.32 Variation of coal ash emissivity with temperature: △, ▲ = Rawdon (CRC 902), ○, ● = Ackton Hall (CRC 502).[76]

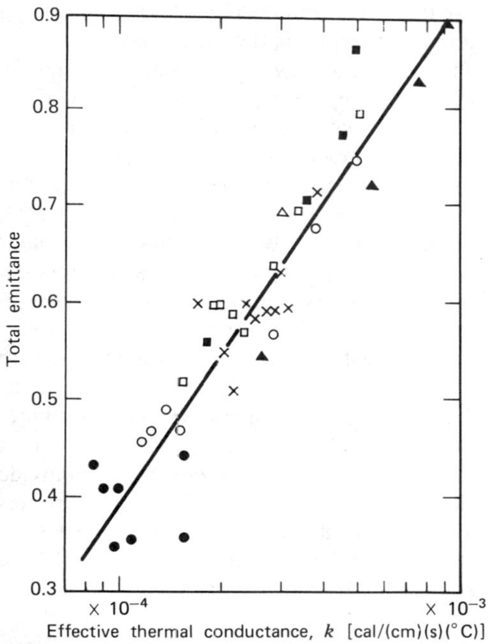

Figure 21.33 Variation of thermal conductance with thermal emittance at surface temperature 500°C: □ = furnace deposits, △ = primary superheater deposits, × = precipitator and cyclone dusts, ● = laboratory ashes, ○ = "colorless" synthetic slags, ▲ = "green" synthetic slags, ■ = "amber" synthetic slags.[78]

may be discharged as slag at 1200°C, the corresponding heat loss would be 1% for a coal containing about 22% ash. It is generally assumed that molten slag when tapped will represent a loss of roughly 650–800 Btu/lb, but this does not include additional losses that may occur from the latent heat of fusion of solid phases in the slag.

Hoy lists earlier Russian data on the heat content of coal ash slags, as shown in Table 21.9.

21.4 BEHAVIOR OF ASH IN BOILER FURNACES

As a rough estimate, some 40 million tons of coal ash passed through central station boiler furnaces in the United States during 1975. This noncombustible residue from steam generation is a major cause of outages, and it is

[78] "Radiant Emittance and Thermal Conductance of Boiler Deposits," *Coal Res. CSIRO,* No. 42, 6–8 (December 1970).

Table 21.9 Heat Content of Coal Ash Slags[70]

Temperature (°C)	Heat Content (Btu/lb)[a]
1100	475
1200	550
1300	630
1400	710
1500	790
1600	865
1700	935

[a] These figures should be used with caution.

the limiting factor in establishing boiler furnace volume for a given steam output. Since almost all coal for generating electricity is now burned in pulverized form, clinkering is no longer a problem, but it has been replaced by deleterious effects on heat flux and on gas flow patterns caused by ash deposits on wall tubes, superheaters, reheaters, economizers, and air heaters. Coal ash also leads to external corrosion, but although interaction between the inorganic material in coal and the boiler might be a serious problem because the alkalis and the sulfur in coal can combine under easily attained conditions, significant wastage is the exception in properly designed and operated coal-fired boilers. Industrial consumption of coal for steam generation in chain grate, underfeed, and spreader stokers amounts to about a tenth of that in central station power plants; although ash is troublesome here, it does not lead to problems of the magnitude faced by the public utilities, with their larger furnaces, higher heat release rates, and greater demand for reliability.

21.4.1 Pulverized-Coal-Fired Furnaces

No major changes in pulverized-coal firing have occurred over the past two decades except that boiler furnaces continue to increase in size.[79] Of 41 contracts for boiler furnaces let in 1975 by investor-owned electric utilities, half were rated at 600 MW or more. The largest was 1281 MW. Supercritical pressure remains popular, with eight of these units operating at pressures of 3500 psi

[79] "1976 Annual Plant Design Report," *Power,* 32 pp. (November 1976).

or higher. Nevertheless, for reasons of reliability, the trend in pressure is moving downward toward 2600 psi. Steam temperature, both superheat and reheat, is almost universally 538/538°C (1000°F/1000°F). Two stations now are using the two boilers per turbine approach to obtain maximum reliability, allowing continued operation of the turbine at half-load in case an outage occurs with one boiler. Mechanical problems continue to be troublesome, but such difficulties related to coal ash as excessive wall slagging and superheater plugging still lead to too many unscheduled outages. External corrosion also causes outages, but 85% of all unscheduled outages result from other causes.

Despite engineering advances aimed at increasing reliability and lowering heat rates, our basic knowledge of the behavior of coal ash in boiler furnaces is still not widely applied. There have been only gradual advances in our understanding of the ways in which ash and slag particles become attached to boiler surfaces. For example, we have yet to determine experimentally which components of the mineral matter in coal are mainly responsible for fouling. Yet significant advances have been made in coping with ash and slag problems although much work remains to be done. Indexes have been developed that relate fouling troubles to the sodium content of coal but these are qualitative at best. Efforts are now under way to identify the mineral species in coal that enhance fouling, but these field and laboratory investigations will not be completed until the 1980s.

21.4.1.1 *Formation of Fireside Deposits*

Accumulation of ash on wall tubes and on convection tube banks occurs by a variety of processes, including both diffusion and impaction. Diffusion may involve both molecular size particles that essentially follow gas laws, and larger particles that become involved in turbulent flow. Impaction accounts for deposits of much larger particles, with their momentum received from the main gas stream sufficient to penetrate boundary layers or turbulence near a surface. None of these processes has been investigated satisfactorily by experimental methods under actual furnace conditions. Hence essentially no progress has yet been made in minimizing fouling by applying theoretical concepts. The advances that have been made depend on empirical design parameters developed by experience with operating units and on large-scale flow models.

Much of the work done on fouling has involved the alkalis (see Section 21.1.3), since sodium has been identified as a major contributor to objectionable ash deposits. Silicon also has been found to contribute to fouling, by the formation of a cloud of SiO_2, through heating silicate minerals in admixture with carbon at temperatures as high as 1750°C.[80] At this temperature carbon reduces SiO_2 to SiO, which has a much higher vapor pressure than SiO_2. This vaporized monoxide then reverts to SiO_2 as oxygen becomes available from excess combustion air. The result is the submicrometer particles of SiO_2 sometimes found in inner layers of furnace deposits.

The growing importance of low-rank Western coals in the United States has led to increased attention to the fouling characteristics of lignite. Because lignite ash often contains appreciable amounts of CaO and Na_2O, the ash has a low fusion temperature, leading to serious fouling problems. As a result, boiler furnaces for lignite must be larger, tube spacing must be more generous, and more wall and soot blowers must be provided. Larger induced-draft fans also must be available to allow for the added draft loss caused by deposits in the convective tube banks, compared with furnaces burning bituminous coal.

Table 21.10 illustrates how the composition of the deposits from lignite may change at varying locations in a boiler circuit.[81] Most noticeable is the pronounced gain in both Na_2O and K_2O on the reheater, apparently by condensation, explaining in large part why this ash caused serious fouling problems in the experiment. The high CaO content of the coal ash contributed to this problem by forming $CaSO_4$, which can serve as a bonding agent for fly ash particles. Even if no segre-

[80] M.-Th. Mackowsky, "The Mineral Constitution of Coal, etc.," in *The Mechanism of Corrosion by Fuel Impurities,* H. R. Johnson and D. J. Littler, Eds., Butterworths Scientific Publications, London, 1963, pp. 80–89.

[81] R. J. Wagner, "Fireside Deposits in Lignite-Fired Boilers," *1965 Lignite Symp.,* April 29, 1965.

Table 21.10 Observed Segregation of Ash Constituents in a Lignite-Fired Boiler Furnace at Hoot Lake[81]

Recalculated on SO_3- and TiO_2-free basis.

	Coal Ash	Reheater	Primary Superheater	Fly Ash
SiO_2	29.1	16.4	20.7	20.7
Al_2O_3	15.5	14.1	20.4	12.4
Fe_2O_3	13.8	13.7	13.5	13.8
C_aO	27.4	27.6	28.2	34.3
M_gO	9.9	9.3	6.4	7.3
Na_2O	2.9	12.2	6.4	7.6
K_2O	1.4	6.7	4.4	3.9
	100.0	100.0	100.0	100.0

gation of alkalis had occurred, this ash still would have been troublesome because of bonding by $CaSO_4$. High-lime, high-alkali lignite ash generally is expected to cause problems with fouling, but not enough experience is available to make firm predictions.

Figure 21.34 illustrates three stages of deposit formation on superheater elements. Stage I consists of a uniform thin layer of particles smaller than 10 μm, uniformly distributed on the tube surface, presumably by diffusion and condensation. In Stage II low-melting particles of mineral matter (e.g., feldspars) impinge on the Stage I surface and are captured. These form a strongly adhering deposit that is difficult to remove with soot blowers. In Stage III all particles hitting the sticky surface adhere to it, and there is a rapid accumulation of material. As the figure shows, soot blowing can be expected to remove the Stage III deposit, but the sticky surface of Stage II remains to accumulate additional deposits.

The temperature of the gas stream, hence in part the temperature of the ash particles, strongly influences the rate of deposition of ash. Figure 21.35 shows that the deposition rate is low at gas temperatures below about 1000°C but increases rapidly as the temperature rises. This is in agreement with the general premise that melting to produce a liquid phase begins at about 1000°C (1832F) for components of most coal-ash systems.

21.4.1.2 *Effect of Deposits on Boiler Performance*

As Section 21.3.2.3 pointed out, the thermal conductivity of coal ash slags is a complex function of the stage of melting of the ash particles on a heat-receiving surface, which in turn depends on ash composition and furnace conditions. Generally it is not possible to predict such actions from first principles. Instead, the design of a boiler furnace is based on engineering judgment supported by past experience with burning similar coals in furnaces of the same general geometry. Radical changes in the coal—for example, using lignite in place of bituminous coal—

[82] R. D. Winship and F. Bender, "Ash Deposition Research on Canadian Lignites," *1969 Lignite Symp.*, U.S. Bur. Mines and Univ. of North Dakota, Grand Forks, N.D.

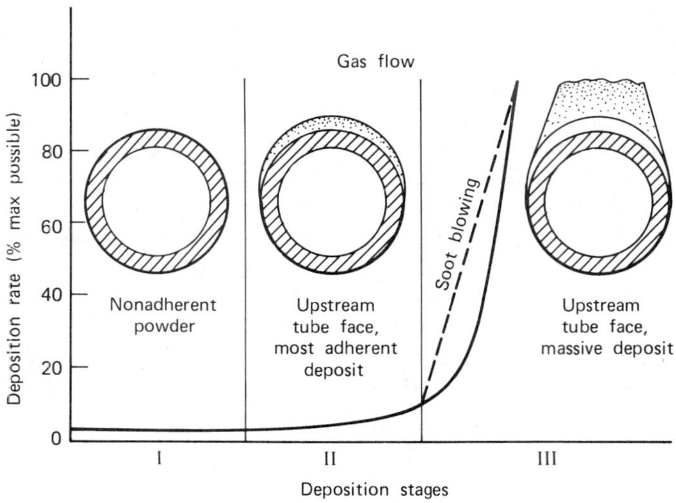

Figure 21.34 Deposits formation with lignite.[82]

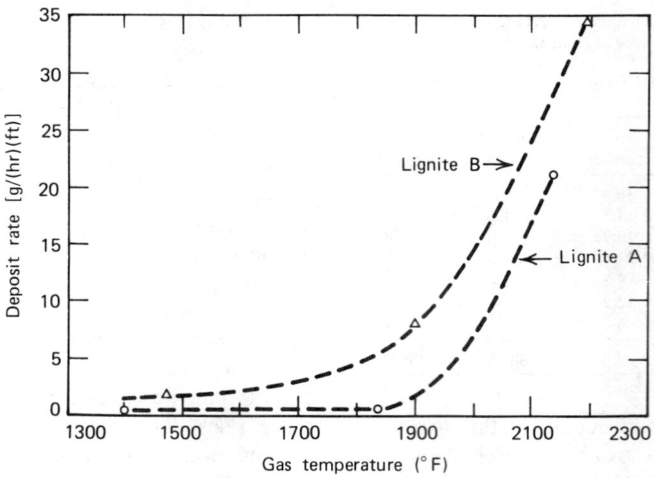

Figure 21.35 Effect of gas temperature on rate of deposit formation with lignite; 4-hr probe exposures.[82]

cause entirely new problems in ash behavior. The cyclone furnace, quite different from pulverized-coal-fired units, required many years of field experience before ash behavior could be predicted.

Even when no new concept is involved, problems with furnace performance can occur because of ash characteristics. Typical is the Keystone experience,[83] where furnace slagging, formation of massive deposits near burners, and slag falls prevented operation at the rated capacity of 900 MW. Even with 50% excess air to lower furnace temperature, availability deteriorated because of fly ash erosion and an imbalance between superheater and reheater surfaces, so that continuous operating levels had to be lowered to 740 and 780 MW in the two units. There were several options to improve availability, with all but one based on changes in the furnace and the boiler circuits. That one, cleaning the coal to improve its quality, is still under study.

By costly redesign of the economizer arrangement and by adding additional superheater surface, availability was improved so that the units now can deliver 855 MW with a boiler efficiency of 89.20%, even with 30% excess air. The main point is that this unit was designed on the basis of information available in 1963, when large steam generators were first being built in the 900–1000 MW size, based on an extrapolation from 500–600 MW design criteria. Mechanical engineers were able to modify that early design sufficiently to compensate for shortcomings in the only available coal, since Keystone fuel comes from a captive mine. No acceptable improvement in coal quality with regard to fouling has yet been achieved here by conventional coal preparation techniques.

Furnace performance may eventually be predicted from basic knowledge of coal characteristics and ash properties.

21.4.1.3 *High-Temperature External Corrosion*

No significant advances in preventing external corrosion, other than by limiting metal temperature to the 540°C level, have been made since the 1940s, when the alkali iron trisulfates were first identified as a major cause of trouble. Sir Harold Hartley's comments at the Marchwood Conference in 1963 still hold: " . . . knowledge of the mineralogical constitution of the fuels, and of the behavior of the minerals during combustion, [is] of vital importance. . . . " It is being recognized now that these are still major problems, which will, it is hoped, be resolved within the next decade. In the meantime, demands for increased efficiency in the use of fuels point to the need for early solution of problems with external corrosion.

Most advances since 1958 have been concerned with the role of sulfur in external corrosion, and mainly with the formation of

[83] J. L. C. Bachofer, Jr., E. R. Cathcart, and E. H. Smith, "Reliability Improvement at the Keystone Generating Station," paper presented at ASME–IEEE Power Generation Conf., Buffalo, N.Y., September 19–22, 1976.

SO_3 at the high levels necessary to form the trisulfates (see Section 21.2.5). Although flame-produced SO_3 seldom exceeds about 40 ppm in large boiler furnaces, the stagnant layer of flue gas beneath slag deposits in contact with catalytically active surfaces can easily reach SO_3 levels approaching 2500 ppm with high-sulfur coal. Even with coals containing as little as 0.3% sulfur, enough SO_3 can be present to form the trisulfates. Coal preparation and chemical treatment of coal to lower the sulfur content will not eliminate metal wastage unless the sulfur is decreased to well below 0.3%. More effective would be the selective removal of sodium and potassium, which is not considered economically feasible at this time. Adjustment of the calcium and magnesium content of the coal ash by coal preparation can have a marked effect on superheater and reheater wastage.[8]

Fundamental studies of corrosion reactions have been aided by the use of radioactive ^{35}S as a tracer, as noted in Section 21.2.5. Using ^{35}S, the Battelle team[84] was able to show that 30 ppm SO_3 added to a gas stream containing O_2 and 5.8% SO_2 increased the reactivity 4.2 times on an Fe_2O_3 surface. An oxidized steel substrate showed a reactivity gain more than twice that of Fe_2O_3; when Na_2SO_4 was applied to an oxidized steel substrate the reactivity ratio increased to 24.9. Hence adding SO_3 in an amount no greater than that in boiler furnace atmospheres increased the rate of reaction between the atmosphere and a Na_2SO_4-coated oxidized steel surface almost 25 times above that which would occur when only SO_2 and O_2 were present.

Figure 21.36 shows the reactivity gain as a function of temperature for a mixture of Fe_2O_3 and Na_2SO_4. The reactivity with 30 ppm SO_3 present again is much greater than that with 5.8% SO_2 alone. Peak reactivity occurred at about 650°C, compared with about 700°C for the well-known "bell-shaped curve." Even at 540°C, there is a distinctly greater formation of complex sulfates with small amounts of SO_3 present initially. This

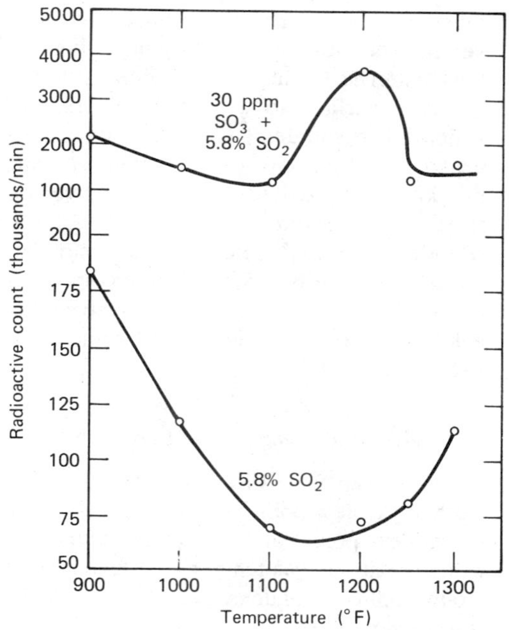

Figure 21.36 Reactivity of SO_2 and SO_3 with a mixture of Fe_2O_3 and Na_2SO_4 as a function of temperature.[84]

work confirmed that SO_3 is much more reactive than $SO_2 + O_2$ on oxidized steel surfaces, particularly when Na_2SO_4 or K_2SO_4 is also present.

Figure 21.37 summarizes the probable sequence of reactions that leads to external corrosion by the trisulfate route, below 650°C, and by pyrosulfates, below 480°C.

In boiler furnaces, metal wastage of wall

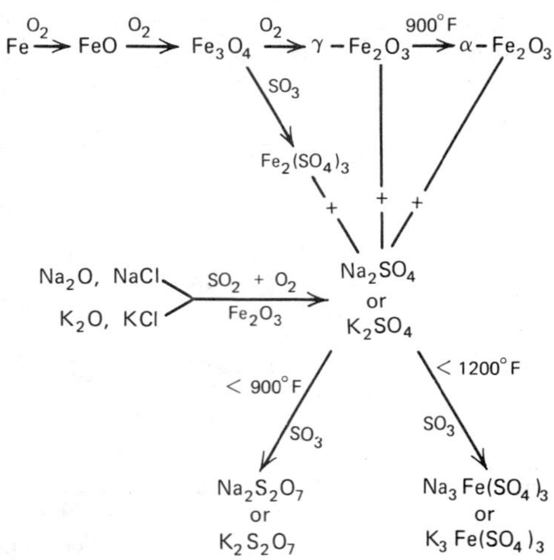

Figure 21.37 Reaction sequence of iron oxides and alkalis with sulfur oxides.[84]

[84] H. H. Krause, A. Levy, and W. T. Reid, "Radioactive Sulfur Oxide Studies of External Corrosion Reactions on Surfaces," ASME Paper No. 68-WA/CD-2, 1968. [Published as a pamphlet by Fuels Division, at Winter annual meeting of ASME in 1976; does not appear in any ASME *Transaction*.]

tubes can be controlled adequately by preventing slag adherence, or by adjusting airflow patterns to dilute the sulfur oxides at the tube surface to a level below that at which the trisulfates can form. With superheater and reheater sections, however, air dilution is not possible and external corrosion remains troublesome. Much research activity recently has been directed toward eliminating wastage in tube banks by finding corrosion-resistant alloys. An economically attractive solution is not yet available but some laminates offer promise.

21.4.1.4 *Low-Temperature Corrosion*

Because even very low levels of SO_3 raise the flue gas dew point appreciably above the water dew point, air heaters must operate with an exit flue gas temperature higher than around 160°C if condensation of sulfuric acid is to be prevented.

Figure 21.38 summarizes a widely accepted relationship between SO_3 level and acid dew point. The experimental points, obtained by measuring the partial pressure of H_2SO_4 at different temperatures, agree closely with calculated values.[85]

The importance of these data relates to the exceedingly low levels of SO_3 required to approach the water dew point. Even with 0.1 ppm SO_3, the acid dew point is still 93°C when only 5% H_2O is present in the flue gas. Chemical reaction to capture SO_3 (e.g., by added limestone or dolomite) or by the presence of slightly reducing flue gas conditions is necessary to achieve such low concentrations of SO_3. Large boiler systems cannot generally reach these levels; therefore an efficiency penalty is accepted by discharging the flue gas to the stack well above the acid dew point. The use of flue gas scrubbers may eliminate this acid dew point problem. It is worth noting, however, that attempts to meet environmental restrictions by the chemical cleaning of coal to reduce the sulfur content to a point below the levels of conventional coal-cleaning systems still are likely to put enough SO_3 into the flue gas to raise the acid dew point significantly.

21.4.1.5 *Composition of Fly Ash*

Fly ash is a heterogeneous mixture of inorganic matter and unburned combustibles, the composition of which is roughly related to the original mineral matter in the coal. This mineral matter, and the inherent ash in the coal, is modified to a varying degree by exposure to temperatures as high as 1650°C for periods probably as long as 50 ms while the inorganic matter passes through the flame front. Interactions between particles continue to occur for the next 2–5 s as collisions take place between the particles while they are passing through the furnace and the associated heat transfer sections of the boiler, with the particle temperature decreasing from 1650°C to the stack temperature of about 160°C.

Generally the chemical composition of fly ash does not differ appreciably from that of

[85] W. T. Reid, *External Corrosion and Deposits: Boilers and Gas Turbines,* Elsevier Publishing Company, New York, 1971, p. 27.

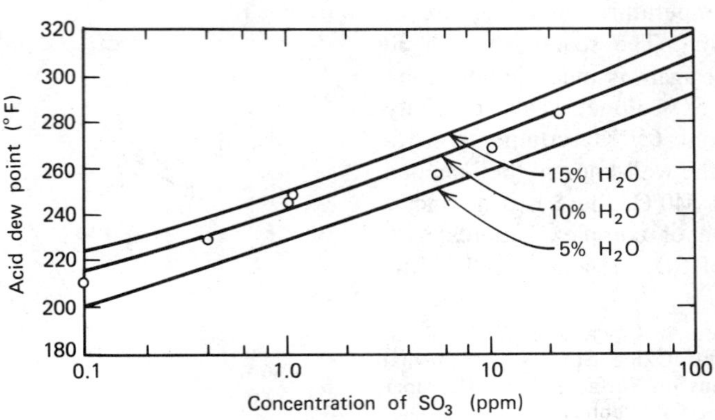

Figure 21.38 Dew point of SO_3 in flue gas[85]: curves, calculated by Müller; open points, experimental data of Lisle and Sensenbaugh, as cited in Ref. 85.

the ash in the coal. Segregation can occur if droplets of mineral matter are captured preferentially by slag deposits already on heat-receiving surfaces, so that there can be a trend for the fluxes Fe_2O_3, CaO, and MgO to be somewhat lower in fly ash than in bottom ash. Alkalis, because of their relatively high volatility, are expected to be somewhat higher in fly ash. But these are mainly surmises, and no procedure exists whereby the composition of fly ash can be predicted based on the composition of the mineral matter in the coal as fired, or the method of burning.

Cenospheres in fly ash pose special problems in ash ponds, where these hollow microspheres of essentially silica-alumina glass float on the surface instead of settling like normal fly ash. There are no major differences in the composition of the cenospheres compared with the original coal ash[86] except that the CaO content may be lower. Figure 21.39 shows that the tendency to form cenospheres increases as the Fe_2O_3 content of the coal ash is greater. This may be related to iron-catalyzed gas evolution; the residual gas within the cenospheres is mainly composed of CO_2 and N_2.

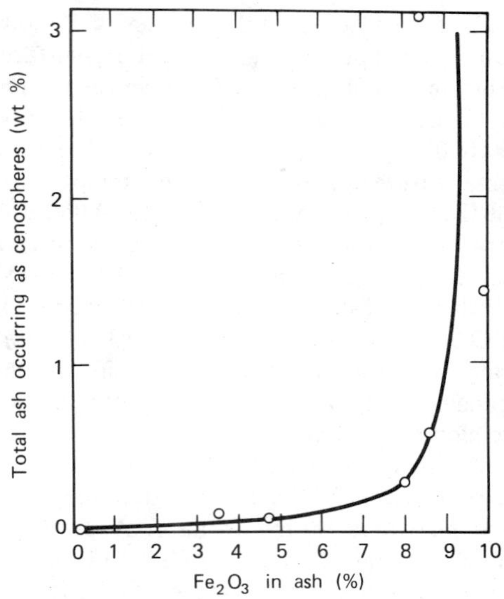

Figure 21.39 Dependence of cenosphere formation on iron oxide in ash.[86]

21.4.1.6 *Effect of Ash on Furnace Design and Performance*

As has been noted many times, a major factor affecting furnace performance in large coal-burning utility steam generators is the inorganic matter in the fuel. Not only does this material deposit on heat transfer surfaces and in flue gas passages, it also provides the reactants and the environment leading to external corrosion. Most problems with ash, however, have to do with its effect on heat transfer by thermally insulating furnace wall tubes and convective tube banks. Accumulation of slag on furnace wall tubes, for example, can decrease the heat transfer rate to the tube surface to one-fourth its normal value, thereby leading to excessively high temperatures for the flue gas leaving the furnace and to inordinately high steam temperature from the superheater. Tilting burners, flue gas recirculation, and attemperation are ways of controlling steam temperature when wall slagging is excessive.

Furnace designs differ among the major boiler builders with respect to such features as corner firing or wall firing, cyclone furnaces, and position of burners, superheaters, and reheaters. Furnace geometry, however, does not change radically. Heat release rates based on the area of heat-receiving surfaces also do not differ to a great extent, since they must be established to keep furnace temperatures below the point at which ash accumulations on furnace walls become troublesome. Elaborate empirical relationships have been established to predict furnace heat absorption; of necessity, however, they have not included the effects of ash and slag on the emissivity and temperatures of the furnace walls,[87] nor have they included the effects of heat transfer through the slag layer.

Each boiler builder has his own criteria to allow for the effects of ash characteristics; these are based on good engineering judgment and accumulated experience with boilers of similar size, and on expected performance. One manufacturer[88] proportions his furnaces for Eastern coals with a cone fusion initial deformation temperature above 1315°C

[86] E. Raask, "Cenospheres in Pulverized-Fuel Ash," *J. Inst. Fuel*, **41,** 339–344 (September 1968).

[87] G. R. Fryling, *Combustion Engineering,* Combustion Engineering, Inc., New York, 1966, p. 6-30.

[88] *Steam. Its Generation and Use,* Babcock and Wilcox Company, New York, 1963, p. 18-5.

to give furnace heat release rates of 100,000 Btu/hr-ft^2 of projected heat-receiving surface. For coals such as those from western Pennsylvania, the heat release rate is down to 90,000 Btu/hr-ft^2. For Midwest coals with a cone fusion initial deformation temperature of 1200°C, it is further lowered to 80,000 Btu/hr-ft^2. And for some Illinois coals with high alkali content, the limit is 43,000 Btu/hr-ft^2. With fuel oil, the furnace heat release rate is 135,000 Btu/hr-ft^2, or more than 3 times the rate with the worst coal, attesting to the penalty that must be paid for the fouling tendency of the ash in coal.

Another broad parameter, also based on experience, is the heat release rate related to the plan area of the furnace. A preferred level is about 1.8×10^6 Btu/hr-ft^2 of furnace cross section near the burners. Badly fouling coals may cause problems even when this value is reduced to 1.5×10^6, and most coals apparently will be troublesome if this figure exceeds about 2.4×10^6. These are very flexible yardsticks; they attest to the difficulties in accurately assessing the role of ash characteristics in furnace design.

The deliberate raising of the heat release rate to increase furnace temperature so that the molten slag accumulated was an answer to many problems early in the history of pulverized-coal firing. But because droplets of slag caused excessive fouling as the hot gases moved through the furnace, furnace heat release rates had to be decreased and the dry bottom furnace largely took over. Nevertheless, cyclone furnaces with a very high recovery of ash as molten slag have been widely used. Predicting slag behavior as a guide to coal selection for cyclone furnaces, based on a maximum viscosity of 250 poises at furnace temperatures, depends on furnace characteristics as well as on the nature of the slag. Maintaining a free flow of slag requires an adequately high gas temperature, which can be calculated by methods devised by BCURA in 1964.[89] Although slagging boiler furnaces are not currently receiving much attention because of high NO_x emission, their ability to use low-fusion Western coals could make them more attractive in the future.

[89] A. G. Roberts and D. M. Wilkins, "The Slagging Behavior of Coals in Cyclone-fired Boilers. Part I. A Method of Predicting Slagging Behavior," *J. Inst. Fuel,* **37**(285), 449–454 (October 1964).

21.4.1.7 *Additives*

The role of additives in the operation of oil-fired boiler furnaces is well established. It is known that controlled amounts of MgO added to the fuel oil minimize problems with vanadium, decrease the catalytic formation of SO_3 on boiler surfaces, prevent formation of corrosive deposits, increase the sintering temperature of deposits, change the emissivity of furnace surfaces, and neutralize acids normally condensing on cool surfaces, such as in air preheaters. In addition, some additives contain manganese compounds as a combustion catalyst.

The ash content of most fuel oils is less than 0.1%; hence the addition of modest amounts of materials such as MgO can play a significant role in oil-fired boilers. With coal firing, however, where the ash content is seldom less than 10%, reasonably small amounts of additives are ineffective because the burden of solids in the flue gas stream is already roughly 100 times that of fuel oil. Hence although additives are available on the market to modify coal ash fusion characteristics, they are not widely used because the quantity necessary for effective action can be excessive.

Some idea of the amount of additive required can be judged from the effect on silica percentage, shown to be a good measure of coal ash slag viscosity. Since viscosity increases with an increase in silica percentage (see Section 21.3.1), how much silica must be added to a typical coal ash slag to increase its viscosity from 100 to 1000 poises at 1430°C? The silica percentage for 100 poises is 67; for 1000 poises it is 82. To make this change requires adding 72 lb of SiO_2 per 100 lb of coal ash, or for a coal with 10% ash, about 7% of the weight of the coal—140 lb of SiO_2—must be added per ton of coal.

To lower the viscosity, fluxes such as limestone or iron ore can be added. A coal ash slag having a viscosity of 500 poises at 1430°C with a silica percentage of 78 can be fluxed with limestone to achieve a viscosity of 50 poises, well within the fluidity range for easy slag removal, by lowering the silica percentage to 63. This would require adding 17 lb of CaO, or about 30 lb of $CaCO_3$ per 100 lb of coal ash. Or, again for a coal with 10% ash, this would mean adding 60 lb of limestone per ton of coal. It is evident that much less additive is required to lower viscosity than to increase it.

The addition of fluxes to decrease the viscosity of coal ash slags was tried in the 1930s when slag tap furnaces were being installed, but it was found preferable to switch to a coal with a lower silica percentage. Recently, with the substitution of low-sulfur coals for high-sulfur Eastern bituminous coals, operators of cyclone furnaces have had trouble maintaining adequate slag fluidity, since low-sulfur coals may have a higher silica percentage. One large utility has mixed limestone with the low-sulfur coal burned in its cyclone furnaces with fair success. Ease of tapping the slag was critically dependent on the quantity of limestone added, but addition of only 6% limestone enabled operation at 75% of full load with satisfactory slag flow. The theoretical basis of these tests was sound, but much more experience will be required before all the operational problems are solved for fluxing-type additives.

There was a flurry of activity in England in the late 1960s concerning the use of copper oxychloride to control slag formation in coal-fired boiler furnaces.[90,91] It was observed that very small amounts of copper oxychloride added to the coal being burned at Aberthaw Power Station of the CEGB modified the physical nature of slag deposits on heat-receiving surfaces, so that the usual hard, glassy deposits were converted into a soft, friable deposit easily removed by soot blowing. Cost of the treatment was low, about 0.05 pence per ton.

A similar addition of copper oxychloride by the Electricity Supply Board, Ireland, with peat as the fuel, also reduced external fouling. Addition of copper sulfate produced comparable effects.

Despite these promising early results, the use of copper additives was not extended, presumably because such small amounts of additives did not provide reliable improvement in slag characteristics.

This action of trace amounts of copper in coal ash slags is not readily explained. The copper may provide nuclei in the slag to enhance devitrification as the slag cools, thereby converting the slag from a glass to a brittle nonhomogeneous solid, but this is purely speculation. Further investigations should be made into the role of copper and other heavy metals in the devitrification of coal ash slags and in the possible modifications of slag properties, even though this would raise environmental issues, which would also have to be considered.

21.5 FLUIDIZED–BED COMBUSTION

The only significant new method of burning coal on a large scale devised over the past 30 years is fluidized-bed combustion (FBC) (see Chapter 20, Sections 20.1.2–20.1.4). Using sound basic principles in combustion and ash control, FBC has gained most prominence because it also offers control of SO_2 emission without external systems such as flue gas scrubbers. And since FBC operates at fuel bed temperatures below 1000°C, problems with ash are expected to be minimal and emission of NO_x will be limited to that from fuel nitrogen. Disposal of solid wastes will be troublesome until a practical regeneration system is devised.

21.5.1 Ash Tolerance

Bench and laboratory-scale studies have dominated efforts to develop FBC; pilot and demonstration plants in sizes greater than about 0.5 MW have only recently been planned, built, and operated. As a result, the behavior of ash in FBC systems of practical size is only now being understood. Most early predictions of ash tolerance in FBC have been unreliable because of the lack of knowledge of mixing effects, temperature variations, oxygen activity levels, and interactions of coal ash with limestone in large-scale systems. Yet in England, more than 20,000 hours of testing has been accumulated with FBC units having thermal inputs up to 3.5 MW without turning up any critical operational problems. The Renfrew boiler,[92]

[90] W. G. James and A. H. Fisher, "A New Fuel Additive for Coal-Fired Boilers," *J. Inst. Fuel,* **40**(315), 170–172 (April 1967).

[91] E. F. O'Connor, "Use of Copper Salts for Inhibition of External Fouling of Peat-Fired Boilers," *J. Inst. Fuel,* **43**(358), 449–450 (November 1970).

[92] Unpublished results of B & W Ltd., Great Britain.

rated at 46,000 lb of steam per hour, had operated for more than 4500 hours by mid-1977.

Conceptual designs of FBC units generally assume that coal ash will not agglomerate at temperatures lower than 1000°C and that fuel bed temperatures will not exceed this level, but it may be that neither premise is sound. Coals containing more than 0.5% Na_2O can be expected to be troublesome because they can produce strongly sintered ash at temperatures as low as 760°C, as demonstrated by the Barnhart and Williams sintering test.[93] Also, the assumption that the maximum temperature of individual particles in an FBC fuel bed will not exceed the average temperature may be questioned. Hence even in FBC beds operated at temperatures as low as 815°C, which is about the lowest temperature suitable for acceptable burning rates, the ash in the coal may sinter to produce undesirable agglomeration of ash particles. Such problems have not proved to be serious yet. Nonetheless, our understanding of ash behavior in fluidized fuel beds is still not good enough to assure freedom from ash problems in the future with large FBC systems.

The quantity of ash in coals burned in FBC units is not an important parameter, since fluidized beds consist mostly of noncombustible material so that the coal represents only a small percentage of the entire bed. For use in the FBC system, coals containing as much as 50% ash are acceptable.

At present, ash characteristics are not dominating factors in the development of successful FBC steam generators. Operating experience with the 30-MW Rivesville, West Virginia, unit and with the 200-MW units now being considered will be necessary to learn whether ash properties in FBC systems are important compared to pulverized-coal-fired boiler furnaces.

21.5.2 Erosion-Corrosion

Successful FBC units must withstand erosion and corrosion of heat transfer surfaces immersed in the fluidized bed. To attain and use the high heat transfer coefficients possible with FBC, horizontal tube bundles are preferred, but movement of the hot bed particles past these surfaces at velocities up to about 4 m/s may cause problems directly by erosion, and indirectly by removal of the protective oxide scale on the tube surface, thereby accelerating metal loss. Erosion of immersed tube bundles has been a problem in an experimental pressurized FBC unit at velocities no greater than 10 ft/s.[94] It has not been troublesome universally. How the coal ash is involved here, compared with the much greater proportion of limestone in the bed, has not been clearly defined. Nevertheless, coal ash is a potential source of chemically reactive species that can cause problems with immersed heat transfer surfaces.

The role of the mineral matter in coal in causing metal wastage in FBC has been considered in great detail, but more on a theoretical than an experimental basis so far. The problem arises because heat-absorbing surfaces must be immersed in the fluidized bed for steam generation, superheating, reheating, and economizer heating, as well as for control of temperature in the combustion zone. Although it had been assumed earlier that the low fuel bed temperatures would eliminate fireside problems with immersed tube banks, it appears from early test data that some tube materials are attacked rapidly.[95] The corrosion problem here is complex and poorly understood.

Alkali iron trisulfates probably are not involved, as in conventional pulverized-coal-fired boilers, because the great excess of CaO and MgO in FBC essentially eliminates SO_3 in the furnace atmosphere. Chlorides in coal may be troublesome, but most coals in the United States contain little or no chlorine. Low oxygen activity in FBC fuel beds can lead to new problems because the usual protective oxide films may not develop on metal surfaces, and alternately reducing and oxidizing conditions bring up a wholly new

[93] F. G. Ely and D. H. Barnhart, in *Chemistry of Coal Utilization,* Supplementary Volume, H. H. Lowry, Ed., John Wiley and Sons, New York, 1963, p. 843.

[94] M. S. Nutkis, "Operation and Performance of the Pressurized FBC Miniplant," *Proc., 4th Int. Conf. Fluid-Bed Comb.,* U.S. Energy Research and Development Administration, Washington, D.C., December 9–11, 1975, pp. 221–238.

[95] J. Stringer and S. Ehrlich, "High Temperature Corrosion in Fluidized Bed Combustors," ASME Paper No. 76-WA/CS-4, 1976, 11 pp. [Published as a pamphlet by Fuels Division, at Winter annual meeting of ASME in 1976; does not appear in any ASME transaction.]

environment for heat-receiving tube banks. Coal ash will contribute to these problems, but in a manner not yet predictable.

21.5.3 Interaction of Ash with Limestone or Dolomite

Ash tolerance in FBC systems is complicated further by the great excess of CaO and MgO surrounding coal particles in FBC as the coal burns and the ash is released. No method is available yet to predict how this will affect ash agglomeration. Both CaO and MgO are potent fluxes for coal ash, lowering the viscosity markedly until the silica percentage drops below about 30, at which point the fluxed coal ash slag has a viscosity of one poise at 1425°C. At higher CaO levels, with more than about 50% CaO in the molten slag, "apparent" viscosity increases, probably because of the presence of an unreacted solid phase of CaO.

It is not clear how extensively the available CaO and MgO in a fluidized bed will react with the coal ash as the ash is released by combustion. The fluxing ability of CaO and MgO appears to be a potential source of trouble. On the other hand, the great surface area of the calcined bed and the high preponderance of CaO and MgO compared to the coal ash, in a weight ratio as great as 200 : 1, may lead to the elimination of coal ash as a troublesome component in FBC. Again, not until larger FBC steam generators are available can the potential problems with coal ash be identified and evaluated.

21.5.4 Agglomerating Fluidized-Bed Systems

Most FBC systems are designed to operate with a fuel bed temperature below 1000°C, in part to permit capture of SO_2 but also to minimize problems with ash fusion. Some FBC units, such as those for gasification, are intended to have fuel bed temperatures up to about 1200°C so that the coal ash is deliberately agglomerated to form solidified slag particles of manageable size. These hot balls of solidified slag then can be physically moved from the combustor to a gasification bed, where the sensible heat in the slag provides the thermal input for the carbon-steam reaction.

The usual parameters for evaluating ash fusion characteristics—silica percentage, temperature of critical viscosity, and ASTM cone fusion temperatures—will be applicable here in choosing coals of specified ash composition to control agglomeration at fuel bed temperatures. It is not possible to predict the behavior of coal ash in such unconventional FBC systems. That ability will come only after extensive trials in agglomerating fluidized fuel beds.

21.6 BEHAVIOR OF ASH DURING GASIFICATION

Ash problems in gasification have been troublesome since the earliest days of the fixed-bed gas producers, which used a variety of grate designs to handle and remove clinkers. These problems were comparable to clinkering troubles in stokers, and they were controlled by limiting fuel bed temperatures or by choosing high ash fusion coals.

Slagging gasifiers, both fixed bed and suspension, permit higher operating temperatures and greater throughput than is obtainable when ash must be kept from melting. Hoy and his associates have summarized the factors involved in good slagging practice in an experimental fixed-bed gasifier.[70] Except for the reducing conditions in a gasifier, slag behavior is much like that in a slag tap, pulverized-coal-fired boiler furnace, and the same parameters of ash composition and viscosity apply. A possible difference is the limited retention of iron compounds in the resultant slag, since the reducing conditions in a gasifier may lead to half the iron oxides normally in slag being converted to metallic iron, thereby lost to the slag as a flux.

Slag viscosity for good tapping normally must be less than 250 poises for cyclone boiler furnaces, but a viscosity less than 50 poises seems necessary for a slagging fixed-bed gasifier. Ease of tapping depends on tap hole location and configuration, as well as on ash composition, temperature, and slag volume.

Coal ash also has a deleterious effect on refractory linings used in some gasifiers.[96] High-alumina castable linings and preformed

[96] N. S. Raymon and L. Y. Sadler, II, *Refractory Lining Materials for Coal Gasifiers,* U.S. Bur. Mines Inf. Circ. No. 8721, 1976, 22 pp.

alumina shapes have been used, but attack by potassium compounds in the coal mineral matter has been observed. Broadly, resistance to alkalis is inversely proportional to the alumina content of the refractory. In general, alkali-refractory reactions will be troublesome.

One early experimental slagging gasifier, designed to operate at pressures up to 600 psi, preferably with lignite and lignite char, demonstrated the importance of the viscosity-temperature relationships of the coal ash slag.[97] The ideal slag should behave as a Newtonian fluid over a wide temperature range and should have a relatively low viscosity. In these tests, fluxing ash from Disco char with blast furnace slag in the ratio of 1 : 3 yielded a slag containing 4.1% Fe_2O_3, 29.8% CaO, and 5.8% MgO. This slag was essentially Newtonian with a viscosity at 1425°C of 7 poises. Without the flux, ash from the Disco char contained 18.5% Fe_2O_3, 3.3% CaO, and 1.0% MgO, and had a viscosity at 1425°C of 110 poises.

[97] G. H. Gronhovd, A. E. Harak, W. R. Kube, and W. H. Oppelt, *Design and Initial Operation of a Slagging, Fixed Bed, Pressure Gasification Pilot Plant,* U.S. Bur. Mines Rep. Invest. No. 6085, 1962, 50 pp.

Figure 21.40*a* shows that both slags, curves *A* and *B*, were essentially Newtonian even down to 1150°C (~ 2100°F). But the much lower viscosity of slag *B* permitted more satisfactory operation.

The importance of Newtonian flow was indicated when lignite and lignite char were gasified. Here, as shown in Fig. 21.40*b*, Newtonian behavior occurred over limited temperature intervals. The rapid rise in "apparent" viscosity as the slag cooled indicated that the fluid slag was changing into a pseudoplastic solid. Even when these lignite ashes were fluxed with blast furnace slag, Curves *D* and *F*, with a viscosity at 1400°C (2550°F) of 6.5 and 3.5 poises, respectively, satisfactory operation was not possible because the

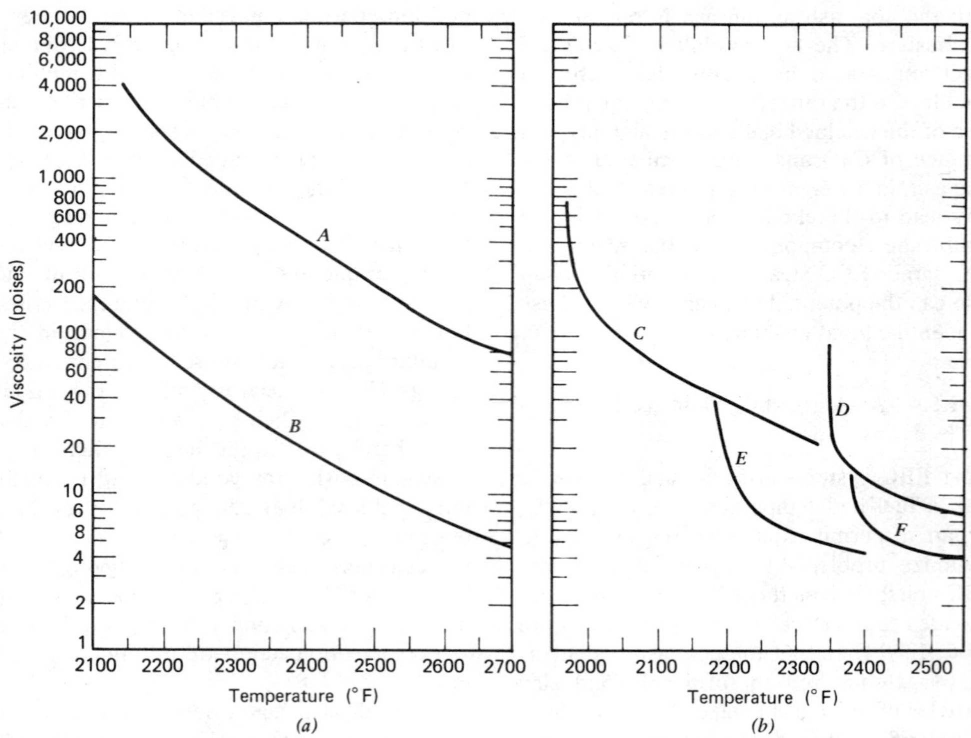

Figure 21.40 Measured viscosity-temperature relationships for various coal ashes and for mixtures of coal ashes and blast furnace slag used in a slagging gasifier. Curve *A*, Disco ash; curve *B*, Disco ash + 3 lb of blast furnace slag per pound of Disco ash; curve *C*, Baukol-Noonan lignite ash + 3 lb of blast furnace slag per pound of lignite ash; curve *E*, Dickinson lignite char ash; curve *F*, Dickinson lignite char ash + 3 lb of blast furnace slag per pound of lignite ash.[97]

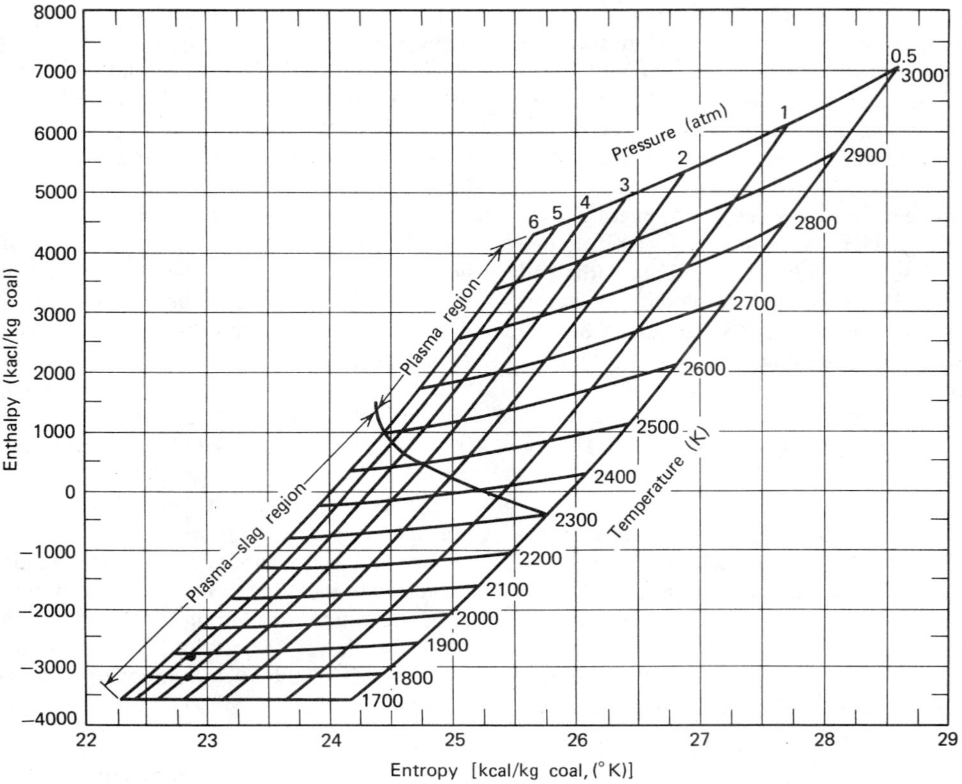

Figure 21.41 Mollier diagram for Belmont County, Ohio, Pittsburgh Seam, No. 8 bituminous coal, 5% excess air, 0.7 g-mole K_2CO_3/per kilogram of coal.[106]

slag froze in the taphole. Such behavior can be explained on the basis of slag composition, since the lignite ash already contained 19.1% and the lignite char 24.9% CaO + MgO. Adding blast furnace slag high in CaO led to an excessive quantity of CaO + MgO in the slag so that a solid phase separated at about 1300°C (2400°F), causing the slag to freeze.

The important consideration here is that experimental verification demonstrated that successful slag tap operation will depend on having a slag with a low temperature of critical viscosity (T_{cv}). Viscosity is important in establishing the *rate* of flow, but T_{cv} is equally significant in determining whether the slag will continue to flow as it cools on leaving the hearth.

More recent work by the same investigators[98] showed that initiating the flow of slag through the taphole was critical on startup. But once begun with the aid of additional blast furnace slag, continuous slag flow could usually be maintained with lignite ash alone. Recycling product slag to increase the slag burden was tried and rejected. Slag flow rate during operation was variable, particularly when blast furnace slag was used as a flux, presumably because of poor mixing in the fuel bed. Few tests lasted more than 10 hr because slag eventually froze in the taphole; attempts to devise a mechanical poker or to heat the taphole were only partly successful. Larger gasifiers with a more favorable surface:volume ratio would be less troubled by freezing slag.

The Disco ash contained 15.1% Fe_2O_3, and the lignite ashes contained 6.7 and 6.4% Fe_2O_3. This led to problems with the formation of metallic iron under the highly reducing conditions in the gasifier, about a fourth of the Fe_2O_3 in the feed being converted to iron, which accumulated on the hearth bottom in the taphole area. Iron oxide is a flux for coal

[98] G. H. Gronhovd, A. E. Harak, M. M. Fegley, and D. E. Severson, *Slagging Fixed-Bed Gasification of North Dakota Lignite at Pressures to 400 psig,* U.S. Bur. Mines Rep. Invest. No. 7408, 1970, 40 pp.

ash slags; hence any reduction to metallic iron represents a loss of flux and an increase in slag viscosity. This in turn could mean that slagging gasifiers using coals with low-viscosity ash attributable to high Fe_2O_3 content would need to have CaO added as a flux in order to tap satisfactorily.

Problems with the refractory were severe in this gasifier. Several types of refractory were tried, but even SiC could not withstand flowing slag in the taphole vicinity. A water-cooled taphole eliminated slag attack but led to problems in tapping.

21.7 BEHAVIOR OF ASH IN MHD COMBUSTORS

Interest in MHD energy conversion systems has varied widely over the past quarter-century. The expected advantages of a high overall thermal efficiency when teamed with a conventional steam cycle are balanced against extremely difficult materials, design, and operational problems associated with combustion at temperatures as high as 2800°C. Because of shortages of natural gas and clean petroleum-based fuels, and since MHD performance improves as the carbon/hydrogen ratio of the hot gas stream increases, coal is now the preferred energy source for MHD generators. Coal combustion under MHD conditions poses no great difficulties, but the mineral matter in coal leads to new slag behavior patterns not previously faced in lower temperature coal-burning systems.

Coal ash in MHD channels modifies performance in two ways: by a change in the electrical insulation characteristics of the MHD channel, evidenced by changes in the breakdown voltage between adjacent electrode modules in the presence of slag; and by decreasing the near-electrode voltage drop so as to increase the size and number of arcs, thereby increasing the total area of arc spots in the presence of slag and hence minimizing electrode erosion.[99] Thus coal ash is not necessarily an objectionable constituent in MHD systems. Using char in place of coal, which is advantageous because of the still higher carbon/hydrogen ratio, will influence MHD generator design but will not affect the ash behavior appreciably. Other coal-derived fuels with low ash content would behave differently. A two-stage, cyclone-type combustor burning char, with removal of a large proportion of the ash in the char as molten slag, has been proposed as an MHD combustor.[100] Although a strong case was made by Young et al. and by others for a low-ash system, it is not certain that coal ash slag is really objectionable in an MHD channel; hence the current viewpoint of designing combustors and channels to benefit from any improvement in electrical characteristics when coal-ash slag is present.

Effects of coal ash slag in MHD systems are currently receiving much attention. No fewer than eight papers related to slag behavior were included in the 1975 International Conference on Magnetohydrodynamic Electrical Power Generation,[101] with another eight papers in the 1976 Symposium on Engineering Aspects of MHD.[102]

21.7.1 Rate of Slag Accumulation

Two approaches have been used in dealing with slag in MHD systems: use of the slag as a protective coating on the electrode walls, and removal of as much of the slag as possible in the combustor—the "cyclone" technique—so that a hot refractory-lined channel can minimize condensation of slag. Materials problems are severe with the hot procedure, not only from the slag but from the potassium seed as well. Oxides such as Y_2O_3, MgO, $SrZrO_3$, and Y_2O_3-stabilized ZrO_2 have been

[99] A. Ye. Sheindlin, *The Influence of Ash on the Operation of an MHD Generator Channel in the U-O$_2$ Facility,* Nat. Acad. Sci., U.S.S.R., Moscow, 1974 (ERDA-66 translation).

[100] W. E. Young, S. Way, T. C. Tsu, D. Q. Hoover, and N. P. Cochran, "Energy Systems. The MHD Combination," *Mech. Eng.*, **89**, 49–57 (November 1967).

[101] *6th Int. Conf. Magnetohydrodynamic Electrical Power Generation,* Washington, D.C., June 9–13, 1975. Energy Research and Development Administration, National Bureau of Standards, Office of Naval Research, Electric Power Research Institute, in cooperation with International Atomic Energy Agency and the Nuclear Energy Agency of the Organization for Economic Cooperation and Development.

[102] 15th Symp. Engineering Aspects of Magnetohydrodynamics, Philadelphia, May 24–26, 1976, University of Pennsylvania. [Refers to a series of eight papers.]

found most satisfactory at temperatures up to 1700°C but fail rapidly at 1800°C.[103]

Because of such materials problems, attributable largely to the slag, much attention is being given to "cold" channel walls, where the gas side surface is made of water-cooled metal parts, typically copper or aluminum, with refractory spacers between adjacent modules.

An experimental program with cooled electrodes,[104] burning slurries of coal or ash with a liquid fuel such as methanol, has demonstrated slag deposition rate with a synthetic ash, a Pittsburgh coal, and a Montana coal. The channel in which these tests were made had water-cooled transversely grooved copper surfaces on one side wall and the top wall, maintained at 125°C. The bottom wall was grooved Type 304 stainless steel, with a surface temperature of about 750°C. The other side wall was smooth magnesia with a surface temperature of about 1600°C. Grooves in the metal surfaces were packed with castable ceramic to provide a bond for the molten slag.

Accumulation of slag on these channel walls required 15–20 min for total slag equilibrium, with coverage of the walls, by slag, varying from 75 to 100%. These slags appeared to be highly fluid, caused in part by the fluxing action of the added potassium seed. Ripples in the slag surface induced by the high-velocity gas stream varied and seemed to affect slag transport, but their formation, as well as occasional streams of molten slag with much more mobility than the bulk slag deposit, was not predictable. Ceramic-filled grooves on cooled metal electrodes were important in slag attachment, since none of the slags adhered to surfaces colder than 800°C. Hence the design of the wall surface, the life of ceramic inserts, and interactions between slag and ceramic will all be important in achieving slag-coated MHD channels with practical lifetimes.

[103] B. R. Rossing, J. A. Dilmore, and H. D. Smith, "Slag-Refractory Behavior Under Hot Wall Generator Conditions," *Proc. 15th Symp. Eng. Aspects of MHD,* Philadelphia, May 24–26, 1976, Paper No. 1.2A.

[104] J. K. Koester, M. E. Rodgers, and R. H. Eustis, *In-Channel Observations on Coal Slag,* Stanford Univ., EPRI Contract No. RP 468-1, 1976.

Reaction between coal ash and added potassium seed has received considerable attention, both because the flow characteristics of the slag are changed by the added flux and because the captured potassium is not available in the plasma phase to improve electrical conductivity. Unless a high degree of rejection of seed by slag is achieved in the combustor, keeping seed losses to an acceptable level in coal-fired systems is expected to be a major problem.[105]

21.7.2 Electrical Characteristics

During combustion in an MHD system, the temperature is high enough to ensure that much of the ash is vaporized, later condensing to form a liquid phase. This condensation and the reaction with potassium added as seed may be more important to the operation of MHD processes than the thermodynamic and physical properties of the combustion products. The slag "dew point" and the electrical conductivity of the gas phase are included in calculations that have been made for seven typical coals of varying rank, under the assumptions that the system is at chemical equilibrium, the gas is ideal, and the liquid slag phase behaves as an ideal solution.[106]

Figure 21.41 is a Mollier diagram for one of these coals, showing the region where the coal ash is completely vaporized and the region where a condensed slag is present. The boundary between these two regions varies with the rank of the coal and the ash composition, tending toward higher temperatures for the lower rank coals.

Electrical properties of the system are sig-

[105] E. R. Plante, C. D. Olson, and T. Negas, "Interaction of K_2O with Slag in Open Cycle, Coal-Fired MHD," *6th Int. Conf. Magnetohydrodynamic Electrical Power Generation,* Washington, D.C., June 9–13, 1975, pp. 211–218. Energy Research and Development Administration, National Bureau of Standards, Office of Naval Research, Electric Power Research Institute, in cooperation with International Atomic Energy Agency and Nuclear Energy Agency of the Organization for Economic Cooperation and Development.

[106] H. F. Feldman, M. A. Leonardo, W. H. Simons, and D. Bienstock, *Thermodynamic, Electrical, Physical, and Compositional Properties of Seeded Coal Combustion Products,* U.S. Bur. Mines Bull. No. 655, 1970, 289 pp.

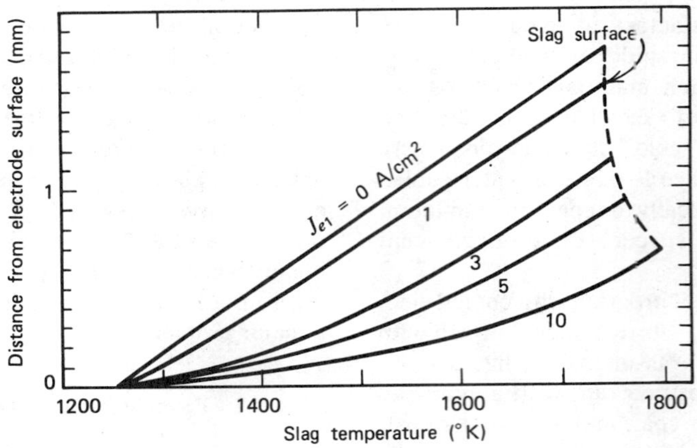

Figure 21.42 Slag temperature profile as function of current density.[104]

nificantly affected by slag, and the slag properties also change because of thermal effects resulting from differences in the electrical conductivity of the slag layers on electrodes. With cooled electrodes, as shown by the EPRI-sponsored studies at Stanford,[104] hot spots indicating locally high current flow increased the slag surface temperature, resulting in lowered viscosity and a region of highly fluid slag downstream of the spot. With electrodes held between 700 and 1100°C, such hot spots were larger and less intense, and they tended to disintegrate the slag coating on the electrode. Analyses have been made based on slag flow properties, electrical conductivity, and expected temperature profiles.

Figure 21.42 shows the likely slag temperature for different current densities, from which slag flow characteristics can be predicted if the chemical composition of the slag is known. Figure 21.43 shows how these temperature profiles affect slag flow rates. In each case, joule heating of the slag contributes to changes in flow and electrical characteristics of the slag layer.

Figure 21.44 indicates how the electrical characteristics of slag affect the stability of such electrode systems. At the anode a stable discharge mode results at high current densities if the electrode temperature is high. At the cathode the occurrence of thermionic emission will lead to instability, suggesting that the electrode will behave like a metal surface with an adsorbed layer of potassium.

Other work[99] also has shown that the high temperature of a slag layer on a cooled electrode sharply lowers breakdown voltage, and that in the presence of a slag layer there is no characteristic concentration of hot spots as with slag-free electrodes. There is a con-

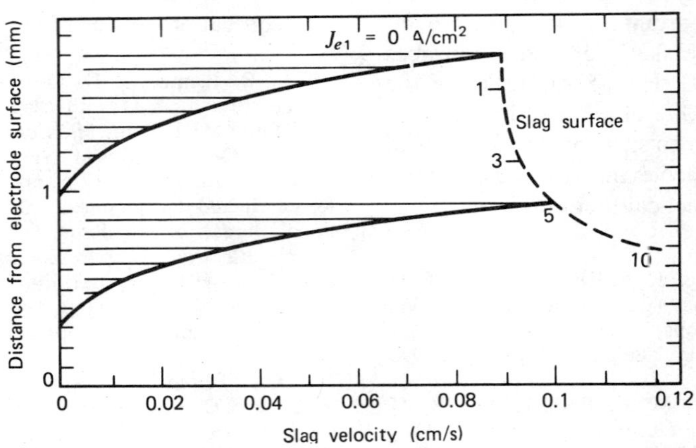

Figure 21.43 Slag velocity profiles as function of current density.[109]

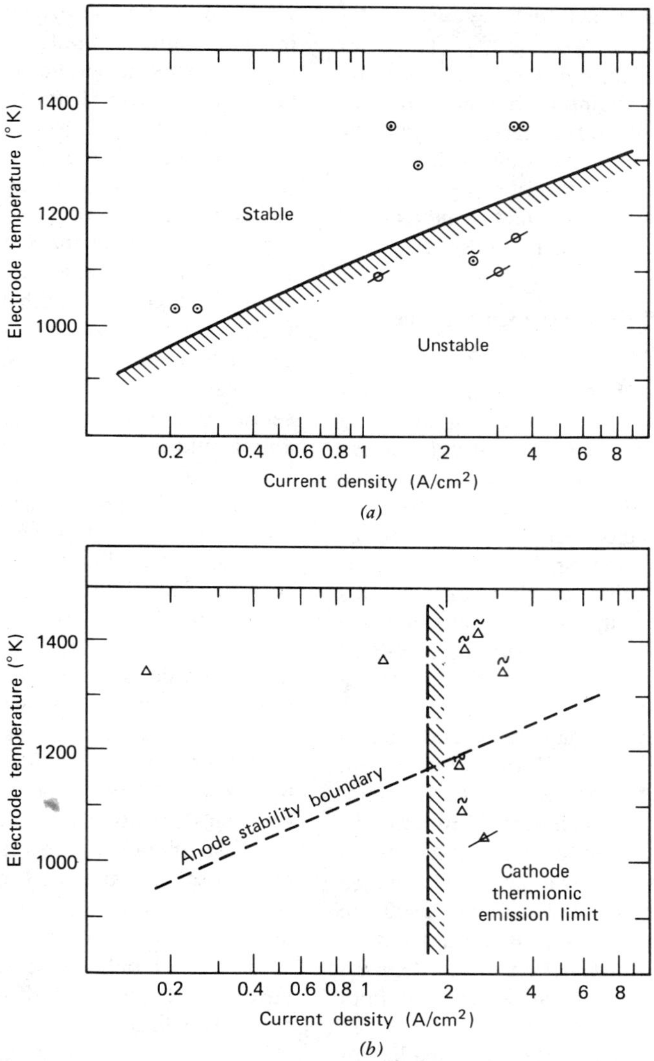

Figure 21.44 Slag electric discharge stability map for MHD electrodes: (*a*) anode and (*b*) cathode. ⊙, △ = stable; Õ, $\tilde{\triangle}$ = quasi-stable; ○, △ = unstable.[104]

tinuous change in the number of spots, their location and size, and the character of the slag layer.

21.8 BEHAVIOR OF ASH IN ADVANCED CYCLES

Despite the current expanding interest in energy conservation and in more efficient energy conversion systems, little progress has been made in lowering the heat rate in modern large coal-fired power plants. In fact, the "design" heat rate for fossil-fired plants in 1976 showed a definite upward trend compared to 1972,[107] with only 70% of the plants in 1976 designed for a heat rate less than 10,000 Btu/kWhr (34% overall thermal efficiency), compared with 92% in 1972. Part of this decline can be blamed on pollution control systems and on the rapidly escalating costs of heat recovery systems. But its main significance is the realistic approach being taken by the utilities in their search for the lowest incremental costs of electrical gener-

[107] Annual Plant Design Reports, *Power,* November 1972 and 1976.

ation by balancing fuel costs against capital investment.

Combined cycles, however, are not being ignored on the assumption that modern technology might provide the materials and techniques to link two generating cycles into a more efficient pair than either cycle alone. As yet, that expectation has been realized only when burning oil or natural gas.

21.8.1 Direct Coal-Fired Combustion Turbines

The outstanding success of combustion turbines in aircraft led to industrial applications and to the early realization that combustion turbines required "clean" fuels to achieve satisfactory performance records. Distillate fuels and natural gas were suitable, but residual fuel oil was marginal at best. But for widest applicability, combustion turbines must be able to burn coal or a coal-derived fuel to compete with pulverized-coal-fired steam generators on overall cost, performance, and reliability.

The earliest significant tests of a coal-burning gas turbine in the United States were conducted in 1953 and 1955 by the Locomotive Development Committee of Bituminous Coal Research.[108] In the later tests, about 1400 hr of operation was achieved over 83 days, at an average load of about 2900 hp; but the development was finally abandoned because of the accumulation of ash deposits in the combustor and localized wear on both stator and rotor blades.

At about the same time, in Canada, an experimental coal-burning, exhaust-heated gas turbine was operated,[109] using a cyclone furnace to supply hot coal combustion products to a tubular heat exchanger, which in turn heated pressurized air going to the gas turbine. Feasibility of the system was demonstrated. Coal ash was no more troublesome here than in any cyclone furnace and was not a major factor in stopping further work.

A coal-burning gas turbine of 1400 hp was operated experimentally in Australia in 1963.[110] It used pulverized brown coal burned in a conventional tubular combustor, with 48 unit cyclones in parallel to remove particulates, mainly ash, from the hot gas stream going to the turbine. The solids were smaller than 40 μm, with the cleaned gases entering the turbine carrying about 3.5×10^{-4} lb of solids per pound of gas. Without the ash separator, wear on the second-stage turbine stator blades exceeded 2 mg/hr-cm^2, an excessively high rate; with the separator the wear rate was reduced by a factor of 70. Based on these tests, compared with operation using oil, it was concluded that the service life of the turbine burning coal probably would not exceed 3000 hr.

In a further effort to determine whether direct-fired, coal-burning combustion turbines might find a place in increasing the efficiency of coal-burning power plants, the Bureau of Mines in the mid-1960s tested the same turbine used by the Locomotive Development Committee a decade earlier.[111] A new blade design was used to try to reduce erosion, but metal loss was still excessive, and it was estimated that the rotor blades would probably last 20,000 hr and the stator blades would have to be replaced after 5000 hr. Ash deposits also showed the presence of alkali metal sulfates, and much of the metal loss was attributed to corrosion rather than to erosion. Although the Bureau of Mines hoped to do further research and development on open-cycle, coal-burning combustion turbines, that effort apparently has not materialized.

The problem of mineral matter in coal provides the major stumbling block to the development of a successful combustion turbine to operate with direct coal firing. It is not now possible to foresee when technology will advance to the point where ash in coal can be removed so completely that direct firing of a turbine is feasible. Closed-cycle gas turbines, using coal as the energy source, appear promising, even though the problem with alloys in the required heat exchangers

[108] J. I. Yellott, P. R. Broadley, and W. H. Meyer, Annual Report, Locomotive Development Committee, Dunkirk, N.Y., September 15, 1955.

[109] D. L. Mordell and R. W. Foster-Pegg, "Tests of an Experimental Coal Burning Turbine," ASME Paper No. 55-A-158, 1955, 15 pp.

[110] J. C. Wisdom, "Coal-Burning Gas Turbine," *Mech. Eng.*, **85**, 57 (July 1963).

[111] J. Smith, D. C. Strimbeck, N. H. Coates, and J. P. McGee, "Bureau of Mines Progress in Developing Open and Closed-Cycle Coal-Burning Gas Turbine Power Plants," *Trans. ASME, J. Eng. Power, Ser. A,* **88,** 313–322 (October 1966).

is formidable. Pressurized FBC is being evaluated in considerable detail for such a cycle.

21.8.2 Coal-Derived Fuels for Combustion Turbines

The liquefaction of coal eventually may produce ash-free liquid fuels suitable for combustion turbines. Gasification systems currently being developed may also one day serve open-cycle combustion turbine systems. Although combustion characteristics of such fuels are important, their main requirement, other than the cost, is that they be free of inorganic matter. This probably would pose no problem with liquid fuels, such as methanol, or a distillate hydrocarbon, but it places a stringent restriction on gasification systems. Also, coal-treating processes such as solvent coal refining leave far too much ash in the product to be suitable for a combustion turbine, although work now under way is aimed at that eventual use. Solvent coal refining leading to a distillate product would be suitable for open-cycle turbines, but fuel costs probably would be high.

Fuels oils currently burned in gas turbines are strictly limited in the amount of inorganic matter they may contain. According to ASTM D2880,[112] No. 4 GT fuel oil may be used if it contains up to 500 ppm vanadium, but not more than 10 ppm sodium and potassium, or 10 ppm calcium. Additives must be used if the temperature exceeds 1200°F (649°C). For No. 1 GT to No. 3 GT, the upper sodium-plus-potassium limit is 5 ppm. Calcium limit for No. 1 GT is 5 ppm, and for No. 2 GT and No. 3 GT it is 10 ppm.

Similar limits have not been placed on other possible fuels for combustion turbines, but it is likely that the permissible upper limits for sodium, potassium, and calcium will not change appreciably. On this basis, much research and development will be needed to arrive at coal-derived fuels other than distillates, methanol, or fuels sufficiently free of metal contaminants at the ppm level to serve as the fuel for combustion turbines. Again, this problem will be transferred from the gas turbine to the heat exchanger in closed-cycle systems. That problem does not have an easy solution either.

[112] ASTM Standards, 1975, Part 24, Gas Turbine Fuel Oils.

CHAPTER 22

Control of Pollution from Combustion Processes

A.V. SLACK

A. V. Slack *is president of SAS Corporation, a consulting firm in gas cleaning to control emissions of sulfur dioxide, particulates, and nitrogen oxides. Educated at the University of Tennessee, he worked with the Tennessee Valley Authority from 1941 to 1973, when he entered the consulting field. At TVA he served in several capacities, the last one as Chief Chemical Engineer with major responsibility for TVA's gas cleaning research and development program. Over the past seven years Mr. Slack has advised the leading U.S. utilities on gas cleaning process selection and process design. He is the author or coauthor of two books and 38 papers and articles.*

22.1 INTRODUCTION

Combustion processes are a major source of emissions of waste materials regarded as undesirable additions to the environment, both from the standpoint of quantity and of variety of pollutants. This is particularly true for the combustion of coal. Control of pollutant emissions during combustion becomes quite complex when all the environmental considerations are brought into play.

The ash produced by the burning of coal pollutes the atmosphere if released with the combustion gas (hereafter, simply "gas") and may pollute groundwater or water courses if removed from the gas and discarded in land disposal areas. Moreover, even with the most effective collection techniques available, a small part of the ash escapes. Unfortunately, it is the smallest particles that are not removed from the gas, and these have the most adverse effect on health. Reduction in fine-particle emission, although not yet required by regulations except in a few instances, has become a major goal of governmental regulatory agencies.

The potential for water pollution is considered to be significant because of the heavy metal content of the ash. In the past 5 years or so, there has been much concern about leaching of such compounds into groundwater, to the extent that containment to isolate the material from the environment is already required in some areas.

Coal also contains sulfur, which is converted to sulfur oxides (SO_x) during combustion. Pollution abatement agencies have been concerned with sulfur oxide emission from combustion sources for 50 years or more, but it is only in the past decade that a major drive to reduce emissions has been launched. The U.S. Congress passed amendments to the Clean Air Act in 1971 and gave the Environmental Protection Agency (EPA) the task of enforcing them. The standards that have been set, with regard to both quantity emitted and ambient concentration, have led many industries either to shift to a low-sulfur fuel or to install equipment to remove the sulfur compounds from the gas.

Of the sulfur oxides emitted, the bulk (usually well over 90%) is sulfur dioxide (SO_2) and the remainder sulfur trioxide (SO_3). Sulfur dioxide itself is considered harmful to health, but it may be even more harmful after emission, when it is oxidized to SO_3, which combines with water vapor to form acid mist or with basic compounds to form sulfates. EPA has claimed that these acidic compounds are even more harmful than SO_2 and that the level of SO_2 concentration now acceptable under the regulations may still be too high: moreover, a high degree of emission reduction may be required to reduce the general sulfate concentration over wide areas. This is a subject of great controversy, and many do not agree that sulfate is a hazard or that regulations should be adopted to deal with it. Nevertheless, the prospect of future sulfate regulations affects industry planning with respect to the control of sulfur oxides.

The nitrogen oxides NO and NO_2 (NO_x) make up the third major class of pollutants encountered in coal combustion. They are formed both from nitrogen compounds in the coal and from the nitrogen in the combustion air. Environmental regulations are not as well defined for NO_x as they are for SO_2 and particulates. In the United States, for example, particulates must be removed from the combustion gas in all large boiler operations and SO_2 must be removed in some operations, but so far there are no installations in which NO_x is separated from the gas. This has been done only in Japan, where the concentration of combustion sources is so high in some areas that health authorities are concerned about the ambient concentrations of NO_x. The regulations are so rigorous in some instances that facilities to remove NO_x from combustion gas have been installed.[1] Most of the problem with NO_x emissions,

[1] J. Ando, *SO_2 and NO_x Removal Technology in Japan—1976,* Japan Management Association, 1976, 61 pp.

however, involves combustion of gas or oil; for although a considerable amount of coal is used as a fuel in Japan, coal-burning plants are located in the less congested areas.

In the United States the main approach to the reduction of NO_x emissions during coal combustion is to alter the combustion in such a way that a smaller quantity of NO_x is produced (Chapter 20). This may involve gas recycling, staged combustion, less excess air, and other variations from the standard methods of operating a boiler. Moreover, the emphasis by environmental authorities on reducing NO_x emission has meant that utilities generally favor the tangentially fired, pulverized-fuel boilers over the cyclone type. The more standard wall-fired, pulverized-fuel boilers are in an intermediate position with respect to NO_x production.

Several other pollutants are produced during the combustion of coal. Most are present in the gas in much lower concentrations than SO_2 and NO_x, but have received increasing attention from regulatory authorities because of their potential for pollution. This particularly applies to the so-called "minor elements" or "trace metals": mercury, selenium, boron, and several others. These trace metals may cause harm in the ambient air when they escape from the boiler (mainly in the very fine particulates, where they tend to concentrate) or in water supplies when leached from collected ash or scrubber sludge.

Chlorides and fluorides are emitted in gaseous form, but in concentrations so small that they have little significance as pollutants. The main problem posed is the possible interference with removal of other pollutants from the gas. For example, chloride may reduce removal efficiency and cause corrosion in wet processes for removing SO_2 from the gas.

In addition, some of the metals in the coal may also volatilize and have a polluting effect that is separate from that of the ash. Since very little is known about this area, the effect of any such volatilization is usually considered along with the general problem of ash.

The problem of pollution from coal combustion is centered in the heavily industrialized areas of the world, particularly in the United States and in West Germany. In both countries the large domestic coal deposits, the relatively low cost of mining, and problems in the supply of alternative fuels has made coal a basic material for production of heat and power. The situation is different in other countries; in Japan the deposits are small and spotty; in England the high cost of mining and the discovery of North Sea oil has shifted emphasis to other fuels; and in France and the U.S.S.R. other fuels have been found more desirable.

The main use of coal in the United States and Germany is by the utilities. There is some use, of course, in industrial boilers and even in residential heating (quite minor), but the utility consumption is so overwhelming that any discussion of pollution control must deal mainly with the large boilers operated at utility power-generating stations.

Control of pollution from coal combustion is quite complex because of the interactions between pollutants and the effect of these interactions on the design and operation of control equipment. For example, the efficiency of an electrostatic precipitator for particulate removal is greatly affected by the SO_x content of the gas; the trend toward use of tangentially fired boilers as a means of reducing NO_x formation works against the use of cyclone boilers to reduce particulate emission. The equipment used to remove SO_2 by wet processes can also be used to remove particulates, but there are complexities in the process that often make it preferable to remove SO_2 and particulates separately. When using dry processes to remove SO_2, the particulates, as a rule, must be removed separately. The location of the particulate collection equipment in the system depends on the type of SO_2 removal process. It may or may not be desirable to combine NO_x and SO_2 removal, depending on several process factors. All this has complicated design and made it difficult for operators of coal-fired boilers to select the optimum combination for control of all pollutants.

22.2 PARTICULATE MATTER

22.2.1 Characteristics and Quantity

The particulate matter that evolves during coal combustion is made up of smoke (unburned solid combustible matter) and ash (mineral matter from the coal). In modern boilers the smoke is such a minor constituent that it is usually considered a part of the general problem of reducing particulate emissions to the atmosphere, rather than as a separate issue. In any event, in most areas, regulations governing particulate emissions are so rigorous with regard to both amount emitted and opacity of the stack "plume" that the collection equipment must remove smoke as well as ash.

It is also possible to have a particulate emission of acid mist—fine droplets of sulfuric acid formed by reaction of emitted SO_3 with moisture in the gas and atmosphere. This is normally encountered in oil combustion, however, and seldom if ever occurs in the burning of coal.

Not all the ash leaves the boiler with the gas, since part remains molten in the boiler long enough to flow to the bottom, where it either solidifies or is tapped out and becomes bottom ash. The amount that leaves in this way varies with type of boiler (Chapter 20) and, to some extent, with type of coal. The main difference is between pulverized-coal and cyclone boilers. The cyclone is constructed to operate at relatively high temperatures and uses coal of relatively low ash fusion temperature, with the result that the bulk of the ash, roughly 85–90%, remains molten in the furnace and flows out the bottom.[2] In contrast, pulverized-fuel furnaces produce only 15–40% of the ash as bottom ash.

Wet bottom furnaces of the pulverized-fuel type (slag tap), popular at one time for burning coal with low ash fusion temperatures, also divert much of the ash (~ 50%) to the furnace bottom. Although some of these are still operating, they are relatively small and are not very significant in the overall pollution problem. They have generally been replaced with cyclone furnaces, which have several operating advantages. Because the cyclones produce relatively large quantities of NO_x, however, they have fallen into disfavor; several remain in operation, but few if any new power stations use this type of unit.

The net result of these changes in boiler practice is that the dry bottom, pulverized-fuel type has gained in prominence. This results in maximum evolution of the ash into the gas stream, from which it must be removed. The dust loading in the gas varies with the coal composition (see Chapter 7), but usually is on the order of 4–6 gr/scf (9.2 to 13.7 g/m^3).[3] For cyclone boilers the amount is much less, on the order of 1–2 gr/scf (2.3–4.6 g/m^3). Stoker boilers, still in use in small installations, also retain most of the ash in the boiler rather than letting it escape with the gas.

Thus the amount of dust to be removed from the gas from pulverized-fuel boilers (the type most widely used) is quite large. For a 1000-MW boiler burning coal containing 15% ash,* the ash loading in the gas is about 1200 tons/day, of which more than 99% must be removed to meet the particulate regulations generally applicable in the United States. This is a major assignment for both designers and operators, since the collection equipment must be very large and efficient and must be operated very carefully.

The EPA emission regulation for large boilers is that not more than 0.1 lb of particulates per million Btu of fuel burned can remain in the gas (40 CFR 60.40 et seq.), which corresponds to roughly 0.05 gr/scf (0.1 g/m^3) for a typical coal and for typical operating conditions. Since residual dust produces a visible stack plume, there is also a plume opacity regulation (40 CFR 60.62) based on visual observation from which a Ringelmann number for smoke density is obtained. On a scale of 1 to 10, the standard calls for a maximum Ringelmann number of 2. These regulations apply only to new plants. State regulations for existing plants vary widely, with some requiring emission rates as low as 0.05 lb/10^6 Btu (which may also be applicable to new stations) and others allowing considerably higher rates of emission than EPA's 0.1 lb/10^6 Btu.

The properties of the ash (Chapters 7 and 21) are quite important in the design of removal equipment. Particle size is particularly significant, since the finer particles are the most difficult to remove by any type of collection equipment. The size range depends on several factors, including type of boiler, boiler temperature, ash melting point, slag

[2] *Chemistry of Coal Utilization,* Supplementary Volume, H. H. Lowry, Ed., John Wiley and Sons, New York, 1963.

[3] A. V. Slack and G. A. Hollinden, *Sulfur Dioxide Removal from Waste Gases,* Noyes Data Corporation, Park Ridge, N.J., 1975.

*In this chapter all percentages except those expressing figures of merit are given as percent by weight, unless otherwise stated.

viscosity, use of additives for changing the fusion point, and possibly the original size of mineral inclusions in the coal. Under the microscope, ash appears as small spheres—usually with a smooth surface but sometimes with small protuberances that seem to make the ash much more erosive in downstream equipment. The spheres are formed, of course, by melting of the ash, followed by the formation and solidification of drops. There are a number of factors that are significant in determining the size range of the solidified particles.

Because the size range varies, it is difficult to present one that may be considered typical. Table 22.1 gives the range for an Eastern coal burned in a pulverized-fuel boiler.[4] Most of the particles are smaller than 1.0 μm, but the weight fraction in this range is quite small.

A large percentage of the particles produced in cyclone boilers are in the smaller size range, whereas stokers perform in the opposite manner. Fluidized-bed boilers, in which the ash does not melt, are unique with regard to size range and shape of particles. Few data are available on this, however, because such boilers have not been built in commercial sizes.

The composition of the ash is quite significant to collector performance when an electrostatic precipitator is used for ash removal. This is the method most widely practiced. The content of alkali metal compounds appears to be especially important in increasing conductivity.[5] On the other hand, CaO, MgO, Fe_2O_3, Al_2O_3, and SiO_2 generally give higher resistivity.[6]

In addition to the large boilers used by utilities, there are many smaller ones of the industrial type. For these, the ash characteristics are more variable; the ash may not be fused, combustion is often less complete, and the ash may have a higher carbon content.

[4] *Sulfur Oxide Removal from Power Plant Stack Gas: Use of Limestone in Wet-Scrubbing Process*, Tennessee Valley Authority, 1969.

Table 22.1 Typical Size Range of Fly Ash[4]

Particle Size (μm)[a]	Weight in Fly Ash (%)	Number of Particles per 100 g
0.5	0.13	8.1×10^{11}
1	0.37	2.9×10^{11}
2	1.5	1.5×10^{11}
3	2.5	7.4×10^{10}
4	3.5	4.3×10^{10}
5	1.5	9.5×10^{9}
6	3.5	12.9×10^{9}
7	3.5	8.1×10^{9}
8	4.5	7.0×10^{9}
9	4.0	4.4×10^{9}
10	5.0	4.0×10^{9}
20	20.0	2.0×10^{9}
30	12.0	3.5×10^{8}
40	8.0	1.0×10^{8}
50	8.0	5.0×10^{7}
70	9.0	1.7×10^{7}
100	3.0	2.4×10^{6}
+100[b]	10.0	2.9×10^{6}

[a] Assumes particles smaller than 0.5 μm to be 0.5 μm in diameter; particles smaller than 1 μm and larger than 0.5 μm to be 1 μm in diameter, etc., to 100 μm.

[b] Assumes all +100– μm particles to be 150 μm in size.

22.2.2 General Principles of Ash Collection

Each of the basic approaches to ash removal from gas involves one of the following principles: gravity, inertia, electrostatic attraction, wet scrubbing, and filtration. Of these, gravity is seldom a factor except incidentally from fallout of large particles during travel through the boiler equipment train. The principle of inertia, as exemplified by the cyclone (mechanical) collectors, once generally used in utility boilers as the sole collection device, is not efficient enough in removing the smaller particles to meet emission regulations. It is primarily used for precollection, to reduce the size of the subsequent main collector.

Electrostatic precipitation is by far the most widely used collection method. Low pressure drop, high collection efficiency, and dependable operation make it the standard choice for large boilers and quite often for smaller ones. Because of the rigorous emission limits in most areas, precipitator design is quite complex, particularly when low-sul-

[5] Roy E. Bickelhaupt, "Surface Reactivity and the Chemical Composition of Fly Ash," *Proc. Symp. Electrostatic Precipitators for the Control of Fine Particles,* January 1965, EPA-650 2-75-016.

[6] J. Stanley Selle, Philip H. Tufte, and Gordon H. Gronhovd, "A Study of the Electrical Resistivity of Fly Ashes from Low-Sulfur Western Coals Using Various Methods," paper presented at 65th Annual Meeting of the Air Pollution Control Association, June 18–22, 1972.

fur coal is used to meet the SO_2 emission regulations. Unfortunately, precipitators are normally operated in a temperature range in which particle resistivity is quite high, requiring adsorption of SO_3 to make the ash adequately conductive. However with low sulfur coal there is not much SO_3 in the gas. The situation can be corrected by making the precipitator larger, which is expensive, or by adding SO_3 to the gas stream. The latter approach involves several problems but is a growing practice. Another approach is to locate the unit between the economizer and air heater so that the operating temperature is about 700°F; at this temperature the ash is adequately conductive without SO_3, but the precipitator must be large because of the greater gas volume at the high temperature.

The problems with low-sulfur coal have led some to the use of wet scrubbing to remove particulates, in which case ash resistivity is not a consideration. This method is based on the principle of inertia, since the particle must impinge on a liquid drop with enough force to stick. In contrast to cyclones, however, a wet scrubber can be designed with high efficiency for particles in the smaller size range.

The problems with wet systems, including corrosion and scaling, have given impetus to the development and use of gas filtration in cases of high resistivity, which means that the precipitators must be large. The gas is passed through fabric bags in a cyclical manner, with the gas flow reversed periodically to shake the dust out. There is a major trend in the utility industry to employ bag filter systems, although there are several problems here not encountered in precipitator operation.

22.2.3 Removal Methods

22.2.3.1 *Cyclonic Separation*

The simplest and most economical method for removing particulates is that of the cyclone collector, which has been used for a long time. The equipment is relatively small and the pressure drop incurred is not high enough to be unacceptable; the only trouble is that even with multicyclones, the efficiency in removing fine particles is too low to meet emission regulations. For this reason, cyclone collectors are seldom considered for new boilers, and even in older plants where they are already installed, the cyclones are often removed when the dust collection system is upgraded to meet modern emission regulations.

Cyclones may still have a place, however, when used in conjunction with wet scrubbers. Although precipitators and bag filters can accept the full dust loading without difficulty, the problems in operating wet scrubbers are aggravated when all the dust as well as all the SO_2 must be removed by the scrubber. Thus placement of a cyclone before the scrubber may be an economical and effective combination. For example, a cyclone is used ahead of the scrubber at the Reid Gardner station of the Nevada Power Company,[7] and this arrangement has been proposed by engineering firms for other utilities.

Cyclonic collectors are based on the principle of whirling the gas stream to impart centrifugal force to the particles, which throws them out against the wall, from which they are collected. A small diameter is desirable because the forces that collect particles on the wall increase as the diameter decreases. Thus it is customary to use many small units rather than a single large unit. For example, at Reid Gardner station, each mechanical collector contains 500 small cyclones.

The efficiency of cyclonic collectors depends mainly on the pressure drop and diameter. The Reid Gardner units have an efficiency of 75–80% based on the inlet weight loading [~ 3 gr/scf (6.9 g/m^3), from coal containing 8–10% ash].

Various theories have been developed for predicting the performance of a cyclone. The theory that apparently gives the best correlation with actual data uses two functional relationships, one depending on cyclone geometry and the other on gas and particle characteristics.[8] In general, the efficiency increases at higher particle size ranges, particle density, vortex strength in the cyclone core,

[7] Personal communication, Nevada Power Company, 1976.

[8] D. Leith and W. Lieth, "Collection Efficiency of Cyclone Type Particle Collectors, A New Theoretical Approach," in Am. Inst. Chem. Eng. Symp. Ser., *Air—1971*, 1972.

and cyclone length—and at lower levels of gas temperature and cyclone diameter.

22.2.3.2 *Electrostatic Precipitation*

Removal of particulates by electrostatic precipitation has several advantages for large boilers, including low pressure drop, no drop in efficiency when boiler output is reduced (efficiency actually increases because of slower gas flow), and high efficiency for fine particle removal. For these reasons precipitators have replaced the cyclonic collectors, which were common as late as the 1940–1950 period, as the standard method for particulate collection.

Although cyclone-precipitator combinations were popular at the time when the growing stringency of particulate emission regulations made it unacceptable to use cyclones as the sole collection device, it became evident that the reduction in precipitator size brought about by using cyclones to first remove large particles was not enough to offset the pressure drop and cost of the cyclone. A large precipitator is required anyhow, because of the slow migration velocity of fine particles to the collection surfaces. This means that not much extra capacity is needed to remove the larger particles. For this reason standard practice is to install a precipitator capable of removing all the dust. This requires low gas velocity and an extended retention time for the fine particules to migrate to the collector plates.

Theory. Electrostatic precipitation is a very complicated phenomenon, not fully understood by research people in the field. The basic principle is that of suspending electrodes (wires or plates) in the gas stream and imposing a potential difference of 40,000–60,000 V between these electrodes and a grounded collection surface (tubes or plates). Ionization of the gas by the high potential causes the particles to pick up a charge. They are thus attracted to the collection area. The particles become charged because they are bombarded by negative ions as they pass through the corona discharge from the electrodes. The collector plates are vibrated (rapped) to shake off the collected particulates into a hopper, located below.

The discharge electrode is usually negative ("negative corona") and the collection electrode positive, which allows a higher voltage differential before disruptive sparking occurs. The reason for this is not fully understood, but it is suggested that sparking begins at the positive electrode. This would explain why the collection electrode is positive, since if the discharge electrode were positive, the high field strength there would promote sparking.

The basic relationship used in the design of precipitators is the Deutsch equation

$$E = 1 - \exp\frac{A}{V}W$$

where E = fractional collection efficiency
A = collection surface area (ft²)
V = gas flow rate (10^3ft³/min)
W = effective migration velocity of particles (cm/s)

(22.1)

For the very high efficiencies required in modern practice, better correlation with actual data is obtained by applying a factor to this equation,[9] making it:

$$E = 1 - \exp\frac{-A}{V}W_K \qquad (22.2)$$

where K is usually about 0.5. This is a rough correction for particle size and other random effects.

Such an equation, however, cannot take into account the several nonideal conditions involved, including nonuniform gas flow, reentrainment of dust during rapping, and "sneakage" (bypassing of gas around the electrified sections). Gooch and Francis (Southern Research Institute) have developed a mathematical model for making corrections and obtaining a better prediction of performance.[10] The procedure is as follows:

1. Migration velocities for representative particle sizes are calculated as a function

[9] Alan B. Walker, "Experience with Hot Precipitators for Fly Ash Collection in Electric Utilities," paper presented at American Power Conf., Chicago, April 29–May 1, 1974.

[10] J. P. Gooch and N. L. Francis, "A Theoretically Based Mathematical Model for Calculation of Electrostatic Precipitator Performance," *Proc. Symp. Electrostatic Precipitators for the Control of Fine Particles,* January 1975, EPA-650 2-75-016.

of electrical conditions for increments of length in the direction of gas flow through the precipitator.

2. The Deutsch equation is used to calculate the number of particles in each size band (the particle size distribution is represented by a histogram). These bands are collected in each increment of length.
3. The collection efficiency with respect to each representative particle size is determined by summing over the length of the precipitator. The overall mass collection efficiency is obtained by adding together the masses of particles collected in each size band of the histogram.

Precipitator Design. Several parameters must be considered in the design of a precipitator. One of the more important is *gas velocity.* The velocity must be low enough for a small particle entering the unit near the discharge electrode to move sidewise to the gas flow, then all the way to the collection surface, before it is carried out of the unit with the gas. The usual velocity range is 4–8 ft/s.

Collection surface area is an even more basic parameter because of its effect on precipitator cost. Since large precipitators are normally fitted with plates rather than tubes as collectors, this is usually called "plate area." The area can range from 100 to 900 ft^2 per 1000 ft^3/min of gas. Plate area and gas velocity are interrelated, because gas velocity is a function of precipitator cross-sectional area, whereas plate area is determined by precipitator height and number and length of plates. Spacing between the plates is usually 8–12 in., and the length-to-height ratio is 0.5–1.5.

Corona current ranges from 5 to 70 $\mu A/ft^2$ and *corona power* from 50 to 500 W per 1000 ft^3/min.

The type of *discharge electrode* differs among designers. In American practice the wire type is normally used, but in Europe a rigid frame is favored. The latter has also been somewhat accepted in the United States.

Particle Resistivity. Much of the variation in design parameters is due to differences in the resistivity of the ash particles, a phenomenon that is not well understood despite its major effect on precipitator cost and efficiency. The subject has been extensively treated by White.[11]

The migration velocity of particles declines rapidly when the resistivity is higher than about 2×10^{10} Ω-cm. This effect is attributed to back corona, which is caused by a voltage drop in the resistive dust layer that has collected, and has the effect of reducing the voltage that can be maintained by the discharge electrodes. The space charge, on which particle charging depends, may also be reduced.

A major factor affecting resistivity is the temperature at which the precipitator is operated. Unfortunately the usual operating temperature is the worst that could be selected over the range available in a boiler train. Precipitators are normally placed after the air heater, at which point the gas temperature is in the range of 275–300°F—as low as possible to get good energy recovery but high enough to avoid acid condensation from the SO_3 present in the gas. As this level, resistivity is relatively high, as shown in Fig. 22.1.

The sulfur content of the coal is another major consideration, as indicated in Fig. 22.1. With a sulfur content of 2.5–3.0%, the resistivity is well within the limit of about 2×10^{10} Ω-cm at which migration velocity begins to decline. At 1.5–3.0% sulfur the problem with resistivity is marginal for a normal plate area, and satisfactory performance can perhaps be achieved in the lower range of the sulfur content at 275°F. For even lower sulfur contents, the resistivity is much too high in the 275–300°F range to achieve good performance, except with a very large plate area.

It should be noted that these curves are only approximate, since characteristics of the ash, especially moisture and sodium content, also have an effect. Tests on ash from various coals have shown a range of 10^8–10^{13} Ω-cm resistivity at 300°F.

Hot Precipitators. Since low-sulfur coal (1.0% S) has such a high resistivity at the usual 275–300°F operating temperature for

[11] H. J. White, "Resistivity Problems in Electrostatic Precipitation," *J. Air Pollut. Control. Assoc.*, **24**(4) (April 1974).

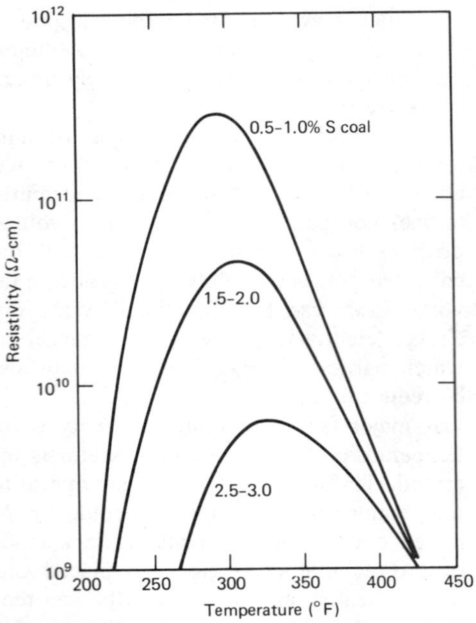

Figure 22.1 The Effects of gas temperature and the sulfur content of coal on ash resistivity.[12]

precipitators, an obvious solution is to operate at a higher temperature. In a power boiler train this requires placing the unit ahead of the air heater, where the temperature is around 700°F.

The "hot precipitator" practice came into use in the early 1970s. It has grown rapidly, not only in the West where almost all coal is low in sulfur, but also in the East, where utilities started using low-sulfur coal, mainly from the West, to comply with the SO_2 emission regulations.

Although the resistivity problem is reduced or eliminated by operating precipitators at a higher temperature, there are several offsetting factors. One is the large size required because of the increased gas volume at the higher temperature, which is about 1.7 times greater than the volume at 300°F. Other problems are frame warpage at the higher temperature, heavier insulation requirement, and greater difficulty in doing maintenance work. Because of these difficulties, some utilities have chosen "cold" precipitators and built them large enough (up to 900 ft² per 1000 ft³/min) to obtain adequate performance even at the high resistivities associated with low-sulfur coal.

Proponents of the "hot-side" approach argue that possible future changes in ash characteristics (as compared to the design coal) are less likely to upset precipitator performance when the higher operating temperature is used. Coals vary so widely in this respect, however, that even hot precipitators may not perform adequately if the ash characteristics vary widely from those of the design coal.

Operating experience with hot precipitators has been reviewed by Walker.[13] The number of boilers equipped with hot precipitators was reported to be 145, with a total capacity of about 52,000 MW. For Eastern bituminous coals, performance was generally good; performance specifications were almost always met, and since the units operated at temperatures far above the SO_3 dew point, operational reliability was good.

Performance was not as good for the 21 units operated with Western coal. The ash tends to foul equipment more than that from Eastern coal, and the problem of resistivity is more complex. More attention must be paid to the proper selection of electrode materials and to cleaning of the discharge electrodes.

The main difference between the Western and Eastern coals was the higher CaO and MgO content of the Western coal, as well as the higher sodium ion content. The higher the CaO-MgO content, the more severe the fouling problems. The best results were obtained with a high ratio of Na_2O to CaO MgO. A hard, thin deposit on the electrodes seemed to be the main problem, causing suppression and maldistribution of the corona. The presence of P_2O_5 in the ash was also an adverse factor, resulting in a higher requirement for power input and increased plume opacity.

Thus hot precipitators are not the complete answer to the problems in particulate collection with low-sulfur coal, but they are quite helpful if ash characteristics are favorable.

Low-Temperature Operation. To avoid the

[12] A. B. Walker, "Experience with Hot Precipitators for Fly Ash Collection in Electric Utilities," paper presented at American Power Conf., Chicago, April 29–May 1, 1974.

[13] A. B. Walker, "Operating Experience with Hot Precipitators on Western Low-Sulfur Coals," presented at American Power Conf., Chicago, April 18–20, 1977.

undesirable resistivity peak at 300°F, it is also possible to operate the precipitator at lower temperatures. However the temperature must be dropped considerably, to 220°F or lower, to get below the desired maximum resistivity of 10^{10} Ω-cm. At this temperature SO_3 condensation and equipment corrosion become a problem, even with the 1–3 ppm SO_3 produced in burning low-sulfur coal.

It has not been entirely proved, however, that low-temperature operation is not feasible. One of the Western utilities made tests with low-sulfur coal at 220°F.[14] The resistivity dropped from more than 10^{11} (at 300°F) to 10^{10} Ω-cm, the collection efficiency was increased considerably, and no corrosion was noted. The main drawback to general acceptance of low-temperature operation apparently is the fear that the danger of corrosion is too great.

Precipitator Reliability. The very low emission limits imposed by regulatory agencies have made it difficult to keep precipitators operating reliably at the required performance level. Problems such as gas maldistribution, rapping reentrainment, and "sneakage" have become critical, whereas they were not a problem with the regulations of the past, which were more lenient (or absent altogether).

Proper sizing of the precipitator is a very important consideration, a phase of design that has become much more sophisticated with the advent of requirements for high removal efficiency. Extensive investigation of the relationships between process operating variables and performance, plus the gathering of field data that allow comparison of predicted and actual performance, have made it possible to set up a central computer data bank.[15] By using this data bank, the best size of the precipitator for high removal efficiency can be chosen with greater certainty than was possible before.

Division of the precipitator into sections—sectionalization—also improves reliability, especially with regard to the electrical system. Efficiency generally increases as the system approaches a power supply for each electrode. Although this is not feasible, the number of electrical sections can be increased somewhat, within economic limits. This not only allows, to some extent, for variations in temperature and dust loading, but also makes it possible to shut a section down for maintenance while the rest of the precipitator remains onstream. Modern precipitators are divided into as many as 12 sections for this purpose, and a spare section or two is often included.

Modern design also pays more attention to the uniformity of gas flow, since flow uniformity is quite important in maintaining efficiency. This applies not only to the original design but also to the system after dust deposits have developed, which may change the flow patterns.[16] It is common practice to model a precipitator system thoroughly—upstream and downstream—before the design is finalized. The effect of the dust-receiving hoppers underneath the electrodes on flow patterns is especially important, as are the contours of dust deposits likely to form during operation.

Of the several problems that interfere with reliable performance, failure of the discharge electrodes due to electrical erosion (arcing) is the main one.[17] Trouble with the rapping system is next, failures in the rapping system third, and fatigue corrosion failure of collector plates is in fourth place.

Retrofitting. One of the main problems brought about by the more rigorous regulations promulgated in the 1970s was that of improving the efficiency of existing precipitators to comply with the new regulations. Not only is the retrofitting expensive, but interference with operation of the main plant can be a major problem, particularly in the utility industry, where the power must be produced as needed. This usually means that it is not feasible to make the precipitator larger.

The usual practice has been to build a new precipitator, either one that will operate at full capacity to replace the old unit, or a partial unit to be operated in series, to be ducted into the main system when complete. This is an operation requiring a minimum

[14] Personal communication, 1976.

[15] R. L. Bump, *Chem. Eng.*, 129 (January 17, 1977).

[16] *Electr. World*, 93 (April 15, 1976).

[17] APCA TC-1 Particulate Committee, *J. Air Pollut. Control. Assoc.*, **26**(11), 1061 (November 1976).

amount of down time for the system. The new unit is sometimes located to one side of the existing precipitator and is sometimes "double-decked" over it.

In other cases it has been possible to add electrical sections, improve gas distribution, or generally remove constraints on efficiency to bring the unit up to an efficiency that meets the regulation. Ash conditioning or precharging (see next two sections) are also possibilites for increasing efficiency to an adequate degree, without installing additional capacity.

Fly Ash Conditioning. Since SO_3 in the gas and Na^+ in the ash improve precipitator performance, these materials are routinely added when normal operation does not produce enough of them. Other additives have also been used, including moisture, ammonia, and various proprietary mixtures.

The main conditioning agent is SO_3, which is added as such, as sulfuric acid, or by burning sulfur, oxidizing the resulting SO_2 to SO_3, and injecting it into the gas stream. The SO_3 is adsorbed on the dust, which makes it more conductive. This lowers the resistivity of the collected dust layer to the corona current passing from the discharge electrode to the collector plate, which in turn prevents sparking and back corona.[18] The objective is to add just enough SO_3 to make the gas comparable to that from burning high-sulfur coal.

This appears to be a very acceptable solution to the problem of precipitator performance associated with low-sulfur coal. Although the sulfur required is expensive, the overall annualized cost is much lower than for installing additional precipitator capacity, either as a very large unit for a new installation or for adding an extra unit in a retrofit. However it is difficult to ensure the adequate mixing of the very small amount added—on the order of 20–50 ppm—before the gas reaches the precipitator. The SO_3 is normally added just after the air heater (to minimize air heater corrosion); thus it has only a limited mixing time.

Moreover, regulatory authorities generally frown on the practice because not all the added SO_3 is adsorbed on the dust particles. There is some evidence that as much as 30% passes out of the stack.[19] Although this is a small amount, far below that prohibited by any regulation, environmental authorities generally resist the deliberate addition of a pollutant to the waste gas.

Addition of Na^+ is not yet a fully commercial practice, although at least one utility injects Na_2CO_3 into the gas stream to improve precipitator performance. This was an emergency measure to offset the failure of the precipitator to meet the required removal efficiency. The practice is probably too expensive to be adopted widely. Addition of Na^+ to the coal might be more effective but would be likely to cause problems in the boiler.

Water has also been added to the gas stream as a "conditioning" agent. Although the addition was fairly effective, a large amount was required and there may be adverse effects; thus it is unlikely to be developed into standard practice.

Ammonia has been used in a few instances. Results have varied and the mechanism is not well understood.[20] General commercial acceptance seems unlikely.

Additives such as these have been used almost entirely to increase the efficiency of existing, inadequate precipitators. There are few, if any, cases in which conditioning was incorporated into the design of a new precipitator to attain the required performance.

Precharging. One of the older concepts in electrostatic dust collection involves charging the particles in one vessel and collecting them in another. It has generally been more economical, however, to perform both functions in the same chamber, as in standard precipitators.

This concept has received more attention in the 1970s because of the need to upgrade existing precipitators to meet the more stringent regulations. A new design has been tested in which a plate containing a spaced array of several venturis is placed at the front

[18] R. S. Atkins and D. H. Klipstein, Techn. Bull. Research—Cottrell.

[19] D. C. Drehmel and D. E. Blake, *J. Air Pollut. Control. Assoc.*, **26**(12), 1141 (December 1976).

[20] E. B. Dismuckes, "Conditioning of Fly Ash with Ammonia," *Proc. Symp. Electrostatic Precipitators for the Control of Fine Particles*, January 1975, EPA-650 2-75-016.

of a precipitator.[21] Located in the throat of each venturi is a disk, which acts as the cathode in charging the particles that are passing through. The venturi anodes are around 12 in. in diameter, and the disk cathodes 4–5 in.

The assembly, called an "ionizer," is claimed to be capable of maintaining a stable, very high intensity corona discharge that charges the particles to a higher level than is possible in a conventional precipitator. As a result, the efficiency of an existing precipitator can be increased significantly and new precipitators can be built in smaller sizes. The device appears to be particularly applicable to low-sulfur coal.

The ionizer concept appears promising, but further test data are required to evaluate the economics.

22.2.3.3 *Fabric Filtration*

The trend toward use of low-sulfur coal as a means of meeting SO_2 regulations has also promoted use of fabric filtration. Utilities have been slow to use this technique because acid condensation in the filter bags can wet the ash and cause plugging of the fabric. Moreover, any increase in temperature in the air heater can ruin the bags because the glass fiber normally used cannot stand high temperatures.

Bag filters have usually been operated at a temperature about 20°F higher than that which is normal for precipitators.[22] Specifications issued by utilities for new baghouses have generally reflected this increase in temperature. The higher temperature reduces the hazard of acid condensation but there may not be enough danger to warrant the resulting loss in energy efficiency. As the utility industry becomes more familiar with this type of dust collection, perhaps the operating temperature will be lowered.

Bag life under utility operating conditions has also not been well proved. Accidental tears can be a problem, and the pressure drop is higher than for precipitators.

On the positive side, ash characteristics have little or no effect on bag filter operation, thus ensuring the possible use of bag filters irrespective of future changes in the coal supply. Rapping reentrainment is avoided, and sneakage is not a factor in bag filter operation. Other drawbacks of precipitators are also avoided, including wire breakage, possibility of maldistribution of gas, frame warpage in hot precipitators, and the tendency to lose efficiency with continued operation. Annualized operating cost can also be lower when the ash characteristics make the collection of the ash in precipitators difficult.

Perhaps the main advantage of fabric filters is their intrinsic ability to remove very fine particles without any extreme design precautions. This property has attracted the attention of regulatory authorities, who in some cases have favored bag filters over precipitators.

There are various design possibilities. The most popular one in the United States consists of a number of long bags of relatively small diameter, with the gas passing up into the bags and out into the bag chamber. The unit is sectionalized, so that the gas flow in a section can be periodically reversed to collapse the bags and drop the dust out of the bottom. The bags are sometimes preconditioned by injecting finely ground limestone. Only a dozen or so boilers have been fitted with bags in the United States, but several other boilers designed to be fitted with bags were under construction in 1977.

Both the theory and practice of fabric filtration have been thoroughly covered in a symposium sponsored by EPA.[23]

22.2.3.4 *Wet Scrubbing*

The troubles encountered when trying to use precipitators with low-sulfur coal have led several utilities to try wet scrubbing. Here there is the outstanding advantage, not shared by either precipitators or bag filters, that both particulates and SO_2 can be removed by the same equipment, thus resulting in great savings in operating costs.

Theory. The theory behind particulate collection by wet scrubbing methods has been covered extensively in EPA symposia and

[21] EPRI FP-291 Project No. 386-1, October 1976.

[22] M. J. Hobson, "Review of Baghouse Systems for Boiler Plants," *J. Air Pollut. Control. Assoc.*, **26**(1) (January, 1976).

[23] *Proc. Symp. Use of Fabric Filters for the Control of Submicron Particulates,* April 8–10, 1974, EPA-650 2-74.043.

surveys.[24–27] Collection results from several interactions between the gas and liquid phases. No single mathematical model has been developed that will adequately predict performance for the wide range of scrubber types in use. The design equations generally used are empirical. Scrubbers have been classified according to unit mechanism of collection (Table 22.2).

Using the unit mechanism approach to predict scrubber efficiency requires analysis of transport phenomena such as diffusion (molecular and turbulent), inertial impaction, interception, electrostatic attraction, diffusiophoresis, thermophoresis, and condensation. The subject is quite complicated and is not yet well enough developed to be a useful design tool.

Practice. Wet scrubbers have been used for a long time to collect dust from various industrial operations. In coal combustion, however, utilities and other large boiler operators have avoided their use because of (1) the relatively high pressure drop required to remove small particles, (2) corrosion of the wetted surfaces, (3) carrythrough of mist and subsequent problems with downstream equipment, (4) reduced gas buoyancy resulting from the cooling, and (5) wetting of the ash, which interferes with its recovery for sale (where this is feasible).

Use of particulate scrubbers in the power industry began in the early 1970s when the growing stringency of emission regulations made cyclones no longer adequate. To avoid the very high cost of precipitators for low-sulfur coal, companies such as Pennsylvania Power and Light and Arizona Public Service tried to economize by installing venturi scrubbers, using water as the scrubbing medium. Although such scrubbers have been used successfully in the steel and other industries, many problems arose in applying the technology to boilers. The main problems were scaling and corrosion, both caused by absorption of SO_2 in the scrubbing medium. The SO_2 caused the pH to decrease and reacted with Ca^{2+} leached from the ash to form gypsum scale ($CaSO_4 \cdot 2H_2O$) by oxidation of calcium sulfite. These problems were so severe at the Four Corners station of Arizona Public Service that the operating cost was extremely high, probably much higher than it would have been for precipitators or bag filters.

Other utilities have installed scrubbers for particulate removal only and encountered the same troubles. The practice developed of adding lime to the circulating liquor to raise the pH, as a means of reducing the corrosion and scaling. This was successful but expensive, and it converted the scrubbers to SO_2 removal, even though this was not originally intended.

Several of the particulate scrubbers have been or will be converted to achieve full SO_2 removal, to meet new local regulations. It seems unlikely that any more scrubbers will be installed for particulate removal alone (where the coal is so low in sulfur that SO_2 removal is not needed).

The main place for scrubbers in particulate removal is for high-sulfur coal, where an SO_2 removal system must be installed anyway, and economy can be achieved by designing the scrubbers for the double function of removing particulates as well as SO_2. Some of the major systems in the United States are of this type, including Mansfield 1 and 2 (Ohio Edison; two 825-MW boilers), La Cygne (Kansas City Power and Light; 825-MW), Sherburne County (Northern States Power; two 680-MW), and Colstrip (Montana Power; two 360-MW).

Although the double-function system costs less, the various troubles encountered have convinced many utilities that it is safer to install an efficient precipitator for particulate

Table 22.2 Unit Mechanisms of Particulate Collection by Wet Scrubbing[26]

Scrubber Type	Unit Mechanisms
Venturi	Liquid jets, sheets, drops
Sieve plate	Gas jets and bubbles
Impingement plate	Gas jets, jet impingement, liquid drops, sheets
Mobile packing	Gas jets, liquid jets, drops, sheets
Packed column	Liquid sheets and drops, curved liquid sheets, gas jets, bubbles

[24] L. J. Shannon, *Control Technology for Fine Particulate Emissions,* May 1974, EPA-650 2-74-027.

[25] *Proc. EPA-APT Fine Particle Scrubber Symp.* May 28–30, 1974, EPA Contract No. 68-02-1328.

[26] *Wet Scrubber System Study,* APT, Inc., July 1972; PB-213016.

[27] A. V. Slack, "Scrubber Survey: A Lime-Limestone Trend," *Electr. World,* October 1, 1974.

removal and follow it with an SO_2 scrubber.[27] The main problem is that when both SO_2 and particulates are collected, the large amount of solids accumulating in the scrubber interferes with operating the scrubber in such a way as to minimize scaling (see subsequent discussion). There are indications, however, that when Mg^{2+} is carried in the circulating slurry the system can be operated without scaling, even with the full particulate loading collected in the scrubber. If this proves to be true, there may well be a trend toward particulate removal by scrubbing.

The scrubber type most often used when particulate removal is required is a venturi, in which liquid is sprayed into the gas. The mixture is then passed through a constricted area to produce a high differential velocity between gas and liquid and to break up the liquid into very fine drops. Efficient particulate removal can be obtained, even for fine particles, but the pressure drop is relatively high and there must be some way of changing the throat cross-sectional area (usually a "plumb bob" arrangement) so that an adequate pressure drop can be maintained at low gas flow. The venturi also has the drawback of poor mass transfer capability for SO_2 removal.

The only other scrubber type used commercially for the sole purpose of removing particulates from boiler gas is the mobile bed. Plastic or rubber balls are used as packing and are retained between grids so that the bed is violently agitated and fluidized by the upward flow of gas. The extreme turbulence generated gives efficient particulate removal at relatively low pressure drop. This type of scrubber, usually called TCA (turbulent contact absorbent), also has good mass transfer capability for SO_2 removal.

When removing both particulates and SO_2, it is not clear which scrubber type gives the best compromise between the differential particle-droplet velocity needed for particulate removal and the rapid gas-liquor transfer rate needed for SO_2 removal. Various tests have indicated, however, that most scrubber types will give adequate particulate collection if they are designed for a high efficiency of SO_2 removal.[28]

[28] *EPA Alkali Scrubbing Test Facility: Limestone Wet Scrubbing Test Results,* January 1974, EPA-650 2-74-010.

22.2.4 The Fine Particulate Problem

Environmental regulations for particulates are usually based on two parameters—bulk removal to a certain level (0.1 lb/10^6 Btu for new plants) and opacity of the plume. Such regulations can be met even if the number of particles escaping is larger than the number removed. The plume opacity regulation controls this to some extent, since fine particles contribute more to opacity than large ones. Even so, most of the fine particulates can escape without violating regulations.

Environmental authorities are placing more emphasis on fine particulates because they are considered more harmful to health than the larger particles. At least one state (New Mexico) has already adopted a regulation requiring a limit not only on total emissions but also on particles less than 2μm in size. EPA is expected to eventually follow suit at the federal level.

The technology for removing such fine particulates is not well developed. One approach is to make conventional particulate removal equipment more efficient. For example, if a precipitator is designed for 99.8% removal efficiency instead of 99.3%, many of the particulates smaller than 2μm will be removed. The drawbacks are high cost, difficulty in maintaining such high efficiency, and the problem of determining, by measurement methods, just what efficiency has been reached.

Bag filters and wet scrubbers can also be made efficient for fine particulate removal, but with the same drawbacks as above—except that efficiency is probably easier to maintain. The bag filters may be the most promising, since they are less subject to some of the factors that affect efficiency of the other types (e.g., maldistribution of gas). The performance of filters when removing very fine particles is not well understood, but it appears that the properties of the dust are important, including size distribution, shape, surface characteristics, electrostatics, chemical reactivity, adhesiveness, and hygroscopicity. The fabric is also important; little is known about the electrical, geometric, adhesive, and mechanical properties of the filter "surface" (dust-fabric interface). More development in this area is needed.

Numerous devices have been proposed

with special features for removing fine particulates without incurring the drawbacks of conventional collection equipment. Space does not permit full description of these, and none are very far advanced. A partial list follows.

1. *High-pressure nozzle*. Water droplets are generated by a nozzle and injected into the gas stream at high velocity. Such devices in tests have removed 99.9% or more of the dust less than 3 μm in size (mean diameter, 1.5 μm). Power requirement, however, is very high.
2. *Wetted-mesh filter*. Water is sprayed into the gas and the mixture passed through a filter made of stainless steel in knitted form (void fraction, 0.983). Collection efficiency was 99.95% for 2-μm particles (11-in. filter depth).
3. *Condensation scrubbers*. Water vapor is condensed from the gas to form a liquid envelope around the particles, thus increasing their size.
4. *Charged droplet scrubbers*. The drops are charged so that particles will be attracted to them.
5. *Electrified filters*. The filter is charged to attract particles.
6. *Sonic agglomeration*. Sound waves are used to agglomerate fine particles.

22.2.5 The Economics of Particulate Removal

The cost of removing particulates varies widely, depending on the situation. The two extremes can be illustrated as follows.

1. A *retrofit cold precipitator,* which is installed when the fuel supply has been shifted from high- to low-sulfur coal, and the particulate emission standard is low (0.05 lb/10^6 Btu in some areas). Because of the very high plate area required (600 ft^2 per 1000 ft^3/min or higher) and the extra expense of retrofitting, such a unit would be very expensive. Operating cost would also be relatively high because of the high electrical requirement, which can approach 2% of a station's capacity.

2. *Combined removal of particulates and SO_2 in a wet scrubber.* If SO_2 removal is required because high-sulfur coal is burned, it costs little extra to include the removal of particulates, unless the emission regulation is so strict that it requires a high pressure drop. The SO_2 scrubbing is complicated somewhat by the dust, but probably not enough to have much effect on cost.

Most methods of particulate removal are somewhere between these two extremes.

If regulations are adopted in the future concerning the removal of fine particulates, the cost of removal will be greatly increased.

22.3 SULFUR OXIDES

22.3.1 Scope of Problem

The sulfur oxides have received more attention than any other pollutant, except perhaps dust. Medical authorities generally agree that the ambient concentrations of SO_2 resulting from industrial operations and natural sources can become high enough to be a hazard to health, but they do not necessarily agree about the highest acceptable concentration level. In the United States, EPA has set the limits at 80 μg/m^3 as an annual average, and 365 μg/m^3 as a 24-hr average.

The ambient standard can be met either by limiting emissions or by dispersing the gas so that it will be adequately diluted with air before it comes back to the ground. Utilities and other larger boiler operators have generally favored the latter approach because it can be accomplished at relatively low cost with high stacks (plus occasional reduction in output when weather conditions are adverse to dispersion). EPA has generally opposed this, on the basis that even if adequately dispersed, the SO_2 is harmful because it is oxidized to acid sulfates, which are harmful im much lower concentrations that is SO_2. There is much disagreement on this point; EPA admits not having enough data to set a sulfate standard. It has, however, set SO_2 emission standards for new boilers and has Congress agreed with this in framing the 1977 amendments to the Clean Air Act. The new source standard (late 1977) is 1.2 lb of SO_2/10^6 Btu.

Thus different utilities have varying degrees of flexibility with respect to their options concerning SO_2 removal. For some there can be a choice between low-sulfur, conforming coal (coal that meets a given standard) and the use of scrubbers. Others, especially in Western states such as Wyoming, must employ scrubbers because no coal is available anywhere that has a sulfur content low enough to meet the particular state standard. Finally, in some Eastern areas low-sulfur coal is too expensive because of the long shipping distance, and it may not be available anyway.

If high-sulfur coal is available nearby, it is usually less expensive to use this coal than to ship low-sulfur coal from a greater distance (low-sulfur coal usually comes from Montana and Wyoming, or from eastern Kentucky and West Virginia). Thus the main economic factor in choosing between conforming coal and scrubbers is the annualized scrubber cost versus incremental coal cost. For existing boilers there are additional problems, since the lower sulfur content has an adverse effect on precipitator operation and the different coal characteristics may cause fouling of the boiler and poor combustion.

The amount of sulfur oxides generated by combustion processes is quite large, estimated to exceed 35 million tons in 1970 and expected to increase by more than threefold by the year 2000. EPA estimates that more than half of all the "man-made" SO_2 comes from electric power plants. For individual plants, use of a typical 3.5% sulfur coal results in an emission exceeding 600 tons of SO_2 per day from a 1000-MW boiler.

Most of the sulfur in the coal is oxidized to SO_2 and is emitted with the combustion gas. Some is retained in the ash, usually only 5% or so, but much more when the ash contains a high percentage of free lime. A small percentage is oxidized to SO_3—the extent depending on the amount of excess oxygen (usually higher in older boilers), boiler temperature, and other factors. The usual range is 0.5–5% oxidation.

The problems with obtaining and using coal that conforms to SO_2 standards have made scrubbers the choice for numerous utility and industrial boilers. Design of scrubbers for very large boilers is a major undertaking; for a 1000-MW boiler the quantity of gas to be scrubbed is quite large, on the order of 125,000 tons/day. In addition, scaling and corrosion have presented formidable problems in design.

22.3.2 Throwaway Operation—Dry Processes

Most of the processes for removing sulfur oxides from waste gases involve contacting the gas with a material that combines with the sulfur oxides, thus converting them to a liquid or solid that can be removed. Either chemical reaction or adsorption may take place. In either case, if the product is to be discarded (throwaway process), the added material must be quite inexpensive. Only three materials have been seriously considered for throwaway processes, including limestone, water, and naturally occurring sodium salts. Water has the disadvantage that SO_2 is not very soluble, thus requiring large volumes. Use of the sodium salts is limited geographically. For these reasons most of the emphasis in throwaway process development has been on the use of limestone.

The methods for using limestone can be divided into two major classes, dry and wet systems. In the dry system the limestone (as is, or calcined to produce lime) is injected into the gas and recovered after reaction with the sulfur oxides. Hence equipment is required both for injecting into the gas and for removal.

22.3.2.1 Chemistry and Kinetics

Limestone injection is complicated by the three different types of reactions that must take place—calcination, absorption, and oxidation. The main reactions are as follows:

$$CaCO_3 \rightarrow CaO + CO_2 \tag{22.3}$$

$$CaO + SO_2 \rightarrow CaSO_3 \tag{22.4}$$

$$CaSO_3 + \tfrac{1}{2}O_2 \rightarrow CaSO_4 \tag{22.5}$$

Calcination is necessary for good SO_2 absorption because the reaction of SO_2 with $CaCO_3$ is extremely slow at the temperature below which $CaCO_3$ is stable (~1400°F). Thus limestone must be injected at the point at which the temperature is high enough (on the order of 1500–2100°F) to give an adequate calcination rate.

Sulfation and oxidation appear to occur simultaneously in the boiler. About 1800°F is the optimum temperature, above which the lime tends to "dead burn" and become unreactive because of the decrease in internal pore area. At lower temperatures the reaction is too slow.

The kinetics of SO_2 absorption in lime are quite complicated. Pertinent factors include retention time in the boiler, temperature, inlet SO_2 concentration, lime particle diameter, sulfate loading, the Arrhenius parameters (frequency factor and activation energy), excess lime present, degree of dead burning, surface area of the lime, oxygen and moisture contents in the gas, and pore diameter in the lime particles. The general mechanism is stepwise, as follows.

1. Calcination produces a pore structure whose nature depends mainly on limestone type and on temperature.
2. The SO_2 reacts rapidly with the fresh pore surface in the initial phase, during which the chemical reaction rate is the controlling resistance.
3. Further absorption proceeds by a combination of two types of diffusion: diffusion into pores so small that the diffusion rate is controlling, and diffusion of SO_2 through the surface layer of $CaSO_4$ formed during the initial phase.

The ideal pore size appears to be 0.2–0.3μm. Smaller pores become blocked rapidly, and larger pores do not have as much surface area, which is important because it is desirable to have as much absorption as possible take place during the initial phase.

22.3.2.2 *Conventional Boilers*

Some of the earliest work on SO_2 removal was done by injecting ground limestone into conventional boilers. The process appeared to be attractive because little investment would be required.

In the earlier tests the limestone was merely mixed with the coal or injected along with it; but it soon became evident that dead burning at the high temperature made the lime too unreactive. The limestone was then injected at a point where the temperature level was around 2000–2100°F. Results were fairly promising in German, Czechoslovakian, and Japanese tests,[29] and EPA funded a major program in the United States to test the process at the Shawnee station of the Tennessee Valley Authority (TVA).[3] An extensive program was conducted using a 175-MW boiler, but results were poor. It was difficult to achieve good distribution of the limestone over the boiler cross section, and the retention time in the zone of effective temperature (high enough to get an adequate absorption rate but not so high as to cause dead burning) was too short for SO_2 to penetrate very far into the particles. Only about 40% removal could be obtained, even with injection of 3 times the stoichiometric amount of limestone. Plugging in the tube area and loss of precipitator efficiency were also major problems.

The TVA–EPA program ran from 1970 to 1973 and then was terminated as unsuccessful. It is not likely that there will be any further application to conventional boilers.

22.3.2.3 *Fluidized-Bed Combustion*

Combustion of coal in a fluidized bed (see Chapters 20 and 21: Sections 20.1.1–20.1.4, 20.3.2, 20.3.3, 21.5) is a process being developed in England and the United States.[30] Its objective is to operate with the boiler tubes in the bed, which gives better heat transfer and allows use of a smaller boiler. Although not likely to be available commercially for a decade or so (if at all), the process appears to have some promise for operation in a combined-cycle mode (with the combustion chamber of the boiler operated under pressure).

One of the advantages claimed is that SO_2 removal can be carried out in an economical way by injecting limestone into the bed. Compared to conventional boilers, the retention time in the bed is much longer; therefore the efficiency of limestone utilization should be higher.

[29] Personal communications from numerous organizations during several tours of the utility industry in Japan, 1970–1977.

[30] *Proc. Fluidized Combustion Conf.*, Institute of Fuel, London, September 1975.

It is not clear whether SO_2 removal carried out in this way would be efficient enough to meet emission regulations in the United States. The results in England, where most of the test work has been done, have not been impressive in this respect. Although the lime retention time is extended, the SO_2 retention time is not; thus there is the same problem of attaining adequate transfer of SO_2 into the lime particles. Furthermore, pressurized operation appears to be more attractive than atmospheric operation, but this makes it more difficult to calcine the limestone because of the higher partial pressure of CO_2. The operating temperature in the fluidized bed is not very high anyway—1500–1650°F, which is not much higher than the temperature at which CaO absorbs CO_2 under conventional boiler conditions.

Nevertheless, fluidized-bed combustion has the significant advantages of lower NO_x formation and the ability to burn low-grade coals that would be difficult to use in a conventional pulverized-fuel boiler. If the removal efficiency can be increased to a level adequate to meet the standards, these advantages would go far toward offsetting the large excess of limestone (and the resulting large solid waste tonnage) likely to be required for SO_2 removal.

22.3.3 Throwaway Operation—Wet Scrubbing

Use of the wet scrubbing operation with lime or limestone as the absorbent is by far the most popular method for removing SO_2 from waste gases. A high degree of removal can be attained with good utilization of the absorbent. Despite several unresolved design and operating problems, the advantages are such that companies operating large boilers seldom give much attention in planning to any process other than lime or limestone scrubbing.

22.3.3.1 *History*

The earliest work on the wet scrubbing process was done in England,[3] where the well-known Battersea and Bankside (London) scrubbers were installed several decades ago. The process was based on water scrubbing, with much of the Thames River diverted to run through the scrubbers. It was found necessary to add lime or chalk, however, to prevent the system from reaching a low pH.

The extreme cooling of the gas resulted in an environmental problem associated with the drooping plume. Some of the scrubbers are still operating, but the installations are considered obsolete.

The next effort in England was the Howden-ICI process, in which lime or chalk slurry was used in a closed-loop system (no effluent to a water course, as at Battersea and Bankside). The method was installed at the Fulham station and operated a few years until it was closed down during World War II.

In the United States pilot-plant work on the wet scrubbing process was done by TVA in the early 1950s, by Wisconsin Electric in 1965, and by Detroit Edison in 1966. Detroit Edison's work resulted in two full-scale systems, one at Union Electric (St. Louis) and one at Kansas Power and Light (Lawrence). Both of these, involving 125–140 MW boilers, were started up late in 1968. This can be considered the beginning of modern lime-limestone scrubbing technology.

Several lime-limestone systems have also been installed in Japan. Although most are associated with oil-fired boilers, the experience gained is applicable, at least in part, to the coal-fired boilers in other parts of the world.

22.3.3.2 *Chemistry and Kinetics*

The chemistry of lime-limestone scrubbing is complex and is not fully understood. The overall reaction is CaO or $CaCO_3$ plus SO_2 to form $CaSO_3$, the same as the intermediate reaction in dry absorption, but with only limited oxidation to $CaSO_4$ ($CaSO_4 \cdot 2H_2O$ in this case). The actual reaction path to these products is complicated, involving gas-liquid reactions, both ionic and nonionic reactions in the liquid phase, and liquid-solid reactions. The system has been studied by Lowell,[31] who assumed that the key species are calcium and magnesium from the lime; CO_2, SO_2, SO_3, O_2, and NO_x from the gas, and sodium, potassium, chlorine, and others from

[31] P. S. Lowell, *A Theoretical Description of the Limestone Injection—Wet Scrubbing Process*, 1970, 1; PB-193029.

the ash. These are assumed to interact, giving 41 neutral and ionic species in solution and 7 solid species. The ionic reactions (22 in all) that result in the formation of ion pairs from the individual ions were then studied; examples include:

$$Ca^{2+} + HCO_3^- \rightarrow CaHCO_3^+ \quad (22.6)$$

$$H^+ + HSO_3^- \rightarrow H_2SO_3^0 \quad (22.7)$$

$$Mg^{2+} + SO_4^{2-} \rightarrow MgSO_4^0 \quad (22.8)$$

Equilibrium constants were determined for these reactions, activities calculated, further equations developed to cover ionic reactions and solid-liquid equilibria, and mass balance equations set up. The final result was 36 equations with 36 unknowns, for which a computer program was developed to solve the equations for any given input of the major species. This program is useful not only for elucidating the chemistry but also for calculating slurry or solution composition in a given situation. However it is limited in that it gives only equilibrium data.

One feature of the chemistry of lime-limestone scrubbing is that calcium sulfite and sulfate can become highly supersaturated. It can be shown that there is a critical level of sulfate supersaturation, about 1.3–1.4 times the concentration at saturation, above which nucleation suddenly accelerates. This can result in scaling of scrubber surfaces. Sulfite, however, can supersaturate to much higher levels without incurring this effect.

To absorb a given amount of SO_2 from the gas, a sufficient amount of basic species must be present in the liquor entering the scrubber. Basic species are those that act as bases in the first ionization step of sulfurous acid, such as OH^-, $SO_3{}^{2-}$, HCO_3^-, $CO_3{}^{2-}$, $CaOH^+$, $CaHCO_3^+$, $CaSO_3$ (1), $MgSO_3$ (1). These species can react to hold SO_2 in solution. The factors affecting the concentration of basic species have been studied by Hatfield and Slack.[32] In tests with lime and limestone, the species present in highest concentration were HCO_3^-, $CaSO_3$ (1), $MgSO_3$ (1), $CaHCO_3^+$, and $MgCO_3$ (1). The concentrations are greatly increased when magnesium is added, either by use of dolomitic lime or by adding a soluble magnesium compound during limestone scrubbing.[33] Chloride, however, has the opposite effect.

The main kinetic factors involved in lime-limestone scrubbing are as follows:

1. Diffusion to and through the gas film at the liquid surface.
2. Rate of SO_2 dissolution and hydration to H_2SO_3, H^+, and HSO_3^-.
3. Diffusion of H_2SO_3 and ions through the liquid film and into the interior of the droplet.
4. Dissolution rate of $CaCO_3$.
5. Reaction rate of Ca^{2+} with acidic ions to precipitate $CaSO_3$.

In limestone scrubbing, the rate of $CaCO_3$ dissolution is especially important because limestone reacts slowly with a weak acid such as H_2SO_3. The reaction rate of lime is much faster, and as a result lime systems can generally be operated at a lower slurry recirculation rate than limestone systems. Few data, however, are available on these rate factors. It is known that the rate increases directly with inlet SO_2 concentration, amount of excess absorbent, liquid-gas ratio, absorbent surface area (fine grinding of limestone), and gas velocity; but data on the interactions and relationships between these are scarce.

The precipitation of the $CaSO_4$ formed by sulfite oxidation is an important part of lime-limestone scrubbing chemistry. As noted earlier, if the sulfate supersaturation exceeds the concentration at saturation by 1.3–1.4, rapid nucleation and scaling are likely to occur. One factor that tends to reduce or eliminate supersaturation is coprecipitation of $CaSO_4$ in the calcium sulfite crystals, a phenomenon that is not completely understood but can be quite helpful in avoiding scaling by gypsum. It has been found that if not more than 20–30% of the sulfur is converted to the sulfate form in the scrubber (the critical level depending on several factors), this amount of sulfate will coprecipitate with the sulfite and keep the solution from becoming saturated with respect to sulfate.

[32] J. D. Hatfield and A. V. Slack, "Lime-Limestone Scrubbing: Factors Affecting the Concentration of Sulfur Dioxide-Adsorbing Species in Solution," in Am. Chem. Soc. Adv. Chem. Ser. No. 139, 1975.

[33] J. M. Potts, A. V. Slack, J. D. Hatfield, *2nd Int. Lime-Limestone Wet-Scrubbing Symposium*, EPA, November 8–12, 1971.

22.3.3.3 *Direct Use of Lime or Limestone*

Lime or limestone can be used in scrubbing operations in two ways—directly and indirectly. As the name implies, in direct use the absorbent is added to the circulating slurry and comes into direct contact with the gas. In indirect methods the absorbent consists of a clear solution of a material with an affinity for SO_2, and the scrubber effluent solution is treated with lime or limestone outside the scrubber to precipitate calcium sulfate and regenerate the absorbent.

Direct processes all have the same general flowsheet (Fig. 22.2). The makeup lime or limestone is added to, or upstream from, a tank in the main scrubber circulation loop. This tank serves as a reaction tank, or just before it, in which the absorbent can react with HSO^{-}_3 formed in the scrubber and crystallization of the product solids can occur without causing scaling on the inside of the scrubber. If lime is used it is slaked (hydrated) before addition. Limestone when used must be ground to a fairly small particle size.

The scrubbed gas is passed through a mist eliminator to knock out mist droplets entrained in the gas stream and is then usually reheated to restore buoyancy and prevent condensation in the stack. Part of the product slurry is bled off from the hold tank and sent to a pond, where the solids settle out. In some cases it may be more economical to remove the water by settling (thickening), plus filtration in some instances, before the solids are transported to the disposal area.

There are numerous variations of the simplified system shown in Fig. 22.2, including no preremoval of dust, locating the fan after the scrubber, use of a prescrubber (usually a venturi) to humidify and cool the gas, downflow of gas in the scrubber, use of a horizontal scrubber (horizontal flow of gas), absence of hold tank, mist eliminator in slanted or horizontal position, mist eliminator outside the scrubber, no reheat, and special treatment of the separated solids (sludge) to stabilize them (increase strength and reduce leachability).

Choice of Absorbent. The choice between lime and limestone is often difficult, with several tradeoffs that must be evaluated in making the decision. In general, limestone is cheaper but not as reactive or as easy to use effectively in the scrubber.

With respect to the economics of limestone, although the quarry cost is relatively low, shipping cost per unit of Ca^{2+} is higher than for lime because of the weight loss in calcining. Thus in Western areas, where deposits are spotty and the limestone must be shipped long distances, lime is often cheaper per unit of Ca^{2+}.

In addition, limestone must be ground to a very fine size to give an adequate dissolution rate, whereas crushing to lump size is all that is required for calcination. There is additional expense beyond calcination, however, because of the slaking required. The slaker is expensive but probably costs less than grinding equipment.

The water used for preparing the two absorbents is another consideration; slaking (needed when using lime) requires fairly pure water, whereas grinding (limestone) can be

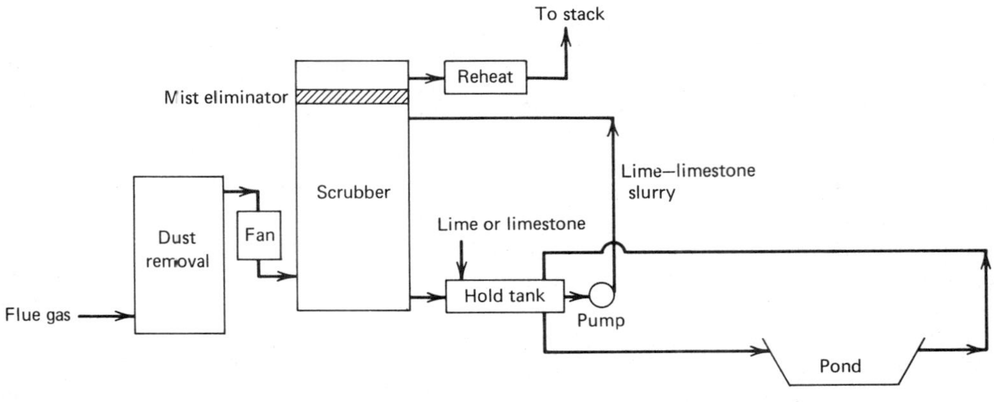

Figure 22.2 Simplified flowsheet of lime-limestone scrubbing process.

done with recycled supernatant liquor from the pond, with cooling tower blowdown, or any other wastewater. This can have a significant effect on the water balance, which gives limestone an advantage because of the limited amount of water that can be added to the system in any scrubbing process.

The lower reactivity of limestone can be offset by recirculating more slurry, grinding to finer particle size, use of two scrubbing stages, or providing more contact time in the scrubber. All these possibilities are expensive but may be justified by the lower cost of limestone.

The use of limestone is subject to question when very high removal efficiency is required, as in areas where the sulfur content of the coal is high and emission limits are low. In such cases lime may be the best choice, or it may even be worthwhile to employ an indirect system, in which a highly reactive absorbent can be used. Nevertheless, there is evidence that techniques such as two-stage scrubbing or sparged scrubbing can give high removal efficiency with limestone.

The content of impurities is another consideration, especially with respect to magnesium. Limestones vary widely in magnesium content; some contain very little (high-calcium limestone), but others (dolomitic type) contain varying proportions of dolomite ($CaCO_3 \cdot MgCO_3$), and thus can contain 25% or more $MgCO_3$. This is undesirable when the limestone is used without calcining because dolomite reacts more slowly than limestone with weak acids. If calcined, however, the resulting MgO is more reactive than CaO and also helps the operation in other ways, as is discussed later. Dolomitic lime is now being used intentionally, rather than simply because it happens to be the type most readily available.

The reactivity of limestone may also be affected by crystallographic and other properties of the finely ground particles. Some types of particles appear to be more porous or more easily broken down in the slurry; however, there is no good system for classifying the various types with respect to this property.

The tradeoffs balance out to the point that both absorbents are used. There is not a significant difference in the number of installations that use one type of absorbent or the other.

Deposition of Soft Solids. Since the recirculating slurry contains a fairly high content of solids, usually in the range of 5–15%, there is often a problem caused by deposition on equipment surfaces of "soft" solids (called "soft" because the mechanism involves sticking or silting as opposed to formation of a hard scale). The main remedy is to avoid stagnant areas and horizontal surfaces as much as possible.

A point of particular concern is the "wet-dry interface," where the gas first encounters liquid as it moves through the system and the drying effect can produce a solid buildup. The interface in single-scrubber systems is normally in the duct leading into the scrubber, as shown in Fig. 22.2, but in some cases it is located in a venturi prescrubber used to humidify the gas. Soot blowing (jets of air, liquor, or steam) is often used to remove such deposits.

The mist eliminator is also subject to soft solids deposition because the drops caught on the baffles tend to stick. Careful washing is required to keep removing such deposits without incurring wash water carrythrough.

Scaling. Crystallization of calcium sulfite or sulfate on scrubber surfaces was the main problem in early work with lime and limestone, and it continues to be a major consideration, although progress has been made. The main problem, as noted earlier, is that the formation of $CaSO_4 \cdot 2H_2O$, which occurs as the slurry moves down through the scrubber, can increase supersaturation to the point that there is a sudden increase in nucleation and consequent crystallization on surfaces.

The early work in England[3] established three basic methods for avoiding this effect:

1. Circulate a large amount of slurry so that the amount of SO_2 picked up per unit volume of solution will not cause the critical limit of supersaturation to be increased beyond the danger point.
2. Provide enough delay time outside the scrubber for reactions started in the scrubber to be completed, for the added lime or limestone to react fully, and for calcium sulfate and sulfite to crystallize from solution—thereby reducing supersaturation to a minimum in the solution returned to the scrubber.
3. Carry enough crystals in the slurry to

provide a crystal surface so that the dissolved salts will crystallize on that surface instead of on scrubber surfaces.

These methods worked well in the British tests and have been effective elsewhere when the quantitative limits set by the British workers were followed. The liquid-gas ratio (gallons of slurry circulated per 1000 ft^3 of gas) should be around 40 for lime and 65 for limestone, although lower levels may be acceptable if offsetting increases are made in the other two factors mentioned above. Delay time outside the scrubber is affected by other considerations but should be in the range of 5–10 min. Crystal content in the slurry should be around 3% each of calcium sulfate and sulfite.

One problem in maintaining high crystal content is that if ash is also collected in the scrubber, it will take up part of the allowable solids concentration—which is usually held at about a maximum of 15% to keep the slurry from becoming too thick. The 15% level, however, allows adequate crystal content unless the ash-SO_2 ratio is unusually high.

As mentioned in Section 22.2.3.2, much of the sulfate will coprecipitate with the sulfite, thus providing another means of minimizing sulfate scaling. The system can be operated below the sulfate saturation point (no supersaturation at all) if oxidation of sulfite to sulfate can be kept within limits. Whether this is possible depends largely on the ratio of oxygen to SO_2 in the gas. If it is high, as when low-sulfur coal is burned or leaks occur in the air heaters, oxidation may be beyond the 20–30% limit and gypsum will precipitate. About all the designer can do is to minimize excess oxygen in the gas and the exposure of the slurry to air.

Sulfite scaling can also be a problem but is not as serious as sulfate scaling, because the sulfite can supersaturate to high levels without catastrophic nucleation. Such scaling has been encountered in limestone scrubbing when an excess of absorbent was used.

Scaling can occur in systems using lime if the pH in the returned scrubber liquid is too high, in which case carbonate or sulfite is thrown out of solution when it encounters the gas and the solution in the scrubber.

The amount of fresh water added in the scrubber loop is a major consideration, since it dilutes the solution, and thus reduces supersaturation. If the system is "closed loop" (i.e., no liquid is discharged), the water addition is limited to that required for replacing both water lost by humidifying the gas stream and that remaining with the sludge in the pond. Many systems, however, discharge supernatant liquor from the pond to watercourses; thus they can use more fresh water in the scrubber loop.

A more recent addition to the methods for reducing scaling involves carrying a fairly high Mg^{2+} content in the circulating slurry. This way, much of the SO_2 is absorbed in the scrubber as $Mg(HSO_3)_2$, which has a much lower tendency to cause scaling than $CaSO_3 \cdot 0.5H_2O$ or $CaSO_4 \cdot 2H_2O$. The $Mg(HSO_3)_2$ reacts with lime or limestone in the hold tank to precipitate the calcium compounds.

The use of Mg^{2+} to prevent scaling is not new. For example, it has been the practice for many years to reduce scaling of equipment during the neutralization of waste acids by using dolomitic rather than high-calcium lime.

Scrubber Design and Performance. There is a difficult choice to make between the simpler and less expensive types of scrubbers, which are less prone to plugging, and the types of scrubbers with internal elements that promote mass transfer but also cost more, are more susceptible to plugging, and are more difficult to maintain.[34] The usual fixed-bed type of packing gives good mass transfer but is more subject to plugging and scaling than other types. High gas velocity is also helpful, but in vertical scrubbers it is likely that too much mist will be carried into the mist eliminator.

Since limestone dissolves at a relatively slow rate, it is helpful to have a large holdup of slurry in the scrubber to provide a large limestone surface area in contact with the solution. Scrubbers differ in the amount of holdup provided.

In some installations, moreover, the scrubbers must remove part or all of the dust in addition to the SO_2. This requires a scrubber with some pressure drop and even more resistance to plugging.

Unlike most scrubber installations, those

[34] A. V. Slack, "Comparison of Conventional Stack Gas Scrubbing Processes," paper presented at 3rd Int. Conf. Coal Gasification and Liquefaction, University of Pittsburgh, August 3–5, 1976.

attached to sources of SO_2 emission, such as power plants and smelters, are subject to widely varying gas flows. This can be compensated for by installing several scrubbers on each boiler and shutting off one or more if the gas flow drops. The operation is simpler, however, if all the scrubbers can be left on line and the flow through each be reduced, which requires a type that can operate efficiently at a gas velocity less than that for which it was designed.

To achieve the lowest investment cost, it is desirable to make the scrubbers quite large, but the cross section may become so large that problems arise in attaining uniform gas and liquor flow. Some scrubber types perform much better in this respect than do others.

No one type of scrubber gives the best performance for each of these criteria, and designers have varied widely in their choices. The major types are shown in Fig. 22.3, in simplified form.

Several fixed-packing scrubbers (packed tower) are in use, particularly in Japan. An open packing (cooling tower type) is used. Absorption efficiency is good, but even with low-sulfur coal and open-loop operation, scaling in the packing has been a problem.

Mobile packing has the advantage of good mass transfer, with little chance for plugging

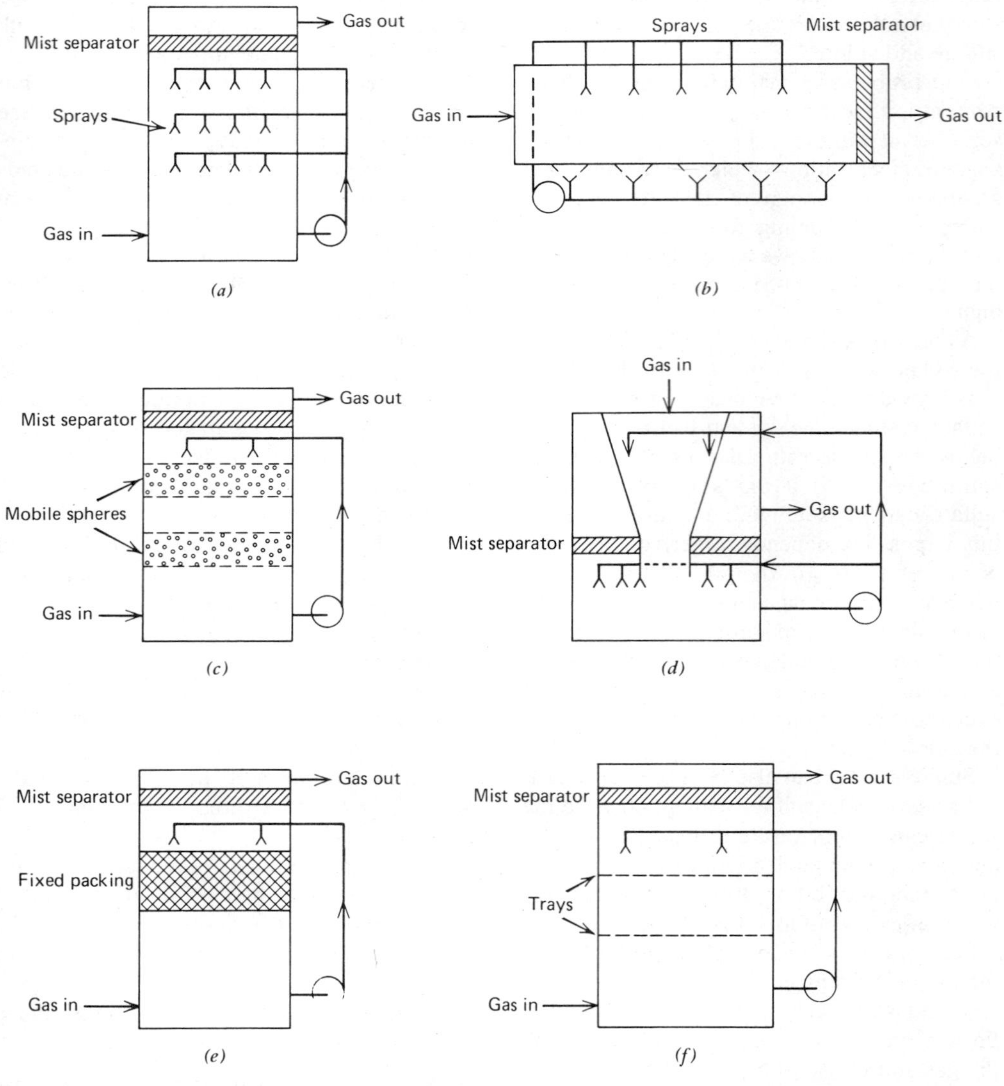

Figure 22.3 Six scrubber types used in lime-limestone scrubbing: (*a*) spray type (vertical), (*b*) spray type (horizontal), (*c*) mobile bed, (*d*) venturi, (*e*) packed tower, and (*f*) sieve tray.

because the movement of the spheres keeps solids from adhering. Several units have been installed at smelters in the United States and Japan, and systems are under construction at utilities in the United States. Good removal of both SO_2 and particulates has been achieved in various pilot-plant tests and in the EPA–TVA test facility at Shawnee.[35] The main drawback is ball wear, but more resistant spheres (plastic and rubber) are being developed.

The vertical spray tower gives maximum protection against plugging and ability to adjust to the load (turndown) is good, but mass transfer is relatively poor. Nozzle wear may also be a problem because there is a greater need to break up the slurry into fine drops in spray towers than in scrubbers, which depend on internal elements to attain good gas-slurry contact. The horizontal spray chamber has the advantage that it can be operated at above 20 ft/s, whereas the vertical towers generally are operated at not more than 8 ft/s. The higher velocity not only reduces scrubber size and cost but also apparently increases mass transfer to the point that it offsets the lack of packing or other internal elements. Nozzle wear is a problem, and more area is required than for vertical scrubbers.

The venturi scrubber has no internal elements, velocity is high, and there is no nozzle wear. Turndown, however, requires a variable throat mechanism, and retention time is so short that two venturis in series have been required for adequate SO_2 removal. Performance can be improved by use of additives (e.g., MgO) and sprays in the mist knockout section, making a single venturi adequate unless a very high removal efficiency is required.

In the sieve-tray type of scrubber the gas flowing up through the holes holds a body of slurry on the plate. Violent agitation provides good mass transfer. However the plates can become plugged, and there is some doubt about turndown capability.

It should be noted that work now being done on most of these scrubber types may eliminate some of these drawbacks.

[35] *EPA Alkalysized Scrubbing Test Facility, Advanced Program,* 3rd Progress Rep., September 1977, EPA-600 7-77-105.

Mist Separation. Probably the most difficult problem in lime-limestone scrubbing is removing mist from the gas and at the same time avoiding scaling of the mist eliminator surfaces. Since chevrons (zigzag baffles) are relatively inexpensive, have a low pressure drop, and take up little room, they are used in most lime-limestone scrubbing installations. The usual arrangement is to install the chevron at the top of the scrubber in a horizontal position, perpendicular to the gas flow (see Fig. 22.3). At least three "passes" (individual baffles, each turning the gas to a new direction) are normally used, and there may be two or more "banks" (sets of baffles) with some distance between them.

The main drawback to the horizontal position is that the gas flowing up tends to keep separated mist in the chevron rather than allowing it to drain back into the scrubber. As a result, mist can remain on the surface of the chevron and be reentrained into the gas. Tests have shown that much better performance is obtained if the slats are slanted longitudinally or are set vertically, so that the mist can drain off. This can be accomplished either by placing the unit in a horizontal duct after the scrubber or by using a slanted arrangement in the top of the scrubber, similar to the rafters of a house.

If a horizontal chevron is used, the gas velocity is usually limited in commercial practice to about 8 ft/s to avoid excessive hangup and reentrainment, whereas tests have shown that 10 ft/s is necessary to achieve adequate impact of the gas on the baffles and that up to 20 ft/s is feasible if the unit is in a vertical position. Such high velocity gives much better efficiency but requires "hooks" or pockets on the trailing edges of the baffles to catch collected liquid pushed forward by the gas and thus prevent reentrainment.

The cross-flow scrubber is especially suited to vertical chevron installation because the unit can be mounted at the end of the scrubber without any duct bend. The bend required for a vertical scrubber causes trouble because soft solids ("mud") sticking in it may change the gas flow pattern enough to result in nonuniform gas flow through the chevron.

Cyclonic chamber collectors are also effective, even at high gas velocities, but have been used in only two or three SO_2 scrubbing installations.[3] They are relatively large and usually more expensive than the other types.

In addition to "mud" sticking on the baffle surfaces, there is also the problem of the liquor remaining on the baffles long enough for salts (especially calcium sulfate) to crystallize out of solution and cause scaling. This is particularly true for horizontal chevrons, since the liquid tends to be held in the baffles by the force of the gas flowing up, and the separator then becomes a miniature scrubber, absorbing residual SO_2 from the gas and reacting it with residual lime or limestone in the collected droplets. Sulfate supersaturation in the drops can thus increase rapidly and result in scaling. The tendency to plug is increased by the action of the scale crystals knitting the "mud" solids together and forming a relatively hard mass.

Tests in the EPA–TVA test program at Shawnee[36] showed that mist separator scaling can be significantly reduced by adjusting scrubber conditions to attain high absorbent utilization. Since little unreacted absorbent would then reach the separator, formation of calcium sulfate would be limited.

The presence of Mg^{2+} in the scrubbing slurry is also quite helpful in preventing scaling, just as in the scrubber itself. The few installations that have used dolomitic lime have had little trouble with mist eliminator scaling.

Even with these improvements, mist eliminators must be washed to avoid "mud" accumulation. Adequate washing is not easy to accomplish. It is difficult to get the wash liquid to all areas with sufficient intensity, not only because of structural considerations but also because there may not be enough of the proper type of wash liquid available. If recycled clear liquor from the pond or thickener is used, there is an almost unlimited amount of liquor available; but using this liquor for washing usually causes trouble because it is already saturated with calcium sulfate. If Mg^{2+} is carried in the system, the problem is much less severe.

Fresh water has been used for washing, and equipment has been installed to catch the wash so that it can be recycled, thus increasing the volume of wash liquor. Such equipment is subject to scaling and sedimentation in some installations.

There has been some indication that washing for only a few minutes per day is adequate if the wash is heavy and intensive. With this approach, fresh water can be used because the total volume required is small. This, plus use of Mg^{2+}, appears to be the best way of solving the problem of plugging in the mist separator.

Reheating. The flue gas is cooled down from the air heater or precipitator outlet temperature (275–300°F) to about 130°F by humidification in the scrubbers. This is undesirable from the environmental standpoint because the plume comes back to ground level more quickly than it would if it were not cooled; thus it is not diluted with as much air. This would not be significant if all the pollutants were removed, but such high efficiency is seldom if ever accomplished. As a result, a given reduction in emission, say 80%, actually means that there is less than an 80% reduction in ambient concentration.

In addition, the saturated gas can cause moisture to condense, with adverse effects on downstream equipment, since the condensate film absorbs residual SO_2 and quickly becomes acidic. Corrosion resistant materials must be used unless reheating is provided.

Practice with respect to the degree of reheating varies widely. In Japan the gas is normally reheated to a high level, approaching air heater outlet temperature.[3] The objective is to avoid a visible plume of condensed water vapor in the crowded industrial areas. In the United States there are several instances of the other extreme—no reheating at all.[3] The usual practice in the United States, however, is to reheat to about 50°F, enough to avoid downstream condensation in equipment.

One of the potential problems of not reheating at all is "rainout" of condensate around the stack. There are varying opinions on this issue; some who operate without reheating claim no trouble, but others have had rainout, particularly during very cold weather. Various means to reduce the problem have been proposed and a few used, including low stack velocity, barriers on the stack interior to prevent condensate from "crawling" up the wall, heavier stack insu-

[36] M. Epstein et al. *Proc. Symp. Flue Gas Desulfurization,* New Orleans, March 1976, EPA 600 2-76-136a.

lation, and passing warm gas up the annular space between the stack and the lining.[3]

There is also wide variation in the type of equipment used for reheating. The approaches include (1) tubed heaters in the gas stream using steam or hot water from the boiler as the heating medium, (2) heating air by using steam or hot water from the boiler and injecting the hot air into the gas, and (3) direct firing of oil or gas with injection of the hot combustion gas into the scrubbed gas stream.

The equipment most often used involves the tubed heaters in the main gas stream. Most installations have had corrosion problems with this equipment because any mist passing through the mist separators impinges on the reheater tubes and results in corrosion, especially under the solid deposits that form when the drops evaporate. Remedial measures include use of high-nickel alloys (expensive and sometimes not effective), recirculating heated gas through the tubed heater, and frequent soot blowing to remove deposits (probably the most effective and most economical). In gas recirculation a portion of the main gas stream (say, at 200°F) is passed through a tubed heater to heat it to a much higher temperature; then it is injected back into the main stream to heat the 120°F gas up to 200°F. This method should be effective, since the mist would be evaporated before it reached the tubes, but it has not been tried in large boilers.

Air heating is expensive because the air must be heated as well as the gas. Gas or oil reheating is not feasible in many cases because the fuel either is not available or is too expensive.

Reheating is quite expensive because of the energy requiremtnt, no matter what system is used. The cost could be reduced by introducing heat exchange between the inlet and outlet scrubber gas, but no acceptable system for doing this has been devised.

Sludge Disposal. The quantity of waste solids generated when using high-sulfur coal as the fuel is quite large, on the order of 400 lb/ton of coal for 4% sulfur. If dewatering is accomplished by simple settling, there will be about 2 lb water/lb of solids, giving 1200 lb/ton of coal or about 6000 tons/day of wet sludge for a 1000-MW boiler.

In comparison, the ash tonnage for coal containing 15% ash is about the same as the dry sludge tonnage above.

The usual practice is simply to use a disposal pond, as shown in Fig. 22.2. Since the solids settle out fairly rapidly, it is usually not difficult to obtain a clear supernatant liquor for recycling. The main problem is the very high cost of constructing the large disposal pond required; moreover, lining may be necessary to meet environmental regulations on water pollution. The size of the pond required depends on the projected operating life of the boiler and the level of operation, both of which can vary widely.

If the pond must be placed at some distance from the boiler, say a mile or more, it may be more economical to operate a thickener at the plant and pump a 30–35% slurry to the pond. Without the thickener, the scrubber would be operated at a fairly high content of slurry solids, say 15%, to reduce pumping cost to and from the pond.

Some environmental authorities, especially EPA, have questioned the acceptability of simple disposal ponds on the basis that the sludge never becomes structurally stable in storage, thus making the land unusable in the future when the pond is full. They also question it because pond liners do not fully assure that leaching will not occur. Instead, they favor a stabilization treatment that chemically converts the sludge into a form that will support weight, thus making the disposal area suitable for normal use in the future. The treatment would also reduce leachability of the material to an acceptable level.

Several stabilization processes have been tested, but only two have reached the commercial stage for SO_2 scrubber sludge. One involves mixing dry fly ash and lime with the sludge, after thickening or filtration to give partial dewatering, to produce a wet solid that can be piled or transported immediately to a landfill area. In time reactions take place between the three constituents to produce a strong solid with low leachability. Since the product can be used as landfill, no expensive pond is necessary.

The other process is a "wet" type, involving addition of a hardening agent (the main constituent is finely ground blast furnace slag) to thickened sludge and pumping the mixture to a disposal pond. There the sludge settles, as in normal pond disposal, and in time hardens to the consistency of clay. This method incurs the cost of the pond but transport by pumping should be less expensive than hauling by truck or other solids transport equipment.

It is not clear which of these two methods is the more economical. Local conditions probably affect the choice, but not enough data have been developed to evaluate the tradeoffs. It is also not clear whether stabilization will achieve the degree of nonleachability desired by the environmental authorities. EPA has an extensive research program underway to resolve this problem.

Another approach is to completely oxidize the calcium sulfite to gypsum (as compared to the partial oxidation that occurs incidentally in the scrubber), since gypsum produced in other industries (e.g., phosphoric acid production) is discarded without any apparent environmental hazard. Moreover, since the gypsum crystals grow to a larger size than the sulfite crystals, dewatering should be less expensive even if stabilization is required. It is not yet clear whether disposal of the unstabilized gypsum will be acceptable for the very large quantities involved. Several development projects in full-scale scrubber installations are being undertaken by engineering firms.

Water Balance. The problems brought about by the limitation on fresh water addition in closed-loop systems were mentioned earlier. The allowable losses (for closed-loop, environmentally acceptable operation) are evaporation in the scrubber, mist carried through the mist eliminator, water of hydration in the product solids ($CaSO_3 \cdot \frac{1}{2}H_2O$ and $CaSO_4 \cdot 2H_2O$), and water remaining with the solids after disposal. The amount of water remaining with the solids varies most widely because the different processes call for varying degrees of dewatering.

The Japanese scrubber systems have the tightest water balances. The sulfite is usually oxidized and the resulting gypsum used in cement and wallboard manufacture. Centrifuges are used to make the gypsum dry enough for sale; at the resulting 10–15% water content, only a relatively small amount of water leaves the system with the solids. (In most Japanese plants, however, some liquor discharge is necessary to purge impurities such as chloride.)

The water balance for disposal in ponds is also affected by evaporation from the pond and by rainfall into the pond. These vary widely depending on pond geometry, climate, and degree of mixing between rainwater and sludge liquor. There is also a difference between a diked pond and one constructed by damming off a small valley or ravine: for the latter, the watershed area is larger than the pond area—thus more rainwater is collected. The rainwater can, of course, take the place of fresh water added to the scrubber loop, but is not as good as fresh water if it mixes very much with the saturated pond liquor. It seems likely that fairly clean water can be obtained by pumping from the pond surface, because of stratification. However if the rainfall gain is larger than the allowable makeup, there must be some discharge from the system and closed-loop operation is infeasible.

Construction Materials. The main problem in selecting the best construction materials for lime-limestone systems is that the slurry causes erosion, which tends to remove the protective layer of corrosion product that normally builds up on metal surfaces. Plastic or rubber coatings are therefore favored. Erosion can still cause trouble, however, if a stream of slurry impinges with some force on a surface.

Another problem with metals, particularly stainless steels, results from the chloride accumulation in the slurry. Some coals contain up to 0.2% or more chloride, which can produce a concentration of several thousand ppm in the scrubber liquor before the bleedstream brings the system to steady state. In addition, cooling tower blowdown is used in some installations as makeup water, in which case the chloride concentration can reach a very high level.

There is a wide variety of construction materials available for the different parts of the system, but the following are most commonly used: rubber-lined pumps, scrubbers with fiber-reinforced plastic linings, scrubber internal elements of high-nickel alloy, nozzles of silicon carbide, mist eliminator of fiber-reinforced plastic (FRP), reheater of high-nickel alloy or mild steel (depending on mode of reheating), fans with a rubber-lined housing, if wet, plus Inconel blade assembly, rubber-lined piping, tanks of mild steel or FRP lining, and stacks lined with FRP, unless a very high degree of reheating is employed.

22.3.3.4 *Indirect Use of Limestone*

The indirect use of limestone (see Section 22.3.3.3) has the advantage that the erosion, "mud" formation, and scaling generally encountered in direct use of lime-limestone are avoided. The alkali and other clear solutions employed in the indirect method simplify scrubber operation to a great extent. The main drawback is at the back end of the system because after calcium sulfite and sulfate are precipitated, the solids must be separated (usually by a filter) and the cake washed thoroughly to minimize loss of absorbent. Some loss occurs, however, because a portion of the solution must be purged to rid the system of impurities such as chlorides.

A simplified flowsheet for the indirect method is given in Fig. 22.4. There are many variations: collection of ash as well as SO_2 in the scrubber, scrubbers of various kinds, addition of an oxidizing vessel in the scrubber loop, use of a centrifuge rather than a filter, and oxidation of the calcium sulfite after precipitation.

The indirect method is generally called "double alkali" in commercial practice because most of the processes use an alkali as the absorbent in the scrubber. Since the method has not attained much commercial significance except in Japan (on oil-fired boilers), only a brief summary is given here.

Basic Chemistry. The chemistry varies according to process type. For sodium-based double alkali systems, the most common type, the reactions are as follows:

$$Na_2SO_3 + H_2O \rightarrow 2NaHSO_3 \qquad (22.9)$$

$$2NaOH + SO_2 \rightarrow Na_2SO_3 + H_2O \qquad (22.10)$$

$$2Na_2SO_3 + O_2 \rightarrow 2Na_2SO_4 \qquad (22.11)$$

$$2NaHSO_3 + Ca(OH)_2 \rightarrow Na_2SO_3 + CaSO_3 + 2H_2O \qquad (22.12)$$

$$Na_2SO_4 + Ca(OH)_2 + 2H_2O \rightarrow 2NaOH + CaSO_4{\cdot}2H_2O \qquad (22.13)$$

Equation 22.12 also holds for limestone, but Equation 22.13 does not proceed very far with limestone as the reactant.

Since some oxidation in the scrubber is unavoidable, Reaction 22.13 is desirable as a means of regenerating the absorbent. Unfortunately, the reaction does not proceed very well, even with lime, unless a very dilute reaction system is used. This "dilute mode" of operation was the type used initially, but required such large equipment that dilute operation was essentially abandoned.

For operation with concentrated solutions, the coprecipitation of sulfate with sulfite must be depended on, as in direct lime-limestone operation (described in Section 22.2.3.2). This became the standard operating procedure, following demonstration in a 20-MW test facility.[37]

Sodium-Based Processes. The sodium-based processes available in the United States are much the same; all operate in the concentrated mode (0.4–1.0 *M* sulfite-bisulfite) and all use lime. For operation at the 1.0 *M* level, as in one of the processes, not all the sulfate is coprecipitated; therefore some must be allowed to escape with the filter cake to purge it from the system. This is undesirable, from the standpoints of both absorbent cost and environmental considerations, but these

[37] *Evaluation of Three MW Prototype Flue Gas Desulfurization Processes,* EPRI FP-713, March 1978.

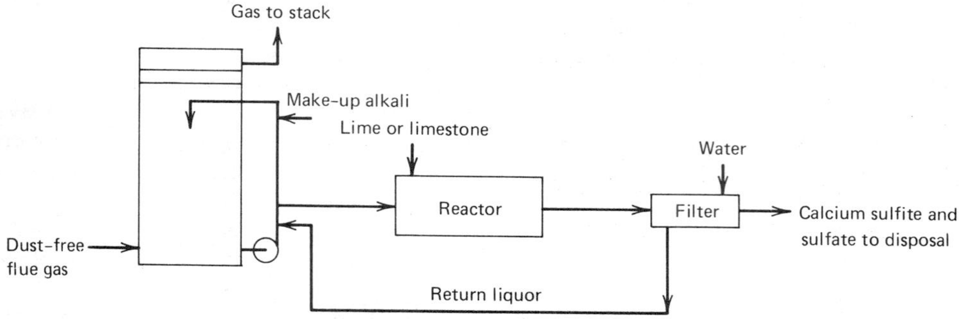

Figure 22.4 Simplified flowsheet for the indirect use of lime-limestone scrubbing.

drawbacks are offset by the smaller equipment size and resulting lower capital cost.

Sulfate purging is not a problem for processes operating at the 0.4–0.5 *M* level, but the purge needed to remove chloride (a relatively small amount) results in some loss of sodium. One process avoids this by using a prescrubber to absorb the chloride in water, thus keeping it out of the sodium loop.

The sodium-based processes are in the first stage of commercial acceptance in the United States. Their main advantage is avoiding the problems brought about by the direct contact of lime or limestone with the gas in the scrubber. If these problems are resolved, however, the complications associated with the double-alkali system would make the use of indirect methods unnecessary.

In Japan sodium-based, double-alkali systems have been installed in several utility and industrial boilers. The usual practice is to use limestone as the regenerant, and in all cases the calcium sulfite is oxidized to gypsum for later sale. The processes are somewhat complicated because of the oxidation and because a separate loop is used to regenerate sodium sulfate, according to the reaction:

$$Na_2SO_4 + 2CaSO_3\cdot\frac{1}{2}H_2O + H_2SO_4 + 3H_2O \rightarrow 2CaSO_4\cdot 2H_2O + 2NaHSO_3 \quad (22.14)$$

Acid-Based Processes. One of the Japanese indirect processes[3] uses water as the absorbent (actually dilute H_2SO_4, but this is incidental), with ferric sulfate carried in the solution to oxidize H_2SO_3 to H_2SO_4. The solution is circulated through an oxidizer tower to complete the oxidation. The weak acid (2–5%) formed in the scrubber-oxidizer loop is bled off and treated with limestone to precipitate gypsum. This process is used in several plants in Japan, one of which has a 350-MW scrubber ($\sim 10^6$ ft^3/min), the largest in the world.

Other Absorbents. Another Japanese process,[3] employed in a small boiler installation and in a sulfuric acid plant, uses basic aluminum sulfate as the absorbent and limestone as the regenerant.

$$Al_2(SO_4)_3\cdot Al_2O_3 + 3SO_2 \rightarrow Al_2(SO_4)_3 + Al_2(SO_3)_3 \quad (22.15)$$

$$Al_2(SO_4)_3 + Al_2(SO_3)_3 + \frac{3}{2}O_2 \rightarrow 2Al_2(SO_4)_3 \quad (22.16)$$

$$2Al_2(SO_4)_3 + 3CaCO_3 + 2H_2O \rightarrow Al_2(SO_4)_3\cdot Al_2O_3 + 3CaSO_4\cdot 2H_2O + 3CO_2 \quad (22.17)$$

Ammonia-based, double-alkali processes have been tested in Japan, France, and the United States, but no full systems have been built. The reactions are similar to those of sodium-based processes except that ammonium sulfate is easier to regenerate than sodium sulfate.

Carbon is used as the absorbent in a large plant in Japan. The resulting H_2SO_4 is washed out of the carbon and treated with limestone to precipitate gypsum, which is used in the construction industry.

22.3.3.5 *Alkali Scrubbing*

The only throwaway process other than lime-limestone scrubbing that has achieved any prominence is the use of naturally occurring sodium carbonate in desert areas. In the western part of the United States such an absorbent is used in a full-scale installation (Reid Gardner station, Nevada Power). The product sodium sulfite–sulfate solution is removed to a disposal pond; in this particular desert area (northern Nevada), the water evaporates to give a solid deposit that is environmentally acceptable.

Such a system is quite economical but is limited, of course, to situations where the sodium carbonate is cheap and the climate is very dry.

22.3.4 Recovery Processes

Because it is obviously more desirable to recover the sulfur from combustion gases as a useful product than to throw it away, there has been a major effort on the part of developers to develop economical recovery processes. Unfortunately, this has been more difficult than expected. Recovery methods are intrinsically more complicated than throwaway methods. Both are very expensive, and any return from the sale of the recovered sulfur does not do much to tip the scales in favor of recovery. This does not mean that it may not be better to produce a salable product in special situations. However there is some question of whether recovery will ever replace lime-limestone scrubbing as the standard SO_2 removal technology.

Recovery processes have the same advantage as double alkali in the scrubbing step; in fact, scrubbing is much the same in both types of process. Both avoid the problems with erosion and solids deposition that are so troublesome in lime-limestone scrubbing.

Recovery methods cover a much wider range of process types than do the throwaway approaches. There are dry and wet processes, some involve oxidation and some reduction, numerous types of absorbents are used, some can tolerate dust and some cannot, and the products can be sulfur, sulfuric acid, liquid SO_2, or ammonium sulfate.

Most recovery processes are still at the pilot-plant or demonstration stage; only one has reached commercial status in the United States, but three or four have been installed in Japan.

22.3.4.1 *Alkali Scrubbing*

Sulfite-Bisulfite Equilibrium. Alkali absorbents have an outstanding advantage when used in recovery processes because of the unique properties of the sulfite-bisulfite equilibrium. In the scrubber sulfite is largely converted to bisulfite by countercurrent operation, and the bisulfite can then be regenerated by heating to a temperature not much higher than the absorption temperature. This is not feasible with most absorbents because if the reaction with SO_2 gives a product with low SO_2 vapor pressure, which is necessary for good removal, the product is too stable to release the SO_2 except at high temperature. The use of the alkali sulfite-bisulfite combination reduces the temperature differential between absorption and regeneration. The reactions are quite simple:

$$M_2SO_3 + SO_2 + H_2O \rightarrow 2MHSO_3 \quad (22.18)$$

$$2MHSO_3 \rightarrow SO_2 + M_2SO_3 + H_2O \quad (22.19)$$

where M represents an alkali such as sodium or ammonium.

The main drawback is that it is difficult to drive the equilibrium to a high degree of bisulfite conversion in the scrubber. To do this requires several stages of scrubbing, with liquor recirculation in each stage. This task is complicated by the need to generate a high concentration of dissolved salts in the scrubber effluent solution, to minimize the energy requirement for regeneration. The high bisulfite content also reduces pH, and thus increases corrosion.

Bisulfite regeneration by heating is a convenient method; however other methods, including acidulation and reduction, are also useful because of the relative instability of the bisulfite.

Sodium-Based Scrubbing. The only recovery process that has attained any commercial prominence is thermal regeneration based on sulfite scrubbing, as discussed above.[38] The gas is scrubbed with recycled Na_2SO_3 solution in a sieve tray tower to give an effluent solution containing about 20 wt % of $NaHSO_3$, 7% Na_2SO_3, and 4–5% Na_2SO_4. The solution is pumped to a double-effect vacuum evaporator system where the water is boiled off, the $NaHSO_3$ decomposed to evolve SO_2, and the resulting Na_2SO_3 crystallized. The crystals are then dissolved and recycled.

The crystallization of Na_2SO_3 is a unique feature of the process. Removal of the sulfite from the liquid phase eliminates its detrimental effect on SO_2 vapor pressure, which should be kept high to get good SO_2 evolution.

The evolved SO_2 stream, after cooling to condense and remove moisture, has an SO_2 content of about 90%. It can be used as feed to a sulfuric acid plant or reduced to elemental sulfur in a Claus-type or other sulfur-producing system.

There are about a dozen installations using this process in Japan, mainly in conjunction with industrial boilers. In the United States one is under construction at the San Juan station (Public Service of New Mexico; low-sulfur coal) and another (high-sulfur coal) has been built as a joint EPA–utility demonstration venture at the Mitchell station (Northern Indiana Public Service) in northern Indiana. In both cases the SO_2 will be converted to elemental sulfur, whereas in Japan H_2SO_4 is the usual product.

One of the main drawbacks to the thermal regeneration process in sulfite scrubbing (thermal scrubbing) is that there is no commercial method for regenerating Na_2SO_4. It

[38] R. I. Pedroso, *Proc. Symp. Flue Gas Desulfurization*, New Orleans, March 1976, EPA 600 2-76-136.

is necessary to remove it in a purge stream and then either discharge the purge or crystallize out the sulfate and sell it. In Japan the general practice is to oxidize the sulfite to sulfate (to prevent oxygen demand) and discharge the oxidized liquor to the sea. This would not be feasible in the United States; it is likely that the sulfate will be separated by crystallization and an attempt made to sell it. The market, however, is limited.

Another approach in Japan, used mainly with small boilers and various industrial processes, is to scrub with sodium carbonate or caustic soda solution in a nonregenerative process, with the product sodium sulfite solution sold to the pulp and paper industry. Again, the market is relatively small.

Treatment with Hydrogen Sulfide. The U.S. Bureau of Mines[3] has worked on a process in which the solution from a sodium-based scrubber is treated with H_2S to produce elemental sulfur directly. Citric acid is carried in the loop to buffer the pH at the low level (3.0–3.7) needed for the H_2S reaction. The chemistry is complex and not fully understood; the following reactions have been proposed (Ct = citrate):

$$(NaO)_3Ct + 3SO_2 + 3H_2O \rightarrow 3NaHSO_3 + Ct(OH)_3 \quad (22.20)$$

$$6NaHSO_3 + 11H_2S + 2Ct(OH)_2 \rightarrow 17S + 2(NaO)_3Ct + 16H_2O \quad (22.21)$$

$$2NaHSO_3 + S \rightarrow Na_2S_2O_3 + H_2O + SO_2 \quad (22.22)$$

$$6Na_2S_2O_3 + 4Ct(OH)_2 \rightarrow 4(NaO)_3Ct + 7S + 5SO_2 + 4H_2O \quad (22.23)$$

In a variation of the process, phosphate is used instead of citrate as the buffering agent.

The main drawback is that two-thirds of the product sulfur must be converted to H_2S and recycled. This not only increases equipment size but also, in the processes commercially available for converting S to H_2S, requires natural gas as the reducing agent. Since gas is usually not available, it will be necessary to develop methods for carrying out the reduction with other materials. The process will be tested on an industrial boiler as a joint government-industry demonstration project.[3] Aside from that, only pilot-plant tests have been made.

Treatment with Other Reducing Agents. Sodium sulfite can also be reduced directly with coke or coal, which should be more economical than using natural gas or other gaseous reducing agents. Reactions are as follows:

$$Na_2CO_3 + SO_2 \rightarrow Na_2SO_3 + CO_2 \quad (22.24)$$

$$Na_2SO_3 + \frac{3}{2}C \rightarrow Na_2S + \frac{3}{2}CO_2 \quad (22.25)$$

$$Na_2S + H_2O + CO_2 \rightarrow Na_2CO_3 + H_2S \quad (22.26)$$

$$H_2S + \frac{3}{2}O_2 \rightarrow S + H_2O \quad (22.27)$$

The scrubber in this case is a spray dryer, which gives solid Na_2SO_3 as the product. The reduction is carried out at high temperature with the Na_2SO_3 and Na_2S in molten form.

The main advantage is that carbon can be used directly as the reducing agent (Reaction 22.25) without first having to convert the carbon to hydrogen and CO. If bituminous coal can be used as the reducing agent instead of coke, the cost of raw materials should be significantly lower.

The process will be employed at a northeastern utility as a joint utility–EPA demonstration project.

Ammonia-Based Scrubbing. A great deal of research effort has been undertaken during the past few decades to develop recovery methods based on ammonia as the absorbent. Ammonia is cheaper than sodium compounds and has certain other advantages.

A drawback common to all ammonia-based processes is the tendency to form fumes in the scrubber. When gaseous ammonia contacts SO_2, the extremely fine solid particles of ammonium sulfite formed can produce a dense white fume that is not removed in the scrubber. Since ammonia is volatile, it is difficult to prevent some of it from entering the gas phase and reacting there with SO_2.

Theoretical calculations and small-scale tests indicate that it should be possible to control solution composition and temperature at levels that would avoid fume formation. It has not been demonstrated, however, that this can be accomplished in full-scale operation.

The main commercial application of ammonia scrubbing has been in Japan,[1] where two large scrubbing systems are under construction for use on iron ore sinter gas. The resulting NH_4HSO_3 neutralized with NH_3 to convert all the SO_3^{2-} to $(NH_4)_2SO_3$, the solution passed through an oxidizer to convert sulfite to sulfate, and the sulfate crystallized to yield fertilizer-grade ammonium sulfate.

This may be a promising recovery process for United States utilities. The system is relatively simple and therefore has minimum capital cost. The extent of the market for ammonium sulfate is the main consideration. For utilities located on a navigable waterway, it seems likely that a sizable tonnage could be sold at the price required to recover the ammonia cost. However the market is more limited than that for sulfur or sulfuric acid.

Other ammonia-based processes are aimed at elemental sulfur or sulfuric acid as the product. In France, for example, thermal stripping has been tested on an intermediate scale (30 MW) with an oil-fired boiler.[39] In this case the ammonia is volatilized along with the SO_2 and both are passed through a catalytic converter, where synthesis gas (H_2 + CO) is added to convert two-thirds of the SO_2 to H_2S. The mixture than travels to a "wet" Claus system, where the H_2S and SO_2 react to form elemental sulfur in the molten form, which separates from the Claus liquid and is tapped off. The ammonia is not affected by this process and is recycled. (See Chapter 18.) The process performed well in the demonstration tests but has not been operated on a full scale.

TVA and EPA have conducted a joint pilot-plant project[3] for several years in which the SO_2 is absorbed by ammonia, and the sulfite-bisulfite solution treated with molten ammonium bisulfate to decompose the sulfites and evolve SO_2. The resulting $(NH_4)_2SO_4$ is heated to an elevated temperature to regenerate bisulfate; the ammonia that has been evolved is recycled:

$$NH_3 + H_2O + SO_2 \rightarrow NH_4HSO_3 \qquad (22.28)$$

$$NH_4HSO_3 + NH_4HSO_4 \rightarrow SO_2 + (NH_4)_2SO_4 + H_2O \qquad (22.29)$$

$$(NH_4)_2SO_4 \rightarrow NH_4HSO_4 + NH_3 \qquad (22.30)$$

The rich stream of SO_2 can be converted into sulfur or sulfuric acid.

Potassium-Based Scrubbing. The only potassium-based process that has reached any significance is that in which potassium sulfite formed in the scrubber is converted to potassium thiosulfate and then reacted with CO in an autoclave to give H_2S and K_2CO_3 for recycling[40]:

$$2KHCO_3 + 2SO_2 \rightarrow 2KHSO_3 + 2CO_2 \qquad (22.31)$$

$$2KHSO_3 + KHS \rightarrow \frac{3}{2}K_2S_2O_3 + \frac{3}{2}H_2O \qquad (22.32)$$

$$\frac{3}{2}K_2S_2O_3 + 6CO + \frac{7}{2}H_2O \rightarrow 2H_2S + KHS + 2KHCO_3 + 4CO_2 \qquad (22.33)$$

$$2H_2S + O_2 \rightarrow 2S + 2H_2O \qquad (22.34)$$

The method has been tested only on a small pilot-plant scale.

22.3.4.2 *Magnesia Scrubbing*

Processes involving magnesia as the absorbent have been studied widely and are used in three commercial installations in Japan.[1] The process is similar to that based on sodium in that $MgSO_3$ absorbs SO_2 to give $Mg(HSO_3)_2$. The latter is precipitated with MgO and the resulting $MgSO_3$ calcined to evolve SO_2 and to regenerate MgO.

$$MgSO_3 + SO_2 + H_2O \rightarrow Mg(HSO_3)_2 \qquad (22.35)$$

$$Mg(HSO_3)_2 + MgO \rightarrow 2MgSO_3 + H_2O \qquad (22.36)$$

$$MgSO_3 \rightarrow SO_2 + MgO \qquad (22.37)$$

Part of the $MgSO_3$ from Equation 22.36 is recycled around the scrubber loop to provide absorbent.

The process has been tested on coal-fired boilers (100 MW or more) in two joint EPA-utility demonstration projects in the United States. Two of the Japanese installations employing magnesia scrubbing are small (25–30 MW equivalent, on a H_2SO_4 plant and copper smelter), but the third is an oil-fired boiler of 162 MW.

22.3.4.3 *Carbon Adsorption*

The cost of reheating in all the wet processes (both throwaway and by-product recovery) is so high that much effort has been expended on developing dry processes. One of the leading dry processes is carbon adsorption, which has been tested in demonstration units

[39] Personal communication.

[40] R. T. Struck, E. Gorin, and W. E. Clark, *Proc. Symp. Flue Gas Desulfurization,* New Orleans, March 1976, EPA 600 2-76-13.6.

in West Germany, the United States, and Japan. Carbon has the unique property that it not only adsorbs SO_2 but also O_2 and H_2O, and then catalyzes the reaction of these to form H_2SO_4 in the carbon pores. All this must take place at a fairly low temperature (175–300°F) because adsorption drops off at higher temperatures. The installations in the United States and Germany are of a demonstration size, 20–45 MW.[3] In Japan there are three commercial systems, ranging from 53 to 150 MW; all three are used with utility or industrial boilers.

There are two general types of procedure: in one the carbon is held in the gas stream until loaded with SO_2 and then moved to another vessel for regeneration, and in the other the carbon stays in place and the gas flow is switched from vessel to vessel. It is not clear which procedure is best.

The practice also differs with respect to the type of regeneration used. The problem is to get the sulfuric acid out of the carbon pores. This is accomplished either by washing it out or by heating up the carbon to a temperature at which reduction of the H_2SO_4 (by some of the carbon) to SO_2 takes place. The washing gives a weak sulfuric acid that is not worth much as a product, while the reduction consumes part of the expensive carbon. The evolved SO_2, although diluted with CO_2, is suitable for conversion either to acid (concentrated) or sulfur.

Another approach, tested on a pilot-plant scale in the United States,[3] is to react the loaded carbon with H_2S to reduce the H_2SO_4 to sulfur, after which part of the sulfur is converted to H_2S (for recycling) by reaction with hydrogen. The hot hydrogen volatilizes and sweeps the sulfur out of the carbon pores, partly as H_2S and partly as elemental sulfur; the sulfur is condensed as product while the H_2S goes on to reduce the H_2SO_4. All this is carried out in a series of fluidized beds.

22.3.4.4 *Catalytic Oxidation*

The boiler train can be converted, in effect, to a sulfuric acid plant by inserting a vanadium catalyst to convert the SO_2 to SO_3. As the gas cools, the SO_3 condenses as sulfuric acid, which is separated and removed as product.

This simple system has many advantages, including no need for raw materials, equipment with few moving parts, low labor requirement, no need for reheating, and production of fairly strong sulfuric acid. Among the drawbacks are high capital cost (because the large gas volume at the elevated operating temperature requires large equipment), need for a hot precipitator (which must be used even with high-sulfur coal because the catalyst cannot tolerate much dust), and a product acid that is not as concentrated as the usual commercial grade.

The process was tested in a joint EPA-utility demonstration project at an Illinois utility.[3] There were so many mechanical problems that the unit was shut down; further operation is unlikely.

22.3.4.5 *Metal Oxide Absorption*

Some of the metal oxides are good SO_2 absorbents at elevated temperature, and the oxide can be regenerated by reduction or by treatment with another reactant. Numerous oxides and process variations have been tested; the one that has shown the most progress is copper oxide absorption. In this process copper oxide absorbs SO_2 at an elevated temperature and is converted to $CuSO_4$, which is reacted with H_2 at the same temperature to evolve SO_2. The SO_2 is converted either to sulfur or sulfuric acid.

A 40-MW industrial boiler using the copper oxide process has been operating in Japan for several years. The gas is switched from one reactor vessel to another, to achieve an alternating cycle between absorption and regeneration.

This is a dry process and has the advantage that it can be adapted to NO_x removal (see subsequent discussion). However the equipment is large, and the hydrogen required for reaction with $CuSO_4$ is expensive.

22.3.5 The Economics of SO_2 Removal

As noted earlier, the economic tradeoffs in SO_2 removal are so complicated that it will be some time before economic optimization is achieved. Even when optimized, however, flue gas desulfurization will be much more expensive than originally was contemplated.

Estimating the costs for capital and operating expenses is a fruitless exercise in view of continuing inflation. The best that can be done is to list the basic principles that are generally applicable.

1. Control of stack emissions, including both SO_2 and particulates (because they are interrelated), accounts for roughly 20–25% of the capital cost of a new power station.
2. Annualized operating cost, including capital charges, often equals the delivered cost of the coal.
3. The main choice in terms of economics is between scrubbing and using low-sulfur coal. Low-sulfur coal usually costs much more to deliver than high-sulfur coal, and particulate removal is more expensive. These additional costs must be balanced against the cost of scrubbing.
4. Low-sulfur coal is usually favored when it is necessary to meet new source performance standards (1.2 lb $SO_2/10^6$ Btu), in the areas nearest the sources of such coal. As the distance increases and local supplies of high-sulfur coal become available, as in Illinois and western Kentucky, scrubbers may be the better choice.
5. High-sulfur coal can in some cases be converted to conforming coal by washing; but since part of the sulfur is in a form that does not wash out, this usually applies only to existing plants located where state regulations are lenient.
6. Coal can also be treated chemically to reduce the sulfur content. Such methods are in too early a stage of development for economic evaluation.
7. For small boilers the cost of scrubbers is so high that low-sulfur coal is usually favored.
8. Retrofit scrubbers usually cost 10–20% more than scrubbers in new installations.
9. Spare scrubbers increase the cost but are needed for reliability.
10. The cost of recovery compared to that of throwaway processes is so high that few recovery processes have been adopted.
11. Some of the recovered SO_2 could be converted to $(NH_4)_2SO_4$, which could be a good recovery product. By offering the by-product $(NH_4)_2SO_4$ at a reduced price, it should be possible to market a fair tonnage if the plant location is favorable.
12. Among the various lime-limestone scrubbing processes, those with the simplest internal elements in the scrubber should cost the least, unless very high degrees of removal are required.
13. Double-alkali systems may be no more expensive than lime-limestone systems if both involve dewatering by filtration.
14. Removing both particulates and SO_2 by the same equipment is definitely cheaper than removing them separately, but it may impair reliability to some extent.

TVA published cost estimates in 1976 that give some perspective on flue gas desulfurization cost. For a 500-MW boiler burning 3.5% coal, the estimated capital costs, in dollars per kilowatt, are as follows:

Lime	61.1
Limestone	68.4
Magnesia	71.7
Sodium (thermal stripping)	84.8
Catalytic oxidation	109.6

Annualized costs including capital charges and return for product (mills/kW-hr) are as follows:

Lime	3.65
Limestone	3.41
Magnesia	3.23
Sodium	4.99
Catalytic oxidation	3.04

It should be noted that these estimates involved numerous assumptions and that site-specific conditions at a given installation could make the comparison quite different. Moreover, these estimated costs assume that construction begins in 1975; inflation continually increases both the capital requirement and the annualized cost.

22.4 NITROGEN OXIDES

22.4.1 Scope of the Problem

There are several differences between controlling nitrogen oxide emissions from coal combustion and controlling emissions of particulates or sulfur oxides. In the first place, a smaller amount of nitrogen oxides is emitted; for example, a typical 1000-MW boiler burning 3.5% sulfur coal (12% ash) will emit around 900 tons of ash, 600 tons of SO_2, and 100 tons of NO_x per day. In addition, there is not much that can be done about the amount of ash and sulfur involved; both are present in the coal and must come out as waste products either in the gas stream or in the bottom ash, whereas much of the NO_x comes from the air and is produced by the reaction of nitrogen and oxygen at the high combustion temperature. Thus the amount of NO_x can vary widely with boiler conditions.

EPA set new source regulations in 1971 for NO_x as well as other pollutants (40 CFR 60.40 et seq.). The limit is 0.7 lb/10^6 Btu (about 500 ppm for an average coal), which is more rigorous on a weight basis than the limit of 1.2 lb/10^6 Btu for SO_2, but more lenient on a volume (molar) basis, since about 25% more NO_x, by volume, can be emitted. The guidelines for ambient standards are not as clear, with EPA changing emphasis from time to time with respect to changing state regulations. It is expected that until 1985 NO_x will remain a local problem in five areas of the United States—Los Angeles, Chicago, New York City, metropolitan Denver, and the McLean-Mercer-Oliver County area in North Dakota.[41]

Thus the utilities have not been pressured to do much about controlling NO_x emissions other than to obtain some guarantee on NO_x production when buying new boilers. This refers to the federal regulations; state and local regulations, as for SO_2, may be more stringent. In New Mexico, for example, the new source emission standard is 0.45 lb/10^6 Btu, and in New York City it is 100 ppm (equivalent to about 0.15 lb/10^6 Btu for an average coal).[41]

State regulations for existing boilers vary widely. About three-fourths of the states have no such regulations at all, but some have very stringent regulations—for example, 0.7 lb/10^6 Btu in New Mexico.

The amount of NO_x emitted from boilers varies so widely that it is difficult to arrive at a typical value. Emission is generally a direct function of flame temperature, excess air, nitrogen content in the coal, percentage of boiler load, and rate of gas cooling. In oil firing the presence of metals in the oil can have a catalytic effect on NO_x formation, but this does not seem to be significant when burning coal. Over the range of relevant parameters, the emission usually ranges from 400 to 1000 ppm by volume, and averages about 700 ppm (roughly 0.6–1.4 lb/10^6 Btu). About 90% of the NO_x is in the form of NO.

Flame temperature is mainly a function of burner and boiler design (Section 22.3.2.1). The nitrogen content of coal varies over a considerable range, roughly 0.6–2.2%. However not all the nitrogen in the coal is necessarily converted to NO_x; under some conditions much of it is converted to molecular nitrogen. A larger part appears as NO_x when the excess air is increased and a hotter, more turbulent flame is produced. NO_x emission drops off at lower boiler loads, presumably because of lower temperature. Rate of gas cooling is not usually a significant factor, since it is determined by the boiler design. Reducing excess oxygen is helpful but may result in incomplete combustion.

In the United States the general approach to reducing NO_x emissions has been to alter the combustion conditions. There are no plans to install any sort of gas treatment system. In Japan, however, increasing restrictions, particularly at the local level, have resulted in installation of numerous gas treatment units, both of the wet (scrubbing) and dry (catalytic decomposition) types.[1] Practically all these units are on gas or oil-fired boilers or various industrial operations.

22.4.2 Combustion Control to Reduce Formation of NO_x

22.4.2.1 *Boiler or Burner Type*

The main types of boilers are cyclone, tangential, and the standard "wall-fired" multiburner. Of these, the cyclone generally produces the highest NO_x concentrations because of the very hot, turbulent flame. The tangential is at the other extreme, with a long, "lazy" flame that swirls around the boiler; tangential boilers often meet the 0.7 lb/10^6 Btu standard without any changes or adjustments. The wall-fired boiler gives NO_x emissions that are in between the other two.

The superior performance of the tangen-

[41] K. Juris, *Proc. NO_x Control Technology Seminar,* February 1976, EPRI SR-39.

tially fired boiler with respect to NO_x emissions has led to its increased use. In addition, the wall-fired type can be designed with special provisions to reduce NO_x formation and meet the standard.

Specially designed burners are also offered to reduce NO_x formation. The general approach is to maximize uniformity of mixing and combustion, to avoid hot spots, which tend to generate large amounts of NO_x.

22.4.2.2 *Staged Combustion*

In staged combustion the amount of air introduced into the burner is less than stoichiometric and the remainder of the air is added into the boiler through separate ports. This is also termed "off-stoichiometric" firing. The advantage is that the reduced amount of oxygen in the burner tends to react more with the fuel than with the nitrogen in the air, and the remainder of the oxygen reacts in an area where the temperature is lower. As a result, there is an overall reduction in NO_x formation.

Tests by several utilities have shown that staged combustion is effective in reducing NO_x emission by approximately 30%, which is usually enough to meet the 0.7 lb/10^6 Btu standard. For example, in TVA tests on a front-fired boiler,[42] emission was reduced from 590 to 375 ppm by a combination of staged combustion and low excess air (2% O_2), a reduction of 36%. However there can be offsetting problems. In the TVA tests the CO levels in the gas were high (500 ppm by volume and higher), the carbon in the ash was increased, boiler efficiency was reduced, and tube corrosion was significantly increased.

Although the concentration of vapor phase organic compounds was not measured, it is likely that this also increased in the TVA tests. Such materials are of considerable concern to health authorities but do not seem to be present in unacceptable concentrations when combustion efficiency is high, as it normally is in utility boilers.

In tests on a tangentially fired boiler,[43] staged combustion gave a maximum reduction of 20–30%. The resulting emission was about 0.4 lb/10^6 Btu, well below the 0.7 lb/10^6 Btu limit.

The technique of reducing excess air has much the same effect as staged combustion and is often used in existing boilers, although poor combustion can result. Both air feed reduction and staged combustion can also produce reducing conditions on the tube walls, with the resulting danger of slagging. This can be limited or avoided by introducing some air near the walls.

The use of staged combustion appears to be the most promising route to reducing NO_x formation. This approach has been widely adopted in Japan, both with oil- and coal-fired boilers. For coal the NO_x concentration is reduced to 300 ppm or lower, and for oil it is usually reduced to less than 200 ppm.[44]

The effect of combustion conditions on NO_x formation is too complicated to be presented in detail here. EPA has sponsored numerous studies on the subject; a list of the EPA reports is given below.

22.4.2.3 *Gas Recirculation*

NO_x formation can also be reduced by recirculating some of the combustion gas and introducing it into the burner combustion air, thereby cooling the flame. The method is not very effective, as shown in tests carried out by the Electric Power Research Institute (EPRI)[45]; about a 15% reduction in NO_x emission was attained by recirculating 15% of the gas. Moreover, when gas recirculation was combined with staged combustion, recirculation gave an added percentage reduction of only 6%.

22.4.3 *Decomposition with Ammonia*

One of the more promising methods for decreasing NO_x emissions is to reduce the oxides to molecular nitrogen. Various reducing agents can be used, but the problem is in obtaining selectivity; since it is not practical to operate without excess oxygen in the gas,

[42] G. A. Hollinden, N. P. Moore, and R. L. Zielke, *Proc. NO_x Control Technology Seminar,* February 1976, EPRI SR-39.

[43] W. W. Habelt and B. M. Howell, *Proc. NO_x Control Technology Seminar,* February 1976, EPRI SR-39.

[44] Personal communication, Jumpei Ando, 1977.

[45] R. E. Thompson, M. W. McElroy, and R. C. Carr, *Proc. NO_x Control Technology Seminar,* February 1976, EPRI SR-39.

EPA Number	Report
APTD 1164	*Systematic Field Study of NO_x Emissions Control Methods for Utility Boilers* (Esso, December 1971, PB 210-739)
EPA 650 2-73-005	*Program for Reduction of NO_x from Tangential Fired Boilers* (Combustion Engineering, August 1973, PB 226-547)
EPA 650 2-73-021	*Proceedings, Coal Combustion Seminar,* June 19–20, 1973 (IERL RTP, 9-73, PB 224-210)
EPA 650 2-74-002a	*Effects of Design and Operating Variables on NO_x from Coal-Fired Furnaces, Phase I* (Babcock and Wilcox, January 1974, PE 229-986)
EPA 650 2-74-038	*Combustion Control of Pollutants from Multiburner Coal-Fired Systems* (U.S. Bureau of Mines, May 1974, PB 233-037)
EPA 650 2-74-066	*Field Testing: Application of Combustion Modifications to Control NO_x Emissions from Utility Boilers* (Exxon, June 1974, PB 237-344 AS)
EPA 650 2-75-012	*Analysis of Test Data for NO_x Control in Gas- and Oil-Fired Utility Boilers* (Aerospace, January 1975, PB 241-918)
EPA 650 2-74-0026	*Effects of Design and Operating Variables on NO_x from Coal-Fired Furnaces, Phase II* (Babcock and Wilcox, February, 1975, PB 241-283 AS)
EPA 650 2-73-005a	*Program for Reduction of NO_x from Tangential Coal-Fired Boilers, Phase IIa* (Combustion Engineering, August 1975, PB 246-889 AS)
EPA 650 2-73-046	*NO_x Combustion Control Methods and Costs for Stationary Sources* (Aerotherm, September 1975, PB 246-750 AS)
EPA 600 2-76-152a,b,c	*Proceedings of the Stationary Source Combustion Symposium,* Vols. I–III (IERL RTP, June 1976, PB 256-320 AS)

the reducing agent must react preferentially with the NO_x rather than the O_2.

It was discovered that ammonia has such selectivity under the proper operating conditions. As a result, in Japan, which has the only commercial installations that use this process, ammonia has become the standard reducing agent for NO_x.[29]

22.4.3.1 *Noncatalytic Reduction*

Ammonia can be used either with or without a catalyst. The main advantage to operating without a catalyst (direct use) is to eliminate the investment required for the catalyst and the reactor vessel. The problems with direct use include inefficient ammonia utilization (at the high temperature required for the reduction, part of the ammonia decomposes without reacting with NO_x) and the adverse effect of the temperature variation that occurs as a boiler is turned up or down.

Laboratory studies[46] on the relationship between temperature and the extent of reaction of NO_x show that a peak is attained between 875 and 975°C (1600–1785°F). For a gas containing 230 ppm NO_x, injection of 2 moles of NH_3 per mole of NO_x results in a conversion of 80–90% of the NO_x. This does not, however, take into account the presence of unreacted ammonia in the gas, which is undesirable because it is itself a pollutant. Moreover, it causes air heater plugging through the formation of ammonium bisulfate. Reducing the amount of excess ammonia injected obviously lowers the residual concentration in the gas, but also reduces the NO_x conversion because part of the ammonia decomposes directly to nitrogen. Thus if the $NH_3 : NO_x$ mole ratio is reduced from 2.0 to 1.0, conversion drops from 80–90% to 50–60%.

Although maximum NO_x conversion can be obtained over the range of 875–975°C, the lower part of this temperature range cannot be used because the concentration of residual ammonia is too high. Only at temperatures of about 1000°C and higher is a minimum ammonia concentration obtained (~ 20 ppm); below 1000°C the concentration goes up very rapidly. For example, at 950°C (with an $NH_3 : NO_x$ ratio of 2.0), the residual ammonia is about 150 ppm—much too high to be tolerated.

Thus to get maximum NO_x conversion and minimum residual ammonia, it is necessary to operate in an extremely narrow temperature zone; at higher temperatures the amount of residual ammonia is decreased but the amount of unreacted NO_x increases rapidly, whereas the opposite occurs at lower temperatures. Below 875°C the NO_x concentration also begins to increase rapidly. The situation is depicted in Fig. 22.5, which is typical for the ammonia injection process.

This situation would not be as bad if the temperature at a given point of injection

[46] R. K. Lyon and J. P. Longwell, *Proc. NO_x Control Technology Seminar,* February 1976, EPRI SR-39.

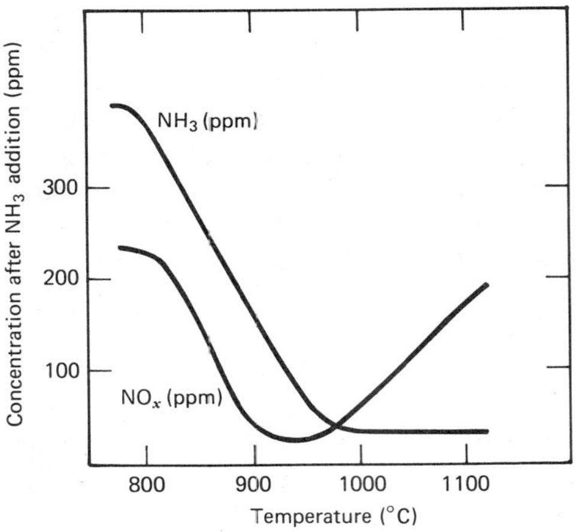

Figure 22.5 The effect of temperature on concentration of NH_3 and NO_x.

stayed constant, but such uniformity is not feasible in a utility boiler. Thus if the injector is located at the point where the temperature is about 1000°C at full load (a little high for optimum NO_x removal but on the safe side with respect to residual ammonia), the temperature will drop at reduced load, giving an excessively high residual ammonia concentration downstream.

One approach to this problem is to use two injection points, one at the optimum temperature for full load and the other located upstream and at the optimum temperature for a reduced load. As the load drops, injection is shifted to the upstream point. The difficulty, of course, is that the temperature at either point must be high enough to be on the safe side with respect to residual ammonia in the gas. The resulting loss in NO_x removal efficiency caused by operating at a temperature somewhat removed from the optimum means that much less conversion of NO_x occurs than that obtained in laboratory tests. At a 2.0 $NH_3 : NO_x$ mole ratio, 40–50% conversion can be achieved in full-scale tests, but the concentration of residual ammonia is high. Dropping the ratio to 1.1–1.2 gives about 40% conversion and 10–20 ppm ammonia (depending on load), which seems to be the best that can be obtained in full-scale operation.

It should be possible to have enough injection points that injection would always occur at the optimum or near-optimum temperature, thereby resulting in maximum NO_x conversion and minimum residual ammonia. The main problem, however, is the 20 ppm of residual ammonia that seems to be unavoidable. An ammonia decomposition catalyst could be installed downstream, but if this were done it might be just as well to use an NO_x decomposition catalyst, thus reducing the sensitivity to temperature. Even with multiple injection points, however, the temperature variation due to changes in boiler condition could be a problem, particularly if only two points were used.

The main conclusions concerning the chemistry of noncatalytic reduction seem to be as follows:

1. At an $NH_3 : NO_x$ mole ratio of 1.0, the optimum temperature is 1050°C and the maximum NO_x conversion is about 50%.
2. The reaction of ammonia and NO_x is very fast, most of it occurring in 0.2 s or less. For low NO_x levels ($<$ 200 ppm), there is an optimum residence time at the optimum temperature; for higher concentrations (in the range tested), the longer the retention time, the better.
3. The degree of conversion is affected by the type of fuel because of the water vapor concentration. Conversion increases linearly with water vapor content in the gas.
4. Hydrogen reduces the optimum temperature for conversion by about 65°C at an $H_2 : NO_x$ mole ratio of 1.0. The hydrogen does not affect the rate of conversion.
5. Ammonia added to an NO_x-free gas begins to decompose at about 800°C. At 1000°C as much as 40% decomposition can occur in 0.1 s.
6. NH_2 generated by NH_3 decomposition has a beneficial effect. Thus H_2 reduces the optimum temperature because it forms more NH_2.

22.4.3.2 *Catalytic Reduction*

Because kinetic considerations require that noncatalytic reduction with ammonia be carried out at such a high temperature that ammonia decomposition becomes a problem, use of a catalyst is indicated to lower the optimum temperature to a level where decomposition does not occur to any great extent. If this is done, NO_x removal of better than 90% can be attained. With a catalyst, a temperature level of 350–450°C is satisfactory. Thus "selective catalytic reduction" (SCR) has become the favored process in Japan. Several problems, however, cast doubt on the viability of the process.

1. *Poisoning by SO_3.* The SO_3 present in the flue gases can be a problem with high-sulfur oil or coal, because it attacks the alumina often used as a catalyst support. Another approach is to use a catalyst that does not have alumina as the support; most catalyst producers have developed this type of product; for example, titania is used in some catalysts.

2. *Dust plugging.* High-sulfur oil can produce enough dust to plug the catalyst in a relatively short time. For a 2% sulfur oil, a dust loading of 0.05–0.08 g/scf is not unusual, which amounts to 5 or 6 tons/day for a 500-MW boiler.

The first installations in Japan used the SCR process to treat "clean gas," such as that produced by burning natural gas or LPG. Since little or no dust is produced from such fuels, fixed-bed catalysts were acceptable and gave good service. The next step in the development of the SCR process was to use low-sulfur oil; here there were few problems with poisoning by SO_3, but catalyst blinding was troublesome, and moving-bed systems had to be developed in which the catalyst was removed from the reaction chamber, cleaned, and returned. These systems were generally satisfactory but troublesome to operate.

A further modification was the development of a catalyst manufactured in such a way that the gas, in effect, flows by it rather than through it. Various geometrical shapes are used—plates, pipes, rings, thin beds simulating plates—all aimed at the parallel flow of the gas past the catalyst surface, through passages large enough to avoid any plugging. Molecular diffusion of the NO_x and NH_3 gives adequate contact, but the dust tends to travel with the gas and has relatively little contact with the catalyst. As a result, blinding of the catalyst surface is minimized.

Such a catalyst is applicable to coal-fired as well as oil-fired boilers and has been tested in Japan for use with coal.

3. *Temperature variation.* The problem of temperature change with load is not as great with a catalyst as it is without one, but it still requires attention. The reasons for this are complex, involving factors such as lower NO_x removal efficiency when the temperature is increased, possible trouble at elevated temperature with catalyzed ammonia decomposition and SO_3 formation, deposition of ammonium sulfate or bisulfate on the catalyst at low temperature, and greater poisoning of metal oxides by SO_3 at low temperature levels.

4. *Residual ammonia.* Both the catalytic and noncatalytic methods have the problem of residual ammonia. This is because all the ammonia will not react, except perhaps with a less than stoichiometric feed, in which case adequate NO_x conversion may not be achieved.

In catalytic conversion, most systems have about 10 ppm ammonia in the gas downstream from the catalyst. An ammonia decomposition catalyst can be used, but this approach seems to have lost favor for various reasons. If the ammonia is left in the gas, the resulting ammonium bisulfate formed downstream is corrosive while it is in the liquid form and deposits on surfaces when the gas has cooled enough for it to solidify. Any ammonia passing through the air heater (or when there is not enough SO_3 to react with it) either becomes an atmospheric pollutant or, when a scrubber is used to remove SO_2, becomes a pollutant of water.

Catalyst Shape. Various shapes are used for catalysts, including standard granular, cylindrical pellets, plates, tubes, and "honeycomb." The last three types involve parallel flow; the cylinders are sometimes stacked to provide parallel flow passages and are sometimes dumped like a granular catalyst. The parallel flow types are more applicable in the United States; although packed beds are acceptable under some circumstances in Japan, they are not likely to be suitable for coal-fired boilers. It would be necessary either to install a hot precipitator before the catalyst or to install the catalyst after a cold precipitator and reheat the gas. Neither option is attractive for use with high-sulfur coal. For low-sulfur coal, some utilities prefer cold precipitators or baghouses to hot precipitators; in such cases the gas must be reheated.

The plate type of reactor consists of a base with the catalyst impregnated into the surface. The plates are made into modules about one meter square and are used in the quantity necessary to give the required surface area. The length of the catalyst assembly, in the direction of gas flow, depends on the degree of removal required, the gas velocity needed to get adequate "mass transfer," and erosion by the dust carried by the gas. The velocity is normally less than 10 m/s.

The preferred configuration is downflow, but in most cases this increases the cost because it requires extra ductwork. Horizontal flow can be used, but there is the problem of the dust settling out, especially with the high dust loadings encountered in power plants. This would be particularly troublesome for a reactor designed for full load, during periods of reduced boiler load. A dust removal hopper would be required in this situation; on the other hand, it might be difficult to fit a long reactor (which can be as long as 8 m or more) into the boiler in a vertical position, which would make horizontal flow the necessary choice.

The "honeycomb" types have various designs to give maximum surface area per unit volume. A typical honeycomb arrangement is shown in Fig. 22.6. Other reactors have a symmetrical internal structure, typically in the form of hexagonal passages.

The parallel flow type of reactor, such as the honeycomb, permits a high linear velocity, as much as 8 times higher than that of the granular catalyst, therefore has a much smaller reactor cross section. The high velocity gives better "mass transfer" because of the lower gas film resistance. Test data show a higher allowable space velocity for the honeycomb catalyst than for the granular catalyst; thus a smaller reactor can be used. Overall cost of the honeycomb arrangement is higher because the prefabricated honeycomb catalyst is more expensive than the more commonly used catalyst granules, and the cost of labor for installing a honeycomb is higher because stacking is required. However utilities are interested in the honeycomb because a reactor with a smaller cross section is easier to install.

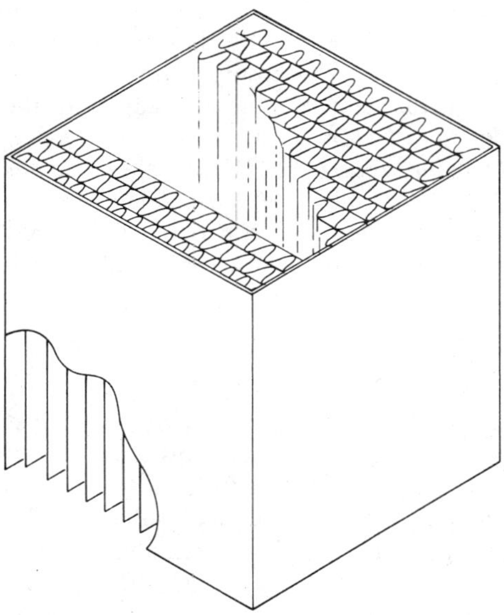

Figure 22.6 Honeycomb-type reactor.

The parallel flow type of reactor is not used solely for the conversion of NO_x; one was installed on a small oil-fired boiler (Japan) in 1967 to remove SO_2 from the gas. The reactor consists of an assembly of "plate" sections through which the gas passes in a direction parallel to the "plates." The arrangement is similar to the plate reactor except that the "plates" are actually very thin sections containing a granular catalyst and held in place by stainless steel wire mesh.

The catalyst seems to be adequately resistant to dust. However there apparently have been no tests made at the 5–10 gr loadings encountered in coal-fired operation. As noted earlier, the plate and honeycomb types can operate at high loadings without any particular difficulty; however fine dust that settles on the catalyst tends to be swept off the smooth surfaces, whereas there is no gas flow to sweep out dust from the interstices of the granular catalyst inside the wire mesh envelopes.

The tube type of catalyst does not seem to be as popular as the others. The tubes are usually 2–3 ft long and 7–28 mm (0.28–1.10 in.) I.D.

Catalyst Composition. One of the main research efforts has been to develop a catalyst support that is resistant to SO_3. The various materials used have the following order of resistance: $TiO_2 = SiO_2 > \alpha\text{-}Al_2O_3 > \eta\text{-}Al_2O_3 > \gamma\text{-}Al_2O_3$. TiO_2 and SiO_2 seem to have been adopted as the standard support materials.

Metal supports are also used, although the

ceramic bases listed above are said to be better because they are more porous and thus provide more catalytic surface after impregnation with the active catalyst material. The main problems are erosion (which may require a reduction in gas velocity) and breakage.

The adverse effect of SO_3 on the alumina catalyst support results from clogging of the Al_2O_3 pores by sulfate formation, thus preventing the gas from contacting catalyst impregnated in the pores. Titania is resistant to chloride as well as SO_3.

Various materials are deposited on the support to serve as the active catalyst. The precise composition of these materials is usually proprietary information, but vanadium is known to be one of the better catalysts. Iron oxide is sometimes used.

Ammonia Leakage. Incomplete reaction of the ammonia (ammonia leakage) is probably the main drawback to catalytic NO_x reduction. The resulting problems of air heater corrosion and plugging are very serious problems to a utility.

By keeping the $NH_3 : NO_x$ mole ratio close to 1.0 and using an active catalyst, the residual ammonia can be held to a level below 10 ppm. However the temperature must be kept up to ensure that there are no bisulfate deposits on the catalyst surface. For operation with a reduced load, the technique of bypassing the economizer to keep the temperature up is favored. This would be recommended for any boiler installation using coal or high-sulfur oil; some are already being operated in this manner.

There is some indication that the high dust loading with coal-fired operation may actually be helpful in scouring bisulfate deposits from the air heater. It is not clear, however, whether this would be enough or whether soot blowing and periodic washing would still be needed. Sandblasting has been used in some installations.

To combat corrosion, an enamel coating can be used in the air heater at the point where bisulfate deposition is expected to occur. Such protection is needed because corrosion of the carbon steel by bisulfate has occurred with even less than 10 ppm ammonia.

One possibility for resolving the ammonia leakage problem is to use an ammonia decomposition catalyst downstream. The type of catalyst used is similar to that for NO_x conversion—Cr_2O-Al_2O_3 seems to be the best—but only about half as much is required. Although the residual ammonia can almost be eliminated, the method is not usually favored because of the additional investment.

The ammonia can be either an air or water pollutant. Since SO_2 scrubbing is likely to be required on new U.S. boilers in the future, there will be the problem of water pollution by ammonia caught in the scrubber.

SO_3 Formation. Some catalysts oxidize much of the SO_2 to SO_3, which can further aggravate the bisulfate problem. Ammonia inhibits the oxidation; without it, greater quantities of SO_3 would be formed. It is claimed that certain catalyst types can minimize this effect, resulting in a level of oxidation as low as 1%, but there seems to be no information available on the properties that inhibit oxidation.

22.4.4 Wet Scrubbing

Although there has been much research on a small scale in the United States and Europe concerning the removal of NO_x by wet scrubbing, again the only commercial application has been in Japan.[3] A major problem is the inertness of NO, making it very difficult to find an absorbent that will react with it. Developers have found it difficult to choose between direct NO absorption, oxidation to NO_2 before absorption, and adjusting the inlet $NO : NO_2$ ratio to an equimolar proportion before absorption. Of the eight processes in Japan that might be considered to be in a leading position, three use the direct NO approach, four oxidize to NO_2, and one adjusts to the equimolar ratio.

22.4.4.1 *Direct NO Absorption*

Direct absorption of NO requires a catalyst and a very large absorption tower because of the slow mass transfer. Like most wet NO_x processes, the method uses sulfite as the absorbent, thus making it essential to combine removal of SO_2 and NO_x. One of the Japanese wet scrubbing processes involves the following absorption reaction (using ferrous chelate as the catalyst):

$$SO_2 + Na_2SO_3 + H_2O \rightarrow 2NaHSO_3 \quad (22.38)$$

$$NO + 3NaHSO_3 \rightarrow NH(NaSO_3)_2 + \frac{1}{2}Na_2S_2O_6 + H_2O \quad (22.39)$$

The sodium imidodisulfonate is then decomposed by treatment with lime, H_2SO_4, and $CaNO_3$ in a complicated series of reactions. The final products are $CaSO_4 \cdot 2H_2O$, Na_2SO_4, and N_2. Thus the process is a double-alkali type that also removes NO.

In a similar method, ammonia is used as the absorbent, ferrous chelate is the catalyst, and imidodisulfonate is formed. However a different decomposition sequence is followed:

$$SO_2 + (NH_4)_2SO_3 + H_2O \rightarrow 2NH_4HSO_3 \quad (22.40)$$

$$NO + 3NH_4HSO_3 \rightarrow NH(NH_4SO_3)_2 + \frac{1}{2}(NH_4)_2S_2O_6 + H_2O \quad (22.41)$$

$$NH(NH_4SO_3)_2 + 2H_2O \rightarrow (NH_4)_2SO_4 + NH_4HSO_4 \quad (22.42)$$

$$NH_4HSO_4 + NH_3 \rightarrow (NH_4)_2SO_4 \quad (22.43)$$

$$(NH_4)_2S_2O_6 \rightarrow (NH_4)_2SO_4 + SO_2 \quad (22.44)$$

The last three reactions are carried out in a hydrolysis step at low pH, followed by neutralization with ammonia.

These methods require a high ratio of SO_2 to NO_x, to provide enough bisulfite for the NO absorption reaction to occur. Thus they are mainly applicable to high-sulfur coal. Neither has been tested beyond the pilot-plant scale.

22.4.4.2 *Oxidation to NO₂*

Nitrogen oxide absorption is much easier to accomplish if the NO is first oxidized to NO_2, which can be done by adding a strong oxidizing agent, such as ozone or chlorine dioxide, to the gas before it enters the absorber.

Processes are being developed in which NO oxidation is combined with lime scrubbing, with a catalyst (probably an iron compound) carried in the slurry. The chemistry of the process is not very clearly understood, but is postulated to be as follows:

$$2NO_2 + 5CaSO_3 + Ca(HSO_3)_2 + 11H_2O \rightarrow Ca(NH_2SO_3)_2 + 5CaSO_4 \cdot 2H_2O \quad (22.45)$$

$$Ca(NH_2SO_3)_2 + Ca(OH)_2 + 4H_2O \rightarrow 2NH_3 + 2CaSO_4 \cdot 2H_2O \quad (22.46)$$

There is also a considerable amount of NO_2 decomposition in the scrubber:

$$2NO_2 + 4CaSO_3 + 8H_2O \rightarrow N_2 + 4CaSO_4 \cdot 2H_2O \quad (22.47)$$

The oxidation approach has much in its favor, since most of the capital investment, except for ozone generation, can be charged to SO_2 removal. However use of either ozone or chlorine dioxide is quite expensive and makes NO_x removal a costly operation. Oxidation processes have been used with a few small boilers in Japan.

22.4.4.3 *Equimolar NO : NO₂*

The amount of expensive oxidant required can be reduced by oxidizing just enough NO to give an equimolar NO : NO_2 ratio, which results in better absorption than with NO alone, because of nitrite formation:

$$NO + NO_2 + H_2O \rightarrow 2HNO_2 \quad (22.48)$$

A pilot plant has been operated in Japan, in which this principle is integrated into a magnesia-scrubbing process. Part of the magnesium nitrite [$Mg(NO_2)_2$] formed is treated with H_2SO_4 to evolve NO_2 for recycling and adjusting to the equimolar NO : NO_x ratio. The remainder, plus the $MgSO_3$ formed by SO_2 absorption, is treated in a complicated series of reactions to give calcium nitrate and gypsum as the products.

Although absorption is made easier than with NO alone, it does not appear to be as good as with NO_2 alone. Furthermore, calcium nitrate is not an especially desirable product.

22.4.5 The Economics of NO_x Removal

The inert nature of NO makes removal processes very expensive and therefore favors methods, such as modification of the combustion process, that reduce formation of NO_x. The economics of combustion modification are not very clear; the tangentially fired boiler presumably is no more expensive than others and appears to be capable of meeting the 0.7 lb/10^6 Btu standard in most cases. A limited amount of overfire air addition seems adequate in the more difficult situations.

For boilers with a higher flame temperature, where relatively extreme combinations of two-stage combustion, gas recirculation, and reduced excess air must be used to meet the standard, there is considerable question of how much the operating cost is increased by insufficient combustion, boiler derating, and boiler corrosion. This is particularly true for existing boilers, where it is more difficult to approach the optimum design than in new installations.

Thus the overall cost of combustion modification is not well established, but it is clear that removal methods are much more expensive. In the United States removal methods probably will not be used to meet emission standards unless the standard is reduced to the point where it cannot be met by combustion modification. This is a strong possibility. If removal is employed, staged combustion would also probably be used to reduce overall cost.

The noncatalytic use of ammonia to remove NO_x is probably the cheapest removal method, if it can be made to work. The estimated cost, which can vary widely with cost of ammonia and other factors, ranges from 0.75 to 1.5 mills/kW-hr. However as noted earlier, noncatalytic reduction probably does not remove more than about half the NO_x; therefore this is not the "best available control technology" favored by regulatory agencies.

Catalytic reduction by ammonia and wet scrubbing methods is much more expensive. There has not been enough experience to be certain about the economics of the process, but estimates in Japan[29] are about 3.75 mills/kW-hr, which is about the same order of magnitude as for SO_2 scrubbing. This estimate is for the separate removal of NO_x and SO_2. If the two are combined, as in the CuO process for SO_2, the cost for NO_x removal is reduced to a minimum—little more than the cost of the ammonia at an 0.8 $NH_3 : NO_x$ mole ratio.

The cost of wet systems is even less well established. Again, combining removal of NO_x and SO_2 saves money, but if an oxidizing agent is used the cost of oxidant production alone can be as high as that for processes that involve reduction with ammonia. Direct absorption of NO eliminates the oxidant expense but requires a much larger scrubber and a higher pressure drop. The available data are not extensive enough to permit making an economic comparison.

CHAPTER 23

Fundamentals of Coal Gasification

JAMES L. JOHNSON*

James L. Johnson *This chapter is the final publication of Dr. James L. Johnson, whose career was cut short by his untimely death in November 1977. Dr. Johnson's 20 years' work with the Institute of Gas Technology, in Chicago, concentrated on measuring and correlating the chemical kinetics of gas-solid reactions. He intensively studied the oxidation and reduction of iron oxides, reactions of mercaptans with iron oxides, carbon deposition and oxidations on iron ores, and, most significantly, coal gasification. His achievements in the latter field gained international recognition. He studied the gasification of all types of coal of major commercial significance over a broad range of temperatures, pressures, and gas compositions. The work included characterizing coal and char reactivity, based on physical properties and catalyst content of the solids.*

* Deceased.

23.1 INTRODUCTION

The subject of coal gasification covers the conversion of coals to light gases, condensable liquids and tars, and solid products, in the presence of reactive gaseous atmospheres. Thus gasification is distinct from other forms of coal conversion such as pyrolysis or carbonization, which occur in inert gas atmospheres, or liquefaction, which occurs in a liquid medium. Since the early 1960s research efforts to develop the technology for converting coals into gases have been greatly accelerated, particularly in the United States, because of projected shortages of other economically competitive fuels for use as energy sources. These expanded efforts have been focused on reaction systems relevant to the conversion of coals into relatively low-energy gases suitable for power generation and other industrial uses, as well as into high-energy gases that can be economically transported over long distances and used as a substitute for natural gas.

This chapter discusses fundamental aspects of coal gasification that have been developed in the many investigations conducted during the past 15–20 years in the areas of thermodynamics, physics, and chemical kinetics. Because this subject area is very broad, selection of the specific topics for discussion has necessarily been subjective and has depended on individual interests, particularly those concerning physical processes involved in coal gasification systems, for which a vast literature exists. Discussion of the thermodynamics and kinetics of coal gasification, however, extends to coal char gasification reactions as well as to the initial coal gasification processes, in which volatile matter is evolved and coal char is formed in reactive atmospheres. Elucidation of the kinetic behavior during the initial coal gasification stages is particularly important to understanding the conversion of coal to methane at elevated pressures.

23.2 THERMODYNAMICS OF GASIFICATION

Thermodynamic considerations related to the gasification of coals and coal chars are important to the theoretical evaluation of limiting performance characteristics for individual process concepts and to the practical design of experimental or commercial reactor systems. For these purposes thermochemical properties of coal gasification systems must be known or estimated, to permit a definition of heat effects and equilibrium or pseudoequilibrium behavior. Measurement of the thermochemical properties of coals and coal chars, however, is very difficult because of the complex and heterogeneous nature of such materials, compared to graphite. Furthermore, there is uncertainty about the accuracy and even the meaning of much of the available experimental information.

The following discussion focuses on equilibrium activity coefficients, specific heats, and heats of carbonization and hydrogenation reactions, as they apply to coal and coal char gasification systems.

23.2.1 Equilibrium Effects

To determine the equilibrium characteristics of coal or coal char gasification systems, it is necessary to define the standard free energies of formation of the reactants and products. (The standard heats and entropies of formation can also be used.) For the major gaseous species involved in such systems, this information is available in a variety of tabulations and correlations. For coals and coal chars, however, accurate experimental information is lacking, and relatively crude techniques have been used to estimate free energies of formation or equivalent parameters.

One technique for estimating coal thermodynamic properties has been to extrapolate or interpolate from the properties of pure, solid aromatic compounds.[1] It is not easy, however, to evaluate the uncertainties that result from using this technique, not even taking into consideration the complex heterogeneous character of coal as well as its amorphous nature. The use of the estimated free energies of formation of coal itself in thermodynamic analyses of coal gasification systems is perhaps most applicable for estimating overall potential performance characteristics of gasification concepts by accounting for second-law constraints.

Some investigators have interpreted experimental yields obtained during the gasification of coals or coal chars in terms of pseudoequilibrium constants, or equivalently, by assigning "activity" coefficients to the reacting carbon. This has been done when the investigator is interested in the methane yields resulting from gasification with hydrogen or with synthesis gas mixtures, according to the following reaction:

$$C + 2H_2 \rightleftarrows CH_4 \qquad (23.1)$$

The equilibrium constant for this reaction is generally expressed by:

$$K_{CH_4} = \frac{P_{CH_4}}{a_c P_{H_2}^2} \text{ or } K'_{CH_4} = a_c K_{CH_4} = \frac{P_{CH_4}}{P_{H_2}^2} \qquad (23.2)$$

where P_{H_2}, P_{CH_4} = partial pressure of H_2 and CH_4, respectively, (atm)
a_c = "activity" of a particular carbon, referred to graphite (for graphite $a_c = 1$)

Wen et al.[2] evaluated experimental values of $P_{CH_4}/P_{H_2}^2$ obtained at apparent pseudoequilibrium conditions during the gasification, with hydrogen, of low-temperature bituminous coal char. Gasification occurred at elevated pressure in a circulating gas system at 705°C. Values of K'_{CH_4}, reported as a function of total carbon gasified, decreased from about 0.8 atm^{-1} at zero carbon conversion to about 0.12 atm^{-1} at carbon conversion fractions approaching unity. Based on Equation 23.2, this corresponds to a range of apparent carbon activity, a_c, from 7 to 1.

Similarly, Squires[3] compiled plots of values of $P_{CH_4}/P_{H_2}^2$ from data obtained in integral steam-oxygen gasification of coals and coal chars; these data show an apparent systematic trend in which values of a_c are generally greater than unity but decrease with increasing pressure, approaching unity at total pressures above 50 atm. Although apparent values of a_c are relatively independent of temperature from 760 to 900°C, values as high as $a_c = 20$ are indicated for data obtained at 1 atm total pressure.

There are several possible explanations for the experimental values of a_c greater than unity obtained in coal and coal char gasification systems. Squires[3] suggests that amorphous carbons have excessive free energies compared to graphites. He also suggests that in gases containing steam as well as hydrogen, the steam activates carbons for gasification with hydrogen. Blackwood and McCarthy,[4] however, have argued that for gasification of carbons in gases containing both steam and hydrogen, it is not valid to interpret apparent carbon activities solely on the basis of the reaction given by Equation 23.1, since methane can be formed by other reactions involving both steam and hydrogen. As indicated by Blackwood and McCarthy, for integral fluidized-bed gasification of brown coal with hydrogen at elevated pressures, Birch et al.[5] found that $P_{CH_4}/P_{H_2}^2$ ratios in product gases never exceed equilibrium values where $a_c = 1$. Blackwood and McCarthy also noted that the equilibrium measurements of hydrogen-methane mixtures made by Browning and Emmett,[6] using a carbon formed from iron carbide decom-

[1] O. H. Krikorian, U.S. Atomic Energy Commission Rep. No. UCID-16587; NTIS, 1972.

[2] C. Y. Wen, O. C. Abraham, and A. T. Talwalkar, *Prepr., Am. Chem. Soc., Div. Fuel Chem.*, **10**(4), 168–185 (1966); Am. Chem. Soc. Adv. Chem. Ser. No. 69, 1967, pp. 251–271.

[3] A. M. Squires, *Trans. Inst. Chem. Eng., London,* **39,** 3–9 (1961).

[4] J. D. Blackwood and D. J. McCarthy, *Aust. J. Chem.,* **19,** 797–812 (1966).

[5] T. J. Birch, K. R. Hall, and R. W. Urie, *J. Inst. Fuel,* **33,** 422–435 (1960).

[6] L. C. Browning and P. H. Emmett, *J. Am. Chem. Soc.,* **73,** 581–583 (1951).

position, corresponded closely to equilibrium values obtained for graphite gasification in hydrogen ($a_c = 1$), over a wide range of temperatures. Such evidence prompted Blackwood and McCarthy to suggest that coal char gasification systems are generally bounded by equilibria in which carbon activities are approximately the same as for graphite.

Johnson[7] has also shown that results such as those interpreted by Squires to reflect values of a_c greater than unity can be reasonably simulated with a kinetic model accounting for the reactions of steam as well as hydrogen with carbon. Significantly, Johnson's kinetic expressions all involve equilibrium terms in which coal char activities are assumed to be the same as those of graphite. Such analyses emphasize the possible dangers of interpreting thermodynamic activities based on the characteristics of a reaction subsystem when the total reaction system is not in equilibrium.

Another assumption sometimes made in estimating the thermodynamic characteristics of coal chars is that the higher heats of formation of such materials with respect to graphite are directly related to higher free energies of formation. This assumption can have significant consequences. If, for example, an amorphous carbon having a heat of formation of 2.6 kcal/mole is assumed to have an increased free energy of formation of 2.6 kcal/mole relative to graphite (independent of temperature), then values of a_c close to 3.4 would result at 800°C. This approach does not recognize, however, that amorphous carbons would also be expected to have higher entropies than graphite, which could either partially or wholly compensate for the higher heats of formation.

For experimental systems in which coal devolatilization reactions are occurring in addition to the subsequent gasification of coal char, apparent high values of a_c can also result. This occurs because coal decomposition reactions are generally irreversible, and the methane yields obtained are a result of relative kinetic factors rather than equilibrium or pseudoequilibrium characteristics, particularly in the presence of hydrogen at elevated pressures. This is discussed in a later section of this chapter concerned with initial stage coal gasification kinetics. Such a consideration, however, suggests that apparent equilibrium constants, such as those obtained by Wen et al.[2] using a low-temperature coal char containing significant volatile matter, may reflect kinetic behavior that depends on the specific reaction conditions employed and therefore should be applied or extrapolated with caution. In any case, it is apparent that at this time there is no general correlation defining effective char-carbon activity coefficients in terms of temperature, pressure, gas composition, the reaction that is occurring, and any known property of the solid phase.

The gas composition obtained under equilibrium conditions for the gasification of coal chars can be at least qualitatively estimated by assuming that the solid carbon has the same free energy of formation as that of graphite. Based on the discussion above, it is quite possible that such an assumption may be quantitatively applicable to a wide range of conditions relevant to coal char gasification systems. Such equilibrium calculations have been made by a number of authors, including von Fredersdorff and Elliott.[8]

For the graphite-hydrogen-methane system, equilibrium gas composition for the reaction:

$$C + 2H_2 \rightleftarrows CH_4 \qquad (23.1)$$

can be expressed only as a function of temperature and total pressure. Equilibrium methane mole fractions for this system (Fig. 23.1) are based on computations using thermodynamic data from the JANAF Thermochemical Tables,[9] in which gas phase species are assumed to be ideal gases. As indicated, equilibrium methane mole fractions decrease with increasing temperature and decreasing pressure.

For the graphite–carbon dioxide–carbon monoxide system, equilibrium gas compositions for the reaction:

$$C + CO_2 \rightleftarrows 2CO \qquad (23.3)$$

[7] J. L. Johnson, in Am. Chem. Soc. Adv. Chem. Ser. No. 131, 1974, pp. 145–178.

[8] C. G. von Fredersdorff and M. A. Elliott, in *Chemistry of Coal Utilization,* Supplementary Volume, H. H. Lowry, Ed., John Wiley and Sons, New York, 1963, pp. 892–1022.

[9] D. R. Stull and H. Prophet, Eds., *JANAF Thermochemical Tables,* 2nd ed., for U.S. Dept. of Commerce, Government Printing Office, Washington, D.C., 1971, 1141 pp.

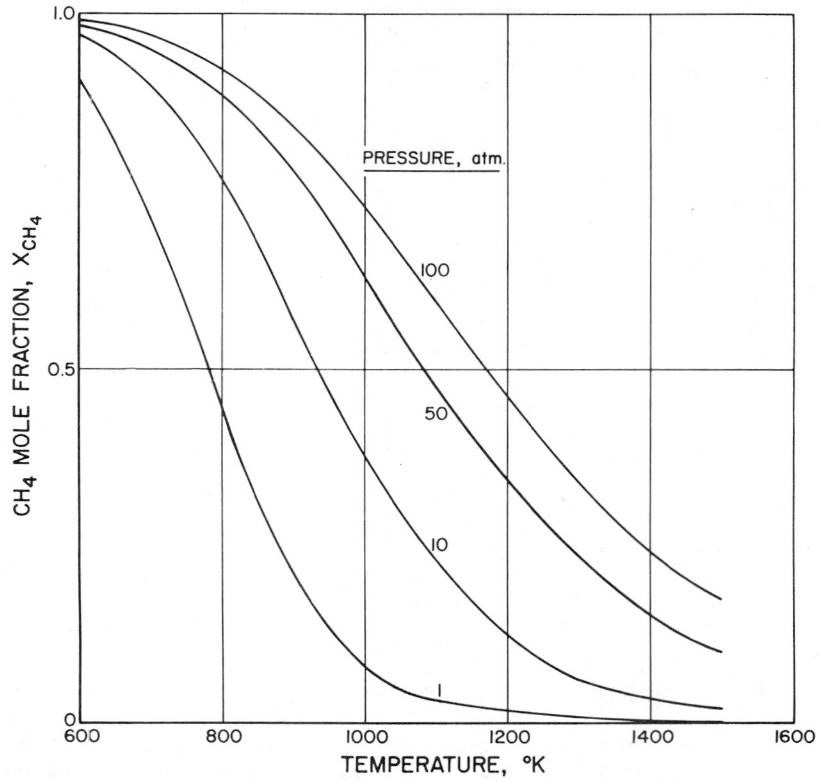

Figure 23.1 Equilibrium characteristics for the graphite hydrogen-methane system.

are also a function only of temperature and total pressure. Equilibrium carbon monoxide mole fractions for this system (Fig. 23.2) increase with increasing temperature and decreasing pressure, a trend opposite to that exhibited for the methane mole fraction in the graphite-hydrogen-methane system.

In a carbon-hydrogen-oxygen gasification system in which all three components are present, it is usually assumed that the following species exist in the gas phase: CO, CO_2, H_2, H_2O, and CH_4. The equilibrium composition for this system can be expressed as a function of temperature, total pressure, and the H/O ratio (atom/atom) in the gas phase. For purposes of calculation, it can be assumed that the equilibrium reactions are comprised of those given by Equations 23.1 and 23.3, as well as the water-gas shift reaction:

$$CO + H_2O \rightleftarrows CO_2 + H_2 \qquad (23.4)$$

This selection of reactions is of course arbitrary, since a variety of equivalent combinations of reactions could be chosen for equilibrium calculations. An example of equilibrium trends in this more complex system is shown in Fig. 23.3 for an H/O ratio of unity. These results are based on calculations made by Baron et al.,[10] who also used the JANAF Thermochemical Tables and assumed ideal gas properties. The trends exhibited in Fig. 23.3 for an H/O ratio of 1 can be more generally characterized as shown in Table 23.1.

The arrows in this table indicate whether a variable increases (↑), decreases (↓), or initially increases, reaches a maximum, and then decreases (↷). For example, the carbon monoxide mode fraction increases with increasing temperature, but decreases with increasing pressure and increasing H/O ratio. The carbon dioxide mole fraction reaches a maximum value with increasing temperature, increases with increasing pressure, and decreases with increasing H/O ratio. The relative amount of gas phase carbon as methane, given by $X_{CH_4}/(X_{CO} + X_{CO_2} + X_{CH_4})$, increas-

[10] R. E. Baron, J. H. Porter, and O. H. Hammond, Jr., *Chemical Equilibria in Carbon-Hydrogen-Oxygen Systems,* MIT Press, Cambridge, Mass., 1976.

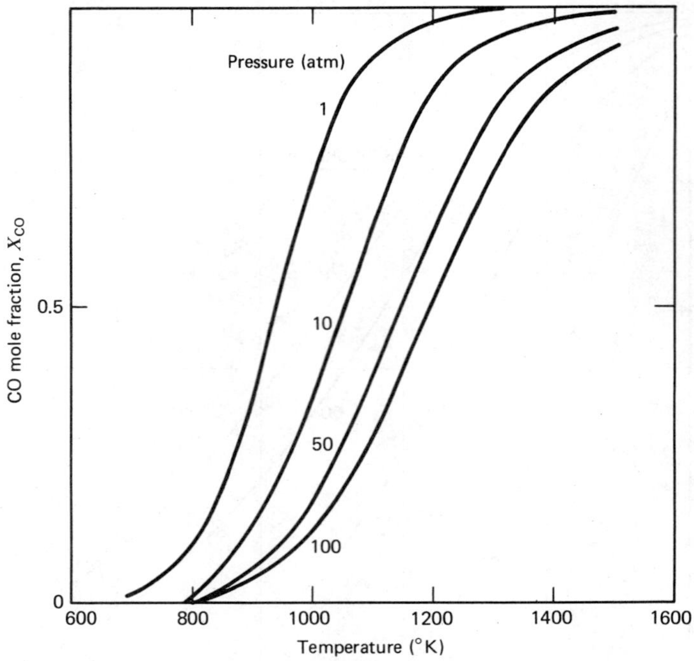

Figure 23.2 Equilibrium characteristics for the graphite–carbon monoxide–carbon dioxide system.

es with both increasing pressure and increasing H/O ratio. The effect of increasing the temperature on this variable depends on the H/O ratio. Relative methane yields have a maximum value for H/O values less than 2, although they decrease continuously with increasing temperature for H/O values equal to or greater than 2.

The equilibrium characteristics shown in Figs. 23.1–23.3 are all constructed on the assumption that the solid carbon phase has the same thermodynamic properties as graphite. For the case of the solid carbon having a greater activity than graphite (i.e., a greater free energy of formation), the equilibrium constants for the reaction given by Equations 23.1 and 23.3 can be more generally represented by:

$$K_{CH_4} = \frac{X_{CH_4}}{X_{H_2}{}^2(Pa_c)} \tag{23.5}$$

and

$$K_{CO} = \frac{X_{CO}{}^2}{X_{CO_2}}\left(\frac{P}{a_c}\right) \tag{23.6}$$

where P = total pressure (atm)
X_i = mole fraction of i species
K_{CH_4}, K_{CO} = equilibrium constants for Equations 23.1 and 23.3, respectively

In Equations 23.5 and 23.6, the equilibrium constants are the same as those for the gasification of graphite, where, by definition, a_c = 1. For values of a_c other than unity, the equilibrium characteristic for the C-H_2-CH_4 and C-CO-CO_2 systems can still be computed from equilibrium characteristics where a_C = 1, by defining:

$$P' = Pa_c \tag{23.7}$$

$$P'' = \frac{P}{a_c} \tag{23.8}$$

If, for example, P = 10 atm and a_c = 5, the equilibrium methane mole fraction can be determined from Fig. 23.2 by using an effective pressure of $P' = 10 \times 5 = 50$ atm.

Increasing the value of a_c generally leads both to increased methane and carbon monoxide mole fractions in each of the simple systems considered. The situation in the C-H-O system is more complex, and the generalized effects of increasing a_c have not been evaluated in this analysis. However an example has been computed for the case in which P = 100 atm, T = 1100 K, the H/O ratio = 1.0, and the a_c has values of 1, 2, and 4. The results of the calculation are given in Table 23.2. They indicate that under these conditions, not only do methane and carbon

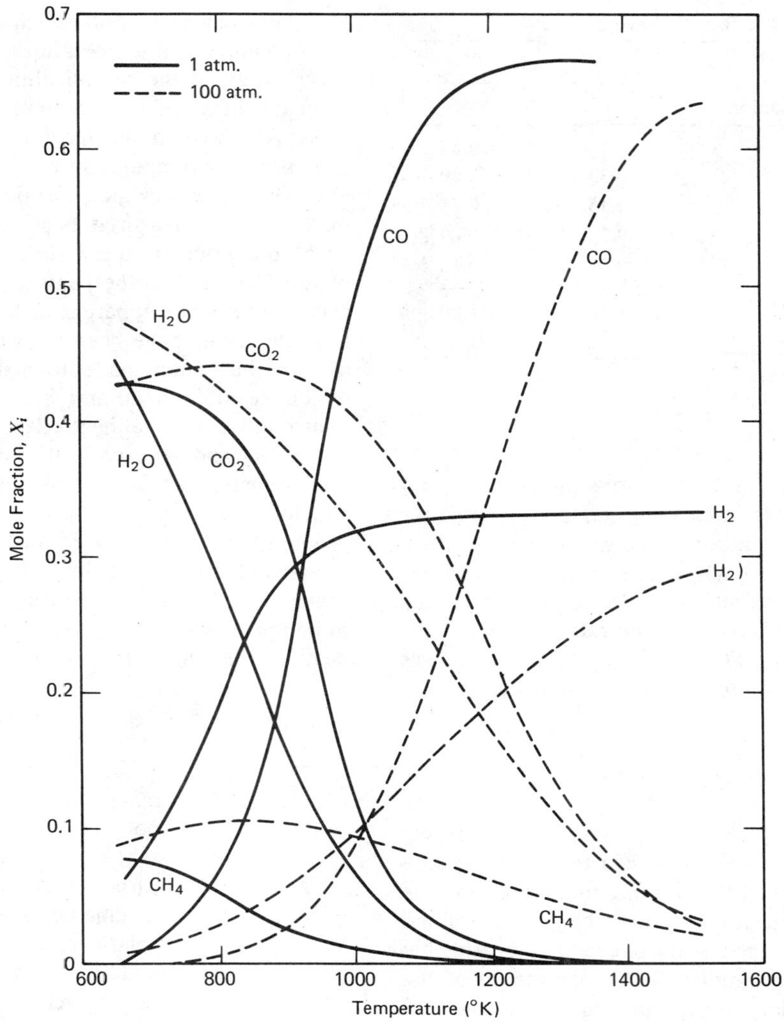

Figure 23.3 Equilibrium characteristics for the carbon-hydrogen-oxygen system (H/O = 1 g-atom/g-atom); solid curves, 1 atm; dashed curves, 100 atm.

Table 23.1 Equilibrium Trends for C-H-O System (H/O = 1 g-atom/g-atom)

Mole Fraction, X_i	Temperature ↑	Pressure ↑	H/O Ratio ↑
X_{CO}	↑	↓	↓
X_{CO_2}		↑	↓
X_{H_2}	↑	↓	↑
X_{H_2O}	↓	↑	↷
X_{CH_4}	↷	↑	↑
$\left[\dfrac{X_{CH_4}}{X_{CH_4}+X_{CO}+X_{CO_2}}\right]$	↷* ↓ **	↑	↑

* H/O < 2.
** H/O ≥ 2.

Table 23.2 Effect of Varying a_c

	a_c		
Mole Fraction	1	2	4
X_{CO}	0.19	0.27	0.35
X_{CO_2}	0.33	0.32	0.29
X_{H_2}	0.15	0.13	0.11
X_{H_2O}	0.25	0.16	0.09
X_{CH_4}	0.08	0.12	0.16
Total	1.00	1.00	1.00
$\frac{X_{CH_4}}{(X_{CO} + X_{CO_2} + X_{CH_4}}$	0.13	0.17	0.20
$\frac{(X_{CO} + X_{CO_2} + X_{CH_4})}{X_{CO} + 2X_{CO_2} + X_{H_2O}}$	0.54	0.67	0.79

monoxide mole fractions individually increase with increasing a_c, but the relative amount of carbon in the gas present as methane, $X_{CH_4}/(X_{CO} + X_{CO_2} + X_{CH_4})$ also increases with increasing a_c. In addition, the total carbon to oxygen in the ratio of gas $(X_{CO} + X_{CO_2} + X_{CH_4})/(X_{CO} + 2X_{CO_2} + X_{H_2O})$ increases with increasing a_c, which indicates greater gasification yields.

23.2.2 Heat Effects

The effects of heat on coal gasification systems can be expressed in terms of the sensible heats of the reacting substances and the heats of reaction. Specific heats of coals and coal chars, necessary to evaluate the sensible heat contributions, have been measured using a variety of experimental techniques.[11–23]

For the most part, studies with coals have been conducted at temperatures below about 400°C to avoid the complications that result from the heats of coal decomposition reactions. Although studies have also been made over a higher temperature range using coals and coal chars containing volatile matter,[11,23–27] measured heat effects under such conditions are due both to the sensible heat effects and to the heats of reaction. In such cases either apparent or average pyroheat capacities have been reported, or some attempt has been made to distinguish between sensible heat and heat of reaction contributions to total heat effects.

A novel and apparently useful correlation for estimating actual specific heats of coal and its solid conversion products has been proposed by Kirov,[28] based on an analysis of selected literature data. In his correlation the specific heat of a coal or coal char is assumed to be the weighted summation of the specific heats of the major coal components, that is:

$$C_p = X_v C_v + X_a C_a + X_w C_w + X_c C_c \qquad (23.9)$$

where

X_v, X_a, X_w, X_c = mass fractions of the potential volatile matter, ash, water, and fixed carbon, respectively

C_v, C_a, C_w, C_c = specific heats of potential volatile matter, ash, water, and fixed carbon, respectively (cal/g-°C)

C_p = average specific heat of coal or coal char (cal/g-°C)

[11] A. A. Agroskin and E. I. Goncharov, *Coke Chem. (USSR)*, **11**, 16–20 (1965).
[12] J. D. Batchelor, R. M. Yavorsky, and E. Gorin, *J. Chem. Eng. Data*, **4**, 241–246 (1959).
[13] J. S. Binford, Jr., J. M. Strohmenger, and T. H. Hebert, *J. Phys. Chem.*, **71**, 2404-2408 (1967).
[14] J. D. Clendenin, K. M. Barclay, H. J. Donald, D. W. Gilmore, and C. C. Wright, *Trans 7th Annu. Anthracite Conf. Lehigh Univ.*, Bethlehem, Pa., 1949, pp. 12–67.
[15] G. Coles, *J. Soc. Chem. Ind.*, **42**, 435–439 (1923).
[16] V. W. Fritz and H. Moser, *Feuerungstechnik*, **5**, 97–107 (1940).
[17] M. Gomez, J. B. Gayle, and A. R. Taylor, Jr., U.S. Bur. Mines Rep. Invest. No. 6607, 1965, 45 pp.
[18] V. V. Kazmina, *Coke Chem. (USSR)*, **11**, 26–29 (1965).
[19] C. L. Mantell, *Industrial Carbon*, 2nd ed., D. Van Nostrand, New York, 1947, p. 432.
[20] H. C. Porter and G. B. Taylor, *Ind. Eng. Chem.*, **5**, 289–293 (1913).
[21] E. Terres, H. Dahne, B. Nandi, C. Scheidel, K. Trappa, and P. Rauth, *Brennstoff-Chemie*, **37**, 366–377 (1956).
[22] E. Terres, H. Dahne, B. Nandi, C. Scheidel, and K. Trappe, *Brennstoff-Chemie*, **37**, 269–277 (1956).
[23] A. K. Voloshin, I. V. Virozub, V. V. Kazmina, and M. Y. Kurbatova, *Coke Chem. (USSR)*, **3**, 17–20 (1962).
[24] P. H. Dewey and D. R. Harper, III, *J. Res. Natl. Bur. Stand.*, **21**, 457–474 (1938).
[25] R. J. Morlock, A. C. Naso, and J. R. Cameron, *Symp. Sci. Tech. Coal*, Canada Dept. Energy, Mines, and Resources, Mines Branch, Ottawa, 1967, pp. 127–131.
[26] A. L. Lee, *Prepr., Am. Chem. Soc., Div. Fuel Chem.*, **12** (3), 19–31 (1968).
[27] O. P. Mahajan, A. Tomita, and P. L. Walker, Jr., *Fuel*, **55**, 63–69 (1975).
[28] N. Y. Kirov, *BCURA, Month. Bull.*, **29**(2), 33–57 (1965).

The specific heats are represented as functions of temperature with the exception of C_w, which is assumed to be constant at 1 cal/g-°C. Correlations proposed for the ash and fixed carbon fractions are:

$$C_a = 0.18 + 1.4 \times 10^{-4} t' \tag{23.10}$$

$$C_c = 0.165 + 6.8 \times 10^{-4} t' - 4.2 \times 10^{-7} (t')^2 \tag{23.11}$$

where t' = temperature (°C)

Kirov assumed that the total potential volatile matter of a coal or coal char is divided into two types: primary volatile matter (type 1), which is evolved at relatively low temperatures, and secondary volatile matter (type 2), which is evolved at higher temperatures. The specific heat for potential primary volatile matter, C_{v_1}, was approximated by the specific heat reported by Cragoe[29] for a low-temperature tar of specific gravity 1.04 in the liquid phase:

$$C_{v_1} = 0.395 + 8.1 \times 10^{-4} t' \tag{23.12}$$

The specific heat of the potential secondary volatile matter was represented by:

$$C_{v_2} = 0.71 + 6.1 \times 10^{-4} t' \tag{23.13}$$

Equation 23.13 was derived from the formulation by Clendenin et al.[30] of the specific heats of anthracites as a function of temperature and volatile matter, extrapolated for the case of 100% volatile matter.

Kirov defined the relative amounts of potential primary and secondary volatile matter in a coal solid by assuming that for the total volatile matter mass fraction (V_T) of 0.10 or less on an maf basis, all volatile matter is secondary volatile matter. For total volatile matter mass fractions greater than 0.10 (maf), the secondary volatile matter mass fraction, X_{v_2}, is assumed to be equal to $0.10(1 - X_a - X_w)$, with the remaining volatile matter existing as primary volatile matter (X_{v_1}). These assumptions are described by the following expressions:

[29] C. S. Cragoe, U.S. Natl. Bur. Stand. Misc. Publ. No. 97, 1929.

[30] J. D. Clendenin, K. M. Barclay, H. J. Donald, D. W. Gilmore, and C. C. Wright, Pennsylvania State Univ., Miner. Ind. Exp. Sta., Tech. Paper No. 160, 1949, pp. 11–67.

$$V_T = \frac{X_{v_1} + X_{v_2}}{1 - X_a - X_w} \tag{23.14}$$

For:

$$V_T \leqslant \frac{0.1}{X_{v_1}} = 0 \qquad \text{and} \qquad X_{v_2} = (1 - X_a - X_w)\, V_T \tag{23.15}$$

For:

$$V_T > \frac{0.1}{X_{v_1}} = (V_T - 0.10)(1 - X_a - X_w)$$
$$X_{v_2} = 0.10\,(1 - X_a - X_w) \tag{23.16}$$

Kirov compared the predictions of his total correlation with experimental data available in 1965 for mean specific heats of different coals over the temperature range of 20° to 100°C. Here, mean specific heat, $\overline{C}_p$, is defined as:

$$\overline{C}_p = \frac{\int_{t_0}^{t'_f} C_p \, dt'}{t'_f - t'_0} \tag{23.17}$$

where t'_0 = initial coal temperature
t'_f = final coal temperature

For coals ranging from approximately 0 to 50% volatile matter, agreement with the correlation was obtained to within about ±7%.

In a recent study at the Institute of Gas Technology (IGT),[31] Kirov's correlation was compared to the mean specific heats of coals and coal chars at temperatures of up to 300°C, and good agreement was found for coals ranging from about 0 to 50% volatile matter. In this IGT study mean specific heats of low-volatile coal char at 25–600°C were also found to be accurately predicted by Kirov's correlation. Since Kirov's correlation accounts only for specific heats of ungasified components in coals or coal chars, there is little basis to evaluate its applicability at higher temperatures, where significant thermal decomposition can occur, leading to reaction heat effects.

Although Kirov suggests that net reaction heat effects, with the exception of water

[31] Institute of Gas Technology, *Preparation of a Coal Conversion Systems Technical Data Book*, Final Report, October 31, 1974–April 30, 1976, U.S. ERDA Rep. No. FE-1730-21; NTIS, 1976.

vaporization, are negligible under carbonization conditions, some measurements of the total heats of coking have indicated that estimates based on Kirov's correlation are low by more than 15%.[25] Assuming that specific heats of carbonizing coal solids are reasonably described by his correlation, this would suggest that net reaction heat effects during carbonization are somewhat endothermic. There is, however, no real agreement in the literature concerning the endothermicity or exothermicity of reaction heat effects during coking; this is apparently because of the complications in distinguishing between relatively small heat of reaction effects and sensible heat requirements.

Net heat effects during the carbonization of coals and coal chars of varying volatile content were reported by Lee,[26] based on results of an experimental calorimeter study. With the technique employed, measurements were made of total heat requirements for the transition of feed solids at 21°C to carbonized solid residue and gasified products at calorimeter temperature, t'. Results were reported in terms of:

$$\overline{C}_{pm} \equiv \frac{Q}{t' - 21} \qquad (23.18)$$

where $\overline{C}_{pm}$ = pyroheat capacity (cal/g-°C)
Q = total heat requirement (cal/g)
t' = temperature (°C)

Values of $\overline{C}_{pm}$ obtained at final temperatures of 300–800°C for feed solids having a volatile matter content ranging from approximately 3 to 43 wt % (dry) were correlated with the following expression (modified to adjust units):

$$\overline{C}_{pm} = 0.174 + 1.98 \times 10^{-4}t' + (0.33 + 5\ 49 \times 10^{-4}t')V \qquad (23.19)$$

where t' = final temperature (°C)
V = volatile matter of feed coal or coal char, weight fraction, dry

In the development of Equation 23.19, the effect of the ash content in the feed solids used, which varied from 9 to 25% (dry), was not considered, nor was an adjustment made for the approximately 1% moisture in the feed solids. In effect, Equation 23.19 reflects the total heat, given by $Q = \overline{C}_{pm}(t' - 21)$, that is required to heat the feed solids from 21°C to t', plus the heat required for water vaporization and any reactions that occur in the system, including thermal decomposition, polymerization, and secondary gas phase reactions. Since total solids conversion is not a parameter in Equation 23.19, it is important to note that measurements were made under conditions in which solids were brought to calorimeter temperature in about 5–10 min and maintained at temperature for about 90 min. This is significant because product yields could vary for other heatup rates and residence times, leading to different heat effects.

A plot of sensible heat effects predicted by Kirov's correlations[28] and of total heat effects predicted by Lee's correlation[26] for the carbonization of coals or coal chars with 10, 30, and 50% volatile matter (maf) is shown in Fig. 23.4. Values of $\overline{C}_p$ were computed from Kirov's correlation under the assumption that $X_a = 0.17$ and $X_w = 0.01$, which is consistent with the average concentrations of these components in the coal chars used by Lee to develop Equation 23.19. Figure 23.4 shows that Lee's correlation for coals of 30 and 50% volatile matter generally predicts greater heat requirements than Kirov's correlation. If these differences are attributed to endothermic heats of reaction during carbonization, then at 800°C and from 30 to 50% feed volatile matter, heat of reaction values will vary from 65 to 125 cal/g of feed coal solids (maf). Although the basis used to compute these values is subject to considerable uncertainty, these heat effects are relatively small compared to the heats of reaction of graphite with various gases, as shown below*:

Reaction	Heat of Reaction at 800°C (cal/g carbon gasified)
$C + O_2 \longrightarrow CO_2$	−7857
$C + 2H_2 \longrightarrow CH_4$	−1798
$C + H_2O \longrightarrow CO + H_2$	2705
$C + CO_2 \longrightarrow 2CO_2$	3382

* Thermodynamic data from *JANAF Thermochemical Tables,* 2nd ed., D. R. Stull and H. Prophet, Eds. for U.S. Dept. of Commerce, Washington, D.C., 1973, 1141 pp.

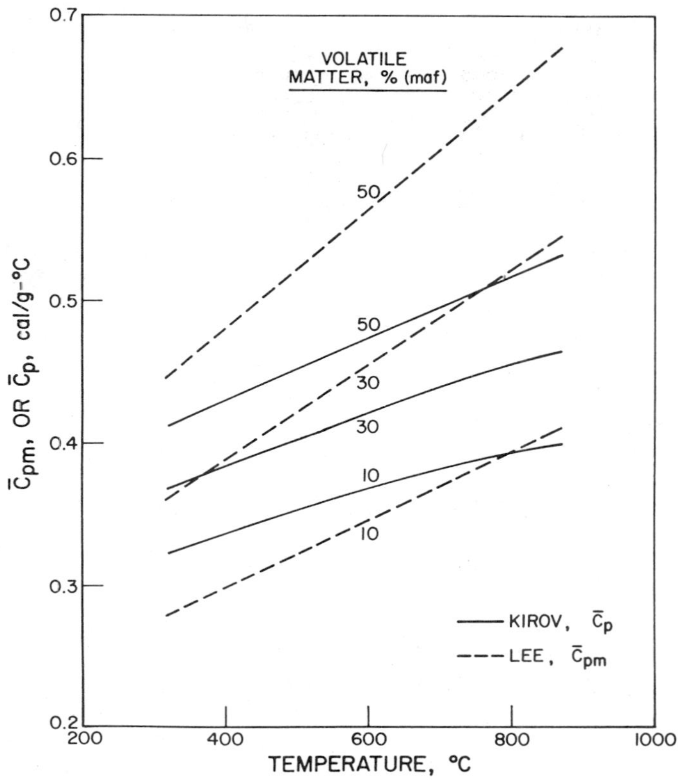

Figure 23.4 Comparison of correlations of Kirov[28] and Lee[26] for carbonization heat effects.

Mahajan et al.[32] have also reported carbonization heats of a variety of coals, based on differential scanning calorimetry (DSC) studies. Figure 23.5 shows total heats of reaction determined as a function of feed volatile matter for coals heated to 580°C at 10°C/min, in helium at 55 atm. With this treatment incomplete devolatilization was observed for all coals tested by Mahajan et al., and the ratio of total weight loss to initial volatile matter ranged from about 0.5 to 0.75. Also shown in Fig. 23.5 are carbonization heats computed from the differences in heat requirements predicted by Kirov's and Lee's correlations at 580°C for concentrations of 30 and 50% feed volatile matter (maf). As is apparent, endothermic values obtained from Kirov's and Lee's correlations are consistent only with the largest endothermic heats of reaction reported by Mahajan et al. The general scatter in the data of Mahajan et al. probably emphasizes the complications and uncertainties that exist in obtaining accurate and well-characterized heat of reaction data in coal carbonization and gasification systems. The proper distinction between sensible heat and reaction heat effects represents

[32] O. P. Mahajan, A. Tomita, J. R. Nelson, and P. L. Walker, Jr., *Prepr., Am. Chem. Soc., Div. Fuel Chem.*, **20** (3), 19–22 (1975).

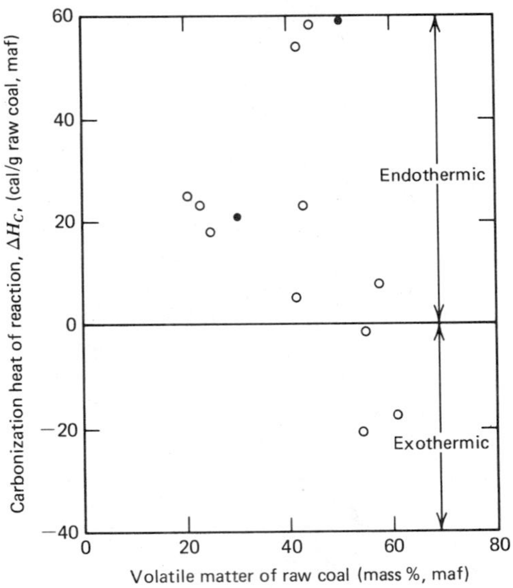

Figure 23.5 Effect of coal rank on carbonization heat of reaction: open points, data of Mahajan et al.[32]; solid points, estimated from Fig. 23.4.

only one of the many difficulties involved in quantitatively interpreting results of calorimetry studies as well as DSC or differential thermal analysis (DTA) studies.

Efforts have also been made to evaluate heat effects during the gasification of coals and coal chars in hydrogen at elevated pressures. In such a reactive gaseous environment, total heat measurements comprise sensible heat effects, heats of reaction due to thermal pyrolysis, heats of reaction due to direct gas-coal or coal-char interactions, and heats of reaction of secondary gas phase interactions. Lee et al.[33] have reported heat of reaction measurements for bituminous coal and coal chars gasified in hydrogen, which were based on calorimetry studies at 68 atm and temperatures of 705 and 816°C. This study employed the same experimental procedure used to evaluate carbonization heat effects,[26] with reaction products retained in the calorimeter for periods of up to 300 min. In his study with hydrogen, Lee also measured total solid conversion and computed net exothermic heats of reaction per mass of coal gasified, $-\Delta H_r$, by adjusting total heat effects according to:

$$-\Delta H_r = \frac{-Q + \overline{C}_{pm}(t' - 21)}{\Delta m} \qquad (23.20)$$

where Q = total heat input (cal/g feed solids)
$\overline{C}_{pm}$ = pyroheat capacity, calculated from Equation 23.19 (cal/g-°C feed solids)
Δm = fraction of feed coal gasified (g/g feed solids)

One interpretation of Equation 23.20 is that the kinetics and thermal characteristics of carbonization depend only on temperature, even in a reactive atmosphere where additional reaction processes occur. Using this simplifying assumption, it can be seen that $-\Delta H_r \Delta m$ reflects the net heat of reaction of the additional reaction processes that occur in the hydrogen environment, such as carbon gasification to form methane ($C + 2H_2 \rightarrow CH_4$) and the secondary hydrogenation of gasified volatile matter.

Figure 23.6 shows values of $-\Delta H_r$ obtained by Lee et al.[33] as a function of the fraction of coal solids reacted. Although temperature does not appear to have a significant effect on values of $-\Delta H_r$, the results do indicate a trend in which $-\Delta H_r$ decreases with increasing volatile matter in the feed solids. Although Lee has suggested that this trend might be due to the effects of increased endothermic pyrolysis with the higher volatile materials, his conclusion appears to conflict with the use of the pyroheat capacity in adjusting total heat measurements, which ideally should compensate for pyrolysis heat effects.

Lee's results can also be interpreted by referring net heat effects only to the fixed carbon gasified in Lee's calorimeter experiments and not to the total solid conversion, including volatile matter gasified. This approach assumes that "fixed carbon" and "volatile matter" are independent components in coal solids, a simplifying assumption that is not valid over a wide range of conditions. Let $-\Delta H_r' = \Delta H_r \Delta m$ = net exothermic heat of reaction in calories per gram (of feed solids). Also assume that the net heat of reaction, $-\Delta H_r'$, is the sum of the heat of reaction of fixed carbon hydrogenation plus the heat of hydrogenation of gasified volatile matter:

$$-\Delta H_r' = h_f \Delta F + h_v \Delta V \qquad (23.21)$$

or

$$-\Delta H_r' = h_f \Delta F + h_v(\Delta m - \Delta F) \qquad (23.22)$$

where ΔF = fixed carbon gasified (g/g feed solids)
$\Delta V = \Delta m - \Delta F$ = volatile matter gasified (g/g feed solids)
h_f, h_v = average heats of reaction of fixed carbon gasification and of the gas phase hydrogenation of volatile matter, respectively (cal/g)

In addition, it can be expected that under the conditions of Lee's experiments, with residence times of up to 300 min, most of the pyrolysis hydrocarbons other than heavy tars will be hydrogenated to methane.[34] Equation 23.22 can be rearranged to yield:

[33] A. L. Lee, H. L. Feldkirchner, F. C. Schora, and J. J. Henry, *Ind. Eng. Chem., Process Des. Dev.*, **7**, 244–249 (1968).

[34] H. N. Woebcke, L. E. Chambers, and P. S. Virk, *Prepr., Am. Chem. Soc., Div. Fuel Chem.*, **18**(2), 97–132 (1973).

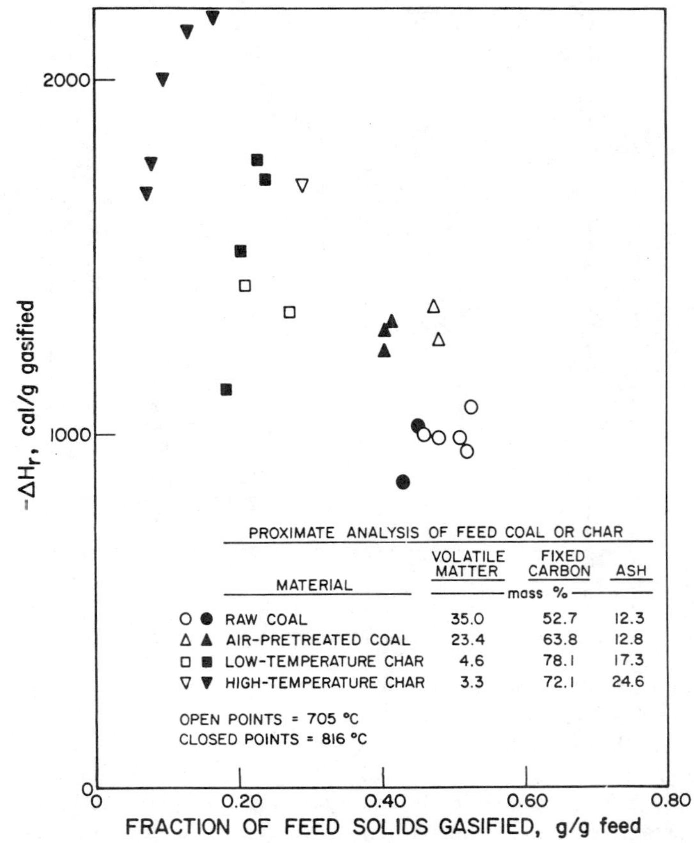

Figure 23.6 Effect of solid conversion on hydrogenation heat of reaction.[33]

$$-\Delta H_r'' = \frac{-\Delta H_r'}{\Delta F} = \frac{-\Delta H_r \Delta m}{\Delta F}$$

$$= h_f + h_v\left(\frac{\Delta m}{\Delta F} - 1\right) \quad (23.23)$$

Thus a plot of $-\Delta H_r''$ versus $\Delta m/\Delta F - 1$ should yield a straight line with a y-axis intercept equal to h_f and a slope equal to h_v. Such a plot (Fig. 23.7) is based on calculations made with Lee's data, in which ΔF values were computed from reported proximate analyses of solid feed and residues.

In plotting the line shown in Fig. 23.7, it was arbitrarily assumed that for zero volatile matter gasified ($\Delta m = \Delta F$), h_f is equal to the heat of the hydrogen gasification of graphite. Based on this assumption, a slope of h_v = 650 cal/g was obtained. At best, this value of h_v is applicable only to the bituminous coal and coal chars used by Lee, since the distribution of volatile products would vary for different coal types. In addition, since not all the volatile matter can be hydrogenated, h_v does not represent an average heat of reaction per mass of volatile matter actually hydrogenated, but rather the heat of reaction per mass of total volatile matter gasified.

A more detailed representation of h_v is:

$$h_v = \sum_i X_i h_{v_i} \quad (23.24)$$

where X_i = mass fraction of species i in gasified volatiles, (g/g volatiles gasified)

h_{v_i} = heat of hydrogenation for conversion of species i to methane (cal/g species i)

Equation 23.24 shows that the value of h_v depends on the distribution of the species gasified during devolatization. For example, it can be assumed that the raw bituminous coal used by Lee (see Fig. 23.6 for proximate analysis) has the distribution of volatiles shown in Table 23.3. Based on this simplified distribution of volatile products, the value of h_v computed from the equation is 650 cal/g of total volatile matter. Although this example was obviously contrived, a reasonable vari-

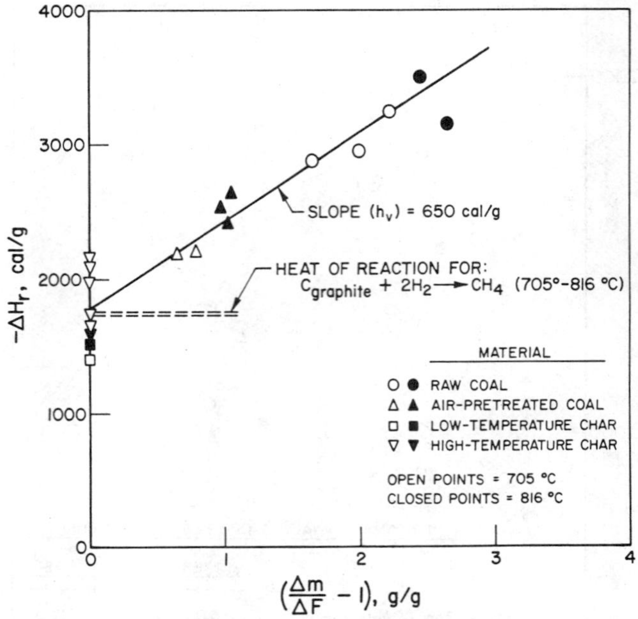

Figure 23.7 Correlation of hydrogenation heats of reaction.

ation in the assumed distribution that is consistent with experimentally observed devolatilization yields for high-volatile coals would not result in a greatly different value for h_v.

The heat of reaction of the fixed carbon in coal chars with reactive gases has been treated in different ways by different investigators. For example, the assumed value of h_f in Fig. 23.7 is consistent with the assumption that the standard heat of formation of carbon in coal chars is the same as that of graphite. This simplifying assumption is most commonly employed in calculating the heats of reaction in coal char gasification systems. It has also been proposed, however, that the carbon in coal chars is amorphous, and that by analogy with other types of amorphous carbon, coal char carbon should have a higher heat of formation than graphite.

Stephen,[35] is computing thermodynamic characteristics of subbituminous coal char gasification systems, assumed that coal char carbon has a standard heat of formation that is 2.6 kcal/g-atom C (216 cal/g C) larger than that of graphite. This value is consistent with the extreme values for standard heats of formation of amorphous carbon at 298 K that have been reported in the literature[36] and with values of standard heats of formation of

[35] D. R. Stephens, U.S. Atomic Energy Commission Rep. No. 16094; NTIS, 1972.

[36] O. A. Hougen and K. M. Watson, *Industrial Chemical Calculations,* 2nd ed., John Wiley and Sons, New York, 1936, 487 pp.

Table 23.3 Distribution of Volatiles for a Raw Bituminous Coal

Coal Component Devolatized	g/g Total (C + H + O)	Volatile Species	g/g Total (C + H + O)	Heat of Hydrogenation to CH_4 (cal/g volatile species)
Carbon	0.67	CO	0.15	121
Hydrogen	0.14	H_2O	0.12	
Oxygen	0.19	CH_4	0.17	
		C_2H_6	0.13	592
		C_6H_6	0.30	1850
		H_2	0.04	
		C(tar)	0.10	
		Total	1.01	

carbon that can be deduced from some of the generalized heat of combustion formulas expressed as a function of coal composition.

Accurate determination of the standard heat of formation of a raw coal is particularly difficult. Although such a value can be theoretically computed from experimental determinations of the heats of carbonization, such a procedure is complicated by the need to know or estimate carbonization product yields and the thermodynamic properties of these products. As indicated previously, this information is neither complete nor well defined, even for a few coals. Therefore the more usual procedure in computing heat effects for gasification reactions in which coal is converted to coal char is to make use of the heats of combustion, as determined experimentally or taken from one of the many published general correlations.[37,38] The main disadvantage of this procedure, however, is the error that can result from computing a relatively small number as the difference of two much larger numbers that are of limited accuracy. In addition, very little information is available for estimating heats of combustion of partially devolatilized coal residues as a function of composition and preparation conditions—and such information would be of value in detailed analyses of coal gasification systems.

23.3 PHYSICS OF GASIFICATION SYSTEMS

Both physical and chemical processes are important in coal and coal char gasification. In a limited sense, physical processes can be defined as those that affect the practical operability of an integral conversion system as well as the conditions under which chemical reactions occur. Such processes include those that affect mass transfer, heat transfer, pressure drop, other coal hydrodynamic behavior, and physical properties of solids undergoing gasification, solids agglomeration, and attrition. Much of the information necessary to quantitatively characterize certain of these processes, as they pertain to coal gasification systems as well as to other types of systems, is available in a voluminous literature, including textbooks and handbooks. Although most of this information is not amenable to review within the scope of this chapter, some areas that are particularly important to modern coal gasification concepts have received considerable attention in recent years and are discussed here. Two of these areas are physical processes related to fluidized-bed systems, and the characterization of coal and coal char structural variations during gasification. Characterization of the physical processes that occur in fluidized-bed contacting systems is important to the design of most modern gasification systems currently being developed; moreover, coal and coal char structural characteristics are important parameters that affect the behavior of any contacting system, including fluidized beds. Although fixed beds and suspension systems are also of interest in coal gasification, the physical characterization of such systems has been described in various reviews.[8,39,40]

23.3.1 Characteristics of Gas-Fluidized Beds

Aggregative gas–fluidized beds can be generated by passing a gas up through a bed of relatively fine particles. At low flow rates the bed remains stationary (i.e., fixed bed) as the gas passes through voids between the particles. As the gas flow rate increases, the pressure drop across individual particles and across the total bed of solids also increases. Ideally, a specific flow rate is eventually reached at which the particles are just suspended in the upward-flowing gas, and the pressure drop through any section of the bed is approximately equal to the weight of particles in that section. At this point, the net vertical components of compressive force between adjacent particles disappear, and the bed is considered to be incipiently fluidized, or at minimum fluidization. The superficial gas velocity at this condition, referred to the empty tube plug flow value, is defined as the minimum fluidization velocity, V_{mf}. The av-

[37] D. W. Van Krevelen, *Coal,* Elsevier, Amsterdam, 1961, 514 pp.

[38] J. C. Janka and R. Malhotra, U.S. Office Coal Res., R & D Rep. No. 22—Interim Rep. No. 7 (PB 235 370); NTIS, 1971.

[39] F. A. Zenz and D. F. Othmer, *Fluidization and Fluid-Particle Systems,* Reinhold Publishing Corp., New York, 1960, 513 pp.

[40] M. A. Field, D. S. Gill, D. B. Morgan, and P. G. W. Hawksley, *Combustion of Pulverized Coal,* BCURA, Leatherhead, Surrey, England, 1967, 413 pp.

erage interstitial gas velocity in the fluidized bed, V_f, is:

$$V_f = \frac{V_{mf}}{\epsilon_{mf}} \qquad (23.25)$$

where ϵ_{mf} is the average void fraction in the bed at minimum fluidization.

As the gas velocity is increased beyond minimum fluidization, the total pressure drop remains relatively constant; but gas bubbles tend to form and pass up through the bed, as depicted in Fig. 23.8. There are two major regions of solids density in a bubbling fluidized bed, usually referred to as the emulsion phase (high solids density) and the bubble phase (low solids density). It is often assumed, as a first approximation, that all gas in excess of that needed to bring the bed to incipient fluidization passes through the bed as bubbles, while the emulsion phase remains at minimum fluidizing conditions.

Bubbles formed at the bottom of a bed do not stay the same size as they pass upward through the bed, although this is often assumed in analyses of fluidized-bed behavior. Instead, bubbles can grow by coalescing with other bubbles or by splitting under conditions of instability. The rate of bubble growth generally increases with increasing bubble diameter. A particularly important effect of bubble interactions in the bed is the agitation and relatively large solids mixing rates that result. This feature is particularly desirable in coal gasification systems, in which control is simplified by the approximately isothermal conditions that result from large solids circulation rates, even with the highly exothermic and endothermic reactions that can occur.

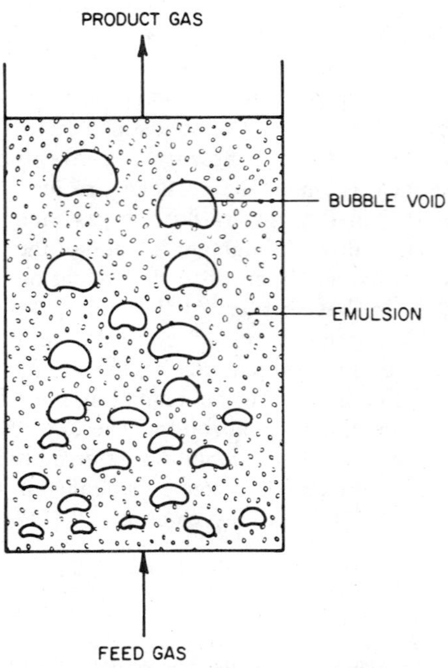

Figure 23.8 Schematic diagram of bubbling fluidized bed.

Other advantages of the fluidized bed for coal gasification include relatively high gas-solids heat transfer rates and the ease of controlling the processes due to the fluidlike flow of particles, which facilitates large-scale operation. Potential disadvantages of the fluidized bed include gas bypassing of the solids, nonuniform solids residence times for continuous solids feed and withdrawal, gaseous backmixing, solids attrition and agglomeration, and solids elutriation. Solids elutriation becomes a problem particularly when a relatively large distribution of particle sizes is used, which causes fine particles to be carried out of the bed with the product gas.

A major objective in elucidating the behavior of fluid beds is to understand the movement of gases and solids. A significant breakthrough in clarifying these processes was made by Davidson and Harrison[41] in 1963. Their model was later extended by Murray,[42] with an alternative treatment given by Jackson.[43] Based on these models, it is now understood that when the bubble velocity is greater than the interstitial gas velocity in the emulsion phase, there is a convective circulation of gas from the top of the bubble void into the emulsion phase. The gas is then swept around and returned to the bubble void near its lower extremity. The region around the bubble, through which this circulation occurs, is called the bubble cloud, as schematically depicted in Fig. 23.9. Significantly, a turbulent mixing zone just below the bubble void causes particles of solids to be drawn up behind the rising void and to form a wake, also shown in Fig. 23.9. Solids in this wake move up with the bubble void, but are also continuously exchanged with emulsion solids, thus resulting in a primary

[41] J. F. Davidson and D. Harrison, *Fluidised Particles,* Cambridge University Press, New York, 1963, 155 pp.

[42] J. D. Murray, in Am. Inst. Chem. Eng. Chem. Eng. Prog. Symp. Ser., Vol. 62, 1966, pp. 71–82.

[43] R. Jackson, *Trans. Inst. Chem. Eng.*, **41**, 13–28 (1962).

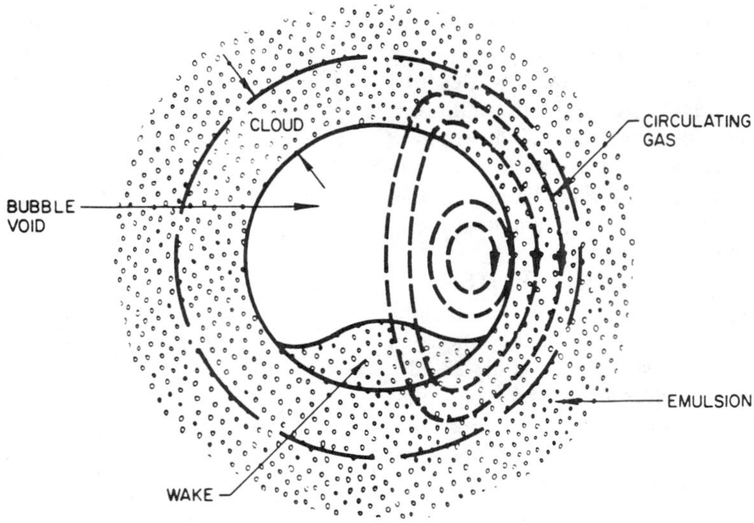

Figure 23.9 Simplified diagram of bubble region.

mechanism for solids mixing in a vigorously bubbling bed. Convective movement of the solids also occurs by displacement of the emulsion phase by rising bubbles. Solids in the wake carried to the surface of the bed fall back and lie on top of the bed, leading to a net downward mass flow of solids in the emulsion phase of a batch solids system to compensate for the upward mass flow of wake solids. In systems where solids are continuously fed and withdrawn from the bed, the direction and mass flow rate of emulsion solids depends on a number of factors, including solids feed and withdrawal locations.

As the ratio of bubble velocity to interstitial emulsion gas velocity decreases, the cloud thickness increases until, eventually, the bubble velocity is lower than the interstitial emulsion gas velocity. At this point, in addition to a circulating pattern of gas flow between bubble void and emulsion, there is also a flow of emulsion gas into the bottom and out of the top of the bubble void. This "through-flow" gas does not directly circulate back to the bubble void. Diffusive-type gas transfer can also occur between bubble void and cloud, and between cloud and emulsion, as another mechanism for net transfer of gas from bubble void to emulsion.

Despite the significant advances made in the last decade or two in elucidating some of the complex processes involved in fluidized-bed technology, there are still important gaps in information and theory that prevent a priori prediction of fluid bed behavior. Recent reviews have indicated the need for more information on bubble inception, the effects of gas distributor design, the phenomenon of bubble coalescence, other bubble interactions, and bubble distribution in large beds.[44–48] Idealized models of varying sophistication have, however, been proposed to describe behavior in fluidized beds for a limited range of conditions. A consideration of the different types of models proposed is therefore of interest in evaluating available techniques for designing coal gasification fluid bed reactors.

23.3.1.1 *Fluidized-Bed Models*

The earliest models proposed to describe the gas-solid contacting configurations in fluidized beds did not reflect the existence of bubbles. Such models idealized the system by assuming isothermal conditions and by various combinations of assumptions with respect to perfect backmixed or plug flow of

[44] J. R. Grace, in Am. Inst. Chem. Eng. Chem. Eng. Prog. Symp. Ser., Vol. 67, No. 116, 1971 *p. 159*.

[45] D. L. Pyle, in Am. Chem. Soc. Adv. Chem. Ser., No. 109, 1972, pp. 106–130.

[46] P. N. Rowe, in *Proc. 5th European–2nd Int. Symp. Chem. Reaction Eng.*, J. M. H. Fortuin, Ed., Elsevier, Amsterdam, 1972, pp. A9-1–14.

[47] C. Chavarie and J. R. Grace, *Ind. Eng. Chem. Fundam.*, **14,** 75 (1975).

[48] J. F. Davidson and D. Harrison, *Fluidization,* Academic Press, London, 1971.

solids and perfect backmixed or plug flow of gas. Although such models (which are still extensively used) can adequately describe experimental results for limited conditions, a variety of models has also been proposed that directly or indirectly recognize the multiphase nature of bubbling fluidized beds and account for a number of mass and heat transfer mechanisms.

These multiphase models can be broadly classified into two major categories: simple two-phase models and bubbling-bed models. For simple two-phase models (Fig. 23.10), the fluidized bed is considered to be composed of two parallel, single-phase regions, one of dense solids concentration and one with dilute or zero solids concentration. Individual gas streams pass through each of these phases, and there is also a cross-flow of gases between phases. With this type of model, the dense solids concentration region is associated with the emulsion phase, and the lean solids concentration region with the bubble phase.

A summary of the assumptions made in different studies to characterize simple two-phase models is shown in Table 23.4, which was adapted from very thorough surveys made by Grace[44] and by Rowe.[46] In most of the models considered in Table 23.4, parallel gas flow occurs in the emulsion and bubble phases, although in some models net gas flow in the emulsion phase is assumed to be zero. In the bubble phase, plug flow is usually assumed, whereas assumptions concerning gas flow patterns in the emulsion phase have

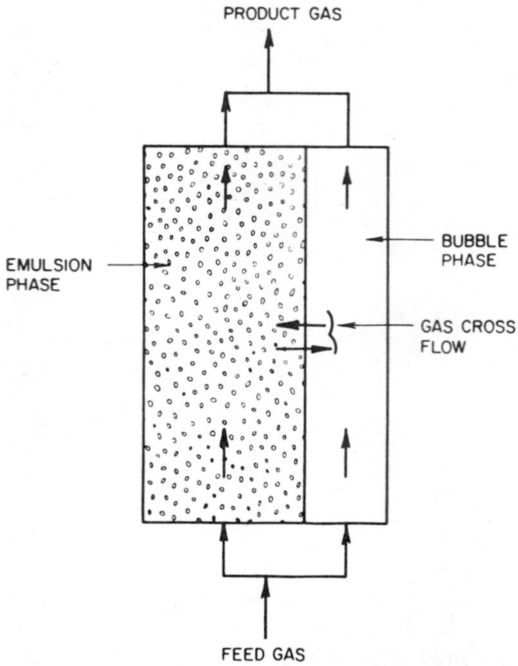

Figure 23.10 Schematic diagram of simple two-phase model.

varied from plug flow to perfect mixing, with finite backmixing represented by both axial dispersion and tanks in series models. Although some of the simple two-phase models account for the presence of solids in the bubble phase, relatively little consideration has been given to solids mixing patterns or residence time distributions, since most of the interest has been in gas-solid catalytic reactions and not in heterogeneous gas-solids

Table 23.4 Characteristics of Simple Two-Phase Models

Bubble Gas Flow[a]			Solids in Bubble Phase		Gas Flow in Emulsion Phase					
$Q_e = 0$	$Q_b = Q_0 - Q_{mf}$	Not Stated	Yes (Dilute)	No	Plug Flow	Perfect Mixing	Axial Dispersion	Tanks in Series	Down Flow	Ref.
—	×	—	—	×	×	×	—	—	—	49
—	×	—	×	—	×	—	—	—	—	50
×	—	—	×	—	×	×	—	—	—	51
—	×	—	—	×	—	—	×	—	—	52
—	—	×	×	—	—	—	—	—	×	53
—	×	—	—	×	×	—	—	—	—	54
×	—	—	—	×	—	—	×	—	—	55
—	×	—	—	×	—	—	—	×	—	56
—	×	—	×	--	×	×	—	—	—	57
—	×	—	—	×	—	—	×	—	—	58
—	—	×	—	×	×	—	—	—	×	59

[a] Q_{mf} = gas flow rate at minimum fluidization
Q_0 = total gas flow rate
Q_b = gas flow rate in bubble phase
Q_e = gas flow rate in emulsion phase

reactions, which are important in coal and coal char gasification.

Simple two-phase models provide a useful engineering approach for designing or scaling up fluidized-bed reactors, based on experimental evaluation of the parameters inherent in a given model. Because of the level of empiricism involved in using this type of model, however, extrapolation of laboratory data to design large-scale reactors in commercial coal gasification systems necessarily involves a high degree of uncertainty in projecting performance. Thus for coal gasification systems, there may be an advantage to using the more complex bubbling-bed models for scaleup, since these models, based at least in part on characterization of fundamental fluid mechanical processes, make it possible to account for phenomena that may occur on a large scale but cannot be easily anticipated from empirical descriptions of laboratory data.

Grace[44] and Rowe[46] have also made thorough reviews of the more complex bubbling-bed models depicted in Figs. 23.8 and 23.9, and the assumptions made in the various proposed models are summarized in Table 23.5, adapted from the work of these authors. The models are all based on the properties of single rising bubbles, often using results from independent theoretical or experimental studies to account for bubble rise velocity, cloud formation, through-flow of gas in bubbles, wake characteristics, and so on. For most of the models listed in Table 23.5, unknown parameters can be selected to fit experimental reaction data. Although some of the parameters can be at least approximately determined on the basis of fundamental considerations, as indicated previously, there is not enough information to accurately predict bubble size distributions and bubble interaction effects. For this reason an adjustable parameter that is often selected to fit experimental results is the effective bubble diameter.

In most experimental studies of fluidized beds, models have been evaluated in catalytic reactor systems by measuring the concentrations of the reactants and products in the inlet and outlet gas streams. However Chavarie and Grace[47] have recently compared the predictions of some of the bubbling-bed models not only with total reaction yields but also with internal gas concentration gradients within the bubble and emulsion phases. These authors made a detailed experimental investigation of the catalytic decomposition of ozone in a two-dimensional fluidized bed using porous alumina particles. The fluidized-bed models considered were those of Orcutt et al.,[60] Partridge and Rowe,[63] Kunii and Levenspiel,[66] and Kato and Wen.[67] Of the four, Kunii and Levenspiel's model was found to best fit the internal gas concentration profiles and net gas yield for the range of experimental conditions investigated. This was true even when the model-derived values of parameters were modified to improve agreement. It is interesting to note that measured mean bubble sizes halfway up the fluidized bed used by Chavarie and Grace were surprisingly close to calculated effective bubble sizes necessary to optimize the Kunii and Levenspiel model. However the analysis made was for a two-dimensional bed, and radial variations in properties were not considered in comparative evaluations. Thus it is uncertain how meaningful this analysis is in terms of large-scale three-dimensional fluidized beds.

23.3.1.2 *General Properties of Fluidized Beds*

Quantitative empirical correlations and data tabulations useful in the design of fluidized-bed systems have been obtained in various investigations. Such correlations are valuable

[49] C. Y. Shen and H. F. Johnstone, *AIChE J.*, **1**, 349–354 (1955).

[50] J. F. Mathis and C. C. Watson, *AIChE J.*, **2**, 518–524 (1956).

[51] W. K. Lewis, E. R. Gilliland, and W. Glass, *AIChE J.*, **5**, 419 (1959).

[52] W. G. May, Am. Inst. Chem. Eng. Chem. Eng. Prog. Symp. Ser., Vol. 55, No. 12, 1959, pp. 49–56.

[53] K. P. Lanneau, *Inst. Chem. Eng. Trans.*, **38**, 125–143 (1960).

[54] A. Gomezplata and W. W. Shuster, *AIChE J.*, **6**, 454–459 (1960).

[55] J. J. van Deemter, *Chem. Eng. Sci.*, **13**, 143–154 (1961).

[56] T. Mamuro and I. Muchi, *Int. Chem. Eng.*, **5**, 732–736 (1965).

[57] H. Kobayashi, F. Arai, T. Chiba, and Y. Tanaka, *Kagaku Kogaku*, **33**, 274–280 (1969).

[58] J. P. Mireur and K. B. Bischoff, *AIChE J.*, **13**, 839–845 (1967).

[59] J. J. van Deemter, in *Proc. Int. Symp. Fluidization*, A. A. H. Drinkenburg, Ed., Netherlands University Press, Amsterdam, 1967, pp. 334–347.

Table 23.5 Characteristics of Bubbling-Bed Models

Bubble Gas Flow[a]		Bubble Phases			Gas Flow in Emulsion Phase			
$Q_b = Q_0 - Q_{mf}$	$Q_b = Q_0 - Q_{mf} - Q_t$	Simple Empty	Empty + Cloud (1 phase)	Empty + Cloud (2 phases)	Plug Flow	Perfect Mixing	Axial Dispersion	Tanks in Series
×	—	×	—	—	×	×	—	—
×	—	×	—	—	×	×	—	—
×	—	—	×	—	×	—	—	—
×	—	—	×	—	×	—	—	—
×	—	—	×	—	×	—	—	—
×	—	×	×	—	×	×	×	—
×	—	×	—	—	×	—	—	—
×	×	—	—	×	—	—	×	—
—	—	—	×	—	—	—	—	×
×	—	—	×	—	×	—	—	—

[a] Q_{mf} = gas flow rate at minimum fluidization
Q_0 = total gas flow rate
Q_b = gas flow rate in bubble phase
Q_e = gas flow rate in emulsion phase
Q_t = "through-flow" gas rate in bubble
[b] 0. = no mass transfer resistance
Int. = intermediate mass transfer resistance
∞ = infinite mass transfer resistance (no transfer)
[c] Net transfer rate between bubble and emulsion kept as correlating parameter.

both for estimating gross bed behavior and for providing a basis for estimating more detailed properties necessary to apply certain of the fluidized-bed models discussed above.

Particle Shape Factors. Shape factors for different nonspherical coal particles are often needed for the application of empirical correlations to estimate properties in fluidized beds. The shape factor, ϕ, is defined as the ratio of the surface area of a sphere to the surface area of a particle, both having the same volume. Thus the shape factor, ϕ, for a sphere is equal to unity, and for any other shaped particle, $0 < \phi < 1$. Table 23.6 lists shape factors determined for a variety of coals ranging in rank from anthracite to subbituminous. The results of Austin et al.[69] are of particular interest in that shape factors were found to be relatively independent of particle size for particle diameters ranging from about 40 to 1200 μm. These investigators also found that the shape factor varied systematically with coal volatile matter, with maximum sphericity at a volatile matter content of about 17.9% maf.

Bed Voidage at Minimum Fluidization. The bed voidage fraction at minimum fluidization, ϵ_{mf}, is also an important parameter in designing fluidized beds, particularly since it is often assumed that the properties of the emulsion phase, even in bubbling beds, are the same as those at minimum fluidization. The value of ϵ_{mf} has been found both to decrease,[72,73] and to increase[74-77] with in-

[60] J. C. Orcutt, J. F. Davidson, and R. L. Pigford, in Am. Inst. Chem. Eng. Chem. Eng. Prog. Symp. Ser., Vol. 58, No. 38, 1962, pp. 1–15.

[61] P. N. Rowe, Am. Inst. Chem. Eng. Chem. Eng. Prog., Symp. Ser., Vol. 60, No. 3, 1964, pp. 75–80.

[62] P.L. Rose, Ph.D. thesis, Cambridge Univ., 1965.

[63] B. A. Partridge and P. N. Rowe, *Trans. Inst. Chem. Eng.*, **44**, 335–348 (1966).

[64] F. D. Toor and P. H. Calderbank, in *Proc. Int. Symp. Fluidization,* A. A. H. Drinkenburg, Ed., Netherlands University Press, Amsterdam, 1967, pp. 373–392.

[65] S. Hovmand and J. F. Davidson, *Trans Inst. Chem. Eng.*, **46**, 190–203 (1968).

[66] D. Kunii and O. Levenspiel, *Fluidization Engineering,* John Wiley and Sons, New York, 1969, 534 pp.

[67] K. Kato and C. Y. Wen, *Chem. Eng. Sci.*, **24**, 1351–1369 (1969).

[68] T. Chiba and H. Kobayashi, *Chem. Eng. Sci.*, **25**, 1375–1385 (1970).

[69] L. G. Austin, R. P. Gardner, and P. L. Walker, Jr., *Fuel,* **42**, 319–323 (1963).

[70] M. Leva, M. Weintraub, M. Grummer, and M. Pollchik, *Ind. Eng. Chem.*, **41**, 1206 (1949).

[71] R. H. Perry and C. H. Chilton, *Chemical Engineers' Handbook,* 5th ed., McGraw-Hill Book Company, New York, 1973.

[72] O. P. Agarwal and J. A. Storrow, *Chem. Ind. (London)*, 278–286 (1951).

Gas Exchange Resistance[b]		Bubble Detail						
		Model						
Bubble-Cloud Interface	Cloud-Emulsion Interface	Davidson	Murray	Through-Flow	Mixing	Wake	Growth	Ref.
Int.	0	—	—	—	×	—	—	60
Int.	0	—	—	—	×	—	—	41
0	∞	×	—	—	×	×	—	61
0, int.	∞	×	—	—	×	×	—	62
0	Int.	—	×	—	×	×	×	63
Int.	0, int, ∞	×	—	—	×	—	—	64
Int.	Int.	—	—	—	×	—	—	65
Int.	Int.	—	—	—	×	×	—	66
Correlating[c]		×	—	×	×	—	×	67
Correlating[a]		—	×	—	×	×	—	68

creasing particle size for coal-related materials. Correlations of this parameter with other potentially significant factors such as shape factors, particle densities, and particle

[73] M. Leva, M. Weintraub, M. Grummer, M. Pollchik, and H. H. Storch, U.S. Bur. Mines Bull. No. 504, 1951, 149 pp.

[74] C. van Heerden, A. P. P. Nobel, and D. W. Van Krevelen, *Chem. Eng. Sci.*, **1**, 37–49 (1951).

[75] J. F. Jones, R. T. Eddinger, and L. Seglin, in *Proc. Int. Symp. Fluidization*, A. A. H. Drinkenburg, Ed., Netherlands University Press, Amsterdam, 1967, pp. 676–687.

[76] G. P. Curran and E. Gorin, U.S. Office Coal Res. R & D Rep. No., 16—Interim Rep. No. 3, Book 1 (PB 184718); NTIS, 1970.

[77] H. F. Feldmann, K. Kiang, and P. Yavorsky, *Prepr., Am. Chem. Soc., Div. Fuel Chem.*, **15**(3), 62 (1971).

Table 23.6 Shape Factors of Different Coals

Material	Volatile Matter (%, maf)	Shape Factor, ϕ	Ref.
St. Nicholas, Pa.	4.5	0.64	69
Lorree Raven, Pa.	5.4	0.61	69
Dorrance, Pa.	5.8	0.59	69
Treverton, Pa.	9.0	0.69	69
Upper Freeport, Md.	17.9	0.83	69
Upper Kittanning, Pa.	18.0	0.78	69
Pratt, Ala.	29.2	0.75	69
Lower Kittanning, Pa.	30.2	0.71	69
Upper Freeport, Pa.	33.0	0.70	69
Cobb, Ala.	33.1	0.68	69
Pittsburgh, Pa.	37.4	0.65	69
Glen Mary, Tex.	39.4	0.68	69
Pittsburgh, Pa.	39.5	0.65	69
Pittsburgh, Pa.	40.0	0.62	69
Pittsburgh No. 8, Ohio	42.4	0.62	69
Illinois No. 6, Ill.	45.4	0.62	69
Anthracite	—	0.63	70
Bituminous	—	0.63	66
Coal dust (natural up to ⅜ in.)	—	0.65	71
Coal dust (pulverized)	—	0.73	71

size distributions, are not available. Kunii and Levenspiel[66] have suggested that since ϵ_{mf} is only slightly larger than voidage in a packed bed, values can be estimated from random packing data or can be measured experimentally.

Minimum Fluidization Velocity. A variety of correlations proposed to predict minimum fluidization velocities has been reviewed by Leva,[78] Wen and Yu,[79] and Frantz.[80] Many of the correlations proposed are based on the definition that minimum fluidization occurs when the total drag force across the bed (pressure drop × cross-sectional area of tube) is equal to the weight of solids in the bed. Wen and Yu[79] and Kunii and Levenspiel[66] have adopted this procedure, using the fixed-bed pressure drop correlation of Ergun,[81] to develop the expression:

$$\frac{1.75}{\phi\epsilon_{mf}^3}\left(\frac{D_p V_{mf}\rho_g}{\mu}\right)^2 + \frac{150(1-\epsilon_{mf})}{\phi^2\epsilon_{mf}^3}\left(\frac{D_p V_{mf}\rho_g}{\mu}\right) = \frac{D_p^3\rho_g(\rho_s-\rho_g)g}{\mu^2} \tag{23.26}$$

where D_p = average particle diameter (cm)
ρ_g = gas density (g/cm³)
ρ_s = solids particle density (g/cm³)
μ = gas viscosity (g/cm-s)
g = gravitational constant (cm/s²)

Although the fixed-bed pressure drop equation developed by Ergun was for uniformly sized solids, it, as well as the related minimum fluidization equation given above, is often applied to systems with a particle size distribution based on the approximation:

$$\frac{1}{D_p} = \sum_i \frac{X_i}{d_{p_i}} \tag{23.27}$$

where X_i = weight fraction
d_{p_i} = average particle size of each weight fraction (cm)

For cases where ϵ_{mf} and ϕ are unknown, Wen and Yu[79] recommend the following modification of Equation 23.26. They found that for a wide variety of systems:

$$\frac{1}{\phi\epsilon_{mf}^3} \simeq 14 \quad \text{and} \quad \frac{1-\epsilon_{mf}}{\phi^2\epsilon_{mf}^3} \simeq 11 \tag{23.28}$$

and that substitution of these expressions in Equation 23.26 gives:

$$V_{mf} = \frac{\mu}{D_p\rho_g}\left\{\left[(33.7)^2 + 0.0408\frac{D_p^3\rho_g(\rho_s-\rho_g)g}{\mu^2}\right]^{1/2} - 33.7\right\} \tag{23.29}$$

Although Wen and Yu found Equation 23.29 to be in reasonable agreement with a wide variety of experimental data, particularly at relatively low pressures, Knowlton[81] found a significant disagreement between predictions of this expression and results obtained on fluidization of coals and related materials at pressures up to 70 atm. In a study conducted at the Institute of Gas Technology to prepare a technical coal data book,[31] it was proposed to use a set of values for the parameters $1/\phi\epsilon_{mf}^3$ and $(1-\epsilon_{mf})/\phi^2\epsilon_{mf}^3$ different from those recommended by Wen and Yu. This proposal came after analysis of fluidization data obtained with coal-related materials. The IGT values:

$$\frac{1}{\phi\epsilon_{mf}^3} = 8.81 \quad \text{and} \quad \frac{1-\epsilon_{mf}}{\phi^2\epsilon_{mf}^3} = 5.19 \tag{23.30}$$

when substituted into Equation 23.26, lead to:

$$V_{mf} = \frac{\mu}{D_p\rho_g}\left\{\left[(25.25)^2 + 0.0651\frac{D_p^3\rho_g(\rho_s-\rho_g)g}{\mu^2}\right]^{1/2} - 25.25\right\} \tag{23.31}$$

Equation 23.31 was found to agree with minimum fluidization velocities obtained in the studies described in Table 23.7, with a standard deviation of 30%.

The ratio of minimum fluidization veloci-

[78] M. Leva, *Fluidization*, McGraw-Hill Book Company, New York, 1959, 327 pp.
[79] C. Y. Wen and Y. H. Yu, *Am. Inst. Chem. Eng.*, **12**, 610–612 (1966).
[80] J. F. Frantz, *Chem. Eng.*, **69**(19), 161–178 (1962).
[81] S. Ergun, in Am. Inst. Chem. Eng. Chem. Eng. Prog. Symp. Ser., Vol. 48, 1952, pp. 89–94.

Table 23.7 Sources of Data for Minimum Fluidization Velocity Correlation[a]

Bed Diameter (cm)	Fluidized Solids	Particle Diameter (cm)	Particle Density (g/cm^3)	Fluidizing Medium	Operating Pressure (atm)	Ref.
4.2	Coal	0.031–0.056	1.31–1.33	Air	1	72
8.5	Coke	0.009–0.066	1.80	H_2, air, CO_2, Ar	1	74
2.5, 5.1	Lignite-char, dolomite, periclase	0.007–0.044	0.82–3.56	N_2H_2, CO_2	1	76
9.4	Char	0.013–0.287	0.26–0.37	CO_2	1	77
6.4	Siderite	0.012–0.021	3.93	Air, CO_2, H_2, Freon	1–69	83
29.2	Coal, char, lignite, siderite	0.024–0.029	1.17–3.93	N_2	1–69	82

[a] All data obtained at 21–27°C.

ties predicted by Equation 23.31 to velocities predicted by Equation 23.29 ranges from about 2 at low Reynolds numbers to about 1.25 at high Reynolds numbers. These variations may be due in part to differences in the range of conditions and types of materials considered in development of the two correlations.

Bed Expansion. When superficial velocities of the gas entering a fluidized bed are increased beyond the point required for minimum fluidization, the bubbles that form and pass through the bed lead to bed expansion. Therefore it is of interest to determine the extent of this expansion as a function of operating conditions, to facilitate design of fluidized beds. Experimental determination of bed expansion, however, is not a straightforward procedure. This is because gas bubbles that erupt from the surface of a fluidized bed tend to throw solid particles up above the bed; some of these are carried out of the reactor (elutriation), and some fall back into the bed. As a result, there is not necessarily a sharp demarcation between the top of a fluidized bed and the lean gas phase above the bed, which can lead to differences in the interpretation of bed expansion measurements.

Numerous correlations for bed expansion in fluidized beds have been proposed, and some studies have found that bed expansion tends to decrease with increasing tube diameter.[39,84,85] One empirical correlation proposed in the IGT study for the coal data book,[32] and based on results obtained primarily with coal-related solids, is as follows:

for $D_T \leqslant 6.35$ cm:

$$\frac{L_f}{L_{mf}} = 1 + \frac{0.437\,(V - V_{mf})^{0.570}\,\rho_g^{0.083}}{\rho_s^{0.166}V_{mf}^{0.063}D_T^{0.445}} \qquad (23.32)$$

for $D_T > 6.35$ cm:

$$\frac{L_f}{L_{mf}} = 1 + \frac{1.95\,(V - V_{mf})^{0.738}D_p^{1.006}\rho_s^{0.376}}{V_{mf}^{0.937}\rho_g^{0.125}} \qquad (23.33)$$

where L_{mf} = bed height at minimum fluidization (cm)
L_f = bed height (cm)
ρ_s = average particle density of solids (g/cm^3)
ρ_g = average gas density (g/cm^3)
D_T = tube diameter (cm)
V = superficial feed gas velocity referred to empty tube plug-flow value (cm/s)

Note that for $D_T \leqslant 6.35$ cm, the expansion ratio L_f/L_{mf} decreases with increasing tube diameter, whereas for $D_T > 6.35$ cm, the expansion ratio is independent of the tube

[82] T. M. Knowlton, Paper No. 9b, presented at 67th Annual Meeting AIChE, Washington, D.C., December 1–5, 1974.

[83] P. Tarman, D. Punwani, M. Bush, and A. Talwalkar, U.S. Office Coal Res. R & D Rep. No. 95—Interim Rep. No. 1 (PB 237363); NTIS, 1975.

[84] P. J. Bakker and P.M. Heertjes, *Br. Chem. Eng.*, **3**, 240–245 (1958).

[85] P. J. Bakker and P.M. Heertjes, *Chem. Eng. Sci.*, **12**, 260–271 (1960).

diameter. Equations 23.32 and 23.33 were fitted to 90% of the data obtained in the studies described in Table 23.8 including high-pressure data, with a maximum deviation of 12%.

Solids Circulation. In a fluidized bed, gross solids are circulated upward primarily by being transported in the wake of rising bubbles, and downward by compensating flow in the emulsion phase. Talmor and Benenati[88] have studied the movement of solids in bubbling-air fluidized beds by a tracer technique, which was cation-exchange resin in hydrogen and sodium forms. Beginning with a layered bed of tracer and nontracer, they fluidized the bed with air for varying lengths of time and determined the amount of tracer solid that crossed the original interface. The solids circulation flux, J, was determined from the experimental results based on a differential equation given by Katz and Zenz,[89] and was correlated with the following expression for the range of conditions employed:

$$J = 0.785\,(V - V_{mf})\exp(-66.3D_p) \qquad (23.34)$$

where J = solids circulation flux referred to tube cross-sectional area (g/cm²-s)
D_p = average particle diameter (cm)

[86] Bituminous Coal Research Inc., U.S. ERDA Rep. No. FE-1527-1; NTIS, 1975.
[87] C. E. Fink, J. D. Sudbury, and G. P. Curran, Paper No. 15d, presented at 77th National Meeting, AIChE, Pittsburgh, June 2–5, 1974.
[88] E. Talmor and R. F. Benenati, *Am. Inst. Chem. Eng. J.*, **9**, 536–540 (1963).
[89] S. Katz and F. A. Zenz, *Hydrocarbon Process. Pet. Refiner*, **33**, 203 (1954).

Talmor and Benenati[88] showed that Equation 23.34 correlates the data of Koble[90] and of Leva and Grummer.[91] Woollard and Potter[92] subsequently demonstrated that Equation 23.34 also fits the data of Bart[93] and of Vakrushev and Erokin,[94] with the result that the correlation is applicable on the following range of parameters:

Bed diameter	3.2–10.2 cm
Particle diameter	0.0067–0.065 cm
Particle density	0.75–2.65 g/cm³
Size distribution	Mixed and uniform
Solids	Microspheres, round sand, sharp sand, silica gel, ion-exchange resins, cracking catalyst, coke

Thus as an empirical representation, Equation 23.34 appears to describe circulation data obtained over a relatively wide range of conditions, at least at atmospheric pressure. Kunii and Levenspiel[66] have suggested, however, that in view of the complicated interactions in fluidized beds, the relatively simplistic basis used for deducing solids fluxes from experimental tracer studies should limit the application of Equation 23.34 to first-order approximations.

[90] R. A. Koble, Thesis, West Virginia University, Morgantown, 1952.
[91] M. Leva and M. Grummer, Am. Inst. Chem. Eng. Chem. Eng. Prog., Symp. Ser., Vol. 48, 1952, pp. 307–313.
[92] I. N. M. Woolard and O. E. Potter, *Am. Inst. Chem. Eng. J.*, **14**, 388–389 (1968).
[93] R. Bart, Ph.D. thesis, Massachusetts Institute of Technology, Cambridge, 1950.
[94] I. A. Vakhrushev and G. S. Erokhin, *Int. Chem. Eng.*, **3**, 333–338 (1963).

Table 23.8 Sources of Data for Bed Expansion Correlation[a]

Bed Diameter (cm)	Fluidized Solids	Particle Diameter (cm)	Particle Density (g/cm)	Fluidizing Medium	Operating Pressure (atm)	Range of V/V_{mf}	Range of L_f/L_{mf}	Ref.
2.5, 5.1	Lignite, char, dolomite, periclase	0.007–0.044	0.82–3.56	N_2, H_2, CO_2	1	1–20.5	1–2	76
5.1	Char	0.013	0.37	CO_2	1	1–12.0	1–1.53	77
6.4, 14.0, 29.2	Siderite	0.007–0.036	3.93	Air, Freon, N_2	1–69	1–10.5	1–1.49	83
29.2	Lignite, coal, char, siderite	0.024–0.029	1.17–3.93	N_2	1–69	1–2.6	1–1.12	82
25.4	Coal	0.005–0.020	1.39	Air	1	1.3–40.2	1–1.63	86
30.5	Char, limestone	0.046–0.216	1.36–2.64	Air	1	1.2–7.3	1–2.15	87

[a] All data obtained at 21–27°C.

In addition to the net upward and downward solids fluxes in fluidized beds, there is a cross-flow between wake and emulsion solids. Kunii and Levenspiel, in developing their fluidized-bed model, recommended the correlation of Yoshida and Kunii[95] to describe this cross-flow; that is:

$$K^* = \frac{\text{volume of solids transferred between the wake to the emulsion}}{\text{(volume of bubble) (time)}} \approx \frac{3(1-\epsilon_{mf})V_{mf}}{(1-\delta)\epsilon_{mf}\,D_b} \qquad (23.35)$$

where δ = volume fraction of bubbles in bed
D_b = characteristic bubble diameter

23.3.2 Pore Structure Characteristics of Coals and Coal Chars

Variations in the pore structure of coal and coal char particles that occur during gasification are important in considering hydrodynamic aspects of integral gas-solid contacting systems as well as in evaluating gaseous mass transfer behavior and available reactive surface areas within individual particles. Coals and coal chars exhibit complex and unique pore structures, with a wide distribution of pore sizes. Systematic analysis of coals and coal-derived solids is complicated by the heterogeneity of both the organic and mineral constituents of such substances. Thus the total pore size distribution for a "whole" coal represents the sum of the distribution characteristics of relatively homogeneous organic constituents (macerals), the distribution characteristics of mineral constituents, and the pore volume contributions of interface regions between the various phases present.

Anderson et al.,[96] in a pioneering study of the pore structures of a number of American coals, found total porosities ranging from 2.5 to 18% and observed that a significant fraction of pores had openings (diameters) of 4.9–5.6 Å. In a later study Gan et al.[97,98] characterized the pore structure of a wide range of American coals in terms of porosities divided into three pore size ranges: (1) micropores, which are accessible through effective pore diameters of less than 12 Å; (2) transitional pores, with diameters from 12 to 300 Å; and (3) macropores, with diameters greater than 300 Å. It was also found in this study that the micropores can represent a significant fraction of pore volume.

Walker and Hippo[99] have discussed some of the fundamental aspects relevant to characterizing the pore structure of chars and their precursor coals, based largely on the detailed X-ray study of Cartz and Hirsch.[100] They pointed out that coals are composed of aromatic and hydroaromatic layers, terminated at their edges by various functional groups and cross-linked by different functional groups. The average size of the layers and the number of layers that are aligned tend to increase with increasing coal rank. Poorer alignment between packets of layers produces internal porosity, which makes coal a microporous material. Upon heating, coals lose volatile matter, primarily from the periphery of the layers. Some coals pass through a softening or plastic stage before solidifying into a coke or char. The extent of softening, which does not occur to a significant extent with low- or high-rank coals, is thought to be largely dependent on the concentration and thermal stability of cross-linking groups. The micropore structure of the precursor coal is preserved in chars that are not heated to too high a temperature; in fact, microporosity can become even more accessible in chars because of the loss of volatile matter. At very high temperatures, however, microporosity is rapidly lost because of thermal breakage of the cross-links between planar regions in the char, allowing improved alignment of these regions.

With respect to their structure, Walker and Hippo envision chars as being composed of

[95] K. Yoshida and D. Kunii, *J. Chem. Eng. Jap.*, **1**, 11–16 (1968).
[96] R. B. Anderson, W. K. Hall, J. A. Lecky, and K. C. Stein, *J. Phys. Chem.*, **60**, 1548–1558 (1956).
[97] H. Gan, S. P. Nandi, and P. L. Walker, Jr., U.S. Office Coal Res. R & D Rep. No. 61—Interim Rep. No. 4 (PB 233994); NTIS, 1972.
[98] H. Gan, S. P. Nandi, and P. L. Walker, Jr., *Fuel*, **51**, 272–277 (1972).
[99] P. L. Walker, Jr., and E. J. Hippo, *Prepr., Am. Chem. Soc., Div. Fuel Chem.*, **20**(3), 45–51 (1975).
[100] L. Cartz and P. G. Hirsch, *Trans. R. Soc. (London)*, **A252**, 557 (1960).

small trigonally bonded planar regions, terminated by far fewer heteroatomic functional groups than are present in the original coal, and still containing substantial cross-linking, although less than in the original coal. They also suggest that the exact structure of a char produced from a given coal will depend on a number of variables such as coal pretreatment, coal particle size, heating rate, maximum temperature, time at maximum temperature, gas composition, and total pressure during heating. Based on differences in the distribution of pore volume with respect to pore size (diameter), chars from coals of different rank would be expected to exhibit differing degrees of inhibition to internal mass transfer effects during gasification. This is related to the fact that the internal surface area of coals and chars exists almost solely in the micropores, where gaseous diffusion is slow. Diffusion is, in fact, an activated process for pore diameters less than about 5 Å, as shown by Walker et al.[101] Walker and Hippo proposed that to efficiently use the active sites located in the micropores for reaction, a large number of feeder pores (macro and transitional), to which the micropores are connected, are required to limit diffusion length within micropores. Diffusion through feeder pores to the mouth of micropores would be reasonably rapid; therefore gaseous reactant concentrations at the mouths of micropores could closely approach reactant concentrations at the exterior particle surface. Furthermore since the results of Gan et al.[97,98] have shown that lower rank coals tend to have a greater percentage of the pore volume in larger pores than do higher rank coals, it can be expected that chars derived from lower rank coals would have a larger feeder pore system and would be less limited with respect to mass transport during their gasification.

Because of the complex pore structure of coals and coal chars, many experimental techniques have been employed to characterize their physical properties in terms of solids densities, total pore volumes, pore volume distributions, and internal surface areas. Useful information has been obtained by density measurements and sorption studies using different fluids at various temperatures.

23.3.2.1 *Solids Densities of Coals and Coal Chars*

Helium displacement is used most often to determine solids densities or true densities, since the helium is not easily adsorbed on the internal surfaces of coal at room temperature, and it is believed to penetrate pore opening diameters greater than 3–4 Å.[102,103] With some highly adsorptive carbons, some adsorption of helium has been found at low temperatures.[104,105] For this type of material it is recommended that helium density measurements be made at elevated temperatures of about 300°C.[106]

Coal and coal char true densities measured by helium displacement, and expressed on an maf basis, have shown a strong correlation with the hydrogen content of the material.[31,97,107] In the IGT study for the coal data book,[31] the following correlation was developed to describe the relationship between the densities reported in the literature[96,97,102,107,108–115] for a variety of coals,

[101] P. L. Walker, Jr., L. G. Austin, and S. P. Nandi, *Chemistry and Physics of Carbon,* Vol. 2, Marcel Dekker, New York, 1960, pp. 257–371.

[102] R. E. Franklin, *Fuel,* **27,** 46–49 (1948).

[103] J. G. Hirshfelder, C. F. Curtis, and R. B. Bird, *Molecular Theory of Gases and Liquids,* John Wiley and Sons, New York, 1954, p. 1127.

[104] F. A. P. Maggs, P. H. Schwabe, and J. H. Williams, *Nature* (*London*), **186,** 956–958 (1960).

[105] K. A. Kini and W. O. Stacy, *Carbon* (*Oxford*), **1,** 17–24 (1963).

[106] R. L. Bond, *Porous Carbon Solids,* Academic Press, New York, 1967, p. 99.

[107] S. Ergun and M. Mentser, *Fuel,* **39,** 509–510 (1960).

[108] P. L. Agarwal, *Proc. Symp. Nature Coal,* Central Fuel Research Institute, Jealgora, India, 1959, pp. 121–128.

[109] R. L. Bond and D. H. T. Spencer, *Proc. Soc. Chem. Ind., London Chem. Eng. Group,* 231–251 (1958).

[110] H. N. M. Dormans, F. J. Huntjens, and D. W. Van Krevelen, *Fuel,* **36,** 321–339 (1957).

[111] R. E. Franklin, *Trans. Faraday Soc.,* **45,** 668–682 (1949).

[112] S. Fujii and H. Tsuboi, *Fuel,* **46,** 361–366 (1967).

[113] S. Fujii and T. Shinbata, *Nenryo Kyokai-shi* (*Fuel Soc. Jap. J.*), **45,** 868–875 (1966).

[114] Y. Osawa, H. Sugimura, Y. Kawaski, and S. Fujii, *Nenryo Kyokai-shi* (*Fuel Soc. Jap. J.*), **49,** 31–39 (1970).

[115] H. Sugimura, Y. Osawa, M. Hatami, H. Sato, and H. Honda, *Nenryo Kyokai-shi* (*Fuel Soc. Jap. J.*), **45,** 618–630 (1966).

carbonized coals, high-temperature cokes, and maceral concentrates, including exinite, fusinite, and vitrinite:

$$V_{He} = \frac{1}{\rho_{He}} = 0.4397 + 0.1223\ Y_H - 0.01715\ Y_H^2 + 0.001077\ Y_H^3 \qquad (23.36)$$

where Y_H = hydrogen content of coal or char (mass %, maf)
V_{He} = specific volume in helium (cm^3/g, maf)
ρ_{He} = "true" density in helium (g/cm^3, maf)

This correlation, applied to data with Y_H ranging from about 0 to 7.7, represents specific volumes or densities of the organic solids on an maf basis. In view of the arbitrary form of the equation, extrapolation should be avoided. To obtain the specific volume of the solids, including ash, the following relationship can be applied:

$$V'_{He} = V_{He}(1 - X_A) + \frac{X_A}{\rho_A} \qquad (23.37)$$

where X_A = ash mass fraction
ρ_A = true density of ash (g/cm^3)
V'_{He} = specific volume of coal or coal char solids (cm_3 /g)

An average value for ash density of about 2.7 g/cm^3 is often used to represent coal density data on an ash-free basis.[97,116] Of course, experimental values of ash or mineral densities are preferred, particularly if there is a large concentration of high-density substances such as pyrites or marcasite.[31] Equation 23.36 was developed on the basis of a set of 190 data points, with a standard deviation of 0.012 cm^3/g over the whole range. Figure 23.11 compares the correlation with some of the data considered in its development.

23.3.2.2 *Pore Volumes of Coals and Coal Chars*

Density measurements obtained by mercury displacement have been used in conjunction

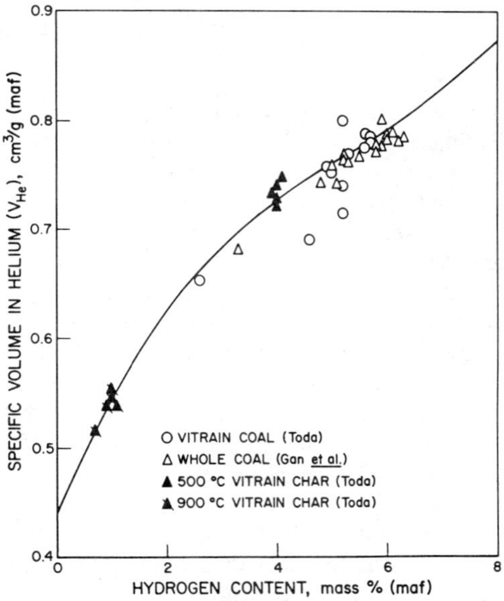

Figure 23.11 Correlation of true specific volume of coals and coal-derived solids as a function of hydrogen content.[97,117]

with helium density measurements to obtain information on the total pore volumes and pore volume distributions of coals and chars. When a porous solid material is immersed in mercury at a given pressure P, the mercury will penetrate all pores with an entry diameter greater than some minimum value, D_m, depending on the pressure. For circular pores the relationship between minimum pore diameter and pressure is usually computed from the Washburn equation[118]:

$$D_m = -\frac{\sigma \cos \theta}{P} \qquad (23.38)$$

where D_m = minimum pore diameter (for pores having circular entries)
P = pressure
σ = surface tension of mercury
θ = contact angle between mercury and the material forming the opening

Using reasonable values of σ and θ at room temperature, Equation 23.38 reduces to:

[116] A. M. Wandless and J. C. Macrae, *Fuel*, **13**, 4 (1934).

[117] Y. Toda, *Study on Pore Structure of Coals and Their Carbonized Products*, Rep. No. 5, National Research Institute for Pollution and Resources, Kawaguchi, Japan, 1973, 112 pp. (in Japanese).

[118] E. W. Washburn, *Proc. Natl. Acad. Sci. (U.S.)*, **7**, 115 (1921).

$$D_m = \frac{120.800}{P} \qquad (23.39)$$

for D_m in angstroms and P in atmospheres. Using Equation 23.39, measurements made over a pressure range of 1 atm (often used to estimate total porosities) to as high as 1000 atm (used in some studies of macropore distributions) will result in a range of values of D_m from 120,800 to 121 Å. The specific volume measurements, V_{Hg}, made by mercury displacement, are equal to the sum of the specific volume of solids plus the volume of pores accessible through openings with a diameter less than D_m. Thus assuming that helium displacement measurements give the true specific volume of the solids, the void volume accessible through pore openings having diameters less than D_m can be determined from the expression:

$$V_{D_m} = V_{\text{Hg}} - V_{\text{He}} \qquad (23.40)$$

The total pore volume between two values for the pore-opening diameter can be determined without helium density measurements by measuring the difference in mercury penetration into the solid between the two pressure levels corresponding to the two pore-opening diameters. Some caution is necessary, however, in interpreting mercury displacement results in terms of pore volumes, at low and high hydrostatic pressures. At low pressures some mercury penetration can be due to the filling of voids between particles. The pressure at which interparticle filling can occur increases with decreasing particle size.[97] It has been recommended that the limiting pressures for interparticle filling be determined experimentally with nonporous materials of the same particle size as the porous solids.[97] Interparticle filling may also be accounted for by calibration. In addition, at very high pressures the compressibility of the coal or coal char material must be considered, or else apparent pore volumes will be too high.[37]

Total pore volumes of a variety of American coals and of Japanese and American vitrains are shown in Fig. 23.12. The results are represented on an ash-free basis and assume that the entire pore volume is associated with the organic material, not with the ash. The American coals used by Gan et al.[97] (315-μm average particle diameter) varied in

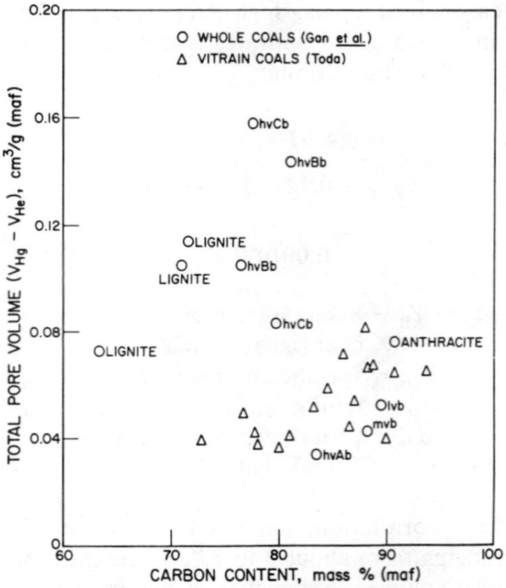

Figure 23.12 Total pore volumes of coals measured by mercury and helium displacement.[97,117]

vitrinite content from about 62 to 96 vol %, and in ash content from about 2 to 25 mass %. The mercury densities used by Gan et al. were obtained at about 4 atm total pressure; thus according to Equations 23.37 and 23.39, the total pore volumes computed are for pore regions accessible through openings (D_m) less than about 3 μm. The vitrains used by Toda[117] (420-μm average particle diameter) varied in ash content only from about 1 to 5 mass %, and mercury densities were apparently obtained at 1 atm. The minimum pore diameter penetrated by mercury at this pressure is about 12 μm.

The results shown in Fig. 23.12 indicate that the total pore volumes of the vitrains and whole coals are similar only if the total carbon content is greater than 83%. At lower ranks the total pore volume for the whole coal is substantially greater than those for the vitrains over corresponding ranges of carbon content. This appears to be the case for coals with a rank of high-volatile B bituminous and lower. Thus it is not possible to accurately predict trends for the lower rank coals solely as a function of carbon content, and a significant fraction of the pore volume of the whole coals probably results from their more heterogeneous composition, compared to vitrains of low ash content.

Pore volumes in the regions accessible through openings having diameters (D_m)

smaller than those corresponding to the results illustrated in Fig. 23.12 were also determined by Gan et al.[97] and by Toda.[117] These results, shown in Fig. 23.13, were based on mercury porosimetry measurements at maximum pressures of 10,000–15,000 psi. In the study by Gan et al., estimates of pore volume at lower values of D_m (12–300 Å) were based on the interpretation of nitrogen adsorption isotherms at 77 K. The results shown in Fig. 23.13 suggest that a substantial portion of the high total pore volumes in whole bituminous coals exists in pores having entry diameters between 12 and 300 Å.

Variations in total pore volumes during carbonization have been determined by Toda[117] for American and Japanese vitrains. Pore volumes obtained on carbonization at 900°C are compared to original pore volumes in Table 23.9. These data indicate that char pore volumes, referred to the original coal mass, show an increase compared to the original coal if the initial carbon content is 77.8 or 81.1%, but are similar or smaller than pore volumes of the original coals if the initial carbon content is 86.2, 87.2, or 89.8%.

These and other results have been interpreted by Toda to indicate that because of net particle shrinking, the pore volumes in macro and transitional pores of caking coals show a net decrease after high-temperature carbonization. This was not observed for lower rank noncaking coals. However, Toda found that at intermediate carbonization stages of caking coals (400–600°C), pore volumes changed drastically, passing through a maximum and then decreasing at higher carbonization temperatures. This was explained by the appearance and disappearance of a

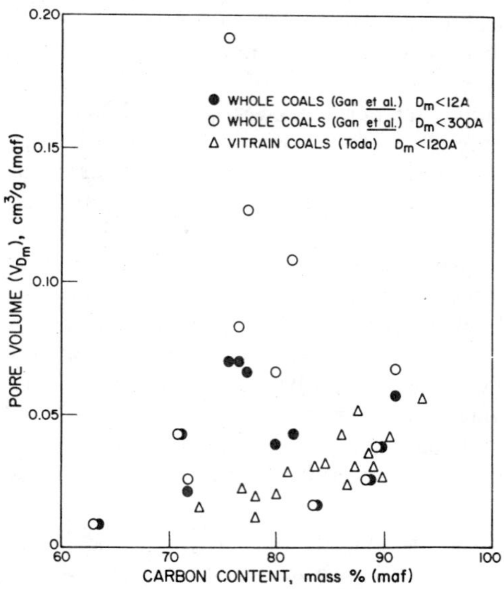

Figure 23.13 Pore volumes of coals in regions accessible through openings having different entry diameters.[97,117]

"bubble" structure of the caking coal in the plastic stage of carbonization.

In Toda's study coals were heated to elevated temperatures at 3°C/min, a relatively low rate, then maintained at temperature for 15 min. This is of interest because the total pore volume variations that occur with bituminous coals during heat treatment generally must be considered to be partly a function of the time-temperature history, as suggested by Walker and Hippo.[99] With bituminous coals, the evolution of volatile matter during plasticization leads to an initial expansion of individual particles and to the intermediate formation of a bubble structure, reflected in

Table 23.9 Comparison of Coal and Carbonized Char Pore Volumes

		Pore Volume (maf)		
		Coal	Char	
Initial Carbon Content of Vitrain (%, maf)	Char Weight Loss Fraction (g/g coal, maf)	cm³/g Coal	cm³/g Char	cm³/g Original Coal
77.8	0.41	0.042	0.127	0.075
81.1	0.41	0.041	0.222	0.131
86.2	0.33	0.071	0.063	0.042
87.2	0.24	0.054	0.045	0.034
89.8	0.32	0.039	0.056	0.038

Source: Derived from Ref. 117.

an increase in pore volume of the macro and, to some extent, transitional pores. In Toda's study the exposure of carbonized chars to higher temperatures for extended times apparently led to subsequent particle shrinkage. In other studies, where bituminous coals have been heated to relatively high temperatures at rapid rates (in dilute solid phase transport or free-fall systems), then quenched after a few seconds or less, the expanded bubble structure of the resultant chars can be retained. Feldmann et al.,[119] studying the hydrogasification of bituminous coals at elevated pressure in a free-fall system, found that the gross volume of char particles had increased severalfold relative to the original coal particle volumes, for particle residence times of a few seconds or less. Thus it is currently recognized that bituminous coal char particles can have widely different pore structures, depending on the initial heat treatment conditions. Detailed analyses of the complex physical and chemical processes that affect the formation of these structures have only recently been initiated.[120]

When precarbonized coal chars are subsequently gasified in reactive atmospheres, total pore volumes generally increase with increasing carbon conversion levels. Some smoothed results obtained for the gasification of chars derived from various coals are shown in Fig. 23.14, and pertinent details of the studies represented are given in Table

[119] H. F. Feldmann, J. A. Mima, and P. M. Yavorsky, Am. Chem. Soc. Adv. Chem. Ser. No. 131, 1974, pp. 108–125.

[120] P. C. Lewellen, Ph.D. thesis, Massachusetts Institute of Technology, Cambridge, 1976.

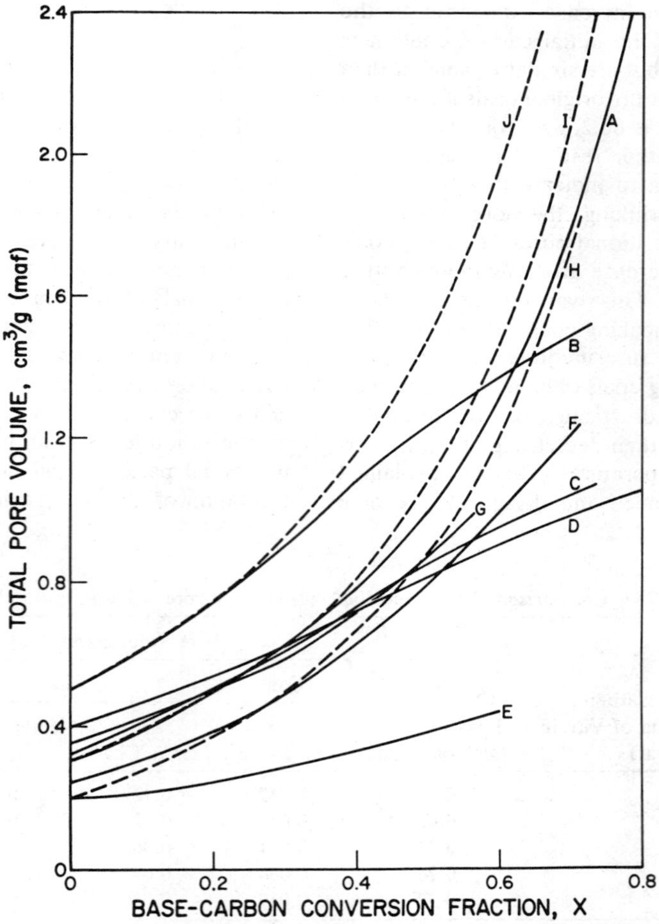

Figure 23.14 Variations in total pore volumes of coal chars as a function of carbon conversion level during gasification in reactive atmospheres.

23.10. Pore volume variations in Fig. 23.14 are given as a function of the base carbon conversion fraction, X, which represents the fraction of carbon in the initial precarbonized char that is gasified during subsequent treatment in the gasifying medium. For most practical purposes this conversion term is directly equivalent to other conversion terms applied to coal char gasification systems, such as carbon "burnoff" and char weight loss fraction.

Although most of the curves shown in Fig. 23.14 are approximately similar (taking into account differences in the pore volumes of the precarbonized chars), the increases in pore volumes at higher conversions, with one exception (curve A), are significantly less than the volumes originally occupied by the components of the solids that were gasified. This can be illustrated by calculating the pore volume variation that would occur if the external particle volume dimensions remained constant during gasification and pore volumes increased in direct proportion to the solids gasified. The following expression represents this behavior as a function of initial pore volume, true solids density, and carbon conversion fraction:

$$V_p = \frac{V_p^0 + X/\rho_{\mathrm{He}}}{1 - X} \qquad (23.41)$$

where V_p^0 = total pore volume of carbonized char (cm^3/g carbonized char)
X = base carbon conversion fraction of char (g carbon gasified/g carbon in carbonized char)
ρ_{He} = true density of char (g/cm^3)
V_p = total pore volume of gasified char (cm^3/g gasified char)

Curves H, I, and J in Fig. 23.14 were calculated from Equation 23.41 for initial pore volumes of 0.2, 0.3, and 0.5 cm^3/g, assuming a reasonable true solids density of 2.1 g/cm^3. Although most of the experimental curves are reasonably consistent with the calculated curves, up to carbon conversions of 0.4–0.5, experimentally obtained pore volume increases at higher conversions are usually less than the calculated increases, which suggests possible particle shrinkage. The curves depicting Johnson's[121] results (A–E) represent total pore volumes with entry diameters less than 120 μm. Johnson verified particle shrinkage with anthracite, high-volatile bituminous char, and lignite char by direct measurement of the individual particle dimensions before and after gasification in hydrogen at 35 atm and 927°C. In a series of tests a few particles from each of the different chars were photographed in several orientations under optically calibrated conditions. The particles were then gasified in a thermobalance apparatus to yield carbon conversion fractions from 0.7 to 0.8 and were photographed again. Detailed examination of all the photographs showed significant reduction in particle volumes consistent with estimates made from mercury displacement measurements. The photographic evidence indicated that the fraction of volume reduction was independent of initial particle diameter in the range from 200 to 800 μm, and that external topological characteristics remained un-

[121] J. L. Johnson, *Prepr., Am. Chem. Soc., Div. Fuel Chem.*, **20**(4), 85–101 (1975).
[122] N. Torikai and P. L. Walker, Jr., U.S. Office of Coal Res. SROCR-3 Report, 1968.
[123] E. J. Hippo, Ph.D thesis, Pennsylvania State University, University Park, 1977.

Table 23.10 Conditions for Studies Illustrated in Fig. 23.14

			Carbonization Conditions		Gasification Conditions		
Curve (Fig. 23.14)	Ref.	Coal Type	Temperature (°C)	Gas	Temperature (°C)	Pressure (atm)	Gas
A	121	Lignite	927	N_2	927	35	H_2-H_2O
B	121	hvAb (air-pretreated)	927	N_2	927	35	H_2-H_2O
C	121	Lignite	927	N_2	927	35	H_2
D	121	Subbituminous A	927	N_2	927	35	H_2
E	121	Anthracite	927	N_2	927	35	H_2-H_2O
F	122	hvAb	950	N_2	900	1	CO_2
G	123	Lignite	700	N_2	650	1	H_2O
H	Calculated	$V_p^0 = 0.2$	—	—	—	—	—
I	Calculated	$V_p^0 = 0.3$	—	—	—	—	—
J	Calculated	$V_p^0 = 0.5$	—	—	—	—	—

changed except for a decrease in size. This indicated that the observed shrinkage occurred throughout the individual particles and was not the result of preferential gasification of external char surface layers, nor was it caused by particle attrition or decomposition.

In Johnson's study the rates of gasification of lignite in a steam-hydrogen mixture (pore volume characteristics shown by curve *A*, Fig. 23.14) were 7 times greater than the largest gasification rate obtained with the chars shown in curves *B–E*. Since curve *A* was the only curve that showed approximate agreement with calculated pore volume variations for the case of no particle shrinkage, this indicated that dynamic modifications—which could occur within the coal structure during gasification, altering gross dimensions—are inhibited when gasification rates are sufficiently large.

For particles gasified in a fluidized bed rather than a fixed-bed system, such as that used by Johnson, particle attrition could contribute to some extent to decreasing internal particle void volumes. Results shown in Fig. 23.14 for particles gasified in a fluidized bed (curves *F* and *G*), however, do not exhibit characteristics that suggest significant attrition in addition to particle shrinkage. Dutta et al.[124] have reported that coal char particles gasified in carbon dioxide tend to disintegrate at high levels of carbon conversion fraction (> 0.8). Such disintegration of coal char particles would be favored by high gasification rates, which inhibit particle shrinkage and instead result in highly porous and friable structures at high conversion levels.

Pore volume distributions determined by mercury porosimetry were reported by Stacy

[124] S. Dutta, C. Y. Wen, and R. J. Belt, *Prepr., Am. Chem. Soc., Div. Fuel Chem.*, **20**(3), 103–114 (1975).

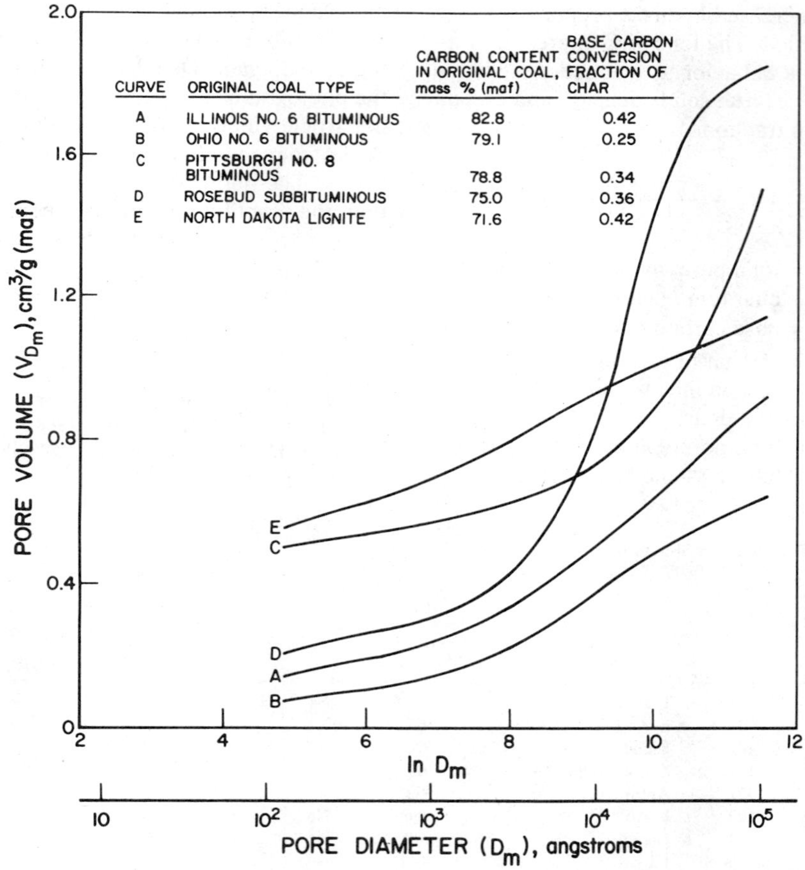

Figure 23.15 Pore volume distributions of coal chars produced from integral fluid bed gasification experiments in process development studies.

and Walker[125] for a variety of coal char residues resulting from gasification in integral systems, in different process development studies. Some of their smoothed results are shown in Fig. 23.15 for char residues with base carbon conversions from 0.25 to 0.42. The results obtained by these investigators, for a total pore diameter range of 130–100,000 Å, indicate that there are relatively small pore volumes for pore diameter openings between 130 and about 2000 Å, compared to pore volumes for the interval between 2000 and 100,000 Å. Significant differences in pore volume are apparent for the various chars with pore entry diameters of less than 130 Å, and these differences cannot be accounted for simply by differences in the rank of the coal from which the char was derived.

Pore volume distributions obtained by Johnson[121] for different coal chars gasified in hydrogen and in steam-hydrogen mixtures in a thermobalance apparatus are shown in Fig. 23.16. Average particle sizes of the initial coals used in this study were 600 μm. Pore volume evaluations were made by mercury porosimetry for values of D_m greater than 120 Å and by interpretation of nitrogen adsorption isotherms for values of D_m between about 13 and 120 Å.

The results shown in Fig. 23.16 indicate that for most of the chars, the regions with pore entry diameters between about 55 and 400 Å contribute little to total pore volume, with major variations in pore volume occurring outside this range. However the pore volume distribution of lignite that has been rapidly gasified in a steam-hydrogen mixture (curve *E*) shows that pore volume varies continuously over the total range of pore diameters examined. Total pore volumes obtained with this lignite at different base carbon conversion levels were described in Fig. 23.14 (curve *A*). Comparison of curves *E* and *D* (Fig. 23.16), which illustrate results obtained with the same lignite gasified in steam-hydrogen mixtures and hydrogen alone, indicates that the much more rapid gasification rates observed with steam-hydrogen mixtures contribute to increased pore volumes

[125] W. O. Stacy and P. L. Walker, Jr., U.S. Office Coal Res. R & D Rep. No. 61 Interim Rep. No. 3 (PB 233996); NTIS, 1973.

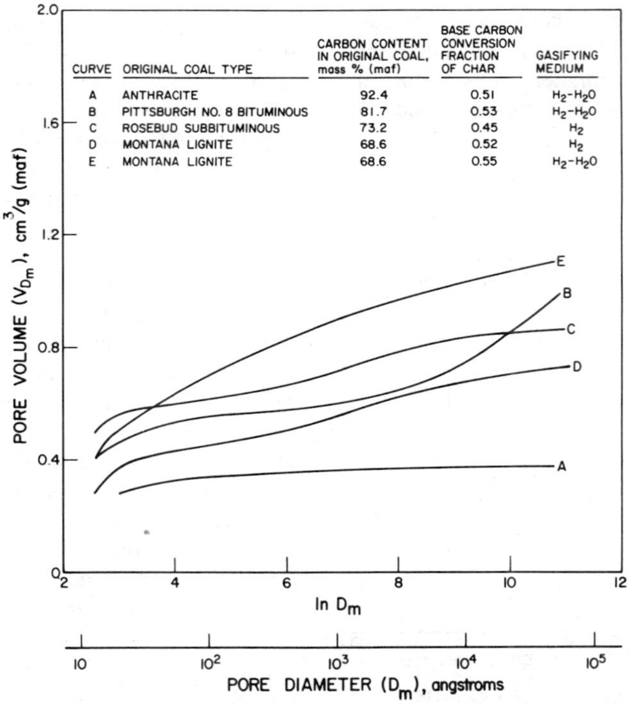

Figure 23.16 Pore volume distributions of coal chars produced from gasification in a thermobalance apparatus at 927°C, 35 atm.

mainly in the pores having entry diameters greater than 13 Å, that is, in transitional pores and macropores rather than in micropores.

Char pore volumes for values of D_m less than 130 Å and for the range 130–50,000 Å are represented in Figs. 23.17 and 23.18 as a function of the rank of the coal from which the char was derived, based on results obtained by Stacy and Walker[125] (Fig. 23.15) and by Johnson[121] (Fig. 23.16). In both Figs. 23.17 and 23.18 there are relatively slight systematic trends in pore volume variations with respect to the carbon content of the raw coals, apparently due to unexplained variances in physical and chemical structures of the raw coals and to differences in the conditions of char formation.

Johnson has found a more systematic relationship between pore volume and carbon conversion values for D_m less than 55 Å, as shown in Fig. 23.19. On the basis of results such as those shown in that figure, Johnson has suggested that when gasification rates are not too great, coal char shrinkage occurs during gasification primarily because of contraction of the microporous volume through crystallite reorientation, as reflected by the lack of a significant increase in micropore volumes at base carbon conversion fractions greater than about 0.5. He concluded that macropores at these higher levels of conversion also shrink, but in a manner analogous to the shrinkage of cavities in a metallic solid undergoing thermal contraction.

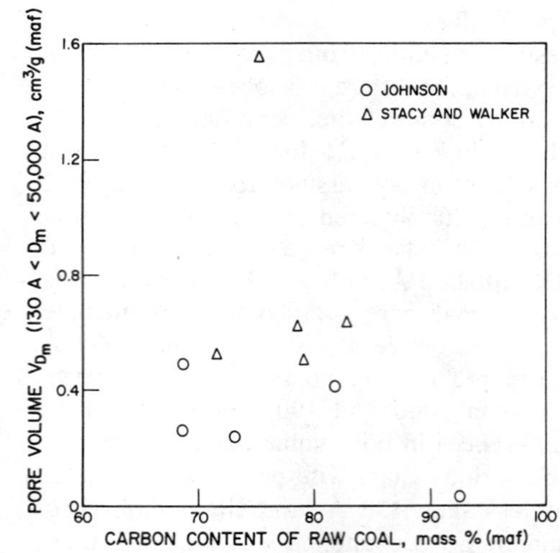

Figure 23.18 Pore volumes of partially gasified coal chars in regions accessible through openings having diameters between 130 and 50,000 Å.[12,125]

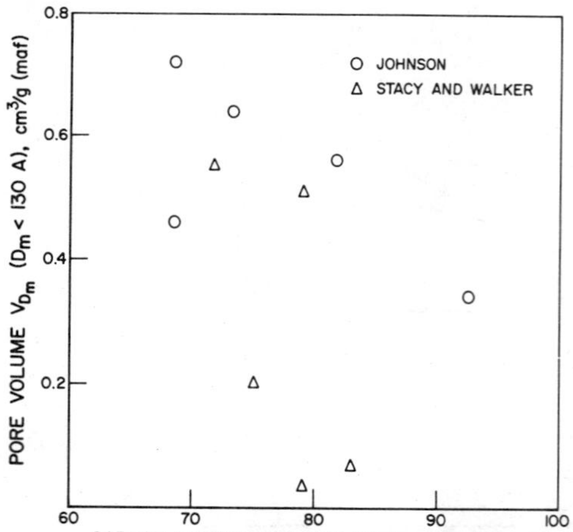

Figure 23.17 Pore volumes of partially gasified coal chars in regions accessible through openings having diameters less than 130 Å.[121,125]

23.3.2.3 *Internal Surface Areas of Coals and Coal Chars*

The internal surface area of coals and of carbonized and gasified chars has been measured in a variety of studies to obtain information on variations in structural characteristics during gasification. Although different experimental techniques have been used to estimate the internal surface area of coal-derived solids, the most widely used technique has been analysis of gas adsorption isotherms obtained in nitrogen or carbon dioxide. Nitrogen isotherms have usually been measured at 77 K and interpreted in terms of the Brunauer, Emmett, and Teller (BET)[126] equation to obtain surface areas. The relative adsorption pressures usually considered in application of the BET equation range from 0.05 to 0.3. Adsorption isotherms obtained in carbon dioxide are usually measured at 298 K at relative adsorption pressures less than 0.05, and they are interpreted in terms of the Dubinin-Polanyi equation[127] to compute surface areas.

[126] S. Brunauer, P. H. Emmett, and E. Teller, *J. Am. Chem. Soc.*, **60,** 309 (1938).

[127] T. G. Lamond and H. Marsh, *Carbon (Oxford)*, **1,** 281 (1964).

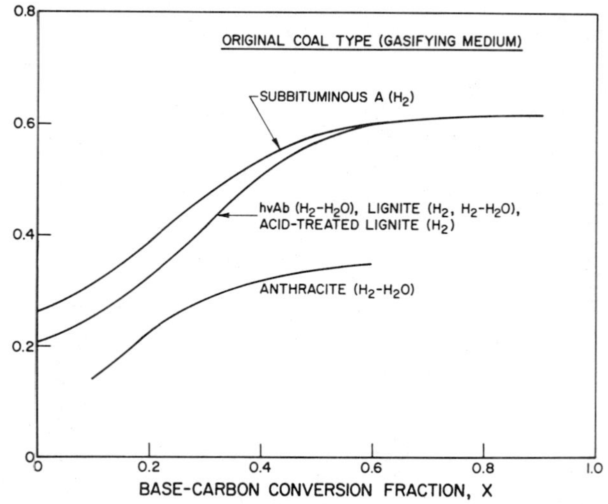

Figure 23.19 Effect of carbon conversion level on pore volumes of partially gasified chars in regions accessible through openings having diameters less than 55 Å.[121]

Computations of the surface areas of coal-derived solids based on nitrogen and carbon dioxide adsorption isotherms are not often in agreement; for reasons described below, greater significance is usually attributed to values obtained in carbon dioxide.[97,121,128–131] Penetration of nitrogen into the micropore structure of coals or carbonized coal chars is apparently severely limited by slow, activated-diffusion processes at 77 K, so that the low values of apparent surface areas at most reflect areas of macropores and transitional pores. For partially gasified coal chars, which have more open micropore structures, capillary condensation of nitrogen can lead to unreasonably high values of apparent surface area.[132]

Adsorption isotherms obtained with carbon dioxide at the higher temperature of 298 K facilitate activated diffusion into micropore structures, and capillary condensation is inhibited by the low relative adsorption pressures. There is, however, some question about whether carbon dioxide adsorption isotherms should be interpreted in terms of micropore volume rather than micropore surface area. Although it has been argued that carbon dioxide adsorption on a carbon surface should be limited to a monolayer thickness because of the strong adsorption potential of carbon dioxide molecules with carbon surfaces as well as high intermolecular attraction in the adsorption plane,[97,128,130] it has also been suggested that in micropores of molecular dimensions, adsorption forces from opposing surfaces overlap and the concept of monolayer filling of the pores is invalid.[133] Because of this controversy, values for both the surface area and the volume have been reported in the literature, as interpreted from carbon dioxide adsorption isotherms in terms of the Dubinin-Polanyi equation. The surface area and volume values are, however, convertible to each other from assumed values of the density of the adsorbed carbon dioxide phase and of the equivalent surface coverage area per carbon dioxide molecule. For convenience, the following discussion considers only surface area values. These are based either on directly reported values or on calculated values obtained by converting reported micropore volumes to equivalent surface areas, assuming

[128] D. H. Everett, *Proc. Chem. Soc., London,* 38–53 (1957).

[129] H. Marsh and T. Siemieniewsska, *Fuel,* **44,** 355 (1965).

[130] H. Marsh and W. F. K. Wynne-Jones, *Carbon (Oxford),* **1,** 281 (1964).

[131] P. L. Walker, Jr., and R. L. Patel, *Fuel,* **49,** 91 (1970).

[132] D. H. T. Spencer and R. L. Bond, in Am. Chem. Soc. Adv. Chem. Ser. No. 55, 1966, pp. 724–728.

[133] M. M. Dubinin, in *Chemistry and Physics of Carbon,* Vol. 2, P. L. Walker, Jr., Ed., Marcel Dekker, New York, 1966, p. 51.

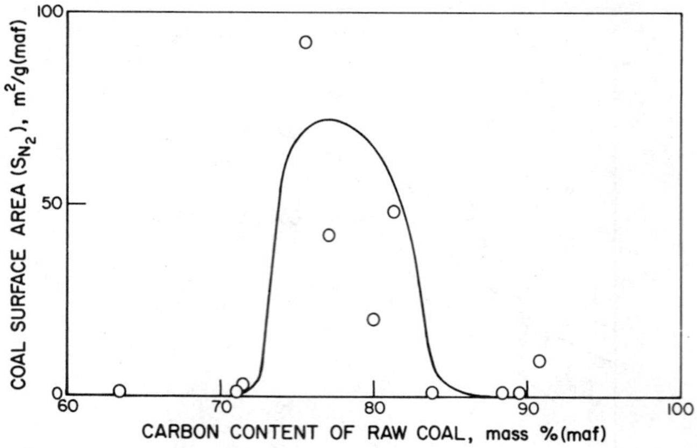

Figure 23.20 Variation with carbon content of N, surface areas of coals.[97]

an adsorbed carbon dioxide density of 1.03 g/cm³ and an equivalent surface coverage of 25.3 Å²/molecule.

Nitrogen surface areas, S_{N_2}, determined by Gan et al.[76] for a variety of American whole coals (Fig. 23.20), are relatively negligible for low- or high-rank coals, whereas values up to 93 m²/g (maf) were obtained for coals in the bituminous range. Carbon dioxide surface areas, S_{CO_2}, for these same coals are shown in Fig. 23.21, along with values of S_{CO_2} computed from Toda's[117] reported micropore volumes (determined by carbon dioxide adsorption) for a variety of American and Japanese vitrains. Values of S_{CO_2} are generally much higher than values of S_{N_2}, and tend to follow a reverse trend with respect to the relationship between surface area and rank, compared with values of S_{N_2}. It is of interest that values of S_{CO_2} obtained by Gan et al.[97] and by Toda[117] are relatively consistent over the total range of overlapping rank (~ 72–90% C in Fig. 23.21), whereas total pore volumes of the same coals and vitrains (Fig. 23.12) are substantially different for carbon contents less than about 83%. Since almost all the

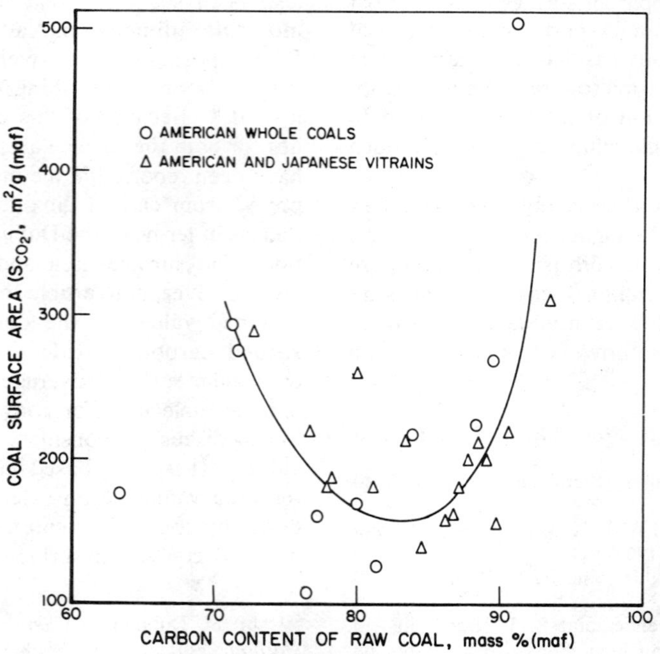

Figure 23.21 Variation with carbon content of CO_2 surface areas of coals.[97]

internal surface area of coals exists in micropores, these results suggest a somewhat systematic relationship between micropore internal surface area and coal rank, in contrast with the significant randomness in macropore and transitional pore structures that can occur in coals of comparable rank, arising from effects of organic and mineral heterogeneities.

Values of S_{CO_2} obtained with carbonized coal chars prepared under comparable conditions are shown in Fig. 23.22 as a function of the carbon content of the raw coals. This plot shows a relatively systematic decrease in S_{CO_2} with increasing coal rank, compared to the behavior obtained for raw coals (reproduced from Fig. 23.21).

The values of S_{CO_2} for carbonized chars (Fig. 23.22) are referred both to the char mass and to the original coal mass. Assuming no attrition or agglomeration during carbonization, the values of S_{CO_2} referred to the coal mass can be considered to be proportional to the surface area on a per particle basis. On this basis, values for the surface area are closer to those obtained with the raw coal, and in fact are less than raw coal values for carbon contents greater than about 85%. Because of the predominant effect of the removal of volatile matter, the overall results described may correspond to a net increase

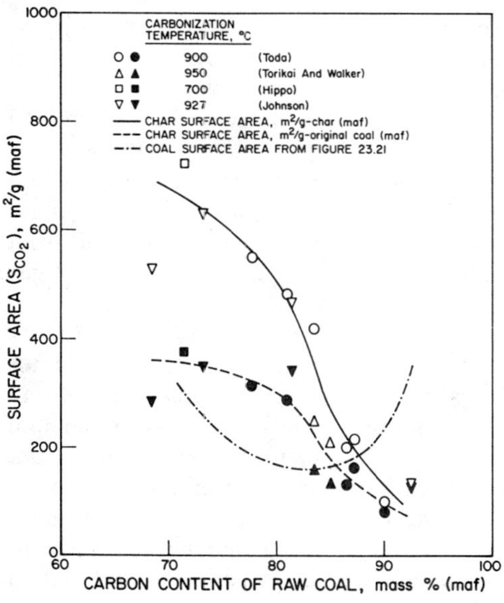

Figure 23.22 Variation with carbon content of the raw coals of CO_2 surface areas of carbonized coal chars.[117,121–123]

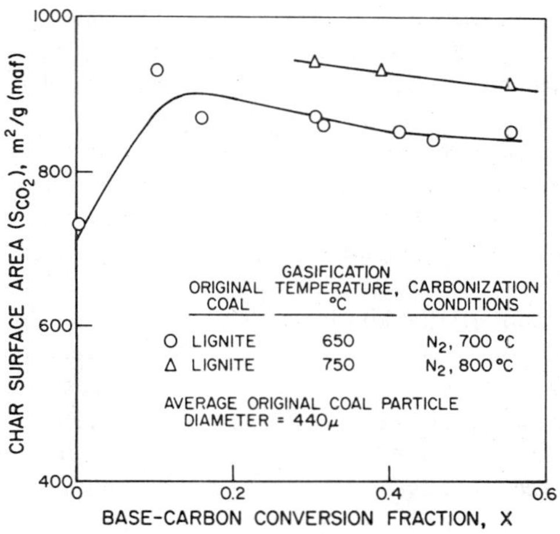

Figure 23.23 Effect of carbon conversion level on CO_2 surface area of lignite coal chars gasified with steam in a fluid bed at 1 atm.[123]

in available surface area during carbonization of coals with carbon contents less than 85%. For coals having carbon contents greater than 85%, however, less volatile matter is evolved; and the net effect of carbonization under these conditions is a decrease in surface area, due possibly to shrinkage or closure of pore-opening regions induced by the heat treatment.

In some studies coal char surface area variations during the course of gasification in reactive atmospheres have been determined as a function of the carbon conversion level. Results obtained by Hippo[123] for the gasification of Montana lignite in steam, at 1 atm, are shown in Fig. 23.23. A fluidized-bed reactor was used in this study, and chars were initially prepared in nitrogen. The results shown in Fig. 23.23 indicate an initial increase in surface area (S_{CO_2}) for carbon conversions less than 10–20%, with a relatively slight decline in surface area thereafter for carbon conversions up to 56%. Hippo has suggested that the increase in surface area is due to a reduction in pore restrictions during initial gasification stages, but that the pore walls collapse in later stages, leading to the joining of pores and the resultant decrease in surface areas.

Torikai and Walker[122] studied surface area variations of coal chars that were prepared by carbonization of hvbA coals and were then gasified in carbon dioxide, in a fluidized-bed reactor at 1 atm. In this study two coal lithotypes obtained from the same seam were used: one, a relatively low vitrinite material classified as a durain, and the other, a high-vitrinite material classified as a clarain. The raw coals were pretreated in air at a low temperature to reduce agglomeration tendencies during the subsequent carbonization. Surface area variations (S_{CO_2}) of partially gasified chars obtained with these samples are shown in Fig. 23.24. The trends indicated are qualitatively similar to those obtained by Hippo, with surface areas initially increasing with increasing carbon conversion, reaching a maximum at intermediate carbon conversion, and then decreasing with further increases in carbon conversion. Maximum surface areas increase with decreasing gasification temperature from 1000 to 850°C, although the variation is relatively slight from 900 to 850°C. At still lower temperatures it would be expected that this trend would reverse for chars carbonized at gasification temperatures, because of the effect of residual volatile matter in inhibiting gas penetration into pores during the measurement of adsorption isotherms.

Such inhibition may account for the somewhat higher apparent surface areas obtained by Hippo for lignite chars resulting from gasification at 750°C (carbonization temperature = 800° C), compared to surface areas obtained for lignite chars resulting from gasification at 650°C (carbonization temperature = 700°C). The decrease in surface area at higher carbon conversions shown by the data of Torikai and Walker is relatively slight at the gasification temperature of 850°C, but quite severe at the gasification temperature of 900°C, indicating the sensitivity of char structural characteristics to temperature.

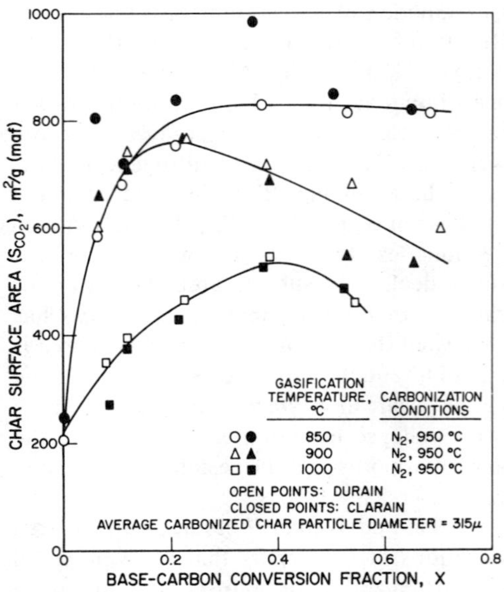

Figure 23.24 Effect of carbon conversion level on CO_2 surface areas of bituminous coal chars gasified in a fluid bed at 1 atm.[122]

Although Torikai and Walker obtained similar results with chars from two different lithotypes of the same rank, surface area characteristics during gasification would be expected to depend on initial coal properties as well as conditions of carbonization and gasification.

Johnson[121] studied surface area variations for a variety of coal chars gasified in hydrogen and steam-hydrogen mixtures, in a thermobalance apparatus, at 35 atm. Johnson's results (Fig. 23.25) showed only slight decreases in surface areas during gasification at 927°C for chars derived from anthracite, high-volatile bituminous coal, and lignite. Only with char obtained from a subbituminous A coal did the surface areas decrease with carbon conversion to an extent comparable to that observed by Torikai and Walker for bituminous char gasified at 900°C. In Johnson's study similar surface area trends were obtained for the gasification of individual chars in hydrogen and in steam-hydrogen mixtures. In addition, no significant effects on lignite char surface area characteristics were observed when initial acid pretreatment was employed, even though chars derived from acid-treated lignite were much less reactive than chars derived from untreated lignite. Significantly, Johnson also found essentially identical char surface area characteristics during the gasification of chars formed by initial carbonization in nitrogen and chars formed directly in the reactive gas atmosphere.

The maximum surface areas exhibited by the curves drawn in Figs. 23.23–23.25 for chars gasified at temperatures from 650 to 927°C, are plotted in Fig. 23.26 as a function of the carbon content of the raw coal. Considering the variations in gasification conditions and original coal types used in these studies, the maximum char surface areas are surprisingly similar for original coal carbon contents from 68 to 85%. The anthracite char surface area, however, is significantly less than the area of lower rank chars. The dashed

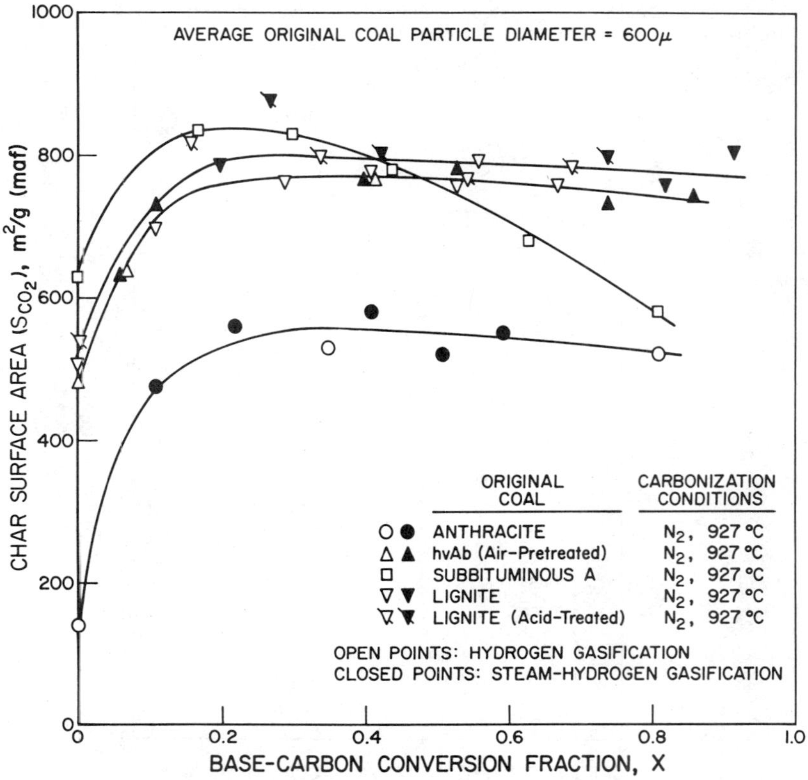

Figure 23.25 Effect of carbon conversion level on CO^2 surface areas of different coal chars gasified in hydrogen and steam-hydrogen mixtures in a thermobalance apparatus at 927°C, 35 atm.[121]

curve in Fig. 23.26 represents a correlation proposed by Johnson[121] to define relative gasification reactivities of coal chars as a function of original coal carbon content (see Fig. 23.50). Although there appears to be a possible relationship between variations in char surface areas and variations in gasification reactivities for chars derived from coals having carbon contents greater than about 82%, chars derived from lower rank coals exhibit higher reactivities than can be attributed solely to variations in surface areas.

23.4 KINETICS OF GASIFICATION REACTIONS

Rational design of coal gasifiers requires the best possible understanding of fundamental chemical and physical processes that occur: (1) during the transition of coal to coal char and (2) during the subsequent gasification of the coal char itself. The following sections focus primarily on studies relevant to these two main reaction stages, in which attempts have been made to eliminate or separately account for mass and heat transfer effects with the hope of characterizing intrinsic chemical behavior in the experimental gasification systems employed. It should be kept in mind, however, that it is not often evident to what extent this objective has been achieved in individual studies, and that as suggested by Ergun and Mentser,[134] it is likely that some of the wide variation in kinetic data reported in the literature is due in large measure to differences in experimental systems and to difficulties in making a quantitatively complete evaluation of physical factors.

23.4.1 Initial Stage Coal Gasification Kinetics

Gaseous hydrocarbons produced during the initial coal gasification stages are extremely

[134] S. Ergun and M. Mentser, in *Chemistry and Physics of Carbon,* P. L. Walker, Jr., Ed., Vol. 1, Marcel Dekker, New York, 1965, pp. 204–263.

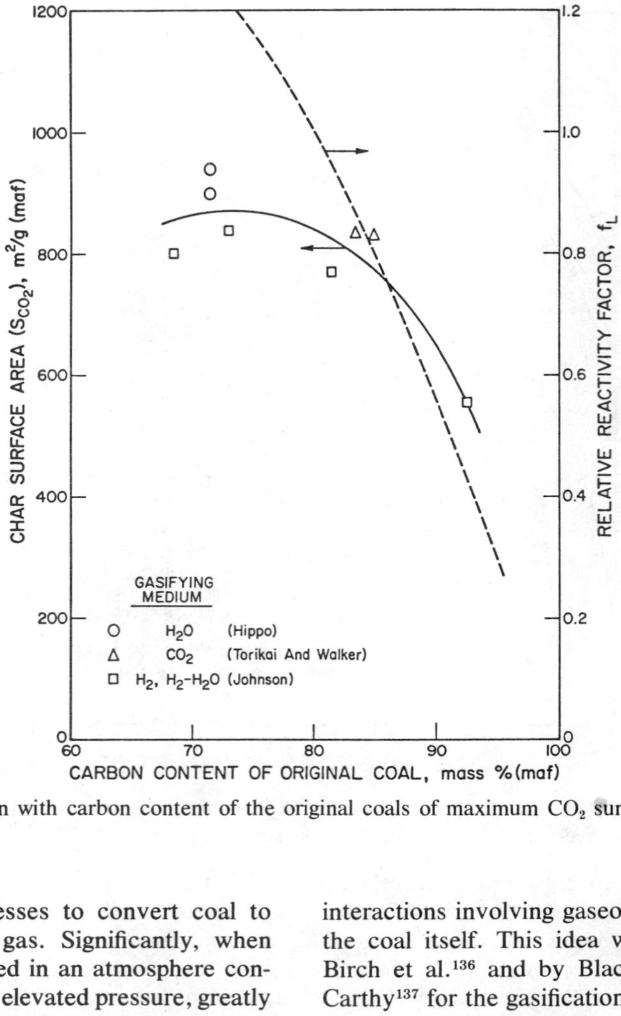

Figure 23.26 Variation with carbon content of the original coals of maximum CO_2 surface areas of coal chars.[121–123]

important in processes to convert coal to substitute natural gas. Significantly, when raw coals are heated in an atmosphere containing hydrogen at elevated pressure, greatly increased gaseous hydrocarbon yields are obtained, compared to those obtained when devolatilization occurs in an inert gas atmosphere. This high reactivity, however, is transient, lasting only for seconds or less at elevated temperatures corresponding to the time required for conversion of the coal to the relatively much less reactive char.

Because of its importance, this phenomenon, first observed by Dent[135] in the 1940s, has been studied in a variety of experimental investigations, particularly during the last 15 years. In some of the earliest work during this period, it was usually believed that these excessive gaseous hydrocarbon yields, particularly of methane, resulted from hydrogenation or hydrogenolysis of the evolved volatile matter and not from heterogeneous interactions involving gaseous hydrogen and the coal itself. This idea was proposed by Birch et al.[136] and by Blackwood and McCarthy[137] for the gasification of a brown coal in a fluidized bed, and later by Feldkirchner and Linden[138] as well as by Wen and Huebler[139] for the gasification of bituminous coal in a fixed bed. Moseley and Paterson,[140,141] however, were the first investigators to propose that although excessive methane yields resulted in part from hydrogenation of volatile matter, a significant fraction did indeed result from heterogeneous coal-hydrogen re-

[135] F. J. Dent, *Gas J.,* **244,** 502–505 (1944).

[136] T. J. Birch, K. R. Hall, and R. W. Urie, *J. Inst. Fuel,* **33,** 422–435 (1960).

[137] J. D. Blackwood and D. J. McCarthy, *Aust. J. Chem.,* **19,** 797–813 (1966).

[138] H. L. Feldkirchner and H. R. Linden, *Ind. Eng. Chem., Process Des. Dev.,* **2,** 153–162 (1963).

[139] C. Y. Wen and J. Huebler, *Ind. Eng. Chem. Process Des. Dev.,* **4,** 142–147 (1965).

[140] F. Moseley and D. Paterson, *J. Inst. Fuel,* **38,** 378–391 (1965).

[141] F. Moseley and D. Paterson, *J. Inst. Fuel,* **40,** 523–530 (1967).

action. Subsequently other investigators have also usually agreed that heterogeneous interactions play an important role in initial methane formation reactions. There is, however, considerable variation in the mechanisms proposed and the kinetic representations made to explain experimentally observed behavior, a diversity of opinion that is understandable in view of the complexity of the reaction processes that can occur during initial coal gasification stages, as well as the difficulty in accurately characterizing experimental parameters during extremely short reaction times.

Some of the overall reactions that can occur during the initial coal gasification stages are simply described in Fig. 23.27. In an inert gas atmosphere the devolatilization reactions represented are similar to the pyrolysis model proposed by Van Krevelen,[37] whereby coal pyrolysis is divided into three overall processes. Initially the coal undergoes a kind of depolymerization reaction, which leads to the formation of a metastable intermediate product. The product then undergoes cracking and recondensation to result in the evolution of primary gases and oils, yielding semichar. After the rate of devolatilization has passed a maximum, the third reaction becomes important, in which the semichar is converted to char primarily through the evolution of hydrogen.

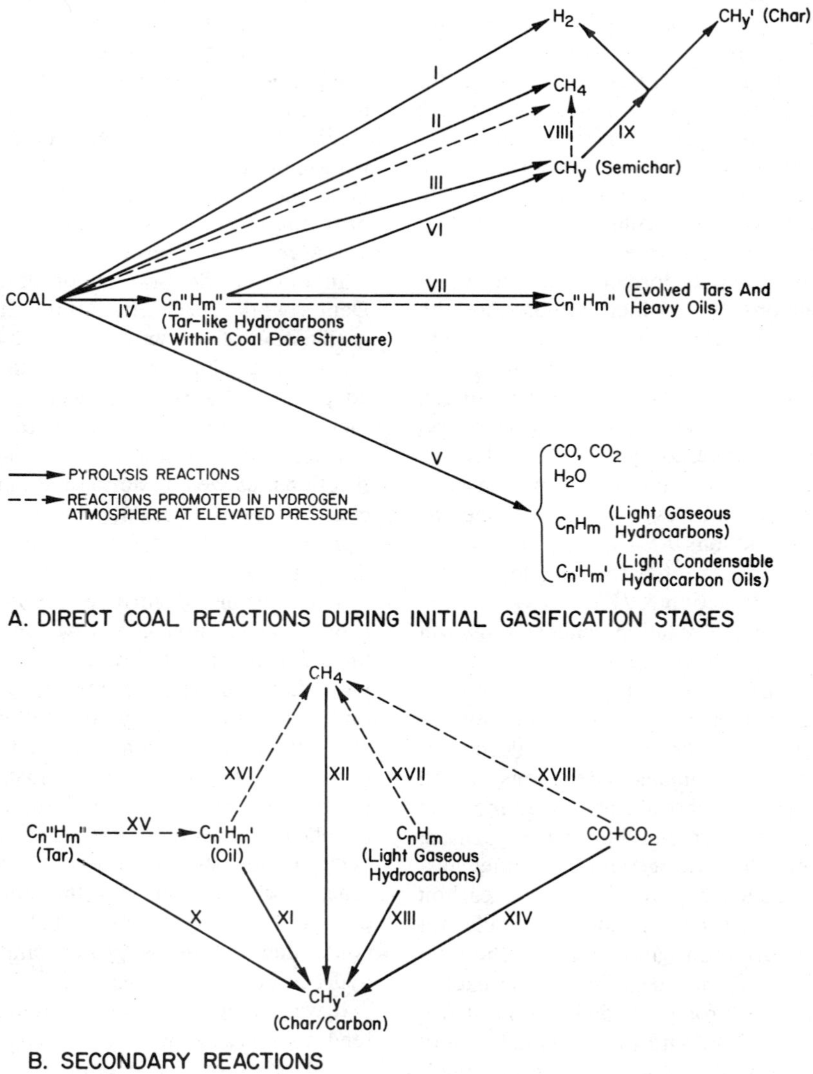

Figure 23.27 Simplified reaction scheme.

In Fig. 23.27*a* primary devolatilization in an inert atmosphere is illustrated by Reactions I–VIII, which involve the evolution of coal oxygen as carbon oxides and water, in addition to the formation of other gaseous hydrocarbons, heavy oils and tar, and some gaseous hydrogen. The solid product of this reaction stage is semichar. Secondary devolatilization is illustrated by Reaction IX, in which semichar is converted to char through the evolution of hydrogen. In a hydrogen atmosphere at elevated pressure, Reaction VII, giving higher yields of evolved oils and tars, is promoted over the internal cracking of oils and tars (Reaction VI). Methane formation from the direct reaction of gaseous hydrogen and coal is indicated by Reaction II, which occurs concurrently with primary devolatilization reactions. Reaction VIII, on the other hand, illustrates another path for the formation of methane in hydrogen, which results from heterogeneous interaction during secondary devolatilization concurrent with conversion of semichar to char, along with associated hydrogen evolution.

Figure 23.27*b* illustrates the secondary reactions of products formed from the direct coal reactions. In an inert atmosphere the various hydrocarbon products and carbon oxides primarily are cracked to form carbon or char (Reactions X–XIV). In a hydrogen atmosphere, however, initial hydrogenation products can undergo hydrogen enrichment, leading to the formation of light oils, gaseous hydrocarbons, and eventually methane, as shown in Reactions XV–XVII. The possibility of methanation of carbon oxides is also included, as Reaction XVIII.

In terms of the reaction scheme shown in Fig. 23.27, excessive methane or other light gaseous hydrocarbon yields can result during the initial coal gasification stages in a hydrogen atmosphere from (1) direct hydrogenation of coal or semichar (Reactions II and VIII), which also should result in an increase in the amount of coal-carbon that is gasified; or (2) from the hydrogenation of other gaseous hydrocarbons, oils, tars, and carbon oxides, which, for the most part, would not directly contribute to an increase in the total amount of coal-carbon gasified. One exception would be through a combination of Reactions VII, XV, and XVI, which would increase the total amount of carbon gasified, as well as the methane yield.

A variety of investigations has been conducted to study initial reactions during coal gasification in hydrogen-containing gases. The range of conditions employed in some of these investigations is given in Table 23.11. Three types of experimental system were used in these studies: fixed coal bed–flowing gas, fluidized coal bed, and dilute solid phase transport, or free-fall systems. In addition to differences in the coal type and the range of conditions employed, there are differences in the methods for characterizing experimental results. In some of the fixed-bed studies gasification behavior was characterized on the basis of total coal weight loss or coal-carbon conversion.[7,144–146] In such cases, although little information is directly available describing gasified product yields or secondary reactions, the data do reflect the net gas-solid interactions leading to enhanced carbon gasification rates during initial reaction stages. In other studies more complete information has been obtained, including yields of major gasified species or species groups, thus providing a basis for interpretation of individual reactions that occur.

In one of the first major studies since Dent's work, Birch et al.[136] investigated the gasification of brown coal with hydrogen in a continuous fluidized-bed system, at elevated pressures. Results showed a substantial increase in methane yield with increasing hydrogen pressure during the initial stage reactions, as well as during the much slower char-hydrogen reaction, which reaction was significant at the relatively long char residence times employed.

The authors described the gasification process as occurring in two fairly well-defined stages, an initial rapid stage associated with devolatilization processes, and a second, much slower stage associated with the char-hydrogen reaction. The first stage was considered to be essentially instantaneous, relative to the amount of time required for significant char-hydrogen reaction. This is because the first-stage reaction involves rapid reaction of hydrogen with the more reactive parts of the coal structure such as oxygen-containing functional groups and aliphatic hydrocarbon side chains.

It was found that although methane yields tend to increase with increasing hydrogen partial pressure, the maximum yield during the initial stage is stoichiometrically limited,

Table 23.11 Nominal Conditions for Investigations of Initial Coal Gasification Reactions in Hydrogen-Containing Gases

Coal Type	Primary Particle Diameter Range (μm)	Feed Gases	Temperature Range (°C)	Pressure Range (atm)	Residence Time at Temperature (s)		Solids Heatup Rate (°C/s)	Ref.
					Solids	Gas		
Fixed-Bed Studies								
Low-temperature bituminous coal char	420–1190	H_2, H_2O, H_2-H_2O	900–1150	70	0–1500	—	~20	142
Pittsburgh high-volatile bituminous coal	250–590	H_2	800, 1200	5–70	0–1800	<1	~10	143
Pittsburgh high-volatile bituminous (air pretreated) coal	420–840	N_2, H_2, H_2O, H_2-H_2O, H_2-CH_4	450–900	10–70	60–3600	—	~10	7
Pittsburgh high-volatile bituminous coal, Montana lignite	50–80	H_2, He, N_2, H_2-He	400–1100	0.001–70	0–20	—	60–10^4	144–146
Illinois high-volatile bituminous coal	<40	H_2, N_2	650–1000	100	2–65	0.2–23	20–1400	147, 148
North Dakota lignite	40	H_2	550–850	10–280	0.3–10	0.3–10	~10^4	149
Fluidized-Bed Study								
Brown coal	300–760	H_2, H_2-N_2, synthesis gas	500–950	2–40	900–6500	6	—	136
Dilute Solid Phase Transport Studies								
British high-volatile bituminous coal	100–150	H_2	840–930	50–500	0.5–5	60–120	—	141
Pittsburgh high-volatile bituminous coal	150–300	H_2-CH_4	730	100–200	1–2	120–240	—	150
North Dakota lignite	70–150	Synthesis gas	940–970	70–85	2–10	2–10	—	151
Pittsburgh high-volatile bituminous coal, Illinois high-volatile bituminous coal, North Dakota lignite	70–300	H_2, H_2-CH_4	650–900	35–140	1–2	30–200	—	119, 152
Montana lignite, Montana subbituminous coal, North Dakota lignite	70–90	H_2, He, H_2-He	450–850	20–50	5–15	5–15	30–~10^4	153–155
North Dakota lignite, New Mexico subbituminous coal	<50, <150	H_2, Ar	700	70–100	15–40	10–70	—	156

and is independent of hydrogen pressure. For the brown coal used, this maximum yield was estimated at about 28% conversion of the carbon feed to methane and 12% to other volatiles. The authors pointed out that the amount of solid carbon remaining after completion of the first reaction stage corresponds fairly closely to that known to be present in aromatic ring structures in brown coal.

A similar explanation was postulated by Blackwood and McCarthy,[137] who later analyzed the data of Birch et al. in greater detail, using an extrapolation procedure to distinguish between first-stage and second-stage methane yields. Blackwood and McCarthy concluded that at temperatures above 700°C, methane yields obtained during the first-stage reaction are dependent only on hydrogen partial pressure up to the maximum yield possible, which was obtained at a hydrogen partial pressure of 30–40 atm (Fig. 23.28).

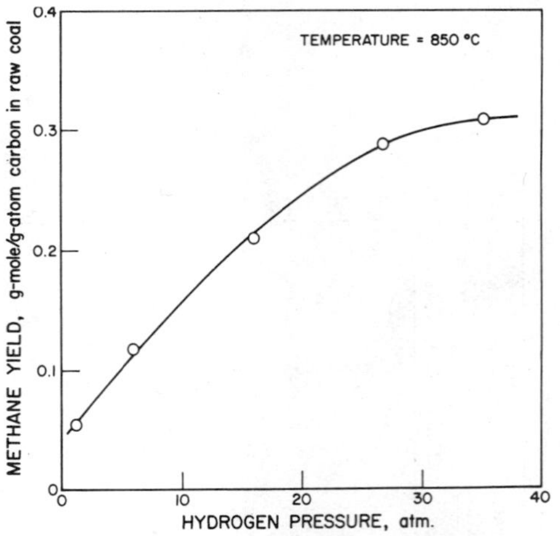

Figure 23.28 Effect of hydrogen pressure on first-stage methane yields obtained for gasification of brown coal in a fluid bed.[137]

The model proposed by Blackwood and McCarthy suggests that there is a fixed fraction of reactive carbon in the raw coal structure that can potentially be converted to methane, once sufficient time has elapsed to complete devolatilization processes. During the main devolatilization process the hydrogenation reactions leading to methane formation compete with certain of the pyrolysis processes that tend to form carbon oxides, heavier hydrocarbons, and char or fixed carbon. Within the context of the simplified reaction scheme shown in Fig. 23.27*a*, high-pressure hydrogen tends to favor Reaction II in competition with Reactions IV and V.

A two-stage gasification model has also been proposed by Wen and Huebler[139] in analyzing data obtained by Feldkirchner and Huebler[142] for the gasification of lignite and a low-temperature bituminous coal char in hydrogen. Based on the improved time resolution of the data obtained by Feldkirchner and Huebler with a fixed-bed flowing gas reactor and differential gas conversion, Wen and Huebler proposed the following correlation to account for methane yields obtained during the first-stage reaction in hydrogen:

$$\frac{dY_{CH_4}}{dt} = k(f - X)\, P_{H_2} \qquad (23.42)$$

where X = total carbon gasified (g/g feed carbon)
Y_{CH_4} = methane yield (g/g feed carbon)
f = fraction of feed carbon that can react by first-stage reactions
P_{H_2} = hydrogen partial pressure (atm)

[142] H. L. Feldkirchner and J. Huebler, *Ind. Eng. Chem. Process Des. Dev.*, **4**, 134–142 (1965).
[143] R. W. Hiteshue, S. Friedman, and R. Madden, U.S. Bur. Mines, Rep. Invest. No. 6376, 1964.
[144] D. B. Anthony, Thesis, Massachusetts Institute of Technology, Cambridge, 1974.
[145] D. B. Anthony, J. B. Howard, H. C. Hottel, and H. P. Meissner, *Fuel*, **55**, 121–128 (1976).
[146] D. B. Anthony and J. B. Howard, *Am. Inst. Chem. Eng. J.*, **22**, 625–656 (1976).
[147] R. A. Graff, S. Dobner, and A. M. Squires, *Prepr., Am. Chem. Soc., Div. Fuel Chem.*, **20**(3), 23–32 (1975).
[148] R. A. Graff, S. Dobner, and A. M. Squires, *Fuel*, **55**, 109–115 (1976).
[149] F. B. Growcock and D. R. MacKenzie, *Fuel* **55**, 349–354 (1976).
[150] P. S. Lewis, S. Friedman, and R. W. Hiteshue, in Am. Chem. Soc. Adv. Chem. Ser. No. 69, 1967, pp. 50–63.
[151] R. A. Glenn, E. E. Donath, and R. J. Grace, in Am. Chem. Soc. Adv. Chem. Ser. No. 69, 1967, pp. 81–103.
[152] H. F. Feldmann, W. H. Simons, J. A. Mima, and R. W. Hiteshue, *Prepr., Am. Chem. Soc., Div. Fuel Chem.*, **14**(4), Pt. 2, 1–25 (1970).
[153] J. L. Johnson, *Prepr., Am. Chem. Soc., Div. Fuel Chem.*, **20**(3), 61–87 (1975).
[154] J. L. Johnson, *Prepr., Am. Chem. Soc., Div. Fuel Chem.*, **22**(1), 17–37 (1977).
[155] J. L. Johnson, paper presented at Joint U.S.–U.S.S.R. Symp. Coal Gasification/Liquefaction, Moscow, October 1976.
[156] M. Steinberg and P. Fallon, paper presented at 68th Annual Meeting, American Institute of Chemical Engineers, Los Angeles, November 16–20, 1975.

t = time (s)
k = rate constant (5.6×10^{-4} atm^{-1} s^{-1})

The variable f is a function of coal type and varies slightly with temperature from 800 to 950°C.

In the correlation of Equation 23.42, Wen and Huebler suggested values at 927°C of f = 0.25 for low-temperature bituminous coal char and f = 0.48 for lignite. It should be noted that the term $f - X$ represents the fraction of potentially volatile carbon still remaining in the reacting coal particle at any time. Thus Wen and Huebler's model proposes that methane yields obtained during the first-stage reaction result from direct interaction of hydrogen with reactive groupings remaining in the coal, a process that competes with pyrolysis reactions involving these groups. In this sense the model is essentially identical to that proposed by Blackwood and McCarthy, although Wen and Huebler's formulation indicates that only the gasified product yield is affected by gaseous hydrogen, not the total amount of carbon gasified during the first-stage reaction.

Moseley and Paterson[140,141] concluded that the high yields of methane obtained during the initial gasification stages in hydrogen resulted not only from the hydrogenation of a limited amount of coal volatile matter, or potential volatile matter, but also from direct hydrogasification of an active intermediate that would otherwise form relatively unreactive char or fixed carbon. These conclusions were based on results obtained in dilute phase transport reactor systems for the gasification of low-temperature bituminous char and noncaking, high-volatile bituminous coal, with feed hydrogen pressures ranging up to 500 atm.

Results such as those shown in Fig. 23.29, which were obtained with a time resolution (0.5–5 s) greatly improved over that existing in previous studies, showed that at sufficiently high hydrogen pressures, substantial fractions of potential fixed carbon can be gasified in a few seconds. At an average product gas hydrogen partial pressure of about 350 atm with a feed hydrogen pressure of 500 atm, almost complete conversion of the feed coal carbon was achieved at 900–925°C.

Moseley and Paterson postulated that at elevated temperatures the hydrogenation of

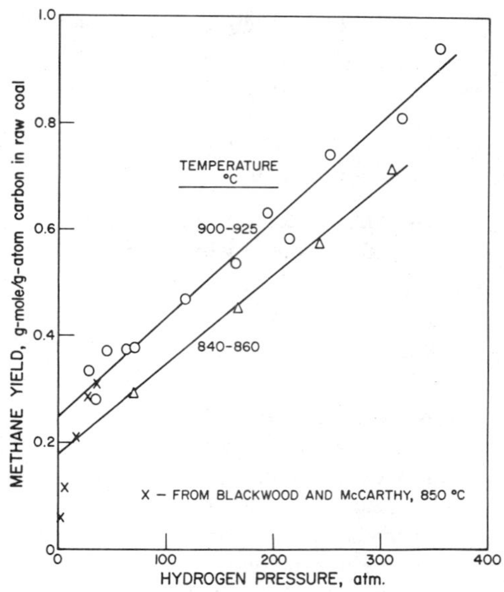

Figure 23.29 Effect of hydrogen pressure on methane yields obtained from gasification of noncaking bituminous coal in a dilute solid phase transport reactor.[137,140,141]

evolved volatile matter to form methane is essentially complete at a hydrogen pressure of less than 100 atm. At higher pressures methane yield increases linearly with hydrogen partial pressure, solely because of the gasification of potential fixed carbon in the intermediate coal structure. According to this description, the lines drawn in plots like Fig. 23.29 should curve downward at low hydrogen partial pressures as a result of incomplete hydrogenation of volatile matter. A possible example of this effect is shown in Fig. 23.29 by Blackwood and McCarthy's representation of methane yields obtained during the first-stage reaction in Birch's study, at a hydrogen partial pressure of less than 40 atm. Although Blackwood and McCarthy interpreted the indicated trends as corresponding to a leveling off of methane yields with increasing partial pressure, the data of Moseley and Paterson show that yields above around 50 atm are approximately proportional to hydrogen partial pressure.

Moseley and Paterson have suggested that when the hydrogen reaction with volatile matter is complete, methane formation is due solely to the attack by hydrogen on transient, unstable structures in the pyrolyzing coal char, and that this reaction competes effectively with normal thermal stabilization processes leading to unreactive carbon residues. In a formalized model, these investigators have proposed the following overall reaction:

$$C_0(\text{coal}) \xrightarrow{K_A} V \text{ (volatiles)} + I \text{ (active intermediate)} \quad (23.43)$$

$$\xrightarrow{K_B} \text{cross-linked, less active structure, not easily gasified by hydrogen}$$

$$\downarrow K \; (+\, 2n\text{H}_2)$$

$$I + nCH_4$$

where K, K_A, and K_B are rate constants.

The initial reaction corresponds to a purely thermal decomposition-polymerization process that is not influenced by hydrogen pressure and leads to the formation of volatiles and an active intermediate in the char structure. The active intermediate promotes rapid hydrogenation of the coal char structure to form methane in competition with a rapid deactivation reaction leading to the formation of unreactive cross-linked structures, with associated loss of reactive intermediates. The formation of reactive intermediates is considered to be rate limiting, and it is assumed that no net consumption of the active intermediate occurs in the hydrogenation step. With this model, and assuming first-order reaction processes, the following rate equation was derived to describe the rate of methane formation under isothermal conditions:

$$\frac{dY_{CH_4}}{dt} = \frac{KK_A C_0 e^{-K_A t}}{K_B} P_{H_2} M \quad (23.44)$$

where C_0 = initial concentration of structures in the coal capable of forming the active intermediate

M = effective char concentration

Moseley and Paterson suggest that the model above represents a drastically simplified picture of a complex process and that not only will C_0, K_A, and K_B depend on the original coal and the temperature of hydrogenation, but also K_A and K_B are unlikely to remain strictly constant at any one temperature, since devolatilization and linking require larger activation energies to overcome the hindrance of previous cross-linking. The solid carbon concentration, M, in Equation 23.44 is not considered to change significantly with time compared to changes in the concentration of the active intermediate. This is equivalent to assuming that the solid carbon is always in excess kinetically, even when a large portion of it has been gasified. Although this assumption is consistent with the straight lines drawn in Fig. 23.29, it would be reasonable to expect increasing downward curvature at levels of carbon conversion approaching unity.

The model proposed by Moseley and Paterson, then, suggests that the formation of methane during the initial gasification of coal in hydrogen results in part from hydrogenation of volatile matter—probably through a secondary reaction such as that described in Fig. 23.27*b*, but also through heterogeneous hydrogenation of the potential fixed carbon in the devolatilizing coal structure. Their suggestion that the main agents of activation for the heterogeneous hydrogenation may be groups that normally yield hydrogen under pyrolysis conditions indicates that Reaction VIII in Fig. 23.27*a* occurs instead of Reaction II. Since secondary stage coal-hydrogen evolution is usually thought to occur after primary devolatilization reactions leading to semichar,[37] the results of Moseley and Paterson can be interpreted as concluding that the rapid-rate heterogeneous reaction by methane formation also occurs after primary devolatilization.

Zahradnik and Glenn[157] later proposed a model somewhat similar to that of Moseley and Paterson, based on analysis of experimental studies conducted at Bituminous Coal Research, Inc. In this model primary devolatilization reactions are considered to go to completion very rapidly—in milliseconds at temperatures above 950°C—prior to any significant reaction between gaseous hydrogen and active intermediates formed from the primary devolatilization. Active intermediates then can either react with hydrogen to form methane or deactivate to form inactive char. These processes are shown schematically as follows:

$$\text{coal} \rightarrow C^{**} + CH_4 + (CO, CO_2, H_2O, \text{etc.}) \quad (23.45)$$

$$C^{**} \xrightarrow{k_1} C^{*} \quad (23.46)$$

[157] R. L. Zahradnik and R. A. Glenn, *Fuel,* **50,** 77–90 (1971).

$$C^{**} \xrightarrow[H_2]{k_2} CH_4 \qquad (23.47)$$

where C^{**} = active intermediate produced directly from coal when coal is thermally depolymerized

C^{*} = inactive char produced by reaction of the active intermediate with itself

k_1, k_2 = rate constants

One of the main characteristics of the model is that the active intermediate is consumed in the methane formation step, contrary to Moseley and Paterson's assumption that there is no net consumption of the active intermediate in this reaction step. Zahradnik and Glenn developed the following correlation, based on their model, to describe total methane yields after complete conversion of the active intermediate either to methane or to inactive char:

$$Y_{CH_4} = \frac{Y^0_{CH_4} + bP_{H_2}}{1 + bP_{H_2}} \qquad (23.48)$$

where $Y^0_{CH_4}$ = methane yield resulting only from devolatilization

b = kinetic parameter, proportional to the ratio of rate constants for methanation and deactivation of the active intermediate

Kinetic parameters in these equations were determined, to describe experimental results obtained for the gasification in hydrogen of several coals in dilute phase transport reactor systems. Among the results to be described are certain of the results obtained by Moseley and Paterson. Examples of the values obtained are given in Table 23.12.

[158] Bituminous Coal Research, Inc., U.S. Office Coal Res. R & D Rep. No. 20—Final Report (PB 235530); NTIS, 1970.

Interestingly, the range of methane yields from pyrolysis and hydrogenation of volatiles given above is considerably lower than the corresponding yields of 0.2–0.25 obtained by Moseley and Paterson for the hydrogenation of volatile matter. This difference apparently arises because Moseley and Paterson used a linear extrapolation with hydrogen pressure to evaluate these yields, whereas Zahradnik and Glenn's model accounts for the downward curvature of yields with decreasing hydrogen pressure. Assuming that the parameter b, equal to the ratio k_2/k_1, is dependent only on temperature and not on coal type, Zahradnik and Glenn determined an activation energy of approximately 15 kcal/g-mole, representing the difference in activation energies between methanation and the deactivation process.

One particularly important aspect of the models and correlations proposed by Moseley and Paterson, as well as by Zahradnik and Glenn, is that they successfully use only the hydrogen partial pressure, rather than the total gas composition, to characterize the experimental data. From Moseley and Paterson's data and correlations, this implies that the methane partial pressure does not have a significant effect on reaction kinetics for methane concentrations of up to 80% in the product gases. Similarly, for the Bituminous Coal Research, Inc., data, obtained at lower hydrogen partial pressures, the significant concentrations of carbon monoxide, carbon dioxide, and methane were not considered to affect hydrogenation reactivity.

The initial gasification stages of coal were also studied by Johnson[7] and Pyrcioch et al.,[159] using a thermobalance apparatus, to

[159] E. J. Pyrcioch, H. L. Feldkirchner, C. L. Tsaros, J. L. Johnson, W. G. Bair, B. S. Lee, F. C. Schora, J. Huebler, and H. R. Linden, IGT Res. Bull., No. 39, 1972, 225 pp.

Table 23.12 Kinetic Parameters by Zahradnik and Glenn[157]

Coal	Temperature (°C)	$Y^0_{CH_4}$	b (atm^{-1})	Data Source Ref.
North Dakota lignite	950	0.07	0.0067	151
Elkol subbituminous	950	0.08	0.0060	158
High-volatile bituminous	900–925	0.08	0.0060	141
High-volatile bituminous	850	0.07	0.0035	141
High-volatile bituminous	725	0.09	0.0017	150

measure reaction behavior in terms of the gasification rates of individual coal components under conditions of differential gas conversion. A typical result with this type of experimental system that illustrates the high reactivity exhibited by coals during initial reaction stages in hydrogen is shown in Fig. 23.30. An important effect of using this thermobalance apparatus is that after the coal sample is lowered into the reactor environment, the initial reactions involving devolatilization and methane formation are completed within the first few minutes required for the sample to heat up to reactor temperature. Figure 23.30 illustrates several results: curve *A* shows that for simple pyrolysis conditions, evolution of volatile matter is completed within the first few minutes and the total weight loss is about equal to the volatile matter of the feed coal, as determined by proximate analysis. When the coal char residue obtained after pyrolysis is exposed to hydrogen at elevated pressures (curve *B*), relatively slow gasification to form methane occurs. However when the feed coal is directly exposed to a hydrogen atmosphere without prior devolatilization in nitrogen (curve *C*), the total weight loss during the first few minutes significantly exceeds the weight loss obtained under pyrolysis conditions. This excessive weight loss corresponds to the additional carbon gasified primarily to "rapid-rate" methane, with yields increasing with increasing hydrogen pressure (curve *D*).

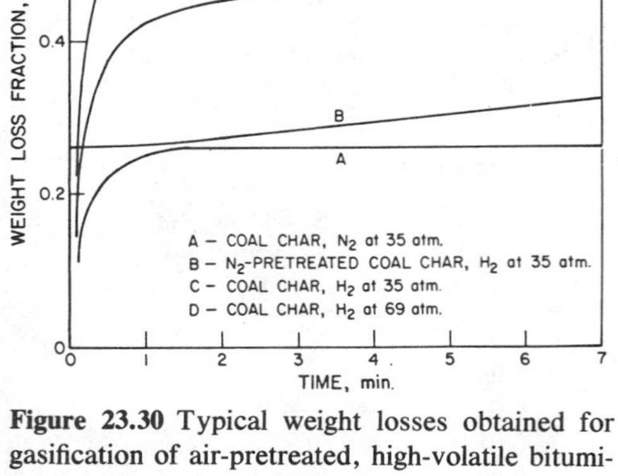

Figure 23.30 Typical weight losses obtained for gasification of air-pretreated, high-volatile bituminous coal in a thermobalance apparatus.[7]

Johnson correlated the excessive carbon conversion due to rapid-rate methane formation at final reaction temperatures above 800°C (hydrogen partial pressures up to 70 atm) with the following expression:

$$X = 1 - \exp(-f_R 0.0092 P_{H_2}) \qquad (23.49)$$

where X = base carbon gasified per base carbon in feed coal

f_R = reactivity factor used to characterize coal type ($f_R \sim 1.0$ for high-volatile bituminous coals)

P_{H_2} = hydrogen partial pressure (atm)

(Base carbon is defined as the carbon in the "fixed carbon" fraction, determined by standard proximate analyses.) As with correlations developed by other investigators, Equation 23.49 was successfully applied not only to systems using pure hydrogen but also to systems with hydrogen-methane mixtures and hydrogen-steam mixtures; this implies that the rapid-rate phenomenon occurs only in the presence of gases containing hydrogen and is not affected by other gaseous species.

Equation 23.49 was developed to describe yields obtained above 800°C, at which temperature devolatilization and those processes leading to rapid-rate methane formation are completed in relatively short times. For gasification at lower temperatures, Johnson proposed a more detailed model and correlation to describe methane yields. The model proposed was similar to that of Moseley and Paterson, but it included the formation of active intermediates, which occurs by an infinite set of parallel reactions having different activation energies. With this assumption, the correlation proposed by Moseley and Paterson was modified by replacing the term $K_A e^{-K_A t}$ in Equation 23.44 by the following term:

$$\int_0^\infty K_A e^{-K_A t} f(E)\, dE \qquad (23.50)$$

where

$$K_A = K_A^0 e^{-E/RT} \qquad (23.51)$$

The distribution function $f(E)$ is expressed by:

$$f(E) = \frac{4\alpha^{3/2}(E - E_1)^2}{\sqrt{\pi}} \exp[-\alpha(E - E_1)^2] \quad \text{for} \quad E \geqslant E_1 \qquad (23.52)$$

and

$$f(E) = 0 \qquad \text{for} \quad E < E_1 \qquad (23.53)$$

With this formulation, results such as those shown in Figs. 23.31 and 23.32 were described with the following evaluation of kinetic parameters:

$$E_1 = 18.1 \text{ kcal/g-mole} \qquad (23.54)$$

$$\alpha = 0.111 \text{ (kcal/g-mole)}^{-2} \qquad (23.55)$$

$$K_A^0 = 150 \text{ s}^{-1} \qquad (23.56)$$

$$\frac{C_0 K}{K_B} = 0.0092 f_R \text{ (atm}^{-1}) \qquad (23.57)$$

In describing these curves, additional correlations were developed to define weight loss contributions due to devolatilization reactions only. The evaluations above were based on data obtained for coal residence times greater than 180 s and should not be extrapolated to shorter residence times, where they are not applicable.

A study of the gasification of lignite and bituminous coal at short residence times (up to 20 s) has been reported by Anthony et al.,[144–146] using total coal weight loss as a measure of kinetic behavior. In this study small samples of coal were sandwiched in a steel mesh screen that could be rapidly heated to a desired temperature and held there

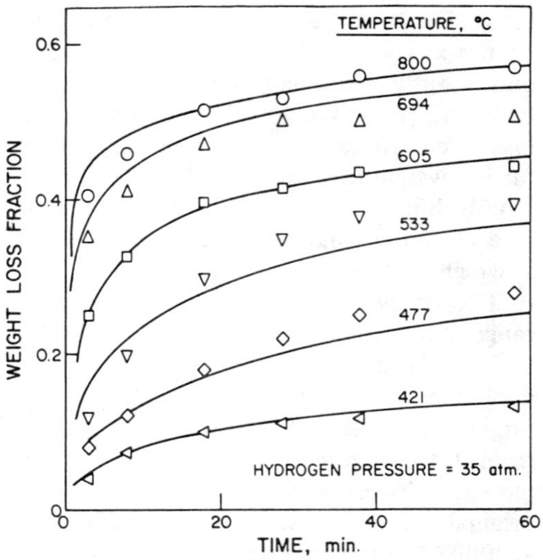

Figure 23.31 Effects of temperature and residence time on weight loss obtained for hydrogen gasification of air-pretreated, high-volatile bituminous coal in a thermobalance apparatus.[159]

for a desired time. The following parameters were measured: coal weight loss, as a function of heating rate (65–100,000°C/s), coal residence times at temperature (0–20 s), final temperature (400–1100°C), total pressure (0.001–70 atm), hydrogen partial pressure (0–70 atm), and coal particle size (70–1000 μm). Tests were conducted using hydrogen, helium, and hydrogen-helium mixtures. It was observed that under these conditions, total weight loss increases significantly with

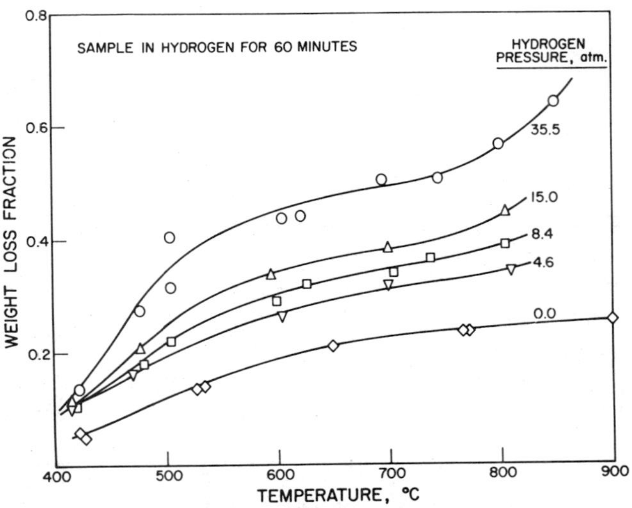

Figure 23.32 Effects of pressure and temperature on gasification of air-pretreated, high-volatile bituminous coal in a thermobalance apparatus.[159]

increasing hydrogen partial pressure above 10 atm and with increasing temperature, but only slightly with increasing heatup rate. At temperatures above 900°C maximum weight loss occurred in 2–3 s with no significant further weight loss up to 20–30 s.

With bituminous coal, at a final temperature of 1000°C and residence times of 5–20 s, weight losses in helium decreased from 54 to 37% of the original amount of coal, in a range of pressures from 0.001 to 70 atm (Fig. 23.33). Weight losses in hydrogen reached a minimum of approximately 48–50% of the original coal weight in the pressure range from 1 to 10 atm, then rose with further increases in pressure to about 60% of the original coal weight at 70 atm. Significantly, at 1000°C these variations in total weight loss are due almost solely to variations in the amount of carbon gasified, since the evolution of other coal components (oxygen, hydrogen, etc.) would be essentially complete.

In explaining these and other results obtained in this study, Anthony et al. proposed that two types of volatiles are formed within coal particles due to thermal activation processes: nonreactive species, and species that are extremely reactive, particularly in the initial stages of decomposition, when free radicals are likely to exist. The nonreactive volatiles are all evolved from the particle; but the reactive volatiles either evolve directly from the particle and are stabilized and rendered inactive by reaction with gaseous hydrogen, or they evolve from the particle and polymerize or crack on hot coal surfaces, thereby decreasing the volatile yields. The authors further postulated that gaseous hydrogen can also react directly with the coal for a transient period concurrent with the formation of reactive volatiles to produce—presumably—stable, light hydrocarbon gases such as methane. Quantitative characterization of this model was represented in terms of final yields after complete conversion of unreactive char as well as in terms of the kinetics describing intermediate reaction stages.

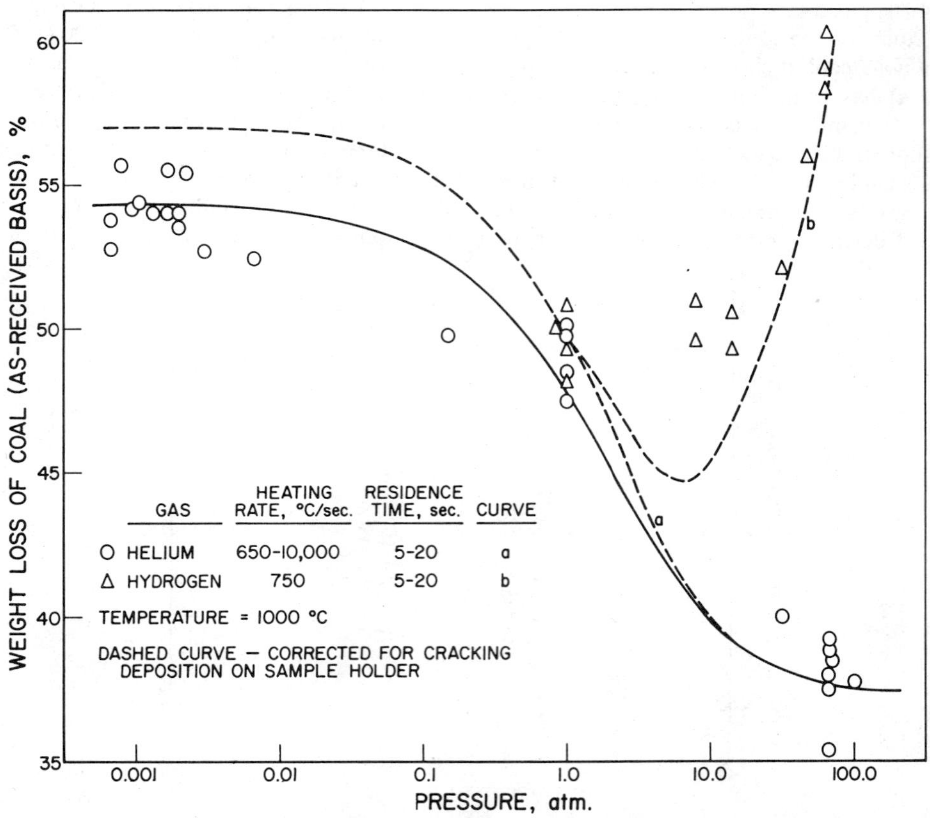

Figure 23.33 Effects of pressure on weight loss of high-volatile bituminous coal heated in hydrogen and helium atmospheres.[144]

The nomenclature in Table 23.13 defines the different classes of volatiles formed or evolved in the model proposed by Anthony et al. (Chapter 12). Using this notation, a material balance performed on the reactive volatiles within the coal particle leads to the following expression:

$$\frac{dV'_r}{dt} = \frac{k_c}{P} C + k_1 C + k_2 P_{H_2} C + \frac{dC}{dt} \quad (23.58)$$

where C = concentration of reactive volatiles in the particle
k_c = overall coefficient for mass transfer of reactive volatiles out of the particle
k_1 = overall rate constant for the deposition of reactive volatiles within the particle
k_2 = overall rate constant for the stabilization of reactive volatiles within the particle, by gaseous hydrogen
$\frac{dV'_r}{dt}$ = rate of formation of reactive volatiles
P = total pressure
P_{H_2} = hydrogen partial pressure

Assuming that the derivative dC/dt is approximately 0 and that the ratios of rate constants and transfer coefficients are independent of temperature, the total yield of evolved reactive volatiles, for reaction times approaching infinity, is given by:

$$V^*_r = \frac{V^{**}_r \left(\frac{1}{P} + \frac{k_2}{k_c} P_{H_2} \right)}{\frac{1}{P} + \frac{k_2}{k_c} P_{H_2} + \frac{k_1}{k_c}} \quad (23.59)$$

Table 23.13 Nomenclature for the Model of Anthony et al.[146]

V'_r	=	reactive volatiles formed up to time t
V^{**}_r	=	reactive volatiles formed up to $t = \infty$ (potential ultimate yield)
V^*_r	=	reactive volatiles lost from particles up to time $t = \infty$ (ultimate yield)
V^{**}_{nr}	=	nonreactive volatiles formed and lost from particles up to $t = \infty$ (ultimate yield)
V	=	total volatiles lost from particles up to time t
V^*	=	total volatiles (reactive and nonreactive) lost from particle up to time $t = \infty$

As indicated by Equation 23.59, this model assumes that the total yield of evolved reactive volatiles is independent of the time-temperature history, but that it increases with increasing hydrogen partial pressure due to stabilization of reactive volatiles and decreases with increasing total pressure due to increased diffusion resistance. In the model the total weight loss is the sum of the nonreactive volatiles evolved, the reactive volatiles evolved, and the amount of carbon that is directly gasified with hydrogen concurrent with the formation of reactive volatiles. The weight loss fraction attributed to the latter process is assumed to be expressed by the term $k_3 P_{H_2}$, thus:

$$V^* = V^{**}_{nr} + V^*_r + k_3 P_{H_2} \quad (23.60)$$

Analysis of data obtained in pure helium, in pure hydrogen, and in hydrogen-helium mixtures led to the following evaluations (in Equations 23.59 and 23.60) of kinetic parameters, which were made to describe final weight losses obtained with bituminous coal at 1000°C, over the total range of conditions employed:

$$V^{**}_r = 0.20 \quad (23.61)$$

$$V^{**}_{nr} = 0.372 \quad (23.62)$$

$$\frac{k_2}{k_c} = 0.02 \text{ atm}^{-2} \quad (23.63)$$

$$\frac{k_1}{k_c} = 0.56 \text{ atm}^{-1} \quad (23.64)$$

$$k_3 = 0.001 \text{ atm}^{-1} \quad (23.65)$$

The rate constant k_3 was not actually determined from the data developed by Anthony[122] but was estimated from the high-pressure data of Moseley and Paterson.[140] This assumption has a strong effect on the evaluation of the parameter k_2/k_c, thus strongly affects the relative amounts of evolved carbon that are attributed to (1) hydrogenation of the reactive volatiles within the particle, which is stoichiometrically limited and is likely to lead to heavy hydrocarbon products, and (2) direct hydrogenation of the coal, a process likely to lead to light gaseous hydrocarbons.[157]

As discussed previously, Zahradnik and

Glenn[157] and Zahradnik and Grace[160] used a different form of correlation to describe the methane yields obtained in Moseley and Paterson's study, so that the carbon conversion due to the direct reaction of coal and hydrogen is given by:

$$Y_{CH_4} = Y^0_{CH_4} + \frac{(1 - Y^0_{CH_4})bP_{H_2}}{1 + bP_{H_2}} \qquad (23.66)$$

where $Y^0_{CH_4} = 0.08$
$b = 0.006\ \text{atm}^{-1}$

This expression can be converted into the weight loss fraction for the bituminous coal used by Anthony et al. by assuming a reasonable value of feed carbon in the raw coal of 0.7 g per gram of coal, yielding the expression:

$$\left(\frac{\Delta W}{W_0}\right)_r = \frac{0.00386\,P_{H_2}}{1 + 0.006\,P_{H_2}} \qquad (23.67)$$

where the left-hand side is the weight fraction of the raw coal lost by direct coal hydrogenation. If it is assumed that the expression above represents the contribution of the direct reaction of coal and hydrogen in Anthony's experiments rather than the term $k_3 P_{H_2}$ and that $k_2/k_c = 0$ rather than 0.02, as assumed by Anthony et al., then in terms of yields due to coal-hydrogen reaction and yields due to stabilization of reactive volatiles, there is a significantly different interpretation of total weight loss. This is shown in Fig. 23.34, where Case I refers to the evaluation of parameters by Anthony et al. and Case II refers to the use of Zahradnik and Glenn's formulation for direct reaction of coal and hydrogen, with $k_2 = 0$. Although the total weight loss of Case II is somewhat less than that obtained by Anthony et al., this could be explained partly by the fact that the evaluation of b by Zahradnik and Glenn, using Equation 23.66, was based on the hydrogen pressure in the feed gas. Since there were substantial average concentrations of methane in Moseley and Paterson's tests, use of average or product gas hydrogen pressures would result in an increased value of b, which would tend to increase the computed total weight loss for Case II (Fig. 23.34). In any event it is apparent that the quantitative distribution of products evolved during the gasification of hydrogen, interpreted with respect to the model proposed by Anthony et

[160] R. L. Zahradnik and R. J. Grace, in Am. Chem. Soc. Adv. Chem. Ser. No. 131, 1974, pp. 126–144.

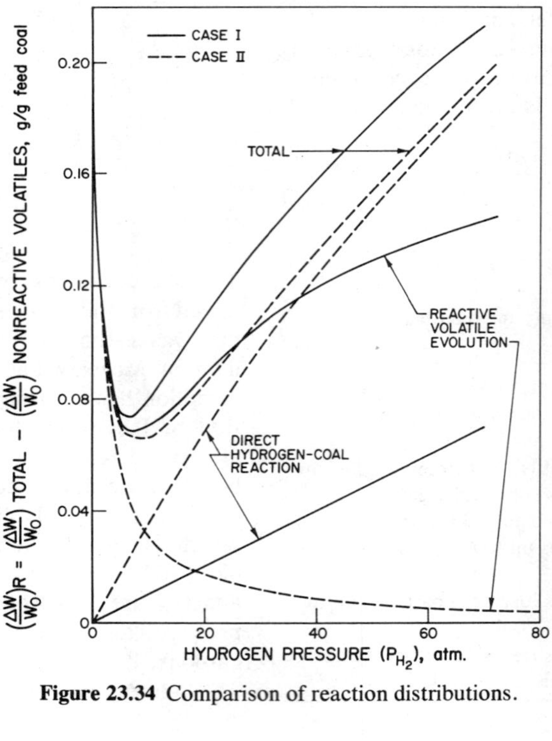

Figure 23.34 Comparison of reaction distributions.

al., should be treated as tentative, based on the experimental evidence considered. The qualitative features of this model, in terms of the reaction scheme depicted in Fig. 23.27*a*, indicate that tars and oils are evolved by Reactions IV and VII, with production of gaseous hydrogen favoring Reaction VII over the competing decomposition Reaction VIII. Light hydrocarbon production is favored by gaseous hydrogen via Reaction II.

The rates of formation of nonreactive and reactive volatiles have also been quantitatively described by Anthony et al. (Chapter 12), assuming first-order kinetics but using a distribution of activation energies for the reacting species. The specific formulation proposed leads to the following expression to define total coal weight loss at any time during the hydrogenation process:

$$V = V_{nr}^{**}\left\{1 - C_1 \int_0^{E_1} f(E) \exp -\left[\int_0^t k_0^0 e^{-E/RT}\, dt\right] dE\right\} + \left(V_r^* + k_3 P_{H_2}\right)\left\{1 - C_2 \int_{E_1}^{\infty} f(E) \exp -\left[\int_0^t k_0^0 e^{-E/RT} dt\right] dE\right\} \quad (23.68)$$

where $f(E)$ = activation energy distribution function (Gaussian)

$$= \left[(2\pi)^{1/2}\sigma\right]^{-1} \exp \frac{-(E - E_0)^2}{2\sigma^2} \quad (23.69)$$

$$C_1 = \left[\int_0^{E_1} f(E)\, dE\right]^{-1} \quad (23.70)$$

$$C_2 = \left[\int_{E_1}^{\infty} f(E)\, dE\right] \quad (23.71)$$

Equation 23.70 was fitted to the data of Anthony et al., using the values for limiting yields indicated above and the following values for other parameters:

$k_0^0 = 1.67 \times 10^{13}\ s^{-1}$

E_0 (mean activation energy) = 54.8 kcal/g-mole

σ (standard deviation) = 17.2 kcal/g-mole

E_1 (minimum activation energy for formation of reactive volatiles) = 61.4 kcal/g-mole

Results obtained by Anthony et al. for the gasification of lignite showed that although weight loss increased with increasing hydrogen partial pressure, weight loss did not decrease with increasing total pressure, as was observed with bituminous coals at pressures up to 10 atm. This implies that with lignite, either no reactive volatiles capable of repolymerizing on coal surfaces are formed, or the diffusion of the reactive volatiles that are formed is extremely rapid, even at elevated pressures. Although both explanations are likely, diffusion resistance is supported by the fact that gaseous diffusion with bituminous coals would be particularly inhibited during the partial liquefaction that occurs at intermediate devolatilization stages, a phenomenon that does not occur with lignites.

The effects of diffusion inhibition on initial coal gasification behavior have received relatively little attention in studies conducted so far. In one series of experiments by Anthony et al. (in hydrogen at 1000°C and 69 atm), total weight loss obtained with bituminous coal for residence times of 5–20 s was found to increase from 44 to 59% of the original weight of the coal as the diameter of the original coal particle decreased from 1000 to 70 μm. This was interpreted as reflecting an increased internal mass transfer resistance with increased particle size, which inhibited diffusion of external hydrogen into the coal particle interstices, thus reducing the effective hydrogen partial pressure. There is a possible complication in the qualitative interpretation of the results of Anthony et al., however, because of the use of a fixed bed of bituminous coal particles, which can agglomerate during devolatilization and significantly increase effective particle sizes. For example, with lignite, a nonagglomerating "coal," Anthony et al. found no effect of particle size on weight loss.

Moseley and Paterson,[141] using a nonagglomerating coal, also obtained some indication of enhanced gasification yields with decreased particle sizes. However interpretation of these results was complicated by considerable data scatter; in varying particle size from 60 to 200 μm, moreover, it was necessary to vary heating rate and coal residence times as well. The possibility that larger particles may not reach reactor temperatures in the short residence times characteristic of transport reactor systems could well be a significant factor affecting Moseley and Paterson's results.

This effect may also apply to results obtained by Hill et al.[161] for the catalyzed gasification of a nonagglomerating coal in hydrogen at 137 atm, in a dilute phase transport reactor system. In that investigation, conducted at 515°C, total coal conversion to liquids and gases decreased from 91 to 29% on an maf basis when average particle sizes were increased from 40 to 420 μm. Even though the use of catalysts may drastically alter the gasification process, it is unlikely that the severe effect of particle size observed by Hill et al. is due solely to mass transfer inhibition. Thus it is uncertain at this time what causes particle size to have an effect during the initial stages of gasification in hydrogen, and future detailed studies using more refined experimental techniques are required to distinguish among possible alternatives.

Johnson recently studied the initial gasification stages of low-rank coals in a dilute phase transport reactor.[153–155] This study is distinct from an earlier investigation using a thermobalance apparatus, discussed previously. The transport reactor consists of a 200-ft-long helical coiled tube that is $^1/_{16}$ in. in diameter. The reactor tube itself is the heating element, with electrodes attached at various points along the length of the tube to permit the establishment of desired temperature profiles and therefore to preset time-temperature histories for the gas-solid streams passing through the coil. With this system the gasification kinetics of lignite and subbituminous coal were studied in hydrogen, helium, and hydrogen-helium mixtures as a function of pressure (18–52 atm), gas composition, temperature (500–850°C), controlled rates of heat up to 80°C/s, and coal gas residence times (5–15 s) when the reactor was operated isothermally.

Typical experimental results of this study, expressed as yields of gasification products and as conversions of individual coal components are shown in Figs. 23.35 and 23.36. Based on such results, Johnson described the initial gasification of low-rank coals with a two-stage model similar to the models proposed by Zahradnik and Glenn and by Moseley and Paterson. The first stage involves primary devolatilization (pyrolysis) processes related to thermally activated reactions that lead to evolution of the bulk of the volatile matter. As shown in Fig. 23.35 for the gasification of lignite in hydrogen at a relatively slow heatup rate, the first-stage devolatilization is initiated below 500°C. Coal-oxygen is evolved primarily as carbon dioxide, carbon monoxide, and water, with a small amount observed in the phenolic oils.

Evolution of carbon dioxide occurs below 500°C, apparently resulting from the decomposition of carboxyl functional groups in the coal. Decomposition of other oxygenated functional groups leads to the evolution of both carbon monoxide and water, with evolution of the total available oxygen in the coal being completed at about 700°C. This is shown more clearly in Fig. 23.36. Although carbon dioxide yields begin to decrease above 650°C, this was attributed solely to a secondary reaction occurring in the reactor tube, probably catalyzed by the reactor walls. Evolution of C_3^+ gaseous hydrocarbons as well as formation of oils and tars occur below 550°C, although yields of this species group decrease above 650°C because of the hydrogenation to methane and ethane. Total yields of oils and tars remained relatively constant up to 850°C, although the composition of this fraction changed substantially with increasing temperature, becoming lighter. Conversion of the heavier oils to benzene is initiated at about 650°C, and at 850°C approximately one-half the total oil-tar fraction consists of benzene.

Johnson considers the first-stage devolatilization reaction to be complete when evolution of the available coal-oxygen is completed. For the results shown in Fig. 23.36, this occurs at about 700°C. The solid product of this first-stage reaction is a semichar consisting primarily of carbon and hydrogen. Under pyrolysis conditions in an inert atmosphere, this semichar undergoes second-stage reaction by evolution of the coal-hydrogen, leading to the formation of relatively unreactive char. In the presence of gaseous hydrogen at elevated pressures, however, the evolution of coal-hydrogen is accompanied by coal-carbon gasification to form methane and ethane. This behavior was interpreted by

[161] G. R. Hill, L. L. Anderson, W. H. Wiser, S. A. Qadar, R. E. Wood, and D. M. Bodily, U.S. Office Coal Res. R & D Rep. No. 18—Final Report (PB 224748); NTIS, 1970.

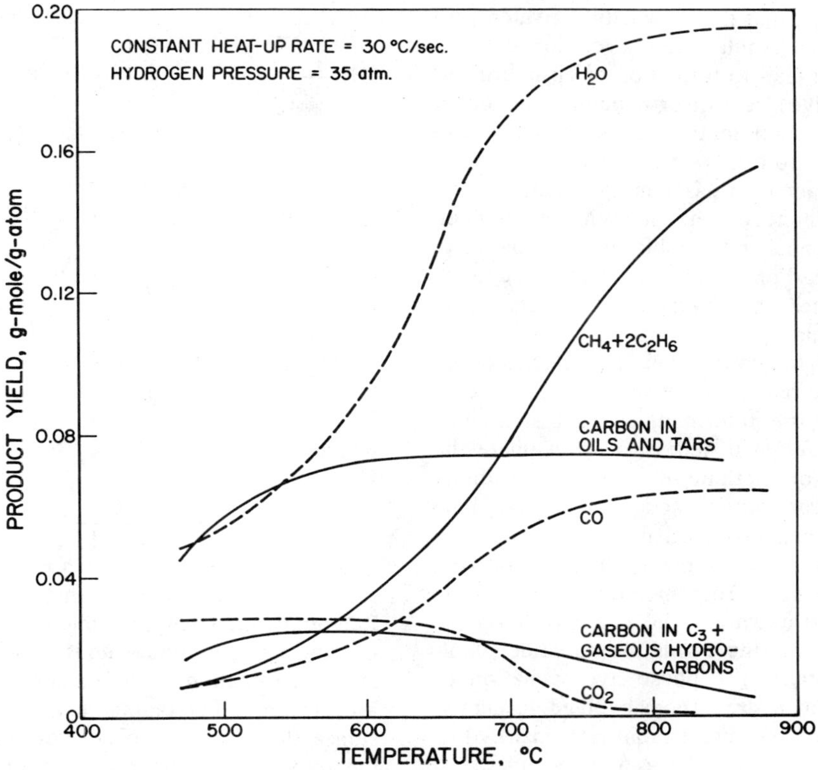

Figure 23.35 Typical product yields for gasified Montana lignite in hydrogen.[153]

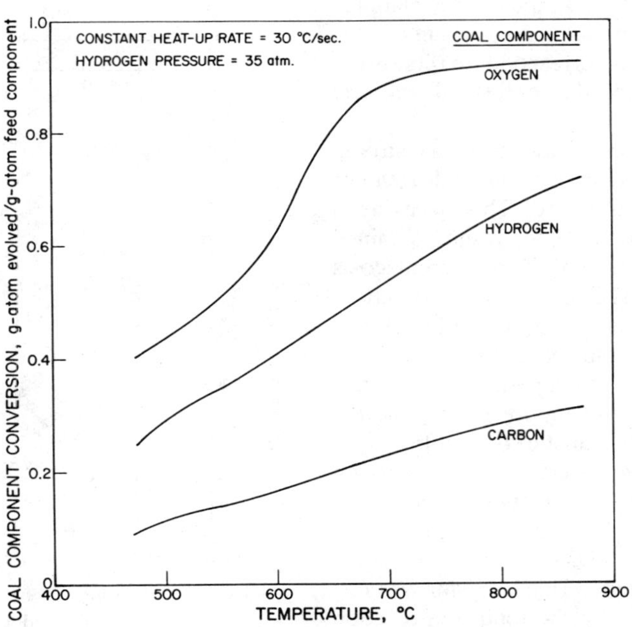

Figure 23.36 Typical coal component conversion during gasification of Montana lignite in hydrogen.[153]

postulating that the thermally activated processes that result in the evolution of coal-hydrogen lead to formation of an intermediate reactive structure or structure component in the carbon matrix. This structure can deactivate to form a stable char structure, or it can react with gaseous hydrogen to form both methane and ethane. When conversion of the semichar to stable char is complete, further carbon gasification occurs at much lower rates characteristic of coal-char gasification kinetics.

To isolate quantitatively the kinetics of the rapid-rate reactions between gaseous hydrogen and the intermediate coal structure, Johnson found it necessary to estimate the fraction of methane-plus-ethane products formed from other reactions. This was done by assuming that total methane-plus-ethane yields obtained in the gasification of coal result primarily from three overall processes: thermally activated coal decomposition reactions occurring during the initial devolatilization stage, gas phase hydrogenation of C_3^+ gaseous hydrocarbons evolved during the first stage, and direct rapid-rate coal hydrogenation occurring during the second stage. With a number of approximations, generalized kinetic correlations were proposed to describe the coal decomposition reactions and the gas-phase hydrogenation reactions as functions of residence time, temperature, and coal type. With these simplified correlations, total observed methane-plus-ethane yields were adjusted to correspond to first-order estimates of yields due only to direct coal hydrogenation.

Adjusted methane yields correlated strongly with evolution of the coal-hydrogen for the different coals tested. This behavior is illustrated in Fig. 23.37 for results obtained from the gasification of four different coals with hydrogen. The stoichiometric relationship between coal-hydrogen evolution and adjusted methane-plus-ethane yields depended only on the hydrogen partial pressure and not on the time-temperature history for the different coals. The kinetics of coal-hydrogen evolution, however, were dependent only on time-temperature history, not on gas composition or pressure, for a given coal. This is illustrated in Fig. 23.38 for results obtained with Montana lignite. This behavior was formally represented by the following reaction scheme:

Semichar — Active Intermediate

$$CH_x^0 \xrightarrow[\text{thermally activated coal-hydrogen evolution}]{k_0} CH_y^* \quad \text{Reaction/A}$$

$$CH_y^* \xrightarrow{k_1} \text{inactive char } (1 - m)CH_y^0$$

Reaction B

$$CH_y^* \xrightarrow[\text{(gaseous hydrogen)}]{k_2} m\left[zCH_4 + \frac{1-z}{2}C_2H_6\right]$$

where

$$m = \frac{k_2}{k_1 + k_2} \qquad (23.72)$$

This model is similar to that proposed by Zahradnik and Glenn, in which it is assumed that the active intermediate is consumed during hydrogenation. It differs somewhat from Moseley and Paterson's model, which assumes that there is no net consumption of the active intermediate in the hydrogenation step and that the intermediate is lost through thermal deactivation.

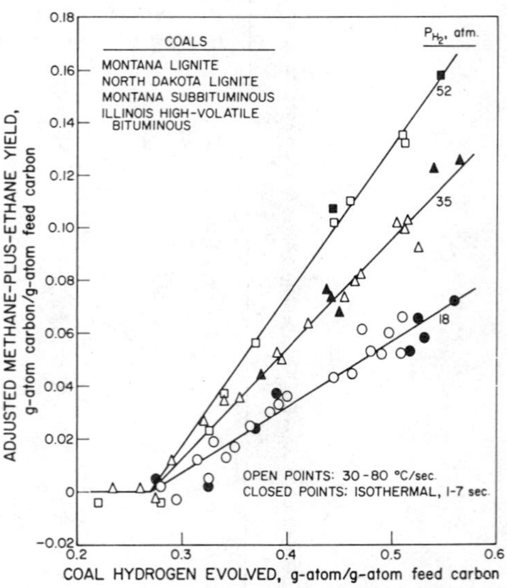

Figure 23.37 Relationship between adjusted methane-plus-ethane yields and coal-hydrogen evolution for different coals.[154]

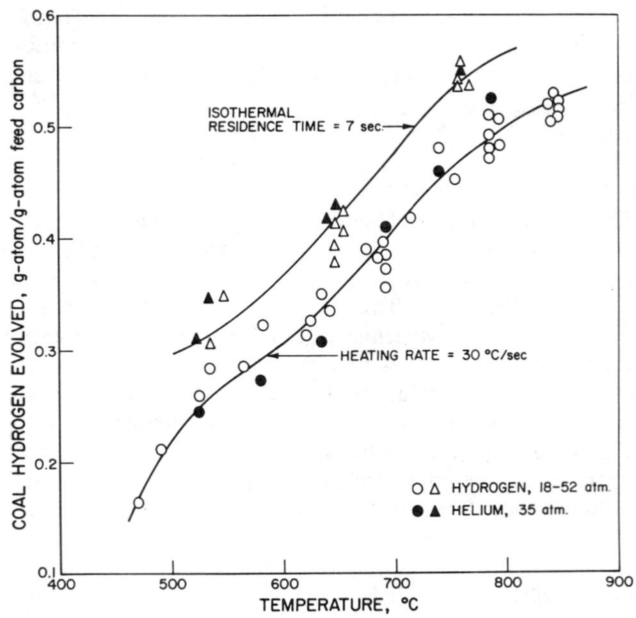

Figure 23.38 Effect of temperature on coal-hydrogen evolution during gasification of Montana lignite in hydrogen and in helium.[155]

Although in describing the rapid-rate degasification step it is important theoretically to distinguish whether the active intermediate is or is not consumed in the hydrogenation reaction, this distinction is difficult to make on the basis of existing experimental evidence, since similar forms of correlations can be derived using either premise to describe gasification behavior under conventional conditions.

A particularly important feature of Johnson's model compared to others is that it specifically characterizes the active intermediate as a structure resulting from the evolution of coal-hydrogen during second-stage devolatilization. This concept was previously proposed by Moseley and Paterson in stating that reactive groupings, which normally yield hydrogen during hydrogasification, are probably the main agents of activation with respect to the rapid-rate methane formation reaction. In Johnson's model the rate of Reaction B was assumed to be very rapid relative to Reaction A, and the ratio k_2/k_1 was assumed to be independent of temperature. This ratio, however, was found to be directly proportional to the hydrogen partial pressure, apparently reflecting an effect of this partial pressure on the hydrogenation step rather than on the deactivation step. With the assumptions made, the total methane-plus-ethane yield from direct hydrogenation, after completion of deactivation of the active intermediate, is independent of the time-temperature history and depends only on hydrogen partial pressure, according to the following relationship:

$$Y''_{CH_4} = \lambda m \qquad (23.73)$$

$$m = \frac{k_2/k_1}{1 + k_2/k_1} \qquad (24.74)$$

$$k_2 = k_2^0 P_{H_2} \qquad (23.75)$$

where λ = total carbon that forms semichar (g-atom/g-atom carbon in feed coal)

Y''_{CH_4} = carbon in methane-plus-ethane formed in rapid-rate hydrogenation after completion of deactivation of the active intermediate (g-atom/g-atom carbon in feed coal)

The form of this expression is very similar to that proposed by Zahradnik and Glenn. Prior to complete deactivation of the active intermediate, the yield of methane-plus-ethane, Y'_{CH_4}, is described by the following more general expression:

$$Y'_{CH_4} = m\lambda f \qquad (23.76)$$

where f = fraction of initial semichar that has been reacted by means of Reaction A

To describe the kinetics of coal-hydrogen evolution that reflect the conversion of semichar to char, Johnson assumed that Reaction A was first order and dependent only on the time-temperature history. This overall reaction was considered to involve thermally activated bond-rupturing processes, with a distribution of activation energies to account for the heterogeneous characteristics of the reacting solids. For simplicity, a continuous uniform distribution of activation energies was assumed, leading to the expression:

$$1 - f = \int_{E_0}^{E_1} f(E)\left[\exp\left(-k_0^0 \int_0^t e^{-E/RT}dt\right)\right]dE \qquad (23.77)$$

where $f(E) = (E_1 - E_0)^{-1}$
$k_0 = k_0^0 e^{-E/RT}$

The parameters in Johnson's model were quantitatively evaluated, based on data obtained from the gasification in hydrogen of a Montana lignite, a Montana subbituminous coal, and a North Dakota lignite, as well as on data reported by Feldmann et al.[119] for the gasification of an Illinois No. 6 high-volatile bituminous coal in methane-hydrogen mixtures. The following evaluations were made to describe results obtained with these coals:

$$k_0^0 = 1.97 \times 10^{10}\ s^{-1}$$

$$E_0 = 40.8\ \text{kcal/g-mole}$$

$$E_1 = 62.9\ \text{kcal/g-mole}$$

$$\lambda = 0.8\ \text{g-atom/g-atom feed carbon}$$

$$\frac{k_2^0}{k_1} = 0.0083\ \text{atm}^{-1}$$

These evaluations describe results obtained at coal-gas heatup rates of 30 and 80°C/s, and at isothermal residence times of 1–7 s.

Johnson also tested the possible applicability of his model to describe rapid-rate methane formation kinetics with a variety of other bituminous coals, based on data reported in the literature that were obtained in investigations employing dilute solid phase reactor systems. These data, obtained over a wide range of conditions, are shown in Fig. 23.39. The values of coal and gas residence times indicated in Fig. 23.39 were reported directly or were deduced from other reported information. Since a significant conversion of feed gases occurred in many of the studies, the hydrogen pressure selected for correlation was the hydrogen partial pressure in the reactor product gases.

In interpreting the data shown in Fig. 23.39, Johnson computed methane-plus-ethane yields corresponding only to direct coal-hydrogen reactions by subtracting the values estimated by his correlations predicting pyrolysis yields and yields from secondary hydrogenation of C_3^+ gaseous hydrocarbons from total methane-plus-ethane yields. Next, it was assumed that the values of Y'_{CH_4} were in agreement with Equation 23.80, $Y'_{CH_4} = m\lambda f$. Values of m were then computed by dividing Y'_{CH_4} by the values of λ and f predicted by his model, in which λ = 0.8, and f was computed from Equation 23.81 for the appropriate temperature and coal residence times reported. The computed values of m, as a function of hydrogen partial pressure, are compared in Fig. 23.40 with Johnson's correlation describing results obtained with subbituminous coals and lignite. In spite of the reasonable consistency shown for hydrogen partial pressures up to 120 atm, Johnson points out that certain of Moseley and Paterson's data at higher pressures, from 200 to 350 atm, show higher values of m than were predicted by the correlation, which suggests that this model may not be applicable in this pressure range.

Most of the data shown in Fig. 23.40 were obtained with gases containing substantial concentrations of methane, carbon oxides, and steam, suggesting that these species have at best only secondary effects on the rapid-rate methane reaction. As discussed previously, this is consistent with the fact that correlations proposed by various investigators to describe this reaction have been expressed in terms of hydrogen partial pressure only, independent of the concentration of other individual gaseous species.

An exception to this is the correlation of Anthony et al.[145] describing the evolution of reactive volatiles, in which the total carbon gasified also depends on total pressure due to the proposed effect of diffusion inhibition.

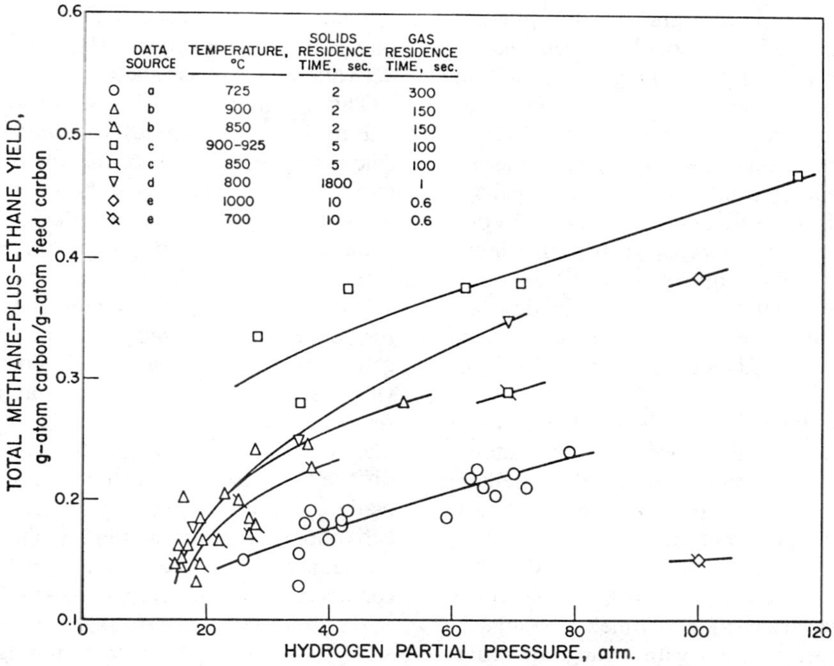

Figure 23.39 Total methane-plus-ethane yields obtained in different studies on gasification of bituminous coals with hydrogen. Data sources: a = Ref. 150, b = Ref. 119, c = Ref. 141, d = Ref. 143, e = Ref. 148.

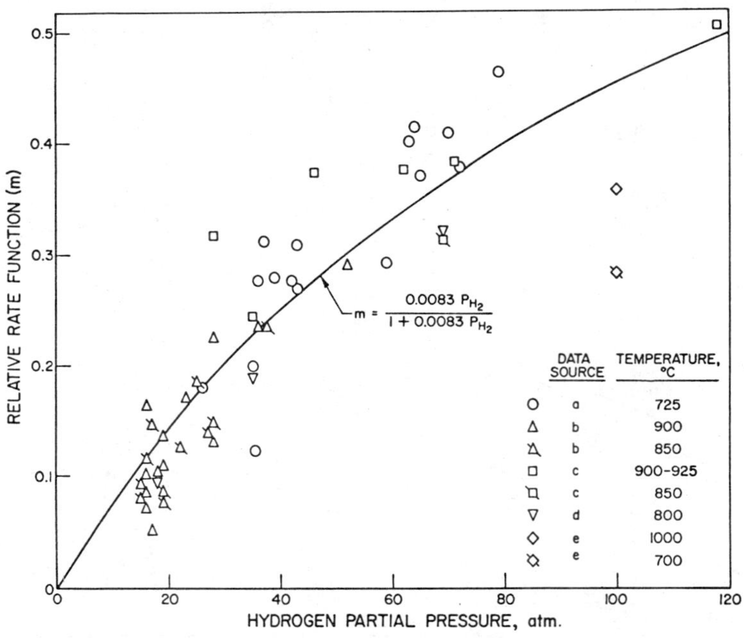

Figure 23.40 Correlation of methane-plus-ethane yields obtained in various investigations of high-volatile bituminous coal gasification in hydrogen. Data sources: a = Ref. 150, b = Ref. 119, c = Ref. 141, d = Ref. 143, e = Ref. 148.

Nevertheless, it appears that for practical purposes, the direct coal-hydrogen reaction is irreversible under the ranges of conditions that have been employed in most investigations. This, of course, does not preclude the possibility of secondary reactions (either homogeneous or catalyzed by reactor walls or by the coal itself) that may alter product methane yields, as reported by some investigators.[160] The contribution of secondary reactions under the complex conditions existing in most of the experimental systems used, however, is an area in need of future study.

The assumption of first-order kinetics with a range of activation energies to characterize the controlling step in rapid-rate methane formation appears to be a convenient technique to approximate the complex phenomena that must actually be involved in this reaction. It is important to note, however, that in representing experimental results by a first-order reaction with a range of activation energies, caution should be exercised in interpreting the fundamental processes based on numerical evaluation of the kinetic parameters. For example, in describing the kinetics of coal-hydrogen evolution by a first-order reaction, with a constant activation energy distribution ranging from E_0 to E_1, Johnson has shown the existence of a large covariance between the preexponential factor and the activation energy parameters.[154,155]

This is illustrated in Fig. 23.41, where variations of the values of conversion, *f*, calculated using values of the kinetic parameters recommended by Johnson are compared to values of *f* calculated using a significantly distinct set of kinetic parameters. Nevertheless, very similar characteristics result over a wide range of temperatures and residence times. A much greater range of estimates of kinetic parameters could result when describing data obtained at a single residence time over a wide range of temperatures. It is thus apparent that significantly different numerical evaluations could be made by different investigators assuming an activation energy distribution to characterize the same first-order reaction, even if the data considered are apparently consistent.

As discussed previously, Anthony et al.[144–146] proposed that a rate-limiting step in the initial stage of coal hydrogasification is the formation of reactive volatiles—a first-order reaction with a distribution of activation energies—within the coal structure. These investigators represented the activation energy distribution by a function equivalent to the high-energy tail of a Gaussian

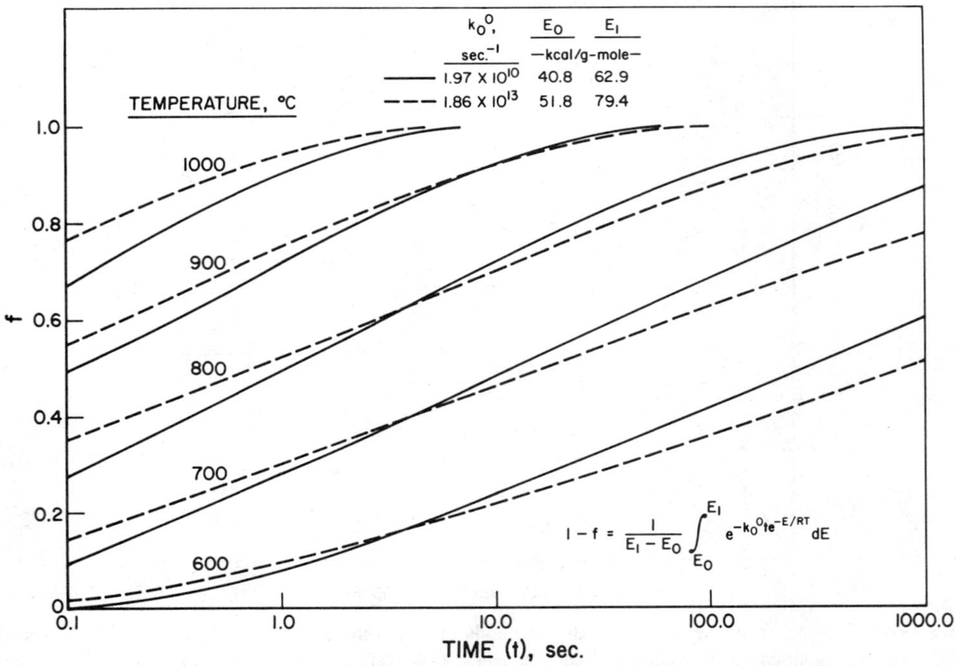

Figure 23.41 Comparison of model characteristics with different sets of kinetic parameters.

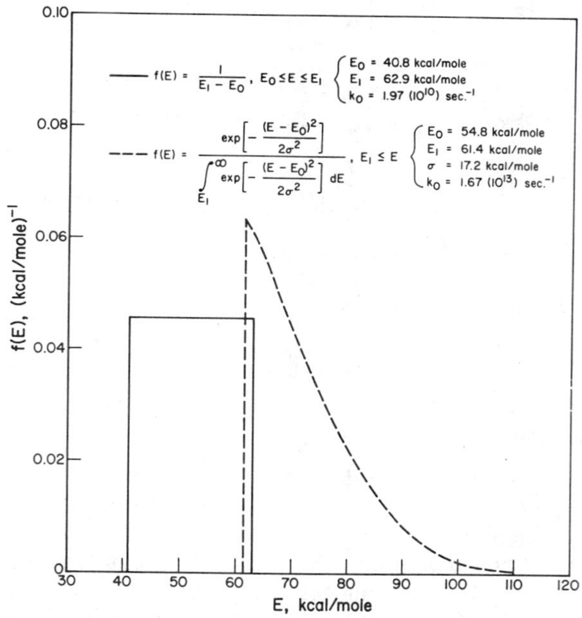

Figure 23.42 Comparison of activation energy distribution functions.

distribution. In Fig. 23.42 this unusual function is compared to the constant distribution function proposed by Johnson to describe a possibly equivalent reaction step, the evolution of coal-hydrogen.

Because of differences in the empirical forms used by these investigators, in addition to differences in the activation energy ranges, it is of interest to compare the predictions of the models proposed to describe carbon gasification yields during coal hydrogasification. The model of Anthony et al. (Anthony's model) describes coal weight loss due to the evolution of reactive volatiles and direct coal-hydrogen reaction as a function of hydrogen pressure and time-temperature history. If it is assumed that 95% of this weight loss is due to carbon gasification and that the feed coal has a carbon content of 70%, then predicted weight losses can be translated into carbon conversion fractions, not including the carbon in the nonreactive volatiles gasified. If it is further assumed that this carbon conversion fraction corresponds to rapid-rate methane formation, then the predictions of Anthony's model can be directly compared to the predictions of Johnson's model describing methane formation due solely to coal-hydrogen reaction.

This comparison is shown in Fig. 23.43 as a function of residence time for gasification in hydrogen at 20 and 50 atm and at 900°C. Although residence times up to 100 s are shown, the data in both investigations considered were generally obtained at coal residence times less than about 20 s. Nevertheless, both models predict that relatively little conversion occurs after 20 s at 900°C. In addition, the magnitude of conversion predicted by the two models is similar, particularly at the lower hydrogen pressure, in spite of the differences in the models and correlations applied. This is also true at other temperatures, as shown in Fig. 23.44, which compares the predictions as a function of temperature at a residence time of 10 s.

The controlling reaction stages in the models proposed by Anthony and by Johnson are characterized by relatively high activation energies, greater than 40 kcal/g-mole. In other investigations simpler kinetic models have been proposed to describe rapid-rate methane formation, which have led to relatively low activation energies. Feldmann,[119] for example, proposed a model in which rapid-rate methane formation rates are directly proportional to the hydrogen partial pressure and to the concentration of solid carbon participating in the reaction. This model, like those proposed by Moseley and Paterson and by Zahradnik and Glenn, assumes that at elevated temperatures devolatilization occurs essentially instantaneously, followed by the slower but still rapid reaction

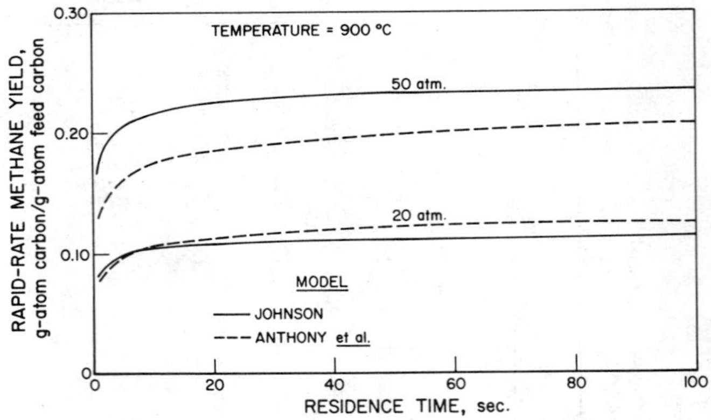

Figure 23.43 Comparison of predicted methane yields as a function of residence time and hydrogen pressure, based on different models.[144–146,153–155]

of hydrogen and coal to form methane. Feldmann's model does not, however, consider deactivation of the reactant intermediate.

In analyzing data obtained at the U.S. Bureau of Mines using Feldmann's model (characterized by a single first-order rate constant), Gray et al.[162] have obtained an activation energy of 15.1 kcal/g-mole. A similarly low activation energy of 15 kcal/g-mole was determined by Zahradnik and Glenn for coal hydrogenation in a dilute phase flow reactor. Although these relatively low activation energies have been interpreted as representing the difference between activation energies for gasification and deactivation of reactive intermediates, or as indicating the effects of mass transfer limitations, such values are not inconsistent with the much higher activation energies proposed in the models that employ ranges of activation energies. If, for example, values of f predicted from Johnson's model at a single residence time are plotted against reciprocal temperature, the slope of the line can be interpreted as an overall activation energy according to the expressions:

$$\frac{df}{dt} = k(1 - f) \tag{23.78}$$

$$k = k^0 e^{-E/RT} \tag{23.79}$$

$$-\ln(1 - f) = kt = k^0 t e^{-E/RT} \tag{23.80}$$

[162] J. A. Gray, P. J. Donatelli, and P. M. Yavorsky, *Prepr., Am. Chem. Soc., Div. Fuel Chem.*, **20**(4), 103–145 (1975).

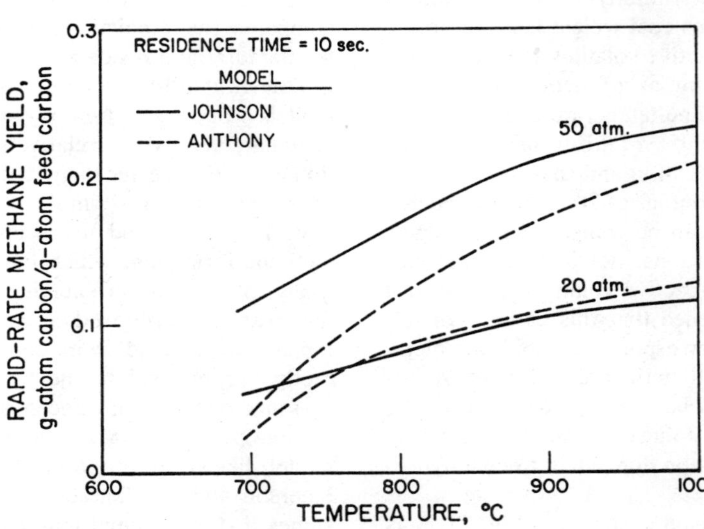

Figure 23.44 Comparison of predicted methane yields as a function of temperature and hydrogen pressure, based on different kinetic models.[144–146,153–155]

$$\ln[-\ln(1 - f)] = \ln k^0 t - \frac{E}{RT} \qquad (23.81)$$

Such Arrhenius plots are shown in Fig. 23.45 for residence times of 1, 10, and 100 s, resulting in activation energies ranging from 13.7 to 15.6 kcal/g-mole. This range of values is quite consistent with the apparent activation energies reported by Zahradnik and Glenn and by Gray, based on data obtained at similar coal residence times.

23.4.2 Coal Char Gasification Kinetics

Although the initial coal gasification stage is completed in seconds or less at elevated temperatures, the subsequent gasification of the coal chars produced during this initial reaction stage is much slower, requiring minutes or hours to obtain significant conversions under practical conditions. Since reactor design criteria and volume requirements for commercial gasifiers are largely dependent on coal char reactivity, the kinetics of coal char gasification systems have been extensively investigated.

One advantage of studying coal char gasification kinetics, compared to initial coal reaction kinetics, is that it is relatively easy to conduct experiments under conditions of differential gas conversion, whereby the composition of gas in contact with the coal char is known and remains constant. This greatly facilitates kinetic analysis of the experimental results. This procedure has usually been applied to fixed-bed flowing gas systems, which have been used in the majority of experimental studies to obtain detailed kinetic information about gasification kinetics of coal chars.

A few studies have also begun to use fluidized beds to obtain differential gasification rates, although this technique has been criticized because of the potential heterogeneous characteristics of such a system.[3] The fluidized bed, however, does have the advantage of good circulation of solids, leading to uniform bed temperatures. With fixed beds, temperature control is a greater problem because of the strong heat effects that can occur during coal char gasification, an effect that can lead, when undefined, to apparently inconsistent results obtained in different experimental studies. Dilute phase solid transport systems have not been employed to study coal char reactions because of the negligible carbon conversions that would occur during the short residence times for solids characteristic of such systems.

Other complications that must be considered in interpreting data obtained from coal char gasification studies are related not only to the different coals used but also to differences in the procedures employed in preparing chars from these coals. For example, the term "coal char" is usually defined as the residue resulting from any heat treatment of coal. If, however, the pretreatment temperature and residence times are sufficiently low, the resultant coal char can exhibit some rapid-rate methane formation reactivity during subsequent gasification. In the discussions that follow the term "coal char" is intended to refer primarily to residues that have been subjected to preparation conditions sufficiently severe that negligible rapid-rate activity remains. As will be discussed, even with this limitation, pretreatment conditions can still affect the subsequent reactivity of coal chars during gasification.

The following coal char gasification reactions have received the greatest attention in recent years because of the importance of

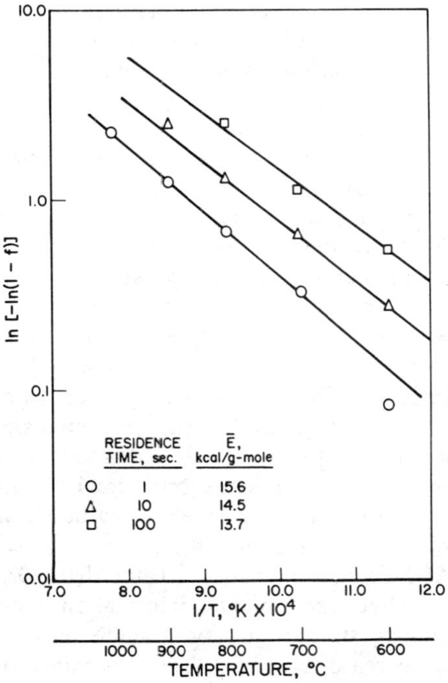

Figure 23.45 Arrhenius plots for the constant residence time conversion function.

developing commercial processes to convert coal to high- or low-energy gases:

$$C + 2H_2 <==> CH_4 \qquad (23.1)$$

$$C + CO_2 <==> 2CO \qquad (23.3)$$

$$C + H_2O <==> CO + H_2 \qquad (23.82)$$

The carbon-hydrogen reaction (Equation 23.1) has been studied almost solely at elevated pressures because of the strongly favorable effect of pressure on this reaction and the almost negligible rates obtained at atmospheric pressure. This reaction is particularly important to the production of high-energy gas, whereas the reactions given by Equations 23.3 and 23.82 are important to the production of both high- and low-energy gases. Although Equations 23.1 and 23.3 can each occur alone in gasification systems, roughly according to the stoichiometry indicated, the steam-carbon reaction (Equation 23.82) cannot generally occur alone under conditions of integral gas conversion. Because of the water-gas shift reaction:

$$H_2O + CO <==> CO_2 + H_2 \qquad (23.4)$$

carbon dioxide can be formed even with a pure steam feed gas and can react competitively with the steam-carbon reaction according to Equation 23.3 at low or high pressures. Since hydrogen is a product of the steam-carbon reaction, as well as the water-gas shift reaction, methane can also be formed in steam-carbon systems, at sufficiently high pressures, by a reaction such as Equation 23.1.

The representations given by Equations 23.1, 23.3, and 23.82 are oversimplified for the purpose of illustration, since the hydrogen present in coal chars will alter reaction stoichiometry in real systems. Furthermore, the reactions describe overall stoichiometry in limiting cases only and do not imply fundamental reaction steps. In gases containing both steam and hydrogen, for example, methane could conceivably be formed by the direct interaction of carbon, steam, and hydrogen by means of a reaction such as:

$$2C + H_2 + H_2O <==> CO + CH_4 \qquad (23.83)$$

A variety of other reactions can also be postulated to occur in coal char gasification systems, which, in the absence of molecular oxygen, consist of some stoichiometric combination of Equations 23.1, 23.3, and 23.82. In complex gasification systems containing steam, hydrogen, and carbon oxides, however, most experiments have determined overall reaction stoichiometry only. Although individual reactions have been variously postulated to rationalize this type of result to aid in kinetic formulations, such postulated reaction schemes are likely to be more arbitrary than those that could be obtained with more sophisticated techniques, such as the use of isotopic tracers or reactions with atomic gaseous species.

Gasification reactions have usually been studied under isothermal conditions, often with differential gas conversion. To a lesser extent, experiments have also been made under conditions of constant heatup rate in order to obtain information on temperature effects in a single experiment. In most studies, the carbon gasification rate, W, has been defined by the expression:

$$W = \frac{dX/dt}{1 - X} \qquad (23.84)$$

where X = carbon conversion fraction initially present (moles/mole carbon initially present)
t = time

Using this expression, in which W is sometimes referred to as the "specific" carbon gasification rate, implies that if absolute carbon gasification rates (moles of carbon gasified per unit time) under constant environmental conditions are proportional to the amount of carbon present at any time, then W will remain constant with increasing conversion, X. This has frequently been found to be true, at least for carbon conversion fractions up to about 80% (although in a number of studies some transient behavior has been observed at levels of carbon conversion less than about 10%).

At carbon conversions greater than 80%, the limited experimental information available suggests that specific gasification rates may decrease somewhat. The approximation of relatively constant specific gasification rates over a large range of carbon conversions is at best applicable only to the gasification of chars in the absence of significant catalytic effects from natural or added impurities. The activity of catalysts present can vary substantially during the course of gasification, even under constant environmental conditions, leading to significant variations in specific gasification rates. Devolatilization of fixed beds of bituminous coals can also subsequently lead to variations in the specific gasification rates of chars formed, due to the agglomerating properties of this type of coal. Agglomeration of coal particles can occur during devolatilization, leading to the possible inhibition of gaseous diffusion during the subsequent gasification. When this occurs, initial specific gasification rates are relatively low, but they increase with increasing conversion as restrictions are removed and the internal char surface becomes more available to the reacting atmosphere.

The following discussion of experimental work on char gasification focuses on three main reaction systems:

1. The char-hydrogen-methane system, which is relevant to the reaction given by Equation 23.1.
2. The char–carbon dioxide–carbon monoxide system, relevant to Equation 23.3.
3. The char–hydrogen–steam–carbon monoxide–carbon dioxide–methane system, relevant to all the gasification reactions considered, including the steam-carbon reaction (Equation 23.82).

Although this discussion focuses primarily on the behavior of coal chars, pertinent results obtained with other carbon solids are also considered.

23.4.2.1 *The Coal Char–Hydrogen–Methane System*

For the gasification of coal chars and other carbons in pure hydrogen, specific carbon gasification rates have been correlated with two main forms of equation:

$$W = \frac{k_H P_{H_2}^2}{1 + K_H P_{H_2}} \qquad (23.85)$$

(see Refs. 7, 137, 159, 163, 164), or

$$W = k_H^* P_{H_2} \qquad (23.86)$$

(see Refs. 137, 164–169)

where k_H, K_H, k_H^* = kinetic parameters that depend on temperature and the nature of the char, often expressed by an Arrhenius form of expression

P_{H_2} = hydrogen partial pressure

At sufficiently high hydrogen partial pressures, Equation 23.85 reduces to Equation 23.86, with:

$$k_H^* = \frac{k_H}{K_H} \qquad (23.87)$$

Based on either of these correlation forms, gasification rates increase significantly with increasing hydrogen partial pressure.

J. D. Blackwood and co-workers have proposed a mechanism to rationalize these correlation forms, based on results of several fixed-bed studies conducted from 1959 to 1967, using coconut char, wood char, brown coal, and Australian bituminous coal chars.[137,164–167] The following fundamental reaction steps are assumed:

Step 1. $C_f + H_2 \underset{k_1'}{\overset{k_1}{\rightleftarrows}} C(H_2)$

Step 2. $C(H_2) + H_2 \underset{k_2'}{\overset{k_2}{\rightleftarrows}} CH_4 + C_f$

[163] C. W. Zielke and E. Gorin, *Ind. Eng. Chem.*, **47**, 820–825 (1955).
[164] S. Sastry, M. M. Mallikarjunan, and R. Vaidyeswaran, *Brennstoff-Chemie*, **50**, 363–367 (1969).
[165] J. D. Blackwood, D. J. McCarthy, and B. D. Cullis, *Aust. J. Chem.*, **20**, 2525–2528 (1967).
[166] J. D. Blackwood, B. D. Cullis, and D. J. McCarthy, *Aust. J. Chem.*, **20**, 1561–1570 (1967).
[167] J. D. Blackwood, *Aust. J. Chem.*, **12**, 14–28 (1959).
[168] A. Tomita, O. P. Mahajan, and P. L. Walker, Jr., *Prepr., Am. Chem. Soc., Div. Fuel Chem.*, **20**(3), 99–102 (1975).
[169] F. Moseley and D. Paterson, *J. Inst. Fuel*, **38**, 13–23 (1965).

where C_f = active center on carbon surface
$C(H_2)$ = surface complex formed by the chemisorption of hydrogen or by decomposition of methane at an active center

According to this mechanism, gaseous hydrogen chemisorbed at an active center on the carbon surface can react with hydrogen to form methane and to regenerate an active center. The individual reaction steps are reversible. Under steady state conditions, the following rate equation can be derived based on steps 1 and 2:

$$W = \frac{dX/dt}{1-X} = \frac{n_T \dfrac{k_1 k_2}{k_1'} P_{H_2}^2 \left(1 - \dfrac{k_1' k_2' P_{CH_4}}{k_1 k_2 P_{H_2}^2}\right)}{1 + \dfrac{k_1 + k_2}{k_1'} P_{H_2} + \dfrac{k_2'}{k_1'} P_{CH_4}} \tag{23.88}$$

where n_T = total number of active centers per mass of carbon

Blackwood and McCarthy suggest that variations in the total number of active centers, n_T, contribute primarily to differences in the reactivity of different coal chars, implying that the relative behavior of different coal chars should be similar.[137] He has also considered a different set of reaction steps in place of step 1 to describe the hydrogen chemisorption process:[170]

Step 3. $C_f + H_2 \underset{k_3'}{\overset{k_3}{\rightleftharpoons}} (CH)\,(H)$

Step 4. $(CH)(H) \underset{k_4'}{\overset{k_4}{\rightleftharpoons}} C(H)_2$

Steps 3 and 4 are based on the postulate that the main function of active sites on the carbon surface is to split the hydrogen molecule, so that one hydrogen atom becomes attached to the active center and the other hydrogen atom reacts either with the original active site or with another active site to form a —CH_2 group. The formation of hydrogen atoms in the chemisorption of hydrogen on carbon surfaces was proposed in 1935 by Barrer,[171] who suggested that these atoms are chemisorbed on adjacent dual sites. It was further shown that the chemisorption and desorption of hydrogen at a carbon surface were first-order processes that could be expressed by a Langmuir type of adsorption isotherm. However Blackwood argues that if hydrogen is adsorbed on dual sites, the desorption would be a second-order process, contrary to experimental results, unless the fraction of surface covered by the two hydrogen atoms could be regarded as fixed—no longer available for chemisorption but able to be reoccupied by the breakdown of a hydrogen molecule.

This requirement for the formation of surface hydrogen atoms as a prerequisite to the formation of methane is supported, although not necessarily proved, by experiments in which various carbons have been reacted directly with hydrogen atoms produced in microwave or radiofrequency fields. Such studies, usually conducted at pressures less than 1 mm Hg and at temperatures below 300°C, have shown gasification rates equivalent to those obtained in molecular hydrogen at elevated pressures and at temperatures several hundred degrees higher.

Although Blackwood had earlier favored the mechanism given by steps 3 and 4 as an explanation for the formation of —CH_2 groups on the carbon surface,[170] later he suggested that step 1 was a more likely path[137] and also proposed alternative explanations to characterize the active sites responsible for the chemisorption step. In earlier studies he found a strong correlation between gasification rate and the concentration of residual oxygen in the carbon.[167] He suggested that this oxygen, combined as chromene or benzpyran structures, was responsible for activating certain sites in the carbon crystallite. Blackwood and McCarthy later suggested that some atoms at the edges of or at defects in the carbon structure may be chemically unsaturated, able to act as centers for initiating molecular dissociation or redox reactions, and that they represent free or unpaired electrons.[137]

Zielke and Gorin[163] have proposed a somewhat different reaction mechanism, which

[170] J. D. Blackwood, *Aust. J. Chem.*, **15**, 397–408 (1962).

[171] R. M. Barrer, *Proc. R. Soc. (London)*, **A149**, 253–269 (1935).

they used to develop a kinetic correlation in the form of Equation 23.85 to describe gasification of a Disco char, derived from commercial processing of a Pittsburgh Seam bituminous coal. This mechanism is based on the hydrogenation of exposed edges in the carbon lattice that contain hydrogen atoms. The first step proposed is:

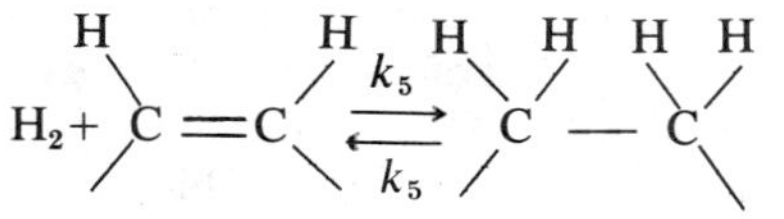

(23.89)

which is similar to the reaction given by step 1. Subsequent reactions in the model involve the progressive hydrogenation of the —CH_2 groups to —CH_3, and then to CH_4, with the regeneration of a new

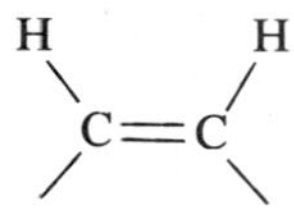

grouping for each one consumed. In deriving Equation 23.85, it was assumed that the hydrogenation of the —CH_3 group, which is effectively equivalent to the reaction given by step 2, is rapid compared to other reaction steps.

Based on existing evidence, Equation 23.88 appears to be the best general kinetic representation of experimental results currently available for steady state gasification conditions where specific gasification rates are relatively constant during the course of conversion. Some simplification of this expression, however, has been assumed in various applications. The $(k_2'/k_1')P_{CH_4}$ term in the denominator is usually either ignored or assumed to be negligible compared to other terms. In addition, the $(k_1 + k_2)/k_1'$ term in the denominator has been found to be suitably described by a single Arrhenius expression to account for temperature effects. Blackwood and McCarthy have suggested that this probably indicates that the rate constant k_2 is significantly larger than k_1 under most conditions.[116] They further suggest that under most practical gasification conditions, the term $k_2 P_{H_2}/k_1'$ is large compared to unity, and that the rate expression reduces to:

$$W = \frac{dX/dt}{1 - X}$$

$$= n_T(k_1)P_{H_2}\left(1 - \frac{k_1'k_2'}{k_1k_2}\frac{P_{CH_4}}{P_{H_2}^2}\right) \quad (23.90)$$

This in turn reduces to Equation 23.86 for the gasification in pure hydrogen, where

$$k_H^* = n_T k_1 \quad (23.91)$$

Other investigators, however, have found reaction orders between 1 and 2 with respect to hydrogen pressure, at pressures greater than 1 atm, that were best described by Equation 23.85, at least for pressures less than 20 atm and temperatures up to 900°C.[7,159,163,164] Johnson[7] correlated hydrogen gasification data for air-pretreated bituminous coal char with an expression of the form given by Equation 23.85, with the following evaluations of the parameters k_H and K_H as a function of temperature:

$$k_H = \exp\left[2.6741 - \frac{27170}{RT}\right] \quad (23.92)$$

$$K_H = \exp\left[-10.452 + \frac{22050}{RT}\right] \quad (23.93)$$

where the units are as follows:

k_H min^{-1} atm^{-2}
K_H atm^{-1}
R cal-mole^{-1}-K^{-1}
T kelvins

The evaluation above of k_H is restricted to bituminous coal chars prepared at the same temperature as the subsequent gasification in hydrogen; it is based on results obtained at temperatures from 800 to 1050°C with hydrogen pressures up to 70 atm.

Figure 23.46 compares the effect of hydrogen pressure on specific gasification rates, as predicted by Johnson's correlation at 927°C,[7] to rates obtained by Zielke and Gorin[163] for the gasification of Disco char and to rates interpreted from Blackwood's data for the gasification of brown coal char.[137] In representing Blackwood's results, obtained at 750, 850, and 950°C and at 25 atm pressure, it was assumed that the gasification rates are directly proportional to hydrogen pressure (Equation 23.86) and that the activation en-

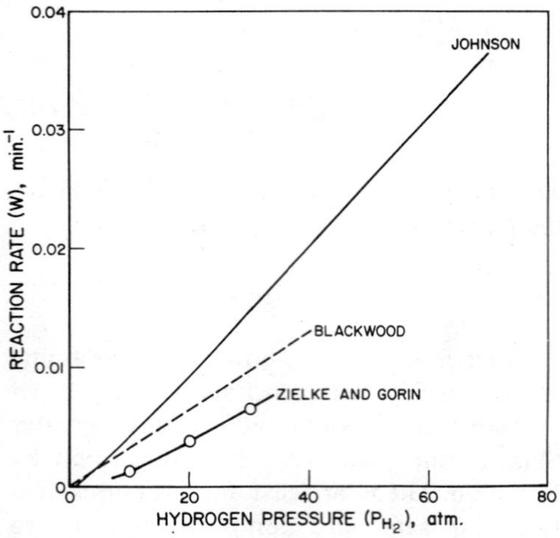

Figure 23.46 Comparison of hydrogen pressure effects on specific gasification rates.[7,137,163]

ergy for the rate constant k_H^* is about 30 kcal/mole, as suggested by these investigators. Although the different curves represent different reactivity levels for the coal chars used, it can be seen that the nonlinearity exhibited by Johnson's correlation and Zielke and Gorin's data is relatively unimportant, except at low pressure, and that a reasonable approximation of the reactivities of different coal chars can be made by assuming the form of correlation given in Equation 23.86.

Figure 23.47 presents a more general comparison of rate constants k_H^* determined from data obtained or correlations proposed in several investigations of the char-hydrogen reaction. Pertinent information relevant to these investigations is given in Table 23.14. The range of values of k_H^* in Fig. 23.47, from Johnson's correlation at a given temperature, exists because his correlation is of the form given in Equation 23.85 and therefore:

[172] J. D. Blackwood, *Aust. J. Appl. Sci.*, **13**, 199–206 (1962).

[173] H. Jűntgen, K. H. Van Heek, and J. Klein, paper presented at Gordon Res. Conf. Coal Sci., New Hampton School, New Hampton, N.H., July 1973.

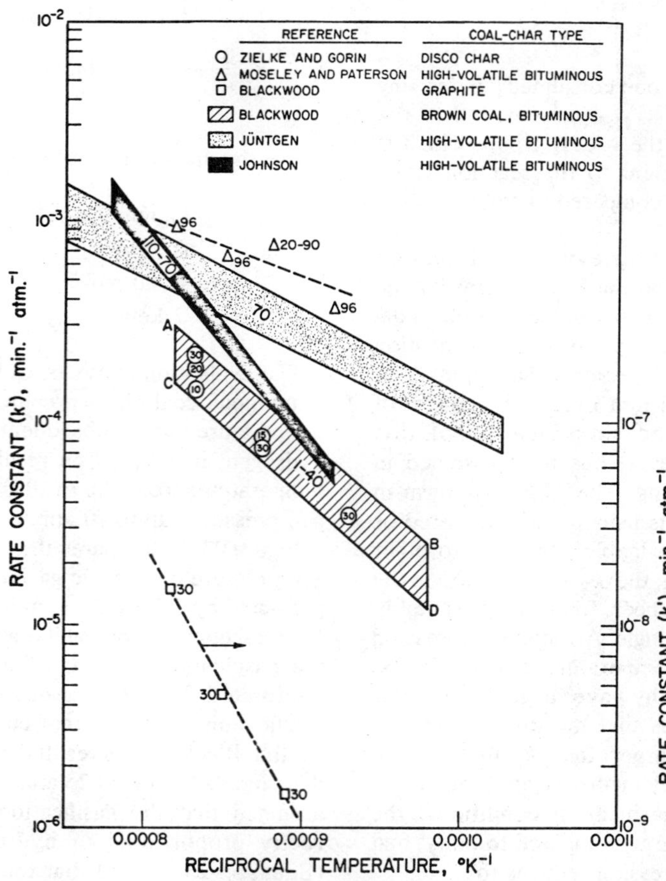

Figure 23.47 Comparison of rate constants for char gasification in hydrogen.[7,137,140,163,165,167,172,173]

Table 23.14 Conditions for Investigations of the Char-Hydrogen Reaction

Coal Type	Primary Particle Diameter Range (μm)	Temperature Range (°C)	Pressure Range (atm)	Maximum Carbon Conversion Fraction	Char Preparation Conditions	Ref.
High-volatile bituminous	150–300	815–950	20–100	~0.3	700°C N_2	169
High-volatile bituminous	—	Ambient–1100°C[a]	70	~0.90	—[a]	173
High-volatile bituminous	420–840	815–1040	10–70	0.90	Reaction temp N_2	7
Disco char	150–230	815–930	10–30	0.80	Commercial	163
Graphite (reactor grade)	1100–2200	850–950	30	—	Commercial	172
Brown coal	300–760	750–950	25	~0.5	750°C N_2	137
Brown coal	300–760	750–950	25	—	Reaction temp N_2	166
Bituminous	1100–2200	750–950	20	—	Reaction temp N_2	165
Coconut char	1100–2200	650–870	2–40	0.10	950°C N_2	167

[a] Coal heated at constant rate of 10°C/min in hydrogen.

$$k_H^* = \frac{k_H P_{H_2}}{1 + K_H P_{H_2}} \qquad (23.94)$$

This range of values tends to decrease with decreasing temperature, since K_H increases with decreasing temperature.

The range of values indicated for Blackwood's results reflects the use of different carbon solids including wood char, coconut char, brown coal char, and bituminous coal chars. Rate constants obtained with these different materials vary by a factor of approximately 2 at a given temperature and are generally less than values predicted by Johnson's correlation at temperatures greater than about 850°C. This difference may be partly due to the effects of coal rank on the reactivity of coal chars for gasification with hydrogen, as shown for an extreme case in Fig. 23.47 for results obtained with graphite. Values of k_H^* given in Fig. 23.47 from Jüntgen et al.[173] were based on correlations derived for the gasification in hydrogen of a German high-volatile bituminous coal in hydrogen exposed to a constant heatup rate (10°C/min). The range of values shown corresponds to two different consecutive reaction stages of char gasification postulated by these authors, with a different rate constant applicable to each stage.

Moseley and Paterson's[169] values of k_H^* were determined for the gasification of a British noncaking high-volatile bituminous coal char in a unique experimental system. In this system a horizontal bed of coal char, through which hydrogen was continually flowing, was repeatedly and rapidly heated to reaction temperature (~15°C/s), maintained at temperature for 15 s, then rapidly cooled by water quenching of the reactor tube. Reaction rate determinations were made assuming a 15-s residence time at temperature. The values of k_H^* obtained with this procedure are higher than any other values shown in Fig. 23.47. A possible explanation for the apparently high values of k_H^* obtained by Moseley and Paterson as well as those reported by Jüntgen et al. for temperatures below 950°C is discussed subsequently.

The effect of methane partial pressure on the gasification kinetics of coal chars in hydrogen-methane mixtures is important in any real system. Neglecting the denominator term containing methane partial pressure in Equation 23.88, the effect of methane on gasification rates appears in a term that represents thermodynamic reversibility. In this sense, the term

$$K_{1,2} \left(= \frac{k_1' k_2'}{k_1 k_2} \right)$$

can be considered an equilibrium constant for the coal char–hydrogen-methane system. Significantly, and perhaps fortuitously, different investigators, including Zielke and

Gorin,[163] Johnson,[7] and Blackwood and McCarthy,[137] have successfully correlated experimental results obtained with coal chars by evaluating $K_{1,2}$ as the equilibrium constant for the graphite-hydrogen-methane system. Blackwood therefore suggested that this equilibrium constant is not affected by char type. Although others have reported results obtained from the gasification of coal chars in gases containing steam as well as hydrogen that indicate methane concentrations in excess of values computed from the graphite-hydrogen-methane equilibrium constant,[3] such results do not necessarily indicate excess carbon-free energies, since competitive reactions involving steam can lead to methane formation. Also, with integral systems that use coal rather than coal char as the feed solids, methane can result from pyrolysis as well as rapid-rate methane formation, both of which are irreversible processes.

Effects of Coal Char Preparation Temperatures. There are two major causes of variations in the reactivity of different coal chars during gasification with hydrogen. One relates to the inherent properties of the coals from which the chars were derived, and the other to the conditions employed in preparing the chars, particularly temperature. It has usually been found that char reactivity decreases with increasing preparation temperature, relative to the temperature of subsequent gasification in hydrogen.

Blackwood et al.[165] have conducted a systematic study to characterize quantitatively the effect of preparation temperature on the reactivity of brown coal chars gasified in hydrogen at 25 atm. Results of this study are shown in Fig. 23.48, which plots values of the rate constant k_H^* (defined in Equation 23.86) as a function of preparation and gasification temperatures. In no case shown in Fig. 23.48, in the range of gasification temperatures from 700 to 950°C and char preparation temperatures from 750 to 950°C, was the char gasified at a temperature higher than its temperature of preparation. Values of k_H^* shown in Fig. 23.48 increase systematically with decreasing preparation temperature and at constant gasification temperature, and they increase with increasing gasification temperature at constant pretreatment temperatures.

Blackwood et al. correlated these results with the following expression:

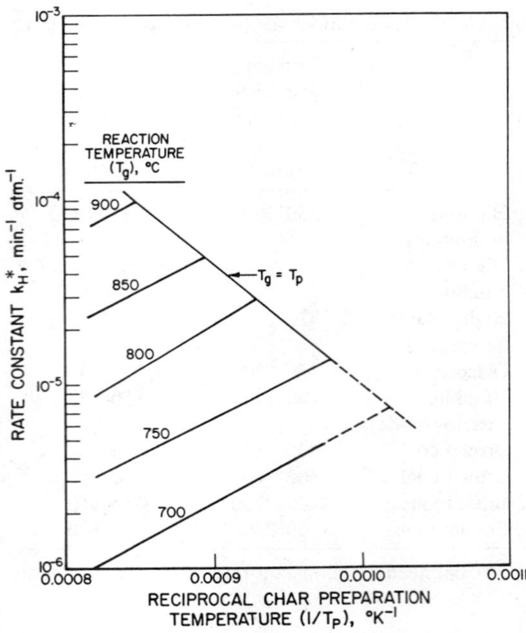

Figure 23.48 Effect of char preparation temperature on rate constant for gasification in hydrogen.[162]

$$k_H^* = k_{HO}^* \exp\left[\frac{E_P}{RT_P}\right] \exp\left[-\frac{E_g}{RT_g}\right] \quad (23.95)$$

where T_g = gasification temperature
T_P = preparation temperature ($T_P \geqslant T_g$)
k_{HO}^*, E_P, E_g = constants

For preparation temperatures equal to or less than the gasification temperatures, the value of T_P used in Equation 23.95 is equal to the gasification temperature, T_g. In rationalizing these results, Blackwood et al. proposed that the term exp $[E_P/T_P]$ reflects the effect of increasing temperature in decreasing the reactivity by the loss of free electrons due to condensation of the carbon structure into more stable condensed-ring systems, and that previously unstabilized bonds are stabilized in the process. This may be brought about partly through the loss of char hydrogen, which exposes free valences that act as electron traps. This process is thought to be irreversible and to be solely a function of temperature, independent of time after a short period (possibly on the order of seconds). A value of E_P = 20 kcal/mole was determined by Blackwood et al. to describe the results given in Fig. 23.48.

For gasification under conditions where $T_g = T_P$, Blackwood et al. previously determined a total activation energy for k_H^* of 30 kcal/mole for different coal chars. From Equation 23.95, then:

$$E_g - E_P = 30 \text{ kcal/mole} \qquad (23.96)$$

and

$$E_P = 20 \text{ kcal/mole} \qquad (23.97)$$

so that

$$E_g = 50 \text{ kcal/mole} \qquad (23.98)$$

Based on Equation 23.90, the value of E_g corresponds to the activation energy of the rate constant k_1, whereas the value of E_P reflects the temperature coefficient pertinent to the concentration of active sites defined by the term n_T.

Johnson[7] has also examined the effect of pretreatment temperatures from 927 to 1100°C on the rates of gasification of high-volatile bituminous coal char in hydrogen at 927°C and 35 atm. Johnson interpreted his results in terms of Equation 23.85 and attributed pretreatment effects to decreases in the rate constant k_H. Assuming that Equation 23.88 is applicable, k_H can generally be expressed by:

$$k_H = k_1 n_T \frac{k_2}{k_1} \qquad (23.99)$$

Johnson assumed that the variations in k_H with pretreatment temperatures were due to variations in n_T and employing the same correlation form used by Blackwood (Equation 23.95), determined a value of $E_P = 9.340$ kcal/mole, about one-half the value determined by Blackwood for brown coal char. According to Johnson's correlation, the overall activation energy of the term $k_1 n_T (k_2/k_1)$, where $T_P = T_g$, is 27.170 kcal/mole, and the net activation energy for the (k_2/k_1) term is 22.05 kcal/mole. Thus the net activation energy for the term $k_1 n_T$ is 27.17 + 22.05 = 49.22 kcal/mole = $E_g - E_P$. For a value of E_P = 9.34 kcal/mole, then, E_g = 58.6 kcal/mole, compared to the value of E_g = 50 kcal/mole, which was based on Blackwood's results with brown coal char.

The differences in the two values of E_P determined by Blackwood[166] and by Johnson[7] may be due to the different coal chars employed or to the different ranges of char pretreatment temperatures used (Blackwood: 750–950°C; Johnson: 927–1100°C). If the difference is due to differences in the pretreatment temperature, the two sets of results can be made consistent by using a different empirical form of expression to describe the relative effects of pretreatment temperature. For example, if the term Ae^{E_P/T_P} is normalized to a value of unity at a temperature of 950°C by determining the appropriate values of A for both values of E^P, then a continuous temperature function can be generated, as shown in Fig. 23.49. Obviously the normalized temperature function cannot be described fully with a term of the form Ae^{E_P/T_P}, using a single value of E^P over the total range of temperatures considered, but it could be described by a variety of other empirical temperature functions. The curve shown in Fig. 23.49 was computed using the function $(T - T_0)^{-1}$, normalized to a value of unity at 950°C. Although other, more theoretically meaningful forms (which probably require additional parameters) could also be applied,

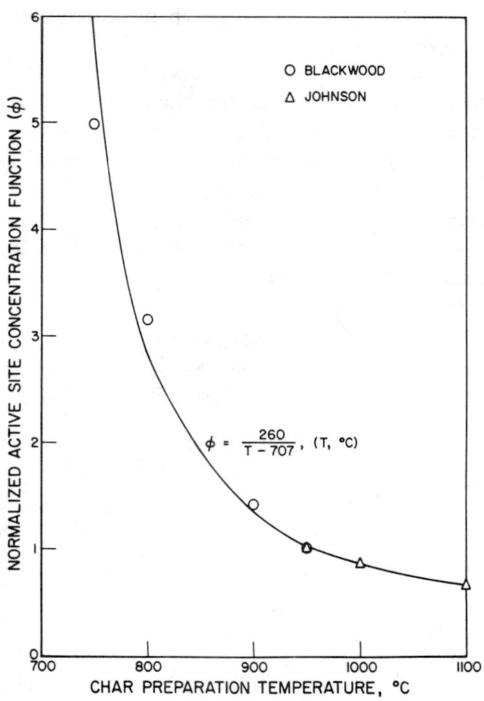

Figure 23.49 Correlation of active site function with char preparation temperature.[7,166]

it is apparent that additional experimental information would be useful.

The total concentration of active sites in coal chars is a function of both the total char surface area and the concentration of active sites per unit surface area, as expressed by the relationship:

$$n_T = S\lambda \tag{23.100}$$

where S = total internal char surface area (m^2/mole carbon)
λ = concentration of active sites (mole/m^2)

Thus a decrease in n_T with increasing pretreatment temperature can potentially result from decreasing S and/or decreasing λ. However unpublished results obtained at the Institute of Gas Technology, given in Table 23.15, show no decrease in the surface area (measured in carbon dioxide) of a lignite char pretreated in nitrogen at temperatures of 927–1100°C and gasified in hydrogen at 927°C. This is consistent with the failure of true graphitization of coal chars to occur at these temperature levels. Thus the results in Table 23.15 suggest that the decrease in the reactivity of chars formed at higher temperatures is due to a decrease in the concentration of active sites.

Johnson[121] has also observed that the internal surface areas do not change during the major course of coal char conversion when specific gasification rates remain essentially constant, under constant environmental conditions. This implies that both λ and S remain constant with most coal chars, at least up to carbon conversions of approximately 80–90%. One subbituminous coal char tested by Johnson, however, did exhibit continuously decreasing specific gasification rates with increasing carbon conversion; in this case total surface area was found to decrease almost in direct proportion to these rates. Although no reason for these decreasing rates and surface areas was found, it is possible that this particular coal char contained more than one type of carbon, each with a significantly different internal specific surface area. If values of λ were the same for the different carbons, the carbon types having larger surface areas would react preferentially, resulting in decreasing total gasification rates and total internal surface areas with increasing carbon conversion.

From the available evidence, it is probable that decreases in char reactivity resulting from increased pretreatment temperature are due to a decrease in the concentration of active sites, which can be expressed over limited temperature ranges, by the expression:

$$\lambda = \lambda_0 \exp \frac{-E_P}{T_P} \tag{23.101}$$

where λ_0 is a constant that depends on the nature of the particular coal used in char preparation. If Equation 23.101 applies, the rate of char gasification in hydrogen can be expressed as:

$$W = \frac{dX/dt}{1-X} = \frac{k_1^0(S\lambda_0)\exp(E_P/RT_P)\exp(-E_g/RT_g)P_{H_2}^2}{P_{H_2} + k_1'/k_2} \tag{23.102}$$

Table 23.15 Effect of Devolatilization Temperature on the Internal Surface Area of Montana Lignite Char Subsequently Gasified in Hydrogen at 927°C

Devolatilization in nitrogen, 1 atm, 60 min; char gasification in hydrogen, 35 atm, 30 min.

Devolatilization Temperature (°C)	Base Carbon Conversion Fraction of Char	Char Surface Area: m^2/g residue	Char Surface Area: m^2/g residue (ash free)
927	0.53	740	907
983	0.48	757	905
1038	0.40	774	914
1094	0.30	792	915

Table 23.16 Comparison of Kinetic Parameters

Correlation Ref.	$k_1^0(S\lambda_0)$ (atm^{-1} min^{-1})	Energy (kcal/g-mole) E_P	E_g	$\frac{k_1}{k_2}$ (atm)
7	5.02×10^5	9.4	58.6	$\exp\left[10.452 - \frac{22{,}050}{RT}\right]$
137	37.0	20.0	50.0	—

Equation 23.102 considers the effect of pretreatment temperature, reaction temperature, and hydrogen partial pressure. With respect to the previous discussion, the term $S\lambda_0$ is a property of the feed coal used. A comparison of the parameters in Equation 23.102 as determined from Johnson's results with coal char from high-volatile bituminous coal,[7] and Blackwood and McCarthy's[137] results with coal char derived from brown coal is summarized in Table 23.16. The ratio of values of the rate constant, k_H^*, predicted by Blackwood's correlation relative to values predicted by Johnson's correlation in the overlapping range of gasification temperatures shown in Fig. 23.47 (line CD for Blackwood) ranges from approximately 0.3 to 0.6, increasing with decreasing temperature. In another study reported by Blackwood,[166] in which higher values of k_H^* were obtained with brown coal char (line AB in Fig. 23.47), the ratio ranges from about 0.6 to 1.1.

The Inherent Reactivity of Different Coal Chars. Johnson has determined the gasification reactivity, in hydrogen, of a large number of coal chars derived from different coal and coal maceral concentrates, varying in rank from anthracite to lignite.[121] Specific gasification rates were obtained at 927°C and 35 atm for coal chars 400–800 μm in diameter, prepared in nitrogen at 927°C (i.e., $T_P = T_g$). Results were expressed in terms of a relative reactivity factor, f_L, defined as:

$$f_L = \frac{W}{W_s} \qquad (23.103)$$

where W_s represents the specific gasification rate obtained with a reference coal char. Under the gasification conditions considered ($W_s = 0.0174$ min^{-1}), the values of f_L obtained for all coal chars tested, except those derived from untreated lignite, were correlated as a function of carbon content, Y, in the raw coals, as shown in Fig. 23.50. The curve drawn in Fig. 23.50 was computed from the expression:

$$f_L = 6.2Y(1 - Y) \quad \text{for} \quad 0.67 \leq Y \leq 0.93 \qquad (23.103a)$$

A similar study was reported by Tomita et al.,[168] who prepared coal chars (40 × 100 mesh) in helium at 1000°C and determined specific gasification rates in hydrogen at 900°C and 27 atm. Relative reactivity factors (Equation 23.103) for these gasification rates can be evaluated by defining W_s on the basis of the kinetic parameters given in Table 23.16 for Johnson's correlation. The values of f_L computed from data obtained with chars derived from anthracite to subbituminous coals are compared to values calculated using Equation 23.103 in Fig. 23.51. As indicated, Tomita's data are quite scattered and show only an approximate trend toward the curve of Equation 23.103.

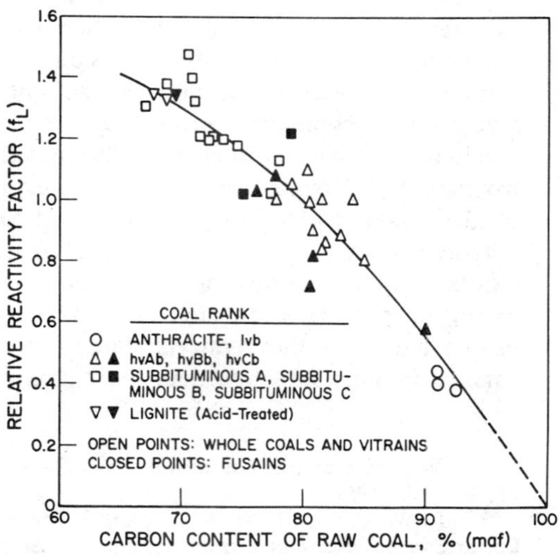

Figure 23.50 Correlation of reactivity factors.[121]

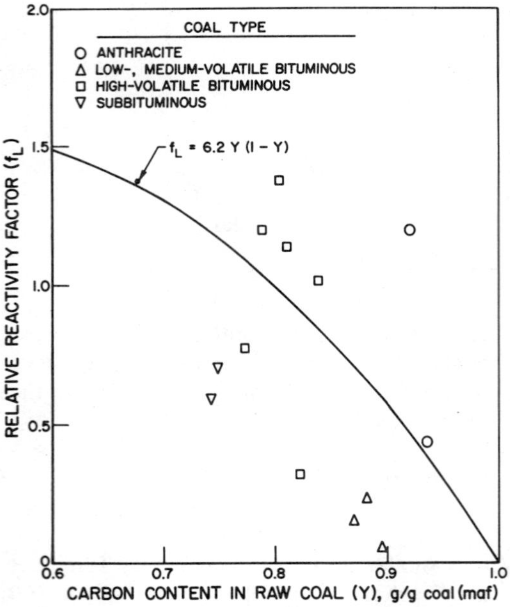

Figure 23.51 Comparison of relative reactivity factors from Tomita et al.[1168] with correlation proposed by Johnson.[121]

Transient Effects. Transient effects have been observed during the early stages of coal char gasification in hydrogen in which the specific gasification rates either decrease or increase to a steady state value.[7,159,167,174] Agglomeration in the char bed during devolatilization of the bituminous coal can be one cause of the initial increase in gasification rates of the coal char. As discussed previously, this can result in a significant increase in the effective particle size, leading to mass transfer inhibition. Initial gasification can open up additional paths for diffusion, resulting in increased rates, until the char surface becomes completely accessible.

Initially increasing rates can also result from catalytic effects. For example, Walker et al.[175] have pointed out that if a carbon initially contains a catalytic component in an oxidized state, gasification of the char in hydrogen could be accompanied by reduction of the oxide. If the reduced state of the catalyst is more effective than the oxidized state, gasification rates will initially increase until reduction of the catalyst is complete. Conversely, if the oxidized state is more effective, initial gasification rates will decrease.

Another explanation for decreasing gasification rates during the early reaction stages has been proposed by Blackwood.[137,167] He suggests that when a coal char is prepared in an inert atmosphere, the char surface becomes saturated with $—CH_2$ complexes. Subsequently, when exposed to a hydrogen atmosphere, these complexes can react relatively rapidly with hydrogen, according to the reaction given by step 1 of the mechanism proposed by Blackwood and co-workers[137,165–167] (see Section 23.4.2), to produce a surge of methane, as the chemisorbed $—CH_2$ complexes approach the lower steady state concentrations characteristic of reaction conditions. Blackwood suggests that the removal of excess $—CH_2$ complexes is rapid, occurring in less than a few minutes at practical char gasification temperatures above 700°C. Removal may even occur in seconds, since gas mixing in experimental reactors can mask the time effect of an impulse of methane. Blackwood suggests that this effect can also be seen in the data obtained by Moseley and Paterson[169] with bituminous coal char.

An additional explanation for the decreasing rates during early gasification stages is related to the rapid-rate methane reaction discussed previously. This can occur if the char preparation conditions are such that complete deactivation of the intermediate responsible for rapid-rate reaction has not occurred. Conditions favoring this effect would be low temperatures and a short residence time during char formation.

Some of the transient effects that can result during coal char gasification with hydrogen under unsteady state conditions can be examined in more detail in terms of the mechanism described by steps 1 and 2 discussed in Section 23.4.2. The steady state gasification rate predicted by this mechanism is given by Equation 23.88, and the steady state concentration of the chemisorbed $—CH_2$ complex $(n^*)_{SS}$ is given by:

$$(n^*)_{SS} = \frac{(k_1 P_{H_2} + k_2' P_{CH_4})\, n_T}{k_1' + (k_1 + k_2) P_{H_2} + k_2' P_{CH_4}} \tag{23.104}$$

[174] C. W. Zielke and E. Gorin, *Ind. Eng. Chem.*, **49**, 396–403 (1957).
[175] P. L. Walker, Jr., L. G. Austin, and F. Rusinko, Jr., in *Advances in Catalysis,* Vol. 11, Academic Press, New York, 1959, pp. 133–221.

In the absence of methane, this simplifies to:

$$(n^*)_{SS} = \frac{k_1 P_{H_2} n_T}{k'_1 + (k_1 + k_2) P_{H_2}} \qquad (23.105)$$

and for the case of $k_2 \gg k_1$ and $k_2 P_{H_2} \gg k'_1$, then:

$$(n^*)_{SS} = \frac{k_1}{k_2} n_T \qquad (23.106)$$

Equation 23.105 leads to the steady state rate expression given by Equation 23.85, and Equation 23.106 leads to the steady state rate expression given by Equation 23.86.

When the —CH_2 surface complex concentration, n^*, is not at steady state under given conditions of temperature and hydrogen pressure, the following equation applies for gasification in hydrogen:

$$\frac{dn^*}{dt} = k_1 n_T P_{H_2} - [k'_1 + (k_1 + k_2) P_{H_2}] n^* \qquad (23.107)$$

and when

$$k_2 \gg k_1 \quad \text{and} \quad k_2 P_{H_2} \gg k'_1 \qquad (23.108)$$

$$\frac{dn^*}{dt} = k_1 n_T P_{H_2} - k_2 P_{H_2} n^* \qquad (23.109)$$

and

$$W = \frac{dX/dt}{1 - X} = k_2 P_{H_2} n^* \qquad (23.110)$$

Equations 23.109 and 23.110 bring up some interesting points. One involves the situation of a coal char that is first exposed to a hydrogen atmosphere at a relatively low temperature such that steady state gasification rates are negligible but the char surface can become saturated with respect to —CH_2 complex. The char is then directly exposed to a high-temperature hydrogen environment. Blackwood has experimentally shown that with this type of treatment, exposure of the char to high temperatures will result in an initial burst of methane formed at a rapid rate, followed by an approach to the steady state rate characteristic of the temperature and pressure used.[137] This behavior can be explained in terms of Equations 23.109 and 23.110 as follows. Integrating Equation 23.109 at constant temperature and pressure yields:

$$n^*(T, t) = \frac{k_1 n_T}{k_2} + \left\{n_0^* - \frac{k_1 n_T}{k_2}\right\} \exp -k_2 P_{H_2} t \qquad (23.111)$$

where n_0^* is the concentration of —CH_2 complex at time $t = 0$. The steady state concentration of —CH_2, $(n^*)_{SS}$, at temperature T is given by Equation 23.106. Thus:

$$n^*(T, t) = (n^*)_{SS} + (n_0^* - n^*_{SS}) \exp -k_2 P_{H_2} t \qquad (23.112)$$

Substituting Equation 23.112 into Equation 23.110 and integrating with respect to time yields:

$$-\ln \frac{1 - X}{1 - X_0} = k_2 P_{H_2} [(n^*)_{SS}\, t + \frac{n_0^* - n_{SS}^*}{k_2 P_{H_2}} (1 - \exp -k_2 P_{H_2} t)] \qquad (23.113)$$

or, at sufficiently long times, when $\exp -k_2 P_{H_2} t \simeq 0$:

$$-\ln \frac{1 - X}{1 - X_0} = k_2 P_{H_2} [(n^*)_{SS}\, t] + (n_0^* - n^*_{SS}) \qquad (23.114)$$

Equation 23.114 shows an excess of conversion during time t, which is reflected by the term $(n_0^* - n_{SS}^*)$, relative to conversion that would be obtained under steady state conditions, that is:

$$k_2 P_{H_2} n_{SS}^* t, \qquad \text{if } n_0^* > n_{SS}^*$$

This condition of excess is satisfied by the initial treatment of the char at a lower temperature, since steady state concentrations of the surface complex decrease with increasing temperature. Moseley and Paterson[169] have observed this "burst" effect, and Blackwood[137] has shown that the excess methane formed by the treatment described above can be continuously repeated.

If instead of being cycled between two temperatures at constant pressure, the char is cycled between two pressures at constant temperature, there are no transient effects in cases where Equation 23.106 can be applied, since n_{SS}^* depends only on temperature. Moseley and Paterson[169] observed this type of behavior when the hydrogen pressure is suddenly increased from 42 to 96 atm. Although the rate was higher at 96 atm, no transient effects occurred. For conditions under which Equation 23.105 is applicable, an increase in hydrogen pressure would result in an interval in which gasification rates increased to the steady state value at the higher pressure, since $(n^*)_{SS}$ increases somewhat with increasing pressure. At pressures above around 10 atm, however, this effect would be small.

Dent[176] has shown that coal chars subjected to continuously increasing temperatures show a higher reactivity during hydrogen gasification than would be expected on the basis of rates obtained isothermally. This result can be predicted from Equations 23.109 and 23.110. If, for example, the concentration of the —CH_2 complex is at a steady state value at some temperature T_0, and the temperature is increased at a uniform rate, the value of n^* will always be greater than n^*_{SS}, since n^*_{SS} decreases with increasing temperature. Thus the steady state rate given by

$$W_{SS} = k_2 P_{H_2} n^*_{SS}$$

will always be less than the actual rate, given by $W = k_2 P_{H_2} n^*$.

This example illustrates a possible danger in integrating rate expressions obtained under steady state conditions to describe behavior under unsteady state conditions. Conversely, with respect to the char-hydrogen reaction, it indicates that rate constants interpreted directly in terms of an expression such as Equation 23.86 from data obtained in experiments with constant heatup rates could be too high. This would give an incorrect activation energy. This is one possible explanation for the relatively high rate constants reported by Jüntgen (see Fig. 23.47), which were determined from experiments with constant heatup rates. The high rate constants shown in Fig. 23.47, deduced from Moseley's experiments under assumed steady state conditions, may also be excessive because of the experimental procedure employed in cyclically heating and cooling the coal char in hydrogen.

23.4.2.2 *The Coal Char–Carbon Dioxide–Carbon Monoxide System*

The reaction of coal chars and other carbon solids with carbon dioxide:

$$CO_2 + C = 2CO \qquad (23.3)$$

has been studied for many years because of the interest in both gasification and combustion systems. Studies prior to the 1960s, which have been reviewed by von Fredersdorff and Elliott,[8] were largely conducted at atmospheric and subatmospheric pressures, using relatively purified carbon solids rather than actual coal chars. An excellent review of more recent studies has been prepared by Mentser and Ergun.[177] Although experimental work during the last 15 years has still focused primarily on the use of purified carbons at low pressure, increasing attention has been paid to gasification at elevated pressures.

Significant differences between the gasification of carbons with carbon dioxide and the gasification with hydrogen have been found. At atmospheric conditions, for example, gasification rates in pure carbon dioxide are generally greater than rates in pure hydrogen by a factor of 10^2 to 10^3. However, although gasification rates in hydrogen increase almost in direct proportion to pressure over a wide pressure range, rates in carbon dioxide have usually been found to increase with pressure only at relatively low pressure levels and then to plateau at pressures less than approximately 10 atm. Such an effect with carbon dioxide is reflected in the following kinetic correlation, which is frequently employed to describe the gasification kinetics of carbon in carbon dioxide and carbon monoxide mixtures, at conditions reasonably removed from equilibrium for this reaction:

$$W = \frac{dX/dt}{1 - X} = \frac{k_c P_{CO_2}}{1 + K_{c,1} P_{CO} + K_{c,2} P_{CO_2}} \qquad (23.115)$$

[176] F. J. Dent, W. H. Blackburn, and H. C. Millett, *Trans. Inst. Gas Eng.*, **88**, 150–217 (1938).

[177] M. Mentser and S. Ergun, U.S. Bur. Mines Bull. No. 644, 1973, 42 pp.

where k_c, $K_{c,1}$, $K_{c,2}$ = kinetic parameters that depend on temperature and nature of char

P_{CO}, P_{CO_2} = carbon monoxide and carbon dioxide partial pressures

Equation 23.115 shows that gasification rates decrease with increasing carbon monoxide partial pressure because of an inhibiting effect that is not directly related to thermodynamic reversibility. This contrasts to the behavior observed in carbon gasification with hydrogen, in which the effects of methane are usually reflected only in terms of thermodynamic reversibility, not as a poisoning agent. At a sufficiently high total pressure, Equation 23.115 simplifies to the following expression, which has been used in some studies to analyze experimental results:

$$W = \frac{k_c^*}{1 + K_c^* P_{CO} / P_{CO_2}} \qquad (23.116)$$

$$\text{where } k_c^* = \frac{k_c}{K_{c,2}} \qquad (23.117)$$

$$K_c^* = \frac{K_{c,1}}{K_{c,2}} \qquad (23.118)$$

Although Equation 23.115 has been rationalized with a variety of mechanisms, the following reaction steps have been postulated most often in recent analyses[177]:

Step 5. $C_f + CO_2 \underset{k_5'}{\overset{k_5}{\rightleftarrows}} C(O) + CO$

Step 6. $C(O) \underset{k_6'}{\overset{k_6}{\rightleftarrows}} CO + C_f$

Step 5 (the first four steps were shown in Section 23.4.2) represents an oxygen exchange reaction in which a carbon dioxide molecule dissociates at an active site, C_f, on the carbon surface, releasing a molecule of carbon monoxide and forming a solid carbon-oxygen complex, C(O). This reaction, considered to be reversible, results only in the exchange of oxygen with the solid and does not result in the gasification of solid carbon. The actual carbon gasification occurs in step 6, with the dissociation of the carbon-oxygen surface complex from the bulk carbon matrix, leading to the formation of carbon monoxide and to the generation of a new active center on the solid carbon surface.

Assuming that the total number of active sites plus carbon-oxygen complexes is constant and assuming further that under steady state conditions the net rate of formation of carbon-oxygen complex is zero, the following rate expression has been derived, based on steps 5 and 6:

$$W = \frac{k_5\, n_t\, P_{CO_2}\left[1 - \dfrac{k_5' k_6' P_{CO}^2}{k_5 k_6 P_{CO_2}}\right]}{1 + \left(\dfrac{k_6'}{k_6} + \dfrac{k_5'}{k_6}\right) P_{CO} + \dfrac{k_3}{k_6} P_{CO_2}} \qquad (23.119)$$

where n_t = concentration of total free active sites plus oxygen-carbon complexes (moles/mole solid carbon)

Equation 23.119 is analogous to Equation 23.115 with the identities:

$$k_c = k_5\, n_t \qquad (23.120)$$

$$K_{c,1} = \frac{k_6'}{k_6} + \frac{k_5'}{k_6} \qquad (23.121)$$

$$K_{c,2} = \frac{k_5}{k_6} \qquad (23.122)$$

To satisfy thermodynamic equilibrium requirements, the term $k_5' k_6' / k_5 k_6$ in Equation 23.119 should be equal to the equilibrium constant, $K_{5,6}$, for the overall reaction given by Equation 23.3, that is:

$$\frac{k_5 k_6}{k_5' k_6'} = K_{5,6} \qquad (23.123)$$

Under the conditions employed in most experimental studies, this term is usually so large that reversibility effects of carbon monoxide are swamped by the poisoning effect of carbon monoxide.

The parameters in the reduced form of Equation 23.116 are related to Equation 23.119 by the following expressions:

$$k_c^* = k_6\, n_t \qquad (23.124)$$

$$K_c^* = \frac{k_6'}{k_5} + \frac{k_5'}{k_5} \qquad (23.125)$$

For the condition in which $k_6' \ll k_5'$ (which often occurs), the term:

$$K_c^* \simeq \frac{k_5'}{k_5} \qquad (23.126)$$

represents the equilibrium constant for the oxygen exchange reaction, step 5. The temperature coefficient for this term, when expressed by the Gibbs-Helmholtz equation, represents the heat of reaction for this step.

The rate expressions described by Equations 23.115 or 23.119 are based not only on results of studies that determined total gasification rate as a function of temperature, pressure, and gas composition, but also on studies using isotopic gases and/or solid carbons. The use of isotopes to study the mechanism of the carbon–carbon dioxide reaction has been a particularly useful tool, although not applicable to the study of the stoichiometrically simpler carbon-hydrogen system. Perhaps the strongest evidence for the validity of oxygen exchange as an essential reaction step is that similar evaluations of the oxygen exchange equilibrium constant, K_5 $(= k_5/k_5')$, have been made from indirect results obtained from kinetic measurements of the overall gasification reactions and by direct measurement using isotopic techniques.

Different investigators, however, have postulated reaction steps in addition to or in place of steps 5 and 6. For example, Blackwood and Ingeme,[178] have proposed a mechanism different from that expressed by steps 5 and 6 to account for results obtained upon gasification of a coconut char in carbon dioxide–carbon monoxide mixtures, at pressures up to 40 atm. This study is particularly significant because it represents one of the few reported in the literature for carbon gasification with carbon dioxide at elevated pressures. Blackwood and Ingeme postulated the following sequence:

Step 5 (forward). $C_f + CO_2 \xrightarrow{k_5} CO + C(O)$

Step 6 (forward). $C(O) \xrightarrow{k_6} CO + C_f$

Step 7. $CO + C_f \underset{k_7}{\overset{k_7}{\rightleftarrows}} C(CO)$

[178] J. D. Blackwood and A. J. Ingeme, *Aust. J. Chem.*, **13**, 194–209 (1960).

Step 8 (forward).

$$CO_2 + C(CO) \xrightarrow{k_8} 2CO + C(O)$$

Step 9 (forward).

$$CO + C(CO) \xrightarrow{k_9} CO_2 + 2C_f$$

Although over the years a number of other investigations have also postulated mechanisms for carbon gasification by carbon dioxide that include reaction steps involving chemisorbed carbon monoxide, such studies represent a minority view. As discussed recently by Mentser and Ergun,[177] there are many more studies that have provided various types of evidence to support the thesis that carbon monoxide retards reaction not through chemisorption but by the reverse oxygen exchange reaction (step 5, reverse), which decreases the number of surface oxygen complexes that can react by step 6 (forward) to gasify carbon.

Perhaps the most significant part of the mechanism proposed by Blackwood and Ingeme is in the rate equation derived from it which describes results obtained at elevated pressures. Based on a number of simplifying approximations, the following equation was proposed:

$$W = \frac{\left(k_5 P_{CO_2} + \frac{k_5 k_8}{k_7'} P_{CO_2}^2\right) n_t}{1 + \frac{k_7}{k_7'} P_{CO} + \frac{k_5}{k_6} P_{CO_2}} \qquad (23.127)$$

With the exception of the term $P_{CO_2}^2$ in the numerator of Equation 23.127, the form of expression is the same as that of Equation 23.115, which is frequently used to describe results of atmospheric pressure studies. As shown by the numerical evaluation of Blackwood and Ingeme's results, obtained on gasification of coconut charcoal from 790 to 870°C at a total pressure up to 40 atm (Table 23.17), the term

$$\frac{k_5 k_8}{k_7'} P_{CO_2}^2 n_t$$

can be neglected at carbon dioxide partial pressures less than 1 atm. At sufficiently high pressures, however, Equation 23.127 predicts that at constant gas composition, gasification rates will become proportional to total pressure, that is,

Table 23.17 Numerical Values for Constants of the Rate Equation[178]

Temperature (°C)	$k_5 n_t$ ($min^{-1}\ atm^{-1} \times 10^4$)	$\left(\frac{k_5 k_8}{k_7'}\right) n_t$ ($min^{-1}\ atm^{-2} \times 10^4$)	$\frac{k_7}{k_7'}$ (atm^{-1})	$\frac{k_5}{k_6}$ (atm^{-1})
790	0.62	0.048	13.7	0.23
830	2.76	0.24	17.1	0.17
870	7.80	0.84	12.7	0.31

$$W \simeq \frac{n_t \frac{k_5 k_8}{k_7'} P_{CO_2}^2}{\frac{k_7}{k_7'} P_{CO} + \frac{k_5}{k_6} P_{CO_2}} \tag{23.128}$$

rather than becoming independent of total pressure, as predicted by Equation 23.115. Unfortunately, little evidence is available in the literature to support the large effects of pressure predicted by Blackwood and Ingeme's correlation, particularly at elevated pressures.

In an unpublished study conducted at the Institute of Gas Technology, results obtained for the gasification of bituminous coal char in carbon monoxide and carbon dioxide mixtures, at 2–35 atm and 850–1000°C, were correlated with an expression of the form given by Equation 23.119. That equation predicts that gasification rates will eventually become independent of pressure, under conditions reasonably removed from equilibrium. The IGT study made the following evaluations, based on gasification of an hvAb coal char prepared by devolatilizing the raw coal in nitrogen at the same temperature used for subsequent gasification:

$$k_c = k_5 n_t = 3.56 \times 10^3 \exp \frac{-28{,}430}{RT} \tag{23.129}$$

$$K_{c,1} = \frac{k_6'}{k_6} + \frac{k_5'}{k_6} = 0.15 \exp \frac{6400}{RT} \tag{23.130}$$

$$K_{c,2} = \frac{k_5}{k_6} = 1.04 \times 10^{-7} \exp \frac{36{,}500}{RT} \tag{23.131}$$

where k_c $min^{-1}\ atm^{-1}$,
$K_{c,1}$ atm^{-1}
$K_{c,2}$ atm^{-1}.

The effects of total pressure on gasification rates, as predicted by Blackwood and Ingeme's correlation,[178] are contrasted in Table 23.18 to those determined in the IGT study, which involved gasification in carbon dioxide and in a carbon dioxide and a carbon monoxide mixture at 870°C. Although Blackwood and Ingeme's correlation predicts continually increasing rates with increasing pressure, the IGT correlation indicates that total pressure has relatively little effect at pressures greater than 10 atm. A trend similar to that suggested by the IGT correlation was also reported by Fuchs and Yavorsky,[179] who studied the kinetics of the gasification of bituminous coal chars in carbon dioxide under conditions of differential gas conversion. These investigators found that at 750 and 800°C, total pressure had no effect on gasification rates at pressures of 18–35 atm. Since it seems unreasonable to attribute the significant differences in the effects of total pressure solely to the different carbon solids used in the few studies conducted at elevated pressure, it appears that additional research would be useful in elucidating the mechanism of the carbon (coal char)–carbon dioxide reaction at high pressures.

Quite recently, Yergey and Lampe[180] also proposed a reaction mechanism different from that expressed by steps 5 and 6, based on experimental results observed for the gasification of an isotopic mixture of purified

[179] W. Fuchs and P. M. Yavorsky, *Prepr., Am. Chem. Soc., Div. Fuel Chem.*, **20**(3), 115–133 (1975).

[180] A. L. Yergey and F. W. Lampe, *Fuel*, **53**, 280–281 (1974).

Table 23.18 Comparison of Pressure Effects on Gasification Rates[178]

Pressure, P (atm)	Specific Rate Ratio, $\frac{W(P)}{W}(1\ atm)$: CO_2 Only, IGT	CO_2 Only, Blackwood	10% CO, 90% CO_2, IGT	10% CO, 90% CO_2, Blackwood
1	1.00	1.00	1.00	1.00
10	1.80	5.98	1.71	2.78
30	1.91	14.56	1.80	5.75
∞	1.98	∞	1.90	∞

Table 23.19 Comparison of Kinetic Parameters from Different Investigators

Curve (Figs. 23.52–23.54)	Carbon Type	Temperature (°C)	Pressure (atm)	Data Source or Ref.
A	Subbituminous coal char	600–790	1	182
B	High-volatile bituminous coal char (Illinois Seam)	750–900	18–35	179
C_1	High-volatile bituminous coal char (Pittsburgh Seam)	880–1080	1	183
C_2	High-volatile bituminous coal char (Illinois Seam)	850–1080	1	183
D	High-volatile bituminous coal char (Pittsburgh Seam)	870–980	1–35	IGT
E	Spheron No. 6 carbon black	750–850	1	177
F	Graphite	—	—	184
G	Coke	—	—	185
H	SP-1 Graphite	1050–1200	1	186
I	SP-1 Graphite	900–1000	1	187
J	Coke	900–1000, 1000–1200	1	188
K	Coconut charcoal	790–870	2–40	178
L_1	Activated carbon	700–1000	1	189
L_2	Activated graphite	900–1150	1	189
L_3	Ceylon graphite	1000–1400	1	189
M_1	Coke	800–1100	1	190
M_2	Anthracite	800–1100	1	190
N	Coconut charcoal	730–830	1	191

Kinetic Parameters

Curve	$k_5 n_t = A_5 e^{-E_5/RT}$ (min^{-1} atm^{-1})	E_5 (kcal/g-mole)	$k'_5 n_t = A'_5 e^{-E'_5/RT}$ A'_5 (min^{-1} atm^{-1})	E'_5 (kcal/g-mole)	$k_6 n_t = A_6 e^{-E_6/RT}$ A_6 (min^{-1} atm^{-1})	E_6 (kcal/g-mole)	$K_5 = \frac{k_5}{k'_5} = \frac{A_5}{A'_5} \exp \frac{-E_5 - E'_5}{RT}$ A_6/A'_5 (min^{-1} atm^{-1})	$E_5 - E'_5$ (kcal/g-mole)
A	—	—	—	—	—	—	—	—
B	—	—	—	—	6.8×10^8	54.5	—	—
C_1	—	—	—	—	—	—	—	—
C_2	—	—	—	—	—	—	—	—
D	3.56×10^3	28.4	5.02×10^9	58.6	3.42×10^{10}	65.0	7.09×10^{-7}	−30.2
E	1.54×10^6	53.0	4.48×10^3	36.0	5.15×10^7	58.0	3.44×10^2	17.0
F	—	41	—	—	—	124	—	—
G	—	61.7	—	27.5	—	67.8	—	34.2
H	—	103.5	—	79.5	—	72.5	—	24.0
I	—	99.0	—	74.0	—	87.0	—	25.0
J	—	—	—	—	8.82×10^8, 1.43×10^3	65.3, 31.5	1.72×10^4	27.0
K	2.7×10^{11}	76.0	1.63×10^{13}	76.0	1.13×10^{12}	76.0	1.66×10^{-2}	0.0
L_1	—	—	—	—	8.58×10^8	59.0	4.15×10^3	23.0
L_2	—	—	—	—	8.28×10^7	59.0	4.15×10^3	23.0
L_3	—	—	—	—	1.55×10^7	59.0	4.15×10^3	23.0
M_1	8.3×10^6	47.6	5.52×10^5	38.9	3.95×10^7	53.9	15.0	8.7
M_2	2.6×10^4	32.5	6.76×10^4	32.2	1.47×10^7	49.1	0.39	0.3
N	7.6×10^9	58.8	3.01×10^{-5}	−16.8	2.39×10^3	28.7	2.62×10^{14}	75.6
Median value	—	53	—	39–59	—	59	—	23

carbons (^{12}C and ^{13}C) with carbon dioxide ($^{12}CO_2$):

Step 10. $C_f + CO_2 \underset{k'_{10}}{\overset{k_{10}}{\rightleftharpoons}} C(CO_2)$

Step 11. $C(CO_2) \underset{k'_{11}}{\overset{k_{11}}{\rightleftharpoons}} 2CO + C_f$

Marsh and Taylor,[181] however, subsequently argued that the conclusions made by Yergey and Lampe were invalid because they misinterpreted their experimental results, primarily because they used a solid carbon that was not degassed prior to reaction with carbon dioxide. It should also be noted that reaction steps 10 and 11 do not lead to a kinetic expression of the form given by Equation 23.122 or 23.126, but instead lead to an expression with a denominator containing a P_{CO}^2 term in place of a P_{CO} term, contrary to the vast majority of kinetic studies made.

Since the existing evidence favors the mechanism given by steps 5 and 6, it is of interest to compare the numerical evaluations of kinetic parameters determined from data obtained by different investigators. Such a comparison, in terms of the notation given in Equation 23.119, is shown in Table 23.19. Results in this table were prepared with the common assumption that under the reaction conditions experimentally employed, the rate of step 6 (reverse) is slow compared to step 5 (reverse). Thus:

$$\frac{k'_5}{k_6} \ll \frac{k'_6}{k_6} \quad (23.132)$$

and, from Equation 23.122,

$$K_{c,1} = \frac{k'_6}{k_6} \quad (23.133)$$

The results in Table 23.19 include the evaluation of earlier work that was summarized by von Fredersdorff and Elliott,[8] and some of the more recent results summarized by Mentser and Ergun.[177] The values of kinetic parameters shown either were reported directly by individual investigators or were estimated from reported, graphic, or tabular information reduced to a common basis. As pointed out by Mentser and Ergun, individual rate constants cannot be determined directly from reported data without knowing the concentration of active sites, n_t. Thus the apparent activation energies of the individual terms may be affected by changes in the concentration of active sites with changing temperature. As can be seen, these studies show a relatively wide range of values of parameters, as shown in Equation 23.119. This is partly due, of course, to the different carbonaceous solids used and to the different experimental procedures and methods of data analyses employed.

Graphic representation of the terms K_5 and $k_6 n_t$ are shown in Figs. 23.52 and 23.53, respectively. In spite of the range of values for differences in activation energy ($E_5 - E'_5$) for the parameter K_5, the values of K_5 predicted from the correlation obtained using different carbons show a surprisingly similar overall trend (Fig. 23.52), reasonably well supported by evaluations made from Ergun's data[189]; that is,

$$K_5 = 4.15 \times 10^{13} \exp(-23{,}000/RT)$$

This similarity gives some credence to the supposition that K_5 is in fact an equilibrium constant representative of the oxygen exchange reaction (step 5) and is relatively independent of carbon type.

If this is true, the values of $E_5 - E'_5 \simeq 23$ kcal/g-mole can be considered as the heat of reaction, ΔH, for the oxygen exchange re-

[181] H. Marsh and D. W. Taylor, *Fuel*, **54**, 218–219 (1975).
[182] R. W. Taylor and D. W. Bowen, Lawrence Livermore Lab. Rep., No. UCRL 52002; NTIS, 1976.
[183] S. Dutta, C. Y. Wen, and R. J. Belt, *Prepr., Am. Chem. Soc., Div. Fuel Chem.*, **20**(3), 103–114 (1975).
[184] H. J. von Grabke, *Ber. Bunsenges. Phys. Chem.*, **70**, 664–674 (1966).
[185] H. C. Hottel, G. C. Williams, and P. C. Wu, *Prepr., Am. Chem. Soc., Div. Fuel Chem.*, **10**(3), 58–71 (1966).
[186] D. L. Biederman, Ph.D. thesis, Pennsylvania State University, University Park, 1965, 105 pp.
[187] J. F. Strange, Ph.D. thesis, Pennsylvania State University, University Park, 1964, 120 pp.
[188] S. Ergun, U.S. Bur. Mines Bull. No. 598, 1962, 38 pp.
[189] S. Ergun, *J. Phys. Chem.*, **60**, 480–485 (1956).
[190] W. K. Lewis, E. R. Gilliland, and G. T. McBride, Jr., *Ind. Eng. Chem.*, **41**, 1213–1226 (1949).
[191] J. Gadsby, F. J. Long, P. Sleightholm, and K. W. Sykes, *Proc. R. Soc. (London)*, **A193**, 357–376 (1948).

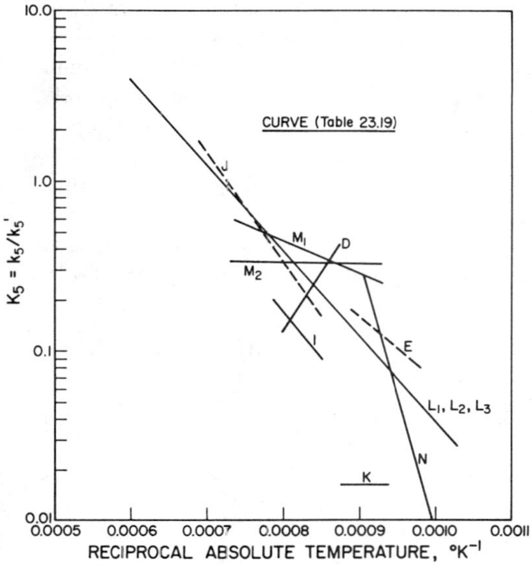

Figure 23.52 Oxygen exchange equilibrium constant as determined by various investigators.

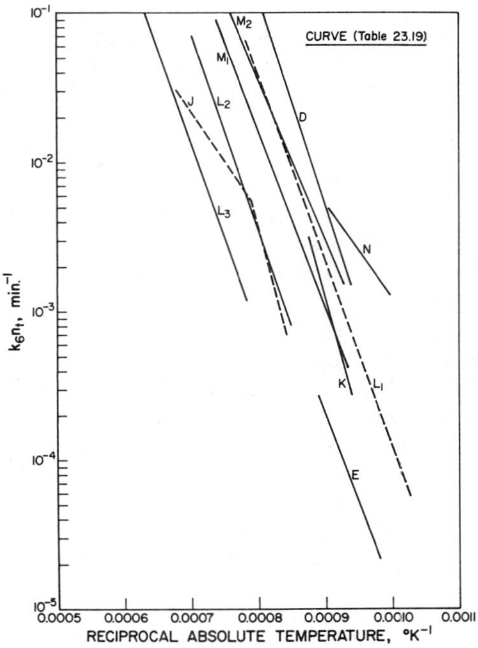

Figure 23.53 Composite rate constant compiled from various investigations.

action (step 5). This magnitude of ΔH has been interpreted by some investigators to imply certain characteristics of the nature of carbon-oxygen bonding in the solid complex, —C(O), formed in the oxygen exchange reaction. According to Mentser and Ergun,[177] for example, the probable values of ΔH obtained in recent investigations suggest that the bonding between oxygen and carbon in the surface complex is, on the average, intermediate between that of true single bonds and double bonds, although somewhat closer to double bonds. Other investigators have reached similar conclusions.

Values of $k_6\,n_t$, as shown in Fig. 23.53, do exhibit significant differences for the various carbon solids. Although the overall activation energies for different materials are comparable, clustered about a median value of 59 kcal/g-mole, the preexponential factors vary by a factor of about 100–200. In general, the variations in the term $k_6\,n_t$ are due to possible variations both in k_6 and n_t. If, however, it is assumed that k_6 is a fundamental rate constant, similar for different carbon solids, major differences in the magnitude of values of $k_6\,n_t$ for different materials can be attributed to differences in values of the concentration of active sites, n_t. The value of this term would be expected to decrease generally with increasing solid preparation or reaction temperature for any material, as suggested previously in the discussion of the carbon-hydrogen reaction and also as indicated by Duval's theory of the neutralization of active sites in carbon solids.[192,193] For coal chars, n_t would also be expected to decrease with increasing coal rank. These general trends are reasonably substantiated by the approximate order of the values of $k_6\,n_t$, exhibited in Fig. 23.53 for curves with similar slopes, that is:

$$D > M_2, L_1 > K, M_1 > L_2, J, E > L_3 \tag{23.134}$$

The order for the carbon solids referred to by each curve is high-volatile bituminous coal char (Pittsburgh Seam) > anthracite, activated carbon > coconut charcoal, coke (M_1) > activated graphite, coke (J), Spheron 6 carbon black > Ceylon graphite.

Another way to consider the relative reactivities of different carbon solids in the carbon–carbon dioxide reaction is to compare gasification rates under a standard condition or range of conditions. This type of procedure, in part, avoids some of the problems

[192] X. Duval, *J. Chim. Phys.*, **47,** 339–347 (1950).
[193] X. Duval, *Ann. Chim.*, **10,** 903–967 (1955).

associated with the accurate evaluation of individual kinetic parameters and reflects overall reaction rate measurements to a greater extent. In Fig. 23.54, values of the specific reaction rate, *W*, are given as a function of temperature at 1 atm, in pure carbon dioxide, based on the use of kinetic parameters given in Table 23.19 for the different carbon solids. Figure 23.54 also includes rate values computed from two investigations with coal chars, in which results were correlated with empirical rate expressions that differed from the form given by Equation 23.126. Each of these studies was conducted at atmospheric pressure with pure CO_2, under conditions in which carbon monoxide concentrations were very small. In one study conducted with subbituminous coal char, reported by Taylor and Bowen (curve *A*),[182] specific gasification rates were correlated with the expression:

$$W = 4.2 \times 10^7 \exp \frac{-45{,}000}{RT} \qquad (23.135)$$

(W in min^{-1}) for temperatures from 600 to 790°C, under conditions in which the chars were produced at reaction temperature. Taylor and Bowen report that specific gasification rates for chars prepared from 6-mm-diameter subbituminous coal are relatively constant, at least for char conversions up to 90%.

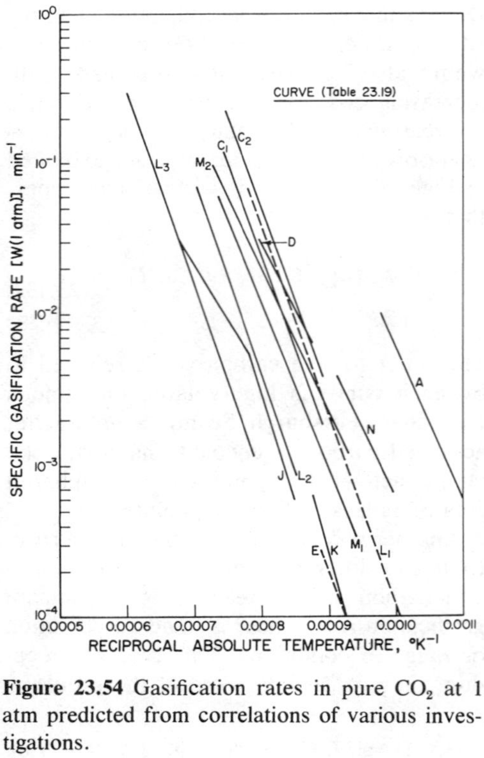

Figure 23.54 Gasification rates in pure CO_2 at 1 atm predicted from correlations of various investigations.

Dutta et al.[183] expressed gasification rates of chars derived from high-volatile bituminous coals in carbon dioxide, at 1 atm and 850–1080°C, with a correlation of the form:

$$W = ak_vC \qquad (23.136)$$

where k_v = rate constant
C = carbon dioxide concentration
a = parameter that reflects the relative available internal char surface area

This correlation was used to describe rates obtained with minus 35, plus 60 mesh chars obtained from Illinois and Pittsburgh Seam coals. The parameter a decreased with carbon conversion for some of the chars tested (particularly those from the Pittsburgh Seam coal) can be considered to be proportional to the total concentration of active sites, n_t, defined previously. Thus variations in a can be due both to a variation in the available char surface area and a variation in the active sites per unit of available surface area. Variations in available surface area would be more likely to occur during gasification under isothermal conditions. For gasification at 1 atm total pressure, initial gasification rates, where a = 1.0 by definition, can be expressed by:

$$W = \frac{k_v}{RT} \qquad (23.137)$$

based on Equation 23.136. Average values of kinetic parameters for the term k_v/RT, for the two types of char used by Dutta, can be deduced from the reported information:

Pittsburgh Seam

$$W = \frac{k_v}{RT} = (0.61 \pm 0.13) \times 10^9 \exp \frac{-59{,}260}{RT} \qquad (23.138)$$

Table 23.20 Comparison of Specific Gasification Rates from Fig. 23.54 (900°C)

Curve (Table 23.19)	Carbon Type	Rate, W (min^{-1})
A	Subbituminous coal char	0.23
C_2	High-volatile bituminous coal char (Illinois)	0.012
N	Coconut char	0.012
D	High-volatile bituminous coal char (Pittsburgh)	0.011
L_1	Activated carbon	0.009
M_2	Anthracite	0.007
C_1	High-volatile bituminous coal char (Pittsburgh)	0.006
M_1	Coke	0.003
K	Coconut char	0.002
L_2	Activated graphite	0.0009
E	Carbon black	0.0009
J	Coke	0.0006
L_3	Ceylon graphite	0.0002

Illinois Seam

$$W = \frac{k_v}{RT} = (1.21 \pm 0.25) \times 10^9 \exp \frac{-59{,}260}{RT} \qquad (23.139)$$

where W is in min^{-1}. It should be noted that the deviations indicated above are not based on reported evaluations, but were derived by averaging reported parameters for different samples and using the single reported activation energy, based on concentration units, to express the deviations in units of pressure, over the range of temperatures employed.

Evaluations of specific gasification rates at 1 atm are shown in Fig. 23.54 as a function of temperature. Similar to the trend exhibited by values of $k_6 n_t$, values of W generally show a wide variation, but with less of a specific trend for medium-reactivity carbons. It is convenient to evaluate the order of carbon reactivities by comparing their interpolated or extrapolated rates at 900°C (Table 23.20), although too great an extrapolation beyond reported temperature ranges may be meaningless.

The values given in Table 23.20 again show the expected trends, with some exceptions. The gasification rate for subbituminous char (A) appears to be unusually high, indicating the possibility of catalysis or initial stage reaction effects, since actual rate measurements were made at relatively low temperatures. Also, the rates for the two coconut charcoals (N and K) are different for reasons that are not easily apparent. The higher value (N, 0.012 min^{-1}) is comparable to values obtained with bituminous coals, whereas the low value (K, 0.002 min^{-1}) is comparable to values obtained with high-temperature coke. This low value may be particularly suspect because Blackwood and Ingeme's correlation,[178] used to evaluate the low rate, predicts gasification rates at higher pressures that are comparable to those of bituminous coals. This is illustrated in Fig. 23.55, which also includes rates reported by Fuchs and Yavorsky[148] for the gasification of high-volatile bituminous coal char in pure carbon dioxide at 18–35 atm and 750–900°C.

Specific gasification rates in carbon dioxide, at 1 atm and 900°C, have also been determined in studies conducted at Pennsylvania State University[99,194,195] for chars de-

[194] W. Spackman, A. Davis, P. L. Walker, Jr., H. L. Lovell, R. H. Essenhigh, F. J. Vastola, and P. H. Given, U.S. ERDA Rep. No. FE-2031-1; NTIS, 1975.

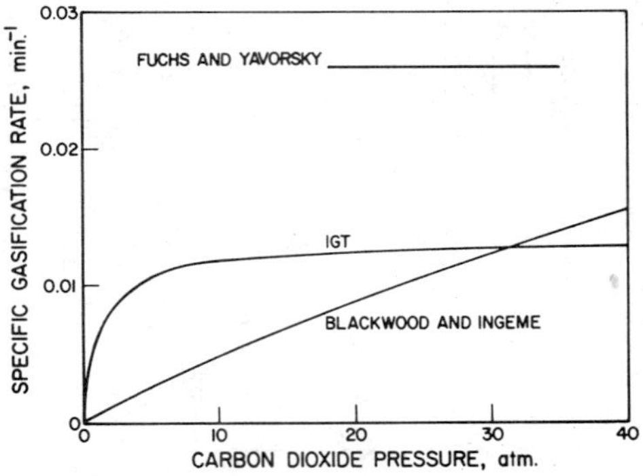

Figure 23.55 Effect of pressure on gasification in pure CO_2 at 870°C.[148,178]

Table 23.21 Specific Gasification Rates of Coal Chars in Carbon Dioxide at 1 atm and 900°C: 40 × 100 Mesh Chars Prepared in Nitrogen at 1000°C

Coal Type	Carbon Content in Raw Coal (% maf)	Specific Gasification Rate (min^{-1})
Lignite (North Dakota)	63.3	0.060–0.093
Lignite (Montana)	70.7	0.058–0.105
Lignite (North Dakota)	71.2	0.045–0.195
Lignite (Texas)	74.3	0.020–0.032
Subbituminous A (Wyoming)	74.3	0.022–0.028
Subbituminous C (Wyoming)	74.8	0.055–0.077
High-volatile B (Illinois)	77.3	0.003–0.025
High-volatile C (Illinois)	78.8	0.010–0.011
High-volatile B (Illinois)	80.1	0.006–0.022
High-volatile C (Indiana)	81.3	0.010
High-volatile A (West Virginia)	82.3	0.0013–0.0032
High-volatile A (Kentucky)	83.8	0.0033–0.0072
Low-volatile (Pennsylvania)	88.2	0.0012–0.0037
Anthracite (Pennsylvania)	91.9	0.0022–0.0046

rived from coals ranging widely in rank from anthracite to lignite. The selection of these values in Table 23.21 shows that gasification rates can be reasonably correlated with rank, which is not inconsistent with the results given in Table 23.20.

Figure 23.56 plots the rates given in Table 23.21 as a function of carbon content in the raw coals. Results for lignites are not included because of the strong possibility of significant, inherent catalytic effects with these materials. The curve drawn in Fig. 23.56 represents the relative reactivity factor correlation developed by Johnson[122] to describe reactivity variations in the gasification of coal chars with hydrogen (Equation 23.79). In spite of the scatter of the data apparent in Fig. 23.56, the results do suggest some relationship between active sites, which is important in the carbon-hydrogen reaction and the carbon dioxide–carbon reaction. This relationship is also indicated from a study reported by Blackwood[196] and by Blackwood and Ingeme,[178] who examined the effect of char preparation temperature on wood char gasification rates obtained in carbon dioxide at 1.2 atm and at 650°C. Results from this study, showing the effects of char preparation temperatures from 750 to 1150°C on gasification rates, are illustrated in Fig. 23.57.

The curve in Fig. 23.57, which is reproduced from Fig. 23.49, describes the relative effects of pretreatment temperatures on char-hydrogen gasification rates, based on the analysis of Blackwood et al.[166] and Johnson's corrections[7] and interpreted in terms of the effect of increasing temperature, on decreasing active site concentration in the char-hy-

[195] E. Hippo and P. L. Walker, Jr., *Fuel*, **54**, 245–248 (1975).

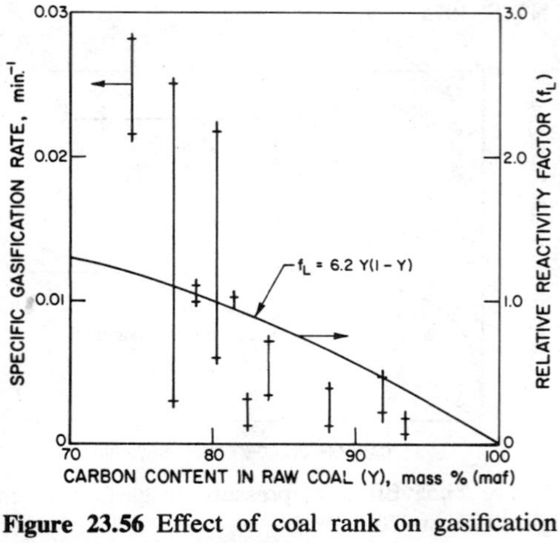

Figure 23.56 Effect of coal rank on gasification rate in CO_2 at 1 atm and 900°C.[194,195]

[196] J. D. Blackwood, *Coke Gas*, **22**, 293–297 (1960).

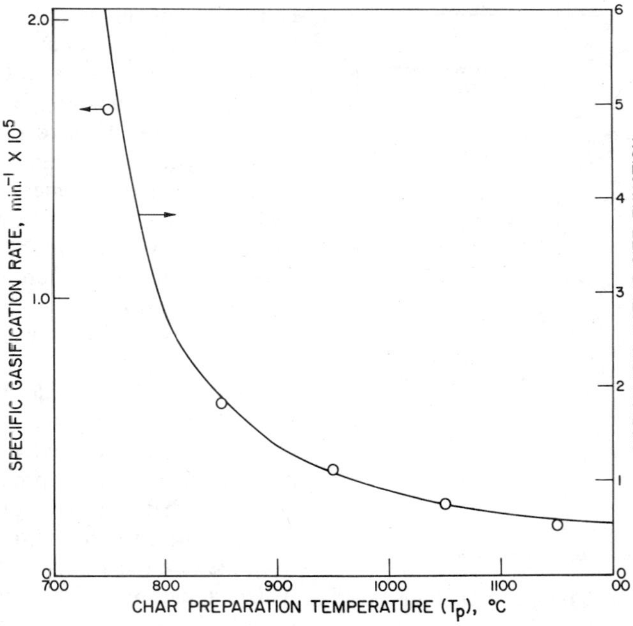

Figure 23.57 Effect of preparation temperature on gasification rates of wood chars in CO_2 at 1.2 atm and 650°C.[178]

drogen reaction. The close correspondence of this curve to the data, however, strongly corroborates the evidence discussed previously that there is a direct relationship between the active sites involved in the carbon-hydrogen reaction and in the carbon–carbon dioxide reaction, although it does not prove that the sites participating in these two reactions are identical.

As indicated by Equation 23.119, an increase in the carbon monoxide partial pressure decreases the gasification rate. The relative decrease in gasification rates with increasing carbon monoxide mole fraction, z, at constant temperature and at total pressure, P, can be expressed by the relationship:

$$\frac{W(z)}{W(z=0)} = \left(1 + \frac{\dfrac{k_5'}{k_6} zP}{1 + \dfrac{k_5}{k_6}(1-z)P}\right)^{-1} \quad (23.140)$$

Quantitative examples of the relative effects of the carbon monoxide mole fraction on gasification rates, at 1 atm pressure and at 800 and 900°C, are shown in Figs. 23.58 and 23.59. These examples were calculated from Equation 23.147, using selected data given in Table 23.19. Although the use of different kinetic parameters results in differences in the magnitude of the relative decrease in gasification rates with increasing temperature, the results of most studies predict that the effect of carbon monoxide will diminish with increasing gasification temperature. The general requirements for this trend to occur can be deduced by examining Equation

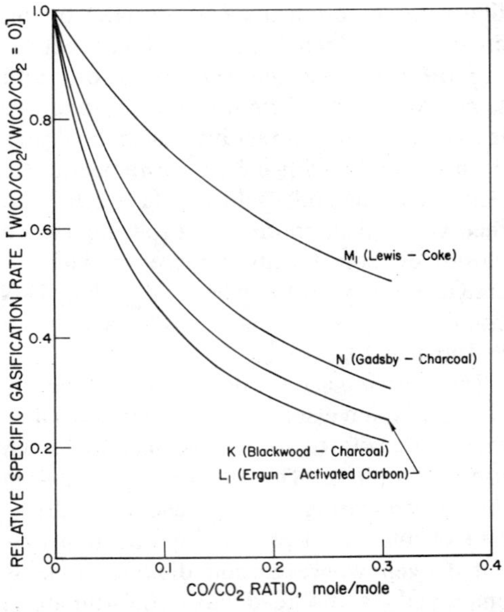

Figure 23.58 Effect of CO concentration on gasification rates at 1 atm and 800°C.

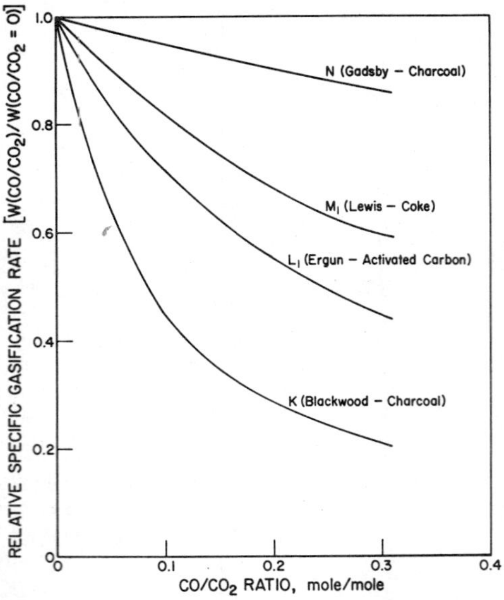

Figure 23.59 Effect of CO concentration on gasification rates at 1 atm and 900°C.

23.14. Under conditions in which $E_6 > E_5 > E'_5$, the denominator term in Equation 23.14 will always increase with increasing temperature, which results in decreasing relative gasification rates. This will also occur at pressures other than 1 atm. The inequality above is satisfied for results given in Table 23.19 for cases *E*, *G*, M_1, and M_2. The situation is more complex for most other cases in Table 23.19, since relative gasification rates theoretically can increase or decrease with increasing temperature, depending on pressure, gas composition, and temperature level. The one exception to this is case *K*, for which relative gasification rates do not change with increasing temperature.. For conditions under which $k_5(1 - z)P \gg k_6$, relative gasification rates at a specific carbon monoxide concentration increase with increasing temperature when $E_5 > E'_5$. This condition exists for almost all the cases given in Table 23.19.

Many carbonaceous materials, particularly coal chars, have some hydrogen in the solids that can gasify to form molecular hydrogen in the gas phase. The effect of hydrogen on the carbon–carbon dioxide reaction is therefore of interest in practical gasification systems, even where carbon dioxide alone is employed as the feed gas. Unfortunately, relatively little work has been done to define quantitatively the effects of hydrogen concentration on this reaction, and the qualitative conclusions from available work are not consistent. Whereas investigations by Long and Sykes,[197] Gadsby et al.,[191] and Blakely and Overholser,[198] have indicated inhibition due to hydrogen at atmospheric pressure, Blackwood and Ingeme[178] found that 10% hydrogen at 10 and 20 atm had no effect. Although von Fredersdorff and Elliott[8] have suggested that the different results may be due to differences in the pressure levels, it is apparent that this area needs further investigation.

Equation 23.119 was derived using reaction steps 5 and 6 for steady state conditions. Certain types of behavior under nonsteady state conditions can be predicted in a manner similar to that discussed previously with respect to the carbon-hydrogen reaction. One example of a nonsteady state effect is that in which a carbon sample undergoing steady state gasification in a carbon dioxide–carbon monoxide mixture is instantaneously exposed to a higher or lower temperature. For this case an equation similar to Equation 23.114 can be derived to describe the relationship between carbon conversion and the amount of time after steady state is achieved at the new temperature:

$$-\ln\frac{1-X}{1-X_0} = k_6\, n_{SS}^{**}\, t + \frac{k_6\,(n_0^{**} - n_{SS}^{**})}{k_5 f_{CO_2} + k'_5 P_{CO} + k_6} \qquad (23.141)$$

where n_{SS}^{**} = steady state concentration of the oxygen complex at the new temperature

$$= \frac{n_t\, k_5 P_{CO_2}}{k_6 + k_5 P_{CO_2} + k'_5 P_{CO}} = \frac{n_T (k_5/k_6) P_{CO_2}}{1 + (k_5/k_6) P_{CO_2} + (k'_5/k_6) P_{CO}} \qquad (23.142)$$

n_0^{**} = steady state concentration of the oxygen complex at the initial temperature

[197] F. J. Long and K. W. Sykes, *Proc. R. Soc. (London)*, **A193**, 377–399 (1948).
[198] J. P. Blakely and L. G. Overholser, *Carbon (Oxford)*, 3, 269–275 (1965).

It is apparent from Equation 23.141 that at the new temperature there will be either an excess or a deficiency in carbon conversion relative to that predicted by applying the steady state rate equation, depending on the sign of the term $(n_0^{**} - n_{SS}^{**})$. For conditions under which the steady state concentration of the oxygen complex increases with increasing temperature, a temperature increase results in a deficiency in conversion, and a temperature decrease results in an excess of carbon conversion. This is the opposite of what was predicted for the carbon-hydrogen reaction. If the steady state concentration of the oxygen complex decreases with increasing temperature, a behavior similar to that predicted for the carbon-hydrogen reaction will occur.

23.4.2.3 *The Coal Char–Synthesis Gas System*

There are many complexities involved in the determination of reaction kinetics for the gasification of carbonaceous solids in systems containing the components carbon, hydrogen, and oxygen. This is mainly because the reaction stoichiometry in such systems is usually a function of reaction conditions rather than being approximately fixed, as is the case in the relatively simpler carbon-oxygen and carbon-hydrogen systems. Under some limiting conditions reaction stoichiometry in carbon-hydrogen-oxygen systems is simplified because of the nature of the relative kinetics of the gasification reaction that occurs. At lower pressures, for example, methane formation is inhibited, and gasification occurs by some combination of the following reactions:

$$CO_2 + C \rightleftarrows 2CO \qquad (23.3)$$

$$H_2O + C \rightleftarrows CO + H_2 \qquad (23.82)$$

$$H_2O + CO \rightleftarrows CO_2 + H_2 \qquad (23.4)$$

In addition, at lower pressures—where steam and hydrogen alone predominate in the gas—the only reaction that occurs is that given by Equation 23.82, and reaction stoichiometry is fixed.

At elevated pressures, however, methane becomes an increasingly more important gasification product through overall reactions of the types:

$$2H_2 + C \rightleftarrows CH_4 \qquad (23.1)$$

and

$$H_2 + H_2O + 2C \rightleftarrows CO + CH_4 \qquad (23.143)$$

Under such conditions, in which both methane and carbon oxides are the primary products of gasification, there are greater variations in the models or mechanisms proposed by different investigators (as well as the kinetic evaluations pertinent to these models) compared to evaluations made for the simpler gasification systems. In this sense the few kinetic correlations developed to describe carbon gasification kinetics at elevated pressures with synthesis gases, or even with steam and hydrogen alone, are largely empirical, with relatively little theoretical basis from a fundamental, mechanistic point of view. This does not, of course, preclude the practical application of these correlations to gasifier design, but it does caution against using them for reaction conditions that have not been verified experimentally.

A wide variety of rate expressions has been used to correlate the gasification rates of carbon to form carbon oxides in steam-hydrogen mixtures, under various reaction conditions. Some of these are given below:

$$W = k_{OH} P_{H_2O} \qquad (23.144)$$

$$W = \frac{k_{OH} P_{H_2O}}{1 + K_{OH,1} P_{H_2} + K_{OH,2} P_{H_2O}} \qquad (23.145)$$

$$W = k_{OH}\left(\frac{P_{H_2O}}{P_{H_2}}\right)^{1/2} \qquad (23.146)$$

$$W = \frac{k_{OH} P_{H_2O}^2}{1 + K_{OH,1} P_{H_2} + K_{OH,2} P_{H_2O}} \qquad (23.147)$$

$$W = \frac{k_{OH} P_{H_2O}}{(1 + K_{OH,1} P_{H_2} + K_{OH,2} P_{H_2O})^2} \qquad (23.148)$$

$$W = \frac{k_{OH} P_{H_2O} + k'_{OH} P_{H_2} P_{H_2O} + k''_{OH} P_{H_2O}^2}{(1 + K_{OH,1} P_{H_2} + K_{OH,2} P_{H_2O})} \qquad (23.149)$$

where k_{OH}, k'_{OH}, k''_{OH}, $K_{OH,1}$, and $K_{OH,2}$ are kinetic parameters, and P_{H_2} and P_{H_2O} are hydrogen and steam partial pressures, respectively.

Except for Equation 23.144, all the foregoing rate expressions predict that hydrogen will inhibit gasification rates. Although some investigators have proposed Equation 23.144 on the basis of the limited available kinetic evidence, this equation can be applied only over a very narrow range of temperature, pressure, and gas composition.

Equation 23.145 has been used by several investigators to describe the kinetics of gasification at atmospheric and subatmospheric pressures, over a wide range of temperatures. The studies using this expression for the gasification of cokes, graphites, and charcoals were made primarily in earlier investigations, and have been reviewed by von Fredersdorff and Elliott.[8] They have shown, however, that kinetic parameters determined in different studies can show a wide variation. For the studies considered, the activation energy for the k_{OH} term was found to vary from 13 to 62 kcal/g-mole; for $K_{OH,1}$, from 0 to −75 kcal/g-mole; and for $K_{OH,2}$, from 20 to −83 kcal/g-mole. In spite of these variations, which result from the strong covariance existing in quantitative evaluations of individual kinetic parameters, actual rate determinations show more consistent trends. This is illustrated in Fig. 23.60,[199–201] which shows the rates of gasification in pure steam predicted by several correlations, at a pressure of 1 atm. Included in Fig. 23.60 are more recent results obtained for the gasification of coal chars, as well as results of earlier studies with less reactive carbons.

One of the mechanisms proposed to describe the kinetics of the steam-carbon reaction is based on adsorption of steam and hydrogen on active sites.[199,202] This mechanism assumes the following sequence:

Step 13. $H_2O + C_f \underset{k_{13}}{\overset{k_{13}}{\rightleftarrows}} C(H_2O)$

Step 14. $C(H_2O) \underset{k_{14}}{\overset{k_{14}}{\rightleftarrows}} CO + H_2 + C_f$

[199] J. Gadsby, F. R. S. Hinshelwood, and K. W. Sykes, *Proc. R. Soc. (London)*, **A187**, 129–151 (1946).

[200] J. D. Blackwood and F. McGrory, *Aust. J. Chem.*, **11**, 16–33 (1958).

[201] G. P. Curran, C. E. Fink, and E. Gorin, *Prepr., Am. Chem. Soc., Div. Fuel Chem.*, **12**(3), 62–82 (1968).

[202] H. F. Johnstone, C. Y. Chen, and D. S. Scott, *Ind. Eng. Chem.*, **44**, 1564–1569 (1952).

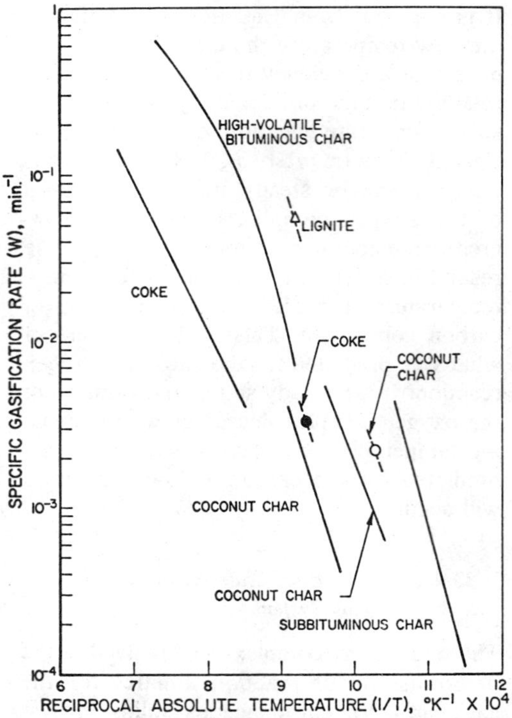

Figure 23.60 Comparison of carbon gasification rates in pure steam at 1 atm. Data sources: a = Ref. 7, b = Ref. 188, c = Ref. 201, d = Ref. 197, e,f = Ref. 199, g = Ref. 182.

Step 1. $H_2 + C \underset{k_1}{\overset{k_1}{\rightleftarrows}} C(H_2)$

In the presence of hydrogen and steam alone, this mechanism leads to a rate expression of the form given in Equation 23.145, where:

$$k_{OH} = \frac{k_{13}\,k_{14}\,n_t}{k'_{13} + k_{14}} \quad (23.150)$$

$$K_{OH,1} = \frac{k_1}{k'_1} \quad (23.151)$$

$$K_{OH,2} = \frac{k_{13}}{(k'_{13} = k_{14})} \quad (23.152)$$

Long and Sykes[197] have proposed another mechanism, which they interpret to lead to Equation 23.145. They suggest that steam adsorption is accompanied by dissociation into hydroxyl radicals and hydrogen atoms, followed by a surface reaction involving hydrogen exchange. The following sequence was proposed to replace steps 13 and 14:

Step 15.

$$H_2O + C_f \text{ (dual site)} \underset{k_{15}}{\overset{k_{15}}{\rightleftarrows}} C(H)(OH)$$

Step 16. $C(H)(OH) \underset{k_{16}}{\overset{k_{16}}{\rightleftarrows}} C(H_2)(O)$

Step 6. $C(O) \underset{k_6}{\overset{k_6}{\rightleftarrows}} CO + C_f$

This sequence, together with step 1, can lead to Equation 23.145 if it is assumed that step 16 occurs only in the forward direction and that the concentration of dual sites interacting in step 15 is equal to or directly proportional to the total concentration of free active sites. Although the fundamental significance of these assumptions is not readily apparent, the dependence of individual rate steps on first-order surface species concentrations only makes it much easier to derive the overall gasification rate expression. Long and Sykes suggest that the active sites responsible for the dissociative steam chemisorption, which is described by step 15, are carbon atoms that can readily provide a bond approaching the order of unity. These investigators noted that the opposite is true for the reaction of carbon dioxide with carbon, which occurs by the oxygen exchange reaction (step 5) on active sites capable of forming bonds of an order greater than unity. This result has been postulated to explain why some investigators have observed that carbon monoxide and hydrogen inhibit the C-CO_2 reaction, whereas only hydrogen inhibits the carbon-steam reaction, at least with coconut charcoal. Other investigators using coal chars, however, have found that carbon monoxide significantly inhibited the carbon-steam reaction, by carbon monoxide,[7,201] which may indicate that both the nature of the active sites and the reaction mechanism of importance depend on the nature of the carbon used.

The mechanism most widely proposed to describe the gasification of carbon in gases containing steam and hydrogen assumes the following sequence[188,203,204]:

Step 12. $H_2O + C_f \underset{k_{12}}{\overset{k_{12}}{\rightleftarrows}} H_2 + C(O)$

Step 6. $C(O) \underset{k_6}{\overset{k_6}{\rightleftarrows}} CO + C_f$

Step 12 is analogous to the oxygen exchange reaction (step 5) in CO_2 and CO, and step 6 is the same as that assumed to occur in the C-CO_2 reaction. In the presence of steam and hydrogen alone, the sequence above leads to the following rate expression:

$$W = \frac{k_{12}\, n_t\, P_{H_2O}}{1 + (k'_{12}/k_6)\, P_{H_2} + (k_{12}/k_6)\, P_{H_2O}} \tag{23.153}$$

By comparison with Equation 23.145,

$$k_{OH} = k_{12}\, n_t \tag{23.154}$$

$$K_{OH,1} = \frac{k'_{12}}{k_6} \tag{23.155}$$

$$k_{OH,2} = \frac{k_{12}}{k_6} \tag{23.156}$$

When carbon monoxide and carbon dioxide are also present in gas mixtures (in addition to steam and hydrogen) step 5 should also be considered:

Step 5. $CO_2 + C_f \underset{k_5}{\overset{k_5}{\rightleftarrows}} CO + C(O)$

which leads to the total gasification rate expression:

$$W = \frac{k_{12}\, n_t\, P_{H_2O}\left[1 - \dfrac{k'_6\, k'_{12}\, P_{H_2} P_{CO}}{k_6\, k_{12}\; P_{H_2O}}\right] + k_5\, n_t\, P_{CO_2}\left[1 - \dfrac{k'_5\, k'_6\, P^2_{CO}}{k_5\, k_6\, P_{CO_2}}\right]}{1 + \dfrac{k_5}{k_6} P_{CO_2} + \left(\dfrac{k'_5}{k_6} + \dfrac{k'_6}{k_6}\right) P_{CO} + \dfrac{k'_{12}}{k_6} P_{H_2} + \dfrac{k_{12}}{k_6} P_{H_2O}} \tag{23.157}$$

It can be seen that Equation 23.157 reduces to Equation 23.153 if only steam and hydrogen are present, and to Equation 23.119 if

[203] R. F. Strickland-Constable, *J. Chim. Phys.*, **47**, 356–360 (1950).

[204] A. Key, Gas Research Board, Commun. GRB No. 40, London, 1948, 36 pp.

only carbon dioxide and carbon monoxide are present. Thus, in general, Equation 23.157 represents the sum of gasification rates by the carbon-steam (C-H_2O) reactions (first term) and the C-CO_2 reaction (second term).

Ergun[188] used the model containing reaction steps 5, 6, and 12 to analyze kinetic results obtained for the gasification of metallurgical coke in a continuous fixed-bed integral gas conversion system, using various feed gases, at atmospheric pressure. With CO_2 as the feed gas it was assumed that the following reduced form of Equation 23.157 applies under nonequilibrium conditions for the C-CO_2 reaction:

$$W = \frac{k_6\, n_t}{1 + (k_5'/k_5)(P_{CO}/P_{CO_2})} \qquad (23.158)$$

Using this expression is equivalent to assuming that step 5 is in equilibrium and that step 6 is rate limiting. For feed gases containing steam, it was also assumed that step 12 is in equilibrium. The assumption of equilibrium in both oxygen exchange reaction steps 5 and 12 leads to the following result:

$$\frac{k_5' P_{CO}}{k_5 P_{CO_2}} = \frac{k_{12}' P_{H_2}}{k_{12} P_{H_2O}} \qquad (23.159)$$

or

$$\frac{P_{CO_2} P_{H_2}}{P_{CO} P_{H_2O}} = \frac{k_5' k_{12}}{k_5 k_{12}'} = K_{5,12} \qquad (23.160)$$

This implies that the water-gas shift equilibrium should be maintained in the gas phase. With these assumptions, it follows that the rate of gasification in gases containing CO, CO_2, H_2, and H_2O can also be described by the following expressions for the C-H_2O reaction and the C-CO_2 reaction under nonequilibrium conditions:

$$W = \frac{k_6\, n_t}{1 + \dfrac{k_5' P_{CO} + k_{12}' P_{H_2}}{k_5 P_{CO_2} + k_{12} P_{H_2O}}} \qquad (23.161)$$

$$= \frac{k_6\, n_t}{1 + \dfrac{k_{12}' P_{H_2}}{k_{12} P_{H_2O}}} \qquad (23.162)$$

$$= \frac{k_6\, n_t}{1 + \dfrac{k_5' P_{CO}}{k_5 P_{CO_2}}} \qquad (23.158)$$

To analyze results obtained in an experimental system where there was significant conversion of the feed gas, Ergun employed an integrated form of the differential rate expression, based on an assumed gas-solid contacting mode, to derive values for the fundamental kinetic parameters. Ergun's values for the oxygen exchange equilibrium constant, k_{12}/k_{12}', are given in Table 23.22. He found these values to be reasonably consistent with values of the equilibrium constant obtained from gasification in CO/CO_2 mixtures, k_5/k_5' by assuming the following relationship:

$$\frac{k_5}{k_5'} = \frac{k_{12}/k_{12}'}{K_{5,12}} \qquad (23.163)$$

where $K_{5,12}$ is the equilibrium constant for the water-gas shift reaction.

Values of $k_6\, n_t$ determined in experiments that used only steam as the feed gas, based on the assumption that Equation 23.162 is applicable, were found to be about 60% higher than values of $k_6\, n_t$ determined with only CO_2 in the feed gas, using Equation 23.158. The relative effect of temperature on values

Table 23.22 Comparison of Experimental Equilibrium Constants for Oxygen Exchange in CO_2-C Reaction with Values Calculated from H_2O-C Reaction[177,178]

	H_2O-C Reaction			CO_2-C Reaction
Temp. (°C)	$\frac{k_{12}}{k_{12}'}$	$K_{5,12}$	$\frac{k_5}{k_5'} = \frac{k_{12}/k_{12}'}{K_{5,12}}$	$\frac{k_5}{k_5'}$
1000	0.23	0.58	0.40	0.54
1100	0.39	0.47	0.83	1.0
1200	0.80	0.39	2.0	1.8

of $k_6 n_t$ determined with either CO_2 or H_2O in the feed gas was quite similar, however, corresponding to an activation energy of approximately 31 kcal/g-mole at 1000–1200°C.

The higher apparent values of $k_6 n_t$ obtained with steam as the feed gas compared to carbon dioxide as the feed gas were attributed tentatively by Ergun to (1) a polymodal distribution of active sites, with steam capable of reacting with more types than carbon dioxide, or (2) diffusion inhibition effects of various types. Ergun also obtained results with various feed gas mixtures of steam and CO, CO_2, or H_2. Using the kinetic parameters obtained when only steam was present in the feed gas, it is possible to compare predicted integral gasification rates with rates obtained experimentally with the gas mixtures. In general, the predicted rates are greater than the experimental rates with both CO_2-H_2O and CO-H_2O feed gas mixtures, and less than experimental rates with H_2-H_2O mixtures. It was not determined whether these differences are due to the use of an incorrect reaction mechanism, to the assumptions made in developing the rate expressions, or to other factors such as diffusion and/or gas-solid contacting phenomena or a distribution of active site types. These results illustrate the complications of analyzing and correlating data obtained in heterogeneous systems with no fixed stoichiometry.

At elevated pressures the gasification of carbon with gases containing steam becomes even more complicated, because of the formation of methane. Blackwood and McCrory[200] suggested that reaction steps 1, 2, and 17 occur in addition to steps 5, 6, and 12, to account for the synergistic interaction of steam and chemisorbed hydrogen to produce methane:

Step 17. $H_2O + C(H_2) \underset{k'_{17}}{\overset{k_{17}}{\rightleftarrows}} CH_4 + C(O)$

Step 1. $H_2 + C_f \underset{k'_1}{\overset{k_1}{\rightleftarrows}} C(H_2)$

Step 2. $C(H_2) + H_2 \underset{k'_2}{\overset{k_2}{\rightleftarrows}} CH_4 + C_f$

Given this, the overall gasification rates can be described by the following complex expressions:

$$W = W_{CO_x} + W_{CH_4} \qquad (23.164)$$

$$(\phi)W_{CH_4} = (k_6 + k'_5 P_{CO} + k'_{12} P_{H_2} + k_{17} P_{CH_4}) (k_1 k_2 P_{H_2}^2 - k'_1 k'_2 P_{CH_4})\, n_t + k_1 k_{17} P_{H_2} P_{H_2O} (k'_5 P_{CO} + k_6 + k'_{12} P_{H_2})\, n_t - k'_1 k'_{17} P_{CH_4} (k_5 P_{CO_2} + k'_6 P_{CO} + k_{12} P_{H_2O})\, n_t \qquad (23.165)$$

$$(\phi)W_{CO_x} = (k'_1 P_{H_2} + k_2 P_{H_2} + k_{17} P_{H_2O}) [k_6(k_{12} P_{H_2O} + k_5 P_{CO_2}) - k'_6 P_{CO} (k'_{12} P_{H_2} + k'_5 P_{CO})]\, n_t + k_6 k_{17} P_{H_2O} (k_1 P_{H_2} + k'_2 P_{CH_4})\, n_t - k'_6 k'_{17} P_{CO} P_{CH_4}(k' + k_2 P_{H_2})\, n_t \qquad (23.166)$$

$$\phi = (k'_1 + k_2 P_{H_2})(k_5 P_{CO_2} + k'_5 P_{CO} + k_6 + k'_6 P_{CO} + k_{12} P_{H_2O} + k'_{12} P_{H_2}) + (k_1 P_{H_2} + k'_2 P_{CH_4})(k_5 P_{CO} + k_6 + k'_{12} P_{H_2}) + (k_{17} P_{H_2O}) (k_1 P_{H_2} + k'_2 P_{CH_4} + k_5 P_{CO_2} + k'_5 P_{CO} + k_6 + k'_6 P_{CO} + k_{12} P_{H_2O} + k'_{12} P_{H_2}) + (k'_{17} P_{CH_4}) (k_1 P_{H_2} + k'_1 + k_2 P_{H_2} + k'_2 P_{CH_4} + k_5 P_{CO_2} + k'_6 P_{CO} + k_{12} P_{H_2O}) \qquad (23.167)$$

where W_{CH_4} = net gasification rate leading to methane formation (moles C/mole C-min)

W_{CO_x} = net gasification rate leading to CO + CO_2 formation (moles C/mole C-min)

The terms on the right-hand side of Equation 23.165 suggest that carbon gasification leading to net methane formation occurs by four paths equivalent to the following overall reactions:

Steps 1, 2. $2H_2 + C \rightarrow CH_4 \qquad (23.168)$

Steps 1, 12 , 17. $2H_2 + C \rightarrow CH_4 \qquad (23.169)$

Steps 1, 6, 17.
$2C + H_2 + H_2O \rightarrow CO + CH_4 \qquad (23.170)$

Steps 1, 5′, 17.
$C + H_2 + H_2O + CO \rightarrow CH_4 + CO_2 \qquad (23.171)$

Similarly, Equation 23.159 suggests that carbon gasification leading to net carbon oxide formation occurs by four paths equivalent to these overall reactions:

Steps 6, 12. $H_2O + C \rightarrow CO + H_2$ (23.172)

Steps 2′, 6, 17. $H_2O + C \rightarrow CO + H_2$ (23.173)

Steps 5, 6. $CO_2 + C \rightarrow 2CO$ (23.174)

Steps 1, 6, 17. $2C + H_2 + H_2O \rightarrow CO + CH_4$ (23.175)

Equations 23.165 and 23.166 also reduce to certain rate expressions proposed for the simpler reaction systems. For example, in the presence of CO and CO_2 alone, Equation 23.166 reduces to Equation 23.119; in the presence of H_2 and CH_4 alone, Equation 23.165 reduces to Equation 23.88; in the presence of only CO, CO_2, H_2, and H_2O, and neglecting reaction steps 1, 2, and 17, Equation 23.166 reduces to Equation 23.157.

Equations 23.165 and 23.166 define the net rates of carbon gasification to form CH_4 and (CO + CO_2), respectively. To define the net formation rates of all gaseous species, due only to reaction steps 1, 2, 5, 6, 12, and 17, it is necessary to derive an additional expression, depending on the model considered. An example of such an expression, which defines the net rate of CO_2 formation, is:

$$(\phi)W_{CO_2} = (k_1' + k_2 P_{H_2} + k_{17} P_{H_2O}) [k_5' P_{CO}(k_6' P_{CO} + k_{12} P_{H_2O}) - k_5 P_{CO_2} (k_6 + k_{12}' P_{H_2})] n_t + (k_1' + k_2 P_{H_2}) [k_5' k_{17} P_{CO} P_{H_2O} - k_5 k_{17}' P_{CO_2} P_{CH_4}] n_t \quad (23.176)$$

The net rates of formation of CO, H_2, and H_2O can be defined from Equations 23.165, 23.166, and 23.176:

$$W_{CO} = W_{CO_x} - W_{CO_2} \quad (23.177)$$

$$W_{H_2} = W_{CO_x} + W_{CO_2} - W_{CH_4} + \lambda(W_{CO_x} + W_{CH_4}) \quad (23.178)$$

$$W_{H_2O} = -W_{CO_x} - W_{CO_2} \quad (23.179)$$

where λ is the ratio of hydrogen in char to carbon in char (mole/mole).

The expression given above for the differential formation of gaseous species would not, of course, apply to individual cases under conditions of homogeneous reactions occurring in the gas phase, or catalytic reactions occurring possibly influenced by char mineral matter or reactor walls. The water-gas shift reaction (H_2O + CO <====> H_2 + CO_2: Equation 23.4) is an important example of a reaction that could be catalyzed in coal char gasification systems. Although this reaction has often been found to be approximately in equilibrium in gasification systems at temperatures above around 800°C, equilibrium cannot always be assumed, particularly in fluidized beds and under conditions of relatively little steam decomposition.[3] At the present time, however, there is no general way to reliably estimate water-gas shift equilibrium approach ratios as a function of the many kinetic and mass transfer parameters that may affect such ratios. Although simplified empirical correlations have been proposed to define equilibrium approach factors for this reaction in char gasification systems,[3] such correlations cannot be considered generally applicable, and should therefore be used with caution.

It has not been determined whether Equations 23.165, 23.166, and 23.176 are applicable under conditions where all the major gas species are present, particularly at elevated pressures, where carbon gasification to form methane becomes significant. This is, of course, understandable in view of the large number of kinetic parameters involved, which tends to dilute the fundamental significance of numerical evaluations made of such parameters. It is also apparent that even a relatively slight alteration of the mechanism described can significantly alter the form of the resulting rate expressions, particularly if more than one type of active site is postulated or if nonuniform surface properties are important. Even at low pressures, the consideration of nonuniform surfaces with respect to reaction energetics can lead to unusual rate expressions, such as that given by Equation 23.146.[205,206] Because of such complica-

[205] E. M. Cherednik, L. O. Apel'baum, and M. I. Temkin, *Dokl. Akad. Nauk SSSR*, 174(4), 891–894 (1967).

[206] M. I. Temkin, E. M. Cherednik, and L. O. Apel'baum, *Kinet. Catal.* (Engl. transl.), 9(1), Pt. 1, 76–83 (1968).

tions, kinetic data obtained for the gasification of carbons, under conditions of the simultaneous formation of methane and carbon oxides,[3,7,200,201,207–212] have been correlated in some studies by what were intended, or probably should be considered, to be empirical expressions. These give relatively little direct insight into the reaction mechanisms.

In an extensive study conducted at the Consolidation Coal Company,[174,207,208,210] the gasification in steam-hydrogen mixtures of a Disco char derived from a Pittsburgh Seam coal was investigated in a laboratory fluidized bed, at 816–927°C and 1–30 atm. Differential gasification rates determined by an extrapolation procedure were reported by Zielke and Gorin[174] in terms of the following empirical expressions:

$$W_T \text{ (total carbon gasification rate)} = AP^n \tag{23.180}$$

$$W_{CH_4} \text{ (methane formation rate)} = DP^m \tag{23.181}$$

$$W_{CO + CO_2} \text{ (carbon oxide formation rate)} = W_T - W_{CH_4} \tag{23.182}$$

where P is the total pressure and A, D, m, n are functions of total carbon conversion and gas composition at a given temperature but are independent of total pressure. In this study values for the constants A, D, m, and n were represented graphically.

In another study by Johnson[7] empirical kinetic correlations were developed to describe differential coal char gasification rates, based on results obtained from several hundred tests in a thermobalance apparatus. Numerical evaluations of the kinetic parameters in the proposed correlations were determined from results obtained with a variety of bituminous coal chars gasified in H_2-CH_4 mixtures, H_2-H_2O mixtures, and CO-CO_2-H_2-H_2O-CH_4 mixtures, at temperatures from 800 to 1100°C and total pressures from 1 to 70 atm. The formulation of this model with respect to gasification in hydrogen-methane mixtures was discussed previously.

The model assumes that in gas mixtures containing hydrogen and steam, three reversible gasification reactions occur independently of one another.

$$H_2O + C \rightleftarrows CO + H_2 \tag{I}$$

$$2H_2 + C \rightleftarrows CH_4 \tag{II}$$

$$H_2 + H_2O + 2C \rightleftarrows CO + CH_4 \tag{III}$$

It is further assumed in this model that the gasification of carbon by carbon dioxide is negligible under the conditions considered and that carbon dioxide interacts only in the water-gas shift reaction.

Total carbon conversion rate, dx/dt, is described by the following expression:

$$\frac{dx}{dt} = f_L k_T (1 - X)^{2/3} e^{-\alpha X^2} \tag{23.183}$$

where f_L is the relative reactivity factor, which depends on the char type and char pretreatment temperature, and α, k_t are kinetic parameters.

Also

$$k_T = k_{\text{I}} + k_{\text{II}} + k_{\text{III}} \tag{23.184}$$

where k_{I}, k_{II}, k_{III} are the rate constants for the three gasification reactions under consideration. Thus knowledge of k_{I}, k_{II}, and k_{III} not only gives the total gasification rate but also defines reaction stoichiometry.

Although the correlations above indicate that the three gasification reactions occur independently, implying that different active sites are involved, the rates of each of the reactions are assumed to be proportional to the same carbon conversion term, $(1 - X)^{2/3} e^{-\alpha X^2}$. Individual parameters in Equations 23.183 and 23.184 are defined as a function of the temperature and partial pressure of individual gas species, as follows:

$$f_L = f_L^0 \exp - \left[\frac{9340}{R}\left(\frac{1}{T} - \frac{1}{T_P}\right)\right] \tag{23.185}$$

[207] G. E. Goring, G. P. Curran, R. P. Tarbox, and E. Gorin, *Ind. Eng. Chem.*, **44**, 1057–1065 (1952).
[208] G. E. Goring, G. P. Curran, R. P. Tarbox and E. Gorin, *Ind. Eng. Chem.*, **45**, 2856–2891 (1953).
[209] K. H. Van Heek, H. Jüntgen, and W. Peters, *J. Inst. Fuel*, **46**, 249–258 (1973).
[210] G. E. Goring, G. P. Curran, R. P. Tarbox, and E. Gorin, *Ind. Eng. Chem.*, **44**, 1051–1057 (1952).
[211] J. D. Blackwood, *Coke Gas*, **22**, 190–218 (1960).
[212] K. Hedden and G. Mienkina, *Brennstoff-Chemie*, **46**, 366–371 (1965).

$$k_I(\text{min}^{-1}) = \frac{\exp\left(9.0201 - \dfrac{12{,}910}{T}\right)\left(1 - \dfrac{P_{CO}\,P_{H_2}}{P_{H_2O}K_I^E}\right)}{\left[1 + \exp\left(-22.216 + \dfrac{24{,}882}{T}\right)\left(\dfrac{1}{P_{H_2O}} + 16.35\dfrac{P_{H_2}}{P_{H_2O}} + 43.5\dfrac{P_{CO}}{P_{H_2O}}\right)\right]^2} \quad (23.186)$$

$$k_{II} = \frac{P_{H_2}^2 \exp\left(2.6741 - \dfrac{13{,}672}{T}\right)\left(1 - \dfrac{P_{CH_4}}{P_{H_2}^2 K_{II}^E}\right)}{1 + P_{H_2}\exp\left(-10.452 + \dfrac{11{,}098}{T}\right)} \quad (23.187)$$

$$k_{III} = \frac{P_{H_2}^{1/2}P_{H_2O}\exp\left(12.4463 - \dfrac{20{,}043}{T}\right)\left(1 - \dfrac{P_{CH_4}P_{CO}}{P_{H_2}P_{H_2O}K_{III}^E}\right)}{\left[1 + \exp\left(-6.6696 + \dfrac{8443}{T}\right)\left(P_{H_2}^{1/2} + 0.85\,P_{CO} + 18.62\dfrac{P_{CH_4}}{P_{H_2}}\right)\right]^2} \quad (23.188)$$

$$\alpha = \frac{52.7P_{H_2}}{1 + 54.3P_{H_2}} + \frac{0.521P_{H_2}^{1/2}P_{H_2O}}{1 + 0.707P_{H_2O} + 0.50P_{H_2}^{1/2}P_{H_2O}} \quad (23.189)$$

where

$K_I^E, K_{II}^E, K_{III}^E$ = equilibrium constants for reactions I, II, and III, considering carbon as graphite

T = reaction temperature (K)

T_P = maximum temperature to which the char has been exposed prior to gasification (K) (if $T_P < T$, then a value of $T_P = T$ is used)

P_i = partial pressure of the gas phase species (atm)

f_L^0 = relative reactivity factor, which depends on the nature of the particular carbonaceous solid

t = time (min)

As can be seen from the description above, Johnson has proposed that the effect of pretreatment temperature on the rates of gasification in gases containing steam and hydrogen is the same as that proposed for gasification in hydrogen alone. Although fewer different types of coal chars were tested in gases containing steam and hydrogen than were tested in hydrogen alone, Johnson nevertheless suggested that the relative reactivity factor correlation (Fig. 23.50) for gasification in hydrogen also appears to be applicable for gasification in steam-hydrogen mixtures.

In addition to the thermobalance data used to develop the model described, Johnson also considered the differential rate data of Zielke and Gorin[174] for the gasification of Disco char. Zielke and Gorin's data exhibited the constant or decreasing specific gasification rates with increasing carbon conversion that are predicted by Equations 23.183 and 23.189 as a function of gas composition and pressure. A comparison of the predictions of Johnson's correlation with the initial gasification rates ($X = 0$) reported by Zielke and Gorin is shown in Figs. 23.61 and 23.62 for the gasification of Disco char at 917°C. The value of f_L^0 was 0.49. Although Zielke and Gorin's complete data show that relative values of $W_{CH_4} : W_{CO}$ increase somewhat with increasing carbon conversion—contrary to Johnson's model, which predicts that this ratio should not vary under constant environmental conditions—it has been suggested by investigators at the Consolidation Coal Company[213] that this effect may be due to a catalytic reaction downstream of the fluidized-bed reactor, in which some of the carbon monoxide produced in the reactor was converted to methane.

For gasification in pure steam, Johnson's correlation predicts that rates initially in-

[213] G. Curran and E. Gorin, U.S. Office Coal Res. R & D Rep. No. 16—Interim Rep. No. 3, Book 2 (PB 184719); NTIS, 1970.

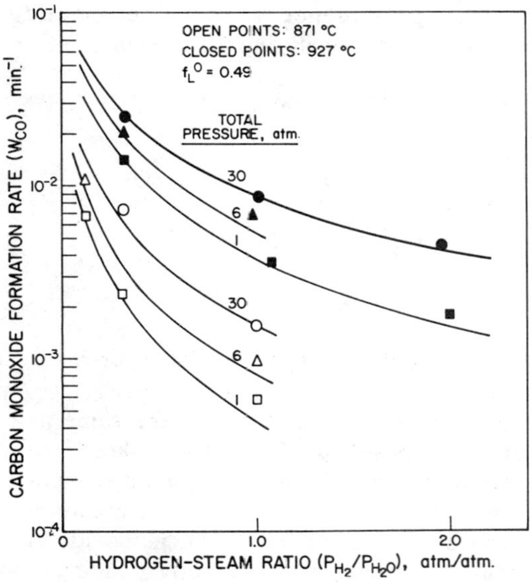

Figure 23.61 Comparison of experimental and calculated CO formation rates for gasification of Disco char in steam-hydrogen mixtures.[7174]

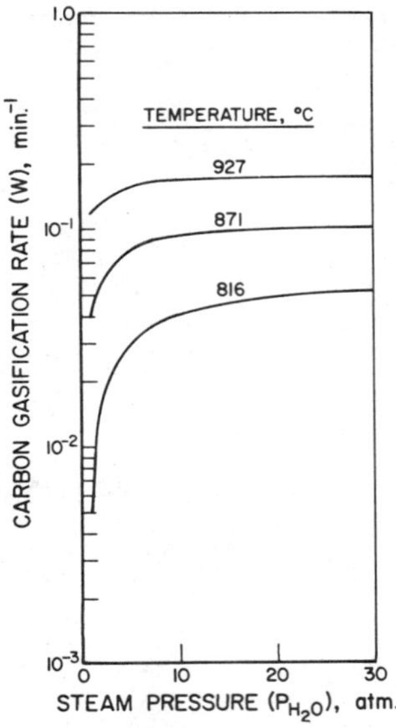

Figure 23.63 Effect of pressure on carbon gasification rate in pure steam ($f_L^0 = 1.0$).[7]

crease with pressure, but eventually become independent of pressure at pressures greater than 10 atm, as shown in Fig. 23.63 (f_L^0 = 1.0). Similar behavior was observed by Van Heek et al.[209] for the gasification of various coals or coal chars in steam, under conditions of constant coal heatup rate. Figure 23.64 compares the initial specific gasification rate in pure steam at 10 atm as predicted by Johnson's correlation[7] (f_L^0 = 1.0) with that predicted by Van Heek's correlation[209] for the gasification of a high-volatile bituminous coal.

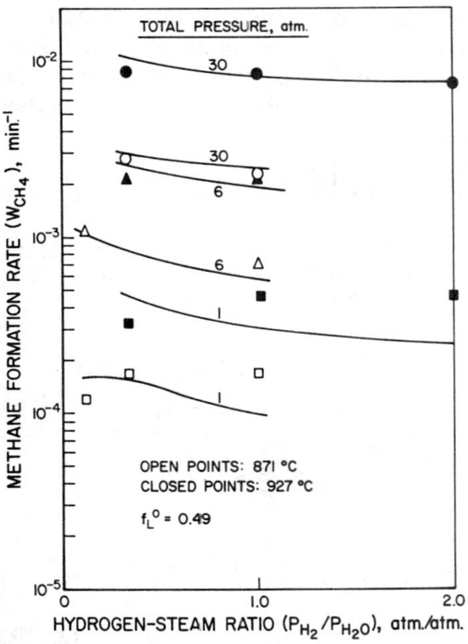

Figure 23.62 Comparison of experimental and calculated methane formation rates for gasification of Disco char in steam-hydrogen mixtures.[7,17.4]

The kinetics of the gasification of coconut charcoal in steam-hydrogen mixtures were studied by Blackwood[211] and by Blackwood and McGrory[200] at total pressures from 1 to 50 atm and at temperatures from 750 to 850°C. Of the carbon gasified in this study at the relatively low hydrogen/steam ratios employed (below about 0.1 mole/mole) less than approximately 20% was converted to methane, with the remaining gas composed of carbon monoxide and carbon dioxide. Blackwood and McGrory suggest that carbon dioxide is a secondary product of the water-gas shift reaction. Relatively large gas flow rates were used in that study, to limit steam decomposition to less than about 6%, and results were interpreted as if the system were a differential reactor. Although total carbon conversion was usually less than about 10%, in some tests specific carbon conversion rates at 830°C were approximately constant for carbon conversions up to 50%.

That study was primarily interested in quantitative evaluations of the kinetics of carbon oxide formation. It was found that at

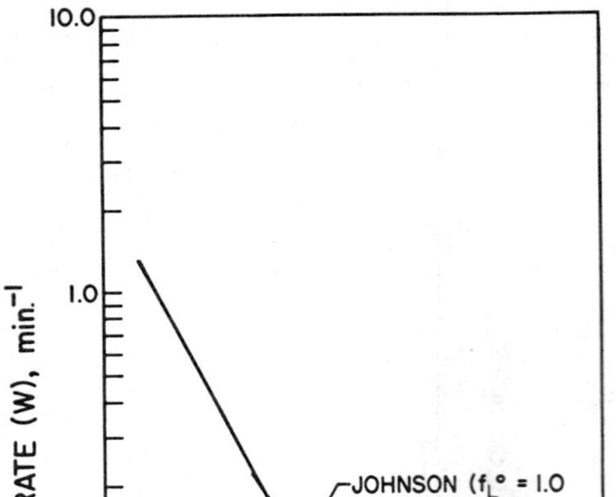

Figure 23.64 Comparison of gasification rates in pure steam at 10 atm.[7,209]

constant steam partial pressure, carbon oxide formation rates decreased significantly with increasing hydrogen partial pressure, although this inhibition by hydrogen was less severe at 830°C than at 750°C. At constant hydrogen partial pressure, rates increased continuously with increasing steam partial pressure. At 830°C and a low hydrogen partial pressure, rates were essentially proportional to the steam partial pressure. At lower temperatures and higher hydrogen partial pressures, the reaction order with respect to steam partial pressure was greater than unity. Methane formation rates at 830°C were found to be relatively independent of the hydrogen partial pressure over the range employed, but increased with steam partial pressure at a reaction order less than unity.

Blackwood and McGrory[200] postulated the following sequence to develop kinetic expressions for analyzing their results:

Step 15. $C_f\,(\text{dual site}) + H_2O \underset{k'_{15}}{\overset{k_{15}}{\rightleftarrows}} C(H)\,(OH)$

Step 16. $C(H)\,(OH) \xrightarrow{k_{16}} C(H_2)\,(O)$

Step 6. $C\,(O) \xrightarrow{k_6} CO$

Step 1. $H_2 + C_f \underset{k_2}{\overset{k_1}{\rightleftarrows}} C(H_2)$

Step 17. $C(H_2) + H_2O \xrightarrow{k_{17}} CH_4 + C(O)$

Reaction steps 15 and 16 are the same as those proposed by Long and Sykes[197] to account for steam adsorption and decomposition on dual sites present on the carbon surface. In this model carbon monoxide is formed by means of step 6, corresponding to the decomposition of the surface carbon-oxygen complex. Methane formation occurs by step 17, postulated to account for the interaction between surface carbon-hydrogen complex and gaseous steam to produce methane and surface oxygen–carbon complex. Although Blackwood and McGrory indicated that methane could also be formed by reaction step 2 [i.e., $C(H_2) + H_2$ <=====> $CH_4 + C_f$], they felt that this reaction would only be important at higher hydrogen partial pressures than those employed in their study.

Based on these reaction steps, the following expression was derived to describe the rate of carbon gasification to form carbon monoxide:

$$W_{CO} = \left[\frac{k_{15}\,k_{16}}{k'_{15}\,k_{16}} P_{H_2O} + \frac{k_1\,k_{17}}{k'_1} P_{H_2} P_{H_2O} + \frac{zk_{15}\,k_{16}\,k_{17}}{k_5\,(k_{16} + k'_{15})} P^2_{H_2O}\right] n_t \Bigg/ \Bigg\{1 + \frac{k_1}{k'_1} P_{H_2} + \left(\frac{k_{17}}{k_5} + \frac{\dfrac{k_{15}\,k_{16}}{k_6} + \dfrac{k_{15}\,k_{16}}{k_5} + k_{15}}{k'_{15} + k_{16}}\right) P_{H_2O} + \left[\frac{2k_{15}\,k_{16}\,k_{17}}{k_5\,k_6\,(k'_{15} + k_{16})} + \frac{k_{15}\,k_{17}}{k_5\,(k_{15} + k_6)}\right] P^2_{H_2O} + \frac{k_1\,k_{17}}{k_5\,k_6} P_{H_2} P_{H_2O}\Bigg\} \tag{23.190}$$

The P_{H_2}, P_{H_2O}, and $P^2_{H_2O}$ terms in the denominator of Equation 23.190 were considered to be negligible, leading to the equation form:

Table 23.23 Values of Kinetic Parameters

Temp. (°C)	k_{OH} (min^{-1} atm^{-1} $\times 10^{-3}$)	k'_{OH} (min^{-1} atm^{-2} $\times 10^{-3}$)	k''_{OH} (min^{-1} atm^{-2} $\times 10^{-5}$)	$K_{OH,1}$ (atm^{-1})	$K_{OH,2}$ (atm^{-1})
750	0.43	0.36	1.80	35	0.06
790	1.50	0.60	1.80	35	0.09
830	4.44	1.26	1.80	35	0.14

$$W_{CO} = \frac{k_{OH} P_{H_2O} + k'_{OH} P_{H_2} P_{H_2O} + k''_{OH} P^2_{H_2O}}{1 + K_{OH,1} P_{H_2} + K_{OH,2} P_{H_2O}} \tag{23.191}$$

Although Blackwood and McGrory did not give an expression to describe methane formation rates, the following equation can be derived from their model:

$$W_{CH_4} = \frac{k'_{OH} P_{H_2} P_{H_2O} + (k''_{OH}/2) P^2_{H_2O}}{1 + K_{OH,1} P_{H_2} + K_{OH,2} P_{H_2O}} \tag{23.192}$$

where the kinetic parameters are defined in the same way as those in Equation 23.191. Blackwood and McGrory evaluated the kinetic parameters in their model by analyzing carbon monoxide formation rates, and obtained the results given in Table 23.23. The apparent activation energies for k_{OH}, k'_{OH}, and $K_{OH,2}$ are 60, 30, and 20 kal/g-mole, respectively.

Experimentally obtained gasification rates[7] at 830°C are compared to values calculated from Blackwood and McGrory's model[200] in Figs. 23.65 and 23.66. In Figure 23.65 the predicted carbon monoxide formation rates are in good agreement with experimental values, which is to be expected because these data were used for the evaluation of kinetic parameters. Figure 23.66 shows less consistency between experimentally determined methane formation rates and values predicted from Equation 23.192 using the kinetic parameters in Table 23.23, which were deduced from carbon monoxide formation data. The model does, however, predict the relative independence of methane formation rates from hydrogen partial pressure over the range studied. It is probable that using a different set of kinetic parameters derived both from carbon oxide and methane formation rates would show better consistency with all the data obtained.

Figures 23.65 and 23.66 also show predict-

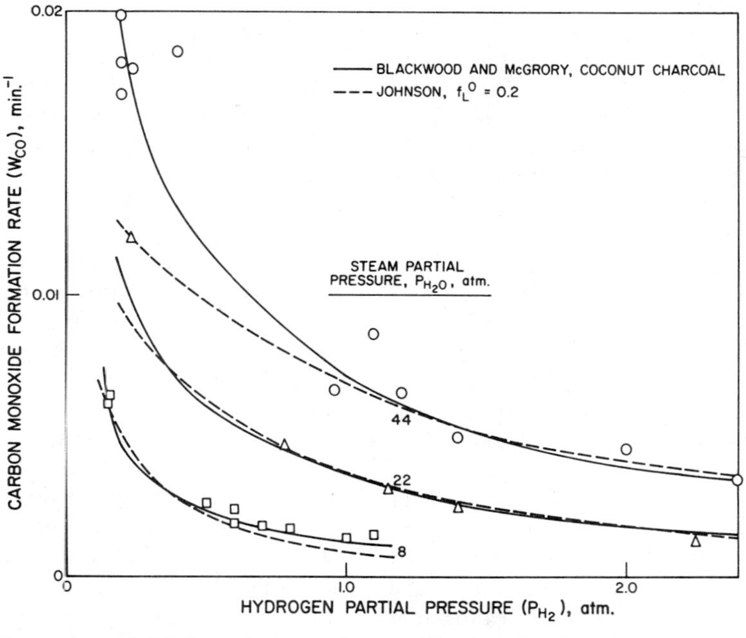

Figure 23.65 Comparison of CO formation rates from gasification of carbons, based on different kinetic correlations.[7,200]

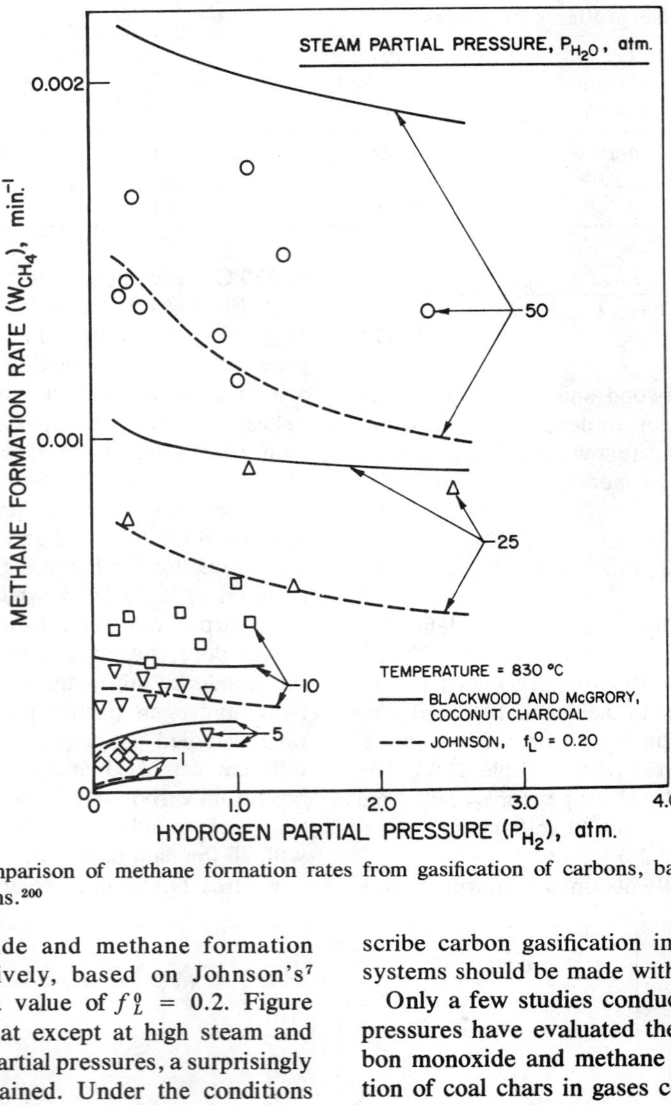

Figure 23.66 Comparison of methane formation rates from gasification of carbons, based on different kinetic correlations.[200]

ed carbon oxide and methane formation rates, respectively, based on Johnson's[7] model, using a value of $f_L^0 = 0.2$. Figure 23.65 shows that except at high steam and low hydrogen partial pressures, a surprisingly good fit is obtained. Under the conditions given in Fig. 23.66, the predictions of Johnson's model for methane formation rates are only approximate, as are those of Blackwood and McGrory's model. Johnson's model does, however, show that the predicted methane formation is relatively insensitive to variations in the hydrogen partial pressure, over the range considered. The general similarity between the predictions of Blackwood and McGrory's model and Johnson's model, as illustrated in Figs. 23.65 and 23.66, is of particular interest in view of the significant differences in the forms of correlations used and in the numerical evaluation of kinetic parameters. This is a direct indication that the theoretical interpretation of kinetic parameters in correlations developed to describe carbon gasification in more complex systems should be made with caution.

Only a few studies conducted at elevated pressures have evaluated the effects of carbon monoxide and methane on the gasification of coal chars in gases containing steam and hydrogen. In Johnson's model carbon monoxide severely inhibits the steam-carbon reaction, to an even greater extent than does hydrogen. This model also predicts that methane formation is inhibited by hydrogen, carbon monoxide, and methane itself, even at conditions far removed from equilibrium. Curran et al.[201] have also studied effects of gas composition on the gasification of Renner Cove lignite char in gases containing steam, hydrogen, carbon monoxide, and methane, at 816°C and at total pressures from 10 to 15 atm. The results of this study were interpreted by assuming that the following two reactions predominate:

$$C + H_2O \Longleftrightarrow CO + H_2 \quad (23.82)$$

$$H_2 + H_2O + 2C <= ==> CO + CH_4 \quad (23.143)$$

The rate-controlling step for Equation 23.82 was assumed to be the reaction between adsorbed H_2O and a free active site on the carbon surface to produce an adsorbed carbon-oxygen complex and an adsorbed carbon-hydrogen complex. This model can be represented by the following sequence:

Step 13. $C_f + H_2O \underset{k_{13}}{\overset{k_{13}}{\rightleftarrows}} C(H_2O)$

Step 1. $H_2 + C \underset{k_1}{\overset{k_1}{\rightleftarrows}} C(H_2)$

Step 6. $C(O) \underset{k_6}{\overset{k_6}{\rightleftarrows}} CO + C_f$

Step 20. $C(H_2O) + C \underset{k_{20}}{\overset{k_{20}}{\rightleftarrows}} C(O) + C(H_2)$

Reaction steps 1, 6, and 13 were assumed to be rapid, therefore in equilibrium.

The rate-controlling step for the reaction given by Equation 23.143 was assumed to be the reaction between adsorbed hydrogen and adsorbed H_2O on adjacent sites to produce adsorbed carbon-oxygen complex and adsorbed methane. The model for this overall reaction can be represented by the sequence:

Step 13. $C_f + H_2O \underset{k_{13}}{\overset{k_{13}}{\rightleftarrows}} C(H_2O)$

Step 1. $C_f + H_2 \underset{k_1}{\overset{k_1}{\rightleftarrows}} C(H_2)$

Step 6. $C(O) \underset{k_6}{\overset{k_6}{\rightleftarrows}} CO + C_f$

Step 19. $C(CH_4) \underset{k_{19}}{\overset{k_{19}}{\rightleftarrows}} CH_4 + C_f$

Step 18. $C(H_2) + C(H_2O) \underset{k_{18}}{\overset{k_{18}}{\rightleftarrows}} C(O) + C(CH_4)$

Reaction steps 1, 6, 13, and 19 were also assumed to be rapid and in equilibrium.

In the models proposed it was assumed that Equations 23.82 and 23.143 occur independently of each other on the carbon surface; that is, different active sites and surface complexes are involved in the two reaction sequences. With these assumptions, the following equations and evaluations of kinetic parameters were derived to describe initial gasification rates ($X = 0$):

$$W_C = \frac{0.0576\,[P_{H_2O} - P_{CO}P_{H_2}/3.26]}{(1 + 0.035\,P_{H_2O} + 0.36\,P_{H_2} + 1.30\,P_{CO})^2} \quad (23.193)$$

$$W_{CH_4} = \frac{0.01564\,[P_{H_2O}P_{H_2} - P_{CH_4}P_{CO}/0.388]}{(1 + 0.19\,P_{H_2O} + 0.16\,P_{H_2} + 1.25\,P_{CO} + 1.18\,P_{CH_4})^2} \quad (23.194)$$

where P_i = partial pressure (atm) (i = CO, H_2, H_2O, CH_4)
W_C = rate of reaction given in Equation 23.82 [moles C gasified/(moles C-min)]
W = rate of methane formation via Equation 23.143 [moles C gasified/(moles C-min)]

From the stoichiometry of Equations 23.82 and 23.143, the net rate of carbon gasification to form carbon monoxide, W_{CO}, is given by:

$$W_{CO} = W_C + W_{CH_4} \quad (23.195)$$

and the total carbon gasification rate, W_T, by:

$$W_T = W_C + 2W_{CH_4} \quad (23.196)$$

The effect of carbon monoxide in strongly inhibiting the rate of reaction given by Equation 23.82 is similar to the predictions of Johnson's model for the same reaction. The correlation of Curran et al. also predicts that carbon monoxide and methane will strongly inhibit the methane formation reaction.

The numerical evaluations given in Equations 23.193 and 23.194 correspond only to initial gasification rates. Different sets of parameters were reported at higher carbon conversion levels, reflecting the finding that total specific gasification rates as well as relative methane and carbon monoxide formation rates varied as a function of carbon conversion level. Such behavior, not usually observed with bituminous coal chars, may be related to variations in catalytic influences during the course of gasification of certain alkali mineral constituents inherent to many lignites.

The general reactivity levels of the Renner Cove lignite char used by Curran et al.[201] and of a typical char derived from a high-volatile bituminous coal can be compared on the basis of the correlations given in Equations 23.193 and 23.194 and the correlation proposed by Johnson ($f_L = 1.0$). Such a comparison is shown in Figs. 23.67 and 23.68. Although the qualitative trends exhibited in Figure 23.67 for the two kinds of char are similar with respect to the gasification of carbon to form carbon monoxide, the lignite char is about 100 times more reactive than the bituminous char. As a later section explains, this difference in reactivity level is due primarily to inherent catalytic influences on lignite char gasification, influences that are not present to a significant extent in the gasification of most bituminous chars.

As shown in Fig. 23.68, the reactivity of the lignite char with respect to methane formation is only about 8 times greater than that of the bituminous char. This reflects the selectivity of the catalyst present in lignite chars in favoring the formation of carbon monoxide and also points out that the kinetic correlations developed for bituminous char gasification are not generally applicable to the gasification of lignite char.

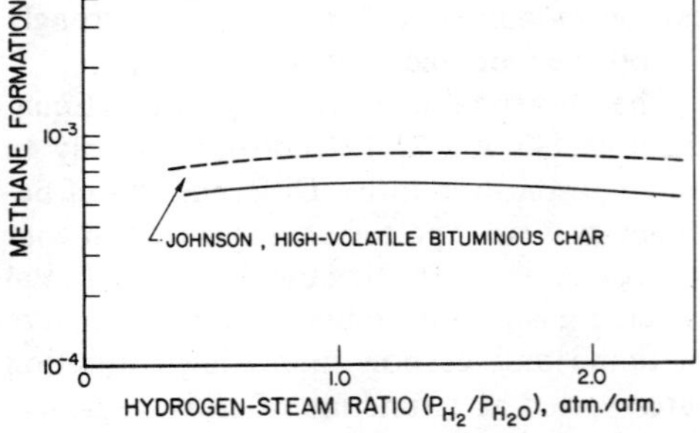

Figure 23.67 Comparison of CO formation during gasification of different coal chars in steam-hydrogen mixtures.[7,201]

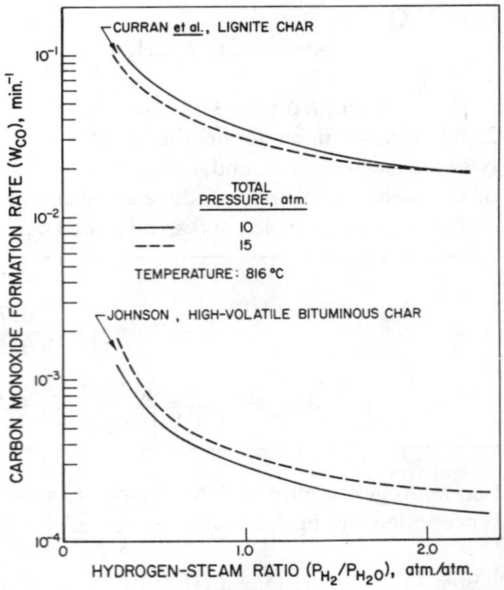

Figure 23.68 Comparison of methane formation rates during gasification of different coal chars in steam-hydrogen mixtures.[7,201]

23.4.2.4 *Particle Size Effects on Char Gasification*

The gasification of porous carbons involves the transport of reactant and product gases between the external bulk gas phase and the internal particle surfaces, where chemical reactions take place. Under conditions where gaseous concentrations within a particle are different from the external gaseous concentrations, mass transport or diffusion processes affect overall gasification rates. When this occurs, overall gasification rates generally tend to decrease with increasing particle size. Gaseous concentration gradients can exist (1) between the bulk gas phase and the external particle surface (because of film diffusion resistances) or (2) between the external particle surface and internal particle surfaces (because of diffusion limitations within the particle pore structures). In either case, where there are limitations on the gaseous diffusion in the gasification of carbon particles, this effect must be considered in interpreting results of laboratory studies and in applying chemical kinetic models to anticipate behavior in large-scale systems.

Film diffusion effects can be reasonably well anticipated on the basis of existing technologies. The more complicated situation is in dealing with internal particle diffusion.

Walker et al.[214] have described this situation in detail by considering diffusion effects on the gasification of purer carbons. Other investigators[174,183,215] have also quantitatively considered effects of intraparticle diffusion on the gasification of coal chars and other carbons, employing various theories[216–218] to account for gaseous mass transport within porous solids. With coal chars, however, the complex pore structures, which change during the course of gasification, present a formidable challenge to theoretical analysis of diffusion behavior.

Walker and co-workers[98,99] have indicated that coal chars have a trimodal pore structure consisting of macropores, transitional or "feeder" pores, and micropores (< 12 Å in diameter). They have also suggested that although gaseous diffusion through macropores and transitional (feeder) pores is quite rapid (a situation in which gaseous concentrations at the mouths of micropores closely approach the concentration at the external particle surface), gaseous diffusion within the micropore system can be quite slow. In fact, for pores or apertures less than about 5 Å, gaseous diffusion is an activated process.[219] The micropore system within coal char particles, however, contains almost all the internal surface area, and therefore almost all the active sites for gasification. As a general principle, then, it has been suggested that coal chars that contain large feeder pore systems tend to show fewer mass transport limitations during gasification. This is because increasing the feeder pore channels decreases the average diffusion length within micropores. In this context it has been shown that chars derived from low-rank coals have larger feeder pore systems than chars derived from high-rank coals.[98,120]

The existence of diffusion limitations in the gasification of a carbon depends both on the chemical reactivity of the carbon and on the pore structure characteristic of the carbon. Thus when chemical reactivity is low because of the inherent nature of the carbon or the reaction conditions employed, more severe diffusion resistance can be tolerated, without affecting gasification kinetics, than would be the case with high chemical reactivity. High chemical reactivity is, of course, favored by high temperatures. This leads to the observation that there is a transition in the controlling mechanism of gas-solid reactions from chemical reaction control to diffusion control that occurs with increasing temperature. This transition is usually accompanied by a decrease in the overall temperature coefficient of reaction.

Because of the complex effects that occur in pore diffusion processes during coal char gasification, quantitative predictions describing the interaction of chemical and diffusion phenomena require, at the very least, a detailed characterization of variations in the char pore structure. Although some information of this type is available, much more is needed. Fortunately, for many of the conditions relevant to coal char gasification systems that have been studied, relatively little effect of particle size on reaction kinetics has been observed. This is indirectly suggested by the large temperature coefficients derived to describe experimental results and the dependence of gasification rates on carbon conversion level, as well as by the more direct comparison of gasification rates obtained with different particle sizes. Van Heek et al.,[209] for example, found that particle sizes from about 0.1 to 1.4 mm had no effect on the gasification of a bituminous coal in steam, at 10 atm and temperatures up to 950°C. The nominal gasification rate at 950°C was about 0.01 min^{-1}. Similarly, for a brown coal, particle size in the same range had no effect at temperatures up to 800°C, where a rate of 0.02 min^{-1} was obtained. Ergun[188] found that particle sizes from approximately 0.2 to 1.0 mm did not affect the gasification of a metallurgical coke in steam at 1 atm and temperatures up to 1200°C. At 1200°C an integral carbon gasification rate of about 0.03 min^{-1} was obtained.

In spite of these results, there is little basis on which to accurately predict—even from

[214] P. L. Walker, Jr., L. G. Austin, and F. Rusinko, Jr., in *Advances in Catalysis,* Vol. 11, Academic Press, New York, 1959, pp. 133–221.

[215] Y. Kunio and D. Kunii, *J. Chem. Eng. Jap.,* **2,** 170–174 (1969).

[216] E. W. Thiele, *Ind. Eng. Chem.,* **31,** 916 (1939).

[217] J. B. Zeldowitsch, *Acta Physicochim. USSR,* **10,** 583 (1939).

[218] A. Wheeler, in *Advances in Catalysis,* Vol. 3, Academic Press, New York, 1951, pp. 249–327.

[219] P. L. Walker, Jr., L. G. Austin, and S. P. Nandi, *Chemistry and Physics of Carbon,* Vol. 2, Marcel Dekker, New York, 1960, pp. 257–371.

an empirical standpoint—conditions under which internal particle diffusion may become important in coal char gasification systems. This is because there is only a limited number of studies in which the particle sizes of coal chars were systematically varied. Although very approximate considerations suggest that chemical reaction control will predominate with most coal chars during gasification (uncatalyzed) at temperatures less than 1000°C and for particle sizes less than about 2 mm, the possibility of diffusion inhibition should be considered at more severe conditions, such as those likely to exist in steam-oxygen gasifiers.

23.4.2.5 *The Use of Catalysts in Coal Char Gasification*

Existing and projected shortages of natural gas in the United States have stimulated extensive research to develop the technology for commercial production of synthetic high-Btu gases of pipeline quality. In this effort coal gasification has been given particular emphasis because of the extensive availability of coal as a natural resource in this country, and several large-scale process development programs are currently being conducted with government and industrial support. The main objective of these programs has been to use modern technologies to develop processes with significantly improved efficiencies, compared to classical European gasification systems. Data available from the different programs currently at various stages of development already indicate that the objectives of improved efficiency and performance will indeed be achieved in the relatively near future. This evidence also indicates, however, that there are still practical limitations to obtaining increased efficiency solely by manipulating process configurations and using modern gas-solid contacting systems, and that these limitations are due primarily to inherent chemistry of coal gasification in the absence of major catalytic influences. Since these limitations are not theoretical or thermodynamic but result instead from the relative and absolute chemical kinetic characteristics, there is considerable interest in the use of catalysts in gasification systems as a means for further significant improvement in process efficiencies.

The primary process objective in the gasification of coal chars to produce methane is to achieve the following reaction stoichiometry:

$$2C + 2H_2O \rightleftharpoons CH_4 + CO_2 \quad (23.197)$$

Unfortunately, this reaction is not feasible in uncatalyzed systems. The actual reaction path corresponds more closely to a three-part synthesis:

$$C + H_2O \rightleftharpoons CO + H_2 \quad (23.82)$$

$$C + 2H_2 \rightleftharpoons CH_4 \quad (23.1)$$

$$CO + H_2O \rightleftharpoons CO_2 + H_2 \quad (23.4)$$

$$\overline{2C + 2H_2O \rightleftharpoons CH_4 + CO_2} \quad (23.197)$$

The reaction given by Equation 23.82 is endothermic and does not normally occur with bituminous coals at reasonable rates below about 800°C. In fact, for most processes, temperatures of about 1000°C or higher are necessary to obtain the required conversions with a reasonable reactor size. This temperature level, however, does not favor the exothermic reaction (Equation 23.1), which is relatively slower than Reaction 23.82 at elevated temperatures. Some processes improve this situation somewhat by using multizoned reactor systems, in which hydrogen generated in a high-temperature zone (by Reaction 23.82) subsequently reacts with carbon in a lower temperature zone to form methane, by Reaction 23.1. But even with this procedure, the overall gasification process is still endothermic. Additional heat to sustain the reactions—usually achieved by further combustion of the char with oxygen—requires increased feed coal and oxygen and increased catalytic methanation to produce a given amount of final methane product, all of which decrease the process thermal efficiencies.

The potential for approaching the stoichiometry indicated by Reaction 23.197 is possible only as a result of the action of catalysts that preferentially enhance either direct formation of methane, as in Reaction 23.1, or indirect formation of methane through a reaction such as Equation 23.198, occurring in the gasifier itself:

$$CO + 3H_2 \rightleftharpoons CH_4 + H_2O \quad (23.198)$$

Thermodynamic considerations, however, dictate that even with optimal catalytic action, substantial improvements in process efficiencies can occur only by operating the gasifier at temperatures much lower than those normally required in the absence of catalysts.

Since the 1920s various investigations of the effects of catalysts on the gasification of carbonaceous solids have had the objective of improving processes for the production of water gas, producer gas, or hydrogen as sources for producing ammonia. More recently, studies have also been conducted in connection with gas-cooled nuclear reactors.[176,220–232] Although most work with these objectives has been conducted at low pressures with cokes, charcoals, and graphite, much of the information obtained, with the large number of catalysts evaluated, has at least qualitative value in assessing the potential effects of catalysts on char gasification for methane production. Catalysts surveyed in the earlier work include metals, metal oxides, metal halides, alkali carbonates, and iron carbonyls, with particular emphasis on K_2CO_3, Na_2CO_3, KCl, NaCl, CaO, and transition metals. Some of the general conclusions are as follows:

1 Relative catalytic effects decrease with increasing gasification temperature.
2 In the gasification process catalysts are generally more effective with gases containing steam than with gases containing hydrogen alone.
3 There is usually an intermediate optimum catalyst concentration, beyond which either negligible or negative effects result.
4 The relative effects of different catalysts can differ at different reaction conditions.
5 The specific methods and conditions used for catalyst impregnation can significantly affect subsequent gasification reactivity.
6 Catalyst impregnation is more effective than physical mixing with the carbon.

With respect to point 6, the method of catalyst addition is very important to the process of coal gasification. The advantages of catalyst impregnation must be economically weighed against the corresponding requirements for catalyst addition operations and catalyst recovery problems, which are likely to be more severe than with physically mixed catalyst-coal systems. The use of physical mixtures of catalysts and coal in laboratory studies presents a particular problem when the results are interpreted in terms of the performance of projected continuous gasifiers, because the transfer mechanism required for the catalyst to interact in the gasification step may depend on the physical nature of the contacting between the catalyst and the coal. Such contacting would be quite different for the fixed beds frequently used in laboratory studies than for the fluidized beds more likely to be used in large-scale systems.

Information that is directly relevant to the catalysis of coal char gasification has been obtained only within the past few years. In one such investigation, conducted at Case Western Reserve University,[233,234] a thermobalance apparatus was used to study the effects of approximately 20 impregnated catalysts on coal char gasification kinetics. In these studies by Gardner et al.[233] and by Wilks et al.,[234] catalytic effects were investigated at 850–950°C and 500–1000 psi for gasification in either hydrogen or steam.

[220] P. L. Walker, Jr., M. Shelef, and R. A. Anderson, *Chemistry and Physics of Carbon,* Vol. 4, Marcel Dekker, New York, 1968, pp. 287–383.
[221] F. J. Dent, W. H. Blackburn, and H. C. Millett, *Trans. Inst. Gas Eng.*, **88,** 150–217 (1938).
[222] D. A. Fox and A. H. White, *Ind. Eng. Chem.*, **23,** 259 (1931).
[223] F. H. Frank and M. Meraikib, *Carbon (Oxford)*, **8,** 423–433 (1970).
[224] V. I. Kislykh and N. V. Shishakov, *Gas Prom.*, **8,** 15–19 (1960).
[225] C. Kroger, *Angew. Chem.*, **52,** 129–139 (1939).
[226] C. Kroger and G. Melhorn, *Brennstoff-Chemie,* **19,** 257–262 (1938).
[227] F. J. Long and K. W. Sykes, *Proc. R. Soc. (London)*, **A215,** 100–110 (1952).
[228] B. Neumann, C. Kroger, and E. Fingas, *Z. Anorg. Allg. Chem.*, **197,** 321–418 (1931).
[229] H. S. Taylor and H. A. Neville, *J. Am. Chem. Soc.*, **43,** 2055–2071 (1921).
[230] A. Tomita and Y. Tamai, *J. Catal.*, **27,** 293–300 (1972).
[231] F. J. Vastola and P. L. Walker, Jr., *J. Chem. Phys.*, **58,** 20–24 (1961).
[232] L. Vignon, *Ann. Chim.*, **15,** 42–60 (1921).
[233] N. Gardner, E. Samuels, and K. Wilks, *Prepr., Am. Chem. Soc., Div. Fuel Chem.*, **18**(2), 43–85 (1973).
[234] K. Wilks, E. Samuels, and N. Gardner, Thesis of K. Wilks, Case Western Reserve University, Cleveland, Ohio, 1974.

Most tests used a 5% catalyst concentration. It was concluded that although a few catalysts are as effective as the alkali metals in promoting gasification rates, none was more active. With 5% $KHCO_3$, gasification rates in hydrogen at 500 psi and 950°C are increased twofold, whereas in steam, under the same conditions, gasification rates are increased about threefold. More recently, data have been obtained indicating that catalyst combinations, such as alkaline earth metals with cobalt molybdate, are even more effective than single catalysts in enhancing gasification rates,[235] suggesting that a study of catalyst combinations would be a promising area for future research.

A thermobalance apparatus was also used at the Institute of Gas Technology to study the catalytic effects of exchangeable cations, such as sodium or calcium, on lignite gasification reactivities.[120] This study was prompted by the knowledge that lignite chars are inherently more reactive than bituminous chars for gasification. When treated in acid (such as HCl), however, the gasification reactivity of the lignite chars is substantially reduced and becomes comparable to the reactivity of the bituminous chars. It has been postulated that the inherently high reactivities of lignites are due to the catalytic influences of sodium or calcium combined with carboxyl functional groups in the organic structure of the lignites. When treated in acid, however, the cation can be removed by the following exchange reaction:

$$\text{—COONa} + H^+ \rightleftharpoons \text{—COOH} + Na^+ \quad (23.199)$$

Since the concentration of carboxyl functional groups decreases significantly with increasing coal rank, this inherent catalytic effect is not significant in bituminous coal chars.

In studies conducted at IGT the reversibility of this exchange reaction was used as a means to systematically investigate the effects of exchangeable cation concentration on lignite char gasification reactivities. Lignites were initially treated in hydrochloric acid to remove inherent calcium or sodium; controlled amounts of calcium or sodium were then added to the lignite by subsequent exchange, through reversal of the reaction above. Experimental results showed that for a 3% sodium concentration, a twentyfold increase in gasification rates was obtained at 760°C and 500 psi in a steam-hydrogen mixture, compared to rates obtained with the lignite when containing no exchangeable cations (acid-treated only).

In general, gasification rates increased with increasing catalyst concentration, at least up to a concentration of 3%. Similar results were obtained with both calcium and sodium, at comparable mass concentrations, for gasification in steam-hydrogen mixtures. Relative catalytic effects decreased with increasing gasification temperature, such that with steam-hydrogen mixtures at 930°C, only a sevenfold increase in gasification rates was obtained with 3% catalyst addition. For gasification in pure hydrogen at 930°C, the catalyst had significantly less effect than with steam-hydrogen mixtures. Sodium was also found to be about 4 times more effective than calcium for gasification in pure hydrogen.

Attempts to add sodium and calcium to bituminous coals were unsuccessful, presumably because of the lack of carboxyl functional groups. It is not known whether the exceptionally large catalytic effects obtained with organically combined sodium and calcium in lignites are due to the particular chemical state of the catalytic complexes or to the greater dispersion that is likely, compared with simple deposition on internal surfaces by solution evaporation. Future studies to determine methods of promoting organic complexes between coal and catalyst should clarify this matter, however.

Some recent studies have been conducted at several laboratories to investigate the gasification yields that can be obtained with physical catalyst-coal mixtures, under conditions of integral gas conversion at elevated pressures. In one such study conducted by the U.S. Bureau of Mines,[236] the effects of physically mixed catalysts were investigated in a continuous, 4-in. diameter, fluidized-bed reactor used in the development of the Synthane process. Additives included 2, 5, and 10% dolomitic limestone, 2 and 5% hydrated

[235] N. Gardner, paper presented at Symp. Catal. Conv. Coal, Pittsburgh, April 1975.

[236] A. J. Forney, W. P. Haynes, S. J. Gasior, and R. F. Kenny, *Prepr., Am. Chem. Soc., Div. Fuel Chem.*, **19**(1), 111–122 (1974).

lime, and 5% lignite ash. For operation at 600 psi with a steam-oxygen feed at 875–950°C, total carbon conversions increased by an average of approximately 10%, with the added materials. This increase in conversion is equivalent to the effect that would be achieved with a decrease in gasifier temperatures of about 50°C, compared to the temperatures required for the same coal throughput without the additives. It was also noted that with the majority of additives tested, no slag formation occurred, suggesting that the gasifier could be operated at higher temperatures, if desired, to achieve higher throughputs.

Another Bureau of Mines study, reported by Haynes et al.,[237] evaluated the effects of approximately 40 different catalysts on yields obtained by the gasification of a high-volatile bituminous coal with steam, at 850°C and 300 psig. In this study 5% catalyst concentrations were used in physical mixtures with the coal charge. A fixed-bed, flowing-gas system was employed, and total gaseous yields were reported for 4-hr test periods. It should be noted that since raw coals were used in this study as well as the one discussed above, total gas yields result from reactions that may occur in both the coal and char gasification stages.

As in earlier studies, Haynes et al. found that the best overall reactivities were achieved with alkali catalysts, particularly K_2CO_3 and KCl. With these materials, total carbon conversions of about 80% were obtained during the 4-hr test period, compared to a reference carbon conversion of less than 50% obtained without any catalyst. For the most part, increase reactivity led to increased formation of carbon oxides and hydrogen, relative to increases in methane formation.

An exception to this trend was obtained with an unactivated Raney nickel catalyst sprayed onto the surface of a metal tubular insert in the coal bed. In this case, total methane yields increased by about 24%, although total carbon conversion only increased by about 10%, indicating that the major influence of the catalyst was on the methanation of the carbon monoxide and hydrogen formed by direct gasification. The sprayed Raney nickel catalyst, however, decreased in reactivity with continued use, and loss of reactivity was estimated at 20 hr of operation. This loss of reactivity under the relatively severe conditions existing in the gasifier was attributed both to exfoliation of the catalyst and to sulfur poisoning, which is currently one of the major problems in the use of nickel-based catalysts in coal gasifiers.

In evaluating the significance of the role played by catalysts in coal gasification systems with gases containing both steam and hydrogen, it is important to distinguish between two major aspects: the increase in total carbon gasification rate, and the relative increase in hydrocarbon production. The latter aspect is particularly important because it contributes to increasing the thermal efficiency of a gasification process, and in many instances is probably an overriding factor in determining the practicality of using catalysts in coal gasifiers. Thus for the Bureau of Mines studies discussed above, it is significant that although with one exception, additives generally tended to increase the total carbon conversion rates, the increased carbon conversion apparently corresponded to a decrease in thermal efficiency for the overall gasification step.

The use of catalysts at lower temperatures has been the subject of several studies conducted at the University of Wyoming[238–240] with the objective of promoting the overall optimum reaction:

$$2C + 2H_2O ==> CH_4 + CO_2 \quad (23.197)$$

In one series of experiments the gasification, with steam, of low-rank coals mixed with K_2CO_3 and commercial nickel catalyst was studied in a fixed-bed, flowing-gas system at 650°C. It was possible with this system, even at low pressure (30 psi), to produce a gas consisting primarily of CO_2 and CH_4, with optimum catalyst concentrations of about 1 g of nickel catalyst per gram of coal and 0.2

[237] W. P. Haynes, S. J. Gasior, and A. J. Forney, *Prepr., Am. Chem. Soc., Div. Fuel Chem.*, **18**(2), 1–28 (1973).

[238] J. L. Cox and L. J. Sealock, Jr., *Prepr., Am. Chem. Soc., Div. Fuel Chem.*, **19**(1), 64–78 (1974).

[239] E. J. Hoffman, J. L. Cox, R. W. Hoffman, J. A. Roberts, and W. G. Willson, *Prepr., Am. Chem. Soc., Div. Fuel Chem.*, **16**(2), 64–67 (1972).

[240] W. G. Willson, L. J. Sealock, Jr., F. C. Hoodmaker, R. W. Hoffman, J. L. Cox, and D. L. Stinson, *Prepr., Am. Chem. Soc., Div. Fuel Chem.*, **18**(2), 29–42 (1973).

to 0.25 g of K_2CO_3 per gram of coal. On a CO_2-free basis, product gases were consistently obtained with heating values of about 850 Btu/scf. At 650°C, these yields were obtained for total test periods of about 7.5 hr and for total carbon conversions of less than 50%. Such results suggest that for acceptable large-scale operation at this low temperature, a much more active catalyst system or coal-catalyst contacting technique would be required.

Two other major problems also inherent to this system are catalyst recovery and loss of catalyst activity. Under optimum reaction conditions only about 80% of the potassium charged was recovered by a 1-hr water wash at room temperature, a relatively serious problem considering the large amounts of K_2CO_3 charged. In addition, relatively large fractions of the coal's sulfur combined with the nickel catalyst during gasification. Although this did not result in noticeable decreases in reactivity during batch charge tests, considering the large amounts of nickel catalysts employed, extrapolation of these effects indicated that replacement of the completely sulfided catalyst would be economically unfeasible, and some means of catalyst regeneration would have to be employed in commercial application. Several approaches to dealing with catalyst poisoning by sulfur have not yet been completely satisfactory, and it has been suggested that ultimately a combination of the use of sulfur scavengers, catalyst regeneration, more powerful sulfur-resistant catalysts, and prior sulfur removal may be required.

The observations above are generally consistent with other recent evaluations,[241–243] which conclude that there is a substantial need for additional research in areas related to the use of catalysts in coal char gasification. Specifically, these needs include the development of new catalysts and catalyst systems, with particular emphasis on detailed investigations of the mechanisms involved in the interactions of catalysts with coal or char gasification processes. Such detailed considerations would ideally involve a much more fundamental characterization of reaction systems than has been done in most of the previous studies. For example, particular attention should be directed toward characterization of the coals and coal chars used, not only in terms of elemental composition or rank type but also in terms of petrographic composition and internal physical properties, such as surface area and pore volume distribution. The dispersion characteristics, chemical state, and physical properties of the catalysts employed should also be determined, not only in the raw coal but also during the course of gasification.

For catalyst systems having the desired operating characteristics, detailed kinetic studies would be needed to provide the information necessary to evaluate fully their application in an integrated commercial system over the range of environmental conditions to which the catalyst would typically be exposed. Such kinetic studies should include not only major reactions but also side reactions and reactions pertinent to catalyst recovery.

[241] J. L. Johnson, *Catal. Rev.—Sci. Eng.*, **14**(1), 131–152 (1976).

[242] H. Katzman, Electric Power Research Institute, Rep. No. EPRI-FF207-0-0, Palo Alto, Calif., 1974.

[243] G. A. Mills, *Prepr., Am. Chem. Soc., Div. Fuel Chem.*, **16**(2), 107–123 (1972).

CHAPTER **24**

Coal Gasification Processes

DENNIS HEBDEN AND HENRY J. F. STROUD

Dennis Hebden, *O.B.E. B.Sc., Ph.D., F. Eng., F.I. Gas E., F. I. Chem. E, has been involved in the development of high-pressure processes for the production of town gas and SNG since he started work for the British gas industry in 1941. He worked for 25 years with Dr. F. J. Dent on the high-pressure slagging and nonslagging gasification of coal, the hydrogenation of coal, the catalytic gasification of petroleum distillates, the hydrogenation of crude, heavy, and distillate oils, and the catalytic synthesis of methane. Dr. Hebden was particularly concerned with the commercial introduction of the CRG and GRH processes in the United Kingdom and overseas. More recently as Programme Director of the Westfield Development Centre in Scotland he has been responsible for the development and demonstration of the British Gas/Lurgi Slagging Gasifier.*

Henry J. F. Stroud, *B.Sc., Ph.D., C.Eng., D.I.C., A.R.C.S., M.I. Gas E.,* ***M. Inst. E.,*** *has worked for the British Gas Corporation at its Midlands Research Station since 1968. Following several years carrying out research on the small-scale development of oil gasification processes, he was appointed in 1973 to take charge of the Process Evaluation Section in the Gas Production Division. He was appointed Head of Process Studies in 1979. Dr. Stroud's responsibilities include the evaluation and development of integrated processes for the gasification of oils and coal. He has written a number of technical papers and patents in this field.*

24.1 GENERAL INTRODUCTION

This chapter deals with processes in which steam, air, oxygen, hydrogen, or carbon dioxide are reacted heterogeneously with coal or char to produce gaseous mixtures comprised primarily of carbon oxides, hydrogen, methane, and steam. Coal gasification is an upgrading process in that it takes a solid, which is difficult to handle and to strip of undesirable constituents, and converts it into a gaseous form that can be readily purified and used directly as a fuel, or further reacted to provide a feedstock for the synthesis of other gaseous or liquid fuels, or chemicals. In the past gasification has fulfilled all these roles; but with an increasing awareness of the need for conservation of natural resources and the environment, coupled with rising oil prices, it has taken on a new significance, particularly in countries that have coal reserves and lack reserves of oil and gas or are depleting them.

Cradled in producer gas and water gas production, coal gasification grew up and developed during a period of increasingly intense competition from oil and gas in both the energy and chemical feedstock fields. Although this situation fostered innovation, many new processes never reached commercial maturity and others, after a short period of application, were abandoned owing to the introduction of cheaper, more flexible liquid and gaseous fuels and feedstocks. In the 1950s and 1960s petroleum dominated the market; a few recognized the transitory nature of this dominance and continued research and development, but no new gasification processes emerged and few coal gasification plants were installed. It was not until the later 1960s and early 1970s, when the United States began to experience natural gas shortages, and the finite nature of fossil fuel reserves and the need for conservation and good resource management were becoming universally accepted, that the significance of the world's coal reserves was fully recognized. This recognition was finally reinforced in 1973–1974 by the oil embargo and by the new attitude of some of the oil-producing countries. These events produced a tremendous surge in the interest in coal utilization, primarily in the fields of gasification and liquefaction, and large sums of money became available from governments for the research, development, and demonstration of new processes and the improvement and recasting of existing ones. This stimulation, evidenced by schemes such as Project Independence and the formation of the Energy Research and Development Administration (ERDA) in the United States, Project Sunshine in Japan, and a comprehensive government-funded energy research and development program in West Germany, has resulted in the proliferation of many new coal conversion processes.

The oil-based industrial expansion, which was initiated during World War II, and the subsequent growth of the oil and chemical industry, led to the development and introduction of many new chemical engineering techniques with the adoption of large integrated plants with large throughputs. These new techniques and engineering philosophies have become the basis of many of the gasification processes now being developed and specially tailored to end products such as substitute natural gas (SNG), synthesis gas, or fuel gas for electricity production.

24.1.1 The Evolution of Gasification Processes

The simplest and oldest gasification process is the air-blown gas producer. Formerly used to provide low-grade fuel gas for furnaces and for power generation in internal combustion engines, it is once again being installed in small units to replace natural gas and oil as a fuel. Air is passed through a fixed bed of screened coal or coke to yield mainly carbon monoxide, some carbon dioxide together with small amounts of hydrogen, methane, and so on, arising from the distillation of the coal. With the gas containing about 50% nitrogen, its calorific value is low, and with a preference for graded noncaking coal or coke, the application of the process is limited. The reactions that occur are:

$$C + O_2 \rightarrow CO_2 \qquad (24.1)$$

$$C + CO_2 \rightarrow 2CO \qquad (24.2)$$

$$2C + O_2 \rightarrow 2CO \qquad (24.3)$$

and they are collectively sufficiently exothermic to sustain reaction with the reactants supplied countercurrently at ambient temperature. High temperatures can be developed in the fuel bed, and when the ash in the fuel is fusible, the endothermic steam-carbon reaction must be imposed by adding steam to the air:

$$C + H_2O \rightarrow CO + H_2 \qquad (24.4)$$

This not only moderates the temperatures, but also yields hydrogen in the product gas. The steam-carbon reaction, however, is in its own right perhaps the most important gasification reaction of all; it not only gasifies carbon but it also splits water to liberate hydrogen. It is the basis of the other main original gasification process—the water gas process—in which the reactions above are carried out alternately in such a way as to allow the production of hydrogen and carbon monoxide without contamination with nitrogen.

The water gas process was once a major source of synthesis and town gas but was based on coke and therefore tended to be associated with carbonization. Coal-based water gas generators required special noncaking coals, and their application was therefore limited. The process is cyclic, involving a gas-making period during which the fuel bed is blown with steam to produce carbon monoxide and hydrogen, followed by an air-blowing period during which the heat to support the steam-carbon reaction is generated in the fuel bed. Because of its cyclic character, a gas holder is required to provide a continuous gas supply.

The advent of the large-scale production of industrial oxygen led to attempts, by admixing oxygen with the steam, to create a thermodynamic balance that would sustain a sufficient level of reaction temperature to allow for the continuous gasification of coal in a single fuel bed. This balance can be illustrated by the following summation of the combustion and steam-carbon reactions:

$$\tfrac{1}{2}C + \tfrac{1}{2}O_2 \rightarrow \tfrac{1}{2}CO_2 \qquad (24.1)$$

$$C + H_2O \rightarrow CO + H_2 \qquad (24.4)$$

$$1\tfrac{1}{2}C + H_2O + \tfrac{1}{2}O_2 \rightarrow \tfrac{1}{2}CO_2 + CO + H_2 \qquad (24.5)$$

Although the use of air separation to provide oxygen gave a product that was now nitrogen-free, some additional carbon dioxide was present. The quest for means of removing this gas led to the idea of operating the gasifier at boiler pressure, compressing the oxygen, and washing out the carbon dioxide with water under pressure. Thus pressure gasification was introduced and with it the production of methane in the fuel bed was discovered.

The Lurgi Company of Germany exploited these features, applying the system to the gasification of lignite, a fuel that was unsuitable for the then-traditional processes of carbonization for town gas manufacture. It was F. J. Dent[1] in England who discovered that the methane was produced by the hydrogenation or hydrogasification of the coal substance.

$$C + 2H_2 \rightarrow CH_4 \qquad (24.6)$$

[1] H. H. Lowry, Ed., *Chemistry of Coal Utilization,* Vol. 2, John Wiley and Sons, New York, 1945, p. 1607; Supplementary Volume, 1963, pp. 900, 996–1005.

He demonstrated the possibility of hydrogenating at high pressure much of the coal to methane at high rates; complete gasification could be achieved but at lower rates. These two events had profound effects on gasification technology. First, methane production was important because it offered the prospect of producing, by the complete gasification of coal in steam and oxygen, a medium-Btu gas suitable for distribution as town gas. Second, operation under pressure was significant because it not only favored methane formation but also offered the possibility of increasing gasifier specific loadings, and at the same time, simplifying and extending the downstream purification and treatment steps while delivering the gas under pressure for synthesis purposes or for admission to a gas transmission system. The Lurgi gasifier remains unique in this field, being the only currently commercially applied process for the gasification of coal under pressure. It is expected to become the source of synthesis gas for the first coal-based SNG plants in the United States, and extra units are presently being installed in South Africa to extend the Sasol oil-from-coal plant.

In most gasifiers the residual mineral matter of the coal was removed as unfused ash or light clinker, and this imposed limitations in output and in operating conditions. Slagging operation based on blast furnace practice in which the ash is removed as a molten liquid was introduced as early as the 1910s, but only for air operation. Little interest was shown in slagging operation on oxygen-blown gasifiers until the 1940s, but the application was limited to one or two cases. Slagging operation at elevated pressure is a relatively recent development, on the threshold of commercial application.

Also aimed at increasing specific output, using steam and oxygen or air and utilizing the reserves of reactive lignite, was the Winkler gasifier developed in Germany in the 1920s. This fluidized gasifier, the first application of fluidization in a commercial process, produced a gas low in methane suitable for ammonia synthesis, for hydrogen production or for electrical power generation.

The technique of fluidization received little further attention in the coal gasification field until the late 1940s, when it was applied experimentally to the control of temperatures and the removal of heat from fuel beds undergoing hydrogenation to produce methane, leading eventually to such process developments as Hygas and Hydrane. At about the same time interest was being shown in its application to the transfer of fuel and the circulation of chemical and inert heat carriers between reaction zones to permit the supply of heat by combustion with air separately from the gasification of carbon in steam to produce nitrogen-free synthesis and fuel gas.

Processes in which the coal is introduced with the gasifying agent, known as fully entrained gasification, first made their appearance in the 1930s, after which there was a marked interest in this system both in the United States and in Germany. Only one process, however, the atmospheric pressure Koppers-Totzek process, reached full commercial status. Some saw full-scale trial, but others were abandoned because of unfavorable economics in competition with oil and natural gas. However recent events have revived interest in them.

24.1.2 The Basic Elements of Gasification

The basic principles, stoichiometry, and thermodynamics of coal gasification are dealt with in detail in Chapter 23 and have been given considerable attention in previous volumes in this series.[2] The essential objective of gasification is to convert coal into a combustible gas with the maximum proportion of the potential heat in the coal appearing as potential heat in the gas. If the gas is to be used for synthesis rather than as a fuel, there is the further requirement that its composition needs to relate to the required product (e.g., whether it be hydrogen, hydrogen and carbon monoxide, a hydrocarbon, an alcohol, etc.). The primary difference therefore between fuel gas and synthesis gas production is that in the latter case, hydrogen has to be produced either by the "splitting" of water by the reaction:

$$C + H_2O \rightarrow CO + H_2 \qquad (24.4)$$

[2] H. H. Lowry, Ed., *Chemistry of Coal Utilization*, Supplementary Volume, John Wiley and Sons, New York, 1963, Ch. 20.

or by the secondary reaction:

$$CO + H_2O \rightarrow CO_2 + H_2 \qquad (24.7)$$

That water is a major reactant in gasification processes is often not fully appreciated; indeed, water supply can be a deciding factor in plant location. As distinct from coal, however, it is a reactant that is resynthesized when the product is combusted.

A secondary difference is the influence the end product may have on the choice of gasification process. In the simple case of hydrogen and carbon monoxide manufacture, one is concerned with the basic endothermic reaction (24.4), supported by the exothermic reaction (24.1) to give the overall reaction (24.5). By supplying extra steam, Reaction 24.7 can be promoted either in the fuel bed or in a separate catalytic conversion stage downstream to produce an increasing proportion of hydrogen. Taken to the ultimate, the product is hydrogen and carbon dioxide:

$$1\tfrac{1}{2}C + H_2O + \tfrac{1}{2}O_2 \rightarrow CO + H_2 + \tfrac{1}{2}CO_2 \qquad (24.5)$$

$$CO + H_2O \rightarrow CO_2 + H_2 \qquad (24.7)$$

$$1\tfrac{1}{2}C + 2H_2O + \tfrac{1}{2}O_2 \rightarrow 1\tfrac{1}{2}CO_2 + 2H_2 \qquad (24.8)$$

Hydrogen as the sole product is obtained by removing the carbon dioxide.

It will be noted that the oxygen in the water combines with the carbon to produce carbon dioxide. Carbon dioxide is the inevitable produce of "splitting" water with carbon or carbonaceous compounds to produce hydrogen, and although carbon is rejected by its removal, in its formation most of the potential heat of the carbon appears in the hydrogen. Thus, although there is little loss of "energy," there is a loss of carbon to a form that is converted by natural processes back to a fossil fuel so slowly that fossil fuels are regarded as nonrenewable.

One of the principal applications of hydrogen production by gasification is in the manufacture of hydrocarbons. The most significant of these is methane, which is the main constituent of natural gas, hence of SNG. The direct route to the production of methane is by coal hydrogenation, Reaction 24.6, the hydrogen being produced by Reaction 24.8 and the source of carbon being the residue from the hydrogenation of the coal which in practice is incomplete. This, the hydrogenation route, can be represented as follows:

gasification and CO shift

$$1\tfrac{1}{2}C + 2H_2O + \tfrac{1}{2}O_2 \rightarrow 2H_2 + 1\tfrac{1}{2}CO_2 \uparrow^{\text{remove}} \qquad (24.8)$$

hydrogenation

$$C + 2H_2 \rightarrow CH_4 \qquad (24.6)$$

$$2\tfrac{1}{2}C + 2H_2O + \tfrac{1}{2}O_2 \rightarrow CH_4\ (+\ 1\tfrac{1}{2}CO_2) \qquad (24.9)$$

An indirect route to methane from coal is first to gasify all the coal in steam by Reaction 24.5 to produce a synthesis gas, which is then reacted over a catalyst to yield methane and carbon dioxide, by Reaction 24.10. This, the synthesis route, can be represented by:

gasification

$$3C + 2H_2O + O_2 \rightarrow 2CO + 2H_2 + \downarrow_{\text{remove}} CO_2$$

synthesis

$$2CO + 2H_2 \rightarrow CH_4 + CO_2 \qquad (24.10)$$

$$3C + 2H_2O + O_2 \rightarrow CH_4\ (+\ 2CO_2) \qquad (24.11)$$

It will be noted that for purposes of explanation, the hydrogenation route (Reaction 24.9) is shown as requiring less oxygen, producing less carbon dioxide, and consuming less carbon or coal than the synthesis route (Reaction 24.11), since it has potentially a higher thermal efficiency.[3] The difference, which is not insignificant, has concentrated SNG process development effort on the hydrogenation route. It also gives the existing high-pressure Lurgi gasification process, in which up to 40% of the methane can be produced in the fuel bed, the advantage over

[3] H. H. Lowry, Ed., *Chemistry of Coal Utilization*, Supplementary Volume, John Wiley and Sons, New York, 1963, p. 995.

the low-pressure processes such as Koppers-Totzek and Winkler, which yield little or no methane and must rely entirely on the synthesis route for SNG manufacture. Even the new SNG processes, however, cannot produce all the methane by the hydrogenation route. As with the Lurgi process, some methane must be synthesized in a catalytic step by reaction of hydrogen and carbon oxides produced in the gasifier.

The cryogenic separation of nitrogen from air produces oxygen for admixture with gasification steam to manufacture nitrogen-free synthesis gas. Oxygen production is very capital intensive and consumes a great deal of power, and attempts have been made to avoid it by using air in process arrangements that keep the products of combustion separate from the products of gasification. Not only is nitrogen then excluded, but also the carbon dioxide associated with combustion. The classic example of such a process is the water gas process with a fixed bed being alternately blown with air and steam. The ICI moving burden process[4] was the first attempt to invert this approach by applying fluidization and fully entrained gasification technology to cycle the fuel bed through two fixed reaction zones, one supplied with steam and the other with air. Variations of this system include the recycling of the agglomerated ash or special refractory particles as heat carriers. Examples of new processes adopting these techniques are the Garrett, Union Carbide–Battelle, Cogas, and Exxon processes. In other concepts heat is supplied directly to the fuel bed by using electrodes or electric heating elements, or by means of heat exchangers immersed in the bed and supplied with hot gases heated in a nuclear reactor. Other direct systems of avoiding oxygen manufacture include supplying the heat by a chemical reaction other than combustion. The CO_2 acceptor process is one such example in which heat is generated by the carbonation of calcium and magnesium oxides in admixture with fuel, the oxides reacting with the carbon dioxide produced by gasification. These proposed systems, which are described later, introduce complications to the process and problems of operation that tend to offset the intended advantages. Except in special circumstances and for the production of low-Btu fuel gas, oxygen is likely to remain the preferred choice for synthesis gas and SNG manufacture.

[4] J. W. R. Rayner, *J. Inst. Fuel,* **25,** 50–59 (1952).

24.1.3 Classification of Gasification Processes

In this supplementary volume gasification processes are mainly classified by contacting mode, with subgroups distinguishing between the status of the processes (i.e., whether it is commercial or experimental) and between the nature of the product gas (i.e., whether it does or does not contain nitrogen).

The principal features of the four main groups are illustrated diagrammatically in Fig. 24.1. For comparison, the common and simple case of gasification in steam and oxygen has been taken, with the gas produced being free from nitrogen. In a fixed-bed gasifier (Fig. 24.1*a*) the coal is supplied in lump form—usually in the range 2–0.25 in. (50–6 mm)—in countercurrent flow to the gasifying agent, with the ash, stripped of carbon, being discharged from the base of the reactor. Both the ash—by heat exchange with the entering gasifying agent—and the product gas—by heat exchange with the entering fuel—leave the bed at modest temperatures. A maximum temperature is generated in the region where the oxygen is consumed. If the temperature is limited with excess steam and maintained below the ash fusion point, the ash is removed "dry" through a grate. Alternatively, the ash can be allowed to melt and be run off as a liquid slag. In a fluidized-bed gasifier (Fig. 24.1*b*) the coal is supplied as a powder usually less than 0.125 in. (3 mm) in size, which is then maintained in a suspended state of continuous random motion by the gasifying agent. Mixing and heat transfer are rapid, resulting in uniformity of composition and temperature throughout the bed. The ash remains intimately mixed with the char, which is discharged, as with the product gas, at full reaction temperature, there being no heat exchange with the entering gasifying agent or fuel. Fuel elutriated from the bed may be separated from the product gas and returned. In the entrained-suspension gasifier (Fig. 24.1*c*) pulverized coal is entrained with the gasifying agent to react in cocurrent flow in a high-temperature flame from which part of the ash may separate as a liquid slag.

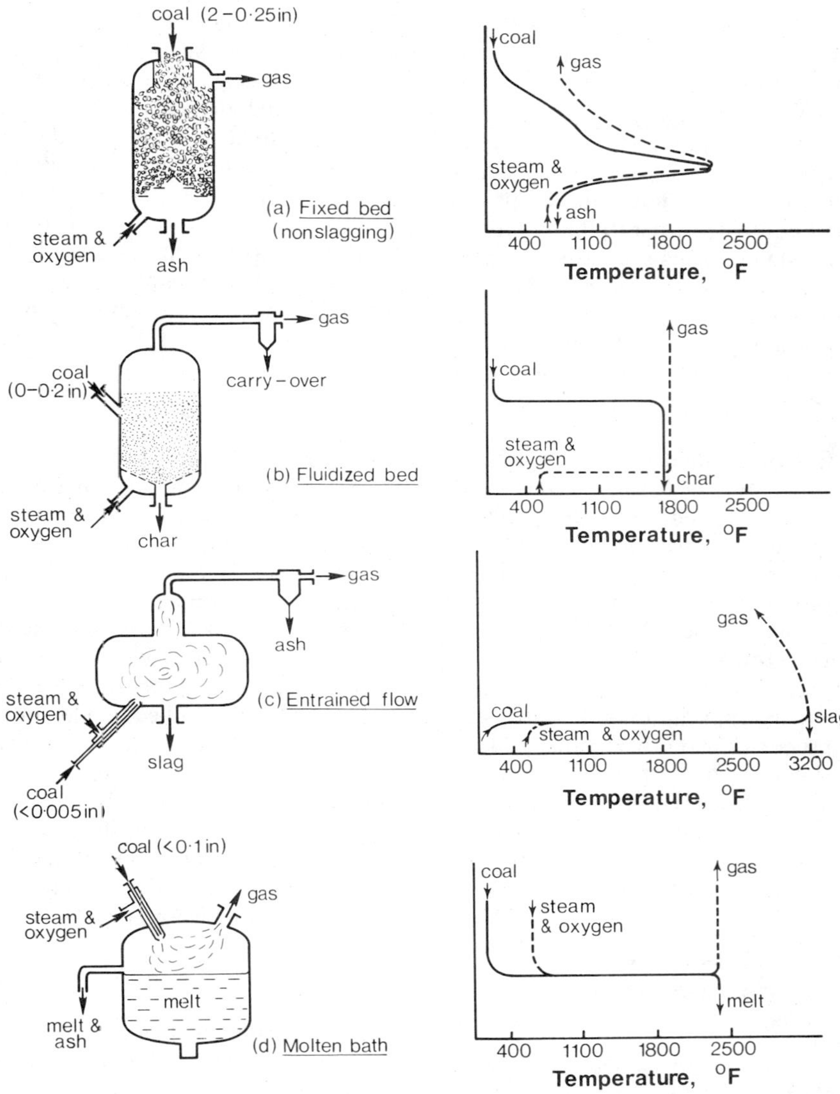

Figure 24.1 Classification and characteristics of principal gasification systems.

Reaction may continue downstream in space provided, but the product gas and the slag leave close to reaction temperature. Ash may be carried forward and separated from the product gas jointly with unreacted fuel. Fuel and gasifying agent are also supplied cocurrently to the molten bath gasifier (Fig. 24.1*d*), the pool being of liquid ash, salt, or metal. The melt is variously claimed to provide heat capacity, a vehicle for fuel particles, and either a reagent-absorbent for sulfur and a catalyst for the steam-carbon reaction, or both. Similar again to the entrained-flow gasifier, the product gas and ash leave at reaction temperature without the opportunity of heat exchange with the entering fuel and gasifying agent.

Fixed-bed gasifiers are typified by their use of lump coal, the presence in the product gas of the volatile constituents such as tar and oils, and the need to make special provisions for handling fine and caking coal. Fluidized-bed operation requires all the coal to be crushed below about 0.125 in. (3 mm), but it offers the opportunity to realize good heat and mass transfer and to convey solids in bulk so that process steps can be designed in which product yield and gasification efficiency can be optimized. Although special provisions must be made to handle caking coal, the product gas is essentially free of tars, oils, and so on. Entrained-flow gasifiers require the coal to be pulverized to less than 0.005 in. (0.12 mm) in size and to be maintained in constant supply with the gasifying agent, since there is no carbon inventory to absorb interruptions. They will accept any coal type but are sensitive to the properties of the ash. Molten baths are similar in many respects to entrained-flow gasifiers, but they can accept a coarser coal size feed and present the possibility of using the melt as a heat transfer vehicle as well as a heat sink. Containment of the melt presents serious problems.

Fixed-bed gasification represents the traditional mode and includes—in the Lurgi gasifier—the only commercially available process operating at high pressure [300–450 psi (20–30 bar)]. The fluidized-bed and entrained modes have only one commercially operating example each, in the Winkler and the Koppers-Totzek gasifiers, respectively, and both are currently available for atmospheric pressure application only. These gasification processes and others either still under development or awaiting commercial application are described in subsequent sections. They have also been conveniently catalogued and their main features summarized by Howard-Smith and Werner.[5]

24.1.4 The Influence of Coal Characteristics on Gasification Processes

Although the means of contacting the gasifying agent with the fuel has a major influence, the process and gasifier design must take into account other factors, of which coal characteristics are perhaps the most important. Gasification units may be on a very large scale, as proposed for SNG manufacture, with a dedicated source of coal, or on a relatively much smaller scale for synthesis or fuel gas production with coal shipped to the site. In the latter case the coal supply may be selected in quality and size and optimized in relation to a gasification process chosen on other grounds.

Freedom of coal choice allows purchasing flexibility and cost optimization, with the ability to change supply to achieve an economic advantage. With a dedicated supply, however, and the desirability of the plant taking the total mine output, the choice of process route becomes more critical both with regard to the gasifier and the downstream treatment plant. The main coal properties bearing on process choice are moisture, ash, volatile matter, caking tendency, fixed carbon, reactivity, ash fusion, and size distribution.

24.1.4.1 *Moisture*

Moisture content varies from less than 5% with the older anthracite and bituminous coals, to more than 40% with the younger low-rank subbituminous coals and lignite or to even higher amounts with peat. Fixed-bed gasifiers can accommodate moistures of up to 35%, provided the ash content is not in excess of about 10%; above these amounts some predrying may be desirable. Fully entrained and fluidized-bed gasifiers require the moisture content to be reduced generally to less than 5% by drying to maintain free-flowing behavior during crushing, conveying, and charging to the gasifier. In the fully entrained system the residual moisture contributes to the gasification steam but requires process heat to evaporate it.

24.1.4.2 *Ash*

Ash is basically inert, but it can convey heat usefully either by its flow countercurrent to the products of gasification and the gasifying agents in fixed-bed systems, or by special provisions in two-stage, fully entrained–fluidized systems such as that of Union Carbide–Battelle. Generally speaking, ash should be at a minimum, since provision must be made for introducing it to and withdrawing it from the system, provisions that increase in complexity and cost with increasing pressure of operation. Ash is present in all coals in amounts varying from 1 or 2% to upward of 50%, but it can be reduced in content by cleaning processes. In fixed-bed systems it accumulates at the base of the fuel bed below the combustion-reduction zone; if unfused it is withdrawn by a mechanical grate, or if slagged it is run off as a liquid through some form of taphole. It is always removed as a

[5] I. Howard-Smith and G. J. Werner, *Coal Conversion Technology,* Noyes Data Corporation, Park Ridge, N.J., 1976.

liquid slag from fully entrained systems, although a small proportion may pass forward with the reacting coal to be separated downstream as fine dust. Fluidized systems present a special problem, since by their nature the ash is maintained intimately mixed with the fuel. Separation here can involve either deliberate sintering and agglomeration of the ash to persuade it to separate, or circulation of some particles from the bed through a fully entrained combustor to melt and separate the ash as a liquid slag. In contrast to fixed-bed operation, the separation of ash from fluidized and fully entrained systems tends to be associated with higher losses of carbon in ash.

Ash constituents must be taken into account when selecting materials of construction, particularly in slagging gasifiers. Ash can vary in character from acidic to basic, and it can contain volatile components and elements such as iron, which may separate or require special provisions for handling. Also, in slagging operation it may be desirable to add reagents such as fluxes to improve the flow characteristics of the slag or to adjust its chemical behavior.

Attempts have been made to use molten ash as a heat transfer agent, but these have failed because of the difficulties of containment, conditioning, and iron separation. Molten salts providing a catalytic potential have been substituted, but heat losses and containment as well as ash contamination still present problems.

24.1.4.3 *Volatile Matter*

Volatile matter refers primarily to the gas, tar, oils, and moisture of decomposition that are distilled or flashed off as coal is heated. The gas—which adds to the products of gasification, therefore can increase the gaseous yield without incurring steam decomposition or oxygen consumption—consists partly of carbon oxides and partly of hydrogen, with some light hydrocarbons and traces of nitrogenous compounds, and so on. At pressures above about 75 psig (5 bar) this hydrogen may be evolved in combination with carbon as methane and perhaps some ethane. The organic liquid products of distillation that survive and can be condensed from the gas from fixed-bed systems are cracked essentially to carbon and hydrogen in fluidized and fully entrained systems operating above 1450–1650°F (800–900°C). Volatile matter contents can vary from less than 5% with the older coals like anthracite to over 50% on the maf basis with the young subbituminous coals and lignites. Its composition, the type of coal, and the conditions under which it is driven off greatly affect the nature of the residual fixed carbon or char that remains.

24.1.4.4 *Caking*

Caking, coking, and swelling occur with certain coals as they are heated and the volatile matter is driven off. Certain high-volatile bituminous coals become fluid during this phase, and the particles not only adhere together but also swell. Others of similar volatile matter content may retain the original shape, with the individual pieces weakly caking together. Yet others, as the moisture and volatile matter are driven off, can disintegrate into small pieces. In general, these coals are referred to respectively as strongly caking and swelling, weakly caking, and noncaking. Such behavior has to be catered for in the gasifier and process design. Some gasifiers may be specially designed to handle caking and swelling coals, but others will require the coal to be pretreated. Although fixed- and fluidized-bed systems tend to be sensitive to strongly caking, swelling, and fusible coals, fully entrained systems can accept coals with any of these properties.

24.1.4.5 *Fixed Carbon*

The fixed carbon is the major component of the residual char remaining after the moisture and volatile matter have been driven off; it is the part of the coal that remains to be gasified in steam, hydrogen, carbon dioxide, air, or oxygen. Its nature, which is important to the performance of the gasifier both physically and chemically, can vary considerably. It can be dense, or light and cellular in structure, it can be hard and brittle, or soft and friable. It can be highly reactive, or very unreactive to steam or hydrogen. These properties depend primarily on the original coal, but they are influenced by the pressure, by the rate at which the coal is heated, and by its final temperature.

24.1.4.6 *Reactivity*

Coals vary in their reactivity to steam and to a lesser extent to hydrogen. Reactive coals, or rather their char or fixed carbon, decompose steam more rapidly and sustain that decomposition down to a lower temperature than do less reactive coals. Reactivity can be measured in the laboratory, but it is difficult to relate to practice because the temperature and the reaction history, which has a strong influence on the nature of the surface, are not easily simulated, nor is it easy to present the surface to the reacting gases under the same conditions that exist in the gasifier.

Reactivity has three important influences. First, it favorably influences methane formation to the advantage of SNG processes. Second, it reduces oxygen consumption by allowing steam decomposition down to a lower temperature. Third, it allows less steam to be used per volume of oxygen or air than with less reactive coals of the same ash fusion temperature, without incurring ash clinkering.

24.1.4.7 *Ash Fusion*

Ash fusion temperature is an approximation of the maximum temperature at which a fuel bed can be operated without incurring fusion, sintering, or clinkering of the ash. Alternatively, it is an indication of the temperature that must be exceeded if the ash is to fuse and be run off as a liquid slag. It is only an approximation, since, for example, ash that is heavily diluted with carbon can be taken in a fluidized bed to temperatures well in excess of the fusion point without incurring sintering or clinkering problems. On the contrary, again in a fluidized bed, ash sintering can occur well below the fusion point when the ash concentration increases above a certain threshold.

In fixed-bed operation, as in the conventional Lurgi gasifier, ash clinkering is prevented by supplying excess steam with the gasifying steam and oxygen or air to limit the maximum temperature generated in the combustion zone to below the ash fusion temperature. Thus the lower the ash fusion temperature, the higher the ratio of steam to oxygen or air that must be used. The more reactive the coal, however, the more rapidly the steam reacts endothermically with carbon to a lower temperature and the less excess steam is required to keep the maximum temperature below a given ash fusion point.

Although ash fusion temperature is relevant to slagging gasifier operation, a more important relationship is that between temperature and viscosity, since it is the flow characteristics of the slag that are critical. Ash can have a low fusion temperature but a flat viscosity temperature curve that may require the addition of fluxes for satisfactory slag removal. Slag viscosity-temperature curves can be determined by direct measurement, but extensive data now exist in the shape of multicomponent phase diagrams and composition-viscosity-temperature correlations, so that the necessary information can be calculated from an ash analysis. One of the widely used correlations is that based on the silica ratio[6] (Chapter 21).

24.1.4.8 *Size Distribution*

The output from a mine varies in size from large pieces to dust. Crushed and screened to a size range suitable for fixed-bed operation [e.g., 2–0.25 in. (50–6 mm)], the proportion of fines smaller than 0.25 in. (6 mm) depends on the method of crushing, the nature of the coal, and the system of mining employed. Strip-mined bituminous coal may contain up to 30% of fines, whereas heavily mechanized deep-mined coal may show 40% or more, and strip-mined lignite can reach as high as 60%. In a fixed-bed gasification plant some of the fines may be consumed in steam and power generation, but not all can be utilized unless special provisions are made such as briquetting, extrusion, or injection to allow them to be supplied to the gasifiers. Developments in fixed-bed gasification are moving toward the acceptance of a greater proportion of fine material in the feed.

The size distribution of the mined coal is less critical with fluidized and fully entrained gasification plants, since in any event all the coal has to be dried and crushed to a size less than 0.125 in. (3 mm). Indeed, the cheaper mining techniques that result in high fines production offer an advantage in this case, since they help offset the high capital and

[6] H. R. Hoy, A. G. Roberts, and D. M. Wilkins, *Behavior of Mineral Matter in Slagging Gasification Processes,* I. G. E. Commun. No. 672; J. I. G. E. 1965, pp. 5, 44.

running costs of the coal preparation plant necessary for powdered-coal processes.

There is not yet a universally or even nationally accepted convention for classifying coal for gasification. A classification system is needed, and certain international organizations such as the International Gas Union and the United Nations Economic Commission for Europe are about to embark on the task of preparing such an instrument.

24.2 FIXED-BED GASIFIERS

24.2.1 Introduction

The term "fixed bed" is used to describe a bed of granular or lump fuel maintained at a nominally constant depth between two fixed extremities. Fuel is supplied at the upper extremity and fuel residues or ash are removed at the lower one, which also provides a support for the bed. Fuel moves slowly down, ideally in plug flow against an ascending stream of gasifying agent. In the strictest sense, this arrangement is a moving bed with fixed extremities, but the term "fixed bed" is in common usage and is the one employed in this book to describe such a system.

The fuel enters the top or upper extremity of the bed and moves down to replace that being consumed by gasification. It descends countercurrent and in heat exchange with the ascending products of gasification through four definable but not distinctly separate zones, illustrated diagrammatically in Fig. 24.2. As it passes through these zones, the fuel is subject to progressively increasing temperature and changing gas composition, until it reaches the combustion zone where it is finally consumed to leave an incombustible residue, the ash. The ash is discharged from the base or lower extremity, either through a grate if it is in the form of an unfused granular material, or through some form of slag tap if the ash has been melted and collected as a liquid slag.

In nonslagging operation the gasifying agent, steam and oxygen or air, enters at the base or lower extremity through the ash zone, where it is preheated by heat exchange before entering the combustion zone, the ash being thereby cooled before discharge

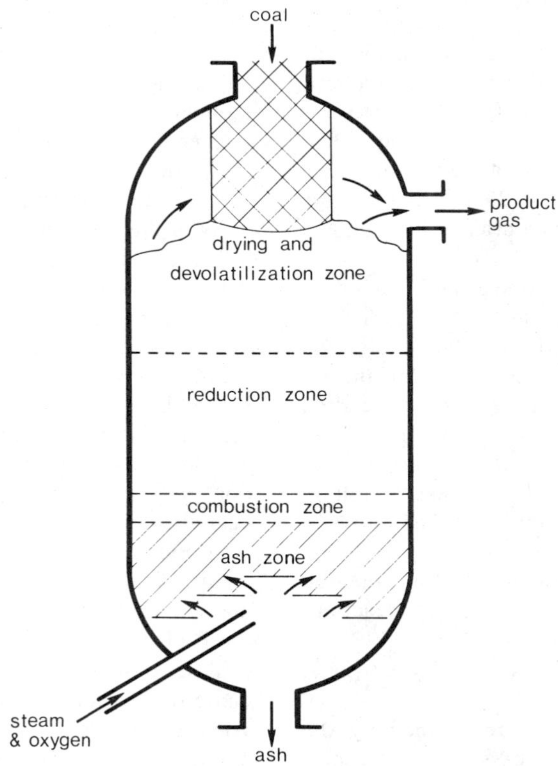

Figure 24.2 The four main reaction zones in a nonslagging fixed-bed gasifier.

through the grate. All the oxygen reacts with carbon in the combustion zone and releases heat

$$C + O_2 \rightarrow CO_2 \qquad (24.1)$$

that, as the temperature rises, is absorbed by the onset of the steam-carbon reaction,

$$C + H_2O \rightarrow CO + H_2 \qquad (24.4)$$

and the carbon dioxide–carbon reaction:

$$C + CO_2 \rightarrow 2CO \qquad (24.2)$$

The temperature rises to a maximum (see Fig. 24.1*a*) that depends on the ratio of steam to oxygen or air and on the reactivity of the fuel to steam. In nonslagging operation the level of the maximum is limited below the fusion point of the ash by supplying extra steam in excess of that required to gasify the carbon; nitrogen in the air, by virtue of its heat capacity, has a similar moderating effect. The effect of reactivity to steam lies in its influence on the relative rates of Reactions 24.1 and 24.4. Thus fuels with low reactivity and low fusion point ash require more excess steam than those with high reactivity and high fusion point or refractory ash. In slagging operation the steam supplied in admixture with the oxygen is only slightly in excess of that required to gasify the carbon. This results in the generation of a zone with temperatures in excess of 2500°F (1375°C) in which the ash melts and drains off as a liquid slag.

Leaving the combustion zone is high temperature unreacted steam admixed with some gasification products (i.e., carbon oxides and hydrogen). This mixture passes into the reduction zone where the steam and carbon dioxide continue to react with carbon (Reactions 24.4 and 24.2) with an absorption of heat and a consequent progressive fall in temperature until a level is reached at which reaction ceases.

Accompanying gasification in the reduction zone is the carbon monoxide shift reaction:

$$CO + H_2O \rightarrow CO_2 + H_2 \qquad (24.7)$$

the extent of which is determined by the relative quantity of excess steam present. This reaction is exothermic and the heat evolved helps to support the steam-carbon reaction. Of particular importance in synthesis gas production for SNG manufacture is the reaction of carbon with hydrogen

$$C + 2H_2 \rightarrow CH_4 \qquad (24.6)$$

which occurs when gasification is carried out under pressure. This strongly exothermic reaction not only contributes heat to support the steam-carbon reaction but also gasifies carbon. Consequently its promotion by operation under pressure reduces the quantity of oxygen required to gasify a given amount of fuel. The reduction depends on fuel reactivity, but is around 20% at 450 psig (30 bar).

The hot products pass from the reduction zone into the drying and devolatization zone, where their sensible heat content pyrolizes or decomposes and carbonizes the coal with the evolution of water vapor and volatile products such as tar and oil together with gases such as carbon oxides, methane, ethane, hydrogen, hydrogen sulfide, and ammonia. These gases and vapors are added to the products from the reduction zone, the residual sensible heat of which finally drives off the moisture contained in the entering fuel. The temperature at which the gases and vapor leave the fuel bed depends on a large number of factors, such as the volatile matter, moisture and ash contents, the reactivity of the fuel, and the ratio of steam to oxygen or air in the gasifying agent.

As in the reduction zone, pressure has an influence on the gases produced in the devolatilization zone. Hydrogen is evolved during coal pyrolysis or decomposition at atmospheric pressure, but at pressures above about 75 psig (5 bar) this hydrogen is evolved in combination with carbon, primarily as methane. At high pressure, hydrogen arising from the steam-carbon reaction in the reduction zone can react with the decomposing coal also to produce methane.

In addition to the evolution of gases and vapors, coal undergoes physical changes during heating in the devolatilization zone, and these changes depend on the nature of the coal. Lignites and subbituminous coal become friable and tend to disintegrate. High-volatile bituminous coals fuse, swell, and eventually carbonize or cake into strong coke, the coal particles having lost their original identity and agglomerated into dense monoliths. Lower rank bituminous coals are less strongly caking and some—even if high in volatile matter—do not fuse or coke, the original particles remaining essentially unagglomerated. The older high-rank coals such as anthracite, as well as cokes and chars produced by pretreatment, do not cake, and they maintain their original size distribution in this zone.

All coals need to be supplied to the drying and devolatilization zone in a manner that prevents size segregation and ensures the even distribution of ascending gases. In addition, coals that swell, fuse, and/or cake must either be pretreated to prevent agglomeration, or, while these changes are occurring, be subjected to mechanical-type stirring or agitating action. This action needs to be designed to break up the coal as it fuses and agglomerates and, if possible, to backmix devolatilized char with the new coal to dilute and weaken the bond between particles. The prime objectives are to prevent bridging and hangups, and to deliver to the reduction zone a uniform flow of fuel particles, the bulk of which are preferably in the size range 2–0.25 in. (50–6 mm).

Although the output of a fixed-bed gasifier is determined by the rate at which the oxygen in the oxygen- (or air-) steam mixture is supplied, it is primarily, but not solely, limited by the quantity of entrained fuel acceptable in the exit product gas. The amount of fuel entrained from a fuel bed of a given size range is dependent on the entrainment velocity, hence on the actual volume of gaseous vapors and unreacted steam leaving the fuel bed. If the quantity of excess or unreacted steam can be reduced and/or the presence of nitrogen can be eliminated by using oxygen instead of air, and/or if the volume of the gases, and so on, can be decreased by operating at higher pressures, higher rates of gas production can be realized without exceeding the entrainment velocity and incurring carryover of fuel. Increasing output is an important advantage of operating under pressure and using oxygen instead of air that is additional to those of promoting methane formation in the fuel bed and producing a nitrogen-free fuel or synthesis gas. It is also one of the principal advantages of operating under slagging conditions when the excess steam otherwise supplied to prevent ash fusion and clinkering is eliminated and the ash is run off as a liquid slag.

Slagging operation, in addition to increasing output, has other advantages of which the reduced steam demand is the most important. The lower ratios of steam to oxygen used result not only in a lower steam consumption per unit of gas produced, but also in the virtual absence of unreacted steam, hence in a lower heat capacity of the products leaving the gasifier. These changes lead to a substantial gain in overall thermal efficiency and to a reduction in the volume of aqueous liquor subsequently condensed from the gas on cooling. Slagging will show to particular advantage with low reactivity coals having a low fusion point ash, that is, with the coals for which a large excess of steam must be present in the gasifying agent to prevent ash clinkering in a conventional nonslagging grate system. The favorable margin is reduced with the high-reactivity, high fusion point ash coals, particularly if these coals have a very high ash content ($> 30\%$). For SNG manufacture, one disadvantage of slagging operation is that the amount of methane produced in the fuel bed is restricted by the high temperatures prevailing in the reduction zone and is lower than that observed with nonslagging operation. Air-blown slagging gasification requires either some oxygen enrichment or substantial air preheating to compensate for the high nitrogen content of the gasifying agent and to form a slag sufficiently mobile to be tapped. No air-blown slagging fixed-bed gasifiers are currently known to be under development.

In general, fixed-bed gasification systems are inherently simple and potentially reliable, and—with the feedstock in countercurrent flow to the products to provide effective heat utilization—they have a high thermal efficiency. They are largely confined to graded fuels in the range of 2.0–0.25 in. (50–6 mm) and have only a limited capability to accept material less than 0.125 in. (3 mm) in size. They yield tars and oils as by-products, and caking coals require special provisions. In contrast to some other systems, gasification is completed in a single reactor with minimally pretreated coal entering at the top and ash stripped of carbon being discharged at the base. Product gas and vapors leave at modest temperatures, conveying a high proportion of the potential heat in the original coal. For purposes of description in the following sections, the various systems are divided into three main groups: earlier, present, and new generation gasifiers. They are further divided into the processes that produce nitrogen-containing gas and those that produce nitrogen-free gas.

24.2.2 Earlier Generation Gasifiers

From the early hand-operated producers and water gas generators, a generation of more sophisticated and mechanized gasifiers was developed, as described and referenced[7,8] in the previous volumes of *Chemistry of Coal Utilization*. These gasifiers operated on air and steam at atmospheric pressure under nonslagging conditions and included such types as the Wellman-Galusha, IFE two-stage, Kerpely, Power Gas, and Marischka producers, and the UGI, Humphreys & Glasgow, Tully, Power Gas, and Gas-Integrale water gas processes. They were limited to closely graded coke or noncaking coal with ash fusion points in excess of 2100°F (1150°C), to gasification rates of about 50 lb of coal/(ft^2)(hr) [244 kg/(m^2)(hr)], and to maximum outputs of about 5 million scf/day (139 × 10^3 m^3/day). With the advent of cheaper industrial oxygen and an increasing demand for synthesis gas, some were converted for operation on oxygen and steam when outputs were raised up to twofold without any serious difficulty. For similar reasons, gasification under slagging conditions was introduced and two commercially successful gasifiers, the Thyssen Galoczy and the Leuna, demonstrated the potential savings that can arise from this mode of operation.

The Lurgi gasifier must be mentioned here, since it was one of the first to adopt oxygen blowing and the first and only one to exploit pressure operation. It has been under continuous development since 1936, and it dominates the fixed-bed gasification field. The Lurgi and a number of the other earlier generation gasifiers that have been modified and/or are being further developed to meet today's commercial requirements are described in Section 24.2.3.

24.2.3 Present Generation Commercial Gasifiers

24.2.3.1 *Present Generation Processes Yielding Nitrogen-Containing Product Gas*

Air-blown processes that have been developed from the earlier generation of fixed-bed gasifiers include the Woodall Duckham two-stage (formerly Gas-Integrale), the Wellman two-stage, the Wilputte, the Riley-Morgan, and the air-blown Lurgi. All have been demonstrated on the full commercial scale, but the last two are not yet in general commercial operation. Published performance data for these gasifiers are summarized in Table 24.1.

The Woodall Duckham Two-Stage Process. The Woodall Duckham two-stage process[9] is based on the Gas-Integrale generator.[10] It comprises a coal carbonization stage above a gasification stage in a single vessel provided with two gas offtakes, so arranged that only sufficient hot gasification products to devolatilize the coal pass through the upper stage, the remainder being withdrawn at the outlet of the reduction zone of the gasification stage. The process is illustrated in Fig. 24.3. Coal is fed to the devolatilization zone via a lock hopper and distributor, the latter additionally providing buffer storage to ensure a continuous flow of coal, hence a constant "top" gas composition. In the refractory-lined devolatilization zone the fuel descends countercurrent to a stream of hot product gas and is dried and devolatilized over a period of several hours. The resultant char or semicoke then passes into the water-jacketed gasification stage, from which the bal-

[7] H. H. Lowry, Ed., *Chemistry of Coal Utilization*, Vol. 2, John Wiley and Sons, New York, 1945, Chs. 36 and 37.

[8] H. H. Lowry, Ed., *Chemistry of Coal Utilization*, Supplementary Volume, John Wiley and Sons, New York, 1963, pp. 949–965.

[9] A. J. Grant, "Applications of the Woodall-Duckham Two-Stage Coal Gasification," paper prepared for presentation at the 3rd Ann. Int. Conf. Coal Conversion, Pittsburgh, August 3–5, 1976.

[10] Bituminous Coal Research Inc. Rep. No. L-156, *Gas Generator Research and Development, Survey and Evaluation, Phase 1*, Vol. 2, March 1965.

Table 24.1 Performance Data for Air-Blown, Commercial-Size Fixed Bed Gasifiers

	Gasifier and Fuel					
	Woodall Duckham	Wilputte	Riley-Morgan			Lurgi
	Bituminous	Bituminous	Anthracite	Bituminous	Bituminous	Subbituminous
Proximate analysis (wt %)						
Volatile matter	32.9	37.3	3.6	30.8	21.4	} 64.4
Fixed carbon	50.6	54.8	84.8	56.6	66.5	
Moisture	11.6	1.9	3.6	5.5	7.1	16.3
Ash	4.9	6.0	8.0	7.1	5.0	19.3
Ultimate analysis (wt %)						
Carbon	—	77.9	83.5	75.6	79.1	49.1
Hydrogen	—	5.5	2.3	5.4	5.2	5.5
Nitrogen	—	1.6	} 5.4	} 11.1	} 10.0	0.9
Oxygen	—	8.4				24.5
Sulfur	—	0.6	0.8	0.8	0.7	0.7
Ash	4.9	6.0	8.0	7.1	5.0	19.3
HHV (Btu/lb)	14,430	15,210	12,930	15,340	15,730	13,450
Operating conditions						
Pressure (psia)	15	15	15	15	15	295
Air/steam weight ratio	—	6.0	—	4.8	—	2.1
Products						
Gas composition (vol %)						
CO	27.8	22.7	22.7	21.0	23.5	17.4
H_2	16.2	16.6	16.6	17.9	16.4	23.3
CH_4	1.7	3.6	0.3	2.0	1.7	5.1
C_nH_m	—	—	—	0.5	0.4	0.7
H_2S	—	—	—	0.2	0.1	0.2
CO_2	6.3	5.9	9.1	8.9	7.3	14.8
N_2	48.0	50.5	51.3	49.5	50.6	38.5
HHV (Btu/scf)	159	163	130	156	153	196
Specific data						
Consumption						
Steam						
lb/lb coal (maf)	—	0.65	—	0.65	—	0.96
lb/10^6 Btu gas	—	48.6	—	63.0	—	93.3
Oxygen						
scf/lb coal (maf)	—	10.43	—	8.66	—	5.44
scf/10^6 Btu gas	—	777	—	836	—	528
Production						
Gas						
scf/lb coal (maf)	68.3	82.3	80.2	66.4	66.1	52.6
Gasification efficiency						
E_1	75.2	72.0	80.6	67.5	64.3	76.3
E_2	—	84.2	80.6	77.8	70.8	—

ance of the product gas is withdrawn through the side gas offtake. The fuel bed rests on a mechanical grate, which extracts the lightly clinkered ash into a lock hopper and distributes the gasifying agent, either steam and air or steam and oxygen. Steam raised in the water jacket is used for gasification.

About half the gas from the reduction zone is drawn off via the side gas offtake. It is known as "clear" gas, since it is free of tar and oils, and it leaves at about 1200°F (650°C), passing through a cyclone to remove dust before being cooled.

The "top" gas leaves at about 250°F

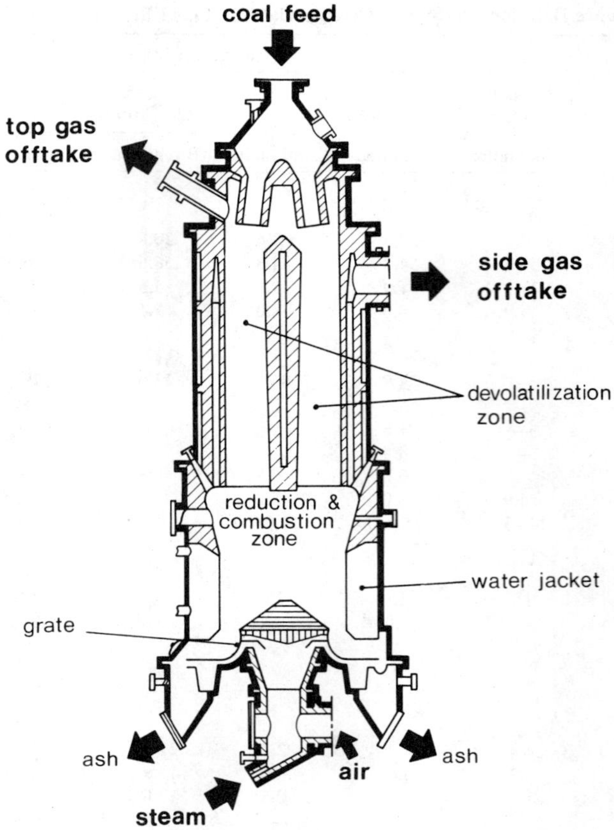

Figure 24.3 The Woodall Duckham synthesis gas generator.[9]

(120°C)[9] and passes to an electrostatic detarrer. Because of the moderate distillation conditions in the gasifier and low water dew point of the gas, the tar is very fluid. It runs easily off the detarrer electrodes and is low in moisture ($<$ 1%). The detarred gas is cooled to condense the light oils and water before being passed to a second electrostatic precipitator to remove tar fog. The oil and water are separated, the aqueous condensate being recycled internally.

If desulfurizing is not necessary, the hot crude gases can be mixed, the hotter "clear" gas revaporizing without cracking the tar in the "top" gas to yield a product gas at about 700°F (370°C), from which soot and pitch will not deposit when transmitted in externally insulated pipes over distances of up to 1000 ft (305 m). The mixed "top" and "clear" gas has a tar-free higher heating value (HHV) of 160–175 Btu/scf (6–6.65 MJ/m^3). Taking account of the tar and oil vapor together with the sensible heat, the total heat content is equivalent to an HHV of 210 Btu/scf (7.97 MJ/m^3).[9,11] When operated on oxygen instead of air, the mixed gas has an HHV of 290 Btu/scf (11.0 MJ/m^3).

Methanation of the low-pressure gas is not considered to be economical, but modified operating procedures have produced a mixed gas containing less hydrocarbon, which is more suitable for ammonia and methanol synthesis. The main advantages of using two stages are the higher quality tar and oil obtained, typically 0.12 lb/100 scf (19.6 g/m^3) of product gas, and the low fines carryover. The process can handle most coals with free-swelling indices less than 2.5 in gasifiers with shaft diameters of 11.8 ft (3.6 m) at throughputs of 80–100 long tons/day (81–102 tonnes/day) producing up to 11.4 million scf/day (317 $\times$ 10^3 m^3/day) of low-Btu gas.

The Wellman Processes. A single-stage

[11] I. Williams, "Fuel Gas Plants for the Process Industries Burning Coal Cleanly, Efficiently," *Process Eng.*, 52–57 (May 1974).

gasification process initially licensed by McDowell-Wellman Company in Ohio, and Wellman-Galusha Company in England has been in commercial use for more than 30 years.[7,8,12]

It utilizes a revolving grate and operates at atmospheric pressure to produce a low-Btu fuel gas of HHV 120–160 Btu/scf (4.56–6.07 MJ/m^3) or, if oxygen is used, a synthesis gas of HHV 270 Btu/scf (10–25 MJ/m^3).[13] Two plants are still operational in the United States. A two-stage gasification process, developed from the single-stage Wellman-Galusha gasifier, was introduced in the early 1960s and is now offered by Wellman Incandescent Ltd. (England). It is illustrated in Fig. 24.4, and since it is very similar to the Woodall Duckham gasifier (Fig. 24.3),[14] it is not described in detail here. The gasifier is claimed to be capable of handling coals with free-swelling indices up to 3 and can be designed to gasify low-grade fuels or fuels such as anthracite beans [0.2–0.75 in. (5–19 mm)]. Gasifiers between 4.5 ft (1.37 m) and 11 ft (3.35 m) diameter[15] are available, producing up to 8.4 million scf/day (233 × 10^3 m^3/day) of clean gas from 72 long tons/day (73 tonnes/day) of coal.[16] Twenty-seven plants have been built to date,[17] but no performance data have been published.

[12] G. W. Hamilton, "Gasification of Solid Fuels," *Cost Eng.*, **8**, 4–11 (July 1963).

[13] I. Howard-Smith and G. J. Werner, *Coal Conversion Technology*, Noyes Data Corporation, Park Ridge, N.J., 1976.

[14] Redrawn from general sales pamphlet distributed by Wellman Gas Engineering, Cornwall Road, Smithwick, Birmingham, England.

[15] Anon., "Wellman Two-Stage Gas Producers," *Heat–Air Cond. J.*, **43**(515) 26–27 (October 1974).

[16] Anon., *Wellman Modern Gas Producer Plants*, Wellman Incandescent Ltd., Birmingham, England (brochure).

[17] A. V. W. Bourton, Wellman Incandescent Ltd., private communication, October 1976.

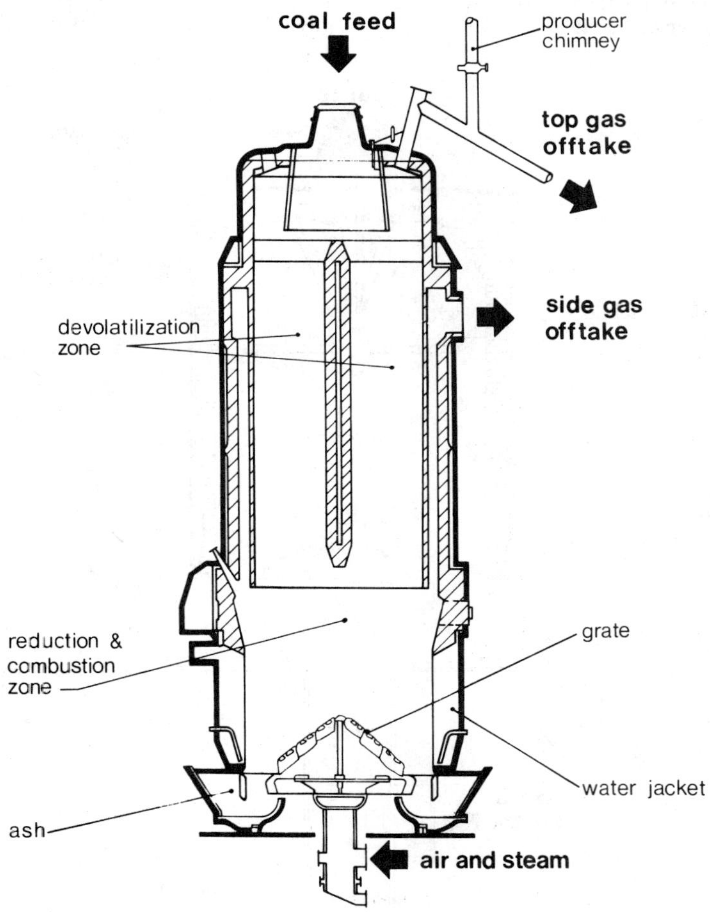

Figure 24.4 The Wellman two-stage producer.[14]

The Wilputte Low-Btu Fuel Gas Process. In 1974 the Wilputte Corporation commenced the development of a low-Btu fuel gas process based on an earlier generation gasifier originally developed from the Semet-Solvay water gas producer and incorporating improvements in coal handling, automatic feeding, ash handling, and so on. The new process[18] employs a single-stage fixed bed, air or oxygen blown, operating at atmospheric pressure under nonslagging conditions and producing gas of 160–170 Btu/scf (6.07–6.45 MJ/m³) on air and 290–300 Btu/scf (11.0–11.38 MJ/m³) on oxygen. It is illustrated in Fig. 24.5.

Coal, sized up to 4 in. (100 mm), is fed to the fuel bed via a revolving drum system and is agitated by means of a rotating rabble arm during drying and distillation to prevent caking and to promote uniform gasification. High temperatures are experienced in the devolatilization zone, which cause cracking of the tar and the formation of pitch and carbonaceous deposits. Deeper fuel beds reduce these temperatures and improve the quality of tar but only marginally decrease fines carryover. The fuel bed is supported on a grate and ash pan assembly that rotates on roller supports; a fixed plow removes the ash from the pan and discharges it into a trough. The gasifier shell is supported independently by columns and is sealed by water, which helps break up the clinker and cools the ash in the ash pan.

The gas leaves the fuel bed at approximately 3 ft/s (0.9 m/s) and then passes to a cyclone dust collector, waste heat boiler, water scrubber, and electrostatic tar precipitator, before final desulfurization by the Holmes Stretford process. It is claimed[18] that the gasifier will handle all types of large coal and is most applicable to the production of fuel gas for industrial use. A 10.3-ft (3.14-m) diameter gasifier will produce 7.2 million scf/day (200 × 10³ m³/day) of 160–170 Btu/scf

[18] G. R. Cooper, Wilputte Corporation, private communication, September 1976.

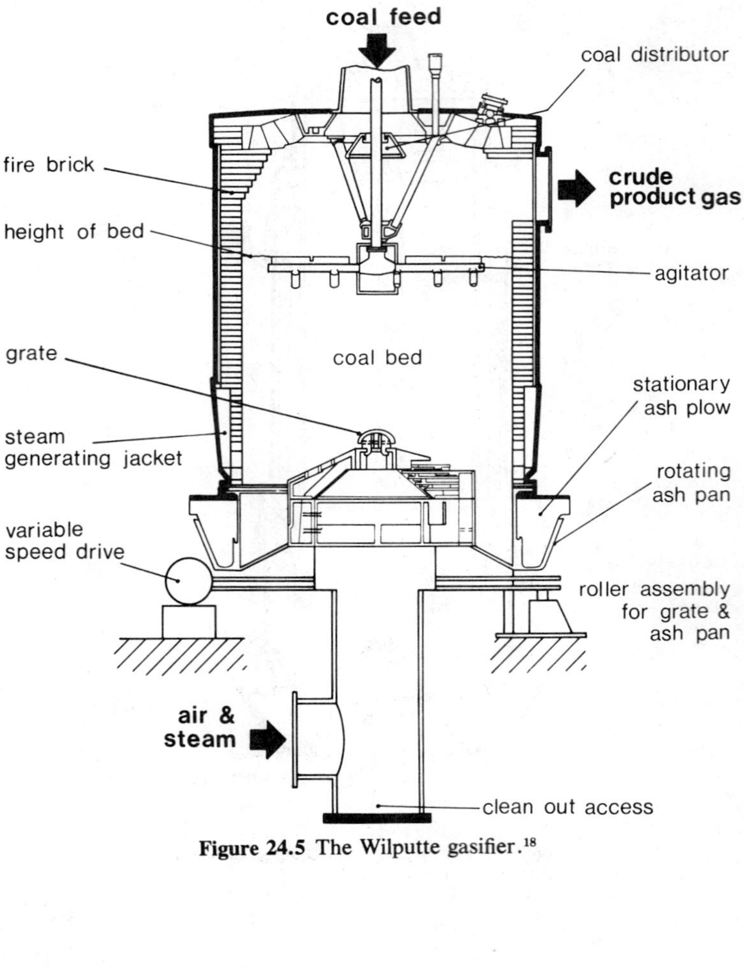

Figure 24.5 The Wilputte gasifier.[18]

(6.07–6.45 MJ/m^3) gas from 60 tons/day (54.4 tonnes/day) of bituminous coal.[18] Published performance data are given in Table 24.1.

The Riley-Morgan Process. The first cylindrical gas producer used in the United States was built by Charles H. Morgan in Worcester, Massachusetts, in 1880. By 1964 the Morgan Construction Company had built and installed more than 9000 gas producers of this type for use in steel, zinc, ceramic, and glass manufacturing plants throughout the world.

In 1971 the Riley Stoker Corporation was granted exclusive manufacturing rights for the international distribution of Morgan's designs and technology. Modifications were made to the design, including (*a*) an increase in the height of the gasifier to allow for the swelling of caking coals, (*b*) a design of opposing water seals to enable the gasifier to be operated at an outlet pressure of 60 in. water gauge (150 mbar) and allow the gas to pass through a cleanup train without the aid of a booster fan, (*c*) the addition of a small lock hopper to permit coal feed against the higher pressure, and (*d*) the addition of a deep-bed agitator or rabble arm based on a Bureau of Mines design for caking coals.[19,20]

This new gasifier[21,22] which is now known as the Riley-Morgan gas producer (Fig. 24.6) comprises a water-jacketed carbon steel shell, refractory-lined at the top and closed at the base by a rotating ash pan, to which is attached a refractory-lined, water-sealed barrel containing the fuel bed. Coal is fed from a drum feeder and lock hopper system and metered into a twin lock hopper arrangement designed to prevent the coal gates from closing against a head of coal. The entering coal is spread evenly on top of the bed by the

[19] R. V. Rahfuse, G. B. Goff, and A. J. Liberatore, *Noncaking Coal Gasified in a Stirred-Bed Producer,* U.S. Bur. Mines Clean Energy Program, Tech. Prog. Rep. No. 77, March 1974.

[20] P. S. Lewis, A. J. Liberatore, and J. P. McGee, *Strongly Caking Coal Gasified in a Stirred-Bed Producer,* U.S. Bur. Mines, Rep. Invest. No. 7644, 1972.

[21] A. H. Rawdon, R. A. Lisauskas, and S. A. Johnson, "Operation of a Commercial Size Riley-Morgan Coal Gasifier," paper presented at American Power Conf., Chicago, April 19–21, 1976.

[22] *Riley-Morgan Coal Gasification System,* Brochure of Riley Stoker Corporation, Worcester, Mass, 1975.

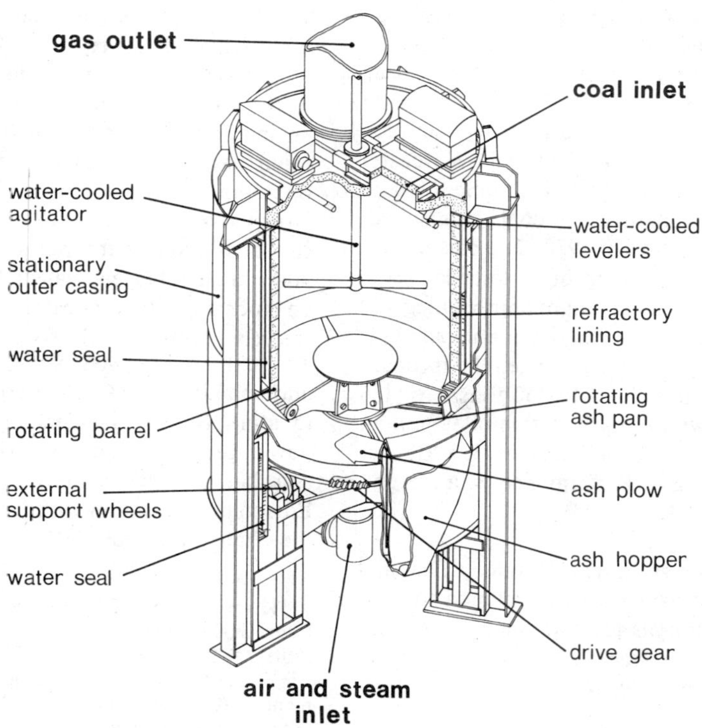

Figure 24.6 The Riley-Morgan gas producer.[22]

rotating fuel bed moving against stationary leveler arms. Air (or oxygen) and steam are introduced into the bottom of the rotating ash pan to be distributed by the bed of ash that supports the fuel bed. The temperature of the gasification zone, typically about 2100°F (1150°C), is maintained below the ash fusion point of the coal. Ash is intermittently removed by means of a plow designed and operated to ensure that all portions of the bed settle uniformly.

The product gas with an HHV of about 150 Btu/scf (5.69 MJ/m³) leaves at a temperature of 1000–1400°F (540–760°C).[23] The heavy cracked tars produced in the top of the bed increase the complexity of the cleanup system. Before sulfur removal, the gas passes through a direct water quench and an indirect cooler, an electrostatic precipitator, and a light oil absorber. The main advantages claimed for the gasifier stem from the use of the water-cooled agitator, which moves vertically as well as rotationally in the bed. It is reported that this action not only breaks up agglomerates formed with caking coal but also creates a fuel bed of uniform permeability, which leads to consistent gas quality and the avoidance of ash clinkering. The developer's program has included the successful gasification of two caking bituminous coals with free-swelling indices of 6 and 8.5.

Although no commercial plant has been built, a full-size 10.5-ft (3.15-m) diameter Riley-Morgan gasifier has been in operation at the Riley Research Center in Worcester since 1975, and a program including trials on all ranks of United States coals was scheduled for completion in 1978. The gasifier has achieved[23] a throughput of 67 tons/day (60.8 tonnes/day) and a maximum approaching 100 tons/day (90.7 tonnes/day) is anticipated. It is expected that by using oxygen the throughput could be increased to 150 tons/day (136 tonnes/day) with the production of 10 million scf/day (278 × 10³ m³/day) of 305 Btu/scf (11.58 MJ/m³) gas. Performance data are given in Table 24.1.

The Air-Blown Lurgi Gasifier. The air-blown Lurgi gasifier was developed to produce a low-Btu gas for combined-cycle electric power generation. It is based on, and is essentially similar to, the oxygen-blown Lurgi gasifier described in Section 24.2.3.2. The main differences are that the operating pressure of about 280 psig (19.3 bar)—being determined by the optimum balance between power recovery and air compression requirements, and so on—is somewhat lower, and a more elaborate crude-gas particulate-removal system is employed. The only gasifiers so far operating in this mode are the Kellerman power station of Steinkohlen Elektrizität AG (STEAG)[24] at Lünen, West Germany, where they are supplying low-Btu gas to a 170-MW electricity generating system.

The STEAG plant at Lünen is illustrated diagrammatically in Fig. 24.7. The gas from the gasifier, after quenching to remove tar and some dust, passes through cyclones to reduce the particulate matter below 0.092 lb/10⁶ scf (1.5 mg/m³). It is then partially expanded in a turbine driving the gasification air compressor before being burnt in a pressurized boiler. The exhaust gases leaving the boiler at 1510°F (820°C) are expanded through a 74-MW gas turbine, and before entering the stack at 320°F (160°C), they pass through a waste heat boiler and boiler feedwater heater. Steam that is raised at 1800 psig (124 bar) and superheated to 970°F (520°C) passes to a 96-MW turbine from which a supply of gasification steam is bled off the high-pressure stage. Initial testing in 1973 revealed problems of high temperatures in the gasifier, inadequate gas cleanup, and an inability to respond to the frequent and rapid load changes associated with peak load operation. These problems have now been largely overcome, and the gasifiers have been run continuously for periods of more than 3000 hr at rates equivalent to 1000 tons of coal per day (907 tonnes/day) on 12–25% ash weakly caking coal (free-swelling index 1.5–3.5) sized between 0 and 1 in. (0–25 mm) and containing 10% of material below 0.2 in. (5 mm). Similar gasifiers are proposed for power generation on certain projected Lurgi coal-based SNG plants.[25] Commonwealth Edison in association with ERDA, EPRI, and the State of Illinois is installing two Lurgi

[23] T. F. Walsh, Riley Stoker Corporation, private communication, September 1976.

[24] K. H. Krieb, "KDV Process (High-Pressure Coal Gasification," Paper No. IGU/B3-76, presented at 13th World Gas Conf., London, 1976.

[25] El Paso Natural Gas Company, 2nd Supplement to Application of El Paso Natural Gas Company for a Certificate of Public Convenience and Necessity, U.S. Federal Power Commission Rep. No. CP 73-131, October 1973.

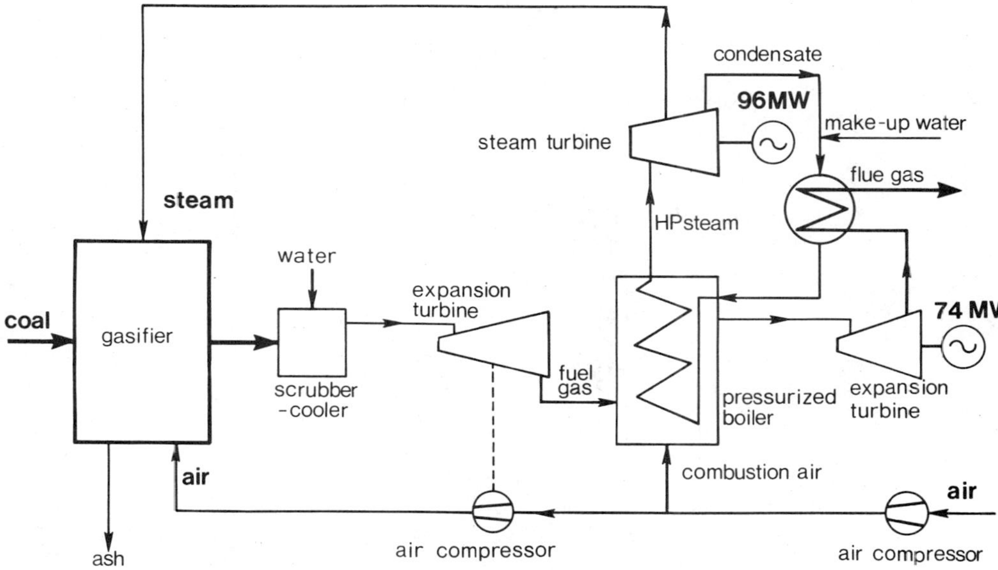

Figure 24.7 The 170-MW power plant in Lünen.[24]

gasifiers (one spare) at Powerton, Illinois, to produce low-Btu gas from 25–30 tons/hr (22.5–27.2 tonnes/hr) of high-sulfur coal to fuel a combined-cycle system generating 25 MW of electrical power using a gas turbine. Performance data predicted for the power generation gasifiers on the projected El Paso SNG plant are listed in Table 24.1.

24.2.3.2 *A Present Generation Process Yielding Nitrogen-Free Product Gas*

All the atmospheric pressure air-blown gasifiers described in Section 24.2.3.1 are available in oxygen-blown versions, and reference should be made to that section for details. It is the high-pressure Lurgi gasifier, however, that dominates the oxygen-blown, fixed-bed gasification scene, since it is the essential part of the only viable process currently available for the production of SNG from coal.

The Lurgi Process. The Lurgi gasifier, the principal features of which are illustrated diagrammatically in Fig. 24.8, has been progressively improved and refined by the developers, Lurgi Kohle und Mineralöltechnik GmbH of Frankfurt,[26–28] West Germany, since the first plant was built in 1936.[8,13] It employs a fixed bed, is steam and oxygen blown, and operates at pressures up to 450 psig (30 bar). Graded coal is supplied via a lock hopper to a distributor, which spreads it uniformly across the surface of the fuel bed. When the coal is weakly or strongly caking a stirrer or agitator is attached to the distributor to break up agglomerates as they form in the devolatilization zone. A typical design of distributor-stirrer is illustrated in Fig. 24.9*a*. The fuel bed is supported on a rotating grate through which the gasifying agent is supplied and ash is withdrawn. Figure 24.9*b* shows the main features of a typical grate. Gases rise up the bed in countercurrent heat exchange and mass transfer with the fuel and leave the top of the reactor together with unreacted steam at a temperature ranging from 570°F (300°C) with lignite to 1200°F (650°C) with low-moisture, low-reactivity bituminous coal.

It is claimed[26] that most types and all ranks of coal can be gasified in present designs of the Lurgi gasifier. Typical performance data for a wide range of feedstocks are summa-

[26] Paul F. H. Rudolph, *The Art of Coal Gasification*, Lurgi Mineralöltechnik GmbH, Frankfurt (Main), Germany.

[27] P. F. H. Rudolph and P. K. Herbert, "Conversion of Coal to High Value Products," paper presented to Symp. Coal Gasification, Liquefaction and Utilization, University of Pittsburgh, August 5–7, 1975.

[28] P. J. H. Rudolph, "The Lurgi Process, The Route to SNG from Coal," paper presented at 4th Synthetic Pipeline Gas Symp., Chicago, 1972.

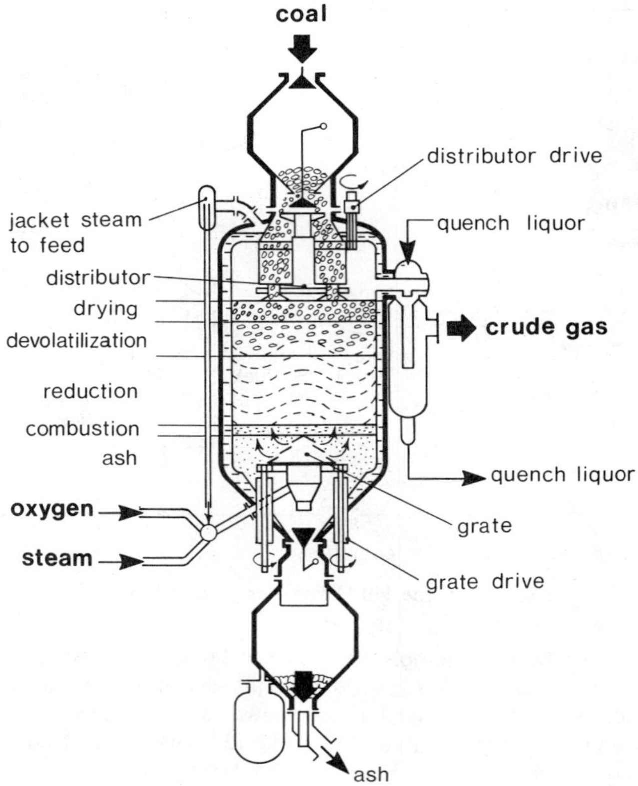

Figure 24.8 The Lurgi gasifier.[27]

rized in Table 24.2. Coal graded between 0.25 and 2.0 in. (6–50 mm) is preferred, but an upper-to-lower size ratio up to 10 is acceptable under certain circumstances. Generally 0.04 in. (1 mm) and 2.0 in. (50 mm) are regarded as lower and upper limits. Ash contents up to 30% by weight are acceptable, but the ratio of ash to fixed carbon normally should not exceed 0.7, although coals with a ratio of 1.2 have been processed successfully. Ash properties and coal reactivity largely determine the ratio of gasification steam to oxygen, the oxygen consumption being mainly dependent on reactivity. Moisture content is not critical, provided it does not reduce the temperature of the gases leaving the fuel bed below the dew point.

[29] British Pat. 1,443,293, Metalgesellschaft AG, July 21, 1976.

[30] British Pat. 1,451,155, Metalgesellschaft AG, September 29, 1976.

[31] British Pat. 1,442,942, Metalgesellschaft AG, July 14, 1976.

[32] British Pat. 1,445,809, Metalgesellschaft AG, August 11, 1976.

[33] Private communication, Lurgi Kohle and Mineralöltechnik GmbH, Frankfurt (Main), Germany, March 1977.

Generally standard boiler plate is used in the construction of the reactor and its associated equipment. Coals with high chlorine content may necessitate the cladding of certain parts of the gasifier to resist corrosion. Heat-resisting steels are used in the grate, and certain valves and internal equipment is to special design. Transport requirements limit shop-fabricated gasifiers to 12.8-ft (3.9-m) I.D., but units of 16-ft (5-m) I.D. are planned. Gasification rates equivalent to 16,500 scf/(ft^2)(hr) [5000 m^3/(m^2)(hr)] can be achieved at a pressure of 400 psig (28 bar), but lower figures are to be expected with coals that disintegrate, swell, or cake in the drying and devolatilization zone.

Some 16 Lurgi plants have been built,[28] involving a total of 65 gasifiers. The largest, which started operation at Sasolburg in South Africa in 1954, produces gas for oil synthesis. It has recently been added to, and a second plant is under construction.[34] Examples of plants built to produce town gas are West-

[34] J. C. Hoogendoorn, "Gas from Coal for Synthesis of Hydrocarbons, Status of Sasol II," paper presented at 9th Synthetic Pipeline Gas Symp., Chicago, 1977.

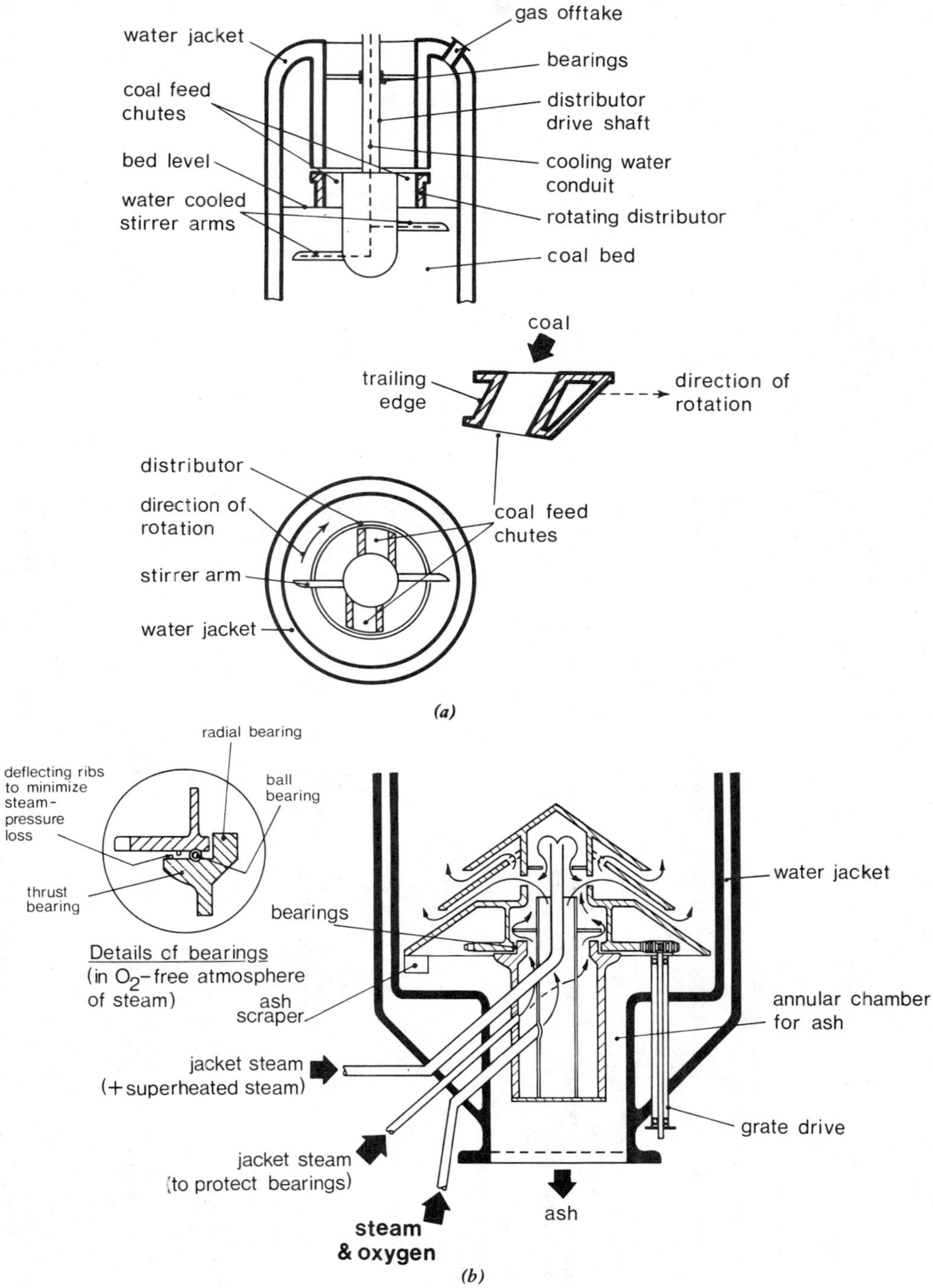

Figure 24.9 Examples of components of the Lurgi gasifier: (*a*) a coal feed system[29,30] and (*b*) a mechanical grate.[30–32]

field, Scotland (1960),[35] Dorsten, West Germany (1955),[36] and Coleshill, England (1963); although all are now out of production, gasifier development programs are active at both Dorsten and Westfield.[37]

[35] T. S. Ricketts, "High-Pressure Coal Gasification Plants in Scotland and Abroad," *J. Inst. Fuel*, **34**(244), 177–188 (May 1961).

[36] H. Weittenhiller, "Pressure Gasification with Oxygen of Solid Fuels in Germany," *J. Inst. Gas Eng.*, **1**(10) 655–669 (1961).

[37] D. Hebden and C. T. Brooks, "Westfield—The

Table 24.2 Typical Operation Results for Oxygen-Blown Lurgi Gasifiers Using Various Coals[33]

	Peat	Lignite		Subbituminous		Bituminous		Anthracite
Origin:	Ireland	Germany	North Dakota	Rio Turbio	Sasol, South Africa	Westfield, Scotland	Dorsten, West Germany	Vietnam
Size (in.):	0.6–1.6	0.04–0.4	0.16–1.20	0.2–1.2	0.2–1.2	0.2–1.2	0.2–1.2	0.2–1.2
Proximate analysis (wt %)								
Volatile matter	57.3	36.8	29.1	30.5	19.7	28.9	29.1	5.7
Fixed carbon	25.2	32.6	30.4	38.7	38.3	43.0	47.6	87.3
Moisture	15.5	26.5	34.3	16.5	5.4	15.7	5.1	2.0
Ash	2.0	4.1	6.2	14.3	36.3	12.4	18.2	5.0
HHV (Btu/lb, (maf))	9,830	11,250	12,010	13,400	12,940	13,810	14,330	15,440
Fischer assay tar (wt %, maf)	15.0	11.6	8.3	14.3	4.3	11.1	13.5	0
Sulfur (wt %)	0.1	0.4	1.1	1.0	0.6	2.1	2.0	0.5
Ash fusion point (°F)	—	2,200	2,280	2,600	2,590	2,550	2,520	2,730
Operating conditions								
Pressure (psia)	295	365	440	440	415	355	325	415
Steam oxygen (molar ratio)	8.7	8.7	9.3	6.7	6.8	5.6	5.3	6.8
Products								
Crude gas composition (vol %)								
CO	17.0	19.1	15.9	20.2	21.4	24.2	24.8	20.3
H_2	34.1	37.2	39.2	38.9	38.4	39.7	38.3	45.3
CH_4	13.6	11.8	10.8	11.7	9.6	9.4	9.3	4.7
C_nH_m	0.6	0.4	1.2	0.4	0.5	1.0	0.6	0.3
H_2S	0.1	0.2	0.4	0.3	0.2	0.8	0.5	0.1
CO_2	33.8	30.7	32.2	28.1	28.9	24.4	25.8	27.5
N_2	0.8	0.5	0.3	0.4	1.0	0.5	0.7	1.3
HHV (Btu/scf)	311.7	307.2	306.2	315.3	297.8	317.0	307.0	264.0
Specific data								
Consumption								
Oxygen								
scf/lb of coal (maf)	2.31	2.91	4.05	4.45	4.95	5.33	6.51	9.50
scf/10^6 Btu gas	322	.326	392	422	446	458	554	720
Steam								
lb/lb of coal (maf)	0.96	1.20	1.79	1.42	1.61	1.42	1.65	3.09
lb/10^6 Btu gas	134	134	173	135	145	122	140	234
Production								
Crude gas								
scf/lb of coal (maf)	23.05	29.08	33.73	33.44	37.23	36.72	38.30	50.01
10^6 Btu/ton (maf)	16.09	20.01	23.14	23.62	24.84	26.07	26.34	29.57
Tar and naphtha (wt %, maf)								

Perhaps the greatest future application for the Lurgi gasification plant will be in the manufacture of synthesis gas for the production of SNG from coal. In the United States, at least five major plants (Table 24.3),[38] each having nominal daily capacities of 250×10^6 scf (7×10^6 m^3) of SNG, are in an advanced planning stage. They are based on Western subbituminous coal and lignite, which, being noncaking and highly reactive, are preferred for Lurgi gasification. The ability of the Lurgi gasifier to handle the strongly caking, less-reactive Eastern U.S. coals has been demonstrated in special trials carried out on Pittsburgh Seam coal at Dorsten in 1963[39] and on a variety of coals at Westfield, Scotland, in 1973–1974.[40,41]

The Westfield trials were sponsored by the American Gas Association (AGA) and were conducted on 4000-ton batches of Illinois Nos. 5 and 6 and Pittsburgh No. 8 bituminous caking coal. A noncaking reactive subbituminous Western U.S. coal from Wyoming (Rosebud) was also included for comparison. One gasifier was isolated from the production plant and fitted with a special design of stirrer to handle the high swelling and strong caking properties of the bituminous coals. The trials were conducted using both graded and simulated run-of-mine coal, but because of limitations in coal supply and restraints on output associated with the plant equipment, operating conditions could not be fully optimized. The trials demonstrated, however, that these coals, of which there are large reserves

Development of Processes for the Production of SNG from Coal," paper presented at 113th Annual General Meeting, Institution of Gas Engineers, Edinburgh, April 1976.

[38] H. Hiller, "The Lurgi Pressure Gasification," Lurgi Information No. 01145/6.75; also *Erdoel Kohle,* **28**(2), (February 1975).

[39] W. H. Oppelt, H. Perry, J. L. MacPherson, and E. J. Vitt, *Lurgi-Gasifier Tests of Pittsburgh Bed Coal,* U.S. Bur. Mines Rep. Invest. No. 6721 (1966).

[40] D. C. Elgin and H. R. Perks, "Results from Trials of American Coals in Lurgi Pressure Gasification Plant at Westfield, Scotland," paper presented at 6th Synthetic Pipeline Gas Symp., Chicago, October 28–30, 1974, pp. 249–268.

[41] "Report on Trials of American Coal in a Lurgi Gasifier at Westfield, Scotland," Woodall-Duckham, Ltd., Crawley, England, U.S. Government Rep. No. FE-105, 1975, 342 pp.

Table 24.3 Lurgi Coal-to-SNG Plants Proposed for the United States[38]

	Company and Site				
	El Paso, Burnham, New Mexico	Wesco, Farmington, New Mexico	Panhandle, Converse County, Wyoming	American Natural Gas, Beulah-Hazen, North Dakota	Northern Natural Gas, Gillette, Wyoming
Fuel:	Subbituminous	Subbituminous	Hard Lignite	Hard Lignite	Hard Lignite
Proximate analysis (wt %)					
Volatile matter	30.7	26.5	32.7	31.9	30.3
Fixed carbon	33.8	35.5	33.7	34.4	35.0
Moisture	16.2	12.4	28.0	25.0	28.0
Ash	19.3	25.6	5.6	8.7	6.7
HHV fuel (Btu/lb, maf)	13,223	13,509	12,719	12,060	12,917
Coal supplied (10^6 tons/year)	7.90	8.10	7.73	7.99	7.56
SNG production (10^6 scf/year)	93,300	89,600	89,600	89,600	82,100
HHV product gas (Btu/scf)	967	970	970	970	985
By-products					
Tar + oil (10^3 tons/year)	450	522	415	300	467
Naphtha (10^3 tons/year)	78	98	62	52	83
Crude phenol (10^3 tons/year)	40	33	41	39	41
NH_3 (10^3 tons/year)	66	65	38	73	52

close to centers of population and industry, could be effectively gasified by the Lurgi process. Some selected results are given in Table 24.4. The results of the trials have been

Table 24.4 Summary of Lurgi Gasification Trials on American Coals at Westfield[31]

	Fuel and Size (in.)			
	Rosebud, 0.25–1.5	Illinois No. 5, 0.5–1.5	Illinois No. 6, 0.5–1.5	Pittsburgh, 0.25–1.5
Proximate analysis (wt %)				
Volatile matter	29.2	35.2	34.7	37.4
Fixed carbon	36.4	44.7	46.0	50.3
Moisture	24.7	12.0	10.2	4.6
Ash	9.7	8.1	9.1	7.7
HHV (Btu/lb, maf)	13,130	14,330	14,210	15,330
Caking index	0	15	15	30
Free-swelling index	0	2–2½	3	7½
Ash-fusion point (°F)				
Oxidizing	2300	2370	2490	2480
Reducing	2150	1940	2100	2140
Total sulfur (wt %, maf)	1.61	3.88	3.44	2.87
Operating conditions				
Pressure (psia)	371	356	350	365
Gas offtake temperatures (°F)	709	1148	1121	1195
Steam/oxygen ratio (molar)	9.1	8.6	9.8	9.8
Products				
Crude gas composition (vol %)				
CO	15.1	17.6	17.3	16.9
H_2	41.1	38.3	39.1	39.4
CH_4	11.2	9.2	9.4	9.0
C_2H_6	0.5	0.5	0.7	0.7
C_nH_m	—	0.3	—	0.1
H_2S	0.5	1.1	1.1	0.8
CO_2	30.4	31.0	31.2	31.5
N_2	1.2	1.5	1.2	1.6
HHV (Btu/scf)	298	291	290	285
Tar and oil (wt % coal, maf)	6.1	6.13	4.21	5.59
Liquor (wt % coal, maf)	148.7	221.2	262.2	296.5
Specific data				
Consumption				
Oxygen				
scf/lb coal (maf)	4.39	6.88	6.64	7.95
scf/10^6 Btu gas	418	620	618	688
Steam				
lb/lb coal (maf)	1.91	2.81	3.11	3.70
lb/10^6 Btu gas	182	253	289	320
Fuel (lb/hr-ft^2 maf)	138	135	127	83.6
Crude gas				
scf/lb coal (maf)	35.2	38.2	37.2	40.5
10^6 Btu/ton (maf)	21.0	22.2	21.6	23.1
10^6 Btu/hr-ft^2	1.45	1.50	1.37	0.96
Gasification efficiencies (%)[a]				
E_1	86.3	81.9	83.8	79.7
E_2	70.1	64.7	62.4	59.2
E_3	79.9	74.3	69.8	63.4

[a] See Appendix.

summarized by Elgin and Perks[40] and are given in full in an official government report.[41]

Other published performance data for Lurgi gasification come from Czechoslovakia,[42] where during 1974 a series of full-scale trials was carried out to evaluate gasifier performance on high-sulfur, high-arsenic lignite. Tests indicated that 81–87% of the sulfur appeared as H_2S in the crude gas, 10% in the ash, and the remainder in the liquid products. Little arsenic passed forward in the gas, the bulk appearing in the ash and tar.

Trials have been carried out on U.S. coals at Sasolburg on behalf of a number of U.S. gas pipeline and oil companies, but the results have not been published. With the support of the West German government. a consortium including Ruhrgas, Ruhrkohle, and Lurgi Kohle und Mineralöltechnik GmbH are to build an experimental gasifier at Dorsten to investigate operation at pressures up to 1500 psig (100 bar) and develop new processes for the treatment of the product gas.

24.2.4 New Generation Experimental Fixed-Bed Gasifiers

A number of American companies have recently proposed new fixed-bed gasification systems designed to increase output and gasification efficiencies, especially with the Eastern U.S. caking coals. Four such processes in the initial development stages are described below. Three of them appear to be of similar design, but their process configurations are not yet fully established, and few experimental data are available.

[42] J. Macak, K. Prasek, and J. Simanek, "Ecological Aspects of Gas Production from High Sulfur Lignite," Paper No. IGU/B7-76, presented at 13th World Gas Conf., London, 1976.

24.2.4.1 *New Generation Processes Yielding Nitrogen-Containing Product Gas*

The M. W. Kellogg Fixed-Bed Gasifier. The process proposed by the M. W. Kellogg Company utilizes a low-pressure [10 psig (0.7 bar)], fixed-bed gasifier with a revolving grate.[43–45] It is claimed to be capable of gasifying coal containing up to 20% fines using air, enriched air, or oxygen mixed with steam to produce gas with an HHV of approximately 300 Btu/scf (11.4 MJ/m³) on oxygen and 135 Btu/scf (5.12 MJ/m³) on air.

The process is illustrated in Fig. 24.10. The gasifier comprises a water-jacketed, refractory-lined shell containing the fuel bed, which is fed with coal via a lock hopper. The fuel bed rests on a variable speed revolving grate, the ash from which is discharged through a lock hopper. The upper part of the fuel bed is agitated by a water-cooled stirrer to break up the coal as it cakes in the devolatilization zone. The product gas passes through a cyclone to remove particulate matter before being cooled, and other by-products are condensed in a waste heat boiler. Some compositions of the gases typically produced with coke, anthracite, and bituminous coal are given in Table 24.5.[46]

The Morgas Process (USBM Stirred Fixed-Bed Gasifier). The Morgas process, initially developed on the pilot scale at the Morgan-

[43] Anon., "Producer Gas from Coal," *Compressed Air,* 16, 17 (October 1974).

[44] Anon., "Conversion of Coal to Gas by Fixed-Bed Process," *Ind. Heat,* **41,** 44, 46, 48 (November 1974).

[45] Anon., "Further Progress in Coal Gasification," *Chem. Eng. News,* **52**(32) 17–18 (August 12, 1974).

[46] Anon., "U.S. Coal-to-Gas Process Is Ready," *Oil Gas J.,* **72**(36), 86, 88 (September 9, 1974).

Table 24.5 Typical Product Gas Compositions for the Kellogg Fixed-Bed Gasifier[46]

Analysis	Coke	Anthracite Coal	Bituminous Coal
Component (% v/v)			
CO	29.0	27.1	28.6
H_2	10.0	16.6	15.0
CH_4	0.7	0.5	2.7
CO_2	3.5	5.0	3.4
N_2	56.8	50.8	50.3
HHV (Btu/scf)	132	146	168

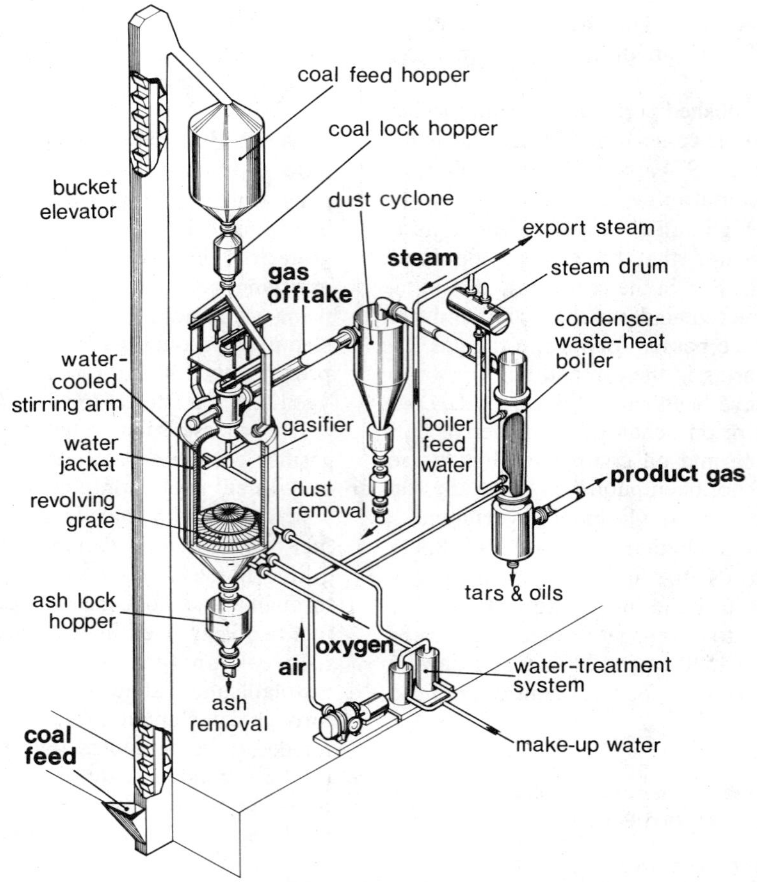

Figure 24.10 The Kellogg low-pressure fixed bed.[43]

town, West Virginia, Research Center of the U.S. Bureau of Mines (USBM) under cosponsorship with the Office of Coal Research (OCR), employs a fixed-bed, single-stage gasifier operating on air at 200 psig (13.8 bar). The Morgantown pilot gasifier,[47] which is 3.5 ft (1.08 m) in diameter and has a throughput of 18 tons/day (16.3 tonnes/day), is illustrated in Fig. 24.11. Its main features are the agitator, which both rotates and reciprocates within the fuel bed, and the twin coal lock hoppers. Otherwise the gasifier is conventional in that the fuel bed is supported on a rotating grate through which ash is withdrawn and superheated steam and air are supplied. Product gas leaves through a gas offtake above the fuel bed and, in the pilot plant, passes through a cyclone to remove dust before being discharged to atmosphere.

Coal is charged alternately through the two lock hoppers to the fuel bed. The stirrer is normally rotated at about 0.5 rpm and moved up and down vertically over a distance of 2 ft (0.61 m) so that it passes any particular

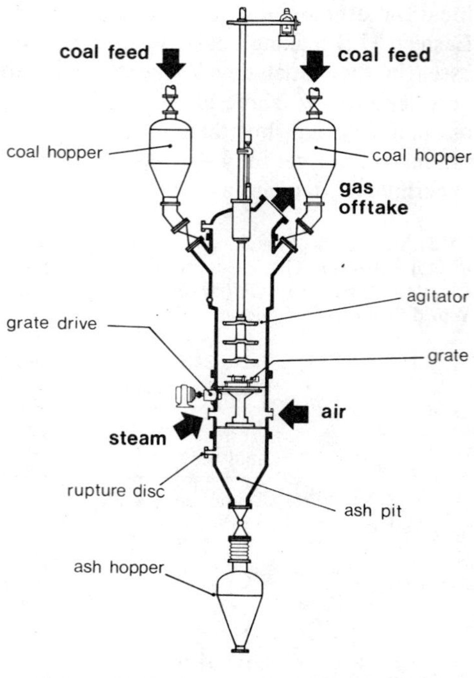

Figure 24.11 The Morgas gas producer.[47]

point once every 15 min. Mechanical details of the stirrer[47] are given in Fig. 24.12. Strongly caking coals classified as high-volatile A bituminous with free-swelling indices greater than 8 (e.g., Pittsburgh, Upper Freeport, and Middle Kittanning), and containing a high proportion of fines, have been successfully gasified.[48] A material balance obtained during tests on a noncaking subbituminous coal is given in Table 24.6. Further development of this gasifier for the utilization of caking coals is being carried out by McDowell-Wellman Engineering Company.[49]

[47] R. V. Rahfuse, G. B. Goff, and A. J. Liberatore, *Noncaking Coal Gasified in a Stirred-Bed Producer,* U.S. Bur. Mines Clean Energy Program, Tech. Prog. Rep. No. 77, March 1974.

[48] P. S. Lewis, A. J. Liberatore, and J. P. McGee, *Strongly Caking Coal Gasified in a Stirred-Bed Producer,* U.S. Bur. Mines Rep. Invest. No. 7644, 1972.

[49] I. Howard-Smith and G. J. Werner, *Coal Conversion Technology,* Noyes Data Corporation, Park Ridge, N.J., 1976, p. 72.

The GEGas Process. The General Electric Company's GEGas process is designed to process all types of coal, including fines, with reduced steam requirements. Most of the recent research work has been directed toward an air-blown gasifier producing a low-Btu gas primarily for use in combined-cycle power plants. The present 3-ft (1-m) I.D. pilot gasifier (Fig. 24.13) is capable of operating at pressures up to 300 psig (20 bar). The main features are a screw-feed system, a stirrer-agitator (based on the design of the Morgas system),[48] and a novel grate design.

Coal introduced through lock hoppers is metered continuously into the gasifier by an auger or screw-type feeder. A three-bladed, water-cooled stirrer, which can travel from the top of the gasifier to the vicinity of the grate, rotates to break up caking coal and remove deposits from the hemispherical dome and gas offtake areas. The grate is equipped with a similar rotating stirrer or agitator paddle that discharges the ash when in a lowered position and breaks up weakly

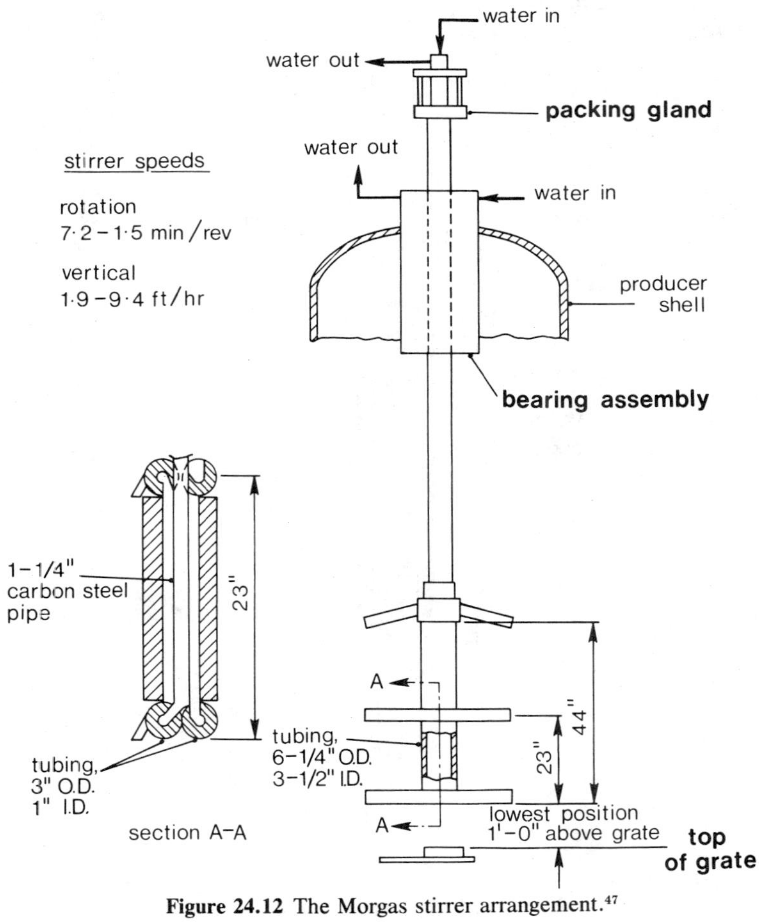

Figure 24.12 The Morgas stirrer arrangement.[47]

Table 24.6 Material Balance for the Morgas Gasifier (lb/hr)[47]
Product gas composition (vol %): CO, 16.0; H_2, 19.0; CH_4, 3.5; C_2H_6, 0.3; H_2S, 0.2; CO_2, 12.6; N_2, 48.4.

	Carbon	Hydrogen	Oxygen	Nitrogen	Sulfur	Ash	Total
Input							
Coal (as received)	757	90.6	243.1	23.8	14.9	360.6	1490
Air			803	2647			3450
Steam		113	897				1010
	757	203.6	1943.1	2670.8	14.9	360.6	5950
Output							
Gas (dry)	722.2	100.4	1212.1	2493.5	11.8		4540
Tar	42.5	3.5	2.5	0.5	1.0		50
Dust	11.5	0.7	0.1	0.3	0.4	12.0	25
Steam		107.4	852.6				960
Residue	29.3	1.9	2.3	1.9	1.0	287.6	324
	805.5	213.9	2069.6	2496.2	14.2	299.6	5899

fused clinker in a raised position. It is claimed that with this grate the ratio of steam to air can be reduced while maintaining satisfactory ash removal.

Pittsburgh No. 8 coal has been successfully gasified in the GEGas-D gasifier, at pressures of 225 and 300 psig (15–20 bar), in runs lasting 40 and 10 hr, respectively. Heat and mass balances obtained in the Pittsburgh coal run are given in Table 24.7. For this difficult coal, continuous agitation of the top 2 ft (600 mm) of bed, with only periodic deep-bed

[50] W. R. Epperly and H. M. Siegel, "Catalytic Coal Gasification for SNG Production," *Proc. 11th Intersociety Energy Conversion Engineering Conf.*, State Line, Nev., September 12–17, 1976.

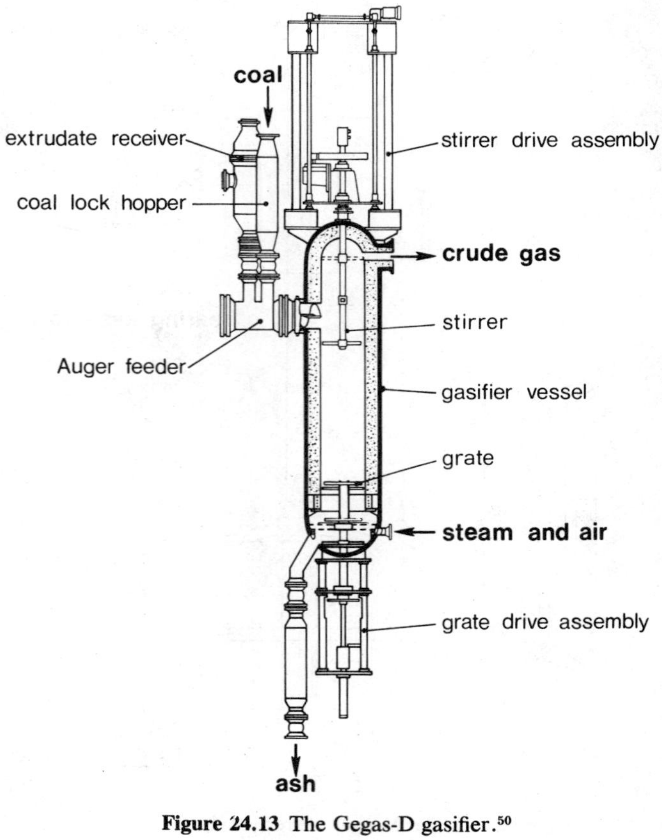

Figure 24.13 The Gegas-D gasifier.[50]

Table 24.7 Material Balance and Heat Balance for the GEGas Gasifier[51]

Product gas composition (vol %) = CO, 23.8; H_2, 17.0; CH_4, 3.2; CO_2, 6.7; N_2, 49.2; O_2, 0.1.

	Material Balance (lb/hr)							Heat Balance (10^6 Btu/hr)			
	Carbon	Hydrogen	Oxygen	Nitrogen	Sulfur	Ash	Total	Temp (°F)	HHV	Sensible + Latent	Total Heat
Input											
Coal	890.0	65.7	97.5	17.8	50.7	198.0	1320.0	70	16.24	—	16.24
Air	—	—	800.0	2670.0	—	—	3470.0	350	—	0.235	0.235
Steam	—	66.0	528.0	—	—	—	594.0	350	—	0.709	0.709
Water to ash—quench	—	0.6	4.4	—	—	—	5.0	70	—	—	—
	890.0	132.3	1429.9	2687.8	50.7	198.0	5388.7	—	16.24	0.944	17.184
Output											
Gas	788.0	95.1	1165.0	1687.8	50.7	—	4786.6	1020	12.315	1.246	13.561
Tar	54.0	3.0	0.5	—	—	—	57.5		0.957	—	0.957
Dust	—	—	—	—	—	—	—	—	—	—	—
Steam	—	30.3	241.0	—	—	—	271.3	—	—	0.320	0.320
Residue	49.0	2.8	23.8	—	—	114	189.6	500	0.661	0.031	0.692
Cooling water	—	—	—	—	—	—	—	90	—	0.661	0.661
	891.0	131.2	1430.3	2687.8	50.7	114	5305.0	—	13.933	2.258	16.191

Table 24.8 Performance Data for *GEGas* Experimental Extrusion Feeders[52]

Barrel Diameter (in.)	Backpressure (atm)	Specific Output [lb/(hr)(ft²)]	Power Consumption [kW(e)-hr/lb]
1	1	2700	0.0575
2	1	9150	0.0075
2	1	4125	0.0150
2	20	4125	0.0160

penetration by the stirrer, gave satisfactory gasification and caused no damage to the internals.

The feeding of fines as tar-bonded extrudates has also been demonstrated. The concept is to separate from the main coal feed the fines, which tend to cause uneven flow in the fuel bed and excess dust carryover, bind them with tar produced in the process, and make extrudates of a size suitable for gasification. These extrudates are then fed to the gasifier with the graded coal.

The proportion of coal that can be made into extrudates economically depends on the cost of power and equipment required. Performance data for an experimental extruder are given in Table 24.8. It is estimated that a commercial gasifier of 12-ft (3.7-m) I.D. would require four 10-in. (254-mm) diameter extruders having a power consumption of 0.005 kW-hr/lb (0.011 kW-hr/kg) of coal mixture.[52]

General Electric has patented and tested an alternative system of handling caking and swelling coals[53,54] involving the use of inert refractory shapes that maintain the fuel bed porous and free flowing and can be recovered and recycled. Experimental work done with 2-in. (50-mm) Raschig rings of silicon carbide has met with limited success and has indicated that dilution of the bed results in lower gas offtake temperatures and reduced times of contact.

24.2.4.2 *New Generation Processes Yielding Nitrogen-Free Product Gas*

The British Gas–Lurgi Slagging Gasifier. The conversion of the Lurgi gasifier to slagging operation was a logical evolutionary step to improve performance and raise specific output, particularly for coals with low fusion point ash and/or low reactivity to steam. Following some earlier work by the Lurgi Company at low pressure, the British Gas Corporation (then the Gas Council) carried out some experiments on slagging operation[55,56] at their Midlands Research Station in Solihull in the mid-1950s, and later (early 1960s) developed a pilot slagging gasifier[57] operating at 300 psig (20 bar) on coal at a throughput of 100 long tons/day (101 tonnes/day). The discovery of natural gas in the North Sea postponed its further development on the full scale in Great Britain until the Continental Oil Company in the United States assembled a consortium of some 15 U.S. gas, pipeline, and oil companies to sponsor its development to commercial status. This development, engineering, and testing commenced in 1974 and was carried out by the British Gas Corporation at their Westfield Development Center in Scotland, with the close collaboration of Lurgi Kohle und Mineralöltechnik GmbH (Frankfurt). Little has been published[37,58] on these more recent

[51] D. E. Woodmansee, *Power Cycles and Fuels Projects,* General Electric Company, Research and Development Center, Schenectady, N.Y., private communication, February 1977.

[52] A. M. Beuche, paper presented at 6th Synthetic Pipeline Gas Symp., Chicago, 1974, pp. 159–186.

[53] British Pat. 1,435,088, General Electric Company, May 12, 1976.

[54] British Pat. 1,435,089, General Electric Company, May 12, 1976.

[55] D. Hebden and R. F. Edge, *Experiments with a Slagging Pressure Gasifier,* Gas Council Research Commun. No. GC 50, November 18–19, 1958.

[56] H. H. Lowry, Ed., *Chemistry of Coal Utilization,* Supplementary Volume. John Wiley and Sons, New York, 1963, pp. 960–963.

[57] D. Hebden, A. G. Horsler, and J. A. Lacey, *Further Experiments with a Slagging Pressure Gasifier,* Gas Council Research Commun. No. GC 112, 1964.

[58] D. Hebden, "High-Pressure Gasification Un-

events, but U.S. ERDA (subsequently a part of DOE) has announced a project to evaluate, engineer, construct, and operate a 60-million-ft^3/day (1.7×10^6-m^3/day) SNG-from-coal demonstration plant[59] based on the British Gas–Lurgi slagging gasifier.

The essential features of the gasifier are illustrated diagrammatically in Fig. 24.14*a*. The upper part resembles a conventional Lurgi gasifier with a coal lock hopper, coal distributor-stirrer, gas offtake, and gas quench. In the lower part the fuel bed rests on a refractory hearth that is surrounded by a number of tuyeres. Steam and oxygen are injected through the tuyeres in a ratio that produces the high temperature required to melt the ash and allow it to accumulate in the base of the hearth. The slag is run off from the hearth into a chamber below, where it is quenched in water and settles out into a lock hopper from which it is discharged to atmosphere as a dense glassy frit.

No details of the slagging gasifier developed at Westfield are available, but the unique system of intermittent slag tapping has been described in some detail,[57] as has the hearth arrangement developed on the pilot plant, which is illustrated in Fig. 24.14*b*. In the pilot plant the fuel bed rests on a conical refractory hearth surrounded by four tuyeres fitted with nozzles, through which steam and oxygen are injected at velocities of between 175 and 260 ft/s (53–79 m/s).

The slag is discharged through the taphole by intermittently adjusting the differential pressure between the quench chamber and the hearth so as to stop, start, or control its flow. It is claimed that the system gives complete control over the rate of slag flow, prevents the formation of stalactites of frozen slag, absorbs changes in ash constituents, and gives time for fluxing reaction to take place, thereby producing a homogeneous slag with uniform and consistent flow characteristics. Slag flow rates of up to 30,000 lb/hr (13,600 kg/hr) were observed through a 1 1/2-in. (38-mm) diameter slag tap. The main difficulties encountered were molten iron attack on the water-cooled metal slag tap and the erosion of the refractory hearth lining by molten slag. Ways of overcoming these difficulties were predicted.

Performance data for two weakly caking British coals and a coke-oven coke tested on the pilot plant are given in Table 24.9, and a heat and material balance for one of the coals is given in Table 24.10.

Comparisons between the conventional dry bottom Lurgi gasifier and the British Gas–Lurgi slagging gasifier indicate the potentially two- to threefold greater output per unit cross-sectional area[58] of the latter over the latest versions of the former proposed for the U.S. SNG market and for oil synthesis in South Africa. The six- to eightfold lower requirement for process steam leads to an 8–10% higher gasification efficiency based on the ratio of the heating value of the crude gas produced to that of the coal required for gasification and gasification steam raising; it also results in lower aqueous liquor production. Although the slagging gasifier shows a less favorable oxygen consumption with the lower rank reactive coals such as lignite, a lower oxygen consumption than is experienced on the conventional Lurgi gasifier has been observed when gasifying the less-reactive bituminous coals, particularly when the ash has a low melting point.[37] More performance data are expected to be published as the ERDA demonstration plant project based on this gasifier progresses.

The Grand Forks Energy Center Slagging Gasifier. The USBM designed and operated a fixed-bed, steam- and oxygen-blown experimental slagging gasifier at Grand Forks, North Dakota, during the period 1952–1965 with special reference to the utilization of Dakota lignites.[60] The facility was mothballed until 1974 when it was recommissioned by ERDA[61] to investigate the slagging fixed-bed gasification potential of lignites and other Western noncaking coals with particular reference to the characteristics of the slag and aqueous effluents produced.

der Slagging Conditions," paper presented at 7th Synthetic Pipeline Gas Symp., Chicago, 1975, pp. 385–399.

[59] Anon., "British Gas's SNG Process to Receive $20 million ERDA Grant," *Eur. Chem. News,* 19 (July 16, 1976).

[60] G. H. Gronhord, A. E. Hook, M. M. Fegley, and D. E. Severson, *Slagging Fixed-Bed Gasification of North Dakota Lignite at Pressures to 400 psig,* U.S. Bur. Mines Rep. Invest. No. 7408, July 1970.

[61] R. C. Ellman and B. C. Johnson, "Slagging Fixed-Bed Gasification at Grand Forks Energy Research Center," paper presented at 8th Synthetic Pipeline Gas Symp., Chicago, October 1976.

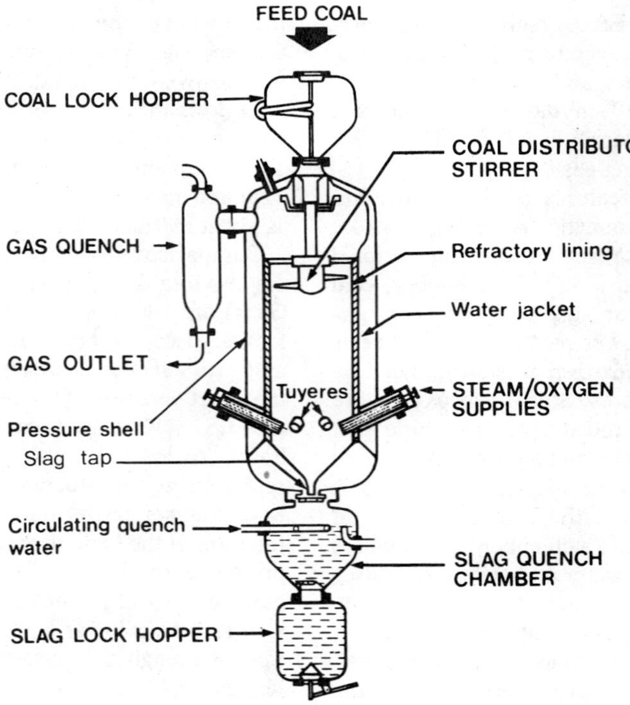

(a)

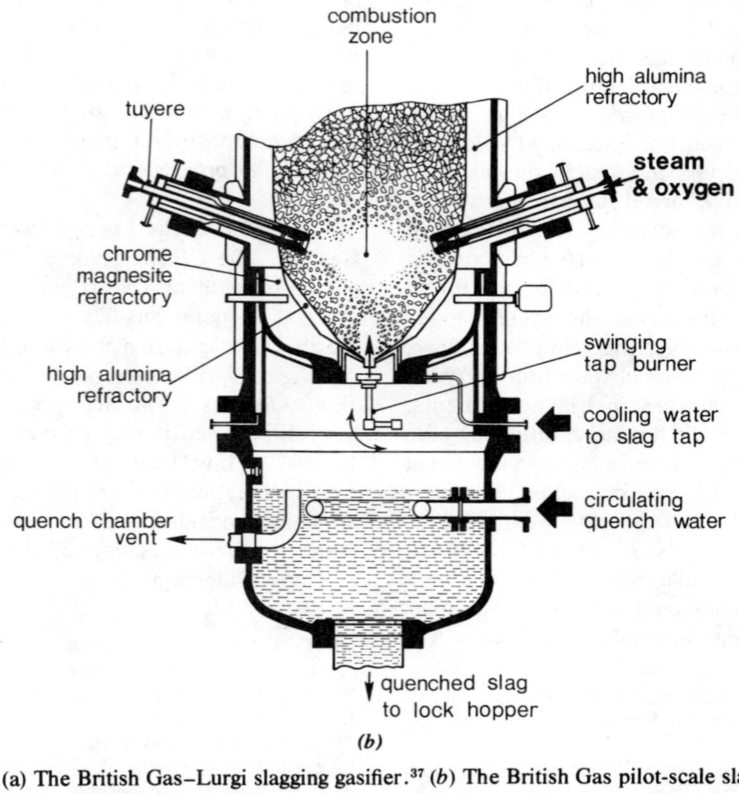

(b)

Figure 24.14 (a) The British Gas–Lurgi slagging gasifier.[37] (*b*) The British Gas pilot-scale slagging gasifier, showing hearth area and slag tapping.[57]

Table 24.9 Performance Data from the Midlands Research Station Slagging Gasifier

	Fuel (origin) and Size (in.)				
	Oven Coke, 1.0–1.5	Donisthorpe, 1.0–1.5	Donisthorpe, 0.5–1.0	Donisthorpe, 1.0–1.5	Newstead, 1.0–2.0
Moisture (as charged)	9.45	12.7	15.1	13.8	12.6
Ash (as charged)	6.85	6.4	9.7	4.85	6.65
Total sulfur (as charged)	1.05	1.25	1.5	1.1	0.6
Ash fusion temperature (flow point) (°F)	2430	2450	2680	2430	2810
Flux type	Blast furnace slag	Blast furnace slag	Dolomite	—	Dolomite
Flux addition (lb/lb ash)	0.83	0.75	0.3	0.0	0.32
Operating conditions					
Pressure (psia)	315	315	315	315	315
Steam/oxygen ratio (molar)	1.17	1.10	1.10	1.10	1.06
Products					
Crude gas composition (vol %, N_2-free basis)					
CO	70.35	61.2	60.05	60.85	60.55
H_2	27.25	28.05	29.25	28.1	28.65
CH_4	0.45	6.7	5.65	6.7	6.4
C_2H_6	0.0	0.95	0.85	1.0	0.85
C_nH_m	0.0	0.45	0.7	0.55	1.05
CO_2	1.85	2.55	3.5	2.7	2.35
O_2	0.1	0.1	0.0	0.1	0.15
HHV (Btu/scf)	319.7	379.6	371.5	380.6	384.7
Tar (wt % of coal, maf)		7.3	7.7	—	6.9
Liquor (wt % of coal, maf)	—	23.2	27.7	—	21.5
Specific data					
Consumption					
Oxygen					
scf/lb coal (maf)	8.27	5.69	6.05	5.69	6.28
scf/10^6 Btu gas	613	488	507	503	505
Steam					
lb/lb coal (maf)	0.453	0.294	0.313	0.292	0.312
lb/10^6 Btu gas	33.2	24.9	26	25.7	24.9
Fuel (lb/hr-ft^2)	516	981	924	1,440	930
Production of crude gas					
scf/lb coal (maf)	42.4	30.7	32.1	29.7	32.3
10^6 Btu/ton coal (maf)	30.4	26.1	26.7	25.3	27.8
10^6 Btu/hr-ft^2	7	11.4	11	16.3	11.5

Table 24.10 Material Balance and Heat Balance for the Midlands Research Station Slagging Gasifier Operating on (1.0–1.5 in.) Donisthorpe Coal

	Material Balance (lb/hr)			Heat Balance (10^6 Btu/hr)		
	Carbon	Hydrogen	Oxygen	HHV	Sensible and Latent	Total
Input						
Coal	5323	373	993	95.8	—	95.8
Moisture	—	121	972	—	—	—
Steam and oxygen	—	226	5106	—	2.85	2.85
Slag tap burner	41	15	378	1.35	—	1.35
	5364	735	7449	97.15	2.85	100.0
Output						
H_2S, NH_3, etc.	—	—	—	0.46	—	0.46
Gas and aqueous liquor	4910	685	7301	80.3	5.55	85.85
Tar	406	38	45	8.44	—	8.44
Dust	48	2	6	0.86	—	0.86
Slag	0.0	0.0	0.0	—	0.80	0.80
Cooling water						
Hearth area	—	—	—	—	0.66	0.66
Gasifier jacket, etc.	—	—	—	—	1.52	1.52
Unaccounted for	0	12	97	1.41	—	1.41
	5364	737	7449	91.47	8.53	100.0

The gasifier, which contains a refractory-insulated fuel bed, 16 in. (405 mm) in diameter, is illustrated in Fig. 24.15*a*, and details of the hearth are given in Fig. 24.15*b*. It is fitted with coal and slag lock hoppers, and the water-cooled shell rated at 600 psig (40 bar) is arranged so that the fuel bed height can be varied between 6 and 15 ft (1.8–4.6 m). The fuel bed rests on the refractory hearth surrounded by four tuyeres. Slag drains continuously from the hearth through a water-cooled taphole; the slag falls into water below, where it is quenched before being discharged through the slag lock hopper. Slag breakers and a burner are provided to remove stalactites as they form, the latter being also used to melt out slag plugs.

Performance data collected in the earlier program and during more recent tests are given in Table 24.11 for two North Dakota lignites. Apart from evaluating the slagging performance of various coals, it is intended to study the effect of moisture content and the injection of fine coal and tar at the tuyeres. No commercial version of the gasifier is currently being considered.

The BCURA and CJB Slagging Gasifier. During the late 1950s and early 1960s the British Coal Utilisation Research Association (BCURA) of Leatherhead, England, carried out a program of research and development of fixed-bed slagging gasification initially under contract to the British Ministry of Power,[62] and later (1962–1964) under the joint sponsorship of the Gas Council, the National Coal Board, and the Central Electricity Generating Board.[63] In both programs the investigation involved work in the laboratory and on experimental gasifiers, but it included close association with a contract awarded by the Ministry of Power to Constructors John Brown (CJB) for the development of a high-pressure slagging gasifier to produce synthesis gas[64] for making oil from coal by the Fischer-Tropsch process. Although no commercial gasifier resulted, a great deal of information was published both on the performance of the CJB gasifier operating on oven coke and on the behavior of the mineral matter in slagging gasification processes.

The CJB gasifier (Fig. 24.16) contained a

[62] H. R. Hoy, A. G. Roberts, and D. M. Wilkins, "Some Investigations into the Gasification of Solid Fuel in a Slagging Fixed-Bed Gasifier," *Proc. Joint Conf. Gasification Processes,* IGE–Institute of Fuel, 1962, C11-C22.

[63] H. R. Hoy, A. G. Roberts, and D. M. Wilkins, *Behavior of Mineral Matters in Slagging Gasification Processes,* IGE Commun. No. 672. JIGE 1965, 5, 44.

[64] S. Masterman and W. A. Peet, "The Development of a Pressurized Slagging Fixed-Bed Gasifier," *Proc. Joint Conf. Gasification Processes,* IGE–Institute of Fuel, 1962, C23-32.

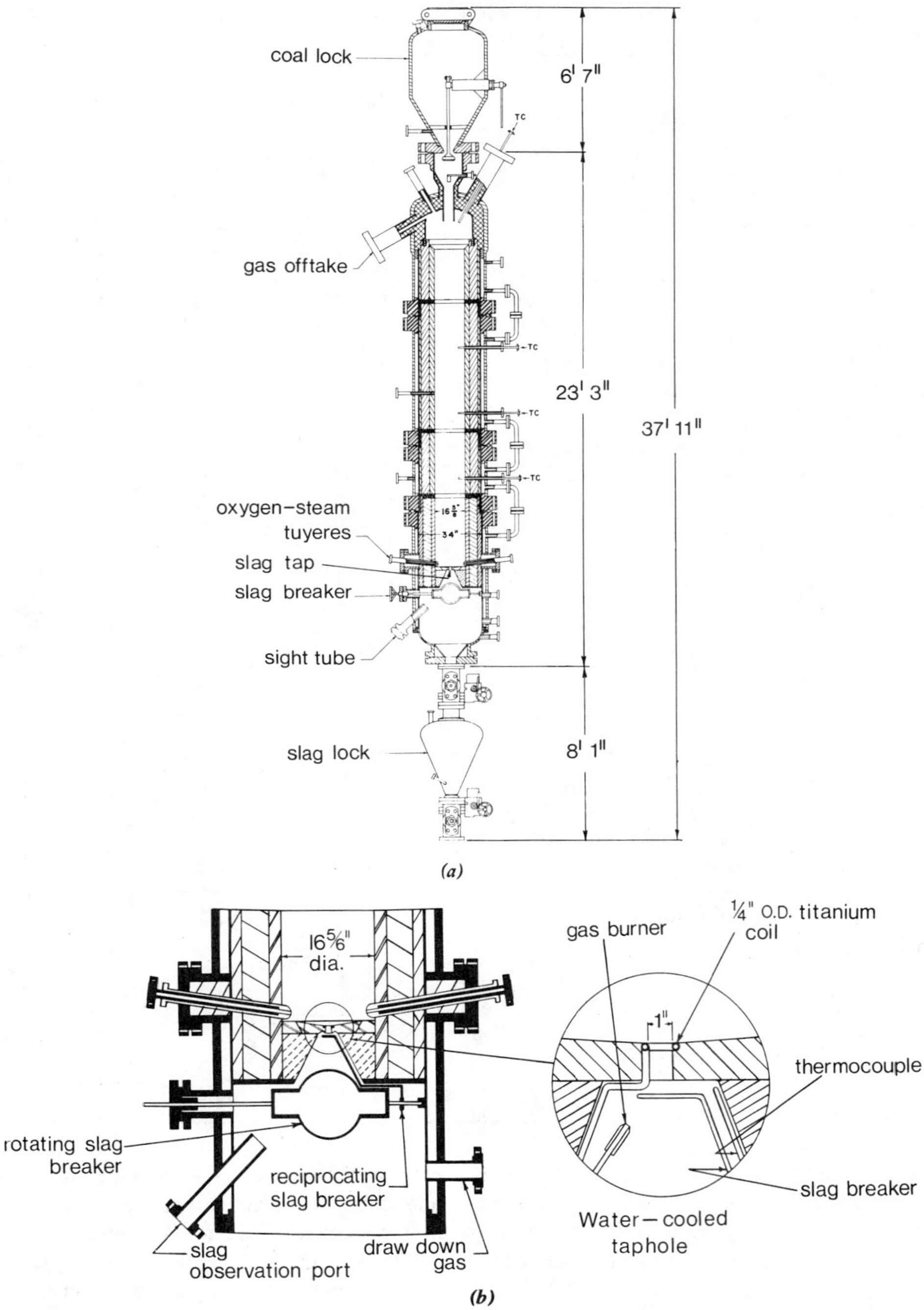

Figure 24.15 (*a*) The Grand Forks slagging fixed-bed gasifier, and (*b*) its hearth and taphole.[61]

fuel bed approximately 4 ft (1.2 m) in diameter and 6 ft (1.8 m) deep supported on a flat hearth and surrounded by a refractory-lined and water-cooled liner suspended in a main pressure shell. Steam and oxygen were injected via two diametrically opposed tuyeres, and slag was withdrawn continuously through a central slag offtake, discharging into water, where it was quenched as a glassy frit prior to removal from the pressure system. The product gas left the gasifier at about 930°F (500°C) via a gas offtake. The fuel used was generally coke-oven coke of various grades and ash content, but some tests were undertaken on a low-temperature char to simulate performance on coal. The fuels were

Table 24.11 Performance Data from the Grand Forks Slagging Gasifier Operating on North Dakota Lignite[61]

	Velva Lignite	Indianhead Lignite
Proximate analysis (wt %)		
Volatile matter	28.8	29.4
Fixed carbon	32.5	35.0
Moisture	35.2	27.0
Ash	3.5	8.6
HHV (Btu/lb, maf)	11,760	11,830
Ash fusion (fluid) temperature (°F)	2,620	2,150
Operating conditions		
Pressure (psia)	415	215
Steam/oxygen ratio (molar)	1.11	1.11
Products		
Crude gas composition (vol %)		
CO	56.0	56.9
H_2	28.8	29.6
CH_4	6.9	5.3
C_2H_6	0.2	0.2
C_nH_m	0.5	0.4
CO_2	7.4	7.5
O_2	0.2	0.1
HHV (Btu/scf)	352	339
Tar (wt %, maf)	5.3	5.0
Specific data		
Consumption		
Oxygen		
scf/lb of coal (maf)	5.77	5.34
scf/10^6 Btu gas	563.8	564
Steam		
lb/lb of coal (maf)	0.305	0.282
lb/10^6 Btu gas	29.8	29.8
Fuel (lb/hr-ft^2)	577	499
Production of crude gas		
scf/lb coal (maf)	29.1	27.95
10^6 Btu/ton coal (maf)	22.9	21.2
10^6 Btu/hr-ft^2	5.9	4.72
Gasification efficiencies		
E_1	87.1	80.1
E_2	94.5	87.1

fluxed with either blast furnace slag or dolomite. Operation was at 45 psig (3 bar); at a nominal oxygen supply rate of 1920 lb/hr (870 kg/hr), the gas output was equivalent to about 2.4×10^6 Btu/(ft^2)(hr) [272.4 MJ/(m^2)(hr)]. The program was terminated in March 1962.

The work by BCURA[63] on slag behavior confirmed the correlation between the silica ratio and the viscosity of the slag with alternative empirical correlations for high iron content and high silica ratio slags. Approximate relationships between temperature and silica ratio at four selected viscosity levels are given in Fig. 24.17. This work also covered slag formation, slag flow behavior, slag quenching and granulation, and slag separation from gases. Slag flow behavior was studied in experimental gasifiers, which included one incorporating a water-cooled grate with pulverized fuel injection through the tuyeres equivalent to 25% of the fuel supplied.

The Chevron Process. The Chevron Research Company has patented a process[65,66]

[65] U.S. Pat. 3,759,677, Chevron Research Company, September 18, 1976.
[66] U.S. Pat. 3,775,072, Chevron Research Company, November 27, 1976.

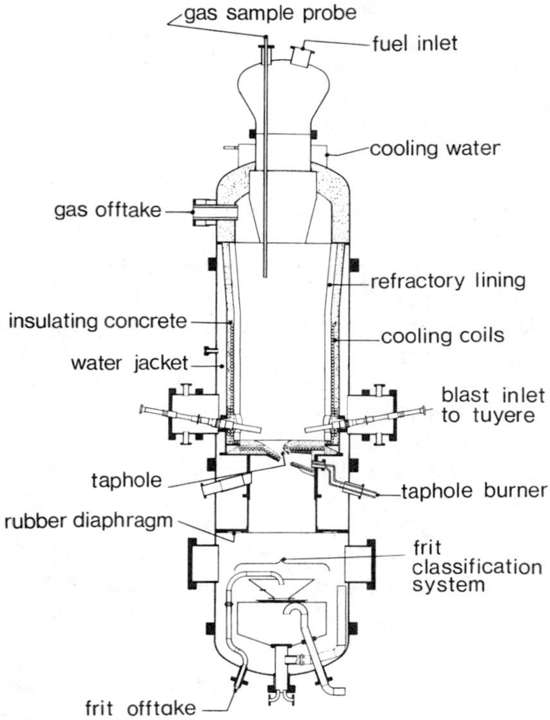

Figure 24.16 The Ministry of Power–CJB pressurized slagging fixed-bed gasifier.[64]

to produce a medium-Btu gas from a wide range of organic feeds such as lignite, organic waste material, and wood, in which the feedstock is reacted with steam at 300–800 psig (20–55 bar) and 1200–1400°F (650–760°C) in the presence of a catalyst (e.g., K_2CO_3).

No details of the gasifier are published beyond reference to a fixed bed with ash removal via a grate. The catalyst (5–15 wt %) is added to the feedstock, removed with the ash, recovered, and recycled. So far only a small laboratory demonstration using lignite appears to have been attempted.

24.2.5 Commentary on Fixed-Bed Gasifiers

Except in special circumstances, low-pressure, air-blown gasifiers are confined in their application to the production of low-Btu fuel gas. If the gas is available under pressure, as in the case of the Lurgi, Gegas, and Morgas

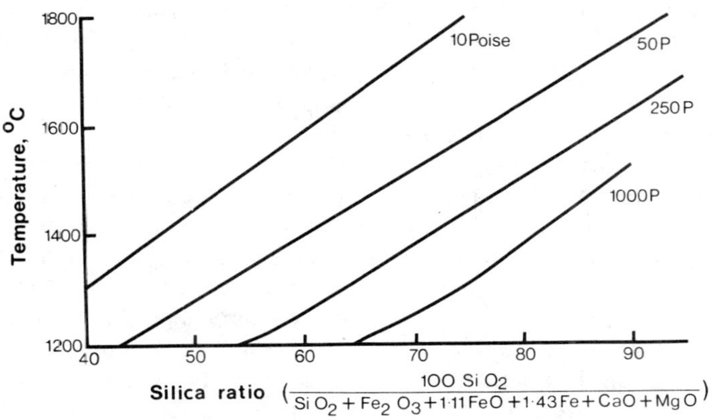

Figure 24.17 The approximate relation between temperature and silica ratio at selected levels of viscosity.[63]

gasifiers, it can with advantage fuel a combined-cycle electric power generating system (see Section 20.3 and "The Air-Blown Lurgi Gasifier," in Section 24.2.3.1). The compression of the nitrogen in the air can be justified only if power recovery is practiced and full advantage taken of the higher pressure to increase output per unit of cross-sectional area.

Apart from the Woodall Duckham and Wilputte two-stage processes and the two purely experimental CJB and Grand Forks slagging gasifiers, all the fixed-bed gasifiers are seeking to extend the range of coals available to them with particular emphasis on caking and swelling coals containing a high proportion of fines. The caking problem is being universally tackled with conventional water-cooled rotating stirrers, agitators, or rabble arms. In some of the experimental gasifiers an additional axial up-and-down movement has been included to increase the depth of agitation down as far as the ash zone. This latter innovation requires sophisticated seals for operation under pressure, but the claims are that the increased mixing allows caking and swelling coals containing high proportions of fines to be successfully gasified without excessive carryover. Whether this system can be scaled up from experimental gasifiers 3 or 4 ft (0.9–1.2 m) in diameter to commercial gasifiers 12–15 ft (3.6–4.5 m) in diameter remains to be seen.

There is evidence that caking and swelling coals can be charged with a high proportion of fines without incurring excessive carryover. As mining operations become more mechanized, the proportion of fines produced will increase, making the ability to gasify them more significant. Briquetting is one solution, and injection at the tuyeres of a slagging gasifier is another; the one is expensive and the other limited. Extrusion direct into a pressure gasifier as being developed by General Electric from EPRI designs has shown some success and is perhaps a more attractive approach. For SNG manufacture, introduction at the top of the fuel bed is to be favored over injection at the tuyeres if the benefit of methane produced during devolatilization is to be preserved.

Coal preparation prior to gasification in a fixed bed is at a minimum in comparison to that required for fluidized-bed and fully entrained systems, in which all the coal not only must be crushed and indeed pulverized for the latter, but also dried to less than 5% moisture to facilitate its transportation. On the other hand, little control can be imposed on fixed-bed systems; and tar, oils, and aqueous liquor are joint products with the gas, adding to the complexity of the downstream cleanup processes. In compensation, however, the countercurrent arrangements, which give rise to the tars, and so on, result in a system that is basically of high thermal efficiency. It is also a system that is still developing and will remain for many years a strong challenge to the new alternative systems now under development and described in the next sections.

24.3 FLUIDIZED-BED PROCESSES

24.3.1 Introduction

Fluidization of a bed of particles occurs when a gas or liquid passes through it at a rate such that the particles are suspended, being maintained in a state of continuous random motion. In coal gasifiers the fluidization is normally achieved by the gasifying medium. Fluidized beds are characterized by their high gas-to-solid heat and mass transfer rates, so that the mixing of solids within the bed approaches that of an ideally mixed reactor, with a resultant uniformity of solids composition and temperature. This makes selective removal of char and ash difficult, and the bed often has to operate at high ash concentrations to avoid excessive carbon loss in the ash. The heat and mass transfer characteristics of the bed can be modified by the choice of the bed geometry and the positioning of any internal baffles, heat transfer elements, and so on.

The particle size distribution of the feedstock must be carefully regulated to satisfy constraints of low entrainment losses and complete solids fluidization, if such features are required. Operation at elevated pressure allows throughputs to be increased without gas velocities becoming such as to cause excessive fines carryover. It also improves the performance of the gas distributor at the base of the bed, since it can operate with a higher allowable pressure drop.

The physical and chemical properties of the coal present several special problems with regard to its gasification within a fluidized bed. For example, during devolatilization coal particles tend to cohere, particularly if the coal is of the caking variety, resulting in loss of good fluidization. This problem is particularly common in the first stages of multistage fluidized-bed systems, which often operate at moderately low temperatures to promote the desired product distribution. In such cases a mildly oxidizing (e.g., steam-air) coal pretreatment stage can be employed to destroy the caking tendency of the coal. However such coal pretreatment reduces the volatile matter contained in the coal feed to the main gasifier, which loss can be undesirable in processes designed to yield liquid and gaseous hydrocarbons. Pretreatment can be avoided if the coal is introduced into a bed operating at a high temperature, since rapid heating and devolatilization coupled with dilution by char prevent particle agglomeration. High temperatures can, however, introduce similar problems if conditions are such that the ash becomes sticky and builds up into agglomerates. Such a condition results in a loss of full fluidization.

These constraints limit the maximum bed temperature, hence the throughput and carbon conversion for a particular coal. Therefore gasification rates attainable in noncatalytic fluidized beds tend to be lower than those for entrained and molten type processes (which operate at higher temperatures), but are rather higher than those for fixed beds. Consequently fluidized beds generally require solids residence times between those of fixed-bed and entrained-flow reactors.

Fluidized beds have been favored in many of the coal gasification projects presently under development, basically because they allow reaction conditions to be tailored to achieve the desired reaction chemistry. Fluidized beds are particularly suited for operation as multibed systems because of the ability to transfer solids between individual stages. This characteristic, together with the inherent attainment of uniform and controlled temperatures (perhaps by using indirect heat transfer) allows great flexibility in the choice of gasifying conditions, hence product distribution. By allowing the use of separate combustion and gasification stages, fluidized beds have largely superseded the older cyclic gasification processes, so that a relatively nitrogen-free crude gas can be continuously produced while retaining air-blown combustion. In some processes this is achieved by the circulation of a heat carrier material between the combustion and gasification stages, thereby providing the heat requirements of endothermic gasification reactions without introducing the nitrogen-containing products of the combustion stage. Fluidized beds are particularly well suited for catalytic gasification by virtue of their good mixing properties and the ease of fresh catalyst addition and spent catalyst removal. They are also preferred when high heat transfer rates are required either internally in the bed (e.g., for cooling exothermic carbon hydrogenation stages), or to provide heat to support endothermic gasification reactions. Corrosion and erosion problems are often less severe than those experienced in molten and entrained-flow gasifiers.

Baffles or weirs may be employed within fluidized beds to reduce mixing, usually with the aim of introducing controlled temperature gradients. By these means, a range of temperatures may be obtained in a single bed to favor, for example, both high-temperature steam-carbon and low-temperature methane-forming reactions. Reduction of the throughput of fluidized beds is not always easily possible, since it can lead to poor fluidization.

The fluidized-bed processes have been subdivided into three groups: Section 24.3.2, "Wholly Fluidized-Bed Processes," broadly as described above; Section 24.3.3, "Hybrid Fluidized-Bed Processes," and Section 24.3.4, "Fluidized-Bed Processes Producing Gas and Oil" in which gas is a major product.

The hybrid fluidized-bed processes involve main gasification stages that combine wholly fluidized beds with other contacting methods, such as dilute phase entrainment, free-fall dilute phase gasification, and controlled ash agglomeration. In the last method conditions near the bottom of the gasifier are deliberately controlled to allow operation at temperatures high enough to result in ash particle agglomeration. This allows selective removal of the larger agglomerates, with a low carbon content, from the bed. Performance data for all fluidized-bed processes, for which such data are available, are summarized for comparison purposes in Section 24.3.2 (Table 24.20).

24.3.2 Wholly Fluidized-Bed Processes

24.3.2.1 *Wholly Fluidized-Bed Processes Yielding a Nitrogen-Free Product Gas*

The CO_2 Acceptor Process. The CO_2 acceptor process being developed by the Conoco Coal Development Company employs two main fluidized-bed reactors—a gasifier and a regenerator—both of which operate at about 150 psig (10.3 bar). Lignite or subbituminous coal is fed to the gasifier operating at 1480–1550°F (800–845°C), where after rapid devolatilization, the fixed carbon is gasified with steam. Heat for the gasification reactions is provided by circulating partially calcined limestone or dolomite (the acceptor) from the regenerator, the heat being supplied mainly by the exothermic reaction:

$$CaO + CO_2 = CaCO_3 \quad \Delta H = -76{,}200 \text{ Btu/lb-mole (at } 77°\text{F)} \quad (24.12)$$

rather than by the sensible heat of the circulating solids (cf. Battelle–Union Carbide, Cogas processes), which does, however, make a lesser contribution. The reaction above is reversed in the regenerator [operating at about 1850°F (1010°C)], where the endothermic heat requirement is provided by combustion of residual char from the gasifier with air. The separate regenerator allows the use of air for combustion without introducing nitrogen into the product gas.

A 40-ton/day pilot plant, situated at Rapid City, South Dakota, has been in operation since early 1972.[67] In this plant feed coal is crushed to approximately 8 mesh (2.4 mm) and simultaneously dried using hot flue gases, typically reducing the total moisture content from 35 to 5% w/w. The dried coal is then preheated by hot furnace flue gas in an atmospheric pressure fluidized bed operating at about 550°F (290°C). The prepared coal is passed to a lock hopper system, utilizing two lock hoppers operating alternately, in which carbon dioxide is used as the pressurizing gas.

Since laboratory-scale studies have shown pretreatment to be unnecessary,[68] preheated coal is fed at pressure directly to the single gasification stage. Steam gasification of the coal occurs, supported by the heat liberated by the calcined acceptor (MgO, CaO) as it reacts with the CO_2 produced. This removal of CO_2 also enhances the exothermic carbon monoxide shift and methanation reactions. A product gas with a H_2/CO volume ratio in excess of 3 is produced, eliminating the need for a separate CO shift stage. The size of the acid gas removal plant is also reduced, the bulk of the CO_2 being rejected in the flue gas from the regenerator.

Figure 24.18 shows the whole process schematically. The fluidized-bed gasifier has two zones. The lower zone, within the smaller diameter section, contains predominantly fluidized acceptor; that of the upper zone, in the larger diameter section, is mainly partially gasified char. Carbonated acceptor leaves the gasifier near the bottom of the fluidized bed and is conveyed through a solids transfer line by a lift gas to the bottom of the regenerator vessel. Partially gasified char is withdrawn from a point near the top of the gasifier fluid bed and is also transferred through a similar lift line to the regenerator to maintain the required carbon feed rate.

In the regenerator the remaining carbon of the char is combusted in air, and regenerated acceptor leaves near the top of the bed to return via a solids transfer line to the gasifier. The acceptor loses reactivity as it circulates between the reactors, so acceptor is continuously purged from the system through the ash removal line leaving the regenerator; fresh makeup acceptor solid is added in the transfer line between the gasifier and regenerator. The removal of spent acceptor and ash from the regenerator is a recently intro-

[67] J. D. Sudbury, "Status of CO_2 Acceptor Process Pilot Plant," paper presented at 4th Synthetic Pipeline Gas Symp., Chicago, October 1972.

[68] C. Fink, G. Curran, and J. Sudbury, "CO_2 Acceptor Process Pilot Plant—1974, Rapid City, South Dakota," paper presented at 6th Synthetic Pipeline Gas Symp., Chicago, October 1974.

[69] C. Fink, "CO_2 Acceptor Process Pilot Plant—1975, Rapid City, South Dakota, paper presented at 7th Synthetic Pipeline Gas Symp., Chicago, October 1975.

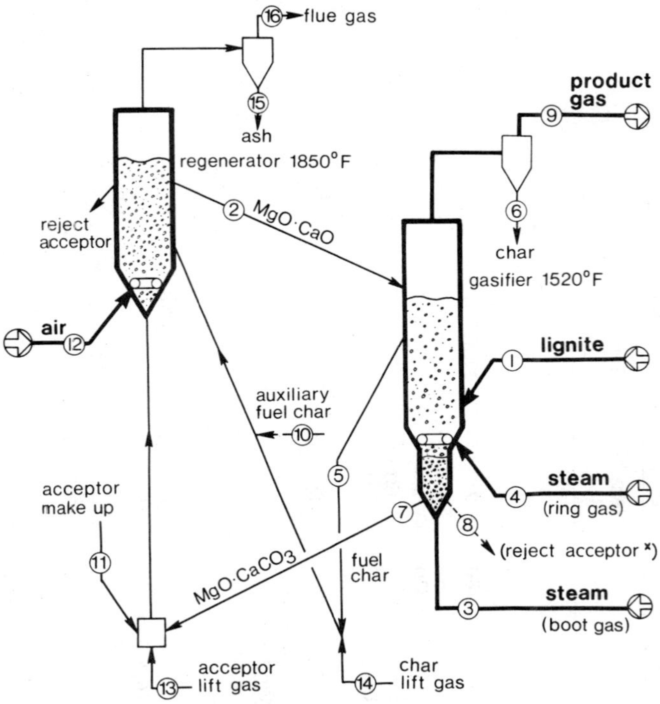

Figure 24.18 Pilot-plant flow diagram of the CO_2 acceptor process.[69]

duced process modification.[70] Formerly this stream was removed from the gasifier through a lock hopper system, but pressure fluctuations introduced by the operation of the lock hoppers adversely affected the gasifier fluidization. Movement of the offtake to the regenerator is claimed to have overcome this problem.[71]

Early pilot-plant tests used lignite char as the gasifier feedstock.[72] During startup, initial heating of the limestone acceptor caused formation of liquids in the regenerator,[68] leading to agglomeration of the acceptor and unreliable circulation. It was found that use of dead-burned dolomite as an acceptor instead of limestone solved this problem. This modification allowed continuous operation of the pilot plant for periods of up to several days, the duration of runs usually being limited by materials problems.[68]

After successful operation with lignite char, Velva (North Dakota) lignite was gasified, but a problem arose[69] because particles too large to be entrained from the regenerator with the char ash, yet too small to be removed with the spent acceptor, accumulated in the upper zone of the gasifier. This adversely affected the stability of the boundary between the zones, and a lock hopper system was added to remove a solids stream from the gasifier char bed.[69] Heat and mass balance data for a run lasting about 5 days, in which this arrangement was used, are given in Tables 24.12 and 24.13.

Early laboratory-scale work showed that acceptor agglomeration would occur at high gasifier steam partial pressures (in excess of 190 psi). So up to Run 26, some gasifier product gas was recycled and mixed with the steam feed to the gasifier. However during

[70] D. C. McCoy, G. P. Curran, and J. D. Sudbury, "CO_2 Acceptor Process Pilot Plant—1976, Rapid City, South Dakota," paper presented at 8th Synthetic Pipeline Gas Symp., Chicago, October 1976.

[71] D. R. Mosher, U. D. Marwig, and J. A. Phinney, "Basic Features of the CO_2 Acceptor Process," paper presented at 162nd National Meeting, American Chemical Society, Washington, D.C. September 1971.

[72] C. Fink, G. Curran, and J. Sudbury, "CO_2 Acceptor Process Pilot Plant—1973," paper presented at 5th Synthetic Pipeline Gas Symp., Chicago, October 1973.

Table 24.12 Heat Balance and Material Balance for the CO_2 Acceptor Process[69]

Run 26: 1-hr basis; system pressure, 150 psig; heat balance datum, H_2O (l)at 60°F.

Stream Number (Fig. 24.18)		Material Balance			Heat Balance (× 10^3 Btu)	
		lb	scf	°F	Sensible Heat	Heat of Combustion
		Gasifier				
	Input					
1	Lignite (*a*)	2,450		400	253.453	27,825.979
2	Acceptor (*h*)	8,795		1840	3,810.364	
3	Boot gas			1430		
	Recycle (*i*)	466	14,077		421.693	
	Steam	1,573	33,082		2,724.700	
4	Ring gas			1505		
	Steam	1,046	22,000		1,855.654	
	Purge			60		
	Recycle (*i*)	106	3,218			
	Inert (*k*)	43	540			
	N_2	79	1,070			
		14,558			9,065.864	27,825.979
	Heats of reaction, CaO + CO_2 = $CaCO_3$	30.07 moles at 76,200 Btu/mole			2,291.273	
					11,357.137	
	Output					
	Char			1495		
5	To regenerator (*b*)	623			325.154	6,881.688
	Reject (*b*)	75			39.143	828.440
6	Char loss from cyclone (*c*)	112			55.418	1006.129
	Acceptor (*g*)			1495		
7	To regenerator	9,749			3,522.511	
8	Reject	317			114.626	

9	Gas			1495		
	Product (*i*)	1,866	56,369		1,777.457	21,404.000
	Recycle (*i*)	572	17,295		545.359	
	Steam	1,220	25,664		2,157.906	
	H_2S	6	69		2.524	43.978
	NH_3	18	398		17.100	173.013
		14,558			8,557.198	30,337.248
	Heats of reaction					
	Heat of combustion	Output minus input			2,511.269	
	$CaO + S \rightarrow CaS + \frac{1}{2}O_2$	0.23 mole at 196,540 Btu/mole			45.373	
	Heat loss				243.295	
					11,357.135	

Analysis	Lignite/Char				
	(*a*)	(*b*)	(*c*)	(*d*)	(*e*)
Moisture (% w/w)	0	0	0	0	3.80
Moisture-free basis (% w/w)					
Hydrogen	4.59	0.83	0.78	0.00	2.20
Carbon	66.16	73.29	59.10	7.38	74.30
Nitrogen	1.00	0.36	0.22	0.00	0.90
Oxygen	20.87	0.00	0.00	0.00	7.10
Sulfur	0.54	0.00	0.00	0.00	1.50
Ash (as oxides)	6.84	18.70	33.05	90.79	14.00
CaS-CaO	0.00	0.07	0.49	1.83	0.00
CO_2	0.00	6.75	6.36	0.00	0.00

Table 24.12 *Continued*

Stream Number (Fig. 24.18)		Material Balance			Heat Balance (× 10³ Btu)	
		lb	scf	°F	Sensible Heat	Heat of Combustion
		Regenerator				
	Input					
5	Char (*b*)	623		1495	325.154	6881.688
10	Auxiliary fuel (*e*)	100		60		1119.690
	Acceptor					
7	From gasifier (*g*)	9.749		1495	3,522.511	
11	Makeup (*f*)	600		60		
12	Air (*l*)	5,733	75,340	830	1,096.142	
	H_2O	21	440		29,587	
13	Acceptor lift (*j*)	4,386	50,857		1,236.221	
	H_2O	14	297		22.485	
14	Char lift gas			725		
	Inert gas (*k*)	553	7,000		89.062	
	Recycle (*j*)	248	2,873		39,683	
	H_2O	1	17		1,087	
	Purge			60		
	Recycle (*j*)	165	1,916			
	Inert (*k*)	18	230			
	N_2	9	119			
		22,220			6,361.932	8001.378
	Heats of reaction					
	Heat of combustion	Output minus input			7,142.227	
	$CaS + \frac{1}{2}O_2 \rightarrow CaO + S$	0.03 mole at 196,540 Btu/ mole			4.923	
					13,509.082	
	Output					
15	Ash (*d*)	143		1840	65.980	150.538
2	Acceptor (*h*)			1840		
	To gasifier	8,795			3,810.364	
15	Overhead	49			21.191	

16	Gas			1840		
	Flue gas (*j*)	8,330	96,559		3,972.083	700.537
	Recycle (*j*)	4,799	55,647		2,289.096	
	Steam	97	2,045		191.242	
	COS	1	7		0.269	3.640
	H_2S	1	6		0.421	4.438
	SO_2	2	14		0.592	
		22,217			10,351.238	859.153
	Heats of reaction					
	$MgCO_3 = MgO + CO_2$				150.040	
	$CaCO_3 = CaO + CO_2$				2,421.108	
	Heat loss				586.712	
					13,509.098	

	Acceptor (% w/w)				Gas (mole %)			
	(*f*)	(*g*)	(*h*)		(*i*)	(*j*)	(*k*)	(*l*)
MgO	22.41	36.02	41.23	CH_4	13.75	0.00	0.00	0.00
CaO	27.01	43.41	49.69	H_2	58.79	0.06	0.00	0.00
Inert	4.93	7.93	9.08	CO	15.47	2.20	0.00	0.00
CO_2	45.65	12.61	0.00	CO_2	9.08	29.35	12.00	0.00
CaS-CaO	0.00	0.03	0.00	N_2	2.91	68.39	88.00	79.00
				O_2	0.00	0.00	0.00	21.00

Table 24.13 Summarized Heat Balance—Run 26[69]

	10^3 Btu/hr	%
Lignite feed, heat of combustion	27,826	100.00
Heats of combustion		
Gasifier Streams		
Product gas	21,404	76.92
Net char[a]	714	2.57
NH_3 and H_2S	217	0.78
Regenerator streams		
Flue gas + H_2S + COS	708	2.54
Overhead ash	151	0.54
Net heats of reaction		
$MgCO_3 = MgO + CO_2$	150	0.54
$CaCO_3 = CaO + CO_2$	130	0.47
$CaO + S = CaS + O$	41	0.15
Sensible heats		
Outlet streams–inlet streams	3,481	12.51
Heat losses		
Gasifier	243	0.87
Regenerator	587	2.11
	27,826	100.00

[a]Net char = heat for combustion of the char losses not compensated for by the use of auxiliary fuel.

Runs 27 and 28 this precaution was shown to be unnecessary for Velva lignite. During Run 28 the use of limestone as an acceptor was proved. After Run 29 the spent acceptor and ash removal point was moved[70] from the bottom of the gasifier to the side of the regenerator to facilitate control of the boundary between the zones in the gasifier. This change reduced the overall process efficiency by about 0.3%, as a result of the higher exit temperature of the stream, but it eliminated the need for the gasifier char removal system, which had been used to prevent buildup of intermediate-sized fines.

Since tests on higher sodium content lignites[70] indicated that acceptor agglomeration may again become a problem (buildup of Na_2O caused the surface of the acceptor to become sticky), product gas recycle to the gasifier was employed again.

The University of Wyoming Catalytic Process. The concept of this process, developed by the National Resources Research Institute at the University of Wyoming, is the direct conversion of coal with steam to a methane-rich gas by the use of catalysts, to achieve both gasification and methanation reactions in a single reactor.

The process has been studied only at the laboratory stage, so reactor configurations and operating parameters have not yet been determined. However outlines of various proposed systems are given in patents assigned to the developers, and it is indicated that some type of fluidized bed will be utilized.

Experiments with a 1-in.-diameter batch reactor were carried out using a number of coals. Steam was passed through a heated bed of coal mixed with a catalyst containing both reduced nickel and potassium carbonate. The data obtained (Table 24.14) show that in the temperature range 1200–1400°F (650–760°C) the catalyst has adequate activity. From the high methane yield it appears that the nickel catalyst is active at these temperatures in spite of the presence of sulfur compounds in the coal, although the gas product contains more methane than had been predicted by equilibrium considerations. It was concluded[74] that although the coal conversion reactions are promoted by the alkali carbonate, the nickel catalyzes methanation and the gasification of liquids that would be produced in its absence. At 1200°F (650°C) and 30 psig a carbon conversion of approximately 60% was achieved, with potassium carbonate and nickel added to the coal feedstock in proportions of 20 and 10%, respectively, of the weight of the coal.

The overall reaction is approximately autothermal, and it is thought that in a full-scale plant the reactions can be supported solely by preheating the feedstream.

Fluidized-Bed Processes Developed by Exxon. In one process developed by Exxon Research and Engineering Company (a subsidiary of the Exxon Corporation) and its affiliate Carter Oil Company, it is claimed that all types of coal can be gasified with steam in a fluidized bed at 1500–1700°F

[73] E. J. Hoffman, J. L. Cox, R. W. Hoffman, J. A. Roberts, and W. G. Wilson, "Behavior of Nickel-Methanation Catalysts in Coal-Steam Reactions," *Prepr., Am. Chem. Soc. Div. Fuel Chem.*, **16**(2), 64–67 (1972).

[74] W. G. Wilson, L. J. Sealock, F. C. Hoodmaker, R. W. Hoffman, D. L. Stinson, and J. L. Cox, "Alkali Carbonate and Nickel Catalysis of Coal-Steam Gasification," in *Coal Gasification,* Am. Chem. Soc. Adv. Chem. Ser. No. 69, 1967, pp. 203–216.

[75] E. J. Hoffman, "The Direct Production of Hydrogen from Coal-Steam Systems," *Prepr., Am. Chem. Soc. Div. Pet. Chem.,* **16**(2), C20–C23 (1971).

Table 24.14 Data for the University of Wyoming Catalytic Process[73–75]

	Run Number			
	274	361	312	356
Coal:	Glenrock[a]	Reynolds[a]	Consol[b]	Minnkota[b]
Catalyst:	Ni-0104	Ni-0104	G-65RS	Ni-0104
Pressure (psi):	250	250	250	250
Reaction temperature (°F):	1200	1400	1350	1400
Gas composition, dry (% v/v)				
CO_2	48.0	46.9	44.2	44.9
H_2	1.8	3.1	8.7	5.1
CH_4	50.2	50.0	47.1	50.0
CH_4 yield (scf/ton)	15,725	13,783	14,795	16,556
Theoretical CH_4 yield (scf/ton)	16,600	16,100	15,900	18,400
Calorific value (Btu/scf, CO_2 free)	977	962	899	939

[a] Subbituminous.
[b] Lignite.

(815–925°C); the necessary heat is provided by a stream of circulating char. Char is withdrawn continuously from the gasifier and reheated in an air-blown external combustor. The heated char, after separation from the flue gas, is returned to the gasifier. A 0.5-ton/day integrated pilot plant has been operated at Baytown, Texas, for some years. Plans for a 500-ton/day pilot plant were announced in early 1974[76] but were deferred, apparently because of the rising costs and competition for financing from other projects.[76]

Further work by Exxon indicated the advantage of using catalysts in coal gasification, and emphasis was given to a fluidized-bed process involving an alkali metal salt (K_2CO_3). This second process[77] combines a relatively low gasification temperature of 1200–1400°F (650–760°C) with recycle to the gasifier of the carbon monoxide and hydrogen recovered from the product gas. Figure 24.19 is a schematic flow diagram of the Exxon process.

The ability of alkali metal salts to increase the gasification rate is well known, and a number of other processes have utilized this effect. The key feature of the process is the regulation of the recycle rate such that the same amount of CO and H_2 enters the gasifier as leaves it. This, in effect, results in the net gaseous product being only CO_2 and CH_4, together with smaller amounts of H_2S and NH_3. This reaction scheme is approximately autothermal.

The product gas is cooled by raising process steam and is then separated into a number of streams: the product methane stream, a carbon dioxide stream that is vented to atmosphere, by-products H_2S and NH_3, and the H_2-CO mixture, which is recycled to the gasifier. Part of this separation is achieved cryogenically. The ash-char residue from the gasifier passes to a catalyst recovery unit where water washing is said to recover most of the catalyst, which can then be recycled.

The rate at which gas has to be recycled is strongly dependent on the operating conditions, and Fig. 24.20 shows its variation with steam conversion at various pressures, at 1300°F (705°C). At the expected commercial operating conditions (viz., 500 psig, 1300°F, and 40–60% steam conversion), a molar recycle rate of about 1–1.5 times the product methane rate is required.[77] Since temperatures below 1300°F (705°C) are necessary to obtain realistic recycle rates and sufficient methane production, the use of catalysts is essential. The addition of the catalyst is said also to reduce the caking characteristics[77] of the coal and to promote the methanation reactions.

Research and development work on this process is being carried out in bench-scale

[76] Anon., "Exxon To Spend $200 million on Coal Conversion," *Oil Gas J.*, 72(11), 45 (March 18, 1974).

[77] W. R. Epperly and H. M. Siegel, "Catalytic Coal Gasification for SNG Production," paper presented at 11th Intersociety Energy Conversion Engineering Conf., State Line, Nev., September 1976.

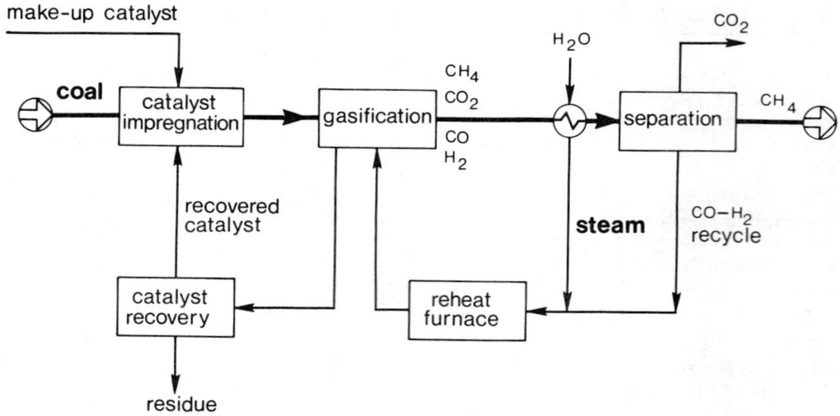

Figure 24.19 Schematic flow plan for the Exxon catalytic coal gasification process.[50]

units that can operate at pressures of up to 1000 psig, and in a small pilot plant of 25-lb/hr coal feed throughput with a maximum operating pressure of 100 psig. Exxon states that catalytic coal gasification has higher thermal efficiencies and results in reduced costs compared with conventional thermal processes.

The British Gas Coal Hydrogasification Process. British Gas proposed a process having two fluidized coal hydrogasification stages. Coal is fed to the first stage, operating at 1470–1560°F(800–850°C) and 750 psig (Fig. 24.21), which promotes the very rapid reaction of hydrogen with part of the volatile matter contained in the coal. Under these conditions the reactions are not limited by the carbon-hydrogen-methane equilibrium and are largely independent of temperature and pressure. In the second stage further relatively slow hydrogenation reactions involving the fixed carbon of the residual char (from the first stage) take place at higher temperatures, typically 1650–1740°F (900–950°C), at a rate that is proportional to the hydrogen partial pressure.[78] The use of the second stage allows the carbon-hydrogen reaction to approach equilibrium without being suppressed by the methane formed by the hydrogasification of volatiles.

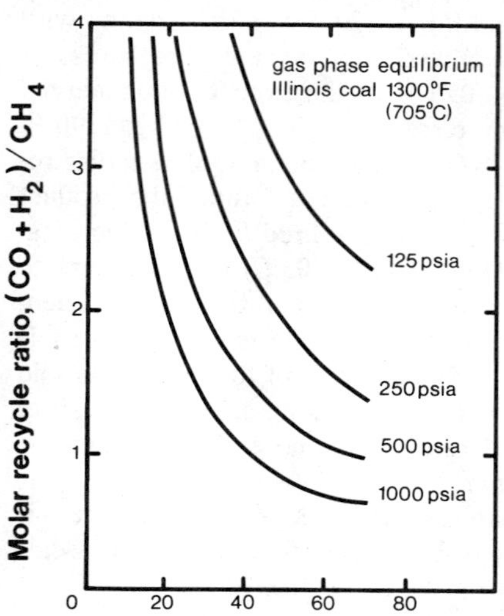

Figure 24.20 The effect of pressure on recycle ratio in the Exxon catalytic gasification process.[50]

Fluidized beds with internal recirculation of solids are proposed for these two stages to aid temperature control and also to discourage agglomeration at the coal-char injection point. The residual char from the second stage is then gasified with steam and oxygen in a further recirculating fluidized bed to produce (after CO shift and CO_2 removal) the hydrogenating gas for the two hydrogasification stages.

Following initial studies of the reaction between hydrogen and coal by Dent et al.,[79,80]

[78] F. Mosely and J. A. Lacey, "The Conversion of Coal to Gases Containing Methane," paper presented at World Power Conf., Tokyo, October 1966.

[79] F. J. Dent, "Hydrogenation of Coal to Gaseous Hydrocarbons," *Proc. Int. Conf. Complete Gasification Mined Coal,* Liège, Belgium, 1954, pp. 113–124.

[80] F. J. Dent, W. H. Blackburn, and H. D. Millett, "Investigation of the Use of Oxygen and High Pressure in Gasification. II. Synthesis of Gaseous Hydrocarbons at High Pressure," IGE Commun. No. 190, 1938; *Trans. Inst. Gas Eng.,* **88,** 150–217 (1938–1939).

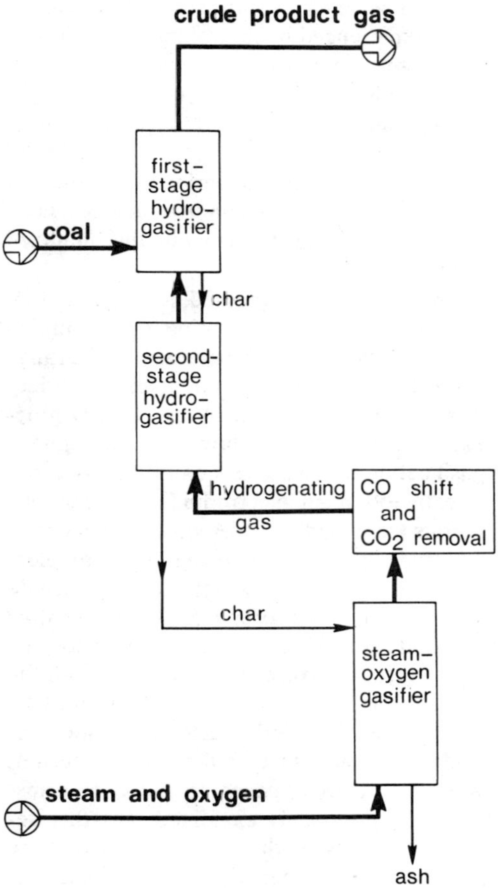

Figure 24.21 The British Gas coal hydrogasification process.

large-scale atmospheric pressure models were used to establish a design for the fluidized-bed hydrogasifiers. This work was followed by the construction of a 34-ft reactor of 3-ft I.D. operating up to 300 psig and 400°F (205°C), which was used to establish the recycle rates required within the recirculating fluidized beds. This work led to the design of a 450-psig, 4-million-scf/day pilot plant to demonstrate the fluidized steam-oxygen gasifier.[81] The plant constructed at the Midlands Research Station, Solihull, England, was sized to gasify about 3360 lb/hr of char at 1920°F (1050°C) using 4750 lb/hr of steam and 2275 lb/hr of oxygen. This plant was in the final stages of construction when work was terminated because of the discovery of large supplies of natural gas in the North Sea in 1965. However the fluidized-bed hydrogasification stage has since been further developed, in a 1-million-scf/day pilot plant on the same site for oil gasification.[80,82]

[81] A. G. Horsler, J. A. Lacey, and B. H. Thompson, "High Pressure Fluidized Beds," *Chem. Eng. Prog.*, **65**(10), 59–64 (October 1969).

Commentary on the Four Wholly Fluidized-Bed Processes Just Described. The CO_2 acceptor process is the only one to have reached the pilot-plant stage. The advantages of the process, all of which should result in reduced processing costs, are as follows:

1. Very good carbon conversions are attainable, so that the ash normally contains less than 1% w/w of the feed carbon.
2. The process uses air for the combustion-regeneration stage, obviating the use of an air separation plant while still producing a nitrogen-free product.
3. The combination of chemical and sensible heat released by the acceptor in the gasifier results in a lower solids recirculation requirement than for processes employing sensible heat transfer alone.
4. Downstream processing of the product gas involves relatively little CO_2 removal and no tar or oil separation, nor is a separate CO shift stage needed before catalytic methanation.

These advantages must be set against the considerable cost of the acceptor replacement to maintain its activity. Furthermore, a complex acceptor solids transfer system using lines with ceramic linings to prevent erosion is required.

The process is confined to the use of lignites and some subbituminous coals, and the drying of these feedstocks, normally needed prior to gasification, imposes a high heat demand.

In contrast, the catalytic process proposed by the University of Wyoming has been demonstrated only at laboratory scale in a 1-in.-I.D. reactor. A process for direct methane production from coal using only steam as the gasifying medium appears very attrac-

[82] B. H. Thompson and H. L. Conway, *The Hydrogasification of Oils to Gaseous Hydrocarbons,* Gas Council Research Commun. No. GC 122, November 1965.

tive, not only in terms of a possible reduced requirement for process heat, but also because of the reduction of subsequent processing of the gas to yield SNG. However the process obviously rests on the discovery of effective catalysts, particularly if high sulfur content coals are to be used, and the costs of such catalysts and their ability to survive continuous regeneration cycles will dictate the possibility of further development of this process.

The process outlined by Exxon to produce a methane-rich gas again relies on the use of a catalyst to enable gasification to take place at lower temperatures with enhanced methane formation. The recycle ratio required is very sensitive to the operating conditions, which have not yet been optimized. However if the process can be demonstrated successfully with acceptable catalyst consumption and operating costs, it should have good development potential.

Demonstration of a successful catalytic gasifier has not yet been achieved, but the claim that catalysts can reduce the caking characteristics of the coal[77] may be significant in future developments.

The Iowa State University Electrofluidic Process. The electrofluidic (electrothermal) process employs a single fluidized bed, in which coal or coal char is gasified with steam. The heat for this endothermic reaction is supplied electrically, utilizing the resistivity of the char between electrodes placed within the bed.

The bench-scale experimental program was undertaken by Iowa State University to demonstrate the general technical feasibility of the electrofluidic reactor and to establish criteria for a commercial-scale design. Initially a number of batch runs were made with a 4-in.-I.D. reactor[83] using coal chars from the FMC Corporation's COED process (see Section 24.3.4.1) and a residue from Pittsburgh Seam coal produced in hydrogasifiers at the Institute of Gas Technology (IGT) (see "The Hygas Processes," in Section 24.3.3.1). Tests at atmospheric pressure were carried out at various temperatures between 1500 and 1960°F (815–1070°C), using various steam flow rates. No unexpected difficulties were experienced in controlling power inputs or reactor temperatures; however problems were encountered with the electrodes. Both copper and stainless steel (type 44b) electrodes were badly worn, became coated with an ashlike material, and seriously overheated (being up to several hundred degrees fahrenheit hotter than the surrounding fluidized bed).

The research progressed to a 12-in.-I.D. continuous reactor operating under atmospheric pressure and temperatures in the range 1300–2000°F (700–1100°C), using a variety of coal chars produced by the COED process. These tests have shown that continuous gasification of coal chars can be carried out effectively in an electrofluidic reactor. Attempts were made to overcome the persisting electrode problem by using different electrode materials, of which silicon carbide showed the most promise. An alternative electrode configuration that greatly increased the electrode surface area in contact with the fluidized bed was also tried. The 12-in.-I.D. reactor was modified to accommodate electrodes arranged so that the flow of current was vertical as opposed to radial or transverse, as previously examined. For this design, the walls of the reactor needed to be electrically nonconducting, so the lining was recast with a high-alumina refractory, causing the I.D. to be reduced to 10 in. (Fig. 24.22). Graphite electrodes were used in this reactor, and the problems of overheating and erosion experienced with the previous reactors were not encountered. This result was due to the negligible contact resistance between the electrodes and the fluidized bed, which previously contributed appreciably to the overall resistance. Results of various runs in the continuous reactors are given in Tables 24.15 and 24.16; the analyses of the feedstock appear in Table 24.17. Catalyzed gasification was also attempted at lower temperatures.

In some experiments the electrical resistance of the bed was either too high or too low to permit adequate heating of the reac-

[83] A. H. Pulsifier, T. M. Knowlton, and T. D. Wheelock, "Coal Char Gasification in an Electrofluid Reactor," *Ind. Eng. Chem. Process Des. Dev. Q.*, **8**, 539–545 (1969).

[84] A. H. Pulsifier and T. D. Wheelock, *Coal Processing by Electrofluidics,* Final Report No. FE-0479-1, to U.S. ERDA, Engineering Research Institute, Iowa State University 1975.

[85] T. D. Wheelock, Iowa State University, private communication, September 1976.

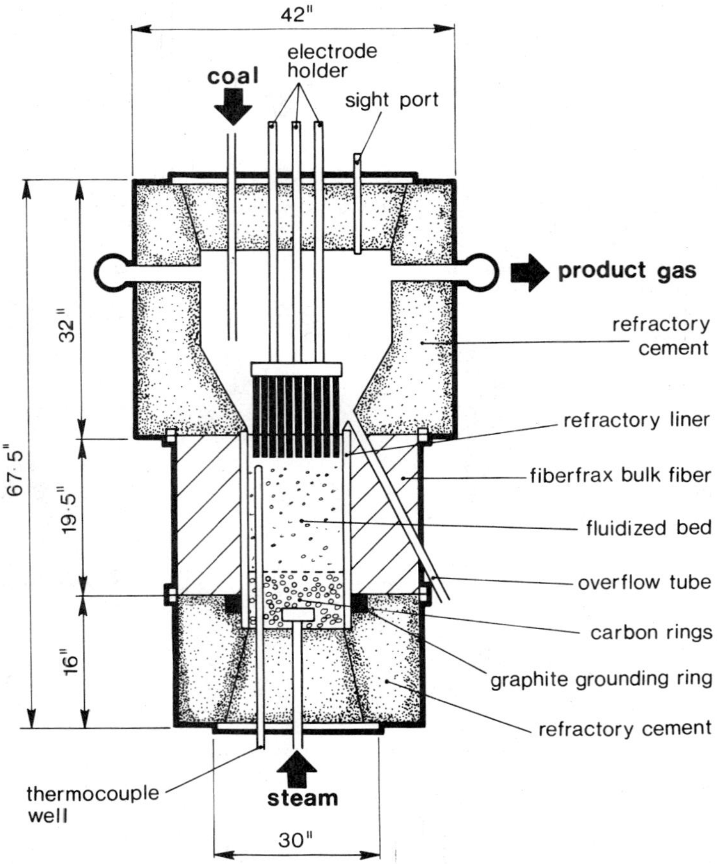

Figure 24.22 A 10-in.-I.D. continuous electrofluidic gasifier.[84]

tor,[86] but a suitable resistivity was obtained by adding char that had been heat treated or had previously passed through the reactor. This procedure may be impractical in a full-scale plant. From these results gas compositions at various temperatures and pressures have been predicted to evaluate commercial applications for the process. The IGT has demonstrated the process as a potential method of hydrogen generation for their hydrogasifier[87] (Fig. 24.23); Table 24.18 compares the predicted and actual results. However this approach has been superseded by alternative methods of hydrogen production because difficulties were encountered with electrodes and high electricity generating costs. The work done by IGT on this process is described below.

The results of this program have enabled a feasibility study to be conducted with a preliminary process design and estimates of capital and operating costs for a full-scale commercial plant of 250-million-scf/day capacity.[87] The process flowsheet would consist of three parallel trains, each as shown in Fig. 24.24 (stream flows are for the combined trains). Table 24.19 summarizes the estimated electric power requirements. These large electric power requirements restrict the development of the process until a more efficient and economic method of power generation is available.[85]

Nuclear-Aided Fluidized-Bed Gasification Processes. A group of companies in West

[86] A. H. Pulsifier and T. D. Wheelock, "Observations of Coal Processing in an Electrofluidic Reactor," *J. Inst. Fuel,* **47**(393), 235–241 (December 1974).

[87] A. H. Pulsifier and T. D. Wheelock, *Coal Processing by Electrofluidics,* Interim Rep. No. 2, Engineering Research Institute, Iowa State University, 1974, published by OCR, U.S. Dept. of Interior, PB 236-588.

Table 24.15 Data from 12-in. Electrofluidic Reactor with Radial Current Flow[85]

	Colestrip Char[a]			Bear Char[a]					
	2-44	2-46	2-47	2-54	2-56	2-57	2-63	2-63	2-67
Run Number:	1,950	1,925	1,850	1,700	1,600	1,700	1,750	1,550	1,500
	8	8	10	11	11	10.7	11	9	11
Bed temperature (°F):	14	15	14	14	11	15	14	10	8
Steam rate (lb/hr):									
Hours gasified:									
Electrode:	Three 1-in. O.D. Silicon Carbide Rods					Three 1.5-in. O.D. Silicon Carbide Rods			Three 1.9-in. O.D. Steel Pipes
Electrode immersion (in.)	8	17	8	10	15	15	12	12	15
Initial bed height (in.)	12	24	16	15	15	15	15	15	15
Initial char weight (lb)	29.8	47.2	28.0	29.0	29.7	30.4	37.5	27.2	37.5
Char added in run (lb)	46.2	54.5	95.2	71.3	54.0	65.3	62.8	57.1	39.7
Char feed rate (lb/hr)	5.3	7.5	7.9	7.1	7.8	8.0	8.9	11.4	7.9
Char overflow (lb)	7.1	15.9	39.3	47.4	30.3	4.5	36.5	37.0	38.8
Bed residue (lb)	30.1	55.1	19.5	26.5	26.1	46.1	21.7	24.5	19.8
Bed residue (% w/w, C)	67.3	69.7	61.4	76.5	75.1	70.5	78.5	NA[c]	83.1
Product gas rate (scf/min)[b]	2.3	4.3	5.5	3.0	3.5	5.2	4.4	3.5	2.2
Product gas composition, dry (% v/v)[d]									
H_2	60	53	60	58	58	62	57	61	62
CO	28	32	32	28	27	28	31	25	20
CO_2	12	15	8	14	15	10	12	14	18
Overall material balance (% input/output)	90	92	96	93	99	98	100	NA[c]	93
Char balance (% input/output)	95	98	91	100	99	96	97	NA[c]	98
Steam balance (% input/output)	78	55	84	88	90	98	103	NA[c]	84

[a] See Table 24.17 for feed material properties.
[b] Average value for entire run.
[c] NA = data not available.
[d] At 32°F and atmospheric pressure.

Table 24.16 Data from 10-in. Electrofluidic Reactor with Longitudinal Current Flow[85]

	Run Number								
	13	14	15	16	17	18	19	20	21
Bed temperature (°F)[a]	1325	1330	1320	1300	1320	1560	1525	1260	1330
Final electrode gap (in.)	11	5	2	3	3	6	2	2	3
Electrode gap range (in.)	11	3–5	2	2–3	3–5	11–3	10–2	10–2	10–3
Steam fed (lb)	155.4	83.8	86.9	102.1	121.3	98.5	106.0	108.7	108.2
Steam feed rate (lb/hr)[a]	11.0	8.6	8.6	8.8	8.8	9.0	9.0	8.8	9.3
Hours of gasification	15	10	10	11.5	14	11	11.7	13	12
Hours solids fed	11	8	8	10	10	9	8.5	11	10
Initial bed weight (lb)	20.5	20.9	20.9	20.9	20.9	22.0	22.0	22.0	22.0
Final bed weight (lb)	22.5	27.1	27.3	20.5	24.9	20.3	9.9	17.2	17.6
Solids fed (lb)	79.6	59.3	43.2	41.9	43.7	62.2	30.2	61.5	55.6
Solids feed rate (lb/hr)[a]	5.7	4.9	5.3	4.2	4.0	7.7	3.5	5.5	5.5
Solids overflow (lb)	61.3	24.7	22.0	28.2	27.8	47.4	10.8	19.8	16.8
Overflow rate (lb/hr)[a]	3.5	3.5	3.5	3.1	2.9	4.9	0.7	2.2	1.3
Feed material[b]	Bear	W Ky	W Ky	W Ky	Bear	Bear	Lignite	Lignite	Lignite
Catalyst (+5% w/w of char feed)	None	None	K_2CO_3	Na_2CO_3	Na_2CO_3	None	None	None	None
Water condensed (lb)	120.8	63.9	62.4	79.1	98.1	74.7	70.1	74.1	71.9
Steam conversion (%)[a]	8	18	25	23	19	28	45	41	38
Product gas rate (scf/hr)[a]	24	49	74	64	52	109	150	160	165
Product gas composition (% v/v)[a]									
H_2	54	66	64	62	61	58	62	61	64
CO	16	5	5	5	5	17	15	12	13
CO_2	27	25	28	31	33	23	21	23	20
CH_4	3	4	3	2	1	2	2	4	3
Overall material balance (% input/output)	86	93	94	97	97	100	NA[c]	87	92
Specific gasification rate[d]	0.021	0.033	0.047	0.053	0.041	NA[c]	NA[c]	NA[c]	0.166

[a] Average over last 5 hr of run.
[b] See Table 24.17; Bear = Bear char; W Ky = Western Kentucky char.
[c] NA = data not available.
[d] Given in pounds of carbon gasified per hour per pound of carbon in the bed.

Table 24.17 Composition of Coal Chars and Lignite Used in Electrofluidic Reactor Tests[84]

Analyses	Colestrip Char	Bear Char	Western Kentucky Char	Lignite
Proximate (wt%)				
Moisture	6.0	1.9	0.7	0.1
Volatile matter	8.6	5.5	9.1	48.3
Fixed carbon	71.7	79.1	74.4	43.5
Ash	13.7	11.6	15.8	8.1
Ultimate (wt%)				
Carbon	78.0	82.3	75.7	64.5
Hydrogen	1.7	1.7	2.1	4.8
Nitrogen	1.1	1.4	—	—
Sulfur	0.5	0.5	—	1.0
Oxygen	4.1	2.5	—	—
Ash	14.6	11.6	15.8	8.1

Germany is collaborating on a project, with sponsorship from the Bundesministerium für Forschung und Technologie, to construct a prototype nuclear-aided coal gasification plant. It was considered in 1976 that a plant with a 3000 MW(t) nuclear energy input could be operational by 1990,[88] consuming 13,200 tons/day of coal to produce 300 million scf/day of SNG and 500 MW of electrical power.

The companies participating in the project are Bergbau-Forschung GmbH (BF), Essen; Gesellschaft für Hochtemperaturreaktortechnik GmbH (GHT), Bensberg; Hochtemperatur-Reaktorbau GmbH (HRE), Mannheim; Kernforschungsanlage Julich Gmbh (KFA), Julich; and Rheinische Braunkohlenwerke AG (RBW), Cologne.

Their particular areas of investigation are shown in Fig. 24.25. This work could lead to the development of several processes, including the steam gasification of coal in a fluidized bed, utilizing heat from a high-temperature, gas-cooled nuclear reactor (HTGR). This reactor supplies the heat for the endothermic gasification reaction, for raising gasification steam, and perhaps for downstream processing and preheating the coal feed.

The heat is transferred from the HTGR to the gasifier using two independent helium transfer circuits (Fig. 24.26). This arrangement effectively isolates the nuclear and gasification sections of the plant, so preventing the possibility of permeation of hydrogen from the gasifier into the core of the reactor and tritium from the nuclear part into the gasifier. Leakage is further lessened by operating the two helium circuits and the fluidized bed at similar pressures (~ 600 psig). Superheated reactant steam fluidizes the bed, in which is immersed a large heat exchanger. Caking coals first pass through a pretreatment step where they are devolatilized by the hot product gas, producing tar and oil byproducts. The resultant char enters at the top of the gasifier. Additional heat requirements for pyrolysis can be provided from the helium stream at the appropriate temperature level. It is thought that noncaking coals can be fed directly into the gasifier. The coal entering at one end of the gasifier moves progressively along to an outlet at the opposite end, where the flow of steam is reduced so that selective fluidization allows separation of the ash from the char and its transfer to a conventional lock hopper discharge system. However fluidized-bed gasifiers used in some other processes have experienced problems with such ash removal techniques, even without the presence of heat transfer tubes within the gasifier.

Gasification studies have progressed from

[88] I. Howard-Smith and G. J. Werner, *Coal Conversion Technology,* Noyes Data Corporation, Park Ridge, N.J., 1976, pp. 102–103.

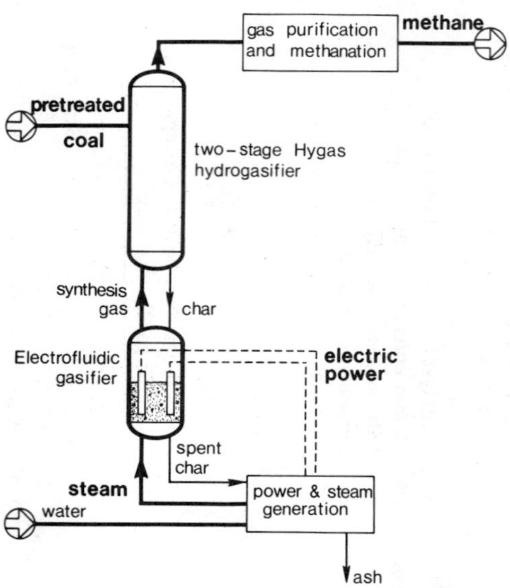

Figure 24.23 Integration of the electrofluidic gasifier with the Hygas process.[87]

[89] H. Jüntgen and K. H. Van Heek, Bergbau-Forschung GmbH, Essen, West Germany, private communication.

Table 24.18 Comparison of Predicted and Observed Gas Compositions from Electrofluidic Reactors[87]

Reactor	Temperature (°F)	Pressure (psig)	Steam Velocity (ft/s)	Steam Conversion (%)	Gas	Composition, Dry Predicted (% v/v)	Observed (% v/v)
4-in. diam (Run 11)	1600	0	0.14	51	CO	23	20
					CO_2	18	20
					H_2	59	58
4-in. diam (Run 17)	1600	0	0.14	58	CO	25	26
					CO_2	16	15
					H_2	59	56
4-in. diam (Run 19)	1600	0	0.41	37	CO	17	24
					CO_2	22	14
					H_2	61	58
IGT (Run EG-46)	1900	1008	0.21	65	CO	34	33
					CO_2	13	12
					H_2	45	46
					CH_4	8	9
IGT (Run EG-48)	1900	1010	0.24	53	CO	28	27
					CO_2	17	14
					H_2	48	51
					CH_4	7	8

a 1-g batch reactor used for kinetic studies:[90] to a small pilot plant of 11 lb/hr producing 635 scf/hr of gas. Typical experimental data using bituminous coal in the small pilot plant are given in Table 24.20. A larger pilot plant, with a capacity of 1540 lb/hr producing up to 24,700 scf/hr of gas, began operation in July 1976. Gasification runs longer than 100 hr have been performed, with the helium in the intermediate circuit being electrically heated to simulate the input of nuclear heat. The main feedstocks under investigation are Rhenish lignite and German hard coal. The throughput of the gasifier is determined by the rates of the endothermic gasification reactions, which are strongly dependent on the heat transfer capacity of the reactor. The rate of heat transfer depends on the overall heat transfer coefficient (U) between the helium stream and the fluidized bed. The estimated value of U, 41 Btu/hr-ft²-°F (200 kcal/

[90] K. H. Van Heek, H. Jüntgen, and W. Peters, "Fundamental Studies on Coal Gasification in the Utilization of Thermal Energy from Nuclear High-Temperature Reactors," *J. Inst. Fuel,* **46,** 249–258 (June 1973).

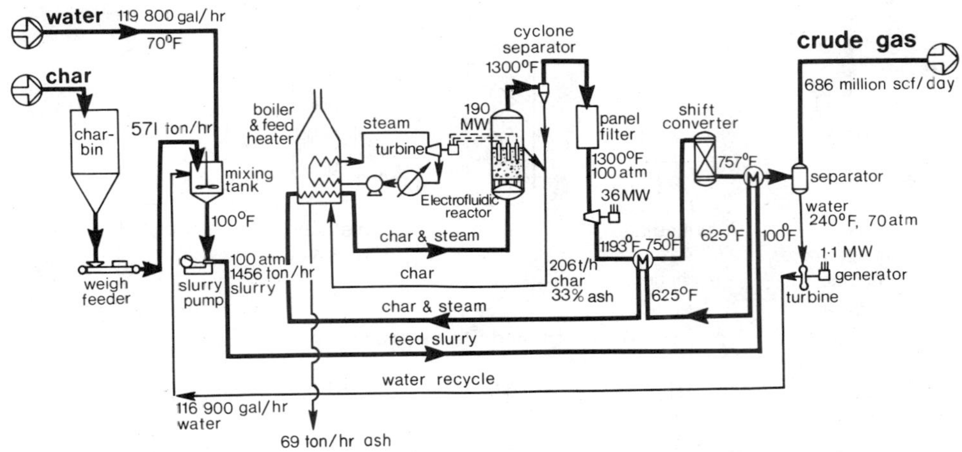

Figure 24.24 The electrofluidic process for the production of 250 million scf/day of SNG from lignite char.[87]

Table 24.19 Electrical Power Requirements for Electrofluidic Coal-to-SNG Plant Producing 250 million scf/day SNG[87]

	Lignite[a]		Bituminous Coal[b]	
Reactor pressure (psig):	1470	2940	1470	2940
Conveyors	135	135	135	135
Char feeders	30	30	30	30
Slurry preparation	75	75	75	75
Make up water pumps	110	110	110	110
Slurry pumps	2,480	5,110	4,480	5,110
Electrofluidic reactors	189,940	141,300	362,353	272,734
Hot carbonate scrubbers	12,000	12,000	12,000	12,000
Total power (kW)	204,770	158,760	379,183	290,194
To operate power plant (12.5%)	25,596	19,845	47,398	36,274
Subtotal	230,366	178,605	426,581	326,468
Less power recovered	−37,050	−101,823	−40,711	−110,891
Net power required (kW)	193,316	76,782	385,870	215,577

[a] Reactor temperature, 1300°F.
[b] Reactor temperature, 1500°F.

m²-hr-°C), is low because the tube walls are exceptionally thick—0.79 in. (20 mm).[91]

The rate of reaction, k, for the steam-carbon reaction,

$$C + H_2O \rightarrow CO + H_2 \tag{24.4}$$

depends on the reactivity of the coal, the size of coal particles, the gasification temperature, and the pressure. Laboratory-scale experiments within the pressure range 145–1015 psia have shown that the rate is independent of the steam partial pressure if this is greater than 145 psia.[91] It was also shown that the coal particle size had no influence when within the range 60 μm–2 mm. Figure 24.27 plots the rate of carbon conversion, as a function of temperature, for various coals. In general, hard coals show a lower reactivity than lignites, so that gasification temperatures have to be at least 270°F (150°C) higher to achieve the same reaction rate.

[91] H. Jüntgen, K. H. Van Heek, R. Overrfeld, and P. P. Feistel, "Kinetics, Heat Transfer and Engineering Aspects of Coal Gasification with Steam Using Nuclear Heat," Paper No. 12, Session III, "The High Temperature Reactor and Process Applications," British Nuclear Energy Society Int. Conf., London, November 1974.

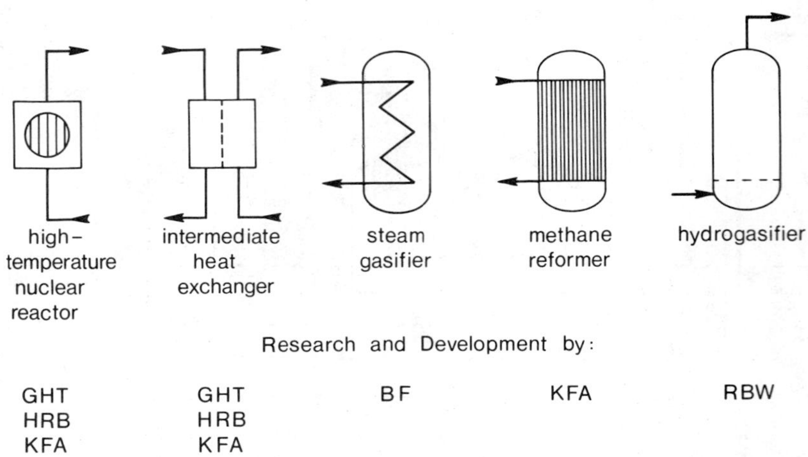

Figure 24.25 Components to be tested in a proto-type nuclear-aided gasification plant (abbreviations spelled out in text).[89]

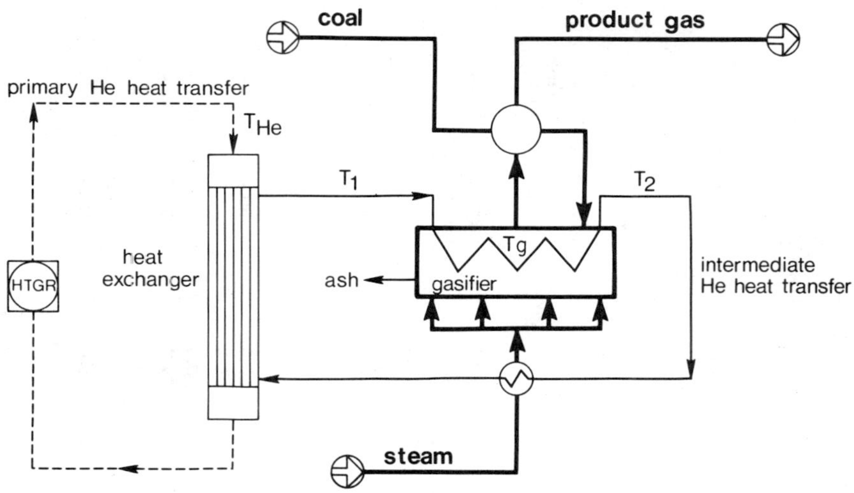

Figure 24.26 The Bergbau-Forschung nuclear-aided steam gasification process.

Conceptual design studies of the gasifier heat transfer system have shown that the difference between the temperatures of the helium leaving the gasifier exchanger (T_2) (Fig. 24.26) and the gasification zone (T_g) should be at least 90°F (50°C) to keep the heat exchange area within reasonable dimensions.[92] For hard coals,[92] assuming an HTGR outlet temperature of 1740°F (950°C) and a 90°F (50°C) temperature difference across the helium-helium exchanger, helium will enter the gasifier at 1650°F (900°C) to give a corresponding gasification temperature of 1420°F (770°C), and a throughput of 52 tons/hr of coal for the conceptual commercial-scale gasifier shown in Fig. 24.28. The reactor, which is a horizontal cylindrical vessel 98 ft long × 23 ft diameter, contains a heat exchanger with a surface area of 43,000 ft².

Investigations are to be carried out to assess the potential increase in throughput attainable by increasing U and k. Thus if U were increased by 50%, the throughput could be increased to 70 tons/hr of coal. This effect could be achieved by a reduction in tube wall thickness, which is possible when the helium temperature is maintained below 1920°F (1050°C). Also, the throughput could be increased from 52 to 58 tons/hr by increasing the density of the fluidized bed by using higher pressures or smaller particles. It is thought that a throughput of up to 80 tons/hr of coal may be possible by combining all these factors.

Fouling of the heat transfer area, ash agglomeration, and temperature control problems that may arise in a large-scale reactor must be investigated further.

It appears from performance data predicted for a commercial-scale gasifier that the amount of coal gasified in the single vessel—approximately 43 tons/hr with helium at 1650°F (900°C)—can be increased by 20, 40, or 80% by raising the helium temperature to 1740, 1830, and 2000°F (950, 1000, and 1100°C), respectively. Although HTGR outlet temperatures up to 1800°F (980°C) may be feasible for plants built in the 1990s, temperatures higher than this appear to be out of the reach of present technology because of materials problems. Nevertheless, higher helium temperatures may be necessary to make this process competitive with other gasification schemes.[92]

The improvement in gasifier capacity resulting from the higher HTGR temperatures will reduce the fixed costs of the gasification section but will probably raise those of the nuclear section. The heat transfer surface accounts for the major portion of the cost of the gasification reactor, and although the fluidized-bed reactor minimizes the area required compared with fixed-bed or entrained-flow reactors, the severe conditions that promote erosion, carburization, embrittlement,

[92] D. Wiegand, K. H. Van Heek, and H. Jüntgen, "The Significance of the HTR-Temperature in the Economic Use of Nuclear Heat for Coal Gasification," Paper No. 13, Section III, "The High Temperature Reactor and Process Applications," British Nuclear Energy Society Int. Conf., London, November 1974.

Table 24.20 Comparison of Performance Data for Fluidized-Bed Processes

Gasification process:	CO_2 Acceptor	University of Wyoming Catalytic	Electrofluidic	Nuclear Aided	Synthane	Hygas: Steam-Oxygen	Hygas: Steam-Iron	Hydrane
Developer and site:	Conoco Coal Development Company, Rapid City, South Dakota	University of Wyoming, Glen Rock	Iowa State University	Bergbau-Forschung, Essen, West Germany	ERDA, Bruceton, Pennsylvania, Rosebud		IGT, Chicago	ERDA, Morgantown, West Virginia
Fuel:	Lignite	Subbituminous	Lignite Char	Low-Volatile Bituminous	Subbituminous	Western Subbituminous	Western Subbituminous	Pittsburgh No. 8
Proximate analysis								—
Volatile matter	—	39.6	3	12.2	36.1	34.3	34.3	—
Fixed carbon	—	36.1	85	68.2	46.0	46.7	46.7	—
Moisture	0	12.2	—	1.2	7.5	11.0	11.0	—
Ash	6.8	12.1	11	18.4	10.4	8.0	8.0	—
Ultimate analysis								—
Carbon	66.2	52.7	—	70.9	61.6	60.4	60.4	—
Hydrogen	4.6	5.1	—	3.3	5.5	5.3	5.3	—
Nitrogen	1.0	0.6	—	1.6	0.9	0.9	0.9	—
Oxygen	20.9	28.6	—	4.0	13.5	24.7	24.7	—
Sulfur	0.5	0.8		1.8	0.6	0.7	0.7	—
Ash	6.8	12.1	—	18.4	10.4	8.0	8.0	—
HHV fuel (Btu/lb)[a]	11,358	9140	—	11,812	10,185	10,245	10,245	—
Steam/coal (lb/lb)[a]	1.07	—	1.14	4.42	1.00	1.10	1.93[b]	
Oxygen/coal (lb/lb)[a]	0.54[b]	0	0	0	~0.32	0.25	0.40[b]	
Air/coal (lb/lb)[a]	2.34	0	0	0	—	—	1.72	
Gasification temperature (°F)	1500	1200–1400	1300	1530	800/1400–1800	570/1200/1700	—	
Gasification pressure (psia)	165	250	1500	600	600–1000	1220	1220	

Product gas analysis,								
dry (% v/v)								—
CO	15.5	—	8.4	13.7	13.2	25.0	14.1	—
H_2	58.8	1.8	22.4	56.2	32.3	29.9	48.1	1000
CH_4	13.7	50.4	31.1	7.8	15.0	17.1	24.1	6.0
CO_2	9.1	48	38.1	22.3	36.3	25.2	8.9	21.7
N_2	2.9	—	—	—	—	—	1.2	68.2
C_nH_m	0	—	—	—	1.6	1.9	2.3	1.7
H_2S	—	—	—	—	1.6	0.3	0.4	1.4
Trace components	—	—	—	—	—	0.6	—	0.4
HHV product gas (Btu/scf)	378	514	413	305	337	385	487	0.6
Tars or oils produced	No	—	—	—	Yes	Yes	No	—
Gasifier outlet temperature (°F)	1495	—	1300	1530	—	600	600	788
Product gas yield (scf/lb)[a]	23.0	15.0	25	47.2	—	12.5	14.4	Yes
Specific gasification rates[c]								1000–1500
10^6 Btu/ft²-hr	—	—	16.7	—	—	—	26.3	—
10^6 Btu/ft³-hr	—	—	0.40	0.14	0.47	—	0.19	—
Carbon conversion (%)[c]	66.0	—	70.0	95.0	75.0[d]	—	97.1	—
Gasification efficiency (%)[c]								—
E_1	77.0	87.7	79.0	—	~57	47.0[e]	68.5[e]	—
E_2	—	—	—	—	~63	—	—	—

[a] Containing moisture and ash indicated in proximate analysis.
[b] Oxygen supplied as air.
[c] See Appendix.
[d] Conceptual data.
[e] Derived value.

Table 24.20 Continued

Gasification process:	Tri-Gas	West Virginia University Two-Stage	Chemically Active Fluidized Bed	Winkler		U-Gas	Westinghouse	Cogas
Developer and site:	BCR, Monroeville, Pennsylvania	West Virginia University	Esso Petroleum Co. Abingdon, England	Davy Powergas, Various Sites		IGT, Chicago	Westinghouse, Pittsburgh	Cogas Development Company, Prince Division, Leatherhead, Surrey, England
Fuel:	Eastern Coal	High-Volatile Bituminous	Illinois No. 6	German Lignite	Lignite	—	Eastern U.S.	Illinois No. 6
Proximate analysis								
Volatile matter	37.9	—	—	—	—	—	—	36.5
Fixed carbon	54.5	—	—	—	—	—	—	51.5
Moisture	1.4	3.4	—	8.0	8.0	5.0	—	0.0
Ash	6.2	—	—	—	—	—	—	12.0
Ultimate analysis								
Carbon	79.2	71.0	—	61.3	54.3	67.8	—	69.0
Hydrogen	5.5	5.5	—	4.7	3.7	5.6	—	5.0
Nitrogen	1.4	0.7	—	0.8	1.7	1.6	—	1.3
Oxygen	6.1	12.8	—	16.3	15.4	5.0	—	8.1
Sulfur	1.5	2.7	—	3.3	1.2	4.2	—	4.0
Ash	6.3	7.3	—	13.8	23.7	15.8	—	12.0
HHV fuel (Btu/lb)[a]	14,090	12,664	—	9490	8030	—	13,600	—
Steam/coal (lb/lb)[a]	0.11	0.29	0	0.12	0.39	0.31	0.5	0.63
Oxygen/coal (lb/lb)[a]	0.86[b]	0.45[b]	—	0.59[b]	0.39	0.71[b]	0.65[b]	0.65[b]+0.32
Air/coal (lb/lb)[a]	3.70	1.91	—	2.51	—	3.07	2.80	2.80
Gasification temperature (°F)	1200/2000/2100	1440/1900	1560	1500–2200	1500–2200	~1900	1600/800/2100	600–1750
Gasification pressure (psia)	250	100	15	~15	~15	115–365	130–120	26

Product gas analysis, dry (% v/v)								
CO	31.3	27.2	12.2	22.5	36.0	23.0	19.2	25.0
H_2	15.9	19.7	13.6	12.6	40.0	14.7	14.4	43.6
CH_4	—	3.9	4.2	0.7	2.5	3.2	2.7	4.2
CO_2	0.5	6.8	9.9	7.7	19.5	6.4	9.3	12.6
N_2	51.7	41.1	59.2	55.7	1.7	52.0	54.3	11.3
C_nH_m	—	0.7	0.9	—	—	—	—	1.9
H_2S	0.2	0.6	—	0.8	0.3	0.7	0.1	1.4
Trace components	0.4	—	—	—	—	—		—
HHV product gas (Btu/scf)	153	207	141	125	272	158	136	—
Tars or oils produced	No	Yes	Yes	No	No	No	No	Yes
Gasifier outlet temperature (°F)	2000	1440	1560	1430–1840	1430–1840	1500–1900	1400–1700	1650
Product gas yield (scf/lb)[a]	74.4	44.8	—	47.0	21.9	—	—	17.4
Specific gasification rates[c]								
10^6 Btu/ft^2-hr	—	—	—	1.78	1.81	6.41	—	—
10^6 Btu/ft^3-hr	—	—	—	—	—	—	—	—
Carbon conversion (%)[c]	95.9	—	—	83.0	81.0	—	—	—
Gasification efficiency (%)[c]								
E_1	80.8	73.0	—	61.9	74.4	—	—	51.0
E_2	—	73.0	—	76.2	88.6	—	—	77.0

[See footnotes on page 1661.]

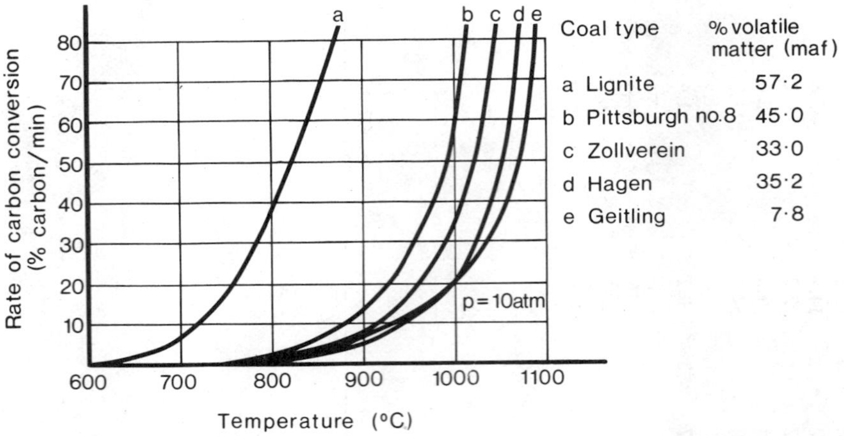

Figure 24.27 Rate of gasification as a function of temperature in a Bergbau-Forschung gasifier.[90]

and sulfidation make the choice of a suitable material a formidable research topic. The pressure shell of the gasifier is constructed from a normal carbon steel and insulated with a ceramic material such as alumina-silica. Such materials have been found suitable for use at temperatures below 1100°F (600°C), but in high-temperature zones pure Al_2O_3 is used because at high temperatures silica is soluble in steam.

[93] W. Peters, in *Coal Processing Technology*, Vol. 2, American Institute of Chemical Engineers, New York, 1975, pp. 72–75.

As a part of the same project, RBW is investigating the hydrogasification of lignite and bituminous coals in a fluidized bed, also using the HTGR as a heat source. A 220-lb/hr pilot plant has recently been designed and constructed to investigate the hydrogasification reaction at pressures between 600 and 1500 psig and temperatures between 1470 and 1650°F (800–900°C).

In this process (Fig. 24.29) the hydrogen required for the hydrogasification reaction would be produced by the catalytic steam reforming of part of the methane product (followed by CO shift and CO_2 removal), with the hot helium stream from the HTGR pro-

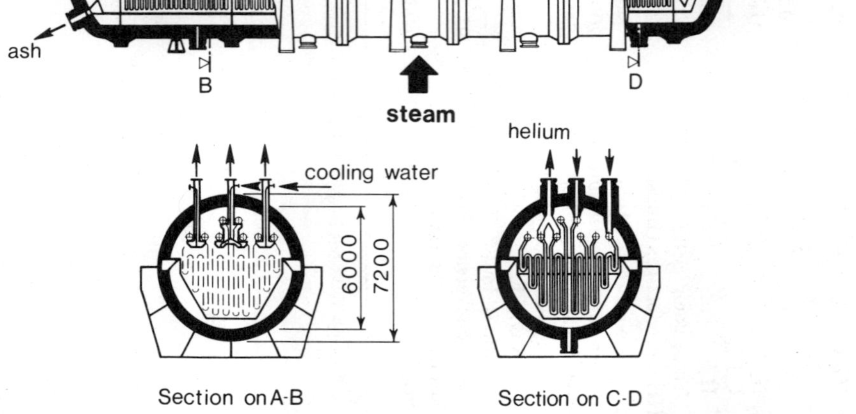

Figure 24.28 The Bergbau-Forschung gasifier; dimensions in millimeters.[93]

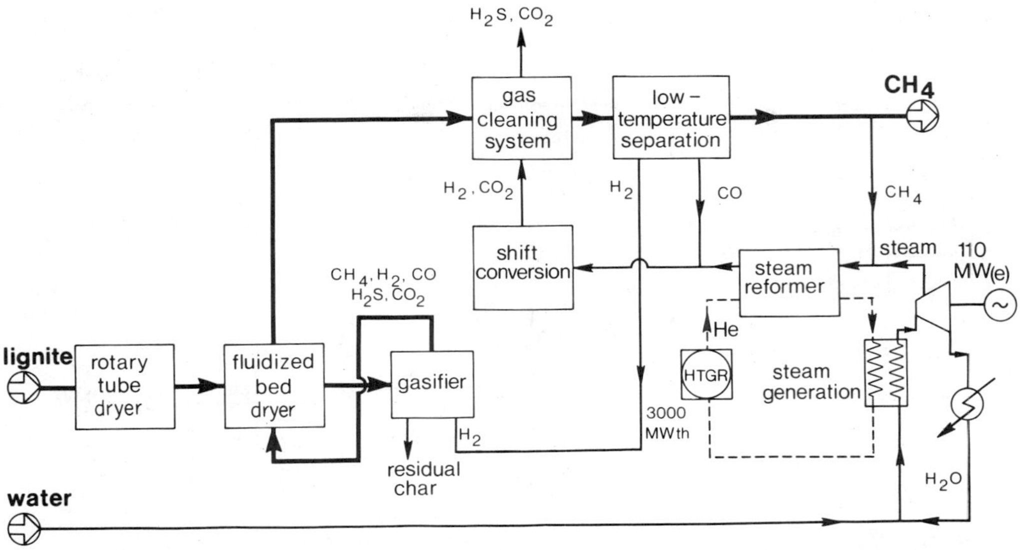

Figure 24.29 The hydrogasification of lignite: steam reforming of methane using nuclear heat.

viding the heat required for the reforming reaction. The residual char remaining from the hydrogasifier is thought to be suitable for further steam-carbon gasification or as a fuel for steam raising.

The reactor, approximately 8 in. (200 mm) I.D. and 26 ft (8 m) high, has an internal cyclone for separation of fines. The objectives of the pilot-plant experiments are the definition of operating parameters, the determination of the properties of the residual char and other by-products, and the design of a suitable feed system.

Predicted performance data for a commercial plant are given in Table 24.21, for which it is assumed that the helium leaves the HTGR at 1740°F (950°C) and is then cooled to 1470°F (800°C) in the reforming stage. The helium is further cooled to 480°F (250°C) to raise steam for use in both the reforming and power-generation stages.

An alternative scheme (Fig. 24.30) is to gasify the coal with a mixture of hydrogen, carbon monoxide, and steam that is preheated to 1650°F (900°C) by the helium stream from the HTGR. This approach would dispense with the steam reformer and the CO shift stage by replacing them with a heat exchanger to preheat the gasifying medium, but the comparatively low hydrogen partial pressure of this medium is likely to increase the size of the gasifier required. The conceptual performance data for this process are also given in Table 24.21.

[94] H. Teggers, "The High Temperature Reactor and Process Applications," Paper No. 10, presented at British Nuclear Energy Society Int. Conf., London, November 1974.

Table 24.21 Performance Data for the Two Rheinische Braunkohlenwerke Hydrogasification Schemes[94]

	Methane Reforming	Preheating Gasifying Agent
Thermal power of nuclear reactor in 1740–480°F (950–250°C) range (MW)	3000	3000
Feed rate of raw lignite (tons/hr)	2200	2500
Products		
Net electric power generation (MW)	110	180
SNG produced		
10^6 scf/hr	13.4	14.5
m^3/hr	(380,000)	(410,000)
Residual char (tons/hr)	270	330

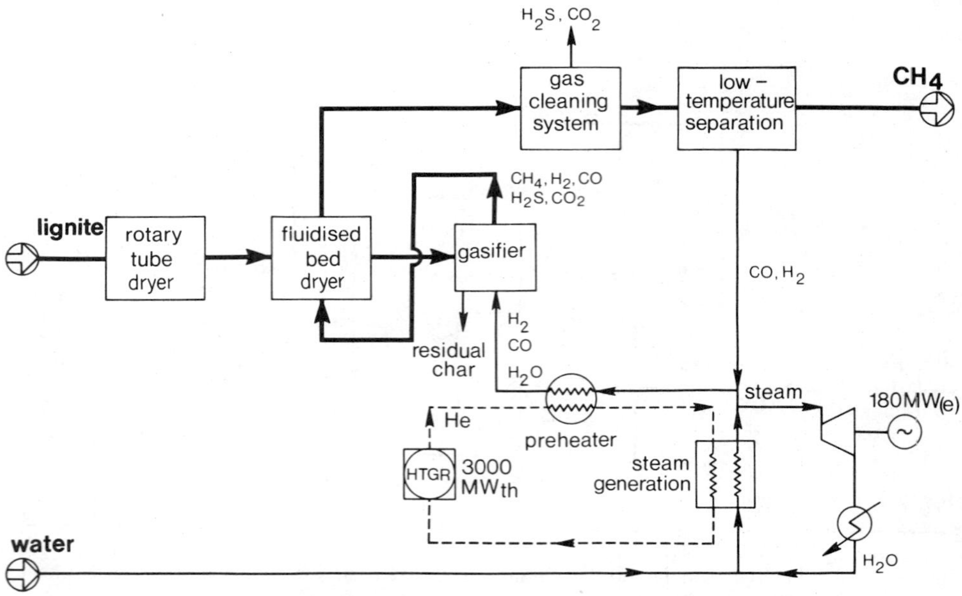

Figure 24.30 The hydrogasification of lignite: pre-heating the gasifying agent by nuclear heat.

The hydrogasification routes to SNG proposed by RBW have some advantages over the steam-carbon gasification route when using an HTGR as a heat source. These include fewer materials problems (since the heat exchangers may operate at lower temperatures in less hostile atmospheres) and the virtual elimination of fouling (since heat is transferred not into the fluidized bed, but only between clean streams). The major advantages are that low-reactivity coals can be hydrogasified without the HTGR having to operate at temperatures in excess of 950°C, and the steam-carbon gasifier, with its associated heat transfer surfaces, is eliminated.

If the process scheme can operate safely using only one helium heat transfer loop, an added advantage is the availability of higher grade heat. The use of a single loop will require careful design of steam boilers and of the preheater, or methane reformer, to ensure its integrity. If an intermediate loop is required, however, the maximum available temperature is reduced and this advantage is lost.

Commentary on the Electrofluidic and Nuclear-Aided Processes. In these processes the heat for the endothermic reactions is supplied from an external source rather than by combustion of part of the feedstock within the gasifier. An oxygen supply is not required, and the absence of the associated compressors facilitates high-pressure operation that together with moderate temperatures, enhances methane formation.

A high steam rate is required in the electrofluidic reactor for adequate fluidization. This promotes the CO shift reaction within the gasifier, so reducing the amount of CO shift required to transform the product gas into SNG. This process does not appear to be economically attractive at present because the high capital and operating costs associated with electricity generation are not offset by the elimination of the oxygen plant. Problems of electrode corrosion appear to have been solved, but the process may be restricted by the electrical properties of the bed, although these can be modified by feed pretreatment or solids recycle.

The major limitation of the BF nuclear-aided process is the strong dependence of the gasification rate on the operating temperature of the HTGR. The currently available temperatures demand long residence times and high steam rates to ensure good carbon conversion. Possibly this problem may be overcome by the use of catalysts or by "superheating" the helium from the HTGR by a conventionally fired heater to provide increased gasification temperatures, or by the adoption of alternative approaches such as the RBW process.

A preliminary evaluation has been made of a process in which both the electrofluidic gasifier and the HTGR are used,[95] so that the gasification rate is independent of the operating temperature of the HTGR. Siting of the gasifier at an electrical power station would allow gas production during periods of low electricity demand, enabling the nuclear power station to operate at high load continuously. The use of the electrofluidic reactor would also avoid the considerable materials problems encountered in the BF gasifier.

24.3.2.2 *Wholly Fluidized-Bed Processes Yielding a Nitrogen-Containing Product Gas*

The Bituminous Coal Research (BCR) Tri-Gas Process. The BCR Tri-Gas process employs three gasification stages producing a tar- and oil-free, low-Btu gas that is suitable for an integrated combined-cycle electric power generation plant. Fig. 24.31 is a schematic diagram of the process.

Sized coal passes from a pressurized lock hopper through a rotary airlock feeder and flows by gravity to the first of three fluidized-bed reactors in series, operating at 250 psig. The operating temperature increases progressively in each reactor from Stage 1 [1200°F, (650°C)] to Stage 3 [2100°F (1150°C)].[96] In Stage 1 the fresh coal is fluidized with hot gas from Stage 3, and it is said that this allows virtually all types of coal to be used. Both the gas, containing volatile tars and oils, and the devolatilized coal from Stage 1 pass to Stage 2, where the main gasification reactions take place.

In the next stage, which operates at 2000°F (1090°C), char is fluidized by the Stage 1 gas together with air and steam. The product gas from Stage 2 passes to a cyclone to allow the return of the entrained solids. The cooled gas is filtered to remove any remaining char fines and purified to give a product with a heating value of approximately 150 Btu/scf.

Char from Stage 2 is gasified by air and steam in the Stage 3 fluidized bed, with the

[95] A. H. Pulsifier and T. D. Wheelock, "Electrofluidic Gasification of Coal with Nuclear Energy," paper presented at American Institute of Chemical Engineers, National Meeting, Chicago, November 1976.

[96] J. T. Stewart and E. K. Diehl, "Fluidized Bed Coal Gasification Process and Equipment Development," paper presented at 3rd Int. Conf. Fluidized Bed Combustion, Hueston Woods State Park, Ohio, October 1972.

[97] E. K. Diehl, Bituminous Coal Research, private communication, August 1977.

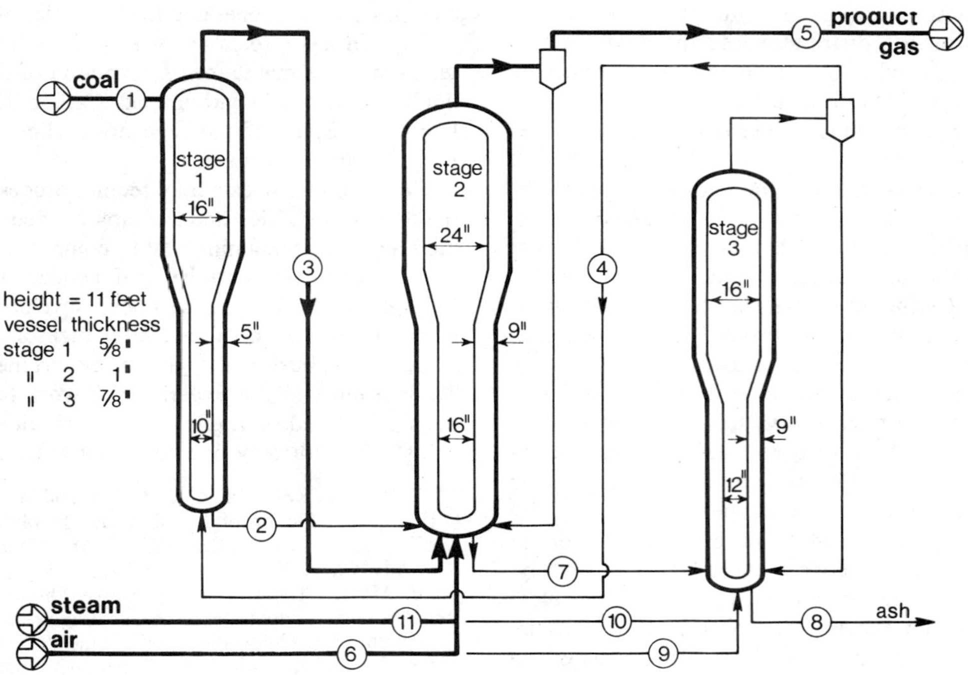

Figure 24.31 Tri-Gas pilot-plant gasifiers[96,97]; material balance data for Streams 1–11 appear in Table 24.22.

resultant ash being removed by lock hopper. The Stage 3 product gas passes via a cyclone (where entrained ash is removed and returned to the Stage 3 bed) to Stage 1.

Bench-scale work began in the late 1960s, and in 1974, under the OCR sponsorship, construction of a 100-lb/hr pilot plant sited at Monroeville, Pennsylvania, commenced. This plant was completed in late 1975 and experimental work started in early 1976. All three stages are positioned to allow solids flow, by gravity, between them. Operating data are not yet available but a design material balance is given in Table 24.22. Additional design data are shown in Table 24.20.

The West Virginia University Two-Stage Process. The two-stage pyrolyzer-gasifier system of West Virginia University consists essentially of a fluidized bed in which coal is pyrolyzed at 1400°F (760°C), and a char gasifier in which the char, with a small amount of raw coal, is completely gasified at 1900°F (1040°C) with air and steam to produce the gas for the pyrolysis section.

Experimental work to date, under OCR sponsorship, has been limited to a 15-in.-diameter fluidized bed, which was used to simulate the pyrolysis section of the process (Fig. 24.32). Natural gas is burned in this experimental unit, and a mixture of the hot combustion gases and extra air passes through a distributor into the fluidized bed. The combustion zone is not a permanent process feature, but is only used in an attempt to simulate the gas from the proposed char gasifier.

Coal is added to the middle section of the vessel, which contains a fluidized bed of sand. The sand provides rapid heat transfer to the fuel particles, which undergo abrasion and grinding to expose constantly fresh reactive surfaces. It also reduces the tendency for the reacting fuel particles to cohere, and it provides a heat sink to dampen any temperature fluctuations. Coal feed rates of up to 60 lb/hr have been employed, and a composition for the crude product gas is given in Fig. 24.33. Conceptual performance data are given in Table 24.20. In spite of the important differences between the commercial pyrolyzer-gasifier shown schematically in Fig. 24.33 and the experimental pyrolyzer, the results from the latter have formed the basis of the design for a conceptual 500-MW power plant.

Material flow rates and gas stream compositions relating to the pyrolyzer-gasifier for this plant are given in Table 24.23. A calculated overall process efficiency of 38.9% for electrical power generation is claimed.[98]

The Esso Chemically Active Fluidized-Bed Process. The Esso Petroleum Company is developing the chemically active fluidized-bed (CAFB) process[99,100] at the Esso Research Centre, Abingdon, England. A schematic diagram of the process is shown in Fig. 24.34. Coal particles in the size range 0–1/2 in. are injected into a shallow bed (2–4 ft deep) of lime particles. The bed is fluidized by air with a superficial velocity of 2–6 ft/s. In addition to acting as a heat transfer agent, the lime combines with the major part of the sulfur in the coal. The air-to-coal ratio to the gasifier is about 30% of that required for complete coal combustion, and at a bed temperature in the range 1560–1740°F (850–950°C), and at about atmospheric pressure, partial oxidation and pyrolysis take place. The gaseous product (see Table 24.20) contains tars, entrained ash, lime, and ungasified coal particles from the bed.

The bed material is circulated continuously through a smaller adjacent air-blown fluidized regenerator at a temperature of 1920–2010°F (1050–1100°C). Here the sulfided limestone is regenerated to CaO, releasing a SO_2-rich gas stream (~ 8–10% SO_2). The temperature is maintained by reaction of the fluidizing air with carbon and CaS in the limestone. Some of the regenerated lime is purged from the system.

The main fuel stream may require processing to remove solids and perhaps condensables and any remaining sulfur compounds, the exact requirements being dependent on its final use. The process has been developed primarily to produce a low-sulfur fuel for an adjacent combustor (e.g., boiler or turbine). To date most of the experimental work has been performed on a pilot plant using an oil feedstock, with only one continuous run on

[98] C. Y. Wen, R. C. Bailie, C. Y. Lin, and W. S. O'Brien, "Production of Low-Btu Gas Involving Coal Pyrolysis and Gasification," in Am. Chem. Soc. Adv. Chem. Ser. No. 131, 1974, pp. 9–28.

[99] G. Moss, "The Chemically Active Fluidized Bed Gasifier," paper presented at Engineering Foundation, Int. Fluidization Conf., June 1975.

[100] R. Newstead, "Fluid Bed Desulfurization," *Oil Gas Firing,* 10–13 (September 1972).

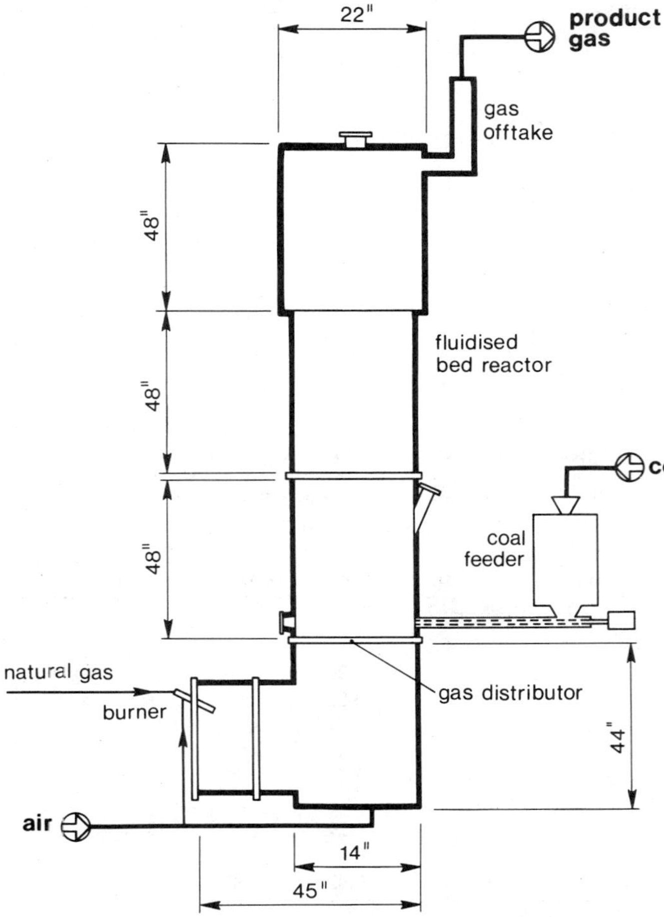

Figure 24.32 The West Virginia University pyrolysis reactor.[98]

coal of 8.5 hr during the summer of 1976. Desulfurization was reported to be very satisfactory during this trial, the main problem being with the coal feed system.

Commentary on Wholly Fluidized-Bed Processes That Produce a Nitrogen-Containing Fuel Gas. Primarily, the three processes above give gases for use in electric power generation. The BCR Tri-Gas process should offer the advantage of being able to gasify both caking and noncaking coals by using the first stage of the process as a pretreatment step. This step also enables the major part of the gasification to be carried out at higher temperatures, resulting in high carbon conversion, elimination of tars and oils, but little production of methane. These effects are reflected in the amount of feedstock combusted (to provide the necessary high-grade heat) being high, which is reflected in the low gasification efficiency (E_1) [for definition see Appendix to this chapter].

The other two processes are at a much earlier stage of development and are therefore difficult to assess. The final process envisaged by West Virginia University has not yet been demonstrated and the experimental pyrolysis gas composition is far from that predicted (from equilibrium data) for the conceptual commercial plant. The use of a low gasification temperature and separate pyrolysis stage should result in the formation of some methane and higher hydrocarbons to give a gas with a HHV of about 200 Btu/scf, which is high in comparison to many other air-blown systems.

The Esso CAFB process may offer a method of combined sulfur removal and gasification to produce a hot fuel stream suitable for direct use. This is advantageous because the gas might otherwise have to be cooled before

Table 24.22 Design Material Balance for BCR Tri-Gas Plant (Fig. 24.31)

	Stream 1: Coal Feed (lb)	Stream 2: Char from Stage 1 (lb)	Stream 3: Stage 1 Gas		Stream 4: Stage 3 Gas		Stream 5: Product Gas	
			mole	% v/v	mole	% v/v	mole	% v/v
Coal (ash free)	93.8	77.0						
Ash	6.2	6.2						
H_2			0.20	4.46	0.20	4.86	3.10	15.80
CO			0.98	21.88	0.98	24.15	6.13	31.23
CO_2			0.01	0.22	0.01	0.14	0.09	0.48
H_2O			0.00	0.00	0.00	0.06	0.10	0.49
N_2			2.58	57.59	2.58	63.65	10.09	51.41
H_2S							0.05	0.24
O_2			0.29	6.47	0.29	7.14		
NH_3							0.07	0.35
Stage 1 volatiles			0.42	9.38				
Total	100.0	83.2	4.48	100.0	4.06	100.0	19.63	100.0
Temperature (°F)	77	1200	1200		2100		2000	
Pressure (psia)			250		250		250	
Gas flow (scf)			28.29		25.66		123.97	
Average molecular weight			28.27		27.05		23.91	

Stream 6: Air to Stage 2		Stream 7: Char from Stage 2	Stream 8: Char from Stage 3	Stream 9: Air to Stage 3		Stream 10: Steam to Stage 3		Stream 11: Steam	
moles	% v/v	(lb)	(lb)	moles	% v/v	mole	% v/v	mole	(% v/v)
		14.8	3.0						
		6.2	6.2						
						0.20	100.0	0.41	100.0
7.54	79.00			2.58	79.00				
2.00	21.00			0.69	21.00				
9.54	100.0	21.0	9.2	3.27	100.0	0.20	100.0	0.41	100.0
1000		2000	2100	1000		1000		1000	
300		250		300		300		300	
60.26				20.66		1.26		2.59	
28.85				28.85		18.02		18.02	

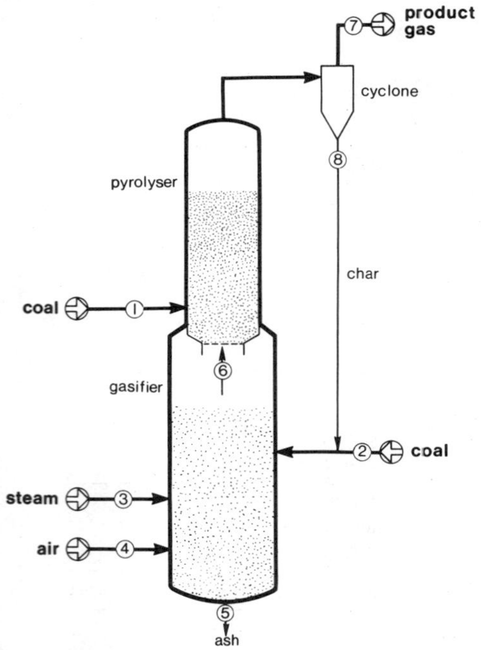

Figure 24.33 The Pyrolyzer-gasifier proposed by West Virginia University; for material flow rates and gas stream composition, see Table 24.23.[98]

sulfur removal, resulting in an efficiency drop. However the SO_2-containing gas from the regenerator will require special treatment. This process differs from others in this section in that it utilizes partial oxidation and pyrolysis of the coal rather than steam gasification. The limited air supply to the gasifier results in CO formation rather than complete oxidation to CO_2, which with the hydrocarbon pyrolysis products gives a product stream with a heating value comparable with that from a steam-air gasifier (cf. "The Westinghouse Fluidized-Bed Process," Section 24.3.3.2).

24.3.3 Hybrid Fluidized-Bed Processes

24.3.3.1 *Hybrid Fluidized-Bed Processes Yielding a Nitrogen-Free Product Gas*

The Battelle–Union Carbide Agglomerating Ash Process. The Battelle–Union Carbide agglomerating ash process has been developed from work conducted by Battelle Memorial Institute in the 1960s on agglomerating combustion. The process consists of two fluidized beds; combustion of coal occurs in one and steam gasification in the other. The heat requirement for the gasification reactions is provided by circulating ash agglomerates between the two stages. Ash agglomeration in the combustor obtained by operating at temperatures approaching the ash fusion point of the coal is a key feature of the process and allows selective removal of relatively large ash lumps. This feature results in a low carbon content ash (typically < 2% w/w) and overcomes the generally low carbon conversion associated with the temperature uniformity and good solids mixing characteristics of conventional fluidized beds.

A pilot plant producing about 1 million scf/day of synthesis gas from 25 tons/day of coal has been constructed.[102] It operates at a relatively low pressure of about 100 psig, chosen as a compromise to achieve acceptable plant capital cost and scaleup risk, since the initial laboratory-scale tests were made at atmospheric pressure. Operation at yet higher pressures is thought to be feasible if commercially necessary.[103]

Figure 24.35 is a diagram of the pilot-plant combustor, gasifier, and associated feed and circulation systems. The coal is reduced to two separate size ranges and dried by a hot-gas-swept roller mill to provide feed at minus 100 mesh (< 0.15 mm) for the combustor and feed at minus 8 plus 100 mesh (0.15–2.4 mm) for the gasifier. Coal pretreatment is probably unnecessary, given the dilution effect of the ash at the proposed high ratio of circulating ash to coal feed (4 : 1). If a coal agglomeration problem exists, however, an atmospheric pressure fluidized-bed stage, operating at about 750°F (400°C) would be used, in which mild oxidation of coal is promoted. It is anticipated that a fluidizing gas consisting largely of recycled off-gas from this stage together with additional air to maintain the oxygen content of the fluidizing gas between 1–5% v/v would be used, with the pretreated coal being conveyed pneumatically to a conventional gasifier lock hopper feed system. Untreated fine coal is conveyed with air into

[101] G. Johnes, Esso Petroleum Company, Esso Research Centre, Abingdon, Oxon, England.

[102] Anon., "Agglomerating Burner Gasification," *Hydrocarbon Process.*, **54**(4), 118 (April 1975).

[103] W. C. Corder and W. M. Goldberger, "Status of Battelle–Union Carbide Process Development Unit Installation," paper presented at 6th Synthetic Pipeline Gas Symp., Chicago, October 1974.

Table 24.23 Material Flow Rates and Gas Stream Compositions for West Virginia University Pyrolyzer-Gasifier (Fig. 24.33)

	Stream 1: Coal to Pyrolyzer	Stream 2: Coal to Gasifier	Stream 3: Steam to Gasifier	Stream 4: Air to Gasifier	Stream 5: Char Ash to Disposal	Stream 6: Gas from Gasifier	Stream 7: Gas from Pyrolyzer	Stream 8: Recycle Char	Experimental Pyrolyzer
ton/hr:	191.8	0	49.9	335	31.7	455	545	103	
lb-mole/hr:	—	—	5,539	23,200	—	35,800	46,800	—	
Gas	Gas Analyses (% v/v)								
H_2						11.63	18.84		4.95
CO						27.87	25.96		2.24
CO_2						5.26	6.50		11.28
CH_4						0.18	3.68		1.79
C_2H_6						—	0.21		0.1
C_2H_4						—	—		1.12
C_2H_2						—	0.44		0.22
H_2S						0.02	0.63		—
N_2						50.90	39.15		77.4
O_2/Ar						—	—		0.9
H_2O						4.14	4.59		—
						100.00	100.00		100.00

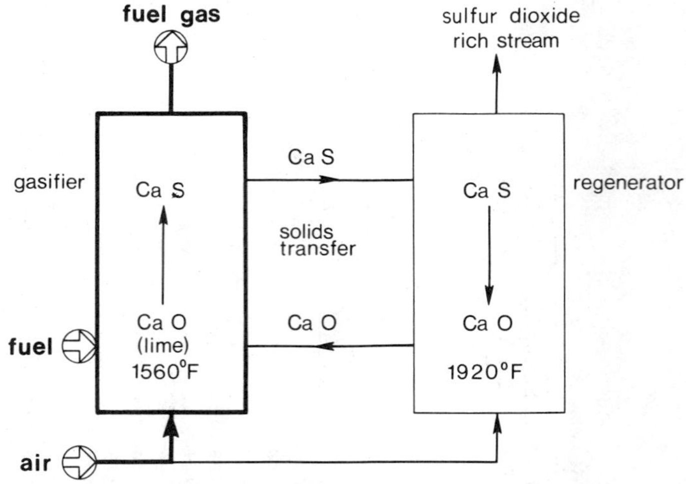

Figure 24.34 The Esso chemically active fluidized-bed process.[101]

the combustor from a lock hopper system through the air distributor plate (see Fig. 24.35), below which additional air enters the plenum chamber. For Eastern bituminous coals, operation of the combustor at a temperature of about 2050°F (1120°C) and a fluidizing velocity of about 7 ft/s[103] should produce the required agglomerating conditions. It is hoped that by maintaining the fluid bed level below the tapered sections of the combustor with the upper section acting as a freeboard, the hot flue gases will be sufficiently free of fly ash to enable direct energy recovery using expansion turbines. The ash agglomerates formed are free-flowing spherical particles, which overflow from the combustor fluid bed through a sidepipe and solids transfer system to the gasifier.

The refractory-lined transfer pipe from the combustor to the gasifier (Fig. 24.36) is sparged with steam and recycled gasifier product gas to strip combustion gases from the interstices of the ash and to help maintain solids fluidization. This avoids compaction of agglomerated ash and minimizes the possibility of transfer line blockage. The uplift leg is more highly sparged so that the effective fluid density can be varied to obtain the desired degree of interstage solids transfer.

Coal is conveyed pneumatically into the gasifier feed inlet situated in the tapered transition zone (Fig. 24.37) and enters at the

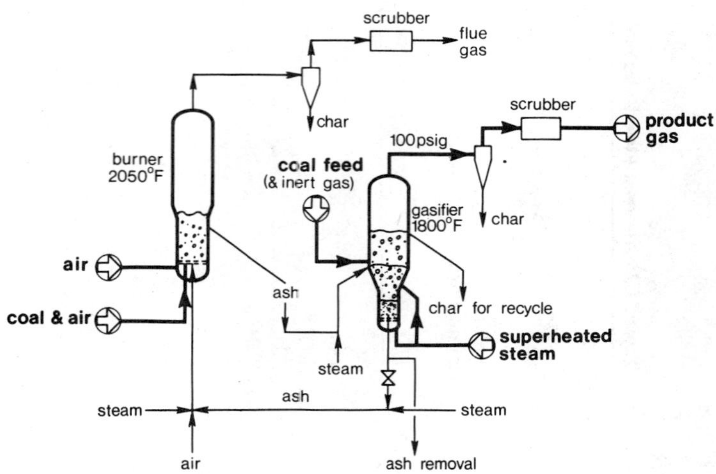

Figure 24.35 The Battelle–Union Carbide agglomerating ash process: burner-gasifier feed and circulation system.[103]

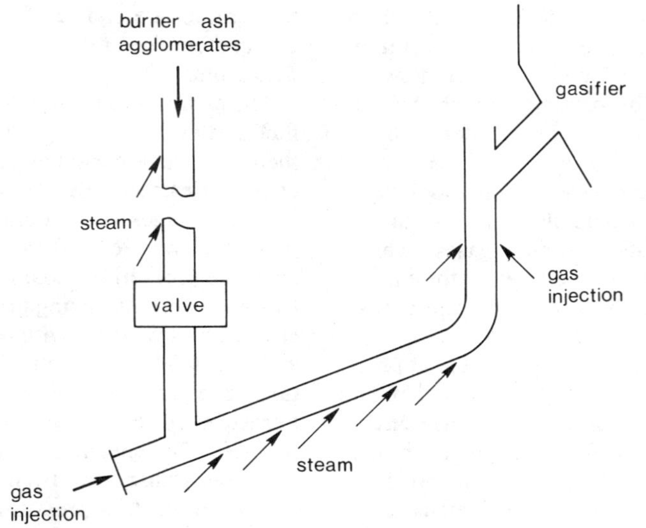

Figure 24.36 The solids transfer system for the Battelle–Union Carbide process.[103]

same vertical level as the circulating ash flow from the burner. The ash agglomerates and char segregate into two zones, separated by an intermediate zone in which the two materials are mixed. The lower zone occupies about two-thirds of the smaller diameter region of the gasifier and is predominantly filled with agglomerated ash particles. It is maintained in a condition of incipient fluidization by superheated steam entering below the fluidized-bed distributor plate; the remaining steam enters through injection points located circumferentially around the transition zone. The intermediate zone, situated approximately within the tapered region, consists of mixed hot ash agglomerates, untreated coal, and char. The upper zone is mainly carbon-rich char and extends up to the char overflow point. The gas velocities[103] (~ 1 ft/s) in the two upper zones are much higher than the minimum for fluidization.

Flow of the agglomerated ash particles is downward through the gasifier; thus their

[104] W. C. Corder, "The Union Carbide–Battelle Coal Gasification Process Development Unit Design," paper presented at 5th Synthetic Pipe Gas Symp., Chicago, October 1973, pp. 107–124.

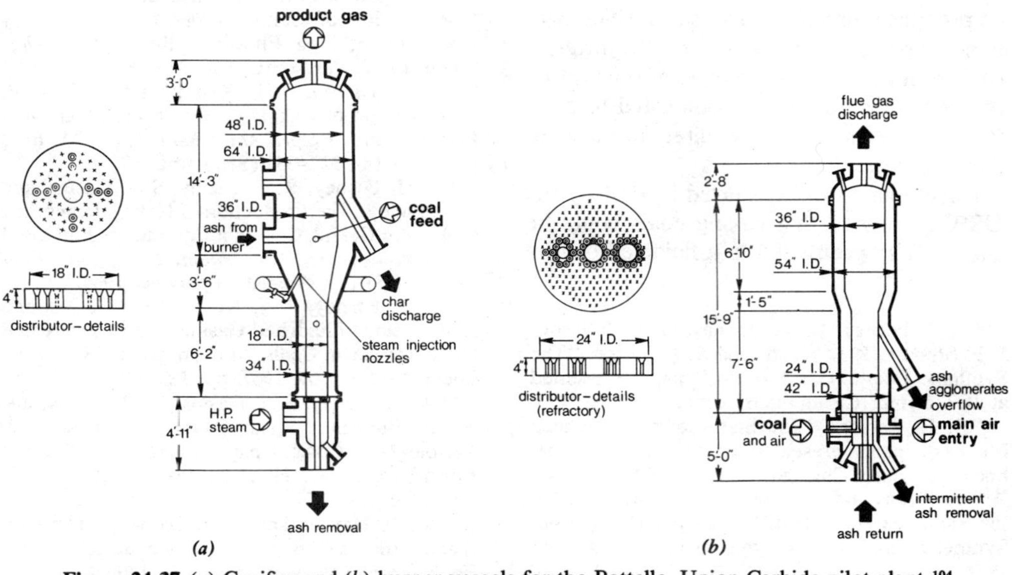

Figure 24.37 (*a*) Gasifier and (*b*) burner vessels for the Battelle–Union Carbide pilot plant.[104]

temperature falls from 2000°F (1090°C) to 1550°F (845°C). They are continuously stripped of carbon by the upward flow of steam and leave by passing through the distributor plate, mostly to be transferred back to the combustor. However sufficient ash is removed from the gasifier through lock hoppers to avoid the accumulation of ash in the system. The gasifier product gases, which are free from tar and oil, pass through a cyclone, for char fines removal, and a gas-scrubbing stage.

No performance data are available at present for this process. Following initial pilot-plant tests[103] to establish that the ash circulation rates required for heat transfer in the gasifier can be obtained, it is planned to demonstrate operation of the combustor in the agglomerating mode before its integration with the gasifier. Tests on materials for the power recovery turbines are also planned.

The Synthane Process. The Synthane process will produce high-Btu gas by gasification, with steam and oxygen, of all types of coal[105] in a fluidized-bed system. Dried, pulverized coal is first pretreated (if it has caking properties) before entering the gasification stage. There it falls through a devolatilization zone into the fluidized bed, where the major part of the gasification occurs. The devolatilization step, together with a high operating pressure of 1000 psig, encourages methane production, so it is claimed that in an SNG process up to about 60% of the methane will be produced directly.[106] The char, which contains about 25–35% w/w[107] of the original carbon in the coal, is removed at the bottom of the fluidized bed and combusted to raise the high-pressure steam required for the complete process.

Development work, started in 1961 by the USBM, showed that caking coals could be successfully pretreated in a fluidized bed by the appropriate choice of fluidizing gas oxygen content, bed temperature, and coal residence time.[108–110]

The earliest gasification tests with a double fluidized-bed system[111] led to the development of a single-vessel arrangement in which coal pretreatment, devolatilization, and gasification stages were combined.[112] In this plant coal was fed (40 lb/hr maximum rate) from the pressurized coal hopper to the pretreater, where its caking properties were destroyed by treatment with oxygen and steam at 750°F (400°C). The fluidized-bed gasification zone was operated at a maximum temperature of between 1500 and 1800°F (815–980°C), depending on the coal type. The char was transported from the gasifier by a screw system to a pressurized lock hopper, and the gas was treated to remove dust, tars, and steam.[111,112]

Table 24.24 shows the results of gasification tests on four different coals: Pittsburgh, Illinois No. 6, Montana Rosebud subbituminous, and North Dakota lignite. Whereas the first two required pretreatment, the other two noncaking coals could have been used directly, but the Montana coal was pretreated in the reported experiment. Progressively higher gasification rates were possible at lower bed temperatures and oxygen feed rates for coals of reducing rank, through experiments 1–4.

[105] A. J. Forney, W. P. Haynes, G. E. Johnson, J. P. Strakey, J. J. Elliott, and S. J. Gasior, "The Synthane Coal-to-Gas Process," paper presented at IGT Symp. Clean Fuels, Chicago, June 1973.

[106] S. E. Carson, "The Status of the Synthane Process," paper presented at 7th Synthetic Pipeline Gas Symp., Chicago, October 1975.

[107] T. Torkas and R. Lewis, "A Pictorial Tour of the Synthane Pilot Plant," paper presented at 6th Synthetic Pipeline Gas Symp., Chicago, October 1974.

[108] S. J. Gasior, A. J. Forney, and J. H. Field, "Destruction of Caking Properties of Coal by Pretreatment in a Fluidized Bed," *Ind. Eng. Chem. Prod. Res. Dev.*, **3** 48–53 (1964).

[109] A. J. Forney, R. F. Kenny, and J. H. Field, "Gasification of Caking Coal in a Free-Fall Fluid Bed Reactor," *Prepr., Am. Chem. Soc., Div. Fuel Chem.*, **11**(4), 322–332 (September 1967).

[110] A. J. Forney, R. F. Kenny, S. J. Gasior, and J. H. Field, "The Production of Nonagglomerating Char from Caking Coal in a Continuous Fluid Bed Reactor," *Prepr., Am. Chem. Soc., Div. Fuel Chem.*, **8**(3) 1–9 (August–September 1964).

[111] A. J. Forney, R. F. Kenny, S. J. Gasior, and J. H. Field, "Fluid Bed Gasification of Pretreated Pittsburgh Seam Coals," in Am. Chem. Soc. Adv. Chem. Ser. No. 69, 1967, pp. 128–140.

[112] A. J. Forney, S. J. Gasior, R. F. Kenny, and W. P. Haynes, "Steam-Oxygen Gasification of Various U.S. Coals," paper presented at 2nd Int. Fluid-Bed Conf., Hueston Woods State Park, Ohio, October 1970.

[113] W. P. Haynes and A. J. Forney, "The Synthane Process," paper presented at IGT Clean Fuels from Coal Symp., Chicago, June 1975.

Table 24.24 Data for Various Fuels from 4-in.-Diameter Synthane Pilot Unit at 600 psig[113]

Fuel:	Pittsburgh	Illinois No. 6	Montana Rosebud	North Dakota Lignite
Experiment:	1	2	3	4
Average maximum temperature (°F)	1814	1760	1629	1587
Coal ash (% w/w)	7.9	14.2	11.6	7.2
Coal moisture (% w/w)	1.5	2.7	10.2	23.0
Average coal feed rate				
lb/hr	17.5	23.7	27.8	37.0
lb/hr-ft^3	30.7	50.4	59.2	78.2
Product gas composition, dry (% v/v)				
H_2	30.3	35.4	37.3	36.2
CO	27.0	12.3	11.4	17.2
CO_2	23.2	35.3	38.0	32.6
CH_4	18.2	13.9	11.3	12.9
C_2H_6	0.9	1.6	1.5	0.8
H_2S	0.4	1.6	0.5	0.3
Product gas yield (scf/lb coal, maf)	18.0	22.3	21.0	19.3
Carbon gasified (%)	64.9	66.5	62.0	59.0
Steam rate (lb/lb coal, maf)	1.0	1.5	1.3	0.6
Oxygen rate (lb/lb coal, maf)	0.29	0.26	0.24	0.14
CH_4 produced in gasifier (%)[a]	53.4	48.6	43.3	46.7

[a] See Appendix.

From the results of this laboratory-scale unit, a pilot plant has been designed by the C. E. Lummus Company.[107,114] Raw coal is fed into a pulverizing and drying system (Fig. 24.38) to give coal finer than 20 mesh (0.84 mm) with low moisture content. This coal falls by gravity, via a weigh hopper of approximately 1 ton capacity, into one of the two lock hoppers, where it is raised to operating pressure (600–1000 psia) using CO_2 by-product from the gas purification stage of the plant. Dual lock hoppers reduce pressurizing gas requirements and allow continuous coal feed into a 5-ton-capacity pressurized feed hopper. From here, high-pressure steam conveys coal continuously to the pretreater. The coal is mildly pretreated with oxygen to prevent particle agglomeration in the feedline. In the main part of the pretreatment stage, the coal is fluidized at about 800°F (425°C) in a flow of steam and oxygen. Coal residence time in the bed can be varied (3–7 min planned) by adjustment of the bed height.[106] For Pittsburgh coal[113] the pretreatment requires about 0.08 lb of oxygen and 0.27 lb of steam per pound of coal feed. Pretreated coal flows by gravity, together with all the gaseous products, into the main gasifier. The pilot-plant main gasifier has been designed to process up to 72 and 144 tons/day of caking and noncaking coal, respectively,[116] corresponding to a 300-fold scaleup from the laboratory plant.

In the main gasifier (Fig. 24.39) the pretreated coal drops 10–20 ft[106,107] through a devolatilization zone in which it is heated from 800 to 1400°F (425–760°C), with any remaining volatile compounds distilling off. The devolatilized coals are fluidized in a 20-ft-deep bed by a stream of steam and oxygen flowing at a velocity of 0.25–0.30 ft/s. The bed temperature increases from about 1400°F (760°C) at the top to 1800°F (980°C) at the

[114] *Coal-to-Gas Prototype Pilot Plant,* The C. E. Lummus Company (New Jersey), U.S. Bur. Mines Contract No. HO 110989, October 13, 1972.

[115] R. Lewis, E. L. Clark, W. P. Haynes, and J. P. Strakey, "Operational Status of the Synthane Pilot Plant," paper presented at 8th Synthetic Pipeline Gas Symp., Chicago, October 1976.

[116] A. J. Forney and W. P. Haynes, "Synthane Process for High-Btu Gas," *Ind. Heating,* 68 (November 1975).

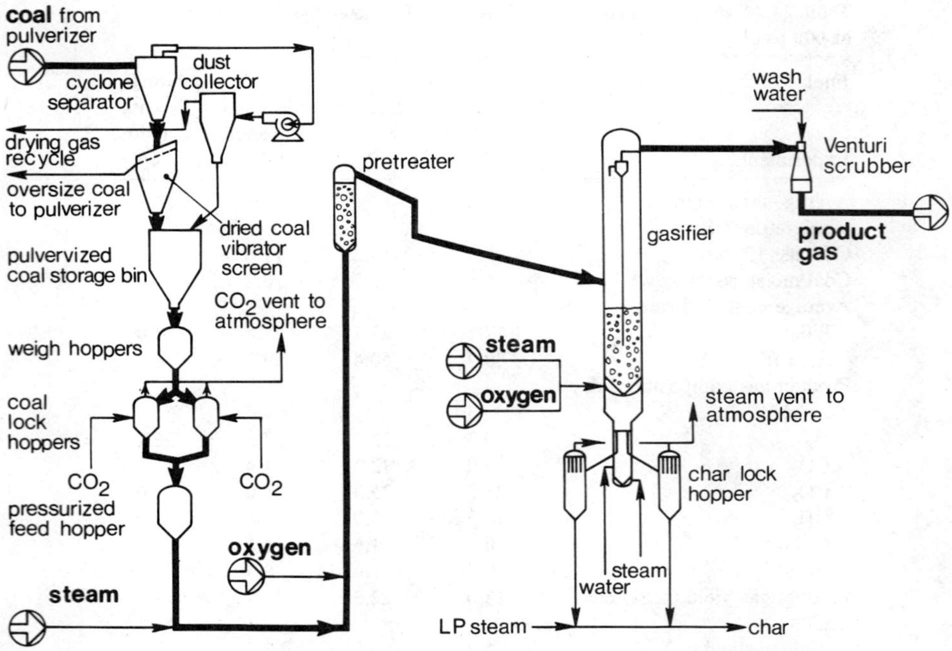

Figure 24.38 The Synthane process.[115]

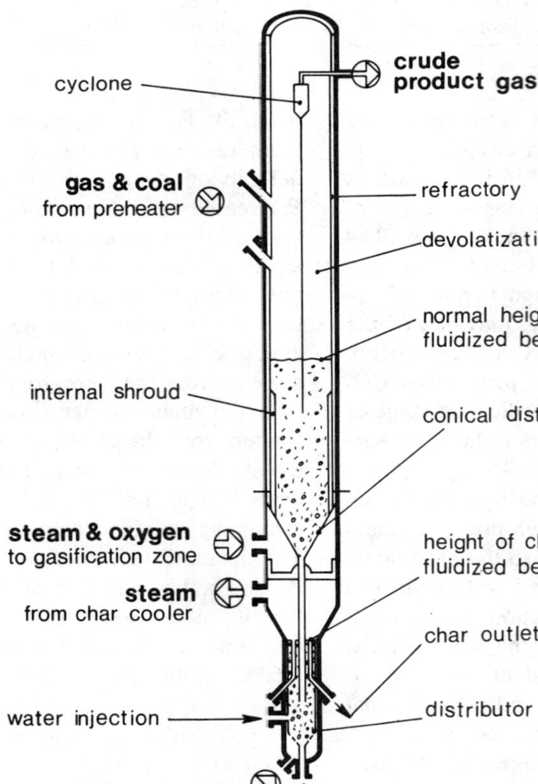

Figure 24.39 The Synthane pilot-plant gasifier.[106,107]

bottom, with mixing within the bed limited, if necessary, by the use of baffles, to obtain the required temperature distribution. The lower temperature at the top of the bed favors methane production, and the maximum bed temperature is kept below 2000°F (1090°C) to avoid ash clinkering. Steam and oxygen enter the bottom of the fluidized bed through a conical distributor.[117] The gasifier does not have a water jacket, and wall thermal insulation is provided by a refractory lining within the gasifier. An internal cyclone at the top of the gasifier allows most of the fines (larger than 50 μm) to be returned to the fluidized bed. The crude gas from the gasifier at 1400°F (760°C) is immediately quenched with water in a venturi scrubber to cool it to about 250°F (120°C). This step also removes tars, oils, and fines escaping the cyclone.

Char at 1800°F (980°C) leaves the gasification zone[113] situated at the bottom of the conical steam-oxygen distributor and passes to a lower fluidized bed maintained at 600°F (315°C) by the quenching action of steam and water sprays, which also act as the fluidizing medium. The high-pressure steam produced is employed in the subsequent CO shift conversion stage. Quenched char overflows (alternately) into two pipes which allow continuous char discharge by the use of a lock hopper system (Fig. 24.38) similar to that used for the coal feed. Since these lock hoppers operate at about 600°F (315°C), condensation of the steam present as the pressurizing medium is not a problem.

The pilot plant built with ERDA sponsorship at Bruceton, Pennsylvania, was commissioned in early 1976.[106] A 38-month test was scheduled to start in July 1976, using Montana Rosebud subbituminous coal, but experimental results were not available at the time of writing. Initial gasification tests were to be at 600 psia, successful completion of which would allow tests at 1000 psia.

Provisions have been made for changing the mixing characteristics of the fluidized bed, hence its temperature gradient, by inserting baffles within the bed, and also for installing shrouds of different diameters to permit different ranges of fluidizing velocity. It is said that the addition to the fluidized bed of materials such as limestone, hydrated lime, and quicklime, can have the desirable effect of increasing both the coal gasification rate and its sintering temperature, and also lowering the sulfur content of the char product.

Problems have been encountered during the commissioning of the pilot plant as a result of ball-valve blockages in the coal feed and char removal lock hopper systems.[106] These valves, together with the gasification zone gas distributor,[117] are two of the major experimental mechanical items of the plant.

The Hygas Processes. There are several Hygas processes designed around the IGT hydrogasifier, which consists of a multistage fluidized-bed system, primarily for high-Btu gas production from caking and noncaking coals. Hydrogenation reactions are encouraged by hydrogasifying the most reactive fraction of the coal at high pressure to produce methane, the residual fraction being used to generate the required hydrogen. In the hydrogasifier the heat of the exothermic coal hydrogenation reactions is used to sustain the endothermic steam-carbon reaction. Thus the steam-carbon reaction assists temperature control of the hydrogasifier and results in additional hydrogen production.

Three alternative methods have been proposed by IGT for producing the gasifying medium for the hydrogasifier: electrothermal gasification, the steam-iron process, and steam-oxygen gasification. Since these three methods produce hydrogenating gases with different hydrogen and carbon monoxide contents, the choice significantly affects the performance of the hydrogasifier section.

Work on forerunners of the Hygas process began at IGT in 1946 and continued under the sole sponsorship of AGA until 1964. During this period much work was done on the fundamentals of reaction kinetics, process yields, and the incorporation of several key process innovations. Studies progressed from the bench-scale batch reactors to small continuous units having throughputs of about 10 lb/hr coal feed. In 1964 the OCR (U.S. Department of the Interior) joined AGA as cosponsors in a 3-year program that ended

[117] J. S. Frith, "Engineering Aspects of the Synthane Coal Gasification Process," paper presented at 2nd Synthetic Fuels from Coal Conf., Oklahoma State University, Stillwater, May 1972.

with a preliminary design for a 80-ton/day Hygas pilot plant. During this period effort was directed toward the testing of various U.S. coals to define suitable gasification conditions, mainly in 100-lb/hr coal units. Construction of a 75-ton/day pilot plant at Chicago was completed in May 1971, the hydrogenating gas being supplied by a natural gas reforming plant. The development of the electrothermal gasifier was suspended in 1974 because of rapidly escalating electricity generating costs, and effort was then shifted to developing the steam-oxygen and steam-iron methods of hydrogen production. Operation of the integrated steam-oxygen process was demonstrated in 1975–1976, during which time the steam-iron facility was constructed on an adjacent site.

A detailed breakdown of the chronology of the development of Hygas processes to date is given in Table 24.25.

In each process alternative the coal is initially crushed and dried simultaneously in a heated roll mill to provide a relatively coarse-sized coal (8/14–100 mesh [2.4/1.4–0.15 mm]). If necessary, the caking tendency of the coal is destroyed by oxidizing the surface of the particles in a fluidized-bed pretreatment reactor. This reactor operates at atmospheric pressure and temperatures of about 800°F (430°C), using a limited air supply as the oxidant, to produce a free-flowing coal. Indirect cooling is employed in this stage to raise high-pressure steam. In a commercial plant the gaseous products of this stage would be combusted to raise high-pressure steam.

The coal, with or without pretreatment, is slurried with hydrogasifier by-product aromatic oil, and pumped to a pressure of 1200 psig prior to feeding to the slurry dryer stage of the hydrogasifier (Fig. 24.40). These slurries can have a solids concentration as high as 45% w/w. A centrifugal pump circulates the slurry at about 50 psig through the suction manifold of a high-pressure reciprocating "mud pump" and the mix tank. The circulation rate is 3–4 times the feed rate to the gasifier, to maintain the coal particles suspended in the low-viscosity aromatic oil throughout the whole system. At present IGT considers this approach to be more effective than lock hoppers, but lock hoppers have not been ruled out for any future commercial plant, provided they can be designed to operate reliably at high pressure (1200 psig). Slurry feed systems have lower energy requirements than lock hoppers, but the latent heat of vaporization of the slurry oil presents a significant heat load to the hydrogasifier.

The hydrogasifier vessel (Fig. 24.40), with four internally connected process stages, is housed within a pressure vessel 132 ft high and 5.5 ft O.D. The coal slurry is sprayed into the slurry dryer stage in which it is fluidized by the gases from the hydrogasifier stage, leaving dry coal at about 570°F (300°C) and under a pressure of 1100 psig.

From a point near the bottom of the drying stage, the solids flow down a 4-in.-diameter pipe to a lift pot, where they mix with hot gases [1700°F (930°C)] from the second hydrogasifier stage. The coal is carried through a 4-in.-I.D. lift line that constitutes the first

Table 24.25 Main Events in the Hygas Development Program

1946	Small-scale development work initiated under AGA sponsorship
1964	OCR and AGA joint sponsorship
1967	OCR contract extended
1968	300-kW Electrothermal unit operational (August)
1971	Construction of 2-MW(e) electrothermal plant started (April); 72 ton/day pilot-plant hydrogasifier completed (May); laboratory-scale tests on steam-oxygen gasification; first coal gasification tests using hydrogen from reformed natural gas (October)
1972	2-MW(e) Electrothermal pilot plant completed (June)
1973	First production of nitrogen-free SNG (June); problem of fines congestion in hydrogasifier solved (October)
1974	Electrothermal program suspended (March); steam-oxygen agglomerating gasifier successfully operated (May)
1975	Test 37, 15-day (final) lignite gasification run
1976	Test 54, 9-day bituminous coal gasification run; Illinois No. 6 coal successfully pretreated to destroy agglomerating properties (June); tests started on Montana Rosebud subbituminous coal (September); steam-iron pilot-plant construction completed (July)

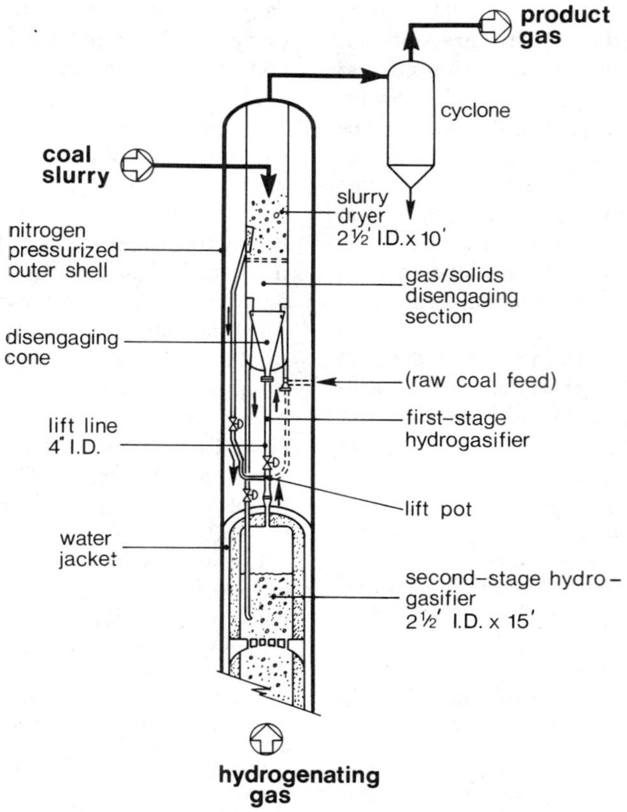

Figure 24.40 The Hygas slurry dryer and first- and second-stage hydrogasifier.[118]

hydrogasifier, where rapid coal hydrogenation occurs in the dilute phase, with the gasification of as much as 20% of the coal feed (within 3–10 s). At the top of the lift line partly reacted char separates from the gas in a conical disengaging section. Two further 4-in.-I.D. transfer pipes are provided near the top of the disengagement cone. The first of these pipes, redundant at present, may be used to recycle disengaged char back to the lift pot as a means of feeding unpretreated caking coals through the slurry dryer to the first hydrogasifier.

The second pipe from the disengaging cone conveys the char down to the second higher temperature hydrogasifier stage, where the char is further gasified in a fluidized bed. Since the hydrogenating gas feed contains hydrogen and steam, both the endothermic steam carbon and exothermic carbon hydrogenation gasification reactions occur. These simultaneous reactions effectively control the temperature of this stage at about 1700°F (925°C). Any tendency for the temperature to exceed this level is counteracted by a corresponding increase in the rate of the steam-carbon reaction.

The three sections that follow treat the different methods of hydrogen production in the Hygas process: the electrothermal, steam-iron, and steam-oxygen methods.

The Electrothermal Method. In the electrothermal gasifier (see also discussion of the electrofluidic process, above), char from the second hydrogasifier stage is further gasified in a bed fluidized by steam. Oxygen is not required by this process, the heat to support the steam-carbon reaction being provided by electrical heating of the bed. In a commercial plant the electric power would be provided by a power generation facility fueled by the residual char from the electrothermal stage.

It was concluded from experiments with a 300-kW unit,[118–122] operating at 1000 psig,

[118] Anon., *HYGAS 1972 to 1974, Pipeline Gas from Coal—Hydrogenation (IGT Hydrogasification Process),* U.S. ERDA R & D Rep. No. 110—Interim Rep. No. 1, OCR Contract No. 14-32-0001-1221, July 1975.

[119] V. J. Kavlick, B. S. Lee, and F. C. Schora,

that bed temperatures in the range 1850–1900°F (1010–1040°C) were optimal. The product gas compositions showed that the CO shift and carbon hydrogenation reactions were near equilibrium in most cases but that the calculated equilibrium temperature for the steam-carbon reaction was 200–300°F (110–165°C) lower than the actual bed temperature. Successful operation of this unit was followed by the installation of a 2.25-MW pilot-plant[118,122] unit adjacent to the Hygas hydrogasifier.

Because of the sharp increase in the cost of electrical power generation, the electrothermal method became the least economically attractive of the three hydrogen-producing methods under development by IGT, and the program was suspended in March 1974.

The Steam-Oxygen Method. A second char-based hydrogen generator under development at IGT uses the steam-oxygen method,[118,122] shown in its integrated form with the hydrogasifier in Fig. 24.41. The hot char from the hydrogasifier is reacted with steam and oxygen in a high-pressure fluidized bed. The reactants are added in a ratio controlled to obtain a suitable gasification temperature, ranging from about 1500°F (815°C)[123] for highly reactive lignite to about 1900°F (1040°C)[122,124] for less reactive bituminous coals and anthracites. The gasification zone temperature must be maintained below the ash-softening point to avoid slagging the ash.

Local ash slagging conditions are avoided by maintaining a sufficiently high fluidizing velocity at the base of the reactor, and using a carefully designed distributor so that rapid

"Electrothermal Char Gasification," in Am. Inst. Chem. Eng. Chem. Eng. Prog. Symp. Ser., Vol. 67, No. 116, 1971, pp. 228–235.

[120] B. S. Lee, E. J. Pyricioch, and F. C. Schora, "Hydrogasification of Coal in High Pressure Fluidized Bed," in Am. Inst. Chem. Eng. Chem. Eng. Prog. Symp. Ser.,Vol. 66, No. 155, 1970, pp. 152–156.

[121] V. J. Kavlick and B. S. Lee, "High Pressure Electrothermal Fluid-Bed Gasification of Coal Char," in Am. Inst. Chem. Eng. Chem. Eng. Prog. Symp. Ser., Vol. 66, No. 105, 1970, pp. 145–151.

[122] Anon., *HYGAS 1964 to 1972 Pipeline Gas from Coal—Hydrogenation,* NTIS Publication No. FE-381-T9-P2, September 1972.

[123] B. S. Lee, "Status of the HYGAS Program," paper presented at 7th Synthetic Pipeline Gas Symp., Chicago October 1975.

[124] L. J. Anastasia, "The HYGAS Process," paper presented at Clean Fuels from Coal Symp., IGT, Chicago, June 1975, pp. 23–27.

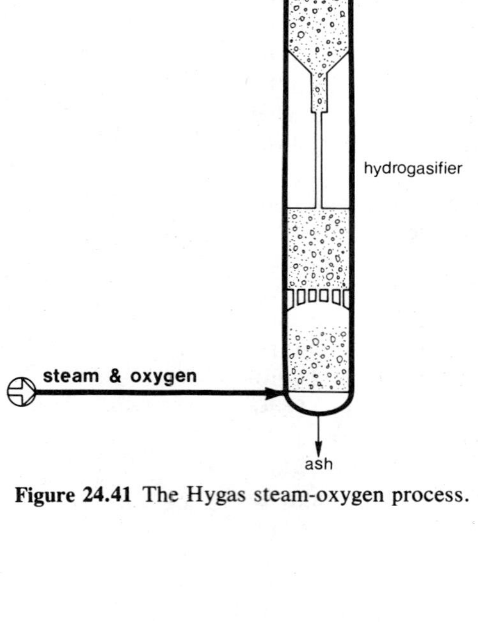

Figure 24.41 The Hygas steam-oxygen process.

initial mixing of the reactants is promoted. Local stagnant areas within the bed may also adversely affect reaction rates and result in combustion of some of the H_2 and CH_4 produced.

The IGT began development of the steam-oxygen gasifier using 2-, 4-, and 6-in.-I.D. high-pressure gasifiers.[122] Results from the last of these[122] showed that for bituminous coal chars an average gasification zone temperature of about 1825°F (995°C) and superficial feed gas velocities of about 0.5 ft/s were appropriate.[125]

The steam-oxygen gasifier was adopted in 1974 as the preferred means of making hydrogen, and design of a steam-oxygen pilot plant facility was initiated. The heat exchange bed beneath the second-stage hydrogasifier (Fig. 24.40) was modified to form the steam-oxygen gasifier.[118] This resulted in a completely integrated Hygas steam-oxygen process contained within one pressure vessel, albeit of a design compromised by the limited space available in the vessel.[126] Also modifications were required[127] to allow a greater residence time in the second hydrogasification stage, to compensate for the reduced hydrogenation rate that results from the lower hydrogen concentrations in the gas from the steam-oxygen gasifier.

Fully integrated operation of the steam-oxygen gasifier and hydrogasifier stages was first demonstrated in April 1975 (Hygas Test 34), during which the plant operated for 20 hr without any external hydrogen supply.[123] One of the most significant Hygas tests reported to date[123] is Test 37, a run of 15 days duration in which 1000 tons of Montana lignite were gasified (see Fig. 24.42; Table 24.26).

Subsequently an Illinois No. 6 bituminous coal[128] requiring oxidative pretreatment was tested. Successful pretreatment without agglomeration problems was achieved in Test 54 by reducing the maximum particle size of the coal feed to 14 mesh (1.4 mm). Figure 24.43 shows the arrangement for Test 54, in which the objective was to demonstrate the coal feed system rather than to maximize carbon conversion; results are given in Table 24.26. In subsequent tests Montana Rosebud subbituminous coal has been handled,[128] although performance data are not yet available.

[125] B. S. Lee, P. B. Tarman, and K. C. Youngblut, "Status of HYGAS, Electrothermal Gasification and Steam-Oxygen Gasification Programs," paper presented at 4th Synthetic Pipeline Gas Symp., Chicago, October 1972.

[126] B. S. Lee and P. B. Tarman, "Status of the HYGAS Program," paper presented at 6th Synthetic Pipeline Gas Symp., Chicago, October 1974.

[127] B. S. Lee, "Status of HYGAS Process—Operating Results," paper presented at 5th Synthetic Pipeline Gas Symp., Chicago, October 1973.

[128] B. S. Lee, "Status of the HYGAS Program," paper presented at 8th Synthetic Pipeline Gas Symp., Chicago, October 1976.

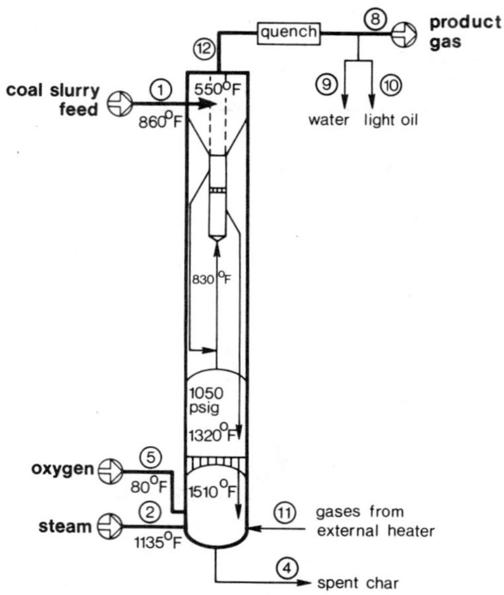

Figure 24.42 Apparatus used for Hygas Test 37 (Montana lignite).[123] Results appear in Table 24.26.

Table 24.26 **Results of Hygas Tests 54 and 37 (Figs. 24.42 and 24.43)**

	Test 54 on Illinois No. 6 Coal					
	Stream 1: Feed Slurry	Stream 2: Steam	Stream 4: Spent Char	Stream 5: Oxygen	Stream 8: Product Gas	Stream 9: Water
Composition of solids (% w/w)						
Carbon	74.11		76.78			
Hydrogen	3.33		1.28			
Oxygen	8.40		1.39			
Nitrogen	1.84		0.60			
Sulfur	2.15		0.62			
Ash	10.17		19.33			
Total (lb/hr, dry)	5,912		2,735			
Total (lb/hr, incl. moisture)	6,100		2,735			
Light oil (lb/hr)	11,388					
Composition of gases (% v/v)						
H_2					26.12	
CO					10.02	
CO_2					27.57	
CH_4					28.14	
C_2H_6					0.40	
H_2S					0.69	
N_2					7.06	
O_2				100		
H_20		100				100
Total (lb-mole/hr)		343		26	137	378

	Test 37 on Montana Lignite						
Stream 10: Light Oil	Stream 1: Feed Slurry	Stream 2: Steam	Stream 4: Spent Char	Stream 5: Oxygen	Stream 8: Product Gas	Stream 9: Water	Stream 10: Light Oil
	61.70		54.30				
	4.58		0.70				
	20.40		3.25				
	0.96		0.27				
	0.75		0.11				
	11.61		41.37				
	5,180		1,274				
	5,730		1,340				
11,993	11,013						11,136
					33.40		
					9.35		
					32.98		
					17.40		
					0.34		
					0.24		
					6.29		
				100			
		100				100	
		314		24	217	256	

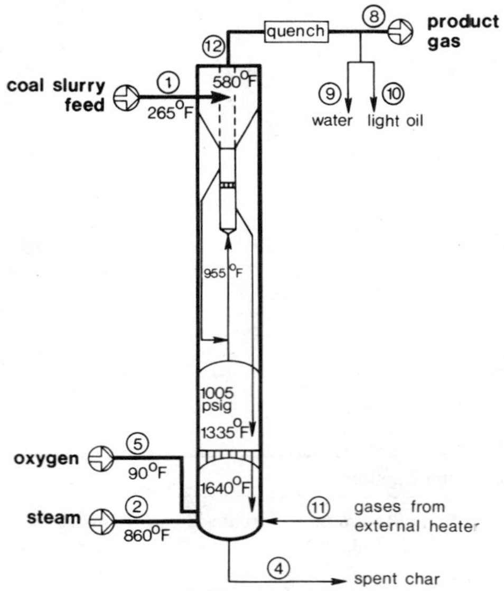

Figure 24.43 Apparatus used for Hygas Test 54 (Illinois No. 6).[128] Results appear in Table 24.26.

To improve overall carbon conversion, IGT has been concurrently developing a low-pressure agglomerating ash steam-oxygen gasifier operating at about 1900°F (1040°C).[118] The gasifier has operated on coke breeze and produced dense agglomerates of size range 1/8–1/4 in.[126] with 5–31% w/w carbon content. However blockage problems encountered in the venturi discharge line[118] appear to have delayed development.

The Steam-Iron Process. The production of hydrogen by the reaction of steam with iron using two fixed beds of iron-containing solids operated cyclically is a very old process that has been largely abandoned. In 1961 IGT started work on a new version of this method of hydrogen production. Instead of cyclically operating fixed beds, an arrangement of fluidized beds with continuous transfer of iron-containing solids between them is used; it operates at pressure so that the hydrogenating gas produced can pass directly into the Hygas hydrogasifier reactor. Although the method was developed primarily as a part of the Hygas process, it could be applied to any process requiring hydrogen.

The conceptual steam-iron process for hydrogen production[129] (Fig. 24.44) consists of three main stages: the producer, the reducer, and the oxidizer. Fuel (hydrogasifier char in the integrated steam-iron process) is fed to the producer vessel, where it reacts with air and steam at temperatures in the range 1900–2000°F (1040–1095°C) to generate a reducing gas with a ratio of ($CO + H_2$) to ($CO_2 + H_2O$) of about 4 : 1.[131] This high ratio should ensure efficient use of the producer gas in the reducer stage. In the reducer this gas converts iron oxides to iron, according to Equations 24.13–24.16.

$$Fe_3O_4 + H_2 \rightarrow 3FeO + H_2O \qquad (24.13)$$

$$Fe_3O_4 + CO \rightarrow 3FeO + CO_2 \qquad (24.14)$$

$$FeO + H_2 \rightarrow Fe + H_2O \qquad (24.15)$$

$$FeO + CO \rightarrow Fe + CO_2 \qquad (24.16)$$

The hot reduced solids are then fed to the oxidizer vessel where, in contact with steam, Reactions 24.13 and 24.15 are reversed to give iron oxide, thereby producing a hydrogen-steam hydrogenating gas. About 5–15%[118] of the iron in the circulating solids stream has to be converted to provide a steam conversion of about 45%, as required for the Hygas process.

The reducer and oxidizer will each consist of two fluidized beds to give higher conversions in each stage and improved utilization of the reducing gas. Figure 24.45 shows schematically the relative positions of these beds. Locating the reducer above the oxidizer allows use of the spent reducing gases to convey the solids to the top of the reactor and leads to a favorable pressure balance. Power recovered by expanding the spent reducer gas may be used to compress the air feed to the producer vessel.

A steam-iron pilot plant for the production of 1.1 million scf/day of hydrogen has been constructed adjacent to the Hygas hydrogasifier pilot plant. Commissioning was started in late 1976.[132] Because of the requirement of independent operation of the steam-iron plant, it has both its own char slurry sys-

[129] F. C. Schora, B. S. Lee, and J. Huebler, "The HYGAS Process," Paper No. IGU/B-73, presented at IGU 12th World Gas Conf., Nice, France, 1973, pp. 1–20.

[130] B. S. Lee, "Development of the Hygas Process for Converting Coal to Synthetic Pipeline Gas," *J. Pet. Technol.*, 1407–1410 (December 1972).

[131] Anon., *Development of the Steam-Iron System for Production of Hydrogen for the HYGAS Process,* OCR R & D Rep. No. 95—Interim Rep. No. 1, OCR Contract No. 14-32-0001-1518, June 1974.

[132] P. B. Tarman, "Status of the Steam-Iron Program," paper presented at 8th Synthetic Pipeline Gas Symp., Chicago, October 1976.

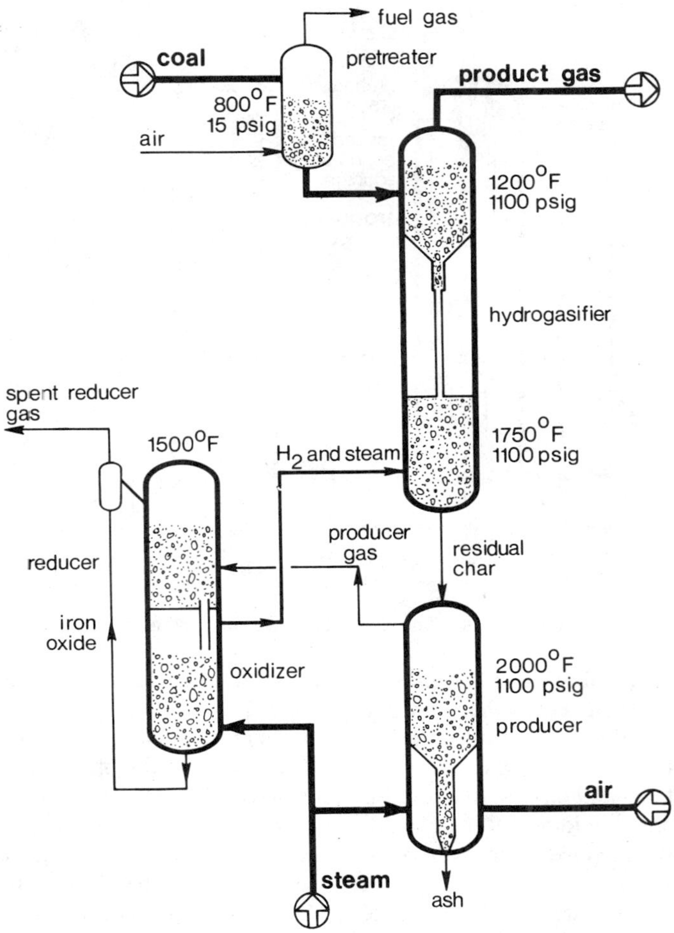

Figure 24.44 Schematic flow diagram of the Hygas steam-iron process.[130]

tem[133] and an additional air-steam-blown fluidized bed situated above the main producer vessel (Fig. 24.45) for char preheating.

Although the IGT steam-iron process has not yet been demonstrated in its integrated form, separate laboratory-scale studies have been made[131] on the producer gas stage and on the fluidizing characteristics of the solids. Producer stage tests on a variety of cokes and chars have resulted in a Husky lignite char being selected for initial pilot-plant runs. This char is about 7 times more reactive than a bituminous char and should allow operation at a bed temperature of about 1900°F (1040°C). A fluidization velocity of 0.8 ft/s has been selected[131] for this stage. Small-scale tests have also been made on the slurry feed system. Table 24.20 shows conceptual performance data for the Hygas steam-iron process; a corresponding mass balance can be seen in Fig. 24.46 and Table 24.27.

The Hydrane Process. The USBM Hydrane process has been designed to maximize direct coal hydrogenation by employing a high-pressure (1000 psig) two-stage hydrogasifier fed with essentially pure hydrogen.

The first hydrogasifier stage consists of a free-fall dilute phase (FDP) reactor operating in the temperature range 1400–1650°F (760–900°C). The coal falls at about 6 ft/s[135] in a cocurrent flow of hydrogen and methane, to produce very porous reactive char particles. In this stage any caking tendency of the coal is lost, and about 20% of the carbon in the feed coal is converted to methane.[136]

[133] Anon., "The Steam-Iron Process, IGT Updates an Old Hydrogen Production Method," *IGT Gas Scope,* No. 34, Winter 1975–1976.

[134] J. W. Loeding, Institute of Gas Technology, private communication, November 1976.

[135] H. F. Feldmann and P. M. Yavorsky, "The Hydrane Process," paper presented at 5th Synthetic Pipeline Gas Symp., Chicago, October 1973.

[136] J. A. Gray and P. M. Yavorsky, "The Hydrane

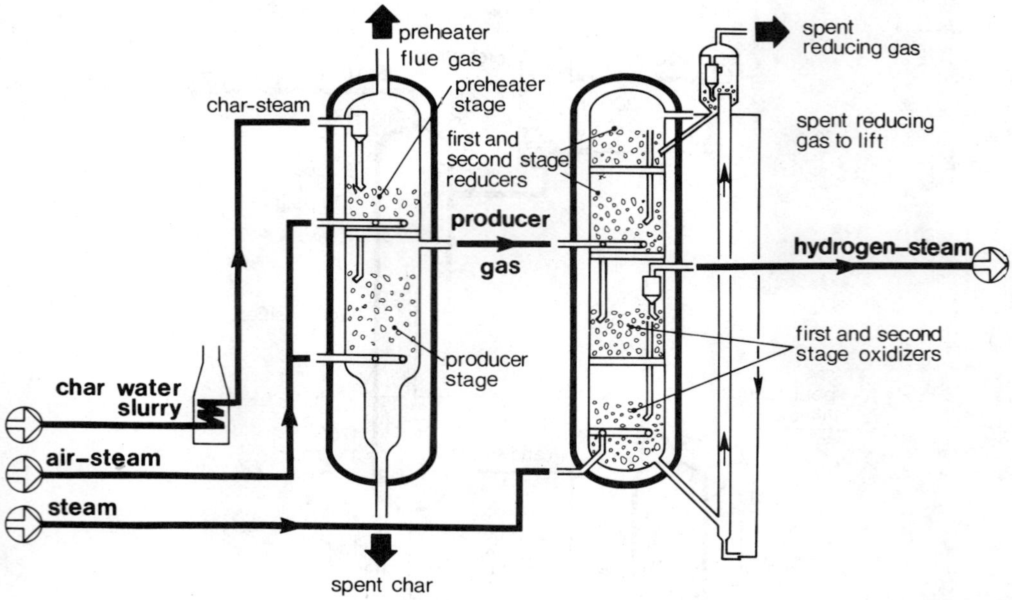

Figure 24.45 The IGT steam-iron pilot plant.[129]

In the second stage about another 25% of the carbon is hydrogenated in the fluidized bed, operating in the temperature range 1650–1800°F (900–980°C). The resultant char is reacted with steam and oxygen in a further fluidized-bed gasifier, yielding, after CO shift and acid gas removal stages, the hydrogenating gas requirement.

It is claimed[136] that more than 90% of the methane content of the final SNG can be produced by direct hydrogenation within the Hydrane gasifier. The hydrogasifier operates with a dry hydrogen feed, so that exothermic hydrogenation reactions predominate. This stage therefore requires considerable cooling, to prevent very high outlet temperatures and possible methane cracking. Temperature control in the fluidized bed may be accomplished by a combination of hydrogen preheat temperature adjustment and heat removal using coils immersed in the bed.[137]

The method of steam-oxygen char gasification has not yet been selected, but it should be capable of operating at 1000 psig to produce a low-carbon ash: the IGT steam-oxygen agglomerating ash process may be suitable.[135]

Integrated operation of the two hydrogasifier stages of the Hydrane process has not

Process," Symp. Clean Fuels from Coal, IGT, Chicago, June 1975.

[137] H. F. Feldmann, C. Y. Wen, W. H. Simons, and P. M. Yavorsky, "Supplemental Pipeline Gas from Coal by the Hydrane Process," paper presented at 71st National Meeting, American Institute of Chemical Engineers, Dallas, Tex., February 1972.

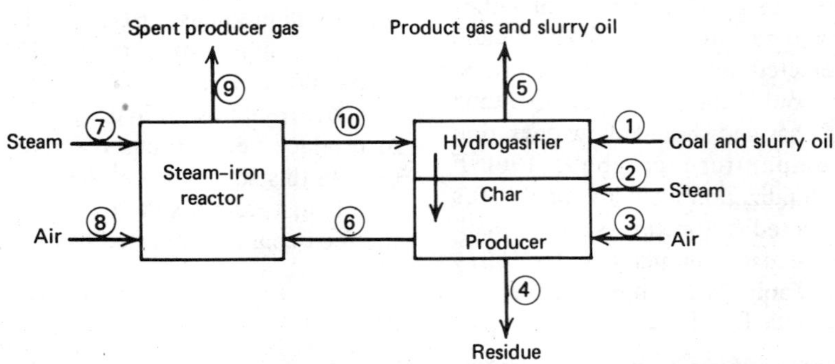

Figure 24.46 The Hygas steam-iron process[134]; for conceptual mass balance Table 24.27.

Table 24.27 Mass Balance Data for the Hygas Steam-Iron Process (see Fig. 24.46)[134]

	Stream 1: Coal and Oil Slurry	Stream 2: Steam	Stream 3: Air	Stream 4: Residue	Stream 5: Product Gas	Stream 6: Producer Gas	Stream 7: Steam	Stream 8: Air	Stream 9: Spent Producer Gas	Stream 10: Hydrogen
Temperature (°F)	270	1050	360	1950	600	1950	570	300	1660	1650
Gas composition (mole/hr)										
CO	—	—	—	—	0.107	0.423	—	—	0.085	—
CO_2	—	—	—	—	0.067	0.106	—	—	0.444	—
H_2	—	—	—	—	0.362	0.243	—	—	0.065	0.409
H_2O	—	0.362	0.01	—	1.397	0.121	1.774	0.003	0.511	1.154
CH_4	—	—	—	—	0.182	0.0175	—	—	0.0175	—
C_2H_6	—	—	—	—	0.017	—	—	—	—	—
NH_3	—	—	—	—	0.007	—	—	—	—	—
H_2S	—	—	—	—	0.0028	0.0007	—	—	0.0007	—
N_2	—	—	—	—	0.009	0.741	—	—	0.943	0.009
Air	—	—	0.938	—	—	—	—	0.256	—	—
Coal and oil compositions (mole/hr)										
Carbon	1.0			0.029	0.035[a]					
Hydrogen	0.406			0.002	0.019[a]					
Oxygen	0.205			—	0.0001[a]					
Nitrogen	0.0054			0.002	—					
Sulfur	0.0036			0.0001	—					
Ash[b]	1.37			1.37	—					
Moisture	0.278			—	—					
Oil[b]	22.744			—	22.744					

[a] Make oil.

[b] In pounds per hour.

yet been demonstrated, since effort has been concentrated on the critical FDP hydrogasifier. These laboratory-scale tests have demonstrated the FDP stage at high pressures[138] (500–3000 psig) in a bench-scale reactor 3 1/4 in. I.D., 5 ft long[136] and using a lock hopper system to feed coal at rates of 5–18 lb/hr. The coal is partially hydrogasified in an equimolar mixture of hydrogen and methane (~ 150 scf/hr) at temperatures in the range 1470–1650°F (800–900°C). It was found that a free-fall distance of 5 ft was sufficient to destroy the caking properties of the coal. The yield of organic liquids was found to decrease as gasifier pressure increased, typically being equivalent to 1–6% of the carbon in the feed. Table 24.28 shows typical results for these tests.

The fluidized-bed hydrogasifier stage has been tested in a very small scale apparatus.[136] A batch charge of 0.02 lb of FDP-stage char product in a reactor 5/16 in. I.D., 32 in. long, was gasified at 100 psig in a stream of hydrogen flowing at 1.3 scf/hr. At 1650°F (900°C) a 52% carbon conversion was obtained within 2 min.

More recently the FDP and fluid bed hydrogasifier stages (in 3 1/4-in.-I.D. reactors) have been operated in combination[136] with char flowing from the first to the second. However complete integration of stream flows between these two stages was not achieved in these experiments, since the gas feed to the FDP stage was provided externally. The methane content of the fluidized-bed stage product gas was lower than desired, because an excessive amount of hydrogen was required to obtain satisfactory bed fluidization. More successful results were achieved when the second stage was replaced by a fixed-bed reactor. This modification allowed lower hydrogen feed flow rates, resulting in greater methane concentrations (up to 46%) in the product gas from this stage. Typical results from these tests (Table 24.29) show that lower methane yields were obtained, as well as levels of oil production higher than were required for balanced operation.

[138] H. F. Feldmann, J. A. Mina, and P. M. Yavorsky, "Pressurized Hydrogasification of Raw Coal in a Dilute Phase Reactor," *Prepr., Am. Chem. Soc., Div. Fuel Chem.*, **18**(1), 182–202 (April 1973).

Table 24.28 Laboratory-Scale Results for the Free-Fall Dilute Phase Stage of the Hydrane Process [136]

	Test 1	Test 2	Test 3	Test 4	Test 5	Test 6
Pressure (psig):	1,000	1,200	1,500	2,000	1,200	1,000
Fuel:	hvAb[a]	hvAb[a]	hvAb[a]	hvAb[a]	hvCb[b]	Lignite[c]
Feed gas/coal ratio (scf/lb):	12.3	12.3	13.1	12.4	14.0	12.3
Feed gas composition, dry (% v/v)						
H_2	49.0	49.2	53.8	49.9	50.1	50.1
CH_4	49.4	48.7	43.4	48.4	47.7	47.8
N_2	1.6	2.1	2.6	1.7	2.2	1.8
CO	0.0	0.0	0.0	0.0	0.0	0.2
CO_2	0.0	0.0	0.2	0.0	0.0	0.0
Carbon conversion (% w/w)	25.0	25.6	24.2	30.0	27.8	32.1
Product gas yield (scf/lb)	13.4	13.3	13.5	13.9	16.9	10.6
Product gas composition, dry (% v/v)						
H_2	22.4	22.7	19.7	18.1	21.9	27.9
CH_4	71.4	72.2	75.2	79.0	72.8	57.5
C_2H_6	0.5	0.1	0.3	0.1	Trace	0.1
CO	3.2	2.3	1.4	0.5	2.4	6.3
CO_2	0.7	0.5	0.8	0.4	0.7	5.9
N_2	1.4	2.3	2.4	1.7	1.9	2.1
H_2S	0.4	0.2	0.1	0.2	0.3	0.1
Crude product gas, HHV (Btu/scf)	817	815	835	863	818	695
CH_4 produced in gasifier (%)[d]	90.8	91.8	92.8	94.2	92.3	86.8

[a] Pittsburgh Seam hvAb.
[b] Illinois hvCb.
[c] North Dakota lignite.
[d] See Appendix.

Table 24.29 Hydrane Laboratory-Scale Results for Hydrogasification of Illinois No. 6 Coal in a Two-Stage Reactor at 1000 psig[136]

	Test 1		Test 2		Test 3	
Reactor zone[a]:	FDP	FB	FDP	FB	FDP	FB
Temperature (°F):	1,560	1,260	1,560	1,480	1,560	1,320
Coal or char rate (lb, dry/hr):	10.51	6.68	10.26	5.08	10.32	5.01
Bed height (in.):	—	0	—	36	—	36
Residence time (min):	—	0	—	10.4	—	10.4
Feed gas rate (scf/hr):	164.4	141.4	181.7	152.0	166.2	150.7
Feed gas composition, dry (% v/v)						
H_2	56.2	99.4	52.0	99.0	50.9	98.6
CH_4	37.2	—	42.1	—	42.8	—
N_2	1.05	0.50	1.10	1.00	1.50	1.30
He	5.45	—	4.70	—	4.70	—
Product gas rate (scf/hr)	168.6	141.4	179.0	124.6	169.8	130.3
Product gas composition, dry (% v/v)						
H_2	34.8	99.4	32.4	54.2	30.1	58.0
CH_4	55.1	—	57.2	43.5	58.0	39.0
Run time (min)	187		193		187	
Solids conversion (% w/w)						
Coal, maf	43.1		60.2		60.4	
Carbon	33.0		50.7		53.6	
Hydrogen	75.4		96.4		93.4	
Sulfur	66.7		74.8		76.3	
Nitrogen	59.4		89.7		86.4	
Oxygen	91.0		99.6		90.0	
Gas yields (scf/lb dry coal)						
CH_4	3.01		7.80		7.58	
C_2H_6	0.11		0.17		0.13	
CO	0.43		0.69		0.80	
CO_2	0.10		0.09		0.11	
H_2	−3.21		−11.63		−10.31	
H_2S	0.04		0.03		0.07	
Oil yield (lb/lb dry coal)	0.048		0.041		0.026	
Mean char particle diameter (mm)	0.433		—		0.397	

[a] FDP = free-fall dilute phase reactor, FB = fixed-bed reactor.

A feasibility study of the Hydrane process has been completed by the Dravo Corporation. A conceptual commercial Hydrane hydrogasifier reactor for producing 125 million scf/day of SNG has multiple FDP reactors feeding a single fluidized-bed char hydrogasifier. This arrangement reduces the specific solids rate (lb/hr-ft^2 reactor cross-sectional area) in the FDP zone to minimize particle agglomeration.

The estimated gas composition for a commercial Hydrane gasifier is shown in Table 24.20.

Commentary on Hybrid Fluidized-Bed Processes Yielding a Nitrogen-Free Product Gas. The processes described employ special contacting methods not found in classical fluidized beds and produce a relatively nitrogen-free product gas intended for upgrading to SNG. As a group they exemplify most of the potential virtues of fluidized-bed coal gasifiers.

In the Battelle–Union Carbide agglomerating ash process, air can be used as the oxidant in the combustor, without introducing nitrogen into the product gas (cf. CO_2 acceptor and Exxon two-stage processes). This feature eliminates the need for an air separation plant and reduces the CO_2 removal requirement when upgrading the product gas to SNG. If commercialized, the process will probably operate at a higher pressure than the pilot plant (100 psig), to increase methane production (so reducing the air requirement) and decrease the sizes of the gasifier and upgrading section. The energy required for air compression is considerable. especially at higher operating pressures, making energy recovery from the combustor products of prime importance. It is said that the agglomerating action in the burner vessel will reduce fly ash carryover in the combustion products and it is hoped to pass this tar- and oil-free gas direct to power recovery turbines without further treatment. Removal of agglomerated ash from the combustor has yet to be demostrated.

The process is unusual in requiring different size ranges of coal for the gasifier and combustor stages. Considerable advantage will accrue if by virtue of the dilution effect of the circulating agglomerates, caking coals can be handled without pretreatment. The solids circulation system is relatively complex and may prove troublesome to operate. Furthermore, the ability to transfer adequate heat from the combustor to the gasifier requires demonstration.

Special features of the Synthane process include the countercurrent free-fall devolatilization zones (contrast the Hydrane process) and the provision for introducing large temperature gradients (e.g., 400°F) within the fluidized bed by the use of baffles. These process features are largely responsible for the high methane formation within the gasifier. This has been shown to range between 48 and 56% v/v of the total methane content of the SNG with various coals at 600 psig (Table 24.24) and is expected to be about 60% at 1000 psig operating pressure. However, the free-fall dilute phase stage does not eliminate the need for pretreatment of caking coals (which tends to reduce direct methane production; cf. Hydrane process).

Conceptual performance data for the gasification of Rosebud coal in the Synthane pilot plant[139] (Table 24.20) and experimental data from the smaller plant (Table 24.24) are in reasonable agreement, apart from the ratios of steam to oxygen. The projected gasification efficiencies [E_1 = 57%, E_2 = 63%, excluding the pretreatment stage (see Appendix)] for the conceptual commercial gasifier are surprisingly low. It appears that the advantages of high methane formation within the gasifier have been largely lost because of the relatively low carbon conversion (75% projected), presumably stemming from limitations on the maximum fluidized-bed temperature (~ 2000°F) (cf. Hydrane process).

The Hygas processes employ a special arrangement for the first-stage hydrogasifier in which the coal is suspended in a high-velocity, upward-flowing, dilute phase bed. Each of the three process alternatives (electrothermal, steam-oxygen, and steam-iron) has potential technical merit.

These three processes produce hydrogenating gases of different compositions (Table 24.30). The hydrogen content is important because it determines the hydrogen partial pressure in the second hydrogasifier stage, high values of which favor rapid methane formation. The electrothermal process is seen to give the most favorable hydrogenating gas on this account.

Table 24.30 also gives the estimated product gas compositions for the three Hygas configurations and shows that the Hygas steam-iron process has the lowest CO_2 content.

The Hygas electrothermal process does not need either oxygen or air supplies to the gasifier. This feature eliminates the need for oxygen or air compressors, required by the other two Hygas processes, and it results in a product gas with a lower CO_2 content. However the penalty of high costs associated with the generation of the required electrical power currently outweighs these advantages.

The Hygas steam-iron process also avoids the use of oxygen, but it does require an air compressor for the producer stage. At the operating pressure of 1100–1200 psig the energy requirement for air compression is very high, and the—albeit expensive—recovery of energy from the spent producer gas is essential if the potentially high thermal efficiency of the process is to be realized. The high ratio of hydrogen to carbon oxides of the product gas means that in producing SNG, little CO_2 removal is necessary (since most of the CO_2 is vented in the spent producer gas). However the process with its numerous separate fluidized beds linked by gas and solid transfer systems is very complex. The steam-iron process itself has not yet been demonstrated, and there may be problems caused by iron and iron oxide particle cohesion. Therefore although the Hygas

[139] A. J. Forney, Pittsburgh Energy Research Center (ERDA), private communication, November 1976.

Table 24.30 Conceptual Gas Compositions for the Hygas Processes

	Electrothermal[129]		Steam-Oxygen[129]		Steam-Iron[134]	
Stage[a]:	1	2	1	2	1	2
Gas composition, dry (% v/v)						
CO	18.7	21.3	15.0	18.0	0.0	5.0
CO_2	10.5	14.4	15.0	18.5	0.0	3.1
H_2	32.2	24.2	23.3	22.8	26.0	16.8
H_2O	33.6	17.1	43.2	24.4	73.4	65.0
CH_4	4.8	19.9	3.4	14.1	—	8.5
C_2H_6	—	0.8	—	0.5	—	0.8
H_2S	0.2	1.3	0.1	0.9	—	0.1
Trace components	—	1.0	—	0.8	0.6	0.7
	100.0	100.0	100.0	100.0	100.0	100.0

[a] 1 = gas leaving hydrogen production stage, 2 = gas leaving hydrogasifier.

steam-iron process may offer the highest overall thermal efficiency of the three options,[129] it has the highest technical risk and requires the most development effort.

The integrated Hygas steam-oxygen process, which has been demonstrated already at pilot-plant scale for both Montana lignite and Illinois coals, is the most likely first Hygas variant to be commercialized. Conceptual commercial performance data for a Western subbituminous coal are shown in Table 24.20. Table 24.31 compares these data with those for Tests 37 and 54, showing that for Western subbituminous coal[134] the projected oxygen requirement is higher than that for the fuels employed in the Montana and Illinois coal tests.

The conceptual and experimental data exhibit carbon conversions, gas yields, and gasification efficiencies (E_1, E_2) that are all comparatively low (Table 24.31). These result from the high carbon content of the rejected char, high carryover of fines (representing about 8% of the coal feed for Test 54), and the appreciable light oils production (accounting for about 15% of the HHV of the coal feed for Test 54). The estimates of E_1 and E_2 exclude the pretreatment stage but include product gas losses associated with the hydrogasifier product gas quench system, part of which may be recoverable.

The carbon conversion attainable is strongly dependent on the temperature of the steam-oxygen fluidized bed. The process has not yet been demonstrated at the bed temperatures proposed for the commercial-scale process, and agglomeration problems may be encountered under these conditions. The fines entrainment problem has been increased by dispensing with the slurry dryer stage freeboard section, a development modification made to overcome the problems due to the accumulation of fines, encountered during early runs.[126] Light oils production during Test 54 was about fourfold that for Test 37 and accounts in part for its lower efficiency (E_1).

In the Hydrane process coal hydrogasification is maximized by employing an FDP stage in addition to a conventional fluidized bed. The use of pure hydrogen should result in high direct methane production within the gasifier, and it is claimed that the caking properties of the coal are destroyed during the FDP stage, eliminating the need for an oxidative coal pretreatment stage (cf. Synthane process). This situation will allow more of the volatile components of the coal to be directly hydrogenated to methane. The method to be used to produce the required hydrogen from the hydrogasifier char has not yet been indicated. The high direct methane formation (~ 90%) should lead to lower overall oxygen (or air) requirements than for most other processes (with a corresponding reduction in CO_2 production), and the process has the potential of attaining a higher overall thermal efficiency than many of its competitors, if the hydrogen requirement can be met efficiently.

Heat transfer problems may be encountered in the hydrogasifiers, particularly the FDP stage, because of the highly exothermic hydrogasification reactions. The limited flow of gasifying agent may also make it difficult to maintain satisfactory fluidization in the

Table 24.31 Comparison of Conceptual and Pilot-Plant Data for the Hygas Steam-Oxygen Process for Various Coal Feedstocks

	Western[134] Subbituminous (Conceptual)	Montana[132] Lignite (Test 37)	Illinois[128] No. 6 (Test 54)
Steam-oxygen fluidized-bed temperature (°F)	1850	1510	1640
Carbon conversion (%)[a]	—	49.3	25.0
Gas yield (scf/lb coal, as-received)	12.5	14.2	10.2
Steam/coal (lb/lb)	1.10	0.84	1.27
Oxygen/coal (lb/lb)	0.25	0.15	0.16
Gasification efficiency (%)[a]			
E_1[b]	47.0	~ 48.6	~ 28.7
E_2[b]	—	~ 52.7	~ 43.5

[a] See Appendix.
[b] Estimated values.

second (fluidized-bed) stage. Also, two separate acid gas removal trains are required by the process, together with extensive CO shift conversion in the hydrogen production stage.

The FDP stage has been operated only on a very small scale, and the feasibility of the integrated process has not yet been demonstrated. Difficulties may also be encountered in obtaining sufficient carbon hydrogenation for balanced operation (i.e., at acceptable overall carbon conversions).

24.3.3.2 *Hybrid Fluidized-Bed Processes Yielding a Nitrogen-Containing Product Gas*

The Winkler Process. The commercial Winkler process,[140] offered by Davy Powergas GmbH, involves a combination of fluidized-bed and partial entrainment gasification.[141] It is, however, a single-stage process in the sense that these gasification zones are in a single chamber. Coal at atmospheric pressure is fluidized by air, oxygen-enriched air, or oxygen and steam. The larger particles are gasified in the fluidized state, while smaller particles in the feed or produced in the bed are entrained and gasified in suspension by injecting further gasifying agent above the fluidized bed to give a gas that is free of tar and oils.

The first commercially operated Winkler plant was built at Leuna, Germany, in 1926. Since then there has been continuous development in the design of the gasifier and waste heat boiler, but the process principles have remained unchanged.

The Winkler gasifier consists of a tall cylindrical vessel (Fig. 24.47) with a lower conical section containing the fluidized bed. The upper section may be 6–10 times the fluidized-bed height. The type of refractory lining used is determined by the properties of the ash present in the feedstock, but a refractory that is resistant to erosion is necessary in the lower part of the gasifier. The carbon steel shell is not cooled.

Coal, of 0–3/8 in. size, is fed to the fluidized bed by means of a screw conveyor. The coal does not have to be dried if it is free from surface moisture and flows freely. It is probable that with caking coals a pneumatic system will be tried in an attempt to overcome blockage problems encountered with the present system. It is said that this will allow gasification of caking coals without bed agglomeration problems, owing to the rapid mixing and dilution of feed coal in the ash-rich fluidized bed.

Oxygen (or air) and steam are injected into the Winkler fluidized bed by means of several rings of nozzles at various heights. The flow rates and compositions of the gasifying media are adjusted such that the temperature in the

[140] K. K. Sud, Davy Powergas GmbH, private communication, October 1976.
[141] H. H. Lowry, Ed., *Chemistry of Coal Utilization,* Supplementary Volume, John Wiley and Sons, New York, 1963, pp. 950, 965–966, 990, 1015, 1087.

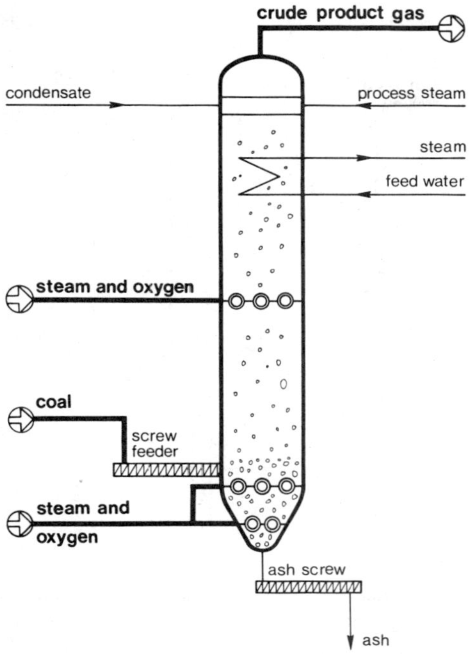

Figure 24.47 The Winkler gasifier.

fluidized bed does not exceed the softening point of the ash. The larger ash-rich particles, being more dense than the coal particles, sink toward the bottom of the fluidized bed and are removed by a screw conveyor. Approximately 30% of the ash leaves from the bottom of the bed, and 70% is entrained in the product-gas stream.

To increase the gasification efficiency and to enable low-reactivity coals to be gasified, the entrained zone of the reactor is operated at temperatures very close to the ash melting point. A radiant boiler is installed near the top of the gasifier in the form of water-cooled tubes along the inner walls, to resolidify any particles of ash that may have become partially molten before they reach the gas offtake. The operating temperatures of the fluidized bed [1470–1830°F (800–1000°C)] and the entrainment zone [1830–2190°F (1000–1200°C)], depend on the reactivity of the feedstock and the ash fusion properties. The product gas is cooled by the radiant boiler and leaves the gasifier some 350–400°F (195–220°C) below the temperature of the entrainment section.

The nature of the coal gasified significantly affects the steam consumption. For an oxygen-blown gasifier the steam/coal weight ratio increases from 0.14 for lignite to 0.61 for subbituminous coal. The HHVs of the gas produced in oxygen-blown and air-blown Winkler gasifiers are 280–300 and 120–125 Btu/scf, respectively. The process can handle a wide spectrum of coals ranging from brown coal to moderately caking coals. It cannot handle coals that have both a very low reactivity and a very low ash melting point [< 1472°F (800°C)]. Some such fuels can, however, be gasified by subsequent developments of the Winkler process, notably the Flesch-Winkler and BASF–Flesch Demag processes.[142]

Performance data for a commercial Winkler plant gasifying lignite are given in Table 24.20.

The minimum and maximum throughputs of a gasifier are limited by the smallest flow required for fluidization and the minimum residence time for gasification, respectively. The quantity of gasifying agent injected into the zone above the bed must be accurately metered in proportion to the amount of unreacted carbon being carried over. If too little is injected, ungasified carbon will be rejected from the gasifier, resulting in a lower gasification efficiency; if too much, potential product gas is unnecessarily combusted.

Prior to 1976, 36 Winkler gasifiers were installed at 16 locations. Currently there are six gasifiers at three installations (Gorzade, Yugoslavia; Kutahya, Turkey; and Madras, India) which are still operating to produce synthesis gas for ammonia production. The largest Winkler gasifier built has an inside diameter of 18 ft and stands approximately 100 ft high. Operated at atmospheric pressure, this air-blown gasifier produces 89.3 million scf/day of low-Btu gas, consuming 38.5 tons/hr of coal. Davy Powergas has completed the process and equipment design work for a pressurized (44 psig) Winkler demonstration plant, which may result in improved gasification efficiency. Gas produced in Winkler gasifiers has been used in low-Btu gas applications, as a synthesis gas for the Fischer-Tropsch process, for ammonia and methanol production, and to produce

[142] H. H. Lowry, Ed., *Chemistry of Coal Utilization,* Supplementary Volume, John Wiley and Sons, New York, 1963, pp. 950, 966.

hydrogen for hydrogenating low-temperature tars and shale oil.

The IGT U-Gas Process. The IGT U-Gas process employs a single-stage, fluidized-bed gasifier to produce low-Btu (150–300 Btu/scf) fuel gas from coal. Crushed, partly dried coal is contacted with steam and air (or oxygen), at 100–350 psig and 1900°F (1040°C). The U-Gas process uses an ash agglomeration technique that continuously and selectively removes a low-carbon ash from the bed, under nonslagging conditions. This technique overcomes the problem of low carbon conversion often associated with fluidized-bed processes. The process is designed to accommodate all coal types, although caking coals require pretreatment to avoid problems in the gasifier.

In this process coal is crushed to 1/4-in. size before being fed to the gasifier, which can accept up to 10% coal fines less than 200 mesh (0.07 mm). When required, pretreatment consists of contacting the coal—introduced through a lock hopper—with air, in a fluidized bed at a temperature of 700–800°F (370–425°C). During this process steam in excess of process requirements is raised in tubes immersed in the bed. Oxidation of the outer surface of the coal particles prevents coal from caking during gasification.

The pretreated coal flows into the gasifier (Fig. 24.48), and the pretreater off-gases, after passing through a cyclone, are introduced into the bottom of the gasifier. It is said that this design will ensure that the tars and oils evolved during pretreatment are gasified. In an alternative gasifier design both the pretreated coal and the gas stream from the pretreater are introduced into the gasifier through a single connecting line, so that these gases mix with the gasifier-bed product gas.

Part of the fluidizing stream enters the gasifier through a grid-type inverted-cone distributor, at a fluidizing velocity of 1–2.5 ft/s.[143] Secondary fluidizing gas flows at high velocity through the ash removal throat. By careful selection of the steam/air ratio of this gas, the temperature within the throat, which is higher than the rest of the bed, can be controlled near the softening point of the ash. Hence the conditions are such that ash agglomeration occurs within the throat until the agglomerates are heavy enough to drop from the fluidized bed. The ash is removed through a lock hopper system and can be collected dry or sent to a settling pond. Residence time of the coal in the gasifier is about 45 min.

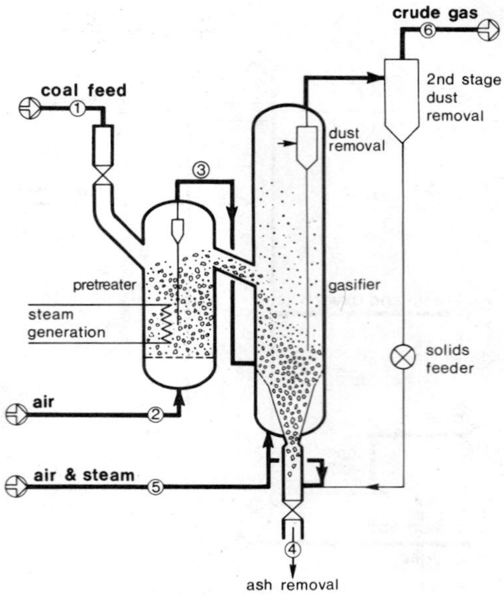

Figure 24.48 Schematic diagram of proposed the U-Gas gasifier.[143] Table 24.32 gives conceptual mass balance data.

Fines are removed from the product gas in a series of cyclones and returned to the gasifier in a stream of air and steam where they are rapidly gasified, the ash so produced agglomerating with that of the bed.[144,145] The conditions in the disengaging space above the bed, which is designed for a gas residence time of 10–15 s, are said[143] to favor cracking of any tars and oils which may have been formed in the bed.

This process has only been demonstrated using an 8-ft-diameter pretreatment reactor and a separate 4-ft-diameter ash agglomerating gasifier, built as part of the Hygas program, both operating at atmospheric pres-

[143] J. W. Loeding and J. G. Patel, "IGT U-Gas Process," paper presented at 67th Annual Meeting, American Institute of Chemical Engineers, Washington, D.C., December 1974.

[144] A. Godel, "A New Combustion Technique," *Eng. Boiler House Rev.*, **71,** 145–153 (1965).

[145] L. Jéquier, L. Longchambon, and G. van de Putte, "The Gasification of Coal Fines," *J. Inst. Fuel,* **33,** 584–591 (1960).

sure. The gasifier differs[146] from that shown in Fig. 24.48 in that the fines are removed by two external cyclones, and the gasifier has operated using air, enriched air, and oxygen. The feeds used were a metallurgical coke, FMC char, and pretreated Illinois No. 6 coal. A conceptual mass balance for the gasifier section of a U-Gas plant producing 587 million scf/day of 161 Btu/scf fuel gas is given in Table 24.32.

One of the major problems with this process will be the control of the temperature at the gasifier throat and distributor. The air/steam ratio of the gas that passes through the throat will need to be much higher than that of the gas passing through the distributor, to promote ash agglomeration; but it must be controlled accurately, to avoid the temperature in the throat becoming so high that the ash is caused to melt. The volume of gas passing up the throat must also be sufficient

[146] J. G. Patel, J. W. Loeding, and K. E. Burnham, "The IGT Low-Btu Gas Process Design and Economics," paper presented at Symp. Comparative Economics of Synthetic Fuel Processes, American Chemical Society Centennial Meeting, New York, April 1976.

Table 24.32 Conceptual Mass Balance for the Gasifier Section of the U-Gas Plant in Fig. 24.48

	Stream 1: Coal from Storage		Stream 3: Pretreated Coal to Gasifier		Stream 4: Gasifier Residue	
Component	lb/hr	% w/w	lb/hr	% w/w	lb/hr	% w/w
Carbon	268,740	67.89	248,235	68.16	4,984	7.11
Hydrogen	22,088	5.58	15,834	4.35	77	0.11
Oxygen	19,594	4.95	18,011	4.95	—	—
Nitrogen	6,254	1.58	6,254	1.72	2,376	3.39
Sulfur	16,665	4.21	13,340	3.66	161	0.23
Ash	62,504	15.79	62,504	17.16	62,504	89.16
Water	20,821					
	416,666		364,178		70,102	

	Stream 2: Air to Pretreater	Stream 5: Air-Steam to Gasifier	Stream 6: Crude Gas	
Component	lb/hr	lb/hr	mole/hr	% w/w
CO			15,525	21.77
CO_2			4,286	6.01
H_2			10,518	13.93
H_2O		128,368	3,916	5.50
CH_4			2,126	2.98
C_2H_6			—	—
H_2S			491	0.69
COS			24	0.03
SO_2			—	—
N_2	253,619	727,075	34,994	49.07
O_2	76,994	220,726	—	—
NH_3			12	0.02
	330,612	1,076,169	71,892	100.00
Temperature (°F)	700		1730	
Pressure (psia)			350	

to entrain the fines recycled to the gasifier. It will be necessary to design the base of the gasifier so that the agglomerates falling from the throat are cooled before they have the opportunity to fuse together. Materials problems are likely to be encountered with the internal cyclone, which must operate at temperatures of up to 2000°F (1090°C) in the product gas.

The Westinghouse Fluidized-Bed Process. The Westinghouse gasification process, which employs a two-stage, pressurized, air-blown, fluidized-bed system, is being developed under a 9-year program, commenced in 1972, sponsored by OCR, Bechtel Corporation, Amax Coal Company, the Public Services of Indiana, and 11 electric power utilities.

In this process (Fig. 24.49) coal from a drying stage is fed through the central draft tube in the devolatilizer-desulfurizer reactor, where the coal and large quantities of recycled solids (char and dolomite sorbent) are carried up by hot gases from the gasifier at velocities greater than 15 ft/s. The recycle solids descend in an annular downcomer at rates up to 100 times the coal feed rate. The high-velocity turbulent gas in the draft tube heats the diluted feed coal rapidly through its plastic stage so that no caking occurs.

Devolatilization of the feed coal takes place at 1400–1700°F (760–925°C),[148] producing methane and higher hydrocarbons in the product gas. Dolomite is added near the top of the bed and mixed with the recirculating solids, partially removing hydrogen sulfide from the product gas. The ungasified coal particles have a size and density that allow them to be entrained in the product gas, from which they are separated using cyclones. Dolomite is withdrawn from the reactor after stripping out the char, either in the transfer line or a special drawoff vessel. The dolomite may then be regenerated.

To facilitate transfer into the devolatilizer-desulfurizer reactor, it is planned to dry the coal with hot recycle product gas in a fluidized bed. The outlet gas from the dryer is cooled to 120°F (50°C) to remove condensables; then it is compressed, heated to about 275°F (135°C), and used to transport the dried coal to the devolatilizer-desulfurizer reactor via a slide valve and standpipe.

The char and fines from the devolatilizer are fed to the gasifier-combustor reactor. In its upper gasification section steam reacts with the char to produce hydrogen and carbon monoxide at temperatures of 1800–2000°F (982–1093°C). In the lower combustor section a high-velocity flow of oxygen or air and steam is directed up through a venturi throat. The temperature in the region of this throat approaches 2100°F (1149°C), causing agglomeration of the ash particles. Above a critical size, the particles are not fluidized and drop down against a countercurrent flow of cooling steam to lock hoppers and quench pots (cf. the U-Gas process).

The Westinghouse gasification system is a combination of process steps that have been successfully operated by others. The recirculating bed was developed by the British Gas Corporation.[81,149–153] The desulfurization step employing fluidized dolomite has been investigated by the Consolidation Coal Company,[154] and Esso Petroleum Company (U.K.).[155] FMC employs multiple fluidized

[147] J. D. Holmgren and L. A. Salvador, "Low-Btu Gas from the Westinghouse System," *Chem. Eng. Prog.*, **71**(4), 87–88 (1975).

[148] D. H. Archer, E. J. Vidt, D. L. Keairns, J. P. Morris, and J. L. Chen, "Coal Gasification for Clean Power Production," paper presented at 2nd Clean Fuels from Coal Symp. IGT, Chicago, June 1975, p. 577.

[149] F. J. Dent, "Methane From Coal," 9th Coal Science Lecture, BCURA, 1960.

[150] F. J. Dent, R. F. Edge, D. Hebden, F. C. Wood, and T. A. Yarwood, "Experiments on the Hydrogenation of Oils to Gaseous Hydrocarbons," *Trans. Inst. Gas Eng.*, **106**, 594–643 (1956–1957).

[151] P. S. Murphy and R. F. Edge, "The Hydrogenation of Oils to Gaseous Hydrocarbons," *J. Inst. Gas Eng.*, **3**, 459–476 (1962).

[152] B. H. Thompson, B. B. Majumdar, and H. L. Conway, "The Hydrogenation of Oils to Gaseous Hydrocarbons," *J. Inst. Gas Eng.*, **6**, 415 (1966).

[153] A. G. Horsler and B. H. Thompson, "Fluidization in the Development of Gas Making Processes," paper presented at Tripartite Chemical Engineering Conf., Montreal, Canada, September 1968.

[154] F. W. Theodore, *Low Sulfur Boiler Fuel Using the Consol CO_2 Acceptor Process,* Rep. No. 2, a feasibility study prepared for OCR; PB-176-910, November 1967.

[155] J. W. T. Craig, G. L. Johnes, G. Moss, J. H. Taylor, and D. E. Tisdall, "Study of Chemically Active Fluid Bed Gasifier for Reduction of Sulfur Oxide Emissions," Esso Research Centre, Abingdon, Oxon, England, Final Rep., OAP Contract No. CPA 70-46, June 1970–March 1972.

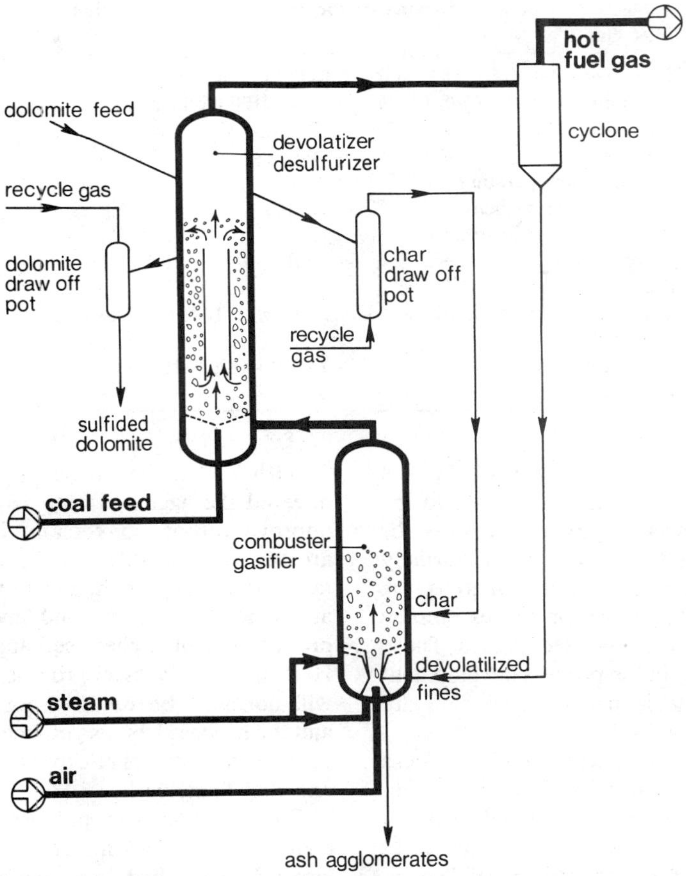

Figure 24.49 The Westinghouse fluidized-bed process.[147]

stages to produce char[156]; and IGT, Battelle,[157] and others[145,158] have used agglomerating fluidized-bed combustors in their processes. The Westinghouse system relies on the satisfactory operation of interacting fluidized beds. If the gas velocity in the draft tube is too low, the solids-carrying capacity of the gas stream may be exceeded, resulting in slugging flow. Maintaining the gas velocity at a level high enough to avoid this condition presents further operating constraints.

[156] J. F. Jones, F. H. Schoemann, J. A. Hamshar, and R. T. Eddinger, "Char Oil Energy Development," prepared for OCR by FMC Corp., Chemical Research and Development Centre, Contract No. 14-01-0001-498, October 1966–June 1971.

[157] W. M. Goldberger, "Collection of Fly Ash in a Self-Agglomerating Fluidized-Bed Coal Burner," ASME Paper No. 67-WA/FU-3, Winter Annual ASME Meeting, Pittsburgh, November 1967.

[158] A. A. Godel and P. Cosar, "The Scale-up of a Fluidized-Bed Combustion System to Utility Boilers," Am. Inst. Chem. Eng. Chem. Eng. Prog. Symp. Ser., Vol. 67, No. 116, 1971, pp. 210–218.

Initial demonstration of the process is being conducted in a 1200-lb/hr pilot plant at Waltz Mills, Pennsylvania. This plant has two refractory-lined, 36-in.-I.D. carbon steel reactors, each containing a 20-in.-diameter fluidized bed, designed for maximum process conditions of 2000°F (1090°C) at 300 psig. The reactors are arranged to function either separately or in combination. The separate operation of the devolatilizer requires a source of "synthetic" gasifier gas, which is produced from the partial combustion at pressure of propane with air, in the presence of CO_2 and steam. Operating parameters for the pilot plant are given in Table 24.33.

Conceptual performance data for a process incorporating an air-blown gasification system, producing a 90–160 Btu/scf gas, are shown in Table 24.20.

Commentary on Hybrid Fluidized-Bed Processes Yielding a Nitrogen-Containing Product Gas. The three processes described

Table 24.33 Operating Parameters for the 1200-lb/hr Westinghouse Pilot Plant[159]

Coal feed rate	300–1500 lb/hr
Dolomite feed rate, Ca/S for 3% S coal	≤ 3/1 (at 1200 lb/hr coal)
Pressure	≤ 300 psig
Maximum temperature	≤ 2200°F
Fluidizing gas velocity through gasifier distributor	≤ 7.5 ft/s
Devolatilizer-desulfurizer, draft tube height	≤ 10 ft
Total bed depth, devolatilizer-desulfurizer	≤ 20 ft (with 10-ft freeboard)
Gasifier bed depth	≤ 13.5 ft (with 10-ft freeboard)
Combustor length	≤ 6.5 ft

are primarily for the production of low-Btu gas, and all utilize special techniques to improve overall carbon conversion. Thus the commercial Winkler process has a partially entrained gasification section, whereas the U-Gas and Westinghouse processes employ agglomerating ash removal techniques. These latter two are still in the process development stage. The available performance data are presented in Table 24.20.

The commercially proved Winkler process produces a low-pressure, tar-free, sulfur-containing product gas and can handle all but strongly caking coals (free-swelling index > 4) without need for pretreatment.[160] The radiant waste heat boiler installed near the top of the gasifier is a relatively new development. Its cooling effect allows operating temperatures very close to the ash-softening temperature in the entrained-gasification zone. The gasifier efficiency (E_1) is lower than that projected for some of the low-Btu fluidized-bed gasification processes presently under development (cf. BCR Tri-Gas, West Virginia University processes). The oxygen-blown Winkler gasifier is not particularly suited for SNG production, since the product gas has an inherently low methane content and is at low pressure. Nevertheless, it produces a medium-Btu gas at a higher gasification efficiency (E_1) than the air-blown process.

[159] D. H. Archer, D. L. Keairns, and E. J. Vidt, "Development of a Fluidized Bed Coal Gasification Process for Electric Power Generation," *Energy Commun.*, 1(2), 115–134 (1975).

[160] *Handbook of Gasifiers and Gas Treatment Systems*, Dravo Corporation, Pittsburgh, February 1976; ERDA Publication No. FE-1772-11.

With the U-Gas process, IGT has pioneered the agglomerating ash technique to improve carbon conversion. However data are not yet available to quantify the advantages of this approach. The process operates at elevated pressures and gives a tar-free product gas of higher methane content and HHV than the Winkler process. Pretreatment will normally be required for caking coals and the process has, as yet, been demonstrated only at atmospheric pressure with coke and char feedstocks.

The Westinghouse process utilizes both controlled ash agglomeration and a recirculating fluidized bed to increase carbon conversion and to allow gasification of caking coals, without prior pretreatment. It is said that the process produces a tar-free, low-sulfur product gas, particularly suitable for application to combined-cycle power generation schemes. The sulfur is removed *in situ* (in the devolatilizer) by the addition of dolomite to the fluidized bed. It is possible that solids- transfer problems may be encountered (e.g., dolomite carryover to the combustor-gasifier).

24.3.4 Fluidized-Bed Processes Producing Gas and Oil

The energy crisis of 1973–1974 brought about a renewed interest not only in processes for the complete gasification of coal, but also those able to produce both gas and oils. These oils could be used as refined fuels or as feedstock for particular chemical processes. Processes for the liquefaction of coal are described in Chapter 28; but four processes that produce a significant proportion of gas as well as oil are discussed below. In contrast to the previous sections of this chapter, the entire process—rather than just the gasification section—must be considered, since the various stages tend to be highly integrated.

24.3.4.1 *The Cogas Process*

The principle of the Cogas process (Fig. 24.50) is the low-pressure, multistage, fluidized-bed pyrolysis of coal to produce oils and gas, followed by steam gasification of the resulting char in a fluidized bed. The heat for the carbon steam gasification reaction, generated in an external, air-blown, slagging, entrained-flow combustor using char fines as fuel, is transferred to the gasifier by a continuously circulating char stream, heated by hot combustion gases. Part of the hot gas from the gasifier passes to the pyrolysis section, where it is used as the fluidizing and heating medium. The remainder of the gas is available as a product stream.

Pyrolysis Section. The pyrolysis section of the process is essentially the COED process, developed by the FMC Corporation under OCR sponsorship, and consists of a series of fluidized-bed reactors operated at low pressure (15–30 psig). This section of the process has been successfully demonstrated at Princeton, New Jersey, in pilot plants that handle 100 lb/hr and 36 tons/day of coal. The work has been widely published under the COED project since 1962,[162–165] and more than 18,000 tons[166] of various coals has been processed through the larger pilot plant in runs of up to 30 days' duration.

In this COED plant coal passes through four pyrolysis stages, each at a successively higher temperature: typically 600, 850, 1000 and 1600°F (315, 450, 540, 870°C). Each stage consists of a bed of fluidized coal char, heated by gas from the subsequent stage and recycled char streams. In each fluidized bed pyrolysis liberates a fraction of the volatile

[161] R. Bloom, Cogas Development Company, private communications, September 1976 and April 1977.

[162] J. F. Jones, "Clean Fuels from Coal for Power Generation," *Energy Sources*, 7(2), 179–202 (1975).

[163] L. J. Scotti, J. F. Jones, L. Ford, and B. D. McMunn, "Multi-Stage Fluidized Bed Pyrolysis of coal at the Project COED Pilot Plant," paper presented at American Institute of Chemical Engineers 77th National Meeting, Pittsburgh, Pennsylvania, June 1974.

[164] FMC Corporation, Princeton, New Jersey, *COED Process Development Unit, Results and Commercial Analysis*, Rep. No. PE 234 166, 1970.

[165] J. F. Jones, *Char Oil Energy Development*, FMC Corporation, Princeton, New Jersey, Rep. No. PB 234 168, December 1971.

[166] J. A. Hamsher, H. D. Terzian, and L. J. Scotti, "Clean Fuels from Coal by the COED Process," paper presented at Environmental Protection Agency Symp., St. Louis, Mo., May 1974.

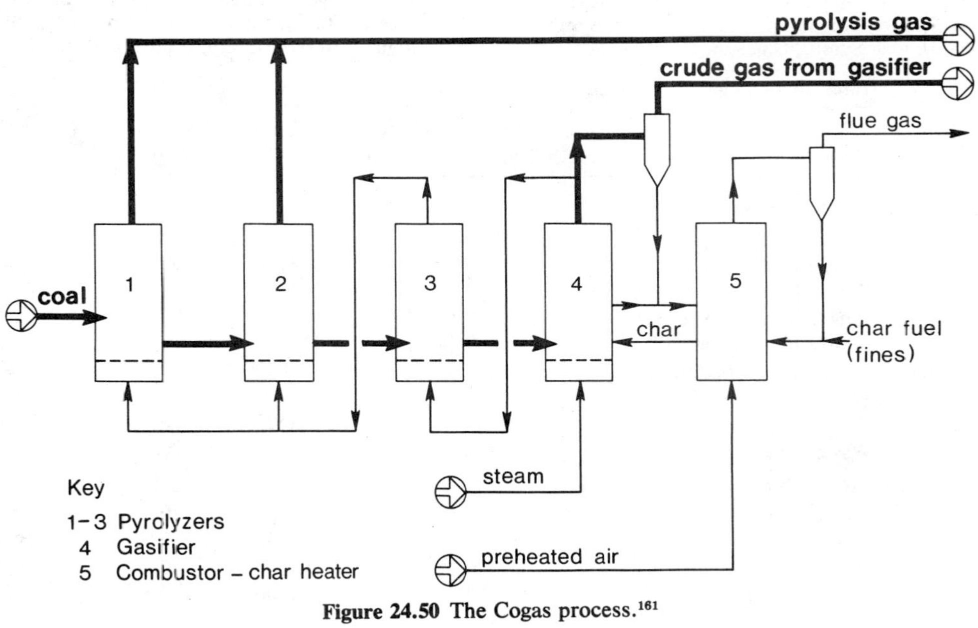

Figure 24.50 The Cogas process.[161]

matter of the coal, with the temperature of each reactor maintained just below that at which the coal would cake and plug the bed. As the coal is partially devolatilized in one reactor, its caking tendency is reduced, and it can be heated further in the next reactor to produce additional volatile products, finally leaving a large quantity of highly reactive char. Some of the liberated volatile material is gaseous under normal conditions and does not condense. However most can be condensed to give a thick, tarry liquid that can be hydrocracked to a marketable oil product. Pyrolysis product compositions are given in the literature.[162–166]

Gasification-Combustion Section. The Cogas Development Company is developing a char gasification-combustion system for integration with the COED pyrolysis process. In the gasification stage steam reacts with the char in a fluidized bed, the heat of reaction being provided by sensible heat transfer from the combustion stage. The reactivity of the COED-type char with steam is an important factor in designing the gasification stage, and a bench-scale, 4-in.-I.D. fluidized-bed batch reactor was used to obtain the data given in Table 24.34.[167]

Two alternative combustor designs were experimentally investigated,[169] each meeting the requirements of using air for combustion, using char fines as fuel, rejecting low-carbon ash, and operating at low pressure. In each case heat was transferred by the continuous circulation of a solid heat carrier, from an entrained-flow combustor to the gasifier. The two designs (Figs. 24.51 and 24.52) differ primarily in the type of solid heat carrier used to transport heat from the combustor to the gasifier, and in the type of combustor system used.

In the first process (Fig. 24.51,) refractory pellets are used to carry the heat to the gasifier. The combustor operates below the ash agglomerating temperature, and entrained ash is separated from the flue gas. The pilot plant at Princeton, New Jersey, was designed for a char feed rate of 2.5 tons/day and operated at a pressure of 50–60 psig to reduce the vessel size.[169] Many of the process objectives were realized in tests in 1974; but attrition of the circulating solid refractory was encountered, and its continuous replacement would have contributed significantly to the cost of the products.[168] An extensive screening program for possible heat carriers was undertaken at bench scale and the most promising was fly ash pellets. This process was not developed further, since at that stage

[167] L. Seglin, L. D. Friedman, and M. E. Sacks, "The Cogas Process for the Gasification of Coal," *Prepr., Am. Chem. Soc., Div. Fuel Chem.*, **19**, 31–34 (1974).

[168] H. R. Hoy and R. T. Eddinger, "Cogas Progress Report," paper presented at 7th Synthetic Pipeline Gas Symp., Chicago, October 1975.

[169] W. A. Paige, "Cogas Status Report," paper presented at 5th Synthetic Fuels from Coal Conf., Oklahoma State University, Stillwater, May 1975.

Table 24.34 Cogas Char-Steam Reactivity Data[168]

Char	Gasification Rate (lb C/hr per lb C in bed)	Temp. (°F)	Pressure (psig)
From pyrolysis			
Western bituminous	0.17	1600	0
Western bituminous	0.21	1600	30
Illinois bituminous (1500°F)	0.10	1600	30
Illinois bituminous (1500°F)	0.37	1700	30
Illinois bituminous (1000°F)	0.24	1600	30
Western subbituminous	0.34	1600	0
Western subbituminous	0.55	1700	0
Western lignite	1.0	1600	0
Western lignite	1.0	1500	30
From other processings			
Western lignite	0.19	1600	0
Eastern bituminous	0.15	1600	0
Western subbituminous	0.19	1600	0
Western subbituminous	0.31	1700	0

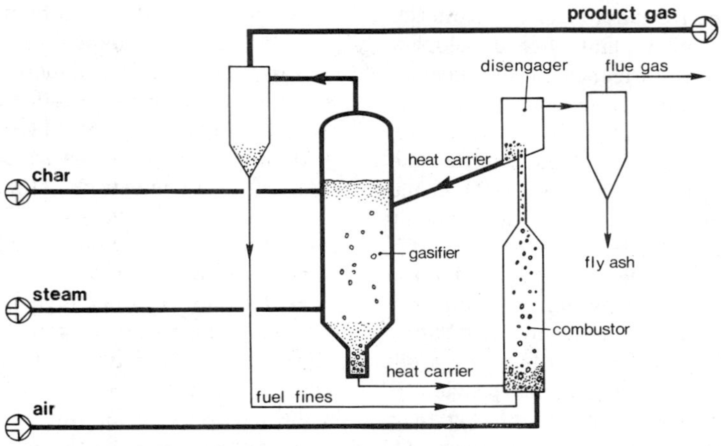

Figure 24.51 The Cogas gasifier-combustor with inert heat carrier.[170]

the results from the alternate process appeared more favorable.

The second process (Fig. 24.52) utilizes char as the heat transfer medium. This circulating char is heated in an entrained-flow lift tube by hot gas from a slagging cyclone combustor and returns to mix with the gasifier bed, providing heat for the gasification reactor. Char fines recovered are consumed in the combustor normally supplied with preheated air. Heat losses in the pilot-plant runs, were high, however, and oxygen-enriched air has been used in some runs. Ash is discharged as a molten slag from the combustor via a water quench system and handled in slurry form. Slagging cyclone combustors have had wide commercial use for coal-fired power-plant boilers; indeed, there are more than 80 such combustors in operation in the United States, although generally of a horizontal design.[169] These units have been constructed up to 30 ft in diameter, well above the scale required for a commercial Cogas plant.

Under contract to Cogas, BCURA, Ltd. has demonstrated this process (Fig. 24.52) in a 50-ton/day of char pilot plant at Leatherhead, Surrey, England. Operation began in February 1974, and more than 600 hr of char circulation and 350 hr of char gasification have been achieved in the course of 12 separate runs.[169]

[170] R. Bloom and R. T. Eddinger, "Status of the Cogas Project," paper presented at 6th Synthetic Pipeline Gas Symp., Chicago, October 1974.

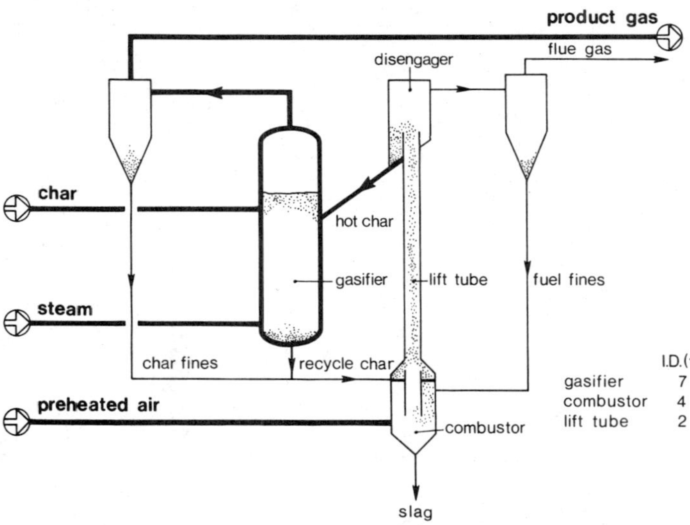

	I.D. (ft)	height (ft)
gasifier	7	30
combustor	4	4
lift tube	2	28

Figure 24.52 The Cogas gasifier-combustor with char heat carrier.[170]

The char used for these tests was from the COED process pilot plant. The developers claim that the performance of the gasifier, combustor, and char circulation systems has met or exceeded expectations.[169] Typical operating data[168] and a heat and material balance for Test 12 are given in Tables 24.35 and 24.36. The char recycle rate is about 30–40 times the char feed rate.

One particular problem experienced in the pilot plant was sulfur emission (as H_2S) from the combustor, necessitating the installation of a Stretford sulfur removal plant to scrub the flue gases. Another potential problem is reaction of carbon in the circulating char with CO_2 in the combustion gas, to form CO. The potential heat of the CO can be recovered by raising steam; but if this is not required by the process, CO formation will reduce the thermal efficiency. The objective of current development work is to investigate the effect of char circulation rate on fines formation, the amount of gas transfer between the combustor and gasifier, the efficiency of the slagging combustor in removing ash, means of avoiding slag buildup in the mixing section and lift tube, the refinement of the design to minimize heat loss, and the widening of the range of operating conditions. Operation at low pressure, though simplifying coal feed and air compression equipment, results in relatively large gasifiers. If SNG is to be made from the product gases, extensive methanation will be necessary and gas compression costs will be high. The operating pressure of a commercial-scale plant is therefore likely to be influenced strongly by both technical and cost factors.

Although the pyrolysis section has been successfully demonstrated, the temperature control systems are very complex, involving many recycle char and gas streams. This complexity will increase in the integrated plant, leading to rather higher capital costs than with single-stage systems. The gas selling price is likely to be heavily dependent on finding a suitable market for the oils produced in the pyrolysis stage. Any necessity to reduce contamination of the oil by coal fines will demand complex oil separation equipment and will further increase capital costs.

A considerable advantage of the Cogas concept resulting in operating and capital cost savings, is that an oxygen plant is not required. Since the heat-generating char combustion is external to the gasifier, the CO_2 it produces leaves in the flue gas and so decreases the size of the CO_2 removal plant required, compared with oxygen-fed gasifiers. Some evidence exists that char from multiple-stage pyrolysis is more reactive toward steam than the char produced from other processes (Table 24.34), a feature that is attributed tentatively to its open structure and high pore volume. The utilization of fines, separated from the gas streams, is

Table 24.35 Performance Data from the Cogas 50-ton/day Char Heat Carrier (BCURA) Plant[168]

	Run 8	Run 9	Run 10	Run 11	Run 12
Operating pressure (psia):	22.1	22.1	21.3	23.5	25.0
Gasifier temperature (°F):	1,700	1,625	1,630	1,590	1,655
Gasifier feed, (lb/hr)[a]:	2,850	3,080	3,490	2,360	2,040
Carbon gasified (lb/hr):	1,040	640	920	560	1,000
Thermal output of product gas (10^6 Btu/hr):	21.0	13.0	17.5	10.4	18.6
Product gas analysis, dry (% v/v)					
H_2	52	56	54	53	52
CO	29	21	22	24	30
CH_4	1	1	1	1	1
CO_2	16	20	21	20	15
N_2	2	2	2	2	2
Air to combustor (lb/s)	4.3	5.0	5.1	4.9	4.5
Cold gas efficiency (%)[b]	53	32	37	31	50
Steam decomposition (%)	86	45	52	—	71

[a] Carbon content of char.

[b] $\dfrac{\text{potential heat of make gas}}{\text{potential heat of char + supplementary fuels}} \times 100\%$.

Table 24.36 Material Balance and Heat Balance for Test 12 on the Cogas Char Heat Carrier (BCURA) Plant[168]

	Carbon (lb/hr)	Oxygen (lb/hr)	Nitrogen (lb/hr)	Hydrogen (lb/hr)	Total (lb/hr)	Higher Heat of Combustion (10^6 Btu/hr)	Sensible Heat (10^6 Btu/hr)	Total (10^6 Btu/hr)
Input								
Char to gasifier	2042	65	28	30	2165	35.64		35.64
Supplementary fuel to combustor	35			10	45	0.98		0.98
Fuel to air preheater	113			17	130			
Natural gas to slag burner	11			4	15	0.33		0.33
Air to combustor		2862	9167	10	12039		2.68	2.68
Oxygen to combustor		1370			1370			
Miscellaneous nitrogen			1251		1251			
Steam		2406		301	2707		3.20	3.20
Gasifier inventory change	313				313		0.18	0.18
Total	2514	6703	10,446	372	20035			43.01
Output								
Product gas	1000	2657	696	308	4661	18.61	4.50	23.11
Flue gas	1449	3954	10,099	99	15601	5.35	7.98	13.33
Flue gas fines	105	1	1	1	108			1.71
Char with slag	10				10	0.16	0.78	0.94
Slag					—		0.12	0.12
Radiation and cooling losses					—		5.58	5.58
Total	2564	6612	10,796	408	20380			44.79
Mass balance, output/input (%)	102.0	98.6	103.4	109.7	101.7			104.1

another achievement of the process; and the use of a slagging combustor to enable a low-carbon slag to be removed is beneficial, although erosion by the slag and flow control of it present some problems.

The successful but separate demonstration of the two sections of the Cogas process has led to ERDA awarding a contract worth $23 million to the Illinois Coal Gasification Group of Chicago for a conceptual design study of a fully integrated Cogas demonstration plant to be sited at Perry County, in southern Illinois. The plant will use 2200 tons/day of coal and produce 18 million scf/day of SNG and 2400 bbl/day of oil.

A conceptual product distribution for a commercial Cogas plant is given in Table 24.37. Table 24.20 shows conceptual performance data for the integrated process, derived from experimental data.

24.3.4.2 *The Stone and Webster Process*

The Stone and Webster process arose from a study undertaken for the State of Oklahoma in May 1971, by General Atomic and Stone and Webster Engineering Corporation[171] on the feasibility of using a high-temperature, gas-cooled reactor (HTGR) to supply heat for the gasification of Oklahoma coals.

This complex multistage process is best understood by referring to Fig. 24.53. In the initial stages coal is processed in the presence of hydrogen and a coal-derived solvent to give a concentrated coal liquid. This solution is catalytically hydrocracked and the products fractionated. The residue fraction passes through a fluidized-bed coking stage. Makeup solvent is produced in both these stages, as are lighter oils, which are then gasified in hydrogen. The gasification products, obtained after treatment and purification stages, are an aromatic liquid and SNG. The hydrogen required for these reactions is produced by the catalytic steam reforming of part of the SNG product, the heat for this reaction being supplied by a circulating helium stream from the HTGR at a temperature currently attainable.

This conceptual route was selected for the feasibility study because, although the heat obtainable from an HTGR is not at a sufficient temperature to support the uncatalyzed steam-carbon reaction at acceptable rates, the temperature [1620°F (880°C)] is high enough to support the catalyzed steam-methane reforming reaction, and hydrogen made from the reformed gas can be used in a variety of coal and oil hydrogenation processes. The hydrogasifier is a fluidized bed of coke particles.

The principal material flows, a heat balance summary, and design data for a conceptual plant, using a HTGR of 10.5 × 10^9 Btu/hr (3000 MW) thermal output are given in Tables 24.38–24.40. This process converts 32,800 tons/day of Oklahoma coal (7% moisture) into 540 million scf/day of SNG having an HHV of 1080 Btu/scf (70% CH_4, 13% H_2, 17% C_NH_M).

This process differs significantly from many of its competitors in that coal is processed as a liquid. Although the result is a very complex process that may be expensive

[171] *Applications Study of a Nuclear Coal Solution Gasification Process for Oklahoma Coal,* Final Report, prepared under Contract No. 10000 for State of Oklahoma, GA-A12068, May 1972.

Table 24.37 Conceptual Product Distribution for a Commercial Cogas plant[161]

Illinois No. 6 Seam coal; feed rate = 28,680 tons/day (including 10% moisture).

Main products	
SNG	263 × 10^6 scf/day (at 950 Btu/scf)
Oils	29,640 bbl/day (22°API)
C_2–C_4 hydrocarbons	13.13 × 10^6 scf/day
By-products	
Ammonia, 25% aqueous	120 tons/day (100% NH_3 basis)
Sulfur	722 tons/day

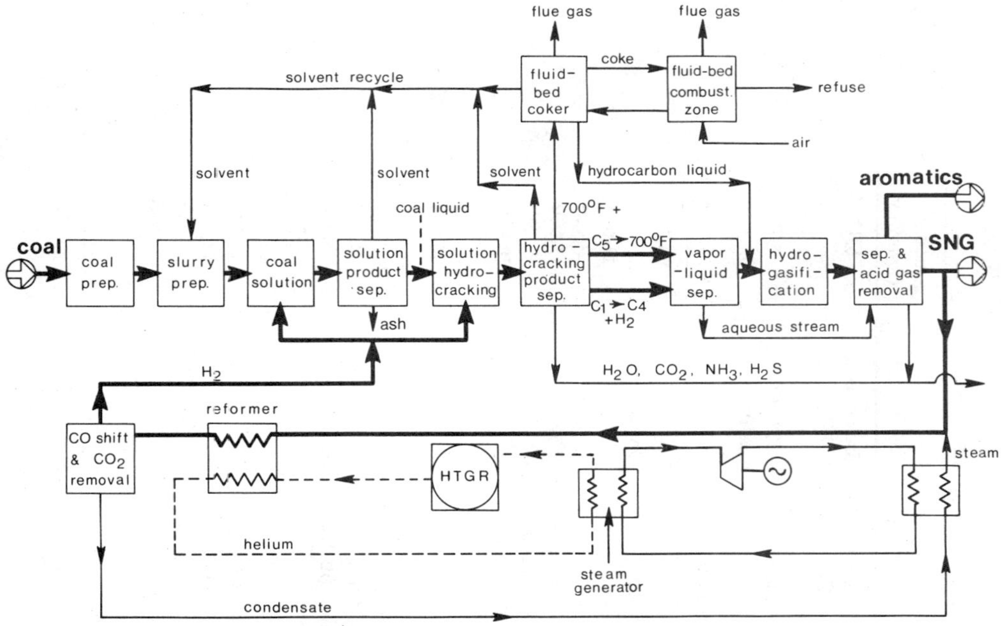

Figure 24.53 The Stone and Webster nuclear-aided process.

and difficult to operate, it is claimed that there are four potential advantages:

1. It offers a means by which agglomerating coals can be handled without the need for preoxidation.
2. The chemical species produced in the coal solvation step are more reactive than those in the parent coal, thus improving the hydrogasification kinetics.
3. Coals with high sulfur content (e.g., Oklahoma coals) can be handled effectively.
4. The process concept is predominantly based on existing processes and equipment.

No suitable market for the low-sulfur aromatic liquid was indicated in the study. The conceptual plant will produce about 21,200 bbl/day of these aromatics, equivalent to approximately 20% of the gross heating value of the gas. If required, they could be converted to gas, but at a loss in thermal efficiency. A loss in the overall process thermal efficiency also results from the production of a high-carbon ash (47% w/w C). However some of this potential heat could be recovered by either air-steam gasification to produce a low-Btu gas or by fluidized combustion to generate steam, although sulfur cleanup of the fuel or the flue gases would be required.

The process is conceptual at present and operating conditions are not finalized. Sections of the process have been tested in bench-scale runs and a two-year joint research and development program by Stone and Webster and General Atomics, costing $740,000, has been undertaken.

24.3.4.3 *The Coalcon Process*

The Coalcon process, being developed jointly by Union Carbide and the Chemical Construction Company (Chemico), originated from work conducted by Union Carbide on liquid phase coal hydrogenation in the 1950s.

Sized, dried coal (see Fig. 24.54) is fed to a fluidized-bed "hydrocarbonizer" operating at moderate temperature and pressure; hydrogen is used as the fluidizing medium. Hydrocarbon vapors and gas from the hydrocarbonizer are quenched, cooled, and fractionated to yield light and heavy oil products. The char removed from the bottom of the reactor is gasified to produce the required hydrogen, any excess char being employed

Table 24.38 Conceptual Material Balance for a Stone and Webster Plant Producing 540 × 10^6 scf/day SNG[171]

Unit	Input	(ton/day)	Output	Amount (tons/day)
Coal solvation unit	Pulverized coal	31,100	Hydrogen + light hydrocarbons	6,120
	Recycle solvent	61,100	($C_1 \rightarrow C_4$)	
	Hydrogen	4,100	Acid gases	3,440
			Light distillate ($C_3 \rightarrow$ 375 °F)	1,720
			Coal liquid (375 °F + ash)	69,920
				81,200
			Recycle solvent	15,100
		96,300		96,300
Solution hydrocracking unit	Output from coal solvation unit	81,200	Hydrogen + light hydrocarbons	
	Hydrogen	4,090	($C_1 \rightarrow C_4$)	11,610
			Acid gases	4,420
			Distillate ($C_3 \rightarrow$ 700 °F)	13,500
			Heavy coal liquid (700 °F +)	6,980
			Recycle solvent	46,000
			Ash	2,780
		85,290		85,290
Fluid bed coker unit	Heavy coal liquids	6,980	Fuel gas	120
	Ash	2,780	Hydrocarbons liquids (700–	
	Air	3,470	1000 °F)	3,230
			Refuse (including ash)	6,020
			Flue gas	3,860
		13,230		13,230
Hydrogasification unit	Hydrogen + light		Product SNG	12,230
	hydrocarbons	11,610	Reformer feed	12,350
	Acid gases	1,430	Light aromatics	3,290
	Distillate ($C_5 \rightarrow$ 700 °F)	13,500	H_2S	840
	Hydrocarbon liquid	3,230	CO_2	550
			NH_3	510
		29,770		29,770

Reformer unit	Reformer feed (SNG)	12,350	Reformed gas	75,500
	Steam	63,150		
		75,500		75,500
Hydrogen production	Reformed gas	75,500	Hydrogen (89 vol %)	8,190
	Makeup water	18,410	Condensate	63,150
			CO_2	22,570
		93,910		93,910
Overall mass balance	Coal	31,100	SNG	12,230
	Air	3,470	Aromatic product	3,290
	Water	18,410	CO_2	23,120
			Refuse	6,020
			Flue gas	3,860
			Water	2,990
			H_2S	840
			NH_3	510
			Fuel gas	120
		52,980		52,980

Table 24.39 Heat Balance for Stone and Webster Process[171]

Exothermic Reactions	10^6 Btu/hr	Temperature of Heat Available (°F)	Endothermic Reactions	10^6 Btu/hr	Temperature of Heat Required (°F)
Hydrocracking	800	< 800	Hydrogen production	4,700	1000–1350
Hydrogasification	2,000	< 1300	Process steam	5,300	555–670
HTGR	10,500	< 1625		500	865–975
			Coal solution	1,500	800
			Solvent distillation	600	750
			Coal drying	100	
			Exchanger inefficiency	600	
Total	13,300		Total	13,300	

as fuel for steam generation. The gaseous products of the hydrocarbonizer pass through purification stages before entering a cryogenic gas separation stage, so that excess hydrogen can be separated and recycled to the hydrocarbonizer. Gas obtained by methanation of a second stream is enriched with a third stream from the cryogenic separation stage, which contains light hydrocarbons, to produce SNG of approximately 1030 Btu/scf.

After conducting small-scale tests in 1- and 10-lb/hr hydrocarbonization units, a 20-ton/day pilot plant was built to acquire the design data necessary for large-scale plant construction. On the basis of these data, a 5000-ton/day coal (Western subbituminous)-to-chemicals plant was designed in preparation for construction in the mid-1960s. However economics moved in favor of oil as a basic raw material and further development work ceased until the early 1970s, when the shortage and increased cost of oil feedstocks again supported the use of coal. Figures 24.55–24.58 include experimental data obtained in the 20-ton/day pilot plant for a Lake de Smet (subbituminous) coal.

On the basis of data for lignite obtained from the 20-ton/day pilot plant and earlier liquid phase hydrogenation work, the Coalcon Company estimated performance data for Eastern bituminous coals and submitted a proposal for a 2790-ton/day Eastern bituminous coal demonstration plant to OCR. Figure 24.59 shows the expected process product distribution. In January 1975 the Coalcon Company was awarded a contract for the design, construction, and operation of such a plant. By 1976, however, the project cost, originally estimated at $237 million, had risen to $440 million. The first phase of the program, costing $20 million, included process design and environmental studies. Initially scheduled for completion in April 1976, the date was extended by more than a year

Table 24.40 Design Data for Commercial Stone and Webster Process[171]

Process Unit (No.)	Size	Design Temperature (°F)	Design Pressure (psig)
Coal liquefaction reactors (3)	14 ft diam × 134 ft	500	2200
Hydrocracking reactors (6)	13 ft diam × 90 ft	1000	2000
Fluid cokers (2)	9 ft → 27 ft diam × 120 ft	1025	70
Fluid coker burners (2)	20½ ft diam × 20 ft	1200	40
Hydrogasification reactors (5)	22 ft 8 in. I.D. × 100 ft	500	1650
Steam-methane reformers (6)	12 ft O.D. × 40 ft bundles	1780	750 (850 × 4½ in. diam tubes)

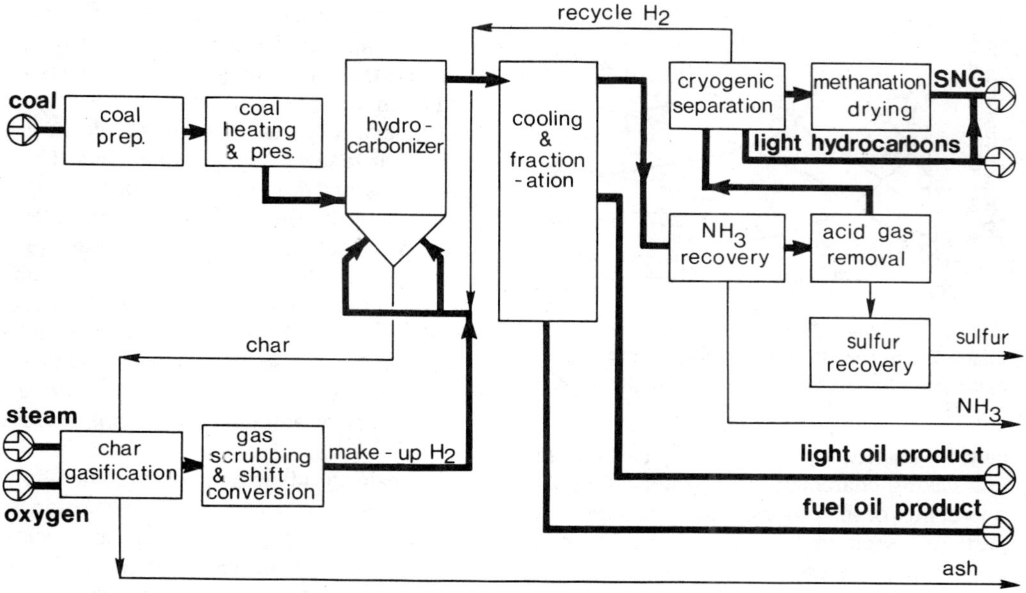

Figure 24.54 The Coalcon hydrocarbonization process.[172,173]

because of difficulties in adapting the hydrocarbonization process to caking coals. Pilot-plant tests continued, with ERDA support, to research the technical problems of the process, and the feasibility of the demonstration plant will be reassessed after the process design and environmental studies have been completed.[175] In 1977 ERDA ceased its funding of this project, although it is said that a caking coal has been hydrocarbonized in the pilot plant without the need for pretreatment.

[172] E. J. Ferretti, "Design Concept for Coal Hydrocarbonization Plant," *Chem. Eng. Prog.*, 72(8), 62–63 (August 1976).
[173] J. R. Martin, "Union Carbide Coalcon Process," paper presented at 2nd Clean Fuels from Coal Symp., IGT, Chicago, June 23–25, 1975, pp. 869–906.
[174] W. D. Morgan, "Coalcon's Demonstration Plant for Clean Fuels," *Chem. Eng. Prog.*, 72(8), 64–65 (August 1976).

24.4 ENTRAINED-FLOW GASIFIERS

24.4.1 Introduction

In entrained-flow gasifiers pulverized fuel reacts as it is carried in the flow of the gasifying medium. Virtually all types of coal

[175] Anon., "ERDA to Reassess Coal Demo Program," *IGT Highlights*, 7(1) (1977).

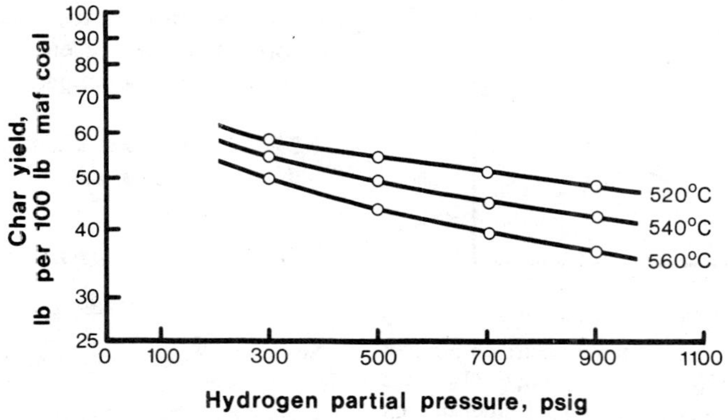

Figure 24.55 The effect of hydrogen partial pressure on char yield.[173]

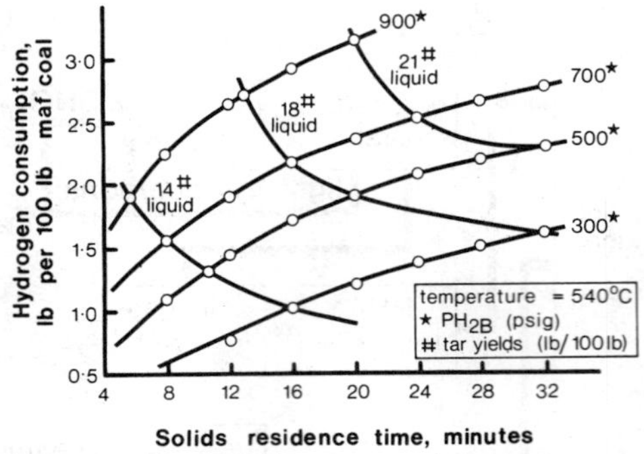

Figure 24.56 The effect of solids residence time on hydrogen consumption: PH_{2B} = H_2 pressure in bed.[173]

can be gasified by this technique. Swelling and caking characteristics do not affect the operability of the process, since each discrete particle is separate from others in the flowing gas stream. The full entrainment of even the very small particles requires relatively high gas velocities, and these lead to high throughputs and reaction times of only a few seconds. Also, diminishing concentration and cooling of the coal toward the exit of the gasifier cause the rate of gasification to decrease, resulting in the formation of char, which is entrained in the product gas.

High carbon conversions can be achieved by two methods. Either the gasifier is operated at moderate temperatures [~ 2000°F (1095°C)] and part of the entrained char is returned to the gasifier, or higher temperatures are applied [> 3000°F (1650°C)], resulting in the ash melting. In some processes both methods are used.

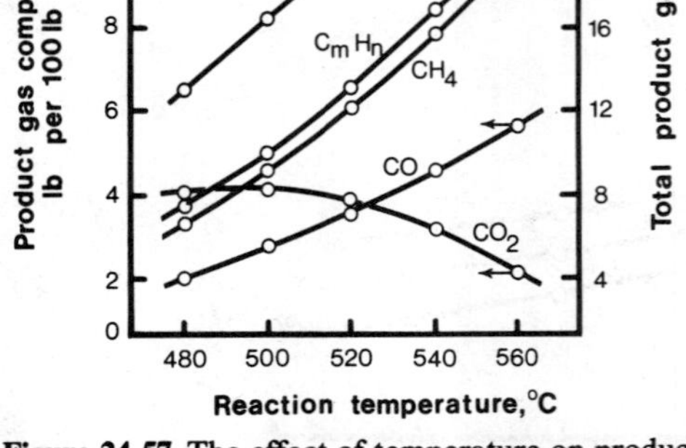

Figure 24.57 The effect of temperature on product gas yield: PH_{2B} = H_2 pressure in bed, GRT_B = gas residence time in bed, GRT_F = gas residence time in freeboard, SRT = solids residence time.[173]

Entrained-flow reactors are simple in design and have few internal components. Their dimensions are determined by particle size, reactivity of the fuel, reaction temperatures, and gas phase velocity. Product gases generally contain little or no tars, oils, or methane, the volatile matter being flashed off and rapidly gasified as the fuel enters the high-temperature reaction zone. The low methane content is an advantage if the gas is required for synthesis of ammonia, methanol, and so on.

The main disadvantages of entrained-flow gasifiers result from the low concentration of fuel in the gasifying medium and the cocurrent flow of the reactants; this arrangement eliminates the possibility of internal heat exchange between the product gas and incoming fuel, resulting in high gas outlet temperatures compared with fixed- and fluidized-bed processes. The need for extensive heat exchange surfaces to recover the sensible heat of the product gases (to enhance an otherwise low thermal efficiency) offsets to some extent the simplicity of the gasifier. The heat recovery problem is complicated by the presence of entrained solids and by the need to avoid molten ash particles contacting heat transfer surfaces. The cost of heat exchange and particulate removal equipment therefore forms a large portion of the capital cost of the gasification section.

Early entrained-flow gasification processes include those of Koppers-Totzek, Texaco, Babcock and Wilcox, Panindco, Ruhrgas, Bianchi, and DuPont, all of which are reported in the first supplementary volume.[176] The processes described below are classified according to the presence or absence of nitrogen in the product gas.

24.4.2 Processes Yielding a Nitrogen-Free Product Gas

24.4.2.1 *The Koppers-Totzek Process*

The Koppers-Totzek process[177] employs a

[176] H. H. Lowry, Ed., *Chemistry of Coal Utilization*, Supplementary Volume, John Wiley and Sons, New York, 1963, Ch. 20.
[177] Ref. 176, p. 973.

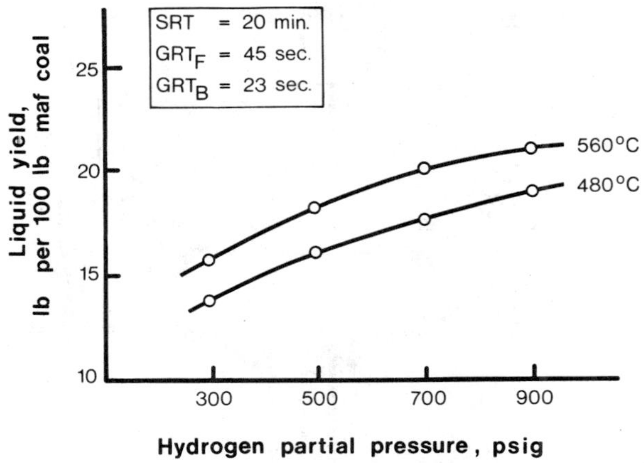

Figure 24.58 The effect of hydrogen partial pressure on liquid yield; symbols as in Fig. 24.57.[173]

single-stage, entrained-flow, ash-slagging gasifier that operates at slightly above atmospheric pressure to produce a synthesis gas from virtually any type of coal. The process has been commercial since 1952, with 49 gasifiers either built or under construction, prior to 1976, in 14 countries.[178] Most of these plants have produced a hydrogen-rich gas for use in ammonia synthesis.

Coal is dried to between 2 and 8% moisture, depending on rank, and pulverized to give more than 80% below 150 mesh (0.1 mm) in a single operation using ball, rod, or roller mills, with hot flue gas (or gasifier product gas combusted with excess air) circulated through the mill. The gas flow dries, entrains, and transports the coal dust to a classifier, where the finer particles continue to a cyclone separator and the oversize particles return to the mill. This system permits the utilization of the entire mine output. The pulverized coal is then fed via a screw feeder at a controlled rate to a mixing head, where it is entrained in a mixture of steam and oxygen. The particles are accelerated through a nozzle at velocities above the speed of flame propagation, to prevent flashback.

The gasifier (Fig. 24.60) has a double-walled shell constructed of boiler-quality plate and is normally lined with a thin layer of refractory. Low-pressure process steam generated in the annulus between the inner and outer shells cools the inner wall below the ash-slagging temperature so that during operation an accumulation of slag provides

[178] J. E. Franzen, Krupp-Koppers GmbH, Research and Development Division, Essen, West Germany, private communication, November 1976.

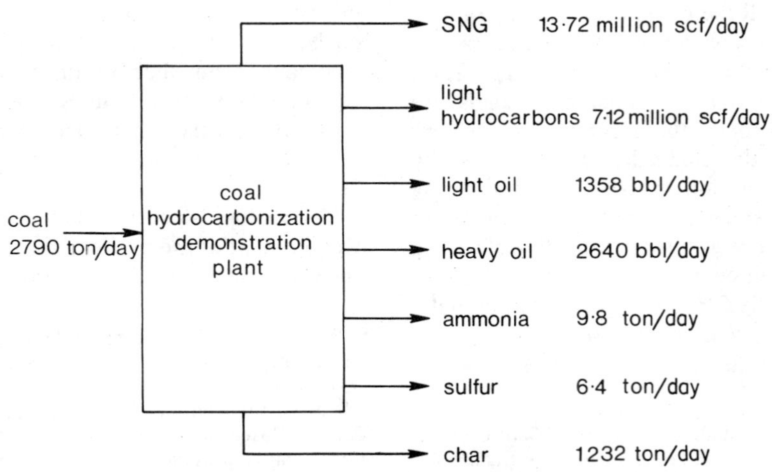

Figure 24.59 The Coalcon demonstration plant product distribution.[173,174]

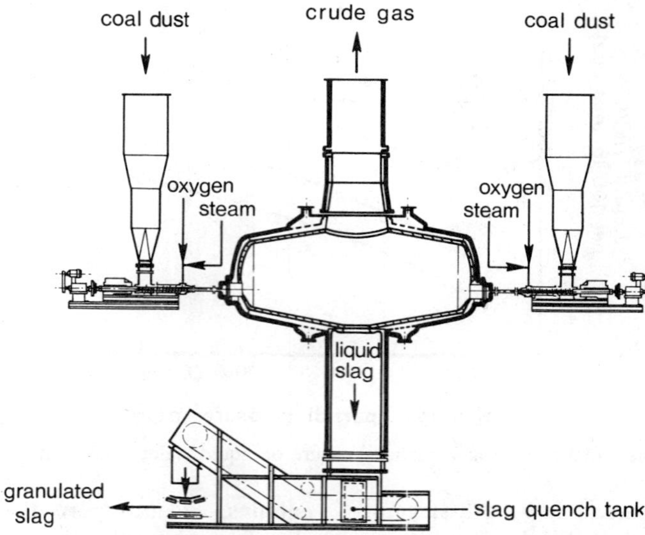

Figure 24.60 The Koppers-Totzek gasifier.[178]

additional protection. Gasifiers have operated satisfactorily with high-ash lignites (at Coruna, Spain) without the need for any refractory lining.

Two burners, each pointing slightly down, are positioned directly opposed, both to improve turbulence and to ensure continuous safe ignition, should one burner become temporarily blocked. This arrangement has two further advantages: projected particles that pass unburned through one flame region are gasified in the opposing flame, and since the highest temperature region is in the central part of the gasifier, refractory problems are minimized.

The reaction temperature in front of the burner port is 3300–3500°F (1825–1925°C), and nearly all the carbon reacts very rapidly with oxygen and steam to produce carbon oxides, hydrogen, and molten slag. Tars, condensable hydrocarbons, or phenols are not formed under these conditions. Decomposition of the coal takes place so rapidly that agglomeration of coal particles during the plastic stage does not occur; therefore any coal can be gasified irrespective of caking property, ash content, or ash fusion temperature.[179] Little CH_4 ($\sim 0.1\%$) is present in the product gas, with all the sulfur in the coal appearing in the gas, 80–90% as H_2S and the remainder as COS.[180] For most coals, more than 50% of the ash flows down the gasifier walls as molten slag and drains into a slag quench tank; the remainder leaves as fly ash entrained in the exit gas.

The quench tank temperatures must be carefully controlled by rapid water circulation, since excessive flash steam rising into the gasifier could cause the downflowing slag to solidify. If the initial deformation, fusion, and fluid temperatures of the ash are in the range of 1900–2400°F (1040–1315°C), there is little or no difficulty in removing slag from the gasifier. If the ash is more refractory, the molten slag can solidify on the gasifier walls and the addition of a fluxing agent is required. Rejection of ash at very high temperatures means that it is inefficient to use high-ash coals, although lignites with up to 40 wt% ash have been successfully gasified on a commercial scale at the Kutahya plant in Turkey.[181] On the other hand, erosion of the refractory lining occurs if the slag is too fluid.

The gasifier is controlled by monitoring the CO_2 content of the product gas at the exit of the spray washer cooler. At normal operating

[179] D. L. Katz et al., *Evaluation of Coal Conversion Processes to Provide Clean Fuels, Part II*, EPRI Rep. No. PB-234 203, February 1974.

[180] V. Ingenhoff, "Koppers-Totzek Coal Dust Gasification Process," *J. Fuel. Soc., Jap.*, **53** (9), 757–761 (1974).

[181] J. E. Franzen and E. K. Goeke, "SNG Production Based on Koppers-Totzek Coal Gasification," paper presented at 6th Synthetic Pipeline Gas Symp., Chicago, October 1974.

temperatures about 11–12% of the carbon is converted to CO_2.[179] When insufficient oxygen is used, indicated by lower CO_2 concentrations, ash clinkers can form because of the lower temperatures. Some physical control of the slag fluidity can therefore be achieved by control of the oxygen supply.

As a result of the endothermic reactions and radiation to the refractory walls, the gas temperature decreases to about 2750°F (1510°C) at the gasifier exit, where it is quenched with water to solidify entrained slag droplets, so preventing their adherence to the tubes of the high-pressure steam boiler. The gas leaving the waste heat boiler at 350–500°F (175–260°C) passes to a refractory-lined spray washer, where about 90% of the heavy particulate matter is removed and the gas is cooled to about 95°F (35°C), reducing entrained solids to 5×10^{-7} lb/scf. Electrostatic precipitators can reduce this entrainment to 1.5×10^{-8} lb/scf, so that the gas can be compressed before sulfur removal.

Gasifiers with two burner heads have progressively increased in capacity from an early design of 2.88 million scf/day to the present 21.6 million scf/day, gasifying more than 394 tons of coal per day.[178] The larger gasifiers are 10–12 ft I.D., tapering down to 6.8 ft at either end, with an overall vessel length of about 25 ft and a volume of approximately 1000 ft^3. Recent development of a gasifier with four burner heads positioned 90° apart (Fig. 24.61) has increased this capacity to 785 tons/day,[182] producing around 43.2 million scf of gas per day.

From a standby condition, gasifiers can be in full production in about 45 min. Variable speed drives on the coal screw feeders permit the operating rate to be reduced to no lower than 60% of design capacity, ensuring adequate feed velocity to prevent flashback. With a four-burner gasifier, a turndown to 30% is possible by shutting off one pair of nozzles completely.[182]

Carbon conversion is a function of coal reactivity, varying between 88% for bituminous coal and 98% for lignite.[180] A major drawback of this process is the high oxygen demand (the capital cost of the oxygen plant is greater than that of the gasifiers). The use of air in place of oxygen is not practical because a high preheat would be required to maintain slagging conditions. Sufficient steam is generated in the waste heat boiler to operate the oxygen plant, but additional steam is required to operate the crude gas compressors. The process suffers economically because its operation at low pressure requires high subsequent compression, especially for ammonia production (2000 psig). A pressurized (440 psia) version of this process, which promises improved economics mainly through reduction of power require-

[182] J. F. Farnsworth et al., "K-T Koppers Commercially Proved Coal and Multiple Fuel Gasifier," *Iron Steel Eng.*, **51**(8), 33–38 (August 1974).

Figure 24.61 The Koppers-Totzek gasifier with four burner heads.[178]

ments for compression, is under development jointly by Krupp-Koppers and Shell Internationale Petroleum Maatschappij.[178] This modification should not appreciably alter the product gas composition. A 130-ton/day demonstration plant, under construction in Germany, is scheduled to be on stream in 1977. Koppers is developing a six-burner gasifier[181] that is expected to have twice the capacity of a four-burner version. Material and heat balances for a four-burner, low-pressure gasifier using bituminous coal (Fig. 24.62) are given in Table 24.41; and a summary of the performance data appears in Table 24.42.

24.4.2.2 *The Occidental Research Corporation Process*

The Occidental Research Corporation (ORC) process (formerly known as the Garrett process) involves low-pressure partial gasification of coal, in which methane and other hydrocarbons are obtained directly by "flash pyrolysis."

The essential features are a very rapid heating and devolatilization of pulverized coal in an entrained-flow reactor in the absence of oxygen, followed by rapid quenching of the gas to prevent further degradation. The yield of either gaseous or liquid products can be optimized by varying the pyrolysis temperature. Part of the residual char produced in the process is combusted with air in a separate char heater and recycled to provide the heat for the pyrolysis; the remainder is a fuel by-product. The major process operating parameter is reactor temperature. Tests have been conducted to determine the distribution and yields of products from a subbituminous coal, in the temperature range 1500–1700°F (815–925°C) (Figs. 24.63–24.65). For a commercial SNG plant, it is clearly desirable to operate the ORC gasifier at above 1700°F (925°C). However the operating temperature must be controlled to prevent ash agglomeration.

A simplified flowsheet of the ORC process is shown in Fig. 24.66. Coal is pulverized and dried to permit pneumatic transportation (by a portion of the product gas) to the entrained-flow gasifier. There it is mixed with a stream of hot recycled char and heated to 1600–1750°F (870–955°C) within 2 s. This rapid heating maximizes the production of methane but necessitates the use of fine particles of coal (70% through 200 mesh/0.074 mm), since the rate of heat transfer is con-

[183] K. H. Mody, Occidental Research Corporation, Laverne, Calif., private communication, October 1976.

[184] R. J. Grace, Bituminous Coal Research, Inc., Monroeville, Pa., private communication, September 1976.

[185] U.S. Pat. 3,544,291, Texaco Inc., April 22, 1968.

[186] R. C. Patterson, "The Combustion Engineering Coal Gasification Program," paper presented at ASME–IEEE Joint Power Generation Conf., Portland, Ore., September–October 1975.

[187] R. A. McCallister, Foster Wheeler Energy Corporation, private communication, September 1976.

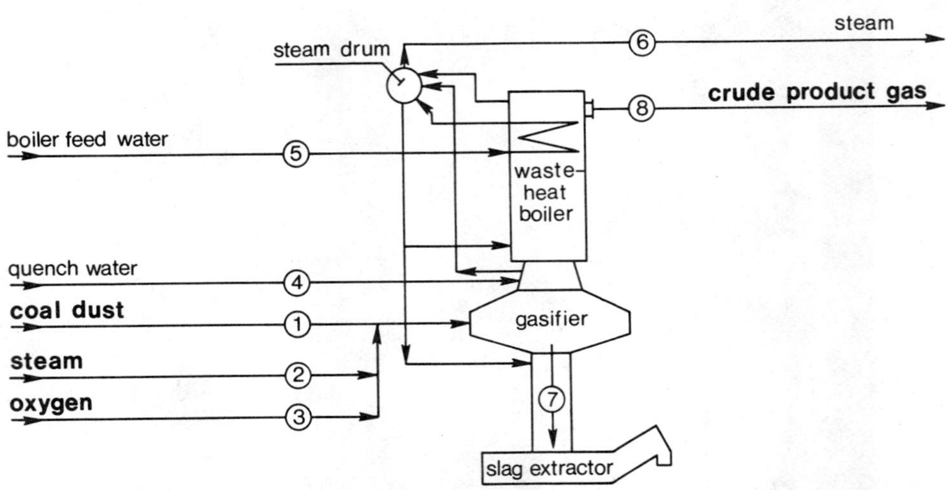

Figure 24.62 Flow diagram of the Koppers-Totzek gasifier.[178] Table 24.41 gives material and heat balance data.

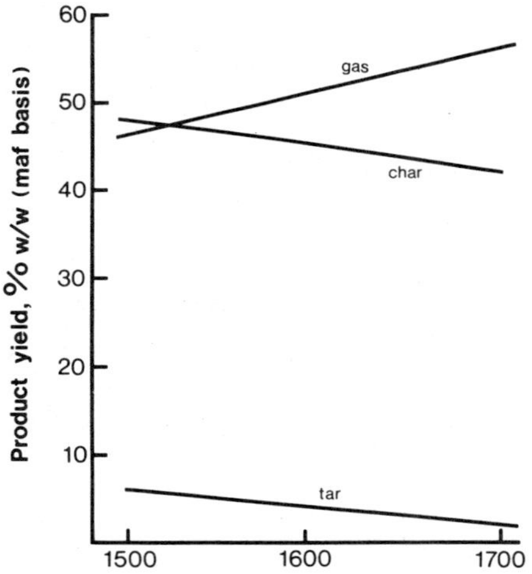

Figure 24.63 The ORC process: product distribution with temperature.[188]

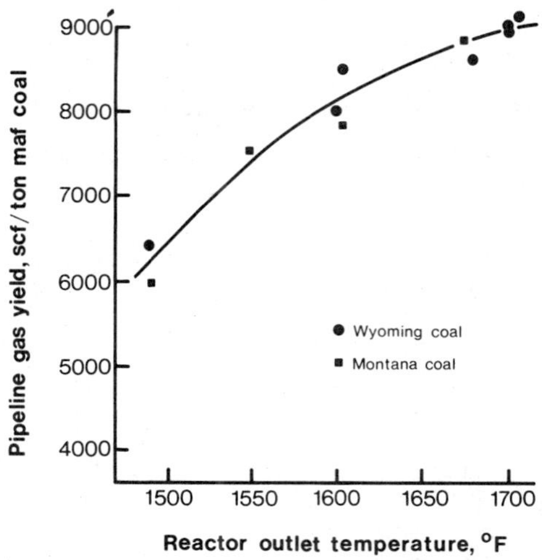

Figure 24.65 The ORC process: pipeline gas yield.[188]

trolled by free convection.[189] With very fine particles, heatup rates in excess of 5000°F/s are reported.[190]

After cyclonic separation of char, the gas from the reactor is rapidly quenched to 1350°F (730°C) to avoid secondary decomposition. Heavy tar is removed from the gas in a series of scrubbers. The condensed tar and any char fines are recycled to the pyrolysis reactor, where the tar is cracked to extinction, producing additional gas and char.

Part of the char is transported to a char heater, where it is heated by controlled combustion in air in an entrained-flow reactor. The short residence time reduces the formation of CO, even though excess carbon is present, because the reaction between carbon dioxide and carbon is relatively slow. The net yield of char is 45% by weight (maf basis) of the products from the process and represents 41% of the combustion heat of the coal.[183]

Small-scale testing of the process has been completed. This work included pyrolysis studies in a 5-lb/hr bench-scale unit, which was indirectly heated, and in a 50-lb/hr pilot plant (see Table 24.42) in which recycle char was used to supply the process heat. In January 1976 a design study of a 250-ton/day pilot plant was begun by the Foster Wheeler Energy Corporation under a 2-year program, jointly financed by ORC and the State of Kentucky. The plant will be designed to process coal primarily at low temperatures to produce liquids, but it will also be capable of operating under high-temperature gasification conditions. It is claimed[183] that the char produced from a plant processing 40,000 tons of coal per day, producing 250 million scf of

[188] H. G. McMath, R. E. Lumpkin, J. R. Longanbach, and A. Sass, "A Pyrolysis Reactor for Coal Conversion," *Chem. Eng. Prog.*, **70(6)**, 72–73 (1974).

[189] N. W. Green, Synthetic Fuels from Coal: The Garrett Process," Paper No. 18, presented at 2nd Clean Fuels from Coal Symp., IGT, Chicago, June 1975, pp. 299–322.

[190] A. Sass, "Garrett's Coal Pyrolysis Process," *Chem. Eng. Prog.*, **70**(1), 72 (January 1974).

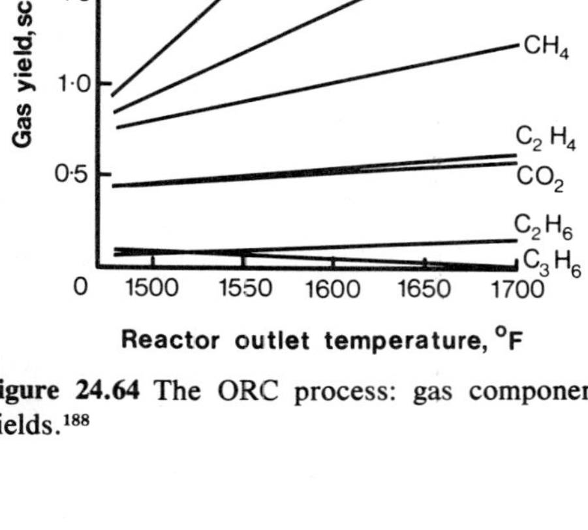

Figure 24.64 The ORC process: gas component yields.[188]

Table 24.41 Material Balance and Heat Balance for the Koppers-Totzek Gasifier (See Fig. 24.62)[178]

Total input, Streams 1–4: 717.954 × 10^6 Btu/hr; total output, Streams, 5–8, 717.954 × 10^6 Btu/hr.

	Stream 1: Coal to Gasifier		Stream 2: Steam to Gasifier		Stream 3: Oxygen to Gasifier		
	lb/hr	% w/w	lb/hr	mole/hr	lb/hr	mole/hr	% w/w
Hydrogen	2,277	3.6					
Carbon	40,172	62.8					
Sulfur	253	0.4					
Nitrogen	887	1.4					
Oxygen	6,769	10.6					
Ash	12,905	20.2					
CO_2							
H_2S							
COS							
CO							
H_2							
CH_4							
N_2					163	5.7	0.4
Ar					922	23.1	1.6
H_2O	638	1.0	6,327	351.1			
HCN							
NH_3							
O_2					45,282	1,415.0	98.0
Total	63,901	100.0	6,327	351.1	46,367	1,443.8	100.0
Temperature (°F)				261		194	
Pressure (psig)				21.8		8.7	
Volume (scf/hr)				124,009		518,986	
Volume (scf/hr)				74,292			
Average molecular weight				18.016		32.113	
HHV (Btu/scf)							
Heat of combustion (10^6 Btu/hr)	686.215						
Enthalpy (10^6 Btu/hr)	0.920			7.302		1.167	
Total heat (10^6 Btu/hr)	687.135			7.302		1.167	

[a] Weight basis.

Stream 4: Quench Water lb/hr	Stream 5: Boiler Feed Water lb/hr	Stream 6: Steam lb/hr	Stream 7: Slag lb/hr	Stream 8: Crude Product Gas			
				lb/hr	moles/hr	% v/v Wet	% v/v Dry
				4,017			
			5,161	7,744			
				12,925	293.7	5.27	6.91
				240	7.0	0.13	0.17
				66	1.1	0.02	0.03
				76,023	2,714.1	48.96	64.29
				2,301	1,141.6	20.52	27.08
				66	4.2	0.07	0.10
				1,115	39.8	0.72	0.94
				799	20.0	0.36	0.48
17,805	115,225	115,225		24,251	1,346.0	23.85	
						130 ppm[a]	
						100 ppm[a]	
17,805	115,225	115,225	5,161	129,547	5,567.5	100.00	100.0
86	216	469	>2,417		320		
217.5	652.5	507.5			0.4		
		2,258,075			1,992,896		
286	1928	104,553			3,170,096		
					21.16		
					333		
					534.074		
0.964	21.386	139.232	4.258		38.644		
0.964	21.386	139.232	4.258		572.718		1.746

Table 24.42 Comparison of Performance Data for the Entrained-Flow Process

Gasification process:	Koppers Totzek[178]	Occidental Research[183]	Bi-Gas[184]	Babcock and Wilcox[197]	Texaco[185]	Combustion Engineering[186]	Foster Wheeler[187]
Developer:	Krupp-Koppers	Occidental Research Corp.	Bituminous Coal Research	Babcock and Wilcox	Texaco Development Corp.	Combustion Engineering	Foster Wheeler
Site:	Germany	California	Pennsylvania	Ohio	California	Connecticut	New Jersey
Status:	Commercial	Experi-mental	Experimental[a]	Commercial scale	Experimental[a]	Experiment-al[a]	Experimental
Fuel:	Black coal	Wyoming subbitumi-nous	Pittsburgh Seam	Pittsburgh Seam	Coke	Bituminous	Bituminous
Proximate analysis							
Volatile matter	—	33.5	30.5	—	—	38.1	33.4
Fixed carbon	—	38.8	58.2	—	—	51.1	50.8
Moisture	—	20.0	2.0	10.0	—	2.4	4.2
Ash	—	7.7	9.3	7.5	—	8.4	11.6
Ultimate analysis							
Carbon	—	51.16	74.3	68.8	—	75.6	66.3
Hydrogen	—	5.62	4.8	5.4	—	5.4	5.5
Nitrogen	—	0.99	1.5	0.9	—	1.4	1.5
Oxygen	—	33.86	7.9	14.9	—	6.6	11.6
Sulfur	—	0.67	2.2	2.4	—	2.6	3.5
Ash	—	7.70	9.3	7.5	—	8.4	11.6
HHV fuel (Btu/lb)[b]	10,716	8604	15,930[c]	—	14,493	13,480	12,303
Steam/coal (lb/lb)[b]	0.10	0	1.20	0.18	0.92	0	—
Oxygen/coal (lb/lb)[b]	0.73	0.28	0.65	0.88	1.17	1.15[d]	—[d]
Air/coal (lb/lb)[b]	—	1.20	0	0	0	4.92	—
Gasification temperature (°F)	3300–3500	1600	1700–3300	3000–3400	2500	3200	1700–3000
Gasification pressure (psia)	15	15	755–1515	315	405	15	385

Product analysis, dry (% v/v)							
CO	64.3	28.6	29.1	59.1	46.5	24.4	28.8
H_2	27.1	24.0	40.0	30.2	33.1	10.6	13.6
CH_4	0.1	17.1	8.9	—	0	—	3.4
CO_2	6.9	17.0	21.1	7.4	19.0	4.1	2.8
N_2	0.9	0.3	0.2	2.5	1.2	60.5	47.8
C_nH_m	—	11.4	—	—	—	—	—
H_2S	0.2	0.6	0.7	0.9	0.1	0.3	0.6
Trace components	0.5	1.0	—	—	0.1	0.1	3.0
HHV product gas (Btu/scf)	296	545	317	293	257	115	174
Tar/oils produced	No	No	No	No	No	No	No
Gasifier outlet temperature (°F)	—	1350	1800	1600–1800	—	1700	1700
Product gas yield (scf/lb)[b]	25.1	6.8	44.0	31.9	41.8	84.3	58.2
Specific gasification rates[e]							
10^6 Btu/ft²-hr	—	166.7 [f]	7.1	—	—	3.57	
10^6 Btu/ft³-hr	0.22	16.7 [f]	0.16	—	0.48	0.03	0.0
Carbon conversion (%)[e]	90	—	—	—	>90	97.5	—
Gasification efficiency (%)[e]							
E_1	69.3	42.8	87.4 [c]	77.6	74.1	71.9	82.3
E_2	69.3	104.8	87.4 [c]	77.6	74.1	71.9	82.3

[a] Conceptual data.
[b] Containing moisture and ash indicated in proximate analysis.
[c] Derived value from Bi-Gas material balance.
[d] Oxygen supplied as air.
[e] See Appendix.
[f] Gasifier volume taken as volume of vessel in which coal is contacted with hot recycle char.

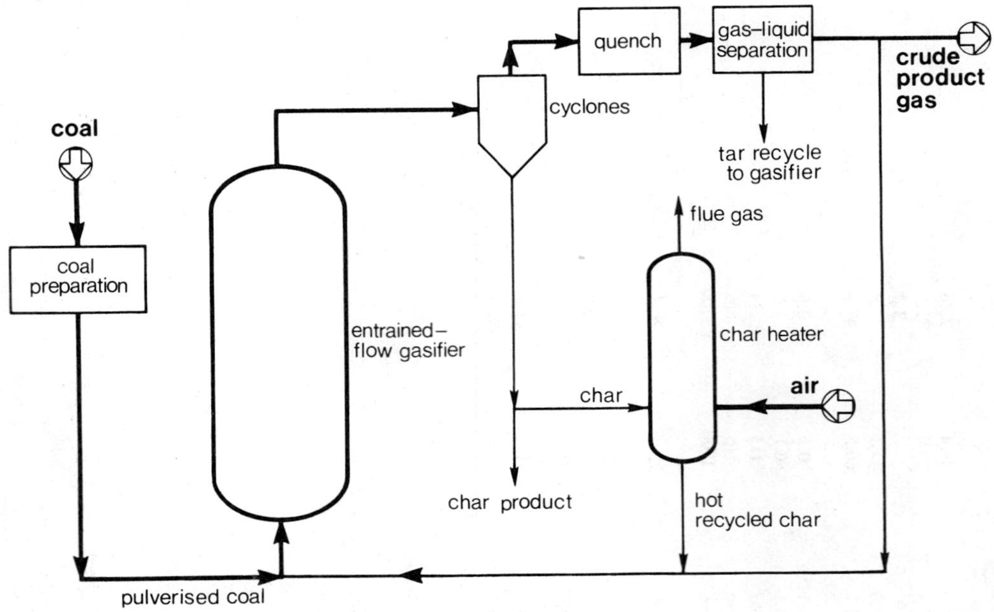

Figure 24.66 Flowsheet for the ORC process.[183]

SNG per day, could fuel a 1300-MW electrical power plant, presently consuming 16,500 tons of coal per day.

24.4.2.3 *The Bi-Gas Process*

The Bi-Gas process uses a two-stage, entrained-flow slagging gasifier (Fig. 24.67). Pulverized coal and steam are fed to the upper section of the gasifier (Stage 2), in which the coal is partially gasified in the presence of the hot gas flowing from the lower section (Stage 1). The residual char entrained in the Stage 2 product is separated

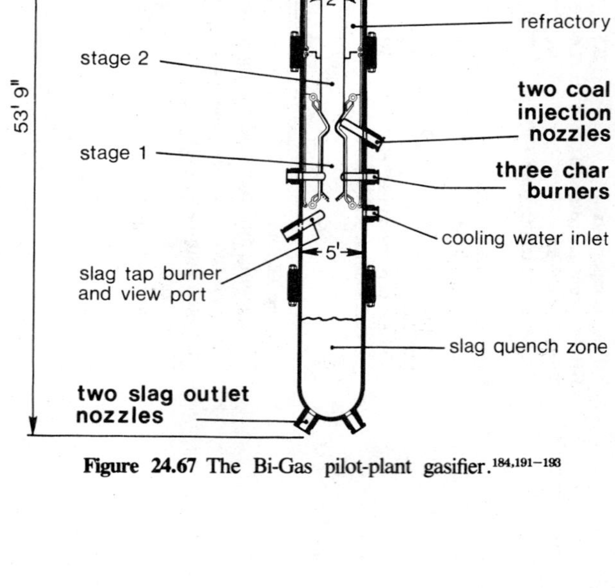

Figure 24.67 The Bi-Gas pilot-plant gasifier.[184,191–193]

using cyclones and sent to the lower stage, where it is gasified under slagging conditions with steam and oxygen, to produce the hot synthesis gas required for Stage 2. Molten ash flows out of the base of Stage 1 into a water quench, from which it is removed using lock hoppers.

Process development has proceeded from batch-autoclave experiments to operation of a 100-lb/hr Stage 2 unit. These tests indicated that at elevated temperatures and pressures, a high methane yield could be obtained (Fig. 24.68), the actual yield being primarily dependent on coal rank, temperature, and hydrogen partial pressure.[194] This work, together with cold-flow dynamic modeling tests,[194] has led to the construction of a 5-ton/hr integrated pilot plant (Fig. 24.69) at Homer City, Pennsylvania. This gasifier comprises an outer pressure shell approximately 5 ft I.D. and 53 ft 9 in. long, containing the reaction stages, water-cooling circuits, and slag quench zone.[195]

[191] W. P. Hegarty and B. E. Moody, "Evaluating the Bi-Gas SNG Process," *Chem. Eng. Prog.*, **69**(3), 37–42 (March 1973).

[192] *Engineering Study and Technical Evaluation of the Bituminous Coal Research, Inc., Two-Stage Super Pressure Gasification Process,* Air Products and Chemicals, Inc., Report to U.S. OCR, R & D Rep. No. 60, Government Printing Office, Washington, D.C., 1970.

[193] R. J. Grace, V. L. Brant, and V. D. Kliewer, "Design of Bi-Gas Pilot Plant," paper presented at 5th Synthetic Pipeline Gas Symp., Chicago, October 1973.

[194] R. J. Grace, "Development of Bi-Gas Process," paper presented at Clean Fuels from Coal Symp., IGT, Chicago, September 1973.

Steam, oxygen, and pulverized coal (70% through 200 mesh) are injected into Stage 2 through nozzles inclined at an angle of about 30° to the horizontal. The nozzles are spaced so that impingement occurs within the gas space, to ensure that the refractory lining does not suffer excessive erosion. These reactants are very rapidly heated to a temperature of approximately 2200°F (1200°C) by the hot gases rising from Stage 1. The coal is rapidly devolatilized, to yield methane and a highly reactive carbon char, then reacts with both hydrogen and steam. The upward velocity of approximately 5–8 ft/s results in residence times of 6–8 s, and the products leave Stage 2 at a temperature of approximately 1700°F (925°C).

The tar- and oil-free crude gas and char leaving the gasifier are then quenched with water to about 650–800°F (345–425°C), so providing most of the steam required for any subsequent shift conversion and also minimizing losses due to the Boudouard reaction. The char is separated from the gas stream in a cyclone and is returned to Stage 1.

Char, oxygen, and steam are injected into Stage 1 through tangentially directed nozzles to produce a vortex flow, sufficient to cause the heavier slag particles to impinge on the walls, down which the slag flows into the slag quenching zone. The temperature is controlled at 2700–3200°F (1480–1760°C) by the relative quantities of steam and oxygen sup-

[195] J. M. Miles, "Status of the Bi-Gas Program. Part I. Pilot-Plant Activities," paper presented at 8th Synthetic Pipeline Gas Symp., Chicago, October 1976.

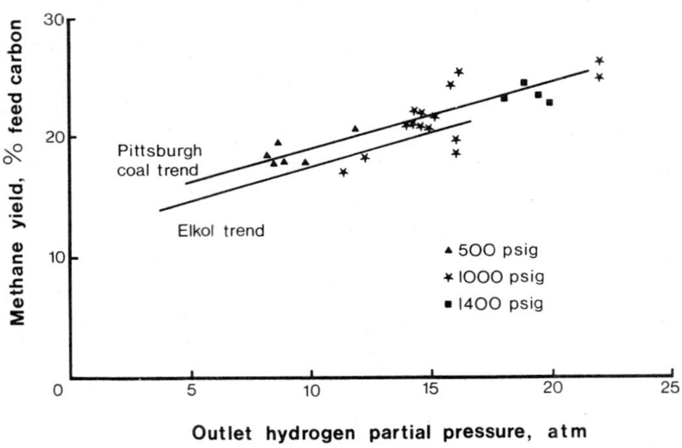

Figure 24.68 The effect of outlet hydrogen partial pressure on methane yield for a 100-lb/hr Bi-Gas Stage 2 unit.[194]

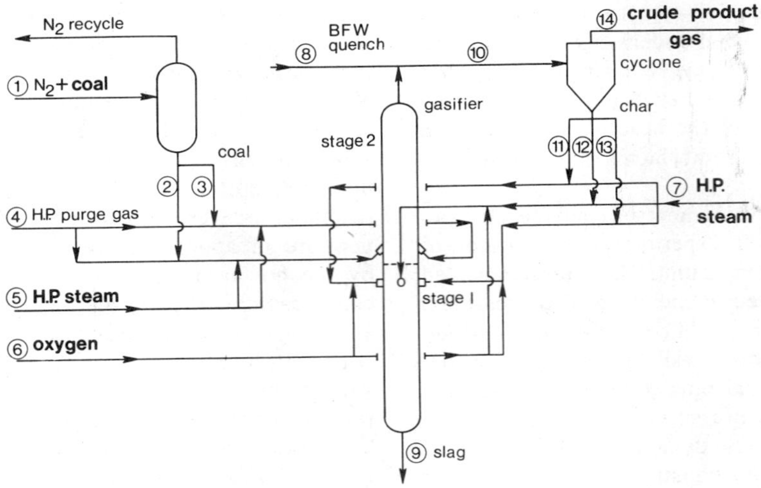

Figure 24.69 Flowsheet for a 5-ton/hr Bi-Gas pilot plant[184]; material balance data appear in Table 24.43.

plied. The upward gas velocity in this stage is approximately 6 ft/s resulting in a residence time of around 2 s. The gas produced, together with any unreacted char and entrained slag, passes through the constriction between the two stages into the upper stage. The gas may be partially quenched with steam between the two stages to solidify any entrained slag. The upper part of Stage 1 is conically shaped, leading to a 9-in. throat. This design encourages the separation of molten slag before the stream enters Stage 2, and also results in a high gas velocity at the throat so that unreacted coal or char does not fall from Stage 2 into Stage 1.

The temperature of the gasifier shell is limited by water-cooled tubes, located between the refractory and the pressure shell of Stages 1 and 2, and by a tube bundle in the refractory of Stage 1.

Design data and a material balance for the 5-ton/hr pilot plant are given in Tables 24.43 and 24.44. Operating conditions for Stage 2 favor methane formation, which, for SNG production, decreases the amount of downstream methanation. It is claimed that this feature lowers the capital and operating costs of the overall process.

24.4.2.4 *The Babcock and Wilcox Process*

The process of the Babcock and Wilcox Company uses a single-stage, two-zone gasifier. In the lower zone coal and recycled char are contacted with oxygen (or air) at 3000–3400°F (1650–1855°C) and molten slag is removed from its base. In the upper zone of the gasifier, product gas is cooled to between 1600 and 1800°F (870–980°C).

Early development work on this process was performed by Babcock and Wilcox in cooperation with the USBM. In 1951 a 36-ton/day pilot plant was erected at the Belle Works of the Dupont Company in West Virginia. It provided data for the construction of a 400-ton/day commercial plant at the same site in 1955. Information on these gasifiers is given in the first Supplementary Volume.[196]

In the early 1960s a 60-ton/day, air-blown pilot plant was constructed by Babcock and Wilcox at Alliance, Ohio, for the study of combined-cycle power generation.

The Babcock and Wilcox gasifier has a vertical, cylindrical pressure vessel of carbon steel (Fig. 24.70), housing a cylindrical "membrane" wall of water-cooled tubes, which extend to the height of the two gasifier zones. In the lower gasification zone these tubes are covered with high-alumina refractory (Fig. 24.71), so that the high temperatures necessary to keep the slag in a fluid form are maintained. In the upper zone max-

[196] H. H. Lowry, Ed., *Chemistry of Coal Utilization*, Supplementary Volume, John Wiley and Sons, New York, 1963, p. 974.

imum cooling of the product gas is obtained by having the tubes exposed.

Pulverized coal, oxygen, and steam are injected through two or more rows of nozzles near the bottom of the lower zone. Centrifugal forces cause the molten slag to impinge on the refractory walls, and it is continuously withdrawn into a quench tank. Char and carbon are separated using cyclones from the product gas, and reinjected into the gasification zone.

The concept of an oxygen-blown 1000-ton/day coal plant is illustrated in Fig. 24.72; and summaries of performance data and material balance data are given in Tables 24.42 and 24.45.

24.4.2.5 *The Texaco Process*

The coal gasification process being developed by Texaco Development Corporation, United Stages, is closely related to the commercially proved Texaco partial oxidation process for oils. Development work on the Texaco coal gasification process was started in 1948 and has involved pilot-plant operation (up to 15 tons/day) on 20 solid fuels, including lignites, bituminous coals, anthracites, coal liquefaction chars, and petroleum cokes.[198] A 100-ton/day demonstration plant, erected at the Morgantown Ordnance Works in West Virginia in 1956, was in operation for 2 years. The results confirmed scaleup criteria and demonstrated the gasifier design and the ash-handling system.

The present Texaco coal gasification process has been simplified and improved as a result of tests using the full-scale heavy-oil partial oxidation unit at the Texaco refinery in Wilmington, California. In the present system the coal is ground in conventional equipment and slurried in oil or water. This slurry is then fed at pressure through a specially developed burner into a refractory-lined gasifier.

Partial combustion with oxygen takes place in the downflow reactor at a pressure of between 100 and 800 psig and a temperature of between 2000 and 2800°F (1095–1540°C). The gas produced is quenched by direct contact with water, and the slag is removed from the bottom of the gasifier by a lock hopper system. Performance data for a pilot plant with a 11.75-ft^3 reactor are included in Table 24.42.

Texaco has patented a high-Btu gasification process. This conceptual process involves the partial oxidation of carbonaceous fuel with oxygen in the presence of carbon monoxide, hydrogen, and steam at pressures of between 300 and 4500 psig and temperatures of between 1900 and 2200°F (1040–1205°C) in a conventional Texaco partial oxidation reactor. The product gas passes through a carbon dioxide removal stage to a cryogenic separation unit. The carbon monoxide and hydrogen are recycled to the gasifier, leaving a product gas rich in methane.

24.4.3 Processes Yielding a Nitrogen-Containing Product Gas

24.4.3.1 *The Combustion Engineering Process*

The Combustion Engineering process for producing low-Btu gas consists essentially of an atmospheric pressure, air-blown, entrained-flow gasifier (Fig. 24.73) with lower combustor and upper reductor reaction sections separated by a reduced cross section throat or diffuser zone.[186] This throat is intended to minimize mixing of reactants between the sections. The design of the gasifier is similar to that of a pulverized-coal-fired boiler, being constructed from fusion-welded, tubular, water-cooled walls supporting an inner refractory lining.

Part of the pulverized coal (70% through 200 mesh) and recycle char, transported by air preheated to 300°F (150°C), is mixed with a further volume of air, preheated to 600°F (315°C); the stoichiometric mixture is fed to the combustor section through nozzles that are arranged tangentially to promote mixing. Molten slag drops into a quench and the hot [3200°F (1760°C)] combustion products flow through the throat to the reductor section. It is thought that about 50% of the slag will be directly quenched, the balance being carried upward with the unconverted carbon.

In the reductor the remainder of the pulverized coal is injected tangentially into the

[197] H. M. Majdeski and M. P. May, Babcock and Wilcox Power Generation Group, Barberton, Ohio, private communication, October 1976.

[198] P. L. Paull, Texaco Development Corporation, private communication, October 1976.

Table 24.43 **Design Material Balance for a 5-ton/hr Bi-Gas Pilot Plant (Fig. 24.69)**[134]

	Stream 1: Coal and Nitrogen		Streams 2 and 3: (each) coal	Stream 4: High-Pressure Purge Gas to Stage II		Stream 5: High-Pressure Steam to Stage II		Stream 6: Oxygen to Stage II	
	lb/hr	mole/hr	(lb/hr)	lb/hr	mole/hr	lb/hr	mole /hr	lb/hr	mole/hr
Coal (maf)	9097		4549						
Char (maf)									
Ash	703		352						
CO_2				1009	22.9				
CO				437	15.6				
H_2				75	37.1				
CH_4				112	7.0				
N_2	223,700	7989						32	1.1
H_2S									
H_2O	17,800	988	100						
Steam						9097	504.9		
O_2								6445	201.4
Total	251,300	8977	5001	1633	82.6	9097	504.9	6477	202.5
				Total in Streams 2–8: 46,100 lb/hr					
Average molecular weight	26.9			19.8		18.0		32.0	
Temperature (°F)	550		475	200		800		60	
Pressure (psig)	1175		1175	1250		1250		1250	
Volume (scf/min)	59,705			521		3189		1280	

Stream 7: High-Pressure Steam to Stage I		Stream 8: Boiler Feed Water Quench	Stream 9: Slag	Stream 10: Crude Gas from Gasifier		Streams 11, 12, 13: (each) Char	Stream 14: Gas from Cyclones	
lb/hr	mole/hr	(lb/hr)	(lb/hr)	lb/hr	(mole/hr)	lb/hr	lb/hr	mole/hr
				8,523		2754	260	
			668	1,152		372	35	
				10,762	244.5		10,762	244.5
				9,442	337.1		9,442	337.1
				937	463.9		937	463.9
				1,649	102.8		1,649	102.8
				62	2.2		62	2.2
				283	8.3		283	8.3
		15,945		22,001	1,220.9		22,001	1,220.9
2947	163.7							
2947	163.7	15,945	668	54,811	2379.7	3126	45,431	2,379.7
				Total out, Streams 14 and 9: 46,000 lb/hr				
18.0				19.0			19.0	
800		230	170	800		800	800	
1850		1150	1150	1,150		1145	1145	
1034				15,032			15,032	

Table 24.44 Design Data for the Bi-Gas Gasifier[184]

Operating pressure, gas side	500–1500 psig
Process gas outlet temperature	1800°F
Shell design pressure	1650 psig
Shell design temperature	650°F
Shell (60 in. I.D., 3⁹/₁₆ in. thick)	2¼ Cr, 1 Mo
Tubes (1½ in. O.D., 0.203 in. thick)	2¼ Cr, 1 Mo
Total gasifier weight	140 tons

hot stream flowing up from the combustion section. Gasification occurs to produce a low-Btu gas containing a small amount of methane with an HHV of 100–125 Btu/scf. At a nominal 2.5-s residence time it is expected that the gases will leave the reductor at 1700°F (930°C),[186] eliminating the need for a water quench. The gases are cooled to about 300°F (150°C) by a heat recovery train before passing to the gas purification stage.

In late 1974 the then OCR, the Electric Power Research Institute, and Combustion Engineering agreed to the further development of this process, and a 5-ton/hr pilot plant is currently (1978) under construction. The gasifier is approximately circular in cross section, with a maximum internal diameter of 10 ft, a dimension set by the dynamic flow modeling requirements necessary to simulate the gas velocity and mixing criteria for a commercial gasifier. The height of the pilot-plant gasifier (150 ft) is set by the predicted residence time necessary for the reactions in the reductor section. Heat transfer losses to the water-cooled walls of a commercial gasifier of this type are predicted to be less than 10% of the total heating value of the coal feed if the inside surface of the welded walls has a thin refractory lining. Heat transfer from the pilot-plant gasifier is therefore being controlled by the use of refractory and insulation materials, to emulate the heat transfer characteristics of a commercial gasifier. Predicted design performance data for the pilot plant (Fig. 24.74) are given in Tables 24.42 and 24.46.

Under the present research program the completion of pilot-plant testing is scheduled for late 1978. Combustion Engineering is negotiating with ERDA and others for the design and construction of a 150-MW dem-

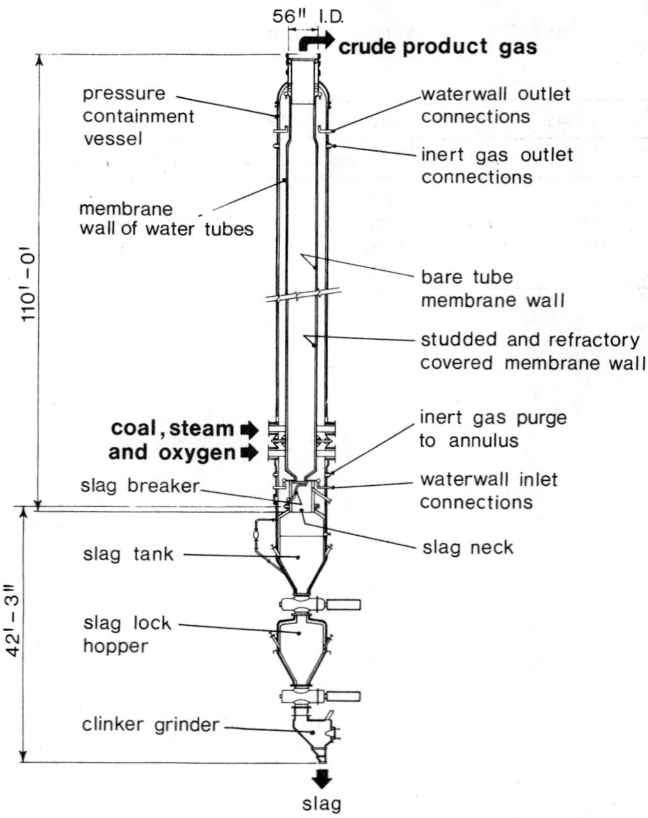

Figure 24.70 The Babcock and Wilcox gasifier.[197]

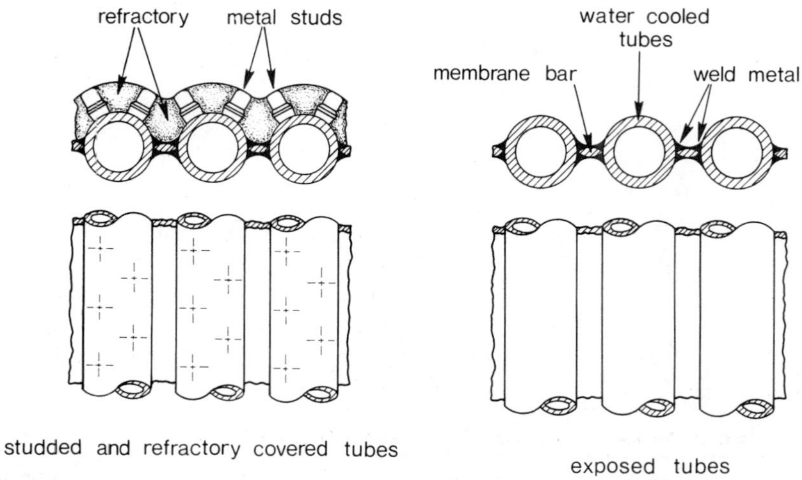

Figure 24.71 Membrane walls for the Babcock and Wilcox gasifier.[197]

onstration plant at an existing electric utility plant. Construction and test operation of the demonstration plant should be completed by the end of 1982. It is considered[199] that the main application of the process will be to make low-Btu gas for electric power plants, but oxygen blowing of the gasifier to produce a 295-Btu/scf fuel gas may also be investigated.

[199] R. C. Patterson, Product Development Manager, Coal Gasification C-E Power Systems, Combustion Engineering, Inc., Windsor, Conn., private communication, November 1976.

24.4.3.2 *The Foster Wheeler Process*

The Foster Wheeler process utilizes an air-blown version of the Bi-Gas gasifier (Section 24.4.2.3) to produce a low-Btu gas (150 Btu/scf) for gas turbine fuel. For this product the gasifier would operate at a pressure of between 200 and 1500 psig and temperatures of between 1700 and 2100°F (925–1150°C) for the upper stage, and 2800 and 3000°F (1540–1650°C) for the lower stage.

Practical development so far has been limited to tests carried out by Bituminous Coal Research, Inc., using the two plants of 5- and 100-lb/hr capacities. In these tests sev-

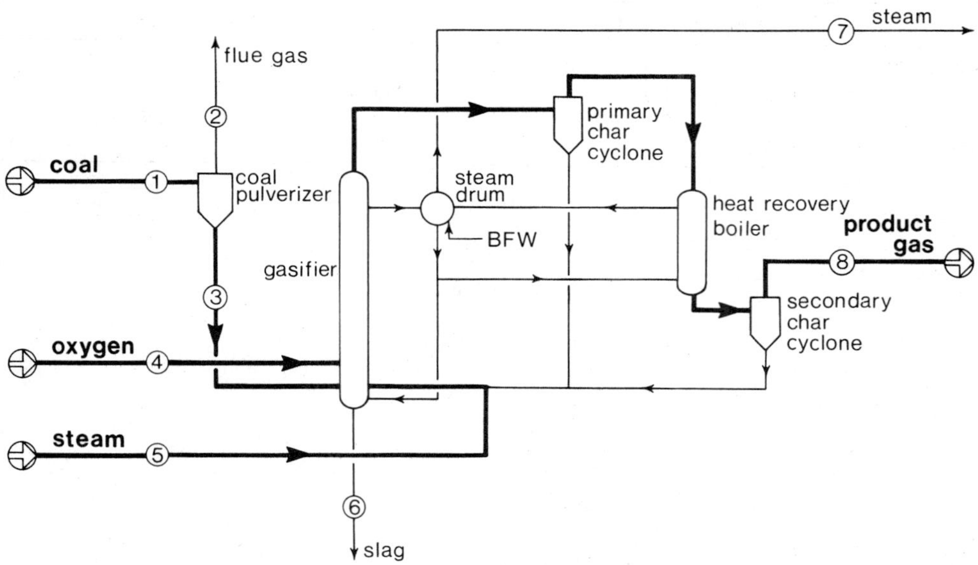

Figure 24.72 Flowsheet for the Babcock and Wilcox process[197]; material balance data appear in Table 24.45.

Table 24.45 Material Balance for the Babcock and Wilcox Process (Fig. 24.72)[147]

	Stream 1: Coal (lb/hr)	Stream 2: Flue Gas From Pulverizer lb/hr	mole/hr	Stream 3: Pulverized Coal to Gasifier lb/hr	mole/hr	Stream 4: Oxygen to Gasifier lb/hr	mole/hr
Coal/Carbon (maf):	68,051	7.2		68,044			
Ash:	6,192	0.7		6,191			
Moisture:	8,249	0.1		1,515			
Total coal:	82,492	8.0		75,750			
CO							
CO_2		4,417	100.4				
H_2							
H_2S + COS		1.3	0.02				
O_2		26,898	840.6			70,528	2,204.0
H_2O		9,246	513.7				
N_2 + Ar		96,538	3,444	2,756	97.9	1,798	45.0
NH_3							
Total	82,492	137,108		78,506		72,326	
Temperature (°F)	77	150		77		77	
Pressure (psig)	0	0.2		200		225	
Flow (scf/min)		30,966				14,206	
Flow (ton/day)	989.9			909.1		868	

Stream 5: Steam to Gasifier		Stream 6: Slag from Gasifier		Stream 7: Steam to Superheater		Stream 8: Crude Gas from Gasifier After Particulate Removal	
lb/hr	mole/hr	lb/hr	mole/hr	lb/hr	mole/hr	lb/hr	mole/hr
		284				6	
		5,015				6	
						114,666	4,095.1
						22,422	509.3
						4,185	2,092.4
						2,169	60.9
15,150	840.9		27.8	215,646	11,969.7	11,937	662.5
						5,291	169.8
						0.7	0.04
15,150		5,299		215,646		160,683	
400		95		491		200	
225		0		615		145	
						47,171	
		69.6					

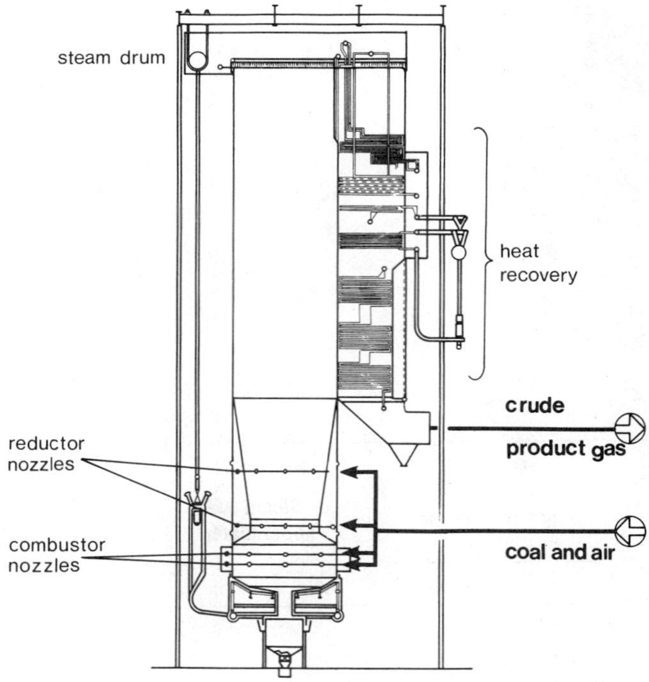

Figure 24.73 Conceptual diagram of a commercial Combustion Engineering gasifier.[186]

eral types of coal, including Pittsburgh Seam No. 8, have been gasified using air. Performance data for the larger pilot plant are given in Table 24.42.

In August 1972 a consortium led by Foster Wheeler Energy Corporation[200] was formed to design and build a 25-ton/hr, low-Btu demonstration plant to be located at Northern States Power Company's Lawrence plant near Sioux Falls, South Dakota. This plant was to operate at 440 psig, using a lock hopper system to feed the pulverized coal (70% through 200 mesh) to the upper reaction stage (although it was anticipated that a steam jet injector will be developed for the commercial plant).[187] It was also planned to construct a hot-flow model with a coal feed rate of 0.75 ton/hr. Information from the Sioux Falls plant will be used to prepare a conceptual design for a 400–1000 MW electric power plant.[201]

[200] I. Howard-Smith and G. J. Werner, *Coal Conversion Technology,* Noyes Data Corporation, Park Ridge, N.J., 1976.

[201] Anon., "Combined-Cycle Gasification Pilot Plant to be Developed by Foster Wheeler," *Chem. Eng. Prog.,* **71**(4), 100 (April 1975).

24.4.4 Commentary on Entrained-Flow Gasifiers

The only fully entrained process yet to have achieved commercial status is that of Koppers-Totzek. Although this process is well proved and the gasifiers have a fairly simple design, it does have a relatively high oxygen demand; but the steam requirement is very low, or zero, since moisture can be present in the fuel. Its main disadvantage is its very low operating pressure. Nevertheless, the main use of the product gas has been for high-pressure ammonia synthesis. The planned extension of the process to high pressures will involve a major redesign of equipment for coal feeding and the removal of ash and char.

The Texaco process has achieved the feeding of a pulverized fuel into a pressurized gasifier by utilizing a coal-liquid slurry. However the resultant additional heat requirement is reflected in the very high oxygen demand.

For SNG manufacture, a disadvantage of both these processes is that they yield virtually no methane. Gases containing significant amounts of methane appear to have

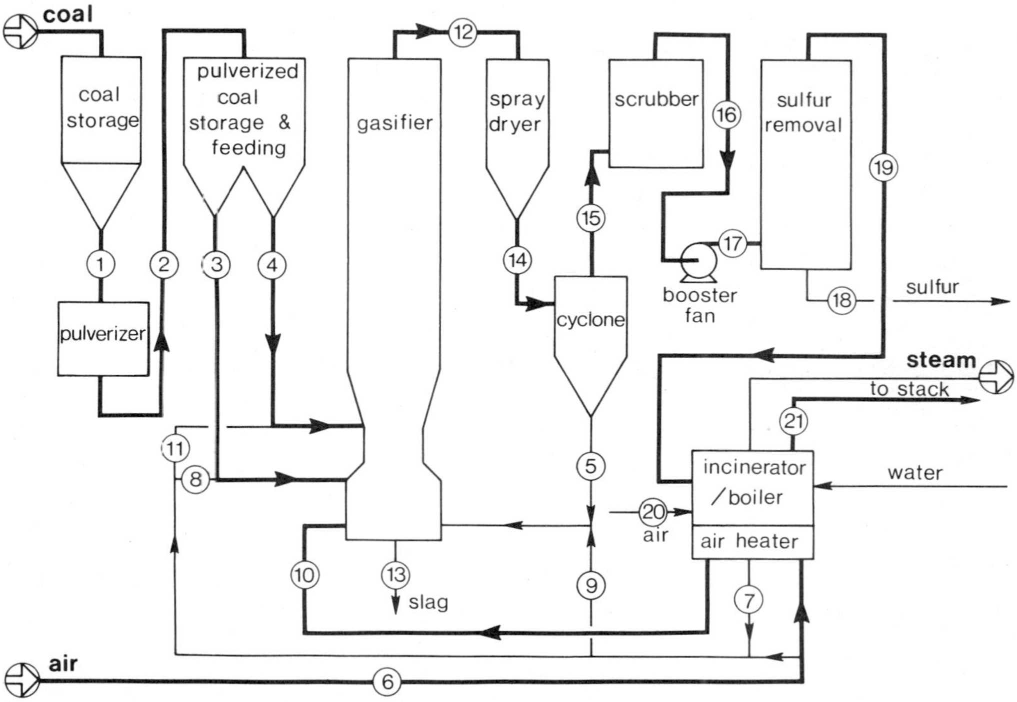

Figure 24.74 Flowsheet for the Combustion Engineering pilot plant[199]; material balance data appear in Table 24.47.

been obtained in the Bi-Gas process, by using two stages.

Conditions chosen to promote methane production in the upper stage are not conducive to good carbon conversions, and so another stage is required that operates under more severe conditions in which the less-reactive carbon can be gasified.

Complete carbon gasification may not be necessary if the char can be consumed in another process, in which event the ORC pyrolysis route provides a possible means of making a gas containing substantial amounts of methane. The efficiency and specific gasification rate data (Table 24.42) for this process cannot be meaningfully compared with other processes because the gasification is incomplete.

The much lower carbon inventory and thermal capacity of entrained-flow gasifiers, compared with the fixed- or fluidized-bed types, causes temperature control and product gas compositions to be much more sensitive to any minor fluctuations in reactant feed rates. Interruption of fuel supply is a particular hazard, which requires protective instrumentation. Although it has been demonstrated that the feed rates can be held sufficiently constant at low pressure in commercial installations (of the Koppers-Totzek process), this has yet to be proved at elevated pressures. However the lower thermal capacity of these units does enable them, within their turndown range, to follow fluctuating load demands more easily.

In all entrained-flow gasifiers the ash entrained in the crude product gas must be cooled below its softening temperature before the gas enters a waste heat recovery train. Often this has been achieved by providing additional residence time within the gasifier to allow either endothermic reactions to proceed or sensible heat to be removed. The height of the Combustion Engineering process gasifier illustrates the consequence of this approach. An alternative approach, used in the Bi-Gas process, is to quench the gas with water, simultaneously providing, in the gas, the steam required for any subsequent carbon monoxide conversion stage,

Table 24.46 Material Balance for Combustion Engineering Pilot Plant (Fig. 24.74)[199]

Stream:	1	2	3	4	5	6	7	8	9
Solids component flows (lb/hr)									
Coal	10,000	9858	3195	6663					
Char					3268				
Ash and carbon									
Sulfur									
Gas component flow (lb/hr)									
CO						0	0	0	0
CO_2						0	0	0	0
H_2						0	0	0	0
H_2O						632	470	62	63
N_2						37,310	27,763	3,636	3,719
H_2S						0	0	0	0
COS						0	0	0	0
O_2						11,238	8,362	1095	1120
SO_2						0	0	0	0
Total flow (lb/hr)	10,000	9858	3195	6663	3268	49,180	36,595	4793	4902
Average molecular weight						28.62	28.62	28.62	28.62
Volumetric flow (acf/min)						11,292	17,515	1473	1506
Temperature (°F)	80	180	160	160	160	80	689	300	300
Pressure (in. w.g.)	0	.9	0	0	0	0	+8.5	+20	+20

10	11	12	13	14	15	16	17	18	19	20	21
		3,268		3,788	520						
			869								
								234			
0	0	15,086		15,086	15,086	15,086	15,086		15,086	0	0
0	0	3973		3,973	3,973	3,973	3,973		3,973	0	27,704
0	0	473		473	473	473	473		473	0	0
377	129	946		3,926	3,926	5,296	2,553		2,612	767	7,603
22,370	7586	37,399		37,399	37,399	37,399	37,399		37,399	45,309	82,708
0	0	259		259	259	259	259		7	0	0
0	0	34		34	34	34	34		34	0	0
6,738	2285	0		0	0	0	0		0	13,648	1,242
0	0	0		0	0	0	0		0	0	51
29,485	10,000	61,438	869	64,938	61,670	62,520	59,777	234	59,584	59,724	119,308
28.62	28.62	25.75		25.22	25.22	25.00	25.45		25.42	28.62	29.50
14,232	3080	21,234		20,404	20,689	19,018	14,059		15,299	13,714	54,089
689	300	300	3000	217	217	118	100	80	100	80	630
+5	+20	−6.5	0	−9.0	−14	−30.5	+29.5	0	+19.5	0	+1

thus eliminating expensive high-temperature heat recovery equipment.

Virtually all types of coal can be gasified in entrained-flow gasifiers, which provide higher specific gasification rates and lower residence times than either fixed- or fluidized-bed processes. The problems associated with coal feed and ash removal, however, still impose major difficulties.

24.5 MOLTEN BATH GASIFIERS

24.5.1 Introduction

Most of the molten bath processes involve the gasification of coal in direct contact with steam and either air or oxygen in a bath of molten slag, metal, or salt. Development of such systems has been encouraged by their potential to overcome the disadvantages of gas-solid reaction systems. High temperatures are normally required to maintain the bath molten, and this favors high reaction rates, hence high throughputs. Gasification is also often enhanced by catalytic properties of the melt. The high thermal capacity of the melt causes rapid heating of the fuel on entry to the gasifier, so that tars and oils are not formed. The rapid release of volatile matter causes the coal to disintegrate, resulting in a large increase in surface area, which in turn also increases the reaction rates. Unlike fixed bed, fluidized-bed, and entrained-flow processes, coal size is not unduly restricted, and the majority of molten bath gasifiers utilize coarsely ground coal, including fines. Advantages of molten bath processes are that strongly caking coals—even with high ash or sulfur contents—can be gasified, and retention of sulfur in the melt can result in the production of an essentially sulfur-free gas stream.

Many of these advantages are attributable to the constant high temperature of the melt, a feature that is also responsible for the main disadvantages of molten bath processes: high heat losses and the problems associated with the containment of the melt and the removal of slag and ash. Severe corrosion problems are encountered with high-temperature molten salts, and particularly with molten metals. Early gasifiers—for example, those used in the various Rummel processes—relied on the cooling of a steel shell to produce a protective crust of slag on the inside of the walls, which reduced the corrosive effect of the melt on the steel. This resulted in very high heat losses from the gasifier. Later designs of gasifiers, used in the Atgas and Kellogg processes, have refractory-lined walls. Although they reduce the heat losses, the durability of these refractories has yet to be proved over long periods of operation.

The material employed for the melt (e.g., slag, iron, molten salts) largely determines the characteristics of each particular process.

1. Slag acts mainly as a heat transfer medium but can also sometimes exert a catalytic effect. For example, ferric oxide in the slag can oxidize carbon:

$$Fe_2O_3 + C \rightarrow 2FeO + CO \qquad (24.17)$$

   The ferrous oxide so produced is reoxidized by the gasifying medium, for example, with oxygen:

$$2FeO + 0.5O_2 \rightarrow Fe_2O_3 \qquad (24.18)$$

2. Iron is used because of its strong affinity for sulfur (as seen by the difficulty in making low-sulfur iron and steel) and because any iron oxide present in the melt will donate oxygen.
3. Molten salts, which are generally less corrosive and have lower melting points than molten metals, can strongly catalyze the basic steam-coal reaction, permitting essentially complete gasification of some coals at relatively low temperatures. Sodium carbonate was the obvious choice, but the use of sodium sulfate has also been suggested, since it provides not only the heat but also the oxygen required for the gasification reactions.

24.5.2 Processes Yielding a Nitrogen-Free Product Gas

24.5.2.1 *The Rummel-Otto Single-Shaft–Saarberg-Otto Processes*

The first pilot-scale (32-in.-I.D.) Rummel-

Otto gasifier was installed by Union Kraftstoff in 1950, based on the inventions of Rummel.[202] It was located at Wesseling, West Germany, and was in operation for several years. A maximum gas output of 3 million scf/day was recorded.

A single-shaft gasifier of 6 ft I.D., which was designed to produce approximately 13 million scf/day of medium-Btu gas from 250 tons/day of brown coal using an oxygen blast, was commissioned on the same site in 1956. Performance data for these gasifiers are given in the first Supplementary Volume.[203]

A 360-psig demonstration-scale gasifier is now under construction as part of an energy research program instituted by the government of West Germany. The gasifier (Fig. 24.75) used in the pressurized Saarberg-Otto version of the process is 4.5 ft I.D. and 50 ft high, consisting of three stages. Stage 1 (at the bottom) is a 6.5-ft-high chamber containing the molten slag bath; Stage 2 is a 20-ft-high, entrained-flow zone; and Stage 3 (at the top) is a gas-cooling section. Stages 1 and 2 have walls that contain cooling tubes.

Dried, pulverized coal (100% through 16 mesh/1 mm) is pneumatically transported by means of an inert gas to a coal bin, from which it is entrained in a recycled gas and injected, with steam and oxygen, into Stage 1 through nozzles (Fig. 24.75). These nozzles are tangentially directed toward the surface of the molten slag bath in Stage 1. Coal is injected in a steam-oxygen mixture such as to induce rotation of the slag bath, which is said to increase the rates of reaction (see Fig. 24.76). The slag passes through a slag tap at the base of the gasifier to be quenched in a water bath, giving granular particles.

Gas from Stage 1 passes through a throat to Stage 2, and as it does, entrained slag droplets and some of the char particles are separated as a result of centrifugal action and travel down the walls back into the slag bath. Entrained char particles are further gasified in Stage 2. The gas then passes to Stage 3, where it is quenched by contact with a recycle gas stream (cooled product gas) to solidify any remaining liquid slag droplets before leaving the gasifier.

[202] R. Rummel, "Gasification in a Slag Bath," *Coke Gas,* **21**(247), 493–501 (December 1959).

[203] H. H. Lowry, Ed., *Chemistry of Coal Utilization,* Supplementary Volume, John Wiley and Sons, New York, 1963, p. 990.

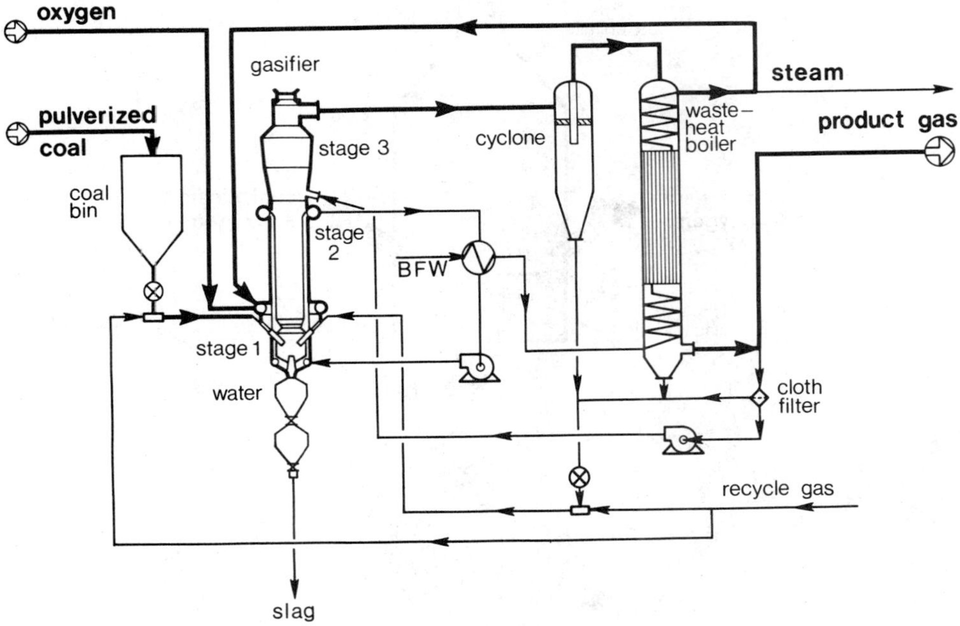

Figure 24.75 The Saarberg-Otto coal gasification process.[204]

The crude product gas, containing slag and char particles amounting to 10–30% of the coal feed, passes through a cyclone and a waste heat boiler. Slag and char particles separated in these units are pneumatically returned to the gasifier by means of recycle gas. The gas from the waste heat boiler is quenched with water before final cooling to about 210°F (100°C), to yield a medium-Btu (280 Btu/scf) gas. It is claimed that the gasifier will accept all types of coal and produce about 17 million scf/day of gas from 220 tons/day of coal. A commercial design of this gasifier is projected that will gasify about 850 tons/day of coal to produce 65 million scf/day of gas.[204]

24.5.2.2 *The Rummel-Otto Double-Shaft Process*

Parallel to the development of the single-shaft gasifier, Rummel, in conjunction with Dr. C. Otto, GmbH, performed experimental work on a small double-shaft gasifier owned by Union Kraftstoff AG at Wesseling in the early 1950s. In this atmospheric pressure process, molten slag at 2800°F (1540°C) is contained within an annular vessel (Fig. 24.77). Air and fuel (pulverized coal) are injected tangentially into the combustion shaft, and steam and coal are injected into the gasification shaft. Vertical partitions, slightly submerged in the slag bath, separate the two shafts so that the combustion and product gases leave separately. The endothermic heat of the gasification reactions is supplied by the heat generated in the combustion shaft and transferred to the gasification shaft by the circumrotating slag.[202] The performance of this double-shaft pilot gasifier was disappointing because the gas produced contained over 60% w/w nitrogen; alleviation of this problem resulted in the gas production rate dropping to an unacceptable level.

In spite of the high nitrogen content of the gas produced at the Wesseling pilot gasifier, the Gas Council (now the British Gas Cor-

[204] *Handbook of Gasifiers and Gas Treatment Systems,* Final Report, prepared by Dravo Corportion, Pittsburgh, Pa., for U.S. ERDA under Contract No. E (49-18) 1772, February 1976 (FE-1772-11).

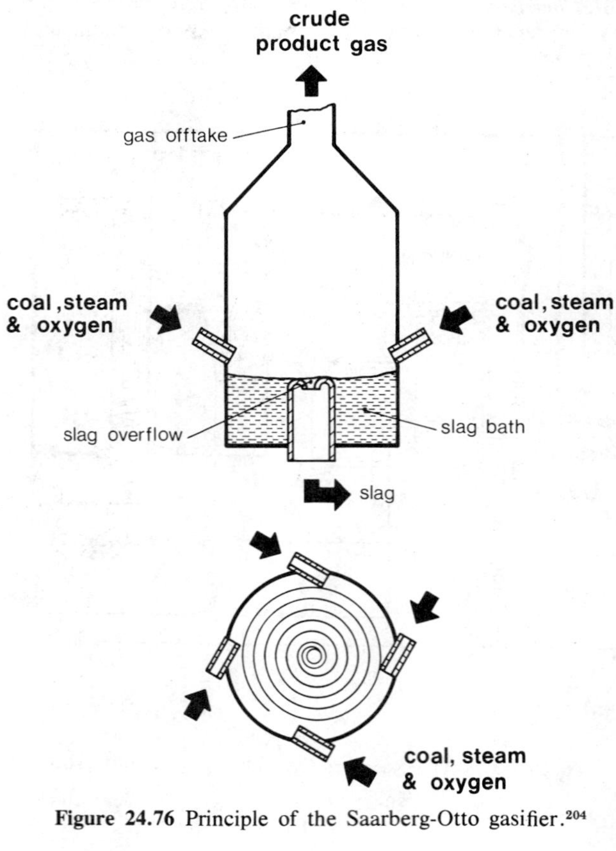

Figure 24.76 Principle of the Saarberg-Otto gasifier.[204]

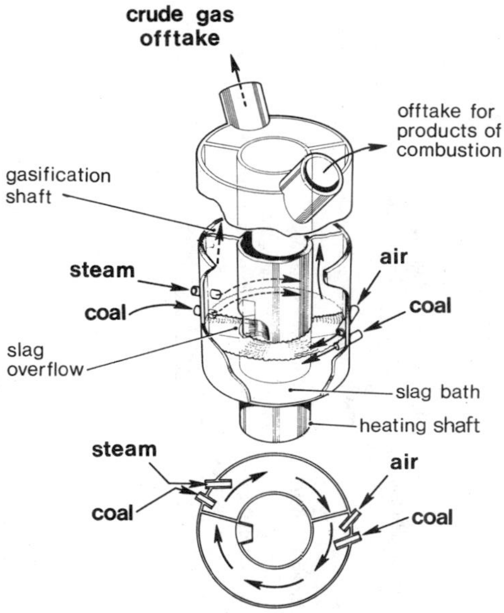

Figure 24.77 The Rummel-Otto double-shaft gasifier.[205]

poration) maintained some interest in the development of the double-shaft gasifier. A 20-ton/day gasifier with an annulus of 90 in. O.D. and 28 in. I.D. was erected at Bromley, England,[206] and was commissioned in 1962. The plant never operated above half its design capacity of 2 million scf/day because performance of the combustion and gasification sections was unsatisfactory. Changes in tuyere positioning resulted in improved combustion performance, but the limited throughput of the gasifier section continued to be a problem. Table 24.47 shows typical plant performance data, from which it is apparent that the process has very low overall thermal efficiencies (E_1 = 15–19%) and produces a product gas of relatively high nitrogen content. The low thermal efficiency was due both to the carryover of unreacted coal fines in the product gas and to the high sensible heat losses from the system. These high heat losses through the gasifier walls were required to maintain a solid slag crust on the walls to inhibit corrosion. The duration of runs never exceeded 4 days because of separation problems in the slag bath. In view of the technical and operating problems of the gasifier, its very low efficiency, and diminishing interest of the Gas Council in coal gasification processes, the project was abandoned in 1964.

24.5.2.3 *The Atgas Process*

Applied Technology Corporation has been developing the Atgas process, in which pulverized coal and limestone are injected with steam and oxygen at 50 psig into a bath of molten iron at a temperature of 2500–2600°F (1370–1425°C). Coal volatiles are released and the remaining carbon dissolves in the iron, where it is gasified to give a product comprising almost solely CO and H_2 (Fig. 24.78). The sulfur in the coal becomes combined with the limestone. This forms part of the slag layer on top of the molten iron. By balancing the coal feed and gasification rates, a steady state carbon concentration in the molten iron can be achieved. It is reported that if this concentration is maintained at a

[205] *Gas Making and Natural Gas,* BP Trading Limited, London, 1972, p. 40.

[206] M. Maccormac and J. Wrobel, *The Gasification of Coal in an Experimental Rummel Double-Shaft Slag Bath Gasifier,* Gas Council, Research Commun. No. GC113, 1964.

[207] J. A. Karnavas, P. L. Larosa, and E. A. Pelczarski, "Atgas—A New Coal Gasification Process to Produce Pipeline Quality Gas," paper presented at American Institute of Chemical Engineers 65th Annual Meeting, New York, November 1972.

Table 24.47 Performance Data for Molten Bath Processes

Gasification:	Rummel-Otto (Single Shaft)	Saarberg-Otto	Atgas	Sun	Kellogg Molten Salt		Atgas (Two-Stage)	Rummel-Otto (Double shaft)
Developer:	Dr. C. Otto & Company	Dr. C. Otto & Company	Applied Technology	Suntech Inc.	Pullman-Kellogg		Applied Technology	Dr. C. Otto & Company
Site:	West Germany	West Germany	Pennsylvania	Philadelphia	Texas		Pennsylvania	Bromley, England
Status:	Commercial	Experimental	Experimental	Conceptual[a]	Conceptual[a]		Experimental	Experimental
Fuel:	Brown coal	Bituminous	Bituminous	—	Kentucky No. 9	Kentucky No. 9	Bituminous	Gedling Smalls
Proximate analysis								
Volatile matter	—	—	—	—	—	—	—	33.2
Fixed carbon	—	—	—	—	—	—	—	51.0
Moisture	—	0	—	—	7.25	7.25	—	5.9
Ash	—	—	—	—	15.37	15.37	—	9.9
Ultimate analysis								
Carbon	—	—	—	—	61.13	61.13	—	—
Hydrogen	—	—	—	—	5.68	5.68	—	—
Nitrogen	—	—	—	—	1.41	1.41	—	—
Oxygen	—	—	—	—	12.13	12.13	—	—
Sulfur	—	—	4.0	—	4.28	4.28	4.0	—
Ash	—	—	—	—	15.37	15.37	—	—
HHV fuel (Btu/lb)[b]	9985	12,500	13,100	—	11,644	11,644	13,100	11,800
Steam/coal (lb/lb)[b]	0.13	0.4	0.11	0.48	1.80	0	0	0.24
Oxygen/coal (lb/lb)[b]	0.59	1.0	0.59	0.855[c]	0.42	0.69[c]	0.85[c]	1.67[c] + 0.20
Air/coal (lb/lb)	0	0	0	3.67	0	2.99	3.64	7.34
Gasification temperature (°F)	2900–3300	2700–3100	2500–2600	1740–2015	1700–1800	1700	2500–2600	2800
Gasification pressure (psia)	15	375	65	15	1200	215	15	15

Product gas analysis dry (% v/v)								
CO	52.0	53.6	64.5	65.0	29.5	29.1	30.0	23.0
H_2	32.1	30.7	35.0	35.0	26.2	13.9	15.0	45.0
CH_4	0.4	0.5	—	—	18.6	1.2	—	2.5
CO_2	13.6	14.0	—	—	23.2	2.6	—	8.5
N_2	1.9	0.7	0.5	—	0.8	53.2	55.0	20.2
C_2H_6	—	—	—	—	—	—	—	0.3
H_2S	—	—	—	—	1.6	—	—	—
Trace components	—	0.5	—	—	0.1	—	—	0.5
HHV product gas (Btu/scf)	275	276	320	322	383	151	145	249
Tars or oils produced	No	No	No	No	No	No	No	No
Gasifier outlet temperature (°F)	1830	1500–1700	—	1922	1700	1700	—	2200
Product gas yield (scf/lb)[b]	29.3	38.2	33.5	—	26.5	58.8	67.4	9.41
Specific gasification rates								
10^6 Btu/ft²-hr[d]	6.93	12.29	2.74	—	14.43[e]	4.06[a]	2.49	0.22
10^6 Btu/ft³-hr[d]	0.24	0.25	—	—	0.22[e]	0.76[e]	—	0.10
Carbon conversion (%)[d]	98.6	—	—	—	97.6[a]	97.7[a]	—	50.0
Gasification efficiency, E, (%)[d]	80.7	85.0	82.0	—	87.3	76.4	74.6	19.8

[a] Conceptual data.
[b] Containing moisture and ash indicated in proximate analysis.
[c] Oxygen supplied as air.
[d] See Appendix.
[e] For Pittsburgh Seam coal.

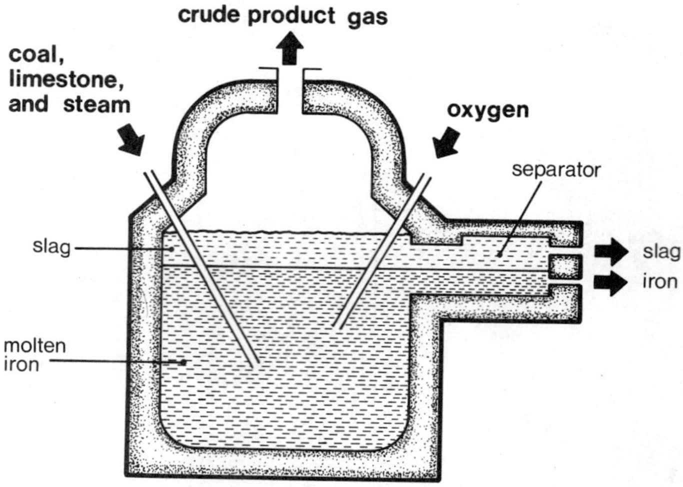

Figure 24.78 The Atgas molten iron gasifier.[207]

relatively high level (3–4%),[208] the oxygen and steam will react preferentially with the carbon without the oxidation of sulfur compounds, so that the gas produced is relatively sulfur free. The concentration of coal in the molten iron is controlled by the depth of its injection, since the solution rate is controlled by mass transfer,[209,210] and sufficient coal particle residence time must be provided. Experimental results show that the fixed carbon content of crushed (less than 1/4 in.) coal is fully soluble at injection depths of 25 in.[208,207] To ensure complete oxygen utilization, the depth need be only 4–5 in. The total depth of the iron melt will be 3–4 ft.

The limestone added with the coal not only reacts with the sulfur from the coal to produce calcium sulfide, it also fluxes the ash constituents of the coal.[207]

The slag is continuously removed and treated with steam to yield elemental sulfur, H_2S, and desulfurized slag, some of which is recycled to make use of its lime content. Sulfur can be recovered as a by-product, and the remaining desulfurized slag could be used as road-building aggregate. High sulfur retention by the slag was thought to be desirable. However bench-scale studies resulted in unacceptably high slag viscosities because sulfur contents were in excess of 14%,[211] and this could present slag removal problems. Consequently it is planned to maintain the sulfur content between 4 and 8% for the optimum flow characteristics.

The gasifier (Fig. 24.78) is a cylindrical vessel, refractory lined, with two oil-cooled injection lances; in one lance coal and limestone are transported pneumatically by steam, and the other is for oxygen. The unit is equipped with a separator for the removal of both slag and excess iron produced from pyrites in the coal.

The largest experimental unit is a batch reactor, 25 in. I.D., of 2 tons capacity, incorporating an induction furnace. The integration of gasification and melt regeneration process has not yet been demonstrated. Further work on the process has been suspended while laboratory results are evaluated.[212]

[208] J. A. Karnavas, P. J. Larosa, and E. A. Pelczarski, "Two-Stage Coal Combustion Process (for SO_2-Free Stack Gas)," in *Recent Advances in Air Pollution Control,* Am. Inst. Chem. Eng. Chem. Eng. Prog. Symp. Ser., Vol. 70, No. 137, 1974, pp. 245–249.

[209] R. J. Leary and E. J. Ostrowski, *Pneumatic Lance Injection of Carbonaceous Solids for Recarburizing Open-Hearth Melts,* U.S. Bur. Mines, Open File Rep., 1963.

[210] R. G. Olsson, V. Kowmp, and T. Penzak, "Fate of Solution of Carbon in Molten Iron Carbon Alloys," paper presented at American Institute of Chemical Engineers Annual Meeting, February 1965.

[211] P. Larosa and R. J. McGarvey, "Fuel Gas from Molten Iron Coal Gasification," paper presented at Clean Fuels from Coal, Symp., IGT, Chicago, September 1973.

[212] Anon., IGU 13th World Gas Conf., London, 1976, Report of Committee B, IGU/B-76, p. 31.

24.5.2.4 *The Sun Gasification Process*

Suntech Inc., a subsidiary of Sun Company, has patented a molten salt gasification process[213] using a bath of molten sodium sulfate, which is said to be less corrosive to the oxide refractories used as reactor linings than carbonate salts.

Coal is introduced into the molten salt and maintained at a temperature between 1740 and 2015°F (950–1100°C) at atmospheric pressure to yield a gas consisting predominantly of carbon monoxide and hydrogen. The composition of the gas can be varied by the degree of steam addition. In conjunction with the steam-carbon reaction, oxidation of the coal also takes place by its reaction with sodium sulfate:

$$Na_2SO_4 + 4C \rightarrow Na_2S + 4CO \qquad (24.19)$$

This exothermic reaction supplies some of the gasification heat requirement. The molten salt circulates continuously through an external oxidizer (Fig. 24.79), where air is injected to regenerate the sodium sulfate and to oxidize unreacted carbon:

$$Na_2S + 2O_2 \rightarrow Na_2SO_4 \qquad (24.20)$$

The hot gases from the oxidizer (N_2 and CO_2) pass through a heat exchanger to preheat the air feed. An internal screen at the base of the gasifier prevents excessive amounts of coal particles from being carried into the oxidizer with the molten salt. Addition of sodium carbonate (~ 5 wt %) to the molten salt allows sulfur removal by reaction with sulfur compounds:

$$C + H_2S + Na_2CO_3 \rightarrow Na_2S + 2CO + H_2O \qquad (24.21)$$

$$4C + SO_2 + Na_2CO_3 \rightarrow Na_2S + 5CO \qquad (24.22)$$

The sodium carbonate can be regenerated by removing a stream of molten salt and treating it with CO_2 and steam:

$$CO_2 + H_2O + Na_2S \rightarrow Na_2CO_3 + H_2S \qquad (24.23)$$

This stream of molten salt is also used to remove the ash that accumulates within the gasifier. No details of experimental equipment or data are available.

24.5.2.5 *Commentary on Molten Bath Processes Yielding a Nitrogen-Free Product Gas*

Of the four experimental gasification systems described above, the Saarberg-Otto process is the nearest to commercialization, since it is a high-pressure version of the already proved Rummel-Otto single-shaft gasifier. A comparison of data for the Saarberg-Otto and the Rummel-Otto single-shaft processes (Table 24.47) shows that the increase in operating pressure results in only minor changes in crude gas composition. The efficiency (E_1) of 85% is to some extent limited by high heat losses through the gasifier walls, which result in a high oxygen requirement.

The Rummel-Otto double-shaft gasifier was an unsuccessful attempt to produce a nitrogen-free product gas using air as the combustion medium. The Atgas process has both lower oxygen and steam demands than the Saarberg-Otto process and typically gives a gas of 320 Btu/scf with an efficiency (E_1) of 82.0%. It must be emphasized, however, that these figures are for a conceptual plant design, and the process has been demonstrated on a small scale only. The major obstacle that must be convincingly surmounted before further development can occur is the problem of melt containment, and this depends on the development of high-durability refractory materials. A low-pressure gasifier is undesirable for SNG production because of the large amount of gas compression required, the relatively low specific gasification rate, and the absence of methane in the product gas.

The Sun gasification process is conceptual, and no experimental work to validate the patent claims (Table 24.46) is planned in the near future.[213,214] The process requires relatively high oxygen and steam rates but does have the major advantage of producing a nitrogen-free gas, although the oxidant is air.

Both the Kellogg and Rockgas processes

[213] U.S. Pat. 3,770,399, Sun Research and Development Co., October 22, 1971.

[214] F. W. Camp, Suntech Inc., Research and Engineering Division, Richardson, Tex., private communication, November 1976.

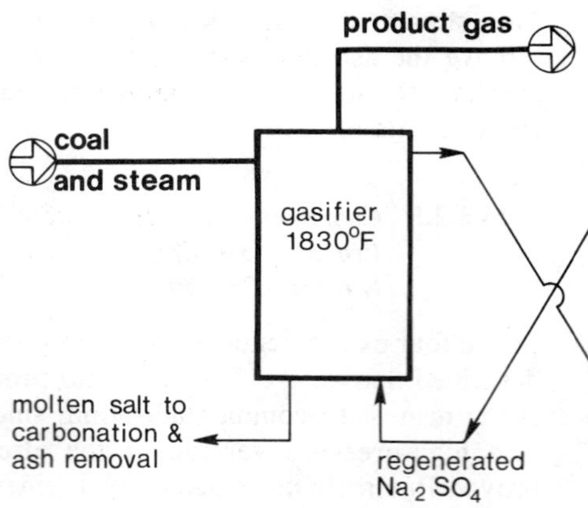

Figure 24.79 Flowsheet of the Sun gasification process.[214]

are also capable of nitrogen-free gas production, but since their main potential appears to be for the production of low-Btu gas for power raising, they are described in the next section.

24.5.3 Processes Yielding a Nitrogen-Containing Product Gas

24.5.3.1 *The Atgas Two-Stage Process*

Applied Technology Corporation's two-stage process for the production of low-Btu gas (145 Btu/scf) utilizes the same gasifier design as the Atgas process (Section 24.5.2.3).[211] There are only three differences in operation: (1) air replaces oxygen for the gasification step, (2) the diluting effect of N_2 in the air eliminates the need to inject steam to moderate the iron temperature: indeed, the combustion air must be preheated to at least 800°F (430°C) to maintain a thermal balance in the gasifier,[208] and (3) the gasifier operates at lower pressures and temperatures [2–3 psig and 2040°F (1115°C),[211] respectively].

To date experimental work has only been undertaken in a 2-ton (batch) capacity plant (also used for the Atgas process).[211] Coal and air have been simultaneously injected into molten iron and the product gas composition determined under steady state conditions. Performance data obtained for bituminous coal are included in Table 24.47.

24.5.3.2 *The Kellogg Molten Salt Process*

Ground coal (< 14 mesh/1.4 mm) is gasified in a pool of molten sodium carbonate in the Kellogg molten salt process. The crude product gas leaves at the top of the gasifier. Ash retained in the molten sodium carbonate is removed by taking a purge stream of melt from the bottom of the gasifier. Kellogg undertook basic bench-scale studies from 1964 to 1967 under OCR sponsorship. The problems of ash chemistry, gasifier mechanical design, melt handling, and materials of construction were studied. Initially molten salt was circulated between separate gasification and combustion stages (operating at 430 psia) as illustrated in Fig. 24.80, so that air could be used for combustion to eliminate the cost of oxygen. Steam was the reactant in the gasification stage. The use of two stages allows a nitrogen-free gas to be produced. However analysis of this system revealed high operating costs and high process heat requirements due to the low methane content of the product gas.

Operation at pressure, leading to an increased methane yield and reduced heat requirement, was then investigated. A pressure of 1200 psia was chosen, and oxygen was used in the combustion stage to reduce the compression requirements. Two stages (in separate vessels) were retained to prevent CO_2, produced during combustion, from mixing with the product gas. Salt circulation was induced by steam injection, as shown in Fig. 24.81.

Further investigation of the chemistry of the process revealed that at the operating temperature of 1800°F (980°C), about 6% of the sodium carbonate entering the gasifier was hydrolyzed by steam:

$$Na_2CO_3 + H_2O \rightarrow 2NaOH + CO_2 \quad (24.24)$$

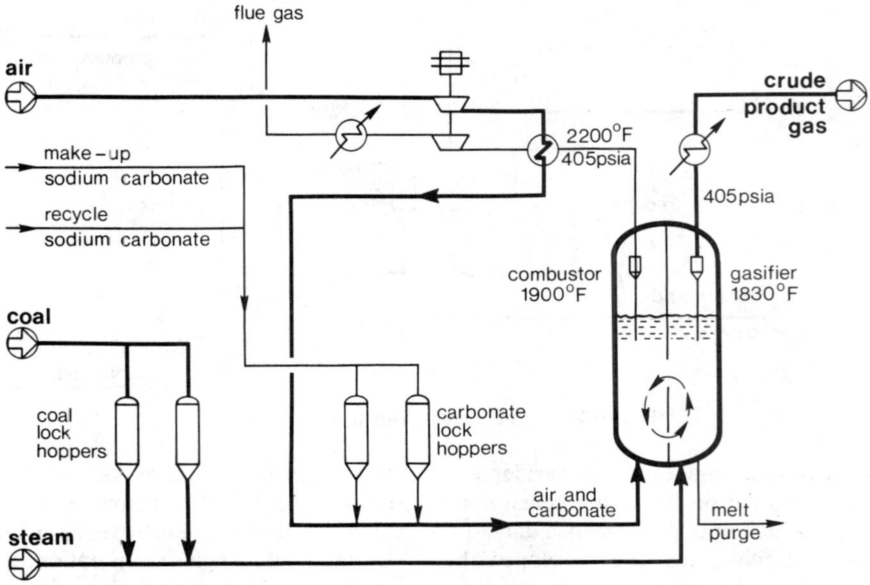

Figure 24.80 The Kellogg gasifier and combustor in a single vessel.[215]

Although this endothermic reaction is reversed in the combustor, it imposes an added heat load on the gasifier and also releases substantial quantities of CO_2 into the product gas, diminishing the expected advantage of using two stages. Potassium carbonate could have been used to minimize these effects, but this material was considered to be too costly. The basic process was modified by eliminating the separate combustor so that oxygen and steam were injected with coal into a single-stage gasifier. The present concept of the process and ash removal system is shown in Figs. 24.82 and 24.83. In 1974 Kellogg demonstrated the suitability of Monofrax "A" (a fused, castable, high-purity alumina refractory) for containment of the melt.

[215] A. E. Cover, W. C. Schreiner, and G. I. Skaperdas, "Kellogg's Coal Gasification Process," *Chem. Eng. Prog.*, **69**(3) 31—36 (March 1973).

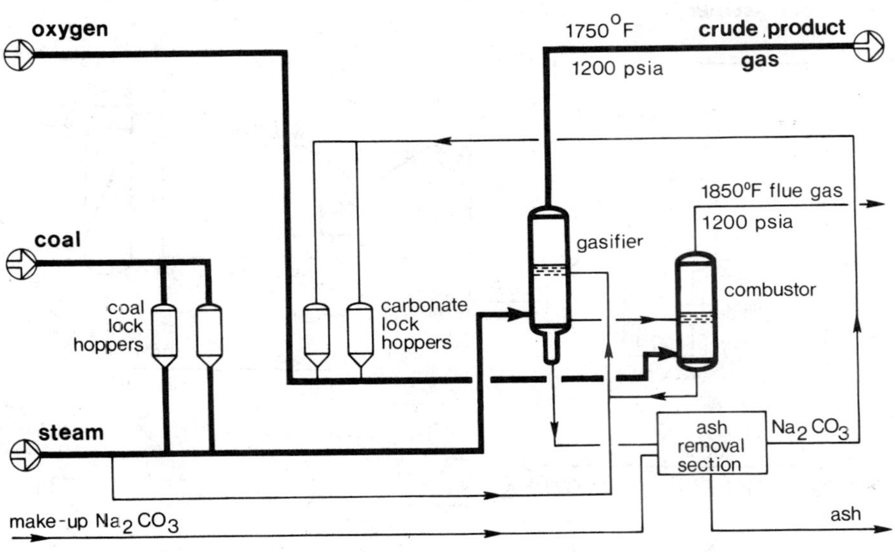

Figure 24.81 The original Kellogg two-vessel gasification system.[215]

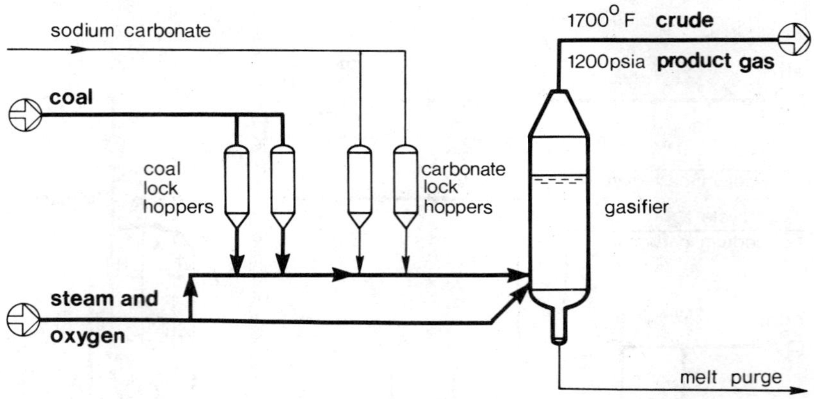

Figure 24.82 The Kellogg single-vessel gasifier.[215]

A mathematical model for the gasifier was prepared from the results of the bench-scale work for prediction of performance data (Table 24.47). Kellogg has also developed the use of air as opposed to steam and oxygen as the gasifying medium.

Dimensions suggested by the developers for conceptual gasifiers, given in Table 24.48, show the large melt volume required for steam-oxygen gasification compared with air gasification. The height of the gasifier is sufficient to allow for expansion and turbulence of the melt. A design drawing for an air-blown molten salt gasifier, with a calculated quiescent bed height of 10 ft, has a design height of 57 ft (Fig. 24.84).

ERDA is sponsoring two major programs for the development of molten salt gasification. One of these is in connection with the design of a complete combined-cycle power plant by the Power Systems Division of United Technologies Corporation. As a subcontractor, Kellogg is designing the integrated power plant components, including a molten salt gasification process, the gas turbine subsystem, and the steam system.

The second program covers the design, construction, and operation of a pilot plant for an air-blown gasifier by the Atomics International Division of Rockwell International.

24.5.3.3 *The Atomics International Rockgas Process*

The Rockgas process is similar to the Kellogg

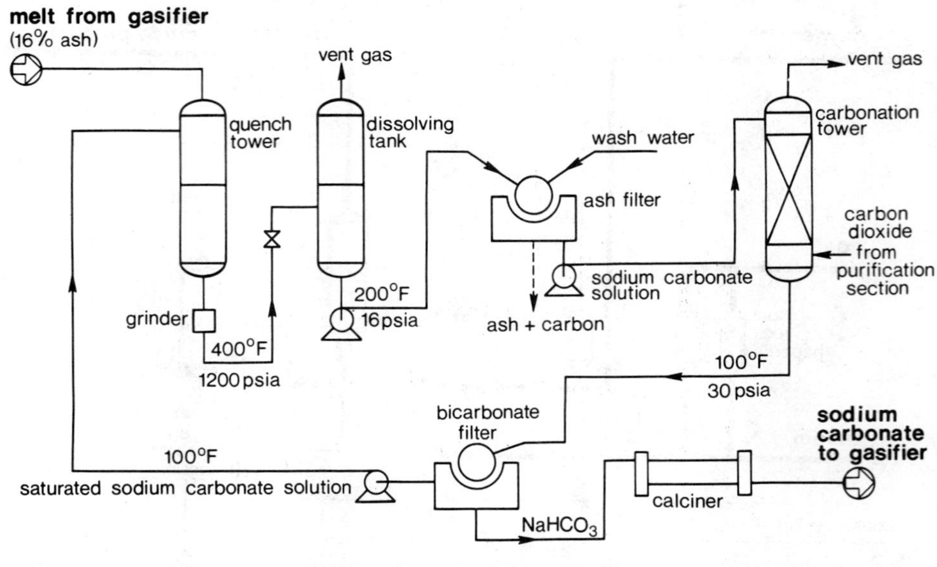

Figure 24.83 Flowsheet of the ash removal section of the Kellogg gasifier.[215]

Table 24.48 Dimensions for Conceptual Kellogg Gasifiers of Thermal Output 1 × 10¹⁰ Btu/hr[216]

Dimension	Steam-Oxygen Gasification	Air Gasification
Melt surface area (ft^2)	693	2,466
Melt volume (ft^3)	45,450	13,100
Melt height (ft)	65.6	5.3
Approximate overall gasifier height (ft)	100	50

process in that coal is gasified by reaction with air in a pool of molten sodium carbonate (Fig. 24.85) to produce a low-Btu gas of low sulfur content. It is being developed by Atomics International Division of Rockwell International and has been demonstrated in a 5-ton/day unit at Santa Susana, California, at pressures of up to 30 psig.[217]

[216] T. W. Holzberger, Pullman Kellogg, Research and Development Center, Houston, Tex., private communication, February 1977.

[217] W. V. Botts, A. L. Kohl, and C. A. Trilling, "Low-Btu Gasification of Coal by Atomics International's Molten Salt Process," paper presented at 11th Intersociety Energy Conversion Engineering Conf., Lake Tahoe, September 1976.

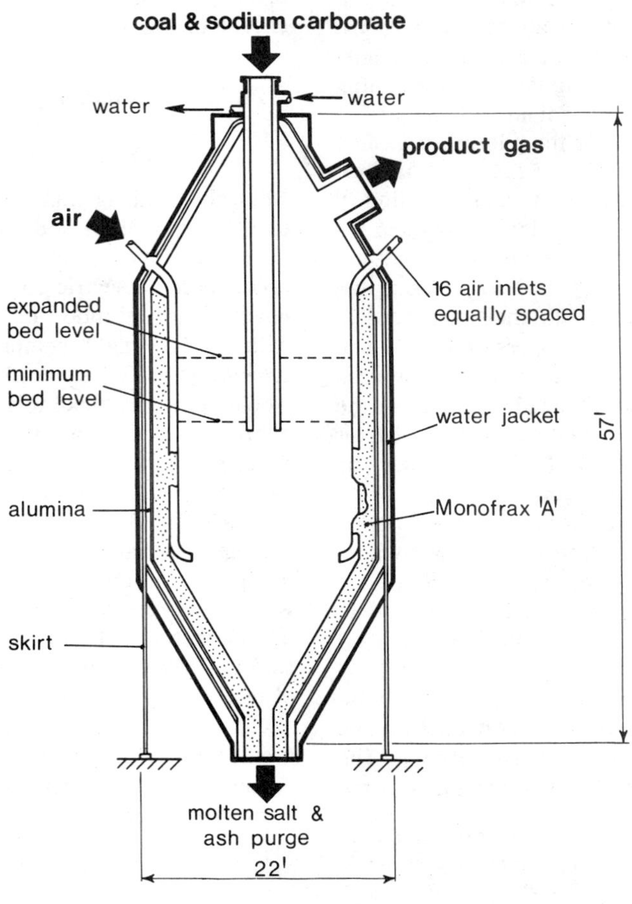

Figure 24.84 Conceptual diagram of a Kellogg molten salt gasifier.[216]

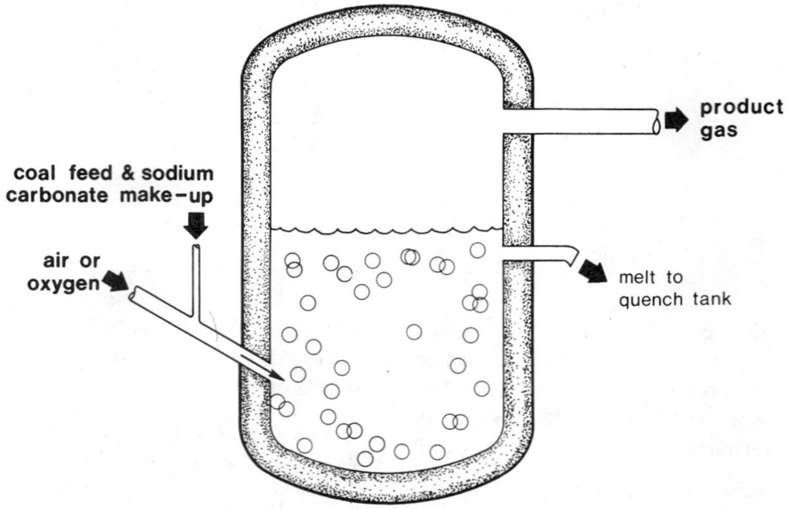

Figure 24.85 The Rockgas molten salt gasification process.

The construction of a pilot plant was scheduled for completion in late 1977. It was designed to operate at about 300 psig and 1800°F (980°C) and to be a completely integrated system including facilities for continuous regeneration of sodium carbonate.[217]

The design calls for the pilot-plant gasifier to be about 36 ft high, 6.5 ft O.D., and lined internally with castable alumina insulation.[217] It should accept 24 tons/day of coal and 9.4 tons/day of sodium carbonate. The solids are to be mixed and brought up to pressure in a mixer-transporter vessel, then conveyed in batches to four feed hoppers. The coal and sodium carbonate mixture is to be conveyed pneumatically from the feed hoppers into the gasifier, below the melt level, with the air feed. All types of coal (< 0.25 in.) may be utilized without pretreatment.

The air-to-coal ratio is chosen to obtain the desired temperature for the melt within a range set by the melting point of sodium carbonate [1570°F (855°C)] and the need to avoid excessive melt vaporization [i.e., 2000°F (1095°C)].[218]

A continuous hot stream of melt is withdrawn from the gasifier, water quenched, and sent to melt regeneration and filtration. The viscosity of the melt, which increases as the concentration of suspended particulates increases, is held at the maximum operationally acceptable value.

[218] C. A. Trilling, "Molten Salt Process for the Gasification of Coal," paper presented at Workshop on Materials Problems and Research Opportunities in Coal Conversion, Ohio State University, April 1974.

24.5.3.4 *Commentary on Molten Bath Processes Yielding a Nitrogen-Containing Gas*

Both the Kellogg and Rockgas processes, employing molten sodium carbonate, are being developed primarily for low-Btu gas production for electric power generation. The much higher retention of sulfur in the melt with air blowing[216] compared with oxygen blowing (> 85% vs. < 15% w/w retention) would appear to make this the more suitable operating mode and may even allow the crude gas to be combusted without the need for further sulfur removal.

It is claimed that the catalytic nature of sodium carbonate enables these two gasifiers to operate at lower temperatures. For the Kellogg oxygen-blown gasifier, this feature in conjunction with the high operating pressure is said to produce a substantial quantity of methane in the crude product gas.[219] This situation reduces the heat required by the gasification reactions and hence the oxygen demand. The low temperature and the less corrosive nature of the sodium carbonate melt, compared with iron, also allows a wider choice of refractory materials.

[219] Anon., "Molten Salt Coal Gasification," *Hydrocarbon Process.*, 125 (April 1975).

The complex and incomplete carbonate regeneration systems present a significant energy demand for the process, and this offsets the thermal advantage of lower gasifier temperatures. Further disadvantages include the loss of Na_2CO_3 in the ash removal system and the possibility of entrained salt in the product gas fouling downstream equipment. However the efficiency and practicality of salt regeneration and ash removal have yet to be demonstrated on a large scale.

The processes described in Section 24.5.2 produce a medium-Btu gas suitable for upgrading to SNG, whereas those described in Section 24.5.3 are primarily for the production of a low-Btu gas. The major difference in operating characteristics is that steam is generally not injected into the air-blown gasifiers, the nitrogen associated with the air serving to moderate the melt temperature. Indeed, that air is normally preheated to above 800°F (425°C) to obtain a thermal balance.

Molten bath gasifiers thus offer yet another approach to gasifying coal to provide either low- or medium-Btu gases. There are many inherent advantages, not the least of which are the potential to provide a hot sulfur-free product gas and the ability to handle strongly caking coals in a wide size range. The high temperature and often corrosive nature of the melts has, however, resulted in notable technical development problems, and only one process, the Rummel-Otto single-shaft, has so far achieved commercial status.

24.6 UNCLASSIFIED PROCESSES

This section discusses processes that either because of the nature of the gasification environment or through lack of information about the gasifier, cannot be classified according to contacting mode.

24.6.1 The Total Energy Process

The conceptual process[220,221] of the Total Energy Corporation produces a methane-rich gas from an integrated gasifier and hydrogasifier (Fig. 24.86). Part of the coal is fed to the gasifier, where it is contacted with carbon dioxide preheated externally to 2100°F (1150°C). The reaction:

$$CO_2 + C \rightarrow 2CO \qquad (24.2)$$

takes place at 100 psia. Oxygen or air is added to maintain a temperature of 2000°F (1095°C).

The desulfurized CO-rich gas is compressed to about 500 psia and passed to a CO shift stage to provide the hydrogen requirement for the process.[220] Carbon dioxide is recovered from the outlet gas in a caustic scrubber, and some is passed to the gasifier, the excess being vented through a gas turbine to produce electricity.

The hydrogen-rich gas is compressed to 1010 psia, heated to 1400°F (760°C), and passed to the hydrogasifier, where it contacts with the remainder of the coal. The temperature in the hydrogasifier is controlled at 1400°F (760°C), not by the usual means of direct steam addition but by an internal heat exchanger in which steam is raised for the CO shift stage. The product gas passes through CO_2 and H_2S removal stages to leave a gas containing 96% methane. The estimated yield based on results from bench-scale apparatus is 12 scf per pound of char.[220]

Gasifier designs have not been described, and there is no indication that development of this process is continuing.

24.6.2 The Columbia University Electric Arc Process

A process that uses a high-intensity arc in which the carbon anode is vaporized at temperatures in the region of 18,000°F (10,000°C) is being developed at Columbia University, in New York. Gas is injected into the arc column by means of a fluid convection cathode (FCC) that consists of an annular nozzle surrounding an electrode (Fig. 24.87).

Laboratory-scale experiments[222] using a carbon anode and various gaseous reactants (e.g., pure hydrogen, an equimolar H_2-CO mixture, and steam) have been conducted, using the apparatus shown in Fig. 24.88. The

[220] U.S. Pat. 3,692,506. Total Energy Corporation, February 13, 1970.

[221] U.S. Pat. 3,635,672, Total Energy Corporation, March 25, 1969.

[222] C. Sheer and S. Korman, "Arc Synthesis of Hydrocarbons" in *Coal Gasification,* Am. Chem. Soc. Adv. Chem. Ser. No. 69, 1967.

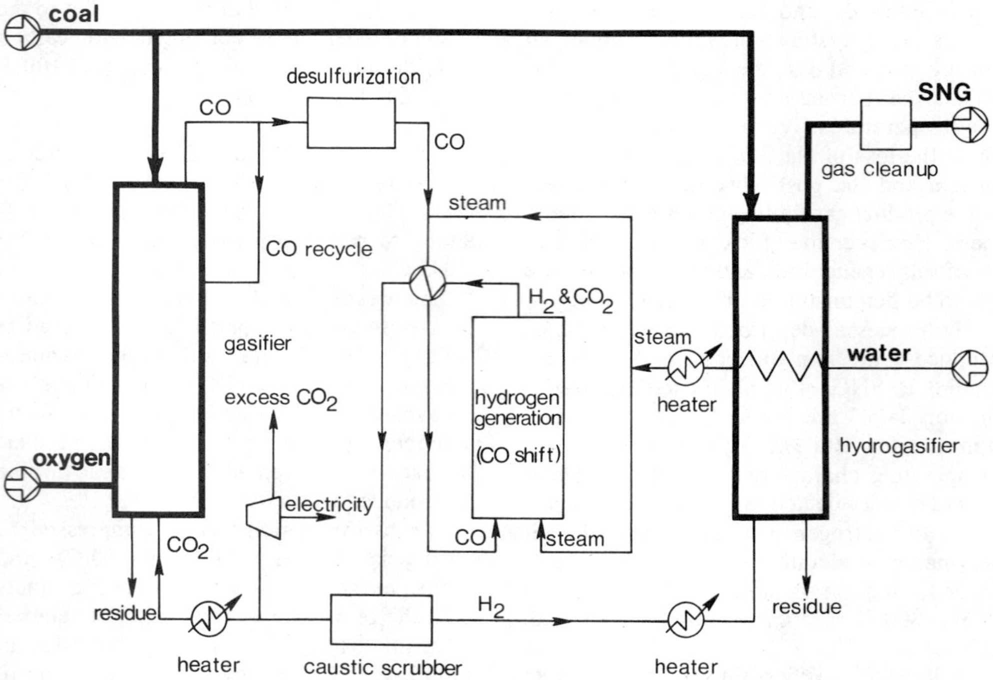

Figure 24.86 The Total Energy Corporation process.[220]

gaseous products are drawn through a tube furnace where residence times and temperatures [up to 1825°F (1000°C)] are varied to control the product distribution. Experimental results are given in the literature.[222]

It is claimed[222] that this technique could be applicable to coal gasification for SNG production, but no experimental work on coal has been reported. The process had been tested on a batch basis with a power input of about 30 kW, under the sponsorship of Consolidated Natural Gas Corporation.

24.7 CONCLUSION

Little space in this chapter has been given to a consideration of the relative ranking of the multifarious processes that have been described. There are several reasons for this approach. First, great caution needs to be exercised in comparing processes that are at different stages of development, as is particularly the case for gasification processes in this period so close to the resurgence in the interest of coal utilization mentioned in the opening section. As the development of a process passes from bench experiments to pilot plant, and on to a commercial-scale unit in a demonstration plant, and finally to full commercial exploitation, it is progressively more possible to quantify its often optimistically claimed advantages and to discover its often latent limitations.

Second, it must be remembered that even the relative ranking of processes is not universal or permanent, and a wide variety of gasification systems is needed to meet in the optimum way special requirements of each application. For instance, the intrinsic properties and size distribution of the economically available coal might govern the best type of contacting mode, especially if the feed contains a large proportion of fines, which would make its direct use in a fixed bed difficult. The final use of the gas is an important factor; for example, the production of methane in the gasifier is desirable for SNG manufacture, but not if a feed gas for ammonia or methanol synthesis is being made. The presence of nitrogen in the gasifier product gas can be particularly troublesome if a unit is producing SNG. The end use of the gas also influences the choice of operating pressure, to minimize gas compression and treatment costs.

The required plant output is another important factor with regard both to the number of gasifiers needed and to general process

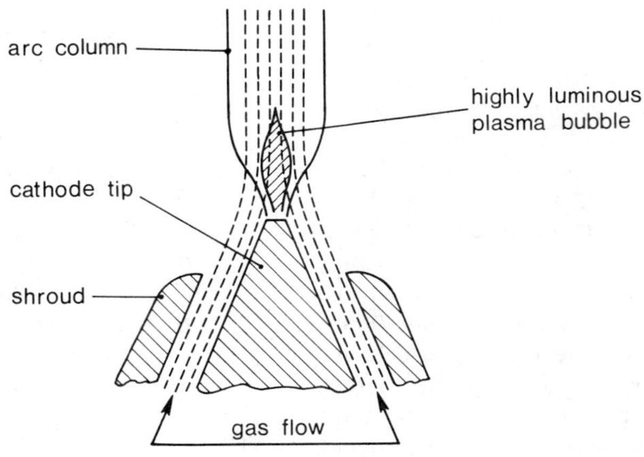

Figure 24.87 Diagram of a fluid convection cathode showing the compressive effect of gas flow on the base of the arc column.[222]

complexity. For instance, simple low-pressure, air-blown, fixed-bed gasifiers might be the best way of producing in small installations a (low CV) fuel gas for use in adjacent combustion equipment, but these might be quite impractical for SNG production. The yields of by-products can be a major disincentive to employing some processes, particularly if they may be difficult to dispose of on a regular basis throughout the expected life of the plant or if their disposal would cause environmental problems. The latter point is particularly pertinent to reactor liquors and gaseous sulfur compounds, which require expensive treatment. Perhaps the last major factors to be considered in selecting a process are its ease of operation (startup time, turndown ratio, etc.) and its flexibility toward variations in coal properties, so allowing some purchasing freedom throughout the life of the plant.

In no one process will the best solutions to all the constraints placed on a plant design be achievable. Therefore an optimum choice must be made normally within the overriding constraint of minimizing the final gas cost. This interplay leads to the third identifiable difficulty in ranking processes: the entire gasification process—including, among other factors, coal preparation and feeding, gasification and gas cooling, gas purification and treatment, char or ash disposal, the supply of all the utilities, and effluent upgrading—must be evaluated to establish the strong and

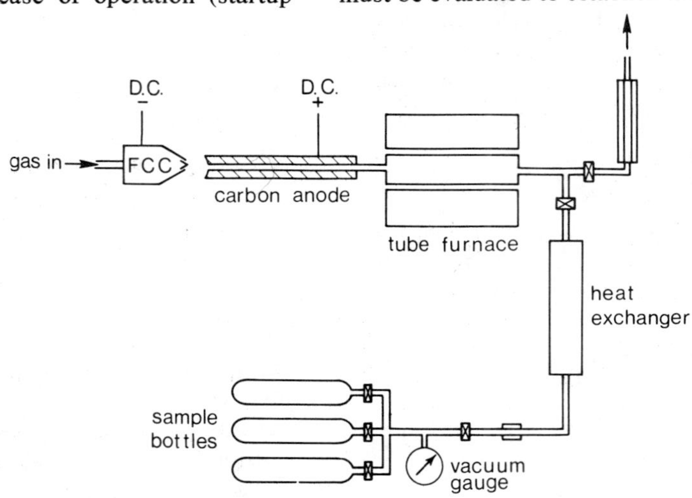

Figure 24.88 Experimental arrangement for the Columbia University electric arc process (FCC = fluid convection cathode).[222]

weak points of the process, and of course its capital cost. Often the gasification section accounts for only a relatively minor part of the capital cost (e.g., 10–15% in a typical coal-based SNG plant). However its careful design is still most important, since it normally has associated with it the greatest technological risk, and its correct choice can often result in the minimization of the complexity, therefore the cost, of the other ancillary stages. For example, an admittedly complex and expensive gasifier, which yet produces a gas with a high methane content with little accompanying tars and phenolic liquor, could be the basis of the cheapest and thermally most efficient route to SNG, because it serves to reduce gas and by-product cleanup costs and avoids expensive and inefficient methanation stages. It is not uncommon in SNG designs for the acid gas removal stages to have a higher capital cost than the gasification stage, so gasifier types that minimize the gas cleanup requirements can score in this way. Capital costs for complex gasification plants need to be critically examined if the design data are derived from small bench or pilot-plant units.

Undoubtedly the growing interest in coal gasification for various applications as a way of upgrading the vast global reserves of coal into environmentally acceptable and highly marketable fuels or chemical feedstocks will encourage the large-scale development and exploitation of many of the processes described in this chapter, and others also, so that valid process comparisons may be more readily made in the future. The development of the newer processes will only enhance the desirability of coal gasification, as already proved by the presently available commercial processes.

ACKNOWLEDGMENT

The authors acknowledge the contribution made by Dr. K. R. Tart and K. R. Wild for compilation of much of the information in this chapter.

APPENDIX: DEFINITIONS OF SPECIAL TERMS USED IN THE TABLES

Gasification Efficiencies

$$E_1 = \frac{H_1R_1}{CR_c} \times 100\%$$

$$E_2 = \frac{H_2R_2}{CR_c} \times 100\%$$

where H_1 = higher heating value of crude product gas excluding condensables (Btu/scf)

H_2 = higher heating value of crude product gas including condensables (Btu/scf)

C = higher heating value of coal feed (Btu/lb)

R_1 = crude gas production rate, excluding condensables (scf/hr)

R_2 = crude gas production rate, including condensables (scf/hr)

R_c = feed rate of coal to gasifier (lb/hr)

Specific Gasification Rates

(*a*) Based on gasifier volume

$$\frac{H_1R_1}{V}$$

where V = volume of gasifier, excluding freeboard, quench, or cooling sections of gasifier (ft^3)

(*b*) Based on gasifier cross section

$$\frac{H_1R_1}{A}$$

where A = maximum cross-sectional area of gasifier shaft (ft^2)

Carbon Conversion

$$C_c = \frac{\text{C content of crude gas including condensables (lb/hr)}}{\text{C content of coal feed (lb/hr)}}$$

Methane Produced in Gasifier

$$\frac{M_1}{M_1 + M_2} \times 100\%$$

where

M_1 = methane produced directly in gasifier (scf/hr)

M_2 = theoretical methane yield following complete conversion to methane of CO, CO_2, C_2H_4, and C_2H_6 in product gas (scf/hr)

Gas Volume

It is not always possible to establish from the literature the exact conditions under which quoted gas volumes are measured; but where possible, standard conditions have been taken as 60°F and 14.696 psia (1 atm) with dry gas. These conditions correspond to an ideal gas molar volume of 379.49 scf/lb-mole.

CHAPTER **25**

Processing of Gasification Products

HOMER E. BENSON

Homer E. Benson *was a member of the technical staff of the Research and Development Branch, Office of Synthetic Fuels, U.S. Bureau of Mines, from 1942 until 1958. His work involved Fischer-Tropsch catalysis and process development, catalytic methanation, and gas purification. The gas purification activities led to the development of the Bureau of Mines hot carbonate process. Mr. Benson worked with the Consolidated Natural Gas System and the Institute of Gas Technology on coal gasification process development. He is currently president of the Benfield Corporation, which is active in purification of synthesis gases, hydrogen, natural gas, and coal-based gases.*

Raw gas made by the gasification of coal with oxygen or air contains many objectionable impurities that must be removed or reduced in concentration to produce a commercially and environmentally acceptable fuel. This chapter discusses the impurities present in the gas, the extent to which they should be removed, and the processes that are available or under development to accomplish this.

25.1 IMPURITIES PRESENT IN COAL-BASED GAS

The effluent gas from steam-oxygen coal gasification consists principally of H_2, CO, CH_4, CO_2, and unreacted steam. The gas will also be diluted with N_2 when air is used as the oxygen source. Although the competing chemical reactions that coal undergoes during the process are complex, they are usually represented as follows. Steam and carbon react to produce H_2 and CO. Some CH_4 is produced by reaction of carbon with H_2 and by thermal cracking of heavy hydrocarbons contained in the coal. The CO_2 is produced by the combustion of carbon with O_2, which supplies the heat needed in the process. Finally the effluent composition is modified by reaction between CO and steam to produce H_2 and CO_2.

Coal contains a large number of components in addition to carbon. These also react or vaporize during gasification so that byproducts appear in the raw gas often as do undesirable contaminants. The volatile matter in coal can appear in the gas as tars, oil, naphtha, phenols, cresols, and so on. Sulfur is converted mainly to H_2S, but the gas will also contain COS, CS_2, mercaptans, and thiophene. The coal's nitrogen emerges in the gas as NH_3, HCN, and sulfocyanide compounds. Halogens are converted to their acids. Coal also contains other reactive or readily vaporized elements. Those that are considered to be a potential hazard include beryllium, arsenic, selenium, cadmium, mercury, and lead.[1]

The concentration of major and minor components in effluent gas is extremely variable, dependent on both the composition of the raw coal and the parameters of the gasification process. It has been noted[2] that the production of most by-products, except sulfur compounds and NH_3, is reduced when (*a*) the heatup rate of fresh coal is high, (*b*) the reaction temperature is high, (*c*) gas residence time is long, and (*d*) gas-solid mixing is vigorous. Differences in these operating conditions are characteristic of entrained, fixed-bed, and fluidized-bed gasifiers. Raw gas compositions that have been reported for each type of gasifier are summarized in the following sections. The information is illustrative only, since the gas made depends on the specific coal and the specific design and operating conditions.

25.1.1 Entrained Gasification

Entrained gasification is typified by high operating temperature and by rapid heating of the coal feed. Examples of entrained gasification include the commercially demonstrated K-T process and the Bi-Gas process being developed by Bituminous Coal Research. Typical effluent gas composition reported by Farnsworth et al.[3] for the K-T process is summarized below. Because of the high-temperature operating conditions, tar, oils, and phenols are not produced but are thermally cracked or hydrogenated to lower molecular weight compounds.

Component	Amount Present After Waste Heat Boiler
CO	37.36 vol %
CO_2	7.13 vol %
CH_4	0.08 vol %
H_2	25.17 vol %
N_2	0.30 vol %
H_2S	0.23 vol %
COS	178 ppm vol
HCN	288 ppm vol
NH_3	0.17 vol %
H_2O	29.19 vol %
Ar	0.32 vol %
SO_2	22 ppm vol
NO	7 ppm vol
Particulates	11.57 gr/scf

25.1.2 Fixed-Bed Gasification

In fixed- or moving-bed gasification coal slowly descends against the rising gas. The raw gas exit temperature is low and the feed coal is heated relatively slowly. Thus volatile matter is distilled from the coal and the quantity of by-products produced from the coal is high. Examples of fixed-bed gasifiers include Lurgi and Wellman Galusha commercially demonstrated units and an experimental pressurized stirred reactor operated by the U.S. Department of Energy (DOE) at Morgantown, West Virginia. The respective concentrations of major components reported for Lurgi,[4] DOE,[5] and Wellman Galusha[4] gasifiers are given below.

	Concentration (vol %)				
	Lurgi		DOE	Wellman Galusha	
Component	Oxygen	Air	Air	Oxygen	Air
CO	9.2	13.3	12.3	29.6	26.0
CO_2	14.7	13.3	10.2	12.3	3.0
CH_4	4.7	5.5	1.7	0.7	2.5
H_2	20.1	19.6	8.6	30.3	13.9
N_2	—	37.5	47.8	1.1	45.6
H_2O	50.2	10.1	19.3	25.3	8.3

[1] Fred L. Robson and Albert J. Giramonti, "The Environmental Impact of Coal-Based Advanced Power Generating Systems," *Symp. Proc.: Environmental Aspects of Fuel Conversion Technology I*, St. Louis, Mo., May 1974.

[2] M. J. Massey, D. V. Nakles, A. J. Forney, and W. P. Haynes, "Role of Gasifier Process Variables in Effluent and Product Gas Production in the Synthane Process." *Symp. Proc.: Environmental Aspects of Fuel Technology II*, Hollywood, Fla., December 1975.

[3] J. Frank Farnsworth, D. Michael Mitsak, and J. F. Kamody, "Clean Environment with K-T Process," *Symp. Proc.: Environmental Aspects of Fuel Conversion Technology I*, St. Louis, Mo., May 1974.

[4] William W. Bodle and C. Kiret Vyas, "Clean Fuels from Coal," *Oil Gas J.*, 73–88 (August 26, 1974).

[5] D. W. Gillmore and A. J. Liberatore, "Pressur-

Sulfur is converted predominantly to H_2S during fixed-bed gasification. Cameron et al.[6] report the following distribution in raw Lurgi gas on a weight basis.

Compound	Total Sulfur (wt %)
H_2S	95.0
COS	2.4
CS_2	0.3
Mercaptans	2.0
Thiophenes	0.3

DOE has reported a higher degree of COS formation amounting to 10–12% of the total sulfur on a weight basis. DOE also found 500–1000 ppm NH_3 and 10 ppm HCN in the raw gas.

Oil (SG $<$ H_2O) and tar (SG $>$ H_2O water) expected in Lurgi effluent has been reported[6,7] as about 2.8% of the coal feed, with about 65% of the total in the heavier tar fraction. Phenols produced amount to about 0.2% of the coal feed. Analysis of the oil and tar fraction from the Morgantown gasifier[8] identified 54 compounds, including basic structures such as phenols, phenanthrene, anthracene, fluoranthene, pyrene, chrysene, picene, pentacene, dibenzofluorene, and dibenzopyrene. The conventional Lurgi gasifier produces a raw gas that is about 50% steam by volume: when cooled, it produces a water stream that must be treated. This problem is greatly minimized with the development of the slagging fixed-bed gasifier, where the undecomposed steam is stated by Hebden to be negligible.[9]

While particulates are blown out of fixed-bed gasifiers, they tend to collect in heavy tar, which is typically condensed and recycled to the gasifier. Thus particulate emissions in the gas are expected to be negligible.[6]

25.1.3 Fluidized-Bed Gasification

Operating characteristics of fluidized gasifiers fall between those of fixed and entrained gasifiers. Therefore raw gases tend to vary in composition between those typical of entrained and fixed-bed systems. Examples of fluid bed gasifiers include the Winkler process that has been used commercially, and the CO_2 acceptor and Synthane processes under development. Both the CO_2 acceptor and Winkler processes are similar to entrained systems in that coal is introduced at the base of the bed and is heated rapidly, and average operating temperature is high. Thus only trace amounts of tars, oils, and phenols are produced. In some modifications of the Winkler gasifier, oxygen is injected above the dense phase bed that further aids in reducing tar formation. In the Synthane process the coal is introduced higher in the bed and is heated slowly by rising gas, and the average temperature is lower. Thus by-products similar to those of fixed-bed gasifiers are produced. This process characteristic also results in a higher methane content than that from other fluidized-bed gasifiers. Major component analyses reported for raw gas[4] are summarized below.

	Concentration (vol %)				
	Winkler		Synthane		
Component	Oxygen	Air	Oxygen	Air	CO_2 Acceptor
CO	25.7	19.0	10.5	5.4	14.1
CO_2	15.8	6.2	18.2	14.9	5.5
CH_4	2.4	0.5	15.4	7.0	17.3
H_2	32.2	11.7	17.5	14.3	44.6
N_2	0.8	51.1	0.5	33.9	0.2
H_2O	23.1	11.5	37.1	23.8	17.1

ized, Stirred, Fixed-Bed Gasification," *Symp. Proc.: Environmental Aspects of Fuel Conversion Technology II*, Hollywood, Fla., December 1975.

[6] D. S. Cameron, C. R. Gibson, and G. A. Hammons, "Environmental Aspects of El Paso's Burnham Coal Gasification Complex," *Symp. Proc.: Environmental Aspects of Fuel Conversion Technology I*, St. Louis, Mo., May 1974.

[7] A. J. Paquette and M. R. Beychok, "Coal Gasification for Clean Energy," *Energy Pipeline Syst.*, 49–51 (March 1974).

[8] *Third Quarterly Progress Report: A Kentucky Coal Utilization Research Program*, Institute for Mining and Minerals Research, University of Kentucky, April 30, 1973, pp. 23–27.

The most complete studies available of raw gas composition and trace element analyses have been conducted by Forney and coworkers around the Synthane gasifier.[10,11] One set of their results shows the following distribution of sulfur compounds on a weight basis.

Compound	Total Sulfur (wt %)
H_2S	94.9
COS	2.6
CS_2	0.2
Mercaptans	0.8
Thiophenes	1.3
SO_2	0.2

Only small quantities of HCN (< 20 ppb) have been reported in Synthane effluent. The oil and tar content has been from 2 to 4 wt % of the coal feed. Analysis of the tar and oil fraction is similar to that reported for the Morgantown fixed-bed gasifier. Analyses of trace element distribution around the Synthane unit indicate that most trace elements are concentrated in the char. The tabulation that follows gives the weight percentages of selected components in the original feed coal that are collected from the vapor in H_2O condensate and tar.

Element	Range in Product Gas (wt % of coal fed)
As	1.3– 4.6
B	26.6–100
Be	0.1– 0.3
Cd	0 – 4.1
Cl	97.9–100
Cr	0.1– 0.3
F	11.5–52.7
Hg	45.1–78.0
Mn	0 – 0.5
Ni	0.2– 0.9
P	0.2– 1.4
Pb	0.2– 2.0
Se	1.4–41.5
U	0 – 0.1
V	0 – 0.1
Zn	0.3– 0.9

These results were obtained at gasifier temperatures of 1600–1700°F, with coal fed at about 750°F. The portion of these elements in the gas would be expected to be higher in entrained or fluidized gasifiers operated at higher temperatures. They would be somewhat lower in the raw effluent from fixed-bed gasifiers.

25.1.4 Product Gas Specifications

Methods of purifying and cleaning raw gas vary not only with the wide variety and varying concentrations of contaminants that might be present, but also with the end use of the gas. The degree of purification required may be fixed by environmental restrictions or by downstream processing requirements.

The final products of coal gasification are often classified as low-, intermediate-, or high-Btu gases. Low-Btu gas has a heating value of 100–200 Btu/scf and is produced by gasification with air rather than oxygen. Typically it is used at the plant site as a boiler fuel or as feed to a turbine in combined power cycles or in peak-load turbine systems. Environmental restrictions for these systems include limits on sulfur, nitrogen compounds (principally NH_3 and HCN, since these are converted to nitrogen oxides during combustion), and particulates. New Mexico standards that would apply to the proposed El Paso Burnham complex are as follows[6]:

Component	Allowed Emission ($lb/10^6$ Btu)	Approx. Conc. in Fuel Gas (190 Btu/scf)
SO_2	0.16	180 ppm S
NO_2	0.20	320 ppm NH_3, HCN[a]
Particulate	0.03	5.7 $lb/10^6$ scf

[a] Assumes that NO_2 is not formed from N_2 in gas or combustion air.

[9] D. Hebden, "High Pressure Gasification Under Slagging Conditions," paper presented at 7th Synthetic Pipeline Gas Symp., Chicago, October 27–29, 1975.

[10] A. J. Forney, W. P. Haynes, Stanley J. Gasior, Glenn E. Johnson, and Joseph P. Strakey, Jr., *Analysis of Tars, Chars, Gases, and Water Found in Effluents from the Synthane Process,* U.S. Bur. Mines Tech. Prog. Rep. No. 76, January 1974.

[11] A. J. Forney, W. P. Haynes, S. J. Gasior, R. M. Kornosky, C. E. Schmidt, and A. G. Sharkey, *Trace Element and Major Component Balances Around the Synthane PDV Gasifier,* U.S. ERDA Rep. No. PERC/TPR-75/1, August 1975.

The maximum allowable concentrations of contaminants in a turbine fuel have been reported as follows by Robson and Giramonti[12]:

Impurity	Maximum Allowable Concentration
Particulate	0.8 lb/10^6 scf
Condensable hydrocarbon	0.5 lb/10^6 scf
Sulfur	0.18 mole %
Metals	
V	0.1–0.2 ppm (wt)
Na + K	0.2–0.6 ppm (wt)
Ca	0.1 ppm (wt)
Pb	0.1 ppm (wt)

Intermediate-Btu gas has a heating value of 300–450 Btu/scf and is produced by gasification with oxygen or by one of the processes that generate an N_2-free gas without using oxygen. Uses typically are similar to those of low-Btu gas, so that similar standards apply. On an absolute basis, however, higher contaminant levels may be permissible because of the higher Btu content.

High-Btu gas has a heating value exceeding 900 Btu/scf and is interchangeable with pipeline-quality natural gas. It is produced by the same gasification methods as intermediate-Btu gas and upgrading by methanation. Purification requirements are fixed by downstream processing equipment. The most stringent standard is removal of sulfur to a value of less than 0.1 ppm to protect methanation catalysts.

In 1965 the Office of Coal Research (OCR) issued the following guideline objectives for development of processes to make high-Btu gas from coal. The gas should be delivered at 1000 psig, it should be interchangeable with natural gas, and it should have the following desirable characteristics[13]:

Heating value	> 900 Btu/scf HHV
CO	< 0.1%
H_2S	< 0.25 gr/100 scf
Total S	<10 gr/100 scf
Inerts	< 5%
CO_2	< 3%
Water	< 7 lb/10^6 scf
Specific gravity	0.59–0.62 recommended
Hydrocarbon dew point	< −40°F at 1000 psig

No poisonous compounds or gum formers

[12] F. L. Robson and A. J. Giramonti, "The Environmental Impact of Coal-Based Advanced Power Generating Systems," *Symp. Proc.: Environmental Aspects of Fuel Conversion Technology I*, St. Louis, Mo., May 1974.

[13] J. P. Strakey, A. J. Forney, and W. P. Haynes, "Methanation in Coal Gasification Processes," paper presented at 2nd Ann. Symp. Coal Gasification, University of Pittsburgh, August 5–7, 1975.

25.2 REMOVAL OF PARTICULATES

The initial purification step for most raw gases consists of cooling with heat recovery and scrubbing or quenching with water. The exact mechanical arrangements and sequence of steps depend on the temperature of the raw gas and on the quantity of condensable by-products in the gas. Systems in use or proposed for the K-T, Winkler, and Lurgi processes are illustrative. In addition, several systems are under development for cleaning gases of particulates while hot.

25.2.1 The K-T Process

In the K-T process[3] gas exits the gasifier at about 3,300°F and a slight positive pressure. It is then directly water quenched to solidify entrained slag droplets and is passed through a waste heat boiler where steam is raised and the gas temperature is reduced to about 350°F. The gas is cleaned and cooled in two stages of venturi scrubbing, where its temperature is lowered to about 95°F, and is then desulfurized. Figure 25.1 is a flowsheet showing the operation.

A typical analysis of K-T gas after the waste heat boiler has already been given. The extent of removal in the water-scrubbing

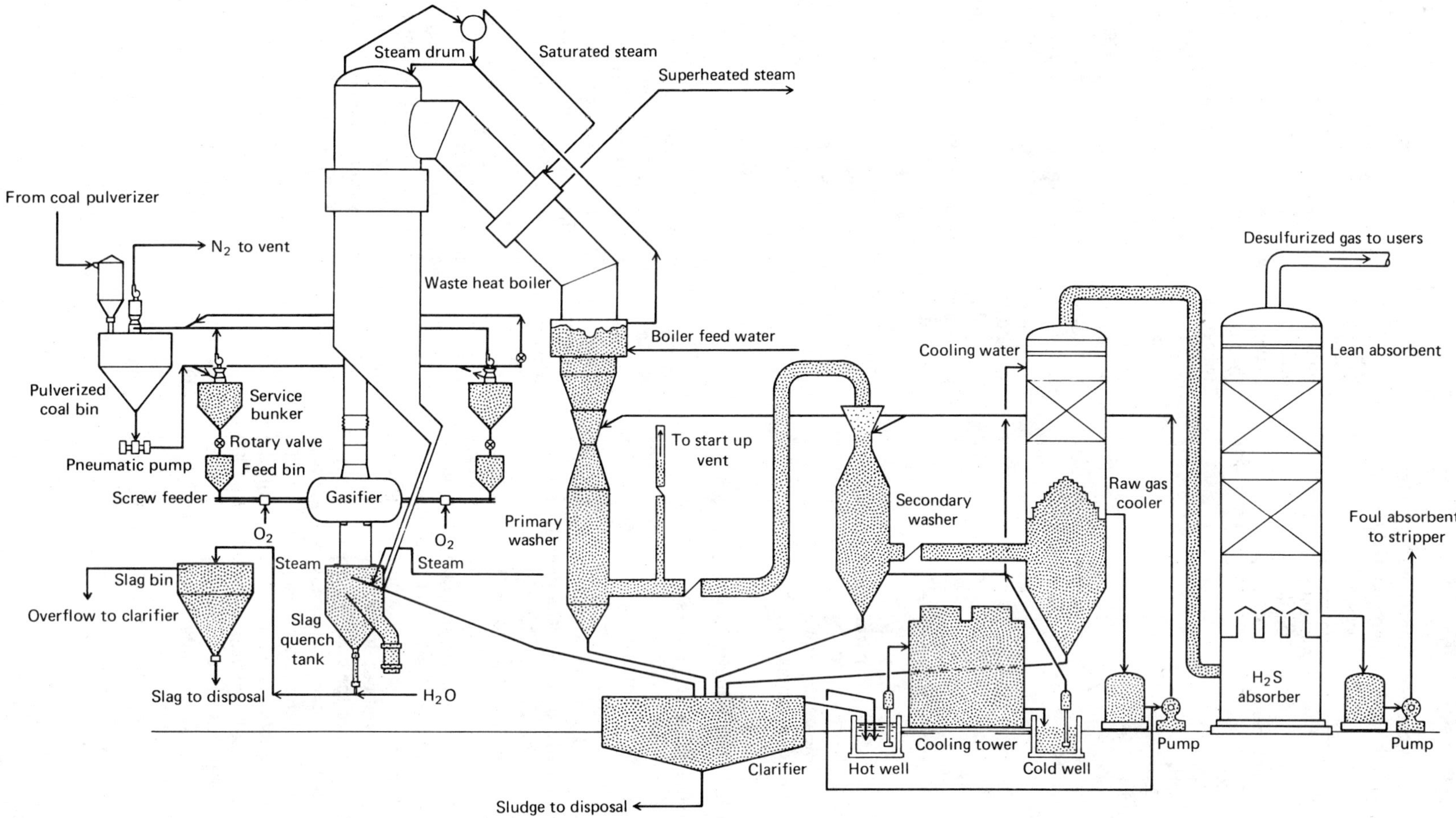

Figure 25.1 The K-T gasification process.

system calculated from data reported by Farnsworth et al.[3] is summarized below:

Component	Amount Removed During Water Quench (%)
CO	—
CO_2	0.3
CH_4	—
H_2	—
N_2	—
H_2S	Trace
COS	Trace
HCN	22
NH_3	2
H_2O	—
Ar	—
SO_2	51
NO	25

More than 99% of particulates are also reported to be removed. No data are available for trace elements, but is is expected that most would be trapped in the water.

25.2.2 The Winkler Process

With the Winkler process[14] 50–75% of the ash in the coal leaves the top of the gasifier entrained in the hot product gas. Because the bed temperature is relatively high (1700–2000°F), no tars or liquids are formed, which greatly simplifies the downstream purification. In Fig. 25.2, showing raw gas cleaning, hot gases leave the top of the gasifier and go to the heat recovery train, where high pressure steam is generated and superheated and boiler feed water is heated. The temperature of the gas leaving the heat recovery train is about 300°F.

About 85% of the char is carried out of the top of the gasifier and is removed in the heat recovery train and downstream cyclone. The venturi scrubber removes the rest of the remaining char down to a level of 1 gr/1000 scf. Cooled gas is desulfurized to a level of 100 ppm with a combination of carbonyl sulfide hydrolysis and Alkazid process to remove the H_2S, which is sent to a Claus unit for recovery of elemental sulfur.

[14] I. N. Banchik, T. K. Subramaniam, and J. H. Marten, "Pressure Reaction Cuts Gasification Costs," *Hydrocarbon Process.*, 121–125 (March 1977).

25.2.3 The Lurgi Process

The initial purification step in Lurgi gasification is somewhat more complex because of pressurized operation and because of the tar and oil content of the gas. Typically gas leaves the gasifier at 800–1000°F, passes through a quench cooler, then through a waste heat boiler, leaving at about 310°F. Tar (SG $> H_2O$), oil (SG $< H_2O$), and condensate are collected at this point and separated. The heavy tar, in which most of the particulate accumulates, is recycled to the gasifier. Two additional stages of indirect gas cooling follow. In the first, intermediate stage, additional tar, oil, and water are collected. In the second stage gas is cooled to 85–100°F and only oil and water are collected. Finally, the gas is scrubbed countercurrently with an oil wash to remove light oils.

In producing high-Btu gas the shift converter is inserted in the process sequence after the waste heat boiler. This arrangement minimizes cooling and reheating of the gas. In addition, some desulfurization of heavy oils occurs over the shift conversion catalyst as well as conversion of COS and CS_2 to H_2S and CO_2.

A typical Lurgi plant produces six different by-product streams.[15] These are tar, tar oil, Rectisol naphtha, crude phenol, ammonia, and sulfur. Ammonia is recovered from the water and sulfur from the acid gas removal process. The other by-product streams require upgrading. A scheme to upgrade these streams to salable products appears in Fig. 25.3. Although the benzene and phenol yield from a single installation would be only a small percentage of the total consumed in the U.S., other products would represent a marketing problem. Alternatively, unwanted by-product chemical streams could be recycled to the gasifier and converted to gas.

25.2.4 Particulate Removal from Hot Gases

The technology employed in the foregoing discussion on cleaning gas by removing solid and liquid particulate matter such as ash, char, tars, and oils involves cooling the gas to ambient temperature and separating the particulates in waste heat boilers, cyclones,

[15] R. Serrurieur, "Prospects for Marketing Coal Gasification By-Products," *Hydrocarbon Process.*, 253–257 (September 1976).

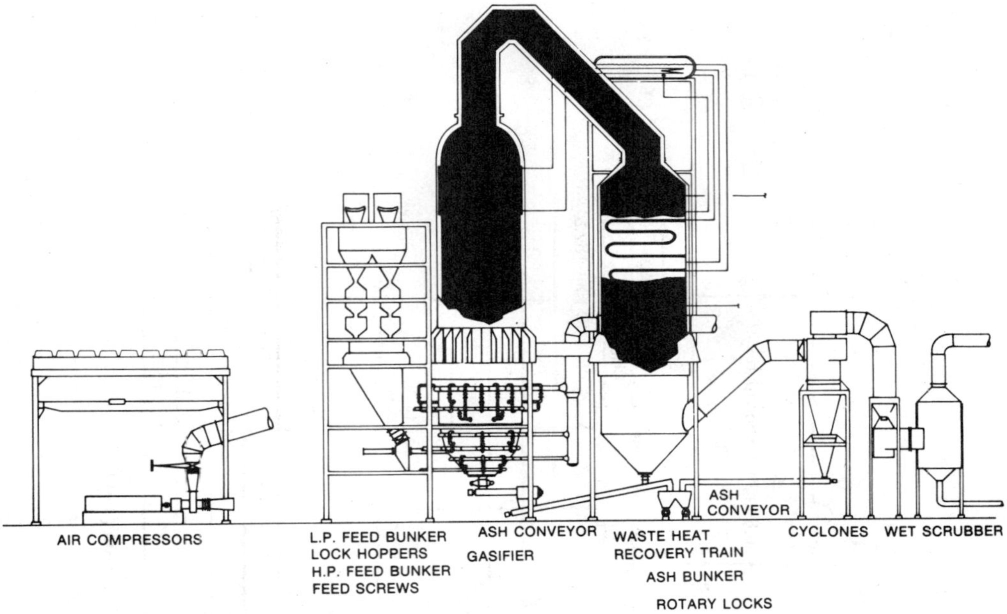

Figure 25.2 The Winkler gasification process.

scrubbers, and washers. In many applications the gas would normally be transmitted some distance and would preferably be cooled and dried.

However in some applications it would be advantageous to clean the gas of particulates while hot, and to remove objectionable sulfur compounds at the elevated temperature. This would avoid the necessity of cooling the gas and reheating for combustion purposes.

If the gas is to be used on the site for process heat or for power generation in combined cycle operation, the potential improvement in thermal efficiency is attractive. It should be noted, however, that state-of-the-art gas turbines cannot utilize high-temperature fuel. Before high-temperature gas purification systems can be used for efficient combined cycles, there must be parallel development of high-temperature control valves, gas turbine combustors, and turbine blades. Desulfurization of hot gases is discussed in Section 25.3.9.

Several hot gas cleaning systems are under development that show promise for use at 300 psi and 1500°F.[16] These are summarized in Table 25.1.

[16] E. R. Zabolotny and R. W. Kuhr, "Hot Gas Purification," *Chem. Eng. Prog.*, 69–74 (October 1976).

Table 25.1 Particulate Removal Systems[16]

Developer	Size Relative to Cyclone	Efficiency at 2-μm Size	Face Velocity (ft/min)	Filter Media
Aerodyne Development Corporation	1	80	—	Cyclone
Brunswick Corporation	0.3 to 2.0	99.9	100–120	Metal fabric
3M Company	3 to 6	—	—	Ceramic fibers
J. P. Stevens	3 to 6	—	—	Silica fibers
American Precision	0.6 to 2.0	99.9	20–40	Electrostatically charged filter media
City College of New York	0.6 to 3.0	99.9	30–45	Panel bed sand filter
Ducon Company	0.3 to 2.0	99	30–90	Panel bed sand filter
Combustion Power Company	3	80	30–100	Moving-pebble bed
Pall Trinity Micro Corporation	5 to 10	98	7–10	Sintered metal
Babcock and Wilcox	2 to 3	99.9	—	Impingement scrubber

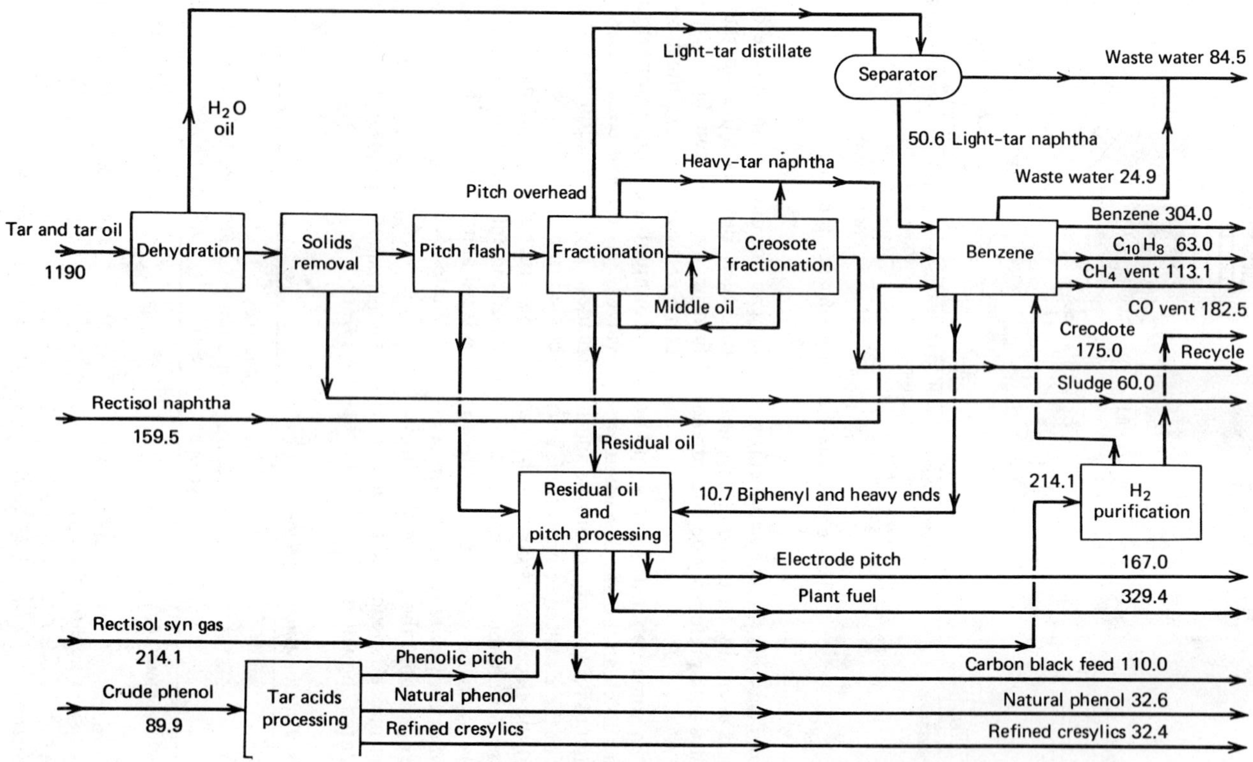

Figure 25.3 Lurgi gasification product streams; all flows in million pounds per year.

The efforts to remove particulates are aimed primarily at protecting gas turbine blades from erosion and corrosion and at meeting environmental regulations. In a report to the OCR, Westinghouse estimated that the total concentration of alkali metals, calcium, lead, and copper may have to be as low as 40 ppb.[16] Other turbine manufacturers expect that blade-cooling techniques will avoid the need for such stringent fuel specifications.

25.3 PURIFICATION OF GASES FROM COAL GASIFICATION

Because of the lack of available commercial technology and the inherent difficulties of sulfur removal at high temperatures close to those of gasification, most engineers familiar with gas-treating technology now believe that sulfur removal should be accomplished at lower temperatures, after water quench. In going from coal to electric power, the gas temperature must be reduced before the gas can be purified. Estimates of the penalty in overall thermal efficiency as a result of this reduction are in the range of 1–4%.[16] However the simplicity and effectiveness of water scrubbing with venturis or in direct contact towers is a distinct advantage. Even with water scrubbing, most of the heat that is in the hot gases after gasification can be recovered by heat exchange; thus the loss in thermal efficiency resulting from cooling the gas to purification temperature is minimized. covered by heat exchange; thus the loss in thermal efficiency resulting from cooling the gas to purification temperature is minimized.

Solids carryover from the gasification operation, as either ash or unburned coal, as well as tars and oils if present, can be removed by the quench operation. In addition, certain chemical impurities in the gas are effectively reduced or eliminated, such as chlorine as HCl or chlorides, HCN, and ammonia, and can be recovered from the water.

The burden of cleaning the raw gas from the gasifier varies greatly depending on the type of gasification employed. Fixed-bed gasifiers such as the Lurgi and fluidized beds such as the Synthane, where the gasifier outlet temperature is relatively low, will produce tars, oils, coal chemicals, and organic sulfur that complicate the cleaning treatment. The entrained gasifiers that have a higher gas outlet temperature, such as the K-T, Texaco, and Bi-Gas units, will have little or no tars and related chemicals, and no difficult-to-remove organic sulfur compounds such as thiophenes and mercaptans.

The gas leaving the water quench operation can be at a temperature from 95 to 250°F. The lower temperature is necessary if the subsequent gas purification step to remove acid gases is one that operates at a relatively low temperature, as is the case for amine-scrubbing processes, physical solvent processes, or the Stretford process. If the H_2S-absorbing process is to be hot carbonate, it is not necessary to cool the gas to below 250°F. If a physical solvent process is being considered, cooling by refrigeration may be advantageous. Absorption at about −40 to −100°F is employed for the Rectisol process. The solubility of impurities is greatly increased by lower temperatures in other physical solvent processes as well.

In addition to the use of liquid absorbent systems for gas purification, sulfur in the gases can be removed by solids such as molecular sieves, active carbon, and iron oxide boxes.

There are currently no commercial processes available for purifying gases at high temperatures. Three processes under development are discussed in Section 25.3.9.

The liquid processes under consideration for purification of coal-based gases generally fall into one of three classes: amines, hot potassium carbonate, or physical solvents.

25.3.1 Amine Gas Purification Processes

Alkanolamines have been used as absorbents for acidic gases for more than 40 years in treating natural gas and refinery vapor streams but have not been employed to any great extent in treating gas made from coal. Their use is considered to be a chemical absorption process, as they are alkaline and abosrb acidic gases such as CO_2 and H_2S quite effectively. A temperature-sensitive reversible chemical reaction provides the absorption-desorption mechanism. Thus absorption is done at ambient temperatures or at temperatures that can be reached economically with air or water coolers.

A number of amines have been used for commercial applications. They include monoethanolamine (MEA), diethanolamine (DEA), triethanolamine (TEA), methydiethanolamine (MDEA), diisopropanolamine (DIPA), and diglycolamine (DGA).[17] Each of these has at least one hydroxyl group (OH), which tends to reduce vapor pressure and increase water solubility, and at least one amino group (NH_2) to provide the alkalinity for acid gas reaction.

With low H_2S and CO_2 partial pressure, the chemistry of the reaction with a primary amine such as MEA can be represented by the following:

$$2RNH_2 + H_2S <====> (RNH_3)_2S \quad (25.1)$$
$$2RNH_2 + CO_2 + H_2O <====> (RNH_3)_2CO_3 \quad (25.2)$$

At high acid gas partial pressures, the reactions above proceed as follows:

$$(RNH_3)_2S + H_2S <====> 2RNH_3HS \quad (25.3)$$
$$(RNH_3)_2CO_3 + CO_2 + H_2O <====> 2RNH_3HCO_3 \quad (25.4)$$

The compounds formed in these reactions have H_2S and CO_2 vapor pressures that increase rapidly with temperature. Consequently, heating the solution to boiling and using stripping steam causes the absorbed gases to be removed from the absorbing solution.

Selection of a particular amine will depend on the application. MEA and DEA will react with CO_2 in a similar manner to H_2S, permitting coabsorption of these two acid gases. The tertiary amines TEA and MDEA react more slowly with CO_2 than do MEA and DEA and are therefore more selective for H_2S removal.

Other sulfur compounds such as carbonyl sulfide (COS) and carbon disulfide (CS_2) are also capable of being absorbed by aqueous amine solutions. However MEA acts irreversibly with COS and CS_2, so that its use is limited to applications with little or none of these compounds present. The same is true of the other primary amine, DGA. The secondary amines DEA and DIPA are less sensitive to degradation by COS and are used when COS and CS_2 are present in small concentrations. Tertiary amines are thought to be even less sensitive, but commercial experience is lacking.

In addition to their possible selectivity as discussed above, the primary advantage of the amines is the capability of lowering residual H_2S and CO_2 concentrations, at reasonable energy consumption, in gases having low partial pressure of these acid gases. Bulk removal of H_2S and CO_2 at higher partial pressures in the gas to be treated, however, does not compare favorably with other purification systems insofar as regeneration heat is concerned.

A recent development in the use of MEA as a means of removing CO_2 is the Ucar process, a development by the Union Carbide Corporation.[18] Solutions of MEA have usually been limited to concentrations of about 15% by weight to avoid the corrosion that occurs when concentrations are greater. Union Carbide has developed a more effective corrosion inhibitor formulation that allows the concentration of the MEA in the solution to be greatly increased without corrosion problems. Inasmuch as the heat consumption for regeneration per gallon of solution circulated is approximately constant, amounting to about 0.8–1.2 lb of steam/per gallon,[19] any additional CO_2 carried by the solution results in a heat savings.

There are also economies in lowered circulation rate and heat consumption with DEA solution when the solution strength is increased. From 1948–1956 when the U.S. Bureau of Mines (USBM) was investigating synthetic fuels processes at Louisiana, Missouri, a concentrated solution of about 35–50% DEA was used to remove the H_2S and CO_2 made from the Koppers-Totzek coal gasification process.[20] Commercialization of the use of higher concentrations of DEA was developed by the Société Nationale des Petroles d'Aquitaine (SNPA), leading to the SNPA-DEA process.[21]

[17] A. L. Kohl and F. C. Riesenfeld, *Gas Purification,* 2nd ed., Gulf Publishing Company, Houston, Tex., pp. 22–32, 1974.

[18] E. N. Hawkes and B. F. Mago, "Stop MEA CO_2 Unit Corrosion," *Hydrocarbon Process.,* 109–112 (August 1971).

[19] "NG/SNG Handbook, Girbotol Process," *Hydrocarbon Process.,* 106 (April 1971).

[20] L. F. Willmott et al., *Performance of a Girbotol Purification Plant at Louisiana, Mo.,* U.S. Bur. Mines Rep. Invest. No. 5196, February 1956.

[21] L. W. Dailey, "Status of SNPA-DEA," *Oil Gas J.,* 136–141 (May 4, 1970).

Amines other than the ethanolamines have been used successfully in gas purification. The Fluor Corporation and the Jefferson Chemical Company have developed the use of aqueous solutions of DGA (Econamine).[22] Concentrations are in the range of 40–60%, resulting in lower circulation rates and regeneration heat when compared to the usual concentrations of the ethanolamines. This amine can also be used for combined acid gas removal and dehydration.

The Sulfinol process[23] developed by the Shell International Petroleum Company uses DIPA along with a physical solvent, sulfolane (tetrahydrothiophene dioxide). This process has been used successfully in many commercial plants for H_2S removal to low levels. The flowsheet is similar to that of other amines, and removal of COS and mercaptans is quite good without solution degradation.

The Adip process[23] also uses DIPA as the alkaline reactant and has been used commercially in 30–40% solutions. It is used primarily to remove H_2S and COS without the degradation of the solution that occurs when MEA is employed with gases that contain COS.

[22] "NG/SNG Handbook, Econamine," *Hydrocarbon Process.*, 101 (April 1971).

[23] E. J. Fisch and J. A. Sykes, Jr., "Synthetic Fuel Gas Purification Using Shell Treating Processes," paper presented at American Chemical Society Meeting, Division of Fuel Chemistry, Dallas, Tex., April 8–13, 1973.

Selective removal of H_2S can be accomplished with solutions of TEA and MDEA.[24] The rate of reaction of CO_2 with these tertiary compounds is quite slow at room temperature, giving improved selectivity with respect to removal of H_2S. This characteristic of the solutions is highly desirable where the acid gas stream is sent to a Claus plant for sulfur recovery.

From an operating standpoint, two primary concerns with respect to amine solutions are corrosion and amine loss. Corrosion can be reasonably minimized with strict control of plant operating procedures, proper selection of construction materials, and use of corrosion inhibitors. Amine loss occurs because of entrainment in the scrubbed gas, vaporization in the absorber and regenerator, and chemical degradation. Vaporization losses can be controlled by addition of capital equipment such as water wash or adsorption systems.

Chemical degradation is the most serious

[24] A. L. Kohl, "Selective H_2S Absorption," *Pet. Process.*, 26–31 (January 1951).

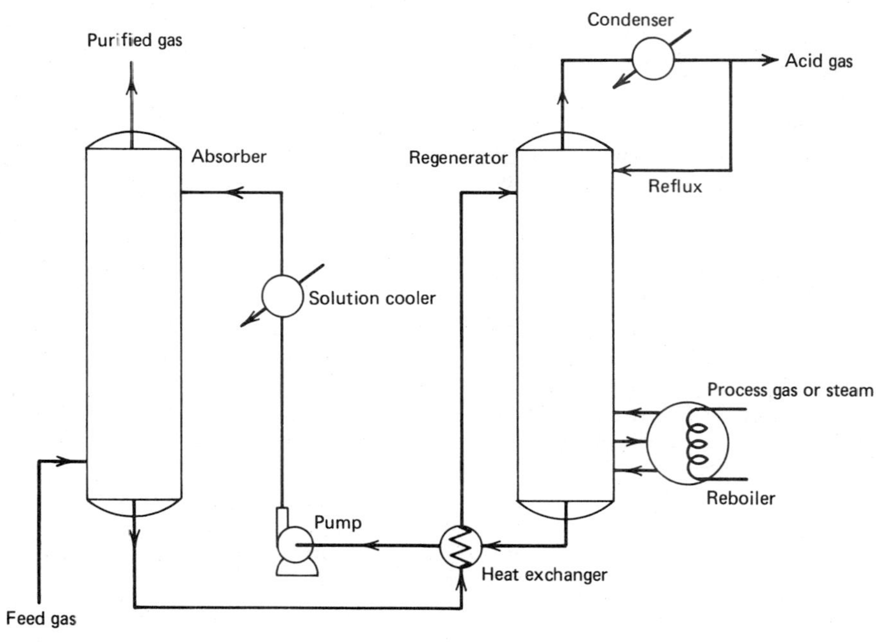

Figure 25.4 Basic amine scrubbing system.[25]

aspect, since irreversible reactions can occur with a wide variety of impurities such as carboxylic acids, hydrogen cyanide, and the COS and CS_2 discussed previously. Oxidative degradation can also occur, adding to the formation of precipitates and sludges. Continual solution reclaiming is frequent, especially with MEA solutions, which adds capital equipment and operating costs to a purification plant.

A basic flow scheme for an amine system is shown in Fig. 25.4.

25.3.2 The Hot Potassium Carbonate Process

The hot potassium carbonate process was developed during the early 1950s at the USBM laboratory at Bruceton, Pennsylvania, as part of the synthetic fuels research program. There is a large amount of CO_2 along with H_2S in gases made from coal, and the cost of the gas purification operation with amine solutions was considered to be quite high because of the large amount of regeneration heat required. The Bureau's investigation with the hot carbonate process showed that the regeneration heat could be decreased to 40–50% of that required for amine solution.[26] In addition, because absorption occurs at the same temperature as regeneration, no heat exchangers are required, and capital investment is reduced. The higher temperature of absorption allows a more concentrated solution, about 30–40%, to be used, thus increasing the pickup of acid gas per gallon of solution circulated.

The chemistry of the reaction of potassium carbonate with H_2S and CO_2 is shown below:

$$K_2CO_3 + CO_2 + H_2O \Longleftrightarrow 2KHCO_3 \quad (25.5)$$

$$K_2CO_3 + H_2S \Longleftrightarrow KHS + KHCO_3 \quad (25.6)$$

Flowsheet arrangements appear in Fig. 25.5. When lower outlet CO_2 or H_2S contents are desired in the purified gas, a small stream at the top of the absorber is slightly cooled. If still lower percentages of CO_2 and H_2S are desired, a two-stage design is used.

The hot potassium carbonate process was commercialized under the name of the Benfield process. The first plants installed were for the removal of CO_2 and H_2S from gases made from the partial oxidation of heavy fuel oil for ammonia synthesis. The composition of the raw gas was about the same as that of gas made from the partial oxidation of coal, and acid gas removal was at about 400 psig. Such a gas contained about 34% CO_2 and 0.7% H_2S. The purified gas had a composition of 0.8% CO_2 and 10 ppm H_2S. Removal of COS and CS_2 by hydrolysis at the higher temperature of absorption is almost complete. In 1960 two Benfield units were installed in the Lurgi coal gasification plant in Westfield, Scotland, and operated for 14 years until the plant was shut down. During the early 1960s the Benfield process was modified to include additives that increased the rate of absorption, further decreased regeneration heat, and decreased capital investment. Table 25.2 summarizes cost and utility requirement comparisons of the MEA, Benfield hot carbonate, and promoted Benfield hot carbonate processes. Some 400 Benfield units have been installed for removal of CO_2 and H_2S from natural gas, partial oxidation gases, ammonia synthesis gas, and hydrogen. A recent development, the Benfield HiPure process,[29] is capable of reducing residual H_2S to less than 1 ppm where this degree of purity is desired. Other hot carbonate processes with different additives have been used in the Catacarb process,[30] by Giammarco Vetrocoke,[31] and by the Lurgi Company.[32]

Where the purification of gases made from coal is concerned, the use of hot potassium

[25] "NG/SNG Handbook," *Hydrocarbon Process.*, 106 (April 1971).

[26] J. H. Field, H. E. Benson, G. E. Johnson, J.S. Tosh, and A. J. Forney, *Pilot Plant Studies of the Hot Potassium Carbonate Process for Removing Carbon Dioxide and Hydrogen Sulfide,* U.S. Bur. Mines Bull. No. 597, 1962.

[27] H. E. Benson and J. H. Field, "New Data for Hot Carbonate Process," *Pet. Refiner,* 127–132 (April 1960).

[28] H. E. Benson, in *Ammonia,* Part 2, Vol. 2, A. V. Slack and G. R. James, Eds., Marcel Dekker, New York, p. 162, 1974.

[29] H. E. Benson and R. W. Parrish, "High Pure Process Removes CO_2/H_2S," *Hydrocarbon Process.*, 81–82 (April 1974).

[30] "Gas Processing Handbook," *Hydrocarbon Process.*, 87 (April 1975).

[31] "Gas Processing Handbook," *Hydrocarbon Process.*, 90 (April 1975).

[32] P. F. H. Rudolph and P. K. Herbert, "Con-

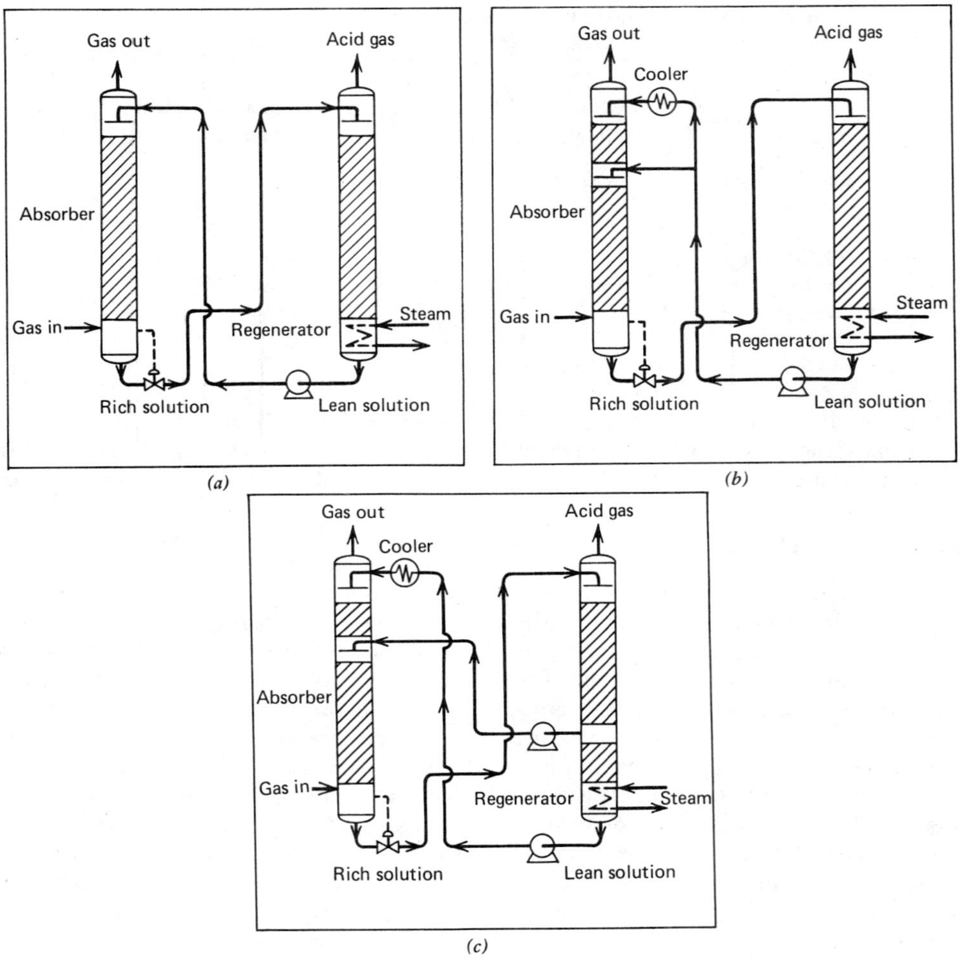

Figure 25.5 Flow arrangements with the Benfield hot potassium carbonate process: (*a*) single stage, (*b*) single stage, split stream, (*c*) two stage.[27]

carbonate process fits well into the usual process conditions. The gasification pressure is usually in the range of 100–400 psig and in some cases as high as 1000 psig, where the hot carbonate process is quite efficient. The preferred method of dedusting and cleaning the gas is by water quench, which results in a water-saturated gas that is at a temperature of 200–260°F after heat exchange. Thus the gas can be fed directly into the absorber without further cooling, which is necessary for processes that operate at lower temperatures.

The raw gases from most coal gasifiers usually have a CO_2 content of 3–12% and an H_2S content of 0.5–1.0%. The preferred method of disposing of the H_2S is to send the acid gas to a Claus process unit, where conversion to elemental sulfur occurs. The percentage of H_2S in the acid gas must be 16% or higher for operation of the conventional Claus process. With some raw gases, if the amount of H_2S is low and the amount of CO_2 is high, some selectivity for the absorption of H_2S is necessary in the absorber. Unless the coal gasified is low-sulfur coal, the Benfield process can be designed with enough selectivity to obtain an acid gas sufficiently high in H_2S to satisfy the requirements of the Claus process.

A Benfield CO_2 and H_2S removal unit has been installed in the U.S. government Syn-

version of Coal to High Value Products," paper presented at Symp. Coal Gasification, Liquefaction and Utilization, University of Pittsburgh, August 5–7, 1975.

Table 25.2 Process Comparisons: MEA-Benfield for Gas Feed Rate of 2.5 × 10^6 scf/hr, 400 psig, 20% CO_2 Purified Gas Content of 1000 ppm CO_2[28]

Item	MEA Solution	Benfield Hot Potassium Carbonate Process	Promoted Benfield Carbonate Process
Relative capital investment	1.00	0.85	0.68
Regeneration heat (10^6 Btu/hr)	136	68	42–60
Solution flow rates (g/hr)	167,000	158,000	125,000
Cooling duty (10^6 Btu/hr)	124	74	58–76

thane coal gasification pilot-plant installation at Bruceton, Pennsylvania, and was undergoing startup in late 1977.

25.3.3 Physical Solvents

Physical solvents in general are well suited to removing large quantities of acid gases from high-pressure gas streams. The higher the feed gas partial pressure of CO_2 and H_2S, the more applicable a physical solvent process becomes. The solvent processes that are available commercially are Selexol, Fluor Solvent, Purisol, and Rectisol. The flowsheets for all these processes are quite similar, the major difference being the chemical compound used for the solvent. All the processes operate close to ambient temperature except the Rectisol process, which operates at much lower temperatures (as low as −100°F). Although scrubbing of gases with water is considered a physical solvent process, the use of organic chemical compounds results in greater efficiency, since the solubility of the acid gases H_2S and CO_2 in organic solvents is several times that in water. Thus the required circulation rate for the recently developed organic solvent processes is a fraction of that required for water, which has been used in the past.

Gas purification by physical solvents has the advantage of requiring little or no heat for regeneration of the solution. Regeneration can be accomplished by pressure reduction of the rich solution, resulting in flashing off of the dissolved acid gases. If a high degree of purification is desired, additional regeneration can be done by stripping with air or an inert gas or by the application of a vacuum to the solvent. Other gaseous components in the treated stream of gas, such as CH_4, CO, or H_2, are also somewhat soluble in the organic solvent. To avoid excessive loss of these valuable gases, the pressure of the spent solvent is usually let down in two or more stages. The flashed gases are then recompressed and sent back to the absorber. The solubility of the higher molecular weight hydrocarbons is much greater than methane, and recovery by flashing is not as effective.

The solubility of H_2S is usually much greater than that of CO_2 in organic solvents. Thus a good degree of selectivity for H_2S removal and concentration in the acid gas stream is possible and desirable when a feed stream to the Claus process is required. The presence of hydrocarbons in the H_2S stream to the Claus process is undesirable because it results in carbon in the recovered sulfur.

Power required for circulation of the solvent can be reduced by expansion of the rich solvent or flashed gases through power recovery turbines. Inasmuch as lower temepratures increase the solubility of gases in the solvent, the gas to be treated should be as cool as possible; sometimes refrigeration is used. The refrigeration is sometimes accomplished by expansion of the flashed gases in power recovery turbines.

The cost of the solvent used for the absorption process is a major consideration both from a first-cost and replacement viewpoint. The physical solvent processes, because they use solvent in nearly pure form rather than in water solution, are costly in this respect. The cost of Selexol solvent, for instance, at $9.00/gal when compared to a 15% MEA solution, is 15 times more expensive.

25.3.4 The Selexol Process

The Selexol process of Allied Chemical Corporation is a physical absorbent process suited to selective removal of H_2S and bulk removal of CO_2. Used commercially in natural gas sweetening and synthesis gas purification,[33] the process can be designed to remove H_2S to 1ppm while keeping CO_2 pickup relatively low.

The Selexol solvent is the dimethyl ether of polyethylene glycol; it is usually used without dilution but can be used with 20% water. Like other physical absorbent processes, it is more efficient at high pressure and low temperature, typically 500–1000 psig pressure and near ambient temperature. When scrubbing to obtain low concentrations of sulfur compounds, the Selexol solvent must be stripped with an inert gas to provide a low H_2S concentration in the lean solvent. A non-oxygen-containing gas is necessary to avoid converting the H_2S to free sulfur. If the H_2S can be vented, surplus nitrogen can be used as the stripping gas; but when the sulfur must be treated further (e.g., in a Claus unit), steam is used to drive the H_2S from the solvent to prevent dilution of the Claus feed. The Selexol process can use a high-pressure flash system to recover most of the desired product gas components, which are then recompressed and put back in with the feed. Heavier hydrocarbons (C_4^+, aromatics, etc.) would be absorbed extensively if present in the feed gas.

Since there are no chemical reactions taking place with trace impurities, impurities that are dissolved in the solvent pass on to the Claus unit. For example, COS, CS_2, NH_3, and HCN are all easily absorbed, as well as H_2S, and these gases can present operating problems in the Claus unit if present in large concentrations. Heavier hydrocarbons, including aromatics, are also absorbed and can produce black sulfur from the Claus unit.

The solvent reportedly has good compatibility with these minor gas impurities, and operating units have not encountered any degradation of the solvent. Besides its high stability, it has a low vapor pressure, which keeps solvent losses very low. The solvent is inherently nonfoaming and nontoxic, and is not known to be corrosive, so carbon steel construction is used throughout for Selexol units. Since the solvent is entirely organic, any trace carryover with the product gas would not be harmful to gas turbines, burners, and other components downstream.

The flowsheet in Fig. 25.6 represents an installation designed for purification of a natural gas at a pressure of about 1000 psig.

The prime advantage of the Selexol process is its selectivity for absorbing H_2S. For removal of the sulfur from a coal gasification gas, it can yield a high H_2S content stream for feeding to a Claus unit, leaving the bulk of the CO_2 for removal in a second unit. High loadings of acid gas per gallon of solvent circulated are possible, when the acid gas is present at high partial pressures. Power recovery from pressure letdown of the circulating solvent is not only helpful in the overall energy balance, but desirable in removing heat from this part of the process.

Like all physical solvent processes, the Selexol process cannot be used economically at low system pressures. Also, it works better at cooler temperatures, and it is necessary to cool the inlet gas to ambient temperature before it is processed. The solvent has an affinity for water and can dehydrate some feed gases, a reaction that can reduce the overall efficiency of operation. On the other hand, if dehydration of the gas is required, this can be done simultaneously with acid gas removal. When removing H_2S to low residual levels, some reboil steam is required for regeneration, but the amount is less than that needed by competitive chemical absorbent systems.

A Selexol unit is being installed in the Homer City, Pennsylvania, pilot-plant installation as part of the coal gasification development program sponsored jointly by the U.S. government and the American Gas Association (AGA).

25.3.5 The Fluor Solvent Process

The process licensed by the Fluor Corporation of Los Angeles uses propylene carbonate, having a boiling point of 464°F as the solvent. The process has been employed commercially for the absorption of CO_2 and CO_2/H_2S mixtures, largely from natural gas where the pressure of the treated gas is relatively high. Table 25.3 summarizes operating data from several Fluor solvents installations. The same general comments made

[33] J. P. Valentine, "New Solvent Process Purifies Crude Coal Acid Gases," *Oil Gas J.*, 60–62 (November 18, 1974).

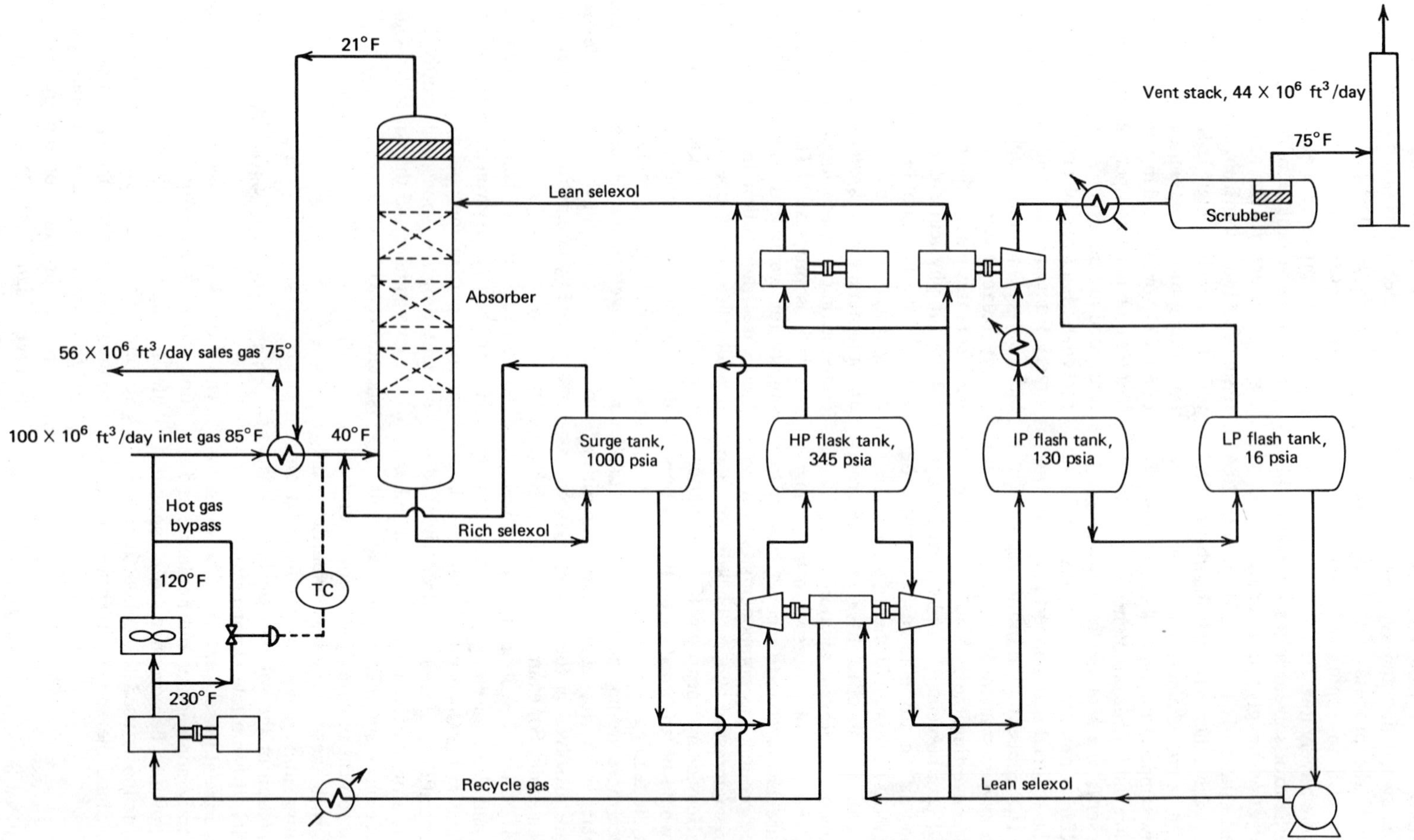

Figure 25.6 Flowsheet of the Selexol process.[34]

Table 25.3 Data from Fluor Solvent Operating Plants[35]

Plant Owner	Plant Location	Design Feed Gas Rate (10^6 scf/day)	Feed Gas Composition		Absorption Pressure (psig)	Acid Gas Partial Pressure in Feed Gas (psi)		Sales Gas Specification	
			CO_2 (%)	H_2S				CO_2(%)	H_2S (gr/100 scf)
El Paso Natural Gas Company	Terrell County, Texas	220	53	3 gr/100 scf	850 (1000 max)	CO_2	432	2	0.25
Standard Oil Company of California	Lost Hills, California	10	17	None	450	CO_2	75	5	
Sinclair Oil & Gas Company	Grand County, Utah	20	22.8	None	800 (1100 max)	CO_2	180	1	
Wintershall AG	Barnstorf, West Germany	28	10 to 30	5–15%	1000	$CO_2 + H_2S$	250–300	0.1	0.8

for other solvent processes apply also to the Fluor solvent process.

25.3.6 The Purisol Process

The Purisol process is a commercially proven physical absorption process for CO_2 and H_2S, licensed by Lurgi of West Germany. The process employs *n*-methyl-2-pyrrolidone (NMP) at ambient temperatures to purify high-pressure gas streams such as natural gas, it is especially suited to selective removal of sulfur compounds.

The NMP solvent is used with dilution. It has a high boiling point (395°F) and a low freezing point (−12°F), is miscible with water, is not subject to foaming, and has good thermal stability and low corrosiveness, allowing all carbon steel construction. Its low vapor pressure under the normal process conditions and its thermal stability assure low losses. The chemical is apparently readily available in the quantities needed for large gas-scrubbing plants.

The flowsheet for this process (Fig. 25.7) shows a typical configuration for scrubbing of CO_2 from a high-pressure hydrogen stream. The absorption step is accomplished in one or two absorbers, depending on whether only bulk CO_2 removal is desired, or H_2S selective removal with a minimum of CO_2. Solvent regeneration is a two- or three-stage pressure reduction flashing for bulk CO_2 removal, followed by stripping with inert gas if a high degree of CO_2 removal is desired, or if the plant is to selectively remove H_2S. The gas from the first pressure reduction flash is compressed and reinjected into the feed gas for recovery of dissolved product gas. Some cooling of the solution is required, and some steam is needed for distilling off water from the solvent.

For purifying gases at high pressure, Purisol has many advantages in its high acid gas loading, ambient temperature operation, and low solvent losses. The process is reported to be capable of purifying to less than 0.1% CO_2, with only a few ppm of H_2S.

As with all physical solvent processes, the losses of product gas are higher than with chemical absorbents, such as amines or potassium carbonate.

There is little information available in the literature on commercial results of sulfur removal with this process, but the potential for selectivity for H_2S with Purisol is higher than with the Rectisol process. Hochgesand[37] gives the following ratios of loadings for H_2S and for CO_2 at equilibrium, which indicate the high potential for selectivity: Purisol, 8.2; Rectisol, 5.1 (−10°C).

25.3.7 The Rectisol Process

The Rectisol process is a development of the Lurgi company in cooperation with Linde AG for the removal of CO_2 and H_2S along with organic sulfur compounds, COS, HCN, aromatic hydrocarbons, and other higher hydrocarbons from synthesis gases.[37] It utilizes low-temperature methanol as a physical solvent in conventional trayed or packed columns.

Solubility of gases in physical solvents increases with decreasing temperature, increasing partial pressure in the feed gas, and decreasing residual content in the regenerated solvent. To take practical advantage of the temperature factor, the Rectisol system operates in a range of −80 to −100°F for the solvent circulated to the absorber column.

It is not possible to provide control of acid gas partial pressure, but because of its effect on solubility the process is most attractive for gases with high acid gas content that are available at high pressure.

Regeneration of rich methanol can be accomplished in three different ways, and some plants use all three. First is flashing the solvent to a lower pressure to free the dissolved gases. Since the solvent undergoes a temperature gain in the absorber, this flashing operation reduces the liquid temperature. The flashed liquid will have an acid gas content determined by the absolute pressure of the system. For example, in a CO_2 system flashing the rich solvent at 16 psia will estab-

[34] D. P. Raney, "Remove Carbon Dioxide with Selexol," *Hydrocarbon Process.*, 73–75 (April 1976).

[35] P. H. Buckingham, "Fluor Solvent Process Plants," *Hydrocarbon Process.*, 113–116 (April 1964).

[36] "Gas Processing Handbook," *Hydrocarbon Process.*, 92 (April 1975).

[37] G. Hochgesand, "Rectisol and Purisol," *Ind. Eng. Chem.*, 37–43 (July 1970).

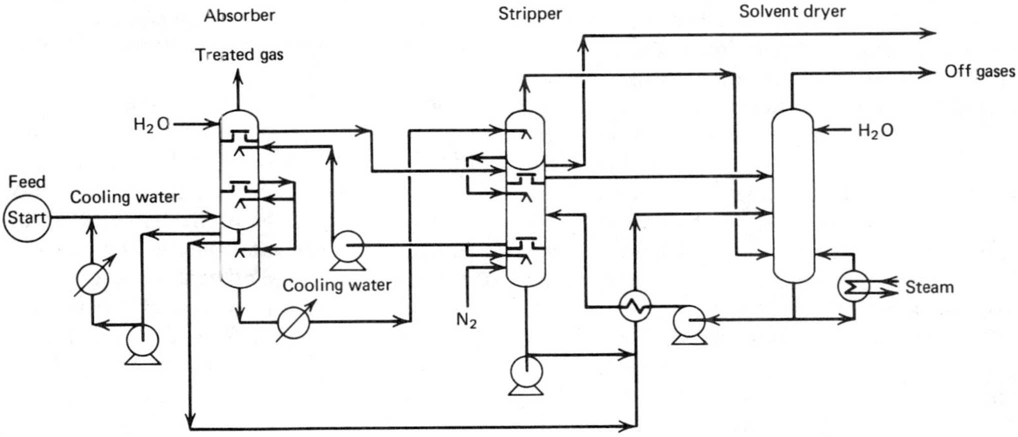

Figure 25.7 Flowsheet of the Purisol process.[36]

lish the CO_2 partial pressure to approximately that value, since the vapor pressure of methanol at the reduced temperature is low in comparison. When the solvent is recycled to the absorber column, this flashing technique then limits the final degree purity in the treated gas to a practical range of 25–30 psia partial pressure of acid gas components. Additional flash stages to a subatmospheric range of about 3–4 psia as a practical economic limit can improve performance in the absorber accordingly.

A second regeneration method is to strip with an inert gas. Since the acid gas partial pressure of the inlet stripping gas is essentially nil, it is possible to adjust the purity of the outlet treated gas to relatively low levels.

A third method is a high-temperature regeneration where the combination of decreasing solubility at increasing temperatures and stripping with solvent vapors produces the lowest residual acid gas content of the three.

The solubility of H_2S in the Rectisol solvent is 5–6 times the solubility of CO_2 at a partial pressure of about 1 atm and a temperature of −15 to −20°F. The solvent can absorb both the H_2S and the CO_2 simultaneously and almost independently of each other. Selective absorption can thus be achieved by using a regenerated solvent still somewhat rich in CO_2 but lean in H_2S.

Figure 25.8 shows a schematic flow diagram for combined desulfurization and decarbonation for purifying a synthesis gas where a sulfur-sensitive shift conversion catalyst is involved.

The process gas is first contacted with a CO_2-rich solvent for H_2S removal before going to the shift reactor. The shifted gas is

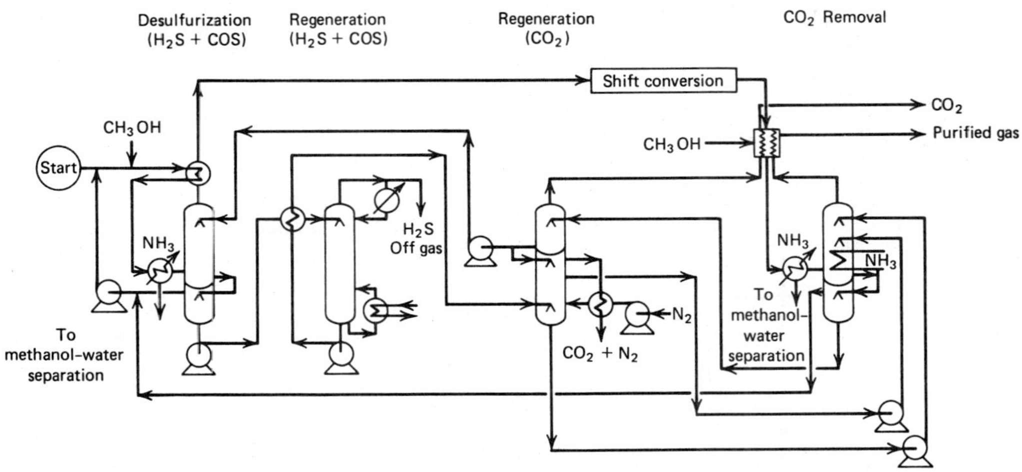

Figure 25.8 Flowsheet of the Rectisol process for the removal of H_2S, COS, and CO_2.[38]

then contacted with thoroughly regenerated solvent to remove CO_2 to low levels.

The rich solvent from each tower is subjected to a controlled flash to return dissolved process gases to the desulfurization column after compression. Then the rich solvent from the first stage is thermally stripped to remove as efficiently as possible all H_2S. It is then directed to an inert gas stripper before being pumped to the top of the second-stage decarbonation.

The rich solvent from this stage receives a second flash treatment and is split, with one CO_2-rich portion going to the desulfurization and the other receiving additional stripping with inert gas.

Selective desulfurization followed by CO_2 removal for a partial oxidation process is shown in the once-through flow diagram in Fig. 25.9.

The process gas first receives a methanol injection to prevent icing as it is heat exchanged with the purified gas and stripped CO_2 and then with ammonia refrigeration. Desulfurization is accomplished in a first column with CO_2-rich solvent from the CO_2-removal column, followed by decarbonation in a second column.

Again, both absorbers are shown with first-stage lean solvent flashing to return soluble gases such as H_2 and CO back to the process stream via compression. The injected methanol-water is separated and is returned off-battery for distillation.

The major difference is that after inert gas stripping of the rich solution from the desulfurization step, any released H_2S reabsorbed for removal in the thermal regeneration. This allows a low-sulfur level discharge of nitrogen and CO_2 to the atmosphere, and concentrates the H_2S to one effluent stream.

Operating with a feed flow of 108 million scf/day (100 million scf/day of H_2 and CO) and at a pressure of 685 psig, the composition of a typical treated gas is as shown on page 1775.[38]

Although the Rectisol process is inherently capable of producing very low residual acid gas contents by utilizing what appear to be basically simple phenomena, it has certain disadvantages compared to other purification processes.

1. The system is quite complex and includes many special low-temperature items of equipment, resulting in high capital investment.
2. Solvent losses are relatively high because of the vapor pressure characteristics of the methanol.

[38] "Gas Processing Handbook," *Hydrocarbon Process.*, 93 (April 1975).

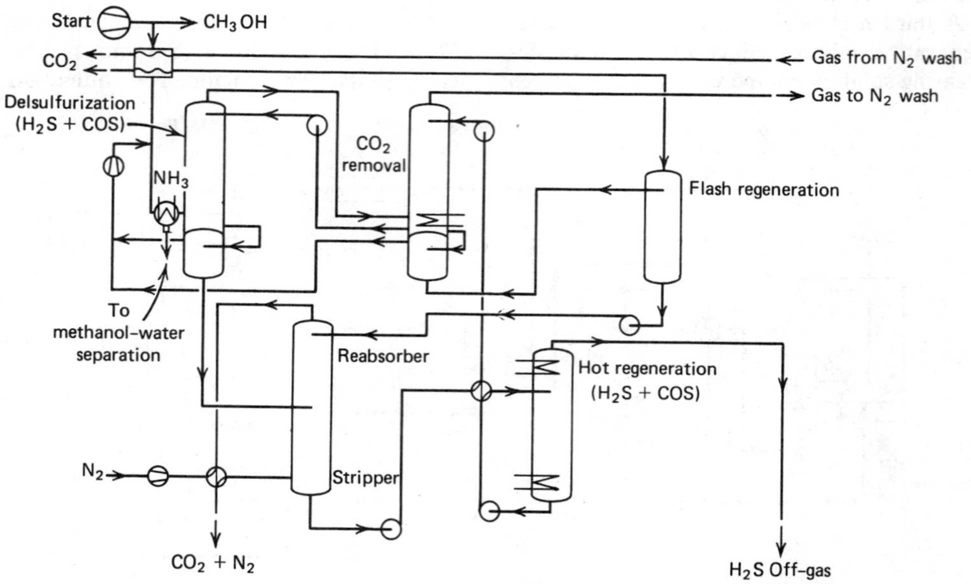

Figure 25.9 Rectisol desulfurization followed by CO_2 removal.[37]

Operation	Feed	Treated Gas	H_2S Off-Gas
Desulfurization			
CO_2 (vol %)	5.3	5.3	57.3
$H_2S + COS$ (vol %)	0.7	<0.1 ppm	40.1
H_2 (vol %)	44.6	45.0	—
CO (vol %)	48.4	48.7	—
N_2 + Ar (vol %)	1.0	1.0	1.6
CO_2 removal			
CO_2 (vol %)	36.1	0.1 (1 ppm possible)	—
$H_2S + COS$ (vol %)	—	—	—
H_2 (vol %)	62.8	98.2	—
CO (vol %)	0.5	0.8	—
N_2 + Ar (vol %)	0.6	0.9	—

3. Losses of process gas components due to solubility in the methanol are high.

25.3.8 Sulfur Removal by Solid Adsorbents at Ambient Temperature

In addition to the use of liquid absorbent systems and hot solids or molten salts discussed subsequently for gas purification, another group of processes should be mentioned. These include the solid adsorbent materials: molecular sieves, active carbon, and iron oxide boxes. Research is also being done on selectively permeable membranes.

25.3.8.1 *Molecular Sieves*

Molecular sieves made by Union Carbide and Davison Chemical Company have been used for removal of CO_2 and H_2S from various gases. In general if the quantity of gas to be treated is relatively large and the percentage of acid gas is high, the use of sieves is not economical, and other bulk removal systems would be favored. However if the acid gas content is relatively low—around 1% or so—the use of sieves could be considered. If the gas quantity is small, even though the percentage of acid gas to be removed is high, again sieves could be considered for this application.

The proper type of molecular sieve can selectively remove H_2S without removing much of the CO_2. These sieves can reduce the H_2S content of the purified gas to 1 ppm or less, an important feature in some applications. However, if 90–95% removal of the H_2S is generally sufficient in the fuel gas being considered, the use of molecular sieves may not be indicated, since other purification processes are better suited for bulk removal of CO_2 and H_2S. Molecular sieves can also be effectively used after a bulk removal process that normally is not suited for complete removal of H_2S. Except for special cases molecular sieves do not fit very well in the purification schemes for low- and intermediate-Btu fuel gases, but are more appropriate in high-Btu gas technology, where the sulfur level must be reduced to meet stringent methanation catalyst requirements.

25.3.8.2 *Active Carbon*

Much of what has been said regarding the use of molecular sieves also applies to the use of active carbon in purification of fuel gases. Active carbon has a much lower capacity for H_2S, but it is capable of adsorption of other organic compounds such as thiophene. Again, no great application is seen for active carbon in purifying fuel gases where complete removal of the sulfur compounds is not required.

25.3.8.3 *Iron Oxide Boxes*

Iron oxide boxes, long used in the old manufactured-gas industry, again could be useful following a bulk removal purification system. Iron oxide boxes are quite effective in reducing residual H_2S to the ppm level, should this be required in a specific installation. Mercaptans, if present, are also removed effectively. However iron oxide boxes would be better applied where bulk removal of H_2S is done economically by another process. Iron oxide boxes do have the distinct advantage of being highly selective; they will remove only sulfur compounds from the gas stream. Carbon oxysulfide, however, is not removed completely by iron oxide. A detailed discussion of iron oxide boxes and other dry oxidation processes used for sulfur removal is given by Kohl and Riesenfeld.[39]

25.3.8.4 *Desulfurization by Membranes*

The General Electric Company laboratories at Schenectady, New York, are researching the use of membranes permeable to H_2S but not to CO_2 to obtain a stream of gas sufficiently high in H_2S to operate a Claus process.[40] Although some CO_2 is expected with the H_2S stream, it is thought that the CO_2 content can be minimized. Operation will be at 100–130°F, with gas made by a new coal gasifier under construction.

25.3.9 Hot Purification of Gases

Hot purification of gas made from coal (i.e., the removal of impurities at gasification temperature) has the theoretical advantage of a higher thermal efficiency in the utilization of the gas. This heat economy is accomplished by keeping the gas at an elevated temperature to remove the sulfur components and avoiding the energy loss and costs associated with heat exchangers if the removal of sulfur and particulates is to be accomplished at lower temperature.

The technology of hot purification is poorly developed and currently is the subject of several investigations. Commercial processes are not yet available. This section reviews the status of three of the hot purification processes under development: the hot iron oxide process (DOE, Morgantown), the Conoco dolomite process (Library, Pennsylvania), and the molten calcium carbonate process (Battelle Northwest).

In addition to these processes, OCR has sponsored a research project with Air Products and Chemicals using a fixed hot bed of limestone and dolomite to remove sulfur from low-Btu gas.[41] The fixed bed continuously lost activity and capacity for H_2S removal; the time required for regeneration was excessively long; and coking on the bed occurred when coal-based producer gas was used. No further work on the project was recommended.

All hot purification processes to remove sulfur would require a particulate removal treatment such as cyclones, filters, or sand beds. This step would be necessary to clean the gas of entrained solid particles such as coal ash before combustion or expansion turbines; otherwise the life of turbine blades would be prohibitively short. Alkali content of the hot gases must also be reduced to a very low value because of corrosive attack on the turbine. It is stated that to be suitable for turbines, the alkali must be reduced to less than 40 ppb. Gasification processes that produce tar would impose an additional burden on a solids removal system.

Mention should be made of the Appleby-Frodingham process that was developed to a large scale in the 1950s at Exeter, England, at a South Western Gas Board installation. A large pilot plant processed 2.5×10^6 ft^3/day of coke-oven gas, and a commercial installation having a capacity of 8×10^6 ft^3/day of coke-oven gas was operated.[42] Desulfurization took place in a fluidized bed of iron oxide at about 650–750°F, and 95–99% desulfurization was obtained.

Regeneration was by air oxidation at 1290–1380°F, with the production of SO_2, which ultimately was converted to H_2SO_4. The losses of iron oxide were quite high, and the production of a fine oxide dust complicated the downstream processing. The plant operated for several years and, generally speaking, established that a fluidized–iron oxide process for desulfurization of the gas was technically feasible. A distinct disadvantage to this process is that the final form of sulfur is sulfuric acid, which does not have as high a value as elemental sulfur.

[39] A. L. Kohl and F. C. Riesenfeld, *Gas Purification,* 2nd ed., Gulf Publishing Company, Houston, Tex., pp. 342–395, 1974.

[40] S. G. Kumura, "Permselective Membrane for Acid Gas Scrubbing from Coal Gas," paper presented at 9th Synthetic Pipeline Gas Symp., Chicago, October 31–November 2, 1977 (published by AGA, 1978).

[41] W. G. O'Brian, and W. A. Grossmans, *Hot Stage Desulfurization of Gasified Coal, Bench-Scale Limestone Fixed-Bed Adsorption and Regeneration Study,* Air Products and Chemicals, Inc., Marcus Hook, Pa., Rep. No. FE-1510T-2 prepared for U.S. ERDA, 1975.

[42] L. Reeve, "Desulfurization of Coke-Oven Gas at Appleby-Frodingham," *J. Inst. Fuel,* **31** 319 (1958).

25.3.9.1 *The Hot Iron Oxide Process*

The Morgantown (West Virginia) Energy Research Center (MERC) is developing a hot purification process capable of operating at 1000–1500°F for the absorption of H_2S on iron solids from hot fuel gases.[43] Regeneration is accomplished with air, preferably below 1500°F, oxidizing the sulfur to SO_2 that subsequently would be converted to sulfuric acid. Reactions involved are as follows:

$$3Fe_2O_3 + 9H_2S \rightarrow 3Fe_2S_3 + 9H_2O \quad (25.7)$$

$$6FeS_{1.5} + 13O_2 \rightarrow 2Fe_3O_4 + 9SO_2 \quad (25.8)$$

$$2Fe_3O_4 + \tfrac{1}{2}O_2 \rightarrow 3Fe_2O_3 \quad (25.9)$$

Reactions are carried out in a fixed bed. Research is being conducted to test fly ash and silica supports for the iron oxide and binding materials to improve physical strength. The effects of water vapor in the fuel gas as well as tars and particulate matter are being studied. The sorbents are put through 30 cycles to test their stability to process conditions. Removal of 92–97% of the H_2S has been obtained.

Equipment used for the laboratory-scale investigation is an absorber 2 in. diameter × 25 in. long, with regeneration being carried out with air in a tube 1 3/8 in. in diameter × 24 in. long. An SO_2 concentration in the effluent gases of 11.5–12.0% has been obtained. A flowsheet of this equipment is shown in Fig. 25.10.

A 16-in.-diameter atmospheric pressure coal gasifier is being operated to provide a source of hot gas for investigating the removal of H_2S by means of a larger iron oxide system. The absorber in this case can handle 9000 scf/hr or producer gas and is 12 in. in diameter × 10 ft long. Regeneration again will be done by air. The equipment is large enough to permit investigation of the problems of temperature control and other important process variables and should allow an economic evaluation of the process.

It appears that temperature of operation and other process conditions are such that development to large scale is feasible. The effect of tars in the hot gases is a problem that must be investigated. The recovery of

[43] E. C. Oldaker, A. M. Poston, Jr., and W. L. Farrior, Jr., *Hydrogen Sulfide Removal from Hot Producer Gas with a Solid Fly Ash–Iron Sorbent,* Rep. No. MERC/TPR-75/2, Morgantown Energy Research Center, Morgantown, W. Va., 1975.

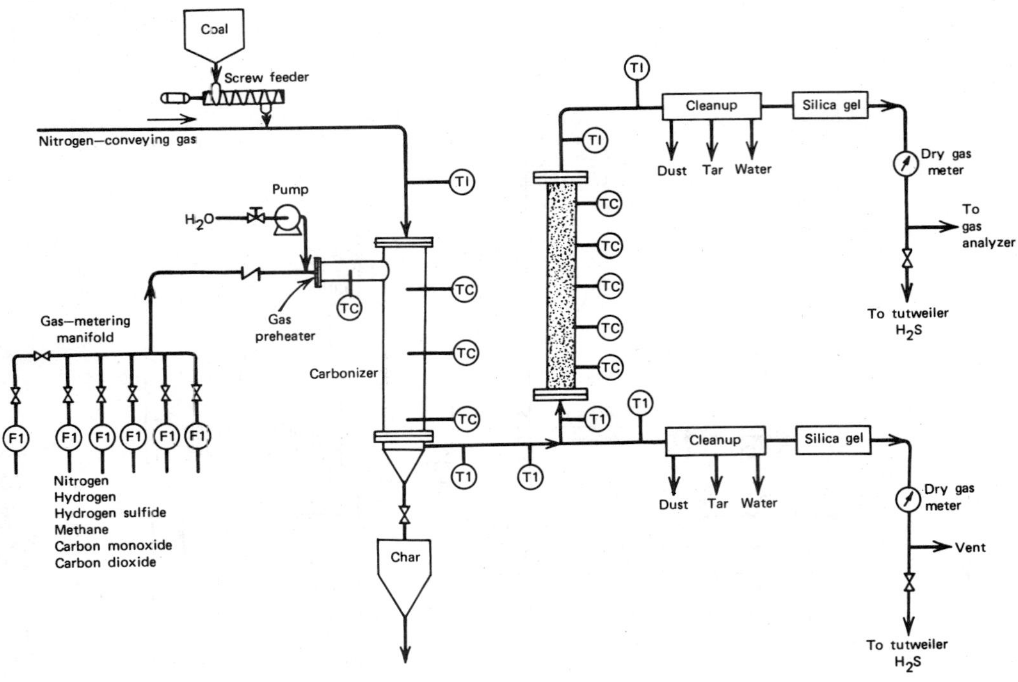

Figure 25.10 Flow diagram of the MERC apparatus for measuring iron solids sorption capacity for H_2S.

the sulfur as SO_2 and subsequently as sulfuric acid is not as attractive as obtaining elemental sulfur as a final product.

25.3.9.2 *The Conoco Dolomite Process*

A process using a hot fluidized bed of dolomite for desulfurization of the hot gas is under development at the Conoco Laboratories at Library, Pennsylvania.[44] The conceptual process is shown in the flowsheet in Fig. 25.11. The hot gas containing H_2S contacts the fluidized dolomite at about 1678°F and 15 atm. The following reaction takes place:

$$H_2S + CaCO_3{\cdot}MgO \rightarrow CaS{\cdot}MgO + H_2O + CO_2 \quad (25.10)$$

The reaction proceeds very well with dolomites but not with limestone. In a gas containing 0.6% H_2S, 9% CO_2, and 13% H_2O, the residual H_2S in the purified gas can be reduced to about 0.02%. The experimental desulfurization bed has a diameter of 3 in. and height of 18 in. The regenerator diameter is 2 in. with a height of 46 in.

The spent solids are lifted by a reactant stream of steam and CO_2 to a fluidized-bed regenerator at about 1300°F, where the reverse reaction takes place. Composition of the effluent gas stream from the overheat of the regenerator is typically 63.9% CO_2, 32.5% H_2O, and 3.6% H_2S. The spend solids contain about 88% CaS compared to 75% CaS in the regenerated solids.

The gas from the regenerator is sent to a liquid phase Claus reactor at 310°F where the Claus reaction takes place:

$$H_2S + \tfrac{1}{2}\,SO_2 \rightarrow 1\tfrac{1}{2}\,S_l + H_2O \quad (25.11)$$

Because of the low concentration of H_2S—only 3.6% in the regenerator off-gas (5.3% dry basis)—it is not possible to operate the conventional vapor phase Claus reaction and get high conversions of the H_2S sulfur. Conversion of the H_2S in the liquid phase Claus

[44] G. P. Curran, J. T. Clancy, B. Pasek, M. Pell, G. D. Rutledge, and E. Gorin, *Production of Clean Fuel Gas from Bituminous Coal,* Consolidation Coal Company Report to EPA, December 1973; NTIS PB 232-695/AS.

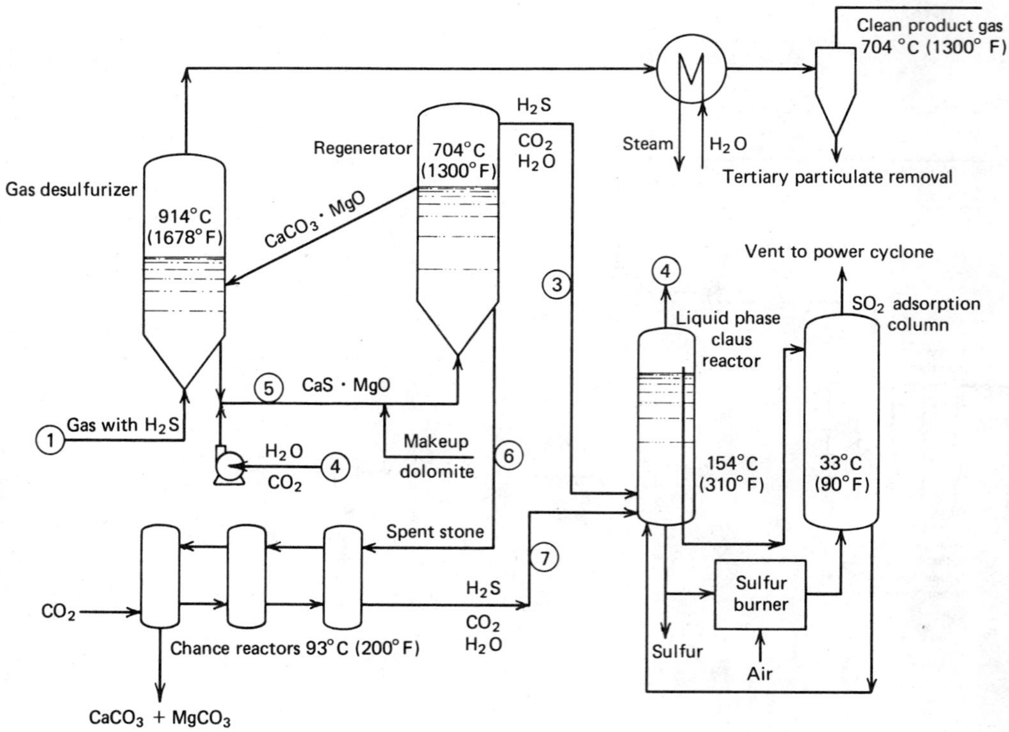

Figure 25.11 The Conoco hot desulfurization process.

reactor is 96%. The effluent gas, mostly CO_2 plus a small amount of H_2S, is recycled to the regenerator.

The solids have been put through more than 50 cycles of regeneration, and attainment of equilibrium has been demonstrated. Makeup rates of fresh dolomite can be expected to be 1–2% of the solids circulation rate.

The spent stone is converted to $CaCO_3$ by the Chance reaction, making it more suitable for disposal:

$$H_2O + CO_2 + CaS \rightarrow H_2S + CaCO_3 \quad (25.12)$$

The H_2S-containing stream is fed to the liquid phase Claus reactor for conversion to elemental sulfur.

The economics of the Conoco dolomite process, based on data collected in 1973, are summarized in Table 25.4.

Table 25.4 Economic Evaluation for Plant Operating in 1980: 70% Plant Operating Factor, 3.4 × 10⁶ Tons of Coal

Item	1975 design
Total investment (includes 9½% escalation and 8½% interest during construction)	$81.2 × 10^6$
Investment per kilowatt at 36% efficiency based on coal HHV	58.0
Direct operating cost	$11.13 × 10^6$
Acceptor at $16/ton	2.27
Interest on W.C. at 8½%	0.23
Capital charges at 18%	14.62
Sulfur credit at $25/tonne	(2.98)
Net annual cost	$25.27 × 10^6$
Cost per ton of feed coal	$7.40
Cost per 10^6 Btu HHV in feed coal	31.0¢
Cost per 10^6 Btu HHV in product gas	40.3¢
Cost per 10^6 Btu HHV plus sensible heat in product gas	38.0¢

In general, the chemistry of the various steps has been fairly well demonstrated in the small-scale unit at the Library laboratories. In many respects the Conoco dolomite process is similar to the CO_2 acceptor process that was operated at Rapid City, South Dakota. Circulating hot fluidized beds of dolomite are common to both processes, so that the basic process concepts have been successfully demonstrated on a large scale. Fines carryover from the fluidized bed into the purified gas stream could be a major problem, especially if the gas is to be used in a turbine.

Makeup requirements of fresh dolomite could be a major cost element, but this can only be defined accurately by operation of larger scale equipment.

25.3.9.3 *The Battelle Northwest Molten Calcium Carbonate Process*

A molten salt process for the removal of H_2S from fuel gases is being developed at the laboratories of Battelle Northwest in the state of Washington.[45] The advantage of a molten salt is that there are no reaction barriers at the gas interface as with solids. The molten salt used is a mixture of sodium, potassium, lithium, and calcium carbonates that is liquid at 500°C and stable at 900°C in an atmosphere of CO_2. Calcium carbonate has been found to be much more active in reaction with H_2S than other carbonates and is considered to be the main reactant:

$$CaCO_3 + H_2S \rightarrow CaS + CO_2 + H_2O \quad (25.13)$$

A pilot plant has been operated at atmospheric pressure using coke in a gasifier to supply the gas. A flowsheet is shown in Fig. 25.12. Ash from the coke is left in the gas. Hot gas from the gas heater is sent to a venturi where it contacts molten salt at about 800°C. The gas-salt mixture flows through a vertical line to a de-entrainer arrangement, where the salt separates and flows downward to a pot furnace and the gas flows upward to a separator filled with 1-in.-diameter alumina balls topped with a sapphire demister to provide separation of the gas and entrained salt.

The pilot plant has been operated briefly with a typical flow of 0.9 gal/min of molten salt at 700°C and 1 psig with a gas flow of 50 scf/min. With an inlet H_2S content of 0.84% H_2S, the outlet H_2S was as low as 5 ppm. Outlet COS at about 0.1% tends to be higher

[45] R. H. Moore et al., "Molten Salt Scrubbing for Removal of Particles and Sulfur from Producer Gas," *Proc. EPA/ERDA Symp. High Temperature Purification/Particle Removal,* September 20–21, 1977.

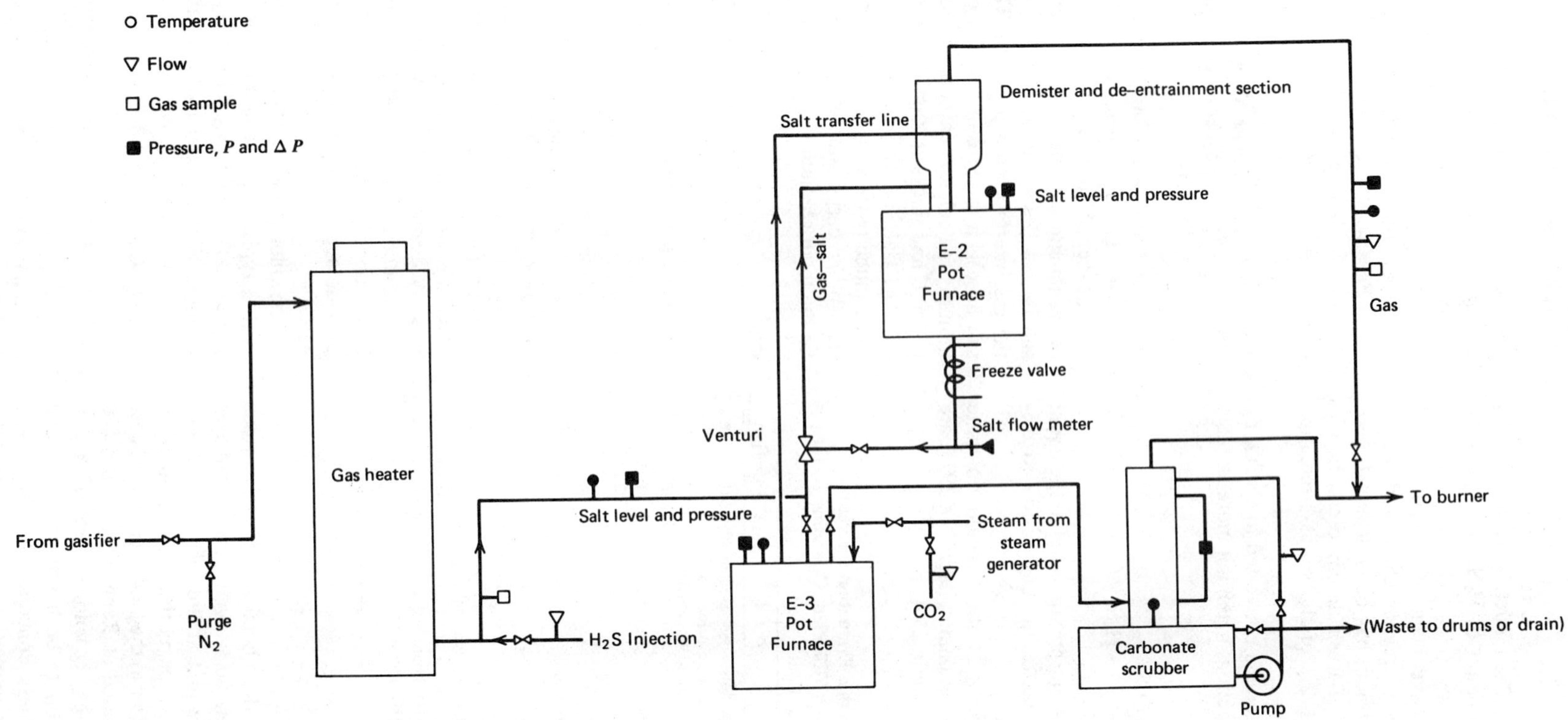

Figure 25.12 The Battelle Northwest molten salt process.

than the inlet COS at 0.02%. Overall sulfur removal from the gas is in the range 95–98%.

Regeneration of the spent salt is done typically with a mixture of 31 lb/hr of steam and 45 scf/hr of CO_2, giving an H_2S concentration of 4.2% in the CO_2/H_2S mixture. Although this off-gas is not sufficiently high in H_2S to provide a suitable feed to a Claus sulfur recovery unit, gases much higher in H_2S content are theoretically possible.

The solids carryover from the pilot-plant unit indicates that much better separation must be accomplished. With a total particle burden of 0.074 gr/scf in the inlet gas, the outlet gas contained 0.0245 gr/scf. About 30% of the solids from the unit were molten salt particles. With respect to the tolerance of a downstream turbine, the alkali carryover is excessively large by about 2 orders of magnitude.

The process entails several inherent problems:

1. Excessive solids and alkali carryover to downstream facilities.
2. Buildup of soluble and insoluble ash in the molten salt.
3. Requirement of a large quantity of CO_2 and steam for regeneration.
4. Corrosion of equipment in contact with the molten salt.
5. Concentrations of H_2S in the H_2S-CO_2 mixture obtained during regeneration may be too low to supply a suitable feed gas to a Claus unit.

25.4 WATER–GAS SHIFT CATALYSTS

Gas produced by coal gasification can be used directly as fuel gas after purification to lower the concentration of sulfur compounds and CO_2. In many types of application, however, it is necessary to reduce the level of CO or to eliminate it altogether. For methanol synthesis or upgrading the gas by catalytic methanation it is desirable to adjust the level of CO to have the preferred ratio of H_2 to CO for subsequent processing.

For hydrogen or ammonia manufacture the CO must be removed nearly completely. This is most conveniently accomplished by use of water-gas shift catalysts, sometimes referred to as CO shift catalysts, where by addition of steam the CO is shifted to an equal volume of the more desirable hydrogen. Carbon dioxide also produced by the shift reaction can be removed to the desired level required for downstream processing. Small amounts of CO and CO_2 that are objectionable as ammonia catalyst poison are generally converted to CH_4 by catalytic methanation.

The gas composition from the K-T coal gasification process,[46] which is being used in some parts of the world as a source of ammonia synthesis gas, is shown below. For comparison purposes, also shown is the gas for ammonia synthesis made from re-forming of natural gas with steam in the primary reformer, and air in the secondary re-former. Considerably more CO in the coal-based gases must be shifted in downstream processing than in the gas from natural gas reforming.

Ammonia Synthesis Gas

	Composition (vol %, dry basis)	
Component	K-T Coal Gasification	Natural Gas Re-Forming
H_2	32.86	56.0
CO	58.68	13.0
CO_2	7.04	8.0
CH_4	Nil	0.5
N_2	1.12	22.5
H_2S	0.28	Nil
COS	0.02	Nil

25.4.1 The Shift Reaction

The upgrading of a coal-based fuel gas to a higher Btu content requires a methanation step that proceeds as follows:

$$3H_2 + CO \rightarrow CH_4 + H_2O \qquad \Delta H_{572^\circ F} = -93{,}240 \text{ Btu/lb-mole} \qquad (25.14)$$

Complete conversion of the CO in this step requires a minimum ratio of H_2 to CO of about 3 : 1. As shown in the K-T gas analy-

[46] J. F. Farnsworth, H. F. Leonard, D. M. Mitsak, and R. Wintrell, "Production of Gas from Coal through a Commercially Proven Process," paper presented at 5th Synthetic Pipeline Gas Symp., Chicago, October 29–31, 1973.

sis, this ratio is 0.56 : 1. Therefore the water-gas shift reaction is necessary to increase the H_2 content while decreasing the CO to achieve the desired methanation ratio.

The classic water-gas shift reaction is:

$$CO + H_2O <====> CO_2 + H_2 \Delta H_{500°F} = -17,006 \text{ Btu/lb-mole} \quad (25.15)$$

The technology of the shift reaction has been highly developed since its very early conception. It is a simple reaction that involves no change in volume; equilibrium is unaffected by changes in pressure, and side reactions are negligible. It is promoted in the presence of various catalysts and actually is one of the earliest industrial applications of catalysis.

As noted in Equation 25.15, the reaction is exothermic and is reversible.

Temperature plays an extremely important part in the various aspects of the CO shift reaction.

The equilibrium constant[47] is expressed as follows:

$$k = \frac{[CO_2][H_2]}{[CO][H_2O]} \quad (25.16)$$

where the brackets indicate molar values.

It may also be related to temperature, as follows:

$$k = e^x \quad (25.17)$$

$$\text{where } x = \frac{1}{RT}\left(11{,}321 - 31.08T + 3T \ln T - 2.8 \times 10^{-4}T^2 - \frac{91{,}500}{T}\right)$$

T = temperature (K)

R = gas constant [cal/(g−mole)(K)]

Table 25.5 gives the heat of reaction and equilibrium constants for a broad range of temperatures.

Selective substitution of temperatures in Equation 25.17 produces equilibrium constants of 149 at 220°C and 4.9 at 500°C. It becomes obvious that the reaction is favored by lowered temperatures.

In an adiabatic reactor, with little or no dissipation of the heat of reaction, the increasing temperature thus tends to reduce the equilibrium constant and to limit the conversion of CO.

As in most catalytic processes, catalyst activity in the shift reaction increases with increasing temperature.[48] However increasing temperature increases the rate of decrease of active surface area, which results in a decline of catalyst activity. Thus it becomes necessary to operate a CO conversion system at the minimum temperature to secure a practical equilibrium constant, yet at a temperature high enough to provide a satisfactory reaction rate, and still maintain it below a maximum temperature to preserve catalyst activity.

Table 25.5 Heat of Reaction and Equilibrium Constant of the Water-Gas Shift Reaction

Temperature (°F)	Heat of Reaction, ΔH (Btu/lb-mole)	Equilibrium Constant, k
200	−17,570	4523
300	−17,410	783.6
400	−17,220	206.8
500	−17,006	72.75
600	−16,777	31.44
700	−16,538	15.89
800	−16,293	9.030
900	−16,044	5.610
1,000	−15,787	3.749
1,100	−15,544	2.653
1,200	−15,299	1.966
1,300	−15,056	1.512
1,400	−14,814	1.202
1,500	−14,575	0.9813
1,600	−14,344	0.8192
1,700	−14,117	0.6970
1,800	−13,892	0.6037
1,900	−13,672	0.5305
2,000	−13,459	0.4712
2,100	−13,246	0.4233
2,200	−13,041	0.3843

25.4.2 Catalysts for CO Conversion

Three basic catalysts have been developed to promote the CO shift reaction, and each is characterized primarily by the temperature range in which it operates as well as by the chemical composition. They have been classified as high-temperature shift (HTS), low-

[47] D. J. Borgars and J. S. Campbell, in *Fertilizer Science and Technology Series,* Vol. 2, Marcel Dekker, New York, 1974.

[48] D. Allen, in *Fertilizer Science and Technology Series,* Vol. 2, Marcel Dekker, New York, 1974.

temperature shift (LTS), and medium-temperature shift (MTS).

Commercial experience with the HTS catalyst is quite extensive, and a great deal is known and can be predicted regarding its performance. The LTS catalyst was introduced in 1962, and its capabilities have been well established. Only a few commercial installations have included an MTS reactor, and most properties and performance data are a result of laboratory and pilot-plant investigations.

25.4.2.1 High-Temperature Shift Catalyst

The first developed was the HTS catalyst, which is a chromia-promoted iron oxide and varies in composition with the manufacturer. A typical analysis as follows:

Iron	56.5–57.5 wt %
Chromium	5.6–6.0 wt %
Carbon	2.0–4.0 wt %
Sulfur	0.07–0.19 wt %

This catalyst finds its best operational capabilities in a range of 350–550°C. Using Equation 25.17, the maximum equilibrium constant at the lower temperature is 20.3, decreasing to 3.4 at the upper level.

It has been found that when the shift reactor exit temperature exceeds 450°C, a reasonable approach to equilibrium is equivalent to a temperature 20–30°C above the exit temperature.[47] Thus if the exit temperature is 525°C, the exit gas composition will correspond to equilibrium at 550°C. Consequently, with a gas of a known composition, the degree of conversion possible with the HTS catalyst can be calculated. Since the CO_2 and H_2 increase and the H_2O decrease is equal to the CO decrease, a simple substitution in Equation 25.16 permits an evaluation of the HTS catalyst for a given application.

Design considerations should take into account the temperature rise due to the heat of reaction to see that it is limited to about 200°C. This ΔT will vary with temperature, but it is approximately 10–11°C per 1% CO converted (wet basis). Maximum CO conversion in a single stage is about 19%. Space velocities increase to about 2500 when the pressure is at 30 atm and inlet CO concentration is 13.0%. [Space velocity as used in this chapter is volume of inlet gas per hour measured at 0°C and 760 mm Hg pressure per volume of catalyst.]

25.4.2.2 *Low-Temperature Shift Catalyst*

The second type of shift catalyst to be developed was the LTS catalyst. This catalyst, consisting of copper oxide and zinc oxide plus aluminum oxide, has the following typical composition[48]:

CuO	15–31 wt %
ZnO	32–62 wt %
A_2O_3	0–40 wt %

The effective temperature range is 200–250°C. With an allowable ΔT of 50°C, the maximum CO conversion rate based on 10–11°C per 1% CO for a single-stage LTS is about 4.7%.

In a recent development a chromia-type LTS catalyst was tested in a commercial plant with good results.[49] The plant was one in which natural gas was re-formed and shifted, and it represents a typical combination of the use of an HTS and LTS shift catalyst combination used in the preparation of ammonia synthesis gas. The flow sequence and the inlet and outlet gas analysis with respect to the shift reactors are shown in Fig. 25.13. The space velocity at which the LTS catalyst is operated is about 2000 hr^{-1} and the outlet gas has the residual CO reduced to 0.15%.

In calculating catalyst volumes using appropriate kinetic rate equations, it may be also necessary to investigate the economics of steam addition. Excess steam drives the reaction to a higher degree of conversion but is effective only up to a certain optimum ratio of steam to dry gas.[50] Larger catalyst volumes could produce the same effect, but the economics of the two amounts should be compared.

[49] R. S. Collard and R. McKintosh, "Low Temperature Shift Regeneration," paper presented at Symp. Safety in Ammonia Plants and Related Facilities, American Institute of Chemical Engineers, Boston, September 7–10, 1975.

[50] A. L. Kohl and F. C. Riesenfeld, *Gas Purification,* 2nd ed., Gulf Publishing Company, Houston, Tex., pp. 634–638, 1974.

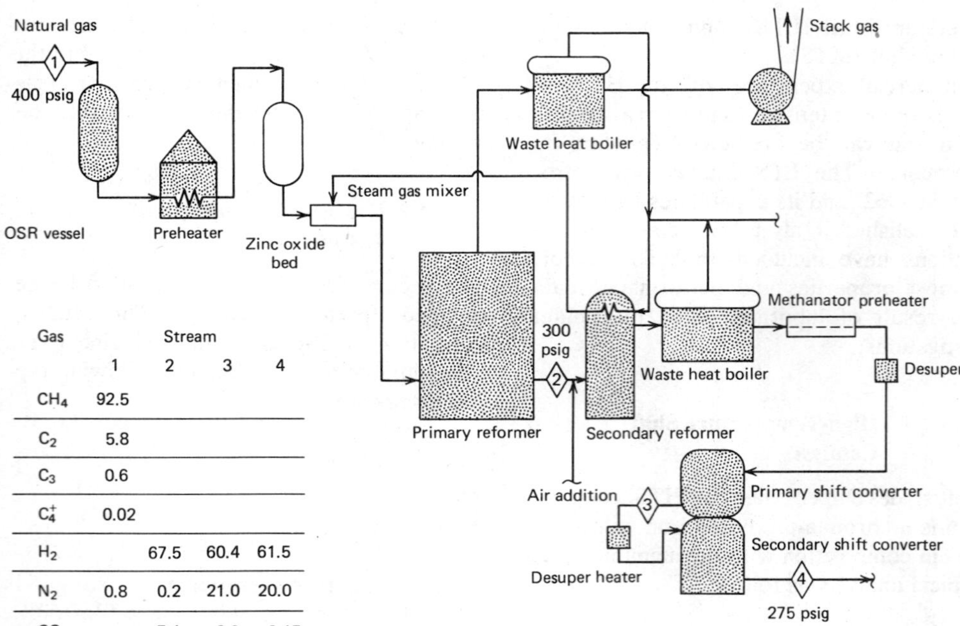

Gas	Stream 1	2	3	4
CH_4	92.5			
C_2	5.8			
C_3	0.6			
C_4^+	0.02			
H_2		67.5	60.4	61.5
N_2	0.8	0.2	21.0	20.0
CO		7.1	2.0	0.15
CO_2	0.2	15.0	16.1	16.9
Ar			0.2	0.2
SG		1.2	1.0	1.2

Figure 25.13 A typical process for the preparation of ammonia synthesis gas from natural gas, and inlet-outlet gas analysis.

In addition to the ratio of steam to dry gas, other variables effect the catalyst activity[47]:

1. Increasing pressure increases catalytic activity, but in the case of the iron base HTS it seems to level off at 350–400 psig at a level of about 4 times that at atmospheric pressure. The LTS activity does not level off at pressures up to 500 psig where activities of 6–8 times the atmospheric pressure activity are realized.[48]
2. Decreasing flow rate or space velocity results in a closer approach to equilibrium and a higher percentage CO conversion.
3. Decreasing catalyst particle size will increase reaction rate at the expense of increasing pressure drop.
4. Catalyst activity is also affected by a variety of extraneous materials that can be found in process streams.

Hydrogen sulfide can be tolerated to a great degree with the HTS catalyst. The percentage of H_2S in the feed gas can be higher at higher temperatures, since deactivation increases at lower temperatures. For example, at temperatures above 410°C there is little effect even at contents up to 4% H_2S. Even at 350°C percentages up to 0.5% H_2S produce only a small effect.[51]

However the LTS catalyst is severely affected at all temperatures in its operating range by H_2S concentrations exceeding about 0.1 ppm. LTS reactors generally have a guard bed of ZnO preceding them.

Phosphates can be an accumulative catalyst poison and can be introduced through boiler water solids. Only clean steam and the use of condensate for water injection should be considered. Most hydrocarbon compounds found in gasification or partial oxidation raw gases have little effect on shift catalysts. Chlorides, which may result from boiler solids, seawater, and water treatment facilities, can poison the LTS catalysts in

[51] Tetsuei Sueyama, in *Fertilizer Science and Technology Series,* Vol. 2, Marcel Dekker, New York, 1974.

concentrations as low as 0.1 ppm.[52] Oxygen, if present, will reoxidize both the HTS and LTS catalysts. Physical blocking of the catalyst can occur if the feed gas contains a significant quantity of finely divided solid particles.

Initial activation of both the HTS and LTS catalysts requires a reduction step that must be carefully performed in accordance with the manufacturer's recommendations. After reduction, both catalysts are extremely pyrophoric and extreme care must be taken to prevent uncontrolled reoxidation.[53] In practice, it has been found that the life of HTS catalysts is in the range of 3–5 years; the useful life of the LTS catalyst is 1–2 years.[47]

25.4.2.3 *Medium-Temperature Shift Catalyst*

A recent development has been a catalyst to fit between the upper limit of the LTS (250°C) and the low practical limit (350°C) of the HTS.[54,55] The so-called MTS catalyst consists of about 3.0–4.0 wt % CoO and 13.0–15.0 wt % MoO_3 supported on alumina. It has an effective temperature range of 260–450°C and is limited to about 18.0% CO conversion in a single stage.

Whereas the HTS and LTS catalysts operate in a reduced state, the MTS catalyst must operate in a sulfided state. It requires a sulfiding step with H_2S for initial activation[54,55] and a presence of more than 20 ppm S in the feed gas to maintain satisfactory operation. Estimated life of the MTS catalyst is 3+ years.[54]

25.4.3 Shift Catalyst Selection

Selection of a CO shift catalyst or a combination of catalysts is a complex problem that depends on many factors of each specific application. There can be a wide variation in the inlet and the outlet CO contents, and the gas may contain impurities that poison the catalysts. The composition of the raw and converted gases plus the rate of conversion determine the volume of catalyst required, but other factors have to be considered in the design and operation of the overall CO shift system[47]:

1. The operating characteristics of the catalysts used, their physical characteristics, their chemical behavior and stability, and the temperature limits of operation.
2. The kinetics of the catalyzed shift reaction.
3. The equipment needed for normal operation, and to start up and shut down in a safe and controlled manner.
4. The ability to provide effective, quick, and economic maintenance.

Selection of a catalyst for the gasification process requires an overall analysis of the application. High conversion of CO for an ammonia synthesis gas, for example, could be achieved in four ways, depending on the economics of the system.

1. Multistage HTS reactors with interstage cooling could provide a high percentage conversion. This could be further modified with interstage CO_2 removal if desired.
2. An HTS reactor could be utilized for bulk CO conversion followed by an LTS reactor. However interstage desulfurization is required, because all LTS catalysts are very quickly deactivated as a result of sulfur poisoning.
3. An HTS reactor could be utilized for bulk CO conversion followed by an MTS reactor, provided there is sufficient sulfur in the gas to maintain MTS catalyst activity.
4. Multistage MTS reactors can be used with interstage cooling as in item 1.

For adjusting the ratio of H_2 to CO for methanation, either single or multistage HTS or MTS reactors can be selected, depending on the specific gas to be treated and the comparative economics.

[52] John F. Lombard, *Hydrocarbon Process.*, **48**(11), 111–116 (1969).

[53] R. Habermehl and K. Atwood, *Prepr., Am. Chem. Soc. Div. Fuel Chem.*, **8**(1) (1964).

[54] Catalyst and Chemicals Inc., Louisville, Ky., Tech. Bull. No. C25-104, and private correspondence.

[55] W. Auer, E. Lorenz, and K. H. Grundler, paper presented at 68th National Meeting, American Institute of Chemical Engineers, Houston, Tex., 1971.

25.5 CATALYTIC METHANATION

A consideration of the production of gas from coal to supplement our natural gas supplies indicates that the advantages of obtaining methane as the final product are many. When compared to a mixture of hydrogen and carbon monoxide made by all coal gasification processes, the higher heating value of CH_4 represents a more concentrated form of energy. This factor makes the transmission of CH_4 in existing pipeline networks and distribution systems the most economical method of transport of coal-based energy. In addition, CH_4 is stored economically in depleted natural gas fields or as liquefied gas for use during high demand periods. Methane does not have the broad explosive limits of H_2 nor the toxic properties of CO.

Although it is possible to obtain a product gas by direct hydrogenation of coal sufficiently high in CH_4 to yield 900 Btu/scf gas (~ 90% CH_4), the required pressure is very high ($\gg$ 1000 psig) and the residence time relatively long. Consequnetly it is currently considered more economical to accept whatever CH_4 is made in the primary coal gasification step at moderate pressures, and after CO_2 and sulfur removal, to react the H_2 and CO to CH_4. The final product gas will then be in the desired range of 900–950 Btu/scf.

The CH_4 yield from the gasification step varies widely with the process used and is largely dependent on the outlet temperature and pressure of the gas leaving the gasifier. The high-temperature processes such as the Koppers-Totzek and Texaco processes produce very little CH_4. The gasifiers that have lower temperatures of the gas effluent and operate at high pressure produce much more CH_4. Thus in the Lurgi fixed-bed gasifier about 40% of the CH_4 is produced in the gasification step and 60% in subsequent catalytic methanation. In the HyGAS process under development by the Institute of Gas Technology, about 66% of the CH_4 is produced by gasification-hydrogenation and the remainder by follow-up methanation.[56]

A mixture of 84% CH_4 and 16% H_2 will have a heating value of 900 Btu/scf. Inasmuch as there will be inerts such as N_2 and CO_2 present, the methane content should be higher than 84% to compensate for the diluting effect of the inerts.

[56] W. G. Bair, D. Peppen, and A. L. Lee, "Methanation of Synthesis Gas," in Am. Chem. Soc. Adv. Chem. Ser. No. 146, 1974, pp. 123–136.

25.5.1 Chemistry and Thermodynamics

The first study of catalytic methanation of H_2 and CO took place in 1902 and used nickel as a catalyst.[57] Iron and cobalt catalysts are also quite active in producing CH_4, although they are best known for producing liquid hydrocarbons in the Fischer-Tropsch synthesis, which operates as lower temperatures.[58]

The reactions that are involved in the catalytic reaction of H_2 and CO to CH_4 are given below with the heat of reaction and equilibrium constant at 700K[59]:

$$CO + 3H_2 \rightleftarrows CH_4 + H_2O \quad \Delta H(700\text{ K}) = -52.69\text{ kcal/g-mole}, \quad k(700\text{ K}) = 3726.5 \tag{25.18}$$

$$CO + H_2O \rightleftarrows CO_2 + H_2 \quad \Delta H(700\text{ K}) = -9.064\text{ kcal/g-mole}, \quad k(700\text{ K}) = 9.01 \tag{25.19}$$

$$CO_2 + 4H_2 \rightleftarrows CH_4 + 2H_2O \quad \Delta H(700\text{ K}) = -43.63\text{ kcal/g-mole}, \quad k(700\text{ K}) = 413.7 \tag{25.20}$$

$$2CO + 2H_2 \rightleftarrows CH_4 + CO_2 \quad \Delta H(700\text{ K}) = -61.75\text{ kcal/g-mole}, \quad k(700\text{ K}) = 33{,}566 \tag{25.21}$$

$$(2n + 1)\,H_2 + nCO \rightleftarrows C_nH_{2n+2} + nH_2O \tag{25.22}$$

$$2n\,H_2 + nCO \rightleftarrows C_nH_{2n} + nH_2O \tag{25.23}$$

$$C_2H_6 + H_2 \rightarrow 2CH_4 \tag{25.24}$$

$$C_3H_8 + 2H_2 \rightarrow 3CH_4 \tag{25.25}$$

Reaction 25.18 is the primary methanation reaction, where a 3 : 1 ratio of H_2 to CO is used. Usually the gas from the gasifier is richer in CO than desired and the ratio is adjusted to 3 : 1 by means of a shift reactor ahead of the methanation reactor in accordance with Reaction 25.19. Reaction 25.19 can also take place in the methanation reactor. It is pertinent that for each 1% CO that reacts

[57] P. Sabatier and J. B. Senderens, *C.R. Acad. Sci.*, 134, 514 (1902).

[58] H. H. Storch, N. Golambic, and R. B. Anderson, *The Fischer-Tropsch and Related Syntheses*, John Wiley and Sons, New York, 1951.

[59] J. P. Strakey, A. J. Forney, and W. P. Haynes, "Methanation in Coal Gasification," paper presented at 2nd Ann. Symp. Coal Gasification, University of Pittsburgh, August 5–7, 1975.

in accordance with Reaction 25.18 there is about a 60°C rise in temperature in an adiabatic-type reactor. To avoid excessive temperature rise, the inlet gas is diluted with hot or cold recycle gas so that the temperature rise is limited to about 100–200°C.

Reaction 25.20 is the methanation of CO_2. It has been noted that this reaction does not begin to take place until the CO concentration is lowered to a few hundred ppm. Reaction 25.21, involving a 1 : 1 ratio of H_2 to CO to make CH_4, can result in carbon deposition; thus this synthesis gas ratio is not preferred. However this 1 : 1 ratio is employed in the methanation reactor used in the RM process (developed by Ralph M. Parsons); steam is added to shift the CO and avoid carbon deposition.[60]

The production of higher hydrocarbons shown in Reactions 25.22 and 25.23 takes place at lower temperatures for most catalysts in accordance with the Fischer-Tropsch synthesis. Although iron catalyst is excellent for liquid production, it promotes carbon deposition at the higher temperatures used for CH_4 synthesis.

Reactions 25.24 and 25.25 show the hydrocracking of higher hydrocarbons to CH_4. Thus any ethane or propane made in the coal gasification reactor would be lost when using the usual methanation catalysts.

A mechanism for methanation of CO and CO_2 was proposed by Vlasenko and Yuzefovich in 1969.[61] It was stated that CO reacts with molecularly adsorbed hydrogen to make CH_4. However CO_2 requires adsorbed hydrogen in the atomic form. Carbon monoxide adsorbs on the surface and poisons it for the adsorption of atomic hydrogen. Thus when CO is present, CO_2 is not methanated, an occurrence consistent with the observation that for many hydrogenation reactions CO acts as a poison.

25.5.2 Catalysts

Methanation catalysts have been reviewed by Mills and Steffgen.[62] They place the order of activity as follows:

$$\text{Ra} > \text{Ir} > \text{Rh} > \text{Ni} > \text{Co} > \text{Os} > \text{Pt} > \text{Fe} > \text{Mo} > \text{Pd} > \text{Ag}$$

Considering availability, cost, and activity, nickel catalyst is the logical choice for methanation, and nickel or nickel-containing mixtures are used in nearly all the methanation reactors currently under development. Precipitated nickel catalyst is the most active form of catalyst, although not as strong physically as other forms. Most commercial catalysts are in pelleted form and contain 25–70% nickel on alumina or kieselguhr base.

A Raney nickel catalyst (42% Ni, 58% Al) is used in the Synthane process. It is leached with NaOH to dissolve aluminum and to leave a spongy skeletal nickel catalyst that is quite active for methanation.

25.5.3 Catalyst Poisons

Gas made from coal contains many trace impurities. Most of these are removed in the water wash and purification system that follows the gasification step. The effect of several potential poisons was tested by Hausberger et al.[63] on nickel catalyst. The compounds tested were hydrogen sulfide, hydrogen chloride, nitrogen dioxide, ethylene, ethane, propane, propylene, methanol, and benzene. At the normal levels that these compounds might be found in gas from coal, only H_2S and HCl were considered to be catalyst poisons. Hydrogen sulfide and other sulfur compounds poison the catalysts by reacting with nickel to form inactive nickel sulfide. Wainwright determined that when COS in the feed gas to a methanation unit was only 0.04 gr/100 scf, the life of the nickel catalyst was more than 40 days.[64] In related experiments where the sulfur content was increased to as high as 1.5 gr/100 scf, the activity of the catalyst dropped sharply after

[60] G. A. White, T. R. Roszkowski, and D. W. Stanbridge, "The R.M. Process," in Am. Chem. Soc. Adv. Chem. Ser. No. 146, 1975.

[61] V. M. Vlasenko and G. E. Yuzfovich, "Mechanism of the Catalytic Hydrogenation of Oxides of Carbon to Methane," *Russ. Chem. Rev.*, **38**(9), 728–739 (1969).

[62] G. A. Mills and F. W. Steffgen, "Catalytic Methanation," *Catal. Rev.*, **8**(2), 159–210 (1973).

[63] A. L. Hausberger, C. B. Knight, and K. Atwood, "Development of Methanation Catalysts for the Synthetic Natural Processes," in Am. Chem. Soc. Adv. Chem. Ser. No. 146, 1975.

[64] H. W. Wainwright, G. C. Egleson, and C. M. Brock, *Laboratory Scale Investigation of Catalytic Conversion of Synthesis Gas to Methane*, U.S. Bur. Mines. Rep. Invest. No. 5046, April 1954.

about 0.58 g of sulfur had been absorbed per 100 g of nickel.

25.5.4 Reactor Types

Catalytic methanation has been used commercially in ammonia and hydrogen plants to convert small amounts of CO or CO_2 (0.1–1.0%) to CH_4. However very little operating experience has been gained on large-scale units installed to methanate high concentrations of H_2 and CO. This lack has led to the installation of several pilot plants to test different types of reactors.

Because of the highly exothermic nature of the methanation reaction—amounting to about 25% of the heat of combustion of the methane formed—considerable care must be taken to avoid catalyst overheating. In the various reactor development programs, the heat of reaction is removed from adiabatic reactors by use of hot gas recycle (DOE), cold gas recycle (HyGAS), or by having several stages of reactors in which a small conversion per pass is obtained (Koppers, RM process).

In 1974 at Westfield, Scotland, Continental Oil Company, the British Gas Corporation, and other companies operated a large pilot unit employing a fixed bed of catalyst cooled by gas recycle using feed gas from a Lurgi coal gasifier.[65] Production of gas from the methanator unit was about 2.1 million scf/day with a heating value of 979 Btu/scf. Because of the proprietary nature of the development, few of the technical details have been published.

A fluidized bed in which heat exchangers in the bed remove the heat of reaction is under development (Bi-Gas). A reactor of flame-sprayed catalyst on tubes in a boiling bath (DOE) is in the pilot-plant stage. In another development the catalyst is dispersed in a mineral oil bath, the oil being circulated through a heat exchanger to remove heat (Chem Systems). It is highly desirable from a thermal efficiency standpoint to convert the reaction heat to high-pressure steam that can be used effectively in the plant. Discussion of these reactor types is detailed in the sections that follow.

[65] J. E. Landers, "Review of Methanation Demonstration at Westfield," paper presented at 6th Synthetic Pipeline Gas Symp., Chicago, October 28–30, 1974.

25.5.5 The U.S. Bureau of Mines Investigations

As part of the synthetic fuels research and development program, the USBM studied the use of iron and nickel methanation catalysts in various types of reactors. Iron catalyst does have the potential advantage of producing higher hydrocarbons, but nickel catalyst at the operating temperatures used produces only methane.

25.5.5.1 *Catalyst Testing*

During the early 1950s[66] a small (0.622 in. I.D. × 12 in. long) catalyst-testing reactor was used, with a boiling Dowtherm jacket for temperature control. Test temperatures ranged from 260 to 400°C and pressure was 300 psig. Catalysts tested were precipitated and impregnated on various supports (e.g., as activated charcoal, kieselguhr, kaolin, alumina, silica gel, and zirconia cement). Fused-type iron oxide (Fe_3O_4), reduced and nitrided, and iron ore reduced with hydrogen were also tested. The most active catalysts were found to be precipitated nickel and Raney nickel extracted with caustic.[67] Carbon deposition was always a problem with iron catalysts and to a lesser extent with nickel. When the ratio of H_2 to CO was increased from 1 : 1 to 2.5 : 1, the carbon deposition was nearly eliminated with nickel catalysts.

25.5.5.2 *Catalyst Deactivation*

Mechanisms of methanation catalyst deactivation have been discussed by Strakey et al.[68] Nickel catalysts are extremely susceptible to sulfur poisoning, and the effect of increasing sulfur concentration has been shown.[69] To avoid poisoning, general practice is to keep the total sulfur concentration below 0.1 ppm. Higher temperature of operation

[66] M. Greyson et al., *Synthesis of Methane*, U.S. Bur. Mines Rep. Invest. No. 5137, July 1955.

[67] W. P. Haynes, J. J. Elliott, and A. J. Forney, "Experience with Methanation Catalysts," preprint of paper presented to American Chemical Society, Division of Fuel Chemistry, Boston, April 10–14, 1972.

[68] J. P. Strakey, A. J. Forney, and W. P. Haynes, "Methanation in Coal Gasification Processes," paper presented at 2nd Ann. Symp. Coal Gasification, University of Pittsburgh, August 5–7, 1975.

[69] H. W. Wainwright, G. C. Egleson, and C. M.

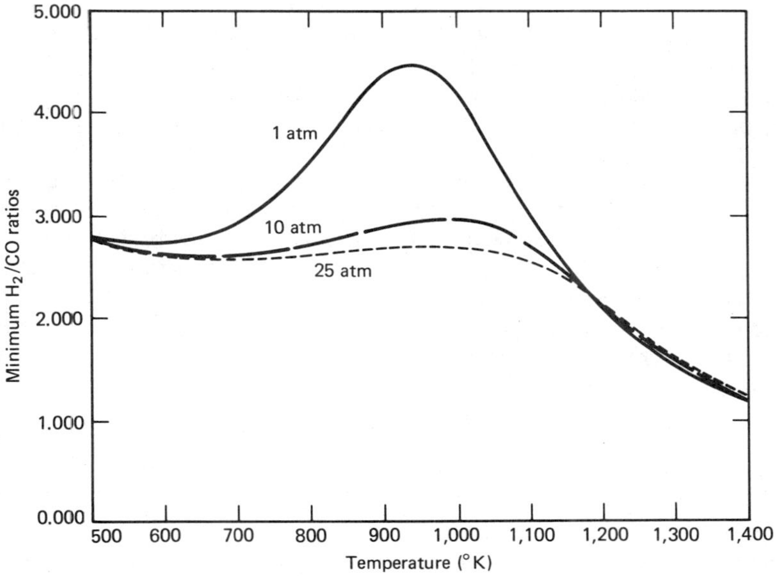

Figure 25.14 Carbon deposition boundaries; deposition occurs in area beneath curve.[66]

causes sintering of the nickel catalysts and a loss of surface area. Nickel carbide can form under synthesis conditions and is catalytically inactive. The carbiding reaction can be reversed by treatment with hydrogen above 250°C, which restores the catalyst activity.Carbon deposition from the decomposition of CO, which can result in catalyst deactivation, is favored at lower temperatures. Carbon formation from CH_4 can result from excessively high temperatures. Figure 25.14 shows carbon deposition boundaries calculated by Greyson et al.[66] Nickel carbonyl is formed at low temperatures and causes loss of the nickel catalyst. It has been found that iron carbonyl can result from the reaction of CO in the synthesis gas at high pressures and low temperatures (100–200°C) with carbon steel piping. The carbonyl decomposes on contact with nickel catalyst, which results in a deposition of iron on the surface. The iron then catalyzes formation of carbon, leading to catalyst fouling.[68]

25.5.5.3 *The Synthane Hot Gas Recycle*

The USBM has developed the hot gas recycle (HGR) reactor in which hot product gas is recycled to the inlet of the reactor diluting the fresh feed gas, thereby lowering the CO concentration and lessening danger of carbon formation. The small pilot-plant reactor was 3 in. in diameter with a bed height of 3 ft when using Raney nickel plates. With an inlet temperature of 300°C, the exit temperature is about 400°C because of the heat of reaction. This temperature range is well suited to the Raney nickel catalyst. The recycle gas volume is about 10 times the volume of fresh feed gas. Because the reactor is adiabatic, the heat of reaction would be recovered as high-pressure steam in a commercial unit. About 30% of the recycle stream is cooled to room temperature so that water is condensed and removed from the system. If this were not done, the water vapor concentration in the reactor would be about 50%, which is undesirable from a methane equilibrium standpoint and could, in addition, cause catalyst oxidation.

Because of the large volume of recycle gas, it was quite important to develop a catalyst form that gives very low pressure drops, thereby reducing horsepower requirements for a recycle gas compressor. Rather than using pelleted or garnular catalysts, therefore, the USBM researchers developed a vertical parallel plate assembly that was flame sprayed with Raney nickel alloy powder. This resulted in a pressure drop through the catalyst bed that was less than 1% of that through a granular bed. Fresh gas space velocities in the range of 2000–3000 have

---

Rock, *Laboratory Scale Investigation of Catalytic Conversion of Synthesis Gas to Methane,* U.S. Bur. Mines Rep. Invest. No. 5046, 1954.

been obtained, corresponding to total gas feed space velocities of 22,000–33,000. The life of the catalyst used in a parallel plate assembly was about 2300 hr with a productivity of 32,000 scf/lb of Raney nickel in a typical experiment.[70]

The hot gas recycle methanation reactor will be tested in DOE 75-ton/day Synthane pilot plant. It will handle about one-eighth of the gas production. The parallel plates are 3/16-in. thick with 3/16-in. spacing between the plates. A flowsheet is shown in Fig. 25.15. Inasmuch as the residual concentration of CO in the product gas leaving the HGR reactor is usually in the range of 0.1–1%, it is necessary to use a cleanup methanation reactor to assure that the concentration of CO in the final product gas is less than 0.1%. This reactor is a small fixed-bed unit with granular packing; it has no provision for recycle inasmuch as temperature rise is relatively low because of the small amount of methane actually made in this final step.

The USBM investigated a combination iron catalyst HGR reactor followed by a fixed-bed Raney nickel catalyst reactor.[71] The hydrocarbon yield of the iron catalyst was only about 40% CH_4, the remainder being higher hydrocarbons including 2–5% liquid oil. When the hot effluent gases from the iron catalyst were passed over the Raney nickel catalyst, the C_2–C_4 hydrocarbons were unaffected, but the C_5 plus material was hydrogenated to CH_4. The resulting product gas on a CO_2-free basis had an HHV as high as 1130 Btu/scf because of the presence of about 9% C_2's and 5% C_3's. With the use of a nickel methanation catalyst alone, the equivalent value is about 950 Btu/scf, since the only hydrocarbon product is CH_4.

[70] W. P. Haynes, A. J. Forney, and H. W. Pennline, "Synthesis of Methane in Hot Gas Recycle Reactor—Pilot-Plant Tests," *Prepr., Am. Chem. Soc., Fuels Div.*, **19**(3), 10–42 (1974).

[71] D. Bienstock, J. H. Field, A. J. Forney, and H. E. Benson, "Synthesis of High-Btu Gas from Carbon Monoxide and Hydrogen Using a Hot Gas Recycle Reactor," paper presented at American Gas Association Operating Section Conf., 1960.

25.5.5.4 *The Synthane Tube Wall Reactor*

The USBM has also developed another type of reactor referred to as the tube wall reactor (TWR), which is built in the form of a heat exchanger with Raney nickel catalyst flame sprayed on the outside of seven 2-in. diameter tubes, 27 in. long. The heat of reaction is transferred through the tube wall to a boiling Dowtherm bath. Operation of this type of reactor has demonstrated catalyst life in excess of 4 months and a productivity of 177,000 scf of methane per pound of catalyst. A single tube reactor of a diameter of 2 in. and length of 14 ft with Raney nickel catalyst sprayed on the inside of the tube has also been operated with good results. The 75-ton/

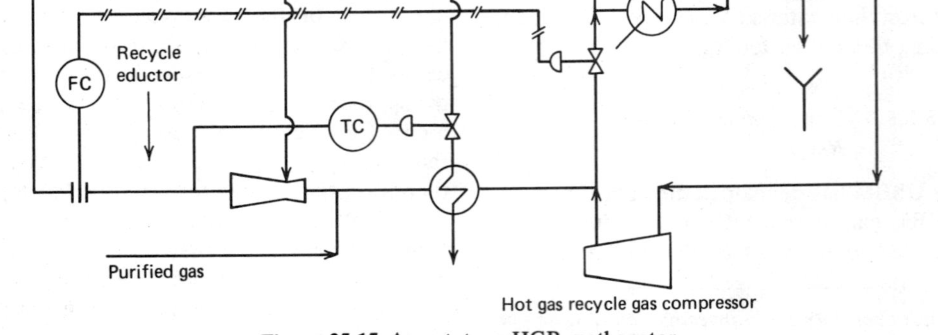

Figure 25.15 A prototype HGR methanator.

day coal gasification Synthane pilot plant also contains a TWR, again designed for methanating about one-eighth of the gasifier output. The tube bundle contains 22 tubes, 2 in. in diameter and 20 ft long. A flowsheet is shown in Fig. 25.16.

25.5.5.5 *The Synthane Hybrid Reactor*

The DOE laboratory at Bruceton, Pennsylvania, has developed a reactor that incorporates some of the features of the TWR and HGR reactors. The uncoated tubes of the TWR are provided with X-shaped inserts that have been sprayed with Raney nickel catalyst. The gas inlet temperature is 300°C and the outlet gas temperature is 400°C. However the recycle gas ratio can be reduced to about 8.5 : 1 rather than the usual 10 : 1 used in the HGR reactor. The main advantage is that the inserts can be easily replaced when the declining catalyst activity justifies it. About 65% of the heat of reaction is removed through the boiling Dowtherm bath and can be recovered.

25.5.6 The Institute of Gas Technology Methanation Process

The Institute of Gas Technology (IGT) has had under development since the 1950s coal gasification with oxygen-steam mixtures and hydrogenation processes to produce a high-Btu gas.[72] This work has evolved into a large pilot plant where pretreated coal char (~ 3 tons/hr) is hydrogenated in a fluidized bed at pressures of 1000–1500 psig and temperatures of 760–982°C.[73] The product gas from the hydrogasification reactor, largely H_2, CO, and CH_4, is cleaned and desulfurized and sent to the catalytic methanation reactor, where the H_2 and CO are reacted to CH_4. About one-third of the total methane is made catalytically, the other two-thirds coming from the hydrogasification reactor.

[72] *HYGAS: 1964 to 1972, Pipeline Gas from Coal Hydrogasification (IGT Hydrogasification Process),* Final Rep. No. HYGAS 381, Washington, D.C.

[73] W. G. Bair, D. Leppin, and A. L. Lee, *Methanation of Synthesis Gas,* Am. Chem. Soc. Adv. Chem. Ser. No. 146, 1974, pp. 123–136.

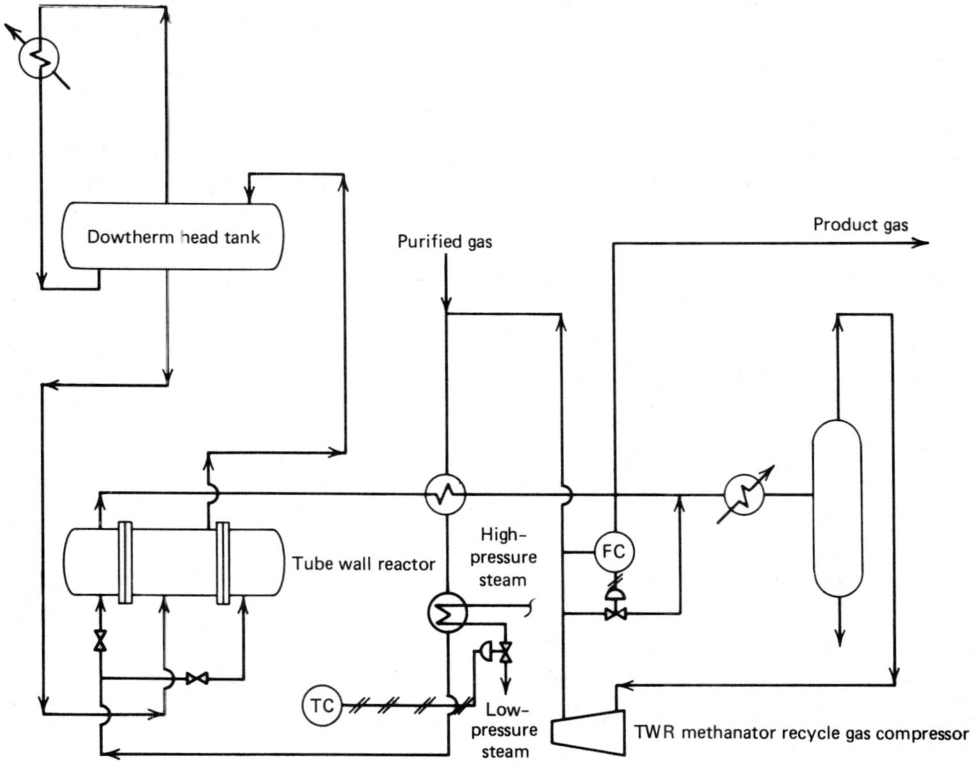

Figure 25.16 A prototype TWR methanator.

25.5.6.1 *The Cold Gas Recycle Methanation Reactor*

IGT has installed and operated as part of the HyGAS pilot plant an adiabatic cold gas recycle reactor system that satisfactorily removes the heat of reaction and controls the temperature of operation. The methanation system uses a feed gas that contains H_2 and CO in a volumetric ratio of about 3 : 1, to yield CH_4 in addition to that made in the hydrogasifier. The reactors are staged with no internal cooling.

The fresh gas is mixed with the recycle gas and preheated to 288°C inlet temperature. The ratio of fresh gas to recycle gas is adjusted so that the mixed gas contains about 4% CO. The outlet temperature is 482°C with the CO content below 0.05%. Before being fed to the second stage, the product gas from the first stage is mixed with additional fresh feed and cold recycle gas, so that the resulting temperature is again about 288°C. The inlet CO concentration is again about 4% and the outlet temperature 482°C. The flow scheme is shown in Fig. 25.17.

The reactors are 2 ft in diameter, with Stage 1 having a bed height of 5 ft and a catalyst volume of 15.7 ft^3. Stage 2 has height of 10 ft with a catalyst volume of 31.4 ft^3. The catalyst used is a commercially available nickel impregnated on kieselguhr (Ni-0104T), made by Harshaw, in the form of ¼ × ¼-in. pellets. The ratio of recycle gas to fresh gas was about 1.48 : 1 for one typical experiment, and when operating with a lignite coal feed the design ratio is 1.67 : 1. In experiments run to date, the inlet gas space velocity ranged from 4397 to 3220 in the first reactor and 2732 to 1720 in the second reactor. At the maximum rate of operation there was a total of 8.8 lb-moles/hr of CO converted to CH_4. A total of more than 500 hr of operation of the catalyst was accumulated with no apparent decline of activity. The heating value of the gas was 916 Btu/scf.

Subsequently the methanation catalyst was poisoned by sulfur and deactivated. The methanation system was then modified to alleviate the problem.[74] The first-stage methanation used was converted to a zinc oxide guard chamber and a new catalyst, a molecular-sieve type made by Union Carbide, was installed in the second-stage methanation vessel.

25.5.7 The Fluidized-Bed Methanation Reactor

Catalytic methanation has been studied in a fluidized bed by several organizations. In this type of reactor the catalyst is in a crushed granular form of about 20-200 mesh size. The

[74] B. S. Lee, "Status of the HYGAS Program," paper presented at 8th Synthetic Pipeline Gas Symp., Chicago, October 18–20, 1976.

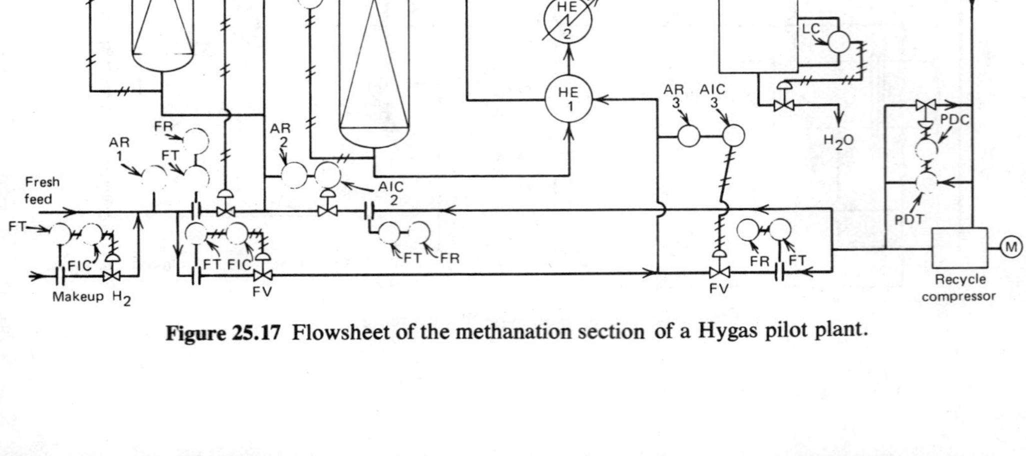

Figure 25.17 Flowsheet of the methanation section of a Hygas pilot plant.

synthesis gas with or without recycle gas is fed to the bottom of the reactor at a velocity sufficient to lift the catalyst and expand the bed. Linear velocities are usually in the range of 0.5–2.0 ft/s, depending on particle size and catalyst density.

For small reactors with a single catalyst tube, cooling and temperature control has been done by means of a jacket containing boiling Dowtherm A. Heat transfer from the catalyst particles to the cooling medium is quite effective and is promoted by the internal circulation of the catalyst to the reactor wall. Larger reactors have employed bayonet tubes in the bed in which coolant is circulated.

25.5.7.1 *Laboratory-Scale Fluidized Bed*

The USBM extended its studies of iron and nickel catalysts in fixed-beds to a fluidized-bed reactor[66] with a diameter of 0.96 in. and a length of 72 in. The reactor was cooled by means of a jacket around the tubular reactor containing Dowtherm A under pressure. Temperatures were in the range of 300–400°C and pressure was 300 psig. Space velocities (settled bed) were in the range of 880–3000 hr^{-1} for iron catalyst. Iron catalyst types were synthetic ammonia catalysts, Alan Wood magnetite, steel shot, and nitrided iron. Even with high ratios of $3H_2 : 1CO$, carbon deposition was excessive and methane content of the product gas was usually less than 30%.

The nickel catalysts were much more active than were the iron catalysts. Space velocities of 10,000 were attained with a catalyst life of 1000 hr. A product gas having a heating value of 900 Btu/scf was obtained. Attrition of the nickel catalyst resulting from the abrasive characteristic of fluidized beds resulted in a loss of as much as 37% of the catalyst in 220 hr of operation.

A particularly active catalyst that performed well in the fluidized bed was a Raney nickel catalyst.[75] The Raney nickel alloy used was 42% nickel and 58% aluminum of 80–230 mesh size. Immersion in NaOH dissolves some of the aluminum, leaving a porous structure. Operation of the catalyst at 300 psig, 370–387°C, and 7700 fresh gas space velocity (settled bed) with a feed gas ratio of 2.94 H_2 : 1 CO, gave a product gas of 950 Btu/scf. After 492 hr the catalyst activity decreased. It was possible to restore the activity for 472 additional hours by extracting the discharged catalyst again with alkali.

25.5.7.2 *The Institute of Gas Technology Fluidized-Bed Plant*

The IGT built and operated a pilot unit capable of converting 3000 scf/hr of synthesis gas to methane.[76] The reaction zone was made of a schedule 40, 6-in.-diameter pipe, 8.5 ft high and jacketed with an 8-in.-diameter pipe. Inside the 6-in. reaction zone were six bayonet heat exchange tubes of 0.75-in., schedule 40 pipe. Dowtherm A was circulated inside the jacket and bayonet tubes to bring the reactor up to temperature. When synthesis began, the Dowtherm served as the heat removal medium.

The longest pilot-plant run lasted 505 hr before fading activity resulted in its termination. Operating pressure ranged from 105 to 152 psi, temperatures from 325 to 450°C, fresh gas space velocity from 5160 to 6320 based on 0.5 ft^3 of catalyst. The catalyst used was 42 wt % Ni and 58 wt % Al of 40–200 mesh size. Treatment of the catalyst with NaOH was used to activate it. The heating value of the product gas on a CO_2-free basis was 902 Btu/scf at 105 hr and dropped gradually to 615 Btu/scf at the end of the run. The yield of methane during the experiment was 471 lb per pound of original catalyst alloy. This yield was only about 25% of the methane production obtained in laboratory catalyst-testing units and indicated difficulty in scaling up of the fluidized reactor.

25.5.7.3 *The Bituminous Coal Research Fluidized Bed*

The Bituminous Coal Research, Inc., laboratories at Monroeville, Pennsylvania, also worked on the development of a fluidized

[75] M. D. Schlesinger, J. J. Demeter, and M. Gregson, "Catalyst for Producing Methane from Hydrogen and Carbon Monoxide" *Ind. Eng. Chem.*, **48**, 68–70 (January 1956).

[76] H. A. Dirksen and H. R. Linden, "Pipeline Gas by Methanation of Synthesis Gas over Raney Nickel Catalyst," *Ind. Eng. Chem.*, **52**(7), 584–589 (July 1960).

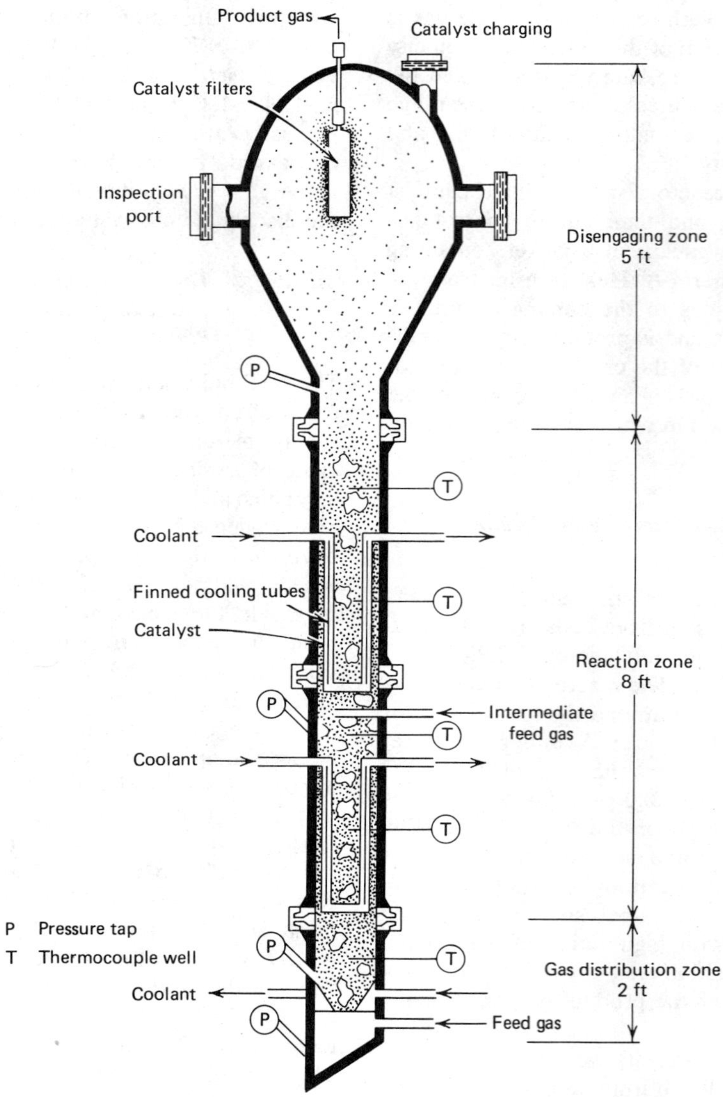

Figure 25.18 Flowsheet of a fluidized-bed methanator.

methanation reactor.[77] This work led to the construction and operation of a 6- in.-I.D., 15-ft-long vessel with a reaction zone of approximately 8 ft (Fig. 25.18). This reactor was called a process and equipment development unit (PEDU). Provisions for removal of the heat of reaction was made by two vertical finned-tube bundles with approximately 33 ft^2 of internal heat transfer area. The heat transfer medium was Dowtherm G.

[77] R. A. Streeter, D. A. Anderson, and J. T. Cobb, Jr., "Status of the Bi-Gas Program: Evaluation of Fluidized Bed Methanation Catalysts," paper presented at 8th Synthetic Pipeline Gas Symp., Chicago, October 18, 1976.

The usual catalyst charge was approximately 50–60 lb, which occupied approximately one cubic foot based on minimum fluidizing velocity. The normal temperature range was from about 426 to 537°C and pressure was 1000 psig. The synthesis gas used was a mixture of 20% CH_4, 60% H_2, and 20% CO.

There was a total of 21 tests reported of durations ranging from less than one day to as long as 10 days. The catalysts tested were largely nickel impregnated on alumina. The most promising was stated to be Ni-CU-Mo on alumina. The usual reason for terminating a run was loss of catalyst activity. Nearly all the catalysts tested exhibited minimum fluidization velocities of 0.01 ft/s. For normal operations, superficial velocities in the range of 0.08–0.18 ft/s were used. Conversions of H_2 and CO as high as 95–97% were obtained, although most results were appreciably lower because of declining catalyst activity.

25.5.8 Methanation with Multiple Reactors

A variation on the use of a gas stream to remove the heat of reaction in adiabatic reactors is the use of several reactors in series with gas intercoolers. In this type of process no cold gas or hot gas recycle is used. A controlled amount of methanation reaction occurs, to give a specified temperature rise in each reactor. The gas is cooled between stages by generation of high-pressure steam. The RM process and the Koppers SNG process employ this type of reactor.

25.5.8.1 *Catalytic Methanation by the RM Process*

The Ralph M. Parsons Company has under development a catalytic methanation process that is capable of methanating a synthesis gas having a 1 : 1 ratio of H_2 to CO and containing 1% methane without use of recycle gas.[78]

The process calls for six adiabatic reactors in series, with heat removal by generation of high-pressure steam between beds. The temperature is progressively lowered, since the decreased temperature favors a high conversion of H_2 and CO to CH_4. The flow arrangement is shown in Fig. 25.19, and resulting gas steam compositions and conditions are listed in Table 25.6. After the bulk methanation in the six reactors, the CO_2 is removed and the residual CO is reduced to less than 0.1% in a single cleanup reactor.

The RM process adds steam to the first stage to retard carbon formation and to effect shift of the CO to CO_2. The efficiency of converting the heat of methanation to usable heat is quite good. About one-third of the total steam produced is consumed for shift conversion, methanation, and CO_2 removal. The remaining two-thirds of the heat is available for export as 1500-psig steam to other sections of the plant.

The process is applicable at pressures as low as 65 psia as well as at higher pressures. The use of low-pressure methanation could improve the economics of low-pressure coal gasification because compression of the reduced volume of methane to high pressure results in energy savings when compared to the compression of the larger volume of the synthesis gas. A pilot plant under construction will operate at a raw gas rate of about 10^6 scf/day. This is equivalent to 5000 space velocity based on wet gas outlet space velocity.

25.5.8.2 *The Koppers SNG Methanation Process*

A SNG methanation process has been developed to follow the generation of synthesis gas by the Koppers-Totzek gasifier.[79] It is a joint development of Heinrich Koppers

[78] G. A. White, T. R. Roszkowski, and D. W. Standbridge, "The R.M. Process," in Am. Chem. Soc. Adv. Chem. Ser. No. 146, 1975.

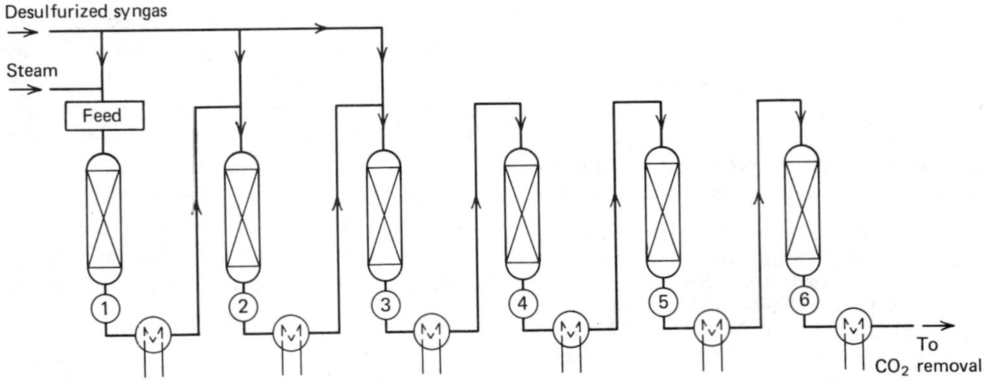

Figure 25.19 Flowsheet for bulk methanation by the RM process.

Table 25.6 Methane Production at 400 psia in the RM Process[78]

	Feed	Outlet					
Reactor No.:	1	1	2	3	4	5	6
Composition (vol %)							
H_2	49.80	54.53	48.07	43.09	36.90	22.86	9.29
CO	49.80	13.97	18.46	20.63	15.25	5.64	.87
CO_2	0.10	25.80	24.04	23.64	29.21	39.90	46.84
CH_4	0.30	5.70	9.43	12.64	18.64	31.60	43.00
	100.00	100.00	100.00	100.00	100.00	100.00	100.00
Steam/gas	1.20	0.88	0.56	0.43	0.50	0.65	0.83
Pressure (psia)	397	387	372	357	342	327	312
Temperature (°F)	900	1424	1434	1423	1322	1119	881

GmbH and Imperial Chemical Industries. Because the gas is CO rich, a shift converter precedes the first methanation reactor. Effluent gas from the shift converter passes without cooling to the methanation reactors, where additional shifting and methanation occur. After three stages of methanation with heat recovery between stages, the gas is cooled to condense the water and is sent to a final stage of methanation. Carbon dioxide is removed from the product gases after methanation. The heating value of the product gas is stated to be 975 Btu/scf.

The flow arrangement is shown in Fig. 25.20; stream analyses appear in Table 25.7. Operating pressure would presumably be in the usual methanation pressure range of 300–400 psi after compression of gases made by atmospheric pressure gasification. The advantages of avoiding compression costs for gases generated at atmospheric pressure have led to the development of a pressurized Kopper-Totzek gasifier at pressures of 15–30 atm. This pressurized version is being presently developed by Heinrich Koppers and Shell International Petroleum, Maatschappij, Netherlands.

25.5.9 The Lurgi–Sasol Methanation Reactor

Semicommercial tests were conducted jointly at Sasolburg, South Africa[80,81] by Lurgi Mineralöltechnik and Sasol using a source gas made from coal gasification for the Fischer-Tropsch synthetic oil plant that has been in operation since the 1950s.

The flow scheme used in the test is shown in Fig. 25.21. Synthesis gas after purification with Rectisol was passed over hot zinc oxide that protected the catalyst from sulfur breakthrough. The reactor type was adiabatic, with the heat of reaction being removed by hot recycle gas in the main methanation reactor. A final methanation reactor further converted the synthesis gas to 972 Btu/scf gas after CO_2 removal to 0.5%, with a reduced CO content of less than 0.1%. Conditions of operation and product gas analyses are shown in Table 25.8. Pressure of operation was 18 kg/cm^2 (gauge). The advantage of retaining CO_2 in the gas during the methanation reaction was said to be that the unreacted hydrogen was decreased, giving a somewhat higher Btu content of the gas after CO_2 removal.

The preferred catalyst was a high nickel on a support provided by BASF. Catalyst life was 4000 hr, with good activity remaining after termination of the test so that an expected life of 16,000 hr was projected.

The presence of trace components before and after methanation is shown in Table 25.9. Reactor D-1 in the table refers to the zinc oxide guard chamber, and its effect on removal of the trace components as well as the methanation reactors can be noted. The only compounds that can be regarded as poisons are chlorine and sulfur. The chlorine was still quite low at 0.02 ppm downstream of the

[79] J. E. Franzen and E. K. Goeke, "SNG Production Based on Koppers-Totzek Coal Gasification," paper presented at 6th Synthetic Pipeline Gas Symp., Chicago, October 28–30, 1974.

[80] K. H. Eisenlohr, F. W. Moeller, and M. Dry, "Effect of Certain Reaction Parameters on Methanation of Coal Gas to SNG," in Am. Chem. Soc. Adv. Chem. Ser. No. 146, 1975.

[81] F. W. Moeller, H. Roberts, and B. Britz, "Methanation of Coal Gas for SNG," *Hydrocarbon Process.*, 69–74 (April 1974).

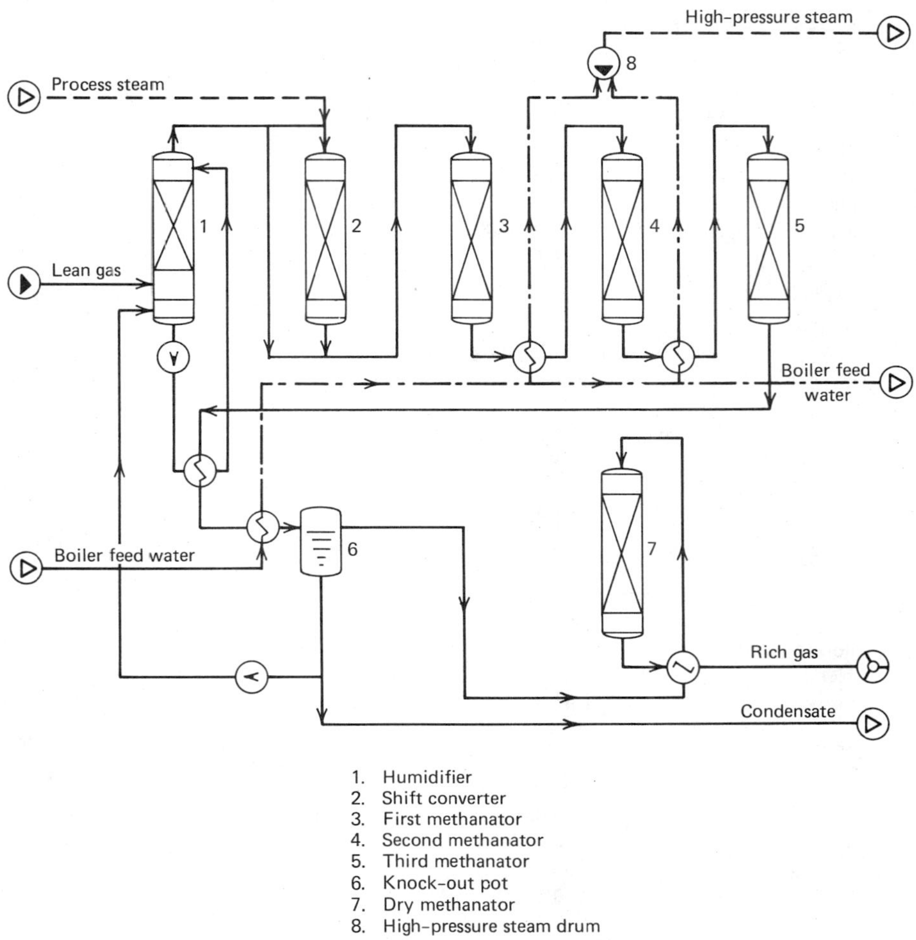

Figure 25.20 Flowsheet for combined shift conversion–methanation in the Koppers-SNG process.

Rectisol unit. No carbon was detected in the catalyst after 4000 hr. Higher hydrocarbons are largely hydrogenated to CH_4.

25.5.10 A Liquid Phase Methanation Reactor

The liquid phase methanation reactor being developed by Chem System Inc.[82,83] is shown in Fig. 25.22. In this reactor the nickel catalyst in granular form is submerged in a mineral oil coolant bath to control reaction temperature. The oil is circulated upward through the reactor along with the fresh synthesis gas, resulting in expansion and fluidization of the catalyst. Catalyst-free oil overflows from the top of the reactor and is cooled to remove the heat of reaction. In a large-scale reactor the heat of reaction would be used to generate high-pressure steam. This reactor type is quite similar to that developed by the USBM for Fischer-Tropsch synthesis of liquid hydrocarbons.[84]

Fresh synthesis gas is fed to the bottom of the reactor. Although a 3 : 1 ratio of H_2 to CO is usually preferred for the synthesis of CH_4, it is possible to use synthesis gas of

[82] M. B. Sherwin, M. E. Frank, and D. G. Blum. "Recent Developments in Liquid Phase Methanation," paper presented at 5th Synthetic Pipeline Gas Symp., Chicago, October 29–31, 1973.

[83] M. E. Frank, M. B. Sherwin, D. B. Blum, and R. L. Mednick, "Liquid Phase Methanation/Shift," paper presented at 8th Synthetic Pipeline Gas Symp., Chicago, October 18–20, 1976.

[84] H. E. Benson, J. H. Field, D. Bienstock, and H. H. Storch, "Oil Circulation Process for the Fischer-Tropsch Synthesis," *Ind. Eng. Chem.*, **46**, 2278 (November 1954)

Table 25.7 Shift Conversion—Methanation[79]

	Raw Coal to the Plant (wt %)
Moisture	8.00
Ash	19.06
Carbon	55.89
Hydrogen	4.16
Sulfur	0.72
Nitrogen	1.06
Oxygen	11.11
	100.00
Calorific value	
Net	5300 kcal/kg = 9540 Btu/lb
Gross	5575 kcal/kg = 10,025 Btu/lb
	SNG Exit Plant (vol %)
CO_2	0.43
CO	0.43
H_2	1.39
N_2 + Ar	2.60
CH_4	95.15
	100.00
Calorific value	
Net	8190 kcal/N-m³ = 876 Btu/scf
Gross	9115 kcal/N-m³ = 975 Btu/scf

Gas Analysis Conversion—Methanation

Gas	Entrance (vol %)	Exit (vol %)
CO_2	9.95	65.58
CO	56.87	0.15
H_2	32.47	0.48
N_2 + Ar	0.61	0.90
CH_4	0.10	32.89
	100.00	100.00

Table 25.8 Lurgi–Sasol Methanation Pilot Plant: Gas Analyses and Process Conditions[81]

		Main Methanator D-2		Final Methanator D-3	
	Synthesis gas	Reactor Inlet	Reactor Outlet	Reactor Inlet	Reactor Outlet
Temperature (°C):	270	300	450	260	315
Gas flow rate, wet (N-m³/hr):	18.2	96.0	89.6	8.2	7.9
Gas composition (vol %)					
CO_2	13.0	19.3	21.5	21.5	21.3
CO	15.5	4.3	0.4	0.4	0.05
H_2	60.1	21.3	7.7	7.7	0.7
CH_4	10.3	53.3	68.4	68.4	75.9
C_2^+	0.2	0.1	0.05	0.05	0.05
N_2/Ar	0.9	1.7	2.0	2.0	2.0
H_2O (v/v dry gas)	0.0	0.37	0.50	0.04	0.08

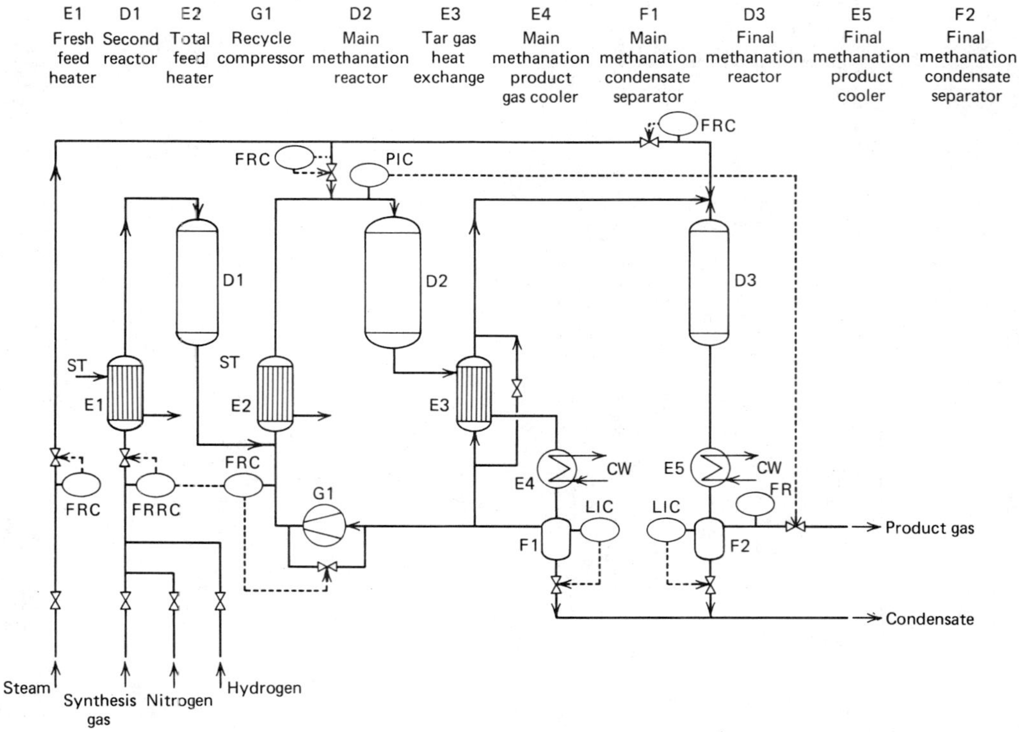

Figure 25.21 Flowsheet of the Lurgi–Sasol methanation pilot plant.[81]

lower ratios. In this case some water can be fed into the base of the reactor to effect a CO shift reaction. Consequently it is claimed that a separate shift reactor ahead of the methanation reactor can be avoided. As in most other methanation reactor systems, a second methanator is required to further upgrade the gas and to lower the CO content to a value less than 0.1%.

The process development unit that was the largest size reactor to be operated had a diameter of 3.62 in. with a height of 7 ft to accommodate bed heights up to 6 ft. Operating pressure ranged from 300 to 1000 psig

Table 25.9 Lurgi–Sasol Methanation Pilot Plant: Effect of Methanation on Trace Components[81]

Component	Synthesis Gas (D-1 off-line)	Synthesis Gas (D-1 on line)	Effluent Gas D-2	Product Gas Downstream D-3
Hydrocarbons (ml/N-m^3)				
C_2H_4	180	180	< 5[a]	< 5[a]
C_2H_6	750	750	50[a]	80[a]
C_3H_6	10	10	< 5[a]	< 5[a]
C_3H_8	14	14	< 5[a]	< 5[a]
Total S (mg/N-m^3)	0.08	0.04	ND[b]	ND[b]
H_2S (mg/N-m^3)	0.03	0.02	ND[b]	ND[b]
NO + NO_2 (ml/N-m^3)	0.1	Not analyzed	<0.003[a]	<0.003[a]
NH_3 (ml/N-m^3)	0.4	0.4	ND[b]	ND[b]
HCN (ml/N-m^3)	0.8	0.8	0.5	0.5
C_1 (ml/N-m^3)	ND[b]	ND[b]	ND[b]	ND[b]
O_2 (ml/N-m^3)	< 20	< 20	ND[b]	ND[b]
CH_3OH (ml/N-m^3)	40	40	ND[b]	ND[b]

[a] Below the detectable limit.
[b] Not detectable.

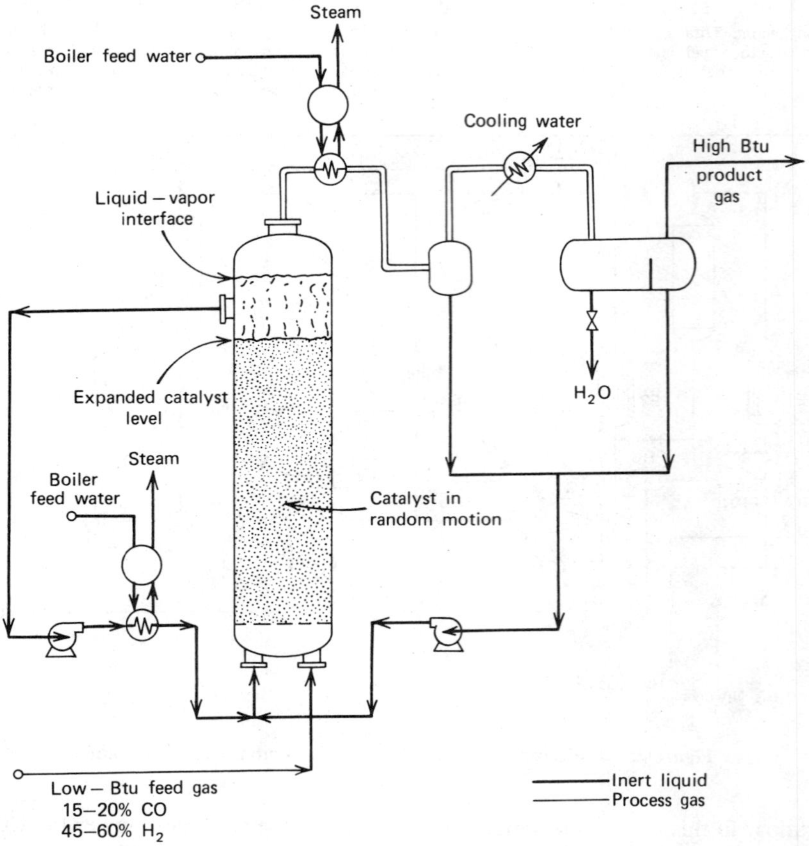

Figure 25.22 The liquid phase methanation process.[81]

with a nominal gas rate of 1500 scf/hr. Operating temperatures were from 300 to 380°C. Catalyst sizes investigated were 1/32–3/16 in. Carbon monoxide conversions in the range of 94–96% were obtained.

Tests were run in the second or polishing reactor (fixed bed) with 20 g of La Porte–Davison catalyst with a simulated product gas from the liquid phase reactor having a typical composition of 8.3% H_2, 2.5% CO, 23.8% CO_2, and 65.4% CH_4. Operation was at 9000–10,000 hr^{-1} space velocity, a pressure of 500 psig, and a temperature of 335–350°C. The catalyst performed well, with no CO being detected in the product gas, although carbon deposition was found in the catalyst.

A large pilot unit has been constructed[82] with a diameter of 2 ft and a height of 15 ft and is capable of processing a synthesis gas feed of 2 million scf/day. It was installed at the HyGAS plant for evaluation.

CHAPTER **26**

Underground Gasification of Coal

THOMAS F. EDGAR AND DAVID W. GREGG

Thomas F. Edgar *has been on the faculty at the University of Texas, Austin, since 1971. He did his undergraduate work in chemical engineering at the University of Kansas and received his Ph.D. in chemical engineering from Princeton University, where he specialized in control theory. He has been involved in a variety of theoretical and experimental coal research activities since 1972. Prof. Edgar is Director of the* In Situ *Lignite Gasification Program at UT-Austin, an interdisciplinary research project sponsored by NSF-RANN, the Department of Energy, the State of Texas, and 10 private companies.*

David W. Gregg *received his B.A. in chemistry from Dartmouth College in 1957 and his Ph.D. in chemical engineering from the University of Illinois in 1961. Upon leaving the University of Illinois, he went to work for Lawrence Livermore Laboratory, where he has been to the present (1979). Dr. Gregg worked from 1962 to 1974 for the laser fusion program. In 1974 he joined the Underground Gasification Program and has been working on the alternative system options for the underground gasification of coal, with special emphasis on the Soviet approach.*

26.1 INTRODUCTION

The underground (*in situ*) gasification of coal (UCG) is a method that combines extraction and conversion of coal into a single step, thus eliminating the need for mining and coal-processing equipment and gasification reactors. It is a process that should be considered as a competitor with shaft mining but not with surface mining, since a high overburden/coal ratio is needed to minimize gas leakage. The successful application of this method would provide a low- to intermediate-Btu gas (100–300 Btu/scf depending on whether air or oxygen-steam is used) that can be readily treated for sulfur compounds and particulates. At the same time, UCG can eliminate many of the health, safety, and environmental problems associated with conventional deep mining of coal. The *in situ* method also has the potential to recover the energy from thick, deep coal deposits and steeply dipping seams, which are not economic to mine using conventional schemes. It can be efficient in utilizing thick seams, since pillars do not have to be left behind to hold up the roof.

It is clear that UCG has near-term commercial potential in the Western world, and it has been used on a commercial scale in the USSR for more than 25 years.[1] A review of the Soviet experience makes it abundantly clear that they have expended far more effort than all other countries combined. The first field tests were performed in 1933, and their efforts peaked in the mid-1960s, with as many as 3000 people employed in testing and operation of the gasification stations. It appears that in the Soviet Union UCG received more attention than conventional gasification in the mid-1960s.

The idea of UCG is not new: it dates back more than a century. Historical reviews of the pre-1965 technology for underground coal gasification have been given by A. D. Little,[2] Wang,[3] Capp, et al.,[4] and Elder.[5] However these four reports contain incomplete information about recent Russian contributions to UCG technology, especially in field testing.

This chapter presents a brief overview of the basic principles of UCG, with special emphasis given to the Soviet technology. However when applicable, information is drawn from other sources as well. We attempt to show not just what has been done and how the systems worked, but also the logic of the successful systems—why they are designed as they are. This overview is divided into 11 sections following the introduction. Since the chemistry of UCG has an important influence on the product gas produced, this subject is covered in the next section. We then review the large body of field results available in the Soviet literature (Section 26.3) and interpret these results in terms of the physicochemical principles presented in Section 26.2.

Section 26.4 analyzes the effects of plastic properties of coal relative to UCG. In Section 26.5 we turn to some of the engineering problems encountered in UCG. Using the Soviet system as a model, specific features of the Soviet design are outlined. One of the most critical steps in the design of a UCG system is the creation of a zone of enhanced permeability between injection and exhaust boreholes; there exist a number of techniques for the "linking" of two boreholes, and these are discussed in detail in Section 26.6.

Although the Soviet technology is applicable to either horizontal or steeply dipping seams, the latter case does involve some design modifications, which are discussed in Section 26.7. Section 26.8 deals with sweep efficiency and thermal efficiency, as well as the operating procedure for maintenance of a consistent product gas heating value. Section 26.9 reviews field tests and commercial installations in the USSR, Europe, Japan, and North America. Specific emphasis is placed on recent tests in the United States and Canada, where there are presently eight field tests being operated by government and university laboratories and private companies.

The results from these field tests are interpretable in a qualitative sense through the discussion in Sections 26.2 and 26.3. However there has been a significant effort recently to provide quantitative predictions of product gas quality and sweep efficiency during UCG via mathematical modeling, and Section 26.10 discusses some of the more successful and comprehensive models. One limitation to the commercial application of UCG is the possibility of environmental problems, especially with respect to groundwater quality. Field and laboratory results dealing with environmental impact are discussed in Section 26.11, and geological factors influencing water quality after a UCG operation are reviewed. The concluding section deals with future field and laboratory research needs for *in situ* conversion of coal to synthetic fuels. Both oxygen-blown UCG systems and the successful application of UCG to Eastern coals would broaden the application of this technology in the United States and abroad; some aspects of the use of *in situ* liquefaction as an extraction-conversion technique are also discussed in Section 26.12.

26.2 THE CHEMISTRY OF UNDERGROUND COAL GASIFICATION

A very extensive treatment of coal gasification processes has been presented in the first

[1] D. W. Gregg, R. W. Hill, and D. U. Olness, Lawrence Livermore Laboratory Rep. No. UCRL-52004, 1976.

[2] A. D. Little, PB 209274, 1972.

[3] C. Wang, Dissertation, Columbia University, New York, 1969.

[4] J. P. Capp, W. L. Rober, and D. W. Simon, U.S. Bur. Mines Inf. Circ. No. 8193, 1963.

[5] J. L. Elder, in *Chemistry of Coal Utilization*, Supplementary Volume, H. H. Lowry, Ed. John Wiley and Sons, New York, 1963, pp. 1023–1040.

Supplementary Volume,[6] and no attempt is made here to reproduce that information. To understand UCG, however, it is essential to present some of that information and interpret it in terms of some of the special problems associated with underground gasification. This section covers coal pyrolysis, the chemistry of char gasification, downstream chemistry, and field results.

26.2.1 Coal Pyrolysis

In both surface and underground coal gasification, coal pyrolysis greatly affects the composition of the product gas. During gasification the coal is usually heated by the hot product gases to a very high temperature in a reducing environment before it is consumed by oxidation reactions. This is the inherent nature of a flame front propagating cocurrently with the gas flow through a rubble zone of coal. Thermal decomposition of the coal releases various chemical products into the gas stream. In conjunction with UCG field tests in the United States, Campbell[7] has studied the pyrolysis of Roland Seam (Wyoming) subbituminous coal at the low heating rates expected for UCG, and his results illustrate the behavior of the coal and its gas evolution during slow heating.

During pyrolysis this coal typically lost about 40–50% of its dry weight as low molecular weight gases, light hydrocarbons, and heavy tars. This result clearly illustrates how influential the pyrolysis products can be on the product gas composition. The samples of coal, consisting of particle sizes ranging from 1.7 to 3.3 mm, were heated at a constant rate of 3.33°C/min, a characteristic heating rate for UCG, and the rate of gas evolution as a function of temperature was measured for various gas species. The results are presented in Fig. 26.1. The temperature dependency of the evolution of the different species of gas is clearly illustrated.

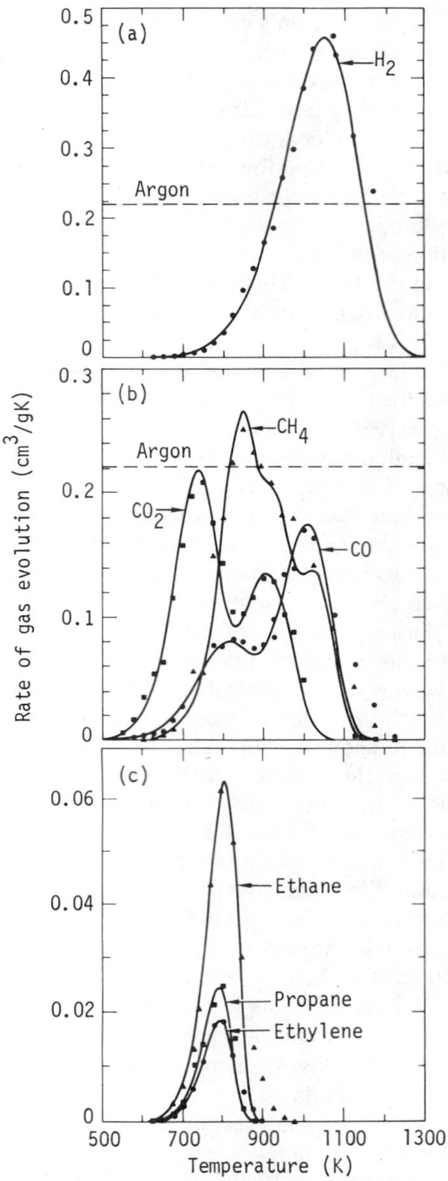

Figure 26.1 The effects of temperature on gas evolution during pyrolysis of Roland Seam coal. The heating rate is 3.33 K/min (3.33°C/min), the carrier gas is argon, and particle size is 1.65–3.35 mm. The rate of gas evolution is relative to a constant flow of carrier gas. In (*a*) and (*b*) one unit on the vertical axis equals 0.737 cm^3/g-min. (*a*) For H_2 evolutions. (*b*) For CO_2, CO, and CH_4 evolution. (*c*) For C_2H_6, C_3H_8, and C_2H_4 evolution. One unit on the vertical axis equals 0.812 cm^3/g-min.

26.2.2 The Chemistry of Char Gasification

An extensive treatment of the thermodynamics of coal gasification is presented by von Fredersdorff and Elliott.[6] However a brief outline of the significant features of the chemistry of char gasification and the impli-

[6] C. G. von Fredersdorff and M. A. Elliott, in *Chemistry of Coal Utilization,* Supplementary Volume, H. H. Lowry, Ed., John Wiley and Sons, New York, 1963, pp. 892–1022.

[7] J. H. Campbell, Lawrence Livermore Laboratory, Rep. No. UCRL-52035, 1976.

cations of these features to the gasification process follows.

In many gasification operations the primary oxidation reactions take place mainly between the char, the end product of pyrolysis (primarily carbon), and oxygen. There are varying degrees of oxidation of the pyrolysis products, but this effect is neglected here. The primary oxidation reactions are

$$C + O_2 \rightarrow CO_2 + 94.05 \text{ kcal/g-mole} \quad (26.1)$$

$$C + \tfrac{1}{2}O_2 \rightarrow CO - 26.42 \text{ kcal/g-mole} \quad (26.2)$$

$$CO + \tfrac{1}{2}O_2 \rightarrow CO_2 + 67.63 \text{ kcal/g-mole} \quad (26.3)$$

In these and subsequent reactions the heats of reaction assume all compounds as gases, carbon as β-graphite, and a temperature of 298.16 K.[6] However these heats vary only slightly with temperature and are not significantly influenced by the fact that char and β-graphite have different heats of formation.[8]

Of these three reactions, only Reaction 26.2 produces a combustible gaseous product. Therefore we need to determine the conditions that will preferentially produce CO instead of CO_2. Figure 26.2 presents the fractional composition of CO and CO_2 in equilibrium with β-graphite at 1 atm of pressure as a function of temperature. It can be seen that to produce CO preferentially, a combustion temperature greater than 1000 K must be maintained.

If char is combusted stoichiometrically in pure oxygen in the absence of heat losses, burn temperatures as high as 3000 K would be expected—far higher than the temperature needed to produce a reaction product of essentially all CO. However such a reaction system is inefficient because a large fraction (~ 25%) of the heat of combustion of the char is expended for further heating of the CO rather than for producing more combustible product. Therefore steam is added to utilize this excess heat of reaction and to produce more combustible gas:

$$H_2O + C \rightarrow CO + H_2 - 31.4 \text{ kcal/g-mole} \quad (26.4)$$

In UCG systems operating below the water table, water is added unavoidably into the reaction zone by leakage from the surrounding formations. In fact, the Soviet experience in the use of air for underground gasification has revealed that the natural water intrusion rate in wet seams can easily exceed the optimum rate, resulting in reduced efficiency of the process because of excessive cooling of the reaction zone.[10] When oxygen is used instead of air, the cost and quantity of oxygen required to gasify a ton of coal is a major concern. In this case the addition of steam not only improves the energy efficiency of the process by utilizing the excess heat, but also minimizes the amount of oxygen required per ton of coal gasified, significantly improving the economics of the process.[6]

A similar means of effectively utilizing the excess combustion energy is accomplished by reducing CO_2 to CO:

$$CO_2 + C \rightarrow 2CO - 41.2 \text{ kcal/g-mole} \quad (26.5)$$

In fact, this reaction routinely occurs in long gasification channels as the CO_2 formed upstream is reduced to CO further downstream.[11]

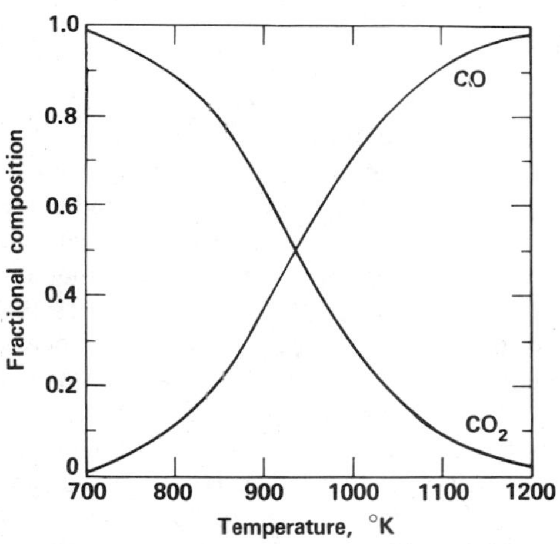

Figure 26.2 The fractional composition of CO and CO_2 in equilibrium with β-graphite at 1 atm of pressure as a function of temperature.[9]

[8] D. R. Stephens, Lawrence Livermore Laboratory, Rep. No. UCID-16094, 1972.

[9] W. K. Lewis, A. H. Radasch, and H. C. Lewis, *Industrial Stoichiometry,* McGraw-Hill Book Company, New York, 1954.

[10] R. I. Antonova, I. S. Garkusha, E. G. Gersheivich, E. V. Kreinin, N. V. Lavrov, D. K. Semenenko, and N. A. Fedorov, *Khim. Tverd. Topl.,* No. 1, 86 (1967); UCRL Transl. No. 10804, 1974.

[11] P. V. Skafa, *Gosudarstvennoe Nauchno-Tekhnicheskoe Izdatel'stvo Literatury po Giornomu Delu,* 1960; UCRL Transl. No. 10880, 1975.

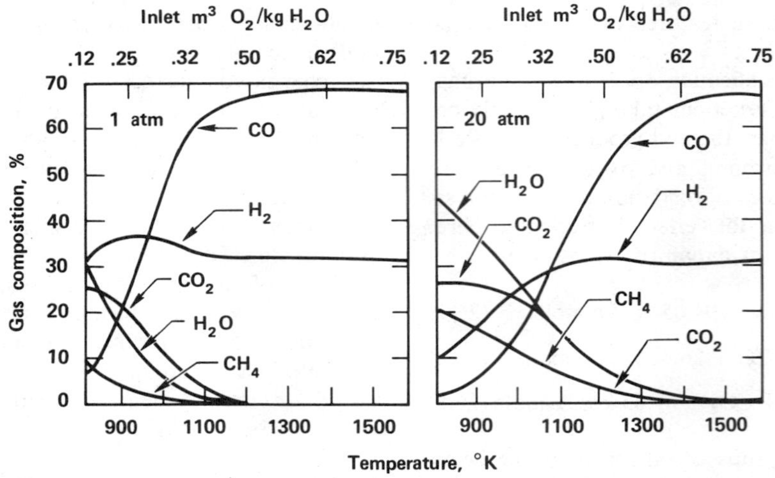

Figure 26.3 Equilibrium gas compositions of the carbon-oxygen-steam system at adiabatic conditions.[6]

The introduction of water (steam) complicates the combustion chemistry.[6] Figure 26.3 presents the calculated equilibrium compositions for a carbon-steam system at adiabatic conditions. It can be seen that CH_4, one of the primary additions to the list of products, increases in concentration as the temperature decreases. However the CO/CO_2 ratio is not significantly altered from the situation where steam is not present (see Fig. 26.2). [All gas analyses are given for water-free mixtures.]

One of the most interesting features of the carbon-oxygen-steam equilibrium calculations (Fig. 26.3) is that throughout the temperature range considered the heating value of the dry product gas does not decrease significantly as the temperature decreases, in contrast to the CO-CO_2-C equilibrium data shown in Fig. 26.2. This is because as the CO and H_2 concentrations fall off with decreasing temperature, the loss of their contribution to the heating value of the product gas is partially offset by the increase in the concentration of CH_4. In practice, at the higher temperature ranges, the compositions calculated assuming thermodynamic equilibrium are fairly accurate for most systems. However as the temperature decreases, the kinetics of reactions become limiting, and in the case of coal gasification the rate of formation of CH_4 is very slow compared to the formation of other components of the system, thus limiting the heating value of the product gas.[6] It is therefore very difficult to predict accurately the optimum operating temperature, thus the optimum oxygen/steam ratio for a specific system design. These factors depend heavily on the degree of formation of CH_4 at low temperatures, which in turn depends on the residence time of the gas as well as the extent to which the formation of CH_4 is catalyzed by other components in the system (ash). In most UCG system designs the amount of CH_4 produced through the char-hydrogen reaction will be small. Hence as shown in Fig. 26.3, the combustion temperature should be kept above 1000 K, since the heating value of the product gas will be derived principally from CO and H_2. Because the details of the design of a UCG system vary significantly, it is not possible to make a generalized systement about the optimum oxygen/steam ratio for all gasification systems. Such an optimum ratio must be determined for each system with its particular set of operating parameters.

Figure 26.3 also reveals that increased pressure will enhance the formation of CH_4. However it is not clear that this effect can be capitalized on for UCG. Soviet experience indicates that subsidence of the overburden, which is unavoidable for a system that removes essentially all the coal, renders the gasification zone so porous that the process must be conducted at the lowest possible pressure to minimize gas losses. This is a major obstacle to the production of a substitute natural gas by UCG. However it might be possible (but it is yet to be shown) for a system to be operated at high pressures at considerable depths where the overburden pressure is sufficient to seal cracks.

26.2.3 Downstream Chemistry

In UCG, as the gas progresses past the combustion zone and continues on its path to the exhaust hole, it inevitably loses heat to the surroundings. The heat loss results not only from conduction into the surrounding formations but also from the vaporization of intruding water in the hot gas stream. As the gas cools its composition tends to shift as described by Figs. 26.2 and 26.3. The primary reactions of concern are:

$$CO + H_2O \rightarrow CO_2 + H_2 + 9.8 \text{ kcal/g-mole} \qquad (26.6)$$

$$2CO \rightarrow CO_2 + C + 41.2 \text{ kcal/g-mole} \qquad (26.7)$$

$$C + 2H_2 \rightarrow CH_4 + 17.9 \text{ kcal/g-mole} \qquad (26.8)$$

It can be seen that Reactions 26.6 and 26.7 reduce the heating value of the product gas, whereas Reaction 26.8 increases it. However in most production systems, as the temperature decreases the reaction rates slow. Equilibrium compositions are not reached, and the final product gas composition is determined primarily by the rates of reaction, not by the equilibrium composition.

Reaction 26.6: $CO + H_2O \rightarrow CO_2 + H_2 +$ 9.8 kcal/g-mole

This reaction, commonly called the water-gas or carbon monoxide shift reaction, reduces the heating value of the product gas in two ways.[6] It exchanges an H_2 molecule for a CO molecule with a corresponding reduction in the heat of combustion of 9.8 kcal/g-mole. It also exchanges a CO_2 molecule for an H_2O molecule. Because H_2O is easily condensed out of the product gas and CO_2 is not so easily removed, the product gas is diluted with an incombustible gas and the specific heating value is further reduced.

The Soviet experience in this field indicates that this reaction can play a major role in establishing the product gas composition[11,12] because it is probably catalyzed by coal surfaces and inorganic salts that are abundant in underground gasification systems.[11,13,14] In some situations Reaction 26.6 has limited the useful length of a gasification channel.[15] If the channels were too long and if there were too much water intrusion, this reaction would shift the product gas composition to one with an uneconomically low heating value.

Reaction 26.7: $2CO \rightarrow CO_2 + C +$ 41.2 kcal/g-mole

This reaction is commonly called the Boudouard reaction.[10] It clearly reduces the heating value of the product gas by substituting a CO_2 molecule for two CO molecules. The reaction in the gas phase is so slow that it usually does not contribute significantly to product gas composition. However this reaction readily takes place at 400–500°C in the presence of iron oxides and metallic iron.[7] The Soviets felt it was important to flash cool the product gas with a water spray at the bottom of the exhaust hole, "freezing" the gas composition before it reached the iron casing. However the spray primarily served to lower the pipe temperature, which preserved the pipe and the pipe-cement seal in the access holes.[9]

Reaction 26.8: $C + 2H_2 \rightarrow CH_4 +$ 17.9 kcal/g-mole

In contrast to Reactions 26.6 and 26.7, this reaction dramatically increases the heating value of the product gas if it occurs to any great extent. However evidence from Soviet operations at Podmoskovia indicate that as the gas cools in a long channel, the CO/CO_2 ratio decreases as expected but the CH_4 concentration does not increase. This observation indicates that in the Soviet process, the rate for the formation of CH_4 is not rapid enough for this reaction to contribute significantly to the heating value of the product gas. This formation rate may not occur at all temperatures, and it is unclear what quantity of the CH_4 found in the product gas results from coal pyrolysis and how much is formed by Reaction 26.8. The equilibrium composition calculations indicate that the CH_4 composition can be increased by increasing the pressure of gasification. However Soviet large-scale field operations leaked too much gas through subsidence cracks to permit operation at high pressure.[11]

[12] E. V. Kreinin and M. Revva, "Kemerovskoe Knizhnoe Izdatel'stvo," 1966; UCRL Transl. No. 10810, 1974.

[13] H. R. Batchelder, R. M. Busche, and W. P. Armstrong, *Ind. Eng. Chem.*, **45**, 1156 (1953).

[14] G. B. Grebenshchikova, *Podzemn. Gazif. Uglei*, No. 2, 54 (1957); UCRL Transl. No. 10900, 1975.

[15] N. V. Lavrov, M. A. Kulakova, S. T. Kazachkova, A. E. Zhirnyi, R. I. Antonova, and A. F. Volk, *Khim. Tverd. Topl.*, No. 1, 73 (1971); UCRL Transl. No. 10922, 1975.

In summary, as heat conduction into the surrounding formations and vaporization of the intruding water cause the downstream gas to cool, Reactions 26.6 and 26.7 progress, reducing the heating value of the product gas. In addition, the rate of formation of CH_4 is not rapid enough to contribute significantly to the product gas heating value. The Soviets thus found that the maximum economic distance between pipes in some cases was limited by the falling off of the heating value due to the water-gas reaction.

26.3 FIELD RESULTS

The underground gasification of coal is a complex physical-chemical process and is difficult to study because it occurs well below the ground surface. The course of the process is influenced by many different factors. These factors can be divided into two basic categories: the chemical process of fuel gas formation and the four-dimensional hydrodynamic interaction between the reactant–product gas phase and the solid coal seam. The interdependence of these factors is only imperfectly known; thus it is difficult to arrive at rigorous scientific or engineering interpretation of the results of gasification. However by analyzing field data, trends can be established and a rough physical-chemical interpretation applied. The quantitative identification and explanation of these trends is essential for establishing engineering methods to control the process and is a necessary initial step that eventually enables the selection of the optimum process conditions.

One Soviet paper[10] presents a survey of laboratory and field data obtained during the decade from 1955 to 1965 at the Yushno-Abinsk station and illustrates trends in the gasification process. All the data are for gasification with air, with product gas compositions given on a water-free basis. This study related the product gas quality to three very important parameters: the blast (oxidant) rate, the water intrusion rate, and the coal seam thickness.

The Soviet system design for steeply dipping beds was essentially a series of parallel channels drilled along the dip of the coal seam from the surface to a manifold connecting the channels at the bottom. The coal seam is combusted from the bottom up, with air injected through a separate set of pipes that are drilled vertically (not through coal) from the surface. The details of the system are discussed later. The observed trends can be generally understood in terms of the coal chemistry discussion presented in the previous section.

The water content in the product gas is a reflection of the rate at which the water intrudes into the gasification zone from the surrounding formation. The water intrusion rate is determined by the permeability of the surrounding formation and the hydrostatic head for the general region; thus it varies slowly with time. For the purposes of this discussion, the water intrusion rate varies so slowly compared to other parameters of concern that it can be considered to be essentially constant with time.

It is interesting to note that the Soviet systems were operated in such a fashion that the gasification channels in the coal were always hot enough to vaporize all intruding water. This is an essential design feature because if the water were not removed, the water level would build up and plug the channel to gas flow. Therefore in the Soviet studies essentially all the intruding water appears in the product gas as water or as products of the reaction between water and coal. Thus it is expected that the water concentration in the product gas essentially will vary inversely with the blast rate. This was found to be true, and data illustrating this effect for different water intrusion rates are presented in Fig. 26.4.

It is difficult to discern the actual axial distribution of water intrusion, that is, water that enters the combustion zone versus water that intrudes downstream of the combustion zone. However in either case the intruding water will cool the product gases and will shift the equilibrium compositions as discussed earlier. Figure 26.5 illustrates the effect of gasification intensity (oxidant injection rate) on the heating value of the product gas for various water intrusion rates. The data show that the heating value always increases with increased gasification intensity and reduced water inflow rate. This is consistent with the chemical equilibrium interpretation of the process that as the system cools, more CO_2 is produced at the expense of CO. However the equilibrium model also predicts that this loss in heating value is offset by the formation of CH_4, a hypothesis

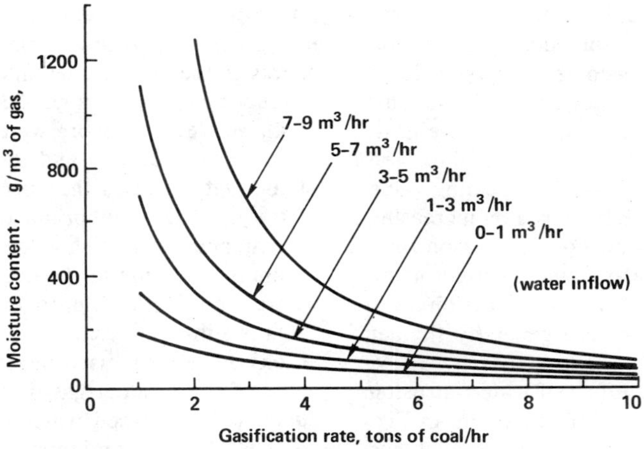

Figure 26.4 Moisture content of the gas as a function of the gasification rate of the coal seam at various water intrusion rates as indicated.

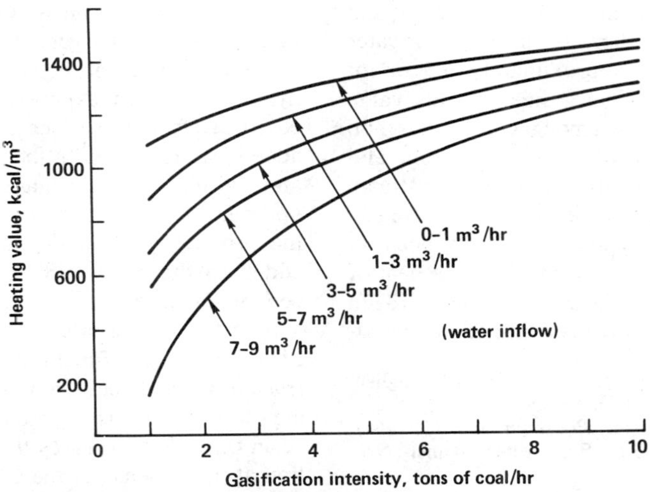

Figure 26.5 The effect of gasification intensity on the calorific value of the gas at various water intrusion rates as indicated.

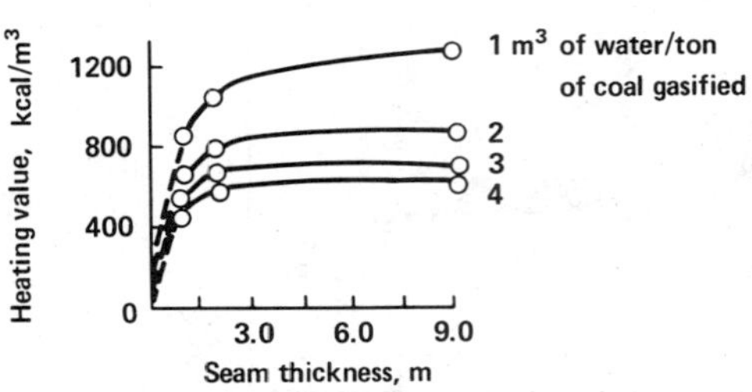

Figure 26.6 The effect of seam thickness and the specific water intrusion on the calorific value of gas obtained by underground gasification. Numbers next to curves are the water intrusion rates.

that is not consistent with data in Fig. 26.6. In this case the reaction rate for CH_4 formation appears to be too slow for any significant contribution to the heating value of the product gas.

The combustion zone and the product gas can also be cooled by thermal conduction to the surrounding formations. Figure 26.6 illustrates the effect of the thickness of the coal seam on the heating value of the product gas as a function of different water intrusion rates. The data in this figure are for a fixed gasification rate of approximately 2 tons of coal per hour. It can be seen that the detrimental effect of the surrounding formation cooling the system becomes very significant as the seam thickness falls below 2 m. This "cutoff" thickness is a function of the gasification intensity and of the fractional oxygen concentration of the blast. The heating value for thinner seams can be raised by increasing either of these factors. For this reason oxygen-enriched air was commonly used at the Soviet gasification station at Lisichansk, where the coal seams are generally thinner than 2 m.[11]

Data showing the effect of water intrusion can be presented in a different form: in Fig. 26.7 the heating value of the product gas appears as a function of the water intrusion per ton of coal gasified for a coal seam thicker than the apparent 2-m "cutoff" thickness. The heating value continuously improves with decreased water intrusion. However these data were taken using air injection, and when air is used there is an optimum water intrusion rate at approximately 0.5 m^3 of water per ton of coal gasified.[16] This value varies considerably among coal seams and depends on the coal moisture content and whether the water intrusion occurs upstream or downstream from the combustion zone.[17] An optimum intrusion rate exists because with no water intrusion, much of the heat of combustion is expended to heat the nitrogen in the air blast. However limited water addition increases the heating value of the product gas by converting some of the thermal energy in the hot nitrogen into a combustible product by char-steam gasification (Reaction 26.4). However as more water is added, the optimum is exceeded and the cooling effect of the water causes the corresponding shift of the CO/CO_2 equilibrium to dominate the system, and the heating value drops. Such an optimum is unique to the use of air for blast where there is a high nitrogen content.

Quite different results are expected when an oxygen blast is used because the heating value of the product gas decreases continuously with increased water intrusion; at no point is there an optimum intrusion rate in terms of the heating value of the product gas. However there are both economic and energy efficiency optima for water intrusion using an oxygen blast. A pure oxygen blast gives the highest heating value gas, but much of the energy of combustion is wasted as sensible heat in the product gas. Its use is also economically inefficient, because an unnecessarily large amount of expensive oxygen is needed to gasify a ton of coal. When water is added it improves both the energy efficiency and the economics of gasification up to an optimum, even though the heating value of the product gas is reduced. Thus steam is added to the oxygen blast to optimize the economics of the system.

Figure 26.8 presents some data on the gas composition as a function of both water intrusion rate and coal seam thickness. Increasing the water intrusion rate or decreasing the seam thickness increases the CO_2 concentration at the expense of the CO concentration.

[16] "Underground Processing of Fuels," English Transl. No. TT 63-11063, 1963.

[17] K. F. Bogoroditskii, *Podzemn. Gazif. Uglei,* No. 1, 65 (1957); U.S. Bur. Mines Transl. No. BM-404.

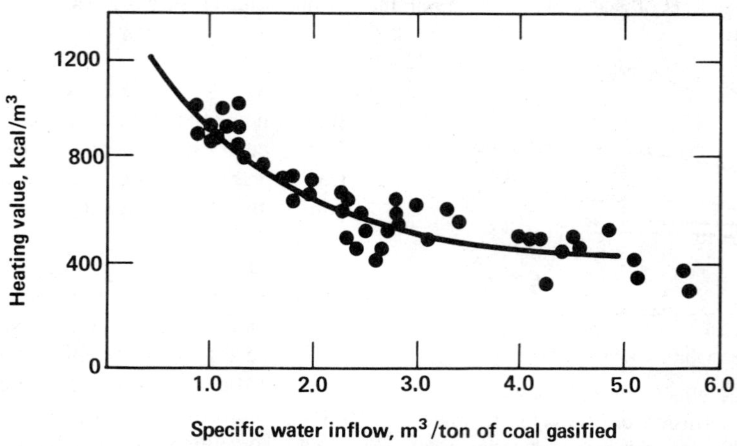

Figure 26.7 The calorific value of the gas as a function of its moisture content.

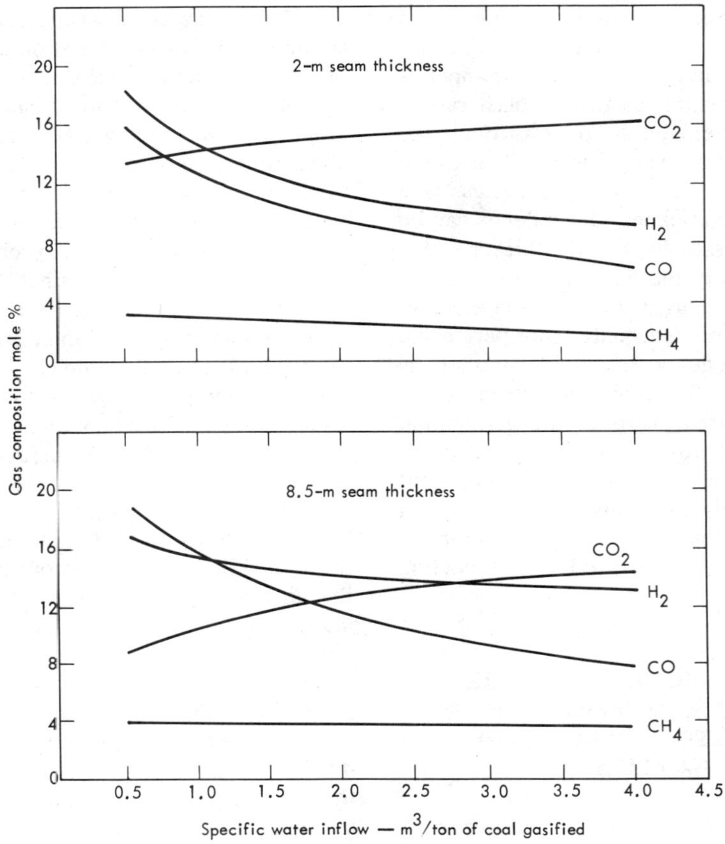

Figure 26.8 The effect of the specific water intrusion on the chemical composition of gas obtained by underground coal gasification.

These observations are consistent with the equilibrium calculations that predict higher CO_2 concentrations at lower temperatures. In addition, the CH_4 concentration is unaffected by lower temperatures, which indicates that the kinetics of formation are slow.

Data taken at both Angren and Yushno-Abinsk (Fig. 26.9) show an optimum blast rate (using air) that is clearly dependent on the thickness of the coal seam.[15,18] We can see that the optimum blast intensity is

[18] N. V. Lavrov, M. A. Kulakov, and S. T. Kazachkova, *Tr. Inst. Proizvod. Opyt*, No. 5, 133 (1971); UCRL Transl. No. 10979, 1975.

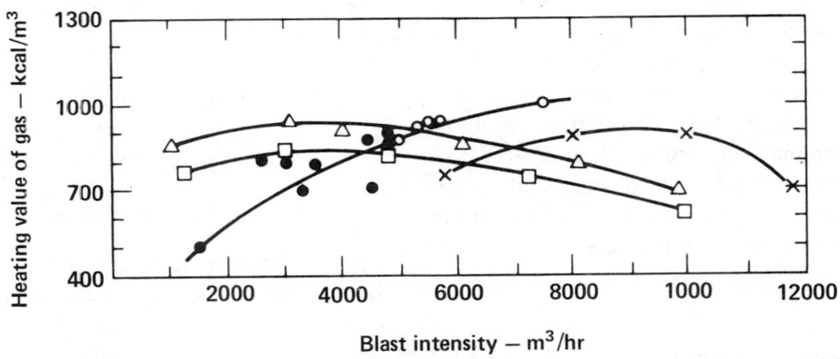

Figure 26.9 The effect of blast intensity on the calorific value of the gas for various seam thicknesses. Line 1 represents a seam thickness of 14–16 m (Angren); line 2, a seam thickness of 5–6 m (Angren); and lines 3 and 4, a seam thickness of approximately 3 m (Yushno-Abinsk).

reached at lower blast rates for thin seams than for thick seams. Because the water intrusion rate into a particular channel is essentially constant, as the air blast rate is increased we expect to pass through the water-air optimum and observe a maximum in the heating value of the product gas (see Fig. 26.9). The thickness dependence can be explained by assuming that the thicker seams have larger water intrusion rates than do the thinner seams. This is a reasonable assumption because coal is usually more permeable than its surrounding formations. If there is water present, the coal seam itself is the major aquifer that supplies the gasification zone with intruding water. Therefore if everything else is constant, thicker coal seams are likely to supply more water than thinner ones, thus will require higher blast rates to maximize the product gas heating value.

26.4 THE IMPORTANCE OF THE PLASTIC PROPERTIES OF COAL TO UNDERGROUND COAL GASIFICATION

A detailed treatment of the plastic properties of coal has been presented by Loison et al.[19] (Chapter 6). Since there are no quantified field data on the performance of a UCG system as a function of the plastic properties of coal, this section merely outlines the magnitude of these properties and the means by which they are likely to influence the operation of a system.

The primary plastic property of concern is whether the coal swells or shrinks as it is heated. The ideal coals for underground gasification do not swell at all. When they are heated they only shrink and fall apart. This is true of most coals with a rank of subbituminous or lower. The shrinking is important, since this ensures that the gasification channels will remain open to gas flow. The breaking up into particles is also important, since it provides a large amount of surface area for the gases to react with the coal.

Swelling (bituminous) coals present problems for underground gasification, the magnitude of which will depend on the degree of swelling. The first problem is that it probably will be impossible to form gasification channels by the technique used most often, countercurrent combustion (discussed later), since the technique depends on the effects of heat on the natural permeability of the coal. Swelling can also present a problem even if the channels are drilled, since large drilled holes can be plugged to gas flow if the swelling is sufficiently great. (Some coals swell up to 300% of original volume.) One possible design solution is to use a combination of drilled holes that are conditioned by a countercurrent combustion step before production gasification is initiated.

If we assume that gasification channels can be developed by some means and held open, there is still another potential problem. Swelling coals generally do not fall apart into many small pieces on heating. They have a plastic phase that tends to "glue" the coal together. Thus in comparison to nonswelling coals, they will generate far less surface area for gasification. It will therefore be more difficult to maintain a good heating quality in the product gas. This potential problem might be moderated somewhat by the tendency of the coal to stop swelling and start to char and shrink above a certain temperature. If this later shrinking phase produces sufficient flaking to generate a high surface area reaction zone, a good quality product gas might be maintainable.

In countercurrent combustion the fire face and the blast move in opposite directions, whereas in cocurrent combustion they move in the same direction. These two methods of gasification are discussed in greater detail later, as well as the above-mentioned gasification design.

26.5 THE FIELD DESIGN OF AN UNDERGROUND COAL GASIFICATION SYSTEM

26.5.1 Problems Encountered[20]

The design of a successful UCG system is a formidable task because we are attempting to carry out a fairly complicated chemical proc-

[19] R. Loison, A. Peytavy, A. F. Boyer, and R. Grillot, in *Chemistry of Coal Utilization,* Supplementary Volume, H. H. Lowry, Ed., John Wiley and Sons, New York, 1963, pp. 150–210.

[20] D. W. Gregg and D. U. Olness, Lawrence Livermore Laboratory Rep. No. UCRL-52107, 1976.

ess with very few adjustable parameters. In a surface plant the dimensions as well as other operating parameters of the reaction vessel can be varied. The rate and nature of the coals fed into the vessel in terms of particle size, ash content, and moisture content can be controlled. In addition, it is possible to blend different types of coal as needed. No such adjustable parameters are available in the design of a UCG system. Despite this, the performance requirements in terms of the quality of the product are determined by the user and remain essentially the same as those for the surface plant.

The underground portion of the system must be engineered to operate in a predictable and controllable manner even though as the combustion zone moves through the coal seam, the seam thickness, ash content, moisture content, natural permeability, and overburden nature change significantly. If most of the coal is removed, the process is further complicated by roof subsidence. Because so little control is possible after the process has been initiated, the basic system design must be inherently stable or insensitive to such changes. The design is also limited by the desire to use no men underground; hence access to the coal is limited to drilled holes. It is somewhat surprising that with all these rigid requirements such a system is possible. However the Soviets have shown with as many as five commercial plants that it is feasible.

The basic Soviet designs for horizontal and dipping coal seams as well as the logic for the designs are presented in this section. The primary distinguishing feature of the design of a system versus that of an individual experiment is that the system must satisfy a large number of criteria simultaneously. A UCG system must be designed to handle all the physical and chemical processes influencing its operation in a manner that optimizes its economics. This in turn may require handling certain problems in a less than optimum manner so that other difficulties are resolved and the overall system is optimized. This section lists some of the problems that must be considered and discusses workable solutions developed in the USSR.

26.5.2 Features of the Soviet System

The Soviet UCG systems for horizontal and steeply dipping beds (essentially the same basic design but at different angles to the surface) are deceptively simple. However it has been demonstrated repeatedly that they can operate successfully in an extremely complex environment: a burning coal seam with a subsiding roof. In spite of the wide variations in the geological conditions that exist among coal seams, the basic system design can be transferred from one coal seam to another with reasonably predictable results. Therefore we examine the design to determine how it handles a number of physical phenomena that must be dealt with in any design for underground gasification of coal.

26.5.2.1 *Gas Leakage*

One design requirement of a UCG process is the minimization of the loss of both the injected blast and the product gas. Because gas losses increase with the square of the pressure, maintaining a gas-tight zone is very difficult. As coal is removed, the roof must eventually subside, especially where a high fraction of the coal is recovered over a large area. Such subsidence forms very permeable cracks to higher formations, sometimes all the way to the surface, allowing gas to escape from the process zone.

The extent of these leakage cracks is unpredictable as the combustion zone moves across a coal seam; therefore it is important to design the system with a minimum sensitivity to such leaks. The only way to accomplish this is to operate the system at the lowest possible pressure and to ensure that the permeability in the coal between the injection and exhaust holes is always much higher than the permeability through subsidence cracks. The Soviets accomplish this by establishing highly permeable paths in the coal (linking) between the injection and exhaust holes before gasification and operating at the lowest pressure that is consistent with production requirements. In some cases they have also sealed the cracks at the surface by bulldozing them full of mud.

26.5.2.2 *Gas Flow Rate at Low Pressure*

A high gas flow rate is necessary to compensate for the water intrusion rate. It is also needed to maximize the eventual channel width, thus the resource recovery for each line of drilled holes as well as for achieving an economically high production rate. This requirement is tightly coupled to the gas leakage problem discussed above because increased flow rates require increased pressure drops. The high gas flow rates can be maintained with minimum leakage only if the permeability between the injection and production holes exceeds some minimum value. The Soviets generally use gas flow rates of 3000–10,000 m^3/hr with driving pressures that do not exceed 2.5 atm.

An additional advantage of these low operating pressures is the possibility of using high-capacity, relatively inexpensive blowers to supply the air, although this is not as important when oxygen-steam is used.

26.5.2.3 *Directional Control of Gas Flow*

Directional control of gas flow is essential for a reproducible system. Only by establishing flow control with the consequent reproducibility can we begin to engineer a system that maximizes resource recovery or optimizes the economics (these two criteria may not lead to the same system). This requires a means of control that is insensitive to variations in geology or coal type.

The Soviets achieve directional control by using highly permeable linkage paths with predetermined spacing at the bottom of the coal seams. These paths render the gasification process insensitive to variations in the natural coal permeability. Thus after obtaining initial scaling data, it is possible to design the required predictable system. The directional control of the gas flow also ensures that the gas will flow into the exhaust pipe instead of spreading uncontrollably throughout the formation.

26.5.2.4 *Surface Area and Permeability*

High permeability is needed for the reasons stated above. The high surface area, consisting of a coal rubble zone without bypass channels, is generally needed for efficient gas-solid reactions. In such reactions the rate of reaction is limited by the available surface area of the coal because this is where the initial reactions must take place. In the Soviet system for horizontal coal seams, the linkage channels are placed along the bottom of the seam so that the flame front undercuts the coal. The coal then falls into the void and forms rubble. In the system for steeply dipping beds, rubble is formed by coal falling into the void created at a bottom manifold, where combustion is initiated.

26.5.2.5 *Liquid Control*

One of the primary effects of liquids (water and pyrolysis tars) on the propagation of the flame front in a horizontal seam (without linkage channels) is to cause the flame front to move preferentially along the top of the coal seam. This occurs because liquids are much more dense than the gases and migrate to the bottom of the seam. The gas flow, thus the flame front, then traverses the top of the seam.

The difficulty that arises is that once a channel has been formed across the top of the seam, the coal ash tends to seal the bottom of the channel, preventing further combustion of coal and resulting in very poor resource use. The Soviet process minimizes this effect by forming the channel at the bottom of the seam during the preparatory linking step. The channels are kept hot enough throughout the gasification process so that any liquids intruding into the region, as well as liquids formed by pyrolysis, are carried out as vapors. The channels thus remain clear to gas flow, with the additional benefit that the removal of the liquids as vapors permits recovery of the economically valuable coal tars.

26.5.2.6 *Phenol Contamination*

The mechanism for total liquid removal described above also removes phenols formed in the pyrolysis of the coal. The phenols are never allowed to condense but are removed almost immediately as vapors. The efficiency of this removal is enhanced by a "steam flush." Because the intruding water is vaporized deeper in the coal seam (farther from the hot channel) than where the phenols are formed, the steam flows into the channels through the phenol forming zone. This effectively flushes out these potential aquifer contaminants into the hot channel, where they then flow to the surface in the gas stream.

Thus a minimum quantity of phenols remains in the ground after gasification, minimizing aquifer contamination.

26.5.2.7 *Access Pipes*

A major portion of the cost of the UCG system is associated with the access pipes. Thus it is important to maximize the spacing and probability of survival of the piping during the gasification process for greatest resource recovery. In addition, subsequent removal of the casings for reuse might be possible and desirable. This importance of maximizing the total resource utilization as well as the quantity of coal obtained per hole dictates the tradeoff between fractional resource recovery and hole spacing.

The Soviet designs place the access holes so that they are out of the (moving) subsiding zone until they are no longer needed. This increases the probability of the survival of the pipes and makes it possible to pull the casing for reuse. The exhaust pipes are also cooled with a spray of water at the bottom of each pipe. This prevents the pipes from separating from the cement that seals them to the formation. However care must be taken not to overcool the gases, or the coal tars could condense in the pipes and plug them. The best operating temperature range was between 100 and 200°C.

It is difficult to prove that the pipe spacing was maximized for the Soviet system. However certain features are essential for a large pipe spacing. A highly permeable channel between the injection and exhaust holes permits high gas flow rates at low pressure drops. The reproducibility of the system also allows the optimization of the pipe spacing.

26.5.2.8 *Coal Seam Thickness*

The Soviets have found no maximum limit in thickness of coal seams for their processes. They have operated in seams up to 20 m thick with no indication that they were approaching a limit. However the Soviets have found that there is a minimum thickness limit. If the seam thickness falls below about 1 m, the heating value of the gas becomes unusably low for air-blown gasification.

26.5.2.9 *No Men Underground*

The present Soviet design requires no men underground during gasification; this was not true of the British design or of earlier Soviet systems. Initially the desirable requirement of no men underground was dictated more by Soviet politics than by technology. To support the principles of Marxism as viewed by Lenin, the goal was to improve the miners' lot by taking them out of the coal mines. This move was intended to prove that a socialistic society was more concerned with the welfare of the worker than were societies based on capitalism. However from a safety standpoint, the requirement of no men underground is an extremely important technological advance because once gasification has been initiated, it is essentially impossible to guarantee that toxic product gases will not leak into adjacent mine workings.

26.5.2.10 *Multilayered Coal Seams*

The Soviet system has been used to sequentially gasify multiple layers of coal, beginning with the top seam and working down.[21] This permits a more complete and economical use of available coal reserves.

26.5.2.11 *Continuous Versus Intermittent Load System*

The Soviet system sweeps in a continuous manner across a coal seam (as opposed to a batch system); thus it can be used for electrical base load demands. However there is also a need for electricity on an intermittent basis to supply peak loads. Under some circumstances the gasification system can be turned on and off in time periods as short as a few hours. Thus the system could prove useful for meeting intermittent electricity demands.

26.5.2.12 *Constant Gas Composition*

The design of the Soviet system permits the maintenance of a product gas composition that is quite constant in its heating value as a function of time. This is primarily due to the hot gases flowing through a continually forming, high surface area rubble zone of coal. (This subject is discussed in detail later.)

[21] J. Koranda, *Sb. Pr. UVP (Ustav Pro Vyzkum Paliv)*, No. 13, 138 (1969); UCRL Transl. No. 11085, 1976.

26.5.2.13 *Reproducibility, Predictability, and Control*

The performance of the process design is reproducible and predictable within reasonable limits from one generator to another within the same coal seam, as well as when transferred from one coal-bearing area to another. Only minor modifications are needed to adapt this design for operation in widely varying coal types. Adequate controls are available to achieve a constant gas quality and high resource recovery; 75% or more of the coal is commonly consumed, with between 50 and 70% of the heating value of the consumed coal recovered as heating value in the product gas.[12,22]

26.5.2.14 *Minimum Sensitivity to Coal Swelling*

The large channels formed in the linking phases minimize the probability of plugging by moderately swelling coals.

26.5.2.15 *Minimum Sensitivity to Flame Front Channeling*

The Soviet system makes no attempt to avoid the natural tendency of flame front channeling[23]; in fact, it encourages it. However the gas quality is insensitive to such channeling because the channels through the coal seam are very long, undercutting the coal and becoming filled with coal rubble.

26.5.2.16 *Simplicity of Design and Operation*

The Soviet system design and operation are exceptionally simple, deceptively so. The only technology required is the drilling of a simple pattern of holes (although slant drilling is not always easy) and the handling of compressed air. The end result is that the gasification process is insensitive to the many uncontrollable physical phenomena that might otherwise influence the process. Such simplicity and insensitivity make the process technically feasible.

[22] A. Gibb, *The Underground Gasification of Coal,* Pitman and Sons, London, 1964.
[23] D. W. Gregg, Lawrence Livermore Laboratory, Rep. No. UCRL-51595, 1974.

26.6 ENHANCED COAL PERMEABILITY

Although gas can be made to flow in coal beds using only naturally occurring permeability, these permeabilities have never been high enough for a large-scale gasification process without initial enhancement. This is not only because the initial permeability of a coal bed is low but also because a cocurrent combustion process tends to plug the coal bed with liquids that condense downstream from the combustion zone. There are a limited number of methods that have been successfully employed to enhance the permeability of a coal bed before gasification, and the nature and limitations of these methods dictate a major portion of the remaining process design.

The four methods that have been employed with varying degrees of success are directional drilling, countercurrent combustion, electrolinking, and hydrofracking followed by countercurrent combustion.[12] The primary feature that characterizes these methods is that they form narrow, highly permeable, approximately cylindrical channels. It is evident how this is achieved by a drilled hole, but it is not so apparent why countercurrent combustion and electrolinking also form these channels. This will be explained in some detail below. However it is important to first emphasize that these proved successful methods for permeability enhancement in coal beds form narrow linear channels of high permeability. They do not create planes of permeability or increase the permeability of the bulk of the coal seam (volume permeability enhancement). Thus the total system design must be compatible with and effectively utilize this feature.

26.6.1 Directional Drilling

Drilling vertical holes down to the coal seam involves no novel technology and is not discussed. The drilling advancement developed by the Soviets for UCG is the technology of drilling along a coal seam at varying angles to provide a gasification channel in the coal. The techniques used for both steeply dipping and horizontal coal beds are described in considerable detail by de Crombrugghe[24] and Arinenkov and Markman.[25] Having tried a number of approaches, the Soviets presently employ downhole electric motors that allow them to drill directionally curved holes with radii of curvature as small as 600 m. The drill is guided with the aid of a downhole compass and a pendulum. The data from the compass and pendulum are transmitted to the surface over the power lines for the drill motor. For steeply dipping coal beds the directionally drilled holes are usually straight but at an appropriate angle to the surface so that the drilled holes remain in the coal seam. However for horizontal beds the drilling is begun at an angle to the surface and is then curved into the coal seam.

Holes with diameters of 0.3 m and as long as 100 m have been drilled in the coal seams by this technique. When fully developed, the penetration rates are expected to reach 3–10 m/hr.[24]

26.6.2 Linking by Countercurrent Combustion

The differences between countercurrent and cocurrent combustion processes are very significant when applied to UCG. Cocurrent combustion is a process by which the flame front and gas flow propagate in the same direction and the rate at which the coal is consumed determines the flame front velocity. As the coal is consumed, the combustion zone moves downstream toward the higher concentrations of coal.

[24] O. de Crombrugghe, *Ann. Mines Belg.*, No. 5, 478 (1959); U.S. Bur. Mines Transl. No. BM-394.
[25] D. M. Arinenkov and L. M. Markman, "Knizhnoe Izdatel'stvo Stalino-Donbass," 1960; UCRL Transl. No. 11007, 1976.

However in countercurrent combustion the flame front moves in the opposite direction to the gas flow. The velocity of the flame front is determined by the rate at which heat is conducted upstream against the gas flow. Coal upstream is heated to its ignition temperature and ignited, thus depriving the remaining downstream coal (char) of oxygen and causing incomplete combustion of the coal reserve. Thus a cocurrent combustion consumes essentially all the combustible material in its path, whereas a countercurrent combustion can pass through a combustible material and consume only a portion of it.

The description above addresses only the differences associated with an idealized one-dimensional combustion process. There are other equally large differences between these two modes of combustion involving their spatial propagation in two and three dimensions. The most significant difference is that countercurrent combustions tend to burn narrow, fixed-diameter channels, whereas cocurrent combustions generally propagate on comparatively broad fronts.[11] The reasons for this difference can be visualized with the air of Fig. 26.10.

Consider the idealized case presented in Fig. 26.10 where a narrow, highly permeable channel of char or ash extends into a coal seam of much lower permeability. If we draw lines for flow from either point *A* to point *C* or from *C* to *A*, we see that these flow lines converge at the tip of the channel. Thus in the region from *A* to *B*, the majority of the gas flow is carried by the channel. In the area from *B* to *C*, the gas flow distributes more uniformly across the bed, independent of the flow direction. If the coal is ignited in the channel in the region of *B* and if the flow direction is from *C* to *A* (countercurrent

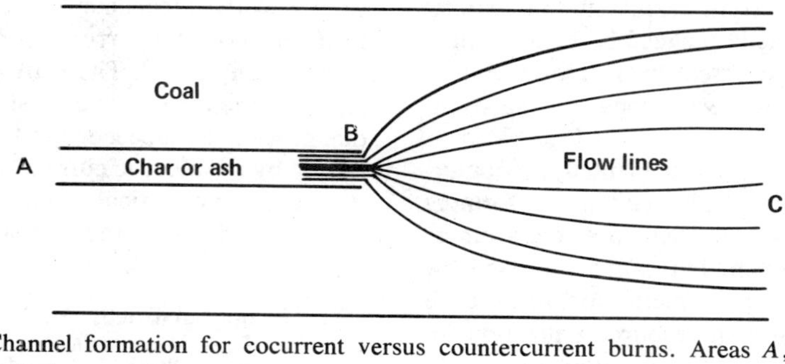

Figure 26.10 Channel formation for cocurrent versus countercurrent burns. Areas *A*, *B*, and *C* are discussed in the text.

combustion), the flame front at the tip of the channel (at *B*) sees a much larger supply of oxygen than will the walls of the channel further toward *A*. Therefore as the flame front at the tip progresses toward *C*, the oxygen-starved channel walls do not significantly expand. This countercurrent combustion process creates a channel of a relatively small and constant diameter. In contrast, if the flow direction is from *A* to *C* (cocurrent combustion), the channel walls will be exposed to the highest concentration of oxygen, supporting combustion on the walls, thus promoting a widening of the channel.

An additional difference between cocurrent and countercurrent burns has considerable importance to UCG. Countercurrent burns are less subject to plugging due to the presence of a condensing liquid front downstream from the flame front than are cocurrent burns. It is easy to see how a cocurrent burn would be very vulnerable to such plugging. If air contact between two pipes drilled into the coal is established where the airflow is conducted by the natural permeability of the coal and ignition is initiated at the air injection pipe, the hot gases carrying water vapor and vaporized tars will cool before reaching the exhaust pipe. The cooling of the gases will cause the water and tars to condense and to plug the small cracks on which the natural permeability of the coal is based.

However if ignition is initiated at the exhaust pipe and a countercurrent burn is carried out, it is possible to maintain high enough temperatures in the combusted channel throughout the process so that the vapors never condense. If the channel does cool enough to allow some condensation, the condensation will occur in the highly permeable channel that is much more difficult to plug. Similar arguments can be made to illustrate that a countercurrent burn would be expected to be less easily plugged by coal swelling than would a cocurrent burn. However it is not clear that either will propagate in a swelling coal.

For the reasons just presented, a countercurrent burn is ideally suited for forming highly permeable channels in a coal seam as a preparatory step, but cannot be used for gasification in the production phase because of the resulting poor resource utilization. In contrast, a cocurrent burn cannot be used to form the initial gasification channels but can be used in the production phase, because it results in high resource utilization. Therefore both burn methods are used in a gasification process: countercurrent combustion as the linking technique, and cocurrent combustion for the production phase.

In practice the Soviets were able to achieve countercurrent combustion linking rates of 1–3 m/day using air.[11] Higher linking velocities could be achieved with oxygen, but the resulting channels were not sufficiently permeable to satisfactorily carry out the following cocurrent burn. Apparently air has an almost optimum oxygen concentration to produce a sufficiently permeable channel at an acceptable channel formation rate.

26.6.3 Electrolinking

A very extensive treatment of the Soviet experience in the use of an electric current to establish a highly permeable channel in coal has been presented by Bondarenko et al.[26] No attempt is made here to reproduce or even summarize that work. However it is useful to explain the electrolinking technique and how, in some circumstances, it can be employed competitively with other linking techniques.

At first glance it appears that it is almost impossible to develop a competitive channel formation technique employing an electric current to char a channel of coal between two access holes. We might expect that the electric power required would far exceed that which would be practical. This would probably be true if the electric current were required to char the entire channel length and width by bringing all the coal to charring temperature at the same time.

However the charring process proceeds in quite a different manner and is in fact very similar to the mechanism of channel formation for countercurrent combustion, as shown in Fig. 26.11. There are certain crucial physical phenomena that must be present for this process to succeed. First, as the coal is heated by the electric current, it must strongly increase in electrical conductivity. Second, the conductivity of the unheated coal must

[26] S. T. Bondarenko, B. K. Brodskaya, S. N. Lyandres, E. A. Meerovich, V. I. Pen'kovskii, and A. D. Reznikov, *Acad. Sci. USSR*, 1959; UCRL Transl. No. 11050, 1976.

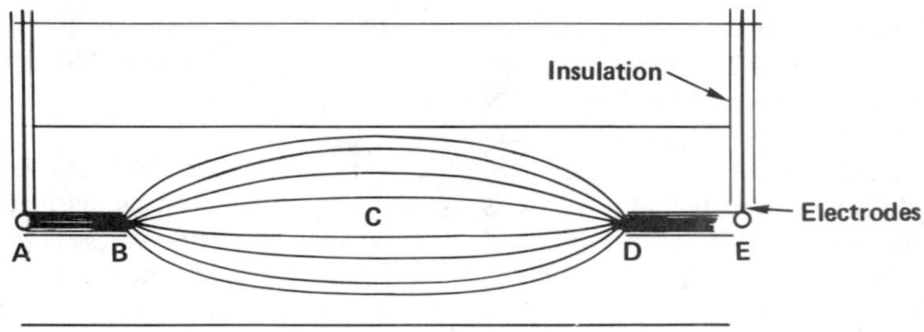

Figure 26.11 Electrolinking process. Areas A–E are discussed in the text.

be reasonably high but still much lower than that of the heated coal.

Figure 26.11 describes the process at some time after the power has been turned on. Highly conducting channels from both electrodes appear and gradually grow toward the center until they meet. This mechanism has been verified experimentally,[26] and the explanations for the progress of this process are similar to those describing channel formation with a countercurrent burn. The channel sections *A–B* and *D–E* are highly conducting. Therefore the current is concentrated into the channels. However the current spreads out over a larger cross section in the unheated coal (region *C*). When this happens, most of the potential drops; therefore most of the energy dissipation occurs at points *B* and *D*, where the highly conducting channels connect with the unheated coal.

Energy dissipation in the channels is minimized by the high conductivity of the channels. Energy dissipation in the unheated coal seam far from the channels (region *C*) is reduced because the large cross-sectional area that carries the current compensates for the relatively low specific conductivity of the unheated coal. The combination of high current density and high resistivity of the unheated coal is found only where the current lines converge on the tips of the highly conducting channels. Thus as the channel tips heat, the coal chars, increasing its conductivity. Each channel is thus extended further into the coal until the two channels meet.

This mechanism of channel formation requires only that the small regions *B* and *D* be heated by the electric current at any one time. This requires far less electric power to overcome heat losses than does a process that heats the entire channel length at once. Thus because of this "point" heating effect, electrolinking is a competitive linking technique.

The efficiency of the electrolinking process clearly depends on the specific conductivities of the channels and the unheated coal. Energy dissipated in either of these zones is wasted. Only the energy that heats the tips of the channels is used constructively. Therefore, depending on the conductivities achieved, which in turn depend on the nature of the particular coal seam, electrolinking may or may not be the most efficient means to form the linkage channels. The Soviets report that electrolinking was slightly more energy efficient than other linking techniques at Podmoskovia, but was significantly less efficient at Angren.[27,28]

The primary problem encountered with electrolinking is that it is very difficult to control. In an idealized and uniform medium, the link proceeds in a predictable manner. However the direction the linking channel follows is strongly influenced by anomalous and unpredictable regions of conductivity that are specific to the local geology. This lack of reliability has forced this technique to be abandoned in many cases even though, had it worked, it would have been the most efficient linking process available.[26]

When this technique is successfully used, the electrolinked channel is usually widened further with a relatively fast countercurrent combustion step before the production phase is initiated.

26.6.4 Hydrofracking

As has been stated, the choice of linking technique is critically dependent on the physical or electrical properties of the coal seam.

[27] N. A. Fedorov, *Podzemn. Gazif. Uglei,* No. 1, 16 (1957); U.S. Bur. Mines Transl. No. BM-432.

[28] S. N. Lyandres and A. D. Reznikov, *Podzemn. Gazif. Uglei,* No. 2, 49 (1958); U.S. Bur. Mines Transl. No. BM-583.

If the electrical conductivities of the coal and char are not correct, electrolinking is not acceptable, and if the natural permeability of the coal seam is too low, countercurrent combustion linking is not a good approach. Directional drilling is one technique that is relatively independent of the coal properties, but in some situations it is too sophisticated an approach to the problem. Therefore in some cases hydrofracking is employed to establish a linking channel. The effectiveness of hydrofracking is independent of both the electrical properties and the natural permeability of the coal. Hydrofracking also requires considerably less sophisticated equipment than is needed for directional drilling but its lack of control is a definite disadvantage.[12]

Essentially, hydrofracking requires the pumping of water down a pipe, into the ground. The water is forced out into the coal formation at the bottom of the pipe and forms cracks. Because cracks tend to close after pumping ceases, sand may be added to the water to prop open the cracks. The effectiveness of the operation is often improved by the addition of a thickening agent to increase the viscosity of the water. This improves the sand-carrying capability of the water and helps to confine the underground flow to a few major cracks.

Many studies on the use of hydrofracking for preparation of a coal bed for gasification have been conducted.[11,12,29] However it appears that this technique has been employed primarily for steeply dipping beds, where it is not essential to ensure that the cracks form at the bottom of the seam or that they form parallel or perpendicular to the bedding planes.

[29] P. P. Klimentov, *Izv. Vyssh. Uchnebn. Zaved—Geol. Razved.*, No. 10, 97 (1964); Transl. No. SLA-73-6002, 1973.

26.7 FIELD DESIGN FOR STEEPLY DIPPING COAL SEAMS

Several approaches to UCG were investigated by the Soviets in the early 1930s, such as explosive fracturing to prepare the bed (the Chamber method) and excavation or drilling of holes from one drift to another (the borehole-producer method).[2,3,5] Poor economics and unstable operation forced these methods to be abandoned.

26.7.1 The Stream Method

The first design that showed promise is commonly called the stream method. This method, as applied to steeply dipping beds, is illustrated in Fig. 26.12.[30,31] The injection and exhaust holes were drilled from the surface along the coal seam and were connected at the bottom by a mined shaft. Air was used for the gasification process. The flame was initiated in the connecting channel and it gradually spread over the entire length. The flow was reversed periodically to approximate a "horizontal" burn. The key feature of this system is that as the coal is consumed, more coal falls into the gasification void. Thus the coal automatically becomes a highly permeable rubble and is fed into the combustion zone. The coal rubble thus formed has the additional essential feature of providing a large surface area on which the oxidation reactions take place. As a result, the product gas composition remains relatively stable as a function of time. In all further design improvements for both steeply dipping and hor-

[30] L. J. Jolley and N. Booth, *Fuel,* **24,** 31 (1945).
[31] V. A. Matveev, *Podzemn. Gazif. Uglei,* No. 2, 4 (1957); National Coal Board Transl. A, No. 1408/SEH, 1957.

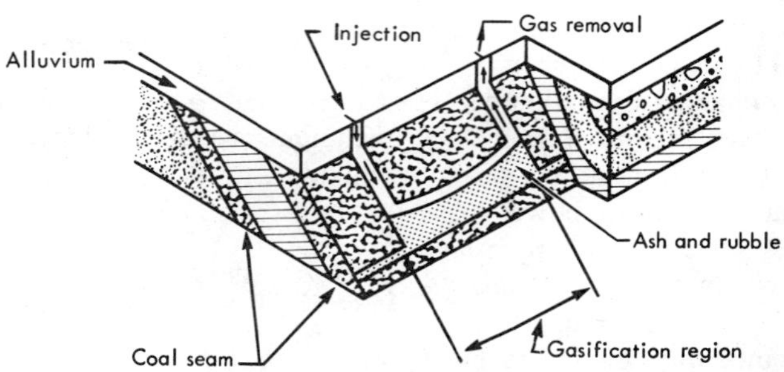

Figure 26.12 The stream method for gasifying coal in steeply dipping coal beds.

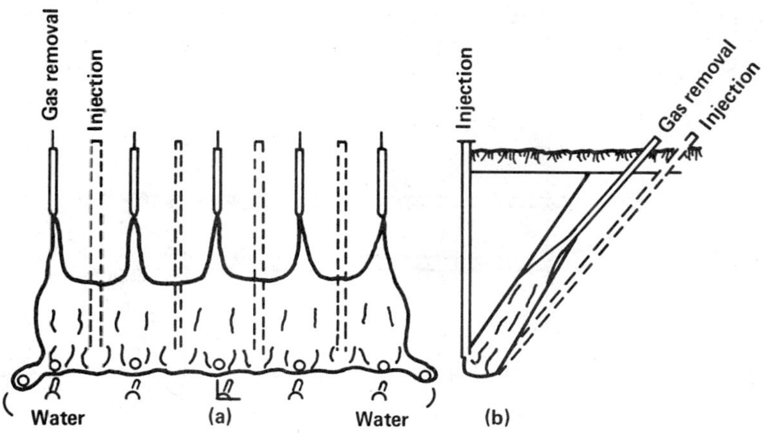

Figure 26.13 Typical diagrams of an underground gas generator for steeply dipping coal beds: (*a*) end view (*b*) plan view.

izontal coal beds, this self-forming rubble feature (due to undercutting the coal) was retained.

26.7.2 Modifications of the Stream Method

Two major problems associated with the stream method as presented in Fig. 26.12 have been resolved in later designs (Fig. 26.13). First, there is a tendency for oxygen (air) to leak through the coal between the two access holes, above the region where combustion takes place. This air either consumes part of the product gas, or introduces a hazardous level of oxygen into the product gas. Second, there is a tendency for the combustion front to channel over the top of coal, resulting in reduced resource recovery and erratic gas quality. In steeply dipping beds these problems have been resolved by drilling cased air injection holes to the bottom manifold through the overburden. In this manner the injection holes are isolated from the exhaust holes. The probability of the flame front channeling over sections of coal is also minimized by permanently fixing the injection point at the bottom of the seam. If the coal seam is thicker than approximately 8 m, it is necessary to drill the injection holes at a slant, under the seam, so that the pipes will not be sheared off by subsidence. When these costly slant holes are used, however, vertical holes are usually drilled to help establish the initial bottom manifold by hydrofracking or countercurrent combustion. The vertical holes are also used in the first stages of gasification.[12,32]

A phenomenon commonly observed with the stream method is illustrated in Fig. 26.14. The pressure drop often is observed to occur in the lower regions, near the injection point in the region of the flame front. Two factors contribute to this phenomenon: a gas expansion at the flame front resulting from the increase in temperature and in molar quantity of gases, and a channel blockage formed by a subsiding roof, restricting the gas flow. If the second factor is the one of importance, not only could the gas flow be restricted in a one-dimensional sense, but severe two-dimensional flow distortion problems could also result. This two-dimensional redirection over a long distance in the coal seam would eventually cause a loss of control over the direction of the gas flow and would result in poor resource recovery. For these reasons, an additional modification has been introduced (Fig. 26.15). A number of additional gas injection holes are drilled along the dip, intersecting the seam at prescribed intervals from the bottom manifold up to the surface. This modification minimizes the pressure drop along the coal seam while retaining spatial control over the gas flow.

An alternative layout using a directionally drilled hole to form the bottom manifold is presented in Fig. 26.16.

[32] S. I. Buyalov, *Ugol*, No. 11, 17 (1938); A. D. Little Transl. No. G-159.

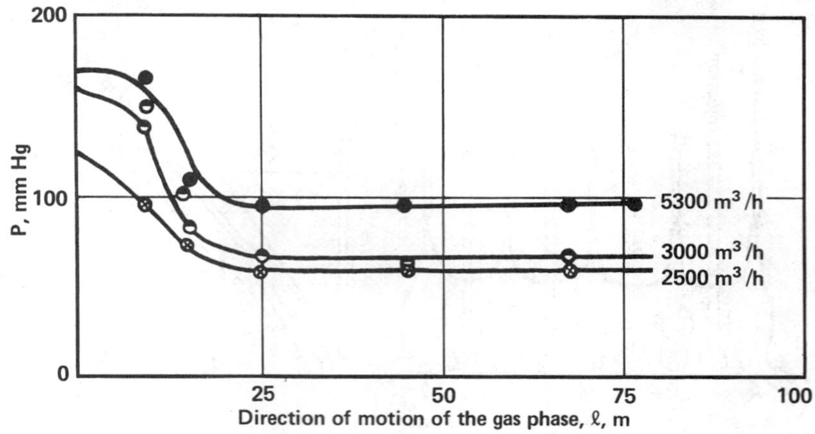

Figure 26.14 Change of pressure of the gas phase along the channel length of gas generator No. 1 of the Yushno-Abinsk station. The values given in cubic meters per hour are injection rates of the blast.

26.8 FIELD DESIGN FOR HORIZONTAL COAL SEAMS

The basic design of the Soviet system for gasifying horizontal coal seams is essentially independent of the linking technique used. Therefore we explain the system design in terms of countercurrent combustion and then generalize to other linking methods. Figure 26.17 shows a typical plan view of the Soviet process. The dotted lines indicate the location of the underground linkage channels formed in the coal by countercurrent combustion. Efforts are made to ensure that these channels are formed at the bottom of the seam. This is accomplished by sealing the access pipes to the coal, two-thirds or more of the way down the coal seam. The injected air is thus forced to enter and the product gas must exit at the bottom of the seam. Thus the linkage channel at least starts and ends at the bottom of the seam, even though it may wander a bit between the pipes.

In Fig. 26.17 the production cocurrent combustion phase is shown in the direction of Row 2 to Row 1. However in practice the Soviets conduct the gasification in both directions; the choice is critically dependent on the exact nature of the roof subsidence. Such subsidence can create leakage paths for

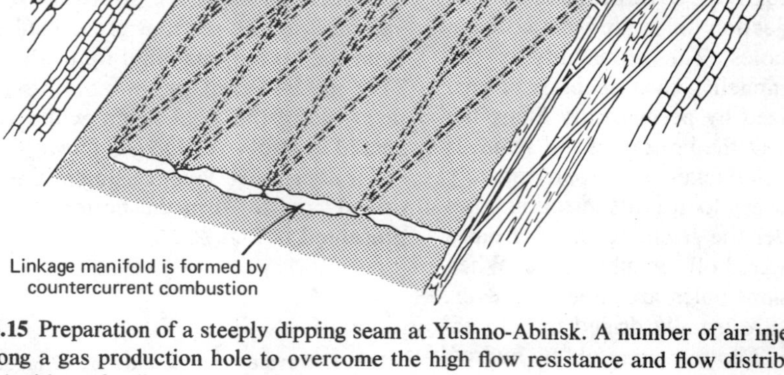

Figure 26.15 Preparation of a steeply dipping seam at Yushno-Abinsk. A number of air injection holes are drilled along a gas production hole to overcome the high flow resistance and flow distribution problems associated with roof collapse.

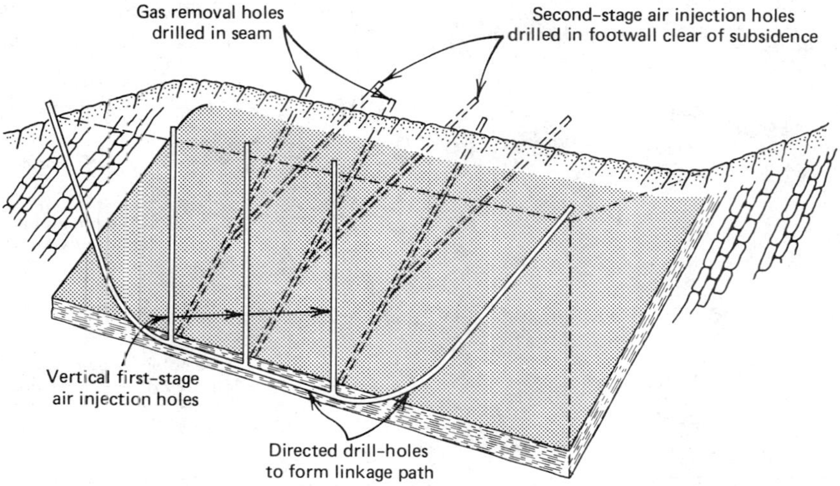

Figure 26.16 Alternative layout using directed drillholes for linkage. The bottom manifold between the gas injection and production holes is formed using directional drilling. This design for steeply dipping beds was first developed at Lisichansk in 1955.

the gas as well as obstruct the gas flow, influencing the best direction for the forward combustion gasification step.

26.8.1 Modifications for Thick Seams

The modifications in the drilling pattern and operational method used to accommodate thick coal seams are shown in Fig. 26.18. The primary problem encountered for thick seams is the tendency of ground subsidence to shear off the access pipes. To prevent this, two modifications are made. First, a lateral withdrawal-injection technique is used. In this case one of the two functions (gas withdrawal or injection) is performed by pipes in a row along the side of the gasification zone. The remaining function is accomplished using

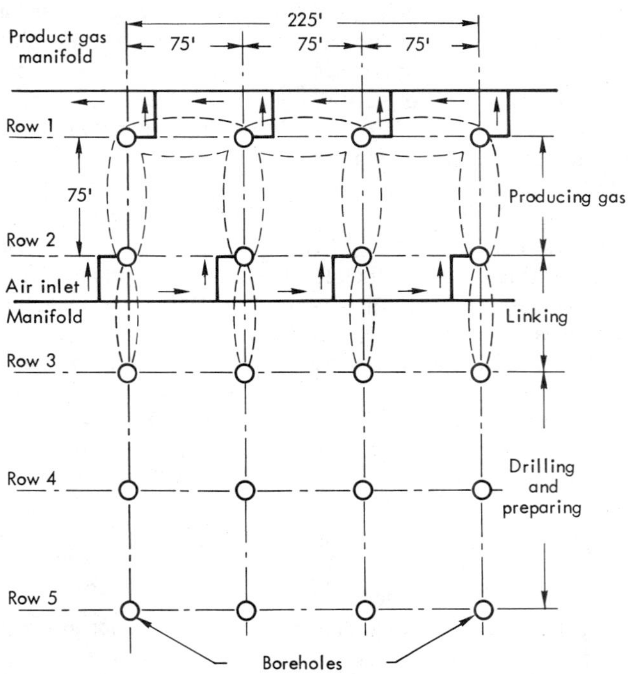

Figure 26.17 Plan view of a shaftless generator for a plant in the Moscow region.

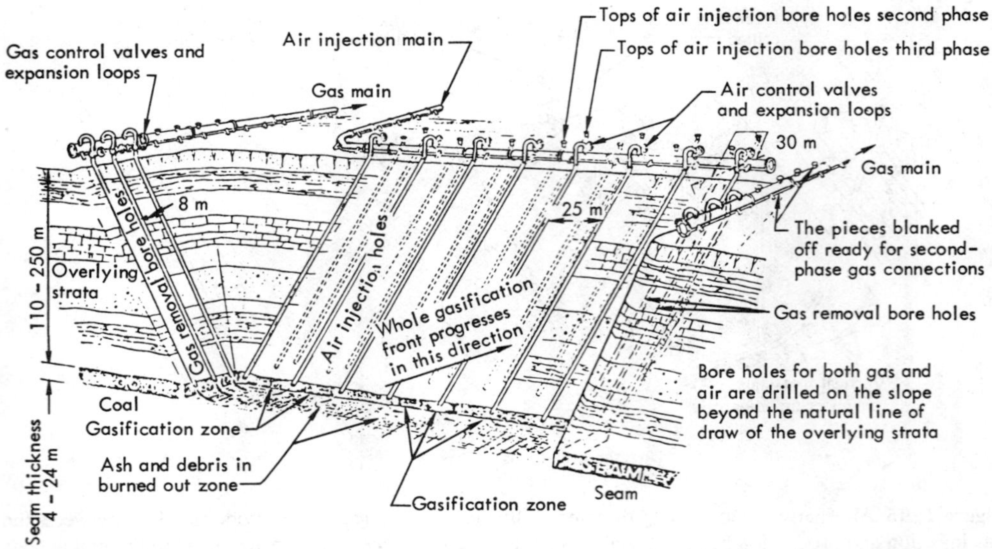

Figure 26.18 Projected layout of boreholes at Angren to suit thick brown coal and heavy subsidence.

pipes in a rectangular pattern, entering the seam directly over the coal to be gasified.

Second, within the framework of this general pattern, all holes are drilled on a slant to keep them out of the subsidence region during their functioning lifetime. When gasification is conducted in this manner, the pipes along the side of the gasification zone are never affected by subsidence, and the pipes in the pattern directly over the coal to be gasified are not affected until they are no longer needed. Therefore it is never necessary to operate with a pipe that has been exposed to any ground movement or subsidence.[22]

Using this technique, a cross-flow component is also added to the gas flow, which helps recover mounds of coal that might otherwise be left between channels. The slant-drilled portion of this design, although fully satisfactory in concept, was not always cost-effective, thus was not always used. In such cases the lateral withdrawal technique was made to work equally well with a combination of vertical pipes and more extended underground linkage channels that substitute for the slant-drilled holes.

26.8.2 Critical Channel Width

The spacing between linkage paths depends on a phenomenon called the critical channel width. If we imagine a channel that is extended indefinitely with a series of drilled holes and operated for a long period of time, we must know how wide the channel will grow. This factor determines the maximum initial spacing allowable between parallel channels for 100% resource recovery. Eventually, the channel will reach a stable width and will expand no further because as the channel widens, the transfer rate of oxygen to the combusting coal faces diminishes, but the conductive and radiative heat losses from the faces remain constant or increase. At a specific width, the critical channel width, it is no longer possible to sustain combustion on the faces, and the channel ceases to widen. The actual critical width is clearly affected by the permeability distribution in the channel after roof subsidence. The critical channel widths for the Soviet gasification stations and for the British experiment at Newman Spinney are presented in Table 26.1. With air as blast, critical channel widths as great as 5–10 times the seam thickness were achieved for thin seams. Widths were further increased by enriching the blast with oxygen. The British employed hole spacings 25 times the seam thickness, with 82% coal recovery.[22,27]

26.8.3 Maintenance of Constant Composition of Product Gas

It is very important to operate the gasification system so that the composition of the product gas remains relatively constant with

Table 26.1 Critical Channel Widths at Soviet and British Coal Gasification Stations

Station	Critical Channel Width (m)	Seam Thickness (m)
Soviet		
Tula	18–36	2–4
Lisichansk	9–27 (air blast)	1–3
	19–57 (oxygen-enriched blast)	1–3
Yushno-Abinsk	14	2
British		
Newman Spinney P-5	14–19	0.6–0.8

time, maintaining the heating value at a desirable level. This is accomplished by ensuring that the hot gases always pass through a zone that is essentially packed with coal rubble. Such a zone has a large amount of coal surface area available for reaction and permits intimate contact between the hot gases and the coal. The downstream tip section of a long channel during gasification of a thick coal seam is illustrated in Fig. 26.19. (The wider upstream portion of the channel with the additional air injection pipes is not shown.)

An essential feature of this region is that as coal is consumed, which occurs initially at the bottom of the seam because of the placement of the linking channel, more coal falls into the growing void. This creates a high surface area rubble bed that can serve as a final reduction zone. Such a zone is highly reactive and, by converting excess CO_2 to CO ($CO_2 + C \rightarrow 2CO$), is responsible for the uniform quality of product gas.

Further upstream the channel may widen to many times the coal seam thickness. In this area coal falling into the growing void at the sides of the channel does not have such an important effect on the product gas composition because of an oxygen bypass at the channel center. A sampling of the upstream gases in the oxidation zone shows low heating values. As long as the tip of each channel has a hot and highly reactive zone of coal rubble through which all the hot gases must

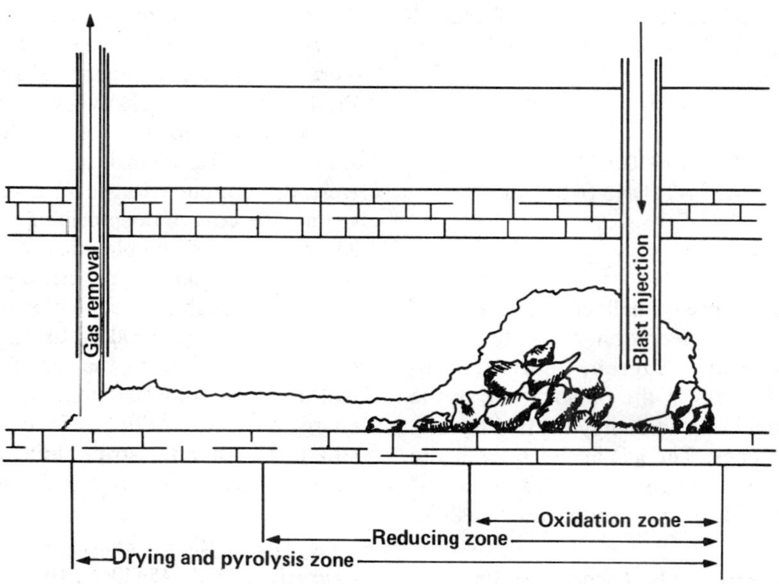

Figure 26.19 Conceptual view of a channel during gasification.

flow, however, the product gas will be properly reduced and will have a high heating value that is uniform with time.

As the flame front advances and the channel width expands (Fig. 26.19), there comes a point when this zone will no longer hold the composition of the gas constant. At this time another section is added to the channel (as illustrated in Fig. 26.17) to maintain the required high surface area coal zone. The gasification process can thus be continued indefinitely along a coal seam with excellent control over the uniformity of the gas composition. Fractional resource recovery can be controlled by proper spacing of the parallel channels.

26.8.4 System Efficiency

There are at least two efficiencies of concern for the underground gasification of coal. The fraction of the heating value of the coal that appears as heating value in the product gas must be maximized. Also, we must consider the overall energy efficiency of the plant, including the conversion efficiency mentioned above and the energy required to maintain the gasification station. The Soviets have well documented the coal-to-gas heating value conversion but report little on the overall plant efficiency. The fraction of coal gasified varied from 50 to 100% depending on the particular gasification station and on the specific nature of the ground subsidence. Of the coal actually gasified, 50–70% of its heating value appeared as heating value in the product gas.[12]

26.9 FIELD AND COMMERCIAL INSTALLATIONS SINCE 1960

26.9.1 USSR

In the USSR three small commercial-scale plants are now in operation; these three sites are at Shatsky, Angren, and Yushno-Abinsk.[33] The gas at the Shatsky site in the Moscow basin is used mainly for fueling low-Btu gas turbines. The gas at the Angren site is used to fuel steam boilers for electrical generation, and the gas produced at Yushno-Abinsk is used for space heating and process heat. Angren, near Tashkent, is a site operating in brown coal. The seams are nearly flat, between 4 and 24 m thick, lying at depths between 40 and 200 m. Exploitation is entirely by vertical holes from the surface. It is reported that gas is produced at a calorific value of 90 ±10 Btu/ft^3 (lower heating value). The lignite has a heating value of 6600 Btu/lb and contains 11% ash.[11] The roof is essentially a gas-impermeable layer of kaolin.[33,34] Extraction efficiency, defined as the fraction of the coal gasified, is claimed to be 65% and the cold gas thermal efficiency of the coal reacted is also 65%. The overall recovery for the process is the product of these two efficiencies, which in this case exceeds 40%. The gas is piped 10–12 miles (15–20 km) to a power station, which is also fired by coal from a nearby mine; the generating plant is about 100 MW in size. Yushno-Abinsk in Southern Siberia operates in steeply inclined bituminous coals. The coal is low ash and has a heating value of about 10,000 Btu/lb. The seams are 3–10 m thick, with a dip of about 10°.[1] Surrounding strata are composed of siltstones and sandstones. Access is by vertical, slanting, and deviated holes.[11] Gas quality varies between 90 and 120 Btu/scf.[11,12] There is no accurate information on current production rates.

In 1963 a total of 1.4 × 10^9 m^3/year of low-Btu gas was produced at four sites in the USSR, with 63% of that total coming from Angren.[33] The combined Podzemgas effort in 1957 involved 3000 workers, of whom 700 were identified as engineers and technicians. Production at Podmoskovia was discontinued in 1963 after depletion of the coal reserve. Activities at Lisichansk in the Donets basin were discontinued for several reasons: thin seams, the cost of deep drilling (approaching 350 m), and the complexities of operating in seams that dipped steeply because of subsidence.[33] Although the economics of the Soviet system of underground gasification are difficult to extrapolate to the terms of Western world, it was reported that the cost on an energy basis for *in situ* gasification was about 40% higher than for strip mining at the same general area (it is not certain what depths

[33] J. L. Dossey, Sandia Laboratories Rep. No. SAND 76-0380, 1976.

[34] K. Heide and K. Eilhauer, *Freiburger Forschungsh., Ser. A,* 45 (1963); UCRL Transl. No. 10756, 1973.

were involved in each case).[11] The major elements of the facilities at each gasification station included a transformer substation for electric power, air compressors, cooling towers, settling tanks for gas liquors, and gas scrubbers.

26.9.2 Europe and Japan

Most of the field testing on UCG in Europe was done prior to 1960, thus was reported in the review by Elder.[5] The major impediment to use of underground gasification in Europe is the threat of subsidence in populated areas. Tests operated in Great Britain by the National Coal Board have been described by Gibb.[22] The final test (No. P5) run by the British at Newman Spinney used the borehole producer method, where four horizontal holes (14-in. diameter) were drilled through the 2.5-ft seam over a distance of 400 ft. Each borehole was separated by 75 ft. A gallery was dug at one end to drill such holes; the British felt that this was a more economical approach for their coals than using vertical or deviated boreholes. A 118-day test resulted in 84% of the coal being gasified, but the heating value of the product gas was quite low, averaging 50 Btu/scf the last 2 months of the test. It should be recognized that the thin coal seam and the method of preparation of the seam were insurmountable obstacles to a successful test, in light of the Soviet data discussed earlier. Field tests have not been resumed in Great Britain since trial P5.

The Eastern European countries also performed some tests on UCG in the 1960s; their approaches were patterned after the Soviet method and were not necessarily geared to generating new results. The test results in Poland have been reported by Revva et al.,[35] and tests in Czechoslovakia have been described by Heide and Eilhauer[34] and Rauk[36].

In Japan some small-scale tests patterned after the British tests for research purposes were performed between 1960 and 1965.[37] A method called the partial blind borehole method was tested, but the product gas heating values were low and underground excavation of galleries was required. Because of these poor results, the tests were short-lived.

[35] M. K. Revva, D. K. Semenenko, and V. G. Smirnov, *Podzemn. Gazif. Uglei,* No. 12, 161 (1964); UCRL Transl. No. 10796, 1975.

[36] Jerzy Rauk, *Underground Gasification of Coal in Czechoslovakia,* in *Przegl. Gorni.,* **22**(5), 216–220 (1966).

[37] T. Yano, T. Kusunoki, M. Sato, and H. Nakazima, *Hokkaido Tanko Gijutsu,* **17,** 365 (1962).

26.9.3 The United States and Canada

During the past decade a revitalization in the interest in underground gasification has occurred in the United States, as indicated by the performance of a number of field tests. The lack of success of underground gasification in Alabama in the 1950s[5] has been a major obstacle to the initiation of field tests since 1965; but the Alabama site was a poor candidate for underground coal gasification, when considered in the context of the Soviet field data presented earlier.

In 1969 Gulf Research and Development conducted a field test near a strip mine in Kentucky.[38] The coal was a 9-ft seam at a depth of 107 ft, and it was fractured by use of air pressure about 30% above overburden pressure. The coal was gasified with a forward combustion without any prior thermal permeability enhancement, and gas with an average heating value of 270 Btu/ft^3 was withdrawn from sample wells. No attempt was made to sustain a reasonable level of gas production; the high heating value indicated a high percentage of coal carbonized versus gasified. It was estimated that the temperature in the combustion zone reached 2500–3000°F. After gasification cement was pumped down the boreholes so that the pattern of gasification could be observed when the coal was uncovered by the strip mine operation. Some samples recovered indicated that the reaction followed both small and large crack patterns in the coal. Other samples were composed of fused rock, clay, and coke.

In 1972 the Laramie Energy Research Center initiated an *in situ* coal gasification experiment in Hanna, Wyoming, to test the feasibility of underground gasification in a deep, thick seam of Western subbituminous coal. The goal of this test was to maximize gas heating value while sustaining stable gas production rates and achieving a high coal utili-

[38] P. Raimondi, P. L. Terwilliger, and L. A. Wilson, *J. Pet. Tech.,* 35 (January 1975).

zation efficiency. Three separate UCG tests on the Hanna No. 1 seam have been conducted from 1973 to 1977. This seam is 9 m thick and dips at an angle of approximately 7°. Depths of the three tests have ranged between 91 and 122 m.

The coal seam was ignited using a propane burner in March 1973, at a depth of 122 m. This test, called Hanna I, occurred over a one-year period; then air injection was stopped. Forward combustion without prior reverse combustion was attempted but was not successful. Thereafter reverse combustion has been used for linking along the bottom of the seam prior to forward combustion. From mid September 1973 through February 1974 an average gas production rate of 1.6×10^6 scf/day of 126 Btu/ft^3 gas was maintained with virtually no gas leakage. Figure 26.20 shows the variability of the heating value over this period. Most of the major fluctuations were due to mechanical problems.[39]

A second Hanna experiment (Hanna II, phase I), initiated in June 1975, was patterned after the Soviet design; over 38 days the heating value averaged 152 Btu/scf (higher heating value), and gas production averaged 2.7×10^6 scf/day. The gasification system was operated at relatively high flow and production rates compared to previous tests, which aided in the high yields.[40,41] The test was terminated before the combustion front reached the production well. One of the significant results from this test was the cold gas thermal efficiency (coal to cold low-Btu gas) of about 85%, which is superior to yields of most conventional gasifiers (e.g., the Lurgi gasifier).

A technique named "line-drive" was tested in 1975–1976 (Hanna II, phases II and III). The goal of this field design was to link two pairs of wells, then perform a gasification sweep over a wide area using the natural permeability of the coal. It was postulated that the gasification sweep in a direction normal to the maximum field permeability would spread out the front. The well pattern consisted of four wells located at the corners of a 60-ft square. After linking to form two parallel paths, the line-drive gasification was attempted. However the natural tendency of the burning process to channel rather than proceed over a broad front made this concept infeasible. Therefore the two sets of linked wells were gasified conventionally. In the first part of the test (phase II) a maximum production rate of 8.5×10^6 scf/day was reached, and the gas heating value ranged between 175 and 185 Btu/scf over the 30-day period. Figure 26.21 shows the variation in flow rates and heating values for this test;[42] the im-

[39] L. A. Schrider, J. W. Jennings, C. F. Brandenburg, and D. D. Fisher, Paper No. SPE 4993, presented to SPE–AIME, Dallas, Tex., 1974.

[40] D. D. Fischer and L. A. Schrider, paper presented at American Institute of Chemical Engineers meeting, Houston, Tex., March 1975.

[41] C. F. Brandenburg, R. P. Reed, R. M. Boyd, D. A. Northrop, and J. W. Jennings, Paper No. SPE 5654, presented to SPE–AIME, Dallas, Tex., 1975.

[42] L. A. Schrider, C. F. Brandenburg, D. D. Fischer, and B. A. Northrop, *Proc. 2nd Annu. Underground Coal Gasification Symp.*, MERC/SP-76/3, 1976, pp. 36–52.

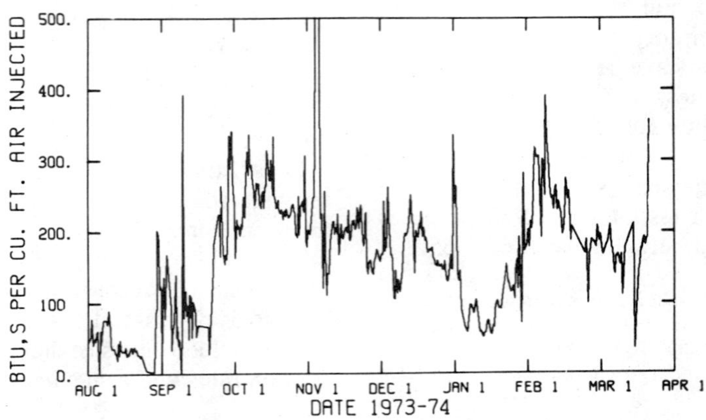

Figure 26.20 Heat value recovered in composite dry gas per standard cubic foot of air injected at the Hanna site (Hanna Test I).

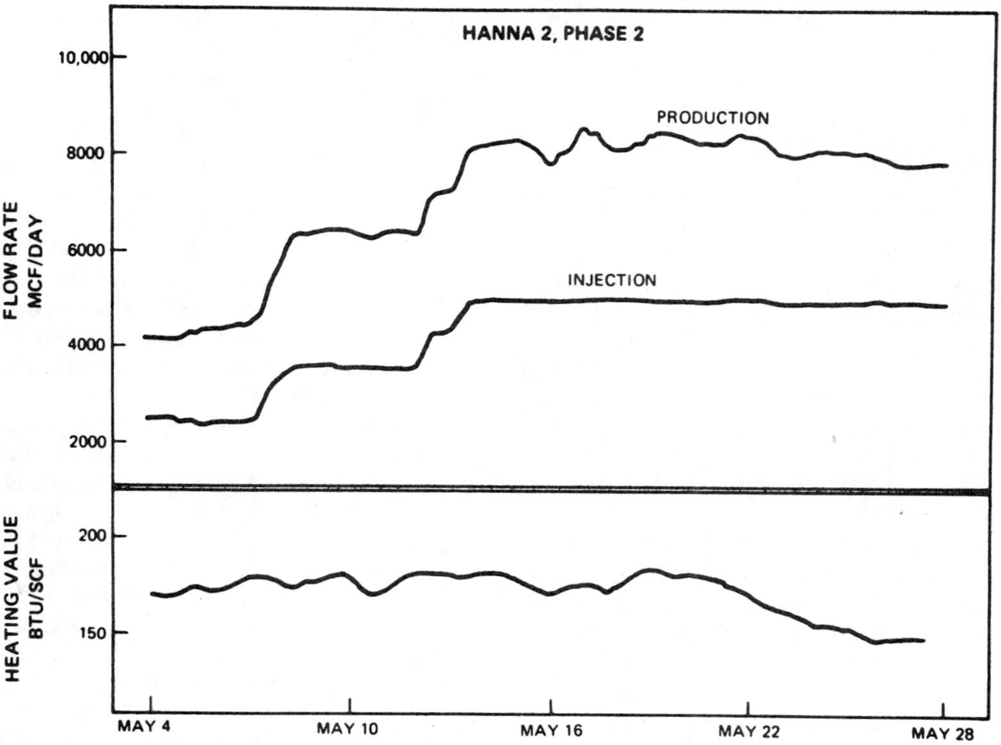

Figure 26.21 Flow rates and heating values for second phase of Hanna Test II.[42]

provement in control of heating value (cf. Fig. 26.20) is noteworthy. The second half of the test (phase III) attained a production level of 11.5×10^6 scf/day with an average heating value of 138 Btu/scf. The lower heating value was due to higher water intrusion rates (even at higher gas flow rates); it is believed that the previous test run nearby caused higher groundwater flows. Tables 26.2 and 26.3 summarize the energy balance data, which are very encouraging, for the Hanna I and II field experiments.[43] Three measures

[43] C. F. Brandenburg, personal communication, 1977.

Table 26.2 Energy Balance Calculations for Hanna II[43]

	Energy (10^9 J)		
	Phase I	Phase II	Phase III
Heating value of dry product Gas	13.3	30.5	43.7
Heating value of liquid by-products	1.0	1.9	3.7
Latent and sensible heat of water vapor	0.1	0.4	4.3
Sensible heat of dry product Gas	1.0	2.0	3.8
Heat losses	1.9	1.6	6.5
	17.3	36.4	62.0
Energy available in coal consumed	17.3	36.4	62.1
Energy used to run process			
Diesel fuel	2.3	7.1	9.4
Electricity	0.2	0.1	0.2
Propane	0.2	0.0	0.9

Table 26.3 Energy Balance Results for Hanna I and II[43]

		Hanna II		
	Hanna I	Phase I	Phase II	Phase III
Energy return ratio	3.1	5.3	4.5	4.5
Overall process efficiency (%)	44.0	71.5	74.3	65.3
Thermal efficiency (%)	52.0	82.7	89.0	76.3

of operating efficiency are used in those tables.

1. *Energy return ratio*. The ratio of total usable energy produced to energy consumed in operating the process; total usable energy is defined as the sum of the heat of combustion of the dry product gas and of the liquid by-products.

2. *Overall process efficiency.* The ratio of total usable energy produced to the sum of energy required to operate the process and energy available in the coal; this includes compressor operations for linking and gasification.

3. *Thermal efficiency.* The ratio of total usable energy produced to total available energy in the coal contacted. Because of the small scale of the tests, the efficiencies are lower than would be realizable for a larger test. A typical gas composition (water-free) is reported in Table 26.4.

One major benefit of the Hanna tests has been the extensive instrumentation designed and installed by Sandia Corporation. Various thermal, acoustic, seismic, and resistivity measurements were taken in the seam, and surface subsidence was monitored using tiltmeters and displacement gauges. Surface resistivity and acoustic measurements were tested for the purposes of locating the combustion front, but these have not demonstrated the same reliability as direct measurement of temperature via thermocouple wells. Figure 26.22 shows the location of thermocouples for one set of boreholes tested during Hanna I; Fig. 26.23 shows the temperature profiles over time. Note that the wells were linked at the bottom (well 1 was completed two-thirds from the top of the seam), yet the combustion front was not entirely predictable in its burning pattern. Thermocouple CCT7 shows a temperature increase before CCT9, which indicates that the linkage path does wander a bit (both horizontally and vertically) between the boreholes. Thermocouple data for Hanna II have indicated the development of multiple linkage paths.[44] The breakthrough of hot gas due to fissures or fractures in the coal seam is also rather unpredictable. The maximum temperature recorded during the 1975–1977 tests was about 2200°F. A detailed assessment of the instrumentation techniques used in Hanna II has been performed by Northrop et al.[44]

Table 26.4 Phase I Product Gas from Hanna II[41]

Gas	Proportion (vol %)
H_2	17.3
CO	14.7
CO_2	12.4
CH_4	3.3
N_2	51.0
Ar	0.6
H_2S	0.1
C_2–H_4	0.6

The initial gasification concept of Lawrence Livermore Laboratory consisted of placing an array of chemical explosives within a thick coal seam (>15 m).[45,46] Using explosive fracturing, a vertical, rubblized zone is formed, yielding in effect a packed-bed reactor. Steam and oxygen are pumped into the bed at high pressure to yield a product with a high percentage of methane. The intermediate-Btu gas then is collected through wells drilled to the bottom of the rubble zone and upgraded on the surface to substitute natural gas or other products.

The site selected for the initial testing of

[44] D. A. Northrop, S. G. Beard, L. C. Bartel, L. W. Beckham, and P. J. Hommert, Sandia Laboratories Rep. No. SAND 77-1072, 1977.

[45] G. H. Higgins, Lawrence Livermore Laboratory Rep. No. UCRL-51217, 1972.

[46] D. R. Stephens, Lawrence Livermore Laboratory Rep. No. UCRL-75494, 1974.

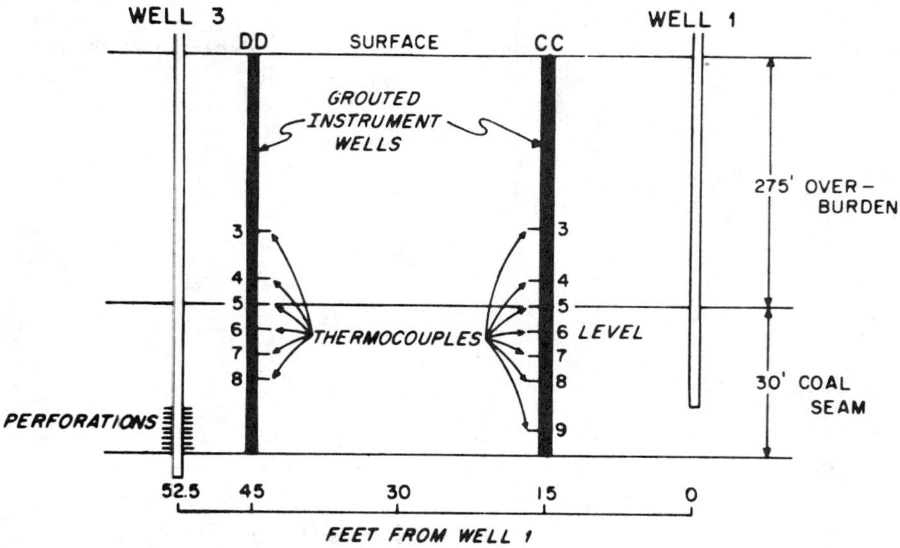

Figure 26.22 Cross section of wells 1, CC, DD, and 3, showing well completions and thermocouple locations.[41]

the explosive fracturing concept with air gasification was in the Powder River basin (Hoe Creek).[47] The test, a two-well fracturing experiment (Hoe Creek I), was conducted in a 7.6-m seam at a depth of 38 m. Two explosive charges 7 m apart were fired simultaneously at the bottom of the coal seam. Examination of cores taken after the blast showed extensive fracturing near the bottom of the seam. However hydrologic and airflow tests led to the conclusion that the explosive fracturing caused the permeability at the bottom of the seam to decrease, because of the large amounts of coal fines produced in the blast, which in turn restricted the flow. The net result was that the most permeable flow paths existed at the top of the seam.

Forward combustion gasification was achieved without any problem of formation plugging. Although the pressure was kept quite low, however, gas losses of 7% occurred. The gas composition, gas heating value, and oxygen utilization were all fairly constant during the course of the experiment, except at the end, as shown in Fig. 26.24. The total water influx into the gasification

[47] R. W. Hill, D. R. Stephens, and C. B. Thorsness, *Proc. 3rd Annu. Underground Coal Conversion Symp.*, Lawrence Livermore Laboratory Rep. Conf. No. 770652, 1977, pp. 29–52.

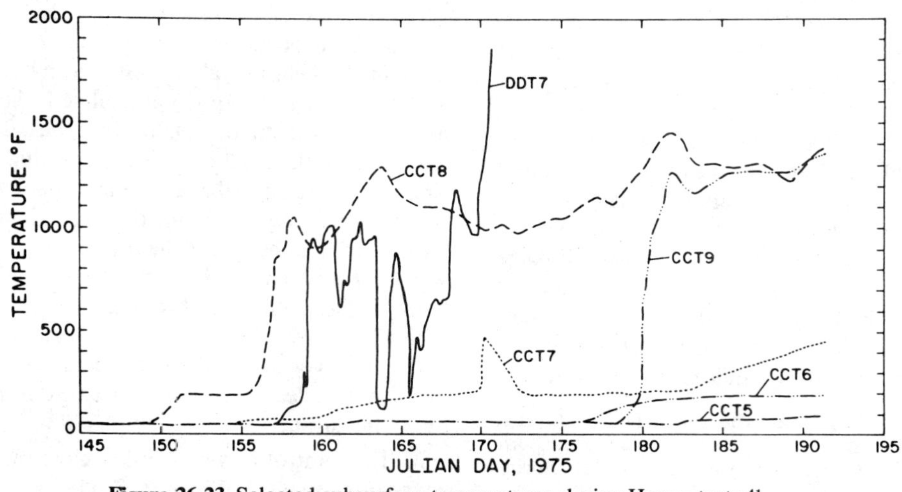

Figure 26.23 Selected subsurface temperatures during Hanna tests.[41]

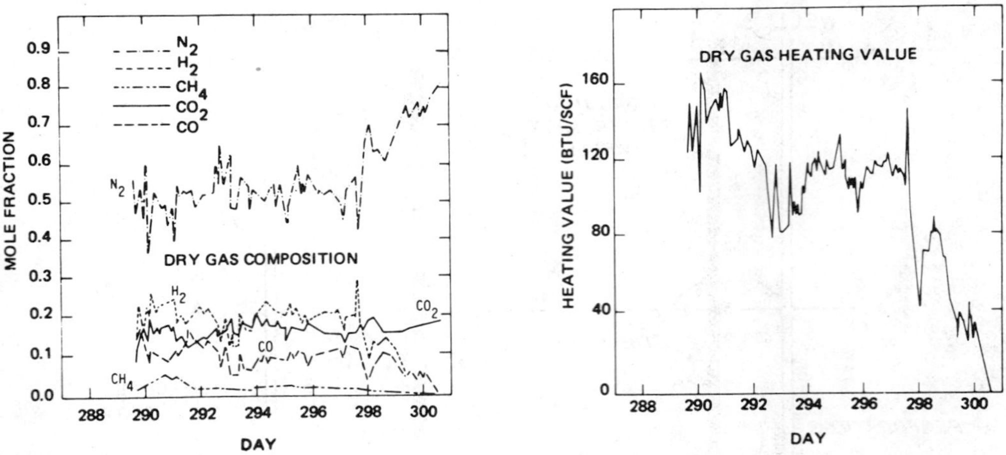

Figure 26.24 Dry gas compositions and heating values for Hoe Creek I test.[47]

region was about 65% of the pregasification level. About 30% of this amount entered the hot zone. Energy recovery in the form of recovered products amounted to 73% of the energy in the consumed coal. There were essentially no heat losses to the subsurface formation, the greatest energy loss being the production of steam. About 16% of the energy production arose from collected tars and oils. Gasification proceeded for 11 days, during which 130 tons was gasified out of an estimated 1000 tons in the fractured region. Analysis of thermocouple data confirmed the override of the gasification zone along the top of the bed, which explains the low resource extraction and the significant variability in product heating value. These observations underscore the importance of the linkage at the bottom of the seam.

A second field test, Hoe Creek II, was operated in late 1977 in another coal seam, although at essentially the same depth and thickness conditions of Hoe Creek I.[48] This test was heavily instrumented for temperatures, pressures, and gas sampling, the latter especially for determining the distribution of gasification-induced contaminants. Test objectives included (1) investigation of reverse combustion in a wet seam (path, linkage rates, specific air consumption, gas quality, water influx); (2) determination of forward air·gasification sweep width, efficiency, gas composition, water influx, gas losses, and pressure and temperature gradients; (3) operation of a three-day steam-oxygen burn; and (4) measurement of air and water quality parameters.

It took 14 days to produce the reverse burn link over the 18-m well spacing. This rate of linking is comparable to that obtained at Hanna. Thermocouple measurements indicated that multiple paths formed through the highly -developed cleat system in the coal. Following this phase, forward gasification using air was initiated and lasted 58 days. The product gas for forward gasification varied from 100 to 150 Btu/Scf. During a 2 day period in the middle of the test, oxygen and steam were injected in place of air, yielding a medium-Btu gas of 250–300 Btu/scf. The data analysis is incomplete but several conclusions are apparent:

1. Override of the coal seam by the combustion front occurred, causing wild fluctuations in the heating value, more severe than in Hoe Creek I. The gradual decline in heating value over time, similar to other field tests that exhibited override, was due to oxygen bypassing and channeling through the rubble zone. The reason for the override was a well casing failure near the top of the seam, which caused gas flow to that point rather than to the bottom of the seam. Financial limitations prevented the drilling of another well. The override of the coal seam resulted in excessive leakage, as much as 30% into overlying strata.

2. The control of water influx gave rather surprising results. Whereas Soviet experi-

[48] R. W. Hill, C. B. Thorsness, D. R. Stevens, D. S. Thompson, and W. R. Aiman, *Proc. 4th Annu. Coal Conversion Symp.*, Sandia Labortories Rep. No. SAND 78-0941, 1978, pp. 19–40.

ence has shown that total water influx has a large effect on heating value (see Fig. 26.7) and the amount of water influx can be controlled by system backpressure, gas heating values for the Hoe Creek test tended to be more or less independent of the total water influx. At the backpressures used, apparently only a fraction of the water actually entered the reaction zone. The excess water intruded downstream of the reaction zone and merely cooled down the gases (and did not promote the water-gas shift reaction). Material balance data indicated that the reacting water was roughly equal to water bound in the coal (about 25% by weight).

3. Oxygen injection was handled successfully and expected heating values were obtained, although channeling of the gas did produce oxygen breakthrough after several days. Somewhat surprisingly, the oxygen burn tended to reduce gas losses; it also attained higher temperatures and gave better resource utilization.

The Morgantown Energy Research Center is concentrating on *in situ* gasification of a West Virginia coal, which is a swelling coal as opposed to the Western shrinking coals. The seam of interest is relatively deep (250 m) and thin (1.8 m) compared to the coal seams tested in Wyoming. The approaches under study in this experiment include the linked vertical well concept as well as a design called the "longwall generator," in which two parallel horizontal holes are directionally drilled (up to 600 ft), then linked; this method is supposed to provide an areal gasification sweep over a wide area.[49] The economic tradeoff in this case is between directional holes and a larger number of vertical holes. Tests on directional drilling for linking were performed but found to be uneconomical; a later field test repeatedly used countercurrent combustion to form a link. As of 1978 other field tests have been operated by Texas Utilities Generating Company on Texas lignite,[50] by Texas A&M University on Texas lignite,[51] and in Alberta, Canada, by the Alberta Research Council.[52] Tests in the design stage include those operated by Gulf Research and Development and Atlantic Richfield. None of the experiments has approached the scope of the Hanna or Hoe Creek tests, but they will provide further data on the application of UCG to other North American coals.

[49] L. D. Strickland, *Proc. 3rd Annu. Underground Coal Conversion Symp.*, Lawrence Livermore Laboratory Rep. Conf. No. 770652, 1977, pp. 53–65.

[50] B. R. Clements, *Proc. 3rd Annu. Underground Coal Conversion Symp.*, Lawrence Livermore Laboratory Rep. Conf. No. 770652, 1977, pp. 81–82.

26.10 REACTOR MODELS FOR UNDERGROUND COAL GASIFICATION

Two basic approaches to modeling the chemistry of UCG systems have been explored: the channel model and the packed-bed model. Both models probably describe the process reasonably accurately in different locations and at different times. Yet the predicted gas compositions for the two models, given a common set of input parameters, differ significantly. These differences have been highlighted in a paper by Gregg and Edgar.[53] As mentioned earlier, the underground gasification system is purposely designed so that at some point in the gasification channel there is a zone that can best be described as a packed bed, with a large amount of surface area available for reactions. When such a zone exists, it will favorably dominate the outcome of the reactions. The minimum length of the packed bed required for acceptable product gas heating values is probably around 2–3 m, such as in a conventional gasifier. Hence the most productive approach to modeling the product composition from a successfully operating and properly designed UCG system is to consider the reaction zone as a packed bed. However the channel model may be more useful for analyzing sweep efficiency.

Packed-bed models for the *in situ* gasifier

[51] J. W. Jennings, R. F. Strickland, and W. D. Von Gonten, *Proc. 3rd Annu. Underground Coal Conversion Symp.*, Lawrence Livermore Laboratory Rep. Conf. No. 770652, 1977, pp. 83–91.

[52] A. A. Roehl, R. A. S. Brown, and E. J. Jensen, *Proc. 3rd Annu. Underground Coal Conversion Symp.*, Lawrence Livermore Laboratory Rep. Conf. No. 770652, 1977, pp. 92–103.

[53] D. W. Gregg and T. F. Edgar, *Am. Inst. Chem. Eng. J.*, 24, 753 (1978).

have been developed by Gunn and Whitman[54] and Thorsness and Rosza.[55] Both have been validated experimentally. The first model has been successfully compared with field data, and the second model has been validated using combustion tube data. Both models represent phenomena both in time and space, although only one spatial dimension is considered. Hence both models take into account a moving reaction zone, but in different ways. The Lawrence Livermore Laboratory (LLL) model[55] is much more detailed than the Laramie Energy Research Center (LERC) model,[54] the latter involving more simplifying assumptions. The more complete model is discussed first; then the simplifying assumptions for the LERC model are identified.

The main function of the LLL model is to predict the reaction front propagation rate and product gas composition for given injection gas flow rates and compositions. Gas and solid temperatures and composition profiles in time and distance are also determined. These results are obtained by solving 13 species mass balances, energy balances on gas and solid phases, and a Darcy flow momentum equation. The differential equations are coupled by source-sink terms involving heat and mass generation from 11 chemical reactions. Reactions include four heterogeneous gas-char reactions, the water-gas shift reaction, coal pyrolysis, coal drying, steam condensation–vaporization, and three gas-phase oxidation reactions. Reaction rates are controlled by chemical kinetics at low temperatures, and they are limited by mass transfer at higher temperatures. Most of the reactions are controlled by equilibrium considerations, except for pyrolysis and oxidation, which are considered to be irreversible. Solid phase heat conduction is included along with gas-solid heat transfer.

The eight gas species include N_2, O_2, H_2O, H_2, CH_4, CO, CO_2, and tar. Tar is a pseudocomponent used to close the stoichiometry for coal pyrolysis. The solid species are coal, char, H_2O(M), H_2O(F), and ash. H_2O(M) is mobile liquid water located in porosity available for flow. H_2O(F) is fixed liquid water present in small pores in the coal; it will not flow out of the coal under the influence of any reasonable pressure gradient—it can be removed only by heating the coal.

To reduce the difficulty in integrating this large set of differential equations, the gas phase equations are assumed to be quasi–steady state (i.e, the gas composition for a given solid temperature profile returns rapidly to the steady state profile). Hence the model can be written as a series of steady state problems evolving in time. Because of the moving reaction zone, however, the solid phase temperature must be integrated over space and time; therefore it remains a partial differential equation. The important equations are listed below.

The gas phase mass balance is

$$\frac{dC_i}{dx} = \frac{1}{v}\left[-C_i\frac{dv}{dx} + \sum_j a_{ij}R_j\right] \qquad (26.9)$$

where C_i is concentration of the ith component, v is velocity, x is axial distance, R_j corresponds to the reaction rate of the jth reaction, and a_{ij} represents the stoichiometric coefficients for the ith component in the jth reaction. Gas phase diffusion is neglected.

The equation for the gas phase velocity is obtained by summing Equation 26.9 over all gas species and replacing the total concentrations using the ideal gas law. This leads to:

$$\frac{dv}{dx} = \frac{-ydP}{P\,dx} + \frac{v\,dT}{T\,dx} + \frac{RT}{P}\left[\sum_i\sum_j a_{ij}R_j\right] \qquad (26.10)$$

where P is gas pressure, T is gas temperature, and R is the gas constant.

The gas phase energy balance leads to the following equation for temperature:

$$\frac{dT}{dx} = -\frac{1}{vc_g}\left[h_T(T - T_s) + H_g\right] \qquad (26.11)$$

where c_g is the heat capacity (gas phase), T_s is the solid temperature, h_T is the heat transfer coefficient (solid-gas), and H_g is the contribution to the gas energy balance due to reaction. Contributions due to pressure-volume work, turbulent heating, and mass diffusion are neglected. It is also assumed that conductive contributions can be adequately

[54] R. D. Gunn and D. L. Whitman, Laramie Energy Research Center Rep. No. LERC/RI-762, 1976.

[55] C. G. Thorsness and R. B. Rosza, *Soc. Pet. Eng. J.*, 105–116 (April 1978).

handled by incorporating them into the solid phase equations.

The solid phase energy balance is used in the following form:

$$c_s \frac{\partial T_s}{\partial t} = \frac{\partial}{\partial x}\left[(1-\phi)k_s \frac{\partial T_s}{\partial x}\right] + h_T(T - T_s) - H_s - v_s \rho_3 c_{s3} \frac{\partial T_s}{\partial x} \quad (26.12)$$

where c_s is the solid heat capacity, ϕ is the porosity, ρ_3 is density of water (component 3), c_{s_3} is the heat capacity of component 3, k_s is the solid thermal conductivity, and v_s is the mobile water velocity; H_s corresponds to heat contributions due to solid phase reactions, and t is time. Bed settling is not considered.

The solid-mass balances are written for all solid species except mobile water as follows:

$$\frac{\partial \rho_i}{\partial t} = \frac{M_i}{1000} \sum_{j=1}^{11} a_{s_{ij}} R_j \quad (26.13)$$

For mobile water, convection is included, but the balance is simplified by assuming that the velocity is constant, which allows liquid water to pass out of the system and not build to unacceptable levels at various points in the system. The balance thus becomes

$$\frac{\partial \rho_3}{\partial t} = -v_s \frac{\partial \rho_3}{\partial x} + 0.018 \sum_{J=1}^{11} a_{s_{ij}} R_j \quad (26.14)$$

in which $a_{s_{ij}}$ is the stoichiometric coefficient of the ith solid component in the jth reaction.

In general, some type of momentum balance should be written for each flowing phase. However the liquid phase balance is circumvented by assuming that all mobile water in the system is moving at a constant velocity, set arbitrarily at one-half the exhaust gas velocity. The momentum balance for the gas phase is replaced by Darcy's law:

$$\frac{dP}{dx} = -\frac{v\mu}{k} \quad (26.15)$$

in which μ is viscosity of the gas and k is the permeability.

The determination of the parameters for chemical reaction rates, heat transfer coefficients, effective thermal conductivity, and gas permeability are nontrivial problems. In this model transport coefficients are estimated using packed-bed correlations, although these correlations are not entirely applicable. Reaction kinetics for pyrolysis reactions are fitted from pyrolysis data using first-order Arrhenius-type expressions.

The LERC model[54] introduces the following simplifications:

1. Constant porosity of the packed bed and constant velocity of the fluid with respect to axial positions are assumed (the velocity of the fluid can be varied at the inlet, though).

2. A constant rate of propagation of the reaction zone is assumed; this allows one to convert the partial differential equations to ordinary differential equations through a variable transformation, which simplifies the integration procedure. This assumption has been validated in combustion tube experimental results.[55]

3. Pyrolysis of the coal is treated as a steady state phenomenon that occurs prior to gasification. No kinetic expressions are used; rather, material balance data provide the pyrolysis gas composition. Hence the pyrolysis products up to 1000°C are additive, with gasification of the remaining char.

4. A homogeneous rather than a heterogeneous model is used, so solid and gas phase temperatures are not explicitly monitored. No gas phase oxidation reactions are included (these do not become very important for the packed-bed system). Reactions are kinetically controlled, rather than being controlled by mass transfer as in the LLL model.

5. All oxygen initially reacts to CO_2, which in turn provides the heat necessary to drive the carbon-steam reaction. Therefore the CO_2 produced in the oxidation zone is "diluted" by the production of CO and H_2 downstream in the reduction zone. Water-gas shift reaction kinetics may be included.

As a result of these assumptions, the LERC model requires integration of 11 nonlinear differential equations, rather than the 17 equations used in the LLL model, but at apparently no great reduction in accuracy. It is felt that these models could be readily applied to conventional fixed-bed gasifiers and would prove to be useful tools for these systems, augmenting existing thermodynamic models.[6] However for rapid order-of-magnitude calculations of product gas heating values, the solution of steady state material and energy balances, such as described by several authors,[6,51] is a satisfactory approach.

Several models have been developed for an open channel undergoing gasification around its walls.[56–58] It is believed that such models may be the key to predicting sweep efficiency in UCG. Qualitatively these models give results similar to the much simpler model for a carbon channel discussed by Thring and Essenhigh.[59] The open-channel models are much more sensitive to velocity, if the oxidation reaction is assumed to be diffusion controlled. They all are based on ordinary differential equations (in space), and steady state conditions hold. The effects for water influx, conductive heat transfer to the overburden and surrounding coal, and both radiative and convective heat losses from the solid to gas phases are included in the models of Dinsmoor et al.[56] and Stewart and Wall.[58] For these two models essentially the same reactions as in the LLL model are used, except in Dinsmoor et al. the effect of pyrolysis is not considered (char or coke is assumed as the reacting medium). For the Magnani–Farouq Ali model,[57] only carbon, oxygen, and carbon oxides are assumed to be present.

For turbulent flow, as the gas velocity in the open channel increases, the overall rate of reaction can decrease because residence time in the tube is decreased; thus the yield of products is reduced and the oxygen content of the exiting gas is increased. On the other hand, near the laminar-turbulent transition point, the rate of reaction can increase markedly with increases in the velocity. Thus for an open channel there will be an optimum velocity (blast rate) that maximizes the rate of reaction because of the combined effects of residence time and mass transfer rates. Such an optimum blast velocity was observed in the models above as well as for the Newman Spinney trial,[22] where horizontal boreholes were drilled for linking purposes.

In the open channel, because of the limited amount of surface area, the occurrence of gas phase oxidation of CO to CO_2 becomes much more significant relative to the heterogeneous reactions. This phenomenon can lower the heating value of the gas produced to less than 100 Btu/scf. To better define the dominant mode of carbon monoxide generation in a channel, Szekely and Maroudas[60] have analyzed the three-zone theory of Thring.[61] Using a classical transport phenomena approach to treat the diffusion of CO from the reacting surface into the gas phase, they performed a simulation for the expected range of diffusion and reaction rate coefficients. The simulations led Szekely and Maroudas to conclude that in the presence of gas phase oxidation, the production of CO in an open channel could proceed only by reduction of CO_2 to CO on the hot carbon surfaces, rather than by primary formation of CO at the reaction surface and "escape" of CO from gas phase oxidation. Gasification channel simulations by Dinsmoor et al.[56] have also demonstrated the unimportance of primary CO production on product gas yields, since all CO formed in the oxidation zone is rapidly oxidized to CO_2 in the gas phase.

Other evidence for gas phase oxidation of gaseous species (CO, H_2) in an open channel is due to Derman et al.[62] In this laboratory study the gas phase temperature rose above the solid phase temperature at some intermediate point axially. This profile is in disagreement with those developed by Magnani and Farouq Ali,[57] which show the gas phase temperature asymptotically approaching but not exceeding the solid temperature because gasphase reactions were omitted from their model.

The growth of a gasification channel can be simulated through the pseudo-steady state assumption. The simulations of Dinsmoor et al.[56] have shown that as the channel

[56] B. Dinsmoor, J. M. Galland, and T. F. Edgar, *J. Pet. Tech.*, 695–704 (May 1978).

[57] C. F. Magnani and S. M. Farouq Ali, *Soc. Pet. Eng. J.*, 425 (October 1975).

[58] I. M. Stewart and T. F. Wall, *16th Int. Symp. Combustion*, Boston, 1976.

[59] M. W. Thring and R. H. Essenhigh, in *Chemistry of Coal Utilization*, Supplementary Volume, H. H. Lowry, Ed., John Wiley and Sons, New York, 1963, pp. 754–772.

[60] J. G. Szekely and N. G. Maroudas, *Trans. Inst. Chem. Eng.*, **44**(1), 3 (1966).

[61] M. W. Thring, *Science of Flames and Furnaces*, Chapman and Hall, London, 1952.

[62] B. M. Derman, N. V. Lavrov, V. A. Nikoleva, and I. L. Farberov, *Underground Processing of Fuels*, U.S. National Science Foundation and Dept. of Interior, TT-63-11063, 1963, pp. 35–39.

grows, the oxidation zone moves toward the end of the tube, with a subsequent decrease in heating value of the gas and any increase in gas temperature due to gas phase oxidation. The open channel ceases to grow at the point where the negative effects of heat losses to the surrounding formation and water influx combine to lower the coal face temperature, which in turn lowers the reaction rate. Eventually, loss of ignition occurs. The ultimate open channel width is related to the Reynolds number in the channel, which affects the rate of reaction. More realistic models incorporating the occurrence of roof collapse may give simulation results that agree to some extent with field data for channel growth.

Warner and Szekely[63] have considered a simple heat transfer model for enlargement of the channel. As the cavity enlarges, more surface area is exposed; the maximum width of the combustion channel is reached when the heat generated by the reaction zone equals the heat losses. They assumed to be applicable, the solution of unsteady state conduction in a semi-infinite slab with heat losses to the overburden only (area proportional to two times the width) which is

$$q = \frac{k_s(T_w - T_\infty)}{\sqrt{\pi \alpha t'}} \tag{26.16}$$

where k is the thermal conductivity, q the heat flux, T_w the burning face temperature, T_∞ the ambient temperature, α the thermal diffusivity, and t' the elapsed time. The rate of heat generation is given by

$$Q = V_c H \rho_c h_c \tag{26.17}$$

where H is the seam thickness, ρ_c the coal density, V_c the linear burning velocity of the coal, and h_c the coal heat of combustion. The burning velocity, which is diffusion controlled, is not constant in this model but is assumed to decrease with increasing cavity size (this step utilizes an empirical constant, K'). The final equation for the maximum half-width, Y, is

$$Y = \frac{\rho_c h_c H}{2k_s(T_w - T_\infty)\, 1.5708}\sqrt{\frac{\pi \alpha V_c}{2K'}} - \frac{0.5708}{1.5708\, K'} \tag{26.18}$$

[63] F. E. Warner and J. Szekely, *Am. Inst. Chem. Eng. I. Ch. E. Symp. Ser.*, No. 2, A 68, 1962.

Note that this result is different from that given by Warner and Szekely[63] because of an integration error in the model of the latter. Therefore this model predicts that the maximum half-width is proportional to the seam thickness (H), coal calorific value $[h_c]$, and the square root of the initial burn velocity $[V_c]$. It is inversely proportional to the heat loss, as given by $T_w - T_\infty$. Simple models such as the one described above offer some insight into the factors affecting cavity size, and more efforts should be made along these lines.

26.11 ENVIRONMENTAL ASPECTS OF UNDERGROUND COAL GASIFICATION

In situ gasification overcomes many of the objectionable environmental features of strip and deep mining. As with any gasification process, the sulfur in the coal is principally converted to hydrogen sulfide, which is readily removed from the product gas stream. Pollution by solid waste is reduced, since inert ash material is left underground, and massive excavation of the overburden is avoided. Surface water pollution problems common to conventional mining are also not encountered. However UCG does present potential water pollution problems in the subsurface environment, because gasification will often be carried out in a freshwater environment.

Indicative of the compounds produced during underground gasification that might be considered to be pollutants is the list of byproducts recovered from the low-Btu gas during *in situ* gasification, shown in Table 26.5. The coal tars consist of phenols, ani-

Table 26.5 By-Products Obtained at Yushno-Abinsk, USSR (Podzemgas), 1963[11]

Product Gas	Yield (g/m^3)
Gas condensate	140–400
Ammonia	2–2.5
Benzene hydrocarbons	1–2
Tar	0.3–5.7
Pyridine bases	0.44–0.67
Hydrogen sulfide	0.3
Naphthalene	0.04–0.09
Acetylene	0.003–0.0100
Hydrocyanic acid	0.007

lines, and quinolines. The phenols are water soluble, but the other organic compounds have rather low water solubility. Therefore it is expected that these compounds will be transmitted into the groundwater regime.

Extensive groundwater sampling and analysis was performed during the Hoe Creek I test described earlier.[64] Twelve wells in the vicinity of the UCG experiment were sampled for water quality before, during, and up to 6 months following gasification. The water samples were analyzed for the presence of 70 inorganic elements and compounds, phenolic materials, dissolved organic compounds, and volatile organics. Table 26.6 shows a simplified summary of data for contaminants found after gasification was completed, both inside and outside the burn zone. Permissible concentrations for public water supplies for selected components are also shown.

The total amount of tars and oils produced from UCG is lower than that predicted by pyrolysis data. This indicates that the high temperatures experienced in the channel during UCG cause many of the heavy organics to form gaseous compounds by cracking or reacting with the steam. The steam flush phenomenon, in which organics in the carbonized zone of the coal face are swept into the channel by evaporation of water in the coal matrix, could be responsible for these results; the actual cause is unknown.

By keeping the channel hot at all points, which occurs naturally, most of the organics should leave in the vapor phase. However carbonized coal and other organics that are not removed in the gas stream remain underground and are subject to leaching by the groundwater, which inevitably intrudes into the burned-out area and eventually fills all the void space. Leaching experiments performed on gasification residues[65] indicate that the raw coal can be a greater source of organic material than the species left after gasification. This is consistent with measurements of water quality in a coal seam prior to gasification, which show a high occurrence of pollutants such as phenol. However the solubilities of coal samples that contain trapped coal tars were not tested.

For the purposes of modeling pollutant dispersion underground, Campbell and Washington have postulated the simplified picture of groundwater pollution shown in Fig. 26.25.[66] A void is left after gasification of the coal seam, and ash remains in the void. A ring of ungasified char surrounds the ash. Outside the char ring is a zone that contains some heavy organics produced during gasification and left underground. After gasification, water slowly enters and fills the void: eventually the natural flow in the aqui-

[64] S. W. Mead, J. H. Campbell, and D. R. Stephens, *Proc. 3rd Annu. Underground Coal Conversion Symp.*, Lawrence Livermore Laboratory Rep. Conf. No. 770652, 1977, pp. 476–489.

[65] M. J. Humenick and C. F. Mattox, *Proc. 3rd Annu. Underground Coal Conversion Symp.*, Lawrence Livermore Laboratory Rep. Conf. No. 770652, 1977, pp. 50–517.

[66] J. H. Campbell and H. Washington, *Proc. 2nd Annu. Underground Coal Gasification Symp.*, MERC/SP-76/3, 1976, pp. 119–132.

Table 26.6 A Simplified Summary of Data for Contaminants That Showed a Large Increase Following Gasification[64]

		Inside Burn Zone		Outside Burn Zone		
Species	Pregasification Value (mg/l)	Concentration (mg/l)	Increase	Concentration (mg/l)	Increase	Permissible Concentration Limit (mg/l)
Phenols	0.001	0.1	100×	500	5×10^5×	0.001
CN^-	0.01	0.4	40×	300	3×10^4×	0.20
NH^+_4	0.5	20	40×	70	100×	0.5
DOC[a]	6	4	—	200	40×	
Br^-	0.1	1.0	—	4	40×	
Pb^{2+}	0.001	0.001	—	0.04	40×	0.05
Ba^{2+}	0.1	—	—	1.0	10×	1.0
K^+	5	60	10×	45	8×	
Li^+	0.03	0.3	10×	0.2	6×	
Mg^{2+}	10	50	5×	60	6×	
SO_4^{2-}	200	2000	10×	1000	5×	250.0
Ca^{2+}	40	600	20×	200	5×	
B^{3+}	0.1	0.7	7×	0.5	5×	1.0

[a] Dissolved organic content.

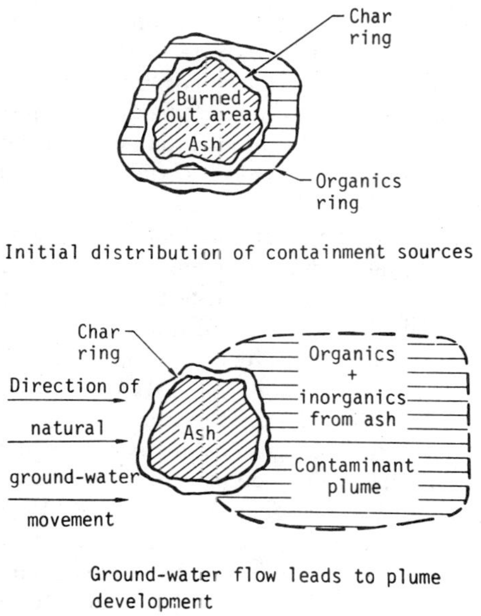

Figure 26.25 (*Top*) Plan view of the postgasification void. (*Bottom*) The plume that develops as a result of the reentry of groundwater.[66]

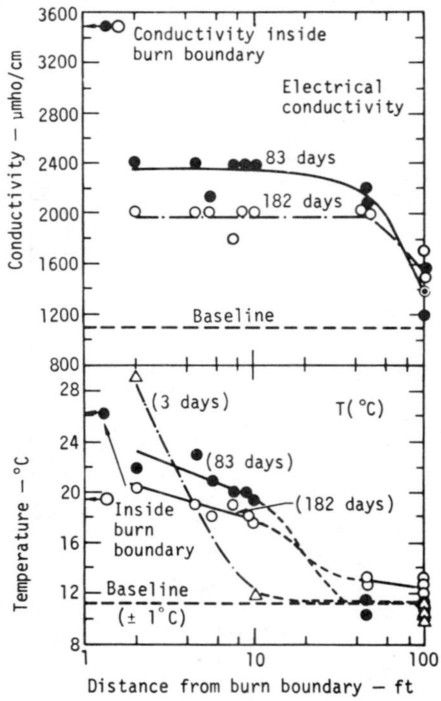

Figure 26.26 Electrical conductivity and temperature as a function of distance from the gasification burn boundary for water samples taken 3, 83, and 182 days following gasification. The arrows near the ordinate indicate the values measured inside the burn boundary. The baseline was determined from pregasification water sampling.[64]

fer is restored, and the pollutants contained near the gasification zone are leached out and carried downstream by the flowing groundwater, thus establishing a pollutant plume. This plume consists of both organic and inorganic species, the latter attributable to coal ash and overburden.

Figure 26.26 shows electrical conductivity and temperature as a function of distance from the gasification burn boundary for Hoe Creek I. The increase in electrical conductivity is a good measure of the ionic species as a result of gasification. The results show, approximately, a twofold increase in conductivity within 10 ft of the burn boundary, with a rapid decrease to nearly baseline values at a distance of 100 ft. Also, there is a slight drop in conductivity between the 83- and 182-day sampling periods; this difference may be an indication of contaminant sorption by the coal. Inside the burn boundary the soluble coal ash species produce a very highly conductive solution (3500 μmho/cm). More details on dispersion of specific ionic species with respect to time and position have also been reported by Mead et al.[64]

The measured postgasification distribution of phenols is reported in Fig. 26.27. The sharp decrease of concentration with respect to distance should be noted, as well as the steady drop in concentration with time. The drop in concentration is presumably due to sorption by the coal. Although dispersion modeling of this phenomenon is still rather primitive because of the lack of field and laboratory data, there does appear to be reasonable agreement between field results and laboratory tests, as discussed by Dalton and Campbell.[67]

Environmental analysis from the Soviet field sites is not nearly as quantitative on details as are the Hoe Creek data reported above. Soviet scientists have reported that the increase in salt loading in the vicinity of the burn zone is roughly tenfold (e.g., from 100 mg/l to about 1000 mg/l). The anions found in conjunction with the cations above included mainly HCO_3^- and SO_4^{2-}, with some HSO_3^- and HS^-.[68] Depending on the particular

[67] V. A. Dalton and J. H. Campbell, *In Situ,* **2,** 295 (1978).

[68] K. F. Bogoroditskii, *Akad. Nauk SSSR,* **23,** 174 (1960); UCRL Transl. No. 10858, 1975.

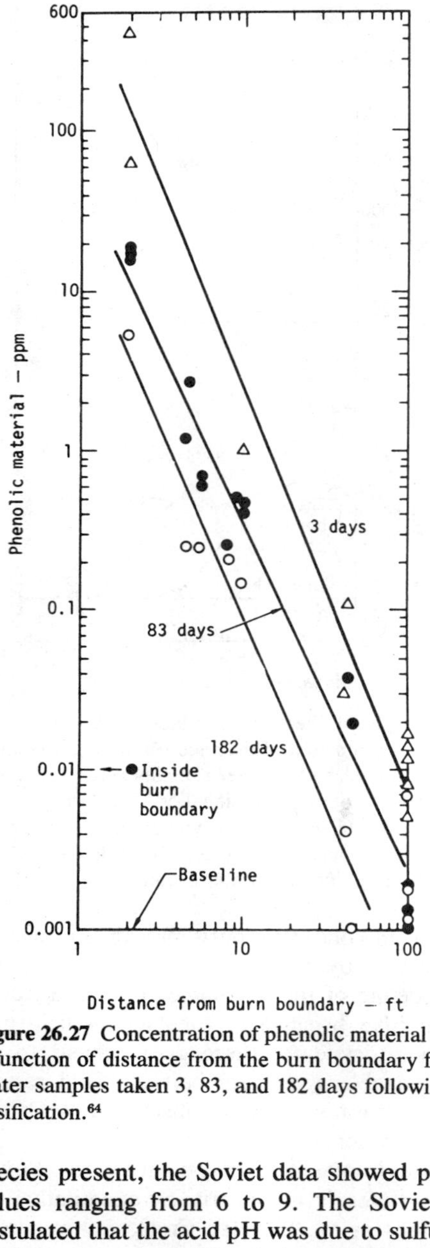

Figure 26.27 Concentration of phenolic material as a function of distance from the burn boundary for water samples taken 3, 83, and 182 days following gasification.[64]

species present, the Soviet data showed pH values ranging from 6 to 9. The Soviets postulated that the acid pH was due to sulfur oxides rather than CO_2; it is possible that such gases could escape from the oxidation zone, but all gaseous sulfur compounds should be reduced downstream from that point. The pH also affects the concentrations of various ionic species because of precipitation.[69] The increase in salt loading may be attributed to both a rise in the salt concentration as water evaporates and temperature rises, and the increase in solubility due to the presence of carbon dioxide. The inorganic species arose from limestones, clays, shales, and sandstones. Kononov[69] has commented that the composition of the groundwater after gasification is not unlike that seen in hot springs in the Soviet Union.

The Soviets also reported that vapors from the gasification process were noted at some distance (100 m away) from the gasification site. This is due to a hydraulic connection underground. Excursions in pH also can occur during gasification.

Water collected from the gasification area immediately after gasification is completed also contains high concentrations of dissolved gases, including H_2, CH_4, CO_2, H_2S, and NH_3. Because of continuing water influx, however, the gas content and mineral content of the water approaches its natural level.[70] Without the presence of appreciable water influx, the formation remains hot, at around 90°C up to one year after gasification. The disturbed region is largely a function of the total area gasified. Rauk[36] reported that the presence of organic species in groundwater after gasification is short-lived, but the mineral compounds and H_2S persisted for a year; the waters reverted to their normal state after 2 years.

The potential of the subsurface environment for natural renovation of polluted groundwater is not surprising, since tests using activated carbon as well as coal have shown a high sorption capacity for organic pollutants such as phenol. Adsorption isotherms for phenols and some inorganic species have been obtained in laboratory tests by Campbell and Washington[66] and Humenick and Mattox.[65] It is also felt that the overburden material may participate in natural renovation, in that gaseous species such as NH_3 and H_2S may be adsorbed by clays.[65]

The monitoring of groundwater was very important at the Soviet gasification stations. Detailed studies of geological structure and hydrogeological conditions were undertaken at each site by professional geohydrologists. Data on water yields and permeability for the coal seam and overburden were usually obtained through pumping tests. The main con-

[69] V. I. Kononov, *Akad. Nauk SSSR; Geol. Inst.*, 35 (1964); UCRL Transl. No. 10812, 1974.

[70] P. P. Klimentov, *Izv. Vyssh. Ucheb. Zaved.—Geol. Razved.*, No. 4, 106 (1963); UCRL Transl. No. 10869, 1975.

cern of the Soviets regarding groundwater was the reduction in thermal efficiency due to water influx, rather than pollution considerations.

Geological research on *in situ* gasification in the United States has focused on characterization of the site in more regional terms,[71] that is, relating hydrologic parameters (transmissivity, hydraulic conductivity and gradient, sand grain size and sorting, and porosity) and water chemistry to the already mapped lithofacies and to outline hydrologically acceptable areas for gasifier sites. This approach identifies regional areas where water productivity (transmissivity) and mobility (hydraulic conductivity) are relatively low. A hydrologically ideal site is characterized by coal seams roofed and floored by thick clay or other impermeable material, a low sand environment, saline waters, and a strike-oriented depositional system. Thus potential sites, on evaluation of exploratory drilling and hydrologic testing, can be compared with the ideal site.

A coal seam roofed and floored by mud is favored for hydrologic and engineering reasons. The impermeable mud seal will minimize groundwater inflow and contain potential pollutants as well as minimize gas leakage. Low values of transmissivity and hydraulic conductivity will correlate with low sand environments. For thick, coarse-grained, areally extensive channel sands, transmissivities typically range from 30 to 2000 $m^3/(day)(m^2)$; conductivities from 1 to 20 $m^3/(day)(m^2)$. Average groundwater velocities, calculated using Darcy's law and divided by porosity, range from 0.3 to 49 m/year. Obviously values at the low end of the ranges cited are desired for aquifers subject to the greatest impact from gasification, that is, those immediately overlying lignite. At the maximum velocity the most mobile toxic species would take 200 years to travel 10 km.

For sands that are thinner, finer grained, more poorly sorted, and less areally extensive, permeabilities are considerably lower, because permeability in unconsolidated sediments is largely a function of grain size and sorting. These conditions create a "local" aquifer, which gives low groundwater productivity and mobility. Lower mobility means that pollutants would be localized, which appears to be desirable. Some formations also have strike-oriented elements, which further restrict groundwater movement. In such strike-oriented sands, which are normal to regional flow movement, water movement is greatly impeded. In dip-oriented systems, water movement is more rapid.

Because any pollution underground will be extremely localized, it is possible that one approach to pollution control will be pumping out of the cavity containing water. Wastewater treatment facilities developed for industrial plants such as coking operations should be applicable here.

The first detailed environmental permit for *in situ* gasification was obtained by Texas Utilities Generating Company from the Texas Water Quality Board. The permit application required the company to supply technical data on geographic setting, geology, hydrology, water chemistry, monitoring procedures, well construction, gasification methodology, surface facilities, and water quality restoration. Geologic and hydrologic data were to be presented in the form of geophysical logs, maps, and cross sections. Geologic data included details on the local stratigraphy, structure, lithology, lignite geometry, water levels, and direction of water movement in every freshwater aquifer affected by the proposed gasifier. Besides major cations and anions, water chemistry included phenols, nitrates, total organic carbon (TOC), temperature, pH, heavy metals, and accuracy assurance techniques. Monitoring of phenols, NH_3, SO_4, H_2S, temperature, and pH was required. Monitoring wells were to be completed in the lignite and adjacent aquifers, 200–400 ft from the burn, and 300 ft apart in the direction of maximum transmissivity and hydraulic conductivity. Casing data, cement program, perforated and/or screened intervals, injection pressures, and logging programs were to be supplied. It should be mentioned that alterations will probably be made in this permit for future tests on UCG. The U.S. Environmental Protection Agency has also promulgated guidelines for any borehole operation, and these will apply to *in situ* gasification when they have been finalized.

[71] T. F. Edgar and W. R. Kaiser, *Geology of Alternate Energy Resources in the South Central United States,* Geological Society of America, Houston, Tex., 1977, pp. 179–192.

26.12 FUTURE RESEARCH ON *IN SITU* CONVERSION OF COAL TO SYNTHETIC FUELS

Current research and development efforts on UCG have benefited tremendously from the abundance of field data from the Soviet research and commercial operations. It should be mentioned, however, that the Soviet efforts in fundamental research on this subject were quite limited. Because of the many process variables involved in UCG, one could be led to the conclusion that operation of and results from this process are site specific and coal specific; however there is much to be gained from smaller scale experiments in which many of the variables can be controlled. Knowledge of how to scale up small-scale experiments to field systems is crucial to the success of such laboratory research.

Before UCG can be economically applied in the Western world, better information on selection and optimization of well spacings must be obtained, since this variable is of paramount importance economically. Well spacing and sweep efficiency are functions of the specific coal and geology at a given site, but improved correlations need to be developed. A better understanding and description of the evolution of the cavity seems to be important here. The successful implementation of an underground system using pure oxygen rather than air is also needed, since much of the interest in coal is for producing a synthesis gas; the high transportation cost of low-Btu gas from air-blown systems is also an incentive to production of intermediate-Btu gas. This effort will require extensive laboratory and field testing before design principles can be established. Very little testing of oxygen-blown systems was done in the USSR.

The research and development work in UCG should also interface with coal research for conventional gasification, since as pointed out earlier, the interpretation of results from *in situ* and conventional gasifiers are related. Hence basic research on gasification and pyrolysis behavior of a range of coals, the effects of temperature on physical and chemical properties of coal, and the interaction of ash chemistry with gasification are some of the important research areas that should be investigated.

The solution of rock mechanical and fluid flow research problems are also needed for the successful application of *in situ* gasification. Research aimed at defining predominant deformation properties, assessing the importance of heterogeneity and anisotropy, determing basic rock strength data for low and high temperature, and better understanding the relationahip between laboratory and field permeabilities is crucial for site analysis prior to gasification. The development of models predicting multiphase fluid flow in the subsurface, roof collapse, and long-term extent of subsidence would be extremely beneficial.

A knowledge of the environmental aspects of UCG is needed to establish the environment feasibility. Much is still unknown about the generation and dispersion of pollutants and their later interactions with the subsurface, such as in natural renovation of polluted groundwater. The procedures for monitoring groundwater pollution are also not well established; for example, should monitoring wells be placed near the burn zone or far away, should monitoring commence during or after gasification, and what are the water quality parameters of significance? Methods and costs of remedial action, if natural cleanup is not feasible, also need to evaluated.

The problem of gasifying Eastern U.S. swelling coals remains to be solved, since initial thermal permeability enhancement, which is applicable to shrinking coals, appears to be impractical. Major breakthroughs in linking technology (e.g., directional drilling) may be the only economical solution. The proximity of Eastern U.S. coals to population centers, which would allow transport of low-Btu gas, and the high sulfur content of these coals, make *in situ* gasification particularly attractive for this region.

Other field problems that should be addressed include gasification of seams that are excessively thick or thin, multiple-seam gasification, combined coal gasification and oil recovery (where applicable), cyclic operation (alternation of steam and air), combined coal and oil shale gasification, and recovery of coal from abandoned coal mines.

The *in situ* extraction of coal in a form other than gas is also of interest. Included under this category are partial mining via carbonization, *in situ* comminution (where solid coal is recovered in a slurry), and *in situ* liquefaction (which may also utilize a slurry approach). The approaches proposed

Table 26.7 Advantages and Disadvantages of *In Situ* Liquefaction Processes[73]

Process	Advantages	Disadvantages
Borehole	Insensitive to plugging by heavy liquid or coal softening; therefore a recycle donor solvent can be used that has a high extraction efficiency at mild conditions	Potential increase of organic liquid inventory underground, with time
	Heat for the reaction is supplied by a donor solvent and synthesis gas	Possible losses of costly donor solvent
		Possible short residence times
		Potentially incomplete resource recovery
	A maximum inlet temperature of 450°C limits materials problems	Possible high drilling costs
		Possible high pumping costs
High-porosity rubble beds with donor solvent	Long residence times, more complete reactions	Potential increase of organic liquid inventory underground, with time
	High resource recovery	Possible donor solvent losses
	Heat for the reaction is supplied by donor solvent and synthesis gas	Potential of plugging bed
		Possible high pumping costs
	A maximum inlet temperature of 450°C limits materials problems	
Aqueous phase liquefaction using high-porosity rubble beds	Insensitive to bed-plugging problems	Must preheat the coal bed; heat balance is critical; possible incomplete reaction of CO + steam at the low (150–1000 psi) pressures that are feasible for *in situ* processes
	Has long residence times	
	Avoids use of donor solvent	
	Avoids inventory of organic liquids underground	Possible high pumping costs in recycling gases
		Possible high materials problems with injecting hot synthesis gas (600^{+}°C
Partial oxidation	Simple, easy to get heat in	The liquids are only a by-product (20% or less) of the gasification process
	Heat losses are not critical	Cocurrent gasification, required for liquids production, may plug the bed

for *in situ* liquefaction follow lines similar to those tested previously for UCG and include (1) blind or other borehole with a donor solvent[72] (similar to the Frasch process for sulfur recovery), (2) explosive fracturing of the bed and liquefaction with a donor solvent, (3) use of water or steam (aqueous phase liquefaction), which produces gas and oil, and (4) partial oxidation, where gasification is used to pyrolyze the coal.

One of the types of donor solvent most actively considered is tetralin, which shows excellent dissolution characteristics. Ammonia and methanol are other possibilities. The solvent is believed to migrate through the naturally occurring cleat system in coal, disrupting the bonding forces present.

All the liquefaction approaches discussed above suffer from the need to operate at high pressures; solvent leakage and subsequent loss constitute a serious economic drawback. Possibly creation of artificial barriers underground could serve to avoid this problem. Uniform contacting of the coal is also a difficult problem that could be solved by fracturing. Potential swelling of the coal creates flow difficulties. Stephens[73] has summarized the advantages and disadvantages of each approach, and these are listed in Table 26.7.

[72] D. R. Skidmore and C. J. Konya, *Prepr., Am. Chem. Soc., Fuel Div.*, April 1973.

[73] D. R. Stephens, Lawrence Livermore Laboratory Rep. No. UCRL-51770, 1975.

CHAPTER 27

Fundamentals of Coal Liquefaction

EVERETT GORIN

Everett Gorin *received his B.S. degree in chemistry in 1931 and his Ph.D. in physical chemistry and chemical physics from Princeton University in 1934. From 1935 to 1938 he held research fellowships at Princeton University and the Karpov Institute of Physical Chemistry. From 1939 to 1946 he was chemical research supervisor at Magnolia Petroleum Company, Dallas, Texas, working on petroleum processing and petrochemicals. In 1947 he was Senior Research Associate at Socony Vacuum Oil Company in Paulsboro, New Jersey. From 1948 until he retired in 1976, Dr. Gorin worked on process research, eventually becoming director of process research and development for Consolidation Coal Company, subsequently Conoco Coal Development Company, where he was in charge of all laboratory, bench-scale and pilot-plant work on coal utilization and coal conversion at the company's facility in Library, Pennsylvania. He is presently a private consultant on coal research and development to Conoco and others.*

27.1 INTRODUCTION

27.1.1 Definition of Coal Liquefaction

Two excellent reviews of the coal liquefaction literature up to the year 1960 have appeared since Volume 2 of *Chemistry of Coal Utilization* was published in 1939. The first is the chapter written by Donath[1] for the first Supplementary Volume edited by H. H. Lowry. The second is a review by Wu and Storch.[2] Both reviews covered the whole gamut of work in coal liquefaction from fundamentals to commercial development. This chapter is concerned with more recent developments, through March 1977, in the fundamental area not covered by the above two reviews.

The term "liquefaction" as used here encompasses conversion of coal in which the major product is a "liquid." The liquid has been rather broadly defined on the basis of solubility in various solvents, and in many cases the major "liquid" product is a solid at room temperature. The extent of liquefaction, or conversion, is universally defined by Equation 27.1:

$$\%\ \text{conversion} = \frac{\text{maf coal in} - \text{maf-insoluble residue out}}{\text{maf coal in}} \times 100 \qquad (27.1)$$

where maf refers to moisture- and ash-free coal.

Gases, water, and distillate oils are produced in addition to the primary heavy liquid

[1] E. E. Donath, in *Chemistry of Coal Utilization,* Supplementary Volume, H. H. Lowry, Ed., John Wiley and Sons, New York, 1963, pp. 1041–1080.

[2] W. R. K. Wu and H. H. Storch, *Hydrogenation of Coal Tar,* U.S. Bur. Mines Bull. No. 633, 1968.

product. The amount of these products increases with severity of the liquefaction conditions.

Unfortunately, the definition of maf-insoluble residue is not universally agreed on and depends on the analytical method used for its determination. This in turn depends on the analytical solvent used. Normally the liquefaction residue is extracted with the solvent in question at its boiling point in a Soxhlet apparatus. Thus the amount of insoluble residue depends on the potency of the solvent chosen. Most investigators[2] in the period prior to 1960 used benzene as the extraction solvent to determine coal conversion. Investigators at Hydrocarbon Research Inc.[3] (HRI) use a modification of this technique that gives a significantly higher conversion, particularly under mild liquefaction conditions. The liquefaction effluent is filtered hot under pressure at about 145°C and the resulting cake is then extracted with benzene.

A more common technique now being used by investigators of coal extraction processes is to define conversion on the basis of the product's solubility in more powerful solvents such as cresol,[4] tetrahydrofuran, quinoline,[5] and pyridine.[6]

The results with the latter group of solvents agree rather closely with one another but give substantially higher conversions than the benzene method.[2] The results are only slightly higher than the conversions reported by the HRI[3] method, since in the latter method most of the cresol-soluble product is in solution in the hot filtrate.

27.1.2 Primary Coal Liquefaction via Solvation

The discussion that follows adopts the second definition and defines primary coal liquefaction as the conversion of coal to a material soluble in cresol or equivalent solvent. The "liquid" product thus produced has been variously defined as extract or solvent-refined coal (SRC) depending on the process used. The term "extract" is used here for all cases and is defined as the nondistillable (at either 240 or 290°C, 1 mm Hg) residue from a primary liquefaction process that is soluble in cresol or equivalent solvent.

Processes such as pyrolysis, hydrocarbonization, and "flash" hydrogenation are excluded from the discussion here. Processes of these types are discussed in other chapters of this volume. The major products from these processes are usually gases and solids rather than liquids, although in some cases liquid yields of up to 35% have been reported.

The liquefaction process discussed here in all cases involves addition of hydrogen to the coal either directly or indirectly by means of a solvent or other hydrogen transfer agents. Because it has been considered necessary from practical considerations, a solvent has been employed in all development projects aimed at primary liquefaction.

A considerable literature exists on the solvent extraction of coals with various pure chemical solvents that do not have hydrogen donor properties, such as ethylene diamine. The work in this field, which was directed primarily toward obtaining a better understanding of coal structure rather than toward establishing a first step in a liquefaction process, is not discussed here. An excellent review of this work up to 1960 was given by Dryden.[7] The major effort in the period prior to 1960[8] in the fundamentals of primary liquefaction was directed toward achieving a better understanding of the structure of coal.

The nature of the extract produced varies widely depending on the severity of the conditions used in the liquefaction process, the hydrogen pressure (if any) imposed, and the amount and type of catalyst employed (if any).

It has been conventional in the past[2,9] to

[3] U.S. Dept. Interior, OCR R & D Rep. No. 26, *Project H-Coal Report on Process Development,* Hydrocarbon Research, Inc., November 1968.

[4] G. P. Curran, R. T. Struck, and E. Gorin, *Ind. Eng. Chem. Process Des. Dev.,* **6,** 166 (1967).

[5] D. L. Kloepper, T. F. Rogers, C. M. Wright, and W. C. Bull, U.S. Dept. Interior, OCR R & D Rep. No. 9, *Solvent Processing of Coal to Produce a De-Ashed Product,* Spencer Chemical Division, Gulf Oil Corp., December 1965.

[6] D. D. Whitehurst and T. O. Mitchell, *Prepr., Am. Chem. Soc., Div. Fuel Chem.,* **21**(5), 127 (1976).

[7] I. G. C. Dryden, in *Chemistry of Coal Utilization,* Supplementary Volume, H. H. Lowry, Ed., John Wiley & Sons, New York, 1963, p. 1142.

[8] D. W. Van Krevelen, *Coal,* Elsevier, Amsterdam, 1961.

[9] M. Farcasiu, T. O. Mitchell, and D. D. Whitehurst, *Prepr., Am. Chem. Soc., Div. Fuel Chem.,* **21**(7), 11 (1976).

characterize the extract in terms of its ultimate analysis and its solubility relationships. Thus the extract conventionally is split into three fractions: a benzene-insoluble fraction more recently termed "asphaltols,"[9] a fraction soluble in benzene but insoluble in hexane termed "asphaltenes," and a fraction soluble in hexane. Investigators from Conoco Coal Development Company,[10] formerly part of Consolidation Coal Company, have used cyclohexane rather than hexane as the splitting solvent. Conoco Coal Development Company hereafter is referred to as CCDC.

The discussion that follows emphasizes the implication of post- 1960 data in providing a rational basis for design and improvements of potential commercial processes.

27.1.3 Primary Coal Liquefaction Without Solvent

Primary coal liquefaction has been achieved by direct hydrogenation in the absence of solvents in laboratory devices,[1,2] usually batch autoclaves. Catalysts are normally required, although in some instances[11] the coal ash itself provides sufficient catalysis.

No continuous operation, however small, has been operated in primary coal liquefaction without a liquid vehicle. A catalyzed primary coal liquefaction system that does not involve solvation is possible in principle, although practical development has not been attempted.

Processes that produce mainly gases, distillate oils, and chars, however, have recently been operated successfully without a vehicle in small continuous units. Entrained-phase hydrogenation with[12,13] and without[14] impregnated catalysts have been operated in this manner.

A small continuous bench-scale unit to be discussed later has also been successfully operated in the CCDC laboratories using molten zinc chloride ($ZnCl_2$) catalyst. A benzene vehicle was used whose only purpose was to transport the coal into the liquefaction zone. This process is not a primary liquefaction process, since the predominant product is a distillate.

27.1.4 Secondary Conversion of Primary Liquids

Secondary conversion of primary liquefaction extracts to produce distillates may be carried out in the same reaction system. Alternatively, the extract may be separated from the organic residue and mineral matter and catalytically hydrocracked in a separate operation. Catalysis is normally required to achieve a high yield of distillate. External catalysts are sometimes added, although in many instances the mineral matter in the coal provides sufficient catalysis.

The addition of disposable quantities of iron or tin catalysts to the liquid phase system was the predominant practice in the period prior to 1960.[1,2] More recently the addition of hydrotreating catalysts of the general type of mixed sulfides of cobalt, molybdenum, and nickel supported on an alumina base has been studied in the form of either an ebullated[3,15] or a fixed[16] bed.

27.2 PRIMARY LIQUEFACTION VIA SOLVATION

27.2.1 General Nature of the Process

This discussion of primary liquefaction is limited to the types of processes in which a catalyst is not added. However, as will be discussed later, the mineral matter in the coal often plays an essential role.

The primary liquefaction process involves the consumption of hydrogen. The solvent appears to play an essential role in transferring the hydrogen from the gas phase to the coal.

[10] *Project Gasoline Pre-Pilot Plant—Phase I Research on CSF Process,* U.S. Dept. Interior, OCR R & D Rep. No. 39, Vol. 2, Part I, Consolidation Coal Company, November 1968.

[11] D. K. Mukherjee and P. B. Choudhury, *Fuel,* **55,** 4 (1976).

[12] S. A. Qader, R. A. Haddadin, L. I. Anderson, and G. R. Hill, *Hydrocarbon Process.* **48**(9), 147 (1969).

[13] R. E. Wood and W. H. Wiser, *Ind. Eng. Chem. Process Des. Dev.,* **15,** 144 (1976).

[14] W. C. Schroeder, *Hydrocarbon Process.* **55**(1), 131 (1976).

[15] C. A. Johnson, M. C. Chervenak, A. G. Comolli, E. S. Johansen, C. C. Kang, and W. Volk, preprint of paper presented at 68th American Institute of Chemical Engineers Meeting, November 1975.

[16] P. M. Yavorsky, S. Akhtar, and S. Friedman, *Chem. Eng. Prog.,* **69**(3), 51 (1973).

The development of primary liquefaction processes has, in general, followed three routes in recent years. The first involves coal extraction[10,17] in the absence of gaseous hydrogen as in the Consol synthetic fuel (CSF) process. Primary liquefaction is achieved by transfer of hydrogen from hydrogenated solvent to the coal as in the original Pott-Broche process.[1,2] The hydrogenated solvent is replenished by supplying makeup distillates produced via catalytic hydrocracking of the extract, by catalytic rehydrogenation of the solvent, or both. The hydrogen transfer reaction appears to be a purely thermal process. Attempts to accelerate the reaction by addition of contact catalysts such as cobalt molybdate on alumina or cracking catalysts such as silica alumina were unsuccessful.[4]

The second method involves conducting the extraction process under hydrogen pressure and without external solvent rehydrogenation. This is the Pittsburg and Midway "solvent-refined coal process" now being developed under the auspices of the U.S. Energy Research and Development Administration (ERDA)[5,18] and the Electric Power Research Institute (EPRI).[19]

Finally, Exxon is developing a process[20] that combines the features of both the processes above: that is, hydrogen pressure is applied along with external catalytic rehydrogenation of the solvent. Strictly speaking, the Exxon process should not be classified as a primary liquefaction process as defined in Section 27.1.2, since the ultimate product is a distillate rather than an extract. However the Exxon extraction step can more readily give a higher yield of extract than distillate and thus may be conducted as a primary liquefaction process, if desired.

Primary liquefaction processes are usually conducted between 375 and 465°C, that is, within the range of which pyrolysis or thermal decomposition of the coal becomes important. Secondary hydrocracking of the primary extract to gases and distillates becomes more important as the temperature is increased, unless it is accompanied by a large decrease in residence time.

It is clear that a number of competing chemical processes are taking place simultaneously under conditions of primary liquefaction. The more important ones are coal pyrolysis, coking, hydrogen transfer from solvent to coal and to extract, solvent rehydrogenation, and direct interaction of molecular hydrogen with coal and extract. The mineral matter of the coal probably plays an important role in the latter two processes.

The following discussion attempts to ascertain the relative role of these various processes insofar as this can be done from the available literature.

27.2.2 Hydrogen Donor Studies with Synthetic Solvents

Fundamental studies of hydrogen donor extraction in the post–1960 period (i.e., in the absence of added hydrogen gas) have been largely concentrated at the University of Utah,[21–25] Exxon Research,[26] and CCDC.[4,10,27,28]

Most investigators used tetralin as the model hydrogen donor. Hill et al.[24] and the

[17] *Pilot Scale Development of the CSF Process,* U.S. Dept. Interior, OCR R & D Rep. No. 39, Vol. 4, Book 3, Consolidation Coal Company, November 1971.

[18] G. R. Pastor, D. J. Keetley, and J. D. Naylor, paper presented at 68th American Institute of Chemical Engineers Meeting, November 1975; *Chem. Eng. Prog.,* **72**(8), 67 (1976).

[19] R. Wolk, N. Stewart, and S. Alpert, *EPRI J.,* 12–16 (May 1976).

[20] L. E. Furlong, E. Effron, L. W. Vernon, and E. L. Wilson, paper presented at the 68th American Institute of Chemical Engineers Meeting, November 1975; *Chem. Eng. Prog.,* **72**(8), 69 (1976).

[21] H. Hariri and G. R. Hill, *Kinetic Studies of Thermal Dissolution of High Volatile Bituminous Coal,* U.S. Dept. Interior Tech. Report to OCR, University of Utah, Contract No. 14-01-0001-271, July 1965.

[22] G. R. Hill, H. Hariri, R. I. Reed, and L. L. Anderson, in Am. Chem. Soc. Adv. Chem. Ser. No. 55, 1966, p. 427.

[23] W. H. Wiser and G. R. Hill, "Kinetic Study of the Thermal Dissolution of High Volatile Bituminous Coal," *Proc. Symp., Science and Technology of Coal,* Mines Br., Dept. of Energy Resources, Ottawa, Canada, 1967, pp. 162–167.

[24] G. R. Hill, L. L. Anderson, W. H. Wiser, S. A. Qader, R. H. Wood, and D. M. Bodily, *Project Western Coal—Conversion of Coal into Liquids,* Final Report to OCR (R & D Rep. No. 18), University of Utah Contract No. 14-01-0001-271, August 1970.

[25] W. H. Wiser, *Fuel,* **47,** 475 (1968).

[26] R. C. Neavel, *Fuel,* **55,** 237 (1976).

[27] E. Gorin, H. E. Lebowitz, C. H. Rice, and R. T. Struck, *Proc. 8th World Petroleum Congr.,* preprints, Session No. PD10 (5), 44 (1971).

[28] Anonymous Technical Summary, *Clean Fuels West,* April 1976.

University of Utah group[25] measured the course of the reaction based on conversion of coal to benzene-soluble product but gave no data on the amount of hydrogen transferred. Neavel[26] followed the reaction based on conversion to both pyridine- and benzene-soluble products. CCDC investigators followed the course of the reaction based on the conversion to cresol-soluble product, although in many instances benzene-soluble yields were also given.

The University of Utah work[24] was conducted in a one-liter stirred autoclave, preheated to reaction temperature, into which the coal-tetralin paste was injected. The course of the reaction was followed by sampling the autoclave contents as a function of time.

Neavel[26] used a microautoclave fabricated from 8-mm tubing that was rapidly heated to reaction temperature by vertical agitation in a preheated sand bath.

Three types of equipment were used by CCDC investigators. Kinetic studies[4] were made in a microautoclave immersed in a fluidized sand bed heater similar to the Exxon technique. Product yield studies were made in a one-liter Parr bomb that was charged; then heated to the extraction temperature at 1.7°C/min. Operability, kinetic, and product studies were also made in a 30-lb/hr continuous bench-scale unit[10,27] previously described.

The University of Utah investigators[24] and the Exxon group[26] used pure tetralin as the solvent. The CCDC group[4,10,27,28] studied not only pure tetralin but also the effect of its dilution with various other compounds such as decane, decalin, xylenol, and methylnaphthalene.

Another method of characterizing the process is the amount of hydrogen transferred to the coal via the reaction,

$$\text{tetralin} + \text{coal} \rightarrow \text{coal-derived products} + \text{naphthalene} \quad (27.2)$$

The University of Utah group[24] did not measure the amount of hydrogen transferred via Reaction 27.2, but Exxon[26] and CCDC investigators did.

The definition of the amount of hydrogen transferred via the Reaction 27.2 appears straightforward; 2 moles of hydrogen are transferred for every mole of naphthalene produced.

CCDC investigators have noted, however, that when donor extraction was conducted with tetralin-methylnaphthalene solvents, the yield of naphthalene was considerably greater than would be obtained via Reaction 27.2. Significant amounts of naphthalene were also obtained by hydrocracking or demethylation of the methylnaphthalenes via Reaction 27.3:

$$\text{tetralin} + 2\ \text{methylnaphthalene} = 3\ \text{naphthalene} + 2\ \text{methane} \quad (27.3)$$

Tetralin losses also occur because of isomerization to methyl-indane,[9] thermal decomposition, and dehydrogenation via Reaction 27.4:

$$\text{tetralin} = \text{naphthalene} + 2H_2 \quad (27.4)$$

These losses via Reaction 27.4 are usually insignificant under mild conditions, but can become appreciable under more severe conditions.

The hydrogen transfer figures quoted below from CCDC work include only the hydrogen transferred to the coal via Reaction 27.2. The tetralin consumed by the other processes is deducted from the total tetralin consumed.

A comparison of Neavel's data[26] with CCDC results using a hvCb coal from Illinois is illustrated in Fig. 27.1. The ultimate analysis of the two coals is similar, as shown in Table 27.1. A major difference is the low ash content of Neavel's coal (3.9 vs. 18.0%). The CCDC fuel is raw run-of-mine coal, whereas the low ash content of Neavel's coal was the result of cleaning at 1.35 specific gravity. This also had the effect of concentrating the more reactive petrographic constituents of the coal (i.e., the vitrinite).

In spite of the foregoing differences, the responses of the two coals in terms of coal conversion to benzene solubles are rather similar, at least at levels of hydrogen transfer up to 1.5 wt %. The CCDC coal shows more efficient use of hydrogen, but this is likely to be a result of a difference in the experimental technique employed. Data from the CCDC work, which employed the same type of apparatus, are also illustrated in Fig. 27.1 but are not directly comparable, since only conversion to cresol solubles was measured. The

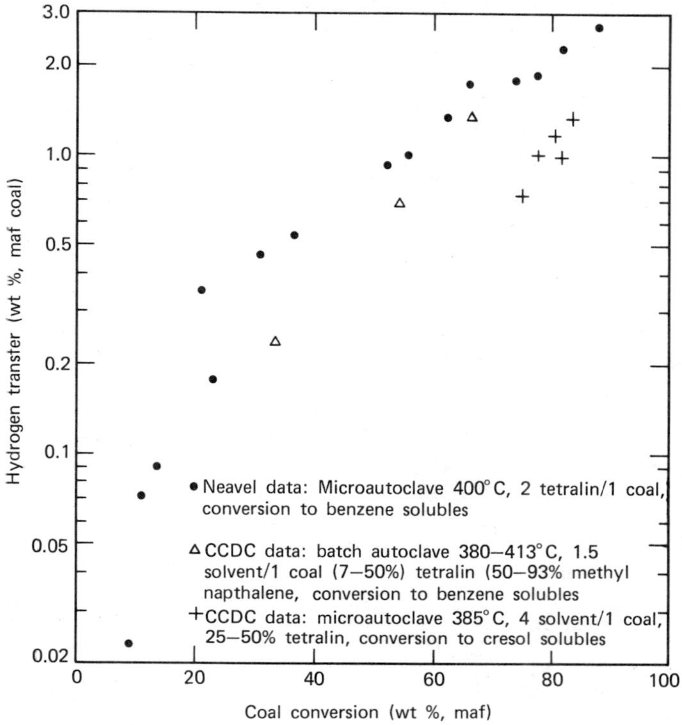

Figure 27.1 Response of hvCb Illinois coal to donor extraction: comparison of Neavel and CCDC data.[4,10,26–28]

data illustrate, of course, that the hydrogen requirement to produce cresol solubles is substantially less than that needed to produce benzene solubles.

The Neavel coal, however, appears to show a much higher extraction rate as well as higher ultimate conversion, as illustrated in Table 27.2. The higher conversions indicate that the particular coal used is much more reactive. As noted above, the beneficiation of the coal also removed some of the relatively unreactive or inert components.

The Spring Canyon, Utah, bituminous coal used by Hill et al.[24] and Wiser[25] has ultimate and proximate analyses similar to the Utah bituminous coal used by CCDC investigators (Table 27.1). The relative reactivity of the two coals is discussed later in more detail, but as the data of Table 27.3 illustrate, there is no great difference in this respect between the two coals. The surprising fact is the high ultimate conversion of 96% to benzene solubles that was achieved with the Spring Canyon coal (i.e., at 450°C and a residence time of 200 min). The CCDC investigators did not reach conversions above 66% to benzene solubles, although extremely severe conditions were not investigated.

27.2.3 Coal Properties in Relation to Hydrogen Donor Liquefaction

It has been conventional in the past to use carbon, Btu, or oxygen content as a measure of coal rank. The lower the carbon content or the higher the oxygen content on an maf basis, the lower is the rank.

Data have been published by both CCDC[27–29] and Exxon[26] investigators concerning the way the rank of coal affects its response to donor extraction.

The ultimate and proximate analyses of some of the coals tested at CCDC laboratories and the yields of extract plus distillate at 1% hydrogen transfer are given in Table 27.1.

The relationship between coal conversion and hydrogen transfer varies somewhat, but

[29] E. Gorin, C. J. Kulik, and H. E. Lebowitz, *Ind. Eng. Chem., Process Des. Dev.*, **16**(1), 95 (1977).

Table 27.1 Properties of Coals and Their Responses (wt%) to Donor Extraction

Mine or area:	Ireland	Browning	Spring Canyon	Hillsboro	Monterey
State:	West Virginia	Utah	Utah	Illinois No. 6	
Seam:	Pittsburgh No. 8	hvCb[a]	hvCb[b]		
Classification:	hvBb[a]			hvCb[a]	hvCb[c]
Ultimate analysis, maf					
Hydrogen	5.55	5.49	6.1	5.45	5.6
Carbon	81.97	80.94	79.8	79.09	78.0
Nitrogen	1.56	1.51	1.7	1.43	1.2
Oxygen	8.74	11.19	11.8	10.88	12.5
Organic sulfur	2.18	0.87	0.7	3.15	2.8
Proximate analysis, mf					
Ash	14.8	7.6	8.4	18.0	3.9
Volatile matter	39.4	41.7	45.7	41.7	45.0
Fixed carbon	45.8	51.2	45.9	40.4	51.1
Results at 1.0% H, transfer					
Total conversion	90	76	—	86	—
Extract + distillate	82	67	—	73	—
Results at 413°C, 60 min, 50% tetralin, 50% methylnaphthalene					
Total conversion	91.2	79.6	—	88.9	—
Conversion to benzene solubles	72.1	66.4	—	67.1	—

[a] Used in CCDC work.
[b] Used in University of Utah work.[21–25]
[c] Used in Exxon work.[26]

not widely, according to the nature of the solvent, the type of experimental unit, and the operating conditions employed. The variation of the relationship above with these factors has been noted previously[27] and also is discussed below.

The relationship derived from the slow heating autoclave experiments will be used to illustrate the influence of coal rank on donor extraction performance. The amount of hydrogen transfer in these experiments was varied by modifying the tetralin content in the tetralin-methylnaphthalene solvent from 7 to 50 wt % and by varying the temperature from 380 to 413°C while holding the nominal residence time constant at 60 min.

The coal conversion versus hydrogen transfer relationships for bituminous coals are illustrated in Fig. 27.2. It is immediately obvious that Pittsburgh Seam high-volatile No. 8 bituminous coal (hvBb) responds best to donor extraction, with Kentucky No. 11 and Illinois No. 6 (hvCb) in relatively close

Table 27.2 Comparison of Extraction Rates: Neavel Data and CCDC Data for an hvCb, Illinois No. 6 Coal[4,10,26–28]

Variable	Neavel Data,[a] Microautoclave	CCDC Data[b] Microautoclave	CCDC Data[b] Batch Autoclave
Solvent	100% Tetralin	50% Tetralin, 50% methyl-naphthalene	50% Tetralin, 50% methyl-naphthalene
Solvent/coal, wt ratio	2	4	1.5
Temperature (°C)	400	413	413.
Time (min)	10	10	—
Conversion (%)	~ 98[c]	78[d]	—
Max conversion, long times (%)	~ 98[c]	~ 86[d]	~ 90[d]

[a] Source not specified.
[b] From Hillsboro Mine.
[c] To pyridine solubles.
[d] To cresol solubles.

Kaiparowits Utah Subbituminous[a]	Black Mesa Arizona Subbituminous[a]	Sweeney Creek Montana Subbituminous[a]	Colstrip Montana Subbituminous[a]	Gillette Wyoming Subbituminous[a]	Glenharold North Dakota Lignite[a]
5.46	5.28	5.01	4.88	5.19	4.70
78.52	78.26	77.38	75.19	74.41	70.88
1.12	1.48	1.12	1.22	1.02	1.45
14.19	14.62	15.92	18.17	19.01	22.61
0.71	0.36	0.57	0.54	0.37	0.36
9.9	5.0	10.5	11.3	9.0	8.1
44.9	47.7	39.8	39.8	46.9	50.6
45.2	47.3	48.9	48.9	44.1	41.3
74	68.5	61	61	67.5	67.0
64	57.0	45	42	51.5	45.5
81.8	84.8	79.0	79.7	85.5	84.1
62.3	68.4	64.0	56.2	69.3	71.2

second place. Browning Mine coal (hvCb) ranks equivalent to or higher than Illinois No. 6 on the basis of carbon content, but it is clear that it responds much more poorly to donor extraction.

The same relationship for subbituminous coals and lignite is illustrated in Fig. 27.3. No clear pattern with regard to rank is apparent and, in fact, the relatively good response of lignite compared to some of the subbituminous coals from Montana is surprising.

Another method of illustrating the effect of rank is shown in Fig. 27.4, where the extract plus distillate yield obtained at 1% hydrogen transfer is plotted against the oxygen content of the feed coal. As might be expected, the yield decreases with increasing oxygen content of the feed coal. It is seen that more than half the feed coals fall on a smooth curve, with the remainder falling well below the curve. On the basis of this criterion, the two poorest are once again the Montana subbituminous coals.

Earlier data[24] based on microautoclave measurements showed that the rates of hydrogen transfer for all coals tested from lignite to Pittsburgh No. 8 were approximately the same. This point is illustrated below for continuous unit results in the section on

Table 27.3 Response of Utah Coals to Donor Extraction

	Coal	
	Spring Canyon	Browning Mine
Investigators	University of Utah[a]	CCDC[b]
Time (min)	60	60+[c]
Temperature (°C)	400	413
Solvent	Tetralin	50% Tetralin, 50% methylnaphthalene
Solvent/coal, wt ratio	10	1.5
Conversion to benzene solubles (wt %)	64	66

[a] References 24 and 25.
[b] Unpublished CCDC work.
[c] Does not include heatup time.

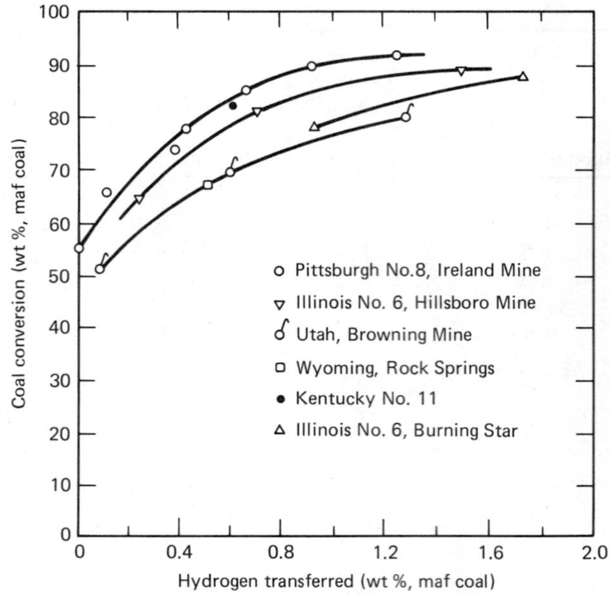

Figure 27.2 Extraction depth versus hydrogen transfer: autoclave results for bituminous coals. (Gorin et al.[27] and unpublished CCDC data.)

kinetics. The result of the batch autoclave data again confirms the near equivalence of hydrogen transfer rates for all the coals above, particularly under the more severe conditions illustrated in Table 27.4. Under these conditions (i.e, 50% tetralin and 413°C), the observed hydrogen transfer for all coals, with one exception, is within 18% of the mean, and no relationship with rank is apparent. The amount of hydrogen transfer is, however, about 30% above the mean for the lowest ranked coal, lignite.

Thus the CCDC data show that the space-time yield of extract in donor extraction, in

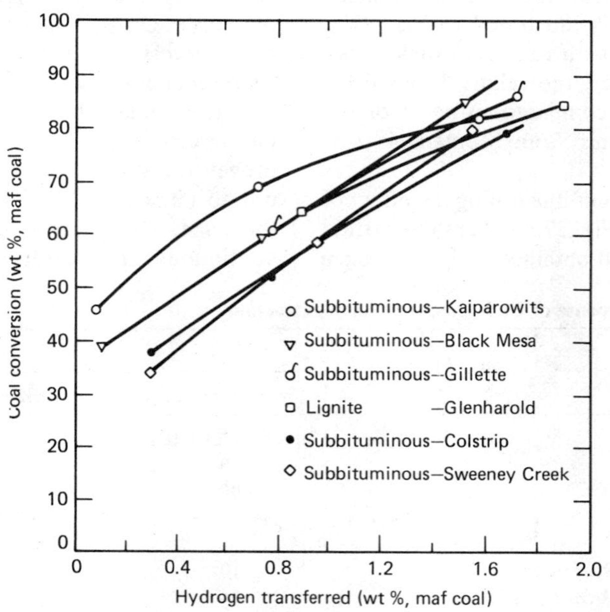

Figure 27.3 Extraction depth versus hydrogen transfer: batch autoclave results for subbituminous and lignite coals. (Gorin et al.[27] and unpublished CCDC data.)

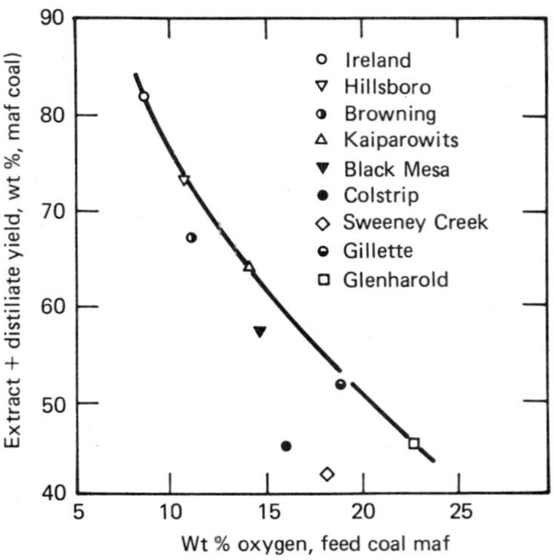

Figure 27.4 Extract yield at 1% hydrogen transfer versus oxygen content of feed coals. Curve drawn through "reactive" coals. (Derived by author from Ref. 27 and unpublished CCDC data.)

general, will decrease with decreasing coal rank. This finding is in direct contrast to previous German data summarized by Storch,[2] which show that under severe sump phase hydrogenation conditions and with German coals, the space-time yield of distillate product increases inversely with the coal rank.

The disagreement here is probably due to the very different processes investigated and the different products. Another contributing factor may be the possibility that the German coals show a reactivity-rank relationship very much different from that of the United States coals.

The response of a number of coals to donor extraction was also investigated by Neavel,[26] using the microautoclave technique. Hydrogenated creosote oil was used as the donor and vehicle. The conversion of the coals to benzene solubles as a function of time is illustrated in Fig. 27.5.

The data of Figure 27.5 show that the rate of conversion to benzene solubles increases with decreasing rank from lvb to lignite, although the increase in conversion rate between hvCb and lignite is slight.

The conclusion above is similar to, but not wholly in accord with, CCDC's data on the effect of rank on rate of conversion to cresol and benzene solubles. Table 27.1, for example, gives the conversion for the various coals ordered according to rank under the following common operating conditions: 413°C, 60 min residence time, and 50% tetralin–50% methylnaphthalene solvent. It is noted that conversion to cresol solubles is highest for the Eastern bituminous coals and lowest for the Montana subbituminous coals. The conversion to benzene solubles fluctuates randomly, but it is highest for the hvBb Pittsburgh Seam coal and lignite. No regular pattern of conversion versus rank is apparent.

The major discrepancy between the CCDC and Exxon data lies in the relative response of the hvBb Pittsburgh Seam coal. The ASTM rank designation is quite arbitrary and the particular coal investigated by CCDC (Ireland Mine) may actually be lower in rank than the Exxon hvBb coals. No ultimate analyses or mine sources are given by Exxon to check this point. At any rate, the Ireland Mine coal is borderline between hvBb and hvCb coal. If the latter classification is used, the major discrepancy disappears.

Consideration of the Exxon and CCDC data indicates that it is difficult to base a prediction of quantitative behavior of a coal in extraction on rank considerations alone. Experimental test of behavior on a given coal is required in practice.

The hvBb Pittsburgh Seam coal (Ireland Mine) responds better to donor extraction in all respects than any other coal tested at CCDC; thus it may be considered to be a "standard" coal. The relative performance of the other coals is poorer by virtue of their higher oxygen content. A theoretical yield and hydrogen consumption were calculated relative to the Ireland Pittsburgh Seam coal at one standard condition as follows: a coal having the same O/C and H/C ratios as Ireland coal was synthesized from the other coals by adding hydrogen and removing oxygen. The pattern of oxygen removal for each coal was assumed to be the same with respect to yields of H_2O and CO_2 as observed in extraction runs with the coal in question. The "synthesized" Ireland coal is then postulated to convert in the same manner as would the real coal with additional hydrogen transfer. The results of these calculations (Table 27.5) indicate how the various coals respond with respect to their ideal performance as a function of rank. Only the Hillsboro (hvCb) coal and lignite closely approach the ideal yield structure. On the basis of this

Table 27.4 Hydrogen Transfer Rates: Effect of Coal Type

Coal	Amount (wt %) Transferred to maf Coals of Different Types Under Different Solvent and Temperature Conditions[a]		
	7wt % Tetralin, 93 wt % Methylnaphthalene At 380°C	25 wt % Tetralin, 75 wt % Methylnaphthalene At 390°C	50 wt % Tetralin, 50 wt % Methylnaphthalene At 413°C
Pittsburgh No. 8, bituminous, Ireland Mine	—	0.90	1.25
Illinois No. 6, bituminous, Hillsboro Mine	0.24	0.70	1.50
Utah, Emery County, bituminous, Browning Mine	0.09	0.52	1.28
Wyoming, bituminous, Rock Springs	—	0.50	—
Utah, subbituminous, Kaiparowits	0.09	0.71	1.57
Arizona, subbituminous, Black Mesa	0.10	0.72	1.52
North Dakota, lignite, Glenharold	—	0.86	1.96
Montana, subbituminous, Colstrip	0.30	0.77	1.64
Montana, subbituminous, Sweeney Creek	0.30	0.95	1.47
Mean	0.19	0.74	1.52
Standard deviation	0.10	0.16	0.22

[a] Common conditions were solvent/mf coal, wt ratio = 1.5, nominal residence time = 60 min.
Source: Unpublished CCDC work.

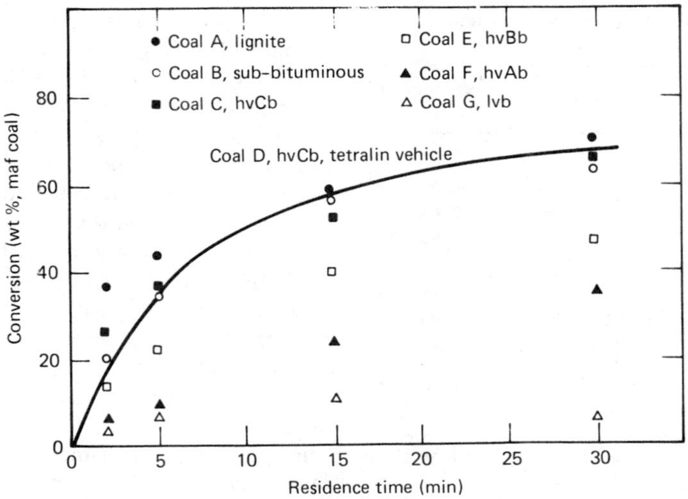

Figure 27.5 Effect of rank on conversion to benzene solubles and gases at 400°C; hydrogenated creosote oil vehicle.[26]

criterion as well, the Montana subbituminous coals again show the poorest yields.

A number of coals listed in Table 27.1 were also analyzed petrographically at the CCDC laboratories by the standard reflectance method used for evaluation of coking coals. Table 27.6 indicates that there is a general qualitative relationship between the vitrinoid content of the coals and their reactivity: high vitrinoid content is synonymous with high reactivity, and vice versa. Reactivity here is defined in relationship to Fig. 27.4. High-reactivity coals are defined as those that fall on the curve, and low-reactivity coals are those that fall below the curve.

27.2.4 Coal Properties in Relation to Liquefaction via Hydroextraction

Primary liquefaction of coal in the presence of both an extraction solvent and imposed hydrogen pressure is called "hydroextraction" in the discussion that follows. This method has been termed the "solvent-refined coal process" by the Pittsburg and Midway Coal Company, which has studied it for a number of years. Four different coals—Pittsburgh Seam, Kentucky bituminous, subbituminous, and lignite—were tested in the original bench-scale unit.[5] Since then, autoclave

Table 27.5 Comparison of Yields with Feed Coals Versus Ireland Coal as Model

	Theoretical Yield[a] Structure			Actual Yield Structure at Same Level of Hydrogen Transfer		
Coal	Conversion	Extract Yield	Hydrogen Transfer	Conversion	Yield	Yield (% Theory)
Ireland	85.0	79.0	0.65	85.0	79.0	100
Hillsboro	85.4	77.0	0.76	82.5	72.5	94
Emery	85.4	76.7	0.79	73.0	64.5	84
Kaiparowits	86.1	73.5	0.85	71.5	61.5	84
Black Mesa	86.1	72.9	0.94	66.5	55.0	75
Colstrip	86.4	71.9	1.13	65.5	49.0	68
Sweeney Creek	86.8	69.5	1.20	68.0	46.5	68
Gillette	87.0	68.1	0.97	66.5	51.0	75
Glenharold	87.6	65.1	1.76	82.5	61.3	94

[a] See text for explanation.
Source of data: Unpublished CCDC work.

Table 27.6 Reactivity Versus Petrographic Composition of Coals

Coals	Vitrinoids (wt %, maf)
High-reactivity coals	
Hillsboro (hvCb)	84
Gillette (subbituminous)	82
Glenharold (lignite)	80
Low-reactivity coals	
Browning Mine (hvCb)	62
Colstrip (subbituminous)	66
Sweeney Creek (subbituminous)	69

Source of data: Unpublished CCDC work.

experiments have been conducted with a number of bituminous coals.[30]

Subsequent pilot-plant work has to date focused on a single Kentucky[18] bituminous coal.

Several different U.S. coals have been evaluated, also, in a 6-ton/day pilot plant at Wilsonville, Alabama; operated by Catalytic Construction Company for Southern Services, Inc., and EPRI.

Bituminous coals from Illinois[31,32] (Illinois No. 6), Kentucky[33] (Kentucky Nos. 9 and 14), West Virginia[34] (Pittsburgh No. 8), and a subbituminous coal from Wyoming[32] have been evaluated.

The same coals as well as a number of others have also been tested in a continuous bench-scale unit by HRI. The unit has a nominal throughput of about 3 lb/hr of slurry. The contactor consists of a 0.814-in. I.D. vertical tube, 117 in. long. The first 22 in. is used for slurry preheat. Provisions were also incorporated for hot recycle of the slurry product, if desired. Detailed description of the equipment and operating results appears in a series of reports to EPRI.[35–38]

Similar coal conversions were achieved in the Wilsonville pilot plant and the HRI bench-scale unit with the same feed coal under nominally the same operating conditions. However, the distillate yields and hydrogen consumption tended to be lower in the pilot plant than in the bench-scale unit under comparable conditions.

CCDC has also evaluated a number of coals for their response to liquefaction in a 30-lb/hr extraction unit previously described.[39] Bituminous coals from Illinois[39] and Utah[40] and a subbituminous coal from Montana[41] have been evaluated.

A more internally self-consistent picture of the relationship between coal properties and hydroextraction performance is obtained by using the bench-scale data and excluding the pilot-plant data.

The HRI and CCDC bench-scale data for a number of coals are summarized in Table 27.7. A number of the coals tested were from the same mine as those evaluated for donor extraction (Table 27.1). Generally, the coals were from different shipments, so their prop-

[30] C. H. Wright, R. E. Perrussel, and G. R. Pastor, *Development of a Process for Producing an Ashless Low-Sulfur Fuel from Coal. Laboratory Studies.* Part I. *Autoclave Experiments,* U.S. Dept. Interior, OCR R & D Rep. No. 53, Interim Rep. No. 6, Vol. 2, Part I, Pittsburg & Midway Mining Company, October 1974.

[31] Solvent-Refined Coal Pilot-Plant Operations, Catalytic Contract No. 34210 with Southern Services, Inc., and EPRI; SRC Tech. Rep. No. 7, *Analysis of Operations, Runs 19 through 40, January 20–August 8, 1975.*

[32] Solvent-Refined Coal Pilot-Plant Operations, Catalytic Contract No. 34210 with Southern Services, Inc., and EPRI; SRC Tech. Rep. No. 8, *Analysis of Operations, Runs 41 through 61, August 8–December 31, 1975.*

[33] Solvent-Refined Coal Pilot-Plant Operations, Catalytic Contract No. 34210 with Southern Services, Inc., and EPRI; SRC Quarterly Tech. Prog. Rep. No. 1, January 1–March 31, 1976.

[34] Solvent-Refined Coal Pilot-Plant Operations, Catalytic Contract No. 34210 with Southern Services, Inc., and EPRI; SRC Quarterly Tech. Prog. Rep. No. 2, April 1–June 30, 1976.

[35] Hydrocarbon Research, Inc., *Liquefaction of Kaiparowits Coal,* Final Report, EPRI 123-2, October 1974.

[36] Hydrocarbon Research, Inc., *Solvent Refining of Illinois No. 6 and Pittsburgh No. 8 Coals,* Final Report, EPRI 389, Vol. 1, June 1975.

[37] Hydrocarbon Research, Inc., *Solvent Refining of Coal,* Final Report, EPRI 123-1-0, October 1974.

[38] Hydrocarbon Research, Inc., *Solvent Refining of Wyodak, Illinois No. 6 (Monterey Mine) and Black Mesa Coals,* Final Report, EPRI 389, Vol. 2, February 1976.

[39] Everett Gorin, C. J. Kulik, and H. E. Lebowitz, *Ind. Eng. Chem. Process Des. Dev.,* **16**(1), 102 (1977).

[40] From unpublished work by CCDC Laboratories.

[41] Conoco Coal Development Company, *Zinc Halide Hydrocracking Process for Distillate Fuels from Coal,* Annual Technical Progress Report for 1975, U.S. ERDA, Rep. No. FE-1743-16, February 15, 1976.

erties should not be considered to be identical.

Data obtained on the various coals by the two different investigators are not strictly comparable because of differences in equipment, operating pressure, and operating technique. The calculated space velocity, for example, does not include the preheater volume, which was especially large (40% of the total contactor volume) for the CCDC runs. The space velocity for the CCDC runs in the data reported in Table 27.7 is reduced by 17% over the previously published figures[39,41] to include the so-called buffer zone. The amount of backmixing or the equivalent number of stages also affects the results. The backmixing factor is unknown but is undoubtedly different in the CCDC and HRI units. Finally, some of HRI data points were obtained with circulation of the hot product slurry to simulate a perfectly mixed reactor. Operation in this mode reduces the contactor efficiency, but the quantitative penalty is not well defined.

The coals in Table 27.7 are arranged in order of increasing oxygen content and, correspondingly, decreasing rank. In spite of the qualifications pointed out above, some qualitative observations are in order.

First of all, the uniquely superior performance of the Pittsburgh Seam coal, previously cited for the donor extraction cases (Table 27.1) is not evident here; indeed, the Kentucky and Illinois coals show superior performance. A higher temperature is required to achieve an equivalent extraction depth, and the distillate and benzene-soluble liquid yields are lower from the Pittsburgh Seam coal. However the Pittsburgh Seam coal used in hydroextraction was from a different mine and was of significantly higher rank, as indicated by its high carbon content, than the Ireland Mine coal used in donor extraction. This likely contributed to its poorer performance. The poorer performance of the higher rank Pittsburgh Seam coal also accords with the previously discussed donor extraction data obtained by Neavel[26] with hydrogenated creosote oil as the donor.

Petrographic analysis of Pittsburgh[31] Seam coal from the same mine (Loveridge) showed it to have a significantly higher fusinite content (7.2 vs. about 3.0% for the Illinois No. 6 coal), which may also have contributed to its poorer performance.

Optimum performance is obvious for the Monterey Illinois No. 6 coal ($\simeq$ 75% C). Especially significant here is the high distillate yield (42% of the maf coal), which is surprising for the low operating pressure of 103 atm. Whether the superior performance here is due to the unique catalytic properties of the coal mineral matter, the reactivity of the coal, or both, cannot be determined from the data at hand. This coal may be from the same mine as the coal cited by Neavel,[26] which showed unusually high reactivity in donor extraction with tetralin as the solvent.

No regular relationship between performance and rank as determined by either conversion or benzene-soluble liquid yield is apparent. It is clear, however, that performance tends to decline as rank decreases below that of the Monterey coal, although not in a regular manner. Again, the decline may in part reflect low catalytic activity of the coal mineral matter.

Two coals stand out as giving poor performance relative to their rank: namely, the Utah bituminous coal (Browning Mine) and the Montana subbituminous coal (Rosebud Seam, Colstrip Mine). As mentioned previously, the same two coals were also relatively poor performers in donor extraction (Table 27.1 and Fig. 27.4).

One coal, Wyodak, appears to be a good performer relative to its rank. The Gillette coal from the same region also showed relatively good performance in donor extraction.

The hydrogen consumption at equivalent depth of conversion is, on the average, somewhat higher in hydroextraction as compared with donor extraction. This may be seen by comparing results for similar coals in Table 27.7 with the conversion hydrogen transfer relationships of Figs. 27.2 and 27.3. The difference may be largely due to hydrogen consumed in hydrocracking of the solvent (which is included in hydroextraction but arbitrarily excluded in donor extraction) and to the higher temperatures used in hydroextraction.

Cudmore[42] investigated, in a batch autoclave, the hydroextraction of a number of Australian coals under initial hydrogen pres-

[42] J. D. Cudmore, *Proc. Symp. Chemicals and Oil from Coal,* Dhanbad, India, 1969; published 1972, Paper No. 47, pp. 438–446, by Central Fuel Research Institute, Dhanbad, India.

Table 27.7 Properties of Coals and Their Response to Hydroextraction

Mine or area: Seam: State or area:	Loveridge Pittsburgh No. 8 West Virginia	Colonial Kentucky No. 14 Kentucky
Carbon (wt %, maf)	85.1	79.7
Oxygen (wt %, maf)	(4.8)[a]	9.9
Operating (wt %, maf) Conditions		
Solvent/coal wt ratio	2.0	3.1
Space velocity, maf coal g/(hr) $(cm)^3$	0.37	0.37
Reactor temperature (°C)	451	441
Reactor pressure (atm, (abs)	104	103
Hydrogen rate (moles/100 g maf coal)	3.0	4.1
Slurry recycle	Yes	No
Investigator	HRI	HRI
Yields (wt %, maf coal)		
Conversion	91	94
Extract	58.9	54.8
Benzene-soluble liquid	55.8	71.4
C_6–535°C distillate	13.6[b]	28.9
C_1–C_5, gas	10.5[c]	7.3
Hydrogen consumption	1.8	2.2
Yield, process solvent	7	19

[a] Oxygen content appears low, possible analytical error.
[b] C_4–535°C distillate, all other values C_4–525°C.
[c] C_1–C_3.
Sources: HRI work = References 35, 36, 37, and 38. CCDC work = References 39,

sure of 69 atm using anthracene oil as the solvent. Coals varying in rank from semianthracite to lignite were tested. Quantitative yield data were not presented, but it is stated that all coals except the semianthracite responded favorably to the extraction process.

27.2.5 Solvent Properties in Relation to Liquefaction

The properties of the solvent can affect the primary liquefaction process in various ways. Unfortunately, published data that could be used to examine this question quantitatively are very limited. The discussion that follows, therefore, is mostly qualitative.

The donor properties of the solvent are obviously of prime importance in donor solvent extraction. The yield structure, under any given set of operating conditions, will depend on the nature and concentration of the donors in the solvent entering and leaving the system. Once the donor concentration has been increased beyond a minimum value, the yield structure becomes relatively independent of a further increase in donor concentration. This matter is discussed in more detail in the section on kinetics.

Quantitative measurement of the donor concentration and of the yield structure as a function of the amount of hydrogen transferred can be made where tetralin or some similar model compound (such as 9, 10-dihydroanthracene) is used as a donor. The donor concentration of natural solvents can now be estimated also by means of recently developed mass spectrometric techniques.[43]

The yield structure in such a system, as a function of the amount of hydrogen transferred, is not completely insensitive to the operating conditions and the nature of the base (nondonor) solvent as the discussion in Section 27.2.2 implies.

The operating method,[27] conditions,[4] and nature of the base solvent do have an effect on the yield structure. The impact is greatest at low conversions and perhaps in the very high conversion region.

The heating rate,[27] for example, affects the yield structure thus: high heating rates such

[43] Thomas Aczel, *Rev. Anal. Chem.*, 1(3), 226 (1972).

Browning	Burning Star Illinois No. 6		Monterey Illinois No. 6	Black Mesa	Kaiparowits	Colstrip Rosebud	Wyodak
Utah	Illinois		Illinois	Arizona	Utah	Montana	Wyoming
81.15	78.8	77.3	74.8	77.7	77.4	74.9	72.9
10.5	11.3	12.5	13.7	15.5	15.8	18.0	20.2
2.0	2.0	3.1	2.0	2.0	2.0	2.0	2.1
0.34	0.43	0.36	0.36	0.36	0.38	0.29	0.48
427	439	440	448	448	427	441	446
117	137	103	103	171	137	120	140
6.6	6.9	4.1	4.5	4.6	4.8	9.3	3.9
No	No	No	Yes	Yes	No	No	Yes
CCDC	CCDC	HRI	HRI	HRI	HRI	CCDC	HRI
79	94	94	94	83	84	69	85
49.4	56.2	59.5	36.5	32.2	46.1	39.2	35.3
54.0	66.8	64.2	73.4	47.6	54.1	33.2	51.4
16.1	20.6	16.5	42.3	30.3	20.6	4.2	28.6
5.6	7.4	9.2	6.5	11.6	8.0	7.0	7.6
2.0	2.2	2.2	3.2	2.9	2.2	2.8	2.9
13	14	10	40	21	15	−3	18

40, and 41.

as those obtained with a microautoclave and continuous unit as compared with the batch autoclave usually, but not always, result in a lower conversion at the same hydrogen input. Data bearing on this question are illustrated in Fig. 27.6.

That the base solvent affects the yield structure, particularly at low conversion levels, is borne out by the data from the continuous bench-scale unit illustrated in Fig. 27.6; when other conditions are equal, a much higher coal conversion is obtained at hydrogen transfer levels below about 0.4 wt % with methylnaphthalene-containing solvents as compared with pure tetralin. The effect of increasing aromaticity of the solvent in increasing conversion at high extraction depths is also illustrated by Run 120 of the continuous bench-scale program[10] at CCDC using Ireland Mine coal. Conversion, at 399°C and 38 min residence time, using tetralin as the feed, increased from 88.7 to 91.1% as the naphthalene content of the recycle solvent increased from 5 to 14%.

At low hydrogen transfer levels the divergence of the microautoclave conversions from the autoclave results may also be largely due to the difference in base solvent; that is, a saturated hydrocarbon was used in the microautoclave versus methylnaphthalene in the autoclave experiments. With Ireland Mine coal, for example, at 385°C, 60 min residence time, and pure methylnaphthalene solvent, a coal conversion of about 55% is achieved, versus about 40% with a saturated solvent such as decalin.

The higher conversions with methylnaphthalene may reflect not only increased solubilization by virtue of its aromaticity but a degree of hydrogen transfer that is not measurable by the usual tetralin-naphthalene couple. Hydrogen transfer can occur by "saturation" of coal-derived free radicals by reaction with the methyl groups of the solvent. This process, of course, leads to coupling or "polymerization" of the methylnaphthalene solvent.

The operating temperature may also affect the yield structure. Lower yields are obtained for the same amount of hydrogen transfer at reduced temperatures.[4]

Neavel[26] has published some interesting results on the microautoclave extraction of Illinois No. 6 coal with naphthalene as the

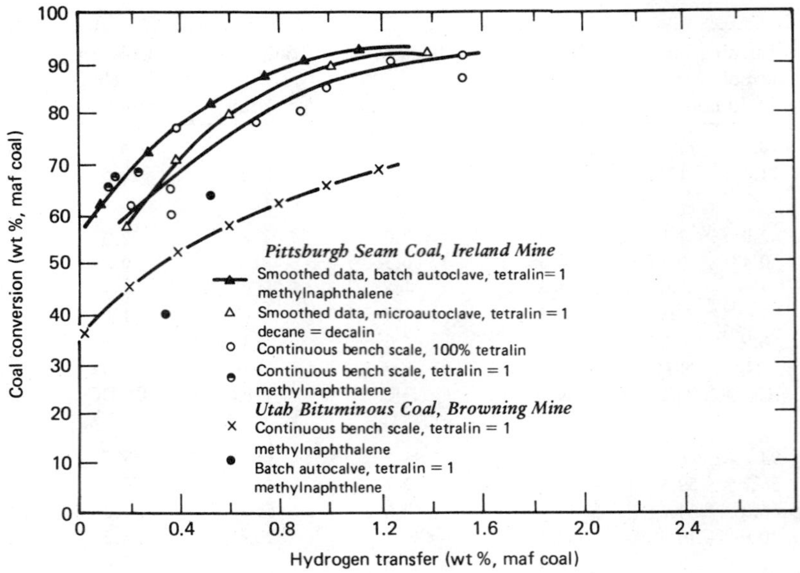

Figure 27.6 Hydrogen transfer versus coal conversion: continuous versus autoclave units. (Pittsburgh data: Ref. 27; Utah data: CCDC, previously unpublished.)

solvent. These results, giving coal conversion versus time, are illustrated in Fig. 27.7, along with previously unpublished CCDC data using methylnaphthalene solvent and an Illinois No. 6 coal from a different mine (Hillsboro). Naphthalene appears to have some stereospecific effect in causing conversion to an unstable soluble product that rapidly transforms to an insoluble product because there is no stabilizing hydrogen. The phenomena above may be absent when methylnaphthalene is the solvent.

The use of high-boiling base solvents (IBP > 240°C) also tends to give improved results as compared with low-boiling synthetic solvents. Such solvents are always natural products and, as such, comprise complex mixtures of ill-defined structure. They also often

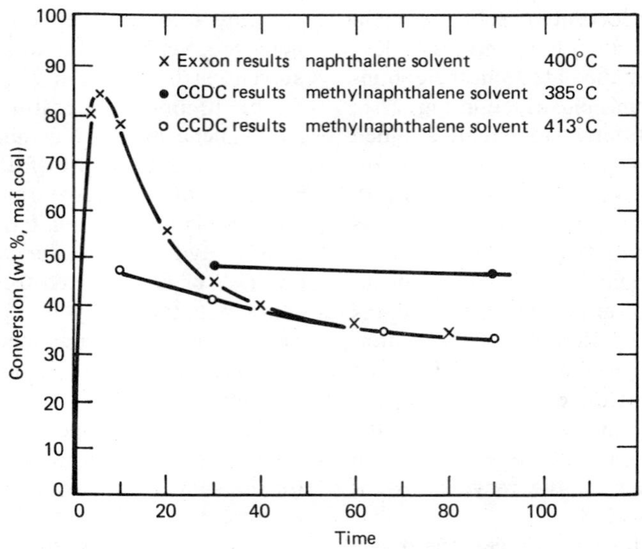

Figure 27.7 Extraction of Illinois No. 6 coal, non-donor solvents. (Neavel[26] and previously unpublished CCDC data.)

posses mild hydrogen donor properties, making quantitative comparison with synthetic solvents difficult.

Hydroextraction operations have always been conducted with complex mixtures rather than with pure compounds. Startup solvents have usually comprised one of the following feedstocks: raw anthracene oil,[35–38] hydrogen-stabilized anthracene oil,[39] catalytically hydrogenated anthracene oil,[35–38] or aromatic petroleum-derived oils.[44] Operation has usually been conducted, where conditions permit, until the displacement of the startup solvent by the steady state, process-derived solvent. As will be shown below, the composition of the steady state solvent is quite different from that of the startup solvent. Hydroextraction results with the startup solvent are, however, similar but not identical to those achieved with the equilibrium solvent. Results from the hydroextraction of Illinois No. 6 coal at CCDC (Burning Star Mine), part of which were previously published,[39] are given in Fig. 27.8, which illustrates the effect of gradual replacement of the startup anthracene oil solvent with natural solvent in operation at 137 atm and 440°C. A slight increase in coal conversion, hydrogen consumption, and production of light products occurs as the solvent composition approaches that of the natural solvent.

Batch autoclave results reported by Whitehurst and Mitchell[45] indicated, however, that the composition of the solvent was critical to achieve good performance not only in donor extraction, which is self-evident, but also in hydroextraction. Whitehurst and Mitchell's results differ from the CCDC results quoted above for several reasons, however. First Whitehurst and Mitchell used a much lower pressure (i.e., 55 atm). Second, the solvent they used, pyridine, is also very volatile, and the extraction temperature employed (427°C) is well above its critical temperature. Finally, pyridine, a nondonor solvent, is incapable of being hydrogenated to produce active donors.

Another consideration is the relationship

[44] R. P. Anderson, paper presented at Symp. Coal Processing, American Institute of Chemical Engineers, Salt Lake City, Utah, August 20, 1974.

[45] D. D. Whitehurst and T. O. Mitchell, *Prepr., Am. Chem. Soc., Div. Fuel Chem.*, **21**(5), 127 (1976).

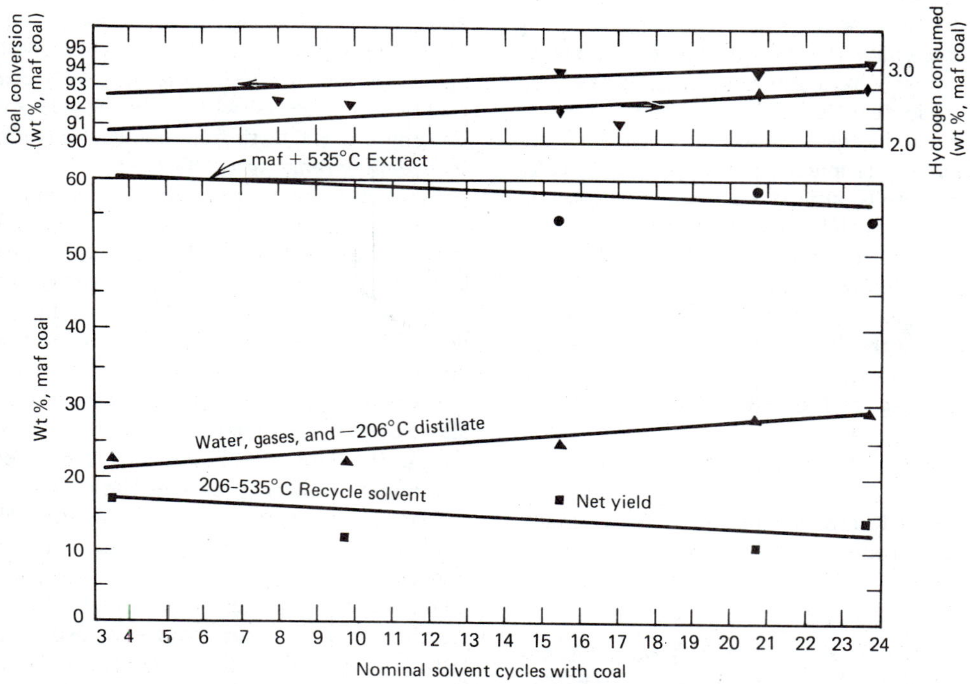

Figure 27.8 Hydroextraction performance of Illinois No. 6 coal at 137 atm and 439°C, and approach to natural solvent: 2 solvent/1 maf coal; 0.43 g maf coal per hour per cubic centimeter; 6.9 moles H_2/100 g maf coal.[39]

between solvent properties and operability in the extraction system. Hydrogen donor properties in the solvent are important, even when the extraction process is conducted under hydrogen pressure, to prevent coking and improve operabiltiy[44] in the preheater. It appears likely[45,46] that in the early stages of the solvation process, the main reaction is transfer of hydrogen from the solvent to the coal with corresponding depletion of the donor content of the solvent. This may be one reason for the importance of catalytic rehydrogenation of the solvent is in the Exxon process[20] before the solvent is reused in hydroextraction.

It should be mentioned that operability was maintained in startup operations in the Cresap pilot plant[17] in both the preheater and extractor with a nondonor extraction solvent (mixed methylnaphthalenes) and without applied hydrogen pressure. Operability was sustained by providing a high degree of turbulence in both units. The operating temperatures, however, were lower at Cresap (400°C) than in the Wilsonville[31–34] and Tacoma[18] pilot plants, where operations are normally conducted at 440°C. Even so, it was noted at Cresap that the extraction proceeded more smoothly when hydrogenated oil was supplied to the solvent.

Operability may also improve by improving the solvency of the solvents. It has been noted in the CCDC bench-scale extractor, operated in the absence of hydrogen, that operability is much improved—that is, coke formation is almost totally absent—when nondonor, high-boiling aromatic oils are used as solvents, as compared with the use of light aromatic solvents such as mixed methylnaphthalenes. The improvement here is ascribed to the improved solubility of extract in the high-boiling oil. The occurrence of the highly viscous transitory[5,46] gel condition is thus minimized.

Similarly, the addition or presence of components such as saturated hydrocarbons,[17] which tend to reduce the solubility of extract, detract from operability of the extraction system.

27.2.6 Natural Solvent Composition, Origin, and Hydrogen Donor Properties

Natural solvent in donor-solvent extraction can be arrived at in many different ways depending on the detailed manner in which the process is operated. Natural solvent is generated from the coal by hydrogen donor cracking, but solvent is destroyed by cracking to lighter components and condensation to residuum oils. For example, condensation may occur by free radical transfer of hydrogen from groups other than six-member naphthenic rings. Experimental data on the amount of solvent polymer produced are not very precise and varied from zero in pilot-plant runs[10,17,27] to 11% of the maf coal in bench-scale runs.

It is clear, however, that in a donor-solvent process the net loss of solvent must be supplied by hydrocracking of at least a part of the extract. This is the method that was used in the development of the CSF process.[10,17,27] The implications of the CCDC bench-scale and pilot-plant data with regard to the solvent balance question were previously presented.[27] Some of these data[27] are reproduced in Table 27.8, which illustrates how solvent balance was achieved. The data above refer to extraction of Ireland Mine coal at 398°C and 30 min residence time to achieve a coal conversion of 78–80 wt %. The recycle and makeup solvent were fractionated such that their boiling range was 230–475°C.

Equilibrium natural solvent was generated by hydrocracking of the extract in an ebullated bed at 440°C and 3500 psig. The distillate generated was recycled until steady state composition was attained. The procedure is described in detail in the aforementioned reports.[10,17]

A closer approach to a true natural steady state solvent was achieved in the bench scale than in the pilot plant because of the limited duration of the latter program. The properties of the solvent were previously reported[10,17] and are reproduced here for the reader's convenience in Table 27.9. The properties given are after recycle of the solvent through extraction. The natural solvent also contained about 3 wt % of the tar acids as determined by caustic extraction. Since proton-type distribution analysis by NMR was

[46] C. H. Wright and R. P. Anderson, "Studies with Kentucky No. 9 Coal in the Solvent-Refined Coal Process," paper presented to Pittsburgh Catalysis Society, April 1975.

Table 27.8 Solvent Balance in Combined Extraction–Hydrogenation[27]

Feed and Products	Scale of Program: Bench Scale	Pilot Plant
Extraction + Low Temperature Carbonization (LTC)		
Feed (kg/100 kg maf coal)		
Solvent	177	362
Fresh solvent	53.0	42.5
Products (wt % maf coal)		
Extraction depth	80.2	78.4
Recovered extract	72.3	71.0
Solvent polymer	11.4	0.0
+475°C LTC tar	1.6	1.6
Total +475°C product to hydro	85.3	72.6
C_5–475°C LTC distillate	1.3	1.3
Hydrogenation + Coking of Residue (85% conversion in hydro)		
Feed (kg/100 kg maf coal)		
325–475°C solvent from extraction	20.2	14.6
+475°C product from extraction	85.3	72.6
Total	105.5	87.2
Products (kg/100 kg maf coal)		
230–475°C hydro distillate to extraction	53.0	42.5
C_5–230°C hydro distillate	26.9	22.9
C_5–475°C coker distillate	5.4	4.6
Spent solvent from extraction	21.4	27.9
C_5–475°C LTC distillate	1.3	1.3
Net yield, C_5–475°C distillate	55.0	56.7

Source of data: Reference 27.

not made, comparison in this respect with natural solvents generated by hydroextraction is not possible.

The hydrogen donor capabilities of the makeup solvent generated by extract hydrocracking is difficult to measure directly. Assessment can be made on the basis of its performance as donor in extraction. Two methods were used and presented[27] previously.

The first method uses the depth of extraction achieved in the continuous bench-scale unit, and the data of Fig. 27.6 to assess the amount of hydrogen actually transferred. As an example of this method, but using an approach somewhat different from that employed previously,[27] the results of successive extraction with 100% hydro oil were used to arrive at a "minimum" value for the hydrogen transfer potential, as shown in Table 27.10. The hydrogen transfer capability derived by this method amounts to a minimum of 1.46 wt % of the solvent or 48 wt % of the hydrogen transfer potential of pure tetralin. The figure above is a "minimum," since obviously the solvent still possesses hydrogen donor properties even after three cycles of extraction. On the other hand, for reasons already outlined, one might expect that less hydrogen transfer would be required for a given level of coal conversion with natural solvent than with tetralin.

The other method outlined previously[27] used a combination of reaction kinetics and the relationship between coal conversion and hydrogen transfer to arrive at a somewhat higher hydrogen transfer potential of the hydro oil of 2.0 wt %, or two-thirds of the value

Table 27.9 Typical Properties of Natural Extraction Recycle Solvent[10]

	Boiling Range	
	230–315°C	315–475°C
Ultimate analysis, (wt %)		
Hydrogen	10.2	9.2
Carbon	89.2	89.9
Nitrogen	0.2	0.3
Oxygen	0.2	0.4
Sulfur	0.2	0.2
Specific gravity at 15°C	0.964	1.021
Viscosity (cSt)		
At 38°C	4.11	25.0
At 100°C	1.36	3.41
PONA analysis[a]		
Paraffins	2.4	1.5
Naphthenes	15.1	10.3
Aromatics	82.5	88.2

TBP Analysis, Whole Oil (vol %)	
IBP	22.5°C
IBP–230°C	4.0
230–275°C	33.5
275–350°C	47.0
350–475°C	15.5

[a] By mass spectrographic analysis.

for tetralin. The actual "true" value is likely to be bracketed between the values of 1.4 and 2.0 wt %.

The solvent balance situation is somewhat simplified when extraction is carried out under imposed hydrogen pressure. It has been shown, for example, that if proper operating conditions[18,31–38,44] are employed, hydroextraction of most coals actually results in generation of a solvent surplus. Thus a balanced operation has been demonstrated where the net products are light distillates and extract, and where the need for an extract hydrocracking step is obviated. This is desirable when the extract itself is a valuable product. The establishment of a solvent surplus means that more solvent range distillates are produced by hydrocracking of the coal or extract than are destroyed by hydrocracking of the solvent. On the other hand, under conditions that exist with at least some of the coals, there is a solvent deficiency.

The definition of the boundary conditions above which a solvent surplus exists is complex and has not been quantified. The major factors are the type of coal and nature of its mineral matter (possible catalytic effect), the hydrogen partial pressure, the space velocity, the temperature, and the boiling range of the recovered and recycle solvent. The nature of the contacting vessel may also have an impact on the solvent balance question, that is, mixed flow versus plug flow as the two extreme situations, as well as isothermal versus rising temperature regimes in the extractors. The criterion for achievement of solvent balance itself, which is somewhat arbitrary, is discussed on the basis of material balance data that give yields of solvent boiling range distillates. The different experimental operations used various boiling ranges in their recycle solvents. The HRI work,[35–38] for example, used recycle distillates that had a normal boiling range of 200–415°C. CCDC[39] work used 240–475, 240–535, and 206–535 boiling ranges. The end points in the vacuum distillates in pilot-plant work at Wilsonville[31–34] and Fort Lewis[18] are about 425°C. The major difference here is that Fort Lewis fractionated the recycle solvent to remove light ends boiling below about 230°C, whereas Wilsonville solvent had an initial boiling point as low as 177°C.

Practical achievement of solvent balance in an actual operation requires a surplus of

Table 27.10 Results of Successive Extraction with 100% Hydro Oil, Continuous Bench-Scale Unit[27]

	Cycle I	Cycle II	Cycle III
Temperature (°C)	398	398	398
Residence time (min)	42	39	42
Solvent/maf coal, wt ratio	2.3	1.73	1.73
Coal conversion (wt %, maf)	87	84	82
Hydrogen transfer (from Fig. 27.6)			
Coal (wt %, maf)	1.1	0.90	0.80
Solvent (wt %)	0.48	0.52	0.46
Cumulative	—	1.00	1.46

solvent (as defined above) because of handling losses and incomplete solvent recovery in vacuum distillation.

Nearly all coals tested by HRI and CCDC under conditions cited in Table 27.7 (operating pressure > 100 atm and temperatures of 427–451°C) generated a surplus of process solvent. The only exception was the run with Colstrip subbituminous coal at 120 atm and 441°C. This observation again points up the refractory nature of the Montana Rosebud Seam subbituminous coal with regard to extraction processing.

The CCDC results illustrated in Fig. 27.8 show that a somewhat lower net yield of process solvent is obtained with natural solvent than with a more thermally stable startup oil such as anthracene oil.

The effect of the process variables on the yield of process solvent was studied in the case of Kentucky No. 14 coal by HRI investigators. A graphic summary of their results[37] is given in Fig. 27.9. A surplus of solvent was generated at all conditions studied. As shown in Fig. 27.9, the process solvent yield increased with increasing pressure in the range 69–171 atm, decreasing space velocity in the range 0.80–0.40 g*(hr) (cm³), and increasing temperature in the range of 427–441°C.

A similar unpublished study was made at CCDC using Illinois No. 6 coal from the Burning Star Mine. The effects of the variables were generally similar to those observed by HRI, except that the effect of temperature was not as well defined. Comparative results given in Table 27.7 show that under similar conditions a higher yield of distillate is obtained from the Kentucky No. 14 coal as compared with the Burning Star Mine coal.

CCDC data[39] showed that with the Illinois No. 6 coal a solvent deficiency was evident when the pressure was reduced to 65.6 atm abs at 441°C, such that replacement of startup anthracene oil with natural solvent could not be achieved. More recent unpublished CCDC data showed that the solvent deficit was eliminated and a surplus achieved at the conditions above by reducing the space velocity from 0.6 to 0.36 g/(hr)(cm³). These results were achieved, however, with anthracene oil, and it remains to be demonstrated

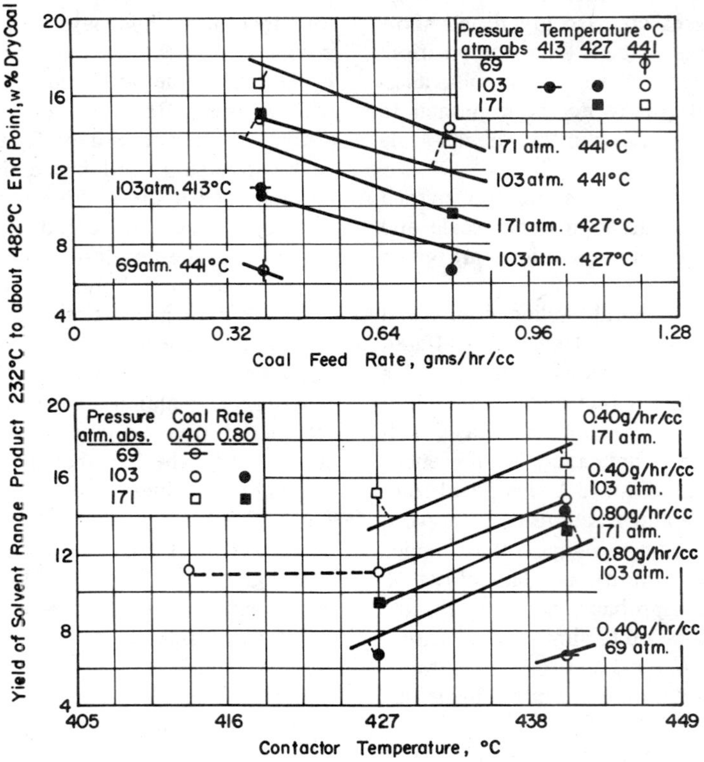

Figure 27.9 Solvent range product yield: noncatalytic solution of Western Kentucky No. 14 coal.[37]

whether a solvent surplus could be achieved at as low a pressure as 66 atm abs with natural solvent.

Wilsonville pilot-plant operations[31–34] also showed that a solvent surplus could be maintained at proper operating conditions with all coals tested to date, namely, Pittsburgh No. 8 and Illinois No. 6 from the Burning Star and Monterey Mines, Wyodak, and Kentucky No. 9 and No. 14 coals from the Colonial Mine. The conditions were similar to those previously demonstrated in the HRI bench-scale programs.[35–38]

Anderson[44] showed how the evolution of the process solvent in the P & M solvent-refined coal process could be followed by analyzing of the solvent with emphasis on use of the proton distribution via NMR analysis. Constancy of the proton distribution on repeated cycling was used as a means of determining when steady state solvent composition had been achieved. An example of the technique was given for bench-scale operations with Kentucky No. 9 coal at 435°C and 103 atm pressure. The IR ratio method[47] was also used to monitor the condition of the solvent. The IR ratio is the ratio of aromatic to hydroaromatic absorbence peak heights. Ultimate analyses are also given and show the same trend to a steady state composition. CCDC investigators[39] used the ultimate analysis method for following the approach to steady state composition in their bench-scale study.

Typical inspections of some startup and steady state solvents from bench-scale and Wilsonville pilot-plant operations are given in Table 27.11.

The choices of initial boiling points and end points of the natural solvents differed widely among the various investigators. This is a reflection of the experimental method of recovering and fractionating the process solvent rather than an indication of differences in the process conditions. A significant fraction of the low-boiling fractions of the solvent (< 230°C) will be in the vapor phase under usual dissolver conditions. The 60-wt % overhead point is probably more significant, and it seems to be quite similar in all cases.

The variation in the quantity of the tetralin-naphthalene fraction also simply reflects different experimental procedures. This cut was partially stripped out in the CCDC work by the relatively large hydrogen stream in the hot separator, thus was partially removed from the recycle solvent.

The other inspections, such as ultimate analysis, phenol content, and proton distribution, are quite similar irrespective of the coal used, the nature of the startup solvent, or whether it was generated in bench-scale or pilot-plant equipment.

The proton distribution via NMR illustrates the increase in H_{ar} and decrease in H_γ when a petroleum-based startup solvent is used. The reverse trend is noted when one starts with a highly aromatic solvent such as anthracene oil.

The oxygen and phenolic contents are higher and the hydrogen content lower than in the natural donor solvent composition given in Table 27.9. This difference is a natural consequence of the generation of natural solvent in the latter case via catalytic hydrogenation and hydrocracking with hydrofining-type catalysts.

On the basis of their work at HRI,[48] Kang et al. have recently attempted to correlate process solvent composition with its performance in the hydroextraction process using inspection data as shown in Table 27.11. In particular, they concluded from the work with Wyodak, Black Mesa, and Illinois No. 6 coals that improved performance, higher coal conversion, and prevention of coking are achieved if the solvent has good donor properties, and this sometimes can be correlated with the tetralin-naphthalene ratio.

Both Kang et al.[48] and Anderson[44] indicate that the donor properties of the solvent can be related to the H_β content. A quantitative relationship is obviously not possible, since β-hydrogen in aliphatic side chains and five-membered rings is a poor donor. This is illustrated by the fact that the petroleum oil (a nondonor solvent) used by Anderson (cf. Table 27.11) has a much higher H_β content than does the steady state solvent. Starting with a given base solvent of low H_β content, the change of H_β content is likely to be qualitatively related to the donor properties of the solvent.

[47] C. H. Wright and D. E. Severson, *Prepr., Am. Chem. Soc., Div. Fuel Chem.*, **16**,(2), 68 (1972).

[48] C. C. Kang, Gavanon Nongbri, and Norman Stewart, *Prepr., Am. Chem. Soc., Div. Fuel Chem.*, **21**,(5), 19 (1976).

Table 27.11 Inspections on Startup and Natural Solvents Hydroextraction

	Colonial Mine Kentucky No. 9			Burning Star Illinois No. 6		Monterey Illinois No. 6	Arizona Black Mesa
Operating pressure (atm):	—	103	115	—	137	103	171
Space velocity [g maf coal/(hr)(cm³)]:	—	0.34	0.38	—	0.43	0.36	0.36
Contactor temperature (°C):	—	435	454	—	439	448	448
Starter solvent:	Gulf FS-120	Gulf FS-120	Recycle Solvent	Stabilized anthracene oil		Pilot plant solvent[a]	Pilot plant solvent[a]
Reference:	44	44	31	29	29	38	38
Solvent analysis:	Startup	Process	Process	Startup	Process	Process	Process
Boiling Point (°C)							
IBP	N.D.	N.D.	172	206	206	173	182
20 wt %	N.D.	N.D.	207	275	240	230	221
40 wt %	N.D.	N.D.	242	318	270	250	243
60 wt %	N.D.	N.D.	258	335	300	278	280
EP	N.D.	N.D.	+448	535	535	—	416
Ultimate Analysis (wt %)							
Hydrogen	7.7	7.0	7.9	5.9	7.8	7.8	7.9
Carbon	89.5	88.7	88.5	91.5	88.4	87.6	87.6
Nitrogen	N.D.	N.D.	0.7	0.9	0.7	N.D.	0.6
Oxygen[b]	N.D.	N.D.	2.7	1.3	2.7	N.D.	3.8
Sulfur	2.2	(2.0)	0.2	0.4	0.4	N.D.	0.1
Tetralin + naphthalene (wt %)	N.D.	N.D.	—	4.5	5.9	14.0	13.4
Tetralin-Naphthalene (wt %)	N.D.	N.D.	—	0.11	0.29	0.5	0.23
Phenols (wt %)	N.D.	N.D.	—	N.D.	31	24	22
Proton distribution (¹H NMR)							
H_{ar}	29.7	46.5	N.D.	77.1[c]	N.D.	48.0	50.3
$H\alpha$	29.7	29.8	N.D.	18.9[c]	N.D.	27.0	23.7
$H\beta$	31.4	18.8	N.D.	34.4[c]	N.D.	18.3	20.5
$H\gamma$	9.2	4.9	N.D.	0.6[c]	N.D.	16.7	5.5

[a] From Wilsonville pilot plant (cf. Ref. 31).
[b] By difference.

HRI[38] also reports PONA analyses via mass spectrometry on some of their recycle solvents. These analyses, likewise, do not shed much light on the hydrogen donor properties of these materials.

CCDC laboratories have used the microautoclave method of coal extraction[4,26] as a quick qualitative test for the donor properties of natural solvents. Until better analytical methods for characterizing the solvent are developed, further development of the microautoclave technique appears to offer promise for the above purpose. Previously unpublished CCDC data using this technique appear in Table 27.12.

27.2.7 Relationship Between Donor and Hydroextraction

The data of Table 27.12, based on Burning Star Mine coal, are pertinent in establishing the relationship between the two types of process. The data clearly indicate that nearly the same coal conversion can be achieved in donor as in hydroextraction by either increasing the solvent-coal ratio to a very high value such as 8 : 1 or by externally rehydrogenating the solvent. Thus it is clear that one of the functions of the applied hydrogen pressure in hydroextraction is to prevent undue depletion of the donor content of the solvent by rehydrogenation of the solvent *in situ,* catalyzed in part by the mineral matter of the coal.

This function of the mineral matter is discussed in more detail in a later section of this chapter.

Another function of the hydrogen pressure is to suppress loss of hydrogen from the donor solvent by catalytic dehydrogenation. The latter process in donor extraction again

Table 27.12 Microautoclave Evaluation of the Donor Properties of Hydroextraction Solvents

Conditions: temperature, 441°C; residence time, 20 min; Illinois No. 6 Coal from the Burning Star Mine.

	Microautoclave Conditions and Results		Hydroextraction Conditions and Results			
Solvent	Solvent/Coal, wt ratio	Coal Conversion (wt %, maf)	Pressure (atm)	Temperature (°C)	Space Velocity [g maf Coal/(hr)(cm³)]	Coal Conversion (wt %, maf)
Methylnaphthalene	3	39.6	—	—	—	—
Stabilized anthracene oil to hydroextraction[a]	3	45.3	65.6	441	0.51	76–81
Stabilized anthracene oil to hydroextraction[a]	8	81.0	65.6	441	0.51	76–81
Process solvent after hydroextraction[b]	3	51.6	65.6	439	0.51	77–81
Process solvent to hydroextraction[c]	3	80.7	65.6	441	0.51	82–84
Process solvent from hydroextraction	3	60.1	65.6	441	0.51	82–84
Stabilized anthracene oil, hydrogenated[d]	3	91.0	—	—	—	—

[a] Stabilized with hydrogen under 65.6 atm at 441°C and LHSV of 2; polymer and light ends removed by fractionation.
[b] After 14 cycles of extraction with stabilized anthracene oil.
[c] Solvent hydrogenated over fixed-bed hydrogen donor solvent catalyst: hydrogen absorbed = 0.53 wt % of feed oil.
[d] Solvent hydrogenated over fixed-bed hydrogen donor solvent catalyst: hydrogen absorbed = 2.2 wt % of feed oil.
Source of data: Unpublished CCDC work.

is dependent on the catalytic properties of the coal ash. The loss of hydrogen by dehydrogenation in donor extraction for iron-rich coals is usually negligible until conditions of severity are imposed such that the amount of hydrogen transferred to the maf coal exceeds about 1.0 wt %. Donor extraction of iron-lean coals (i.e., most subbituminous coals and lignite) can be carried up to about 1.5 wt % hydrogen transfer before loss of hydrogen via dehydrogenation becomes significant.

Another limitation on donor extraction is the limited hydrogen transfer capacity of natural solvents after catalytic hydrogenation: approximately 1.5–2.0 wt %. Since the donor capacity cannot be entirely depleted in extraction, the practical donor capacity is likely to be about 1.0 wt % of the solvent fed.

From the considerations above, it is clear that a pure donor extraction process is basically most adaptable to use where the maximum amount of hydrogen transfer involved is around 1.0–1.5 wt % maf coal. Moderately high conversions can be obtained under these conditions (1.0% hydrogen transfer) from bituminous coals (i.e., 75–90 wt %), but in the case of subbituminous coals and lignites the conversions (at 1.5% hydrogen transfer) are lower (i.e., in the range 70–84 wt %). Specific data for 1.0% hydrogen transfer with a number of coals are given in Table 27.1.

Hydrocracking of the extract will also be required in most cases to generate makeup distillates to preserve the solvent balance. Donor extraction is thus most suitable, as in the CSF process, where the extract is hydrocracked to distillate and the extraction residue is used to generate hydrogen and as a source of fuel for the overall process.

It is generally believed, as stated previously, that in the early stages of the hydroextraction process (up to perhaps 1% hydrogen consumption) there is very little consumption of gaseous hydrogen, and in the main the hydrogen is supplied by depletion of the donor content of the solvent. Thus operation under these moderate conversion conditions, if they should prove to be desirable, is obviously the regime of donor extraction. Hydroextraction is clearly the more attractive process when maximum coal conversion is the desired objective.

If hydroextraction under mild conditions indeed follows a donor mechanism, a lower hydrogen consumption would be estimated for donor extraction at the same conversion level. This is primarily because the hydrogen transfer–conversion relationship of Figs. 27.2 and 27.3 excludes hydrogen consumed via hydrocracking of the solvent, which is automatically included in the hydroextraction results. Comparison is difficult because few hydroextraction runs have been conducted under mild conditions and because different coal samples, albeit from the same mine, were used.

A few data points with different coals are given in Table 27.13, which illustrates the point just made. The subbituminous coals generally show very much higher hydrogen consumptions than those predicted from donor extraction. This is because of the relatively severe conditions employed, with resulting higher yields of gas and light distillate.

27.2.8 Operability Limit in Hydroextraction

A limiting set of conditions of severity exists in hydroextraction, should these conditions be exceeded; coke will form, and the result will be inoperability. The boundary conditions under which coke formation occurs may be defined as those above which the coal conversion drops. The drop in coal conversion that occurs when the temperature is increased actually indicates formation of cresol insolubles via coking. Thus long-term operation in this regime is not possible.

The limiting conditions are a function of the solvent-to-coal ratio, donor content of the solvent, hydrogen partial pressure, and temperature. Qualitatively, the interaction between the variables is understood; the maximum operable temperature will tend to increase with increasing solvent-to-coal ratio and increasing donor content of the solvent. The latter, of course, also increases with increasing catalytic activity of the coal minerals and increasing hydrogen partial pressure. Quantitative relationships of a generalized nature have not been outlined to determine the operability limits.

As illustration of the interaction between the variables, unpublished CCDC data show that extract itself undergoes massive coking at 446°C and 240 atm of hydrogen pressure in the absence of both a solvent and a catalyst.

Inoperability with coal due to coking has been experienced in Wilsonville pilot-plant operations with Illinois No. 6 coals. Discussion of the coking of Illinois No. 6 coal from the Monterey Mine at Wilsonville and of HRI bench-scale results bearing on this question has been given by Kang et al.[48]

It is clear from HRI bench-scale data at 103 atm, for example, that increasing the operating temperature from 446 to 462°C causes the onset of coking. This is reflected by a decrease in coal conversion from 94.0 to 89.5 wt % as well as by plugging of the contactor with coke.

Examination of the Wilsonville deposits has been made by investigators from Pennsylvania State University.[49] They report that

[49] *Characterization of Mineral Matter in Coals and Coal Liquefaction Residue,* Annual Report, Pennsylvania State University Report to EPRI, RP 366-1, June 1977.

Table 27.13 Hydrogen Consumption Under "Mild" Conditions: Hydroextraction Versus Donor Extraction

Mine or area: Seam: State:	Burning Star Illinois No. 6 Illinois	Kaiparowits Utah	Black Mesa Arizona	Colstrip Rosebud Montana
Reference	29	35	38	[a]
Hydroextraction conditions and results				
Temperature (°C)	441	427	448	441
Pressure (atm)	65.6	137	171	120
Weight percent maf coal				
Conversion	77.8	84	83	69
Hydrogen consumption	1.4	2.2	2.9	2.8
Donor extraction				
Hydrogen consumption at same conversion (cf. Figs. 27.2 and 27.3)	0.95	1.7	1.5	1.35

[a] Unpublished CCDC work.

the fine-grained material is composed of insoluble particles surrounded by a layer of anisotropic carbon.

The onset of coking at a rate that would be troublesome is essentially a kinetic question. The only stable hydrocarbon, in the thermodynamic sense, under usual hydroextraction conditions, is methane. The pressure of methane in equilibrium with graphite calculated from thermodynamic data[50] and hydrogen at the rather extreme conditions for hydroextraction of 527°C and a hydrogen partial pressure of 6.7 atm, is 63.3 atm. All normally liquid hydrocarbons are thermodynamically unstable with respect to decomposition to graphite and hydrogen at temperatures above 325 K. The thermodynamically favored reaction is thus complete hydrocracking to produce methane as the end product. Thus we must ask how fast coke is formed by pyrolytic processes and how rapidly the coke is converted to methane by reaction with hydrogen. Data to supply answers to these questions are lacking.

27.3 THE ROLE OF HYDROGEN IN PRIMARY LIQUEFACTION

27.3.1 Thermodynamics: Solvent Rehydrogenation

One objective in the primary liquefaction of coal is to maintain the donor activity of the solvent by hydrogenation of polynuclear aromatics to their partially hydrogenated derivatives, or hydroaromatic forms. Severe hydrogenation to form the saturated or perhydrogenated derivatives is undesirable. This is because the perhydrogenated compounds are relatively poor donors, as shown below, and because they also reduce the solubility of the extract in the process solvent.

The thermodynamic relationships underlying this process can be discussed on the basis of measured equilibria involving model compounds. Equilibria data have been presented by Frye[51] and by Frye and Weitkamp[52] for a number of these compounds.

Equilibria were measured in the following systems: naphthalene-hydrogen, phenanthrene-hydrogen, biphenyl-hydrogen, indene-hydrogen, acenaphthene-hydrogen, and fluorene-hydrogen. Equations for calculation of the equilibrium constants as a function of the absolute temperature for the various reactions are given in Table 27.14.

These equilibrium constants refer to the vapor phase reaction. Fugacities of hydrogen, rather than partial pressures, are used in calculating the equilibrium composition. Frye[51] suggests using the Gilliland equation:

$$f_{H_2} = P_{H_2}(1 + 3.3 \times 10^{-4} P_{H_2}) \qquad (27.5)$$

Gorin calculated the vapor phase equilibria for the naphthalene system using Equation 27.5. The results are given in Fig. 27.10, which covers the range of temperatures and hydrogen partial pressures most normally used in hydroextraction.

It is immediately apparent that hydrogenation equilibria are not closely approached. This can be seen by a comparison between the experimental value for the tetralin-naphthalene ratio for the HRI run with Illinois No. 6 coal from the Monterey Mine and the equilibrium value from Fig. 27.10 as shown below:

Approximate hydrogen partial pressure at contactor outlet 70 atm

Tetralin/naphthalene ratio

Experimental value at 448°C: hydroextraction, Monterey coal 0.5

Equilibrium value at 448°C and 70 atm H_2 1.8

Even lower values for the tetralin-naphthalene ratios were obtained in the runs with Burning Star and Black Mesa coals: 0.29 and 0.23, respectively, in spite of the significantly higher pressures used. In these cases the deviation from equilibrium is obviously even greater.

Potgeiter[53] studied the hydroextraction in

[50] D. D. Wagman, J. E. Kilpatrick, W. J. Taylor, K. S. Pitzer, and F. D. Rossini, U.S. Natl. Bur. Standards Res. Paper No. RP 1634, 1945.

[51] C. G. Frye, *J. Chem. Eng. Data*, 7(4), 592 (1962).

[52] C. G. Frye and A. W. Weitkamp, *J. Chem. Eng. Data*, **14**(3), 372 (1969).

[53] H. G. J. Potgeiter, *Fuel*, **52**, 134 (1973).

Table 27.14 Equilibria and Heats of Hydrogenation for Polynuclear Aromatics[44,45]

Reaction	$-\Delta H$(kcal/g-mole H_2)	Log Kp
Naphthalene → tetralin	14.9	$6,460/T - 12.40$
Naphthalene → *cis*-decalin	15.2	$16,500/T - 32.88$
cis-Decalin → *trans*-decalin	—	$695/T - 0.301$
Biphenyl → cyclohexyl benzene	18.0	$11,750/T - 21.71$
Biphenyl → cyclohexyl cyclohexane	18.0	$23,500/T - 44.08$
Phenanthrene → dihydro-P	12.0	$2,600/T - 6.11$
Phenanthrene → tetrahydro-P	15.0	$6,565/T - 13.25$
Phenanthrene → *sym*-octahydro-P	15.0	$13,030/T - 26.38$
Phenanthrene → perhydro (lowest b.p.)	15.0	$23,190/T - 46.49$
Acenaphthene → tetrahydroacenaphthene	13.4	$5,850/T - 11.76$
Acenaphthene → perhydroacenaphthene	11.8	$12,874/T - 27.87$
Fluorene → *cis*-hexahydrofluorene	14.0	$9,242/T - 19.00$
Fluorene → perhydrofluorene (all isomers)	13.6	$17,830/T - 38.66$

batch autoclaves of a subbituminous South African coal at 360–410°C and pressures from 143 to 212 bars; 1% of either cobalt oxide–potassium oxide or $SnCl_2$ was added as a catalyst. The purpose of the experiments was to study the kinetics of tetralin conversion. When the results reported by Potgeiter are interpreted via the equilibrium data discussed above, it is obvious that little hydrogenation of the naphthalene produced in the reaction was taking place. Large amounts of naphthalene were produced, apparently by the donor reaction of tetralin with coal. The tetralin-naphthalene ratios in the products were very much lower than the equilibrium values for naphthalene hydrogenation at the conditions employed.

Previously unreported batch autoclave data from CCDC laboratories also permit one to estimate the approach to equilibrium in donor extraction in the reverse direction (i.e., in the dehydrogenation of tetralin to naphthalene). Some of these data are summarized in Table 27.15, where the experimentally observed partial pressure of H_2 is compared with equilibrium values associated with the

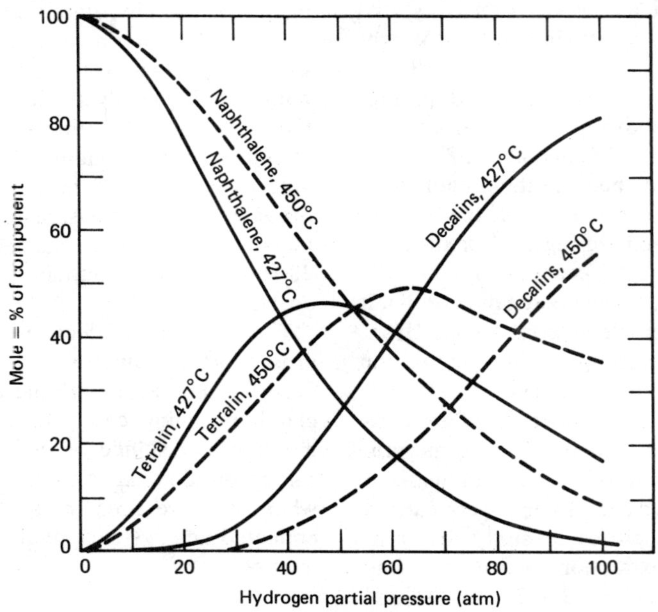

Figure 27.10 Equilibrium composition in the system hydrogen–naphthalene–tetralin–(*cis* + *trans*) decalin. (Calculated by author from data of Frye.[51])

Table 27.15 Batch Donor Extraction Approach to Equilibrium Tetralin Dehydrogenation: Initial Solvent, 50% Tetralin, 50% Methylnaphthalene

Coal	Temperature (°C)	Residence Time (min)	Product Mole Ratio, T/N[a]	Partial Hydrogen Pressure (atm.)[b]	Equilibrium Hydrogen Pressure (atm.)
Ireland Mine	413	60	0.38	15	19
Hillsboro	413	60	0.34	12	18
Burning Star	427	30	0.50	10	27
Kaiparowits	413	60	0.52	2	22
Colstrip	413	60	0.35	1	18
Glenharold	413	60	0.61	4	24

[a] Tetralin to naphthalene.
[b] Solubility in solvent neglected—maximum value.
Source of data: Unpublished of CCDC work.

tetralin-naphthalene ratio in the product. It is noted that in all cases the partial pressure is substantially lower than the equilibrium hydrogen partial pressure. The closest approach is with the high-sulfur, iron-rich, Ireland Mine coal.

The least hydrogen, relative to the equilibrium value, is evolved in the donor extraction of subbituminous coals and lignite. These coals generally are low in iron content. Thus the foregoing data on the approach to tetralin equilibrium from both sides (i.e., from the hydrogen-rich and hydrogen-deficient sides) point to iron in coal ash as a mild catalyst for this reaction. The catalytic action of iron minerals in coal ash was extensively discussed previously by Mukherjee[11] and Wright and Anderson.[46]

The equilibrium data illustrated in Fig. 27.10 show that above hydrogen partial pressures of 60 atm at 427°C and above 83 atm at 450°C, the decalins become the predominant species at equilibrium. Data on the amount of decalins formed during hydroextraction are meager, but it appears likely that the approach to equilibrium is even more difficult than in the naphthalene-tetralin case. Iron, apparently, is a very poor catalyst for the conversion of tetralin to decalin.

The poor catalytic properties of iron, as just mentioned, are illustrated by some previously unpublished CCDC data summarized in Table 27.16. The experiments were carried out in batch autoclaves to study the iron-catalyzed hydroextraction of Ireland Mine coal. A mixture of tetralin (25 wt %) and methylnaphthalene (75 wt %) was used as the solvent. The compositions of the feed and recovered solvent are shown in Table 27.16. The hydrogen partial pressure, though not accurately known in the experiments of Table 27.16, was approximately 200 atm. The equilibrium under these conditions would call for nearly complete conversion to decalin. The equilibrium ratios of tetralin to naphthalene at a hydrogen fugacity of 200 atm are given in Table 27.16 for comparison with the experimental values. The approach to equilibrium in hydrogenation to tetralin is much closer than for hydrogenation to decalin, but the values are still far removed from equilibrium. It must be concluded that iron catalysts are fairly selective for hydrogenation of naphthalene to tetralin rather than to decalin.

Vapor phase hydrogenation of naphthalene over tungsten–nickel sulfide and molybdenum disulfide catalysts yields fairly substantial hydrogenation but not equilibrium values at relatively mild conditions. Thus at a space velocity of 0.51 kg/(l)(hr) at 298°C and 50 atm pressure, with a tungsten–nickel sulfide catalyst,[54] the product contained 6.3, 63.7, and 30% naphthalene, tetralin, and decalin, respectively.

The formation of decalin and other perhydrogenated polynuclear aromatics does not occur readily, even with the use of commercial hydrofining catalysts such as cobalt–molybdenum sulfide or nickel–molybdenum sulfide on alumina. This is particularly true when they are used in the liquid phase in ebullated bed systems involving coal or coal extract.

[54] N. V. Shavolina, D. F. Kasatkin, and V. I. Karzhev, *Koks Khim.*, No. 11, 45 (1959).

Table 27.16 Recovered Solvent Compositions for Autoclave Hydroextraction of Ireland Mine Coal

Conditions: 1% Fe impregnated on coal as $FeSO_4$; residence time, 60 min; solvent/dry coal, wt ratio = 1.0

		Recovered Solvents (wt %)		
Temperature (°C):		400	420	440
Total hot pressure (atm):	Feed Solvent (wt %)	259	273	280
Dimethylnaphthalenes	3.7	0.8	0.6	0.5
(α + β)-Methylnaphthalenes	69.4	54.9	44.2	35.7
(α + β)-Methyltetralins	—	12.8	19.5	20.4
Naphthalene	—	5.3	8.0	13.7
Tetralin	26.8	25.7	26.2	25.5
cis + *trans*-Decalins	0.1	0.3	0.8	2.4
Light ends	—	0.2	0.7	1.8
Tetralin-naphthalene, experimental		4.9	3.3	1.9
Tetralin-naphthalene, equilibrium at f_{H_2} = 200 atm		63	33	18

Source of data: Unpublished CCDC work.

The natural solvent produced in these systems not only has a low content of saturates (cf. Table 27.9), but the ratio of tetralin to naphthalene of the recovered solvent from pilot-plant operation[14] of about 1.6 is well below the equilibrium value considering the severe conditions employed (i.e., 240 atm at 440°C).

Experimental equilibrium data on the hydrogenation of alkylated naphthalenes to the corresponding alkyl tetralins are lacking. Zakharenko[55] used the available thermodynamic data in the naphthalene-tetralin-decalin system to calculate the free energy of formation of these compounds, both from the elements and from methylene groups. The data above were then combined with the approximate method of Ciola[56] to estimate the effect of adding an additional methylene group.

Zakharenko[55] presented results of his calculations on the methylnaphthalene-methyltetralin-methyldecalin system at only one temperature, 527°C. Under these conditions, his results indicate that the methyltetralin-methylnaphthalene equilibrium ratio is higher than the corresponding tetralin-naphthalene ratio. The reverse is true for the equilibrium ratio of methyldecalin to methyltetralin. If the results are valid, they would indicate that from the equilibrium point of view, methyl substitution would permit formation of a higher concentration of active donors. The experimental results of Table 27.15, accordingly, indicate that the tetralin-methylnaphthalene ratio was much further from equilibrium than the tetralin-naphthalene ratio. In other words, hydrogenation of methylnaphthalene is kinetically more difficult than naphthalene. Direct experimental data are required, however, before Zakharenko's calculations may be validated. Bondi,[57] for example, arrived at the opposite conclusion: the equilibrium in the hydrogenation of methylnaphthalene is less favorable than that of naphthalene. Bondi used another semiempirical method of estimating the equilibrium constants. Frye,[51] for example, found that the equilibria in the hydrogenation of phenanthrene calculated by Bondi[57] were seriously in error.

The equilibria in the formation of active donors from the other polynuclear hydrocarbon investigated by Frye[51] and Frye and Weitkamp[52] are less favorable than in the tetralin-naphthalene system. This is illustrated by the calculations summarized in Table 27.17 under common conditions of 450°C and

[55] V. A. Zakharenko, *Khim. Tekhnol. Smol. Term. Pererab. Tverd. Topl. Akad. Nauk, SSR*, 65 (1965).

[56] R. Ciola, *Ind. Eng. Chem.*, **49**, 1789 (1957).

[57] A. Bondi, *Pet. Refiner*, **38**(2), 161 (1959).

Table 27.17 Equilibrium Ratios of Donors at 450°C and 70 atm Hydrogen Partial Pressure: Based on Table 27.14

Polynuclear Aromatics	Donors	Equilibrium Mole Ratio, $\left(\frac{\Sigma \text{ donors}}{\text{aromatic}}\right)$
Naphthalene	Tetrahydro	1.76
Biphenyl	Hexahydro	1.28
Acenaphthene	Tetrahydro	1.10
Phenanthrene	Dihydro + tetrahydro + *sym* octahydro + *asym* octahydro (*cis* + *trans*)	0.72
Fluorene	*cis* + *trans*-Hexahydro	0.28

70 atm partial pressure using the equilibrium constants calculated from the equations summarized in Table 27.14. Heats of hydrogenation for the various reactions are also given in Table 27.14. It is seen that average heat evolution in the hydrogenation of aromatic rings is about 15 kcal/mole H_2.

Wiser et al.[58] investigated the hydrogenation of anthracene in batch autoclaves at 220–435°C and 103 atm total hot pressure catalyzed by nickel–tungsten sulfide. Equilibrium data are not available for anthracene hydrogenation, but they should be similar to those for phenanthrene.

The system is controlled kinetically, not by equilibrium.[58] Dihydroanthracene appears to be the primary product, as indicated by the rate data of the same investigators,[58] illustrated in Fig. 27.11. Equilibrium may be approached in the anthracene-dihydroanthracene reaction after extended periods.

Anthracene is completely consumed after a sufficiently long residence time at all temperatures: 160 min is required at 220°C. In no case was any perhydroanthracene noted in the product; that is, other than cracked products, the products consisted entirely of a mixture of the dihydro, tetrahydro, and octahydro derivatives.

Huang et al.[59] found a very dissimilar pic-

[58] W. H. Wiser, Surjit Singh, S. A. Qader, and G. R. Hill, *Ind. Eng. Chem. Prod. Res. Dev.*, **9**, 350 (1970).

[59] C. S. Huang, K. C. Wang, and H. W. Haynes, *Prepr., Am. Chem. Soc., Div. Fuel Chem.*, **21**,(5), 228, 1976).

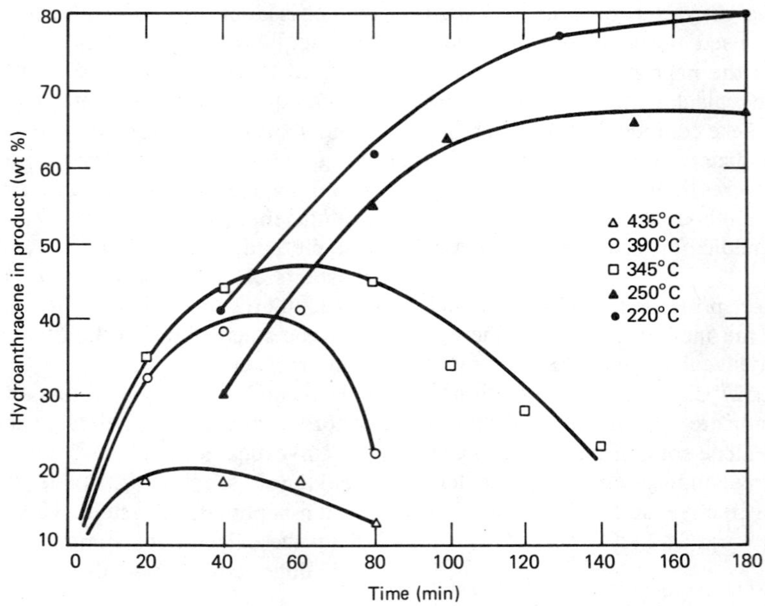

Figure 27.11 Rate data for the formation of 9,10-dihydroanthracene during anthracene hydrogenation.[58]

ture in the hydrogenation of phenanthrene. They used quite a different system: a fixed-bed flow system at 137 atm pressure over a cobalt molybdate on alumina catalyst. Their results indicate, in contrast to those of Wiser et al., that perhydrogenation occurred readily but very little dihydrophenanthrene was formed. Octahydro- and perhydrophenanthrene were the major products below 430°C. Above this temperature extensive ring opening occurred, such that the closest approach to equilibrium at a liquid hourly space velocity (LHSV) of 2 was obtained at about 410°C. A comparison of their results with equilibrium using Frye's data is shown below:

	Experimental	Equilibrium
Hydrogen partial pressure (atm)	137	137
Temperature (°C)	375–427 (Nonisothermal)	413
Amount in Product (mole %)		
Phenanthrene	0.8	0.8
Dihydrophenanthrene	—	0.5
Tetrahydrophenanthrene	1.6	2.5
Octahydrophenanthrene	27.7	9.6
Perhydrophenanthrene	69.9	86.6

27.3.2 Hydrogen Solubility and Vapor-Liquid Equilibria

The solubility of hydrogen in the process liquid and the vapor-liquid equilibria involving the other species present are important for a better understanding of the nature of the process, for elucidation of the reaction kinetics in the system, and for engineering design of the liquefaction systems.

Hydrogen solubility data are available for a number of pure hydrocarbons, although in many cases both the temperatures and pressures used to obtain them were well below the range of practical interest for coal liquefaction systems. Generally speaking, Henry's law is valid at any given temperature; that is, the solubility expressed as mass of hydrogen dissolved per unit mass of liquid is proportional to the hydrogen partial pressure. Thermodynamically, a more exact definition is that the mole fraction of dissolved hydrogen in the liquid is proportional to the hydrogen fugacity. Cukov and Prausnitz[60] used the latter more exact method of reporting solubility data for hydrogen at temperatures in the range of 290–475 K and pressures to 125 atm in hexadecane, bicyclohexyl, and diphenylmethane. Low-temperature (up to 100°C) and low-pressure (up to about 1.25 atm) solubility data of hydrogen in decalin, benzene, cyclohexane, phenol, and cyclohexanol were reported by Kruyer and Nobel.[61] High-pressure hydrogen solubility data have been reported in hexane and benzene[62,63] at temperatures up to about 250°C and pressures up to about 600 atm.

Data at relatively high temperatures (up to about 427°C) and pressures (up to 270 atm) are available also for the following pure compounds: heptane,[64] methylcyclohexane,[64] 2,-2,4-trimethylpentane,[64] tetralin,[10] and mesitylene.[65] Data on the solubility of hydrogen in coal liquids were obtained in Germany[66] in the course of their development of liquefaction. Some of these data are discussed below.

Scattered data have been obtained on the solubility of hydrogen[9] in various liquids from the ebullated bed hydrogenation of coal extract. Rather extensive data have been reported by Guin et al.[67] on the solubility of hydrogen in creosote oil.

[60] P. M. Cukov and J. M. Prausnitz, *J. Phys. Chem.*, **76**(598) (1972).

[61] S. Kruyer and A. P. P. Nobel, *Recueil*, **80**, 1145(1961).

[62] K. C. Chao and J. D. Seader, *Am. Inst. Chem. Eng. J.*, **7**, 598 (1961).

[63] J. F. Connolly, *J. Chem. Phys.*, **36**, 2897 (1962).

[64] S. Peter and K. Reinhartz, *Z. Phys. Chem.*, **24**, 103 (1960).

[65] B. E. Eakin and W. E. Devaney, "Thermodynamics Data and Correlations," in Am. Inst. Chem. Eng. Symp. Ser., Vol. 70, No. 140, 1974, p. 30.

[66] E. E. Donath, consultant, private communication, 1975.

[67] J. A. Guin, A. R. Tarrer, W. S. Pitts, and J. W. Prather, *Prepr., Am. Chem. Soc., Div. Fuel Chem.*, **21**(5), 170 (1976).

Typical data for various pure compounds and natural oils are summarized in Table 27.18, where "pseudo" Henry's law constants (grams of hydrogen dissolved per gram of oil-atm H_2) are given as a measure of solubility.

Some typical qualitative trends are illustrated by the data of Table 27.18. The solubility of hydrogen increases in all cases with temperature. This is a result of the endothermic heat of solution, which is about 1.0–2.0 kcal/mole for most liquids. The solubility, in general, decreases with increasing polarity and aromaticity of the liquid. Finally, solubility decreases with increasing molecular weight. This is especially evident for the data with hydro liquids, where the solubility decreases strongly with increasing content of +475°C residuum.

The general vapor-liquid equilibria for all species present in a liquefaction system have been treated by a number of authors. Such a general correlation was presented, for example, by Chao and Seader.[62] The correlation was based on the use of the thermodynamically exact equation for each species present:

$$K_i = \frac{Y_i}{X_i} = \nu_i^0 \frac{\gamma_i}{\phi_i} \tag{27.6}$$

where

K_i = equilibrium constant of ith component

Y_i = mole fraction in vapor of ith component

X_i = mole fraction in liquid of ith component

ν_i^0 = fugacity coefficient of pure liquid component i

γ_i = activity coefficient of component i in liquid solution

ϕ_i = fugacity coefficient of component i in vapor mixture

The fugacity coefficient ν_i^0 was determined as a function of reduced temperature and pressure by use of the correlating equations of Lyderson et al.[68] and Curl and Pitzer.[69] Special treatment is required for gases such as H_2 and CH_4, which are far above their critical temperature.

[68] A. L. Lyderson, R. A. Greenhorn, and O. A. Hougen, University of Wisconsin Eng. Exp. Sta. Rep. No. 4, Madison, 1955.

[69] R. F. Curl, Jr., and K. S. Pitzer, *Ind. Eng. Chem.*, **50**, 265 (1958).

The activity coefficient in the liquid, γ_i, was correlated by use of Hildebrand[70] equations applicable to regular solutions. Regular solutions are defined as those whose excess entropy is zero. Nonideal behavior is due entirely to the heat of solution.

The fugacity coefficient in the vapor, ϕ_i, is correspondingly determined by the Redlich and Kwong[71] equation of state.

Because of its complexity, the correlation must be programmed on a digital computer for ready application. Unfortunately, it is not valid outside the range in which it was developed, since many of the constants in the correlating equations must be determined empirically to give the best fit to experimental data. Generally, Chao and Seader[62] do not recommend use of the correlation at temperatures above 260°C, pressures above 136 atm, or for liquids whose pseudoreduced temperature exceeds 260°C.

Grayson and Streed[72] extended the Chao and Seader[62] correlation to pressures up to 200 atm and temperatures to 427°C by vapor-liquid experimental data obtained in the hydrocracking of petroleum gas oils. A simplified form, for use without a digital computer, was also presented by Shen.[73]

The Grayson and Streed[72] and Shen[73] correlations for materials boiling above ethylbenzene are directly applicable to petroleum-based systems only and cannot be used to accurately determine K values for the highly aromatic coal liquefaction systems. In the latter system also, water vapor, hydrogen sulfide, and carbon oxides are important components whose K values cannot be estimated within the framework of the correlations.

The Grayson and Streed[72] method has been extended by some engineering companies to include the use of aromatic oils.[74] Their results, however, have not been published.

[70] J. H. Hildebrand and R. L. Scott, *The Solubility of Non-Electrolytes*, 3rd ed., Reinhold, New York, 1950.

[71] O. Redlich and J. N. S. Kwong, *Chem. Rev.*, **44**, 233 (1949).

[72] H. G. Grayson and C. W. Streed, "Vapor-Liquid Equilibria for High-Temperature, High-Pressure Hydrogen-Hydrocarbon Systems," Paper No. 20, presented at 6th World Petroleum Congr., Section VII, Hamburg, 1963.

[73] R. C. Shen, *Proc. Am. Pet. Inst. Sect. III*, **44**, 23 (1964).

[74] M. Sooter, formerly of Catalytic Construction Company, private communication, 1975.

Table 27.18 Solubility of Hydrogen in Various Solvents and Natural Oils, in grams per gram of oil–atm hydrogen × 10⁶

Temp (°C)	n = 1 Hexane[62–63]	Methyl-cyclo-hexane[64]	Decalin[61]	Diphenyl-methane[60]	Tetralin[10]	Creosote Oil[67]	Hydro Liquids[10] (wt % + 475°C residue)			
							37	70	80	100
100	58	—	6	5	—	9	—	—	—	—
200	152	30	—	8	10	10	—	—	—	3
300					14	11	—	—	—	—
370					—	—	26	—	5	—
400					34	14	—	—	—	—
446					—	—	—	10	—	—

CCDC has obtained extensive equilibrium data on the hydrogen-tetralin system.[10] A comparison of the experimental K values with those estimated via the "revised" Grayson and Streed[72] correlation is shown in Fig. 27.12. The Grayson and Streed K values are about 25% lower than the experimental values. This was also noted in recent work by Lawson et al.[75] Grayson and Streed values above 370°C are not shown, since the correlation becomes invalid as the critical temperature of the liquid is approached.

A comparison with German solubility data supplied by Donath[66] is also shown in Fig. 27.12. Here again, the K values calculated by the Grayson and Streed correlation are about 20% below the experimental values.

The K values determined in actual coal and extract hydrogenation systems are available for a few spot data points from Wilsonville pilot-plant[31] and CCDC[10] bench-scale operations. More extensive vapor-liquid equilbria data were obtained in other unpublished CCDC bench-scale hydrogenation work but are not presently available.

The K values for hydrogen and other components were determined at only two temperatures at Wilsonville: about 325 and 65°C, respectively. Assuming that the product ($K \times P$) should be approximately constant, where P is the system pressure, the Wilsonville values average about 100% above the Grayson and Streed[72] values for heavy oil as given in Fig. 27.12.

Two data points from CCDC bench-scale work[10] give K values for hydrogen compared with Grayson and Streed at 370°C as follows:

K Value Source	Pressure (atm)	Hydro Oil (wt % residuum)		
		0	37	80
Grayson-Streed (Fig. 27.12)	200	2.4	—	—
CCDC bench scale	238	—	2.1	5.5

Fairly good agreement is obtained with the hydro oil, which has a low residuum content. Thus both CCDC[10] and Wilsonville[31] data tend to confirm the relatively low solubility, discussed above, of hydrogen in hydro liquids high in extract content.

The K values for the other components were also obtained in the Wilsonville[31] and CCDC[10] tests but are of doubtful accuracy. The K values for the "heavy" distillate cuts (b.p. > ~ 260°C) are in approximate agreement with the Grayson and Streed values for the corresponding petroleum fractions.

It is obvious that many more experimental data are required to better define the vapor-liquid equilibria in coal liquefaction systems.

27.3.3 Liquefaction via Gaseous Hydrogen

It is generally agreed that the predominant reaction, at least in the early stages of liquefaction, is transfer of hydrogen from the solvent to the coal. This is evidenced,[6] for example, by depletion of the solvent donor content even when hydrogen pressure is applied in the early stages of the reaction. Liquefaction of coal can be achieved, how-

[75] C. C. Lawson, K. D. Lui, J. J. Simmick, H. M. Lin, and K. C. Chao, "Vapor-Liquid Equilibriums in the Systems H_2 Diphenylmethane and H_2 Tetralin to 425°C and 120 atm," Paper No. 16e, presented at American Institute of Chemical Engineers, 69th Meeting, Chicago, November 29, 1976.

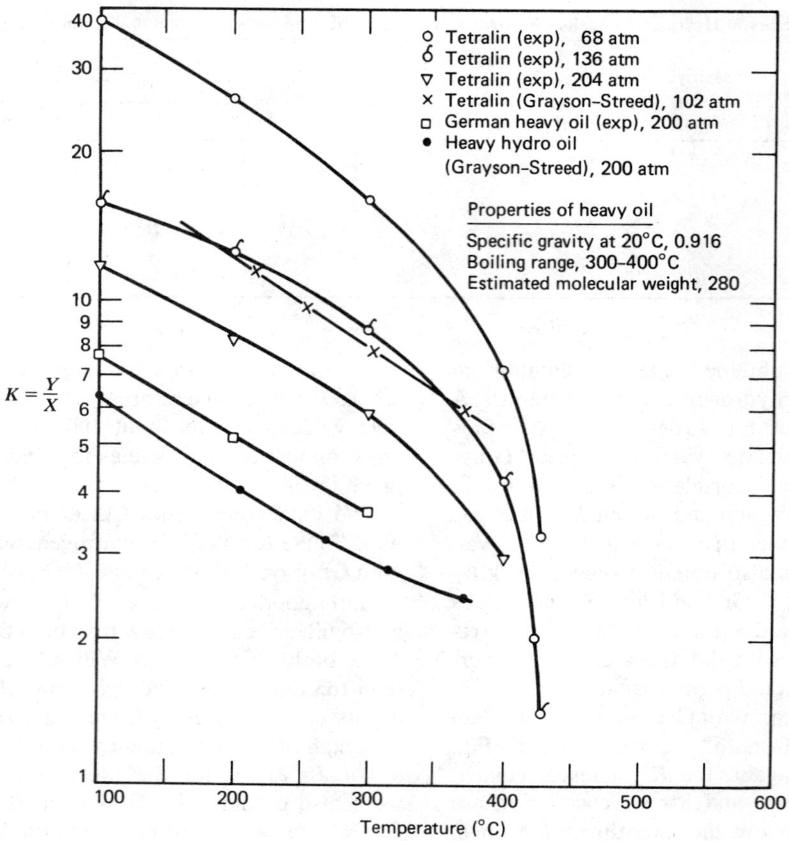

Figure 27.12 Hydrogen vapor–liquid equilibria for hydrogen–tetralin and hydrogen–hydro oil systems. Comparison of experimental data with Grayson-Streed correlation. Grayson-Streed calculations supplied by M. Sooter.[74] (From tetralin data.[10])

ever, even in the absence of a solvent, either as a result of "activation" of the hydrogen[11] by catalytic components of the ash or by addition of catalysts.[76] The question of whether a direct thermal reaction (noncatalytic) of hydrogen with coal can take place has not been satisfactorily answered.

The free radical mechanism to be discussed later, if accepted, would make the direct thermal reaction unlikely because of the high activation energy required for reaction of coal-derived free radicals with hydrogen. This is a consequence of the high bond strength in hydrogen (102 kcal/mole) as compared with the relatively low bond strength in the activated C—H bond of a good donor such as tetralin.

Some previously unpublished CCDC autoclave experiments with Ireland Mine coal with and without impressed hydrogen pressure, bearing on this question, are summarized in Table 27.19. Perhaps the most significant result is the high coal conversion of 92.5% obtained with the weak donor solvent dimethyldecalin when the autoclave was pressurized with hydrogen to 171 atm.

Dimethyldecalin, in the absence of hydrogen when tested as a solvent, gave a conversion of only 42 wt %. Thus it is clear that given sufficient imposed hydrogen pressure, high coal conversions can be achieved even in the absence of a donor solvent. This does not prove, however, that hydrogen reacts thermally with the coal; it may simply point to a reaction catalyzed by coal ash, similar to the results of Mukherjee[11] previously cited. Table 27.19 also compares results using tetralin as solvent with and without added hydrogen pressure. The addition of hydrogen increases the total hydrogen consumption by about 20% and decreases the amount trans-

[76] P. H. Given, D. C. Cronauer, W. Spackman, H. L. Lovell, A. Davis, and B. Biswas, *Fuel*, **54**, 34 (1975).

Table 27.19 Batch Autoclave Experiments with Ireland Mine Coal Comparing Donor Versus Hydroextraction: 60 min Residence Time at Temperature

Temperature (°C):	400	400	400	441	441
Total hot pressure (atm):	171	~35	171	~70	171
Hydrogen added:	Yes	No	Yes	No	Yes
Solvent:	DMD[a]	Tetralin	Tetralin	Tetralin	Tetralin
Solvent/maf coal, wt ratio:	1	1	3	3	1
Weight percent maf coal					
C_4 + distillate	6.8	14.2	12.1	18.4	22.7
Extract	81.4	73.9	75.9	68.8	67.4
Coal conversion	92.5	92.5	92.4	92.6	96.9
Hydrogen consumed	1.13	1.36	1.70	2.48	2.96
Hydrogen transferred (wt %)		1.36	0.71	2.48	1.43
Hydrogen reacted (wt %)		—	0.99	—	1.53

[a] Dimethyl decalin.
Source: Unpublished CCDC work.

ferred by about 50%. The latter may signify rehydrogenation of the naphthalene in situ. Otherwise, the overall results are quite similar with and without the addition of hydrogen.

Another difference was noted in comparing donor extraction with hydroextraction of Illinois No. 6 coal from the Burning Star Mine: namely, 94% coal conversion was readily achieved in hydroextraction at 103 atm and 441°C. Such high coal conversions were never achieved via donor extraction with synthetic donor solvents in the absence of hydrogen. The highest conversion (91%) was achieved with hydrogenated anthracene oil. No such difficulty was apparent with another Illinois No. 6 coal in results reported by Neavel.[26] Very high coal conversions, approaching 100%, were obtained with tetralin in donor extraction of this coal. The Neavel[26] coal was, however, highly cleaned, so that unreactive components were rejected with the mineral matter.

The experience with Burning Star coal may indicate a catalytic reaction between hydrogen and the insolubles to increase the net conversion.

In their study of the hydroextraction of Polish coals with anthracene oil and hydrogenated anthracene oil, Rank et al.[77,78] concluded that hydrogen gas reacted directly with free radicals produced by depolymerization of the coal. The evidence cited was that yields of products were independent of the applied hydrogen pressure with the hydrogenated anthracene oil, but increased with applied hydrogen pressure with the unhydrogenated oil. These data can be interpreted in many other ways, however, and do not prove the investigators' basic hypothesis.

Whitehurst and Mitchell,[45] in runs with Kentucky No. 14 coal at 1.3 min residence time and 413°C using a synthetic donor solvent, compared the effect of impressed hydrogen pressure on conversion. Conversion decreased from 78 to 65% when hydrogen pressure was decreased from 88 to 0 atm.

As a general conclusion, high coal conversions are more readily obtained when hydrogen pressure is applied. It is probable that the bulk of the hydrogen reacts through the donor solvent as an intermediary, but direct reaction with hydrogen catalyzed by coal ash is important in some cases.

27.4 PRIMARY LIQUEFACTION KINETICS

27.4.1 Mechanistic Considerations

An excellent review of the theory of coal extraction has been presented by Roy and Lahiri.[79] Coal can be considered as a gel that

[77] J. Rank, A. Rutkowski, and M. Rutkowski, *Pr. Nauk Inst. Chem. Technol. Nafty Wegla Politech. Wroclaw,* **11**, 43–53 (1972).

[78] J. Rank and M. Rutkowski, *Koks, Smola, Gaz,* **18**(3), 77–80 (1973).

[79] J. Roy and A. Lahiri, *Proc. Symp. Chemicals and Oil from Coal,* Dhanbad, India, December 1969, Paper No. 48, pp. 447–455, published by Central Fuel Research Institute, Dhanbad, India, 1972.

contains soluble molecules fixed by van der Waals forces in an insoluble coal matrix. The use of nondonor extraction solvents, of which methylnaphthalene and unhydrogenated anthracene oil are typical, can extract the soluble molecules without extrinsic addition of hydrogen. This can be accomplished by application of relatively high temperatures (i.e., 350–425°C), which serves to swell the gel coal matrix and also aids in overcoming the van der Waals forces holding the soluble molecules in place. Thus coal conversions in the range of 30–50 wt % of the maf coal can be achieved (cf. Table 27.11, Fig. 27.7, and Ref. 10) without use of hydrogen, although the possibility cannot be excluded that significant internal hydrogen transfer takes place within the coal itself. The conversion with nondonor solvents, in general, increases with coal rank from subbituminous to bituminous, with maximum yields achieved with hvBb Pittsburgh Seam coal.

Anderson et al.[80] showed that removal of part of the soluble fraction with tetralin can be accelerated at low temperature (47–87°C) by use of ultrasonic radiation. No hydrogen transfer from the tetralin takes place at these low temperatures.

Higher coal conversion requires rupturing or depolymerization of the insoluble coal matrix, termed γ-gel phase by Roy and Lahiri.[79] These workers, and many others, consider this γ phase to be a three-dimensional polymer. Bond rupture can be accomplished either by heterolytic or homolytic dissociation.

Heterolytic dissociation has been achieved by use of electrophilic reagents. Neuworth et al.,[81–83] for example, used BF_3-phenol as an electrophilic reagent. Their work is especially interesting because it proved the presence of easily ruptured methylene bridges in the coal structure. Under normal coal liquefaction conditions, however, the polarity of the solvent is probably too low to permit a heterolytic dissociation mechanism. Homolytic dissociation to yield free radicals is likely to be the predominant mechanism. The kinetics development presented below is based largely on this point of view.

Squires[84] has also discussed the mechanism of donor liquefaction. He presents the view that part of the coal may be converted directly to oils and asphaltenes without intermediate conversion via the usual sequence:

$$\text{coal} \rightarrow \text{cresol solubles} \rightarrow \text{asphaltenes} \rightarrow \text{oil}$$

Supporting experimental evidence has been presented by Yoshida et al.[85] based on autoclave hydrogenation of Japanese coals with iron catalysts.

27.4.2 Semiempirical Rate Models: Donor Extraction

27.4.2.1 *Second-Order Kinetics*

Wiser et al.[25,86] and Hill et al.[21,22] studied the kinetics of pyrolysis and solution of a Utah bituminous coal in tetralin. Both the pyrolysis and donor extraction processes could be partially fitted by a second-order rate model followed by a final, slow, first-order rate. The donor extraction second-order rate equation was written as follows:

$$\frac{dx}{dt} = k_2'(a - x)(b - x) \qquad (27.7)$$

where x is fractional coal conversion to benzene solubles, a is the maximum weight fraction of coal extracted at the temperature in question, b is 1, and k_2' is the second-order rate constant. Wiser[25] discussed the kinetics in terms of the Eyring absolute rate theory and showed that k_2' had an enthalpy of activation of 28.8 kcal/mole.

Inspection of Equation 27.7, however, reveals that it cannot represent a second-order

[80] L. L. Anderson, M. Y. Shifai, and G. R. Hill, *Fuel,* **53,** 32 (1974).

[81] L. A. Heredy, A. E. Kostyo, and M. B. Neuworth, in *Coal Science,* Am. Chem. Soc. Adv. Chem. Ser. No. 55, 1966, p. 493.

[82] L. A. Heredy and M. B. Neuworth, *Fuel,* **41,** 221 (1962).

[83] L. A. Heredy, A. E. Kostyo, and M. B. Neuworth, *Fuel,* **44,** 43 (1964).

[84] A. M. Squires, "Reaction Paths in Donor Liquefaction," paper presented at Conf. Coal Gasification–Liquefaction, US–USSR Trade and Economic Council, Inc., Moscow, October 11–14 1976; published in *Appl. Energy,* **4,** 161–185 (1978).

[85] R. Yoshida, Y. Maekawa, T. Ishii, and G. Takeya, *Fuel,* **55,** 337 (1976).

[86] W. H. Wiser, L. L. Anderson, S. A. Qader, and G. R. Hill, *J. Appl. Chem. Biotechnol.,* **21,** 82 (1971).

reaction involving the separate reactants tetralin and coal. This is because in Wiser's experiments a very large excess of tetralin was used; the tetralin-coal weight ratio was 10 : 1. No analyses were presented to follow the change in solvent composition with time, but assuming a reasonable maximum hydrogen transfer of 2.5 wt % of the mf coal, the tetralin concentration could only decrease from 100 to 95% at the end of a typical run. Thus a better way of looking at Wiser's results is that they involve a complex series of first-order decomposition reactions of the coal, where the first-order rate constant k is a function of the coal conversion and can be approximated by the expression

$$k = k_2'(1 - x)$$

CCDC workers[17] have similarly correlated kinetic data on coal extraction by a "pseudo" second-order equation. Equation 27.8 presents their results in a slightly different manner, that is, using Wiser's nomenclature:

$$\frac{dx}{dt} = k_2' (a - x)^2 \qquad (27.8)$$

A series of microautoclave experiments to elucidate the kinetics of donor extraction was carried out by Curran et al.[4] The tetralin concentration in the solvent was varied to test whether a second-order reaction between coal and tetralin could adequately represent the results. It was shown on this basis that the process did not have the characteristics of a true second-order reaction.

Furthermore,[4] it was shown that when a reactive donor was employed, the amount of hydrogen transfer was the same, independent of the donor employed.

27.4.2.2 *First-Order Representation*

The facts just discussed are consistent with the view, proposed by Wiser,[25] that the rate of coal conversion and hydrogen transfer is determined by the rate of thermal decomposition of the coal.

On this basis, the kinetics of coal conversion were treated[4] as occurring via a series of independent first-order reactions of varying rates. The different rates, on the free radical model, could be interpreted as representing the rupture of homopolar bonds of varying strength in the coal matrix.

The rate of coal pyrolysis was represented by Equation 27.9:

$$C_T - C = \sum_{i=1}^{n} C_i \exp -K_i t \qquad (27.9)$$

where C_i is the amount (wt %) of the coal that decomposes with the characteristic rate constant K_i. Where the pyrolysis is carried out in the presence of a high concentration of active donor such as tetralin, all the free radicals produced via coal pyrolysis are stabilized by the hydrogen transfer process. Now C_T is defined as the maximum coal conversion and C is the actual conversion.

The hydrogen transfer rates were treated by an analogous equation:

$$H_T - H = \sum_{i=1}^{n} h_i \exp -k_i t \qquad (27.10)$$

where H_T is the maximum amount of hydrogen that can be transferred (wt % of the maf coal), H is the actual amount of hydrogen transferred, and h_i is the amount (wt %) of the maf coal that accepts hydrogen with the characteristic rate constant h_i.

Sufficiently accurate fit to the experimental data was achieved by simply dividing the coal into two fractions, γ and $1 - \gamma$, which decompose by a fast and by a slow reaction, respectively. The integrated forms of Equations 27.9 and 27.10 are:

$$\frac{C_T - C}{C_T} = \gamma \exp -K_1 t + (1 - \gamma) \exp -K_2 t \qquad (27.11)$$

$$\frac{H_T - H}{H_T} = \gamma_1 \exp -k_1 t + (1 - \gamma_1) \exp -k_2 t \qquad (27.12)$$

where γ, correspondingly, represents the fraction of the free radicals generated by coal pyrolysis at the high rate, and $1 - \gamma$ represents the fraction produced at the low rate. The values of the parameters above were determined as follows, for the Ireland Mine coal, from the Pittsburgh Seam:

$$C_T = 0.91$$

$$\gamma = 0.34$$

$$H_T = 2.6$$

$$\gamma_1 = 0.12$$

$$K_1 = 2.5 \times 10^{10} \exp\left(\frac{-30{,}000}{RT}\right)$$

$$K_2 = 5.1 \times 10^{11} \exp\left(\frac{-38{,}250}{RT}\right)$$

$$k_1 = 4 \times 10^{12} \exp\left(\frac{-40{,}000}{RT}\right)$$

$$k_2 = 9.1 \times 10^{12} \exp\left(\frac{-45{,}500}{RT}\right)$$

The activation energy for the high coal conversion rate of 30 kcal/mole is low for rupture of covalent bonds, and may represent energy required to break the van der Waals forces holding the soluble coal molecules in the matrix. The entropy of activation calculated by Eyring's absolute rate theory had a very high negative value of −20.7 eu. This may be rationalized as meaning that the surrounding coal molecules must be oriented in a very specific manner to permit escape into the surrounding solvent medium.

The fast rate process, therefore, likely represents the first part of the coal dissolution process that takes place without transfer of hydrogen from the solvent.

The activation energies for the slow rates in the range of 38–45 kcal/mole are also low for rupture of most covalent bonds. It may signify that highly resonating stabilized free radicals are being formed in the bond rupturing process. Even so, the activation energies are significantly higher than those calculated by Wiser.[25] Direct comparison is not possible, however, because different coals, different models, and differing measures of conversion were used.

CCDC workers have only presented detailed kinetics for a single coal, namely, the Pittsburgh Seam, Ireland Mine coal.

As was mentioned previously, however, it is a rather curious fact that the hydrogen transfer rates for most coals are quite similar. This is illustrated in Fig. 27.13, which presents microautoclave rate data for a number of coals.

The application of the rate equations to continuous flow systems has also been discussed previously.[27] The equations above are valid for batch runs or continuous runs in a plug-flow reactor. In the latter case the residence time, t, is defined in Equation 27.13:

$$t = \frac{\text{reactor inventory (cm}^3\text{)}}{\text{product withdrawal rate (cm}^3\text{/min)}} \tag{27.13}$$

Where true plug flow is not maintained, it is easy to show that the equations when applied to a continuous system, should be converted to the form:

$$\frac{C_T - C}{C_T} = \frac{\gamma}{(K_1t/n + 1)^n} + \frac{1-\gamma}{(K_2t/n + 1)^n} \tag{27.14}$$

$$\frac{H_T - H}{H_T} = \frac{\gamma_1}{(k_1t/n + 1)^n} + \frac{1-\gamma_1}{(k_2t/n + 1)^n} \tag{27.15}$$

where n is the number of perfectly mixed stages.

The validity of the equations was checked out in the continuous bench-scale, single-stage extractor ($n = 1$). The results are shown in Fig. 27.14, and the agreement is generally good.

Means for modifying the equations, for the case in which effect of donor concentration becomes important, were also outlined previously.[4,27] The free radical model was used as a basis for developing the semiempirical equations.

Equations thus modified were applied to continuous extraction results with a Utah bituminous coal. Satisfactory representation of the experimental results was again achieved[29] by application of the model, assuming the same rate constants applied to this coal as for the Ireland Mine case.

27.4.2.3 *Free Radical Mechanism*

The set of rate equations just given is consistent with a free radical mechanism as had been proposed not only by Curran et al.[4] but also by Wiser[25] and others. The rather good fit of the equations to experimental data does not in itself prove that the mechanism is valid; they should be regarded primarily as a convenient semiempirical method of correlating the data. They should be useful also for scaleup purposes and engineering design.

More direct evidence for a free radical mechanism is the formation of dihydronaphthalene as an intermediate. This was proposed by Curran,[4] but no evidence was presented for its formation. Fortunately, Neavel[26] has subsequently shown that dihydronaphthalene is indeed formed as an intermediate in donor extraction with tetralin.

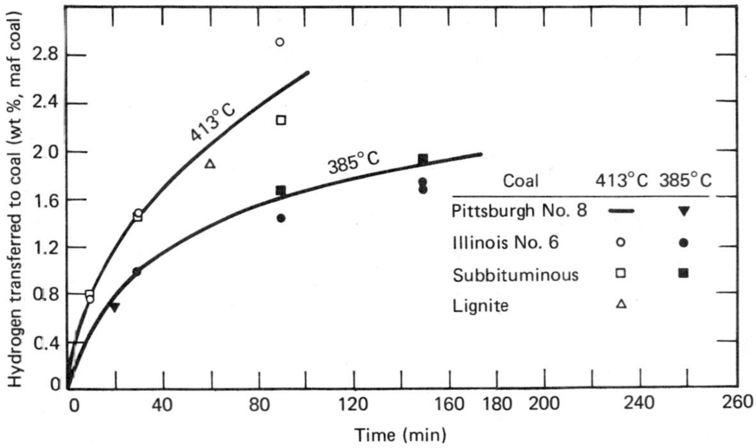

Figure 27.13 Hydrogen transfer rates.[27]

27.4.3 Kinetic Relationships: Donor Extraction and Pyrolysis

Wiser[25] was one of the first to discuss the kinetic relationship between donor extraction and pyrolysis in some detail. The kinetics of both processes were shown to be similar in that both could be represented for the major course of the reaction as a "pseudo" second-order reaction. The magnitude of the rates was also shown to be similar when expressed as a fraction of total conversion as a function of time. The relationship above, reproduced from Wiser,[25] is illustrated in Fig. 27.15.

Neavel,[87] in a somewhat similar vein, discussed the relationship between the development of coal plasticity during pyrolysis and liquefaction.

Brown and Waters[88] had earlier suggested that bitumens of low molecular weight serve as solvating and hydrogen donor agents during coal pyrolysis.

Fitzgerald[89] developed kinetic equations to

[87] R. C. Neavel, *Proc. Coal Agglomeration Conversion Symp.*, West Virginia University, Morgantown, May 5, 1975; published April 1976.

[88] H. R. Brown and P. L. Waters, *Fuel*, **45**, 17, 41 (1966).

[89] P. Fitzgerald, *Fuel*, **35**, 178 (1956).

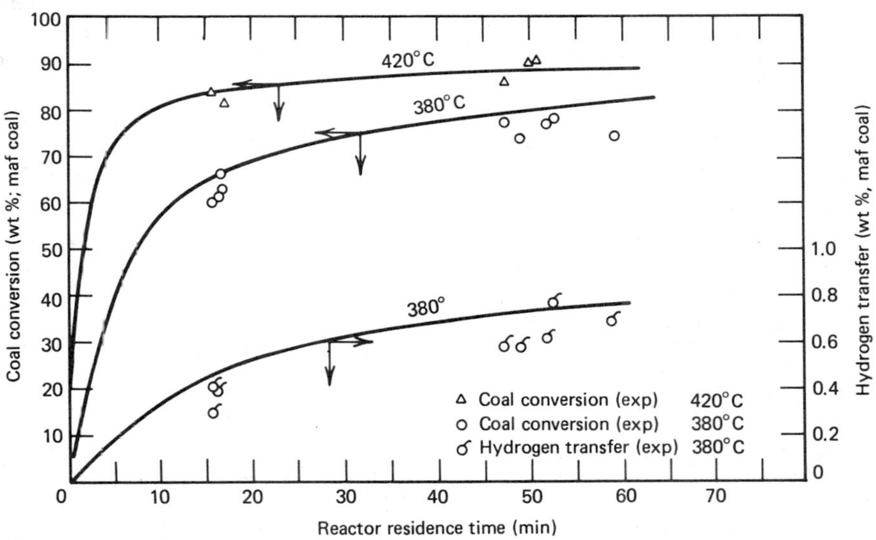

Figure 27.14 Kinetics of continuous extraction calculated versus observed rates for pure tetralin solvent.[27]

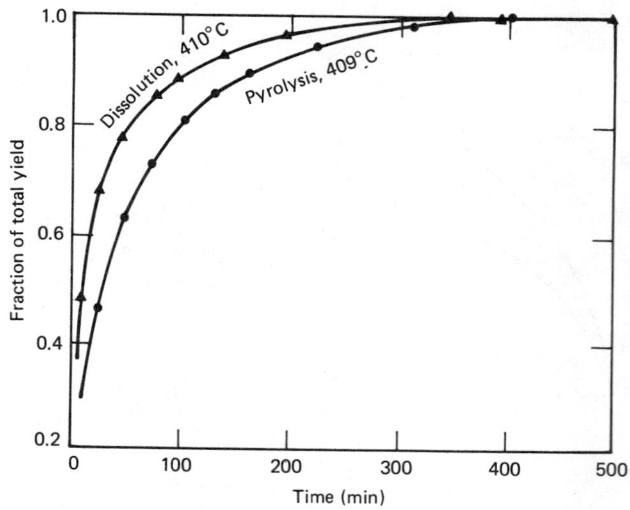

Figure 27.15 Comparison of fraction yield versus time.[25]

define the development and decay of coal fluidity, as measured experimentally on a Gieseler plastometer. The model used involved two consecutive first-order reactions, the first corresponding to coal pyrolysis to produce fluid coal and the second to the decomposition of the fluid coal to semicoke. A typical fluidity-time relationship for a British coking coal, reproduced from Fitzgerald,[89] is shown in Fig. 27.16.

Fitzgerald's equation for defining the rate constants from the fluidity measurements is:

$$\frac{d \ln \phi}{dt} = -\lambda \frac{k \exp(-kt) - k' \exp(-k't)}{\exp(-kt) - \exp(-k't)} \qquad (27.16)$$

where ϕ is the fluidity in arbitrary units, λ is a constant equal to 2.5, and k' and k are the rate constants for development and decay of fluidity, respectively. The rate constant k can be determined from the slope of the right-hand side of the fluidity curve, since at long times Equation 27.16 reduces to:

$$\frac{d \ln \phi}{dt} = -\lambda k \qquad (27.17)$$

Thus k' can be determined after k has been defined.

Fitzgerald's method was applied at CCDC to fluidity data on a Pittsburgh Seam coal from the Loveridge Mine. This work, previously unpublished, resulted in the following expression for the rate constant k' for development of fluidity:

$$k' = 2.3 \times 10^{11} \exp \frac{-38{,}600}{RT} \qquad (27.18)$$

Equation 27.18 is very similar to the slow rate constant discussed above[4] (see discussion of Equations 27.11 and 27.12) for coal dissolution K_2 with almost identical activation energy. The absolute rate is about one-half that previously given[4] for K_2. The comparison is well within the accuracy for determination of k', which is not good. However it does illustrate again at least semiquantitatively the close kinetic relationship between coal pyrolysis and donor extraction.

27.4.4 Kinetic Relationships: Donor Extraction and Hydroextraction

The kinetics of hydroextraction follows the same relationship as donor extraction with the following revision. The kinetic equations must be modified to account for the decrease in concentration of donors as compared with pure tetralin. It was shown, for example, that in a batch reactor[4,27] the effect of tetralin concentration on the rate of hydrogen trans-

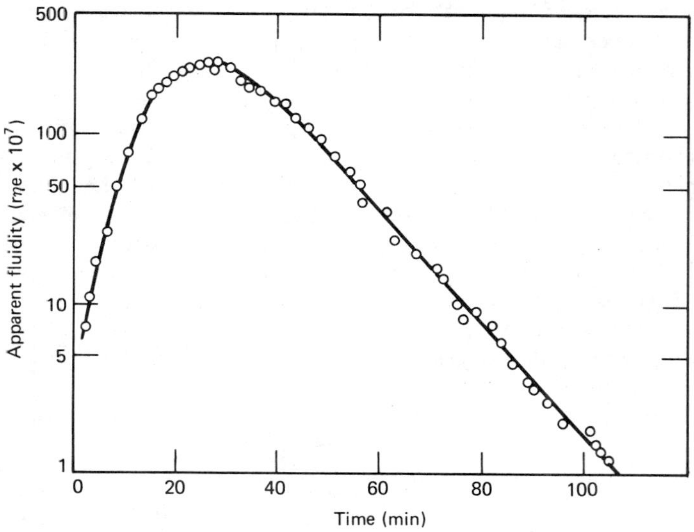

Figure 27.16 Apparent fluidity as a function of time at 407°C for a typical coking coal.[89]

fer could be described by the semiempirical equation:

$$\frac{H_T - H}{H_T} = \frac{B/T - B}{1 + B/T} + \frac{1 + B}{1 + B/T}\left[\gamma_1 \exp -k_1 t + (1 - \gamma_1) \exp -k_2 t\right] \quad (27.19)$$

where all quantities except B and T are the same as described previously, B is an empirical constant equal to 0.75, and T is the weight fraction of tetralin in the extraction solvent.

Equation 27.19 is useful for correlation of results with synthetic solvents. The coal conversion is then estimated from the calculated hydrogen transfer by use of either Fig. 27.2 or Fig. 27.3.

When natural solvents are employed, the equivalent tetralin concentration is unknown. In principle, Equation 27.19 could be used to estimate the equivalent tetralin concentration from actual experimental data on hydrogen consumption. It is clear, therefore, that the reaction rate under hydroextraction conditions will also be a function of the hydrogen partial pressure, since this along with the catalytic effect of the coal ash will determine the actual steady state concentration of donors. Thus the overall kinetics in the case of hydroextraction can be quite complex.

Whitehurst and Mitchell[45] measured the rate of hydrogen transfer at 427°C and 89 atm at short residence times with Kentucky No. 9 coal, using synthetic solvents containing 43 wt % tetralin under applied hydrogen pressure. If the same rate constants developed for Pittsburgh Seam coal[4] (shown following discussion of Equations 27.11 and 27.12) and Equation 27.19 are used, one can obtain a comparison between the calculated and observed results. A further assumption had to be made here, namely, that Kentucky coal follows the same relationship between coal conversion and hydrogen transfer given for Ireland Mine coal in Fig. 27.2. The single experimental point shown for Kentucky coal confirms this assumption. The kinetic comparison is given below:

	0.50 min		1.3 min		40 min	
	Exp	Calc	Exp	Calc	Exp	Calc
Conversion (wt %)	50	64	78	72	92.5	92
Hydrogen transfer (wt %)	0.13	0.13	0.34	0.26	0.89	1.4

In general, the agreement is quite satisfactory in view of the necessary assumptions. The low experimental figure for hydrogen transfer as compared with the calculated value at 40 min indicates that significant rehydrogenation of naphthalene may be occurring at the longer times.

Wen and Han[90] have correlated the batch rate data of the Pittsburg and Midway Coal Company[91] and of the University of Utah[92] on the rate of dissolution of Illinois No. 6 coal and Kentucky No. 9 coal in anthracene oil solvent with impressed hydrogen pressure by an empirical "first-order" equation.

The equation is written in the form:

$$r_a = k_0 \exp \frac{-E}{RT} \exp (0.000684 P_{H_2}) C_{s0} (1 - X) \frac{C}{S} \quad (27.20)$$

or, for a constant hydrogen pressure;

$$r_a = k' \left(\frac{C_{s0}}{C_{a0}} \right) \frac{C}{S} (1 - X) \quad (27.21)$$

where

r_a = rate of dissolution (g/hr-cm^3)

C_{s0} = weight fraction of organics in the untreated coal

C_{a0} = grams of ash-free coal per cubic centimeter of reactor volume (= 0.245 g/cm^3 for the Utah experiments)

X = fractional coal conversion

$\frac{C}{S}$ = coal/solvent wt ratio

P_{H_2} = hydrogen pressure (psi)

It is also interesting to compare the rate constant of Wen and Han[90] with those for donor extraction. Again, use must be made of the rate constants for Pittsburgh Seam coal. One difference is immediately apparent: the activation energies are very low in Wen and Han's correlation (i.e., 11 and 4.5 kcal/mole for the Illinois and Kentucky coals, respectively). Another difference is the magnitude of the rate constant, which is much smaller, as is illustrated below for the Illinois case at 427°C. The equivalent rate constant on the CCDC basis is calculated by means of the equation:

$$K = k' \left[\frac{C_{s0}}{C_{a0}} \right] \left[\frac{C}{S} \right] \quad (27.22)$$

	K (min^{-1}) Values Calculated from Kinetics	
P_{H_2} (atm)	Wen and Han Hydroextraction	Curran et al. Slow Constant K_2 Donor Extraction 100% Tetralin
0	0.016	0.574
68	0.031	
136	0.062	

The hydroextraction rate is an order of magnitude lower than the rate reported by Whitehurst and Mitchell[45] and that calculated from donor extraction. It is clear that the difficulty is the low concentration of donors in the anthracene oil solvent. The approximate linear dependence on hydrogen pressure above 68 atm is an indirect measure of the rate of production of donors by solvent rehydrogenation. Guin et al.[67] also found a first-order dependence of extraction rate on hydrogen pressure in their kinetic studies of the hydroextraction of Kentucky coal with creosote oil as solvent.

Kinetic analyses have seldom been undertaken in continuous hydroextraction. An exception is recent pilot operations at Wilsonville at 113 atm with Kentucky coal.[34] The considerable volume of the preheater plus transfer line was taken into account by assuming plug flow through the preheater and mixed flow in the dissolver to arrive at Equation 27.23:

$$X_{AF} = X_{AM} \frac{1 - \exp -K_P \tau_P}{1 + K_\beta \tau_\beta} \quad (27.23)$$

where

X_{AF} = final conversion

X_{AM} = maximum possible conversion

[90] C. Y. Wen and K. W. Han, *Prepr., Am. Chem. Soc., Div. Fuel Chem.*, **20**(1), 216 (1975).

[91] Pittsburg and Midway Coal Company, Monthly Prog. Reps. Nos. 82–95, Solvent-Refined Coal Project, OCR Contract No. 14-01-0001-496, July 1973–August 1974.

[92] *Hydrogenation of Western Coal in Dilute Phase Under Medium Pressures for Production of Liquids and Gases,* University of Utah, Quarterly Prog. Rep., OCR Contract No. 14-32-0001-1200, January–March 1974.

K_P = first-order rate constant in preheater

τ_P = residence time in preheater

K_β = first-order rate constant in dissolver

τ_β = residence time in dissolver

and K_β corresponds to the slow rate constant K_2 in donor extraction.

A number of simplifying assumptions were necessary in deriving and using Equation 27.23; for example, the assumption of isothermality in the preheater. Also, the final conversions (X_{AF}) were uniformly high, usually more than 90%, making it difficult to obtain accurate kinetic constants. It is instructive, nevertheless, to compare the Wilsonville rate constant with that calculated from donor extraction (Pittsburgh Seam coal) at the same average temperature of 446°C:

K_P (from Wilsonville)	=	$0.22\ \text{min}^{-1}$
K_2 (from donor extraction)	=	$1.20\ \text{min}^{-1}$ (100% tetralin)

Again, the same pattern emerges: the conversion rate is lower in hydroextraction, presumably because of the lower donor concentration.

In conclusion, the kinetics of hydroextraction is not inconsistent with the kinetics of donor extraction; rather, the former is more complex. Consideration must be given to the rate of formation of donors by solvent rehydrogenation and the level of donors present in the solvent during the course of the process.

27.4.5 The Role of Mass Transfer

The role of mass transfer in determining the rate of hydrogen consumption in hydroextraction has not been quantitatively investigated. Guin et al.[67] have shown that mass transfer is not rate controlling in laboratory batch-stirred autoclaves.

Literature correlations on the rate of mass transfer are utilized along with pilot-plant data[31] in the discussion below to investigate this question. The rate-determining process in mass transfer is considered to be the rate of transport of hydrogen from the liquid surface to the bulk liquid. The rate may then be represented by the equation:

$$R = K_V (C_e - C) = 3.6 \times 10^3 k_L A (C_e - C) \qquad (27.24)$$

where

R = the rate of hydrogen consumption experimentally measured (moles/hr-unit volume)

K_V = volumetric mass transfer coefficient (hr^{-1})

k_L = mass transfer coefficient (cm/s)

A = bubble surface area (cm^2/cm^3) = $6\,h/D_p$ for spherical bubbles

h = gas holdup, fraction of volume

D_P = average bubble diameter (cm)

C_e = concentration of hydrogen (moles/unit volume in equilibrium with hydrogen in gas phase)

C = concentration of hydrogen in bulk liquid

It is assumed in this treatment that the rate of reaction of hydrogen in the liquid phase is so slow that no enhancement of the rate of absorption need be considered.

The estimation of the role of mass transfer follows the treatment given in the report on the CSF process.[10] This is done by estimating the concentration of hydrogen in the bulk liquid phase as the percentage of its equilibrium value:

$$\text{percentage saturation} = \frac{C}{C_e} \times 100 \qquad (27.25)$$

The calculation of the mass transfer rates requires estimation of the parameters in Equations 27.24 and 27.25 (i.e., h, C_e, K_L, and D_P). These parameters can be only roughly estimated with our present knowledge. It is also necessary to use the physical properties characteristic of the liquid phase in estimating these quantities; the effect of solids is neglected. The presence of fine solids has been reported to effect an increase in interfacial area in bubble columns.[93]

The fraction voids, h, were estimated by the method of Hughmark.[94] These were cor-

[93] M. M. Sharma and R. A. Mashelkar, Inst. Chem. Eng., Symp. Ser. No. 28, 1968, p. 10.

[94] G. A. Hughmark, *Ind. Eng. Chem., Process Des. Dev.* **6**, 218 (1967).

rected to take account of the buoyancy effect of the gas by the method previously used in the CSF process:

$$h = h' \frac{\rho_L}{(\rho_L - \rho_g)} \qquad (27.26)$$

where

h' = fraction voids calculated by Hughmark procedure

ρ_L = liquid density (g/cm³)

ρ_g = gas density (g/cm³)

The mass transfer coefficient was estimated by two methods. The first uses the penetration theory for individual rigid bubbles, as discussed previously,[10] that is,

$$K_V = 3.6 \times 10^3 \left[\frac{D_{H_2} U_s}{D_P}\right]^{1/2} \left(\frac{6h}{D_P}\right) \qquad (27.27)$$

where

D_{H_2} = diffusion coefficient of hydrogen in liquid

(cm²-s⁻¹) = 4.4 × 10⁻⁴ from the Wilke-Chang equation[95]

U_s = slip velocity (cm/s) = difference between gas and liquid velocities

The second uses the method of Hughmark, which employs the empirical equation below, applicable to bubble swarms:

$$N_{Sh} = \left(\frac{k_L D_p}{D_{H_2}}\right) = 2 + 0.0187 \left[\left(\frac{D_p U_s}{\nu}\right)^{0.484} \left(\frac{\nu}{D_{H_2}}\right)^{0.339} \left(\frac{D_p g^{1/3}}{D_{H_2}^{2/3}}\right)^{0.072}\right] \qquad (27.28)$$

where ν = kinematic viscosity (cm²-s⁻¹)

The application of Equation 27.28 requires the use of the proper parameters to apply to the actual runs. The following values, based on prior determinations,[10] were used: $\rho_L = 0.75$, $\nu = 4 \times 10^{-3}$. Values for ρ_g were calculated from Wilsonville data. Correlations presented in the Wilsonville reports[31] were used to estimate the partial pressure of hydrogen and C_e. The value of C_e was determined as the dissolved hydrogen in equilibrium with the logarithmic mean of the inlet and outlet hydrogen partial pressures.

The largest single uncertainty resides in the value for the average bubble size, for which no experimental data exist in the hydrogenation system. Hughmark[94] recommends a value of $D_p = 0.635$ cm based on the measurements of Towell et al.[96] in aqueous systems, who found that the mean bubble diameter was independent of gas rate. This value of D_p was used for the calculations based on the Wilsonville operation, as summarized in Table 27.20.

The calculations in Table 27.20 are, of course, approximate, but should be of the proper order of magnitude. The values for K_V are about twice as large when calculated by penetration theory for rigid bubbles as compared with the Hughmark method. The approach to saturation of the liquid with hydrogen is any case in always 85% or greater. Directionally, the mass transfer rate tends to become more limiting; that is, the approach to saturation decreases as the pressure increases. The reason for this is the decreasing upflow gas velocity and the corresponding decrease in the value of h, thus the area for mass transfer used in Wilsonville runs.

It may be concluded that under conditions of the Wilsonville pilot-plant operations, mass transfer rate is, in general, not controlling. An extension to higher temperature, shorter residence time, and higher pressure operation where faster reaction rates are involved can lead to a situation where mass transfer rates are partially controlling.

However proper design of the extraction vessel, particularly on a commercial scale, can always prevent mass transfer from being the rate-controlling process. An example to illustrate this is given below. In the situation considered, the donor levels in the solvent are sustained to achieve maximum extraction rates for a Pittsburgh Seam coal using the kinetic equations developed in Section 27.4.2. The donor levels can be maintained in principle by adequate *in situ* catalysis of the solvent rehydrogenation process. Two cases are considered, with hydrogen consumption of 1.5 wt % of the maf coal at 450°C. One case corresponds to a higher rate

[95] C. R. Wilke and Pin Chang, *Am. Inst. Chem. Eng. J.*, **1**, 264 (1955).

[96] G. D. Towell, B. S. Strand, and G. H. Ackerman, Am. Inst. Chem. Eng. Chem. Eng. Symp. Ser. No. 10, 1965.

Table 27.20 Mass Transfer Rates: Wilsonville Pilot-Plant Runs[31] With Illinois No. 6 Coal

Run No.	Hydrogen Partial Pressure (atm) Outlet	LM[a]	Hydrogen Reacted (wt %, maf coal)	Temp (°C)	h^b	K_V (hr^{-1})[c] Calc from Ref. 87	Calc from Penetration Theory[10]	Saturation of Liquid with Hydrogen (%) Est from Ref. 87	Est from Penetration Theory[10]
20C	34	36	1.5	439	0.092	190	400	92	96
22A	50	56	2.3	438	0.055	110	230	88	94
23B	77	84	2.3	446	0.030	75	150	85	93

[a] LM = logarithmic mean = $\dfrac{(^PH_2)\text{ in} - (^PH_2)\text{ out}}{\ln\left[\dfrac{(^PH_2)\text{ in}}{(^PH_2)\text{ out}}\right]}$

[b] Fraction gas bubble volume.

[c] K_V = volumetric mass transfer coefficient = $k_L A$.

because plug flow of liquid is assumed; the other assumes the probably more realistic case of perfect mixing in the liquid phase. The value of hydrogen consumption (1.5 wt %) was chosen because it represents the minimum level at which deep extraction plus an acceptable sulfur content in the product extract can be achieved.

Mass transfer rates adequate to sustain these hydrogen consumption rates are achieved by using a high superficial gas velocity. The mass transfer rates in an acceptable vessel design for the operation above are listed in Table 27.21. It is clear that acceptable mass transfer rates can be achieved.

It should be noted that the fraction voids calculated by the Hughmark[94] method for the cases given in Tables 27.20 and 27.21 are much lower than would be estimated from the German hydro data cited in the report on the CSF process.[10] A probable explanation is that both the gas velocity and the density were much higher in the prior German work, but no figures are available to check this point.

The practical limitation thus is not the rate of mass transfer, but the rate of solvent rehydrogenation, which in most hydroextraction operations is not sufficient to maintain a high donor level in the solvent. The result is a decrease in overall reaction rate.

Table 27.21 Vessel Design and Mass Transfer Rates for High Extraction Rates

Conditions: Solvent/maf coal, wt ratio = 2; temperature, 450°C.

	Plug Flow	Mixed Flow
Log mean hydrogen pressure (atm)	50	50
Hydrogen consumption (wt %, maf coal)	1.54	1.50
Liquid residence time (min)	4	5.75
Superficial gas velocity (m/s)	0.167	0.112
Superficial liquid velocity (m/s)	0.0466	0.0314
Gas density (g/cm^3)	0.062	0.062
Estimated fraction voids	0.37	0.305
K_V (hr^{-1})[a]	1090	800
Saturation of liquid with hydrogen (%)	97	96
Extraction vessel dimensions		
Area [m^2/(1000 kg maf coal)/(hr)]	0.024	0.035
Height (m)	17.8	15.6

[a] Estimated from Ref. 94.

27.5 MISCELLANEOUS PRIMARY LIQUEFACTION AGENTS

27.5.1 Carbon Monoxide and Water in Primary Liquefaction

In 1921 Fischer and Schrader[97] showed that brown coal could be readily liquefied by use of carbon monoxide and water. This method of coal liquefaction was not pursued further until recently. Appell, Wender, and other workers[98,99] at the Pittsburgh Energy Research Center (PERC) showed in autoclave experiments that high yields of a benzene-soluble oil could be obtained by treating low-rank coals such as lignite by this method in the presence of 1 : 1 α-naphthol-phenanthrene as a solvent. Low-rank coals were shown to react more readily with CO-H_2O than with hydrogen under comparable conditions.[100,101] Most experiments were conducted at fairly high total pressures (270–340 atm) and moderate temperatures (380°C). Under these conditions, in the presence of good solvent, as much as 89% conversion of lignite to benzene solubles could be achieved in 10 min residence time. The conversion to benzene solubles is likely overstated. The filtrate due to the technique used undoubtedly contains cresol soluble but benzene insoluble products.

It was shown[100] that the advantages of CO-H_2O as compared with hydrogen were less with higher rank coals and became insignificant with bituminous coals. The same relatively higher reactivity with CO-H_2O of higher oxygen content coals and organic materials such as peat, cellulose, and lignin was also observed by Ueda et al.[102] and by Oelert et al.[103,104]

Unpublished autoclave data from CCDC laboratories showed that with some subbituminous coals, specifically Colstrip, no advantage at the same total hot pressure was apparent in use of CO-H_2O versus pure hydrogen. These experiments were conducted at 380°C, with methylnaphthalene as solvent, one hour nominal residence time, and a relatively low total hot pressure of 132 atm. Low conversions to cresol solubles were obtained: 33.0 and 27.5% with pure hydrogen and CO-H_2O, respectively. These results indicate that the advantage of the CO-H_2O method may depend also on having the proper alkalinity in the coal ash. The Colstrip coal mentioned above has an unusually low sodium content (0.003 wt % of the feed coal).[41]

The relative advantages of CO-H_2O versus pure hydrogen[105] were also shown to decrease with increasing temperature.

It has been shown also that high conversions[104,105] can be obtained without a solvent, but significantly higher pressures are required. The nature of the solvent strongly influences results. Particularly good results are achieved by use of heterocyclic nitrogen bases such as isoquinoline as a solvent: 93% conversion of North Dakota lignite to benzene solubles was obtained at 400°C and 103 atm initial (cold) pressure of CO in 15 min. Poorer results are obtained with solvents such as anthracene oil unless significantly longer reaction times are used; 79 and 93% conversion of North Dakota lignite to benzene solubles is obtained at 380°C and about 250 atm hot pressure with anthracene oil at 15 and 120 min residence time, respectively. Addition of isoquinoline to the anthracene oil improves conversion.

Nitrogen bases used as part of the solvent[105] may be regarded as catalysts for the CO-H_2O reaction. The naturally occurring sodium and possibly the calcium in lignite[105,106] may also be regarded as catalysts.

---

[97] F. Fischer and H. Schrader, *Brennstoff-Chemie,* **2,** 287 (1921).

[98] H. R. Appell and I. Wender, *Prepr., Am. Chem. Soc., Div. Fuel Chem.,* **12**(3), 220 (1968).

[99] H. R. Appell, I. Wender, and R. D. Miller, *Chem. Ind. (London),* No. 47, 1703 (1969).

[100] H. R. Appell, I. Wender, and R. D. Miller, *Proc. Symp. Chem. Oil from Coal,* Central Fuel Research Institute, Dhanbad, India, December 6–8, 1969, pp. 151–155; published 1972, by Central Fuel Research Institute.

[101] H. R. Appell, I. Wender, and R. D. Miller, *Prepr., Am. Chem. Soc., Div. Fuel Chem.,* **13**(4), 39 (1969).

[102] S. Ueda, S. Yokoyama, Y. Nakata, Y. Hasegawa, Y. Maekawa, Y. Yoshida, and G. Takeya, *Nenryo Kyokai-shi,* **53**(572), 977 (1974).

[103] H. H. Oelert, R. Siekmann, and K. U. Islam, *Erdoel Kohle,* **28,** 290 (1975).

[104] H. H. Oelert and R. Siekmann, *Fuel,* **55**(1), 39 (1976).

[105] H. R. Appell, R. D. Miller, and I. Wender, paper presented at American Chemical Society, Division of Fuel Chemistry, 163rd National Meeting, Boston, April 10–14, 1972 (no preprint).

[106] *Technology of Lignite Coals* (in two parts) U.S. Bur. Mines Inf. Circ. No. 7691, 1954, p. 54.

Sodium carbonate[107] has also been shown to be a catalyst for the conversion of various organic wastes to oil by use of the CO-H_2O method. Similarly, Fischer[108] showed many years ago that sodium bicarbonate was a good catalyst for the liquefaction of German brown coal with hydrogen.

27.5.2 Synthesis Gas in Primary Liquefaction

The use of both pure carbon monoxide and CO-H_2 mixtures has been investigated for liquefaction. Aside from any possible improvements in operating conditions and yield structure as compared with pure hydrogen, there are other potential commercial advantages, such as use of cheaper synthesis gas and the possibility of eliminating thermal drying of the feed coal. Synthesis gas is also a much more practical feed gas than pure CO.

Studies of liquefaction of North Dakota lignite with synthesis gas have been conducted in batch autoclaves by Appell et al.[109] In tests with synthesis gas, operation at temperatures higher than those used with pure CO along with pure hydrogen in the gas was found to be beneficial in reducing the viscosity of the oil product. Both these factors were favorable in that they increased the amount of hydrocracking, compared with the use of pure CO.

The direct utilization of hydrogen increased with increasing temperature but even at temperatures up to 460°C, with a 1 : 1 ratio of CO to H_2 feed gas, more hydrogen was produced by water-gas shift than was consumed by reaction.

Conversion of the lignite, even at 420°C, increased with increasing CO content of the feed gas. This is illustrated by data from Appell[109] reproduced in Fig. 27.17.

The use of synthesis gas for lignite liquefaction has been investigated by Del Bel et al.[110] in a small (4-lb/hr) continuous unit. A 1-liter autoclave was used as the reactor, which was agitated at 1500 rpm by a turbine-type impeller. Residence times of 1–2 hr, pressures of 136–272 atm, temperature from 400 to 450°C, and feed gas compositions ranging from pure CO to 3 : 1 (H_2/CO) were investigated. Anthracene oil was used as solvent.

The same qualitative trends were observed with respect to gas compositions and temperatures as in the batch autoclave work.[100] Hydrogen-rich gases gave improved results relative to CO-rich gases as the temperature was increased. This relationship is illustrated in Fig. 27.18. No data with pure hydrogen were presented.

The major difference between the autoclave and continuous unit results was the much slower conversion rate in the latter case. For example, with a 0.7 : 1 ratio of CO to H_2, 2 hr residence time,[110] 425°C, and 273 atm pressure, lignite conversion was 80% in the continuous unit. Under comparable conditions, but with only 1 hr residence time, 98% conversion was obtained in autoclave experiments.[109]

The difference is undoubtedly due to the favorable effect on kinetics of the slow heating rate in the autoclave experiments, as well as the unfavorable effect of the mixed-flow regime in the continuous unit. The relative contribution of each factor cannot be elucidated.

A larger-scale continuous unit (15 lb/hr) has been operated at PERC,[111] although results have not been published yet. The unit is operated analogously to the Synthoil unit, with fuel oil product obtained by centrifugation of the whole reactor effluent. Part of the centrifuge overflow is used as pasting oil for the process. Solvent surplus[111] has been demonstrated by operation in this manner with a 3 : 1 ratio of H_2 to CO feed gas, at 450°C, 0.37 kg coal/(hr)(l) space velocity, and 237 atm pressure.

A larger process development unit (PDU) (50 lb/hr coal feed) has been constructed and put in operation at the Engineering Experi-

[107] H. R. Appell, Y. C. Fu, S. Friedman, P. M. Yavorsky, and I. Wender, *Converting Organic Wastes to Oil, A Replenishable Energy Source*, U.S. Bur. Mines Rep. Invest. No. 7560, 1971.

[108] Franz Fischer, *Conversion of Coal to Oils*, D. Van Nostrand Company, New York, 1925.

[109] H. R. Appell, E. C. Moroni, and R. D. Miller, *Prepr., Am. Chem. Soc., Div. Fuel Chem.*, **20**(1), 58 (1975).

[110] Elsio Del Bel, S. Friedman, P. M. Yavorsky, and I. Wender, paper presented at American Institute of Chemical Engineers Meeting, Houston, Tex., March 16, 1975. Also, *Coal Processing Technology—A Chemical Engineering Progress Technical Manual*, Vol. 2, 1975, p. 104.

[111] Elsio Del Bel, U.S. ERDA, Pittsburgh Energy Research Center, private communication, 1976.

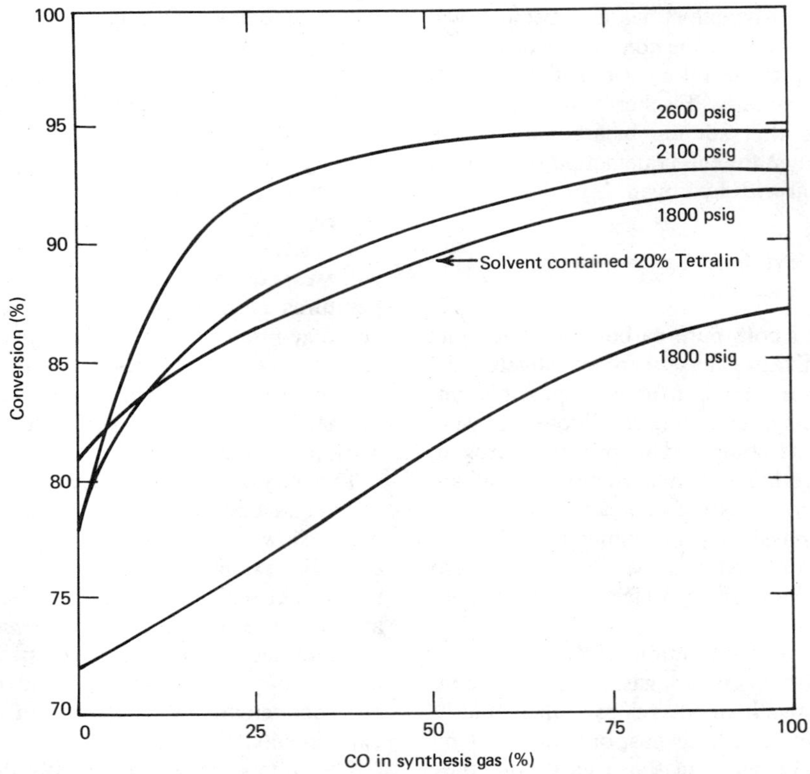

Figure 27.17 Lignite conversion versus composition of synthesis gas at 420°C, 1 hr residence time; ratio of maf lignite to anthracene oil solvent to water, 3 : 6 : 1.[105] (Donor extraction data: Ref. 27; Wilsonville data: Ref. 31.)

ment Station of the University of North Dakota.[112] The method of operation, unlike that at PERC,[111] is patterned after the Pittsburg and Midway solvent-refined oil process, with the difference that synthesis gas rather than pure hydrogen is used as feed gas. The product from the operation is a pyridine-soluble extract rather than fuel oil. Anthracene oil is used as startup solvent, but at this writing operations had not been continued long enough to demonstrate self-sufficient operation with natural solvent.

Other differences from the PERC work[110,111] are operation at lower pressures (103–170 atm) and the use of gas-sparged instead of mechanically agitated vessels. The other operating ranges explored so far are dissolver temperature (367–490°C), solvent-coal weight ratio (1.3 : 2.0), and slurry residence times (20–90 min). Most operations were conducted with an 1 : 1 ratio of H_2 to CO in the feed gas. Recent runs have shown better solvent recovery at a ratio of 3, as well as better hydrogen utilization.

The PDU results with lignite[112] again indicate significantly lower conversion rates than in batch autoclaves.[109] Conversions increased with increases in both temperature and pressure, but decreased with increasing space velocity. Operating temperatures at 170 atm, however, must be maintained below 480°C to prevent coke deposition. The maximum practical conversion (to pyridine solubles) is in the range 85–90% at 170 atm and a space velocity of about 0.5 kg of lignite per hour per liter.

Actual generation of natural solvent has not so far been achieved. Yield data indicate that to generate a solvent surplus, operations must be conducted at pressures around 170 atm or higher.

27.5.3 Mechanism of Reaction

The mechanism of the reaction of CO and

[112] D. E. Severson, principal investigator, *Project Lignite,* Report to U.S. ERDA, Contract No. E(49-18)-1224, October 1976.

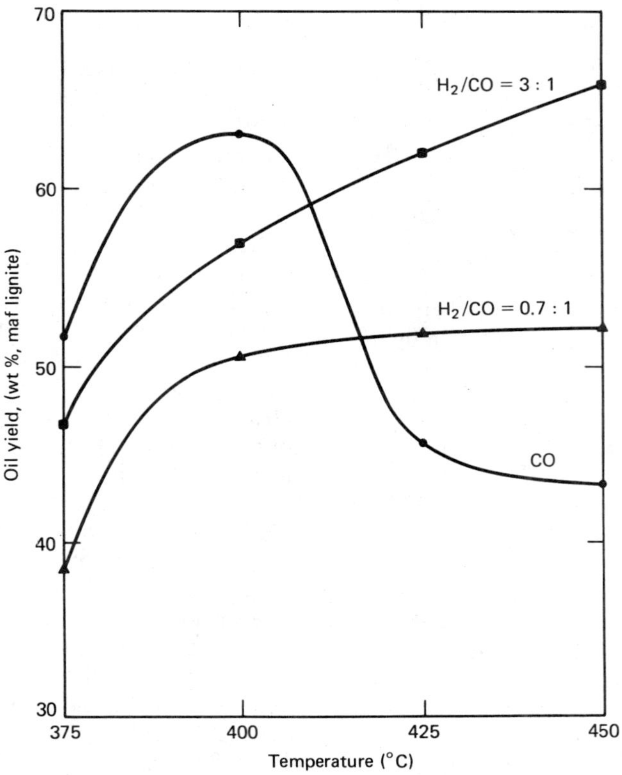

Figure 27.18 Oil yield versus temperature and gas composition at 4000 psig pressure, 2.0 hr residence time.[110]

synthesis gas with low-rank coals has been discussed by Appell et al.[105,109] They attribute the high reactivity of these coals, in part, to the presence of carbonyl groups. Using Na_2CO_3 as a catalyst, experiments with model compounds such as acetophenone showed that reduction to the alcohol was achieved more readily with CO than with hydrogen.

The use of an alkaline catalyst or solvent containing basic nitrogen appears to be essential to achieve the CO-H_2O reaction. Lignites usually are rich in alkaline components, thus have the catalyst naturally present. Fischer[108] remarked on the use of alkaline catalysts and also showed that alkali formates can be used as a source of active hydrogen for conversion of German brown coal.

Appell et al.[107,109] agree that formates formed by reaction of CO and H_2O with alkalis or nitrogen bases are essential intermediates in the process:

$$Na_2CO_3 + 2CO + H_2O = 2NaOOCH + CO_2 \quad (27.29)$$

Equilibrium pressures in the reactions:

$$2KOOCH = K_2CO_3 + CO + H_2 \quad (27.30)$$

$$2KOOCH + CO_2 = K_2CO_3 + 2CO + H_2O \quad (27.31)$$

where $K_1 = P_{CO} \cdot P_{H_2}$ have been reported by Yavorsky[113] and Zielke.[114]

Under conditions where the only phases present are the gas phase and molten potassium formate and solid potassium carbonate, the equilibrium pressure

$$P_T = P_{CO} + P_{CO_2} + P_{H_2} + P_{H_2O} \quad (27.32)$$

is uniquely determined by the above-given equilibrium constant K_1, the equilibrium con-

[113] P. M. Yavorsky, N. J. Mazzocco, G. D. Rutledge, and Everett Gorin, *Environ. Sci. Technol.*, **4**, 757 (1970).

[114] C. W. Zielke, W. A. Rosenhoover, and Everett Gorin, *Prepr., Am. Chem. Soc., Div. Fuel Chem.*, **21**, (7), 163 (1976).

stant for the water-gas shift reaction

$$K_w = \frac{P_{CO_2} \cdot P_{H_2}}{P_{CO} \cdot P_{H_2O}} \tag{27.33}$$

and the parameters

$$r_c = \frac{P_{CO_2}}{P_{CO}} \quad \text{and} \quad r_H = \frac{P_{H_2}}{P_{CO}} \tag{27.34}$$

The equilibrium pressure in atmospheres, P_T, as can be readily shown, is then given by the expression:

$$P_T = \frac{\sqrt{K_1}}{K_w} \left[r_c + K_w \left(1 + \frac{r_c}{r_H} + \frac{1}{r_H} \right) \right] \left(r_H \right)^{1/2} \tag{27.35}$$

The equilibrium pressure is a minimum when no water is present (i.e., $r_c = 0$) and is further minimized when $r_H = 1$. Unpublished data from CCDC laboratories confirm that the total dissociation pressure over molten potassium formate does indeed rise on the addition of water.

Illustrative examples of the equilibrium pressures in the potassium system, using the foregoing equilibrium data[113] and Equation 27.35, are given below for two temperatures (380 and 427°C) for $r_H = 1$ and $r_c = 0$ and 1:

Temp (°C)	Equilibrium Pressure Only, Molten KOOCH		
	r_c	r_H	P_T (atm)
380	0	1	96
380	1	1	149
427	0	1	204
427	1	1	317

Qualitatively, the equilibrium calculations agree with the trends noted in the experimental data: that the major advantages of using gases containing CO are to be expected in the lower temperature ranges (i.e., below about 425°C), unless high pressures are employed.

The above-calculated equilibrium pressures cannot be given absolute significance because one is, in fact, dealing with the sodium instead of the potassium system. Another factor is that formate can be produced by the noncatalytic reaction:

$$M_2CO_3 + 2CO + H_2O = 2MOOCH + CO_2 \tag{27.36}$$

at lower pressures[114] than that which corresponds to final equilibrium. In other words, formate can be formed as a reactive intermediate under conditions where it is not thermodynamically stable. Based on this premise, the addition or presence of catalytic agents that accelerate formate decomposition would have an unfavorable impact on the process. The primary condition that must be satisfied here is that water-gas shift equilibrium not be established. Also, it can be shown that the low-pressure advantage can be preserved only if a relatively small fraction of the CO feed is utilized.

If these concepts of Fischer[108] and Appell et al.[109] are correct, molten potassium formate should be a good donor for coal extraction. Unpublished work was carried out to test this hypothesis at CCDC laboratories. Conversions of 70 and 80 wt %, respectively (of Colstrip subbituminous coal and Glenharold lignite to cresol solubles), were obtained in autoclave experiments at 380°C. A 1-hr residence time was used with 4 : 1 weight ratio of 93 wt % aqueous KOOCH to coal in both cases. The total pressure measured, 117 atm, interestingly enough was in agreement with the equilibrium calculations cited above. As already noted, conversion of Colstrip coal with added hydrogen or CO-H_2O under similar conditions but with no formate was very low: 27–33 wt %.

27.5.4 Other Hydrogen Donors

The formate donor system discussed above probably operates by a nonradical, possibly heterolytic, mechanism. A number of other donors besides tetralin are capable of acting as free radical–type donors. Hydroaromatics such as dihydrophenanthrenes and cyclohexylphenol, which like tetralin have activated carbon-hydrogen bonds, are all good donors as was shown previously.[4] Hydroaromatics with five-membered rings, such as indane,[4] are not only less active than tetralin but are undesirable, since the product—indene—polymerizes readily.

Saturated polycyclic hydrocarbons, such

as decalin and perhydrophenanthrene,[4] were also shown to act as weak donors.

Secondary alcohols such as cyclohexanol[4] and isopropanol[4] were shown to be donors as reactive as tetralin. Their use is of academic interest, since it shows that the only criterion to be satisfied is that the potential donor contain active hydrogen.

Even methanol,[115] a primary alcohol, has been shown to have weak hydrogen donor properties at high temperatures and pressures (600°C and 95 atm) in the vapor phase.

More extensive data have been published recently by Ross and Blessing[116] on the use of isopropanol as a hydrogen donor. They made the interesting observation that metal alkoxides catalyze liquefaction by means of isopropanol.

27.6 PRODUCTS OF PRIMARY LIQUEFACTION

27.6.1 Yield Structure and Product Characterization

The gross yield structure and hydrogen consumption as a function of reaction conditions and coal type were discussed previously. The major product in primary liquefaction is a nondistillable liquid or extract. The imposition of more severe conditions of temperature, space velocity, and pressure results in production of larger quantities of hydrocarbon gases and distillate oils. These are formed mostly be secondary hydrocracking of the primary extract, thus to a large extent are not the primary liquefaction products. Significant quantities of gases and distillates must, however, be produced in any practical hydroextraction process. This follows from the necessity of generating enough distillate oil in the process to stay in solvent balance.

The characterization of the liquefaction products from coal by advanced instrumental methods has been discussed in Section 27.4. More detailed data on the distillate oils boiling in the recycle solvent range were given in Sections 27.2.5 and 27.2.6.

[115] W. J. Huffman, D. N. Garner, and H. W. Parker, *Coal Processing Technology—A Chemical Engineering Progress Technical Manual,* Vol. 2, 1975, p. 76.

[116] D. S. Ross and J. E. Blessing, *Prepr., Am. Chem. Soc., Div. Fuel Chem.,* 22(2), 208 (1977).

A very detailed analysis of the liquefaction products from the Synthoil process was given in a report by Aczel et al.[117] Their publication did not appear in time for its content to be analyzed and summarized here.

This discussion therefore is limited to a summary of the knowledge regarding the major product of primary liquefaction, that is, the nondistillable extract. The characterization of extracts was originally limited to a determination of the following gross properties[9,17]: ultimate analysis, solvent fractionation, molecular weight, ash content, ash composition, and ash size consist. The ash distribution and composition of extracts will be discussed in Section 27.7.

The solvent fractionation analysis was conducted to divide the extract into three fractions, as follows. The benzene-insoluble fraction is the fraction that is soluble in cresol at its normal boiling point of 165°C, but insoluble in boiling benzene. The asphaltenes are defined as the fraction that is soluble in benzene but insoluble in *n*-hexane (room temperature). CCDC investigators used cyclohexane as the solvent instead of *n*-hexane. The oil fraction is defined as the fraction of extract that is soluble in hexane.

The molecular weight, the heteroatom content, particularly oxygen, and the C/H ratio increase in the progression oil–asphaltenes–benzene insolubles. The relationship is illustrated by analytical results reported by Burk and Kutta[118] for extract produced by hydroextraction of Illinois No. 6 coal in the Wilsonville pilot plant. The data are summarized in Table 27.22. Burk and Kutta, incidentally, use the term "preasphaltenes" to describe the benzene-insoluble fraction. The operating conditions unfortunately are not specified. The Wilsonville plant operated with Burning Star coal over a wide range of pressures (70–174 atm), but with a relatively constant dissolver temperature (~ 445°C). The high content of benzene insolubles and the low oil content are not consistent with other published data[39] on extracts from hydroextraction of Burning Star coal.

[117] T. Aczel, R. B. Williams, R. J. Pancirov, and J. H. Karchmer, Exxon Research and Engineering Company, Report to ERDA No. FE-8007, February 1977.

[118] E. H. Burk, Jr. and H. W. Kutta, Paper No. 7, Session II, Coal Chemistry Workshop, Stanford Research Institute, 1976.

Table 27.22 Analytical Results of Extract Fractionation[118]

SRC Product	Distribution (wt %)	Net H/C × 100[a]	Weight Percent: Carbon	Hydrogen	Oxygen	Nitrogen	Organic Sulfur	Total	Mean Molecular Weight
Burning Star Mine Coal (maf)[31]	—	5.09	80.13	5.82	10.27	1.41	2.37	100.00	—
Extract	—	5.36	86.90	5.75	5.00	2.00	0.67	100.32	—
Benzene insolubles	62.8	4.85	85.74	5.39	5.86	2.15	0.66	99.80	870
Asphaltenes	27.0	6.78	86.84	6.88	4.78	1.65	0.60	100.75	565
Oils	10.2	7.37	88.30	7.10	3.00	0.87	0.48	99.75	350
	100.0								

[a] $100 \dfrac{H - \frac{0}{8} - \frac{3N}{14} - \frac{S}{16}}{C}$

The data illustrate the trends noted above. The net ratio of hydrogen to carbon given in Table 27.22 is an indicator of the hydrogen requirement for conversion to hydrocarbon fuels; the higher the ratio, the lower are the hydrogen requirements. The net hydrogen ratio of the benzene insolubles is even less than that of the feed coal.

A more elaborate solvent fractionation procedure has been used by Schwager and Yen.[119] The benzene-insoluble fraction is further fractionated with CS_2 to yield a carboid fraction (CS_2-insoluble) and a carbene fraction (CS_2-soluble). The pentane-soluble fraction is divided into propane insolubles (resin) and propane solubles (oil). The breakdown of extract products from the Wilsonville pilot plant as given by Schwager and Yen[119] is shown in Table 27.23. The extract was produced by the solvent-refined coal hydroextraction process, but operating conditions were not specified, nor was the feed coal.

The pentane-soluble fraction can be split into a more diverse series of fractions by use of liquid chromatography. Ruberto and Jewell and their co-workers,[120,121] for example, devised a scheme to split the pentane solubles into resins, monoaromatics, di-plus triaromatics, polyaromatics, *n*-paraffins, and other saturates. Structural information on the various aromatic fractions are then obtained by a combination of IR, UV, NMR, and mass spectrometry.

The benzene-insoluble fraction, in spite of its hydrogen deficiency, contains a highly reactive component. The fraction that is highly reactive appears to depend on the specific coal used and conditions under which it was derived. Nearly complete conversion to asphaltenes and oils was readily achieved by donor extraction of a Utah[21] and some Illinois No. 6 coals.[26] CCDC investigators[39] found, however, that it was difficult to reduce the benzene-insoluble yield in hydroextraction of Illinois No. 6 Burning Star coal below 10% of the maf feed coal unless severe hydroextraction conditions were employed. A similarly refractory component of the benzene insolubles was noted by CCDC investigators[4,10,27] in donor extraction of Pittsburgh No. 8 coal from the Ireland Mine. Autoclave results to illustrate this point are given in Fig. 27.19, where the yields of benzene insolubles, asphaltenes, and oil are plotted as a function of severity of extraction. The index of severity used in Fig. 27.19 is the amount of hydrogen transferred from tetralin to the Ireland Mine coal.

Benzene insolubles constitute a somewhat artificial entity defined by the analytical procedure used. The conversion sequence, maf cresol insolubles <=> benzene insolubles <=> asphaltenes, is reversed under many operating conditions. Treatment of asphaltenes with tetralin, for example, results in production of some benzene insolubles (unpublished CCDC data). The benzene insolubles produced by such a reverse process are undoubtedly more refractory than the virgin material.

The benzene insolubles also are the least soluble in the extraction solvent. They tend to come out of solution[29,35–38] either when the extraction effluent is cooled or when a deashing solvent is added.[29,39] The precipitation of the benzene insolubles is important in deashing processes, since they cause ag-

[119] I. Schwager and T. K. Yen, *Prepr., Am. Chem. Soc., Div. Fuel Chem.*, **21**(5), 199 (1976).

[120] R. G. Ruberto and D. M. Jewell, "Characterization of Coal Liquids," paper presented at National Science Foundation Workshop, Greenup, Ky., August 21–23, 1974.

[121] R. G. Ruberto, D. M. Jewell, R. K. Jensen, and D. C. Cronauer, *Prepr., Am. Chem. Soc., Div. Fuel Chem.*, **19**(2), 258 (1973).

Table 27.23 Fractionation Analysis of an Extract from the Wilsonville Pilot Plant[119]

	Percent of Total	Analysis (wt %)						Molecular Weight
		Carbon	Hydrogen	Nitrogen	Sulfur	Oxygen	Ash	
Extract		88.71	5.53	1.26	0.19	4.31	0.29	—
Oil	4	90.99	6.94	0.41	0.57	1.09	0.25	—
Resin	15	89.88	6.64	0.84	0.00	2.64	0.05	—
Asphaltene	45	88.79	5.61	1.25	0.12	4.23	0.78	747
Carbene	2	89.79	5.03	1.11	0.26	3.83	0.40	—
Carboid	34	87.08	4.70	0.96	0.34	6.92	0.71	—

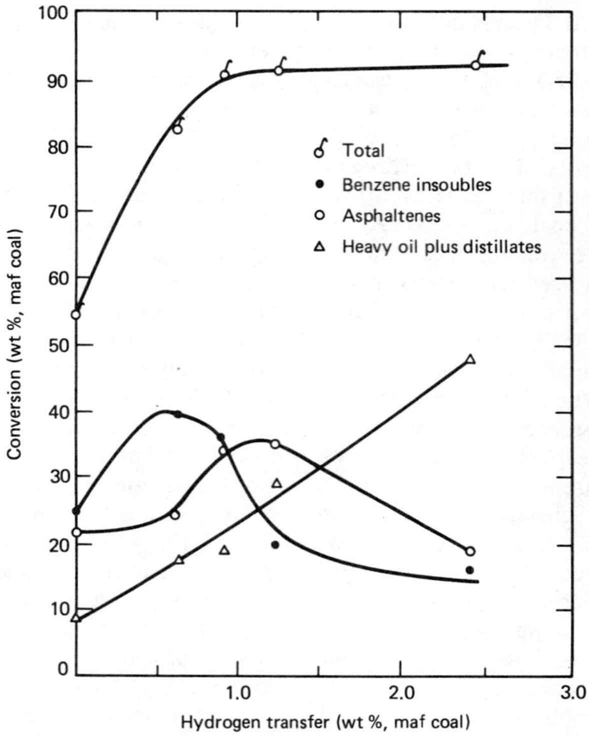

Figure 27.19 Product distribution in donor extraction for Ireland Mine coal. (Unpublished CCDC work.)

glomeration of the undissolved coal particles[29,39] thus facilitating their separation.

A still more extensive fractionation procedure termed SESC has been developed by Mobil Oil investigators[9,122] using liquid chromatography with a silica gel substrate. Ten different elution solvents with polarity varying from *n*-hexane to pyridine–10% methanol–20% water are used to produce the corresponding fractions. The procedure used by the Mobil investigators is aimed at fractionating the coal extract into groups of compounds having similar functionality. Each fraction was then extensively analyzed using a variety of chemical and instrumental methods including ^{1}H and ^{13}C NMR.

Table 27.24 gives the procedure and the major functionality associated with each fraction. The percentage of each fraction from an extract derived from hydroextraction of Kentucky No. 14 coal at the Wilsonville pilot plant is also shown. The Mobil report[122] gives an extensive discussion of probable chemical structures in the various fractions. NMR measurements indicate that most fractions contain a high percentage of aromatic carbon present in complex ring structures.

An important conclusion from the Mobil work is that the original coal and initial solubilized product are less aromatic than the final product. The quantity of condensed aromatic structures containing more than two rings, in particular, increases strongly with conversion.

Another interesting conclusion is that the average molecular weight of the individual SESC fractions did not change significantly with residence time, although that of the whole extract was reduced.

The concept with regard to the aromaticity of coal[123] is undergoing change again. Recent data suggest that the aromatic content of bituminous coals is quite high; more than 75% of the carbon is present in aromatic structures.

High-resolution, low-voltage mass spectro-

[122] Mobil Research & Development Corporation, *Nature and Origin of Asphaltenes in Processed Coals,* Ann. Rep. No. RP 410-1, EPRI AF-252, February 1976.

[123] D. I. Vanderhart and H. L. Retcofsky, *Fuel,* **55,** 202 (1976).

Table 27.24 The Mobil Oil SESC Fractionation Procedure[122]

Classical Description	Fraction	Weight Percent of Extract[a]	Elution Solvent	Major Compounds	Molecular Weight
Oils	1	0.4	Hexane	Saturates	
	2	15.0	Hexane–15% benzene	Aromatics	
	3	30.0	Chloroform	Polar aromatics; nonbasic nitrogen, oxygen, and sulfur heterocyclics	325
Asphaltenes	4	10.2	Chloroform–10 % Et_2O	Simple phenols	
	5	10.1	Et_2O–3% EtOH	Basic nitrogen heterocyclics	~400
	6	4.1	MeOH	Highly functional molecules (>10 wt % heteroatoms)	
Multifunctional compounds	7	6.4	$CHCl_3$–3% EtOH	Polyphenols	
	8	10.2	THF	Increasing oxygen content and increasing basicity of nitrogen	715
	9	8.5	Pyridine		
	10	5.1	—	Noneluted, unknown materials	

[a] From hydroextraction of Kentucky No. 14 coal.

metry has been a useful tool in investigating the most volatile fractions of the extract, principally the oil fraction.[43,124]

A technique[125] for volatilization of relatively high molecular weight extract components into the mass spectrometer has been developed by investigators at Stanford Research Institute. This permits a molecular weight distribution profile to be obtained. The data presented, however, were only for fuel oil products from the H-Coal process. The relatively low molecular weight of the material is surprising; the molecular weight of the most abundant species was in the 400 range.

An interesting new development in the chemistry of asphaltenes is the finding by Sternberg et al.[126] that these compounds can be separated into a basic and a neutral (or acidic) component by treatment with anhydrous HC1. The foregoing procedure for separation of asphaltenes into two fractions has been incorporated into a standard analytical inspection method.[127]

27.6.2 Heteroatom Removal

An important objective in producing nonpolluting boiler fuel from primary liquefaction is the removal of sulfur and nitrogen from the coal. It is clear from the data presented above that on an maf basis, the extract contains as much or more nitrogen than does the feed coal. Sulfur is, however, removed to a very considerable extent.

The removal of sulfur parallels the input of hydrogen. The Wilsonville pilot-plant data, for example, have been correlated[32] by the expression:

$$\text{sulfur ratio} = \frac{S}{S_0} = 0.78\, e^{-KH} \qquad (27.37)$$

where S_0 is the organic sulfur of the feed

[124] Thomas Aczel, J. G. Foster, and J. H. Karchmer, *Prepr., Am. Chem. Soc., Div. Fuel Chem.*, **13**(1), 8 (1969).

[125] M. E. Scoknich, G. A. St. John, and M. Anbar, Paper No. 9, presented at Coal Chemistry Workshop, Session II, Stanford Research Institute, August 1976.

[126] H. W. Sternberg, R. Raymond, and F. K. Schweighardt, *Science*, **188**, 49 (1975).

[127] H. W. Sternberg, *Prepr., Am. Chem. Soc., Div. Pet. Chem.*, **21**(7), 1 (1976).

coal, S is the sulfur content of the extract product, K is a constant, and H is the hydrogen consumption (wt % of the maf feed coal). Typical values of K are given below:

Coal	Temp (°C)	K
Illinois No. 6, Burning Star	443	0.31
Pittsburgh No. 8, Loveridge	454	0.13
Pittsburgh No. 8, Loveridge	446	0.18

These correlations are only valid, however, for the limited range of hydrogen consumption observed in the pilot-plant runs, that is, 2.2–3.3 wt % for the Loveridge coal and 1.5–2.5 wt % for the Burning Star coal.

Comparison of the efficiency of sulfur elimination via donor extraction was made by plotting (Fig. 27.20) the sulfur ratio obtained in continuous donor extraction[27] of Ireland Mine coal with the Wilsonville correlation[32] for Loveridge coal. The data suggest that sulfur elimination is equally good for donor and for hydroextraction. It should be stated, however, that the hydrogen lost by solvent dehydrogenation was not counted as hydrogen consumed in the donor extraction runs.

Previously unpublished CCDC results with Burning Star coal offer perhaps a more direct correlation, as shown below, at 440–443°C:

Sulfur in Extract (wt %)	Hydrogen Consumption (wt %, maf coal)	Process
1.19	1.3	Batch donor extraction
1.10	1.3	Wilsonville correlation, hydroextraction
0.90	2.5	Batch donor extraction
0.76	2.5	Wilsonville correlation, hydroextraction

Somewhat better desulfurization is apparent here via the hydroextraction route. Even so, the results suggest that the major desulfurization in the solvent-refined coal process occurs via thermal hydrocracking rather than by catalytic hydrofining.

The foregoing discussion implies that a large part of the organic sulfur in coal is present in easily removable aliphatic forms rather than in refractory heterocyclic forms. This view does not appear to be consistent with Mobil's[122] conclusion regarding the structure of the sulfur compounds.

Koltz et al.[128] have studied the kinetics of sulfur elimination from Kentucky No. 9 coal in batch hydroextraction equipment using anthracene oil as a solvent. The data fit a "pseudo" first-order reaction where the first-order rate constant decreases linearly with conversion, that is,

$$\frac{dx}{dt} = k_0(1 - ax)(1 - x) \qquad (27.38)$$

where k_0 and a are constants and x is the fractional sulfur removal. This treatment is equivalent to the statement that the residual sulfur becomes increasingly refractory with increasing conversion. It is thus analogous to the kinetics of extraction as discussed above.

Oxygen elimination is another topic that has received attention. Mobil[122] investigators have noted a parallelism between oxygen and sulfur removal in hydroextraction.

Studies of the mechanism of oxygen elimination and of the structural distribution of oxygen in the products formed in pressure hydrogenolysis of Japanese coals at 400°C have been reported by Hasegawa et al.[129,130] They find that at short residence times, the benzene-insoluble–pyridine-soluble fraction has a higher oxygen content than the feed coal. They conclude that there are two types of oxygen structure present, one of high and the other of low reactivity.

Chemical analytical procedures[130] were developed to determine the different types of

[128] R. C. Koltz, R. M. Baldwin, R. L. Bain, J. O. Golden, and J. H. Gary, *Prepr., Am. Chem. Soc., Div. Fuel Chem.*, **21**(5), 96 (1976).

[129] Y. Hasegawa, Y. Nakata, and Y. Yoshida, *Nenryo Kyokai-shi*, **53**, 987 (1974).

[130] Y. Hasegawa, Sh. Ueda, Sh. Yokoyama, Y. Yoshida, and Y. Maekawa, *Nenryo Kyokai-shi*, **53**, 995 (1974).

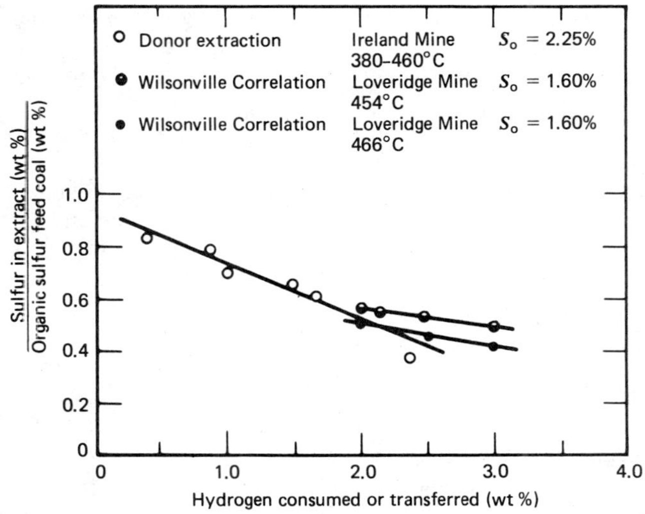

Figure 27.20 Comparison of sulfur elimination from Pittsburgh No. 8 coal: continuous donor extraction versus hydroextraction.

oxygen present. It was concluded that the ether oxygen is the most abundant form and that during hydrogenolysis the ether bonds are broken and phenolic hydroxyl groups are produced. The phenolic and alcoholic hydroxyl groups are gradually removed as the hydrogenolysis reaction progresses.

Cronauer and Ruberto[131] have undertaken a study for EPRI to determine the mechanism of oxygen elimination from coal during liquefaction.

[131] D. C. Cronauer and R. G. Ruberto, Task I Report, *Literature Survey for EPRI,* Contract No. RP 713-1, June 1976.

The form of elimination of oxygen, particularly with low-rank coals, is of some economic importance. The larger the percentage of oxygen eliminated as CO_2 instead of water, the lower is the hydrogen consumption. The data from a number of coals subjected to donor and hydroextraction were examined from this point of view and are summarized in Table 27.25. The donor runs selected were from the batch autoclave runs discussed in Section 27.2.3. The hydroextraction runs were from continuous bench-scale operations as cited in Table 27.25. It is obviously impossible to compare results under identical conditions. Comparative runs were selected

Table 27.25 Oxygen Elimination: Donor Extraction Versus Hydroextraction and Hydrogenation

Fuel	Oxygen (wt %, maf)	Hydrogen Consumed (%, maf)	Oxygen in Gas: As H_2O (%)	Oxygen in Gas: As CO + CO_2 (%)	Process[a]	Source of Data
Burning Star,	12.2	2.5	50	50	D	—[b]
Illinois No. 6		2.7	78	22	H	Ref. 39
Kaiparowits	14.2	1.6	60	40	D	—[b]
		2.2	68	32	H	Ref. 35
Black Mesa	14.6	1.6	46	54	D	—[b]
		2.6	66	34	H	Ref. 38
Colstrip	15.9	1.6	50	50	D	—[b]
		2.3	92	8	H	Ref. 41
Gillette	19.0	1.7	50	50	D	—[b]
Wyodak	20.2	1.7	56	44	H	Ref. 38
Wyodak	20.2	4.4	72	28	HC	Ref. 3
Glenharold lignite	22.6	1.7	57	43	D	—[b]

[a] D = batch donor extraction; H = continuous hydroextraction; HC = H-Coal.
[b] Unpublished CCDC data.

to correspond as closely as possible in hydrogen consumption.

Two conclusions are immediately apparent. The oxygen distribution in donor extraction varies in no regular manner with the type of coal, but on the average about 50% is eliminated as $CO + CO_2$. The continuous hydroextraction results show an even wider fluctuation with respect to the coal processed. A higher percentage of oxygen is eliminated on the average as water in hydroextraction. The difference with some coals is not significant, however.

An additional data point is indicated for processing of Wyodak coal via the H-Coal process. The catalytic treatment results in a higher percentage of oxygen eliminated as water, as might be expected.

Wolk et al.[132] have recently reviewed the available data on removal of nitrogen and sulfur during the liquefaction of coal. They show that the nitrogen removal is only slightly enhanced in liquefaction systems that use active hydrofining-type catalysts as compared with noncatalytic systems. The nitrogen content decreases sharply with decreasing molecular weight or boiling range of the products. The residuum or extract fraction is actually richer in nitrogen on an maf basis than the feed coal. The data of Burk and Kutta[118] (Table 27.22), and the Mobil data,[122] show the same trend of decreasing nitrogen content with molecular weight within the extract itself.

27.7 COAL MINERAL MATTER IN PRIMARY LIQUEFACTION

27.7.1 Types of Mineral Matter in Coal and Distribution in Products

The occurrence and analysis of mineral matter in coal are extensively discussed by Gluscoter et al., Chapter 7. The discussion here is focused on its distribution in the products, and Section 27.7.3 deals with their effect on secondary processing of the extract. These topics were discussed in some depth in two talks[133,134] given before the Gordon Research Conference on Coal Science.

The mineral matter in coal occurs in two broad categories. The first is the mineral matter present as discrete particles not directly associated with the organic matter. Some of the more abundant species here are the various clay minerals, iron pyrite, quartz, gypsum, and calcite. Pyrite is a particularly abundant mineral in the high-sulfur bituminous coals.

The second main category, the organically bound ash, has two main subcategories. The first consists of inorganic salts of organic acids present in the coal. These are sometimes referred to as humates. The second category comprises organic chelates of refractory oxides such as TiO_2, Al_2O_3, and SiO_2, as well as trace elements such as scandium.[135]

The inorganic salts usually comprise the major portion of the alkali and alkaline earth metals present in the coal. It has been shown that the metals associated with coal acids are ion exchangeable[136,137] and readily extractable with mineral acids.[138] CCDC[134] investigators, for example, showed that water saturated with sulfur dioxide at atmospheric pressure (i.e., sulfurous acid) is an efficient extractant for these metals. The ash composition of coals of various rank and the percentage removal of various oxides with sulfurous acid[134] after 60 min residence time at ambient temperature are shown in Table 27.26.

[132] R. H. Wolk, N. C. Stewart, and H. F. Silver, *Prepr., Am. Chem. Soc., Div. Fuel Chem.*, **20**(2), 116 (1975).

[133] Everett Gorin, "The Effect of Coal-Derived Minerals on the Activity of Hydrogenation Catalysts," paper presented at Gordon Conf. Coal Science, New Hampton, N.H., July 2, 1969.

[134] Everett Gorin, "Characterization of Coal Ash and Its Influence on Coal Conversion Processes," papers presented at Gordon Conf. Coal Science, New Hampton, N.H., July 24, 1975.

[135] P. M. Yavorsky and E. Gorin, *Utilization of Radioactive Isotopes in Coal Process Research*, unclassified report to U.S. Atomic Energy Commission No. NYC-9852, under Contract No. AT(30-1)-2350, May 31, 1962.

[136] *Technology and Use of Lignite, Proc: Bureau of Mines–University of North Dakota Symp.*, Grand Forks, N. Dak., April 27–28, 1967; U.S. Bur. Mines Inf. Circ. No. 8376, 1968.

[137] *Technology and Use of Lignite, Proc.: Bureau of Mines–University of North Dakota Symp.*, Grand Forks, N. Dak., May 1–2, 1969; U.S. Bur. Mines Inf. Circ. No. 8471, 1970.

[138] C. H. Wright and D. E. Severson, *Prepr., Am. Chem. Soc., Div. Fuel Chem.*, **16**(2), 68 (1972).

Table 27.26 Ash Composition of Coals and Extraction with Sulfurous Acid

Coal:	Hillsboro	Browning Mine	Colstrip	South Gillette	Glenharold
Mesh size:	−100	−100	−14	−14	−14
Type	Illinois bituminous	Utah bituminous	Montana subbituminous	Wyoming subbituminous	North Dakota Lignite
			Extraction (% removed)		
Total ash	19	37	36	39	62
Na_2O	82	8	69	96	98
K_2O	4	7	33	34	68
CaO	98	92	84	82	85
MgO	11	78	91	85	88
Fe_2O_3	9	23	11	63	92
		Ash Composition of Feed Coal (wt % of Ash, oxidized and SO_3-free basis)			
Na_2O	1.6	3.3	0.2	1.5	10.5
K_2O	1.7	0.2	0.1	0.6	0.9
CaO	7.1	18.6	15.4	22.0	36.4
MgO	0.9	5.0	6.0	3.6	7.1
Fe_2O_3	23.8	5.6	4.1	4.1	8.2

Source of data: Ref 134.

The data illustrate the ease of extraction of the alkali and alkaline earth metals, and their tendency to become an increasingly important component of the mineral matter as the coal rank decreases. A notable exception is potassium: the major portion of the potassium is apparently tied up with clay minerals such as illite in a form that is not readily extractable.

Very little iron is extracted in the case of most coals. Notable exceptions are the Wyoming subbituminous coal and lignite, from which a high portion of the iron was extracted. The iron in these latter two cases apparently is either largely present as an organic salt or exists in some readily extractable inorganic form.

Other evidence for the organic nature of the alkali and alkaline earth metals[134] is their concentration in the pure coal fraction, 1.35 specific gravity, on float-sink separation.

Not much is known about the chemical form of the chelated metal oxides or even the amount that is present.

The mineral matter in the extract derived by primary liquefaction may be characterized both by chemical analysis and by ash size distribution. Published data on the questions above are mostly available from extracts produced via donor extraction.[133]

The mineral matter[133] in extract was classified according to size by filtration in cresol solution through Whatman No. 1 filter paper (+1.2 μm) followed by a series of cellulosic Millipore filters that have a controlled pore opening. The majority of the −1.2-μm ash passed through the finest Millipore paper (i.e., 0.01 μm), and this material is probably present as soluble organometallic compounds.

The quantities and chemical composition of extract ash as compared with the coal ash are given in Table 27.27. The particular extract shown was derived by batch extraction of Ireland Mine coal with tetralin at 380°C.

The ash compositions are given in terms of the oxides and normalized to exclude SO_3 and minor elements such as boron and phosphorus.

It is readily noted that the composition of the coarse extract ash is substantially equivalent in chemical composition to that of the coal ash. The fine or "soluble" extract ash is, however, of quite different composition. The outstanding difference is the twentyfold increase in TiO_2 concentration and the fourfold increase in CaO concentration.

The titania enrichment is likely due to the presence of organotitanium chelated compounds of unknown structure. The CaO and alkali metals are likely present as soluble salts of complex organic acids, perhaps largely as phenolates, as evidenced[10] by their ease of removal by water extraction. The water-extracted fine extract ash contains more than 30% of titania. Other organometallic chelates

Table 27.27 Chemical Composition (wt %) of Coal and Extract Ash

	Ireland Mine Coal	Filtered Extract	
Size:	All > 1 μm	+ 1.2 μm ash	< 0.01 μm ash
Total ash (wt %):	13.6	0.034	0.103
Chemical Composition:			
$Na_2O + K_2O$	2.3	2.8	6.0
CaO	5.4	5.8	22.9
MgO	1.2	4.4	3.5
Fe_2O_3	23.7	14.2	5.3
Al_2O_3	16.9	15.6	17.5
SiO_2	49.3	55.4	14.6
TiO_2	1.2	1.8	30.2

Source of data: Reference 133.

of SiO_2, and perhaps Al_2O_3, are also present in all likelihood.

These chelates of TiO_2, SiO_2, and Al_2O_3 are very refractory, and no method of breaking them down and removing the metals, short of extraction with aqueous HF, was found.

Some further data on size distribution and chemical composition of extract ash are given in Table 27.28. The effectiveness of water washing in removing alkali metals and calcium is again illustrated. About 50% of the −0.01-μm ash is removed. The solvent used and conditions of extraction have an effect on the total ash content, but no gross effect is apparent regarding chemical composition of the ash.

The two extracts illustrated are derived from Ireland Mine coal in a 30-lb/hr pre-pilot-scale continuous unit. In the one case tetralin was used; the other employed "equilibrium" natural solvent derived from the CSF process, where solvent derived from extract hydrogenation is recycled through the extraction step.[10,27]

The water washing step has been previously described[10,27] and has been demonstrated on a continuous pre-pilot scale.

A similar concentration of titania in the −0.01-μm extract ash was also observed in

Table 27.28 Size Consist and Chemical Composition (wt %) of Extract Ash

	Tetralin Extract		Natural Solvent Extract	
	Unwashed	Washed	Unwashed	Washed
> 1.2 μm:	0.02	0.020	0.03	0.02
0.01–1.2 μm:	0.02	0.005	0.01	0.02
< 0.01 μm:	0.06	0.035	0.12	0.06
Total ash (wt %)	0.10	0.060	0.17	0.08
Chemical Composition, Total: Ash				
$Na_2O + K_2O$	3.0	0.7	1.6	1.6
CaO	24.3	5.3	18.2	2.2
MgO	1.9	1.4	1.8	0.9
Fe_2O_3	7.0	8.5	10.0	12.1
Al_2O_3	24.9	42.9 (Al_2O_3 + SiO_2)	26.6	23.4
SiO_2	17.2		8.0	19.6
TiO_2	21.7	36.2	33.8	40.2

Source of data: Reference 10.

CCDC work[134] on the hydroextraction of Illinois No. 6 coal at 950°F and 2000 psig.

The distribution of extract-soluble (i.e., -0.01-μm) ash by molecular weight was determined[133] by selective solvent precipitation. The molecular weight distribution of the organic extract was determined by two methods: by gel permeation chromatography and by a cryoscopic method. Fairly good agreement was obtained between the two methods.

The distribution of the extract and soluble ash by molecular weight is shown in Fig. 27.21. The ash is concentrated in the high molecular weight portion of the extract. More than 90% of the ash is concentrated in the "heaviest" 50% of the extract, mostly in the benzene insolubles.

27.7.2 Mineral Residues

The chemical composition of the ash in the residues from donor or hydroextraction is substantially equivalent to that of the mineral matter in the feed coal, as is evident from the discussion above. The mineral matter present in the extract in soluble form obviously cannot be removed from the extract by purely mechanical means such as filtration.

The primary property of interest for evaluation of process techniques for removal of mineral residues from extraction effluents is the size distribution of the residue. A secondary property of interest is the extract solubility behavior, as was discussed in Section 27.6.1.

The mean size of the extraction residue can be quite a bit larger than that of the mineral matter in the feed coal. Walker et al.[49] at Pennsylvania State University are investigating this problem, and some results have been reported.

The Walker group[49] has been developing computerized scanning electron microscope (SEM) techniques to determine the size distribution of the various mineral components in coal. Data are given, for example, for a number of the feed coals used for the HRI[35–38] bench-scale and Wilsonville[31,32] pilot-plant work on hydroextraction. The median size of the mineral matter in the Kaiparowits coal determined by the SEM technique[49] is only about 1 μm. The median size of the pyridine-extracted residues from hydroextraction of this coal[35] determined by sedimentation[49] techniques is considerably higher (i.e., ~ 10 μm or more).

It is clear from the work cited that the mineral matter is not completely liberated from the organic part of the coal during hydroextraction. Agglomeration of liberated mineral particles can also occur because of repolymerization and coking of extracted organic matter at the surface of mineral particles.[49]

The complex behavior of the various con-

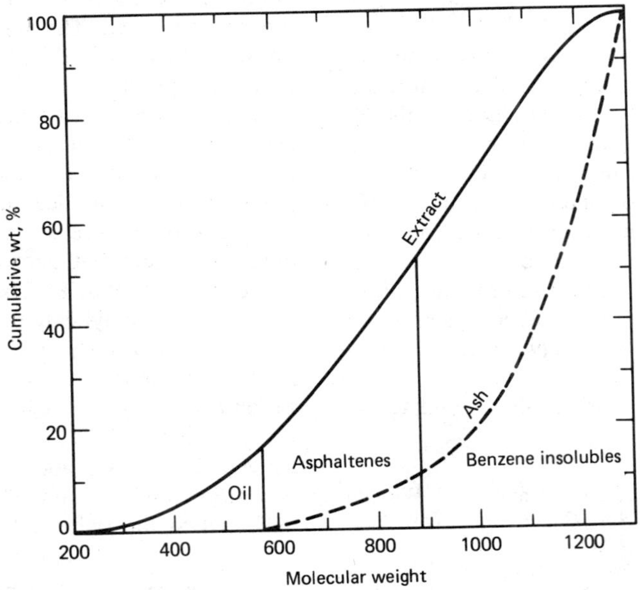

Figure 27.21 Distribution of "soluble" ash (-0.01 μm) and extract by molecular weight.[131,133]

stituents of the coals undergoing liquefaction was discussed in detail from the petrographic viewpoint by Walker et al.[49] and also in papers by Mitchell et al.[139] and Davis.[140]

The size distribution of extraction residues as affected by the nature of the coal, type of process, and conditions employed, was also discussed by CCDC workers.[29,39] The principal generalities arrived at are more or less in accord with the data and considerations advanced by the Penn State workers.[49,139] The mean residue particle size decreases with coal rank and with the severity of liquefaction conditions employed. The depth of conversion to cresol solubles, for example, is one measure of the severity of the liquefaction conditions. Thus the difficulty of residue separation by gravity settling or solvent deashing increases as the process changes in the sequence donor extraction to hydroextraction to catalytic liquefaction.

Some doubt is cast on the CCDC conclusions by particle size distribution data reported by Auburn University investigators[67] on filter feed produced in the Wilsonville pilot plant. The size distribution is not only quite different from CCDC results with the same coal (Burning Star), but also varies in no systematic way with the different coals used. More data are required to clarify the situation.

27.7.3 Effect of Mineral Matter on Catalysts

The mineral matter in extract is of importance when the extract is subjected to catalytic hydrotreating either *in situ* as in the H-Coal and Synthoil processes or after separation from the residue, as in the CSF process.

It is only the organometallics or soluble ash that can permanently damage the catalyst. Particulate ash cannot enter catalyst pores, although it can cause operating problems, particularly in fixed-bed systems, by deposition in the voids between catalyst particles.

The deactivation of cobalt molybdate catalysts in the H-Coal process was discussed by Kang and Johanson.[141] Typical data showing the rate of deactivation in the process had previously been presented by Johnson et al.[15] The three causes discussed by Kang were carbon deposition, sintering by steam, and deposition of organometallics.

Gorin[133] also discussed catalyst deactivation in the CSF process for the same three reasons.

The relative importance of the three causes of deactivation was generally agreed on by both sets of workers. Carbon deposition causes an initial catalyst deactivation, but this occurs rapidly and the carbon level on the catalyst quickly reaches a steady state level and changes only very slowly with further processing time. Sintering by steam generated in the process also causes a slow catalyst deactivation and can be accelerated by using incompletely dried coals.[141] It is agreed, however, that the principal cause of catalyst deactivation is the deposition of metals in the catalyst pores. Synergistic effects also exist among the three sources of catalyst deactivation.

One very sensitive measure of catalyst activity is the sulfur level of the hydrotreated residuum. The greater importance of ash deactivation is illustrated by CCDC results[134] (Fig. 27.22) in a continuous 1-lb/hr, bench-scale, ebullated-bed extract hydrotreating unit. Two extracts produced by continuous donor extraction of Ireland Mine coal with tetralin were used. The first was deashed by filtration, then washed with acidulated water to yield a product extract with 0.045 wt % ash. The second was extensively deashed by adding a third step: washing with hydrofluoric acid. This latter treatment caused some polymerization and degradation of the extract; that is, it increased the benzene-insoluble component but decreased the soluble ash content to 0.01%. The catalyst was a beaded nickel molybdate on alumina catalyst (UOP S-6) previously described.[10]

The much more rapid decline in desulfurization activity of the higher ash feedstock is evident from Fig. 27.22. The major cause of activity decline is again shown to be metals deposition on the catalyst by the activity indicated to the right of the broken line in

[139] G. D. Mitchell, Alan Davis, and William Spackman, *Prepr., Am. Chem. Soc., Div. Fuel Chem.*, **21**(5), 189 (1976).

[140] Alan Davis, Paper No. 2, presented at Coal Chemistry Workshop, Session I, Stanford Research Institute, 1976.

[141] C. C. Kang and E. S. Johanson, *Prepr., Am. Chem. Soc., Div. Fuel Chem.*, **21**(5), 32 (1976).

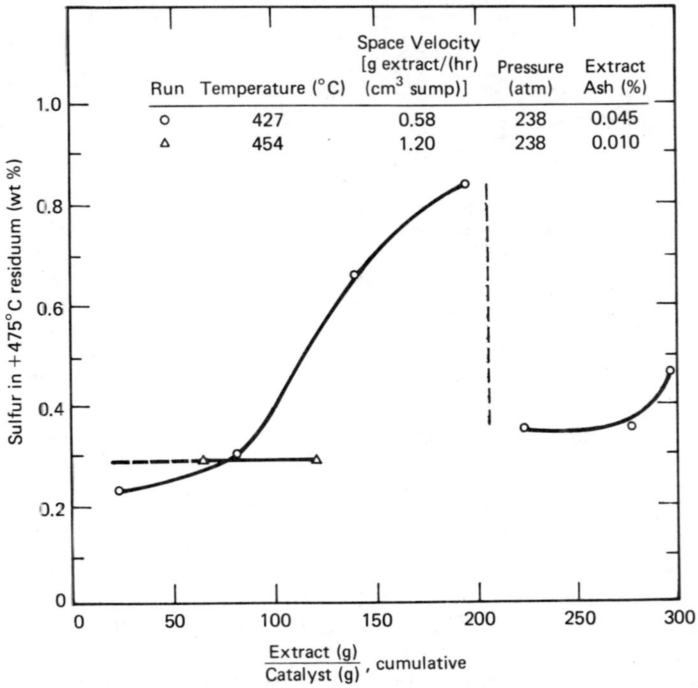

Figure 27.22 Rate of decline of desulfurization activity: ebullated-bed Ni-Mo-Al_2O_3 catalyst.[134]

Fig. 27.22. The catalyst at this point was reactivated, without burning off the carbon, by abrading off the ash-rich outer layer of the catalyst.

Titanium is an excellent indicator to define the metals deposition profile on catalysts, both because of the ease of its detection and because it is one of the principal elements present in the coal as an organic chelate.

HRI[141] and CCDC[134] investigators showed a concentration of titania in the catalyst when hydrotreating bituminous coals or extracts derived therefrom in ebullated beds.

Distribution of titania in the beaded catalyst used by CCDC[133] workers was studied by an electron beam microprobe technique. It was shown that the titania is relatively concentrated at the surface. This phenomenon is apparently the reason that catalyst activity is in part readily restored by abrasion of the surface.

Kang and Johanson[141] report very little titania deposition, however, when a Wyoming subbituminous coal is processed. The rate of deactivation appears to be higher for subbituminous coals than for bituminous coals[3] when they are treated in the H-Coal process. This is to be expected, in view not only of the higher alkali–alkaline earth metals content of subbituminous coals, but also because of the larger amounts of steam evolved when they are hydrogenated.

It is well known that alkali and alkaline earth metals are active poisons for hydrogenation catalysts. Furthermore, a synergistic effect appears to exist between poisoning by alkaline earth metals and deactivation by steam.[133]

Another interesting observation made by CCDC[133] workers is that the more active and more selective the catalyst, the more efficient is the capture or deposition of the soluble metals in the catalyst. Figure 27.23 illustrates this point with results from bench-scale, ebullated-bed hydrogenation of extract with UOP S-6 catalyst. The abscissa shows the percentage of titanium in the feed deposited on the catalyst. Total conversion on the ordinate refers to conversion to distillates plus gas. Selectivity is measured by the yield of C_1–C_3 hydrocarbon gases divided by conversion.

Other studies showed that the rate of catalyst activity decline not only is a function of the total soluble ash, but increases with the proportion of alkali and alkaline earth oxides in the soluble ash.

The implication of these findings to catalyst

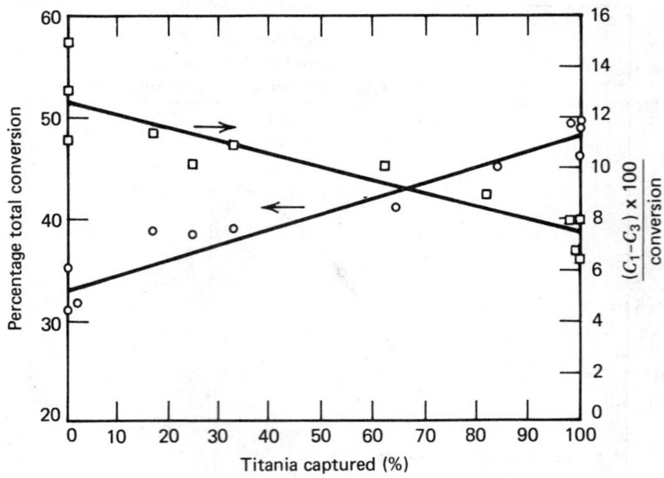

Figure 27.23 Efficiency of titania capture by catalyst at 427°C, 190 atm H_2 pressure: ebullated-bed Ni-Mo-Al_2O_3 catalyst UOP S-6. Space velocity = 3 g of extract per g of catalyst per hour or 0.7 g of extract per cubic centimeter sump per hour.[133]

life in hydrotreating of extracts is clear. The rate of decline, in the high activity range near that of fresh catalyst, is proportional to the soluble ash content of the feed. The proportionality constant is higher the higher the alkali–alkaline earth metals content of the feed. The soluble ash content of extracts is generally too high, usually about 0.1 wt %, to permit catalyst replacement at a high enough rate to maintain a close approximation to fresh activity. In practical systems such as H-Coal or in secondary ebullated-bed treatment of extracts, one must be content with operation at a catalyst activity level that is only a fraction of that of fresh catalyst.

27.8 CATALYSIS IN LIQUEFACTION

27.8.1 Noncatalytic Systems

The liquefaction of coal under hydrogen pressure, as will be shown below, always involves catalytic effects of the mineral matter. However coal liquefaction by means of donor extraction—that is, in the absence of hydrogen pressure—is likely a noncatalytic process.[4] Mineral matter plays a secondary role in donor extraction in that it can cause undesirable catalytic dehydrogenation of the solvent.

The hydrogen transfer to the coal that can be achieved is limited in practice by the number of points of easy thermal rupture of the polymeric coal molecule. Hydrogen donor extraction, from the practical point of view, ceases after the coal has been broker down into its basic polynuclear units, which have thermal stabilities comparable to that of the donor solvent itself. The maximum amount of "practical" hydrogen transfer that can be effected is thus around 2.5–3.0 wt % of the maf coal.[4] The exact amount, of course, is a function of the particular coal being processed.

Donor extraction is thus capable of effecting nearly complete conversion of the cresol-soluble–benzene-insoluble components of extracted coal to asphaltenes plus oil[4,21,26] but is incapable of completing the conversion of asphaltenes to oil and distillate. Catalytic hydrogenation is required here to create new points susceptible to thermal rupture in the asphaltene molecule.

27.8.2 Catalysis by Mineral Matter

The liquefaction of coal by hydroextraction in the absence of added catalysts, however, involves catalysis by the mineral matter of the coal. *In situ* hydrogenation of the solvent is one reaction catalyzed by coal mineral matter. Direct evidence of the catalytic activity of extraction residue for anthracene oil hydrogenation was presented, for example, by Wright and Severson.[138] An iron-rich residue from the hydroextraction of Kentucky

No. 11 hvBb coal containing 5.5% iron was fed as a slurry in anthracene oil to a bench-scale continuous unit. The increase in hydrogen content of the oil was found to be proportional to the solids content in the effluent slurry, as illustrated in Fig. 27.24. The iron in the residue is present largely as iron sulfide to which the catalytic activity was attributed, since the catalytic activity of iron sulfide is well known.[142]

The order of magnitude of the catalysis was shown to be consistent with the rate of uptake of hydrogen by high-sulfur bituminous coals in the solvent-refined coal process.

The sodium catalysis of the liquefaction of lignite by carbon monoxide–water was discussed previously.[104] Wright and Severson[138] also performed experiments that yielded further proof regarding the catalytic effect of the naturally occurring sodium in lignite.

In a later publication Wright and Anderson[46] discussed further the catalytic effect of iron in coal minerals in the "solvent-refined coal process." They showed that the coal minerals, and particularly the iron-rich minerals, tended to concentrate in the dissolver in their continuous bench-scale unit; thus possibly their catalytic action is intensified.

Tarrer et al.[143] also investigated the catalytic action of coal minerals in various batch autoclave studies. The catalytic activity of iron-rich residues from coal extraction and of other minerals such as pyrite was demonstrated in the hydrogenation of creosote oil. Catalytic activity was also observed in other minerals that may occur in coal ash but are free of iron, such as kaolinite, although the activity was lower in magnitude.

Recycle of an iron-rich mineral residue from hydroextraction of Kentucky No. 9 coal[143] was also shown to accelerate the rate of hydrogen uptake in autoclave hydroextraction. Similarly, removal of iron-rich minerals by high-intensity magnetic separation was shown to adversely effect the conversion in subsequent hydroextraction.

The catalytic effect of mineral matter in liquefaction of an Indian coal from North Assam was also demonstrated by Mukherjee and Choudhury.[11] Their experiments were conducted without a liquid vehicle. The hydrogenation was conducted at 400°C for 3 hr with an initial cold hydrogen pressure of 96.9 atm.

The amount of mineral matter was varied by use of various specific gravity fractions derived via float-sink separation. Very low mineral contents were obtained by further treatment of 1.30 specific gravity float fraction with hydrochloric and hydrofluoric acids. The conversion to benzene solubles increased dramatically with increasing mineral content of the charge, as illustrated in Fig. 27.25.

[142] W. Kawa and R. W. Hiteshue, U.S. Bur. Mines Rep. Invest. No. 6179, 1963.
[143] A. R. Tarrer, J. A. Guin, W. S. Pitts, J. P. Henley, J. W. Prather, and G. A. Styles, *Prepr., Am. Chem. Soc., Div. Fuel Chem.*, **21**(5), 59 (1976).

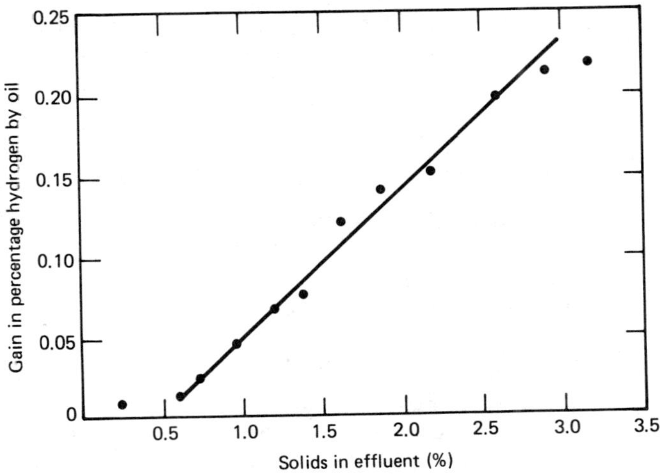

Figure 27.24 Catalytic effect of coal hydro residue solids in anthracene oil hydrogenation at 425°C, 69 atm; liquid hourly space velocity = 0.6.[138]

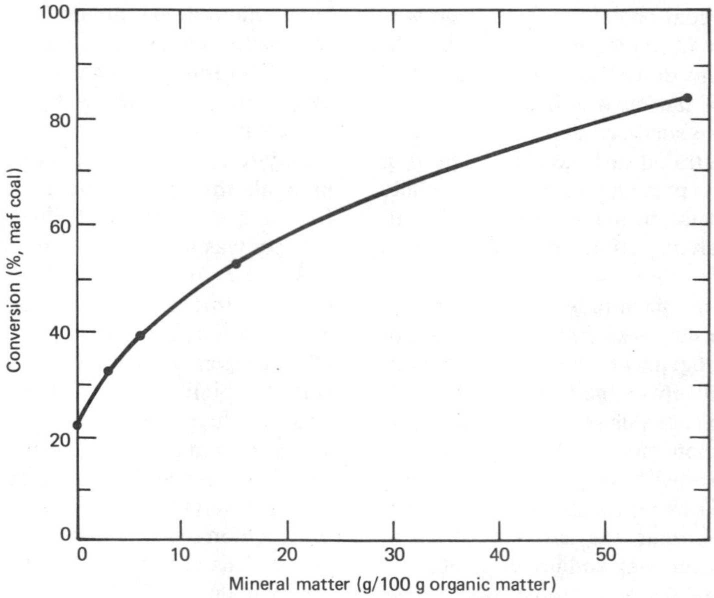

Figure 27.25 Relation between mineral matter content and conversion hydrogenation of North Assam coal.[11]

The mineral matter of North Assam coal used in the study just cited contained 28.8% Fe_2O_3 and 0.60% TiO_2 on an oxidized basis.

The coal conversion plotted[11] against the Fe and Ti contents of the charge gives similar correlations. This in itself does not prove, as Mukherjee and Choudhury[11] suggest, that both iron and titanium are catalysts. It is reasonable to assume, in view of the prior literature, that the major catalytic component is iron sulfide. Mild catalytic effect of titanium was demonstrated by impregnation of the feed coal with freshly precipitated titanium hydroxide. Kaolinite, by a similar method, was also demonstrated to be mildly catalytic.

The overwhelming importance of catalysis in coal liquefaction was further illustrated by petrographic analysis of the high ash charge stocks in experiments[11] just discussed.

The float-sink fractions enriched in mineral matter were also enriched in the refractory petrographic component, inertinite. In spite of this, better conversion was obtained than with the low-ash materials enriched with the more reactive component, vitrinite.

The high degree of liquefaction obtained without a vehicle shows that with adequate catalysis it is not necessary to use a donor solvent. This finding, of course, is not new; it is discussed in more detail below.

Wright and Severson[138] proposed that the rate-limiting factor in hydroextraction is the rate of hydrogenation of the solvent. The latter in turn is proportional to the amount of active catalyst present in the dissolver. The inventory of iron-rich minerals in continuous hydroextraction units has not generally been measured. It should, however, bear some relationship to the iron content of the feed coal. HRI[35–38] and CCDC[39] bench-scale data on hydroextraction indicate no simple relationship between the iron content or rank of the feed coal and the hydrogen consumption. The conditions under which the various coals were tested varied as follows, so that direct comparison is difficult:

Pressure	103–170 atm
Temperature	427–450°C
Space velocity	0.4–0.5 g/(hr)(cm^3)

Even so, the hydrogen consumption in all cases fell within a narrow range of 2–3 wt % of the maf coal, with one exception, namely, the particularly reactive Illinois No. 6 coal from the Monterey Mine, where the hydrogen consumption was 4 wt %.

The iron content of the different feed coals, expressed as Fe_2O_3, varied over a very wide range (0.25–2.30 wt %). As a matter of fact, two Pittsburgh Seam coals, containing 1.0 and 1.3 wt % Fe_2O_3, respectively, were tested by HRI[36] under identical conditions. The hydrogen consumptions, however, were in the reverse order, (i.e., 2.7 and 2.0 wt % of the maf coal, respectively).

The discussion above shows that the catalytic effect of coal ash in hydroextraction is rather complex. The catalysis involves not only solvent hydrogenation, but also to some extent direct hydrogenation of the coal. The major catalytic effect is due to iron sulfides, but other minerals apparently also contribute to the catalysis. The particle size distribution of the mineral matter, particularly of the pyrite, is also likely to be of importance.

27.8.3 Systems with Contact Catalysts

The older coal hydrogenation art generally used small amounts of catalysts in liquid phase hydrogenation which were not recovered and reused. The catalysts employed were usually iron, either added as red mud or sprayed on the coal as iron sulfate.[1] Metal halides[1] were used, as well as tin catalysts, employed in very small quantities and promoted by addition of halogen compounds.

A number of development projects are in progress in the United States in which the catalyst added is of the generic type of the hydrotreating or hydrofining catalysts used in the petroleum industry. The catalyst is used in the form of fairly coarse particles—approximately $^1/_{16}$ in. diameter or greater—and is employed as an ebullated bed[15] as in the H-Coal process or as a turbulent fixed bed as in the Synthoil[16] process.

The catalytic action here probably does not involve the coal directly, but rather the solvent and the primary liquefaction products. In all likelihood the coal is already in solution in the Synthoil process before contact is made with the catalyst bed. This assumption follows from the considerable residence time provided to the coal slurry after preheating and before it contacts the catalyst bed. The coal in the H-Coal process is also probably dissolved to a large extent in the vehicle before it contacts the ebullated bed. The catalyst in both the H-Coal and Synthoil processes may be regarded as primarily serving to rehydrogenate the solvent and to effect hydrogenolysis of the extract.

The analysis of such systems would be expedited if more information on the kinetics of extract hydrocracking were available. The factors effecting the hydrocracking of extract in the H-Coal process are complex and include temperature, hydrogen partial pressure, type of catalyst, catalyst age and metals deposition thereon, extract conversion, nature of coal feed, extract concentration, and catalyst concentration in grams per cubic centimeter of sump.

Limited data on the kinetics of hydrocracking[17] are available from CCDC bench-scale data. These were obtained in an ebullated bed unit using an extract from Ireland Mine coal that had been deashed by washing with hydrofluoric and hydrochloric acid. The data were obtained under the following conditions:

Pressures	239 and 409 atm
Temperatures	427, 441, and 454°C
Extract space velocity	0.27, 0.57, 1.20 g/(hr)(cm^3)
Feed ratio, tetralin to extract	0.39
Hydrogen rate	6.5 moles per 100 g of extract
Catalyst concentration	0.23 g/cm^3 sump
Catalyst	$^1/_{16}$-in. UOP S-6

The hydrocracking kinetics was fairly well represented over this operating range by a "pseudo" first-order equation where the first-order rate constant decreases in proportion with conversion of the extract to distillate and gas. Since the system is a perfectly mixed one, the appropriate rate equation for the system is:

$$\frac{F}{V} = k\,E\,\frac{1-\alpha}{\alpha} \tag{27.39}$$

where

F/V = space velocity (g extract/hr per cm^3 sump)

k = "pseudo" first-order rate constant (hr^{-1})

E = extract concentration (g/cm³ sump), assumed constant for conditions above

α = conversion of extract to distillate plus gas

The experimental data were fit to Equation 27.39 by the method of least squares to yield the equations:

$$\ln(kE) = \frac{-16{,}700}{T} + 23.0 \text{ at 239 atm or } H_2 \text{ partial pressure of 197 atm} \quad (27.40)$$

$$\ln(kE) = \frac{-20{,}300}{T} + 28.4 \text{ at 409 atm or } H_2 \text{ partial pressure of 340 atm} \quad (27.41)$$

where T is the temperature (K).

It was noted in other CCDC work that the rate of hydrocracking was relatively insensitive to the catalyst concentration. This observation strongly implies that thermal pyrolysis of hydrogenated extract plays a significant role in the overall mechanism of hydrocracking in ebullated-bed systems.

The extent of hydrocracking observed over the range of operating conditions given varied between 26 and 85 wt % of the feed extract. Because the rate expressions are empirical, they cannot be used outside the range of the data.

The rates apply to a relatively fresh catalyst with a very low metals content due to the deashing process used. The deashing process did, however, have the effect of decreasing the reactivity of the feedstock.

27.8.4 Iron Catalysts

The addition of iron catalysts in autoclave experiments increases the degree of hydrogenation, even with coals that are rich in iron-bearing minerals. It is not certain, however, that this behavior would also be observed in continuous systems, where concentration of the iron minerals in the hydro reactors tends to occur.

Mukherjee et al.[144] made an extensive study of iron catalysis in batch autoclave studies of the liquefaction in the 200–300 atm pressure range of an Indian coal (North Assam). Iron was added either by impregnation of the coal with ferric chloride or ferrous sulfate or by addition of hydrated ferric oxide, with and without added sulfur. The activity of the iron catalyst systems was compared with the use of tin and ammonium molybdate impregnated on the coal as catalysts.

The optimum method of applying the iron catalyst was found to be precipitation of ferric hydroxide on the coal, combined with the addition of free sulfur. The most active iron catalysts were found to be comparable to impregnated ammonium molybdate in activity.

X-Ray studies indicated that the promotional effect of sulfur may be attributed to the formation of ferromagnetic sulfides having the formula $FeS_{(1+x)}$ where x lies between 0.04 and 0.10.

An interesting sidelight to these studies is that with the best iron catalysts, as good or better results were obtained when no vehicle was used as compared with runs with a vehicle. Ishii and Yoshida and their co-workers[85,145,146] studied the hydrogenation of Japanese coals in autoclave experiments with red mud as a catalyst. Anthracene oil was used as the vehicle. The principal interest of these studies was in elucidating the kinetics and mechanism of the process. The kinetics of hydrogenolysis were found to be consistent with the dual reaction paths:

$$\text{coal} \begin{cases} \nearrow \text{oil} \\ \searrow \text{asphaltenes} \rightarrow \text{oil} \end{cases}$$

All can be represented by a first-order kinetics. The rate constants, however, vary with the rank of the coal processed.

The foregoing picture is somewhat at variance with that formerly presented on United States coals by Weller et al.,[147] who represented coal conversion as proceeding through sequential first-order reactions,

[144] D. K. Mukherjee, J. K. Sama, P. E. Choudhury, and A. Lahiri, *Proc. Symp. Chemicals and Oil from Coal,* Paper No. 8, Central Fuel Research Institute, Dhanbad, India, December 6–8, 1969; published 1972 by Central Fuel Research Institute.

[145] T. Ishii, Y. Maekawa, and G. Takeya, *Chem. Eng., Jap.,* **29,** 988 (1965).

[146] R. Yoshida, Y. Maekawa, T. Ishii, and G. Takeya, *J. Chem. Soc., Jap. Chem. Ind. Chem.,* 1885 (1972).

[147] S. Weller, M. G. Pelipetz, and S. Friedman, *Ind. Eng. Chem.,* **43,** 1572 (1951).

coal → asphaltenes → oil

The view presented by the Japanese investigators is also in accord with that put forth by Squires.[84]

27.8.5 Coal Rank and Petrographic Composition in Catalysis

The subjects of coal rank and petrographic composition in catalysis were addressed in a joint effort by Gulf Research and Pennsylvania State University.[76,148]

Two batch autoclave techniques were employed in the study. The first, termed the Penn State procedure, used no vehicle; the catalyst comprised 1% ammonium molybdate impregnated on the coal. Experimental conditions were 400°C, 3 hr residence time, and an initial hot pressure of 180 atm. Liquefaction was followed by measurement of conversion to benzene solubles.

The second autoclave procedure used anthracene oil as a vehicle and a proprietary catalyst from Gulf. Test conditions were 385 and 400°C, 1 hr residence time, and constant total pressure of 238 atm. Conversion was followed by a procedure similar to that used by HRI[3] and was essentially equivalent to the cresol-soluble basis.

A wide variety of coals of varying rank and geological age, and from different regions of the United States, were tested. Molecular parameters based on proton NMR measurements were also given for the asphaltene fraction.

The high-volatile A, B, and C bituminous coals gave generally the best conversion[148] by both the Gulf and Penn State procedures, but within this group no relationship between liquefaction response and coal rank or geological age was evident.

Two bituminous coals from the State of Washington gave an especially favorable response to liquefaction, but both coals were exceptionally high in mineral matter, which may be relevant.

Subbituminous coals and medium- and low-volatile bituminous coals responded relatively poorly. The response became poorer in the order named above.

[148] P. H. Given, D. C. Cronauer, William Spackman, H. L. Lovell, A. Davis, and B. Biswas, *Fuel*, **54**, 40 (1975).

The response of lignites in this study was quite specific.[148] A North Dakota lignite sample was acid extracted to yield a sodium-free lignite, and part of the product was treated with sodium chloride to yield a sodium-enriched lignite. Both the sodium-free and the sodium-enriched lignites gave exceptionally high conversions, 96 and 97% respectively, by the Gulf procedure. The sodium-enriched lignite gave an oil of much lower viscosity and benzene-insoluble content than the sodium-free lignite. The asphaltenes from the sodium-rich lignite were also unusual in that they showed by NMR analysis an unusually high content of monocyclic aromatics. These results point to a favorable catalytic effect of sodium.

Studies on the liquefaction behavior of a number of coals[148] were also made wherein various macerals were concentrated by float-sink and other techniques. Reactive macerals such as vitrinite were relatively concentrated in the low specific gravity float fractions, whereas unreactive macerals such as fusinite were concentrated in the sink fractions.

Fair correlation between liquefaction behavior and content of reactive macerals was obtained for the case of a subbituminous coal from New Mexico (Fig. 27.26) and an Elkhorn hvAb coal from Kentucky (Fig. 27.27). The reactive macerals in the latter case are defined as the sum of the vitrinites, pseudovitrinites, and sporinites.

The lignite case was quite different. No correlation of liquefaction behavior with maceral type was found. The best results were obtained with a high-ash, fusinite-rich, 1.40 specific gravity, sink sample. The ash in this case concentrated the iron-rich minerals, principally pyrite. The "reverse" behavior here may be due to the catalytic effect of the iron: A similar result was found, as discussed above,[11] in liquefaction of North Assam coal.

27.8.6 Molten Salt Catalysis

Relatively small quantities of various metal halides ($ZnCl_2$, $SnCl_2$, $NiCl_2$, $FeCl_2$, etc.), impregnated on coal, were used as "throw-away catalysts" in bench-scale, pilot-plant, and commercial coal liquefaction operations in Germany, Great Britain, Japan, and the United States for many years.[1]

The use of massive quantities of molten halide catalysts of the Lewis acid type to

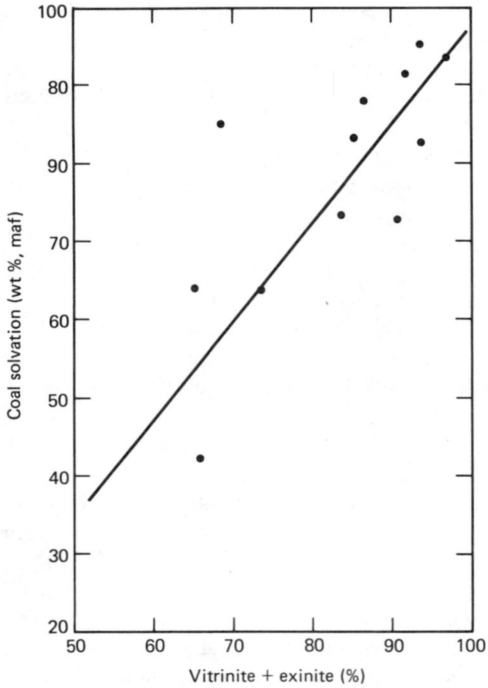

Figure 27.26 Coal solvation versus reactive macerals: New Mexico subbituminous coal from the Blue Seam McKinley Mine.[148]

catalyze coal liquefaction was investigated more recently by workers at CCDC[149–155] and at Shell.[156,157]

[149] Consolidation Coal Company, R & D Rep. No. 39, to U.S. Dept. Interior OCR, under Contract No. 14-01-0001-310—Interim Rep. No. 2, *Research on Zinc Chloride Catalyst for Converting Coal to Gasoline—Phase I—Hydrocracking of Coal and Extract with Zinc Chloride,* Vol. 3, Book 1, Consolidation Coal Company, March 1969.

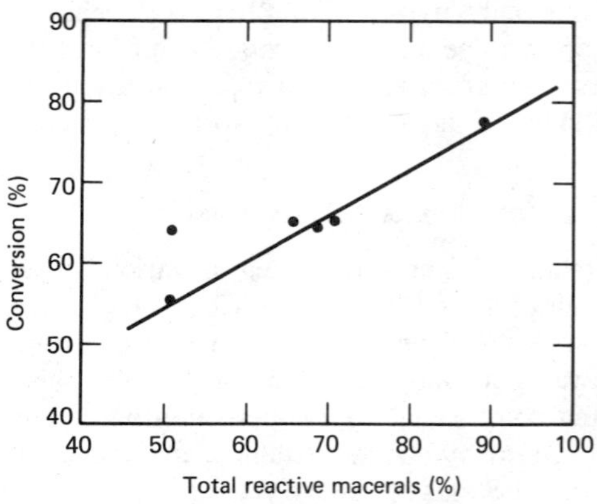

Figure 27.27 Solvation of lithotypes from the Elkhorn No. 3 Seam, hvAb Kentucky coal.[148]

The Shell workers[156,157] investigated the activity of a large number of metal halides of the Lewis acid type for the hydrocracking of Illinois No. 6 coals. The CCDC workers focused their attention principally on the use of zinc chloride, for reasons to be made clear below. A large number of coals were investigated, varying in rank from lignite to hvAb[155] in batch autoclaves.

Kawa et al.[158] also investigated the hydrocracking in autoclaves of Pittsburgh Seam hvAb coal and asphaltenes derived from this coal by mild hydrogenation. They used massive quantities of the various zinc halides as well as nickel, cadmium, iron, and tin.

Continuous unit operations in the earlier CCDC work were confined to evaluation of the hydrocracking of extract derived from Ireland Mine coal. Kinetic expressions[149,151–154] were derived similar to those described for coal extraction,[4,39] as discussed in Section 27.4. These satisfactorily correlated the experimental data.

A summary of the status of the zinc chloride hydrocracking work prior to 1975 was presented by Gorin.[159]

[150] Consolidation Coal Company, R & D Rep. No. 39, to U.S. Dept. Interior OCR, under Contract No. 14-01-0001-310—Interim Rep. No. 2, *Pre–Pilot Plant Research on the CSF Process—Phase I—Regeneration of Zinc Chloride Catalyst,* Vol. 3, Book 2, Consolidation Coal Company, March 1969.

[151] C. W. Zielke, R. T. Struck, J. M. Evans, C. P. Costanza, and E. Gorin, *Ind. Eng. Chem., Process Des. Dev.,* **5,** 151 (1966).

[152] C. W. Zielke, R. T. Struck, J. M. Evans, C. P. Costanza, and E. Gorin, *Ind. Eng. Chem., Process Des. Dev.,* **5,** 158 (1966).

[153] R. T. Struck, W. E. Clark, P. J. Dudt, W. A. Rosenhoover, C. W. Zielke, and E. Gorin, *Ind. Eng. Chem. Process Des. Dev.,* **8,** 546 (1969).

[154] C. W. Zielke, R. T. Struck, and E. Gorin, *Ind. Eng. Chem., Process Des. Dev.,* **8,** 552 (1969).

[155] C. W. Zielke, W. A. Rosenhoover, and E. Gorin, *Prepr., Am. Chem. Soc., Div. Fuel Chem.,* **19**(2), 306 (1974); "Shale Oil, Tar Sands, and Related Fuel Sources," in Am. Chem. Soc. Adv. Chem. Ser. No. 151, 1976, pp. 153–166.

[156] M. M. Wald, British Pat. 1,310,283, assigned to Shell Oil Company, March 14, 1973.

[157] T. E. Kiovsky, U.S. Pat. 3,764,515, assigned to Shell Oil Company, October 9, 1973.

[158] W. Kawa, H. F. Feldmann, and R. W. Hiteshue, *Prepr., Am. Chem. Soc., Div. Fuel Chem.,* **14**(4), 19 (1970).

[159] Everett Gorin, "Hydrocracking of Coal by Zinc Chloride Catalysts," paper presented at Pittsburgh Catalysis Society Meeting, Pittsburgh, April 1975.

Work is now in progress to demonstrate the zinc chloride hydrocracking process on a larger process development scale. A continuous bench-scale unit has also been put in service and successfully operated using subbituminous coal as feedstock.[160,161]

The mechanism of the reaction and the nature of the products obtained are entirely different when molten halides, which are strong Lewis acids, are used in hydrocracking of coal instead of conventional catalysts.

Fundamental studies have been conducted on the hydrogenation and hydrocracking of simple model compounds in strong "superacid" systems. Brouwer and Hogeveen[162] and Olah et al.,[163] for example, have observed that hydrocracking of hydrocarbons occurs under very mild conditions. Ipatieff and Schmerling[164] showed that the HCl-$AlCl_3$ system catalyzes the hydrocracking of heptane at 150°C and 100 atm of hydrogen.

Siskin[165] and Wristers[166] have shown that the hydrogenation of benzene to hexanes is catalyzed under very mild conditions (i.e., 50°C, 16 atm) by HF-TaF_5, HBr-$AlBr_3$, and HF-NbF_5 catalysts. Tertiary hydrocarbons are apparently required as "cocatalysts" in the conversion above.

Ross and Low[167] are investigating the use of massive quantities of strong Lewis acid–Brønsted acid systems such as HX-AlX_3-H_2, for conversion of coal to THF- and pyridine-soluble material at very low temperatures (i.e., 195–210°C).

The liquefaction process undoubtedly proceeds via a heterolytic rather than a homolytic bond-splitting mechanism. It is probable that a carbonium ion[164–166] rather than a free radical mechanism is involved.

The reaction rate is much higher, and the reaction proceeds at much lower temperatures[162,163] with this type of catalyst system.

These characteristics are obtained only when massive quantities of the halides are used. This is because, as discussed below, the catalysts are rapidly poisoned by NH_3 generated in the hydrocracking process. The sulfur in the feed liberated as H_2S also acts to destroy the catalyst by the chemical reaction:

$$MCl_2 + H_2S = MS + 2HCl \qquad (27.42)$$

The nature of the product obtained in hydrocracking of coal or extract depends on the activity and relative quantity of the metal halide used. This is illustrated in Table 27.29, where the results of prior evaluation of various metal halides by Shell,[156,157] CCDC,[149,150] and other investigators are qualitatively summarized.

The most active halides (i.e., the aluminum halides) are too active for liquefaction, since the principal products are light hydrocarbons.

The next group of halides is highly specific for production of light liquids boiling in the gasoline range.

An apparent reason for this is the specificity of $ZnCl_2$, for example, toward hydrogenation and ring splitting of polynuclear versus single ring aromatics. Zinc chloride is almost completely inert toward benzene, but it readily hydrogenates and hydrocracks polynuclear aromatics as typified by naphthalene, pyrene, and so on.

Another characteristic of the $ZnCl_2$ group of halides is the very high ratio of iso- to *n*-paraffins produced. This is a well-known characteristic of the dual-function type of contact hydrocracking catalyst.

The more active aluminum halides produce a lower ratio of iso- to *n*-paraffins, apparently because they are sufficiently active to effect isomerization, thus producing an equilibrium distribution.

The less active catalysts such as $SnCl_2$ produce a heavy oil product more typical of the coal liquefaction products obtained with metal sulfide catalysts. The ratio of isoparaffins to *n*-paraffins is also low, which is

[160] *Zinc Halide Hydrocracking Process for Distillate Fuels from Coal,* U.S. ERDA Contract No. E(49-18)-1743, Quarterly Reports.

[161] R. T. Struck, C. W. Zielke, and E. Gorin, *Summary of Work on Zinc Chloride Liquefaction of Coal,* U.S. ERDA Rep. No. FE-1743-30, November 1, 1976.

[162] D. Brouwer and H. Hogeveen, *Prog. Phys. Org. Chem.,* **9,** 179 (1972).

[163] G. Olah, Y. Halpern, J. Shen, and Y. Mo, *J. Am. Chem. Soc.,* **95,** 4960 (1973).

[164] V. Ipatieff and L. Schmerling, *Ind. Eng. Chem.,* **40,** 2354 (1948).

[165] M. Siskin, *J. Am. Chem. Soc.,* **96,** 3641 (1974).

[166] J. Wristers, *J. Am. Chem. Soc.,* **97,** 4312 (1975).

[167] D. S. Ross and J. Y. Low, Quarterly Reps. Nos. 2 and 3, Stanford Research Institute, to U.S. ERDA, Contract No. E(49-18)-2202, May 18 and August 15, 1976.

Table 27.29 Relative Activity of Metal Halide Catalysts for Coal Hydrocracking

	Most Active	Highly Active	Intermediate Activity	Low Activity
	Al	Sb, As, Bi Zn Ga Ti	Sn Hg	Cu Pb
iC_4/nC_4	Approximate equilibrium	Above equilibrium	Below equilibrium	—
Principal products	Light paraffins and tar	Gasoline	Heavy oil	—

Source: Reference 159

again a reflection of their low hydrocracking activity.

When inhibitors are added to $ZnCl_2$ catalyst, or the ratio of catalyst to coal is reduced, its hydrocracking activity is reduced, and the type of product distribution moves in the direction of those produced by the less active halides. This is presumably true of the other active halides as well.

One characteristic of the metal halide hydrocracking catalyst is that in all cases an undesirable hydrogen-deficient, charlike residue is produced as a by-product, along with the hydrogen-rich hydrocracked products. The former material appears in the form of a catalyst-hydrocarbonaceous complex that is uniformly dispersed in the molten halide. The production of this undesirable by-product may be minimized by operating at high hydrogen pressures and/or relatively low temperatures.

Development of the molten halide catalyst system is presently concentrating on the use of zinc chloride catalyst relative to other halides for the following reasons:

1. It is of relatively low cost compared with most of the other halides.
2. It has a higher chemical stability toward hydrolysis effected by steam liberated in the hydrocracking of coal.
3. It has exceptionally good specificity for production of high-octane gasoline.
4. It is more easily recovered from the spent catalyst than most other halides.

Another interesting facet of the zinc chloride catalyst system relative to a more conventional system is that all distillates and oils that are readily recovered from the catalyst phase are very low in nitrogen content. This characteristic includes not only the lighter distillates but heavier oils in the fuel oil range as well.

The $ZnCl_2$ hydrocracking process and its ramifications are discussed more fully in Chapter 28.

CHAPTER **28**

Liquefaction Processes

SEYMOUR B. ALPERT AND RONALD H. WOLK

Seymour B. Alpert *is Technical Director of the Advanced Fossil Power Systems Division of the Electric Power Research Institute, Palo Alto, California. His industrial career included associations with Stanford Research Institute, Chem Systems, and Hydrocarbon Research, Inc. Mr. Alpert has supervised technical development of process technology for petroleum desulfurization and coal conversion.*

Ronald H. Wolk *is Department Director of the Advanced Fossil Power Systems Department of the Electric Power Research Institute, Palo Alto, California. Formerly, when he was associated with Hydrocarbon Research, Inc, Mr. Wolk carried out process development work on coal conversion, residual oil desulfurization, and hydrocracking.*

28.1 GENERAL CONSIDERATIONS

Since the mid-1970s the field of coal liquefaction has become a lively one that has evolved from "benign neglect" in the early 1960s to one in which an actively competitive community is involved in a wide array of scientific, technical, and administrative activities. The major objective is the development of technology that lays the foundation for a synthetic oil industry to supplement petroleum from natural sources. Commercial use of this technology should be possible by the mid-1980s.

A major effort has been launched to demonstrate the operability of integrated processes and the ability to engineer components for the production of liquids from coal. In addition, small-scale research is under way to seek routes to lower-cost second-generation processes. Suitable levels of funding are allowing plants in the range of 250–600 tons/day of coal capacity to be built that will provide an engineering data base for commercialization of process technology for converting coal to clean liquids and solid products. These products can be used for fueling transportation vehicles, providing clean liquid (and solid) fuels for power generation and space heating, and as raw materials for chemicals.

In the past 5 years the amount of new engineering information on the production of liquids from coal, coupled with an increased understanding of the basic chemistry and technology of coal liquefaction processes, has formed the basis for ultimate realization of a synthetic oil industry in the United States. Not all the information has yet been fully digested by the technical community, nor has full advantage been taken of new concepts obtained from research now being carried out. However the field of coal liquefaction is now established in the United States because large budgets are available to carry out tests of competitive process routes in large pilot-plant equipment. Each of these "spectaculars" will represent a cost of about $250 million, which will support the investment in the plant, a test program, and the engineering design of full-scale plants. These full-scale plants will be capable of providing up to 100,000 bbl/day of liquid fuels. Such commercial plants will be capital intensive, but their products should find a ready market when world oil production rate ceases to grow and perhaps declines.

This chapter provides a summary of coal liquefaction work in progress through the end of 1977. It is likely that over the long run some of these new processes will fail, for reasons not yet readily apparent. However, even failures produce useful data and bring experience and new facts to the technical community.

We know that the routes currently being pursued are not the ultimate solution to the lowest cost production of liquids from coal. New processes will probably use inventions now only dimly perceived. Paths now being explored will appear in retrospect to be primitive approximations for accomplishing the objective of low-cost clean liquids from coal. There is no proof, unfortunately, that new lower cost routes will be found. However, the thrust of work in several areas in this field provides firm bases for potential realization of new approaches. These bases include the following:

- Chemical understanding of coal dissolution and the structure of coal itself.
- Large-scale units that prove engineering designs.
- Engineering of equipment, subcomponents, and process steps that simplify the process schemes.
- The large array of experienced talent now at work.
- Catalyst development by experienced industrial teams.

Process improvements will probably be evolutionary in scope rather than revolutionary. The results should provide a basis for significantly lowering the costs of synthetic liquid fuels which in 1977 were estimated at \$25–\$30/bbl.

The historical procedure for developing coal liquefaction technology has involved a series of increasingly larger experimental operations. The purpose of bench-scale unit operations (at a scale of 10–100 lb of coal/day) is to demonstrate process operability, determine product yield structures and product inspections, provide kinetic information and catalyst aging data, and suggest the ultimate process flowsheet. Work at this scale has been completed on a variety of coals that are of commercial interest for many process concepts. The next step has been a larger pilot operated in an integrated continuous manner, which further firms up the process flowsheet. With equipment of this size, tests are performed on a variety of coals on a larger scale, to uncover process problems caused by scaleup of equipment size. A further objective is to obtain larger samples of products for combustion testing and utilization. These larger pilot plants have a coal feed capacity of 1–10 tons/day, but they still use noncommercial-type equipment that is not markedly different from that used in bench-scale units. If the larger scale test work results in process questions requiring further research for resolution, the bench-scale sized unit is often used to solve the problem, since work at that scale costs less.

The final development step incorporates all the findings from smaller equipment into a large pilot plant that more or less "freezes" the flowsheet at a scale of about 250 tons/day of coal.

At this scale of operation the flowsheet of the process is essentially fixed. Small-size commercial equipment and components are used to verify the engineering data and mechanical features required for operation and construction of commercial plants. If process problems are encountered, the smaller bench-scale equipment may again be used to simulate more expensive tests in large equipment and to sort out alternative solutions to process difficulties. Redundancy of commercial equipment is often employed in such large plants where there are no proved "off-the-shelf" components or where a major extrapolation of design practice is necessary.

A process research and development program has five main objectives:

- A correlation and data summary of the commercially useful background technical data showing limits where data were obtained in small equipment.
- A definition of operable and inoperable combinations of process variables that define safe operating areas.
- A design handbook and an operating manual for commercial plants based on large-scale, pilot-plant tests with continual checking of observations at each step for consistency.
- A skilled team of specialists, which represents an organization with detailed knowledge of the practice developed, to serve as consultants for commercial plants.
- Perhaps the most important product: an enthusiastic management with enough entrepreneurial insight, imagination, and resources to carry new technology to commercial reality.

Process development is not a schedule of discrete events that stretches from an inventor's concept to commercial plants employing an array of skills. Rather, it is three-dimensional spiral of knowledge that expands with added information from many engineering disciplines. At every step in the process development, continual cross-checks and reconfirmation are necessary to achieve commercial success in the marketplace.

28.2 TECHNICAL CONSIDERATIONS

There are four generic methods of producing liquids from coal:

- Pyrolysis and hydrocarbonization.
- Extraction by solvents.
- Catalytic liquefaction.
- Noncatalytic liquefaction.

This chapter describes process routes for the latter two methods. Pyrolysis and hydrocarbonization, and extraction of coal using solvents from external sources, are discussed in Chapters 12 and 13.

Coal is a solid material consisting primarily of carbon and hydrogen with significant amounts of inorganic mineral matter as finely divided constitutents and adsorbed moisture. In addition, coal contains varying quantities of heteroatoms (oxygen, nitrogen, sulfur) and alkali metals and trace elements that may be organically bound within the coal itself. The actual types of ring structure that constitute structural building blocks in coal have not been elucidated with certainty.

Compared with petroleum, coal is hydrogen deficient, containing only about one-half as much hydrogen and far more nitrogen, oxygen, and ash (Table 28.1). The sulfur content of coal varies widely, depending on the precise source of the material. Even different samples from a given mine can vary markedly, not only in the sulfur content but also in their ash content and adsorbed water. The sulfur in coal is both organically bound and in the form of iron pyrites (FeS_2). Under the conditions used in coal liquefaction process, the pyrites is not reduced completely to metallic iron. Because of equilibrium considerations, it is reduced to an intermediate compound with FeS as an approximate formula. This compound must be physically removed from coal conversion products. Organic sulfur in coal is fairly reactive, and substantial amounts of it are conveniently converted to hydrogen sulfide even when added heterogeneous catalysts are not present in the reactor system. Oxygen in coal is readily converted to water or carbon oxides. The formation of water is a major expense in that hydrogen, which is expensive to produce, is consumed. However, oxygen compounds (e.g., phenols) in the product are noxious and can cause difficulties with product storage stability. Nitrogen, in general, has been found to be relatively difficult to remove compared to sulfur and oxygen. Only a relatively small portion of nitrogen in coal is converted to ammonia unless severe conditions are used.

Table 28.1 Comparison of Typical Petroleum and Coal Analyses (wt %, mf)

	Coal		Petroleum
	Illinois	Wyoming	Texas Crude
Carbon	71.5	69.3	84.9
Hydrogen	4.5	4.3	10.8
Sulfur	3.6	0.5	1.6
Oxygen	5.9	14.5	2.1
Nitrogen	1.2	1.0	0.5
Ash	13.3	10.4	0.1
	100.0	100.0	100.0

To convert coal to liquids, process technology has been developed that requires a hydrogen environment at pressures in the range of 500–5000 psi and temperatures that are in the range of 400–550°C. Three basic approaches to coal liquefaction are relatively advanced. These three approaches use different techniques to contact coal with hydrogen at elevated temperature and pressure. The approaches are briefly identified as follows:

1. Noncatalytic solvent extraction methods that while processing coal as a slurry, simultaneously regenerate the donor properties of the solvent *in situ*.
2. Catalytic techniques that use supported catalysts in special reactor systems.
3. Donor-solvent systems that separate distillate fractions from the slurry product that is noncatalytically extracted, and separately hydrogenate the distillates to impart donor properties and tailor the chemical species desired in the solvent.

The major coal liquefaction pilot-plant programs in the United States are summarized in Table 28.2.

Coal is a sensitive material that begins to dissolve on being heated much above 250°C in the presence of aromatic hydrocarbons. In the dissolved state it begins to react with

Table 28.2 Major Operating Pilot Plants That Liquefy Coal

Process[a]	Status	Location	Coal Capacity (tons/day)	Product	Source of Funds[b]
SRC[a]	Operational	Wilsonville, Ala.	6	Solid fuel	DOE, EPRI
SRC and SRC II	Operational	Tacoma, Wash.	50	Solid fuel, liquid fuel oil	DOE
Consol: CSF	Operational	Cresap, W. Va.	206	Extract, distillates	DOE
H-Coal	Operational	Trenton, N.J.	3	Distillates, fuel oil	DOE, EPRI, Amoco, Ashland, Conoco, Mobil, State of Kentucky
	Under construction	Catlettsburg, Ky.	600	Distillates, fuel oil	DOE, EPRI, Amoco, Ashland, Conoco, Mobil, State of Kentucky
Exxon: EDS	Operational	Baytown, Tex.	1	Distillates, distillate fuel oil	DOE, EPRI, Exxon, Phillips
	Under construction	Baytown, Tex.	250	Distillates, fuel oil	DOE, EPRI, Exxon, Phillips
Gulf: CCL	Operational	Harmarville, Pa.	1	Fuel oil	Gulf
Synthoil	Operational	Bruceton, Pa.	0.5	Fuel oil	DOE
	Abandoned		8	Fuel oil	DOE

[a] SRC = solvent-refined coal process.
[b] DOE = U.S. Department of Energy, EPRI = Electric Power Research Institute, Palo Alto, Calif.

molecular hydrogen that is dissolved in the liquid phase. Workers at the U.S. Bureau of Mines,[1] Exxon,[2] and especially at Mobil Oil Corporation[3] have elegantly elucidated the coal dissolution phenomena using sophisticated techniques. This recent work provides new insights for process engineers coping with the conversion of coal. Not all the carbon in coal is easily dissolved and converted to more desirable low molecular weight species.

The liquid used to slurry the coal and carry it into the reactor is known to play an important part in the dissolution process. Its properties must be specified and controlled. Understanding of the reactions of coal from the time it is first heated in the presence of solvent and hydrogen has important process implications.

The undissolved solids and inorganic species in coal can accumlate in reactor systems as discrete particles, precipitate out on reactor surfaces, or deposit on the surface of active catalysts. In reactor systems accreted carbonaceous materials can accumulate and partially fill the reactor space. The underlying mechanism of accumulation of solids in reactor systems and catalyst poisoning is only partially understood, although operation in larger scale equipment in recent years has provided new information that has helped the understanding of the phenomena.

To produce clean solids or liquids from coal, hydrogen is added in the conversion step. The mildest treatment of coal, the solvent-refined coal (SRC) process, requires about 2 wt % hydrogen based on coal to produce the low-sulfur solid product desired. As much as 8 wt % hydrogen can be consumed in other processes to produce lighter, lower boiling liquid products that have lower sulfur, oxygen, and nitrogen contents. Some initial work has been carried out to obtain hydrogen production information from coal or carbonaceous residues from processes, but much remains to be done to firm up the important step to low-cost hydrogen that is required to improve coal liquefaction economics.

The ash constituents in coal have several effects, and again much work still needs to be done to completely describe the interac-

[1] M. G. Pelipetz, E. M. Kuhn, E. S. Friedman, and H. H. Storch, *Ind. Eng. Chem.*, **40**, 1259 (1948); S. Weller, E. L. Clark, and M. G. Pelipetz, *Ind. Eng. Chem.*, **42**, 334 (1950); S. Weller, M. G. Pelipetz, and E. S. Friedman, *Ind. Eng. Chem.*, **43**, 1572 (1951).

[2] R. C. Neavel, *Fuel*, **55**, 237 (1976).

[3] D. D. Whitehurst, M. Farcasiu, and T. O. Mitchell, *The Nature and Origin of Asphaltenes in Processed Coals*, EPRI AF-252, Research Project No. 410-1 Annual Report, February 1976.

tion of coal ash constituents with the process. The ash has a weak catalytic effect that most probably originates from the iron compounds present. However, the ash also presents a macro surface for deposition of coke and coke precursors so that process operability may be helped by its presence. Understanding of the role of ash, however, is predominantly based on empirical observations. A complete chemical understanding is required to allow proper operation of large-scale equipment on which process scaleup ultimately will be based.

Catalysts of varying formulations that are usually supported have been found useful in selectively carrying out desired reactions at less severe conditions than would be required without the presence of catalysts. The catalyst aids in maximizing the yield of liquid products instead of gaseous products or coke by promoting hydrogenation reactions. However, catalyst usage generally represents an added cost and a complicating factor for a process.

Finally, the regimes of mixtures of solid, liquid, and gaseous phases, as they affect the design of reactors, pumps, valves, and other equipment used in coal liquefaction processes, need further elucidation before full-scale equipment can be operated with surety.

In the development of a process technology for converting coal into clean fuels, several generic considerations must be dealt with. The seven major categories are outlined next.

1. Coal is a solid material that needs to be introduced into a pressure system. In general, receiving, drying, and crushing coal are standard steps practiced in other industries that can be adapted for coal liquefaction applications. Pulverized coal burning power plants reliably handle 20,000 tons/day of coal crushed to about 200 mesh. However the drying of coal to moisture levels much below 5% represents some uncertainty. Some questions regarding chemical changes in the coal during drying have implications in terms of process chemistry and costs. In addition, since dry coal is pyrophoric, it can spontaneously ignite in storage.

The use of liquids to slurry dry coal is generally practiced. The liquids suspend the ground coal and permit the mixture to be pumped. Although laboratory-size equipment for pumping coal slurries is available, commercial plants will require a large number of slurry pumps of unknown reliability for this service. Finding improved methods for pumping coal slurries is an objective of the current engineering development program. As in other parts of coal liquefaction process plants, a "balance" or compromise is made. Historically more dilute slurries have reduced operating difficulties. However a concentrated slurry feed may be more desirable from the standpoint of the size of all subsequent equipment that must be used. Much of the "downstream" equipment size is set by the specified coal-slurry mixture. One of the more costly plant sections, for example, is represented by the atmospheric and vacuum distillation facilities. If large quantities of slurrying oil are used, expensive duplicate towers may have to be operated to recover the slurry oil recycled in the process.

2. In most of the processing routes hydrogen is reacted with coal at elevated pressures and temperatures, which results in coal dissolution. Primary products are reduced in molecular weight by further reaction with hydrogen. Heteroatoms are transformed into compounds that allow simple proved process steps to accomplish their removal from gaseous streams. Hydrogen generation from the least valuable solid carbonaceous residues is preferable. However, the generation of hydrogen from such materials represents a relatively high degree of uncertainty as compared with conventional steam reforming of gaseous hydrocarbons or light naphtha components.

3. Liquefaction of coal is an exothermic reaction that presents a reactor control problem. Reactor quench techniques and backmixing of coal reactants and products in the liquefaction reactor are methods being used to control heat evolved. However, the engineering of adequate control systems for commercial-sized reactors remains to be accomplished. Reactor systems that are employed in coal liquefaction systems include the following:

a. Open, unhindered vessels in which coal slurried in oil is contacted with hydrogen. Depending on the velocities used, the ash in the coal fed to the system can accumulate in the reactor systems and may serve as a catalyst. In these reactor systems the liquid upflow velocity is generally maintained at a low level in an attempt to approach a plug-flow system (i.e, system in which there is a minimum of mixing).

b. Ebullated-bed reactor systems have been employed in the H-Coal system. An ebullated reactor system contains a fluidized catalyst bed expanded over its settled height, but one that does not overflow from the top of the reactor. Solids in the form of small particles of unreacted coal or ash pass through the catalytic reactor system. Considerable mixing occurs in ebullated-bed systems.

c. Fixed-bed catalytic reactor systems are used in several coal liquefaction processes. In such systems high-velocity gas streams propel coal particles through the catalyst beds.

Although such systems have been applied at small scale (< 12 in. diameter), they have not yet been operated in large-diameter systems, and uncertainty exists regarding ultimate scaleup.

4. Ash, unreacted coal, and condensed polymeric reaction products that are not liquid at process conditions need to be removed from the final product in equipment that can be scaled up for commercial plants. The following methods are being pursued:

a. Rotating mechanical pressure filters that use a precoat of selected diatomaceous earth as a filter aid to increase filtration efficiency.

b. Cyclone separators that separate the ash-containing stream into an ash-rich stream and a relatively ash-free stream.

c. Continuous centrifuges.

d. Distillation of solids containing product fractions in specially designed towers.

e. Fluidized-bed coking that cokes a portion of the heaviest liquids (vacuum distillation bottoms) and incorporates the ash and other fine solids in the coke.

f. Solvent precipitation that precipitates finely divided solid matter.

g. Combinations of various techniques.

Data from several laboratories indicate that mechanical devices have formidable scaleup and economic problems that stem from their low capacity and poor reliability. Experience with solvent precipitation and distillation equipment has been favorable, and these two techniques will be used in large pilot plants that are under construction. Fluid coking of vacuum distillation bottoms is also being actively studied.

5. Reactor designs must take into account hydrodynamic phenomena, erosion problems, and equilibrium composition of reactor contents of multicomponent three-phase systems. Complete data are not available to permit large-scale reactor system designs to be made with confidence at this time.

6. For large-scale plants, special equipment will need to be developed for particular applications. Reliable pressure letdown valves and level controllers are examples of such equipment.

7. Particular attention to heating and cooling process streams in process equipment represents a challenge to design engineers. Ill-defined hydrogenation reactions that occur with coal and its products can foul heat exchanger surfaces or lead to deposition of coke on walls.

28.3 PROCESS DESCRIPTIONS

28.3.1 Solvent-Refining Processes

28.3.1.1 *Solvent-Refined Coal (SRC I)*

Introduction. The SRC process is perhaps the least complex of the current process schemes. The purpose of the SRC process is to produce an environmentally acceptable solid fuel from coal. The coal would not be environmentally acceptable if conventionally burned. Hydrogenation of the coal in the SRC process takes place at elevated pressure and temperature in the presence of hydrogen. Catalysts other than the minerals contained in the coal are not used.

The Office of Coal Research (OCR) was established as a part of the U.S. Department of the Interior in 1960. In 1962 it contracted for SRC process development work with the Spencer Chemical Company, which became a part of the Pittsburg and Midway Coal Mining Company (P&M). Later P&M became part of the Gulf Oil Corporation. The SRC work was to develop a coal liquefaction process based on the older Pott-Broche technology, developed in Germany before World War II (see Chapter 19, Vol. 1, Chapter 28,

Vol. 2 of this work; also Chapters 6 and 23 of the first Supplementary Volume). In that process coal was extracted with externally generated hydrogenated coal tar and product oil, at elevated temperature and pressure in the absence of gaseous hydrogen. Hydrogen transfer from the tar or oil to the coal was sufficient for solubilization. However, sufficient oil to slurry fresh coal could not be recovered directly from the process; thus the use of externally generated coal tar was required.

Results in bench-scale equipment at P&M's Merriam, Kansas, laboratories indicated that the use of gaseous hydrogen aided in the production of sufficient quantities of solvent. This was confirmed during operation of a 50-lb/hr continuous unit in 1965.[4]

Further development of the technology was limited to bench-scale operations during the late 1960s and early 1970s. A design for a 50-ton/day plant was prepared under an OCR contract with Stearns-Roger.[5] This plant was built at Fort Lewis, Washington, by Rust Engineering. Operations, beginning in 1974, have been carried out by Gulf. A flowsheet of the plant is given in Fig. 28.1.

In an independent effort that built on the basis of P&M's technical development, Southern Services Inc. obtained the support of the Edison Electric Institute (EEI) for the design, construction, and operation of a 6-ton/day SRC plant at Wilsonville, Alabama. Catalytic, Inc., was retained as contractor. EEI provided two-thirds of the funds for the project and Southern Services provided one-third. With the formation of the Electric Power Research Institute (EPRI), funding responsibility was transferred to it from EEI. Joint funding by EPRI and Southern Services continued until the end of 1975. Funding for 1976 and 1977 was provided by the U.S. Energy Research and Development Administration [ERDA, now a part of the federal Department of Energy (DOE)] and EPRI on a two-thirds–one-third basis. A flowsheet of the Wilsonville facility is given in Fig. 28.2.

Process Description. Ground and dried coal is fed to the reactor system as a slurry. The transport liquid, or process solvent, is a distillate fraction recovered from the coal hydrogenation product. The solvent performs vital functions in the dissolution process besides serving as a transport medium. Among these are to serve as the hydrogen donor and solvent for heavy liquefaction products.

The coal-solvent slurry is pumped to system pressure and mixed with hydrogen. The three-phase feed mixture is typically heated to near the desired reaction temperature in a preheater before injection into a reaction vessel. The preheater residence time is short compared to the reactor residence time. The total reaction time is typically 20–60 min.

The mechanism or stages of coal solution are not well defined. The coal apparently first swells and softens, and becomes a viscous gel very rapidly in the preheater. In the preheating process bituminous coal is substantially transformed into pyridine- or cresol-soluble material. This change typically takes 2–4 min at temperatures above 600°F.

The substantially "dissolved" coal passes into the reactor, where dissolution proceeds further, hydrocracking of the previously dissolved coal fractions takes place, and organic sulfur reacts to form hydrogen sulfide. The reactions that take place during dissolution are as follows:

1. Hydrogen transfer from solvent to coal.
2. Coal molecule cleavage due to heat and hydrogen transfer.
3. Further hydrogenation of fragments—CH,—SH,—O—,—N—,—C—C.

After leaving the reactor, the product is cooled to 500–600°F. A vapor liquid separation is made at this temperature, primarily to separate the stream rich in hydrogen and light liquids from that containing process solvent, heavy product, and unconverted coal and ash.

At the 50-ton/day Fort Lewis SRC plant, the separator underflow is flashed and those bottoms are sent to a pair of rotary precoat filters. The filters have surface areas of 40

[4] D. L. Kloepper, T. F. Rogers, C. H. Wright, and W. C. Bull, *Solvent Refining of Coal to Produce a Deashed Product,* R & D Rep. No. 9, Final Report prepared for U.S. Dept. Interior OCR, by Pittsburg and Midway Coal Mining Company, under Contract No. 14-01-0001-496, issued December 1965.

[5] G. E. Chenoweth, "Progress Report of Pilot Plant Work on the Solvent Refined Coal Process," paper presented at 2nd Ann. Conf. Coal Gasification and Liquefaction, University of Pittsburgh, August 5–7, 1975.

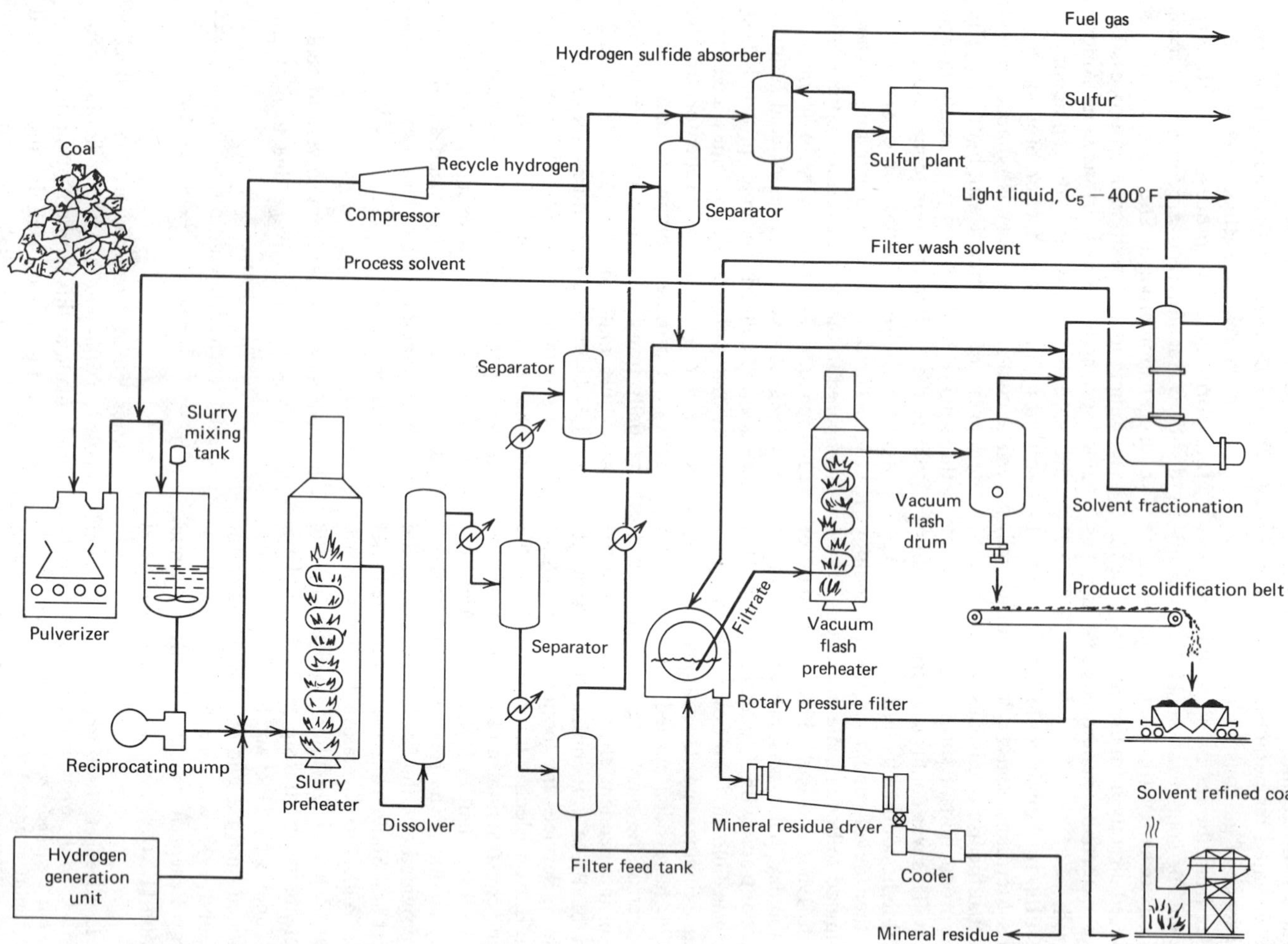

Figure 28.1 Flowsheet for the solvent-refined coal process, Fort Lewis, Washington.[6]

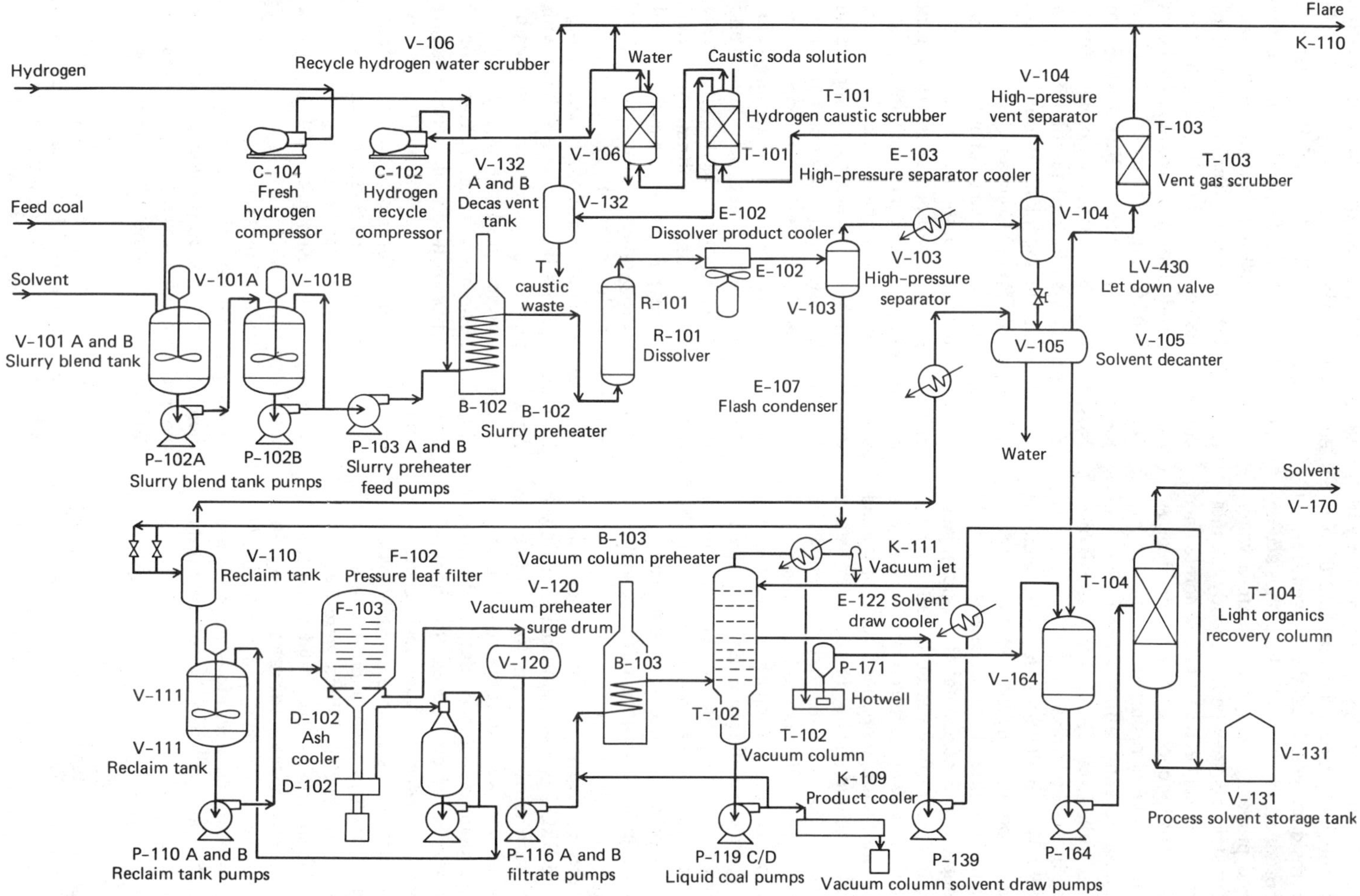

Figure 28.2 Flowsheet for SRC I, Wilsonville, Alabama.[7]

and 80 ft², respectively. Filtrate is sent to a vacuum distillation tower for recovery of wash solvent, process solvent, and the vacuum residue, which is the solvent-refined coal product. Product solidification is accomplished primarily by spraying onto a water-cooled stainless steel belt. Recently the molten material has been cast into concrete lined shallow pits for cooling and ultimate breakup. The filter cake is reslurried and passed to a horizontal rotating kiln, where the slurrying oil is distilled off. Dried filter cake is produced in this operation.

At the Wilsonville 6-ton/day SRC plant the separator bottoms stream is sent to a precoat pressure leaf filter. Filter cake was originally discharged to a horizontal, Dowtherm-jacketed, scraped surface vessel. Because of coke formation, however, this arrangement was never satisfactorily operated for a sustained period. Vacuum drying in the filter vessel proved to be a satisfactory technique for recovering solvent and producing a dry filter cake that could be discharged from the system. Product solidification has been accomplished by spraying the molten SRC into a pool of liquid water and more recently by spraying onto a chilled vibrating plate. The latter technique produces a dry product.

Operating Experience. Before the end of 1977 all SRC-I operations at the Fort Lewis plant were carried out on Kentucky coal. Initial efforts were devoted to attempts to develop material balanced data to determine the response of the process to changes in temperature, pressure, and coal space rate. This objective was compromised by losses of liquids from the system during precoating and filtration cycles. Filters were frequently taken off-stream for repair and modification, as were the systems for transferring filter cake to the rotary kiln and the kiln itself. Material-balanced operations have been reported for periods when the solid separation system was not used (Table 28.3). Through all these difficulties the reaction section equipment worked well.

Late in 1975 ERDA concluded that the major effort for the facility should be the production of 3000 tons of SRC, which met the federal New Source Performance Standards (NSPS) of 0.96 wt % S and 0.16 wt % ash. This production run was completed in November 1976. The feed coal throughout the run was a mixture of Kentucky Seam Nos. 9 and 14 from P&M's Colonial Mine. This product was burned in July 1977 in a 22-MW boiler at the Plant Mitchell Station of the Georgia Power Company.

With its smaller size and inherent flexibility, the Wilsonville unit was utilized to generate process yield and operability data from a variety of major domestic coals, investigate some new processing concepts, and to collect data on a number of different solid separation techniques. Yields and operating conditions are presented in Table 28.4, for operations on Kentucky Nos. 9 and 14; Pittsburgh No. 8, Illinois No. 6 (Burning Star Mine), and Illinois No. 6 (Monterey Mine) bituminous coals, and Wyoming subbituminous coal. Satisfactory operation was achieved with all coals, with the product at least meeting the NSPS for sulfur content. All conditions, however, were not operationally satisfactory. Low hydrogen partial pressure, solvent with depleted donor capability, and coal weathering were suspected of causing some of the difficulties. The phenomenon of formation of large quantities of fluidizable reactor solids was well documented at Wilsonville.[9] In general, the quantity formed is directly proportional to the amount of exchangeable calcium ions present in the feed coal; Wyoming coal is the highest and Kentucky coal the lowest in this regard.

Coordination of the Wilsonville and Fort Lewis programs in 1976 led to the development of a kinetic interpretation of the role of the preheater and reactor on conversion. Additional insights into the role of solvent

[6] B. K. Schmid, "The Solvent Refined Coal Process," in *Encyclopedia of Chemical Processing and Design,* Vol. 2, Marcel Dekker, New York, 1978.

[7] H. E. Lewis, W. H. Weber, G. B. Usnick, and W. R. Hollenack, *Solvent Refined Coal (SRC) Process Operation of SRC Pilot Plant at Wilsonville, Alabama,* Quarterly Tech. Rep., Progress for period July–September 1976, FE-2270-10 Dist. Category UE 90d, p. 124. Catalytic, Inc. for U.S. ERDA, publ. Wilsonville, Ala.

[8] G. R. Pastor, D. J. Keetly, and J. D. Naylor, "Operation of the SRC Pilot Plant," Paper No. 41.b, presented at American Institute of Chemical Engineers Meeting, Los Angeles, November 16–20, 1975.

[9] R. Wolk, N. Stewart, and Seymour Alpert, "Solvent Refining for Clean Coal Combustion," *EPRI J.,* **1**(4), 12–16 (May 1976).

Table 28.3 Reaction Area Process Variable Studies: SRC Pilot Plant, Fort Lewis, Washington[8]

	Run 1	Run 2	Run 3	Run 4	Run 5	Run 6	Run 7	Run 8	Run 9
Coal feed (lb)/(hr)(ft^3), [maf coal basis]:	49.5	29.0	75.2	77.9	97.6	98.5	98.8	79.7	91.0
Solvent/coal wt ratio (dehydrated coal basis):	1.66	1.57	1.65	1.57	1.60	1.59	1.59	1.55	1.54
Hydrogen feed ($\times 10^3$ scf/ton maf coal):	14.	21.	18.	18.	14.	13.	13.	17.	15.
Dissolver pressure (psig):	1528.	1489.	1496.	1499.	1488.	1483.	1471.	1484.	1478.
Dissolver outlet temperature (°F):	851.	848.	870.	844.	849.	870.	850.	846.	859.
Results (% mf coal)									
Hydrogen consumed	2.5	2.7	2.1	2.0	2.0	2.0	1.9	2.4	2.1
C_1 yield	2.2	2.3	2.8	1.5	1.3	2.2	1.4	1.4	2.1
C_2 yield	1.3	1.5	1.6	0.9	0.8	1.3	1.0	1.0	1.2
C_3 yield	1.1	1.2	1.6	0.9	0.6	1.0	0.9	0.9	1.1
C_4 yield	0.8	0.5	2.0	0.4	0.2	0.4	0.4	0.4	0.5
CO yield	0.1	0.2	0.1	0.0	0.0	0.1	0.0	0.0	0.1
CO_2 yield	1.2	1.1	1.2	1.0	0.9	1.1	1.0	0.8	1.0
H_2S yield	2.0	1.3	2.6	1.1	1.1	1.3	1.8	1.4	1.3
H_2O yield	4.9	5.3	7.3	3.6	6.8	5.3	5.0	5.3	4.6
Light oil yield	6.6	10.2	16.2	5.4	2.2	6.8	4.2	11.2	6.3
Wash solvent yield	8.0	11.4	−3.1	−1.1	6.7	2.6	5.5	−1.6	3.4
Process solvent yield	0.1	−3.9	−3.9	9.2	−1.5	−2.6	−0.4	3.6	0.7
SRC yield	58.3	57.6	59.8	64.0	67.3	65.9	65.6	63.0	64.2
Mineral residue	15.9	14.1	16.2	15.2	15.6	16.7	15.5	14.9	15.6
SRC Sulfur (wt %)									
Mean	0.91	0.57	0.72	0.85	0.88	0.78	0.77	0.66	0.7
High	0.94	0.69	0.86	0.90	0.92	0.86	0.94	0.70	0.76
Low	0.87	0.40	0.63	0.80	0.82	0.69	0.62	0.61	0.64

Table 28.4 Wilsonville SRC Operating Data[a]

Mine:	Colonial	Loveridge	Burning Star	Monterey	Belle Ayr
Seam:	Kentucky Nos. 9 and 14	Pittsburgh No. 8	Illinois No. 6	Illinois No. 6	Roland Smith
State:	Kentucky	Pennsylvania	Illinois	Illinois	Wyoming
Feed coal sulfur content (%)	3.1	2.6	3.1	4.4	0.7
Process parameters					
Temperature (°F)	800–850	855	820	855	855
Pressure (psig)	1500–2400	1700	1800	2400	2400
Feed rate [lb/(hr) (ft³) reactor]	25–50	25	23	25	25
Coal conversion (%)	91–95	91	90	95	85
SRC output (%)					
Yield	55–65	69	63	54	45
Sulfur content	0.8	0.9	0.9	0.95	0.1

boiling range in determining conversion and desulfurization were also developed.

28.3.1.2 *Solvent-Refined Coal (SRC II)*

Introduction. Solvent-refined coal process technology has been modified to eliminate a number of processing steps and to produce an all-distillate, low-sulfur fuel oil from coal. This modification of solvent refining is designated as SRC II.[10] Figure 28.3 is a diagram of the SRC II process. The heavier product fractions containing ash are recycled to the dissolver system to reduce their molecular weight. The fuel oil product includes only distillate material recovered by fractionation of the product. The processing steps that have been eliminated from the SRC I process are as follows:

- Filtration to accomplish solids-liquid separation.
- Mineral residue drying to recover occluded liquid product.
- Solidification of the product.

Recycling of part of the product containing mineral residues allows more time at rela-

[10] B. K. Schmid and D. M. Jackson, "The SRC-II Processes," paper presented at 3rd Annu. Conf. Coal Gasification and Liquefaction, University of Pittsburgh, August 3–5, 1976.

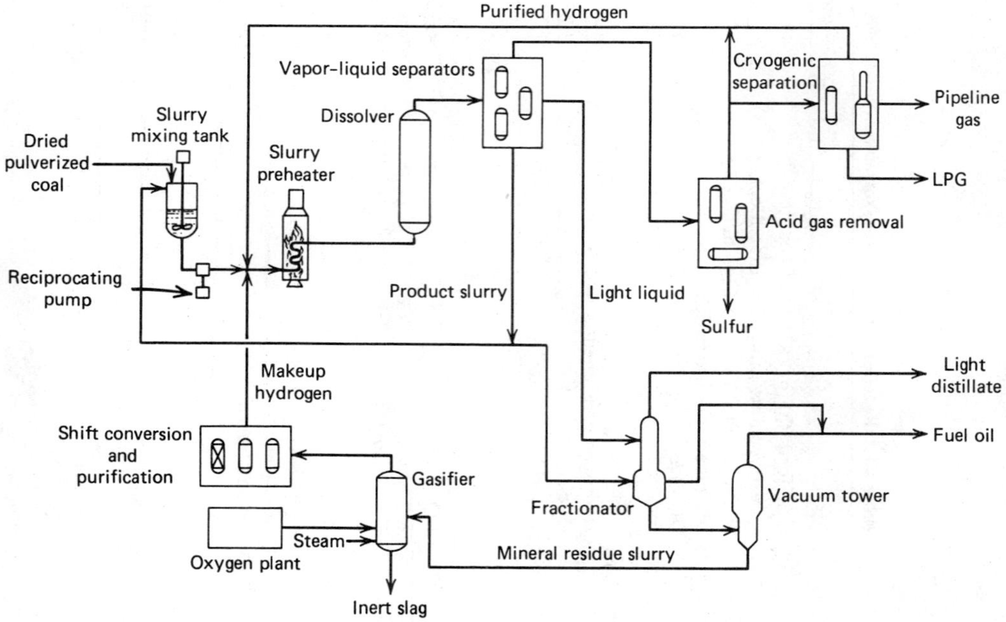

Figure 28.3 Flowsheet for the SRC II process.

tively severe conditions, which further hydrocracks heavier fractions and also permits a buildup of mildly catalytic ash components in the reactor system.[11] The severity of the process is sufficient to provide just enough vacuum residue for producing the hydrogen required for the process (by gasification of the heavy residue with steam and oxygen). As compared with a yield on dry coal feed of about 65 wt % of the 850°F+ fraction in the SRC I process, the target yield of this fraction in the SRC II process is about 20 wt %.

Process Description. Coal is pulverized and dried, then mixed with recycled slurry solvent from the process. The slurry mixture is pumped to reaction pressure of about 140 atm and, together with hydrogen, preheated to temperatures of about 700–750°F. The mixture is then passed to the dissolver, where the exothermic heat of reaction increases the temperature of the reactants to a temperature in the range of 820–870°F. Hydrogen at a lower temperature is injected at various points in the reactor to control the reaction temperature.

The stability of the dissolver system in a commercial plant would be dependent on the backmixing of cool slurry feed and cool hydrogen to dissipate the exothermic heat. Backmixing and uniform feed distribution across the cross section of the reactor are requisites for proper temperature control.

The dissolver effluent is separated into vapor and liquid phase fractions. The overhead vapor stream is cooled in a series of heat exchangers and separators. The condensed liquid is distilled in a fractionation section. The gaseous products are purified of hydrogen sulfide and carbon dioxide. Hydrogen-rich gas is recycled to the reactor together with makeup hydrogen.

The liquid phase product containing solids serves to act as solvent for the SRC II process. Part of this stream is recycled to the slurry preparation equipment. The net yield of liquid phase product is distilled in a product fractionation system to recover the low-sulfur fuel oil product. A portion of these distillates may also be recycled to help slurry the coal feed.

[11] C. H. Wright and D. E. Severson, "Experimental Evidence for Catalytic Activity in Coal Minerals" *Prepr., Am. Chem. Soc., Div. Fuel Chem.*, **16**(2), 68–92 (1972).

The nondistillable residue product from a vacuum tower that contains unconverted coal solids and ash serves as feed to the hydrogen generation section. An oxygen-blown gasifier is used to convert this material, which contains about 30% ash as mineral matter, to hydrogen. Such gasification facilities are not in operation at pilot-plant sites, but samples have been tested in gasification equipment at other locations.

Yields obtained from small-scale laboratory tests, which used Kentucky bituminous coal in the test work, are shown in Table 28.5, and Table 28.6 gives data on inspections of the fuel oil and light distillate products.

Status. Laboratory work has been carried out by the Gulf Mineral Resources Company, a subsidiary of Gulf Oil Corporation. The Tacoma SRC II runs carried out in late 1977 have been successful on both Kentucky Nos. 9 and 14 and Illinois No. 6 coals at a scale of about 30 tons/day. Sustained operations were achieved for periods of about 60 days. The featured recycle of the ash-containing heavy fractions may result in complexities in that the solids in the reactor will increase in concentration. In practicing solvent refining, agglomeration of the mineral matter has been observed.[12] Recycling mineral matter may

[12] Anon., "Characterization of Mineral Matter in Coals and Coal Liquefaction Residues," EPRI Project No. RP366-1, Quarterly Rep., Pennsylvania State University, University Park, June 1976–August 1976.

Table 28.5 SRC II Process Yields for Western Kentucky Bituminous Coal, 4% Sulfur[10]

Yields (wt %, mf coal)		
Light gases, C_1–C_4	16.6	
Liquids		
Naphtha, C_5–380°F		11.4
Mid-distillates, 380–480°F		9.5
Heavy distillates, 480–850°F		22.8
Total liquids	43.7	
Residue, 850°F	20.2	
Carbonaceous solids	3.7	
Ash	9.9	
Hydrogen sulfide	2.3	
Carbon oxides	1.1	
Water	7.2	
	104.7	

Hydrogen consumption-4.7 wt %

Table 28.6 Inspections on SRC II Products[10]

	Fuel Oil	Light Distillate
Gravity (°API)	5.0	39.0
Boiling range, nominal (°F)	400–900	100–400
Molecular weight	230	
Flash point (°F)	168	
Viscosity (SUS at 100°F)	50	
Heating value (Btu/lb)	17,284	
Elemental analysis (wt %)		
Carbon	87.0	84.0
Hydrogen	7.9	11.5
Sulfur	0.3	0.2
Nitrogen	0.9	0.4
Oxygen	3.9	3.9
	100.0	100.0

exacerbate growth of mineral matter in the reactor system.

The enhanced catalytic activity of mineral matter in the reactor system is representative of the coals on which the SRC II process has been tested. However the variability of mineral matter from coal to coal, as well as the variability even in the same seam of coal, may result in process complexity with respect to maintaining the severe hydrocracking and high degree of sulfur removal desired. However the SRC-II process has made rapid progress and is relatively advanced compared to other routes to liquids from coal.

Design of a 6000-ton/day unit is being undertaken, and preliminary cost data are available.[13] Serious consideration is being given for construction of a 6000-ton/day SRC plant. Funds for support of such a facility have been proposed in authorization legislation before the Congress. The Department of Energy is supporting initial design of a 6000-ton/day plant.

28.3.1.3 *Costeam*

Introduction. The Pittsburgh Energy Research Center (PERC) has carried out research work in autoclaves and in continuous small units to produce liquid products useful as low-sulfur boiler fuel from lignites and other low-rank coals.[14] A feature of the Costeam process is the use of crude synthesis gas containing about 50–60% carbon monoxide, 30–50% hydrogen, and some carbon dioxide, rather than hydrogen, which is more conventionally used.[15,16] Water naturally present in high-moisture coal is converted to steam.

The water-gas shift of raw synthesis gas (carbon monoxide plus hydrogen) used in production of hydrogen is eliminated as a processing step. However, the cost of the water-gas shift of raw synthesis gas is relatively low.

The mineral components in the low-rank coals usually provide "weak" catalyst activity. In the application of the process to lignites, the alkali and alkaline earth metals react with carbon monoxide to produce formates, which are in turn decomposed to carbonates during hydrogenative reactions. The formates function to reduce the carbonyl groups in low-rank coals and thereby minimize the cross-linking that leads to high molecular weight species and consequently char formation. The Costeam process has the following features:

- Use of low-rank coals that are poor feedstocks for conventional liquefaction methods.
- Economic advantages from using low-cost coal and synthesis gas.
- Absence of added catalyst.
- Production of a fuel oil product low in sulfur (usually less than 0.3 wt %).

Under Costeam processing conditions, pyrites and other iron compounds are converted to catalytically active iron materials that serve as mild hydrogenation catalysts, thus utilizing the hydrogen in the synthesis gas. For coals that are low in iron, additional pyrite or other low-cost, iron-containing material may be added to improve utilization of the hydrogen.

Severe conditions are used; the pressure is about 4000 psig. A temperature of about 450°C (850°F) is required to effect dissolution

[13] Anon., *Steag 1975 Annual Report,* Steag, AG, Essen, West Germany, p. 20.

[14] E. Del Bel, S. Friedman, and P. Yavorsky, "The Liquefaction of Lignite by the COSTEAM Process," paper presented at American Institute of Chemical Engineers Meeting, Houston, Tex., March 16–20, 1975.

[15] H. Appell, E. Morcni, and E. Miller, "COSTEAM Liquefaction of Lignite, Pittsburgh Energy Research Center, Pittsburgh, Pa.," *Prepr., Am. Chem. Soc., Div. Fuel Chem.,* **20**(1), 58–65 (1975).

[16] H. Appell, I. Wender, and R. Miller, "Solubilization of Low-Rank Coals with Carbon Monoxide and Water," *Chem. Ind. (London),* No. 47 (November 22, 1969).

of lignite. Such operating conditions are a potential drawback for the process, since long residence times and high pressures are costly and present major engineering problems.

Process Description. A continuous laboratory unit that has a capacity of about 10 lb/hr of coal has been in operation for about 2 years at the PERC. Ground coal is slurried with recycled product oil from the process. The slurried coal is pumped to unit pressure (4000 psig) and mixed with synthesis gas (25,000–75,000 scf/ton of coal). The reactants are heated to 400–450°C and maintained at conditions for extended periods (1–2 hr residence time) to effect liquefaction. Gaseous and liquid phase products are separated in a series of receivers. The liquid product, which contains some solid carbonaceous material and mineral matter, is centrifuged. The semisolid centrifuge residue, which contains 40–50% oil, can be pyrolyzed to recover product oil. The solid carbonaceous residue (plus some fresh coal) could be used in a gasifier to produce synthesis gas. In the present laboratory unit the reactor systems that have been investigated include a mechanically agitated reactor, packed tubular reactors (3 in. I.D.; 42 in. long), and helical coil reactors operated upflow or downflow.

The configuration of the reactor is under study. In addition, the research program is investigating continous centrifugation as well as other solid spearation techniques. PERC plans to continue the Costeam research program to form the basis for a 10-ton/day pilot plant and an associated test program.

28.3.1.4 *Solvent-Refined Lignite*

Introduction. The production of liquids from lignite has been investigated at the University of North Dakota. The process involves contacting lignite slurried with recycle solvent and synthesis gas in a noncatalytic reactor. Unconverted lignite and ash are separated from the solvent-refined lignite product by countercurrent decantation with toluene. Further upgrading of the solvent-refined lignite in a catalytic hydrogenation second stage is being considered as a means of producing final products.

From 1965 to 1970 batch autoclave runs were carried out at the University of North Dakota with support from the Great Northern Railway. The OCR sponsored the design and construction of a 50 lb/hr PDU, which was completed in late 1974 and started up in early 1975.[17]

Process Description. Figure 28.4 is a schematic flow diagram of the process. Lignite that has been dried and pulverized is slurried with solvent recovered from the process, and the slurry is pumped to 1500–2500 psi, mixed with a synthesis gas blend, and preheated to 700–850°F. After preheating the mixture enters a reactor at a rate that allows an average residence time of 20–90 min. Light product gases and liquids are recovered by flash separation and distillation. Process solvent is recovered, along with additional light oil in vacuum distillation and fractionation towers.

Bottoms from the vacuum still are then processed in a system operating at approximately 450°F and 350 psi, where the solvent-refined lignite product is contacted with toluene. Unconverted coal, ash, and some heavy oil are precipitated out. Solvent-refined lignite product is recovered from the overflow by distillation.

Operating Experience. A major problem of chloride stress corrosion cracking of the 316 stainless steel parts was solved by the use of Incoloy 800 instead of 316 stainless steel parts. Whether the origin of this problem derives from lignite or is an artifact of one of the cleaning procedures has not been resolved.

Satisfactory operation of the toluene solid separations system had not been achieved before the latter part of 1976. In addition, major difficulties have been experienced with the pumping system that transfers the vacuum bottoms to the decantation system and with accumulation of ash in the settling tower. Another difficulty encountered has been the formation of spherical agglomerates in the reactor.

The range of process variables investigated is given in Table 28.7.

Sustained plant operation in solvent bal-

[17] *Project Lignite Process Development Unit—PDU Operations Through September 1976—Analysis and Interpretation,* U.S. ERDA Contract No. E(49-18)-1224, October 1976.

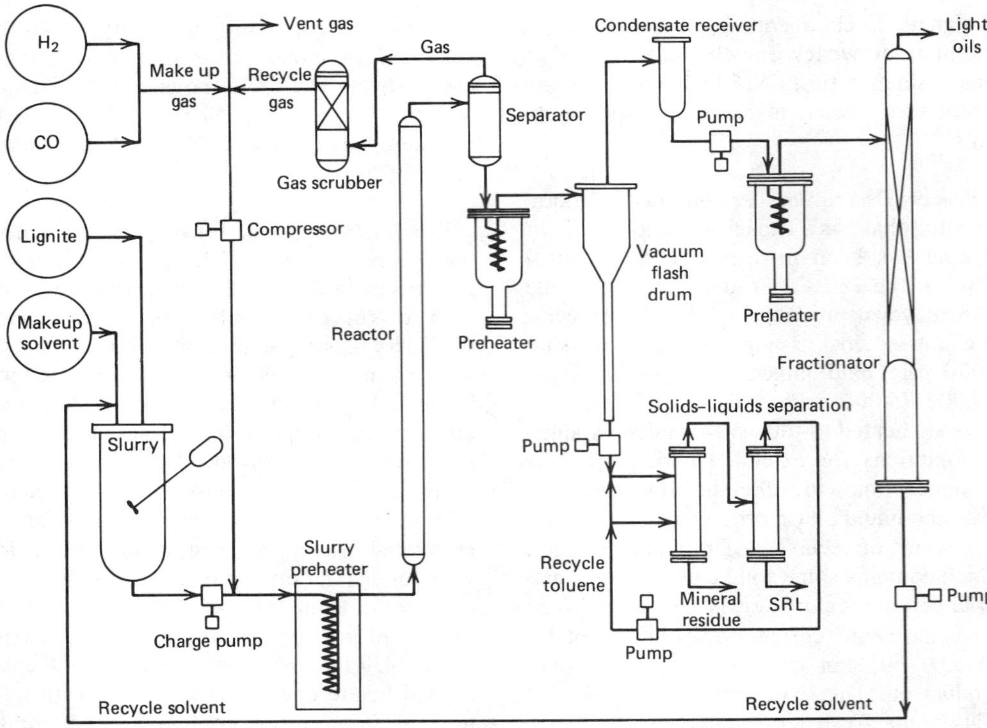

Figure 28.4 Flow diagram of solvent-refined lignite process development unit.

Table 28.7 Range of Process Variables for Solvent-Refined Lignite[17]

Variable	Range
Pressure	1500–2500 psig
Temperature	
At preheater exit	692–768°F
At dissolver exit	683–916°F
Lignite feed rate	14–54 lb/hr
Dissolver volume	0.72–2.86 ft³
Gas feed rate	7.4–20.7 scf/lb lignite
Liquid (slurry) hourly space velocity	0.66–2.8 hr/l
Gas hourly space velocity	139–600 hr/l
Solvent/coal, wt ratio	1.33–1.99
H_2/CO, mole ratio in feed gas	0.33–3.0

Ultimate Analysis of Materials (wt %)

	Coal as Charged	Solvent	Vacuum Bottoms	Deashed SRL
Carbon	46.54	85.76	79.18	87.98
Hydrogen	6.43	8.64	5.11	5.68
Nitrogen	0.59	0.26	1.02	1.13
Sulfur	0.61	1.70	1.03	0.80
Oxygen	39.55	3.64	3.98	4.42
Ash	6.28	0	9.68	0

ance was rarely achieved and was limited to a narrow range of conditions. Total pressures of about 2500 psi with a high ratio of H_2 to CO are required for satisfactory operations that are in solvent balance. Increased temperature results in increased conversion, water consumption, and net gas yield. At 2500 psig total pressure, 817°F dissolver outlet temperature, 1.04 liquid hourly space velocity (LHSV), and a solvent-lignite ratio of 1.62, the yield of solvent-refined lignite was 56.6%, with 5.1% of light oil. Net gas yields of about 25% would be consistent with these results. Typical data are presented in Table 28.7.

28.3.2 Catalytic Processes

28.3.2.1 *The H-Coal Process*

Introduction. The H-Coal process features an ebullated-bed catalytic reactor system. Total liquid yields from the reaction section, including liquids boiling above 1000°F, are usually 3–4 bbl/ton of coal. Reaction conditions are normally in the range of 825–875°F, 1500–3500 psig total pressure with coal space rates of 20–100 lb/hr per cubic foot of reactor. Boiling range distribution of products is primarily a function of space rate, temperature, and coal type.

Initial development of the process began in 1963.[18] The first experiments were carried out by feeding a mixture of petroleum-derived fuel oil and coal to a bench-scale H-Oil reactor. Many of the basic H-Coal process concepts were derived from the H-Oil process developed by Hydrocarbon Research, Inc. (HRI), beginning in 1955. That process was commercialized in 1962 with a 2500-bbl/day unit at the Cities Services refinery in Lake Charles, Louisiana. Two large H-Oil units were operating in 1976 with a total throughput of about 80,000 bbl/day. The first H-Coal experiment was stimulated by the organization in 1962 of OCR. A contract for 3 years of process development was entered into by HRI and OCR in February 1965. This contract was terminated in 1967, for the convenience of the government.

Bench-scale work was started in 1965 in continuous units having capacities of about 25 lb/day. Operations in a 3-ton/day unit were initiated in March 1966.[18]

Between 1968 and 1971 development work on the process was supported by the Atlantic Richfield Company.[19] In 1971 a sponsoring group consisting of Atlantic Richfield, Exxon Research and Engineering, Ashland Oil Company, Consolidation Coal Company, Gulf Mineral Resources, and Sun Oil Company was formed. This group sponsored further development of the program until 1973. At that time Exxon Research and Engineering, Consolidation Coal, Gulf Mineral Resources, and Sun Oil terminated support; but Standard Oil of Indiana (Amoco) joined the program.[19] Later that year EPRI joined the sponsoring group.

In September 1974 HRI entered into a contract with OCR to carry out PDU operations that would confirm the basis of design for a nominal 600-ton/day pilot plant as well as perform a site selection survey and a detailed process and mechanical design for that plant. When ERDA was organized it took over the functions of OCR.[20] The site selected was adjacent to the Ashland Oil refinery in Catlettsburg, Kentucky. Between the beginning of work under this contract and pilot-plant groundbreaking on December 15, 1976, Arco dropped out of the program, Shell Oil joined and dropped out, Sun rejoined and dropped out, Mobil joined, and Conoco Coal Development Company joined. The Commonwealth of Kentucky agreed to provide financial support after the site was selected.

In summary, those participating in the project—along with the Department of Energy (DOE) of which ERDA became a part—are EPRI, Amoco Oil, Ashland, Mobil, Conoco, and the Commonwealth of Kentucky. Other companies that provided support at one time

[18] K. C. Hellwig, M. C. Chervenak, E. S. Johanson, H. H. Stotler, and R. Wolk, *"Convert Coal to Liquid Fuels with H-Coal,"* paper presented at American Institute of Chemical Engineers 62nd Meeting, Salt Lake City, Utah, May 21–24, 1967.

[19] C. A. Johnson, M. C. Chervenak, E. S. Johanson, and R. H. Wolk, *"Scale-Up Factors in the H-Coal Process,"* paper presented at American Institute of Chemical Engineers Meeting, New York, November 26–30, 1972.

[20] Harry de Looper and Leon H. Lehman, *"The Importance of the Kentucky H-Coal Pilot Plant," Proc. 4th Energy Resource Conf.*, University of Kentucky, Louisville, January 7–8, 1976.

or another included Exxon Research and Engineering, Gulf Mineral Resources, Sun Oil, Shell Oil, and Atlantic Richfield.

Process Description. The prominent feature of this process that distinguishes it from all other coal liquefaction processes is the ebullated-bed reactor. The Catlettsburg reactor design is shown in Fig. 28.5. The cylindrical section of the reactor above the distributor tray contains a bed of particulate catalyst. Most prominently mentioned in a number of patents and articles as the preferred catalyst is a 1/16-in. cobalt molybdate on alumina extrudate.[21,22] This catalyst bed

[21] E. S. Johanson, S. C. Schuman, H. H. Stotler, and R. H. Wolk, "Coal Conversion Process," U.S. Pat. 3,519,553, July 7, 1970.

[22] E. S. Johanson, "Hydrogenation of Coal Using Unreduced Catalyst," U.S. Pat. 3,769,198, October 30, 1973.

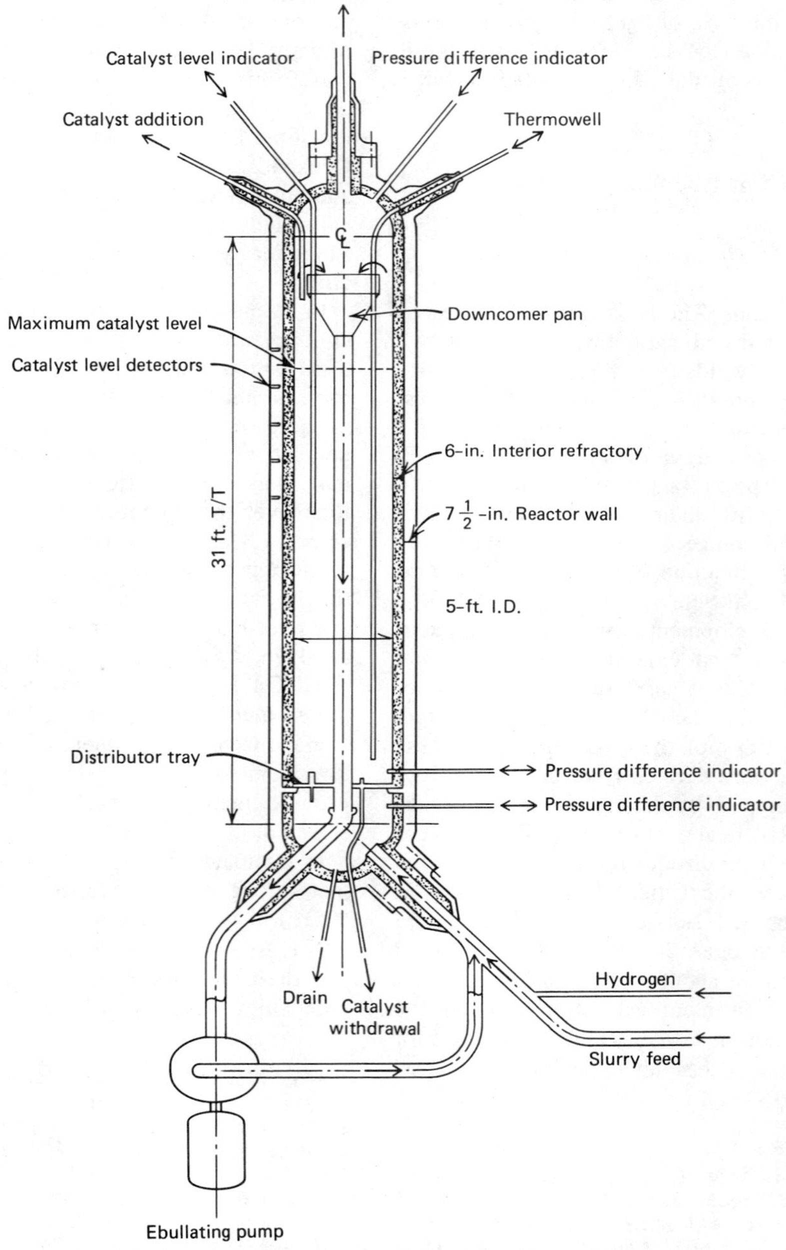

Figure 28.5 Design of the ebullated-bed reactor at Catlettsburg, Kentucky.

is fluidized primarily by the upflowing stream of recycle oil collected near the top of the reactor and recycled to the plenum chamber beneath the distributor tray. In this zone the recycle oil is mixed with the incoming coal-slurry-oil mixture and hydrogen-rich gas. Recycle of oil is required because the velocity of feed slurry and hydrogen-rich gas would not be sufficient by itself to fluidize the bed of catalyst. Normal expansion of the catalyst bed is about 40% over the settled height of the original catalyst charge.[23] The expanded catalyst bed allows ash and unconverted coal to pass between the fluidized catalyst particles. Recirculation of quantities of reactor liquid, which are large relative to the coal-slurry-oil mixture fed to the reactor, keeps the reactor temperature essentially isothermal. The reaction of coal and hydrogen is a highly exothermic reaction, so that there is a potential for a reactor inlet to outlet temperature differential of 150–300°F. Short flow eddies within the catalyst bed also help to minimize temperature differentials.

One major advantage of the ebullated-bed reactor system is the ability to maintain the activity of the catalyst at a constant desired level. This is done by withdrawing small amounts of essentially equilibrium catalyst and adding fresh catalyst to the reactor periodically. This simplifies operation by maintaining product quality and yield distribution at essentially constant levels.

Net liquid and vapor products are withdrawn from the reactor separately and are cooled and depressurized through independent circuits. In the currently favored mode of operation, heavy or residual oil is recycled to the reactor to increase its conversion to lower boiling products.[20] Hydroclones are utilized to separate the solids-containing reactor liquid stream into a high-solids stream and a low-solids stream. The low-solids stream is recirculated to the slurry mixing vessel to minimize the amount of distillate oil required to slurry fresh coal. Since the feed to the hydroclones has been flashed previously, it is enriched in residual oil relative to the reactor liquid effluent. This results in a higher concentration of residual oil in the reactor, which aids in its conversion to lower boiling products. Hydroclone underflow is ultimately vacuum distilled to recover additional distillate slurry oil. This recovery is necessary when operating the process at relatively low coal feed rates to maximize distillate oil production. Figure 28.6 is a simplified flowsheet of the 3-ton/day PDU at HRI's Trenton, New Jersey, laboratory.[24]

[23] C. A. Johnson and J. Y. Livingston, Hydrocarbon Research, Inc., New York, "H-Coal: How Near to Commercialization?" paper presented at University of Pittsburgh, School of Engineering, Symp. Coal Gasification and Liquefaction: Best Prospects for Commercialization, August 6–8, 1974.

[24] C. A. Johnson, H. H. Stotler, and William Volk, "H-Coal Process for Producing Liquid Hydrocarbons," paper presented at AIME Meeting, Dallas, Tex., February 1974.

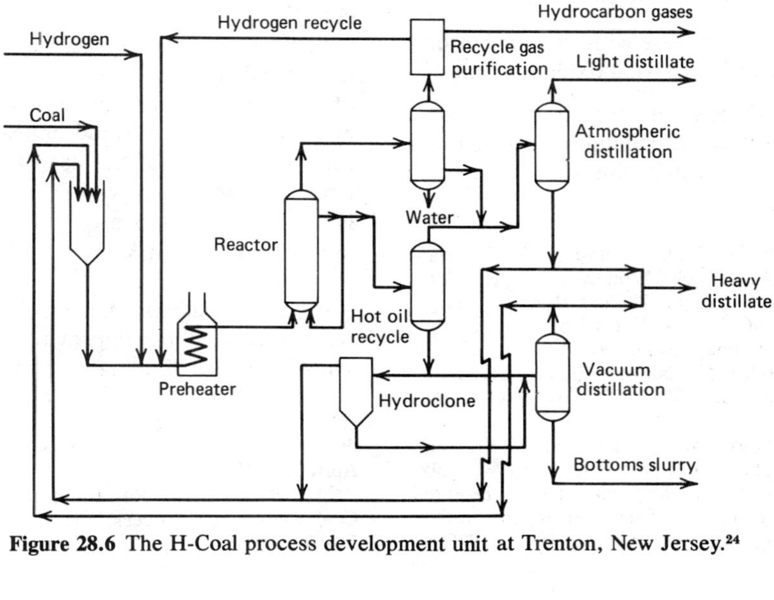

Figure 28.6 The H-Coal process development unit at Trenton, New Jersey.[24]

Early conceptions of the process considered coking the vacuum bottoms stream to recover additional distillable oils.[18] Reportedly only about 35% of the maximum recoverable oil was recovered by coking.[25] More recent publications indicate that partial oxidation of these bottoms in a steam-oxygen-blown gasifier to produce a CO-H_2 mixture is preferable. This approach allows some by-product hydrocarbon fuel gas to be exported instead of steam reforming all light gas products to meet internal plant hydrogen and fuel gas requirements.

When the process is operated at lower severity by means of increasing coal throughput, the product objective is a low-sulfur heavy fuel oil.[26] In this case the ratio of heavy nondistillable oil to unconverted coal and ash is too high for vacuum distillation to be utilized as the method of solids rejection. Experimental work has been carried out on centrifugation, magnetic separation, filtration, and solvent precipitation to obtain data on these various methods of solids rejection.[27] Selection of a Lummus solvent deashing system for the Catlettsburg pilot plant indicates that the latter approach is currently viewed as the preferred system.

Simplified process flowsheets for the following conceptual designs are given in Fig. 28.7.

1. An all-distillate case with fluid coking of vacuum bottoms; in this configuration hydrogen is produced by feeding fluid coke and coal to a Lurgi gasifier.
2. An all-distillate case with vacuum bottoms fed to a Texaco partial oxidation unit.
3. A fuel oil case with the bottoms product from a solvent precipitation unit fed to a Texaco partial oxidation unit to generate synthesis gas.

[25] C. A. Johnson, M. C. Chervenak, E. S. Johanson, H. H. Stotler, Olaf Winter, and R. H. Wolk, "Present Status of the H-Coal Process," paper presented at IGT Clean Fuels from Coal Symp., Chicago, September 10–14, 1973.

[26] H. H. Stotler, M. Caldron, and C. A. Johnson, "Low Sulfur Fuel Oil From Coal," U.S. Pat. 3,617,474.

[27] C. A. Johnson, M. C. Chervenak, A. G. Comolli, E. S. Johanson, C. C. Kang, and W. Volk, "H-Coal Process Development," paper presented at American Institute of Chemical Engineers, 68th Meeting, Los Angeles, November 16–20, 1975.

Solids-Liquid Separation. Vacuum distillation is the preferred method of solids separation when the process is being operated in the so-called syncrude or all-distillate mode. Vacuum bottoms containing all the ash and unconverted coal is sent to the partial oxidation unit for synthesis gas production. In the gasification step the ash is converted to a molten slag, which is quenched in water and ultimately rejected from the system in the form of coarse granules.

Several solids separation options have been investigated in the fuel oil mode of operation. Among the earliest was rotary precoat filtration. Rates achieved with diluted vacuum still bottoms were in the range of 100–150 lb/(hr)(ft^2) of submerged filtration area. With corrections to total machine filter area, the rate corresponds to about 40–60 lb/(hr)(ft^2) of total filter area. This rate was deemed too low to be viable commercially.[27]

A patent described magnetic separation as a potenial method for removing solids from H-Coal liquids.[28] However, later results reported that this approach had a very limited machine capacity, did not remove inorganic sulfur (in the form of FeS_x) to a level low enough to allow the products to meet the EPA NSPS, and resulted in a magnetic material that was difficult to remove from the matrix.[27]

Centrifuge tests indicated that satisfactory ash levels could be achieved by two stages of centrifugal separation.[29] Generalized centrifuge correlations were developed for syncrude mode products from Illinois No. 6 coal.[27]

Solvent washing of vacuum bottoms was mentioned in several early patents as a means of recovering heavy liquids from vacuum bottoms.[21,29] This method appears to be an early exploration of the use of solvent precipitation. It was reported in 1975 that an HRI in-house program on solvent precipitation was underway.[27]

Process Variations. A number of variations were suggested for improving the process. These included limiting reactor liquid com-

[28] M. C. Chervenak, "Solids Removal from Hydrogenated Coal Liquids," U.S. Pat. 3,725,241 April 3, 1973.

[29] H. H. Stotler, "Residuum Recovery From Coal Conversion Process," U.S. Pat. 3,519,554, July 7, 1970.

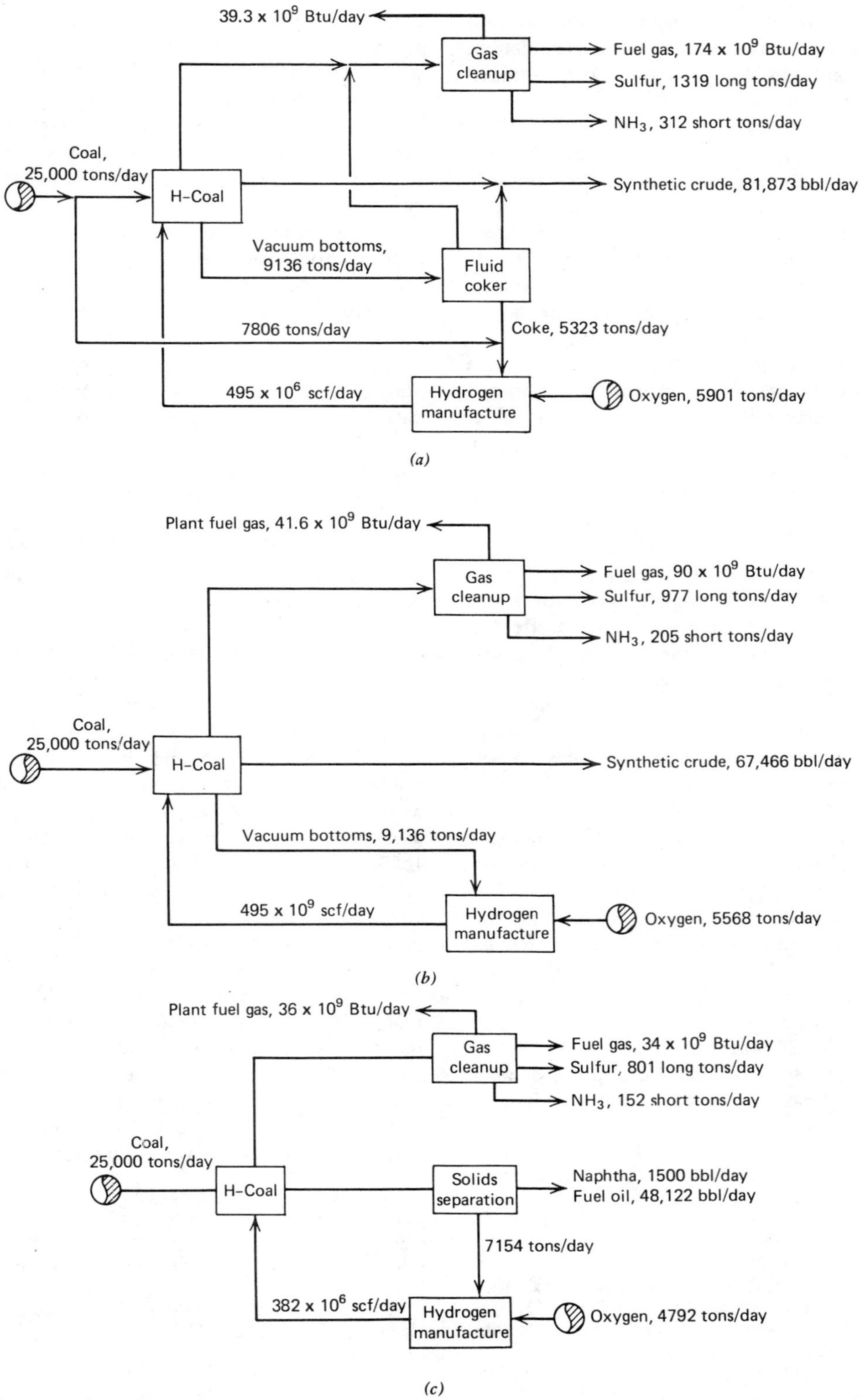

Figure 28.7 Flow schemes for the H-Coal process (*a*) synthetic crude from Illinois coal, case I, (*b*) synthetic crude from Illinois coal, case II, and (*c*) low-sulfur fuel from Illinois coal.[24]

position to 30–45% residuum and 10–25% solids.[30] This, it is claimed, enables ebullation of catalyst in the reactor to be maintained under better control.

The use of a two-stage system with a low-pressure (500–600 psi hydrogen partial pressure) dissolving zone, and a higher pressure ebullated-bed catalytic hydrogenation zone for solvent hydrogenation was suggested. The first reactor contained particulate solids to improve contacting. Hydrogen flow through the slurry preheater was mentioned as a means of reducing fouling.[31]

Utilization of ash and solids formed in the process as catalyst was mentioned as a means of converting Pittsburgh coal at a level of more than 80%.[32] Lower hydrogen consumption and heavier products were obtained compared to normal cobalt molybdate catalyst.

When processing low-rank coal, the use of two reactors in series with interstage separation of gases was suggested as a means of increasing CO_2 yield. This had the effect of reducing hydrogen consumption, since less hydrogen was rejected as water.[33]

A predissolver upstream of a normal H-Coal reactor was suggested as a means of utilizing coarser coal and improving the performance of the H-Coal reactor.[34]

Catalyst. Many of the available references disclose the use of cobalt molybdate on alumina catalyst,[21,26] which has been described as commercially available. Sizes most frequently mentioned are extrudates ¼–¹/₁₆ in. in diameter. A calculated catalyst age distribution for a catalyst replacement rate of 1.0 lb of catalyst per ton of coal processed indicates that 63% of the catalyst has an effective age of less than 2000 lb of coal processed per pound of catalyst. Since the initial activity of fresh catalyst is greater than that of partially aged catalyst, the effective catalyst age would be less than about 1000 lb of coal processed per pound of catalyst.[27]

Spent cobalt molybdate catalyst from operations on Illinois No. 6 coal contained 3.0% titanium, 0.5% calcium, and 0.5% iron. The carbon level was about 16% on this catalyst.[35]

Utilization of two reactors in series, with spent catalyst from the second reactor being rejected to the first reactor, was suggested as a means of minimizing titanium deposition on catalyst in the second stage.[36]

Higher liquid yields were claimed when a startup procedure was used in which the catalyst temperature in the presence of hydrogen was limited to 200–400°F before startup oil was brought into contact with it.[22]

Physical losses due to attrition have been reported as being at most on the order of 5% of the initial inventory in nonreplacement operations.[27] This is lower than the normal catalyst replacement rate.

Experimental Experience. At least 12 coals have been run in H-Coal bench-scale units. These include five bituminous coals [Illinois No. 6 (Burning Star and Monterey), Pittsburgh No. 8, Colorado, and Utah], four subbituminous (Wyodak, Kaiparowits, Black Mesa, and Big Horn), and three lignites (Texas, North Dakota, and Australia). A total of 1657 unit days was claimed, with more than 1100 of these on Illinois No. 6 and Wyodak coals. Operations in the 3-ton/day PDU have been limited to Illinois No. 6 and Wyodak runs and totaled 174 unit days through 1974.[23] Since that time several hundred additional PDU operating days have been logged on Illinois No. 6 and Wyodak coals.

Yield distributions have been reported on Illinois No. 6 and Wyoming operations in a number of papers.[18,19,37] Black Mesa and Kaiparowits H-Coal data were summarized in

[30] R. H. Wolk and E. S. Johanson, "H-Coal Process Slurry Oil System," U.S. Pat. 3,540,995, November 17, 1970.

[31] C. A. Johnson, E. S. Johanson, and R. H. Wolk, "Low Pressure Hydrogenation of Coal," U.S. Pat. 3,607,719.

[32] R. H. Wolk, E. S. Johanson, and S. B. Alpert, "Coal Hydrogenation," U.S. Pat. 3,617,465.

[33] E. S. Johanson, "Hydrogenation of Low Rank Coal," U.S. Pat. 3,700,584, October 24, 1972.

[34] R. H. Wolk, "Coal Hydrogenation Using Pretreatment Reactor," U.S. Pat. 3,791,957, February 12, 1974.

[35] R. H. Wolk, M. C. Chervenak, C. A. Johnson and E. S. Johanson, "The H-Coal Process," paper presented at National Science Foundation Workshop, Ohio State University, Columbus, April 16–18, 1974.

[36] C. A. Johnson, "Two-Stage Countercurrent Hydrogenation of Coal," U.S. Pat. 3,679,573, July 25, 1972.

[37] E. T. Layng and Katherine C. Hellwig, "Liquid Fuels from Coal by the H-Coal Process," *Min. Congr. J.*, **55**(4), 62–67 (April 1969).

Table 28.8 Experimental Results from H-Coal Operations on a Variety of Coals

	Illinois No. 16,[18] River King	Wyoming Subbituminous C,[18] Smith Seam	Australian Brown Coal[39]	Wyodak[38]	Black Mesa[38]	Kaiparowits[38]
Proximate analysis, mf						
Ash (%)	11.60	8.7	8.3	8.1	4.84	6.8
Volatile matter (%)	38.11	44.0	49.0	46.7	43.2	42.3
Fixed carbon (%)	50.29	47.3	42.7	45.2	52.0	50.9
Ultimate analysis (%)						
Carbon	70.28	68.9	62.3	68.8	71.9	70.7
Hydrogen	4.89	4.3	4.5	5.2	5.2	5.0
Nitrogen	0.90	1.0	0.5	0.9	1.2	0.9
Sulfur	3.13	0.6	1.2	0.7	0.5	0.5
Ash	11.60	8.7	8.3	8.1	4.8	6.8
Oxygen	9.2	16.5	23.2	16.3	17.2	16.2
Yields (wt %, maf)						
C_1–C_3	10.2	10.8	7.2			
Liquid	71.0	56.1	62.3			
Unconverted char	10.7	18.6	7.1			
H_2S, H_2O, NH_3	14.0	12.9	19.1			
CO, CO_2		6.6	9.5			
Yields of fractions, bbls/ton, (wt %, maf coal)						
C_4–400°F	1.35 (18.5)	1.35 (18.5)				
400–680°F	1.70 (22.5)	0.95 (15.5)	{ 53.2	{ 3.73	{ 2.92	{ 3.27
680–975°F	0.67 (12.7)	0.62 (11.1)				
975° F+	0.55 (12.3)	0.52 (11.0)	9.0	0.25	0.67	.58
Total	4.27 (71.0)	3.44 (56.1)	3.96 (62.2)	3.98	3.59	3.85
Gravity (°API)						
C_4–400°F	49.2	50				
400–680°F	21.1	21				
680–975°F	8.3	4				
975°F+	−20	−16				
Sulfur content (wt %)						
C_4–400°F	0.099	<0.07				
400–680°F	0.16	<0.07				
680–975°F	0.10	<0.07				
975°F+						
Nitrogen content (wt %)						
C_4–400°F	0.10	0.2				
400–680°F	0.17	0.3				
680–975°F	0.41	0.6				
975°F+						

1973.[38] A paper presented at the Eighth World Petroleum Congress reported yields on Australian brown coal.[39] These data have been summarized in Table 28.8.

Development Issues. The large pilot plant at Catlettsburg, Kentucky, will address many questions that must be resolved before H-Coal technology can be commercialized. Among the parts of the reaction system to be evaluated are the reactor distributor design, catalyst level detection system, and reactor internals. Commercial reactors will be 2–3 times the diameter of the 5-ft diameter pilot-plant reactor that is to be operated at the Kentucky site.

Heat exchange systems, fired heaters to preheat slurries, coal-handling techniques, slurry pumping, pressure letdown, and solid-liquid separation systems will also be tested at this site.

Other important aspects of the pilot-plant program have the following goals:

- Confirmation of scaleup concepts from operations in bench-scale equipment and an 8.5-in.-diameter PDU with a capacity of 3 tons/day.
- Development and establishment of the mechanical integrity of specialized equipment.
- Demonstration of overall plant performance.
- Discovery and description of "new problems areas," if any.
- Provision of data required for optimization to minimize commercial plant costs.
- Provision of large samples of liquid products.[25]

28.3.2.2 *The Lummus Clean Fuel from Coal (CFFC) Process*

Introduction. The Lummus Corporation, a division of Combustion Engineering, Inc., is developing processing technology for liquefying coal. Early work carried out in the 1960s involved extraction of coal by a technique that did not use hydrogen or a ctalytic system, but used a hydrotreated solvent to extract and dissolve the coal. Work on this method was abandoned because of difficulties in solids-liquid separation by filtration techniques and because of unfavorable economics. Several patents were issued for the thermal approach pursued.[40]

Since that time laboratory work in continuous units has been carried out to develop the Lummus Corporation's CFFC process. One objective was to develop a process that provided a heavy coal liquid product similar in physical characteristics to a No. 6 fuel oil derived from petroleum sources. A further objective was to efficiently utilize hydrogen to produce the fuel oil product.

The CFFC process features a catalytic hydroliquefaction system using molecular hydrogen generated from gasification of rejected ash-containing residues obtained from the Lummus proprietary ash separation method.

The key features of the CFFC process are as follows:

- Catalytic hydrodesulfurization to produce a heavy fuel oil product containing less than 0.5 wt % sulfur.
- An ash separation technique to produce a clean fuel product containing less than 0.1 wt % ash.

Process Description. A block flow diagram of the process and a diagram of the ash separation method appear in Figs. 28.8 and 28.9, respectively. In the CFFC process crushed and dried coal is slurried with recycled aromatic solvent generated in the process. The slurry is contacted in the catalytic hydrodesulfurization section with hydrogen at elevated pressures and temperatures. No details of the reaction section have been released, although the patent literature cites

[38] C. A. Johnson, K. C. Hellwig, E. S. Johanson, R. H. Wolk, and H. H. Stotler, "H-Coal: Conversion of Western Coals," *Trans AIME, Soc. Min. Eng.*, **1**(254), 235 (September 1973).

[39] C. A. Johnson, K. C. Hellwig, E. S. Johanson, H. H. Stotler, M. C. Chervenak, and E. T. Layng, "Production of Gasoline from Australian Brown Coal by the H-Coal Process," *Proc. 8th World Petroleum Congr.*, Moscow, June 1971, Applied Science Publishers Ltd. Essex, England.

[40] W. J. Bloomer et al., U.S. Pat 3,109,803, November 5, 1963; W. J. Bloomer et al., U.S. Pat. 3,375,188, March 26, 1968; M. C. Sze and G. J. Snell, U.S. Pat. 3,974,073, August 10, 1976; W. J. Bloomer et al., U.S. Pat. 3,379,638, April 23, 1968; W. J. Bloomer et al., British Pat. 1,075,240, November 26, 1963; W. J. Bloomer et al., British Pat. 1,122,510, December 22, 1966.

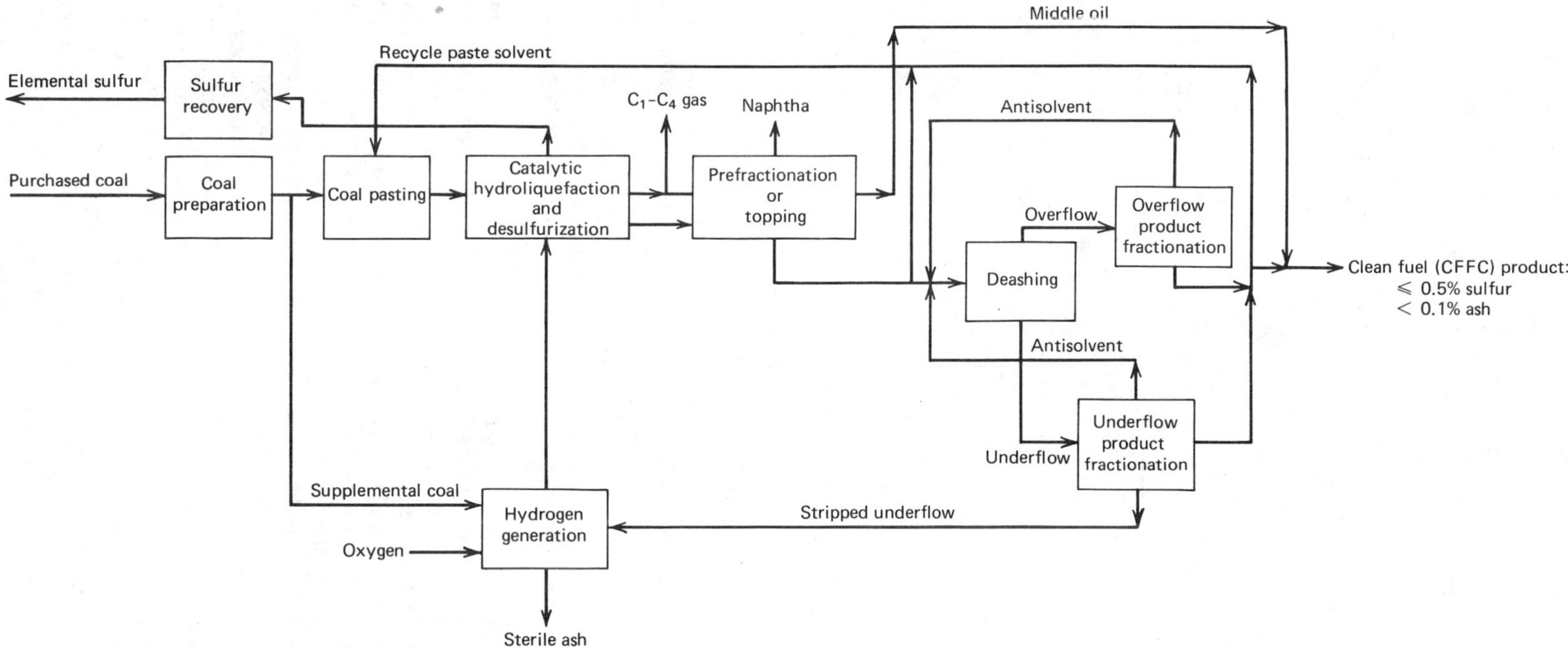

Figure 28.8 Schematic block flow diagram of the Lummus CFFC process.[41]

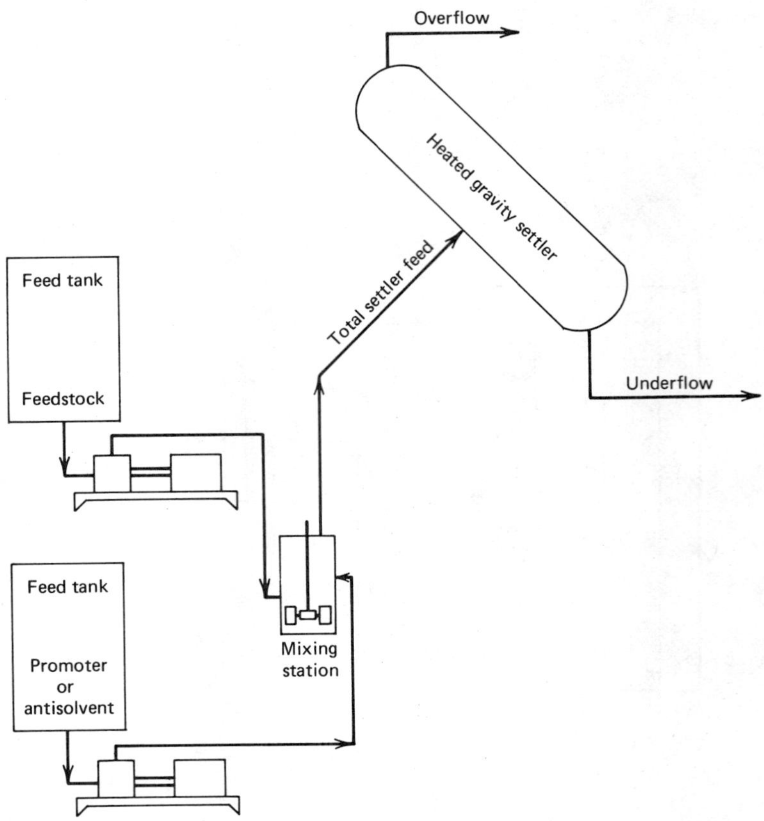

Figure 28.9 Deashing pilot-plant flow diagram for the CFFC process.[42]

the use of an expanded ebullated bed of spheroidal catalyst contained in an upflow reactor.[43] Recent work has been directed at developing catalysts with enhanced activity for reducing the nitrogen content of the fuel oil product[44] from about 1 wt % to less than 0.5 wt %.

The coal that is solubilized in the reaction section is let down in pressure, and distillates boiling below 600°F are recovered by fractionation. For Illinois No. 6 coal fed to the process, about 95% of the maf coal is converted to a liquid product (quinoline solubles), and about 75% of the sulfur is converted to hydrogen sulfide. In addition, 80–85 wt % of the coal is in the form of benzene-soluble asphaltenes and oils. Portions of the oxygen and nitrogen in the coal are converted to water and ammonia, respectively. Hydrogen consumption is about 2.4 wt %. This hydrogen consumption is about the same as that in solvent refining, which produces a solid product.

The residual product containing ash is mixed with a distillate antisolvent, to accomplish agglomeration of ash-containing residues. The antisolvent used in the process is a narrow boiling range fraction (425–500°F). It is usually a kerosine fraction derived from a petroleum source (or alternately generated in the process by a separate hydrogenation of the aromatic kerosine). It is used to promote rapid settling of ash-containing residues. The antisolvent has a Watson-Nelson characterization factor of at least 9.75. The antisolvent is not consumed in the process but is recovered by fractionation and used over again.

The mixture of antisolvent and topped (nominally boiling higher than 600°F) product is separated into an ash-free stream that

[41] G. J. Snell, private communication.

[42] G. J. Snell and A. A. Simone, *The Applications of Lummus Anti-Solvent Deashing Technology to a Solvent Refined Coal Solution,* EPRI AF-234, September 1976.

[43] M. C. Sze and G. J. Snell, U.S. Pat. 3,932,266, January 13, 1976.

[44] A. A. Simone, "Clean Fuel from Coal Process," *Combust. Mag.*, 15–19 (May 1976).

overflows from an inclined settler and an underflow of ash-enriched product. About 62% of the 600°F+ product is the low-ash overflow product and 38% is the high-ash underflow product. The gravity settler operates at a temperature in the 300–600°F range and at a modest pressure of about 100 psig, which is sufficient to maintain the mixture in the liquid phase.[45]

The fuel oil product is a blend composed of low-boiling middle distillates (400–600°F), clean overflow from the settler, and distillate product stripped from the underflow stream.

Status. The CFFC process has been applied to a variety of U.S. bituminous coals, a subbituminous coal, and a foreign brown coal. In all cases the fuel oil product had an ash content of less than 0.1 wt %.[46,47] The CFFC process, by adding hydrotreating and hydrocracking steps, can upgrade the product to a synthetic crude.[43] Typical CFFC product data with an Illinois No. 6 coal feed are shown in Table 28.9.

The process is in an early stage of development with respect to scaleup of the technology to larger scale equipment. In 1977 ERDA announced plans to support additional laboratory development work to complete testing of the catalyst.

The Lummus approach to ash removal has been applied to removing solids from a solvent-refined coal product produced in a 6-ton/day pilot plant.[42] This deashing route appears to represent an economically attractive and more reliable alternative to filters or centrifuges for other processes. It has been selected for use in the H-Coal pilot plant to be constructed in Kentucky.

28.3.2.3 *Synthoil*

Introduction. Since 1967 PERC has carried out research and development on direct catalytic conversion of high-sulfur coals to low-sulfur liquid products, using a novel contacting method to convert slurried coal.[48,49] A feature of the process is the turbulent condition that results from the recycled hydrogen and selected geometry of the reactor. The high velocity from the hydrogen flow creates turbulence in the packed catalyst bed and maintains operability by violently propelling the coal slurry upward through the packed bed. It is also claimed that turbulence keeps the surface of the catalyst clean by attrition and promotes mass transfer of the hydrogen into the slurry mixture.

Table 28.9 CFFC Data on Illinois No. 6 Coal

	Coal	Clean Fuel Product
Inspections (wt %)		
Carbon	66.3	88.8
Hydrogen	4.8	6.9
Sulfur	3.9	0.34
Water	3.0	—
Nitrogen	1.2	1.2
Oxygen	11.1	2.8
Ash	9.7	0.05
	100.0	100.0
Specific gravity at 170/60°F		1.100
Pour point		65°F
Flash point		215°F
Heating value		16,900 Btu/lb
Viscosity at 140°F		5,400 SUS
Viscosity at 180°F		864 SUS
Distillation		
IBP		342°F
10%		469°F
30%		538°F
50%		790°F
67%		900°F
Trace metals (ppm)		
Sodium		5.7 ppm
Vanadium		1.0 ppm
Chlorine		0.025 ppm

Source: Modified from Ref. 44.

Process Description. Coal is slurried with a portion of the product oil in about 2 : 1 ratio of product oil to coal. The slurried coal is then mixed with hydrogen and preheated to reaction conditions. Most of the work has been carried out at about 250 atm, and some at 150 atm. Reaction time in the preheater is about 15 min. In the preheater substantial

[45] M. C. Sze and G. J. Snell, U.S. Pat. 3,852,182, 3,856,675, December 3 and December 24, 1974.

[46] M. C. Sze and G. J. Snell, "Application of the C-E Lummus Deashing Technique to Coal Liquefaction Processes," ECHEMA, Frankfurt, West Germany, June 20–26, 1976.

[47] M. C. Sze and G. J. Snell, "A New Process for Removing Ash from Coal Liquefied by Hydrogeneration," *Proc. American Power Conf.*, **37**, 315 (1975).

[48] P. M. Yavorsky, S. Akhtar, and S. Friedman, "Converting Coal into Non-Polluting Fuel Oil," *Chem. Eng. Prog.*, **69**(3), 31–32 (March 1973).

[49] P. M. Yavorsky, "ERDA Coal Liquefaction Program for Synthetic Fuel Oil Development," paper presented at Symp. Air Quality Management in the Electric Power Industry, Austin, Tex., January 18–30, 1976.

conversion of the coal is accomplished at a high velocity that results from circulating hydrogen. A reactor packed with ⅛ × ⅛ in. catalyst, composed of cobalt molybdate on silica alumina, is used to remove organic sulfur from the product. Contact time in the reactor is 2–5 min.

In the reactor, sulfur is converted to hydrogen sulfide, and desulfurization and hydrocracking are accomplished to produce a low-viscosity, low-sulfur product in the form of a liquid.[50–52] To remove ash from the product oil, centrifugation is practiced. The practicality of using centrifuges in large-scale equipment is doubtful, and other approaches—including filtration, solvent precipitation, and combinations—are being investigated at PERC.[53]

The dimensions of various Synthoil units that have been operated or are being built are summarized below:

Coal Capacity (lb/day)	Reactor Diameter (in.)	Length (ft)	Status
48	0.3	68	Operational
400	1	14	Operational
16,000	4	4 × 15	Under construction

Although construction was begun on the 8-ton/day pilot plant, plans for operating it were deferred, and the project has been terminated.

A flowsheet of the Synthoil process is shown in Fig. 28.10.

Status. In the development of the process 30-day runs were carried out on Kentucky, West Virginia, Ohio, Indiana, and Pennsylvania coals, which were converted to fuel oil products having a sulfur content of about 0.3%. The ash content of the products, although initially about 0.1 wt %, increased to about 2 wt % during extended runs[54] because of changes in viscosity of the product. (The Environmental Protection Agency's specifications for particulate matter in fuel oil used in utility applications allow about 0.1 wt % ash matter in the final product.) The increase in ash content of the Synthoil product results from difficulty in centrifuging the product produced with relatively aged catalyst. The life of the catalyst in the Synthoil process is of concern. The University of Delaware, Sandia Laboratories, and Battelle Memorial Institute are carrying out investigations to reduce the rather high catalyst usage rates associated with the Synthoil process.[55,56] Reportedly, in a 0.5-ton/day unit operation, catalyst deactivation was severe after 500 hr at conditions of 4000 psi and 840°F.[57]

Table 28.10 summarizes Synthoil laboratory data on Kentucky coal. Areas of concern in the Synthoil process include the following:

1. Reactor pressure drop (10 psi/ft) that could be troublesome because of catalyst strength considerations.
2. Slurry preheating requirements that appear impractical to scale up.
3. Solids separation.

[50] S. Akhtar, S. Friedman, and P. M. Yavorsky, "Process for Hydrodesulfurization of Coal in a Turbulent Flow Fixed Bed Reactor," in Am. Inst. Chem. Eng. Symp. Ser., Vol. 70, No. 137, 1974, pp. 106–113.

[51] P. M. Yavorsky, S. Akhtar, and S. Friedman, "Process Developments: Fixed Bed Catalysis of Coal to Fuel Oil," Am. Inst. Chem. Eng. Symp. Ser., Vol. 70, No. 137, 1974, pp. 101–105.

[52] P. M. Yavorsky, S. Akhtar, J. Lacey, M. Weintraub, and A. Reznik, "The Synthoil Process," *Chem. Eng. Prog.*, **71**(4), 79–80 (April 1975).

[53] J. O. Newman and S. Akhtar, "Coagulation and Filtration of Solids from Liquefied Coal Synthoil Process," paper presented at 172nd American Chemical Society Meeting, San Francisco, August 29–September 3, 1976.

[54] H. Huang and J. Fischer, *Laboratory Studies for Separation of Solids from Synthoil Gross Product,* Argonne National Laboratory, Argonne, Ill., U.S. ERDA Contract No. W-31-109 Eng. -38, November 1976.

[55] W. Kawa, W. Friedman, W. Frank, and P. Yavorsky, "Evaluation of Catalysts for Hydrodesulfurization and Liquefaction of Coal," *Prepr., Am. Chem. Soc., Div. Fuel Chem.*, **19**(1), 192–206 (1974).

[56] J. Stanulonis, B. Gates, and J. Olson, "Catalyst Aging in a Process for Liquefaction and Hydrodesulfurization of Coal," *Am. Inst.. Chem. Eng. J.*, **22**(3), 577 (May 1976).

[57] J. C. Davis, "Coal Liquefaction Shooting for Better Catalysts," *Chem. Eng. News,* **83**(16), 77–79 (July 19, 1976).

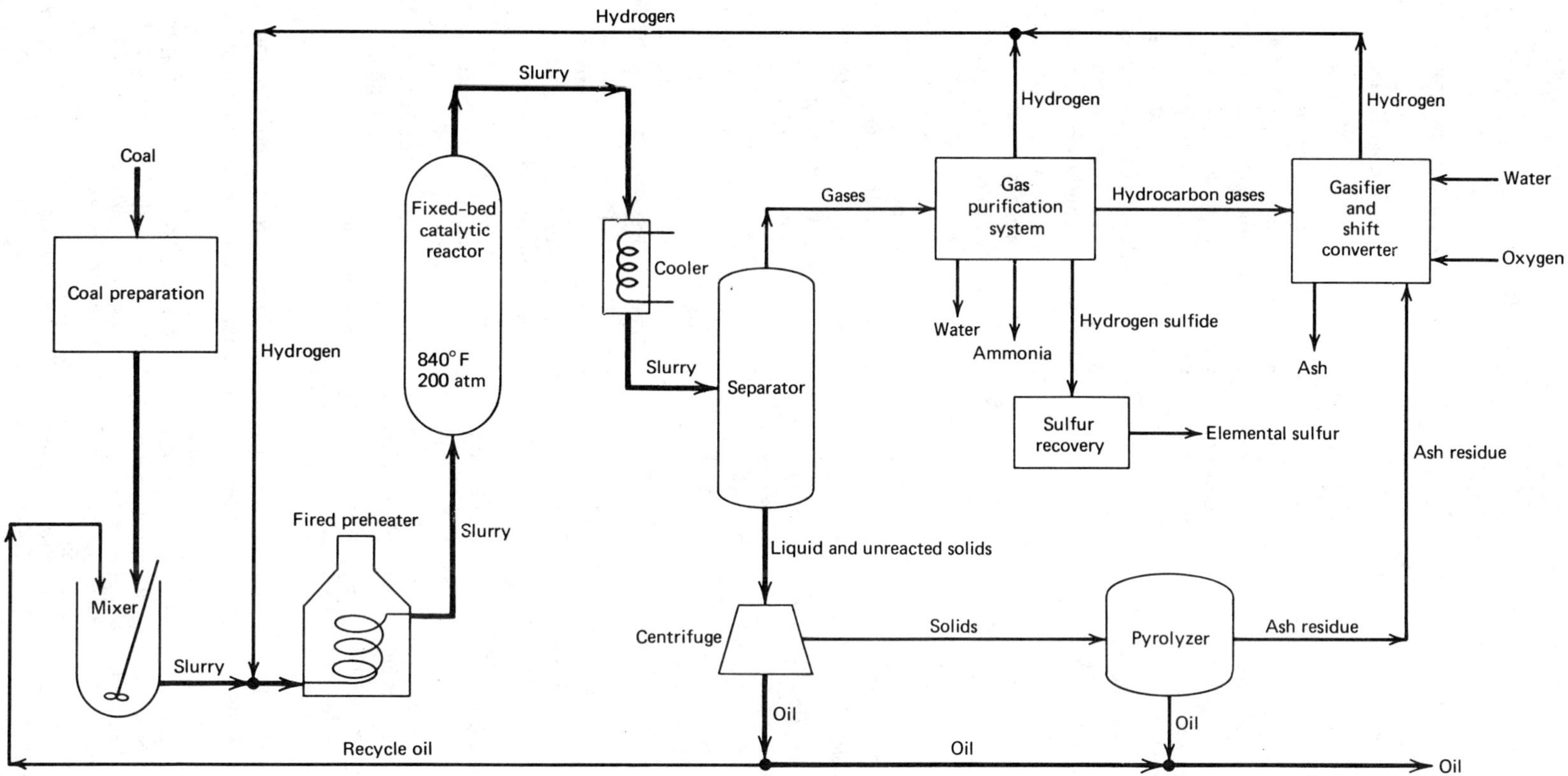

Figure 28.10 Flowsheet for the Synthoil process.

Table 28.10 Summary of Synthoil Data on Kentucky Oil

Pressure	4,000 psig
Temperature	450°C
Hydrogen rate	160,000 scf/ton
Slurry oil/coal ratio	1.85
Space velocity	20 lb coal/(hr)(ft³)
Sulfur in feed coal	5.5 wt %
Sulfur in product oil	0.3 wt %
Ash in product oil	0.2 wt %
Calorific value of product oil	17,400 Btu/lb
Yield of product oil	3.3 bbl/ton

4. Engineering design of the reactor system to provide temperature control at elevated pressures and to avoid plugging of the catalyst bed.[58]
5. Catalyst performance.
6. Process engineering of required heaters and compressors in commercial equipment.
7. Utilization of raw products that are unstable when stored.[59,60]

28.3.2.4 *The Gulf Catalytic Coal Liquids (CCL) Process*

Gulf, using proprietary technology, is carrying out development of a direct catalytic route to liquids from coal. All the work has been supported by Gulf Oil Company. The process involves the use of a fixed catalyst bed of special design, which is intended to avoid deposition of carbonaceous materials. The proprietary catalyst was claimed to have a prolonged activity and to be tolerant of metallic compounds such as iron and titanium that are present in coal. The major objective is the production of low-sulfur fuel oil from high-sulfur Eastern coal. About 3 bbl of liquids is produced per ton of coal. The patent literature[61] reveals radially placed baskets that contain particulate catalyst.

A 1-ton/day pilot plant has been in operation for 2 years. The pilot plant, located at Harmarville, Pennsylvania, will provide design data for a larger plant.[62]

28.3.3 Donor Solvent Processes

28.3.3.1 *The Exxon Donor Solvent (EDS) Process*

Introduction. Exxon Research and Engineering Company (ER & E) has been doing research on coal liquefaction since 1966. The technical feasibility of the Exxon donor solvent (EDS) coal liquefaction process was established in small, continuous pilot plants. A 0.5-ton/day integrated liquefaction pilot plant was operated for several years prior to 1970. More recently smaller pilot plants have been operating to test process improvements and obtain process variable data. A 1.0-ton/day integrated pilot plant began operation in June 1975. Laboratory and engineering studies carried out in 1976 confirmed the viability of the EDS process, showed that it can be applied to a wide range of coal feeds, and identified several potential process improvements. Extensive engineering studies continued to show that the cost of EDS coal liquids should be the same or lower than liquids of similar quality from other coal liquefaction processes.

Process Description. The EDS process converts coal by an indirect catalytic liquefaction technique.[63] In the process a distillate stream is hydrogenated under specially controlled conditions by more or less conventional catalytic hydrogenation methods to provide a "carrier" of hydrogen to the dissolver section. In the dissolver section some of the hydrogen in the recycled distillate is released or "donated" to the coal dissolution fragments. The chief features of the process are as follows:

1. In the primary liquefaction step coal is liquefied in a noncatalytic tubular plug-flow

[58] P. M. Yavorsky, Pittsburgh Energy Research Center, personal communication, November 5, 1976.

[59] F. S. Karn, F. A. Braun, and R. Friedel, *Aging Studies of the Synthoil Product*. Part I. *Viscosity Changes,* Pittsburgh Energy Research Center, Bruceton, Pa,, PERC/TDS-76/1, May 1976.

[60] H. W. Sternberg, R. Raphael, and S. Akhtar, "Synthoil Process and Product Analysis," Am. Chem. Soc. Symp. Ser. No. 20, *Hydrocracking and Hydrotreating,* 1975, pp. 111–122.

[61] South Africa Pats. 05793X/04, 2A7502-628.

[62] S. Chung, *Gulf Catalytic Coal Liquids (CCL) Process, Materials Problems and Research Opportunities in Coal Conversion,* Corrison Center, Ohio State University, Columbus, 1974 pp. 263–276.

[63] L. E. Furlong, E. Effron, L. W. Vernon, and E. L. Wilson, "The Exxon Donor Solvent Process," *Chem. Eng. Prog.*, 72(8), 69 (1976).

reactor in the presence of molecular hydrogen and a hydrogen-rich donor solvent. The liquefaction reactor operates at 800–880°F and at a pressure 1700–2300 psi.

2. The hydrogen donor solvent is a distillate cut that is separated from the liquefaction reactor product and is catalytically hydrogenated to restore its donor capability. The recycle solvent also serves as the slurry medium for the fresh coal, which is then pumped through a preheat furnace to the primary liquefaction system. Although the hydrogenation of the solvent is specially controlled, it is accomplished by conventional petroleum industry equipment that operates at moderate conditions of temperature and pressure. The use of the specially hydrogenated solvent results in higher yields and more stable plant operation than are obtained with unhydrogenated solvent.

3. The total product containing solids is separated by distillation steps into gaseous fuels, naphtha, distillates, and a solids-containing vacuum bottoms.

4. The vacuum bottoms from the distillation section is coked in a fluid coking plant to produce additional liquids.

5. The coke is gasified in a processing step closely integrated with the fluid coking step to produce fuel gas. The combined fluid coking–gasification operation is an application of Exxon's Flexicoking process. Alternatively, hydrogen is produced by steam re-forming of the light gaseous hydrocarbons produced in the liquefaction system or by partial oxidation of a portion of the vacuum bottoms.

The EDS process has been structured to accomplish the following objectives:

1. Maximization of the use of existing petroleum industry process knowledge and experience.

2. Decoupling of the coal dissolution and solvent rehydrogenation steps, thereby minimizing problems and segregating them into discrete sets.

3. Avoiding contact of catalyst with coal minerals and very heavy coal liquids, thereby obtaining long catalyst life.

4. Avoiding the use of mechanical solids-liquid separation methods (such as filters or centrifuges) by using atmospheric and vacuum distillation to fractionate the product.

5. Allowing the process to operate in a "balanced" way that is self-sufficient in hydrogen and in the fuel required for process heat. The products are all distillates; and ammonia, elemental sulfur, and ash are essentially the only by-products.

6. Being able to operate the process to provide flexibility with respect to the relative yields of naphtha or fuels useful for power generation (i.e., turbine fuels and low-sulfur fuel oil).

A flowsheet of the process is shown in Figure 28.11. The process has been operated on Midwestern bituminous coal, Western subbituminous coal, and lignites. On Illinois No. 6 bituminous coal, close to 3 bbl of distillates (C_4–1050°F) is produced per ton of dry coal.

Depending on the disposition of the final products, they can be upgraded to produce specification grade fuels. Table 28.11 shows typical inspections of raw and hydrotreated products from the EDS process produced by operations on Illinois bituminous coal.

The EDS process was chosen by Exxon as the preferred process sequence after about 8 years of exploratory, predevelopment, and evaluation work on a wide variety of possible processes. Table 28.12 shows the range of process conditions that were studied during this period (1966–1973).

Development Status and Plans. Exxon has defined and initiated a 6.5-year (1976-mid 1982), fully integrated liquefaction development program designed to bring the EDS process to commercial readiness. The program incorporates bench-scale research, small pilot-plant operations, engineering studies, and design, construction, and 2.5 years operation of a 250-ton/day pilot plant.

The 250-ton/day large pilot plant is the minimum size needed for confident scaleup of the critical process and mechanical features of the EDS process. A basic design specification for this plant was completed by Exxon in late 1975. Construction was scheduled to start by mid-1978 (after completion of detailed engineering design), to meet the development plan schedule. The 2.5-year operating period will include demonstration of the process, testing of process improvements, and operation on different coals. The total development program also includes re-

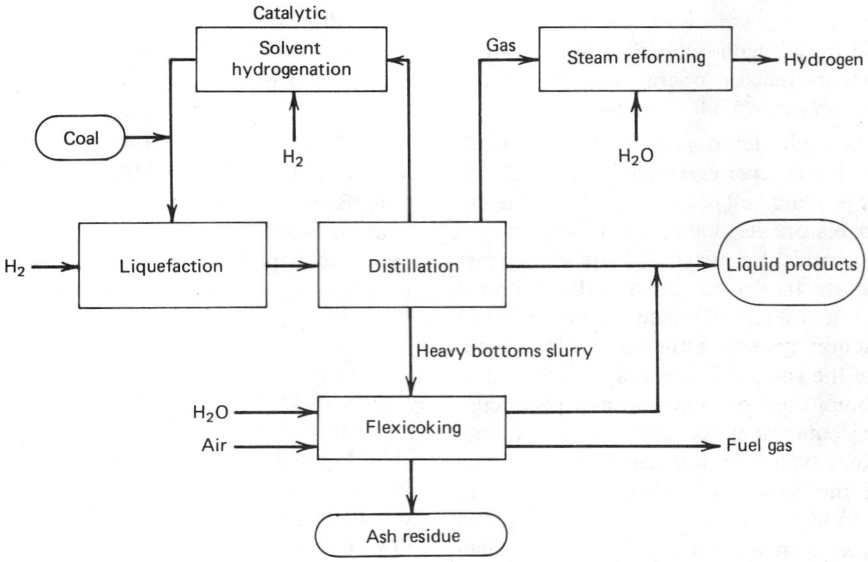

Figure 28.11 Flowsheet for the Exxon donor solvent process.

search and development work on the production of fuel gas and hydrogen from the liquefaction residue.

The combined liquefaction and fuel gas–hydrogen program is estimated to cost $240 million for the period 1976 through mid-1982. The program goal is to develop a flexible coal liquefaction process having good operability and reliability. The results will provide the basis for designing full-scale, 30,000–60,000-bbl/day commercial plants with normal technical risk.

[64] H. M. Siegel, Exxon Research and Engineering Company, B. T. Fant, Baytown, Tex., private communication, April 13, 1977.

The cost of the longer term development program will be shared by Exxon, DOE, EPRI, and Phillips Petroleum Company, Atlantic Richfield Corporation, and the Japan Coal Liquefaction Development Company.

The total development program encompasses five phases, as shown in Fig. 28.12.

Work done in Phases I and II was funded by Exxon alone. Phase I, predevelopment work, extended from 1966 through 1973 and led to the selection of the EDS process configuration. Phase II—research, planning, and design—covered 1974 and 1975 and concluded with the basic process design and definitive cost estimate of the 250-ton/day large pilot plant.

Table 28.11 Typical Inspections on Coal and Products from the Exxon Donor Solvent Process[64]

	Illinois Bituminous Coal	Heavy Naphtha		Fuel Oil, 200°C+	
		Raw Liquid	Hydrotreated Liquid	Raw Liquid	Hydrotreated Liquid
Density (g/cm³)		0.87	0.80	1.08	1.01
Elemental analysis (wt %, mf)					
Carbon	69.7	85.6	86.8	89.4	90.8
Hydrogen	5.1	10.9	12.9	7.7	8.6
Oxygen	9.5	2.8	0.2	1.8	0.3
Nitrogen	1.8	0.2	0.05	0.6	0.25
Sulfur	4.2	0.5	0.005	0.5	0.05
Ash	9.7				
Paraffins (wt %)			10		3
Naphthenes + aromatics (wt %)			90		97
Nominal boiling range (°C)		70/200	70/200	200/540	200/540
HHV (Btu/lb)		18,300	19,300	17,100	18,100

Table 28.12 Range of Process Conditions Studied, 1966–1973

Coal type:	Bituminous, subbituminous
Reactor configuration:	Stirred tank, tubular plug flow, tubular with internal and external recirculation, ebullating bed (catalytic)
Recycle solvent:	Hydrogenated, unhydrogenated
Solids separation:	Distillation, centrifugation, filtration, hydroclones, coking
Liquefaction	
Temperature	370–480°C
Pressure	20–170 atm
Residence times	15–240 min
Solvent/coal ratio	1.2–4.0
Hydrogen rates	0–17 wt % (mf coal)
Solvent Hydrogenation	
Temperature	260–450°C
Pressure	80–210 atm
Space velocity	0.2–4.0 lb oil/(hr)(lb catalyst)
Catalyst sources	Commercial, proprietary

Phase III is the detailed engineering of the large pilot plant, Phase IV is construction, and Phase V is the operation of this plant. The integrated laboratory and engineering research and development program spans all the phases.

The integrated laboratory and engineering research and development effort is concentrated in the following major areas:

1. Development of process correlations for predicting product yields and qualities from

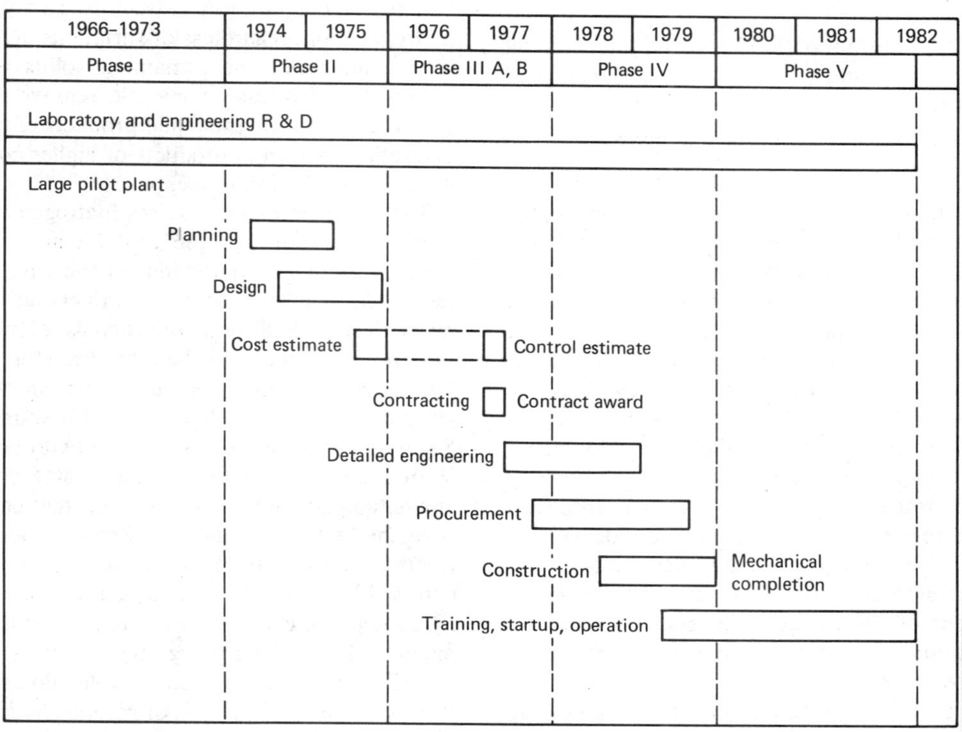

Figure 28.12 Development schedule for the EDS coal liquefaction processes.

operating conditions and coal properties using the 1.0-ton/day integrated liquefaction pilot plant, two smaller integrated pilot plants, and bench-scale equipment.

2. Evaluation of improvements to reduce plant and product cost.

3. Selection of preferred operating conditions and internal process configuration for Flexicoking of vacuum bottoms and gasifying the coke to produce fuel gas or hydrogen. This work will utilize several coking pilot plants ranging up to 2 bbl/day in feed capacity.

4. Testing and evaluation of product quality.

5. Engineering design and cost studies for full-scale plants to establish preferred configurations of process flow and equipment sequencing and to firm up estimates of plant and product cost.

6. Engineering testing and evaluation of critical equipment, development of equipment scaleup criteria, testing of construction materials, and measurement of physical and thermodynamic data on process fluids. The critical equipment includes the slurry preheat furnaces, the liquefaction reactors, the vacuum distillation tower, and other key facilities.

7. Solution of process and mechanical problems encountered with the large pilot plant.

8. Analytical laboratory support.

The laboratory and pilot-plant work on the project is being carried out at Baytown, Texas, and the engineering evaluations and engineering technology programs are being done at Florham Park, New Jersey. The site for the 250-ton/day large pilot plant is adjacent to Exxon's Baytown refinery. This site provides a number of cost, logistical, and timing advantages for the large pilot plant.

Continuous operations in the 1.0-ton/day integrated pilot plant have shown reproducible results and provide high confidence that the 250-ton/day pilot plant can be built and operated successfully to complete development of the scaleup and engineering information needed for designing commercial-sized plants.

A large staff of research and engineering specialists throughout the Exxon organization is engaged in the overall EDS development program, and the program appears to have a high probability for technical success.

28.3.3.2 *The Consol Synthetic Fuels (CSF) Process*

Introduction. The Consol synthetic fuels (CSF) process was conceived by Consolidation Coal Company. From 1963 to 1972 the development effort was sponsored by the OCR. A flowsheet of the process is given in Fig. 28.13.

Process Description. The CSF process produces synthetic crude (or fuel oil) from coal. The distinguishing feature of the process is that it is carried out in two stages: an extraction stage wherein coal is partially converted to extract and ash and nonliquefied coal are removed from that extract, followed by a second stage that involves catalytic hydrogenation of the extract. Hydrogen for the process is generated from gasification of char derived from carbonization of the unextracted residue from the first stage.

The fundamental objective for selecting the multistage flowsheet configuration is the desire to feed a uniform extract to catalytic hydrogenation, regardless of variations in the feed coal. Since the remaining solids are larger, it is also less complex to remove the solids and ash from the liquid products of the extraction than from products of higher conversion liquefaction processes.

The extraction step requires hydrogen addition. The nature of the coal dictates the conditions for the extraction. In the case of the Pittsburgh Seam coal, and in general for many Eastern U.S. bituminous coals, extraction may be accomplished by transfer of hydrogen from the vehicle solvent in the absence of gaseous hydrogen. Pittsburgh Seam coal requires relatively little hydrogen. With about 0.6% hydrogen added, 80% coal conversion is realized. The solvent that contains the hydrogen donor is taken as a sidestream from the extract hydrogenation. Western subbituminous coals require more hydrogen than can readily be supplied in this manner. It has been suggested[66] that as in the SRC process, the proper application for these coals is the inclusion of gaseous hydrogen in the extractor.

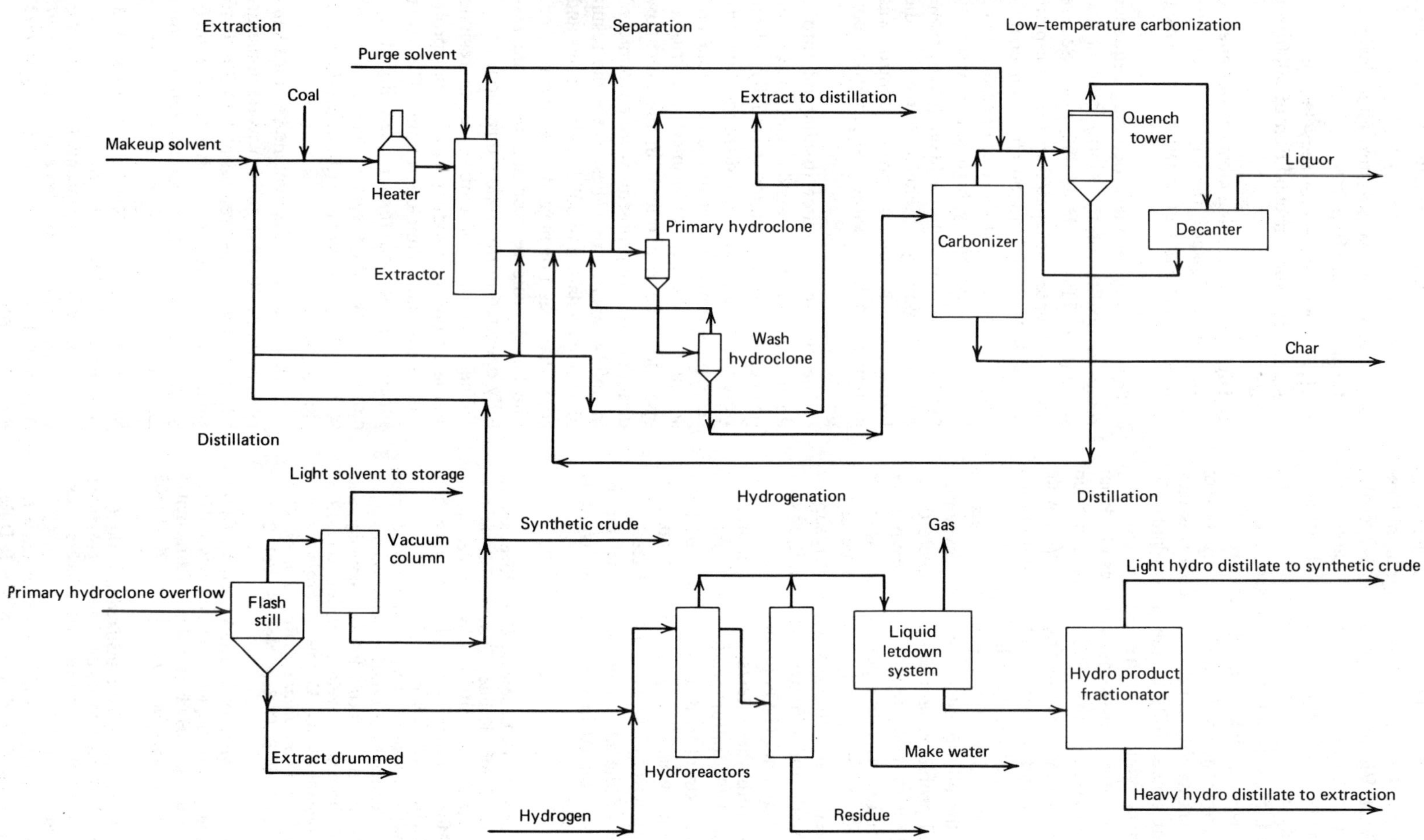

Figure 28.13 Flowsheet for the CSF process.[65]

Pilot-Plant Experience. Most of the work done in the OCR-sponsored program at Cresap, West Virginia, at a 20-ton/day scale, utilized Pittsburgh Seam No. 8 coal. All sections of the integrated plant did not successfully operate, and the project was terminated. Most problems were associated primarily with a lack of mechanical reliability of key components. Continuous extraction of highly caking Pittsburgh Seam coal was carried out successfully at the pilot plant. Kinetic data, yield structure, and process operability were consistent with results from the bench-scale units. These data had been previously developed during tests on a wide variety of bituminous and subbituminous coals as part of the overall program.[65–68]

Hydrocracking of the extract was also a part of the pilot-plant scheme. However mechanical problems precluded complete demonstration of this step at the Cresap pilot plant, and satisfactory performance was not achieved. Subsequently HRI successfully converted coal extract made at Cresap in bench-scale units under the conditions of their commercial H-Oil process.

Data from these tests show that a 90% yield of 0.1% sulfur product was obtained from the extract feed, which contained 1.8% sulfur that was all organic. Hydrogen consumption was 5–7 wt % of extract feed.

Some of the key nonconventional equipment items tested in the pilot plant were as follows[69–71]:

1. A staged, stirred extractor vessel designed for 800°F and 150 psig.
2. A rotary precoat pressure filter designed for 650°F and 150 psig.
3. Hydrogenation reactor recycle pumps designed for suction condition of 5000 psig and 850°F.
4. High-pressure piping closures designed to remain leakproof through temperature cycles from ambient to 850°F at 5000 psig.

The first step in the processing sequence required heating of the slurry of coal and solvent to 715–775°F at 400 psig. Severe problems were anticipated in the pilot-plant preheater because Pittsburgh Seam coal is very highly caking at these temperatures and because an extremely high slurry viscosity develops as coal solution takes place. However no coking problems occurred in the furnace coil as long as proper control was maintained over pressure and tube wall temperature.

The furnace product discharged into a downflow, agitated, staged extractor. Coking problems were encountered initially. These were resolved by adding a disproportionately large horsepower mixer in the feed zone and scraping at the gas-liquid interface. Prior to operation of the extractor in the downflow direction, severe problems were encountered in upflow operations. These were the result of an accumulation of solid agglomerates and coke that was not purged from the reactor. The bench-scale reactor had been successfully operated in the upflow direction.

The separation of carbonaceous residue and ash from the product was a major problem in the pilot-plant operations. The most successful method tested used liquid cyclones. A two-stage system was capable of 90% extract recovery with a product ash below 1 wt %.

Carbonization of the cyclone bottoms slurry containing 40 wt % solids was demonstrated. The slurry was treated at 800–900°F to simultaneously recover solvent, essentially by distillation, and to produce additional dis-

[65] E. Gorin, H. E. Lebowitz, C. H. Rice, and R. T. Struck, "Synthetic Fuels Process," Paper No. 5, Panel Discussion 10, 8th World Petroleum Cong., Moscow, June 1971; Applied Science Publishers Ltd., Essex, England.

[66] *Technical Summary Clean Fuels—West,* prepared by Conoco Coal Development Company for EPRI, Southern California Edison Company, and Mobil Research and Development Company, 1976.

[67] Consolidation Coal Company, *Project Gasoline Pre-Pilot Plant Phase I Research on CSF Process,* OCR R & D Rep. No. 39, Vol. 2, PE-234126/AS.

[68] H. E. Lebowitz, D. J. Kulik, and E. Gorin, "Deashing of Coal Liquefaction Products via Partial Deasphalting. I. Hydrogen Donor Extraction," *Prepr., Am. Chem. Soc., Div. Fuels,* **20**(1), 105 (1975).

[69] J. A. Phinney, "Coal Liquefaction at the Cresap, West Virginia Pilot Plant," paper presented to American Institute of Chemical Engineers, Pittsburgh, June 2–5, 1974.

[70] Consolidation Coal Company, *Pilot Scale Development of the CSF Process,* OCR R & D Rep. No. 39, Interim Rep. No. 6, Vol. 4, Book 3, November 1971; PB-234 131/AS.

[71] Consolidation Coal Company, *Development of CSF Coal Liquefaction Process—Final Report,* OCR R & D Rep. No. 39, Vol. 5, August 1973; PB-234 132/AS.

tillates by coking the combined residue and extract. The carbonizer was fluidized by a mixture of air and recycle gas; partial combustion of char provided sufficient heat to keep the system in balance. Hydroclone underflow was injected at the upper part of the fluidized bed, or alternatively into a lift-leg that forced a high circulation of char past the injection point.

Hydrogenation of the extract (residual portion of hydroclone overflow) was attempted in an ebullated-bed reactor system at 3500 psig and 800°F. Repetitive mechanical problems included the following:

1. Failure of canned reactor liquid recycle pumps.
2. Leakage of reactor contents through piping closures.
3. Failure of packing on reciprocating pumps feeding hot extract to the reactors.

The Cresap pilot plant is being reactivated and substantially revised to function as an equipment test center. This is being done by Fluor Corporation for ERDA.

28.3.4 Miscellaneous Processes

28.3.4.1 *The Zinc Chloride–Catalyzed Coal Liquefaction Process*

Introduction. The use of metal halides for coal hydrogenation has been previously reported in the literature.[72] This work involved relatively small amounts (5%) of the salt impregnated on the coal. Halogen-promoted tin catalysts were reported to be most active.

The use of molten $ZnCl_2$ as a catalyst for coal liquefaction was developed by Consolidation Coal Company in the process of searching for a more active hydrocracking and hydrofining catalyst.[73] The chlorides tested were $ZnCl_2$, $ZnBr_2$, $AlBr_3$, and $SnCl_2$ (1 : 1 catalyst/feedstock or greater by weight).[73–76] The first three were superior to $SnCl_2$ for hydrocracking pyrene, coal extracts as obtained, for example, from the CSF process, and unconverted residue from extract hydrocracking.

Process Description. The molten $ZnCl_2$-catalyzed system is the basis of a unique hydrocracking process. Coal extract of the type produced by the CSF process or solvent refining, as well as other high molecular weight hydrocarbons such as petroleum residues, tar sands, oil, or shale oil, can be used as feedstock. Hydrocarbon, hydrogen, and molten salt are fed separately to a reactor at 625–850°F and pressures of up to 5000 psig. The hydrocarbon reacts rapidly in the molten pool, evolving product vapors that are recovered overhead. Unconverted feed, heavy oil, ash, and "spent" catalyst flow from the reactor as a suspension. A hydrogen-rich gas is separated from the overhead vapors and recycled; light hydrocarbon product is recovered. The gas is free of H_2S and NH_3, which are retained in the molten salt phase as ZnS and $ZnCl_2 \cdot xNH_3$.

The salt is regenerated by feeding the spent melt into a fluidized sand bed at 1700–1800°F, wherein $ZnCl_2$ is vaporized, organics burned to CO_2 and H_2O, ZnS converted to $ZnCl_2$ and SO_2 (proper HCl partial pressure must be maintained), and $ZnCl_2 \cdot xNH_3$ is converted to N_2 and H_2O. Ash is separated as a solid from the gas before the $ZnCl_2$ is condensed. The $ZnCl_2$ is then recovered and the remaining gases treated to recover HCl and SO_2 before discharge.

A flow diagram of the process is given in Fig. 28.14.

An alternative regeneration is also possible in which oxygen amounting to 45% of the stoichiometric requirement is fed to the com-

[72] H. H. Storch, "Hydrogenation of Coal and Tar," in *Chemistry of Coal Utilization*, Vol. 2, H. H. Lowry, Ed., John Wiley and Sons, New York, 1945, pp. 1750–1796.

[73] Consolidation Coal Company, *Research on Zinc Chloride Catalyst for Converting Coal to Gasoline—Phase I—Hydrocracking of Coal and Extract with Zinc Chloride*, OCR R & D Rep. No. 39 under Contract No. 14-01-0001-310; Interim Rep. No. 2, Vol. 3, Book 1, 1968.

[74] Consolidation Coal Company, *Pre-Pilot Plant Research on the CSF Process—Phase I—Regeneration of Zinc Chloride Catalyst*, OCR R & D Rep. No. 39, Interim Rep. No. 2, Vol. 3, Book 2, 1968.

[75] Consolidation Coal Company, *Summary Report on Project Gasoline*, OCR R & D Rep. No. Interim Rep. No. 5, 1969.

[76] C. W. Zielke, R. T. Struck, J. M. Evans, C. P. Costanza, and E. Gorin, "Molten Salt Catalysts for Hydrocracking of Polynuclear Hydrocarbons," *Ind. Eng. Chem. Process Des. Dev.*, **5**, 151–157 (1966).

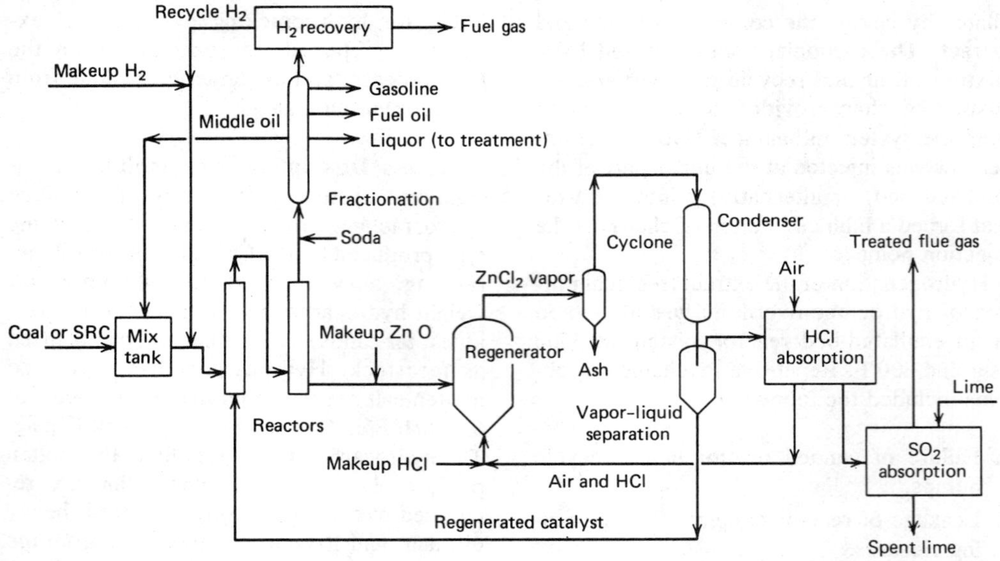

Figure 28.14 Flowsheet for the molten zinc chloride salt process for coal liquefaction.[77]

bustor. In this case fuel gas is a product (in place of steam from heat recovery in the total combustion option), and ZnS remains with the ash for later recovery.

The $ZnCl_2$ catalyst provides the following desirable characteristics: very rapid reaction rates, high yield of gasoline-range naphtha, low yield of hydrocarbon gases, and a high ratio of iso- to normal paraffins resulting in about 90 RON gasoline without reforming (properties of gasoline are given in Table 28.13).

Status. The process was studied during 1963–1968 under sponsorship of OCR. A 2-lb/hr bench unit was operated on coal extract from the CSF process.

Table 28.13 Typical Properties of Gasoline Produced by the Molten Zinc Chloride Process[77]

Nitrogen	100 ppm
Oxygen	2000 ppm
Sulfur	200 ppm
PONA	
Aromatics	24.5
Olefins	—
Naphthenes	34.2
Paraffins	41.3
i C_4/n C_4	9.0
Yields, single pass	
(C_4–200°C)/(C_4–400°C), wt ratio	0.85
Research octane number	90[a,b]

[a] Primary gasoline C_5–180°C.
[b] C_5–200°C.

Considerable effort was devoted to catalyst regeneration. In general, the spent melt contains ZnS, ZnO, $ZnCl_2 \cdot xNH_3$, $ZnCl_2 \cdot yNH_4Cl$, ash, and unreacted organic material. These are present as a single, well-dispersed phase that has no tendency to separate on standing.

Vapor phase combustion was demonstrated on a 5-lb/hr continous unit (3 in. I.D.). Conditions explored include 14–46 psia, 1800–1900°F, 0.16–2.67 ft/s superficial velocity, and 0.4–6.3 s vapor residence time. Throughputs up to 347 lb/hr-ft³ were demonstrated.

Related unit operations were studied that would (1) extract unconverted heavy oils from spent melt with middle oil, (2) scrub HCl from overhead gases using melt containing $ZnCl_2 \cdot xNH_3$ or ZnO, (3) recover NH_3 by thermal decomposition of $ZnCl_2$ complexes, and (4) liberate ZnS from melt by HCl addition.

Zielke[78] and Gorin[77] reported on batch hydrocracking of feeds including lignite, subbituminous, Illinois No. 6, and Pittsburgh

[77] R. T. Struck, C. W. Zielke, and E. Gorin, *Zinc Chloride Coal Liquefaction Process*, Report for U.S. ERDA under Contract No. E(49-18)-1743, Conoco Coal Development Company, Library, Pa., November 1, 1976.

[78] C. W. Zielke, W. A. Rosenhoover, and E. Gorin, *Prepr., Am. Chem. Soc., Div. Fuel Chem.*, **19**(2), 306 (1974), in Am. Chem. Soc. Adv. Chem. Ser. No. 151, 1976, p. 153.

No. 8 coals. Interestingly, each coal was found to be as reactive as its particular extract. Western coals responded nearly as well as Eastern bituminous coals. Yields for two feeds are given in Table 28.14.

An added concern when utilizing coal feed, as opposed to substantially ash-free extract, was the reactions of mineral matter constituents with molten $ZnCl_2$ during regeneration. This would lead to consumption of the catalyst and a buildup of impurites in the circulating melt. Work on reactions occurring during $ZnCl_2$ regeneration was carried out in a 5-lb/hr continous unit under conditions of complete combustion. Four principal conclusions were drawn:

1. Hydrolysis could be prevented by maintaining 5% HCl in the regenerator vapor.
2. Sodium and potassium in the feed coal will react to form their chlorides.
3. Calcium and magnesium do not react, probably because silicates form in preference to chlorides.
4. Some ferrous chloride forms, but not to the extent indicated by equilibrium.

The reactions of sodium and potassium probably contraindicate the use of lignites as a feedstock. The buildup of alkali chlorides would consume chlorine, requiring an appreciable catalyst purge. The iron reactions are probably not critical, since under the most adverse conditions chemical equilibrium would be established in the circulating melt.

The development of the $ZnCl_2$-catalyzed process was resumed in 1975 under joint

Table 28.14 Continuous Zinc Chloride Liquefaction of Coal and Coal Extract

	Coal Extract from hvAb Coal, Pittsburgh No. 8 Seam	Subbituminous Coal, Rosebud Seam
Percent of maf coal in feed:	71.3[a]	100
	Reaction Conditions	
Temperature (°C)	413	413
Hydrogen partial pressure (psig)	2730	3000
$ZnCl_2$/mf feed, wt ratio	1.05	1.52
Solvent/mf feed,[b] wt ratio	0.07	2.33
Average liquid residence time (hr)	1.8	1.6
Hydrogen feed rate (scf/lb organics)	33	32
	Yields (wt % of 475°C+ feed)	
Methane	0.8	0.8
Ethane	2.1	1.1
Propane	12.0	5.0
Isobutane	18.2	6.9
Normal butane	2.5	0.7
$CO + CO_2$	0.0	2.5
C_2O	3.1	13.4
C_5–200°C naphtha	45.4	42.0
200–475°C distillate	7.2	11.2
+475°C	2.7	6.6
MEK-soluble residue[c]	5.5	8.2
MEK-insoluble residue[c]	3.3	7.2
N, O, S, Cl, to catalyst	6.4	2.8
	109.3	108.4
Conversion to −475°C (wt %, maf of feed)[b]	88.5	78.0
H_2 consumed (wt %, maf of feed[b]	9.3	8.4

[a] Extract produced with tetralin solvent at 750°F, with a nominal residence time of 17 min.
[b] "Feed" refers to material boiling above 475°C.
[c] MEK = methyl ethyl ketone.

sponsorship of Conoco Coal Development Company, Shell Development Company, and ERDA. Direct continuous hydrocracking of bituminous and subbituminous coal on the 2-lb/hr unit has been demonstrated (at a catalyst to coal weight ratio of 1 : 1) as has regeneration of the spent melt produced.[79,80] It has been shown that the potential overall zinc loss is negligible. If $ZnCl_2$ is recovered from the ash by secondary processing, 99.5–99.9% overall recovery is feasible on each cycle. Without secondary processing the recovery is 98.5–99.5%.

A 100-lb/hr PDU is being constructed to demonstrate the integrated operation and mechanical details of this process. The major objectives include the following:

1. Establish integrated hydrocracking and regeneration yields at equilibrium to determine steady state yield structure, the steady state composition of the circulating catalyst stream (i.e., ash contamination), the effect of such yield structure and composition on hydrocracking activity, and recovery of zinc and chlorine with an integrated system.

2. Examine the operability and performance of a gas-sparged bubble column reactor as compared to a mechanically agitated reactor used in bench-scale testing.

3. Define construction material for the hydrogenation section.

4. Define materials and mechanical design for the regenerator. This would include determination of the appropriate construction materials for bed grid, nozzles, and probes, definition of acceptable refractory materials, and determination of materials requirements for the zinc chloride condenser.

5. Demonstrate efficient recovery of HCl.

6. Demonstrate liquid-liquid extraction of heavy oil from spent melt using indigenous middle oil.

[79] Conoco Coal Development Company, *Zinc. Halide Hydrocracking Process for Distillate Fuels from Coal,* Ann. Tech. Prog. Rep. No. FE-1743-16, for 1975, U.S. ERDA, February 15, 1976.

[80] Conoco Coal Development Company, *Zinc. Halide Hydrocracking Process for Distillate Fuels from Coal'' Quarterly Tech. Prog. Rep. No. FE 1743-24 for May 1–July 1, 1976, U.S. ERDA, August 15, 1976.*

28.3.4.2 *The Kerr-McGee Coal Liquefaction Process*

Introduction. The work done by Kerr-McGee Corporation in the field of coal conversion has been concentrated on separating solids from various coal liquefaction products.[81] The approach involves the treatment of liquefaction products with certain ''deashing'' solvents near the critical conditions of the solvents.

In January 1970, Kerr-McGee Corporation filed patent applications (now U.S. 3,607,716 and 3,607,717) covering the treatment of coal liquefaction products with certain light organic solvents under elevated temperature and pressure conditions to separate the liquefaction products into a plurality of fractions. That work was an extension of experimental and design technology Kerr-McGee had developed for the pentane deasphalting of petroleum residua. The pentane deasphalting process known as the ROSE® (residuum oil supercritical extraction) process was operated for 6 years in a 750-bbl/day pilot plant.

From 1972 to 1977 Kerr-McGee's experimental work was directed toward developing techniques for separating lower molecular weight organic components from higher molecular weight materials such as asphaltenes, undissolved coal, and mineral matter, using solvents near their critical conditions. Feed materials studied included petroleum residues, tall oil pitch, and coal liquefaction products.

To ensure the availability of an adequate supply of coal liquefaction products to permit the treatment of the products with a number of solvents under conditions in which the solvent was in the critical range, a coal liquefaction program was undertaken. The liquefaction of Illinois No. 6 bituminous and Wyodak subbituminous coal were studied at various temperatures and pressures, using both donor solvents and anthracene oils. Both carbon monoxide and hydrogen were used as the overpressure gases.

Initially batch autoclave tests were used to develop and confirm the concept of coal deashing and fractionation via the treatment

[81] T. W. Clapper and R. A. Baldwin, Kerr McGee Corporation, private communication to N. C. Stewart, EPRI, December 9, 1976.

of the liquefaction products with the solvents under their critical conditions. Subsequently continuous bench-scale units were constructed and used to study the effect of variations in temperature, pressure, and solvent-solute relationships. The first unit was a 600-g/hr critical solvent separation unit. Following successful operation of that unit, a 15-lb/hr unit was built and has been operated successfully for 3 years. A 2.5-ton/day pilot plant began operation in June 1977.

Process Description. Figure 28.15 is a conceptual flowsheet illustrating a continuous process for this method of deashing liquefied coal. In the deashing step ash-containing still bottoms from the liquefaction process are mixed with the deashing solvent. The ratio of deashing solvent to soluble coal is 2 or more to 1. The mixture is heated to a temperature greater than 400°F. On advancing to the deashing-fractionation vessel, the mixture's temperature is increased until the density of the deashing solvent is less than about 0.55 g/ml.

The temperature and pressure conditions selected for operation of the deashing vessel are such that the density of the deashing solvent is low enough to cause the precipitation of a small amount of the heavier coal liquefaction products. A heavy ash-containing phase settles rapidly to the bottom of the deashing vessel. In comparison to published settling rates[82] for somewhat related "solvent precipitation–mineral matter agglomeration" processes, the ash-settling rate in the Kerr-McGee process is substantially faster.

The heavy phase contains mineral ash and insoluble coal fluxed with solvent and a small amount of the soluble coal liquefaction products. The viscosity of this phase is relatively low; therefore it may be withdrawn readily at a controlled rate into a recovery chamber.

By varying the conditions in the deashing vessel, it is possible to vary the yield and ash content of the ash concentrate. If the temperature is increased, the total amount of ash concentrate increases because additional soluble coal is rejected (into the ash concentrate) with a consequent reduction in the percentage of mineral matter. Thus the properties of the ash concentrate can be con-

[82] E. Gorin, C. J. Kulik, and H. E. Lebowitz, *Prepr., Am. Chem. Soc., Div. Fuel Chem.*, **20**(1), 105 (1975).

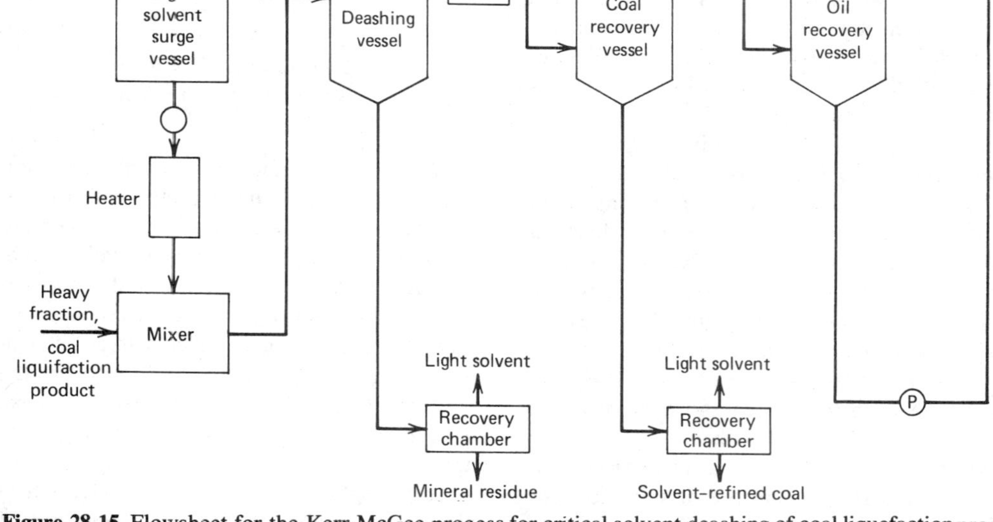

Figure 28.15 Flowsheet for the Kerr-McGee process for critical solvent deashing of coal liquefaction products.

trolled to provide a feed material to a hydrogen generation unit.

The ash concentrate is removed from the recovery chamber as a dry, easy-to-handle powder. This material appears to be a suitable feed for a gasification process. It may contain 20 to more than 40 wt % ash (oxidized), depending on the deashing conditions and the type of material being deashed. Any deashing solvent accompanying the discharged ash concentrate is flashed and recycled for further deashing service.

The solution of soluble coal and deashing solvent formed in the upper part of the deashing vessel is withdrawn, but at a controlled rate that provides sufficient residence time in the vessel to assure adequate settling of the insoluble materials. The settling occurs very rapidly, generally in less than one minute, because the solution has a very low viscosity. Furthermore, the solution is substantially free of insoluble material and does not require filtering or centrifuging to remove the last traces of solids.

The solution leaving the top of the deashing vessel is passed through a heater where it is heated to a substantially higher temperature before being introduced into the coal recovery vessel. The deashing solvent density existing in the coal recovery vessel is substantially less than that existing in the deashing vessel, because the operating temperature is higher and the pressure is slightly lower as a result of the drop in line and heater pressure. This change in the solvent density is large enough to cause a fluidlike phase of heavy deashed coal products to separate from a solvent-rich lighter phase. The heavy phase is withdrawn to a chamber where substantially all the contained deashing solvent is flashed and recycled for further deashing service. The residue remaining after flashing the solvent is a fluid mass of deashed coal, which is easily removed from the chamber.

If desired, there can be several deashed coal recovery subsections, each operating at a temperature somewhat above that in the previous subsection, to permit the recovery of a multiplicity of fractions of differing physical properties. All fractions contain less than 0.16 wt % ash (oxidized) and generally less than 0.1 wt % ash (oxidized). The sulfur content is dependent on that of the raw coal fed to the dissolver and the dissolution conditions.

The solvent-rich light phase contains the deashing solvent and some coal-derived oils. This phase is introduced into an oil recovery vessel. By changing the temperature or pressure or both in that vessel, the oils are recovered and the deashing solvent is recycled for further deashing service. The recovered oils are recycled to liquefaction to dissolve additional coal.

By varying the solvent, the temperature, the retention time, and the solvent-to-feed ratio, a wide spectrum of feed materials can be processed to yield a variety of products. The process appears to be useful for separating ash and undissolved coal from coal liquefaction products that have been prepared by liquefaction processes in which only minimum quantities of hydrogen are utilized.

Status. The data in Table 28.15 are based on continuous bench-scale unit operations on vacuum still bottoms provided by the Wilsonville SRC pilot plant. Those still bottoms were produced in the pilot plant from a mixture of Kentucky No. 9 and No. 14 coals.

Both aliphatic and aromatic solvents operating near their critical conditions have been used as the deashing solvent. For the treatment of coal liquefaction products, the aromatic solvents (i.e., benzene, toluene, xylenes, cresols, and pyridines) seem to provide the maximum deashing power.

Ash-containing tower bottoms obtained from noncatalytic conversion of subbituminous, Illinois No. 6, Kentucky, and Pittsburg No. 8 coals and from catalytically converted Illinois No. 6 coal have been deashed successfully with the Kerr-McGee process.

28.3.4.3 *Coal Liquefaction Research Activity Outside the United States*

Limited coal liquefaction work has been initiated outside the United States with laboratory or bench-scale programs at government laboratories and universities in Poland, Japan, Australia, Great Britain, and South Africa. Such laboratory activities are relatively modest, since the cost of coal itself is relatively high compared to coal costs in the United States or the large resource base needed for a synthetic fuels industry does not exist.

An exception is the interest in German in direct conversion of coal by adaptation of technologies used during World War II. The

Table 28.15 Kerr-McGee Critical Solvent Deashing Process Analysis[a] of Feed and Products[81]

	Feed Ash Containing Still Bottoms (%)	Products Ash Concentrate (%)	Products Deashed Coal (%)
Analysis (wt %)			
Carbon	76.8	59.4	85.4
Hydrogen	5.4	3.3	6.3
Nitrogen	1.2	1.0	1.3
Sulfur	1.6	3.6	0.7
Ash	10.8	32.7	0.08
Oxygen (by difference)	4.2	—	6.22
	100.0	100.0	100.0
Heat content[b]	~14,300 Btu/lb	10,800 Btu/lb	~15,900 Btu/lb

Physical Properties of Ash Concentrate Powder	
Bulk density	31 lb/ft³
Particle density	97 lb/ft³
Angle of repose[c]	40–45°

Size Distribution[d] (Ash Concentrate)

Tyler Screen Mesh Size	Percent Weight
+ 28	9.0
−28 + 48	13.6
−48 + 65	8.2
−65 + 100	8.4
−100 + 150	8.5
−150 + 200	9.6
−200 + 270	7.3
−270 + 400	8.1
−400	27.3
	100.0

[a] Data represent bench-scale deashing results for a feed obtained from the Wilsonville SRC pilot plant and processed under one set of operating conditions. Product yields and properties may vary considerably depending on the source of the feed and the operating conditions employed in the process.

[b] Calculated by Dulong's formula.

[c] Material is readily fluidized.

[d] Ash was uniform in concentration throughout the various size fractions.

desired products are fuels and aromatic chemicals.

The IG-Nue process uses metal salts impregnated or precipitated on coal (e.g., iron sulfate), which serves as a "throwaway" catalyst. During the World War II German liquefaction processes operated at a pressure of about 700 atm, whereas the IG-Nue process operates at 300 atm. Few details are available on the processing scheme. Research and development work was performed by Bergbau-Forschung. A scaleup of laboratory work is planned with support by the Northern Westphalia government. The planned unit would be at the 100–200 ton/day scale and would be carried out by Ruhrkohle Chemie at a site in Germany that has not yet been selected. Designs and cost estimates have been prepared and the total project cost is estimated at about $100 million (U.S.) including 2 years of operation.[83] The

[83] Private communication with Dip. Ing. Christian Circkle, Dip. Ing., Ministerrat, Ministerium für Wirtschaft, Düsseldorf, Germany, and Seymour B. Alpert, October 3, 1977.

cost relative to comparable projects at comparable scale in the United States seems low.

28.4 GENERIC PROBLEMS AND SUBPROCESSES

The majority of the processes discussed in the previous section consist of different combinations of a limited number of unit processes and unit operations. Alternate choices have been made by process developers because of differing perceptions of the value of process reliability, process operability, economics, and product objectives. Each group had to deal with a number of generic problems in developing its own particular process, as discussed in detail in the sections that follow:

1. Catalyst deactivation.
2. Characteristics of solid residues.
3. Solids separation approaches.
4. Process engineering.
5. Hydrogen generation.
6. Mechanical equipment

28.4.1 Catalyst Life Considerations

Although the detailed mechanism of heterogeneous catalytic coal liquefaction is not precisely known, the chemical reactions taking place involve the solubilization and hydrogenation of the coal molecules by a hydrogen transfer solvent. This is followed by migration of the molecules through the liquid phase to the catalyst particle, where further hydrocracking, hydrodesulfurization, hydrodenitrogenation, and so on, take place. In addition, rehydrogenation of the solvent must be simultaneously achieved. Therefore a good coal liquefaction catalyst must have the proper balance of cracking activity and hydrogenation activity. In the case of a particulate catalyst, the required activity levels are achieved by dispersing the hydrogenation components, usually mixtures of the oxides of cobalt, nickel, molybdenum, and/or tungsten, on a porous, high surface area support, often alumina or silica-alumina. The latter support is commonly employed to impart a degree of cracking activity.

The achievement of the proper activity and selectivity depends on mass transport of reactants into and products out of the pores of the catalyst. If the intrinsic, or chemical, rate is high compared to the diffusion rate, utilization of the catalyst is inefficient. Results from studies[84,85] on the effects of both pore size and particle size on liquefaction catalyst activity indicate that coal dissolution and coal-asphaltene conversion as well as desulfurization are pore-diffusion limited. Consequently particle size, particle shape, pore size, and pore size distribution—parameters that relate to both geometric and internal surface area—are important to proper functioning of the catalyst.

Further constraints are placed on catalyst physical properties by reactor type, fluid dynamics, and other mechanical considerations. Cylindrical or spherical catalyst particles of 1/16–1/8 in. diameter are employed in various direct coal liquefaction processes now under development.[86] In addition to adequate initial catalytic activity, there is a critical requirement that the catalyst age in a way that permits the maintenance of adequate performance in the environment of the coal liquefaction reactor. Catalyst particles become deactivated during coal liquefaction as they interact with both the organic and inorganic components in coal. These components interact with the catalyst in a complex manner to reduce its capability for carrying out the various required functions. To provide a conceptual framework for understanding the deactivation of coal liquefaction catalysts, it is convenient to break down the deactivation phenomenon into four general categories: coke deposition, heavy liquid deposition, metals poisoning, and sintering.

28.4.1.1 *Coke Deposition*

"Coke" refers to a hydrogen-deficient carbonaceous solid that is insoluble either in the reactor medium or in solvents such as benzene or pyridine, which can dissolve most of the liquid product. This definition distinguishes coke from heavy coal liquids termed

[84] J. Brooks et al., *Catalyst Development for Coal Liquefaction* EPRI Rep. No. AF-574, 1977.
[85] Y. K. Yen, R. E. Furlani, and S. W. Weller, *Ind. Eng. Chem., Prod. Res. Dev.*, **15**(1), 24 (1976).
[86] J. Brooks et al., *Catalyst Development for Coal Liquefaction,* Task I Report, Literature Search, EPRI Contract No. RP 408-1, July 8, 1975.

variously as preasphaltenes[87] and from asphaltenes that are solids at ambient temperature. Catalyst deactivation caused by coke deposition can result from laydown of coke on the internal surface, pore plugging that either reduces or prevents access of reactant coal liquid molecules to the internal catalytic sites, or deposition of coke on the external or geometric surface. In the latter case the coke on the external surface is not necessarily an actual poison in itself, but it can result in circumstances such as uneven flow distribution and high pressure drop in a fixed-bed reactor, or defluidization from particle agglomeration in an ebullated-bed reactor.

Kovach and Bennett[88] studied the mechanisms of catalyst deactivation during the liquefaction of high-ash West Kentucky No. 11 bituminous coal. The catalyst used was a 1/16-in. Nalcomo 471 Co-Mo/SiO_2-Al_2O_3 extrudate. Deactivation was achieved by successive contacting of the catalyst with a coal–donor solvent slurry in a high-temperature, high hydrogen pressure batch autoclave. After each autoclave run the catalyst was separated from the product. Its residual activity was then measured by determining the relative rate of hydrogenating the bicyclic aromatics (naphthalenes), contained in a new charge of solvent, to the corresponding tetralin forms.

Two types of deactivation were observed:

1 A rapid loss in activity (activity drops to 10–20% of fresh catalyst activity after processing only 20–40 lb coal per pound of catalyst) due to rapid coke deposition, which could be reversed by burning off this contaminant in air.

2 A slower deactivation due to the deposition of metals from the feed coals. This slow deactivation portion of the curve (Fig. 28.16) was a function of the amount of coal processed. After each air regeneration the catalyst rapidly deactivated. The process is plotted in Fig. 28.17, which shows the relation between relative catalyst activity and catalyst age. The rapid deactivation occurred when metals contamination was less than 0.1 wt % while carbon loadings were 15–18 wt %. Therefore the 40–50% reduction in surface area and the 45–55% reduction in pore volume accompanying the rapid deactivation are primarily due to coke deposition.

Kang and Johanson reported a similar rapid decrease in total hydrogen consumption, and an increase in the sulfur content of the fuel oil product with the quantity of Illinois No. 6 Burning Star Mine bituminous proc-

[87] H. W. Sternberg, *Science,* **188,** 49 (1975).

[88] S. M. Kovach and J. Bennett, *Prepr., Am. Chem. Soc., Div. Fuel Chem.,* **20**(1), 143 (1975); Fig. 2, p. 155; Fig. 3, p. 156; Fig. 4, p. 157.

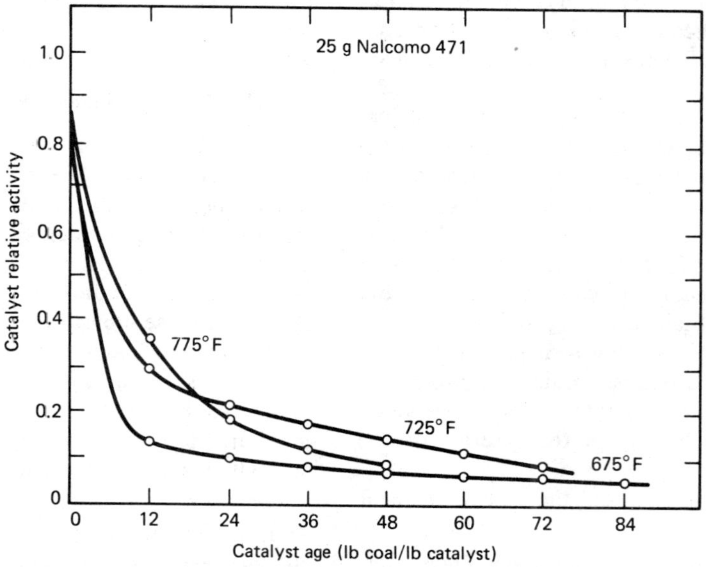

Figure 28.16 Catalyst deactivation as a function of catalyst age.

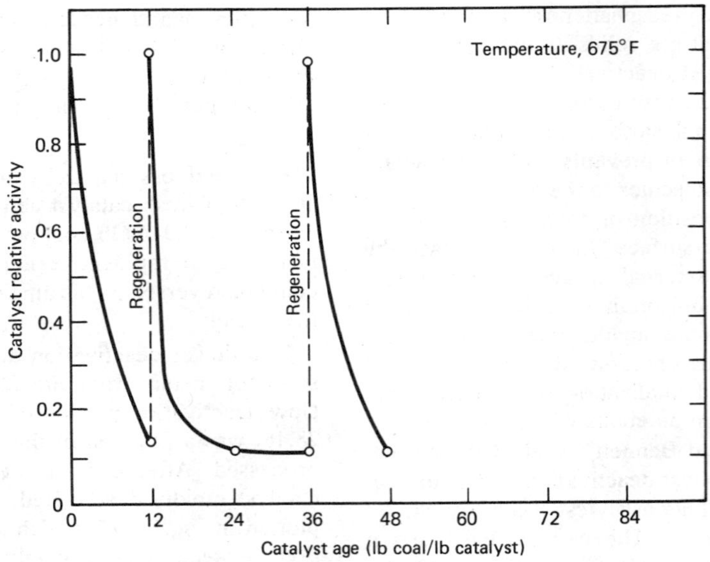

Figure 28.17 Effect of regeneration on catalyst activity.

essed.[89] This work was done in an ebullating-bed, H-Coal pilot-plant liquefaction run employing a 1/16-in. Cyanamid HDS-1442A Co-Mo/Al_2O_3 extrudate. Other data on carbon deposition onto this catalyst during the liquefaction of a Pittsburgh Seam bituminous coal indicate that little additional coke is deposited after approximately 100 lb of coal per pound of catalyst has been processed. This suggests that initial rapid deactivation is due to coke deposition. These investigators also reported that experiments carried out with spent catalyst after regeneration (air burnoff) indicate that coke deposition contributes significantly to initial catalyst deactivation.

They also report no consistent trend between the magnitude of coke deposit and the rank of feed coal, although the lower rank subbituminous coals and lignites tend toward the lower end of the observed 10–35 wt % coke ranges. Coke deposition appears to be more affected by process conditions, with high rates obtained during low-pressure operation (1000 psi hydrogen partial pressure or less). Coke deposition under these conditions appears to result from dehydrogenation of the coal liquids in the interior of the catalyst particle.

Little is known about the actual chemical mechanism of coke deposition on catalysts during coal liquefaction. Evidence suggests that components of the so-called preasphaltene fraction of coal liquid are major coke precursors during thermal degradation in the absence of a catalyst.[90] However, the rapid initial coke buildup, followed by a leveling off of coke accumulation, could be accounted for by the uncontrolled hydrocracking of polynuclear aromatic hydrocarbons with subsequent deactivation of the cracking sites on the catalyst surface rather than by the thermal cracking of the highly polar preasphaltenes.

28.4.1.2 *Heavy Liquid Deposition*

A second category of deactivation is heavy liquid deposition. These high molecular weight species are distinguished from coke in that they have some degree of mobility and solubility in the reaction medium and can react to form lower molecular weight species. However, these molecules may form coke in a hydrogen-deficient environment. Deactivation by these preasphaltenes and asphaltenes results from either a site-shielding mechanism in which the molecules are strongly adsorbed on the internal surface or

[89] C. C. Kang and E. S. Johanson, *Prepr., Am. Chem. Soc., Div. Fuel Chem.*, **21**(5), 32 (1976).

[90] D. D. Whitehurst, M. Farcasiu, and T. Mitchell, *The Nature and Origin of Asphaltenes in Processed Coals*, EPRI Rep. No. AF-252 (1976).

a pore-plugging mechanism in which their size precludes or inhibits intraparticle diffusion of either reactant and/or product molecules. Although deposition of these heavy liquids on the external catalyst surface could conceivably preclude access to the interior pores, their mobility and solubility in the reactor medium would prevent this from resulting in irreversible poisoning unless these species subsequently degrade to coke.

Although there would appear to be logical reasons to postulate catalyst deactivation due to either site shielding or pore plugging by heavy coal liquid molecules, no studies of this phenomenon have been reported. This is probably due to the difficulties of experimental measurement, either during or after the liquefaction reaction. However, it is known[84,86] that a large fraction of the pore volume in catalysts used in coal liquefaction is in pores of less than 100 Å (0.01 μm) in diameter. In contrast to petroleum asphaltenes having molecular weights in the range of 1000–25,000,[91] the molecular weights of the majority of coal asphaltenes and preasphaltenes have been shown to be in the range of 300–900,[90] small enough to penetrate much of the catalyst pore volume but large enough to encounter difficulty in diffusing in and out of pores of less than 100 Å in diameter.

28.4.1.3 *Metals Poisoning*

Metals that can affect catalyst life (e.g., Ca, Fe, Ti, Na) may either originate from metal salts, metallorganic compounds, and ion-exchangeable clay metals, or be present as mineral solids (e.g., SiO_2, FeS_x, $CaCO_3$) in the feed coal. Again deactivation may result from site shielding through adsorption of organometallic compounds or chemical reaction between the "metals" and the catalytic surface species, pore plugging by minerals formed *in situ* or present as finely divided (micrometer or submicrometer size) coal minerals, and "metals" deposition on the external or geometric surface of the catalyst particle.

Kang and Johanson[89] cite typical analyses of coal ash composition and major catalyst contaminants during typical H-Coal processing of both Wyodak subbituminous and Illinois No. 6 bituminous coal (Tables 28.16 and 28.17). The relative levels of metal contaminants on the catalyst are not in the ratio characteristic of ash composition of the coal. Bituminous coal, such as Pittsburgh Seam and Illinois No. 6, deposits appreciable amounts of titanium and low amounts of calcium and sodium; subbituminous coal such as Wyodak usually deposits higher amounts of sodium and similar amounts of calcium in comparison with these bituminous coals. These investigations report that about one-twentieth of the titanium present in Illinois No. 6 bituminous coal deposits on the catalyst. However titanium deposition increases with the amount of coal processed, as does catalyst deactivation.

Lebowitz reported that the most sensitive parameters indicating catalyst deactivation during the hydrogenation of low-ash, bituminous coal extract are the sulfur and the nitrogen contents of the vacuum residuum

[91] K. H. Altgelt and E. Hirsch, *Prepr., Am. Chem. Soc., Div. Pet. Chem.*, **15**(2), A262 (1970).

Table 28.16 Composition of Coal Ash (wt %)[89]

Coal:	Wyodak	Illinois No. 6
Ash (wt %):	9.73	9.95
Components of Ash		
Phosphorus pentoxide	0.61	0.06
Silica	33.64	45.74
Ferric oxide	3.80	16.18
Alumina	17.08	18.95
Titania	1.16	0.95
Calcium oxide	18.50	9.50
Magnesium oxide	4.70	0.94
Sulfur trioxide	18.04	4.52
Potassium oxide	0.71	1.79
Sodium oxide	0.80	0.43
Undetermined	0.95	0.94

Table 28.17 Analysis (wt %) of Contaminants on Used Catalysts[89]

Coal feed:	Wyodak	Illinois No. 6
Catalyst age (lb coal/lb catalyst):	781	2644
Oil-Free Catalyst		
Carbon	13.67	10.9
Sulfur	5.13	4.6
Iron	1.73	1.5
Titanium	0.03	5.11
Sodium	1.46	0.42
Calcium	1.04	0.24

fraction at constant yield of distillates.[92] Increases in the sulfur and nitrogen in this fraction were found to correlate well with the quantity of "soluble" ash to which the catalyst is exposed. Removal of "soluble" ash from the coal extract proportionally increased catalyst life.

"Soluble" ash is defined as that ash in the coal extract feed that passes through a 0.45-μm filter. In fact, more than 75% of this ash passes through a 0.01-μm filter. Since the "soluble" ash contains on the order of 10–15% titanium as compared to around 0.1% in the feed coal used to obtain the extract, the "soluble" titanium and other "soluble" ash components are either present as thermally stable organotitanium compounds or as extremely finely divided salts or minerals. That distillate yield was unaffected while desulfurization and denitrogenation of the large residuum molecules decreased with increased titanium deposition indicates that the titanium does not prevent entrance of the smaller molecules into the catalyst particles where they are subsequently hydrocracked.

Stanulonis et al.[93] employed a scanning electron microscope (SEM) equipped with an energy dispersive X-ray analyzer (EDAX) to investigate the deposition of inorganic contaminants on aged Harshaw 0402T Co-Mo/SiO_2-Al_2O_3 catalysts removed from the upstream and downstream ends of a fixed-bed Synthoil reactor. Figures 28.18 and 28.19 depict the relative concentrations of various elements as a function of distance from the exterior surface of the catalyst particles taken from the upstream and downstream reactor locations, respectively. The upstream catalyst had a thin (varying from 0 to 90 μm) inorganic crust composed mainly of coarse rodlike crystals of ferrous sulfide (covering 50–70% of the exterior surface) apparently growing out of large fissures in the catalysts. Traces of silicon and titanium accumulated in the pore mouths. In contrast, the inorganic crust on the downstream catalyst consisted of alumina and silica, which covered from 10–20% of the exterior surface. The ferrous sulfide crust was entirely absent. However large quantities of titanium were deposited in

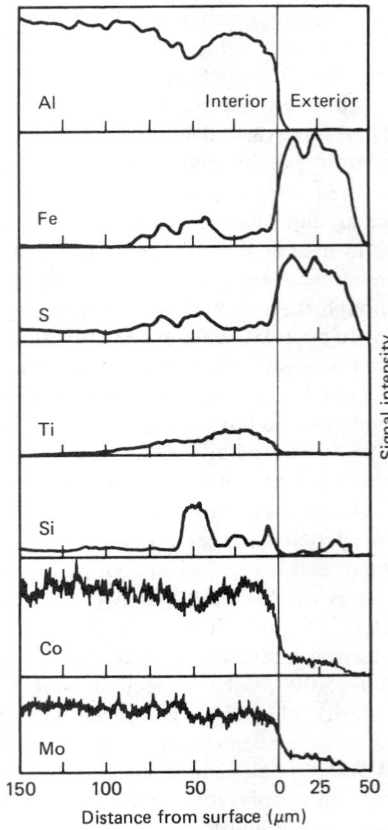

Figure 28.18 Electron microprobe concentration profiles of a catalyst particle from the upstream end of the reactor. The aluminum profile serves to identify the exterior catalyst surface and shows variations in the interior, in part because the signals have not been corrected for varying local absorption of X rays. The iron concentration is approximately proportional to the sulfur concentration in the crust, with the sulfur-iron ratio being much higher in the interior of the catalyst, which is a sulfide in operation. The irregularities indicate variations in the macropore structure. Titanium was deposited primarily in the interior of the catalyst, and the concentration profile shows the presence of finely dispersed titanium. The irregular silicon profile suggests the incorporation of mineral constituents of the coal within the macropores. The cobalt and molybdenum profiles show large variations inside the catalyst, which may be attributed to low dispersion of these components. These two elements penetrate into the organic crust.[93]

the pore mouths up to a depth of 150 μm. This was about the same depth of penetration reported by Kang and Johanson.[89]

Although the ferrous sulfide deposits prevented their deposition in the pore mouths

[92] H. E. Lebowitz, based on speech at Gordon Conf. Fuel Science, New Hampton, N.H., June 1975.

[93] J. J. Stanulonis, B. C. Gates, and J. H. Olson, *Am. Inst. Chem. Eng. J.*, **22**(3), 576 (1976).

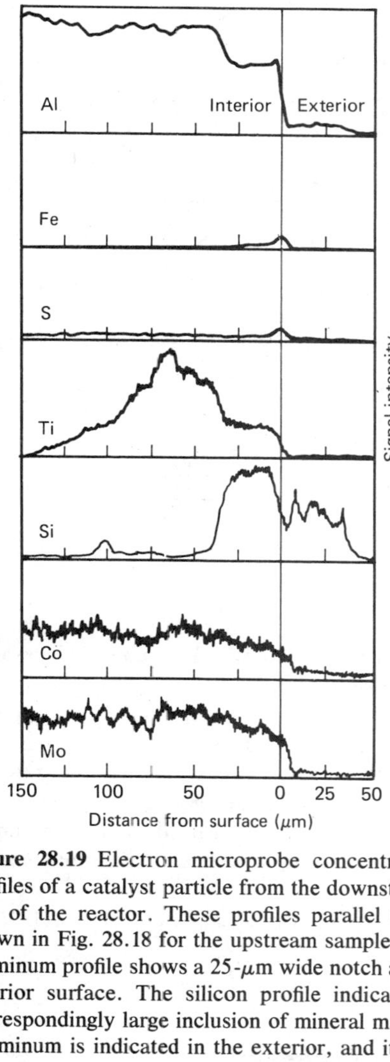

Figure 28.19 Electron microprobe concentration profiles of a catalyst particle from the downstream end of the reactor. These profiles parallel those shown in Fig. 28.18 for the upstream sample. The aluminum profile shows a 25-μm wide notch at the interior surface. The silicon profile indicates a correspondingly large inclusion of mineral matter. Aluminum is indicated in the exterior, and it also is attributed to mineral matter. The iron and sulfur profiles indicate formation of FeS_x in small quantities. The interior sulfur concentration is above the background. The titanium concentration is much greater for the downstream catalyst than for the upstream catalyst. The notch at the surface corresponds to that indicated by the aluminum profile. The silicon profile is irregular, suggesting that the mineral components were incorporated as 1 /10 μm particles. The cobalt and molybdenum profiles show that there was transport of catalytically active material toward the exterior of the crust.[93]

upstream, it would appear that organotitanium compounds were able to penetrate into the pore structure of the downstream catalyst particles and prevent the entrance of the reactant molecules. That deactivation by titanium results from the plugging of pores rather than through chemical reaction with the active sites on the interior surface is indicated by the relative effect of impregnation (onto the interior surface of the catalyst) of individual metal salts on the solvent hydrogenation activity employed by Kovach and Bennett.[88] As shown in Fig. 28.20, the alkali salts (Na, Ca, Mg) gave the highest degree of deactivation, whereas the acidic components (P, Ti, Si) caused little or no deactivation at loadings significantly in excess of those reported for catalyst deactivated during coal liquefaction. These results would indicate that the alkali metals deactive through chemical reaction with active sites on the interior surface, but titanium does not.

Pore size distributions shown in Fig. 28.21 indicate that the largest reduction in pore volume occurs in pores less than 60 μm in diameter.[93] That most of this micropore reduction is due to coke laydown is demonstrated by data showing that most of the lost micropore volume in a spent H-Coal catalyst can be restored by burning off the coke.[94] Brooks et al.[84] ground a deactivated 1/16-in. H-Coal extrudate catalyst to 60 × 100 mesh size and compared its ability to liquefy Illinois No. 6 bituminous coal to benzene-soluble liquids with that of the 60 × 100 mesh fresh catalyst. The activities were nearly identical, further indicating that deactivation of the extrudate during H-Coal processing is largely due to processes occurring at the exterior surface, such as pore mouth plugging.

28.4.1.4 *Sintering*

The fourth category of catalyst deactivation is sintering. This deactivation could result either from sintering of the refractory oxide support, resulting in a substantial loss of internal surface area, or from sintering of the finely dispersed catalytic metal oxides and/or sulfides, resulting in loss of "active" surface area. Such sintering of these types might be induced by either thermal or chemical action. Sintering of the support could also result in physical degradation of the catalyst particle, which would severely hinder its functioning in a liquefaction reactor.

Little is known about the sintering of sup-

[94] C. A. Battista, Hydrocarbon Research, Inc., private communication, September 17, 1976.

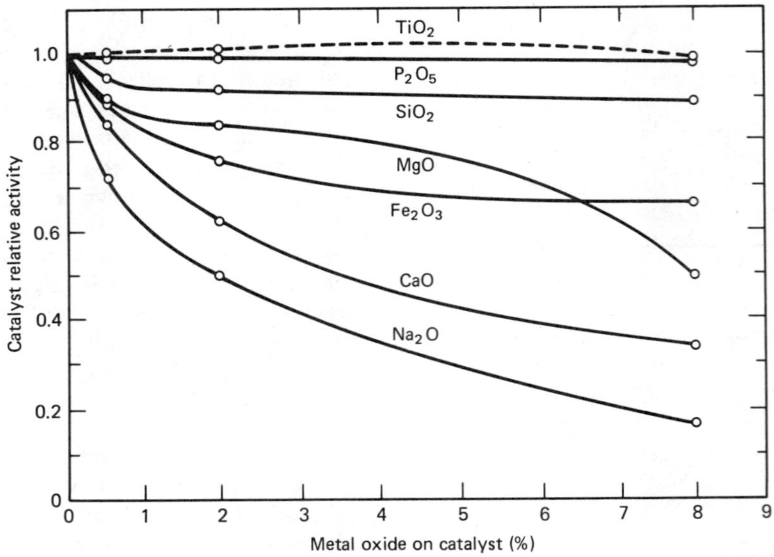

Figure 28.20 Single ash component by impregnation.

ported metal oxides and/or sulfide crystals that might take place under coal liquefaction conditions. However one of the problems unique to the hydrogenation of coal is related to its high oxygen content. This is especially true of lower rank subbituminous coal and lignite. Much of the oxygen is converted to water, which is in the form of high-temperature steam at liquefaction conditions.

Rapid loss of surface area has been shown[95] to occur when a large surface area nickel on γ-alumina catalyst is exposed to a 9 : 1 mixture of steam and hydrogen as compared to its exposure to hydrogen gas. Increased temperature enhances the rate of surface area decline. Kang and Johanson[89] have examined fresh, used, and air-regenerated Co-Mo/Al_2O_3 catalyst after their use in

[95] A. Williams et al., *J. Catal.*, **24,** 352 (1972).

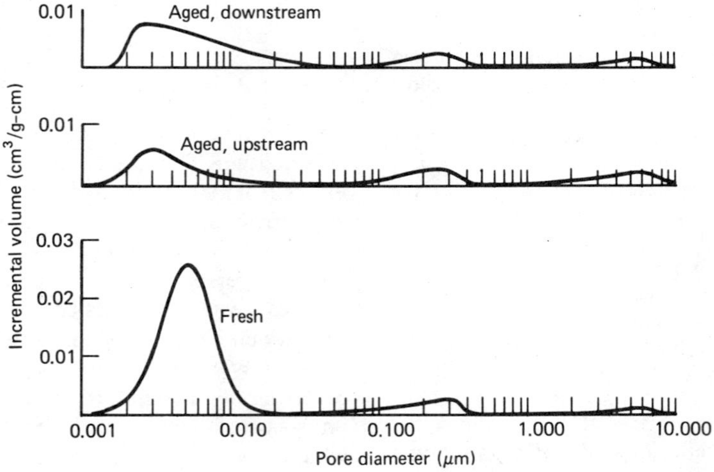

Figure 28.21 Pore size distributions of the fresh and aged catalysts. Three kinds of pores are indicated, micropores in the range 2–8 nm, macropores in the range 0.1–0.4 μm, and cracks of about 5 μm. Most of the area of the original catalyst was in the micropores. The upstream catalyst sample shows a greater loss in micropore volume than the downstream sample, and the micropore size distribution is shifted to lower pore sizes. The pore volume of cracks was greater in the aged catalysts, especially the upstream sample.[93]

H-Coal processing. No sintering of the catalyst was detected after an extended 30-day run on Illinois No. 6 bituminous coal where the H_2O/H_2 ratio in the reactor outlet gas was 1 : 8. In contrast, sintering, as measured by loss in surface area, was quite pronounced after a 49-day run with a "wet" subbituminous Wyodak coal containing 20 wt % moisture. Here the H_2O/H_2 ratio in the outlet gas was 1 : 2. Since the rate of deactivation decreased when the Wyodak coal was dried to 2 wt % moisture, the data suggest that steam sintering of the γ-alumina support may be responsible for the rapid catalyst deactivation during the liquefaction of the "wet" Wyodak coal.

28.4.2 Nature of Solids Formed During Coal Liquefaction

To gain some insight into the nature of the solids formed during the hydrogenative liquefaction of coal, it is necessary to consider certain aspects of the coal itself. Coal is a highly heterogeneous material consisting of two major phases, one organic and the other inorganic. The organic phase can be subdivided into component macerals, and the inorganic phase is composed of a diverse mixture of minerals. During liquefaction the reactive organic components are converted into gaseous, liquid, and solid (i.e., coke) products. The degree of conversion is a complex function of reactant composition and process conditions. It is ultimately limited by both the inherent insolubility of the mineral matter and the insolubility of certain nonreactive macerals. Also contributing to the makeup of the solid residue are the minerals that are formed *in situ* as a result of reactions undergone by the ion-exchangeable clay minerals and organometallic compounds found in coals.

28.4.2.1 *The Organic Nature of Liquefaction Solids*

Although it is not within the scope of this discussion to consider the detailed chemical and physical transformations that the various macerals undergo, it is instructive to consider the behavior of the various organic coal constituents during liquefaction, to obtain a better understanding of the makeup of the resultant solid residue. A residue classification and the dissociation mechanisms that link the organic residue components to their original macerals have been proposed by Mitchell et al.[96] A summary is presented in Table 28.18. Vitrinite and exinite are reactive for transformation to liquid products, whereas fusinite, semifusinite, and micrinite either are substantially unaltered or are transformed to various residue components.

Maceral analyses of various coals and their resultant residues are shown in Tables 28.19 and 28.20, respectively. The analyses are seen to vary from sample to sample for both a given feed coal and its residues. Pyrite is included in the coal analyses, and pyrrhotite, clay, carbonate, and quartz in the residue analyses because they can be distinguished optically. The various organic components share this property.

[96] G. D. Mitchell, A. Davis, and W. Spackman, *Prepr., Am. Chem. Soc., Div. Fuel Chem.*, **21**(5), 189 (1976).

Table 28.18 Behavior of Coal Organic Constituents During Liquefaction[96]

Maceral Precursor	Mechanisms of Reaction		Organic Residue Components
Vitrinite	Slightly altered vitrinite (contracted and/or swollen)	Semicoke	Unreacted vitrinite
	Granular residue		Granular residue
	Vitroplast (high molecular weight) → cenosphere		Cenosphere
	Vitroplast (intermediate molecular weight) → vitroplast		Vitroplast
	Hydrogenated product (repolymerized) → liquid-crystal ⇄ mesophase → semicoke		Semicoke (anisotropic)
Exinite	Fractional contribution to submicrometer granular material		Granular residue
Semifusinite	Possible fractional contribution to liquid product, mechanism similar to vitroplast		Semifusinite, semicoke, or vitroplast
Fusinite	No observable mechanism		Fusinite
Macrinite	Possible fractional contribution to liquid product, mechanism unknown		Unknown
Micrinite	Mechanism unknown		Unknown

Table 28.19 Maceral Analysis (Vol. %) of Feed Coals from Various Seams[97]

Sample No.	Seam	Vitrinite	Pseudovitrinite	Fusinite	Semifusinite	Macro-Micrinite	Granular Micrinite	Exinite	Residue	Pyrrhotite
177-96, Pds. 1–60	Western Kentucky No. 14	71.3	0.6	2.9	5.0	3.3	12.5	1.7	0.4	1.8
177-96, Pds. 6–12	Western Kentucky No. 14	38.4	0.2	2.7	3.9	1.0	1.8	1.0	0.4	0.7
177-96, Pds. 61–66	Illinois No. 6	88.8	0.0	4.8	1.2	0.0	1.1	1.1	0.2	1.9
177-100, Pds. 1–5	Kaiparowits	92.0	0.1	4.0	8.6	0.2	4.1	1.7	0.3	0.0
177-101, Pds. 1–6	Kaiparowits	78.5	0.4	4.2	9.0	0.7	4.7	1.1	1.1	0.0
177-101, Pds. 7–11	Kaiparowits	85.9	0.0	2.9	4.2	1.0	1.3	1.0	1.3	0.0
130-69, LO–7	Illinois No. 6	90.2	0.0	3.3	3.3	0.5	1.0	1.7	0.0	NC
HRI3663, LO–18	Illinois No. 6	93.5	0.0	1.7	1.8	0.7	0.9	1.4	0.0	NC

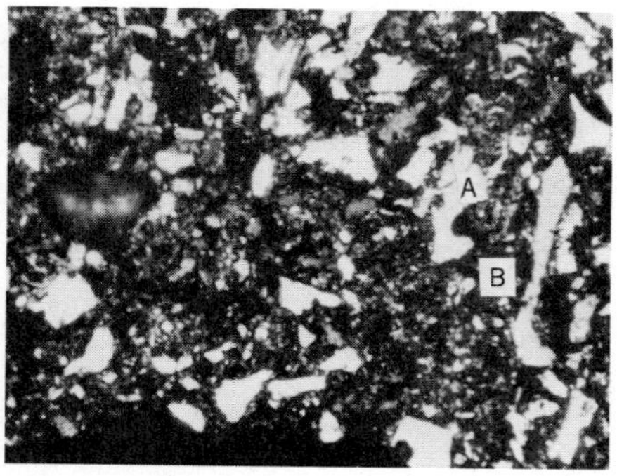

Figure 28.22 Micrograph of HRI 177-101-4B pyridine extracted filter cake from H-Coal processing of Kaiparowits coal (A-fusinite, B-calcite). × 625.

Figure 28.22 is a micrograph of a residue from the H-Coal process, in which particles of fusinite and calcite can be readily distinguished because of their size and reflectance characteristics.[96] As seen, the most abundant maceral in coal is vitrinite, and the most abundant residue component is granular residue. Granular residue is described[98] as a submicrometer granular material derived from several sources. The present inability to determine routinely the origin or nature of individual grains of this residue type using the petrographic microscope causes the individual components to be lumped into this single category. However only about 10% of granular residue is organic, and the remainder is presumed to be finely divided clay particles. Next in abundance is vitroplast, which is described as a spherical plastic or once-plastic thermal degradation product of vitrinite.[96] Another major organic residue component is semicoke, possibly resulting from dehydrogenation, repolymerization, or carbonization reactions occurring during liquefaction. The higher concentration of semicoke in the SRC as compared with H-Coal residues may result from processing differences or, possibly, from coke contamination of the SRC process feed coal during a specific grinding operation.[97] The remaining organic residue components either are present as discrete particles or are intimately associated with one or more of the major organic components and/or the inorganic components. This association is discussed in detail later.

Although determined to be present in significant amounts only in the two Kaiparowits coal residue analyses given in Table 28.20, mesophase is important because of the role it plays in the formation of large agglomerated particles that build up in the SRC reactor during the liquefaction of bituminous coals. Mesophase is a transient intermediate stage between a unique, vitrinite-derived fluid system (liquid crystals) of high molecular weight and a solid anisotrophic carbon such as semicoke. Mesophase in liquefaction residues appears to have grown around a central "inert" particle such as calcite. The role of mesophase in the formation of large reactor solids is discussed in more detail later.

28.4.2.2 *Nature of Mineral Matter in Liquefaction Solids*

The mineral composition of coal is highly complex, as can be seen from Table 28.21, which lists the major and minor types identified. Different coals contain different combinations of these minerals, as well as different proportions. However, the most common types are clays (kaolinite, illite, montmorillonite), quartz, pyrite, siderite, calcite, mixed carbonates (ankerite, dolomite), gypsum, and rutile.

Although not present in coal as minerals, a number of metals (e.g., Na, Ca, Ti) are present as either carboxylates or humates. Without going into a detailed discussion of this topic, it can be stated that the presence of organometallic compounds and ion-exchangeable metal ions in certain clays is important. This is because the crystallization of minerals formed from these species during liquefaction contributes to both the chemical composition and the size of the agglomerated particles formed. In turn, this affects their residence time in a reactor and, ultimately, the solid separation operations.

Inorganic materials in liquefaction residues can be classified as follows[48]:

1. Those remaining physically and chemically unaltered during processing. The most important species in this category is quartz.

[97] P. L. Walker, Jr., W. Spackman, P. H. Given, E. W. White, A. Davis, and R. J. Jenkins, *Characterization of Mineral Matter in Coals and Coal Liquefaction Residues,* EPRI Rep. No. RP366-1, December 1975.

[98] A. Davis, "The Petrographic Composition of Coals: Its Importance in Liquefaction Processes," Preprints, 1976 Coal Chemistry Workshop, Stanford Research Institute, August 26 and 27, 1976.

Table 28.20 Residue Analyses (vol %) of Feed Coals from Several Seams[97]

Sample No.	Feed	Process	Granular Residue	Semi-Coke	Meso-phase	Ceno-sphere	Vitro-plast
177-96-47B	Western Kentucky	SRC	71	2	0	0	7
177-96-60B	Western Kentucky	SRC	63	7	0	0.1	5
177-96-66A	Illinois No. 6	SRC	65	4	0	0.3	7
177-100-6A	Kaiparowits	SRC	52	12	0	0.3	21
177-100-8B	Kaiparowits	SRC	51	10	3	0.3	24
177-101-4B	Kaiparowits	H-Coal	56	5	1	0.1	29
130-69-6A	Illinois No. 6	H-Coal	58	1	0	4	15
130-69-16A	Illinois No. 6	H-Coal	56	0.4	0	1	17
130-69-26A	Illinois No. 6	H-Coal	54	4	0	0	22
130-69-35A	Illinois No. 6	H-Coal	53	4	0	0	25

Table 28.21 Minerals Found in Coals[99]

Silica minerals

Quartz (trigonal), SiO_2
Tridymite (orthorhombic), SiO_2
Cristobalite (tetragonal), SiO_2

Chlorite $(Mg, Al, Fe)_{12}[(Si, Al)_8O_{20}](OH)_{16}$

Serpentine $Mg_3[Si_2O_5](OH)_4$

Clay minerals

Kaolinite group, $Al_4[Si_4O_{10}](OH)_8$
Illite, $K_{1-1.5}Al_4\ [Si_{7-6.5}Al_{1-1.5}O_{20}](OH)_4$
Montmorillonite group, $(\frac{1}{2}\ Ca, Na)_{0.7}(Al, Mg, Fe)_4[(Si, Al)_8O_{20}]\ (OH)_4 \cdot nH_2O$
Vermiculite, $(Mg, Ca)_{0.7}(Mg, Fe^{3+}, Al)_{6.0}[(Al, Si)_8O_{20}]\ (OH)_4 \cdot 8H_2O$

Feldspar group

Alkali feldspars, $(K, Na)\ [AlSi_3O_8]$
Plagioclase, $Na[AlSi_3O_8] - Ca[Al_2Si_2O_8]$
Celsian, $B\ [Al_2Si_2O_8]$

Sulfates

Gypsum, $CaSO_4 \cdot H_2O$
Anhydrite, $CaSO_4$
Hemihydrate, $CaSO_4 \cdot \frac{1}{2}\ H_2O$ (bassanite)
Barytes, $BaSO_4$
Celestine, $SrSO_4$

Sulfides

Pyrite (cubic), FeS_2
Marcasite (orthorhombic), FeS_2
Pyrrhotite, Fe_7S_8–FeS
Chalcopyrite, $CuFeS_2$
Sphalerite, ZnS

Carbonates

Ankerite, $Ca(Mg, Fe^{2+}Mn)\ (CO_3)_2$
Calcite (trigonal); $CaCO_3$
Aragonite (orthorhombic); $CaCO_3$
Magnesite, $MgCO_3$
Rhodochrosite, $MnCO_3$
Siderite, $FeCO_3$
Dolomite, $CaMg(CO_3)_2$
Strontianite, $SrCO_3$
Witherite, $BaCO_3$

Rutile TiO_2

Untreated								
Vitrinite	Pseudo-vitrinite	Fusinite	Semi-fusinite	Pyrrhotite	Clay	Carbonate	Quartz	Other
4	2	1	4	9	0	0.5	0.3	0
2	2	2	5	12	0	0.3	0.4	0
3	2	4	5	8	0	2	1	0
2	0	2	3	2	0.1	6	0.3	0
1	0	3	3	0.4	0	4	0.5	0
1	0	2	1	0.3	0	4	0.5	0
1	1	2	3	8	0.4	5	1	0
0.1	0.1	2	4	10	1	8	1	0
0.2	0.5	3	2	11	0.2	3	0.4	0
0	0.2	3	3	10	0.1	2	1	0

2. Those undergoing extensive physical degradation that produces extremely small particles. Typical of these are the clay minerals, which have been reduced so much that they cannot be recognized with petrographic microscopes. The carbonate minerals also undergo the same procedure to a limited degree.
3. Those undergoing major chemical and crystallographic transformations. Of major importance here is pyrite (FeS_2), which is readily reduced to pyrrhotite ($Fe_{1-x}S$) during liquefaction.
4. Those formed *in situ* during processing. The most significant here are $CaCO_3$, TiO_2, and possibly $CaSO_4$.

Comparable quantitative data on the mineralogical analyses of feed coals and their resultant liquefaction residues are limited. Only recently have X-ray and IR techniques, in combination with low-temperature ashing (LTA), been developed to analyze directly for most of the major minerals present.[97] These techniques must be supplemented by other techniques to ensure accuracy. One of the most complete side-by-side mineralogical analyses of LTA ash from coal and from the SRC process residue is shown in Table 28.22 for Illinois No. 6 coal. Roughly speaking, most of the mineral species remain unchanged chemically during liquefaction. However there is reason to believe that the amounts of gypsum are too high and those of calcite too low, because of some degree of chemical alteration taking place during LTA. The exact values for the iron-containing species (pyrite, pyrrhotite, and siderite) are somewhat in doubt because of the difficulty in obtaining a precise determination of pyrrhotite content. However the application of Mössbauer spectroscopy to the determination of pyrite and pyrrhotite content appears to be possible.

Another technique, computer evaluation of scanning electron microscope images (CESEMI), promises to provide additional information not only on the various minerals present in coals and their residues, but also on the size distribution of individual minerals and their close association with other minerals.[97] However further refining of this technique and its interpretation is required before routine use is indicated. CESEMI also has the potential of distinguishing between pyrite and pyrrhotite in residues.

In summary, it must be said that the direct quantitative analysis of the minerals in coal liquefaction residue is far from routine, with

[99] W. A. Deer, R. A. Howie, and J. Zussman, *An Introduction to Rock Forming Minerals,* Longmans, Green and Company, Ltd., London, 1966.

Table 28.22 Mineralogical Analyses of Ash Produced From Illinois No. 6 (Burning Star) Feed Coal and Its SRC Residue[97]

Mineral	Illinois No. 6 Coal (wt % on LTA)	SRC Residue (wt % on LTA)
Kaolinite	17.0	25.0
Other clays	27.5	22.0
Quartz	11.0	11.1
Pyrite	15.0	(1)[a]
Pyrrhotite ($Fe_{0.905}S$)	—	9.0
Siderite	6.7	(6.7)
Gypsum	14.3	12.4
Dolomite	1.4	3.0
Calcite	Trace	Trace
Rutile	0.35	0.48

[a] From pyritic sulfur.

most information being obtained through analyses of the sulfur forms present and the metal oxides resulting from high-temperature ashing.

28.4.2.3 *Particle Size of Liquefaction Solids*

Coals fed to most liquefaction processes now under development have been ground to pass through at least a 60 U.S. mesh size screen, and some grinds call for most of the coal feed to pass a 200 U.S. mesh screen. During any liquefaction process resulting in conversion of a major fraction of the organic component to a liquid product, analyses have shown that particle sizes (exclusive of particles produced under conditions where massive coke formation occurs) range from the order of a fraction of an inch down through subsieve and even to as small as submicrometer size. Since measurements cannot be made of the solid particles as they exist, however, either as separate entities or as agglomerates dispersed in the coal liquid, a separation must be performed prior to analysis. Depending on the nature of the method employed, the solids are then either analyzed as a dry powder or else dispersed in some suitable liquid.

It is in the laboratory or process-scale separation, drying, and/or dispersion steps that the particles undergo possible size reduction and/or agglomeration. For this reason, and also because different methods at times yield widely different size distributions for the same sample, it is difficult to decide what size particles are actually being handled in a given process solids separation step. Knowledge of particle size is important for solids separation, whether the process is based on filtration, sedimentation, centrifugation, or even magnetic approaches. It is not within the scope of this discussion to examine further the techniques and complications involved in the determination of the particle size of liquefaction solids. However it is important to be aware of the limitations of the measurements.

Although particle size distributions can be on a number, volume, or weight basis, those based on weight usually have the most practical important when being evaluated in conjunction with a process solid separation step, and weight-based distributions are cited in the examples given below. Particles much longer than about 70 μm do not normally constitute a major proportion (except in the case of the intermittent carryover of highly agglomerated solids retained in the liquefaction reactor) of the solids normally suspended in the liquid product. In any event, these can be measured by standard screening techniques. As part of a study of solids separation by solvent precipitation, Gorin et al.[82] employed a Coulter counter to measure the particle size distribution of cresol-insoluble solids from the H-Coal process and from hydroextractions carried out on Illinois coals. Semilog plots are shown in Fig. 28.23. The lack of a straight-line relationship indicates a possible bimodal size distribution. However 50 wt % of the particles measures fall below about 7–10 μm (equivalent diameter), with 90 wt % in sizes smaller than about 25 μm.

Figure 28.23 may be compared with the semilog distribution(weight percent undersize in this case) plots given in Figs. 28.24 and 28.25 for H-Coal and SRC residues from Kaiparowits coal. Although the measurement techniques differ, in each case the methods (including the Coulter counter) involve dispersing the solids in a liquid medium prior to analysis. In the case of the H-Coal solids, the 50% diameter ranges from 2 to 10 μm depending on method, whereas with the SRC solids it ranges from 3 to 4 μm. Despite the difference in coals, processes, and methods, in all cases about 90% of the particles had equivalent spherical diameters less than about 25 μm. To what extent agglomeration has shifted the size distribution above about 1 μm is not known. However a particle number distribution carried out on a transmission electron micrograph (Fig. 28.26) of the Kaiparowits H-Coal solid previously dispersed in acetone yields a number average diameter below 0.5 μm, indicating that submicrometer particles are indeed present in this solid residue.

28.4.2.4 *Reactor Solids*

Distinct from the normal solid residues suspended in the coal liquid product are what may be broadly termed "reactor solids." These accumulate in the reactor and are found after cooling the reactor at the end of an operation. The particles have a size mainly

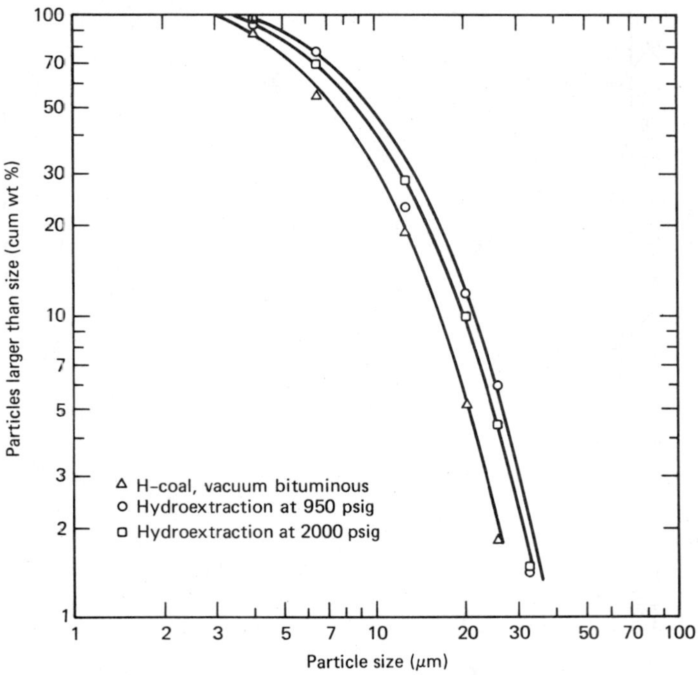

Figure 28.23 Particle size analysis for cresol insolubles, Illinois coals.

in the 50–200 mesh (297–74 μm) range.[101] They result from an incompletely understood process (or processes) whereby solid materials either present in the feed coal or formed in liquefaction reactors are retained in the reactor.[100–102] Subsequently they undergo further transformation and/or agglomeration

[100] H. E. Lewis, W. H. Weber, G. B. Usnick, W. R. Hollenack, and H. W. Hooks, *Solvent Refined Coal, SRC Process, Operation of SRC Pilot Plant at Wilsonville, Alabama,* Quarterly Tech. Prog. Rep. No. FE-2270-15, for October–December 1976, Catalytic, Inc., Wilsonville, Ala.

[101] H. E. Lewis, W. H. Weber, G. B. Usnick,

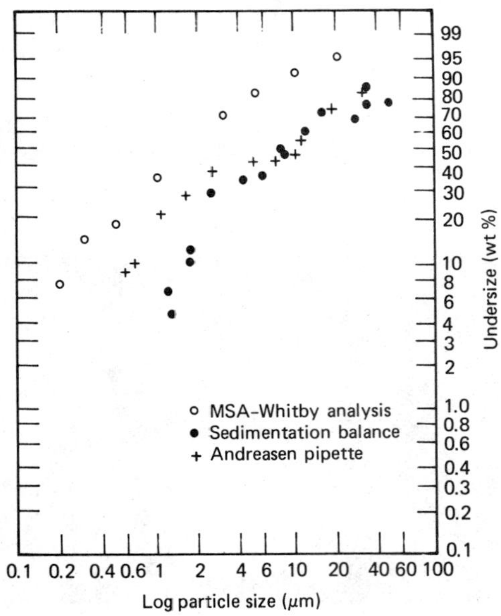

Figure 28.24 Particle size analysis for H-Coal residue from Kaiparowits coal.[97]

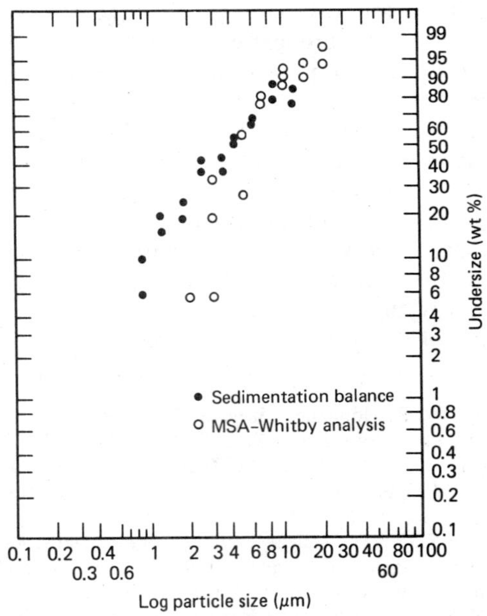

Figure 28.25 Particle size analysis for SRC residue from Kaiparowits coal.[97]

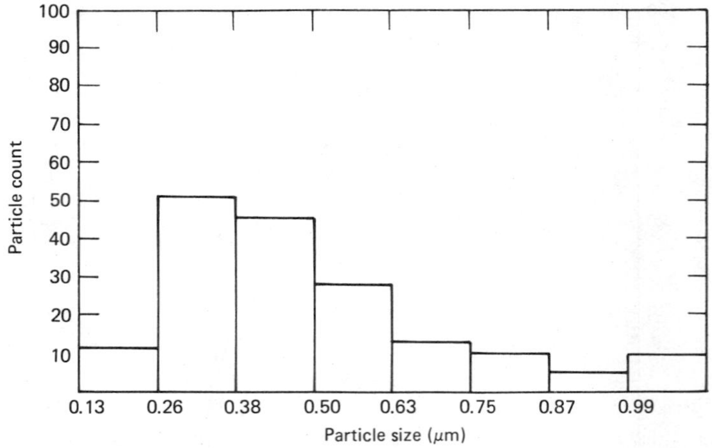

Figure 28.26 Particle size analysis by transmission electron micrography of H-Coal residue from Kaiparowits coal.[97]

with other particles. Although some of these dissolved solids are carried out of the reactor, a large inventory builds up and eventually requires either purging or flushing to avoid process difficulties. Another form of reactor solid formation is reactor wall scale, which is encountered primarily in noncatalytic conversion.[102] This hard wall scale, composed mostly of calcium carbonate along with some iron sulfides, could cause severe processing problems. Since wall scale is retained, it is not an important consideration in requirements for solid separation processes and is not discussed further here.

There are two somewhat different forms of reactor solid brought about by two distinct agglomeration mechanisms. The first type is loosely labeled "coking." In this case a central "inert" particle (e.g., coke breeze contaminant, mineral matter, or residue components) is surrounded by a concentrically deposited mesophase shell, as depicted in Figs. 28.27 and 28.28. The particle adheres to other mesophase-coated particles, forming an agglomerate that is too large to be carried out of the reactor under normal flow conditions. In time, the mesophase is transformed to semicoke. This is the predominant mechanism occurring during the processing of bituminous coals. There is an indication that these agglomerates are weakly bound and that they can be dispersed with increased agitation.

The second type of solids formation is associated with the *in situ* formation of calcium carbonate (from the calcium salts of humic and carboxylic acids) during the liquefaction of lower rank, high-calcium content subbituminous coal and lignite. The formation of this type of reactor solid has been reported for SRC processing of subbituminous Wyoming coal[100] and the liquefaction of German lignites.[103] In the latter case the

and W. R. Hollenack, *Solvent Refined Coal, SRC Process, Operation of SRC Pilot Plant at Wilsonville, Alabama,* Quarterly Tech. Prog. Rep. No. FE-2270-10, for July–September 1976, Catalytic, Inc., Wilsonville, Ala.

[102] *EDS Coal Liquefaction Process Development, Phase IIIA,* Quarterly Tech. Rep. for January–June, 1976, under U.S. ERDA Contract No. E(49-181-2353); Exxon Research and Engineering Company, B. T. Fant, Baytown, Tex., October 1976.

[103] W. Kroenig, *Katalytische Druckhydrierung von Kohlen, Teeren und Mineraloelen,* Springer-Verlag, Berlin, (1950).

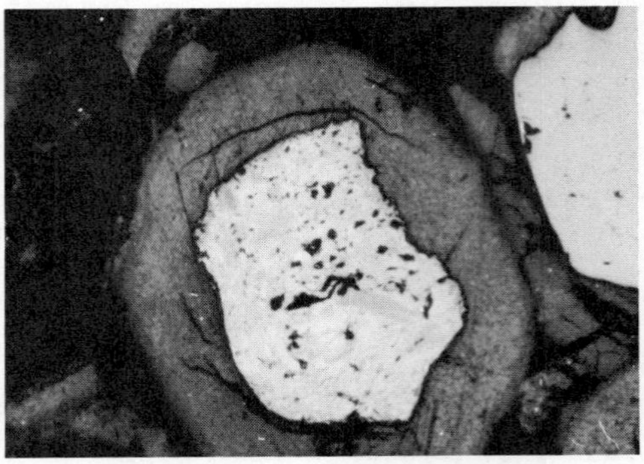

Figure 28.27 Semicoke nucleus surrounded by anisotropic coke derived from mesophase.

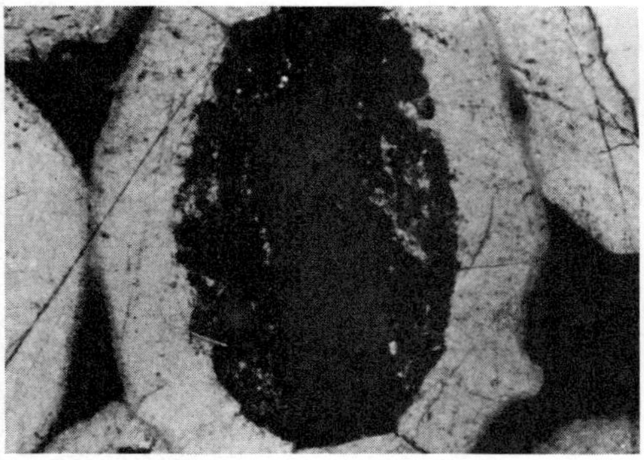

Figure 28.28 Calcite nucleus surrounded by anisotropic coke derived from mesophase.

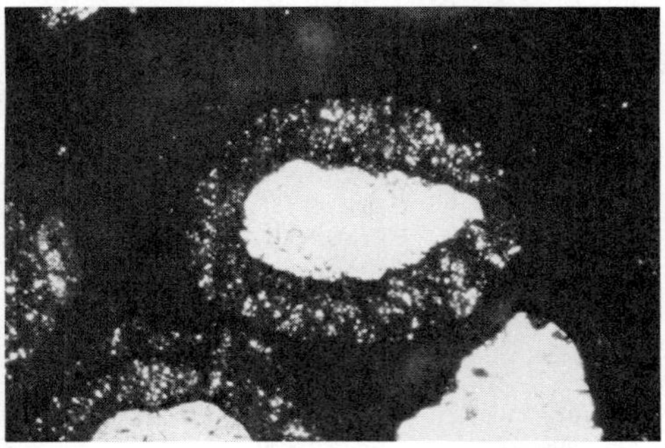

Figure 28.29 Photomicrograph of a particle of reactor solid material recovered after Wyodak coal processing, depicting central particle, a coke breeze contaminant, surrounded by shell of dispersed organic-inorganic matter in a calcium carbonate matrix. × 250.

solids were referred to as "caviar." Here a shell composed primarily of dispersed fine-grained inorganic and organic matter, with a calcium carbonate matrix, is deposited onto a central inert particle (again, semicoke, mineral matter, or residue components). Growth of calcium carbonate "bridges" between primary particles appears to be the major cause of particle agglomeration. Figure 28.29 depicts a typical primary particle. Because of the close association of pyrrhotite and calcium carbonate in the shell, as depicted in Fig. 28.30 and reported by Exxon[102] for the $CaCO_3$-rich wall scale deposits, vaterite (a form of calcium carbonate) formation may be catalyzed by that iron sulfide.

The chemical composition of minerals found in SRC reactor solids[7] and the major and minor mineral components as a function of coal type are listed in Tables 28.23 and 28.24. Comparisons of the analyses[100,101] of the feed coals and resultant reactor solids for these three coals, as shown in Table 28.25, confirm the findings given in Table 28.24. Furthermore, the concentration of calcium in the reactor solids from these three coals follows the trend in their respective concentrations of exchangeable calcium (i.e., calcium that reacts and forms a calcium mineral) as shown in Table 28.26. The low concentration of calcium in the solids from the preheater effluent of the processes that used Illinois No. 6 Monterey bituminous coal (Table 28.25) indicates that formation of calcium compounds required a long reaction time. The extent to which the organocalcium compounds are converted to mineral form would be expected to be influenced by process conditions.

28.4.3 Solids Separation Processes

28.4.3.1 *Solvent Precipitation*

The particulate residues (unconverted coal and ash) from coal liquefaction vary in their intrinsic size distribution as a function of coal liquefaction conditions and coal type. Some examples of characterizations have been previously cited; a large proportion of particles smaller than 10 μm is usually present. In general, the products of coal liquefaction are dispersed in a recycle vehicle, which is an excellent solvent for even the heaviest residual portions of the dissolved coal. This solvent is desirable because it increases the liquid yield and prevents coke formation in the liquefier. In most cases, because of the size and density of the residue particles, a relatively stable suspension is formed in the

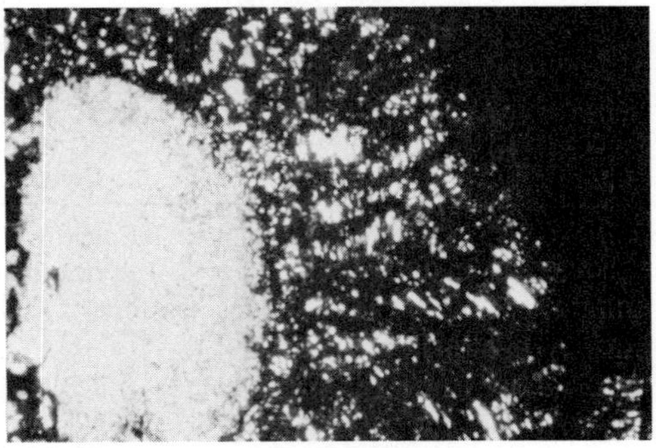

Figure 28.30 Photomicrograph of a particle of reactor solid material recovered after Wyodak coal processing, showing radial arrangement of pyrrhotite in a carbonate shell. × 625.

Table 28.23 Chemical Compositions of Minerals Found in Reactor Solids[97]

Mineralogical Name	Chemical Composition
Pyrrhotite	$Fe_{1-x}S$
Pyrite	FeS_2
Troilite	FeS
Gypsum	$CaSO_4 \cdot 2H_2O$
Bassanite	$CaSO_4 \cdot \frac{1}{2}H_2O$
Anhydrite	$CaSO_4$
Calcite	$CaCO_3$
Aragonite	$CaCO_3$ Different crystallographic structures
Vaterite	$CaCO_3$
Dolomite	$CaMg(CO_3)_2$
Siderite	$FeCO_3$
Quartz	SiO_2
Rutile	TiO_2
Kaolinite	$Al_2(Si_2O_5)(OH)_4$
Illite	$K_{1-1.5}Al_4[Si_{7-6.5}Al_{1-1.5}O_{20}](OH)_4$
Montmorillonite	$(Al, Mg)_2Si_4O_{10}(OH)_2$
Muscovite	$KAl_2(AlSi_3)O_{10}(OH)_2$

reactor effluent; the settling rate is very slow. The basis for the solvent precipitation technique is that the intrinsic particles tend to agglomerate when the heaviest coal liquefaction products are precipitated from solution. This precipitation may be effected by temperature reduction and/or addition of a precipitation solvent (antisolvent) to the coal liquefaction products.

In the case of mild liquefaction of Pittsburgh No. 8 coal by extraction in the CSF process, it was shown that particles can be made to settle from the effluent slurry at a rate greater than 2 in./min.[104] This rate is higher than the Stokes terminal velocity of 75% of the intrinsic particles present. In this case the precipitation of 3–8% of the residuum product was effected by simple cooling to 500–600°F. It was shown that the precipitated material was principally benzene insoluble. When similar experiments were conducted with Western subbituminous coals, little separation or precipitation occurred on cooling. Products from more severe liquefaction processes, such as the SRC process under hydrogen pressure and direct catalytic processes, are also of the latter type; little precipitation or agglomeration takes place on cooling alone. It has been roughly determined that the degree of precipitation is related to the fraction of benzene insolubles present in the particular product.

It was well established in the early patent literature (1930s or earlier) that the addition of solvents with an aliphatic character to coal-derived hydrocarbons containing asphaltic material and solids, affects precipitation of some asphaltic material together with agglomeration of the asphaltic material and the solids.[105–109] Much of the recent work in this field has been done at Consolidation Coal Company[109,110–112] (now Conoco Coal Devel-

[104] E. Gorin, C. J. Kulik, and H. E. Lebowitz, "Deashing of Coal Liquefaction Products via Partial Deasphalting. I–Hydrogen Donor Extraction Effluents," *Prepr., Am. Chem. Soc., Div. Fuel Chem.*, **20**(1), 79 (1975).

[105] William H. Hill, Koppers Company, U.S. Pat. 1,875,458, September 6, 1932.

[106] Harold J. Rose, Koppers Company, U.S. Pat. 1,875,502, September 6, 1932.

[107] Mathias Pier et al., I. G. Farbenindustrie, U.S. Pat. 1.933,226, March 5, 1935.

[108] Karl Schoenenmann, I.G. Farbenindustrie, U.S. Pat. 2,060,447, November 10, 1936.

[109] Mathias Pier et al., I.G. Standard Company, Linden, N.J., U.S. Pat, 2,144,409, January 17, 1939

[110] E. Gorin, C. J. Kulik, and H. E. Lebowitz, "Deashing of Coal Liquefaction Products via Partial Deasphalting. II. Hydrogenation and Hydroextraction Effluents," *Prepr., Am. Chem. Soc., Div. Fuel Chem.*, **20**(1), 105 (1975).

Table 28.24 Constituents of Reactor Solids from Wilsonville SRC I Operation as Determined by Qualitative X-Ray Examination

Feed Coal	Major Constituents	Minor Constituents
Amax (Wyodak)	Calcite	Pyrrhotite, vaterite, troilite
Western Kentucky, Nos. 9 and 14	Pyrrhotite, quartz, anhydrite	Rutile
Illinois No. 6	Anhydrite, quartz, calcite, pyrrhotite	Kaolinite, montmorillonite, illite, aragonite, siderite, gypsum, pyrite, troilite, rutile

Source: Modified from Ref. 97.

Table 28.25 Typical Analyses of Coals and Reactor Solids

	Kentucky Nos. 9 and 14, Colonial Mine, Pittsburg & Midway Coal Mining Company		Illinois No. 6, Monterey Mine, Monterey Coal Company			Roland Smith Seam, Belle Ayr Mine, Amax Coal Company, Wyoming	
	Feed Coal	Reactor Solids	Feed Coal	Preheater Outlet Solids	Reactor Solids	Feed Coal	Reactor Solids
Ultimate analysis (wt %)							
Carbon	72.94	29.7	64.80	81.21	22.31	67.95	17.6
Hydrogen	5.27	0.6	5.39	6.76	2.68	4.94	0.5
Nitrogen	1.14	0.2	1.20	1.04	0.19	1.06	0.2
Chlorine	0.13	1.3	0.07	0.05	0.99	0.01	<0.1
Sulfur	3.07	9.3	4.25	1.69	3.54	0.62	2.5
Ash	8.93	56.5	10.15	4.62	69.94	7.77	63.5
Oxygen	8.52	2.4	14.14	4.63	2.35	17.65	15.3
Sulfur forms (wt %)							
Sulfide, total	—	7.4	—	0.13	3.4	—	1.8
Pyritic	1.12	0.6	1.14	<0.01	0.28	0.06	0.6
Sulfate	0.10	0.6	0.19	<0.01	0.15	0.03	<0.1
Organic	1.85	0.7	2.92	1.56	0.0	0.53	0.1
Carbon dioxide (% of solids)	—	0.2	—	0.02	8.98	—	58.5
Mineral analysis of ash, (wt %)							
Sodium oxide	0.3	0.2	1.3	0.8	0.9	1.4	0.3
Potassium oxide	2.3	0.3	1.8	2.6	1.4	0.2	0.2
Lime	2.7	4.6	3.1	3.3	25.8	15.4	53.9
Magnesia	0.5	0.5	0.8	0.8	0.8	3.7	2.8
Ferric oxide	24.8	39.3	18.2	17.9	14.7	5.8	9.7
Titania	1.0	2.2	1.2	1.5	6.5	1.0	0.8
Phosphorus pentoxide	0.2	0.2	0.3	0.4	0.7	0.9	0.9
Silica	41.1	10.8	47.3	51.5	26.2	50.4	11.8
Alumina	23.7	9.8	21.7	16.9	8.5	17.5	10.4
Sulfur trioxide	2.8	4.9	4.1	4.3	14.3	3.5	9.7
Undetermined	0.6	3.2	0.2	—	0.2	0.2	0.1

Source: Modified from Refs. 100, 101.

opment Company) and the Lummus Company.[42,45,46]

When the separation requires an antisolvent addition, the general flowscheme is as follows:

1. A topped feedstock is prepared in such a way that the boiling range of the feedstock does not overlap that of the antisolvent.
2. Antisolvent and feed are mixed at an appropriate temperature, generally 400–650°F.
3. Solids separation is carried out in a device such as a liquid cyclone or a gravity settler. (It has also been noted that antisolvent addition is beneficial to filtration.[112])
4. A relatively solids-free overflow (filtrate) is separated from a solids-rich underflow (cake).

[111] E. Gorin, C. J. Kulik, and H. E. Lebowitz, Consolidation Coal Company, U.S. Pat. 3,791,956, 1974.

[112] E. Gorin, Consolidation Coal Company, U.S. Pat. 3,018,241, January 23, 1962.

Table 28.26 Amount of Ion-Exchangeable Calcium in Selected Coals

Source of Feed Coal	Total Ca^{2+} (wt %, mf)	Exchangeable Ca^{2+} (wt %, mf)
Wyodak	1.03	0.86
North Dakota lignite	1.84	1.76
Illinois No. 6, Monterey	0.59	0.44
Illinois No. 6, Burning Star	0.89	0.53
Western Kentucky, Nos. 9 and 14	0.21	0.21

Source: Modified from Ref. 97.

5. Antisolvent is fractionated from the products above and is usually recylced.
6. The antisolvent-free products are further processed, as is desirable in the particular process.

It is practical to recover a very low-ash product from settlers using precipitation techniques. It has been possible to produce products with an ash content similar to that obtained with pressure filters. For instance, SRC has been produced with 0.1% ash or less.[42,110]

Since it is desirable to be able to fractionate the antisolvent from the liquefaction recycle solvent, a solvent boiling range within the limits of 350–500°F is usually chosen. This also provides a reasonably low vapor pressure at solids separation conditions (usually 400–650°F). The amount of antisolvent to be added generally depends on the following factors: type of antisolvent, type of coal liquefaction products, operating conditions, and desired separation velocity.

It is possible to get high liquid recoveries. The precipitated residuum is not recoverable with the clarified product and leaves with the underflow. In an SRC-type process this amounts to about 5–15% of the SRC product.[104,110] Higher recoveries require a multistage system. The ultimate recovery that is chosen represents an economic balance that depends largely on the value of the underflow as an internal stream. The system is suitable for multiple-stage settlers (washing)[111] and/or combination with carbonization.[42] The economics of solvent precipitation may represent a substantial improvement over filtration.[113]

28.4.3.2 *Filtration*

Pressure precoat filtration has been used successfully to produce a low-ash; solvent-refined coal. The production of SRC containing less than 0.16 wt % ash has been achieved on the pilot-plant scale, using a pressure precoat rotary drum and pressure precoat horizontal leaf filters.

The Fort Lewis, Washington, SRC plant has utilized two rotary precoat filters having areas of 40 and 80 ft² respectively, to process a filter feed derived from about 50 tons/day of feed coal entering the reactor section. At the Wilsonville, Alabama, 6-ton/day SRC plant, a 50-ft² pressure leaf filter has been utilized. Nominal specifications for these machines are includes in Table 28.27.

Reactor effluent that has undergone some vaporization to remove light ends is normally filtered near its bubble point (450–550°F) for minimum viscosity. Filtrate viscosity is 0.5–1.0 cp at these conditions. The undis-

[113] J. D. Batchelor and C. Shih, "Solid-Liquid Separation in Coal Liquefaction Streams," paper presented at 68th American Institute of Chemical Engineers Meeting, Los Angeles, November 1975.

Table 28.27 Filter Characteristics of Two SRC Plants

	Fort Lewis, Washington	Wilsonville, Alabama
Plant capacity	50 tons/day	6 tons/day
Filter type	Pressure rotary drum (Goslin Birmingham)	Horizontal pressure leaf (Funda)
Filter area	One each, 80 ft²[a]	54 ft²[b]
	One each, 40 ft²	—
Operating pressure	100 psi	
Operating temperature	450–500°F (insulated)	550°F (Dowtherm jacketed)
Pressure drop	20 psi normal	15–75 psi
Cake removal	Pressurized slurry	Batch-mechanical-gravity
Precoat	3 in.	0.19 lb/ft²
Filter aid usage[c]	0.01–0.08 lb precoat/lb solids (0.03 typical)	0.06 lb precoat/lb solids
Cycle time	24–48 hr	60–90 min
Tray spacing	Not applicable	1.6 in.[b] (maximum cake, 1 in.)

[a] Includes excess capacity with 90% operating factor.
[b] Modified in 1976 to 3.2-in. spacing to study thicker cakes.
[c] Typical of reported results only. Optimized cycles for specific installations may vary widely.

solved coal and liberated mineral matter, are very fine, normally less than 1 μm. Commercial filter aids such as diatomaceous earth are adequate precoat media. The filter cakes built up during filtration are highly flow resistant and compressible. The filter cake that is normally formed contains valuable process solvent and SRC, which must be recovered before the mineral residue is rejected. Therefore the cake formed is washed with a lower boiling process-generated "wash" solvent that can be recovered by flashing under noncoking conditions.

The pressure rotary drum filter, developed from earlier vacuum filter designs, is a device that is suitable for essentially continuous thin-cake operation. Cake washing and partial drying can be achieved as the drum rotates. The pressure designs used have had limited success, however, because mechanical limitations restrict filtration efficiency, limit machine onstream time, and increase maintenance costs.

Problems with the cake-shaving doctor knife have limited the potential advantage of reduced cake resistance achieved by rapid drum rotation and precise knife advance. Drum submergence is limited to 35% by the necessity to keep the liquid level well below the shaft bearings.

Support screen life is short and variable. Development of small holes in the support screens produces unacceptably low product clarity and is a significant maintenance problem. Improved methods of support screen fastening have been reported by the plant operators.

Steady transfer of shaved and blown cake from the filter is also a problem. Operability is maintained by rewetting (slurrying) the cake before transfer from the pressurized case. The added wash solvent imposes an extra load on the cake drier.

The pressure filter with multiple horizontal leaves has proved to be reliable at Wilsonville, and a 2-in. cake thickness has been achieved. The leaf filter is a batch operation that operates on an automated time cycle. During the filtration cycle the cake is formed, washed, gas stripped, and vacuum dried. The dried mineral residue cake is rejected to a cooler before it is stored. The leaf design is more reliable and mechanically simpler, but filter aid usage is higher than with the rotary filter.

28.4.3.3 *Magnetic Separation*

Within the last 10 years, practical methods have been devised to employ high magnetic field intensity and high field gradient in large-scale applications. Using these developments, magnetism can be applied on a commercial scale to remove micrometer-sized, feebly magnetic materials. The high gradient magnetic separation (HGMS) technology was developed for wet benficiation of kaolin clay. The magnetic filter is composed of an electromagnet surrounding a magnetic filter bed. The bed is an open structure packed with fine strands of filamentary materials such as stainless steel. Oder has extensively reviewed the development of this technology and its potential application to coal liquefaction.[114,115]

The HGMS technique is capable of removing mineral sulfur constituents (FeS_x) from coal liquefaction products. The most straightforward application of this technology is the SRC process, wherein a relatively low-sulfur product would be produced. This product, however, would still contain most of the coal ash.

Tests on coal liquefaction products have been conducted on small-scale continuous separators.[115–118] These tests have shown that reductions of 75–90% of FeS_x in the feed material are practical.

28.4.3.4 *Centrifuges*

A great deal of experience has been accumulated with centrifuges in bench-scale and small pilot-plant operations. Typical data ob-

[114] Robin R. Oder, "High Gradient Magnetic Separation and Applications," *IEEE Trans. Magn.*, **MAG-12,**(5) (September 1976).

[115] Robin R. Oder, "Magnetic Desulfurization of Coal Liquids: Conceptual Process Design and Cost Estimation," *IEEE Trans. Magn.*, **MAG-12**(5), (September 1976).

[116] E. Maxwell, D. R. Kelland, and I. Y. Akoto, "High Gradient Magnetic Separation of Mineral Particles from SRC," *IEEE Trans. Magn.*, **MAG-12**(5) (September 1976).

[117] C. J. Lin, Y. A. Liu, D. L. Vives, M. J. Oak, G. E. Crow, and E. L. Huffman, "Pilot Scale Studies of Sulfur and Ash Removal from Coals by HGMS," *IEEE Trans. Magn.*, **MAG-12**(5) (September 1976).

[118] Hydrocarbon Research Inc., *Solvent Refining of Wyodak, Illinois 6, and Black Mesa Coals,* Part IV, Task 7, EPRI 389, February 1976.

Table 28.28 Summary of Centrifuge Operating Data

Process Development Effort	Ref.	Type of Centrifuge Used	Temperature of Operation (°F)	Feed Rate Range (gal/min)	Ash Content of Feed (wt %)	Ash Content of Overflow (wt %)	Ratio of Ash in Overflow to Ash in Feed (lb/lb)
CSF, Cresap, West Virginia, 1968	119	Bird solid bowl, 6 in.		6	No data		
H-Coal, Trenton, New Jersey, 1975	27	Sharples horizontal bowl, P-660	300	0.4–4	3–20	0.7–10	0.1–0.4
	27	Sharples horizontal bowl, P-850	500	0.4–4	3–20	0.7–10	0.1–0.4
SRC, Wilsonville, Alabama, 1976	120	Dorr Oliver Merco PC disk bowl	350				
	120	Sharples horizontal bowl, P-850	500	0.5–0.8	2	0.2	0.02
				0.05–3	3–10	0.1–0.3	0.1
Synthoil, Bruceton, Pa., 1974	121	Sharples Supercentrifuge 4 in. I.D. × 30 in. long		Batch		0.1–0.7	

tained at the CSF (Cresap, W. Va.),[119] H-Coal (Trenton, N.J.),[27] SRC (Wilsonville, Ala.),[120] and Synthoil (Bruceton, Pa.)[121] process locations are summarized in Table 28.28. Horizontal, vertical, and disk nozzle machines have been utilized with modest success. The amount of ash remaining in the clarified overflow has usually represented about 10% of the ash in the feed. Seal problems have plagued the operations at Cresap and Wilsonville when operations at projected commercial operating conditions were attempted. Nozzle plugging on disk machines was encountered at both HRI and Wilsonville. This experience indicates that if centrifuges are to play a part in the future of liquefaction, test machines capable of sustained operation must be developed. Specific capabilities must include erosion resistance and the means for withdrawing cakes with solids concentrations of 60% or more. There is some possibility that high-capacity machines may be developed for commercial operation.

28.4.3.5 *Hydroclones*

Hydroclones, or liquid cyclones, have been utilized to accomplish partial separation of ash contained in the liquid product from coal liquefaction systems. The degree of separation obtained is sufficient to enable the hydroclone overflow to be used as slurrying oil for fresh coal feed. If techniques are developed to allow coal-slurry-oil mixing and pumping at 500–600°F, substantial cost savings and thermal efficiency increases could be obtained. Hydroclone underflows, representing a concentrated solids stream, relative to the hydroclone feed, are normally sent to the next stage of solids separation. Vacuum distillation has been mentioned as the next stage for H-Coal syncrude operations and the Gulf SRC II flowsheet. Filtration and centrifugation have followed hydroclones at the Wilsonville SRC plant. In H-Coal fuel oil operations, hydroclone underflow has been fed to a solvent precipitation system.

The degree of separation achieved in a hydroclone is a function of the diameter of the hydroclone, the viscosity and density of the liquid, and the density and particle size of the solid. Ash underflow to ash overflow concentration ratios normally are in the range of 1.5–3 to 1, depending on operating parameters. Normally 20–40% of the feed volume is withdrawn as the hydroclone underflow stream.

Erosion is a problem that is unresolved. Stainless steel does not have adequate erosion resistance. Ceramic units were utilized successfully at the Cresap CSF plant and the H-Coal unit in Trenton. Wilsonville tests have been of limited duration.

One of the major technical questions to be resolved is the potential buildup of fine solids in the hydroclone overflow stream, which is recycled to the slurry mixing step. Particles below a certain minimum effective diameter do not segregate between the overflow and underflow. Since they are not preferentially purged along with the larger solids, they will increase in concentration in the hydroclone overflow stream used as a slurrying oil.

Hydroclones were utilized at the 20-ton/day CSF process pilot plant operated at Cresap, West Virginia, by Consolidation Coal Company.[122] Three configurations of 3-in. hydroclones were tried (Fig. 28.31). The feed for these units was a slurry prepared from Pittsburgh Seam No. 8 coal at extractor conditions of 725–775°F and 400 psig in a mechanically stirred vessel. With the two-stage operations, 90% of the extract with an ash content of less than 1% was routinely obtained.[122]

28.4.4 Gasification of Liquefaction Plant Residues

Liquefaction plants are normally designed to internally supply their own hydrogen and

[119] *Coal Liquefaction Design Practices Manual,* EPRI Rep. No. AF 199, Project 409-1 Final Report, July 1976.

[120] H. E. Lewis, *Solvent Refined Coal (SRC) Process Operation of SRC Pilot Plant at Wilsonville, Alabama,* Quarterly Tech. Prog. Rep., No. FE-2270-3, April–June 1976; publ. by Catalytic, Inc., Wilsonville, Ala., August 1976.

[121] Sayed Akhtar, Nestor J. Mazzocco, Murray Weintraub, and Paul M. Yavorsky, "Synthoil Process for Converting Coal to Nonpolluting Fuel Oil," paper presented at 4th Synthetic Fuel from Coal Conf. Oklahoma State University, Stillwater, May 6–7, 1974.

[122] John A. Phinney, "Coal Liquefaction at eh Cresap, W. Va., Pilot Plant" paper presented at Coal Conversion Pilot Plant Symp., Area 3a, Committee Solid Fuel Technology, Pittsburgh, June 2–5, 1974.

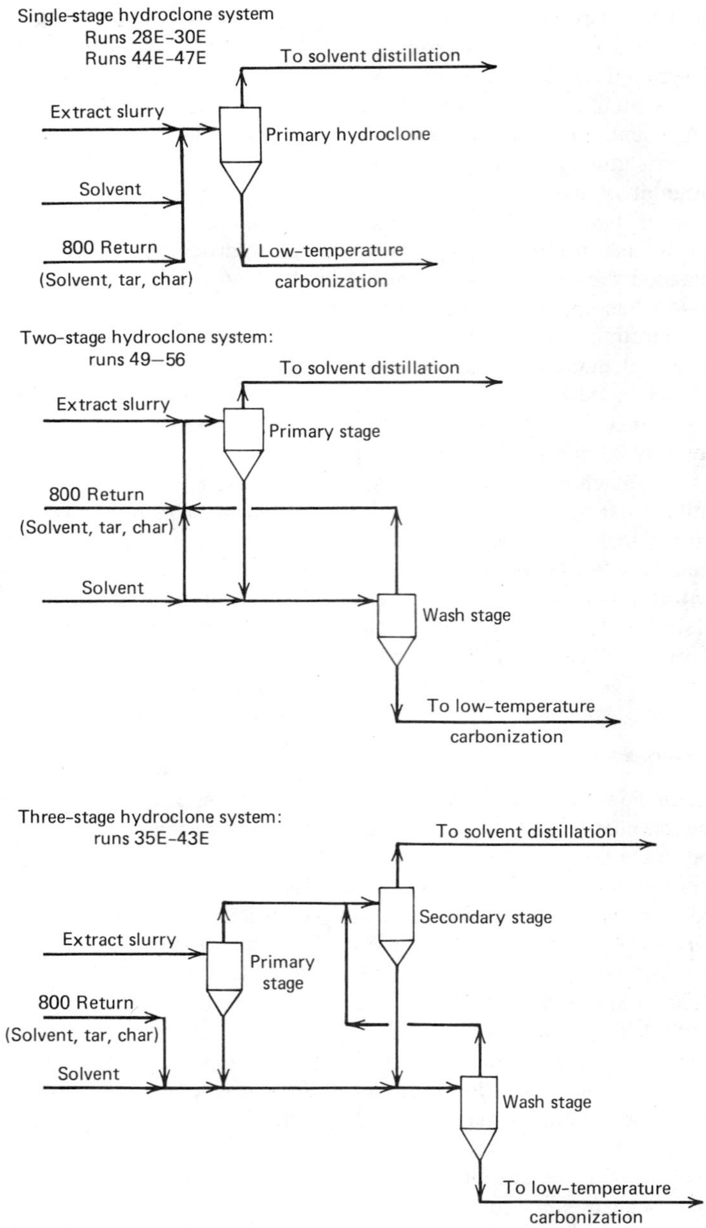

Figure 28.31 Flowsheet of hydroclone system for residue separation at the 20-ton/day CSF process pilot plant at Cresap, West Virginia.

fuel gas requirements. The conventional approach involves gasification of residues that contain unconverted coal and ash. For the SRC process it has been suggested that filter cake be fed to a Koppers-Totzek gasifier. A Texaco unit has been suggested for pumpable vacuum tower residues from H-Coal operations in the syncrude or all distillate mode. Flexicoking has been proposed as the most appropriate process for EDS vacuum tower residues. Depending on individual plant designs, additional hydrogen and fuel gas requirements are met by steam reforming of light hydrocarbon products or gasification of raw coal.

Koppers-Totzek entrained gasification involves the reaction of residue (and/or coal), steam, and oxygen at atmospheric pressure and at an equilibrium temperature of about 2700°F (see Chapter 24). Finely divided solids, normally 70% passing 200 mesh, are fed into a mixing head by a screw feeder. In the mixing head a combined stream of oxygen and steam entrains the pulverized solid feed. To ensure that no flashback occurs, this mixture is delivered to the reaction zone at a velocity that is above the speed of flame propagation. The product gas from this operation is normally essentially all hydrogen, carbon monoxide, and carbon dioxide, with small amounts of nitrogen, hydrogen sulfide, and carbonyl sulfide. Ash is liquified in the high-temperature zone. About half the ash flows down the walls of the vessel into a slag quench tank. The remainder is carried from the gasifier, entrained in the gaseous product, and ultimately removed in a water wash stream. Typical concentrations of the major gas stream components are 36 mole % hydrogen, 56 mole % carbon monoxide, and 8 mole % carbon dioxide.

One disadvantage of the Koppers-Totzek system is that the product gas must be compressed at some point for further processing and ultimately for use in the liquefaction reactor. It is, however, a convenient system for injecting solids into a reaction zone and is particularly useful when the residue is a filter cake, since no slurrying medium is required.

Another approach has been adopted for use in the H-Coal process when it is operated in the syncrude mode. Vacuum distillation is utilized to concentrate all the ash, unconverted coal, and heavy liquid into a pumpable stream. This stream is then injected along with steam and oxygen into a Texaco partial oxidation unit operating at elevated pressure. Test results have been reported on the gasification of residues from Illinois No. 6 and Wyoming coal at about 400 psig and 2340–2570°F.[123] Typical gas compositions were 38 mole % hydrogen, 57 mole % carbon monoxide, and 5 mole % carbon dioxide. Operation at higher pressures would be even more desirable.

In the EDS process a vacuum bottoms stream similar to that utilized in the H-Coal process is fed to a fluid bed coker and additional liquid product is recovered from this operation. Coke is then gasified with steam and air in a fluid bed. Low-Btu gas produced from this step is used as fuel gas within the process. Hydrogen is made from steam reforming of light gaseous products.

[123] *Hydrogen Production from Coal Liquefaction Residues,* EPRI Report No. AF-233, Project 714-1, December 1976.

28.4.5 Process Engineering Considerations

Coal liquefaction plants will be complex plants and costly in terms of investment per barrel of output. Economies of scale dictate that commercial plants will probably have a capacity of 60,000 bbl/day or larger. Investments at this size level for totally integrated plants, including power production, will be of the order to $15,000–$25,000 (1976 dollars) per barrel of daily capacity, for a first plant. To ensure reliable operation, first plants are likely to have a liberal sparing philosophy, extremely conservative design standards, and nonoptimized flowsheets. Reductions of 20–50% in investments for installed equipment for future plants should be possible.

Plant complexity derives in part from the requirement that the operation handle large quantities of solids. Furthermore, many streams must undergo phase changes. Heat exchange service is difficult, since multiphase streams that may be thermally unstable must be handled in multipass exchanger systems. Recycle streams, such as slurrying oil and hydrogen for reaction, are large in comparison to net flows through the plant. As a result, careful and conservative engineering must be carried out to design a reliable plant.

For hydrogenation-based processes, the plant typically can be broken down into the following sections:

Section	Percentage of Investment
Coal receiving, handling, and drying	5
Liquefaction, solids separation	30
Hydrogen and fuel gas manufacture, including gas cleanup	30
Power generation	15
Product storage and handling, and general offsites	20
	100

Except for the liquefaction section, conventional electric utility or petroleum refining industry experience is applicable to design and operating practices. Even in the liquefaction section, the majority of unit operations and processes has its origin in petroleum-refining operations. These include the use of hydrogenation reactor conditions within the current range of temperatures and pressures used for hydrocracking and hydrodesulfurization, distillation, gas compression and purification, high-pressure heat exchange, and fired preheat furnaces.

The degree of process integration that can be achieved is dependent in part on the amount of risk that an investor is willing to take in return for investment reduction. With a high degree of integration, requiring dependence of one unit on another, the failure of a single unit may shut down the entire plant. Since plants will be large for economic reasons, multiple liquefaction reactor trains will be used. Each of these trains will have a capacity of 10,000–30,000 bbl/day. Choices must be made about how the coal is supplied to each reactor system. Coal will most likely be supplied from multiple coal grinding and drying equipment units that are not individually linked to a specific reactor. Slurrying oil, however, will probably come from a single large storage tank. Cooling of reactor effluent streams will be most effectively carried out within a given train.

Certain activities themselves represent an integrated series of processes that can be handled by a specific plant section, and these will not be serially linked to individual liquefaction reactor trains. Examples would be gas cleanup, hydrogen generation, coking, steam-methane reforming, product handling, and storage.

On the other hand, within the liquefaction section other processing activities would be best handled in sequence with each reactor train. These would particularly include feed slurry pumping, feed preheat, and reactor effluent heat exchange.

The question of maximizing thermal efficiency by extensive feed product heat exchange must be addressed. Since the hydrogenation reaction is exothermic, utilization of reactor effluent to reject high-temperature heat to reactor feed streams is desirable. However slurry-slurry heat exchange in a temperature range where both slurries are undergoing chemical reaction may prove to be undesirable because of inherent low heat transfer coefficients and the possibility of surface fouling. Generation of high-pressure steam as an alternative may be desirable if a use can be found for the steam.

28.5 MECHANICAL EQUIPMENT

28.5.1 Pressure Reduction Valves

One of the major operational problems faced in the development of coal liquefaction processes based on slurry hydrogenation has been erosion of the valve trim in the pressure reduction valves used to solids-containing streams leaving the high-pressure liquefaction sections. Conventional practice has been to utilize tungsten carbide parts in valves that are installed in manifolds that contain spare valves. In this way valves can be replaced without shutting down the rest of the plant.

To fabricate tungsten carbide into usable shapes, it is necessary to utilize binders such as cobalt and chromium to hold the tungsten carbide crystals together. These binder materials are much softer and are attacked by the erosive slurry. Failure usually proceeds by removal of the binder materials, which hold the tungsten carbide crystals together. No new materials have been developed in the last 25 years that demonstrate an improvement over tungsten carbide in this service.

The rate of erosion of a given set of valve trim is a function of the amount of slurry passed through, the concentration of solids, the type of solids, the pressure drop, the amount of dissolved and entrained gases in the liquid slurry, and the angle of contact between the slurry and the valve part.

The most significant approach to improving valve trim life is to utilize two or more stages of pressure reduction so that the velocity of the flashing gases in the system is reduced. Other approaches that may be useful are to minimize the amount of entrained gas in the slurry entering the valve and to optimize the angle of contact between slurry and part. Development of improved materials obviously holds the highest potential of all.

28.5.2 High-Pressure Feed Pumps

Reciprocating plunger pumps have been used to pressurize feed slurries to the hydrogenation reactors in all process development efforts in this field to date, whether at the bench scale or the PDU scale. It is most likely that this type of pump will be utilized in any first-generation commercial liquefaction plant, but it does suffer from several deficiencies. These include low volume capacity per machine, and relatively high wear and required maintenance on shafts, packing, seats, and checks. Some improvements have been noted recently in shaft and packing life resulting from the availability of newer materials. The use of tungsten carbide pump seats and checks has decreased maintenance in this area. Improvements are obviously desired in obtaining new solutions and alternatives to the reciprocating plunger pumps of conventional design.

28.5.3 Preheater

Experience at the Wilsonville and Fort Lewis SRC plants has proved the feasibility of radiant-fired helical coil heaters in coal slurry service. The design and experience with heat fluxes of 5000–10,000 Btu/(hr)(ft^2) in the two installations are moderate compared to conventional refinery heater installations, which are designed for fluxes of 10,000–20,000 Btu/(hr)(ft^2). Even this is a significant advance over the German experience with all convection heaters and the overly complicated steam jackets used at Louisiana, Missouri.

The heater utilized at the Cresap, West Virginia, CSF plant had a horizontal hairpin design with individually fired sections. This design has an advantage in that it can be scaled up. Helical coils, which are convenient for small pilot-plant and process development units, cannot be scaled up for use in commercial liquefaction plants.

Wilsonville data indicate an apparent high film resistance toward the end of the helical coil at temperatures above 600°F. This apparent resistance rises sharply and falls before the coil outlet point is reached. The problem of high apparent resistance to heat transfer has been followed closely at Wilsonville. The rapid rise in resistance begins at about 600°F, peaks out at about 750°F, and essentially disappears at around 800–825°F. The terminology "apparent heat transfer resistance" is used because the actual heat transfer rate can be determined only when reliable heat of reaction data have been determined. The phenomenon is unique to coal dissolution and is generally accepted to be a result of the formation and subsequent resolution of a viscous gel. This phenomenon alone is a challenging, unresolved obstacle to radiant furnace design.

In the future, as part of the effort to optimize preheater designs, data should be obtained in the following areas:

1. An experimental design is necessary that will explore the maximum allowable velocity. Wilsonville and Tacoma operate in the 2–6 ft/s range because of the fear of erosion at higher velocities. The Germans apparently were successful with operations at 25–30 ft/s.

2. An experimental hairpin design should be developed that has at least two independently fired sections, so that the 500–600°F and 600–700°F ranges can be separated and studied.

3. Heaters should be instrumented so that good heat balances are possible and heat transfer and pressure drop characteristics can be studied in reasonable tube increments.

4. The heats of reaction involved should be determined, and it should be ascertained whether they are significant.

5. The advantages and disadvantages of utilizing process hydrogen in preheaters should be identified. This is not purely a heater problem but includes fundamental process chemistry considerations. A furnace coil is essentially plug flow. Therefore significant exothermic heat of reaction is accompanied by a temperature increase. Limiting the amount of hydrogen available limits the extent of the exothermic reaction. However hydrogen in the coil may prevent coking and significantly affect fluid viscosity.

6. Methods of controlling mass flow distribution of the three phase mixture entering a multipass preheater must be developed.

28.5.4 Mineral Residue Drying

When filtration is utilized as the method for solid separation, solvent recovery from the separated residue is essential. Unsatisfactory results have been obtained with the rotary-fired kiln at Fort Lewis and the scraped jacketed vessel at Wilsonville. Coke formation in these vessels has been frequent and has required the addition of internal coke cleanout facilities. It is not certain whether the integrity or quality of the solvent can be maintained as it passes through this severe treatment. The most successful operation has been vacuum drying in place on the Funda filter leaves at Wilsonville. Drying on the leaves slightly reduces filter capacity in that it slightly lengthens the overall cycle.

A major unsolved problem associated with ash drying on a rotary precoat filter is the transport of pressure letdown of filter cake from a filter to a dryer.

28.5.5 Product Solidification

Product solidification has been achieved at Wilsonville by direct heat transfer and indirect heat transfer. Direct heat transfer by water quenching is a successful operation. The product of water quenching, however, has the undesirable properties of retained moisture and low bulk density. The indirect approach at Wilsonville uses a water-jacketed plate coil. This equipment is basically a batch device, which means that the capacity of individual units is low. The development of the plate coil and the associated operating cycle has been quite successful on the scale of the Wilsonville plant. Because of the limited capacity of single units, however, it is not suitable for large-capacity plants.

The Sandvik water-cooled continuous belt used at Tacoma has inherent mechanical problems that limit the size of individual units. The whole concept of indirect heat transfer has limitations that result from the inherent physical properties of SRC.

Large-scale casting, such as that used in Frasch sulfur plants, has some possibility, but the fumes present an unevaluated problem.

ACKNOWLEDGMENTS

This chapter was possible only because Norman Stewart, Howard Lebowitz, and William Rovesti prepared sections of the work, and their contribution is acknowledged. Ms J. Gluckman's editorial contribution is acknowledged, as well as the work of a number of reviewers who commented on the subject matter at the request of L. Swaab, of the Exxon Research and Engineering Corporation. Special review efforts and suggestions by H. Siegel of Exxon Research and Engineering, and Bruce Schmid of Gulf Oil Corporation are also acknowledged.

CHAPTER **29**

Processing Coal Liquefaction Products

BILLY L. CRYNES

Billy L. Crynes *is professor and head of the School of Chemical Engineering at Oklahoma State University, Stillwater, Oklahoma. He has been conducting research on catalysts for coal liquids for the past 10 years. This work has been a natural carryover from his experience with heavy petroleum stocks. Dr. Crynes's other coal-related activities date back to his pilot-plant work in industry in the early 1960s and more recently to his activities with the Electric Power Research Institute and Oak Ridge National Laboratory.*

29.1 INTRODUCTION

29.1.1 Scope and Objective

The primary conversion of coal to liquid and gaseous products has been covered in Chapters 24 and 28. This chapter presents the work done since 1963 on secondary chemical processing or upgrading of liquids obtained from coal liquefaction processes. The discussion is confined to chemical (as opposed to physical) treatment and covers the liquids that are derived from liquefaction. Secondary processing of product liquids derived from gasification processes such as methanol, Fischer-Tropsch liquids, and related process liquids, are discussed in Chapter 30. The *in situ* liquefaction of coal and the subsequent treatment of the derived liquids are also excluded from consideration here, because the data available are limited.

The liquids discussed in this chapter include both major and secondary liquids produced from solvent extraction, hydrogenation, and carbonization schemes, as well as hybrid concepts that incorporate more than one conversion method. The type of liquid feedstocks and their characteristics, the processing conditions, the chemicals and catalysts used, the types of reactor vessel, and the nature of the chemical changes are all discussed.

Most of the data reported in this chapter have been obtained from results published since 1963, when the first Supplementary Volume of *Chemistry of Coal Utilization* was published.[1] Most of the material covers work presented through 1976; a few references that became available in 1977 are also briefly mentioned. The literature references and data presentation are not inclusive, but the more important and appropriate works are presented, to cover the entire topic adequately. In the first Supplementary Volume[1] the type of material considered in this chapter was distributed in the chapters discussing low- and high-temperature tar, light oils from carbonization, and hydrogenation of coal and tar.[2] Much of the coverage in these areas was given to liquid characterizations.

29.1.2 Definitions

Table 29.1 presents for the nonspecialist reader some definitions and abbreviations that are used throughout this chapter. These definitions do not necessarily represent stan-

[1] H. H. Lowry, Ed., *Chemistry of Coal Utilization*, Supplementary Volume, John Wiley and Sons, New York, 1963.

[2] In Ref. 1, Chs. 13–15, 22.

Table 29.1 Definitions and Abbreviations

Benzol	A general term for light liquids derived from coal carbonization, usually boiling below 200°C and rich in aromatics
BTX	Benzene, toluene, xylene
Direct hydrogenation liquids	Liquids derived from coal by direct coal-hydrogen contact; no initial solvent added but catalysts may be impregnated on the coal
FCC	Fluidized catalytic cracker
IBP	Initial boiling point
Gasoline range	Describing a material usually boiling over the range of C_5–200°C
HDS, HDN, HDO	Hydrodesulfurization, hydrodenitrogenation, hydrodeoxygenation
Light oils	Usually the lower boiling liquids obtained from coal carbonization or coking, generally boiling below 200°C and rich in BTX
Middle oils	Liquids generally boiling between 200 and 325°C
PNA	Polynuclear aromatics
SRC, SRL	Solvent-refined coal, solvent-refined lignite
Syncrudes	Coal-derived liquids especially directed to serving as a petroleum refining feedstock
WHSV, LHSV, LHST, WHST, VHSV	Measures of catalyst processing capacity or time: they stand for, respectively, weight hourly space velocity (weight of feed oil per hour per weight of catalyst), liquid hourly space velocity (volume of feed oil per hour per volume of catalyst), liquid hourly space time (volume of catalyst per volume of feed oil per hour), weight hourly space time (weight of catalyst per weight of feed oil per hour), volume hourly space velocity (same as LHSV)

dard terms: they are meant only to be consistent within this chapter.

29.2 PROCESSES PRODUCING COAL LIQUIDS

A wide variety of processes has been used to produce coal liquids, and many are now being developed, to obtain coal-based liquids as either a primary product or a coproduct. The nature and characteristics of the liquid to be further processed are greatly influenced by the techniques used for its generation, which in turn are influenced by the properties of the widely varied ranks, types, and fields of coal that can serve as feedstocks to the many processing schemes. The relevant properties of coal, and representative processes, are therefore briefly examined here as an orientation to the discussion of characterization and upgrading that is the principal topic of this chapter. More detailed discussions of specific processes are found in Chapter 28, and rather extensive guides to the production of liquids from coal as taken from the United States patent literature are available.[3,4] Also of interest are the reviews by Wu and Storch[5] and Hawk and Hiteshue.[6] More recent bibliographical surveys have been made by the U.S. Energy Research and Development Administration (ERDA).[7,8]

29.2.1 Coal Properties

Coal is an exceedingly complex substance. Considerable effort has been devoted to the study of its properties, its reactivity, and its structural model; these are reviewed in other chapters of this book. The models that are developed must account for both the macrophysical properties and the chemical elemental analyses. In spite of extensive efforts directed toward coal characterization, no one can yet specify what a "coal molecule" is. Indeed, coal found in all its widely ranging types cannot be considered to have a specific molecular structure. Nevertheless, efforts are made to represent coal as some sort of molecular model.

One such suggested model (Fig. 29.1) reveals the complexity of the aromatic, polynuclear forms incorporating heteroatoms. The models are not entirely planar, rather, they are three-dimensional, entangled assemblies. Regardless of whether they are representative, such models do illustrate the complexity of the basic starting material from which all coal liquids are derived. Any coal-based liquid that must be further processed is related both to the original coal source and its character, and to the time-temperature-pressure processing sequence used to produce that liquid.

Table 29.2 shows that the coal from which the liquids are obtained is a rather hydrogen-deficient substance with a hydrogen carbon atom ratio of 0.7 : 0.8. The ratios shown for fluids indicate that this atom ratio must be increased to 1.4 for syncrude liquids, or even to 4 if methane is an objective. Other substances are listed for reference.

An informative relationship between various coals and comparative substances is illustrated in Fig. 29.2, which shows the ranges and the relationship between carbon, hydrogen, and oxygen. Coals fall in the 5–7 wt % (maf) hydrogen range and the 60–95 wt % (maf) carbon range. For liquid products, not only must the hydrogen/carbon atom ratio be increased, but also the ash and heteroatoms must be extracted.

The sulfur content of coals ranges widely from less than 0.4 to more than 6.0 wt %. Coals from the various coal-producing regions of the United States are typically in the 0.7–3.0 wt % (maf) range. The sulfur may be present in the pyritic, sulfate, or organic form. The nitrogen content of coals ranges from about 0.2 to almost 3.0 wt % (maf) and is present entirely in the organic form. The oxygen content is an important parameter for characterizing coals and may range from a low of about 3 wt % (maf) for anthracite to about 30–35 wt % (maf) for lignite, brown coal, and peat. The quantity of ash identified with coal is functionally related partly to the mining and preparation techniques. The ash

[3] F. W. Richardson, *Oil from Coal,* Noyes Data Corporation, Park Ridge, N.J., 1975.
[4] G. K. Goldman, *Liquid Fuel from Coal,* Noyes Data Corporation, Park Ridge, N.J., 1972.
[5] W. R. K. Wu and H. H. Storch, *U.S. Bur. Mines* Bull. No. 633, 1968.
[6] C. O. Hawk and R. N. Hiteshue, *U.S. Bur. Mines* Bull. No. 622, 1965.
[7] *Coal Processing,* Technical Information Center, U.S. ERDA, TID 3349, October 1974.
[8] *Coal Processing, Production, and Properties,* Technical Information Center, U.S. ERDA, TID-3349-S1P1 and S1P2, September 1976.

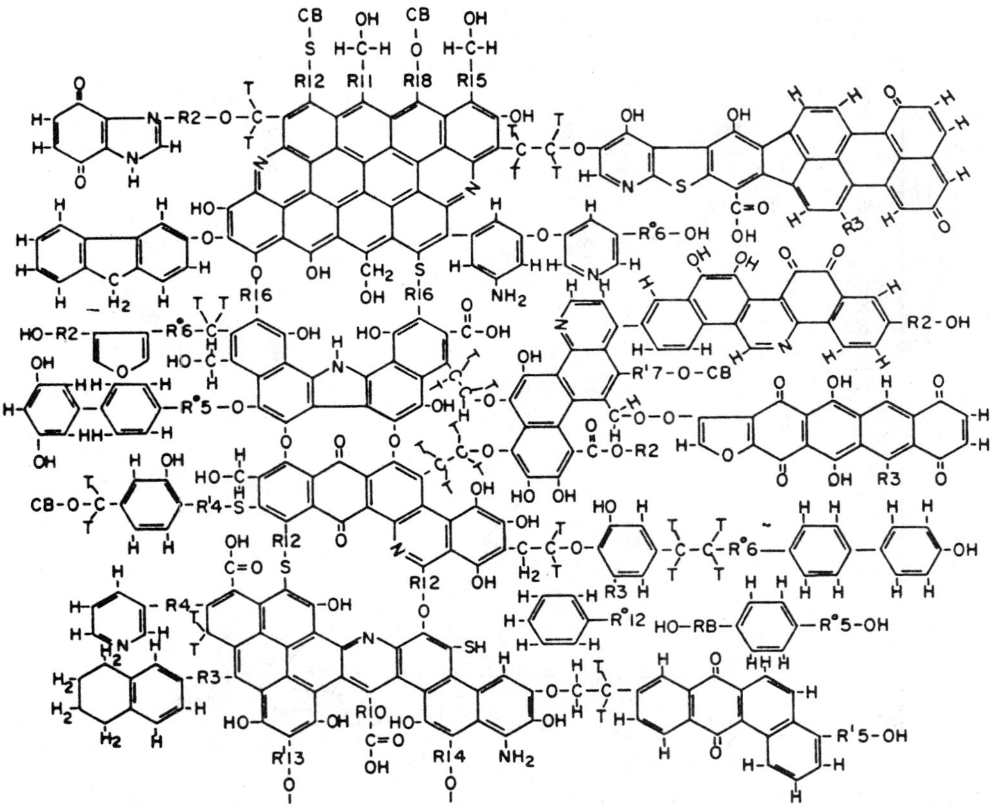

Figure 29.1 A hypothetical coal model.[9]

content of much of the coal mined in the United States ranges from 6 to 20 wt % (mf). Coals that have a low ash content (3–4 wt %) are used to produce metallurgical coke. Some coals, however, might contain as much as 30 wt % ash. In any scheme for producing coal liquids, at least a portion of the heterospecies (S, N, and O) remains in the processing steps and is found in the liquids. Much of the processing of coal-derived liquids is concerned with removal of these heterospecies.

[9] M. G. Huntington, U.S. Pat. 3,244,615, Pyrochem Corporation, April 5, 1966.

Table 29.2 Comparison of Hydrogen/Carbon Atom Ratios

Material	H/C Atom Ratio
Coal	0.7–0.8
Asphaltenes	1.2
Petroleum Crude	1.4
Gasoline	1.6
Toluene	1.14
Methane	4.0

29.2.2 Processing Methods

Methods for deriving liquids from coal can be classified in two general categories: carbon rejection and hydrogen addition. The carbon rejection processing schemes are merely the pyrolysis or carbonization of coal, which for years has been used to produce coke and chemical by-products. Carbonization is the destructive distillation of coal in the absence of air or oxygen. Processes based on this technology are still being developed. Carbonization processes are usually classified into two or three categories, according to the temperature of operations, as shown in Table 29.3.

Table 29.3 Carbonization Ranges

Temperature Classification	Final Temperature (°F)
Low	840–1300
Medium	1300–1650
High	> 1650

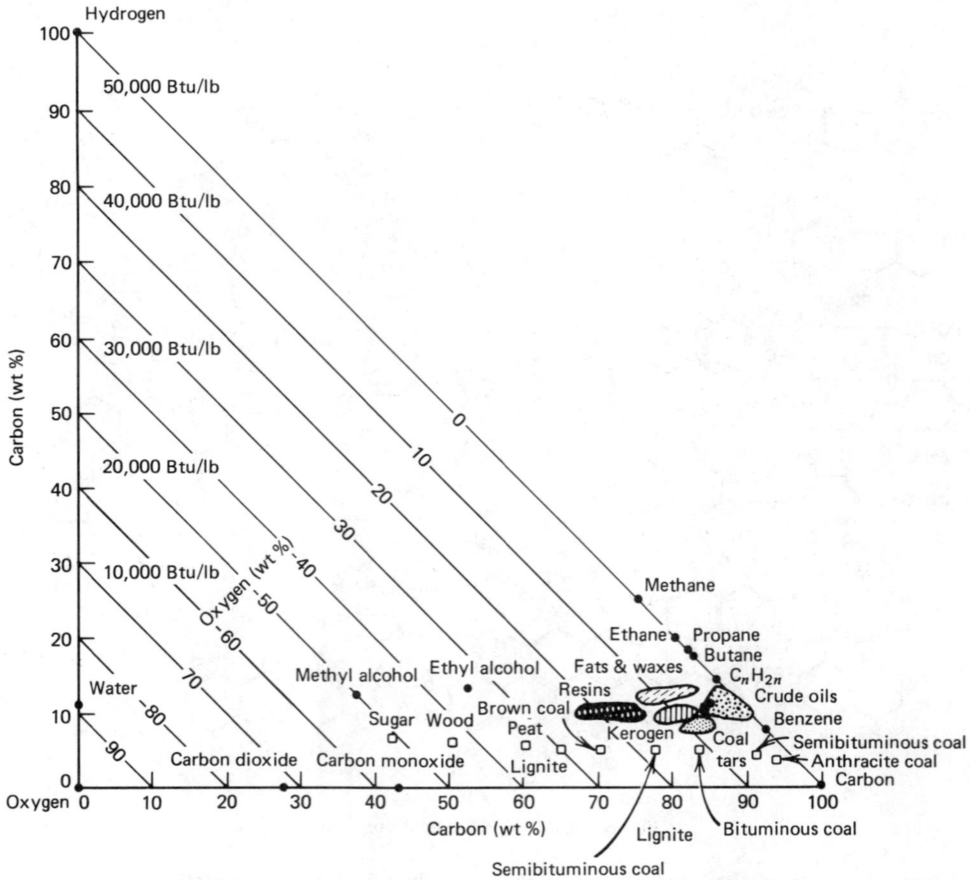

Figure 29.2 The carbon, hydrogen, and oxygen contents of fossil fuels.[10]

The temperature of operation depends on the process objectives (production of liquid fuels, or the type and yield of coke that is desired). In certain classifications the low- and medium-temperature ranges are categorized together as low-temperature carbonization.

The second major category of processing methods, hydrogen addition, incorporates a rather broad range of techniques to obtain a liquid with an increased hydrogen-carbon atom ratio. Direct contact of coal with hydrogen, the use of a solvent as a hydrogen transfer medium, and coal contact with gaseous mixtures are used with a wide variety of process configurations. Some schemes incorporate both basic concepts of liquefaction—carbon rejection and hydrogen addition.

The processing concepts presented below are selected as representative of the range of the process sources that generate the coal liquids.

29.2.2.1 *Carbon Rejection*

When coal is subjected to high temperatures in the absence of air, both gases and liquids evolve as the solid coal undergoes a rather complex sequence of physical and chemical changes. A process (other than the classic methods of coal coking practiced by the steel industries) that typifies the carbon rejection concept is the Char Oil Energy Development (COED) process, developed by the Food Machinery Corporation (FMC).[11,12] This con-

[10] G. A. Mills, "New Liquid Fuels from Coal: Progress in the ERDA Program," paper presented at Gulf Coast Lignite Conf., Austin, Tex., June 2, 1976.

[11] R. C. Merrill, L. J. Scotti, L. Ford, and D. J. Domina, "The Production of Clean Fuels from Eastern Coals by the COED Process," paper presented at American Institute of Chemical Engi-

cept has undergone much developmental work at the scale of 36 tons/day and lower. Dried coal at 375°F (191°C) is processed through four successive stages, operating over the range of 600–1600°F. Product gases and recycled char furnish the heat for the second and third stages. Oils are condensed from the second stage, then hydrogenated at about 3000 psi and 770–800°F. More than half the dry coal fed to the process is converted to char—a major disadvantage of the carbon rejection concept. On the other hand, char produced under conditions that give it high gasification activity is a desirable product.

Some other carbonization processes that yield liquids, at least as coproducts, are listed in Table 29.4. All these processes produce large quantities of char or coke (carbon rejection), operate within the 800–1600°F temperature range, and have relatively high gas yields. In fact, the oil or liquefaction yields may be secondary. Most of the schemes operate at essentially atmospheric pressure, with some designed for up to 1000 psi. Contact times range widely from less than 2 s to more than 2 hr.

neers, 79th National Meeting, Houston, Tex., March 18, 1975.

[12] J. F. Jones, N. J. Brunsvold, H. D. Terzian, L. J. Scotti, F. H. Schoemann, R. C. Merrill, J. D. Alcantara, D. J. Domina, S. J. Romelczyk, and L. Ford, *Char Oil Energy Development,* Vol. 1, *Final Report for Period August 18, 1971, through June 30, 1975,* Rep. No. AFE-1212-T-9, for U.S. ERDA by FMC Corporation, September 1975.

29.2.2.2 *Hydrogen Addition*

Almost all processes that operate on the principle of hydrogen addition use a liquid solvent of some type. These may or may not involve the direct use of hydrogen—and they may be either catalyzed or uncatalyzed. In fact, even when direct coal hydrogenation is attempted by contact between solid coal and hydrogen gas, a "self-liquid" forms immediately, and the subsequent changes can be considered as taking place through the agency of a solvent (i.e., solvent-present changes). When coal is contacted with coal-derived solvents, and when sufficient hydrogen can be transferred from the solvent to the reacting coal, nearly all the organic matter can be "dissolved." The solvent may he saturated with hydrogen (hydrogen gas present) or prehydrogenated outside the dissolver-reactor. A catalyst may or may not be added to the system. In any case the mineral matter of the coal itself is suspected of having catalyzing effects. Three processes that illustrate the hydrogen addition liquefaction concepts are: the Consol Synthetic Fuel (CSF), solvent-refined coal (SRC), and H-Coal processes.

The coal solution process without direct hydrogen addition is represented by the CSF process,[13,14] the precursor of which was Pro-

[13] *Final Report: Development of CSF Coal Liquefaction Processes,* Vol. 5, U.S. Dept. Interior OCR R & D Rep. No. 39, April 1973.

[14] *Zinc Halide Hydrocracking Process for Distillate Fuels from Coal,* Quarterly Rep. No. FE-1743-20, by Conoco Coal Development Company for U.S. ERDA, May 14, 1976.

Table 29.4 Carbonization Processes

Process	Developer	Concept
COED	FMC	Staged beds, 600–1600°F, 5–10 psig
Clean coke	U.S. Steel Corporation	Combined carbonization, 1200–1400°F, 100 psig, and extractive hydrogenation
Hydrocarbonization	Union Carbide Corporation	Fluidized bed, 1050°F, 550 psig
Cogas	Consortium	COED concept plus a char gasification step
Toscoal	Oil Shale Corporation	Low-temperature carbonization; same retort used for oil shale, 800–1000°F, 1.0 atm
Garrett	Occidental Petroleum Corporation	Entrained-bed flash pyrolysis, 1100°F, 1.0 atm

ject Gasoline, developed by the Consolidation Coal Company.[15] Depending on configuration and severity of operation, the products could range from syncrude to low-sulfur fuel oil. The process consists of slurrying crushed coal in a hydrogen donor solvent (process-derived), then processing the slurry at 700–800°F at 400–500 psig. More than 70% of the coal is dissolved. Hydroclone separation yields a solid concentrate and a filtrate that is further treated to remove ash. The solid concentrate is heated to recover solvent and a char. The extract is hydrogenated at about 800°F and 3000–4000 psig to yield a syncrude or distillate and a recycle solvent. Since the donor solvent is derived from the hydrogenation step, it is rich in hydrogen; consequently no hydrogen need be added to the dissolution reactor.

The SRC process being developed by the Pittsburg and Midway Coal Mining Company is illustrative of coal liquefaction with both hydrogen and solvent present, but with no catalyst addition.[16–21] (The ash mineral matter may, however, act as a catalyst.) The primary objective of the process is to produce a low-ash, low sulfur boiler fuel (solid form); a newer concept, SRC II, yields a liquid. In the SRC process coal is slurried with a process-derived solvent at about 1000 psig and 450°F, with hydrogen present. The coal is dissolved in the preheaters and reactors at 1500–2000 psig and 800–900°F. From 90 to 95% of the organic matter in the coal is dissolved. From the recovery section, a spectrum of products is obtained, including a recycle solvent. The SRC II modification incorporates a slurry recycle to the reactor that provides for an increased ash inventory. Other liquefaction schemes operating with hydrogen and solvent contacting coal but with no direct catalyst addition include the Exxon donor solvent process,[22,23] the solvent-refined lignite (SRL) process,[24] and the solvent extraction scheme of Universal Oil Products.[25] Many more process configurations have been conceived.

Of the catalyzed liquefaction processes that have been given major emphasis in the United States, Synthoil and H-Coal are foremost. These two schemes are similar in that they use a hydrotreating catalyst to liquefy coal in the presence of a solvent and hydrogen. They differ significantly in their contacting method. The H-Coal concept[26–29] slurries coal in a process-derived solvent, with hydrogen and catalyst being fed to an ebullated-bed reactor operating at 850°F and 2200–2700 psig. The reaction takes place continuously in the liquid phase, and the products may range from syncrude to lighter fuel oil. The Synthoil process, in contrast,

[15] *Final Report of the Advisory Committee on Project Gasoline,* U.S. Dept. Interior OCR R & D Rep. No. 62, January 1971.

[16] *Solvent Processing of Coal to Produce a Deashed Product,* U.S. Dept. of Interior OCR R & D Rep. No. 53, December 1965.

[17] *Economic Evaluation of a Process to Produce Ashless, Low-Sulfur Fuel from Coal,* U.S. Dept. Interior OCR R & D Rep. No. 53, June 1970.

[18] V. L. Brant and B. K. Schmid, "Design of a Pilot Plant to Produce Ashless, Low-Sulfur Fuel from Coal," paper presented at American Institute of Chemical Engineers, 65th National Meeting, Cleveland, Ohio, May 4–7, 1969.

[19] V. L. Brant and B. K. Schmid, *Chem. Eng. Prog.,* **65**(12), 55 (1969).

[20] W. C. Bull, L. G. Stevenson, D. L. Kloepper, and T. F. Rogers, U.S. Pat. 3,341,447, Gulf Oil Corporation and U.S. Secretary Interior, September 12, 1967.

[21] *Solvent Refined Coal (SRC) Process,* Quarterly Rep. No. FE-496-109, by Pittsburg & Midway Coal Mining Company for U.S. ERDA, July–September, 1975.

[22] L. E. Furlong, E. Effron, L. W. Vernon, and E. L. Wilson, "Coal Processing: The Exxon Donor Solvent Process," *Chem. Eng. Prog.,* **72**(8), 69 (August 1976).

[23] L. E. Furlong, E. Effron, L. W. Vernon, and E. L. Wilson, "Coal Liquefaction by the Exxon Donor Solvent Process," paper presented at American Institute of Chemical Engineers, 68th National Meeting, Los Angeles, November 18, 1975.

[24] D. E. Severson, *Project Lignite,* Quarterly Tech. Prog. Rep. No. 9, April–June 1976, and Annual Summary Fiscal Year 1976, FE-1224-61, by University of North Dakota for U.S. ERDA, October 29, 1976.

[25] W. K. T. Gleim, R. S. Corey, F. J. Riedl, and G. R. Sunagel, U.S. Pat. 3,598,718, Universal Oil Products Company, August 10, 1971.

[26] S. P. Alpert, "Ebullated Beds for Coal Hydrogenation," paper presented at Synthetic Fuels from Coal Conf., Oklahoma State University, Stillwater, May 14, 1970.

[27] *Evaluation of Project H-Coal,* U.S. Dept. Interior OCR R & D Rep. No. 32, December 1967.

[28] *Project H-Coal: Report on Process Development,* U.S. Dept. Interior OCR R & D Rep. No. 26, Final Report, November 1968.

[29] *Commercial Process Evaluation of the H-Coal Hydrogenation Process,* U.S. Dept. Interior OCR R & D Rep. No. 26—Interim Rep. No. 1, February 1967.

operates with hydrogen and coal slurry passing through a packed catalyst bed under extremely turbulent conditions. It operates at essentially the same temperature as the H-Coal process (850°F) but at somewhat higher pressures (2000–4000 psi).[30,31] A low-sulfur fuel oil is the specified product. Other catalyzed liquefaction concepts include Gulf Oil Corporation's Catalytic Coal Liquids (CCL) process[32] and the liquefaction process of the Lummus Company.[33]

The two basic approaches to coal liquefaction—carbon rejection and hydrogen addition—have been illustrated by the processes discussed. The possible variations, combinations, and modifications of the specific steps to bring about the production of liquids are almost limitless. The liquids produced may then be further treated in a number of ways; among several possible secondary treatments are hydrotreatment to remove sulfur, nitrogen, and oxygen, hydrocracking to reduce molecular weight and increase hydrogen content, and coking, dealkylation, isomerization, and halogenation. Before these specific techniques for processing coal liquids can be usefully discussed, typical properties of selected coal liquids need to be presented.

29.3 PROPERTIES OF COAL LIQUIDS

The task of making any general statements about the characteristics of coal-derived liquids is complicated by the wide variations found in the reactive properties of the various coal types. Using the same process under the same operating conditions, the properties found in the liquid product derived from processing a lignite might differ from those of a product derived from a high-volatile bituminous coal. Wide variations in liquid products will also, of course, arise from using the same coal feedstock but in different processes—for example, in a carbonization as opposed to a solvent-refining scheme. Even more subtle considerations, such as changes in liquid properties during storage, make generalization difficult. Nevertheless the attempt to classify and generalize some coal liquid properties has the value of indicating the areas that have to be considered for secondary processing. Following are some specific illustrations of the properties of liquids derived from coal liquefaction processes now being developed.

29.3.1 Process Liquids

29.3.1.1 *Project COED*

The FMC Corporation's Project COED is probably one of the most documented processes developed in the United States. A large number of coals have been processed in the pilot-plant development of this staged carbonization concept. Properties before hydrotreating of the raw pyrolysis oil obtained in processing Western Kentucky and Pittsburgh coals along with the ultimate analysis of the two coals are shown in Table 29.5. Nominal process operating conditions for both coals were as follows:

temperature—Stage 1, 450°F; Stage 2, 735, 765°F; Stage 3, 850, 885°F; Stage 4, 1200, 1285°F; and pressure—5.5–8.0 psig.

Note that the concentrations of hetero atoms in the raw pyrolysis oils are still comparatively high. Consideration must also be given to the concentration of ash if hydrotreating operations are to follow. In fact, FMC has used a guard chamber to protect the catalysts in a subsequent hydrotreating operation.[34] This operation is discussed in Section 29.3.1 on processing coal liquids. For those seeking more detail on COED operations and the resultant liquid characteristics, a number of reports exist.[11,12,34–46]

[30] S. Akhtar, S. Friedman, and P. M. Yavorsky, "Process Hydrodeusulfurization of Coal in a Turbulent-Flow Fixed-Bed Reactor," paper presented at American Institute of Chemical Engineers, 71st National Meeting, Dallas, Texas, February 20–23, 1972.

[31] S. Akhtar, S. Friedman, and P. M. Yavorsky, "Low-Sulfur Liquid Fuels from Coal," *Prepr., Am. Chem. Soc., Div. Fuel Chem.*, **16**(1), (1972).

[32] S. W. Chun, "Gulf Catalytic Coal Liquids (CCL) Process," paper presented at National Science Foundation Workshop, Ohio State University, Materials Problems and Research Opportunities in Coal Conversion, April 16, 1974.

[33] G. J. Snell, U.S. Pat. 3,852,183, Lummus Company, December 3, 1974.

[34] L. J. Scotti, R. C. Merrill, B. D. McMunn, S. J. Romelczyk, D. J. Domina, L. Ford, H. D. Terzian, and J. F. Jones, *Char Oil Energy Development,* Interim Rep. No. 5, Period of Operation July 1973–June 1974, FE-1212-5, by FMC Corporation for U.S. ERDA, August 1975.

Table 29.5 Ultimate Analysis of the Coals Fed and the Raw Pyrolysis Oils Produced in the COED Process[11]

	Western Kentucky Feedstock		Pittsburgh Feedstock	
	Coal	Raw Pyrolysis Oil	Coal	Raw Pyrolysis Oil
Ultimate analysis (wt %, mf)				
Carbon	71.3	82.3	73.8	82.9
Hydrogen	4.9	7.5	5.3	7.3
Nitrogen	1.6	1.2	1.2	0.9
Sulfur	3.1	1.5	3.9	2.3
Oxygen	8.6	6.3	7.4	5.9
Ash	10.5	1.2	8.4	0.7
API gravity of oil at 15.6°C		−6.4		−7.4

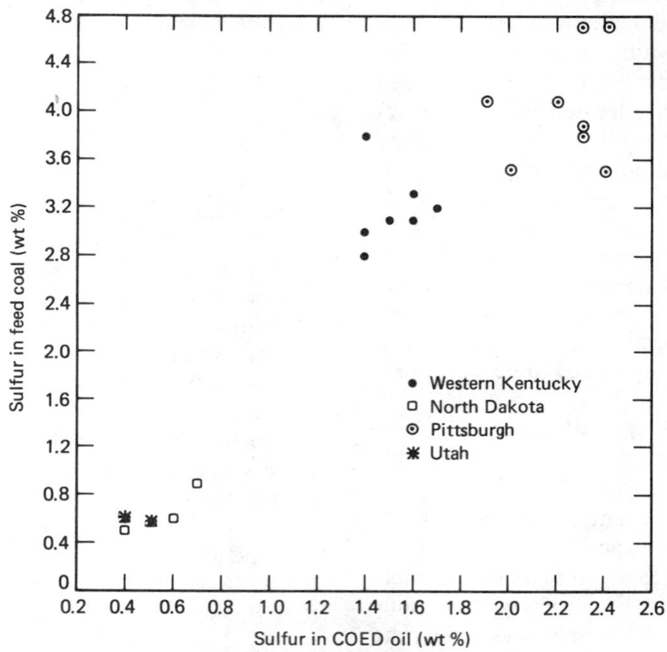

Figure 29.3 The sulfur contents of coal and COED oil.

[35] *Char Oil Energy Development: The Desulfurization of COED Char,* Part III, U.S. Dept. Interior OCR R & D Rep. No. 56—Interim Rep. No. 2, January 1971.

[36] *Char Oil Energy Development Project COED,* U.S. Dept. Interior OCR R & D Rep. No. 11, February 1967.

[37] M. R. Schmid, B. D. McMunn, J. F. Jones, and R. T. Eddinger, "Fluidized-Bed Pyrolysis of Strongly Caking Coals," paper presented at American Institute of Chemical Engineers, 65th National Meeting, Cleveland, Ohio, May 4–7, 1969.

[38] C. A. Gray, M. E. Sacks, and R. T. Eddinger, "Hydrodesulfurization of Coal Chars," paper presented at American Chemical Society 158th National Meeting, New York, September 1969.

[39] J. F. Jones, R. T. Eddinger, and L. Seglin, *Chem. Eng. Prog.*, **62**,(2), 73 (1966).

[40] R. T. Eddinger, "Char Oil Energy Development," paper presented at Synthetic Fuels from Coal Conf., Oklahoma State University, Stillwater, May 14, 1970.

[41] H. E. Jacobs, J. F. Jones, and R. T. Eddinger, *Ind. Eng. Chem. Process Des. Dev.*, **10**(4), 558 (1971).

[42] J. J. Johns, J. F. Jones, and B. C. McMunn, "Hydrogenated COED Oil," paper presented at American Chemical Society Division of Fuel Chemistry, 163rd National Meeting, Boston, April 10–14, 1972.

[43] *Char Oil Energy Development,* U.S. Dept. Interior OCR R & D Rep. No. 56, Final Report, December 3, 1971.

[44] J. F. Jones, *Char Oil Energy Development,* U.S. Dept. Interior OCR R & D Rep. No. 72—Interim Rep. No. 1, 1972.

Figures 29.3–29.6 show the various heteroatom contents of raw COED oils as a function of the corresponding heteroatom content of the parent coal. These data were taken from a range of coal types and under various processing conditions. They indicate that the choice of coal used to produce a liquid can, to some extent, control the concentration of certain heteroatoms that must ultimately be removed by hydrotreatment. In particular, the sulfur content of the product oil depends strongly on the sulfur content of the coal feed (Fig. 29.3). The nitrogen content in the oil, however, is not nearly so sensitive to its concentration in the feed (Fig. 29.4); the same is true for ash and oxygen (Figs. 29.5 and 29.6). Ash particularly would be expected to be a matter of entrainment or carry-through, a physical rather than chemical design problem.

29.3.1.2 *The CSF Process*

The CSF process being developed by the Conoco Coal Development Company has also been well documented, either through the patents of Consolidation Coal Company or through the work contracted with the United States government.[13–15,47–59] Various

[45] S. N. Rothman, D. McNeil, M. I. Greene, L. J. Scotti, and J. F. Jones, "Application of COED Process Coal-Derived Liquids in a Petroleum Refinery," *Proc. 9th Intersociety Energy Conversions Engineering Conf.*, San Francisco, August 26–30, 1974.

[46] Coal Technology Program, Quarterly Prog. Rep. No. 1, for Period Ending December 31, 1974, Oak Ridge National Laboratory Rep. No. ORNL-5026, April 1975.

[47] *Project Gasoline Pre–Pilot Plant: Phase 1, Research on CSF Process*, Vol. 2, U.S. Dept. Interior OCR R & D Rep. No. 39—Interim Rep. No. 1, November 1968.

[48] *Research on Zinc Chloride Catalyst for Converting Coal to Gasoline: Phase I, Hydrocracking of Coal and Extract with Zinc Chloride,* Book 1, Vol. 3; and *Pre–Pilot Plant Research on the CSF Process: Phase I, Regeneration of Zinc Chloride Catalyst,* Book 2, Vol. 3, U.S. Dept. Interior OCR R & D Rep. No. 39—Interim Rep. No. 2, March 1969.

[49] *Pilot Plant Research and Development: Phase II, Design and Construction of the CSF Pilot Plant, Cresap, West Virginia,* Book 1, Vol. 4, U.S. Dept. Interior OCR R & D Rep. No. 39—Interim Rep. No. 3, March 1969.

[50] *Start-up and Initial Operations at Cresap Pilot*

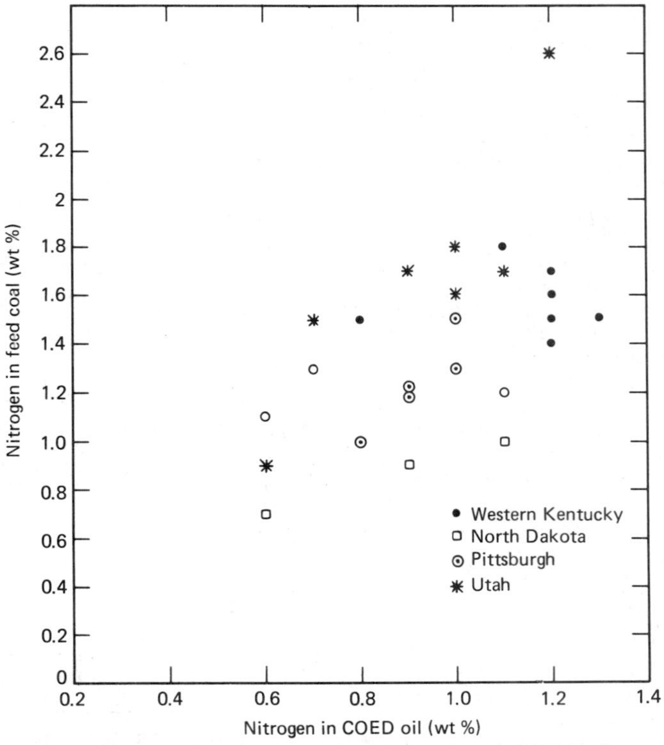

Figure 29.4 The nitrogen contents of coal and COED oil.

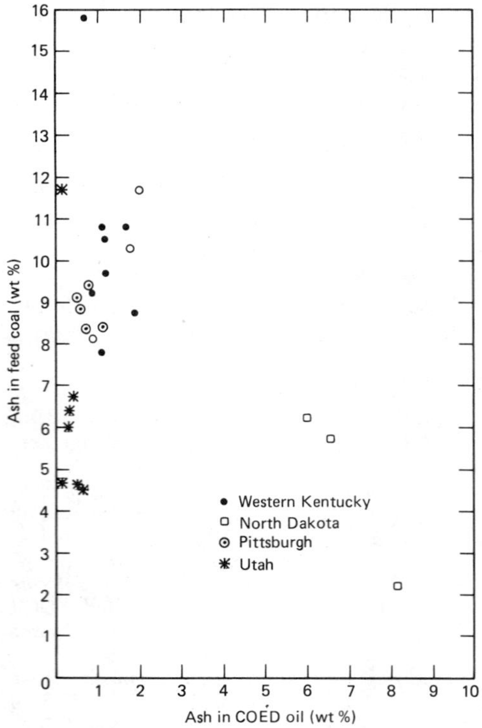

Figure 29.5 The ash contents of coal and COED oil.

process alternatives may yield a syncrude or a low-sulfur fuel oil. The heavy extract fluid from the dissolver-reactor (700–800°F, 400–500 psig) is hydrocracked to generate a recycle solvent and a synthetic crude that could be further hydrotreated. Some typical properties of such a syncrude are listed in Table 29.6; a typical extract might also be similar to that shown in Table 29.6. It should be recognized, of course, that the values shown depend on coal type, the "depth of extraction," and other variables. The sulfur and nitrogen levels, though remaining rather high in the extract, are significantly reduced following hydrotreating to yield a syncrude. Any further processing of the extract will have to accommodate the rather high hetero-atom levels, including ash.

Table 29.6 Typical CSF Syncrude and Extract Properties from Processing Pittsburgh Coal

Syncrude	
API gravity (15.6°C/15.6°C)	21.0
Sulfur	0.09 wt %
Nitrogen	0.17 wt %
Bromine number	7
ASTM Distillation	
Fraction	Temperature (°C)
IBP	94
10%	142
30%	202
50%	231
70%	256
90%	294
EP	344
Extract	
Constituent	Ultimate Analysis (wt %)
Hydrogen	6.2
Carbon	86.4
Nitrogen	1.5
Oxygen	4.2
Sulfur	1.6
Ash	0.13
Oils	27
Asphaltenes	39
Benzene insolubles	34

Source: Modified from Ref. 47.

Plant, Book 2, Vol. 4, U.S. Dept. Interior OCR R & D Rep. No. 39—Interim Rep. No. 4, December 1969.

[51] *Summary Report on Project Gasoline,* Vol. 4, U.S. Dept. Interior OCR R & D Rep. No. 39—Interim Rep. No. 5, March 1969.

[52] *1968 Feasibility Report: Consol Synthetic Fuel Process,* U.S. Dept. Interior OCR R & D Rep. No. 45—Interim Rep. No. 1, 1968.

[53] *Addendum to 1968 Feasibility Report: Consol Synthetic Fuel Process,* U.S. Dept. Interior OCR R & D Rep. No. 45, April 1969.

[54] *1969 Feasibility Report: Consol Synthetic Fuel Process, Synthetic Crude Production,* U.S. Dept. Interior OCR R & D Rep. No. 45—Interim Rep. No. 2, July 1969.

[55] *1970 Final Report: Consol Synthetic Fuel Process,* U.S. Dept. Interior OCR R & D Rep. No. 45, Final Report, July 1970.

[56] *Engineering Evaluation of Project Gasoline, Consol Synthetic Fuel Process,* U.S. Dept. Interior OCR R & D Rep. No. 59, by Foster Wheeler Corporation, 1970.

[57] *Pilot Scale Development of the CSF Process,* Book 3, Vol. 4, U.S. Dept. Interior OCR R & D Rep. No. 39, August 1971.

[58] E. Gorin and H. E. Lebowitz, "The Removal of Sulfur and Mineral Matter from Coal," paper presented at Joint Meeting, American Institute of Chemical Engineers, and Canadian Society for Chemical Engineering, Vancouver, B.C., September 9–12, 1973.

[59] E. Gorin, U.S. Pat. 3,143,489, Consolidation Coal Company, August 4, 1964.

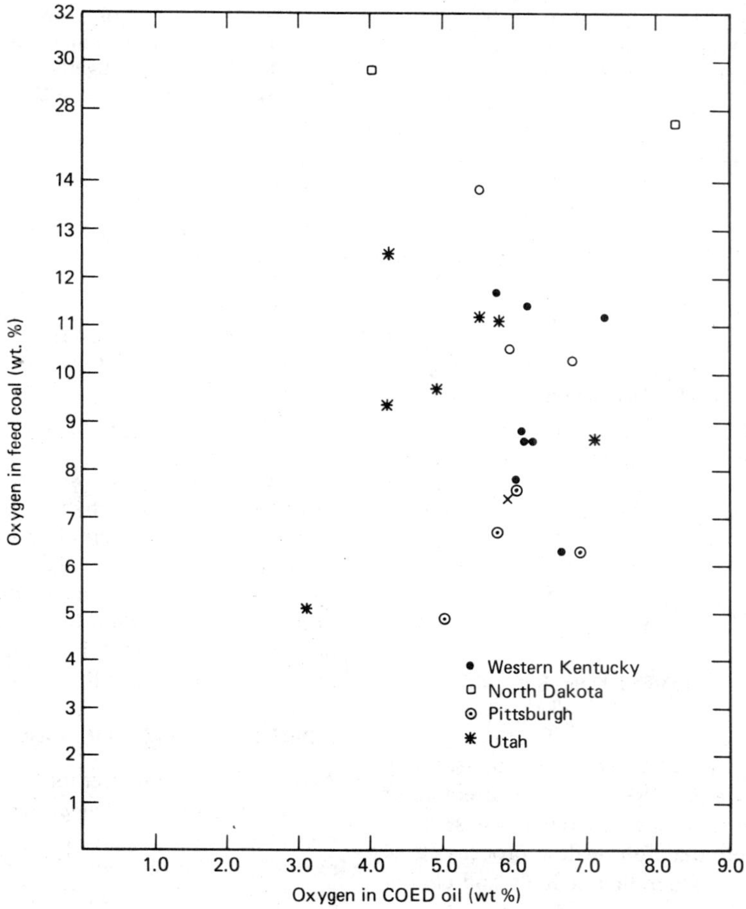

Figure 29.6 The oxygen contents of coal and COED oil.

29.3.1.3 *The SRC Process*

Representative data for a product from an SRC process are those shown in Table 29.7 for the SRC from the 6-ton/day pilot plant at Wilsonville, Alabama.[61] This plant uses essentially the same concept as the 50-ton/day plant of the Pittsburg and Midway Coal Mining Company.[62] The table gives the analysis of the SRC product, obtained from an Illinois No. 6 coal with 3.22 wt % sulfur, 1.42 wt % nitrogen, and a 8 wt % ash, (based on weight of total product). Although the nitrogen still remains rather high, the sulfur and ash have been significantly reduced. The aromatics and saturates fractions of this product have been subjected to detailed gas chromatographic and mass spectrometric analyses.[63] The aromatics were only 2.4 wt % of the sample. A portion of these aromatics (11.5%) was labeled pyrene-fluoranthene. The saturates (1.5 wt %) were predominantly alkanes and one- to four-ring cycloalkanes. The asphaltenes (benzene solubles, hexane insolu-

[60] E. J. Greskovich, *Chemical Characterization, Handling, and Refining of SRC to Liquid Fuels,* Quarterly Prog. Rep. No. FE-2003-15, for July–September, 1976, by Air Products and Chemicals, Inc., for U.S. ERDA, October 25, 1976.

[61] H. E. Lewis, *Solvent Refined Coal (SRC) Process. Operation of Solvent Refined Coal Pilot Plant at Wilsonville, Alabama,* Quarterly Rep. No. FE-2270-3, by Catalytic, Inc., for U.S. ERDA, August 1976.

[62] *Solvent Refined Coal (SRC) Process,* Quarterly Rep. No. FE-496-109, by Pittsburg and Midway Coal Mining Company, for U.S. ERDA, July–September 1975.

[63] E. J. Greskovich, *Chemical Characterization, Handling, and Refining of SRC to Liquid Fuels,* Quarterly Rep. No. FE-2002-6, by Air Products and Chemicals, Inc., for U.S. ERDA, January 5, 1976.

Table 29.7 Typical SRC Product from Processing of Illinois No. 6 Coal; Density (20°C/4°C) = 1.225

Constituent	Ultimate Analysis (wt %)
Carbon	80.0
Hydrogen	5.6
Nitrogen	2.0
Oxygen	3.6
Sulfur	0.6
Ash	0.06
Moisture	0.23
ASTM Distillation	
Distillate	14.3 wt %
Bottoms	86.3 wt %
Fraction	Temperature (°C)
IBP	224
5 vol %	406
10 vol %	474
15 vol %	503

Source: Modified from Ref. 60.

bles) made up 94.9 wt % of the sample. Processes yielding SRC usually direct their products to uses such as boiler fuels; however the potential for further processing of the SRC to produce lighter fuels and chemicals is also being considered.

29.3.1.4 *The Synthoil Process*

The Synthoil process developed by the U.S. Bureau of Mines is under study at the 10-ton/day scale and smaller. Table 29.8 presents data on some properties of a Synthoil product from an Indiana No. 5 coal: 3.4 wt % sulfur, 1.4 wt % nitrogen, 7.9 wt % oxygen, and 8.9 wt % ash. The operating conditions were 450°C, 2000 psig, and 30 wt % coal in oil feed slurry. In a sample of Synthoil product from a Kentucky high-volatile A coal, the structures of the hydrocarbon fraction were predominantly of the pyrene and phenanthrene type, with an average of two aromatic rings per molecule.[64] In a detailed analysis of the type of sulfur species remaining in a Synthoil product,[65] 14 organic sulfur species were identified, 13 of which were derivatives of thiophene. Sulfur species known to be in coal such as thiol, thioether, disulfide, and γ-thiopyrene simply do not survive the liquefaction process. As with petroleum feedstocks, dibenzothiophene and benzothiophene are quite resistant to desulfurization.

Table 29.8 Typical Gross Oil Product from Processing Indiana No. 5 Coal in the Synthoil Process: Specific Gravity at 15.6°C/15.6°C = 1.18

Component	Wt %
Carbon	89.8
Hydrogen	6.3
Sulfur	0.42
Nitrogen	0.8
Ash	2.7
Distillable below 200°C	0.0
Distillable below 355°C	24.0

Source: Modified from Ref. 31.

29.3.1.5 *The H-Coal Process*

Another important method of catalytic liquefaction is the H-Coal process. Table 29.9 presents typical product analyses from processing an Illinois No. 6 coal (nominally 3.0–3.5 wt % S, 1.5 wt % N, and 7–9 wt % ash).[66,67] The sulfur, nitrogen, and ash concentrations are, respectively, significantly lower than those of the feed coal, a typical outcome of most catalytic processes. About 60 vol % of the product oil boils below 482°C. Rather extensive product analyses are presented in Refs. 66 and 67.

29.3.1.6 *The CCL Process*

The CCL process developed by Gulf Oil Corporation is primarily directed toward the

[64] A. Korosi, H. N. Woebcke, and P. S. Virk, "Pyrolysis of a Hydrogenated Coal Liquid," *Prepr, Am. Chem. Soc., Div. Fuel Chem.*, **21**(6), 190 (1976).

[65] S. Akhtar, A. G. Sharkey, J. L. Shultz, and P. M. Yavorsky, "Organic Sulfur Compounds in Coal Hydrogenation Products," paper presented at American Chemical Society 167th National Meeting, Los Angeles, March 31–April 5, 1974.

[66] R. B. Callen, J. G. Bendoraitis, C. A. Simpson, and S. E. Volz, "Upgrading Coal Liquids to Gas Turbine Fuels. 1. Analytical Characterization of Coal Liquids," *Ind. Eng. Chem. Prod. Res. Dev.*, **15**(4), 222 (1976).

[67] J. G. Bendoraitis, A. V. Cabal, R. B. Callen, T. R. Stein, and S. E. Voltz, "Upgrading of Coal Liquids for Use as Power Generation Fuels," Rep. No. EPRI 361-1, Phase 1, January 1976.

Table 29.9 Representative Properties of H-Coal Liquid from Illinois No. 6 Coal: Specific Gravity at 15.6°C/15.6°C = 1.184

Component	Ultimate Analysis (wt %)
Carbon	89.0
Hydrogen	7.9
Oxygen	2.1
Nitrogen	0.77
Sulfur	0.42
Ash	0.01

Simulated Distillation

Fraction	Temperature (°C)
IBP	250
10 vol %	298
20 vol %	327
50 vol %	404
60 vol %	463
70 vol %	> 517

Source: Modified from Ref. 66.

production of clean-burning liquid fuels for power plants; however certain distillate products of the process are also attractive feedstocks for upgrading to higher value fuels and chemicals. The comparative data available for liquids obtained from CCL processing of bituminous and subbituminous coals do not include the processing details or the coal types. The liquids are compared after distillation into naphtha (IBP, 204°C), middle oil (204–343°C), gas oil (343–441°C), and bottoms (+441°C). The analyses of the naphthas (Table 29.10) indicate the presence of significant quantities of heteroatoms, especially oxygen. Each naphtha has about the same proportion of saturates (60 wt %) and aromatics (40 wt %). These naphthas are highly naphthenic, and the aromatic fractions are composed primarily of monoaromatics with very few three-ring compounds. The phenolic stream of the subbituminous liquid has a large percentage of pure phenol (28 wt %). Upgrading these naphthas for heteroatom removal will certainly increase their value as reforming stocks.

About 42 wt % of the CCL liquids fall within the middle distillate range. Some of their properties are also given in Table 29.10. The nitrogen and oxygen contents are rather high, and the oil from subbituminous coal contains a larger proportion of aromatics. The catalytic hydrogenation process is manifested in the significant quantities of hydroaromatics in the middle oils. Following hydrotreatment to remove sulfur and nitrogen, these oils are expected to have properties falling between No. 2 and No. 4 petro-

Table 29.10 Characterization of Fractions from Gulf CCL Liquids

Properties	Bituminous Coal Naphtha	Subbituminous Coal Naphtha	Bituminous Middle Distillate	Subbituminous Middle Distillate	Bituminous Coal Gas Oil	Subbituminous Coal Gas Oil
Gravity (°API)	41.1	39.2	14.0	13.2	−5	−5
Carbon (wt %)	85.78	85.96	89.03	89.25	90.86	90.06
Hydrogen (wt %)	12.70	11.98	9.47	9.55	7.65	8.35
Nitrogen (wt %)	0.34	0.20	0.36	0.35	0.40	0.56
Oxygen (wt %)	0.90	1.61	1.04	1.75	0.99	0.92
Sulfur (wt %)	0.21	0.25	0.10	0.10	0.19	0.17
			Saturates (wt %)			
Fraction	59.0	60.0	24.0	14.0	8.9	13.4
Paraffins	19.6	8.0	16.2	3.6	1.7	6.4
Noncondensed cyclo	34.8	48.7	5.2	6.6	8.6	3.4
Condensed dicyclo	4.4	3.2	1.8	2.7	2.1	2.4
Condensed tricyclo	0.2	0.1	0.8	1.1	0.5	1.2
			Aromatics (wt %)			
Fraction	41.0	40.0	76.0	85.0	91.1	86.6
Alkylbenzenes	15.6	16.7	2.2	4.1	1.4	1.0
Tetralins	13.4	12.7	12.2	18.7	1.7	1.2
Naphthalenes	2.6	1.9	6.3	8.6	0.0	0.0
Phenanthrenes	1.0	0.0	7.0	9.6	0.0	7.1
Hydrophenanthrenes	1.0	0.4	27.4	23.7	6.5	16.6
Hydropyrenes	0.2	0.0	17.8	6.7	45.3	34.7
Phenolics	6.7	7.5	1.9	9.2	—	—
Benzopyrenes	—	—	—	—	4.3	1.9
Chrysenes	—	—	—	—	8.6	4.2
Pyrenes	—	—	—	—	26.5	19.7

Source: Modified from Ref. 32.

leum fuel oils. The gas oil cuts given in Table 29.10 reveal their high aromaticity, with four-ring systems prevailing, especially in the oil from bituminous coal.

29.3.1.7 *The EDS Process*

A typical liquid derived from the EDS process[22] is shown in Table 29.11. The process has been designed to yield naphthas and fuel oils. The liquids shown are to be hydrotreated. An Illinois No. 6 coal containing 9.6 wt % ash, 4.2 wt % sulfur, 1.8 wt % nitrogen, and 9.4 wt % oxygen (mf) was processed to yield these liquids. No operating conditions are available for these data.

29.3.1.8 *Liquids from Carbonization Processes*

Representative characteristics of liquids obtained from carbonization processes, especially low- and high-temperature tars, are discussed in Chapters 16 and 17; two examples relevant to the present discussion are given here. One involves an interesting study comparing the detailed properties of low-temperature tars obtained from two different carbonization reactors—a static bed and an entrained bed.[68] The product liquids were separated into various fractions, and the fraction containing the tar bases revealed significant differences in the types of compound present. The phenols were very similar to one another. However the neutral oils from the static bed were more aliphatic and were especially rich in lower boiling olefins. This study illustrates that variations in liquid products can result from such subtleties as reactor-contacting method.

Another carbonization liquid worthy of mention is raw anthracene oil, which has often been used as a test and process startup solvent in liquefaction processes. Raw and chilled anthracene oils are coal-derived liquids that often serve as solvent in development work on solvent extraction processes. The oils, which are derived from the carbonization of coal, contain very little anthracene, despite the name. A raw anthracene oil has been extensively studied and analyzed in detail in hydrotreating catalyst development work.[69–71] The two-, three- and four-ring aromatics dominate the hydrocarbons, with phenanthrene (and its alkylates) accounting for 15.4 wt % of the raw anthracene oil. Four-ring compounds are mostly pyrene and fluoranthene. The five-ring compounds were

[68] T. P. Maher, "The Liquid Products from the Low-Temperature Carbonization of Two Bituminous Coals," *Fuel*, **47**, 359 (1968).

[69] S. E. Scheppele, G. J. Greenwood, and B. L. Crynes, "Mass Spectrometric Analysis of the Hydrocarbons from Coal Liquids and from the Products Obtained upon Their Hydrogenation-Hydrogenolysis," *Proc. 23rd Annu. Conf. Mass Spectrometry and Allied Topics*, Houston, Tex., May 25–30, 1975.

[70] P. L. Grizzle, T. D. Marriott, N. B. Perreira, and C. S. Hsu, "Mass Spectrometric Analysis of Coal Liquids and Their Hydrogenation-Hydrogenolysis Products," paper presented at 20th Annu. Conf. Analytical Chemistry and Environmental Technology, Gatlinburg, Tenn., October 12–14, 1976.

[71] G. J. Greenwood, P. L. Grizzle, T. D. Marriott, and N. B. Perreira, "Mass Spectrometric Analysis of Coal Liquids and Their Hydrogenation-Hydrogenolysis Products," *Proc. 24th Annu. Conf. Mass Spectrometry and Allied Topics*, San Diego, Calif., May 9–13, 1976.

Table 29.11 Analysis of Raw Liquid from the Exxon Donor Solvent Process When Processing Illinois No. 6 Coal

Normal boiling range:	Heavy Naphtha 70–200°C	200°C + Fuel Oil 200–540°C
Ultimate analysis (wt %)		
Carbon	85.6	89.4
Hydrogen	10.9	7.7
Oxygen	2.82	1.83
Nitrogen	0.21	0.66
Sulfur	0.47	0.41

Source: Modified from Ref. 22.

designated benzopyrene. Hydroaromatics account for at most 6 wt % of the anthracene oil, and in all, the hydrocarbons make up 65 wt % of the oil. The principal sulfur compounds are of the benzothiophene and dibenzothiophene type. The phenolic-type compounds contain one to four rings with up to three hydroxyl groups. The nitrogen compounds that were identified are listed in Table 29.12. Carbazoles and azafluorenes, quinolines, azaphenanthrenes and/or acridines, and azapyrenes and/or azafluoranthenes represent the predominant nitrogen-containing compounds. The ultimate analysis of the raw anthracene oil is shown in Table 29.13. The oil has little or no ash, a total nitrogen content of 1.06 wt %, and about 0.47 wt % sulfur.

29.3.2 Comparison of Coal Liquids

In the past few years, as emphasis has turned to replacing or supplementing petroleum stocks with coal liquids, several excellent studies have been made comparing various liquid types. One recent study has compared solvent refined coal, H-Coal product, Synthoil product, and an El Palito No. 6 fuel oil.[66,67] Table 29.14 presents both physical and chemical data for comparison. The SRC and the H-Coal products were derived from an Illinois No. 6 coal, the Synthoil product from a Pittsburgh Seam coal. Data on operating conditions producing these coal liquids were not available.

The table provides some information of particular relevance to the possibilities for further processing the coal liquids. As is typical in primary liquid from coal, the carbon content is greater and the hydrogen content much lower than that of the petroleum stock. This is to be expected, since coal liquids are much more condensed than petroleum liquids and have a higher degree of aromaticity, as shown in Table 29.14. Figure 29.7 plots the aromaticity versus the C/H atom ratio of these liquids and some petroleum stocks. In a hydrogenation-hydrogenolysis process for upgrading these liquids, this figure suggests the quantities of hydrogen required to produce petroleumlike stocks from coal liquids. The alternative technique for lowering the aromaticity is carbon rejection (carbonization); if total liquid yield is paramount, however, this alternative may not be viable.

The heteroatoms, oxygen and nitrogen, are

Table 29.12 Identified Nitrogen-Containing Species in Anthracene Oil

Empirical Formula	Possible Structural Type	Weight Percent in Feedstock[a]
C_8H_7N	Indole	0.24
$C_{12}H_9N$	Carbazole	1.84
$C_{14}H_9N$	Benzocarbazole	0.25
$C_{14}H_{11}N$	Dihydrobenzocarbazole	0.19
$C_{16}H_{11}N$	Naphthylindanol	0.35
C_9H_7NO	Hydroxyazanaphthalene	0.48
$C_{13}H_9NO$	Hydroxyazaphenanthrene	0.12
$C_8H_{11}N$	Pyridines	0.07
C_8H_9N	Azaindane	0.45
C_9H_7N	Quinoline	2.22
$C_{11}H_9N$	Azaacenaphthene	0.31
$C_{12}H_9N$	Azafluorene	0.78
$C_{13}H_9N$	Azaphenanthrene, acridine	2.72
$C_{14}H_9N$	Cyclopentaazaphenanthrene	0.54
$C_{15}H_9N$	Azapyrene, azaflouranthene	1.42
$C_{17}H_{11}N$	Azabenzoanthracene	0.37
$C_{19}H_{11}N$	Azabenzopyrene	0.03
$C_{11}H_7NO$	Azadibenzophenanthrene	0.16
$C_{13}H_7NO$	—	0.06
$C_{14}H_9NO$	Benzoazanaphthalenfuran	0.05

[a] Weight percent represents the sum of the parent molecule and its alkylated homologues.

Table 29.13 Properties of Raw Anthracene Oil: API Gravity at 60°F = −7

Component	Ultimate Analysis (wt %)
Carbon	90.65
Hydrogen	5.76
Sulfur	0.48
Nitrogen	1.06
Oxygen[a]	2.05
Ash	Nil

Distillation	
Fraction	Temperature (°C)
Initial[b]	193
10 vol %	232
30 vol %	297
50 vol %	343
70 vol %	371
90 vol %	435

Composition of Various Separation Fractions	
Component	Weight Percent
Acid	14.9
Base	9.5
Neutral nitrogens	1.7
Hydrocarbons + ethers	73.9

[a] By difference.
[b] Normal boiling point determined from ASTM D1160 data.

present in higher concentrations in the coal liquids than in the petroleum liquids. In the two catalytic processes (H-Coal, Synthoil) the concentration of the heteroatoms is lower than in the solvent-refining process, as might be expected (see Table 29.14). The sulfur levels of the coal liquids are lower than those of the fuel oil. The predominant ash components found in the SRC are iron (100 ppm) (all ppms are by weight), titanium (130 ppm), and sodium (100 ppm). In the H-Coal liquid are found iron (20 ppm) and titanium (80 ppm). The Synthoil liquid, especially high in total ash, contained predominantly silicon (1348 ppm), iron (375 ppm), aluminum (886 ppm), titanium (150 ppm), and potassium (116 ppm). The silicon and aluminum suggest that catalyst materials have been carried into the product. The El Palito oil contained vanadium (275 ppm), titanium (78 ppm) and nickel (59 ppm). Other trace metals were, of course, present in each liquid, but not in significant quantities. The rather high nitrogen levels found in these coal liquids present a real processing problem and will eventually demand more attention in upgrading the liquids than desulfurization. The nitrogen compounds of the coal liquids are essentially basic (40–70 wt % vs. about half that much for the typical petroleum liquid). The oxygen levels, though relatively easily reduced in hydrotreating, will exact a significant penalty in hydrogen consumption. The oxygen and nitrogen compounds are found in much higher concentrations in the high-boiling (asphaltene) fractions, whereas sulfur is more uniformly distributed throughout the liquid-boiling fractions. Iron and titanium are the most prevalent common metals found in these and other coal liquids. These metals carry over from the parent coal, not from catalyst residuals.

In terms of boiling ranges, the two catalytic processes yield liquids not unlike the No. 6 fuel oil. As might be expected, the SRC, produced as it is under lower hydrogenation conditions and with a different objective, is a better match for petroleum vacuum residuum.

Gradient elution chromatography revealed that the coal liquids contained significantly more asphaltenes than did the petroleum stock; the SRC had an especially high asphaltene content. The saturated fractions of the coal liquids are essentially naphthenic compounds containing one to six rings.

A comparison of an Indian coal liquid (Baragolai, Assam) with an Indian petroleum crude (Nahorkatiya) points out further similarities and differences. The coal liquid was produced from a coal with 3.2 wt % ash by a solvent extraction process using ammonium molybdate as a catalyst, at 400°C and 1565 psig cold autoclave pressure. The properties are compared in Table 29.15. The asphaltene level, higher than in the coal liquid of the syncrude, results in correspondingly high values of the C/H ratio, viscosity, and density. Both liquids were judged to have satisfactory naphtha cuts for reforming for gasoline, provided the concentration of heteroatoms was reduced first. The middle distillates were reported to be similar.

Other workers have compared the properties of coal liquids derived from a catalytic versus a noncatalytic conversion process.

Table 29.14 Comparative Properties of Various Coal Liquids and No. 6 Fuel Oil

Constituent	SRC	H-Coal	Synthoil	Petroleum No. 6 Fuel Oil
		Analysis (wt %)		
Carbon	87.9	89.0	87.6	86.4
Hydrogen	5.7	7.9	8.0	11.2
Oxygen	3.5	2.1	2.1	0.3
Nitrogen	1.7	0.77	0.97	0.41
Sulfur	0.57	0.42	0.43	1.96
Ash	0.01	0.02	0.68	
		Simulated Distillation		
Fraction		Temperature (°C)		
IBP	—	250	222	175
15 vol %	510	312	264	—
20 vol %	>510	327	279	379
50 vol %	>510	404	379	478
70 vol %	>510	>517	>477	>532
90 vol %	>510	>517	>477	>532
EP	>510	>517	>477	>532
Aromaticities (atom %)	77	63	61	24
C/H atom ratio	1.29	0.94	0.92	0.45

Source: Modified from Ref. 66.

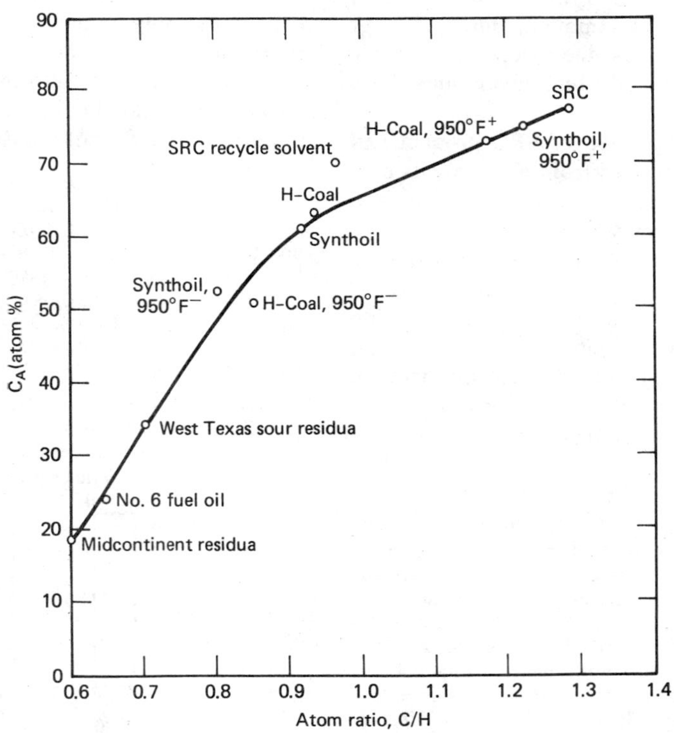

Figure 29.7 The relation between the atom percent of aromatic carbon and the atom ratio of carbon to hydrogen.[67]

Table 29.15 Comparison of an Indian Coal Liquid and Petroleum Crude

Properties	Coal Syncrude[a]	Nahorkatija Crude[a]
Analysis (wt %)		
Carbon	85.8	86.7
Hydrogen	9.7	12.5
Sulfur	0.67	0.27
Nitrogen	0.71	0.05
Ash	0.012	0.015
C/H wt ratio	8.8	6.9
API specific gravity at 15.6°C/15.5°C	9.5	29.5
Viscosity at 38°C (cSt)	60.95	5.8

[a] Difference between sum of percentage of constituents shown and 100 not specified in reference—Ed.
Source: Modified from Ref. 72.

Sternberg[73] has emphasized the differences in asphaltene contents and their influence on liquid viscosity. The oil-asphaltene levels for SRC and Synthoil liquids are shown in Fig. 29.8 under unidentified coal type and processing conditions. The SRC shows nearly 4 times as much asphaltene and preasphaltene. The effect of asphaltene on liquid viscosity is shown in Fig. 29.9. From the standpoint of liquid handling and moving during further upgrading processes, the increased viscosity caused by high asphaltene levels must be a consideration.

Another comparison of SRC product and H-Coal oils produced from Kaiparowits coal (1.07 wt % N, 0.24 wt % S, 14 wt % O, and 7.9 wt % ash) reveals the greater extent of hydrogenation resulting from the catalytic process. The SRC was produced at 427°C and 2000 psig, and the H-Coal operating conditions were specified only as low and high severity.[74] The liquid properties listed in Table 29.16 show the lower heteroatom levels and viscosities resulting from more extensive hydrogenation.

[72] D. K. Mukherjee, S. K. Mukherjee, A. N. Basu, and A. Lahiri, "Characterization of Oils Derived from the Hydrogenation of Upper Assam Coal and Their Comparison with Nahorkatiya Petroleum," *Indian J. Technol.*, **2,** 363 (August 1973).

[73] H. W. Sternberg, R. Raymond, and F. K. Schweighardt, "The Nature of Coal Liquefaction Products," paper presented at American Chemical Society, Petroleum Chemistry Division, 170th National Meeting, Chicago, August 24–29, 1975.

[74] Hydrocarbon Research, Inc., *Liquefaction of Kaiparowits Coals,* EPRI Rep. No. 123-2, Final Report, October 1977.

Comparisons of COED oils with petroleum crudes and other coal-derived oils disclose differences characteristic of carbonization-type oils. Table 29.17 compares the general types of hydrocarbons in a COED syncrude with those in petroleum crudes.[75] The coal liquid is significantly lower in paraffins than even the naphthenic petroleum crude, and correspondingly much higher in aromatics, especially polycyclics. These component properties are best seen in Figs. 29.10 and 29.11, which present information on several coals, a Seacoke liquid and a West Texas crude.[76,77] The Seacoke liquid was obtained

[75] H. J. West, "Nonfuel Chemicals from Coal," *Ind. Eng. Chem.*, **62**(11), 28 (November 1970).

[76] *Project Seacoke,* Vol. 1 with Appendices A, B, and C, U.S. Dept. Interior OCR R & D Rep. No. 29, Final Report, January 1970.

[77] *Catalytic Hydrotreating of Coal-Derived Liquids: Project Seacoke, Phase II,* OCR, U.S. Dept. Interior OCR R & D Rep. No. 29—Interim Rep. No. 1, June 1967.

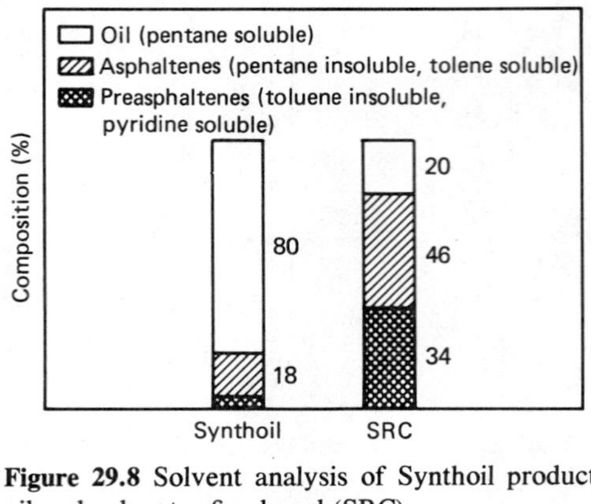

Figure 29.8 Solvent analysis of Synthoil product oil and solvent-refined coal (SRC).

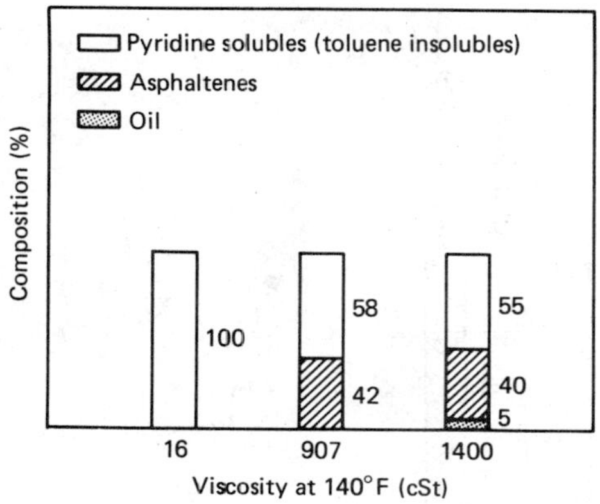

Figure 29.9 Effects of pyridine solubles (toluene insolubles) and asphaltenes on viscosity.

in essentially the same way as the COED product except that the feedstock was a combination of petroleum residuum and coal. The COED product from Utah coal and, of course, the petroleum crude, show greater paraffins content than the others. The lower boiling Seacoke fraction is especially high in monocycloparaffins. As can be seen, the properties of the feed coal do influence the product component distribution.

Table 29.18 shows the relatively lower polyaromatic content and relatively higher content of saturates plus monoaromatics in the COED carbonization oil compared to the catalytic oils from Synthoil and H-Coal process.[79] The di- and triaromatics are very similar in all four products.

Another recent comparison of COED and Synthoil liquids (Table 29.19) essentially confirms the results in Table 29.18, except for the polyaromatics fraction. Again, the importance of coal type is evident.[80] The COED oil listed in Table 29.19 had undergone hydrogenation following the staged carbonization and separation steps. The Synthoil liquid has an estimated 70 wt % material as aromatics (including aromatics in the residuum) in comparison with about 50 wt % for the other two liquids. Two- to three-ring systems are dominant in all three liquids. More details

[78] I. Schwager and T. F. Yen, "Preliminary Examination of Coal Liquefaction Products," *Prepr. Am. Chem. Soc., Div. Fuel Chem.*, **21**(5), 199 (1976).

[79] *Coal Technology Program Annual Interim Report for Fiscal Year Ending June 30, 1975,* Oak Ridge National Laboratory Rep. No. ORNL-5069, October, 1975.

[80] J. E. Dooley and C. J. Thompson, "The Analysis of Liquids from Coal Conversion Processes," *Am. Chem. Soc. Prep. Div. Fuel Chem.*, **21**(5), 243 (1976).

Table 29.16 Properties of SRC and H-Coal Products

Property	SRC	H-Coal: Low Severity	H-Coal: High Severity
Sulfur (wt %)	0.25	0.10	0.11
Nitrogen (wt %)	1.15	1.18	1.05
Ash (wt %)	0.31	0.15	0.06
Viscosity at 177°C (cP)	3690	26	17
API gravity[a]	−15	−9	−10

[a] Reference temperature not specified—Ed.

Source: Modified from Ref. 74.

Table 29.17 Comparison of Typical COED Syncrude with Petroleum Crudes

Hydrocarbons	COED Syncrude	Naphthenic Petroleum Crude	Paraffinic Petroleum Crude
Paraffins (wt %)	13.5	40–46	70–75
Naphthenes (wt %)	57.8	47	19–23
Alkyl benzenes (wt %)	7.9 }	7–13	7–12
Polycyclic and thioaromatics (wt %)	20.8 }		

Source: Modified from Ref. 75.

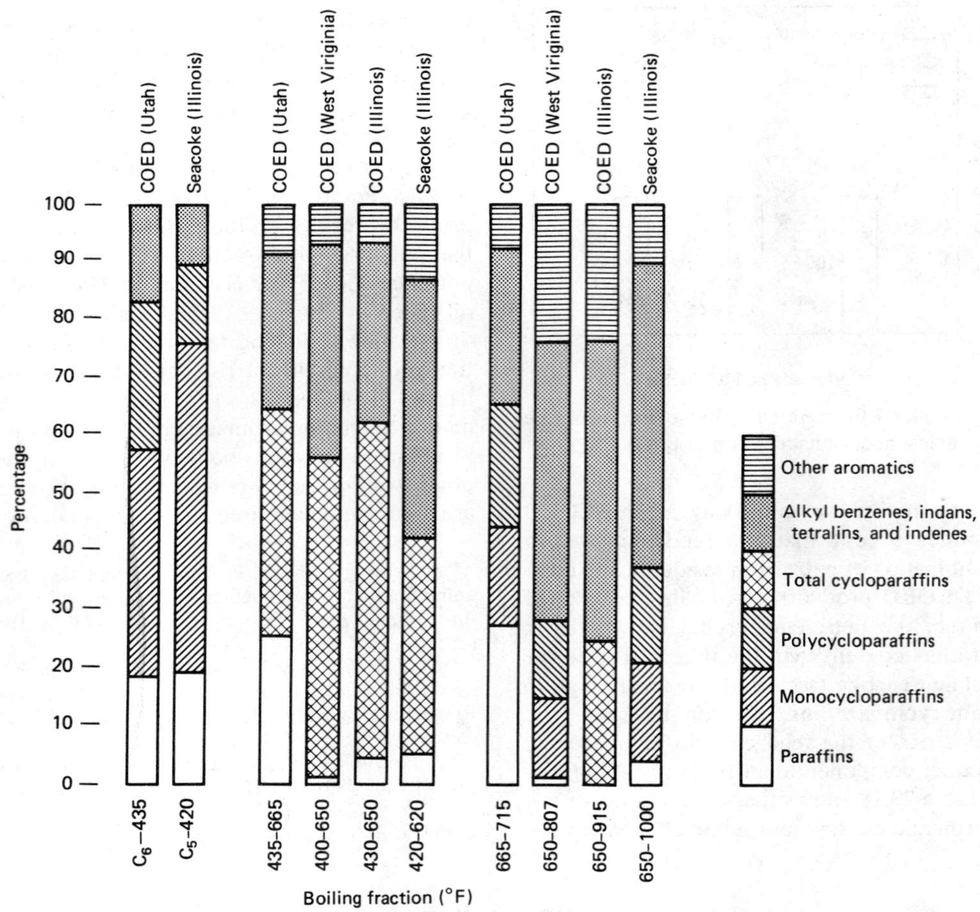

Figure 29.10 Compositions of COED syncrude and Seacoke.[76]

on the distillate fractions of the syncrudes of Table 29.19 are available.[81]

Additional data comparing COED products with Synthoil, H-Coal, and solvent-refining products are available, and since 1974 the

[81] P. W. Woodward, G. P. Sturm, Jr., J. W. Vogh, A. A. Holmes, and J. E. Dooley, "Compositional Analysis of Synthoil from West Virginia Coal," Rep. No. BERC/RI-76/s, Bartlesville Energy Research Center, Bartlesville, Okla., for U.S. ERDA, January 1976.

Table 29.18 Comparison of COED Oil and Oils from Catalytic Processes[a]

Component	COED	Synthoil	H-Coal: Western Subbituminous	H-Coal: Eastern Bituminous
Saturates (wt %)	17.8	4.3	45.5	4.2
Monoaromatics (wt %)	32.8	6.4	18.4	12.4
Di plus triaromatics (wt %)	11.8	13.6	12.8	15.8
Poly- plus polar aromatics (wt %)	23.6	76.6	26.8	42.2

[a] Deviation of sum of percentage of components from 100 not specified in reference—Ed.

Source: Modified from Ref. 78.

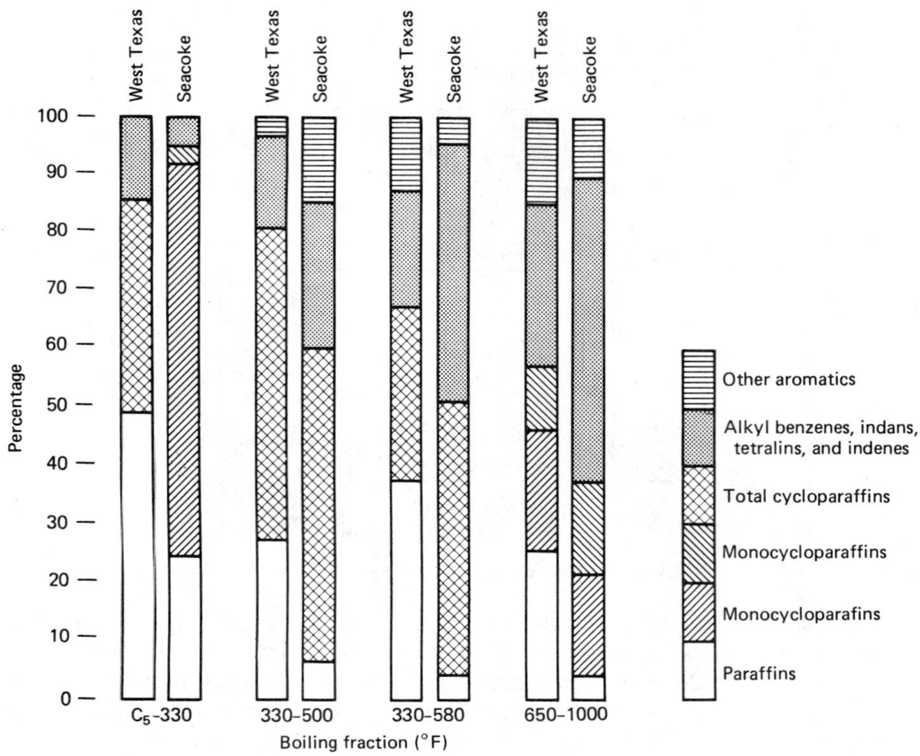

Figure 29.11 Compositions of West Texas crude and Seacoke.[77]

Table 29.19 Comparison of Liquids from the COED and Synthoil Processes[80]

	COED Syncrude (Utah Coal)	COED Syncrude (Western Kentucky Coal)	Synthoil (West Virginia Coal)
Sulfur (wt %):	0.05	0.08	0.42
Nitrogen (wt %):	0.48	0.23	0.79
		Distillates (wt %)	
< 200°C	13.3	21.0	4.4
200–370°C	45.4	54.2	42.6
370–540°C	40.3	24.2	27.3
540°C + residuum	Nil	Nil	25.7
		Major Components (wt %)	
Saturates	30.1	33.4	10.6
Monoaromatics	20.1	33.6	14.2
Diaromatics	14.4	12.8	15.5
Polyaromatics	14.5	7.9	14.6
Heteroatomics	1.5	0.3	2.2
Acids	8.2	3.1	10.4
Bases	1.3	1.0	2.8
Residuum	None	None	25.7
Other	10.0	7.8	3.9

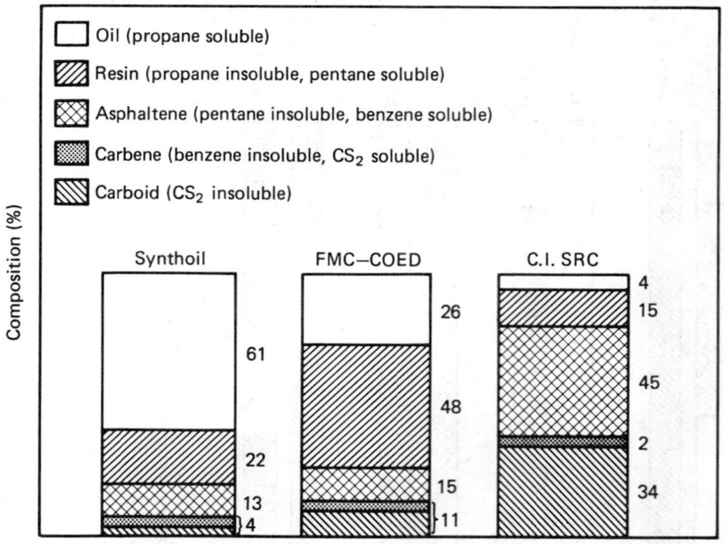

Figure 29.12 Solvent fractionation analysis of coal liquid products.

number of detailed analyses and comparisons of coal liquids published has increased markedly.[74,78,79,82–85] Table 29.20 gives a detailed analysis of liquid fractions for heteroatoms of Synthoil, COED, and SRC liquids. This table presents data on the sulfur, nitrogen, oxygen, and ash for the total liquids, oils (propane soluble), resins (propane insoluble, pentane soluble), asphaltenes (pentane insoluble, benzene soluble), carbenes (benzene insoluble, CS_2 soluble), and carboids (CS_2 insoluble). Again, the higher concentrations of heteroatoms are found in the "heavier" fractions. The percentages of the individual fractions present in the three products are shown in Fig. 29.12. The total quantity of each heteroatom in the individual fraction per 100g of each liquid is given by the numbers in parentheses in Table 29.20. The figure reveals that the asphaltene content of the SRC liquid is about 3 times that of the other two liquids. The more highly hydrogenated Synthoil liquid has by far the highest oil content.

An example of the way in which both the coal type and the process itself dictate the nature of the coal liquid to be upgraded is given in Table 29.21, which presents four different Utah coal liquids produced from the same Utah coal.[86] The low-temperature carbonization oil and the pyrolysis oil have similar properties, as would be expected. The solvent extraction liquid is markedly different, containing far more asphaltenes and residuals, as well as sulfur and nitrogen. The hydrogenation liquid, produced by direct hydrogenation of coal impregnated with a catalyst, has the lowest content of high-boiling oils and heteroatoms.

Some processes tend to mask the initial coal properties and produce similar liquid products from different coals.

This seems to be true for solvent-refining or extraction schemes. The properties of solvent-refined coal and solvent-refined lig-

[82] B. R. Rodger and S. Katz, *Supporting Research and Development on Separations Technology, Phase II Report: Scouting Experiments,* Oak Ridge National Laboratory Rep. No. ORNL-TM-4968, October 1975.

[83] R. G. Ruberto and D. M. Jewell, "Characterization of Coal Liquids," paper presented at Symp. Supplemental Fuels from Coal: New Analytical Needs, 26th Pittsburgh Conf., Cleveland, Ohio, March 3–7, 1975.

[84] Y. Maekawa, S. Veda, Y. Hasegawa, Y. Nakata, S. Yokoyama, and Y. Yoshida, "Chemical Structure Analyses of Coal Hydrogenation Products—Distribution of Structural Parameters and Application of ^{13}C NMR to Structural Analysis," *Prepr. Am. Chem. Soc., Div. Fuel Chem.,* **20**(3) (1975).

[85] J. E. Schiller, "Composition of Coal Liquefaction Products," *Hydrocarbon Process.,* 147 (January 1977).

[86] S. A. Qader, S. Singh, and G. R. Hill, "Quality of Synthetic Gasoline from Coal," *Prepr. Am. Chem. Soc., Div. Fuel Chem.,* **16**(1), 36 (1972).

Table 29.20 Comparison of Synthoil, COED, and SRC Products Showing Heteroatom Content of Liquid Fractions (wt %, maf)[78]

Component	Total Liquid	Oil[a]	Resin[a]	Asphaltene[a]	Carbene[a]	Carboid[a]
			Synthoil			
Nitrogen	0.94	0.60 (0.37)	1.30 (0.29)	1.63 (0.21)	1.72 (0.03)	1.64 (0.03)
Sulfur	0.10	0.43 (0.26)	0.14 (0.03)	0.66 (0.09)	0.74 (0.01)	2.07 (0.04)
Oxygen	3.26	1.65 (1.01)	3.52 (0.77)	3.93 (0.51)	3.64 (0.07)	2.28 (0.05)
Ash	0.69	0.02 (0.01)	0.30 (0.07)	0.48 (0.06)	0.56 (0.01)	8.80 (0.18)
			COED			
Nitrogen	1.09	0.42 (0.11)	1.06 (0.51)	1.70 (0.26)	1.69 (0.03)	1.95 (0.18)
Sulfur	1.11	1.08 (0.28)	1.26 (0.61)	2.58 (0.39)	1.31 (0.03)	1.85 (0.17)
Oxygen	7.08	2.87 (0.75)	7.18 (3.45)	7.11 (1.07)	9.82 (0.20)	12.08 (1.09)
Ash	0.28	0.01 (0.003)	0.39 (0.19)	0.79 (0.12)	0.11 (0.002)	1.31 (0.12)
			SRC			
Nitrogen	1.26	0.41 (0.02)	0.84 (0.38)	1.25 (0.56)	1.11 (0.02)	0.96 (0.33)
Sulfur	0.19	0.57 (0.02)	0.00 (0.00)	0.12 (0.05)	0.26 (0.005)	0.34 (0.12)
Oxygen	4.31	1.09 (0.04)	2.64 (1.19)	4.23 (1.90)	3.83 (0.08)	6.92 (2.35)
Ash	0.29	0.25 (0.01)	0.05 (0.02)	0.78 (0.35)	0.40 (0.008)	0.71 (0.24)

[a] Numbers in parentheses indicate total quantity of heteroatoms in 100 g of liquid.

nite might be more similar than those of the coals from which they were obtained.[87–89]

[87] N. Woolsey, R. Baltisberger, K. Klabunde, V. Stenberg, and R. Kaba, "Solvent Refined Lignite and Solvent Refined Coal Structural Features. Computerized Methodology Development," *Prepr. Am. Chem. Soc., Div. Fuel Chem.*, **21**(7), 33 (1976).

[88] B. K. Schmid, "Processes for Coal Liquefaction and the Production of Low Sulfur Boiler Fuels," paper presented at Synthetic Fuels from Coal Conf., Oklahoma State University, Stillwater, May 3–4, 1971.

29.3.3 Summary of Properties of Coal Liquids

Despite the difficulties of attempting to generalize the properties of coal liquids, some points emerge that can give guidance on the

[89] P. H. Given, W. Spackman, A. Davis, D. C. Cronauer, H. L. Lovell, and B. Biswas, *Dependence of Coal Liquefaction Behavior on Coal Characteristics*, U.S. Dept. Interior OCR R & D Rep. No. 61—Interim Rep. No. 9, by Pennsylvania State University, June 14, 1974.

Table 29.21 Coal Liquids Obtained from Processing a Utah Coal by Different Process Techniques

Liquids	Low-Temperature Carbonization	Pyrolysis	Solvent Extraction	Hydrogenation
Benzene insolubles (wt %)	Nil	2.0	43.0	Nil
Asphaltenes (wt %)	19.0	16.0	25.0	15.0
Light oil, up to 200°C (vol %)	5.0	7.0	3.0	18.0
Residual oil, +370°C (vol %)	27.0	30.0	50.0	25.0
Sulfur (wt %)	0.82	0.98	1.15	0.64
Nitrogen (wt %)	0.91	0.86	1.04	0.76
Oxygen (wt %)	3.8	4.5	8.0	2.8
Tar acids, 200–370°C oil (vol %)	26.0	24.0	—	15.0
Hydrocarbon Analysis of Neutral Oil, 200–370°C (vol %)				
Saturates	29.0	25.0		24.0
Olefins	15.0	17.0		11.0
Aromatics	56.0	58.0		65.0

Source: Modified from Ref. 86.

characterization of coal-derived liquids suitable for further processing or upgrading.

The heteroatom content of coal-derived liquids can be high enough to require reduction during further processing steps. The sulfur content will range from 0.05 to as much as 2.5 wt % but is generally about 0.3–0.7 wt %. This sulfur is found mostly in the form of benzo-dibenzothiophene derivatives and can be relatively uniformly distributed throughout the liquid, but tends to be in relatively high concentrations in the high-boiling constituents. Nitrogen concentrations can range from 0.2 to 2.0 wt % and are comparatively rather high, typically in the range of 0.9–1.1 wt %. The nitrogen heteroatoms might be found as carbazoles, quinolines, azaphenanthrenes, acridines, azapyrenes, and azafluoranthenes. The concentration of oxygen in coal liquids might range from 1.5 to more than 7 wt % depending on the coal type and process. The typical oxygen concentration of 4 or 5 wt % does not present as many problems for further processing as do concentrations of sulfur and especially nitrogen; however the hydrogen consumption entailed by the oxygen levels is a consideration in hydroprocessing operations. Since the ash content of the liquids depends more on physical separation methods, if any, than any other factor, some liquids will contain essentially no ash and others may contain more than 3 wt % ash. This ash can cause serious problems in further processing operations involving catalysts and in certain fuel uses, for instance, in turbines. Although petroleum liquids, especially residuum, might contain relatively high levels of iron, nickel, and vanadium, the coal liquids contain a much broader variety of elements, especially iron, titanium, silicon, and aluminum, some of which are carryovers from catalysts.

Figures 29.13 and 29.14 present interesting data collected from several coals using catalytic (H-coal, Gulf CCL, Synthoil) and noncatalytic processes (SRC, Wilsonville), respectively.[90] The concentration of sulfur is relatively uniform throughout the boiling range of liquids from subbituminous coals,

[90] R. H. Wolk, N. C. Stewart, and H. F. Silver, "Review of Desulfurization and Denitrogenation in Coal Liquefaction," *Prepr. Am. Chem. Soc., Div. Fuel Chem.*, **20**(2), 116 (1975).

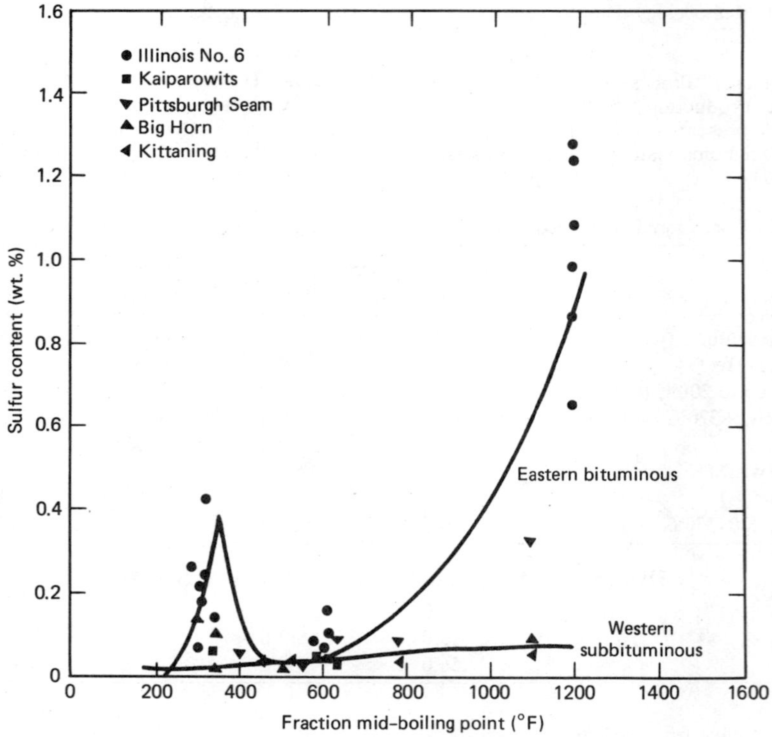

Figure 29.13 The sulfur contents of coal liquids from catalytic operations.[90]

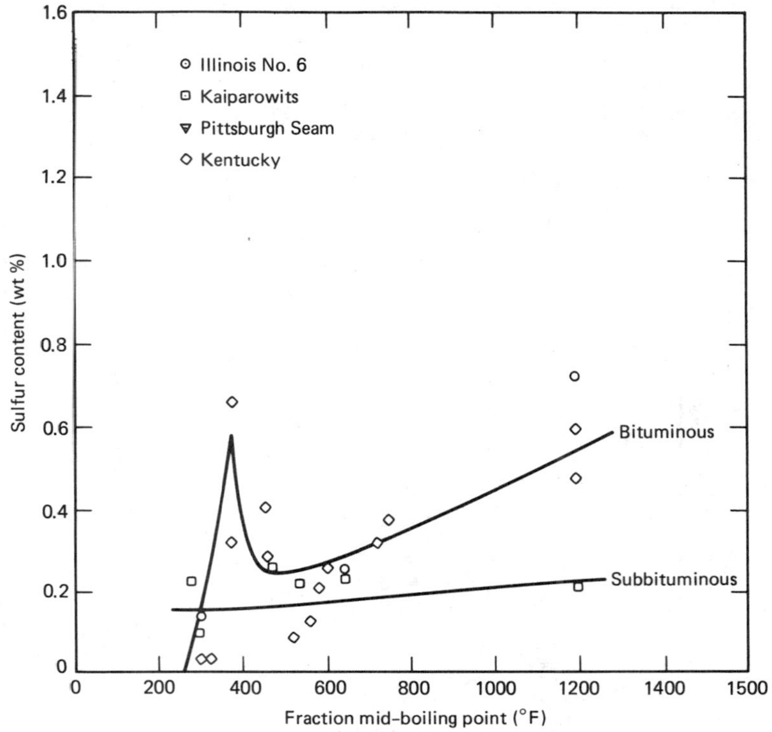

Figure 29.14 The sulfur contents of coal liquids from noncatalytic operations.[90]

but it increases significantly in the residual-type fractions of the liquids from bituminous coals. For the catalytic processes, the low-boiling fractions are particularly low in sulfur, whereas the undistillable fraction is resistant to catalytic sulfur removal. Similar results (Figs. 29.15 and 29.16) are shown for nitrogen data obtained from the same systems. The nitrogen concentration increases uniformly with the increasing boiling point of the liquid fraction in both catalytic and noncatalytic processes. However the proportion of undistillable material from noncatalytic systems is much greater than from catalytic systems and the total nitrogen levels are correspondingly higher.

The quantity of asphaltenes present in coal liquids significantly affects chemical and physical properties. The C/H atom ratios and the aromaticity increase and the density and the viscosity rise as the concentration of asphaltenes increases. Solvent-refined coals are particularly high in asphaltene content. The asphaltenes appear to be monofunctional species with an acidic character from a phenolic-type hydroxyl or an acidic nitrogen as in pyrrole. Sulfur compounds are to be found in asphaltenes and may be rather difficult to remove. The asphaltenes may range from 300 to 1000 in molecular weight.[91,92]

Solvent-refined products from coal or lignite have similar characteristics and seem to be relatively insensitive to the type of feed coal. The nature of solvent used in the solvent extraction process can have some influence on the characteristics of the product.[93]

The hydrocarbons in coal liquids comprise a wide range of compounds, usually containing one to six rings. Alkyl groups seem to be generally no greater than four carbons long, although much longer groups have been identified. Hydroaromatics are more numerous in liquids obtained from catalytic processes, in which more hydrogen can be transferred.

[91] D. D. Whitehurst, M. Farcasiu, and T. O. Mitchell, "The Nature and Origin of Asphaltenes in Processed Oil," Rep. No. EPRI AF-252, by Mobil Research and Development Corporation, February 1976.

[92] M. Farcasiu, T. O. Mitchell, and D. D. Whitehurst, "On the Chemical Nature of the Benzene Insoluble Components of Solvent Refined Coals," *Prepr. Am. Chem. Soc., Div. Fuel Chem.*, **21**(7), 11 (1976).

[93] C. C. Kang, G. Nongbri, and N. Stewart, "The Role of Solvent in the Solvent Refined Coal Process," *Prepr. Am. Chem. Soc., Div. Fuel Chem.*, **21**(5), 19 (1976).

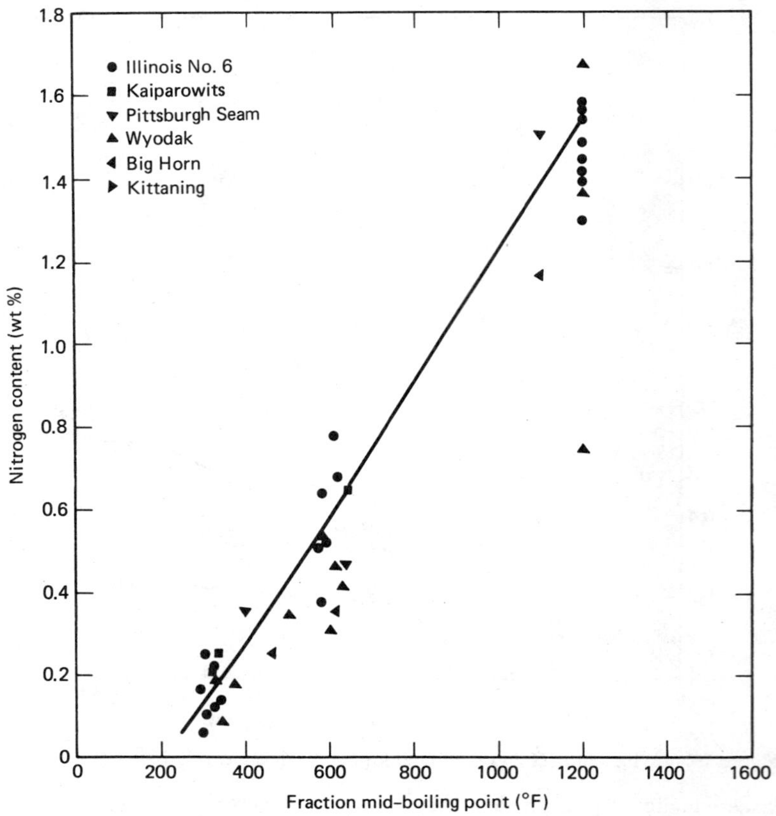

Figure 29.15 The nitrogen contents of coal liquids from catalytic operations.[90]

Pyrolysis liquids can be rather complex; a simple species is seldom present in concentrations greater than 1% in low-temperature carbonization tars. High-temperature tars may have some species present in concentrations of 5%. Oils from low-temperature coking are not as aromatic as those from high-temperature coking.[94] The type of feed coal has less influence on the properties of high-temperature tar than on those of low-temperature tar. Table 29.22 lists fundamental compounds that carry heteroatoms in low-temperature tars.[95] As earlier discussion indicated, generally the higher molecular weight compounds show greatest resistance to heteroatom removal.

Obviously there are many alternatives for processing such complex feedstocks as coal-derived liquids. This complexity also, however, gives rise to many practical problems in trying to achieve the potential products from coal liquids. Many studies have been made, and patents abound with reports on how to transform coal liquids into more valuable chemicals or fuels. The next section addresses some of the significant efforts given to processing coal-derived liquids.

29.4 PROCESSING COAL LIQUIDS

During the last 12–14 years a full range of coal liquids, from benzene-rich light oils derived from coke ovens to solvent-refined coal or extracts, has been studied for various processing techniques. The investigations have included hydrotreating to remove heteroatoms, and mild hydrogenation, hydro-

[94] J. F. Jones, P. L. Barrick, and J. H. Blake, "Pyrolysis of Oils Derived from Coal and Oil Shale in a Fluidized Bed of Catalysts," paper presented at American Chemical Society, Division of Petroleum Chemistry, 151st National Meeting, Pittsburgh, March 23–26, 1966.

[95] S. A. Qader, W. H. Wiser, and G. R. Hill, "Kinetics of the Hydroremoval of Sulfur, Oxygen, and Nitrogen from a Low Temperature Coal Tar," *Ind. Eng Chem. Process Des. Dev.*, 7(3), 390 (July 1968).

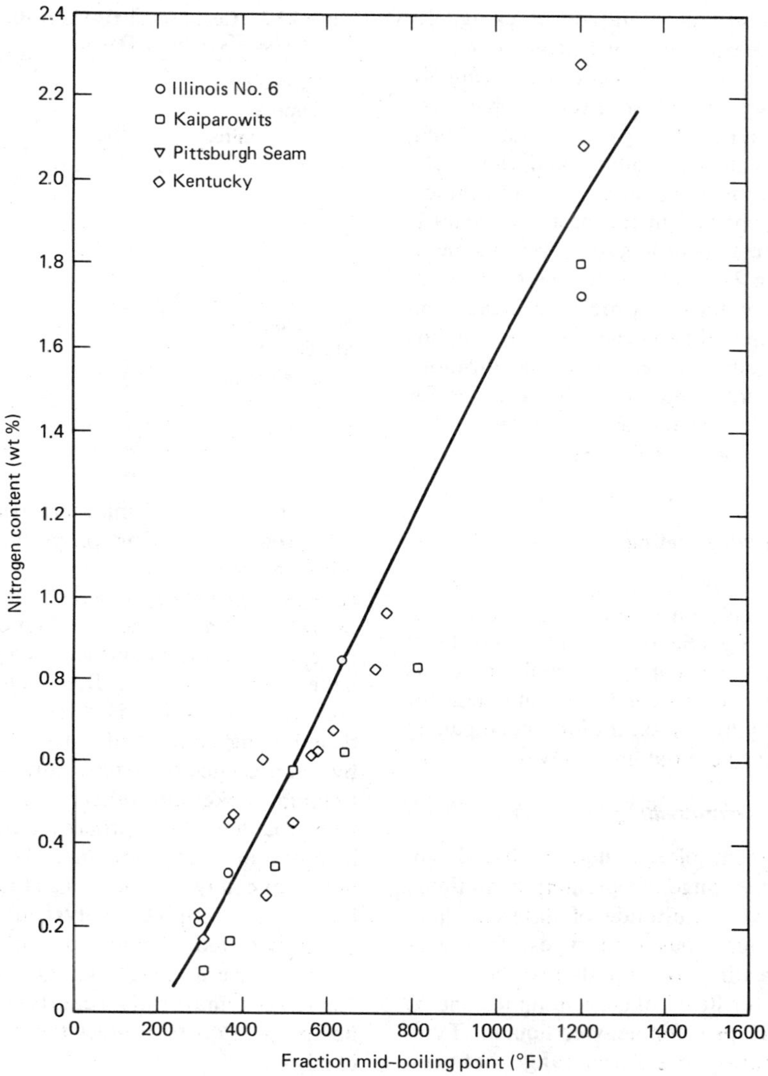

Figure 29.16 The nitrogen contents of coal liquids from noncatalytic operations.[90]

Table 29.22 Heteroatom Species in Low-Temperature Tars

Sulfur Compounds	Oxygen Compounds	Nitrogen Compounds
Sulfides	Phenols	Pyridines
Disulfides	Indenols	Cyclopentanopyridines
Mercaptans	Phenylphenols	Phenylpyridines
Thiophenes	Naphthols	Quinolines
Benzothiophenes	Fluorenols	Benzoquinolines
Dibenzothiophenes	Indanones	Anilines
	Carboxylic acids	Benzylanilines
	Asphaltic and resinous bodies	Naphthylamines
		Pyrroles
		Indoles
		Carbazoles

Source: Modified from Ref. 95.

cracking, reforming, catalytic cracking, dealkylation, deasphalting, and steam pyrolysis. Compatibility of products with petroleum liquids was also determined. Product objectives have been syncrudes, gasoline, diesel fuel, jet fuel, gas turbine liquids, and chemicals.

The material on some of the more important work reported in the past few years is presented here according to processing methods. Section 29.4.1 discusses the hydrotreating of coal liquids, a process critically important to upgrading because the coal liquids usually contain relatively high concentrations of heteroatoms, which must be removed for environmental reasons and to protect rather sensitive catalysts used in subsequent processing.

29.4.1 Hydrotreating

The technical basis for hydrotreating coal liquids is simply an extension of the hydrotreating concepts developed in the petroleum industry. Hydrotreating of coal liquids is done under reaction conditions sufficient for only mild hydrogen saturation accompanied by significant heteroatom removal.

29.4.1.1 *Hydrotreating Summary*

Hydrotreatment of coal-derived liquids involves a wide range of operating conditions, evaluation of a multitude of different catalysts, and numerous feed types. Pressures for hydrotreating coal liquids may be as low as 250 psig for light coke-oven liquids and as high as 4500 psig for heavier liquids. Typically, operating pressures range between 1500 and 2500 psig. Temperatures range just as broadly from 300 to 500°C, with space times from less than 0.5 hr to nearly 5.0 hr. The quantity of gas made or the extent of hydrocracking and coking with most liquids seems to increase drastically at about 425°C. Thus if light hydrotreatment is to be achieved with limited cracking, temperatures must generally be below 425°C or significantly higher pressures must be used. Even so, there will be marked increases in hydrogen consumption.

Some of the many catalysts that have been evaluated for hydrotreating coal liquids are listed in Table 29.23. Of these, the petroleum hydrofining catalysts such as Co-Mo-alumina and Ni-Mo-alumina are favored at present. When a significant nitrogen removal is desired, the Ni-Mo-alumina catalyst would be preferred. The support properties of the catalyst as well as the active metals in the catalyst can be significant for hydrotreating operations. Pore size distribution, average pore size, and total surface area of catalysts have been shown to affect heteroatom removal. Catalyst life is directly affected by deactivating compounds in coal liquids and by the presence of deactivators such as iron, titanium, coke, and other materials. Liquids from coal have been processed for estimated loadings of 4 to more than 11 barrels per pound of catalyst. Processing studies seldom, however, determine catalyst life. One successfully tested technique to prolong catalyst life is to use a guard chamber packed with scavenger materials (catalyst supports, flexirings, etc.) to protect the hydrotreating catalysts.

Table 29.23 Coal-Liquid Hydrotreating Catalyst Types That Have Been Tested

Ni-Mo-alumina
Pt-alumina
Co-Mo-alumina
Na_2CO_3
ZnO
Fe_2O_3
Fe_2S
WS_2
Dicobalt-octacarbonyl
Ni-kieselguhr
Mo-Cr-Zn
Cr_2O_3-alumina
Iodine
Ni-W-alumina

Nitrogen is much more difficult to remove than sulfur and oxygen. The multiple sequential and parallel reactions first of hydrogenation, then of cracking and/or isomerization, followed by further hydrogenation, in addition to the inhibition of catalyst sites by the product, are responsible for a relatively high resistance to removal of nitrogen species in coal.

Coal liquids and their fractions can be hydrotreated to yield reformer feedstock, excellent gasoline blending cuts, material rich in BTX, jet fuel, and gas turbine fuel. In meeting specifications, reducing nitrogen levels to protect sensitive catalysts is especially difficult, and for jet fuels meeting density and smoke points causes some problems. Two stages of hydrotreatment are often necessary.

Table 29.24 Properties of COED Oils Fed to Hydrotreater and Treatment Conditions

Source of Oil	Components (%) Carbon	Hydrogen	Oxygen	Nitrogen	Sulfur	Density (° API)	H/C Atom Ratio
Utah A Seam	83.3	8.5	6.8	1.1	0.3	−3.5	1.22
Illinois No. 6	80.1	7.2	9.1	1.1	2.5	−13.1	1.08
Pittsburgh Seam	83.3	7.1	6.6	1.2	1.8	−12.3	1.02
Venezuelan crude	86.0	12.6	—	0.2	1.2	32.7	1.75

Range of Study, Nominal	
Temperature	343–454°C
Pressure	2000–4500 psig
WHSV	0.3–3.0 lb/(hr)(lb)

Source: Modified from Ref. 41.

29.4.1.2 *Hydrotreating Studies*

Hydrotreatment of COED Oil in Catalyst Testing Units. One coal-derived liquid, COED oil, has been extensively assessed for hydrotreatment. An integral part of the COED process was a hydrotreater reactor in which the raw pyrolysis oils were upgraded to syncrude. This hydrogenation of the raw oils is discussed below.

The hydrotreater in the COED pilot plant had a capacity of 30 bbl/day. Some of the preliminary studies for this unit were conducted in a 100-cm³ catalyst testing unit.[41,96] The work in the catalyst testing unit was performed on filtered raw pyrolysis oil using an American Cyanamid catalyst, HDS-3A. This catalyst has 3.1 wt % NiO + 18.4 wt % MoO_3 on alumina, with a pore volume of 0.56 cm³/g and a surface area of 200 m²/g. The properties of the feed oil and the range of operating conditions are shown in Table 29.24. The data that follow were all taken at 3000-psig pressure[41]; temperature variations within the catalyst bed were rather large (average bed temperature 11°C below maximum). The WHSV required at 382°C and 3000 psig to achieve 80% heteroatom removal from the coal liquids is shown in Table 29.25. Of the three oils listed, the oil from Pittsburgh coal was much more difficult to hydrotreat. In general, the high-sulfur oil from Illinois coal was relatively less resistant to sulfur removal, and the Utah oil was by far the easiest to denitrogenate under hydrotreatment conditions.

The quantity of hydrogen required for saturation to achieve complete heteroatom removal and conversion to all cycloparaffins is shown in Table 29.25, along with the hydrogen required to yield a 25° API product. The hydrogen requirements are much greater for the oils from Illinois and Pittsburgh coals. Figure 29.17 compares the distillation curves for the feeds and products. There is a significant shift between feed and product oils, and little residuum (oils boiling at 538°C) remained in the oils.

[96] J. F. Jones, M. R. Schmid, M. E. Sacks, Y. C. Chen, C. A. Gray, and R. T. Eddinger, *Char Oil Energy Development,* Rep. No. PCR-534, Appendices, for January–October 1966, by FMC Corporation for U.S. Dept. Interior OCR, January 1967.

Table 29.25 Process Conditions and Requirements for Hydrotreating Oils Listed at 382°C, 3000 psi

Oil Type	WHSV for 80% Removal [lb/(hr)(lb)] Nitrogen	Sulfur	Oxygen	Hydrogen Required (scf/bbl) For Saturation	To Produce 25° API Oil
Utah	1.2	1.8	1.8	4600	2700
Illinois	0.8	2.6	1.5	6200	4050
Pittsburgh	0.4	1.1	0.6	6400	4550

Source: Modified from Ref. 41.

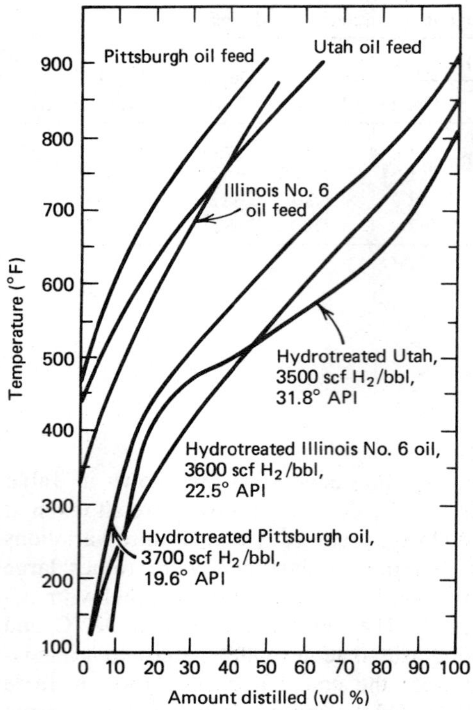

Figure 29.17 Distillation curves for feed and products of COED oil.

Although no data on catalyst life were obtained, the life was anticipated to be proportional to the inverse ratio of the gas make. The gas make (C_3's and lower) was similar for all three oils, remaining at about 1.0 wt % up to about 415°C. At that temperature the gas yield increased rapidly to more than 10 wt % at 427°C.

Hydrotreatment of COED Oils in a 30-bbl/day Pilot Plant. The recommendations for operation of the pilot plant for upgrading raw COED oil were 2000–3000 psig, 400–427°C, and LHSV of 0.8–1.0hr^{-1} with the HDS-3A catalyst.[97] The hydrotreater reactor consisted of a guard chamber holding about 3.2 ft^3 of packing followed by two sections containing, respectively, 6.3 and 9.4 ft^3 of catalyst. The first few experimental runs in the 30-bbl/day pilot plant proved the value of using a guard chamber in front of the catalyst bed to lengthen the life of the catalyst.[98] Results from developmental and production runs in the pilot-plant reactor illustrate performance in a relatively large-scale unit.[12,34] Typical feed oils from various coals are shown in Table 29.26.

The oil from Illinois coal was tested with the HDS-3/3A catalyst mentioned above. At conditions of 399–421°C, 2500 and 3000 psig, and WHSV of 0.22 to 0.54 hr^{-1} the sulfur was readily removed (90–98%); but nitrogen was more difficult to remove, with removal percentages ranging from 65 to 97%, usually only 71–85%. From 73–99% of oxygen, usually 80–88% was removed, and there was essentially no ash in the products.

The oil from Utah coal was tested at 393–429°C, 1650–3000 psig, and WHSV of 0.2–0.55 hr^{-1}. The percentage sulfur removal (69–97%, usually 80–90%), was not as great

[97] J. F. Jones, M. R. Schmid, M. E. Sacks, Y. C. Chen, C. A. Gray, and R. T. Eddinger, *Char Oil Energy Development,* Rep. No. PCR-534 for January–October 1966, by FMC Corporation, for U.S. Dept. of Interior OCR, January 1967.

[98] J. F. Jones, F. H. Schoemann, J. A. Hamsar, B. D. McMunn, L. J. Scotti, and R. T. Eddinger, *Char Oil Energy Development,* U.S. Dept. Interior OCR R & D Rep. No. 73—Interim Rep. No. 1, period of July 1971–June 1972, by the FMC Corporation.

Table 29.26 Ultimate Analyses (wt %) of 30-bbl/day Hydrotreater Feedstocks for 30-bbl/-day Pilot Plant

Constituent	Illinois Oil[a]	Utah Oil[b]	Western Kentucky Oil
Carbon	81	83	82
Hydrogen	7.7	8.6	8.0
Oxygen	7.8	6.2	6.8
Nitrogen	1.08	1.3	0.99
Sulfur	2.44	0.66	1.49
Ash	0.15	0.09	0.09

[a] API gravity = −4.
[b] API gravity = −3.

Source: Modified from Ref. 34.

as for the oil from Illinois coal. The range of both oxygen and nitrogen removal was around 80–90%. The study of the effect of pressure on this feed oil showed that the API gravity of the product was relatively insensitive to pressure increases. At the lower pressure the gravity was 17–21° API and was 30° API for pressures of 2300–2400 psig. Pressure increases from 1400–1800 psig improved hydrodesulfurization markedly, raising percentage removal from 70 to 90%; but once these removal levels were achieved, little was gained by further increases in pressure. In these tests both oxygen and nitrogen removal increased, respectively, from 75 to 100% and from 80 to 95%, as pressure increased from 1400 to 2400 psig.

The oil from Western Kentucky coal was, in general, more difficult to hydrotreat than the oil from Utah coal. For WHSV of 0.26–0.27 hr^{-1} and temperatures of 406–417°C, the API gravity of the resultant syncrude increased as pressure increased from 1400 to 2600 psig. Over the same pressure range, the percentage of sulfur removed increased from 80 to 95% for this relatively high-sulfur oil. The removal percentage for nitrogen also increased from 70% to about 93%, and for oxygen from 82 to 95%. Other reports of the work on Western Kentucky and Colorado high-volatile B coal oils are available.[11,42]

A very important part of the work on the 30-bbl/day unit was to demonstrate the use of a guard chamber as a technique for removing metals and solids from the raw COED oils, thereby significantly extending the useful life of the catalyst. Typically, the filtered raw oil from the COED pyrolysis reactors contains 0.1 wt % solids as fed to the hydrotreaters. These solids consist of very fine char and quinoline-insoluble hydrocarbons, which are high in metals, especially iron. Several materials and combinations of materials have been tested for use in the guard chamber: Kock stainless steel flexirings, alumina tablets, spheres and pellets, HDS-3 catalyst, cobalt-molybdenum-alumina catalysts, and others.

The performances of some of the systems tested are shown in Fig. 29.18 in terms of the total barrels of oil processed before the guard chamber became plugged.[12] The use of catalysts in the guard chamber gave significantly better performance than the use of alumina

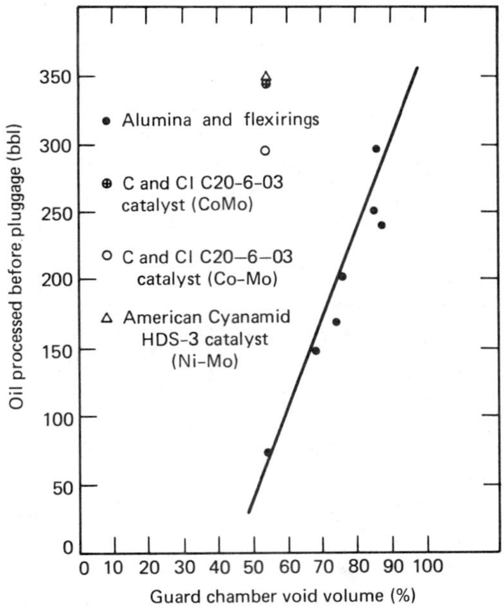

Figure 29.18 Oil processed before pluggage through hydrotreating the guard chamber.

and flexirings. This high performance by the catalysts was probably caused by their ability to accommodate more coke within the pellets and to hydrogenate coke precursors.

Detailed analysis of deposits in the guard chamber showed that coke and iron predominated, with coke ranging from 0.5 to 1.75 lb per pound of packing, and iron from 0.002 to 0.195 lb/lb. These were typical ranges for these deposits depending on feed, time on stream, and operating conditions. The solids removal was about 0.2–0.7 lb/bbl. Solids analysis of the reaction catalysts downstream from the guard chamber indicated iron carrythrough, especially in the first one-third of the reactor bed (0.20 lb of Fe deposited per pound of bed), with little at bed exit. Coke was relatively uniformly distributed throughout at 1.0–1.6 lb coke deposited per pound of bed.

Further modifications, such as using alternating sections of catalyst and alumina layers in the first section of the hydrotreater, followed by the HDS-3 catalyst in the second section, resulted in better temperature control and slower catalyst deactivation. A design capacity for the catalysts appears to be 4–5 barrels of oil per pound of catalyst.

Atlantic Richfield Tests of Different Refining Techniques on COED Oils and Syncrudes. In tests by Atlantic Richfield to assess various refining techniques applied to COED oils, certain fractions were hydrotreated. To prepare feedstocks for reforming operations, two naphtha fractions were first hydrotreated.[99] The naphthas (Table 29.27) were derived from an Illinois coal liquid that contained only 0.07 wt % sulfur, 0.03 wt % nitrogen, and 0.6 wt % oxygen. Since the paraffin content of the naphthas was low, they were potentially good reformer stocks. The naphthas were treated in a 1-in. (diameter) × 20-in. reactor, using an American Cyanamid HDS-3A catalyst (3.1 wt % NiO, 14.8 wt % MoO_3, 200 m^2/g). Results at 750 psig, 399°C, and 1–4 hr^{-1} WHSV indicated that sulfur and nitrogen levels were still too high for use with bimetallic reforming catalysts, but that the use of platinum catalysts might be feasible. Operations at 1000 psig, 399°C, and 2.0 hr^{-1} WHSV produced reformer feeds that generally contained sulfur at 0–0.4 ppm and nitrogen at 4–7 ppm. The naphtha having the lower end point consumed 40 scf/bbl of hydrogen, the higher 80 scf/bbl, but overall there was little saturation and hydrocracking.

A mid-distillate (199–274°C) jet fuel fraction from the oil from Illinois coal was hydrotreated in an attempt to meet density, aromatics, and smoke point specifications. A first-stage treatment over the HDS-3A catalyst at 371°C, 2000 psig, and 2 hr^{-1} WHSV was followed by a second-stage treatment with a platinum catalyst at 260–316°C, 1000–2000 psia, and 1.0 hr^{-1} WHSV. The first-stage treatment with the HDS-3A catalyst was insufficient to meet jet fuel specifications because of the high cycloparaffins (69 vol %). The second stage brought the distillate to the point where it met specifications for smoke point and aromatic saturation was complete; density, however, remained too high. Typical hydrogen consumption in the first stage was 800 scf/bbl; in the second stage it was 670 scf/bbl. A general summary of the overall potential of a COED oil from

[99] S. J. Wachtel, T. A. Collins, J. P. Gallagher, G. V. Lampadarios, M. G. Robison, and J. O. Siemssen, *Evaluation of Hydrotreated COED Syncrude—Final Report,* Memorandum No. MI.73-1, Atlantic Richfield Company, January 29, 1973.

Table 29.27 Properties of COED Naphtha Feedstocks

	Low End Point Naphtha	High End Point Naphtha
Distillation		
Fraction	Temperature (°C)	
IBP	93	94
10%	109	109
50%	135	140
90%	170	177
EP	191	202
Properties		
Elemental analysis (ppm)		
Sulfur	100	90
Nitrogen	87	128
Hydrocarbon type (vol %)		
Paraffins	7.1	7.1
Monocycloparaffins	60.2	58.9
Dicycloparaffins	11.2	12.3
Tricycloparaffins	0.1	0.2
Alkylbenzenes	19.9	19.8
Indanes-tetralins	1.5	1.7
Naphthalene	0	0

Source: Modified from Ref. 99.

Table 29.28 Properties of Feed to Hydrotreater and Properties of Product Fractions

		Fractions				
	Feed	IBP–300°F	300–550°F	550–650°F	650°F+	Vacuum Bottoms
API gravity:	31.0	50.3	27.5	14.5	5.4	—
Property (wt %)			H-Coal			
Total liquid	100	25	61.4	8.1	5.2	—
Hydrogen	11.8	13.3	11.2	10.0	9.0	5.75
Carbon	87.1	86.6	87.2	89.2	89.2	74.39
Sulfur	0.170	0.129	0.073	0.110	0.199	1.93
Nitrogen	0.172	0.066	0.175	0.165	0.504	1.16
Oxygen	0.48	0.61	1.00	0.49	1.34	4.60
Ash	0.0	—	—	—	—	14.62
API gravity:	18.2	47.8	25.2	18.0	16.5	9.6
Property (wt %)			COED Oil			
Total Syncrude	100	4.0	32.5	18.0	27.0	18.5
Hydrogen	10.94	14.8	11.23	10.58	10.86	10.09
Carbon	87.48	85.2	86.45	87.9	88.10	87.54
Sulfur	0.023	0.027	0.021	0.016	0.022	0.02
Nitrogen	0.385	0.068	0.370	0.373	0.340	0.428
Oxygen	0.98	0.34	1.31	0.86	0.74	1.02
Ash	0.05	—	—	<0.001	<0.02	0.03

Source: Modified from Ref. 100.

Illinois coal refinery feed was made by Rothman et al. in 1974.[45]

The Atlantic Richfield Company also conducted a broader study on syncrudes from shale oil and coal stocks to determine technical problems in converting these to aviation turbine fuel.[100] The two coal syncrudes were obtained from the H-Coal and COED processes. A portion of each of these two syncrudes was separated into various fractions (Table 29.28), which were then further processed. The H-Coal liquid was hydrotreated at two levels of severity to meet yield and hydrogen specifications. During the two hydrotreatments the total pressures were maintained at 2000 psig, with a WHSV in the range of 0.45–0.50 hr^{-1} and a hydrogen rate of 8000 scf/bbl. For the low-severity treatment the starting and final temperatures were 343 and 349°C, respectively. For the high-severity treatment these temperatures were 368 and 374°C. The increase in temperature from start to finish of an experiment reflects the loss in catalyst activity, which was compensated for by raising the reaction temperature.

The same experiments were again conducted with the American Cyanamid HDS-3A catalyst (1.59 mm). The reactor was 5.08 cm diameter × 61 cm long and packed with 500 g of catalyst diluted with tabular alumina on a gradient of 3 : 1 (alumina-catalyst) at the inlet down to 1 : 8 at the reactor outlet. The results for the H-Coal hydrotreatment are shown in Table 29.29. The desired fuel specifications of aromatic content (20 and 40 vol %), hydrogen content (12.75 and 13.5 wt %), and sulfur and nitrogen levels were met, along with other requirements.

The COED oil from Utah coal was given somewhat more attention in this study.[100] The limiting factor seemed to be the same as for the H-Coal liquid: that of achieving the required levels of hydrogen in the product. The reactor and catalyst configuration were the same as were used with the H-Coal oil. The

[100] A. C. Antoine and J. P. Gallagher, "Synthesis and Analysis of Jet Fuels from Shale Oil and Coal Syncrudes," paper presented at American Institute of Chemical Engineers, 82nd National Meeting, Atlantic City, N.J., August 29–September 1, 1976.

IBP–288°C fraction was hydrotreated at both low- and high-severity conditions at 2000 psig and 8000 scf/bbl hydrogen rate. The high-severity experiment was conducted with the catalyst fresh, at an initial temperature of 368°C and a final temperature of 374°C, at a WHSV of 0.50 hr^{-1}. The low-severity run, conducted over the same catalyst charge, was started at 343°C, 0.80 hr^{-1}, and terminated at 349°C, 0.65 hr^{-1}. The rates of loss of catalyst activity were observed to be about 11 and 8% per day for the high- and low-severity conditions, respectively. The results (Table 29.30) indicate that, again, the desired specifications were generally met, but with significant loss in catalyst activity.

A heavier COED fraction of 288–427°C was hydrotreated at the "most severe conditions practicable," 385°C, 2000 psig, and 0.70 hr^{-1}. Whereas no indication was given of the rate of catalyst deactivation, the changes from feed to product were (1) API gravity, 17.2–22.9, (2) hydrogen content, 10.6–11.9 wt %, (3) sulfur content, 0.025–0.002 wt %, and (4) nitrogen content, 0.358–0.0237 wt %.

This study indicated in general that with these coal liquids and using distillation and hydrotreating (and hydrocracking), fuel specifications approximating those of JP-4, JP-5 and diesel No. 2 can be met. However achieving the desired degree of hydrogenation and severe catalyst deactivation are noted as significant problem areas.

Potential of Synthoil Liquids for Producing Jet Fuel. The results of another program recently reported, though still incomplete, suggest more pessimistic conclusions about the potential for producing JP-4 and Jet A specification fuels from coal by distillation and hydrotreating.[101] Only limited results were obtained on a Synthoil liquid using the HDS-3A catalyst at 371°C, 1500 psig, and 0.81–0.95 hr^{-1}, LHSV. The feedstock was a fraction boiling up to 293°C from the Synthoil process. The aromatic content of the products with respect to the feedstock was reduced by 40–50%, sulfur by 96–98%; hydrogen was increased by 8–11%. No indication of the activity state or conditioning of the catalyst was given.

Treatment of COED Oils to Produce Gas Turbine Fuels. A more recent detailed study was made on the preparation of a gas turbine engine fuel by hydrotreating distillates from COED oils.[102,103] The feedstock properties (Table 29.31) were obtained by atmospheric and vacuum distillation to remove the light and heavy ends. The hydrotreatments were made in two stages, using the HDS-3A catalyst mentioned above for some tests and the HDS-9A (3.1 wt % NiO + 18.4 wt % MoO_3, 150 m^2/g) for others. The first-stage reactor was 3 in. in diameter and could contain 10,000 cm^3 (7470 g) of catalyst, although somewhat less was used. First-stage conditions were nominally 366°C, 1500 psig, and 4000 H_2 scf/bbl at LHSVs of 0.15 and 0.2 hr^{-1}. The desired level for aromatics was 20–25 vol % in the products. The catalyst decayed at about 1% per day (rise in aromatics) to yield 35–40% aromatics after 20 days on stream with the oil from a Western Kentucky coal. To maintain aromatics specification while using the oil from the Utah coal, the reaction temperature was increased, and after 17 days on stream, a temperature of 383°C was reached.

Second-stage hydrogenation was achieved in a reactor packed with an American Cyanamid platinum on alumina catalyst (PHF-4) distributed in about four equal zones, each of 1220 g interspersed with inert packing. The feedstocks for the second-stage treatment (products from stage 1) are shown in Table 29.32. Stage 2 treatment conditions were 3000 psig, 0.75 hr^{-1} LHSV, 4000 scf/bbl, and 279°C for the oil from Utah coal and 271°C for the oil from Kentucky coal. Although platinum catalysts are rather suscep-

[101] C. D. Kalfadelis and H. Shaw, "A Pilot Plant Study of Jet Fuel Production," paper presented at American Institute of Chemical Engineers, 82nd National Meeting, Atlantic City, N.J., August 29–September 1, 1976.

[102] F. S. Eisen, *Preparation of Gas Turbine Engine Fuel from Synthetic Crude Oil Derived from Coal, Phase II, Final Report,* AD/A-007 923, U.S. Navy Contract No. N00014-74-C-0568, Mod. P00001, by Suntech, Inc. (Sun Oil Company), February 6, 1975.

[103] F. S. Eisen and J. D. Tice, "Preparation of Gas Turbine Engine Fuel from Synthetic Crude Oil Derived from Coal," paper presented at American Institute of Chemical Engineers 82nd National Meeting, Atlantic City, N.J., August 29–September 1, 1976.

Table 29.29 Properties of Low-Yield H-Coal Products

	Low Severity				High Severity			
Property	IBP–650°F	250–653°F	250–550°F	IBP–550°F	IBP–650°F	250–650°F	250–550°F	IBP–550°F
Specific gravity	0.8493	0.8654	0.8565	0.8413	0.8338	0.8488	0.8468	0.8314
Flash point °F	—	102	96	—	—	102	106	—
Freezing point (°F)	−8	−32	−68	−79	(0)	−17	−54	−86
Viscosity at 239 K (−30°F) (cSt)	—	16.99	6.785	5.162	9.757	15.91	9.102	6.264
Smoke point (mm)	14	15	15	16	24	21	24	25
Hydrocarbon type (vol %)								
Aromatics	29.7	33.8	30.9	26.3	5.9	6.7	5.8	5.5
Olefins	1.2	1.8	1.4	1.2	1.3	1.4	1.0	0.9
Naphthalenes	0.54	0.66	0.31	0.27	0.064	0.077	0.065	0.055
Elemental analysis (wt %)								
Hydrogen	12.73	12.47	12.64	12.79	13.56	13.26	13.31	13.73
Nitrogen	0.0005	0.0006	0.0006	0.0001	< 0.0001	< 0.0001	< 0.0001	0.0001
Total sulfur	0.0004	0.0004	0.0001	0.0001	0.0005	0.0005	0.0001	0.0001
Mercaptans	—	—	—	—	—	—	—	—
Oxygen	0.11	0.09	0.06	0.10	0.04	0.06	0.03	0.04

Source: Modified from Ref. 100.

Table 29.30 Properties of High-Yield COED Products

	High Severity				Low Severity			
Property	IBP–650°F	250–650°F	250–550°F	IBP–550°F	IBP–650°F	250–650°F	250–550°F	IBP–550°F
Specific gravity	0.8358	0.8586	0.8493	0.8270	0.8255	0.8458	0.8368	0.8165
Flash point (°F)	—	114	104	—	—	104	106	—
Freezing point (°F)	−23	+2	−62	−70	−23	−30	−64	−72
Viscosity at 239 K (−30°F) (cSt)	—[a]	—[a]	9.851	5.586	—[a]	19.25	9.676	5.565
Smoke point (mm)	16	14	16	19	20	20	24	27
Hydrogen type (vol %)								
Aromatics	22.4	28.5	25.2	20.1	9.3	11.6	7.2	5.4
Olefins	0.6	1.2	0.8	0.5	0.7	1.1	0.9	0.5
Naphthalenes	0.68	0.86	0.38	0.31	0.49	0.62	0.13	0.11
Elemental analysis (wt %)								
Hydrogen	13.07	12.88	12.96	13.24	13.6	13.44	13.63	13.69
Nitrogen	0.0003	0.0002	0.0002	0.0002	0.0002	0.0002	0.0002	0.0002
Total sulfur	0.0003	0.0001	0.0001	0.0001	0.0003	0.0003	0.0001	0.0001
Mercaptans	—	—	—	—	—	—	—	—
Oxygen	0.03	0.03	0.03	0.03	0.05	0.05	0.05	0.03

[a] Solid.

Source: Modified from Ref. 100.

Table 29.31 Properties of COED Oils Fed to First-Stage Hydrogenation Reactor

	Type of Oil		
	Phase I, Utah	Phase I, Western Kentucky	Phase II, Western Kentucky
Crude (wt %):	43.2	51.0	15.7
API Gravity (D287):	24.2	25.3	30.0
	ASTM D86 Distillation		
Fraction		*Temperature (°C)*	
IBP	183	160	154
5%	201	171	164
10%	213	181	170
30%	236	209	182
50%	256	243	194
70%	276	269	208
90%	296	293	234
95%	303	300	252
EP	307	307	262
	Properties		
Smoke point (mm)	9	9	10
Flash point (°C)	82.2	41	47
Freeze point (°C)	−46	−46	−59
Elemental analysis			
Carbon (wt %)	87.2	87.6	87.2
Hydrogen (wt %)	11.1	11.2	11.6
Oxygen (wt %)	1.4	1.0	1.0
Nitrogen (wt %)	0.3	0.2	0.2
Sulfur (ppm)	241	135	33
Hydrocarbon type (vol %)			
Paraffins	14.5	6.7	8.0
Naphthenes	20.2	27.4	37.9
Olefins	3.1	1.0	2.3
Aromatics	62.2	64.9	51.8
Metals (ppm)	61	105	—

tible to sulfur and nitrogen poisoning, no problems were encountered over a three-week experimental run. The products resulting from the second-stage hydrotreatment (Table 29.33) contained less than 5 vol % aromatics.

The specifications for the finished product were met by distillation to remove light and heavy ends. Most of the current JP-5 specifications could be met, but meeting smoke points specifications is difficult for fuels containing more than 20 vol % aromatics. First-stage catalyst deactivation was quite severe, and preferred operation would have to be at 3000 psig as in a production run, rather than 1500 psig, as in stage 1 tests. Some additional studies related to this work further characterize the products.[104–107]

[104] R. N. Hazlett, J. M. Hall, and J. Solash, "Properties and Composition of Jet Fuels Derived from Alternate Energy Sources Part I. Background and *N*- Alkane Content," *Prepr. Am. Chem. Soc., Div. Fuel Chem.*, **21**(6), 219 (1976).

[105] J. Solash and R. F. Taylor, "Characterization of Aromatic Fractions from Non-Petroleum Derived JP-5 Type Fuels," *Prepr. Am. Chem. Soc., Div. Fuel Chem.*, **21**(6), (1976).

[106] C. J. Nowack, J. Solash, and R. J. Delfosse, "Test and Evaluation of Coal Derived JP-5 Type Fuels," paper presented at American Institute of Chemical Engineers 82nd National Meeting, Atlantic City, N.J., August 29–September 1, 1976.

[107] J. Solash and R. F. Taylor, "Characterization of Aromatics in Non-Petroleum JP-5 Type Fuels," paper presented at American Institute of Chemical Engineers 82nd National Meeting, Atlantic City, N.J., August 29–September 1, 1976.

Table 29.32 Properties of Feedstocks Produced from COED Oils for Second-Stage Hydrogenation

	Type of Coal	
	Utah	Western Kentucky
API gravity:	33.1	31.2
ASTM D86 Distillation		
Fraction	*Temperature (°C)*	
IBP	142	143
5%	164	160
10%	181	169
50%	235	231
90%	279	288
95%	292	300
EP	303	310
Properties		
Elemental analysis		
Carbon (wt %)	87.3	87.1
Hydrogen (wt %)	12.7	12.8
Oxygen (ppm)	173	400
Nitrogen (ppm)	16	73
Sulfur (ppm)	9	4
Product analysis (vol %)		
Aromatics	29.3	33.9
Olefins	0.7	0.7
Saturate	70.0	65.4
Metals (ppm)[a]	85	112

[a] Determined by emission spectroscopy.

Source: Modified from Ref. 102.

Use of Iodine as a Hydrotreatment Catalyst. One final catalyst is worthy of mention for hydrotreating not only COED oils but also coal tars in general. Iodine, when added to coal liquids at about 0.6–1.0 wt % of liquid, can hydrotreat in the ranges of 393–449°C and 1000–3000 psig. One advantage suggested for the use of iodine is that gases with relatively low hydrogen content can be used: as low as 25 mole %.[108] Iodine might also help nitrogen removal, but this has not been established. Large-scale use of iodine would be hindered by excessive corrosion and by the recovery of iodine from product streams.[109]

Evaluation of Catalysts for Hydrotreating Coal-Derived Liquids. Coal liquids were hydrotreated in a study of the production of high-temperature gas turbine fuels.[67,110,111] A well-designed shaker bomb reactor was used, with excellent temperature control and few heatup and cooldown effects. Typical reaction conditions ranged from 371 to 427°C at 2000 psig and for reaction times of up to 2 hr. In one series of tests a solvent-refined coal extract was hydrotreated with a 2 : 1 ratio of recycle solvent to extract, to assess the performance of several commercial catalysts. These catalysts and their properties are listed in Table 29.34, and the properties of the feedstocks and products are given in Table 29.35. Reaction conditions for this catalyst-screening series were 427°C, 2000 psig, 1.0 hr^{-1}, and 20 : 1 weight ratio of hydrocarbon to catalyst. Based on removal of heteroatoms, the Ketjen 153-S catalyst was superior. Greater saturation and reduction to fractions boiling below 454°C resulted from use of the HDS-3A catalyst. The performance of the tungsten catalyst was especially poor in these tests. The effects of pore size, surface area, metals concentration, and metals types cannot be distinguished, nor were they intended to be. On the basis of its high degree of hydrogenating ability, the HDS-3A catalyst was used in subsequent tests.

These additional tests used the SRC recycle solvent and extract at weight ratios of 2 : 1 and 1 : 1 at 2000 psig hydrogen pressure, 399°C, and 0.5–2.0 hr^{-1}, both with and without catalysts (20 : 1 hydrocarbon/catalyst). These studies revealed the following points:

1. Hydrogenation is significant with the HDS-3A catalyst (440–825 scf H_2/bbl) and without a catalyst (350 scf H_2/bbl). The recycle solvent apparently serves as a good hydrogen transfer agent.
2. Almost no nitrogen removal was achieved, but 60 and 30 wt % sulfur and

[108] L. D. Friedman and R. T. Eddinger, U.S. Pat. 3,453,202, FMC Corporation, July 1, 1969.

[109] R. T. Eddinger and L. D. Friedman, "Hydrogenation of Low-Temperature Coal Tars Using Iodine as Catalyst," *Fuel,* **47**(4), 320 (July 1968).

[110] A. V. Cabal, S. E. Voltz, and T. R. Stein, "Upgrading Coal Liquids to Gas Turbine Fuels. 2. Compatibility of Coal Liquids with Petroleum Fuels," *Ind. Eng. Chem. Prod. Res. Dev.,* **16**(1), 58 (1977).

[111] T. R. Stein, S. E. Voltz, and R. B. Callen, "Upgrading Coal Liquids to Gas Turbine Fuels. 3. Exploratory Process Studies," *Ind. Eng. Chem. Prod. Res. Dev.,* **16**(1), 58 (1977).

Table 29.33 Properties of Products from Second-Stage Hydrogenation of Liquids Derived from COED Oils

	Type of Coal	
Chargestock aromatics (vol. %):	Utah 29.3	Western Kentucky 33.9
Product Analysis (vol %)		
Aromatics	3.6	3.6
Olefins	1.4	0.8
Saturates	95.0	95.6
ASTM D86 Distillation		
Fraction	*Temperature (°C)*	
IBP	139	143
5%	164	157
10%	180	167
20%	188	182
30%	213	199
50%	232	228
70%	252	254
80%	263	267
90%	279	282
95%	293	293
EP	304	302

Source: Modified from Ref. 102.

oxygen removal were achieved, respectively. Recall that for turbine fuels the levels of sulfur and nitrogen are indicated to be about 0.5 to 1.0 wt % respectively as fixed by the available NO_x codes.

3. Decreasing the ratio of solvent to extract decreases the sulfur removal. The hard-to-remove sulfur species were in the heavier fractions of the extract, but the hard-to-remove oxygen compounds were found in the solvent fraction.

A feedstock was prepared from a fluidized catalytic cracker (FCC) clarified oil using a

Table 29.34 Assessment of Performance of Hydroprocessing Catalysts[67]

	Catalysts[a]					
	3A	618X	9A	1441A	153-S	86
Physical properties						
Total pore volume (cm³/g)	0.54	0.60	0.43	0.57	0.52	0.28
Real density (g/cm³)	3.41	3.60	3.23	3.67	3.55	3.20
Particle density (g/cm³)	1.20	1.14	1.35	1.19	1.24	1.70
Surface area (m²/g)	173	140	145	323	439	198
Average pore diameter (Å)	125	171	119	71	50	56
Chemical composition (wt %)						
Nickel	2.9	2.7	2.6	—	3.1	7.1
MoO_3	12.8	14.8	12.1	13.8	15.1	—
CoO	—	—	—	2.8	—	—
Tungsten	—	—	—	—	—	17.5

[a] 3A = American Cyanamid HDS-3A, 618X = Harshaw 618X, 9A = American Cyanamid HDS-9A, 1441A = American Cyanamid HDS-1441A, 153-S = Ketjen 153-S, 86 = American Cyanamid HDS-86.

Table 29.35 Properties of Liquid Products from Catalyst Evaluation Tests[67]

		Catalysts (see note ,[a] Table 29.34)					
Liquid Product Analysis	Charge[a]	3A	618X	9A	1441A	153-S	86
Specific gravity at 15.6°C	1.068	1.075	1.073	1.081	1.082	1.072	1.094
Elemental analysis (wt %)							
Hydrogen	6.82	8.01	7.71	7.61	7.50	7.71	7.08
Sulfur	0.36	0.14	0.13	0.16	0.18	0.13	0.29
Nitrogen	1.01	0.96	0.85	0.85	0.81	0.89	0.99
Oxygen	4.09	2.72	3.29	2.87	2.54	2.83	3.3
Simulated distillation (wt %)							
IBP–454°C	69.9	78.2	78.0	73.2	69.7	72.2	—[b]
454°C+	30.1	21.8	22.0	26.8	30.3	27.8	—[b]
Heteroatom removal (wt %)							
Sulfur		63	65	55	50	65	19
Oxygen		34	20	30	40	31	19
Nitrogen		6	17	18	22	14	2
Total hydrogen consumption (bbl)		1058	822	759	702	825	262
Hydrogen utilization (wt % total)							
H_2S		0.9	1.2	1.1	1.1	1.2	1.1
NH_3		0.9	3.1	3.4	4.6	2.4	1.7
H_2O		11.7	9.0	14.5	20.8	13.8	25.3
C_{1+}–C_5 gases		15.8	19.7	22.9	26.1	19.2	29.4
C_6 + liquid		70.7	67.0	58.1	47.5	73.4	42.5

[a] Ratio of recycle solvent to SRC, 2 : 1.
[b] Not determined.

2 : 1 FCC oil to SRC extract ratio. The HDS-3A catalyst was used at 399 and 427°C, 2000 psig, and 1.0–2.0 hr. The feedstock had 6.98 wt % hydrogen, 1.02 wt % sulfur, 0.79 wt % nitrogen, and 1.61 wt % oxygen, with 58.5 wt % material boiling below 538°C and 41.3 wt % above 538°C. Desulfurization was 23 wt % and deoxygenation was 47 wt % at the least severe conditions of 399°C and 1.0 hr, whereas they were, respectively, 59 and 54 wt % at the more severe conditions of 427°C and 2.0 hr. The sulfur was not removed uniformly, but varied from 90 wt % in polar resins to 78 wt % for asphaltenes and 46 wt % in the polynuclear aromatic (PNA) oils. Virtually no nitrogen was removed under any conditions. In general, the sulfur was easier to remove from the heavier fractions—a marked contrast to petroleum residua. Similarly, the oxygen was more easily removed from the heavier fractions.

Liquids from H-Coal, Synthoil, and SRC were each tested in the shaker bomb reactor at 427°C, 2000 psig, 1.0 and 2.0 hr, and with a 5 : 1 hydrocarbon-catalyst weight ratio. The feedstock properties are shown in Table 29.36, along with their respective product properties at the two conditions tested. Under the most severe conditions, the asphaltenes were reduced from 22.7 to 6.8 wt % in the product from the H-Coal. In searching for the fractions most resistant to heteroatom removal, experiments with the H-Coal stock revealed that sulfur was rather uniformly removed from all fractions and nitrogen and oxygen were more easily removed from the heavier fractions. Little hydrocracking of this stock occurred. The Synthoil liquid appeared to have properties similar to those of the H-Coal oil, but presented more problems for heteroatom removal. A possible reason for this problem is the relative uniformity of distribution of sulfur throughout the Synthoil fractions, whereas in the H-Coal liquid there is more sulfur in the relatively easily treated higher boiling fractions. The solvent-refined coal was hydrodesulfurized about as easily as the other two liquids; it was more difficult, however, to remove oxygen and especially nitrogen. Significantly greater quantities of hydrogen could be transferred to the SRC stocks than to the other two.

Brief mention should be made of another study in which screening tests of catalysts

Table 29.36 Properties of Feedstocks and of Products Hydrotreated to Yield Turbine Fuels[67]

	Process Used to Prepare Feedstock								
	H-Coal			Synthoil			SRC		
Reaction time (hr): Liquid Product Analysis	Charge	1.0	2.0	Charge	1.0	2.0	Charge	1.0	2.0
Specific gravity at 15.6°C	1.10	1.079	1.050	1.10	1.057	1.060	1.184	1.184	1.184
Elemental analysis (wt %)									
Hydrogen	7.94	8.38	8.66	7.97	8.65	8.44	5.35	5.86	6.24
Sulfur	0.42	0.08	0.03	0.43	0.11	0.10	0.43	0.10	0.07
Nitrogen	0.77	0.66	0.47	0.97	0.65	0.72	1.85	1.75	1.47
Oxygen	2.12	1.33	0.60	2.08	1.00	1.07	4.17	3.00	2.27
Simulated distillation (wt %)									
IBP–454°C	58.5	58.8	70.5	62.8	67.4	66.0			
454°C+	41.5	41.2	29.5	37.2	32.6	34.0			
Hydrogen consumption (scf/bbl)		298	750		538	445		843	1396
Yield (wt %)									
H_2S		0.36	0.41		0.35	0.36		0.35	0.39
NH_3		0.18	0.40		0.44	0.37		0.20	0.56
H_2O		0.98	1.75		1.29	1.22		1.44	2.28
C_1		0.65	1.11		0.55	1.00		0.96	1.62
C_2		0.43	0.71		0.47	0.70		0.69	1.09
C_3		0.31	0.46		0.54	0.48		0.56	0.83
C_4		0.16	0.19		0.34	0.20		0.35	0.41
C_5		0.06	0.05		0.12	0.05		0.11	0.08
C_6 + liquid		93.62	93.97		93.18	92.87		96.42	94.54
Coke		3.66	1.98		3.46	3.36			
		100.41	101.03		100.74	100.61		101.08	101.80
Heteroatom removal (wt %)									
Sulfur		82	93		77	79		78	85
Nitrogen		20	43		38	31		9	25
Oxygen		37	72		52	49		31	49

for hydrotreating solvent-refined coals and lignites were made.[112–114] Usual test conditions in a batch autoclave were at 500–2500 psig and 450°C for about 2.0 hr. Catalyst materials such as Na_2CO_3, ZnO, Fe_2O_3, Fe_2S, Co-Mo-alumina, Ni-Mo-alumina, and WS_2, among others, were tested. The hydrofining catalysts gave superior performance, and similar conversions were obtained with SRC and SRL.

Hydrotreatment of British Coal Extracts. A study of British coal extracts revealed favorable results from hydrotreatment using a standard Co-Mo-alumina catalyst.[115] Batch studies were made with catalyst-feed ratios of 1 : 4 to 1 : 20, with solvents and extracts as depicted in Table 29.37. Under conditions of 450°C, 2060 psig, and 4 hr, the H/C atom ratio in the hydrogenated filtrate was 0.90 compared to 0.62 in the initial extract and 0.77 in the solvent oil. As is typical, the sulfur was significantly easier to remove than were oxygen and nitrogen. The low-rank coal extracts were the easiest to hydrogenate.

Dicobalt Octacarbonyl as a Hydrotreatment Catalyst. Dicobalt octacarbonyl, $Co_2(CO)_8$, has been tested on a coal tar to determine whether carbonyls would promote the addition of hydrogen to certain points in the polynuclear aromatics of coal liquids.[116] The idea is to place hydrogen at points that control, limit, or reduce coking and at the same time reduce the total hydrogen consumption. As an example, hydrogen addition at the 9,10 positions of anthracene should limit its high coking tendency. A set of results is shown in Table 29.38 comparing conditions for the carbonyl and a Co-Mo-alumina catalyst. The results show that the carbonyl can partially hydrogenate and can do this at a significantly lower temperature; however a marked in-

[112] D. E. Severson, *Project Lignite Quarterly Report No. 7,* Rep. No. FE-1224-52, University of North Dakota, for U.S. ERDA, January 22, 1976.

[113] D. E. Severson, *Project Lignite Quarterly Technical Progress Report No. 6,* Rep. No. FE-1224-49, University of North Dakota, for U.S. ERDA, December 22, 1975.

[114] D. E. Severson, *Project Lignite Quarterly Technical Progress Report No. 5 and Fiscal Year 1975 Annual Summary,* Rep. No. FE-1224-48, University of North Dakota, for U.S. ERDA, December 3, 1975.

[115] J. S. Harrison, "Research Activities on Coal Liquefaction in the UK," paper presented at American Chemical Society, Division of Petroleum Chemistry, 171st National Meeting, New York, April 4–9, 1976.

[116] ARCO Chemical Company, *Catalytic Hydrotreating of Coal-Derived Liquids—Project Seacoke Phase II Final Report,* for U.S. Dept. Interior OCR, August 1963–December 1966.

Table 29.37 Analyses of British Solvent Oil, Coals, and Coal Extracts

Properties	Recycle Oil	Beynon High-Rank Coal	Annesley Low-Rank Coal	Beynon Extract	Annesley Extract
Elemental analysis (wt %, maf)					
Carbon	90.8	90.0	84.2	90.5	89.3
Hydrogen	5.8	4.7	5.4	4.7	5.0
Nitrogen	0.95	1.4	1.9	1.6	1.75
Oxygen	1.4	2.9	7.6	2.1	2.9
Sulfur	1.00	1.55	0.65	0.9	0.55
H/C atomic ratio	0.77	0.62	0.76	0.62	0.67
Proximate analysis					
Volatile matter (%, maf)				46.8	48.5
Ash (%, as received)				0.4	0.2
Insolubility (%, as received) in:					
Cyclohexane				66	N.D.
Chloroform				52	49
Benzene				50	55
Quinoline				13	2.5
Asphaltene content				12	25

Source: Modified from Ref. 115.

Table 29.38 Composition of Feed and Products in Hydrotreatment with Dicobalt Octacarbonyl and Cobalt-Molybdenum Catalysts

	Feed	$CO_2(CO)_8$	Co-Mo
Temperature (°C):		204	370
Pressure (psi):		5000 (1 : 1, Co/H_2)	3000 (H_2)
		16	3
Elemental analysis (wt %)			
Carbon	82.6	83.5	85.1
Hydrogen	7.3	9.4	8.8
Nitrogen	1.4	1.3	1.3
Oxygen	8.1	5.6	4.5
Sulfur	0.6	0.2	0.3
Content (wt % increase in hydrogen)		28.8	20.5

Source: Modified from Ref. 116.

crease in time of reaction is required. In general the hydrofining catalyst is able to add greater quantities of hydrogen. No additional work was recommended for use of the carbonyls on coal liquids.

Hydrotreatment of Recycle Solvents. In coal liquefaction schemes a process-derived solvent may be recycled for use in the primary conversion step. Often this recycle solvent will require a relatively mild hydrotreatment to increase its liquefaction capability. However one particular study on Wyodak coal revealed little difference in liquefaction properties of a solvent "as received" and in its "mildly and severely" hydrotreated form.[117] A liquid recycle solvent from the 6-ton/day liquefaction pilot plant at Wilsonville, Alabama, was used as the reference liquid and subjected to hydrotreatment. The mild treatment was achieved with Co-Mo-alumina catalyst (12.5 wt % MoO_3, 3.5 wt % CoO, Nalcomo 471) at 2000 psig hydrogen pressure (cold, batch autoclave), 371°C for 1 hr. The severe hydrotreatment was conducted with the same catalyst, but at 3000 psig and 427°C. The properties of the initial solvent and its hydrotreated products are given in Table 29.39. Little heteroatom removal was achieved at the milder conditions, and even at the more severe operation, only 31% nitrogen removal was achieved. The concentration of tetralin, which is an excellent hydrogen transfer agent for liquefaction, increased from 2.3 to 2.7 and 6.1 wt % after mild and severe hydrogenation, respectively. Much of the remaining nitrogen was in the form of quinoline compounds, which ranged from 63 to 45 wt % of the total nitrogen for the mild hydrogenation and 29 wt % for the severe hydrogenation.

The U.S. Bureau of Mines Study of Hydrotreatment of Tars and Other Coal-Derived Liquids. Coke-oven liquids, pyrolysis tars or low- and high-temperature carbonization tars represent another general class of coal liquids that have been hydrotreated, some in commercial plants. An excellent historical review of previous technical developments in coal-tar hydrogenation is available.[118] Published in 1968, this report covers the period from around 1920 to the mid-1950s. This review is mentioned for those who wish to assess these earlier developments.

Mention has been made earlier of attempts to produce jet fuel (JP-5) from certain coal liquids. Another such study is that made by the U.S. Bureau of Mines.[119] With a low-temperature carbonization tar, a five-step sequence was used: (1) feed distillation, (2) hydrodesulfurization over a Co-Mo-alumina catalyst at 400°C and 2500 psig, (3) vacuum

[117] H. F. Silver and R. J. Hurtubise, *Effects of Solvent Characteristics on Wyodak Coal Liquefaction,* Annu. Tech. Prog. Rep. for May 1976–April 1977, University of Wyoming, for U.S. ERDA, May 15, 1977.

[118] W. R. K. Wu and H. H. Storch, *Hydrogenation of Coal and Tar,* U.S. Bur. Mines Bull. No. 633, 1968.

[119] C. O. Hawk, M. D. Schlesinger, P. Dobransky, and R. W. Hiteshue, *Hydrorefining Coal-Oils to Fuels for Supersonic Aircraft,* U.S. Bur. Mines Rep. Invest. No. 6655, 1965.

Table 29.39 Effect of Hydrotreatment of Recycle Solvent from the SRC Process

	Solvent		
Specific gravity at 15.6°C/15.6C:	As Received from Wilsonville 1.023	Hydrogenated: 371°C, 2000 psig 1.015	Hydrogenated: 427°C, 3000 psig 0.991
	Distillation[a,b]		
Fraction (vol %)		*Temperature (°C)*	
0–5	129–218	93–177	93–149
10–30	224–243	216–238	213–232
50	271	268	254
70–90	310–379	302–354	282–332
95–100	410–	402–	371–
End point	97	97	98
	427[c]	427[c]	402[c]
Temperature (°C)		*Fraction (wt %)*	
177–	2.1	3.6	6.0
177–260	41.0	37.8	46.7
260–343	37.9	40.6	36.7
343–427	14.5	13.9	9.3
427+	4.5	4.1	1.3
	Properties		
Ultimate Analysis (wt %)			
Carbon	87.14	86.81	85.83
Hydrogen	7.94	8.54	9.99
Nitrogen	0.49	0.45	0.34
Sulfur	0.13	0.12	0.04
Oxygen[d]	4.30	4.08	3.80
Tetralin (wt %)	2.27	2.74	6.12
Naphthalene (wt %)	7.30	7.06	6.36

[a] Atmospheric distillation to 177°C, vacuum distillation at 5 mm to end point.
[b] Front end of last-listed solvent partially removed by atmospheric distillation.
[c] At end point.
[d] By difference.

Source: Modified from Ref. 117.

distillation, (4) hydrogenation over a nickel-kieselguhr catalyst at 300°C and 2500 psig, and (5) atmospheric distillation. The first-step hydrogenation was conducted in a 0.97-in. I.D. reactor holding 100 ml of the Co-Mo-alumina catalyst (Girdler G-35B). The liquid feedstock properties are shown in Table 29.40. In considering the products from hydrodesulfurization (Table 29.41), note that the goal of 0.020 wt % sulfur was nearly achieved both in the total product and in the 180–280°C fraction, after alkali washing. In this hydrotreating step the H/C ratio increased from 1.35 to 1.67 and the saturates increased from 33 to 54%. No catalyst deactivation was noted over 182 hr of operation.

The 180–280°C cut from the desulfurization step shown in Table 29.41 was then processed in the saturation step at 300°C and 2500 psig, over a nickel catalyst (Harshaw Ni-0104T). The products from this treatment are shown in Table 29.42. Both the sulfur and nitrogen levels were satisfactorily low: 0.005 and 0.01 wt %, respectively, in the total product. In the overall process, the total aromatics were reduced from 67 vol % in the feed to 46 vol % after desulfurization to 5.0 vol % after saturation. In the jet fuel cut itself (190–241°C) the aromatics were 7.1 vol %.

In this same study a middle oil (78–327°C) obtained from hydrogenation of a Wyoming high-volatile bituminous coal at 465°C and

Table 29.40 Properties of Tar and Mid-Distillate Oil Used in Bureau of Mines Tests

	Tar	Middle Oil
Gravity (°API):	17.8	15.5
Specific gravity at 16°C/16°C:	0.9478	0.9626
ASTM Distillation		
Fraction (vol %)	*Temperature (°C)*	
IBP	87	78
10	172	163
30	200	210
50	215	247
70	229	276
90	272	309
95	297	323
EP	310	327
	Distillation Results (%)	
Recovery	98.3	97.0
Residue	0.8	1.5
Loss	0.9	1.5
Analyses		
Ultimate analysis (wt %)		
Carbon	83.8	87.1
Hydrogen	9.4	9.5
Nitrogen	0.3	1.0
Sulfur	1.0	0.1
Oxygen (by difference)	5.5	2.3
Sulfur, lamp method (titrimetric wt %)	1.1	0.08
H/C atomic ratio	1.35	1.31
FIA analysis (vol %)		
Saturates	33.4	63.6
Aromatics plus oxygenated compounds	66.6	36.4
Mass spectrometric analysis (vol %)		
Alkanes	12.7	
Noncondensed cycloalkanes	11.6	
Condensed cycloalkanes	9.1	
Aromatics	50.9	
Oxygenated compounds	15.7	

Source: Modified from Ref. 119.

2000 psig was also subjected to the two-step desulfurization-saturation process. Typical products are given in Tables 29.41 and 29.42. There was less hydrogenation in the desulfurization step than with the low-temperature tar oil. The H/C atom ratio increased from 1.31 to 1.39. There was 95% hydrodenitrogenation, and the nitrogen in the 180–280°C cut was somewhat high for the nickel catalyst. However the nickel saturation catalyst was not deactivated over the relatively short duration of the experiment. The final jet fuel cut of the middle oil (200–256°C) consisted almost entirely of cycloalkanes. The similarity between the condensed and noncondensed ring cycloalkanes of the middle and tar oil is a striking feature—61.2 and 24.1 vol %, respectively, for middle oil, and 66.2 and 28.2 vol % for tar oil.

A few attempts were made in this study to produce jet fuel in a single step, but results were unsatisfactory when using Co-Mo, WS_2, and Mo-Cr-Zn type catalysts. Temperatures ranged from 400 to 550°C, at 2500 psig, and 1.0 LHSV.

Effect of Catalyst Properties on Removal of Sulfur and Nitrogen from Coal-Derived Liquids. One set of studies has been devoted to tailoring support properties of catalysts, especially for hydrotreating coal-derived liquids for sulfur and nitrogen removal.[120–123] Two carbonization liquids have been studied along with others. These two liquids were an FMC oil from project COED and a raw anthracene oil derived from coal coking. The latter was used as a liquid analogous to a recycle solvent used in coal extraction processes. Some properties of the two liquids are shown in Table 29.43. Catalyst support properties such as pore size and distribution, pore volume, and total surface area were determined. A trickle-flow reactor containing 10–30 g of catalyst was used. The range of conditions investigated included 316–454°C, 500–2000 psig, and 0.5 to 3.0 hr, VHST. The studies were restricted to Co-Mo-alumina catalysts, both commercially available and laboratory prepared. Table 29.44 shows the

[120] M. C. Sooter and B. L. Crynes, "Preliminary Assessment of Catalyst Pore Size Effects on Sulfur Removal from a Coal Derived Liquid," *Ind. Eng. Chem. Process Des. Dev.*, **14**(3), 199 (September 1975).

[121] M. Mushtaq and B. L. Crynes, "Effect of Presulfiding on Coal Liquid Hydrodesulfurization Catalysts," paper preprinted and presented at American Institute of Chemical Engineers 81st National Meeting, Kansas City, Mo., April 11–14, 1976.

[122] R. Sivasubramanian and B. L. Crynes, "Hydrogenation of a Coal Liquid in a Trickle Bed Reactor," paper preprinted and presented at the American Institute of Chemical Engineers 69th Annual Meeting, Chicago, November 28–December 2, 1976.

[123] B. L. Crynes, *Development of a Process for Producing an Ashless, Low-Sulfur Fuel from Coal*, R & D Rep. No. 53—Interim Rep. No. 19, Vol. 4, Part 10, Final Report of Coal Liquids Catalyst Work performed at Oklahoma State University, Rep. No. FE-496-T10, by Pittsburg & Midway Coal Company for U.S. ERDA, 1977.

Table 29.41 Hydrodesulfurization of Tar and Middle Oil Products in U.S. Bureau of Mines Tests

	Low-Temperature Tar Fraction		Middle Oil	
	Total Oil Product[a]	180–280°C Cut[b]	Total Oil Product[a]	180–280°C Cut[b]
Recovery (wt %):	93.2	42.6	97.2	42.6
Gravity:	34.3	30.1	22.0	21.9
Specific gravity at 16°C/16°C:	0.8534	0.8756	0.9218	0.9224
ASTM Distillation				
Fraction (vol %)		*Temperature (°C)*		
IBP	78	142	91	158
10	119	168	139	196
30	144	193		
50	183	210	231	226
70	214	226		
90	255	249	302	255
EP	290	275	327	265
		Results (%)		
Recovery	97.5	98.0		
Residue	1.8	1.5		
Loss	0.7	0.5		
Analyses				
Ultimate analysis (wt %)				
Carbon	87.3	88.4	88.9	89.3
Hydrogen	11.8	11.6	10.3	10.2
Nitrogen	0.06	0.03	0.037	0.05
Sulfur, lamp method (titimetric wt %)	0.025	0.022	0.02	0.013
H/C atomic ratio	1.62	1.57	1.39	1.37
FIA analysis (vol %)				
Saturates	54.0	—		
Aromatics plus oxygenated compounds	46.0	—		

[a] Before alkali wash.
[b] After alkali wash.

Source: Modified from Ref. 119.

properties of these catalysts. Micropore radii of the catalysts ranged from 26 to 50 Å; two catalysts were bidispersed; total surface ranged from 200 to 300 m²/g.

A few results of this set of studies are summarized here. Figures 29.19 and 29.20 compare the sulfur and nitrogen removal achieved for three feedstocks. Typically, the rate of sulfur removal is rapid during the first 0.5–1.0 VHST, but there is relatively little removal thereafter. These feedstocks might be considered as having a number of easily removed sulfur-containing compounds and a group of more unreactive compounds that are affected very little by increasing space times. Comparison of the results presented in Figs. 29.19 and 29.20 shows that nitrogen removal from coal liquids is very much more difficult to achieve with cobalt-molybdenum-type catalysts than is sulfur removal. This is especially true of the heavier coal liquids such as FMC oil and the PAMCO composite oil. In fact, the heavy liquids such as FMC oil have a greater tendency to cause premature reactor shutdown through coking and/or condensation, especially as reactor conditions approach 427°C.

A series of vendor-supplied Co-Mo-alumi-

Table 29.42 Properties of Saturation Products from U.S. Bureau of Mines Tests

	Middle Oil		Tar Oil	
	Total Oil Product	200–256°C Cut	Total Oil Product	190–241°C Cut
Gravity (°API):	34.1	31.5	38.7	35.8
Specific gravity at 16°C/16°C:	0.854	0.866	0.831	0.845
ASTM Distillation				
Fraction (vol %)	*Temperature (°C)*			
IBP	130	200	137	190
10	181	207	165	207
30	199	211		
50	210	216	201	214
70	222	224		
90	243	236	242	230
95	253	242		
End point	261	256	270	241
	Results (%)			
Recovery	98.5	98.0		
Residue	1.5	1.0		
Loss	0	1.0		
Analyses				
Ultimate analysis (wt %)				
Carbon	86.2	86.6	86.6	86.4
Hydrogen	13.6	13.6	14.0	13.7
Nitrogen			0.01	0.01
Sulfur, lamp method (titimetric wt %)	0.004	0.001	0.005	0.001
Yield (wt % feed to desulfurization)	—	30.7	—	22.1
Yield (wt % total saturated product oil)	—	69.4	—	51.1
H/C atomic ratio	1.88	1.87	1.94	1.90
FIA analysis (vol %)				
Saturates			95.0	92.9
Aromatics plus oxygenated compounds			5.0	7.1
Mass spectrometric analysis (vol %)				
Alkanes	—	5.4	—	2.4
Noncondensed cycloalkanes	—	24.1	—	28.2
Condensed cycloalkanes	—	69.2	—	66.0
Monoaromatics	—		—	3.3
Naphthalenes	—		—	0.1

Source: Modified from Ref. 119.

na catalysts, with average pore radii ranging from 25 to 50 Å, were tested at various conditions (Fig. 29.21), using the raw anthracene oil. For desulfurization, catalysts with a pore radius of 25 Å were not as satisfactory as those with pore radii of 33–50 Å. Few differences were noted for denitrogenation for the 33–50 Å catalysts. Even the differences at the 25 Å level could not be considered significant. Other Co-Mo-alumina catalysts prepared in the laboratory from Ketjen support material and with average pore radii from 33 to 38 Å were tested with raw anthracene oil. In contrast to the observations above, the catalysts of the small pore radii were more effective for nitrogen removal, but no effect was noted for desulfurization. When the Ketjen support catalysts were com-

Table 29.43 Properties of Feedstocks Used in Tests of Hydrotreating Catalysts

Component (wt %)	Raw Anthracene Oil[a]	FMC Oil[b]	PAMCO Composite Oil	Synthoil Liquid
Carbon	90.65	83.0	88.2	88.0
Hydrogen	5.76	8.35	7.34	8.07
Sulfur	0.48	0.35	1.2	1.06
Nitrogen	1.06	1.13	1.3	1.19
Oxygen[c]	2.2	7.15	1.6	
Ash	Nil	Nil	0.1–0.2	3.4

[a] API gravity at 60°F (15.5°C), −7.
[b] API gravity at 60°F (15.5°C), −4.5.
[c] By difference.

pared on a surface area basis, all catalysts were equally effective. Hence it is more important to control and tailor total catalyst surface area for these catalysts than simply pore radii. Catalysts with bimodal pore distributions showed no advantage for desulfurization or denitrogenation of raw anthracene oil over the monodispersed catalysts. The same observation was noted for processing an FMC oil (project COED).[120–123]

In activity tests made by operating at a fixed set of conditions for up to 200 hr of continuous oil-on-catalyst contacting, no loss in desulfurization activity was noted with the raw anthracene oil or with the FMC oil for any of the catalysts tested. No significant loss in denitrogenation activity was noted with the raw anthracene oil. However significant loss in denitrogenation activity resulted while processing FMC oil and PAMCO composite oil. Figure 29.22 presents an interesting comparison of the effect of reaction temperature on the loss of catalyst activity for denitrogenation of the FMC oil. After about 120 hr of oil-catalyst contact, operating temperature had comparatively little effect of nitrogen content of product oils. No difference was noted between a bidispersed and a monodispersed catalyst with respect to rates of loss or activity. Loss rates while processing the PAMCO oil and the Synthoil liquid are shown in Table 29.45. For both feeds, significant loss in activity results after a relatively short exposure of the catalyst to oil under processing conditions.

Comparison of Co-Mo-alumina and Ni-W-alumina catalysts in the hydrogenation of a low-temperature tar, from a 25-ton/day pilot

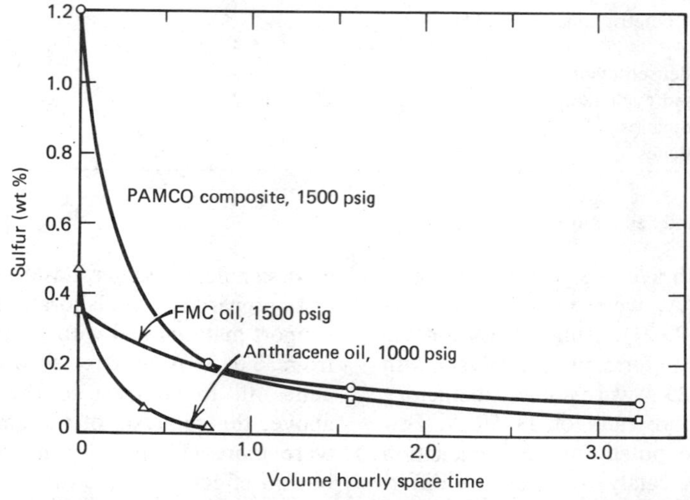

Figure 29.19 Feedstock comparisons: sulfur removal at 700°F, Nalcomo 474 catalyst.

Table 29.44 Catalyst Properties

	Catalysts from Vendors						Supports Impregnated in Oklahoma State University Laboratory				
	Nalcomo 474	N-A	N-B	Nalcomo Sphericat 474	Harshaw HT-400	N-C	Ketjen 007-1.5E	Steam treated Ketjen 007-1.5E	Ketjen 000-3P	Conoco Catapal HP-20	Norton SA6175
	Chemical Composition (wt %, except as shown)										
MoO_3	12.5	12.5	12.5	12.5	15.0	12.5	12.5	12.5	12.5	12.5	12.5
CoO	3.5	3.5	3.5	3.5	3.0	3.5	3.5	3.5	3.5	3.5	3.5
Na_2O	0.08	0.08	0.08				0.03	0.03	0.033	40 ppm	0.02
Fe	0.03	0.03	0.03				—	—	—	50 ppm	0.075
SiO_2	1.5	1.5	1.5				—	—	0.88	0.008	0.180
SO_4							0.5	0.5	0.028	—	
TiO_2										0.12	
	Physical Properties										
Surface area m^2/g	240	298	303	287	234	242	291	230	199	218	220
Pore volume (cm^3/g)	0.463	0.441	0.556	0.997	0.431	0.525	0.503	0.540	0.625	0.875	0.75 0.95
Most frequent pore radius (Å)	34	26	26	30,800[a]	50	47	31	38	36,160[a]	40,190[a]	100

[a] Micropores and macropores.

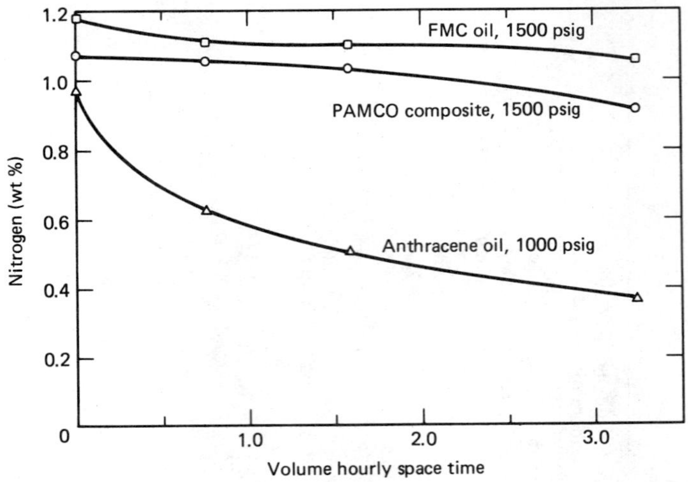

Figure 29.20 Feedstock comparisons: nitrogen removal at 700°F, Nalcomo 474 catalyst.

plant revealed that the two catalysts removed similar amounts of sulfur and tar acid.[124] At the lower temperatures of the study, total range of 350–480°C, the Ni-W catalyst was superior for removing nitrogen. The feedstock was obtained from an Indian (Kothagudem) noncaking coal and its boiling range was 200–356°C, with tar acids, 30.2 vol %; and bases, 4.4 wt %. The liquid had 0.415 wt % sulfur, 0.844 wt % nitrogen, 25.0 vol % saturates, 17.6 vol % olefins, and 57.4 vol % aromatics. The studies were conducted at

[124] A. Mirzo, A. V. Ramaswamy, M. A. Masood, S. A. Qader, and R. Vaidyeswaran, "Catalytic Hydrogenation of Low Temperature Tar Fractions," *Indian J. Technol.*, **3,** 57 (February 1965).

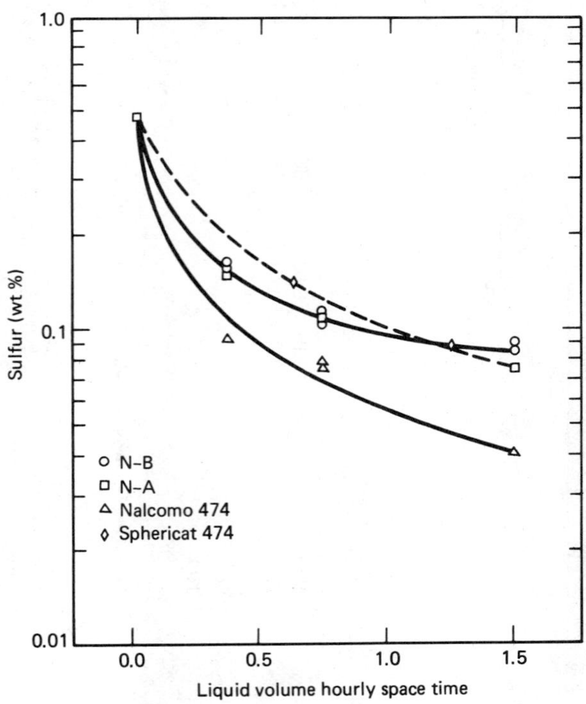

Figure 29.21 Comparison of desulfurization results for four catalysts at 650°F and 1000 psig; catalyst size, 8–10 mesh; hydrogen rate, 1500 scf/bbl.

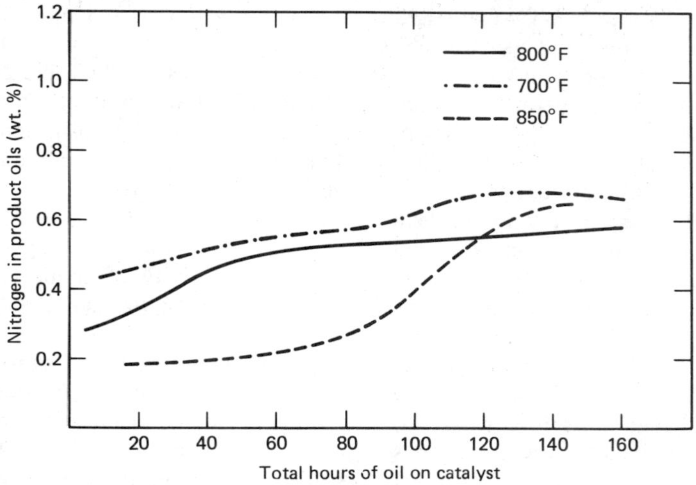

Figure 29.22 Comparison of catalyst (Spherical 474) decay rates: FMC feed oil, 1500 psig, 2.99 hr space time.

995 and 1420 psig at a VHSV of 1.0 hr^{-1}. The tar bases increased in the low-temperature range, as temperature increased, but decreased at higher temperatures. The total nitrogen content decreased throughout the entire temperature range, suggesting that nonbasic species, such as indole or carbazol, are first converted to basic amines, then denitrogenated. The neutral oil derived from the tar feed revealed more aromatic saturation than the entire feed, which suggests that the tar acids, bases, and other compounds inhibit hydrogenation. Overall, little hydrogenation of aromatics was achieved at the test conditions, and temperatures of 450–480°C were required to achieve essentially complete removal of oxygen, nitrogen and sulfur.

Table 29.45 Loss in Catalyst Activity for Two Feedstocks

Conditions for PAMCO composite oil: 800°F, 1500 psig, Nalcomo 474 catalyst, space time = 1.6 hr.

Conditions for Synthoil Liquid; 700°F, 1000 psig, Corning monolithic catalyst, space time = 1.11 hr.

Hours on Oil	DS (%)[a]	DN (%)[b]
PAMCO Composite Oil		
10	85	51
30	78	42
60	73	34
Synthoil Liquid		
40	38	18
130	27	16

[a] Desulfurization.
[b] Denitrogenation.

Another batch study of low-temperature carbonization tar, using pure WS_2 (220 m^2/g) as a catalyst, required 500°C for complete heteroatom removal at 1500 psig, 2.0 hr.[95] The aromatics content of the resultant gasoline product fraction remained steady at 50 vol % from 300 to 375°C, but rose to about 75 vol % at 500°C. At 400°C and 2.0 hr, the conversion to gasoline rose from 12% at 500 psig to almost 40% at 2000 psig. The removal by hydrotreating of sulfur, oxygen, and nitrogen was represented by first-order reactions having activation energies of such a magnitude as to suggest that chemical and not physical processes were the controlling factors in these hydrotreating studies. The feedstock used was derived by carbonizing at 550°C a high-volatile bituminous coal from Utah. The tar acids and bases were 30.2 and 2.5 vol %, respectively, with 0.834 wt % sulfur, 0.398 wt % nitrogen, and 6.31 wt % oxygen. The boiling range was 200–325°C with neutral oil containing 28.3, 18.4, and 53.3 vol % saturates, olefins, and aromatics, respectively.

Hydrotreatment of Light Oils. One more type of coal liquid that must be briefly mentioned is comprised of the relatively light oils that are by-products from coal coking. These liquids are especially rich in BTX and might be used for fuel blending, but usually they are of interest for their chemical value.

One light oil from low-temperature carbonization of an Indian coal (650°C, Kothagudem) was successfully hydrotreated with a Co-Mo-alumina catalyst for removal of sulfur, oxygen, and nitrogen, and for diolefin saturation.[125] The reactor system consisted of a continuous flow, packed-bed reactor containing 60 ml of catalyst. Test conditions were 300–480°C, 213–995 psig, and a VHSV of 0.5–2.0 hr^{-1}. The light-oil (benzol) feedstock contained 0.597 wt % sulfur and 0.176 wt % nitrogen, had a boiling range of 76–180°C, and contained hydrocarbon types of 20.6 vol % saturates, 27.7 vol % olefins, and 51.8 vol % aromatics. This material proved to be easily desulfurized; it was somewhat harder to saturate olefins, and still harder to denitrogenate. Between 400 and 450°C and at 427 psig, 1.0 hr^{-1} VHSV, the sulfur and nitrogen removals were essentially complete. Above 400°C there was no further reduction in olefins in the product as space time increased. The refined benzol (400°C, 711 psig, 1.1 hr^{-1} VHSV product) contained volume percentages of 35.12% BTX (of which 13.97% was toluene), 7.37% *n*-octane, 4.91% *n*-decane, 5.89% 2,7-dimethyl octane, 4.44% 1-methyl-3-ethyl benzene, and 3.68% 3-methylnonane. These conditions were considered optimum for hydrotreating this light coal oil.

The coke-oven plant light oils are especially rich in benzene and similar compounds and can be treated to produce high-purity benzene. Such hydrotreating and purification steps to remove paraffins, naphthenes, olefins, and thiophenes can be achieved by treatment with a catalyst such as Cr_2O_3 on alumina.[126–128] The paraffins and naphthenes are hydrocracked to species with lower molecular weight, and thiophenes can be reduced to as low as 0.5 ppm. Some hydrodealkylation of toluene and xylene to benzene also results. In Table 29.46 a representative feed and product are presented for conditions of 700–900 psig and 400–500°C. The benzene concentration increases to 90.3 vol % from 72.6 vol %, the paraffins are significantly reduced, and thiophene, initially at 0.97 vol %, is nil. A catalyst capacity for acceptable activity of about 11 barrels per pound of catalyst was suggested.

[125] S. A. Qader and M. G. Krishna, "Hydrorefining of Low Temperature Benzole," *Indian J. Technol.*, **3**, 125 (April 1965).

[126] J. J. Donovan and A. H. Weiss, U.S. Pat. 3,081,259, Air Products and Chemicals, Inc., March 12, 1963.

[127] M. O. Tarhan and L. H. Windsor, "Coke Oven Oils Yield High Quality Benzene," *Chem. Eng. Prog.*, **62**(2), 67 (February 1966).

[128] A. K. Logwinuk, L. Friedman, and A. H. Weiss, "Upgrading Coke Oven Aromatics," *Ind. Eng. Chem.*, **56**(4), 20 (April 1964).

Hydrotreatment of Pure Compounds. Several hydrotreating studies have been made on pure compounds meant to be representative of the compounds found in coal liquids. Such studies are usually directed toward determination of mechanisms and kinetics, and although important they are beyond the scope of this discussion dealing with complex liquid products. A few of the studies of pure compounds are referenced here for those who wish to investigate further.[129–131]

Another study worthy of comment, but available too late to be fully incorporated in

[129] J. R. Katzer, B. C. Gates, J. H. Olson, H. Kwart, and A. B. Stiles, *Kinetics and Mechanism of Desulfurization and Denitrogenation of Coal-Derived Liquids,* Quarterly Reps. Nos. FE-2028-1 to-6, University of Delaware, for U.S. ERDA, September 30, 1975–July 15, 1976.

[130] C. S. Huang, K. C. Wang, and H. W. Haynes, "Hydrogenation of Phenanthrene over a Commercial Cobalt Molybdenum Sulfide Catalyst Under Severe Reaction Conditions," *Prepr. Am. Chem. Soc., Div. Fuel Chem.*, **21**(5), 228 (1976).

[131] L. D. Rollman, "Competitive Hydrogenation of Model Heterocycles and Polynuclear Aromatics," *Prepr. Am. Chem. Soc., Div. Fuel Chem.*, **21**(7), 59 (1976).

Table 29.46 Composition of Feed and Products in Light Oil Processing

Component	Feed (Vol %)	Product (Vol %)
Benzene	72.59	90.34
Toluene	16.31	6.36
C_8 aromatics	5.38	0.51
C_9 aromatics	0.83	0.06
Styrene	2.16	—
Indene	0.40	—
Naphthalene	0.02	0.05
Biphenyl	—	1.29
Paraffins	1.39	0.06
Thiophene	0.97	—

Source: Modified from Ref. 126.

this review, was directed to hydroprocessing SRC extract in an ebullated-bed reactor.[132] The first part of this work involves pilot-plant studies to determine solvent-SRC ratio, temperature, catalyst aging, and other conditions useful for processing SRC at various severity levels, ranging from desulfurization to extensive hydrocracking. The second part includes engineering studies defining requirements for adapting a commercial ebullated-bed reactor to process mixtures of SRC and solvent. Some additional recent work has addressed characterization of the processibility, by conventional petroleum hydrorefining technology, of SRC, Synthoil, and H-Coal liquids.[133]

29.4.2 Hydrocracking

The distinction between hydrocracking and "general" hydrotreating discussed in Section 29.4.1 is not always clear. The term "hydrocracking" as used here refers to the catalytic processes in which the molecular weight of the feedstock is reduced and heteroatoms are removed concurrently by hydrogenation.

29.4.2.1 *Hydrocracking Summary*

Several interesting conclusions can be drawn from results of studies of coal-derived liquids presented on the following pages. Such liquids are more difficult to hydrocrack than are petroleum feedstocks, principally because of the high heterocyclic content, especially of nitrogen compounds, and the high level of polynuclear aromatics. Hydrofining and dual-function catalysts are used in hydrocracking at typical reaction conditions of 350–500°C and 1500–2500 psig. In addition to those types of catalyst, relatively high concentrations of fused metal halides ($ZnCl_2$) are good hydrocracking catalysts. In processing heavier feeds, such as extracts, the properties of the catalyst support are important. High porosity, macropores, and large average pore diameters are desirable so that activity can be maintained in the presence of accumulations of ash and coke. The incorporation in a catalyst of components, such as silica, that catalyze cracking reactions, may be advantageous for some but not all liquids.

Not only are the coal liquids more difficult to hydrocrack than petroleum stocks, as indicated by the much slower rates of reaction of coal liquids; in addition, coke buildup on the catalyst can be double that for petroleum stocks. A pretreatment by hydrofining can assist hydrocracking, and in some instances may yield a liquid that behaves like the petroleum stocks. Hydrogen consumption in hydrocracking coal liquids can be twice that for petroleum stocks treated under identical conditions.

The naphthas and gasolines from coal liquids are especially high in aromatics and, with some mild reforming, lead to good stock for further processing. Extraction and dealkylation make these hydrocrackates excellent sources of BTX and pure aromatics. Direct coal hydrogenation liquids, extracts, and solvent-refined products are usually higher in aromatics than are liquids produced by coking, pyrolysis, or carbonization.

Because of the complexity of the reaction mechanisms involved in hydrocracking, the complex mixture of compounds comprising coal liquids, often a simple first-order rate expression is used to follow the course of the reactions. Such empirical or semiempirical models must be used with caution. The consecutive and parallel reactions of cracking, isomerization, and hydrogenation can be presented by a simple nth-order reaction only over a relatively narrow set of operating conditions. It has been suggested that the cracking step for coal liquids is the rate-limiting factor. If this is true, the need to protect the acidic cracking sites from the relatively high poison levels in coal liquids is apparent.

29.4.2.2 *Hydrocracking Studies*

Tests of Catalysts in Batch Autoclaves with Screening Liquids. The objective of hydrocracking is to remove heteroatoms and convert the feedstock to higher quality fuels (gasoline through middle distillates) with high selectivity (little coke or gas formation). This

[132] E. D. Wysocki, *Commercial Scale Ebullated-Bed Hydroprocessing of Solvent Refined Coal (SRC) Extract,* Monthly Prog. Reps. Nos. FE-2038-1 to -10, by Cities Service Research and Development Company for U.S. ERDA, July 1976–March 1977.

[133] A. J. deRossett, *Characterization of Coal Liquids,* Intermediate Reps. Nos. FE-2010-01 to -08, by Universal Oil Products Inc., for U.S. ERDA February 1976–November 1976.

is accomplished in a catalytic process operating at elevated temperature and pressure.

The extracts obtained from Consol's CSF process have been subjected to a wide range of tests to assess appropriate hydrocracking catalysts. Early studies were conducted in batch autoclaves at about 427°C, 4200 psig, with residence times of 1 hr.[134] A standard or reference extract was used in addition to a solvent-refined extract and a high-temperature tar. The properties of these three feeds are shown in Table 29.47, along with the properties of the products, the conversion, and the hydrogen consumption in tests using the Universal Oil Products UOP-S-6 catalyst. The nitrogen levels of the feeds were around 1.0 and 1.5 wt %, the sulfur levels were 0.3–1.8 wt %, and oxygen was about 1.0–7 wt %. The tar liquid contained much lower concentrations of heteroatoms and benzene insolubles. The asphaltene levels were about the same in the three liquids. The data on the products are discussed in the next section.

A number of refining-type, commercial catalysts were assessed using the reference extract. These are listed in Table 29.48, along with their properties. The catalyst properties incorporate active metal types, alumina support with and without silica, high- and low-pore supports, and average pore diameters ranging from 63 to 166 Å.

Some of the properties of the resultant product are shown in Table 29.49.

Batch Autoclave Tests with the Standard Extract and with a UOP-S-6 Catalyst. Of the various catalysts tested in these and other studies, the refining-type catalysts containing Co-Mo or Co-Mo-Ni on alumina proved to work best. The preferred type of catalyst used subsequently in additional tests was a commercial type, UOP-S-6, chosen for both its activity and handling characteristics in an

[134] *Consolidation Coal Company, Project Gasoline Pre-Pilot Plant Phase I Research on CSF Process,* Vol. 2, Parts 1 and 2, U.S. Dept. Interior OCR R & D Rep. No. 39, for August 1963–June 1968.

Table 29.47 Properties of Feeds and Products in CSF Catalyst-Screening Tests

	Standard Extract	Solvent-Refined Extract	High-Temperature Extract
	Properties of Feed (wt %)		
Hydrogen	6.14	5.03	5.73
Carbon	83.73	87.62	91.30
Nitrogen	1.46	1.49	0.95
Oxygen	6.87	4.75	1.17
Sulfur	1.80	1.11	0.31
Benzene insolubles	47	37	16
Asphaltenes	37	31	37
Oils	16	32	47
Ash	0.18	0.52	0.54
	Conversion and Properties of Extract		
Conversion (wt %, maf feed)	61	49	6.4 (of +400°C in feed)
$\frac{(C_1\text{–}C_3)\ 100}{\text{conversion}}$	6.2	6.1	3.1
C_5–200°C (392°F) distillate	24.0	13.7	7
200–400°C (392–752°F) distillate	28.7	30.3	50.0
+400°C (752°F) residue	36.9	45.4	42.6
Hydrogen used (wt %)	6.0	5.3	3.5
Benzene insolubles (wt %)	1.1	0.0	—
Asphaltenes (wt %)	0.5	14.3	—
Oils (wt %)	98.4	85.7	—

Source: Modified from Ref. 134.

Table 29.48 Properties of Catalysts Used in CSF Tests[a,b]

	Catalysts										
	A	B	C	D	E	F	G	H	I	J	K
	Elemental Analysis (wt %)										
Cobalt (as metal)	2.3	—	—	—	2.3	0.5	0.2	2.3	2.5	2.6	0.04
Molybdenum	5.1	—	6.2	6.3	6.9	10.2	10.9	7.6	7.9	8.7	7.8
Nickel	—	3.6[c]	5.7	5.7	—	7.4	2.0	—	—	—	4.2
Silicon	—	—	4.0	—	—	—	—	0.07	0.06	0.14	1.7
	Properties										
Pore volume (cm^3/g)	0.744	0.615	0.765	0.845	0.520	0.517	0.532	0.786	0.727	0.974	0.894
Surface area (m^2/g)	275	227	313	310	313	329	180	290	193	235	303
Average pore diameter (Å)	108	106	98	109	66	63	116	108	151	166	118

[a] Properties of catalyst S-6 not available.
[b] Test also included catalyst U-S-6: used catalyst, 0.06% ash, oxygen regenerated, processed 1580 lb extract per pound of catalyst.
[c] Also contained 2.2% tungsten.

Source: Modified from Ref. 134.

Table 29.49 Data on Products Obtained from Standard Extract in CSF Catalyst Tests

Reaction conditions: temperature = 800°F, pressure = 4200 psig, residence time = 60 min, catalyst/extract wt ratio = 0.30.

	Catalysts												
Property	A	B	C	D	E	F	G	H	I	J	K	S-6	U-S-6
Conversion (wt %, maf feed)	57.0	43.0	62.2	59.4	51.3	56.7	48.7	50.4	51.4	54.9	57.4	61.0	51.2
$\frac{(C_1\text{–}C_3)100}{\text{conversion}}$	6.8	9.1	6.1	4.7	9.5	7.0	8.8	8.3	7.9	7.7	7.8	6.2	9.2
C_5–200°C (392°F) distillate	23.7	17.6	24.4	21.8	19.2	25.5	19.8	20.9	21.7	20.1	23.4	24.0	20.8
200–400°C (392–752°F) distillate	24.2	16.3	29.5	29.9	21.8	22.2	19.1	20.1	20.3	25.3	24.3	28.7	20.8
+400°C (752°F) residue	40.8	54.8	35.2	38.0	43.1	39.6	48.9	47.6	46.7	43.1	40.9	36.9	45.4
Hydrogen used (wt %, maf feed)	5.4	4.6	6.1	5.7	5.1	5.7	5.6	5.7	5.8	5.9	5.6	6.0	4.3

Source: Modified from Ref. 134.

ebullating reactor. Catalysts of high porosity and those containing macropores were recommended for the standard extract because they can accommodate a high molecular weight feedstock and maintain activity in the presence of accumulated ash. In fact, the support properties such as pore volume, pore distribution, and average diameter proved to be more important variables than high content of active metals. Catalysts promoted with phosphoric acid, HCl, and supports with silica revealed little advantage because the rather high levels of nitrogen in the feed resulted in ammonia poisoning. When the UOP-S-6 catalyst was promoted with $ZnCl_2$, however, conversion was significantly higher. Those aged catalysts that were analyzed for ash deposits contained titanium, calcium, and other metal oxides near the surface of the catalyst.

The results of the tests, although affected by batch autoclave heatup and cooldown times, are significant in indicating the advantages of tailoring catalyst support properties for coal liquids.

Batch Autoclave Tests with CSF Screening Liquids and a UOP-S-6 Catalyst. The results of comparative tests with the three feedstocks are shown in Table 29.47. The results with the solvent-refined extract and the standard extract were similar, except that the asphaltene content in the product from solvent-refined extract was significantly higher than that from the standard liquid. The pyrolysis tar was not as easily hydrocracked as were the two other feedstocks.

In most cases and in most fractions the sulfur level was well below 0.2 wt %. The nitrogen levels were somewhat higher, especially in the residue fractions where the level approached 1.0 wt %. The pyrolysis tar was much more resistant to heteroatom removal. These catalyst-screening experiments, though informative, should be considered to be preliminary in that they showed only initial catalyst activity levels.

Tests in a Continuous Fluidized-Bed Reactor. Additional studies of hydrocracking of CSF extract in a continuous fluidized-bed reactor essentially confirmed the results obtained using hydrofining catalysts. The basic limitation of the catalysts is their low cracking activity, and the addition of acidic cracking agents is unlikely to solve the problem because of the ammonia poisoning from nitrogen compounds in the feed.[135]

Batch Autoclave Tests Using Fused Metal Halide Catalysts. Another approach, using large quantities of fused metal halides as catalysts, was assessed. Table 29.50 compares the results of hydrocracking using the refining catalysts described above with results using $ZnCl_2$ melt as a catalyst for an extract residue with a boiling point above 400°C. The conversion of +400°C material was significantly higher with $ZnCl_2$, even at the lower reaction temperature (427 vs. 441°C). The selectivity (lower gas make as measured by the C_1–C_3 produced) was significantly improved with the halide. The percentage of distillate oils within the gasoline range (C_5–200°C) was 91% for $ZnCl_2$ but only 34% for the hydrofining catalyst. However massive amounts of halide are required to "overwhelm" the poisoning action of the nitrogen compounds of the feed. Further studies with pure compounds (pyrene) support the need for large quantities of halide. The catalyst-extract weight ratio required is about 0.3, with significant hydrocracking improvements for ratios up to 1.0.

A rather comprehensive study has been made of several variables in the $ZnCl_2$ system.[136] The range of operation incorporates 329–441°C, 1000–4200 psig, $ZnCl_2$-extract weight ratio of 0.2–3.0, ZnO-$ZnCl_2$ weight ratio of 0.02–0.09, with residence times of 15 to 360 min. The ZnO is used to react with HCl that is produced by reaction of H_2S with $ZnCl_2$. The effects of pressure and temperature on the conversion of the standard extract are shown in Fig. 29.23. Note that an upper limit of 85–95% conversion is not readily exceeded. The naphtha yield (gasoline range) and selectivity (C_1–C_3 make) for a fixed set of operating conditions are shown in Fig.

[135] C. W. Zielke, R. T. Struck, J. M. Evans, C. P. Costanza, and E. Gorin, "Molten Salt Catalysts for Hydrocracking of Polynuclear Hydrocarbons," *Ind. Eng. Chem. Process Des. Dev.,* **4**(2), 151 (April 1966).

[136] *Consolidation Coal Company, Research on Zinc Chloride Catalyst for Converting Coal to Gasoline, Phase I. Hydrocracking of Coal and Extract with Zinc Chloride.* Vol. 3, Book 1, U.S. Dept. Interior OCR R & D Rep. No. 39, August 1963–June 1968.

Table 29.50 Comparison of Results from Hydrofining and $ZnCl_2$ Catalysts

Composition of Feedstock (wt %)		+400°C Residue
Hydrogen		6.82
Carbon		90.79
Nitrogen		1.17
Oxygen		1.07
Sulfur		0.15
Benzene insolubles		3
Asphaltenes		30
Oil		67
	Catalyst Data	
Property	Co-Mo-Ni (0.1, 6.8, 3.8%) on Silica-Alumina	$ZnCl_2$ Melt
Pressure (psig)	4200	4200
Temperature (°C)	441	427
Time (min)	60	60
Catalyst/feed, wt ratio	0.3	1.00
Conversion (wt % feed)	29.6	77.6
$\frac{(C_1\text{–}C_3)100}{\text{conversion}}$	13.4	9.5
Distillates (C_5–150°C)	7.0	52.5
(150–200°C)	1.7	10.4
(200–400°C)	16.9	6.3
MEK-soluble +400°C residue	67.5	19.0
Hydrogen consumption (wt % feed)	2.64	6.82

Source: Modified from Ref. 136.

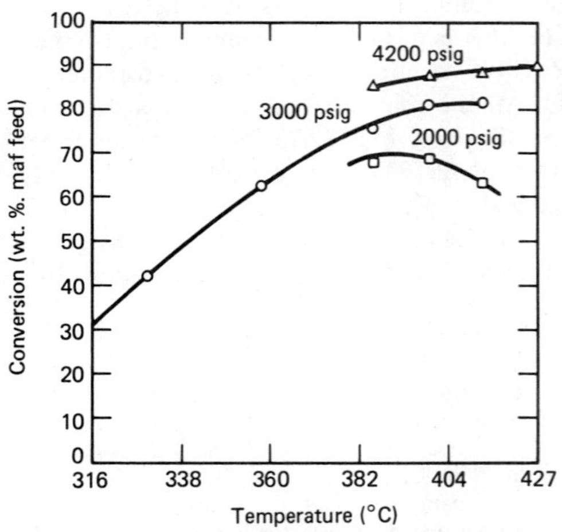

Figure 29.23 Effect of temperature and total hot pressure on extract conversion: feed, standard extract; residence time at temperature, 60 min; weight ratio of zinc chloride to extract, 1.0.

29.24. The yields are especially high for temperatures above 393°C. The selectivity remains favorable throughout the range shown.

Some of the problems with excess halide hydrocracking include the following:

1. Massive amounts of salts to be handled.
2. Poisons consisting of zinc chloride–ammonium and zinc chloride–ammonium chloride double salts tend to collect in the catalyst melt because of high nitrogen concentrations in the coal liquids.
3. Zinc oxide is required to react with HCl that is produced from H_2S reactions.
4. Large quantities of catalyst must be regenerated to remove nitrogen, sulfur, and carbon impurities.
5. Water must be guarded against to prevent excessive metal corrosion.

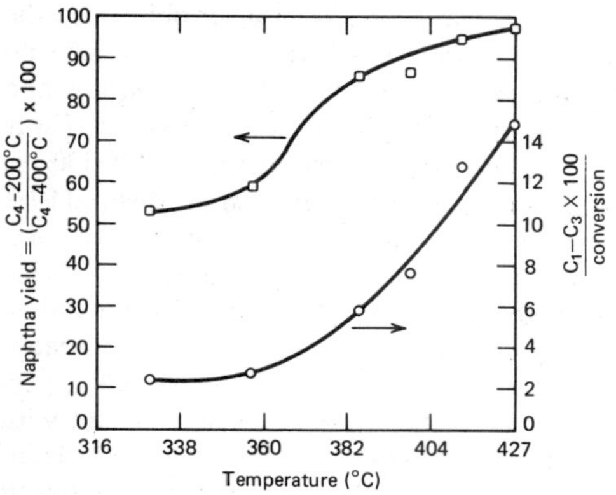

Figure 29.24 Effect of temperature on selectivity and the fraction of the distillate boiling in the naphtha range: feed, standard extract; total hot pressure, 3000 psig; residence time at temperature, 60 min; weight ratio of zinc chloride to extract, 1.0.

Table 29.51 Comparative Study of Properties of Feedstocks Used in Prehydrotreating

Property	Unhydrofined	Hydrofined
Carbon (wt %)	89.84	89.30
Hydrogen (wt %)	8.33	10.45
Sulfur (ppm)	520	—
Nitrogen (ppm)	3400	5
Specific gravity	1.006	0.940
204°C−(wt %)	6.0	18.6
Sulfur (ppm)	280	—
Nitrogen (ppm)	2800	—
Paraffins (vol %)	0.0	2.4
Naphthenes (vol %)	26.5	50.1
Aromatics (vol %)	73.0	47.5
204°C+ (wt %)	94.0	81.4
Sulfur (ppm)	460	—
Nitrogen (ppm)	3500	—
Saturates (vol %)	2.7	22.8
Aromatics (vol %)	97.3	77.2

Source: Modified from Ref. 138.

A semiempirical, first-order conversion model can be used to represent the $ZnCl_2$ hydrocracking.[137] The model consists of three parts: a nonconvertible fraction, a rapid first-order rate, and a slow first-order rate. The size of the nonconvertible fraction ranges between 7 and 14% of the feed extract and depends on the operating conditions. About 75% of the extract hydrocracks through the fast reaction, which is 40 times more rapid than the slow reaction. The fast reaction has an activation energy of 35 kcal/g-mole.

Pretreatment of Coal Extracts. The advantage, chemical if not economic, of prehydrotreating a coal extract before hydrocracking has been demonstrated.[138] The properties of two feeds used in this comparative study are shown in Table 29.51. The hydrofined feedstock was doctored with pure compounds to raise the concentrations of nitrogen and of sulfur to equal those of the unhydrofined stock. The hydrotreating significantly reduced the level of concentration of aromatics and of heteroatoms and increased the hydrogen content. Apparently much of the +204°C aromatic material was converted to −204°C hydroaromatics and naphthenics.

A first-order model, not mechanistic, was used to quantify the effects of pretreatment by hydrofining. The range of conditions reported in hydrocracking included 343–413°C, 1500 and 3000 psig, with an unidentified catalyst. The rate constants in the equation

$$-\frac{d\,(+204°\text{C})}{dt} = k(+204°\text{C})$$

were found to be 0.25 and 0.46 hr^{-1} for the unhydrofined and hydrofined feeds, respectively, at 1500 psig and 399°C. Even this pretreated coal liquid was more difficult to hydrocrack than a comparable petroleum stock. Doubling the pressure to 3000 psig resulted in a threefold increase in the rate constant for the hydrofined feed compared with a twofold increase for the unhydrofined feed. Unfortunately the hydrofining conditions were not specified. Site poisoning caused by the high concentration of nitrogen in the untreated feed cannot be the reason for the lower rate of reaction, because the treated feed was spiked to the same nitrogen level. Prior treatment is clearly beneficial to hydrocracking, but the economic justification remains open to question.

Tests have demonstrated that standard hydrocracking catalysts are rather sensitive to nitrogen compounds and high polyaromatics (four or five rings) in coal liquids even after prehydrofining has been used.[99] A gas oil

[137] R. T. Struck, W. E. Clark, P. J. Dudt, W. A. Rosenhoover, C. W. Zielke, and E. Gorin, "Kinetics of Hydrocracking of Coal Extract with Molten Zinc Chloride Catalysts in Batch and Continuous Systems," *Ind. Eng. Chem. Process Des. Dev.*, **8**(4), 546 (October 1969).

[138] W. T. House, "Catalytic Hydrocracking of Coal Liquids," *Prepr. Am. Chem. Soc., Div. Fuel Chem.*, **14**(4), 68 (1970).

(+343°C) derived from a COED pyrolysis liquid was hydrotreated at 399°C, 2000 psig, 0.33 hr^{-1} WHSV using an American Cyanamid HDS-3A catalyst (14.8 wt % MoO_3–3.1 wt % NiO) to yield an essentially sulfur-free liquid, but with 42 ppm total nitrogen (25 ppm basic). When this liquid was hydrocracked over an unidentified catalyst (commercial), precipitous deactivation resulted. Even when the feedstock was reduced to include only the 343–427°C fraction (84% of the original feed), using rather high hydrogen feed rates (11,000 scf/bbl), the hydrocracking catalyst was at one-third of its original activity after only 170 hr of catalyst-oil contact. Some of the feedstock properties, together with hydrocracking conditions and results, are shown in Table 29.52. At 371°C, 1 hr^{-1} WHSV, the conversion of the 343–427°C fraction to −199°C material is complete. However interpreting the data on time and temperature is complicated by steady decay in catalyst activity. The aromatic content of the product fractions increased in proportion to the loss in activity.

Rapid decay in catalyst activity during hydrocracking was also reported to have occurred when processing a somewhat broader range of COED oil, 288–427°C.[100] A double-reactor system was used consisting of a first reactor operating with a hydrotreating catalyst at 371°C, 2000 psig, 1.00 hr^{-1} WHSV, and 10,000 scf/bbl hydrogen rate. The second reactor, receiving the entire effluent from the first reactor, contained a hydrocracking catalyst and was maintained initially at 334°C and finally at 358°C, 2000 psig, 1.00 hr^{-1}, and 10,000 scf/bbl. Since the two catalysts were listed as proprietary, no details were available. The feedstock and some product yields are shown in Table 29.53. Over the course of the run of 330 hr, the total conversion fell from the desired 97% to 74%, even though the hydrocracking reactor temperature was increased by 24°C. The yields also reflect this significant catalyst decay. The unusually high degree of deactivation in comparison with the excellent performance of the catalyst system tested with petroleum stocks illustrates the significant difference in catalyst performance between processing petroleum- and coal-based stocks.

Hydrocracking a Feedstock Produced by the Direct Hydrogenation of Coal. The production of gasoline and diesel fuel by hydrocracking of coal-derived oil produced from fluid bed hydrogenation of a high-volatile bituminous Utah coal has been demonstrated.[139] The crude was produced at 510°C and 2000 psig, using stannous chloride as the catalyst. The feed properties are shown in Table 29.54, along with the two types of hydrocracking catalyst that were used. The laboratory-scale reactor had an inside diameter of 0.75 in., a length of 40 in., and contained about 50 cm^3 of catalyst.

The effect on product yields of temperatures in the range of 375–500°C is shown in Fig. 29.25 for 1500 psig and 1.0 hr^{-1} space

[139] S. A. Qader and G. R. Hill, "Fuels by Hydrocracking Oil from Coal," *Hydrocarbon Process.*, 141 (March 1969).

Table 29.52 Results of Hydrocracking Tests on COED Gas Oil

Feed: Property	Fraction						
	+343°C		343–427°C		+427°C		
Yield (wt %)	100		84		16		
Nitrogen (ppm)	42		23		123		
Polyaromatics (total)	15.4		12.8		28.4		
Polyaromatics (4 and 5 ring)	5.9		3.6		17.5		
Temperature (°C)	374	374	371	382	382	382	374
Pressure (psig)	1500	1500	1500	1500	1500	1500	1500
WHSV (hr^{-1})	1.14	0.945	0.993	1.02	2.0	3.18	1.02
Conversion (wt %)							
To −199°C	48.1	70.5	100	100	41.6	21.2	25.8[a]
To −82°C	21.6	34	68.9	78	20.1	9.6	8.7[a]

[a] Reflects extent of catalyst decay compared to first entry.

Source: Modified from Ref. 99.

Table 29.53 Results of Tests on the Decay in Activity of Hydrocracking Catalyst

Properties of feedstock—COED fraction: boiling range = 288–427°C, API gravity = 17.1.

Component	Weight Percent
Hydrogen	10.75
Carbon	88.02
Sulfur	0.026
Nitrogen	0.353
Oxygen	0.76
Ash	0.012

Yields and Conversions

Sample (increasing time)	1	2	3	4	5
Conversion (%)	96.6	79.3	79.9	80.2	73.6
Hydrocracking temperature (°C)	334	335	340	360	353
Yield (wt % feed)					
149–288°C	40.3	32.6	28.5	24.8	18.9
38–343°C	88.8	75.3	74.9	69.3	66.4

Source: Modified from Ref. 100.

velocity. The maximum gasoline yield exceeded 60 vol % and the gas yield approached 6 vol %. In this study the material boiling to 200°C was designated gasoline, the neutral fraction from 200 to 360°C was diesel fuel, and the remaining was residue. Under other conditions tested, the maximum gasoline yield was 77 vol %, and at higher yields higher research octane number and low sulfur and nitrogen content were achieved. The gasoline obtained from hydrocracking this coal-derived oil might still require a light reforming treatment to improve it for use.

The two catalysts, NiS + WS_2 and CoO + MoO_3, gave essentially the same gasoline and gas yields, although the yields with the NiS

Table 29.54 Properties of Hydrocracking Feedstock from Direct Coal Hydrogenation for Two Catalysts

Hydrocracking catalysts used: 6% NiS + 19% WS_2 on silica-alumina (200 m²/g) and 3% CoO + 12% MoO_3 on silica-alumina (200 m²/g); pellets 0.083 - in. diameter, 0.125 in. high.

Specific gravity at 25°C	0.988
Sulfur	0.317 wt %
Nitrogen	0.615 wt %
Tar acids	19.0 vol %
Tar bases	1.5 wt %
Distillation from 76°C, IBP	
Up to 200°C	22 vol %
Up to 360°C	80 vol %
Residue	20 vol %
Composition of neutral oil, 76–360°C	
Saturates	42.0 vol %
Olefins	5.0 vol %
Aromatics	53.0 vol %

Source: Modified from Ref. 139.

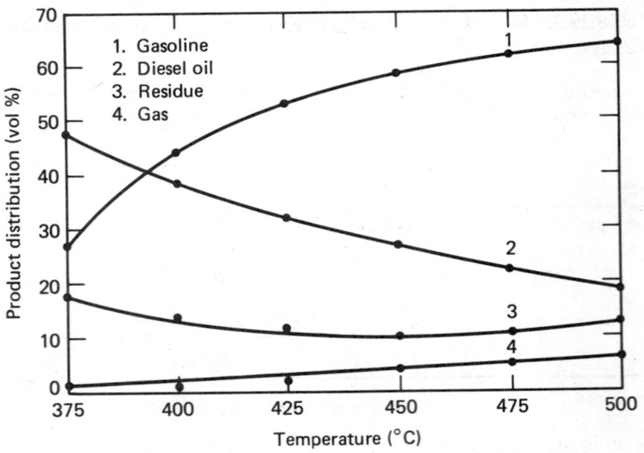

Figure 29.25 Effect of temperature on product distribution at 1500 psig and 1.0 space velocity.[139]

+ WS_2 catalyst were somewhat higher. The compositions of the gasolines produced by the two catalysts (Fig. 29.26) are similar except for the isoparaffins and *n*-paraffins. Apparently the NiS + WS_2 catalyst gave greater isomerization activity. The heteroatom removal (Fig. 29.27) reveals that essentially all sulfur can be removed by both catalysts, but the greater hydrogenating power of nickel is manifested by the increased denitrogenation with the NiS + WS_2 catalyst. The CoO + MoO_3 catalyst gave better tar

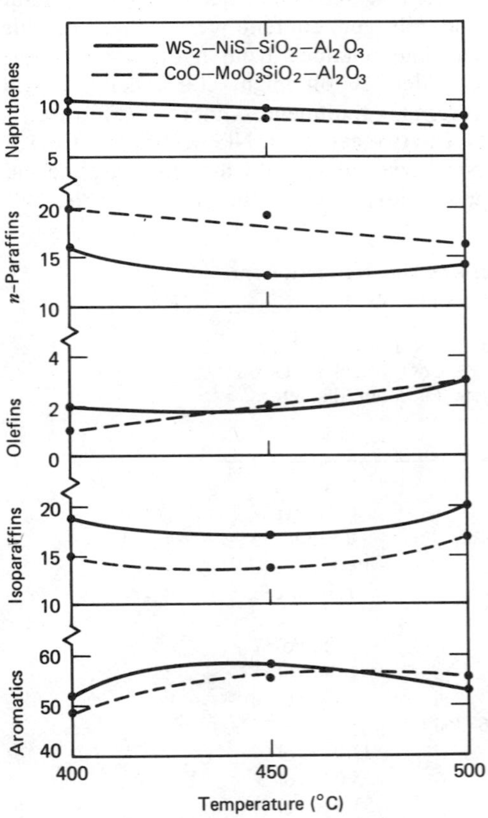

Figure 29.26 Comparison of gasoline compositions for sulfide and oxide catalysts.[139]

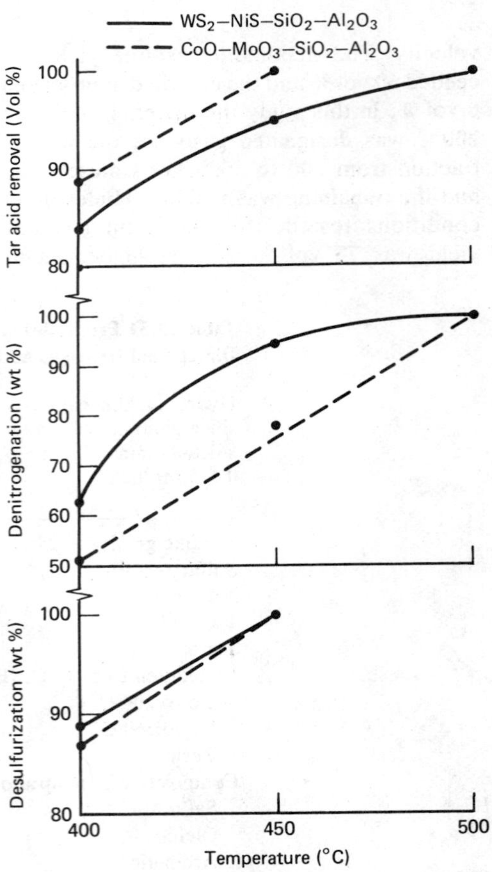

Figure 29.27 Effect of temperature on conversion of heterocyclic compounds.[139]

acid removal. Coking problems occurred, and catalyst regeneration was required when coke levels reached 5 wt %. A once-regenerated catalyst had about the same activity as fresh catalysts. An estimated 100 volumes of oil from coal hydrogenation might be processed per volume of catalyst. In general, this type of oil is harder to hydrocrack because of the higher levels of oxygenated compounds and aromatics. Polynuclear compounds adversely affect the activity and life of the catalyst.

29.4.2.3 *Hydrocracking of Tars*

The hydrocracking studies of coal extracts just discussed were made subsequent to the extensive work done on upgrading by hydrocracking tars or oils produced in a pyrolysis coking operation. COED oils discussed under "Pretreatment of Coal Extracts" in Section 29.4.2.2 are also, of course, liquids produced in a pyrolysis or carbonization process. However the COED process comprises several pyrolysis stages; therefore the liquids produced have been exposed to conditions different from those existing during coking or single-stage pyrolysis. Work on hydrocracking tars produced by coking or carbonization is summarized below.

Low-Temperature Tar from the Carbonization of a High-Volatile Bituminous Coal from Utah. A low-temperature tar obtained in the carbonization of a high-volatile bituminous coal (Utah) at 550°C was hydrocracked for gasoline production using dual-functional, refinery, and conventional cracking catalysts.[140] The study was performed in a one-liter batch autoclave with 100 cm³ of tar and 15 g of catalyst at different combinations of temperature, time, and pressure. The feed oil properties and some of those of the catalysts used are shown in Table 29.55. The volume-percent distribution obtained at 1500 psig as a function of temperature (the reaction time varied) is shown in Fig. 29.28. At temperatures from 350 to 500°C the diesel oil and residue fractions decreased as the gasoline fraction and gas make increased. The sulfur and nitrogen removals were 65 and 30%, respectively, at 350°C, 1500 psig, and 100 and 85%, respectively, at 450°C. The pressure effects from 1000 to 3000 psig are shown in Fig. 29.29. The rate of gasoline formation was high from 1000 to 1500 psig

[140] S. A. Qader and G. R. Hill, "Catalytic Hydrocracking—Hydrocracking of a Low-Temperature Coal Tar," *Ind. Eng. Chem. Process Des. Dev.*, 8(4), 450 (October 1969).

Table 29.55 Properties of Feed Oil and Catalysts Used in Hydrocracking Tests of Low-Temperature Tar

Specific gravity at 25°C	0.989
Distillation at 89°C, IBP	
To 200°C	15 vol %
50% Distillate	283°C
Pitch point	360°C
Pitch	26 vol %
Tar acids	27.0 vol %
Tar bases	2.0 vol %
Sulfur	0.846 wt %
Nitrogen	0.644 wt %
Neutral tar to 360°C	
Saturates	28.0 vol %
Olefins	15.0 vol %
Aromatics	57.0 vol %

Catalysts Used

Type		Composition	Surface Area (m^2/g)
Dual functional	–	6% NiS + 19% WS_2 on silica-alumina	212
Refining	–	6% NiS + 19% WS_2 on alumina	210
Cracking	–	84% silica + 16% alumina	200

Source: Modified from Ref. 140.

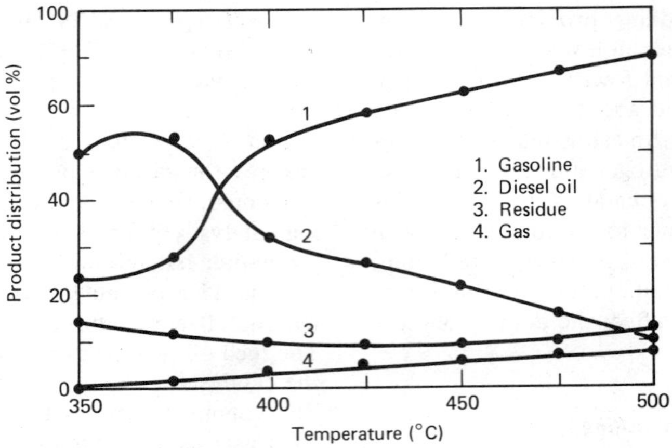

Figure 29.28 Effect of temperature on product distribution (dual-functional catalyst).[140]

and from 2500 to 3000 psig but somewhat slower at intermediate pressures. The aromatic content of the gasoline formed also increased with pressure up to about 2500 psig, but decreased thereafter. In contrast to this, naphthenics in the gasoline increased slowly to 2500 psig and rapidly thereafter. The sulfur removal followed pressure increases linearly, but nitrogen removal was much more responsive at high pressures. Nitrogen compounds simply require much greater hydrogenation for removal.

Using the yields of gasoline and gas as measures of hydrocracking activity, the dual-functional catalyst was superior, followed by the refinery type and then the conventional cracking catalyst. A higher yield and quality of diesel oil were obtained with the refinery-type catalyst. The dual-functional and the refinery catalyst were the most effective for heteroatom removal, as expected.

In 6 hr of reaction time at 475°C and 1500–3000 psig, about 1.0 wt % coke (catalyst weight) was deposited on the dual-function catalyst. However the data from such relatively short times provide few design data for projecting catalyst life in longer operations.

The hydrocracking of tars produced in pyrolysis is a complicated process of parallel and series cracking steps, hydrogenation, and isomerization. Mechanisms have been proposed that incorporate hydrocarbon cracking by way of a carbonium ion scheme, and sulfur, oxygen, and nitrogen removal through radical reaction sequences.[141] Dual-functional catalysts are capable of supplying acidic sites for cracking and isomerization (silica) and hydrogenation sites for hydrogen transfer (Co, Mo, Ni species).

First-order reactions were used to represent the results of hydrocracking a tar similar to that shown in Table 29.55, except that the −200°C material had been removed.[141] For first-order rates of gasoline formation, and oxygen, nitrogen, and sulfur removal, the constants shown below adequately repre-

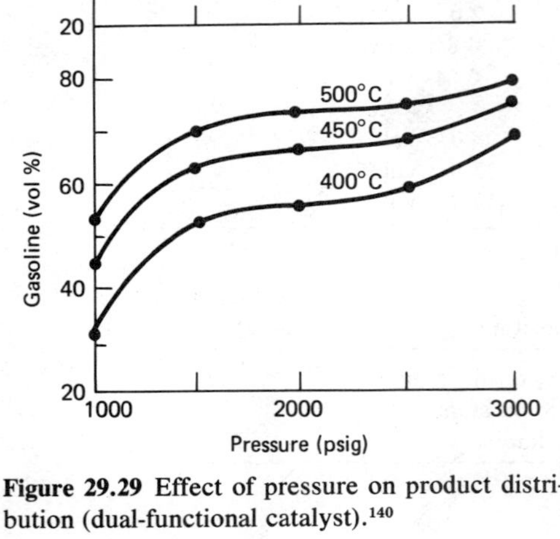

Figure 29.29 Effect of pressure on product distribution (dual-functional catalyst).[140]

[141] S. A. Qader and G. R. Hill, "Catalytic Hydroprocessing of Low Temperature Coal Tar," *Ind. Eng. Chem. Process Des. Dev.*, 8(4), 456 (October 1969).

sented the data at 1500 psig and 400–500°C using the dual-functional catalyst shown in Table 29.55.

$$k_{\text{gasoline}} = 0.157 \times 10^6 \exp \frac{-17{,}600}{RT} \text{hr}^{-1}$$

$$k_S = 0.213 \times 10^5 \exp \frac{-14{,}500}{RT} \text{hr}^{-1}$$

$$k_O = 0.361 \times 10^5 \exp \frac{-13{,}600}{RT} \text{hr}^{-1}$$

$$k_N = 0.474 \times 10^5 \exp \frac{-15{,}900}{RT} \text{hr}^{-1}$$

The results of this study suggest that hydrocracking of the low-temperature tar is controlled by rate of cracking of the carbon-carbon bonds or carbon-sulfur, carbon-nitrogen, or carbon-oxygen bonds, and possibly also by isomerization reactions. The cracking power of catalysts is so important that the disastrous effects of high nitrogen concentrations are understandable. The acidic sites of the catalyst cannot withstand the relatively high concentration of ammonia and other inhibitors. The rate of hydrocracking over the catalyst appears to be controlled by chemical reaction steps, not by physical or mass transport steps; this was not, however, shown conclusively by direct experimentation.

Low-Temperature Tar from the Carbonization of a Lignite from North Dakota. Low-temperature tars from North Dakota lignite yield more and better quality gasoline by using a refinery-type catalyst (Co-Mo-alumina) than by using a zinc chromite on alumina catalyst.[142] Both the raw tar and a creosote oil were processed; the feed properties and catalysts are shown in Table 29.56. The tests were run at 3000 psi in a continuous flow reactor with 150 Cm³ of catalyst. Some of the results of this study are shown in Table 29.57. The heteroatom removal was significantly greater with the Co-Mo catalyst than with the zinc chromite catalyst. Moreover, use of the zinc catalyst yielded less gasoline, and the products were less hydrogenated. The neutral gasoline fractions from the experiments with the zinc catalyst contained large amounts of olefins, but small amounts of naphthenics. The three neutral gasolines from the experiment with the Co-Mo catalyst were high in cyclic hydrocarbons, and the ratio of aromatics to naphthenes was highest from the highest reaction temperatures.

The deposition of coke on these catalysts was quite low, but the active life of the catalyst could not be determined. Table 29.58 makes an interesting comparison of these hydrocracked coal stocks with petroleum stocks for reforming feeds. The low heteroatom content and the high aromatic and naphthene content of the coal liquids lead to good reforming stocks.

Table 29.56 Properties of Feed and Catalysts Used in Tests to Produce Gasoline from Lignite Tar

	Crude Tar	Creosote Oil
Specific gravity at 20°C/20°C:	1.042	1.036
Composition (wt %)		
Benzene insolubles	1.13	—
Ash	0.25	—
Carbon	80.29	80.01
Hydrogen	8.94	8.88
Nitrogen	0.56	0.36
Sulfur	0.90	0.77
Oxygen	9.31	9.98
Distillation		
Fraction	Weight Percent	
1–170°C	2.05	
170–200°C	0.62	
200–210°C	0.71	
210–235°C	3.64	8.70
235–270°C	12.00	30.45
270°C–decomposition	60.81	48.90
Residue	16.65[a]	10.00[b]
Loss	3.51	1.95
Distillate	Volume Percent	
Acids	32.0	42.5
Bases	1.0	3.0
Neutral oil	67.0	54.5
Neutral oil		
Olefines	15.0	—
Aromatics	53.5	—
Paraffins	31.5	—

Catalysts Used

1. 2.3 wt% cobalt + 15.5 wt% molybdenum on alumina.
2. 37.8 wt% zinc oxide + 24.7 wt% chromic oxide on alumina.

[a] At 370°C.
[b] At 359°C.

Source: Modified from Ref. 142.

Topped Tar from the High-Temperature Carbonization of Coal. Other investigations by the U.S. Bureau of Mines of hydrocrack-

[142] H. C. Carpenter, P. L. Cottingham, C. M. Frost, and W. W. Fowkes, *Hydrocracking Low-Temperature Tar from a North Dakota Lignite,* U.S. Bur. Mines Rep. Invest. No. 6237, 1963.

Table 29.57 Results of Tests on the Production of Gasoline from Lignite Tar

	Catalyst and Feedstocks			Catalyst and Feedstocks	
	Cobalt Molybdate			Zinc Chromite	
	Creosote Oil	Total Tar	Total Tar	Total Tar	Total Tar
Temperature (°C):	477	452	394	449	478
Pressure (psig):	3000	3000	3000	3000	3000
Hydrogen feed rate (scf/bbl):	6000	6000	6000	6000	3000
Space velocity (hr^{-1}):	1.0	1.0	1.0	1.0	2.0
Recoveries					
Liquid product (wt %)	78.1	82.2	94.7	90.5	90.6
Gasoline (wt %)	46.8	33.4	24.9	18.9	21.6
Recycle oil (wt %)	31.3	48.8	69.8	71.6	69.0
Catalyst deposit (wt %)	0.5	0.7	0.7	0.9	0.9
Liquid product (vol %)	94.9	98.7	110.2	98.2	97.5
Gasoline (vol %)	59.3	43.0	31.6	22.8	26.4
Recycle oil (vol %)	35.6	55.7	78.6	75.4	71.1
Liquid-product properties					
Specific gravity at 16°C	0.948	0.8684	0.8958	0.9603	0.9684
Sulfur (wt %)	0.08	0.14	0.12	0.47	0.54
Nitrogen (wt %)	0.14	0.09	0.20	0.55	0.57
Hydrogen (wt %)	11.76	11.68	11.61	9.95	9.53
Carbon (wt %)	87.82	85.55	85.52	82.55	82.51
Oxygen, by difference (wt %)	0.20	2.54	2.55	6.48	6.85
C/H, wt ratio	7.47	7.32	7.37	8.30	8.66
Gasoline properties					
Specific gravity at 16°C	0.8174	0.8099	0.8230	0.8656	0.8529
Sulfur (wt %)	0.03	0.02	0.04	0.35	0.30
Nitrogen (wt %)	0.07	0.05	0.11	0.24	0.20
Hydrocarbons in neutral gasoline (vol %)					
Paraffins	20	14	15	28	41
Naphthenes	34	54	59	11	6
Olefins	3	1	3	31	23
Aromatics	43	31	23	30	30

Source: Modified from Ref. 142.

ing and hydrotreating catalysts attempted to duplicate some of the earlier German catalysts.[143] Since this work is mostly of historical interest, no details are presented here.

A typical patent illustrating the interest of those who produce large quantities of coal tar (e.g., the steel industry) emphasizes the use of refining-type catalysts.[144] Topped tar (+300°C fraction) from destructive distillation of coal to form coke was hydrocracked over a Co-Mo catalyst at 3000 psig and 400°C. The catalyst contained 3% CoO + 15% MoO_3 on alumina and had a surface area of 320 m^2/g. For a reaction time of 4 hr, about 73% of the +300°C material was converted to the −300°C fraction.

[143] C. O. Hawk, M. D. Schlesinger, H. H. Ginsberg, R. W. Hiteshue, *Vapor-Phase Hydrogenation of Oils Derived from Coal and Oil Shale,* U.S. Bur. Mines Rep. Invest. No. 6548, 1964.

[144] R. A. Flinn, Emmaus, and M. O. Tarhan, U.S. Pat. 3,503,872, Bethlehem Steel Corporation, March 31, 1970.

Hydrocracking of Coal Tars for Chemicals. Hydrocracking operations are not necessarily always directed toward gasoline production. The rising prices of petroleum and natural gas stocks and their correspondingly decreased availability are redirecting interest to production of chemicals from coal liquids. The trend is by no means new: our modern petrochemical industry was an outgrowth of the coal chemicals industry. A brief illustra-

Table 29.58 Comparison of Tests on Lignite Tar and Petroleum Reforming Stocks

	Coal Liquids			Petroleum Reforming Stocks		
Property	Exp 1[a]	Exp 2[b]	Exp 3[c]	Midcont-inent	Gulf Coast	Kuwait
Yield (vol %)	—	—	—	100	100	100
Gravity (API)	40.4	43.2	—	55.0	—	58.1
Paraffins (wt %)	15	14	5	40	41	67
Olefins (wt %)	3	1	0	Trace	Trace	Trace
Naphthenes (wt %)	59	54	60	52	43	22
Aromatics (wt %)	23	31	35	8	16	11
ASTM distillation at 760 mm						
IBP (°C)	50.6	53.9	77.8	94.1	92.9	101
10 %	97.8	74.4	103	106		113
50 %	141	127	133	130	136	
90 %	188	184	173	168	164	133
EP (°C)	204	211	184	197	193	168
Sulfur (wt %)	0.01	0.01	0.01	0.012	—	177
Nitrogen (wt %)	None	None	None	—	—	0.04
Research octane number: clear	82	80	—	48	52	—

[a] Gasoline from experiment 3 after removing 7.7 vol % tar acids.
[b] Gasoline from experiment 2 after removing 1.6 vol % tar acids.
[c] Gasoline after removing 88 vol % tar acids from experiment 2.

Source: Modified from Ref. 142.

tion of work on the production of chemicals from coal liquids is provided by a current study by Dow Chemical Company.[145] The data from the study, which are still somewhat imprecise, show typical test conditions as 488–529°C, 1500 and 2500 psig, 0.6–1.5 hr^{-1} LHSV, and hydrogen consumption of 20,000 scf/bbl. The catalyst used was a commercial type, Harshaw HT-400E, which is 3.0 wt % CoO, 15.0 wt % MoO_3, with a surface area of 220 m^2/g. The feedstock was a straight-run, middle-distillate coal oil (182–360°C) with 0.74 wt % sulfur and 0.28 wt % nitrogen. Typical C_1–C_9 yields ranged from about 40 to 60 wt %. The paraffin content (10–13 wt %) of the C_5–C_9 liquids was relatively insensitive to operating conditions. The naphthenes and aromatics of the C_5–C_9 liquids ranged between 20 and 36 wt % and 25 and 33 wt %, respectively. The study is preliminary; its intention is to assess conditions under which coal liquefaction products might be processed to produce useful aliphatic and aromatic chemicals. Hydrocracking is one of several processing steps being assessed in the study.

Liquids Produced from the Same Type of Coal Processed in Three Different Ways. The nature and quality of products obtained from hydrocracking depend not only on the coal type but also on the process used to produce the coal liquids. The properties of three coal liquids derived from three different treatments of the same coal are shown in Table 29.59, along with the properties of their hydrocracked gasoline products.[86] The oil derived from the hydrogenation process was highest in aromatics and yielded gasoline with the highest level of aromatics. The detailed composition of the aromatics listed in the table shows the potential for extraction and dealkylation to produce BTX or pure aromatics.

Comparison of Products Produced in Hydrocracking Coal Liquids and Petroleum Fractions. The behaviors of petroleum- and coal-derived stocks during hydrocracking have been compared.[146] Three petroleum and two coal liquids were hydrocracked in a laboratory reactor containing 60 cm^3 of catalyst, 6 wt % NiS + 19 wt % WS_2 on silica-alumina

[145] B. C. Peters, *Chemicals from Coal,* Quarterly Tech. Prog. Rep. No. FE-1534-33, by Dow Chemical Company for U.S. ERDA, July 15, 1976.

[146] S. A. Qader and G. R. Hill, "Hydrocracking of Petroleum and Coal Oils," *Ind. Eng. Chem. Process Des. Dev.,* 8(4) 462 (October 1969).

Table 29.59 Summary of Coal Oils Derived by Different Treatments of the Same Coal

Components	Low-Temperature Carbonization	Pyrolysis	Hydrogenation
	Properties of Coal Oils		
Benzene insolubles (wt %)	Nil	2.0	Nil
Asphaltenes (wt %)	19.0	16.0	15.0
Light Oil (up to 200°C) (vol %)	5.0	7.0	18.0
Residual Oil, +370°C (vol %)	27.0	30.0	25.0
Sulfur (wt %)	0.82	0.98	0.64
Nitrogen (wt %)	0.91	0.86	0.76
Oxygen (wt %)	3.8	4.5	2.8
Tar acids, (200–370°C oil, wt%)	26.0	24.0	15.0
Hydrocarbon analysis of neutral oil, 200–370°C (vol %)			
Saturates	29.0	25.0	24.0
Olefins	15.0	17.0	11.0
Aromatics	56.0	58.0	65.0

Composition of Product Coal Gasolines[a]

	Feedstock		
Composition (vol %)	Low-Temperature Tar	Pyrolysis	Hydrogenation
Saturates	30.0	26.0	13.0
Olefins	1.0	1.0	1.0
Aromatics	69.0	73.0	86.0

Analysis of Gasoline Saturates

	Gasoline from		
Analysis (vol %)	Low-Temperature Tar	Pyrolysis Tar	Hydrogenation Oil
n-Paraffins			
C_6	3.2	1.7	0.9
C_7	5.9	6.6	7.4
C_8	15.2	14.1	12.7
C_9	14.1	13.2	15.5
C_{10}	3.8	4.5	2.1
C_{11}	13.8	14.2	12.2
	56.0	54.3	50.8
Isoparaffins			
C_6	3.1	1.9	2.8
C_7	11.2	8.3	10.1
C_8	10.1	12.0	16.2
C_9	9.8	13.3	10.0
	33.2	35.5	39.1
Cycloparaffins			
C_6	4.1	3.2	1.7
C_7	6.7	7.0	8.4
	10.8	10.2	10.1

Table 29.59 Continued

Analysis of Gasoline Aromatics

Analysis (vol %)	Gasoline from Low-Temperature Tar	Pyrolysis Tar	Hydrogenation
Benzene	33.9	30.7	40.4
Toluene	18.2	20.2	14.8
Ethylbenzene	4.5	2.8	5.5
Xylenes	25.8	27.2	19.4
Propylbenzens	2.5	3.5	1.6
Butylbenzenes	9.0	7.0	11.3
Indanes	4.1	5.1	4.5
Unidentified	2.0	3.5	2.5

[a] Conditions: 480°C, 2000 psi, 1.0 space velocity.

Source: Modified from Ref. 86.

(200 m^2/g). Tests were made at 350–500°C, and 1500 and 2000 psig. The hydrogen feed rate was about 2800 scf/bbl. The feedstock properties are shown in Table 29.60. The material boiling to 200°C was designated naphtha; at 500°C the naphtha yields for petroleum oils were 77–79 vol % and 79–80 vol % at 1500 and 2000 psig, respectively, as compared to 60–63 vol % and 77–79 vol % for coal oils. The coal oils simply require a greater hydrogen pressure to hydrogenate their higher aromatics and heteroatoms. This is shown too by the hydrogen consumption at maximum naphtha yields (80 vol %), which are 800–1000 scf/bbl for petroleum oils and 1500–1600 scf/bbl for coal oils.

Typical compositions of naphthas are shown in Fig. 29.30. The *n*-paraffins are about the same for the two feed types; however the aromatics are much higher for the coal oils. The naphthalenes and isoparaffins are lower in the naphtha from the coal oils. The sulfur and nitrogen levels of the naphtha product from the coal were somewhat higher than in the naphtha from the petroleum oils.

The rate of catalyst coking when processing coal oils is about twice that for petroleum oils. At 480°C, 2000 psig, and one space velocity, the catalyst using coal liquids accumulated 5 wt % coke after processing about 160 volumes of oil per volume of catalyst. When the coal liquids were subject-

Table 29.60 Comparison of Hydrocracking Petroleum and Coal Liquids Used in Properties of Feeds

Properties	Petroleum Stocks: Stove Oil	Diesel Oil	Gas Oil	Coal Stocks: Oil Fraction	Low-Temperature Tar Fraction
API gravity	38.2	36.5	32.4	21.5	23.4
Sulfur (wt %)	0.24	0.22	0.38	0.62	0.84
Nitrogen (wt %)	0.18	0.16	0.14	0.42	0.58
Oxygen (wt %)	Nil	Nil	Nil	3.6	4.8
Distillation					
IBP (°C)	200	250	350	200	200
50%	250	302	430	238	260
EP (°C)	300	380	490	370	360
Hydrocarbon analysis (vol %)					
Saturates	80	80	81	38	28
Aromatics	20	20	19	57	54
Olefins	Nil	Nil	Nil	5	18

Source: Modified from Ref. 146.

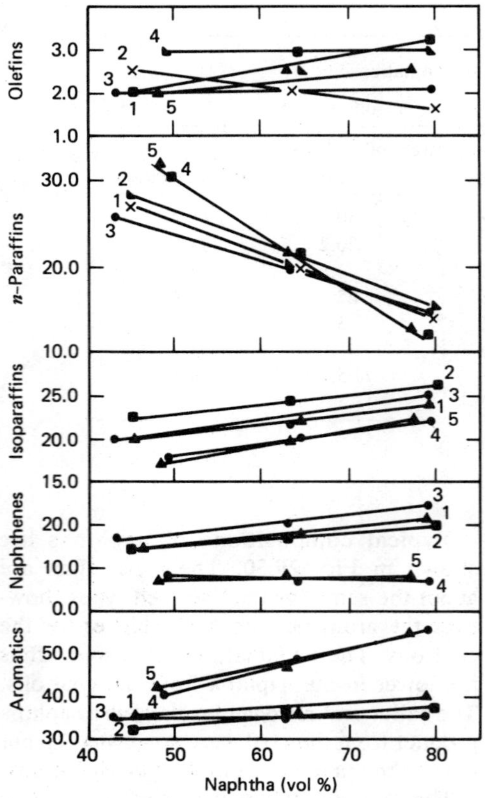

Figure 29.30 Composition of naphtha at differenı levels of formation: 1, stove oil; 2, diesel oil; 3, gas oil; 4, low-temperature tar; 5, coal oil.[146]

ed to a hydrofining pretreatment before hydrocracking, the coking rate was essentially the same as that for processing the petroleum oils. Thus with the cracking step a controlling feature, the pretreatment simply protects the acidic sites, which would otherwise be more readily poisoned by coke or nitrogen compounds.

Simple, first-order kinetics of hydrocracking of petroleum gas oil and a coal oil yielded activation energies of 14,300 and 15,500 cal/g-mole, respectively. The larger activation energy for the coal liquid is therefore simply a greater "resistance" to reaction, but it also means a greater sensitivity or responsiveness to temperature changes.

Liquid Products from Coal Conversion Processes Under Development. The objective of one program worthy of note is to screen a multitude of catalysts to determine their effectiveness for hydrocracking and heteroatom removal from liquids obtained from developing processes such as PAMCO, H-Coal, COED, and Synthoil.[147] Many catalysts have been tested in both laboratory batch and continuous flow reactors under typical hydroprocessing conditions. These catalysts usually include nickel molybdate and sulfide, nickel tungstate, and other combinations on various carriers, prepared both commercially and in the laboratory. The study has presented few general conclusions thus far, except to indicate that the catalysts containing nickel in some state appear to be generally favored for the hydroprocessing of the coal liquids.

Studies of Pure Compounds. Several studies have been made to determine the hydrocracking behavior of pure compounds. The few that are briefly summarized here are directed toward the definition and screening of catalysts.

Several catalysts were assessed, in standard screening tests using pyrene at 427°C, 4200 psig, for 1 hr.[134] Refining type, dual-functional, and metal halides were used in the study. Under the conditions tested the pyrene was readily converted. When cracking promoters such as silica, hydrogen fluoride, and metal halides were added to the supported refining catalysts, hydrocracking activity was increased. Zinc chloride was a strong catalyst when used alone or added separately with a refining catalyst. Even $ZnCl_2$ on alumina showed hydrocracking activity, but experienced high coking rates because the catalyst lacks the capability for hydrogenation.

Pyrene has also been hydrocracked over H-Zeolon + WS_2 and silica-alumina-WS_2 catalysts at 1800 psig, 350–550°C, and up to 60 min.[148] The extent of cracking varied between the two catalysts, but the trends were similar. The silica-alumina catalyst gave better conversion, but higher coke yields (5–23 wt %

[147] L. Berg and F. P. McCandless, *Catalytic Hydrogenation of Coal-Derived Liquids,* Interim Reps. Nos. FE-2034-6 and -5, by Montana State University, for U.S. ERDA, December 1976–February 1977, September 1976–February 1977.

[148] W. H. Wiser, *Hydrogenation of Western Coal in Dilute Phase Under Medium Pressure for Production of Liquids and Gases,* Final Rep. No. FE-1200-1, by University of Utah, for U.S. ERDA, December 1975.

vs. 5–10 wt %). Pore diffusion problems were suspected with the H-Zeolon catalysts, and comparison with results from a similar but much larger pore diameter support confirmed mass transfer effects. Reaction orders of 0.5–2.0 in pyrene concentration represented the data equally well. The sequence proposed for the processing steps is hydrogenation of pyrene, followed by cracking; then hydrogenation of the cracked products.

The hydrocracking patterns of naphthalene, anthracene, and pyrene were found to be essentially the same over several different catalysts.[149] The test conditions were 450–500°C and 500–1250 psig, using multifunctional catalysts of the sulfides and oxides of nickel, cobalt, and molybdenum; tungsten with silica-alumina; and molecular sieves. Sequential first-order reaction steps were used to model the reactions. The general sequence of steps includes the following:

1. Hydrogenation of aromatic rings.
2. Isomerization to structures favoring cracking.
3. Cracking to smaller species.
4. Hydrogenation and cracking of smaller products.

The rate of hydrocracking of four-ring structures is not nearly as rapid as for the smaller, two-ring compounds.

29.4.3 Catalytic Cracking

Few details have been published on the processing of coal liquids by catalytic cracking. The two studies reported here suggest that catalytic cracking is more susceptible to coke and gas formation for coal pyrolysis oils than for petroleum stocks. The coal oils were markedly less favorable to cracking than comparable petroleum oils.

A COED +343°C gas oil was blended with a petroleum stock and after hydrotreating was subjected to catalytic cracking, using a catalyst that contained zeolite and was conditioned or equilibrated in a commercial petroleum unit.[99] Nominal catalyst properties are given in Table 29.61, and some properties of the various feedstocks are presented in Table 29.62. The petroleum stock was much higher in paraffins, whereas the coal oils contained more aromatics and less total hydrogen. The unhydrofined coal oil was especially high in polynuclear aromatics. A small fluidized-bed reactor could not be used because it was too easily plugged when processing the "as-distilled" stock. Most comparative testing was done in a special, fixed-bed microreactor using a blend of petroleum and coal liquids. Partial results are shown in Table 29.63 for reactor temperatures of about 500–520°C.

Table 29.61 Cracking Catalyst Properties

Property	Value
Al_2O_3	45 wt %
Cerium	0.45 wt %
Total rare earth	1.0 wt %
Nickel	0.038 wt %
Vanadium	0.040 wt %
Iron	0.65 wt %
Volatile matter	0.6 wt %
Surface area	60 m^2/g

Source: Modified from Ref. 99.

Conversion to −200°C material was 6–9% greater from processing petroleum liquid than from processing COED oils. The coal liquids yielded less gasoline and more coke and gas make. Maximum gasoline yields were achieved at about 62% conversion. Surprisingly, the results obtained with the hydrotreated oil were similar to those with the blend, suggesting that hydrofining had the same effect as blending 75% petroleum liquid with 25% of the coal liquids. Since fluid catalytic cracking units are designed on the basis of coke burnoff rate, the yields from the three liquids are compared at a constant coke yield. The conversions and selectivities of the coal oils appear to be markedly less favorable than those of the petroleum gas oil. Again, these data were taken in a packed-bed microreactor because the laboratory-scale, fluidized-bed reactor was susceptible to plugging when operated with coal liquids. However plugging in the small-scale reactor does not necessarily mean that a commercial fluidized catalytic cracking unit would be susceptible to plugging problems.

Catalytic cracking of a low-temperature

[149] S. A. Qader and G. R. Hill, "Development of Catalysts for the Hydrocracking of Polynuclear Aromatic Hydrocarbons," *Prepr. Am. Chem. Soc., Div. Fuel Chem.*, **16**(2), 93 (1972).

Table 29.62 Feeds Used in Catalytic Cracking Tests

	Midcontinent Gas Oil	As Distilled, 343°C+ COED	Blend[a]	Hydrotreated, 343°C+ COED
		Analyses		
API gravity	26.1	11.2	22.4	15.3
Sulfur (wt %)	0.902	0.07	0.694	0.001
Nitrogen (ppm)	1300	816	1050	181
Basic nitrogen (ppm)	430	492	330	84
Hydrogen (wt %)	12.68	10.14	11.0	11.20
Oxygen (wt %)	—	0.25		0.08
Carbon (wt %)	86.18	89.69		—
Carbon residue (con) (%)	—	1.13		—
Nickel (ppm)	—	1.8		—
Vanadium (ppm)	—	2.1		—
Paraffins (wt %)[b]	27.7	10.9	23.5	9.7
Monocycloparaffins (wt %)	26.7	11.3	22.8	16.1
Polycycloparaffins (wt %)	5.5	7.7	6.1	12.8
Monoaromatics (wt %)	24.7	33.0	26.8	44.7
Diaromatics (wt %)	8.7	13.7	10.0	9.5
Triaromatics (wt %)	3.6	8.9	4.9	2.0
Tetraaromatics (wt %)	1.7	9.6	3.6	3.5
Benzanthracene + 5 rings (wt %)	1.4	4.9	2.3	1.7
Total polynuclear aromatics (wt %)	15.4	37.1	20.8	16.7
		ASTM Distillation		
Fraction		Temperature (°C)		
IBP	236	272		221
10%	319	374		346
30%	370	382		359
50%	416	313		373
70%	465	416		395
90%	552	466		444
95%		498		476

[a] Contains 25% as-distilled 343°C + COED + 75% midcontinent gas oil.
[b] Paraffinic or cycloparaffinic portions of the aromatics molecules are included as wt % aromatics; paraffinic side chains on cycloparaffins are included as wt % cycloparaffins.

Source: Modified from Ref. 99.

pyrolysis oil derived from a subbituminous coal was demonstrated to be viable in a laboratory-scale fluidized bed.[150] Two cracking catalysts, natural clay–Filtrol and silica alumina, were assessed over a temperature range of 460–538°C at WHSV of 1.0 hr^{-1}. The synthetic silica-alumina catalyst was able to promote cracking at lower temperatures than the Filtrol. A cracking temperature of 510°C resulted in the highest naphtha yields, and more than half the oxygen in the feed oil was removed. Naphtha was taken as the light-end liquids boiling below 210°C. Some of the results and the few feedstock properties stated are shown in Table 29.64.

[150] J. F. Jones, P. L. Barrick, and J. H. Blake, "Pyrolysis of Oils Derived from Coal and Oil Shale in a Fluidized Bed of Catalysts," paper presented at American Chemical Society, Division of Petroleum at Chemistry, 151st National Meeting, Pittsburgh, March 23–26, 1966.

29.4.4 Steam Pyrolysis

Information is as sparse for steam pyrolysis or steam cracking of coal liquids as it is for catalytic cracking. Such information as has been gathered does, however, lead to a significant conclusion that prehydrogenation of coal-derived oils is necessary before they are steam cracked if "petrochemicals" are desired.

Table 29.63 Catalytic Cracking Results

Conversion or Yield	Midcontinent Oil	Hydrotreated COED Oil, +343°C	Blend
Conversion at 12 WHSV (–200°C) (vol %)	65.8	58.8	58.8
Yields at 62 vol % conversion			
C_5–220°C (vol %)	54.5	50.7	50.7
Coke (wt %)	2.7	5.5	4.4
C_4 (vol %)	11.2	11.2	11.5
C_3 + lighter (wt %)	5.0	5.5	5.6
Conversion (constant coke make) (vol %)	74.2	58.2	60.4
Yields (constant coke make)			
C_5–220°C (vol %)	60.4	49.6	50.4
C_4 (vol %)	15.4	10.3	11.2
C_3 + lighter (wt %)	7.7	4.9	5.3

Source: Modified from Ref. 99.

The results of a study of the use of a steam-cracking coil with feeds of H-Coal naphtha, H-Coal mid-distillate, and a COED mid-distillate are summarized here.[151] No details regarding the reaction or the feedstock were stated. Table 29.65 presents the limited data available. The H-Coal naphtha yielded significantly greater quantities of olefins and benzene, both desirable in chemical processing of coal liquids. The mid-distillates, especially COED, yielded about 50% fuel stock. This rather high yield indicates that a pre-hydrocracking step would be desirable or necessary if petrochemical production is the objective.

[151] B. C. Peters, *Chemicals from Coal,* Quarterly Tech. Prog. Rep. No. FE-1534-T-5, by Dow Chemical Company for U.S. ERDA, October 29, 1975.

A study was conducted comparing steam cracking of a Synthoil liquid with a petroleum stock and a representative, pure compound—decahydronaphthalene.[64] The tubular reactor was controlled to simulate the temperature-time history of a commercial reactor. Conditions were as follows: exit temperature, 875°C, exit pressure, 27 psig, time of contact, 0.3 s; and steam-hydrocarbon weight ratio, 1.0. Table 29.66 presents the results. The gravity, distillation, and elemental analyses are similar for the coal and the petroleum feedstocks; however the coal oil essentially consists entirely of alicyclics compared to a broad, even distribution from paraffins to aromatics in the petroleum oil. The coal oil has three-ring alicyclics without side chains. Benzene ring type compounds represent both aromatics groups. The decahydronaphthal-

Table 29.64 Catalytic Cracking of Pyrolysis Coal Oil

Properties	Catalyst		
	Silica-alumina	Filtrol	Filtrol
Temperature (°C)	460	460	510
Liquid product yield (wt %)	86.0	69.2	62.3
Naphtha (wt %)			
In feed	11.2	11.2	11.2
In product	26.1	36.0	40.8
Total yield	22.4	24.9	29.9
Net yield	11.2	13.7	18.7

Source: Modified from Ref. 150.

Table 29.65 Steam Pyrolysis Results: Yields in Weight Percent of Products[a]

Product	H-Coal Naphtha, IBP–177°C	H-Coal Mid-Distillate, 177–343°C	COED Mid-Distillate, 177–343°C
Hydrogen	0.8	0.7	0.9
Methane	16.3	12.1	11.8
Ethylene	23.4	14.4	13.9
Propylene	8.5	5.9	6.5
Butadiene	3.0	1.5	1.3
Benzene[b]	23.6	18.5	9.9
Fuel	24.4	46.9	55.8

[a] Calculated yields assuming recycle gas extinction and hydrodealkylation of toluene and C_8 aromatics.
[b] Includes toluene and xylene equivalents.

Source: Modified from Ref. 151.

ene was chosen as representative of a typical coal compound.

Table 29.66 reveals that the model compound and coal liquid produced about the same ethylene yield, but less than that from the petroleum stock. The coal liquid yielded about twice as much BTX as did the petroleum. The model compound was a good choice of an analogous pure compound in that it gave an excellent match of results with the coal liquid. Mechanistic studies of these results suggest that hydroaromatic species in the coal oil are pyrolyzed to polynuclear aromatics, indicating that coal oils should be fully hydrogenated to alicyclics to prevent partially hydrogenated compounds, hydroaromatics, from reverting to polynuclear compounds. The coal liquid pyrolyzed in this study had first been hydrogenated to provide the large percentage of alicyclics in the pyrolysis feedstock. The conclusion to be drawn is that coal liquids should be carefully and completely hydrogenated to reduce the aromatics to near zero, thereby increasing the yield of olefins upon pyrolysis.

An oil obtained from zinc chloride hydrocracking of a Utah coal followed by hydrogenation of the −360°C fraction was used as a feedstock for steam pyrolysis in a small laboratory reactor.[152] Pyrolysis was performed at 700°C, at near atmospheric pressure. The feed and products shown in Table 29.67 reveal that 24 wt % of the feedstock is converted to ethylene and propylene and that 51 wt % is BTX. Little other information is available.

29.4.5 Reforming

Liquids, whether coal or petroleum, require pretreatment before they can be processed over reforming catalysts, especially bimetallics. Although few data are available for assessing the reforming potential of coal-derived naphthas, those that exist suggest that reforming can be achieved. The preferable catalyst is probably the platinum on alumina rather than the bimetallic catalysts, because of the severity of prehydrotreatment of feeds necessary to achieve the very low sulfur and nitrogen levels for bimetallic catalysts. In addition, the use of bimetallic catalysts is questionable when processing liquids with a very high aromatics content. The rate of aging of platinum catalysts is not yet clearly established, although they may age more slowly than when processing some petroleum naphthas.

Evaluation of reforming of naphthas obtained from a COED oil from Illinois No. 6 coal revealed excellent aging characteristics for a platinum catalyst.[99] This catalyst was a 1/16-in. diameter extrudate with 0.35 wt % platinum and 0.35 wt % chlorine with the balance of η and γ aluminas. Its total pore volume and surface area were 0.57 cm^3/g and 430 m^2/g, respectively. A bimetallic catalyst would offer no advantage because even after prehydrotreating, the naphthas still contain rather high sulfur and nitrogen concentrations. Reforming feedstocks are shown in Table 29.68, along with a midcontinent petroleum naphtha for comparison. The COED naphthas were treated at 399°C, 1000 psig, and 2 hr^{-1} WHSV, over an HDS-3A catalyst to reduce their sulfur and nitrogen concentrations from ranges of 93–180 and 87–128 ppm, respectively. The reformer feedstocks contained less than 1.0 ppm sulfur and 1 or 2 ppm nitrogen. In the coal liquid feeds the naphthenes and aromatics were significantly higher than in the petroleum naphtha.

In catalyst aging tests at 482°C for the light naphtha and 510°C for the other two stocks, 300 psig, 4 hr^{-1} WHSV, and a 3 : 1 hydrogen-hydrocarbon mole ratio, the coal liquids proved to age the catalysts significantly more

[152] S. A. Qader, "Catalytic Hydrocracking of Coal for Chemicals and Clean Fuels," paper presented at American Institute of Chemical Engineers 82nd National Meeting, Atlantic City, N.J., August 29–September 1, 1976.

Table 29.66 Summary of Results of Steam Cracking Tests on Various Liquids

	Feedstock Sample		
	Hydrogenated Coal Liquid	Light Gas Oil Petroleum Distillate	Decahydro-naphthalene
Specific gravity at 15°C/15°C:	0.895	0.828	0.883
Hydrogen (wt %):	12.76	13.70	13.12
Carbon (wt %):	87.14	86.16	86.88
Hydrocarbon group types (wt %)			
n-Paraffins	0.0	23.50	0.0
i-Paraffins	1.0	33.40	0.0
Alicyclics	83.1	23.00	100.0
Aromatics	12.8	20.10	0.0
	Distillation		
Fraction (vol %)		*Temperature (°C)*	
0 (IBP)	184	193	187
10	245	238	
30	268	271	
50	285	293	191
70	304	318	
90	333	351	
100 (EP)	348	389	196
Product Yields (wt %)			
Hydrogen	1.03	0.60	1.05
Methane	11.66	12.00	10.35
Acetylene	0.49	0.39	0.54
Ethylene	18.83	22.66	19.36
Ethane	2.40	3.96	2.86
Propadiene, propyne	0.58	0.41	0.29
Propylene	7.83	13.02	6.57
Propane	0.21	0.34	0.20
1,3-Butadiene	4.68	4.82	4.65
Butenes	1.63	3.12	1.41
C_5 olefins, diolefins	2.56	2.72	2.24
Benzene	12.15	6.71	17.97
Toluene	6.76	3.51	10.34
C_8 aromatics	3.43	2.11	4.88
Other C_6–C_8	0.48	0.50	0.42
C9+ to 205 C	2.82	2.61	5.21
Pyrolysis fuel oil (−205 C)	22.46	20.51	11.66
Total	100.00	100.00	100.00

Source: Modified from Ref. 64.

slowly than the petroleum naphtha. For corrected initial octane numbers (RONC) of 100, the aging rates in units of octane number change per 100 hr of operation (RONC/100 hr) were −0.24 for the lighter coal naphtha, −2.2 for the heavier coal naphtha, and −8.4 for the petroleum naphtha. These activity decay rates for coal naphthas are significantly lower than for the petroleum naphthas, and they indicate that catalyst deactivation would certainly not be a limiting factor.

Studies of reforming process variables were made over a temperature range of 449–482°C, at 300 psig, 4 hr^{-1} WHSV, and at a 7 : 1 mole ratio of hydrogen to hydrocarbon. Typical results are given in Table 29.69 for 482°C operation. The reformate yields are rather high, especially considering that the

Table 29.67 Results of Steam Cracking of an Oil Produced from Utah Coal

Feed properties	
Gravity	18 °API
Carbon	88.89 wt %
Hydrogen	11.11 wt %
Product distribution (wt %)	
Hydrogen	0.5
Methane	8.0
Acetylene	0.2
Ethylene	15.0
Propylene	9.0
Butanes	4.3
BTX	50.7
Naphthalenes	12.3
BTX analysis (vol %)	
Benzene	51
Toluene	20
Ethyl benzene	8
Xylenes	21

Source: Modified from Ref. 152.

midcontinent naphtha yield is only about 74%. Hydrogen production is higher for the coal liquids at around 1400–2000 scf/bbl compared to only 1100–1400 scf/bbl for the petroleum liquid. The improvement in the coal reformate results from converting naphthenes (73 vol % in the feed vs. 5 vol % in the reformate) to aromatics (20 vol % feed vs. 83 vol % products).

Another study of reforming COED naphthas, prehydrotreated, revealed satisfactory results.[145] These naphthas were essentially −177°C cuts; partial properties are shown in Table 29.70, along with products. Hydrotreated naphthas from straight-run COED oil and from hydrocrackate were assessed at conditions of 510–520°C, 250 and 500 psig, and 2.0 hr^{-1} LHSV, using a platinum reforming catalyst, American Cyanamid PHF-4. The feestocks contained about 6 ppm nitrogen and 2 ppm sulfur. The aromatics increased from around 30 vol % in the feeds

Table 29.68 Properties of Reforming Feedstocks

	Midcontinent Naphtha[a]	Hydrotreated COED, 82–188°C	Hydrotreated COED, 82–199°C
	ASTM Distillation		
Fraction		Temperature (°C)	
IBP	93.3	93.3	94.4
5	104	105	104
10%	109	109	109
20%	113	115	117
30%	115	121	123
40%	124	128	131
50%	128	135	140
60%	134	144	148
70%	147	153	158
80%	157	161	167
90%	169	170	177
95%	177	177	182
EP	200	191	202
	Properties		
Carbon (%)	85.40	86.64	86.73
Hydrogen (%)	14.60	13.36	13.27
Sulfur (ppm)	1	< 1	< 1
Nitrogen (ppm)	2	1	2
Paraffins (vol %)	47.8	7.1	7.1
Naphthenes (vol %)	42.8	73.7	72.7
Aromatics (vol %)	9.4	19.2	20.2
RONC	46.4	68	69

[a] Gravity = 55.6 °API.

Source: Modified from Ref. 99.

Table 29.69 Results of Reforming Test of Naphthas Derived from Coal Liquids

	Naphthas	
Properties	Light COED	Heavy COED
C_3 and lower (wt %)	4.87	4.69
C_4 (wt %)	0.36	0.50
C_5 and over (wt %)	94.77	94.81
C_5 and over (vol %)	89.0	89.3
Hydrogen produced (scf/bbl)	1960	1858
Reformate (vol %)		
Paraffins	11.1	11.5
Naphthene	5.4	5.8
Aromatics	83.5	82.7

Source: Modified from Ref. 99.

Table 29.70 Properties of Feedstocks and Products from Reforming Tests

Properties	Feed	Product	Feed	Product
Temperature (°C)		513		520
Pressure (psig)		250		500
LHSV (hr^{-1})		2		1.9
Hydrogen produced (scf/bbl)		1160		980
C_1–C_3 (vol %)		7.5		6.2
C_4–C_5 (vol %)		5.2	0.4	5.8
C_6–C_9 (vol %)				
Paraffin (vol %)	10.9	4.0	13.3	6.4
Naphthene (vol %)	50.1	0.9	35.7	2.2
Aromatic (vol %)	28.7	73.4	33.4	70.2
C_{10} and over (vol %)	10.3	6.0	17.2	6.6

Source: Modified from Ref. 145.

to more than 70 vol % in the products, again almost entirely from the naphthenes. Operation at a pressure of 250 psig revealed rather rapid catalyst aging in contrast to the results with COED oil just discussed and to operation at 500 psig (see Table 29.70). No specific times and aging tests were available over the duration of reforming the straight-run naphtha; however the aromatics content fell from 73 to 53 vol %, indicating rapid decay rates, probably over 100 or 150 hr of operation.

CHAPTER 30

Conversion of Coal and Gases Produced from Coal Into Fuels, Chemicals, and Other Products

HARALD JÜNTGEN, JÜRGEN KLEIN, KARL KNOBLAUCH, HANS–JÜRGEN SCHRÖTER, AND JOACHIM SCHULZE

Harald Jüntgen *is Director of Steinkohlenbergbauverein, Prokurist of Bergbau-Forschung GmbH, Prokurist of Bergwerksverband GmbH, Essen, and apl. professor of University of Essen. Dr. Jüntgen's principal areas of work are research in coal gasification and other coal conversion processes, and research and industrial application of adsorption processes based on carbonaceous adsorbents.*

Jürgen Klein *is Manager of Bergbau-Forschung GmbH, Essen. Dr. Klein directs basic research on the coal conversion and adsorption processes.*

Karl Knoblauch *is Manager of Bergbau-Forschung GmbH, Essen. Dr. Knoblauch's areas of interest are environmental techniques in coal processing and adsorptive gas separation processes.*

Hans-Jürgen Schröter *is a scientist with Bergbau-Forschung GmbH, Essen. Dr. Schröter performs research on coal gasification.*

Joachim Schulze *is professor of chemical engineering economics at the Technical University of Berlin, Institute of Technical Chemistry. Dr. Schulze's main fields of activity are project engineering and cost estimation of chemical processes, especially for coal conversion.*

30.1 INTRODUCTION

The most important nonfuel use of coal is in the reduction of iron ores, mainly by coke and in some special new processes by coal directly. The main process for the production of iron is still the blast furnace. The worldwide coke consumption for this purpose is estimated to be 250×10^9 kg, with consumption in the European community of approximately 44×10^9 kg, in Japan 42×10^9 kg, and in the United States 39×10^9 kg. Since coking coal is scarce and therefore relatively expensive, the tendency to use other reducing agents to modify the processes for producing raw iron is growing. One possibility is to use in the blast furnace other supplemental fuels, such as coal, oil, or gas, to reduce the specific coke consumption for the production of one ton of iron. In this way coke consumption has decreased from approximately 0.7 kg per kilogram of pig iron in 1960 to approximately 0.5 kg/kg in 1975. Another possibility is to manufacture coke from materials other than coking coal or from mixtures of coking and other coal, and to use these products in the conventional blast furnace. In this connection the development of formcoke processes is very important (Chapters 11, 14, and 15). Test runs in large blast furnaces have shown that formcoke is essentially suitable for use in blast furnaces. Thus the construction of first plants for the manufacture of formcoke on a technical scale is under way. A third very different possibility is to develop new processes for reducing iron ores in which coal or reducing gases can be used directly for the reduction to produce sponge iron. Many so-called direct reduction processes have been suggested, and some of these have been put into operation. In this field at present 63% of the installed plant capacity is based on reduction by gases and the remainder on reduction by coal. It is expected that the proportion of gas-using processes will rise in the near future. Since in some regions of the world natural gas is not available today or will not be available in the future, coal as feedstock for the production of reducing gas may lead to a higher indirect use of coal for direct reduction processes. The coal used in 1975 for the production of sponge iron can be roughly estimated at 2.8×19^9 kg. Plants have been built principally in Brazil, South Africa, and New Zealand; however interest in these processes is also increasing in the United States, Canada, and Japan.

Compared with the use of coke and coal as reducing agents for iron ores, the consumption of such materials for the manufacture of nonferrous metals, metal oxides, or chemical products is not very significant. For example, in the European community in 1974, 0.9×10^9 kg of coke was used in the nonferrous metal industry, compared with 44.8×10^9 kg for the reduction of iron ores. Additional coal is used in the metallurgical industry for the carburization of iron. In some cases (e.g., in the production of aluminum) the reduction of metal oxides to the metals is done by electrolysis. The electrodes used in such processes are made from binder pitch and pitch coke. Coal as coke can be used as a reducing agent in the chemical industry (e.g., for the production of phosphorus and phosphoric acid from phosphate rocks) and for the manufacture of some sulfites or sulfides from the corresponding sulfates (e.g., NaS and Bas).

A very different use of coal is as a feedstock for the chemical industry. As demonstrated previously—for example, in Germany in the 1930s—most chemicals can be produced from coal, water, and calcium. Today oil products (e.g., naphtha) and to some extent natural gas are the most important feedstocks used in the chemical industry worldwide. But in some countries that have no oil or natural gas resources, but do have inexpensive coal, the chemical industry is still partly based on coal. In South Africa the Fischer-Tropsch process, which uses synthesis gas made from coal by gasification is important not only for the production of chemicals but also for the manufacture of gasoline. Coal gasification to produce synthesis gas is important to the ammonia and consequently to the fertilizer industry in India, Pakistan, South Africa, Turkey, Greece, Finland, Japan, and other countries. It can be estimated from the operating gasification plants that worldwide approximately 4×10^9 kg/year CE (coal equivalent) (1 kgCE = 29.3 $\times 10^3$ kJ = 27.8 $\times 10^3$ Btu) are used for the production of 6×10^6 kg NH_3/day and 3.2×10^9 kg of South African coal for the production of town gas and gasoline. The consumption of coal for these purposes will rise significantly if in 1982 the Sasol II plant goes into operation in South Africa. Also, in South Africa and in some other countries acetylene made from coal via CaC_2 is converted to polyvinylchloride, crotonaldehyde, and chlorinated hydrocarbons. Approximately 2.7×10^9 kg of coal as coke was converted to CaC_2 in 1975; approximately 50% of this was processed into acetylene (C_2H_2). The other part was used for the production of $Ca(CN)_2$, an important product for the fertilizer industry.

Furthermore, products from coke-oven plants are used in the chemical industry worldwide, especially in the United States, West Germany, and Japan. Thus coke-oven gas can be processed via low-temperature separation or adsorption processes to yield pure hydrogen. Furthermore, benzene, phenols, NH_3, $(NH_4)_2SO_4$, S, or H_2SO_4, and, as a special product NH_4SCN, are produced by treatment of coke-oven gas or condensates (Chapter 18). The tar from the coke-oven plants is a very important source of aromatics such as naphthalene, which is oxidized to phthalic anhydride, anthracene, and so on. A special fraction of the tar distillate is used as feedstock to make carbon black. Pitch is important as binder for briquettes (Chapter 11) and other products such as carbon electrodes, and it can be converted into pitch coke, which is used as raw material for carbon electrodes and artificial graphite. These applications, however, have to compete with petroleum coke, which is produced worldwide in greater quantities.

In view of rising oil prices since 1973,

research and development on the use of coal as a feedstock for the chemical industry is intensifying. Several routes to process coal into basic chemicals are being considered. The production of chemicals from coke-oven plants is strongly dependent on coke production. Thus an expansion of the production of chemicals from this source is possible only if the demand for coke increases. Therefore consideration is being given to devolatilization of coal before burning, to recover gaseous and liquid products and to obtain coke or char with a sulfur content lower than that of the original coal. The gases and liquids can be converted to chemicals if desired, and the coke or char can be burned without flue gas desulfurization. Suitable processes for this pretreatment such as fluid bed carbonization or shock heating with solid heat carriers have been demonstrated on a small scale (Chapters 12 and 13).

Furthermore gasification of coal to make synthesis gas gives great flexibility in the range of products that can be made in the future. Thus methanol plants based on coal are being considered even in some highly industrialized countries such as the United States. Synthesis gas could also be an important source for the production of hydrogen. Furthermore, new developments in the field of more selective catalysts (e.g., in the highly efficient Mobil Oil process for converting methanol into a hydrocarbon mixture having high octane number and suitable for use as gasoline) give the hope that synthesis gas can be used for the production of other basic chemicals such as ethylene (C_2H_4), which could compete with ethylene from naphtha if oil prices rise further.

The hydrogenation of coal was also used in Germany in the 1930s and 1940s to convert coal into a gasoline with high octane number. At present, however, hydrogenation plants are not operating worldwide. But research, especially in the United States, the United Kingdom, West Germany, and Japan, is underway to elucidate the role that hydrogenation and extraction of coal could have in the future to provide feedstocks for the chemical industry that would be as universal as crude oil.

The current production of C_2H_2 from CaC_2 made from coal offers a different route for processing coal into basic chemical products. That use of coal could be expanded by developing a technical process for direct conversion of coal into C_2H_2 using extremely high temperatures and heating rates, as in electric arcs or plasma jets. Research is underway, but no pilot plant has been operated yet to demonstrate these processes.

At present it is very difficult to decide which route for transforming coal to chemicals would have the greatest chance in the future. The basic differences between these four approaches are as follows:

1. The coke-oven plant route is undesirable because of the strong relationship to coke production and is not discussed further.

2. The direct conversion of coal into C_2H_2 would provide the most universal, basic feedstock with a very high energy of formation. Thus most chemicals could be produced directly from that hydrocarbon. The disadvantages are high energy consumption for the conversion of coal into C_2H_2 and the production of by-products such as carbon black.

3. The main constituents of synthesis gas, H_2 and CO, have a high potential for the production of many important chemicals. Moreover, synthesis gas could be of great importance as a source of H_2. The main advantages of this route are that technical processes are now available to produce synthesis gas on a large industrial scale, and that the synthesis of distinct desirable products can be done with high efficiency. However if products containing no oxygen are synthesized, the part of the coal that has to be oxidized by pure oxygen to CO_2 to provide the heat of the strongly endothermic steam-carbon reaction is lost, and this could be a disadvantage.

4. There is the hope that hydrocarbons and especially aromatics could be produced more economically by direct hydrogenation of coal rather than by the route using synthesis gas. In this connection, processes in which coal is only extracted could be important. However the technical problems of refining the crude hydrogenation product from coal have not yet been solved. Also, questions of the best source of the hydrogen needed and of the uses of the residues are open.

Other special uses of coal are in the direct production of carbonaceous products such as graphite, carbon electrodes, plastic materials, carbon fibers, carbon black, fillers, and adsorbents such as activated carbon and molecular sieves. There are direct routes for the production of adsorbents, fillers, and plastic materials from coal. In the case of carbon fibers, carbon black, carbon electrodes, and graphite the route goes via special fractions of coal-tar distillates.

In coal conversion processes such as burning or gasification of coal, by-products such as coal ash are produced. The so-called fly ash coming out of dry deashing processes can be used as raw material for bricks. Similarly, the slag product of liquid deashing is used in the building industry. At present in industrial countries such as the United States, Japan, and West Germany, flue gas desulfurization plants are operating or are under construction. The most important products from these plants are wastes containing mixtures of $CaSO_3$, $CaSO_4$, and water or gypsum. In the future, elemental sulfur is expected to be the final product.

30.2 USE OF COAL AND COKE FOR REDUCTION

30.2.1 Reduction of Iron Ore

Bituminous coal is still an extremely important raw material for the steel industry. Blast furnaces produce 99% of the world's production of iron, and about 90% use metallurgical coke as a reducing agent and fuel.[1]

In view of the scarcity of coking coal, great efforts are being undertaken to reduce dependence on it. New methods of coke-making that permit the use of coals other than coking coals are being developed, and other reduction processes and additives are being tested.[2,3]

Two routes for the production of steel are presently indicated and are shown in Fig. 30.1. Basically they differ in how the oxygen content changes as the ore is converted to steel. In the blast furnace process the ore is first reduced to pig iron, which is converted into steel in a top-charged oxygen converter and finally is cast into ingots by use of a ladle. In the direct reduction process the ore is reduced to iron sponge in the direct reduction furnace, then melted in the electric furnace, and also cast. The discussion that follows investigates the coke and coal consumption, respectively, of these two routes.

30.2.1.1 *The Blast Furnace*

For a long time the largest use of coal has been in the classic metallurgy of iron ore reduction in the blast furnace, which uses more than 80% of the worldwide coke production. As noted earlier, from 1960 to 1975 the specific coke consumption for the production of pig iron in the blast furnace decreased from about 0.7 to about 0.5 kg/kg pig iron. In spite of this, the coke consumption during these years remained nearly constant, at about 250×10^9 kg/year, since the pig iron production increased from 400×10^9 to 660×10^9 kg/year.

The manufacture of metallurgical coke for use in the blast furnace is described in detail in Chapter 14 (also see Chapter 15). Here we cover only the processes that are important for evaluating the specific coke consumption in the use of coke for reduction. In addition, however, the coke properties that permit optimum use in the blast furnace are elucidated.

Characteristics of Coke and Ores

Coke Function. In the blast furnace, coke is used not only for increasing the carbon content of the iron but also as fuel, reducing agent, and support for the burden. Coke can fulfill these roles only if it has certain required chemical and physical properties.[5,6] The function of the coke has three principal aspects, as follows:

1. *It serves as fuel.* Combustion of coke in front in the hot blast produces a gas that ascends from the tuyere zone and transmits heat to the burden. It thus supplies the heat

[1] H. Messerschmidt, in *Bergbau—Rohstoffe—Energie,* Vol. 10, Verlag Glückauf, Essen, 1976, pp. 12–41.

[2] W. Peters and J. Lehmann, in *Bergbau—Rohstoffe—Energie,* Vol. 10, Verlag Glückauf, Essen, 1976, pp. 42–56.

[3] P. Speich, in *Bergbau—Rohstoffe—Energie,* Vol. 10, Verlag Glückauf, Essen, 1976, pp. 90–118.

[4] L. v. Bogdandy, *Stahl Eisen,* **92**(22), 1069–1084 (1972).

[5] L. v. Bogdandy and H. J. Engell, *The Reduction of Iron Ores,* Springer Verlag, Berlin–Verlag Stahleisen GmbH, Düsseldorf, 1967.

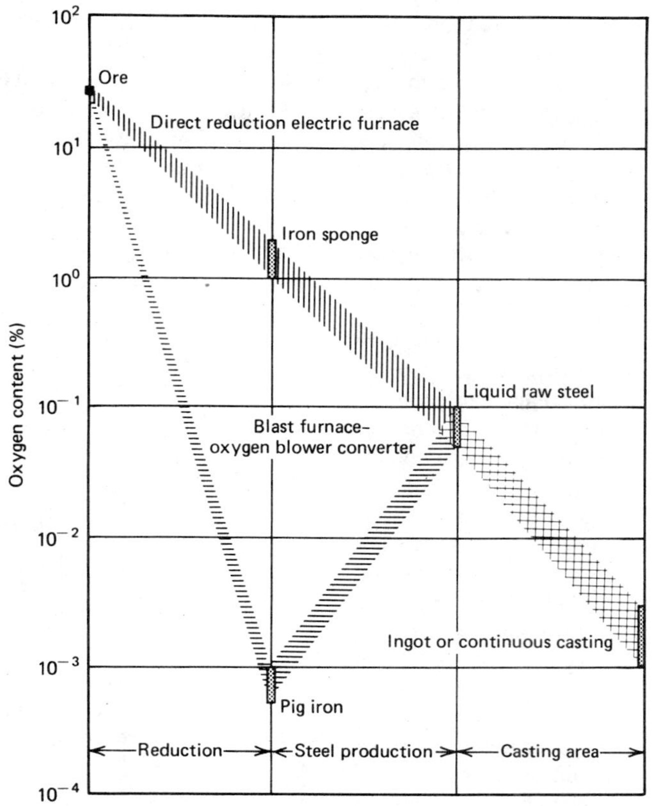

Figure 30.1 Change in oxygen concentration in the production of steel.[4]

required by all endothermic reactions for the preheating and melting of the feed materials. Because of the Boudouard equilibrium (see Equation 30.11), carbon is only oxidized to CO.

$$C + \tfrac{1}{2} O_2 \dashrightarrow CO \quad \Delta H_{298\,K} = -9204 \text{ kJ/kg-°C} \qquad (30.1)$$

2. *It is used as a reducing agent.* Carbon monoxide as well as some hydrogen contained in the gas formed in the tuyere zone are used for reduction according to the following equations:

$$Fe_nO_m + mCO = nFe + mCO_2 \qquad (30.2)$$

$$Fe_nO_m + mH_2 = nFe + mH_2O \qquad (30.3)$$

whereby the iron oxides contained in the ore are reduced to metallic iron.

3. *It gives support to the burden.* Because of its excellent strength at high temperature, the coke assures gas passage through the blast furnace and thereby makes the process possible, especially in those zones where all other components are softened or molten. Since the properties of coke are important for fulfilling these different requirements, the most important properties are listed in Table 30.1 and subsequently discussed in detail.

Coke Quality.[6,7] In the blast furnace process, in addition to the small content of hydrogen, only the carbon content of the coke is used. Therefore a high carbon content and a low content of moisture, ash, and sulfur are desired. It is important to have uniformity of these properties, with only small variation, since otherwise the heat balance of the blast furnace is disturbed and changes in the pig iron analysis occur. Variations of the mois-

[6] G. Winzer and K. Engel, *Glueckauf,* **108**(24), 1124–1131 (1972).

[7] M. Ottow, "Iron," in *Ullmanns Encyklopädie der technischen Chemie,* 4th Ed., Vol. 10, Munich, 1975, pp. 384 ff.

ture content are usually concurrent with variations in both the strength and size consist of the coke. They are therefore liable to lead to disturbances in the passage of the gases. The ash content influences the input of heat and the composition of the slag, and therefore may change the pig iron analysis. A low sulfur content is important not only for the production of steel but also for the further use of the slag and for the use of coke breeze in sintering plants.

The size consist of the coke and its strength are of decisive importance to its role as support for the burden. To achieve even gas flow, a large portion of large-size coke is required. On the other hand, the size consist of the coke must mesh with the small size ore and sinter. Worldwide, in large blast furnaces, size fractions from 20 to 80 mm are used. The content of lumps above 50 mm should be 50–55%. Individual coke particles should be without fissures and as close to cubical as possible. Coke sizes above 80 mm are useful only if they are entirely free of fissures. This usually cannot be achieved without stabilization.[8] To achieve low coke consumption and to avoid breakdowns, high strength of the coke is indispensable. The evaluation of the coke strength must take into consideration the size that is used.[9] The tumbler test,[6] which is standardized in individual countries, is used for this purpose. Examples of national standards include DIN 51 717 (German Institute of Standardization); ISO/R 556-1967 (International Organization for Standardization); AFNOR-NF M 03/023/1962 (French Association of Standardization); ASTM D294-1964 (American Society of Testing and Materials); and JIS K 2151-1960 (Japanese Industrial Standard Committee).

Other standards for determining the properties of metallurgical coke include:

Ash content	DIN 51 719; ISO/R 158-1960; ISO/TC 27 N 874-1966
Sulfur content	DIN 51 724; ISO/R 926-1969
Moisture content	DIN 51 718; ISO/R 579-1967
Ash fusion behavior	DIN 51 730/ISO 540-1974
Reducibility	ISO/TC 102

The reactivity of the coke is also important to the specific coke consumption. It should conform to the reducibility of the ore and in general should be low. Several proposals have been made[10,11] for determining the reactivity of coke, and ISO has promulgated the draft standard ISO/TC 27/WG 8-N175.

Properties of the Feed Ore.[7] Pretreatment of the ore, which removes the less valuable compounds in the iron oxide, thus decreases

[8] H. Echterhoff, *Brennstoff-Chemie*, **49**(10), 294–302 (1968).

[9] H. v. Ende and K. Grebe, *Stahl Eisen*, **92**(7), 303–314 (1972).

[10] K. Hedden, *Chem.-Ing.-Techn.*, **30**, 125–132 (1958).

[11] G. Heynert, W. Zischhale, and E. Schurmann, *Stahl Eisen*, **80**, 981–990 (1960).

Table 30.1 Characteristics of Blast Furnace Coke

	Metallurgical Coke		
Property	I > 80 mm	II > 60 mm	III > 40 mm
Tumbler strength (%)[a]	85	85	85
Abradability, M_{10} (%)[a]	6	6	6
Moisture (% of raw product)	1.5 ±0.5	2.5 ±0.5	3.5 ±0.5
Ash (%)	8.5 ±0.3	8.5 ±0.3	8.5 ±0.3
Sulfur (% of dry product)	0.9 ±0.03	0.9 ±0.03	0.9 ±0.03
Volatile matter (% of dry product)	< 1	< 1	< 1
Alkali (% of dry product)[b]	0.26	0.26	0.26
Bulk density (kg/m^3)			438

[a] According to DIN 51 717, conforming with ISO/R—1967.

[b] Content of Na_2O and K_2O determined in different plants.

the amount of ballast that passes the reactor, is important for every reduction reactor. Therefore preparation processes have gained in importance. Of the great number of ore preparation processes, three have become most important, and treat 95% of all ores[12]:

1. Humphrey separator for the preparation of hematitic ores.
2. Low field intensity magnetic separator for the preparation of magnetite ores.
3. Flotation for the separation of the phosphorus-containing apatite and the taconites from magnetite and hematite.

The concentrate obtained in the preparation plant usually cannot be used directly for reduction but must be further treated by suitable means. Lump ore must be brought to the desired size by crushing and screening, and it may be used directly or after renewed agglomeration in the reduction reactor. Since increasing quantities of ore concentrates are being used, a greater proportion of the ore must be sintered or pelletized. The proportion of sinter and pellets in the total ore feed was about 70% worldwide in 1973. This is mainly because of the better metallurgical properties of these particles in comparison with the original ore. Several tests have been developed to determine the suitability of ores for processing in the blast furnace. They determine the reduction behavior at low and high temperatures.[5,13,14]

Heat and Material Balance of the Reduction Process.[5,7,15] Iron occurs in the form of the following oxides: Fe_2O_3 (hematite), Fe_3O_4 (magnetite), and FeO (wustite). Assuming stoichiometric conversion in the reduction to 1 kg of iron, the following amounts of oxygen must be removed: 0.286 kg in the wustite stage, 0.382 kg in the magnetite stage, and 0.429 kg in the hematite stage.

The general reduction equation:

$$Fe_nO_m + m(CO, H_2) \text{----} > nFe + m(CO_2, H_2O) \quad (30.4)$$

can be given separately for the three iron oxides:

$$3Fe_2O_3 + CO = 2Fe_3O_4 + CO_2 \quad (30.5)$$

$$Fe_2O_3 + CO = 2FeO + CO_2 \quad (30.6)$$

$$\Delta H_{298\,K} = +38.1 \text{ kJ/kg } Fe_2O_3$$

$$Fe_3O_4 + CO = 3FeO + CO_2 \quad (30.7)$$

$$\Delta H_{298\,K} = +149.5 \text{ kJ/kg } Fe_3O_4$$

$$FeO + CO = Fe + CO_2 \quad (30.8)$$

$$\Delta H_{298\,K} = -229.4 \text{ kJ/kg FeO}$$

The wustite, which is especially important for iron ore reduction, may occur with different values for structural disorder, *y*, which means that Equation 30.8 reads:

$$Fe_{1-y}O + \text{---->} (1-y)Fe + CO_2$$

$$\text{for } y = 0.04, \ldots, 0.125 \quad (30.9)$$

where *y* indicates empty spaces in the iron lattice of the wustite, which corresponds to an excess of oxygen.[5] In the blast furnace CO is formed by combustion of the fuel coke according to:

$$C + ½ O_2 \text{---->} CO \quad (30.10)$$

and at temperatures above 900°C, according to the Boudouard reaction:

$$C + CO_2 <====> 2CO \quad (30.11)$$

in which the CO_2 required for the Boudouard reaction is obtained by reduction of the iron oxide according to Equations 30.5–30.8.

The reduction reactions occur at temperatures between 500 and 1300°C and depend on the local concentration of CO in the shaft gas. The relation between temperature, CO and CO_2 concentrations, and the degree of oxidation of the iron establishes the so-called Baur-Glaessner diagram, which is plotted in Fig. 30.2 in modified form. It shows the areas of stability of the iron oxides and the iron in

[12] A. Stieler, "Magnetscheidung," in *Ullmanns Encyklopädie der technischen Chemie,* 4th ed., Vol. 2, Munich, 1972, pp. 86 ff.

[13] R. A. Oliver, I. Iona Steel Inst., London, 205.2 1967, pp. 1131–1135.

[14] A. K. Yakimoff, *South East Asia Iron Steel Inst. Q.,* 1(3), 12–26 (1972).

[15] F. Oeters, "Eisen und Stahl," in *Winnacker-Küchler, Chemische Technologie,* 3rd ed., Vol. 6, Munich, 1973, pp. 368 ff; H. Hoff, O. Heerhaber and O. Meyer, in *Winnacker-Küchler, Chemische Technologie,* 2nd ed., Vol. 5, Munich, 1961, pp. 374 ff.

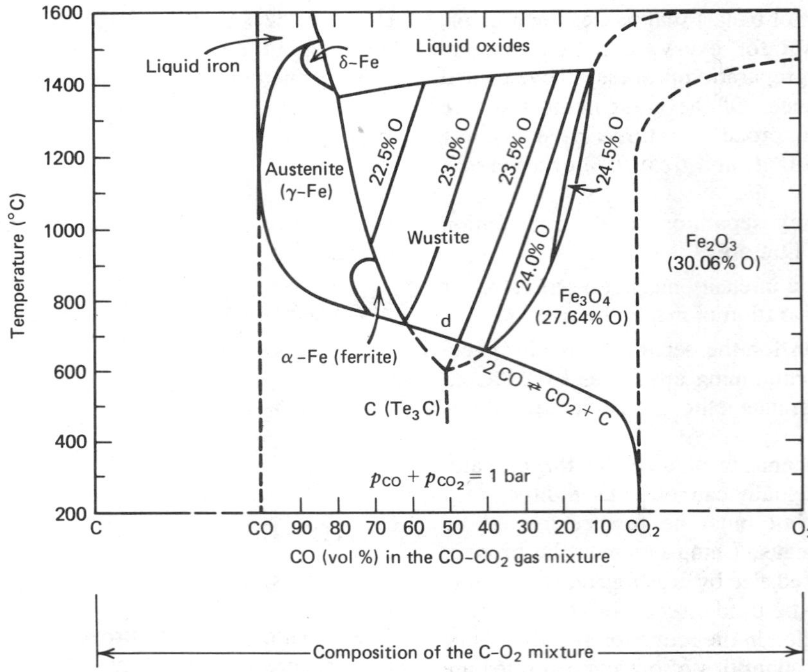

Figure 30.2 Baur-Glaessner diagram with Boudouard curve.[7]

relation to temperature and composition of the CO-CO_2 mixture.[7] In addition, the equilibrium conditions for the Boudouard reaction are given. The countercurrent principle used in the blast furnace leads first to the reduction of the higher oxides at low temperatures. The CO content of the gas is sufficient to achieve this reduction. For the reduction of wustite to iron, there is a limit to the CO_2 content, which at 710°C must be below 42% and at 1000°C below 30%, to achieve reduction. At 710°C the Boudouard equilibrium line crosses the wustite line, so that thermodynamically the CO_2 formed by the wustite reduction may be reduced by the carbon to CO, according to Equation 30.11. Kinetic effects may prevent establishment of equilibrium at this temperature if the metallurgical coke has low reactivity. With increasing temperature, the equilibrium of the Boudouard reaction favors CO formation. Thus above 1000°C, the CO_2 content in the gas is nearly zero; therefore a complete reduction of the wustite to iron is assured in every case.

Figure 30.3, which shows the changes in temperature and gas composition in the blast furnace, can be used to give a simplified description of the reactions that are actually occurring as the burden descends in the blast furnace. The feed material, which consists of coke and the so-called burden rounds (ore and fluxing additives or stone), is heated in the upper part of the furnace by the ascending gas. At temperatures below 500°C the moisture, hydrate water, and CO_2 from decomposed carbonates ($MgCO_3$ ----> MgO + CO_2) are removed. Thus the gas is cooled and leaves the furnace at 150–250°C. In the middle part of the furnace, at temperatures between 500 and 800°C, the reduction of iron oxides according to Equations 30.5–30.8 begins. This so-called indirect reduction is slightly exothermic if the reduction leads to iron itself, although some of the individual reactions (Equations 30.6 and 30.7) are endothermic. The heat that is developed is sufficient to compensate for the heat loss from the furnace. If the burden contains limestone it is decomposed in this zone ($CaCO_3$ ----> CaO + CO_2), for which the heat consumption is 4034 kJ/kg CO_2. This increase in CO_2 content must be compensated by additional CO. Since the Boudouard reaction does not occur for kinetic reasons, this additional CO requirement increases the heat consumption and leads to an increased coke consumption for the decomposition of the limestone (~ 0.8–1.0 kg coke/kg limestone). Thus in this zone of the furnace the

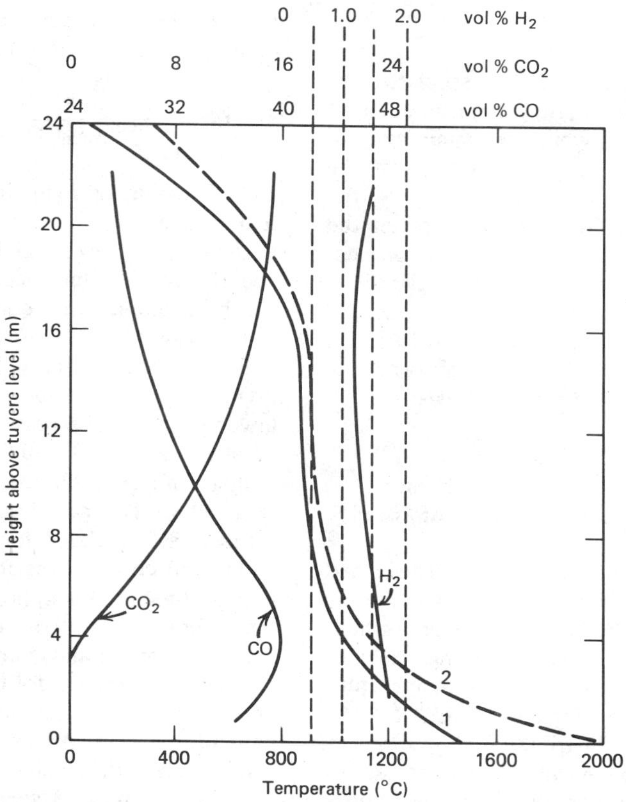

Figure 30.3 Temperature and gas composition in the blast furnace: curve 1, burden temperature; curve 2, gas temperature.[7]

temperature of the gas and burden changes comparatively slowly, while the CO and CO_2 contents change significantly.

In the temperature range 800–1000°C the indirect reduction of wustite is predominant (Equation 30.8), since the higher iron oxides are already almost completely reduced to Fe or FeO. According to Fig. 30.2, thermodynamic considerations require a CO concentration in the gas of 65–70% by volume. However in practice (see Fig. 30.3) this value is not reached. Therefore only a part of the wustite can be reduced according to this exothermic reaction. The Boudouard reaction becomes significant above 950°C, according to Equation 30.11; therefore the CO_2 content is decreased and that of CO increased. Thus in the temperature range 1000–1200°C the iron ore can be completely reduced according to the summary equation:

$$FeO + C = Fe + CO$$
$$\Delta H_{298\ K} = +217.1\ kJ/kg\ FeO \qquad (30.12)$$

This summary equation, which consists of Equations 30.8 and 30.11, is called direct reduction. A significant conversion of CO_2 to CO by the Boudouard reaction occurs only above 950°C, although according to Fig. 30.2 it is thermodynamically possible above 710°C. This is caused by the desired low reactivity of the coke toward CO_2. Thus a greater portion of the oxygen content of the ore is removed by indirect reduction and the gas is better utilized.

Because of the high heat consumption of the direct reduction (~ 4000 kJ/kg pig iron), the ascending gas is cooled substantially as the burden is heated. The composition of the gas is barely changed. The carbon used in the direct reduction is called reduction carbon. Above a temperature of 1300°C the reduction of other components (MnO, P_2O_5, SiO_2) begins and ends when the burden becomes liquid. This begins in the lower part of the direct reduction zone. At 1300°C the iron starts to react with carbon from the coke (carburizing):

$$3Fe + C \dashrightarrow Fe_3C \qquad (30.13)$$

which melts because of the associated decrease in the melting point. Pig iron and slag flow into the hearth while the remaining coke is burned by the hot blast in front of the tuyeres (Equation 30.10). This portion of the coke is called the fuel coke. It supplies the overall heat requirement of the blast furnace.

The total heat requirement of the blast furnace amounts to about 18×10^3 kJ/kg pig iron and must be supplied by the fuel. On the average, 9.2×10^3 kJ/kg is absorbed by the burden accounted for as follows:

40% = heat of reduction
7% = enthalpy of pig iron and slag
5% = amount consumed by heat losses

The remaining 48% of the total heat is tied up as chemically fixed heat (~ 43%) and as sensible heat in the blast furnace gas, which leaves the furnace with a heating value of about 3350 kJ/m³ at standard temperature and pressure (15% CO_2, 25–30% CO, 5% H_2, 50–55% N_2). This distribution shows that to produce iron economically, it is important to reduce the CO content of the gas by using it for reduction and for forming CO_2. A high utilization of the gas requires easy reducibility of the ore and uniform passage of the gas through the column. A low reactivity of the coke with CO_2 moves the beginning of the Boudouard reaction to high temperatures and leads to a greater proportion of indirect reduction, which contributes to the utilization of the gas. A complete conversion of CO into CO_2 is impossible on the basis of the material balance and the reaction equilibrium, since at least two-thirds of the oxygen in the ore must be removed from wustite.

The reaction sequences in the blast furnace are limited by the conditions of material and heat exchange and by the equilibrium position of the iron oxide versus the CO-CO_2 concentration in the gas. Material and heat balances can be used to determine the reaction sequence, thus to permit favorable selection of operating conditions for the blast furnace.

The material balance includes, among others, the following items: (1) amount of oxygen in the feed iron oxide, (2) amount of CO_2 in the blast furnace gas (produced by the removal of oxygen from the ore), (3) proportion of direct and indirect reduction, (4) use factor η in the blast furnace gas:

$$\eta_{CO} = \frac{\%CO_2}{\%CO_2 + \%CO} \qquad (30.14)$$

and (5) extent of reduction and quantity of fuel carbon.

Using the known stoichiometric relations and the equilibrium conditions, these items can be combined into equations for the material balance.

For the heat balance, the blast furnace is divided into two regions: the upper and the lower furnace. The boundary is in the temperature range 800–900°C, in which the enthalpies of gas and burden change very little (Fig. 30.3). The heat balance for the lower furnace is formulated for the portion above 900°C and contains the following items: enthalpy of the hot blast, heat of combustion of the carbon, enthalpy of the pig iron, the slag derived from ore and from coke, heat losses, and heat requirement for the direct reduction of the iron oxide and the elements accompanying the iron. An amount of heat, QN, can be obtained, which is the heat content of the lower furnace. Using the heat balance, QN supplies the total carbon consumption, which includes fuel carbon consumption.

The balance for the upper furnace is prepared in a similar manner. To connect material and heat balance, the heat balance of the lower furnace is usually used. If QN and the hot blast temperature are constant, one degree of freedom remains for the possible reaction sequence in the furnace, which is the utilization of the gas or the proportion of indirect reduction. Both values depend on the reduction rate and can be determined during the operation by analysis of the blast furnace gas. On the basis of such balance models, the operating conditions of the blast furnace are selected.

Development of the Specific Coke Consumption. One basis of the factors affecting the reduction of coke consumption is knowledge of the heat balance in the blast furnace. In modern blast furnaces with large hearth diameters, the heat losses through the wall are very small (~ 3%). Since 48% of the total heat is contained as specific heat or as chemically bonded heat in the blast furnace gas, measures for reduction of coke consumption necessarily require better use of the heat contained in the blast furnace gas or decreasing its amount by improved operating conditions in the blast furnace. The latter goal can be attained by selecting the reducibility of the ore and the reactivity of the coke in such a manner that in the zone of indirect reduction, the CO is used as far as possible. In addition to an increase in the proportion of the indirect reduction of wustite, the contribution of endothermic side reactions (e.g., the decomposition of limestone and the melting of additives) must be kept as low as possible. In the past few years a number of measures have been applied to reduce considerably the specific coke consumption. The most important measures are summarized as follows:

1. Increasing the hearth diameter is a simple construction method to decrease the radiation heat losses and fuel consumption. Since the pig iron output of a blast furnace depends on the amount of fuel that can be gasified per unit of the cross section, reduction of the specific fuel consumption leads directly to an increase in the capacity of the furnace and improves the economics of the process by reducing the capital investment costs. From 1960 to 1975 hearth diameters have increased from an average of 7–8 m to more than 14 m, which led to an increase in the pig iron production of one blast furnace from about 2×10^6 kg/day to more than 10^7 kg/day.[6,16,17]

[16] H. Wysocki, *Stahl Eisen,* **96**(4), 141–145 (1976).

Next to the increase in the size of the furnace, the following measures have yielded considerable fuel reductions:

2. The blast temperature can be increased to more than 1200°C by increasing the proportion of gas used for heating the blast. This leads to a better utilization of the off-gas heat of the furnace, so that overall a fuel reduction of 2% is achieved per 100°C increase of the blast temperature.

3. Improvement of the feed preparation leads to better distribution of the gas and reduction of the ore, which increases the utilization of the gas and the capacity of the furnace. In this category also belong the crushing and screening of all lump ore and the recycling of materials to 8–25 mm. Fine ores are sintered or pelletized. Figure 30.4 shows the change in the composition of the blast furnace feed for low-phosphorus pig iron in West Germany from 1960 to 1975.

4. Variations in the physical and chemical properties of the feed materials can be decreased.

5. The gas pressure in the blast furnace can be increased, making the gas distribution more uniform and leading to a better approximation of the theoretical limit of gas utilization.

6. Oil, gas, or coal can be blown in.[18–20]

[17] L. v. Bogdandy and D. Terlaak, *Stahl Eisen,* **88**(21), 1139–1144 (1968).

[18] T. Ikeshima, "Reduction of the Coke Requirement and New Coking Processes Using Non-Caking Coals," paper presented at Annual Meeting and Conference, Osaka, Japan, 1976.

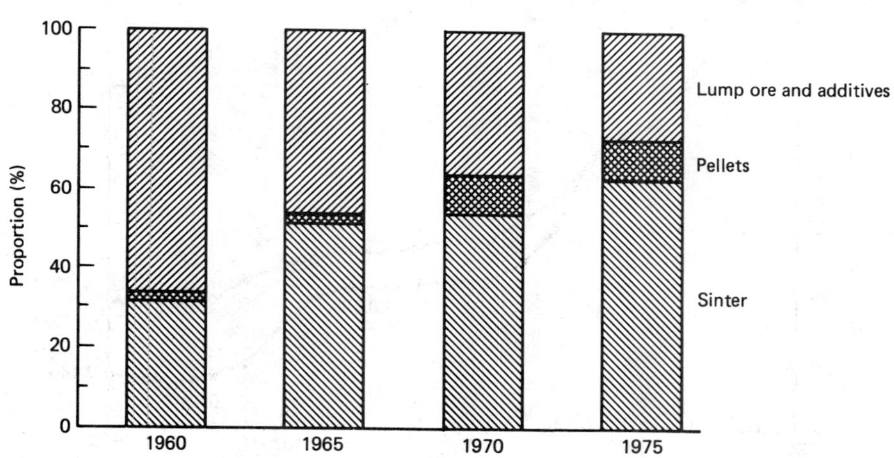

Figure 30.4 Composition of the feed materials for low-phosphorus pig iron in West Germany.[6]

The use of oil has been most successful. The optimum amount at present is about 0.07–0.09 kg/kg pig iron, which can be increased to 0.15 kg/kg if additives are used. This measure decreases the consumption of expensive coke for the melting work.

7. Oxygen can be added to the blast, which leads to a decrease in the amount of gas and in the nitrogen concentration in the shaft and blast furnace gas.

8. Prereduced feed materials can be used. This possibility is in competition with the feeding of a reducing gas to the lower part of the shaft.[21–23]

It is interesting to illustrate this development of reduced coke consumption for pig iron production using Japan as an example.[18] Figure 30.5 shows that between 1961 and 1964 the first major step brought a reduction from 0.6 to 0.5 kg/kg. The most important measures to achieve this improvement were the use of oil, increase of the blast temperature, increase of the blast furnace pressure, and increase of the proportion of sinter and pellets. The second major step occurred between 1968 and 1972 and brought a large decrease: from 0.502 to 0.438 kg/kg. The main reasons for this change were a further increase of the blast temperature and the proportion of sinter, an increase of the backpressure, and an increase of the volume of the blast furnace. Coke consumption in Japan has remained essentially constant since 1972.

An overview of the trend in coke and total energy consumption in pig iron production in the blast furance in the Western industrial countries is given in Table 30.2.

The use of mathematical kinetic models for blast furnace operation permits calculation of the possibility of approaching coke consumption of the thermodynamic minimum.[20,25] Comparison of the calculated values with the operating results shows that in 1975 coke consumption in most blast furnaces was considerably higher than the theoretical minimum coke consumption.

Formcoke as an Alternative to the Classic Metallurgical Coke. The consumption of coking coal for the manufacture of metallurgical coke used in the production of pig iron will increase in the next few years, especially in the "young" steel countries. Despite the

[19] F. W. Luerssen, "Developments for the Reduction of the Requirement of Metallurgical Coal in the U.S.A.," paper presented at Annual Meeting and Conference, Osaka, Japan, 1976.

[20] L. v. Bogdandy, G. Lange, and P. Heinrich, *Stahl Eisen*, **88**(22), 1177–1188 (1968).

[21] J. A. Peart and F. J. Pearce, *J. Met.*, **17**, 1396–1400 (1965).

[22] P. L. Woolf, *J. Met.*, **18**, 99–103 (1966).

[23] F. Moji, S. Otabe, and K. Ohmori, "Injection of Reducing Gas into Blast Furnance," paper presented at Int. Conf. Science Technol. Iron Steel, Tokyo, 1970.

[24] E. Umene, "Energy and Steel," paper presented at Annual Meeting and Conference, Osaka, Japan, 1976.

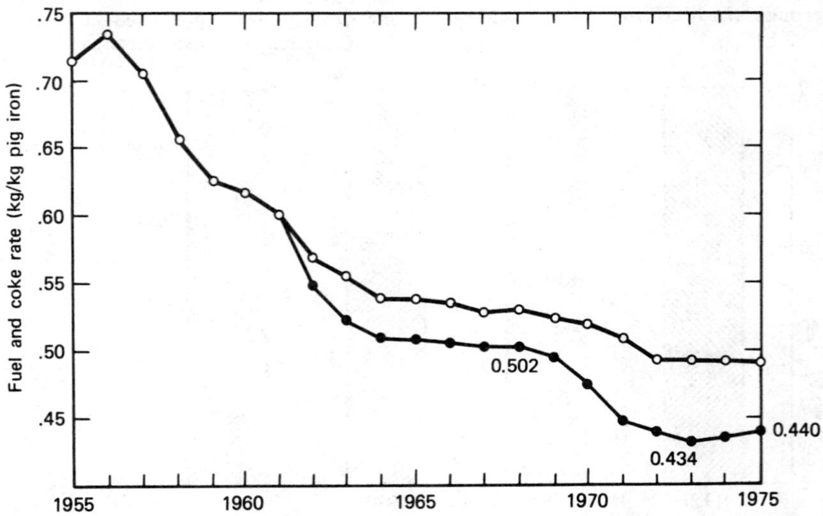

Figure 30.5 Trends in productivity, fuel rate, and coke rate of all blast furnaces in Japan.[18]

Table 30.2 Trend of Coke and Energy Consumption (per Kilogram of Hot Metal) in Ironmaking[24]

	1960		1973	
Country	Coke (kg)	10^3 kJ	Coke (kg)	10^3 kJ
Belgium	0.850	16.8	0.555 (0.580 with oil)	16.9
France	0.970	20.5	0.560 (0.620 with oil)	20.1
Germany	0.825	18.7	0.495 (0.585 with oil)	16.9
Japan	0.615	16.2	0.435 (0.490 with oil)	16.1
Luxembourg	1.060	31.1	0.595 (0.680 with oil)	21.0
United Kingdom	0.825	19.2	0.575 (0.605 with oil)	18.7
United States	0.771	22.6	0.602 (0.617 with oil)	18.2

increased requirement for coking coal, however, worldwide there is only a limited amount of such coal available. Therefore all coal and steel producers are increasing their efforts to develop new technology for the manufacture of blast furnace coke and to obtain a long-range solution to the problems imposed by the limited availability of coking coal.

On the one hand, new methods of conventional discontinuous coke-oven technology are being tried out, to broaden the base of coals suitable for coking; and on the other hand, continuous processes are being developed for the manfacture of formcoke. The object of these processes is to manufacture coke with very little or almost none of the usual coking coal.[18,19,26–28]

Details of formcoke production are given in Chapters 11, 14, and 15. Only the principle is indicated here, and some properties of the product are described.

Of about 20 different proposed processes for formcoke manufacture, only 11 have reached a degree of development that is expected to lead to commercial application (Table 30.3). The main difference from the slot-oven process is the preparation of the desired lump size of the coke by a briquetting process that precedes the final coking. This briquetting can be done using two different process principles: (1) conventional briquetting with the steps of mixing the different coals, drying, grinding, addition of binder, kneading of the briquetting mixture, and pressing; and (2) two-component hot briquetting. In the second process the largest portion of the feed coal (75%) is carbonized at low temperature and the hot char is mixed with a coal (25%) that has softening properties. Heat exchange between these components softens the binder coal, and the plastic mixture is briquetted on a double-roll press at a temperature between 420 and 520°C.

This briquetting process makes it possible to reduce the quantity of binder coal (coking coal) to a minimum of 25%. By adding about 6–9% tar obtained during low-temperature carbonization to the hot briquetting mixture, it is possible to reduce the quality requirements for the binder coal or to reduce the amount of coking coal in the blend.[26] All bituminous coals are suitable for low-temperature carbonization as long as their ash and sulfur contents meet the requirements.

Large-scale experiments have shown that the quality of formcoke made by the processes listed in Table 30.3 is suitable for use in the blast furnace. For testing in the blast

[25] R. Wartmann, *Stahl Eisen,* **95**(10), 442–454 (1975).

[26] W. Peters, "New Technologies for the Production of Blast Furnace Coke Which Widen the Coking Coal Basis," paper presented at Annual Meeting and Conference, Osaka, Japan, 1976.

[27] O. Därmann, S. Henkel, and K. D. Haverkamp, *Stahl Eisen,* **90**(19), 1009–1012 (1970).

[28] E. Ahland, J. Lehmann, W. Peters, and J. Langhoff, "The Manufacture of Metallurgical Formed Coke According to the BFL-Hot-Briqueting Process," Symp. American Chemical Society, Division of Fuel Chemistry, Chicago, August 1975.

Table 30.3 Production of Formcoke in Various Demonstration Plants, Status: 1976[26]

Process	Plant Location	Briquetting Capacity (10^3 kg/day, unless given otherwise)
INIEX	Belgium	100
ICEM	Rumania	100
HBNPC	Denain, France	150–170
Auscoke[a]	Port Kanbla, Australia	100
DKS	Kokura, Japan (1971)	250 (180×10^6 kg until 1976; in the planning stage, 2.3×10^6 kg/day)
Sapozhnikov	Charkow, USSR (1955)	300 (about beginning 1985: 1.5×10^9 kg/year)
BFL	Bottrop, West Germany (1967)	300
BFL	Scunthorpe, England (1976–1977)	650
Ancit	Alsdorf, West Germany (1966–1967)	250
CRIJ	Japan	2×10^3 kg/hr
NIPR	Japan	2
Consol–BNR	Library, Pennsylvania	30
Consol–BNR	Sparrows Point, Maryland	450
FMC[b]	Kemmerer, Wyoming (1960)	250

[a] Not in operation at present.
[b] At present mainly used in phosphorus furnaces.

furnace, the main emphasis was on the strength values and grain structure; the reactivity and the content of volatile matter played only a secondary role in the evaluation. Table 30.4 compares blast furnace tests with formcoke and conventional coke. The briquetting process determines the suitability of formcoke. So far, there has been no breakthrough in the large-scale application of any one of the formcoke processes.

The main factors impeding the adoption of formcoke processes are the following. The development of such processes requires a very long time, therefore a lot of money. The applicability of the formcoke for the blast furnace has to be proved in long-term tests

Table 30.4 Blast Furnace Tests with Formcoke[26]

	Formcoke[a] and Conventional Coke (CC) Test Pairs					
Properties	ANCIT	CC	BFL	CC	HBNPC	CC
Proportion formcoke (%)	50		100		25	
Proportion of CC (%)	50	100		100	75	100
Blast furnace diameter (m)	8		6.8		9.5	
Blast furnace working volume (m^3)	1376		764		1900	
Production rate (tons/day)	1714	1697	1348	1352	3658	3446
Blast volume (N-m^3/min)	1605	1605	1346	1344	2811	2720
Blast temperature (°C)	1146	1139	1048	1071	1010	1070
Blast pressure (bar)	2.42	2.33	2.07	1.91	2.2	2.1
Flue dust (kg/ton, HM)	20	14	49	23	9	10
Permeability index, *P/V*	0.598	0.523	N.D.	N.D.	0.478	0.459
Burden						
Sinter (%)	100	100	74	74	84	80
Pellets (%)			Nil	Nil	7	6
Ore (%)			26	26	9	14
Coke rate (kg dry/ton, HM)	505	454	575	525	405	404
Oil rate (kg dry/ton, HM)	107	106	61	70	81	88
Total fuel rate (kg dry/ton, HM)	612	560	636	595	486	492
Top temperature (°C)	102	111	107	229	163	180

[a] See Table 30.3 and Chapter 11.

in technical plants. Therefore the capacity of a production pilot plant must be in the order of magnitude of about 2×10^6 kg/day. Because the new technology gives no stimulus to process engineering, only to using noncoking coals, the coal industry is reluctant to make large investments. On the other hand, the recent recession in the iron and steel industry caused the shutdown of capacities for producing metallurgical coke, so that from this point of view there is at present only slight chance for large-scale application of formcoke.

30.2.1.2 *Direct Reduction (Reduction at Low Temperature)*

Principle and Present Commercial Use. Processes for the reduction of iron ore outside the blast furnace are called direct reduction processes. In these processes the iron ore is reduced in the solid state at temperatures that are below the softening point of the reactants (low-temperature reduction).[5,7,15]

The oxygen content of the ore is removed at temperatures up to 1100°C by reducing gases such as CO and H_2 or by solid carbon, according to the following equations. The commercial processes attempt to release from the furnace only gaseous products that have low heating values, such as CO_2 and H_2O, to keep the heat requirement low.[29,30]

$$Fe_2O_3 + 3CO \text{ ----> } 2Fe + 3CO_2 \qquad H_{298\text{ K}} = -251 \text{ kJ/kg Fe} \qquad (30.15)$$

$$Fe_2O_3 + 3H_2 \text{ ----> } 2Fe + 3H_2O \qquad H_{298\text{ K}} = +426 \text{ kJ/kg Fe} \qquad (30.16)$$

$$Fe_2O_3 + 3C \text{ ----> } 2Fe + 3CO \qquad H_{298\text{ K}} = +4377 \text{ kJ/kg Fe} \qquad (30.17)$$

$$Fe_2O_3 + 1\ 1/2C \text{ ----> } 2Fe + 1\ 1/2CO_2 \qquad H_{298\text{ K}} = +2065 \text{ kJ/kg Fe} \qquad (30.18)$$

It is characteristic of direct reduction that the reduction process ends at the point at which, on the scale of the reduction potentials, the potential of the metallic iron is reached.[29] This selective reduction of the iron oxide avoids reaction of other components of the ore, such as oxides of phosphorus, manganese, and silicon, as well as carburizing of the iron, which leads to great purity of the product. The product obtained is called iron sponge; it is quite porous because of the removal of the oxygen. The shape and size of the sponge correspond to those of the feed ore.

The development of direct reduction processes stems from the economic requirement to use a greater variety of ores with different chemical composition, and especially different reducing agents, for iron production. From the great variety of proposed processes, only a small number have been developed to commercial scale. They can be divided into two groups depending on their reducing agent: reduction by gases, and reduction by solid agents.[30,31]

The actual process design leads to the following combination of reducing agents and reactors: coal is being used predominantly in traveling-bed and rotary kilns, and gas is used in shaft retorts or fluidized-bed reactors.

The present status of commercialization (Table 30.5) indicates that in 1976 the capacity for all plants was about 8×10^9 kg/year. In 37% of these plants with a combined capacity of 2.9×10^9 kg/year, coal is used as the reducing agent. Thus direct reduction contributes about 1.5% to the total production of iron. In view of the direct reduction processes that are in the planning stage, which predominantly will be using gas as a reducing agent, an increase to 3.5% of total production is expected in the next few years.

Since only the use of coal as a reducing agent is discussed at this point, the gas reduction processes are not treated in detail. They are indirectly of interest, however, since the raw material situation in the future will necessitate gasification of coal to produce reducing gas. This need will lead to an increased use of coal for the manufacture of reducing gas for iron production,[31] an area discussed in detail in Section 30.3.3.10.

Description of Solid Reducing Agent Processes. Traveling-bed (Höganäs) processes have been in commercial use in Sweden since

[29] G. Meyer, *Iron Steel Int.*, 103 (April 1976).
[30] W. Peters and H.-D. Beyer, *Glueckauf*, **107**(21), 811–820 (1971).
[31] U. Kalla and R. Steffen, *Stahl Eisen*, **96**(14), 645 ff. (1976).

Table 30.5 Status of Commercialization (1976) of Various Direct Reduction Processes[31]

Process	Type of Reactor	Plants: Location	Plants: Design Capacity (10^3 kg/year)	Plants: Coal Consumption (10^3 kg/year)
		Reduction with Solid Agents		
Höganäs	Traveling bed	Höganäs, Sweden	162,000	39,000
		Oxklesuca, Sweden	35,000	8,000
		Riverton,	70,000	17,000
Kinglor-Metor	Shaft	Buttno, Italy	8,000	2,500
SL/RN	Rotary kiln	Glenbroch, New Zealand	150,000	42,800
		Porto Alegre, Brazil	65,000	23,000
		Wilbauh, South Africa	800,000	228,500
		Casa Grande, Arizona	65,000	23,000
		Redlake, Canada	350,000	123,000
		Chiverbote, Peru	100,000	35,000
Krupp	Rotary kiln	Bendu, South Africa	150,000	52,000
Koho	Rotary kiln	Musoran-W, Japan	36,000	12,600[a]
Kawasaki	Rotary kiln	Chiba-W, Japan	75,000	24,000[a]
		Kurashibi-W, Japan	240,000	72,000[a]
Nipponkokan	Rotary kiln	Fukuyawa, Japan	350,000	100,000
SDR	Rotary kiln	Wakayama-W, Japan	160,000	26,400
SPM	Rotary kiln	Koshima-W, Japan	100,000	26,000
			2,856,000 (37%)	855,300
		Reduction with Gases		
Hyl	Retort	Monterrey, Mexico	75,000	
		Monterrey, Mexico	190,000	
		Monterrey, Mexico	375,000	
		Puebla, Mexico	225,000	
		Veracruz, Mexico	190,000	
		Bahia, Brazil	225,000	
Wiberg	Shaft furnace	Sanviken, Sweden	25,000	
		Hofors	25,000	
		Yasugi, Japan	10,000	
Midrex	Shaft furnace	Portland,	400,000	
		Georgetown,	400,000	
		Hamburg, West Germany	400,000	
		Contrecomy, Canada	400,000	
Purofer	Shaft furnace	Oberhausen, West Germany	165,000	
Armco	Shaft furnace	Houston, Texas	330,000	
Novalfer	Fluidized-bed reactor	Toulouse, France	18,000	
HIB	Fluidized-bed reactor	Puerto Ordaz, Venezuela	1,000,000	
PIOR	Fluidized-bed reactor	Matanzas, Cuba	400,000	
			4,853,000 (63%)	

[a] Estimated.

1911.[32–34] Because of high investment and operating costs and low specific throughput, no further installation is expected. The feed materials (ore with Fe content above 70% and carbon-coke breeze) are charged together with limestone into ceramic retorts, which travel through a tunnel kiln at 1200°C.

In rotary kiln processes the ore and the fuel-reducing agent travel either concurrently

[32] Wilberg Höganäs, *Iron Coal,* **28,** 1270 (1954).
[33] I. Kalning, *Tech. Geme.,* **12**(8), 334 ff. (1964).
[34] *Direct Reduction of Iron Ore: Bibliographic Study,* Verlag Stahleisen GmbH, Düsseldorf, 1976, p. 335.

or countercurrently through the heated rotary kiln. While passing through the kiln, the ore is reduced to iron sponge. The process is very flexible with respect to size consist and composition of the iron ore. Pelletizing of the feed permits the use of fine ores or even dust removed from gases from blast furnace, sinter, and converter plants.[35–38] The most important representatives of this group are the Krupp iron sponge process[39,40] and the SL/RN[41,42] process developed by Steel-Lurgi–Republic Steel National Leed. Furthermore, there are some Japanese developments such as the Kawasaki,[43] Koho,[44] SDR,[45] and SPM processes.[46] The SPM (Sumitomo Metal Industry) process attempts to use fine dusts without pelletizing, since under controlled conditions, agglomeration and reduction occur simultaneously in the rotary kiln.

The Kinglor-Metor process is the only solid reducing agent process of commercial interest to employ a shaft furnace.[47] Lump ore or pellets are mixed with coke or coal and lime and reduced to iron sponges as they pass through a heated rectangular chamber.

[35] G. Meyer, K.-H. Vopel, and W. Janssen, *Stahl Eisen,* **96**(24), 1228 ff. (1976).

[36] G. Reuter and H. Serbent, *Development and Present Status of the SL/RN Process,* pamphlet of Lurgi Chemie und Hüttentechnik GmbH, September 1976.

[37] K. Sugasawa, Y. Yamada, Sh. Watanabe, K. Kato, K. Masuda, Y. Sato, and T. Kawabata, *Stahl Eisen,* **96**(24), 1239 ff. (1976).

[38] H. Maczek, H. Rellermeyer, G. Kossek, and H. Serbent, *Stahl Eisen,* **96**(24), 1233 ff. (1976).

[39] K.-H. Ulrich, "Direct Reduction in Comparison with the Classical Method of Used Production," paper presented at Fedemetal Direct Reduction Seminar, Bogotá, Columbia, November 1976 (pamphlets of Krupp-Industrie- und Stahlbau).

[40] *Direct Reduction of Iron Ore, Bibliographic Study,* Verlag Stahleisen GmbH, Düsseldorf, 1976, p. 185.

[41] W. Janke and H. Garbe, Metallgesellschaft AG, Frankfurt, Bull. No. 12.

[42] *Direct Reduction of Iron Ore: Bibliographic Study,* Verlag Stahleisen GmbH, Düsseldorf, 1976, p. 255.

[43] *Direct Reduction of Iron Ore: Bibliographic Study,* Verlag Stahleisen GmbH, Düsseldorf, 1976, p. 153.

[44] *Direct Reduction of Iron Ore: Bibliographic Study,* Verlag Stahleisen GmbH, Düsseldorf, 1976, p. 174.

[45] *Direct Reduction of Iron Ore: Bibliographic Study,* Verlag Stahleisen GmbH, Düsseldorf, 1976, p. 242.

[46] *Direct Reduction of Iron Ore: Bibliographic Study,* Verlag Stahleisen GmbH, Düsseldorf, 1976, p. 270.

[47] *Direct Reduction of Iron Ore: Bibliographic Study,* Verlag Stahleisen GmbH, Düsseldorf, 1976, p. 164.

Characterization of Feed Materials and Products. Table 30.6 gives characteristic data of commercially important solid fuel reducing processes.

Iron-Containing Raw Materials. A great variety of ores may be used, since naturally occurring fine and lump ores may be supplemented by pelletizing processes, which may produce mixtures of diverse ore components such as hematite, magnetite, limonite, and iron-containing dusts from blast furnace, sinter, and converter plants.

Reducing Agents. In addition to coal of all ranks (anthracite to lignite), fine coke, coke breeze, and steel mill dust may be used. The following characteristics are important for the selection of the reducing agent: reactivity toward CO_2, content of volatile matter, sulfur and ash content, and ash softening point.[29,48,49]

An increase in reactivity may lead to an increase in the throughput, since the rate of the process depends on the gasification velocity of the carbon, according to the Boudouard reaction, with the CO_2 obtained by reduction of the ore.

When high-volatile coals are used, they produce off-gases having a high content of chemically bound heat, which increases the heat consumption per ton of iron. In this case the use of the waste heat becomes important.

If the ash has an acidic composition, the sulfur content of the coal should not exceed 1.5%, to ensure that an iron sponge with a low sulfur content is obtained.

To avoid deposit formation, the ash content should be below 20% and the softening point should be at least 100°C above the maximum temperature reached by the reactants. In rotary kiln processes, furthermore, the bulk density and the size consist of the

[48] G. Reuter and W. Schnabel, "Charcoal—New Primary Energy Source for Steelmaking," paper presented at ILAFA Coal Seminar, Mexico City, July 1976 (Lurgi pamphlet).

[49] G. Meyer and U. Bongers, "Reduction by Solid Fuels with the KRUPP-Sponge Iron Process," paper presented at ILAFA Seminar, Direct Reduction, Porto Alegre, Brazil, May 1975.

Table 30.6 Processes of Great Commercial Importance for Direct Reduction with Solid Carbon Reducing Agents[134]

					Specific Consumptions		
Process	Type of Reactor	Reducing Agent	Iron Ore	Product	Reducing Agent (kg/kg)	Process Heat (10^3 kJ/kg)	Use
Höganäs	Traveling bed	Coke breeze	Fine ores, concentrates, Fe > 70%	Iron sponge, shapes, cubes (98% Fe)	0.240	~ 16.7	Powder metallurgy, special dusts
Kawasaki	Rotary kiln	Coke breeze	Blast furnace, sinter, LD dust	Reduced pellets (74% Fe)	0.300–0.500	16.6–17.6	Blast furnace, sintering plant
Kinglor-Metor	Shaft furnace	Metallurgical coke	Coals, lump ore, sinter ore, pellets (6.5–25 mm)	Iron sponge	0.310	15.9	Steel mill, foundry
Koho	Rotary kiln	Coke from blast furnace dust	Steel mill dust	Pellets (~ 70% Fe)	No data	~ 12.5	Blast furnace, sintering plant
Krupp iron sponge	Rotary kiln	Bituminous and brown coals	Lump ore, pellets, fine ore	Iron sponge, pellets (90% Fe)	~ 0.3	15.9–16.7	Electro-oven, D-converter
SDR	Rotary kiln	Fine coke	Blast furnace, sinter, LD dust	Pellets (76% Fe)	0.165[a]	~ 5.9	Blast furnace
SL/RN	Rotary kiln	Coke breeze Bituminous coals Brown coals	Lump ore (3–20 mm) Fine ore (0.04–3 mm) Pellets (8–16 mm)	Iron sponge pellets (> 90% Fe)	0.310–0.396	No data	Electro-oven, blast furnace
SPM	Rotary kiln	Fine coke from blast furnace, anthracite	Blast furnace, LD dust	Iron sponge (75% Fe)	0.160[a]	4.2	Blast furnace, sintering plant

[a] Because of the carbon content of the ore dust, the reducing agent requirement is small.

coal must correspond to those of the ore, to avoid segregation of ore and reducing agent.

The Product and Its Use. Ore and reducing agent are selected and the reducing process conducted so as to attain the degree of metalization required for further processing. In most cases an electric furnace is used for the conversion to steel (degree of metalization > 90%). In cases of exceptionally high degrees of metalization (> 98%), the product is used in powder metallurgy and for the manufacture of special steels. If the iron content is especially low, the material is used in the blast furnace or a sintering plant.

In contrast to hot metal, the iron sponge consists of many particles that can be shipped over long distances and stored for extended periods of time if they are protected from reoxidation.[48,50]

Specific Consumption.[34,51–54] The average consumption is 0.3 kg of carbon per kilogram of iron sponge. Depending on the kind and properties of the reducing agent and the desired degree of metalization, values between 0.15 and 0.45 kg/kg are found. The process heat requirement is between 8.4 and 16.7×10^3 kJ/kg, the total energy requirement is given as $20.9–22.2 \times 10^3$ kJ/kg sponge iron.

30.2.2 Other Uses of Coal and Coke in the Metallurgical Industry

As in the case of iron, nearly all metallic materials and commodities require a melting process in their preparation, which is followed by a shaping process before the final finishing. Melting and shaping are done in the foundry. For special products it is necessary to add certain materials during melting or casting, to carburize, desulfurize, or change concentrations of components to permit the achievement of the required properties of the product. In most cases the manufacture of cast iron (gray, malleable, hard-cast iron) entails the use of the cupola, which similar to the blast furnace, is a countercurrent shaft furnace.[55] For the production of a fault-free casting, the composition of the burden is important; thus the action of the individual components accompanying the iron need to be considered. In addition, proper operating conditions of remelting in the cupola must be employed. The coke used in the cupola (foundry coke) has the following functions: (1) to supply the heat needed for the melting by combustion of the coke, (2) to serve as a fire-resistant, supporting skeleton, thus to promote uniform gas distribution during melting; (3) to assure uniform flow of the liquid iron; and (4) to introduce the required amount of carbon into the cast iron.

Foundry coke must therefore have the following properties: high strength, uniform large lump size (80–100 mm diameter), high carbon content (> 85%), and low ash (< 10%) and sulfur (< 1%) content.

The coke consumption of 0.05–0.12 kg/kg cast iron depends on the quality of the casting and the design of the cupola. For special products a high-carbon coke (HCC) is being used, which decreases coke consumption and leads to higher melting throughputs.[55–57]

In the past few years coke consumption has been decreased by the following measures: introduction of the secondary-blast process in various cupolas (United States, West Germany, France, England), use of natural gas as fuel (e.g., France, East Germany), and oxygen enrichment of the blast (~ 2%) during melting.[58]

Table 30.7 gives cast iron production for 1974, as well as the estimated coke consumption.

Besides the cupola, special electric furnaces are used for the melting process (e.g., arc, graphite rod, induction, and high-frequency furnaces). Because of high operating costs, they are used predominantly for special qualities, especially in nonferrous metal

[50] W. D. Liestmann, *Stahl Eisen,* **96**(3), 97 ff. (1976).

[51] O. Schwirten, "Ideas for the Heat Recovery in Rotary Kiln Plants," paper presented at ILAFA Seminar, Iron Sponge Production, Porto Alegre, Brazil, May 1975.

[52] E. Förster and P. Kehl, *VGB Kraftwerkstechnik,* **55**(11), 697 ff. (1975).

[53] L. v. Bogdandy and S. Henkel, in *Bergbau—Rohstoffe—Energie,* Vol. 10, Verlag Glückauf, Essen, 1976, pp. 57 ff.

[54] K. Meyer, *Stahl Eisen,* **95**(10,) 72 (1975).

[55] W. Donike and C. A. Eber, *Schmelzen und Gattieren in der Eisengiesserei,* 2nd ed., Fachverlag Schiele und Schon GmbH, Berlin, 1965.

[56] Hütte, *Taschenbuch für Eisenhüttenleute,* pp. 703 ff.

[57] *Giesserei-Lexikon,* Fachverlag Schiele und Schon GmbH, Berlin, 1976 ed.

[58] A. Dahlmann, *Giesserei,* **62**(12), 329 ff. (1975).

Table 30.7 1974 Production of Iron, Steel, and Malleable Castings and Total Consumption of Coke (10^6 kg)[57]

	Iron Castings	Cast Iron with Spherical Graphite	Steel Castings	Malleable Castings	Total Production	Coke Consumption
Europe[a]	12,445	1935	1,433	710	16,523	1700
United States	12,211	1998	1,896	829	16,934	1700
World	50,838	5700	10,511	3113	70,162	7000

[a] Excluding Eastern Bloc countries.

foundries. In the arc and graphite rod furnaces, carbon is consumed because of burnup of the electrodes, which for iron castings amounts to about 5 g/kg (see Section 30.4.1).

30.2.3 Conversion of Nonferrous Ores into Metal Oxides or Metals

Carbon is used in many forms (coal, coke, electrodes) to obtain pure metals or metal oxides. Thus, for example, pure aluminum is obtained by electrolysis of its molten oxide. Aluminum oxides (electrocorundum and alumina) are produced from bauxite in rotary kilns with coke. Titanium oxides are produced from ilmenite, a titanium–iron ore mixture, by reduction of the iron oxide. Zinc oxide is produced by reduction of a mixture of the ore (franklinite), with subsequent reoxidation of the metallic zinc. These processes are described briefly next.

30.2.3.1 *Alumina*

For the electrolytic production of aluminum from aluminum oxide, large amounts of commercial carbon are used in preparing the carbon electrodes. Addition of cryolite reduces the operating temperature of the electrolysis process to 950°C. The electrolysis cell or pot has a carbon lining used as a cathode, whereas the carbon anodes enter from the top into the melt. As aluminum oxide decomposes into metallic aluminum and oxygen, the oxygen reacts with the carbon anodes, forming a gas mixture consisting of about 20% carbon dioxide and 80% carbon monoxide. Prefabricated anodes as well as the continuous self-calcining Söderberg electrodes are used. The solid component used in the manufacture of the carbon anodes is predominantly petroleum coke; electrode pitch, which is used almost exclusively as binder, is obtained by thermal treatment of coal-tar pitch. The ratio of coke to pitch is 5 : 1 for the prefabricated anodes and 2.5 : 1 for the Söderberg anodes. The two electrode types are used today in about equal quantities for the manufacture of aluminum.[59]

The material for the carbon cathodes can be prepared from a mixture similar to the Söderberg mix or from prefabricated carbon blocks. It does not take part in the chemical reaction and has a life span of about 2–3 years, therefore does not cause a significant usage of carbon. The consumption of anode material is larger and amounts to 0.5–0.55 kg/kg aluminum. The feed material for the Söderberg anodes contains about 30% electrode binder pitch; the anode material for the prefabricated block anodes, 16% electrode binder pitch. Thus the specific consumption figures when using Söderberg anodes are 0.15 kg of electrode binder pitch and 0.4 kg of electrode coke per kilogram of aluminum, and for prefabricated anodes, 0.1 kg of electrode binder pitch and somewhat more than 0.4 kg of electrode coke per kilogram of aluminum.[59]

The development of aluminum production in the world and in Western Europe is shown in Fig. 30.6. The production of aluminum in 1975 required 1.5×10^9 kg of electrode binder pitch and 5×10^9 kg of electrode coke. The electrode binder pitch is almost exclusively obtained from bituminous coal-tar pitch. In the manufacture of electrode coke, only about 10% of the total is derived from the coking of bituminous coal-tar pitch; the greatest portion is supplied by petroleum coke.[60]

In West Germany the major part of commercial carbon production goes to electrode binder pitch and pitch coke for the aluminum industry and graphite production; the smaller

[59] G. Collin and W. Gemmeke, "Bituminous Coal Tar Pitch and Aluminum," *Erdoel Kohle, Erdgas, Petrochem.*, **30**, 25 (1977).

[60] G. Collin, "Technical and Economic Aspects of Tar Chemistry," *Erdoel Kohle, Erdgas, Petrochem.*, **29**, 159 (1976).

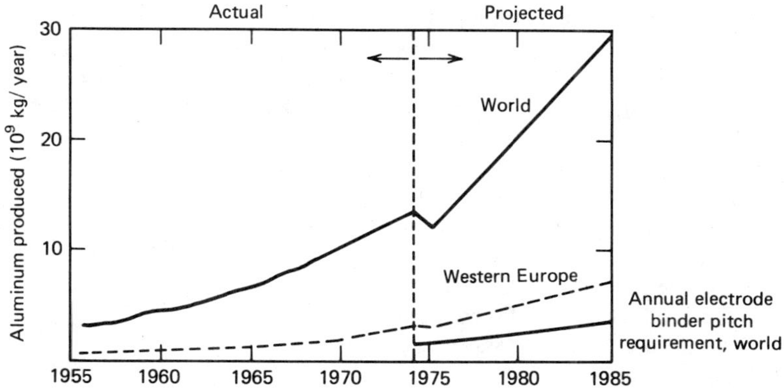

Figure 30.6 Development of aluminum production in the world and in Western Europe with an extrapolation until 1985 and the predicted electrode binder pitch consumption in the world.[60]

part is used for the production of carbon black. Since 1960 the production of electrode binder pitch and pitch coke from bituminous coal tar has increased from 200×10^6 to 450×10^6 kg/year (Fig. 30.7).

30.2.3.2 *Fused Alumina*

Bauxite is calcined in rotary kilns at temperatures above 1000°C to produce electrocorundum or fused alumina for use as an abrasive material. It is then fused in an electric arc furnace with carbon electrodes. Unit consumption figures are 0.09 kg of coke or anthracite and 0.02–0.03 kg of electrode carbon per kilogram of fused alumina. The additional electrical energy requirements are 3–4 kW-hr per kilogram of fused alumina. The production in Canada, which was 210×10^6 kg of fused alumina in 1957, remained constant for several years. Such abrasives are manufactured in several countries and marketed under different brand names.[61–63]

30.2.3.3 *Titanium Dioxide*

Titanium ores are processed either with sulfuric acid or with chlorine via titanium tetrachloride to titanium dioxide, which is used in large quantities as a white pigment.

Ilmenite ($FeTiO_2$) is used as a raw material in the sulfuric acid process; this mineral is in ample supply but has a limited TiO_2 content of about 35–55% by weight, with significant

[61] R. Tertian and D. Papée, "Fused Aluminas," in *Kirk-Othmer Encyclopaedia of Chemical Technology,* 2nd ed., Vol. 2, New York, 1967, p. 56.

[62] C. Burbott, "Elektrokorund," in *Ullmanns Encyklopädie der technischen Chemie,* 3rd ed., Vol. 6, Munich, 1955, p. 237.

[63] P. Haasemann, "Chlorierung von Rutil und TiO_2-Konzentrat," in *Ullmanns Encyklopädie der technischen Chemie,* 3rd ed., Vol. 17, Munich, 1966, p. 419.

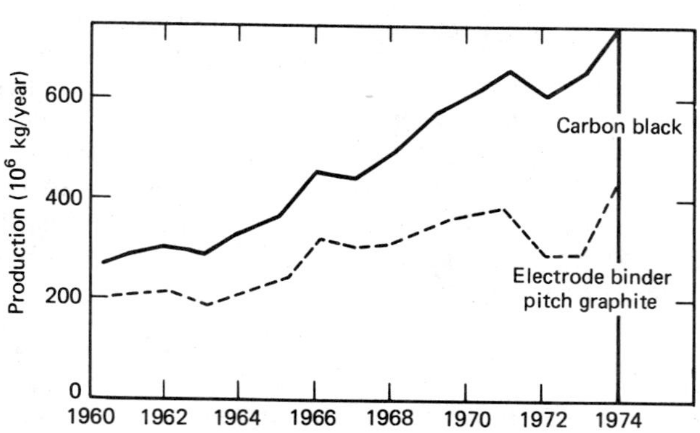

Figure 30.7 Development of the production of commercial carbon in Germany.[60]

impurities, especially iron. Consequently a large fraction of the sulfuric acid input is used for the conversion of iron and the production of waste materials (3–4 kg of iron sulfate hydrates and 8 kg of diluted sulfuric acid and wastewater per kilogram of TiO_2). On the other hand, the process using chlorine is restricted to the processing of rutile, which contains 90–97% TiO_2 but is available in the world only in limited quantities. Therefore processes for the production of TiO_2 concentrates from ilmenite are attracting more and more commercial interest.

To this process group belongs the reduction in metallurgical kilns of iron ores with coal to the metal. The Sorel process of Quebec Iron and Titanium Corporation uses electric arc furnaces with graphite electrodes for ilmenite reduction. The processing of ilmenite with 35 wt % TiO_2 and 40% iron yields in addition to steel a slag that is rich in titanium (72% TiO_2, 9% FeO, 6% Al_2O_3, 6% SiO_2, 1% CaO, and 5% MgO). In 1960 in Canada, 350 $\times 10^6$ kg of titanium slag was produced by that process, for further conversion to TiO_2 for use in the pigments industry.[63] Much less sulfuric acid is required to process this slag than is needed to process ilmenite. Unit consumption figures, related to a titanium slag with 65% TiO_2 are as follows: 0.25–0.3 kg of coke, 2.4–2.6 kg of ilmenite, and 7.1–8.3 kW-hr.[64] To produce a slag with 70% TiO_2, the consumption figures are 0.33 kg of coke and 2.28 kg of ilmenite.[65]

In the chlorine process rutile, coke, and chlorine react to produce titanium tetrachloride, which is subsequently converted to TiO_2 and to chlorine, which is recovered and recycled:

$$3TiO_2 + 4C + 6Cl_2 \dashrightarrow 3TiCl_4 + 2CO + 2CO_2 \qquad (30.19)$$

$$TiCl_4 + O_2 \dashrightarrow TiO_2 + 2Cl_2 \qquad (30.20)$$

The output of 1 kg of TiO_2 (98%) requires 0.25 kg of coke, 1.125 kg of rutile (95%), 0.15 kg of chlorine, and 0.45 kg of oxygen.[66] Titanium tetrachloride is also the starting material for the production of metallic titanium.

In 1973, 700 $\times 10^6$ kg of TiO_2 was manufactured in the United States alone. By far the largest share of world production is based on ilmenite processing with sulfuric acid. However some new plants are using the chlorine process. Its more extensive use is hampered only by the shortage of rutile.[67]

30.2.3.4 *Zinc Oxide and Magnesium Chloride*

The production of zinc oxide by the American or direct process consumes 3–5 kg of anthracite or coke as a reducing agent per kilogram of ZnO. Using franklinite with 20% ZnO content requires 5.25 kg of coal. The electrical energy consumption is 0.400 kW-hr/kg ZnO. The process is similar to metallic zinc production. Ground and concentrated zinc ore is mixed with fine-grain coal and briquetted. In the reducing kiln, ZnO is converted to metallic zinc by solid carbon or carbon monoxide, which is formed in the process at temperatures of 900–1000°C. Vaporized zinc is reoxidized to ZnO outside the furnace with air.[68]

The French or indirect process starts with zinc metal, which is vaporized in a retort in the presence of carbon monoxide. Oxidation to ZnO and CO_2 occurs in a special combustion chamber. Raw material requirements are 0.87 $\times 10^3$ kg of zinc and 0.65 $\times 10^3$ kg of anthracite or coke per thousand kilograms of zinc oxide.[68]

In 1973 in the United States, production of zinc oxide for different uses (elastomers, paint, varnishes, etc.) was 225 $\times 10^6$ kg. The direct and indirect processes shared this production figure about equally.

Dry magnesium chloride was formerly produced quite frequently by chlorination of magnesium oxide in the presence of coal, $MgCl_2$ having been used primarily as an intermediate for the production of magnesium metal.[69] Magnesium oxide could be formed by calcination of magnesite. This requires 0.5 kg of coal for calcination and 0.4 kg of coal

[64] C. K. Stoddard et al., U.S. Bur. Mines Rep. Invest. No. 4750, 1950.

[65] A. W. Knoerr, *Eng. Min. J.*, **153**(3), 72 (1952).

[66] W. L. Faith et al., *Industrial Chemicals*, 4th ed., New York, 1975, p. 817.

[67] K. H. Schmidt, *Chem. Ind.*, **25**, 652 (1973).

[68] W. L. Faith et al., *Industrial Chemicals*, 4th ed., New York, 1975, p. 882.

[69] R. Ritter and G. Jaekel, "Magnesium," *Winnacker-Küchler, Chemische Technologie*, Vol. 6, Munich, 1973, p. 120.

as reducing agent per kilogram of $MgCl_2$.[70,71] This process, however, is no longer economical. Furthermore magnesium metal is no longer produced by reduction with coal.

30.2.4 Reduction of Nonmetallic Compounds

30.2.4.1 *Sodium Sulfide*

Sodium sulfide may be produced by the reduction of sodium sulfate with coal, preferably anthracite. The reactions are:

$$Na_2SO_4 + 4C \longrightarrow Na_2S + 4CO \qquad (30.21)$$

$$Na_2SO_4 + 2C \longrightarrow Na_2S + 2CO_2 \qquad (30.22)$$

The required sodium sulfate or salt cake is formed mostly by conversion of NaCl with sulfuric acid. In addition to heat requirements, 0.45 kg of coal is needed to reduce 1 kg of salt cake.[72] During recent years more and more sodium sulfide has been produced as a by-product of other processes, so that the reduction process with coal is said to be, for example, no longer important in the United States.[73] Sodium sulfide is used extensively in the leather industry for the removal of hair from hide. It is also used in the manufacture of rayon as an agent to desulfurize viscose, in the manufacture of sulfur dyes, in engraving and lithography, and as a reducing agent in the production of amino compounds.

30.2.4.2 *Sodium Sulfate*

Sodium sulfate is used to some extent in the manufacture of glass that goes into lower quality products. When sodium sulfate is used in this manner, the addition of some coal in an empirically determined weight ratio is necessary for the reduction of the salt cake to sodium silicate. One possible reaction is:

$$Na_2SO_4 + SiO_2 + C \longrightarrow Na_2SiO_3 + SO_2 + CO \qquad (30.23)$$

The SO_2 evolved creates some environmental problems.

In one case, for the production of window glass plates, the use of 3 wt % salt cake in the raw material mixture and the addition of 4.2 kg of coal per 100 kg of salt cake was specified.[74] These input ratios vary considerably in glassmaking practice.

30.2.4.3 *Barium Sulfide*

Barium sulfide is made from barite by a reduction process with 0.2 kg of coal per kilogram of barite, exclusive of heating requirements[75]:

$$BaSO_4 + 2C \longrightarrow BaS + 2CO_2 \qquad (30.24)$$

The process takes place in rotary kilns at temperatures of 1000–1200°C with retention times of 90–120 min.

Barium sulfide serves as an intermediate in the manufacture of lithopone, a white pigment consisting of zinc sulfide and barium sulfate, as well as all barium chemical compounds. The volume of barite processing in the chemical industry, however, has decreased in the recent past. In the United States in 1969, 160×10^6 kg barite was used; in 1972 this figure had dropped to 96×10^6 kg. Worldwide use in 1972 was about 350×10^6 kg.

30.2.4.4 *Calcium, Magnesium, and Iron Sulfates*

In several countries the process involving the reduction of calcium sulfate (anhydrite) with coke for the production of sulfuric acid and cement is operated on the commercial scale (United Kingdom, East Germany, Austria, Poland, Brazil, Chile). The decomposition can be described by the following equation:

[70] E. R. Riegel, *Industrial Chemistry,* 5th ed., New York, 1949, p. 352.

[71] R. N. Shreve, *The Chemical Process Industries,* 1st ed., New York, 1945, p. 309.

[72] E. Dönges et al., "Der Schwefel und seine organischen Verbindungen," in *Winnacker-Küchler, Chemische Technologie,* 3rd ed., Vol. 2, Munich, 1970, p. 76.

[73] J. S. Scone, "Sodium Sulfides," in *Kirk-Othmer Encyclopedia of Chemical Technology,* 2nd ed., Vol. 18, New York, 1969, p. 510.

[74] R. Günther, H. Loffler, and W. A. Weyl, "Glas," in *Ullmanns Encyklopädie der technischen Chemie,* 3rd ed., Vol. 8, Munich, 1957, pp. 133 ff.

[75] L. Preisman, "Barium Compounds," in *Kirk-Othmer Encyclopedia of Chemical Technology,* 2nd ed., Vol. 3, New York, 1964, p. 96.

$$2CaSO_4 + C \text{----} > 2CaO + 2SO_2 + CO_2 \quad (30.25)$$

In 1969 total production from all plants in the world amounted to 5×10^6 kg of sulfuric acid per day.[76] The outlook for this process is favorable, since it can use waste calcium sulfate, which is formed in huge amounts during the wet process of phosphoric acid manufacture from phosphate ores with sulfuric acid, and also that formed in some processes for flue gas desulfurization (see Section 30.5.2 and Chapter 22). Unit consumption figures for the process developed in East Germany have been stated as follows for 1 kg of sulfuric acid (100%) and the attendant production of 0.95 kg of cement: 0.13 kg of coke; 1.63 kg of anhydrite (95% $CaSo_4$), 1.84 kg of waste calcium sulfate (94.5% $CaSO_4$), or 2.18 kg of natural gypsum (75% $CaSO_4$); 0.08 kg of sand (94% SiO_2); 0.27 kg of clay (25% Al_2O_3, 51% SiO_2); 0.3 kg of heating oil; and 0.2 kW-hr electricity.[77,78]

In a similar process in East Germany magnesium sulfate or kieserite ($MgSO_4{'}H_2O$) is converted with coke in a multiple hearth roaster, recovering magnesium oxide (MgO) as a by-product:

$$2MgSO_4 + C \text{---->} 2MgO + 2SO_2 + CO_2 \quad (30.26)$$

In this process carbon monoxide could also be used effectively as a reducing agent.[79]

Iron sulfate is decomposed to remove iron sulfate wastes from the processing of ilmenite with sulfuric acid (see Section 30.2.3.3) and from waste pickle liquors in steel manufacture. Waste acids from processing titanium ores contain 9–15 wt % sulfuric acid and 4–10% iron sulfate. The solution is evaporated and iron sulfate is dehydrated to the monohydrate ($FeSO_4{'}H_2O$). This material is decomposed in a separate furnace together with coke or mixtures of coke and pyrite:

$$2FeSO_4 + C \text{---->} Fe_2O_3 + 2SO_2 + CO \quad (30.27)$$

[76] *Chemieanlagen/Verfahren (CAV)*, **12**, 37 (1969).

[77] *Sulphur*, No. 86, 31 (January–February 1970).

[78] W. Gosch, *Verfahrenstechnik*, **7**, 319 (1973).

[79] *Inorganic Technical Processes*, F. Matthes and G. Wehner, Eds., Leipzig, 1964, p. 155.

The gas formed contains around 7 vol % sulfur dioxide, which is converted to sulfuric acid. Iron oxides are sent to the steel mills.[80] Coal requirements are 0.1 kg per kilogram of iron sulfate.[81]

The utilization of large waste gas streams with low concentration of SO_2 is sometimes not economical for the production of sulfuric acid, so that the production of elemental sulfur is preferred. This is possible by increasing the concentration of SO_2 to 25–30% using an adsorption process. A gas of this SO_2 content may be reduced with coke or coal to elementary sulfur[82,83] according to the following reaction:

$$SO_2 + C \text{---->} CO_2 + S \quad (30.28)$$

In a new process using a countercurrent moving-bed reactor, anthracite is used as the reducing agent.

To produce carbon disulfide (CS_2) from the elements carbon and sulfur in a retort, wood charcoal is normally used because of its greater reactivity. However char from low-temperature pyrolysis of lignite or coal also has been used successfully in that process.[84]

30.2.4.5 *Phosphorus*

In the electric furnace process, calcium phosphate is reduced with coke together with the addition of sand for slag formation.[85,86] The coke is converted into carbon monoxide in the off-gas, which may be used as fuel. The simplified reaction is:

[80] T. S. Harrer, "Sulfuric Acid and Sulphur Trioxide," in *Kirk-Othmer Encyclopedia of Chemical Technology*, 2nd ed., Vol. 19, New York, 1969, p. 475.

[81] F. J. Bartholomew, *Chem. Eng.*, **57**(8), 118 (1950).

[82] E. Dönges et al., "Gewinnung von Schwefel aus Schewefeldioxid," in *Winnacker-Küchler, Chemische Technologie*, 3rd ed., Vol. 2, Munich, 1970, p. 9.

[83] J. J. Strum, W. F. Bischoff, R. E. Rush, K. Knoblauch, and K. Goldschmidt, paper presented at EPA Flue Gas Desulfurization Symp., New Orleans, March 8–11, 1976.

[84] A. Hake and R. Leitner, "Schwefel Kohlenstoff," in *Ullmanns Encyklopädie der technischen Chemie*, 3rd ed., Vol. 15, Munich, 1964, pp. 533 ff.

[85] R. N. Shreve, *The Chemical Process Industries*, 3rd ed., New York, 1967, p. 281.

[86] *Phosphoric Acid*, A. V. Slack, Ed., Vol. 2, New York, 1968, p. 927.

$$2Ca_3(PO_4)_2 + 10C + 6SiO_2 \dashrightarrow 4P + 6CaSiO_3 + 10CO \quad (30.29)$$

The consumption figures per kilogram of yellow elementary phosphorus are 1.325 kg of coke, 12 kg of phosphate ore (depending on composition), 2.225 kg of sand, 0.025 kg of electrode carbon, and 14.3 kW-hr.[87] Most of the elementary phosphorus production in the United States (~ 90%) is processed to phosphoric acid. Integrating both processes, the following consumption figures per kilogram of phosphoric acid (100%) are obtained: 0.44 kg of coke, 2.5 kg of phosphate ore (70% BPL), 1 kg of sand (SiO_2), 7–11 × 10^{-3} kg of electrode carbon, and 4.5 kW-hr.[87]

To get a furnace feed with a good and evenly distributed gas flow in the past, only coarse-grained phosphate ores were used; however today fine-grained material is also used because agglomeration processes have been developed that provide the necessary enlargement of the particles. The furnace has a steel shell for the crucible. The floor of the crucible is lined with carbon blocks. The sidewalls are made of carbon blocks or of smaller carbon bricks to a point far enough up to be well out of the reaction zone. Above the sidewall, refractory is used up to the roof line. Good water cooling of the crucible prolongs considerably the onstream time of the carbon lining. The roof of the furnace is made of a special heat-resistant alumina cement. The electrodes, the charging lines for raw materials, and the discharge ducts for the gas are inserted through the roof. Molten slag and ferrophosphorus are discharged through tapholes near the bottom. Normally the electric furnaces are operated with three-phase current.

The gas leaves the furnace at a temperature of between 200 and 400°C and contains about 400 g of vaporized phosphorus as well as 40 g of dust per cubic meter of dry gas at 0°C and 760 mm Hg (m_n^3). The concentration of carbon monoxide is up to 85% by volume. Dust is removed by electrostatic precipitators. Elementary phosphorus is then condensed by injection of water. Small quantities of residual phosphorus are finally reclaimed by low-temperature condensation. Purity of phosphorus is usually better than 99.7%. Söderberg-type as well as prebaked electrodes are used, the Söderberg electrodes being employed almost exclusively outside the United States.[88]

Söderberg electrodes have the following advantages: (1) the specific, end-product-related costs of the electrode consumption in the process are lower because the lower price of the electrode materials more than compensates for the higher unit consumption of electrode materials, which may amount to up to 3 times that of prebaked electrodes, and (2) inventory and handling costs are lower.

On the other hand, some disadvantages of the Söderberg electrodes compared with prebaked electrodes should be mentioned: process control is much more critical, breakage of a Söderberg electrode would lead to a shutdown of the furnace, and the continuous feeding of the electrode materials is critical and must be kept under strict control.[89]

The maximum size of the electrodes is of considerable technical and economic importance because it limits the maximum capacity of the furnace. The maximum available diameters of the prebaked electrodes and of the Söderberg-type electrodes are 1.4 and 1.58 m, respectively. A further increase of these maximum dimensions in the future can be expected. The large-size prebaked electrodes would require a lowering of the current density, which limits the capacity. The current density of Söderberg electrodes does not vary significantly with the diameter.

The capacity of phosphorus furnaces was raised to about 65 MW per unit in the early 1970s.

In 1973 in the United States, about 480 × 10^6 kg of elemental phosphorus was produced; this would have corresponded to a coke consumption of 650 × 10^6 kg. In 1965 world capacity for elemental phosphorus was estimated to be 900 × 10^6 kg.[90] Economic prospects for the electric furnace process are

[87] W. L. Faith et al. in *Industrial Chemicals*, 4th ed., New York, 1975, pp. 928, 940.

[88] H. Harnisch and H.-H. Saenger, "Der Phosphor und seine anorganischen Verbindungen," in *Winnacker-Küchler, Chemische Technologie*, Vol. 1, *Anorganische Technologie*, 3rd ed., Munich, 1970, p. 344.

[89] H. W. Bryant et al., "Phosphorus Plant Design, New Trends," *Ind. Eng. Chem.*, **62**(4), 8 (1970).

[90] R. van Wazer, "Phosphorus and the Phosphides," in *Kirk-Othmer Encyclopedia of Chemical Technology*, 2nd ed., Vol. 15, New York, 1968, p. 276.

not very good because of the rising cost of electrical energy. Its competition against the wet process (i.e., conversion of calcium phosphate with sulfuric acid[91]) is turning more and more in favor of the latter. Purifying wet process phosphoric acid has progressed to a point at which the high-purity specification of the electric furnace acid may even be reached. In 1972 in the United States wet process phosphoric acid production had climbed to 5×10^9 kg, whereas the thermal process acid was stagnating at about 850×10^6 kg or was even slightly decreasing. The electric furnace acid is mostly used for technical purposes, especially for sodium polyphosphates in detergents. Only a very small fraction is used for phosphate fertilizers, which are made almost entirely from wet process acid.

30.3 COAL AND COKE AS FEEDSTOCK FOR CHEMICAL PRODUCTS

30.3.1 Use of Coke-Oven Products

Coal carbonization, especially at high temperature, is a very important process for utilizing coal, as described in detail in Chapter 14. Now, however, coal carbonization is examined as a source of feedstocks for the chemical industry (Chapter 18) and for uses other than supplying coke to the iron and steel industry (Section 30.2.1). Formerly, during the coal-based age of the chemical industry, coal carbonization even had a central position. However coal chemicals are a by-product of the coking process, and therefore their production is linked to the demand for coke by the steel industry. The growth rate in coke production, therefore in the production of coal chemicals, was much behind that of the chemical industry; consequently that industry turned to petroleum as a source of raw materials.

Coke-oven gas, which comprises approximately 12 wt % yield from the input of dry coal, was often separated at low temperature to obtain the components, hydrogen (which has a concentration around 55–60% by volume), carbon monoxide, methane, or even ethylene for further processing. At present this technique is used only in some older plants (e.g., in Yugoslavia). At this time it cannot be stated with certainty which of the older plants, built in the 1950s, are still operating.

Instead of using this low-temperature process for separating the coke-oven gas into five or six pure gas components, a modified and much simpler separation technique is presently being used. This so-called regenerator process, which is considered a more economical version, has only two end components. One fraction is pure hydrogen, and the residual fraction with all the other gaseous components is usable as a heating gas.[93]

A corresponding technique has been used in the following plants[94]:

1. At the U.S. Steel Corporation in Clairton, Pennsylvania, 300,000 m_n^3/hr coke-oven gas is separated into hydrogen, a heating gas, ethylene, and benzene; the attendant NH_3 production is 1000×10^3 kg/day.
2. At Houillières du Bassin du Nord et du Pas de Calais in Mazingarbe, France, 51,000 m_n^3/hr coke-oven gas is separated into hydrogen, methane, and ethylene, with an attendant production of 380×10^3 kg NH_3/day.
3. Ensidesa at Aviles, Spain, produces 184×10^3 kg NH_3/day from an original 27,000 m_n^3/hr coke-oven gas.

The recovery of pure hydrogen from coke-oven gas by means of pressure swing adsorption is in the developmental stage and will provide a further cost reduction, especially in medium-size hydrogen recovery units below 15,000 m^3/hr. Pilot plants are operating at coke-oven plants of Saarbergwerke AG and Ruhrkohle AG, and in connection with the hydrogenation pilot plant of Bergbau-Forschung GmbH, Essen, West Germany.[95]

The H_2S and NH_3 in the coke-oven gas are mostly converted to ammonium sulfate. The yield is about 1% of the dry coal input. The fraction of crude light oil, which is obtained in about the same yield, is a valuable raw

[91] T. P. Hignett and M. M. Striplin, *Chem. Eng. Prog.*, **63**(5), 85 (1967).

[92] G. Collin, *Erdoel Kohle, Erdgas, Petrochem.*, **29**, 159 (1976).

[93] Communication Linde AG Hoellriegelskreuth.

[94] Linde AG, pamphlet, May 1977.

[95] Bergbau-Forschung GmbH, pamphlet, 1977.

material for pure aromatics, especially benzene. The chemical utilization of many fractions and components of the coal tar[92] induced the origin of many chemical specialties (pharmaceuticals, pesticides, dyestuffs, plastics, etc.). Coal-tar yield is around 3% of dry coal input. Coke as the solid product of coal pyrolysis may be used as a chemical feedstock by processing it to calcium carbide and acetylene, thus developing an important route to aliphatic chemical intermediates (see Section 30.3.4). Finally, the condensates of coal pyrolysis may be utilized for their phenolic compounds.

Even now, in an age when petrochemistry still prevails, the chemical uses of the products of coal carbonization (e.g., benzene, ammonium sulfate, and some coal-tar products) are still significant. It is estimated that world steel production will grow from the present 350×10^9 kg to 500×10^9 kg in 1985, which would increase the quantity of byproducts.[92,96] However the preconditions for this increase in by-product production are no further reduction of the specific coke consumption per ton of steel production (Section 30.2.1.1), no further increase of direct reduction processes (Section 30.2.1.2), and construction of plants for recovering the byproducts.

In West Germany 370×10^6 kg of crude light oil was processed into 260×10^6 kg of pure benzene with a yield of approximately 70% in 1974. The residual fractions were used as gasoline and solvents. Compared to the production of coal tar (1.35×10^9 kg), the benzene production was about 20%. The additional benzene production based on petrochemical feedstocks in West Germany was 750×10^6 kg in 1974; as a result, the fraction of coal-based benzene amounted to approximately 25% of the overall production. In other countries the fraction of coal-based benzene production was significantly smaller. For example, in the United States the portion of benzene produced in coke-oven plants (275×10^6 kg) was only 5% of the overall production (5055×10^6 kg) in 1974. Worldwide, coal-based benzene production could be increased to about 3×10^9 kg (20% of the tar production of about 16×10^9 kg) if all possible sources were used, as they are in West Germany. The actual production, however, is still smaller.

The available quantity of coal tar from coal carbonization is now 16×10^9 kg/year. In Germany about 500 different chemical products are manufactured from coal tar. For several aromatic compounds, coal tar up to now has been the exclusive source of raw material. As the production of anthracene from coal tar was not sufficient to satisfy the increasing demand, new synthetic processes have been developed.

Naphthalene is the individual compound present in the highest concentration of 10 wt % of the tar. Tar oils are used for carbon black production (Section 30.4.5). Pitch is produced from coal-tar distillation with a yield of around 50 wt %. It is a valuable raw material for carbon electrode production (Section 30.4.1).[92]

It is usually difficult to justify the application of low- or medium-temperature processes of coal carbonization economically, except under special local conditions of cheap coal supplies and high cost of competitive products from petroleum or natural gas (India, for instance, is building low-temperature coal carbonization plants to produce a smokeless domestic fuel). The economics of low- or medium-temperature carbonization processes may change in the future because of the increased output of liquid hydrocarbons from such processes. The modern Lurgi-Ruhrgas process should be mentioned in this context.[97] The combination of low-temperature coal carbonization followed by upgrading of the products by hydrogenation may well gain economic advantages compared with the direct hydrogenation of coal, mainly because of the reduced hydrogen consumption, which is always a critical economic factor. Further combined processes are under development (e.g., for the prepyrolysis of coal before burning). Here, however, some other considerations are involved: the simultaneous production of gas and electricity, and the decrease of the sulfur content of coal.

30.3.2 Coal Hydrogenation

Coal hydrogenation for the production of liquid hydrocarbons is treated in detail in Chapters 27–29, and we now review the

[96] *Chem. Week,* 11 (February 12, 1975).

[97] H. Döring et al., *Erdoel Kohle, Erdgas, Petrochem.,* **28,** 225 (1975).

derivation of chemical intermediates from coal hydrogenation.

The original industrial application of coal hydrogenation, before and during World War II in Germany, using the Bergius-Pier process, aimed at the production of gasoline. This objective also dominates the present developmental efforts. Besides producing gasoline, the high-boiling hydrocarbon fractions from coal hydrogenation have also attracted much interest as a possible substitute for petroleum oil fractions. These fuel oils might be produced from coal with reduced hydrogen consumption [solvent-refined coal (SRC) processes].

Moreover, hydrogenation of coal may become a valuable source of aromatic hydrocarbons as raw materials for the chemical industry. Since benzene, toluene, and the xylenes are by-products of coal carbonization, the yield of those products is fixed and the quantities produced are by far insufficient to satisfy the market demand (Section 30.3.1). Almost the entire production of aromatics is now derived from petrochemical feedstocks. In 1975 the production figures in the United States, Western Europe, and Japan were 12.3×10^9 kg of benzene, 5.7×10^9 kg of toluene, and 5.1×10^9 kg of mixed xylenes.

To obtain a high yield of low-boiling aromatic hydrocarbons, hydrogenation must be extensive in the coal liquefaction process. Bituminous coal gives fractions consisting predominantly of ring compounds in the liquid products, which may be converted to aromatic compounds by a successive reforming step. This was done in the original German coal liquefaction plants (Bergius-Pier process), when aviation gasoline was upgraded by adding highly aromatic compounds. Also, medium-boiling hydrocarbon fractions (middle oil) were hydrogenated with catalysts at 400°C in gas phase to yield a maximum output of gasoline. These fuels contained up to 59 wt % naphthenic and 10 wt % aromatic compounds. They were treated in a reforming process (DHD process) developed in 1939 just for that purpose. The total output after the reforming step was up to 88 wt % with respect to the gasoline input, and its aromatics concentration was increased to 50 wt %.[98] In addition, highly aromatic gasolines were obtained by direct catalytic gas phase hydrogenation at 500°C. Presently available reforming processes for the dehydroisomerization with noble metal catalysts are much more effective, so that the results will be improved considerably. Separation processes for the isolation of the individual aromatics are also available from petrochemical manufacturing (extraction and rectification) processes.

Recent process developments in coal hydrogenation could also be used for aromatics production. The experimental hydrogenation of Illinois No. 6 coal according to the H-Coal process of Hydrocarbon Research, Inc., gave liquid fractions with relatively high concentrations of naphthenic and aromatic compounds. The gasoline fraction C_4—477 K contained 17.5 wt % alkylbenzenes, 51.3 wt % saturated naphthenes, and 11.2 wt % unsaturated naphthenes.[99]

This type of research has already been investigated in the earlier development programs on coal liquefaction by the U.S. Bureau of Mines (USBM).[100–102] The observed concentration and composition of the naphthenic compounds in the hydrotreated gasoline fraction and the originally formed aromatics permit one to assume that by application of platforming as the reforming process, the following final distribution of aromatics would result: benzene, 25 wt %, + toluene, 40 wt % + C_8-aromatics, 35 wt % = 100 wt % of all aromatics.

To streamline the aromatics output according to market demand, a considerable fraction of the toluene output should be converted to benzene by hydrodealkylation. From the published data an overall yield of around 25 wt % C_6–C_8-aromatics related to the maf coal input can be calculated.

Research work on coal hydrogenation at the University of Utah also aimed at the production of chemical raw materials[103,104]

[98] *Ullmanns Encyklopädie der technischen Chemie,* 3rd ed., Vol. 10, Munich, 1958, p. 531.

[99] K. C. Hellwig et al., *Brennstoff-Chemie,* **50** 263 (1969).

[100] J. Feldman and M. Orchin, "Composition of Gasoline from Coal Hydrogenation," *Ind. Eng. Chem.,* **44,** 2852 (1952).

[101] E. E. Donath, "Fuels and Chemicals from Coal Hydrogenation," *Ind. Eng. Chem.,* **46,** 2032 (1954).

[102] "Coal Hydrogenation—More Cost Data," *Chem. Eng.,* **59**(6), 159 (1952).

[103] S. A. Qader et al., "Coal Can Also Yield Liquid Fuels," *Hydrocarbon Process.,* **48**(9), 147 (1968).

using $ZnCl_2$, $SnCl_2$, and $(NH_4)_2MoO_4$ as catalyst materials for the coal hydrogenation process. To increase the output of chemical raw materials, the processing steps of hydrotreating, hydrocracking, and solvent extraction of aromatics were included with a "BTX from coal" process. It was shown that 33 wt % yield of C_6–C_{9+}-aromatics could be obtained related to the input of maf coal. Within the output structure of the valuable products, the BTX aromatics would comprise around 40 wt %, the other products being fuel gas, motor gasoline, and fuel oil.

In 1967 the U.S. Office of Coal Research (OCR) launched a comprehensive evaluation of the technical and economic prospects of the production of chemical raw materials from coal hydrogenation.[105] The design concept was based on the COED and H-Coal processes for coal hydrogenation and several processing steps for the primary products. Material balances and cost estimates were developed for two basic plant capacities with 50,000 and 100,000 bbl/day gasoline production and subsequent degrees of substitution of the fuels output through the production of chemicals. The profitability of a commercial plant would be increased considerably by the conversion of 25% of the gasoline production to BTX aromatics, and even more so by the design concept of an all-chemical refinery with production of naphthalene, anthracene and other condensed aromatics, and tar acids, and the production of ethylene, propylene, butadiene, and butenes through the cracking of gaseous hydrocarbons. Approximately 50% of the valuable products of such an all-chemical refinery would be chemicals, the remainder being fuels (gasoline, fuel gas, fuel oil, coke).

For further treatment of coal hydrogenation, see Chapter 28. Chapter 29 discusses the processing of primary products.

30.3.3 Manufacture and Utilization of Synthesis Gas

30.3.3.1 *The Competitive Position of Coal-Based Synthesis Gas*

A mixture of CO and H_2 can be used for many chemical syntheses, of which the syntheses of NH_3 and methanol and the Fischer-Tropsch synthesis are the most important. But other products are or could be synthesized from this gas from the thermodynamic point of view.[106] Synthesis gas is also a source for pure hydrogen and pure CO. With coal as feedstock, synthesis gas or hydrogen can be produced from coke-oven gas as described in Section 30.3.1 or by the gasification of coal or coke using the heterogeneous steam carbon reaction (see Chapter 24). Presently synthesis gas from coal gasification is generally not competitive with methane reforming using natural gas as feedstock, nor with naphtha reforming. Even partial oxidation of oil may still have an economic advantage. At present coal gasification processes are used only for the production of ammonia and for the manufacture of hydrocarbons by Fischer-Tropsch synthesis. Ammonia-producing plants based on coal are operating in countries with cheap lignite or coal resources such as Greece, Turkey, India, Pakistan, Zambia, and the Republic of South Africa. Recently in some of these countries several large coal-based ammonia plants were built with a daily capacity of 10^6 kg of NH_3. The Fischer-Tropsch synthesis is practiced on a large industrial scale in South Africa, where a large new plant is under construction to produce 2.1×10^9 kg/year of salable products (mostly liquid motor fuels). The manufacture and processing of synthesis gas involves the following steps: (1) gasification of coal with steam and oxygen; (2) compression of raw gas, in the case of low-pressure gasification processes; (3) purification of raw gas and adjustment of the necessary ratio of H_2 to CO; and (4) processing of the clean synthesis gas to the final product. These steps have been discussed in more detail in Chapters 24 and 25.

30.3.3.2 *Preparation and Purification of Coal-Based Synthesis Gas by Commercially Available Processes*

For the production of synthesis gas by coal gasification, only three processes are commercially applied today: the Winkler, the Lurgi pressure, and the Koppers-Totzek gas-

[104] S. A. Qader, "Process Yields: BTX from Coal for 25¢/gal." *Oil Gas J.*, 58 (July 22, 1974).

[105] *Chemical By-Products from Coal,* U.S. Office of Coal Research, R & D Rep. No. 43, November 1968.

[106] J. D. Loosen, IEA Working Party on Coal Technology, meeting on Basic Coal Science, January 27–28, 1977.

ification processes. In all these processes coal or coke is gasified by steam and oxygen, resulting in a gas containing CO, H_2, CO_2, and CH_4 as the main constituents; only in the case of the Lurgi pressure-gasification process is methane produced in higher concentrations.

The mechanism of gasification is discussed in more detail in Chapters 23 and 24 and is not covered here. This material duplicates some of the information presented in Chapter 24; however the discussion is in relation to the use of the processes for synthesis gas production.

Coal fed into a hot inert atmosphere is decomposed, resulting in the pyrolysis reaction:

$$\text{coal} \dashrightarrow \text{coke} + \text{liquids} + \text{gases} \quad (30.30)$$

The type of process determines which secondary reactions take place with the liquid or gaseous pyrolysis products. In the case of countercurrent processes such as the Lurgi process, they leave the reaction vessel and can be recovered in the purification steps. When cocurrent processes are used, these primary products can react with steam and oxygen, resulting in additional quantities of CO, CO_2, and H_2 according to reactions 30.35 and 30.36, discussed later. The carbon content of the coke formed during the pyrolysis reaction can react with the steam according to the well-known highly endothermic heterogeneous steam-carbon reaction:

$$C + H_2O \text{ ----> } CO + H_2 \qquad \Delta H_{298\ K} = +118.0 \text{ kJ/mole} \quad (30.31)$$

Also the following reactions, with lower significance for the total gasification process, take place. The equilibrium of the homogeneous shift reaction

$$CO + H_2O \text{ ----> } CO_2 + H_2 \qquad \Delta H_{298\ K} = -42.3 \text{ kJ/mole} \quad (30.32)$$

is reached at lower temperatures only if the coal ash contains suitable catalytic components. The Boudouard reaction

$$C + CO_2 \text{ ----> } 2CO \qquad \Delta H_{298\ K} = +160.8 \text{ kJ/mole} \quad (30.33)$$

becomes significant only at high temperatures from the thermodynamic and kinetic points of view; and the direct formation of methane according to reaction 30.34

$$C + 2H_2 \text{ ----> } CH_4 \qquad \Delta H_{298\ K} = -87.5 \text{ kJ/mole} \quad (30.34)$$

has some importance only in higher pressure ranges. On the other hand, methane formed according to Reaction 30.30 or 30.34 can react with steam according to the following reactions:

$$CH_4 + H_2O \text{ ----> } CO + 3H_2 \qquad \Delta H_{298\ K} = +206.0 \text{ kJ/mole} \quad (30.35)$$

or

$$CH_4 + 2H_2O \text{ ----> } CO_2 + 4H_2 \qquad \Delta H_{298\ K} = +167.0 \text{ kJ/mole} \quad (30.36)$$

To provide the heat of reaction required for the endothermic reactions 30.31, 30.33, 30.35, and 30.36, part of the coal has to be burned with oxygen, according to reactions 30.37 and 30.38:

$$C + O_2 \text{ ----> } CO_2 \qquad \Delta H_{298\ K} = -406.1 \text{ kJ/mole} \quad (30.37)$$

$$C + \tfrac{1}{2} O_2 \text{ ----> } CO \qquad \Delta H_{298\ K} = -123.1 \text{ kJ/mole} \quad (30.38)$$

For the production of synthesis gas, only the quantity of CO + H_2 from the coal can be used directly, and the CO/H_2 ratio can be adjusted by the shift reaction (reaction 30.32). CO_2 cannot be used without production of further hydrogen because of the water-gas shift equilibrium. CH_4 can be converted into CO and H_2 by the methane-reforming reaction 30.35, which can be performed after the recovery of the methane from the pure gas. In this way the processes producing a maximum of CO + H_2 in the raw gas are especially advantageous for the production of synthesis gas, whereas the processes leading to a high methane content in the raw gas are suitable for the production of substitute natural gas (SNG) and town gas or for a combination of synthesis gas and town gas production, as performed at the Sasol plant.

The oldest of the modern processes is the Winkler process. It was used for the first time in 1926 for the production of fuel and water gas at Leuna, Germany. According to the engineering contractor (Davy Powergas GmbH, Cologne), 18 plants have been built to date, with 38 gasifier units. Of these, eight have been onstream up to the present time, and three (in Yugoslavia, Turkey, and India) in the Western world.[107] The main feedstock has been lignite (e.g., in East Germany, West Germany, India, and Turkey), but bituminous coal has also been used in Spain and Japan. Some characteristic properties of the feed coal are given in Table 30.8.

The Winkler generator is essentially a brick-lined cylindrical shaft, the lower part of which is drawn to a truncated cone. The coal fed into the gasifier is gasified with steam and oxygen first in a fluidized bed in the lower conical part of the apparatus. After size reduction by gasification and abrasion, the particles are entrained and then gasified further in a suspension zone above the fluidized bed. A major part of the gasification residue leaves as fly ash. The gasification temperature depends on the reactivity of the coal gasified (e.g., 800°C for brown coal or 1100°C for bituminous coal in Japan).[108] The fusion characteristics of the mineral matter are very important. Severe difficulties are reported with the South Arcot lignite as feedstock.[108,109] In this lignite the presence of lumps of marcasite (FeS_2) lowers the fusion point of the ash to about 1000°C and results in frequent slagging of the generator. This problem has been solved by segregating the marcasite-containing heavier particles in the lower part of the gasifier below the oxygen nozzles, where the temperatures are lower. Some typical operating data of Winkler plants with lignite from West Germany,[110] lignite from Turkey, and a young bituminous coal from Spain are summarized in Table 30.9. As these figures indicate, the $CO + H_2$ content and the H_2/CO ratio are dependent on the kind of coal and the operating conditions, and they fall between the corresponding figures of the two other gasification processes under consideration. The throughput of one generator is reported to be 20,000 m_n^3/hr.[108] This process has been operated only under atmospheric pressure until now. The possi-

[107] Davy Powergas GmbH, private communication, Cologne.

[108] W. Flesch and S. Yagneswaran, *Chem. Age India,* **27**(9) (1976).

[109] C. V. S. Ratnam, *Fert. News,* April 1974.

[110] W. A. Flesch and S. Velling, *Erdoel Kohle, Erdgas, Petrochem.,* **15**, 710 (1962).

Table 30.8 Coal Properties Required by the Different Commercial Gasification Processes

	Process		
Properties Required	Winkler Fluidized Bed	Lurgi Fixed Bed	Koppers-Totzek Entrained Bed
Caking properties	No limitation for noncaking to medium caking coal	No limitation for noncaking to weakly caking coal	No limitations
Particle size	0–8 mm	Particles between 6 and 40 mm required; no fines	No limitations, but grinding to 90% < 0.1 mm (bituminous coal) or 80–85% < 0.1 mm (lignite) required
Ash content	Up to 50%	Up to 40%	Up to 40% by weight of dry coal admitted
Ash quality	Sintering point markedly beyond gasification temperature	Sintering point beyond temperature in combustion zone	Melting point of ash distinctly below gasification temperature
Moisture	Up to 8%	Up to 40%	Drying required to 1–2% (bituminous coal) or 8–10% (brown coal) before grinding

Table 30.9 Examples for Product Gas Composition and Coal and Oxygen Consumption for Different Gasification Processes

	Analysis of Coal, Gasified				Composition of Raw Gas			Coal Consumption[a]	Oxygen Consumption
Process	Carbon (%)	Oxygen (%)	Hydrogen (%)	Ash + Moisture (%)	CH_4 (%)	H_2 + CO (%)	H_2/CO	(kg coal, maf/m_n^3 CO + H_2)	(m_n^3 O_2/m_n^3 synthesis gas)
Winkler[107]	59.0	22.4	4.3	13.6	1.8	83	0.73	0.67	0.28
	54.3	15.5	3.7	23.7	2.5	76	1.13	0.75	0.28
	57.9	6.9	3.7	29.6	2.0	75	1.43	0.78	0.30
Lurgi[111]	50.8	8.8	2.8	35.9	12.1	58	2.21	0.82 (0.68)[b]	0.30 (0.26)[b]
Koppers-	40.3	15.7	3.1	38.0	0.1	85	0.52	0.78	0.39
Totzek[112]	69.7	12.2	5.2	10.9	0.1	87	0.59	0.54	0.39

[a] Only for gasification without steam or energy supply.
[b] Related to purified gas, containing CH_4 + H_2 + CO.

bility of running the Winkler generator under pressures up to 15 bar is under investigation.[113]

The Lurgi process, according to the Lurgi Mineralöltechnik GmbH,[111] has been proved in 14 commercial plants with a total of 65 gasifiers. The first plants (1936) were built for the manufacture of town gas from brown coal. Two plants have been in operation in Pakistan and Korea for NH_3 production, and one large plant consisting of 17 gasifiers is onstream in South Africa, producing 400,000 m_n^3/hr clean synthesis gas for conversion to liquid fuels, town gas, and NH_3. Current construction at that plant will result in a threefold increase in gas production capacity.

As shown in Table 30.8, coals with low caking characteristics are suitable for the Lurgi process. It was reported, however, that bituminous coals, even those with stronger caking properties,[114] can now also be gasified in Lurgi gasifiers (Chapter 24). Fractions with a particle size between 6 and 40 mm are optimal. Furthermore, the fusion characteristics of the mineral matter are extremely important; the fusion temperature of the ash should be higher than 1000°C. Compared to the other gasification processes, this process has the advantage of working under pressures up to 30 bar.

The Lurgi process operates as a countercurrent gasifier. Coal is fed at the top of the gasifier and is gasified as it moves slowly down. The reacting gases, steam, and oxygen are introduced at the bottom of the vessel and flow up. The ash is removed dry at the bottom. In this way there are three reaction zones in the gasifier; different in temperature, carbon and ash content of the solid, and composition of the gas phase, but overlapping one another. At the bottom, free oxygen reacts preferentially with coke to form CO_2 and CO, resulting in a small high-temperature zone (900–1200°C). The steam and the formed CO and CO_2 transfer the heat into the larger gasification zone with temperatures between 900 and 700°C, in which mainly the gasification of coke to H_2, CO, CO_2, and CH_4 takes place. In the pyrolysis zone at the top of the gasifier, the steam and the gases produced in the zones below heat up the fresh coal with the attendant formation of typical pyrolysis products at temperatures between 500 and 700°C. All product gases leave the reactor at about 500°C. From the typical operating data given in Table 30.9 it can be seen that the H_2 + CO content of the product gas is smaller than that of the product gases of the other gasification processes and that the H_2/CO ratio is very high. The purified gas, however, also contains 12.1% CH_4, which either can be converted into synthesis gas by methane reforming or can be used as town gas or SNG. Tar is also produced, and phenols and other products can be obtained from the gas liquor. In the Sasol plant the output per gasifier with an internal diameter of 3.60 m is 46,000 m_n^3/hr raw gas with a gasifier availability of 10.8 out of 13. For the new type of gasifier with an internal diameter of 4 m, an output of 50,000 m_n^3/hr raw gas is expected.[115]

In the Koppers-Totzek process, which has been in commercial use since 1950, dry pulverized coal is gasified at very high temper-

[111] Communication Lurgi Mineralöl GmbH, Frankfurt.
[112] H. Staege, *Erdoel Erdgas Z.*, **92,** 381 (1976).
[113] Anver Bögner (1976).
[114] H. Hiller, *Erdoel Kohle, Erdgas, Petrochem.*, **28,** 75 (1975).
[115] J. C. Hoogendoorn, *Gas Waerme Int.*, **25**(6), 283 (1976).

atures with oxygen and a small amount of steam in a gasifier, in which reactants are fed through two or four burners. In the future, eight or even more burners may be used to feed the reactants. The residence time of the coal dust in the reaction chamber is only a few seconds. To date, gasifiers have been operated at atmospheric pressure; however, a pressurized version is under development by a joint venture of Krupp-Koppers and Shell. According to Table 30.8, every kind of coal or coke can be gasified in the Koppers-Totzek process. Because the ash is removed as a liquid slag, the fusion temperature of the ash has to be below the gasification temperature. As shown in Table 30.9, the $CO + H_2$ content of the produced raw gas is very high and nearly independent of the kind of coal gasified. The H_2/CO ratio, however, is smaller than in the other gasification processes. According to the different heating values of the feedstocks, the consumption of lignite for the production of one cubic meter of gas, at standard temperature and pressure, is higher.[112] As Table 30.9 also indicates the oxygen consumption of the Koppers-Totzek process is significantly higher than that of the other processes, a consequence of the high gasification temperature. Twenty plants with 49 gasifier units have been constructed in Finland, Spain, Greece, Turkey, Zambia, the Republic of South Africa, India, and Thailand. Twelve are still in operation, among them some large plants with an output of 2.1×10^6 m_n^3/day, corresponding to an NH_3 capacity of 10^6 kg/day.

Table 30.10 summarizes the synthesis gas capacities, the daily production, and the type and annual consumption of coal for all operating coal gasification plants in the Western world, according to Franzen[116] and to the suppliers of the plants. The coal consumption could only be roughly estimated from the average consumption for the production of 10^3 kg of product, including energy consumption, under the assumption that all plants operate 333 days/year. For the production of NH_3, a coal consumption of 58.5×10^3 kJ/kg NH_3 has been assumed.

It can be seen from Table 30.10 that under these assumptions, approximately 4×10^9 kgCE/year is consumed by the production of NH_3 from coal in the Western world; 27% of this coal consumption is lignite, 73% bituminous coal. The fractions of the coal consumed in the individual gasification processes are Koppers-Totzek 86.7%, Winkler 9.3%, Lurgi 4.0%. The latter process, however, is used on a large scale for the production of hydrocarbons using the Fischer-Tropsch synthesis. The coal consumption of the Sasol plant after its enlargement to a synthesis capacity of 400,000 m_n^3/hr can be estimated to be 2.56×10^9 kgCE/year, assuming a coal consumption of 0.8 kg/m_n^3 synthesis gas. For 1982, a larger unit Sasol II is scheduled, with an estimated coal consumption of about 7.0×10^9 kgCE per year. The Lurgi gasification process will be used here also.

The purification of the raw gas coming out of the gasifier, which depends on the synthesis processes that follow, is discussed in more detail in connection with the different processes later on. The general aims of the gas purification plant are removal of dust from the cooled raw gas (if necessary the removal of tar), and the removal of phenol, NH_3, sulfur compounds, CO_2, HCN, and CH_4. In the case of low-pressure gasification processes, a compression after dust removal is necessary. There exist numerous purification processes for the single cleaning steps, which are summarized in Table 30.11. Furthermore, the H_2/CO ratio must be adjusted as required for the production of different final products, by conversion of CO into H_2 by the shift reaction. The various gas purification processes have been discussed in detail in Chapters 18 and 25. The water-gas shift was discussed in Chapter 20. None of these discussions is repeated here. The consumption of synthesis gas per thousand kilograms of product and its composition are given in Table 30.12 for different final products.

30.3.3.3 *Ammonia Synthesis*

Ammonia is produced by the reaction:

$$3H_2 + N_2 \longrightarrow 2NH_3 \qquad \Delta H_{298\,K} = -91.9 \text{ KJ/mole} \quad (30.39)$$

This process, used commercially since 1913, is most important for nitrogen fixation. The

[116] J. E. Franzen, Technical Communication Krupp Research, Rep. No. 35, Booklet 1, p. 183, 1977.

Table 30.10 Application of Conventional Gasification Processes in the Western World for Production of Synthesis Gas and Its Conversion to NH_3 and Hydrocarbons

				Capacity		Coal Consumption		
Product	Process	Number of Plants	Number of Gasifiers	Synthesis Gas (m_n^3/day)	Product (10^3 kg/day)	Type	Annual Consumption (10^3 kgCE)[a]	Countries
NH_3	Winkler	3	6	1,200,000[b]	530	Brown coal	353,000[c]	Yugoslavia, India, Spain, Turkey
	Lurgi	2	5	510,000[b]	230	Hard coal	153,000[c]	Pakistan, Korea
	Koppers -Totzek	6	14	2,270,000[b]	1030	Brown coal	686,000[c]	Zambia, South Africa, India, Greece, etc.
		6	14	8,590,000[b]	3920	Hard coal	2,611,000[c]	
				13,010,000[b]	5710		3,803,000[c]	
Hydrocarbons, NH_3, town gas	Lurgi	1	17	9,600,000[d,e]		Hard coal	2,560,000[e,f]	South Africa, Sasol I
	Lurgi	1[g]	39[g]	26,400,000[d,g]		Hard coal	7,040,000[g]	South Africa, Sasol II[g]

[a] 1 kgCE (coal equivalent) = 29.3×10^3 kJ = 27.8×10^3 Btu.
[b] $H_2 + CO$.
[c] Calculated from the NH_3 capacity under the assumptions of 8000 hr/year (= 333 days/year) and a coal consumption of 58.5×10^3 kJ/kg NH_3.
[d] $H_2 + CO + CH_4$.
[e] After plant enlargement to 17 gasifiers.
[f] Calculated under the assumption of 0.8×10^3 kgCE consumption/m_n^3.
[g] Scheduled for 1982.

Table 30.11 Processes for Purification and Treatment of Synthesis Gas[112]

Process Type	Process Name	Substances Removed	Washing Agent or Catalyst
Adsorption		traces of H_2S + S org	Activated carbon
		CO_2 + traces of H_2O	Molecular sieves
		CO_2 + traces of H_2O	Silica
Physical absorption		$CO_2 + H_2S$	Water under pressure
	Selexol	$CO_2 + H_2S$ + COS	Dimethylether or polyethylene glycol
	Rectisol	$CO_2 + H_2S$ + S org	Methanol
	Purisol	$CO_2 + H_2S$	*N*-Methylpyrrolidone
	Fluor solvent	$CO_2 + H_2S$	Propylene carbonate
Condensation		CO + Ar + CH_4	Liquid nitrogen
Chemical absorption	Girbitol	$CO_2 + H_2S$	ethanolamine (mono-, di-, tri-,)
	Shell-Adip	$CO_2 + H_2S$ + COS	Diisopropanolamine
	Alkazid DIK	H_2S	Dimethylglycine
	Alkazid M	$CO_2 + H_2S$	*N*-Methylamine
		$CO_2 + H_2S$	Potassium carbonate
	Benfield	$CO_2 + H_2S$ + COS	Potassium carbonate with additive
	Giammarco-Vetrocoke	H_2S	Sodium thioarsenate
	Shell Sulfinol	$CO_2 + H_2S$ + COS	Tetrahydrothiophene-dioxide + diisopropanolamine
	Stretford	H_2S	Sodium carbonate–anthraquinone
		CO	Copper formate–copper chloride
		Traces of CO_2	Ammonia water
		Traces of CO_2	Sodium hydroxide
		Sulfur compounds	Zinc oxide
		H_2S	Bog iron ore–luxmasse
Catalytic conversion		CO	Iron–copper-zinc catalyst
		CO; CO_2	Nickel catalyst
		Organic sulfur compounds	Cobalt molybdenum catalyst

contribution of by-product ammonia or ammonium sulfate from coke-oven operation and the nitrogen fixation via calcium carbide and calcium cyanamide are of minor importance. World ammonia capacity in 1961–1962 had scarcely reached 20 × 10^9 kg of NH_3, but it had climbed to 81.5 × 10^9 kg by 1975–1976. The capacity forecast for

[117] H. Staege, *Gas Waerme Int.*, June 1976.

Table 30.12 Synthesis Gas Consumption for Different Processes[117]

Process	Theoretical Gas Consumption ($m_n^3/10^3$ kg product)	Composition of the Gas: H_2 (vol %)	CO (vol %)	N_2 (vol %)	H_2/CO	Final Product
Ammonia synthesis	2735	75.0	—	25.0	—	NH_3
Methanol synthesis	2300	67.8	32.1	—	2.12	CH_3OH
Fischer-Tropsch synthesis	6000	33.3–66.7	66.7–33.3	—	0.50–2.00	F-T primary product
Oxosynthesis	1200	50.0	50.0	—	1.00	Oxoalcohols
Direct reduction of Fe_2O_3	675	33.3–66.7	66.7–33.3	—	0.50–2.00	Fe (92%)

1977–1978 was 98 × 10^9 kg of NH_3 worldwide.[118] Most ammonia plants operating in the world today are using natural gas as raw material, whereas the relative share of coal has declined constantly. In 1961–1962 the share of coke-oven gas and coal for ammonia production was still 18%, but it was reduced to 9% in 1971–1972. In the same time the share of natural gas climbed from 50 to 63%. The remaining ammonia capacity was based on petroleum products as raw material.[119]

Nevertheless, just recently several new ammonia plants based on coal gasification were constructed, especially in the countries that are interested in the utilization of their own coal deposits and must avoid excessive imports of raw materials in exchange for foreign currencies. As shown in Section 30.3.3.2, the annual NH_3 capacity (1976) in the Western world based on coal gasification can be estimated at 1.9 × 10^9 kg. In addition, the new regenerator process of low-temperature separation of coke-oven gas is the basis of an NH_3 capacity of approximately 0.55 × 10^9 kg (Section 30.3.1). To expand their foodstuffs production, many countries will have to enlarge their ammonia production facilities so that greater supplies of fertilizer can be made available for agriculture. Approximately three-quarters of ammonia production is used for fertilizers, and the remainder is spread over a wide variety of technical uses.

The current optimum production capacity of single-stream ammonia plants is around 10^6 kg/day and even more. In the early 1960s the technology of ammonia synthesis had been automated, and considerable gains in productivity had been made by single-train design, application of turbo compressors, and fully integrated energy systems. Synthesis pressure of modern plants is in the range of 300–400 bar.

Theoretical stoichiometric consumption of hydrogen is 1975 $m_n^3/10^3$ kg of NH_3; the practical requirement in most efficient new plants is 2100 $m_n^3/10^3$ kg. The consumption of coal-based ammonia plants constructed in 1970–1974 lies between 2150 and 2220 $m_n^3/10^3$ kg of NH_3.[116] A typical flowsheet of an NH_3 plant based on coal (Modderfontein, South Africa)[116] is given in Fig. 30.8; corresponding gas compositions (vol 1%) at various points in the plant are as follows:

	Process Stage				
	I	II	III	IV	V
Pressure (bar):	0.86	28.9	49.2	47.1	45.1
Temperature (°C):	32	28	35	−56	23
Gas					
CO_2	13.4	11.3	41.3	—	—
H_2S + COS	0.6	—	—	—	—
CO	54.5	56.0	3.0	5.0	—
H_2	30.0	31.0	54.6	93.1	75.0
N_2	1.0	1.0	0.6	1.1	25.0
CH_4	0.1	0.1	0.1	0.1	—

The gas leaving the gasifier (six gasifiers with two burners each; Stage I) is cooled in a heat exchanger producing high-pressure steam and dedusted in an electrical precipitator (dedusting to 0.5 mg/m_n^3). After compression to 30 bar, H_2S, COS, and other sulfur-containing compounds are removed using a washing process with methanol at low temperatures, the remaining concentration being at 2 ppm (Stage II). After a further compression to 53 bar, the shift conversion of CO to CO_2 and more hydrogen takes place (Stage III), followed by a second methanol washing process to remove the CO_2 formed (Stage IV). The remaining content of CO and CH_4 (~ 7%) is removed by a low-temperature washing process with liquid nitrogen, and the required H_2/N_2 ratio is adjusted at the same time (Stage V). After a compression of the gas mixture to 220 bar, the NH_3 synthesis takes place.

The coal consumption figures for NH_3 production given in the literature differ depending on the gasification process and the coal quality used, and depending on whether the coal consumption is related to gasification alone, to gasification plus steam production, or to the overall energy and coal consumption of the whole plant. According to a recent evaluation, the total energy and raw material consumption of a modern NH_3 plant would be 48.5 × 10^3 kJ/kg NH_3 when using pressure gasification of coal, whereas gasification of coal at atmospheric pressure would result in an overall consumption of 58.5 × 10^3 kJ/kg NH_3.[120] According to another reference[117] the

[118] *Nitrogen,* No. 101, 34 (1976).
[119] *Nitrogen,* No. 79, 5 (1972).

[120] K. J. Mundo, *Chem. Ing. Techn.,* **45,** 632 (1973).

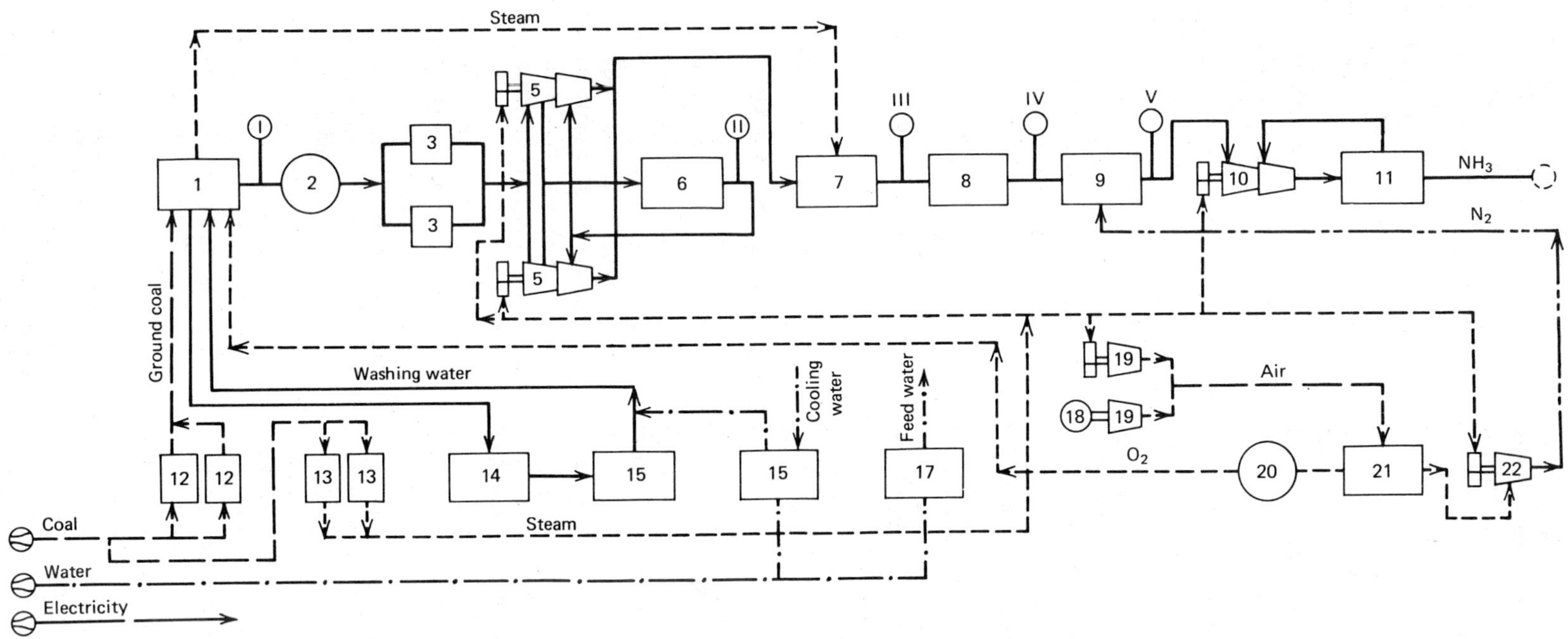

Figure 30.8 Flowsheet of an ammonia plant in Modderfontein, South Africa, based on coal gasification by the Koppers-Totzek process.[116] Components Numbered as follows: 1, gasifier; 2, raw gas storage vessel; 3, electrical dust preciitator; 4, turbine; 5, raw gas compressor; 6, sulfur removal, 7, shift reactor; 8, CO_2 removal; 9, liquid nitrogen low/temperature washing process; 10, synthesis gas compressor; 11, NH_3 synthesis; 12,coal preparation; 13, boiler; 14, settling tank; 15, cleaning water system; 16, cooling water system; 17, feed water preparation; 18, motor-driven air compressor; 19, steam-driven air compressor; 20, ox ygen storage vessel; 21, air seperation plant; 22, nitrogen compressor.

coal consumption per kilogram of NH_3 is 51.9×10^3 kJ, using the Koppers-Totzek process at atmospheric pressure. This would mean a consumption ranging between 1.66 and 2 kg of coal per kilogram of ammonia produced (lower heating value of coal H_u = 29.3 kJ/kg). Within integrated plants most of the energy requirements are covered by waste heat utilization, so that only a small additional amount of steam must be raised separately.

30.3.3.4 *Manufacture of Hydrogen*

In 1974 production of hydrogen in the Western world was 270×10^9 m_n^3. Of this amount, 212×10^9 m_n^3 or 79% was made from petroleum and natural gas. Hydrogen production from coal and coke had dropped to 44×10^9 m_n^3 or 16%, including the utilization of hydrogen for NH_3 production. The remaining 14×10^9 m_n^3 or 5% of the total production volume came from electrochemical processes and other sources.[121] The consumption pattern of hydrogen shows predominant uses in the chemical industry, especially for ammonia manufacture. In 1974 ammonia's share of hydrogen consumption in the world was 54%. Methanol synthesis and several other chemical processes (oxosynthesis, hydrogenation processes, hardening of liquid fats, etc.) together amounted to only 8%. However hydrogen supplies to these diversified chemical processes, which yield products of high value, are also of primary importance. Oil refineries used 22%, mainly for hydrocracking of gas-oil fractions and for hydrotreating processes (sulfur removal). Regional and temporary imbalances of production and consumption led to the use of 16% as fuel.

To convert synthesis gas or mixtures of CO and H_2 from coal gasification into pure hydrogen, it is necessary to desulfurize and shift convert the carbon monoxide with steam into more hydrogen and carbon dioxide, followed by the removal of CO_2 formed. Furthermore, residual CO may be removed by absorption in a copper-containing solution, or, in case of very low CO concentrations, by methanation. Adsorptive purification processes might also be applied (pressure swing adsorption, or PSA), which provide high-purity hydrogen (99 vol % and higher).[122,123] Sometimes it is desired to separate both components—H_2 and CO, to a high degree of purity—from synthesis gas for further chemical processing. In this case hydrogen (90 vol % concentration) is extracted by liquid methane in a low-temperature separation scheme. In an adjacent fractionation CO is separated from the methane.[124] The coal consumed in producing hydrogen is 700 kg/1000 m_n^3 H_2 (20.5×10^6 kJ/1000 m_n^3 H_2).

It is also possible to produce hydrogen from coal by separating it from coke-oven gas, which contains approximately 55% hydrogen by volume. Two processes are available for doing this: low-temperature separation (Chapter 18) and adsorption using suitable adsorbents (e.g., molecular sieves combined with a PSA technique). More details are given in Section 30.3.1.

30.3.3.5 *Methanol and Its Processing to Gasoline*

A very important use of synthesis gas is the conversion to methanol according to:

$$CO + 2H_2 \dashrightarrow CH_3OH \qquad \Delta H_{298\,K} = -91.9 \text{ kJ/mole} \tag{30.40}$$

Production of methanol in the Western world climbed in 1974 to approximately 7.4×10^9 kg, and in the United States alone to 3.1×10^9 kg. The annual production capacity was around 9×10^9 kg in the world. Presently manufacturing processes are based on petrochemical raw materials (natural gas, naphtha, gas oil, etc.).

Instead of the older, high-pressure synthesis (BASF or Union Kraftstoff–Wesseling process, which employs 350 bar, 350°C, and $ZnO-Cr_2O_3$-based catalysts) the modern low- or medium-pressure processes are dominating nearly all new plant installations. The leading low-pressure process of ICI already accounts for one-third of the world production. With the new catalysts based on Cu-Zn-Al, milder reaction conditions are possible (100 bar, 250°C). However these catalysts

[121] G. Kaske, *Chem.-Ing.-Tech.*, **48,** 95 (1976).

[122] E. Supp and H. Jockel, *Erdoel Kohle, Erdgas, Petrochem.*, **29,** 117 (1976).

[123] W. H. Scholz, *Proc. 8th World Petroleum Congr.*, Moscow, 1970.

[124] W. Förg, *GWF-Gas/Erdgas*, **113,** 538 (1972).

require very high degrees of purity of the synthesis gas, especially with respect to chlorine and sulfur compounds.[125–128] The similar low-pressure process of Lurgi[129,130] uses even lower pressures, in the range of 40–50 bar, whereas the medium-pressure processes are operated in the range of 100–250 bar, and mostly with copper-containing catalysts (Mitsui-Topsoe, Vulcan, Pritchard, Catalyst and Chemicals, Inc.). Methanol from the new processes is extremely pure (99.99 wt % methanol from the ICI process). Modern methanol plants are equipped with turbo compressors and have single-stream capacities up to $1–1.5 \times 10^6$ kg/day. Because it is generally assumed that the use of coal for chemical processes will grow in the future, the production of methanol via synthesis gas from coal has already been investigated thoroughly. By integrating and optimizing the gasification and synthesis sections to lowest raw material and energy consumption, production figures on the order of 43.9×10^6 kJ or 1.5×10^3 kg of bituminous coal (standardized to a lower heating value of bituminous coal of 29.3 kJ/kg) per thousand kilograms of methanol can be reached. A pressure gasification process is preferred. In case of a considerable output of liquids and methane during gasification (e.g., the Lurgi process), the coproduction of those byproducts would give better overall thermal efficiency than their recycling.[131] The Koppers-Totzek gasification process, operating at atmospheric pressure, can also be used for methanol production. The overall consumption of coal per thousand kilograms of methanol in this case is 51.8×10^3 kJ/kg.[132] There is no difficulty in adjusting the required H_2/CO ratio of around 2.2 in the methanol synthesis by using the shift reaction. The purity requirements of the synthesis gas for the low-pressure methanol synthesis are exceptionally high, so that a suitable purification process must be provided (e.g., the Rectisol process).

About 50% of the methanol is processed to produce formaldehyde. This product can be converted into the important chemical intermediate ethylene glycol using synthesis gas, a process that was operated in a commercial plant by Dupont until quite recently. Now Union Carbide Corporation is developing a direct synthesis of ethylene glycol from synthesis gas, which in the future could compete with the petrochemical manufacture via ethylene.[133]

Methanol is also used as a solvent, for the production of dimethyl terephthalate (DMT), as raw material for polyester fibers, for methyl methacrylate (plastics), methylamines, methylhalogenides, and many other chemical building blocks. Methanol could also gain importance as a substrate for fermentation to single-cell proteins (SCP). For SCP production, methanol would have some advantages over the competing raw materials methane and *n*-paraffins, including easier biological degradation, reduced oxygen consumption and energy requirements for the stirred fermenting reactors, and good mixing behavior with nutritient salt solutions.

In the energy sector methanol has proved to be suitable as a motor fuel.[134–137] The new methanol applications would require large-scale plant installations with annual plant capacities up to 10^{10} kg/year of methanol.[138]

According to the Mobil process,[139,140] which is still being developed, it may be economical in the future to produce gasoline

[125] J. W. Woolcock, *Chem. Process Eng.*, **49** (6), 69 (1968).

[126] D. H. Bolton, *Chem.-Ing.-Tech.*, **41**, 129 (1969).

[127] D. H. Bolton and D. Hanson, *Chem. Eng.*, **76**(22.9), 154 (1969).

[128] D. D. Mehta and D. E. Ross, *Hydrocarbon Process.*, **49**(11), 83 (1970).

[129] H. Hiller and F. Marschner, *Hydrocarbon Process.*, **49**(9), 281 (1970).

[130] E. Supp, *Chem. Technol.*, **3**, 430 (1973).

[131] P. F. H. Rudolph and P. K. Herbert, Symp. Coal Gasification, Liquefaction and Utilization, University of Pittsburgh, May 7–8, 1975.

[132] Communication Krupp-Koppers GmbH, Essen.

[133] A. M. Brownstein, *Chem. Eng. News*, **53**, 10 (March 31, 1975).

[134] M. Schaffrath, *Erdoel Kohle, Erdgas, Petrochem.*, **29**, 64 (1976).

[135] *Chem. Eng. News*, **53**, 12 (December 1, 1975).

[136] W. Hampton and M. R. Iammartino, *Chem. Eng*. **82**, 58 (July 21, 1975).

[137] G. A. Mills and B. M. Harney, *Chem. Tech.*, **4**, 26 (1974).

[138] D. P. Burke, *Chem. Week*, 33 (September 24, 1975).

[139] J. J. Wise and A. H. Silvestri, *Oil Gas J.*, 140 (November 22, 1976).

[140] S. L. Meisel, C. H. McCullough, C. H. Lechthaler, and P. B. Weisz, *Chem. Tech.*, 86 (February 1976).

synthetically via methanol, the latter being manufactured predominantly from coal. Chemically, the Mobil process incorporates a dehydration of the methanol to the dimethylether, which is subsequently converted to hydrocarbons. The overall equation may be written as:

$$x CH_3OH \dashrightarrow (CH_2)x + x H_2O \quad (30.41)$$

By using a new shape-selective zeolite catalyst with a unique channel structure for this reaction, the hydrocarbons are predominantly in the gasoline boiling range (C_4–C_{10}) and the gasoline is chemically conventional, with unleaded research octane numbers (RON) of 90–100. By comparison, RONs of Fischer-Tropsch gasoline as a primary product are much lower, because of the mostly aliphatic and unbranched nature of the hydrocarbons, so that further processing and blending with other products is necessary to upgrade the RON.

Typical process conditions and yields of the gasoline-from-methanol process are shown in Table 30.13. The total yield of hydrocarbons from crude methanol with 17% water is 36.2 wt %. Purification of the methanol by distillation is not necessary. With pure methanol the yield would be 43.6 wt %. Assuming a specific coal consumption of 1.5 kg (H_u = 29.3 × 10^3 kJ/kg coal) per kilogram of methanol, the coal consumption related to the hydrocarbons would be 3.4 kg/kg. This is a lower consumption figure than had ever been experienced with Fischer-Tropsch synthesis, which would require a coal input of at least about 4 kg/kg of hydrocarbons. It is remarkable that about 80% of the hydrocarbons produced are in the C_{5+} gasoline fraction. Since part of the low-boiling hydrocarbons may be converted into valuable gasoline by alkylation, the total yield of gasoline with this production pattern is even higher. As Table 30.13 shows, 85% gasoline, 13.6% LP gas, and 1.4% fuel gas may be obtained as end products.

Table 30.13 Operating Conditions of the Mobil Process and Yields from Crude Methanol[139,140]

Temperature (°C)	
Inlet	360
Outlet	415
Reaction pressure (bar)	21
Recycle ratio, mole/mole	9
(WHSV)[a]	2
Yields (wt % of charge)[b]	
Methanol, ether	0.0
Hydrocarbons	36.2
Water	63.2
CO, CO_2	0.3
Coke, other	0.3
	100.0
Hydrocarbon product (wt %)	
Light gas	1.4
Propane	5.5
Propene	0.2
i-Butane	8.6
n-Butane	3.3
Butenes	1.1
C_{5+}–gasoline	79.9
	100.0
Gasoline, including alkylate (wt %)	85.0
LP gas (wt %)	13.6
Fuel gas (wt %)	1.4
	100.0

[a] Weight hourly space velocity, the weight of limiting reactant fed per hour per unit weight of catalyst in the reactor.

[b] Crude methanol with 17 wt % water.

30.3.3.6 *Fischer-Tropsch Synthesis*

The Fischer-Tropsch process produces from synthesis gas mostly linear aliphatic hydrocarbons within a wide range of carbon numbers from gaseous products to high-boiling paraffins. Gases rich in H_2 or CO may be converted in the process as indicated by the following reactions:

$$CO + 2H_2 \dashrightarrow {-CH_2-} + H_2O \qquad \Delta H_{298\,K} - 164.7 \text{ kJ/mole}$$
$$2CO + H_2 \dashrightarrow {-CH_2-} + CO_2 \qquad \Delta H_{298\,K} = -204.4 \text{ kJ/mole} \quad (30.42)$$

The modern process versions of the Fischer-Tropsch synthesis all use iron catalysts. Up to now Fischer-Tropsch plants, mostly constructed during World War II, had a total capacity of around 1.5 × 10^9 kg/year. At present only one plant is in operation, Sasol I in South Africa, with a capacity of 270 × 10^6 kg/year (1970) and consuming 3.6 × 10^9 kg of coal. On the basis of the very cheap South African bituminous coal, a second plant, Sasol II, with a capacity of 1.5 × 10^9 kg/year is under construction. Most of the earlier plants were installed for the production of synthetic gasoline during World War II in Germany. However since that time Fischer-Tropsch plants were also designed for the production of coal chemicals. This will become important in the future.

Presently, three types of technically proved processes are available.[141–144] The Arge gas phase process of Ruhrchemie-Lurgi with fixed-bed catalysts was installed in the Sasol I plant and has operated there since 1955; it is preferred for the production of long-chain linear paraffins. The fresh gas from Lurgi pressure gasification has an H_2/CO ratio of 1.8 and is mixed with recycled tail gas in a ratio of 1 : 2. The pressure of the Fischer-Tropsch reaction is 25 bar, the temperature is 220–240°C.

The other commercial process at Sasol I is an entrained-bed reactor system that was invented by Kellogg and further developed by Sasol Corporation. The medium-pressure range is about the same as mentioned before; however the reaction temperatures are elevated to 310–340°C and a high H_2/CO ratio of around 6 is used. It is adjusted by mixing the fresh gas from Lurgi pressure gasification with recycled tail gas. Under these reaction conditions the synthesis produces primarily hydrocarbons in the gasoline boiling range. Also substantial quantities of oxygenated hydrocarbons are produced as by-products. With the entrained-bed system highest single-reactor capacities could be realized. This process alone will be used in the new Sasol II installation.

The liquid phase Fischer-Tropsch process (Rheinpreussen-Koppers) incorporates a bubble column reactor. The fine-grain precipitated iron catalysts are suspended in a fraction of high-boiling paraffins derived from the process itself.[143] The liquid phase Fischer-Tropsch process uses CO-rich gases with a CO/H_2 ratio of 1.5. A large pilot plant was operated with a synthesis pressure of 12 bar and a temperature of around 250°C. By variation of catalyst composition and reaction conditions, the process can achieve great flexibility with respect to the preferred boiling range of the hydrocarbons produced and their olefin contents. In the various hydrocarbon products the olefin contents may reach 90 wt % and more.

Fischer-Tropsch primary products may be processed to a large number of chemical derivatives. By means of the oxosynthesis, the olefins of the C_6–C_{10} hydrocarbons may be converted with synthesis gas to plasticizer alcohols. The olefins with higher carbon numbers (up to C_{18}) may be processed in the same way to detergent fatty alcohols. Generally Fischer-Tropsch products are straight chained. So the linear paraffins in the gas-oil range (C_{11}–C_{18}) are valuable raw materials for detergent manufacture. They could be used for the production of linear alkylbenzenesulfonates (LABS) as well as for paraffin sulfonates. The highest boiling paraffin waxes have a great number of applications in chemical specialties.[145]

To substitute the coal-based Fischer-Tropsch synthesis for the currently most important petrochemical processes, there would have to be an increase in the output of low-boiling olefins in the range of C_2–C_4, especially ethylene and propylene. Current developmental work is aimed at these targets. Because these gaseous products are diluted in the tail gas of the process, the separation by low-temperature or special absorption techniques is somewhat complicated and costly. Likewise, the primary production of a raw material by the Fischer-Tropsch process, which would have to undergo a thermal cracking operation in a subsequent step, is also not a favorable alternative route.[146]

With integrated Fischer-Tropsch plants (gasification, synthesis gas purification, Fischer-Tropsch synthesis, separation, and downstream processing of primary products) the overall coal consumption may be around 5 kg of bituminous coal per kilogram of C_{2+} primary products. Variations are due to different processes and reaction conditions.

As an example of an operating coal-based Fischer-Tropsch plant, Fig. 30.9 shows a simplified flowsheet of the Sasol I plant (sta-

[141] H. Pichler and A. Hector, "Carbon Monoxide–Hydrogen Reactions," in *Kirk-Othmer Encyclopedia of Chemical Technology,* 2nd ed., Vol. 4, New York, 1964, p. 446.

[142] J. Schulze, *Chem.-Ing.-Tech.,* **46,** 925 (1974).

[143] H. Kölbel et al., *Erdoel Kohle,* **9,** 153, 228 (1956).

[144] D. D. Frohning and B. Cornils, *Hydrocarbon Process.,* **53,**(11), 115 (1974).

[145] J. Schulze, *Chem.-Ing.-Tech.,* **46,** 976 (1974).

[146] G. M. Drissel, paper presented at Symp. Comparative Economics for Synfuels Processing, American Chemical Society, New York, April 8, 1976.

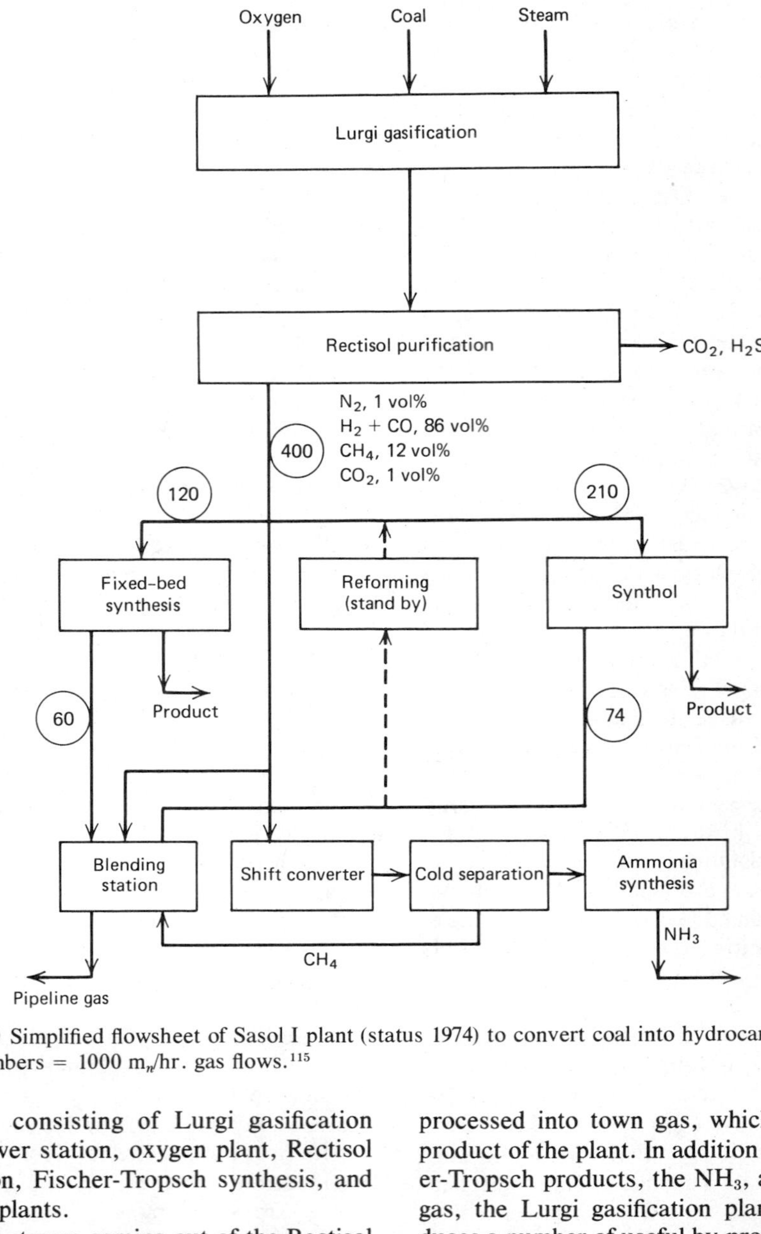

Figure 30.9 Simplified flowsheet of Sasol I plant (status 1974) to convert coal into hydrocarbons and gas; circled numbers = 1000 m_n/hr. gas flows.[115]

tus 1974) consisting of Lurgi gasification plant, power station, oxygen plant, Rectisol purification, Fischer-Tropsch synthesis, and ammonia plants.

The gas stream coming out of the Rectisol purification is divided into three streams. The first is converted by the fixed-bed process into the products shown in Fig. 30.10 and the second by the entrained-bed Synthol process into the range of products shown in Fig. 30.11.

In addition, the third gas stream is converted into NH_3 by shift conversion and cold separation with nitrogen as pretreatment. The separated methane is mixed in a blending station with the tail gases coming from the two Fischer-Tropsch synthesis units and processed into town gas, which is a major product of the plant. In addition to the Fischer-Tropsch products, the NH_3, and the town gas, the Lurgi gasification plant itself produces a number of useful by-products including benzene, aromatic solvents, ammonia, and phenols.

The Sasol II plant, with a total capacity of approximately 2.1 × 10^9 kg of salable products per year, is also based on Lurgi gasification in combination with the entrained-bed Synthol process only. The main factors in favor of the Lurgi gasifier are that gas can be produced at a pressure suitable for Fischer-Tropsch synthesis, which avoids the compression of raw gas, and that on the scale planned, the by-products could be further

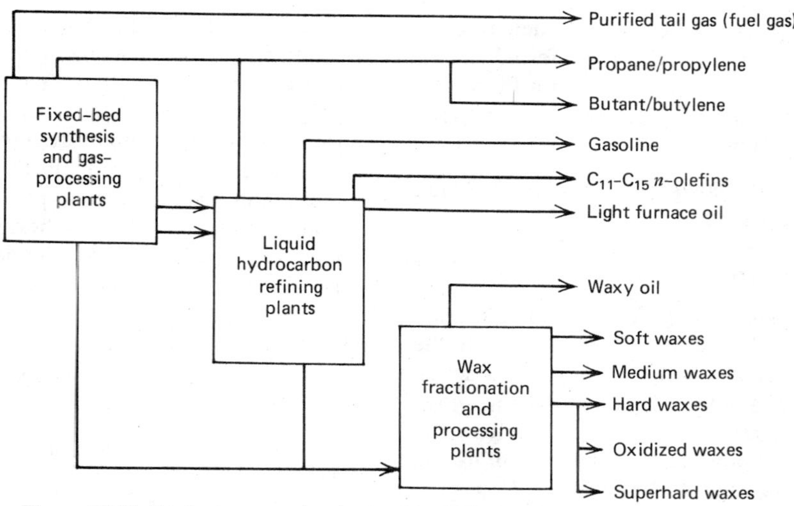

Figure 30.10 Product processing in fixed-bed Fischer/ Tropsch synthesis.[115]

processed in economically sized plants, thus improving the economy of the whole operation.

The reasons for the choice of the Synthol process for hydrocarbon production were as follows[115,147]:

1. In the Synthol reactors removal of reaction heat is easily controlled by the amount of catalyst circulating and by the additional coolers in the reactor system.
2. The system lends itself to high gas loads and a high production per reactor unit.
3. In view of the nature of the entrained system, operating conditions are chosen such that the product selectivity is mainly aimed at hydrocarbons in the gasoline range, and production of higher hydrocarbons is avoided. This makes the Synthol reactor an excellent motor fuel producer, and the lighter olefins such as propylene and butylene can easily be converted into gasoline.
4. The production of oxygenated chemicals such as alcohols and ketones is an additional benefit, and these valuable chemi-

[147] J. C. Hoogendoorn, "Gas from Coal for Synthesis of Hydrocarbons," paper presented at American Institute of Mining Engineers 23rd Annual Meeting, 1974.

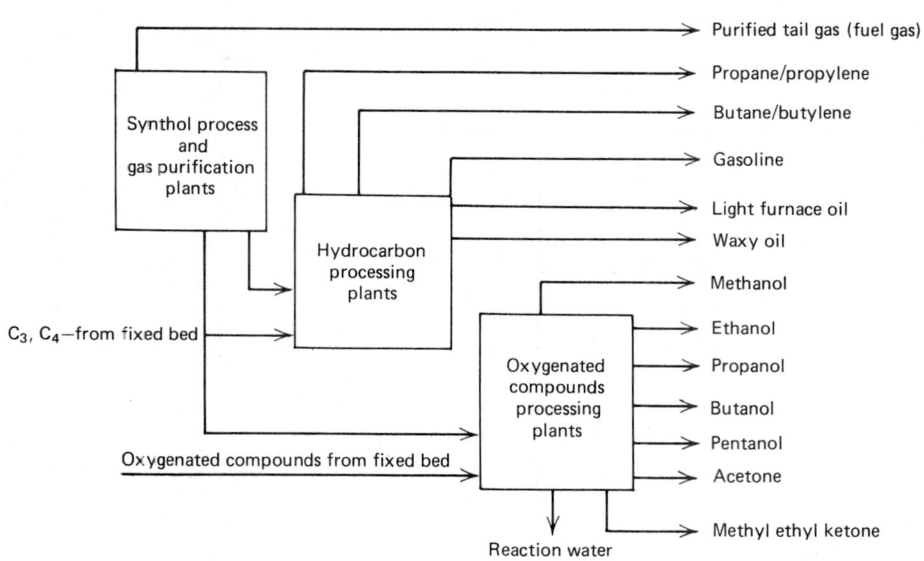

Figure 30.11 Synthol entrained-bed Fischer-Tropsch synthesis, for the product flow diagram.[115]

cals can either be sold as such where a market exists or, if it is preferred to limit the range of products to motor fuels, they can be converted into the corresponding hydrocarbons.

A simplified flowsheet of the Sasol II plant is given in Fig. 30.12.

30.3.3.7 *Oxosynthesis*

Oxosynthesis or hydroformylation is the addition of CO and H_2 to olefins with the formation of aldehydes in the first reaction step, the aldehydes subsequently being mostly hydrogenated to alcohols. Propylene, for instance, is converted to a mixture of *n*- and *i*-ibutyraldehyde (~ 75 and 25 wt %):

The reaction is catalyzed by cobalt compounds at 200–300 bar and 140–180°C. The chain length range of the reacting olefins is C_2–C_{20}. The oxoalcohols C_4–C_6 or their respective ester compounds are mostly used as solvents. The ester products of the C_8–C_{12} oxoalcohols are important plasticizers. A volume product is 2-ethyl hexanol (2EH), which is formed by aldol condensation of butyraldehyde to 2-ethyl hexenal and its subsequent hydrogenation. The oxoalcohols above C_{12} are mainly used in detergent manufacture and for several chemical specialties.[148–151]

[148] T. Mori, *Chem. Econ. Eng. Rev.*, 3(5), 47 (1971).

$$CH_3CH = CH_2 + CO + H_2 \begin{cases} \dashrightarrow CH_3CH_2CH_2CHO \\ \dashrightarrow CH_3CH(CHO)CH_3 \end{cases} \qquad (30.43)$$

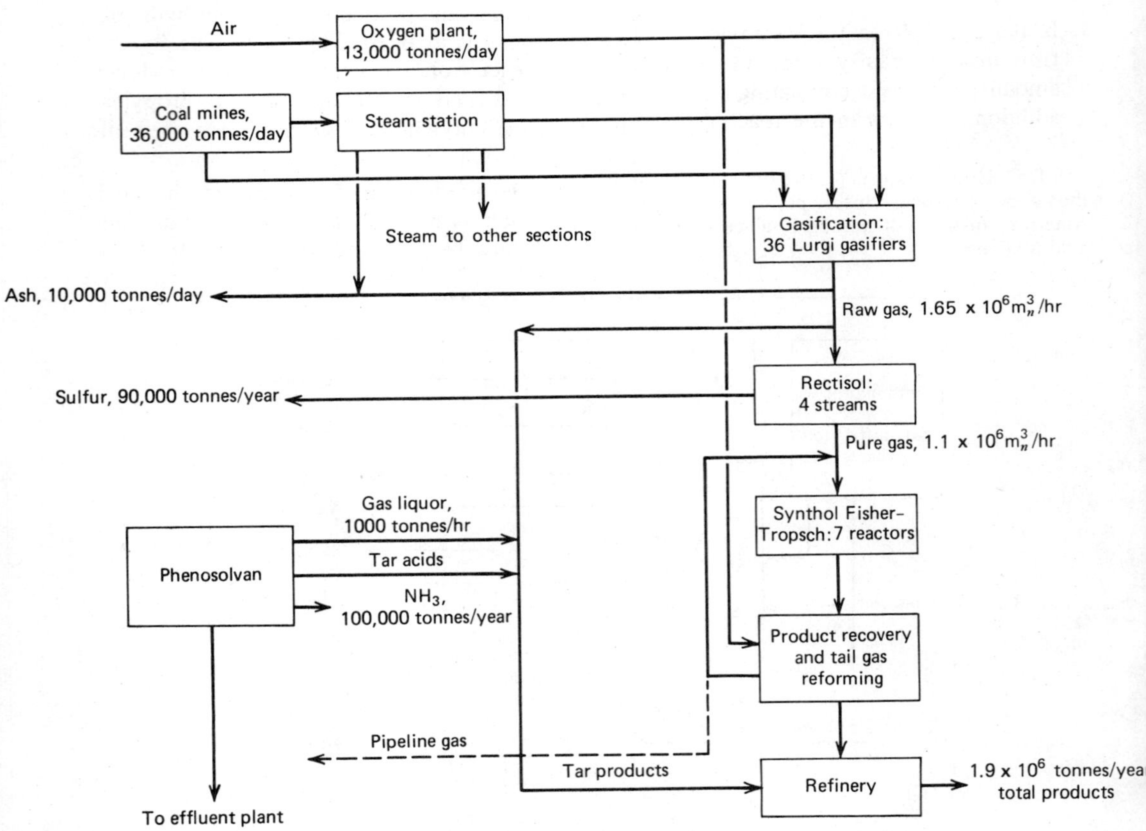

Figure 30.12 Schematic block flow diagram for Sasol II.[115]

The production of all these alcohols via synthesis gas has become a very economical process route, so that production figures could grow considerably. In 1974, world capacity of oxoalcohols was estimated at 3.5×10^9 kg. Synthesis gas for the oxosynthesis as well as hydrogen for the subsequent hydrogenation processes could easily be derived from coal. Coal could even be the exclusive raw material as soon as the olefins are economically available from coal (i.e., by Fischer-Tropsch synthesis: see Section 30.3.3.6). However the very important 2EH may also be produced from coal via the following route (which at present is not competitive): calcium carbide to acetylene (Section 30.3.4) to acetaldehyde to butyraldehyde.

Similar to oxosynthesis is the homologization reaction, that is, the reaction of alcohols with synthesis gas under chain elongation by one $—CH_2—$ group, and the formation of the higher homologous alcohols or aldehydes. For instance, ethanol is produced from methanol and synthesis gas:

$$CH_3OH + CO + 2H_2 \dashrightarrow CH_3CH_2OH + H_2O \quad (30.44)$$

By-products from this reaction are propanol, butanol, methyl acetate, ethyl acetate, and others. In this way commercially important oxygenated hydrocarbon compounds could be produced via synthesis gas and coal, and also their respective low-boiling olefins by dehydration of the oxygenated compounds (e.g., ethylene from ethanol, etc.). The possible shift of the chemical industry from petrochemistry to coal chemistry in the future could make these process routes very important.[152]

30.3.3.8 *Carbon Monoxide*

Pure carbon monoxide may be separated from synthesis gas by removing the other gas components (hydrogen, methane, and nitrogen). Gasification of coke with air gives low-Btu gas with limited CO concentration. High concentrations of CO are obtained by modern partial oxidation processes of coke with oxygen.[153] Also, carbon monoxide containing off-gas streams may be utilized (from carbide furnaces and blast furnaces).

Carbon monoxide is separated from gas mixtures by low-temperature techniques or by absorption processes with selective solvents. For a long time, copper-salt solutions were used for the purification of synthesis gas or hydrogen after the water-gas shift reaction. A newly developed process with solutions of $CuCl_2$ and $AlCl_3$ in toluene (Cosorb process) promises cost reductions.[154,155]

The processing of pure carbon monoxide in the chemical industry is less extensive than is the use of synthesis gas. However several carbonylation reactions are commercially applied.

The conversion of acetylene with CO in the presence of water or alcohols forms acrylic acid or the respective acrylic esters (Reppe process). This process has been in use for a long time, and only quite recently was replaced by the oxidation of propylene:

$$HC \equiv CH + CO + H_2O\,(ROH) \dashrightarrow CH_2{=}CHCOOH(R) \quad (30.45)$$

The manufacture of propionic acid from ethylene and carbon monoxide using nickel tetracarbonyl catalyst is one of the carbonylation reactions of olefins practiced on a commercial scale (BASF):

$$H_2C = CH_2 + CO + H_2O \dashrightarrow CH_3CH_2COOH \quad (30.46)$$

$$\Delta H_{298\,K} = -158.8 \text{ kJ/mole}$$

The conversion of olefins with CO and water in the presence of proton-donating catalysts (mineral acids) forms mixtures of isomeric and branched carboxylic acids (Koch synthesis). In this way trimethylacetic acid or pivalic acid is manufactured from isobutene and carbon monoxide;

[149] G. Dümbgen and D. Neubauer, *Chem.-Ing.-Tech.*, **41,** 974 (1969).
[150] H. Weber and J. Falbe, *Ind. Eng. Chem.*, **62**(4), 33 (1970).
[151] H. Weber et al., *Hydrocarbon Process.*, **55**(4), 127 (1976).
[152] P. H. Spitz, paper presented at American Chemical Society, Centennial Meeting, New York, April 8, 1976.
[153] K. W. Foo and I. Shortland, *Hydrocarbon Process.*, **55**(5), 149 (1976).
[154] D. J. Haase and D. J. Walker, *Chem. Eng. Prog.*, **70**(5), 74 (1974).
[155] D. J. Haase, *Chem. Eng.*, **82,** 52 (August 4, 1975).

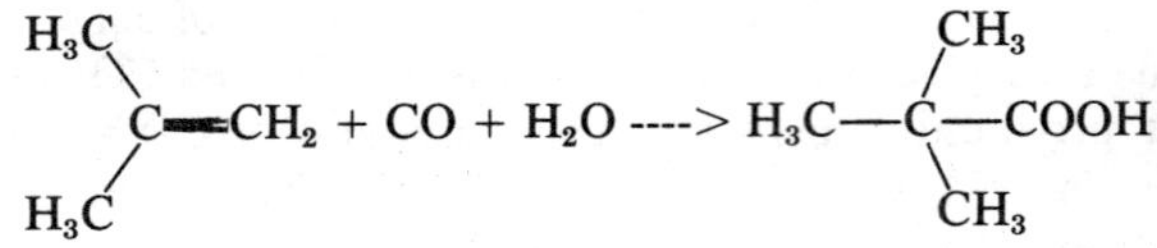

(30.47)

The respective higher olefins are used for the production of C_6–C_{11} carboxylic acids [sold under the brand names Versatic acids (Shell) and Neo Acids (Exxon)].

Of increasing economic importance is the manufacture of acetic acid by methanol carbonylation:

$$CH_3OH + CO \dashrightarrow CH_3COOH \quad \Delta H_{298\ K} = -137.9\ \text{kJ/mole} \tag{30.48}$$

The original high-pressure process of BASF is operated at 680 bar and 250°C with CoI_2 catalysts in the liquid phase.[156] The new process of Monsanto uses catalysts with the very active rhodium metal and requires only very low pressures.[157,158]

In a direct synthesis of formic acid, carbon monoxide is catalytically hydrogenated at moderate pressures or is converted with alcohols to the respective esters:

$$CO + H_2O\ (ROH) \dashrightarrow HCOOH\ (R) \tag{30.49}$$

Methanol is mostly used as an esterification component.

The direct conversion of CO with chlorine over charcoal to phosgene is commercially important:

$$CO + Cl_2 \dashrightarrow COCl_2 \quad \Delta H_{298\ K} = -107.4\ \text{kJ/mole} \tag{30.50}$$

Phosgene is an intermediate chemical for the manufacture of isocyanates, which are used for polyurethane plastics, polycarbonate plastics, and several chemical derivatives.

All processes involving carbon monoxide chemistry could well be based on coal with respect to production of CO itself and with respect to the other reactants (methanol, acetylene, olefins) as well.

[156] F. Ellwood, *Chem. Eng.*, **76**, 148 (May 19, 1969).

[157] J. F. Roth et al., *Chem. Tech.*, **1**, 600 (1971).

[158] H. D. Grove, *Hydrocarbon Process.*, **51**(11), 76 (1972).

30.3.3.9 *Methane*

Production of methane from synthesis gas according to:

$$CO + 3H_2 \dashrightarrow CH_4 + H_2O \quad \Delta H_{298\ K} = -206.1\ \text{kJ/mole} \tag{30.51}$$

has been commercialized on a limited scale; that is, the removal of low concentrations of carbon monoxide in the purification of ammonia synthesis gas has been achieved. Sometimes coke-oven gas has been treated in this way, or the residual amounts of CO are eliminated by the methanation reaction after CO shift conversion when synthesis gas is to be transformed into hydrogen. Methane could not be produced economically from synthesis gas as long as low-cost natural gas resources were available. Actually the reverse is true, because natural gas is economically converted into synthesis gas in many parts of the world. If in the future we have to use more and more coal as an energy resource and a raw material resource for the chemical industry, the methanation reaction might well be of considerable interest (Chapter 25). All the coal gasification processes in which the gasifier produces high concentrations of methane as a primary gas component might have economic advantages.

When the methanation reaction is to be combined with Lurgi pressure gasification of coal, a multistage process with some recycling of product gas is recommended for better temperature control, according to recent development work. Deterioration of the nickel catalyst must be avoided by very effective gas purification methods (Rectisol process) and by suppressing carbon deposits that result from the thermal decomposition of methane and from the Boudouard reaction. The addition of some CO_2 to the reacting system favors reducing the final carbon monoxide concentration to less than 1 vol %.[159]

The new interest in SNG production has intensified the development of new coal gasification processes[160–166] in which both gasification with steam and hydrogasification take place (Chapter 24). Several project studies and profitability appraisals have been based on large, fully integrated SNG plants with capacities of between 2×10^9 and 3×10^9 m_n^3 methane/year. With the application of Lurgi pressure gasification of bituminous coal, the consumption is approximately 1.8×10^6 kJ or 61.4 kg coal/10^6 kJ lower heating value in the methane. This equals 2.04 kg coal/m_n^3 methane or 2.84 kg coal/kg methane (lower heating value of coal $H_u = 29.3 \times 10^3$ kJ/kg).

Within the chemical consumption pattern of methane, processing to synthesis gas and acetylene would no longer be viable in the case of coal-derived methane, because synthesis gas is an intermediate precursor and acetylene might be manufactured from coal more economically by other routes (via calcium carbide, or by the AvCo process; Sections 30.3.4 and 30.3.5). Some other chemical outlets, however, probably would remain, such as the methane chlorides and other halogenated compounds, hydrocyanic acid, or the production of carbon black from methane.[167]

30.3.3.10 *Synthesis Gas as a Reducing Agent*

Synthesis gas as a direct reducing agent has already gained some economic importance in the iron and steel industry. The economic incentive for substituting it for the expensive coke in the conventional blast furnace process arises mainly from the availability of cheap natural gas resources, which might be converted at low cost to synthesis gas as a reducing agent. Natural gas as a raw material for synthesis gas provides the opportunity for adjusting the CO/H_2 ratio immediately to the requirements of the reduction process; furthermore, special gas purification is unnecessary. However coal might be used for iron ore reduction and other technical reduction processes by gasifying it and using the synthesis gas produced as a reducing agent instead of coal in direct reduction processes or coke in the conventional blast furnace process. The dependence on high-quality coking coal resources and the expensive process of coal carbonization might then be reduced (see also Sections 30.2.1 and 30.2.1.2).

Within the commercially proved conventional autothermal gasification processes, the Winkler process, Lurgi pressure gasification, and Koppers-Totzek gasification of coal are applicable to the production of gas as a reducing agent; however some gas treatment processes are necessary. The relatively high methane concentration in the gas from Lurgi pressure gasification can be lowered by secondary reforming. To remove CO_2 and H_2S in an absorption step, the gas must be cooled and then reheated to around 800°C for the reduction process after the absorption of the sour gas components. With a gasification process at atmospheric pressure, such as the Koppers-Totzek process up to now, a gas compression step would be required to compensate for the pressure drop during gas purification and in the direct reduction process. A pressure gasification process might be favorably combined with a reduction process under elevated pressure; however suitable pressure processes have not yet been developed. It has been reported, however, that there is technically proved equipment for feeding the coal at pressures up to 4 bar into the Winkler gasifier.[168] Also, processes for dust removal and absorption of sulfur compounds at high temperature would be required to make the combination of coal gasification and direct reduction more attractive

[159] F. W. Moeller, H. Roberts, and B. Britz, *Hydrocarbon Process.*, **53**(4), 69 (1974).
[160] S. A. Bresler and J. D. Ireland, *Chem. Eng.*, **79**, 94, (October 16, 1972).
[161] G. E. Klingman and R. P. Schaaf, *Hydrocarbon Process.*, **51**(4), 97 (1972).
[162] W. Bodle and K. C. Vyas, *Oil Gas J.*, 73 (August 26, 1974),
[163] R. H. Willis, *Gas Eng. Manage.*, 211 (1976).
[164] P. F. H. Rudolph, *Chem. Econ. Eng. Rev.*, **5**(10), 8 (1973).
[165] H. Jüntgen et al., *Chem.-Ing.-Tech.*, **46**, 937 (1974).
[166] H. Teggers, *GWF-Gas/Erdgas*, **115**, 532 (1974).
[167] S. A. Miller, *Br. Chem. Eng.*, **7**, 656 (1962).
[168] Communication of Davy Powergas GmbH, Cologne.

than its competitor, the direct reduction of iron ore using coal in rotating drums. Additionally, the compressed gas could be sent to an expansion turbine, to generate the electrical energy required in the process.[169]

To reduce 10^3 kg of iron from 1.43×10^3 kg of Fe_2O_3 with 100% oxygen elimination from the ore, theoretically 602 m_n^3 carbon monoxide or hydrogen would be needed. In the sponge iron process, the oxygen content of the ore is reduced practically only by 90%; the theoretical consumption would be lowered to 540 m_n^3 $CO + H_2$ per thousand kilograms of iron. The practical utilization of the reducing gas, however, ranges only between 30 and 40%, depending on several influencing factors such as type of reduction process used, composition of the reducing gas, degree of oxidation of the entering reducing gas, reduction temperature, and residence time of the gas in the reduction reactor. For the process of direct iron ore reduction, a total energy consumption of 16.7×10^3 kJ/kg sponge iron has been calculated, which equals 0.57 kg of bituminous coal per kilogram of sponge iron. Energy or coke consumption of the conventional blast furnace process is $14.6–17.6 \times 10^3$ kJ/kg pig iron or 0.5–0.6 kg coke/kg pig iron[169,170] (Section 30.2.1.1).

30.3.4 Manufacture and Processing of Calcium Carbide

30.3.4.1 *Manufacture of Calcium Carbide*

Calcium carbide is made from coke or coal and lime in an electric arc reduction furnace at around 2100°C[132,171]:

$$3C + CaO_{melt} \dashrightarrow CaC_{2melt} + CO \qquad \Delta H_{298\ K} = +464\ \text{kJ/mole} \qquad (30.52)$$

The (Söderberg) electrodes are deeply immersed into the burden, in which a multiplicity of small arcs is formed. The bulk of the electrical energy is used in the melt below the electrodes. Modern furnaces are fully enclosed; thus the recovery of 100% of the CO formed during reduction is achieved. Furnaces are now being built to accommodate an energy requirement of 65 MW.

The main sources of carbon are oven coke and anthracite. In North America and Western Europe metallurgical coke is predominantly used. Anthracite is used in the USSR and South Africa. The lime is usually screened, dust-free, calcified limestone. The main quality requirements are chemical purity and size consist, since these two factors determine energy consumption and smoothness of furnace operation.

In large modern furnaces, the following are consumption figures for 1 kg (300 liters) of carbide: 0.95 kg of lime (94% CaO), 0.55 kg of coke (dry, 10% ash), 0.02 kg of electrode material, and 3.1 kW-hr.

The term "1 kg (300 liters) carbide" refers to the fact that one kilogram of carbide produces on conversion 300 liters of acetylene (at 15°C and 1.013 bar). This description corresponds (cf. Section 30.3.4.2) to a carbide of 80.5% purity.

The reaction enthalpy for the formation of calcium carbide according to equation 30.52 is 464 kJ/mole at 298 K, corresponding to about 2.0 kW-hr/kg. In practice, for the 300-liter commercial carbide, about 3.0–3.3 kW-hr/kg is required.

Worldwide the production of carbide shows a decreasing trend. The total production in 1976 was estimated to be about 5×10^9 kg/year. Available production figures for individual countries are shown for the years 1972–1975 in Table 30.14.

The three main uses of calcium carbide are listed below.

1. Production of acetylene:

$$CaC_2 + 2H_2O \dashrightarrow C_2H_2 + Ca(OH)_2 \qquad (30.53)$$

2. Reaction of calcium carbide with nitrogen to form calcium cyanamide:

$$CaC_2 + N_2 \dashrightarrow CaCN_2 + C \qquad (30.54)$$

3. Sulfur removal from iron in steelmaking:

$$(CaC_2) + S_{Fe} \dashrightarrow (CaS) + 2C_{Fe} \qquad (30.55)$$

[169] F. Rose, *Stahl Eisen,* **95,** 1012 (1975).
[170] K. Tippmer, *Erdoel Kohle, Erdgas, Petrochem.,* **29,** 153 (1976).
[171] A. G. Scobie, "Calcium Carbide," in *Kirk-Othmer Encyclopedia of Chemical Technology,* 2nd ed., Vol. 4, New York, 1964, p. 100.

Table 30.14 Production of Calcium Carbide

	Production (10^6 kg)			
Country	1972	1973	1974	1975
West Germany	640	604	644	485
East Germany	1332	1353	1342	
France	144	116	129	
Great Britain and Northern Ireland	72	72	14	
Spain	212	167		
Italy	117	93	59	
Norway	104	105		
Sweden	32	31		
Yugoslavia	77	53	64	
Poland	550	580		
Rumania	304	312		
Czechoslovakia	160			
USSR	809			
United States	448	271	294	
Japan	635	577	631	

For the production of 1 kg of acetylene, 3–3.3 kg of carbide is required; and for the production of 1 kg of nitrogen bound in calcium cyanamide, about 3.2 kg of carbide is required. The sulfur removal from iron requires 0.005–0.01 kg of carbide per kilogram of iron, depending on the sulfur content of the hot metal.

The use of calcium carbide differs greatly from country to country; therefore the relevance of the average figures is very limited. In 1974 in the West Germany about 290 × 10^6 kg of CaC_2 (48% of the total production) was converted to calcium cyanamide, and 240 × 10^6 kg (37%) was converted to acetylene. Worldwide, however, only 10% of the carbide production was used for the conversion to calcium cyanamide.

30.3.4.2 *Conversion of Calcium Carbide into Acetylene*

Carbide reacts with water according to the equation

$$CaC_2 + 2H_2O \dashrightarrow C_2H_2 + Ca(OH)_2$$

$$\Delta H_{298\text{ K}} = -125 \text{ kJ/mole} \qquad (30.53)$$

to form acetylene and hydrated lime.[172,173]

[172] H. Beller and J. M. Wilkinson, Jr., "Acetylene," in *Kirk-Othmer Encyclopedia of Chemical Technology,* 2nd ed., Vol. 1, New York, 1963, p. 174.

[173] H. Behringer et al., "Acetylen," in *Ullmanns Encyklopädie der technischen Chemie,* 4th ed., Vol. 7, Munich, 1974, p. 64.

The reaction is highly exothermic. Pure carbide produces 372.66 liters of acetylene at 15°C and 1.013 bar. Commercial carbide produces about 260–300 liters.

For the production of small amounts of acetylene (e.g., for welding purposes), wet converters are being used that require an excess of water. The calcium hydroxide is obtained as a sludge containing 10–20% calcium hydroxide. In use are low-pressure converters with a maximum operating pressure of 0.2 bar and medium-pressure converters with a maximum operating pressure of 1.5 bar above atmospheric pressure.

Large amounts of acetylene for chemical synthesis are preferentially produced in dry converters in which the carbide is mixed with only enough water to ensure the chemical reaction and the removal of the heat. The calcium hydroxide is obtained in dry converters as a dry, easily flowing powder. Dry converters have a maximum production of 3000 m^3/hr acetylene, corresponding to a throughput of about 12 × 10^3 kg/hr carbide.

In addition to its use as chemical raw material (cf. Section 30.3.5) acetylene is important for the production of very hot flames (e.g., for welding and cutting purposes).

30.3.4.3 *Conversion of Calcium Carbide into Calcium Cyanamide*

Another chemical use of calcium carbide besides processing to acetylene is its conversion with nitrogen to calcium cyanamide at a temperature above 1000°C:

$$CaC_2 + N_2 \dashrightarrow CaCN_2 + C \quad (30.54)$$

$$\Delta H_{298\,K} = -289 \text{ kJ/mole}$$

This is the oldest method of nitrogen fixation, which came into use even before ammonia synthesis.

Commercial processes are the batchwise operation in the Frank-Caro kiln, the channel kiln process (Polzeniusz-Krauss), and the rotary kiln process.[174] The practical consumption is 3.2 kg of commercial-grade calcium carbide (78 wt % CaC_2) per kilogram of fixed nitrogen in the product. The commercial-grade calcium cyanamide is supplied with up to 65 wt % $CaCN_2$; the fixed nitrogen contents may amount to up to 23%. Assuming a coke consumption (carbon electrodes included) of 0.75 kg/kg of calcium carbide, there would be a unit consumption of 2.4 kg of coke per kilogram of fixed nitrogen.

Calcium cyanamide is mainly used as a fertilizer, but the additional herbicide effects are also important. Therefore a limited market could be preserved in spite of the higher cost of the product compared to ammonia. A smaller fraction of the output is further processed to chemical derivatives, such as cyanamide, dicyanamide and melamine, thiourea, and guanidines.

In 1974 world production of calcium cyanamide was around 200×10^6 kg N. To that figure, the contribution of West Germany alone was 90×10^6 kg N; Japan had a production of 65×10^6 kg, Belgium 10×10^6 kg, and Spain 9×10^6 kg. Other producing countries are France, Norway, Sweden, Korea, Taiwan, USSR, and Poland.[175]

[174] H. Höger, "Calciumcyanamid," in *Ullmanns Encyklopädie der technischen Chemie,* 4th ed., Vol. 9, Munich, 1975, p. 85.

30.3.5 Direct Conversion of Coal into Acetylene

In addition to the conversion of carbide, acetylene may be produced by the direct conversion of hydrocarbons at high temperature. Several commercial processes are being used with liquid and gaseous hydrocarbons as raw materials. There is no commercial process using coal as raw material, although the theoretical energy requirement is favorable. The enthalpy of formation for acetylene at 298 K, using methane as the raw material, is 385 kJ/mole C_2H_2; for oil as a raw material, only about 260 kJ/mole C_2H_2 (octane: 279 kJ/mole C_2H_2; nonane: 278 kJ/mole C_2H_2); and for benzene, only 210 kJ/mole C_2H_2. Since aromatic structures are prevalent in coal, the enthalpy of conversion of coal to acetylene is about as low as for benzene. The theoretical energy requirement for the production of acetylene consists of three components: the heat required for the heating of the raw materials, the heat of vaporization of the raw material, and the reaction enthalpy for the conversion of the raw material into acetylene. Table 30.15 presents approximate values of these three components for 100% conversion of carbon to acetylene at 1600 K. Thus aromatic feedstocks such as coal and

[175] J. R. McAdam and F. C. Schaefer, "Cyanamides," in *Kirk-Othmer Encyclopedia of Chemical Technology,* 2nd ed., Vol. 6, New York, 1965, p. 553.

[176] W. R. Ladner and J. O. H. Newman, paper presented at IEA meeting, Stoke Orchard, January 27–28, 1977.

Table 30.15 Theoretical Energy Requirements for Conversion to Acetylene[176]

	Energy Required (kW-hr/kg acetylene)			
Feedstock	Sensible Heat to Raise Temperature from 298 to 1600 K	Heat of Vaporization (K)	Heat of Reaction at 298 K	Total
Coal–coal extract $(CH_{0.8})_n$	0.9	0.1	2.3	3.3
Oil, C_9H_{20}	1.1	0.1	3.1	4.3
Natural gas, CH_4	2.2	0.0	4.0	6.2

coal extract have an advantage over oil and natural gas.

Acetylene has a postivie free enthalpy of formation, ΔG, therefore is metastable at room temperature. Above about 700 K the decomposition rate becomes measurable. Above 1600 K the ΔG for acetylene is smaller than for all other hydrocarbons, but still is positive. In this temperature range, acetylene—although the most stable hydrocarbon—still decomposes into the elements hydrogen and carbon, the latter being obtained as carbon black. The equilibrium concentration of acetylene at 1600 K is still negligible and reaches the maximum at about 3300 K[177,178] if the H/C ratio is 1 : 1. The presence of oxygen or nitrogen shifts the maximum only to a small extent,[179–181] but it changes the composition of the gas, which in the presence of nitrogen contains HCN.

The pyrolysis of hydrocarbons—including coal—occurs above 1600 K in consecutive reactions, with acetylene as an intermediate. For the production of acetylene, it is not favorable to select a long reaction time that permits reaching equilibrium; indeed, the reaction must be interrupted by quenching at the point at which the concentration of acetylene reaches a maximum. The residence time depends on the temperature, and its order of magnitude is milliseconds. The extreme reaction conditions required for the production of acetylene can be achieved in plasma pyrolysis. The plasma is produced by subjecting a flow of gas to an electric arc. Plasma conditions ionize monatomic gases and ionize and dissociate polyatomic gases. The enthalpy of the plasma is used to pyrolyze the raw material. Coal, as a fine powder in a carrier gas, is injected either into the plasma or directly into the electric arc. Instead of coal, a coal extract may be used, which can be melted before injection into the plasma. The product gases must be quenched after a short but precisely maintained reaction time to avoid decomposition of the acetylene. There is no commercial plant that produces acetylene from coal by plasma pyrolysis. However a series of investigations has been made in this field. Knapsack-Griesheim AG[182] used brown coal or coke in a hydrogen-plasma stream and reached an acetylene concentration of 18 vol %, but energy consumption was high. Jimeson[183] used a high-volatile coal (hvAb) and a plasma consisting of argon with 12% hydrogen. He obtained an acetylene yield of 8%, based on the weight of the maf coal.

Ladner and co-workers[184–187] used plasmas consisting of argon, argon-hydrogen, and hydrogen and nitrogen. Coal with a particle size below 50 μm was blown into the plasma stream. Coal with a volatile matter content of 36% gave a maximum carbon conversion to acetylene of 20% in the argon plasma and 40% in a plasma consisting of argon with 10% hydrogen. In pure hydrogen the acetylene yield was lower than in the argon-hydrogen mixture. A spinning plasma improved the yields. In a nitrogen plasma acetylene was replaced by hydrogen cyanide as the main product. The conversion of coal to acetylene and HCN, respectively, depends linearly on the volatile matter content.

Ammann and co-workers[188,189] used a consumable anode consisting of pulverized coal. In an argon plasma Graves and co-workers[190,191] achieved an acetylene yield of

[177] R. F. Baddour and J. L. Blanchet, *Ind. Eng. Chem. Process Des. Dev.*, **3**, 258 (1964).

[178] D. M. L. Griffiths and H. A. Standing, in *Coal Science*, Am. Chem. Soc., Adv. Chem. Ser. No. 55, 1966, p. 666.

[179] D. Tremblay and S. Kaliaguine, *Ind. Eng. Chem. Process Des. Dev.*, **11**, 265 (1972).

[180] Yu. I. Krasnokutskii, S. N. Ganz, and V. D. Parkomenko, *Zh. Prikl. Khim.*, **46**, 2425 (1973).

[181] H.-J. Spangenberg, I. Börger, and W. Schirmer, *Z. Phys. Chem., Leipzig*, **255**, 1 (1974).

[182] Knapsack-Griesheim AG, British Pat. 831 522, 1960.

[183] R. M. Jimeson, *Min. Eng.*, **15**, 52 (November 1963).

[184] R. L. Bond, I. F. Galbraith, W. R. Ladner, and G. I. T. McConnel, *Nature*, **200**, 1313 (1963).

[185] T. Aust, W. R. Ladner, and G. I. T. McConnel, Paper No. 7, presented at 6th Int. Conf. Coal Science, Münster, 1965.

[186] L. R. Bond, W. R. Ladner, and G. I. T. McConnel, *Fuel*, **45**, 381 (1966).

[187] L. R. Bond, W. R. Ladner, and G. I. T. McConnel, in *Coal Science*, Am. Chem. Soc. Adv. Chem. Ser. No. 55, 1966, p. 650.

[188] P. R. Ammann, R. F. Baddour, M. M. Johnson, T. W. Mix, and R. S. Timmins, *Chem. Eng. Prog.*, **60**, 52 (1964).

[189] P. R. Ammann, R. F. Baddour, and T. W. Mix, U.S. Patent 3,384,467, 1968.

[190] R. D. Graves, W. Kawa, and R. W. Hiteshue, *Ind. Eng. Chem. Process Des. Dev.*, **5**, 59 (1966).

[191] R. D. Graves, W. Kawa, and P. S. Lewis, *Prepr., Am. Chem. Soc., Div. Fuel Chem.*, No. 8, 118 (1964).

15 wt % based on maf coal. However the process required 200 kW-hr/kg of acetylene.

Kawana and co-workers[192–195] achieved not only higher yields (52 wt %) but also lower energy consumption (66 kW-hr/kg). Newman and co-workers[196] obtained from coal in a hydrogen plasma a maximum acetylene yield of 46% with an energy consumption of 88 kW-hr/kg. Several investigators[197–199] used a fluidized bed of aluminum for quenching the products of the plasma pyrolysis. Ladner and Wheatley[200] investigated acetylene manufacture from a coal extract in a submerged arc reactor.

Nicholson and Littlewood[201,202] used an argon-hydrogen plasma and achieved an acetylene yield of 74% from coke; the apparently very high energy requirement is not being disclosed. Sheer and Korman[203] designed a specially shaped fluid convection cathode.

Chakravartty and co-workers[204] attempted to optimize the reaction conditions of coal pyrolysis in an argon plasma. They found that the acetylene yield depends linearly on the volatile matter content of the coal. They achieved a maximum acetylene yield of 20% on a maf basis with a coal containing 46.6% volatile matter.

So far, only Krukonis and co-workers[205–209] have succeeded in optimizing acetylene production from coal in a hydrogen plasma in a way that appears to be economically attractive. The economic attractiveness of this process was confirmed by an independent study.[210,211] In the so-called AvCo arc coal process (Fig. 30.13), dry, finely pulverized coal is converted into acetylene by injection into a hydrogen arc that uses direct current. The arc is stabilized by rotation in a magnetic field. Part of the product gas is recycled and used as a quenching medium. The feed gas for the electric arc and the carrier gas for the coal are also recovered from the product gas. The acetylene yield is 30% of the coal feed. The specific energy requirement is less than 10 kW-hr per kilogram of acetylene, and the concentration of acetylene in the product gas reaches 15%. If naphtha is used as a quenching medium instead of the gas, ethylene is obtained as a coproduct. The economics of the process can be improved by this means.

The AvCo arc coal process was tested in a pilot plant with a maximum electric power consumption of 180 kW and a maximum coal throughput of 55 kg/hr.[209] In the planning stage is a plant with a capacity of 136×10^6 kg/year, which corresponds to a consumption of 450×10^6 kg/year coal with a moisture content of 10% and a total electric power consumption in the arc reactors of 167 MW.[210]

There are other possibilities for achieving

[192] Y. Kawana, M. Makino, and T. Kimura, *Kogyo Kagaku Zasshi,* **69,** 1144 (1966).

[193] Y. Kawana and M. Makino, *Kogyo Kagaku Zasshi,* **70,** 1657 (1967).

[194] Y. Kawana, M. Makino, and T. Kimura, *Int. Chem. Eng.,* **7,** 359 (1967).

[195] Y. Kawana, *Chem. Econ. Eng. Rev.,* **4,** 13 (1972).

[196] J. O. H. Newman, A. J. T. Coldrick, P. L. Evans, T. J. Kempton, D. G. O'Brien, and B. Woods, paper presented at 7th Int. Coal Science Meeting, Prague, 1968.

[197] W. M. Goldberger and S. H. Oxley, *Am. Inst. Chem. Eng. J., Chem. Eng. Res. Dev.,* **9,** 778 (1963).

[198] S. Bal, A. Musialski, and R. Swierczek, *Koks Smola Gaz,* **16,** 96 (1971).

[199] S. Bal, A. Musialski, and R. Swierczek, *Koks Smola Gaz,* **16,** 123 (1971).

[200] W. R. Ladner and R. Wheatley, *Fuel,* **50,** 443 (1971).

[201] R. Nicholson and K. Littlewood, *Nature* **236,** 397 (1972).

[202] K. Littlewood, *Proc. Symp. Chem. Oil, Coal,* Central Fuel Research Institute, Dhanbad, India, 1972, p. 517.

[203] C. Sheer, and S. Korman, in *Coal Gasification,* L. G. Massey, Ed., Am. Chem. Soc. Adv. Chem. Ser. No. 131, 1974, p. 42.

[204] S. C. Chakravartty, D. Dutta, and A. Lahiri, *Fuel,* **55,** 43 (1955).

[205] V. J. Krukonis and T. Schoenberg, paper presented at 7th Int. Coal Science Meeting, Prague, 1968.

[206] R. E. Gannon, V. J. Krukonis, and T. Schoenberg, *Ind. Eng. Chem. Prod. Res. Dev.,* **9,** 343 (1970).

[207] V. J. Krukonis, R. E. Gannon, and T. Schoenberg, paper presented at 19th Canadian Chem. Eng. Conf., Edmonton, Alberta, 1969.

[208] V. J. Krukonis, R. E. Gannon, and M. Modell, in *Coal Gasification,* L. G. Massey, Ed., Am. Chem. Soc. Adv. Chem. Ser. No. 131, 1974, p. 29.

[209] R. E. Gannon and V. Krukonis, U.S. Dept. Interior OCR R & D Rep. No. 34, Contract No. 14-01-0001-493, 1972.

[210] Blaw-Knox Chemical Plants, Inc., *300 MM lbs/yr Acetylene Plant—Avco Arc-Coal Process,* U.S. Dept. Interior OCR R & D Rep. No. 67, Contract No. 14-32-0001-1215; BKC 1661, November 30, 1971.

[211] J. H. Mashey, Paper No. 40, presented at 168th National Meeting, American Chemical Society, Division of Industrial Engineering Chemistry, Atlantic City, N.J., September 1974.

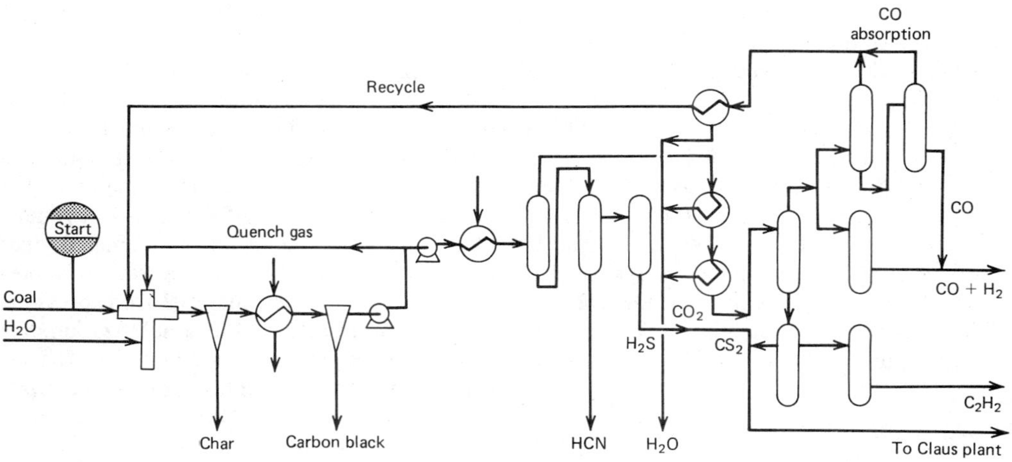

Figure 30.13 Flowsheet for the Avcoarc coal process.[211]

the high heating rates and temperatures that are necessary for acetylene production from coal.[212] Several authors[213–215] used a simple, conventional experimental arrangement by dropping finely ground coal through an electrically heated tubular furnace or blowing it through with a carrier gas.

Others[216] used the action of shock waves to achieve high temperatures. A microwave discharge, too, has been used for the pyrolysis of coal, whereby comparatively low temperatures (e.g., 973 K) were sufficient for the formation of C_2H_2, HCN, and so on.[217–220] Other heating methods include light pulses,[221,222] arc image furnaces,[223] and lasers.[224–232]

In testing a process similar to the Japanese Kureha process for the petrochemical manufacture of acetylene, a maximum acetylene yield of 6% was obtained. In this process finely ground coal was blown into a hot stream of gas obtained by the combustion of hydrogen and oxygen.[233]

The petrochemical industry produces organic raw materials such as ethylene and propylene by cracking of naphtha. Before this development occurred, chemical products such as acetaldehyde, acetic acid, vinylchloride, acrylonitrile, butadiene, vinylester, vinyl acetate and vinyl ether, and chloroprene were made on a large scale from acetylene. These materials today are produced predominantly from olefins, and acet-

[212] W. R. Ladner, *BCURA Month. Bull.*, **28,** 281 (1964).

[213] C. K. Kröger and K. S. Rao, *Brennstoff-Chemie,* **46,** 129 (1965).

[214] H. Pichler, K. H. Lauer, W. Ripperger, and M. Tayar, *Brennstoff-Chemie,* **49,** 377 (1968).

[215] M. Ihnatowicz, *Pr. Glown. Inst. Gorn.*, No. 522 (1971).

[216] E. T. Woodburn, R. C. Everson, and A. R. M. Kirk, *Fuel,* **53,** 38–46 (1974).

[217] Y. C. Fu and B. D. Blaustein, *Fuel,* **47,** 463 (1968).

[218] Y. C. Fu and B. D. Blaustein, *Ind. Eng. Chem. Process Des. Dev.,* **8,** 257 (1969).

[219] Y. C. Fu, B. D. Blaustein, and A. G. Sharkey, Jr., *Fuel,* **51,** 308 (1972).

[220] R. Durmosch, K. Ul. Islam, and H.-H. Oelert, *Chem.-Ing.-Tech.,* **46,** 961 (1974).

[221] E. Rau and L. Seglin, *Fuel,* **43,** 147 (1964).

[222] C. O. Hawk, M. D. Schlesinger, and R. W. Hiteshue, U.S. Bur. Mines Rep. Invest. No. 6264, 1963.

[223] E. Rau and R. T. Eddinger, *Fuel,* **43,** 246 (1964).

[224] A. G. Sharkey, J. L. Schultz, and R. A. Friedel, *Nature,* **202,** 988 (1964).

[225] F. S. Karn, R. A. Friedel, and A. G. Sharkey, *Carbon* (Oxford), **5,** 25 (1967).

[226] J. L. Shultz and A. G. Sharkey, *Carbon* (Oxford), **5,** 57 (1967).

[227] F. S. Karn and A. G. Sharkey, *Fuel,* **47,** 193 (1968).

[228] F. S. Karn, R. A. Friedel, and A. G. Sharkey, *Fuel,* **48,** 297 (1969).

[229] F. S. Karn, R. A. Friedel, and A. G. Sharkey, *Fuel,* **51,** 113 (1972).

[230] W. K. Joy, W. R. Ladner, and E. Pritchard, *Nature,* **217,** 640 (1968).

[231] A. F. Granger, W. K. Joy, and W. R. Ladner, paper presented at 7th Int. Coal Science Meeting, Prague, 1968.

[232] W. K. Joy, W. R. Ladner, and E. Pritchard, *Fuel,* **49,** 26–38 (1970).

[233] R. L. Coates, C. L, Chen, and B. J. Pope, in *Coal Gasification,* L. G. Massey, Ed., Am. Chem. Soc. Adv. Chem. Ser. No. 131, 1974, p. 92.

ylene is used only in comparatively small quantities for the production of vinylchloride, acrylic acid, vinyl acetate, butanediol, tetrahydrofuran, and others.

Energy and chemical production in South Africa is still based on 80% indigenous coal and only 20% imported petroleum. There the production of PVC using acetylene is being increased. In addition to PVC, acetylene is used to produce chlorinated solvents (trichlorethylene and perchlorethylene), acetaldehyde, acetic acid, calcium acetate, crotonaldehyde, and vinyl acetate.[234]

30.3.6 Manufacture of Other Carbides

Among the other carbides, silicon carbide—with a world production of about 600×10^6 kg—is the most important. Uranium and thorium carbides are promising nuclear fuels. Basically, for the manufacture of these carbides, in addition to petroleum coke, graphite, and carbon black, carbon from coal (oven coke, anthracite, pitch coke) could also be used. However the literature does not contain specific data for this possibility. Sections 30.3.6.1 and 30.3.6.2 briefly summarize the production, use, and properties of these materials.

30.3.6.1 *Diamondlike and Saltlike Structured Carbides*[235–237]

Silicon Carbide. Silicon carbide is produced in electric resistance furnaces (Acheson process) used with sand, low-ash oven coke, petroleum coke, tar coke, or pitch coke or anthracite as raw materials. The reaction occurs in two stages:

$$SiO_2 + 2C \dashrightarrow Si + 2CO \qquad (30.56)$$

$$Si + C \dashrightarrow SiC \qquad (30.57)$$

The feed materials are comminuted to a size of 2–5 mm and made more permeable by the addition of sawdust. Furthermore, sodium chloride is added. These materials are charged in the furnace in a manner that causes the formation of a center—a cylindrical nucleus of about 50-cm diameter—that consists of granulated coke, which is predominantly recycled material from previous charges and because of its graphitization, has an improved electrical conductivity. The reaction is strictly a reaction of solid materials and begins on the outside of the nucleus at about 1700–1800°C. It usually occurs at about 2040°C. The nucleus temporarily reaches a temperature of more than 2500°C. Instead of sodium chloride, scrap from chlorinated hydrocarbons (e.g., PVC) may be added.[238] This technique permits chlorination of 90% (sodium chloride, 85%) of the mineral impurities in the carbonaceous material and permits their removal by evaporation. The energy consumption is 7.5–12.5 kW-hr/kg SiC in furnaces with up to 1500 kW electric power consumption. Raw material consumption for one kilogram of SiC is 1.3–1.5 kg of petroleum coke or corresponding carbonaceous material, and 1.7–2.0 kg of quartz sand.

Silicon carbide has the following technically important properties: very high hardness (Mohs hardness, 9.6), high heat resistance and heat conductivity and resistance to temperature changes, high-temperature bending strength, chemical resistivity, and high electric conductivity. The dissociation pressure reaches 1.013 bar at 2425 ±40°C.

The main uses of these materials are as abrasives, as refractory raw materials, as heating elements for electric resistance furnaces, as semiconductor components, and in the metallurgy of iron and steel as reducing agents.

The following yearly production capacities existed early in 1974[235]: Western Europe, 160×10^6 kg; Comecon countries, 250×10^6 kg; United States and Canada, 160×10^6 kg; other countries, 30×10^6 kg; total, 600×10^6 kg.

[234] J. C. Hoogendoorn, "Conversion of Coal into Fuels and Chemicals in South Africa," *Mitt. Sasol Werke, Sasolberg, S. Afr.*

[235] H. Fuchs, *Chem.-Ing.-Tech.*, **46,** 139 (1974).

[236] M. C. Parche, "Silicon Carbide," in *Kirk-Othmer Encyclopedia of Chemical Technology,* 2nd ed., Vol. 4, New York, 1964, pp. 114–132.

[237] F. Ritter and J. Krause, "Carbid, Kalkstickstoff und Siliciumcarbid," *Winnacker-Küchler Chemische Technologie,* 3rd ed., Vol. 1, Munich, 1970, pp. 450–464.

[238] Eikichi Ueki (Shawa Denko K.K.) Jap. Kokai 73 19,499 (Cl.15 N11) March 10, 1973, Appl. 71 53,855, July 21, 1971; *C.A.*, **79,** 440049 (1973).

Boron Carbide. Boron carbide is produced in an electric resistance furnace with silicon carbide resistance elements; it is made from a mixture of molten boric acid, petroleum coke, wood charcoal, or graphite at about 2600°C.[239] Boron carbide may also be produced in an electric arc process by melting together anhydrous boric acid with graphite or coke powder,[240] in a two-stage process according to[241]:

$$2B_2O_3 + 7C \dashrightarrow B_4C + 6CO \qquad (30.58)$$

or according to a Japanese process at 650–830°C by reduction of B_2O_3 with carbon or magnesium in the presence of sulfates.[242]

The properties of B_4C are as follows: it has a higher hardness than SiC (Mohs, > 9), a high melting point (2450°C), and high resistance to corrosion, especially toward acids.

Boron carbide has the following applications: (1) as a loose grinding powder or as a powder paste for grinding operations on hard materials, especially hard metals, (2) as compact, dense shapes for the combat of wear and tear (e.g., for sandblasting nozzles, draw plates, thread guides, gauges, mortars and pestles, filters, contact points, ignition electrodes, and (3) in nuclear reactors as a neutron-absorbing agent.

Beryllium Carbide (Saltlike Carbide). Beryllium carbide is produced by the direct reaction of solid or liquid beryllium with carbon, or by the reaction of beryllium oxide with finely divided carbon at 2080°C.[243,244] Beryllium carbide has a melting point of 2400°C and may be used in the construction of nuclear reactors.

[239] G. Schmutzler and J. Kircheisen, *Chem. Tech. (Leipzig),* **19,** 488 (1967). American Potash and Chemical Company, DAS 1,164,997, 1960; N. Morimoto, *Mine. J.,* **2,** 1 (1956); U.S. Borax, U.S. Patents 2,774,070, 2,785,952, 1953.

[240] U.S. Borax, U.S. Patent 3,408,155, 1955; Solvay, German Patent 1,945,744, 1969; Treibacker Chem. Werke, DAS 1,044,783, 1956; Borax Cons. Ltd., German Pat. 356,824, 1921.

[241] E. Asseo, Netherlands Patent 291,917, 1963.

[242] *Chem. Eng. News,* **47,** (21), 29 (1969).

[243] W. W. Beaver and D. W. Lillie, in *Reactor Handbook,* C. R. Tipton, Jr., Ed., Interscience, New York, 1960, p. 897.

[244] W. W. Beaver, in *The Metal Beryllium,* D. W. White, J. E. Burke, Eds., American Society for Metals, Cleveland, 1955.

30.3.6.2 *Commercial Heavy Metal Carbides*[245]

The commercial heavy metal carbides include the following materials: WC, TiC, TaC, NbC (principal carbides); Cr_3C_2, Ma_2C, VC, HfC, ZrC (supplemental carbides); uranium carbides (UC, U_2C_3, UC_2); and thorium carbides (ThC, ThC_2).

The principal carbides are manufactured by carburizing pulverized metals, metal hydrides, or the oxides, in the form of loose or compacted mixtures, with solid carbon, especially low-ash carbon black, in a hydrogen atmosphere and usually in carbon tube resistance furnaces (Tamman furnace) at temperatures between 1300 and 2300°C.

Their properties include a high melting point, resistance to acids, and stability when heated in hydrogen to high temperatures.

Tungsten carbide is the main hardness-producing component in all commercial hard metals. Titanium carbide in the form of mixed crystals with WC and TaC (NbC) is used in all steel-cutting tools and universal tools. TaC is a component of practically all hard metals. From 20–50% of the TaC content can be replaced by the cheaper NbC without deterioration of quality.

Worldwide production is as follows: WC, $8\text{–}10 \times 10^6$ kg/year; TiC, $0.7\text{–}1.0 \times 10^6$ kg/year; TaC, $0.15\text{–}0.2 \times 10^6$ kg/year.

The supplemental carbides are produced essentially in the same manner as the principal carbides. They are additives to improve the properties of hard metals, except for ZrC, which has little importance as a hard metal. This carbide plays an important role as an intermediate in the production of zirconium chloride or zirconium metal (Kroll process).[246] It is produced from pure ZrO_2, and carbon black, coke, or graphite.

Uranium carbide and thorium carbide are produced from mixtures of metal, metal hydride, or oxide powders with graphite, at around 1800–2200°C in carbon tube resistance furnaces in a hydrogen atmosphere, or in high-frequency heated vacuum furnaces, or by electric arc melting. Their further con-

[245] F. Benesoosky, "Carbide," in *Ullmanns Encyklopädie der technischen Chemie,* 4th ed., Vol. 9, Munich, 1975, pp. 122–136.

[246] W. Völkel, "Zirkonium und -Verbindungen," in *Ullmanns Encyklopädie der technischen Chemie,* 3rd ed., Vol. 19, Munich, 1969, p. 181.

version is pyrophoric and sensitive to moisture, hence requires a protective atmosphere. Carbides of ^{235}U are used as nuclear fuels for graphite-moderated, gas-cooled, high-temperature reactors. The thorium carbides in the form of mixed crystals or mixtures with UC or UC_2 are used as nuclear fuels for breeder reactors.

30.3.7 Manufacture of Carbon Dioxide and Carbonates

Combustion of coal as well as of other solid, liquid, or gaseous fuels produces flue gases that contain CO_2, which can be recovered either as the main product or as the byproduct. Large quantities of CO_2 are produced by the shift reaction of synthesis gas (i.e., the conversion of CO with steam to hydrogen and CO_2). Carbon dioxide is also obtained during the decomposition of limestone or magnesite. If coke is used as a fuel instead of hydrocarbons, a gas is obtained with a high CO_2 content (often as high as 42% by volume). Such a gas is highly desirable in several chemical processes—for example, in the production of sodium carbonate by the Solvay process. When natural gas or fuel oil is used, the CO_2 content usually is only 28–34% by volume.[247] The remaining CO_2 requirement in the manufacture of sodium carbonate is supplied by the calcination of the sodium hydrocarbonate, whereby a gas with a CO_2 content of 90–95% is obtained.

For the manufacture of sodium carbonate (98.8% Na_2CO_3), 0.095–0.11 kg of coke is required per kilogram in the lime kiln, and 0.3–0.34 kg of coal for the calcination and the production of energy if coal is exclusively used for this purpose.[248] The calcium oxide obtained simultaneously is used for the recovery of the ammonia in the process.

In the manufacture of sugar, calcium oxide and CO_2 from the decomposition of limestone are used simultaneously for the purification of the juice. For the decomposition of the limestone, 0.02–0.04 kg of coke is required per kilogram of sugar when using sugar beets. In addition, about 0.4–0.6 kg of coal per kilogram of sugar is needed to supply the required energy.[249]

Depending on the composition and purity of the gas, as well as the purity requirements of the product, different gas purification and absorption processes are used to recover CO_2 from flue gases containing 10–11 vol % of that gas. The production of 1 kg of dry ice using a flue gas containing 18% CO_2 requires 0.85 kg of coke or 1 kg of coal and 0.286 kW-hr.[250]

Total CO_2 production is considerable. In 1973 in the United States 980 × 10^6 kg of CO_2 in liquid and gaseous form and 275 × 10^6 kg in solid form (as dry ice) were produced.

The world production of sodium carbonate in 1970 was about 20 × 10^9 kg, of which 5.5% was produced from naturally occurring soda (Trona). In the United States the production of sodium carbonate by the Solvay process is decreasing in favor of the naturally occurring sodium carbonate.[251] Solvay production figures in the United States were 3.18 × 10^9 kg in 1974, and 2.54 × 10^9 kg in 1975. Capacity figures (U.S.) for 1976 were 1.91 × 10^9 kg/year for the Solvay process and 6.7 × 10^9 kg/year for the natural NA_2CO_3 process.

30.4 USE OF COAL AND RELATED PRODUCTS AS CARBONACEOUS MATERIALS

Coal and related products, especially pulverized coke and coal-tar pitch, are important feedstocks for the production of solid carbonaceous materials that are characterized by a high carbon content, special properties, and very different applications. Table 30.16 presents a list of the materials described in this section and shows the worldwide annual consumption of coal, coke, and pitch for

[247] Z. Rant, *Manufacture of Sodium Carbonate Using the Solvay Process,* Stuttgart, 1968, p. 72.
[248] H. W. Schmidt and H. Roethe, "Natriumcarbonat und natriumhydroxid," in *Ullmanns Encyklopädie der technischen Chemie,* 3rd ed., Vol. 12, Munich, 1960, p. 643.
[249] H. Schiwek, "Zuckergewinnung aus Zuckerrüben," in *Ullmanns Encyklopädie der technischen Chemie,* 3rd ed., Vol. 19, Munich, 1969, p. 207.
[250] R. N. Shreve, *The Chemical Process Industries,* 2nd ed., New York, 1956, p. 127.
[251] *Chem. Ind.,* **28**(12), 725 (1976).

Table 30.16 Uses of Coal and Related Products as Carbonaceous Materials

Carbonaceous Material	State of Development	Coal or Coal-Related Products Used	Worldwide Consumption (kg/year)	Main Applications
Carbon and graphite electrodes	Commercial use	Coal-tar pitch Pitch coke Anthracite Coke (pulverized)	1.5×10^6 (1976) $0.5–0.6 \times 10^6$ (1976) $0.2–0.25 \times 10^6$ (1976) $0.2–0.3 \times 10^6$ (1976)	Electrodes for aluminum production, electric steel furnaces, carbide furnaces, alkaline chloride electrolysis
Carbon bricks	Commercial use	Anthracite Coke (pulverized) Coal-tar pitch	$50–60 \times 10^3$ (1976) $10–15 \times 10^3$ (1976)	Blast furnace
Carbon fibers	Under development	Coal-tar pitch	—	Space and aviation industry, hobby and sport equipment
Plastic materials	Under development	Low-rank coal	—	Construction materials in buildings
Carbon black	Commercial use	Coal-tar pitch	0.5×10^6 (1974)	Tires, printing colors,
Fillers	—	Anthracite, bituminous coal	Unknown	Additives in rubber industry, plastics
Adsorbents	Commercial use	Bituminous coal, anthracite	$0.2–0.25 \times 10^6$ (1974)	Drinking and waste water treatment, gas cleaning and gas separation, decolorization

producing these products. These figures are indeed relatively small compared with the total consumption of coal, but the products have a very high value for the various uses. For all products described here, greatly increasing consumption figures are expected. For example, the demand for carbon and graphite electrodes will increase in the expanding worldwide aluminum industry. The production of adsorbents from bituminous coal and anthracite will be increased by the requirements of environmental protection. Carbon fibers from pitch are cheaper than the known polyacrylonitrile fibers used especially for military purposes. Therefore they may open the way for broad industrial production of fibers for various applications. Furthermore, the use of carbon black and fillers in the rubber industry, especially for tires, is growing.

30.4.1 Carbon and Graphite Electrodes

30.4.1.1 *Fields of Application*

Owing to their resistance to corrosion, their extremely favorable properties at high temperatures, and their good electrical conductivity, carbon electrodes are largely used in electric steel furnaces, carbide and reduction furnaces, and plants for electrolysis in fluid state or in aqueous solution.[252–255]

In electric steel furnaces graphite electrodes with diameters up to 600 mm and lengths up to 2400 mm can be used. To guarantee continuous operation of the furnace and full use, new electrodes are fitted to the ones in use by male-female thread connections. From 4.5 to 6.5 kg of graphite electrodes is used per thousand kilograms of steel. Approximately 600×10^6 kg of graphite electrodes is used in the world's electric steel furnaces per year. In the carbide furnace (see also Section 30.3.4.1) the limestone-coal burden is heated up to a reaction temperature of approximately 2273 K by direct flow of current through graphite electrodes. The specific electrode consumption varies between 0.01 and 0.03 kg/kg of CaC_2. In addition, "on the job" baking electrodes, the so-called S ̈derberg electrodes, are very frequently used in carbide furnaces. These electrodes are made up of steel shells filled with layers

[252] E. Dönges and O. Vohler, "Elektrographit und Kunstkohle," in *Winnacker-Küchler Chemische Technologie,* 3rd ed., Munich, 1969, p. 465.

[253] C. L. Mantell, *Carbon and Graphite Handbook,* John Wiley and Sons, New York 1968.

[254] B. W. Gonser and H. H. Hausner, *Ammerican Modern Materials,* Vol. 4, *Carbon and Graphite,* Academic Press, New York, 1964.

[255] F. Schieber and D. Zöllner, *Chem.-Ing.-Tech.,* **49,** 335 (1977).

of green electrode material consisting of solid matter and binder. In the lower zones of the shell the binder carbonizes, thus constitutes a solid body. The green blend is refilled from the top according to the speed of consumption. Both kinds of electrode are also used in furnaces for phosphorus production (see also Section 30.2.4) as well as for such other electrothermal reduction processes as zinc, ferroalloys, and corundum (see also Section 30.2.3). Worldwide approximately 200–300 $\times 10^6$ kg/year of carbon electrodes is estimated to be consumed in all electrothermal processes (CaC_2, Zn, Fe, and corundum).

A particularly important and strongly expanding field of activity involving carbon electrodes is the electrolytic production of aluminum from molten cryolite (see also Section 30.2.3). In this process the anode carbon is consumed by oxidation; thus the loss of anode material, amounting to 0.45 kg/kg of aluminum, is very high. For this kind of production Söderberg anodes are used or, to a large extent, finished block anodes. The cathode is the bottom of the electrolysis cell and consists of a group of carbon blocks. The cathode consumption here is low. In world aluminum production, approximately 10^8 kg/year of carbon is estimated to be consumed in cathodes, in comparison to more than 5×10^9 kg/year in anodes. Besides aluminum, alkaline metals and alkaline earth metals are also produced in molten state electrolysis (e.g., Li, Na, K, Mg). The consumption of electrode material for these metals is, however, negligible in contrast to the aluminum production.

Alkaline chloride electrolysis is the most important of the electrolytic processes in aqueous solution (for production of chlorine, caustic soda, bleaching lye, and chlorate). In the course of this process, oxygen is generated on the anodes simultaneously with chlorine. The oxygen attacks the anode, thus causing an excessive consumption by the formation of CO and CO_2. This means a consumption of approximately 1.8–2 kg of anode material per thousand kilograms of chlorine in the case of NaCl solutions, and 3–4 kg of graphite anodes per thousand kilograms of chlorine in the case of KCl solutions. Anode consumption is increased by higher pH values and sulfate concentrations. Electrographite material of utmost purity is used for the anodes, at which the chlorine is generated. The cathode, where hydrogen is formed, consists of metal. The graphite anodes are consumed relatively slowly by oxidation in the presence of oxygen-containing anions or by decline of chlorine ions in the electrolyte. The carbon consumption is further increased by gas bubbles, causing erosion. In addition to good electrical conductivity, the anodes must exhibit high mechanical strength, good particle-cementing properties, and minimum porosity. By additional impregnation, the pores can be completely closed and the anode surface can be made smooth. New developments competing with carbon electrodes aim at the use of dimensionally stable metallic anodes for chlorine and alkali electrolysis (e.g., anodes made from titanium and coated with precious metals).

30.4.1.2 *Consumption Data*

The use of bituminous coal and bituminous coal products is mainly reserved for production of carbon and graphite electrodes intended for purposes as listed below. The consumption figures stated have been taken mostly from literature and were updated for 1976 on the basis of information issued by the industry.[252,256–263]

Coal-tar pitch is used as binder and impregnating agent for all electrodes, and the main quantities being used for anodes (90%) are supplied to aluminum production plants. The worldwide consumption for this purpose is approximately 1.5×10^9 kg (i.e., approximately 18% of the world's total tar pitch production). Given the expanding aluminum industry, the demand for electrode pitch for 1985 is projected as 3.5×10^9 kg. It can be satisfied completely by coal-tar pitch.[262]

Pitch coke as an alternative to petroleum coke is used for anodes in alkali chloride

[256] B. E. A. Thomas, *Bitumen-Teere-Asphalte-Peche*, **7**, 279 (1971).

[257] H. G. Frank and G. Collin: *Steinkohlenteer*, Springer-Verlag, Berlin, Heidelberg, New York, 1968.

[258] W. Gemmeke, "Technologie und Analytik von Elektrodenpechen," paper presented at 2nd Int. Carbon Conf., Baden-Baden, Germany, 1976, preprint p. 370.

[259] F. Schieber (Fa. Conrady, Nürnberg, Germany), private communication.

[260] W. Gemmeke (Verkaufsvereinigung für Teererzeugnisse, Essen, Germany), private communication.

[261] G. Pietzka and E. Wege (Fa. Sigri, Meitingen, Germany) private communication.

[262] R. Oberkobusch, *Erdoel Kohle*, **28**, 558 (1975).

[263] G. Collin, *Erdoel Kohle*, **29**, 159 (1976).

electrolysis and in aluminum production; annual worldwide pitch coke consumption ranges approximately between 500 and 600 $\times 10^6$ kg compared to the consumption of 5–6 $\times 10^9$ kg of petroleum coke (for the same fields of application).

Anthracite is used in electrodes of carbide furnaces and in cathodes of aluminum furnaces. Annual anthracite consumption (worldwide) for this purpose is approximately 200–250 $\times 10^6$ kg (i.e., approximately 2–3% of the world's annual anthracite production).

Coke from bituminous coal in pulverized form is used mainly in carbonizing and graphitizing furnaces and, in smaller quantities, for the manufacture of cathodes for aluminum production. The world's annual consumption of coke made from bituminous coal for these purposes is estimated to range between 200 and 300 $\times 10^6$ kg.

30.4.1.3 *Processing of Coal and Related Products into Electrodes*

Coal-Tar Pitch. Coal-tar pitch is by far the most preferred binding agent for production of carbon and graphite electrodes. Its share in electrode production ranges between 18 and 32%. It is used to increase the plasticity of the green blends and to cement the solid matter (i.e., coke and graphite). In the course of the subsequent carbonizing or graphitizing process, the binder is carbonized. Thus solid links consisting of the carbon from the binder are established between the individual solid particles. The significance of pitch as an impregnating agent is less, since the impregnating agent merely serves to reduce electrode porosity and to increase density and strength of the material.

When used as electrode pitch, the coal-tar pitch must satisfy specific requirements that generally cannot be met by normal pitch qualities.[264] Electrode pitch must be a thermoplastic material having a very good wetting effect on the coke particles at high temperatures, and consequently exhibiting optimum cementing properties on carbonization. The hardness of pitch must be chosen such that the green, shaped bodies can be transported and stored at ambient temperature without difficulty. The most important point, however, is a high coke yield, so that after carbonization and graphitization the solid particles are linked by many carbon bridges. In certain cases special additional requirements with respect to ash content and, more and more frequently, to the sulfur content of the pitch, are specified.

[264] K. J. Hüttinger, *Erdoel Kohle,* **23,** 559 (1970).

For the choice of electrode pitch qualities, therefore, softening point, viscosity, and coke yield are important. Table 30.17 shows some characteristic data for pitch that is well suited for use as a binding agent for electrodes.[252] For impregnating pitches, lower softening points (323–333 K) are preferred, to facilitate impregnation at low temperatures. The coke yield is somewhat lower (i.e., 55–60% of original pitch).

Definite pitch fractions with certain average molecular weights are critical in obtaining the required special properties. The molecular weights are determined by laboratory methods of fractionating using certain solvents.[256,257] The electrode pitch qualities achieved by these methods differ from those of normal pitches in having a higher content of components that are soluble or insoluble in certain solvents, especially benzene (or toluene) and quinoline (or anthracene). These components (tar resins) behave differently with respect to the rheological properties of the pitch affecting the wetting and cementing of the solid particles (coke and graphite) during carbonization. For example, quinoline insolubles (QIs) increase coke yield but decrease cementing capacity, whereas the benzene insolubles (BIs) are characterized by good cementing capacity but a smaller coke yield.[256] The interdependences in this respect, however, are not yet known exactly, and further research is being conducted in this field.[258,265–267] Great attention is given to

[265] H. Tillmanns, H. Pauls, and G. Pietzka, "Einfluss der Art und Menge des Chinolinunlöslichen, paper presented at 2nd Int. Carbon Conf., Baden-Baden, Germany, 1976, preprint p. 374.

[266] T. Miyake, K. Ueda, and T. Nakamura, "The Cherry-T-Process," paper presented at 2nd Int. Carbon Conf., Baden-Baden, Germany, 1976, preprint p. 378.

[267] K. J. Hüttinger, *Erdoel Kohle,* **26,** 21 (1973).

Table 30.17 Properties of Typical Pitches Used as Binders for Electrodes

Krämer-Sarnow Softening Point (K)	Viscosity (poises) at 423 K	Coke Yield (%) in Annular Kiln
333–348	2.5–4.5	60–70
353–368	12–45	68–75

the influence during carbonization of the insoluble components on the course of the mesophase formation, which is monitored by a variety of methods, including thermogravimetric methods, polarization microscopy, and electron scan microscopy (see also "Pitch Coke" in Section 30.4.1.3).[264,267–271] The QIs, also called α-resins, are subdivided into α_1-resins (soot- and cokelike particles with a grain size of 10–100 μm) and α_2-resins (high molecular weight aromatic resins formed by polymerization).[258,265–267]

With the current state of technology, investigators are seeking a certain ratio of QI to BI components that will constitute the best possible blend for the purpose or kind of finish envisaged, (i.e, binding or impregnating agent, electrodes to be used in the aluminum or steel industry, finished electrodes, or Söderberg anodes as used in aluminum furnaces).[258] Typical limiting values are shown in Table 30.18.

The difference between BIs and QIs ranges between 18 and 23% for the binder pitch qualities and between 12 and 15% for the impregnating pitches. The components insoluble in benzene and soluble in quinoline are called β-resins; they are medium molecular weight aromatic compounds that are subject to mesophase formation in the same way as α_2-resins.[258,265,266]

A low QI content is important for impregnating pitch to avoid difficulties with soaking of the electrodes (formation of filter cake).[259]

In most instances petroleum pitches contain significantly less QIs than tar pitches, and their aromaticity is lower as well. For this reason the coke yield on carbonization and the mechanical strength obtained are lower. We can assume that the petroleum pitch qualities produce thinner carbon bridges between the solid particles than the tar pitches.[259] Therefore tar pitch is preferred to petroleum pitches as binder because of the higher carbon content of the former and the resulting higher density (~ 1.32 g/cm^3) and higher mechanical strength after carbonization and graphitization.

In recent years reports were received about attempts to add condensation agents to the binder, such as sulfur or organic chlorine combinations (e.g., benzene dichloride or carbon tetrachloride) to increase the coke yield.[252,272] In large-scale operation, however, these methods apparently were not successful either with tar pitch or with petroleum pitch.

The basic material for full-scale tar pitch production for electrodes is generally the normal pitch obtained by continuous distillation of coal tar produced in coke-oven plants. By mere thermal condensation and polymerization, a pitch with higher softening point and increased tar resin content is obtained from the normal pitch.[273] It is important that the components required for electrode pitch are already present in the coal tar to an adequate extent, in their definite or in their preliminary form. The thermal condensation of pitch components is influenced by pressure, temperature, and residence time. In practice, condensation is carried out at temperatures ranging from 648 to 693 K, and

[268] E. Fitzer, *Ber. Dtsch. Keram. Ges.*, **49**, 141 1972).

[269] J. L. White, G. L. Guthrie, and J. O. Gaschner, *Carbon (Oxford)*, **5**, 517 (1967).

[270] R. W. Wallouch, H. N. Murty, and E. A. Heintz, *Carbon (Oxford)*, **10**, 729 (1972).

[271] J. Mochida, K. Kudo, N. Fukuda, K. Takeshita, and R. Takahashi, *Carbon (Oxford)*, **13**, 135 (1975).

[272] "Pitch Binder for Carbon Electrodes," in *Industrial Carbon and Graphite Conference*, London, September 24–26, 1957, A. Darney, Ed. (publ. 1958).

[273] W. Gemmeke and G. Collin, "The Characteristics of Binder Pitches," paper presented at 103rd AIME Annual Meeting, Dallas, Tex., February 24–28, 1974.

Table 30.18 Values for the Content of Quinoline Insolubles (QI) and Benzene Insolubles (BI) for Different Electrode Tar-Pitch Blends as Used in Aluminum Production[253,257,260]

	Pitch as Binder for		Pitch as
Component	Prebaked Anodes	Söderberg Anodes	Impregnating Agent
QI (%)	8–15	8–12	4
BI (%)	28–36	26–34	15–20

in most cases at slight pressure. The process is actually controlled by variation of the residence time.

Full-scale processes for production of electrode pitch have been developed by Rütgerswerke–Teerverwertung AG, Koppers Company, Inc., and the Société pour l'Exploitation des Procédés Ab-der Halden (Proabd process).[257] Most of the pitch for electrodes is produced in these plants.

Recently reports were received about a process (Cherry-T process) developed in 1972–1974 by Mitsubishi Heavy Industry, Ltd., and Osaka Gas Company, Ltd., in Japan.[266] By this process coal tar is not distilled in the classic way. The oil remains in the tar; and in a special reactor about which no detailed information is yet available, polycondensation takes place at high pressures and temperatures, thus producing the binder pitch. The oil distillate produced in this process is also converted to pitch in a second reactor. Because of its composition, this pitch is regarded as an impregnating pitch.

The by-products of the Cherry-T process consist only of some light aromatic oils. Regarding the present-day importance attached to by-products from coal, no definite assessment is yet possible with respect to the economic significance of tar-oil conversion to pitch.

Pitch Coke. Pitch coke as an alternative to petroleum coke is mostly used for the manufacture of anodes for aluminum fusion electrolysis and for alkaline chloride electrolysis.

The quantity of the pitch coke used in aluminum production is only of secondary importance when compared with petroleum coke. Here the decision whether to use pitch coke or petroleum coke is predominantly controlled by availability, by the proximity of coke producer and coke consumer, and by the price.[260]

The sulfur content of the pitch coke, which is generally lower than that of petroleum coke, can be regarded as an advantage (Table 30.19). This advantage could become increasingly important with regard to the more stringent requirements of pollution control, especially for large aluminum plants. Attempts to reduce the sulfur content of the carbon electrodes are also made to avoid corrosion of the iron contacts in the electrolysis cells. During the carbonization process the majority of the sulfur remains in the coke up to a temperature of 1673 K and is expelled only in the temperature range from 1723 to 1773 K by very sudden cracking of heterocyclic C—S bond, resulting in a "puffing effect." In production of graphite anodes (e.g., for alkaline chloride electrolysis) this "puffing effect" must be suppressed by the addition of alkaline metals and iron.[274]

The use of pitch coke qualities for the manufacture of anodes for alkaline chloride electrolysis is more significant from the technical point of view, because of its very low vanadium content (Table 30.19). Vanadium effects a reduction of the overvoltage and may result in the formation of explosive

[274] E. Fitzer, "Development and Application of New and Modified Residues," paper presented at World Petroleum Congr., Tokyo, 1975.

Table 30.19 Properties of Pitch Coke and Petroleum Coke[252,253,257,274]

Property	Pitch Coke	Normal Petroleum Coke	Premium Petroleum Coke
Density after calcination at 1573 K (g/cm^3)	2.07–2.11	2.12–2.15	2.15–2.17
Graphitizing[a]	~ 2.5	1.88–2.2	1.45–5.65
Coefficient of thermal expansion after graphitization at 2973 K (10^{-6}/K)	8.0–10.5	5.5–8.0	4.5–5.5
Ash (%)	0.2–0.4	0.1–0.3	0.05–0.15
Sulfur (%)	0.2–0.6	0.6–0.8	0.2–1.0
Vanadium (ppm)	1–5	Up to 100	

[a] Empirical scale per G. Pietzka and H. Pauls (not yet published).[252]

mixtures of hydrogen and chlorine. Vanadium contents exceeding 20 ppm in the graphite anodes should be avoided whenever possible.[274]

After all, the higher hardness and strength of pitch coke in comparison with most petroleum coke properties can be regarded as an advantage for the former.

The great disadvantage of pitch coke properties compared to petroleum coke properties is to be found in their less favorable graphitizing characteristics. In the present context this applies, above all, to the anodes for alkaline chloride electrolysis; on the other hand, in the future special pitch coke properties might find new applications in high-temperature reactors because of their isotropic character.

The reason for the difference between graphitizing properties of petroleum and coke pitches may be found above all in the difference in behavior during mesophase formation (see preceding section). The mesophase is formed by parallel orientation of planar polyaromatic groups in the temperature range between 623 and 723 K. This phenomenon can be compared to the crystalline nucleus formation, the α-resin particles acting as nuclei. In the range between 723 and 823 K a further agglomeration occurs, resulting in the formation of larger surfaces.[274–278] It is obvious that this preliminary orientation has a decisive influence on the later graphitizing process. The mesophase formation is controlled by the kind and composition of raw materials, above all by the content of insoluble components and by the ash content. A lower content of insolubles and ash components encourages mesophase formation,[267,274] and thus improves the graphitizing properties of the coke produced.

The basic material for pitch coke production is purified hard pitch (impurities $<$ 0.3%); hard pitch has a softening point above 423 K.[257] Carbonization of hard pitch is presently carried out mostly by the Koppers process. Liquid hard pitch from heated container cars is fed discontinuously at approximately 593 K into coke ovens operating at approximately normal pressure. To avoid, as far as possible, premature formation of coke during charging of the ovens, superheated steam can be injected into the ovens at that time. Tars that may be present are separated by condensation, and 55% of the gas produced (80% H_2) is used for heating the coke ovens. Maximum coking temperature is 1473 K. Coking time of a 10–15 $\times$ 10^3 kg charge is approximately 17 hr. The pitch coke is then discharged (red-hot) to a covered quenching car for quenching and cooling by water spray.[257] A similar process was developed by Dr. C. Otto Company in Germany.

Pitch cokes with improved graphitizing properties can be obtained by coking pitch with an especially low content of QIs under special conditions. Such hard pitch can be obtained, for example, from normal pitch by careful low-temperature distillation. Quinoline insolubles can be removed by separation (dissolution, centrifugal process, and filtration). Tests on this procedure, carried out especially in Germany, the United States, and the United Kingdom,[257,274–276] yielded a coke having premium qualities. However these methods have not yet been applied in large scale production.

It was found that the inferior graphitizing properties of pitch coke compared with petroleum coke can be imputed not only to the conditions of the raw material used but also to the kind of coking process: for example, a premature formation of coke in the coke ovens turned out to be a disadvantage.[252,266,279] Consequently development work in recent years was aimed at the "delayed coking" of coal-tar pitch.

The decisive factor in the delyaed coking process is the positive influence of a longer stay in the temperature range below 773 K on mesophase formation, and the inhibited evaporation of low-boiling components under pressure.[274] These influences are obviously favorable to the preliminary orientation in the mesophase, which is necessary for later graphitization.

[275] I. Romey, "Carbon from Purified Coal-Tar Pitch," paper presented at 12th Biennial Conf. Carbon, Pittsburgh, July 28–August 1, 1975.

[276] J. Owen, "Carbons from Coal" (Plenary Lecture), presented at 12th Biennial Conf. Carbon, Pittsburgh, July 28–August 1, 1975.

[277] P. L. Walker, Jr., *Carbon (Oxford)*, **10**, 369 (1972).

[278] J. D. Brooks and G. H. Taylor, in *Chemistry and Physics of Carbon*, Vol. 4, P. L. Walker, Jr., Ed., Marcel Dekker, New York, 1968.

[279] E. Fitzer, K. Müller, and W. Schäfer, in *Chemistry and Physics of Carbon,* Vol. 7, P. L. Walker, Jr., Ed., Marcel Dekker, New York, 1971, p. 237.

In Japan the first production plant for the delayed coking process using coal-tar pitch exclusively as feedstock was built by the Lummus Company (U.S.) and Nittetsu Chemical Industries (Japan) in the period 1968–1974. This plant produces pitch coke with considerably improved graphitizing properties and with low contents of sulfur and ash. A complete production cycle including charging, coking, discharging, and quenching takes 48 hr.[280]

Another possibility for producing coke from coal with improved graphitizing properties has been considered by the National Coal Board (NCB) in England.[276] This process is based on a coal extract instead of pitch. As in the German Pott-Broche process, finely ground and dried coal is dissolved at increased temperature and under pressure in a solvent oil (anthracene). The mineral components and undissolved coal are removed by filtration, and after preheating the extract is fed to a delayed coking reactor. The coke obtained reportedly exhibits good graphitizing properties, and during graphitization there is no puffing effect because of sudden sulfur separation. Production-scale testing of coke from such an extract has not yet been done.

Anthracite and Coke from the Carbonization of Bituminous Coal. Bituminous coal had only limited application as a solid component (filler) in the beginning of the development of electrodes. Together with anthracite it was used for the manufacture of electrodes for electric steel furnaces.[253] The low coke yield and high content of volatile matter and ash, as well as unsatisfactory graphitizing properties, were the reasons for replacement of this material by pitch coke, and later on by petroleum coke. A variety of publications dealing very thoroughly with coking of coal also very frequently treat the range of high temperatures causing graphitization. However these publications are in most cases of merely academic significance.[281,282]

Coke from coal is used as an auxiliary material for the production of all carbon and graphite electrodes.[252] It is used for covering and backing of carbon bodies in carbonizing and graphitizing ovens. Except for specified grain size, dryness, and purity (maximum ash content 10%), no special requirements are set for this material. It is still preferred to other kinds of coke because of its lower price.

Coke is used in smaller quantities for the manufacture of electrodes for carbide furnaces and for cathodes of aluminum furnaces. In the last few years the use of cokes from lignite by the electrode industry was also proposed.[283,284] However anthracite is a more important material in these fields of application. Among all coal properties, anthracite shows the highest carbon content and a sufficient electric conductivity for the above-mentioned purposes.[252,253,285,286] A decisive prerequisite for the use of anthracite, however, is a calcination[252,253] and, in some cases, preparation treatment for reducing the ash content.

For the last-mentioned purposes, anthracite cannot simply be replaced by petroleum coke. The extremely narrow-pored structure of anthracite results in a higher packing density and greater stability with respect to shape and dimensions than can be obtained with petroleum coke, and this means that cathodes made from anthracite suffer less, for example, in the aluminum furnace, from attacks by penetrating sodium vapors from the molten cryolite (swelling effect). Not all kinds of anthracite can be used; nonswelling anthracite with stable structure is preferred. Cathodes made from anthracite can be used for more than 5 years in aluminum furnaces.[259,261]

In the past attempts were made mainly in the United States to use anthracite generally as a replacement for petroleum coke in electrode production, especially for aluminum furnaces. Tests have shown, however, that for sufficient reduction of the ash content,

[280] "Technischer Kurzbericht," *Erdoel Kohle,* **22,** 433 (1969).; Firmenprospekt Lummus/Nittetsu.

[281] P. Chiche, S. Durif, and S. Prégermain, *Fuel,* **44,** 5 (1965).

[282] J. Maire and J. Mering, in *Chemistry and Physics of Carbon,* Vol. 6, P. L. Walker, Jr., and P. L. Thrower, Eds., Marcel Dekker, New York, 1973.

[283] F. H. Franke, B. Wenzel, and R. Kurtz, *Braunkohle,* **11,** 357 (1975).

[284] P. Speich, *Braunkohle,* press, 1977.

[285] A. Oberlin and G. Terrierl, *Carbon (Oxford),* **13,** 367 (1975).

[286] W. Rieger, paper presented at 3rd Conf. Industrial Carbons and Graphite, London, April 14–17, 1970.

anthracite either had to be heated to 3073 K [287] or purified at lower temperatures (1773–2273 K) in chlorine atmosphere.[288,289] However up to the present such processes are less economical than the use of petroleum coke.[289]

30.4.2 Carbon Bricks

The ever-increasing requirements in blast furnace operation, along with larger hearth diameters and boosted productivity, resulted in replacing ceramic material by carbon bricks for the lining of certain zones of blast furnaces.[175,290–293] Since 1930, the bottoms and hearths of blast furnaces have mostly been lined with carbon bricks. These carbon bricks, however, also proved their value when used for blast furnace runout through linings and molds. Furthermore, carbon bricks are used for lining of the bottom, hearth, and walls of furnaces for production of aluminum, copper, phosphorus, carbide, and so on.[291,294]

The manufacturing principle for carbon bricks is nearly identical to that for other synthetic carbon products (e.g., electrodes: Section 30.4.1).[290,293,295] Anthracite, coke from bituminous coal, pitch coke, and petroleum coke, with coal-tar pitch as a binding and impregnating agent, are used as raw materials. Anthracite and different kinds of coke in general are calcined in rotary kilns at temperatures of approximately 1573 K. During this process the volatile matter content is reduced to a value below 0.2%, and a uniform, basic material with defined shrinking properties is obtained. After grinding the coke in roll grinders or hammermills, and after screening, the dried feedstock and the binder are blended at temperatures ranging from 373 to 423 K, then kneaded and eventually shaped by extrusion, pressing, or jarring molding processes. The green shapes are then carbonized in round downdraft kilns or single-chamber kilns according to exactly predetermined carbonization curves and at final temperatures of 1273 K (to max 1573 K). The pitch coke produced by calcination of the binder establishes a rigid linking of the coke particles.

The carbon bricks obtained as just described can be treated with the tools generally used in the metalworking and woodworking industries. When using these bricks as lining material for blast furnaces, anthracite and coke are chosen as the main components, constituting more than 75% of the raw materials used. However the blending ratio may vary considerably with the individual manufacturer. According to moderate estimates, the annual anthracite and coke requirement for the total world production of lining material for blast furnaces amounts to 50–60 $\times$ 10^6 kg. To this quantity a small proportion of coke powder must be added, which is used for covering purposes in the kilns and for support of the bricks during the softening phase of the binding agent.

The preference given to coke from bituminous coal and anthracite in the production of carbon bricks is due, above all, to the more competitive prices when compared with those for pitch coke and petroleum coke, which can be also used. However the use of anthracites having a high density and a low ash content is recommended. Since anthracites with such properties are rare, in certain market situations semianthracite is used instead. In any case the structure of anthracite guarantees a higher stability of shape and a reduced porosity of the carbon bricks.

Because of the low graphitization properties of anthracite and coke, compared with pitch coke and petroleum coke, as well as the kind of forming process—mostly pressing or jarring molding—the carbon bricks are to a great extent isotropic.[175] For the intended use as described above, this does not constitute a disadvantage. The special features of carbon bricks are a dense and homogeneous structure with little porosity, high strength values, good heat conductivity, very good

[287] I. Geller and P. L. Walker, Jr., in *Proc. 5th Carbon Conf.* Vol. 2, Pergamon Press, Oxford, 1963, p. 471.

[288] G. R. Imperial, Thesis, Pennsylvania State University, 1962.

[289] P. L. Walker, Jr., W. O. Stacy, and E. Wege, *Carbon (Oxford),* **4,** 129 (1966).

[290] K. Etzel, H. Rosswurm, and G. Wilhelmi, *Tonind.-Ztg.,* **96,** 167 (1972).

[291] J. H. Chesters, *Ironmaking Steelmaking Q.,* **1**(3), 134 (1974).

[292] J. L. Evans and G. M. Workmann, *J. Iron Steel Inst.,* 264 (April 1973).

[293] R. B. Snow, *Iron Steel Eng.,* 37 (January 1976).

[294] O. Rubisch and O. Vohler, *HDT-Vortragsveröffentlichung, Essen,* **206,** 44 (1969).

[295] C. L. Mantell, *Carbon and Graphite Handbook,* John Wiley and Sons, New York, 1968.

resistance to temperature changes, and high resistance to corrosive slags and liquid metal.[252,293] Table 30.20 contains typical values measured on carbon bricks made from anthracite and coke from bituminous coal and used for blast furnace lining.

The properties of carbon bricks made from anthracite and coke are adequate for the use described, especially for blast furnaces. The present development work, however, aims at their use for lining of additional zones of the blast furnace. Besides the bottom and the hearth, they are intended, above all, for the bosh and the lower zone of the stack.[290,296] In these zones, until now, neither fire clay bricks, sillimanite bricks, high-alumina bricks, nor the carbon bricks described above have given satisfactory results.[290] The attacks by oxidation, slag, alkaline vapors, and erosion are serious. So far, laboratory research and research on blast furnaces in operation have shown that only graphitized materials can be used for lining these zones. For the manufacture of those graphitized products the standard raw material is not anthracite or coke from bituminous coal, but petroleum coke and, to a smaller extent, pitch coke.[289,297] Graphitizing is carried out at temperatures above 2773 K.

30.4.3 Carbon Fibers

The carbon fibers made available so far have been produced by high-temperature (1973–2673 K) treatment of cellulose (rayon) or polyacrylonitrile (PAN) fibers. These fibers are characterized by an extremely favorable ratio of Young's modulus and strength to density. They are stiffer than all other materials of the same density. They exhibit very good thermal resistance as well as very positive sliding and lubricating properties. They are electrically conductive and highly corrosion resistant. Because of their low strength against notch impact stress, the carbon fibers are mostly embedded into a matrix of synthetic resins or metals. Such compound materials are mechanically 2 or even 3 times as strong as comparable metals (e.g., steel, aluminum, or titanium).

In the past carbon fibers were used only for military purposes or in space technology. They were used specifically for the construction of motors, helicopter rotor blades, and different components of missiles.

Because of their extremely good properties, such materials may come to be used in much broader fields of technology. The main barrier to large-scale production, however, has been the high price of carbon fibers, which are costly principally because they are made from comparatively expensive raw materials. From the technical point of view the relatively low carbon content of these raw materials and the considerable shrinkage resulting during carbonization can be regarded as an additional disadvantage.

Hence from the mid-1960s, increased efforts were aimed at the production of carbon fibers from cheap asphalt, petroleum pitch, or coal-tar pitch. The processes described here are based on coal-tar pitch. After minor modifications, however, they can be applied to all other raw materials similar to pitch. The special problem in processing coal-tar pitch, as compared to rayon or PAN, is the presence of an insoluble mesophase component that must be removed before spinning. If this component is not removed, particles of it create points of tension in the carbon thread, greatly reducing its strength. Besides, nodes can be formed on the surface of the threads, making the threads unusable for certain applications. Furthermore, the softening range of coal-tar pitches is lower than the disintegration temperature of pitch, and this temperature range must be passed through under suitable conditions, generally by careful oxidation to prevent the threads from baking together.

[296] F. Reinhart, *Glas-Email-Keramo-Technik,* **24,** 321 (1973).
[297] G. König, C. Weidemüller, and G. Pietzko, *Stahl Eisen,* **91,** 63 (1971).

Table 30.20 Properties of Lining Bricks for Blast Furnaces[175] and Manufacturers' Data (uncited)

Carbon	95–97%
Bulk density	1.55 g/cm³
Real density	1.90 g/cm³
Porosity	13–17%
Young's modulus	$0.9\text{–}3 \times 10^5$ N/cm²
Flexural strength	80–150 N/cm²
Compression strength	350–500 N/cm²
Coefficient of thermal expansion at 293–473 K	$4\text{–}6 \times 10^{-6}$ K^{-1}
Thermal conductivity	8–16 W/m-K

The first industrially applicable process using coal-tar pitch as raw material was developed in Japan by Kureha Kagaku Company. The process is carried out in the following steps.[298–301]

1. Dry distillation of coal-tar pitch in a nitrogen atmosphere at approximately 653 K and/or in vacuum at approximately 543 K.
2. Spinning of the molten distillation product at temperatures between 473 and 643 K through a die and by means of a drawoff system.
3. Oxidation of threads in atmospheric air at a heatup rate of approximately 274.5 K/min up to a temperature below the spinning temperature, but not higher than 533 K.
4. Carbonization and graphitization of the threads made unmeltable in phase 3.

By distillation (step 1) a pitch is obtained that contains virtually no ash; it has a carbon content ranging between 91 and 95%, an H/C mole ratio of 0.47 up to 0.49, and an average molecular weight greater than 400. Only if these conditions are met can the material be spun at high speed in the molten state. If the carbon content is greater than 95%, the addition of chloroform insolubles to the feedstock is recommended for the pretreatment described. The feedstock does not necessarily need to be of the same kind of pitch; that is, blends made up from different kinds of pitch may also be used.

During the spinning process (step 2) a tension is applied to the threads, which has a decisive influence on the longitudinal arrangement of crystalline structures during the subsequent coking and graphitizing process. It was found that[302–304] this influence of tension is more effective with pitch fibers than with the known PAN fibers.

In preparation for the subsequent coking process, the oxidation (step 3) of the raw threads obtained from the spinning makes these threads unmeltable. The oxidation can also be carried out in an atmosphere of ozone or pure oxygen instead of atmospheric air, but correspondingly reduced temperatures must be used.[298] The threads can be also made unmeltable by adding to the pitch a "hardening agent" in the form of peroxide (e.g., barium peroxide) or metal chlorides (e.g., chlorides af aluminum, iron, or zinc) and by keeping this blend for a certain time at temperatures ranging from 423 to 623 K, in inert gas atmosphere.[298]

The threads made unmeltable are carbonized in the usual way in the last step at temperatures below 1273 K or graphitized at higher temperatures exceeding 2273 K. It was found that for many fields of application the less expensive carbonization is adequate.[305,306]

All other processes using coal-tar pitch as basic material for the production of fibers employ the process steps described above. They differ only in the way they are technically carried out, or in some additional details.

The Great Lakes Carbon Corporation, New York,[307] has developed a production process for carbon fibers in which, instead of the distillation step, ground pitch particles smaller than 0.15 mm are partially dissolved in an aromatic solvent (e.g., benzene), to remove the low-boiling, low molecular weight components, and the resulting particles are extracted by a solubilizing agent (quinoline). After evaporation of this agent, a pitch is obtained that is free of quinoline-insoluble components and can be spun and further processed in the usual way.

Another means of pretreatment has been developed in Great Britain by the NCB.[302,303,305,306,308,309] Finely ground coal is

[298] DT-OS 1,669,471, U.S. Patent 3,629,379.
[299] DT-OS 1,669,486.
[300] DT-AS 2,027,385.
[301] T. Araki and S. Gomi, *App. Polym. Symp.*, **9**, 331 (1969).
[302] M. A. A. Jarro, W. R. Ladner, T. D. Rantell, and W. F. Wyss, "The Properties of Coal-Based Carbon Fibers," paper presented at 2nd Int. Carbon Conf., Baden-Baden, Germany, June 27–July 2, 1976.
[303] I. Romey, "Production of Carbon Fibers from Coal-Tar Pitch," paper presented at 2nd Int. Carbon Conf., Baden-Baden, Germany, June 27–July 2, 1976.
[304] D. J. Johnson, I. Tomizuka, and O. Watanabe, *Carbon (Oxford)*, **13**, 529 (1975).
[305] M. A. A. Jarro and W. R. Ladner, "Carbon Fibers," paper presented at 4th Int. Carbon and Graphite Conf., London, September 23–27, 1974.
[306] H. F. Volk, "Low Cost Carbon Fibers from Pitch," paper presented at 4th Int. Carbon and Graphite Conf., London, September 23–27, 1974.
[307] DT-OS 2,153,567.
[308] J. Owen, "Carbon from Coal," paper presented at 12th Biennial Conf. Carbon, Pittsburgh, July 28–August 1, 1975.

dissolved under pressure at temperatures of approximately 673 K to yield an oil with a high boiling point. The coal and mineral particles that are not dissolved are filtered off. After subsequent distillation, a spinnable extract is obtained. The kind of coal has an effect on the properties of the fibers produced in the way just described only if coking is carried out at approximately 1273 K. Strain graphitization has turned out to be very favorable in the NCB process. The threads are pulled through the graphitizing furnace by two reels turning at different speeds.

The process developed by Bergbau-Forschung, Germany,[303,309,310] differs basically from the processes described above in the method of cleaning the pitch—namely, by filtration—and in the oxidizing treatment. During the spinning process the threads are powdered with an oxidizing agent impregnated with carbon dust. In this way carbonization and graphitization of fibers can be obtained without difficulty.

In certain applications additional advantages can be realized if the process of Kanegafuchi Boseki, Tokyo, is used.[311] These pitch fibers—extruded through a common die—consist of one component of the molten pitch, the other of a synthetic polymer with thermoplastic properties (e.g., polyimide, polyester, polyolefine, copolymer); both components of the thread are closely linked longitudinally.

In addition, the process developed by the Union Carbide Corporation for the production of carbon fibers from pitch should be mentioned.[306] In this process pitch is spun during the mesophase at approximately 673 K; thus an additional oxidizing treatment is not required.

Table 30.21 shows some results of measurements of tensile strength and Young's modulus of carbon fibers. From this table it can be seen that these fibers cover an extraordinarily large range of values. Furthermore, the effect of tensile stress during graphitization (strain graphitization as practiced, for example, in the NCB process) is readily apparent. As to the mechanical strength, the carbon fibers match the PAN fiber values very well. The tensile strength values for PAN fibers range generally between 1 and 2 GN/m^2 and the values for the Young's modulus between 100 and 200 GN/m^2.[308,309]

Sometimes values for electric resistivity and the density of carbon fibers are also indicated in the literature (e.g., 80 Ω-m for the graphitized NCB fibers[302,303]). Union Carbide indicates a value of 1.95 g/cm^3 for the density of their fibers.[306]

Porosity measurements show a pore volume in the pore radii range of 10^2–10^4 Å, a value that can be neglected. Also in the micropore range (< 20 Å) no significant pore volume has been found.[302,303]

Inspections by electron microscopy,[302–304]

[309] I. Romey, "Carbon from Purified Coal-Tar Pitch," paper presented at 12th Biennial Conf. Carbon, Pittsburgh, July 28–August 1, 1975.

[310] DT-Patent 2,419,659.

[311] DT-OS 2,038,949.

Table 30.21 Properties of Pitch-Based Carbon Fibers from the Literature

Developer	Diameter (μm)	Tensile Strength (GN/m^2)	Young's Modulus (GN/m^2)	Remarks, References
Kurcha Magaku Company, Japan	20–40	0.28–0.77 1.39	13–32 99	Unstretched Stretched 304, 312
Union Carbide Corporation, United States	6–7	1.55–2.15	205–620	306
National Coal Board, United Kingdom	6-30	0.8 1.7–2.4	45 300–670	Carbonized Strain graphitized 302, 303–306, 308, 310
Bergbau-Forschung GmbH, West Germany	10–50	0.2–1.0	1–35	303

X-ray investigations,[304] and infrared spectrometric measurements[312] were made to determine the structure of the carbon fibers. It was found that carbonized pitch fibers and fibers produced without application of significant tensile stress show a structure containing only a few graphitic aligned formations. Such fibers can still be regarded as isotropic. With more extensive high-temperature treatment and increased tensile stress, short-length crystal formation occurs along the fiber axis so that fibers of that kind are to be regarded as anisotropic. The size of the crystals and their formation vary within a wide range, and this variation probably explains the wide range of mechanical properties of tar-pitch fibers as compared with PAN fibers.[304]

Owing to their special properties and lower price, as compared to cellulose and PAN fibers, the coal-pitch fibers are suitable for many fields of application. Besides the civil aircraft industry, mainly the fields of hobby and sports equipment seem to be very promising to further development of applications. Carbon fibers can be used for the production of sailing and motor boats, gliders, skis, golf clubs and tennis rackets, archery equipment, and fishing rods. Owing to the low weight per unit volume, use of this material can also be made in the automobile industry (e.g., for manufacture of hoods, roofs, and bumpers). Fiber-reinforced rubber is extremely suitable for beltings of all kinds. Eventually fiber-reinforced compound material can be used for a variety of purposes in the construction industry. Carbon fibers made from pitch are throught to have good prospects for the future.[306,313]

30.4.4 Plastic Materials from Coal

Attempts to produce plastic materials from coal date back to the 1930s. Work was resumed after 1960, and since that time promising results have been obtained.

The trials were started with low-rank coals whose content of fusible components permits them to be plasticized in the temperature range between 623 and 723 K, then used for binding and cementing nonfusible inert materials (e.g., kaolin).[314] Furthermore, the binding components of coal were increased by additives (e.g., multifunctional alcohols, aldehyde, polyvalent and aromatic dienes, as well as heterocyclics). Probably those compounds have a certain softening effect[315] on the coal or an effect of linking the molecule clusters of coal, with possible separation of their functional groups by condensation. This condensation may be increased by simultaneous addition of condensation agents.[316] This cluster linking effect may also be achieved by increasing the proportion of functional groups contained therein, by slight oxidation or chlorination.[315]

A blend consisting of 80–85 wt % of coal and 15–20 wt % of modification agents (i.e., polymers and elastomers) turned out to be particularly successful. The following materials can be used as modification agents: polyethylene, polypropylene, polyvinyl chloride, polystyrene, acrylonitrile, natural rubber, and butadiene polymers.[317] The classification of polymers and elastomers as modification agents indicates that no gluing is achieved by their addition, but that a reaction between the coal particles and the additives takes place during pressure and heat treatment. This reaction cannot yet be explained in detail.

Production is carried out in the following phases: grinding of coal, homogeneous addition of polymers and elastomers (powder or emulsion), and pressing (application of pressure and heat).

Pressing can be done by means of discontinuously working pressing units. The pressing temperature ranges from 373 to 523 K. If pressing is to be carried out in a continuous press as used in plastic industries (i.e., extruders or die-cast machines), the coal proportion is reduced to approximately 65–70%, and a softening agent is added as well. The material made up as just described exhibits a behavior similar to that of thermoplastics and may be used for production of foils, hoses, and a variety of shapes, on the same machinery used for thermoplastics. The

[312] S. Otani, K. Yamada, T. Koitabashi, and A. Yokoyama, *Carbon (Oxford)*, **4**, 425 (1966).

[313] P. L. Walker, Jr., *Carbon (Oxford)*, **10**, 369 (1972).

[314] I. Romey, *Erdoel Kohle*, **26**, 449 (1973).

[315] C. Kröger, I. Romey, and G. Falkenhain, *Erdoel Kohle*, **24**, 99 (1971).

[316] DT-Patent 1,796,124.

[317] DT-AS 2,116,417.

process temperatures range from 393 to 463 K (i.e., within the limits regarded as normal). Table 30.22 shows the most important properties of the materials made from coal according to the semitechnical-scale work of Bergbau-Forschung GmbH.

The materials made from coal can be sawed, nailed, drilled, machined, or glued and their surfaces veneered and painted. The material itself, however, cannot be colored.

The favorable properties described above offer these materials made from coal a good chance on the market, especially in construction work, where they can be used for the production of tiles, flagstones, and so on.

30.4.5 Carbon Black and Fillers

30.4.5.1 *Carbon Black*

Carbon black is formed by incomplete combustion of many organic substances: solid, liquid, or gaseous. The generic term "carbon black" refers to a group of industrial products consisting of furnace blacks, channel blacks, thermal blacks, and lamp blacks,[318,319] They are materials composed essentially of elemental carbon in the form of near-spherical particles of colloid sizes, coalesced mainly into particle aggregates obtained by partial combustion or thermal decomposition of hydrocarbons. Carbon blacks are used as rubber reinforcing agents (95%) and as coloring material in printing inks, paints and lacquers, plastics, chemicals, paper, and miscellaneous applications.

[318] J. B. Donnet and A. Voet, *Carbon Black, Physics, Chemistry and Elastomer Reinforcement,* Marcel Dekker, New York and Basel, 1976.

[319] W. R. Smith and D. C. Bean, "Carbon Black," in *Kirk-Othmer Encyclopedia of Chemical Technology,* 2nd ed., Vol. 4, New York, 1964, p. 243.

The oil furnace process is by far the most important commercial process for carbon black production. The feedstock used in the oil furnace process must fulfill a number of requirements, such as high content of aromatics; low content of asphaltenes, free carbon, sulfur, alkaline metal compounds, water, and sediment; and limited boiling range.

The aromatic content is the most important criterion, since it largely controls the all-important yield. Asphaltenes, the toluene-soluble, hexane-insoluble fraction, are relatively nonvolatile and carbonize easily to form highly undersirable gritty material. Maximum allowable quantities are 5–8%, although occasionally feedstocks with a somewhat higher asphaltene content are used. Ash and sulfur accumulate in the black and must be avoided. Although feedstocks have generally been derived from refinery cracking operations, the major part of the feedstock used at present is obtained from the bottoms of petroleum fractions cracked under conditions of high severity. Petroleum extracts are also used.

The requirements mentioned above are also fulfilled by fully aromatic tar oils produced by West Germany's coal-tar refining industry. Using the furnace process with incomplete combustion of this feedstock, yields of up to 60% can be achieved. This technique is used mostly in West Germany, where, consequently, the process above represents one of the biggest market outlets for tar oil. The decreasing use of tar oil products for road paving and briquetting of coal can be offset in this way. From 1960 to 1974 carbon black production from tar oils was increased from 70×10^6 kg to 300×10^6

Table 30.22 Technical Data on Plastic Materials Made from Coal[314]

Compression strength	Up to 10^4 N/cm^2
Flexural strength	Up to 2×10^3 N/cm^2
Tensile strength	Up to 9×10^2 N/cm^2
Impact strength	52 N-cm/cm^2
Notch impact test	32 N-cm/cm^2
Young's modulus	1.3×10^5 N/cm^2
Shore hardness	90–95
Apparent density	1.3 g/cm^3
Coefficient of thermal conductivity	1.55 kJ/m-hr-K
Dc conductivity	$2.10^{-14}\ \Omega^{-1}\ cm^{-1}$

kg.[60,320] This situation is different in other countries, including the United States, where carbon black is produced mainly from natural gas and petroleum fractions.

With the tar oil as feedstock, all types of rubber blacks characterized by the ASTM classification[321] can be produced. They range from those having the largest particle size, SRF (semireinforcing furnace)—surface area measured by nitrogen adsorption of 17–33 m^2/g and particle size of 61–100 nm—to those having the smallest particle size, SAF (superabrasion furnace)—surface area 125–155 m^2/g and particle size 11–19 nm.[322] Also, color blacks with surface areas up to 700 m^2/g and average particle size down to 10 nm can be produced for various applications.

30.4.5.2 *Fillers*

As shown in Section 30.4.5.1, various grades of carbon black are used as reinforcing fillers for rubber blends. In the past 15 years many efforts have been made to substitute finely ground coal for carbon black. The results of these recent investigations show that by grinding low-volatile coal down to a particle size range of 5–10 μm, a filler may be produced that does not attain the desired properties of highly reinforcing carbon blacks such as high abrasion furnace (HAF) or intermediate super abrasion furnace (ISAF) carbon blacks, although it can be added to these grades of carbon black in a ratio up to 10% without significantly affecting the quality of several rubber blends.[323,324] Anthracite and coke are suitable feedstocks. The properties of these fillers can be improved by not admitting any oxygen to the material from the beginning of the grinding process up to the proportioning in the rubber blend. The prevention of contact with oxygen can be achieved, for example, by grinding the feedstock in an atmosphere of nitrogen or under steam in a jet mill, or by making up master batches. Another possibility is to coat the filler, after grinding with suitable material (e.g., zinc stearate) in a nonoxidizing atmosphere. Such a coating agent can easily be removed when necessary by evaporation or sublimation. In the case of coke as feedstock, the elasticity modulus of rubber blends can be controlled by the volatile matter content of the coke used. Thus the elasticity modulus decreases with increasing volatile matter content of the coke.

Ground coal and coke may also be used as filler for plastic material (e.g., polyethylene and polypropylene). However in these cases the requirements with respect to particle size and the chemical state of the surface (prevention of oxygen) are not as stringent as in the case of rubber.

An incomplete list of references on the properties of ground coal used as filler and on compound materials from coal and polymers may be found in Gorlov et al.[325]

In several cases investigators attempted to produce carbon black from coal by coal pyrolysis, involving rapid heatup to high temperatures so that the volatile matter is converted to carbon black by decomposition.[326–331] Such processes, however, are not yet economically viable.

30.4.6 Carbonaceous Adsorbents Made from Coal

30.4.6.1 *General Remarks*

Carbonaceous adsorbents are substances consisting mainly of carbon with high poros-

[320] R. Oberkobusch, *Erdoel Kohle, Erdgas, Petrochem. verein. Brennst.-Chem.*, **28**, 558 (1975).

[321] ASTM-D2159, proposed nomenclature issued 1966.

[322] *Degussa Rubber Blacks*, 1976 Edition.

[323] W. Mueller, S. Palinchak, and P. Stickney, *Rubber Age*, **94**, 255 (1963).

[324] W. Mueller, S. Palinchak, and P. Stickney, *Rubber Age*, **94**, 407 (1964).

[325] E. G. Gorlov, S. R. Zymmerov, and J. M. Pauskin, *Khim. Tverd. Topl.*, **11**(1), 3–16 (1977).

[326] G. E. Johnson, W. A. Decker, A. J. Forney, and J. H. Field, paper presented at Chemicals from Coal Symp., American Chemical Society, New York, September 1969.

[327] G. E. Johnson, W. A. Decker, A. J. Forney, and J. H. Field, *Rubber World*, **156**, 63 (1967).

[328] G. E. Johnson, W. A. Decker, A. J. Forney, and J. H. Field, *Ind. Eng. Chem. Prod. Res. Dev.*, **9**, 382 (1970).

[329] I. F. Bogdanov, L. P. Gilyazetdinov, M. L. Mishchenko, and I. L. Faberov, *Khim. Tverd. Topl.*, No. 4, 121 (1968).

[330] M. L. Mishchenko, I. F. Bogdanov, I. L. Faberov, I. M. Yurkovskii, L. P. Gilyazetdinov, and V. I. Gugaev, *Khim. Tverd. Topl.*, No. 2, 139 (1969).

[331] I. F. Bogdanov, L. P. Gilyazetdinov, V. I. Gugaev, V. P. Zuev, and M. L. Mishchenko, *Khim. Tverd. Topl.*, No. 6, 116 (1969).

ity and extensive inner surface. Their chemical structure allows them to preferentially adsorb organic materials and other nonpolar compounds from the gas or liquid phase. Because of these properties they have been used for many decades for the purification of gases, the separation of gas mixtures, the purification of exhausted air, especially the recovery of solvents, the decolorization of solutions, and the purification of water. In the last decade carbonaceous absorbents found greater use in the solution of problems of keeping the air clean, and in the purification of water before and after use.

The production of carbonaceous adsorbents in the United States in 1940 was 18 × 10^6 kg; it had doubled by 1950 to 36 × 10^6 kg,[332] and in 1974 it was 127 × 10^6 kg.[333] Of this amount, 52 × 10^6 kg was produced in the form of granules and 75 × 10^6 kg as powder. Recently the proportion of granulated or shaped active carbons increased more rapidly than that of powdered material. This situation is caused by the greater ease of regeneration of the granulated material. Western Europe follows the United States as the second largest producer, with 100 × 10^6 kg in 1974. Japan, with 40 × 10^6 kg in 1974, was the third largest producer of carbonaceous adsorbents. In Western Europe and in Japan 75 and 65% of the adsorbents, respectively, were powders. Furthermore, in 1974 about 20 × 10^6 kg of carbonaceous adsorbents was produced in South America and Asia. Production figures for the Eastern bloc countries are not known.

In the United States three of the seven producers had 87% of the total capacity, in Western Europe four out of ten companies produced 85%, and in Japan in 1974 twenty companies supplied the market. Raw materials for the production of carbonaceous adsorbents include currently grown and fossil carbonaceous materials, especially wood, peat, brown coal, bituminous coal, coconut shells, pits from peaches and other fruit, petroleum-based residues, and pulp mill residues (black ash). The classical raw materials were wood, peat, and coconut shells. During World War II parallel developments in Germany and the United States led to processes based on bituminous coal. The rapidly increasing demand for carbonaceous adsorbents in industrial countries made it important to have a broad raw material base. Therefore bituminous coal is being used increasingly as a raw material. We estimate that at present 200–250 × 10^6 kg/year of bituminous coal is converted into carbonaceous adsorbents.

The different carbonaceous adsorbents (Fig. 30.14) are defined as follows.

In addition to the system of macropores, active carbons have a large adsorption pore volume of more than 30 cm^3/100 g; the average pore diameter of this micro- and mesopore system is between 5 and 20 Å. Different degrees of activation intensities may make it possible to change both values to a great extent, to obtain materials with suitable adsorption properties. Increasing adsorption pore volume increases the average pore radius, while the bulk density, which is often used for a first characterization, decreases. Active carbons with low bulk density are assumed to have "large pores," and active carbons with high bulk density are assumed to have "small pores."

By comparison, carbonaceous molecular sieves have very small pores, which have about the same diameter as the molecules to be separated by adsoprtion. The adsorption pore volume is usally below 30 cm^3/100 g.

[332] J. J. Schanz, Jr., and R. H. Parry, "The Activated Carbon Industry," *Ind. Eng. Chem.*, **54**, 24 (1968).

[333] "Environmental Protection Promotes Active Carbon Business," *Chem. Ind.*, **26**(12), 808–809 (1974).

[334] H. Jüntgen, "New Uses for Carbonaceous Adsorbents," paper presented at 2nd Int. Carbon Conf., Baden-Baden, Germany, 1976.

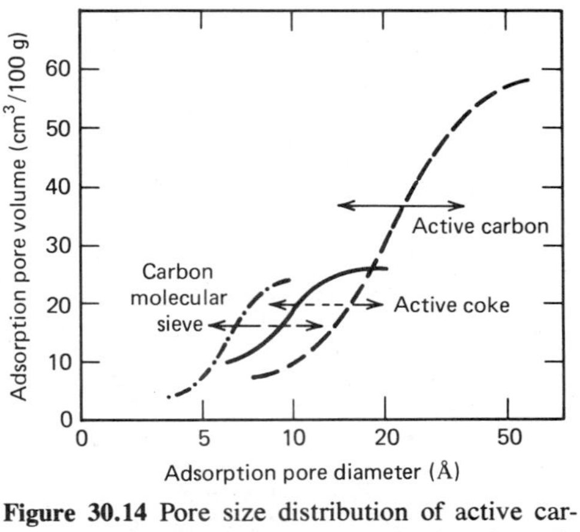

Figure 30.14 Pore size distribution of active carbon, active coke,and carbon (Molecular sieve).[334]

One characteristic property is their ability to separate permanent gases (e.g., oxygen from nitrogen) or mixtures of hydrocarbons (e.g, olefins from paraffins, or *n*-paraffins from isoparaffins). This ability to separate the gases or hydrocarbons is usually a result of the different diffusion behavior of the adsorbed materials in the narrow submicropores.

Active cokes with a pore volume of about 20 cm^3/100 g are preferentially used in Germany for the desulfurization of flue gas using thermal reactivation. They have a high adsorption capacity for SO_2, high resistance to abrasion, and high ignition temperatures, and low resistance to the flow of gases. When used for the adsorption of SO_2 with thermal regeneration, they change their pore structure, which increases the capacity for adsorbing SO_2.[335]

30.4.6.2 *Production Processes*

The principle of manufacturing carbonaceous adsorbents is the selective removal of some groups of compounds from the suitable carbon-containing material without reducing the particle size. These removed compounds are replaced by well-defined pores. Chemical activation or gas activation may be used. The former method is used only when wood is the raw material; activating compounds are $ZnCl_2$ or H_3PO_4. Gas activation is used with all raw materials: wood, peat, brown coal, bituminous coal, and petroleum coke; the activating media most often used are H_2O and CO_2.[336]

If suitable coals are selected as the raw material, and an appropriate pretreatment such as oxidation or modification of the carbonization process is applied, coal can be converted by gas activation into carbonaceous adsorbents with properties that are identical or comparable to the absorbents made from wood, peat coke, or coconut shells.[337] The feed coals must have the following properties:

1. Low ash content, since the percentage of mineral matter in the course of activation is increased and the adsorption capacity decreased.
2. Formation of cokes that are sufficiently reactive toward steam and CO_2 and have a crystallite structure that contains faults and edge atoms that on activation, lead to narrow micropores. Only noncaking bituminous coals have these properties by nature. However they can be obtained when caking coals are pretreated by oxidation with air below the ignition point, or are very slowly carbonized. To obtain an effective macropore system, the coal is ground to a size below 0.1 mm and the fine coal is then briquetted with suitable binders such as coal-tar pitch.

The major raw material for the production of active carbon from coal is bituminous coal, but brown coal and anthracite are also used. The possibility of using brown coal was tested in Germany in the laboratory and on a small technical scale.[338]

When anthracite serves as the raw material, route 1 of Fig. 30.15 is used; this means that after removal of the mineral matter, the anthracite is screened to the desired size consist, then carbonized and activated with steam or carbon dioxide. The active carbon obtained by activation is further screened. For bituminous coal as the raw material, the literature indicates essentially three processing procedures for the preparation of active carbon. Basically similar or identical processing steps are used, but in a different sequence or under different conditions. According to route 2 in Fig. 30.15, the coal is pulverized and then briquetted with binder at high pressure. A very slow heating rate is used for the carbonization of the briquettes, which are then activated with steam. Next the activated briquettes are comminuted and separated into granulated active carbon of the desired size range. If pulverized active carbons are to be produced, the active carbon obtained in granulated form is comminuted. The comminution may also be done before or after carbonization.

Manufacturing route 3 uses drying and oxidation of the granules before carboniza-

[335] H. Jüntgen, K. Knoblauch, and D. Zündorf, "Large-Scale Testing of Active Coke from Bituminous Coal for Adsorptive Desulfurization of Exhaust Gases with Integrated Regeneration," *Chem.-Ing.-Tech.*, **45,** 1148–1151 (1973).

[336] M. Smisek and S. Cerny, *Active Carbon,* Elsevier Publishing Company, New York, 1970.

[337] H. Jüntgen, "Physikalishe Eigenschaften und Reaktionsverhalten der Steinkohle," in *Winnacker-Küchler, Chemische Technologie,* 3rd ed., Vol. 3, Munich, 1971, p. 14.

[338] H. Franke, R. Jäger, and M. Meraikib, "Adsorption Cokes from Brown Coal for Air Purification," *Chem.-Ing.-Tech.*, **18,** 1017 (1971).

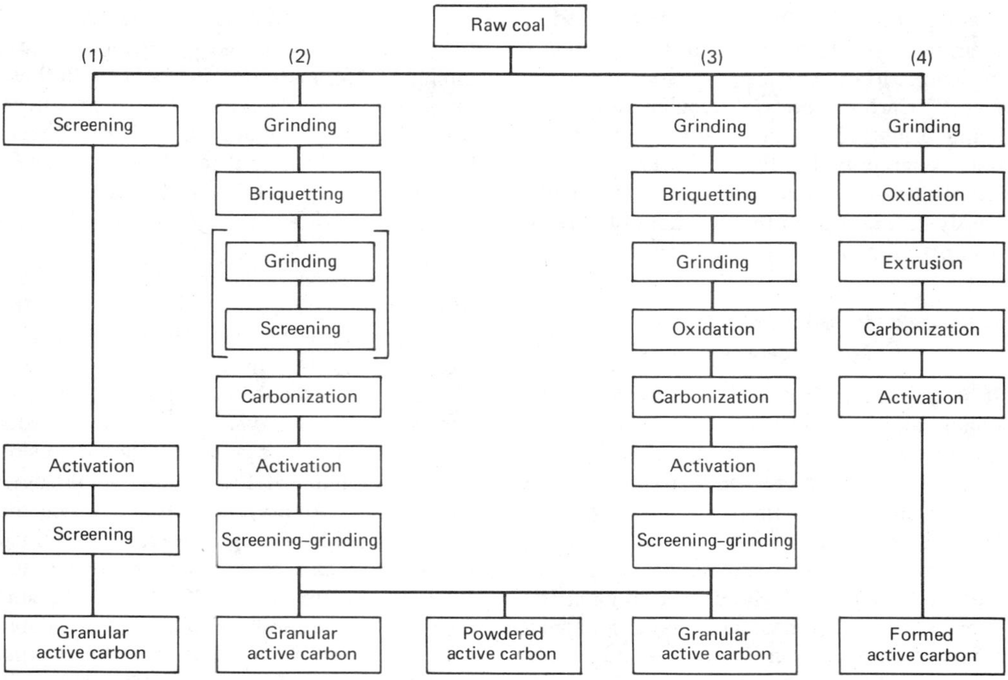

Figure 30.15 Different process routes for the production of activated carbon from coal.

tion. After carbonization and activation, granules or pulverized active carbons are obtained. In manufacturing route 4 the bituminous coal is oxidized immediately after grinding and then is shaped with a binder by extrusion into cylindrical shapes of uniform diameter. These shapes are carbonized and activated, so that active carbon of a uniform particle size is obtained. For the individual processing steps—that is, for oxidation, carbonization, and activation—specific equipment of different character is used (e.g., fluidized-bed kilns, rotary kilns, shaft kilns, and traveling-grate and rotating-grate kilns).

Published typical processes for the manufacture of active carbon from bituminous coal are those of the following companies: Pittsburgh Activated Carbon, Westvaco, Sutcliff and Speakman, and the Bergwerksverband GmbH. The processes of Pittsburgh Activated Carbon[339] and of Sutcliff and Speakmann[340] correspond to route 2, the process of Westvaco is similar to route 3,[341] and the Bergwerksverband GmbH process[342] corresponds to route 4. The process of Sutcliff and Speakman differs from the others in that it uses a single unit with a traveling bed for carbonization and activation.[343] Active carbon from anthracite is made according to route 1 on a small scale in England.[344] In the United States extensive laboratory results have been published.[345]

Active cokes are produced by the Bergwerksverband GmbH using a specific carbonization process for extruded shapes. For the production of carbonaceous molecular sieves, two processes are known. Feed materials such as anthracite, which after carbonization have very narrow pores, are activated by a limited partial gasification process.[346] Other pore systems obtained by

[339] *Manufacture of Active Coal;* open for public inspection, No. 2,361,097; German Patent Office.

[340] Pamphlet by Sutcliff and Speakman.

[341] *Process for the Oxidation of Coal;* open for public inspection, No. 2,425,547; German Patent Office; *Process for the Manufacture of Nonagglomerating Coal;* open for public inspection, No. 2,425,548; German Patent Office.

[342] H. Jüntgen, "Preparation of Specific Pore Systems from Carbonaceous Materials," *Ber. Bunsenges. Phys. Chem.*, **79**(9), 747–748 (1975).

[343] F. R. Houghton and J. Wildmann, "Manufacture and Uses of Activated Carbon," *Chem. Process Eng.*, May 1971.

[344] J. Owen, lecture to 12th Biennial Conf. on Carbon, Pittsburgh, July 29, 1975.

[345] M. Kawahata and P. L. Walker, Jr., "Activated Carbon from Anthracite," *Anthracite Conf. Proc.*

[346] R. L. Patel, S. P. Nandi, and P. L. Walker, Jr., "Molecular Sieve Characteristics of Slightly Activated Anthracite," *Fuel,* **51,** 47–51 (1972).

shaping and carbonization can be changed in their pore structure by depositing products of hydrocarbon cracking at the inner surface.[342] Carbonaceous molecular sieves are being produced commercially from bituminous coal only by the Bergwerksverband GmbH process. Their properties are significantly different from the product obtained in Japan[347] from wood charcoal.

30.4.6.3 *Properties of Adsorbents Made from Coal*

It is important that active carbons, active cokes, and molecular sieves have a ratio of macro- or transport pores to micro- or adsorption pores (micro-submicro pores) that on the one hand permits rapid transport of the adsorbate to the inner surface, and on the other hand has a sufficiently large inner surface to obtain a high adsorption capacity. Molecular sieves must also have a narrow range of pore sizes in the micro- and submicropore volume, since these materials selectively separate mixtures of adsorbates by differences in the diffusion rate. For the use of adsorbents in commerical adsorption processes, a high mechanical hardness and abrasion resistance is required. The saturated active carbons, active cokes, and molecular sieves must be regenerated for reuse. Therefore a small desorption energy is desired. Furthermore great stability of the carbon structure is needed if reactivation should become necessary under more severe conditions than normally are used in regeneration, to keep carbon loss by gasification at a minimum. Commercial adsorbents with these properties can theoretically be produced from different raw materials. Adsorbents can be produced from coal that have practically all the properties of adsorbents available commercially.

Active Carbons. The inner surface of active carbons is usually determined by the Brunauer-Emmett-Teller (BET) method. This technique involves measuring the volume of nitrogen gas adsorbed at liquid nitrogen temperatures (78 K) at various pressures. From this isotherm it is possible to calculate, using the BET equation, the volume of nitrogen adsorbed at monolayer coverage, hence the area covered by nitrogen. Areas determined in this way can be taken to represent the total surface area. Table 30.23 gives the surface area of several commercially available active carbons, determined according to this method and produced from different raw materials. It can be seen that the properties of active carbons made from bituminous coal correspond to those of active carbons made from other raw materials. It must be emphasized that this test method can give only a crude characterization of active carbons. It

[347] K. Kawazoe, M. Suzuki, and K. Chakava, *Ind. Chem. Eng. Japan,* **7,** 151 (1974).

[348] Pamphlet, Sutcliff and Speakman Co. Ltd.
[349] Pamphlet, Chemviron, S.A.
[350] Pamphlet, Bergwerksverband GmbH.
[351] J. C. Abram, "The Characteristics of Activated Carbon," *Papers Proc. Water Research Association Conf.,* University of Reading, April 3–5, 1973.
[352] Pamphlet, Thomas Ness Ltd.

Table 30.23 Surface Areas of Activated Carbons

Name of Product	Origin	Area (m^2/g)	Ref.
203 A Sutcliff and Speakman	Coal	600–700	348
208 A Sutcliff and Speakman	Coal	1150–1250	348
Pittsburgh CAL	Coal	1000–1100	349
Pittsburgh SGL	Coal	950–1050	349
Pittsburgh Filtrasorb 400	Coal	1000–1100	349
BKA Bergwerksverband	Coal	1300–1400	350
BKC Bergwerksverband	Coal	1200–1300	350
D40/3 Bergwerksverband	Coal	1000–1100	350
Davco S 51	Lignite	500–550	351
Anthrasorb	Anthracite	600–1200	352
203 C Sutcliff and Speakman	Coconut shell	700–800	348
Nuchar Aqua	Pulp mill residue	550–650	351
Nuchar C	Pulp mill residue	1050–1100	351
Davco KE	Wood	980–1000	351

gives no information about the actual pore structure, since the conditions of the BET equations are not fulfilled. Furthermore, the adsorbed molecules encountered in the gas phase and especially in liquid phase operation are usually much larger than the nitrogen molecule. As a result, the actual area available for adsorption could be considerably smaller than the total surface area. Therefore the total surface area is not sufficient for characterizing an adsorbent. In addition, the size distriubtion of the molecules to be adsorbed must be known.

The pore structure of an adsorbent is characterized by transport pores with a diameter of about 500 Å, by mesopores with diameters below 500 Å down to 20 Å, by micropores with diameters below 20 Å to 8 Å, and by pores of less than 8 Å, which are called submicropores. The size of the total pore system is determined using the helium and mercury density. Helium, because of its small molecular size, enters the submicropores, but under the conditions used the mercury does not enter the pore system at all, not even the transport pores. For the measurement of the pore distribution in the macropore area and a great portion of the mesopore area, mercury porosimetry is used.[353,354] The mesopores and part of the micropores are measured by the adsorption of vapors using capillary condensation for the evaluation, by means of the Kelvin equation.[355] The micro- and submicropores can be measured by the adsorption of methanol.[356]

Figure 30.16 shows pore volume distributions for active carbons made from coal, measured according to the methods above.[348–351] The figure shows that the pore volumes of these active carbons are distributed over a wide range of pore diameters. Active carbons made from bituminous coal characteristically have a larger micro- and submicropore volume than do those made from lignite. The pore volume distribution in the adsorption and transport pore range can be influenced by the manufacturing conditions, especially the degree of activation (ratio of partially gasified carbon to the originally available carbon).

Carbonaceous adsorbents differ characteristically from other adsorbents by their behavior toward water vapor adsorption. Figure 30.17 plots the water adsorption versus the water vapor pressure for a carbonaceous molecular sieve, a silica gel, a zeolitic molecular sieve, and a reactive carbon.[357,358] It can be seen that the carbon-containing adsor-

[353] IUPAC Classifikation 1972.

[354] W. Pruss, "Determination of Pore Size and Pore Distribution in Coal and Coke," *Brennstoff-Chemie,* **42,** 157–160 (1961).

[355] S. Gregg and K. S. W. Sing, *Adsorption, Surface Area and Porosity,* Academic Press, New York, 1967, pp. 124–128, 200–211.

[356] H. Jüntgen and H. Seewald, "Characterization of Pore Structure of Microporous Carbonaceous Adsorbents," *Ber. Bunsenges. Phys. Chem.,* **79,** 734–738 (1975).

[357] Herbert Kroll, "Molecular Sieves—Modern Adsorbents," *Chem. Tech.* **26**(7) (July 1974).

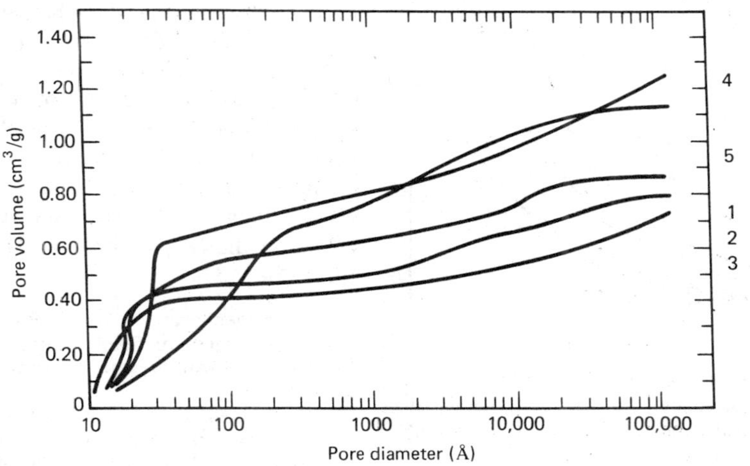

Figure 30.16 Pore volume distribution and diameter for various active carbons: 1, 2 = Pittsburgh carbon; 3, 4 = Bergwerks Verband; 5 = Atlas Powder Company; 1–4, butuminous coals; 5, liginte base. 348–351

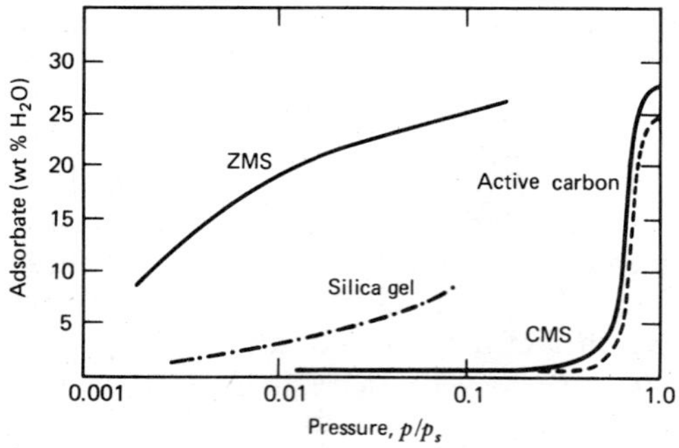

Figure 30.17 Water vapor isotherms for a zeolite molecular sieve(ZMS),[357] a silica gel[357], a carbon aceous molecular sieve. $(CMS)_3$[358] and a reactive carbon.

bents, up to a water vapor partial pressure lower than 0.6 p/p_s (where p = partial pressure and p_s = partial pressure at saturation) barely adsorb water in comparison to the silica gel and the zeolite. The amounts of adsorbed water become similar only at higher water vapor pressures.

Active carbons made from different raw materials show great differences in their activity toward gasification media such as H_2O, O_2, and CO_2. Figure 30.18 shows the effect of temperature[359] on the reactivity, r, toward steam of six different active carbons: three made from wood charcoal, one made from coconut shells, and two made from bituminous coal. It shows that the reactivity of the active carbons made from wood or coconut shells is considerably higher than the reactivity of active carbons made from bituminous coal.

This property of active carbons from bituminous coal is especially important for use in processes such as the purification of drinking water and wastewater, in which the active carbon after the adsorption has to be reactivated with steam.[360]

The mechanical properties such as hardness and resistance to abrasion of adsorbents made from coal are similar to those of adsorbents made from other raw materials. As a whole, it can be stated that the hardness of adsorbents that are only slightly activated is higher than that of higly activated materials.

Active Cokes. The properties of active cokes that are used for the removal of SO_2 from flue gases with thermal regeneration change during extended use. This effect is caused by the carbon consumption due to the decomposition of the H_2SO_4, which is the form in which sulfur dioxide is adsorbed. As illustrated in Figs. 30.19–30.21, changes of the pore volume, the shock hardness, and the SO_2 adsorption, respectively, depend on the number of regeneration cycles in commercial use.[335] As Fig. 30.19 shows, the total pore volume increases from an initial 32 cm³/100 g after 180 recycles to 58 cm³/100 g; up to the 50th recycle the changes are very marked, but they decrease after that. Correspondingly, the shock hardness (Fig. 30.20) decreases with the number of recycles from an initial value of 95 wt %, until a constant value of 85 wt % is reached. With the increase in pore volume, the adsorption properties change, and this is shown by an increase in the SO_2 adsorption capacity (Fig. 30.21).

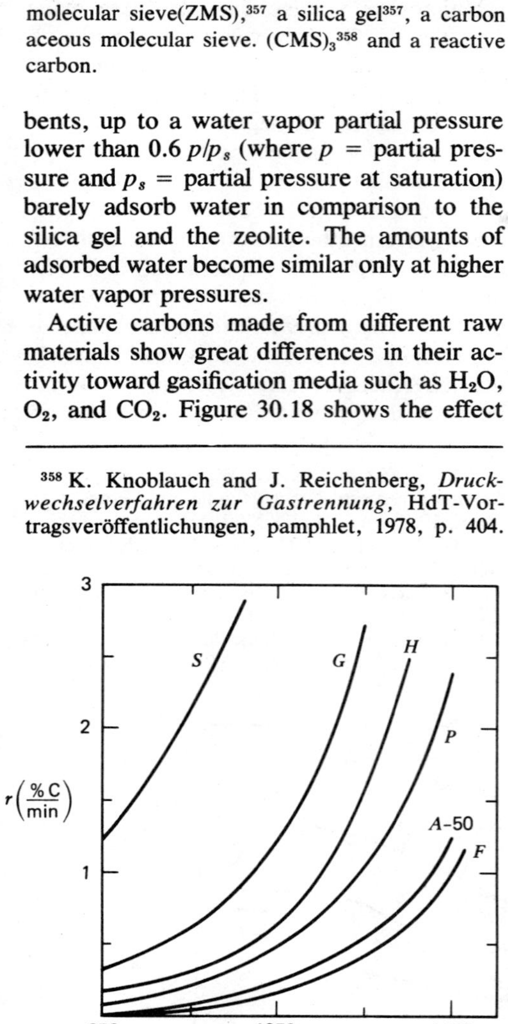

Figure 30.18 Reactivity of active carbons made from various raw materials: S,G,H, = activated carbons produced from wood chars; P, activated carbon made from coconut chars; A,-SO; F = activated carbon produced from bituminous coal.[359]

[358] K. Knoblauch and J. Reichenberg, *Druckwechselverfahren zur Gastrennung*, HdT-Vortragsveröffentlichungen, pamphlet, 1978, p. 404.

[359] J. Reichenberger, "The Mechanisms of Thermal Regeneration of Spent Active Carbons in a Fluidized Bed with Steam," Dissertation, Aachen, 1974.

[360] H. Jüntgen, K. Knoblauch, and H. J. Schröter, "Diffusion of Gas Mixtures Through Constriction as the Rate Determining Steps of Diffusion in Pores of Carbonaceous Materials," *Ber. Bunsenges. Phys. Chem.*, 79(9) (1975).

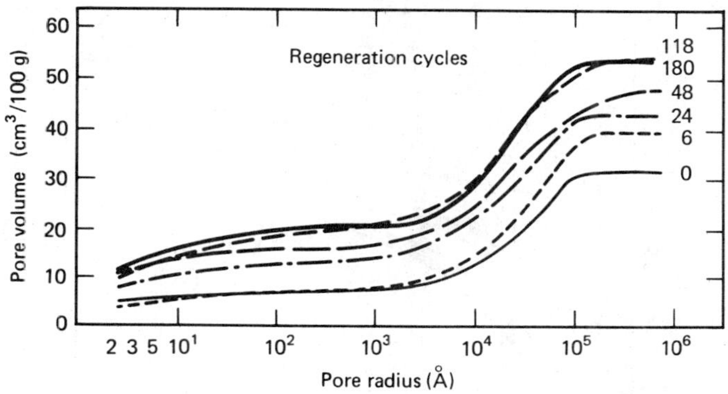

Figure 30.19 Pore volume distribution measured by benzine adsorption and mercury porosimetry after increasing number of regeneration cycles.[335]

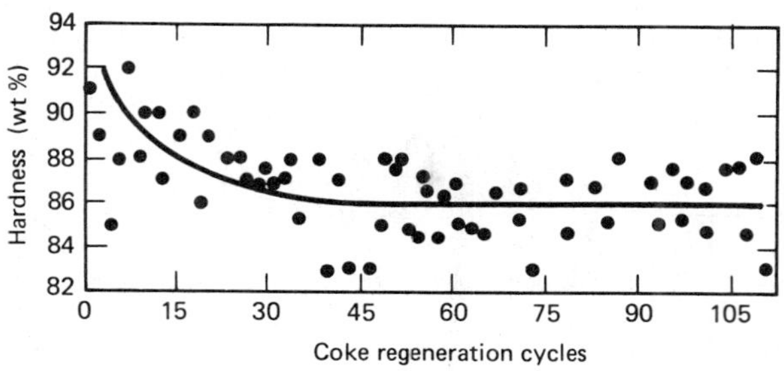

Figure 30.20 Shock hardness (wt%>0.5 mm) of active coke after increasing number of generation cycles in the demonstration plant.[335]

Carbonaceous Molecular Sieves. The carbon-containing molecular sieves, which separate adsorbates because of the differences in the diffusion rates in the pore system, have an especially developed micro- and submicropore volume. Pores of a radius smaller than 5 Å are observed and cannot be measured by the classic method of adsorption of benzene but can be characterized by adsorption of methanol. The separation efficiency of different molecular sieves can be described by measuring the diffusion coefficient of different adsorbents. Figure 30.22 shows the measured diffusion coefficient of different adsorbents for carbonaceous molecular sieves produced by the Bergwerksverband GmbH. The molecular sieves can be characterized by their average adsorption pore diameter, which is determined by the methanol isotherms.[360] The figure shows graphically the relation between the pore structure and the diffusion parameters of different gases and permits one to draw conclusions about the separation efficiency of the different materials.

P. L. Walker prepared products by the activation of anthracite.[361] They can be char-

[361] R. L. Patel and P. L. Walker, "Preparation of Molecular Sieve Materials from Anthracite," U.S. Dept. Interior OCR R & D Rep. No. 61—Interim Rep. No. 6.

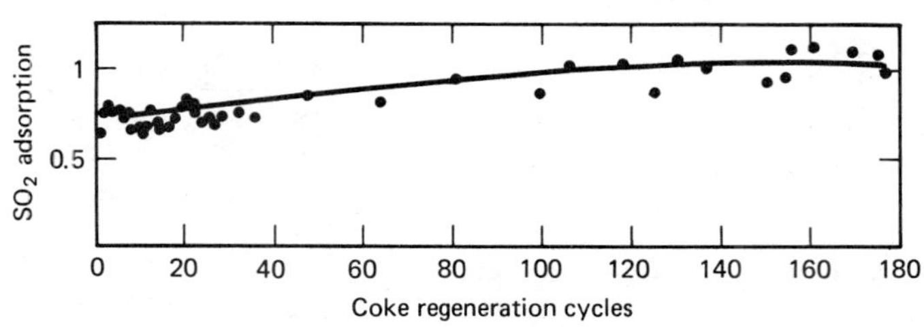

Figure 30.21 Sulfur dioxide adsorption in the demonstration plant under standard conditions as a fraction of the adsorption after coke regeneration cycles. [360]

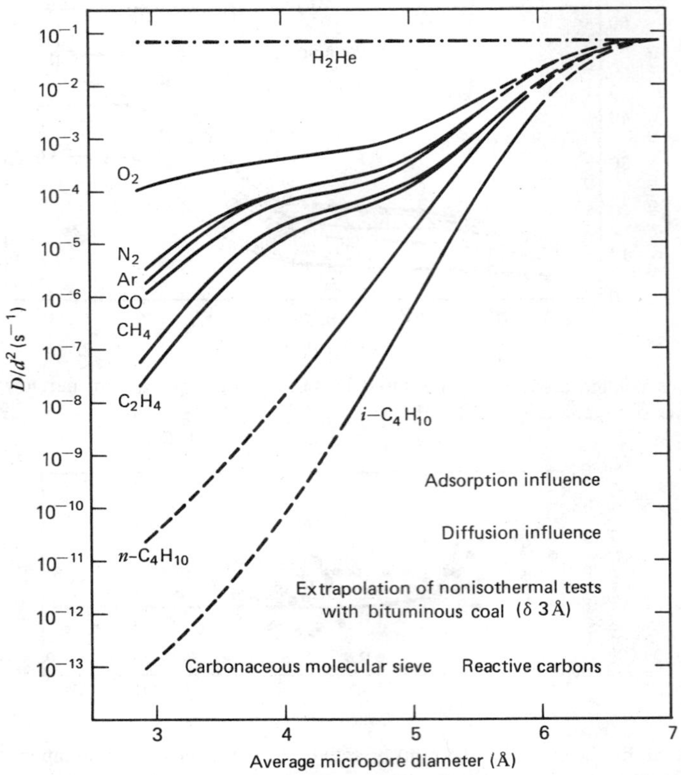

Figure 30.22 The diffusion parameter D/d^2 for molecular sieves from bituminous coal.[360]

acterized by determining the diffusion coefficient of methane. As the degree of activation increased, larger diffusion coefficients were measured. Anthracite with 6.9% burnoff shows a diffusion coefficient for methane of 0.95×10^{-9} cm²/s, the anthracite with 8.5% burnoff 0.85×10^{-8} cm²/s, and the anthracite with 9.1% burnoff 0.109×10^{-6} cm²/s. These molecular sieves made from anthracite are comparable to molecular sieves made by the Bergwerksverband Gmbh, which have an average diameter of adsorption pores below 3.5 Å.

30.4.6.4 *Applications*

Activated Carbon and Activated Char. In 1972 the largest of active carbon consumed about 30% of the total for decolorizing sugar and syrup using pulverized active carbon (Table 30.24).[362] Another large area of consumption (23% of the total) was the treatment

[362] *Chemical Economics Handbook,* Stanford Research Institute, 1972.

[363] H. Jüntgen, "New Applications for Carbonaceous Adsorbents," *Carbon 1977,* Vol. 15, pp. 273–282.

Table 30.24 Application of Active Carbon: 1972[363]

Type of Use	Proportion (%)	Condition of Activated Carbon
Decolorization of sugar and syrup	30	Pulverized
Treatment of water, especially drinking water	23	Pulverized granules
Adsorption of industrial gases, including solvent recovery	12	Granules
Adsorption of gasoline vapors from automobiles	7	Granules
Pharmaceuticals	5	Pulverized
Purification of fats, oils, foodstuffs	4	Pulverized
Miscellaneous	19	Pulverized granules

of water, mainly drinking water. In this case also, pulverized active carbon was mainly used. The next largest use was for the adsorption and separation of industrial gases, especially the recovery of solvents. In this case only granulated active carbon is used. Finally, a large market has been developed, especially in the United States, for granulated active carbons for the adsorption of gasoline vapors from automobile tanks. Further applications, mostly of pulverized active carbons, are medicinal uses and the purification of fat, oil, and food materials.

The growth rates of the various applications are different; for instance, the decolorizing of sugar and syrup has a slow growth rate, but other fields of application are more promising. Thus it is to be expected that purification of air for the removal of organic materials, a process that is already being used for the recovery of solvents, will increase in importance in the future. In some countries governmental regulations make it necessary to remove certain substances present in small concentrations, although in cost-benefit terms removal is very expensive. Special applications are the separation of odoriferous materials and of radioactive gases from nuclear plants. Furthermore it is expected that the separation of sulfur compounds such as H_2S, SO_2, and CS_2 from gases will increase in importance.

Large-scale uses can be foreseen for active cokes, especially in the separation of SO_2. The use of active carbons for the purification of drinking water will also increase, since the amount of surface water used for the preparation of drinking water is increasing, as well as the concentration of organic materials in this surface water. The development of new techniques for the regeneration of saturated active carbon will increase the use of granulated active carbon at the expense of the use of pulverized active carbon. An area of increasing importance is the adsorptive treatment of wastewater for the main purpose of removing the organic materials that are resistant to removal by biological treatment. Molecular sieves made from bituminous coal seem to find commercial users for the separation of gases. The use of carbon-containing adsorbents in industry is described in the literature and in reports. Therefore known uses are only treated briefly in this chapter.

Active Carbon and Active Coke for Gas Purification. For gas purification processes such as solvent recovery, the separation of sulfur-containing compounds, the separation of radioactive gases, and the separation and decomposition of nitrogen oxides, adsorption processes with active carbon are being used.[364–366] Active carbons made from bituminous coal are of the same quality as active carbons from other raw materials. Since the use of bituminous coal for the production of active carbon increases from year to year, the proportion of coal-derived adsorbents will also increase.

[364] H. Krill and H. Wirth, "Gas Purification with Materials Based on Coal," *Chem.-Ing.-Tech.* **46,** 757–762 (1974).

[365] H. Jüntgen, "Current Status and Development of Gas Purification," *Tech. Mitt.,* **70**(1) (January 1977).

[366] K. Storp, "Desulfurization of Exhaust Gases and Spent Air Through Adsorption and Catalysis of Active Carbon," *Dechema Monogr.,* **64,** 91–102 (1970).

Table 30.25 Adsorption Processes for the Removal of Organic Substances[365]

Process	Adsorbate	Adsorbent	Adsorption Temperature (K)	Type of Reactor	Regeneration	Status
Solvent recovery	Water insoluble (e.g. toluene, xylene, gasoline, hexane, chlorinated hydrocarbons)	Active carbon	293–333	Fixed bed	Steaming, condensation, phase separation	Commercial
	Partially water soluble (ethyl acetate)	Active carbon	293–333	Fixed bed	Steaming, enriching condensation, distillation	Commercial
	Water soluble (ethanol, isopropanol, acetone)	Active carbon	293–333	Fixed bed	Steaming, distillation	Commercial
Adsox	Organic substances in low concentration, odoriferous material	Active carbon	293–333	Fixed-bed, loop adsorber	Inert gas, 800°C	Commercial

Table 30.26 Adsorption Processes for the Retention and Separation of Radioactive Gases and Vapors[363]

Process	Adsorbate	Adsorbent	Adsorption Temperature (K)	Type of Reactor	Regeneration	Status
Radioactive iodine separation	^{129}I and ^{131}I (isotope exchange)	Active carbon, fine pores	293–323	Fixed bed	—	Commercial
Retardation of radioactive noble gases	Krypton-xenon	Active carbon, fine pores	293–323	Fixed bed, long columns	—	Commercial
Separation of ^{85}Kr	^{85}Kr	Active carbon, fine pores	293–323	Fixed bed	Pressure variation, flushing	Laboratory experiments

Adsorption processes for the separation of organic materials are used worldwide in plants that have a gas throughput of up to 350,000 m^3/hr. Usually several parallel fixed-bed adsorbers are used, which are alternatively regenerated. Table 30.25 gives an overview of the diversity of uses. As shown in Table 30.26, active carbon adsorption processes are also used for retaining and separating radioactive gases and vapors. The separation of radioiodine and the retardation of radioactive noble gases is already being used commercially, whereas the separation of the long-lived ^{85}Kr has been tested only on the laboratory scale.

Table 30.27 shows in detail the multifaceted nature of the adsorption process for the removal of sulfur-containing components. Processes for the separation for H_2S and CS_2 from air and of H_2S and SO_2 from off-gases from Claus plants or other chemical plants are already in commercial use. By comparison, the adsorption processes for SO_2 removal from flue gases and off-gases of sintering plants are in the testing stage in large prototype plants.[367,368] These adsorption processes for SO_2 separation use operating temperatures above 373 K and therefore do not require cooling of the flue gases and the energy-consuming reheating, which are needed for the wet absorptive desulfurization processes. The adsorption processes using activated cokes for SO_2 removal may be regenerated to produce sulfuric acid, liquid SO_2, or elemental sulfur. These final products may be sold, or in the case of elemental

[367] K. Knoblauch and R. Noack, "Operating Experience with the Bergbau-Forschung Process for Flue Gas Desulfurization," *Staub-Reinhaltung Luft,* **36**(7), 318–323 (1976).

[368] K. Knoblauch, H. Jüntgen, and W. Peters, "The Bergbau-Forschung Process for the Desulfurization of the Flue Gas," paper presented at 4th Int. Clean Air Congr., Tokyo, May 1977.

Table 30.27 Processes for the Sorption of Sulfur-Containing Materials[363]

Process	Adsorbate	Adsorbent	Adsorption Temperature (K)	Type of Reactor	Regeneration	Status
Sulfosorbon	H_2S and CS_2 from air	Wide-pore active carbon + iodine, fine-pore active carbon	293–323	Fixed bed	Steaming extraction	Commercial
Thiocarbon	H_2S and CS_2 from air	Wide-pore active carbon, low in iron	293–323	Fixed bed	Steaming extraction	Commercial
Sulfreen	H_2S and SO_2 from Claus off-gas	Wide-pore active carbon + aluminum silicate or aluminum oxide	293–323	Fixed bed	Inert gas at 723 K	Commercial
Sulfacid	SO_2 from chemical plants	Active carbon + metal catalyst	353	Fixed bed	Spraying with water	Commercial
Hitachi	SO_2 from flue gas	Active carbon	373	Fixed bed	Scrubbing with water	Prototype, 400,000 m^3/hr
Sumitomo	SO_2 from flue gas	Active carbon	373–403	Traveling bed	Hot gas at 623 K	Prototype, 170,000 m^3/hr
BF	SO_2 from flue gas	Active coke	373–403	Traveling bed	Hot sand at 923 K	Prototype, 150,000 m^3/hr
Westwaco	SO_2 from flue gas	Active carbon	373	Rotating bed	Hydrogen at 623 K	Prototype, 30,000 m^3/hr

sulfur, stored without danger to the environment.

An adsorption process is in the development stage for the catalytic oxidation of NO to NO_2 with active carbon with adjoining NO_2 absorption, and for the reduction of NO_2 to N_2 and H_2O with ammonia, using impregnated active carbon[369,370] (Chapter 22).

Application of Carbonaceous Molecular Sieves. Separation of gases by means of carbonaceous molecular sieves is at the beginning of technical development. In pilot plants and the first technical plants, the Bergwerksverband GmbH used molecular sieves for the separation of air into nitrogen and oxygen and for the recovery of highly concentrated hydrogen from hydrogen-containing gases, especially coke-oven gas.[350,371,372]

Application of Activated Carbon for Water Purification. Active carbons are used in powder form or as granules or extruded shapes for the purification of drinking water and effluent water, especially for the separation of organic molecules. Two different process types are being used: one for the purification of drinking water[373] with a low content of organic impurities, and the other for the so-called tertiary purification of effluents leaving biological purification plants,[374] from which small amounts of organic material also need to be removed. The adsorption beds containing active carbon may be used for several weeks or months. The adsorption plant consists of several parallel adsorbers, through which flows the water to be purified. These are connected to the regeneration plant by hydraulically operated lines. As necessary, a filter in the adsorption process may be taken offstream and the spent active carbon removed from the adsorber to be regenerated in a captive or contract regeneration plant.

In the purification of aqueous effluents, much higher concentrations of organic materials are present and the active carbon filters are exhausted in a few days or possibly hours. In this case moving-bed active carbon filters are used, which are directly coupled with the regeneration plant.[375] An overview of the status of the commercial plants for the purification of drinking water and aqueous effluents is given in Ref. 376.

30.5 BY-PRODUCTS OF COAL-BURNING PROCESSES

30.5.1 Fly Ash and Slag

Large, modern, coal-burning boilers usually fire pulverized coal, which contains 10–40% particles smaller than 90 μm. The components of the fuel that cannot be burned or gasified are obtained as ash. In the so-called slagging boilers, the mineral matter of the coal is obtained to a large extent in molten form on the floor of the furnace. The remainder is finely divided fly ash, which is removed with a high efficiency either in mechanical separators or in electrostatic precipitators. Worldwide, dry bottom boilers—units in which the mineral matter is not molten—are most frequently used. Unburned particles of the coal are present in the flue gas and are removed from it. The ash that is present in dust form and carried out with the flue gas is called fly ash. The slag that is obtained in liquid form and quenched in water is called granulated boiler slag. The use of slagging boilers is limited to locations with suitable

[369] W. Becker, K. Knoblauch, and H. Jüntgen, "The Removal of Nitroxides with Carbonaceous Adsorbents," DGMK paper presented at 4th Meeting of OGEW/DGMK. Salzburg, October 4–6, 1976.

[370] Jumpei Ando, *SO_2 and NO_x Removal Technology,* Environmental Technical Information Center, Japan Management Association, 1976.

[371] K. Knoblauch, J. Reichenberger, and H. Jüntgen, "Oxygen and Molecular Sieve Cokes for the Manufacture of Nitrogen," *Gas Erdgas,* **116**(9), 382–386 (1975).

[372] K. Knoblauch, J. Reichenberger, and H. Jüntgen, "Oxygen and Nitrogen," *Erdoel Kohle,* **28**(9), 426–428 (1975).

[373] H. Sontheimer, paper presented at 7th Meeting for Exchange of Results of Special Questions of Water Adsorption Technology. Publications of the Engler-Bunte-Institut, University of Karlsruhe, Booklet No. 9, 1975. Translation of reports on special problems of water technology, *Adsorption,* Vol. 9, EPA-600/9-76-030, December 1976.

[374] P. Roberts and W. Stumm. "Sewage Treatment with Active Carbon," *Gas-Wasser/Abwasser,* **54,** 74–86 (1974).

[375] H. Jüntgen and J. Klein, "Purification of Wastewater from Coking and Coal Gasification Plants Using Activated Carbon," *Energy Sources,* **2**(4), 311–330 (1976).

[376] "Activated Carbon in Water Treatment," *Papers Proc. Water Research Association Conf.* University of Reading, April 3–5, 1973.

coal, and the number in comparison to the dry bottom boilers is very small. Therefore slag is obtained more rarely, and if the term "ash" is mentioned in the literature, it includes fly ash.

Similar ashes are obtained during gasification of coal. Here, too, there are processes involving the removal of liquid slag and dry bottom ash removal, as well as processes that produce fly ash. At present the amount of ash obtained by coal gasification is negligible; therefore the pertinent literature exclusively treats the use of ash from coal burning.

In general, only raw materials of uniform composition can be utilized. The first step in ash utilization is therefore a partial preparation. It consists of subjecting fly ashes to a homogenizing or classifying process, while molten slags are crushed and screened.

Low-cost pretreatment is necessary to keep the cost of raw material ash low and to retain its price advantage over other materials.

For the use of ash, the carbon content is especially important. It depends on the coal, the construction of the boiler, and the boiler operation. Molten ashes (i.e., slags) usually contain small amounts or only traces of unburned components. Fly ash may have a varying content of unburned material, depending on the design of the boiler, the operation of the boiler, and the throughput. The carbon content of fly ash also depends on the size of the fly ash particles and can range from 1 to 37%. Figure 30.23 gives a typical example of a bituminous coal-fired utility boiler: here the loss on ingnition, which can be considered to be a measure of the carbon content, is seen to increase with increasing particle size. The larger particles are not completely burned. Small particles (< 63 μm), which comprise 78% of the total, have a carbon content that is smaller than the average content of the total fly ash. Therefore screening or mechnaical classifying of the fly ash permits separation into fractions with different carbon content.

The utilization of the ash also critically depends on the composition of the mineral matter. The main components are Al_2O_3, SiO_2, Fe_2O_3, CaO, and MgO, as well as alkali oxides. Figure 30.24, a diagram for the composition of bituminous coal and brown coal ashes in West Germany, using the three-component system of Al_2O_3, SiO_2, and CaO, gives a comparison with blast furnace slag and several building materials. Table 30.28 shows that the composition of German bituminous coal ash brings it into the area of clays, fireclay, and burned stone, whereas brown coal ash, because of the high lime content, is located more in the area of cement. The simple differentiating of brown and bituminous coal ashes through their chemical decomposition is not possible worldwide, since brown coal and lignite ashes are found that are chemically similar to German bituminous coal ash. The silica con-

[377] H. Erythropel, in *Yearbook of Steam Production Technology,* 2nd ed., Vulkan-Verlag, Essen, 1972–1973, p. 182.

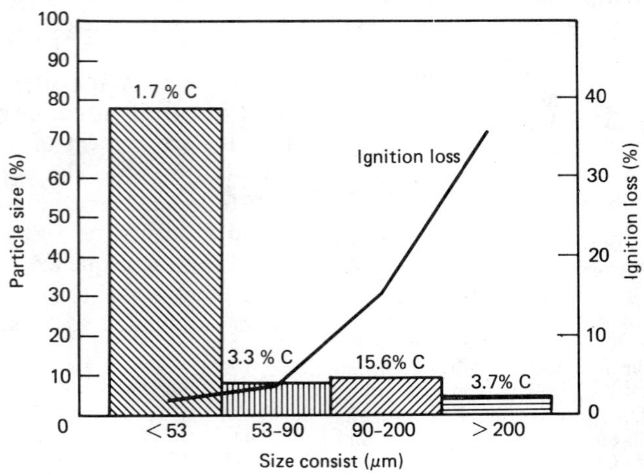

Figure 30.23 Ignition loss of individual particle size groups of a fly ash with an average carbon content of 4.56%.[377]

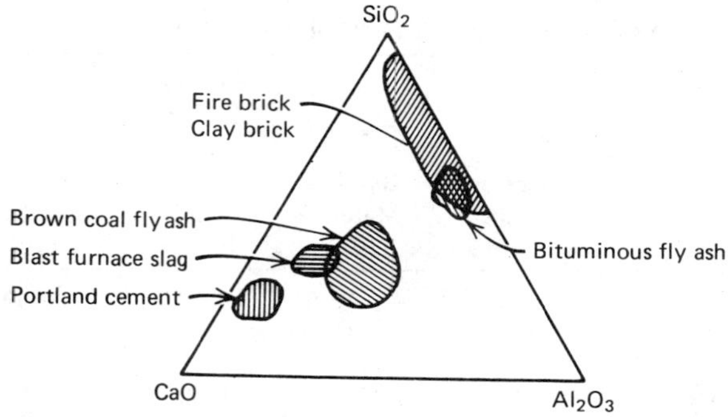

Figure 30.24 Composition of bituminous coal and brown coal ashes in West Germany; see Table 30.28 for chemical analyses.[378]

tent of brown coal ashes can also vary to a great extent depending on the mining method, which may bring in different amounts of sand and other minerals. Brown coal ash may contain up to 80% sand, which must be separated because it is a harmful component in utilization.

The main uses of brown coal and bituminous coal ash are in the manufacture of cement or hydraulic binders, as building material, and as gravel in highway building, depending on the ratio of the main chemical components SiO_2, Al_2O_3, and CaO and the above-mentioned carbon content.

The production of bituminous coal and brown coal ash and their use is shown in Tables 30.29 and 30.30,[379,380] for some important industrial countries. In the European countries about equal amounts of bituminous coal and brown coal ash are used, but in general the portion of ash that is economically usable is found to be considerably higher for bituminous coal than for brown coal.

The utilization of brown coal fly ash is limited, since it has varying physical and chemical properties and furthermore is needed as material to fill in nearby surface mines.

Brown coal ash is used mainly as cement, whereas bituminous coal ash is predominantly used in the building sector. Bituminous coal ash and fly ash are used as additives to mortar, grout, and concrete because of their pozzolanic properties.[381] Pozzolans are materials that have no cementing properties in their natural state but on addition of lime and water, attain properties similar to those of cement. These contribute to the strength. A mixture of fly ash can thus improve the handling properties and the flow properties when pumped, and it may result in a lower temperature increase on setting. Concrete that is made with fly ash addition has lower porosity, low specific gravity, and lower permeability to water, gas, and air.[382] Such concrete has higher resistance to sulfate and other corrosive water. In some countries (especially in France), a cement is produced that is mixed in the factory with up to 40% fly ash.[381] The strength of these cements after aging is higher than that of the cement used so far; however the initial strength can be considerably lower.

Fly ash is also used in large quantities as a filling material in highway building. For this purpose the fly ash may be mixed with granulate, medium, and coarse sand to work as a supporting layer of material. For building

[378] H. Erythropel, *VGB Kraftwerkstechnik*, **55**, 297 (1975).

[379] Economic Commission for Europe, Genf 1975, *Production and Use of Ash in 1974*, EP/GE No. 3/R 17.

[380] *Ash at Work*, Vol. 7, No. 3, 1975, p. 4. Published by National Ash Association, Washington, D.C.

[381] E. G. Berber, *J. Inst. Fuel*, 4 (January 1970).

[382] H. Erythropel, *Mitt. VGB*, **50**, 193 (1970).

Table 30.28 Chemical Analysis of Bituminous Coal and Brown Coal Ashes[378]

Component	Bituminous Coal (%)	Brown Coal (%)
SiO_2	45–55	20–70
Al_2O_3	20–30	3–15
Fe_2O_3	10–20	2–25
CaO	2–5	10–45
MgO	1–3	0.5–10
SO_3	0.5–3	2–15
K_2O	1–5	1–3
Na_2O	0.5–2	—

Table 30.29 Production[a] and Utilization of Bituminous Coal Ash: 1974[380]

Country	Production (10^6 kg)	Cement or Hydraulic Binder (10^6 kg)	Building Materials (10^6 kg)	Highway Building, Dams, and Filler (10^6 kg)	Miscellaneous (10^6 kg)	Amount Utilized (%)
Belgium	723	66.5	95.6	9.3	18.7	26.3
Bulgaria	1,500	—	15	—	5	1.33
West Germany	6,000	350	1800	1400	500	67.5
East Germany	225	31	—	—	51.3	36.6
Finland	462	12.9	—	366.4	8.8	84
France	3,082	1298.5	83.1	749	96.4	72.5
Italy	120	—	6	25	—	26
Poland	7,122	671	1120	669	259	38
Rumania	1,336	98	1	—	3	7.5
Turkey	197	—	5	—	—	—
Hungary	750	81	—	—	—	10.8
Ukraine SSR	13,769	97.8	441.6	147.6	513	8.7
Yugoslavia	20	—	—	—	—	—
Total	35,326	2706.7	3562.3	3366.3	1455.2	31
United States[b]	59,500	1100	200	2600	5300	15.5
Total	94,836	3806.7	3762.3	5966.3	6755.2	21

[a] According to definition of ECE: total ash from pulverized coal, combustion derived from the ash content of the coal.
[b] According to the National Ash Association and the Edison Electric Institute.[381]

bituminous highways, fly ash with a carbon content above 5% is often economically superior to the fillers that have been used thus far, such as limestone.[378] The suitability of fly ash as a component of the asphalt mixture is determined by a special test.

Fly ash with a high carbon content (> 10%) is in great demand in the steel and metallurgical industries as foundry power; therefore it finds no other use. Fly ashes contain 1–5% cenospheres, which are hollow but strong spheres with a density between

Table 30.30 Production and Utilization of Brown Coal Ash: 1974[380]

Country	Production[a] (10^6 kg)	Cement or Hydraulic Binder (10^6 kg)	Building Material (10^6 kg)	Highway, Dams, and Filler (10^6 kg)	Miscellaneous (10^6 kg)	Amount Utilized (%)
Bulgaria	2,500	—	100	55	10	6.6
West Germany	7,600	120	10	250	100	6.3
East Germany	13,115	461	286	39.5	787	12
France	433	106	—	87	0.1	44.6
Italy	360	27	—	9	—	10
Poland	4,939	—	7	186	34	4.6
Rumania	3,806	90	61	11	16	4.4
Turkey	417	0.5	0.5	—	—	0.25
Hungary	4,195	82	118	—	—	4.77
Yugoslavia	3,354	—	190	190	—	5.96
	40,719	886.5	772.5	827.5	947.1	8.4

[a] According to the ECE definition: total ash from pulverized coal combustion derived from the ash content of the coal.

0.4 and 0.6 g/cm^3. They can be easily separated because their density is lower than that of water, especially in the wet ash removal units. They can be converted to fire brick of high quality.[381,382]

Molten granulates can be converted by comminution, (e.g., in an impact mill) into a material that is similar to sand. Since the resulting material contains only small amounts of free crystalline silica, it does not cause silicosis and can be used to replace quartz sand for sandblasting. After preparation, the ash may also be used as an additive to building concrete, for lime or cement bonded stone, and in highway building.[382]

30.5.2 By-Products of Flue Gas Desulfurization

Depending on the process, the sulfur dioxide removed from flue gas can be obtained in the form of a variety of chemical compounds (Chapter 22). These can be converted into final products that are salable, or used simply as fill. Such compounds may be a sludge that contains $CaSO_3$ in addition to gypsum. As salable products, gypsum, sulfuric acid, liquid SO_2, and elemental sulfur may be obtained. Table 30.31 shows the annual quantities of these different products obtained in a 600-MW power plant fired with bituminous coal containing 2% sulfur. The plant removes 80% of the sulfur, and operates 4000 hr/year. The table also shows the most important flue gas sulfur removal processes that yield these products. In the different countries of the world, different interests exist for the manufacture and use of individual products. Japan is interested in the manufacture of gypsum, since it does not have any natural gypsum deposits. In other countries that have gypsum deposits, gypsum from the flue gas sulfur removal plants is not desired, cannot find uses, and therefore must be used as fill.

The sale of sulfuric acid and liquid SO_2 depends on the proximity of chemical plants that use these products and can be reached with low freight cost. Most favorable are the sales possibilities for elemental sulfur, which is produced in the smallest amount and in all industrial countries can be transported over long distances in large tank cars, in liquid form, at low cost.

With regard to the cost of sulfur removal from flue gas, at present the throwaway processes are superior to regenerative processes as long as there is the possibility of disposing of the throwaway material in the vicinity of the power plant.[383] Thus in the United States lime- and limestone-scrubbing processes contribute 80% to all flue gas desulfurization and produce gypsum or sludge as final product. For the throwaway processes, not only is the amount of sludge of importance but also its structure, its components being mainly calcium sulfate and sulfite. The thixotropic properties of the sludge make mechanical drying difficult. Removal of moisture before use as fill is necessary, since after dumping

[383] H. Jüntgen, VDI Rep. No. 267, 1976, p. 17.

Table 30.31 Products of Flue Gas Sulfur Removal in a 600-MW Power Plant Operating 4000 hr/year and Using 2% Sulfur Bituminous Coal, Heating Value 29.3 × 10^3 kJ/kg and 80% Removal Efficiency; Yields Depend on Process Used

Product	Amounts (10^3 kg/year)	Process
Sulfur	128,000	Bergbau-Forschung, IFP, Wellman-Lord, SFGD
Sulfuric acid (100%)	390,100	Wellman-Lord, Sumitomo, Monsanto, Lurgi-Sulfacid, Bergbau-Forschung
Sulfur dioxide	256,000	Wellman-Lord, Chemico, Sumitomo, Bergbau-Forschung
Gypsum ($CaSO_4 \cdot 2H_2O$)	692,000	Chiyoda-, Hölter, Mitsubishi-Jecco, Chemico
Sludge (e.g., 70% $CaSO_3$ ½ H_2O) 15% $CaSO_4 \cdot 2H_2O$ 15% $CaCO_3$	688,000	Bischoff, Bahco, Combustion Engineering, Chemico

only surface drying occurs and solidification of the bulk is hindered.

Currently attempts are being made to improve the solidification properties by changes in the sulfate-sulfite ratio or by the use of additives. Meanwhile in the United States it has become possible to convert partially dewatered sludge into materials strong enough to be used in the building sector and in highway building. Then, however, the throw-away processes lose their cost advantage.[384] For densely populated and highly industrialized countries that have no indigenous supply of raw materials, salable sulfur products are desirable, since there is not sufficient room to dispose of the sludges. If the resulting products have a market, one should not expect that they can be sold at high prices.

As a salable product, sulfur has the best possibilities because it can be stored and, if worse comes to worst, deposited. To keep the cost of storing and transporting sulfuric acid low, it is important to have consumers in the vicinity, with uniform consumption.

Similarly, the sale of liquid SO_2 is not expected to cause problems. For gypsum, uniform qualities are required. Since natural gypsum has different setting properties, it cannot be replaced without added cost by chemical (flue gas desulfurization) gypsum. Therefore a direct further conversion (e.g., gypsum board) is attempted.

[384] W. Brocke, N. Haug, H. Pettelkan, and W. Schenkel, VDI-Rep. No. 267, 1976, p. 123.

CHAPTER 31

Environmental, Health and Safety Implications of Increased Coal Utilization*

CARL W. GEHRS, DAVID S. SHRINER, AND STEPHEN E. HERBES

ELIAHU J. SALMON AND HARRY PERRY

Carl W. Gehrs *received his B.S. degree from Concordia Teachers College, River Forest, Illinois, and his Ph.D. in Zoology from the University of Oklahoma. Dr. Gehrs's research interests center on potential environmental risks of advanced fossil energy systems, with emphasis on organic contaminants and predictive capabilities regarding ecological effects and environmental fate of these contaminants. Dr. Gehrs served as coordinator for the development of the U.S. Energy Research and Development Administration's health and environmental Balanced Program Plan for Coal Conversion. He served as chairperson for the Water Quality, Quantity and Aquatic Ecology sections of DOE/EPA/HEW Interagency Committee on Energy Workshops on Oil Shale and Coal Conversion. He is the author of more than 50 publications and papers on the environmental effects of advanced fossil energy technologies and aquatic ecology. Currently, Dr. Gehrs is Manager of the Advanced Fossil Energy Program, Environmental Sciences Division, Oak Ridge National Laboratory.*

David S. Shriner *holds an undergraduate degree in forestry and graduate degrees in plant pathology and has been on the staff of the Terrestrial Ecology Section, Environmental Sciences Division, at Oak Ridge National Laboratory since 1974. His principal research interests are in the areas of air pollution effects on vegetation, the physiology of stress in plants, and the biogeochemical cycling of elements in terrestrial ecosystems. Dr. Shriner has written approximately 30 papers in the general area of effects on vegetation of air pollutants that are related to fossil fuels.*

Stephen E. Herbes *received his B.S. in chemistry from Muhlenberg College, Allentown, Pennsylvania, in 1970, and his Ph.D. in environmental health from the University of Michigan in 1974. He has been a*

*This chapter represents the combination of the separate efforts of the two teams of authors named.

Research Associate with the Environmental Sciences Division, Oak Ridge National Laboratory, since 1974. Dr. Herbes has authored 20 papers and presentations on the aquatic transport of trace organic contaminants from coal conversion.

Eliahu J. Salmon *received his B.S. degree at Utah State University in 1951 and his Ph.D. in public health and industrial hygiene at the University of Michigan in 1964. He is presently the senior staff officer for a National Academy of Sciences (NAS) committee that is evaluating the Department of Energy research on health effects of ionizing radiation. Before joining the NAS Division of Medical Sciences, Dr. Salmon was with Resources for the Future where, as Senior Research Associate, he participated in a project on the evaluation of energy alternatives for the future and the impacts on health and the environment of various fossil fuel alternatives. From 1974 to 1976 he served as Executive Secretary of the Committee on Energy and Environment within the Commission on Natural Resources. Dr. Salmon has been senior lecturer in Environmental Sciences at the Israel Institute of Technology, Visiting Professor of Ecology at Tel-Aviv University, Permanent Consultant in Radiation, Occupational, and Environmental Health, Ministry of Health in Israel, and senior staff member of the units of Environmental Pollution and Radiation Health at the World Health Organization in Geneva.*

Harry Perry *is a private energy consultant, employed part-time by Resources for the Future. He was employed by the U.S. Bureau of Mines 1946 to 1968 in various capacities involving fossil fuel research. He also served as Science Advisor to the Assistant Secretary for Mineral Resources (Department of Interior), Senior Specialist for Energy at the Library of Congress, and Special Assistant to the Director of the Licensing, Atomic Energy Commission. Mr. Perry is the author of more than 200 articles and books on coal research, energy activities, and environmental impacts of coal production and use. He has received the Distinguished Service Medal of the Department of Interior and the British Coal Utilisation Research Association Coal Science Medal.*

31.1 INTRODUCTION

Neither the first two-volume edition of the *Chemistry of Coal Utilization* published in 1945 nor the supplemental volume published in 1963 contained a comprehensive discussion of implications of the coal system for the environment, human health and safety. When these topics were covered at all, the discussions appeared in the individual chapters on technology where significant impacts on health and the environment were an aspect of the particular technology. This volume continues the practice of individual chapter discussions but adds to them the present integrative chapter. Of course, the range of subjects discussed in the review is so broad that a critical analysis of all topics has not been attempted.

The development of the coal industry has been one of the most important features of the Industrial Revolution. During the earliest phases of that revolution, any machine or industrial process that worked was an achievement. Under these conditions health and environmental hazards associated with industrial processes generally, and coal processing in particular, were accepted as a matter of necessity and tolerated in the interest of industrial and economic progress.

As knowledge expanded in all phases of biology and engineering from the late nineteenth century on, more enlightened attitudes developed gradually about the environmental and health consequences of industrialization.

In the coal industry injuries and loss of life in mining accidents led first to public concern, then to regulations regarding mine safety. Surface mining is safer than underground mining, but it produces unsightly disturbance of the landscape and heaps of waste materials near the mines. This too became a matter of public concern and eventually led to regulations requiring the reclamation of land disturbed by mining. As the harmful effects of toxic gases and particulate emissions from combustion of coal became more obvious, public concern about pollution of the air and water increased, and eventually regulations were issued pertaining to the concentration of hazardous substances that could be permitted in air and water.

At present three principles of environmental protection have become accepted by leaders in industry and in society generally:

1. The air, water, soils, plants, and animals of the earth are public resources. Thus the emission of any substance into these resources is a privilege to be granted by society, not a right of any new or existing industry.
2. Every gaseous, aerosol, or particulate substance that is emitted into the atmosphere ultimately returns to the biosphere, where it is deposited by meteorological processes onto the soil vegetation or into the surface water systems of the earth.
3. In assessing the costs and benefits associated with any industrial process, a total material balance approach is necessary.

These three principles of environmental protection have special relevance to the coal industry. The ultimate fate of every constituent (including trace constituents) moved, removed, or produced in the process of using coal is coming to be recognized and dealt with explicitly.

Given the inadequacy of presently available data, it is still difficult to apply the total material balance approach. More complete information is needed on the concentrations, quantities, and ultimate fate of emissions released during coal processing.

Underground coal mining is a hazardous occupation, and its accident and fatality rates remain among the highest of any industry. Uncontrolled combustion of coal causes serious air pollution problems, the more obvious of which—smoke and particulates—were recognized in England as early as 1272. Concern about the particulate problem in Pittsburgh in 1915 led to extensive investigation of its causes and methods of abating it.

In the decade of the 1960s there was a rapid development of concern about occupational health and safety, the health and safety of the general public exposed to man-made pollutants, and risks and adverse impacts imposed on the air, water, and land. The concern was in part a result of greater impacts on the environment resulting from the need to meet the demands of a growing, more affluent population, and in part a recognition that the human and environmental resources described above are limited and must be protected.

Thus efforts were initiated to minimize the harmful side effects of economic growth. In the United States the Occupational Health and Safety Act,[1] the Clean Air Act,[2] the Clean Water Act,[3] the National Environmental Policy Act,[4] the Toxic Substances Control Act,[5] and the Resource Conservation and Recovery Act[6] were enacted. These laws are intended to minimize the harmful effects of industrial activities, including those of the coal system. In addition, the coal activities that can have serious adverse impacts were given special attention through passage of the Federal Coal Mine Health and Safety Act[7] and the Surface Mining Control and Reclamation Act.[8] Similar measures were adopted in some other countries where production and use of coal are important. (Although some other countries have enacted legislation to minimize the adverse effects of coal production and use, this chapter deals mainly with U.S. efforts to control adverse impacts on the environment and on health and safety.)

Despite these legislative efforts, the greatly increased use of coal currently anticipated in the United States and other countries may be expected to aggravate certain environmental, health and safety problems. Even under strict controls, large amounts of pollutants will be emitted, and even the most advanced regulations will only reduce emissions, not eliminate them. Surface-mine reclamation may in some cases be only partially successful. In addition, environmental and health and safety regulations will impose on coal production and utilization constraints that may limit the total amount of coal actually supplied and consumed and the mix of coal technologies that eventually penetrate the marketplace.

For these reasons it is useful to describe various impacts associated with the coal system, so that the total effect of their actions and interactions can be considered and evaluated comprehensively.

This chapter reviews the key impacts on the environment, health and safety from major categories of the coal system. All phases of the coal system and their potential impacts are described and assessed within the framework of the complete system. The main phases evaluated include cleaning and processing, transportation, and use of coal and coal-derived products.

To reduce the complexity of the chapter, only major impacts are discussed. The decision of whether a potential impact is major was based on an assessment of its magnitude, its importance relative to other impacts of the system, and the extent of any special public concern about it. In each section general impacts characteristic of coal processes are assessed first, and the impacts that depend on regional and local siting of coal activities are considered later.

It should be noted that many uncertainties are involved in the evaluation of potential impacts. Where sufficient data were available, the impacts were quantified; but in most cases, because of limited knowledge, we could do no more than identify the potential impact.

[1] Occupational Safety and Health Act of 1970; 29 USC 651.

[2] Public Law 91-604 (1970), Clean Air Amendments of 1970; 42 USC 1857 *et seq.*, 1975.

[3] Public Law 92-500 (1972), Federal Water Pollution Control Act Amendments of 1972; 33 USC 1251, 1975.

[4] Public Law 91-190 (1970), National Environmental Policy Act of 1969; 42 USC 4321 *et seq.*

[5] Toxic Substances Control Act of 1976; 15 USCA 2601, 1976.

[6] Public Law 94-580 (1976), Resource Conservation and Recovery Act of 1976.

[7] Public Law 91-173 (1969), Federal Coal Mine Health and Safety Act of 1969; 30 USC 801 *et seq.*, 1970.

[8] Public Law 95-87 (1977), Surface Mining Control and Reclamation Act of 1977; 91 Stat. 445 (30 USC Sec. 1201 *et seq.*).

On the basis of the assessments that follow in this chapter, it is possible to evaluate information about external costs to be expected from various coal systems, costs that affect the quality of life but normally are not covered fully by the price paid for coal. Effects of regulatory actions, control technologies, and other protective strategies to mitigate potentially harmful effects of coal activities can also be evaluated, as can the major tradeoffs involving the environment and health presented by various options among alternative coal technologies.

In addition to direct effects on the environment and human health from the construction and operation of coal facilities, certain indirect effects also should be considered. These include social and economic changes caused by increased coal production and utilization, such as manpower requirements and the influence on the quality of community life, needs for housing and services, local taxes, revenues, rate of inflation, and standard of living. The secondary impacts, which can be positive or negative, in some instances could be more significant than the primary consequences of the coal activities themselves.

In the discussions that follow (except for the discussion of air pollution emissions) one segment of the coal system is conspicuous by its absence, that is, the ultimate use of fuels produced by coal conversion processes leading to substitute natural gas or synthetic liquid fuel. The segment has not been considered completely because no commercial plants exist for the production of synthetic fuels, and the only data are from products made in small pilot plants, which may not be typical of commercial plants when such facilities are constructed. Clearly, however, this segment of the coal fuel system must be investigated before a complete assessment can be made of the impacts of synthetic fuels on the environment, health, and safety.

31.2 OVERVIEW OF THE COAL SYSTEM

31.2.1 General

Evaluation of implications for the environment and health of anticipated increased coal utilization requires understanding of all the activities involved in the various coal technologies. Each technology involves different processes, products, and end uses. and different impacts may be expected from each technology or its components. All technologies, however, share common elements and exert similar impacts, and all involve the following sequence of activities: exploration, extraction, beneficiation, transportation, distribution, and finally, utilization of coal and its by-products.

Figure 31.1 summarizes the major activities involved in the coal technologies in use as of 1976.

31.2.2 Exploration

Exploration for coal determines whether a particular area contains resources that can be economically extracted and marketed. Exploration usually involves two stages: identification of the existence of sufficient coal of quality adequate to support extraction at competitive cost; and a second, more extensive program to collect and evaluate data necessary for development of an extraction plan.

31.2.3 Extraction

Extraction involves mining the coal by underground or surface methods. Choice of a mining method is determined by depth and thickness of the seam, size of deposit, characteristics of the overburden, and the geology and hydrology of the site. Commonly, coal at depths less than 150 ft is mined by surface methods.

As of 1975, in the Eastern portions of the United States, 54% of coal was from underground mines, whereas in the West 87% was from surface mines.[9] A shift to surface mining is taking place in the United States because of the great increase in mining of Western coal. In the years beyond 2000, however, a

[9] National Coal Association, *Coal Facts 1978–1979,* Washington, D.C., 1978.

[10] U.S. Energy Research and Development Administration, *Comparative Risk-Cost Benefit Study of Alternative Sources of Electrical Energy,* Rep. No. WASH-1224, December 1974.

[11] Executive Office of the President, Council on Environmental Quality, *Energy Alternatives: A Comparative Analysis,* Washington, D.C., 1975.

[12] The MITRE Corporation, *Annual Environmental Analysis Report,* MTR-7626, Vol. 1, McLean, Va.; prepared for U.S. ERDA, September 1977.

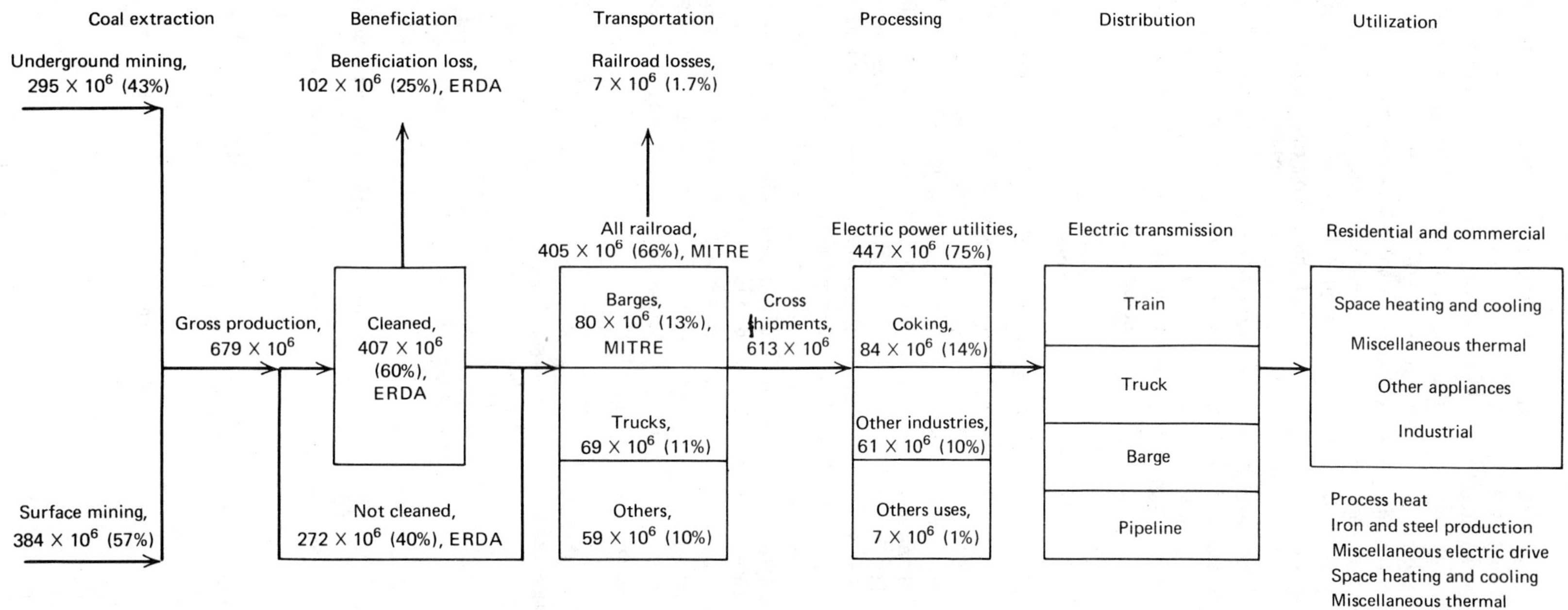

Figure 31.1 Summary of coal systems in the United States. Number other than percentages represent annual tons of bituminous coal; unless otherwise indicated, all numbers are obtained or derived from *Coal Facts*[9] and apply to 1976. Numbers identified as "ERDA" and "MITRE" are derived from Refs. 10 and 12, respectively. The "railroad losses" data are based on the assumptions that about two-thirds of coal transported by rail is by mixed trains with 2% coal losses and that about one-third is transported by unit train with 1% loss.[11]

shift back to underground mining can be expected because 77–91% of the nation's coal reserves are inaccessible to the surface mining methods available under current technology.[13]

There are two major types of surface mining methods,* namely, contour and area mining.[11] Contour mining is generally used in hilly or mountainous terrain and area mining is used in flat terrain.

In underground mining there are again two major methods: room-and-pillar mining and longwall mining.[11] Room-and-pillar was used for about 96% of all underground mining carried out in the United States in 1976.[9] Longwall mining methods are extensively used in Europe, but only to a limited extent in the United States.

31.2.4 Beneficiation

Prior to transportation or use, mined coal is often subjected to some degree of sizing and cleaning called beneficiation, which upgrades coal by increasing its heating values through removal of noncombustible materials. Beneficiation lowers transportation costs and reduces problems of waste disposal for the user. Removal of some of the sulfur and mineral matter, for example, can reduce sulfur dioxide and particulate emissions during combustion. Cleaning also increases the uniformity of the physical and chemical properties of coal.

In the United States it is estimated that on the average, about 20–25% of raw coal is rejected during the cleaning process.

31.2.5 Transportation

Transportation involves moving coal from the mine to the beneficiation plant (when one is used) or to the facility using the coal. Chapter 9 gives a detailed discussion of coal transportation systems.

[13] National Research Council, *Implications of Environmental Regulations for Energy Production and Consumption,* National Academy of Sciences, Washington, D.C., 1977.

*Some coal is surface mined by open-pit mining, where the coal seams are thick; auger mining (drilling into the coal seam at an exposed coal face) is used when the ratio of the thickness of the overburden to cost in contour mining becomes uneconomic.

Table 31.1 summarizes the mix of current modes of transportation in the United States.[9] Railroads carry about two-thirds of all the coal produced in the United States, although some of it may have been transported at some stage by other modes.

Currently the more efficient unit trains are used in the United States to carry about one-third of all coal shipped by rail, and their use is expected to increase. Other major modes of transporting coal are barges and trucks. Slurry pipelines, which transport coal with water, are expected to increase in number as more tons of coal are moved and transport distances increase.

31.2.6 Utilization

The major uses of coal are for electricity generation, coking, industrial process heat, and industrial, commercial, and residential heating. Of these, electricity generation is by far the largest. In 1976 in the United States, about 75% of the coal consumed was burned in conventional utility burners[9] and about 46% of the electricity produced was generated from coal.[14] The next largest users are industrial consumers, who use it largely for coke production, for process and space heating, and for steam generation. In 1976 coking accounted for 14% of the coal used; industrial uses in manufacturing and mining accounted

[14] U.S. Congress, House Subcommittee on Energy and Power of the Committee on Interstate and Foreign Commerce, *Energy Information Handbook. Data on Energy Resources, Reserves, Production, Consumption and Prices;* prepared by Congressional Research Service of the Library of Congress; 95 Congr., 1 Sess., committee print.

Table 31.1 Coal Transportation Modes in the United States

Mode of Transportation	Percentage of Total Coal Transportation: 1975
Rail (total)	66
Mixed trains	44
Unit trains	22
Barges	13
Other	21

Source: Derived from Ref. 9.

for about 10%.[15] The only new markets of coal consumption that may develop are coal gasification, liquefaction, and coal for chemical feedstocks.

All these technologies are discussed in detail in earlier chapters of this book; here they are referred to as coal conversion.

31.2.7 Distribution to Consumers

Distribution involves conveying the supply of coal, electricity, or coal products and by-products to end users. The electrical distribution system consists of subtransmission lines from the plant, primary and secondary distribution lines, and service lines to residential and commercial customers.

Coal and its solid products and by-products are usually distributed by railroads, trucks, barges, or ships (the latter mostly in exporting). Liquid and gaseous products produced from coal are distributed by pipelines, rail, or trucks. As a means of regulating supply, storage is provided at points of production and collection.

31.3 FACTORS INFLUENCING THE IMPACTS OF THE COAL SYSTEM ON THE ENVIRONMENT, HEALTH AND SAFETY

31.3.1 General

Various factors, acting singly or in combination, influence the nature and size of potential impacts of coal utilization on the environment and health. The factors may be grouped in three general categories: those related to the coal system, those related to the siting of the coal activities, and those related to legislative and technical controls used to mitigate potential impacts.

The factors associated with the coal system include efficiencies of the total system and of its different components in using resources; requirements of water, land, material, and manpower resources; types and rates of air emissions, water effluents, and solid wastes; and occupational and general public health and safety risks inherent in the different coal system activities and processes.

Critical factors related to siting of the coal activities include location relative to the mine, the coal markets, and the centers of population; the density and other characteristics of the population at the site; variations over time in the types and concentrations of environmental pollutants and chemically reactive species already existing in the area of the facilities; regional precipitation and ecosystem patterns; and the potential availability of an infrastructure of support services.

The main factors related to strategies to control or mitigate impacts include types of regulation, guidelines, and incentives, and the nature of the control technology that is used.

Table 31.2 compares the performance of different coal technologies in producing equivalent amounts of useful energy. It is important to be aware of the numerous uncertainties, the uneven data base, and the large number of assumptions that underlie Table 31.2. Some of the data are based on the best available measurements derived from commercial coal plants, and other data are scaled from operating experiences with pilot plants, or, to a lesser degree, extrapolated from data on operations in similar industrial processes.

The comparisons in Table 31.2 include measurements of impacts of different ways of using coal in space heating. This end use was selected because residential space heating accounts for a significant portion of energy demand, and it uses all three major forms of energy (i.e., electricity, oil, and natural gas).

The size of the unit system for the technologies compared in the table is based on useful energy flow from a representative commercial high-Btu coal gasification plant producing 66.5×10^{12} Btu/year of useful energy for which a large number of cost estimates has been published. The other unit coal energy systems in Table 31.2 have been sized to produce the same amount of useful energy per year.

In the case of coal conversion technologies, the unit systems represent approximately the expected size of the commercial plant of the future. For the electric power and combined-cycle unit systems, the plants are

[15] Energy Information Administration, Annual Report to Congress, Vol. 2, *Statistics and Trends in Energy Supply, Demand and Prices,* DOE/EIA CC 36/3, 1977.

Table 31.2 Important Factors of Some Coal Energy Technologies Based on Unit Model Systems[16]

	3390-MWe Electric Power Plant with Scrubber Producing 2540 MWe of Electricity	3390-MWe Integrated Low-Btu–Combined Cycle Electric Power Plant Producing 2540 MWe of Electricity	278×10^6 scf/day High-Btu Coal Gasification Plant Producing daily 250×10^6 scf/day 1000-Btu/scf Gas	58.0×10^3 bbl/day Coal Liquefaction Plant Producing 52.2×10^3 bbl/day of 5.6×10^6 Btu/bbl Oil
Load factor[11,17] (% annually)	75	75	90	90
Energy efficiencies (%)				
Coal transport[11,18]	97.5	99	97.5	97.5
Coal preparation and storage[19]	95.5	95.5	95.5	95.5
Thermal conversion[11,20]	35	37	60	65
Transmission or distribution[11,21]	92	92	97.1	98.8
Residential space heating[11]	95	95	75	63
Overall efficiency	28.5	30.6	40.7	37.7
Annual energy (10^{12} Btu/year)				
Input into plant	217	206	152	164
Output from plant	76.1	76.1	91.3	107
Useful energy	66.5	66.5	66.5	66.5
Annual coal requirements (10^6 tons/year of 25×10^6 Btu/ton coal)				
Mined	9.34	8.70	6.53	7.06
Combusted or converted	8.70	8.22	6.08	6.57
Plant resource requirements				
Water evaporated (10^9 gal/year)[22]	7.9	7.9	8.4	9.8
Land (10^3 acres/life of plant)[23]	2.4	1.4	0.81	1.1
Manpower				
Peak construction (people)[24,25]	2,500	2,600	4,700	2,700
Man-years (construction)[23]	10,000	10,500	10,800	8,700
Operation (people)[23]	440	480	590	600
System resource requirements				
Water evaporated (10^9 gal/year)[22]	9.4	9.4	9.3	10.8
Land (10^3 acres/life of plant)[23,a]	12.1–16.6	10.4–14.7	7.6–10.7	8.4–11.9
Manpower				
Man-years (construction)[23,a]	10,000–11,300	11,200–11,700	11,400–11,700	9,300–9,700
Operation (people)[23,a]	1,030–1,560	1,030–1,520	1,000–1,370	1,040–1,450
Annual air emissions (10^3 tons/year)[b]				
SO_2	88.2	8.4	6.2	16.3
NO_x	85.7	41.0	8.1	8.8
Particulates	12.3	2.4	1.7	4.2
HC	1.3	3.6	0.5	1.2
CO	4.5	0.9	1.5	2.3
Liquid effluents		No discharges		
Annual solid wastes (10^6 tons/year)				
Plant[17,23]	2.3	1.5	0.8	0.6
System[23,a]	2.5–3.1	1.7–2.2	0.9–1.3	0.7–1.2

[a] The range is based on 56 and 90% of coal mined by surface mining; only underground coal is beneficiated.
[b] See Table 31.3.

[16] E. J. Salmon, "Health and Environmental Implications of Various Energy Systems," draft report prepared for National Energy Strategies Project (Resources for the Future, Washington, D.C.), 1978.

[17] U.S. Energy Research and Development Administration, *Market Oriented Program Planning Study (MOPPS). Assessments of Environmental, Health, and Safety Constraints on Selected Energy Technologies,* final draft, September 1977.

[18] Hittman Associates, *Environmental Impacts, Efficiency, and Cost of Energy Supply and End Use,* Final Report, Vol. 1, Columbia, Md., 1974.

[19] Brookhaven National Laboratory, *Sourcebook for Energy Assessment,* M. Beller, Ed., BNL-50483. Upton, N.Y., December 1975.

[20] National Research Council, Ad Hoc Panel on Liquefaction of Coal, Committee on Processing and Utilization of Fossil Fuels, *Assessment of Technology for the Liquefaction of Coal,* 2 vols., National Academy of Sciences, Washington, D.C., 1978.

[21] H. Perry, *Conservation of Energy,* commissioned by Library of Congress, Congressional Research Service, for Committee on Interior and Insular Affairs, U.S. Senate, 92nd Cong., serial 92-18, 1972.

[22] J. Harte and M. El-Gasseir, "Energy and Water," *Science,* **199,** 623–634 (February 10, 1978).

[23] Radian Corporation, *The Coal Resource System,* final draft, April 1976; Office of Planning and Evaluation, U.S. ERDA.

significantly larger than those that have been built. Therefore the electric plant unit should be viewed as being made up of several smaller plants operating at the same site.

The air emissions referred to in Tables 31.2 and 31.3 include oxides of sulfur and nitrogen, particulates, hydrocarbons, and carbon monoxide. The quantities are those emitted from the complete coal energy system, including the consumer's use of electricity or fuel derived from the coal. Table 31.3 presents more detail concerning the major sources of emissions.

The values of emissions shown in Tables 31.2 and 31.3 are based on the assumptions that the coal has sulfur and ash contents of 3.0 and 10.0%, respectively, and a heating value of 25 × 10^6 Btu/ton, and that the efficiency of the sulfur oxide removal control equipment is 85%, and emissions of particulates and NO_2 meet the 1970 New Source Performance Standards (NSPS).

[24] Science and Public Policy Program of the University of Oklahoma, and Radian Corporation, *Energy from the West: A Progress Report of a Technology Assessment of Western Energy Resource Development,* 3 vols., prepared for U.S. ERDA, July 1977.

[25] M. Carrasso et al. *The Energy Supply Planning Model,* Vol. 1, Bechtel Corporation, San Francisco, 1975.

31.3.2 Coal System Factors

31.3.2.1 *Resource Utilization Efficiencies*

One of the most important factors influencing impacts of a coal system on the environment and health is the efficiency of the system in using the coal resource. The lower the utilization efficiency, the greater the amount of coal that needs to be mined, transported, cleaned, and processed to provide equivalent amounts of useful energy or products. Greater requirements for coal to supply energy to the consumer will produce proportionately greater impacts on the environment and health from mining and transportation of the coal, which account for a significant proportion of the impacts of coal use. The types and quantities of pollutants and wastes and the safety risks from remaining phases of the coal system are also inversely related to coal utilization efficiency.

Efficiency in resource use varies considerably among coal systems. For example, a system that includes an electrical power plant with flue gas desulfurization has an overall energy efficiency of 30% compared to 40.7% for a high-Btu coal gasification system.[16]

Table 31.3 Air Emissions for Some Coal Energy Technologies Based on Unit Model Plants

	Annual Air Emissions (10^3 tons/year)				
Technology	SO_2	NO_x	Particulates	HC	CO
Electricity from coal with flue gas desulfurization	88.2[a]	85.7[a]	12.3[a]	1.3[b]	4.5[b]
Integrated low-Btu gasification–combined cycle power plant					
Gasification	8.39[c]	8.1[c]	1.17[c]	0.27[b]	0.91[b]
Electricity[b]	0.05	32.9	1.26	3.371	0.03
	8.44	41.0	2.43	3.64	0.94
High-Btu gasification					
Gasification	6.18[c]	6.00[c]	0.86[c]	0.20[b]	0.67[b]
Residential gas use[b]	0.02	2.14	0.82	0.34	0.86
	6.20	8.14	1.68	0.54	1.53
Coal liquefaction					
Liquefaction	4.4[a]	4.18[a]	0.40[a]	0.06[a]	0.42[d]
Residential oil use[b]	11.9	4.57	3.81	1.14	1.90
	16.3	8.75	4.21	1.20	2.32

[a] Derived from Ref. 17.
[b] Derived from Ref. 19.
[c] Based on data from Refs. 17 and 19. In addition, it is assumed that most of the emissions during gasification result from combustion of the coal, that about 10% of the coal is thus consumed, and that the energy efficiency of low-Btu gasification is about 83%.
[d] Derived from Ref. 23.

31.3.2.2 *Resource Requirements*

General. The four major resource requirements that influence the potential impacts of increased coal use on the environment and health are coal, water, land, and manpower. Other resources, such as capital, other materials, and other energy inputs, are required but have less impact on the ecosystem. In addition to these direct resource requirements, there are extensive potential impacts associated with the use of resources. Thus the ''land requirement'' of a coal facility refers to the land within the boundaries of the facility necessary to support activities directly or indirectly relating to coal. These activities, however, may in turn have additional impacts on land that may extend far beyond the site, such as disturbances caused by changes in runoff or river flow patterns.[25]

Estimation of resource requirements is further complicated because the resources are not all permanently consumed. For example, water is used for cooling and then returned to the river, and land may be used simultaneously for electricity transmission and for farming. When an energy facility is dismantled, the people who provided the manpower and the land the facility used become available again for other uses, even though some rehabilitation of the site may be necessary.

Coal. Two major factors affect the quantity of coal needed to produce a fixed amount of useful energy, coal products, or by-products: the composition of the coal (discussed under ''Feedstock Characteristics'' in Section 31.3.2.3) and the overall efficiency of coal utilization, which depends on the coal technology employed (discussed in the preceding section). Economic and technological considerations primarily determine the grade of resources used or the efficiency of utilization achieved, but considerations of health and safety may also be contributing factors–for example, when it is decided to require use of low-sulfur coal.

Water. An adequate and suitable water supply is an important requirement in developing coal as a source of energy. In the future, limited availability of water could become a significant constraint. Not only are the nationwide water requirements of the coal industry expected to grow rapidly, but a greater proportion of coal mining and processing activities is anticipated in arid regions of the United States. At the same time, the demand for water for many other uses is also increasing.

Information on water requirements is often contradictory. A distinction is made between water consumed and water withdrawn. Consumed water (e.g., water that is required to produce process hydrogen) is no longer available for further use. Heavily polluted water is also usually considered to be consumed because it is not reuseable without extensive treatment. In contrast, water that has been warmed only slightly when it is used for cooling, then returned to the original water body, is considered to be withdrawn water. Unless otherwise noted, the following discussion refers to consumed water.

The major uses of water in the coal system include the following:

1. Dust suppression during mining.
2. Waste disposal.
3. Reclamation of mined land. The reclamation of land from a surface mine producing 12 million tons of coal annually has been estimated to consume about 730–1100 million gallons of water per year for 5 years.[23] The figure represents about 10–15% of the total water consumption of the coal energy system.
4. Beneficiation of coal. Beneficiation would consume about 900 million gallons of water in cleaning the 12 million tons of coal referred to above.[23]
5. Transportation of coal in slurry pipelines. About 12 million tons of water/year (2.8 billion gallons of water/year) would be required to transport 12 million tons of coal. (Some of this water may be reused after proper treatment.[23])
6. Processing of coal to produce synthetic fuels or chemical feedstocks. During coal conversion water is consumed as process water and steam in the chemical reactions, including the production of hydrogen to upgrade the fuels. This consumption is estimated at about 15% of the cooling water withdrawal requirements.[23]
7. Disposal of wastes produced during electricity generation and coke and synthetic fuel production.

Table 31.4 summarizes estimates of the amount of water that will be consumed by electricity generation and in the conversion of coal to gaseous and liquid fuels. (More recent estimates of consumption of water are lower by a factor of up to 3 for coal conversion technologies.[26,27]) The table shows that most of the water is consumed during the energy conversion phase of the various coal technologies, when coal is converted to electricity or fuels. Most of the consumed water evaporates during cooling.[23] The other major potential consumers of water in the coal system are reclamation of land after surface mining and transportation of coal in slurry pipelines. The quantity of water consumed by the various coal technologies varies according to processes employed, water use practices, and characteristics of the coal, particularly its hydrogen-to-carbon ratio and moisture content.

Two basic kinds of water problem are associated with future increases of coal utilization. One arises from the geographical location of the coal activities, where clear

[26] R. F. Probstein and H. Gold, *Water in Synthetic Fuel Production—The Technology and Alternatives,* MIT Press, Cambridge, Mass., 1978.
[27] H. Gold and D. J. Goldstein, *Water-Related Environmental Effects in Fuel Conversion,* Vol. 1, *Summary;* Vol. 2, *Appendices,* EPA 600/7-78-197a, October 1978.

Table 31.4 Estimates of Consumption of Water (10^{10} gal/quad[a] of electricity or synthetic fuel product for the various categories of use) by Energy Technologies for Generating Electricity or Producing Synthetic Fuels[b]

Fuel	Mining[c]	Reclamation[d]	Transport by Slurry Pipeline[e]	Conversion	Associated Urban	Total with Slurry Pipelines	Total Without Slurry Pipelines
Electricity generation							
Eastern coal							
Surface mined	0.18	0.0–1.6	3.0	5.1–39[f]	0.50	8.8–44	5.8–41
Deep mined	0.41	0.0	3.0	5.1–39[f]	0.50	9.0–43	6.0–40
Western coal							
Surface mined	0.18–0.37	0.19–7.5	2.9–6.0	5.1–39[f]	0.50	8.9–53	6.0–47
Deep mined	0.41–0.54	0.0	2.9–6.0	5.1–39[f]	0.50	8.9–46	6.0–40
Low-Btu gas							
Eastern coal							
Surface mined	0.078–0.097	0.0–0.83	1.2–1.6	2.3–16	0.50	4.1–19	2.9–17
Deep mined	0.17–0.22	0.0	1.2–1.6	2.3–16	0.50	4.2–18	3.0–17
Western coal							
Surface mined	0.078–0.19	0.078–3.9	1.2–3.1	2.3–16	0.50	4.2–24	3.0–21
Deep mined	0.17–0.28	0.0	1.2–3.1	2.3–16	0.50	4.2–20	3.0–17
High-Btu gas							
Eastern coal							
Surface mined	0.097–0.12	0.0–1.0	1.6–1.9	2.3–16	1.4	5.4–20	3.8–19
Deep mined	0.22–0.26	0.0	1.6–1.9	2.3–16	1.4	5.5–20	3.9–18
Western coal							
Surface mined	0.097–0.24	0.10–4.7	1.6–3.9	2.3–16	1.4	5.5–26	3.9–22
Deep mined	0.22–0.33	0.0	1.6–3.9	2.3–16	1.4	5.5–22	3.9–18
Coal liquefaction							
Eastern coal							
Surface mined	0.086–0.16	0.0–1.3	1.4–2.6	3.1–21	0.80	5.4–26	4.0–23
Deep mined	0.19–0.36	0.0	1.4–2.6	3.1–21	0.80	5.5–25	4.1–22
Western coal							
Surface mined	0.086–0.31	0.089–6.4	1.4–5.3	3.1–21	0.80	5.5–34	4.1–29
Deep mined	0.19–0.47	0.0	1.4–5.3	3.1–21	0.80	5.5–28	4.1–22

[a] 1 quad = energy from, for example, 10^{12} ft³ of natural gas.
[b] Calculations are based on coal energy content of 24, 19, and 12×10^6 Btu per ton of bituminous coal, subbituminous coal, and lignite, on electrical generation efficiency of 35%; and coal conversion efficiencies of 67–85, 55–67 and 41–75% for low- and high-Btu gasification and liquefaction, respectively.
[c] In the East surface and deep mining consume 64 and 140 gal of water per million Btu of coal mined. In the West consumption of water is 64–130 and 140–190 gal per million Btu of coal, for deep and surface mining respectively.
[d] Slurry pipelines consume 10.6 and 10.3–21.2 gal of water per million Btu of coal mined in the East and the West, respectively.
[e] Slurry water could be returned in a parallel pipeline for reuse, thereby reducing water consumption to a value much lower than that shown here.
[f] Water requirements for coal electricity generation are based on 35% generation efficiency and on 17% of the waste heat dissipated directly to the atmosphere in the form of hot stack gases. The cooling water consumption requirements range between 5.6 and 11×10^{10} gallons per quad of electric output for once-through cooling with no storage; 14 to 42×10^{10} for once-through cooling with reservoir storage to meet backup storage requirements of 1000-MWe-size plants for 90 days; and 11 to 17×10^{10} for wet cooling tower.

Source: Derived from Ref. 22.

differences can be seen between regions with adequate rainfall, and the drier (e.g., western) locations. Problems of the second kind are associated with the quantitatively greater demand for water created by coal processes, particularly coal conversion to energy, fuels, and feedstocks.

It would be possible to decrease the amount of water consumed by the coal industry by minimizing evaporation and by reclaiming water after use. Consumption would then be limited to the need to replace water unavoidably lost in the system and water chemically consumed during coal conversion. But such intensive water conservation practices are not usually followed, because in most locations the savings in water cannot currently be justified on either economic or environmental grounds.

Availability of water in the future may be a problem in some regions. One approach to the problem is to conserve the available supply through better water management; the other is to attempt to increase the supply. Available water supplies can be increased by developing new water resources. For example, local unused waters, such as those underground, can be tapped, although care must be taken to avoid excessive withdrawal of underground water that may cause an intrusion of salt water or subsidence of the ground surface. Surface or underground waters that are too saline for direct use may be desalinated, and polluted water may be cleaned. Cooling and waste disposal may also be achieved with seawater whenever it is readily available.

Land. Four activities of the coal system require substantial amounts of land: mining (surface or underground); electrical power generation, conversion, or processing; distribution of the coal, electricity, fuels, and products; and disposal of wastes from mining, beneficiation, and processing. It is expected that wastes will usually be disposed of at or near the production site.

Land requirements are of two types: fixed amounts of land for facilities and associated activities, and fluctuating amounts of land in annual increments as it is removed from use for surface mining or for disposal of wastes. The product of the annual increments of land use multiplied by the life of the facility is usually added to the fixed land requirement to provide an estimate of the total land use of each energy system.

The predominant use of land by coal activities is in mining. It is estimated that about 80% of the total area used (excluding partial uses of land for such needs as electricity transmission lines or rights of way) will be for mining.[23,24,28] The other main uses of land are for energy conversion facilities and for the associated disposal of solid wastes.

The potential problems imposed by the aggregate land demand of increased coal utilization appear to be minimal, since only a small fraction of 1% of the land area of the United States will be required for coal activities.[12] The situation will be different in countries with smaller land areas.

Manpower. There are usually two distinct phases of manpower requirements, which differ in numbers of workers and kinds of skill needed. The first phase involves construction of energy facilities and service infrastructures. This phase usually lasts 4–7 years, with labor requirements peaking toward the end of the period. A longer operational phase follows, lasting 20–40 years, during which workers are required for the operation and maintenance of the facilities. The number of workers needed does not fluctuate significantly during this phase.

Estimates of needs of manpower and skills and timetables of employment have been compiled by several authors.[23–25,28,29] An adequate potential labor force, including engineers, technicians, and miners, has been forecast for both coal activities projected for 1985 and for continuing expansion in coal utilization.[30]

[28] U.S. Department of Energy, *Energy Supply Projections (1985 and 2000) and Emerging Energy Technology Characteristics,* Washington, D.C., January 1978.

[29] U.S. Department of the Interior, *Project Independence: Potential Future Role of Oil Shale: Prospects and Constraints,* Interagency Task Force on Oil Shale, 1974.

[30] MITRE Corporation, *Actions to Increase the Use of Coal: To Date to 1990: Proc. Symp.* sponsored by METREK Division, November 7–9, 1977; published March 1978.

31.3.2.3 *Types and Rates of Emissions of Pollutants and Solid Wastes*

General. The following generalizations apply to the various coal technologies discussed in this chapter.

1. Air emissions with important implications for health and the environment occur primarily during fuel combustion in electricity generation, conversion of coal to liquids and gases, and combustion of fuels by end users. In conventional combustion facilities the contaminants include gaseous oxides of sulfur and nitrogen and, to a lesser extent, hydrocarbons, particulate materials, and volatilized trace elements. The known or suspected effect on the environment and on health of several of these classes of pollutants is the major reason for the development of more advanced combustion processes (e.g., fluidized-bed combustion) and conversion processes, in which most of the pollutants are captured and removed rather than being released to the environment.

2. The principal liquid effluents arise from washing coal during beneficiation, from quenching waters used to cool hot products and ashes, and from condensates and blowdown waters from boilers, cooling, and scrubbing systems. In the case of electricity generation or coal conversion, waste heat is also released into water effluents. A policy of zero water discharge from energy facilities appears feasible and desirable in cases where particularly troublesome effluents are formed. Indirect water effluents that may be leached at solid waste disposal sites present a potential problem that needs further evaluation. Additionally there are problems of acid mine drainage, particularly from abandoned underground mines in the East, and of increased salinity of surface and underground waters caused by Western coal mining.

3. Solid wastes are associated with extraction and beneficiation, and with generation of electricity from coal, and conversion of coal to liquid and gaseous fuels. The largest quantities of solid wastes associated with coal systems are produced during beneficiation. They usually account for about two-thirds of total wastes, excluding the overburden removed during surface mining. The other major source of solid wastes is the energy facility, which produces large quantities of coal ash, fly ash, and sludges.

The major system factors that influence the quantities and rates of emissions and wastes from the coal system are the characteristics of the coal feed, the type of process employed (including both design and operational factors), the efficiency of the process, and the emission control technology used.

Feedstock Characteristics and Types of Process Factor. Coal type also influences the quantitative and qualitative characteristics of contaminants from coal combustion facilities. The most important fuel variables are the percentage of sulfur, the ratio of organic to inorganic sulfur, the percentage of nitrogen, the percentage of volatile matter, moisture, and ash content, the heating value of the coal (Btu/lb), and the range of sizes of the coal particles.[31,32] The effluent concentration of particulate matter varies with the ash content of the coal. The density of particulate matter depends on the carbon content of the ash.[32,33]

The sulfur content of some major coal deposits ranges from 0.14 to 1.2% for Wyoming coals, and from 0.6 to 4% for West Virginia coals. Ash content ranges from 1.9 to 11% for some Virginia coals, and from 5 to 30% for western Kentucky coals. Heating values range from 18 to 24 million Btu/ton for some Arizona coals, and from 29 to 30 million for some Tennessee coals.[34]

The average trace elements concentration of different coals also exhibits wide regional variations. (See Chapter 7 for ranges of impurities in different coals.)

[31] National Research Council, *Air Quality and Stationary Source Emission Control,* NAS/NAE/NRC Rep. No. 94-4 to Committee on Public Works. U.S. Senate, 94th Congr., 1st Sess., 1975.

[32] A. V. Slack and G. A. Hollinden, *Sulfur Dioxide Removal from Waste Gases,* 2nd ed., Noyes Data Corporation, Park Ridge, N.J., 1975, pp. 15–23.

[33] U.S. Environmental Protection Agency, *National Air Quality and Emissions Trends Report, 1975,* EPA-450/1-76-002, 1976.

[34] Nuclear Regulatory Commission, *The Environmental Effects of Using Coal for Generating Electricity,* prepared by Argonne National Laboratory, Chicago, June 1977.

Table 31.5 Estimated Emissions from an Uncontrolled 1000-MW Coal-Fired Power Plant and Emissions Allowable Under Federal New Source Performance Standards[36,37]

Assumptions: 10% ash, 3% sulfur, 11,200 Btu/lb, 9×10^6 Btu, 1 MW-hr.

		Emissions: Particulates			Sulfur Dioxide			Nitrogen Oxides		
Type of Furnace	Prevalence (%)	lb/hr	lb/ton coal	lb/10^6 Btu	lb/hr	lb/ton coal	lb/10^6 Btu	lb/hr	lb/ton coal	lb/10^6 Btu
Pulverized										
Wet bottom	13.4	52,200	130	5.80	45,900	114	5.1	12,060	30	1.34
Dry bottom	73.8	68,400	170	7.60	45,900	114	5.1	7,227	18	0.80
Cyclone	9.5	8,010	20	0.89	45,900	114	5.1	22,050	55	2.45
Other	3.3									
Allowable emissions under NSPS		900	2.24	0.10	10,800	26.9	1.2	6,300	15.7	0.70
Required efficiency (for most prevalent type)		98.7%			76.5%				13 %	
BACT[a] efficiency		99.5%[b]			90%[c]				24. 50%[d]	

[a] Best available control technology.
[b] Ref. 41
[c] Gartrell.[80]
[d] Gartrell[80]—short-term tests with staged firing; this is not BACT yet.

Process Characteristics

Process Types. Differences in pollutants and solid wastes emitted by the same coal technology are primarily influenced by process variables. Among these are the amount of excess oxygen or air used, the type of catalyst employed, the temperature and pressure at which the process operates, the configuration of reaction vessels, the feed entry geometry, the rate of cooling of decomposition products, the gas velocity, the method used for recycling residuals, and the type of control equipment employed[35] (Chapter 22).

Table 31.5 summarizes the influence of boiler design on major emissions from a typical coal combustion power plant. Table 31.6 illustrates how variations in temperature and pressure affect residual characteristics of coal gasification processes. For example, fixed-bed systems in which the gas leaves the reactor at low temperatures tend to be "dirty" (i.e., the process produces such contaminants as oils, tars, and phenols) compared to high-temperature entrained systems, which produce "clean gas."

Table 31.7 provides some quantitative estimates of the influence of process type on projected major atmospheric emissions from selected coal gas and liquid conversion plants in comparison to fossil fuel combustion units.

Feed entry geometry (such as alteration of the point of fresh coal injection) may significantly influence the character of effluents. Variations in process technology can have a major effect on the concentration of pollutants in process streams. Dramatic reductions in the amount of phenol produced have been reported[42] for a high-Btu gasification plant when coal is injected deeper into a fluidized bed, as well as reductions of up to

[35] W. S. Smith and C. W. Gruber, *Atmospheric Emissions from Coal Combustion—An Inventory Guide,* U.S. Public Health Service Bull. No. 999-AP-24, 1966.

[36] C. Cowherd, Jr., M. Marcus, C. M. Gunther, and J. L. Spigarelli, *Hazardous Emission Characterization of Utility Boilers,* EPA-650/1-75-66, 1975.

[37] W. L. Polglase, "Fossil and Nuclear Power Options and Related Air Pollution Considerations," *Remarks, Proc. Symp. Power Generation and Air Quality,* Triangle Universities Consortium on Air Pollution (TUCAP), Raleigh, N.C., 1974.

[38] S. M. Hossain, J. W. Mitchell, and A. B. Cherry, *Control Technology Development for Products/By-Products of Coal Conversion Systems,* Catalytic, Inc., Philadelphia 1977.

[39] U.S. Energy Research and Development Administration, *Synthetic Fuels Commercialization Program,* draft environmental statement, ERDA-1547, 1975.

[40] D. S. Shriner, S. B. McLaughlin, and C. F. Baes, Jr., *Character and Transformation of Pollutants from Major Fossil Fuel Energy Sources,* ORNL/TM-5919, Oak Ridge National Laboratory: Oak Ridge, Tenn., 1977, 37 pp.

[41] U.S. Environmental Protection Agency, *Compilation of Air Pollutant Emission Factors,* 2nd ed., AP-42, 1973, 153 pp.

86% in tar production when the injection of lignite is shifted from free-fall to shallow-bed feeding in a gasification reactor.

Scale of Operations. Discussions of the Risk/Impact Panel of the National Academy of Sciences Committee on Nuclear and Alternative Energy Systems[43] emphasized that impacts from coal activities do not vary proportionately with the scale of operations or the level of stress. This is because there appear to be stress levels below which no lasting impacts can be detected on organisms or communities.

On the basis of energy output, commercial coal gasification plants are expected to be about 3–4 times as large as electric power plants, and liquefaction plants are expected to be even larger. On the basis of size alone, these plants might have disproportionately greater ecological and social impact at their sites, but this will depend on the nature, type, and quantity of pollutants, and on the efficiency of the control technology employed.

31.3.3 Siting Factors

The location of energy activities relative to the site of extraction of the resource is particularly important in the case of impacts of coal technologies on health. The distance the bulky coal is transported will influence the number of deaths and injuries to be expected during transportation, which is one of the system's major impacts on health. Transportation is also a source of air pollutants, and the quantity of emissions increases with the transportation distance. Proximity of energy conversion facilities to the mining site would make it easier to return some solid wastes to the coal mine for burial, thus alleviating the solid waste disposal problem.

Because of the high cost of electricity transmission, we may expect electricity generating plants to be located closer to consumers than plants for high-Btu gasification or liquefaction of coal.[44] Consequently the environmental impacts of electricity generation on land use may be greater than those of the other conversion processes.

The following site characteristics are important with respect to impacts of coal activities on the environment, health, and socioeconomic factors.

1. Population characteristics, which include the size, density, and age distribution of the population at risk. Density of the site population affects the number of people that may be exposed to toxic emissions. Age distribution affects the degree of damage that the site population can be expected to experience from exposure to pollutants. Community size affects the ease with which the capacity of services can be expanded without disruption and deleterious social impacts.

2. Variations over time of the types and concentrations of environmental pollutants and chemically reactive species from other sources existing in the vicinity of the plant. Residents of the Middle Atlantic and northeastern regions of the United States and the Ruhr area in West Germany are expected to experience greater damage to health than would residents in other areas for an equal increment of pollution. This is because background levels of both emitted pollutants and transformation products damaging to health are already greater in those regions. Thus the regional population tends to have a smaller reserve capacity for accommodating further intakes without damage.

3. Regional climatic patterns, both the total amount of rainfall and its distribution throughout the year. These become critical factors in arid regions where water is scarce. Little rainfall and short rainy seasons adversely influence the ecological impacts relative to the reclamation potential of the site.

4. Existence of fragile ecosystems that are more easily damaged by pollutants and more difficult to reestablish than others. Desert and tundra areas are examples of systems that recover only slowly from perturbation and therefore need careful attention during the planning of facilities.

[42] D. V. Nakles and M. J. Massey, "Gasification Process/Environmental Characterization from Pilot Plant Data," in *3rd Symp. Proc., Environmental Aspects of Fuel Conversion Technology,* Hollywood, Fla., September 13–16, 1977, published as EPA 600/7-78-063, April 1978.

[43] National Research Council, *Risks and Impacts of Energy Production and Use,* Committee on Nuclear and Alternative Energy Sources, Risk and Impact Panel, 1977 (to be published).

[44] U.S. Senate, Conservation of Energy, Committee on Interior and Insular Affairs, Ser. No. 9218, 1972, p. 31.

Table 31.6 Comparison[a] of Products and By-Products from Various Coal Gasification Processes[38]

	Low-Temperature Fixed Bed				Intermediate-Temperature Fluidized Bed			
	Low Pressure	Intermediate Pressure			Low Pressure	Intermediate Pressure	High Pressure	
Products and By-Products	Wellman-Galusha	Lurgi	BGC/Lurgi Slagging Gasifier	Pressurized Stirred Fixed Bed, Morgantown	Winkler	U-Gas	Synthane	Hygas
High-Btu gas—SNG	P	P	P	—	P	—	P	P
Low (Intermediate)-Btu gas	P	P	P	P	P	P	—	—
H_2S–acid gas/sulfur	P	P	P	P	P	P	P	P
Ammonia	P	P	P	P	N.A.	P	P	P
Phenols	P	P	P	P	Neg.	Neg.	P	P
Naphthas and benzenes	N.A.	P	P	—	—	N.A.	P	P
Tar oils and light oils	P	P	P	P	—	Neg.	P	P
Tars	—	P	P	P	—	Neg.	P	—
Char and unreacted coal	P	—	—	P	P	P	P	P
Ash and slag	—	P	P	P	—	—	—	—

[a] Key: P = product or by-product present in recoverable quantities
Neg. = negligible or small amounts present
N.A. = information not available, not complete, or not reported at this time
[b] Intermediate temperature, low pressure.
[c] Intermediate temperature, intermediate pressure.

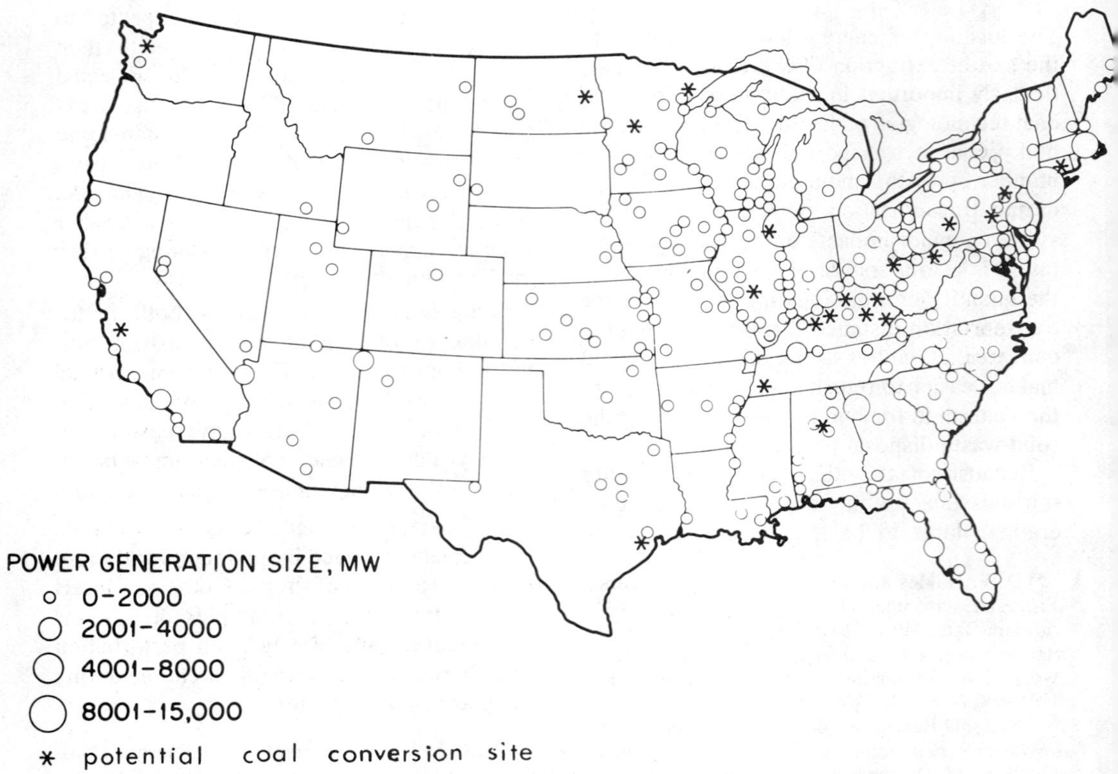

Figure 31.2 Location of major coal- and oil-fired power units, 1971.[45]

High-Temperature Entrained Bed		Dolomite Acceptor		Coal Pyrolysis	
Low Pressure	High Pressure	Intermediate Temperature, Intermediate Pressure		Entrained Bed[b]	Fluid Bed[c]
Koppers - Totzek	Bi-Gas	CO_2 Acceptor	Westinghouse Advanced Gasifier	Battelle Ash Agglomerating	Garrett's Coal Gasification
P	P	P	—	P	P
P	—	—	P	P	P
P	P	P	P	P	P
Neg	P	P	N.A.	P	N.A.
—	—	N.A.	—	—	—
—	—	N.A.	—	—	—
—	—	N.A.	—	—	—
—	—	N.A.	—	—	P
—	—	P	—	—	P
P	P	—	P	P	—

A major factor in determining whether gaseous emissions have adverse regional effects on the environment and health is the spatial distribution of sources. The spatial distribution of major coal- and oil-fired steam electric generating facilities in the United States as of 1971 is shown in Fig. 31.2, which makes three important points: (1) the large number of generating units in the eastern United States as compared with the western part of the country (~ 80% of the U.S. electric generating capacity derived from fossil fuels is in the eastern half of the country, and 50% of the total is in the northeastern states), (2) the high density of these sources in the eastern United States, and (3) the relatively few areas remote (> 100 km) from sources of pollutants. The longest distance of any point in the eastern United States from some fossil-fuel-fired electric power plant is on the order of 175 km.[40] Site selection, therefore, will be an important factor in determining the relative contributions to regional air quality from future coal combustion and coal conversion facilities. In attempting to limit impacts of gas emissions from plants utilizing coal, Western Europe, Japan, and other densely populated industrial countries will have even less choice in plant site selection.

[45] U.S. Environmental Protection Agency, Sulfate briefing for regional administrators. Draft of Report presented February, 1975, New York. 76 pp.

Two counteracting factors will be operative in site selection for coal-fired facilities in the United States. Over much of the northeastern United States the addition of minor quantities of sulfur oxides to the atmosphere may cause the air quality of a region to be unacceptable under the significant risk provisions of the Energy Supply and Environmental Coordination Act of 1974.[46] This would suggest the desirability of development in the West. However the nonsignificant deterioration provisions of the National Clean Air Act are designed to prevent the deterioration of air quality in regions in which the concentrations of criteria pollutants (currently only sulfur dioxide and total suspended particulates) are below the limits specified in the National Ambient Air Quality Standards.[47] This could limit the amount of development in areas of the western United States where the density of sources and total emissions of sulfur dioxide are currently relatively low.[40]

In addition to the emission density, factors that strongly influence the transport and transformation of pollutants from steam elec-

[46] Public Law 93-319, Energy Supply and Environmental Coordination Act of 1974, U.S. Congress, 15 USC 791-798; 42 USC 1857b-1 *et seq.*, amended by Energy Policy and Conservation Act (1975), Public Law 94-163; 15 USC 792, 796.

[47] K. E. Yeager, *The Impact of Air Quality Regulations on the Electric Power Industry,* EPRI draft, Electric Power Research Institute, Palo Alto, Calif., 1978.

Table 31.7 Influence of Process Type on Projected Major Emissions from Different Coal Conversion Plants and Comparison to Fossil Fuel Combustion Units[39–41]

Process Type[a]	Total Particulates[b] (tons/year)	Sulfur Oxides (tons/year)	Nitrogen Oxides (tons/year)	Hydro-carbons (tons/year)	Carbon Monoxide (tons/year)	Alde-hydes (tons/year)	Control[c]	Fuel Consumed (10^6 tons/year)
Fischer-Tropsch[d]	2110	29,010	14,570	320	1080	5.4	C	8.6
Liquefaction (other than F-T)[d]	810	6620	3880	80	340	0.7	C	5.2
SRC[d]	570	3830	3340	65	310	0.3	C	5.4
High-Btu, fixed-bed gasification[d]	770	11,910	5340	120	395	2.0	C	5.3
Low-Btu gasification	600	9240	4140	92	310	1.5	C	4.1
Carbonization								
By-product coking				—[e]				1.8
Unloading	360	—	—	—		—	U	—
Charging	1350	18	27	2250	540	—	U	—
Coking cycle	90	—	9	1350	540	—	U	—
Discharging	540	—	—	180	63	—	U	—
Quenching	810	—	—	—	—	—	U	—
Underfiring	—	3600	—	—	—	—	U	—
Coal, direct combustion[f, g]	6000[h]	102,000	24,000	240	600	5.1[i]	U	2.4
Oil, direct combustion[g]	2300[h]	54,000	24,000	736	9	N.D.[j]	U	1.7
Natural gas, direct combustion[g]	465[h]	12	13,000	—	—	N.D.[j]	U	1.5

[a] Based on 328 operating days a year (90% stream factor).
[b] Eastern interior coal, Henderson County, Kentucky.
[c] C = controlled emissions; U = uncontrolled emissions (except as noted in *h*).
[d] Annual controlled air emissions from a unit plant (output equal to product required to support a 1000-MW electric power plant operating at 33% overall thermal efficiency).
[e] HC as methane.
[f] Coal type not specified (~2.25% S, 10% ash).
[g] 1000-MW capacity.
[h] About 97% control.
[i] Estimated.
[j] Not determined.

tric plants are regional meteorological conditions (predominantly wind flow patterns, frequency of air stagnation events, and air temperature and humidity) and the presence of chemically reactive pollutants such as ozone and other photochemical oxidants. Conditions that most strongly influence the interchange and diffusion of pollutants in the atmosphere include primary wind speed and direction, and the frequency, persistence, and variation in height of stable layers of air.[48] Avoidance of regions where air stagnation is a frequent occurrence would be helpful in the selection of future sites for combustion and conversion facilities.

31.3.4 Control Technology

Control technologies exist that can reduce emission of pollutants and effluents that enter the air and water streams. Tables 31.8 and 31.9 summarize the applicability and status of these pollution control technologies.

Emissions, effluents, or wastes from the various coal activities may be reduced by any one or by a combination of the following strategies:

1. Changing the processes so that smaller amounts of pollutants or less hazardous ones are created.

2. Containing the harmful materials both within the plant and within wastes.

3. Decreasing the amount of wastes by utilizing these materials as by-products.

4. Removing some or all of the harmful materials from the air, water, or by-product residual streams by means of emission control equipment.

Process changes such as reducing excess air during combustion, staged combustion, greater gas recycling, and lower peak flame temperature of the boiler are used successfully to reduce emissions of oxides of nitrogen.[50] Combustion modification techniques, such as fluidized-bed combustion, have been proposed as a means of reducing emissions of sulfur oxides. Operational and design modifications can also influence the rates at which phenols, tars, and tar oils are emitted (see "Process Types" in Section 31.3.2.3). More research on process changes useful for reducing emissions is needed (1) to improve the thermal efficiency of such energy systems as power plants and coal gasification or liquefaction plants; such improvements would decrease the emissions and thermal pollution per unit of useful products; (2) to improve techniques of minimizing water consumption and improving wastewater cleanup and enhancing its reuse; and (3) to increase the removal of trace elements and reduce the quantities of volatile trace elements carried from the reactor vessel into the cleanup systems during coal conversion.

31.3.4.1 *Control of Effluents Discharged to the Atmosphere*

The technology used to control air pollutants (SO_x, NO_x, and particulates) resulting from combustion is discussed in detail in Chapters 18 and 22. The Claus process, which converts some sulfur compounds to easily controllable elemental sulfur, can be used to control the sulfur compounds produced during gasification and liquefaction. It is a commercial, dry oxidation process and has removal efficiencies of between 90 and 95%* for H_2S, COS, CS_2, and organic sulfur compounds (R–SH).[51] The process also removes about 90% of the hydrocarbons in the gas stream.

The competing Stretford process (Chapter 18) is best used for gas streams where H_2S is

[48] "Atmospheric Areas of the United States," *Fed. Reg.*, **33** (10), 548 (1968).

[49] U.S. Department of Energy–Environmental Protection Agency, *Energy/Environmental Fact Book,* EPA 600/9-77-041, 1977.

[50] H. M. Braunstein, E. D. Copenhaver, and H. A. Pfuderer, Eds., *Environmental Health and Control Aspects of Coal Conversion: An Information Overview,* Vol. 2, Oak Ridge National Laboratory ORNL/EIS-94, Oak Ridge, Tenn., April 1977.

* The addition of a tail gas cleanup unit to a Claus plant increases recovery of sulfur to over 99%.

[51] U.S. Environmental Protection Agency, *Environmental Assessment Data Base for Low/Medium-Btu Gasification Technology,* Vol. 1, *Technical Discussion;* Vol. 2, Appendices A–F; Interagency Publ. No. EPA 600/7-77-125a,b, prepared by Radian Corporation, Austin, Tex., 1977.

Table 31.8 Control Technologies[49]

Extraction		Processing-Conversion		Generation	
				Power Plants	
				Conventional	
				Sulfur oxides →	scrubbers, fluidized-bed clean or cleaned coal[a]
Underground coal mining		*Physical coal cleaning*			
				Nitrogen oxides →	scrubbers, combustion modification, flue gas treatment
Acid drainage →	contain or neutralize	Air particles →	dust control		
Solid waste →	well-managed landfill	High-sulfur solid waste →	recover sulfur	Fly ash and smoke →	cyclones, baghouses, electrostatic precipitators, scrubbers
		Runoff solids →	contain		
Surface coal mining		Acid drainage →	contain or neutralize		
				Solid waste (ash) →	well-managed landfill
Runoff solids, Acid drainage, Sediments, Air particles, Solid waste →	control or treatment of runoff, well-managed mine, restoration and revegetation	*Gasification/ liquefaction*[a]			
		Runoff solids →	physical-chemical treatment	*Industry*	
		Organic wastes, Toxins, Carcinogens →	physical-chemical-biochemical treatment	Sulfur oxides →	fluidized-bed, clean or cleaned coal,[a] scrubbers
				Nitrogen oxides →	combustion modification
		Solid waste →	landfill	Fly ash and smoke →	electrostatic precipitators, baghouses, cyclones, scrubbers
		Waste heat →	cooling or reuse		
		Fugitive or accidental release of toxins, carcinogens →	design to eliminate	Solid waste (ash) →	landfill
				Commercial and residential	
				Sulfur oxides →	low-sulfur fuel
				Nitrogen oxides →	combustion modification
				Fly ash and air particles →	electrostatic precipitators
				Solid waste (ash) →	landfill

[a]Technologies not yet available.

Table 31.9 Environmental Problems and Control Technologies[49]

Pollutants and Control Technology	Pollutant Reduction Efficiency (%)	Time Frame of Applicability		
		Utility	Industrial	Residential and Commercial
For SO_1				
Flue gas desulfurization	80–95	Current (new and existing plants)	Current for large-scale installations only—about 1980	Not applicable
Physical coal cleaning	20–40	Current (limited availability)	Current (very limited availability)	Current (very limited availability)
Chemical coal cleaning	10–60	Post–1980	Post–1980	Post–1980 (limited applicability)
Use of low-sulfur coal	12–30	Current (limited availability)	Current (limited availability)	Current (limited availability and applicability)
Fluidized-bed combustion (with chemical sorbent)	80–90	Post–1980 (widely applicable)	Current (or very near-term)	Not fully evaluated
Coal gasification				
Low Btu	90–95	Post–1980 (more applicable to new units)	Applicable only to largest units	Not applicable
High Btu	90–95	Post–1980 (new and existing units; high costs)	Post–1980 (widely applicable; high costs)	Post–1980 (widely applicable; high costs)
Coal liquefaction	90–95	Post–1985 (new and existing units)	Post–1985 (probably applicable to larger units only)	Under evaluation
For NO_x				
Combustion modification	20–80	Current (applicable to certain types of units only)	Current (widely applicable for larger units)	Partially applicable (under evaluation)
Flue gas denitrification	60–95	Post–1985	Post–1985	Not applicable
For particulate matter				
Inertial devices (cyclones)	98	Current (widely applicable)	Current for larger installations	Not required
Electrostatic precipitators	>99	Current (widely applicable)	Current for larger installations	Not required
Wet scrubbers	80–98	Current	Current	Not required

the preponderant compound and only small amounts of organic sulfur compounds are to be removed. Several other gas cleanup processes convert sulfur compounds to a nonusable waste that requires disposal. Their H_2S removal efficiencies exceed 99%, and their carbonyl sulfide–carbon disulfide removal exceeds 98%.

Afterburners and adsorbers are used to control the hydrocarbon emissions produced in coal conversion plants. Afterburners burn the hydrocarbons, changing them into carbon dioxide and water, with conversion efficiencies of more than 99%.[51] If the hydrocarbon content of the gas stream is too low to support combustion, supplemental fuel can be added. Adsorbers using activated carbon or other sorbents that remove the hydrocarbons from the gas streams at efficiencies of more than 99.9%[51] can also be used. The hydrocarbons are then released by thermal regeneration of the sorbent and are either partially recovered by condensation or burned.

Although control technologies that can reduce carbon monoxide emissions exist, they are considered unnecessary for electric power and coal conversion facilities. Since carbon monoxide is emitted primarily from motor vehicles, rigid control of stationary sources appears to be unwarranted at this time.

Coarse particulates from loading, unloading, conveyor belt operation, beneficiation, coke-oven operations, and stockpiling generally can be controlled effectively (Chapters 9, 10, and 14).

Several major strategies may be employed to control sulfur emissions produced by the combustion of coal. Coal cleaning (physical separation by differences in density) reduces sulfur content by an average 40% for coals high in inorganic sulfur; however coals high in organic sulfur are not affected by physical separation methods. The use of low-sulfur, low-ash products from such processes as solvent refining may be a promising source of clean-burning fuels derived from coal. Flue gas desulfurization, by means of limestone scrubbing or other new scrubbing techniques currently being developed, will be required in the United States on all new coal-burning power plants in the near future.

In the case of oxides of nitrogen, the nitrogen content of the coal contributes a portion of the total emission of nitrogen oxides (the amount depends on the nitrogen content of the coal) from a coal-fired utility boiler; the remainder is contributed by thermal oxidation of atmospheric nitrogen during combustion. The moisture and nitrogen content of the coal, the burner design, the furnace design, and the operating conditions are major factors controlling the relative contributions of fuel nitrogen and thermal fixation.[31] Data on fuel combustion from stationary sources collected in 1972 account for 50 percent of the total emission of man-made nitrogen oxides.[52] In most countries a much larger share comes from auto exhausts. Even with large increases in coal utilization, it is unlikely that significant reductions in the emissions of nitrogen oxides from the stacks of stationary source coal utilization plants would have a major influence on the levels of nitrogen oxide in the atmosphere in urban areas. Power plants in rural areas could be an important contributor to nitrogen oxide ambient levels.

31.3.4.2 *Control of Water Pollutants*

Technologies exist that can substantially purify polluted liquid effluents. Acid or alkaline drainage from mines may be contained and neutralized prior to release into the environment. Coal-cleaning processes introduce suspended particles and other pollutants into the washwater; these pollutants can be largely removed by sedimentation prior to discharge. The other major sources of liquid effluents from coal facilities are the quenching water used to cool hot products and ashes, the condensates from various processes, and the blowdown water from boilers and cooling and scrubbing systems. The most troublesome wastewater streams produced during coal carbonization and conversion are those with high levels of organic and inorganic contaminants.

Many small-volume wastewater streams may be diluted in combustion ash sluice water, which is discarded after sedimentation in ash ponds. Even in properly operated

[52] Commission on Natural Resources, *Air Quality and Stationary Source Emission Control,* published for Committee on Public Works, Ser. No. 94-4, Government Printing Office, Washington, D.C., March 1975, p. 738.

ponds it is common to have difficulty in meeting EPA discharge criteria for suspended solids levels, particularly in plants that use slurries to remove fly ash to combined settling ponds. Recycling or skimming the slurry water to remove cenospheres (floating ash) may then be required. High concentrations of iron, copper, and other trace metals in boiler washwater effluent can be reduced by precipitation in ash ponds after neutralization or by evaporation.[53]

Currently used procedures for treating coking wastewater effluent fall into five general categories (Chapter 18): physical, physicochemical, chemical oxidation, chemical precipitation, and biological. A comprehensive review of available methods, including effective concentrations and costs, has appeared elsewhere.[54] Generally, treatment procedures have been designed to concentrate on the removal of phenols (and their chlorination products) because of the exceedingly low taste and odor thresholds maintained for these substances. Solvent extraction and biological oxidation are the most widely used treatment procedures because of their effectiveness and low cost.

Benzene, benzene-octanol mixtures, diisopropyl ether, tricresyl phosphate, and light oil have been used as solvents to remove phenols[55–58]; reductions of 97–99% in the phenol concentrations have been observed.[59] Removal efficiencies decrease rapidly as phenol concentrations decrease, although greater removal can be obtained by increasing the ratio of extractant to water,[57] or by increasing the number of stages.

Even after this treatment, the concentrations of phenol generally exceed EPA discharge criteria by several hundredfold. For this reason, further treatment of the wastewater is needed. The usual method is biological oxidation. In the 1970s activated sludge and trickling filter processes for treatment of coking wastewater have been perfected to achieve removal efficiencies of 99.9%.

U.S. EPA guidelines for the best practical control technology currently available require phenol removal of 99.97% for a wastewater flow of 150 gal per ton of coke (Table 31.10). The Bethlehem Steel Corporation's activated sludge facility at Bethle-

[53] T.- Y. J. Chu, R. J. Ruane, and G. R. Steiner, "Characteristics of Wastewater Discharges from Coal-Fired Power Plants," paper presented at 31st Ann. Purdue Industrial Waste Conf., West Lafayette, Ind., May 4–6, 1976.

[54] H. R. Eisenhauer, "Dephenolization of Water and Wastewater," *Water Pollut. Control,* **106,** 34, 38–41 (1968).

[55] A. S. Pierce and S. E. Punt, "Biological Treatment of Liquid Toxic Wastes, Part I," *Effluent Water Treat. J.,* **15**(32), 33, 35, 37–39 (1975).

[56] G. Gutzeit, E. J. Lyons, and D. C. MacLean, "Treatment of Phenolic Wastes," *Ind. Wastes,* **4**(4), 57–60, (1959).

[57] N. H. Kirchgessner, "Phenol Recovery by Use of Isopropyl Ether," *Sew. Ind. Wastes,* **30**(2), 191–198 (1958).

[58] H. A. Kjellman, "Operation of a Phenol Recovery Plant," *Sew. Ind. Wastes,* **27**(7), 854–859 (1955).

[59] R. G. Bond, C. P. Straub, and R. Prober, Eds., *Handbook of Environmental Control,* Vol. 4, *Wastewater: Treatment and Disposal,* CRC Press, Cleveland, Ohio, 1974.

[60] C. E. Adams, R. M. Stein, and W. W. Eckenfelder, Jr., "Treatment of Two Coke Plant Wastewaters to Meet Guideline Criteria," *Proc. 29th Industrial Waste Conf.,* Part 2, pp. 864–880, 1975.

Table 31.10 EPA Guidelines for By-Products in Coke Wastewater[a] and Estimated Efficiency Requirements for Treating Wastewater[60]

By-Product	Amount (lb/ton coke)	Concentration (mg/1)[b]	Efficiency (%)
Cyanide	0.0002	0.16	96
Phenol	0.0004	0.32	99.97
Ammonia	0.0084	6.7	92
BOD	0.0166	13.2	99
Sulfide	0.0002	0.16	—
Oil and grease	0.0084	6.7	74
Suspended solids	0.0084	6.7	97

[a] ph = 6.0–9.0

[b] Estimate based on wastewater flow of 150 gal/ton of coke produced.

hem, Pennsylvania, removes phenols with an efficiency of nearly 99.99%.[61] A bibliography of recent advances in biological and extractive dephenolization has appeared.[62].

Other physical, physicochemical, and chemical oxidation treatment procedures have been proposed (chlorination, ozonation, and activated carbon); but, in general, either their removal efficiencies are low or their costs are excessively high.

Although phenols, thiocyanates, and thiosulfates are readily oxidized by biological treatment, removal of other contaminants is far less complete. Fluorides may be precipitated as calcium fluoride by use of lime for pH adjustment in the ammonia still.[60] In pilot-plant studies removal of cyanide averaged about 78% for both activated-sludge and aerated-lagoon processes.[60] Activated carbon, chlorination, and ferric or ferrous sulfate precipitation have been proposed or used to reduce cyanide concentrations; costs ranged from $0.60 to $1.80 per million gallons (in 1968 dollars) for precipitation and from $9.50 to $12 for activated carbon and chlorination.

Ammonia concentrations increased in biological pilot-plant studies, apparently because of degradation of organic nitrogen compounds to ammonia.[60] Although ion exchange has been used for ammonia removal, constant fouling by iron cyanide complexes discouraged long-term use.[63] Adams et al.[60] concluded that ammonia "may be the most difficult parameter to control in the treatment of coke plant wastewater." Ammonia concentrations in wastewater streams of some coal gasification and liquefaction processes could be great enough to justify removal and sale of the ammonia.

Because conversion processes consume large quantities of water, most designs provide for complete reuse of wastewater streams after treatment. Of 10 proposed processes analyzed by EPA on a commercial-scale basis, only two included planned wastewater discharges[64,65]; the other eight proposed to evaporate the wastewater completely, leaving a solid waste residue for disposal. Gasification processes that were examined proposed to return treated wastewater either to the raw water supply[66,67] or directly to cooling[68,69] or scrubbing loops,[70] or to use wastewater for quenching and slurrying ashes prior to their return to the mine.[68,71] The liquefaction plants propose to recycle product-separator water to the reactor,[65,72] or, after treatment, to the cooling loop to offset evaporative losses.[69] However, concentrations of salts will rapidly accumulate in such closed systems, making it almost certainly necessary to remove the salts and to either evaporate or discharge a fraction of the recycling wastewater stream and replace it with fresh water. Moreover, injection of product-separator water into conversion reactors may not be practicable because of corrosion or incomplete pyrolysis of organic contaminants within the reactor.[72]

[61] P. D. Kostenbader and J. W. Flecksteiner, "Biological Oxidation of Coke Plant Weak Ammonia Liquor," *J. Water Pollut. Control Fed.*, **41**(2), 199–207 (1969).

[62] D. G. Brinn, *A Bibliography on the Pollution Aspects of Coke Ovens*, NTIS-SM/TN/1/25, 1973.

[63] W. G. Cousins and A. E. Mindler, "Tertiary Treatment of Weak Ammonia Liquor," *J. Water Pollut. Control Fed.*, **44**(4), 607–618 (1972).

[64] C. E. Jahnig, *Evaluation of Pollution Control in Fossil Fuel Conversion Processes. Gasification*, Section 6, *HYGAS Process*, EPA-650/2-74-009-h, 1975.

[65] C. E. Jahnig, *Evaluation of Pollution Control in Fossil Fuel Conversion Processes. Liquefaction*, Section 2, *SRC Process*, EPA-650/2-74-009-f, 1975.

[66] C. E. Jahnig and E. M. Magee, *Evaluation of Pollution Control in Fossil Fuel Conversion Processes. Gasification*, Section 1, *CO_2 Acceptor Process*, EPA-650/1-74-009-d, 1974.

[67] C. D. Kalfedelis and E. M. Magee, *Evaluation of Pollution Control in Fossil Fuel Conversion Processes. Gasification*, Section 1, *Synthane Process*, EPA-650/2-74-009-b, 1974.

[68] H. Shaw and E. M. Magee, *Evaluation of Pollution Control in Fossil Fuel Conversion Processes, Gasification*, Section 1, *Lurgi Process*, EPA-650/2-74-009-c, 1974.

[69] C. E. Jahnig, *Evaluation of Pollution Control in Fossil Fuel Conversion Processes. Liquefaction*, Section 3, *H-Coal Process*, EPA-650/1-74-009-m, 1975.

[70] C. E. Jahnig, *Evaluation of Pollution Control in Fossil Fuel Conversion Processes. Gasification*, Section 5, *Bi-Gas Process*. EPA-650/2-74-009-g, 1975.

[71] E. M. Magee, C. E. Jahnig, and H. Shaw, *Evaluation of Pollution Control in Fossil Fuel Conversion Processes. Gasification*, Section 1, *Koppers-Totzek Process*, EPA-650/2-74-009-a, 1974.

[72] C. D. Kalfadelis and E. M. Magee, *Evaluation of Pollution Control in Fossil Fuel Conversion Processes. Liquefaction*, Section 1, *COED Process*, EPA 650/2-74-009-e, 1975.

The discharges of treated water necessary to prevent buildup of solids for several commercial (1.0×10^4 tons of maf coal per day) gasification plants have been estimated at $6.0–6.7 \times 10^6$ l/day.[64,73] The estimated discharge of treated wastewater by one proposed liquefaction process is 5.8×10^6 l/day.[65] Unless effluent flows are reduced or eliminated by use of evaporation ponds or other means, commercial coal conversion plants probably will discharge significant amounts of aqueous effluents into receiving waters.

Some form of wastewater treatment to remove phenols will be mandatory, since a 20,000-ton/day coal liquefaction plant would otherwise discharge 5000 times the EPA limit for a coke plant of equal size, even if a large fraction of the product water flow is recycled. Moreover, in the United States, commercial coal conversion plants will be subject to federal NSPS, which require "best adequately demonstrated technology" for effluent cleanup. Physical and chemical treatment techniques for coal conversion facilities will probably have phenol removal efficiencies similar to those of coking wastewater treatment facilities. The most likely treatment procedure, therefore, is biological treatment preceded by either dilution or solvent extraction to reduce the concentration of phenols to concentrations nontoxic to the treatment microorganisms. Treated water will then be largely recycled, a fraction being discharged (either to evaporation ponds or directly to receiving waters) to prevent buildup of solids.

Although wastewater treatment may effectively eliminate phenols from coal conversion effluents, removal of other classes of compounds may be less complete (Table 31.11). Solvent extraction is not completely effective in removing aromatic bases,[75] possibly because of pH effects. Moreover, pyridine, aniline, quinoline, and their alkylated derivatives, which may be present in the aqueous streams of coal conversion plants, often pass virtually unchanged through treatment plants.[76] Monocyclic aromatic hydrocarbons are relatively easily degraded, but branched-chain alkyl substitution greatly increases their resistance to breakdown.[77]

Reported efficiencies for removal of polycyclic aromatic hydrocarbons vary from 30 to 96% during sedimentation and biological treatment. Generally removal efficiencies decrease as the concentration of these compounds decreases. This is also true for solvent extraction.[78] Virtually nothing is known of the efficiency of removal of heterocyclic sulfur compounds, but presumably they be-

[73] C. E. Jahnig, *Evaluation of Pollution Control in Fossil Fuel Conversion Processes. Gasification,* Section 7, *U-Gas Process,* EPA-650/2-74-009-i, 1975.

[74] S. E. Herbes, G. R. Southworth, and C. W. Gehrs, "Organic Contaminants in Aqueous Coal Conversion Effluents: Environmental Consequences and Research Priorities," in *Trace Substances in Environmental Health, A Symposium, University of Missouri at Columbia,* D. D. Hemphill, Ed., 1976, pp. 295–303.

[75] J. Horacek and K. Repiska, *Hutn. Listy,* **29,** 580–582 (1974) (in Czech); *C.A.,* **81,** 158307 (1975).

[76] L. S. Bark, R. L. Cooper, and K. C. Wheatstone, "The Determination of Organic Bases in Carbonization Effluents," *Water Res.,* **6,** 117–126 (1972).

[77] G. W. Malaney and R. E. McKinney, "Oxidative Abilities of Benzene-Acclimated Activated Sludge," *Water Serv. Wks,* **113,** 303–309 (1966).

[78] J. Borneff and H. Kunte, *Arch. Hyg. Ber.,* **151,** 202–210 (1967) (in German).

Table 31.11 Concentrations and Wastewater Treatment Efficiencies of Major Constituents of Coal Conversion Effluents

Constituent	Anticipated Effluent Concentration Range (mg/l)	Wastewater Treatment Removal Efficiency (%)	Expected Levels in Final Effluent (mg/l)
Phenols	10,000	99.9+	< 1
Aromatic amines	100–1000	30–50	50–700
Monoaromatic hydrocarbons	10–100	90+	1–10
Thiophenes	10	?	1–10
Polycyclic hydrocarbons	0.10–1	30–80	0.02–0.7

Source: Adapted from Ref. 74.

have similarly to polycyclic aromatic hydrocarbons, although oxidation to sulfones may occur under aerobic conditions.

If coal conversion effluents are treated in a biological treatment unit (possibly preceded by solvent extraction), some general predictions about the composition of the treated wastewater are possible. Phenols will be reduced to concentrations of a few milligrams per liter or less, and aromatic amines may constitute quantitatively the largest class of compounds. Monocyclic aromatic and aliphatic hydrocarbons may occur in concentrations of about 10 mg/l. Polycyclic aromatic hydrocarbons and sulfur compounds will probably occur at concentrations on the order of 0.1–1 mg/l. Coal conversion plant wastewater will contain organic contaminants at the concentrations indicated above. Wastewater flows will probably average 10–15%[73] of total sour water flow, unless the wastewater streams are not evaporated to dryness.

31.3.4.3 *Control, Containment, and Use of Wastes*

Fugitive emissions of hazardous materials during combustion and conversion of coal and processing of the fuels produced from coal can be important potential sources of environmental contamination. They may be minimized by good maintenance of pump seals, joints, and flanges and other plant equipment. Development of improved seals on pumps and on other mechanical devices, and releasing vent gases only after they have undergone treatment to remove hazardous materials, would also reduce emissions.

Many potential wastes from coal conversion processes may be used as by-products. These wastes include large quantities of sulfur, ammonia, phenol, naphtha, tars, and tar oils, which may be used as chemical or petrochemical feedstocks.[38]

Dispersion of trace elements from solid wastes and sludges can be minimized by improved management of the wastes. Sludges can be dewatered by evaporation in lined ponds that reduce potential for leaching into the soil. Solid wastes can be disposed of in coal mines where leaching can be prevented, or they can be buried in earth that is then revegetated.

A combination of strategies for controlling or eliminating pollutants is possible. For example, contaminated wastewater from coal conversion processes may be used to cool ash or slag removed from the process. The hot ash or slag acts as a source of heat to evaporate the water, solidifying the contaminants and making them considerably less mobile. Of course suitable containment and controls are necessary during the quenching processes to safeguard against emission of the hazardous pollutants.

Removing pollutants from one medium (e.g., air) may simply transfer them to other media (e.g., solid wastes and water), creating a different set of potential problems. For example, if sulfur is removed during coal conversion and disposed of with the ash as elemental sulfur, water percolating through the solids may be acidified, increasing its capability to mobilize pollutants present in the wastes. In contrast, wastes resulting from the removal of sulfur by limestone scrubbing (alkaline) would tend to immobilize the trace elements present in other solid residues with which they might be discarded.

31.4 EFFLUENT CHARACTERISTICS (AIR, WATER, SOLID)

31.4.1 Effluents Discharged to the Atmosphere

A wide variety of factors influence the flux of contaminants from the coal system into the atmosphere (Table 31.12). (See also ex-

[79] V. H. Sussman, "Process Emissions and Their Control: Mineral Product Industries," in *Air Pollution,* 3rd ed., Vol. 4, *Engineering Control of Air Pollution,* A. C. Stern, Ed., Academic Press, New York, 1977, pp. 705–735.

[80] F. E. Gartrell, "Process Emissions and Their Control: Power Generation," in *Air Pollution,* 3rd ed., Vol. 4, *Engineering Control of Air Pollution,* A. C. Stern, Ed., Academic Press, New York, 1977, Ch. 12.

[81] J. E. Sickles, II, W. C. Eaton, L. A. Ripperton, and R. S. Wright, *Literature Survey of Emissions Associated with Emerging Energy Technologies,* EPA 600/7-77-104, 1977.

[82] R. B. Engdahl and R. E. Barrett, "Process Emissions and Their Control: Fuels and Their Utilization," in *Air Pollution,* 3rd ed., Vol. 4, *Engineering Control of Air Pollution,* A. C. Stern, Ed., Academic Press, New York, 1977, pp. 380–425.

tensive discussion in Chapter 14 of discharges to the atmosphere and their control in high-temperature carbonization.) As an example of the magnitude of the flux, burning coal for generating electricity currently contributes approximately 60% of all sulfur oxide emissions[84] and approximately 25% of all nitrogen oxides emitted from all anthropogenic sources in the United States.

Particulate matter emitted from the combustion of coal is composed of mineral (ash from coal), some unburned combustible matter (smoke), and aerosols generated as transformation products (e.g., sulfates). Sulfur dioxide (with a small fraction of sulfur trioxide) and nitrogen oxides are formed from the sulfur and nitrogen compounds in the coal and, in the case of nitrogen oxides, from nitrogen in the air. If emissions are uncontrolled, particulate matter has the greatest emitted weight and nitrogen oxides have the smallest. The reverse is true with respect to controlled emissions, because the removal efficiencies of control equipment are highest for particulates and lowest for oxides of nitrogen.

Sulfur compounds are also present in the raw medium- and low-Btu gases produced by coal conversion. In addition to hydrogen sulfide, other gaseous sulfur compounds, mainly carbon oxysulfide, carbon disulfide, and mercaptans, are found in these gases.

Other important pollutants that may need to be controlled to some degree are trace elements and trace organic compounds, hydrocarbons, and carbon monoxide.

Trace elements and trace organic compounds are elements and compounds present in concentrations of 0.1% or less. They are associated with deteriorating quality of the environment and health. The main potential source of emissions of trace elements from electricity generation and coal conversion plants is the mineral matter in the coal. The potential emissions of trace organic compounds come mainly from the coal combustion and conversion reactions.

[83] B. A. Steiner, "Process Emissions and Their Control: Ferrous Metallurgical Operations," in *Air Pollution,* 3rd ed., Vol. 4, *Engineering Control of Air Pollution,* A. C. Stern, Ed., Academic Press, New York, 1977, pp. 89–930.

[84] U.S. Environmental Protection Agency, *Monitoring and Air Quality Trends Report—1974,* EPA-450/1-76-001, 1976.

Most of the trace elements are retained in the solid wastes. However some elements or their compounds can be volatilized, as indicated by sampling air emissions from power plants.[85] Small concentrations of polynuclear aromatic hydrocarbons have also been measured in flue gases from coal-fired boilers,[23] in vent gases, and in fugitive emissions at coal conversion plants.[51,85]

Little information is available on the potential emissions of trace pollutants from coal combustion or conversion processes. Some control of the emissions is possible through existing control technology, modifications in the design of processes, and changes in operating procedures.[51,85] More data are needed to evaluate carefully the control options and to improve them, if necessary.

Coal transport, storage, handling, preparation, and carbonization may produce significant emissions in localized areas, but they are generally not major contributors to regional-scale pollution. The emissions consist mainly of particulate matter, and the air pollutants come from burning refuse piles and from drying coal after it has been beneficiated. Coking operations also emit hydrocarbons and hydrogen sulfide.

Developing technologies to control air pollutants resulting from combustion and conversion is a major concern. Most of the coal that is used is burned directly, and large volumes of sulfur and nitrogen oxides and particulates along with smaller quantities of hydrocarbons, carbon monoxide, and trace elements are emitted in the flue gas.

Although no commercial coal to gas or oil conversion plants exist in the United States (and only a few exist in the rest of the world), projected emissions of nonmethane hydrocarbons, carbon monoxide, and oxides of nitrogen from coal conversion sources are somewhat smaller than the actual emissions from coal combustion units producing the same amount of useful energy for the consumer. Emissions of sulfur dioxide and particulates from gasification facilities may be an order of magnitude less than those from combustion units (Table 31.7).

[85] U.S. Energy Research and Development Administration, *Effects of Trace Contaminants from Coal Combustion,* ERDA-77-64, *Proceedings of Work-Shop,* R. I. Van Hook and W. D. Shults, Eds., Knoxville, Tenn., 1977.

Table 31.12 Summary of Atmospheric Emissions from the Coal Fuel Cycle

	Pollutants Emitted	Relative Quantities	Control Technology Available	Remarks	Ref.
		Transportation Step			
	Particles	Highly variable depending on size fractionation of material transported	Wet sprays		79
		Storage and Handling Step			
	Particulates	May be locally heavy, generally small areas affected (e.g., on site)	Wet sprays		79
	Volatile hydrocarbons		Enclosure with local exhaust ventilization		
		Preparation Step			
Cleaning	Fine particulates	Generally localized	Inertial collectors, wet scrubbers		79
Drying	Fine particulates, SO_2, NO_x, CO		Wet scrubbing		
Burning refuse	H_2S, SO_2	20–50% of raw coal	Prevent ignition or extinguish		
		Utilization Step			
Combustion	Particulates		Electrostatic precipitation, wet scrubbers tall stacks, low-sulfur coal, FGD, low excess air, low overfire air, staged firing		80
	SO_2	70% of U.S. total			
	NO_x	25% of U.S. total			

Carbonization	Volatilized hydrocarbons	Variable	Aspiration systems, water scrubbers, sequential charging		
	Coal fines, chars, tars, coke fines		Secondary collectors		
	H_2S, HCN		Sulfur recovery		
Conversion	H_2S, COS, SO_x, CS_2, NO_x, HF, HCl, HCN, CH_3SH, C_2H_5SH, CH_4, C_2–C_4+ paraffins, C_5+ aromatics, olefins, particulates, plus pollutants from first three steps, above	Unknown	Electrostatic precipitation, wet scrubbers, incineration, sulfur recovery	Presently not possible to make good qualitative or quantitative distinction between contributions of various conversion options	81
		Product Utilization Step			
Gasification	NO_x, SO_2, CO, HC	Unknown	Research needed for NO_x	NO_x a potential problem	82
Liquefaction	NO_x, SO_2, CO, HC				
Coke	CO, SO_x, HC		Electrostatic precipitation, wet scrubbers		83

31.4.2 Aqueous Effluents

The types and quantities of materials that might reach the aquatic environment are dependent on the source of coal and the method of transport of emissions. Eastern coals generally have higher levels of acid-forming substances (e.g., pyrites) than do Western coals (Chapter 7). Expanded use of coal slurry pipelines for transporting coal is currently under consideration. The major potential environmental problem is separating the coal from the water at the coal's destination.[86]

Runoff from coal piles (found in all parts of the coal system) results in polluted wastewater, but few data are available concerning the quantitative and qualitative characteristics of the leachate. (Nichols[87] found only nine installations in the United States for which such data existed.) Such data are necessary to design and to construct environmentally acceptable coal stockpiles.

On the basis of these limited data, it has been found that the characteristics of the leachate from coal piles are highly variable and depend on the properties of the coal, the rainfall, the topography, and the drainage of the site. Additionally, the volume, configuration, and age of the stockpile and the intensity of precipitation affect the characteristics of the leachate. Microbial and chemical activities that mobilize various elements and compounds in an aging stockpile, although poorly understood, also affect the composition of the leachate.

Iron is often the major contaminant of the leachate, giving rise to problems similar to acid mine drainage. Various trace elements and heavy metals (e.g., chromium, mercury, and magnesium) are also sometimes present in quantities sufficiently great to create a potential hazard.

Another source of aqueous effluents is runoff from refuse and ash disposal. It is estimated that more than 3×10^8 tons of refuse material has been accumulated in 3000–5000 refuse dumps in the United States.[88] This material usually contains heavy metals and pyrite, and leaching and runoff from exposed refuse sites have been major sources of acid mine drainage and related environmental problems.[88,89]

Leaching of ash and slag and the water used for handling ashes as slurries also creates water pollution problems. The slurries are ultimately pumped into settling ponds. The average flows through the ash-settling ponds of 10 coal-fired power plants ranged from 1.5×10^7 to 2.0×10^8 l/day.[53]

The chemical characteristics of the ash and the discharged ash slurry water vary widely. Generally, discharges are characterized by high pH (often above 11), although acidic discharges also occur. The mean total of dissolved solids in discharges from 10 plants ranged from 197 to 452 mg/l, consisting largely of sulfate and carbonate salts. Mean concentrations of suspended solids at nine of the plants ranged from 13 to 42 mg/l, although intermittent concentrations as high as 431 mg/l were reported, largely because of the presence of floating hollow-sphere ashes (cenospheres) not removed by settling.[53]

Several studies have been conducted to determine the partitioning of trace elements between vapor releases, fly ash, and bottom ash.[90–94] A composite of their results is shown in Table 31.13. Settling ponds remove fly ash with low efficiency; therefore the elements most likely to reach the aquatic

[86] W. S. Gray and P. F. Mason, "Slurry Pipelines: What the Coal Man Should Know in the Planning Stage," *Coal Age*, National Coal Association, Washington, D.C., 1975.

[87] C. R. Nichols, *Development Document for Effluent Limitations Guidelines and New Source Performance Standards for the Steam Electric Power Generating Point Source Category,* EPA, 1974.

[88] E. M. Wewerka, J. M. Williams, P. L. Wanke, and J. D. Olsen, *Environmental Contamination from Trace Elements in Coal Preparation Wastes: A Literature Review and Assessment,* EPA-600/7-76-007; ERDA LA-6600-MS, 1976, 61 pp.

[89] J. F. Martin, (1974) *Coal Refuse Disposal in the Eastern United States. News of Environmental Research in Cincinnati, EPA, Cincinnati, Ohio, 1974, 3 pp.*

[90] D. H. Klein, H. A. Andren, J. A. Carter, J. F. Emery, C. Feldman, W. Fulkerson, W. S. Lyon, J. C. Ogle, Y. Talmi, R. I. Van Hook, and N. Bolton, "Pathways of Thirty-Seven Trace Elements Through a Coal-Fired Plant," *Environ. Sci. Technol.,* **9,** 973–979 (1975).

[91] R. E. Lee, Jr., H. L. Crist, A. E. Riley, K. E. Macleod, "Concentration and Size of Trace Metal Emissions from a Power Plant, a Steel Plant, and a Cotton Gin," *Environ. Sci. Technol.*, **9,** 643–647 (1975).

environment are those that are concentrated in fly ash.

The amounts of trace elements mobilized by coal combustion, estimated on the basis of the data of van Hook and Shults[85] and Klein et al.,[90,96] were compared with estimates of mobilization from natural weathering.[97,98] The results (Table 31.14) show arsenic, cadmium, cobalt, copper, iron, lead, molybdenum, nickel, scandium, selenium, uranium, and zinc exhibiting mobilization rates that are 10% of weathering rates or higher.

Few data are available that qualitatively or quantitatively characterize the dissolved organic constituents in wastewater from coal combustion. Because of the high temperatures and completeness of oxidation in large boilers, there is little reason to expect high concentrations of organic matter in ash slurries. What data exist[99] show no increase in the total amount of organic carbon in river water used to slurry ash. The use of limestone slurries to remove sulfur dioxide may increase the quantities of organic matter in wastewaters because the low temperature (< 80°C) at which the slurries contact the flue gas facilitates their absorption on the limestone particles.[100]

During carbonization a highly contaminated wastewater, "weak ammonia liquor," is produced in the effluent of the coke gas scrubber. This wastewater is generated at a rate of approximately 90–100 liters per thousand kilograms of dry coal processed (Chapter 18).[60] Weak ammonia liquor may constitute less than 50% of the total wastewater, but it contributes the bulk of some of the highly objectionable contaminants such as phenols, ammonia, and cyanide (Table 31.15). Additional significant fluxes of contaminants occur in light-oil separator liquor (oil and grease) and final cooler liquor (ammonia, cyanide, and phenols). Process conditions may significantly alter effluent concentrations: raising the oven temperature from 950 to 1050°C has reduced by 60% the phenol content of the weak ammonia liquor.[101] Procedures used to remove ammonia also alter other contaminant concentrations. The more recently developed direct and semidirect processes, which use sulfuric acid treatment, not only remove ammonia

[92] R. L. Davison, D. F. S. Natusch, J. R. Wallace, and C. A. Evans, Jr., "Trace Elements in Fly Ash—Dependence of Concentration on Particle Size," *Environ. Sci. Technol.*, **8,** 1107–1113 (1974).

[93] S. Oglesby, Jr., *A Survey of Technical Information Related to Fine Particle Control,* EPRI 259, Electric Power Research Institute, Palo Alto, Calif., 1975.

[94] J. W. Kaakinen, R. M. Jorden, M. H. Lawasoni, and R. F. West, "Trace Element Behavior in Coal-Fired Power Plants," *Environ. Sci. Technol.*, **9,** 862–869 (1975).

[95] S. S. Ray and F. G. Parker, *Characterization of Ash from Coal-Fired Power Plants,* EPA-600/7-77-010, 1977.

[96] D. H. Klein, A. W. Andren, and N. E. Bolton, "Trace Element Discharge from Coal Combustion for Power Production," *Water Air Soil Pollut.*, **5,** 71–77 (1975).

[97] K. K. Turekian, in *Impingement of Man on the Oceans,* D. W. Hood, Ed., John Wiley and Sons, New York, 1971, 738 pp.

[98] E. D. Goldberg, W. S. Broecker, M. G. Gross, and K. K. Juekian, *Radioactivity in the Marine Environment,* National Academy of Sciences, Washington, D.C., 1971, 137 pp.

[99] S. E. Herbes, Environmental Sciences Division, Oak Ridge National Laboratory, Oak Ridge, Tenn., unpublished data.

[100] D. F. S. Natusch, Dept. of Chemistry, Colorado State University, private communication to S. E. Herbes.

[101] H. H. Black, G. N. McDermott, C. Henderson, W. A. Moore, and H. R. Pahren, "Industrial Wastes Guide—By-Product Coke Industry," *Sew. Ind. Wastes,* **29**(1), 53–75 (1957).

Table 31.13 Partition of Elements by Their Tendencies for Distribution in Coal Combustion Residues[95]

Group I: Elements Concentrated Approximately Equally in Bottom Ash and Fly Ash						
Al	Ce	Fe	La	Ro	Sm	Tn
Ba	Co	Hf	Mg	Sc	Sr	Ti
Ca	Eu	K	Mn	Si	Ta	
Group II: Elements Preferentially Concentrated in Fly Ash						
As	Ga	Sb				
Cd	Mo	S				
Cu	Po	Zn				
Group III: Elements Tending To Be Discharged to Atmosphere as Vapors						
Hg	Cl					
	Br					

Table 31.14 Annual U.S. Discharge of Trace Elements from Coal Combustion[85,96]

Element	Flows (10^3 tons/year) Discharged Slag and Fly Ash	Weathering Mobilization	Annual Flow (% of weathering mobilization)
Al	3640	64,000	5.7
As	1.72	8	21
Ba	22.7	480	4.7
Br	0.1	34	0.3
Ca	1510	34,000	4.4
Cd	0.16	0.6	27
Co	1	7.3	13.7
Cr	6.2	90	6.9
Cs	0.38	5.4	7
Cu	2.9	28	10
Fe	3780	35,000	10.8
Hg	0.005	0.12	4
K	536	16,000	3.4
Mg	420	11,000	3.8
Mn	11.7	770	1.5
Mo	3.1	2.7	115
Na	243	15,000	1.6
Ni	5.6	36	15.6
Pb	1.7	13	13
Rb	5.3	90	5.9
Sb	0.17	5	3.4
Sc	0.77	6.3	12.2
Se	0.66	0.5	132
Si	8020	30,500	2.6
Sr	8	370	2.2
Th	0.73	11	6.6
Ti	176	4,500	3.9
U	0.76	0.95	80
V	9.9	90	11
Zn	26.8	75	36

Table 31.15 Composition of Weak Ammonia Liquor (after steam stripping) and Total Wastewater at One Coking Plant[60]

Constituent	Concentration (mg/l)[a] Weak Ammonia Liquor, 2.7×10^5 l/day Flow, pH 10.9	Total Wastewater, 5.4×10^5 l/day Flow
Suspended solids	122	252
BOD	4565	1800
COD	6652	2663
Oil and grease	34	25.3
Phenolics	2800	1135
Total Kjeldahl nitrogen	179	256
NH_3, total	27	86
Cl^-	2375	800
F^-	93	37
CN^-	5.1	4.0
SO_3^{2-}	70[b]	—
SCN^-	0[b]	—
S^{2-}	30[b]	—

[a] Unless otherwise noted.
[b] Ref. 59.

from the liquor more efficiently, they also reduce cyanide and sulfide to far lower concentrations than the older indirect process.[55]

Major flows of contaminated wastewater from coal conversion plants are expected to be a "sour water," which consists of the water used to scrub the gas produced or product-separator liquor (in liquefaction processes). Estimates of sour water stream flows for a plant converting 20,000 tons of coal per day range from 7.3 to 31×10^6 l/day for gasification processes, and from 2.7 to 7.1×10^6 l/day for liquefaction processes (Table 31.16). Total concentrations of organic carbon are far greater in most samples of coal conversion wastewater than in combustion effluents. Although wastewater from high-temperature gasification processes such as Koppers-Totzek may contain only traces

Table 31.16 Estimated Quantities of Sour Water Produced in Commercial Coal Conversion Plants[a]

Process	Stream Source	Sour Water (10^6 l/day)	Ref.
Gasification			
Bi-Gas	Gas scrubber	9.5	66
CO_2 acceptor	Gas scrubber	7.3	62
Lurgi	Gas scrubber	18.	64
Koppers-Totzek	Gas scrubber	31.	67
Synthane	Gas scrubber	28.	63
U-Gas	Gas scrubber	8.3	69
Hygas	Gas scrubber	12.	60
Liquefaction			
COED	Gas condensate hydrotreatment	6.8	68
H-Coal	Gas condensate product separator	7.1	65
SRC	Product separator	2.7	61

[a] Based on 20,000-ton/day (maf) coal consumption.

of organic matter,[102] measurements of total organic carbon in effluents from several other bench- and pilot-scale liquefaction and low-temperature gasification processes ranged from 2600 to 11,600 mg/l.[103] The total amount of contaminated water from conversion plants will be much greater than that from coal combustion plants.

More than 80 major ultraviolet-absorbing compounds have been qualitatively separated from a sample of liquefaction product water. The aqueous streams from both liquefaction[104] and gasification reflect the product oil, gas composition, and water solubilities of the various constituents. Thus phenols, which make up only a small percentage of the product streams of several liquefaction processes, are more water soluble than most organic compounds and phenols partition into the product water. Therefore the largest fraction of dissolved organic matter in most sour water streams from coal conversion processes is phenolic; up to 6000 mg/l of total phenols has been detected in scrubber liquor from low-temperature gasification processes.[104] (High-temperature entrained processes are an exception; scrubber water in those processes contains only traces of phenols.) Monoaromatic phenols (phenol, cresols, and xylenols) are major constituents of both low-temperature, gas scrubber liquor and bench-scale liquefaction product water, totaling 5170 and 6000 mg/l, respectively, in several sample analyses. Lesser amounts of dihydric phenols (catechol, resorcinol, and hydroquinone) and naphthols have also been detected.[104] Three- and four-ring phenols are present in some synthetic crudes,[105] and trace amounts may occur in aqueous effluents.

Aromatic amines and azaarenes (e.g., quinoline and carbazole) are major contaminants in some samples of wastewater from coal conversion processes, although their concentrations are about an order of magnitude below those of phenols. Between 15 and 20 individual compounds totaling between 150 and 300 mg/l have been detected in product water samples.[106]

Monoaromatic amines (including aniline, pyridine, and methylated derivatives), the less water-soluble, two-ring bases, and traces of three-ring and larger azaarenes may be present in low concentrations in wastewater.

Monoaromatic hydrocarbons (benzene, toluene, and xylene) will probably be present at concentrations of a few milligrams per liter in wastewater. Aliphatic hydrocarbons have been detected in coal-conversion wastewater at concentrations ranging from 2.3 mg/l for *n*-octane to 1.02 mg/l for *n*-heptadecane.[107]

As many as 24 two-, three-, and four-ring compounds have been identified in product water at concentrations of from 5 mg/l for naphthalene to 0.05–0.6 mg/l for the larger compounds.[107] Benzo(*a*)pyrene was measured at 0.009 mg/l in another sample.[104] To date, analyses suggest that the larger (three–five ring) compounds are present at roughly saturation levels.

[102] F. Glaser, A. Hershaft, and R. Shaw, *Emissions from Processes Producing Clean Fuels,* Booz-Allen Applied Research, Bethesda, Md., BA-9075-015, 1974.

[103] Pittsburgh Energy Research Center Quarterly Report, PERC/ATR-75/1, 1975.

[104] W. D. Shults, Ed., *Chemical and Biological Examination of Coal-Derived Materials,* Oak Ridge National Laboratory ONRL/NSF/EATC-18, Oak Ridge, Tenn., 1976.

[105] A. J. Forney, W. P. Haynes, S. J. Gasior, G. E. Johnson, and J. P. Strakey, Jr., "Analysis of Tars, Chars, Gases, and Water found in Effluents from the Synthane Process," *Symp. Proc.: Environmental Aspects of Fuel Conversion Technology* St. Louis, Mo., May 1974, EPA-650/2-74-118.

[106] T. Ho and S. E. Herbes, Environmental Sciences Division, Oak Ridge National Laboratory, Oak Ridge, Tenn., unpublished data.

[107] M. R. Petersen, J. S. Fruchter, and J. C. Laul, *Characterization of Substances in Products, Effluents, and Wastes from Synthetic Fuel Production Tests,* Battelle Northwest Laboratories Quarterly Rep. No. 2131, 1976.

Sulfur-containing polycyclic aromatic compounds have also been detected; 15 individual compounds were present in one sample of product water.[104] Four three-ring aromatic sulfur compounds were present at concentrations up to 1.1 mg/l in another wastewater sample.[107] Low molecular weight organic acids have been detected in wastewater from the EDS process.[108]

Aqueous wastes are produced intermittently in boiler cleaning operations and during boiler blowdown, but these wastes are created no matter which fuel is used for generating steam.

31.4.3 Solid Wastes and Residues

Large quantities of solid and semisolid materials result from the preparation and consumption of coal and ultimately require disposal. Ash content of coal ranges from 2 to 30% by weight,[109] and future coal needs will probably be met by using coals having higher ash contents. Currently the great majority of this residue is disposed of in spoil banks or is returned to mines; in the United States only about 14–16% of this residue is used productively,[110] and this is generally typical of the experience in other countries (Chapter 30).

The major solid wastes produced during coal beneficiation and utilization for generating electricity are ash from coal combustion, residues from coal and wastewater treatment, particulates from emission control and product cleanup processes, spent catalysts (if not regenerated), and filter media.

Some of these wastes may be sold as byproducts after cleaning. The volume of sludges can be reduced by drying in incinerators, by pyrolysis, or by biological oxidation in an oxygen-deficient atmosphere. (For further details, see Ref. 48.) Sludges and solid wastes can also undergo chemical fixation, which consists of reacting chemicals with them and pumping the resulting mixture to a landfill, where solidification occurs in several days or weeks. The resultant solids can be used as landfill, sold (e.g., as a soil conditioner), or stored at the production site.

Changes in federal requirements[2] are affecting both the quantities and composition of materials requiring disposal (e.g., residue from limestone scrubbers must be disposed of), and both coal gasification and coal liquefaction processes will create residues containing varying quantities of organic matter.

Recent U.S. legislation[111] requires industries to determine the potential hazards of solid materials before they are disposed of in landfills. It is estimated that materials judged to be hazardous will cost approximately an order of magnitude more to dispose of than the same quantity of nonhazardous materials.[110]

If all the trace elements in coal ashes were assumed to be mobilized into the water, most, if not all, ashes would be classified as hazardous, requiring expensive disposal techniques. Fortunately, not all materials present in residues and ash are leachable, and even those that are mobilized leach at different rates and to different extents.

The principal solid residues produced by conventional combustion facilities are fly ash and bottom ash. Fly ash makes up 10–85% of the total ash, depending on the type of facility,[112] and ranges from 0.5 to 200 μm in diameter. Bottom ash is a coarser, larger, heavier material. Several investigations have been conducted to evaluate the partitioning of trace elements between bottom ash and fly ash (Table 31.13).

Although the major potential environmental impact of residues and ash associated with coal preparation results from the leachability of trace contaminants and their infiltration into groundwater and surface water, some concern centers on resuspension of collected fly ash in the atmosphere during disposal. The latter interest arises not only

[108] *Coal Liquefaction Process Development, Phase III*[A], Fuel Technical Prog. Rep. for Period January 1, 1976–June 30, 1977, U.S. ERDA Rep. No. 2853-20.

[109] *Mineral Matter and Trace Elements in U.S. Coal,* U.S. Dept. Interior OCR R & D Rep. No. 61, Pittsburgh, 1972.

[110] H. M. Braunstein, Ed., *Environmental and Health Aspects of Disposal of Solid Wastes from Coal Conversion:* An Information Assessment, Oak Ridge National Laboratory, ORNL-5361, Oak Ridge, Tenn., 1978.

[111] Resource Conservation and Recovery Act of 1976, PL 94-580, 1976.

[112] *Steam-Electric Plant Air and Water Quality Control Data,* Federal Power Commission Publ. No. FPC-S-229, 1973.

because some fly ash contains smaller, potentially respirable particles, but also because some trace elements are more highly concentrated on these particles.[113]

Also, if sulfur is produced that is not sold, it will require disposal. Local soil acidification has resulted from windblown sulfur deposited on farmland. It has been reported[114] that the leachate resulting from acidified farmland could mobilize trace elements present, thereby increasing the potential hazard of these materials.

Few experimental data describing solid wastes produced by coal conversion facilities are available. Conceptual designs show ash, slag, filter cake, and sulfur as materials ultimately requiring disposal. The major differences between ash from conversion processes and from combustion are expected to be the presence of organic matter in residues from coal conversion and the existence of the elements in a reduced rather than oxidized form. The method of treating these residues before disposal will also be different. If solid materials from conversion processes were recycled to retrieve heating value before disposal, the organic matter in the solid residues would be reduced and the element in the ash oxidized.

Tamura[115] used ignition weight loss of residues from the Lurgi facility in Westfield, Scotland, as an initial estimate of organic content. Weight losses ranged from 0.8 to 10% for ashes from three types of U.S. coals. Although atmospheric moisture may have infiltrated the samples, resulting in a biased determination, it is possible that organic matter was present in the residue, perhaps because the gasifer was not operated in a steady state condition.

No data have been published on the chemical form of components in any ash from coal conversion processes,[110] and an understanding of how these contaminants modify the environment is also lacking. Collecting data necessary to arrive at such an understanding is critical. Such data will not only allow design and implementation of successful landfills, but they will aid in decreasing the ill effects on the environment of solid wastes resulting from coal conversion processes. Until then these wastes should be disposed of as if they could release harmful materials into the environment.

[113] D. F. S. Natusch, J. R. Wallace, and C. A. Evans, Jr., "Toxic Trace Elements: Preferential Concentration in Respirable Particles," *Science*, **183**, 202–204 (1977).

[114] J. B. Mudd and T. T. Kozlowski, Eds., *Responses of Plants to Air Pollution*, Academic Press, New York, 1975, 383 pp.

[115] T. Tamura, *Solid Waste Characterization*, Coal Technology Progress Quarterly Prog. Rep. for period ending June 30, 1977, ORNL-5327, Oak Ridge National Laboratory, Oak Ridge, Tenn.

31.5 IMPACTS ON THE ENVIRONMENT

31.5.1 General

Coal utilization affects terrestrial, aquatic, and atmospheric environments. Impacts on the terrestrial environment include disruption of land, aquifers, and natural drainage contours, soil erosion, flooding, subsidence, and contamination of land surfaces. Impacts on the aquatic environment include disruption of surface and underground aquatic systems, and thermal, particulate, and chemical contamination of waters. Impacts on the atmospheric environment include deterioration of air quality by particulate and gaseous emissions, and modification of the climate. The various impacts on air, land, and water can result in damage to plant and animal species and to their habitats, as well as modification or elimination of ecosystems, damage to property or food resources, and modification of the aesthetic environment.

Acid rain has caused significant perturbations in aquatic ecosystems. Experiments have indicated that acid rain may affect terrestrial ecosystems as well. Little is known, however, about the potential effects of the various hydrocarbons that might be released into the environment in connection with coal conversion or about their ultimate fate in the environment.

This section describes the types of potential terrestrial and aquatic effects associated with increased use of coal.

31.5.2 Global

Potentially the increased use of coal (and other fossil fuels) may modify the global climate by increasing the concentration of carbon dioxide in the atmosphere. Since carbon dioxide concentrations have been increasing over the past 100 years, a recent report by the National Academy of Sciences[116] concluded that "the climatic effects of carbon dioxide release may be the primary limiting factor on energy production from fossil fuels over the next few centuries."

There is no consensus about the long-term effects of carbon dioxide on the weather, and climatologists even disagree about whether a significant buildup of carbon dioxide in the atmosphere will result in global warming or cooling (although most experts expect a warming to occur as CO_2 levels increase), the extent of climatic changes, and their distribution among geographical latitudes. There is also controversy about how climatic modifications might alter global food production and about the possibility of coastal areas being flooded if a warming trend melts glacial ice.

Large increases in the amount of carbon dioxide in the atmosphere could both help and harm agricultural systems. Among the factors that may be speculated on now are (1) the effects of increased carbon dioxide on plant metabolism, (2) higher annual average temperatures, (3) spatial shifts in agroclimatic regions, especially through changes in precipitation patterns, (4) the possibilities of greater or lesser climatic variability from year to year in different regions, and (5) the effects of possible increased cloudiness on the growth of crops.[117]

The many topics in this area that require additional research are discussed in Section 31.6.2. Critical among these are the partitioning of carbon dioxide among the biosphere, oceans, and atmosphere; the sensitivity of climate to disturbances in the radiation balance; world population growth and related energy demands; and possible uses of agricultural and water resources to mitigate the effects of climatic change.

[116] National Research Council, "Overview and Recommendations," in *Energy and Climate. A Report by the Geophysics Study Committee, Assembly of Mathematical and Physical Sciences,* National Academy of Sciences, Washington, D.C., 1977, pp. 1–31.

[117] C. F. Baes, Jr., H. E. Goeller, J. S. Olson, and R. M. Rotty, "Carbon Dioxide and Climate: The Uncontrolled Experiment," *Am. Sci.,* **65,** 310–320 (1977).

31.5.3 Regional

The major combustion effluents currently recognized as affecting terrestrial ecosystems are sulfur dioxide, oxides of nitrogen, particulate matter, and ozone produced as a secondary pollutant through the photochemical reaction between oxides of nitrogen and hydrocarbons. Reviews of the literature about the responses of plants to air pollution have been compiled by Mudd and Kozlowski,[114] Ziegler,[118] Heck,[119] Heagle,[120] Treshow,[121] and others.

Gaseous mixtures of sulfur oxides, ozone, and oxides of nitrogen may affect both small local areas and large regional ones. A considerable amount of data[122] indicates that very low concentrations (0.05–0.15 ppm) of the pollutants may act synergistically on vegetation. The high density of electric power plants in the eastern United States and the emissions from urban and industrial sources of pollution make the widespread occurrence of phytotoxic combinations of pollutants particularly likely.[40]

The most significant potential harm to a region would be caused by (1) dry and wet deposition of gaseous and particulate pollutants created by the combustion of fossil fuels and (2) thermal pollution coming from energy parks.

The major effects of dry and wet deposition of some pollutants on the environment are changes in the nutrient content of soils and aquatic systems, changes in the populations and activities of microorganisms, and changes in yields of various plant and animal species due to a combination of direct and indirect modifications of environmental conditions.

[118] I. Ziegler, "The Effect of Air-Polluting Gases on Plant Metabolism," *Environ. Qual. Saf.,* **2,** 182–208 (1973).

[119] W. W. Heck, "Factors Influencing Expression of Oxidant Damage to Plants," *Annu. Rev. Phytopathol,* **6,** 165–188 (1968).

[120] A. S. Heagle, "Interactions Between Air Pollutants and Plant Parasites," *Annu. Rev. Phytopathol.,* **11,** 365–388 (1973).

[121] M. Treshow, "The Impact of Air Pollutants on Plant Populations," *Phytopathology,* **58,** 1108–1113 (1968).

[122] R. A. Reinert, A. S. Heagle, and W. W. Heck, "Plant Responses to Pollutant Combinations," in *Responses of Plants to Air Pollution,* J. B. Mudd and T. T. Kozlowski, Eds., Academic Press, New York, 1975, 383 pp.

Plants are often more susceptible than animals to gaseous air pollutants. Some species of fish (e.g., trout and salmon) are also quite sensitive. Animals in general, however, are about an order of magnitude less sensitive than plants to acute exposures of sulfur dioxide, ozone, and nitrogen oxide.

Other major alterations of the environment that affect the welfare of man are decreases in agricultural and forest yields, decreases in freshwater fish production, corrosion or deterioration of materials such as steel and paints, and decrease of visibility.

The effects of burning fossil fuel on the environment of a region are derived from many sources within an area of thousands of square miles because the pollutants may be transported for hundreds of miles. Sulfur oxides, particulates, or sulfates are often used as indicators of the potential harm resulting from the various pollutants. In the discussion that follows, sulfur dioxide is used as the indicator.

Ramsay,[123] using U.S. data, attempted to estimate some of the economic effects of dry and wet deposition of pollutants. The costs included damage to paints and metals (\$0.044–\$0.088/lb of SO_2) and agricultural damage (\$0.0059–\$0.012/lb of SO_2). Table 31.17, based on these estimates of damage, summarizes property and agricultural damage caused by unit coal plants (see Section 31.3.1).

[123] W. Ramsay, *Unpaid Costs of Electrical Energy: Health and Environmental Impacts from Coal and Nuclear Power,* preliminary draft, Resources for the Future, Washington, D.C., 1978.

Concentrating electric generating plants in energy parks would have some economic and social advantages. However an increase in the size of generating capacity at one site produces more than a simple arithmetic increase in the amount of disruption to the environment. Whereas one energy facility may cause barely noticeable harm, a number of plants could cause major damage to the ecological balance of a locality or region, particularly with respect to thermal pollution. The great quantities of heat that would be released by a large energy park could seriously damage a river system or could cause major atmospheric perturbations.

One of the most extensively studied regional effects of atmospheric contamination is acid rain. The literature about acid rainfall in the eastern United States has been reviewed recently by Likens.[124] The acidity of precipitation in this region has increased greatly during the past two or three decades. Annual average acidity has increased from a normal pH of 5.6 (rainfall in equilibrium with ambient carbon dioxide is weakly acidic) to pH 4.5 or below (hydrogen ion concentration approximately 13 times as high) over sub-

[124] G. E. Likens, "Acid Precipitation," *Chem. Eng. News,* **54**(48), 29–44 (November 26, 1976).

Table 31.17 Damage from Sulfur Dioxide[a] from Unit Plants[16,b]

		Annual Cost of Damage ($\$10^6$)			
Stage of Fuel Cycle	Type of Damage	Coal-Electric	Low-Btu Gas-Coal Conversion, Combined Cycles	High-Btu Coal Gasification	Coal Liquefaction
Generation or conversion	Property	7.8–16	0.74–1.5	0.54–1.1	0.38–0.77
	Crops	1.1–2.1	0.10–0.20	0.07–0.15	0.05–0.10
Utilization	Property	0	0	0	1.0–2.1
	Crops	0	0	0	0.14–0.28
Total fuel cycle	Property	7.8–16	0.74–1.5	0.54–1.1	1.4–2.9
	Crops	1.1–2.1	0.10–0.20	0.07–0.15	0.19–0.38

[a] Based on data of emissions (Table 31.7) and estimates of damage[123]:

$$\text{property damage} = \$88{,}000\text{–}\$176{,}000/1000 \text{ tons } SO_2$$
$$\text{crop damage} = \$12{,}000\text{–}\$24{,}000/1000 \text{ tons } SO_2$$

[b] A unit plant is defined as the useful energy flow from a representative commercial high-Btu coal gasification plant producing 66.5×10^{12} Btu/year.

stantial areas of the East. Precipitation data reviewed by Cogbill and Likens[125] indicated that the area affected by this high acidity had increased from approximately 2000 square miles in 1955 and 1956 to approximately 5500 square miles in 1972. About 1500 square miles showed pH levels of 4.3 or lower (hydrogen ion concentration 20 or more times normal). Individual storms having acidity approximately 100 times normal (pH 3.6 or lower) are now being reported with increasing frequency at widely separated sampling stations in the eastern United States.[125]

Two studies, one in Europe[126] and one in the northeastern United States,[127] reported decreased productivity of forest systems during the past few decades; however a clear cause-and-effect relationship with acid rain could not be established. Analysis of the significance of acid rain for forest and agricultural systems is still only speculative.

Likens and Borman[128] have reviewed the known effects of acid rain on plant-soil systems. Controlled laboratory and field studies have demonstrated decreased productivity, necrosis of leaves, nutrient loss from foliage and soil, weathering of leaf surfaces, and altered pathogenicity of plant parasites. Acid rain has a high potential for producing environmental stress through additive or synergistic action in combination with other factors, such as gaseous pollutants.

In almost all the laboratory experiments, rather high levels of acidity have been used (pH 3.2 and below). However recent preliminary data[129] have shown that yields of kidney beans are not reduced by episodic application of rainfall of pH 4.0 unless the plants are costressed by periodic exposure to ozone. Evans and Raynor[130] reported that the reproductive processes of ferns may be affected by pH levels as high as 5.2. Jacobson and Van Leuken[131] found necrosis of the foliage of certain herbaceous species with single applications of rainfall ranging from a 1-min duration at pH 2.6 up to 9 hr at a pH of 3.4.

An analysis of sulfur movement through a forested watershed in east Tennessee[132] indicated that approximately 35% of the deposited sulfur was being accumulated in the soil and vegetation of that deciduous forest. It is not known whether the effects of continued or accelerated additions of sulfur to this forest system will result in acidification of soil, disruption of normal mineral cycles, or loss of productivity.

Although acid rains cannot be attributed solely to coal combustion, the large increase in coal utilization expected in the United States during the next several decades would result in further decreases in the pH of rainfall if strict control technology is not required.

Evidence is mounting that we are already nearing amounts of acid input through precipitation that may eventually cause diminished productivity of forest and agricultural ecosystems. Jacobson and Van Leuken[131] concluded that "current levels of acidity present little margin of safety for foliar injury to susceptible plant species and, with increasing emissions of pollutants which contribute to the free hydrogen ion concentration of rain, there is substantial risk of surpassing the threshold for foliar effects in the future." The provisions of the Clean Air Act Amendments of 1977, when fully implemented, should help slow the rate of spread of the acid rain problem.

[125] C. V. Cogbill and G. E. Likens, "Acid Precipitation in the Northeastern United States," *Water Resour. Res.*, **10**, 1133 (1974).

[126] B. Bolin, Ed., "Air Pollution Across National Boundaries: The Impact on the Environment of Sulfur in Air and Precipitation, in *Report of the Swedish Preparatory Committee for the U.N. Conference on the Human Environment,* Kungl. Boktryckeriet P.A. Norstedt et Soner, Stockholm, 1971, 96 pp.

[127] R. H. Whittaker, F. H. Bormann, G. E. Likens, and T. G. Siccama, "The Hubbard Brook Ecosystem Study: Forest Biomass and Production," *Ecol. Monogr.*, **44,** 233–252 (1974).

[128] G. E. Likens and F. H. Bormann, "Acid Rain: A Serious Regional Environmental Problem," *Science,* **184** 1176–1179 (1974).

[129] D. S. Shriner, "Interactions Between Acidic Precipitation and SO_2 or O_3: Effects on Plant Response," *Phytopathol. News,* **12,** 153 (1978).

[130] L. S. Evans and G. S. Raynor, Acid Rain Research Program Annual Progress Report, Brookhaven National Laboratory, BNL 50575, September 1976, 67 pp.

[131] J. S. Jacobson and P. Van Leuken, "Effects of Acidic Precipitation on Vegetation," *Proc. 4th Int. Clean Air Congr.*, Toyko, May 16–20, 1977, pp. 124–127.

[132] D. S. Shriner and G. S. Henderson, "Sulfur Distribution and Cycling in a Deciduous Forest Watershed," *J. Environ. Qual.* (in press).

31.5.4 Local

31.5.4.1 *Terrestrial Ecosystems*

Terrestrial ecosystems will be affected by virtually all phases of the coal fuel system. Unless effectively controlled, however, gaseous emissions, primarily sulfur and nitrogen oxides from combustion and conversion of coal, will represent the most important sources of harm to these terrestrial systems, both quantitatively and qualitatively. Since plants tend to be very sensitive to air pollution, this section is primarily concerned with phytotoxicity.

Erosion of overburden exposed by surface mining and not properly reclaimed can lead to decreases in plant yields, destruction of wildlife habitats, and disruption of current land use. Erosion can also increase sediment loading in streams and can affect groundwater. Erosion damage is prevented by controlling runoff through such practices as mulching and minimizing slopes. The 1977 Federal Surface Mining Act requires land reclamation for both surface mining and the surface effects of underground mining. If the act is properly enforced, it should prevent most, if not all, future damage, but its provisions will be difficult to administer properly in water-short areas.

Coal mining also causes subsidence, or a settling of the ground surface after coal extraction. Effects of subsidence can include destruction of highways, bridges, and water and gas lines in developed areas, and modification of drainage patterns in undeveloped regions. The latter may result in creation or depletion of swamplands.

Various methods are available to minimize hazards associated with potential future subsidence. In areas where structures are already developed on the surface, mining can be conducted to prevent future subsidence. Currently undeveloped areas can be required to remain undeveloped until subsidence and stabilization have been achieved.

Coal storage, transportation, and preparation produce dust; however the dust is generally confined to relatively small areas surrounding a plant and railroad rights of way, and it should not significantly degrade land areas off site. Localized changes may occur in soil pH and in concentrations of potentially toxic trace elements; the discussion of acid mine drainage (Section 31.5.4.2) can be reviewed for further information on this subject.

Emissions from coal carbonization include hydrocarbon oils, which have a variety of effects on plants. Many hydrocarbon molecules appear to damage cell membranes by "dissolving" the plasma membrane, opening it up by displacing lipid molecules, increasing membrane permeability, and permitting the movement of the cellular contents into intercellular[133] spaces, resulting in loss of cell turgor.

Currier and People[134] exposed barley and carrots to hydrocarbon vapors and found that the toxicity of the vapors increases in the order: hexane hexene cyclohexane cyclohexene benzene. Currier[135] concluded that increasing the number of methyl groups promotes penetration. Toxicity appeared to be inversely correlated with water solubility. Polycyclic aromatic compounds may cause chronic rather than acute injury to plants because they penetrate more slowly into the plasma membrane.[136]

Kozyukina and Obraztsova,[137] in investigating the effects of coke-oven gases (including volatile sulfur compounds and phenols) on the amino acid dynamics of vegetation, concluded that when changes are induced in the leaves of trees and shrubs exposed to coke-oven gases, the order of nitrogen exchange changes. The content of free and bound amino acids (proline, serine, glycine, threonine, and valine) may be associated with the resistance mechanisms of the plants, possibly resulting in increased protein synthesis under stress conditions.

[133] J. M. Baker, "The Effects of Oils on Plants," *Environ. Pollut.*, **1**, 27–44 (1970).

[134] H. B. Currier and S. A. Peoples, "Phytotoxicity of Hydrocarbons," *Hilgardia*, **23**, 15–173 (1954).

[135] H. B. Currier, "Herbicidal Properties of Benzene and Certain Methyl Derivatives," *Hilgardia*, **20**, 383–406 (1951).

[136] J. Van Overbeek and R. Blondeau, "Mode of Action of Phytotoxic Oils," *Weeds*, **3**, 55–65 (1954).

[137] Zh. T. Kozyukina and V. I. Obraztsova, "Dynamics of Free and Bound Amino Acids in Leaves of Trees and Shrubs Under Conditions of Coke Oven Gas Production," *Ukr. Bot. Zh.*, **30**(3), 332–339 (1973); transl. ORNL-tr-2948, Office of Language Services, Oak Ridge National Laboratory, Oak Ridge, Tenn.

The toxicity to plants of some compounds present in coal-tar vapors, such as anthracene and acridine, has long been recognized.[138,139] Halbwachs and Hlawatsch[140] were able to show that photolabile compounds in coal-tar vapors are oxidized under the influence of short-wave solar radiation and become phytotoxic only in this form. Kronberger[138] concluded that the differences in phytotoxicity between vapors derived from soft coal pitch and asphaltic bitumen were due to their relative concentrations of polynuclear aromatic hydrocarbons, the primary toxic agents. No correlation was found between concentrations of 3,4-benzopyrene in carrots and lettuce and concentrations in soil when the plants were grown in soil contaminated with very low (ppm) concentrations of the compound.[141]

A few reports suggest that endogenous synthesis of benzo(*a*)pyrene occurs, but there is no agreement on this.[142–147] In carefully controlled studies designed to exclude atmospheric contamination of plant materials,[148] no proof was found for the biosynthesis of polycyclic hydrocarbons in lettuce, soybean, rye, or tobacco plants.

The quantity of solid wastes produced by coal combustion plants is expected to increase because of the flue gas desulfurization sludges generated by processes used to prevent sulfur oxide emissions. Potential local adverse consequences of disposing of these wastes are the diversion of land from other uses and the contamination of ground and surface waters.

Some trace elements found in coal are toxic, even at fairly low concentrations, to both terrestrial and aquatic animals and to plant species. Other trace elements are essential to plant growth, but they are toxic in amounts only slightly higher than those required for growth. These types of toxic contaminant are not degraded in the environment, so they persist for extended periods. The effect on animals and plants is cumulative and will often manifest itself only after years of exposure.

Exposures to trace elements are usually local. The elements concentrate mainly in coal ash, fly ash, sludges from desulfurization systems, or particles trapped by other control systems. Trace elements can be leached from the solid waste by runoff and seepage waters that come in contact with them. Most of the trace elements in leachates are fixed and retained in soils,[34] a condition that would minimize their migration to groundwater. However some of the trace ions can be taken up by plants or moved with water that is evaporated from upper layers of the soil. Available information is insufficient for estimating the rates of release of trace contaminants from soil and solid wastes. The HEW Advisory Committee on Health and Environmental Effects of Increased Coal Utiliza-

[138] W. Kronberger, "Effect of Vapors from Asphaltic Bitumen on Plants," *Eur. J. Forest Pathol.*, **5**(1), 21–34 (1975); transl. ORNL-tr-4112, Office of Language Services, Oak Ridge National Laboratory, Oak Ridge, Tenn.

[139] R. Ewert, "Die Ermittelung der in den Teer dampfen enhaltenen pflanzenschädlichen Bestandteile und die Unterscheidung ihrer Wirkung von Anderen Akuten Rauch beschädigungen der Pflanzen," *Landwrtsch. Jahrb., Berlin,* **50,** 693–832 (1916).

[140] G. Halbwachs and H. Hlawatsch, "Photooxidation als Ursache von Pflanzen-schädigungen durch Teerdampfe," *Naturwissenschaften,* **55,** 90 (1968).

[141] R. Sieggried, "Effect of Garbage Compost on the 3,4-Benzpyrene Content of Carrots and Head Lettuce," *Naturwissenschaften,* **62,** 300 (1975); transl. ORNL-tr-4124, Office of Language Services, Oak Ridge National Laboratory, Oak Ridge, Tenn.

[142] W. Gräf and H. Diehl, "Über der naturbedingten Normalpegel kanzerogener polyzyklisher Aromate und seine Ursache (Concerning the Naturally Caused Normal Level of Carcinogenic Polycyclic Aromatics and Its Cause)," *Arch. Hyg.*, **150**(4), 49–59 (1966).

[143] J. Borneff, F. Selenka, H. Kunte, and A. Maximos, "Die Synthese von 3,4-Benzpyren und anderen polyzyklischen aromatischen Kohlenwasserstoffen in Pflanzen (The Synthesis of 3,4-Benzopyrene and Other Polycyclic, Aromatic Hydrocarbons in Plants)," *Arch. Hyg.*, **152**(3), 279–282 (1968).

[144] W. Gräf and W. Nowak, *Arch. Hyg.*, **150,** 513 (1966).

[145] W. Gräf, *Anz. Schaedlingsk.*, **39,** 152 (1966).

[146] K. Wulf, *Naturwiss. Rundsch.*, **21,** 69 (1968).

[147] K. H. Wagner, E. Wagner-Hering, and K. Buchhaupt, *Pflanzenernaehr. Bodenk.*, **123,** 186 (1969).

[148] G. Grimmer and D. Duvel, "Untersuchungen zur endogenen Bildung von polyzyklischen Kohlenwasserstoffen in höheren Pflanzen, 8, Mitt.: Cancerogene Kohlenwasserstoffe in der Umgebung des Menschen (Investigation of Biosynthetic Formation of Polycyclic Hydrocarbons in Higher Plants)," *Z. Naturforsch.*, **256,** 1171–1175 (1970).

tion[149] recommended that further information be collected and evaluated on the environmental movement, availability, and uptake of trace elements such as cadmium, fluorine, cobalt, and nickel, which have potentially high biological impact (Section 31.4).

Since few data about the operation of coal conversion facilities are currently available, answering questions about potential effects on the environment will require close cooperation at the earliest opportunity between environmental scientists and technologists. Such cooperation should make it possible to avoid both expensive pollution abatement add-ons and deleterious effects on the environment. Moreover, the restraints that the NSPSs and state implementation plans are expected to place on effluents from coal conversion plants should prevent significant, acute injury to terrestrial ecosystems.

The interactions of sulfur oxide, nitrogen oxide, and ozone with some of the less well studied pollutants from conversion plants, such as hydrogen sulfide, carbonyl sulfide, carbon disulfide, and hydrocarbons, are still unknown. The probable effects include decreased rates of photosynthesis, increased rates of respiration, enzyme inhibition, and changes in membrane permeability.

Many of the compounds that might be emitted from coal conversion facilities may be eliminated from consideration as phytotoxicants on the basis of their expected emission concentration and known range of phytotoxicity. For example, stack concentrations of hydrogen sulfide are not expected to exceed 10 ppm. Hydrogen sulfide is known to cause measurable negative effects on plant growth and yield at concentrations in the range from 0.5–1 ppm, but it is unlikely to reach this concentration because of effects on humans. If reasonable stack-to-ground dilution ratios are achieved, hydrogen sulfide from stack effluents would not be expected to reach phytotoxic levels. Table 31.18 summarizes similar relationships for other effluent gases of potential ecological hazard. (Benzene may reach transitory concentrations as high as 2000 ppm, but 10,000 ppm is needed to cause phytotoxicity.)

Kotcon et al.[150] indicate that exposure of kidney beans to 0.20 ppm of carbonyl sulfide (one of the potential effluents from conversion plants) for one hour resulted in significant increases in the transpiration rate of the plants. Exposure to 0.30 ppm resulted in significant decreases in the photosynthetic rate, and exposure to concentrations up to 6 ppm resulted in total inhibition of photosynthesis. Chronic exposures to 0.30 ppm for 4 hr/day for 16 days resulted in no significant reduction in yield, however.

A recent assessment of the effects of car-

[149] "Report of the HEW Advisory Committee on Health and Environmental Effects of Increased Coal Utilization," *Fed. Reg.*, **43**(10).

[150] J. B. Kotcon, *Assessment of Effects of Gas Effluents*, Coal Technology Program Quarterly Prog. Rep. for period ending December 31, 1976, ORNL Rep. No. 5252, 1976.

Table 31.18 An Assessment of Potential Phytotoxicity of Selected Coal Conversion Effluents

Effluent Gas	(A) Expected Stack Concentrations to Achieve Standards (ppm)	(A^1) Potential Fugitive Concentrations (ppm)	(B) Reasonable Range to Expect Phytotoxicity (ppm)	(C) Concentration Factor A or A^1/B[a]		Expected Phytotoxicant (without synergism)[b]	
				A	A^1	A	A^1
CS_2, COS	90		0.30	300			
H_2S	10	200–10,000	0.3–1,000	33	667	No	Possible
NH_3	25		20–50	0.5–1		No	
HCN	10		1,000	0.1		No	
BTX[c]		10–2,000	10,000		0.001–0.20		No

[a] Concentration factors greater than 1 indicate phytotoxic concentrations at a source, prior to atmospheric dilution.

[b] Behavior relative to phytotoxicity of mixtures of most of these compounds is unknown. It is reasonable to assume that some of these pollutants and "regional background" pollutants such as SO_2, O_3, and NO_x may interact additively or synergistically to become phytotoxic at concentrations below those presented in column B.

[c] Benzene, toluene, xylene.

bonyl sulfide[151] on the environment tentatively recommends limiting long-term average concentrations to 400 mg/m^3 (equivalent to 0.16 ppm). The limit would appear to be adequate for protecting vegetation as well.

Some useful insight into the potential problems associated with hydrocarbon pollutants can be derived from existing studies of related classes of compounds from other sources (e.g., petroleum and coking industries).

31.5.4.2 *Aquatic Ecosystems*

The impact of coal mining on the environment has been intensively researched and well documented. Past perturbations have included large-scale removal of vegetation and destruction of soil structure, loss of valuable fish and wildlife habitats, changes in hydrology, including pollution of underground and surface waters, and severe erosion and sedimentation, which have contributed to landslides and flooding. Studies in southern Appalachia have shown that erosion from untreated surface-mined slopes (contour mining) can be more than 1000 times that found on adjacent unmined lands.[152] Uncontrolled runoff from unreclaimed slopes may flow directly into natural drainageways, causing flooding, increased turbidity, and degradation or destruction of aquatic habitats.[153] The exposure of previously undisturbed soil and rock may change the chemical composition of the groundwater and runoff.

Perhaps the most damaging waterborne contaminant from coal mining is the acid produced from exposure of the iron sulfide present in some coal and overburdens. The acid directly affects the biota and also mobilizes heavy metals, allowing them to enter downstream waterways. In the United States, acid mine drainage, which occurs mainly in Appalachia and parts of Indiana, Illinois, and western Kentucky, is of widespread concern. Alkaline drainage in Western coalfields in the United States presents a hazard to the biota in the immediate receiving waters, but it does not result in increased mobilization of heavy metals from coal and the surrounding overburden.

Acid mine drainage is associated with both underground and surface mining, although the former is greater in extent and more difficult to treat, since the subterranean waters flowing through the coalfields often do not result in point-source discharges. In the United States the Surface Mining Control and Reclamation Act of 1977[8] requires treatment of water leaving the mine; thus the problem of acid mine drainage should be minimized in future mining.

Combustion of coal produces ash that must be disposed of. Improperly dredged ash ponds or ponds with leaky dikes may permit downstream sedimentation and iron precipitation that will adversely affect bottom-dwelling organisms. Extremely acidic or alkaline water overflowing from ash ponds will put aquatic organisms under stress in the vicinity of the overflow before dilution with the receiving stream neutralizes the effluent.

To what extent do trace metals in ash slurry water adversely affect the aquatic biota? The mean concentrations of most heavy metals (except iron and selenium) in 10 ash pond effluents were below EPA limits for drinking water.[53] Downstream organisms will be chronically exposed if disposal sites are not managed properly. Sublethal effects may include reduction in growth rates, reproductive impairment, reduced ability to avoid predation, and other behavioral aberrations.

The current understanding of trace metal toxicity is not adequate to permit assessment of the long-term effects of ash pond effluents. Potential toxic effects may be modified by pH, alkalinity, water hardness, dissolved oxygen, complexing capacity, and other factors in receiving waters.[154] Moreover, trace element toxicities may interact in an additive, antagonistic, or synergistic manner[155,156];

[151] T. O. Peyton, R. V. Steele, and W. R. Mabey, *Carbon Disulfide, Carbonyl Sulfide: Literature Review and Environmental Assessment;* NTIS PB 257 947, 1976, 57 pp.

[152] C. R. Collier, R. J. Pickering, and J. J. Musser, *Influence of Strip Mining on the Hydrologic Environment of Parts of Beaver Creek Basin, Kentucky, 1955–1965,* U.S. Geol. Surv. Prof. Paper No. 427-C, 1970.

[153] U.S. Environmental Protection Agency, *Erosion and Sediment Control: Surface Mining in the Eastern U.S.,* EPA-625/3-76-006, 1976, 102 pp.

[154] G. M. Hendry, K. Baakrud, M. Laake, and G. Raddum, "Acid Precipitation: Some Hydrologic Changes," *Ambio,* **5,** 224–227 (1976).

[155] J. B. Sprague, "Review Paper: Measurement of Pollutant Toxicity to Fish, I. Bioassay Methods for Acute Toxicity," *Water Res.,* **3,** 793–821 (1969).

present knowledge of their potential interactions is inadequate. Further studies to permit evaluation of the potential for long-term environmental hazards have been recommended by the Committee on Health and Environmental Effects of Increased Coal Utilization.[149]

The primary contaminants in coking wastewaters are residual phenols, ammonia, cyanide, sulfide, thiosulfate, and thiocyanate (Section 31.4.2). The acute toxic concentration ranges reported for fish are summarized in Table 31.19. Acute toxicities of phenols for fish and other aquatic organisms have been extensively studied, and several comprehensive reviews have appeared.[157,158] Coarse fish, such as goldfish, carp, and roach, are generally only a tenth to a fifth as sensitive to phenols as trout and salmon, which are killed at concentrations of 3–5 mg/l. Sensitivities decrease with age of the fish and with rises in oxygen tension and water temperature. In general, both alkylated phenol derivatives (cresols and xylenols) and polycyclic phenols are less toxic than phenol.

[156] J. B. Sprague, "Review Paper: Measurement of Pollutant Toxicity to Fish. III," "Sublethal Effects and Safe Concentrations," *Water Res.*, **5**, 245–266 (1971).

[157] J. E. McKee and H. W. Wolf, *Water Quality Criteria*, 2nd ed., Water Resources Control Board, State of California, Publ. No. 3-A, 1963.

[158] European Fisheries Advisory Commission Working Party on Water Quality Criteria for European Freshwater Fish, "Phenols," *Water Res.*, **7**, 929–941 (1973).

Table 31.19 Toxicity to Fish of Substances in Wastewater Discharged from Coke Plants[157]

Toxicant	Toxic Concentration Range (mg/l)[a]
Phenol	5.0–333[b]
NH_3 (unprotonated)	2.0 –24.4[c]
CN^-	0.05–0.18
CNS^-	> 1800
S^{2-}	0.86–25
$S_2O_3^{2-}$	> 600

[a] Fish deaths reported in the literature in tests of greater than 24-hr duration.

[b] Of 21 values reported, 18 were equal to or less than 28.9 mg/l.

[c] Of 11 values reported, 8 were equal to or less than 6.25 mg/l.

Hydroquinone, however, is reported to be 20 times as toxic as phenol.[157]

Ammonia in the unprotonated form is highly toxic to fish; a toxic threshold concentration of 2.5 mg/l in the pH range of 7.4–8.5 is often given.[157] Because free ammonia is several hundred times as toxic as the ammonium ion, increases in pH above 8 may increase the toxicity because of the shift of the equilibrium toward the unprotonated form. Similarly both sulfide and cyanide are far more toxic as nonionic species (hydrogen sulfide and hydrogen cyanide) and for that reason are most toxic in neutral or slightly acidic waters.

Thiocyanate and thiosulfate ions are insufficiently toxic to be of concern in most coke effluents. Several studies[157,159] have confirmed that the toxicity of partially treated carbonization effluents can be predicted with reasonable accuracy on the basis of the concentration of phenols and ammonia alone. As increasingly efficient procedures for treating wastewater for phenol and ammonia come into use, the effects of minor constituents may become more important.

Carbonization effluents also contain low concentrations of nitrogen-containing heterocyclic and polycyclic aromatic hydrocarbons, and other organic compounds that are less efficiently removed than phenols.

Many constituents of the effluents from coal conversion processes are similar to those of coking discharges. In the future coal conversion may surpass coking in volume and potential for pollution.

Samples of raw product water taken from both liquefaction and low-temperature gasification processes have been tested and found to be lethal to test fish and zooplankton at dilutions of several hundred to a thousand.[160] In solvent-extracted fractions the preponderant toxicity was due to weak acids (phenols), as was expected from the reported LC_{50} values of phenols for aquatic organisms (generally 4–40 mg/l).[161]

[159] D. W. M. Herbert, "The Toxicity to Rainbow Trout of Spent Still Liquors from the Distillation of Coal," *Ann. Appl. Biol.*, **50**, 755–777 (1962).

[160] S. I. Auerbach, Environmental Sciences Division Annual Report, period ending September 31, 1977, Rep. No. ORNL 5294, Oak Ridge National Laboratory, Oak Ridge, Tenn.

Although aromatics containing nitrogen are expected to constitute a significant fraction of the dissolved organic contaminants in some coal conversion wastewaters, their toxicities are approximately a tenth of those of the corresponding phenols. The concentrations of some aromatic amines may approach toxic amounts, but biologically treated coking effluents appear to be below the toxic thresholds for most of the individual constituents.[76]

The toxic thresholds of aliphatic and monocyclic aromatic hydrocarbons for fish are substantially higher than the concentrations expected in wastewater discharges. However naphthalenic hydrocarbons are more toxic than monocyclics and have been cited as the most toxic water-soluble fraction of petroleum.[162] Identifiable alkylnaphthalene compounds totaled 12 mg/l in one sample of product water from liquefaction.[105] Treatment and dilution will be necessary to reduce the concentrations in wastewater discharges below the amounts that cause acute toxicity in fish (1.5–25 mg/l).[157]

Because wastewater dilution, treatment, and recycling appear to be mandatory for coal conversion facilities, releases of toxic effluents causing immediate death to fish and other aquatic organisms probably will not occur. Of possible concern, however, are chronic, subacute stresses. Phenols at sublethal concentrations cause decreased reproduction in fish.[157] The water-soluble fraction of petroleum (consisting primarily of monocyclic and bicyclic aromatic hydrocarbons) affects the behavior of some aquatic invertebrates at concentrations 1–2 orders of magnitude below lethal concentrations[162–164]; many similar trace organics may be present in discharges from coal conversion plants. The concentrations of four metals in one sample of coal conversion product water were reported to be within the toxic range to fish,[165] although treatment could reduce the concentrations of trace metals (and trace organics) to subtoxic levels. Because little is known of the potential sublethal effects of complex mixtures of trace contaminants on aquatic organisms, only field studies at large-scale coal conversion plants can demonstrate whether significant ecological problems will occur.

The effects of trace contaminants in coal conversion wastewaters should be considered in conjunction with those of other industrial discharges that may enter the same waterways. Because coal conversion products are more easily and economically transported than coal, most development strategies foresee construction of coal conversion facilities in coal mining regions. Potential sites in the eastern United States will generally be located on highly industrialized water systems, which are already under stress from effluents. Western coal regions are less highly developed. The increase in environmental stress caused by similar quantities of effluent contaminants in the two regions may thus differ greatly. In the six-state Ohio River watershed, the impact of six new 20,000-ton/day coal conversion plants complying with existing discharge regulations has been estimated to be comparable in magnitude to the total impact of all the coking plants in the region[166]; but that magnitude is only a small fraction of the contaminants emitted by the existing petroleum refineries, which include one-fifth of the total U.S. refining capacity. The effluents from the same six plants in the relatively undeveloped West could have a greater relative impact, although the ecological effects of the two scenarios cannot be accurately assessed at present.

[161] C. W. Gehrs, Technology Program Quarterly Prog. Rep. for period ending September 30, 1975, Oak Ridge National Laboratory Rep. No. ORNL TM 5093, Oak Ridge, Tenn., pp. 61–62.

[162] J. W. Anderson, J. M. Neff, and S. R. Petrocelli, "Sublethal Effects of Oil, Heavy Metals, and PCBs on Marine Organisms," in *Survival in Toxic Environments,* M. Khan, M. A. Q. and M. P. Bederka, Eds., Academic Press, New York, 1974, pp. 83–122.

[163] S. F. Moore and R. L. Dwyer, "Effects of Oil on Marine Organisms: A Critical Assessment of Published Data," *Water Res.,* **8,** 819–827 (1974).

[164] S. M. Jacobson and D. B. Boylan, "Effect of Seawater Soluble Fraction of Kerosene on Chemotaxis in a Marine Snail, *Nassavis obsoletus,*" *Nature,* **241,** 213–215 (1973).

[165] S. G. Hildebrand, R. M. Cushman, and J. A. Carter, "The Potential Toxicity and Bioaccumulation in Aquatic Systems of Trace Elements Present in Aqueous Coal Conversion Effluents," in *Trace Substances in Environmental Health, A Symposium,* University of Missouri, Columbia, D. D. Hemphill, Ed., 1976.

[166] E. S. Rubin and F. C. McMichael, "Impact of Regulation on Coal Conversion Plants," *Environ. Sci. Technol.,* **9,** 112–117 (1975).

31.6 IMPACTS ON HEALTH AND SAFETY

31.6.1 General

Many potential effects on health and safety are associated with increased use of coal. They range from immediate death to premature death from impaired health or aggravation of existing disease. They can also reduce the quality of life and lead to deferred losses of life from genetic defects created in earlier generations by photochemical oxidants or radiation. In addition, there are nonfatal injuries and illnesses that range in severity from immediate and serious cases, sometimes accompanied by permanent disability or the need for hospitalization, to delayed illnesses associated only with stress and manifested mainly as discomforts.

It should be noted that at the present levels of air quality the overt adverse health effects that were readily detectable in the 1950s and 1960s are not demonstrable. If air pollution is having an adverse impact on health, it is far more subtle and difficult to detect, and may possibly even be marginal in terms of other health hazards. The intent of Congress with respect to the meaning of the term "adverse health effect" is not clear, but present interpretation of the law is that even marginal effects should be eliminated. Continuation of this interpretation could have far-reaching implications for the energy technologies that are expected to be introduced in the future.

The information available for assessing deleterious effects of different coal technologies on health is insufficient, uneven, and difficult to evaluate. Data on some effects are considerably more reliable than others. A common practice is to indicate the degree of confidence associated with reported effects by providing ranges of figures rather than a single value. The ranges may be as small as a few percent for such effects as accidental deaths from coal mining, or as large as 2 orders of magnitude or more for premature deaths of members of the general public caused by emissions of air pollutants.[167]

[167] C. L. Comar and L. A. Sagan, "Health Effects of Energy Production and Conversion," in *Annual Review of Energy,* Vol. 1, Jack Hollander, Ed., Annual Reviews Inc., Palo Alto, Calif., 1976.

The reasons for differences in the data base are related to many factors: the straightforwardness of the cause-and-effect relationship between the impact and its effect on health and safety, the amount of information available about the kinds and quantities of emissions (particularly from new processes), and the lack of good information about the responses of people to low concentrations of many of the pollutants.

The most serious uncertainty in the data used to estimate impacts on health is information needed to estimate the number of premature deaths and disabilities resulting from various lung diseases associated with coal mining. For example, there is an average delay of 14 years between the beginning of exposure to coal mine dust and the appearance of clinical symptoms of coal workers' pneumoconiosis (CWP).[168] Thus old cases of CWP and clinical symptoms of the disease will continue to surface for many years, even if current exposures to coal dust are controlled.

The potential effects of coal technologies on health and safety can be classified according to the phase of the coal system with which they are associated, how widespread they are (global, regional, or local), and whether they affect occupational workers only or members of the general public as well.

31.6.2 Global

The major potential global effect of increased use of coal is the possibility that emissions of carbon dioxide may cause the modification of our climate. The effects of modifying world climate are discussed in Section 31.5.2; other effects on the health of the earth's inhabitants are discussed here.

If significant long-term climatic changes were to result from increased use of coal, the indirect effects on human health and welfare would depend to a large extent on human adaptation to the changes and on the state and distribution of agricultural practices and

[168] National Research Council, Committee on Mineral Resources and the Environment, *Mineral Resources and the Environment,* Supplementary Report, *Coal Workers Pneumoconiosis—Medical Considerations, Some Social Implications,* Report of the Committee, National Academy of Sciences, Washington, D.C., 1976.

habitation. It is not yet possible to determine whether there will be risks to human health associated with changes in the amount of carbon dioxide in the atmosphere or how extensive such changes will be. It is necessary first to be able to predict the global requirements for different fuels, the distribution of carbon dioxide among the various reservoirs in the carbon cycle, the global and regional impacts of variations in carbon dioxide on climate, and the sociopolitical and environmental impacts that may be caused by climatic change, and to develop methods of reducing the amounts of carbon dioxide released by energy production, and methods of influencing the distribution of carbon dioxide among the various reservoirs in the carbon cycle, to decrease the quantity in the atmosphere.[149,169,170]

31.6.3 Regional

The regional impacts of the various coal technologies on human health include (1) deaths and disabilities from respiratory diseases associated with inhalation of pollutants, including both emissions into the air from combustion processes and the transformation products produced during transport of pollutants by air, and (2) accidental deaths and injuries associated with the mining and transportation of coal.

A third category of impact that may affect the public health consists of deaths and disablements from cancer and genetic damage associated with trace contaminants that may be emitted during the combustion of synthetic liquids from coal. Better data are needed to assess the probability and extent of such contamination before its importance can be judged. Studies are currently underway[171] for comparing the carcinogenicities of synthetic and natural crude oils, but thus far they permit only limited conclusions concerning projection of comparative risks. Before sound estimates of the increased cancer risks, if any, of synthetic fuels can be made, further data will be needed on the composition of future synthetic fuels, emissions of trace contaminants and exposure to them, and dose-response relationships between exposures and impacts.

31.6.3.1 *Effects on Health from Combustion Products and Their Transformation Derivatives*

Emissions of combustion products from the various coal technologies are expected mainly from the processes of electricity generation and coal conversion and from the ultimate use of the fuels. The emissions, in turn, give rise to many transformation products formed largely by photochemical reactions in the atmosphere.

There are hundreds, if not thousands, of combustion and transformation products that may have adverse effects on health, acting singly or in combination with others. This section references a wide range of estimates of these potential effects but does not come to any conclusions about whether or to what extent these effects will actually occur. The severity of the potential impact depends on the characteristics of the pollutant (both its chemical and physical properties, and including variations in its concentration over time), conditions of the exposure (duration, environmental conditions such as temperature and humidity during exposure, and the presence of other chemical species or physical agents that influence the impacts), and such characteristics of the exposed population as its state of health, the nutritional status of exposed subjects, and the presence of susceptible groups (e.g., the elderly or the young).

Currently there is little information about exposures to most potentially harmful pollutants. Even less is known about the relation of these substances to disease.[172] Major research on health effects has been done in relation to only six pollutants that are derived mainly from combustion of fossil fuels or

[169] U.S. Congress, House Committee on Science and Technology, *Environmental Challenges of the President's Energy Plan: Implications for Research and Development,* Report of the Subcommittee on Environment and the Atmosphere, 95th Congr., 2 Sess., October 1977.

[170] National Academy of Sciences, *Energy and Climate: Studies in Geophysics,* Washington, D.C., 1977.

[171] J. M. Holland, research sponsored by U.S. Dept. of Energy, personal communication.

[172] U.S. Department of Health, Education and Welfare, *Human Health and the Environment—Some Research Needs,* DHEW Publ. No. NIH 77-1277, Report by PHS/NIH/NIEHS, Washington, D.C., 1977.

from atmospheric transformation processes: oxides of sulfur, nitrogen dioxide, particulate matter, carbon monoxide, photochemical oxidants (including ozone), and nonmethane hydrocarbons.[168,173–176] In the United States these pollutants are referred to as "criteria" or "regulated" air pollutants. Existing data about them have led to preliminary National Ambient Air Quality Standards. The basis for establishing the health standards for regulated air pollutants is open to question. Although we recognize its importance, resolution is beyond the scope of this chapter.

There are even greater uncertainties about the health implications of a number of unregulated pollutants such as organic aerosols, acid sulfates, and nitrate particles. Because of the absence of health data on these pollutants, in the discussion that follows the three criteria pollutants—SO_2, NO_2, and particulates—are used as the pollutant index, even though they are not a good index for the aerosol-type pollutants.

Studies of health effects of air pollutants include laboratory studies of animals, studies of occupational groups, and epidemiological surveys of the general population. The scope of this section precludes a review of all the important studies. Instead, the discussion is limited to issues or pollutants believed to be most critical. The references are limited to important reviews that evaluate and refer to many of the separate studies. The studies themselves contain many weaknesses that limit their application.

The main problem associated with attempts to extrapolate to man the results from laboratory studies of animals relates to physiological differences between man and other species. In addition, most animal studies are designed for short-term, high exposures, so that there are further problems in extrapolating the results to long-term, low-level exposures of humans. Long-term, low-dose experiments on animals are not only considerably more expensive than short-term, high-dose experiments, but they are also limited by the shorter life spans of the laboratory animals as compared with man. It is also extremely difficult to extrapolate from the fixed concentrations of single pollutants to which animals are usually exposed to the fluctuating levels of combined exposures to several pollutants that are typical for humans. Laboratory studies using human subjects could eliminate some of the extrapolation problems, but they would raise ethical and practical problems.[173,175]

Extrapolation to the general population from studies of occupational groups also is of limited value. The conditions of exposure differ for occupational groups and for the general exposed population in respect to the mix of pollutants, their concentrations, the environmental conditions during exposure (e.g., temperature and humidity), and the duration of exposure (8 hr/day, 5 days/week). Furthermore, the working population is not representative of the general population, since it does not include such susceptible groups as the elderly and the very young. As a result, if an effect is not demonstrated with a study of occupational groups, it cannot be concluded that there will be no effect in the general population. If an effect is demonstrated, a greater effect in the general population can be expected. Nevertheless, such studies are important because they can provide information on amounts of intake and elimination of pollutants, and they should lead to better estimates of the consequences for humans of mixed pollutants in the air.[173]

Epidemiologic studies of the general population also have serious limitations. They usually cannot demonstrate a cause-and-effect relation between adverse health effects and exposure to a single pollutant but can only indicate that a relationship exists between effects and a complex of pollutants.[177] Supporting experimental studies would be necessary to evaluate the toxicity of the individual components of the complex, but it is unlikely that such studies will be successful

[173] B. G. Ferris, "Health Effects of Exposure to Low Levels of Regulated Air Pollutants: A Critical Review," *J. Air Pollut. Control Assoc.*, **28**(5), 482–487 (May 1978).

[174] National Research Council, *Air Quality and Stationary Source Emission Control,* Ref. 31.

[175] "Health Effects of Exposure to Low Levels of Regulated Air Pollutants—Discussion Papers," *J. Air Pollut. Control Assoc.*, **28**(9), 883–894 (September 1978).

[176] World Health Organization, *Environmental Health Criteria for Sulfur Oxides and Suspended Particulate Matter,* Geneva, Switzerland, 1978.

[177] L. B. Lave and E. P. Seskin, *Air Pollution and Human Health,* published for Resources for the Future by John Hopkins University Press, Baltimore, 1977.

in the near future. It is expected that sulfur oxide levels will continue to be used as the surrogate for the mixture of pollutants. There are also limitations with respect to estimating the dose, measuring the effects, and correcting for variations in the environment, and in the socioeconomic and racial makeup of the exposed population. Serious problems with the interpretation of the results are often caused by competing risks, such as cigarette smoking, which may completely mask the effects of chronic exposures to low levels of air pollution.[173,176] In general, it is useful to combine information from studies of all three kinds to improve the evaluation of impacts on health.

Various reviews of the effects on health of the six regulated air pollutants indicate that there are acute deleterious health effects only at concentrations greater than and usually in excess of the limits currently recommended by the U.S. National Ambient Air Quality Standards.[173,176,178–180] With few exceptions, there are no manifest impacts at concentrations equal to or lower than those recommended by the standards.[173] No proof of excess mortality of persons exposed to the usual ambient levels of the regulated pollutants can be inferred from a critical review of the literature.[173] Studies of acute episodes of air pollution, such as those that occurred in the Meuse Valley; Donora, Pennsylvania; London; and New York City, have clearly shown excess mortality.[173] However when an attempt is made to interpret the results in terms of effects of specific pollutants or to extrapolate them to the more modest day-to-day variations in concentration of specific pollutants, serious difficulties are encountered.[181–183] Evaluation of several acute episodes has shown correlations between sulfur dioxide, total suspended particulate counts, or sulfates and excess mortality, morbidity, or respiratory symptoms.[173] The chronic respiratory illnesses or aggravation of symptoms associated with these pollutants include increased incidence or aggravation of emphysema, asthma, bronchitis, and various cardiovascular symptoms. On the other hand, Schimmel[182] evaluated the daily mortality data of New York City during 1973–1976 and concluded that little, if any, effect of pollution could be observed. His study indicated that temperature changes are the main factor affecting day-to-day variations in mortality.

With our current knowledge it is impractical to attempt to assess separately the potential impacts on health of each air pollutant to which the population may be exposed. Moreover, since simultaneous exposure to a number of pollutants may produce combined effects that are different from the sum of the separate effects, it is better to consider pollutants together. One suggested solution to this problem is to use a single pollutant as an index of all the others and to calculate the potential health impacts from this exposure. The rate of emission of the index pollutant from a particular energy plant can be calculated from the characteristics of the energy plant, including fuel and type of emission control equipment in use. An air pollution model can then be developed to take into account the transport of the index pollutant, its atmospheric reactions, dosages or exposure of specified populations, and their responses to the doses. It should be realized,

[178] R. D. Stewart, "The Effect of Carbon Monoxide on Humans," *J. Occup. Med.*, **18**(5), 304–309 (1975).

[179] National Research Council, *Airborne Particles,* Report by Subcommittee on Airborne Particles, Committee on Medical and Biological Effects of Environmental Pollutants, Assembly of Life Sciences, National Academy of Sciences, University Park Press, Baltimore, 1977; *Nitrogen Oxides,* Report by Committee on Medical and Biological Effects of Environmental Pollutants, Assembly of Life Sciences, National Academy of Sciences, Washington, D.C., 1977; *Ozone and Other Photochemical Oxidants,* Report by Committee on Medical and Biological Effects of Environmental Pollutants, Assembly of Life Sciences, National Academy of Sciences, Washington, D.C., 1977; *Carbon Monoxide,* Report by Committee on Medical and Biological Effect of Environmental Pollutants, Assembly of Life Sciences, National Academy of Sciences, Washington, D.C., 1977.

[180] National Research Council, *Sulfur Oxides,* Report by Committee on Sulfur Oxides, Board on Toxicology and Environmental Health Hazards, Assembly of Life Sciences. National Academy of Sciences, Washington, D.C., 1978.

[181] H. Schimmel and T. J. Murawski, "The Relation of Air Pollution to Mortality," *J. Occup. Med.*, **18**(5), 316–333 (1975).

[182] H. Schimmel, "Evidence for Possible Acute Health Effects of Ambient Air Pollution from Time Series Analysis: Methodological Questions and Some New Results Based on New York City Daily Mortality 1973–1976," paper presented at VDI colloquium on Oxidized Sulphur Compounds, Augsberg, May 1978.

[183] R. E. Waller, Discussion, see Ref. 175.

however, that errors are associated with each of these assumptions that, together, lead to great uncertainties and broad ranges of estimated impact.

Usually sulfur dioxide, its sulfate equivalent, or particulate air pollution has been used as the index pollutant.[174,177,184,185] These three pollutants are in fact interrelated. Most of the sulfate particles found in ambient air are derived from conversion of the primary product, sulfur dioxide, in the air. Sulfate particles can also account for a large proportion of the respirable particulates in the air in some circumstances.[186] It is not now known whether these particulates, particularly the finer sizes, have health consequences similar to those particulates emitted by the source of the air pollutants, or what type of technologies would have to be developed for effective control.

Estimates of expected deaths and illnesses from the combustion and transformation products of coal vary greatly among different authors. The variations can be 2 orders of magnitude or more, and they reflect the great uncertainties involved in the assessments. A number of detailed evaluations of the various studies have been made with the objective of determining the causes of the wide variations.[174,177,184,185,187]

[184] L. D. Hamilton, *Health Effects of Air Pollution*, Brookhaven National Laboratory, Upton, N.Y., for U.S. ERDA, 1975.

[185] W. A. Buehring, W. K. Foell and R. L. Keeney, *Energy/Environment Management: Application of Decision Analysis*, Res. Rep. No. RR-76-14, International Institute for Applied System Analysis, Laxenburg, Austria, May 1976.

[186] Greenfield, Attaway and Tyler, Inc., *An Evaluation of CHESS; New York Asthma Data, 1970–1971*, Vol. I, EPRI, EA-450, San Rafael, Calif., May 1977.

[187] See Ref. 172.

31.6.3.2 *Impacts of Coal Transportation*

The transportation of coal is the other major source of regional impacts on health. Transportation by train accounts for about two-thirds of all current coal transportation in the United States (see Section 31.2.5).

Rates of accidental death and injury resulting from transporting coal are estimated in Table 31.20.

The estimates of rail deaths and injuries have two components: (1) deaths and injuries associated with collisions of trains and motor vehicles at rail grade crossings, and (2) deaths and injuries of rail passengers. The rates of accidental death and injury from transportation accidents depend on the number of grade-level crossings, the presence of automatic safety equipment at intersections and on trains, and maintenance of the rail bed.

Deaths and injuries from train accidents have been apportioned to coal rail transportation in two ways. One is based on miles traveled, the other on ton-miles.

Estimates of annual fatalities for round-trip deliveries of coal to a 1000-MW power plant consuming 1.9 million tons of coal per year have been made on a train-mile basis.[34] (Only unit coal trains carrying 10,000 tons of coal each are assumed. This is a conservative assumption, because mixed trains carry about two-thirds of the coal transported by trains today. They also carry considerably less coal than unit trains, and they involve greater mileage for the transportation of a

[188] C. P. Bliss, P. Clifford, G. Goldgraben, E. Graf-Webster, and K. Krickenberger, *Accidents and Unscheduled Events Associated with Non-Nuclear Energy Resources and Technology*, prepared by MITRE Corporation for U.S. Environmental Protection Agency, EPA/600/7-77-016, Washington, D.C., February 1977.

Table 31.20 Rates of Fatalities and Injuries Resulting from Various Transportation Modes[188]

Transportation Mode	Fatalities (cases/10^{12} Btu)^a	Injuries (cases/10^{12} Btu)
Rail	0.062–0.075	0.60–1.00
Trucking	0.032	0.1
Barge	0.0019	0.0032
Slurry pipeline	0.0019	0.0032

^{a} There is about 10^{12} Btu in 40,000 tons of bituminous coal.

million tons of coal.) The number of fatalities per year ranges from 0.11 for a plant located in the Eastern Interior Region and supplied by southern Illinois coal (a round-trip distance of 200 miles) to 1.45 for plants located in the Eastern Interior and supplied by Wyoming coal (1900 miles round-trip). If we assume an average of 600 miles per round trip and the portion of general public deaths as 90% of all accidental deaths,[189] we derive an annual fatality rate of 0.15 per million tons of coal.

Other authors assume that accidental impacts are calculated on the basis of ton-miles.[19,188,190] Their estimates of general public accidental deaths per million tons of coal (assuming 25×10^6 Btu per ton of coal) are about 1.5 and 15, values about 10 times greater than those calculated using train-miles.

31.6.4 Local

Based on current knowledge the main potential local impacts on health and safety are those associated with coal mining.[191,192] These account for a predominant fraction of occupational deaths, injuries, and damage to health from the various coal technologies.

There are several other local impacts on workers in coal technologies. In order of importance, based on current knowledge, they consist of accidental deaths and injuries from construction and operation of energy conversion facilities, transportation of coal, and coal preparation and storage. If adequate controls are not devised, workers in coal conversion plants will face increased risk of death and disablement from cancer. Such controls would involve engineering factors, industrial hygiene monitoring, medical surveillance, personal hygiene, work practices, and worker health education.

31.6.4.1 *Mining*

Effects on Health. Several respiratory diseases, including chronic bronchitis, coal workers' pneumoconiosis (CWP), and various obstructive airway diseases are aggregated under the legal definition used by the U.S. Social Security Administration to ascertain eligibility for social security benefits to victims of black lung disease.[168] Chronic bronchitis and obstructive airway diseases are almost 3 times as common as CWP among coal miners,[193] but CWP is a diagnosable disorder of the respiratory system that is clearly attributable to the prolonged inhalation of coal dust; thus it is a better index of the respiratory impacts that can be attributed to working with coal.

The prevalence of CWP increases in direct relation to exposure to coal dust. The incidence is greatest among underground coal miners with duties at the face of the coal seam, where dust pollution counts are highest. The lowest prevalence of CWP is found among surface coal miners and underground coal miners whose duties are mainly on the surface.[168] It is therefore inferred that control of respirable coal dust concentrations in mines is the most effective means of reducing the number of new CWP cases. This conclusion is supported by the British experience of limiting coal dust concentrations to 2 mg per cubic meter of air. At these concentrations the prevalence of CWP among British underground miners has dropped to 3% of the work force, and these workers show only mild symptoms of the disease.

The incidence and character of CWP are influenced by local environmental factors, regional differences in prevalence, delays between commencement of exposure and diagnosis of the disease, and severity of the disease. The reasons for the differences are not known, but variations in the chemical

[189] W. Baldewicz, G. Haddock, Y. Lee, R. Whitley, and V. Denny, *Historical Perspectives on Risk for Large-Scale Technological Systems,* University of California at Los Angeles, Rep. No. UCLA-ENG-7485, November 1974.

[190] W. A. Sevian and P. D. Moskowitz, *Energy System Network Simulator,* Brookhaven National Laboratory, Biomedical and Environmental Assessment Division, Upton, N.Y., September 1978.

[191] National Research Council, *Mineral Resources and the Environment,* report by Committee on Mineral Resources and the Environment, Commission on Natural Resources, National Academy of Sciences–National Research Council, 1975.

[192] W. H. Walton, Ed., *Inhaled Particles,* Vol. 4, Pergamon Press, New York, 1977.

[193] National Institute for Occupational Safety and Health, *Occupational Safety and Health Implications of Increased Coal Utilization* (draft), prepared for President's Committee on Health and Ecological Effects of Increased Coal Utilization, Washington, D.C., November 4, 1977.

composition and size distribution of the coal dust produced in different coal seams may be responsible.

Estimates of the prevalence of CWP in the United States vary widely among reporting sources. According to the National Institute of Occupational Safety and Health, between 10 and 15% of underground coal miners have the disease in some form.[193] The Committee on Mineral Resources and the Environment of the National Academy of Sciences, however, estimates that the disease is considerably more prevalent and that it involves nearly 30% of underground miners.[168] Sixty-five new cases were reported for 1976 per million tons for underground mining and 4.0 cases per million tons for surface mining.[193]

Miners in the United States disabled by CWP are compensated under provisions of the Federal Coal Mine Health and Safety Act of 1969[7] and the Black Lung Benefits Act of 1972.[194] Payments of benefits to miners and their dependents totaled $1.7 billion from 1970 to 1973, reaching almost $1 billion in 1973 alone.[168] The rapid increase in cost was due to provisions in the 1972 law that no miner could be refused compensation solely on the basis of a normal roentgenogram.[191]

The U.S. Coal Mine Health and Safety Acts of 1969 and 1977 placed restrictions on the permissible respirable dust concentration in coal mines, and the restrictions are expected to reduce the incidence of CWP. However projections of future rates of CWP are uncertain.[168,193] A significant decline in the number of cases per million tons of coal mined would occur if the respirable dust controls specified in the Safety Act were rigidly implemented, and if the shift from underground to surface coal mining were to continue.

Although a survey of 223 coal mines in the United States indicated that none had concentrations of radioactive materials exceeding those permitted by current radiation protection guides and standards,[195] some coal seams, especially in the West, contain high concentrations of uranium.

[194] Federal Black Lung Benefits Act, 1972.

[195] R. L. Rock, G. Svilar, R. T. Beckman, and D. D. Roop, *Evaluation of Radioactive Aerosols in United States Underground Coal Mines*, Mining Enforcement and Safety Administration Inf. Rep. No. 1025, 1975.

Coal Mining Accidents. The mining of coal is the largest source of accidental death and injury to workers associated with any part of the coal energy system. Underground coal mining is not as safe as surface mining; there were 0.48 death and 37.11 injuries per million man-hours for underground mining during 1975, as compared with 0.35 and 18.10 for surface mining.[9] Because of this and because of its higher productivity, surface coal mining is about one-fourth as hazardous with respect to accidental deaths per million tons of coal mined and about one-sixth as hazardous per million tons mined with respect to injuries. In 1975 about 200 accidental deaths were associated with coal mining, of which about 170 occurred among underground miners.[191]

The number of accidental deaths and injuries per million tons of coal from underground mining have declined during the last 50 years. During the period from 1949 to 1969 the decline in accidental rates per million tons of coal produced was largely due to a tripling of productivity. From 1970 to 1976 the decline was attributable to major improvements in safety conditions. The output per man-day during the 1970–1976 period declined by about 40%, whereas a 60% improvement occurred in safety in the mines.[9]

In the case of surface mining, the number of accidental deaths and injuries per million tons of coal produced has shown little improvement during the last decade. Improvements in safety have been largely offset by decreased output per man-day.[9]

31.6.4.2 *Construction of Facilities*

A significant local impact on the occupational health and safety of workers involves construction deaths and injuries. The impact is, however, considerably smaller than that of mining: 7.4 deaths per million man-years and 30,000 disabling but nonfatal injuries per million man-years occur in the construction industry. Since the predominant manpower requirements are those for construction of the energy conversion facilities,[23,25] the majority of occupational deaths and injuries are associated with this phase of the coal system.

The other major additional manpower requirements for coal systems are those for construction of the mining facilities. They account for less than 5% of the manpower requirements of the energy facilities, based on the requirements for coal over the lifetime of the facility.[23,25]

31.6.4.3 *Carbonization*

Occupational Hazards

Cancer. Cancer has been the primary health concern of workers in carbonization plants (see Chapter 17, Section 17.5). The concern is based on two types of study:

1. Epidemiologic studies, which have indicated an increased risk of cancer to workers routinely exposed to the products of coal, such as workers in coal gas, coking, and coal hydrogenation industries, street pavers, and asphalters.
2. Laboratory studies to determine the carcinogenicity of various coal products and the chemicals in them, using animals, bacteria, and human cell culture systems.

Redmond[197] attempted to correlate the risk from lung cancer within a coke plant with work area and duration of employment (Tables 31.21 and 31.22 and Chapter 18, Section 18.1.5).

The study covered about 60,000 steel workers employed at seven plants. It found that men who worked for 5 years or more in areas where exposure to coke-oven emissions was greatest (i.e., the top of the ovens) experienced about 900% excess lung cancer mortality compared with other steel workers. As expected, the excess cancer mortality rates were found to be considerably lower or even nonsignificant for coke workers at cleaner localities and/or those who had shorter periods of exposure.

In another study Lloyd[199] calculated the relative risk of death from different causes among 2048 full-time coke-oven workers according to race and work station in the plant. The study extended from 1953 to 1961 (Table 31.23). Lloyd found that the mortality for all coke-oven workers from malignant neo-

[196] U.S. Energy Research and Development Administration, *Carcinogens Relating to Coal Conversion Processes,* Rep. No. FE-2213-1, prepared by TRW Systems, Washington, D.C., June 1976.

[197] C. K. Redmond, A. Ciocco, J. W. Lloyd, and H. W. Rush, "Long Term Mortality of Steel Workers. VI. Mortality from Malignant Neoplasms Among Coke Oven Workers," *J. Occup. Med.,* **14,** 621–629 (1972).

[198] S. Mazumdar, C. Redmond, W. Sollecito, and N. Sussman, "An Epidemiological Study of Exposure to Coal Tar Pitch Volatiles Among Coke Oven Workers," *J. Air Pollut. Control Assoc.,* **25**(4), 382–389 (1977).

[199] J. W. Lloyd, "Long-Term Mortality Study of Steelworkers. V. Respiratory Cancer in Coke Plant Workers," *J. Occup. Med.,* **13**(2), 53–67 (1971).

Table 31.21 Observed and Expected Deaths and Relative Risks for Malignant Neoplasms for Coke-Oven Workers:

	Total Oven			5 Years or More, Coke Oven		
Cause of Death	Observed Deaths	Expected Deaths	Relative Risk[a]	Observed Deaths	Expected Deaths	Relative Risk[a]
Malignant neoplasms, all sites	145	122.1	1.34[b]	114	82.6	1.69[b]
Bronchus, lung, and trachea	69	41.5	2.85[b]	55	28.0	3.48[b]
Other respiratory system	1	2.4	*d*	1	1.8	*d*
Buccal and pharynx	3	3.4	*d*	1	2.2	*d*
Digestive organs and peritoneum	31	37.0	0.77	25	26.0	0.95
Prostate	12	8.8	1.64	11	7.4	1.83
Other genital organs	0	0.4	*d*	0	0.3	*d*
Kidney	8	2.6	7.49[b]	5	1.6	5.69
Bladder and other urinary organs	1	1.5	*d*	1	1.3	*d*
Skin	1	1.5	*d*	1	0.8	*d*
Leukemias and lymphomas	9	10.8	0.76	5	6.0	0.77
Other and unspecified sites	10	12.1	0.69	9	7.3	1.46

[a] Significance of relative risk based on summary chi-square with 1 degree of freedom.

[b] 1% level.

[c] 5% level.

[d] Fewer than 5 deaths.

plasms of the lung was 2.5 times the expected rate. The risk was 6.5 times for full topside workers compared to that for men employed only at the side of the oven (standard mortality ratio 818 vs. 125). The mortality of side oven workers from all other causes did not differ from that expected.

Mazumdar[198] et al. were able to develop a relationship between exposure to coal-tar pitch volatiles (CTPV) and excess cancer risk to exposed coke workers (Table 31.24). They applied Redmond's risk factors to exposure data developed by the Pennsylvania Department of Health at 10 coke-oven installations.[200] The data revealed considerable differences in exposure to CTPV for the various coke-oven jobs. Exposures varied from 3.23 mg/m^3 for the high average exposure to 0.50 mg/m^3 for the low average. It was concluded that both amount and length of exposure are related to development of cancer, particularly lung cancer. The period from first exposure to CTPV until death varied from 10 to 40 years and averaged 25 years. No excess cancer incidence has been observed for workers exposed for an average of 30 years to the established threshold limit of 0.2 mg of CTPV per cubic meter.

Another study measured CTPV and benzo(*a*)pyrene profiles in a coke manufacturing plant.[201] The results were similar to those reported by Fannick et al.[200] CTPV concentrations in more contaminated areas varied between 0.0313 and 1.67 mg/m^3, and for the cleaner areas, between 0.0072 and 0.322 mg/m^3. The benzo(*a*)pyrene concentrations on top of the oven were 100–1000 times the typical levels found in urban atmospheres.

Identification of Carcinogens. Bituminous coals contain between 14 and 50% (on a dry basis) volatile material and anthracite coals contain 2–14% (dry basis).[34] The volatile matter includes many organic compounds that are potential carcinogens, among them aromatic and heterocyclic compounds.

Many researchers have attempted to identify the potentially carcinogenic substances in coal, in its products, and in pollutants from coal processing. A wide range of potential carcinogens has been listed, and their carcinogenicity has been postulated on the basis of animal experiments, epidemiological studies, or recognition of relationships between the chemical structure of substances and their carcinogenicity. The most important

[200] N. Fannick, L. T. Gonshor, and J. Shockley, Jr., "Exposure to Coal Tar Pitch Volatiles at Coke Ovens," *Am. Ind. Hyg. Assoc. J.*, **33,** 461–462 (1972).

[201] J. O. Jackson, P. O. Warner, and T. F. Mooney, Jr., "Profiles of Benzo (*a*) pyrene and Coal Tar Pitch Volatiles at and in the Immediate Vicinity of a Coke Oven Battery," *Am. Ind. Hyg. Assoc. J.*, **35,** 276–281 (1974).

1951–1966[197]

5 years or More, Full-Time Topside			5 years or More, Topside and Side Oven Exposure			5 years or More, Side Oven, Never Topside		
Observed Deaths	Expected Deaths	Relative Risk[a]	Observed Deaths	Expected Deaths	Relative Risk[a]	Observed Deaths	Expected Deaths	Relative Risk[a]
41	20.0	2.74[b]	27	18.9	1.51	46	35.5	1.40
25	7.4	6.87[b]	15	5.5	3.22[b]	15	8.7	2.10[a]
0	0.5	*d*	0	0.4	*d*	1	1.0	*d*
0	0.5	*d*	1	1.0	*d*	0	0.8	*d*
3	5.9	0.46	4	6.3	0.62	18	14.2	1.37
4	2.0	*d*	0	1.1	*d*	7	3.9	2.11
0	0.1	*d*	0	0.1	*d*	0	0.2	*d*
0	0.1	*d*	3	0.4	*d*	2	0.7	*d*
0	0.3	*d*	0	0.3	*d*	1	0.8	*d*
0	0.1	*d*	0	0.3	*d*	1	0.3	*d*
2	1.4	*d*	3	2.1	*d*	0	2.3	*d*
7	1.7	12.37[b]	1	1.3	*d*	1	2.6	*d*

(continued on next page)

Table 31.21 Continued

Cause of Death	<5 years, Coke Oven Observed Deaths	Expected Deaths	Relative Risk[a]
Malignant neoplasms, all sites	31	38.6	0.74
Bronchus, lung, and trachea	14	9.9	1.70
Other respiratory system	0	1.0	*d*
Buccal and pharynx	2	1.4	*d*
Digestive organs and peritoneum	6	12.2	0.42
Prostate	1	1.4	*d*
Other genital organs	0	0.2	*d*
Kidney	3	0.7	*d*
Bladder and other urinary organs	0	0.4	*d*
Skin	0	0.8	*d*
Leukemias and lymphomas	4	5.3	0.68
Other and unspecified sites	1	5.2	*d*

potential carcinogens appear to be the polycyclic aromatic hydrocarbons (PAH). Other important groups are various types of organic compounds, such as the polycyclic azaheterocyclic compounds, aromatic amines and benzenes, and the inorganic trace elements.

Coal tars produced at higher temperatures are considerably more carcinogenic than among workers at gas plants using coal in England.[203] The study includes 11,500 workers who were involved regularly or periodically in the carbonization part of gas plants for more than 5 years.

The group of carbonization workers who were most heavily exposed to coal and to coal products showed a standardized annual death rate from bronchitis of 2.86 versus 1.36 those produced at lower temperatures.[196,202] Only small amounts of carcinogenic material are produced by destructive distillation of coal below 450°C (842°F). The formation of carcinogenic material increases rapidly between 450 and 550°C (842–958°F) coking range and continues to increase at lower rates up to 1250°C (2282°F). In 1973 an attempt was made to correlate coking temper-

[202] National Institute for Occupational Safety and Health, *Criteria for a Recommended Standard: Occupational Exposure to Coke Oven Emissions,* Cincinnati, Ohio, 1973.

[203] R. Doll, R. E. W. Fisher, E. J. Gammon, W. Gunn, G. O. Hughes, F. H. Tyrer, and W. Wilson, "Mortality of Gasworkers with Special Reference to Cancers of the Lung and Bladder, Chronic Bronchitis, and Pneumoconiosis," *Br. J. Ind. Med.*, **22,** 1–12 (1965).

Table 31.22 Observed and Expected Deaths and Relative Risks for Selected Causes of Deaths for Coke-Oven Workers[a]: 1951–1966[197]

Cause of Death	5 years or More, Coke Oven Observed Deaths	Expected Deaths	Relative Risk[b]	< 5 years, Coke Oven Observed Deaths	Expected Deaths	Relative Risk[b]
All causes	383	365.9	1.08	206	235.2	0.83[c]
Malignant neoplasms—lung, bronchus, and trachea	55	28.0	3.48[b]	14	9.9	1.70
Malignant neoplasms—genitourinary organs	17	10.6	2.01[c]	4	2.6	*e*
Other malignant neoplasms	42	44.0	0.93	13	26.0	0.40[d]
Tuberculosis of the respiratory system	5	5.0	1.00	3	2.9	*e*
Other diseases of the respiratory system	20	16.8	1.20	10	11.0	0.88
Cardiovascular renal diseases	173	186.2	0.90	81	96.4	0.79
Accidents	20	21.9	0.88	42	38.6	1.12
All other causes	51	53.4	0.94	39	47.6	0.77

[a] Coke-oven workers employed during 1951–1955 at 10 non-Allegheny County plants and coke-oven workers employed during 1953 at two Allegheny County steel plants.

[b] Significance of relative risk based on summary chi-square with 1 degree of freedom.

[c] 5% level.

[d] 1% level.

[e] Fewer than 5 deaths.

ature with incidence of lung cancer in operators.[202] Table 31.25 presents this correlation.

Other Health Impairments. Data are available on increased mortality from all causes for non-gas-plant workers in England and Wales. These workers also exhibited a death rate of 1.19 from other respiratory diseases versus 1.14 for the general population. For the 10 groups of diseases considered, this occupational group also experienced the following increased death rates compared to those for the general public: cancer of lung, 3.06 versus 1.81 for the general population; cancer of bladder, 0.31 versus 0.14; cancer of skin and scrotum, 0.07 versus 0.02; all causes, 17.21 versus 16.35. The data are similar to those of Lloyd,[199] Redmond,[197] and Mazumdar et. al.,[198] which were discussed in the preceding section. It is interesting to note that the death rates of this occupational group were lower than those of non-gas-plant workers in England and Wales for the following causes: pneumoconiosis, 0.05 versus 0.07; arteriosclerotic and other heart disease, 3.34 versus 4.48; other cancers, 1.96 versus 2.29; other causes, 3.59 versus 4.31. In the case of accidental deaths, this group had a slightly higher rate: 0.76 versus 0.72.

The occupational group that had lighter or intermittent exposure to coal products showed the same death rates from respiratory diseases as the general population.

The mortality data for coke-oven workers[197–199] also indicate greater relative risk from diseases of the respiratory system. Among those who worked 5 years or more at the coke oven, the relative risk of death from tuberculosis of the respiratory system is 1.0; from other diseases of the respiratory system the relative risk is 1.28. All other relative risks (except cancer) are somewhat lower than for non-coke-oven workers.

Hazards to the General Population. In the United States coke ovens are one of the major sources of emissions of benzo(*a*)pyrene (BaP), accounting in 1972 for 170 tons of emissions, or about 20% of the nationwide BaP emissions from all sources.[204] About 15 million people are exposed each year to coke-oven emissions.[205]

[204] B. E. Suta, *Human Population Exposures to Coke Oven Atmospheric Emissions,* U.S. Environmental Protection Agency, Rep. No. 27, 1977.

The cancer risk to the population living near coke ovens or gas works has been estimated by extrapolation from the data of excess cancer incidence and exposure to BaP and benzene-soluble organics among coke-oven workers.[205] The estimates of excess cancer risk to members of the general public are based on linear, nonthreshold dose-response assumptions. The exposure of people living in the vicinity of coke plants is estimated to be of the order of 1% of those for lightly exposed workers. No significant excess cancer incidence has been observed among workers at these lower levels of exposure, so that there may be no effect at all on the general population.

31.6.4.4 *Combustion*

Only limited data are available on morbidity and mortality among workers in coal-burning electric power plants or on their exposure to gaseous or aerosol emissions from coal combustion. No statistically significant measurements have been reported in the literature on increased mortality or morbidity. Occasionally concentrations of combustion products and hydrocarbons are monitored. Some measurements indicate intermittent peak values higher than the threshold limit values recommended by the National Institute for Occupational Safety and Health, but whether these exposures are sufficient to impair health is a matter of controversy.[193]

Organic Contaminants. The main organic contaminants of concern that are formed during the combustion of coal include polynuclear aromatic hydrocarbons and polynuclear azaheterocyclic compounds. These contaminants may reach the environment by emission into the air during normal coal combustion processes, by runoff and seepage from inadequate waste management and storage operations, and through use of solid waste by-products for landfill and soil conditioning.

Contamination from any of these sources can be controlled by containment of wastes, use of control technologies to reduce envi-

[205] U.S. Environmental Protection Agency, Carcinogenic Assessment Group, Roy E. Albert, Chairman, *Preliminary Report on Population Risk to Ambient Coke Oven Exposures,* 1978.

Table 31.23 Number Employed, Observed and Expected Deaths, and Standardized Mortality Ratios (SMR) Prior Years[199]

		All Causes			Malignant Neoplasms of Lung		
Work Area and Race	Number Employed	Observed Deaths	Expected Deaths	SMR	Observed Deaths	Expected Deaths	SMR
Total coke oven	2048	184	173.1	106	31	12.3	252[a]
White	993	80	80.2	100	8	4.7	170
Nonwhite	1055	104	92.9	112	23	7.6	303[a]
Side oven[c]	1431	106	114.6	92	10	8.0	125
White	606	45	47.2	95	5	2.7	185
Nonwhite	825	61	67.3	91	5	5.4	93
Partial topside[d]	315	24	26.3	91	2	1.7	
White	307	22	25.1	88	2	1.6	
Nonwhite	8	2	1.2		0	0.1	
Full topside	302	54	32.2	168[a]	19	2.6	731[a]
White	80	13	7.8	167	1	0.5	
Nonwhite	222	41	24.4	168[a]	18	2.2	818[a]

[a] 1% level.
[b] Includes two respiratory cancer deaths other than lung.
[c] Never partial or full topside.
[d] Never full topside.

Table 31.24 Mean Levels of Coal-Tar Pitch Volatiles at Coke-Oven Plants (mg/m³)[198]

Coke Plant Exposure Group, by Job	Plant[a]											
	A	B	C	D	E	F	G	H	I	J	Total	Average
Topside												
Lidman	2.06	3.14	2.02	2.06	3.34	4.11	1.95	6.27	1.88	3.23	3.23	
	(3)	(6)	(6)	(9)	(3)	(5)	(5)	(10)	(5)	(7)	(59)	
Tar chaser				1.18	11.59	3.32	3.31	1.62	2.34	1.81	3.14	
				(2)	(2)	(3)	(7)	(3)	(6)	(4)	(27)	
Larry car	2.73	1.62	8.33	2.73	2.76	3.91	1.11	4.97	2.62	2.50	3.04	
operator	(1)	(4)	(2)	(4)	(3)	(7)	(4)	(3)	(5)	(8)	(41)	3.15
Side oven 1												
Luterman			3.53		2.62		2.18				2.57	
			(4)		(4)		(10)				(18)	
Machine		2.29	1.99			0.94	3.43	3.90	1.18	3.11	2.43	
operator, coke side		(1)	(2)			(2)	(3)	(4)	(5)	(2)	(19)	
Benchman,	0.84	1.42		0.61	0.85	0.70		1.24	4.44	2.09	1.91	
pusher side	(8)	(2)		(1)	(1)	(5)		(4)	(8)	(2)	(31)	
Benchman, coke	0.76	1.58		0.73		0.76			1.07	2.35	1.08	
side	(2)	(2)		(4)		(5)			(3)	(2)	(18)	1.99
Side oven 2												
Heater	1.12	0.61	0.65	1.39	0.61	1.38	1.40	1.11	0.78	1.05	1.07	
	(2)	(2)	(4)	(5)	(2)	(7)	(4)	(3)	(6)	(4)	(39)	
Miscellaneous	0.74		0.24		1.35		0.57		2.15	1.06	0.93	
	(2)		(2)		(1)		(3)		(1)	(9)	(18)	
Quenching car	0.19	0.35	0.83	0.00	0.76	0.34	0.25	0.77	0.27	2.67	0.94	
operator	(2)	(2)	(2)	(1)	(1)	(3)	(2)	(4)	(1)	(5)	(23)	
Pusher machine	0.56	0.94	0.15	0.57		0.46	0.37	0.22	0.53	0.31	0.50	
operator	(2)	(5)	(2)	(2)		(3)	(2)	(3)	(4)	(3)	(26)	0.88
Total	1.02	1.68	2.16	1.55	3.30	1.98	1.95	3.24	2.06	2.02	2.08	
	(22)	(24)	(24)	(28)	(17)	(40)	(40)	(34)	(44)	(46)	(319)	

[a] Numbers in parentheses indicate number of samples.

for Selected Causes by Race, 1953–1961: Men Employed in Subdivisions of the Coke Oven in 1953 and

Other Malignant Neoplasms			Heart Disease			All Other Causes		
Observed Deaths	Expected Deaths	SMR	Observed Deaths	Expected Deaths	SMR	Observed Deaths	Expected Deaths	SMR
27[b]	26.9	100	49	61.5	80	77	72.4	106
12	12.2	98	29	33.9	86	31	29.4	105
15[b]	14.8	101	20	27.6	72	46	43.0	107
13[b]	17.5	74	34	40.0	85	49	49.1	100
6	7.3	82	18	19.9	90	16	17.4	92
7[b]	10.3	68	16	20.1	80	33	31.6	104
6	4.0	150	5	10.9	46	11	9.7	113
5	3.8	132	5	10.5	48	10	9.2	109
1	0.2		0	0.3		1	0.5	
8	5.4	148	10	10.6	94	17	13.6	125
1	1.2		6	3.5	171	5	2.8	179
7	4.3	163	4	7.2	56	12	10.8	111

ronmental pollution, and limitations on uses of contaminated products and by-products.

There is insufficient knowledge to quantify or even to fully evaluate qualitatively the potential impacts on health from trace organic contaminants. Only limited information is available on potential air emissions and the resulting human intakes of contaminants from coal systems. Information about potential exposures associated with contamination of water supplies and agricultural products from runoff and seepage waters has not yet appeared in the literature. Furthermore, complications in the evaluation of health impacts—particularly carcinogenicity—arise because they depend not only on levels of exposure to trace contaminants but on complicated interactions among carcinogens, cocarcinogens, carcinogen initiators, and various other organic agents and components.[204] As a result, characterization of the emitted pollutants is not sufficient for estimating the potential impacts on health. Further research is needed to understand the transformation processes and the exposures of sensitive target organisms to the transformation products.

However, preliminary estimates of the potential impacts on health from emission of various polynuclear aromatic hydrocarbons from coal combustion processes can be made

Table 31.25 Temperature Range of Carbonization Chambers and Excess of Lung Cancers Reported[196]

Type of Carbonizing Chamber	Temperature Range (°C)	Excess of Lung Cancers Reported (%)
Vertical retorts	400–500	27
Horizontal retorts	900–1100	83
Coke ovens	1200–1400	255
Japanese gas generators	1500	800

by using BaP or benzene-soluble organics as surrogates.[205,206] Data are available on emissions and atmospheric concentrations of BaP and benzene-soluble organics,[204,207,208] and relatively good correlations have been made between their atmospheric concentrations and those of various polynuclear aromatic compounds.[204]

The major sources of BaP emissions are reported to be home coal and wood furnaces, home wood fireplaces, small residential and industrial boilers, vehicular emissions, and various industrial processes, such as coke production. BaP emissions from coal-burning power plants in the United States were estimated to be less than 1 ton in 1972, or only about 0.33% of the estimated national emission of BaP.[209] Based on coke-oven estimates ("Hazards to the General Population," Section 31.6.4.3), the excess annual cancer risk associated with a 1000-MWe coal-burning electric power plant is estimated at about 0.005.

Inorganic Contaminants. Conclusions similar to those for organic constituents can be reached with respect to potential health impacts from exposure to inorganic contaminants from coal combustion. The major source of releases of inorganic materials from coal is their mineral content, which is made up almost completely of silicon, calcium, aluminum, and iron and also includes almost all the known elements, but usually in much smaller quantities (Table 31.26). Some of these inorganic constituents of coal can be carcinogenic. These include the metallic elements arsenic, beryllium, cadmium, cobalt, chromium, nickel, and lead, and the nonmetal selenium.

Table 31.26 Minor and Trace Elements in Coal and Earth's Crust[210]

	Concentration (ppm)		
Element	Coal (mean)	Coal (range)	Earth's Crust (mean)
Arsenic	14.0	0.5–93.0	1.8
Beryllium	1.6	0.2–4.0	2.8
Cadmium	2.5	0.1–65.0	0.2
Chromium	13.8	4.0–54.0	100.0
Cobalt	9.6	1.0–43.0	25.0
Iron	1.9	0.3–4.3	5600.0
Lead	34.8	4.0–218.0	12.5
Nickel	21.1	3.0–80.0	75.0
Selenium	2.1	0.5–7.7	0.05

The large ranges in the amounts of trace elements present in various coals are noteworthy. Variations in composition occur even for coal samples derived from the same seam. Because of this, large errors can be made in projecting potential emissions of trace elements from a coal-utilizing plant and the resulting impacts on health from these emissions.

Other inorganic contaminants that are found in some coals include the trace elements mercury and the radionuclides ^{235}U, ^{238}U, ^{232}Th, and various associated radioactive decay products. These contaminants can reach the environment before, during, or after coal combustion. They enter the human body through inhalation of contaminated air and through ingestion of contaminated water or food. The main sources of contaminated water include runoff and seepage from coal mining and storage and from coal waste piles; air emissions result from fires in coal and waste storage piles and from the combustion of coal.

Trace element emissions, including radioactive elements, from coal combustion in power plants are generally considered an unimportant source of exposure to trace elements.[149,211,212] The main sources of human

[206] E. J. Salmon, *Health and Environmental Implications of Various Energy Systems,* draft report prepared for National Energy Strategies Project, Resources for the Future, Washington, D.C., 1978.

[207] University of California, Lawrence Livermore Laboratory, *Environmental Effects of Energy Production and Utilization in the U.S.,* Vol. 1, *Sources, Trends and Costs of Control,* compiled and edited by H. W. Newkird, Berkeley, Calif., May 1, 1976.

[208] National Research Council, *Particulate Polycyclic Organic Matter,* Report by Committee on Biologic Effects of Atmospheric Pollutants, Division of Medical Sciences, National Research Council, Washington, D.C., 1972.

[209] U.S. Environmental Protection Agency, "Preferred Standards Path Report for Polycyclic Organic Matter," 1974, reported in Ref. 204.

[210] M. R. Kornreich, *Coal Conversion Processes: Potential Carcinogenic Risk,* MTR-7155, MITRE Corporation, McLean, Va., March 1976.

[211] H. L. Falk, "Health Effects of Coal Mining and Combustion: Carcinogens and Cofactors" (draft), Paper No. 7 prepared for President's Committee on Health and Ecological Effects of Increased Coal Utilization, November 4, 1977.

[212] U.S. Energy Research and Development Administration, *Fossil Energy Technical Program Summary Book, FY 1977,* February 15, 1977.

exposure to trace inorganic contaminants are not coal but smelters, incinerators, industrial processes, and automobiles.

31.6.4.5 *Conversion*

A report on the incidence of cancer among workers in coal conversion plants was published by Sexton.[213,214] His information pertains to a coal liquefaction pilot plant at Institute, West Virginia. The plant was in operation from 1952 until the mid-1950s, employed 259 men, and produced 300 tons of coal hydrogenation products per day. Because of awareness of the cancer risks involved, an industrial hygiene program was conducted. It included employee education to use personal hygiene procedures of barrier cream, protective clothing, daily changes of all clothing, and daily inspection of the skin. Sexton[214] reported that between 1954 and 1959, 42 cases of skin lesions were diagnosed as precursors of skin cancer. This is 16–37 times the rates found in the populations of West Virginia and the United States as a whole.

The periods of exposure of the men to the chemicals varied from several months to several years, since some of them had been exposed to similar chemicals in previous laboratory work. The effectiveness of the industrial hygiene program is questionable because spot checks conducted on workers during the operation of the plant indicated skin contamination in spite of the cleansing procedures and protective measures. Only limited conclusions may be drawn from these data, namely, that exposure to coal liquefaction products can induce skin cancer. No correlation between exposure and incidence of skin cancer has been shown, nor are the data conclusive concerning the incidence of other kinds of cancer.

More recent data on experience at Exxon Research and Engineering with exposures of 966 man-years to coal hydrogenation in liquid products showed three cases of basal cell carcinomas. However it is difficult to draw firm conclusions on their significance because of the small size of the population exposed.

Table 31.27, which lists carcinogens in coal conversion process streams, demonstrates the range of potential carcinogenic pollutants; but more information on the relative hazards of the pollutants, the emission rates, and projected exposures is necessary before the potential impacts on health can be evaluated.

Larger emissions of polynuclear aromatic hydrocarbons are expected from coal conversion processes than from coal-burning power plants. However the contributions are expected to remain small[205,206] relative to other sources. This is a provisional conclusion that needs to be confirmed after data from operating coal conversion facilities have been obtained.

Because of the similarity of the liquid products from coal conversion plants and from coke-oven plants, emissions of polynuclear aromatics from all these sources are expected to be similar.[206] Coal conversion plants may produce greater emissions of polynuclear aromatics than equivalent power plants, but their expected location in areas of low population density will reduce the potential exposures. Also, emissions are expected to be easier to manage because of the large size of the plants and use of rigid controls. However more information concerning emissions, exposures, and the impacts of the contaminants on health is needed to confirm this preliminary conclusion.

There is insufficient knowledge to evaluate the potential exposures associated with contamination of water and food by trace elements and with water runoff and seepage from solid wastes from coal conversion plants.

31.7 COMPARISON OF THE IMPACTS OF ALTERNATIVE COAL TECHNOLOGIES

An evaluation and comparison of the potential impacts on health of various coal technologies is useful in understanding the relative importance of the impacts.

The key impacts on health from the various coal technologies discussed in this chapter

[213] R. J. Sexton, "The Hazards to Health in the Hydrogenation of Coal. I. An Introductory Statement on General Information, Process Description, and a Definition of the Problem," *Arch. Environ. Health,* **1** 181–186 (1960).

[214] R. J. Sexton, "The Hazards to Health in the Hydrogenation of Coal. IV. The Control Program and the Clinical Effects," *Arch. Environ. Health,* **1,** 208–231 (1960).

Table 31.27 Carcinogens of Concern in Coal Liquefaction and Gasification Process Streams[215]

Chemical Classification	Compound	Phase	Remarks
Amines	Diethylamines	Gas	Potential source of nitrosamines from reaction with nitrogen oxides
	Methylethylamines	Gas	
Heterocyclics	Pyridines	Gas/liquid	
	Pyrroles	Gas/liquid	
Hydrocarbons	Benzene	Gas/liquid	
Phenols	Cresols	Gas/liquid	May be promoters
	Alkyl cresols	Gas/liquid	
Polynuclears	Anthracene	Gas	
	BaP[a]	Gas	
	Chrysene	Gas	
	Benzo(*a*)anthracene	Gas	
	Benzo(*a*)anthrone	Gas	
	Dibenzo(*a*,*l*)pyrene	Gas	
	Dibenzo(*a*,*n*)pyrene	Gas	
	Dibenzo(*a*,*i*)pyrene	Gas	
	Indeno(1,2,3–*c*,*d*)pyrene	Gas	
	Benzoacridine	Gas	
Trace elements	Nickel	Gas	
	Beryllium	Gas	
	Cadmium[a]	Gas	
	Arsenic[a]	Liquid	
Organometallics	Nickel carbonyl	Gas	
Fine particulates	Sulfur particulates	Gas	May be a cocarcinogen
	Coke	Gas	
	Coal dust	Gas	

[a] Potential mutagens or teratogens, based on Teratogen/Mutagen Sulfide, *Registry of Toxic Effects of Chemical Substances,* National Institute for Occupational Safety and Health.

are summarized in Table 31.28. The impacts are compared on the basis of production of an equivalent quantity of useful energy for space heating. (For further information about characteristics and choices of various unit coal technologies, see Section 31.2.) The information available for the assessment of health impacts of different coal technologies is difficult to evaluate. The effects on health of different coal processes cannot be compared just on the basis of equal numbers of projected deaths, injuries, or illnesses. Data on some impacts are considerably more reliable than others, and such data should receive more weight when comparisons are made.

The most reliable information is that about accidental deaths from mining coal. At the other extreme of the data base are the potential impacts on health from environmental pollutants emitted by coal technologies that are still under development, such as impacts from conversion of coal to gaseous and liquid fuels. The information is deficient with respect to the kinds and quantities of potential emissions, their environmental transport, the resulting exposures of and intakes by sensitive individuals among the exposed population, the expected dose-response relationships, and the complications that may arise when exposure to a combination of pollutants takes place.

Between these extremes of uncertainty are many degrees of comprehensiveness and reliability of data. Rail transport of coal is an important source of deaths and injuries to the general public from accidents at grade

[215] K. Bridbord, *Carcinogenic and Mutagenic Risks Associated with Fossil Fuels,* National Institute of Occupational Safety and Health, Rockville, Md., 1976.

Table 31.28 Health Impacts from Coal Unit Systems (number of cases per unit of useful energy to consumer) During the Five Stages of the Fuel Cycle[206]

Health Effect	Electric	Low-Btu Gasification–Combined Cycle Electric	High-Btu Gasification	Liquefaction
		Mining		
Occupational				
CWP deaths	2.4–10.9	2.4–10.6	1.8–8.0	1.9–8.6
CWP disease	95–290	93–280	70–210	76–230
Accidental deaths	1.0–1.8	1.0–1.7	0.7–1.3	0.8–1.4
Accidental injuries	46–82	43–76	34–57	35–62
		Transportation		
Occupational				
Accidental deaths	0.2–0.6	0.2–0.6	0.1–0.4	0.2–0.5
Accidental injuries	20–60	20–60	10–40	20–50
General public				
Accidental death	1.9–5.8	1.8–5.4	1.3–4.0	1.5–4.4
Accidental injuries	5–16	5–15	4–11	4–12
		Coal Preparation		
Occupational				
Accidental deaths	0.08	0.08	0.06	0.06
Accidental injuries	8.7	8.1	6.1	6.6
		Generation or Conversion plus Utilization		
Occupational				
Accidental deaths	0.1	0.6	0.5	0.2
Accidental injuries	3	20	15	3
General public				
Disease deaths	0–58	0–5.5	0–4.1	0–11
Asthma	860–86,000	82–8200	61–6100	160–16,000
		Distribution of Fuels		
General public				
Accidental deaths	—	—	0.004	0.007
Accidental injuries	—	—	1.3	0.4

crossings. Comprehensive statistical data exist but only with respect to general rail freight accidents; that is, the type of cargo involved is not identified. Data are available on safety risks to workers in the construction industry. These may be multiplied by the estimated number of man-days needed to construct coal facilities, to derive the expected number of deaths and injuries likely to be associated with the construction phase.

Serious errors are involved in attempting to estimate rates of delayed death and disability from various lung diseases associated with coal mining (see Section 31.6.4.1). Another error involves the projection of CWP rates, which depends on estimating the future ratio of surface to underground mining, the regional distribution of mines, and the amounts of exposure to coal dust. Although a significant decline in the occurrence of new cases of CWP is expected in the future because of new requirements to maintain low dust levels and because of the continuing shift to surface mining, it is doubtful whether the number of reportable cases of the disease will decline in the near future. Old cases of CWP will continue to be reported for many years even if current exposures to coal dust are controlled.

Ramsay[123] has compared various models that estimate the adverse health impacts of air pollution emissions from coal-burning power plants on adults and children. He used emissions of sulfur oxides as an index to predict impacts from all the air pollutants derived from power plants. His reported range of impact values is used in Table 31.28 to derive estimates of impacts on health from air pollutants associated with electricity generation and coal conversion. In the case of deaths, the lower value of zero is chosen to reflect the great uncertainties involved in the assessments. In the case of morbidity, variation by a factor of about 100 between the lower and upper values indicates the considerable uncertainties in these estimates. If the upper value turns out to be correct, it would mean that a large number of people would be adversely affected.

When the potential local impacts of the various coal technologies are compared, those associated with the mining of coal are predominant. In fact, they are large enough to mask all the other local occupational impacts. In general, their magnitudes depend mainly on the overall energy efficiencies of the coal system. High-Btu coal gasification, which has the highest overall energy efficiency of all the technologies discussed (40.7%) is associated with an impact only about two-thirds as great as the corresponding plant for the conversion of coal to electricity (efficiency 28.5%). Coal liquefaction involves slightly greater overall impacts on health than gasification, and a low-Btu gas–combined cycle electric power plant has a small advantage in impacts over a direct-fired coal-electricity plant.

These results could change significantly if plants used different mixes of coal in their methods of mining. Since most of the mining impacts are associated with underground mining, the larger the proportion of underground coal in the feed supply, the greater will be the occupational impacts.

The key potential regional impacts on the health of the general public from coal electric and coal conversion plants are those from the transportation of coal and from emission of combustion products. It is difficult to determine which of these two kinds of impact is the more important. There are large uncertainties associated with estimates of disease, death, and disability attributed to emission of combustion products. The lower part of the range of the impacts may well deserve more credence than the higher. The average shortening of life from transportation fatalities is estimated to be 40 years, as compared with a few days to a few years for air pollution. The spread must be given weight when accidental deaths from transportation are compared with those from air emissions.

The regional impacts on health from a unit coal conversion system are considerably smaller than those from a unit coal electric system (see Section 31.3). The main reason is that most of the impacts on health from environmental pollutants are associated with combustion products. Only about 10–20% of the coal used in the plant is burned directly in coal conversion processes. The synthetic fuels produced from coal conversion are largely cleaned of sulfur before use. Thus considerably smaller amounts of sulfur dioxide are emitted from coal conversion systems than from the useful energy supplied from a coal-electricity system. However the synthetic oil and gas products are normally burned closer to a larger number of people than coal in electric power plants, and this partly offsets the advantage just noted. An equivalent amount of total pollution from a conversion system therefore will have greater overall impact on health, since more people are exposed to the pollutants.

The annual number of occupational deaths associated with the various coal technologies per unit of useful energy (66.5×10^{12} Btu/year) is approximately 3–13 or about 180–780 per year for all the coal consumed in the United States as of 1978. This compares with a total of about 12,600 occupational deaths reported for the United States during 1976.

The annual number of deaths per unit of useful energy among members of the general public is estimated at 2–60 for the coal-electricity system and 1–15 for the others, or about 90–2400 per year for all useful energy consumed in the United States in 1978. This compares with approximately one million deaths in the United States each year from all causes.

31.8 SUMMARY

This chapter has reviewed the environmental, health and safety implications of the expanding U.S. reliance on coal as an energy source, suggesting some of the problems that may result from mining, transportation, and utilization of the coal and from the disposal of waste and effluents. Possible environmental effects include changes in surface water and groundwater movement as a result of mining and waste disposal, infiltration of groundwater and surface water by hazardous contaminants resulting from solid waste disposal, acid rain, increased mobilization of trace elements from coal, and release of organic contaminants produced by coal conversion. Undesirable health effects may result for those segments of the occupational work force that work near certain components of the fuel system, and for the general population because of the release of oxides of sulfur and nitrogen, particulates, and toxic organic compounds during production, transport, or utilization of coal and its products.

Although the preceding discussion has allowed us to identify the types of environmental and health problems that might arise as a result of increased coal utilization, quantification is uncertain. The uncertainty results from a lack of the data necessary to provide an understanding of the types of material being produced and/or released, their transport and fate in the environment, the kinetics of the processes that lower or increase their quantities, and the pathways to man and other species.

The research that we have summarized is professionally sound and valuable work, and it provides specific information based on the research approach chosen by its authors. However the researchers have not all used the same methodology and experimental design; thus the data are often not comparable. Furthermore, because research by individuals has not been comprehensive, gaps exist in the data. These circumstances suggest the need for a major coordinated effort involving all the groups interested in the environmental, health and safety aspects of coal utilization. Such an effort would require identifying the issues, defining and initiating the needed research, and setting realistic goals and time frames for obtaining the data. The development and execution of such an integrated, orderly research plan is no small task. It is, however, important. To make rational decisions about the use of coal as a future energy source, we must have accurate estimates of its costs and benefits, including an environmental, health, and safety prospectus.

Finally, if we are to rely on increased use of coal to supply future energy needs without allowing the quality of our environment to deteriorate, effluents and emissions must be carefully controlled. Environmental and health research activities must supply the data necessary not only for risk assessment but also for identifying the substances whose emissions must be reduced.

ACKNOWLEDGMENTS

The production of a chapter covering, in depth, the breadth of environmental, safety, and health aspects of coal utilization required expertise greater than that possessed by the six authors. Numerous people have been involved in the writing, critical reviewing, and editing of this material. We appreciate it all. Most notably, we acknowledge the contributions of Drs. J. S. Huckabee and R. J. Haynes, of the Environmental Sciences Division of Oak Ridge National Laboratory, for the scientific content of the sections concerning aquatic effects of coal combustion and environmental effects of coal mining, respectively. The contribution of E. J. Salmon is based on a large-scale effort conducted by Resources for the Future for the National Energy Strategies Project during 1977–1978.

Glossary

AAS	Atomic absorption spectroscopy
ACGIH	American Conference of Government and Industrial Hygienists, Cincinnati, Ohio
ADA	2, 7-Anthraquinone disulfonic acid
AERE	Atomic Energy Research Establishment, Great Britain
AFNOR	Association Française de Normalisation (French Standardization Organization)
AIChE	American Institute of Chemical Engineers
AIME	American Institute of Mining Engineers; now American Institute of Mining, Metallurgical, and Petroleum Engineers
AISI	American Iron and Steel Institute
Amoco	Standard Oil of Indiana
ANCIT	Formcoke process of Eschweiler Bergwerks-Verein (West Germany)
ANSI	American National Standards Institute, New York (formerly United States of America Standards Institute)
AOAC	Association of Official Analytical Chemists
Arco	Atlantic Richfield Company
ASA	Active surface area
ASCE	American Society of Construction Engineers
ASTM	American Society for Testing and Materials, Philadelphia
BaP	Benzo(*a*)pyrene
BASF	Badische Anilin und Soda Fabrik
BCR	Bituminous Coal Research
BCURA	British Coal Utilisation Research Association, Leatherhead, Surrey, England
BET	Brunauer, Emmett, and Teller (adsorption equation developers)
BF	Bergbau-Forschung GmbH
BFL	Bergbau-Forschung-Lurgi
BHP	Broken Hill Proprietary Co. Ltd. (Australia)
BIs	Benzene insoluble
BM–AGA	U.S. Bureau of Mines–American Gas Association
BOD	Biological oxygen demand
BPCA	Benzenepolycarboxylic acid
BS	British Standard
BSI	British Standard Institution
BTX	Benzene, toluene, xylene
CA	Chemical Abstracts
CAB	Critical air blast
CAD	Continuous ash discharge
CAFB	Chemically active fluidized bed
CAS	Cyanogen, ammonia, and sulfur
CBC	Carbon burning cell
CCDC	Conoco Coal Development Company
CCL	Catalytic coal liquids
CE	Coal equivalent (1 kgCE = 29.3×10^3 kJ = 27.8×10^3 Btu)
CEC	Carbonisation et Entreprise Ceramique
CEGB	Central Electricity Generating Board, Great Britain
CERCHAR	Centre d'Études et Recherches des Charbonnages de France
CESEMI	Computer-evaluated scanning electron microscope images
CFFC	Clean fuel from coal
CFR	Continuous flow reactor
CJB	Constructors John Brown
CNRM	Centre National des Recherches Metallurgiques, Liège
COD	Chemical oxygen demand

COED	Char, oil, energy development
COMA	Coke-Oven Managers' Association, Yorkshire, England
COMECON	Organization for Eastern Europe
COMINPECC	Comission Interprofessionelle des Essais des Charbons et des Cokes
Consol	Consolidation Coal Company
CORMA	Coke reactivity measurement apparatus
CORMA	Committee on Metallurgical Research, Norway
c.p.	Crystallization point
CP	Centipoise
CST	Centistokes
CRE	Coal Research Establishment of the National Coal Board (Stoke Orchard, U.K.)
CRIJ	Coal Research Institute of Japan
CSF	Consol synthetic fuel
CS-SRT	Cities Service-Short Residence Time
CSIRO	Commonwealth Scientific and Industrial Research Organization, Australia
DAP	Diammonium phosphate
DGA	Diglycolamine
DKS	Didier-Keihan-Sumitomo, Japan
DIN	Deutsches Institut für Normung (German Institute for Standardization)
DIPA	Diisopropanolamine
D–LM	Dwight–Lloyd McWane
dmmf	Dry, mineral matter free
DMT	Dimethyl terephthalate
DOE	Department of Energy, United States
DSC	Differential scanning calorimetry
DTGA	Derivative thermogravimetric analysis
DTA	Differential thermal analysis
DTG	Differential thermogravimetric
DWT	Dead weight tons
EBV	Eschweiler Bergwerks-Verein
ECE	Economic Commission for Europe, United Nations
ECSC	European Coal and Steel Community
EDAS	Energy dispersive X-ray analyzer
EDA	Exxon donor solvent
EGKS	Europäische Gemeinschaft für Kohle und Stahl
EEC	European Economic Communities
EEI	Edison Electric Institute
2EH	2-Ethyl hexanol
EHV	Extra high voltage
EP	End point
EPA	Environmental Protection Agency, United States
EPRI	Electric Power Research Institute, Palo Alto, Calif.
ESCA	Electron spectroscopy for chemical analysis
ERDA	Energy Research and Development Administration; now a part of DOE
ER & E	Exxon Research and Engineering Company
ESR	Electron spin resonance
EVT	Equiviscous temperature
FB	Fluidized bed
FBC	Fluidized-bed combustion
FC	Fixed carbon
FCC	Fluid catalytic cracking
Fcc	Fluid convection cathode
FCL	Formed Coke Ltd., Great Britain
FDP	Free-fall dilute phase
FGD	Flue gas desulfurization
FIA	Fluorescent indicator analysis
f.i.d.	Flame ionization detector
FMC	Food Machinery and Chemical Corporation
FSI	Free swelling index
GC	Gas chromotography
GHT	Gesellschaft für Hochtemperaturreaktor technik GmbH
GPC	Gel permeation chromatography
GOST	Gosudarstvenny Obshesousny Standart (Russian Standards)
GT	Gas turbine
H_{al}/C_{al}	Ratio of aliphatic hydrogen to aliphatic carbon
Har/Hal	Ratio of aromatic hydrogen to aliphatic hydrogen

Haru/C	H/C ratio for the unsubstituted aromatic nucleus
HBNPC	Houillères du Bassin du Nord et du Pas de Calais
HCC	High-carbon coke
HGI	Hardgrove Grindability Index
HGR	Hot gas recycle
HGMS	High-gradient magnetic separation
HHV	Higher heating value
HMPA	Hexamethylphosphoramide
H_n	Net (lower) heating value
HPLC	High-pressure liquid chromatography
HRE	Hochtemperatur-Reaktor Bau, GmbH Julich, West Germany
HRI	Hydrocarbon Research Incorporated
H_s	Hydrogen saturates
HTGR	High-temperature, gas-cooled reactor
HTS	High-temperature shift
HTT	Heat treatment temperature
hvAb	High-volatile A bituminous
hvb	High-volatile bituminous
hvBb	High-volatile B bituminous
hvCb	High-volatile C bituminous
Hy	*y*-Butyrolactone
IBC	Inert binding capacity
IBP	Initial boiling point
ICEM	Institute of Energetics of the Rumanian Academy of Sciences
IChPW	Institute of Chemical Coal Processing, Poland
ICI	Imperial Chemical Industries, Great Britain
I.D.	Inside diameter
IFP	Institut Français du Petrole
IGI	(Sapozhnikov)
IGT	Institute of Gas Technology, Chicago
IISI	International Iron and Steel Institute (Brussels, Belgium)
IITRI	Illinois Institute of Technology Research Institute, Chicago
ILAFA	Institudo Latinoamericano del Fierro y el Aciero
IMMA	Ion micropobe mass analyzer
INAA	Instrumental neutron activation analysis
INICHAR	Institut National de l'Industrie Charbonniere
INIEX	Institute National Des Industries Extractives (Belgium)
IR	infrared
ISCOR	Iron and Steel Industrial Corporation
ISO	International Organization for Standardization
ISRM	International Society for Rock Mechanics
JANAF	Joint Army Navy Air Force-Thermochemical data bank
JIS	Japanese Industrial Standard
JP-4	Jet fuel specification designation
JP-5	Jet fuel specification designation
KEG	Kommission der Europäischen Gemeinschaften
KFA	Kernforschunganlage Julich GmbH
K and S	Krämer and Sarnow softening point
K-T	Koppers-Totzek
LAH	Lithium aluminum hydride
LAB	Linear alkylbenzenesulfonate
LERC	Laramie (Wyoming) Energy Research Center, now Laramie Energy Technology Center
L-H	Langmuir-Hinshelwood
LHSV	Liquid hourly space velocity
LLL	Lawrence Livermore Laboratory
LR	Lurgi-Ruhrgas
Long ton	2240 lb
LPG	Liquefied petroleum gases
LTA	Low-temperature ashing
LTC	Low-temperature carbonization
LTS	Low-temperature shift
lvb	Low-volatile bituminous
maf	Moisture- and ash free
mEq	Milliequivalent(s)
MERC	Morgantown (West Virginia) Energy Research Center, now Energy Technology Center
meV	Millielectron volt(s)
mf	moisture free

MHD Magnetohydrodynamic(s)
MOF Miyazu, Okuyama, and Fukuyama (authors) of a paper on evaluation of coal blends)
m_n^3 cubic meters of dry gas at 0°C and 760 mmHg
MS Mass spectrometry
MTS Medium temperature shift
N Newton (IN = 1kg-m/S^2)
NAA–RC Neutron activation analysis–radiochemical separations
NAPCA National Air Pollution Control Association, United States
NBS National Bureau of Standards, United States
NCB National Coal Board, Great Britain
NIPR National Institute for Pollution and Resources, Japan
NKK Nippon Kokan, Japan
NMP *N*-Methyl-l-pyrrolidone
NMR Nuclear magnetic resonance
NRDC National Research Development Corporation
NTIS National Technical Information Service, Springfield, Va.
NSPS New Source Performance Standards (of U.S. EPA)
OCR Office of Coal Research (of U.S. Department of the Interior)
OE Optical emission spectrography
OP Oxidation parameter
ORC Occidental Research Corporation
ORNL Oak Ridge National Laboratory, Oak Ridge, Tenn.
OSHA Occupational Safety and Health Administration, United States
PAH Polynuclear aromatic hydrocarbon
PAMCO Pittsburg and Midway Company
p.c. Pulverized coal
PDU Process development unit
PER Pope, Evans, and Robbins
PERC Pittsburgh Energy Research Center, now (Pittsburgh Energy Technology Center)
PFBCC pressurized fluidized-bed, combined cycle
PIXE Proton-induced x-ray emission
PLOT Porous layer tubular
P & M Pittsburg and Midway Coal Mining Company
PMR Proton magnetic resonance
PONA Paraffins, olefins, naphthenes and aromatics
PNA Polynuclear aromatic
ppb Parts per billion
ppm vol Parts per million by volume
PSA Pressure swing absorption
psia Pounds per square inch absolute
psig Pounds per square inch gauge
PSR Perfectly stirred reactor
PTSA p-Toluene sulfonic acid
PV Permanganate value
PVC Polyvinyl chloride
Quad 1×10^{15} BTU
QI Quinoline insoluble
(R) Reverse reaction
RASA Reacting fraction of the active surface area
R and B Ring and ball
RBW Rheinische Braunkohlenwerke AG,
RHS Right-hand side (of an equation)
RMP Ralph M. Parsons
RON Research octane number
RONC Rate of octane number change
ROSE Residual oil supercritical extraction
SAF Superabrasion furnace
SB Special bitumen
SCOT Shell Claus off-gas treating
SCP Single cell proteins
SEM Scanning electron microscope
SG Specific gravity
Sh Sherwood number
SMRE Safety in Mines Research Establishment, Great Britain
SMTTP Standard Methods of Testing Tar and Its Products
SNG Substitute natural gas
s.p. Softening point
sp^2 Hybridization of one *s* and two *p* valence electrons into

	three equivalent energy levels
sp^3	Hybridization one *s* and three *p* valence electrons into four equivalent energy levels
SPM	Sumitomo Metal Industries, Japan
SRC	Solvent-refined coal
SRF	Semireforming furnace
SRL	Solvent-refined lignite
SSMS	Spark source mass spectrometry
STEAG	Steinkohlen Elektrizität, AG
STP	Standard temperature and pressure: 0°C, 760 mm Hg
STV	Standard tar viscometer
SUS	Saybott universal second
T	Tesla (1T = 1 Wb/m^2 = 10^4 G)
T_{cv}	Temperature of critical viscosity
TEA	Triethanolamine
TEMS	Thermal emission mass spectrometry
TGA	Thermogravimetric analysis
THF	Tetrahydrofuran
TI	Toluene insoluble
TLC	Thin-layer chromatography
TOC	Total organic carbon
Ton	short ton, 2000 lb
Ton	(long) = 2240 lb
Tonne	Metric ton, 2204.62 lb
TSA	Total surface area
TWR	Tube wall reactor
UASA	Unreacting fraction of active surface area
UCC	Union Carbide Corporation, New York
UCG	Underground gasification of coal
UNIDO	United Nations Industrial Development Organization
UOP	Universal Oil Products
USBM	U.S. Bureau of Mines (Department of the Interior)
USS	United States Steel Corporation
UV	ultraviolet
VGB	Technische Vereinigung der Grosskraftwerksbetrieber e.V.
VM	Volatile matter
WHSV	weight hourly space velocity
XPS	X-Ray photoelectron spectroscopy
XRF	X-Ray fluorescence
ZMS	Zeolitic molecular sieve

Author Index

Subject Index